Volume

0.7646

m^3

35.315

220

6.229

10^3

4.546

10^6

IMP Gallon

US Gallon

1.201

3.785

Liter

231

61.024

10^3

cm^3

16.387

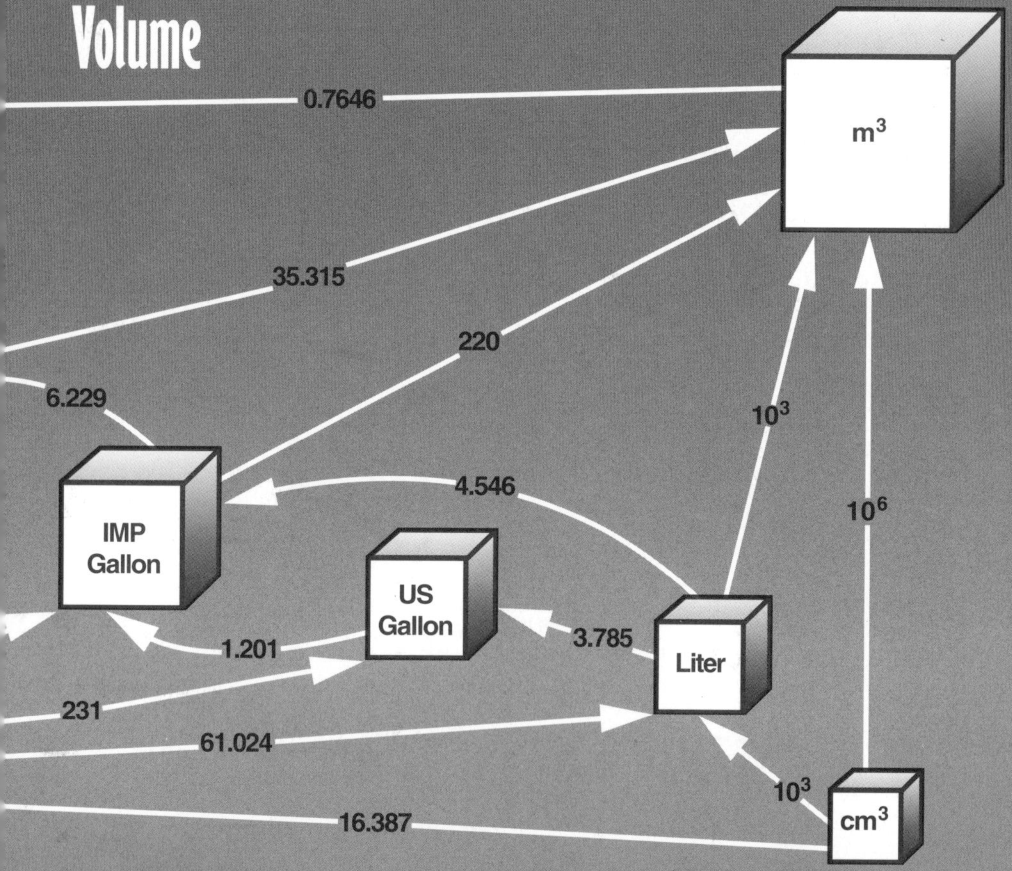

kW = kilowatt
lb = pound
M = mega
m = meter
mm = millimeter
MPa = megapascal
N = newton
oz = ounce

Pa = pascal
PS = pferdestarke (metric hp)
s = seconds
US = United States (gallon)
W = watt
yd = yard

Machinery's Handbook

A REFERENCE BOOK
FOR THE MECHANICAL ENGINEER, DESIGNER,
MANUFACTURING ENGINEER, DRAFTSMAN,
TOOLMAKER, AND MACHINIST

26th Edition Machinery's Handbook

BY ERIK OBERG, FRANKLIN D. JONES,
HOLBROOK L. HORTON, AND HENRY H. RYFFELL

CHRISTOPHER J. MCCAULEY, EDITOR
RICCARDO HEALD, ASSOCIATE EDITOR
MUHAMMED IQBAL HUSSAIN, ASSOCIATE EDITOR

2000

INDUSTRIAL PRESS INC.

NEW YORK

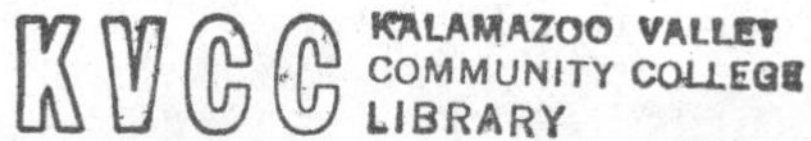

Library of Congress Cataloging-in-Publication Data

Oberg, Erik, 1881—1951

Machinery's Handbook.

2640 p.

Includes index.

I. Mechanical engineering—Handbook, manuals, etc.

I. Jones, Franklin Day, 1879-1967

II. Horton, Holbrook Lynedon, 1907-

III. Ryffel, Henry H. I920- IV. Title.

TJ151.0245 2000 621.8'0212 72-622276

ISBN 0-8311-2625-6 (Thumb Indexed 11.7 x 17.8 cm)

ISBN 0-8311-2635-3 (Thumb Indexed 17.8 x 25.4 cm)

LC card number 72-622276

INDUSTRIAL PRESS, INC.
200 Madison Avenue
New York, New York 10016-4078

MACHINERY'S HANDBOOK
26th Edition
First Printing

PREFACE

For more than 85 years of continuous publication, *Machinery's Handbook* has served as the principal reference in design and manufacturing facilities, and in colleges throughout the world. The editors' objective continues to be that of making the *Handbook* a practical tool to be used in the same way that other kinds of tools are used, to make or repair products of high quality, at the lowest cost, and in the shortest time possible.

Reference works such as *Machinery's Handbook* cannot carry the same information in successive editions if they are to justify the claim that new or updated material is always presented. The editors of such a book must move with the times, keeping a finger on the pulse of manufacturing industry to learn what subjects have less, and what have more, usefulness to the majority of users. At the same time, material that is of proven worth must continue to be included if the *Handbook* is to provide for the needs of disciplines that do not develop as fast as, for instance, the numerical control field. Thus, it remains a difficult task to select suitable material from the almost limitless supply of data pertaining to the manufacturing and mechanical engineering fields, and to provide for the needs of design and production departments in all sizes of manufacturing plants and workshops, as well as those of jobbing shops, trade schools, and technical schools.

The editors rely to some extent on conversations with users of the *Handbook*, and on postcards and other written communications from *Handbook* users, for guidance on which topics should be introduced, revised, lengthened, shortened, or omitted. In response to users' suggestions, in recent years material on logarithms, trigonometry, and other topics was restored, and in this edition sine-bar tables have finally been restored after numerous requests for this topic. Also at the request of users, in 1997 the first ever large-print or "desktop" edition of the *Handbook* was published, followed in 1998 by the publication of *Machinery's Handbook CD-ROM* including several hundred additional pages of material restored from earlier editions.

Regular users of the *Handbook* should be able to identify some of the many changes embodied in the present edition. "Old style" numerals, in continuous use since the first edition, and occasionally a source of confusion for readers, have been replaced by a modern numeral style. The entire text of this edition, including all the tables and equations, has been reset, and a great many of the numerous figures have been redrawn. The addition of 80 pages brings the total length of the book to 2640 pages.

The 26th edition of the *Handbook* contains significant format changes and major revisions of existing content, as well as new material on a variety of topics including: aerodynamic lubrication, high speed machining, grinding feeds and speeds, machining econometrics, metalworking fluids, ISO surface texture, pipe welding, geometric dimensioning and tolerancing, gearing, and EDM.

Other subjects in the *Handbook* that are new or have been revised, expanded, or updated are: graphic descriptions of functions of angles, imaginary and complex numbers, complex coordinate systems, contour milling, weight of piles, Ohm's law, binary multiples, force on inclined planes, and measurement over pins.

Those users involved in aspects of machining and grinding will be interested in the new topics *MACHINING ECONOMETRICS* and *GRINDING FEEDS AND SPEEDS*, presented in the *Machining* section. The core of all manufacturing methods start with the cutting edge and the metal removal process, and improving the control of the machining process is a major component in order to achieve a **Lean chain** of manufacturing events. These sections describe the means that are necessary to get metal cutting processes under control and how to properly evaluate the decision making.

A major goal of the editors is to make the *Handbook* easier to use. The 26th edition of the *Handbook* continues to incorporate the time-saving thumb tabs, much requested by users in the past. The table of contents pages beginning each major section, first introduced for the 25th edition, have proven very useful to readers. Consequently, the number of contents pages has been increased to several pages each for many of the larger sections, more thoroughly reflecting the contents of these sections. In the present edition, the Plastics section, formerly a separate thumb tab, has been incorporated into the Properties of Materials section. A major task in assembling this edition has been the expansion and reorganization of the index. For the first time, most of the many Standards referenced in the *Handbook* are now included in the index.

The American Standards Association was reconstituted in August 1969 as the United States of America Standards Institute, and standards that had been approved as American Standards were designated as USA Standards. In October 1969, the name was changed to the American National Standards Institute. Thus, the designation of present standards is ANSI instead of ASA or USAS. Standards originally adopted by the American Standards Association and not revised are still referred to in the *Handbook* by the designation ASA.

ANSI Standards are copyrighted by the American National Standards Institute, West 42nd Street, New York, NY 10017, from whom current copies may be purchased. Many of the American National Standards Institute (ANSI) Standards that deal with mechanical engineering, extracts from which are included in the *Handbook*, are produced by the American Society of Mechanical Engineers (ASME), and we are grateful for their permission to quote extracts and to update the information contained in the standards, based on the revisions regularly carried out by the ASME. Information regarding current editions of any of these Standards can be obtained from ASME International, Three Park Avenue, New York, NY 10016.

Users who call possible defects to the attention of the editors, or the omission of some matter that is considered to be of general value, often render a service to the entire manufacturing field. We desire to increase the usefulness of the *Handbook* as far as possible, so all criticisms and suggestions about revisions, omissions, or inclusion of new material are welcome.

Christopher J. McCauley, Editor

CONTENTS

ACKNOWLEDGMENTS

The editors would like to acknowledge all those who contributed ideas and suggestions to the *Handbook*. Several individuals and companies, in particular, contributed substantial amounts of information to this edition.

Dr. Bertil Colding, Colding International Corp., Lansing, Michigan provided extensive material on grinding speeds, feeds, depths of cut, and tool life for a wide range of materials. He also provided practical information on machining econometrics, including tool wear and tool life and machining cost relationships.

Dr. Bruce Harding, Director and Professor of MET, Perdue University contributed information on GD&T as well material on the differences between ISO and ANSI surface finish symbology.

Mr. Edward Craig of WeldTrain International and ABB Flexible Automation, Inc. contributed information on pipe welding.

Mr. Sydney Kravitz provided valuable information on the weight of piles for different materials.

Dr. T. A. Stolarski, Professor at Brunel University, provided detailed explanations and equations on aerodynamic lubrication.

Mr. Alec Stokes provided much new material that was incorporated into gearing sections including highpoint gears, British spur and helical gearing, addendum modification to involute spur and helical gears, and hypoid gears.

Mr. Richard Pohanish contributed material on metal working fluids.

Hansvedt Industries provided a detailed listing of EDM terms.

Mr. Matthew Radcliff supplied data on wood screw pilot hole sizes.

Mr. Robert H. Green, as editor emeritus, contributed much useful, well organized material to this edition. He also provided invaluable practical guidance to the editorial staff during the *Handbook*'s compilation.

Finally, Industrial Press is extremely fortunate that Mr. Henry H. Ryffel, author and editor of *Machinery's Handbook*, continues to be deeply involved with the *Handbook*. Henry's ideas, suggestions, and vision are deeply appreciated by everyone who worked on this book.

MATHEMATICS

NUMBERS, FRACTIONS, AND DECIMALS

ALGEBRA AND EQUATIONS

GEOMETRY

SOLUTION OF TRIANGLES

LOGARITHMS

NUMBERS, FRACTIONS, AND DECIMALS

Mathematical Signs and Commonly Used Abbreviations

Sign	Meaning	
$+$	Plus (sign of addition)	
$+$	Positive	
$-$	Minus (sign of subtraction)	
$-$	Negative	
$\pm (\mp)$	Plus or minus (minus or plus)	
$\times$	Multiplied by (multiplication sign)	
$\cdot$	Multiplied by (multiplication sign)	
$\div$	Divided by (division sign)	
$/$	Divided by (division sign)	
$:$	Is to (in proportion)	
$=$	Equals	
$\neq$	Is not equal to	
$\equiv$	Is identical to	
$::$	Equals (in proportion)	
$\cong$	Approximately equals	
$\approx$		
$>$	Greater than	
$<$	Less than	
$\geq$	Greater than or equal to	
$\leq$	Less than or equal to	
$\rightarrow$	Approaches as a limit	
$\propto$	Varies directly as	
$\therefore$	Therefore	
$\sqrt{\ }$	Square root	
$\sqrt[3]{\ }$	Cube root	
$\sqrt[4]{\ }$	4th root	
$\sqrt[n]{\ }$	*n*th root	
a^2	*a* squared (2nd power of *a*)	
a^3	*a* cubed (3rd power of *a*)	
a^4	4th power of *a*	
a^n	*n*th power of *a*	
a^{-n}	$1 \div a^n$	
$\frac{1}{n}$	Reciprocal value of *n*	
log	Logarithm	
$\log_e$	Natural or Napierian logarithm	
ln	Natural or Napierian logarithm	
e	Base of natural logarithms (2.71828)	
lim	Limit value (of an expression)	
∞	Infinity	
α	Alpha	
β	Beta	commonly used to denote angles
γ	Gamma	
θ	Theta	
ϕ	Phi	
μ	Mu (coefficient of friction)	

Sign	Meaning
π	Pi (3.1416)
Σ	Sigma (sign of summation)
ω	Omega (angles measured in radians)
g	Acceleration due to gravity (32.16 ft. per sec. per sec.)
i (or j)	Imaginary quantity $(\sqrt{-1})$
sin	Sine
cos	Cosine
tan	Tangent
cot	Cotangent
sec	Secant
csc	Cosecant
vers	Versed sine
covers	Coversed sine
$\sin^{-1} a$ arcsin a	Arc the sine of which is *a*
$(\sin a)^{-1}$	Reciprocal of sin *a* ($1 \div \sin a$)
$\sin^n x$	*n*th power of sin *x*
$\sinh x$	Hyperbolic sine of *x*
$\cosh x$	Hyperbolic cosine of *x*
Δ	Delta (increment of)
δ	Delta (variation of)
d	Differential (in calculus)
∂	Partial differentiation (in calculus)
$\int$	Integral (in calculus)
$\int_b^a$	Integral between the limits *a* and *b*
!	$5! = 1 \times 2 \times 3 \times 4 \times 5$ (Factorial)
$\angle$	Angle
∟	Right angle
$\perp$	Perpendicular to
△	Triangle
○	Circle
▱	Parallelogram
$^\circ$	Degree (circular arc or temperature)
$'$	Minutes or feet
$''$	Seconds or inches
a'	*a* prime
a''	*a* double prime
a_1	*a* sub one
a_2	*a* sub two
a_n	*a* sub *n*
( )	Parentheses
[ ]	Brackets
{ }	Braces

Prime Numbers and Factors of Numbers

The *factors* of a given number are those numbers which when multiplied together give a product equal to that number; thus, 2 and 3 are factors of 6; and 5 and 7 are factors of 35.

A *prime number* is one which has no factors except itself and 1. Thus, 3, 5, 7, 11, etc., are prime numbers. A factor which is a prime number is called a *prime factor*.

The accompanying "Prime Number and Factor Tables" give the smallest prime factor of all odd numbers from 1 to 9600, and can be used for finding all the factors for numbers up to this limit. For example, find the factors of 931. In the column headed "900" and in the line indicated by "31" in the left-hand column, the smallest prime factor is found to be 7. As this leaves another factor 133 (since $931 \div 7 = 133$), find the smallest prime factor of this number. In the column headed "100" and in the line "33", this is found to be 7, leaving a factor 19. This latter is a prime number; hence, the factors of 931 are $7 \times 7 \times 19$. Where no factor is given for a number in the factor table, it indicates that the number is a prime number.

The last page of the tables lists all prime numbers from 9551 through 18691; and can be used to identify quickly all unfactorable numbers in that range.

For factoring, the following general rules will be found useful:

2 is a factor of any number the right-hand figure of which is an even number or 0. Thus, $28 = 2 \times 14$, and $210 = 2 \times 105$.

3 is a factor of any number the sum of the figures of which is evenly divisible by 3. Thus, 3 is a factor of 1869, because $1 + 8 + 6 + 9 = 24 \div 3 = 8$.

4 is a factor of any number the two right-hand figures of which, considered as one number, are evenly divisible by 4. Thus, 1844 has a factor 4, because $44 \div 4 = 11$.

5 is a factor of any number the right-hand figure of which is 0 or 5. Thus, $85 = 5 \times 17$; $70 = 5 \times 14$.

Tables of prime numbers and factors of numbers are particularly useful for calculations involving change-gear ratios for compound gearing, dividing heads, gear-generating machines, and mechanical designs having gear trains.

Example 1: A set of four gears is required in a mechanical design to provide an overall gear ratio of $4104 \div 1200$. Furthermore, no gear in the set is to have more than 120 teeth or less than 24 teeth. Determine the tooth numbers.

First, as explained previously, the factors of 4104 are determined to be: $2 \times 2 \times 2 \times 3 \times 3 \times 57 = 4104$. Next, the factors of 1200 are determined: $2 \times 2 \times 2 \times 2 \times 5 \times 5 \times 3 = 1200$. Therefore $\frac{4104}{1200} = \frac{2 \times 2 \times 2 \times 3 \times 3 \times 57}{2 \times 2 \times 2 \times 2 \times 5 \times 5 \times 3} = \frac{72 \times 57}{24 \times 50}$. If the factors had been combined differently, say, to give $\frac{72 \times 57}{16 \times 75}$, then the 16-tooth gear in the denominator would not satisfy the requirement of no less than 24 teeth.

Example 2: Factor the number 25078 into two numbers neither of which is larger than 200.

The first factor of 25078 is obviously 2, leaving $25078 \div 2 = 12539$ to be factored further. However, from the last table, *Prime Numbers from 9551 to 18691*, it is seen that 12539 is a prime number; therefore, no solution exists.

Prime Number and Factor Table for 1 to 1199

From To	0 100	100 200	200 300	300 400	400 500	500 600	600 700	700 800	800 900	900 1000	1000 1100	1100 1200
1	P	P	3	7	P	3	P	P	3	17	7	3
3	P	P	7	3	13	P	3	19	11	3	17	P
5	P	3	5	5	3	5	5	3	5	5	3	5
7	P	P	3	P	11	3	P	7	3	P	19	3
9	3	P	11	3	P	P	3	P	P	3	P	P
11	P	3	P	P	3	7	13	3	P	P	3	11
13	P	P	3	P	7	3	P	23	3	11	P	3
15	3	5	5	3	5	5	3	5	5	3	5	5
17	P	3	7	P	3	11	P	3	19	7	3	P
19	P	7	3	11	P	3	P	P	3	P	P	3
21	3	11	13	3	P	P	3	7	P	3	P	19
23	P	3	P	17	3	P	7	3	P	13	3	P
25	5	5	3	5	5	3	5	5	3	5	5	3
27	3	P	P	3	7	17	3	P	P	3	13	7
29	P	3	P	7	3	23	17	3	P	P	3	P
31	P	P	3	P	P	3	P	17	3	7	P	3
33	3	7	P	3	P	13	3	P	7	3	P	11
35	5	3	5	5	3	5	5	3	5	5	3	5
37	P	P	3	P	19	3	7	11	3	P	17	3
39	3	P	P	3	P	7	3	P	P	3	P	17
41	P	3	P	11	3	P	P	3	29	P	3	7
43	P	11	3	7	P	3	P	P	3	23	7	3
45	3	5	5	3	5	5	3	5	5	3	5	5
47	P	3	13	P	3	P	P	3	7	P	3	31
49	7	P	3	P	P	3	11	7	3	13	P	3
51	3	P	P	3	11	19	3	P	23	3	P	P
53	P	3	11	P	3	7	P	3	P	P	3	P
55	5	5	3	5	5	3	5	5	3	5	5	3
57	3	P	P	3	P	P	3	P	P	3	7	13
59	P	3	7	P	3	13	P	3	P	7	3	19
61	P	7	3	19	P	3	P	P	3	31	P	3
63	3	P	P	3	P	P	3	7	P	3	P	P
65	5	3	5	5	3	5	5	3	5	5	3	5
67	P	P	3	P	P	3	23	13	3	P	11	3
69	3	13	P	3	7	P	3	P	11	3	P	7
71	P	3	P	7	3	P	11	3	13	P	3	P
73	P	P	3	P	11	3	P	P	3	7	29	3
75	3	5	5	3	5	5	3	5	5	3	5	5
77	7	3	P	13	3	P	P	3	P	P	3	11
79	P	P	3	P	P	3	7	19	3	11	13	3
81	3	P	P	3	13	7	3	11	P	3	23	P
83	P	3	P	P	3	11	P	3	P	P	3	7
85	5	5	3	5	5	3	5	5	3	5	5	3
87	3	11	7	3	P	P	3	P	P	3	P	P
89	P	3	17	P	3	19	13	3	7	23	3	29
91	7	P	3	17	P	3	P	7	3	P	P	3
93	3	P	P	3	17	P	3	13	19	3	P	P
95	5	3	5	5	3	5	5	3	5	5	3	5
97	P	P	3	P	7	3	17	P	3	P	P	3
99	3	P	13	3	P	P	3	17	29	3	7	11

Prime Number and Factor Table for 1201 to 2399

From To	1200 1300	1300 1400	1400 1500	1500 1600	1600 1700	1700 1800	1800 1900	1900 2000	2000 2100	2100 2200	2200 2300	2300 2400
1	P	P	3	19	P	3	P	P	3	11	31	3
3	3	P	23	3	7	13	3	11	P	3	P	7
5	5	3	5	5	3	5	5	3	5	5	3	5
7	17	P	3	11	P	3	13	P	3	7	P	3
9	3	7	P	3	P	P	3	23	7	3	47	P
11	7	3	17	P	3	29	P	3	P	P	3	P
13	P	13	3	17	P	3	7	P	3	P	P	3
15	3	5	5	3	5	5	3	5	5	3	5	5
17	P	3	13	37	3	17	23	3	P	29	3	7
19	23	P	3	7	P	3	17	19	3	13	7	3
21	3	P	7	3	P	P	3	17	43	3	P	11
23	P	3	P	P	3	P	P	3	7	11	3	23
25	5	5	3	5	5	3	5	5	3	5	5	3
27	3	P	P	3	P	11	3	41	P	3	17	13
29	P	3	P	11	3	7	31	3	P	P	3	17
31	P	11	3	P	7	3	P	P	3	P	23	3
33	3	31	P	3	23	P	3	P	19	3	7	P
35	5	3	5	5	3	5	5	3	5	5	3	5
37	P	7	3	29	P	3	11	13	3	P	P	3
39	3	13	P	3	11	37	3	7	P	3	P	P
41	17	3	11	23	3	P	7	3	13	P	3	P
43	11	17	3	P	31	3	19	29	3	P	P	3
45	3	5	5	3	5	5	3	5	5	3	5	5
47	29	3	P	7	3	P	P	3	23	19	3	P
49	P	19	3	P	17	3	43	P	3	7	13	3
51	3	7	P	3	13	17	3	P	7	3	P	P
53	7	3	P	P	3	P	17	3	P	P	3	13
55	5	5	3	5	5	3	5	5	3	5	5	3
57	3	23	31	3	P	7	3	19	11	3	37	P
59	P	3	P	P	3	P	11	3	29	17	3	7
61	13	P	3	7	11	3	P	37	3	P	7	3
63	3	29	7	3	P	41	3	13	P	3	31	17
65	5	3	5	5	3	5	5	3	5	5	3	5
67	7	P	3	P	P	3	P	7	3	11	P	3
69	3	37	13	3	P	29	3	11	P	3	P	23
71	31	3	P	P	3	7	P	3	19	13	3	P
73	19	P	3	11	7	3	P	P	3	41	P	3
75	3	5	5	3	5	5	3	5	5	3	5	5
77	P	3	7	19	3	P	P	3	31	7	3	P
79	P	7	3	P	23	3	P	P	3	P	43	3
81	3	P	P	3	41	13	3	7	P	3	P	P
83	P	3	P	P	3	P	7	3	P	37	3	P
85	5	5	3	5	5	3	5	5	3	5	5	3
87	3	19	P	3	7	P	3	P	P	3	P	7
89	P	3	P	7	3	P	P	3	P	11	3	P
91	P	13	3	37	19	3	31	11	3	7	29	3
93	3	7	P	3	P	11	3	P	7	3	P	P
95	5	3	5	5	3	5	5	3	5	5	3	5
97	P	11	3	P	P	3	7	P	3	13	P	3
99	3	P	P	3	P	7	3	P	P	3	11	P

Prime Number and Factor Table for 2401 to 3599

From To	2400 2500	2500 2600	2600 2700	2700 2800	2800 2900	2900 3000	3000 3100	3100 3200	3200 3300	3300 3400	3400 3500	3500 3600
1	7	41	3	37	P	3	P	7	3	P	19	3
3	3	P	19	3	P	P	3	29	P	3	41	31
5	5	3	5	5	3	5	5	3	5	5	3	5
7	29	23	3	P	7	3	31	13	3	P	P	3
9	3	13	P	3	53	P	3	P	P	3	7	11
11	P	3	7	P	3	41	P	3	13	7	3	P
13	19	7	3	P	29	3	23	11	3	P	P	3
15	3	5	5	3	5	5	3	5	5	3	5	5
17	P	3	P	11	3	P	7	3	P	31	3	P
19	41	11	3	P	P	3	P	P	3	P	13	3
21	3	P	P	3	7	23	3	P	P	3	11	7
23	P	3	43	7	3	37	P	3	11	P	3	13
25	5	5	3	5	5	3	5	5	3	5	5	3
27	3	7	37	3	11	P	3	53	7	3	23	P
29	7	3	11	P	3	29	13	3	P	P	3	P
31	11	P	3	P	19	3	7	31	3	P	47	3
33	3	17	P	3	P	7	3	13	53	3	P	P
35	5	3	5	5	3	5	5	3	5	5	3	5
37	P	43	3	7	P	3	P	P	3	47	7	3
39	3	P	7	3	17	P	3	43	41	3	19	P
41	P	3	19	P	3	17	P	3	7	13	3	P
43	7	P	3	13	P	3	17	7	3	P	11	3
45	3	5	5	3	5	5	3	5	5	3	5	5
47	P	3	P	41	3	7	11	3	17	P	3	P
49	31	P	3	P	7	3	P	47	3	17	P	3
51	3	P	11	3	P	13	3	23	P	3	7	53
53	11	3	7	P	3	P	43	3	P	7	3	11
55	5	5	3	5	5	3	5	5	3	5	5	3
57	3	P	P	3	P	P	3	7	P	3	P	P
59	P	3	P	31	3	11	7	3	P	P	3	P
61	23	13	3	11	P	3	P	29	3	P	P	3
63	3	11	P	3	7	P	3	P	13	3	P	7
65	5	3	5	5	3	5	5	3	5	5	3	5
67	P	17	3	P	47	3	P	P	3	7	P	3
69	3	7	17	3	19	P	3	P	7	3	P	43
71	7	3	P	17	3	P	37	3	P	P	3	P
73	P	31	3	47	13	3	7	19	3	P	23	3
75	3	5	5	3	5	5	3	5	5	3	5	5
77	P	3	P	P	3	13	17	3	29	11	3	7
79	37	P	3	7	P	3	P	11	3	31	7	3
81	3	29	7	3	43	11	3	P	17	3	59	P
83	13	3	P	11	3	19	P	3	7	17	3	P
85	5	5	3	5	5	3	5	5	3	5	5	3
87	3	13	P	3	P	29	3	P	19	3	11	17
89	19	3	P	P	3	7	P	3	11	P	3	37
91	47	P	3	P	7	3	11	P	3	P	P	3
93	3	P	P	3	11	41	3	31	37	3	7	P
95	5	3	5	5	3	5	5	3	5	5	3	5
97	11	7	3	P	P	3	19	23	3	43	13	3
99	3	23	P	3	13	P	3	7	P	3	P	59

Prime Number and Factor Table for 3601 to 4799

From To	3600 3700	3700 3800	3800 3900	3900 4000	4000 4100	4100 4200	4200 4300	4300 4400	4400 4500	4500 4600	4600 4700	4700 4800
1	13	P	3	47	P	3	P	11	3	7	43	3
3	3	7	P	3	P	11	3	13	7	3	P	P
5	5	3	5	5	3	5	5	3	5	5	3	5
7	P	11	3	P	P	3	7	59	3	P	17	3
9	3	P	13	3	19	7	3	31	P	3	11	17
11	23	3	37	P	3	P	P	3	11	13	3	7
13	P	47	3	7	P	3	11	19	3	P	7	3
15	3	5	5	3	5	5	3	5	5	3	5	5
17	P	3	11	P	3	23	P	3	7	P	3	53
19	7	P	3	P	P	3	P	7	3	P	31	3
21	3	61	P	3	P	13	3	29	P	3	P	P
23	P	3	P	P	3	7	41	3	P	P	3	P
25	5	5	3	5	5	3	5	5	3	5	5	3
27	3	P	43	3	P	P	3	P	19	3	7	29
29	19	3	7	P	3	P	P	3	43	7	3	P
31	P	7	3	P	29	3	P	61	3	23	11	3
33	3	P	P	3	37	P	3	7	11	3	41	P
35	5	3	5	5	3	5	5	3	5	5	3	5
37	P	37	3	31	11	3	19	P	3	13	P	3
39	3	P	11	3	7	P	3	P	23	3	P	7
41	11	3	23	7	3	41	P	3	P	19	3	11
43	P	19	3	P	13	3	P	43	3	7	P	3
45	3	5	5	3	5	5	3	5	5	3	5	5
47	7	3	P	P	3	11	31	3	P	P	3	47
49	41	23	3	11	P	3	7	P	3	P	P	3
51	3	11	P	3	P	7	3	19	P	3	P	P
53	13	3	P	59	3	P	P	3	61	29	3	7
55	5	5	3	5	5	3	5	5	3	5	5	3
57	3	13	7	3	P	P	3	P	P	3	P	67
59	P	3	17	37	3	P	P	3	7	47	3	P
61	7	P	3	17	31	3	P	7	3	P	59	3
63	3	53	P	3	17	23	3	P	P	3	P	11
65	5	3	5	5	3	5	5	3	5	5	3	5
67	19	P	3	P	7	3	17	11	3	P	13	3
69	3	P	53	3	13	11	3	17	41	3	7	19
71	P	3	7	11	3	43	P	3	17	7	3	13
73	P	7	3	29	P	3	P	P	3	17	P	3
75	3	5	5	3	5	5	3	5	5	3	5	5
77	P	3	P	41	3	P	7	3	11	23	3	17
79	13	P	3	23	P	3	11	29	3	19	P	3
81	3	19	P	3	7	37	3	13	P	3	31	7
83	29	3	11	7	3	47	P	3	P	P	3	P
85	5	5	3	5	5	3	5	5	3	5	5	3
87	3	7	13	3	61	53	3	41	7	3	43	P
89	7	3	P	P	3	59	P	3	67	13	3	P
91	P	17	3	13	P	3	7	P	3	P	P	3
93	3	P	17	3	P	7	3	23	P	3	13	P
95	5	3	5	5	3	5	5	3	5	5	3	5
97	P	P	3	7	17	3	P	P	3	P	7	3
99	3	29	7	3	P	13	3	53	11	3	37	P

Prime Number and Factor Table for 4801 to 5999

From To	4800 4900	4900 5000	5000 5100	5100 5200	5200 5300	5300 5400	5400 5500	5500 5600	5600 5700	5700 5800	5800 5900	5900 6000
1	P	13	3	P	7	3	11	P	3	P	P	3
3	3	P	P	3	11	P	3	P	13	3	7	P
5	5	3	5	5	3	5	5	3	5	5	3	5
7	11	7	3	P	41	3	P	P	3	13	P	3
9	3	P	P	3	P	P	3	7	71	3	37	19
11	17	3	P	19	3	47	7	3	31	P	3	23
13	P	17	3	P	13	3	P	37	3	29	P	3
15	3	5	5	3	5	5	3	5	5	3	5	5
17	P	3	29	7	3	13	P	3	41	P	3	61
19	61	P	3	P	17	3	P	P	3	7	11	3
21	3	7	P	3	23	17	3	P	7	3	P	31
23	7	3	P	47	3	P	11	3	P	59	3	P
25	5	5	3	5	5	3	5	5	3	5	5	3
27	3	13	11	3	P	7	3	P	17	3	P	P
29	11	3	47	23	3	73	61	3	13	17	3	7
31	P	P	3	7	P	3	P	P	3	11	7	3
33	3	P	7	3	P	P	3	11	43	3	19	17
35	5	3	5	5	3	5	5	3	5	5	3	5
37	7	P	3	11	P	3	P	7	3	P	13	3
39	3	11	P	3	13	19	3	29	P	3	P	P
41	47	3	71	53	3	7	P	3	P	P	3	13
43	29	P	3	37	7	3	P	23	3	P	P	3
45	3	5	5	3	5	5	3	5	5	3	5	5
47	37	3	7	P	3	P	13	3	P	7	3	19
49	13	7	3	19	29	3	P	31	3	P	P	3
51	3	P	P	3	59	P	3	7	P	3	P	11
53	23	3	31	P	3	53	7	3	P	11	3	P
55	5	5	3	5	5	3	5	5	3	5	5	3
57	3	P	13	3	7	11	3	P	P	3	P	7
59	43	3	P	7	3	23	53	3	P	13	3	59
61	P	11	3	13	P	3	43	67	3	7	P	3
63	3	7	61	3	19	31	3	P	7	3	11	67
65	5	3	5	5	3	5	5	3	5	5	3	5
67	31	P	3	P	23	3	7	19	3	73	P	3
69	3	P	37	3	11	7	3	P	P	3	P	47
71	P	3	11	P	3	41	P	3	53	29	3	7
73	11	P	3	7	P	3	13	P	3	23	7	3
75	3	5	5	3	5	5	3	5	5	3	5	5
77	P	3	P	31	3	19	P	3	7	53	3	43
79	7	13	3	P	P	3	P	7	3	P	P	3
81	3	17	P	3	P	P	3	P	13	3	P	P
83	19	3	13	71	3	7	P	3	P	P	3	31
85	5	5	3	5	5	3	5	5	3	5	5	3
87	3	P	P	3	17	P	3	37	11	3	7	P
89	P	3	7	P	3	17	11	3	P	7	3	53
91	67	7	3	29	11	3	17	P	3	P	43	3
93	3	P	11	3	67	P	3	7	P	3	71	13
95	5	3	5	5	3	5	5	3	5	5	3	5
97	59	19	3	P	P	3	23	29	3	11	P	3
99	3	P	P	3	7	P	3	11	41	3	17	7

Prime Number and Factor Table for 6001 to 7199

From To	6000 6100	6100 6200	6200 6300	6300 6400	6400 6500	6500 6600	6600 6700	6700 6800	6800 6900	6900 7000	7000 7100	7100 7200
1	17	P	3	P	37	3	7	P	3	67	P	3
3	3	17	P	3	19	7	3	P	P	3	47	P
5	5	3	5	5	3	5	5	3	5	5	3	5
7	P	31	3	7	43	3	P	19	3	P	7	3
9	3	41	7	3	13	23	3	P	11	3	43	P
11	P	3	P	P	3	17	11	3	7	P	3	13
13	7	P	3	59	11	3	17	7	3	31	P	3
15	3	5	5	3	5	5	3	5	5	3	5	5
17	11	3	P	P	3	7	13	3	17	P	3	11
19	13	29	3	71	7	3	P	P	3	11	P	3
21	3	P	P	3	P	P	3	11	19	3	7	P
23	19	3	7	P	3	11	37	3	P	7	3	17
25	5	5	3	5	5	3	5	5	3	5	5	3
27	3	11	13	3	P	61	3	7	P	3	P	P
29	P	3	P	P	3	P	7	3	P	13	3	P
31	37	P	3	13	59	3	19	53	3	29	79	3
33	3	P	23	3	7	47	3	P	P	3	13	7
35	5	3	5	5	3	5	5	3	5	5	3	5
37	P	17	3	P	41	3	P	P	3	7	31	3
39	3	7	17	3	47	13	3	23	7	3	P	11
41	7	3	79	17	3	31	29	3	P	11	3	37
43	P	P	3	P	17	3	7	11	3	53	P	3
45	3	5	5	3	5	5	3	5	5	3	5	5
47	P	3	P	11	3	P	17	3	41	P	3	7
49	23	11	3	7	P	3	61	17	3	P	7	3
51	3	P	7	3	P	P	3	43	13	3	11	P
53	P	3	13	P	3	P	P	3	7	17	3	23
55	5	5	3	5	5	3	5	5	3	5	5	3
57	3	47	P	3	11	79	3	29	P	3	P	17
59	73	3	11	P	3	7	P	3	19	P	3	P
61	11	61	3	P	7	3	P	P	3	P	23	3
63	3	P	P	3	23	P	3	P	P	3	7	13
65	5	3	5	5	3	5	5	3	5	5	3	5
67	P	7	3	P	29	3	59	67	3	P	37	3
69	3	31	P	3	P	P	3	7	P	3	P	67
71	13	3	P	23	3	P	7	3	P	P	3	71
73	P	P	3	P	P	3	P	13	3	19	11	3
75	3	5	5	3	5	5	3	5	5	3	5	5
77	59	3	P	7	3	P	11	3	13	P	3	P
79	P	37	3	P	11	3	P	P	3	7	P	3
81	3	7	11	3	P	P	3	P	7	3	73	43
83	7	3	61	13	3	29	41	3	P	P	3	11
85	5	5	3	5	5	3	5	5	3	5	5	3
87	3	23	P	3	13	7	3	11	71	3	19	P
89	P	3	19	P	3	11	P	3	83	29	3	7
91	P	41	3	7	P	3	P	P	3	P	7	3
93	3	11	7	3	43	19	3	P	61	3	41	P
95	5	3	5	5	3	5	5	3	5	5	3	5
97	7	P	3	P	73	3	37	7	3	P	47	3
99	3	P	P	3	67	P	3	13	P	3	31	23

Prime Number and Factor Table for 7201 to 8399

From To	7200 7300	7300 7400	7400 7500	7500 7600	7600 7700	7700 7800	7800 7900	7900 8000	8000 8100	8100 8200	8200 8300	8300 8400
1	19	7	3	13	11	3	29	P	3	P	59	3
3	3	67	11	3	P	P	3	7	53	3	13	19
5	5	3	5	5	3	5	5	3	5	5	3	5
7	P	P	3	P	P	3	37	P	3	11	29	3
9	3	P	31	3	7	13	3	11	P	3	P	7
11	P	3	P	7	3	11	73	3	P	P	3	P
13	P	71	3	11	23	3	13	41	3	7	43	3
15	3	5	5	3	5	5	3	5	5	3	5	5
17	7	3	P	P	3	P	P	3	P	P	3	P
19	P	13	3	73	19	3	7	P	3	23	P	3
21	3	P	41	3	P	7	3	89	13	3	P	53
23	31	3	13	P	3	P	P	3	71	P	3	7
25	5	5	3	5	5	3	5	5	3	5	5	3
27	3	17	7	3	29	P	3	P	23	3	19	11
29	P	3	17	P	3	59	P	3	7	11	3	P
31	7	P	3	17	13	3	41	7	3	47	P	3
33	3	P	P	3	17	11	3	P	29	3	P	13
35	5	3	5	5	3	5	5	3	5	5	3	5
37	P	11	3	P	7	3	17	P	3	79	P	3
39	3	41	43	3	P	71	3	17	P	3	7	31
41	13	3	7	P	3	P	P	3	11	7	3	19
43	P	7	3	19	P	3	11	13	3	17	P	3
45	3	5	5	3	5	5	3	5	5	3	5	5
47	P	3	11	P	3	61	7	3	13	P	3	17
49	11	P	3	P	P	3	47	P	3	29	73	3
51	3	P	P	3	7	23	3	P	83	3	37	7
53	P	3	29	7	3	P	P	3	P	31	3	P
55	5	5	3	5	5	3	5	5	3	5	5	3
57	3	7	P	3	13	P	3	73	7	3	23	61
59	7	3	P	P	3	P	29	3	P	41	3	13
61	53	17	3	P	47	3	7	19	3	P	11	3
63	3	37	17	3	79	7	3	P	11	3	P	P
65	5	3	5	5	3	5	5	3	5	5	3	5
67	13	53	3	7	11	3	P	31	3	P	7	3
69	3	P	7	3	P	17	3	13	P	3	P	P
71	11	3	31	67	3	19	17	3	7	P	3	11
73	7	73	3	P	P	3	P	7	3	11	P	3
75	3	5	5	3	5	5	3	5	5	3	5	5
77	19	3	P	P	3	7	P	3	41	13	3	P
79	29	47	3	11	7	3	P	79	3	P	17	3
81	3	11	P	3	P	31	3	23	P	3	7	17
83	P	3	7	P	3	43	P	3	59	7	3	83
85	5	5	3	5	5	3	5	5	3	5	5	3
87	3	83	P	3	P	13	3	7	P	3	P	P
89	37	3	P	P	3	P	7	3	P	19	3	P
91	23	19	3	P	P	3	13	61	3	P	P	3
93	3	P	59	3	7	P	3	P	P	3	P	7
95	5	3	5	5	3	5	5	3	5	5	3	5
97	P	13	3	71	43	3	53	11	3	7	P	3
99	3	7	P	3	P	11	3	19	7	3	43	37

Prime Number and Factor Table for 8401 to 9599

From To	8400 8500	8500 8600	8600 8700	8700 8800	8800 8900	8900 9000	9000 9100	9100 9200	9200 9300	9300 9400	9400 9500	9500 9600
1	31	P	3	7	13	3	P	19	3	71	7	3
3	3	11	7	3	P	29	3	P	P	3	P	13
5	5	3	5	5	3	5	5	3	5	5	3	5
7	7	47	3	P	P	3	P	7	3	41	23	3
9	3	67	P	3	23	59	3	P	P	3	97	37
11	13	3	79	31	3	7	P	3	61	P	3	P
13	47	P	3	P	7	3	P	13	3	67	P	3
15	3	5	5	3	5	5	3	5	5	3	5	5
17	19	3	7	23	3	37	71	3	13	7	3	31
19	P	7	3	P	P	3	29	11	3	P	P	3
21	3	P	37	3	P	11	3	7	P	3	P	P
23	P	3	P	11	3	P	7	3	23	P	3	89
25	5	5	3	5	5	3	5	5	3	5	5	3
27	3	P	P	3	7	79	3	P	P	3	11	7
29	P	3	P	7	3	P	P	3	11	19	3	13
31	P	19	3	P	P	3	11	23	3	7	P	3
33	3	7	89	3	11	P	3	P	7	3	P	P
35	5	3	5	5	3	5	5	3	5	5	3	5
37	11	P	3	P	P	3	7	P	3	P	P	3
39	3	P	53	3	P	7	3	13	P	3	P	P
41	23	3	P	P	3	P	P	3	P	P	3	7
43	P	P	3	7	37	3	P	41	3	P	7	3
45	3	5	5	3	5	5	3	5	5	3	5	5
47	P	3	P	P	3	23	83	3	7	13	3	P
49	7	83	3	13	P	3	P	7	3	P	11	3
51	3	17	41	3	53	P	3	P	11	3	13	P
53	79	3	17	P	3	7	11	3	19	47	3	41
55	5	5	3	5	5	3	5	5	3	5	5	3
57	3	43	11	3	17	13	3	P	P	3	7	19
59	11	3	7	19	3	17	P	3	47	7	3	11
61	P	7	3	P	P	3	13	P	3	11	P	3
63	3	P	P	3	P	P	3	7	59	3	P	73
65	5	3	5	5	3	5	5	3	5	5	3	5
67	P	13	3	11	P	3	P	89	3	17	P	3
69	3	11	P	3	7	P	3	53	13	3	17	7
71	43	3	13	7	3	P	47	3	73	P	3	17
73	37	P	3	31	19	3	43	P	3	7	P	3
75	3	5	5	3	5	5	3	5	5	3	5	5
77	7	3	P	67	3	47	29	3	P	P	3	61
79	61	23	3	P	13	3	7	67	3	83	P	3
81	3	P	P	3	83	7	3	P	P	3	19	11
83	17	3	19	P	3	13	31	3	P	11	3	7
85	5	5	3	5	5	3	5	5	3	5	5	3
87	3	31	7	3	P	11	3	P	37	3	53	P
89	13	3	P	11	3	89	61	3	7	41	3	43
91	7	11	3	59	17	3	P	7	3	P	P	3
93	3	13	P	3	P	17	3	29	P	3	11	53
95	5	3	5	5	3	5	5	3	5	5	3	5
97	29	P	3	19	7	3	11	17	3	P	P	3
99	3	P	P	3	11	P	3	P	17	3	7	29

Prime Numbers from 9551 to 18691

9551	10181	10853	11497	12157	12763	13417	14071	14747	15361	16001	16693	17387	18043
9587	10193	10859	11503	12161	12781	13421	14081	14753	15373	16007	16699	17389	18047
9601	10211	10861	11519	12163	12791	13441	14083	14759	15377	16033	16703	17393	18049
9613	10223	10867	11527	12197	12799	13451	14087	14767	15383	16057	16729	17401	18059
9619	10243	10883	11549	12203	12809	13457	14107	14771	15391	16061	16741	17417	18061
9623	10247	10889	11551	12211	12821	13463	14143	14779	15401	16063	16747	17419	18077
9629	10253	10891	11579	12227	12823	13469	14149	14783	15413	16067	16759	17431	18089
9631	10259	10903	11587	12239	12829	13477	14153	14797	15427	16069	16763	17443	18097
9643	10267	10909	11593	12241	12841	13487	14159	14813	15439	16073	16787	17449	18119
9649	10271	10937	11597	12251	12853	13499	14173	14821	15443	16087	16811	17467	18121
9661	10273	10939	11617	12253	12889	13513	14177	14827	15451	16091	16823	17471	18127
9677	10289	10949	11621	12263	12893	13523	14197	14831	15461	16097	16829	17477	18131
9679	10301	10957	11633	12269	12899	13537	14207	14843	15467	16103	16831	17483	18133
9689	10303	10973	11657	12277	12907	13553	14221	14851	15473	16111	16843	17489	18143
9697	10313	10979	11677	12281	12911	13567	14243	14867	15493	16127	16871	17491	18149
9719	10321	10987	11681	12289	12917	13577	14249	14869	15497	16139	16879	17497	18169
9721	10331	10993	11689	12301	12919	13591	14251	14879	15511	16141	16883	17509	18181
9733	10333	11003	11699	12323	12923	13597	14281	14887	15527	16183	16889	17519	18191
9739	10337	11027	11701	12329	12941	13613	14293	14891	15541	16187	16901	17539	18199
9743	10343	11047	11717	12343	12953	13619	14303	14897	15551	16189	16903	17551	18211
9749	10357	11057	11719	12347	12959	13627	14321	14923	15559	16193	16921	17569	18217
9767	10369	11059	11731	12373	12967	13633	14323	14929	15569	16217	16927	17573	18223
9769	10391	11069	11743	12377	12973	13649	14327	14939	15581	16223	16931	17579	18229
9781	10399	11071	11777	12379	12979	13669	14341	14947	15583	16229	16937	17581	18233
9787	10427	11083	11779	12391	12983	13679	14347	14951	15601	16231	16943	17597	18251
9791	10429	11087	11783	12401	13001	13681	14369	14957	15607	16249	16963	17599	18253
9803	10433	11093	11789	12409	13003	13687	14387	14969	15619	16253	16979	17609	18257
9811	10453	11113	11801	12413	13007	13691	14389	14983	15629	16267	16981	17623	18269
9817	10457	11117	11807	12421	13009	13693	14401	15013	15641	16273	16987	17627	18287
9829	10459	11119	11813	12433	13033	13697	14407	15017	15643	16301	16993	17657	18289
9833	10463	11131	11821	12437	13037	13709	14411	15031	15647	16319	17011	17659	18301
9839	10477	11149	11827	12451	13043	13711	14419	15053	15649	16333	17021	17669	18307
9851	10487	11159	11831	12457	13049	13721	14423	15061	15661	16339	17027	17681	18311
9857	10499	11161	11833	12473	13063	13723	14431	15073	15667	16349	17029	17683	18313
9859	10501	11171	11839	12479	13093	13729	14437	15077	15671	16361	17033	17707	18329
9871	10513	11173	11863	12487	13099	13751	14447	15083	15679	16363	17041	17713	18341
9883	10529	11177	11867	12491	13103	13757	14449	15091	15683	16369	17047	17729	18353
9887	10531	11197	11887	12497	13109	13759	14461	15101	15727	16381	17053	17737	18367
9901	10559	11213	11897	12503	13121	13763	14479	15107	15731	16411	17077	17747	18371
9907	10567	11239	11903	12511	13127	13781	14489	15121	15733	16417	17093	17749	18379
9923	10589	11243	11909	12517	13147	13789	14503	15131	15737	16421	17099	17761	18397
9929	10597	11251	11923	12527	13151	13799	14519	15137	15739	16427	17107	17783	18401
9931	10601	11257	11927	12539	13159	13807	14533	15139	15749	16433	17117	17789	18413
9941	10607	11261	11933	12541	13163	13829	14537	15149	15761	16447	17123	17791	18427
9949	10613	11273	11939	12547	13171	13831	14543	15161	15767	16451	17137	17807	18433
9967	10627	11279	11941	12553	13177	13841	14549	15173	15773	16453	17159	17827	18439
9973	10631	11287	11953	12569	13183	13859	14551	15187	15787	16477	17167	17837	18443
10007	10639	11299	11959	12577	13187	13873	14557	15193	15791	16481	17183	17839	18451
10009	10651	11311	11969	12583	13217	13877	14561	15199	15797	16487	17189	17851	18457
10037	10657	11317	11971	12589	13219	13879	14563	15217	15803	16493	17191	17863	18461
10039	10663	11321	11981	12601	13229	13883	14591	15227	15809	16519	17203	17881	18481
10061	10667	11329	11987	12611	13241	13901	14593	15233	15817	16529	17207	17891	18493
10067	10687	11351	12007	12613	13249	13903	14621	15241	15823	16547	17209	17903	18503
10069	10691	11353	12011	12619	13259	13907	14627	15259	15859	16553	17231	17909	18517
10079	10709	11369	12037	12637	13267	13913	14629	15263	15877	16561	17239	17911	18521
10091	10711	11383	12041	12641	13291	13921	14633	15269	15881	16567	17257	17921	18523
10093	10723	11393	12043	12647	13297	13931	14639	15271	15887	16573	17291	17923	18539
10099	10729	11399	12049	12653	13309	13933	14653	15277	15889	16603	17293	17929	18541
10103	10733	11411	12071	12659	13313	13963	14657	15287	15901	16607	17299	17939	18553
10111	10739	11423	12073	12671	13327	13967	14669	15289	15907	16619	17317	17957	18583
10133	10753	11437	12097	12689	13331	13997	14683	15299	15913	16631	17321	17959	18587
10139	10771	11443	12101	12697	13337	13999	14699	15307	15919	16633	17327	17971	18593
10141	10781	11447	12107	12703	13339	14009	14713	15313	15923	16649	17333	17977	18617
10151	10789	11467	12109	12713	13367	14011	14717	15319	15937	16651	17341	17981	18637
10159	10799	11471	12113	12721	13381	14029	14723	15329	15959	16657	17351	17987	18661
10163	10831	11483	12119	12739	13397	14033	14731	15331	15971	16661	17359	17989	18671
10169	10837	11489	12143	12743	13399	14051	14737	15349	15973	16673	17377	18013	18679
10177	10847	11491	12149	12757	13411	14057	14741	15359	15991	16691	17383	18041	18691

Continued and Conjugate Fractions

Continued Fractions.—In dealing with a cumbersome fraction, or one which does not have satisfactory factors, it may be possible to substitute some other, approximately equal, fraction which is simpler or which can be factored satisfactorily. Continued fractions provide a means of computing a series of fractions each of which is a closer approximation to the original fraction than the one preceding it in the series.

A continued fraction is a proper fraction (one whose numerator is smaller than its denominator) expressed in the form

$$\frac{N}{D} = \cfrac{1}{D_1 + \cfrac{1}{D_2 + \cfrac{1}{D_3 + \ldots}}}$$

It is convenient to write the above expression as

$$\frac{N}{D} = \frac{1}{D_1+}\frac{1}{D_2+}\frac{1}{D_3+}\frac{1}{D_4+}\cdots$$

The continued fraction is produced from a proper fraction N/D by dividing the numerator N both into itself and into the denominator D. Dividing the numerator into itself gives a result of 1; dividing the numerator into the denominator gives a whole number D_1 plus a remainder fraction R_1. The process is then repeated on the remainder fraction R_1 to obtain D_2 and R_2; then D_3, R_3, etc., until a remainder of zero results. As an example, using N/D = 2153/9277,

$$\frac{2153}{9277} = \frac{2153 \div 2153}{9277 \div 2153} = \cfrac{1}{4 + \cfrac{665}{2153}} = \frac{1}{D_1 + R_1}$$

$$R_1 = \frac{665}{2153} = \cfrac{1}{3 + \cfrac{158}{665}} = \frac{1}{D_2 + R_2} \quad \text{etc.}$$

from which it may be seen that $D_1 = 4$, $R_1 = 665/2153$; $D_2 = 3$, $R_2 = 158/665$; and, continuing as was explained previously, it would be found that: $D_3 = 4$, $R_3 = 33/158$; …; $D_9 = 2$, $R_9 = 0$. The complete set of continued fraction elements representing 2153/9277 may then be written as

$$\frac{2153}{9277} = \frac{1}{4+}\frac{1}{3+}\frac{1}{4+}\frac{1}{4+}\frac{1}{1+}\frac{1}{3+}\frac{1}{1+}\frac{1}{2+}\frac{1}{2}$$

$$D_1 \ldots\ldots\ldots D_5 \ldots\ldots\ldots D_9$$

By following a simple procedure, together with a table organized similar to the one below for the fraction 2153/9277, the denominators D_1, D_2, … of the elements of a continued fraction may be used to calculate a series of fractions, each of which is a successively closer approximation, called a *convergent*, to the original fraction N/D.

1) The first row of the table contains column numbers numbered from 1 through 2 plus the number of elements, 2 + 9 = 11 in this example.

2) The second row contains the denominators of the continued fraction elements in sequence but beginning in column 3 instead of column 1 because columns 1 and 2 must be blank in this procedure.

3) The third row contains the convergents to the original fraction as they are calculated and entered. Note that the fractions 1/0 and 0/1 have been inserted into columns 1 and 2. These are two arbitrary convergents, the first equal to infinity, the second to zero, which are used to facilitate the calculations.

4) The convergent in column 3 is now calculated. To find the numerator, multiply the denominator in column 3 by the numerator of the convergent in column 2 and add the numerator of the convergent in column 1. Thus, $4 \times 0 + 1 = 1$.

5) The denominator of the convergent in column 3 is found by multiplying the denominator in column 3 by the denominator of the convergent in column 2 and adding the denominator of the convergent in column 1. Thus, $4 \times 1 + 0 = 4$, and the convergent in column 3 is then ¼ as shown in the table.

6) Finding the remaining successive convergents can be reduced to using the simple equation

$$\mathrm{CONVERGENT}_n = \frac{(D_n)(\mathrm{NUM}_{n-1}) + \mathrm{NUM}_{n-2}}{(D_n)(\mathrm{DEN}_{n-1}) + \mathrm{DEN}_{n-2}}$$

in which n = column number in the table; D_n = denominator in column n; NUM_{n-1} and NUM_{n-2} are numerators and DEN_{n-1} and DEN_{n-2} are denominators of the convergents in the columns indicated by their subscripts; and $\mathrm{CONVERGENT}_n$ is the convergent in column n.

Convergents of the Continued Fraction for 2153⁄9277

Column Number, n	1	2	3	4	5	6	7	8	9	10	11
Denominator, D_n	—	—	4	3	4	4	1	3	1	2	2
$\mathrm{Convergent}_n$	$\frac{1}{0}$	$\frac{0}{1}$	$\frac{1}{4}$	$\frac{3}{13}$	$\frac{13}{56}$	$\frac{55}{237}$	$\frac{68}{293}$	$\frac{259}{1116}$	$\frac{327}{1409}$	$\frac{913}{3934}$	$\frac{2153}{9277}$

Notes: The decimal values of the successive convergents in the table are alternately larger and smaller than the value of the original fraction 2153⁄9277. If the last convergent in the table has the same value as the original fraction 2153⁄9277, then *all* of the other calculated convergents are correct.

Conjugate Fractions.—In addition to finding approximate ratios by the use of continued fractions and logarithms of ratios, conjugate fractions may be used for the same purpose, independently, or in combination with the other methods.

Two fractions a/b and c/d are said to be conjugate if $ad - bc = \pm 1$. Examples of such pairs are: 0/1 and 1/1; 1/2 and 1/1; and 9/10 and 8⁄9. Also, *every successive pair of the convergents of a continued fraction are conjugate.* Conjugate fractions have certain properties that are useful for solving ratio problems:

1) No fraction between two conjugate fractions a/b and c/d can have a denominator smaller than either b or d.

2) A new fraction, e/f, conjugate to both fractions of a given pair of conjugate fractions, a/b and c/d, and lying between them, may be created by adding respective numerators, $a + c$, and denominators, $b + d$, so that $e/f = (a + c)/(b + d)$.

3) The denominator $f = b + d$ of the new fraction e/f is the smallest of any possible fraction lying between a/b and c/d. Thus, 17/19 is conjugate to both 8⁄9 and 9/10 and no fraction with denominator smaller than 19 lies between them. This property is important if it is desired to minimize the size of the factors of the ratio to be found.

The following example shows the steps to approximate a ratio for a set of gears to any desired degree of accuracy within the limits established for the allowable size of the factors in the ratio.

Example: Find a set of four change gears, ab/cd, to approximate the ratio 2.105399 accurate to within ± 0.0001; no gear is to have more than 120 teeth.

Step 1. Convert the given ratio R to a number r between 0 and 1 by taking its reciprocal: $1/R = 1/2.105399 = 0.4749693 = r$.

Step 2. Select a pair of conjugate fractions a/b and c/d that bracket r. The pair a/b = 0⁄1 and c/d = 1⁄1, for example, will bracket 0.4749693.

Step 3. Add the respective numerators and denominators of the conjugates 0/1 and 1/1 to create a new conjugate e/f between 0 and 1: $e/f = (a+c)/(b+d) = (0+1)/(1+1) = 1/2$.

Step 4. Since 0.4749693 lies between 0/1 and 1/2, e/f must also be between 0/1 and 1/2: $e/f = (0+1)/(1+2) = 1/3$.

Step 5. Since 0.4749693 now lies between 1/3 and 1/2, e/f must also be between 1/3 and 1/2: $e/f = (1+1)/(3+2) = 2/5$.

Step 6. Continuing as above to obtain successively closer approximations of e/f to 0.4749693, and using a handheld calculator and a scratch pad to facilitate the process, the fractions below, each of which has factors less than 120, were determined:

Fraction	Numerator Factors	Denominator Factors	Error
19/40	19	$2 \times 2 \times 2 \times 5$	+.000031
28/59	$2 \times 2 \times 7$	59	−.00039
47/99	47	$3 \times 3 \times 11$	−.00022
104/219	3×41	7×37	−.000066
142/299	2×71	13×23	−.000053
161/339	7×23	3×113	−.000043
218/459	2×109	$3 \times 3 \times 3 \times 17$	−.000024
256/539	$2 \times 2 \times 2 \times 2 \times 2 \times 2 \times 2 \times 2$	$7 \times 7 \times 11$	−.000016
370/779	$2 \times 5 \times 37$	19×41	−.0000014
759/1598	$3 \times 11 \times 23$	$2 \times 17 \times 47$	−.00000059

Factors for the numerators and denominators of the fractions shown above were found with the aid of the Prime Numbers and Factors tables beginning on page 4. Since in Step 1 the desired ratio of 2.105399 was converted to its reciprocal 0.4749693, all of the above fractions should be inverted. Note also that the last fraction, 759/1598, when inverted to become 1598/759, is in error from the desired value by approximately one-half the amount obtained by trial and error using earlier methods.

Using Continued Fraction Convergents as Conjugates.—Since successive convergents of a continued fraction are also conjugate, they may be used to find a series of additional fractions in between themselves. As an example, the successive convergents 55/237 and 68/293 from the table of convergents for 2153/9277 on page 14 will be used to demonstrate the process for finding the first few in-between ratios.

Desired Fraction N/D = 2153/9277 = 0.2320793

	a/b	e/f	c/d
(1)	55/237 = .2320675	[a]123/530 = .2320755 error = −.0000039	68/293 = .2320819
(2)	123/530 = .2320755	191/823 = .2320778 error = −.0000016	68/293 = .2320819
(3)	191/823 = .2320778	[a]259/1116 = .2320789 error = −.0000005	68/293 = .2320819
(4)	259/1116 = .2320789	327/1409 = .2320795 error = +.0000002	68/293 = .2320819
(5)	259/1116 = .2320789	586/2525 = .2320792 error = − .0000001	327/1409 = .2320795
(6)	586/2525 = .2320792	913/3934 = .2320793 error = − .0000000	327/1409 = .2320795

[a] Only these ratios had suitable factors below 120.

Step 1. Check the convergents for conjugateness: $55 \times 293 - 237 \times 68 = 16115 - 16116 = -1$ proving the pair to be conjugate.

Step 2. Set up a table as shown on the next page. The leftmost column of line (1) contains the convergent of lowest value, a/b; the rightmost the higher value, c/d; and the center column the derived value e/f found by adding the respective numerators and denominators of a/b and c/d. The error or difference between e/f and the desired value N/D, error = $N/D - e/f$, is also shown.

Step 3. On line (2), the process used on line (1) is repeated with the e/f value from line (1) becoming the new value of a/b while the c/d value remains unchanged. Had the error in e/f

been + instead of −, then *e*/*f* would have been the new *c*/*d* value and *a*/*b* would be unchanged.

Step 4. The process is continued until, as seen on line (4), the error changes sign to + from the previous −. When this occurs, the *e*/*f* value becomes the *c*/*d* value on the next line instead of *a*/*b* as previously and the *a*/*b* value remains unchanged.

Positive and Negative Numbers

The degrees on a thermometer scale extending upward from the zero point may be called *positive* and may be preceded by a plus sign; thus +5 degrees means 5 degrees above zero. The degrees below zero may be called *negative* and may be preceded by a minus sign; thus − 5 degrees means 5 degrees below zero. In the same way, the ordinary numbers 1, 2, 3, etc., which are larger than 0, are called positive numbers; but numbers can be conceived of as extending in the other direction from 0, numbers that, in fact, are less than 0, and these are called negative. As these numbers must be expressed by the same figures as the positive numbers they are designated by a minus sign placed before them, thus: (−3). A negative number should always be enclosed within parentheses whenever it is written in line with other numbers; for example: $17 + (-13) - 3 \times (-0.76)$.

Negative numbers are most commonly met with in the use of logarithms and natural trigonometric functions. The following rules govern calculations with negative numbers.

A negative number can be added to a positive number by subtracting its numerical value from the positive number.

Example: $4 + (-3) = 4 - 3 = 1$.

A negative number can be subtracted from a positive number by adding its numerical value to the positive number.

Example: $4 - (-3) = 4 + 3 = 7$.

A negative number can be added to a negative number by adding the numerical values and making the sum negative.

Example: $(-4) + (-3) = -7$.

A negative number can be subtracted from a larger negative number by subtracting the numerical values and making the difference negative.

Example: $(-4) - (-3) = -1$.

A negative number can be subtracted from a smaller negative number by subtracting the numerical values and making the difference positive.

Example: $(-3) - (-4) = 1$.

If in a subtraction the number to be subtracted is larger than the number from which it is to be subtracted, the calculation can be carried out by subtracting the smaller number from the larger, and indicating that the remainder is negative.

Example: $3-5 = -(5-3) = -2$.

When a positive number is to be multiplied or divided by a negative numbers, multiply or divide the numerical values as usual; the product or quotient, respectively,is negative. The same rule is true if a negative number is multiplied or divided by a positive number.

Examples: $4 \times (-3) = -12 \quad (-4) \times 3 = -12$

$15 \div (-3) = -5 \quad (-15) \div 3 = -5$

When two negative numbers are to be multiplied by each other, the product is positive. When a negative number is divided by a negative number, the quotient is positive.

Examples: $(-4) \times (-3) = 12$; $(-4) \div (-3) = 1.333$.

The two last rules are often expressed for memorizing as follows: "Equal signs make plus, unequal signs make minus."

Powers, Roots, and Reciprocals

The *square* of a number (or quantity) is the product of that number multiplied by itself. Thus, the square of 9 is $9 \times 9 = 81$. The square of a number is indicated by the *exponent* (^{2}), thus: $9^2 = 9 \times 9 = 81$.

The *cube* or *third power* of a number is the product obtained by using that number as a factor three times. Thus, the cube of 4 is $4 \times 4 \times 4 = 64$, and is written 4^3.

If a number is used as a factor four or five times, respectively, the product is the fourth or fifth power. Thus, $3^4 = 3 \times 3 \times 3 \times 3 = 81$, and $2^5 = 2 \times 2 \times 2 \times 2 \times 2 = 32$. A number can be raised to any power by using it as a factor the required number of times.

The *square root* of a given number is that number which, when multiplied by itself, will give a product equal to the given number. The square root of 16 (written $\sqrt{16}$) equals 4, because $4 \times 4 = 16$.

The *cube root* of a given number is that number which, when used as a factor three times, will give a product equal to the given number. Thus, the cube root of 64 (written $\sqrt[3]{64}$) equals 4, because $4 \times 4 \times 4 = 64$.

The fourth, fifth, etc., roots of a given number are those numbers which when used as factors four, five, etc., times, will give as a product the given number. Thus, $\sqrt[4]{16} = 2$, because $2 \times 2 \times 2 \times 2 = 16$.

In some formulas, there may be such expressions as $(a^2)^3$ and $a^{3/2}$. The first of these, $(a^2)^3$, means that the number a is first to be squared, a^2, and the result then cubed to give a^6. Thus, $(a^2)^3$ is equivalent to a^6 which is obtained by *multiplying* the exponents 2 and 3. Similarly, $a^{3/2}$ may be interpreted as the cube of the square root of a, $(\sqrt{a})^3$, or $(a^{1/2})^3$, so that, for example, $16^{3/2} = (\sqrt{16})^3 = 64$.

The multiplications required for raising numbers to powers and the extracting of roots are greatly facilitated by the use of logarithms. Extracting the square root and cube root by the regular arithmetical methods is a slow and cumbersome operation, and any roots can be more rapidly found by using logarithms.

When the power to which a number is to be raised is not an integer, say 1.62, the use of either logarithms or a scientific calculator becomes the only practical means of solution.

The *reciprocal* R of a number N is obtained by dividing 1 by the number; $R = 1/N$. Reciprocals are useful in some calculations because they avoid the use of negative characteristics as in calculations with logarithms and in trigonometry. In trigonometry, the values *cosecant, secant*, and *cotangent* are often used for convenience and are the reciprocals of the *sine, cosine*, and *tangent*, respectively (see page 83). The reciprocal of a fraction, for instance $\frac{3}{4}$, is the fraction inverted, since $1 \div \frac{3}{4} = 1 \times \frac{4}{3} = \frac{4}{3}$.

Powers of Ten Notation

Powers of ten notation is used to simplify calculations and ensure accuracy, particularly with respect to the position of decimal points, and also simplifies the expression of numbers which are so large or so small as to be unwieldy. For example, the metric (SI) pressure unit pascal is equivalent to 0.00000986923 atmosphere or 0.0001450377 pound/inch^2. In powers of ten notation, these figures are 9.86923×10^{-6} atmosphere and 1.450377×10^{-4} pound/inch^2. The notation also facilitates adaptation of numbers for electronic data processing and computer readout.

Expressing Numbers in Powers of Ten Notation.—In this system of notation, every number is expressed by two factors, one of which is some integer from 1 to 9 followed by a decimal and the other is some power of 10.

Thus, 10,000 is expressed as 1.0000×10^4 and 10,463 as 1.0463×10^4. The number 43 is expressed as 4.3×10 and 568 is expressed. as 5.68×10^2.

In the case of decimals, the number 0.0001, which as a fraction is $\frac{1}{10{,}000}$, is expressed as 1×10^{-4} and 0.0001463 is expressed as 1.463×10^{-4}. The decimal 0.498 is expressed as 4.98×10^{-1} and 0.03146 is expressed as 3.146×10^{-2}.

Rules for Converting Any Number to Powers of Ten Notation.—Any number can be converted to the powers of ten notation by means of one of two rules.

Rule 1: If the number is a whole number or a whole number and a decimal so that it has digits to the left of the decimal point, the decimal point is moved a sufficient number of places to the *left* to bring it to the immediate right of the first digit. With the decimal point shifted to this position, the number so written comprises the *first* factor when written in powers of ten notation.

The number of places that the decimal point is moved to the left to bring it immediately to the right of the first digit is the *positive* index or power of 10 that comprises the *second* factor when written in powers of ten notation.

Thus, to write 4639 in this notation, the decimal point is moved three places to the left giving the two factors: 4.639×10^3. Similarly,

$$431.412 = 4.31412 \times 10^2$$

$$986388 = 9.86388 \times 10^5$$

Rule 2: If the number is a decimal, i.e., it has digits entirely to the right of the decimal point, then the decimal point is moved a sufficient number of places to the *right* to bring it immediately to the right of the first digit. With the decimal point shifted to this position, the number so written comprises the *first* factor when written in powers of ten notation.

The number of places that the decimal point is moved to the *right* to bring it immediately to the right of the first digit is the *negative* index or power of 10 that follows the number when written in powers of ten notation.

Thus, to bring the decimal point in 0.005721 to the immediate right of the first digit, which is 5, it must be moved *three* places to the right, giving the two factors: 5.721×10^{-3}. Similarly,

$$0.469 = 4.69 \times 10^{-1}$$

$$0.0000516 = 5.16 \times 10^{-5}$$

Multiplying Numbers Written in Powers of Ten Notation.—When multiplying two numbers written in the powers of ten notation together, the procedure is as follows:

1) Multiply the first factor of one number by the first factor of the other to obtain the first factor of the product.

2) Add the index of the second factor (which is some power of 10) of one number to the index of the second factor of the other number to obtain the index of the second factor (which is some power of 10) in the product. Thus:

$$(4.31 \times 10^{-2}) \times (9.0125 \times 10) =$$
$$(4.31 \times 9.0125) \times 10^{-2+1} = 38.844 \times 10^{-1}$$

$$(5.986 \times 10^4) \times (4.375 \times 10^3) =$$
$$(5.986 \times 4.375) \times 10^{4+3} = 26.189 \times 10^7$$

In the preceding calculations, neither of the results shown are in the conventional powers of ten form since the first factor in each has two digits. In the conventional powers of ten notation, the results would be

$$38.844 \times 10^{-1} = 3.884 \times 10^{0} = 3.884$$

since $10^0 = 1$, and

$$26.189 \times 10^{7} = 2.619 \times 10^{8}$$

in each case rounding off the first factor to three decimal places.

When multiplying several numbers written in this notation together, the procedure is the same. All of the first factors are multiplied together to get the first factor of the product and all of the indices of the respective powers of ten are added together, taking into account their respective signs, to get the index of the second factor of the product. Thus, $(4.02 \times 10^{-3}) \times (3.987 \times 10) \times (4.863 \times 10^{5}) = (4.02 \times 3.987 \times 4.863) \times (10^{-3+1+5}) = 77.94 \times 10^{3} = 7.79 \times 10^{4}$ rounding off the first factor to two decimal places.

Dividing Numbers Written in Powers of Ten Notation.—When dividing one number by another when both are written in this notation, the procedure is as follows:

1) Divide the first factor of the dividend by the first factor of the divisor to get the first factor of the quotient.

2) Subtract the index of the second factor of the divisor from the index of the second factor of the dividend, taking into account their respective signs, to get the index of the second factor of the quotient. Thus:

$$(4.31 \times 10^{-2}) \div (9.0125 \times 10) =$$

$$(4.31 \div 9.0125) \times (10^{-2-1}) = 0.4782 \times 10^{-3} = 4.782 \times 10^{-4}$$

It can be seen that this system of notation is helpful where several numbers of different magnitudes are to be multiplied and divided.

Example: Find the quotient of $\dfrac{250 \times 4698 \times 0.00039}{43678 \times 0.002 \times 0.0147}$

Solution: Changing all these numbers to powers of ten notation and performing the operations indicated:

$$\frac{(2.5 \times 10^{2}) \times (4.698 \times 10^{3}) \times (3.9 \times 10^{-4})}{(4.3678 \times 10^{4}) \times (2 \times 10^{-3}) \times (1.47 \times 10^{-2})} =$$

$$= \frac{(2.5 \times 4.698 \times 3.9)(10^{2+3-4})}{(4.3678 \times 2 \times 1.47)(10^{4-3-2})} = \frac{45.8055 \times 10}{12.8413 \times 10^{-1}}$$

$$= 3.5670 \times 10^{1-(-1)}$$

$$= 3.5670 \times 10^{2}$$

$$= 356.70$$

ALGEBRA AND EQUATIONS

Rearrangement and Transposition of Terms in Formulas

A formula is a rule for a calculation expressed by using letters and signs instead of writing out the rule in words; by this means, it is possible to condense, in a very small space, the essentials of long and cumbersome rules. The letters used in formulas simply stand in place of the figures that are to be substituted when solving a specific problem.

As an example, the formula for the horsepower transmitted by belting may be written

$$P = \frac{SVW}{33,000}$$

where P = horsepower transmitted; S = working stress of belt per inch of width in pounds; V = velocity of belt in feet per minute; and, W = width of belt in inches.

If the working stress S, the velocity V, and the width W are known, the horsepower can be found directly from this formula by inserting the given values. Assume $S = 33$; $V = 600$; and $W = 5$. Then

$$P = \frac{33 \times 600 \times 5}{33,000} = 3$$

Assume that the horsepower P, the stress S, and the velocity V are known, and that the width of belt, W, is to be found. The formula must then be rearranged so that the symbol W will be on one side of the equals sign and all the known quantities on the other. The rearranged formula is as follows:

$$\frac{P \times 33,000}{SV} = W$$

The quantities (S and V) that were in the numerator on the right side of the equals sign are moved to the denominator on the left side, and "33,000," which was in the denominator on the right side of the equals sign, is moved to the numerator on the other side. Symbols that are not part of a fraction, like "P" in the formula first given, are to be considered as being numerators (having the denominator 1).

Thus, any formula of the form $A = B/C$ can be rearranged as follows:

$$A \times C = B \quad \text{and} \quad C = \frac{B}{A}$$

Suppose a formula to be of the form

$$A = \frac{B \times C}{D}$$

Then

$$D = \frac{B \times C}{A} \qquad \frac{A \times D}{C} = B \qquad \frac{A \times D}{B} = C$$

The method given is only directly applicable when all the quantities in the numerator or denominator are standing independently or are *factors of a product*. If connected by + or − signs, the entire numerator or denominator must be moved as a unit, thus,

Given:

$$\frac{B + C}{A} = \frac{D + E}{F} \qquad \text{to solve for } F$$

then

$$\frac{F}{A} = \frac{D + E}{B + C}$$

and

$$F = \frac{A(D + E)}{B + C}$$

A quantity preceded by a + or – sign can be transposed to the opposite side of the equals sign by changing its sign; if the sign is +, change it to – on the other side; if it is –, change it to +. This process is called *tranposition* of terms.

Example: $B + C = A - D$ then
$$A = B + C + D$$
$$B = A - D - C$$
$$C = A - D - B$$

Sequence of Performing Arithmetic Operations

When several numbers or quantities in a formula are connected by signs indicating that additions, subtractions, multiplications, and divisions are to be made, the multiplications and divisions should be carried out first, in the sequence in which they appear, before the additions or subtractions are performed.

Example:
$$10 + 26 \times 7 - 2 = 10 + 182 - 2 = 190$$
$$18 \div 6 + 15 \times 3 = 3 + 45 = 48$$
$$12 + 14 \div 2 - 4 = 12 + 7 - 4 = 15$$

When it is required that certain additions and subtractions should precede multiplications and divisions, use is made of parentheses ( ) and brackets [ ]. These signs indicate that the calculation inside the parentheses or brackets should be carried out completely by itself before the remaining calculations are commenced. If one bracket is placed inside another, the one inside is first calculated.

Example:
$$(6 - 2) \times 5 + 8 = 4 \times 5 + 8 = 20 + 8 = 28$$
$$6 \times (4 + 7) \div 22 = 6 \times 11 \div 22 = 66 \div 22 = 3$$
$$2 + [10 \times 6(8 + 2) - 4] \times 2 = 2 + [10 \times 6 \times 10 - 4] \times 2$$
$$= 2 + [600 - 4] \times 2 = 2 + 596 \times 2 = 2 + 1192 = 1194$$

The parentheses are considered as a sign of multiplication; for example, $6(8 + 2) = 6 \times (8 + 2)$.

The line or bar between the numerator and denominator in a fractional expression is to be considered as a division sign. For example,

$$\frac{12 + 16 + 22}{10} = (12 + 16 + 22) \div 10 = 50 \div 10 = 5$$

In formulas, the multiplication sign (×) is often left out between symbols or letters, the values of which are to be multiplied. Thus,

$$AB = A \times B \quad \text{and} \quad \frac{ABC}{D} = (A \times B \times C) \div D$$

Ratio and Proportion

The *ratio* between two quantities is the quotient obtained by dividing the first quantity by the second. For example, the ratio between 3 and 12 is ¼, and the ratio between 12 and 3 is 4. Ratio is generally indicated by the sign (:); thus, 12 : 3 indicates the ratio of 12 to 3.

A *reciprocal*, or *inverse* ratio, is the opposite of the original ratio. Thus, the inverse ratio of 5 : 7 is 7 : 5.

In a *compound* ratio, each term is the product of the corresponding terms in two or more simple ratios. Thus, when

$$8:2 = 4 \qquad 9:3 = 3 \qquad 10:5 = 2$$

then the compound ratio is

$$8 \times 9 \times 10 : 2 \times 3 \times 5 = 4 \times 3 \times 2$$
$$720:30 = 24$$

Proportion is the equality of ratios. Thus,

$$6:3 = 10:5 \qquad \text{or} \qquad 6:3::10:5$$

The first and last terms in a proportion are called the *extremes;* the second and third, the *means*. The product of the extremes is equal to the product of the means. Thus,

$$25:2 = 100:8 \qquad \text{and} \qquad 25 \times 8 = 2 \times 100$$

If three terms in a proportion are known, the remaining term may be found by the following rules:

The first term is equal to the product of the second and third terms, divided by the fourth.

The second term is equal to the product of the first and fourth terms, divided by the third.

The third term is equal to the product of the first and fourth terms, divided by the second.

The fourth term is equal to the product of the second and third terms, divided by the first.

Example: Let x be the term to be found, then,

$$x : 12 = 3.5 : 21 \qquad x = \frac{12 \times 3.5}{21} = \frac{42}{21} = 2$$

$$\tfrac{1}{4} : x = 14 : 42 \qquad x = \frac{\tfrac{1}{4} \times 42}{14} = \frac{1}{4} \times 3 = \frac{3}{4}$$

$$5 : 9 = x : 63 \qquad x = \frac{5 \times 63}{9} = \frac{315}{9} = 35$$

$$\tfrac{1}{4} : \tfrac{7}{8} = 4 : x \qquad x = \frac{\tfrac{7}{8} \times 4}{\tfrac{1}{4}} = \frac{3\tfrac{1}{2}}{\tfrac{1}{4}} = 14$$

If the second and third terms are the same, that number is the *mean proportional* between the other two. Thus, 8 : 4 = 4 : 2, and 4 is the mean proportional between 8 and 2. The mean proportional between two numbers may be found by multiplying the numbers together and extracting the square root of the product. Thus, the mean proportional between 3 and 12 is found as follows:

$$3 \times 12 = 36 \qquad \text{and} \qquad \sqrt{36} = 6$$

which is the mean proportional.

Practical Examples Involving Simple Proportion.—If it takes 18 days to assemble 4 lathes, how long would it take to assemble 14 lathes?

Let the number of days to be found be x. Then write out the proportion as follows:

$$4:18 = 14:x$$

(lathes : days = lathes : days)

Now find the fourth term by the rule given:

$$x = \frac{18 \times 14}{4} = 63 \text{ days}$$

Thirty-four linear feet of bar stock are required for the blanks for 100 clamping bolts. How many feet of stock would be required for 912 bolts?

Let x = total length of stock required for 912 bolts.

$$34:100 = x:912$$

(feet : bolts = feet : bolts)

Then, the third term $x = (34 \times 912)/100 = 310$ feet, approximately.

Inverse Proportion.—In an inverse proportion, as one of the items involved *increases*, the corresponding item in the proportion *decreases*, or vice versa. For example, a factory employing 270 men completes a given number of typewriters weekly, the number of working hours being 44 per week. How many men would be required for the same production if the working hours were reduced to 40 per week?

The time per week is in an inverse proportion to the number of men employed; the shorter the time, the more men. The inverse proportion is written:

$$270 : x = 40 : 44$$

(men, 44-hour basis: men, 40-hour basis = time, 40-hour basis: time, 44-hour basis) Thus

$$\frac{270}{x} = \frac{40}{44} \quad \text{and} \quad x = \frac{270 \times 44}{40} = 297 \text{ men}$$

Problems Involving Both Simple and Inverse Proportions.—If two groups of data are related both by direct (simple) and inverse proportions among the various quantities, then a simple mathematical relation that may be used in solving problems is as follows:

$$\frac{\text{Product of all directly proportional items in first group}}{\text{Product of all inversely proportional items in first group}}$$
$$= \frac{\text{Product of all directly proportional items in second group}}{\text{Product of all inversely proportional items in second group}}$$

Example: If a man capable of turning 65 studs in a day of 10 hours is paid $6.50 per hour, how much per hour ought a man be paid who turns 72 studs in a 9-hour day, if compensated in the same proportion?

The first group of data in this problem consists of the number of hours worked by the first man, his hourly wage, and the number of studs which he produces per day; the second group contains similar data for the second man except for his unknown hourly wage, which may be indicated by x.

The labor cost per stud, as may be seen, is directly proportional to the number of hours worked and the hourly wage. These quantities, therefore, are used in the numerators of the fractions in the formula. The labor cost per stud is inversely proportional to the number of studs produced per day. (The greater the number of studs produced in a given time the less the cost per stud.) The numbers of studs per day, therefore, are placed in the denominators of the fractions in the formula. Thus,

$$\frac{10 \times 6.50}{65} = \frac{9 \times x}{72}$$

$$x = \frac{10 \times 6.50 \times 72}{65 \times 9} = \$8.00 \text{ per hour}$$

Percentage

If out of 100 pieces made, 12 do not pass inspection, it is said that 12 per cent (12 of the hundred) are rejected. If a quantity of steel is bought for $100 and sold for $140, the profit is 28.6 per cent of the selling price.

The per cent of gain or loss is found by dividing the amount of gain or loss by the *original* number of which the percentage is wanted, and multiplying the quotient by 100.

Example:—Out of a total output of 280 castings a day, 30 castings are, on an average, rejected. What is the percentage of bad castings?

$$\frac{30}{280} \times 100 = 10.7 \text{ per cent}$$

If by a new process 100 pieces can be made in the same time as 60 could formerly be made, what is the gain in output of the new process over the old, expressed in per cent?

Original number, 60; gain 100 – 60 = 40. Hence,

$$\frac{40}{60} \times 100 = 66.7 \text{ per cent}$$

Care should be taken always to use the original number, or the number of which the percentage is wanted, as the divisor in all percentage calculations. In the example just given, it is the percentage of gain over the old output 60 that is wanted and not the percentage with relation to the new output too. Mistakes are often made by overlooking this important point.

Interest

Interest is money paid for the use of money lent for a certain time. *Simple* interest is the interest paid on the principal (money lent) only. When simple interest that is due is not paid, and its amount is added to the interest-bearing principal, the interest calculated on this new principal is called *compound* interest. The compounding of the interest into the principal may take place yearly or more often, according to circumstances.

Interest Formulas.—The symbols used in the formulas to calculate various types of interest are:

P = principal or amount of money lent

I = nominal annual interest rate stated as a percentage, i.e., 10 per cent per annum

I_e = effective annual interest rate when interest is compounded more often than once a year (see *Nominal vs. Effective Interest Rates*)

i = nominal annual interest rate per cent expressed as a decimal, i.e., if I = 12 per cent, then i = 12/100 = 0.12

n = number of annual interest periods

m = number of interest compounding periods in one year

S = a sum of money at the end of n interest periods from the present date that is equivalent to P with added interest i

R = the payment at the end of each period in a uniform series of payments continuing for n periods, the entire series equivalent to P at interest rate i

Note: The exact amount of interest for one day is 1/365 of the interest for one year. Banks, however, customarily take the year as composed of 12 months of 30 days, making a total of 360 days to a year. This method is also used for home-mortgage-type payments, so that the interest rate per month is 30/360 = 1/12 of the annual interest rate. For example, if I is a 12 per cent per annum nominal interest rate, then for a 30-day period, the interest rate is (12 × 1/12) = 1.0 per cent per month. The decimal rate per month is then 1.0/100 = 0.01.

Simple Interest.—The formulas for simple interest are:

$$\text{Interest for } n \text{ years} = Pin$$

$$\text{Total amount after } n \text{ years, } S = P + Pin$$

Example: For \$250 that has been lent for three years at 6 per cent simple interest: P = 250; I = 6; i = I/100 = 0.06; n = 3.

$$S = 250 + (250 \times 0.06 \times 3) = 250 + 45 = \$295$$

Compound Interest.—The following formulas apply when compound interest is to be computed and assuming that the interest is compounded annually.

$$S = P(1+i)^n$$
$$P = S/(1+i)^n$$
$$i = (S/P)^{1/n} - 1$$
$$n = (\log S - \log P)/\log(1+i)$$

Example: At 10 per cent interest compounded annually for 10 years, a principal amount P of $1000 becomes a sum S of

$$S = 1000(1 + 10/100)^{10} = \$2{,}593.74$$

If a sum $S = \$2593.74$ is to be accumulated, beginning with a principal $P = \$1{,}000$ over a period $n = 10$ years, the interest rate i to accomplish this would have to be $i = (2593.74/1000)^{1/10} - 1 = 0.09999$, which rounds to 0.1, or 10 per cent.

For a principal $P = \$500$ to become $S = \$1{,}000$ at 6 per cent interest compounded annually, the number of years n would have to be

$$n = (\log 1000 - \log 500)/\log(1 + 0.06)$$
$$= (3 - 2.69897)/0.025306 = 11.9 \text{ years}$$

To triple the principal $P = \$500$ to become $S = \$1{,}500$, the number of years would have to be

$$n = (\log 1500 - \log 500)/\log(1 + 0.06)$$
$$= (3.17609 - 2.69897)/0.025306 = 18.85 \text{ years}$$

Interest Compounded More Often Than Annually.—If interest is payable m times a year, it will be computed m times during each year, or nm times during n years. The rate for each compounding period will be i/m if i is the nominal annual decimal interest rate. Therefore, at the end of n years, the amount S will be: $S = P(1 + i/m)^{nm}$.

As an example, if $P = \$1{,}000$; n is 5 years, the interest payable quarterly, and the annual rate is 6 per cent, then $n = 5$; $m = 4$; $i = 0.06$; $i/m = 0.06/4 = 0.015$; and $nm = 5 \times 4 = 20$, so that

$$S = 1000(1 + 0.015)^{20} = \$1{,}346.86$$

Nominal vs. Effective Interest Rates.—Deposits in savings banks, automobile loans, interest on bonds, and many other transactions of this type involve computation of interest due and payable more often than once a year. For such instances, there is a difference between the *nominal* annual interest rate stated to be the cost of borrowed money and the *effective* rate that is actually charged.

For example, a loan with interest charged at 1 per cent per month is described as having an interest rate of 12 per cent per annum. To be precise, this rate should be stated as being a *nominal* 12 per cent per annum compounded monthly; the actual or *effective* rate for monthly payments is 12.7 per cent. For quarterly compounding, the effective rate would be 12.6 per cent:

$$I_e = (1 + I/m)^m - 1$$

In this formula, I_e is the effective annual rate, I is the nominal annual rate, and m is the number of times per year the money is compounded.

Example: For a nominal per annum rate of 12 per cent, with monthly compounding, the effective per annum rate is

$$I_e = (1 + 0.12/12)^{12} - 1 = 0.1268 = 12.7 \text{ per cent effective per annum rate}$$

Example: Same as before but with quarterly compounding:

$$I_e = (1 + 0.12/4)^4 - 1 = 0.1255 = 12.6 \text{ per cent effective per annum rate}$$

Example: Same as before but with annual compounding.

$$I_e = (1 + 0.12/1)^1 - 1 = 0.12 = 12 \text{ per cent effective per annum rate}$$

This last example shows that for once-a-year-compounding, the nominal and effective per annum rates are identical.

Finding Unknown Interest Rates.—If a single payment of P dollars is to produce a sum of S dollars after n annual compounding periods, the per annum decimal interest rate is found using:

$$i = \sqrt[n]{\frac{S}{P}} - 1$$

Present Value and Discount.—The present value or present worth P of a given amount S is the amount P that, when placed at interest i for a given time n, will produce the given amount S.

$$\text{At simple interest, } P = S/(1 + ni)$$

$$\text{At compound interest, } P = S/(1 + i)^n$$

The *true discount* D is the difference between S and P: $D = S - P$.

These formulas are for an annual interest rate. If interest is payable other than annually, modify the formulas as indicated in the formulas in the section *Interest Compounded More Often Than Annually.*

Example: Required the present value and discount of $500 due in six months at 6 per cent simple interest. Here, $S = 500$; $n = 6/12 = 0.5$ year; $i = 0.06$. Then, $P = 500/(1 + 0.5 \times 0.06)$ = $485.44.

Required the sum that, placed at 5 per cent compound interest, will in three years produce $5,000. Here, $S = 5000$; $i = 0.05$; $n = 3$. Then,

$$P = 5000/(1 + 0.05)^3 = \$4{,}319.19$$

Annuities.—An annuity is a fixed sum paid at regular intervals. In the formulas that follow, yearly payments are assumed. It is customary to calculate annuities on the basis of compound interest. If an annuity A is to be paid out for n consecutive years, the interest rate being i, then the present value P of the annuity is

$$P = A\frac{(1 + i)^n - 1}{i(1 + i)^n}$$

Example: If an annuity of $200 is to be paid for 10 years, what is the present amount of money that needs to be deposited if the interest is 5 per cent. Here, $A = 200$; $i = 0.05$; $n = 10$:

$$P = 200\frac{(1 + 0.05)^{10} - 1}{0.05(1 + 0.05)^{10}} = \$1{,}544.35$$

The annuity a principal P drawing interest at the rate i will give for a period of n years is

$$A = P\frac{i(1 + i)^n}{(1 + i)^n - 1}$$

Example: A sum of $10,000 is placed at 4 per cent. What is the amount of the annuity payable for 20 years out of this sum: Here, $P = 10000$; $i = 0.04$; $n = 20$:

$$A = 10{,}000\frac{0.04(1 + 0.04)^{20}}{(1 + 0.04)^{20} - 1} = \$735.82$$

If at the *beginning* of each year a sum A is set aside at an interest rate i, the total value S of the sum set aside, with interest, at the end of n years, will be

$$S = A\frac{(1+i)[(1+i)^n - 1]}{i}$$

If at the *end* of each year a sum A is set aside at an interest rate i, then the total value S of the principal, with interest, at the end of n years will be

$$S = A\frac{(1+i)^n - 1}{i}$$

If a principal P is increased or decreased by a sum A at the end of each year, then the value of the principal after n years will be

$$S = P(1+i)^n \pm A\frac{(1+i)^n - 1}{i}$$

If the sum A by which the principal P is decreased each year is greater than the total yearly interest on the principal, then the principal, with the accumulated interest, will be entirely used up in n years:

$$n = \frac{\log A - \log(A - iP)}{\log(1+i)}$$

Sinking Funds.—Amortization is "the extinction of debt, usually by means of a sinking fund." The sinking fund is created by a fixed investment R placed each year at compound interest for a term of years n, and is therefore an annuity of sufficient size to produce at the end of the term of years the amount S necessary for the repayment of the principal of the debt, or to provide a definite sum for other purposes. Then,

$$S = R\frac{(1+i)^n - 1}{i} \qquad \text{and} \qquad R = S\frac{i}{(1+i)^n - 1}$$

Example: If \$2,000 is invested annually for 10 years at 4 per cent compound interest, as a sinking fund, what would be the total amount of the fund at the expiration of the term? Here, $R = 2000$; $n = 10$; $i = 0.04$:

$$S = 2000\frac{(1 + 0.04)^{10} - 1}{0.04} = \$24{,}012.21$$

Evaluating Investments in Industrial Assets

Investment in industrial assets such as machine tools, processing equipment, and other means of production may not be attractive unless the cost of such investment can be recovered with interest. The interest, or *rate of return*, should be equal to, or greater than, some specified minimum rate for each of such investments. Three methods used in analyzing prospective investments are

1) *Annual cost* of the investment at a specified minimum acceptable rate of return used as the interest rate.

2) *Present worth*, using as an interest rate a specified minimum acceptable rate of return.

3) *Prospective rate of return* compared to a specified minimum acceptable rate.

Annual Cost Method.—In the annual cost method, comparisons are made among alternative investment plans. If the annual costs in any investment plan form a non-uniform series of disbursements from year to year, a much used method for reducing *all* comparisons to an equivalent basis is as follows: Take each of the annual disbursements and use the *present worth* method developed in the next section to bring all annual costs down to a common present worth date, usually called "year 0." When this has been done, each present worth is converted to an equivalent uniform annual series of disbursements using the applicable formulas from Table 1.

Table 1. Summary of Useful Interest Formulas

The meanings of the symbols P, R, S, L, i, n, and F are as follows:

P = Principal sum of money at the present time. Also the present worth of a future payment in a series of equal payments

R = Single payment in a series of n equal payments made at the end of each interest period

S = A sum, after n interest period, equal to the compound amount of a principal sum, P, or the sum of the compound amounts of the payments, R, at interest rate i

i = Nominal annual interest rate expressed as a decimal

n = Number of interest periods, usually annual

L = Salvage value of an asset at the end of its projected useful life

$F = (1 + i)^n$ for which values are tabulated in Table 2

To Find	Given	Formula	
Simple interest, Sum	P, find S	$S = P(1 + ni)$	(1)
Single payment, Compound-amount	P, find S	$S = PF$	(2)
Single payment, Present-worth	S, find P	$P = \frac{S}{F}$	(3)
Equal-payment series, Compound-amount	R, find S	$S = R\frac{(F-1)}{i}$	(4)
Equal-payment series, Sinking-fund	S, find R	$R = S\frac{i}{(F-1)}$	(5)
Equal-payment series, Present-worth	R, find P	$P = R\frac{(F-1)}{iF}$	(6)
Equal-payment series, Capital-recovery	P, find R	$R = P\frac{iF}{(F-1)}$	(7)
Equal-payment series with salvage value, Capital-recovery	P and L, find R	$R = (P-L)\frac{iF}{(F-1)} + Li$	(8)

Example 1 (Annual Cost Calculations): An investment of $15,000 is being considered to reduce labor and labor-associated costs in a materials handling operation from $8,200 a year to $3,300. This operation is expected to be used for 10 years before being changed or discontinued entirely. In addition to the initial investment of $15,000 and the annual cost of $3,300 for labor, there are additional annual costs for power, maintenance, insurance, and property taxes of $1,800 associated with the revised operation. Based on comparisons of annual costs, should the $15,000 investment be made or the present operation continued?

The present annual cost of the operation is $8,200 for labor and labor-associated costs. The proposed operation has an annual cost of $3,300 for labor and labor extras plus $1,800 for additional power, maintenance, insurance, and taxes, plus the annual cost of recovering the initial investment of $15,000 at some interest rate (minimum acceptable rate of return).

Assuming that 10 per cent would be an acceptable rate of return on this investment over a period of 10 years, the annual amount to be recovered on the initial investment would be $15,000 multiplied by the capital recovery factor calculated using Formula (7) in Table 1. From Table 2, the factor F for 10 per cent and 10 years is seen to be 2.594.

Putting this value into Formula (7) gives:

$$R = P\frac{iF}{F-1} = 15{,}000\frac{0.1 \times 2.594}{2.594 - 1} = \$2{,}442$$

Adding this amount to the $5,100 annual cost associated with the investment ($3,300+ $1,800 = $5,100) gives a total annual cost of $7,542, which is less than the present annual cost of $8,200. Thus, the investment is justified unless there are other considerations such as the effects of income taxes, salvage values, expected life, uncertainty about the required rate of return, changes in the cost of borrowed funds, and others.

A tabulation of annual costs of alternative plans A, B, C, etc., is a good way to compare costs item by item. For Example 1:

	Item	Plan A	Plan B
1	Labor and labor extras	$8,200.00	$3,300.00
2	Annual cost of $15,000 investment using Formula (7), Table 2		2,442.00
3	Power		400.00
4	Maintenance		1,100.00
5	Property taxes and insurance		300.00
	Total annual cost	$8,200.00	$7,542.00

Example 2 (Annual Cost Considering Salvage Value): If in Example 1 the salvage value of the equipment installed was $5,000 at the end of 10 years, what effect does this have on the annual cost of the proposed investment of $15,000?

The only item in the annual cost of Example 1 that will be affected is the capital recovery amount of $2,442. The following formula gives the amount of annual capital recovery when salvage value is considered:

$$R = (P-L)\frac{iF}{(F-1)} + Li = (15{,}000 - 5{,}000)\frac{0.1 \times 2.594}{2.594 - 1} + 5{,}000 \times 0.1 = \$2{,}127$$

Adding this amount to the $5,100 annual cost determined previously gives a total annual cost of $7,227, which is $315 less than the previous annual cost of $7,542 for the proposed investment.

Present Worth Method.—A present worth calculation may be described as the discounting of a future payment, or a series of payments, to a cash value on the present date based on a selected interest rate. The present date is referred to as "day 0," or "year 0." Initial costs are already at zero date (present worth date), so no factors need be applied to initial, or "up-front," costs. On the other hand, if salvage values are to be considered, these must be reduced to present worth and *subtracted* from the present worth of the initial investment.

The present worths of each of the alternative investments are then compared to find the lowest cost alternative. The present worth of the lowest cost alternative may then be converted to a uniform series of annual costs and these annual costs compared with an existing, in-place, annual cost. Present worth calculations are often referred to as *discounted cash flow* because this term describes both the data required and the method of calculation. *Cash flow* refers to the requirement that data must be supplied in the form of amounts and dates of receipts and payouts, and *discounted* refers to the calculation of the present worth of each of one or more future payments. The rate of return used in such calculations should be the minimum attractive interest rate before taxes.

Example 3 (Present Worth Calculation): Present worths are calculated as of the zero date of the payments being compared. Up-front (zero-day) disbursements are already at their present worth and no interest factors should be applied to them; the present worths of salvage values are *subtracted* to get the present worth of the net disbursements because the present worth of a salvage value, in effect, reduces the amount of required initial disbursements.

A) Find the present value of a salvage value of $1000 from the sale of equipment after 10 years if the expected rate of return (interest rate) is 10 per cent? Using Formula (3) from Table 1 and the value of $F = 2.594$ for $n = 10$ years and $i = 0.1$ from Table 2, the present worth $P = 1000/2.594 = \$385.51$.

B) In Example 1, the annual cost of an investment of $15,000 at 10 per cent over 10 years was $2,442. Convert this annual outlay back to its present worth.

Using Formula (6) and the value 2.594 from Table 2, $P = 2{,}442 \times (2.594 - 1)/(0.1 \times 2.594) = \$15{,}005$, which rounds to \$15,000.

Table 2. Values of $F = (1 + i)^n$ for Selected Rates of Interest, i, and Number of Annual Interest Periods, n

Number of Years, n	Annual Interest Rate, i, Expressed as a Decimal											
	0.050	0.060	0.070	0.080	0.090	0.100	0.110	0.120	0.130	0.140	0.150	0.160
	Factor $F = (1 + i)^n$											
1	1.050	1.060	1.070	1.080	1.090	1.100	1.110	1.120	1.130	1.140	1.150	1.160
2	1.103	1.124	1.145	1.166	1.188	1.210	1.232	1.254	1.277	1.300	1.323	1.346
3	1.158	1.191	1.225	1.260	1.295	1.331	1.368	1.405	1.443	1.482	1.521	1.561
4	1.216	1.262	1.311	1.360	1.412	1.464	1.518	1.574	1.630	1.689	1.749	1.811
5	1.276	1.338	1.403	1.469	1.539	1.611	1.685	1.762	1.842	1.925	2.011	2.100
6	1.340	1.419	1.501	1.587	1.677	1.772	1.870	1.974	2.082	2.195	2.313	2.436
7	1.407	1.504	1.606	1.714	1.828	1.949	2.076	2.211	2.353	2.502	2.660	2.826
8	1.477	1.594	1.718	1.851	1.993	2.144	2.305	2.476	2.658	2.853	3.059	3.278
9	1.551	1.689	1.838	1.999	2.172	2.358	2.558	2.773	3.004	3.252	3.518	3.803
10	1.629	1.791	1.967	2.159	2.367	2.594	2.839	3.106	3.395	3.707	4.046	4.411
11	1.710	1.898	2.105	2.332	2.580	2.853	3.152	3.479	3.836	4.226	4.652	5.117
12	1.796	2.012	2.252	2.518	2.813	3.138	3.498	3.896	4.335	4.818	5.350	5.936
13	1.886	2.133	2.410	2.720	3.066	3.452	3.883	4.363	4.898	5.492	6.153	6.886
14	1.980	2.261	2.579	2.937	3.342	3.797	4.310	4.887	5.535	6.261	7.076	7.988
15	2.079	2.397	2.759	3.172	3.642	4.177	4.785	5.474	6.254	7.138	8.137	9.266
16	2.183	2.540	2.952	3.426	3.970	4.595	5.311	6.130	7.067	8.137	9.358	10.748
17	2.292	2.693	3.159	3.700	4.328	5.054	5.895	6.866	7.986	9.276	10.761	12.468
18	2.407	2.854	3.380	3.996	4.717	5.560	6.544	7.690	9.024	10.575	12.375	14.463
19	2.527	3.026	3.617	4.316	5.142	6.116	7.263	8.613	10.197	12.056	14.232	16.777
20	2.653	3.207	3.870	4.661	5.604	6.727	8.062	9.646	11.523	13.743	16.367	19.461
21	2.786	3.400	4.141	5.034	6.109	7.400	8.949	10.804	13.021	15.668	18.822	22.574
22	2.925	3.604	4.430	5.437	6.659	8.140	9.934	12.100	14.714	17.861	21.645	26.186
23	3.072	3.820	4.741	5.871	7.258	8.954	11.026	13.552	16.627	20.362	24.891	30.376
24	3.225	4.049	5.072	6.341	7.911	9.850	12.239	15.179	18.788	23.212	28.625	35.236
25	3.386	4.292	5.427	6.848	8.623	10.835	13.585	17.000	21.231	26.462	32.919	40.874
26	3.556	4.549	5.807	7.396	9.399	11.918	15.080	19.040	23.991	30.167	37.857	47.414
27	3.733	4.822	6.214	7.988	10.245	13.110	16.739	21.325	27.109	34.390	43.535	55.000
28	3.920	5.112	6.649	8.267	11.167	14.421	18.580	23.884	30.633	39.204	50.066	63.800
29	4.116	5.418	7.114	9.317	12.172	15.863	20.624	26.750	34.616	44.693	57.575	74.009
30	4.322	5.743	7.612	10.063	13.268	17.449	22.892	29.960	39.116	50.950	66.212	85.850

Prospective Rate of Return Method (Discounted Cash Flow).—This method of calculating the prospective return on an investment has variously been called the *discounted cash flow method, the Investor's Method*, the *Profitability Index*, and the *interest rate of return*, but *discounted cash flow* is the most common terminology in industry.

The term "discounted cash flow" is most descriptive of the process because "cash flow" describes the amounts and dates of the receipts and disbursements and "discounted" refers to the calculation of present worth. Calculating the present worth of future payments is often described as discounting the payments to the present "zero" date.

The process is best illustrated by an example. The data in Table 3 are a tabulation of the cash flows by amount and year for two different plans of investment, the fourth column showing the differences in cash flows for each year and the differences in the totals. This type of information could equally well represent a comparison involving a choice between two different machine tools, investments in rental properties, or any other situation where the difference between two or more alternative investments are to be evaluated.

The differences in cash flows in the fourth column of Table 3 consist of a disbursement of \$15,000 at date zero and receipts of \$3,000 per year for 10 years. The rate of return *on the net cash flow* can be calculated from these data using the principal that *the rate of return is that interest rate at which the present worth of the net cash flow is zero.*

In this example, the present worth of the *net* cash flow will be 0 if the present worth of the \$15,000 disbursements is numerically equal to the present worth of the uniform annual series of receipts of \$3,000. For the disbursement of \$15,000, the present worth is already \$15,000, because it was made on date 0. The uniform annual series of receipts of \$3,000 a year for 10 years must be converted to present worth using Formula (6) from Table 1. The interest rate needed to calculate the factor F in Formula (6) to get the necessary present worth conversion factor is not known, so that a series of trial-and-error substitutions of assumed interest rates must be made to find the correct present value factor. As a first guess, assume an interest rate of 15 per cent. Then, from Table 2, for 15 per cent and 10 years, $F = 4.046$, and substituting in Formula (6) of Table 1,

$$P = 3000\frac{4.046 - 1}{0.15 \times 4.046} = 15{,}056 \textit{ nearly}$$

so that the present worth of the net cash flow at 15 per cent interest (–\$15,000 + \$15,056 = \$56) is slightly more than 0. If a 16 per cent interest rate is tried, P is calculated as \$14,499, which is too small by \$501 and is about 10 times larger than the previous difference of \$56. By interpolation, the interest rate should be approximately 15.1 per cent (0.1 per cent above 15 per cent, or 0.9 per cent less than 16 per cent). Then $F = (1 + 0.151)^{10} = 4.0809$, and $P = 3{,}000(F - 1)/(0.151 \times 4.0809) = 3{,}000(4.0809 - 1)/(0.151 \times 4.0809) = 14{,}999$, *nearly*. Thus, the present worth of the net cash flow at 15.1 per cent (–\$15,000 + \$14,999 = –\$1) is only slightly less than 0, and the prospective rate of return may be taken as 15.1 per cent.

Table 3. Comparison of Cash Flows for Two Competing Plans

Year	Annual Costs Plan A	Annual Costs Plan B	Net Cash Flow B — A
0		-\$15,000	-\$15,000
1	-\$8,000	-5,000	+3,000
2	-8,000	-5,000	+3,000
3	-8,000	-5,000	+3,000
4	-8,000	-5,000	+3,000
5	-8,000	-5,000	+3,000
6	-8,000	-5,000	+3,000
7	-8,000	-5,000	+3,000
8	-8,000	-5,000	+3,000
9	-8,000	-5,000	+3,000
10	-8,000	-5,000	+3,000
Totals	-\$80,000	-\$65,000	+\$15,000

Principal Algebraic Expressions and Formulas

$$a \times a = aa = a^2$$

$$a \times a \times a = aaa = a^3$$

$$a \times b = ab$$

$$a^2b^2 = (ab)^2$$

$$a^2a^3 = a^{2+3} = a^5$$

$$a^4 \div a^3 = a^{4-3} = a$$

$$a^0 = 1$$

$$a^2 - b^2 = (a+b)(a-b)$$

$$(a+b)^2 = a^2 + 2ab + b^2$$

$$(a-b)^2 = a^2 - 2ab + b^2$$

$$\frac{a^3}{b^3} = \left(\frac{a}{b}\right)^3$$

$$\frac{1}{a^3} = \left(\frac{1}{a}\right)^3 = a^{-3}$$

$$(a^2)^3 = a^{2\times3} = (a^3)^2 = a^6$$

$$a^3 + b^3 = (a+b)(a^2 - ab + b^2)$$

$$a^3 - b^3 = (a-b)(a^2 + ab + b^2)$$

$$(a+b)^3 = a^3 + 3a^2b + 3ab^2 + b^3$$

$$(a-b)^3 = a^3 - 3a^2b + 3ab^2 - b^3$$

$$\sqrt{a} \times \sqrt{a} = a$$

$$\sqrt[3]{ab} = \sqrt[3]{a} \times \sqrt[3]{b}$$

$$\sqrt[3]{a} \times \sqrt[3]{a} \times \sqrt[3]{a} = a$$

$$\sqrt[3]{\frac{a}{b}} = \frac{\sqrt[3]{a}}{\sqrt[3]{b}}$$

$$(\sqrt[3]{a})^3 = a$$

$$\sqrt[3]{a^2} = (\sqrt[3]{a})^2 = a^{2/3}$$

$$\sqrt[3]{\frac{1}{a}} = \frac{1}{\sqrt[3]{a}} = a^{-1/3}$$

$$\sqrt[4]{\sqrt[3]{a}} = \sqrt[4\times3]{a} = \sqrt[3]{\sqrt[4]{a}}$$

$$\sqrt{a} + \sqrt{b} = \sqrt{a + b + 2\sqrt{ab}}$$

When		then	
	$a \times b = x$	then	$\log a + \log b = \log x$
	$a \div b = x$	then	$\log a - \log b = \log x$
	$a^3 = x$	then	$3\log a = \log x$
	$\sqrt[3]{a} = x$	then	$\frac{\log a}{3} = \log x$

Equations

An equation is a statement of equality between two expressions, as $5x = 105$. The unknown quantity in an equation is generally designated by the letter x. If there is more than one unknown quantity, the others are designated by letters also selected at the end of the alphabet, as y, z, u, t, etc.

An equation of the first degree is one which contains the unknown quantity only in the first power, as $3x = 9$. A quadratic equation is one which contains the unknown quantity in the second, but no higher, power, as $x^2 + 3x = 10$.

Solving Equations of the First Degree with One Unknown.—Transpose all the terms containing the unknown x to one side of the equals sign, and all the other terms to the other side. Combine and simplify the expressions as far as possible, and divide both sides by the coefficient of the unknown x. (See the rules given for transposition of formulas.)

Example:

$$22x - 11 = 15x + 10$$

$$22x - 15x = 10 + 11$$

$$7x = 21$$

$$x = 3$$

Solution of Equations of the First Degree with Two Unknowns.—The form of the simplified equations is

$$ax + by = c$$
$$a_1x + b_1y = c_1$$

Then,

$$x = \frac{cb_1 - c_1b}{ab_1 - a_1b} \qquad y = \frac{ac_1 - a_1c}{ab_1 - a_1b}$$

Example:

$$3x + 4y = 17$$
$$5x - 2y = 11$$

$$x = \frac{17 \times (-2) - 11 \times 4}{3 \times (-2) - 5 \times 4} = \frac{-34 - 44}{-6 - 20} = \frac{-78}{-26} = 3$$

The value of y can now be most easily found by inserting the value of x in one of the equations:

$$5 \times 3 - 2y = 11 \qquad 2y = 15 - 11 = 4 \qquad y = 2$$

Solution of Quadratic Equations with One Unknown.—If the form of the equation is $ax^2 + bx + c = 0$, then

$$x = \frac{-b \pm \sqrt{b^2 - 4ac}}{2a}$$

Example: Given the equation, $1x^2 + 6x + 5 = 0$, then $a = 1$, $b = 6$, and $c = 5$.

$$x = \frac{-6 \pm \sqrt{6^2 - 4 \times 1 \times 5}}{2 \times 1} = \frac{(-6) + 4}{2} = -1 \quad \text{or} \quad \frac{(-6) - 4}{2} = -5$$

If the form of the equation is $ax^2 + bx = c$, then

$$x = \frac{-b \pm \sqrt{b^2 + 4ac}}{2a}$$

Example: A right-angle triangle has a hypotenuse 5 inches long and one side which is one inch longer than the other; find the lengths of the two sides.

Let x = one side and $x + 1$ = other side; then $x^2 + (x + 1)^2 = 5^2$ or $x^2 + x^2 + 2x + 1 = 25$; or $2x^2 + 2x = 24$; or $x^2 + x = 12$. Now referring to the basic formula, $ax^2 + bx = c$, we find that $a = 1$, $b = 1$, and $c = 12$; hence,

$$x = \frac{-1 \pm \sqrt{1 + 4 \times 1 \times 12}}{2 \times 1} = \frac{(-1) + 7}{2} = 3 \quad \text{or} \quad x = \frac{(-1) - 7}{2} = -4$$

Since the positive value (3) would apply in this case, the lengths of the two sides are $x = 3$ inches and $x + 1 = 4$ inches.

Cubic Equations.—If the given equation has the form: $x^3 + ax + b = 0$ then

$$x = \left(-\frac{b}{2} + \sqrt{\frac{a^3}{27} + \frac{b^2}{4}}\right)^{1/3} + \left(-\frac{b}{2} - \sqrt{\frac{a^3}{27} + \frac{b^2}{4}}\right)^{1/3}$$

The equation $x^3 + px^2 + qx + r = 0$, may be reduced to the form $x_1{}^3 + ax_1 + b = 0$ by substituting $x_1 - \frac{p}{3}$ for x in the given equation.

Series.—Some hand calculations, as well as computer programs of certain types of mathematical problems, may be facilitated by the use of an appropriate series. For example, in some gear problems, the angle corresponding to a given or calculated involute function is found by using a series together with an iterative procedure such as the Newton-Raphson method described on page 35. The following are those series most commonly used for such purposes. In the series for trigonometric functions, the angles x are in radians (1 radian = 180/π degrees). The expression $\exp(-x^2)$ means that the base e of the natural logarithm system is raised to the $-x^2$ power; $e = 2.7182818$.

(1) $\sin x = x - x^3/3! + x^5/5! - x^7/7! + \cdots$ for all values of x.

(2) $\cos x = 1 - x^2/2! + x^4/4! - x^6/6! + \cdots$ for all values of x.

(3) $\tan x = x + x^3/3 + 2x^5/15 + 17x^7/315 + 62x^9/2835 + \cdots$ for $|x| < \pi/2$.

(4) $\arcsin x = x + x^3/6 + 1 \cdot 3 \cdot x^5/(2 \cdot 4 \cdot 5)$ $+ 1 \cdot 3 \cdot 5 \cdot x^7/(2 \cdot 4 \cdot 6 \cdot 7) + \cdots$ for $|x| \le 1$.

(5) $\arccos x = \pi/2 - \arcsin x$

(6) $\arctan x = x - x^3/3 + x^5/5 - x^7/7 + \cdots$ for $|x| \le 1$.

(7) $e^x = 1 + x + x^2/2! + x^3/3! + \cdots$ for all values of x.

(8) $\exp(-x^2) = 1 - x^2 + x^4/2! - x^6/3! + \cdots$ for all values of x.

(9) $a^x = 1 + x \log_e a + (x \log_e a)^2/2! + (x \log_e a)^3/3! + \cdots$ for all values of x.

(10) $1/(1+x) = 1 - x + x^2 - x^3 + x^4 - \cdots$ for $|x| < 1$.

(11) $1/(1-x) = 1 + x + x^2 + x^3 + x^4 + \cdots$ for $|x| < 1$.

(12) $1/(1+x)^2 = 1 - 2x + 3x^2 - 4x^3 + 5x^4 - \cdots$ for $|x| < 1$.

(13) $1/(1-x)^2 = 1 + 2x + 3x^2 + 4x^3 + 5x^5 + \cdots$ for $|x| < 1$.

(14) $\sqrt{(1+x)} = 1 + x/2 - x^2/(2 \cdot 4) + 1 \cdot 3 \cdot x^3/(2 \cdot 4 \cdot 6)$ $- 1 \cdot 3 \cdot 5 \cdot x^4/(2 \cdot 4 \cdot 6 \cdot 8) - \cdots$ for $|x| < 1$.

(15) $1/(\sqrt{1+x}) = 1 - x/2 + 1 \cdot 3 \cdot x^2/(2 \cdot 4) - 1 \cdot 3 \cdot 5 \cdot x^3/(2 \cdot 4 \cdot 6)$ $+ \cdots$ for $|x| < 1$.

(16) $(a+x)^n = a^n + na^{n-1}x + n(n-1)a^{n-2}x^2/2!$ $+ n(n-1)(n-2)a^{n-3}x^3/3! + \cdots$ for $x^2 < a^2$.

Derivatives of Functions.—The following are formulas for obtaining the derivatives of basic mathematical functions. In these formulas, the letter a denotes a constant; the letter x denotes a variable; and the letters u and v denote functions of the variable x. The expression d/dx means the derivative with respect to x, and as such applies to whatever expression in parentheses follows it. Thus, d/dx (ax) means the derivative with respect to x of the product (ax) of the constant a and the variable x, as given by formula (3).

To simplify the form of the formulas, the symbol D is used to represent d/dx. Thus, D is equivalent to d/dx and other forms as follows:

$$D(ax) = \frac{d(ax)}{dx} = \frac{d}{dx}(ax)$$

1) $D(a) = 0$ 2) $D(x) = 1$ 3) $D(ax) = a \cdot D(x) = a \cdot 1 = a$

4) $D(u+v) = D(u) + D(v)$ *Example:* $D(x^4 + 2x^2) = 4x^3 + 4x$

5) $D(uv) = v \cdot D(u) + u \cdot D(v)$ *Example:* $D(x^2 \cdot ax^3) = ax^3 \cdot 2x + x^2 \cdot 3ax^2 = 5ax^4$

6) $D(u/v) = \dfrac{v \cdot D(u) - u \cdot D(v)}{v^2}$ *Example:* $D(ax^2/\sin x) = (2ax \cdot \sin x - ax^2 \cdot \cos x)/\sin^2 x$

7) $D(x^n) = n \cdot x^{n-1}$ *Example:* $D(5x^7) = 35x^6$

8) $D(e^x) = e^x$

9) $D(a^x) = a^x \cdot \log_e a$ *Example:* $D(11^x) - 11^x \cdot \log_e 11$

10) $D(u^v) = v \cdot u^{v-1} \cdot D(u) + u^v \cdot \log_e u \cdot D(v)$

11) $D(\log_e x) = 1/x$

12) $D(\log_a x) = 1/x \cdot \log_e a = \log_a e/x$

13) $D(\sin x) = \cos x$ *Example:* $D(a \cdot \sin x) = a \cdot \cos x$

14) $D(\cos x) = \sin x$ 15) $D(\tan x) = \sec^2 x$

Solving Numerical Equations Having One Unknown.—The Newton-Raphson method is a procedure for solving various kinds of numerical algebraic and transcendental equations in one unknown. The steps in the procedure are simple and can be used with either a handheld calculator or as a subroutine in a computer program.

Examples of types of equations that can be solved to any desired degree of accuracy by this method are

$$f(x) = x^2 - 101 = 0, \quad f(x) = x^3 - 2x^2 - 5 = 0$$
$$\text{and} \quad f(x) = 2.9x - \cos x - 1 = 0$$

The procedure begins with an estimate, r_1, of the root satisfying the given equation. This estimate is obtained by judgment, inspection, or plotting a rough graph of the equation and observing the value r_1 where the curve crosses the x axis. This value is then used to calculate values $r_2, r_3, \ldots, r_n$ progressively closer to the exact value.

Before continuing, it is necessary to calculate the first derivative. $f'(x)$, of the function. In the above examples, $f'(x)$ is, respectively, $2x$, $3x^2 - 4x$, and $2.9 + \sin x$. These values were found by the methods described in *Derivatives of Functions* on page 34.

In the steps that follow,

r_1 is the first estimate of the value of the root of $f(x) = 0$;

$f(r_1)$ is the value of $f(x)$ for $x = r_1$;

$f'(x)$ is the first derivative of $f(x)$;

$f'(r_1)$ is the value of $f'(x)$ for $x = r_1$.

The second approximation of the root of $f(x) = 0$, r_2, is calculated from

$$r_2 = r_1 - [f(r_1)/f'(r_1)]$$

and, to continue further approximations,

$$r_n = r_{n-1} - [f(r_{n-1})/f'(r_{n-1})]$$

Example: Find the square root of 101 using Newton-Raphson methods. This problem can be restated as an equation to be solved, i.e., $f(x) = x^2 - 101 = 0$

Step 1. By inspection, it is evident that $r_1 = 10$ may be taken as the first approximation of the root of this equation. Then, $f(r_1) = f(10) = 10^2 - 101 = -1$

Step 2. The first derivative, $f'(x)$, of $x^2 - 101$ is $2x$ as stated previously, so that

$$f'(10) = 2(10) = 20.$$

Then,

$$r_2 = r_1 - f(r_1)/f'(r_1) = 10 - (-1)/20 = 10 + 0.05 = 10.05.$$

Check: $10.05^2 = 101.0025$; error = 0.0025

Step 3. The next, better approximation is

$$r_3 = r_2 - [f(r_2)/f'(r_2)] = 10.05 - [f(10.05)/f'(10.05)]$$
$$= 10.05 - [(10.05^2 - 101)/2(10.05)] = 10.049875$$

Check: $10.049875^2 = 100.9999875$; error = 0.0000125

Coordinate Systems

Rectangular, Cartesian Coordinates.—In a Cartesian coordinate system the coordinate axes are perpendicular to one another, and the same unit of length is chosen on the two axes. This rectangular coordinate system is used in the majority of cases.

The general form of an equation of a line in a Cartesian coordinate system is $y = mx + b$, where (x, y) is a point on the line, m is the slope (the rate at which the line is increasing or decreasing), and b is the y coordinate, the y-intercept, of the point $(0, b)$ on the y-axis where the line intersects the axis at $x = 0$.

Another form of the equation of a line is the point-slope form $(y - y_1) = m(x - x_1)$. The slope, m, is defined as a ratio of the change in the y coordinates, $y_2 - y_1$, to the change in the x coordinates, $x_2 - x_1$,

$$m = \frac{\Delta y}{\Delta x} = \frac{y_2 - y_1}{x_2 - x_1}$$

Example 1: Find the general equation of a line passing through the points (3, 2) and (5, 6), and it's intersection point with the y-axis.

First, find the slope using the equation above

$$m = \frac{\Delta y}{\Delta x} = \frac{6 - 2}{5 - 3} = \frac{4}{2} = 2$$

The line has a general form of $y = 2x + b$, and the value of the constant b can be determined by substituting the coordinates of a point on the line into the general form. Using point (3,2), $2 = 2 \times 3 + b$ and rearranging, $b = 2 - 6 = -4$. As a check, using another point on the line, (5,6), yields equivalent results, $y = 6 = 2 \times 5 + b$ and $b = 6 - 10 = -4$.

The equation of the line, therefore, is $y = 2x - 4$, indicating that line $y = 2x - 4$ intersects the y-axis at point (0,–4), the y-intercept.

Example 2: Using the point-slope form find the equation of a line passing through the point (3,2) and having a slope of 2.

$$\begin{aligned}(y - 2) &= 2(x - 3)\\ y &= 2x - 6 + 2\\ y &= 2x - 4\end{aligned}$$

Because the slope, 2, is positive the line is increasing and the line passes through the y-axis at the y-intercept at a value of –4.

Polar Coordinates.— Another coordinate system is determined by a fixed point O, the origin or pole, and a zero direction or axis through it, on which positive lengths can be laid off and measured, as a number line. A point P can be fixed to the zero direction line at a distance r away and then rotated in a positive sense at an angle u. The angle, u, in polar coordinates can take on values from 0° to 360°. A point in polar coordinates takes the form of (u, r).

Changing Coordinate Systems.—For simplicity it may be assumed that the origin on a Cartesian coordinate system coincides with the pole on a polar coordinate system, and it's axis with the x-axis. Then, if point P has polar coordinates of (u, r) and Cartesian coordinates of (x, y), by trigonometry $x = r \times \cos(u)$ and $y = r \times \sin(u)$. By the Pythagorean theorem and trigonometry

$$r = \sqrt{x^2 + y^2} \qquad \theta = \operatorname{atan}\frac{y}{x}$$

Example 1: Convert the cartesian coordinate (3, 2) into polar coordinates.

$$r = \sqrt{3^2 + 2^2} = \sqrt{9+4} = \sqrt{13} = 3.6 \qquad \theta = \text{atan}\frac{2}{3} = 33.78°$$

Therefore the point (3.6, 33.78) is the polar form of the cartesian point (3, 2). Graphically, the polar and Cartesian coordinates are related in the following figure

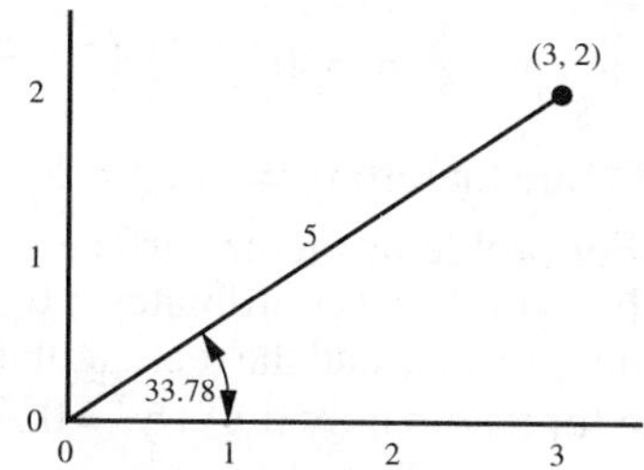

Example 2: Convert the polar form (5, 608) to Cartesian coordinates. By trigonometry, $x = r \times \cos(u)$ and $y = r \times \sin(u)$. Then $x = 5 \cos 608 = 2.5$ and $y = 5 \sin 608 = 4.33$. Therefore, the Cartesian point equivalent is (2.5, 4.33).

Spherical Coordinates.—It is convenient in certain problems, for example, those concerned with spherical surfaces, to introduce non-parallel coordinates. An arbitrary point P in space can be expressed in terms of the distance r between point P and the origin O, the angle ϕ that OP' makes with the x–y plane, and the angle λ that the projection OP' (of the segment OP onto the x–y plane) makes with the positive x-axis.

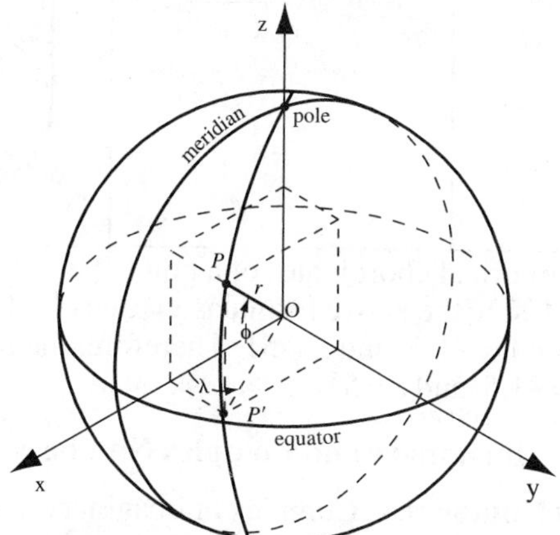

The rectangular coordinates of a point in space can therefore be calculated by the following formulas

Relationship Between Spherical and Rectangular Coordinates

Spherical to Rectangular	Rectangular to Spherical	
$x = \cos\phi\cos\lambda$	$r = \sqrt{x^2 + y^2 + z^2}$	
$y = r\cos\phi\sin\lambda$	$\phi = \text{atan}\frac{z}{\sqrt{x^2 + y^2}}$	(for $x^2 + y^2 \neq 0$)
$z = r\sin\phi$	$\lambda = \text{atan}\frac{y}{x}$	(for $x > 0, y > 0$)
	$\lambda = \pi + \text{atan}\frac{y}{x}$	(for $x < 0$)
	$\lambda = 2\pi + \text{atan}\frac{y}{x}$	(for $x > 0, y < 0$)

Example: What are the spherical coordinates of the point $P(3, 4, -12)$?

$$r = \sqrt{3^2 + (-4)^2 + (-12)^2} = 13$$

$$\phi = \operatorname{atan}\frac{-12}{\sqrt{3^2 + (-4)^2}} = \operatorname{atan}-\frac{12}{5} = -67.38°$$

$$\lambda = 360° + \operatorname{atan}-\frac{4}{3} = 360° - 53.13° = 306.87°$$

The spherical coordinates of P are therefore $r = 13$, $\phi = 267.388$, and $\lambda = 306.878$.

Cylindrical Coordinates.—For problems on the surface of a cylinder it is convenient to use cylindrical coordinates. The cylindrical coordinates r, θ, z, of P coincide with the polar coordinates of the point P′ in the x-y plane and the rectangular z-coordinate of P. This gives the conversion formula. Those for θ hold only if $x^2 + y^2 \neq 0$; θ is undetermined if $x = y = 0$.

Cylindrical to Rectangular	Rectangular to Cylindrical	
$x = r\cos\theta$	$r = \frac{1}{\sqrt{x^2 + y^2}}$	z, P, O, r, θ, x, P′, y
	$\cos\theta = \frac{x}{\sqrt{x^2 + y^2}}$	
$y = r\sin\theta$	$\sin\theta = \frac{y}{\sqrt{x^2 + y^2}}$	
$z = z$	$z = z$	

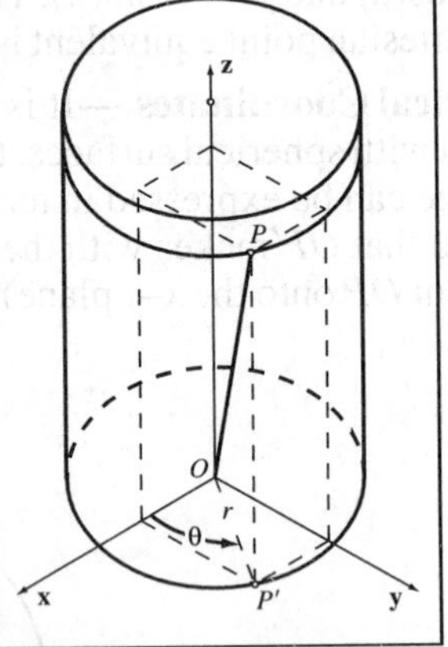

Example: Given the cylindrical coordinates of a point P, $r = 3$, $\theta = -30°$, $z = 51$, find the rectangular coordinates. Using the above formulas $x = 3\cos(-30°) = 3\cos(30°) = 2.598$; $y = 3\sin(-30°) = -3\sin(30°) = -1.5$; and $z = 51$. Therefore, the rectangular coordinates of point P are $x = 2.598$, $y = -1.5$, and $z = 51$.

Imaginary and Complex Numbers

Complex or Imaginary Numbers.—Complex or imaginary numbers represent a class of mathematical objects that are used to simplify certain problems, such as the solution of polynomial equations. The basis of the complex number system is the unit imaginary number i that satisfies the following relations:

$$i^2 = (-i)^2 = -1 \qquad i = \sqrt{-1} \qquad -i = -\sqrt{-1}$$

In electrical engineering and other fields, the unit imaginary number is often represented by j rather than i. However, the meaning of the two terms is identical.

Rectangular or Trigonometric Form: Every complex number, Z, can be written as the sum of a real number and an imaginary number. When expressed as a sum, $Z = a + bi$, the complex number is said to be in rectangular or trigonometric form. The real part of the number is a, and the imaginary portion is bi because it has the imaginary unit assigned to it.

Polar Form: A complex number $Z = a + bi$ can also be expressed in polar form, also known as phasor form. In polar form, the complex number Z is represented by a magnitude r and an angle θ as follows:

$$Z = r\angle\theta$$

where $\angle\theta$ = a direction, the angle whose tangent is $b \div a$, thus $\theta = \operatorname{atan}\frac{b}{a}$; and, r = a magnitude $= \sqrt{a^2 + b^2}$.

A complex number can be plotted on a real-imaginary coordinate system known as the complex plane. The figure below illustrates the relationship between the rectangular coordinates a and b, and the polar coordinates r and θ.

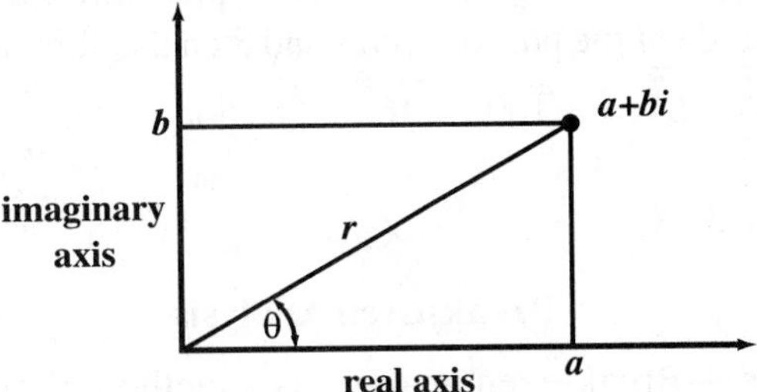

Complex Number in the Complex Plane

The rectangular form can be determined from r and θ as follows:

$$a = r\cos\theta \qquad b = r\sin\theta \qquad a + bi = r\cos\theta + ir\sin\theta = r(\cos\theta + i\sin\theta)$$

The rectangular form can also be written using Euler's Formula:

$$e^{\pm i\theta} = \cos\theta \pm i\sin\theta \qquad \sin\theta = \frac{e^{i\theta} - e^{-i\theta}}{2i} \qquad \cos\theta = \frac{e^{i\theta} + e^{-i\theta}}{2}$$

Complex Conjugate: Complex numbers commonly arise in finding the solution of polynomials. A polynomial of nth degree has n solutions, an even number of which are complex and the rest are real. The complex solutions always appear as complex conjugate pairs in the form $a + bi$ and $a - bi$. The product of these two conjugates, $(a + bi) \times (a - bi) = a^2 + b^2$, is the square of the magnitude r illustrated in the previous figure.

Operations on Complex Numbers

Example 3, Addition: When adding two complex numbers, the real parts and imaginary parts are added separately, the real parts added to real parts and the imaginary to imaginary parts. Thus,

$$(a_1 + ib_1) + (a_2 + ib_2) = (a_1 + a_2) + i(b_1 + b_2)$$

$$(a_1 + ib_1) - (a_2 + ib_2) = (a_1 - a_2) + i(b_1 - b_2)$$

$$(3 + 4i) + (2 + i) = (3 + 2) + (4 + 1)i = 5 + 5i$$

Example 4, Multiplication: Multiplication of two complex numbers requires the use of the imaginary unit, $i^2 = -1$ and the algebraic distributive law.

$$\begin{aligned}(a_1 + ib_1)(a_2 + ib_2) &= a_1a_2 + ia_1b_2 + ia_2b_1 + i^2b_1b_2 \\ &= a_1a_2 + ia_1b_2 + ia_2b_1 - b_1b_2\end{aligned}$$

$$\begin{aligned}(7 + 2i) \times (5 - 3i) &= (7)(5) - (7)(3i) + (2i)(5) - (2i)(3i) \\ &= 35 - 21i + 10i - 6i^2 \\ &= 35 - 21i + 10i - (6)(-1) \\ &= 41 - 11i\end{aligned}$$

Multiplication of two complex numbers, $Z_1 = r_1(\cos\theta_1 + i\sin\theta_1)$ and $Z_2 = r_2(\cos\theta_2 + i\sin\theta_2)$, results in the following:

$$Z_1 \times Z_2 = r_1(\cos\theta_1 + i\sin\theta_1) \times r_2(\cos\theta_2 + i\sin\theta_2) = r_1r_2[\cos(\theta_1 + \theta_2) + i\sin(\theta_1 + \theta_2)]$$

Example 5, Division: Divide the following two complex numbers, 2 + 3*i* and 4 – 5*i*. Dividing complex numbers makes use of the complex conjugate.

$$\frac{2+3i}{4-5i} = \frac{(2+3i)(4+5i)}{(4-5i)(4+5i)} = \frac{8+12i+10i+15i^2}{16+20i-20i-25i^2} = \frac{-7+22i}{16+25} = \left(\frac{-7}{41}\right) + i\left(\frac{22}{41}\right)$$

Example 6: Convert the complex number 8+6*i* into phasor form.

First find the magnitude of the phasor vector and then the direction.

$$\text{magnitude} = \sqrt{8^2+6^2} = \sqrt{100} = 10 \qquad \text{direction} = \text{atan}\frac{6}{8} = \text{atan}0.75 = 36.87°$$

$$\text{phasor} = 10\angle 36.87°$$

Break-Even Analysis

Break-Even Analysis.—Break-even analysis is a method of comparing two or more alternatives to determine which works best. Frequently, cost is the basis of the comparison, with the least expensive alternative being the most desirable. Break-even analysis can be applied in situations such as: to determine if it is more efficient and cost effective to use HSS, carbide, or ceramic tooling; to compare coated versus uncoated carbide tooling; to decide which of several machines should be used to produce a part; or to decide whether to buy a new machine for a particular job or to continue to use an older machine. The techniques used to solve any of these problems are the same; however, the details will be different, depending on the type of comparison being made. The remainder of this section deals with break-even analysis based on comparing the costs of manufacturing a product using different machines.

Choosing a Manufacturing Method: The object of this analysis is to decide which of several machines can produce parts at the lowest cost. In order to compare the cost of producing a part, all the costs involved in making that part must be considered. The cost of manufacturing any number of parts can be expressed as the sum: $C_T = C_F + n \times C_V$, where C_T is the total cost of manufacturing one part, C_F is the sum of the fixed costs of making the parts, *n* is the number of parts made, and C_V is the total variable costs per piece made.

Fixed costs are manufacturing costs that have to be paid whatever the number of parts is produced and usually before any parts can be produced. They include the cost of drafting and CNC part programs, the cost of special tools and equipment required to make the part, and the cost of setting up the machine for the job. Fixed costs are generally one-time charges that occur at the beginning of a job or are recurrent charges that do not depend on the number of pieces made, such as those that might occur each time a job is run again.

Variable costs depend on the number of parts produced and are expressed as the cost per part made. The variable costs include the cost of materials, the cost of machine time, the cost of the labor directly involved in making the part, and the portion of the overhead that is attributable to production of the part. Variable costs can be expressed as: C_V = material cost + machine cost + labor cost + overhead cost. When comparing alternatives, if the same cost is incurred by each alternative, then that cost can be eliminated from the analysis without affecting the result. For example, the cost of material is frequently omitted from a manufacturing analysis if each machine is going to make parts from the same stock and if there is not going to be a significant difference in the amount of scrap produced by each method. The time to produce one part is needed to determine the machine, labor, and overhead costs. The total time expressed in hours per part is $t_T = t_f + t_s$, where t_f equals the floor-to-floor production time for one part and t_s the setup time per part. The setup time, t_s, is the time spent setting up the machine and periodically reconditioning tooling, divided by the number of parts made per setup.

Material cost equals the cost of the materials divided by the number of parts made.

Machine cost is the portion of a machine's total cost that is charged toward the production of each part. It is found by multiplying the machine rate (cost of the machine per hour) by the machine time per part, t_f. The machine hourly rate is calculated by dividing the lifetime costs (including purchase price, insurance, maintenance, etc.) by the estimated lifetime hours of operation of the machine. The total operating hours may be difficult to determine but a reasonable number can be based on experience and dealer information.

Labor costs are the wages paid to people who are directly involved in the manufacture of the part. The labor cost per part is the labor rate per hour multiplied by the time needed to manufacture each part, t_T. Indirect labor, which supports but is not directly involved in the manufacture of the part, is charged as overhead.

Overhead cost is the cost of producing an item that is not directly related to the cost of manufacture. Overhead includes the cost of management and other support personnel, building costs, heating and cooling, and similar expenses. Often, overhead is estimated as a percentage of the largest component cost of producing a part. For example, if direct labor is the largest expense in producing a part, the overhead can be estimated as a percentage of the direct labor costs. On the other hand, if equipment costs are higher, the overhead would be based on a percentage of the machine cost. Depending on the company, typical overhead charges range from about 150 to 800 per cent of the highest variable cost.

Most of the time, the decision to use one machine or another for making parts depends on how many pieces are needed. For example, given three machines *A*, *B*, and *C*, if only a few parts need to be produced, then, in terms of cost, machine *A* might be the best; if hundreds of parts are needed, then machine *B* might be best; and, if thousands of components are to be manufactured, then machine *C* may result in the lowest cost per part. Break-even analysis reveals how many components need to be produced before a particular machine becomes more cost effective than another.

To use break-even analysis, the cost of operating each machine needs to be established. The costs are plotted on a graph as a function of the number of components to be manufactured to learn which machine can make the required parts for the least cost. The following graph is a plot of the fixed and variable costs of producing a quantity of parts on two different machines, *Machine 1* and *Machine 2*. Fixed costs for each machine are plotted on the vertical *cost* axis. Variable costs for each machine are plotted as a line that intersects the cost axis at the fixed cost for each respective machine. The variable cost line is constructed with a slope that is equal to the cost per part, that is, for each part made, the line rises by an amount equal to the cost per part. If the calculations necessary to produce the graph are done carefully, the total cost of producing any quantity of parts can be found from the data plotted on the graph.

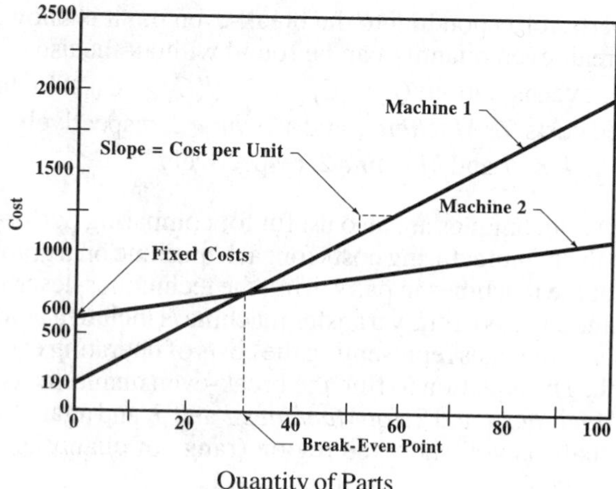

As an example, the graph shown is a comparison of the cost of manufacturing a quantity of a small part on a manually operated milling machine (*Machine 1*) and on a CNC machining center (*Machine 2*). The fixed costs (fixed costs = lead time × lead time rate + setup time × setup rate) for the manual machine are \$190 and the fixed costs for the CNC machine are higher at \$600. The fixed cost for each machine is the starting point of the line representing the cost of manufacturing a quantity of parts with that machine. The variable costs plotted are: \$18 per piece for the manual machine and \$5 per piece for the CNC mill.

The variable costs are calculated using the machine, labor, and overhead costs. The cost of materials is not included because it is assumed that materials cost will be the same for parts made on either machine and there will be no appreciable difference in the amount of scrap generated. The original cost of *Machine 1* (the manual milling machine) is \$19,000 with an estimated operating life of 16,000 hours, so the hourly operating cost is 19,000/16,000 = \$1.20 per hour. The labor rate is \$17 per hour and the overhead is estimated as 1.6 times the labor rate, or \$17 × 1.6 = \$27.20 per hour. The time, t_f, needed to complete each part on *Machine 1* is estimated as 24 minutes (0.4 hour). Therefore, by using *Machine 1*, the variable cost per part excluding material is (1.20 + 17.00 + 27.20) \$/h × 0.4 h/part = \$18 per part. For *Machine 2* (the CNC machining center), the machine cost is calculated at \$3 per hour, which is based on a \$60,000 initial cost (including installation, maintenance, insurance, etc.) and 20,000 hours of estimated lifetime. The cost of labor is \$15 per hour for *Machine 2* and the overhead is again calculated at 1.6 times the labor rate, or \$24 per hour. Each part is estimated to take 7.2 minutes (0.12 h) to make, so the variable cost per part made on *Machine 2* is (3 + 15 + 24) \$/h × 0.12 h/part = \$5 per part.

The lines representing the variable cost of operating each machine intersect at only one point on the graph. The intersection point corresponds to a quantity of parts that can be made by either the CNC or manual machine for the same cost, which is the break-even point. In the figure, the break-even point is 31.5 parts and the cost of those parts is \$757, or about \$24 apiece, excluding materials. The graph shows that if fewer than 32 parts need to be made, the total cost will be lowest if the manual machine is used because the line representing *Machine 1* is lower (representing lower cost) than the line representing *Machine 2*. On the other hand, if more than 31 parts are going to be made, the CNC machine will produce them for a lower cost. It is easy to see that the per piece cost of manufacturing is lower on the CNC machine because the line for *Machine 2* rises at a slower rate than the line for *Machine 1*. For producing only a few parts, the manual machine will make them less expensively than the CNC because the fixed costs are lower, but once the CNC part program has been written, the CNC can also run small batches efficiently because very little setup work is required.

The quantity of parts corresponding to the break-even point is known as the break-even quantity Q_b. The break-even quantity can be found without the use of the graph by using the following break-even equation: $Q_b = (C_{F1} - C_{F2})/(C_{V2} - C_{V1})$. In this equation, the C_{F1} and C_{F2} are the fixed costs for *Machine 1* and *Machine 2*, respectively: C_{V1} and C_{V2} are the variable costs for *Machine 1* and *Machine 2*, respectively.

Break-even analysis techniques are also useful for comparing performance of more than two machines. Plot the manufacturing costs for each machine on a graph as before and then compare the costs of the machines in pairs using the techniques described. For example, if an automatic machine such as a rotary transfer machine is included as *Machine 3* in the preceding analysis, then three lines representing the costs of operating each machine would be plotted on the graph. The equation to find the break-even quantities is applied three times in succession, for *Machines 1* and *2*, for *Machines 1* and *3*, and again for *Machines 2* and *3*. The result of this analysis will show the region (range of quantities) within which each machine is most profitable.

GEOMETRY

Arithmetical Progression

An arithmetical progression is a series of numbers in which each consecutive term differs from the preceding one by a fixed amount called the *common difference*, *d*. Thus, 1, 3, 5, 7, etc., is an arithmetical progression where the difference d is 2. The difference here is *added* to the preceding term, and the progression is called increasing. In the series 13, 10, 7, 4, etc., the difference is (−3), and the progression is called decreasing. In any arithmetical progression (or part of progression), let

a = first term considered

l = last term considered

n = number of terms

d = common difference

S = sum of n terms

Then the general formulas are $l = a + (n-1)d$ and $S = \frac{a+l}{2} \times n$

In these formulas, d is positive in an increasing and negative in a decreasing progression. When any three of the preceding live quantities are given, the other two can be found by the formulas in the accompanying table of arithmetical progression.

Example: In an arithmetical progression, the first term equals 5, and the last term 40. The difference is 7. Find the sum of the progression.

$$S = \frac{a+l}{2d}(l+d-a) = \frac{5+40}{2 \times 7}(40+7-5) = 135$$

Geometrical Progression

A geometrical progression or a geometrical series is a series in which each term is derived by multiplying the preceding term by a constant multiplier called the *ratio*. When the ratio is greater than 1, the progression is increasing; when less than 1, it is decreasing. Thus, 2, 6, 18, 54, etc., is an increasing geometrical progression with a ratio of 3, and 24, 12, 6, etc., is a decreasing progression with a ratio of 1⁄2.

In any geometrical progression (or part of progression), let

a = first term

l = last (or nth) term

n = number of terms

r = ratio of the progression

S = sum of n terms

Then the general formulas are $l = ar^{n-1}$ and $S = \frac{rl - a}{r - 1}$

When any three of the preceding five quantities are given, the other two can be found by the formulas in the accompanying table. For instance, geometrical progressions are used for finding the successive speeds in machine tool drives, and in interest calculations.

Example: The lowest speed of a lathe is 20 rpm. The highest speed is 225 rpm. There are 18 speeds. Find the ratio between successive speeds.

$$\text{Ratio } r = \sqrt[n-1]{\frac{l}{a}} = \sqrt[17]{\frac{225}{20}} = \sqrt[17]{11.25} = 1.153$$

Formulas for Arithmetical Progression

To Find	Given	Use Equation
a	$d \quad l \quad n$	$a = l-(n-1)d$
	$d \quad n \quad S$	$a = \frac{S}{n}-\frac{n-1}{2}\times d$
	$d \quad l \quad S$	$a = \frac{d}{2}\pm\frac{1}{2}\sqrt{(2l+d)^2-8dS}$
	$l \quad n \quad S$	$a = \frac{2S}{n}-l$
d	$a \quad l \quad n$	$d = \frac{l-a}{n-1}$
	$a \quad n \quad S$	$d = \frac{2S-2an}{n(n-1)}$
	$a \quad l \quad S$	$d = \frac{l^2-a^2}{2S-l-a}$
	$l \quad n \quad S$	$d = \frac{2nl-2S}{n(n-1)}$
l	$a \quad d \quad n$	$l = a+(n-1)d$
	$a \quad d \quad S$	$l = -\frac{d}{2}\pm\frac{1}{2}\sqrt{8dS+(2a-d)^2}$
	$a \quad n \quad S$	$l = \frac{2S}{n}-a$
	$d \quad n \quad S$	$l = \frac{S}{n}+\frac{n-1}{2}\times d$
n	$a \quad d \quad l$	$n = 1+\frac{l-a}{d}$
	$a \quad d \quad S$	$n = \frac{d-2a}{2d}\pm\frac{1}{2d}\sqrt{8dS+(2a-d)^2}$
	$a \quad l \quad S$	$n = \frac{2S}{a+l}$
	$d \quad l \quad S$	$n = \frac{2l+d}{2d}\pm\frac{1}{2d}\sqrt{(2l+d)^2-8dS}$
S	$a \quad d \quad n$	$S = \frac{n}{2}[2a+(n-1)d]$
	$a \quad d \quad l$	$S = \frac{a+l}{2}+\frac{l^2-a^2}{2d} = \frac{a+l}{2d}(l+d-a)$
	$a \quad l \quad n$	$S = \frac{n}{2}(a+l)$
	$d \quad l \quad n$	$S = \frac{n}{2}[2l-(n-1)d]$

GEOMETRY

Arithmetical Progression

An arithmetical progression is a series of numbers in which each consecutive term differs from the preceding one by a fixed amount called the *common difference, d*. Thus, 1, 3, 5, 7, etc., is an arithmetical progression where the difference d is 2. The difference here is *added* to the preceding term, and the progression is called increasing. In the series 13, 10, 7, 4, etc., the difference is (−3), and the progression is called decreasing. In any arithmetical progression (or part of progression), let

a = first term considered

l = last term considered

n = number of terms

d = common difference

S = sum of n terms

Then the general formulas are $l = a + (n - 1)d$ and $S = \frac{a+l}{2} \times n$

In these formulas, d is positive in an increasing and negative in a decreasing progression. When any three of the preceding live quantities are given, the other two can be found by the formulas in the accompanying table of arithmetical progression.

Example: In an arithmetical progression, the first term equals 5, and the last term 40. The difference is 7. Find the sum of the progression.

$$S = \frac{a+l}{2d}(l + d - a) = \frac{5+40}{2 \times 7}(40 + 7 - 5) = 135$$

Geometrical Progression

A geometrical progression or a geometrical series is a series in which each term is derived by multiplying the preceding term by a constant multiplier called the *ratio*. When the ratio is greater than 1, the progression is increasing; when less than 1, it is decreasing. Thus, 2, 6, 18, 54, etc., is an increasing geometrical progression with a ratio of 3, and 24, 12, 6, etc., is a decreasing progression with a ratio of 1⁄2.

In any geometrical progression (or part of progression), let

a = first term

l = last (or nth) term

n = number of terms

r = ratio of the progression

S = sum of n terms

Then the general formulas are $l = ar^{n-1}$ and $S = \frac{rl - a}{r - 1}$

When any three of the preceding five quantities are given, the other two can be found by the formulas in the accompanying table. For instance, geometrical progressions are used for finding the successive speeds in machine tool drives, and in interest calculations.

Example: The lowest speed of a lathe is 20 rpm. The highest speed is 225 rpm. There are 18 speeds. Find the ratio between successive speeds.

$$\text{Ratio } r = \sqrt[n-1]{\frac{l}{a}} = \sqrt[17]{\frac{225}{20}} = \sqrt[17]{11.25} = 1.153$$

Formulas for Arithmetical Progression

To Find	Given	Use Equation
a	$d \quad l \quad n$	$a = l - (n-1)d$
	$d \quad n \quad S$	$a = \frac{S}{n} - \frac{n-1}{2} \times d$
	$d \quad l \quad S$	$a = \frac{d}{2} \pm \frac{1}{2}\sqrt{(2l+d)^2 - 8dS}$
	$l \quad n \quad S$	$a = \frac{2S}{n} - l$
d	$a \quad l \quad n$	$d = \frac{l-a}{n-1}$
	$a \quad n \quad S$	$d = \frac{2S - 2an}{n(n-1)}$
	$a \quad l \quad S$	$d = \frac{l^2 - a^2}{2S\; l - a}$
	$l \quad n \quad S$	$d = \frac{2nl - 2S}{n(n-1)}$
l	$a \quad d \quad n$	$l = a + (n-1)d$
	$a \quad d \quad S$	$l = -\frac{d}{2} \pm \frac{1}{2}\sqrt{8dS + (2a-d)^2}$
	$a \quad n \quad S$	$l = \frac{2S}{n} - a$
	$d \quad n \quad S$	$l = \frac{S}{n} + \frac{n-1}{2} \times d$
n	$a \quad d \quad l$	$n = 1 + \frac{l-a}{d}$
	$a \quad d \quad S$	$n = \frac{d-2a}{2d} \pm \frac{1}{2d}\sqrt{8dS + (2a-d)^2}$
	$a \quad l \quad S$	$n = \frac{2S}{a+l}$
	$d \quad l \quad S$	$n = \frac{2l+d}{2d} \pm \frac{1}{2d}\sqrt{(2l+d)^2 - 8dS}$
S	$a \quad d \quad n$	$S = \frac{n}{2}[2a + (n-1)d]$
	$a \quad d \quad l$	$S = \frac{a+l}{2} + \frac{l^2 - a^2}{2d} = \frac{a+l}{2d}(l + d - a)$
	$a \quad l \quad n$	$S = \frac{n}{2}(a+l)$
	$d \quad l \quad n$	$S = \frac{n}{2}[2l - (n-1)d]$

Formulas for Geometrical Progression

To Find	Given			Use Equation
a	l	n	r	$a = \frac{l}{r^{n-1}}$
	n	r	S	$a = \frac{(r-1)S}{r^n - 1}$
	l	r	S	$a = lr - (r-1)S$
	l	n	S	$a(S-a)^{n-1} = l(S-1)^{n-1}$
l	a	n	r	$l = ar^{n-1}$
	a	r	S	$l = \frac{1}{r}[a + (r-1)S]$
	a	n	S	$l(S-l)^{n-1} = a(S-a)^{n-1}$
	n	r	S	$l = \frac{S(r-1)r^{n-1}}{r^n - 1}$
n	a	l	r	$n = \frac{\log l - \log a}{\log r} + 1$
	a	r	S	$n = \frac{\log[a + (r-1)S] - \log a}{\log r}$
	a	l	S	$n = \frac{\log l - \log a}{\log(S-a) - \log(S-l)} + 1$
	l	r	S	$n = \frac{\log l - \log[lr - (r-1)S]}{\log r} + 1$
r	a	l	n	$r = \sqrt[n-1]{\frac{l}{a}}$
	a	n	S	$r^n = \frac{Sr}{a} + \frac{a-S}{a}$
	a	l	S	$r = \frac{S-a}{S-l}$
	l	n	S	$r^n = \frac{Sr^{n-1}}{S-l} - \frac{l}{S-l}$
S	a	n	r	$S = \frac{a(r^n - 1)}{r-1}$
	a	l	r	$S = \frac{lr - a}{r-1}$
	a	l	n	$S = \frac{\sqrt[n-1]{l^n} - \sqrt[n-1]{a^n}}{\sqrt[n-1]{l} - \sqrt[n-1]{a}}$
	l	n	r	$S = \frac{l(r^n - 1)}{(r-1)r^{n-1}}$

Geometrical Propositions

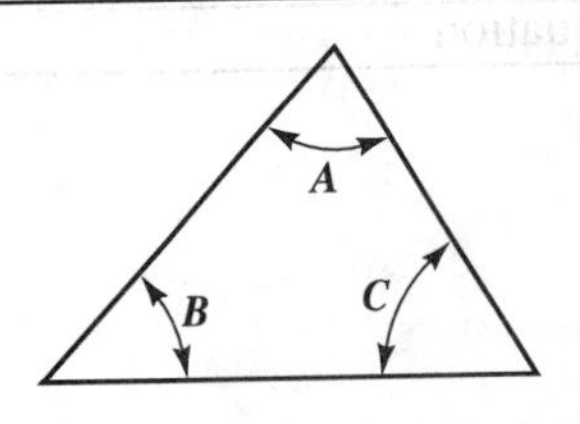

The sum of the three angles in a triangle always equals 180 degrees. Hence, if two angles are known, the third angle can always be found.

$$A + B + C = 180° \qquad A = 180° - (B + C)$$
$$B = 180° - (A + C) \qquad C = 180° - (A + B)$$

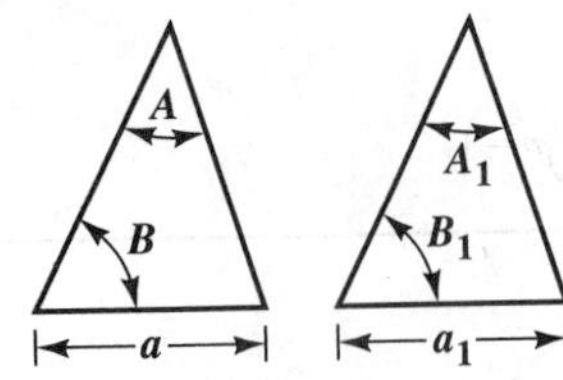

If one side and two angles in one triangle are equal to one side and similarly located angles in another triangle, then the remaining two sides and angle also are equal.

If $a = a_1$, $A = A_1$, and $B = B_1$, then the two other sides and the remaining angle also are equal.

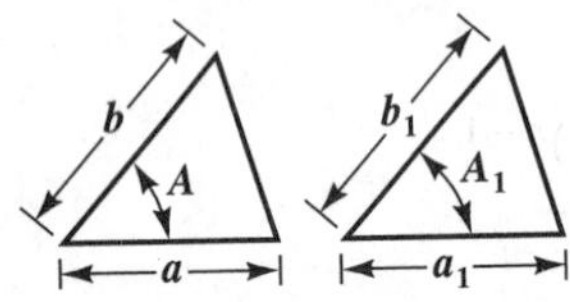

If two sides and the angle between them in one triangle are equal to two sides and a similarly located angle in another triangle, then the remaining side and angles also are equal.

If $a = a_1$, $b = b_1$, and $A = A_1$, then the remaining side and angles also are equal.

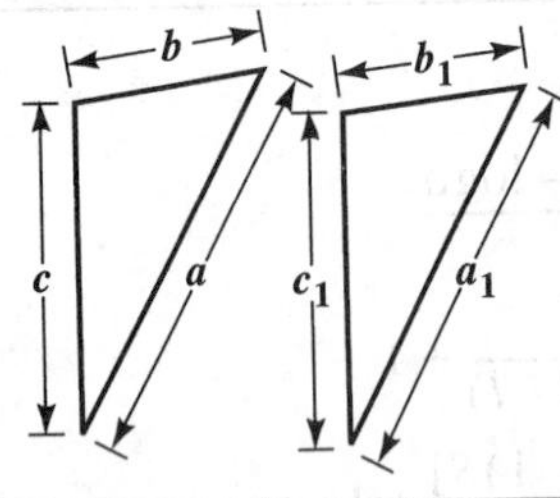

If the three sides in one triangle are equal to the three sides of another triangle, then the angles in the two triangles also are equal.

If $a = a_1$, $b = b_1$, and $c = c_1$, then the angles between the respective sides also are equal.

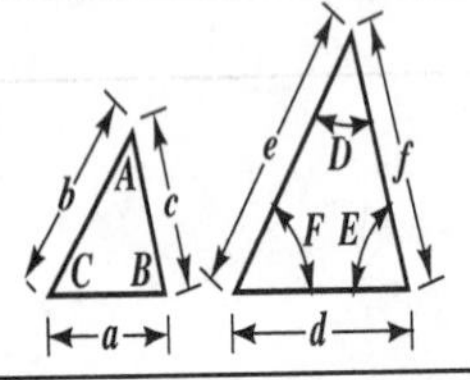

If the three sides of one triangle are proportional to corresponding sides in another triangle, then the triangles are called *similar*, and the angles in the one are equal to the angles in the other.

If $a : b : c = d : e : f$, then $A = D$, $B = E$, and $C = F$.

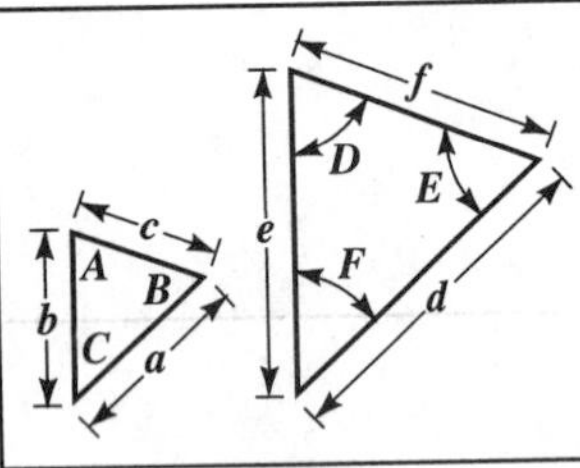

If the angles in one triangle are equal to the angles in another triangle, then the triangles are similar and their corresponding sides are proportional.

If $A = D, B = E$, and $C = F$, then $a : b : c = d : e : f$.

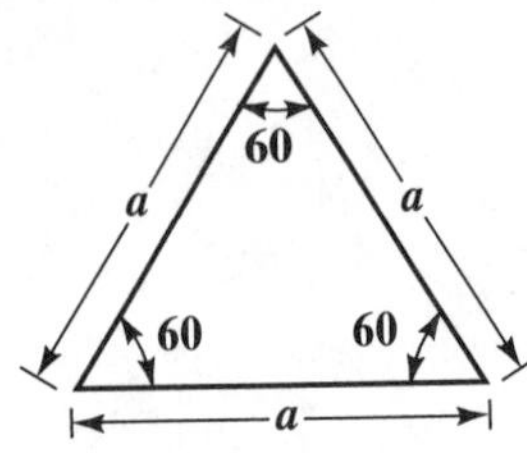

If the three sides in a triangle are equal—that is, if the triangle is *equilateral*—then the three angles also are equal.

Each of the three equal angles in an equilateral triangle is 60 degrees.

If the three angles in a triangle are equal, then the three sides also are equal.

Geometrical Propositions

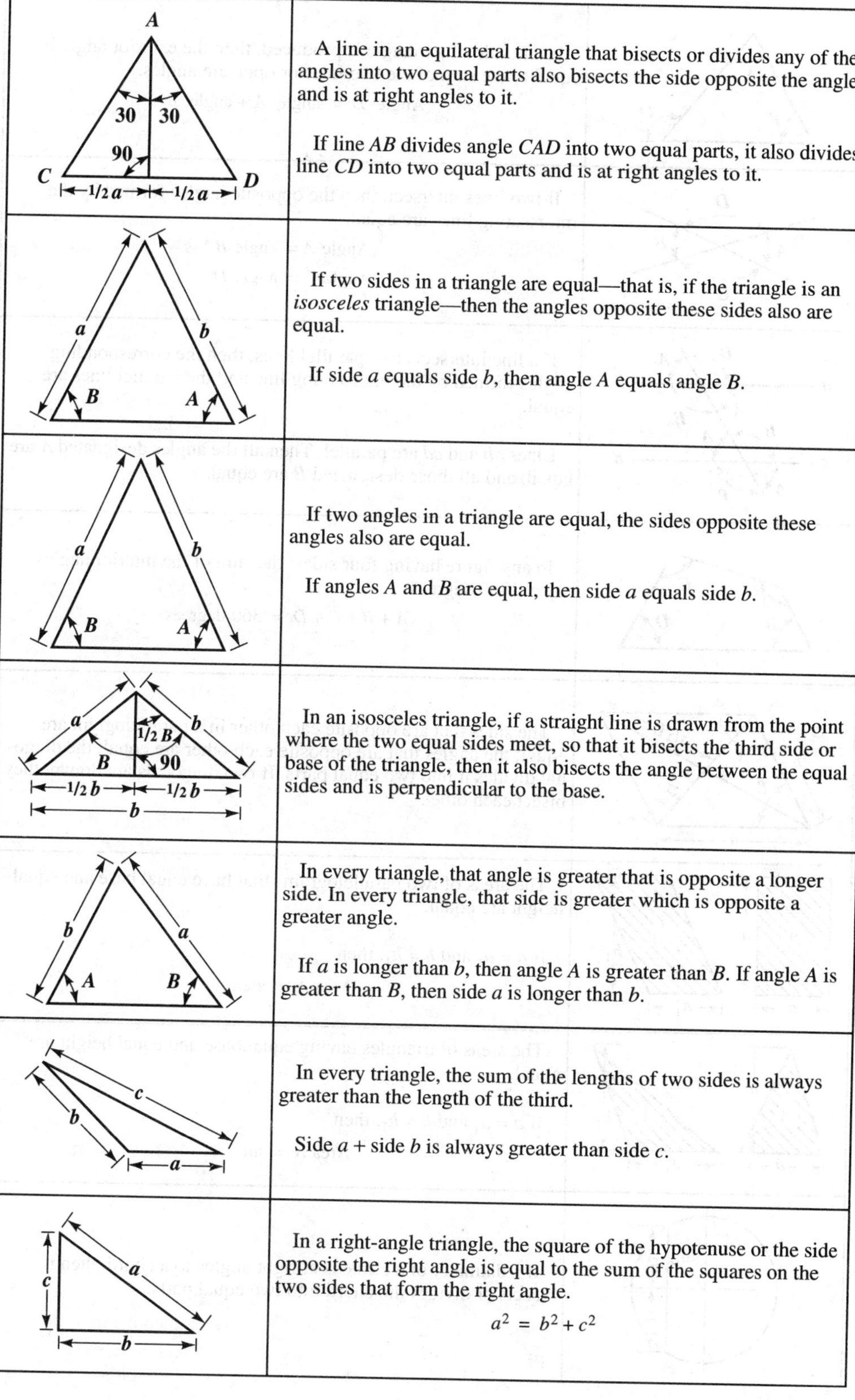

A line in an equilateral triangle that bisects or divides any of the angles into two equal parts also bisects the side opposite the angle and is at right angles to it.

If line *AB* divides angle *CAD* into two equal parts, it also divides line *CD* into two equal parts and is at right angles to it.

If two sides in a triangle are equal—that is, if the triangle is an *isosceles* triangle—then the angles opposite these sides also are equal.

If side *a* equals side *b*, then angle *A* equals angle *B*.

If two angles in a triangle are equal, the sides opposite these angles also are equal.

If angles *A* and *B* are equal, then side *a* equals side *b*.

In an isosceles triangle, if a straight line is drawn from the point where the two equal sides meet, so that it bisects the third side or base of the triangle, then it also bisects the angle between the equal sides and is perpendicular to the base.

In every triangle, that angle is greater that is opposite a longer side. In every triangle, that side is greater which is opposite a greater angle.

If *a* is longer than *b*, then angle *A* is greater than *B*. If angle *A* is greater than *B*, then side *a* is longer than *b*.

In every triangle, the sum of the lengths of two sides is always greater than the length of the third.

Side *a* + side *b* is always greater than side *c*.

In a right-angle triangle, the square of the hypotenuse or the side opposite the right angle is equal to the sum of the squares on the two sides that form the right angle.

$$a^2 = b^2 + c^2$$

Geometrical Propositions

Figure	Proposition
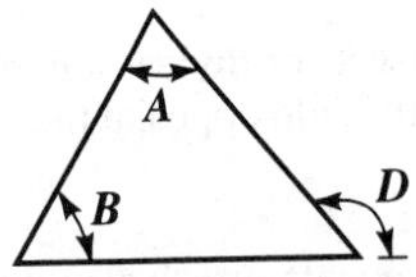	If one side of a triangle is produced, then the exterior angle is equal to the sum of the two interior opposite angles. Angle D = angle A + angle B
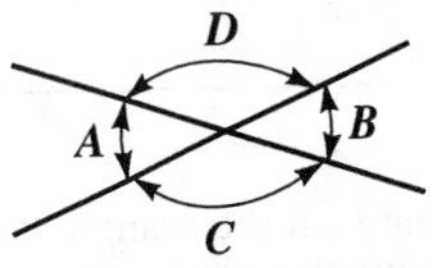	If two lines intersect, then the opposite angles formed by the intersecting lines are equal. Angle A = angle B Angle C = angle D
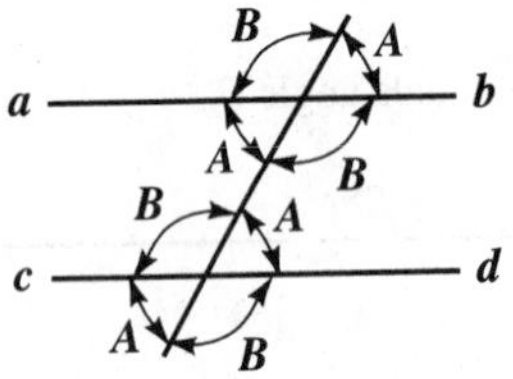	If a line intersects two parallel lines, then the corresponding angles formed by the intersecting line and the parallel lines are equal. Lines *ab* and *cd* are parallel. Then all the angles designated *A* are equal, and all those designated *B* are equal.
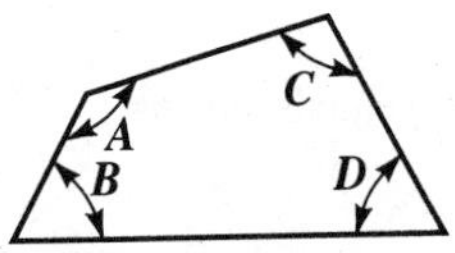	In any figure having four sides, the sum of the interior angles equals 360 degrees. $A + B + C + D$ = 360 degrees
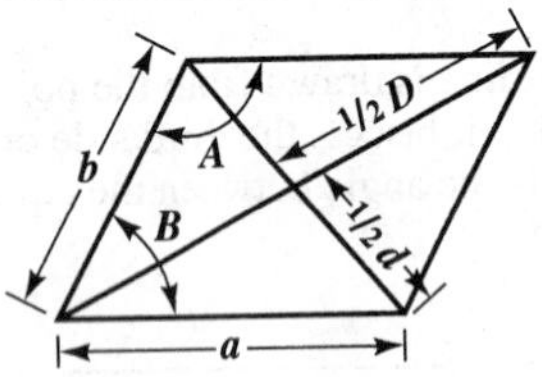	The sides that are opposite each other in a parallelogram are equal; the angles that are opposite each other are equal; the diagonal divides it into two equal parts. If two diagonals are drawn, they bisect each other.
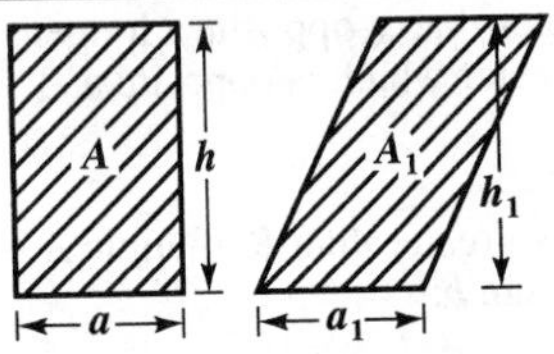	The areas of two parallelograms that have equal base and equal height are equal. If $a = a_1$ and $h = h_1$, then Area A = area A_1
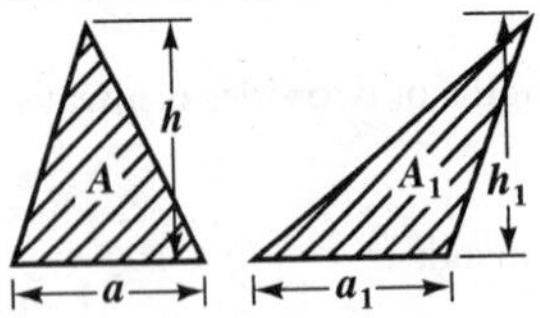	The areas of triangles having equal base and equal height are equal. If $a = a_1$ and $h = h_1$, then Area A = area A_1
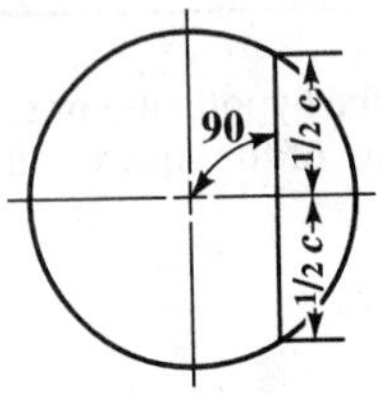	If a diameter of a circle is at right angles to a chord, then it bisects or divides the chord into two equal parts.

Geometrical Propositions

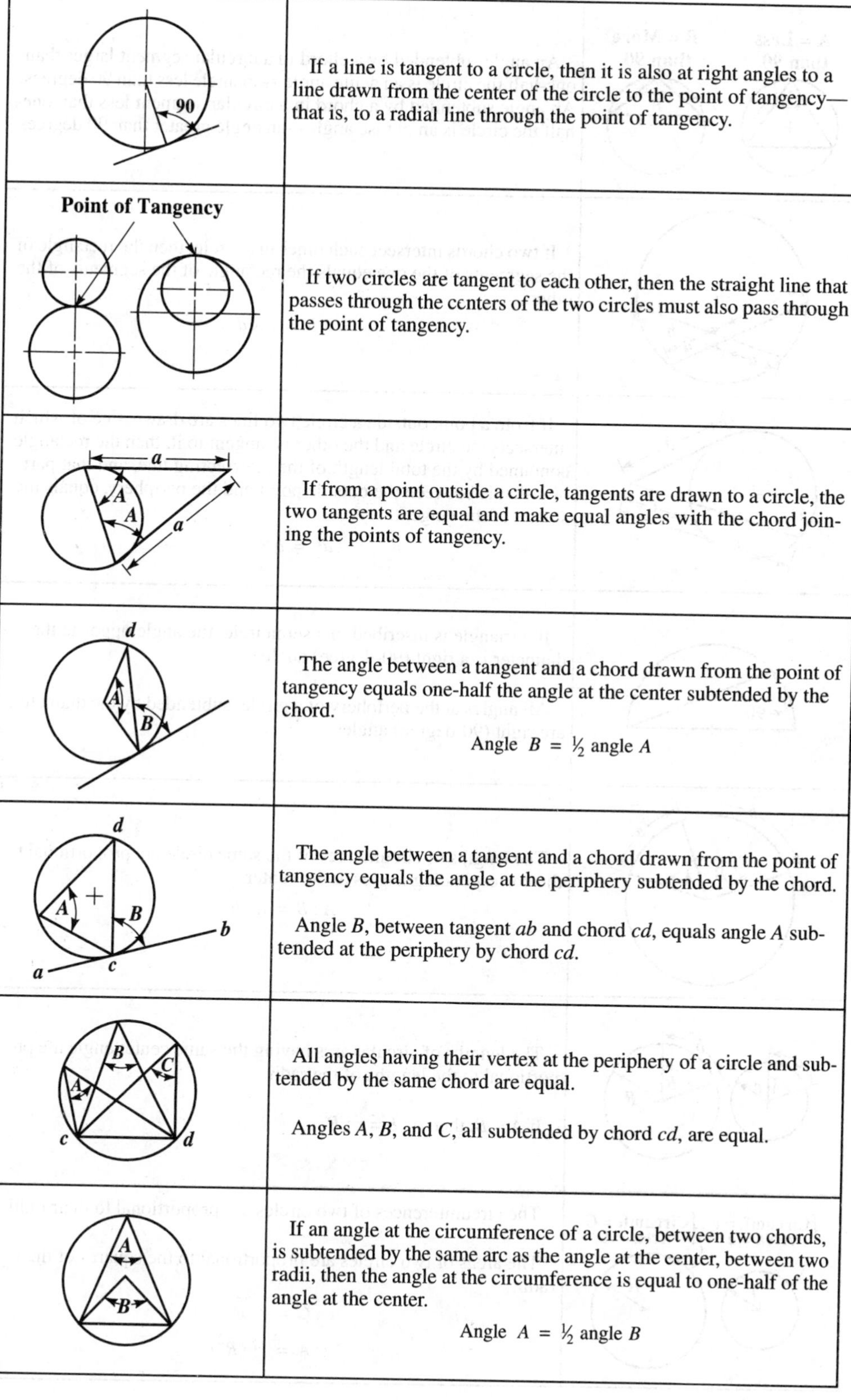

90	If a line is tangent to a circle, then it is also at right angles to a line drawn from the center of the circle to the point of tangency—that is, to a radial line through the point of tangency.
Point of Tangency	If two circles are tangent to each other, then the straight line that passes through the centers of the two circles must also pass through the point of tangency.
a, A, A, a	If from a point outside a circle, tangents are drawn to a circle, the two tangents are equal and make equal angles with the chord joining the points of tangency.
d, A, B	The angle between a tangent and a chord drawn from the point of tangency equals one-half the angle at the center subtended by the chord. Angle $B = \frac{1}{2}$ angle A
d, A, B, a, b, c	The angle between a tangent and a chord drawn from the point of tangency equals the angle at the periphery subtended by the chord. Angle *B*, between tangent *ab* and chord *cd*, equals angle *A* subtended at the periphery by chord *cd*.
A, B, C, c, d	All angles having their vertex at the periphery of a circle and subtended by the same chord are equal. Angles *A*, *B*, and *C*, all subtended by chord *cd*, are equal.
A, B	If an angle at the circumference of a circle, between two chords, is subtended by the same arc as the angle at the center, between two radii, then the angle at the circumference is equal to one-half of the angle at the center. Angle $A = \frac{1}{2}$ angle B

Geometrical Propositions

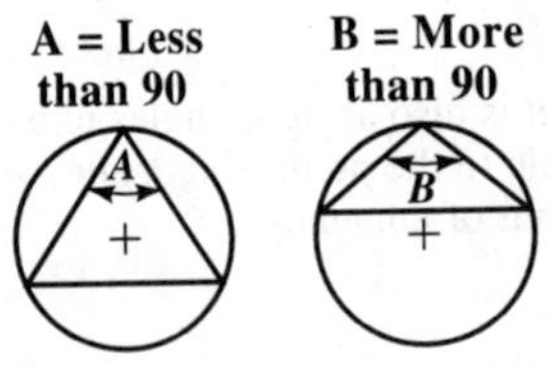 A = Less than 90 B = More than 90	An angle subtended by a chord in a circular segment larger than one-half the circle is an acute angle—an angle less than 90 degrees. An angle subtended by a chord in a circular segment less than one-half the circle is an obtuse angle—an angle greater than 90 degrees.
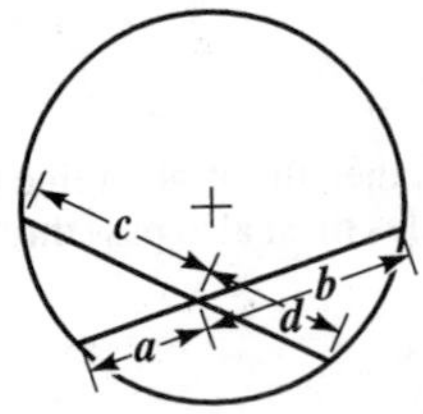	If two chords intersect each other in a circle, then the rectangle of the segments of the one equals the rectangle of the segments of the other. $a \times b = c \times d$
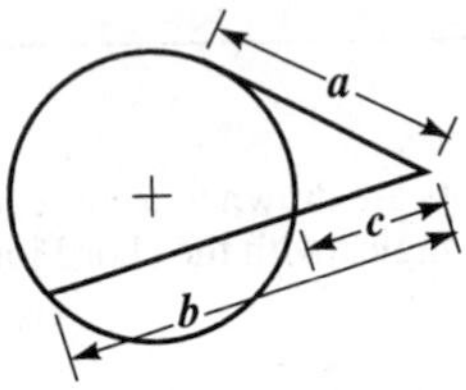	If from a point outside a circle two lines are drawn, one of which intersects the circle and the other is tangent to it, then the rectangle contained by the total length of the intersecting line, and that part of it that is between the outside point and the periphery, equals the square of the tangent. $a^2 = b \times c$
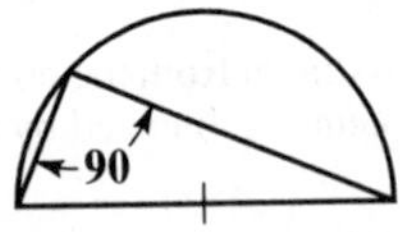 90	If a triangle is inscribed in a semicircle, the angle opposite the diameter is a right (90-degree) angle. All angles at the periphery of a circle, subtended by the diameter, are right (90-degree) angles.
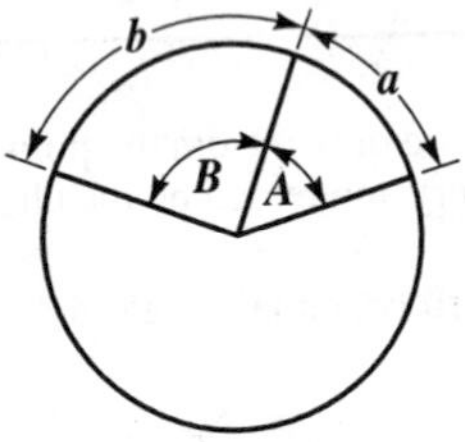	The lengths of circular arcs of the same circle are proportional to the corresponding angles at the center. $A : B = a : b$
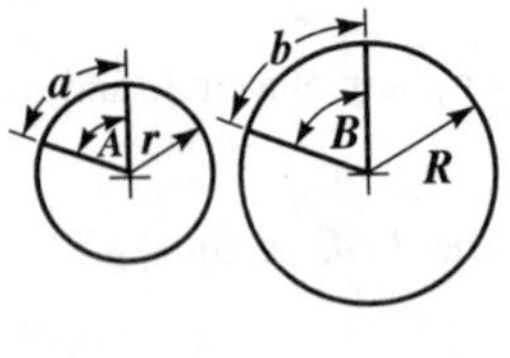	The lengths of circular arcs having the same center angle are proportional to the lengths of the radii. If $A = B$, then $a : b = r : R$.
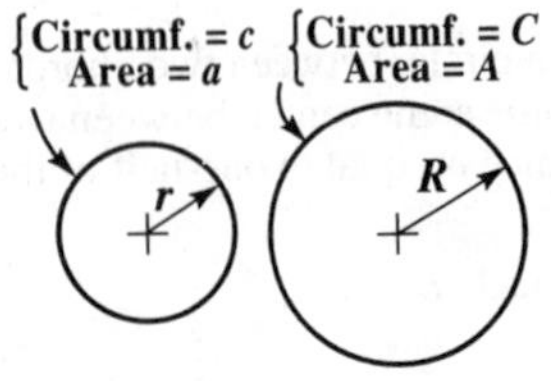 Circumf. = c, Area = a Circumf. = C, Area = A	The circumferences of two circles are proportional to their radii. The areas of two circles are proportional to the squares of their radii. $c : C = r : R$ $a : A = r^2 : R^2$

Geometrical Constructions

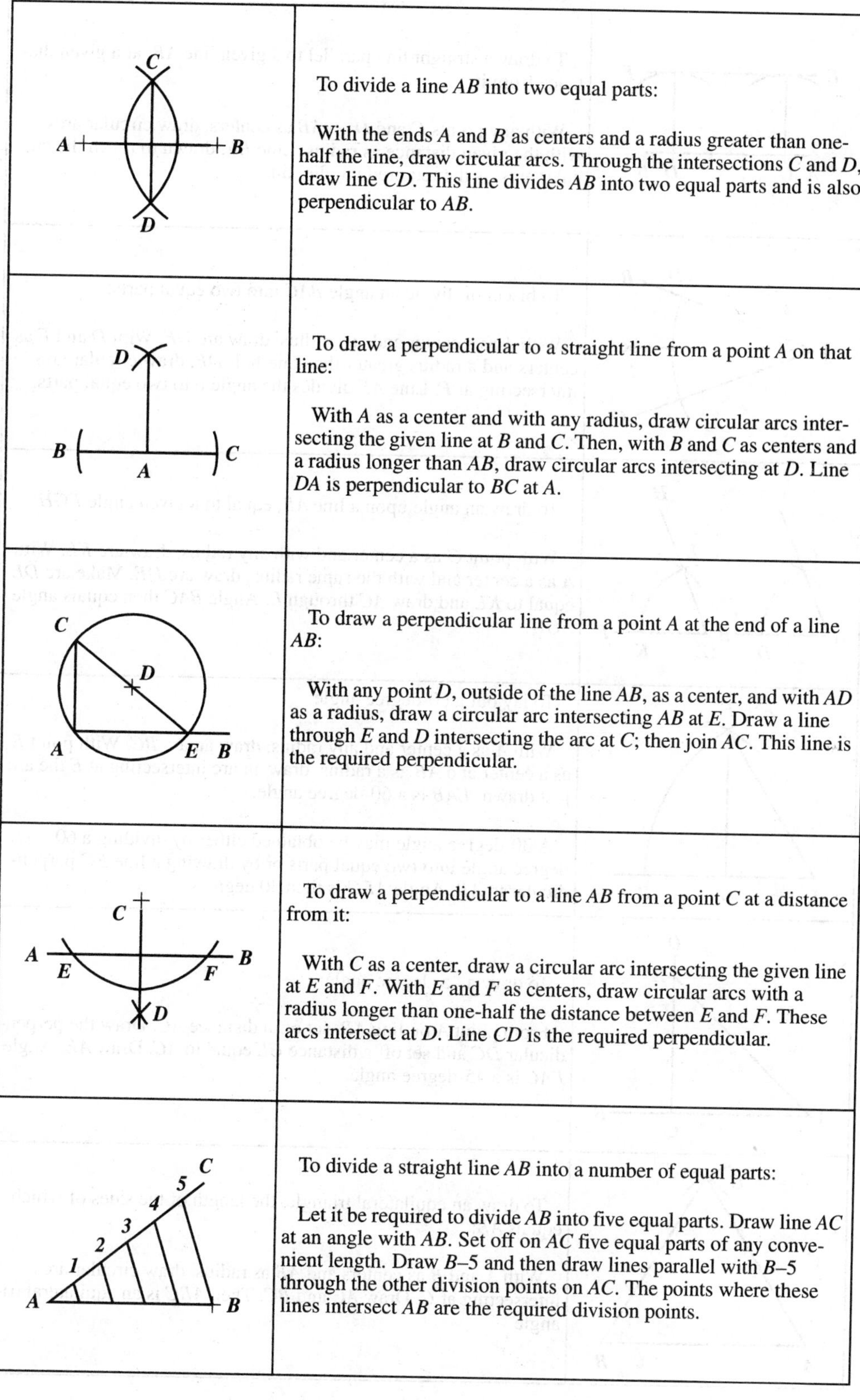

To divide a line *AB* into two equal parts:

With the ends *A* and *B* as centers and a radius greater than one-half the line, draw circular arcs. Through the intersections *C* and *D*, draw line *CD*. This line divides *AB* into two equal parts and is also perpendicular to *AB*.

To draw a perpendicular to a straight line from a point *A* on that line:

With *A* as a center and with any radius, draw circular arcs intersecting the given line at *B* and *C*. Then, with *B* and *C* as centers and a radius longer than *AB*, draw circular arcs intersecting at *D*. Line *DA* is perpendicular to *BC* at *A*.

To draw a perpendicular line from a point *A* at the end of a line *AB*:

With any point *D*, outside of the line *AB*, as a center, and with *AD* as a radius, draw a circular arc intersecting *AB* at *E*. Draw a line through *E* and *D* intersecting the arc at *C*; then join *AC*. This line is the required perpendicular.

To draw a perpendicular to a line *AB* from a point *C* at a distance from it:

With *C* as a center, draw a circular arc intersecting the given line at *E* and *F*. With *E* and *F* as centers, draw circular arcs with a radius longer than one-half the distance between *E* and *F*. These arcs intersect at *D*. Line *CD* is the required perpendicular.

To divide a straight line *AB* into a number of equal parts:

Let it be required to divide *AB* into five equal parts. Draw line *AC* at an angle with *AB*. Set off on *AC* five equal parts of any convenient length. Draw *B*–5 and then draw lines parallel with *B*–5 through the other division points on *AC*. The points where these lines intersect *AB* are the required division points.

Geometrical Constructions

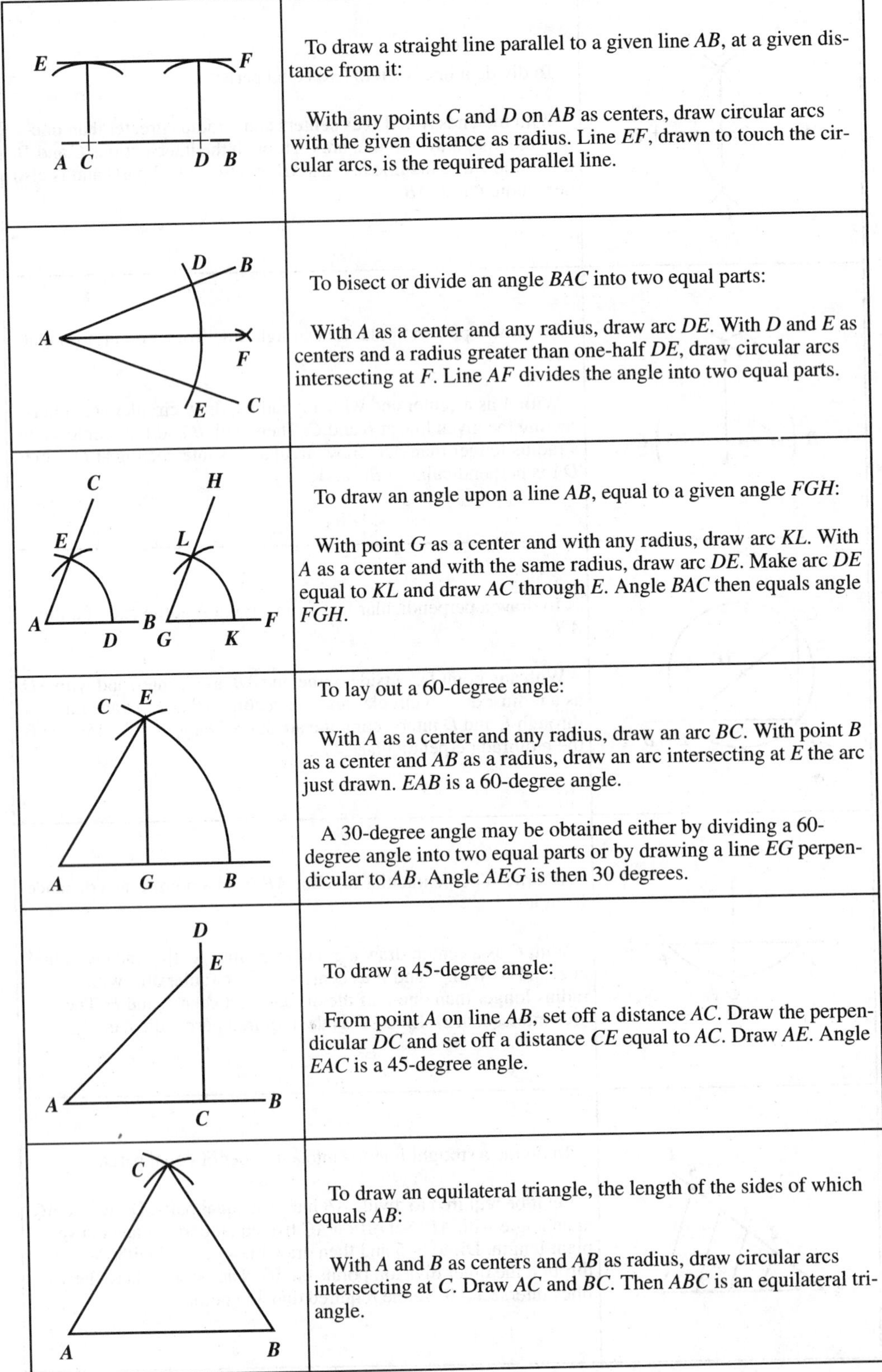

To draw a straight line parallel to a given line *AB*, at a given distance from it:

With any points *C* and *D* on *AB* as centers, draw circular arcs with the given distance as radius. Line *EF*, drawn to touch the circular arcs, is the required parallel line.

To bisect or divide an angle *BAC* into two equal parts:

With *A* as a center and any radius, draw arc *DE*. With *D* and *E* as centers and a radius greater than one-half *DE*, draw circular arcs intersecting at *F*. Line *AF* divides the angle into two equal parts.

To draw an angle upon a line *AB*, equal to a given angle *FGH*:

With point *G* as a center and with any radius, draw arc *KL*. With *A* as a center and with the same radius, draw arc *DE*. Make arc *DE* equal to *KL* and draw *AC* through *E*. Angle *BAC* then equals angle *FGH*.

To lay out a 60-degree angle:

With *A* as a center and any radius, draw an arc *BC*. With point *B* as a center and *AB* as a radius, draw an arc intersecting at *E* the arc just drawn. *EAB* is a 60-degree angle.

A 30-degree angle may be obtained either by dividing a 60-degree angle into two equal parts or by drawing a line *EG* perpendicular to *AB*. Angle *AEG* is then 30 degrees.

To draw a 45-degree angle:

From point *A* on line *AB*, set off a distance *AC*. Draw the perpendicular *DC* and set off a distance *CE* equal to *AC*. Draw *AE*. Angle *EAC* is a 45-degree angle.

To draw an equilateral triangle, the length of the sides of which equals *AB*:

With *A* and *B* as centers and *AB* as radius, draw circular arcs intersecting at *C*. Draw *AC* and *BC*. Then *ABC* is an equilateral triangle.

Geometrical Constructions

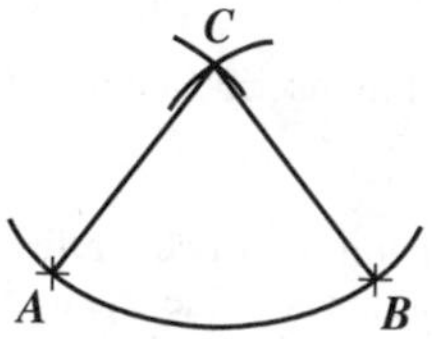	To draw a circular arc with a given radius through two given points *A* and *B*: With *A* and *B* as centers, and the given radius as radius, draw circular arcs intersecting at *C*. With *C* as a center, and the same radius, draw a circular arc through *A* and *B*.
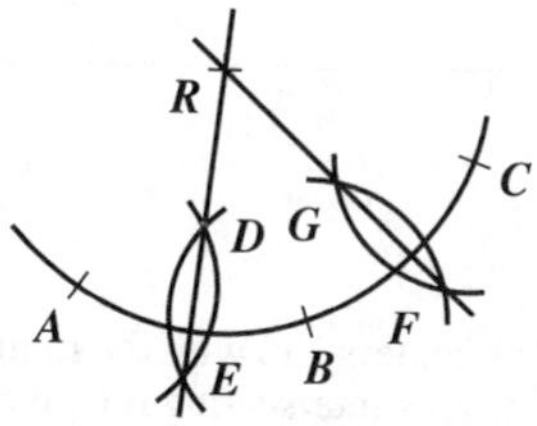	To find the center of a circle or of an arc of a circle: Select three points on the periphery of the circle, as *A*, *B*, and *C*. With each of these points as a center and the same radius, describe arcs intersecting each other. Through the points of intersection, draw lines *DE* and *FG*. Point *H*, where these lines intersect, is the center of the circle.
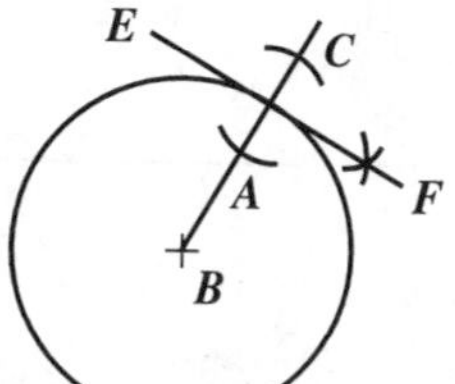	To draw a tangent to a circle from a given point on the circumference: Through the point of tangency *A*, draw a radial line *BC*. At point *A*, draw a line *EF* at right angles to *BC*. This line is the required tangent.
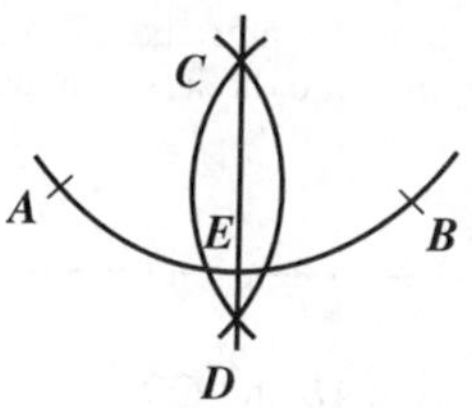	To divide a circular arc *AB* into two equal parts: With *A* and *B* as centers, and a radius larger than half the distance between *A* and *B*, draw circular arcs intersecting at *C* and *D*. Line *CD* divides arc *AB* into two equal parts at *E*.
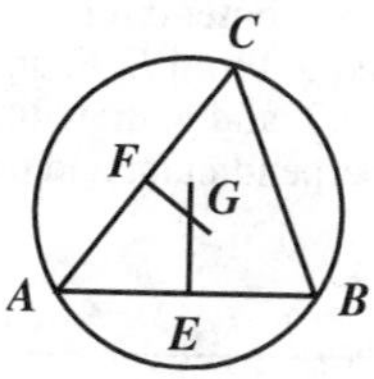	To describe a circle about a triangle: Divide the sides *AB* and *AC* into two equal parts, and from the division points *E* and *F*, draw lines at right angles to the sides. These lines intersect at *G*. With *G* as a center and *GA* as a radius, draw circle *ABC*.
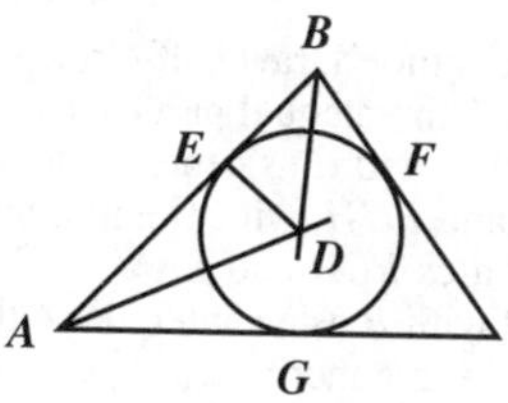	To inscribe a circle in a triangle: Bisect two of the angles, *A* and *B*, by lines intersecting at *D*. From *D*, draw a line *DE* perpendicular to one of the sides, and with *DE* as a radius, draw circle *EFG*.

Geometrical Constructions

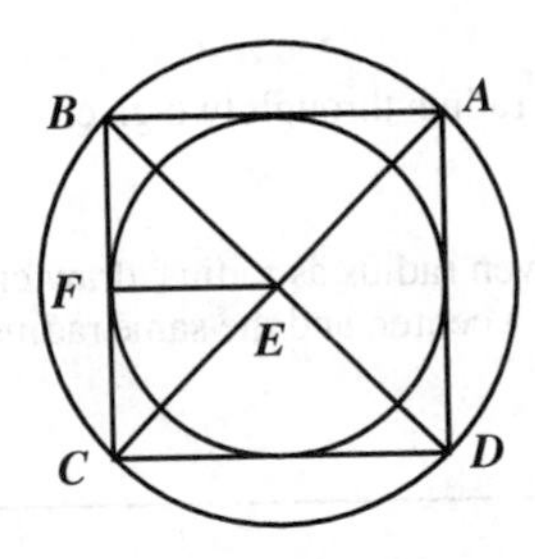	To describe a circle about a square and to inscribe a circle in a square: The centers of both the circumscribed and inscribed circles are located at the point *E*, where the two diagonals of the square intersect. The radius of the circumscribed circle is *AE*, and of the inscribed circle, *EF*.
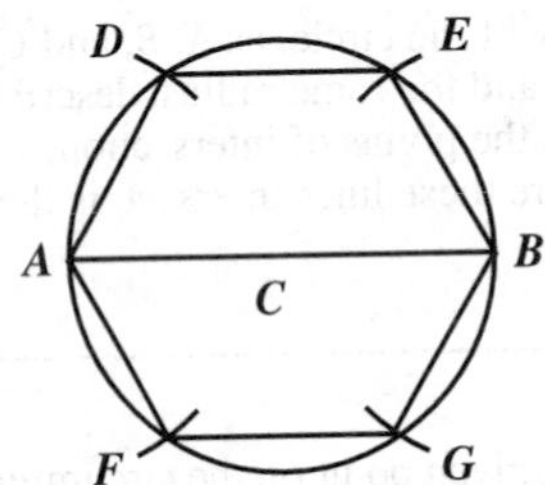	To inscribe a hexagon in a circle: Draw a diameter *AB*. With *A* and *B* as centers and with the radius of the circle as radius, describe circular arcs intersecting the given circle at *D*, *E*, *F*, and *G*. Draw lines *AD*, *DE*, etc., forming the required hexagon.
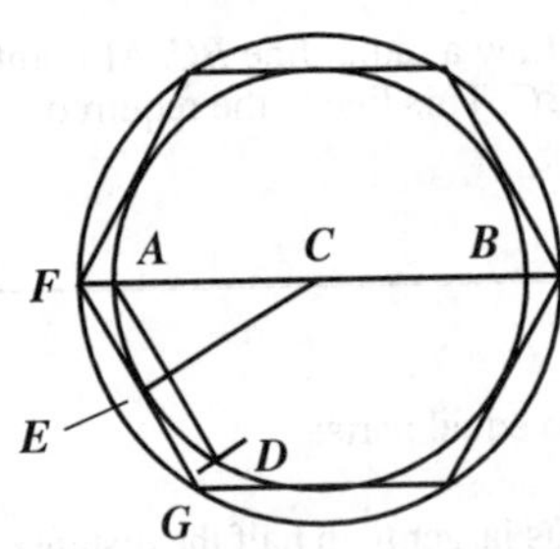	To describe a hexagon about a circle: Draw a diameter *AB*, and with *A* as a center and the radius of the circle as radius, cut the circumference of the given circle at *D*. Join *AD* and bisect it with radius *CE*. Through *E*, draw *FG* parallel to *AD* and intersecting line *AB* at *F*. With *C* as a center and *CF* as radius, draw a circle. Within this circle, inscribe the hexagon as in the preceding problem.
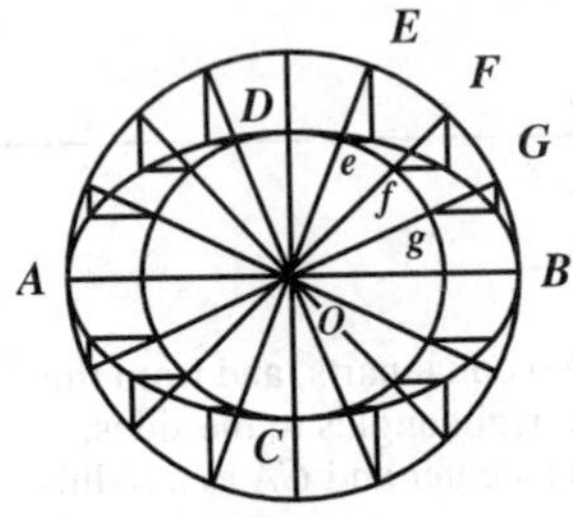	To describe an ellipse with the given axes *AB* and *CD*: Describe circles with *O* as a center and *AB* and *CD* as diameters. From a number of points, *E*, *F*, *G*, etc., on the outer circle, draw radii intersecting the inner circle at *e*, *f*, and *g*. From *E*, *F*, and *G*, draw lines perpendicular to *AB*, and from *e*, *f*, and *g*, draw lines parallel to *AB*. The intersections of these perpendicular and parallel lines are points on the curve of the ellipse.
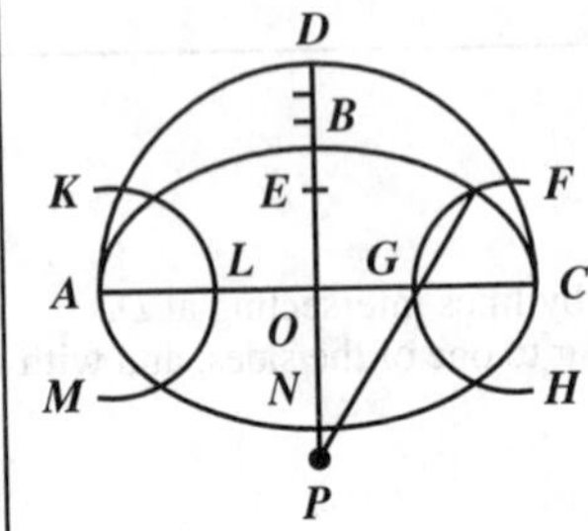	To construct an approximate ellipse by circular arcs: Let *AC* be the major axis and *BN* the minor. Draw half circle *ADC* with *O* as a center. Divide *BD* into three equal parts and set off *BE* equal to one of these parts. With *A* and *C* as centers and *OE* as radius, describe circular arcs *KLM* and *FGH*; with *G* and *L* as centers, and the same radius, describe arcs *FCH* and *KAM*. Through *F* and *G*, drawn line *FP*, and with *P* as a center, draw the arc *FBK*. Arc *HNM* is drawn in the same manner.

Geometrical Constructions

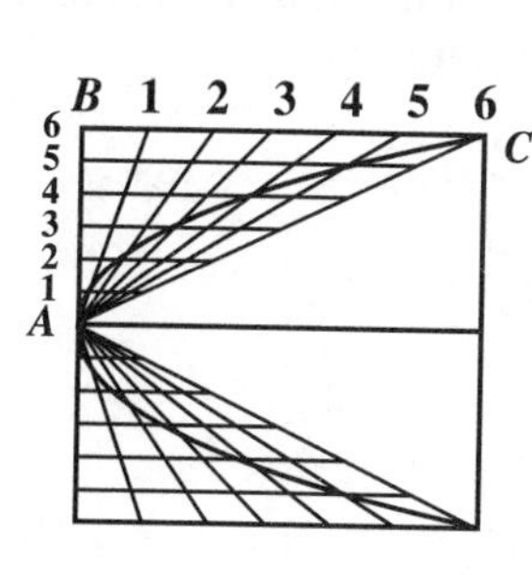	To construct a parabola: Divide line AB into a number of equal parts and divide BC into the same number of parts. From the division points on AB, draw horizontal lines. From the division points on BC, draw lines to point A. The points of intersection between lines drawn from points numbered alike are points on the parabola.
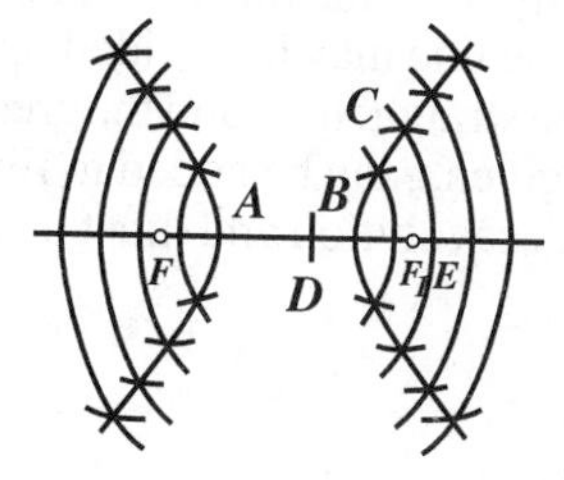	To construct a hyperbola: From focus F, lay off a distance FD equal to the transverse axis, or the distance AB between the two branches of the curve. With F as a center and any distance FE greater than FB as a radius, describe a circular arc. Then with F_1 as a center and DE as a radius, describe arcs intersecting at C and G the arc just described. C and G are points on the hyperbola. Any number of points can be found in a similar manner.
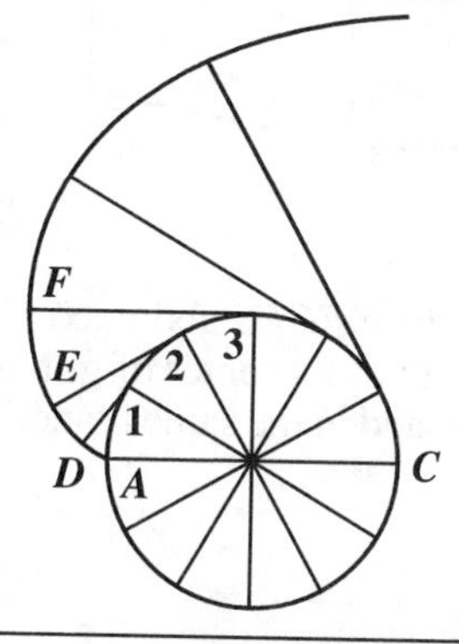	To construct an involute: Divide the circumference of the base circle ABC into a number of equal parts. Through the division points 1, 2, 3, etc., draw tangents to the circle and make the lengths D–1, E–2, F–3, etc., of these tangents equal to the actual length of the arcs A–1, A–2, A–3, etc.
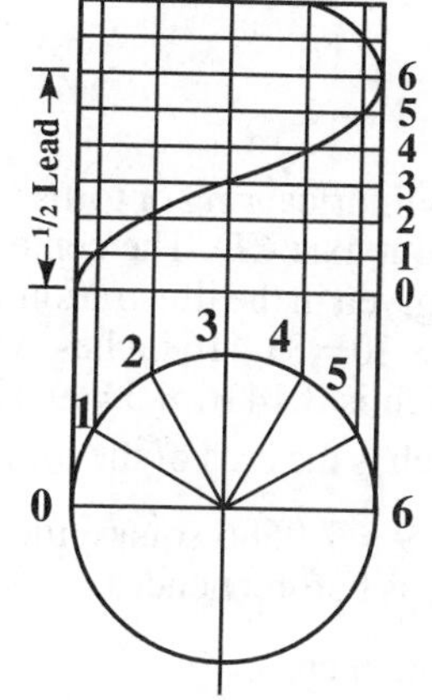	To construct a helix: Divide half the circumference of the cylinder, on the surface of which the helix is to be described, into a number of equal parts. Divide half the lead of the helix into the same number of equal parts. From the division points on the circle representing the cylinder, draw vertical lines, and from the division points on the lead, draw horizontal lines as shown. The intersections between lines numbered alike are points on the helix.

Areas and Volumes

The Prismoidal Formula.—The prismoidal formula is a general formula by which the volume of any prism, pyramid or frustum of a pyramid may be found.

A_1 = area at one end of the body

A_2 = area at the other end

A_m = area of middle section between the two end surfaces

h = height of body

Then, volume V of the body is $V = \frac{h}{6}(A_1 + 4A_m + A_2)$

Pappus or Guldinus Rules.—By means of these rules the area of any surface of revolution and the volume of any solid of revolution may be found. The area of the surface swept out by the revolution of a line ABC (see illustration) about the axis DE equals the length of the line multiplied by the length of the path of its center of gravity, P. If the line is of such a shape that it is difficult to determine its center of gravity, then the line may be divided into a number of short sections, each of which may be considered as a straight line, and the areas swept out by these different sections, as computed by the rule given, may be added to find the total area. The line must lie wholly on one side of the axis of revolution and must be in the same plane.

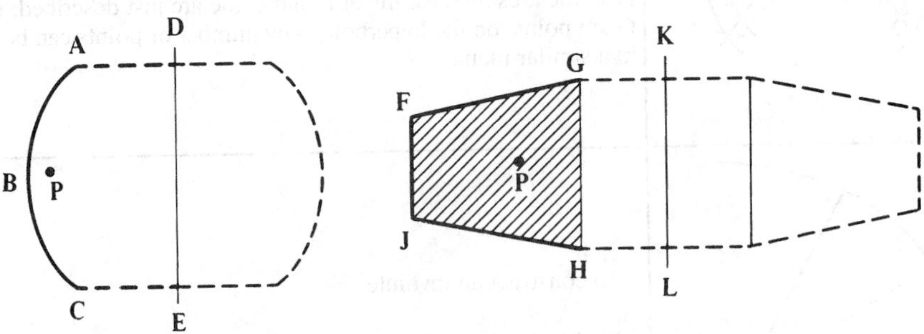

The volume of a solid body formed by the revolution of a surface $FGHJ$ about axis KL equals the area of the surface multiplied by the length of the path of its center of gravity. The surface must lie wholly on one side of the axis of revolution and in the same plane.

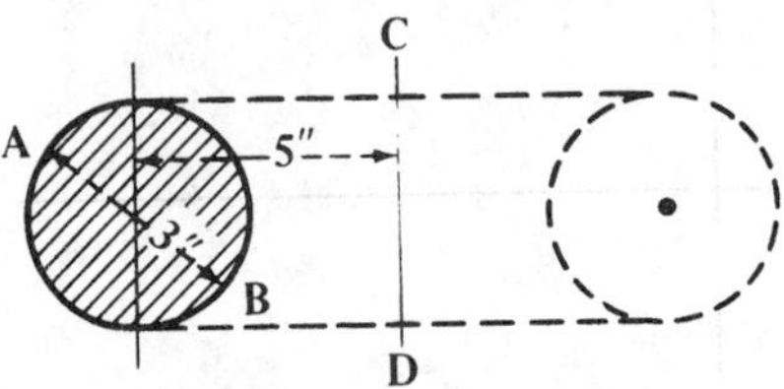

Example: By means of these rules, the area and volume of a cylindrical ring or torus may be found. The torus is formed by a circle AB being rotated about axis CD. The center of gravity of the circle is at its center. Hence, with the dimensions given in the illustration, the length of the path of the center of gravity of the circle is $3.1416 \times 10 = 31.416$ inches. Multiplying by the length of the circumference of the circle, which is $3.1416 \times 3 = 9.4248$ inches, gives $31.416 \times 9.4248 = 296.089$ square inches which is the area of the torus.

The volume equals the area of the circle, which is $0.7854 \times 9 = 7.0686$ square inches, multiplied by the path of the center of gravity, which is 31.416, as before; hence,

$$\text{Volume} = 7.0686 \times 31.416 = 222.067 \text{ cubic inches}$$

Approximate Method for Finding the Area of a Surface of Revolution.—The accompanying illustration is shown in order to give an example of the approximate method based on Guldinus' rule, that can be used for finding the area of a symmetrical body. In the illustration, the dimensions in common fractions are the known dimensions; those in decimals are found by actual measurements on a figure drawn to scale.

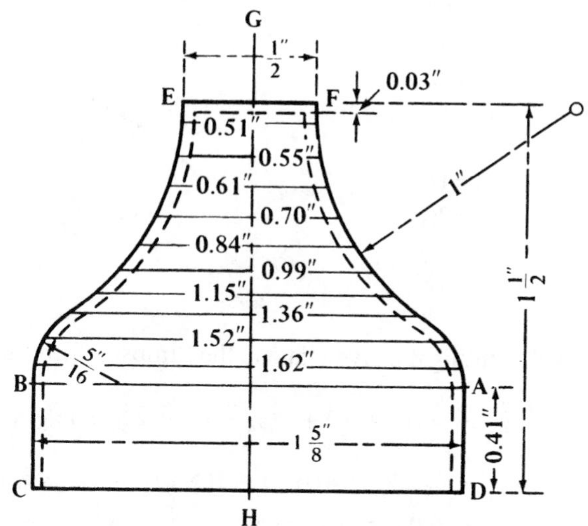

The method for finding the area is as follows: First, separate such areas as are cylindrical, conical, or spherical, as these can be found by exact formulas. In the illustration *ABCD* is a cylinder, the area of the surface of which can be easily found. The top area *EF* is simply a circular area, and can thus be computed separately. The remainder of the surface generated by rotating line *AF* about the axis *GH* is found by the approximate method explained in the previous section. From point *A*, set off equal distances on line *AF*. In the illustration, each division indicated is ⅛ inch long. From the central or middle point of each of these parts draw a line at right angles to the axis of rotation *GH*, measure the length of these lines or diameters (the length of each is given in decimals), add all these lengths together and multiply the sum by the length of one division set off on line *AF* (in this case, ⅛ inch), and multiply this product by π to find the approximate area of the surface of revolution.

In setting off divisions ⅛ inch long along line *AF*, the last division does not reach exactly to point *F*, but only to a point 0.03 inch below it. The part 0.03 inch high at the top of the cup can be considered as a cylinder of ½ inch diameter and 0.03 inch height, the area of the cylindrical surface of which is easily computed. By adding the various surfaces together, the total surface of the cup is found as follows:

Cylinder, 1 ⅝ inch diameter, 0.41 inch high	2.093 square inches
Circle, ½ inch diameter	0.196 square inch
Cylinder, ½ inch diameter, 0.03 inch high	0.047 square inch
Irregular surface	3.868 square inches
Total	6.204 square inches

Area of Plane Surfaces of Irregular Outline.—One of the most useful and accurate methods for determining the approximate area of a plane figure or irregular outline is known as *Simpson's Rule*. In applying Simpson's Rule to find an area the work is done in four steps:

1) Divide the area into an *even* number, N, of parallel strips of equal width W; for example, in the accompanying diagram, the area has been divided into 8 strips of equal width.

2) Label the sides of the strips V_0, V_1, V_2, etc., up to V_N.

3) Measure the heights V_0, V_1, V_2, … , V_N of the sides of the strips.

4) Substitute the heights V_0, V_1, etc., in the following formula to find the area A of the figure:

$$A = \frac{W}{3}[(V_0 + V_N) + 4(V_1 + V_3 + \cdots + V_{N-1}) + 2(V_2 + V_4 + \cdots + V_N$$

Example: The area of the accompanying figure was divided into 8 strips on a full-size drawing and the following data obtained. Calculate the area using Simpson's Rule.

$W = \frac{1}{2}''$
$V_0 = 0''$
$V_1 = \frac{3}{4}''$
$V_2 = 1\frac{1}{4}''$
$V_3 = 1\frac{1}{2}''$
$V_4 = 1\frac{5}{8}''$
$V_5 = 2\frac{1}{4}''$
$V_6 = 2\frac{1}{2}''$
$V_7 = 1\frac{3}{4}''$
$V_8 = \frac{1}{2}''$

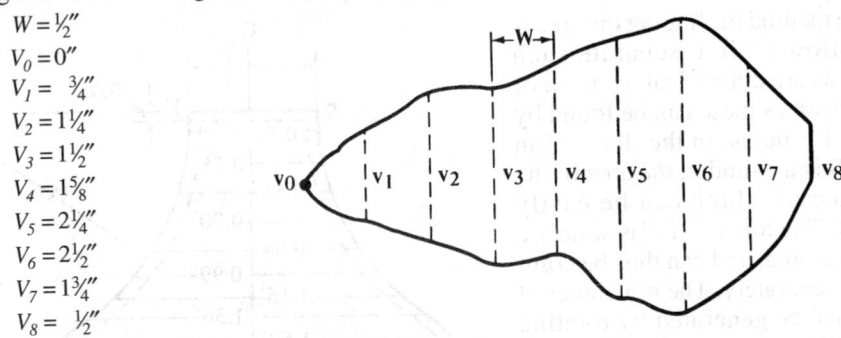

Substituting the given data in the Simpson formula,

$$A = \frac{\frac{1}{2}}{3}[(0 + \tfrac{1}{2}) + 4(\tfrac{3}{4} + 1\tfrac{1}{2} + 2\tfrac{1}{4} + 1\tfrac{3}{4}) + 2(1\tfrac{1}{4} + 1\tfrac{5}{8} + 2\tfrac{1}{2})]$$
$$= \tfrac{1}{6}[(\tfrac{1}{2}) + 4(6\tfrac{1}{4}) + 2(5\tfrac{3}{8})] = \tfrac{1}{6}[36\tfrac{1}{4}]$$
$$= 6.04 \text{ square inches}$$

In applying Simpson's Rule, it should be noted that the larger the number of strips into which the area is divided the more accurate the results obtained.

Areas Enclosed by Cycloidal Curves.—The area between a cycloid and the straight line upon which the generating circle rolls, equals three times the area of the generating circle (see diagram, page 63). The areas between epicycloidal and hypocycloidal curves and the "fixed circle" upon which the generating circle is rolled, may be determined by the following formulas, in which a = radius of the fixed circle upon which the generating circle rolls; b = radius of the generating circle; A = the area for the epicycloidal curve; and A_1 = the area for the hypocycloidal curve.

$$A = \frac{3.1416b^2(3a + 2b)}{a} \qquad A_1 = \frac{3.1416b^2(3a - 2b)}{a}$$

Find the Contents of Cylindrical Tanks at Different Levels.—In conjunction with the table *Segments of Circles for Radius = 1* presented on pages 80 and 81, the following relations can give a close approximation of the liquid contents, at any level, in a cylindrical tank.

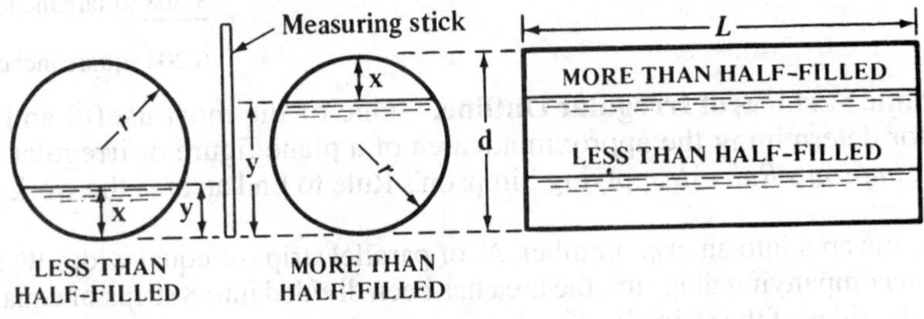

A long measuring rule calibrated in length units or simply a plain stick can be used for measuring contents at a particular level. In turn, the rule or stick can be graduated to serve as a volume gauge for the tank in question. The only requirements are that the cross-section of the tank is circular; the tank's dimensions are known; the gauge rod is inserted vertically through the top center of the tank so that it rests on the exact bottom of the tank; and that consistent English *or* metric units are used throughout the calculations.

$K = Cr^2L$ = Tank Constant (remains the same for any given tank) (1)

$V_T = \pi K$, for a tank that is completely full (2)

$V_s = KA$ (3)

$V = V_s$ when tank is less than half full (4)

$V = V_T - V_s = V_T - KA$, when tank is more than half full (5)

where C = liquid volume conversion factor, the exact value of which depends on the length and liquid volume units being used during measurement: 0.00433 U.S. gal/in^3; 7.48 U.S. gal/ft^3; 0.00360 U.K. gal/in^3; 6.23 U.K. gal/ft^3; 0.001 liter/cm^3; or 1000 liters/m^3

V_T = total volume of liquid tank can hold

V_s = volume formed by segment of circle having depth = x in given tank (see diagram)

V = volume of liquid at particular level in tank

d = diameter of tank; L = length of tank; r = radius of tank (= ½ diameter)

A = segment area of a corresponding unit circle taken from pages 80 or 81

y = actual depth of contents in tank as shown on a gauge rod or stick

x = depth of the segment of a circle to be considered in given tank. As can be seen in above diagram, x is the actual depth of contents (y) when the tank is less than half full, and is the depth of the void ($d - y$) above the contents when the tank is more than half full. From pages 80 and 81 it can also be seen that h, the height of a segment of a corresponding unit circle, is x/r

Example: A tank is 20 feet long and 6 feet in diameter. Convert a long inch-stick into a gauge that is graduated at 1000 and 3000 U.S. gallons.

$$L = 20 \times 12 = 240\text{in.} \qquad r = \tfrac{6}{2} \times 12 = 36\text{in.}$$

From Formula (1): $K = 0.00433(36)^2(240) = 1347$

From Formula (2): $V_T = 3.142 \times 1347 = 4232$ US gal.

The 72-inch mark from the bottom on the inch-stick can be graduated for the rounded full volume "4230"; and the halfway point 36″ for 4230/2 or "2115." It can be seen that the 1000-gal mark would be below the halfway mark. From Formulas (3) and (4):

$A_{1000} = \dfrac{1000}{1347} = 0.7424$ from page 81, h can be interpolated as 0.5724; and $x = y = 36 \times 0.5724 = 20.61$. If the desired level of accuracy permits, interpolation can be omitted by choosing h directly from the table on page 81 for the value of A nearest that calculated above.

Therefore, the 1000-gal mark is graduated 20⅝″ from bottom of rod.

It can be seen that the 3000 mark would be above the halfway mark. Therefore, the circular segment considered is the cross-section of the void space at the top of the tank. From Formulas (3) and (5):

$$A_{3000} = \frac{4230 - 3000}{1347} = 0.9131;\ h = 0.6648;\ x = 36 \times 0.6648 = 23.93''$$

Therefore, the 3000-gal mark is 72.00 − 23.93 = 48.07, or at the 48 1/16″ mark from the bottom.

Areas and Dimensions of Plane Figures

In the following tables are given formulas for the areas of plane figures, together with other formulas relating to their dimensions and properties; the surfaces of solids; and the volumes of solids. The notation used in the formulas is, as far as possible, given in the illustration accompanying them; where this has not been possible, it is given at the beginning of each set of formulas.

Examples are given with each entry, some in English and some in metric units, showing the use of the preceding formula.

Square:

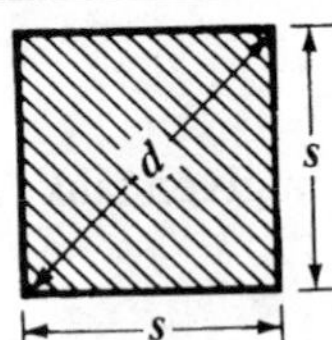

$$\text{Area} = A = s^2 = \tfrac{1}{2}d^2$$
$$s = 0.7071d = \sqrt{A}$$
$$d = 1.414s = 1.414\sqrt{A}$$

Example: Assume that the side *s* of a square is 15 inches. Find the area and the length of the diagonal.

$$\text{Area} = A = s^2 = 15^2 = 225 \text{ square inches}$$
$$\text{Diagonal} = d = 1.414s = 1.414 \times 15 = 21.21 \text{ inches}$$

Example: The area of a square is 625 square inches. Find the length of the side *s* and the diagonal *d*.

$$s = \sqrt{A} = \sqrt{625} = 25 \text{ inches}$$
$$d = 1.414\sqrt{A} = 1.414 \times 25 = 35.35 \text{ inches}$$

Rectangle:

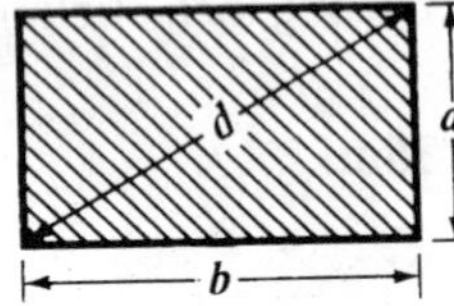

$$\text{Area} = A = ab = a\sqrt{d^2 - a^2} = b\sqrt{d^2 - b^2}$$
$$d = \sqrt{a^2 + b^2}$$
$$a = \sqrt{d^2 - b^2} = A \div b$$
$$a = \sqrt{d^2 - a^2} = A \div a$$

Example: The side *a* of a rectangle is 12 centimeters, and the area 70.5 square centimeters. Find the length of the side *b*, and the diagonal *d*.

$$b = A \div a = 70.5 \div 12 = 5.875 \text{ centimeters}$$
$$d = \sqrt{a^2 + b^2} = \sqrt{12^2 + 5.875^2} = \sqrt{178.516} = 13.361 \text{ centimeters}$$

Example: The sides of a rectangle are 30.5 and 11 centimeters long. Find the area.

$$\text{Area} = A = a \times b = 30.5 \times 11 = 335.5 \text{ square centimeters}$$

Parallelogram:

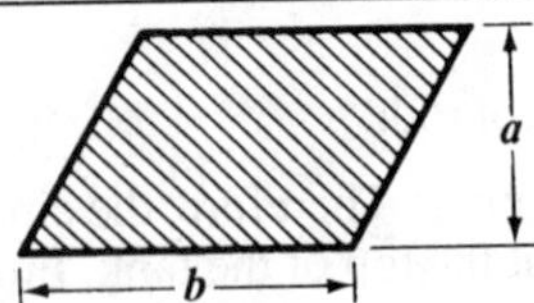

$$\text{Area} = A = ab$$
$$a = A \div b$$
$$b = A \div a$$

Note: The dimension *a* is measured at right angles to line *b*.

Example: The base *b* of a parallelogram is 16 feet. The height *a* is 5.5 feet. Find the area.

$$\text{Area} = A = a \times b = 5.5 \times 16 = 88 \text{ square feet}$$

Example: The area of a parallelogram is 12 square inches. The height is 1.5 inches. Find the length of the base b.

$$b = A \div a = 12 \div 1.5 = 8 \text{ inches}$$

Right-Angled Triangle:

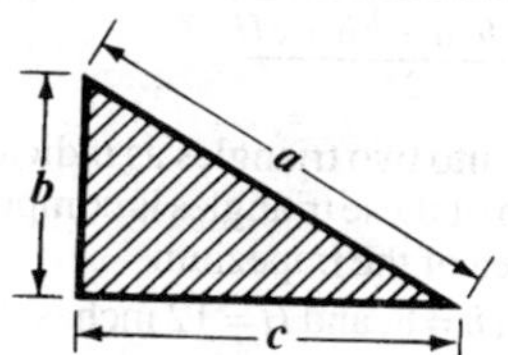

$$\text{Area} = A = \frac{bc}{2}$$

$$a = \sqrt{b^2 + c^2}$$

$$b = \sqrt{a^2 - c^2}$$

$$c = \sqrt{a^2 - b^2}$$

Example: The sides b and c in a right-angled triangle are 6 and 8 inches. Find side a and the area

$$a = \sqrt{b^2 + c^2} = \sqrt{6^2 + 8^2} = \sqrt{36 + 64} = \sqrt{100} = 10 \text{ inches}$$

$$A = \frac{b \times c}{2} = \frac{6 \times 8}{2} = \frac{48}{2} = 24 \text{ square inches}$$

Example: If $a = 10$ and $b = 6$ had been known, but not c, the latter would have been found as follows:

$$c = \sqrt{a^2 - b^2} = \sqrt{10^2 - 6^2} = \sqrt{100 - 36} = \sqrt{64} = 8 \text{ inches}$$

Acute-Angled Triangle:

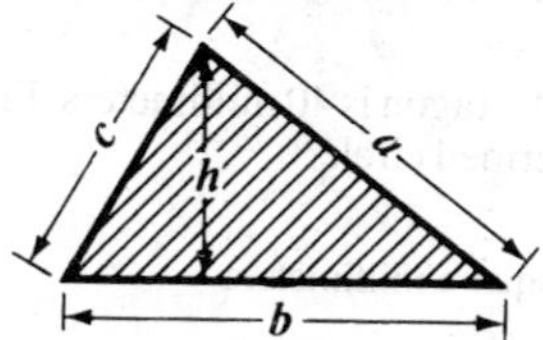

$$\text{Area} = A = \frac{bh}{2} = \frac{b}{2}\sqrt{a^2 - \left(\frac{a^2 + b^2 - c^2}{2b}\right)^2}$$

If $S = \frac{1}{2}(a + b + c)$, then

$$A = \sqrt{S(S-a)(S-b)(S-c)}$$

Example: If $a = 10$, $b = 9$, and $c = 8$ centimeters, what is the area of the triangle?

$$A = \frac{b}{2}\sqrt{a^2 - \left(\frac{a^2 + b^2 - c^2}{2b}\right)^2} = \frac{9}{2}\sqrt{10^2 - \left(\frac{10^2 + 9^2 - 8^2}{2 \times 9}\right)^2} = 4.5\sqrt{100 - \left(\frac{117}{18}\right)^2}$$

$$= 4.5\sqrt{100 - 42.25} = 4.5\sqrt{57.75} = 4.5 \times 7.60 = 34.20 \text{ square centimeters}$$

Obtuse-Angled Triangle:

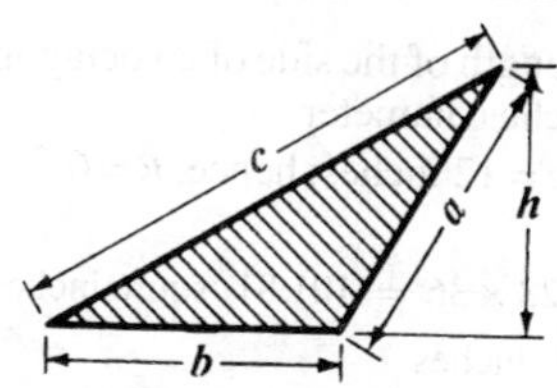

$$\text{Area} = A = \frac{bh}{2} = \frac{b}{2}\sqrt{a^2 - \left(\frac{c^2 - a^2 - b^2}{2b}\right)^2}$$

If $S = \frac{1}{2}(a + b + c)$, then

$$A = \sqrt{S(S-a)(S-b)(S-c)}$$

Example: The side $a = 5$, side $b = 4$, and side $c = 8$ inches. Find the area.

$$S = \tfrac{1}{2}(a + b + c) = \tfrac{1}{2}(5 + 4 + 8) = \tfrac{1}{2} \times 17 = 8.5$$

$$A = \sqrt{S(S-a)(S-b)(S-c)} = \sqrt{8.5(8.5-5)(8.5-4)(8.5-8)}$$

$$= \sqrt{8.5 \times 3.5 \times 4.5 \times 0.5} = \sqrt{66.937} = 8.18 \text{ square inches}$$

Trapezoid:

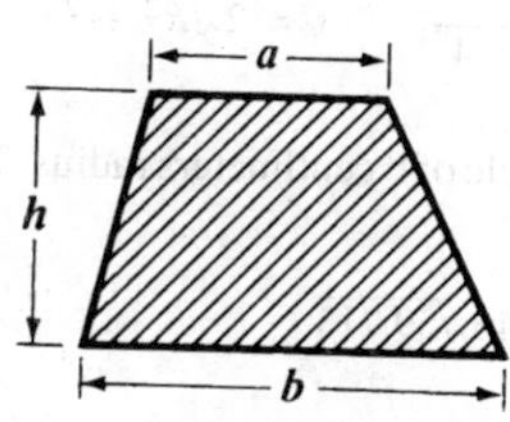

$$\text{Area} = A = \frac{(a + b)h}{2}$$

Note: In Britain, this figure is called a *trapezium* and the one below it is known as a *trapezoid*, the terms being reversed.

Example: Side $a = 23$ meters, side $b = 32$ meters, and height $h = 12$ meters. Find the area.

$$A = \frac{(a + b)h}{2} = \frac{(23 + 32)12}{2} = \frac{55 \times 12}{2} = 330 \text{ square meters}$$

Trapezium:

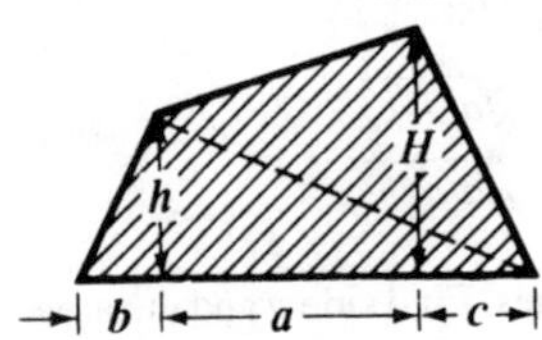

$$\text{Area} = A = \frac{(H+h)a + bh + cH}{2}$$

A trapezium can also be divided into two triangles as indicated by the dashed line. The area of each of these triangles is computed, and the results added to find the area of the trapezium.

Example: Let $a = 10$, $b = 2$, $c = 3$, $h = 8$, and $H = 12$ inches. Find the area.

$$A = \frac{(H+h)a + bh + cH}{2} = \frac{(12+8)10 + 2\times 8 + 3\times 12}{2}$$

$$= \frac{20\times 10 + 16 + 36}{2} = \frac{252}{2} = 126 \text{ square inches}$$

Regular Hexagon:

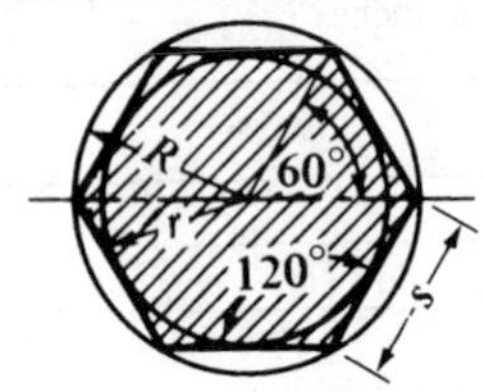

$A = 2.598s^2 = 2.598R^2 = 3.464r^2$

$R = s$ = radius of circumscribed circle = $1.155r$

r = radius of inscribed circle = $0.866s = 0.866R$

$s = R = 1.155r$

Example: The side s of a regular hexagon is 40 millimeters. Find the area and the radius r of the inscribed circle.

$$A = 2.598s^2 = 2.598\times 40^2 = 2.598\times 1600 = 4156.8 \text{ square millimeters}$$

$$r = 0.866s = 0.866\times 40 = 34.64 \text{ millimeters}$$

Example: What is the length of the side of a hexagon that is drawn around a circle of 50 millimeters radius? — Here $r = 50$. Hence, $s = 1.155r = 1.155\times 50 = 57.75$ millimeters

Regular Octagon:

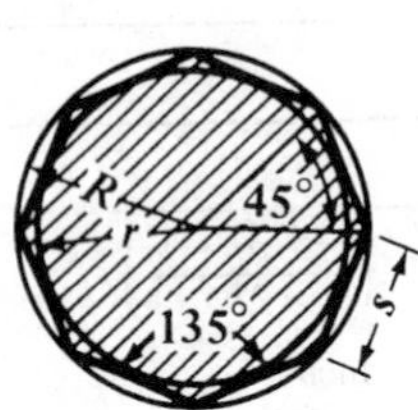

A = area = $4.828s^2 = 2.828R^2 = 3.3\,14r^2$

R = radius of circumscribed circle = $1.307s = 1.082r$

r = radius of inscribed circle = $1.207s = 0.924R$

$s = 0.765R = 0.828r$

Example: Find the area and the length of the side of an octagon that is inscribed in a circle of 12 inches diameter.

Diameter of circumscribed circle = 12 inches; hence, $R = 6$ inches.

$$A = 2.828R^2 = 2.828\times 6^2 = 2.828\times 36 = 101.81 \text{ squre inches}$$

$$s = 0.765R = 0.765\times 6 = 4.590 \text{ inches}$$

Regular Polygon:

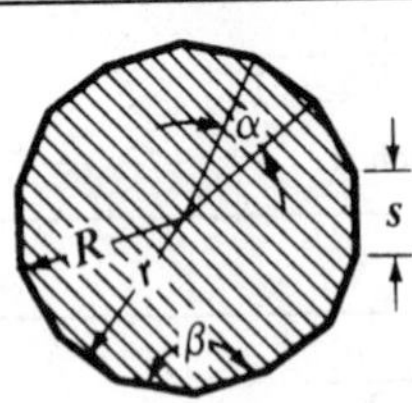

A = area $\qquad n$ = number of sides

$\alpha = 360° \div n \qquad \beta = 180° - \alpha$

$$A = \frac{nsr}{2} = \frac{ns}{2}\sqrt{R^2 - \frac{s^2}{4}}$$

$$R = \sqrt{r^2 + \frac{s^2}{4}} \qquad r = \sqrt{R^2 - \frac{s^2}{4}} \qquad s = 2\sqrt{R^2 - r^2}$$

Example: Find the area of a polygon having 12 sides, inscribed in a circle of 8 centimeters radius. The length of the side s is 4.141 centimeters.

$$A = \frac{ns}{2}\sqrt{R^2 - \frac{s^2}{4}} = \frac{12\times 4.141}{2}\sqrt{8^2 - \frac{4.141^2}{4}} = 24.846\sqrt{59.713}$$

$$= 24.846\times 7.727 = 191.98 \text{ square centimeters}$$

Circle:

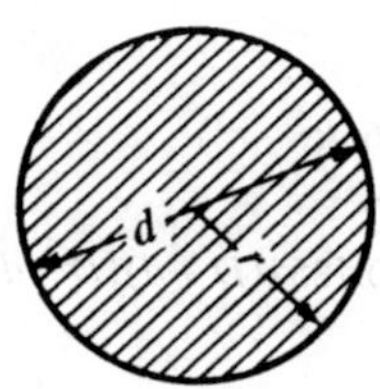

$$\text{Area} = A = \pi r^2 = 3.1416r^2 = 0.7854d^2$$

$$\text{Circumference} = C = 2\pi r = 6.832r = 3.1416d$$

$$r = C \div 6.2832 = \sqrt{A \div 3.1416} = 0.564\sqrt{A}$$

$$d = C \div 3.1416 = \sqrt{A \div 0.7854} = 1.128\sqrt{A}$$

Length of arc for center angle of $1° = 0.008727d$

Length of arc for center angle of $n° = 0.008727nd$

Example: Find the area A and circumference C of a circle with a diameter of $2\frac{3}{4}$ inches.

$$A = 0.7854d^2 = 0.7854 \times 2.75^2 = 0.7854 \times 2.75 \times 2.75 = 5.9396 \text{ square inches}$$

$$C = 3.1416d = 3.1416 \times 2.75 = 8.6394 \text{ inches}$$

Example: The area of a circle is 16.8 square inches. Find its diameter.

$$d = 1.128\sqrt{A} = 1.128\sqrt{16.8} = 1.128 \times 4.099 = 4.624 \text{ inches}$$

Circular Sector:

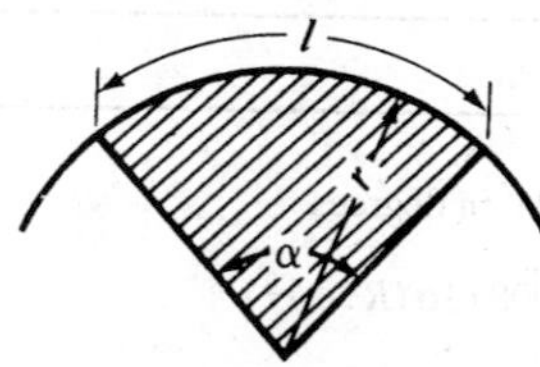

$$\text{Length of arc} = l = \frac{r \times \alpha \times 3.1416}{180} = 0.01745r\alpha = \frac{2A}{r}$$

$$\text{Area} = A = \tfrac{1}{2}rl = 0.008727\alpha r^2$$

$$\text{Angle, in degrees} = \alpha = \frac{57.296\ l}{r} \qquad r = \frac{2A}{l} = \frac{57.296\ l}{\alpha}$$

Example: The radius of a circle is 35 millimeters, and angle α of a sector of the circle is 60 degrees. Find the area of the sector and the length of arc l.

$$A = 0.008727\alpha r^2 = 0.008727 \times 60 \times 35^2 = 641.41 \text{mm}^2 = 6.41 \text{cm}^2$$

$$l = 0.01745r\alpha = 0.01745 \times 35 \times 60 = 36.645 \text{ millimeters}$$

Circular Segment:

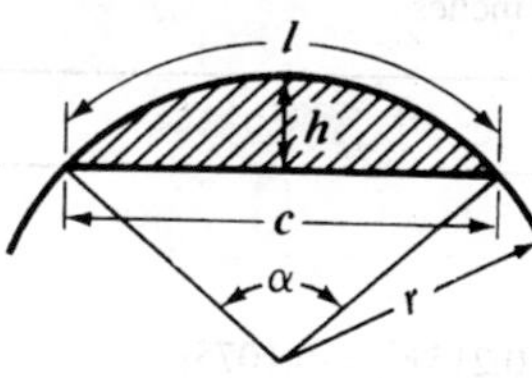

A = area $\quad l$ = length of arc $\quad \alpha$ = angle, in degrees

$$c = 2\sqrt{h(2r-h)} \qquad A = \tfrac{1}{2}[rl - c(r-h)]^*$$

$$r = \frac{c^2 + 4h^2}{8h} \qquad l = 0.01745r\alpha$$

$$h = r - \tfrac{1}{2}\sqrt{4r^2 - c^2} = r[1 - \cos(\alpha/2)] \qquad \alpha = \frac{57.296\ l}{r}$$

Example: The radius r is 60 inches and the height h is 8 inches. Find the length of the chord c.

$$c = 2\sqrt{h(2r-h)} = 2\sqrt{8 \times (2 \times 60 - 8)} = 2\sqrt{896} = 2 \times 29.93 = 59.86 \text{ inches}$$

Example: If $c = 16$, and $h = 6$ inches, what is the radius of the circle of which the segment is a part?

$$r = \frac{c^2 + 4h^2}{8h} = \frac{16^2 + 4 \times 6^2}{8 \times 6} = \frac{256 + 144}{48} = \frac{400}{48} = 8\tfrac{1}{3} \text{ inches}$$

Cycloid:

$$\text{Area} = A = 3\pi r^2 = 9.4248r^2 = 2.3562d^2$$

$$= 3 \times \text{area of generating circle}$$

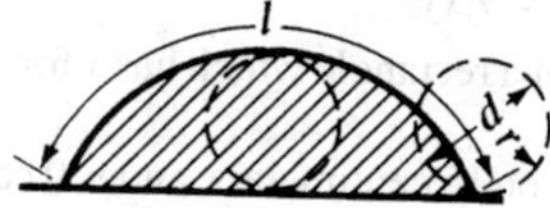

$$\text{Length of cycloid} = l = 8r = 4d$$

Example: The diameter of the generating circle of a cycloid is 6 inches. Find the length l of the cycloidal curve, and the area enclosed between the curve and the base line.

$$l = 4d = 4 \times 6 = 24 \text{ inches}$$

$$A = 2.3562d^2 = 2.3562 \times 6^2 = 84.82 \text{ square inches}$$

Circular Ring:

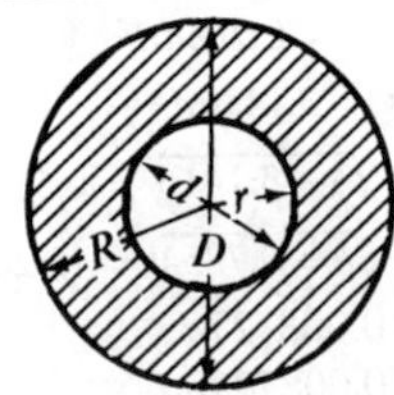

$$\text{Area} = A = \pi(R^2 - r^2) = 3.1416(R^2 - r^2)$$
$$= 3.1416(R + r)(R - r)$$
$$= 0.7845(D^2 - d^2) = 0.7854(D + d)(D - d)$$

Example: Let the outside diameter $D = 12$ centimeters and the inside diameter $d = 8$ centimeters. Find the area of the ring.

$$A = 0.7854(D^2 - d^2) = 0.7854(12^2 - 8^2) = 0.7854(144 - 64) = 0.7854 \times 80$$
$$= 62.83 \text{ square centimeters}$$

By the alternative formula:

$$A = 0.7854(D + d)(D - d) = 0.7854(12 + 8)(12 - 8) = 0.7854 \times 20 \times 4$$
$$= 62.83 \text{ square centimeters}$$

Circular Ring Sector:

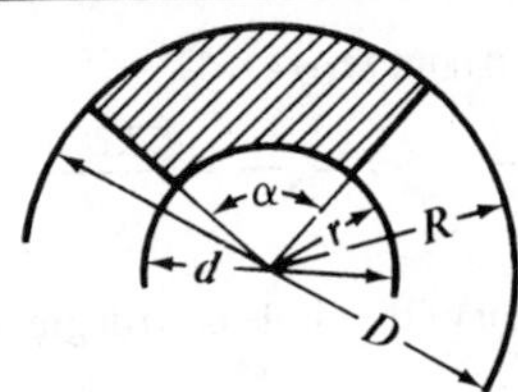

A = area α = angle, in degrees

$$A = \frac{\alpha\pi}{360}(R^2 - r^2) = 0.00873\alpha(R^2 - r^2)$$
$$= \frac{\alpha\pi}{4 \times 360}(D^2 - d^2) = 0.00218\alpha(D^2 - d^2)$$

Example: Find the area, if the outside radius $R = 5$ inches, the inside radius $r = 2$ inches, and $\alpha = 72$ degrees.

$$A = 0.00873\alpha(R^2 - r^2) = 0.00873 \times 72(5^2 - 2^2)$$
$$= 0.6286(25 - 4) = 0.6286 \times 21 = 13.2 \text{ square inches}$$

Spandrel or Fillet:

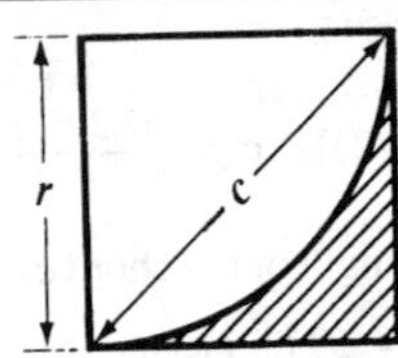

$$\text{Area} = A = r^2 - \frac{\pi r^2}{4} = 0.215r^2 = 0.1075c^2$$

Example: Find the area of a spandrel, the radius of which is 0.7 inch.

$$A = 0.215r^2 = 0.215 \times 0.7^2 = 0.105 \text{ square inch}$$

Example: If chord c were given as 2.2 inches, what would be the area?

$$A = 0.1075c^2 = 0.1075 \times 2.2^2 = 0.520 \text{ square inch}$$

Parabola:

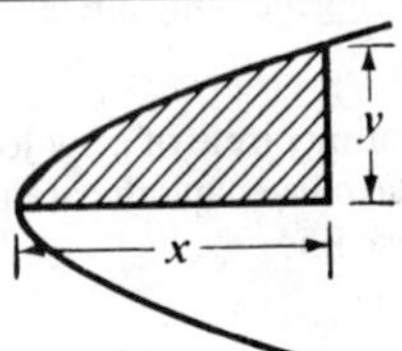

$$\text{Area} = A = \tfrac{2}{3}xy$$

(The area is equal to two-thirds of a rectangle which has x for its base and y for its height.)

Example: Let x in the illustration be 15 centimeters, and y, 9 centimeters. Find the area of the shaded portion of the parabola.

$$A = \tfrac{2}{3} \times xy = \tfrac{2}{3} \times 15 \times 9 = 10 \times 9 = 90 \text{ square centimeters}$$

Parabola:

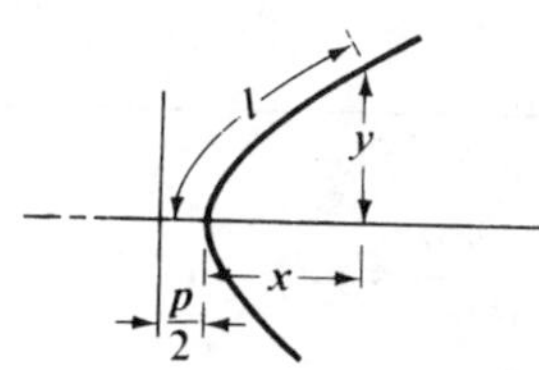

$$l = \text{length of arc} = \frac{p}{2}\left[\sqrt{\frac{2x}{p}\left(1+\frac{2x}{p}\right)} + \ln\left(\sqrt{\frac{2x}{p}} + \sqrt{1+\frac{2x}{p}}\right)\right]$$

When x is small in proportion to y, the following is a close approximation:

$$l = y\left[1 + \frac{2}{3}\left(\frac{x}{y}\right)^2 - \frac{2}{5}\left(\frac{x}{y}\right)^4\right] \text{or } l = \sqrt{y^2 + \frac{4}{3}x^2}$$

Example: If $x = 2$ and $y = 24$ feet, what is the approximate length l of the parabolic curve?

$$l = y\left[1 + \frac{2}{3}\left(\frac{x}{y}\right)^2 - \frac{2}{5}\left(\frac{x}{y}\right)^4\right] = 24\left[1 + \frac{2}{3}\left(\frac{2}{24}\right)^2 - \frac{2}{5}\left(\frac{2}{24}\right)^4\right]$$

$$= 24\left[1 + \frac{2}{3} \times \frac{1}{144} - \frac{2}{5} \times \frac{1}{20{,}736}\right] = 24 \times 1.0046 = 24.11 \text{ feet}$$

Segment of Parabola:

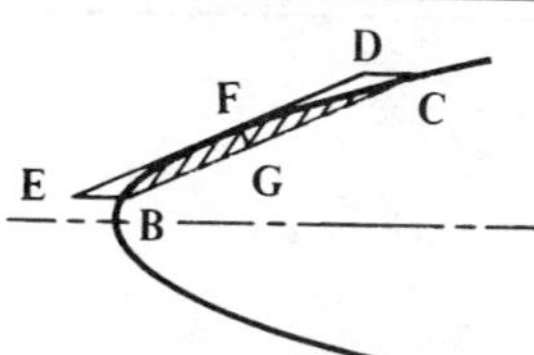

$$\text{Area BFC} = A = \tfrac{2}{3} \text{ area of parallelogram BCDE}$$

If FG is the height of the segment, measured at right angles to BC, then:

$$\text{Area of segment BFC} = \tfrac{2}{3}\text{BC} \times \text{FG}$$

Example: The length of the chord $BC = 19.5$ inches. The distance between lines BC and DE, measured at right angles to BC, is 2.25 inches. This is the height of the segment. Find the area.

$$\text{Area} = A = \tfrac{2}{3}\text{BC} \times \text{FG} = \tfrac{2}{3} \times 19.5 \times 2.25 = 29.25 \text{ square inches}$$

Hyperbola:

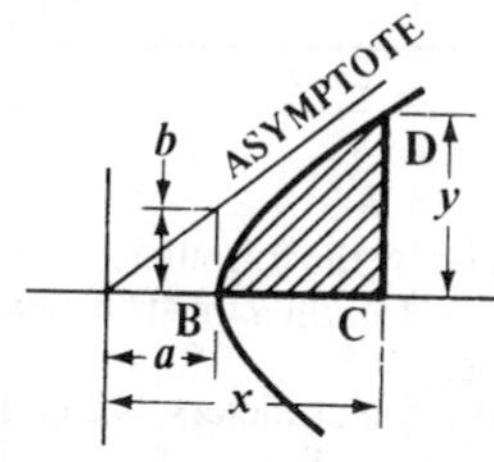

$$\text{Area BCD} = A = \frac{xy}{2} - \frac{ab}{2}\ln\left(\frac{x}{a} + \frac{y}{b}\right)$$

Example: The half-axes a and b are 3 and 2 inches, respectively. Find the area shown shaded in the illustration for $x = 8$ and $y = 5$.

Inserting the known values in the formula:

$$A = \frac{8 \times 5}{2} - \frac{3 \times 2}{2} \times \ln\left(\frac{8}{3} + \frac{5}{2}\right) = 20 - 3 \times \ln 5.167$$

$$= 20 - 3 \times 1.6423 = 20 - 4.927 = 15.073 \text{ square inches}$$

Ellipse:

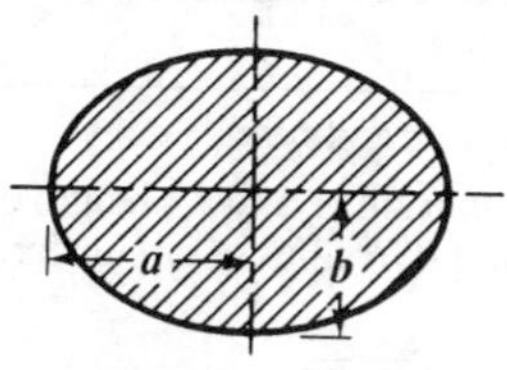

$$\text{Area} = A = \pi ab = 3.1416ab$$

An approximate formula for the perimeter is

$$\text{Perimeter} = P = 3.1416\sqrt{2(a^2 + b^2)}$$

A closer approximation is $P = 3.1416\sqrt{2(a^2 + b^2) - \frac{(a-b)^2}{2.2}}$

Example: The larger or major axis is 200 millimeters. The smaller or minor axis is 150 millimeters. Find the area and the approximate circumference. Here, then, $a = 100$, and $b = 75$.

$$A = 3.1416ab = 3.1416 \times 100 \times 75 = 23{,}562 \text{ square millimeters} = 235.62 \text{ square centimeters}$$

$$P = 3.1416\sqrt{2(a^2 + b^2)} = 3.1416\sqrt{2(100^2 + 75^2)} = 3.1416\sqrt{2 \times 15{,}625}$$

$$= 3.1416\sqrt{31{,}250} = 3.1416 \times 176.78 = 555.37 \text{ millimeters} = (55.537 \text{ centimeters})$$

Volumes of Solids

Cube:

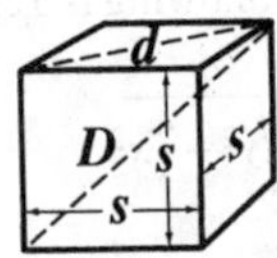

Diagonal of cube face $= d = s\sqrt{2}$

Diagonal of cube $= D = \sqrt{\frac{3d^2}{2}} = s\sqrt{3} = 1.732s$

Volume $= V = s^3$

$s = \sqrt[3]{V}$

Example: The side of a cube equals 9.5 centimeters. Find its volume.

Volume $= V = s^3 = 9.5^3 = 9.5 \times 9.5 \times 9.5 = 857.375$ cubic centimeters

Example: The volume of a cube is 231 cubic centimeters. What is the length of the side?

$s = \sqrt[3]{V} = \sqrt[3]{231} = 6.136$ centimeters

Square Prism:

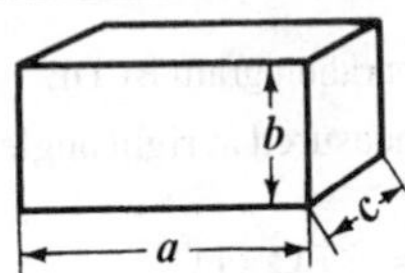

Volume $= V = abc$

$a = \frac{V}{bc}$ $b = \frac{V}{ac}$ $c = \frac{V}{ab}$

Example: In a square prism, $a = 6$, $b = 5$, $c = 4$. Find the volume.

$V = a \times b \times c = 6 \times 5 \times 4 = 120$ cubic inches

Example: How high should a box be made to contain 25 cubic feet, if it is 4 feet long and 2½ feet wide? Here, $a = 4$, $c = 2.5$, and $V = 25$. Then,

$b = \text{depth} = \frac{V}{ac} = \frac{25}{4 \times 2.5} = \frac{25}{10} = 2.5$ feet

Prism:

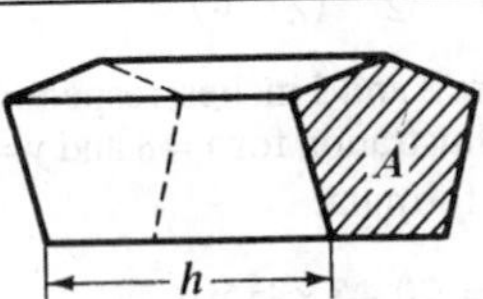

V = volume

A = area of end surface

$V = h \times A$

The area A of the end surface is found by the formulas for areas of plane figures on the preceding pages. Height h must be measured perpendicular to the end surface.

Example: A prism, having for its base a regular hexagon with a side s of 7.5 centimeters, is 25 centimeters high. Find the volume.

Area of hexagon $= A = 2.598s^2 = 2.598 \times 56.25 = 146.14$ square centimeters

Volume of prism $= h \times A = 25 \times 146.14 = 3653.5$ cubic centimeters

Pyramid:

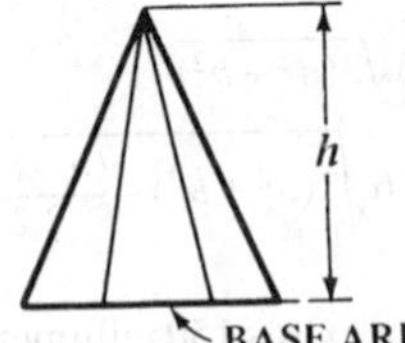

Volume $= V = \frac{1}{3}h \times$ area of base

If the base is a regular polygon with n sides, and s = length of side, r = radius of inscribed circle, and R = radius of circumscribed circle, then:

$$V = \frac{nsrh}{6} = \frac{nsh}{6}\sqrt{R^2 - \frac{s^2}{4}}$$

Example: A pyramid, having a height of 9 feet, has a base formed by a rectangle, the sides of which are 2 and 3 feet, respectively. Find the volume.

Area of base $= 2 \times 3 = 6$ square feet; $h = 9$ feet

Volume $= V = \frac{1}{3}h \times$ area of base $= \frac{1}{3} \times 9 \times 6 = 18$ cubic feet

Frustum of Pyramid:

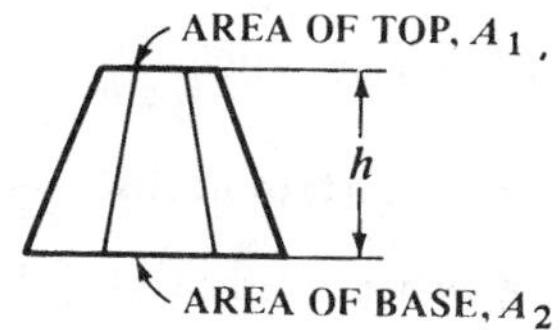

$$\text{Volume} = V = \frac{h}{3}(A_1 + A_2 + \sqrt{A_1 \times A_2})$$

Example: The pyramid in the previous example is cut off 4½ feet from the base, the upper part being removed. The sides of the rectangle forming the top surface of the frustum are, then, 1 and 1½ feet long, respectively. Find the volume of the frustum.

Area of top $= A_1 = 1 \times 1\frac{1}{2} = 1\frac{1}{2}$ sq. ft. Area of base $= A_2 = 2 \times 3 = 6$ sq. ft.

$$V = \frac{4 \cdot 5}{3}(1.5 + 6 + \sqrt{1.5 \times 6}) = 1.5(7.5 + \sqrt{9}) = 1.5 \times 10.5 = 15.75 \text{ cubic feet}$$

Wedge:

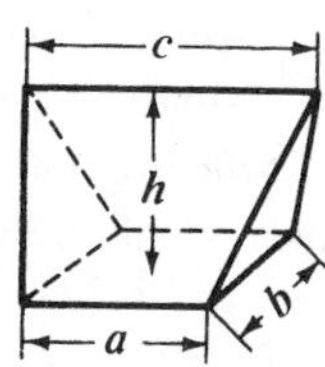

$$\text{Volume} = V = \frac{(2a + c)bh}{6}$$

Example: Let $a = 4$ inches, $b = 3$ inches, and $c = 5$ inches. The height $h = 4.5$ inches. Find the volume.

$$V = \frac{(2a + c)bh}{6} = \frac{(2 \times 4 + 5) \times 3 \times 4.5}{6} = \frac{(8 + 5) \times 13.5}{6}$$

$$= \frac{175.5}{6} = 29.25 \text{ cubic inches}$$

Cylinder:

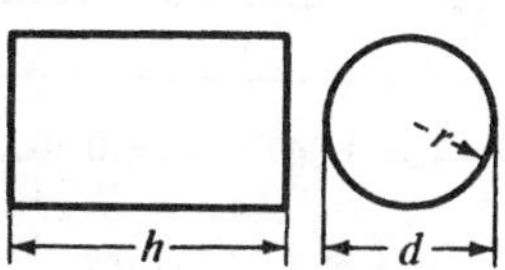

$$\text{Volume} = V = 3.1416r^2h = 0.7854d^2h$$

$$\text{Area of cylindrical surface} = S = 6.2832rh = 3.1416dh$$

Total area A of cylindrical surface and end surfaces:

$$A = 6.2832r(r + h) = 3.1416d(\tfrac{1}{2}d + h)$$

Example: The diameter of a cylinder is 2.5 inches. The length or height is 20 inches. Find the volume and the area of the cylindrical surface S.

$$V = 0.7854d^2h = 0.7854 \times 2.5^2 \times 20 = 0.7854 \times 6.25 \times 20 = 98.17 \text{ cubic inches}$$

$$S = 3.1416dh = 3.1416 \times 2.5 \times 20 = 157.08 \text{ square inches}$$

Portion of Cylinder:

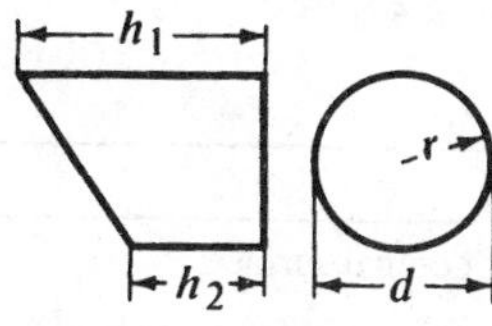

$$\text{Volume} = V = 1.5708r^2(h_1 + h_2)$$

$$= 0.3927d^2(h_1 + h_2)$$

$$\text{Cylindrical surface area} = S = 3.1416r(h_1 + h_2)$$

$$= 1.5708d(h_1 + h_2)$$

Example: A cylinder 125 millimeters in diameter is cut off at an angle, as shown in the illustration. Dimension $h_1 = 150$, and $h_2 = 100$ mm. Find the volume and the area S of the cylindrical surface.

$$V = 0.3927d^2(h_1 + h_2) = 0.3927 \times 125^2 \times (150 + 100)$$

$$= 0.3927 \times 15{,}625 \times 250 = 1{,}533{,}984 \text{ cubic millimeters} = 1534 \text{ cm}^3$$

$$S = 1.5708d(h_1 + h_2) = 1.5708 \times 125 \times 250$$

$$= 49{,}087.5 \text{ square millimeters} = 490.9 \text{ square centimeters}$$

Portion of Cylinder:

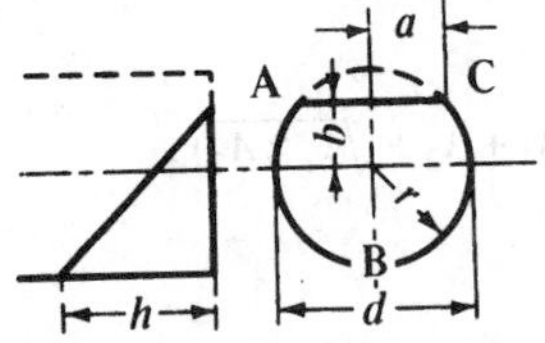

$$\text{Volume} = V = (\tfrac{2}{3}a^3 \pm b \times \text{area ABC})\frac{h}{r \pm b}$$

$$\text{Cylindrical surface area} = S = (ad \pm b \times \text{length of arc ABC})\frac{h}{r \pm b}$$

Use + when base area is larger, and − when base area is less than one-half the base circle.

Example: Find the volume of a cylinder so cut off that line AC passes through the center of the base circle — that is, the base area is a half-circle. The diameter of the cylinder = 5 inches, and the height h = 2 inches.

In this case, $a = 2.5$; $b = 0$; area $ABC = 0.5 \times 0.7854 \times 5^2 = 9.82$; $r = 2.5$.

$$V = \left(\frac{2}{3} \times 2.5^3 + 0 \times 9.82\right)\frac{2}{2.5 + 0} = \frac{2}{3} \times 15.625 \times 0.8 = 8.33 \text{ cubic inches}$$

Hollow Cylinder:

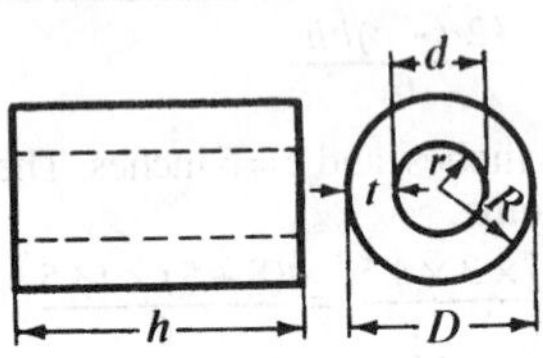

$$\begin{aligned} \text{Volume} = V &= 3.1416h(R^2 - r^2) = 0.7854h(D^2 - d^2) \\ &= 3.1416ht(2R - t) = 3.1416ht(D - t) \\ &= 3.1416ht(2r + t) = 3.1416ht(d + t) \\ &= 3.1416ht(R + r) = 1.5708ht(D + d) \end{aligned}$$

Example: A cylindrical shell, 28 centimeters high, is 36 centimeters in outside diameter, and 4 centimeters thick. Find its volume.

$$\begin{aligned} V = 3.1416ht(D - t) &= 3.1416 \times 28 \times 4(36 - 4) = 3.1416 \times 28 \times 4 \times 32 \\ &= 11{,}259.5 \text{ cubic centimeters} \end{aligned}$$

Cone:

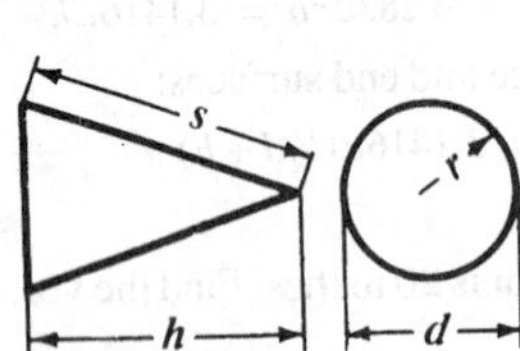

$$\text{Volume} = V = \frac{3.1416r^2h}{3} = 1.0472r^2h = 0.2618d^2h$$

$$\begin{aligned} \text{Conical surface area} = A &= 3.1416r\sqrt{r^2 + h^2} = 3.1416rs \\ &= 1.5708ds \end{aligned}$$

$$s = \sqrt{r^2 + h^2} = \sqrt{\frac{d^2}{4} + h^2}$$

Example: Find the volume and area of the conical surface of a cone, the base of which is a circle of 6 inches diameter, and the height of which is 4 inches.

$$V = 0.2618d^2h = 0.2618 \times 6^2 \times 4 = 0.2618 \times 36 \times 4 = 37.7 \text{ cubic inches}$$

$$\begin{aligned} A &= 3.1416r\sqrt{r^2 + h^2} = 3.1416 \times 3 \times \sqrt{3^2 + 4^2} = 9.4248 \times \sqrt{25} \\ &= 47.124 \text{ square inches} \end{aligned}$$

Frustum of Cone:

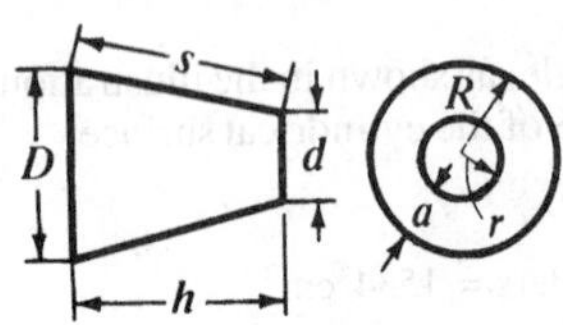

V = volume A = area of conical surface

$$V = 1.0472h(R^2 + Rr + r^2) = 0.2618h(D^2 + Dd + d^2)$$

$$A = 3.1416s(R + r) = 1.5708s(D + d)$$

$$a = R - r \qquad s = \sqrt{a^2 + h^2} = \sqrt{(R - r)^2 + h^2}$$

Example: Find the volume of a frustum of a cone of the following dimensions: $D = 8$ centimeters; $d = 4$ centimeters; $h = 5$ centimeters.

$$\begin{aligned} V &= 0.2618 \times 5(8^2 + 8 \times 4 + 4^2) = 0.2618 \times 5(64 + 32 + 16) \\ &= 0.2618 \times 5 \times 112 = 146.61 \text{ cubic centimeters} \end{aligned}$$

Sphere:

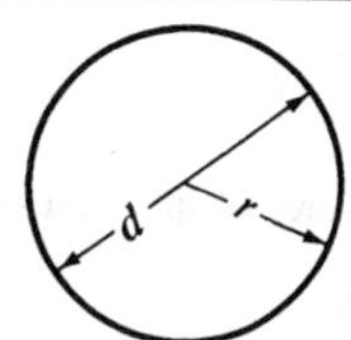

$$\text{Volume} = V = \frac{4\pi r^3}{3} = \frac{\pi d^3}{6} = 4.1888r^3 = 0.5236d^3$$

$$\text{Surface area} = A = 4\pi r^2 = \pi d^2 = 12.5664r^2 = 3.1416d^2$$

$$r = \sqrt[3]{\frac{3V}{4\pi}} = 0.6024\sqrt[3]{V}$$

Example: Find the volume and the surface of a sphere 6.5 centimeters diameter.

$$V = 0.5236d^3 = 0.5236 \times 6.5^3 = 0.5236 \times 6.5 \times 6.5 \times 6.5 = 143.79 \text{ cm}^3$$

$$A = 3.1416d^2 = 3.1416 \times 6.5^2 = 3.1416 \times 6.5 \times 6.5 = 132.73 \text{ cm}^2$$

Example: The volume of a sphere is 64 cubic centimeters. Find its radius.

$$r = 0.6204\sqrt[3]{64} = 0.6204 \times 4 = 2.4816 \text{ centimeters}$$

Spherical Sector:

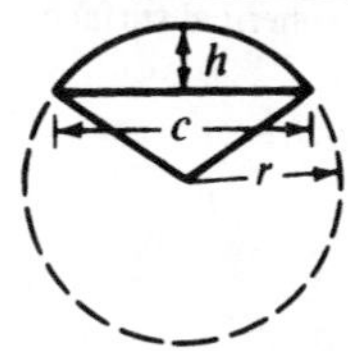

$$V = \frac{2\pi r^2 h}{3} = 2.0944r^2h = \text{Volume}$$

$$A = 3.1416r(2h + \tfrac{1}{2}c)$$

$= $ total area of conical and spherical surface

$$c = 2\sqrt{h(2r-h)}$$

Example: Find the volume of a sector of a sphere 6 inches in diameter, the height h of the sector being 1.5 inch. Also find the length of chord c. Here $r = 3$ and $h = 1.5$.

$$V = 2.0944r^2h = 2.0944 \times 3^2 \times 1.5 = 2.0944 \times 9 \times 1.5 = 28.27 \text{ cubic inches}$$

$$c = 2\sqrt{h(2r-h)} = 2\sqrt{1.5(2 \times 3 - 1.5)} = 2\sqrt{6.75} = 2 \times 2.598 = 5.196 \text{ inches}$$

Spherical Segment:

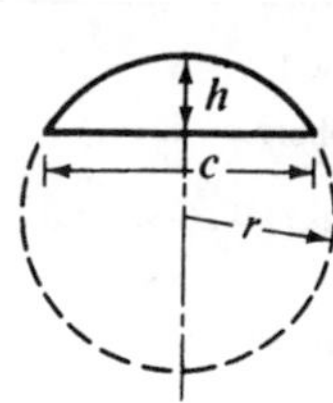

$V = $ volume $\qquad A = $ area of spherical surface

$$V = 3.1416h^2\left(r - \frac{h}{3}\right) = 3.1416h\left(\frac{c^2}{8} + \frac{h^2}{6}\right)$$

$$A = 2\pi rh = 6.2832rh = 3.1416\left(\frac{c^2}{4} + h^2\right)$$

$$c = 2\sqrt{h(2r-h)}; \qquad r = \frac{c^2 + 4h^2}{8h}$$

Example: A segment of a sphere has the following dimensions: $h = 50$ millimeters; $c = 125$ millimeters. Find the volume V and the radius of the sphere of which the segment is a part.

$$V = 3.1416 \times 50 \times \left(\frac{125^2}{8} + \frac{50^2}{6}\right) = 157.08 \times \left(\frac{15{,}625}{8} + \frac{2500}{6}\right) = 372{,}247 \text{ mm}^3 = 372 \text{ cm}^3$$

$$r = \frac{125^2 + 4 \times 50^2}{8 \times 50} = \frac{15{,}625 + 10{,}000}{400} = \frac{25{,}625}{400} = 64 \text{ millimeters}$$

Ellipsoid:

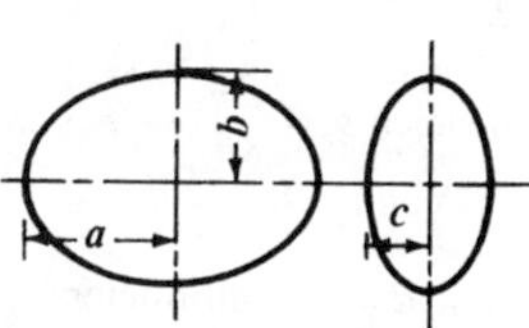

$$\text{Volume} = V = \frac{4\pi}{3}abc = 4.1888abc$$

In an ellipsoid of revolution, or spheroid, where $c = b$:

$$V = 4.1888ab^2$$

Example: Find the volume of a spheroid in which $a = 5$, and $b = c = 1.5$ inches.

$$V = 4.1888 \times 5 \times 1.5^2 = 47.124 \text{ cubic inches}$$

Spherical Zone:

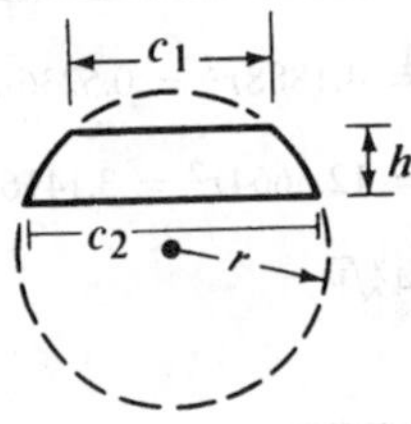

$$\text{Volume} = V = 0.5236h\left(\frac{3c_1^2}{4} + \frac{3c_2^2}{4} + h^2\right)$$

$$A = 2\pi rh = 6.2832rh = \text{area of spherical surface}$$

$$r = \sqrt{\frac{c_2^2}{4} + \left(\frac{c_2^2 - c_1^2 - 4h^2}{8h}\right)^2}$$

Example: In a spherical zone, let $c_1 = 3$; $c_2 = 4$; and $h = 1.5$ inch. Find the volume.

$$V = 0.5236 \times 1.5 \times \left(\frac{3 \times 3^2}{4} + \frac{3 \times 4^2}{4} + 1.5^2\right) = 0.5236 \times 1.5 \times \left(\frac{27}{4} + \frac{48}{4} + 2.25\right) = 16.493 \text{ in}^3$$

Spherical Wedge:

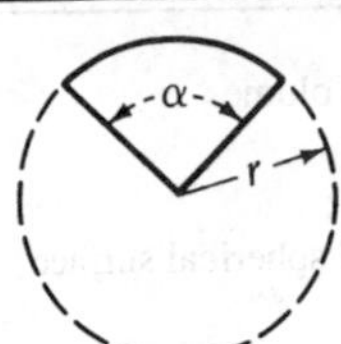

V = volume A = area of spherical surface

α = center angle in degrees

$$V = \frac{\alpha}{360} \times \frac{4\pi r^3}{3} = 0.0116\alpha r^3$$

$$A = \frac{\alpha}{360} \times 4\pi r^2 = 0.0349\alpha r^2$$

Example: Find the area of the spherical surface and the volume of a wedge of a sphere. The diameter of the sphere is 100 millimeters, and the center angle α is 45 degrees.

$$V = 0.0116 \times 45 \times 50^3 = 0.0116 \times 45 \times 125{,}000 = 65{,}250 \text{ mm}^3 = 65.25 \text{ cm}^3$$

$$A = 0.0349 \times 45 \times 50^2 = 3926.25 \text{ square millimeters} = 39.26 \text{ cm}^2$$

Hollow Sphere:

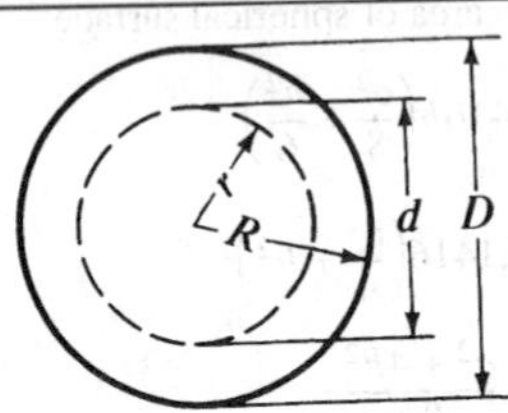

V = volume of material used to make a hollow sphere

$$V = \frac{4\pi}{3}(R^3 - r^3) = 4.1888(R^3 - r^3)$$

$$= \frac{\pi}{6}(D^3 - d^3) = 0.5236(D^3 - d^3)$$

Example: Find the volume of a hollow sphere, 8 inches in outside diameter, with a thickness of material of 1.5 inch.

Here $R = 4$; $r = 4 - 1.5 = 2.5$.

$$V = 4.1888(4^3 - 2.5^3) = 4.1888(64 - 15.625) = 4.1888 \times 48.375 = 202.63 \text{ cubic inches}$$

Paraboloid:

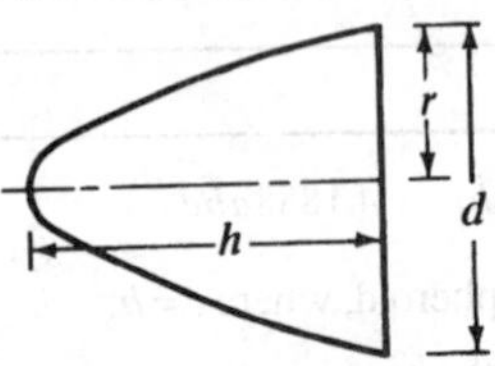

$$\text{Volume} = V = \tfrac{1}{2}\pi r^2 h = 0.3927d^2h$$

$$\text{Area} = A = \frac{2\pi}{3p}\left[\sqrt{\left(\frac{d^2}{4} + p^2\right)^3} - p^3\right]$$

$$\text{in which } p = \frac{d^2}{8h}$$

Example: Find the volume of a paraboloid in which $h = 300$ millimeters and $d = 125$ millimeters.

$$V = 0.3927d^2h = 0.3927 \times 125^2 \times 300 = 1{,}840{,}781 \text{ mm}^3 = 1{,}840.8 \text{ cm}^3$$

Paraboloidal Segment:

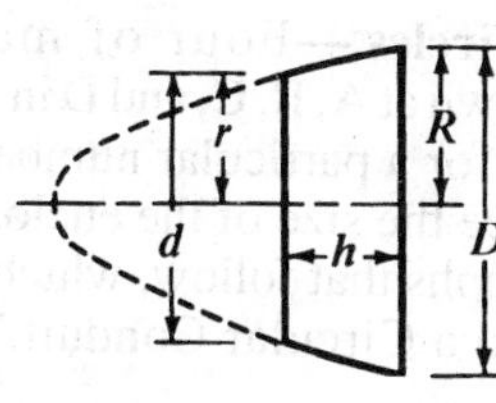

$$\text{Volume} = V = \frac{\pi}{2}h(R^2 + r^2) = 1.5708h(R^2 + r^2)$$
$$= \frac{\pi}{8}h(D^2 + d^2) = 0.3927h(D^2 + d^2)$$

Example: Find the volume of a segment of a paraboloid in which $D = 5$ inches, $d = 3$ inches, and $h = 6$ inches.

$$V = 0.3927h(D^2 + d^2) = 0.3927 \times 6 \times (5^2 + 3^2)$$
$$= 0.3927 \times 6 \times 34 = 80.11 \text{ cubic inches}$$

Torus:

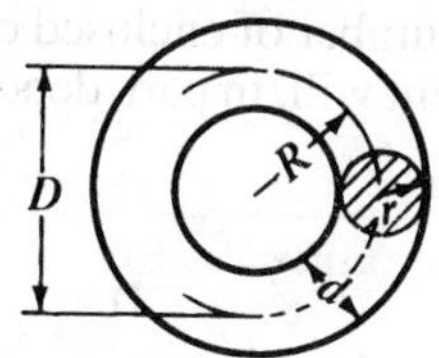

$$\text{Volume} = V = 2\pi^2 Rr^2 = 19.739Rr^2$$
$$= \frac{\pi^2}{4}Dd^2 = 2.4674Dd^2$$

$$\text{Area of surface} = A = 4\pi^2 Rr = 39.478Rr$$
$$= \pi^2 Dd = 9.8696Dd$$

Example: Find the volume and area of surface of a torus in which $d = 1.5$ and $D = 5$ inches.

$$V = 2.4674 \times 5 \times 1.5^2 = 2.4674 \times 5 \times 2.25 = 27.76 \text{ cubic inches}$$
$$A = 9.8696 \times 5 \times 1.5 = 74.022 \text{ square inches}$$

Barrel:

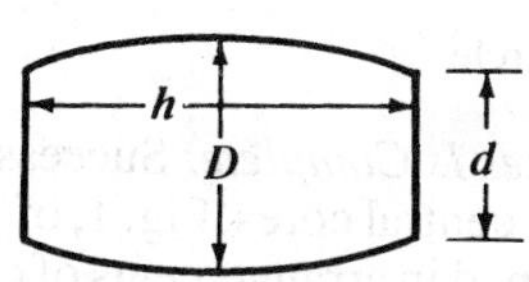

V = approximate volume.

If the sides are bent to the arc of a circle:

$$V = \frac{1}{12}\pi h(2D^2 + d^2) = 0.262h(2D^2 + d^2)$$

If the sides are bent to the arc of a parabola:

$$V = 0.209h(2D^2 + Dd + \tfrac{3}{4}d^2)$$

Example: Find the approximate contents of a barrel, the inside dimensions of which are $D = 60$ centimeters, $d = 50$ centimeters; $h = 120$ centimeters.

$$V = 0.262h(2D^2 + d^2) = 0.262 \times 120 \times (2 \times 60^2 + 50^2)$$
$$= 0.262 \times 120 \times (7200 + 2500) = 0.262 \times 120 \times 9700$$
$$= 304{,}968 \text{ cubic centimeters} = 0.305 \text{ cubic meter}$$

Ratio of Volumes:

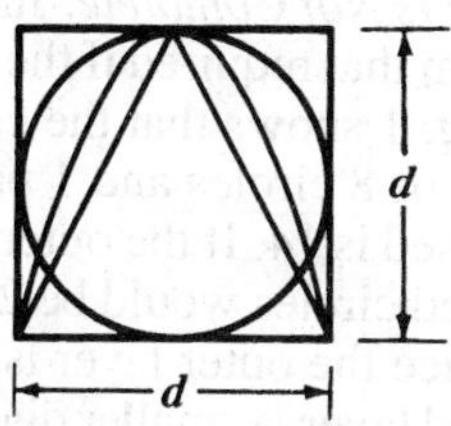

If d = base diameter and height of a cone, a paraboloid and a cylinder, and the diameter of a sphere, then the volumes of these bodies are to each other as follows:

$$\text{Cone:paraboloid:sphere:cylinder} = \tfrac{1}{3} : \tfrac{1}{2} : \tfrac{2}{3} : 1$$

Example: Assume, as an example, that the diameter of the base of a cone, paraboloid, and cylinder is 2 inches, that the height is 2 inches, and that the diameter of a sphere is 2 inches. Then the volumes, written in formula form, are as follows:

$$\underset{\text{Cone}}{\frac{3.1416 \times 2^2 \times 2}{12}} : \underset{\text{Paraboloid}}{\frac{3.1416 \times (2p)^2 \times 2}{8}} : \underset{\text{Sphere}}{\frac{3.1416 \times 2^3}{6}} : \underset{\text{Cylinder}}{\frac{3.1416 \times 2^2 \times 2}{4}} = \tfrac{1}{3} : \tfrac{1}{2} : \tfrac{2}{3} : 1$$

Packing Circles in Circles and Rectangles

Diameter of Circle Enclosing a Given Number of Smaller Circles.—Four of many possible compact arrangements of circles within a circle are shown at A, B, C, and D in Fig. 1. To determine the diameter of the smallest enclosing circle for a particular number of enclosed circles all of the same size, three factors that influence the size of the enclosing circle should be considered. These are discussed in the paragraphs that follow, which are based on the article "How Many Wires Can Be Packed into a Circular Conduit," by Jacques Dutka, *Machinery*, October 1956.

1) *Arrangement of Center or Core Circles:* The four most common arrangements of center or core circles are shown cross-sectioned in Fig. 1. It may seem, offhand, that the "A" pattern would require the smallest enclosing circle for a given number of enclosed circles but this is not always the case since the most compact arrangement will, in part, depend on the number of circles to be enclosed.

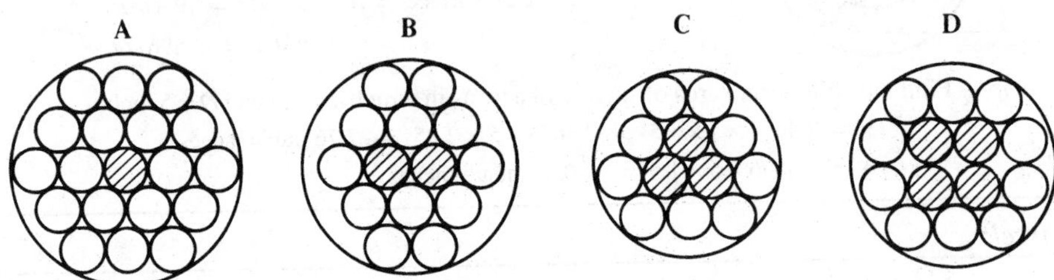

Fig. 1. Arrangements of Circles within a Circle

2) *Diameter of Enclosing Circle When Outer Layer of Circles Is Complete:* Successive, complete "layers" of circles may be placed around each of the central cores, Fig. 1, of 1, 2, 3, or 4 circles as the case may be. The number of circles contained in arrangements of complete "layers" around a central core of circles, as well as the diameter of the enclosing circle, may be obtained using the data in Table 1. Thus, for example, the "A" pattern in Fig. 1 shows, by actual count, a total of 19 circles arranged in two complete "layers" around a central core consisting of one circle; this agrees with the data shown in the left half of Table 1 for $n = 2$.

To determine the diameter of the enclosing circle, the data in the right half of Table 1 is used. Thus, for $n = 2$ and an "A" pattern, the diameter D is 5 times the diameter d of the enclosed circles.

3) *Diameter of Enclosing Circle When Outer Layer of Circles Is Not Complete:* In most cases, it is possible to reduce the size of the enclosing circle from that required if the outer layer were complete. Thus, for example, the "B" pattern in Fig. 1 shows that the central core consisting of 2 circles is surrounded by 1 complete layer of 8 circles and 1 partial, outer layer of 4 circles, so that the total number of circles enclosed is 14. If the outer layer were complete, then (from Table 1) the total number of enclosed circles would be 24 and the diameter of the enclosing circle would be $6d$; however, since the outer layer is composed of only 4 circles out of a possible 14 for a complete second layer, a smaller diameter of enclosing circle may be used. Table 2 shows that for a total of 14 enclosed circles arranged in a "B" pattern with the outer layer of circles incomplete, the diameter for the enclosing circle is $4.606d$.

Table 2 can be used to determine the smallest enclosing circle for a given number of circles to be enclosed by direct comparison of the "A," "B," and "C" columns. For data outside the range of Table 2, use the formulas in Dr. Dutka's article.

Table 1. Number of Circles Contained in Complete Layers of Circles and Diameter of Enclosing Circle (English or metric units)

No. Complete Layers Over Core, n	Number of Circles in Center Pattern							
	1	2	3	4	1	2	3	4
	Arrangement of Circles in Center Pattern (see Fig. 1)							
	"A"	"B"	"C"	"D"	"A"	"B"	"C"	"D"
	Number of Circles, N, Enclosed				Diameter, D, of Enclosing Circle[a]			
0	1	2	3	4	d	$2d$	$2.155d$	$2.414d$
1	7	10	12	14	$3d$	$4d$	$4.055d$	$4.386d$
2	19	24	27	30	$5d$	$6d$	$6.033d$	$6.379d$
3	37	44	48	52	$7d$	$8d$	$8.024d$	$8.375d$
4	61	70	75	80	$9d$	$10d$	$10.018d$	$10.373d$
5	91	102	108	114	$11d$	$12d$	$12.015d$	$12.372d$
n	b	b	b	b	b	b	b	b

[a] Diameter D is given in terms of d, the diameter of the enclosed circles.

[b] For n complete layers over core, the number of enclosed circles N for the "A" center pattern is $3n^2 + 3n + 1$; for "B," $3n^2 + 5n + 2$; for "C," $3n^2 + 6n + 3$; for "D," $3n^2 + 7n + 4$; while the diameter D of the enclosing circle for "A" center pattern is $(2n + 1)d$; for "B," $(2n + 2)d$; for "C," $(1 + 2\sqrt{n^2 + n + \frac{1}{3}})d$ and for "D," $(1 + \sqrt{4n^2 + 5.644n + 2})d$.

Table 2. Factors for Determining Diameter, D, of Smallest Enclosing Circle for Various Numbers, N, of Enclosed Circles (English or metric units)

No. N	Center Circle Pattern "A"	"B"	"C"	No. N	Center Circle Pattern "A"	"B"	"C"	No. N	Center Circle Pattern "A"	"B"	"C"
	Diameter Factor K				Diameter Factor K				Diameter Factor K		
2	3	2	...	34	7.001	7.083	7.111	66	9.718	9.545	9.327
3	3	2.733	2.155	35	7.001	7.245	7.111	67	9.718	9.545	9.327
4	3	2.733	3.310	36	7.001	7.245	7.111	68	9.718	9.545	9.327
5	3	3.646	3.310	37	7.001	7.245	7.430	69	9.718	9.661	9.327
6	3	3.646	3.310	38	7.929	7.245	7.430	70	9.718	9.661	10.019
7	3	3.646	4.056	39	7.929	7.558	7.430	71	9.718	9.889	10.019
8	4.465	3.646	4.056	40	7.929	7.558	7.430	72	9.718	9.889	10.019
9	4.465	4	4.056	41	7.929	7.558	7.430	73	9.718	9.889	10.019
10	4.465	4	4.056	42	7.929	7.558	7.430	74	10.166	9.889	10.019
11	4.465	4.606	4.056	43	7.929	8.001	8.024	75	10.166	10	10.019
12	4.465	4.606	4.056	44	8.212	8.001	8.024	76	10.166	10	10.238
13	4.465	4.606	5.164	45	8.212	8.001	8.024	77	10.166	10.540	10.238
14	5	4.606	5.164	46	8.212	8.001	8.024	78	10.166	10.540	10.238
15	5	5.359	5.164	47	8.212	8.001	8.024	79	10.166	10.540	10.452
16	5	5.359	5.164	48	8.212	8.001	8.024	80	10.166	10.540	10.452
17	5	5.359	5.164	49	8.212	8.550	8.572	81	10.166	10.540	10.452
18	5	5.359	5.164	50	8.212	8.550	8.572	82	10.166	10.540	10.452
19	5	5.583	5.619	51	8.212	8.550	8.572	83	10.166	10.540	10.452
20	6.292	5.583	5.619	52	8.212	8.550	8.572	84	10.166	10.540	10.452
21	6.292	5.583	5.619	53	8.212	8.811	8.572	85	10.166	10.644	10.866
22	6.292	5.583	6.034	54	8.212	8.811	8.572	86	11	10.644	10.866
23	6.292	6.001	6.034	55	8.212	8.811	9.083	87	11	10.644	10.866
24	6.292	6.001	6.034	56	9.001	8.811	9.083	88	11	10.644	10.866
25	6.292	6.197	6.034	57	9.001	8.938	9.083	89	11	10.849	10.866
26	6.292	6.197	6.034	58	9.001	8.938	9.083	90	11	10.849	10.866
27	6.292	6.568	6.034	59	9.001	8.938	9.083	91	11	10.849	11.067
28	6.292	6.568	6.774	60	9.001	8.938	9.083	92	11.393	10.849	11.067
29	6.292	6.568	6.774	61	9.001	9.186	9.083	93	11.393	11.149	11.067
30	6.292	6.568	6.774	62	9.718	9.186	9.083	94	11.393	11.149	11.067
31	6.292	7.083	7.111	63	9.718	9.186	9.083	95	11.393	11.149	11.067
32	7.001	7.083	7.111	64	9.718	9.186	9.327	96	11.393	11.149	11.067
33	7.001	7.083	7.111	65	9.718	9.545	9.327	97	11.393	11.441	11.264

Table 2. *(Continued)* **Factors for Determining Diameter, *D*, of Smallest Enclosing Circle for Various Numbers, *N*, of Enclosed Circles** (English or metric units)

No.	Center Circle Pattern			No.	Center Circle Pattern			No.	Center Circle Pattern		
	"A"	"B"	"C"		"A"	"B"	"C"		"A"	"B"	"C"
N	Diameter Factor *K*			*N*	Diameter Factor *K*			*N*	Diameter Factor *K*		
98	11.584	11.441	11.264	153	14.115	14	14.013	208	16.100	16	16.144
99	11.584	11.441	11.264	154	14.115	14	14.013	209	16.100	16.133	16.144
100	11.584	11.441	11.264	155	14.115	14.077	14.013	210	16.100	16.133	16.144
101	11.584	11.536	11.264	156	14.115	14.077	14.013	211	16.100	16.133	16.144
102	11.584	11.536	11.264	157	14.115	14.077	14.317	212	16.621	16.133	16.144
103	11.584	11.536	12.016	158	14.115	14.077	14.317	213	16.621	16.395	16.144
104	11.584	11.536	12.016	159	14.115	14.229	14.317	214	16.621	16.395	16.276
105	11.584	11.817	12.016	160	14.115	14.229	14.317	215	16.621	16.395	16.276
106	11.584	11.817	12.016	161	14.115	14.229	14.317	216	16.621	16.395	16.276
107	11.584	11.817	12.016	162	14.115	14.229	14.317	217	16.621	16.525	16.276
108	11.584	11.817	12.016	163	14.115	14.454	14.317	218	16.621	16.525	16.276
109	11.584	12	12.016	164	14.857	14.454	14.317	219	16.621	16.525	16.276
110	12.136	12	12.016	165	14.857	14.454	14.317	220	16.621	16.525	16.535
111	12.136	12.270	12.016	166	14.857	14.454	14.317	221	16.621	16.589	16.535
112	12.136	12.270	12.016	167	14.857	14.528	14.317	222	16.621	16.589	16.535
113	12.136	12.270	12.016	168	14.857	14.528	14.317	223	16.621	16.716	16.535
114	12.136	12.270	12.016	169	14.857	14.528	14.614	224	16.875	16.716	16.535
115	12.136	12.358	12.373	170	15	14.528	14.614	225	16.875	16.716	16.535
116	12.136	12.358	12.373	171	15	14.748	14.614	226	16.875	16.716	17.042
117	12.136	12.358	12.373	172	15	14.748	14.614	227	16.875	16.716	17.042
118	12.136	12.358	12.373	173	15	14.748	14.614	228	16.875	16.716	17.042
119	12.136	12.533	12.373	174	15	14.748	14.614	229	16.875	16.716	17.042
120	12.136	12.533	12.373	175	15	14.893	15.048	230	16.875	16.716	17.042
121	12.136	12.533	12.548	176	15	14.893	15.048	231	16.875	17.094	17.042
122	13	12.533	12.548	177	15	14.893	15.048	232	16.875	17.094	17.166
123	13	12.533	12.548	178	15	14.893	15.048	233	16.875	17.094	17.166
124	13	12.533	12.719	179	15	15.107	15.048	234	16.875	17.094	17.166
125	13	12.533	12.719	180	15	15.107	15.048	235	16.875	17.094	17.166
126	13	12.533	12.719	181	15	15.107	15.190	236	17	17.094	17.166
127	13	12.790	12.719	182	15	15.107	15.190	237	17	17.094	17.166
128	13.166	12.790	12.719	183	15	15.178	15.190	238	17	17.094	17.166
129	13.166	12.790	12.719	184	15	15.178	15.190	239	17	17.463	17.166
130	13.166	12.790	13.056	185	15	15.178	15.190	240	17	17.463	17.166
131	13.166	13.125	13.056	186	15	15.178	15.190	241	17	17.463	17.290
132	13.166	13.125	13.056	187	15	15.526	15.469	242	17.371	17.463	17.290
133	13.166	13.125	13.056	188	15.423	15.526	15.469	243	17.371	17.523	17.290
134	13.166	13.125	13.056	189	15.423	15.526	15.469	244	17.371	17.523	17.290
135	13.166	13.125	13.056	190	15.423	15.526	15.469	245	17.371	17.523	17.290
136	13.166	13.125	13.221	191	15.423	15.731	15.469	246	17.371	17.523	17.290
137	13.166	13.289	13.221	192	15.423	15.731	15.469	247	17.371	17.523	17.654
138	13.166	13.289	13.221	193	15.423	15.731	15.743	248	17.371	17.523	17.654
139	13.166	13.289	13.221	194	15.423	15.731	15.743	249	17.371	17.523	17.654
140	13.490	13.289	13.221	195	15.423	15.731	15.743	250	17.371	17.523	17.654
141	13.490	13.530	13.221	196	15.423	15.731	15.743	251	17.371	17.644	17.654
142	13.490	13.530	13.702	197	15.423	15.731	15.743	252	17.371	17.644	17.654
143	13.490	13.530	13.702	198	15.423	15.731	15.743	253	17.371	17.644	17.773
144	13.490	13.530	13.702	199	15.423	15.799	16.012	254	18.089	17.644	17.773
145	13.490	13.768	13.859	200	16.100	15.799	16.012	255	18.089	17.704	17.773
146	13.490	13.768	13.859	201	16.100	15.799	16.012	256	18.089	17.704	17.773
147	13.490	13.768	13.859	202	16.100	15.799	16.012	257	18.089	17.704	17.773
148	13.490	13.768	13.859	203	16.100	15.934	16.012	258	18.089	17.704	17.773
149	13.490	14	13.859	204	16.100	15.934	16.012	259	18.089	17.823	18.010
150	13.490	14	13.859	205	16.100	15.934	16.012	260	18.089	17.823	18.010
151	13.490	14	14.013	206	16.100	15.934	16.012	261	18.089	17.823	18.010
152	14.115	14	14.013	207	16.100	16	16.012	262	18.089	17.823	18.010

The diameter D of the enclosing circle is equal to the diameter factor, K, multiplied by d, the diameter of the enclosed circles, or $D = K \times d$. For example, if the number of circles to be enclosed, N, is 12, and the center circle arrangement is "C," then for $d = 1\frac{1}{2}$ inches, $D = 4.056 \times 1\frac{1}{2} = 6.084$ inches. If $d = 50$ millimeters, then $D = 4.056 \times 50 = 202.9$ millimeters.

Approximate Formula When Number of Enclosed Circles Is Large: When a large number of circles are to be enclosed, the arrangement of the center circles has little effect on the diameter of the enclosing circle. For numbers of circles greater than 10,000, the diameter of the enclosing circle may be calculated within 2 per cent from the formula $D = d(1 + \sqrt{N \div 0.907})$. In this formula, D = diameter of the enclosing circle; d = diameter of the enclosed circles; and N is the number of enclosed circles.

An alternative approach relates the area of each of the same-sized circles to be enclosed to the area of the enclosing circle (or container), as shown in Figs. 1 through 27. The table shows efficient ways for packing various numbers of circles N, from 2 up to 97.

In the table, D = the diameter of each circle to be enclosed, d = the diameter of the enclosing circle or container, and $\Phi = ND^2/d^2$ = ratio of the area of the N circles to the area of the enclosing circle or container, which is the packing efficiency. Cross-hatching in the diagrams indicates loose circles that may need packing constraints.

Data for Numbers of Circles in Circles

N	d/D	Φ	Fig.	N	d/D	Φ	Fig.
2	2.0000	0.500	1	17	4.7920	0.740	15
3	2.1547	0.646	2	18	4.8637	0.761	16
4	2.4142	0.686	3	19	4.8637	0.803	16
5	2.7013	0.685	4	20	5.1223	0.762	17
6	3.0000	0.667	5	21	5.2523	0.761	18
7	3.0000	0.778	5	22	5.4397	0.743	19
8	3.3048	0.733	6	23	5.5452	9.748	20
9	3.6131	0.689	7	24	5.6517	0.751	21
10	3.8130	0.688	8	25	5.7608	0.753	22
11	3.9238	0.714	9	31	6.2915	0.783	23
12	4.0296	0.739	10	37	6.7588	0.810	24
13	4.2361	0.724	11	55	8.2111	0.816	25
14	4.3284	0.747	12	61	8.6613	0.813	26
15	4.5214	0.734	13	97	11.1587	0.779	27
16	4.6154	0.751	14	…	…	…	…

Packing of large numbers of circles, such as the 97 in Fig. 27, may be approached by drawing a triangular pattern of circles, as shown in Fig. 28, which represents three circles near the center of the array. The point of a compass is then placed at A, B, or C, or anywhere within triangle ABC, and the radius of the compass is gradually enlarged until it encompasses the number of circles to be enclosed. As a first approximation of the diameter, $1.14D\sqrt{N}$ may be tried.

Fig. 1. $N = 2$

Fig. 2. $N = 3$

Fig. 3. $N = 4$

Fig. 4. $N = 5$

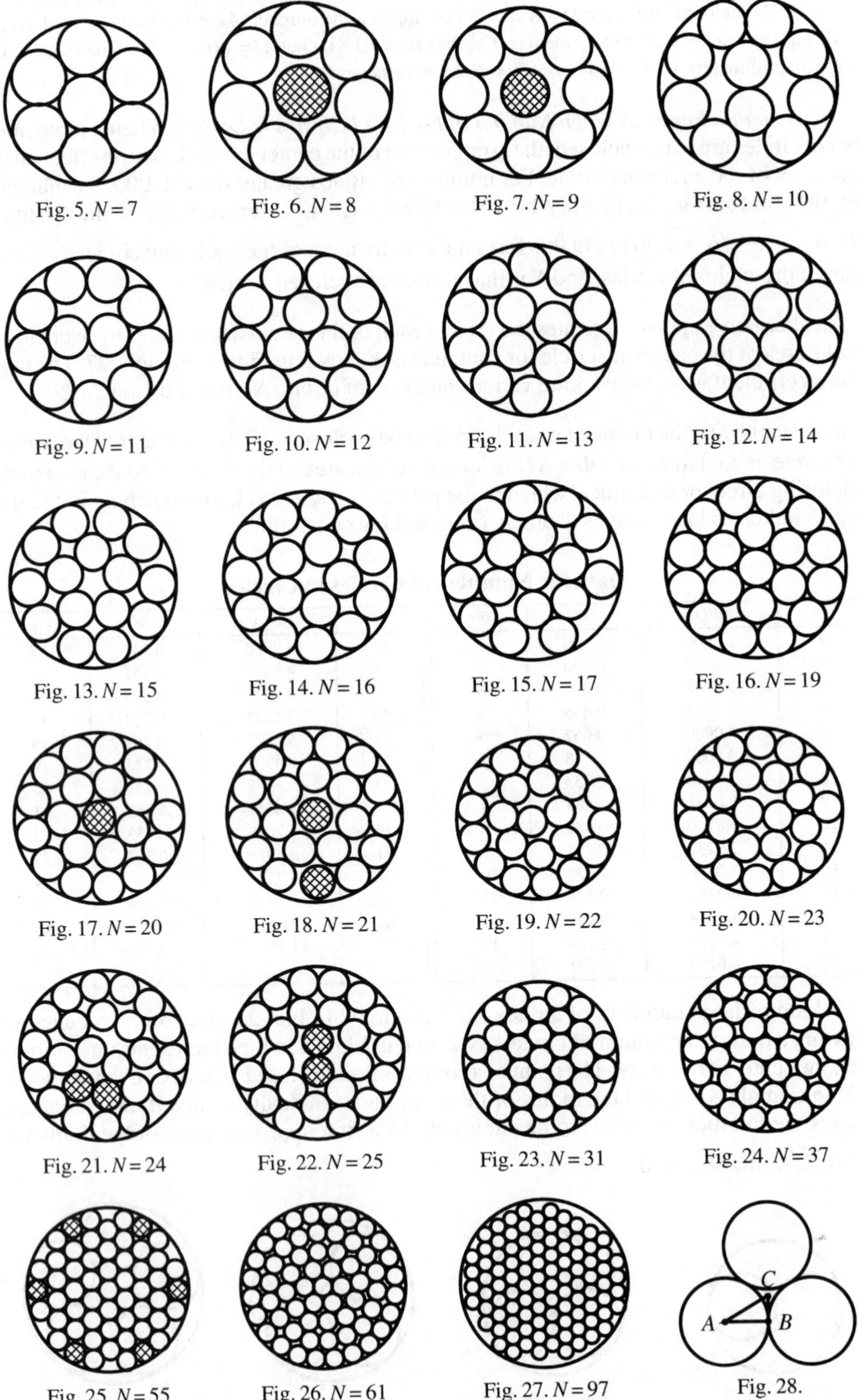

Fig. 5. $N=7$ Fig. 6. $N=8$ Fig. 7. $N=9$ Fig. 8. $N=10$

Fig. 9. $N=11$ Fig. 10. $N=12$ Fig. 11. $N=13$ Fig. 12. $N=14$

Fig. 13. $N=15$ Fig. 14. $N=16$ Fig. 15. $N=17$ Fig. 16. $N=19$

Fig. 17. $N=20$ Fig. 18. $N=21$ Fig. 19. $N=22$ Fig. 20. $N=23$

Fig. 21. $N=24$ Fig. 22. $N=25$ Fig. 23. $N=31$ Fig. 24. $N=37$

Fig. 25. $N=55$ Fig. 26. $N=61$ Fig. 27. $N=97$ Fig. 28.

Circles within Rectangles.—For small numbers N of circles, packing (for instance, of cans) is less vital than for larger numbers and the number will usually govern the decision whether to use a rectangular or a triangular pattern, examples of which are seen in Figs. 29 and 30.

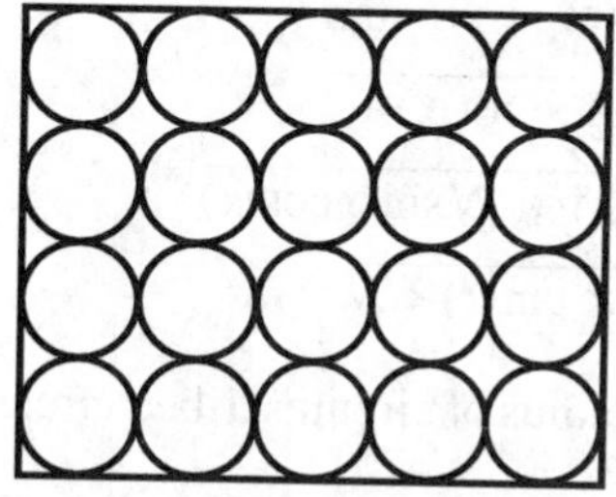

Fig. 29. Rectangular Pattern ($r = 4$, $c = 5$)

Fig. 30. Triangular Pattern ($r = 3$, $c = 7$)

If D is the can diameter and H its height, the arrangement in Fig. 29 will hold 20 circles or cans in a volume of $5D \times 4D \times H = 20D^2 H$. The arrangement in Fig. 30 will pack the same 20 cans into a volume of $7D \times 2.732D \times H = 19.124D^2 H$, a reduction of 4.4 per cent. When the ratio of H/D is less than 1.196 : 1, the rectangular pattern requires less surface area (therefore less material) for the six sides of the box, but for greater ratios, the triangular pattern is better. Some numbers, such as 19, can be accommodated only in a triangular pattern.

The following table shows possible patterns for 3 to 25 cans, where N = number of circles, P = pattern (R rectangular or T triangular), and r and c = numbers of rows and columns, respectively. The final table column shows the most economical application, where V = best volume, S = best surface area (sometimes followed by a condition on H/D). For the rectangular pattern, dimensions of the container are $rD \times cD$, and for the triangular pattern, the dimensions are' $\times [1 + (r + 70(-1))\sqrt{3}/2]$, or $cD^2[1 + 0.866(r - 1)]$.

Numbers of Circles in Rectangular Arrangements

N	P	r	c	Application
3	T	2	2	V, S
4	R	2	2	V, S
5	T	3	2	V, S
6	R	2	3	V, S
7	T	2	4	V, S
8	R	4	2	V, (S, $H/D < 0.732$)
	T	3	3	(S, $H/D > 0.732$)
9	R	3	3	V, S
10	R	5	2	V, (S, $H/D > 1.976$)
	T	4	3	(S, $H/D > 1.976$)
11	T	3	4	V, S
12	R	3	4	V, S
13	T	5	3	(S, $H/D > 0.236$)
	T	2	7	V, (S, $H/D < 0.236$)
14	T	4	4	(S, $H/D > 5.464$)
	T	3	5	V, (S, $H/D < 5.464$)

N	P	r	c	Application
15	R	3	5	(S, $H/D > 0.038$)
	T	2	8	V, (S, $H/D < 0.038$)
16	R	4	4	V, S
17	T	3	6	V, S
18	T	5	4	V, S
19	T	2	10	V, S
20	R	4	5	(S, $H/D > 1.196$)
	T	3	7	V, (S, $H/D < 1.196$)
21	R	3	7	(S, $0.165 < H/D < 0.479$)
	T	6	4	(S, $H/D > 0.479$)
	T	2	11	V, (S, $H/D < 0.165$)
22	T	4	6	V, S
23	T	5	5	(S, $H/D > 0.366$)
	T	3	8	V, (S, $H/D < 0.366$)
24	R	4	6	V, S
25	R	5	5	(S, $H/D > 1.10$)
	T	7	4	(S, $0.113 < H/D < 1.10$)
	T	2	13	V, (S, $H/D < 0.133$)

Formulas and Table for Regular Polygons.—The following formulas and table can be used to calculate the area, length of side, and radii of the inscribed and circumscribed circles of regular polygons (equal sided).

$$A = NS^2 \cot\alpha \div 4 = NR^2 \sin\alpha\cos\alpha = Nr^2 \tan\alpha$$

$$r = R\cos\alpha = (S\cot\alpha) \div 2 = \sqrt{(A \times \cot\alpha) \div N}$$

$$R = S \div (2\sin\alpha) = r \div \cos\alpha = \sqrt{A \div (N\sin\alpha\cos\alpha)}$$

$$S = 2R\sin\alpha = 2r\tan\alpha = 2\sqrt{(A \times \tan\alpha) \div N}$$

where N = number of sides; S = length of side; R = radius of circumscribed circle; r = radius of inscribed circle; A = area of polygon; and, $\alpha = 180° \div N$ = one-half center angle of one side (see *Regular Polygon* on page 62) .

Area, Length of Side, and Inscribed and Circumscribed Radii of Regular Polygons

No. of Sides	$\frac{A}{S^2}$	$\frac{A}{R^2}$	$\frac{A}{r^2}$	$\frac{R}{S}$	$\frac{R}{r}$	$\frac{S}{R}$	$\frac{S}{r}$	$\frac{r}{R}$	$\frac{r}{S}$
3	0.4330	1.2990	5.1962	0.5774	2.0000	1.7321	3.4641	0.5000	0.2887
4	1.0000	2.0000	4.0000	0.7071	1.4142	1.4142	2.0000	0.7071	0.5000
5	1.7205	2.3776	3.6327	0.8507	1.2361	1.1756	1.4531	0.8090	0.6882
6	2.5981	2.5981	3.4641	1.0000	1.1547	1.0000	1.1547	0.8660	0.8660
7	3.6339	2.7364	3.3710	1.1524	1.1099	0.8678	0.9631	0.9010	1.0383
8	4.8284	2.8284	3.3137	1.3066	1.0824	0.7654	0.8284	0.9239	1.2071
9	6.1818	2.8925	3.2757	1.4619	1.0642	0.6840	0.7279	0.9397	1.3737
10	7.6942	2.9389	3.2492	1.6180	1.0515	0.6180	0.6498	0.9511	1.5388
12	11.196	3.0000	3.2154	1.9319	1.0353	0.5176	0.5359	0.9659	1.8660
16	20.109	3.0615	3.1826	2.5629	1.0196	0.3902	0.3978	0.9808	2.5137
20	31.569	3.0902	3.1677	3.1962	1.0125	0.3129	0.3168	0.9877	3.1569
24	45.575	3.1058	3.1597	3.8306	1.0086	0.2611	0.2633	0.9914	3.7979
32	81.225	3.1214	3.1517	5.1011	1.0048	0.1960	0.1970	0.9952	5.0766
48	183.08	3.1326	3.1461	7.6449	1.0021	0.1308	0.1311	0.9979	7.6285
64	325.69	3.1365	3.1441	10.190	1.0012	0.0981	0.0983	0.9988	10.178

Example 1: A regular hexagon is inscribed in a circle of 6 inches diameter. Find the area and the radius of an inscribed circle. Here $R = 3$. From the table, area $A = 2.5981R^2 = 2.5981 \times 9 = 23.3829$ square inches. Radius of inscribed circle, $r = 0.866R = 0.866 \times 3 = 2.598$ inches.

Example 2: An octagon is inscribed in a circle of 100 millimeters diameter. Thus $R = 50$. Find the area and radius of an inscribed circle. $A = 2.8284R^2 = 2.8284 \times 2500 = 7071$ mm^2 = 70.7 cm^2. Radius of inscribed circle, $r = 0.9239R = 09239 \times 50 = 46.195$ mm.

Example 3: Thirty-two bolts are to be equally spaced on the periphery of a bolt-circle, 16 inches in diameter. Find the chordal distance between the bolts. Chordal distance equals the side S of a polygon with 32 sides. $R = 8$. Hence, $S = 0.196R = 0.196 \times 8 = 1.568$ inch.

Example 4: Sixteen bolts are to be equally spaced on the periphery of a bolt-circle, 250 millimeters diameter. Find the chordal distance between the bolts. Chordal distance equals the side S of a polygon with 16 sides. $R = 125$. Thus, $S = 0.3902R = 0.3902 \times 125 = 48.775$ millimeters.

Tabulated Dimensions Of Geometric Shapes

Diameters of Circles and Sides of Squares of Equal Area

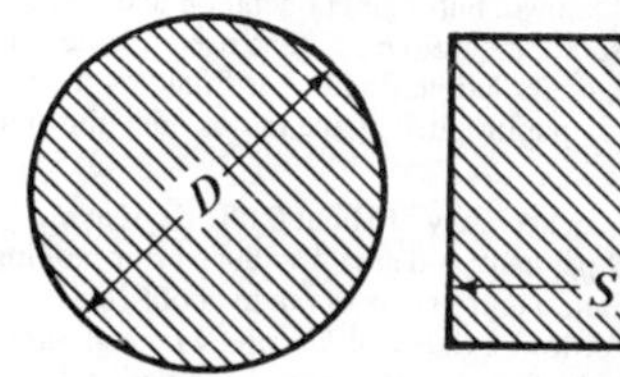

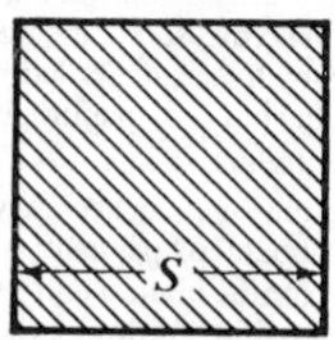

The table below will be found useful for determining the diameter of a circle of an area equal to that of a square, the side of which is known, or for determining the side of a square which has an area equal to that of a circle, the area or diameter of which is known. For example, if the diameter of a circle is 17½ inches, it is found from the table that the side of a square of the same area is 15.51 inches.

Diam. of Circle, *D*	Side of Square, *S*	Area of Circle or Square	Diam. of Circle, *D*	Side of Square, *S*	Area of Circle or Square	Diam. of Circle, *D*	Side of Square, *S*	Area of Circle or Square
½	0.44	0.196	20½	18.17	330.06	40½	35.89	1288.25
1	0.89	0.785	21	18.61	346.36	41	36.34	1320.25
1½	1.33	1.767	21½	19.05	363.05	41½	36.78	1352.65
2	1.77	3.142	22	19.50	380.13	42	37.22	1385.44
2½	2.22	4.909	22½	19.94	397.61	42½	37.66	1418.63
3	2.66	7.069	23	20.38	415.48	43	38.11	1452.20
3½	3.10	9.621	23½	20.83	433.74	43½	38.55	1486.17
4	3.54	12.566	24	21.27	452.39	44	38.99	1520.53
4½	3.99	15.904	24½	21.71	471.44	44½	39.44	1555.28
5	4.43	19.635	25	22.16	490.87	45	39.88	1590.43
5½	4.87	23.758	25½	22.60	510.71	45½	40.32	1625.97
6	5.32	28.274	26	23.04	530.93	46	40.77	1661.90
6½	5.76	33.183	26½	23.49	551.55	46½	41.21	1698.23
7	6.20	38.485	27	23.93	572.56	47	41.65	1734.94
7½	6.65	44.179	27½	24.37	593.96	47½	42.10	1772.05
8	7.09	50.265	28	24.81	615.75	48	42.54	1809.56
8½	7.53	56.745	28½	25.26	637.94	48½	42.98	1847.45
9	7.98	63.617	29	25.70	660.52	49	43.43	1885.74
9½	8.42	70.882	29½	26.14	683.49	49½	43.87	1924.42
10	8.86	78.540	30	26.59	706.86	50	44.31	1963.50
10½	9.31	86.590	30½	27.03	730.62	50½	44.75	2002.96
11	9.75	95.033	31	27.47	754.77	51	45.20	2042.82
11½	10.19	103.87	31½	27.92	779.31	51½	45.64	2083.07
12	10.63	113.10	32	28.36	804.25	52	46.08	2123.72
12½	11.08	122.72	32½	28.80	829.58	52½	46.53	2164.75
13	11.52	132.73	33	29.25	855.30	53	46.97	2206.18
13½	11.96	143.14	33½	29.69	881.41	53½	47.41	2248.01
14	12.41	153.94	34	30.13	907.92	54	47.86	2290.22
14½	12.85	165.13	34½	30.57	934.82	54½	48.30	2332.83
15	13.29	176.71	35	31.02	962.11	55	48.74	2375.83
15½	13.74	188.69	35½	31.46	989.80	55½	49.19	2419.22
16	14.18	201.06	36	31.90	1017.88	56	49.63	2463.01
16½	14.62	213.82	36½	32.35	1046.35	56½	50.07	2507.19
17	15.07	226.98	37	32.79	1075.21	57	50.51	2551.76
17½	15.51	240.53	37½	33.23	1104.47	57½	50.96	2596.72
18	15.95	254.47	38	33.68	1134.11	58	51.40	2642.08
18½	16.40	268.80	38½	34.12	1164.16	58½	51.84	2687.83
19	16.84	283.53	39	34.56	1194.59	59	52.29	2733.97
19½	17.28	298.65	39½	35.01	1225.42	59½	52.73	2780.51
20	17.72	314.16	40	35.45	1256.64	60	53.17	2827.43

Segments of Circles for Radius = 1 (English or metric units)

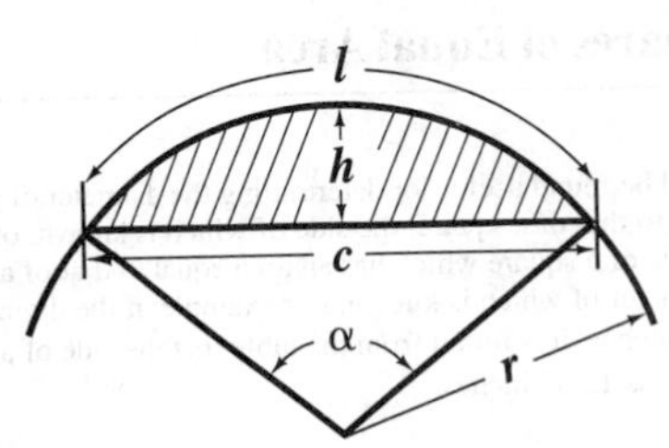

Formulas for segments of circles are given on page 63. When the central angle α and radius *r* are known, the tables on these pages can be used to find the length of arc *l*, height of segment *h*, chord length *c*, and segment area *A*. When angle α and radius *r* are not known, but segment height *h* and chord length *c* are known or can be measured, the ratio *h*/*c* can be used to enter the table and find α, *l*, and *A* by linear interpolation. Radius *r* is found by the formula on page 63. The value of *l* is then multiplied by the radius *r* and the area *A* by r^2, the square of the radius.

Angle α can be found thus with an accuracy of about 0.001 degree; arc length *l* with an error of about 0.02 per cent; and area *A* with an error ranging from about 0.02 per cent for the highest entry value of *h*/*c* to about 1 per cent for values of *h*/*c* of about 0.050. For lower values of *h*/*c*, and where greater accuracy is required, area *A* should be found by the formula on page 63.

θ, Deg.	*l*	*h*	*c*	Area *A*	*h*/*c*	θ, Deg.	*l*	*h*	*c*	Area *A*	*h*/*c*
1	0.01745	0.00004	0.01745	0.0000	0.00218	41	0.71558	0.06333	0.70041	0.0298	0.09041
2	0.03491	0.00015	0.03490	0.0000	0.00436	42	0.73304	0.06642	0.71674	0.0320	0.09267
3	0.05236	0.00034	0.05235	0.0000	0.00655	43	0.75049	0.06958	0.73300	0.0342	0.09493
4	0.06981	0.00061	0.06980	0.0000	0.00873	44	0.76794	0.07282	0.74921	0.0366	0.09719
5	0.08727	0.00095	0.08724	0.0001	0.01091	45	0.78540	0.07612	0.76537	0.0391	0.09946
6	0.10472	0.00137	0.10467	0.0001	0.01309	46	0.80285	0.07950	0.78146	0.0418	0.10173
7	0.12217	0.00187	0.12210	0.0002	0.01528	47	0.82030	0.08294	0.79750	0.0445	0.10400
8	0.13963	0.00244	0.13951	0.0002	0.01746	48	0.83776	0.08645	0.81347	0.0473	0.10628
9	0.15708	0.00308	0.15692	0.0003	0.01965	49	0.85521	0.09004	0.82939	0.0503	0.10856
10	0.17453	0.00381	0.17431	0.0004	0.02183	50	0.87266	0.09369	0.84524	0.0533	0.11085
11	0.19199	0.00460	0.19169	0.0006	0.02402	51	0.89012	0.09741	0.86102	0.0565	0.11314
12	0.20944	0.00548	0.20906	0.0008	0.02620	52	0.90757	0.10121	0.87674	0.0598	0.11543
13	0.22689	0.00643	0.22641	0.0010	0.02839	53	0.92502	0.10507	0.89240	0.0632	0.11773
14	0.24435	0.00745	0.24374	0.0012	0.03058	54	0.94248	0.10899	0.90798	0.0667	0.12004
15	0.26180	0.00856	0.26105	0.0015	0.03277	55	0.95993	0.11299	0.92350	0.0704	0.12235
16	0.27925	0.00973	0.27835	0.0018	0.03496	56	0.97738	0.11705	0.93894	0.0742	0.12466
17	0.29671	0.01098	0.29562	0.0022	0.03716	57	0.99484	0.12118	0.95432	0.0781	0.12698
18	0.31416	0.01231	0.31287	0.0026	0.03935	58	1.01229	0.12538	0.96962	0.0821	0.12931
19	0.33161	0.01371	0.33010	0.0030	0.04155	59	1.02974	0.12964	0.98485	0.0863	0.13164
20	0.34907	0.01519	0.34730	0.0035	0.04374	60	1.04720	0.13397	1.00000	0.0906	0.13397
21	0.36652	0.01675	0.36447	0.0041	0.04594	61	1.06465	0.13837	1.01508	0.0950	0.13632
22	0.38397	0.01837	0.38162	0.0047	0.04814	62	1.08210	0.14283	1.03008	0.0996	0.13866
23	0.40143	0.02008	0.39874	0.0053	0.05035	63	1.09956	0.14736	1.04500	0.1043	0.14101
24	0.41888	0.02185	0.41582	0.0061	0.05255	64	1.11701	0.15195	1.05984	0.1091	0.14337
25	0.43633	0.02370	0.43288	0.0069	0.05476	65	1.13446	0.15661	1.07460	0.1141	0.14574
26	0.45379	0.02563	0.44990	0.0077	0.05697	66	1.15192	0.16133	1.08928	0.1192	0.14811
27	0.47124	0.02763	0.46689	0.0086	0.05918	67	1.16937	0.16611	1.10387	0.1244	0.15048
28	0.48869	0.02970	0.48384	0.0096	0.06139	68	1.18682	0.17096	1.11839	0.1298	0.15287
29	0.50615	0.03185	0.50076	0.0107	0.06361	69	1.20428	0.17587	1.13281	0.1353	0.15525
30	0.52360	0.03407	0.51764	0.0118	0.06583	70	1.22173	0.18085	1.14715	0.1410	0.15765
31	0.54105	0.03637	0.53448	0.0130	0.06805	71	1.23918	0.18588	1.16141	0.1468	0.16005
32	0.55851	0.03874	0.55127	0.0143	0.07027	72	1.25664	0.19098	1.17557	0.1528	0.16246
33	0.57596	0.04118	0.56803	0.0157	0.07250	73	1.27409	0.19614	1.18965	0.1589	0.16488
34	0.59341	0.04370	0.58474	0.0171	0.07473	74	1.29154	0.20136	1.20363	0.1651	0.16730
35	0.61087	0.04628	0.60141	0.0186	0.07696	75	1.30900	0.20665	1.21752	0.1715	0.16973
36	0.62832	0.04894	0.61803	0.0203	0.07919	76	1.32645	0.21199	1.23132	0.1781	0.17216
37	0.64577	0.05168	0.63461	0.0220	0.08143	77	1.34390	0.21739	1.24503	0.1848	0.17461
38	0.66323	0.05448	0.65114	0.0238	0.08367	78	1.36136	0.22285	1.25864	0.1916	0.17706
39	0.68068	0.05736	0.66761	0.0257	0.08592	79	1.37881	0.22838	1.27216	0.1986	0.17952
40	0.69813	0.06031	0.68404	0.0277	0.08816	80	1.39626	0.23396	1.28558	0.2057	0.18199

Segments of Circles for Radius = 1 (English or metric units)

θ, Deg.	*l*	*h*	*c*	*Area A*	*h/c*	θ, Deg.	*l*	*h*	*c*	*Area A*	*h/c*
81	1.41372	0.23959	1.29890	0.2130	0.18446	131	2.28638	0.58531	1.81992	0.7658	0.32161
82	1.43117	0.24529	1.31212	0.2205	0.18694	132	2.30383	0.59326	1.82709	0.7803	0.32470
83	1.44862	0.25104	1.32524	0.2280	0.18943	133	2.32129	0.60125	1.83412	0.7950	0.32781
84	1.46608	0.25686	1.33826	0.2358	0.19193	134	2.33874	0.60927	1.84101	0.8097	0.33094
85	1.48353	0.26272	1.35118	0.2437	0.19444	135	2.35619	0.61732	1.84776	0.8245	0.33409
86	1.50098	0.26865	1.36400	0.2517	0.19696	136	2.37365	0.62539	1.85437	0.8395	0.33725
87	1.51844	0.27463	1.37671	0.2599	0.19948	137	2.39110	0.63350	1.86084	0.8546	0.34044
88	1.53589	0.28066	1.38932	0.2682	0.20201	138	2.40855	0.64163	1.86716	0.8697	0.34364
89	1.55334	0.28675	1.40182	0.2767	0.20456	139	2.42601	0.64979	1.87334	0.8850	0.34686
90	1.57080	0.29289	1.41421	0.2854	0.20711	140	2.44346	0.65798	1.87939	0.9003	0.35010
91	1.58825	0.29909	1.42650	0.2942	0.20967	141	2.46091	0.66619	1.88528	0.9158	0.35337
92	1.60570	0.30534	1.43868	0.3032	0.21224	142	2.47837	0.67443	1.89104	0.9314	0.35665
93	1.62316	0.31165	1.45075	0.3123	0.21482	143	2.49582	0.68270	1.89665	0.9470	0.35995
94	1.64061	0.31800	1.46271	0.3215	0.21741	144	2.51327	0.69098	1.90211	0.9627	0.36327
95	1.65806	0.32441	1.47455	0.3309	0.22001	145	2.53073	0.69929	1.90743	0.9786	0.36662
96	1.67552	0.33087	1.48629	0.3405	0.22261	146	2.54818	0.70763	1.91261	0.9945	0.36998
97	1.69297	0.33738	1.49791	0.3502	0.22523	147	2.56563	0.71598	1.91764	1.0105	0.37337
98	1.71042	0.34394	1.50942	0.3601	0.22786	148	2.58309	0.72436	1.92252	1.0266	0.37678
99	1.72788	0.35055	1.52081	0.3701	0.23050	149	2.60054	0.73276	1.92726	1.0428	0.38021
100	1.74533	0.35721	1.53209	0.3803	0.23315	150	2.61799	0.74118	1.93185	1.0590	0.38366
101	1.76278	0.36392	1.54325	0.3906	0.23582	151	2.63545	0.74962	1.93630	1.0753	0.38714
102	1.78024	0.37068	1.55429	0.4010	0.23849	152	2.65290	0.75808	1.94059	1.0917	0.39064
103	1.79769	0.37749	1.56522	0.4117	0.24117	153	2.67035	0.76655	1.94474	1.1082	0.39417
104	1.81514	0.38434	1.57602	0.4224	0.24387	154	2.68781	0.77505	1.94874	1.1247	0.39772
105	1.83260	0.39124	1.58671	0.4333	0.24657	155	2.70526	0.78356	1.95259	1.1413	0.40129
106	1.85005	0.39818	1.59727	0.4444	0.24929	156	2.72271	0.79209	1.95630	1.1580	0.40489
107	1.86750	0.40518	1.60771	0.4556	0.25202	157	2.74017	0.80063	1.95985	1.1747	0.40852
108	1.88496	0.41221	1.61803	0.4669	0.25476	158	2.75762	0.80919	1.96325	1.1915	0.41217
109	1.90241	0.41930	1.62823	0.4784	0.25752	159	2.77507	0.81776	1.96651	1.2084	0.41585
110	1.91986	0.42642	1.63830	0.4901	0.26028	160	2.79253	0.82635	1.96962	1.2253	0.41955
111	1.93732	0.43359	1.64825	0.5019	0.26306	161	2.80998	0.83495	1.97257	1.2422	0.42328
112	1.95477	0.44081	1.65808	0.5138	0.26585	162	2.82743	0.84357	1.97538	1.2592	0.42704
113	1.97222	0.44806	1.66777	0.5259	0.26866	163	2.84489	0.85219	1.97803	1.2763	0.43083
114	1.98968	0.45536	1.67734	0.5381	0.27148	164	2.86234	0.86083	1.98054	1.2934	0.43464
115	2.00713	0.46270	1.68678	0.5504	0.27431	165	2.87979	0.86947	1.98289	1.3105	0.43849
116	2.02458	0.47008	1.69610	0.5629	0.27715	166	2.89725	0.87813	1.98509	1.3277	0.44236
117	2.04204	0.47750	1.70528	0.5755	0.28001	167	2.91470	0.88680	1.98714	1.3449	0.44627
118	2.05949	0.48496	1.71433	0.5883	0.28289	168	2.93215	0.89547	1.98904	1.3621	0.45020
119	2.07694	0.49246	1.72326	0.6012	0.28577	169	2.94961	0.90415	1.99079	1.3794	0.45417
120	2.09440	0.50000	1.73205	0.6142	0.28868	170	2.96706	0.91284	1.99239	1.3967	0.45817
121	2.11185	0.50758	1.74071	0.6273	0.29159	171	2.98451	0.92154	1.99383	1.4140	0.46220
122	2.12930	0.51519	1.74924	0.6406	0.29452	172	3.00197	0.93024	1.99513	1.4314	0.46626
123	2.14675	0.52284	1.75763	0.6540	0.29747	173	3.01942	0.93895	1.99627	1.4488	0.47035
124	2.16421	0.53053	1.76590	0.6676	0.30043	174	3.03687	0.94766	1.99726	1.4662	0.47448
125	2.18166	0.53825	1.77402	0.6813	0.30341	175	3.05433	0.95638	1.99810	1.4836	0.47865
126	2.19911	0.54601	1.78201	0.6950	0.30640	176	3.07178	0.96510	1.99878	1.5010	0.48284
127	2.21657	0.55380	1.78987	0.7090	0.30941	177	3.08923	0.97382	1.99931	1.5184	0.48708
128	2.23402	0.56163	1.79759	0.7230	0.31243	178	3.10669	0.98255	1.99970	1.5359	0.49135
129	2.25147	0.56949	1.80517	0.7372	0.31548	179	3.12414	0.99127	1.99992	1.5533	0.49566
130	2.26893	0.57738	1.81262	0.7514	0.31854	180	3.14159	1.00000	2.00000	1.5708	0.50000

Distance Across Corners of Squares and Hexagons (English and metric units)

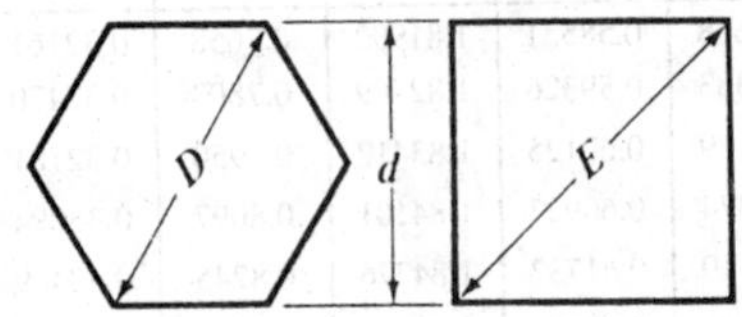

$D = 1.154701d$

$E = 1.414214d$

A desired value not given directly in the table can be obtained by the simple addition of two or more values taken directly from the table. Further values can be obtained by shifting the decimal point.

Example 1: Find D when d = 2 5/16 inches. From the table, 2 = 2.3094, and 5/16 = 0.3608. Therefore, D = 2.3094 + 0.3608 = 2.6702 inches.

Example 2: Find E when d = 20.25 millimeters. From the table, 20 = 28.2843; 0.2 = 0.2828; and 0.05 = 0.0707 (obtained by shifting the decimal point one place to the left at d = 0.5). Thus, E = 28.2843 + 0.2828 + 0.0707 = 28.6378 millimeters.

d	*D*	*E*	*d*	*D*	*E*	*d*	*D*	*E*	*d*	*D*	*E*
1/32	0.0361	0.0442	0.9	1.0392	1.2728	32	36.9504	45.2548	67	77.3650	94.7523
1/16	0.0722	0.0884	29/32	1.0464	1.2816	33	38.1051	46.6691	68	78.5197	96.1666
3/32	0.1083	0.1326	15/16	1.0825	1.3258	34	39.2598	48.0833	69	79.6744	97.5808
0.1	0.1155	0.1414	31/32	1.1186	1.3700	35	40.4145	49.4975	70	80.8291	98.9950
1/8	0.1443	0.1768	1.0	1.1547	1.4142	36	41.5692	50.9117	71	81.9838	100.409
5/32	0.1804	0.2210	2.0	2.3094	2.8284	37	42.7239	52.3259	72	83.1385	101.823
3/16	0.2165	0.2652	3.0	3.4641	4.2426	38	43.8786	53.7401	73	84.2932	103.238
0.2	0.2309	0.2828	4.0	4.6188	5.6569	39	45.0333	55.1543	74	85.4479	104.652
7/32	0.2526	0.3094	5.0	5.7735	7.0711	40	46.1880	56.5686	75	86.6026	106.066
1/4	0.2887	0.3536	6.0	6.9282	8.4853	41	47.3427	57.9828	76	87.7573	107.480
9/32	0.3248	0.3977	7.0	8.0829	9.8995	42	48.4974	59.3970	77	88.9120	108.894
0.3	0.3464	0.4243	8.0	9.2376	11.3137	43	49.6521	60.8112	78	90.0667	110.309
5/16	0.3608	0.4419	9.0	10.3923	12.7279	44	50.8068	62.2254	79	91.2214	111.723
11/32	0.3969	0.4861	10	11.5470	14.1421	45	51.9615	63.6396	80	92.3761	113.137
3/8	0.4330	0.5303	11	12.7017	15.5564	46	53.1162	65.0538	81	93.5308	114.551
0.4	0.4619	0.5657	12	13.8564	16.9706	47	54.2709	66.4681	82	94.6855	115.966
13/32	0.4691	0.5745	13	15.0111	18.3848	48	55.4256	67.8823	83	95.8402	117.380
7/16	0.5052	0.6187	14	16.1658	19.7990	49	56.5803	69.2965	84	96.9949	118.794
15/32	0.5413	0.6629	15	17.3205	21.2132	50	57.7351	70.7107	85	98.1496	120.208
0.5	0.5774	0.7071	16	18.4752	22.6274	51	58.8898	72.1249	86	99.3043	121.622
17/32	0.6134	0.7513	17	19.6299	24.0416	52	60.0445	73.5391	87	100.459	123.037
9/16	0.6495	0.7955	18	20.7846	25.4559	53	61.1992	74.9533	88	101.614	124.451
19/32	0.6856	0.8397	19	21.9393	26.8701	54	62.3539	76.3676	89	102.768	125.865
0.6	0.6928	0.8485	20	23.0940	28.2843	55	63.5086	77.7818	90	103.923	127.279
5/8	0.7217	0.8839	21	24.2487	29.6985	56	64.6633	79.1960	91	105.078	128.693
21/32	0.7578	0.9281	22	25.4034	31.1127	57	65.8180	80.6102	92	106.232	130.108
11/16	0.7939	0.9723	23	26.5581	32.5269	58	66.9727	82.0244	93	107.387	131.522
0.7	0.8083	0.9899	24	27.7128	33.9411	59	68.1274	83.4386	94	108.542	132.936
23/32	0.8299	1.0165	25	28.8675	35.3554	60	69.2821	84.8528	95	109.697	134.350
3/4	0.8660	1.0607	26	30.0222	36.7696	61	70.4368	86.2671	96	110.851	135.765
25/32	0.9021	1.1049	27	31.1769	38.1838	62	71.5915	87.6813	97	112.006	137.179
0.8	0.9238	1.1314	28	32.3316	39.5980	63	72.7462	89.0955	98	113.161	138.593
13/16	0.9382	1.1490	29	33.4863	41.0122	64	73.9009	90.5097	99	114.315	140.007
27/32	0.9743	1.1932	30	34.6410	42.4264	65	75.0556	91.9239	100	115.470	141.421
7/8	1.0104	1.2374	31	35.7957	43.8406	66	76.2103	93.3381	…	…	…

SOLUTION OF TRIANGLES

Any figure bounded by three straight lines is called a triangle. Any one of the three lines may be called the base, and the line drawn from the angle opposite the base at right angles to it is called the height or altitude of the triangle.

If all three sides of a triangle are of equal length, the triangle is called *equilateral*. Each of the three angles in an equilateral triangle equals 60 degrees. If two sides are of equal length, the triangle is an *isosceles* triangle. If one angle is a right or 90-degree angle, the triangle is a *right* or *right-angled* triangle. The side opposite the right angle is called the *hypotenuse*.

If all the angles are less than 90 degrees, the triangle is called an *acute* or *acute-angled* triangle. If one of the angles is larger than 90 degrees, the triangle is called an *obtuse-angled* triangle. Both acute and obtuse-angled triangles are known under the common name of *oblique-angled* triangles. The sum of the three angles in every triangle is 180 degrees.

The sides and angles of any triangle that are not known can be found when: 1) all the three sides; 2) two sides and one angle; and 3) one side and two angles are given.

In other words, if a triangle is considered as consisting of six parts, three angles and three sides, the unknown parts can be determined when any three parts are given, provided at least one of the given parts is a side.

Functions of Angles

For every right triangle, a set of six ratios is defined; each is the length of one side of the triangle divided by the length of another side. The six ratios are the trigonometric (trig) functions sine, cosine, tangent, cosecant, secant, and cotangent (abbreviated sin, cos, tan, csc, sec, and cot). Trig functions are usually expressed in terms of an angle in degree or radian measure, as in $\cos 60° = 0.5$. "Arc" in front of a trig function name, as in arcsin or arccos, means find the angle whose function value is given. For example, $\arcsin 0.5 = 30°$ means that 30° is the angle whose sin is equal to 0.5. Electronic calculators frequently use $\sin^{-1}$, $\cos^{-1}$, and $\tan^{-1}$ to represent the arc functions.

Example: $\tan 53.1° = 1.332$; $\arctan 1.332 = \tan^{-1} 1.332 = 53.1° = 53° 6'$

The *sine* of an angle equals the opposite side divided by the hypotenuse. Hence, $\sin B = b \div c$, and $\sin A = a \div c$.

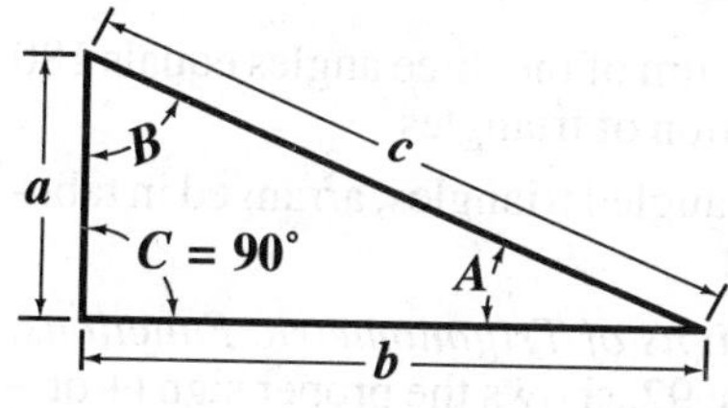

The *cosine* of an angle equals the adjacent side divided by the hypotenuse. Hence, $\cos B = a \div c$, and $\cos A = b \div c$.

The *tangent* of an angle equals the opposite side divided by the adjacent side. Hence, $\tan B = b \div a$, and $\tan A = a \div b$.

The *cotangent* of an angle equals the adjacent side divided by the opposite side. Hence, $\cot B = a \div b$, and $\cot A = b \div a$.

The *secant* of an angle equals the hypotenuse divided by the adjacent side. Hence, $\sec B = c \div a$, and $\sec A = c \div b$.

The *cosecant* of an angle equals the hypotenuse divided by the opposite side. Hence, $\csc B = c \div b$, and $\csc A = c \div a$.

It should be noted that the functions of the angles can be found in this manner only when the triangle is right-angled.

If in a right-angled triangle (see preceding illustration), the lengths of the three sides are represented by a, b, and c, and the angles opposite each of these sides by A, B, and C, then the side c opposite the right angle is the hypotenuse; side b is called the *side adjacent* to angle A and is also the *side opposite* to angle B; side a is the side adjacent to angle B and the side opposite to angle *A*. The meanings of the various functions of angles can be explained with the aid of a right-angled triangle. Note that the cosecant, secant, and cotangent are the reciprocals of, respectively, the sine, cosine, and tangent.

The following relation exists between the angular functions of the two acute angles in a right-angled triangle: The sine of angle *B* equals the cosine of angle *A*; the tangent of angle *B* equals the cotangent of angle *A*, and *vice versa*. The sum of the two acute angles in a right-angled triangle always equals 90 degrees; hence, when one angle is known, the other can easily be found. When any two angles together make 90 degrees, one is called the *complement* of the other, and the sine of the one angle equals the cosine of the other, and the tangent of the one equals the cotangent of the other.

The Law of Sines.—In any triangle, any side is to the sine of the angle opposite that side as any other side is to the sine of the angle opposite that side. If *a*, *b*, and *c* are the sides, and *A*, *B*, and *C* their opposite angles, respectively, then:

$$\frac{a}{\sin A} = \frac{b}{\sin B} = \frac{c}{\sin C}, \quad \text{so that:}$$

$$a = \frac{b \sin A}{\sin B} \quad \text{or} \quad a = \frac{c \sin A}{\sin C}$$

$$b = \frac{a \sin B}{\sin A} \quad \text{or} \quad b = \frac{c \sin B}{\sin C}$$

$$c = \frac{a \sin C}{\sin A} \quad \text{or} \quad c = \frac{b \sin C}{\sin B}$$

The Law of Cosines.—In any triangle, the square of any side is equal to the sum of the squares of the other two sides minus twice their product times the cosine of the included angle; or if *a*, *b* and *c* are the sides and *A*, *B*, and *C* are the opposite angles, respectively, then:

$$a^2 = b^2 + c^2 - 2bc \cos A$$

$$b^2 = a^2 + c^2 - 2ac \cos B$$

$$c^2 = a^2 + b^2 - 2ab \cos C$$

These two laws, together with the proposition that the sum of the three angles equals 180 degrees, are the basis of all formulas relating to the solution of triangles.

Formulas for the solution of right-angled and oblique-angled triangles, arranged in tabular form, are given on the following pages.

Signs of Trigonometric Functions.—The diagram, *Signs of Trigonometric Functions, Fractions of π, and Degree–Radian Conversion* on page 92, shows the proper sign (+ or −) for the trigonometric functions of angles in each of the four quadrants, 0 to 90, 90 to 180, 180 to 270, and 270 to 360 degrees. Thus, the cosine of an angle between 90 and 180 degrees is negative; the sine of the same angle is positive.

Trigonometric Identities.—Trigonometric identities are formulas that show the relationship between different trigonometric functions. They may be used to change the form of some trigonometric expressions to simplify calculations. For example, if a formula has a term, 2sin*A*cos*A*, the equivalent but simpler term sin2*A* may be substituted. The identities that follow may themselves be combined or rearranged in various ways to form new identities.

Basic

$$\tan A = \frac{\sin A}{\cos A} = \frac{1}{\cot A} \qquad \sec A = \frac{1}{\cos A} \qquad \csc A = \frac{1}{\sin A}$$

Negative Angle

$\sin(-A) = -\sin A \qquad \cos(-A) = \cos A \qquad \tan(-A) = -\tan A$

Pythagorean

$\sin^2 A + \cos^2 A = 1 \qquad 1 + \tan^2 A = \sec^2 A \qquad 1 + \cot^2 A = \csc^2 A$

Sum and Difference of Angles

$$\tan(A+B) = \frac{\tan A + \tan B}{1 - \tan A \tan B} \qquad \tan(A-B) = \frac{\tan A - \tan B}{1 + \tan A \tan B}$$

$$\cot(A+B) = \frac{\cot A \cot B - 1}{\cot B + \cot A} \qquad \cot(A-B) = \frac{\cot A \cot B + 1}{\cot B - \cot A}$$

$$\sin(A+B) = \sin A \cos B + \cos A \sin B \qquad \sin(A-B) = \sin A \cos B - \cos A \sin B$$

$$\cos(A+B) = \cos A \cos B - \sin A \sin B \qquad \cos(A-B) = \cos A \cos B - \sin A \sin B$$

Double-Angle

$$\cos 2A = \cos^2 A - \sin^2 A = 2\cos^2 A - 1 = 1 - 2\sin^2 A$$

$$\sin 2A = 2\sin A \cos A \qquad \tan 2A = \frac{2\tan A}{1 - \tan^2 A} = \frac{2}{\cot A - \tan A}$$

Half-Angle

$$\sin \tfrac{1}{2}A = \sqrt{\tfrac{1}{2}(1 - \cos A)} \qquad \cos \tfrac{1}{2}A = \sqrt{\tfrac{1}{2}(1 + \cos A)}$$

$$\tan \tfrac{1}{2}A = \sqrt{\frac{1 - \cos A}{1 + \cos A}} = \frac{1 - \cos A}{\sin A} = \frac{\sin A}{1 + \cos A}$$

Product-to-Sum

$$\sin A \cos B = \tfrac{1}{2}[\sin(A+B) + \sin(A-B)]$$

$$\cos A \cos B = \tfrac{1}{2}[\cos(A+B) + \cos(A-B)]$$

$$\sin A \sin B = \tfrac{1}{2}[\cos(A-B) - \cos(A+B)]$$

$$\tan A \tan B = \frac{\tan A + \tan B}{\cot A + \cot B}$$

Sum and Difference of Functions

$$\sin A + \sin B = 2[\sin \tfrac{1}{2}(A+B) \cos \tfrac{1}{2}(A-B)]$$

$$\sin A - \sin B = 2[\sin \tfrac{1}{2}(A-B) \cos \tfrac{1}{2}(A+B)]$$

$$\cos A + \cos B = 2[\cos \tfrac{1}{2}(A+B) \cos \tfrac{1}{2}(A-B)]$$

$$\cos A - \cos B = -2[\cos \tfrac{1}{2}(A+B) \cos \tfrac{1}{2}(A-B)]$$

$$\tan A + \tan B = \frac{\sin(A+B)}{\cos A \cos B} \qquad \tan A - \tan B = \frac{\sin(A-B)}{\cos A \cos B}$$

$$\cot A + \cot B = \frac{\sin(B+A)}{\sin A \sin B} \qquad \cot A - \cot B = \frac{\sin(B-A)}{\sin A \sin B}$$

Solution of Right-Angled Triangles

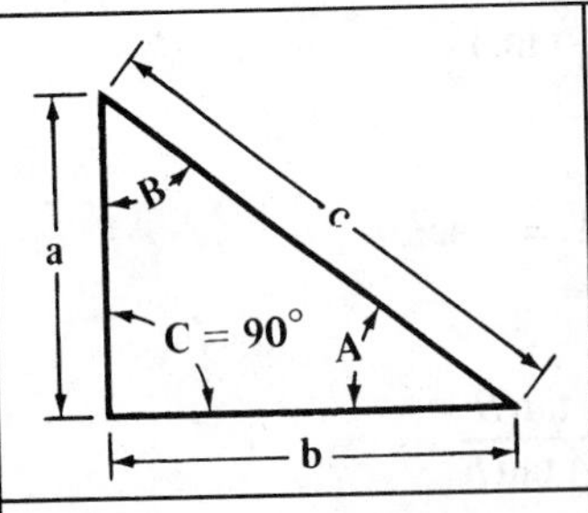

As shown in the illustration, the sides of the right-angled triangle are designated *a* and *b* and the hypotenuse, *c*. The angles opposite each of these sides are designated *A* and *B*, respectively.

Angle *C*, opposite the hypotenuse *c* is the right angle, and is therefore always one of the known quantities.

Sides and Angles Known	Formulas for Sides and Angles to be Found		
Side a; side b	$c = \sqrt{a^2 + b^2}$	$\tan A = \frac{a}{b}$	$B = 90° - A$
Side a; hypotenuse c	$b = \sqrt{c^2 - a^2}$	$\sin A = \frac{a}{c}$	$B = 90° - A$
Side b; hypotenuse c	$a = \sqrt{c^2 - b^2}$	$\sin B = \frac{b}{c}$	$A = 90° - B$
Hypotenuse c; angle B	$b = c \times \sin B$	$a = c \times \cos B$	$A = 90° - B$
Hypotenuse c; angle A	$b = c \times \cos A$	$a = c \times \sin A$	$B = 90° - A$
Side b; angle B	$c = \frac{b}{\sin B}$	$a = b \times \cot B$	$A = 90° - B$
Side b; angle A	$c = \frac{b}{\cos A}$	$a = b \times \tan A$	$B = 90° - A$
Side a; angle B	$c = \frac{a}{\cos B}$	$b = a \times \tan B$	$A = 90° - B$
Side a; angle A	$c = \frac{a}{\sin A}$	$b = a \times \cot A$	$B = 90° - A$

Examples of the Solution of Right-Angled Triangles (English and metric units)

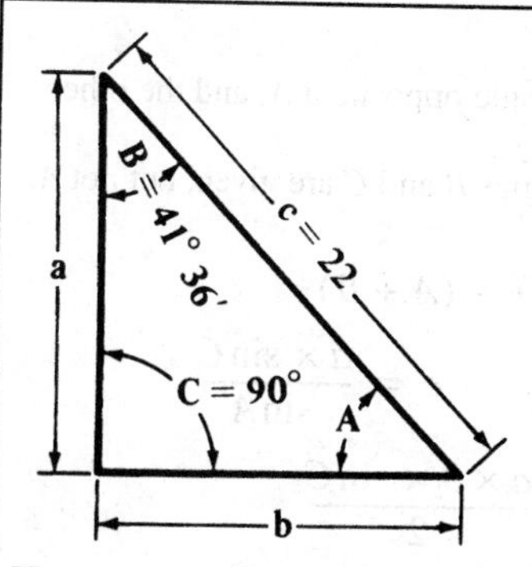 Hypotenuse and one angle known	$c = 22$ inches; $B = 41°\ 36'$. $a = c \times \cos B = 22 \times \cos 41°36' = 22 \times 0.74780$ $= 16.4516$ inches $b = c \times \sin B = 22 \times \sin 41°36' = 22 \times 0.66393$ $= 14.6065$ inches $A = 90° - B = 90° - 41°36' = 48°24'$
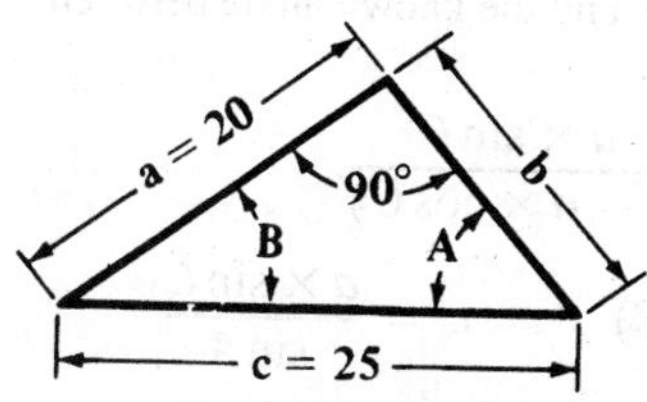 Hypotenuse and one side known	$c = 25$ centimeters; $a = 20$ centimeters. $b = \sqrt{c^2 - a^2} = \sqrt{25^2 - 20^2} - \sqrt{625 - 400}$ $= \sqrt{225} = 15$ centimeters $\sin A = \frac{a}{c} = \frac{20}{25} = 0.8$ Hence, $A = 53°8'$ $B = 90° - A = 90° - 53°8' = 36°52'$
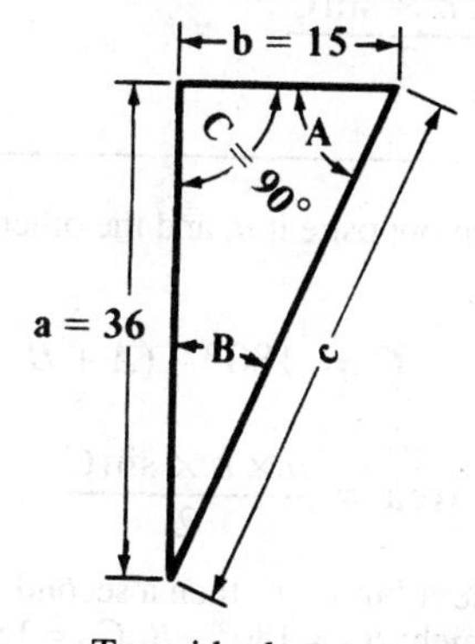 Two sides known	$a = 36$ inches; $b = 15$ inches. $c = \sqrt{a^2 + b^2} = \sqrt{36^2 + 15^2} = \sqrt{1296 + 225}$ $= \sqrt{1521} = 39$ inches $\tan A = \frac{a}{b} = \frac{36}{15} = 2.4$ Hence, $A = 67°23'$ $B = 90° - A = 90° - 67°23' = 22°37'$
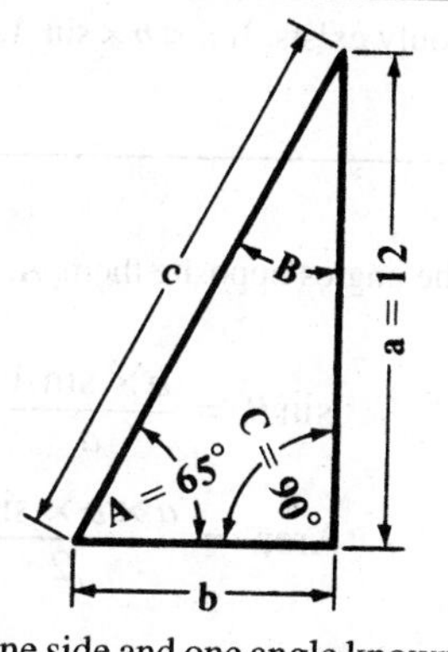 One side and one angle known	$a = 12$ meters; $A = 65°$. $c = \frac{a}{\sin A} = \frac{12}{\sin 65°} = \frac{12}{0.90631} = 13.2405$ meters $b = a \times \cot A = 12 \times \cot 65° = 12 \times 0.46631$ $= 5.5957$ meters $B = 90° - A = 90° - 65° = 25°$

Solution of Oblique-Angled Triangles

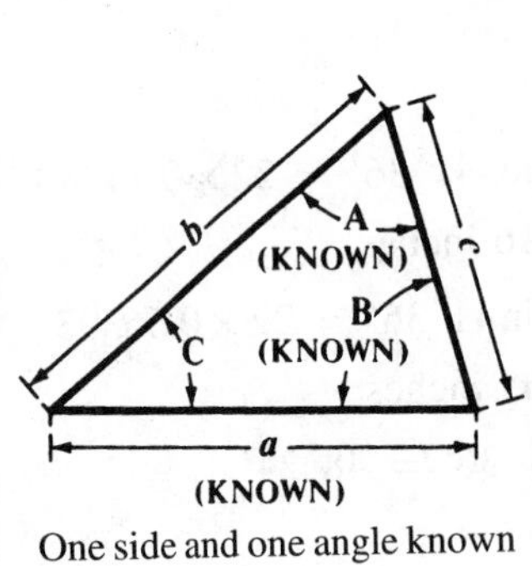 One side and one angle known	Call the known side a, the angle opposite it A, and the other known angle B. Then: $C = 180° - (A + B)$; or if angles B and C are given, but not A, then $A = 180° - (B + C)$. $$C = 180° - (A + B)$$ $$b = \frac{a \times \sin B}{\sin A} \qquad c = \frac{a \times \sin C}{\sin A}$$ $$\text{Area} = \frac{a \times b \times \sin C}{2}$$
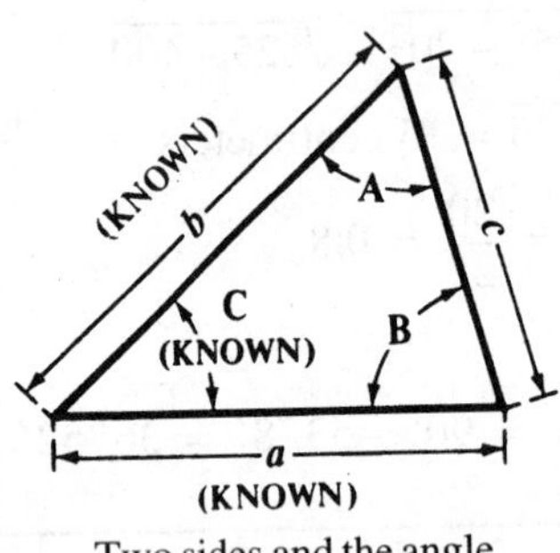 Two sides and the angle between them known	Call the known sides a and b, and the known angle between them C. Then: $$\tan A = \frac{a \times \sin C}{b - (a \times \cos C)}$$ $$B = 180° - (A + C) \qquad c = \frac{a \times \sin C}{\sin A}$$ Side c may also be found directly as below: $$c = \sqrt{a^2 + b^2 - (2ab \times \cos C)}$$ $$\text{Area} = \frac{a \times b \times \sin C}{2}$$
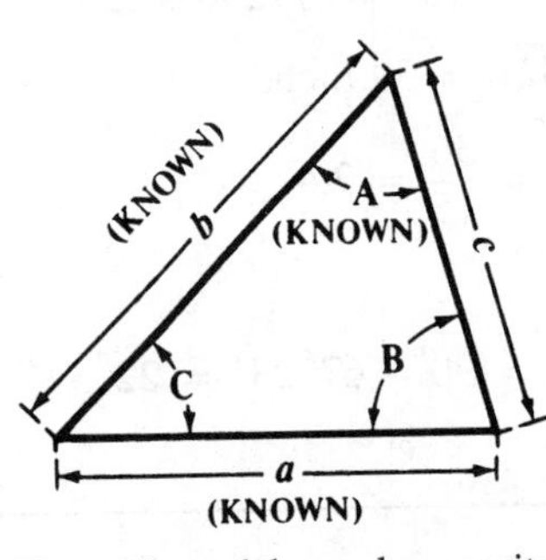 Two sides and the angle opposite one of the sides known	Call the known angle A, the side opposite it a, and the other known side b. Then: $$\sin B = \frac{b \times \sin A}{a} \qquad C = 180° - (A + B$$ $$c = \frac{a \times \sin C}{\sin A} \qquad \text{Area} = \frac{a \times b \times \sin C}{2}$$ If, in the above, angle $B >$ angle A but $<90°$, then a second solution B_2, C_2, c_2 exists for which: $B_2 = 180° - B$; $C_2 = 180° - (A + B_2)$; $c_2 = (a \times \sin C_2) \div \sin A$; area $= (a \times b \times \sin C_2) \div 2$. If $a \geq b$, then the first solution only exists. If $a < b \times \sin A$, then no solution exists.
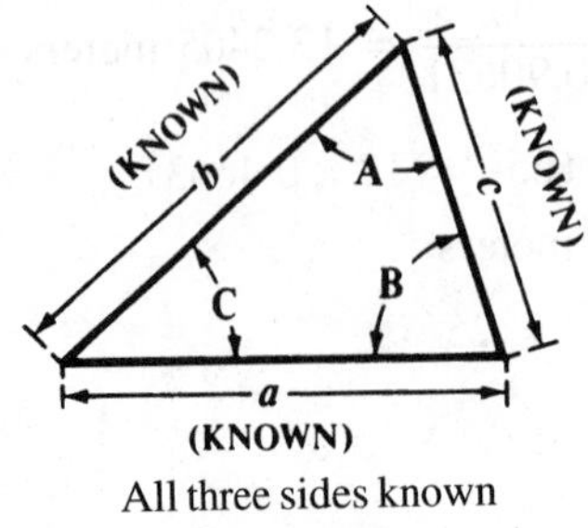 All three sides known	Call the sides a, b, and c, and the angles opposite them, A, B, and C. Then: $$\cos A = \frac{b^2 + c^2 - a^2}{2bc} \qquad \sin B = \frac{b \times \sin A}{a}$$ $$C = 180° - (A + B) \qquad \text{Area} = \frac{a \times b \times \sin}{2}$$

Examples of the Solution of Oblique-Angled Triangles (English and metric units)

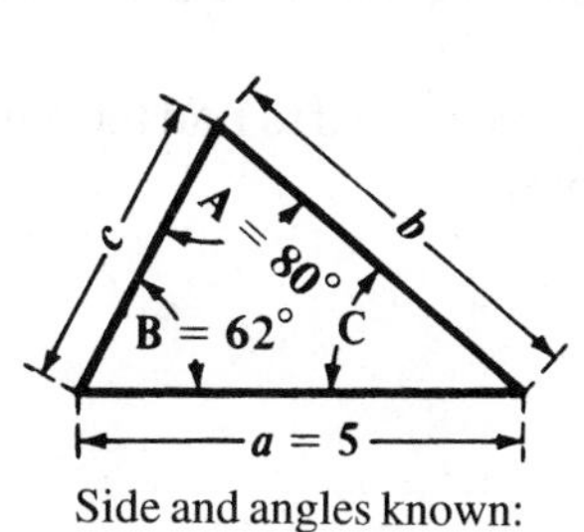

Side and angles known:

$a = 5$ centimeters; $A = 80°$; $B = 62°$

$$C = 180° - (80° + 62°) = 180° - 142° = 38°$$

$$b = \frac{a \times \sin B}{\sin A} = \frac{5 \times \sin 62°}{\sin 80°} = \frac{5 \times 0.88295}{0.98481} = 4.483$$

centimeters

$$c = \frac{a \times \sin C}{\sin A} = \frac{5 \times \sin 38°}{\sin 80°} = \frac{5 \times 0.61566}{0.98481} = 3.126$$

centimeters

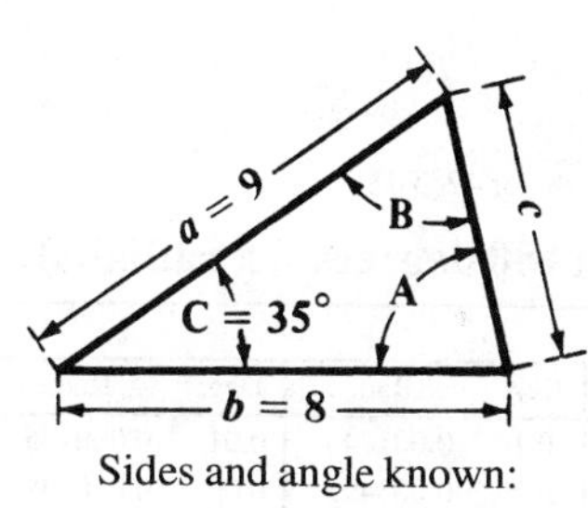

Sides and angle known:

$a = 9$ inches; $b = 8$ inches; $C = 35°$.

$$\tan A = \frac{a \times \sin C}{b - (a \times \cos C)} = \frac{9 \times \sin 35°}{8 - (9 \times \cos 35°)}$$

$$= \frac{9 \times 0.57358}{8 - (9 \times 0.81915)} = \frac{5.16222}{0.62765} = 8.2246$$

Hence, $A = 83°4'$

$$B = 180° - (A + C) = 180° - 118°4' = 61$$

$$c = \frac{a \times \sin C}{\sin A} = \frac{9 \times 0.57358}{0.99269} = 5.2 \text{ inches}$$

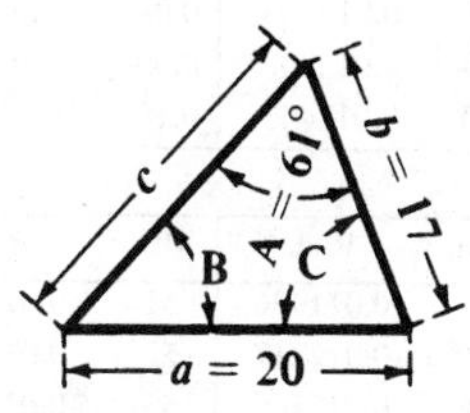

Sides and angle known:

$a = 20$ centimeters; $b = 17$ centimeters; $A = 61°$.

$$\sin B = \frac{b \times \sin A}{a} = \frac{17 \times \sin 61°}{20}$$

$$= \frac{17 \times 0.87462}{20} = 0.74343$$

Hence, $B = 48°1'$

$$C = 180° - (A + B) = 180° - 109°1' = 70°59'$$

$$c = \frac{a \times \sin C}{\sin A} = \frac{20 \times \sin 70°59'}{\sin 61°} = \frac{20 \times 0.94542}{0.87462}$$

$= 21.62$ centimeters

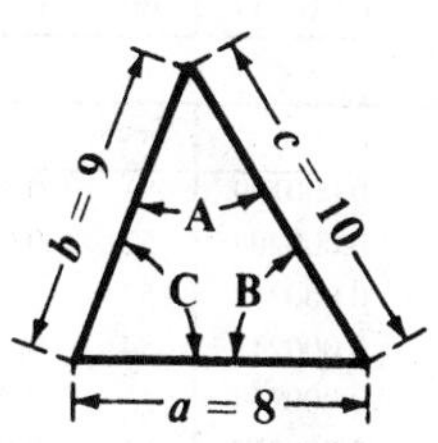

Sides and angle known:

$a = 8$ inches; $b = 9$ inches; $c = 10$ inches.

$$\cos A = \frac{b^2 + c^2 - a^2}{2bc} = \frac{9^2 + 10^2 - 8^2}{2 \times 9 \times 10}$$

$$= \frac{81 + 100 - 64}{180} = \frac{117}{180} = 0.65000$$

Hence, $A = 49°27'$

$$\sin B = \frac{b \times \sin A}{a} = \frac{9 \times 0.75984}{8} = 0.85482$$

Hence, $B = 58°44'$

$$C = 180° - (A + B) = 180° - 108°11' = 71°49'$$

Conversion Tables of Angular Measure.—The accompanying tables of degrees, minutes, and seconds into radians; radians into degrees, minutes, and seconds; radians into degrees and decimals of a degree; and minutes and seconds into decimals of a degree and vice versa facilitate the conversion of measurements.

Example: The Degrees, Minutes, and Seconds into Radians table is used to find the number of radians in 324 degrees, 25 minutes, 13 seconds as follows:

300 degrees	= 5.235988 radians
20 degrees	= 0.349066 radian
4 degrees	= 0.069813 radian
25 minutes	= 0.007272 radian
13 seconds	= 0.000063 radian
324°25′13″	= 5.662202 radians

Example: The Radians into Degrees and Decimals of a Degree, and Radians into Degrees, Minutes and Seconds tables are used to find the number of decimal degrees or degrees, minutes and seconds in 0.734 radian as follows:

0.7 radian	= 40.1070 degrees	0.7 radian	= 40° 6′25″
0.03 radian	= 1.7189 degrees	0.03 radian	= 1°43′8″
0.004 radian	= 0.2292 degree	0.004 radian	= 0°13′45″
0.734 radian	= 42.0551 degrees	0.734 radian	= 41°62′78″ or 42°3′18″

Degrees, Minutes, and Seconds into Radians (Based on 180 degrees = π radians)

Degrees into Radians											
Deg.	Rad.	Deg.	Rad.	Deg.	Rad.	Deg.	Rad.	Deg.	Rad.	Deg.	Rad.
1000	17.453293	100	1.745329	10	0.174533	1	0.017453	0.1	0.001745	0.01	0.000175
2000	34.906585	200	3.490659	20	0.349066	2	0.034907	0.2	0.003491	0.02	0.000349
3000	52.359878	300	5.235988	30	0.523599	3	0.052360	0.3	0.005236	0.03	0.000524
4000	69.813170	400	6.981317	40	0.698132	4	0.069813	0.4	0.006981	0.04	0.000698
5000	87.266463	500	8.726646	50	0.872665	5	0.087266	0.5	0.008727	0.05	0.000873
6000	104.719755	600	10.471976	60	1.047198	6	0.104720	0.6	0.010472	0.06	0.001047
7000	122.173048	700	12.217305	70	1.221730	7	0.122173	0.7	0.012217	0.07	0.001222
8000	139.626340	800	13.962634	80	1.396263	8	0.139626	0.8	0.013963	0.08	0.001396
9000	157.079633	900	15.707963	90	1.570796	9	0.157080	0.9	0.015708	0.09	0.001571
10000	174.532925	1000	17.453293	100	1.745329	10	0.174533	1.0	0.017453	0.10	0.001745
Minutes into Radians											
Min.	Rad.	Min.	Rad.	Min.	Rad.	Min.	Rad.	Min.	Rad.	Min.	Rad.
1	0.000291	11	0.003200	21	0.006109	31	0.009018	41	0.011926	51	0.014835
2	0.000582	12	0.003491	22	0.006400	32	0.009308	42	0.012217	52	0.015126
3	0.000873	13	0.003782	23	0.006690	33	0.009599	43	0.012508	53	0.015417
4	0.001164	14	0.004072	24	0.006981	34	0.009890	44	0.012799	54	0.015708
5	0.001454	15	0.004363	25	0.007272	35	0.010181	45	0.013090	55	0.015999
6	0.001745	16	0.004654	26	0.007563	36	0.010472	46	0.013381	56	0.016290
7	0.002036	17	0.004945	27	0.007854	37	0.010763	47	0.013672	57	0.016581
8	0.002327	18	0.005236	28	0.008145	38	0.011054	48	0.013963	58	0.016872
9	0.002618	19	0.005527	29	0.008436	39	0.011345	49	0.014254	59	0.017162
10	0.002909	20	0.005818	30	0.008727	40	0.011636	50	0.014544	60	0.017453
Seconds into Radians											
Sec.	Rad.	Sec.	Rad.	Sec.	Rad.	Sec.	Rad.	Sec.	Rad.	Sec.	Rad.
1	0.000005	11	0.000053	21	0.000102	31	0.000150	41	0.000199	51	0.000247
2	0.000010	12	0.000058	22	0.000107	32	0.000155	42	0.000204	52	0.000252
3	0.000015	13	0.000063	23	0.000112	33	0.000160	43	0.000208	53	0.000257
4	0.000019	14	0.000068	24	0.000116	34	0.000165	44	0.000213	54	0.000262
5	0.000024	15	0.000073	25	0.000121	35	0.000170	45	0.000218	55	0.000267
6	0.000029	16	0.000078	26	0.000126	36	0.000175	46	0.000223	56	0.000271
7	0.000034	17	0.000082	27	0.000131	37	0.000179	47	0.000228	57	0.000276
8	0.000039	18	0.000087	28	0.000136	38	0.000184	48	0.000233	58	0.000281
9	0.000044	19	0.000092	29	0.000141	39	0.000189	49	0.000238	59	0.000286
10	0.000048	20	0.000097	30	0.000145	40	0.000194	50	0.000242	60	0.000291

Radians into Degrees and Decimals of a Degree
(Based on π radians = 180 degrees)

Rad.	Deg.	Rad.	Deg.	Rad.	Deg.	Rad.	Deg.	Rad.	Deg.	Rad.	Deg.
10	572.9578	1	57.2958	0.1	5.7296	0.01	0.5730	0.001	0.0573	0.0001	0.0057
20	1145.9156	2	114.5916	0.2	11.4592	0.02	1.1459	0.002	0.1146	0.0002	0.0115
30	1718.8734	3	171.8873	0.3	17.1887	0.03	1.7189	0.003	0.1719	0.0003	0.0172
40	2291.8312	4	229.1831	0.4	22.9183	0.04	2.2918	0.004	0.2292	0.0004	0.0229
50	2864.7890	5	286.4789	0.5	28.6479	0.05	2.8648	0.005	0.2865	0.0005	0.0286
60	3437.7468	6	343.7747	0.6	34.3775	0.06	3.4377	0.006	0.3438	0.0006	0.0344
70	4010.7046	7	401.0705	0.7	40.1070	0.07	4.0107	0.007	0.4011	0.0007	0.0401
80	4583.6624	8	458.3662	0.8	45.8366	0.08	4.5837	0.008	0.4584	0.0008	0.0458
90	5156.6202	9	515.6620	0.9	51.5662	0.09	5.1566	0.009	0.5157	0.0009	0.0516
100	5729.5780	10	572.9578	1.0	57.2958	0.10	5.7296	0.010	0.5730	0.0010	0.0573

Radians into Degrees, Minutes, and Seconds
(Based on π radians = 180 degrees)

Rad.	Angle	Rad.	Angle	Rad.	Angle	Rad.	Angle	Rad.	Angle	Rad.	Angle
10	572°57′28″	1	57°17′45″	0.1	5°43′46″	0.01	0°34′23″	0.001	0°3′26″	0.0001	0°0′21″
20	1145°54′56″	2	114°35′30″	0.2	11°27′33″	0.02	1°8′45″	0.002	0°6′53″	0.0002	0°0′41″
30	1718°52′24″	3	171°53′14″	0.3	17°11′19″	0.03	1°43′8″	0.003	0°10′19″	0.0003	0°1′ 2″
40	2291°49′52″	4	229°10′59″	0.4	22°55′6″	0.04	2°17′31″	0.004	0°13′45″	0.0004	0°1′23″
50	2864°47′20″	5	286°28′44″	0.5	28°38′52″	0.05	2°51′53″	0.005	0°17′11″	0.0005	0°1′43″
60	3437°44′48″	6	343°46′29″	0.6	34°22′39″	0.06	3°26′16″	0.006	0°20′38″	0.0006	0°2′ 4″
70	4010°42′16″	7	401°4′14″	0.7	40°6′25″	0.07	4°0′39″	0.007	0°24′4″	0.0007	0°2′24″
80	4583°39′44″	8	458°21′58″	0.8	45°50′12″	0.08	4°35′1″	0.008	0°27′30″	0.0008	0°2′45″
90	5156°37′13″	9	515°39′43″	0.9	51°33′58″	0.09	5°9′24″	0.009	0°30′56″	0.0009	0°3′6″
100	5729°34′41″	10	572°57′28″	1.0	57°17′45″	0.10	5°43′46″	0.010	0°34′23″	0.0010	0°3′26″

Minutes and Seconds into Decimal of a Degree and Vice Versa
(Based on 1 second = 0.00027778 degree)

Minutes into Decimals of a Degree						Seconds into Decimals of a Degree					
Min.	Deg.	Min.	Deg.	Min.	Deg.	Sec.	Deg.	Sec.	Deg.	Sec.	Deg.
1	0.0167	21	0.3500	41	0.6833	1	0.0003	21	0.0058	41	0.0114
2	0.0333	22	0.3667	42	0.7000	2	0.0006	22	0.0061	42	0.0117
3	0.0500	23	0.3833	43	0.7167	3	0.0008	23	0.0064	43	0.0119
4	0.0667	24	0.4000	44	0.7333	4	0.0011	24	0.0067	44	0.0122
5	0.0833	25	0.4167	45	0.7500	5	0.0014	25	0.0069	45	0.0125
6	0.1000	26	0.4333	46	0.7667	6	0.0017	26	0.0072	46	0.0128
7	0.1167	27	0.4500	47	0.7833	7	0.0019	27	0.0075	47	0.0131
8	0.1333	28	0.4667	48	0.8000	8	0.0022	28	0.0078	48	0.0133
9	0.1500	29	0.4833	49	0.8167	9	0.0025	29	0.0081	49	0.0136
10	0.1667	30	0.5000	50	0.8333	10	0.0028	30	0.0083	50	0.0139
11	0.1833	31	0.5167	51	0.8500	11	0.0031	31	0.0086	51	0.0142
12	0.2000	32	0.5333	52	0.8667	12	0.0033	32	0.0089	52	0.0144
13	0.2167	33	0.5500	53	0.8833	13	0.0036	33	0.0092	53	0.0147
14	0.2333	34	0.5667	54	0.9000	14	0.0039	34	0.0094	54	0.0150
15	0.2500	35	0.5833	55	0.9167	15	0.0042	35	0.0097	55	0.0153
16	0.2667	36	0.6000	56	0.9333	16	0.0044	36	0.0100	56	0.0156
17	0.2833	37	0.6167	57	0.9500	17	0.0047	37	0.0103	57	0.0158
18	0.3000	38	0.6333	58	0.9667	18	0.0050	38	0.0106	58	0.0161
19	0.3167	39	0.6500	59	0.9833	19	0.0053	39	0.0108	59	0.0164
20	0.3333	40	0.6667	60	1.0000	20	0.0056	40	0.0111	60	0.0167

Example 1: Convert 11′37″ to decimals of a degree. From the left table, 11′ = 0.1833 degree. From the right table, 37″ = 0.0103 degree. Adding, 11′37″ = 0.1833 + 0.0103 = 0.1936 degree.

Example 2: Convert 0.1234 degree to minutes and seconds. From the left table, 0.1167 degree = 7′. Subtracting 0.1167 from 0.1234 gives 0.0067. From the right table, 0.0067 = 24″ so that 0.1234 = 7′24″.

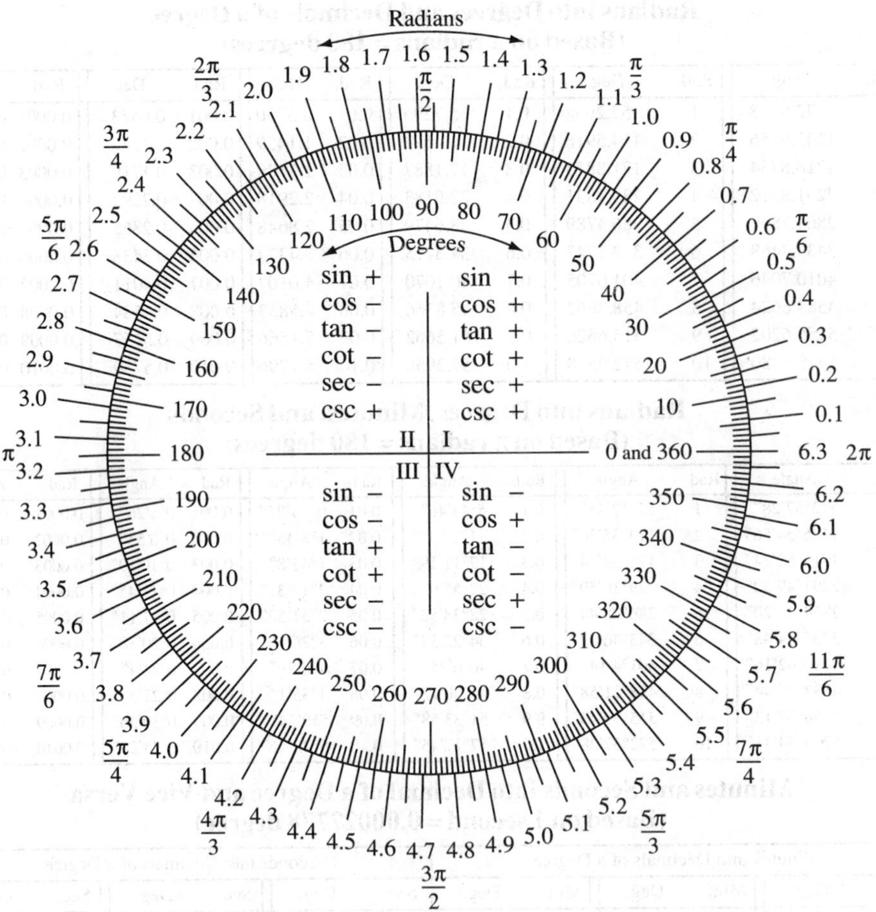

Signs of Trigonometric Functions, Fractions of π, and Degree–Radian Conversion

Graphic Illustrations of the Functions of Angles.—In graphically illustrating the functions of angles, it is assumed that all distances measured in the horizontal direction to the right of line *AB* are positive. Those measured horizontally to the left of *AB* are negative. All distances measured vertically, are positive above line *CD* and negative below it. It can then be readily seen that the sine is positive for all angles less than 180 degrees. For angles larger than 180 degrees, the sine would be measured below *CD*, and is negative. The cosine is positive up to 90 degrees, but for angles larger than 90 but less than 270 degrees, the cosine is measured to the left of line *AB* and is negative.

This table *Useful Relationships Among Angles* that follows is arranged to show directly whether the function of any given angle is positive or negative. It also gives the limits between which the numerical values of the function vary. For example, it will be seen from the table that the cosine of an angle between 90 and 180 degrees is negative, and that its value will be somewhere between 0 and − 1. In the same way, the cotangent of an angle between 180 and 270 degrees is positive and has a value between infinity and 0; in other words, the cotangent for 180 degrees is infinitely large and then the cotangent gradually decreases for increasing angles, so that the cotangent for 270 degrees equals 0.

The sine is positive for all angles up to 180 degrees. The cosine, tangent and cotangent for angles between 90 and 180 degrees, while they have the same numerical values as for angles from 0 to 90 degrees, are negative. These should be preceded by a minus sign; thus tan 123 degrees 20 minutes = −1.5204.

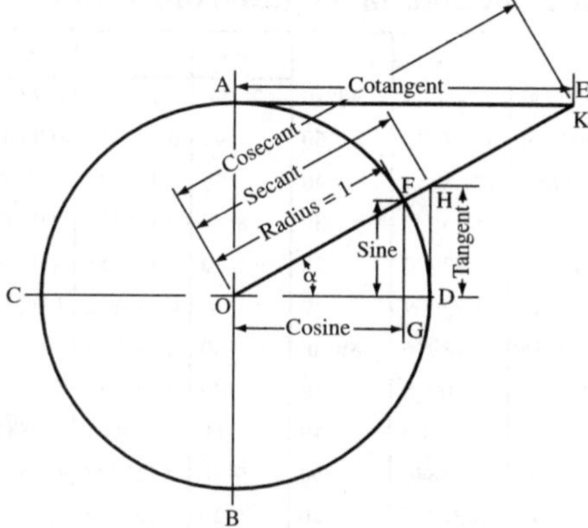

Tables of Trigonometric Functions.—The trigonometric (trig) tables on the following pages give numerical values for sine, cosine, tangent, and cotangent functions of angles from 0 to 90 degrees. Function values for all other angles can be obtained from the tables by applying the rules for signs of trigonometric functions and the useful relationships among angles given in the following. Secant and cosecant functions can be found from sec $A = 1/\cos A$ and csc $A = 1/\sin A$.

The trig tables are divided in half by a double line. The body of each half table consists of four labeled columns of data between columns listing angles. The angles listed to the left of the data increase, moving down the table, and angles listed to the right of the data increase, moving up the table. Labels above the data identify the trig functions corresponding to angles listed in the left column of each half table. Labels below the data correspond to angles listed in the right column of each half table. To find the value of a function for a particular angle, first locate the angle in the table, then find the appropriate function label across the top or bottom row of the table, and find the function value at the intersection of the angle row and label column. Angles opposite each other are complementary angles (i.e., their sum equals 90°) and are related. For example, sin 10° = cos 80° and cos 10° = sin 80°.

All the trig functions of angles between 0° and 90° have positive values. For other angles, consult the chart below to find the sign of the function in the quadrant where the angle is located. To determine trig functions of angles greater than 90° subtract 90, 180, 270, or 360 from the angle to get an angle less than 90° and use Table 1 to find the equivalent first-quadrant function and angle to look up in the trig tables.

Table 1. Useful Relationships Among Angles

Angle Function	θ	−θ	90° ± θ	180° ± θ	270° ± θ	360° ± θ
sin	sin θ	−sin θ	+cos θ	∓sin θ	−cos θ	±sin θ
cos	cos θ	+cos θ	∓sin θ	−cos θ	±sin θ	+cos θ
tan	tan θ	−tan θ	∓cot θ	±tan θ	∓cot θ	±tan θ
cot	cot θ	−cot θ	∓tan θ	±cot θ	∓tan θ	±cot θ
sec	sec θ	+sec θ	∓csc θ	−sec θ	±csc θ	+sec θ
csc	csc θ	−csc θ	+sec θ	∓csc θ	−sec θ	±csc θ

Examples: cos (270° − θ) = −sin θ; tan (90° + θ) = −cot θ.

Example: Find the cosine of 336°40′. The diagram in *Signs of Trigonometric Functions, Fractions of π, and Degree–Radian Conversion* shows that the cosine of every angle in Quadrant IV (270° to 360°) is positive. To find the angle and trig function to use when entering the trig table, subtract 270 from 336 to get cos 336°40′ = cos (270° + 66°40′) and then find the intersection of the cos row and the 270 ± θ column in Table 1. Because cos (270 ± θ) in the fourth quadrant is equal to ± sin θ in the first quadrant, find sin 66°40′ in the trig table. Therefore, cos 336°40′ = sin 66°40′ = 0.918216.

Trigonometric Functions of Angles from 0° to 15° and 75° to 90°

Angle	sin	cos	tan	cot		Angle	sin	cos	tan	cot	
0° 0′	0.000000	1.000000	0.000000	—	**90° 0′**	**7° 30′**	0.130526	0.991445	0.131652	7.595754	**82° 30′**
10	0.002909	0.999996	0.002909	343.7737	**50**	**40**	0.133410	0.991061	0.134613	7.428706	**20**
20	0.005818	0.999983	0.005818	171.8854	**40**	**50**	0.136292	0.990669	0.137576	7.268725	**10**
30	0.008727	0.999962	0.008727	114.5887	**30**	**8° 0′**	0.139173	0.990268	0.140541	7.115370	**82° 0′**
40	0.011635	0.999932	0.011636	85.93979	**20**	**10**	0.142053	0.989859	0.143508	6.968234	**50**
50	0.014544	0.999894	0.014545	68.75009	**10**	**20**	0.144932	0.989442	0.146478	6.826944	**40**
1° 0′	0.017452	0.999848	0.017455	57.28996	**89° 0′**	**30**	0.147809	0.989016	0.149451	6.691156	**30**
10	0.020361	0.999793	0.020365	49.10388	**50**	**40**	0.150686	0.988582	0.152426	6.560554	**20**
20	0.023269	0.999729	0.023275	42.96408	**40**	**50**	0.153561	0.988139	0.155404	6.434843	**10**
30	0.026177	0.999657	0.026186	38.18846	**30**	**9° 0′**	0.156434	0.987688	0.158384	6.313752	**81° 0′**
40	0.029085	0.999577	0.029097	34.36777	**20**	**10**	0.159307	0.987229	0.161368	6.197028	**50**
50	0.031992	0.999488	0.032009	31.24158	**10**	**20**	0.162178	0.986762	0.164354	6.084438	**40**
2° 0′	0.034899	0.999391	0.034921	28.63625	**88° 0′**	**30**	0.165048	0.986286	0.167343	5.975764	**30**
10	0.037806	0.999285	0.037834	26.43160	**50**	**40**	0.167916	0.985801	0.170334	5.870804	**20**
20	0.040713	0.999171	0.040747	24.54176	**40**	**50**	0.170783	0.985309	0.173329	5.769369	**10**
30	0.043619	0.999048	0.043661	22.90377	**30**	**10° 0′**	0.173648	0.984808	0.176327	5.671282	**80° 0′**
40	0.046525	0.998917	0.046576	21.47040	**20**	**10**	0.176512	0.984298	0.179328	5.576379	**50**
50	0.049431	0.998778	0.049491	20.20555	**10**	**20**	0.179375	0.983781	0.182332	5.484505	**40**
3° 0′	0.052336	0.998630	0.052408	19.08114	**87° 0′**	**30**	0.182236	0.983255	0.185339	5.395517	**30**
10	0.055241	0.998473	0.055325	18.07498	**50**	**40**	0.185095	0.982721	0.188349	5.309279	**20**
20	0.058145	0.998308	0.058243	17.16934	**40**	**50**	0.187953	0.982178	0.191363	5.225665	**10**
30	0.061049	0.998135	0.061163	16.34986	**30**	**11° 0′**	0.190809	0.981627	0.194380	5.144554	**79° 0′**
40	0.063952	0.997953	0.064083	15.60478	**20**	**10**	0.193664	0.981068	0.197401	5.065835	**50**
50	0.066854	0.997763	0.067004	14.92442	**10**	**20**	0.196517	0.980500	0.200425	4.989403	**40**
4° 0′	0.069756	0.997564	0.069927	14.30067	**86° 0′**	**30**	0.199368	0.979925	0.203452	4.915157	**30**
10	0.072658	0.997357	0.072851	13.72674	**50**	**40**	0.202218	0.979341	0.206483	4.843005	**20**
20	0.075559	0.997141	0.075775	13.19688	**40**	**50**	0.205065	0.978748	0.209518	4.772857	**10**
30	0.078459	0.996917	0.078702	12.70621	**30**	**12° 0′**	0.207912	0.978148	0.212557	4.704630	**78° 0′**
40	0.081359	0.996685	0.081629	12.25051	**20**	**10**	0.210756	0.977539	0.215599	4.638246	**50**
50	0.084258	0.996444	0.084558	11.82617	**10**	**20**	0.213599	0.976921	0.218645	4.573629	**40**
5° 0′	0.087156	0.996195	0.087489	11.43005	**85° 0′**	**30**	0.216440	0.976296	0.221695	4.510709	**30**
10	0.090053	0.995937	0.090421	11.05943	**50**	**40**	0.219279	0.975662	0.224748	4.449418	**20**
20	0.092950	0.995671	0.093354	10.71191	**40**	**50**	0.222116	0.975020	0.227806	4.389694	**10**
30	0.095846	0.995396	0.096289	10.38540	**30**	**13° 0′**	0.224951	0.974370	0.230868	4.331476	**77° 0′**
40	0.098741	0.995113	0.099226	10.07803	**20**	**10**	0.227784	0.973712	0.233934	4.274707	**50**
50	0.101635	0.994822	0.102164	9.788173	**10**	**20**	0.230616	0.973045	0.237004	4.219332	**40**
6° 0′	0.104528	0.994522	0.105104	9.514364	**84° 0′**	**30**	0.233445	0.972370	0.240079	4.165300	**30**
10	0.107421	0.994214	0.108046	9.255304	**50**	**40**	0.236273	0.971687	0.243157	4.112561	**20**
20	0.110313	0.993897	0.110990	9.009826	**40**	**50**	0.239098	0.970995	0.246241	4.061070	**10**
30	0.113203	0.993572	0.113936	8.776887	**30**	**14° 0′**	0.241922	0.970296	0.249328	4.010781	**76° 0′**
40	0.116093	0.993238	0.116883	8.555547	**20**	**10**	0.244743	0.969588	0.252420	3.961652	**50**
50	0.118982	0.992896	0.119833	8.344956	**10**	**20**	0.247563	0.968872	0.255516	3.913642	**40**
7° 0′	0.121869	0.992546	0.122785	8.144346	**83° 0′**	**30**	0.250380	0.968148	0.258618	3.866713	**30**
10	0.124756	0.992187	0.125738	7.953022	**50**	**40**	0.253195	0.967415	0.261723	3.820828	**20**
20	0.127642	0.991820	0.128694	7.770351	**40**	**50**	0.256008	0.966675	0.264834	3.775952	**10**
7° 30′	0.130526	0.991445	0.131652	7.595754	**82° 30**	**15° 0′**	0.258819	0.965926	0.267949	3.732051	**75° 0′**
	cos	sin	cot	tan	Angle		cos	sin	cot	tan	Angle

For angles 0° to 15° 0′ (angles found in a column to the left of the data), use the column labels at the top of the table; for angles 75° to 90° 0′ (angles found in a column to the right of the data), use the column labels at the bottom of the table.

Trigonometric Functions of Angles from 15° to 30° and 60° to 75°

Angle	sin	cos	tan	cot		Angle	sin	cos	tan	cot	
15° 0′	0.258819	0.965926	0.267949	3.732051	**75° 0′**	**22° 30′**	0.382683	0.923880	0.414214	2.414214	**67° 30**
10	0.261628	0.965169	0.271069	3.689093	**50**	**40**	0.385369	0.922762	0.417626	2.394489	**20**
20	0.264434	0.964404	0.274194	3.647047	**40**	**50**	0.388052	0.921638	0.421046	2.375037	**10**
30	0.267238	0.963630	0.277325	3.605884	**30**	**23° 0′**	0.390731	0.920505	0.424475	2.355852	**67° 0′**
40	0.270040	0.962849	0.280460	3.565575	**20**	**10**	0.393407	0.919364	0.427912	2.336929	**50**
50	0.272840	0.962059	0.283600	3.526094	**10**	**20**	0.396080	0.918216	0.431358	2.318261	**40**
16° 0′	0.275637	0.961262	0.286745	3.487414	**74° 0′**	**30**	0.398749	0.917060	0.434812	2.299843	**30**
10	0.278432	0.960456	0.289896	3.449512	**50**	**40**	0.401415	0.915896	0.438276	2.281669	**20**
20	0.281225	0.959642	0.293052	3.412363	**40**	**50**	0.404078	0.914725	0.441748	2.263736	**10**
30	0.284015	0.958820	0.296213	3.375943	**30**	**24° 0′**	0.406737	0.913545	0.445229	2.246037	**66° 0′**
40	0.286803	0.957990	0.299380	3.340233	**20**	**10**	0.409392	0.912358	0.448719	2.228568	**50**
50	0.289589	0.957151	0.302553	3.305209	**10**	**20**	0.412045	0.911164	0.452218	2.211323	**40**
17° 0′	0.292372	0.956305	0.305731	3.270853	**73° 0′**	**30**	0.414693	0.909961	0.455726	2.194300	**30**
10	0.295152	0.955450	0.308914	3.237144	**50**	**40**	0.417338	0.908751	0.459244	2.177492	**20**
20	0.297930	0.954588	0.312104	3.204064	**40**	**50**	0.419980	0.907533	0.462771	2.160896	**10**
30	0.300706	0.953717	0.315299	3.171595	**30**	**25° 0′**	0.422618	0.906308	0.466308	2.144507	**65° 0′**
40	0.303479	0.952838	0.318500	3.139719	**20**	**10**	0.425253	0.905075	0.469854	2.128321	**50**
50	0.306249	0.951951	0.321707	3.108421	**10**	**20**	0.427884	0.903834	0.473410	2.112335	**40**
18° 0′	0.309017	0.951057	0.324920	3.077684	**72° 0′**	**30**	0.430511	0.902585	0.476976	2.096544	**30**
10	0.311782	0.950154	0.328139	3.047492	**50**	**40**	0.433135	0.901329	0.480551	2.080944	**20**
20	0.314545	0.949243	0.331364	3.017830	**40**	**50**	0.435755	0.900065	0.484137	2.065532	**10**
30	0.317305	0.948324	0.334595	2.988685	**30**	**26° 0′**	0.438371	0.898794	0.487733	2.050304	**64° 0′**
40	0.320062	0.947397	0.337833	2.960042	**20**	**10**	0.440984	0.897515	0.491339	2.035256	**50**
50	0.322816	0.946462	0.341077	2.931888	**10**	**20**	0.443593	0.896229	0.494955	2.020386	**40**
19° 0′	0.325568	0.945519	0.344328	2.904211	**71° 0′**	**30**	0.446198	0.894934	0.498582	2.005690	**30**
10	0.328317	0.944568	0.347585	2.876997	**50**	**40**	0.448799	0.893633	0.502219	1.991164	**20**
20	0.331063	0.943609	0.350848	2.850235	**40**	**50**	0.451397	0.892323	0.505867	1.976805	**10**
30	0.333807	0.942641	0.354119	2.823913	**30**	**27° 0′**	0.453990	0.891007	0.509525	1.962611	**63° 0′**
40	0.336547	0.941666	0.357396	2.798020	**20**	**10**	0.456580	0.889682	0.513195	1.948577	**50**
50	0.339285	0.940684	0.360679	2.772545	**10**	**20**	0.459166	0.888350	0.516875	1.934702	**40**
20° 0′	0.342020	0.939693	0.363970	2.747477	**70° 0′**	**30**	0.461749	0.887011	0.520567	1.920982	**30**
10	0.344752	0.938694	0.367268	2.722808	**50**	**40**	0.464327	0.885664	0.524270	1.907415	**20**
20	0.347481	0.937687	0.370573	2.698525	**40**	**50**	0.466901	0.884309	0.527984	1.893997	**10**
30	0.350207	0.936672	0.373885	2.674621	**30**	**28° 0′**	0.469472	0.882948	0.531709	1.880726	**62° 0′**
40	0.352931	0.935650	0.377204	2.651087	**20**	**10**	0.472038	0.881578	0.535446	1.867600	**50**
50	0.355651	0.934619	0.380530	2.627912	**10**	**20**	0.474600	0.880201	0.539195	1.854616	**40**
21° 0′	0.358368	0.933580	0.383864	2.605089	**69° 0′**	**30**	0.477159	0.878817	0.542956	1.841771	**30**
10	0.361082	0.932534	0.387205	2.582609	**50**	**40**	0.479713	0.877425	0.546728	1.829063	**20**
20	0.363793	0.931480	0.390554	2.560465	**40**	**50**	0.482263	0.876026	0.550513	1.816489	**10**
30	0.366501	0.930418	0.393910	2.538648	**30**	**29° 0′**	0.484810	0.874620	0.554309	1.804048	**61° 0′**
40	0.369206	0.929348	0.397275	2.517151	**20**	**10**	0.487352	0.873206	0.558118	1.791736	**50**
50	0.371908	0.928270	0.400646	2.495966	**10**	**20**	0.489890	0.871784	0.561939	1.779552	**40**
22° 0′	0.374607	0.927184	0.404026	2.475087	**68° 0′**	**30**	0.492424	0.870356	0.565773	1.767494	**30**
10	0.377302	0.926090	0.407414	2.454506	**50**	**40**	0.494953	0.868920	0.569619	1.755559	**20**
20	0.379994	0.924989	0.410810	2.434217	**40**	**50**	0.497479	0.867476	0.573478	1.743745	**10**
22° 30	0.382683	0.923880	0.414214	2.414214	**67° 30**	**30° 0′**	0.500000	0.866025	0.577350	1.732051	**60° 0′**
	cos	sin	cot	tan	Angle		cos	sin	cot	tan	Angle

For angles 15° to 30° 0′ (angles found in a column to the left of the data), use the column labels at the top of the table; for angles 60° to 75° 0′ (angles found in a column to the right of the data), use the column labels at the bottom of the table.

Trigonometric Functions of Angles from 30° to 60°

Angle	sin	cos	tan	cot		Angle	sin	cos	tan	cot	
30° 0′	0.500000	0.866025	0.577350	1.732051	**60° 0′**	**37° 30′**	0.608761	0.793353	0.767327	1.303225	**52° 30′**
10	0.502517	0.864567	0.581235	1.720474	**50**	**40**	0.611067	0.791579	0.771959	1.295406	**20**
20	0.505030	0.863102	0.585134	1.709012	**40**	**50**	0.613367	0.789798	0.776612	1.287645	**10**
30	0.507538	0.861629	0.589045	1.697663	**30**	**38° 0′**	0.615661	0.788011	0.781286	1.279942	**52° 0′**
40	0.510043	0.860149	0.592970	1.686426	**20**	**10**	0.617951	0.786217	0.785981	1.272296	**50**
50	0.512543	0.858662	0.596908	1.675299	**10**	**20**	0.620235	0.784416	0.790697	1.264706	**40**
31° 0′	0.515038	0.857167	0.600861	1.664279	**59° 0′**	**30**	0.622515	0.782608	0.795436	1.257172	**30**
10	0.517529	0.855665	0.604827	1.653366	**50**	**40**	0.624789	0.780794	0.800196	1.249693	**20**
20	0.520016	0.854156	0.608807	1.642558	**40**	**50**	0.627057	0.778973	0.804979	1.242268	**10**
30	0.522499	0.852640	0.612801	1.631852	**30**	**39° 0′**	0.629320	0.777146	0.809784	1.234897	**51° 0′**
40	0.524977	0.851117	0.616809	1.621247	**20**	**10**	0.631578	0.775312	0.814612	1.227579	**50**
50	0.527450	0.849586	0.620832	1.610742	**10**	**20**	0.633831	0.773472	0.819463	1.220312	**40**
32° 0′	0.529919	0.848048	0.624869	1.600335	**58° 0′**	**30**	0.636078	0.771625	0.824336	1.213097	**30**
10	0.532384	0.846503	0.628921	1.590024	**50**	**40**	0.638320	0.769771	0.829234	1.205933	**20**
20	0.534844	0.844951	0.632988	1.579808	**40**	**50**	0.640557	0.767911	0.834155	1.198818	**10**
30	0.537300	0.843391	0.637070	1.569686	**30**	**40° 0′**	0.642788	0.766044	0.839100	1.191754	**50° 0′**
40	0.539751	0.841825	0.641167	1.559655	**20**	**10**	0.645013	0.764171	0.844069	1.184738	**50**
50	0.542197	0.840251	0.645280	1.549715	**10**	**20**	0.647233	0.762292	0.849062	1.177770	**40**
33° 0′	0.544639	0.838671	0.649408	1.539865	**57° 0′**	**30**	0.649448	0.760406	0.854081	1.170850	**30**
10	0.547076	0.837083	0.653551	1.530102	**50**	**40**	0.651657	0.758514	0.859124	1.163976	**20**
20	0.549509	0.835488	0.657710	1.520426	**40**	**50**	0.653861	0.756615	0.864193	1.157149	**10**
30	0.551937	0.833886	0.661886	1.510835	**30**	**41° 0′**	0.656059	0.754710	0.869287	1.150368	**49° 0′**
40	0.554360	0.832277	0.666077	1.501328	**20**	**10**	0.658252	0.752798	0.874407	1.143633	**50**
50	0.556779	0.830661	0.670284	1.491904	**10**	**20**	0.660439	0.750880	0.879553	1.136941	**40**
34° 0′	0.559193	0.829038	0.674509	1.482561	**56° 0′**	**30**	0.662620	0.748956	0.884725	1.130294	**30**
10	0.561602	0.827407	0.678749	1.473298	**50**	**40**	0.664796	0.747025	0.889924	1.123691	**20**
20	0.564007	0.825770	0.683007	1.464115	**40**	**50**	0.666966	0.745088	0.895151	1.117130	**10**
30	0.566406	0.824126	0.687281	1.455009	**30**	**42° 0′**	0.669131	0.743145	0.900404	1.110613	**48° 0′**
40	0.568801	0.822475	0.691572	1.445980	**20**	**10**	0.671289	0.741195	0.905685	1.104137	**50**
50	0.571191	0.820817	0.695881	1.437027	**10**	**20**	0.673443	0.739239	0.910994	1.097702	**40**
35° 0′	0.573576	0.819152	0.700208	1.428148	**55° 0′**	**30**	0.675590	0.737277	0.916331	1.091309	**30**
10	0.575957	0.817480	0.704551	1.419343	**50**	**40**	0.677732	0.735309	0.921697	1.084955	**20**
20	0.578332	0.815801	0.708913	1.410610	**40**	**50**	0.679868	0.733334	0.927091	1.078642	**10**
30	0.580703	0.814116	0.713293	1.401948	**30**	**43° 0′**	0.681998	0.731354	0.932515	1.072369	**47° 0′**
40	0.583069	0.812423	0.717691	1.393357	**20**	**10**	0.684123	0.729367	0.937968	1.066134	**50**
50	0.585429	0.810723	0.722108	1.384835	**10**	**20**	0.686242	0.727374	0.943451	1.059938	**40**
36° 0′	0.587785	0.809017	0.726543	1.376382	**54° 0′**	**30**	0.688355	0.725374	0.948965	1.053780	**30**
10	0.590136	0.807304	0.730996	1.367996	**50**	**40**	0.690462	0.723369	0.954508	1.047660	**20**
20	0.592482	0.805584	0.735469	1.359676	**40**	**50**	0.692563	0.721357	0.960083	1.041577	**10**
30	0.594823	0.803857	0.739961	1.351422	**30**	**44° 0′**	0.694658	0.719340	0.965689	1.035530	**46° 0′**
40	0.597159	0.802123	0.744472	1.343233	**20**	**10**	0.696748	0.717316	0.971326	1.029520	**50**
50	0.599489	0.800383	0.749003	1.335108	**10**	**20**	0.698832	0.715286	0.976996	1.023546	**40**
37° 0′	0.601815	0.798636	0.753554	1.327045	**53° 0′**	**30**	0.700909	0.713250	0.982697	1.017607	**30**
10	0.604136	0.796882	0.758125	1.319044	**50**	**40**	0.702981	0.711209	0.988432	1.011704	**20**
20	0.606451	0.795121	0.762716	1.311105	**40**	**50**	0.705047	0.709161	0.994199	1.005835	**10**
37° 30	0.608761	0.793353	0.767327	1.303225	**52° 30**	**45° 0′**	0.707107	0.707107	1.000000	1.000000	**45° 0′**
	cos	sin	cot	tan	Angle		cos	sin	cot	tan	Angle

For angles 30° to 45° 0′ (angles found in a column to the left of the data), use the column labels at the top of the table; for angles 45° to 60° 0′ (angles found in a column to the right of the data), use the column labels at the bottom of the table.

Using a Calculator to Find Trig Functions.—A scientific calculator is quicker and more accurate than tables for finding trig functions and angles corresponding to trig functions. On scientific calculators, the keys labeled **sin**, **cos**, and **tan** are used to find the common trig functions. The other functions can be found by using the same keys and the **1/x** key, noting that $\csc A = 1/\sin A$, $\sec A = 1/\cos A$, and $\cot A = 1/\tan A$. The specific keystrokes used will vary slightly from one calculator to another. To find the angle corresponding to a given trig function use the keys labeled **sin**$^{-1}$, **cos**$^{-1}$, and **tan**$^{-1}$. On some other calculators, the **sin**, **cos**, and **tan** are used in combination with the **INV,** or inverse, key to find the number corresponding to a given trig function.

If a scientific calculator or computer is not available, tables are the easiest way to find trig values. However, trig function values can be calculated very accurately without a scientific calculator by using the following formulas:

$$\sin A = A - \frac{A^3}{3!} + \frac{A^5}{5!} - \frac{A^7}{7!} \pm \cdots \qquad \cos A = 1 - \frac{A^2}{2!} + \frac{A^4}{4!} - \frac{A^6}{6!} \pm \cdots$$

$$\sin^{-1} A = \frac{1}{2} \times \frac{A^3}{3} + \frac{1}{2} \times \frac{3}{4} \times \frac{A^5}{5} + \cdots \qquad \tan^{-1} A = A - \frac{A^3}{3} + \frac{A^5}{5} - \frac{A^7}{7} \pm \cdots$$

where the angle A is expressed in radians (convert degrees to radians by multiplying degrees by $\pi/180 = 0.0174533$). The three dots at the ends of the formulas indicate that the expression continues with more terms following the sequence established by the first few terms. Generally, calculating just three or four terms of the expression is sufficient for accuracy. In these formulas, a number followed by the symbol **!** is called a factorial (for example, 3! is three factorial). Except for 0!, which is defined as 1, a factorial is found by multiplying together all the integers greater than zero and less than or equal to the factorial number wanted. For example: $3! = 1 \times 2 \times 3 = 6$; $4! = 1 \times 2 \times 3 \times 4 = 24$; $7! = 1 \times 2 \times 3 \times 4 \times 5 \times 6 \times 7 = 5040$; etc.

Versed Sine and Versed Cosine.—These functions are sometimes used in formulas for segments of a circle and may be obtained using the relationships:

$$\text{versed } \sin\theta = 1 - \cos\theta; \text{ versed } \cos\theta = 1 - \sin\theta.$$

Sevolute Functions.—Sevolute functions are used in calculating the form diameter of involute splines. They are computed by subtracting the involute function of an angle from the secant of the angle (1/cosine = secant). Thus, sevolute of 20 degrees = secant of 20 degrees − involute function of 20 degrees = 1.064178 − 0.014904 = 1.049274.

Involute Functions.—Involute functions are used in certain formulas relating to the design and measurement of gear teeth as well as measurement of threads over wires. See, for example, pages 1867 through 1870, 2080, and 2147.

The tables on the following pages provide values of involute functions for angles from 14 to 51 degrees in increments of 1 minute. These involute functions were calculated from the following formulas: Involute of $\theta = \tan\theta - \theta$, for θ in radians, and involute of $\theta = \tan\theta - \pi \times \theta/180$, for θ in degrees.

Example: For an angle of 14 degrees and 10 minutes, the involute function is found as follows: 10 minutes = 10/60 = 0.166666 degrees, 14 + 0.166666 = 14.166666 degree, so that the involute of 14.166666 degrees = tan 14.166666 − $\pi \times$ 14.166666/180 = 0.252420 − 0.247255 = 0.005165. This value is the same as that in Table 1 for 14 degrees and 10 minutes. The same result would be obtained from using the conversion tables beginning on page 90 to convert 14 degrees and 10 minutes to radians and then applying the first of the formulas given above.

Involute Functions for Angles from 14 to 23 Degrees

Minutes	Degrees								
	14	15	16	17	18	19	20	21	22
	Involute Functions								
0	0.004982	0.006150	0.007493	0.009025	0.010760	0.012715	0.014904	0.017345	0.020054
1	0.005000	0.006171	0.007517	0.009052	0.010791	0.012750	0.014943	0.017388	0.020101
2	0.005018	0.006192	0.007541	0.009079	0.010822	0.012784	0.014982	0.017431	0.020149
3	0.005036	0.006213	0.007565	0.009107	0.010853	0.012819	0.015020	0.017474	0.020197
4	0.005055	0.006234	0.007589	0.009134	0.010884	0.012854	0.015059	0.017517	0.020244
5	0.005073	0.006255	0.007613	0.009161	0.010915	0.012888	0.015098	0.017560	0.020292
6	0.005091	0.006276	0.007637	0.009189	0.010946	0.012923	0.015137	0.017603	0.020340
7	0.005110	0.006297	0.007661	0.009216	0.010977	0.012958	0.015176	0.017647	0.020388
8	0.005128	0.006318	0.007686	0.009244	0.011008	0.012993	0.015215	0.017690	0.020436
9	0.005146	0.006340	0.007710	0.009272	0.011039	0.013028	0.015254	0.017734	0.020484
10	0.005165	0.006361	0.007735	0.009299	0.011071	0.013063	0.015293	0.017777	0.020533
11	0.005184	0.006382	0.007759	0.009327	0.011102	0.013098	0.015333	0.017821	0.020581
12	0.005202	0.006404	0.007784	0.009355	0.011133	0.013134	0.015372	0.017865	0.020629
13	0.005221	0.006425	0.007808	0.009383	0.011165	0.013169	0.015411	0.017908	0.020678
14	0.005239	0.006447	0.007833	0.009411	0.011196	0.013204	0.015451	0.017952	0.020726
15	0.005258	0.006469	0.007857	0.009439	0.011228	0.013240	0.015490	0.017996	0.020775
16	0.005277	0.006490	0.007882	0.009467	0.011260	0.013275	0.015530	0.018040	0.020824
17	0.005296	0.006512	0.007907	0.009495	0.011291	0.013311	0.015570	0.018084	0.020873
18	0.005315	0.006534	0.007932	0.009523	0.011323	0.013346	0.015609	0.018129	0.020921
19	0.005334	0.006555	0.007957	0.009552	0.011355	0.013382	0.015649	0.018173	0.020970
20	0.005353	0.006577	0.007982	0.009580	0.011387	0.013418	0.015689	0.018217	0.021019
21	0.005372	0.006599	0.008007	0.009608	0.011419	0.013454	0.015729	0.018262	0.021069
22	0.005391	0.006621	0.008032	0.009637	0.011451	0.013490	0.015769	0.018306	0.021118
23	0.005410	0.006643	0.008057	0.009665	0.011483	0.013526	0.015809	0.018351	0.021167
24	0.005429	0.006665	0.008082	0.009694	0.011515	0.013562	0.015850	0.018395	0.021217
25	0.005448	0.006687	0.008107	0.009722	0.011547	0.013598	0.015890	0.018440	0.021266
26	0.005467	0.006709	0.008133	0.009751	0.011580	0.013634	0.015930	0.018485	0.021316
27	0.005487	0.006732	0.008158	0.009780	0.011612	0.013670	0.015971	0.018530	0.021365
28	0.005506	0.006754	0.008183	0.009808	0.011644	0.013707	0.016011	0.018575	0.021415
29	0.005525	0.006776	0.008209	0.009837	0.011677	0.013743	0.016052	0.018620	0.021465
30	0.005545	0.006799	0.008234	0.009866	0.011709	0.013779	0.016092	0.018665	0.021514
31	0.005564	0.006821	0.008260	0.009895	0.011742	0.013816	0.016133	0.018710	0.021564
32	0.005584	0.006843	0.008285	0.009924	0.011775	0.013852	0.016174	0.018755	0.021614
33	0.005603	0.006866	0.008311	0.009953	0.011807	0.013889	0.016215	0.018800	0.021665
34	0.005623	0.006888	0.008337	0.009982	0.011840	0.013926	0.016255	0.018846	0.021715
35	0.005643	0.006911	0.008362	0.010011	0.011873	0.013963	0.016296	0.018891	0.021765
36	0.005662	0.006934	0.008388	0.010041	0.011906	0.013999	0.016337	0.018937	0.021815
37	0.005682	0.006956	0.008414	0.010070	0.011939	0.014036	0.016379	0.018983	0.021866
38	0.005702	0.006979	0.008440	0.010099	0.011972	0.014073	0.016420	0.019028	0.021916
39	0.005722	0.007002	0.008466	0.010129	0.012005	0.014110	0.016461	0.019074	0.021967
40	0.005742	0.007025	0.008492	0.010158	0.012038	0.014148	0.016502	0.019120	0.022018
41	0.005762	0.007048	0.008518	0.010188	0.012071	0.014185	0.016544	0.019166	0.022068
42	0.005782	0.007071	0.008544	0.010217	0.012105	0.014222	0.016585	0.019212	0.022119
43	0.005802	0.007094	0.008571	0.010247	0.012138	0.014259	0.016627	0.019258	0.022170
44	0.005822	0.007117	0.008597	0.010277	0.012172	0.014297	0.016669	0.019304	0.022221
45	0.005842	0.007140	0.008623	0.010307	0.012205	0.014334	0.016710	0.019350	0.022272
46	0.005862	0.007163	0.008650	0.010336	0.012239	0.014372	0.016752	0.019397	0.022324
47	0.005882	0.007186	0.008676	0.010366	0.012272	0.014409	0.016794	0.019443	0.022375
48	0.005903	0.007209	0.008702	0.010396	0.012306	0.014447	0.016836	0.019490	0.022426
49	0.005923	0.007233	0.008729	0.010426	0.012340	0.014485	0.016878	0.019536	0.022478
50	0.005943	0.007256	0.008756	0.010456	0.012373	0.014523	0.016920	0.019583	0.022529
51	0.005964	0.007280	0.008782	0.010486	0.012407	0.014560	0.016962	0.019630	0.022581
52	0.005984	0.007303	0.008809	0.010517	0.012441	0.014598	0.017004	0.019676	0.022633
53	0.006005	0.007327	0.008836	0.010547	0.012475	0.014636	0.017047	0.019723	0.022684
54	0.006025	0.007350	0.008863	0.010577	0.012509	0.014674	0.017089	0.019770	0.022736
55	0.006046	0.007374	0.008889	0.010608	0.012543	0.014713	0.017132	0.019817	0.022788
56	0.006067	0.007397	0.008916	0.010638	0.012578	0.014751	0.017174	0.019864	0.022840
57	0.006087	0.007421	0.008943	0.010669	0.012612	0.014789	0.017217	0.019912	0.022892
58	0.006108	0.007445	0.008970	0.010699	0.012646	0.014827	0.017259	0.019959	0.022944
59	0.006129	0.007469	0.008998	0.010730	0.012681	0.014866	0.017302	0.020006	0.022997
60	0.006150	0.007493	0.009025	0.010760	0.012715	0.014904	0.017345	0.020054	0.023049

Involute Functions for Angles from 23 to 32 Degrees

	Degrees								
	23	24	25	26	27	28	29	30	31
Minutes	Involute Functions								
0	0.023049	0.026350	0.029975	0.033947	0.038287	0.043017	0.048164	0.053752	0.059809
1	0.023102	0.026407	0.030039	0.034016	0.038362	0.043100	0.048253	0.053849	0.059914
2	0.023154	0.026465	0.030102	0.034086	0.038438	0.043182	0.048343	0.053946	0.060019
3	0.023207	0.026523	0.030166	0.034155	0.038514	0.043264	0.048432	0.054043	0.060124
4	0.023259	0.026581	0.030229	0.034225	0.038590	0.043347	0.048522	0.054140	0.060230
5	0.023312	0.026639	0.030293	0.034294	0.038666	0.043430	0.048612	0.054238	0.060335
6	0.023365	0.026697	0.030357	0.034364	0.038742	0.043513	0.048702	0.054336	0.060441
7	0.023418	0.026756	0.030420	0.034434	0.038818	0.043596	0.048792	0.054433	0.060547
8	0.023471	0.026814	0.030484	0.034504	0.038894	0.043679	0.048883	0.054531	0.060653
9	0.023524	0.026872	0.030549	0.034574	0.038971	0.043762	0.048973	0.054629	0.060759
10	0.023577	0.026931	0.030613	0.034644	0.039047	0.043845	0.049064	0.054728	0.060866
11	0.023631	0.026989	0.030677	0.034714	0.039124	0.043929	0.049154	0.054826	0.060972
12	0.023684	0.027048	0.030741	0.034785	0.039201	0.044012	0.049245	0.054924	0.061079
13	0.023738	0.027107	0.030806	0.034855	0.039278	0.044096	0.049336	0.055023	0.061186
14	0.023791	0.027166	0.030870	0.034926	0.039355	0.044180	0.049427	0.055122	0.061292
15	0.023845	0.027225	0.030935	0.034997	0.039432	0.044264	0.049518	0.055221	0.061400
16	0.023899	0.027284	0.031000	0.035067	0.039509	0.044348	0.049609	0.055320	0.061507
17	0.023952	0.027343	0.031065	0.035138	0.039586	0.044432	0.049701	0.055419	0.061614
18	0.024006	0.027402	0.031130	0.035209	0.039664	0.044516	0.049792	0.055518	0.061721
19	0.024060	0.027462	0.031195	0.035280	0.039741	0.044601	0.049884	0.055617	0.061829
20	0.024114	0.027521	0.031260	0.035352	0.039819	0.044685	0.049976	0.055717	0.061937
21	0.024169	0.027581	0.031325	0.035423	0.039897	0.044770	0.050068	0.055817	0.062045
22	0.024223	0.027640	0.031390	0.035494	0.039974	0.044855	0.050160	0.055916	0.062153
23	0.024277	0.027700	0.031456	0.035566	0.040052	0.044940	0.050252	0.056016	0.062261
24	0.024332	0.027760	0.031521	0.035637	0.040131	0.045024	0.050344	0.056116	0.062369
25	0.024386	0.027820	0.031587	0.035709	0.040209	0.045110	0.050437	0.056217	0.062478
26	0.024441	0.027880	0.031653	0.035781	0.040287	0.045195	0.050529	0.056317	0.062586
27	0.024495	0.027940	0.031718	0.035853	0.040366	0.045280	0.050622	0.056417	0.062695
28	0.024550	0.028000	0.031784	0.035925	0.040444	0.045366	0.050715	0.056518	0.062804
29	0.024605	0.028060	0.031850	0.035997	0.040523	0.045451	0.050808	0.056619	0.062913
30	0.024660	0.028121	0.031917	0.036069	0.040602	0.045537	0.050901	0.056720	0.063022
31	0.024715	0.028181	0.031983	0.036142	0.040680	0.045623	0.050994	0.056821	0.063131
32	0.024770	0.028242	0.032049	0.036214	0.040759	0.045709	0.051087	0.056922	0.063241
33	0.024825	0.028302	0.032116	0.036287	0.040839	0.045795	0.051181	0.057023	0.063350
34	0.024881	0.028363	0.032182	0.036359	0.040918	0.045881	0.051274	0.057124	0.063460
35	0.024936	0.028424	0.032249	0.036432	0.040997	0.045967	0.051368	0.057226	0.063570
36	0.024992	0.028485	0.032315	0.036505	0.041077	0.046054	0.051462	0.057328	0.063680
37	0.025047	0.028546	0.032382	0.036578	0.041156	0.046140	0.051556	0.057429	0.063790
38	0.025103	0.028607	0.032449	0.036651	0.041236	0.046227	0.051650	0.057531	0.063901
39	0.025159	0.028668	0.032516	0.036724	0.041316	0.046313	0.051744	0.057633	0.064011
40	0.025214	0.028729	0.032583	0.036798	0.041395	0.046400	0.051838	0.057736	0.064122
41	0.025270	0.028791	0.032651	0.036871	0.041475	0.046487	0.051933	0.057838	0.064232
42	0.025326	0.028852	0.032718	0.036945	0.041556	0.046575	0.052027	0.057940	0.064343
43	0.025382	0.028914	0.032785	0.037018	0.041636	0.046662	0.052122	0.058043	0.064454
44	0.025439	0.028976	0.032853	0.037092	0.041716	0.046749	0.052217	0.058146	0.064565
45	0.025495	0.029037	0.032920	0.037166	0.041797	0.046837	0.052312	0.058249	0.064677
46	0.025551	0.029099	0.032988	0.037240	0.041877	0.046924	0.052407	0.058352	0.064788
47	0.025608	0.029161	0.033056	0.037314	0.041958	0.047012	0.052502	0.058455	0.064900
48	0.025664	0.029223	0.033124	0.037388	0.042039	0.047100	0.052597	0.058558	0.065012
49	0.025721	0.029285	0.033192	0.037462	0.042120	0.047188	0.052693	0.058662	0.065123
50	0.025778	0.029348	0.033260	0.037537	0.042201	0.047276	0.052788	0.058765	0.065236
51	0.025834	0.029410	0.033328	0.037611	0.042282	0.047364	0.052884	0.058869	0.065348
52	0.025891	0.029472	0.033397	0.037686	0.042363	0.047452	0.052980	.058973	0.065460
53	0.025948	0.029535	0.033465	0.037761	0.042444	0.047541	0.053076	0.059077	0.065573
54	0.026005	0.029598	0.033534	0.037835	0.042526	0.047630	0.053172	0.059181	0.065685
55	0.026062	0.029660	0.033602	0.037910	0.042608	0.047718	0.053268	0.059285	0.065798
56	0.026120	0.029723	0.033671	0.037985	0.042689	0.047807	0.053365	0.059390	0.065911
57	0.026177	0.029786	0.033740	0.038060	0.042771	0.047896	0.053461	0.059494	0.066024
58	0.026235	0.029849	0.033809	0.038136	0.042853	0.047985	0.053558	0.059599	0.066137
59	0.026292	0.029912	0.033878	0.038211	0.042935	0.048074	0.053655	0.059704	0.066251
60	0.026350	0.029975	0.033947	0.038287	0.043017	0.048164	0.053752	0.059809	0.066364

Involute Functions for Angles from 32 to 41 Degrees

	Degrees								
	32	33	34	35	36	37	38	39	40
Minutes	Involute Functions								
0	0.066364	0.073449	0.081097	0.089342	0.098224	0.107782	0.118061	0.129106	0.140968
1	0.066478	0.073572	0.081229	0.089485	0.098378	0.107948	0.118238	0.129297	0.141173
2	0.066591	0.073695	0.081362	0.089628	0.098532	0.108113	0.118416	0.129488	0.141378
3	0.066705	0.073818	0.081494	0.089771	0.098686	0.108279	0.118594	0.129679	0.141584
4	0.066820	0.073941	0.081627	0.089914	0.098840	0.108445	0.118773	0.129870	0.141789
5	0.066934	0.074064	0.081760	0.090058	0.098994	0.108611	0.118951	0.130062	0.141995
6	0.067048	0.074188	0.081894	0.090201	0.099149	0.108777	0.119130	0.130254	0.142201
7	0.067163	0.074312	0.082027	0.090345	0.099303	0.108943	0.119309	0.130446	0.142408
8	0.067277	0.074435	0.082161	0.090489	0.099458	0.109110	0.119488	0.130639	0.142614
9	0.067392	0.074559	0.082294	0.090633	0.099614	0.109277	0.119667	0.130832	0.142821
10	0.067507	0.074684	0.082428	0.090777	0.099769	0.109444	0.119847	0.131025	0.143028
11	0.067622	0.074808	0.082562	0.090922	0.099924	0.109611	0.120027	0.131218	0.143236
12	0.067738	0.074932	0.082697	0.091067	0.100080	0.109779	0.120207	0.131411	0.143443
13	0.067853	0.075057	0.082831	0.091211	0.100236	0.109947	0.120387	0.131605	0.143651
14	0.067969	0.075182	0.082966	0.091356	0.100392	0.110114	0.120567	0.131799	0.143859
15	0.068084	0.075307	0.083101	0.091502	0.100549	0.110283	0.120748	0.131993	0.144068
16	0.068200	0.075432	0.083235	0.091647	0.100705	0.110451	0.120929	0.132187	0.144276
17	0.068316	0.075557	0.083371	0.091793	0.100862	0.110619	0.121110	0.132381	0.144485
18	0.068432	0.075683	0.083506	0.091938	0.101019	0.110788	0.121291	0.132576	0.144694
19	0.068549	0.075808	0.083641	0.092084	0.101176	0.110957	0.121473	0.132771	0.144903
20	0.068665	0.075934	0.083777	0.092230	0.101333	0.111126	0.121655	0.132966	0.145113
21	0.068782	0.076060	0.083913	0.092377	0.101490	0.111295	0.121837	0.133162	0.145323
22	0.068899	0.076186	0.084049	0.092523	0.101648	0.111465	0.122019	0.133358	0.145533
23	0.069016	0.076312	0.084185	0.092670	0.101806	0.111635	0.122201	0.133553	0.145743
24	0.069133	0.076439	0.084321	0.092816	0.101964	0.111805	0.122384	0.133750	0.145954
25	0.069250	0.076565	0.084458	0.092963	0.102122	0.111975	0.122567	0.133946	0.146165
26	0.069367	0.076692	0.084594	0.093111	0.102280	0.112145	0.122750	0.134143	0.146376
27	0.069485	0.076819	0.084731	0.093258	0.102439	0.112316	0.122933	0.134339	0.146587
28	0.069602	0.076946	0.084868	0.093406	0.102598	0.112486	0.123117	0.134537	0.146799
29	0.069720	0.077073	0.085005	0.093553	0.102757	0.112657	0.123300	0.134734	0.147010
30	0.069838	0.077200	0.085142	0.093701	0.102916	0.112829	0.123484	0.134931	0.147222
31	0.069956	0.077328	0.085280	0.093849	0.103075	0.113000	0.123668	0.135129	0.147435
32	0.070075	0.077455	0.085418	0.093998	0.103235	0.113172	0.123853	0.135327	0.147647
33	0.070193	0.077583	0.085555	0.094146	0.103395	0.113343	0.124037	0.135525	0.147860
34	0.070312	0.077711	0.085693	0.094295	0.103555	0.113515	0.124222	0.135724	0.148073
35	0.070430	0.077839	0.085832	0.094443	0.103715	0.113688	0.124407	0.135923	0.148286
36	0.070549	0.077968	0.085970	0.094593	0.103875	0.113860	0.124592	0.136122	0.148500
37	0.070668	0.078096	0.086108	0.094742	0.104036	0.114033	0.124778	0.136321	0.148714
38	0.070788	0.078225	0.086247	0.094891	0.104196	0.114205	0.124964	0.136520	0.148928
39	0.070907	0.078354	0.086386	0.095041	0.104357	0.114378	0.125150	0.136720	0.149142
40	0.071026	0.078483	0.086525	0.095190	0.104518	0.114552	0.125336	0.136920	0.149357
41	0.071146	0.078612	0.086664	0.095340	0.104680	0.114725	0.125522	0.137120	0.149572
42	0.071266	0.078741	0.086804	0.095490	0.104841	0.114899	0.125709	0.137320	0.149787
43	0.071386	0.078871	0.086943	0.095641	0.105003	0.115073	0.125896	0.137521	0.150002
44	0.071506	0.079000	0.087083	0.095791	0.105165	0.115247	0.126083	0.137722	0.150218
45	0.071626	0.079130	0.087223	0.095942	0.105327	0.115421	0.126270	0.137923	0.150434
46	0.071747	0.079260	0.087363	0.096093	0.105489	0.115595	0.126457	0.138124	0.150650
47	0.071867	0.079390	0.087503	0.096244	0.105652	0.115770	0.126645	0.138326	0.150866
48	0.071988	0.079520	0.087644	0.096395	0.105814	0.115945	0.126833	0.138528	0.151083
49	0.072109	0.079651	0.087784	0.096546	0.105977	0.116120	0.127021	0.138730	0.151299
50	0.072230	0.079781	0.087925	0.096698	0.106140	0.116296	0.127209	0.138932	0.151517
51	0.072351	0.079912	0.088066	0.096850	0.106304	0.116471	0.127398	0.139134	0.151734
52	0.072473	0.080043	0.088207	0.097002	0.106467	0.116647	0.127587	0.139337	0.151952
53	0.072594	0.080174	0.088348	0.097154	0.106631	0.116823	0.127776	0.139540	0.152169
54	0.072716	0.080306	0.088490	0.097306	0.106795	0.116999	0.127965	0.139743	0.152388
55	0.072838	0.080437	0.088631	0.097459	0.106959	0.117175	0.128155	0.139947	0.152606
56	0.072960	0.080569	0.088773	0.097611	0.107123	0.117352	0.128344	0.140151	0.152825
57	0.073082	0.080700	0.088915	0.097764	0.107288	0.117529	0.128534	0.140355	0.153044
58	0.073204	0.080832	0.089057	0.097917	0.107452	0.117706	0.128725	0.140559	0.153263
59	0.073326	0.080964	0.089200	0.098071	0.107617	0.117883	0.128915	0.140763	0.153482
60	0.073449	0.081097	0.089342	0.098224	0.107782	0.118061	0.129106	0.140968	0.153702

Involute Functions for Angles from 41 to 50 Degrees

	Degrees								
	41	42	43	44	45	46	47	48	49
Minutes	Involute Functions								
0	0.153702	0.167366	0.182024	0.197744	0.214602	0.232679	0.252064	0.272855	0.295157
1	0.153922	0.167602	0.182277	0.198015	0.214893	0.232991	0.252399	0.273214	0.295542
2	0.154142	0.167838	0.182530	0.198287	0.215184	0.233304	0.252734	0.273573	0.295928
3	0.154362	0.168075	0.182784	0.198559	0.215476	0.233616	0.253069	0.273933	0.296314
4	0.154583	0.168311	0.183038	0.198832	0.215768	0.233930	0.253405	0.274293	0.296701
5	0.154804	0.168548	0.183292	0.199104	0.216061	0.234243	0.253742	0.274654	0.297088
6	0.155025	0.168786	0.183547	0.199377	0.216353	0.234557	0.254078	0.275015	0.297475
7	0.155247	0.169023	0.183801	0.199651	0.216646	0.234871	0.254415	0.275376	0.297863
8	0.155469	0.169261	0.184057	0.199924	0.216940	0.235186	0.254753	0.275738	0.298251
9	0.155691	0.169500	0.184312	0.200198	0.217234	0.235501	0.255091	0.276101	0.298640
10	0.155913	0.169738	0.184568	0.200473	0.217528	0.235816	0.255429	0.276464	0.299029
11	0.156135	0.169977	0.184824	0.200747	0.217822	0.236132	0.255767	0.276827	0.299419
12	0.156358	0.170216	0.185080	0.201022	0.218117	0.236448	0.256106	0.277191	0.299809
13	0.156581	0.170455	0.185337	0.201297	0.218412	0.236765	0.256446	0.277555	0.300200
14	0.156805	0.170695	0.185594	0.201573	0.218708	0.237082	0.256786	0.277919	0.300591
15	0.157028	0.170935	0.185851	0.201849	0.219004	0.237399	0.257126	0.278284	0.300983
16	0.157252	0.171175	0.186109	0.202125	0.219300	0.237717	0.257467	0.278649	0.301375
17	0.157476	0.171415	0.186367	0.202401	0.219596	0.238035	0.257808	0.279015	0.301767
18	0.157701	0.171656	0.186625	0.202678	0.219893	0.238353	0.258149	0.279381	0.302160
19	0.157925	0.171897	0.186883	0.202956	0.220190	0.238672	0.258491	0.279748	0.302553
20	0.158150	0.172138	0.187142	0.203233	0.220488	0.238991	0.258833	0.280115	0.302947
21	0.158375	0.172380	0.187401	0.203511	0.220786	0.239310	0.259176	0.280483	0.303342
22	0.158601	0.172621	0.187661	0.203789	0.221084	0.239630	0.259519	0.280851	0.303736
23	0.158826	0.172864	0.187920	0.204067	0.221383	0.239950	0.259862	0.281219	0.304132
24	0.159052	0.173106	0.188180	0.204346	0.221682	0.240271	0.260206	0.281588	0.304527
25	0.159279	0.173349	0.188440	0.204625	0.221981	0.240592	0.260550	0.281957	0.304924
26	0.159505	0.173592	0.188701	0.204905	0.222281	0.240913	0.260895	0.282327	0.305320
27	0.159732	0.173835	0.188962	0.205185	0.222581	0.241235	0.261240	0.282697	0.305718
28	0.159959	0.174078	0.189223	0.205465	0.222881	0.241557	0.261585	0.283067	0.306115
29	0.160186	0.174322	0.189485	0.205745	0.223182	0.241879	0.261931	0.283438	0.306513
30	0.160414	0.174566	0.189746	0.206026	0.223483	0.242202	0.262277	0.283810	0.306912
31	0.160642	0.174811	0.190009	0.206307	0.223784	0.242525	0.262624	0.284182	0.307311
32	0.160870	0.175055	0.190271	0.206588	0.224086	0.242849	0.262971	0.284554	0.307710
33	0.161098	0.175300	0.190534	0.206870	0.224388	0.243173	0.263318	0.284927	0.308110
34	0.161327	0.175546	0.190797	0.207152	0.224690	0.243497	0.263666	0.285300	0.308511
35	0.161555	0.175791	0.191060	0.207434	0.224993	0.243822	0.264014	0.285673	0.308911
36	0.161785	0.176037	0.191324	0.207717	0.225296	0.244147	0.264363	0.286047	0.309313
37	0.162014	0.176283	0.191588	0.208000	0.225600	0.244472	0.264712	0.286422	0.309715
38	0.162244	0.176529	0.191852	0.208284	0.225904	0.244798	0.265062	0.286797	0.310117
39	0.162474	0.176776	0.192116	0.208567	0.226208	0.245125	0.265412	0.287172	0.310520
40	0.162704	0.177023	0.192381	0.208851	0.226512	0.245451	0.265762	0.287548	0.310923
41	0.162934	0.177270	0.192646	0.209136	0.226817	0.245778	0.266113	0.287924	0.311327
42	0.163165	0.177518	0.192912	0.209420	0.227123	0.246106	0.266464	0.288301	0.311731
43	0.163396	0.177766	0.193178	0.209705	0.227428	0.246433	0.266815	0.288678	0.312136
44	0.163628	0.178014	0.193444	0.209991	0.227734	0.246761	0.267167	0.289056	0.312541
45	0.163859	0.178262	0.193710	0.210276	0.228041	0.247090	0.267520	0.289434	0.312947
46	0.164091	0.178511	0.193977	0.210562	0.228347	0.247419	0.267872	0.289812	0.313353
47	0.164323	0.178760	0.194244	0.210849	0.228654	0.247748	0.268225	0.290191	0.313759
48	0.164556	0.179009	0.194511	0.211136	0.228962	0.248078	0.268579	0.290570	0.314166
49	0.164788	0.179259	0.194779	0.211423	0.229270	0.248408	0.268933	0.290950	0.314574
50	0.165021	0.179509	0.195047	0.211710	0.229578	0.248738	0.269287	0.291330	0.314982
51	0.165254	0.179759	0.195315	0.211998	0.229886	0.249069	0.269642	0.291711	0.315391
52	0.165488	0.180009	0.195584	0.212286	0.230195	0.249400	0.269998	0.292092	0.315800
53	0.165722	0.180260	0.195853	0.212574	0.230504	0.249732	0.270353	0.292474	0.316209
54	0.165956	0.180511	0.196122	0.212863	0.230814	0.250064	0.270709	0.292856	0.316619
55	0.166190	0.180763	0.196392	0.213152	0.231124	0.250396	0.271066	0.293238	0.317029
56	0.166425	0.181014	0.196661	0.213441	0.231434	0.250729	0.271423	0.293621	0.317440
57	0.166660	0.181266	0.196932	0.213731	0.231745	0.251062	0.271780	0.294004	0.317852
58	0.166895	0.181518	0.197202	0.214021	0.232056	0.251396	0.272138	0.294388	0.318264
59	0.167130	0.181771	0.197473	0.214311	0.232367	0.251730	0.272496	0.294772	0.318676
60	0.167366	0.182024	0.197744	0.214602	0.232679	0.252064	0.272855	0.295157	0.319089

LOGARITHMS

Logarithms are used to facilitate and shorten calculations involving multiplication, division, the extraction of roots, and obtaining powers of numbers. The following properties of logarithms are useful in solving problems of this type:

$$\log_c c = 1 \qquad \log_c c^p = p \qquad \log_c 1 = 0$$

$$\log_c(a \times b) = \log_c a + \log_c b \qquad \log_c(a \div b) = \log_c a - \log_c b$$

$$\log_c(a^p) = p\log_c a \qquad \log_c(\sqrt[p]{a}) = 1/p\log_c a$$

The logarithm of a number is defined as the exponent of a base number raised to a power. For example, $\log_{10} 3.162277 = 0.500$ means the logarithm of 3.162277 is equal to 0.500. Another way of expressing the same relationship is $10^{0.500} = 3.162277$, where 10 is the base number and the exponent 0.500 is the logarithm of 3.162277. A common example of a logarithmic expression $10^2 = 100$ means that the base 10 logarithm of 100 is 2, that is, $\log_{10} 100 = 2.00$. There are two standard systems of logarithms in use: the "common" system (base 10) and the so-called "natural" system (base $e = 2.71828+$). Logarithms to base e are frequently written using "ln" instead of "$\log_e$" such as ln 6.1 = 1.808289. Logarithms of a number can be converted between the natural- and common-based systems as follows: $\ln_e A = 2.3026 \times \log_{10} A$ and $\log_{10} A = 0.43430 \times \ln_e A$. Additional information on the use of "natural logarithms" is given at the end of this section.

A logarithm consists of two parts, a whole number and a decimal. The whole number, which may be positive, negative, or zero, is called the characteristic; the decimal is called the mantissa. As a rule, only the decimal or mantissa is given in tables of common logarithms; tables of natural logarithms give both the characteristic and mantissa. The tables given in this section are abbreviated, but very accurate results can be obtained by using the method of interpolation described in *Interpolation from the Tables* that follows. These tables are especially useful for finding logarithms and calculating powers and roots of numbers on calculators without these functions built in.

Evaluating Logarithms

Common Logarithms.—For common logarithms, the characteristic is prefixed to the mantissa according to the following rules: For numbers greater than or equal to 1, the characteristic is one less than the number of places to the left of the decimal point. For example, the characteristic of the logarithm of 237 is 2, and of 2536.5 is 3. For numbers smaller than 1 and greater than 0, the characteristic is negative and its numerical value is one more than the number of zeros immediately to the right of the decimal point. For example, the characteristic of the logarithm of 0.036 is –2, and the characteristic of the logarithm of 0.0006 is –4. The minus sign is usually written over the figure, as in $\overline{2}$ to indicate that the minus sign refers only to the characteristic and not to the mantissa, which is never negative. The logarithm of 0 does not exist.

The table of common logarithms in this section gives the mantissas of the logarithms of numbers from 1 to 10 and from 1.00 to 1.01. When finding the mantissa, the decimal point in a number is disregarded. The mantissa of the logarithms of 2716, 271.6, 27.16, 2.716, or 0.02716, for example, is the same. The tables give directly the mantissas of logarithms of numbers with three figures or less; the logarithms for numbers with four or more figures can be found by interpolation, as described in *Interpolation from the Tables* and illustrated in the examples. All the mantissas in the common logarithmic tables are decimals and the decimal point has been omitted in the table. However, a decimal point should always be put before the mantissa as soon as it is taken from the table. Logarithmic tables are sufficient for many purposes, but electronic calculators and computers are faster, simpler, and more accurate than tables.

To find the common logarithm of a number from the tables, find the left-hand column of the table and follow down to locate the first two figures of the number. Then look at the top row of the table, on the same page, and follow across it to find the third figure of the number. Follow down the column containing this last figure until opposite the row on which the first two figures were found. The number at the intersection of the row and column is the mantissa of the logarithm. If the logarithm of a number with less than three figures is being obtained, add extra zeros to the right of the number so as to obtain three figures. For example, if the mantissa of the logarithm of 6 is required, find the mantissa of 600.

Interpolation from the Tables.—If the logarithm of a number with more than three figures is needed, linear interpolation is a method of using two values from the table to estimate the value of the logarithm desired. To find the logarithm of a number not listed in the tables, find the mantissa corresponding to the first three digits of the given number (disregarding the decimal point and leading zeros) and find the mantissa of the first three digits of the given number plus one. For example, to find the logarithm of 601.2, 60.12, or 0.006012, find the mantissa of 601 and find the mantissa of 602 from the tables. Then subtract the mantissa of the smaller number from the mantissa of the larger number and multiply the result by a decimal number made from the remaining (additional greater than 3) figures of the original number. Add the result to the mantissa of the smaller number. Find the characteristic as described previously.

Example: Find the logarithm of 4032. The characteristic portion of the logarithm found in the manner described before is 3. Find the mantissa by locating 40 in the left-hand column of the logarithmic tables and then follow across the top row of the table to the column headed 3. Follow down the 3 column to the intersection with the 40 row and read the mantissa. The mantissa of the logarithm of 4030 is 0.605305. Because 4032 is between 4030 and 4040, the logarithm of 4032 is the logarithm of 4030 plus two tenths of the difference in the logarithms of 4030 and 4040. Find the mantissa of 4040 and then subtract from it the mantissa of 4030. Multiply the difference obtained by 0.2 and add the result to the mantissa of the logarithm of 4030. Finally, add the characteristic portion of the logarithm. The result is $\log_{10} 4032 = 3 + 0.605305 + 0.2 \times (0.606381 - 0.605305) = 3.60552$.

Finding a Number Whose Logarithm Is Given.—When a logarithm is given and it is required to find the corresponding number, find the number in the body of the table equal to the value of the mantissa. This value may appear in any column 0 to 9. Follow the row on which the mantissa is found across to the left to read the first two digits of the number sought. Read the third digit of the number from the top row of the table by following up the column on which the mantissa is found to the top. If the characteristic of the logarithm is positive, the number of figures to the left of the decimal in the number is one greater than the value of the characteristic. For example, if the figures corresponding to a given mantissa are 376 and the characteristic is 5, then the number sought has six figures to the left of the decimal point and is 376,000. If the characteristic had been $\overline{3}$, then the number sought would have been 0.00376. If the mantissa is not exactly obtainable in the tables, find the mantissa in the table that is nearest to the one given and determine the corresponding number. This procedure usually gives sufficiently accurate results. If more accuracy is required, find the two mantissas in the tables nearest to the mantissa given, one smaller and the other larger. For each of the two mantissas, read the three corresponding digits from the left column and top row to obtain the first three figures of the number as described before. The exact number sought lies between the two numbers found in this manner.

Next: 1) subtract the smaller mantissa from the given mantissa and; and 2) subtract the smaller mantissa from the larger mantissa.

Divide the result of (1) by the result of (2) and add the quotient to the number corresponding to the smaller mantissa.

Example: Find the number whose logarithm is 2.70053. First, find the number closest to the mantissa 70053 in the body of the tables. The closest mantissa listed in the tables is

700704, so read across the table to the left to find the first two digits of the number sought (50) and up the column to find the third digit of the number (2). The characteristic of the logarithm given is 2, so the number sought has three digits to the left of the decimal point. Therefore, the number sought is slightly less than 502 and greater than 501. If greater accuracy is required, find the two mantissas in the table closest to the given mantissa (699838 and 700704). Subtract the smaller mantissa from the mantissa of the given logarithm and divide the result by the smaller mantissa subtracted from the larger mantissa. Add the result to the number corresponding to the smaller mantissa. The resulting answer is 501 + $(700530 - 699838) \div (700704 - 699838) = 501 + 0.79 = 501.79$.

Avoiding the Use of Negative Characteristics.—As previously explained, the logarithm of any number less than 1 has a negative characteristic and a positive mantissa. In many computations, the use of logarithms having negative characteristics is troublesome and frequently a source of error. A simple way to avoid this difficulty is to convert each logarithm having a negative characteristic into an equivalent logarithm having a positive characteristic. This is done according to the following method, which is based on the principle that any number can be simultaneously added to and subtracted from the characteristic of a logarithm without changing its value. Thus: $\log 1 = 0.000000 = 10.000000 - 10$; $\log 0.3 = \bar{1}.47712 = 9.47712 - 10$; $\log 0.000478 = \bar{4}.67943 = 6.67943 - 10$. Usually, 10 to 20 are added to and subtracted from the characteristic, but any convenient number may be so used.

Natural Logarithms.—In certain formulas and in some branches of mathematical analysis, use is made of logarithms (formerly also called Napierian or hyperbolic logarithms). As previously mentioned, the base of this system, $e = 2.7182818284+$, is the limit of certain mathematical series. The logarithm of a number A to the base e is usually written $\log_e A$ or $\ln A$. Tables of natural logarithms for numbers ranging from 1 to 10 and 1.00 to 1.01 are given in this Handbook after the table of common logarithms. To obtain natural logs of numbers less than 1 or greater than 10, proceed as in the following examples: $\log_e 0.239 = \log_e 2.39 - \log_e 10$; $\log_e 0.0239 = \log_e 2.39 - 2\log_e 10$; $\log_e 239 = \log_e 2.39 + 2\log_e 10$; $\log_e 2390 = \log_e 2.39 + 3\log_e 10$, etc.

Using Calculators to Find Logarithms.—Usually, using a scientific calculator is the quickest and most accurate method of finding logarithms and numbers corresponding to given logarithms. On most scientific calculators, the key labeled **log** is used to find common logarithms (base 10) and the key labeled **ln** is used for finding natural logarithms (base e). The keystrokes to find a logarithm will vary slightly from one calculator to another, so specific instructions are not given. To find the number corresponding to a given logarithm: use the key labeled $\mathbf{10^x}$ if a common logarithm is given or use the key labeled $\mathbf{e^x}$ if a natural logarithm is given; calculators without the 10^x or e^x keys may have a key labeled $\mathbf{x^y}$ that can be used by substituting 10 or e (2.718281 …), as required, for x and substituting the logarithm whose corresponding number is sought for y. On some other calculators, the **log** and **ln** keys are used to find common and natural logarithms, and the same keys in combination with the **INV**, or inverse, key are used to find the number corresponding to a given logarithm.

Multiplication by Logarithms.—If two or more numbers are to be multiplied together, find the logarithms of the numbers to be multiplied, and add these logarithms. The sum is the logarithm of the product, and the number corresponding to this logarithm, as found from the logarithmic tables, is the required product.

Example 1: Find the product of $2831 \times 2.692 \times 29.69 \times 19.4$

$$
\begin{aligned}
\log 2831 &= 3.451786 + 0.1 \times (0.453318 - 0.451786) = 3.451939\\
\log 2.692 &= 0.429752 + 0.2 \times (0.431364 - 0.429752) = 0.430074\\
\log 29.69 &= 1.471292 + 0.9 \times (0.472756 - 0.471292) = 1.472610\\
\log 19.4 &= \underline{1.287802}\\
&\ 6.642425
\end{aligned}
$$

The closest number in the table corresponding to the mantissa is 439. The characteristic indicates the number has seven digits to the left of the decimal point; therefore, the product is slightly less than 4,390,000. If a more accurate result is required, interpolate from the table as follows: 438 + (0.642425 – 0.641474) ÷ (0.642465 – 0.641474) = 438.95963. Therefore, the product sought is 4,389,596.

In multiplication problems involving numbers less than 1, the method of avoiding the use of negative characteristics simplifies the addition and tends to reduce the possibility of error.

Example: Find the product of 0.002656 × 155.1 × 0.5833 × 7.968

$$\begin{aligned}
\log\ 0.002656 &= \bar{3}.424228 = 7.424228 - 10\\
\log\ 155.1 &= 2.190611 = 2.190611\\
\log\ 0.5853 &= \bar{1}.767379 = 9.767379 - 10\\
\log\ 7.968 &= 0.901349 = \underline{0.901349}\\
&\qquad 20.283567 - 20 = 0.283567
\end{aligned}$$

Therefore, the product is 1.92. Interpolate for additional accuracy if required.

Division by Logarithms.— When dividing one number by another, subtract the logarithm of the divisor from the logarithm of the dividend; the remainder is the logarithm of the quotient.

Example: Find the quotient of 7658 ÷ 935.3

$$\begin{aligned}
\log\ 7658. &= 3.884115\\
-\log\ 935.3 &= \underline{-2.970951}\\
&\quad 0.913164
\end{aligned}$$

From the tables, 818 + (0.913164 – 0.912753) ÷ (0.913284 – 0.912753) = 818.8. The answer has one digit to the left of the decimal; hence, 7658 ÷ 935.3 = 8.188.

Instead of dividing 7658 by 935.3, the same answer would be obtained if 7658 were multiplied by the reciprocal of 935.3, or 1 ÷ 935.3. To do this by logarithms, the log of 7658 and the log of the reciprocal of 935.3 would be added together.

To find the logarithm of the reciprocal of a number, subtract the log of the number from the log of 1. To do this conveniently, some number, such as 10, is first added to and then subtracted from the characteristic of the log of 1.

Example: Find the log of the reciprocal of 935.3

$$\begin{aligned}
\log 1 &= 0.000000 = 10.000000 - 10\\
-\log 935.3 &= -2.970951 = \underline{-2.970951}\\
\log (1 \div 935.3) &= 7.029049 - 10\\
\log (1 \div 935.3) &= \bar{3}.029049\\
1 \div 935.3 &= 0.001069
\end{aligned}$$

The quotient of 7658 ÷ 935.3 can be found by adding the log of 7658 and the log of the reciprocal of 935.3.

$$\begin{aligned}
\log 7658 &= 3.884115 = 3.884115\\
\log (1 \div 935.3) &= \bar{3}.029049 = \underline{7.0299049 - 10}\\
&= 10.913164 - 10 = 0.91317
\end{aligned}$$

Hence, 7658 ÷ 935.3 = 8.188.

As is readily seen, this method is more cumbersome than the direct method where there is only one factor each in the dividend and divisor. However, it greatly facilitates the solution of problems in division involving several factors in the dividend and the divisor. In such a problem, the logarithm of each factor of the dividend is added to the logarithm of the reciprocal of each factor of the divisor.

Example: Find the quotient of

$$\frac{0.0272 \times 27.1 \times 12.6}{2.371 \times 0.007}$$

$$\begin{array}{llll} \log 0.0272 & = \overline{2}.43457 & = 8.434569 & - 10 \\ \log 27.1 & = 1.432969 & = 1.432969 & \\ \log 12.6 & = 1.100371 & = 1.100371 & \\ \log (1 \div 2.371) & & = 9.625069 & - 10 \\ \log (1 \div 0.007) & & = \underline{2.154902} & \\ & & 22.74788 & - 20 = 2.74788 \end{array}$$

The quotient is $559 + (0.74788 - 0.747412) \div (0.748188 - 0.747412) = 559.6$.

In division problems where the divisor is larger than the dividend, the subtraction of logarithms is facilitated if some number is added to and subtracted from the log of the dividend. (The is the same method used to convert a logarithm with a negative characteristic to an equivalent logarithm with a positive characteristic except that it serves to convert a logarithm with a positive characteristic to one with a larger positive characteristic, but having the same value.)

Example: Find the quotient of $43.2 \div 971.4$

$$\begin{array}{lll} \log\ 43.2 = 1.63584 & = 11.635484 & - 10 \\ -\log\ 971.4 = -2.987397 & = \underline{-2.987397} & \\ & 8.648087 & - 10 = \overline{2}.648087 \end{array}$$

Hence, the quotient of $43.2 \div 971.4$ is 0.044472.

Obtaining the Powers of Numbers.—A number may be raised to any power by simply multiplying the logarithm of the number by the exponent of the number. The product gives the logarithm of the value of the power.

Example 1: Find the value of 6.51^3

$$\log 6.51 = 0.81358$$

$$3 \times 0.81358 = 2.44074$$

The logarithm 2.44074 is the logarithm of 6.51^3. Hence, 6.51^3 equals the number corresponding to this logarithm, as found from the tables, or $6.51^3 = 275.9$.

Example 2: Find the value of $12^{1.29}$

$$\log 12 = 1.07918$$

$$1.29 \times 1.07918 = 1.39214$$

Hence, $12^{1.29} = 24.67$.

Raising a decimal to a decimal power presents a somewhat more difficult problem because of the negative characteristic of the logarithm and the fact that the logarithm must be multiplied by a decimal exponent. The method previously outlined for avoiding the use of negative characteristics is helpful here.

Example 3: Find the value of $0.0813^{0.46}$

$$\log 0.0813 = \bar{2}.91009 = 8.91009 - 10$$

$$\log 0.0813^{0.46} = 0.46 \times (8.91009 - 10) = 4.09864 - 4.6$$

Subtract and add 0.6 to make the characteristic a whole number:

$$\begin{array}{rcl} & 4.09864 - 4.6 \\ & \underline{-0.6 \quad\quad + 0.6} \\ \log 0.0813^{0.46} = & 3.49864 - 4 = \bar{1}.49864 \end{array}$$

Hence, $0.0813^{0.46} = 0.3152$.

Extracting Roots by Logarithms.—Roots of numbers, for example, $\sqrt[5]{37}$, can be extracted easily by means of logarithms. The small (5) in the radical ($\sqrt{\ }$) of the root sign is called the index of the root. Any root of a number may be found by dividing its logarithm by the index of the root; the quotient is the logarithm of the root.

Example 1: Find $\sqrt[3]{276}$

$$\log 276 = 2.44091$$

$$2.44091 \div 3 = 0.81364$$

Hence, $\log \sqrt[3]{276} = 0.81364$ and $\sqrt[3]{276} = 6.511$

Example 2: Find $\sqrt[3]{0.67}$

$$\log 0.67 = \bar{1}.82607$$

Here it is not possible to divide directly, because there is a negative characteristic and a positive mantissa, another instance where the method of avoiding the use of negative characteristics, previously outlined, is helpful. The preferred procedure is to add and subtract some number to the characteristic that is evenly divisible by the index of the root. The root index is 3, so 9 can be added to and subtracted from the characteristic, and the resulting logarithm divided by 3.

$$\log 0.67 = \bar{1}.82607 = 8.82607 - 9$$

$$\log \sqrt[3]{0.67} = \frac{8.82607 - 9}{3} = 2.94202 - 3$$

$$\log \sqrt[3]{0.67} = 2.94202 - 3 = \bar{1}.94202$$

Hence, $\sqrt[3]{0.67} = 0.875$

Example 3: Find $\sqrt[1.7]{0.2}$

$$\log 0.2 = \bar{1}.30103 = 16.30103 - 17$$

$$\log \sqrt[1.7]{0.2} = \frac{16.30103 - 17}{1.7} = 9.58884 - 10 = \bar{1}.58884$$

Hence,

$$\sqrt[1.7]{0.2} = 0.388$$

Table of Logarithms

Table of Common Logarithms

	0	1	2	3	4	5	6	7	8	9
10	000000	004321	008600	012837	017033	021189	025306	029384	033424	037426
11	041393	045323	049218	053078	056905	060698	064458	068186	071882	075547
12	079181	082785	086360	089905	093422	096910	100371	103804	107210	110590
13	113943	117271	120574	123852	127105	130334	133539	136721	139879	143015
14	146128	149219	152288	155336	158362	161368	164353	167317	170262	173186
15	176091	178977	181844	184691	187521	190332	193125	195900	198657	201397
16	204120	206826	209515	212188	214844	217484	220108	222716	225309	227887
17	230449	232996	235528	238046	240549	243038	245513	247973	250420	252853
18	255273	257679	260071	262451	264818	267172	269513	271842	274158	276462
19	278754	281033	283301	285557	287802	290035	292256	294466	296665	298853
20	301030	303196	305351	307496	309630	311754	313867	315970	318063	320146
21	322219	324282	326336	328380	330414	332438	334454	336460	338456	340444
22	342423	344392	346353	348305	350248	352183	354108	356026	357935	359835
23	361728	363612	365488	367356	369216	371068	372912	374748	376577	378398
24	380211	382017	383815	385606	387390	389166	390935	392697	394452	396199
25	397940	399674	401401	403121	404834	406540	408240	409933	411620	413300
26	414973	416641	418301	419956	421604	423246	424882	426511	428135	429752
27	431364	432969	434569	436163	437751	439333	440909	442480	444045	445604
28	447158	448706	450249	451786	453318	454845	456366	457882	459392	460898
29	462398	463893	465383	466868	468347	469822	471292	472756	474216	475671
30	477121	478566	480007	481443	482874	484300	485721	487138	488551	489958
31	491362	492760	494155	495544	496930	498311	499687	501059	502427	503791
32	505150	506505	507856	509203	510545	511883	513218	514548	515874	517196
33	518514	519828	521138	522444	523746	525045	526339	527630	528917	530200
34	531479	532754	534026	535294	536558	537819	539076	540329	541579	542825
35	544068	545307	546543	547775	549003	550228	551450	552668	553883	555094
36	556303	557507	558709	559907	561101	562293	563481	564666	565848	567026
37	568202	569374	570543	571709	572872	574031	575188	576341	577492	578639
38	579784	580925	582063	583199	584331	585461	586587	587711	588832	589950
39	591065	592177	593286	594393	595496	596597	597695	598791	599883	600973
40	602060	603144	604226	605305	606381	607455	608526	609594	610660	611723
41	612784	613842	614897	615950	617000	618048	619093	620136	621176	622214
42	623249	624282	625312	626340	627366	628389	629410	630428	631444	632457
43	633468	634477	635484	636488	637490	638489	639486	640481	641474	642465
44	643453	644439	645422	646404	647383	648360	649335	650308	651278	652246
45	653213	654177	655138	656098	657056	658011	658965	659916	660865	661813
46	662758	663701	664642	665581	666518	667453	668386	669317	670246	671173
47	672098	673021	673942	674861	675778	676694	677607	678518	679428	680336
48	681241	682145	683047	683947	684845	685742	686636	687529	688420	689309
49	690196	691081	691965	692847	693727	694605	695482	696356	697229	698101
50	698970	699838	700704	701568	702431	703291	704151	705008	705864	706718
51	707570	708421	709270	710117	710963	711807	712650	713491	714330	715167
52	716003	716838	717671	718502	719331	720159	720986	721811	722634	723456
53	724276	725095	725912	726727	727541	728354	729165	729974	730782	731589
54	732394	733197	733999	734800	735599	736397	737193	737987	738781	739572
55	740363	741152	741939	742725	743510	744293	745075	745855	746634	747412
56	748188	748963	749736	750508	751279	752048	752816	753583	754348	755112
57	755875	756636	757396	758155	758912	759668	760422	761176	761928	762679
58	763428	764176	764923	765669	766413	767156	767898	768638	769377	770115
59	770852	771587	772322	773055	773786	774517	775246	775974	776701	777427

Table of Common Logarithms

	0	1	2	3	4	5	6	7	8	9
60	778151	778874	779596	780317	781037	781755	782473	783189	783904	784617
61	785330	786041	786751	787460	788168	788875	789581	790285	790988	791691
62	792392	793092	793790	794488	795185	795880	796574	797268	797960	798651
63	799341	800029	800717	801404	802089	802774	803457	804139	804821	805501
64	806180	806858	807535	808211	808886	809560	810233	810904	811575	812245
65	812913	813581	814248	814913	815578	816241	816904	817565	818226	818885
66	819544	820201	820858	821514	822168	822822	823474	824126	824776	825426
67	826075	826723	827369	828015	828660	829304	829947	830589	831230	831870
68	832509	833147	833784	834421	835056	835691	836324	836957	837588	838219
69	838849	839478	840106	840733	841359	841985	842609	843233	843855	844477
70	845098	845718	846337	846955	847573	848189	848805	849419	850033	850646
71	851258	851870	852480	853090	853698	854306	854913	855519	856124	856729
72	857332	857935	858537	859138	859739	860338	860937	861534	862131	862728
73	863323	863917	864511	865104	865696	866287	866878	867467	868056	868644
74	869232	869818	870404	870989	871573	872156	872739	873321	873902	874482
75	875061	875640	876218	876795	877371	877947	878522	879096	879669	880242
76	880814	881385	881955	882525	883093	883661	884229	884795	885361	885926
77	886491	887054	887617	888179	888741	889302	889862	890421	890980	891537
78	892095	892651	893207	893762	894316	894870	895423	895975	896526	897077
79	897627	898176	898725	899273	899821	900367	900913	901458	902003	902547
80	903090	903633	904174	904716	905256	905796	906335	906874	907411	907949
81	908485	909021	909556	910091	910624	911158	911690	912222	912753	913284
82	913814	914343	914872	915400	915927	916454	916980	917506	918030	918555
83	919078	919601	920123	920645	921166	921686	922206	922725	923244	923762
84	924279	924796	925312	925828	926342	926857	927370	927883	928396	928908
85	929419	929930	930440	930949	931458	931966	932474	932981	933487	933993
86	934498	935003	935507	936011	936514	937016	937518	938019	938520	939020
87	939519	940018	940516	941014	941511	942008	942504	943000	943495	943989
88	944483	944976	945469	945961	946452	946943	947434	947924	948413	948902
89	949390	949878	950365	950851	951338	951823	952308	952792	953276	953760
90	954243	954725	955207	955688	956168	956649	957128	957607	958086	958564
91	959041	959518	959995	960471	960946	961421	961895	962369	962843	963316
92	963788	964260	964731	965202	965672	966142	966611	967080	967548	968016
93	968483	968950	969416	969882	970347	970812	971276	971740	972203	972666
94	973128	973590	974051	974512	974972	975432	975891	976350	976808	977266
95	977724	978181	978637	979093	979548	980003	980458	980912	981366	981819
96	982271	982723	983175	983626	984077	984527	984977	985426	985875	986324
97	986772	987219	987666	988113	988559	989005	989450	989895	990339	990783
98	991226	991669	992111	992554	992995	993436	993877	994317	994757	995196
99	995635	996074	996512	996949	997386	997823	998259	998695	999131	999565
100	000000	000434	000868	001301	001734	002166	002598	003029	003461	003891
101	004321	004751	005181	005609	006038	006466	006894	007321	007748	008174
102	008600	009026	009451	009876	010300	010724	011147	011570	011993	012415
103	012837	013259	013680	014100	014521	014940	015360	015779	016197	016616
104	017033	017451	017868	018284	018700	019116	019532	019947	020361	020775
105	021189	021603	022016	022428	022841	023252	023664	024075	024486	024896
106	025306	025715	026125	026533	026942	027350	027757	028164	028571	028978
107	029384	029789	030195	030600	031004	031408	031812	032216	032619	033021
108	033424	033826	034227	034628	035029	035430	035830	036230	036629	037028
109	037426	037825	038223	038620	039017	039414	039811	040207	040602	040998

Table of Natural Logarithms

	0	1	2	3	4	5	6	7	8	9
1.0	0.00000	0.009950	0.019803	0.029559	0.039221	0.048790	0.058269	0.067659	0.076961	0.086178
1.1	0.09531	0.104360	0.113329	0.122218	0.131028	0.139762	0.148420	0.157004	0.165514	0.173953
1.2	0.18232	0.190620	0.198851	0.207014	0.215111	0.223144	0.231112	0.239017	0.246860	0.254642
1.3	0.26236	0.270027	0.277632	0.285179	0.292670	0.300105	0.307485	0.314811	0.322083	0.329304
1.4	0.33647	0.343590	0.350657	0.357674	0.364643	0.371564	0.378436	0.385262	0.392042	0.398776
1.5	0.40546	0.412110	0.418710	0.425268	0.431782	0.438255	0.444686	0.451076	0.457425	0.463734
1.6	0.47000	0.476234	0.482426	0.488580	0.494696	0.500775	0.506818	0.512824	0.518794	0.524729
1.7	0.53062	0.536493	0.542324	0.548121	0.553885	0.559616	0.565314	0.570980	0.576613	0.582216
1.8	0.58778	0.593327	0.598837	0.604316	0.609766	0.615186	0.620576	0.625938	0.631272	0.636577
1.9	0.64185	0.647103	0.652325	0.657520	0.662688	0.667829	0.672944	0.678034	0.683097	0.688135
2.0	0.69314	0.698135	0.703098	0.708036	0.712950	0.717840	0.722706	0.727549	0.732368	0.737164
2.1	0.74193	0.746688	0.751416	0.756122	0.760806	0.765468	0.770108	0.774727	0.779325	0.783902
2.2	0.78845	0.792993	0.797507	0.802002	0.806476	0.810930	0.815365	0.819780	0.824175	0.828552
2.3	0.83290	0.837248	0.841567	0.845868	0.850151	0.854415	0.858662	0.862890	0.867100	0.871293
2.4	0.87546	0.879627	0.883768	0.887891	0.891998	0.896088	0.900161	0.904218	0.908259	0.912283
2.5	0.91629	0.920283	0.924259	0.928219	0.932164	0.936093	0.940007	0.943906	0.947789	0.951658
2.6	0.95551	0.959350	0.963174	0.966984	0.970779	0.974560	0.978326	0.982078	0.985817	0.989541
2.7	0.99325	0.996949	1.000632	1.004302	1.007958	1.011601	1.015231	1.018847	1.022451	1.026042
2.8	1.02961	1.033184	1.036737	1.040277	1.043804	1.047319	1.050822	1.054312	1.057790	1.061257
2.9	1.06471	1.068153	1.071584	1.075002	1.078410	1.081805	1.085189	1.088562	1.091923	1.095273
3.0	1.09861	1.101940	1.105257	1.108563	1.111858	1.115142	1.118415	1.121678	1.124930	1.128171
3.1	1.13140	1.134623	1.137833	1.141033	1.144223	1.147402	1.150572	1.153732	1.156881	1.160021
3.2	1.16315	1.166271	1.169381	1.172482	1.175573	1.178655	1.181727	1.184790	1.187843	1.190888
3.3	1.19392	1.196948	1.199965	1.202972	1.205971	1.208960	1.211941	1.214913	1.217876	1.220830
3.4	1.22377	1.226712	1.229641	1.232560	1.235471	1.238374	1.241269	1.244155	1.247032	1.249902
3.5	1.25276	1.255616	1.258461	1.261298	1.264127	1.266948	1.269761	1.272566	1.275363	1.278152
3.6	1.28093	1.283708	1.286474	1.289233	1.291984	1.294727	1.297463	1.300192	1.302913	1.305626
3.7	1.30833	1.311032	1.313724	1.316408	1.319086	1.321756	1.324419	1.327075	1.329724	1.332366
3.8	1.33500	1.337629	1.340250	1.342865	1.345472	1.348073	1.350667	1.353255	1.355835	1.358409
3.9	1.36097	1.363537	1.366092	1.368639	1.371181	1.373716	1.376244	1.378766	1.381282	1.383791
4.0	1.38629	1.388791	1.391282	1.393766	1.396245	1.398717	1.401183	1.403643	1.406097	1.408545
4.1	1.41098	1.413423	1.415853	1.418277	1.420696	1.423108	1.425515	1.427916	1.430311	1.432701
4.2	1.43508	1.437463	1.439835	1.442202	1.444563	1.446919	1.449269	1.451614	1.453953	1.456287
4.3	1.45861	1.460938	1.463255	1.465568	1.467874	1.470176	1.472472	1.474763	1.477049	1.479329
4.4	1.48160	1.483875	1.486140	1.488400	1.490654	1.492904	1.495149	1.497388	1.499623	1.501853
4.5	1.50407	1.506297	1.508512	1.510722	1.512927	1.515127	1.517323	1.519513	1.521699	1.523880
4.6	1.52605	1.528228	1.530395	1.532557	1.534714	1.536867	1.539015	1.541159	1.543298	1.545433
4.7	1.54756	1.549688	1.551809	1.553925	1.556037	1.558145	1.560248	1.562346	1.564441	1.566530
4.8	1.56861	1.570697	1.572774	1.574846	1.576915	1.578979	1.581038	1.583094	1.585145	1.587192
4.9	1.58923	1.591274	1.593309	1.595339	1.597365	1.599388	1.601406	1.603420	1.605430	1.607436
5.0	1.60943	1.611436	1.613430	1.615420	1.617406	1.619388	1.621366	1.623341	1.625311	1.627278
5.1	1.62924	1.631199	1.633154	1.635106	1.637053	1.638997	1.640937	1.642873	1.644805	1.646734
5.2	1.64865	1.650580	1.652497	1.654411	1.656321	1.658228	1.660131	1.662030	1.663926	1.665818
5.3	1.66770	1.669592	1.671473	1.673351	1.675226	1.677097	1.678964	1.680828	1.682688	1.684545
5.4	1.68639	1.688249	1.690096	1.691939	1.693779	1.695616	1.697449	1.699279	1.701105	1.702928
5.5	1.70474	1.706565	1.708378	1.710188	1.711995	1.713798	1.715598	1.717395	1.719189	1.720979
5.6	1.722767	1.724551	1.726332	1.728109	1.729884	1.731656	1.733424	1.735189	1.736951	1.738710
5.7	1.74046	1.742219	1.743969	1.745716	1.747459	1.749200	1.750937	1.752672	1.754404	1.756132
5.8	1.75785	1.759581	1.761300	1.763017	1.764731	1.766442	1.768150	1.769855	1.771557	1.773256
5.9	1.77495	1.776646	1.778336	1.780024	1.781709	1.783391	1.785070	1.786747	1.788421	1.790091

Table of Natural Logarithms

	0	1	2	3	4	5	6	7	8	9
6.0	1.791759	1.793425	1.795087	1.796747	1.798404	1.800058	1.801710	1.803359	1.805005	1.806648
6.1	1.808289	1.809927	1.811562	1.813195	1.814825	1.816452	1.818077	1.819699	1.821318	1.822935
6.2	1.824549	1.826161	1.827770	1.829376	1.830980	1.832581	1.834180	1.835776	1.837370	1.838961
6.3	1.840550	1.842136	1.843719	1.845300	1.846879	1.848455	1.850028	1.851599	1.853168	1.854734
6.4	1.856298	1.857859	1.859418	1.860975	1.862529	1.864080	1.865629	1.867176	1.868721	1.870263
6.5	1.871802	1.873339	1.874874	1.876407	1.877937	1.879465	1.880991	1.882514	1.884035	1.885553
6.6	1.887070	1.888584	1.890095	1.891605	1.893112	1.894617	1.896119	1.897620	1.899118	1.900614
6.7	1.902108	1.903599	1.905088	1.906575	1.908060	1.909543	1.911023	1.912501	1.913977	1.915451
6.8	1.916923	1.918392	1.919859	1.921325	1.922788	1.924249	1.925707	1.927164	1.928619	1.930071
6.9	1.931521	1.932970	1.934416	1.935860	1.937302	1.938742	1.940179	1.941615	1.943049	1.944481
7.0	1.945910	1.947338	1.948763	1.950187	1.951608	1.953028	1.954445	1.955860	1.957274	1.958685
7.1	1.960095	1.961502	1.962908	1.964311	1.965713	1.967112	1.968510	1.969906	1.971299	1.972691
7.2	1.974081	1.975469	1.976855	1.978239	1.979621	1.981001	1.982380	1.983756	1.985131	1.986504
7.3	1.987874	1.989243	1.990610	1.991976	1.993339	1.994700	1.996060	1.997418	1.998774	2.000128
7.4	2.001480	2.002830	2.004179	2.005526	2.006871	2.008214	2.009555	2.010895	2.012233	2.013569
7.5	2.014903	2.016235	2.017566	2.018895	2.020222	2.021548	2.022871	2.024193	2.025513	2.026832
7.6	2.028148	2.029463	2.030776	2.032088	2.033398	2.034706	2.036012	2.037317	2.038620	2.039921
7.7	2.041220	2.042518	2.043814	2.045109	2.046402	2.047693	2.048982	2.050270	2.051556	2.052841
7.8	2.054124	2.055405	2.056685	2.057963	2.059239	2.060514	2.061787	2.063058	2.064328	2.065596
7.9	2.066863	2.068128	2.069391	2.070653	2.071913	2.073172	2.074429	2.075684	2.076938	2.078191
8.0	2.079442	2.080691	2.081938	2.083185	2.084429	2.085672	2.086914	2.088153	2.089392	2.090629
8.1	2.091864	2.093098	2.094330	2.095561	2.096790	2.098018	2.099244	2.100469	2.101692	2.102914
8.2	2.104134	2.105353	2.106570	2.107786	2.109000	2.110213	2.111425	2.112635	2.113843	2.115050
8.3	2.116256	2.117460	2.118662	2.119863	2.121063	2.122262	2.123458	2.124654	2.125848	2.127041
8.4	2.128232	2.129421	2.130610	2.131797	2.132982	2.134166	2.135349	2.136531	2.137710	2.138889
8.5	2.140066	2.141242	2.142416	2.143589	2.144761	2.145931	2.147100	2.148268	2.149434	2.150599
8.6	2.151762	2.152924	2.154085	2.155245	2.156403	2.157559	2.158715	2.159869	2.161022	2.162173
8.7	2.163323	2.164472	2.165619	2.166765	2.167910	2.169054	2.170196	2.171337	2.172476	2.173615
8.8	2.174752	2.175887	2.177022	2.178155	2.179287	2.180417	2.181547	2.182675	2.183802	2.184927
8.9	2.186051	2.187174	2.188296	2.189416	2.190536	2.191654	2.192770	2.193886	2.195000	2.196113
9.0	2.197225	2.198335	2.199444	2.200552	2.201659	2.202765	2.203869	2.204972	2.206074	2.207175
9.1	2.208274	2.209373	2.210470	2.211566	2.212660	2.213754	2.214846	2.215937	2.217027	2.218116
9.2	2.219203	2.220290	2.221375	2.222459	2.223542	2.224624	2.225704	2.226783	2.227862	2.228939
9.3	2.230014	2.231089	2.232163	2.233235	2.234306	2.235376	2.236445	2.237513	2.238580	2.239645
9.4	2.240710	2.241773	2.242835	2.243896	2.244956	2.246015	2.247072	2.248129	2.249184	2.250239
9.5	2.251292	2.252344	2.253395	2.254445	2.255493	2.256541	2.257588	2.258633	2.259678	2.260721
9.6	2.261763	2.262804	2.263844	2.264883	2.265921	2.266958	2.267994	2.269028	2.270062	2.271094
9.7	2.272126	2.273156	2.274186	2.275214	2.276241	2.277267	2.278292	2.279316	2.280339	2.281361
9.8	2.282382	2.283402	2.284421	2.285439	2.286456	2.287471	2.288486	2.289500	2.290513	2.291524
9.9	2.292535	2.293544	2.294553	2.295560	2.296567	2.297573	2.298577	2.299581	2.300583	2.301585
1.00	0.000000	0.001000	0.001998	0.002996	0.003992	0.004988	0.005982	0.006976	0.007968	0.008960
1.01	0.009950	0.010940	0.011929	0.012916	0.013903	0.014889	0.015873	0.016857	0.017840	0.018822
1.02	0.019803	0.020783	0.021761	0.022739	0.023717	0.024693	0.025668	0.026642	0.027615	0.028587
1.03	0.029559	0.030529	0.031499	0.032467	0.033435	0.034401	0.035367	0.036332	0.037296	0.038259
1.04	0.039221	0.040182	0.041142	0.042101	0.043059	0.044017	0.044973	0.045929	0.046884	0.047837
1.05	0.048790	0.049742	0.050693	0.051643	0.052592	0.053541	0.054488	0.055435	0.056380	0.057325
1.06	0.058269	0.059212	0.060154	0.061095	0.062035	0.062975	0.063913	0.064851	0.065788	0.066724
1.07	0.067659	0.068593	0.069526	0.070458	0.071390	0.072321	0.073250	0.074179	0.075107	0.076035
1.08	0.076961	0.077887	0.078811	0.079735	0.080658	0.081580	0.082501	0.083422	0.084341	0.085260
1.09	0.086178	0.087095	0.088011	0.088926	0.089841	0.090754	0.091667	0.092579	0.093490	0.094401

MECHANICS

MECHANICAL PROPERTIES OF BODIES

MECHANICAL PROPERTIES OF BODIES (cont.)

FLYWHEELS

PENDULUMS

FRICTION

MECHANICS

Throughout the Mechanics section in this Handbook, both English and metric SI data and formulas are given to cover the requirements of working in either system of measurement. Except for the passage entitled *The Use of the Metric SI System in Mechanics Calculations* formulas and text relating exclusively to SI are given in bold face type.

Terms and Definitions

Definitions.—The science of mechanics deals with the effects of forces in causing or preventing motion. *Statics* is the branch of mechanics that deals with bodies in equilibrium, i.e., the forces acting on them cause them to remain at rest or to move with uniform velocity. *Dynamics* is the branch of mechanics that deals with bodies not in equilibrium, i.e., the forces acting on them cause them to move with non-uniform velocity. *Kinetics* is the branch of dynamics that deals with both the forces acting on bodies and the motions that they cause. *Kinematics* is the branch of dynamics that deals only with the motions of bodies without reference to the forces that cause them.

Definitions of certain terms and quantities as used in mechanics follow:

Force may be defined simply as a push or a pull; the push or pull may result from the force of contact between bodies or from a force, such as magnetism or gravitation, in which no direct contact takes place.

Matter is any substance that occupies space; gases, liquids, solids, electrons, atoms, molecules, etc., all fit this definition.

Inertia is the property of matter that causes it to resist any change in its motion or state of rest.

Mass is a measure of the inertia of a body.

Work, in mechanics, is the product of force times distance and is expressed by a combination of units of force and distance, as foot-pounds, inch-pounds, meter-kilograms, etc. **The metric SI unit of work is the joule, which is the work done when the point of application of a force of one newton is displaced through a distance of one meter in the direction of the force.**

Power, in mechanics, is the product of force times distance divided by time; it measures the performance of a given amount of work in a given time. It is the rate of doing work and as such is expressed in foot-pounds per minute, foot-pounds per second, kilogram-meters per second, etc. **The metric SI unit is the watt, which is one joule per second.**

Horsepower is the unit of power that has been adopted for engineering work. One horsepower is equal to 33,000 foot-pounds per minute or 550 foot-pounds per second. The *kilowatt,* used in electrical work, equals 1.34 horsepower; or 1 horsepower equals 0.746 kilowatt. **However, in the metric SI, the term horsepower is not used, and the basic unit of power is the watt. This unit, and the derived units milliwatt and kilowatt, for example, are the same as those used in electrical work.**

Torque or moment of a force is a measure of the tendency of the force to rotate the body upon which it acts about an axis. The magnitude of the moment due to a force acting in a plane perpendicular to some axis is obtained by multiplying the force by the perpendicular distance from the axis to the line of action of the force. (If the axis of rotation is not perpendicular to the plane of the force, then the components of the force in a plane perpendicular to the axis of rotation are used to find the resultant moment of the force by finding the moment of each component and adding these component moments algebraically.) Moment or torque is commonly expressed in pound-feet, pound-inches, kilogram-meters, etc. **The metric SI unit is the newton-meter (N · m).**

Velocity is the time-rate of change of distance and is expressed as distance divided by time, that is, feet per second, miles per hour, centimeters per second, meters per second, etc.

Acceleration is defined as the time-rate of change of velocity and is expressed as velocity divided by time or as distance divided by time squared, that is, in feet per second, per second or feet per second squared; inches per second, per second or inches per second squared; centimeters per second, per second or centimeters per second squared; etc. **The metric SI unit is the meter per second squared.**

Unit Systems.—In mechanics calculations, both *absolute* and *gravitational* systems of units are employed. The fundamental units in absolute systems are *length*, *time*, and *mass*, and from these units, the dimension of force is derived. Two absolute systems which have been in use for many years are the cgs (centimeter-gram-second) and the MKS (meter-kilogram-second) systems. Another system, known as MKSA (meter-kilogram-second-ampere), links the MKS system of units of mechanics with electro magnetic units.

The Conference General des Poids et Mesures (CGPM), which is the body responsible for all international matters concerning the metric system, adopted in 1954 a rationalized and coherent system of units based on the four MKSA units and including the kelvin as the unit of temperature, and the candela as the unit of luminous intensity. In 1960, the CGPM formally named this system the 'Systeme International d'Unites,' for which the abbreviation is SI in all languages. In 1971, the 14th CGPM adopted a seventh base unit, the mole, which is the unit of quantity ("amount of substance"). Further details of the SI are given in the Weights and Measures section, and its application in mechanics calculations, contrasted with the use of the English system, is considered on page 116.

The fundamental units in gravitational systems are *length*, *time*, and *force*, and from these units, the dimension of mass is derived. In the gravitational system most widely used in English measure countries, the units of length, time, and force are, respectively, the foot, the second, and the pound. The corresponding unit of mass, commonly called the *slug*, is equal to 1 pound second^2 per foot and is derived from the formula, $M = W \div g$ in which M = mass in slugs, W = weight in pounds, and g = acceleration due to gravity, commonly taken as 32.16 feet per second^2. A body that weighs 32.16 lbs. on the surface of the earth has, therefore, a mass of one slug.

Many engineering calculations utilize a system of units consisting of the inch, the second, and the pound. The corresponding units of mass are pounds second^2 per inch and the value of g is taken as 386 inches per second^2.

In a gravitational system that has been widely used in metric countries, the units of length, time, and force are, respectively, the meter, the second, and the kilogram. The corresponding units of mass are kilograms second^2 per meter and the value of g is taken as 9.81 meters per second^2.

Acceleration of Gravity *g* Used in Mechanics Formulas.—The acceleration of a freely falling body has been found to vary according to location on the earth's surface as well as with height, the value at the equator being 32.09 feet per second, per second while at the poles it is 32.26 ft/sec^2. In the United States it is customary to regard 32.16 as satisfactory for most practical purposes in engineering calculations.

Standard Pound Force: For use in defining the magnitude of a standard unit of force, known as the *pound force*, a fixed value of 32.1740 ft/sec^2, designated by the symbol g_0, has been adopted by international agreement. As a result of this agreement, whenever the term mass, M, appears in a mechanics formula and the substitution $M = W/g$ is made, use of the standard value g_0 = 32.1740 ft/sec^2 is implied although as stated previously, it is customary to use approximate values for g except where extreme accuracy is required.

American National Standard Letter Symbols for Mechanics and Time-Related Phenomena *ANSI/ASME Y10.3M-1984*

Quantity	Symbol	Quantity	Symbol
Acceleration, angular	α (alpha)	Height	h
Acceleration, due to gravity	g	Inertia, moment of	I or J
Acceleration, linear	a	Inertia, polar (area) moment of[a]	J
Amplitude[a]	A	Inertia, product (area) moment of[a]	I_{xy}
Angle	α (alpha)	Length	L or l
	β (beta)	Load per unit distance[a]	q or w
	γ (gamma)	Load, total[a]	P or W
	θ (theta)	Mass	m
	φ (phi)	Moment of force, including bending moment	M
	ψ (psi)	Neutral axis, distance to extreme fiber from[a]	c
Angle, solid	Ω (omega)	Period	T
Angular frequency	ω (omega)	Poisson's ratio	μ (mu) or ν (nu)
Angular momentum	L	Power	P
Angular velocity	ω (omega)	Pressure, normal force per unit area	p
Arc length	s	Radius	r
Area	A	Revolutions per unit of time	n
Axes, through any point[a]	X-X, Y-Y, or Z-Z	Second moment of area (second axial moment of area)	I_a
Bulk modulus	K	Second polar moment of area	I_p or J
Breadth (width)	b	Section modulus	Z
Coefficient of expansion, linear[a]	α (alpha)	Shear force in beam section[a]	V
Coefficient of friction	μ (mu)	Spring constant (load per unit deflection)[a]	k
Concentrated load (same as force)	F	Statical moment of any area about a given axis[a]	Q
Deflection of beam, max[a]	δ (delta)	Strain, normal	ε (epsilon)
Density	ρ (rho)	Strain, shear	γ (gamma)
Depth	d, δ (delta), or t	Stress, concentration factor[a]	K
Diameter	D or d	Stress, normal	σ (sigma)
Displacement[a]	u, v, w	Stress, shear	τ (tau)
Distance, linear[a]	s	Temperature, absolute[b]	T, or θ (theta)
Eccentricity of application of load[a]	e	Temperature[b]	t, or θ (theta)
Efficiency[a]	η (eta)	Thickness	d, δ (delta), or t
Elasticity, modulus of	E	Time	t
Elasticity, modulus of, in shear	G	Torque	T
Elongation, total[a]	δ (delta)	Velocity, linear	v
Energy, kinetic	E_k, K, T	Volume	V
Energy, potential	E_p, V, or Φ (phi)	Wavelength	λ (lambda)
Factor of safety[a]	N, or n	Weight	W
Force or load, concentrated	F	Weight per unit volume	γ (gamma)
Frequency	f	Work	W
Gyration, radius of[a]	k		

[a] Not specified in Standard

[b] Specified in ANSI Y10.4-1982 (R1988)

The Use of the Metric SI System in Mechanics Calculations.—The SI system is a development of the traditional metric system based on decimal arithmetic; fractions are avoided. For each physical quantity, units of different sizes are formed by multiplying or dividing a single base value by powers of 10. Thus, changes can be made very simply by adding zeros or shifting decimal points. For example, the meter is the basic unit of length; the kilometer is a multiple (1,000 meters); and the millimeter is a sub-multiple (one-thousandth of a meter).

In the older metric system, the simplicity of a series of units linked by powers of 10 is an advantage for plain quantities such as length, but this simplicity is lost as soon as more complex units are encountered. For example, in different branches of science and engineering, energy may appear as the erg, the calorie, the kilogram-meter, the liter-atmosphere, or the horsepower-hour. In contrast, the SI provides only one basic unit for each physical quantity, and universality is thus achieved.

There are seven base-units, and in mechanics calculations three are used, which are for the basic quantities of length, mass, and time, expressed as the meter (m), the kilogram (kg), and the second (s). The other four base-units are the ampere (A) for electric current, the kelvin (K) for thermodynamic temperature, the candela (cd) for luminous intensity, and the mole (mol) for amount of substance.

The SI is a coherent system. A system of units is said to be coherent if the product or quotient of any two unit quantities in the system is the unit of the resultant quantity. For example, in a coherent system in which the foot is a unit of length, the square foot is the unit of area, whereas the acre is not. Further details of the SI, and definitions of the units, are given in the section "*METRIC SYSTEMS OF MEASUREMENT*" at the end of the book.

Other physical quantities are derived from the base-units. For example, the unit of velocity is the meter per second (m/s), which is a combination of the base-units of length and time. The unit of acceleration is the meter per second squared (m/s^2). By applying Newton's second law of motion — force is proportional to mass multiplied by acceleration — the unit of force is obtained, which is the $kg \cdot m/s^2$. This unit is known as the newton, or N. Work, or force times distance, is the $kg \cdot m^2/s^2$, which is the joule, (1 joule = 1 newton-meter) and energy is also expressed in these terms. The abbreviation for joule is J. Power, or work per unit time, is the $kg \cdot m^2/s^3$, which is the watt (1 watt = 1 joule per second = 1 newton-meter per second). The abbreviation for watt is W.

The coherence of SI units has two important advantages. The first, that of uniqueness and therefore universality, has been explained. The second is that it greatly simplifies technical calculations. Equations representing physical principles can be applied without introducing such numbers as 550 in power calculations, which, in the English system of measurement have to be used to convert units. Thus conversion factors largely disappear from calculations carried out in SI units, with a great saving in time and labor.

Mass, weight, force, load: SI is an absolute system (see *Unit Systems* on page 114), and consequently it is necessary to make a clear distinction between mass and weight. The *mass* of a body is a measure of its inertia, whereas the weight of a body is the *force* exerted on it by gravity. In a fixed gravitational field, weight is directly proportional to mass, and the distinction between the two can be easily overlooked. However, if a body is moved to a different gravitational field, for example, that of the moon, its weight alters, but its mass remains unchanged. Since the gravitational field on earth varies from place to place by only a small amount, and weight is proportional to mass, it is practical to use the weight of unit mass as a unit of force, and this procedure is adopted in both the English and older metric systems of measurement. In common usage, they are given the same names, and we say that a mass of 1 pound has a weight of 1 pound. In the former case the pound is being used as a unit of mass, and in the latter case, as a unit of force. This procedure is convenient in some branches of engineering, but leads to confusion in others.

As mentioned earlier, Newton's second law of motion states that force is proportional to mass times acceleration. Because an unsupported body on the earth's surface falls with acceleration g (32 ft/s^2 approximately), the pound (force) is that force which will impart an acceleration of g ft/s^2 to a pound (mass). Similarly, the kilogram (force) is that force which will impart an acceleration of g (9.8 meters per second^2 approximately), to a mass of one kilogram. In the SI, the *newton* is that force which will impart unit acceleration (1 m/s^2) to a mass of one kilogram. It is therefore smaller than the kilogram (force) in the ratio 1:g (about 1:9.8). This fact has important consequences in engineering calculations. The factor g now disappears from a wide range of formulas in dynamics, but appears in many formulas in statics where it was formerly absent. It is however not quite the same g, for reasons which will now be explained.

In the article on page 154, the mass of a body is referred to as M, but it is immediately replaced in subsequent formulas by W/g, where W is the weight in pounds (force), which leads to familiar expressions such as $WV^2 / 2g$ for kinetic energy. In this treatment, the M which appears briefly is really expressed in terms of the slug (page 114), a unit normally used only in aeronautical engineering. In everyday engineers' language, weight and mass are regarded as synonymous and expressions such as $WV^2 / 2g$ are used without pondering the distinction. Nevertheless, on reflection it seems odd that g should appear in a formula which has nothing to do with gravity at all. In fact the g used here is not the true, local value of the acceleration due to gravity, but an arbitrary standard value which has been chosen as part of the definition of the pound (force) and is more properly designated g_o (page 114). Its function is not to indicate the strength of the local gravitational field, but to convert from one unit to another.

In the SI the unit of mass is the *kilogram*, and the unit of force (and therefore weight) is the *newton*.

The following are typical statements in dynamics expressed in SI units:

A force of R newtons acting on a mass of M kilograms produces an acceleration of R/M meters per second^2. The kinetic energy of a mass of M kg moving with velocity V m/s is $\frac{1}{2}MV^2$ kg (m/s)^2 or $\frac{1}{2}MV^2$ joules. The work done by a force of R newtons moving a distance L meters is RL Nm, or RL joules. If this work were converted entirely into kinetic energy we could write $RL = \frac{1}{2}MV^2$ and it is instructive to consider the units. Remembering that the N is the same as the kg · m/s^2, we have (kg · m/s)^2 × m = kg (m/s)^2, which is obviously correct. It will be noted that g does not appear anywhere in these statements.

In contrast, in many branches of engineering where the weight of a body is important, rather than its mass, using SI units, g does appear where formerly it was absent. Thus, if a rope hangs vertically supporting a mass of M kilograms the tension in the rope is Mg N. Here g is the acceleration due to gravity, and its units are m/s^2. The ordinary numerical value of 9.81 will be sufficiently accurate for most purposes on earth. The expression is still valid elsewhere, for example, on the moon, provided the proper value of g is used. The maximum tension the rope can safely withstand (and other similar properties) will also be specified in terms of the newton, so that direct comparison may be made with the tension predicted.

Words like load and weight have to be used with greater care. In everyday language we might say "a lift carries a load of five people of average weight 70 kg," but in precise technical language we say that if the average mass is 70 kg, then the average weight is $70g$ N, and the total load (that is force) on the lift is $350g$ N.

If the lift starts to rise with acceleration a m/s^2, the load becomes $350 (g + a)$ N; both g and a have units of m/s^2, the mass is in kg, so the load is in terms of kg · m/s^2, which is the same as the newton.

Pressure and stress: These quantities are expressed in terms of force per unit area. In the SI the unit is the pascal (Pa), which expressed in terms of SI derived and base units is the newton per meter squared (N/m^2). The pascal is very small—it is only equivalent to 0.15×10^{-3} lb/in^2 — hence the kilopascal (kPa = 1000 pascals), and the megapascal (MPa = 10^6 pascals) may be more convenient multiples in practice. Thus, note: 1 newton per millimeter squared = 1 meganewton per meter squared = 1 megapascal.

In addition to the pascal, the bar, a non-SI unit, is in use in the field of pressure measurement in some countries, including England. Thus, in view of existing practice, the International Committee of Weights and Measures (CIPM) decided in 1969 to retain this unit for a limited time for use with those of SI. The bar = 10^5 pascals and the hectobar = 10^7 pascals.

Force Systems

Scalar and Vector Quantities.—The quantities dealt with in mechanics are of two kinds according to whether magnitude alone or direction as well as magnitude must be known in order to completely specify them. Quantities such as time, volume and density are completely specified when their magnitude is known. Such quantities are called *scalar* quantities. Quantities such as force, velocity, acceleration, moment, and displacement which must, in order to be specified completely, have a specific direction as well as magnitude, are called *vector* quantities.

Graphical Representation of Forces.—A force has three characteristics which, when known, determine it. They are *direction, point of application*, and *magnitude*. The direction of a force is the direction in which it tends to move the body upon which it acts. The point of application is the place on the line of action where the force is applied. Forces may conveniently be represented by straight lines and arrow heads. The arrow head indicates the direction of the force, and the length of the line, its magnitude to any suitable scale. The point of application may be at any point on the line, but it is generally convenient to assume it to be at one end. In the accompanying illustration, a force is supposed to act along line *AB* in a direction from left to right. The length of line *AB* shows the magnitude of the force. If point *A* is the point of application, the force is exerted as a pull, but if point *B* be assumed to be the point of application, it would indicate that the force is exerted as a push.

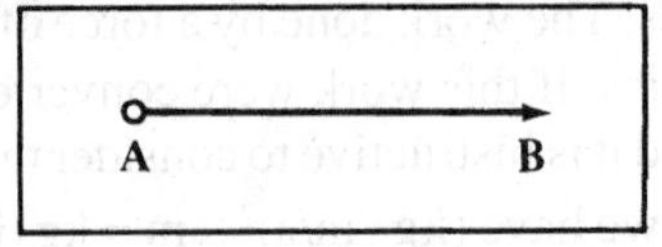

Vector

Velocities, moments, displacements, etc. may similarly be represented and manipulated graphically because they are all of the same class of quantities called vectors. (See *Scalar and Vector Quantities.*)

Addition and Subtraction of Forces: The resultant of two forces applied at the same point and acting in the same direction, is equal to the sum of the forces. For example, if the two forces *AB* and *AC*, one equal to two and the other equal to three pounds, are applied at point *A*, then their resultant *AD* equals the sum of these forces, or five pounds.

Fig. 1. Fig. 2.

If two forces act in opposite directions, then their resultant is equal to their difference, and the direction of the resultant is the same as the direction of the greater of the two forces. For example: *AB* and *AC* are both applied at point *A*; then, if *AB* equals four and *AC* equals six pounds, the resultant *AD* equals two pounds and acts in the direction of *AC*.

Parallelogram of Forces: If two forces applied at a point are represented in magnitude and direction by the adjacent sides of a parallelogram (*AB* and *AC* in Fig. 3), their resultant will be represented in magnitude and direction by the diagonal *AR* drawn from the intersection of the two component forces.

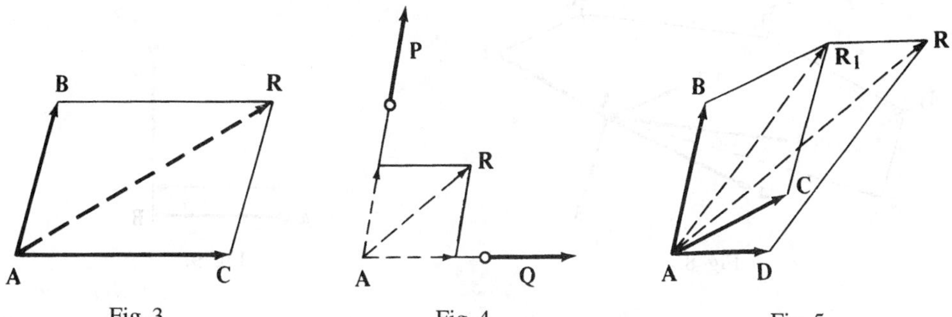

Fig. 3. Fig. 4. Fig. 5.

If two forces *P* and *Q* do not have the same point of application, as in Fig. 4, but the lines indicating their directions intersect, the forces may be imagined as applied at the point of intersection between the lines (as at *A*), and the resultant of the two forces may be found by constructing the parallelogram of forces. Line *AR* shows the direction and magnitude of the resultant, the point of application of which may be assumed to be at any point on line *AR* or its extension.

If the resultant of three or more forces having the same point of application is to be found, as in Fig. 5, first find the resultant of any two of the forces (*AB* and *AC*) and then find the resultant of the resultant just found (AR_1) and the third force (*AD*). If there are more than three forces, continue in this manner until the resultant of all the forces has been found.

Parallel Forces: If two forces are parallel and act in the same direction, as in Fig. 6, then their resultant is parallel to both lines, is located between them, and is equal to the sum of the two components. The point of application of the resultant divides the line joining the points of application of the components inversely as the magnitude of the components. Thus,

$$AB : CE = CD : AD$$

The resultant of two parallel and unequal forces acting in opposite directions, Fig. 7, is parallel to both lines, is located outside of them on the side of the greater of the components, has the same direction as the greater component, and is equal in magnitude to the difference between the two components. The point of application on the line *AC* produced is found from the proportion:

$$AB : CD = CE : AE$$

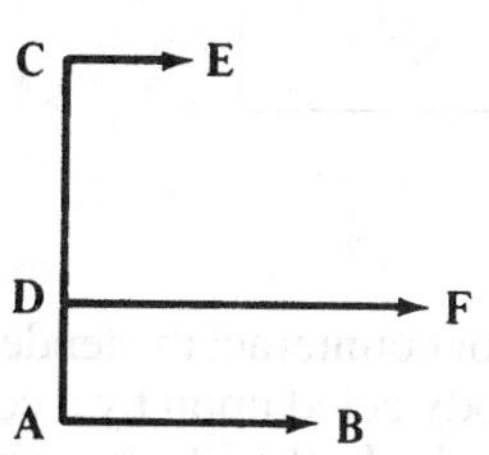

Fig. 6. 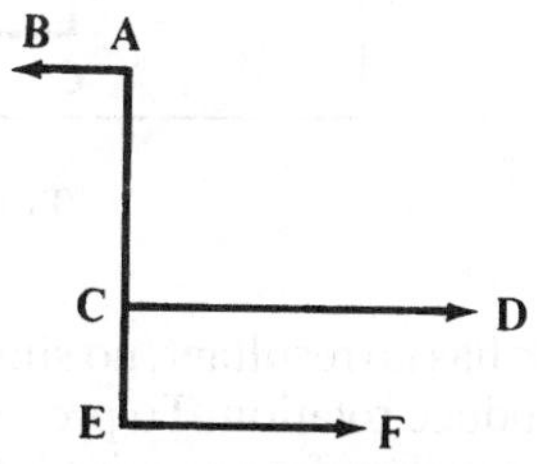 Fig. 7.

Polygon of Forces: When several forces are applied at a point and act in a single plane, Fig. 8, their resultant may be found more simply than by the method just described, as follows: From the extreme end of the line representing the first force, draw a line representing the second force, parallel to it and of the same length and in the direction of the second force. Then through the extreme end of this line draw a line parallel to, and of the same length and direction as the third force, and continue this until all the forces have been thus represented. Then draw a line from the point of application of the forces (as A) to the extreme point (as 5_1) of the line last drawn. This line ($A\ 5_1$) is the resultant of the forces.

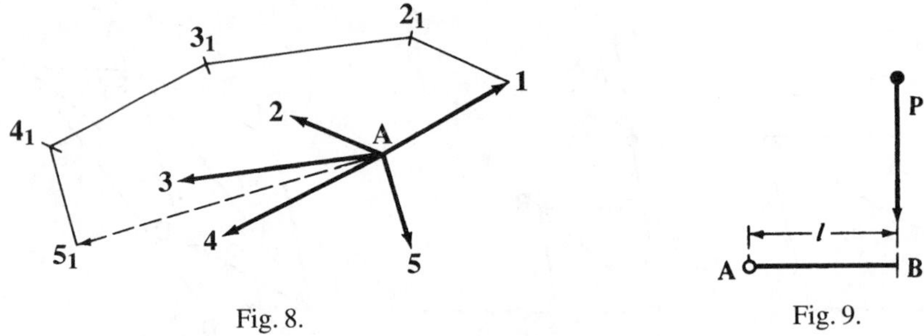

Fig. 8. Fig. 9.

Moment of a Force: The moment of a force with respect to a point is the product of the force multiplied by the perpendicular distance from the given point to the direction of the force. In Fig. 9, the moment of the force P with relation to point A is $P \times AB$. The perpendicular distance AB is called the lever-arm of the force. The moment is the measure of the tendency of the force to produce rotation about the given point, which is termed the center of moments. If the force is measured in pounds and the distance in inches, the moment is expressed in inch-pounds. **In metric SI units, the moment is expressed in newton-meters (N · m), or newton-millimeters (N · mm).**

The moment of the resultant of any number of forces acting together in the same plane is equal to the algebraic sum of the moments of the separate forces.

Couples.—If the forces AB and CD are equal and parallel but act in opposite directions, then the resultant equals 0, or, in other words, the two forces have no resultant and are called a couple. A couple tends to produce rotation. The measure of this tendency is called the moment of the couple and is the product of one of the forces multiplied by the distance between the two.

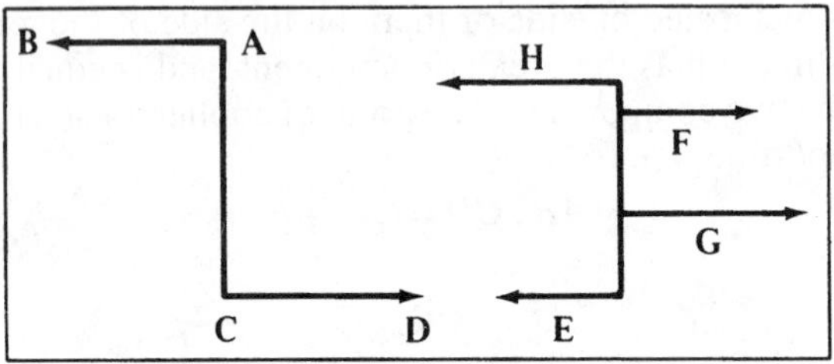

Two Examples of Couples

As a couple has no resultant, no single force can balance or counteract the tendency of the couple to produce rotation. To prevent the rotation of a body acted upon by a couple, two other forces are therefore required, forming a second couple. In the illustration, E and F form one couple and G and H are the balancing couple. The body on which they act is in equilibrium if the moments of the two couples are equal and tend to rotate the body in opposite directions. A couple may also be represented by a vector in the direction of the axis about which the couple acts. The length of the vector, to some scale, represents the magnitude of the couple, and the direction of the vector is that in which a right-hand screw would advance if it were to be rotated by the couple.

Composition of a Single Force and Couple.—A single force and a couple in the same plane or in parallel planes may be replaced by another single force equal and parallel to the first force, at a distance from it equal to the moment of the couple divided by the magnitude of the force. The new single force is located so that the moment of the resultant about the point of application of the original force is of the same sign as the moment of the couple.

In the next figure, with the couple $N-N$ in the position shown, the resultant of $P, -N$, and N is O (which equals P) acting on a line through point c so that $(P-N) \times ac = N \times bc$.

Thus, it follows that,

$$ac = \frac{N(ac + bc)}{P} = \frac{\text{Moment of Couple}}{P}$$

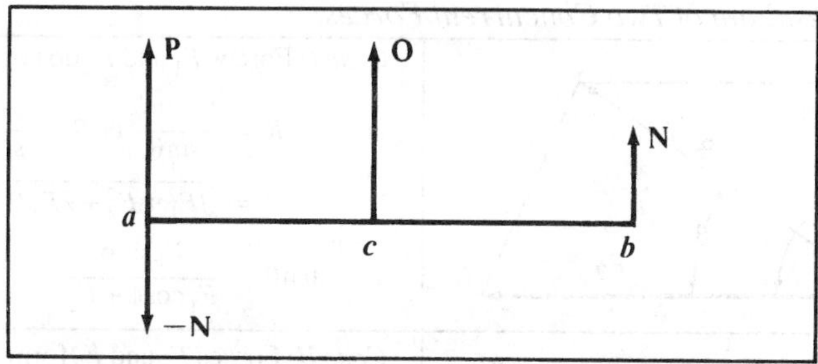

Single Force and Couple Composition

Algebraic Composition and Resolution of Force Systems.—The graphical methods given beginning on page 118 are convenient for solving problems involving force systems in which all of the forces lie in the same plane and only a few forces are involved. If many forces are involved, however, or the forces do not lie in the same plane, it is better to use algebraic methods to avoid complicated space diagrams. Systematic procedures for solving force problems by algebraic methods are outlined beginning on page 121. In connection with the use of these procedures, it is necessary to define several terms applicable to force systems in general.

The single force which produces the same effect upon a body as two or more forces acting together is called their *resultant*. The separate forces which can be so combined are called the *components*. Finding the resultant of two or more forces is called the *composition of forces*, and finding two or more components of a given force, the *resolution of forces*. Forces are said to be *concurrent* when their lines of action can be extended to meet at a common point; forces that are *parallel* are, of course, *nonconcurrent*. Two forces having the same line of action are said to be *collinear*. Two forces equal in magnitude, parallel, and in opposite directions constitute a *couple*. Forces all in the same plane are said to be *coplanar;* if not in the same plane, they are called *noncoplanar* forces.

The *resultant* of a system of forces is the simplest equivalent system that can be determined. It may be a single force, a couple, or a noncoplanar force and a couple. This last type of resultant, a noncoplanar force and a couple, may be replaced, if desired, by two *skewed* forces (forces that are nonconcurrent, nonparallel, and noncoplanar). When the resultant of a system of forces is zero, the system is in equilibrium, that is, the body on which the force system acts remains at rest or continues to move with uniform velocity.

Algebraic Solution of Force Systems—All Forces in the Same Plane

Finding Two Concurrent Components of a Single Force:

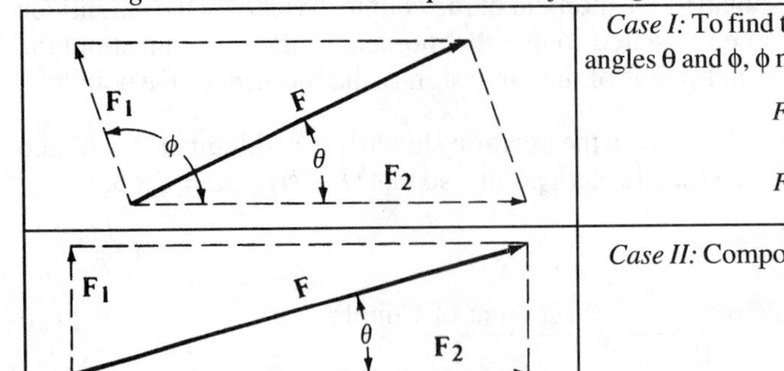

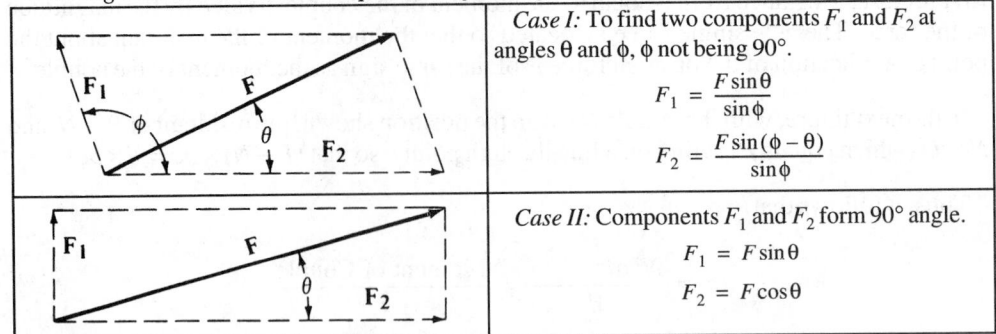

Case I: To find two components F_1 and F_2 at angles θ and φ, φ not being 90°.

$$F_1 = \frac{F \sin\theta}{\sin\phi}$$

$$F_2 = \frac{F \sin(\phi - \theta)}{\sin\phi}$$

Case II: Components F_1 and F_2 form 90° angle.

$$F_1 = F \sin\theta$$

$$F_2 = F \cos\theta$$

Finding the Resultant of Two Concurrent Forces:

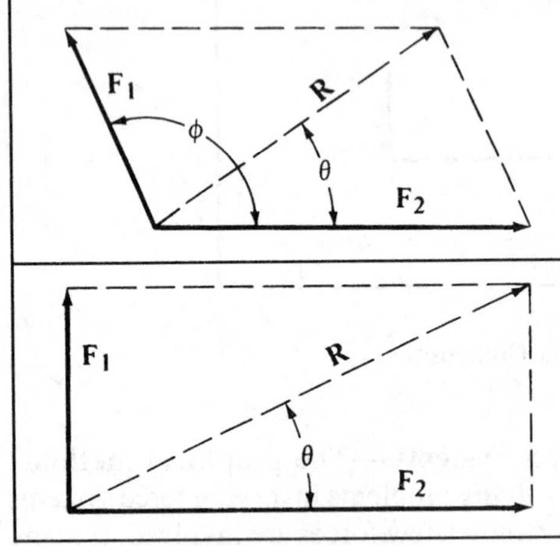

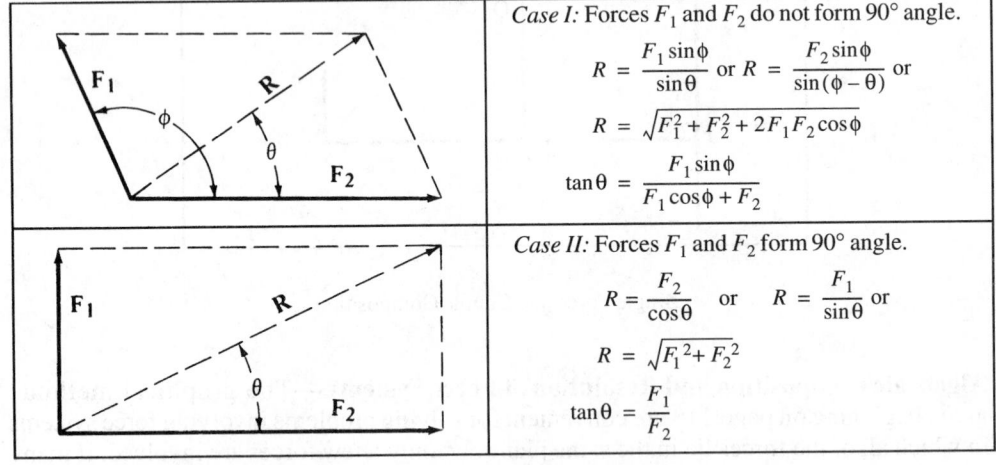

Case I: Forces F_1 and F_2 do not form 90° angle.

$$R = \frac{F_1 \sin\phi}{\sin\theta} \text{ or } R = \frac{F_2 \sin\phi}{\sin(\phi - \theta)} \text{ or}$$

$$R = \sqrt{F_1^2 + F_2^2 + 2F_1F_2\cos\phi}$$

$$\tan\theta = \frac{F_1 \sin\phi}{F_1 \cos\phi + F_2}$$

Case II: Forces F_1 and F_2 form 90° angle.

$$R = \frac{F_2}{\cos\theta} \quad \text{or} \quad R = \frac{F_1}{\sin\theta} \text{ or}$$

$$R = \sqrt{F_1^2 + F_2^2}$$

$$\tan\theta = \frac{F_1}{F_2}$$

Finding the Resultant of Three or More Concurrent Forces:

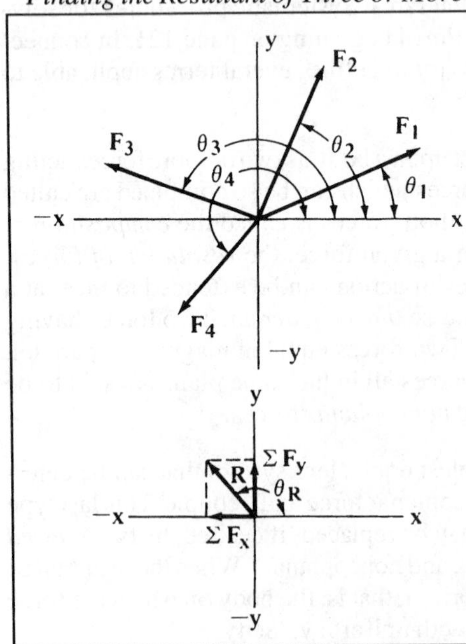

To determine resultant of forces F_1, F_2, F_3, etc. making angles, respectively, of θ_1, θ_2, θ_3, etc. with the *x* axis, find the *x* and *y* components F_x and F_y of each force and arrange in a table similar to that shown below for a system of three forces. Find the algebraic sum of the F_x and F_y components (ΣF_x and ΣF_y) and use these to determine resultant *R*.

Force	F_x	F_y
F_1	$F_1 \cos\theta_1$	$F_1 \sin\theta_1$
F_2	$F_2 \cos\theta_2$	$F_2 \sin\theta_2$
F_3	$F_3 \cos\theta_3$	$F_3 \sin\theta_3$
	$\Sigma\Phi_\xi$	$\Sigma\Phi_\psi$

$$R = \sqrt{(\Sigma F_x)^2 + (\Sigma F_y)^2}$$

$$\cos\theta_R = \frac{\Sigma F_x}{R}$$

$$\text{or } \tan\theta_R = \frac{\Sigma F_y}{\Sigma F_x}$$

Finding a Force and a Couple Which Together are Equivalent to a Single Force:

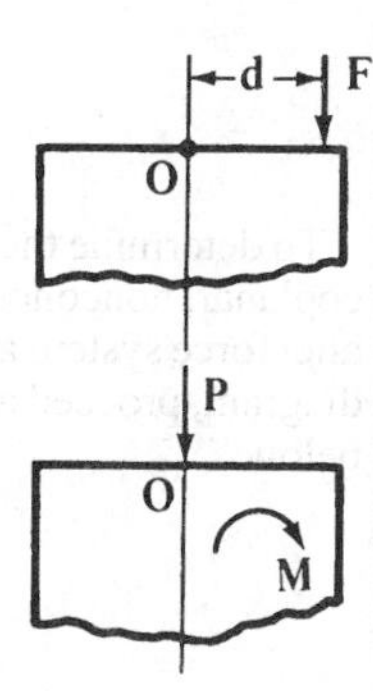

To resolve a single force F into a couple of moment M and a force P passing through any chosen point O at a distance d from the original force F, use the relations

$$P = F$$
$$M = F \times d$$

The moment M must, of course, tend to produce rotation about O in the same direction as the original force. Thus, as seen in the diagram, F tends to produce clockwise rotation; hence M is shown clockwise.

Finding the Resultant of a Single Force and a Couple:

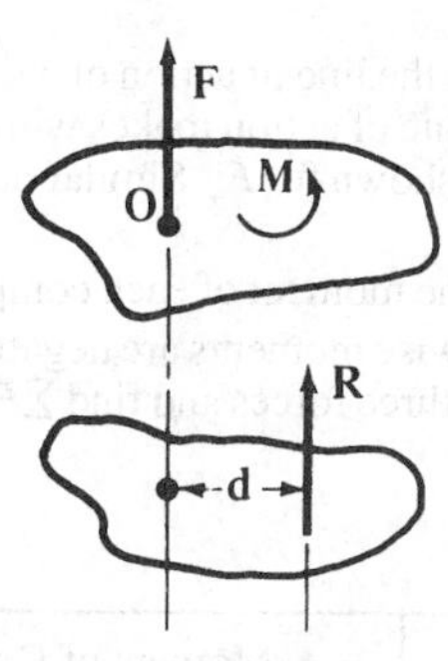

The resultant of a single force F and a couple M is a single force R equal in magnitude and direction to F and parallel to it at a distance d to the left or right of F.

$$R = F$$
$$d = M \div R$$

Resultant R is placed to the left or right of point of application O of the original force F depending on which position will give R the same direction of moment about O as the original couple M.

Finding the Resultant of a System of Parallel Forces:

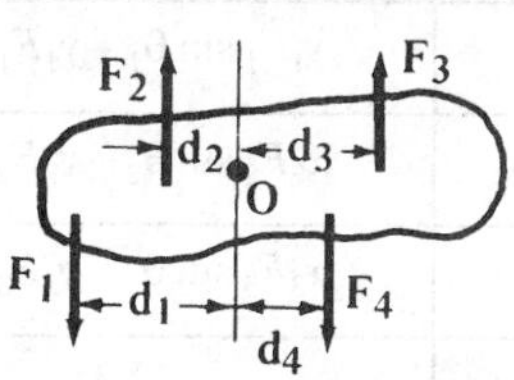

To find the resultant of a system of coplanar parallel forces, proceed as indicated below.

1) Select any convenient point O from which perpendicular distances d_1, d_2, d_3, etc. to parallel forces F_1, F_2, F_3, etc. can be specified or calculated.

2) Find the algebraic sum of all the forces; this will give the magnitude of the resultant of the system.

$$R = \Sigma F = F_1 + F_2 + F_3 + \ldots$$

3) Find the algebraic sum of the moments of the forces about O; clockwise moments may be taken as negative and counterclockwise moments as positive:

$$\Sigma M_O = F_1 d_1 + F_2 d_2 + \ldots$$

4) Calculate the distance d from O to the line of action of resultant R:

$$d = \Sigma M_O \div R$$

This distance is measured to the left or right from O depending on which position will give the moment of R the same direction of rotation about O as the couple ΣM_O, that is, if ΣM_O is negative, then d is left or right of O depending on which direction will make $R \times d$ negative.

Note Concerning Interpretation of Results: If $R = 0$, then the resultant of the system is a couple ΣM_O; if $\Sigma M_O = 0$ then the resultant is a single force R; if both R and $\Sigma M_O = 0$, then the system is in equilibrium.

Finding the Resultant of Forces Not Intersecting at a Common Point:

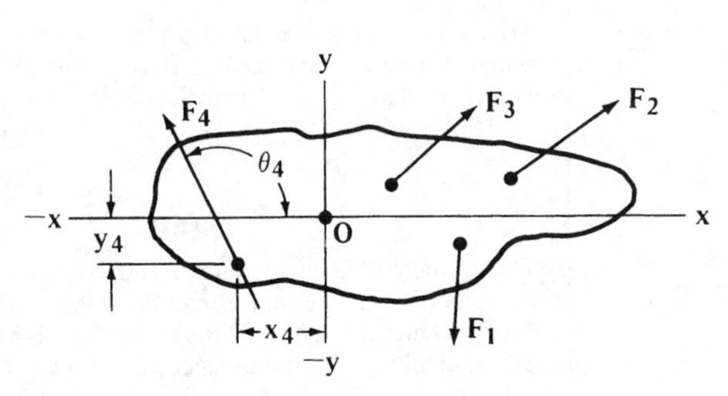

To determine the resultant of a coplanar, nonconcurrent, nonparallel force system as shown in the diagram, proceed as shown below.

1) Draw a set of x and y coordinate axes through any convenient point O in the plane of the forces as shown in the diagram.

2) Determine the x and y coordinates of any convenient point on the line of action of each force and the angle θ, measured in a counterclockwise direction, that each line of action makes with the positive x axis. For example, in the diagram, coordinates x_4, y_4, and θ_4 are shown for F_4. Similar data should be known for each of the forces of the system.

3) Calculate the x and y components (F_x, F_y) of each force and the moment of each component about O. Counterclockwise moments are considered positive and clockwise moments are negative. Tabulate all results in a manner similar to that shown below for a system of three forces and find ΣF_x, ΣF_y, ΣM_O by algebraic addition.

Force	Coordinates of F			Components of F		Moment of F about O
F	x	y	θ	F_x	F_y	$M_O = xF_y - yF_x$
F_1	x_1	y_1	θ_1	$F_1 \cos \theta_1$	$F_1 \sin \theta_1$	$x_1F_1 \sin \theta_1 - y_1F_1 \cos \theta_1$
F_2	x_2	y_2	θ_2	$F_2 \cos \theta_2$	$F_2 \sin \theta_2$	$x_2F_2 \sin \theta_2 - y_2F_2 \cos \theta_2$
F_3	x_3	y_3	θ_3	$F_3 \cos \theta_3$	$F_3 \sin \theta_3$	$x_3F_3 \sin \theta_3 - y_3F_3 \cos \theta_3$
				ΣF_x	ΣF_y	ΣM_O

4. Compute the resultant of the system and the angle θ_R it makes with the x axis by using the formulas:

$$R = \sqrt{(\Sigma F_x)^2 + (\Sigma F_y)^2}$$

$$\cos\theta_R = \Sigma F_x \div R \text{ or } \tan\theta_R = \Sigma F_y \div \Sigma F_x$$

5. Calculate the distance d from O to the line of action of the resultant R:

$$d = \Sigma M_O \div R$$

Distance d is in such direction from O as will make the moment of R about O have the same sign as ΣM_O.

Note Concerning Interpretation of Results: If $R = 0$, then the resultant is a couple ΣM_O; if $\Sigma M_O = 0$, then R passes through O; if both $R = 0$ and $\Sigma M_O = 0$, then the system is in equilibrium.

Example: Find the resultant of three coplanar nonconcurrent forces for which the following data are given.

$$F_1 = 10 \text{ lbs; } x_1 = 5 \text{ in.; } y_1 = -1 \text{ in.; } \theta_1 = 270°$$

$$F_2 = 20 \text{ lbs; } x_2 = 4 \text{ in.; } y_2 = 1.5 \text{ in.; } \theta_2 = 50°$$

$$F_3 = 30 \text{ lbs; } x_3 = 2 \text{ in.; } y_3 = 2 \text{ in.; } \theta_3 = 60°$$

$$F_{x_1} = 10\cos 270° = 10 \times 0 = 0 \text{ lbs.}$$

$$F_{x_2} = 20\cos 50° = 20 \times 0.64279 = 12.86 \text{ lbs.}$$

$$F_{x_3} = 30\cos 60° = 30 \times 0.5000 = 15.00 \text{ lbs.}$$

$$F_{y_1} = 10 \times \sin 270° = 10 \times (-1) = -10.00 \text{ lbs.}$$

$$F_{y_2} = 20 \times \sin 50° = 20 \times 0.76604 = 15.32 \text{ lbs.}$$

$$F_{y_3} = 30 \times \sin 60° = 30 \times 0.86603 = 25.98 \text{ lbs.}$$

$$M_{o_1} = 5 \times (-10) - (-1) \times 0 = -50 \text{ in. lbs.}$$

$$M_{o_2} = 4 \times 15.32 - 1.5 \times 12.86 = 41.99 \text{ in. lbs.}$$

$$M_{o_3} = 2 \times 25.98 - 2 \times 15 = 21.96 \text{ in. lbs.}$$

***Note:* When working in metric SI units, pounds are replaced by newtons (N); inches by meters or millimeters, and inch-pounds by newton-meters (N · m) or newton-millimeters (N · mm).**

Force F	Coordinates of F			Components of F		Moment of F about O
	x	y	θ	F_x	F_y	
$F_1 = 10$	5	−1	270°	0	−10.00	−50.00
$F_2 = 20$	4	1.5	50°	12.86	15.32	41.99
$F_3 = 30$	2	2	60°	15.00	25.98	21.96
				27.86	31.30	13.95

$$R = \sqrt{(27.86)^2 + (31.30)^2}$$

$$= 41.90 \text{ lbs.}$$

$$\tan\theta_R = \frac{31.30}{27.86} = 1.1235$$

$$\theta_R = 48°20'$$

$$d = \frac{13.95}{41.90} = 0.33 \text{ inches}$$

measured as shown on the diagram.

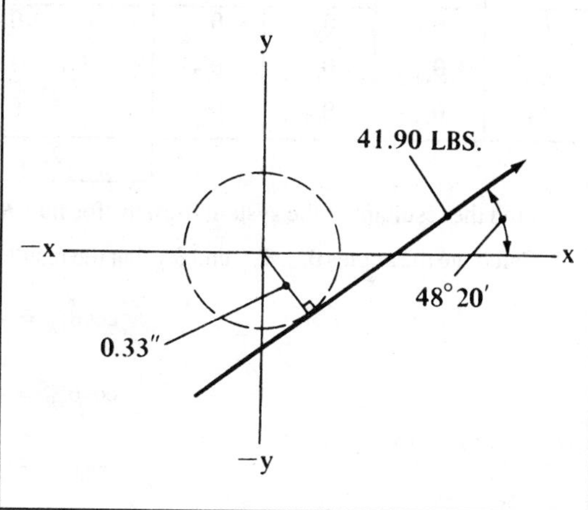

Algebraic Solution of Force Systems — Forces Not in Same Plane

Resolving a Single Force Into Its Three Rectangular Components:

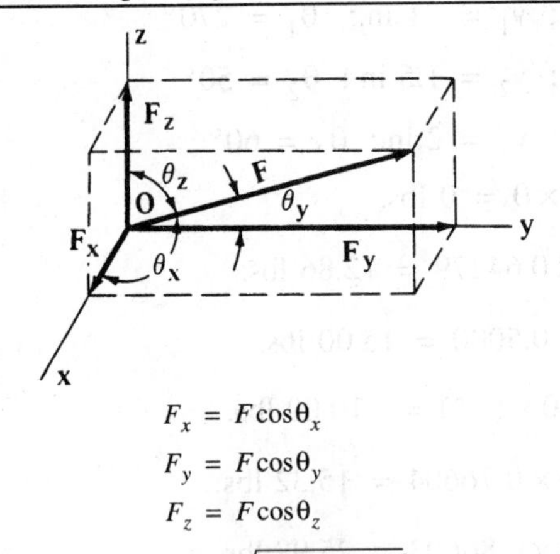

$$F_x = F\cos\theta_x$$
$$F_y = F\cos\theta_y$$
$$F_z = F\cos\theta_z$$
$$F = \sqrt{F_x^2 + F_y^2 + F_z^2}$$

The diagram shows how a force F may be resolved at any point O on its line of action into three concurrent components each of which is perpendicular to the other two.

The x, y, z components F_x, F_y, F_z of force F are determined from the accompanying relations in which θ_x, θ_y, θ_z are the angles that the force F makes with the x, y, z axes.

Finding the Resultant of Any Number of Concurrent Forces:

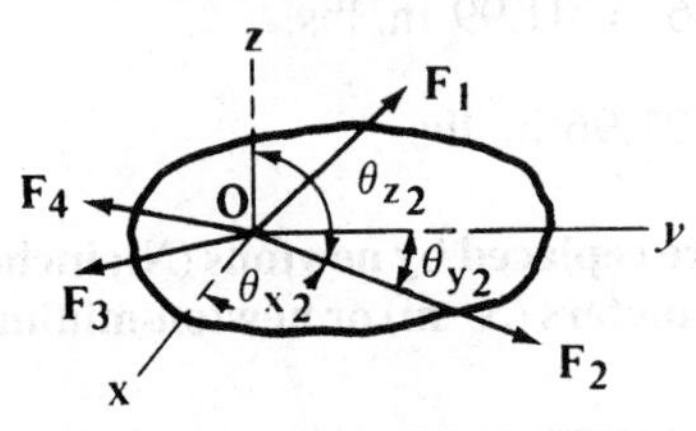

To find the resultant of any number of noncoplanar concurrent forces F_1, F_2, F_3, etc., use the procedure outlined below.

1) Draw a set of x, y, z axes at O, the point of concurrency of the forces. The angles each force makes measured counterclockwise from the positive x, y, and z coordinate axes must be known in addition to the magnitudes of the forces. For force F_2, for example, the angles are θ_{x2}, θ_{y2}, θ_{z2} as indicated on the diagram.

2) Apply the first three formulas given under the heading "Resolving a Single Force Into Its Three Rectangular Components" to each force to find its x, y, and z components. Tabulate these calculations as shown below for a system of three forces. Algebraically add the calculated components to find ΣF_x, ΣF_y, and ΣF_z which are the components of the resultant.

Force	Angles			Components of Forces		
F	θ_x	θ_y	θ_z	F_x	F_y	F_z
F_1	θ_{x1}	θ_{y1}	θ_{z1}	$F_1 \cos\theta_{x1}$	$F_1 \cos\theta_{y1}$	$F_1 \cos\theta_{z1}$
F_2	θ_{x2}	θ_{y2}	θ_{z2}	$F_2 \cos\theta_{x2}$	$F_2 \cos\theta_{y2}$	$F_2 \cos\theta_{z2}$
F_3	θ_{x3}	θ_{y3}	θ_{z3}	$F_3 \cos\theta_{x3}$	$F_3 \cos\theta_{y3}$	$F_3 \cos\theta_{z3}$
				ΣF_x	ΣF_y	ΣF_z

3. Find the resultant of the system from the formula $R = \sqrt{(\Sigma F_x)^2 + (\Sigma F_y)^2 + (\Sigma F_z)^2}$

4. Calculate the angles θ_{xR}, θ_{yR}, and θ_{zR} that the resultant R makes with the respective coordinate axes:

$$\cos\theta_{xR} = \frac{\Sigma F_x}{R}$$
$$\cos\theta_{yR} = \frac{\Sigma F_y}{R}$$
$$\cos\theta_{zR} = \frac{\Sigma F_z}{R}$$

Finding the Resultant of Parallel Forces Not in the Same Plane:

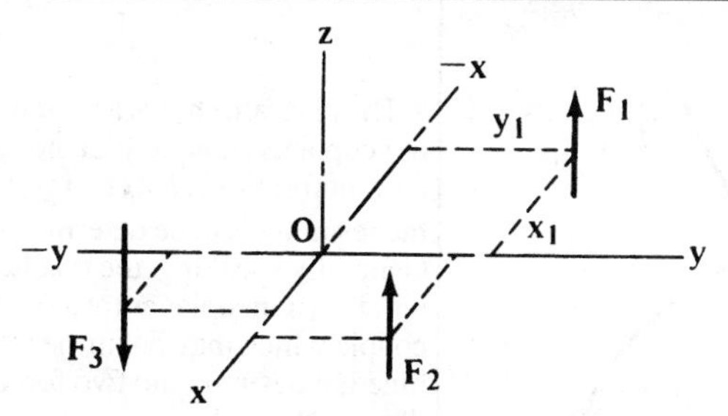

In the diagram, forces F_1, F_2, etc. represent a system of noncoplanar parallel forces. To find the resultant of such systems, use the procedure shown below.

1) Draw a set of *x*, *y*, and *z* coordinate axes through any point *O* in such a way that one of these axes, say the *z* axis, is parallel to the lines of action of the forces. The *x* and *y* axes then will be perpendicular to the forces.

2) Set the distances of each force from the *x* and *y* axes in a table as shown below. For example, x_1 and y_1 are the *x* and *y* distances for F_1 shown in the diagram.

3) Calculate the moment of each force about the *x* and *y* axes and set the results in the table as shown for a system consisting of three forces. The algebraic sums of the moments ΣM_x and ΣM_y are then obtained. (In taking moments about the *x* and *y* axes, assign counterclockwise moments a plus (+) sign and clockwise moments a minus (-) sign. In deciding whether a moment is counterclockwise or clockwise, look from the positive side of the axis in question toward the negative side.)

Force	Coordinates of Force F		Moments M_x and M_y due to F	
F	x	y	M_x	M_y
F_1	x_1	y_1	F_1y_1	F_1x_1
F_2	x_2	y_2	F_2y_2	F_2x_2
F_3	x_3	y_3	F_3y_3	F_3x_3
ΣF			ΣM_x	ΣM_y

4. Find the algebraic sum ΣF of all the forces; this will be the resultant R of the system.

$$R = \Sigma F = F_1 + F_2 + \ldots$$

5. Calculate x_R and y_R, the moment arms of the resultant:

$$x_R = \Sigma M_y \div R$$

$$y_R = \Sigma M_x \div R$$

These moment arms are measured in such direction along the *x* and *y* axes as will give the resultant a moment of the same direction of rotation as ΣM_x and ΣM_y.

Note Concerning Interpretation of Results: If ΣM_x and ΣM_y are both 0, then the resultant is a single force R along the *z* axis; if R is also 0, then the system is in equilibrium. If R is 0 but ΣM_x and ΣM_y are not both 0, then the resultant is a couple

$$M_R = \sqrt{(\Sigma M_x)^2 + (\Sigma M_y)^2}$$

that lies in a plane parallel to the *z* axis and making an angle θ_R measured in a counterclockwise direction from the positive *x* axis and calculated from the following formula:

$$\sin\theta_R = \frac{\Sigma M_x}{M_R}$$

Finding the Resultant of Nonparallel Forces Not Meeting at a Common Point:

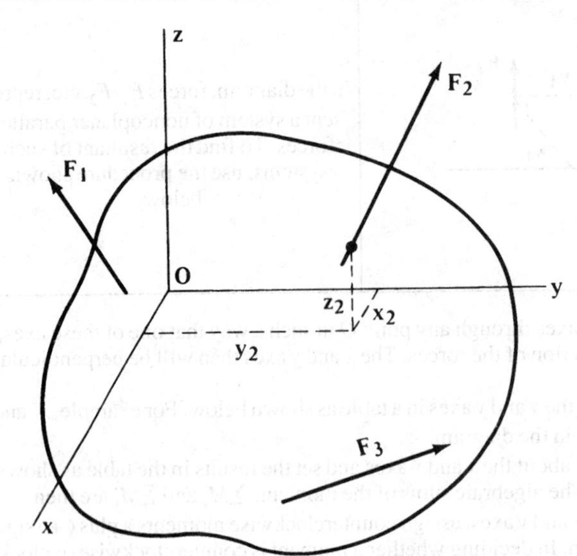

The diagram shows a system of noncoplanar, nonparallel, noncon-current forces F_1, F_2, etc. for which the resultant is to be determined. Generally speaking, the resultant will be a noncoplanar force and a couple which may be further com-bined, if desired, into two forces that are skewed.

This is the most general force sys-tem that can be devised, so each of the other systems so far described represents a special, simpler case of this general force system. The method of solution described below for a system of three forces applies for any number of forces.

1) Select a set of coordinate x, y, and z axes at any desired point O in the body as shown in the diagram.

2) Determine the x, y, and z coordinates of any convenient point on the line of action of each force as shown for F_2. Also determine the angles, θ_x, θ_y, θ_z that each force makes with each coordinate axis. These angles are measured counterclockwise from the positive direction of the x, y, and z axes. The data is tabulated, as shown in the table accompanying Step 3, for convenient use in subsequent calculations.

3) Calculate the x, y, and z components of each force using the formulas given in the accompanying table. Add these components algebraically to get ΣF_x, ΣF_y and ΣF_z which are the components of the resultant, R, given by the formula,

$$R = \sqrt{(\Sigma F_x)^2 + (\Sigma F_y)^2 + (\Sigma F_z)^2}$$

Force	Coordinates of Force F						Components of F		
F	x	y	z	θ_x	θ_y	θ_z	F_x	F_y	F_z
F_1	x_1	y_1	z_1	θ_{x1}	θ_{y1}	θ_{z1}	$F_1 \cos\theta_{x1}$	$F_1 \cos\theta_{y1}$	$F_1 \cos\theta_{z1}$
F_2	x_2	y_2	z_2	θ_{x2}	θ_{y2}	θ_{z2}	$F_2 \cos\theta_{x2}$	$F_2 \cos\theta_{y2}$	$F_2 \cos\theta_{z2}$
F_3	x_3	y_3	z_3	θ_{x3}	θ_{y3}	θ_{z3}	$F_3 \cos\theta_{x3}$	$F_3 \cos\theta_{y3}$	$F_3 \cos\theta_{z3}$
							ΣF_x	ΣF_y	ΣF_z

The resultant force R makes angles of θ_{xR}, θ_{yR}, and θ_{zR} with the x, y, and z axes, respectively, and passes through the selected point O. These angles are determined from the formulas,

$$\cos\theta_{xR} = \Sigma F_x \div R$$

$$\cos\theta_{yR} = \Sigma F_y \div R$$

$$\cos\theta_{zR} = \Sigma F_z \div R$$

4. Calculate the moments M_x, M_y, M_z about x, y, and z axes, respectively, due to the F_x, F_y, and F_z components of each force and set them in tabular form. The formulas to use are given in the accompanying table.

In interpreting moments about the x, y, and z axes, consider counterclockwise moments a plus (+) sign and clockwise moments a minus (-) sign. In deciding whether a moment is counterclockwise or clockwise, look from the positive side of the axis in question toward the negative side.

Force	Moments of Components of $F\ (F_x, F_y, F_z)$ about x, y, z axes		
F	$M_x = yF_z - zF_y$	$M_y = zF_x - xF_z$	$M_z = xF_y - yF_x$
F_1	$M_{x1} = y_1F_{z1} - z_1F_{y1}$	$M_{y1} = z_1F_{x1} - x_1F_{z1}$	$M_{z1} = x_1F_{y1} - y_1F_{x1}$
F_2	$M_{x2} = y_2F_{z2} - z_2F_{y2}$	$M_{y2} = z_2F_{x2} - x_2F_{z2}$	$M_{z2} = x_2F_{y2} - y_2F_{x2}$
F_3	$M_{x3} = y_3F_{z3} - z_3F_{y3}$	$M_{y3} = z_3F_{x3} - x_3F_{z3}$	$M_{z3} = x_3F_{y3} - y_3F_{x3}$
	ΣM_x	ΣM_y	ΣM_z

5. Add the component moments algebraically to get ΣM_x, ΣM_y and ΣM_z which are the components of the resultant couple, M, given by the formula,

$$M = \sqrt{(\Sigma M_x)^2 + (\Sigma M_y)^2 + (\Sigma M_z)^2}$$

The resultant couple M will tend to produce rotation about an axis making angles of β_x, β_y, and β_z with the x, y, z axes, respectively. These angles are determined from the formulas,

$$\cos\beta_x = \frac{\Sigma M_x}{M} \qquad \cos\beta_y = \frac{\Sigma M_y}{M} \qquad \cos\beta_z = \frac{\Sigma M_z}{M}$$

General Method of Locating Resultant When Its Components are Known:

To determine the position of the resultant force of a system of forces, proceed as follows:

From the origin, point O, of a set of coordinate axes x, y, z, lay off on the x axis a length A representing the algebraic sum ΣF_x of the x components of all the forces. From the end of line A lay off a line B representing ΣF_y, the algebraic sum of the y components; this line B is drawn in a direction parallel to the y axis. From the end of line B lay off a line C representing ΣF_z. Finally, draw a line R from O to the end of C; R will be the resultant of the system.

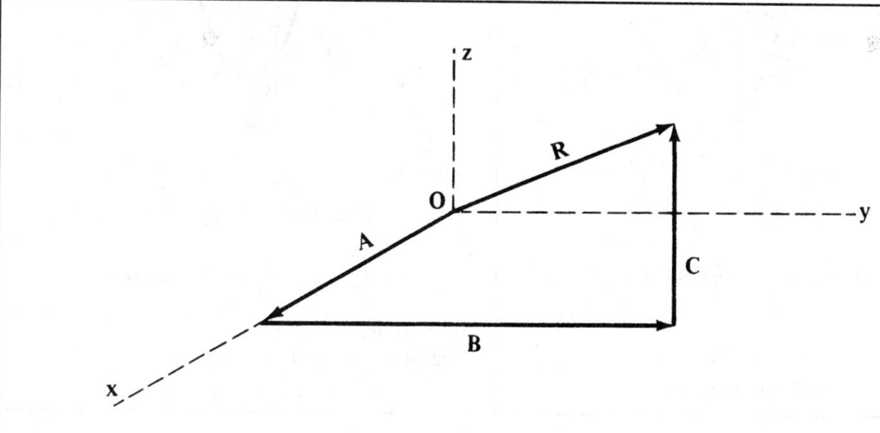

Mechanisms

Inclined Plane—Wedge

W = weight of body

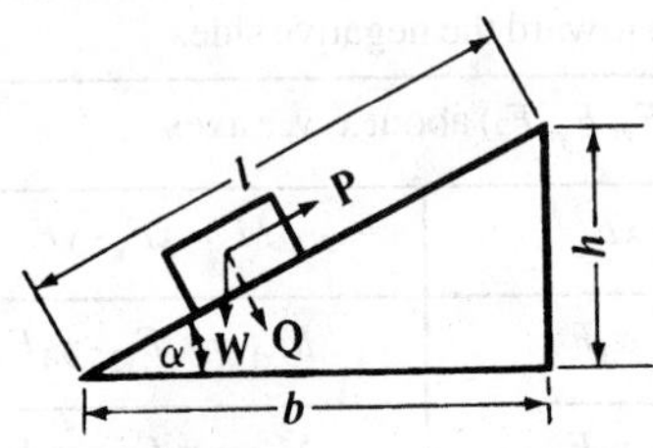

Neglecting friction:

$$P = W \times \frac{h}{l} = W \times \sin\alpha$$

$$W = P \times \frac{l}{h} = \frac{P}{\sin\alpha} = P \times \operatorname{cosec}\alpha$$

$$Q = W \times \frac{b}{l} = W \times \cos\alpha$$

If friction is taken into account, then force P to pull body up is:

$$P = W(\mu\cos\alpha + \sin\alpha)$$

Force P_1 to pull body down is:

$$P_1 = W(\mu\cos\alpha - \sin\alpha)$$

Force P_2 to hold body stationary:

$$P_2 = W(\sin\alpha - \mu\cos\alpha)$$

in which μ is the coefficient of friction.

W = weight of body

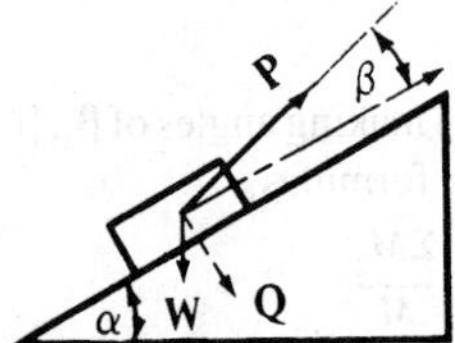

Neglecting friction:

$$P = W \times \frac{\sin\alpha}{\cos\beta}$$

$$W = P \times \frac{\cos\beta}{\sin\alpha}$$

$$Q = W \times \frac{\cos(\alpha + \beta)}{\cos\beta}$$

With friction:

Coefficient of friction $= \mu = \tan\phi$

$$P = W \times \frac{\sin(\alpha + \phi)}{\cos(\beta - \phi)}$$

W = weight of body

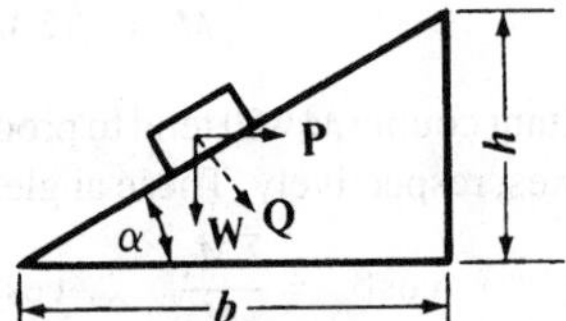

Neglecting friction:

$$P = W \times \frac{h}{b} = W \times \tan\alpha$$

$$W = P \times \frac{b}{h} = P \times \cot\alpha$$

$$Q = \frac{W}{\cos\alpha} = W \times \sec\alpha$$

With friction:

Coefficient of friction $= \mu = \tan\phi$

$$P = W\tan(\alpha + \phi)$$

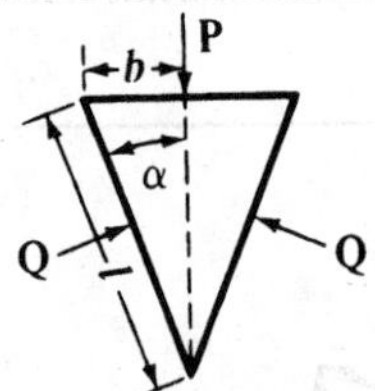

Neglecting friction:

$$P = 2Q \times \frac{b}{l} = 2Q \times \sin\alpha$$

$$Q = P \times \frac{l}{2b} = \frac{1}{2}P \times \operatorname{cosec}\alpha$$

With friction:

Coefficient of friction = μ.

$$P = 2Q(\mu\cos\alpha + \sin\alpha)$$

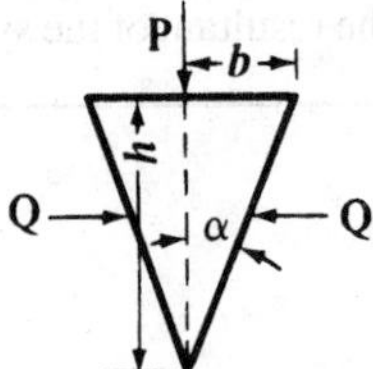

Neglecting friction:

$$P = 2Q \times \frac{b}{h} = 2Q \times \tan\alpha$$

$$Q = P \times \frac{h}{2b} = \frac{1}{2}P \times \cot\alpha$$

With friction:

Coefficient of friction = $\mu = \tan\phi$.

$$P = 2Q\tan(\alpha + \phi)$$

Table of Forces on Inclined Planes

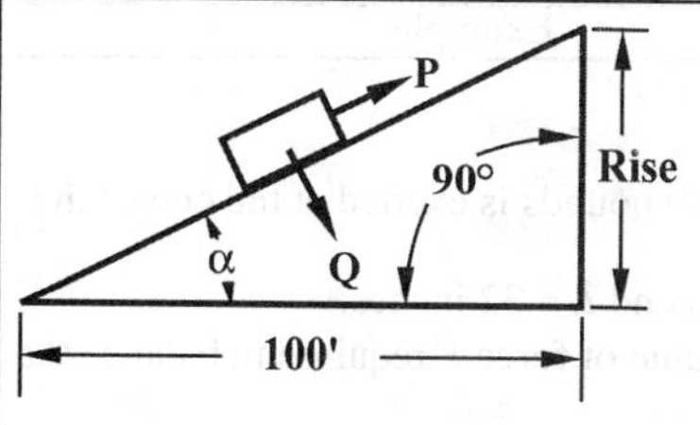

The table below makes it possible to find the force required for moving a body on an inclined plane. The friction on the plane is not taken into account. The column headed "Tension *P* in Cable per Ton of 2000 Pounds" gives the pull in pounds required for moving one ton along the inclined surface. The fourth column gives the perpendicular or normal pressure. If the coefficient of friction is known, the added pull required to overcome friction is thus easily determined:

$$Q \times \text{coefficient of friction} = \text{additional pull required.}$$

Tensions and Pressures in Pounds

Per Cent of Grade. Rise, Ft. per 100 Ft.	Angle α	Tension *P* in Cable per Ton of 2000 Lbs.	Perpendicular Pressure *Q* on Plane per Ton of 2000 Lbs.	Per Cent of Grade. Rise, Ft. Per 100 Ft.	Angle α	Tension *P* in Cable per Ton of 2000 Lbs.	Perpendicular Pressure *Q* on Plane per Ton of 2000 Lbs.
1	0° 35′	20.2	1999.8	39	21° 19′	727.0	1863.0
2	1 9	40.0	1999.4	40	21 49	743.2	1856.6
3	1 44	60.4	1999.0	41	22 18	758.8	1850.4
4	2 18	80.2	1998.2	42	22 47	774.4	1843.8
5	2 52	100.0	1997.4	43	23 17	790.4	1837.0
6	3 27	120.2	1996.2	44	23 45	805.4	1830.6
7	4 1	140.0	1995.0	45	24 14	820.8	1823.6
8	4 35	159.8	1993.6	46	24 43	836.2	1816.6
9	5 9	179.4	1991.8	47	25 11	851.0	1809.8
10	5 43	199.2	1990.0	48	25 39	865.6	1802.8
11	6 17	218.8	1987.8	49	26 7	880.4	1795.6
12	6 51	238.4	1985.6	50	26 34	894.4	1788.8
13	7 25	258.0	1983.2	51	27 2	909.0	1781.4
14	7 59	277.6	1980.6	52	27 29	922.8	1774.2
15	8 32	296.6	1977.8	53	27 56	936.8	1766.8
16	9 6	316.2	1974.8	54	28 23	950.6	1759.4
17	9 39	335.2	1971.6	55	28 49	964.0	1752.2
18	10 13	354.6	1968.2	56	29 15	977.2	1744.8
19	10 46	373.6	1964.6	57	29 41	990.4	1737.4
20	11 19	392.4	1961.0	58	30 7	1003.4	1730.0
21	11 52	411.2	1957.2	59	30 33	1016.4	1722.2
22	12 25	430.0	1953.2	60	30 58	1029.0	1714.8
23	12 58	448.6	1949.0	61	31 23	1041.4	1707.4
24	13 30	466.8	1944.6	62	31 48	1053.8	1699.6
25	14 3	485.4	1940.0	63	32 13	1066.2	1692.0
26	14 35	503.4	1935.4	64	32 38	1078.4	1684.2
27	15 7	521.4	1930.6	65	33 2	1090.2	1676.6
28	15 39	539.4	1925.8	66	33 26	1101.8	1669.0
29	16 11	557.4	1920.6	67	33 50	1113.4	1661.2
30	16 42	574.6	1915.6	68	34 13	1124.6	1653.8
31	17 14	592.4	1910.2	69	34 37	1136.0	1645.8
32	17 45	609.6	1904.6	70	35 0	1147.0	1638.2
33	18 16	626.8	1899.2	71	35 23	1158.0	1630.4
34	18 47	643.8	1893.4	72	35 46	1168.8	1622.8
35	19 18	661.0	1887.6	73	36 8	1179.2	1615.2
36	19 48	677.4	1881.6	74	36 31	1190.0	1607.2
37	20 19	694.4	1875.4	75	36 53	1200.4	1599.6
38	20 49	710.6	1869.4	…	… …	…	…

Levers

Types of Levers	Examples
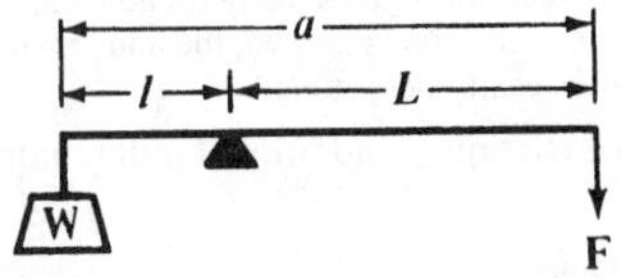 $F:W = l:L$ $F \times L = W \times l$ $F = \frac{W \times l}{L}$ $W = \frac{F \times L}{l}$ $L = \frac{W \times a}{W + F} = \frac{W \times l}{F}$ $l = \frac{F \times a}{W + F} = \frac{F \times L}{W}$	A pull of 80 pounds is exerted at the end of the lever, at *W*; $l = 12$ inches and $L = 32$ inches. Find the value of force *F* required to balance the lever. $F = \frac{80 \times 12}{32} = \frac{960}{32} = 30$ pounds If $F = 20$; $W = 180$; and $l = 3$; how long must *L* be made to secure equilibrium? $L = \frac{180 \times 3}{20} = 27$
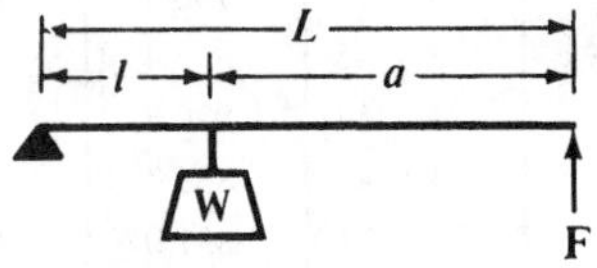 $F:W = l:L$ $F \times L = W \times l$ $F = \frac{W \times l}{L}$ $W = \frac{F \times L}{l}$ $L = \frac{W \times a}{W - F} = \frac{W \times l}{F}$ $l = \frac{F \times a}{W - F} = \frac{F \times L}{W}$	Total length *L* of a lever is 25 inches. A weight of 90 pounds is supported at *W*; *l* is 10 inches. Find the value of *F*. $F = \frac{90 \times 10}{25} = 36$ pounds If $F = 100$ pounds, $W = 2200$ pounds, and $a = 5$ feet, what should *L* equal to secure equilibrium? $L = \frac{2200 \times 5}{2200 - 100} = 5.24$ feet
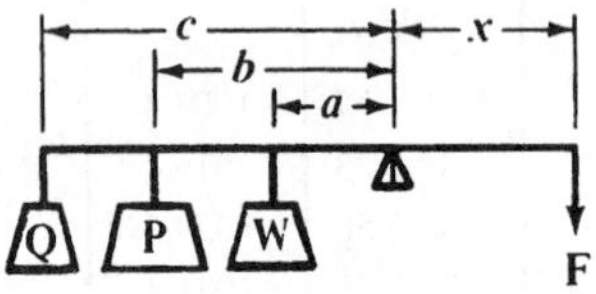 When three or more forces act on lever: $F \times x = W \times a + P \times b + Q \times c$ $x = \frac{W \times a + P \times b + Q \times c}{F}$ $F = \frac{W \times a + P \times b + Q \times c}{x}$	Let $W = 20$, $P = 30$, and $Q = 15$ pounds; $a = 4$, $b = 7$, and $c = 10$ inches. If $x = 6$ inches, find *F*. $F = \frac{20 \times 4 + 30 \times 7 + 15 \times 10}{6} = 73\frac{1}{3}$ lbs Assuming $F = 20$ in the example above, how long must lever arm *x* be made? $x = \frac{20 \times 4 + 30 \times 7 + 15 \times 10}{20} = 22$ ins

The above formulas are valid using metric SI units, with forces expressed in newtons, and lengths in meters. However, it should be noted that the weight of a mass *W* kilograms is equal to a force of *Wg* newtons, where *g* is approximately 9.81 m/s^2. Thus, supposing that in the first example $l = 0.4$ m, $L = 1.2$ m, and $W = 30$ kg, then the weight of *W* is 30*g* newtons, so that the force *F* required to balance the lever is $F = \frac{30g \times 0.4}{1.2} = 10g = 98.1$ **newtons.**

This force could be produced by suspending a mass of 10 kg at *F*.

Toggle-joint.—If arms *ED* and *EH* are of unequal length:

$$P = \frac{Fa}{b}$$

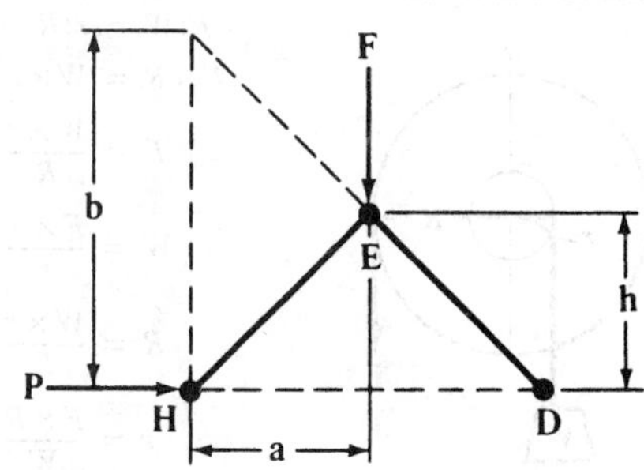

The relation between *P* and *F* changes constantly as *F* moves downward.

If arms *ED* and *EH* are equal:

$$P = \frac{Fa}{2h}$$

A double toggle-joint does not increase the pressure exerted so long as the relative distances moved by *F* and *P* remain the same.

Toggle-joints with Equal Arms

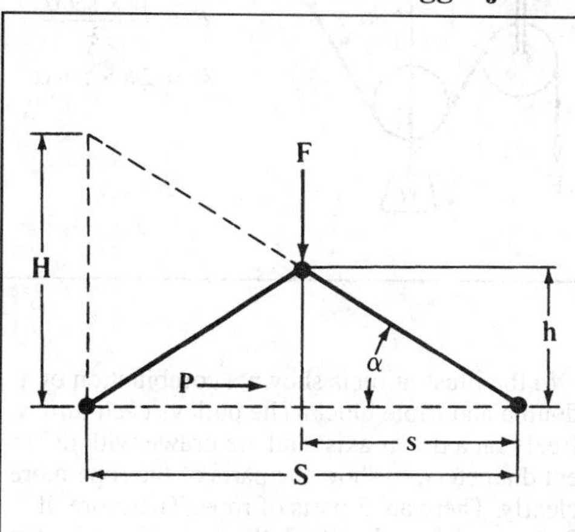

F = force applied
P = resistance
α = given angle

$$2P\sin\alpha = F\cos\alpha$$

$$\frac{P}{F} = \frac{\cos\alpha}{2\sin\alpha} = \text{coefficient}$$

$$P = F \times \text{coefficient}$$

Equivalent expressions (see diagram):

$$P = \frac{FS}{4h} \qquad P = \frac{Fs}{H}$$

To use the table, measure angle α, and find the coefficient in the table corresponding to the angle found. The coefficient is the ratio of the resistance to the force applied, and multiplying the force applied by the coefficient gives the resistance, neglecting friction.

Angle	Coefficient	Angle	Coefficient	Angle	Coefficient	Angle	Coefficient
0° 2′	862	0° 50′	34.4	2° 45′	10.4	8° 0′	3.58
0 4	456	0 55	31.2	2 50	10.1	8 30	3.35
0 6	285	1 0	28.6	3 0	9.54	9 0	3.15
0 8	216	1 10	24.6	3 15	8.81	9 30	2.99
0 10	171	1 15	22.9	3 30	8.17	10 0	2.84
0 12	143	1 20	21.5	3 45	7.63	11 0	2.57
0 14	122	1 30	19.1	4 0	7.25	12 0	2.35
0 15	115	1 40	17.2	4 15	6.73	13 0	2.17
0 16	107	1 45	16.4	4 30	6.35	14 0	2.00
0 18	95.4	1 50	15.6	4 45	6.02	15 0	1.87
0 20	85.8	2 0	14.3	5 0	5.71	16 0	1.74
0 25	68.6	2 10	13.2	5 30	5.19	17 0	1.64
0 30	57.3	2 15	12.7	6 0	4.76	18 0	1.54
0 35	49.1	2 20	12.5	6 30	4.39	19 0	1.45
0 40	42.8	2 30	11.5	7 0	4.07	20 0	1.37
0 45	38.2	2 40	10.7	7 30	3.79	…	…

Wheels and Pulleys

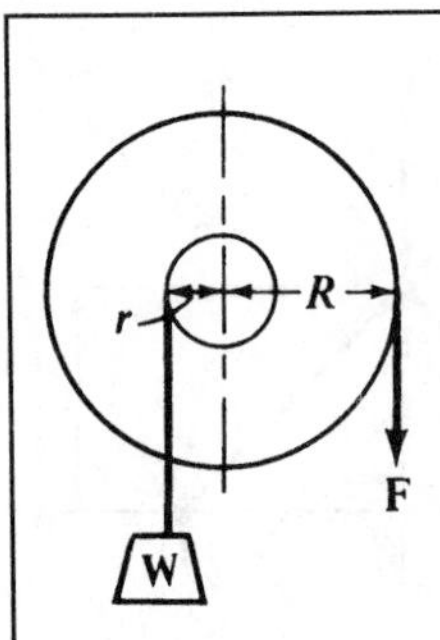

$$F:W = r:R$$
$$F \times R = W \times r$$
$$F = \frac{W \times r}{R}$$
$$W = \frac{F \times R}{r}$$
$$R = \frac{W \times r}{F}$$
$$r = \frac{F \times R}{W}$$

The radius of a drum on which is wound the lifting rope of a windlass is 2 inches. What force will be exerted at the periphery of a gear of 24 inches diameter, mounted on the same shaft as the drum and transmitting power to it, if one ton (2000 pounds) is to be lifted? Here $W = 2000$; $R = 12$; $r = 2$.

$$F = \frac{2000 \times 2}{12} = 333 \text{ pounds}$$

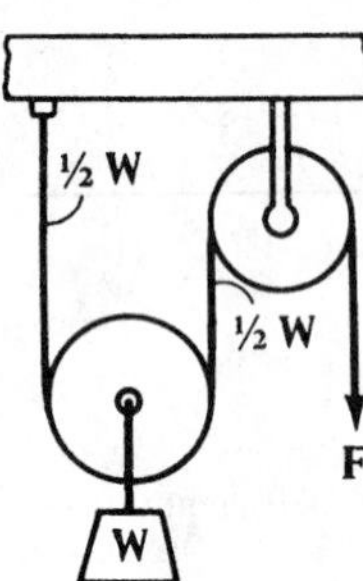

$$F = \tfrac{1}{2}W$$

The velocity with which weight W will be raised equals one-half the velocity of the force applied at F.

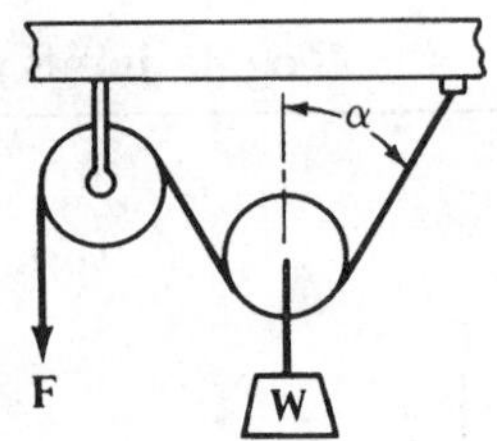

$$F:W = \sec\alpha:2$$
$$F = \frac{W \times \sec\alpha}{2}$$
$$W = 2F \times \cos\alpha$$

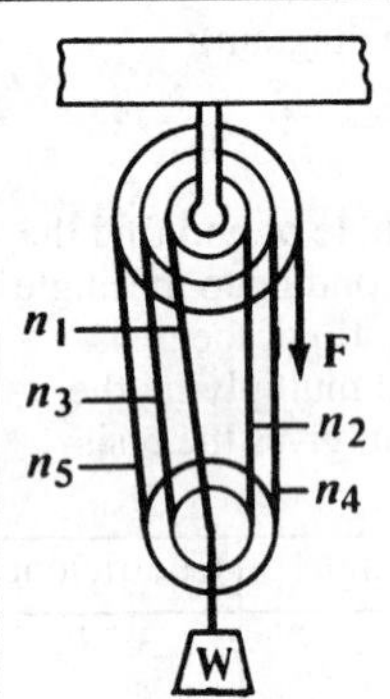

n = number of strands or parts of rope (n_1, n_2, etc.).

$$F = \frac{1}{n} \times W$$

The velocity with which W will be raised equals $\frac{1}{n}$ of the velocity of the force applied at F.

In the illustration is shown a combination of a double and triple block. The pulleys each turn freely on a pin as axis, and are drawn with different diameters, to show the parts of the rope more clearly. There are 5 parts of rope. Therefore, if 200 pounds is to be lifted, the force F required at the end of the rope is:

$$F = \tfrac{1}{5} \times 200 = 40 \text{ pounds}$$

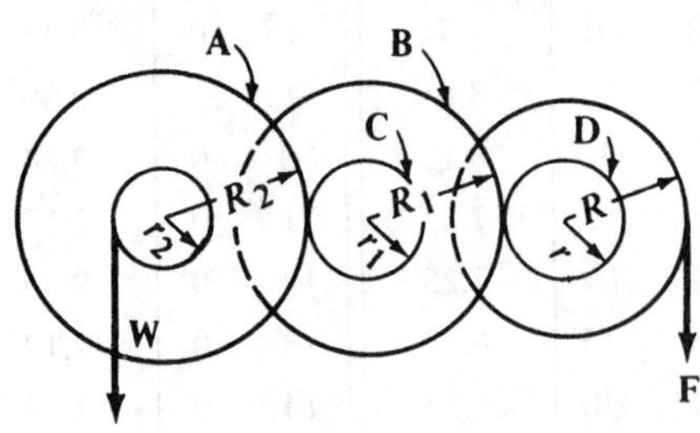

A, B, C and D are the pitch circles of gears.

$$F = \frac{W \times r \times r_1 \times r_2}{R \times R_1 \times R_2}$$

$$W = \frac{F \times R \times R_1 \times R_2}{r \times r_1 \times r_2}$$

Let the pitch diameters of gears A, B, C and D be 30, 28, 12 and 10 inches, respectively. Then $R_2 = 15$; $R_1 = 14$; $r_1 = 6$; and $r = 5$. Let $R = 12$, and $r_2 = 4$. Then the force F required to lift a weight W of 2000 pounds, friction being neglected, is:

$$F = \frac{2000 \times 5 \times 6 \times 4}{12 \times 14 \times 15} = 95 \text{ pounds}$$

Note: The above formulas are valid using metric SI units, with forces expressed in newtons, and lengths in meters or millimeters. (See note on page 132 concerning weight and mass.)

Differential Pulley—Screw

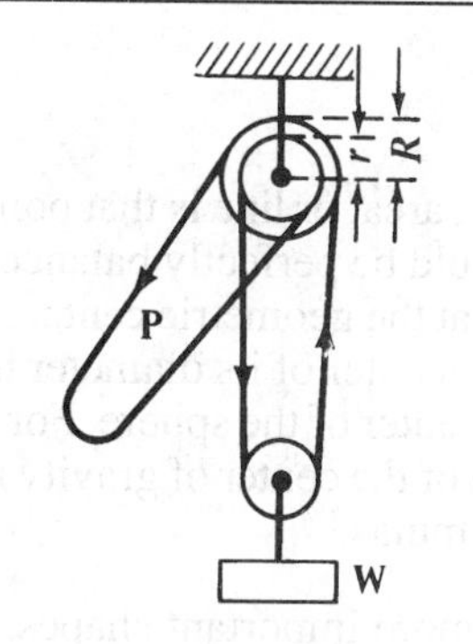	**Differential Pulley.**—In the differential pulley a chain must be used, engaging sprockets, so as to prevent the chain from slipping over the pulley faces. $P \times R = \frac{1}{2}W(R - r)$ $P = \frac{W(R-r)}{2R}$ $W = \frac{2PR}{R-r}$
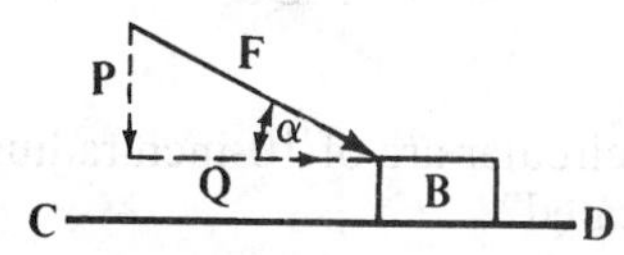	**Force Moving Body on Horizontal Plane.**—*F* tends to move *B* along line *CD*; *Q* is the component which actually moves *B*; *P* is the pressure, due to *F*, of the body on *CD*. $Q = F \times \cos\alpha$ $P = \sqrt{F^2 - Q^2}$
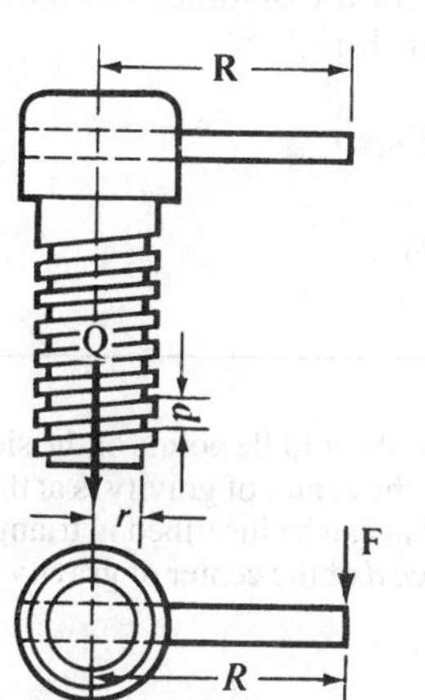	**Screw.**—*F* = force at end of handle or wrench; *R* = lever-arm of *F*; *r* = pitch radius of screw; *p* = lead of thread; *Q* = load. Then, neglecting friction: $F = Q \times \frac{p}{6.2832R}$ $Q = F \times \frac{6.2832R}{p}$ If μ is the coefficient of friction, then: For motion in direction of load *Q* which *assists* it: $F = Q \times \frac{6.2832\mu r - p}{6.2832r + \mu p} \times \frac{r}{R}$ For motion opposite load *Q* which *resists* it: $F = Q \times \frac{p + 6.2832\mu r}{6.2832r - \mu p} \times \frac{r}{R}$

MECHANICAL PROPERTIES OF BODIES

Properties of Bodies

Center of Gravity.—The center of gravity of a body, volume, area, or line is that point at which if the body, volume, area, or line were suspended it would be perfectly balanced in all positions. For symmetrical bodies of uniform material it is at the geometric center. The center of gravity of a uniform round rod, for example, is at the center of its diameter halfway along its length; the center of gravity of a sphere is at the center of the sphere. For solids, areas, and arcs that are not symmetrical, the determination of the center of gravity may be made experimentally or may be calculated by the use of formulas.

The tables that follow give such formulas for some of the more important shapes. For more complicated and unsymmetrical shapes the methods outlined on page 142 may be used.

Example: A piece of wire is bent into the form of a semi-circular arc of 10-inch radius. How far from the center of the arc is the center of gravity located?

Accompanying the third diagram on page 137 is a formula for the distance from the center of gravity of an arc to the center of the arc: $a = 2r \div \pi$. Therefore,

$$a = 2 \times 10 \div 3.1416 = 6.366 \text{ inches}$$

Formulas for Center of Gravity

Triangle:

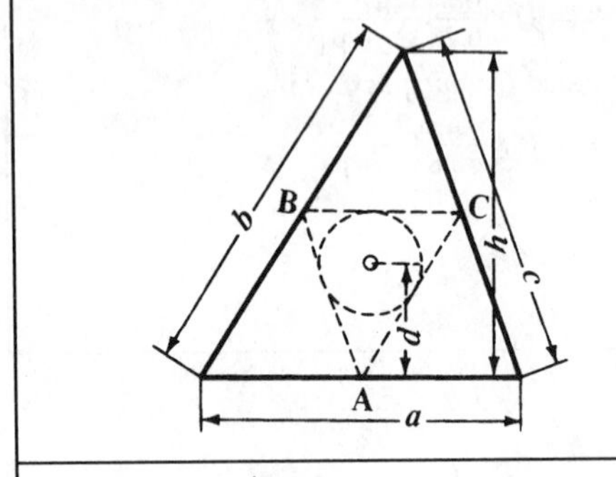

Perimeter

If *A*, *B* and *C* are the middle points of the sides of the triangle, then the center of gravity is at the center of the circle that can be inscribed in triangle *ABC*. The distance *d* of the center of gravity from side *a* is:

$$d = \frac{h(b + c)}{2(a + b + c)}$$

where *h* is the height perpendicular to *a*.

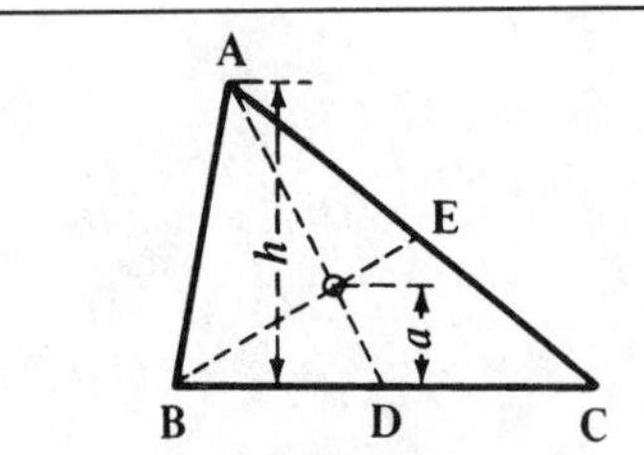

Area

The center of gravity is at the intersection of lines *AD* and *BE*, which bisect the sides *BC* and *AC*. The perpendicular distance from the center of gravity to any one of the sides is equal to one-third the height perpendicular to that side. Hence, $a = h \div 3$.

Perimeter or Area of a Parallelogram :

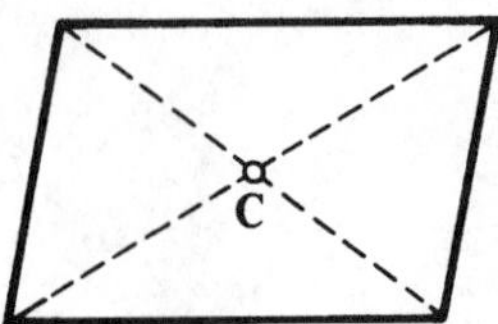

The center of gravity is at the intersection of the diagonals.

Area of Trapezoid:

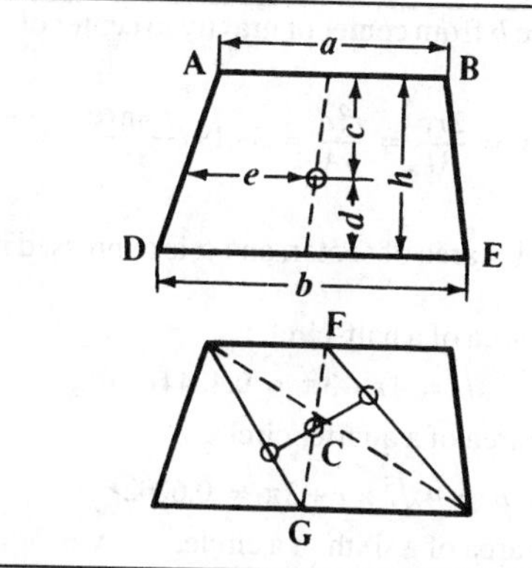

The center of gravity is on the line joining the middle points of parallel lines *AB* and *DE*.

$$c = \frac{h(a+2b)}{3(a+b)} \qquad d = \frac{h(2a+b)}{3(a+b)}$$

$$e = \frac{a^2+ab+b^2}{3(a+b)}$$

The trapezoid can also be divided into two triangles. The center of gravity is at the intersection of the line joining the centers of gravity of the triangles, and the middle line *FG*.

Any Four-sided Figure :

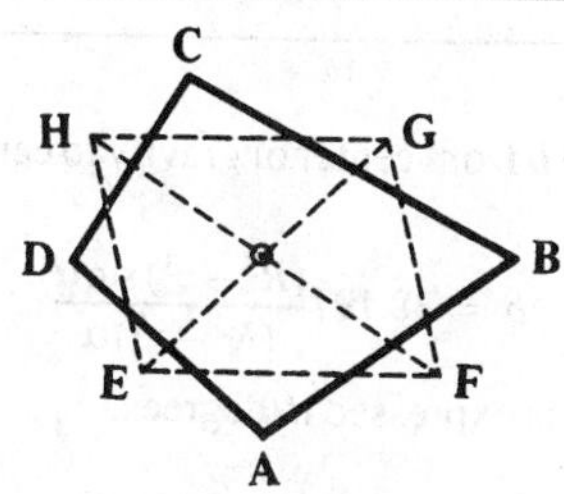

Two cases are possible, as shown in the illustration. To find the center of gravity of the four-sided figure *ABCD*, each of the sides is divided into three equal parts. A line is then drawn through each pair of division points next to the points of intersection *A*, *B*, *C*, and *D* of the sides of the figure. These lines form a parallelogram *EFGH*; the intersection of the diagonals *EG* and *FH* locates center of gravity.

Circular Arc:

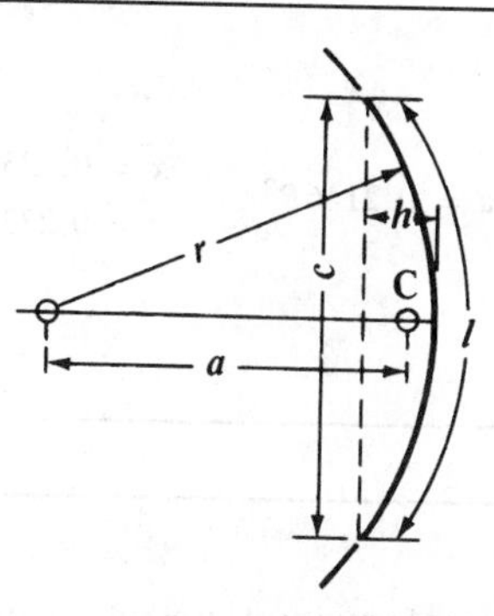

The center of gravity is on the line that bisects the arc, at a distance $a = \frac{r \times c}{l} = \frac{c(c^2+4h^2)}{8lh}$ from the center of the circle.

For an arc equal to one-half the periphery:

$$a = 2r \div \pi = 0.6366r$$

For an arc equal to one-quarter of the periphery:

$$a = 2r\sqrt{2} \div \pi = 0.9003r$$

For an arc equal to one-sixth of the periphery:

$$a = 3r \div \pi = 0.9549r$$

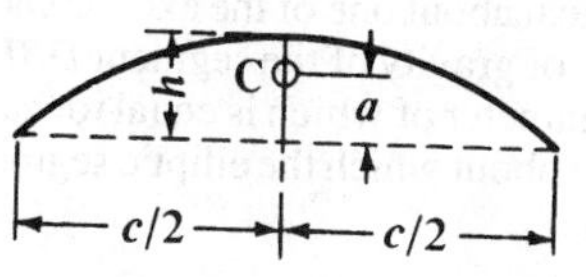

An approximate formula is very nearly exact for all arcs less than one-quarter of the periphery is:

$$a = \tfrac{2}{3}h$$

The error is only about one per cent for a quarter circle, and decreases for smaller arcs.

Circle Segment :

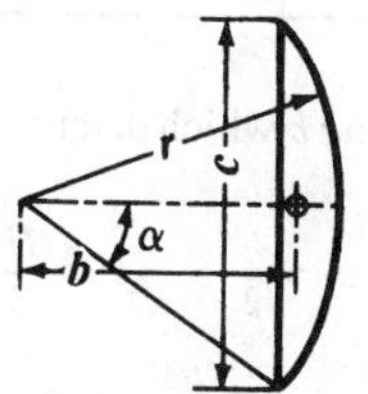

The distance of the center of gravity from the center of the circle is:

$$b = \frac{c^3}{12A} = \tfrac{2}{3} \times \frac{r^3 \sin^3 \alpha}{A}$$

in which A = area of segment.

Circle Sector :

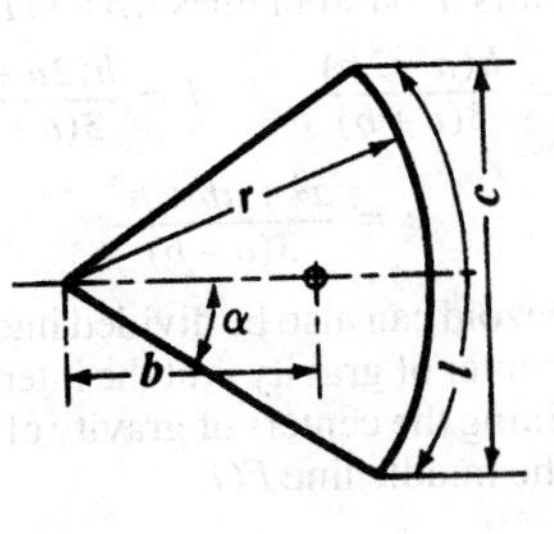

Distance b from center of gravity to center of circle is:

$$b = \frac{2rc}{3l} = \frac{r^2c}{3A} = 38.197\frac{r\sin\alpha}{\alpha}$$

in which A = area of sector, and α is expressed in degrees.

For the area of a half-circle:

$$b = 4r \div 3\pi = 0.4244r$$

For the area of a quarter circle:

$$b = 4\sqrt{2} \times r \div 3\pi = 0.6002r$$

For the area of a sixth of a circle:

$$b = 2r \div \pi = 0.6366r$$

Part of Circle Ring :

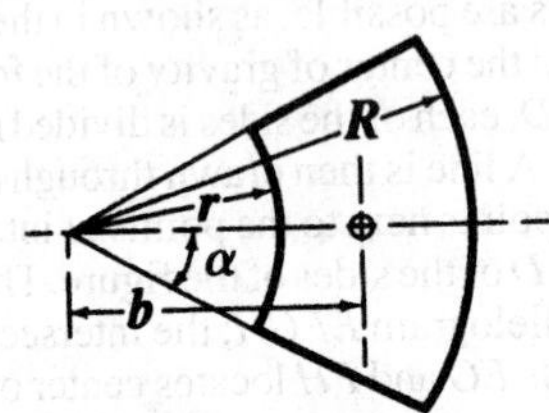

Distance b from center of gravity to center of circle is:

$$b = 38.197\frac{(R^3 - r^3)\sin\alpha}{(R^2 - r^2)\alpha}$$

Angle α is expressed in degrees.

Spandrel or Fillet :

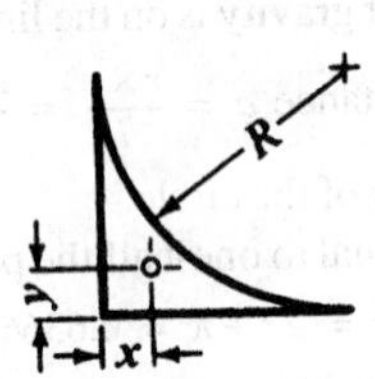

$$\text{Area} = 0.2146R^2 \qquad x = 0.2234R \qquad y = 0.2234R$$

Segment of an Ellipse :

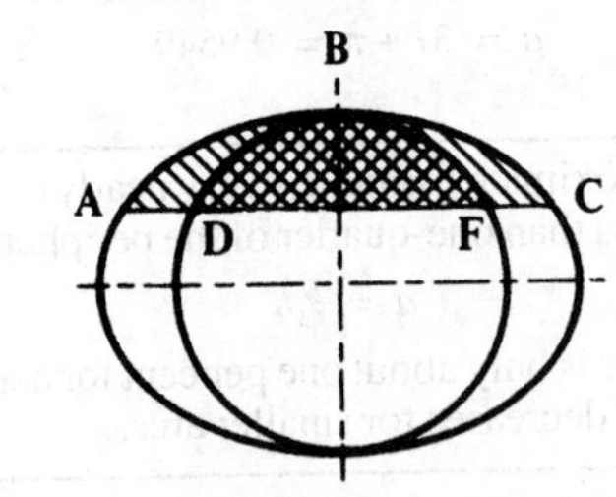

The center of gravity of an elliptic segment ABC, symmetrical about one of the axes, coincides with the center of gravity of the segment DBF of a circle, the diameter of which is equal to that axis of the ellipse about which the elliptic segment is symmetrical.

Spherical Surface of Segments and Zones of Spheres :

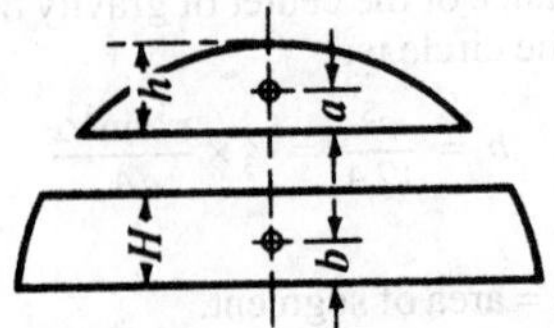

Distances a and b which determine the center of gravity, are:

$$a = \frac{h}{2} \qquad b = \frac{H}{2}$$

Area of a Parabola :

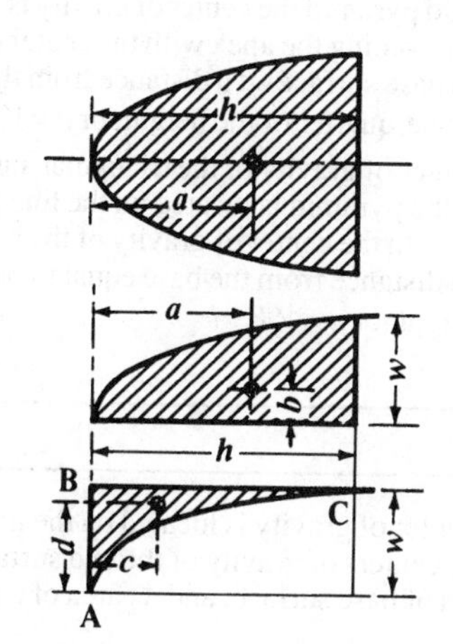

For the complete parabolic area, the center of gravity is on the center line or axis, and

$$a = \frac{3h}{5}$$

For one-half of the parabola:

$$a = \frac{3h}{5} \text{ and } b = \frac{3w}{8}$$

For the complement area ABC:

$$c = 0.3h \text{ and } d = 0.75w$$

Cylinder :

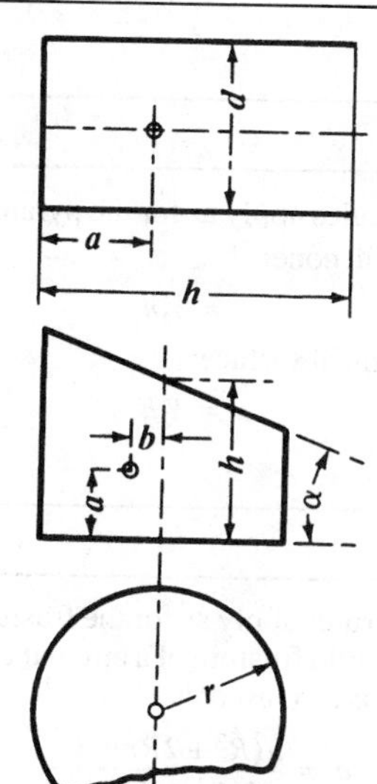

The center of gravity of a solid cylinder (or prism) with parallel end surfaces, is located at the middle of the line that joins the centers of gravity of the end surfaces.

The center of gravity of a cylindrical surface or shell, with the base or end surface in one end, is found from:

$$a = \frac{2h^2}{4h + d}$$

The center of gravity of a cylinder cut off by an inclined plane is located by:

$$a = \frac{h}{2} + \frac{r^2 \tan^2 \alpha}{8h} \qquad b = \frac{r^2 \tan \alpha}{4h}$$

where α is the angle between the obliquely cut off surface and the base surface.

Portion of Cylinder :

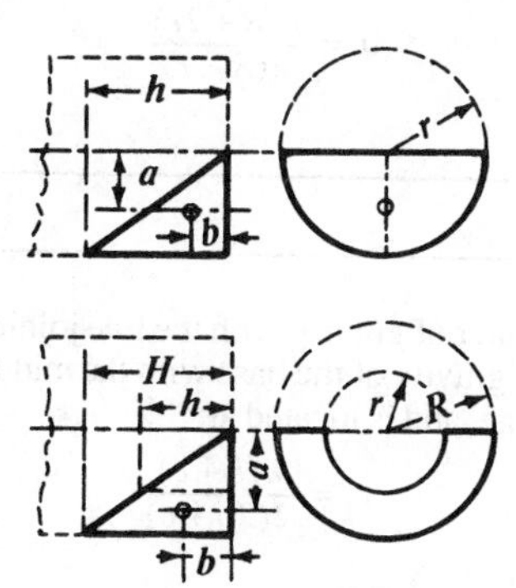

For a solid portion of a cylinder, as shown, the center of gravity is determined by:

$$a = \tfrac{3}{16} \times 3.1416r \qquad b = \tfrac{3}{32} \times 3.1416h$$

For the cylindrical surface only:

$$a = \tfrac{1}{4} \times 3.1416r \qquad b = \tfrac{1}{8} \times 3.1416h$$

If the cylinder is hollow, the center of gravity of the solid shell is found by:

$$a = \tfrac{3}{16} \times 3.1416\frac{R^4 - r^4}{R^3 - r^3}$$

$$b = \tfrac{3}{32} \times 3.1416\frac{H^4 - h^4}{H^3 - h^3}$$

Pyramid:

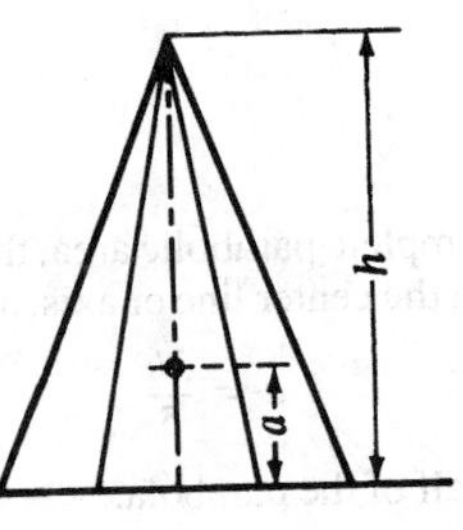	In a solid pyramid the center of gravity is located on the line joining the apex with the center of gravity of the base surface, at a distance from the base equal to one-quarter of the height; or $a = \frac{1}{4}h$. The center of gravity of the triangular surfaces forming the pyramid is located on the line joining the apex with the center of gravity of the base surface, at a distance from the base equal to one-third of the height; or $a = \frac{1}{3}h$.

Frustum of Pyramid:

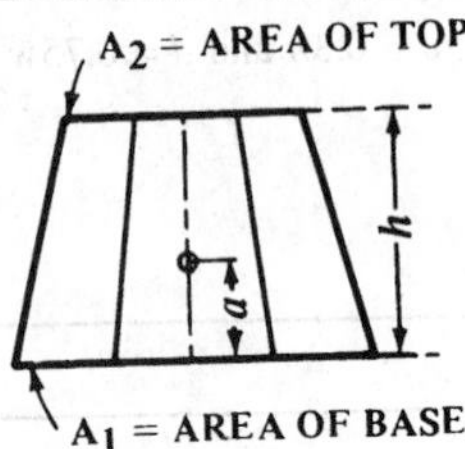	The center of gravity is located on the line that joins the centers of gravity of the end surfaces. If A_1 = area of base surface, and A_2 area of top surface, $$a = \frac{h(A_1 + 2\sqrt{A_1 \times A_2} + 3A_2)}{4(A_1 + \sqrt{A_1 \times A_2} + A_2)}$$

Cone:

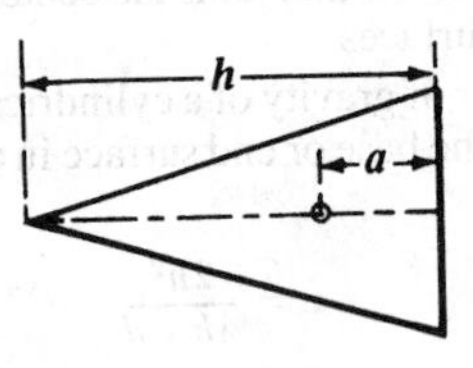	The same rules apply as for the pyramid. For the solid cone: $$a = \frac{1}{4}h$$ For the conical surface: $$a = \frac{1}{3}h$$

Frustum of Cone:

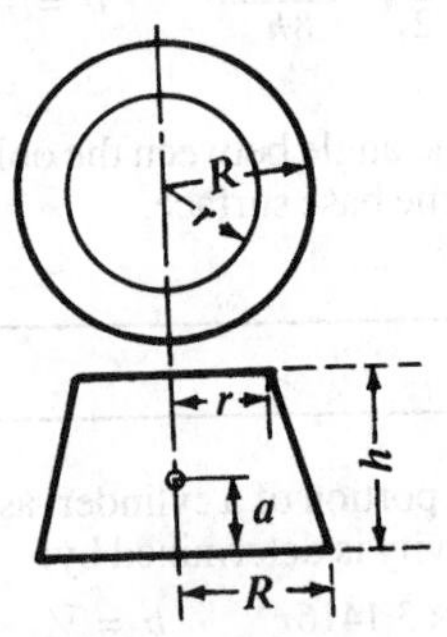	The same rules apply as for the frustum of a pyramid. For a solid frustum of a circular cone the formula below is also used: $$a = \frac{h(R^2 + 2Rr + 3r^2)}{4(R^2 + Rr + r^2)}$$ The location of the center of gravity of the conical surface of a frustum of a cone is determined by: $$a = \frac{h(R + 2r)}{3(R + r)}$$

Wedge:

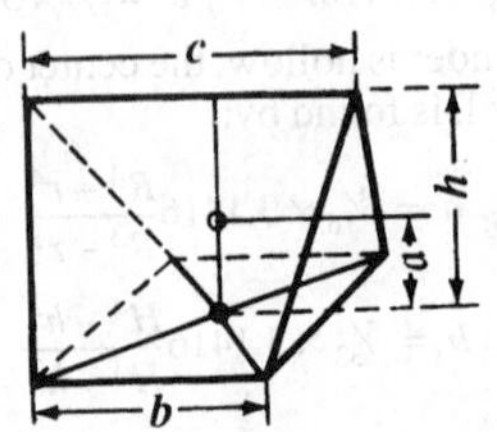	The center of gravity is on the line joining the center of gravity of the base with the middle point of the edge, and is located at: $$a = \frac{h(b + c)}{2(2b + c)}$$

Half of a Hollow Sphere :

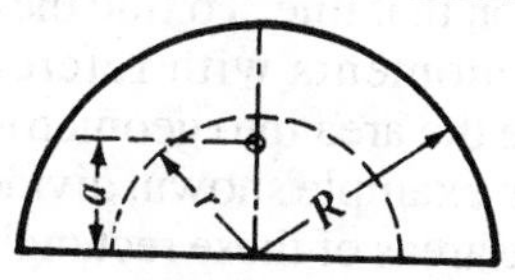

The center of gravity is located at:

$$a = \frac{3(R^4 - r^4)}{8(R^3 - r^3)}$$

Spherical Segment :

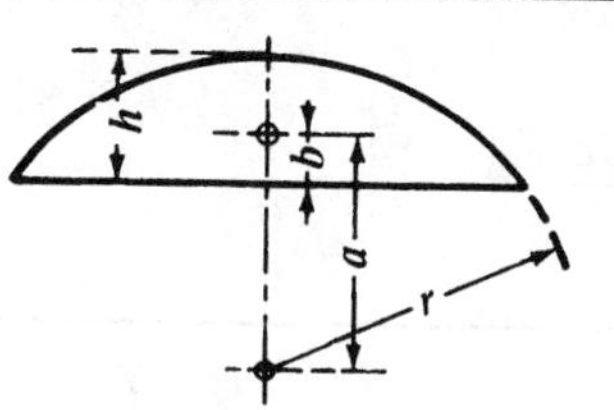

The center of gravity of a solid segment is determined by:

$$a = \frac{3(2r - h)^2}{4(3r - h)}$$

$$b = \frac{h(4r - h)}{4(3r - h)}$$

For a half-sphere, $a = b = \frac{3}{8}r$

Spherical Sector :

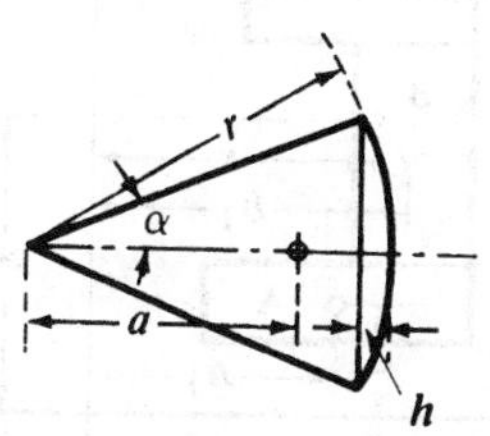

The center of gravity of a solid sector is at:

$$a = \tfrac{3}{8}(1 + \cos\alpha)r = \tfrac{3}{8}(2r - h)$$

Segment of Ellipsoid or Spheroid :

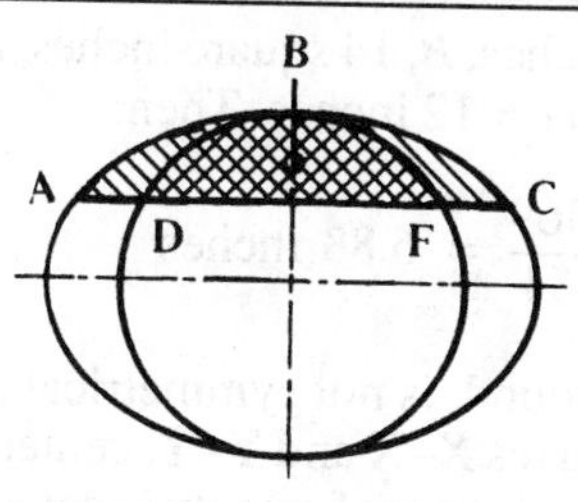

The center of gravity of a solid segment *ABC*, symmetrical about the axis of rotation, coincides with the center of gravity of the segment *DBF* of a sphere, the diameter of which is equal to the axis of rotation of the spheroid.

Paraboloid :

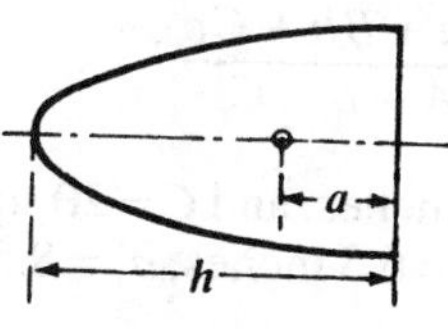

The center of gravity of a solid paraboloid of rotation is at:

$$a = \tfrac{1}{3}h$$

Center of Gravity of Two Bodies :

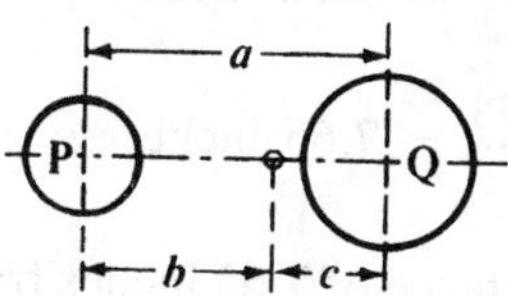

If the weights of the bodies are *P* and *Q*, and the distance between their centers of gravity is *a*, then:

$$b = \frac{Qa}{P + Q} \qquad c = \frac{Pa}{P + Q}$$

Center of Gravity of Figures of any Outline.—If the figure is symmetrical about a center line, as in Fig. 1, the center of gravity will be located on that line. To find the exact location on that line, the simplest method is by taking moments with reference to any convenient axis at right angles to this center line. Divide the area into geometrical figures, the centers of gravity of which can be easily found. In the example shown, divide the figure into three rectangles KLMN, EFGH and OPRS. Call the areas of these rectangles A, B and C, respectively, and find the center of gravity of each. Then select any convenient axis, as X–X, at right angles to the center line Y–Y, and determine distances a, b and c. The distance y of the center of gravity of the complete figure from the axis X–X is then found from the equation:

$$y = \frac{Aa + Bb + Cc}{A + B + C}$$

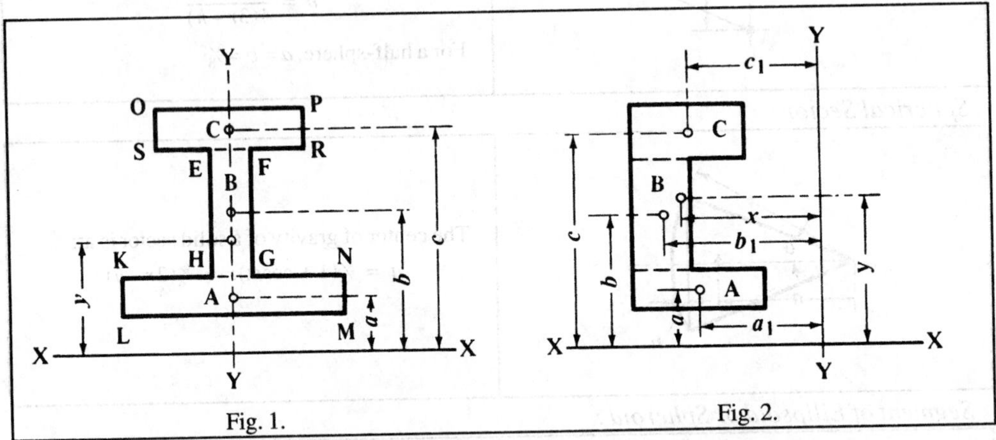

Fig. 1. Fig. 2.

As an example, assume that the area A is 24 square inches, B, 14 square inches, and C, 16 square inches, and that $a = 3$ inches, $b = 7.5$ inches, and $c = 12$ inches. Then:

$$y = \frac{24 \times 3 + 14 \times 7.5 + 16 \times 12}{24 + 14 + 16} = \frac{369}{54} = 6.83 \text{ inches}$$

If the figure, the center of gravity of which is to be found, is not symmetrical about any axis, then moments must be taken with relation to two axes X–X and Y–Y, centers of gravity of which can be easily found, the same as before. The center of gravity is determined by the equations:

$$x = \frac{Aa_1 + Bb_1 + Cc_1}{A + B + C} \qquad y = \frac{Aa + Bb + Cc}{A + B + C}$$

As an example, let $A = 14$ square inches, $B = 18$ square inches, and $C = 20$ square inches. Let $a = 3$ inches, $b = 7$ inches, and $c = 11.5$ inches. Let $a_1 = 6.5$ inches, $b_1 = 8.5$ inches, and $c_1 = 7$ inches. Then:

$$x = \frac{14 \times 6.5 + 18 \times 8.5 + 20 \times 7}{14 + 18 + 20} = \frac{384}{52} = 7.38 \text{ inches}$$

$$y = \frac{14 \times 3 + 18 \times 7 + 20 \times 11.5}{14 + 18 + 20} = \frac{398}{52} = 7.65 \text{ inches}$$

In other words, the center of gravity is located at a distance of 7.65 inches from the axis X–X and 7.38 inches from the axis Y–Y.

Moments of Inertia.—An important property of areas and solid bodies is the moment of inertia. Standard formulas are derived by multiplying elementary particles of area or mass by the squares of their distances from reference axes. Moments of inertia, therefore, depend on the location of reference axes. Values are minimum when these axes pass through the centers of gravity.

Three kinds of moments of inertia occur in engineering formulas:

1) *Moments of inertia of plane area, I,* in which the axis is in the plane of the area, are found in formulas for calculating deflections and stresses in beams. When dimensions are given in inches, the units of I are $inches^4$. A table of formulas for calculating the I of common areas can be found in the *STRENGTH OF MATERIALS* section beginning on page 217.

2) *Polar moments of inertia of plane areas, J,* in which the axis is at right angles to the plane of the area, occur in formulas for the torsional strength of shafting. When dimensions are given in inches, the units of J are $inches^4$. If moments of inertia, I, are known for a plane area with respect to both x and y axes, then the polar moment for the z axis may be calculated using the equation, $J_z = I_x + I_y$

A table of formulas for calculating J for common areas can be found on page 275 in the *SHAFTS* section.

When metric SI units are used, the formulas referred to in (1) and (2) above, are valid if the dimensions are given consistently in meters or millimeters. If meters are used, the units of I and J are in $meters^4$; if millimeters are used, these units are in $millimeters^4$.

3) *Polar moments of inertia of masses,* J_M*, appear in dynamics equations involving rotational motion. J_M bears the same relationship to angular acceleration as mass does to linear acceleration. If units are in the foot-pound-second system, the units of J_M are ft-lbs-sec^2 or slug-ft^2. (1 slug = 1 pound $second^2$ per foot.) If units are in the inch-pound-second system, the units of J_M are inch-lbs-sec^2.

If metric SI values are used, the units of J_M are kilogram-meter squared. Formulas for calculating J_M for various bodies are given beginning on page 144. If the polar moment of inertia J is known for the area of a body of constant cross section, J_M may be calculated using the equation,

$$J_M = \frac{\rho L}{g} J$$

where ρ is the density of the material, L the length of the part, and g the gravitational constant. If dimensions are in the foot-pound-second system, ρ is in lbs per ft^3, L is in ft, g is 32.16 ft per sec^2, and J is in ft^4. If dimensions are in the inch-pound-second system, ρ is in lbs per in^3, L is in inches, g is 386 inches per sec^2, and J is in $inches^4$.

Using metric SI units, the above formula becomes $J_M = \rho L J$, where ρ = the density in $kilograms/meter^3$, L = the length in meters, and J = the polar moment of inertia in $meters^4$. The units of J_M are $kg \cdot m^2$.

* In some books the symbol I denotes the polar moment of inertia of masses; J_M is used in this handbook to avoid confusion with moments of inertia of plane areas.

Formulas for Polar Moment of Inertia of Masses, J_M

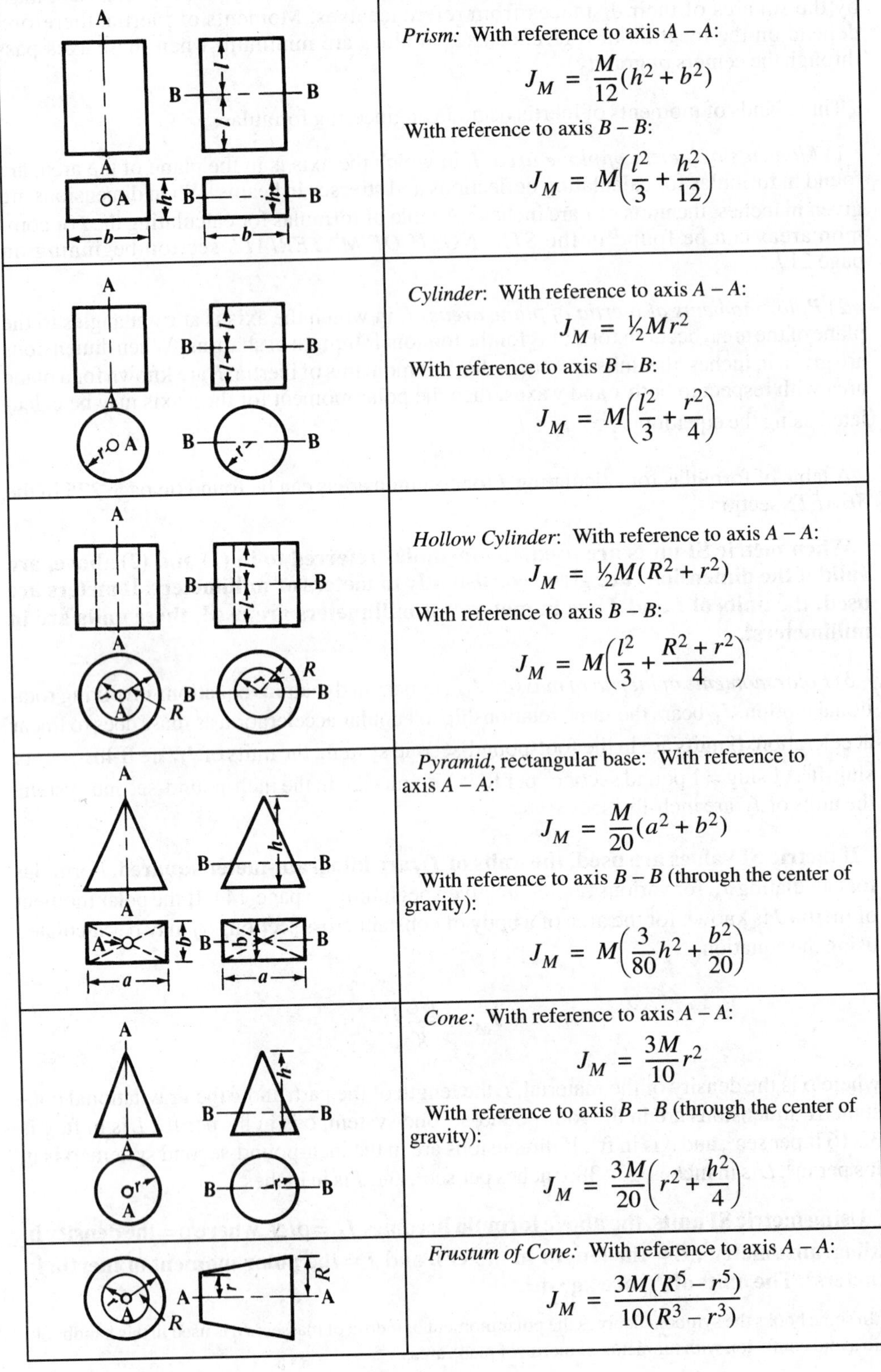

Prism	With reference to axis $A-A$: $$J_M = \frac{M}{12}(h^2 + b^2)$$ With reference to axis $B-B$: $$J_M = M\left(\frac{l^2}{3} + \frac{h^2}{12}\right)$$
Cylinder	With reference to axis $A-A$: $$J_M = \tfrac{1}{2}Mr^2$$ With reference to axis $B-B$: $$J_M = M\left(\frac{l^2}{3} + \frac{r^2}{4}\right)$$
Hollow Cylinder	With reference to axis $A-A$: $$J_M = \tfrac{1}{2}M(R^2 + r^2)$$ With reference to axis $B-B$: $$J_M = M\left(\frac{l^2}{3} + \frac{R^2 + r^2}{4}\right)$$
Pyramid, rectangular base	With reference to axis $A-A$: $$J_M = \frac{M}{20}(a^2 + b^2)$$ With reference to axis $B-B$ (through the center of gravity): $$J_M = M\left(\frac{3}{80}h^2 + \frac{b^2}{20}\right)$$
Cone	With reference to axis $A-A$: $$J_M = \frac{3M}{10}r^2$$ With reference to axis $B-B$ (through the center of gravity): $$J_M = \frac{3M}{20}\left(r^2 + \frac{h^2}{4}\right)$$
Frustum of Cone	With reference to axis $A-A$: $$J_M = \frac{3M(R^5 - r^5)}{10(R^3 - r^3)}$$

Formulas for Polar Moment of Inertia of Masses, J_M

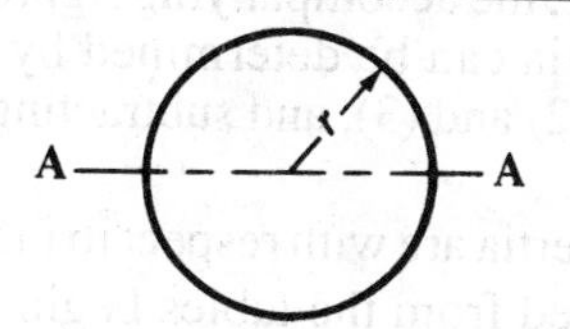	*Sphere*: With reference to any axis through the center: $$J_M = \tfrac{2}{5}Mr^2$$
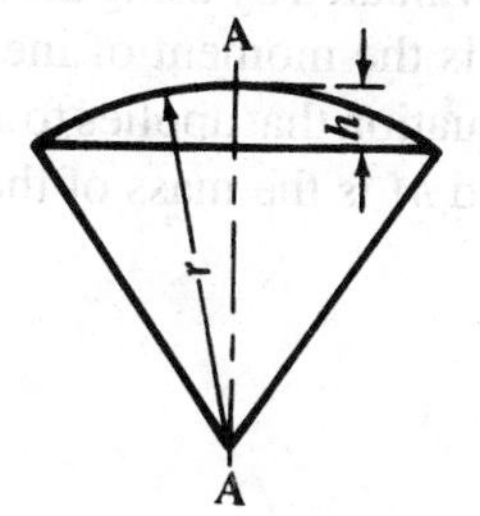	*Spherical Sector*: With reference to axis $A-A$: $$J_M = \frac{M}{5}(3rh - h^2)$$
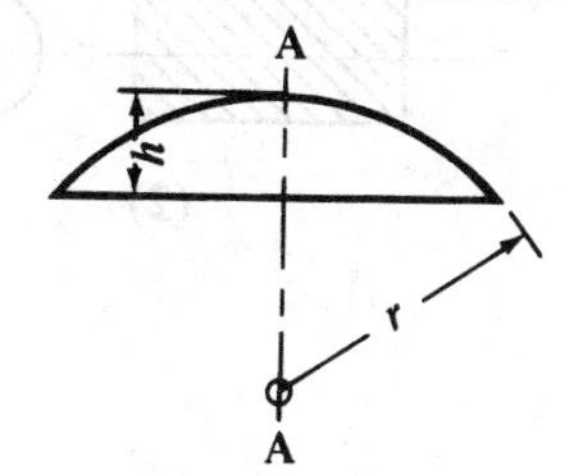	*Spherical Segment*: With reference to axis $A-A$: $$J_M = M\left(r^2 - \frac{3rh}{4} + \frac{3h^2}{20}\right)\frac{2h}{3r-h}$$
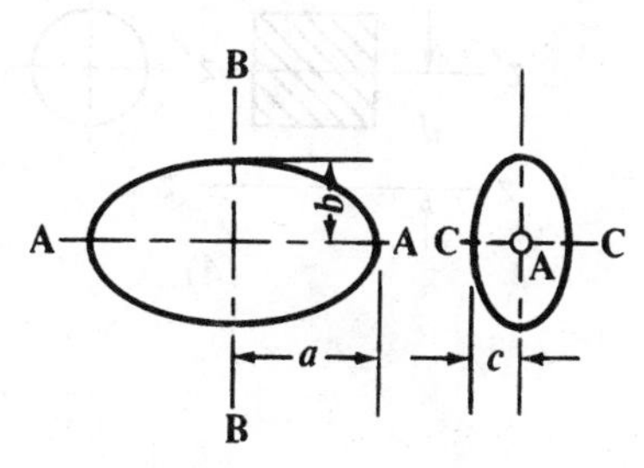	*Ellipsoid*: With reference to axis $A-A$: $$J_M = \frac{M}{5}(b^2 + c^2)$$ With reference to axis $B-B$: $$J_M = \frac{M}{5}(a^2 + c^2)$$ With reference to axis $C-C$: $$J_M = \frac{M}{5}(a^2 + b^2)$$
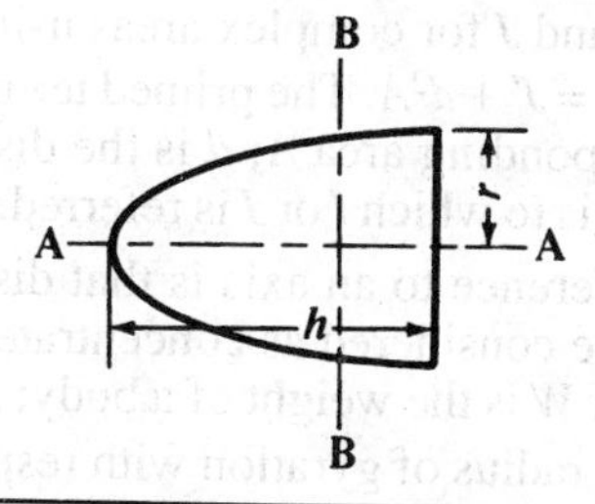	*Paraboloid*: With reference to axis $A-A$: $$J_M = \tfrac{1}{3}Mr^2$$ With reference to axis $B-B$ (through the center of gravity): $$J_M = M\left(\frac{r^2}{6} + \frac{h^2}{18}\right)$$
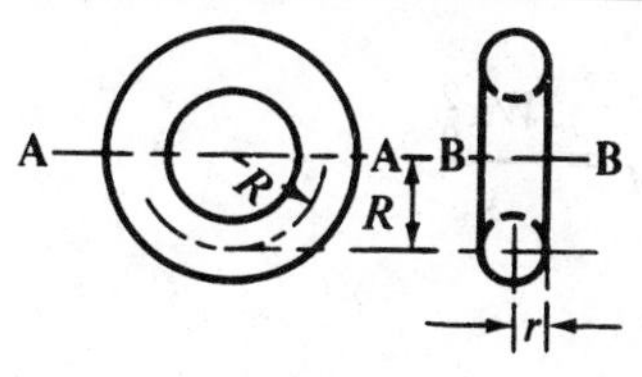	*Torus*: With reference to axis $A-A$: $$J_M = M\left(\frac{R^2}{2} + \frac{5r^2}{8}\right)$$ With reference to axis $B-B$: $$J_M = M(R^2 + \tfrac{3}{4}r^2)$$

Moments of inertia of complex areas and masses may be evaluated by the addition and subtraction of elementary areas and masses. For example, the accompanying figure shows a complex mass at (1); its mass polar moment of inertia can be determined by adding together the moments of inertia of the bodies shown at (2) and (3), and subtracting that at (4).

Thus, $J_{M1} = J_{M2} + J_{M3} - J_{M4}$. All of these moments of inertia are with respect to the axis of rotation $z - z$. Formulas for J_{M2} and J_{M3} can be obtained from the tables beginning on page 144. The moment of inertia for the body at (4) can be evaluated by using the following transfer-axis equation: $J_{M4} = J_{M4}' + d^2M$. The term J_{M4}' is the moment of inertia with respect to axis $z' - z'$; it may be evaluated using the same equation that applies to J_{M2} where d is the distance between the $z - z$ and the $z' - z'$ axes, and M is the mass of the body (= weight in lbs ÷ g).

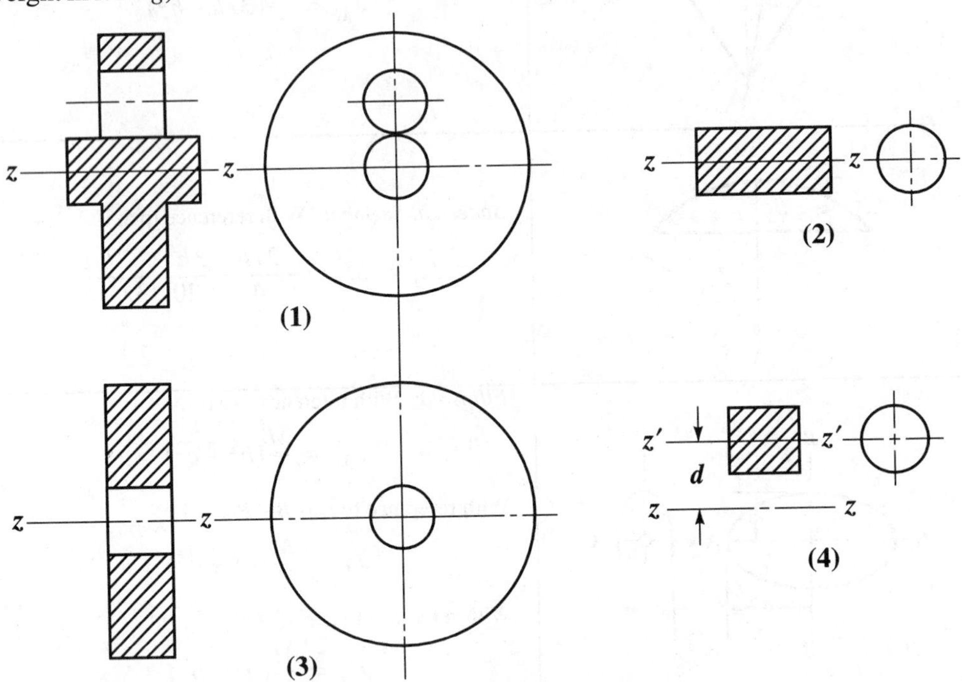

Moments of Inertia of Complex Masses

Similar calculations can be made when calculating I and J for complex areas using the appropriate transfer-axis equations are $I = I' + d^2A$ and $J = J' + d^2A$. The primed term, I' or J', is with respect to the center of gravity of the corresponding area A; d is the distance between the axis through the center of gravity and the axis to which I or J is referred.

Radius of Gyration.—The radius of gyration with reference to an axis is that distance from the axis at which the entire mass of a body may be considered as concentrated, the moment of inertia, meanwhile, remaining unchanged. If W is the weight of a body; J_M, its moment of inertia with respect to some axis; and k_o, the radius of gyration with respect to the same axis, then:

$$k_o = \sqrt{\frac{J_M g}{W}} \qquad \text{and} \qquad J_M = \frac{Wk_o^2}{g}$$

When using metric SI units, the formulas are:

$$\boldsymbol{k_o = \sqrt{\frac{J_M}{M}} \qquad \text{and} \qquad J_M = Mk_o^2}$$

where k_o = the radius of gyration in meters, J_M = kilogram-meter squared, and M = mass in kilograms.

To find the radius of gyration of an area, such as for the cross-section of a beam, divide the moment of inertia of the area by the area and extract the square root.

When the axis, the reference to which the radius of gyration is taken, passes through the center of gravity, the radius of gyration is the least possible and is called the *principal* radius of gyration. If *k* is the radius of gyration with respect to such an axis passing through the center of gravity of a body, then the radius of gyration, k_o, with respect to a parallel axis at a distance *d* from the gravity axis is given by:

$$k_o = \sqrt{k^2 + d^2}$$

Tables of radii of gyration for various bodies and axes follows.

Formulas for Radius of Gyration

Bar of Small Diameter:

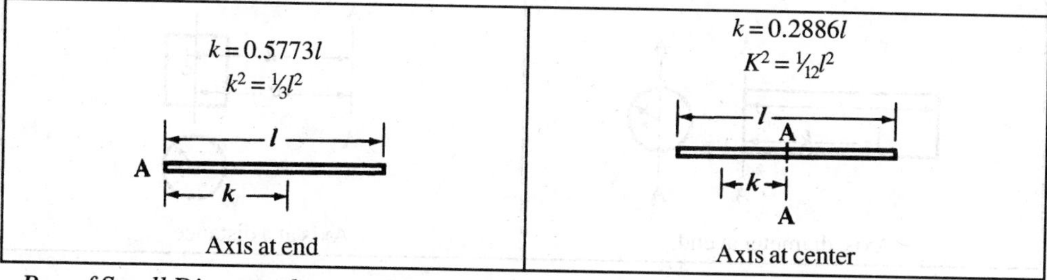

Bar of Small Diameter bent to Circular Shape:

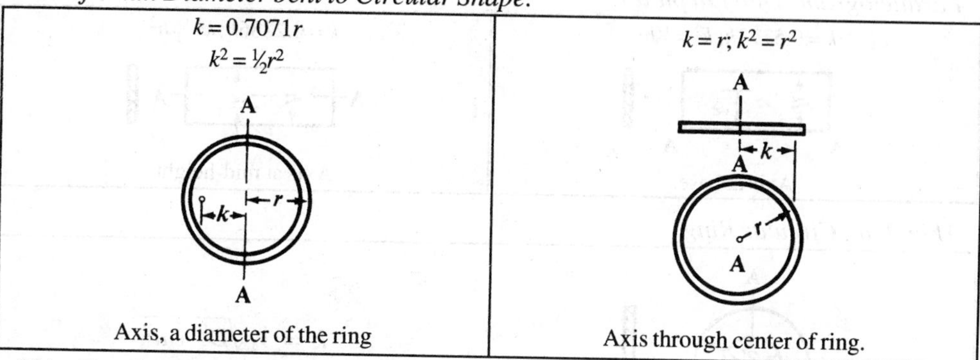

Thin Circular Disk:

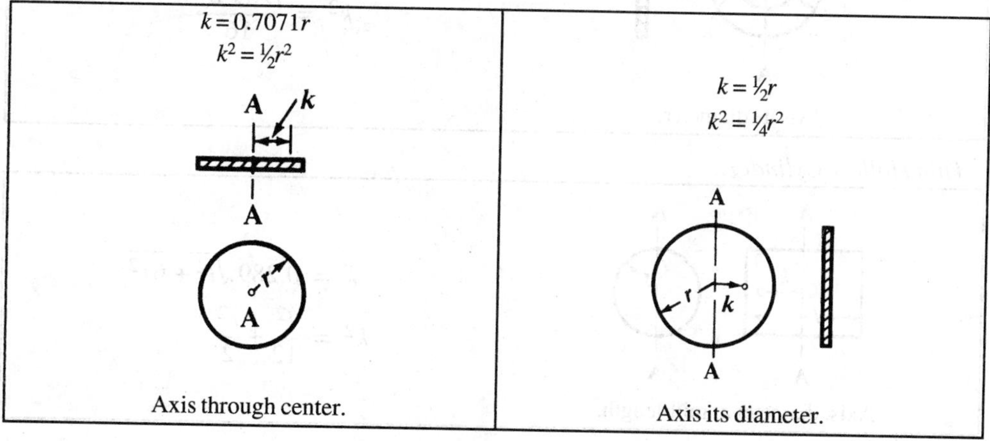

Cylinder:

$k = 0.7071r, \qquad k^2 = \frac{1}{2}r^2$

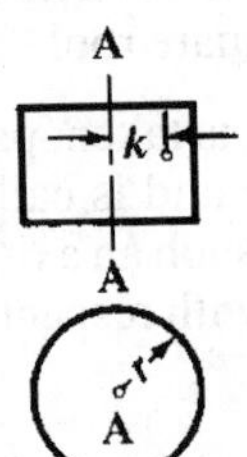

Axis through center.

$k = 0.289\sqrt{l^2 + 3r^2}$

$k^2 = \frac{l^2}{12} + \frac{r^2}{4}$

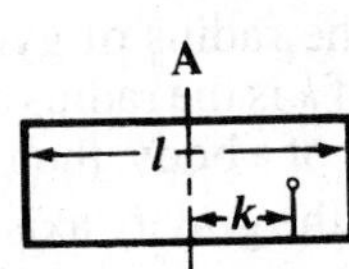

Axis, diameter at mid-length.

$k = 0.289\sqrt{4l^2 + 3r^2}$

$k^2 = \frac{l^2}{3} + \frac{r^2}{4}$

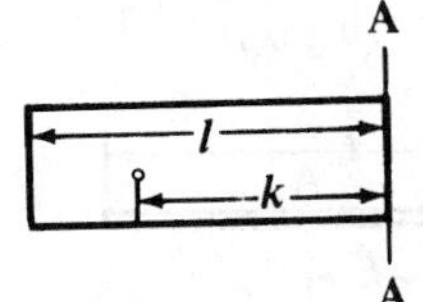

Axis, diameter at end.

$k = \sqrt{a^2 + \frac{1}{2}r^2}$

$k^2 = a^2 + \frac{1}{2}r^2$

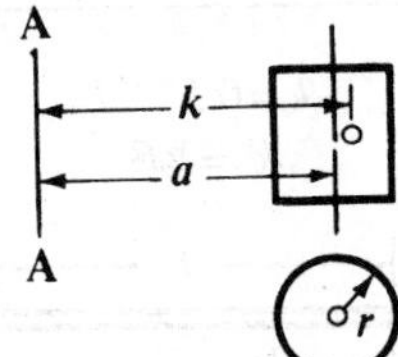

Axis at a distance.

Parallelogram (Thin flat plate):

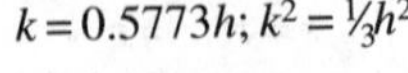

$k = 0.5773h$; $k^2 = \frac{1}{3}h^2$

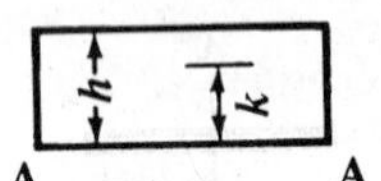

Axis at base.

$k = 0.2886h$; $k^2\ \frac{1}{12}h^2$

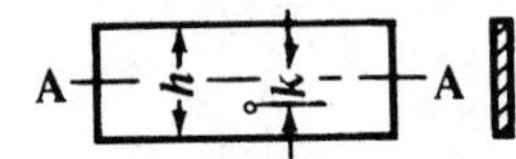

Axis at mid-height.

Thin, Flat, Circular Ring:

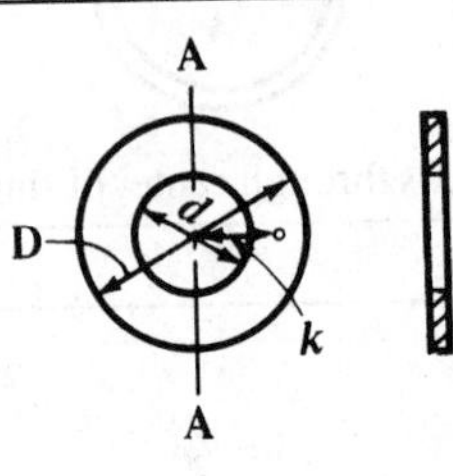

Axis its diameter.

$k = \frac{1}{4}\sqrt{D^2 + d^2}$

$k^2 = \frac{D^2 + d^2}{16}$

Thin Hollow Cylinder.:

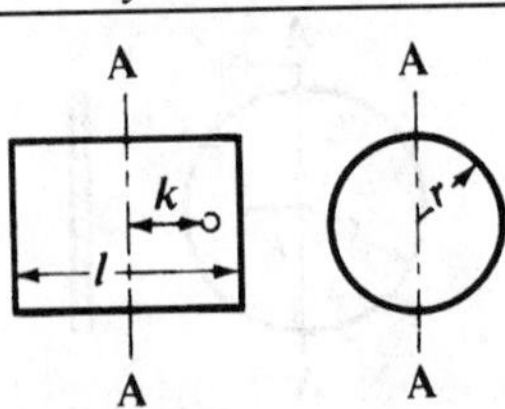

Axis, diameter at mid-length.

$k = 0.289\sqrt{l^2 + 6r^2}$

$k^2 = \frac{l^2}{12} + \frac{r^2}{2}$

Hollow Cylinder.:

$$k = 0.289\sqrt{l^2 + 3(R^2 + r^2)}$$

$$k^2 = \frac{l^2}{12} + \frac{R^2 + r^2}{4}$$

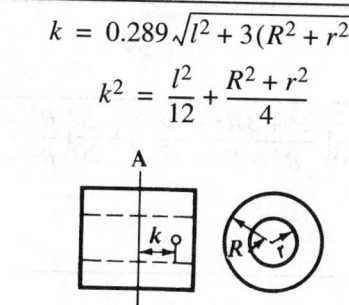

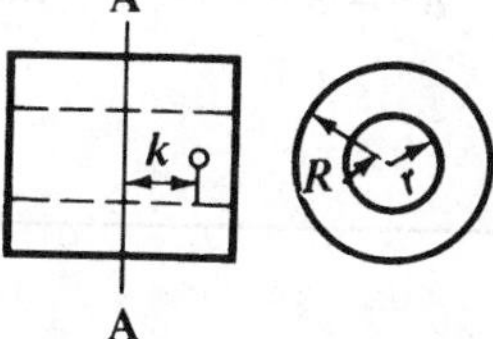

Axis, diameter at mid-length.

$$k = 0.7071\sqrt{R^2 + r^2}$$

$$k^2 = ½(R^2 + r^2)$$

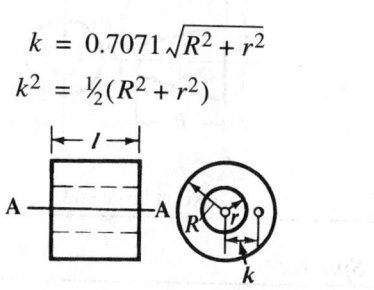

Longitudinal Axis.

Rectangular Prism:

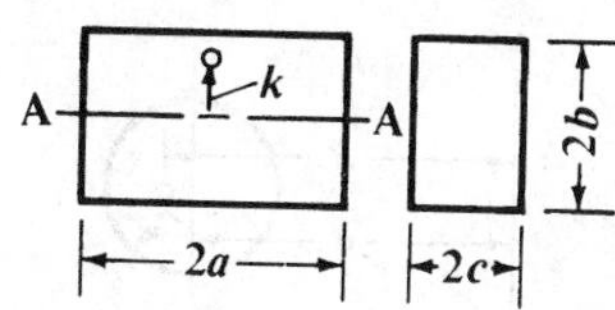

Axis through center.

$$k = 0.577\sqrt{b^2 + c^2}$$

$$k^2 = ⅓(b^2 + c^2)$$

Parallelepiped:

$$k = 0.289\sqrt{4l^2 + b^2}$$

$$k^2 = \frac{4l^2 + b^2}{12}$$

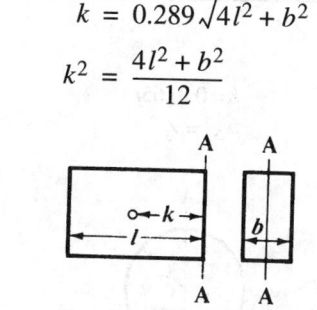

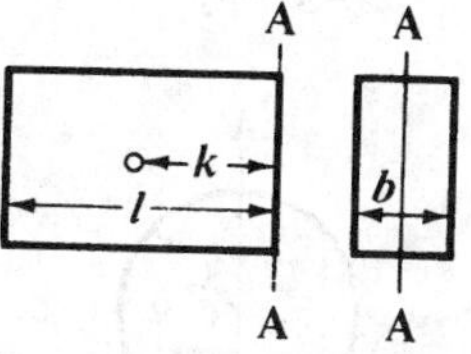

Axis at one end, central.

$$k = \sqrt{\frac{4l^2 + b^2}{12} + a^2 + al}$$

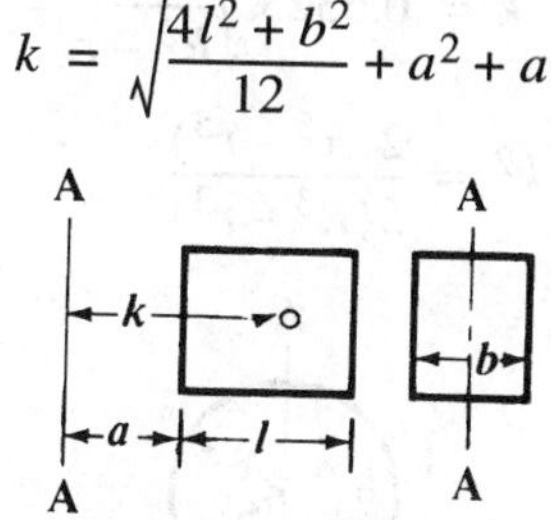

Axis at distance from end.

Cone:

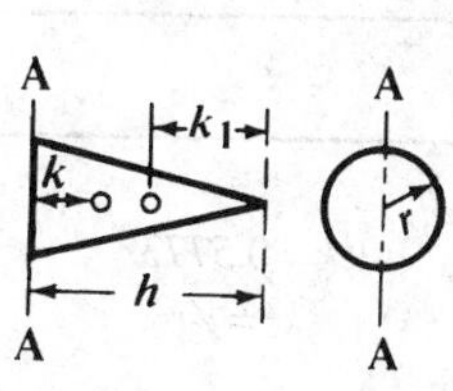

Axis at base.

$$k = \sqrt{\frac{2h^2 + 3r^2}{20}}$$

Axis at apex.

$$k_1 = \sqrt{\frac{12h^2 + 3r^2}{20}}$$

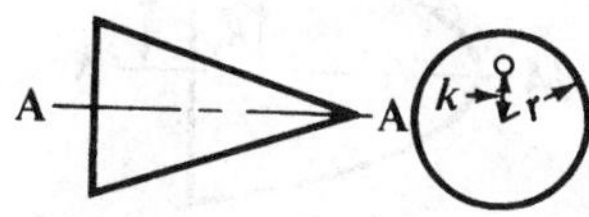

Axis through its center line.

$$k = 0.5477r$$

$$k^2 = 0.3r^2$$

Frustum of Cone:

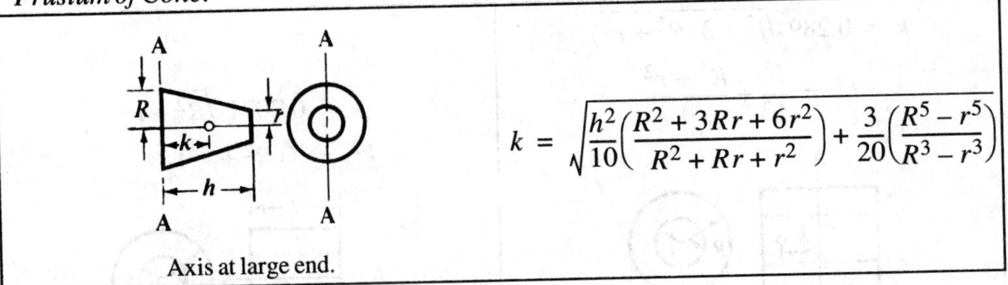

$$k = \sqrt{\frac{h^2}{10}\left(\frac{R^2 + 3Rr + 6r^2}{R^2 + Rr + r^2}\right) + \frac{3}{20}\left(\frac{R^5 - r^5}{R^3 - r^3}\right)}$$

Axis at large end.

Sphere:

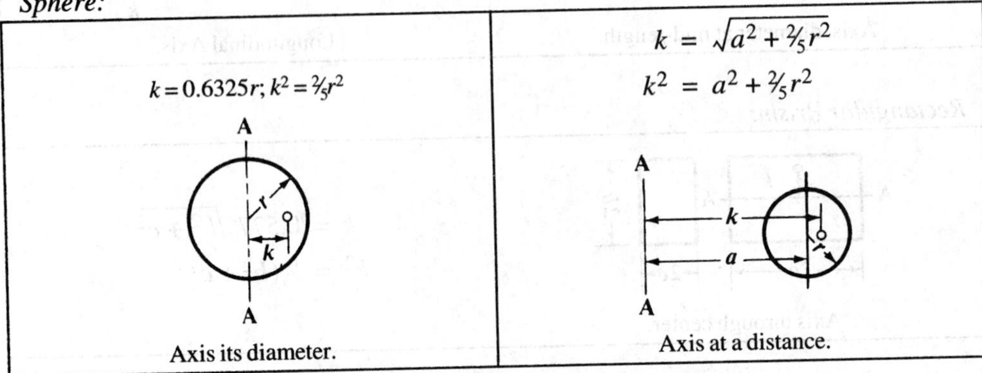

$k = 0.6325r$; $k^2 = \frac{2}{5}r^2$

Axis its diameter.

$$k = \sqrt{a^2 + \tfrac{2}{5}r^2}$$

$$k^2 = a^2 + \tfrac{2}{5}r^2$$

Axis at a distance.

Hollow Sphere and Thin Spherical Shell:

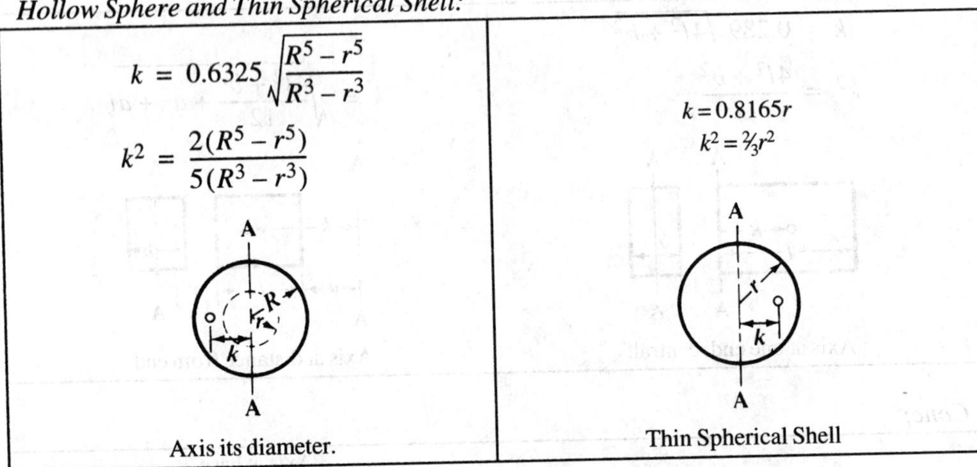

$$k = 0.6325\sqrt{\frac{R^5 - r^5}{R^3 - r^3}}$$

$$k^2 = \frac{2(R^5 - r^5)}{5(R^3 - r^3)}$$

Axis its diameter.

$k = 0.8165r$

$k^2 = \frac{2}{3}r^2$

Thin Spherical Shell

Ellipsoid. and Paraboloid:

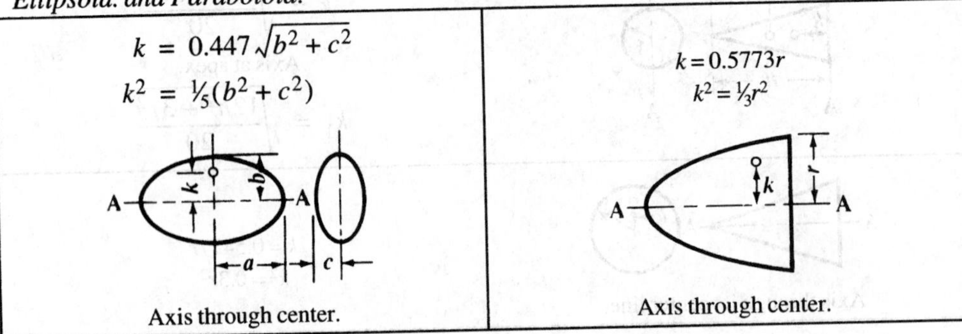

$$k = 0.447\sqrt{b^2 + c^2}$$

$$k^2 = \tfrac{1}{5}(b^2 + c^2)$$

Axis through center.

$k = 0.5773r$

$k^2 = \frac{1}{3}r^2$

Axis through center.

Center and Radius of Oscillation.—If a body oscillates about a horizontal axis which does not pass through its center of gravity, there will be a point on the line drawn from the center of gravity, perpendicular to the axis, the motion of which will be the same as if the whole mass were concentrated at that point. This point is called the *center of oscillation.* The *radius of oscillation* is the distance between the center of oscillation and the point of suspension. In a straight line, or in a bar of small diameter, suspended at one end and oscillating about it, the center of oscillation is at two-thirds the length of the rod from the end by which it is suspended.

When the vibrations are perpendicular to the plane of the figure, and the figure is suspended by the vertex of an angle or its uppermost point, the radius of oscillation of an isosceles triangle is equal to ¾ of the height of the triangle; of a circle, ⅝ of the diameter; of a parabola, 5⁄7 of the height.

If the vibrations are in the plane of the figure, then the radius of oscillation of a circle equals ¾ of the diameter; of a rectangle, suspended at the vertex of one angle, ⅔ of the diagonal.

Center of Percussion.—For a body that moves without rotation, the resultant of all the forces acting on the body passes through the center of gravity. On the other hand, for a body that rotates about some *fixed axis*, the resultant of all the forces acting on it does not pass through the center of gravity of the body but through a point called the *center of percussion*. The center of percussion is useful in determining the position of the resultant in mechanics problems involving angular acceleration of bodies about a fixed axis.

Finding the Center of Percussion when the Radius of Gyration and the Location of the Center of Gravity are Known: The center of percussion lies on a line drawn through the center of rotation and the center of gravity. The distance from the axis of rotation to the center of percussion may be calculated from the following formula

$$q = k_o^2 \div r$$

in which q = distance from the axis of rotation to the center of percussion; k_o = the radius of gyration of the body with respect to the axis of rotation; and r = the distance from the axis of rotation to the center of gravity of the body.

Velocity and Acceleration

Motion is a progressive change of position of a body. Velocity is the rate of motion, that is, the rate of change of position. When the velocity of a body is the same at every moment during which the motion takes place, the latter is called *uniform* motion. When the velocity is variable and constantly increasing, the rate at which it changes is called *acceleration;* that is, acceleration is the rate at which the velocity of a body changes in a unit of time, as the change in feet per second, in one second. When the motion is decreasing instead of increasing, it is called *retarded* motion, and the rate at which the motion is retarded is frequently called the *deceleration*. If the acceleration is uniform, the motion is called *uniformly accelerated* motion. An example of such motion is found in that of falling bodies.

Motion with Constant Velocity.—In the formulas that follow, S = distance moved; V = velocity; t = time of motion, θ = angle of rotation, and ω = angular velocity; the usual units for these quantities are, respectively, feet, feet per second, seconds, radians, and radians per second. Any other consistent set of units may be employed.

Constant Linear Velocity:

$$S = V \times t \qquad V = S \div t \qquad t = S \div V$$

Constant Angular Velocity:

$$\theta = \omega t \qquad \omega = \theta \div t \qquad t = \theta \div \omega$$

Relation between Angular Motion and Linear Motion: The relation between the angular velocity of a rotating body and the linear velocity of a point at a distance r feet from the center of rotation is:

$$V(\text{ft per sec}) = r(\text{ft}) \times \omega(\text{radians per sec})$$

Similarly, the distance moved by the point during rotation through angle θ is:

$$S(\text{ft}) = r(\text{ft}) \times \theta(\text{radians})$$

Linear Motion with Constant Acceleration.—The relations between distance, velocity, and time for linear motion with constant or uniform acceleration are given by the formulas in the accompanying table. In these formulas, the acceleration is assumed to be in the same direction as the initial velocity; hence, if the acceleration in a particular problem should happen to be in a direction opposite that of the initial velocity, then a should be replaced by $-a$. Thus, for example, the formula $V_f = V_o + at$ becomes $V_f = V_o - at$ when a and V_o are opposite in direction.

Linear Motion with Constant Acceleration

To Find	Known	Formula	To Find	Known	Formula
		Motion Uniformly Accelerated From Rest			
S	a, t	$S = \frac{1}{2}at^2$	t	S, V_f	$t = 2S \div V_f$
	Vf, t	$S = \frac{1}{2}V_f t$		S, a	$t = \sqrt{2S \div a}$
	Vf, a	$S = V_f^2 \div 2a$		a, V_f	$t = V_f \div a$
Vf	a, t	$V_f = at$	a	S, t	$a = 2S \div t^2$
	S, t	$V_f = 2S \div t$		S, V	$a = V_f^2 \div 2S$
	a, S	$V_f = \sqrt{2aS}$		Vf, t	$a = V_f \div t$
		Motion Uniformly Accelerated From Initial Velocity V_o			
S	a, t, V_o	$S = V_o t + \frac{1}{2}at^2$	t	Vo, Vf, a	$t = (V_f - V_o) \div a$
	Vo, Vf, t	$S = (V_f + V_o)t \div 2$		Vo, Vf, S	$t = 2S \div (V_f + V_o)$
	Vo, Vf, a	$S = (V_f^2 - V_o^2) \div 2a$	a	Vo, Vf, S	$a = (V_f^2 - V_o^2) \div 2S$
	Vf, a, t	$S = V_f t - \frac{1}{2}at^2$		Vo, Vf, t	$a = (V_f - V_o) \div t$
Vf	Vo, a, t	$V_f = V_o + at$		Vo, S, t	$a = 2(S - V_o t) \div t^2$
	Vo, S, t	$V_f = (2S \div t) - V_o$		Vf, S, t	$a = 2(V_f t - S) \div t^2$
	Vo, a, S	$V_f = \sqrt{V_o^2 + 2aS}$			
	S, a, t	$V_f = (S \div t) + \frac{1}{2}at$			
Vo	Vf, a, S	$V_o = \sqrt{V_f^2 - 2aS}$			
	Vf, S, t	$V_o = (2S \div t) - V_f$			
	Vf, a, t	$V_o = V_f - at$			
	S, a, t	$V_o = (S \div t) - \frac{1}{2}at$			

Meanings of Symbols

S = distance moved in feet
V_f = final velocity, feet per second
V_o = initial velocity, feet per second
a = acceleration, feet per second per second
t = time of acceleration in seconds

Example: A car is moving at 60 mph when the brakes are suddenly locked and the car begins to skid. If it takes 2 seconds to slow the car to 30 mph, at what rate is it being decelerated, how long is it before the car comes to a halt, and how far will it have traveled?

The initial velocity V_o of the car is 60 mph or 88 ft/sec and the acceleration a due to braking is opposite in direction to V_o, since the car is slowed to 30 mph or 44 ft/sec.

Since V_o, V_f, and t are known, a can be determined from the formula

$$a = (V_f - V_o) \div t = (44 - 88) \div 2$$
$$a = -22 \text{ ft/sec}^2$$

The time required to stop the car can be determined from the formula

$$t = (V_f - V_o) \div a = (0 - 88) \div -22$$
$$t = 4 \text{ seconds}$$

The distance traveled by the car is obtained from the formula

$$S = (V_f + V_o)t \div 2 = (0 + 88)4 \div 2$$
$$= 176 \text{ feet}$$

Rotary Motion with Constant Acceleration.—The relations among angle of rotation, angular velocity, and time for rotation with constant or uniform acceleration are given in the accompanying table.

Rotary Motion with Constant Acceleration

To Find	Known	Formula	To Find	Known	Formula
Motion Uniformly Accelerated From Rest					
	α, t	$\theta = \frac{1}{2}\alpha t^2$		θ, ω_f	$t = 2\theta \div \omega_f$
θ	ω_f, t	$\theta = \frac{1}{2}\omega_f t$	t	θ, α	$t = \sqrt{2\theta \div \alpha}$
	ω_f, α	$\theta = \omega_f^2 \div 2\alpha$		α, ω_f	$t = \omega_f \div \alpha$
	α, t	$\omega_f = \alpha t$		θ, t	$\alpha = 2\theta \div t^2$
ω_f	θ, t	$\omega_f = 2\theta \div t$	α	θ, ω_f	$\alpha = \omega_f^2 \div 2\theta$
	α, θ	$\omega_f = \sqrt{2\alpha\theta}$		ω_f, t	$\alpha = \omega_f \div t$
Motion Uniformly Accelerated From Initial Velocity ω_o					
	α, t, ω_o	$\theta = \omega_o t + \frac{1}{2}\alpha t^2$		$\omega_o, \omega_f, \theta$	$\alpha = (\omega_f - \omega_o^2) \div 2\theta$
θ	ω_o, ω_f, t	$\theta = (\omega_f + \omega_o)t \div 2$	α	ω_o, ω_f, t	$\alpha = (\omega_f - \omega_o) \div t$
	$\omega_o, \omega_f, \alpha$	$\theta = (\omega_f^2 - \omega_o^2) \div 2\alpha$		$\omega_o, \theta\ t$	$\alpha = 2(\theta - \omega_o t) \div t^2$
	ω_f, α, t	$\theta = \omega_f t - \frac{1}{2}\alpha t^2$		$\omega_f, \theta\ t$	$\alpha = 2(\omega_f t - \theta) \div t^2$
	ω_o, α, t	$\omega_f = \omega_o + \alpha t$			
ω_f	ω_o, θ, t	$\omega_f = (2\theta \div t) - \omega_o$			
	ω_o, α, θ	$\omega_f = \sqrt{\omega_o^2 + 2\alpha\theta}$			
	θ, α, t	$\omega_f = (\theta \div t) + \frac{1}{2}\alpha t$			
	ω_f, α, θ	$\omega_o = \sqrt{\omega_f^2 - 2\alpha\theta}$			
ω_o	ω_f, θ, t	$\omega_o = (2\theta \div t) - \omega_f$			
	ω_f, α, t	$\omega_o = \omega_f - \alpha t$			
	θ, α, t	$\omega_o = (\theta \div t) - \frac{1}{2}\alpha t$			
t	$\omega_o, \omega_f, \alpha$	$t = (\omega_f - \omega_o) \div \alpha$			
	$\omega_o, \omega_f, \theta$	$t = 2\theta \div (\omega_f + \omega_o)$			

Meanings of Symbols

θ = angle of rotation, radians
ω_f = final angular velocity, radians per second
ω_o = initial angular velocity, radians per second
α = angular acceleration, radians per second, per second
t = time in seconds

1 degree = 0.01745 radians
(See conversion table on page 90)

In these formulas, the acceleration is assumed to be in the same direction as the initial angular velocity; hence, if the acceleration in a particular problem should happen to be in a direction opposite that of the initial angular velocity, then α should be replaced by $-\alpha$. Thus, for example, the formula $\omega_f = \omega_o + \alpha t$ becomes $\omega_f = \omega_o - \alpha t$ when α and ω_o are opposite in direction.

Linear Acceleration of a Point on a Rotating Body: A point on a body rotating about a fixed axis has a linear acceleration a that is the resultant of two component accelerations. The first component is the centripetal or normal acceleration which is directed from the point P toward the axis of rotation; its magnitude is $r\omega^2$ where r is the radius from the axis to the point P and ω is the angular velocity of the body at the time acceleration a is to be determined. The second component of a is the tangential acceleration which is equal to $r\alpha$ where α is the angular acceleration of the body.

The acceleration of point P is the resultant of $r\omega^2$ and $r\alpha$ and is given by the formula

$$a = \sqrt{(r\omega^2)^2 + (r\alpha)^2}$$

When $\alpha = 0$, this formula reduces to: $a = r\omega^2$

Example: A flywheel on a press rotating at 120 rpm is slowed to 102 rpm during a punching operation that requires ¾ second for the punching portion of the cycle. What angular deceleration does the flywheel experience?

From the table on page 187, the angular velocities corresponding to 120 rpm and 102 rpm, respectively, are 12.57 and 10.68 radians per second. Therefore, using the formula

$$\alpha = (\omega_f - \omega_o) \div t$$

$$\alpha = (10.68 - 12.57) \div \tfrac{3}{4} = -1.89 \div \tfrac{3}{4}$$

$$\alpha = -2.52 \text{ radians per second per second}$$

which is, from the table on page 187, – 24 rpm per second. The minus sign in the answer indicates that the acceleration α acts to slow the flywheel, that is, the flywheel is being decelerated.

Force, Work, Energy, and Momentum

Accelerations Resulting from Unbalanced Forces.—In the section describing the resolution and composition of forces it was stated that when the resultant of a system of forces is zero, the system is in equilibrium, that is, the body on which the force system acts remains at rest or continues to move with uniform velocity. If, however, the resultant of a system of forces is not zero, the body on which the forces act will be accelerated in the direction of the unbalanced force. To determine the relation between the unbalanced force and the resulting acceleration, Newton's laws of motion must be applied. These laws may be stated as follows:

First Law: Every body continues in a state of rest or in uniform motion in a straight line, until it is compelled by a force to change its state of rest or motion.

Second Law: Change of motion is proportional to the force applied, and takes place along the straight line in which the force acts. The "force applied" represents the resultant of *all* the forces acting on the body. This law is sometimes worded: An unbalanced force acting on a body causes an acceleration of the body in the direction of the force and of magnitude proportional to the force and inversely proportional to the mass of the body. Stated as a formula, $R = Ma$ where R is the resultant of *all* the forces acting on the body, M is the mass of the body (mass = weight W divided by acceleration due to gravity g), and a is the acceleration of the body resulting from application of force R.

Third Law: To every action there is always an equal reaction, or, in other words, if a force acts to change the state of motion of a body, the body offers a resistance equal and directly opposite to the force.

Newton's second law may be used to calculate linear and angular accelerations of a body produced by unbalanced forces and torques acting on the body; however, it is necessary first to use the methods described under "Composition and Resolution of Forces" to determine the magnitude and direction of the resultant of *all* forces acting on the body. Then, for a body moving with pure translation,

$$R = Ma = \frac{W}{g}a$$

where R is the resultant force in pounds acting on a body weighing W pounds; g is the gravitational constant, usually taken as 32.16 ft/sec^2, approximately; and a is the resulting acceleration in ft/sec^2 of the body due to R and in the same direction as R.

Using metric SI units, the formula is $R = Ma$, where R = force in newtons (N), M = mass in kilograms, and a = acceleration in meters/second squared. It should be noted that the weight of a body of mass M kg is Mg newtons, where g is approximately 9.81 m/s^2.

Free Body Diagram: In order to correctly determine the effect of forces on the motion of a body it is necessary to resort to what is known as a *free body diagram.* This diagram shows 1) the body removed or isolated from contact with all other bodies that exert force on the body and; and 2) *all* the forces acting on the body.

Thus, for example, in Fig. a the block being pulled up the plane is acted upon by certain forces; the free body diagram of this block is shown at Fig. b. Note that all forces acting on the block are indicated. These forces include: 1) the force of gravity (weight); 2) the pull of the cable, P; 3) the normal component, $W \cos \phi$, of the force exerted on the block by the plane; and 4) the friction force, $\mu W \cos \phi$, of the plane on the block.

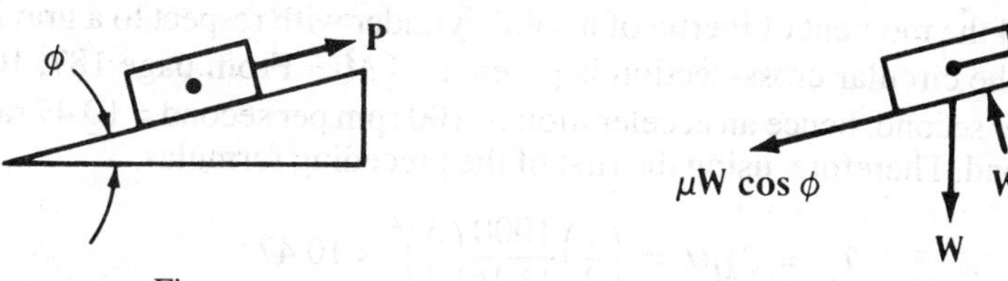

Fig. a. Fig. b.

In preparing a free body diagram, it is important to realize that only those forces exerted *on* the body being considered are shown; forces exerted by the body on other bodies are disregarded. This feature makes the free body diagram an invaluable aid in the solution of problems in mechanics.

Example: A 100-pound body is being hoisted by a winch, the tension in the hoisting cable being kept constant at 110 pounds. At what rate is the body accelerated?

Two forces are acting on the body, its weight, 100 pounds downward, and the pull of the cable, 110 pounds upward. The resultant force R, from a free body diagram, is therefore 110 – 100. Thus, applying Newton's second law,

$$110 - 100 = \frac{100}{32.16}a$$

$$a = \frac{32.16 \times 10}{100} = 3.216 \text{ ft/sec}^2 \text{ upward}$$

It should be noted that since in this problem the resultant force R was positive (110 – 100 = + 10), the acceleration a is also positive, that is, a is in the same direction as R, which is in accord with Newton's second law.

Example using SI metric units: **A body of mass 50 kilograms is being hoistedby a winch, and the tension in the cable is 600 newtons. What is the acceleration? The weight of the 50 kg body is 50*g* newtons, where *g* = approximately 9.81 m/s² (see *Note* on page 163). Applying the formula *R* = *Ma*, the calculation is: (600 – 50*g*) = 50*a*. Thus,**

$$a = \frac{600 - 50g}{50} = \frac{600 - (50 \times 9.81)}{50} = 2.19 \text{ m/s}^2$$

Formulas Relating Torque and Angular Acceleration: For a body rotating about a fixed axis the relation between the unbalanced torque acting to produce rotation and the resulting angular acceleration may be determined from any one of the following formulas, each based on Newton's second law:

$$T_o = J_M \alpha$$

$$T_o = M k_o^2 \alpha$$

$$T_o = \frac{W k_o^2 \alpha}{g} = \frac{W k_o^2 \alpha}{32.16}$$

where T_o is the unbalanced torque in pounds-feet; J_M in ft-lbs-sec² is the moment of inertia of the body about the axis of rotation; k_o in feet is the radius of gyration of the body with respect to the axis of rotation, and α in radians per second, per second is the angular acceleration of the body.

Example: A flywheel has a diameter of 3 feet and weighs 1000 pounds. What torque must be applied, neglecting bearing friction, to accelerate the flywheel at the rate of 100 revolutions per minute, per second?

From page 144 the moment of inertia of a solid cylinder with respect to a gravity axis at right angles to the circular cross-section is given as $\frac{1}{2} Mr^2$. From page 187, 100 rpm = 10.47 radians per second, hence an acceleration of 100 rpm per second = 10.47 radians per second, per second. Therefore, using the first of the preceding formulas,

$$T_o = J_M \alpha = \left(\frac{1}{2}\right)\frac{1000}{32.16}\left(\frac{3}{2}\right)^2 \times 10.47$$

$$= 366 \text{ ft-lbs}$$

Using metric SI units, the formulas are: $T_o = J_M\alpha = Mk_o{}^2\alpha$, where T_o = torque in newton-meters; J_M = the moment of inertia in kg · m², and α = the angular acceleration in radians per second squared.

Example: **A flywheel has a diameter of 1.5 m, and a mass of 800 kg. What torque is needed to produce an angular acceleration of 100 revolutions per minute, per second? As in the preceding example, α = 10.47 rad/s². Thus:**

$$J_M = \tfrac{1}{2}Mr^2 = \tfrac{1}{2} \times 800 \times 0.75^2 = 225 \text{ kg} \cdot \text{m}^2$$

Therefore: $T_o = J_M\alpha = 225 \times 10.47 = 2356$ N · m.

Energy.—A body is said to possess energy when it is capable of doing work or overcoming resistance. The energy may be either mechanical or non-mechanical, the latter including chemical, electrical, thermal, and atomic energy.

Mechanical energy includes *kinetic energy* (energy possessed by a body because of its motion) and *potential energy* (energy possessed by a body because of its position in a field of force and/or its elastic deformation).

Kinetic Energy: The motion of a body may be one of pure translation, pure rotation, or a combination of rotation and translation. By translation is meant motion in which every line in the body remains parallel to its original position throughout the motion, that is, no rotation is associated with the motion of the body.

The kinetic energy of a translating body is given by the formula

$$\text{Kinetic Energy in ft lbs due to translation} = E_{KT} = \tfrac{1}{2}MV^2 = \frac{WV^2}{2g} \qquad \text{(a)}$$

where M = mass of body (= $W \div g$); V = velocity of the center of gravity of the body in feet per second; W = weight of body in pounds; and g = acceleration due to gravity = 32.16 feet per second, per second.

The kinetic energy of a body rotating about a fixed axis O is expressed by the formula:

$$\text{Kinetic Energy in ft lbs due to rotation} = E_{KR} = \tfrac{1}{2}J_{MO}\omega^2 \qquad \text{(b)}$$

where J_{MO} is the moment of inertia of the body about the fixed axis O in pounds-feet-seconds^2, and ω = angular velocity in radians per second.

For a body that is moving with both translation and rotation, the total kinetic energy is given by the following formula as the sum of the kinetic energy due to translation of the center of gravity and the kinetic energy due to rotation about the center of gravity:

$$\begin{aligned}\text{Total Kinetic Energy in ft lbs} &= E_T = \tfrac{1}{2}MV^2 + \tfrac{1}{2}J_{MG}\omega^2 \\ &= \frac{WV^2}{2g} + \tfrac{1}{2}J_{MG}\omega^2 \\ &= \frac{WV^2}{2g} + \frac{1}{2}\frac{Wk^2\omega^2}{g} \qquad \text{(c)} \\ &= \frac{W}{2g}(V^2 + k^2\omega^2)\end{aligned}$$

where J_{MG} is the moment of inertia of the body about its gravity axis in pounds-feet-seconds^2, k is the radius of gyration in feet with respect to an axis through the center of gravity, and the other quantities are as previously defined.

In the metric SI system, energy is expressed as the joule (J). One joule = 1 newton-meter. The kinetic energy of a translating body is given by the formula $E_{KT} = \frac{1}{2}MV^2$, where M = mass in kilograms, and V = velocity in meters per second. Kinetic energy due to rotation is expressed by the formula $E_{KR} = \frac{1}{2}J_{MO}\omega^2$, where J_{MO} = moment of inertia in kg · m^2, and ω = the angular velocity in radians per second. Total kinetic energy $ET = \frac{1}{2}MV^2 + \frac{1}{2}J_{MO}\omega^2$ joules = $\frac{1}{2}M(V^2 + k^2\omega^2)$ joules, where k = radius of gyration in meters.

Potential Energy: The most common example of a body having potential energy because of its position in a field of force is that of a body elevated to some height above the earth. Here the field of force is the gravitational field of the earth and the potential energy E_{PF} of a body weighing W pounds elevated to some height S in feet above the surface of the earth is WS foot-pounds. If the body is permitted to drop from this height its potential energy E_{PF} will be converted to kinetic energy. Thus, after falling through height S the kinetic energy of the body will be WS ft-lbs.

In metric SI units, the potential energy E_{PF} of a body of mass M kilograms elevated to a height of S meters, is MgS joules. After it has fallen a distance S, the kinetic energy gained will thus be MgS joules.

Another type of potential energy is elastic potential energy, such as possessed by a spring that has been compressed or extended. The amount of work in ft lbs done in compressing the spring S feet is equal to $KS^2/2$, where K is the spring constant in pounds per foot. Thus, when the spring is released to act against some resistance, it can perform $KS^2/2$ ft-lbs of work which is the amount of elastic potential energy E_{PE} stored in the spring.

Using metric SI units, the amount of work done in compressing the spring a distance S meters is $KS^2/2$ joules, where K is the spring constant in newtons per meter.

Work Performed by Forces and Couples.—The work U done by a force F in moving an object along some path is the product of the distance S the body is moved and the component $F \cos \alpha$ of the force F in the direction of S.

$$U = FS\cos\alpha$$

where U = work in ft-lbs; S = distance moved in feet; F = force in lbs; and α = angle between line of action of force and the path of S.

If the force is in the same direction as the motion, then $\cos \alpha = \cos 0 = 1$ and this formula reduces to:

$$U = FS$$

Similarly, the work done by a couple T turning an object through an angle θ is:

$$U = T\theta$$

where T = torque of couple in pounds-feet and θ = the angular rotation in radians.

The above formulas can be used with metric SI units: U is in joules; S is in meters; F is in newtons, and T is in newton-meters.

Relation between Work and Energy.—Theoretically, when work is performed on a body and there are no energy losses (such as due to friction, air resistance, etc.), the energy acquired by the body is equal to the work performed on the body; this energy may be either potential, kinetic, or a combination of both.

In actual situations, however, there may be energy losses that must be taken into account. Thus, the relation between work done on a body, energy losses, and the energy acquired by the body can be stated as:

$$\text{Work Performed} - \text{Losses} = \text{Energy Acquired}$$

$$U - \text{Losses} = E_T$$

Example 1: A 12-inch cube of steel weighing 490 pounds is being moved on a horizontal conveyor belt at a speed of 6 miles per hour (8.8 feet per second). What is the kinetic energy of the cube?

Since the block is not rotating, Formula (a) for the kinetic energy of a body moving with pure translation applies:

$$\text{Kinetic Energy} = \frac{WV^2}{2g}$$

$$= \frac{490 \times (8.8)^2}{2 \times 32.16} = 590 \text{ ft-lbs}$$

A similar example using metric SI units is as follows: If a cube of mass 200 kg is being moved on a conveyor belt at a speed of 3 meters per second, what is the kinetic energy of the cube? It is:

$$\textbf{Kinetic Energy} = \tfrac{1}{2}MV^2 = \tfrac{1}{2} \times 200 \times 3^2 = 900 \textbf{ joules}$$

Example 2: If the conveyor in Example 1 is brought to an abrupt stop, how long would it take for the steel block to come to a stop and how far along the belt would it slide before stopping if the coefficient of friction μ between the block and the conveyor belt is 0.2 and the block slides without tipping over?

The only force acting to slow the motion of the block is the friction force between the block and the belt. This force F is equal to the weight of the block, W, multiplied by the coefficient of friction; $F = \mu W = 0.2 \times 490 = 98$ lbs.

The time required to bring the block to a stop can be determined from the impulse-momentum Formula (c) on page 160.

$$R \times t = \frac{W}{g}(V_f - V_o)$$

$$(-98)t = \frac{490}{32.16} \times (0 - 8.8)$$

$$t = \frac{490 \times 8.8}{98 \times 32.16} = 1.37 \text{ seconds}$$

The distance the block slides before stopping can be determined by equating the kinetic energy of the block and the work done by friction in stopping it:

$$\text{Kinetic energy of block}(WV^2/2g) = \text{Work done by friction}(F \times S)$$

$$590 = 98 \times S$$

$$S = \frac{590}{98} = 6.0 \text{ feet}$$

If metric SI units are used, the calculation is as follows (for the cube of 200 kg mass): The friction force = μ multiplied by the weight Mg where g = approximately 9.81 m/s². Thus, $\mu Mg = 0.2 \times 200g = 392.4$ newtons. The time t required to bring the block to a stop is $(-392.4)t = 200(0 - 3)$. Therefore,

$$t = \frac{200 \times 3}{392.4} = 1.53 \text{ seconds}$$

The kinetic energy of the block is equal to the work done by friction, that is $392.4 \times S = 900$ joules. Thus, the distance S which the block moves before stopping is

$$S = \frac{900}{392.4} = 2.29 \text{ meters}$$

Force of a Blow.—A body that weighs W pounds and falls S feet from an initial position of rest is capable of doing WS foot-pounds of work. The work performed during its fall may be, for example, that necessary to drive a pile a distance d into the ground. Neglecting losses in the form of dissipated heat and strain energy, the work done in driving the pile is equal to the product of the impact force acting on the pile and the distance d which the pile is driven. Since the impact force is not accurately known, an average value, called the "average force of the blow," may be assumed. Equating the work done on the pile and the work done by the falling body, which in this case is a pile driver:

$$\text{Average force of blow} \times d = WS$$

or,

$$\text{Average force of blow} = \frac{WS}{d}$$

where, S = total height in feet through which the driver falls, including the distance d that the pile is driven

W = weight of driver in pounds

d = distance in feet which pile is driven

When using metric SI units, it should be noted that a body of mass *M* kilograms has a weight of *Mg* newtons, where g = approximately 9.81 m/s². If the body falls a distance *S* meters, it can do work equal to *MgS* joules. The average force of the blow is *MgS/d* newtons, where *d* is the distance in meters that the pile is driven.

Example: A pile driver weighing 200 pounds strikes the top of the pile after having fallen from a height of 20 feet. It forces the pile into the ground a distance of ½ foot. Before the ram is brought to rest, it will $200 \times (20 + \frac{1}{2}) = 4100$ foot-pounds of work, and as this energy is expended in a distance of one-half foot, the average force of the blow equals $4100 \div \frac{1}{2} =$ 8200 pounds.

A similar example using metric SI units is as follows: A pile driver of mass 100 kilograms falls 10 meters and moves the pile a distance of 0.3 meters. The work done = $100g(10 + 0.3)$ joules, and it is expended in 0.3 meters. Thus, the average force is

$$\frac{100g \times 10.3}{0.3} = 33680 \text{ newtons or } 33.68 \text{ kN}$$

Impulse and Momentum.—The *linear momentum* of a body is defined as the product of the mass *M* of the body and the velocity *V* of the center of gravity of the body:

$$\text{Linear momentum} = MV \text{ or since } M = W \div g$$

$$\text{Linear momentum} = \frac{WV}{g} \qquad \text{(a)}$$

It should be noted that linear momentum is a vector quantity, the momentum being in the same direction as *V*.

Linear impulse: is defined as the product of the resultant *R* of *all* the forces acting on a body and the time *t* that the resultant acts:

$$\text{Linear Impulse} = Rt \qquad \text{(b)}$$

The change in the linear momentum of a body is numerically equal to the linear impulse that causes the change in momentum:

$$\text{Linear Impulse} = \text{change in Linear Momentum}$$

$$Rt = \frac{W}{g}V_f - \frac{W}{g}V_o = \frac{W}{g}(V_f - V_o) \qquad \text{(c)}$$

where V_f, the final velocity of the body after time *t*, and V_o, the initial velocity of the body, are both in the same direction as the applied force *R*. If V_o, and V_f are in opposite directions, then the minus sign in the formula becomes a plus sign.

In metric SI units, the formulas are: Linear Momentum = *MV* kg · m/s, where *M* = mass in kg, and *V* = velocity in meters per second; and Linear Impulse = *Rt* newton-seconds, where *R* = force in newtons, and *t* = time in seconds. In Formula (c) above, *W/g* is replaced by *M* when SI units are used.

Example: A 1000-pound block is pulled up a 2-degree incline by a cable exerting a constant force *F* of 600 pounds. If the coefficient of friction μ between the block and the plane is 0.5, how fast will the block be moving up the plane 10 seconds after the pull is applied?

The resultant force *R* causing the body to be accelerated up the plane is the difference between *F*, the force acting up the plane, and *P*, the force acting to resist motion up the plane. This latter force for a body on a plane is given by the formula at the top of page 130 as $P = W(\mu \cos\alpha + \sin\alpha)$ where α is the angle of the incline.

Thus, $R = F - P = F - W(\mu \cos\alpha + \sin\alpha)$

$$= 600 - 1000(0.5 \cos 2^\circ + \sin 2^\circ)$$

$$= 600 - 1000(0.5 \times 0.99939 + 0.03490)$$

$$= 600 - 535$$

$$R = 65 \text{ pounds.}$$

Formula (c) can now be applied to determine the speed at which the body will be moving up the plane after 10 seconds.

$$Rt = \frac{W}{g}V_f - \frac{W}{g}V_o$$

$$65 \times 10 = \frac{1000}{32.2}V_f - \frac{1000}{32.2} \times 0$$

$$V_f = \frac{65 \times 10 \times 32.2}{1000} = 20.9 \text{ ft per sec}$$

$$= 14.3 \text{ miles per hour}$$

A similar example using metric SI units is as follows: A 500 kg block is pulled up a 2 degree incline by a constant force *F* of 4 kN. The coefficient of friction μ between the block and the plane is 0.5. How fast will the block be moving 10 seconds after the pull is applied?

The resultant force *R* is:

$$R = F - Mg(\mu\cos\alpha + \sin\alpha)$$

$$= 4000 - 500 \times 9.81(0.5 \times 0.99939 + 0.03490)$$

$$= 1378\text{N or } 1.378 \text{ kN}$$

Formula (c) can now be applied to determine the speed at which the body will be moving up the plane after 10 seconds. Replacing *W/g* by *M* in the formula, the calculation is:

$$Rt = MV_f - MV_o$$

$$1378 \times 10 = 500(V_f - 0)$$

$$V_f = \frac{1378 \times 10}{500} = 27.6 \text{ m/s}$$

Angular Impulse and Momentum: In a manner similar to that for linear impulse and moment, the formulas for angular impulse and momentum for a body rotating about a fixed axis are:

$$\text{Angular momentum} = J_M\omega \qquad (a)$$

$$\text{Angular impulse} = T_o t \qquad (b)$$

where J_M is the moment of inertia of the body about the axis of rotation in pounds-feet-seconds^2, ω is the angular velocity in radians per second, T_o, is the torque in pounds-feet about the axis of rotation, and *t* is the time in seconds that T_o, acts.

The change in angular momentum of a body is numerically equal to the angular impulse that causes the change in angular momentum:

$$\text{Angular Impulse} = \text{Change in Angular Momentum}$$

$$T_o t = J_M\omega_f - J_M\omega_o = J_M(\omega_f - \omega_o) \qquad (c)$$

where ω_f and ω_o are the final and initial angular velocities, respectively.

Example: A flywheel having a moment of inertia of 25 lbs-ft-sec^2 is revolving with an angular velocity of 10 radians per second when a constant torque of 20 lbs-ft is applied to reverse its direction of rotation. For what length of time must this constant torque act to stop the flywheel and bring it up to a reverse speed of 5 radians per second?

Applying Formula (c),

$$T_o t = J_M(\omega_f - \omega_o)$$

$$20t = 25(10 - [-5]) = 250 + 125$$

$$t = 375 \div 20 = 18.8 \text{ seconds}$$

A similar example using metric SI units is as follows: A flywheel with a moment of inertia of 20 kilogram-meters² is revolving with an angular velocity of 10 radians per second when a constant torque of 30 newton-meters is applied to reverse its direction of rotation. For what length of time must this constant torque act to stop the flywheel and bring it up to a reverse speed of 5 radians per second? Applying Formula (c), the calculation is:

$$\mathbf{T_o t = J_M(\omega_f - \omega_o),}$$

$$\mathbf{30t = 20(10 - [-5]).}$$

$$\textbf{Thus, } t = \frac{20 \times 15}{30} = \textbf{10 seconds}$$

Formulas for Work and Power.—The formulas in the accompanying table may be used to determine work and power in terms of the applied force and the velocity at the point of application of the force.

Formulas for Work and Power

To Find	Known	Formula	To Find	Known	Formula
S	P, t, F	$S = P \times t \div F$	P	F, V	$P = F \times V$
	K, F	$S = K \div F$		F, S, t	$P = F \times S \div t$
	t, F, hp	$S = 550 \times t \times hp \div F$		K, t	$P = K \div t$
V	P, F	$V = P \div F$		hp	$P = 550 \times hp$
	K, F, t	$V = K \div (F \times t)$	K	F, S	$K = F \times S$
	F, hp	$V = 550 \times hp \div F$		P, t	$K = P \times t$
t	F, S, P	$t = F \times S \div P$		F, V, t	$K = F \times V \times t$
	K, F, V	$t = K \div (F \times V)$		t, hp	$K = 550 \times t \times hp$
	F, S, hp	$t = F \times S \div (550 \times hp)$	hp	F, S, t	$hp = F \times S \div (550 \times t)$
F	P, V	$F = P \div V$		P	$hp = P \div 550$
	K, S	$F = K \div S$		F, V	$hp = F \times V \div 550$
	K, V, t	$F = K \div (V \times t)$		K, t	$hp = K \div (550 \times t)$
	V, hp	$F = 550 \times hp \div V$			

Meanings of Symbols: S = distance in feet; V = constant or average velocity in feet per second; t = time in seconds; F = constant or average force in pounds; P = power in foot-pounds per second; K = work in foot-pounds; and hp = horsepower.

Note: **The metric SI unit of work is the joule (one joule = 1 newton-meter), and the unit of power is the watt (one watt = 1 joule per second = 1 N · m/s). The term horsepower is not used. Thus, those formulas above that involve horsepower and the factor 550 are not applicable when working in SI units. The remaining formulas can be used, and the units are: S = distance**

in meters; V = constant or average velocity in meters per second; t = time in seconds; F = force in newtons; P = power in watts; K = work in joules.

Example: A casting weighing 300 pounds is to be lifted by means of an overhead crane. The casting is lifted 10 feet in 12 seconds. What is the horsepower developed? Here $F = 300$; $S = 10$; $t = 12$.

$$\text{hp} = \frac{F \times S}{550t} = \frac{300 \times 10}{550 \times 12} = 0.45$$

A similar example using metric SI units is as follows: A casting of mass 150 kg is lifted 4 meters in 15 seconds by means of a crane. What is the power? Here $F = 150g$ N, $S = 4$ m, and $t = 15$ s. Thus:

$$\text{Power} = \frac{FS}{t} = \frac{150g \times 4}{15} = \frac{150 \times 9.81 \times 4}{15}$$
$$= 392 \text{ watts or } 0.392 \text{ kW}$$

Centrifugal Force

Centrifugal Force.—When a body rotates about any axis other than one at its center of mass, it exerts an outward radial force called centrifugal force upon the axis or any arm or cord from the axis that restrains it from moving in a straight (tangential) line. In the following formulas:

F = centrifugal force in pounds

W = weight of revolving body in pounds

v = velocity at radius R on body in feet per second

n = number of revolutions per minute

g = acceleration due to gravity = 32.16 feet per second per second

R = perpendicular distance in feet from axis of rotation to center of mass, or for practical use, to center of gravity of revolving body

Note: If a body rotates about its own center of mass, R equals zero and v equals zero. This means that the *resultant* of the centrifugal forces of all the elements of the body is equal to zero or, in other words, no centrifugal force is exerted on the axis of rotation. The centrifugal force of any part or element of such a body is found by the equations given below, where R is the radius to the center of gravity of the part or element. In a flywheel rim, R is the mean radius of the rim because it is the radius to the center of gravity of a thin radial section.

$$F = \frac{Wv^2}{gR} = \frac{Wv^2}{32.16R} = \frac{4WR\pi^2n^2}{60 \times 60g} = \frac{WRn^2}{2933} = 0.000341\,WRn^2$$

$$W = \frac{FRg}{v^2} = \frac{2933F}{Rn^2} \qquad v = \sqrt{\frac{FRg}{W}}$$

$$R = \frac{Wv^2}{Fg} = \frac{2933F}{Wn^2} \qquad n = \sqrt{\frac{2933F}{WR}}$$

(If n is the number of revolutions per second instead of per minute, then $F = 1227WRn^2$.)

If metric SI units are used in the foregoing formulas, W/g is replaced by M, which is the mass in kilograms; F = centrifugal force in newtons; v = velocity in meters per second; n = number of revolutions per minute; and R = the radius in meters. Thus:

$$F = Mv^2/R = \frac{Mn^2(2\pi R^2)}{60^2R} = 0.01097\ MRn^2$$

If the rate of rotation is expressed as n_1 = revolutions per second, then $F = 39.48 MRn_1^2$; if it is expressed as ω radians per second, then $F = MR\omega^2$.

Calculating Centrifugal Force.—In the ordinary formula for centrifugal force, $F = 0.000341\ WRn^2$; the mean radius R of the flywheel or pulley rim is given in feet. For small dimensions, it is more convenient to have the formula in the form:

$$F = 0.2842 \times 10^{-4} Wrn^2$$

in which F = centrifugal force, in pounds; W = weight of rim, in pounds; r = mean radius of rim, in inches; n = number of revolutions per minute.

In this formula let $C = 0.000028416n^2$. This, then, is the centrifugal force of one pound, one inch from the axis. The formula can now be written in the form,

$$F = WrC$$

C is calculated for various values of the revolutions per minute n, and the calculated values of C are given in Table 1. To find the centrifugal force in any given case, simply find the value of C in the table and multiply it by the product of W and r, the four multiplications in the original formula given thus having been reduced to two.

Example: A cast-iron flywheel with a mean rim radius of 9 inches, is rotated at a speed of 800 revolutions per minute. If the weight of the rim is 20 pounds, what is the centrifugal force?

From Table 1, for $n = 800$ revolutions per minute, the value of C is 18.1862.

Thus,

$$F = WrC$$
$$= 20 \times 9 \times 18.1862$$
$$= 3273.52 \text{ pounds}$$

Using metric SI units, $0.01097n^2$ is the centrifugal force acting on a body of 1 kilogram mass rotating at n revolutions per minute at a distance of 1 meter from the axis. If this value is designated C_1, then the centrifugal force of mass M kilograms rotating at this speed at a distance from the axis of R meters, is C_1MR newtons. To simplify calculations, values for C_1 are given in Table 2. If it is required to work in terms of millimeters, the force is $0.001\ C_1MR_1$ newtons, where R_1 is the radius in millimeters.

***Example:* A steel pulley with a mean rim radius of 120 millimeters is rotated at a speed of 1100 revolutions per minute. If the mass of the rim is 5 kilograms, what is the centrifugal force?**

From Table 2, for $n = 1100$ revolutions per minute, the value of C_1 is 13,269.1.

Thus,

$$F = 0.001\ C_1MR_1$$
$$= 0.001 \times 13{,}269.1 \times 5 \times 120$$
$$= 7961.50 \text{ newtons}$$

Table 1. Factors *C* for Calculating Centrifugal Force (English units)

n	*C*	*n*	*C*	*n*	*C*	*n*	*C*
50	0.07104	100	0.28416	470	6.2770	5200	768.369
51	0.07391	101	0.28987	480	6.5470	5300	798.205
52	0.07684	102	0.29564	490	6.8227	5400	828.611
53	0.07982	103	0.30147	500	7.1040	5500	859.584
54	0.08286	104	0.30735	600	10.2298	5600	891.126
55	0.08596	105	0.31328	700	13.9238	5700	923.236
56	0.08911	106	0.31928	800	18.1862	5800	955.914
57	0.09232	107	0.32533	900	23.0170	5900	989.161
58	0.09559	108	0.33144	1000	28.4160	6000	1022.980
59	0.09892	109	0.33761	1100	34.3834	6100	1057.360
60	0.10230	110	0.34383	1200	40.9190	6200	1092.310
61	0.10573	115	0.37580	1300	48.0230	6300	1127.830
62	0.10923	120	0.40921	1400	55.6954	6400	1163.920
63	0.11278	125	0.44400	1500	63.9360	6500	1200.580
64	0.11639	130	0.48023	1600	72.7450	6600	1237.800
65	0.12006	135	0.51788	1700	82.1222	6700	1275.590
66	0.12378	140	0.55695	1800	92.0678	6800	1313.960
67	0.12756	145	0.59744	1900	102.5820	6900	1352.890
68	0.13140	150	0.63936	2000	113.6640	7000	1392.380
69	0.13529	160	0.72745	2100	125.3150	7100	1432.450
70	0.13924	170	0.82122	2200	137.5330	7200	1473.090
71	0.14325	180	0.92067	2300	150.3210	7300	1514.290
72	0.14731	190	1.02590	2400	163.6760	7400	1556.060
73	0.15143	200	1.1367	2500	177.6000	7500	1598.400
74	0.15561	210	1.2531	2600	192.0920	7600	1641.310
75	0.15984	220	1.3753	2700	207.1530	7700	1684.780
76	0.16413	230	1.5032	2800	222.7810	7800	1728.830
77	0.16848	240	1.6358	2900	238.9790	7900	1773.440
78	0.17288	250	1.7760	3000	255.7400	8000	1818.620
79	0.17734	260	1.9209	3100	273.0780	8100	1864.370
80	0.18186	270	2.0715	3200	290.9800	8200	1910.690
81	0.18644	280	2.2278	3300	309.4500	8300	1957.580
82	0.19107	290	2.3898	3400	328.4890	8400	2005.030
83	0.19576	300	2.5574	3500	348.0960	8500	2053.060
84	0.20050	310	2.7308	3600	368.2710	8600	2101.650
85	0.20530	320	2.9098	3700	389.0150	8700	2150.810
86	0.21016	330	3.0945	3800	410.3270	8800	2200.540
87	0.21508	340	3.2849	3900	432.2070	8900	2250.830
88	0.22005	350	3.4809	4000	454.6560	9000	2301.700
89	0.22508	360	3.6823	4100	477.6730	9100	2353.130
90	0.23017	370	3.8901	4200	501.2580	9200	2405.130
91	0.23531	380	4.1032	4300	525.4120	9300	2457.700
92	0.24051	390	4.3220	4400	550.1340	9400	2510.840
93	0.24577	400	4.5466	4500	575.4240	9500	2564.540
94	0.25108	410	4.7767	4600	601.2830	9600	2618.820
95	0.25645	420	5.0126	4700	627.7090	9700	2673.660
96	0.26188	430	5.2541	4800	654.7050	9800	2729.070
97	0.26737	440	5.5013	4900	682.2680	9900	2785.050
98	0.27291	450	5.7542	5000	710.4000	10000	2841.600
99	0.27851	460	6.0128	5100	739.1000	…	…

Table 2. Factors C_1 for Calculating Centrifugal Force (Metric SI units)

n	C_1	n	C_1	n	C_1	n	C_1
50	27.4156	100	109.662	470	2,422.44	5200	296,527
51	28.5232	101	111.867	480	2,526.62	5300	308,041
52	29.6527	102	114.093	490	2,632.99	5400	319,775
53	30.8041	103	116.341	500	2,741.56	5500	331,728
54	31.9775	104	118.611	600	3,947.84	5600	343,901
55	33.1728	105	120.903	700	5,373.45	5700	356,293
56	34.3901	106	123.217	800	7,018.39	5800	368,904
57	35.6293	107	125.552	900	8,882.64	5900	381,734
58	36.8904	108	127.910	1000	10,966.2	6000	394,784
59	38.1734	109	130.290	1100	13,269.1	6100	408,053
60	39.4784	110	132.691	1200	15,791.4	6200	421,542
61	40.8053	115	145.028	1300	18,532.9	6300	435,250
62	42.1542	120	157.914	1400	21,493.8	6400	449,177
63	43.5250	125	171.347	1500	24,674.0	6500	463,323
64	44.9177	130	185.329	1600	28,073.5	6600	477,689
65	46.3323	135	199.860	1700	31,692.4	6700	492,274
66	47.7689	140	214.938	1800	35,530.6	6800	507,078
67	49.2274	145	230.565	1900	39,588.1	6900	522,102
68	50.7078	150	246.740	2000	43,864.9	7000	537,345
69	52.2102	160	280.735	2100	48,361.1	7100	552,808
70	53.7345	170	316.924	2200	53,076.5	7200	568,489
71	55.2808	180	355.306	2300	58,011.3	7300	584,390
72	56.8489	190	395.881	2400	63,165.5	7400	600,511
73	58.4390	200	438.649	2500	68,538.9	7500	616,850
74	60.0511	210	483.611	2600	74,131.7	7600	633,409
75	61.6850	220	530.765	2700	79,943.8	7700	650,188
76	63.3409	230	580.113	2800	85,975.2	7800	667,185
77	65.0188	240	631.655	2900	92,226.0	7900	684,402
78	66.7185	250	685.389	3000	98,696.0	8000	701,839
79	68.4402	260	741.317	3100	105,385	8100	719,494
80	70.1839	270	799.438	3200	112,294	8200	737,369
81	71.9494	280	859.752	3300	119,422	8300	755,463
82	73.7369	290	922.260	3400	126,770	8400	773,777
83	75.5463	300	986.960	3500	134,336	8500	792,310
84	77.3777	310	1,053.85	3600	142,122	8600	811,062
85	79.2310	320	1,122.94	3700	150,128	8700	830,034
86	81.1062	330	1,194.22	3800	158,352	8800	849,225
87	83.0034	340	1,267.70	3900	166,796	8900	868,635
88	84.9225	350	1,343.36	4000	175,460	9000	888,264
89	86.8635	360	1,421.22	4100	184,342	9100	908,113
90	88.8264	370	1,501.28	4200	193,444	9200	928,182
91	90.8113	380	1,583.52	4300	202,766	9300	948,469
92	92.8182	390	1,667.96	4400	212,306	9400	968,976
93	94.8469	400	1,754.60	4500	222,066	9500	989,702
94	96.8976	410	1,843.42	4600	232,045	9600	1,010,650
95	98.9702	420	1,934.44	4700	242,244	9700	1,031,810
96	101.065	430	2,027.66	4800	252,662	9800	1,053,200
97	103.181	440	2,123.06	4900	263,299	9900	1,074,800
98	105.320	450	2,220.66	5000	274,156	10000	1,096,620
99	107.480	460	2,320.45	5100	285,232	…	…

Balancing Rotating Parts

Static Balancing.—There are several methods of testing the standing or static balance of a rotating part. A simple method that is sometimes used for flywheels, etc., is illustrated by the diagram, Fig. 1. An accurate shaft is inserted through the bore of the finished wheel, which is then mounted on carefully leveled "parallels" A. If the wheel is in an unbalanced state, it will turn until the heavy side is downward. When it will stand in any position as the result of counterbalancing and reducing the heavy portions, it is said to be in standing or static balance. Another test which is used for disk-shaped parts is shown in Fig. 2. The disk D is mounted on a vertical arbor attached to an adjustable cross-slide B. The latter is carried by a table C, which is supported by a knife-edged bearing. A pendulum having an adjustable screw-weight W at the lower end is suspended from cross-slide B. To test the static balance of disk D, slide B is adjusted until pointer E of the pendulum coincides with the center of a stationary scale F. Disk D is then turned halfway around without moving the slide, and if the indicator remains stationary, it shows that the disk is in balance for this particular position. The test is then repeated for ten or twelve other positions, and the heavy sides are reduced, usually by drilling out the required amount of metal. Several other devices for testing static balance are designed on this same principle.

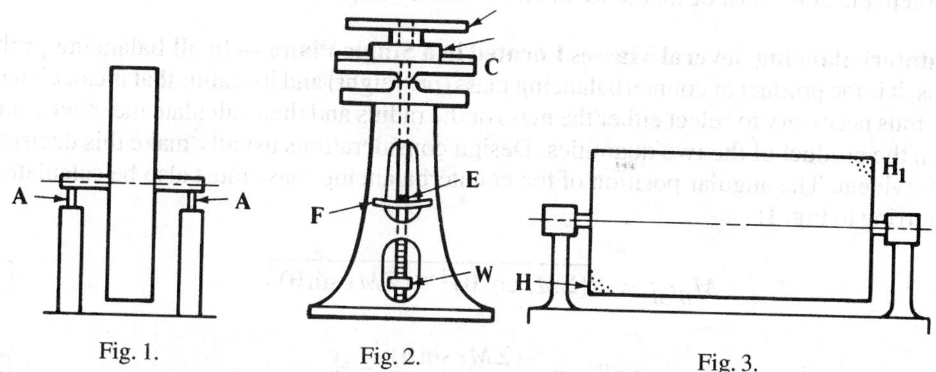

Fig. 1. Fig. 2. Fig. 3.

Running or Dynamic Balance.—A cylindrical body may be in perfect static balance and not be in a balanced state when rotating at high speed. If the part is in the form of a thin disk, static balancing, if carefully done, may be accurate enough for high speeds, but if the rotating part is long in proportion to its diameter, and the unbalanced portions are at opposite ends or in different planes, the balancing must be done so as to counteract the centrifugal force of these heavy parts when they are rotating rapidly. This process is known as a running balance or dynamic balancing. To illustrate, if a heavy section is located at H (Fig. 3), and another correspondingly heavy section at H_1, one may exactly counterbalance the other when the cylinder is stationary, and this static balance may be sufficient for a part rigidly mounted and rotating at a comparatively slow speed; but when the speed is very high, as in turbine rotors, etc., the heavy masses H and H_1, being in different planes, are in an unbalanced state owing to the effect of centrifugal force, which results in excessive strains and injurious vibrations. Theoretically, to obtain a perfect running balance, the exact positions of the heavy sections should be located and the balancing effected either by reducing their weight or by adding counterweights opposite each section and in the same plane at the proper radius; but if the rotating part is rigidly mounted on a stiff shaft, a running balance that is sufficiently accurate for practical purposes can be obtained by means of comparatively few counterbalancing weights located with reference to the unbalanced parts.

Balancing Calculations.—As indicated previously, centrifugal forces caused by an unbalanced mass or masses in a rotating machine member cause additional loads on the bearings which are transmitted to the housing or frame and to other machine members. Such dynamically unbalanced conditions can occur even though static balance (balance at zero speed) exists. Dynamic balance can be achieved by the addition of one or two masses rotating about the same axis and at the same speed as the unbalanced masses. A single unbalanced mass can be balanced by one counterbalancing mass located 180 degrees opposite and in the same plane of rotation as the unbalanced mass, if the product of their respective radii and masses are equal; i.e., $M_1r_1 = M_2r_2$. Two or more unbalanced masses rotating in the same plane can be balanced by a single mass rotating in the same plane, or by two masses rotating about the same axis in two separate planes. Likewise, two or more unbalanced masses rotating in different planes about a common axis can be balanced by two masses rotating about the same axis in separate planes. When the unbalanced masses are in separate planes they may be in static balance but not in dynamic balance; i.e., they may be balanced when not rotating but unbalanced when rotating. If a system is in dynamic balance, it will remain in balance at all speeds, although this is not strictly true at the critical speed of the system. (See *Critical Speeds*.)

In all the equations that follow, the symbol M denotes either mass in kilograms or in slugs, or weight in pounds. Either mass or weight units may be used and the equations may be used with metric or with customary English units without change; however, in a given problem the units must be all metric or all customary English.

Counterbalancing Several Masses Located in a Single Plane.—In all balancing problems, it is the product of counterbalancing mass (or weight) and its radius that is calculated; it is thus necessary to select either the mass or the radius and then calculate the other value from the product of the two quantities. Design considerations usually make this decision self-evident. The angular position of the counterbalancing mass must also be calculated. Referring to Fig. 1:

$$M_B r_B = \sqrt{(\Sigma Mr\cos\theta)^2 + (\Sigma Mr\sin\theta)^2} \tag{1}$$

$$\tan\theta_B = \frac{-(\Sigma Mr\sin\theta)}{-(\Sigma Mr\cos\theta)} = \frac{y}{x} \tag{2}$$

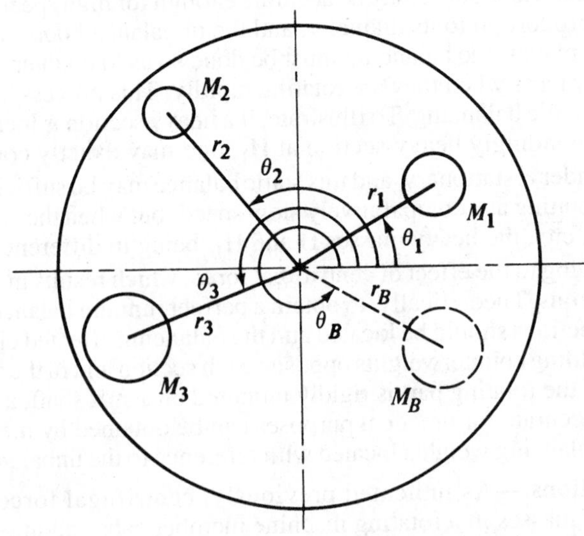

Fig. 1.

Table 1. Relationship of the Signs of the Functions of the Angle with Respect to the Quadrant in Which They Occur

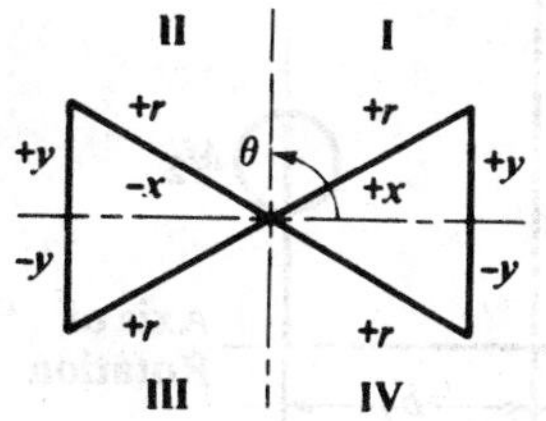	Angle θ			
	0° to 90°	90° to 180°	180° to 270°	270° to 360°
	Signs of the Functions			
tan	$\frac{+y}{+x}$	$\frac{+y}{-x}$	$\frac{-y}{-x}$	$\frac{-y}{+x}$
sine	$\frac{+y}{+r}$	$\frac{+y}{+r}$	$\frac{-y}{+r}$	$\frac{-y}{+r}$
cosine	$\frac{+x}{+r}$	$\frac{-x}{+r}$	$\frac{-x}{+r}$	$\frac{+x}{+r}$

where:

$M_1, M_2, M_3, \ldots M_n$ = any unbalanced mass or weight, kg or lb

M_B = counterbalancing mass or weight, kg or lb

r = radius to center of gravity of any unbalanced mass or weight, mm or inch

r_B = radius to center of gravity of counterbalancing mass or weight, mm or inch

θ = angular position of r of any unbalanced mass or weight, degrees

θ_B = angular position of r_B of counterbalancing mass or weight, degrees

x and y = see Table 1

Table 1 is helpful in finding the angular position of the counterbalancing mass or weight. It indicates the range of the angles within which this angular position occurs by noting the plus and minus signs of the numerator and the denominator of the terms in Equation (2). In a like manner, Table 1 is helpful in determining the *sign* of the sine or cosine functions for angles ranging from 0 to 360 degrees. Balancing problems are usually solved most conveniently by arranging the arithmetical calculations in a tabular form.

Example: Referring to Fig. 1, the particular values of the unbalanced weights have been entered in the table below. Calculate the magnitude of the counterbalancing weight if its radius is to be 10 inches.

M		*r*	θ				
No.	lb.	in.	deg.	cos θ	sin θ	*Mr* cos θ	*Mr* sin θ
1	10	10	30	0.8660	0.5000	86.6	50.0
2	5	20	120	−0.5000	0.8660	−50.0	86.6
3	15	15	200	−0.9397	−0.3420	−211.4	−77.0
						−174.8	59.6
						= Σ*Mr* cos θ	= Σ*Mr* sin θ

$$M_B = \frac{\sqrt{(\Sigma Mr\cos\theta)^2 + (\Sigma Mr\sin\theta)^2}}{r_B} = \frac{\sqrt{(-174.8)^2 + (59.6)^2}}{10}$$

$$M_B = 18.5 \text{ lb}$$

$$\tan\theta_B = \frac{-(\Sigma Mr\sin\theta)}{-(\Sigma Mr\cos\theta)} = \frac{-(59.6)}{-(-174.8)} = \frac{-y}{+x};\quad \theta_B = 341°10'$$

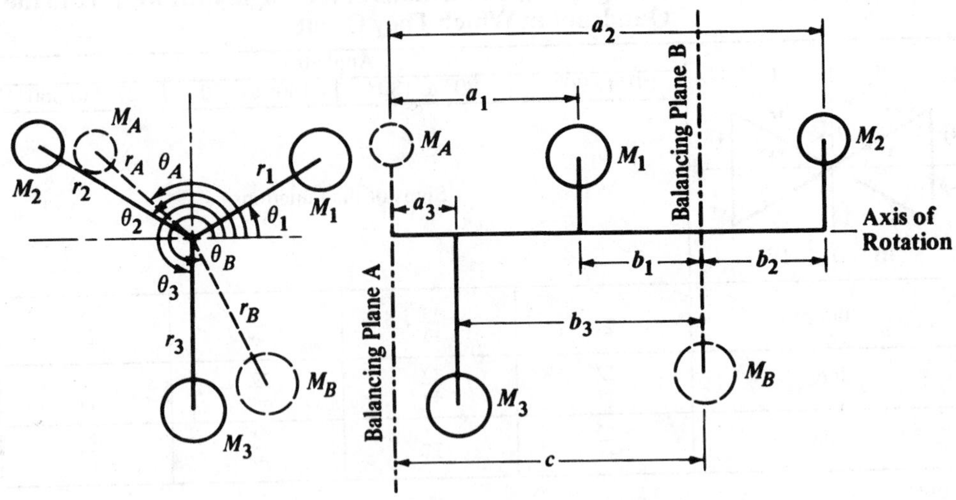

Fig. 2.

Counterbalancing Masses Located in Two or More Planes.—Unbalanced masses or weights rotating about a common axis in two separate planes of rotation form a couple, which must be counterbalanced by masses or weights, also located in two separate planes, call them planes A and B, and rotating about the same common axis (see *Couples*, page 120). In addition, they must be balanced in the direction perpendicular to the axis, as before. Since two counterbalancing masses are required, two separate equations are required to calculate the product of each mass or weight and its radius, and two additional equations are required to calculate the angular positions. The planes A and B selected as balancing planes may be any two planes separated by any convenient distance c, along the axis of rotation. In Fig. 2:

For balancing plane A:

$$M_A r_A = \frac{\sqrt{(\Sigma Mrb\cos\theta)^2 + (\Sigma Mrb\sin\theta)^2}}{c} \qquad (3)$$

$$\tan\theta_A = \frac{-(\Sigma Mrb\sin\theta)}{-(\Sigma Mrb\cos\theta)} = \frac{y}{x} \qquad (4)$$

For balancing plane B:

$$M_B r_B = \frac{\sqrt{(\Sigma Mra\cos\theta)^2 + (\Sigma Mra\sin\theta)^2}}{c} \qquad (5)$$

$$\tan\theta_B = \frac{-(\Sigma Mra\sin\theta)}{-(\Sigma Mra\cos\theta)} = \frac{y}{x} \qquad (6)$$

Where: M_A and M_B are the mass or weight of the counterbalancing masses in the balancing planes A and B, respectively; r_A and r_B are the radii; and θ_A and θ_B are the angular positions of the balancing masses in these planes. M, r, and θ are the mass or weight, radius, and angular positions of the unbalanced masses, with the subscripts defining the particular mass to which the values are assigned. The length c, the distance between the balancing planes, is always a positive value. The axial dimensions, a and b, may be either positive or negative, depending upon their position relative to the balancing plane; for example, in Fig. 2, the dimension b_2 would be negative.

Example: Referring to Fig. 2, a set of values for the masses and dimensions has been selected and put into convenient table form below. The separation of balancing planes, c, is assumed as being 15 inches. If in balancing plane A, the radius of the counterbalancing weight is selected to be 10 inches; calculate the magnitude of the counterbalancing mass and its position. If in balancing plane *B*, the counterbalancing mass is selected to be 10 lb; calculate its radius and position.

For balancing plane *A*:

Plane	*M* lb	*r* in.	θ deg.	Balancing Plane *A* *b* in.	*Mrb*	*Mrb* cos θ	*Mrb* sin θ
1	10	8	30	6	480	415.7	240.0
2	8	10	135	−6	−480	339.4	−339.4
3	12	9	270	12	1296	0.0	−1296.0
A	?	10	?	15[a]	…	755.1	−1395.4
B	10	?	?	0	…	= Σ*Mrb* cos θ	= Σ*Mrb* sin θ

[a] 15 inches = distance *c* between planes *A* and *B*.

$$M_A = \frac{\sqrt{(\Sigma Mrb\cos\theta)^2 + (\Sigma Mrb\sin\theta)^2}}{r_A c} = \frac{\sqrt{(755.1)^2 + (-1395.4)^2}}{10(15)}$$

$$M_A = 10.6 \text{ lb}$$

$$\tan\theta_A = \frac{-(\Sigma Mrb\sin\theta)}{-(\Sigma Mrb\cos\theta)} = \frac{-(-1395.4)}{-(755.1)} = \frac{+y}{-x}$$

$$\theta_A = 118°25'$$

For balancing plane *B*:

Plane	*M* lb	*r* in.	θ deg.	Balancing Plane *B* *a* in.	*Mra*	*Mra* cos θ	*Mra* sin θ
1	10	8	30	9	720	623.5	360.0
2	8	10	135	21	1680	−1187.9	1187.9
3	12	9	270	3	324	0.0	−324.0
A	?	10	?	0	…	−564.4	1223.9
B	10	?	?	15[a]	…	=Σ*Mra* cos θ	=Σ*Mra* sin θ

[a] 15 inches = distance *c* between planes *A* and *B*.

$$r_B = \frac{\sqrt{(\Sigma Mra\cos\theta)^2 + (\Sigma Mra\sin\theta)^2}}{M_B c} = \frac{\sqrt{(-564.4)^2 + (1223.9)^2}}{10(15)}$$

$$= 8.985 \text{ in.}$$

$$\tan\theta_B = \frac{-(\Sigma Mra\sin\theta)}{-(\Sigma Mra\cos\theta)} = \frac{-(1223.9)}{-(-564.4)} = \frac{-y}{+x}$$

$$\theta_B = 294°45'$$

Balancing Lathe Fixtures.—Lathe fixtures rotating at a high speed require balancing. Often it is assumed that the center of gravity of the workpiece and fixture, and of the counterbalancing masses are in the same plane; however, this is not usually the case. Counterbalancing masses are required in two separate planes to prevent excessive vibration or bearing loads at high speeds. Usually a single counterbalancing

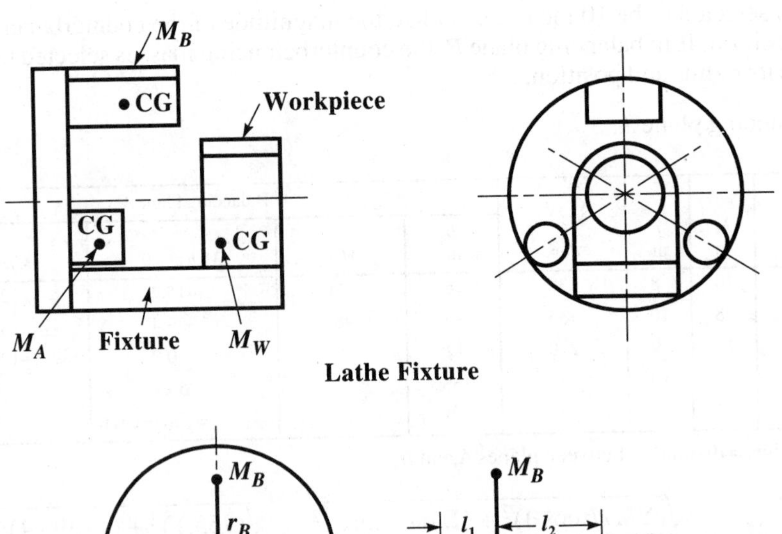

M_B r_B r_A r_A r_W θ M_A M_W M_A θ M_B l_1 l_2 M_A M_W

Schematic View

Fig. 3.

mass is placed in one plane selected to be 180 degrees directly opposite the combined center of gravity of the workpiece and the fixture. Two equal counterbalancing masses are then placed in the second counterbalancing plane, equally spaced on each side of the fixture. Referring to Fig. 3, the two counterbalancing masses M_A and the two angles θ are equal. For the design in this illustration, the following formulas can be used to calculate the magnitude of the counterbalancing masses. Since their angular positions are fixed by the design, they are not calculated.

$$M_B = \frac{M_w r_w (l_1 + l_2)}{r_B l_1} \tag{7}$$

$$M_A = \frac{M_B r_B - M_w r_w}{2 r_A \sin\theta} \tag{8}$$

In these formulas M_w and r_w denote the mass or weight and the radius of the combined center of gravity of the workpiece and the fixture.

In Fig. 3 the combined weight of the workpiece and the fixture is 18.5 lb. The following dimensions were determined from the layout of the fixture and by calculating the centers of gravity: $r_w = 2$ in.; $r_A = 6.25$ in.; $r_B = 6$ in.; $l_1 = 3$ in.; $l_2 = 5$ in.; and $\theta = 30°$. Calculate the weights of the counterbalancing masses.

$$M_B = \frac{M_w r_w (l_1 + l_2)}{r_B l_1} = \frac{18.5 \times 2 \times 8}{6 \times 3} = 16.44 \text{ lb}$$

$$M_A = \frac{M_B r_B - M_w r_w}{2 r_A \sin\theta} = \frac{(16.44 \times 6) - (18.5 \times 2)}{(2 \times 6.25)\sin 30°} = 9.86 \text{ lb (each weight)}$$

FLYWHEELS

Classification of Flywheels.—Flywheels may be classified as *balance wheels* or as *flywheel pulleys*. The object of all flywheels is to equalize the energy exerted and the work done and thereby prevent excessive or sudden changes of speed. The permissible speed variation is an important factor in all flywheel designs. The allowable speed change varies considerably for different classes of machinery; for instance, it is about 1 or 2 per cent in steam engines, while in punching and shearing machinery a speed variation of 20 per cent may be allowed.

The function of a balance wheel is to absorb and equalize energy in case the resistance to motion, or driving power, varies throughout the cycle. Therefore, the rim section is generally quite heavy and is designed with reference to the energy that must be stored in it to prevent excessive speed variations and, with reference to the strength necessary to withstand safely the stresses resulting from the required speed. The rims of most balance wheels are either square or nearly square in section, but flywheel pulleys are commonly made wide to accommodate a belt and relatively thin in a radial direction, although this is not an invariable rule.

Flywheels, in general, may either be formed of a solid or one-piece section, or they may be of sectional construction. Flywheels in diameters up to about eight feet are usually cast solid, the hubs sometimes being divided to relieve cooling stresses. Flywheels ranging from, say, eight feet to fifteen feet in diameter, are commonly cast in half sections, and the larger sizes in several sections, the number of which may equal the number of arms in the wheel. Sectional flywheels may be divided into two general classes. One class includes cast wheels which are formed of sections principally because a solid casting would be too large to transport readily. The second class includes wheels of sectional construction which, by reason of the materials used and the special arrangement of the sections, enables much higher peripheral speeds to be obtained safely than would be possible with ordinary sectional wheels of the type not designed especially for high speeds. Various designs have been built to withstand the extreme stresses encountered in some classes of service. The rims in some designs are laminated, being partly or entirely formed of numerous segment-shaped steel plates. Another type of flywheel, which is superior to an ordinary sectional wheel, has a solid cast-iron rim connected to the hub by disk-shaped steel plates instead of cast spokes.

Steel wheels may be divided into three distinct types, including 1) those having the center and rim built up entirely of steel plates; 2) those having a cast-iron center and steel rim; and 3) those having a cast-steel center and rim formed of steel plates.

Wheels having wire-wound rims have been used to a limited extent when extremely high speeds have been necessary.

When the rim is formed of sections held together by joints it is very important to design these joints properly. The ordinary bolted and flanged rim joints located between the arms average about 20 per cent of the strength of a solid rim and about 25 per cent is the maximum strength obtainable for a joint of this kind. However, by placing the joints at the ends of the arms instead of between them, an efficiency of 50 per cent of the strength of the rim may be obtained, because the joint is not subjected to the outward bending stresses between the arms but is directly supported by the arm, the end of which is secured to the rim just beneath the joint. When the rim sections of heavy balance wheels are held together by steel links shrunk into place, an efficiency of 60 per cent may be obtained; and by using a rim of box or I-section, a link type of joint connection may have an efficiency of 100 per-cent.

Flywheel Calculations

Energy Due to Changes of Velocity.—When a flywheel absorbs energy from a variable driving force, as in a steam engine, the velocity increases; and when this stored energy is given out, the velocity diminishes. When the driven member of a machine encounters a variable resistance in performing its work, as when the punch of a punching machine is passing through a steel plate, the flywheel gives up energy while the punch is at work, and, consequently, the speed of the flywheel is reduced. The total energy that a flywheel would give out if brought to a standstill is given by the formula:

$$E = \frac{Wv^2}{2g} = \frac{Wv^2}{64.32}$$

in which E = total energy of flywheel, in foot-pounds
W = weight of flywheel rim, in pounds
v = velocity at mean radius of flywheel rim, in feet per second
g = acceleration due to gravity = 32.16 ft/s²

If the velocity of a flywheel changes, the energy it will absorb or give up is proportional to the difference between the squares of its initial and final speeds, and is equal to the difference between the energy that it would give out if brought to a full stop and the energy that is still stored in it at the reduced velocity. Hence:

$$E_1 = \frac{Wv_1^{\,2}}{2g} - \frac{Wv_2^{\,2}}{2g} = \frac{W(v_1^2 - v_2^2)}{64.32}$$

in which E_l = energy in foot-pounds that a flywheel will give out while the speed is reduced from v_1 to v_2
W = weight of flywheel rim, in pounds
v_l = velocity at mean radius of flywheel rim before any energy has been given out, in feet per second
v_2 = velocity of flywheel rim at end of period during which the energy has been given out, in feet per second

Ordinarily, the effects of the arms and hub do not enter into flywheel calculations, and only the weight of the rim is considered. In computing the velocity, the mean radius of the rim is commonly used.

Using metric SI units, the formulas are $E = \frac{1}{2}Mv^2$, and $E_1 = \frac{1}{2}M(v_1^{\,2} - v_2^{\,2})$, where E and E_1 are in joules; M = the mass of the rim in kilograms; and v, v_1, and v_2 = velocities in meters per second. Note: In the SI, the unit of mass is the kilogram. If the weight of the flywheel rim is given in kilograms, the value referred to is the mass, M. Should the weight be given in newtons, N, then

$$M = \frac{W(\text{newtons})}{g}$$

where g is approximately 9.81 meters per second squared.

General Procedure in Flywheel Design.—The general method of designing a flywheel is to determine first the value of E_1 or the energy the flywheel must either supply or absorb for a given change in velocity, which, in turn, varies for different classes of service. The mean diameter of the flywheel may be assumed, or it may be fixed within certain limits by the general design of the machine. Ordinarily the speed of the flywheel shaft is known, at least approximately; the values of v_1 and v_2 can then be determined, the latter depending upon the allowable percentage of speed variation. When these values are known, the weight of the rim and the cross-sectional area required to obtain this weight may be computed. The general procedure will be illustrated more in detail by considering the design of flywheels for punching and shearing machinery.

Flywheels for Presses, Punches, Shears, Etc.—In these classes of machinery, the work that the machine performs is of an intermittent nature and is done during a small part of the time required for the driving shaft of the machine to make a complete revolution. To distribute the work of the machine over the entire period of revolution of the driving shaft, a heavy-rimmed flywheel is placed on the shaft, giving the belt an opportunity to perform an almost uniform amount of work during the whole revolution. During the greater part of the revolution of the driving shaft, the belt power is used to accelerate the speed of the flywheel. During the part of the revolution when the work is done, the energy thus stored up in the flywheel is given out at the expense of its velocity. The problem is to determine the weight and cross-sectional area of the rim when the conditions affecting the design of the flywheel are known.

Example: A flywheel is required for a punching machine capable of punching ¾-inch holes through structural steel plates ¾ inch thick. This machine (see accompanying diagram) is of the general type having a belt-driven shaft at the rear which carries a flywheel and a pinion that meshes with a large gear on the main shaft at the top of the machine. It is assumed that the relative speeds of the pinion and large gear are 7 to 1, respectively, and that the slide is to make 30 working strokes per minute. The preliminary layout shows that the flywheel should have a mean diameter (see enlarged detail) of about 30 inches. Find the weight of the flywheel and the remaining rim dimensions.

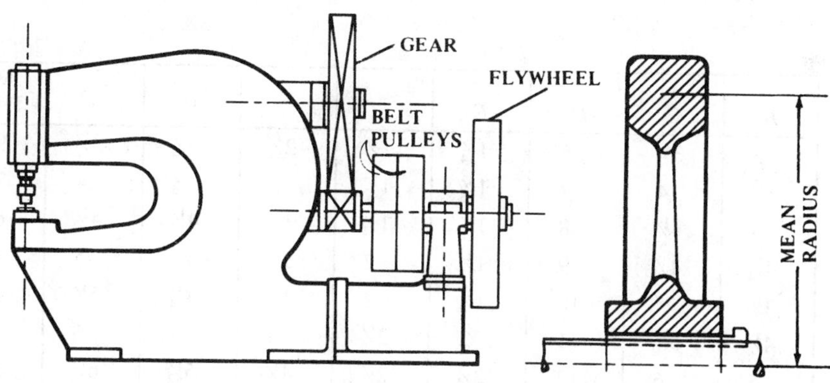

Punch Presss and Flywheel Detail

Energy Supplied by Flywheel: The energy that the flywheel must give up for a given change in velocity, and the weight of rim necessary to supply that energy, must be determined. The maximum force for shearing a ¾-inch hole through ¾-inch structural steel equals approximately the circumference of the hole multiplied by the thickness of the stock multiplied by the tensile strength, which is nearly the same as the shearing resistance of the steel. Thus, in this case, $3.1416 \times 3/4 \times 3/4 \times 60{,}000 = 106{,}000$ pounds. The average force will be much less than the maximum. Some designers assume that the average force is about one-half the maximum, although experiments show that the material is practically sheared off when the punch has entered the sheet a distance equal to about one-third the sheet thickness. On this latter basis, the average energy E_a is 2200 foot-pounds for the example given. Thus:

$$E_a = \frac{106,000 \times 1/3 \times 3/4}{12} = \frac{106,000}{4 \times 12} = 2200 \text{ foot-pounds.}$$

If the efficiency of the machine is taken as 85 per cent, the energy required will equal 2200 0.85 = 2600 foot-pounds nearly. Assume that the energy supplied by the belt while the punch is at work is determined by calculation to equal 175 foot-pounds. Then the flywheel must supply 2600 - 175 = 2425 foot-pounds = E_1.

Dimensions of Flywheels for Punches and Shears

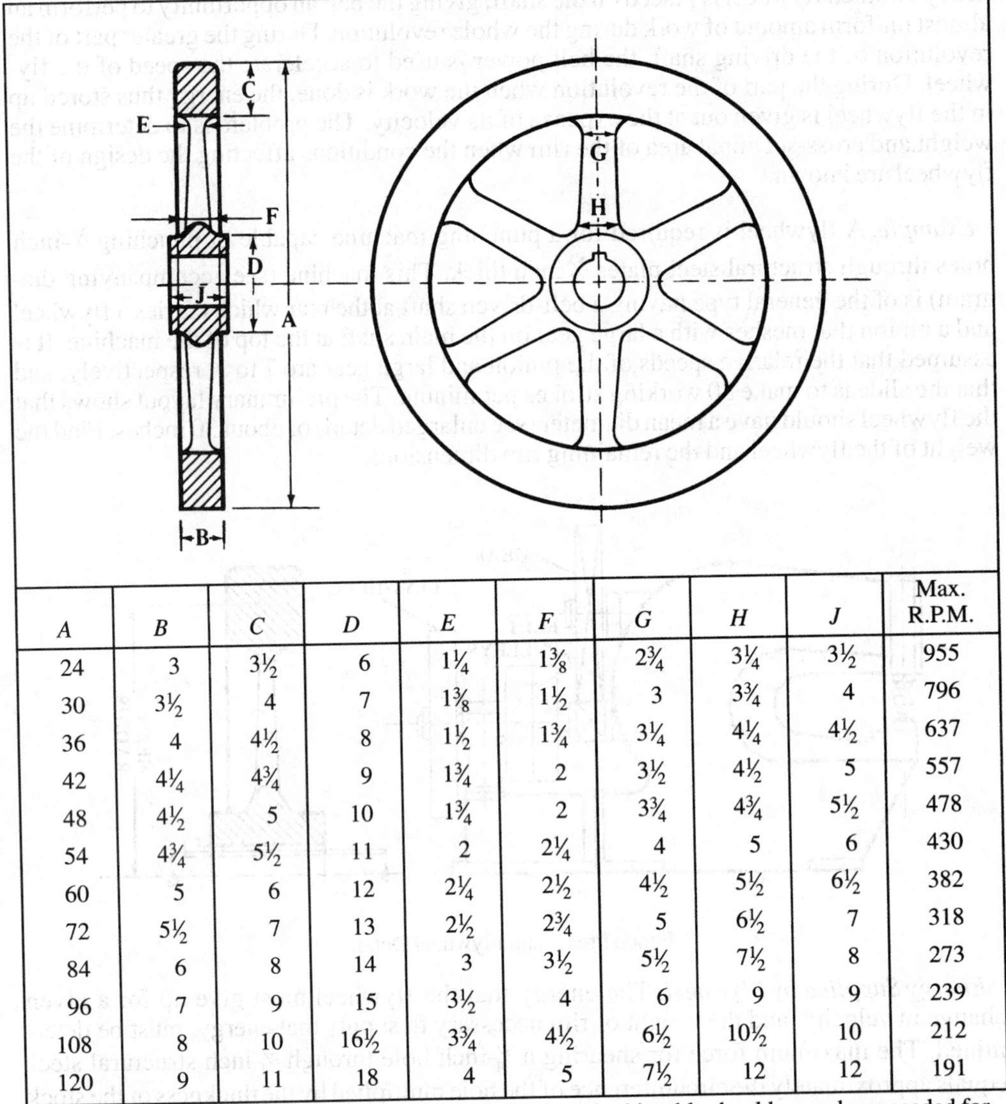

A	*B*	*C*	*D*	*E*	*F*	*G*	*H*	*J*	Max. R.P.M.
24	3	3½	6	1¼	1⅜	2¾	3¼	3½	955
30	3½	4	7	1⅜	1½	3	3¾	4	796
36	4	4½	8	1½	1¾	3¼	4¼	4½	637
42	4¼	4¾	9	1¾	2	3½	4½	5	557
48	4½	5	10	1¾	2	3¾	4¾	5½	478
54	4¾	5½	11	2	2¼	4	5	6	430
60	5	6	12	2¼	2½	4½	5½	6½	382
72	5½	7	13	2½	2¾	5	6½	7	318
84	6	8	14	3	3½	5½	7½	8	273
96	7	9	15	3½	4	6	9	9	239
108	8	10	16½	3¾	4½	6½	10½	10	212
120	9	11	18	4	5	7½	12	12	191

The maximum number of revolutions per minute given in this table should never be exceeded for cast-iron flywheels.

Rim Velocity at Mean Radius: When the mean radius of the flywheel is known, the velocity of the rim at the mean radius, in feet per second, is:

$$v = \frac{2 \times 3.1416 \times R \times n}{60}$$

in which v = velocity at mean radius of flywheel, in feet per second

R = mean radius of flywheel rim, in feet

n = number of revolutions per minute

According to the preliminary layout the mean diameter in this example should be about 30 inches and the driving shaft is to make 210 rpm, hence,

$$v = \frac{2 \times 3.1416 \times 1.25 \times 210}{60} = 27.5 \text{ feet per second}$$

Weight of Flywheel Rim: Assuming that the allowable variation in velocity when punching is about 15 per cent, and values of v_1 and v_2 are respectively 27.5 and 23.4 feet per second ($27.5 \times 0.85 = 23.4$), the weight of a flywheel rim necessary to supply a given amount of energy in foot-pounds while the speed is reduced from v_1 to v_2 would be:

$$W = \frac{E_1 \times 64.32}{v_1^2 - v_2^2} = \frac{2425 \times 64.32}{27.5^2 - 23.4^2} = 750 \text{ pounds}$$

Size of Rim for Given Weight: Since 1 cubic inch of cast iron weighs 0.26 pound, a flywheel rim weighing 750 pounds contains 750 0.26 = 2884 cubic inches. The cross-sectional area of the rim in square inches equals the total number of cubic inches divided by the mean circumference, or 2884 94.25 = 31 square inches nearly, which is approximately the area of a rim 5⅛ inches wide and 6 inches deep.

Simplified Flywheel Calculations.—Calculations for designing the flywheels of punches and shears are simplified by the following formulas and the accompanying table of constants applying to different percentages of speed reduction. In these formulas let:

HP = horsepower required
N = number of strokes per minute
E = total energy required per stroke, in foot-pounds
E_1 = energy given up by flywheel, in foot-pounds
T = time in seconds per stroke
T_1 = time in seconds of actual cut
W = weight of flywheel rim, in pounds
D = mean diameter of flywheel rim, in feet
R = maximum allowable speed of flywheel in revolutions per minute
C and C_1 = speed reduction values as given in table
a = width of flywheel rim
b = depth of flywheel rim
y = ratio of depth to width of rim

$$HP = \frac{EN}{33,000} = \frac{E}{T \times 550} \qquad E_1 = E\left(1 - \frac{T_1}{T}\right)$$

$$W = \frac{E_1}{CD^2R^2} \qquad a = \sqrt{\frac{1.22W}{12Dy}} \qquad b = ay$$

For cast-iron flywheels, with a maximum stress of 1000 pounds per square inch:

$$W = C_1E_1 \qquad R = 1940 \div D$$

Values of C and C_1 in the Previous Formulas

Per Cent Reduction	C	C_1	Per Cent Reduction	C	C_1
2½	0.00000213	0.1250	10	0.00000810	0.0328
5	0.00000426	0.0625	15	0.00001180	0.0225
7½	0.00000617	0.0432	20	0.00001535	0.0173

Example 1: A hot slab shear is required to cut a slab 4 × 15 inches which, at a shearing stress of 6000 pounds per square inch, gives a force between the knives of 360,000 pounds. The total energy required for the cut will then be 360,000 × 4⁄12 = 120,000 foot-pounds. The shear is to make 20 strokes per minute; the actual cutting time is 0.75 second, and the balance of the stroke is 2.25 seconds.

The flywheel is to have a mean diameter of 6 feet 6 inches and is to run at a speed of 200 rpm; the reduction in speed to be 10 per cent per stroke when cutting.

$$HP = \frac{120,000 \times 20}{33,000} = 72.7 \text{ horsepower}$$

$$E_1 = 120,000 \times \left(1 - \frac{0.75}{3}\right) = 90,000 \text{ foot-pounds}$$

$$W = \frac{90,000}{0.0000081 \times 6.5^2 \times 200^2} = 6570 \text{ pounds}$$

Assuming a ratio of 1.22 between depth and width of rim,

$$a = \sqrt{\frac{6570}{12 \times 6.5}} = 9.18 \text{ inches}$$

$$b = 1.22 \times 9.18 = 11.2 \text{ inches}$$

or size of rim, say, $9 \times 11\frac{1}{2}$ inches.

Example 2: Suppose that the flywheel in Example 1 is to be made with a stress due to centrifugal force of 1000 pounds per square inch of rim section.

$$C_1 \text{ for 10 per cent} = 0.0328$$

$$W = 0.0328 \times 90,000 = 2950 \text{ pounds}$$

$$R = \frac{1940}{D} \qquad \text{If } D = 6 \text{ feet,} \qquad R = \frac{1940}{6} = 323 \text{ rpm}$$

Assuming a ratio of 1.22 between depth and width of rim, as before:

$$a = \sqrt{\frac{2950}{12 \times 6}} = 6.4 \text{ inches}$$

$$b = 1.22 \times 6.4 = 7.8 \text{ inches}$$

or size of rim, say, $6\frac{1}{4} \times 8$ inches.

Centrifugal Stresses in Flywheel Rims.—In general, high speed is desirable for flywheels in order to avoid using wheels that are unnecessarily large and heavy. The centrifugal tension or hoop tension stress, that tends to rupture a flywheel rim of given area, depends solely upon the rim velocity and is independent of the rim radius. The bursting velocity of a flywheel, based on hoop stress alone (not considering bending stresses), is related to the tensile stress in the flywheel rim by the following formula which is based on the centrifugal force formula from mechanics.

$$V = \sqrt{10 \times s} \qquad \text{or,} \qquad s = V^2 \div 10$$

where V = velocity of outside circumference of rim in feet per second, and s is the tensile strength of the rim material in pounds per square inch.

For cast iron having a tensile strength of 19,000 pounds per square inch the bursting speed would be:

$$V = \sqrt{10 \times 19,000} = 436 \text{ feet per second}$$

Built-up Flywheels: Flywheels built up of solid disks of rolled steel plate stacked and bolted together on a through shaft have greater speed capacity than other types. The maximum hoop stress is at the bore and is given by the formula,

$$s = 0.0194V^2[4.333 + (d/D)^2]$$

In this formula, s and V are the stress and velocity as previously defined and d and D are the bore and outside diameters, respectively.

Assuming the plates to be of steel having a tensile strength of 60,000 pounds per square inch and a safe working stress of 24,000 pounds per square inch (using a factor of safety of 2.5 on stress or $\sqrt{2.5}$ on speed) and taking the worst condition (when d approaches D), the safe rim speed for this type of flywheel is 500 feet per second or 30,000 feet per minute.

Combined Stresses in Flywheels.—The bending stresses in the rim of a flywheel may exceed the centrifugal (hoop tension) stress predicted by the simple formula $s = V^2 \; 10$ by a considerable amount. By taking into account certain characteristics of flywheels, relatively simple formulas have been developed to determine the stress due to the combined effect of hoop tension and bending stress. Some of the factors that influence the magnitude of the maximum combined stress acting at the rim of a flywheel are:

1) *The number of spokes.* Increasing the number of spokes decreases the rim span between spokes and hence decreases the bending moment. Thus an eight-spoke wheel can be driven to a considerably higher speed before bursting than a six-spoke wheel having the same rim.

2) *The relative thickness of the spokes.* If the spokes were extremely thin, like wires, they could offer little constraint to the rim in expanding to its natural diameter under centrifugal force, and hence would cause little bending stress. Conversely, if the spokes were extremely heavy in proportion to the rim, they would restrain the rim thereby setting up heavy bending stresses at the junctions of the rim and spokes.

3) *The relative thickness of the rim to the diameter.* If the rim is quite thick (i.e., has a large section modulus in proportion to span), its resistance to bending will be great and bending stress small. Conversely, thin rims with a section modulus small in comparison with diameter or span have little resistance to bending, thus are subject to high bending stresses.

4) *Residual stresses.* These include shrinkage stresses, impact stresses, and stresses caused by operating torques and imperfections in the material. Residual stresses are taken into account by the use of a suitable factor of safety. (See *Factors of Safety for Flywheels.*)

The formulas that follow give the maximum combined stress at the rim of fly-wheels having 6, 8, and 10 spokes. These formulas are for flywheels with *rectangular rim sections* and take into account the first three of the four factors listed as influencing the magnitude of the combined stress in flywheels.

For 6 spokes:
$$s = \frac{V^2}{10}\left[1 + \left(\frac{0.56B - 1.81}{3Q + 3.14}\right)Q\right]$$

For 8 spokes:
$$s = \frac{V^2}{10}\left[1 + \left(\frac{0.42B - 2.53}{4Q + 3.14}\right)Q\right]$$

For 10 spokes:
$$s = \frac{V^2}{10}\left[1 + \left(\frac{0.33B - 3.22}{5Q + 3.14}\right)Q\right]$$

In these formulas, s = maximum combined stress in pounds per square inch; Q = ratio of mean spoke cross-section area to rim cross-section area; B = ratio of outside diameter of rim to rim thickness; and V = velocity of flywheel rim in feet per second.

Thickness of Cast Iron Flywheel Rims.—The mathematical analysis of the stresses in flywheel rims is not conclusive owing to the uncertainty of shrinkage stresses in castings or the strength of the joint in sectional wheels. When a flywheel of ordinary design is revolving at high speed, the tendency of the rim is to bend or bow outward between the arms, and the bending stresses may be serious, especially if the rim is wide and thin and the spokes are rather widely spaced. When the rims are thick, this tendency does not need to be considered, but in a thin rim running at high speed, the stress in the middle might become sufficiently great to cause the wheel to fail. The proper thickness of a cast-iron rim to resist this tendency is given for solid rims by Formula (1) and for a jointed rim by Formula (2).

$$t = \frac{0.475d}{n^2\left(\frac{6000}{v^2} - \frac{1}{10}\right)} \quad (1) \qquad t = \frac{0.95d}{n^2\left(\frac{6000}{v^2} - \frac{1}{10}\right)} \quad (2)$$

In these formulas, t = thickness of rim, in inches; d = diameter of flywheel, in inches; n = number of arms; v = peripheral speed, in feet per second.

Factors of Safety for Flywheels.—Cast-iron flywheels are commonly designed with a factor of safety of 10 to 13. A factor of safety of 10 applied to the tensile strength of a flywheel material is equivalent to a factor of safety of $\sqrt{10}$ or 3.16 on the speed of the flywheel because the stress on the rim of a flywheel increases as thdesigned would undergo rim stresses four times as great as at the design speed.

Tables of Safe Speeds for Flywheels.—The accompanying Table 1, prepared by T. C. Rathbone of The Fidelity and Casualty Company of New York, gives general recommendations for safe rim speeds for flywheels of various constructions. Table 2 shows the number of revolutions per minute corresponding to the rim speeds in Table 1.

Table 1. Safe Rim Speeds for Flywheels

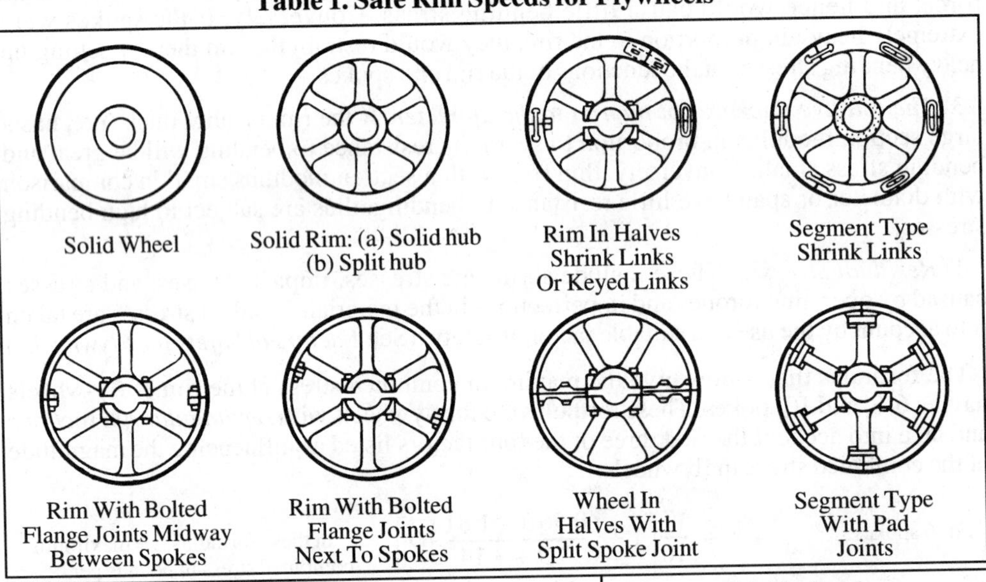

Type of Wheel	Safe Rim Speed	
	Feet per Sec.	Feet per Min.
Solid cast iron (balance wheels—heavy rims)	110	6,600
Solid cast iron (pulley wheels—thin rims)	85	5,100
Wheels with shrink link joints	77.5	4,650
Wheels with pad type joints	70.7	4,240
Wheels with bolted flange joints	50	3,000
Solid cast steel wheels	200	12,000
Wheels built up of stacked steel plates	500	30,000

To find the safe speed in revolutions per minute, divide the safe rim speed in feet per minute by 3.14 times the outside diameter of the flywheel rim in feet. For flywheels up to 15 feet in diameter, see Table 2.

Table 2. Safe Speeds of Rotation for Flywheels

Outside Diameter of Rim (feet)	Safe Rim Speed in Feet per Minute (from Table 1)						
	6,600	5,100	4,650	4,240	3,000	12,000	30,000
	Safe Speed of Rotation in Revolutions per Minute						
1	2100	1623	1480	1350	955	3820	9549
2	1050	812	740	676	478	1910	4775
3	700	541	493	450	318	1273	3183
4	525	406	370	338	239	955	2387
5	420	325	296	270	191	764	1910
6	350	271	247	225	159	637	1592
7	300	232	211	193	136	546	1364
8	263	203	185	169	119	478	1194
9	233	180	164	150	106	424	1061
10	210	162	148	135	96	382	955
11	191	148	135	123	87	347	868
12	175	135	123	113	80	318	796
13	162	125	114	104	73	294	735
14	150	116	106	97	68	273	682
15	140	108	99	90	64	255	637

Safe speeds of rotation are based on safe rim speeds shown in Table 1.

Safe Speed Formulas for Flywheels and Pulleys.—No simple formula can accommodate all the various types and proportions of flywheels and pulleys and at the same time provide a uniform factor of safety for each. Because of considerations of safety, such a formula would penalize the better constructions to accommodate the weaker designs.

One formula that has been used to check the maximum rated operating speed of flywheels and pulleys and which takes into account material properties, construction, rim thickness, and joint efficiencies is the following:

$$N = \frac{CAMEK}{D}$$

In this formula,

N = maximum rated operating speed in revolutions per minute

C = 1.0 for wheels driven by a constant speed electric motor (i.e., a-c squirrel-cage induction motor or a-c synchronous motor, etc.)
0.90 for wheels driven by variable speed motors, engines or turbines where overspeed is not over 110 per cent of rated operating speed

A = 0.90 for 4 arms or spokes
1.00 for 6 arms or spokes
1.08 for 8 arms or spokes
1.50 for disc type

M = 1.00 for cast iron of 20,000 psi tensile strength, or unknown
1.12 for cast iron of 25,000 psi tensile strength
1.22 for cast iron of 30,000 psi tensile strength
1.32 for cast iron of 35,000 psi tensile strength
2.20 for nodular iron of 60,000 psi tensile strength
2.45 for cast steel of 60,000 psi tensile strength
2.75 for plate or forged steel of 60,000 psi tensile strength

E = joint efficiency
1.0 for solid rim
0.85 for link or prison joints
0.75 for split rim — bolted joint at arms

0.70 for split rim — bolted joint between arms

K = 1355 for rim thickness equal to 1 per cent of outside diameter
1650 for rim thickness equal to 2 per cent of outside diameter
1840 for rim thickness equal to 3 per cent of outside diameter
1960 for rim thickness equal to 4 per cent of outside diameter
2040 for rim thickness equal to 5 per cent of outside diameter
2140 for rim thickness equal to 7 per cent of outside diameter
2225 for rim thickness equal to 10 per cent of outside diameter
2310 for rim thickness equal to 15 per cent of outside diameter
2340 for rim thickness equal to 20 per cent of outside diameter

D = outside diameter of rim in feet

A six-spoke solid cast iron balance wheel 8 feet in diameter has a rectangular rim 10 inches thick. What is the safe speed, in revolutions per minute, if driven by a constant speed motor?

In this instance, $C = 1$; $A = 1$; $M = 1$, since tensile strength is unknown; $E = 1$; $K = 2225$ since the rim thickness is approximately 10 per cent of the wheel diameter; and $D = 8$ feet. Thus,

$$N = \frac{1 \times 1 \times 1 \times 2225}{8} = 278 \text{ rpm}$$

(*Note:* This safe speed is slightly greater than the value of 263 rpm obtainable directly from Tables 1 and 2.)

Tests to Determine Flywheel Bursting Speeds.—Tests made by Prof. C. H. Benjamin, to determine the bursting speeds of flywheels, showed the following results:

Cast-iron Wheels with Solid Rims: Cast-iron wheels having solid rims burst at a rim speed of 395 feet per second, corresponding to a centrifugal tension of about 15,600 pounds per square inch.

Wheels with Jointed Rims: Four wheels were tested with joints and bolts inside the rim, using the familiar design ordinarily employed for band wheels, but with the joints located at points one-fourth of the distance from one arm to the next. These locations represent the points of least bending moment, and, consequently, the points at which the deflection due to centrifugal force would be expected to have the least effect. The tests, however, did not bear out this conclusion. The wheels burst at a rim speed of 194 feet per second, corresponding to a centrifugal tension of about 3750 pounds per square inch. These wheels, therefore, were only about one-quarter as strong as the wheels with solid rims, and burst at practically the same speed as wheels in a previous series of tests in which the rim joints were midway between the arms.

Bursting Speed for Link Joints: Another type of wheel with deep rim, fastened together at the joints midway between the arms by links shrunk into recesses, after the manner of flywheels for massive engines, gave much superior results. This wheel burst at a speed of 256 feet per second, indicating a centrifugal tension of about 6600 pounds per square inch.

Wheel having Tie-rods: Tests were made on a band wheel having joints inside the rim, midway between the arms, and in all respects like others of this design previously tested, except that tie-rods were used to connect the joints with the hub. This wheel burst at a speed of 225 feet per second, showing an increase of strength of from 30 to 40 per cent over similar wheels without the tie-rods.

Wheel Rim of I-section: Several wheels of special design, not in common use, were also tested, the one giving the greatest strength being an English wheel, with solid rim of I-section, made of high-grade cast iron and with the rim tied to the hub by steel wire spokes. These spokes were adjusted to have a uniform tension. The wheel gave way at a rim speed of 424 feet per second, which is slightly higher than the speed of rupture of the solid rim wheels with ordinary style of spokes.

Tests on Flywheel of Special Construction: A test was made on a flywheel 49 inches in diameter and weighing about 900 pounds. The rim was 6¾ inches wide and 1⅛ inches thick, and was built of ten segments, the material being cast steel. Each joint was secured by three "prisoners" of an I-section on the outside face, by link prisoners on each edge, and by a dovetailed bronze clamp on the inside, fitting over lugs on the rim. The arms were of phosphor-bronze, twenty in number, ten on each side, and were cros-shaped in section. These arms came midway between the rim joints and were bolted to plane faces on the polygonal hub. The rim was further reinforced by a system of diagonal bracing, each section of the rim being supported at five points on each side, in such a way as to relieve it almost entirely from bending. The braces, like the arms, were of phosphor-bronze, and all bolts and connecting links were of steel. This wheel was designed as a model of a proposed 30-foot flywheel. On account of the excessive air resistance the wheel was enclosed at the sides between sheet-metal disks. This wheel burst at 1775 revolutions per minute or at a linear speed of 372 feet per second. The hub and main spokes of the wheel remained nearly in place, but parts of the rim were found 200 feet away. This sudden failure of the rim casting was unexpected, as it was thought the flange bolts would be the parts to give way first. The tensile strength of the casting at the point of fracture was about four times the strength of the wheel rim at a solid section.

Stresses in Rotating Disks.—When a disk of uniform width is rotated, the maximum stress S_t is tangential and at the bore of the hub, and the tangential stress is always greater than the radial stress at the same point on the disk. If S_t = maximum tangential stress in pounds per sq. in.; w = weight of material, lb. per cu. in.; N = rev. per min.; m = Poisson's ratio = 0.3 for steel; R = outer radius of disk, inches; r = inner radius of disk or radius of bore, inches.

$$S_t = 0.000071wN^2[(3+m)R^2 + (1-m)r^2]$$

Steam Engine Flywheels.—The variable amount of energy during each stroke and the allowable percentage of speed variation are of special importance in designing steam engine flywheels. The earlier the point of cut-off, the greater the variation in energy and the larger the flywheel that will be required. The weight of the reciprocating parts and the length of the connecting-rod also affect the variation. The following formula is used for computing the weight of the flywheel rim:

Let W = weight of rim in pounds
D = mean diameter of rim in feet
N = number of revolutions per minute
$1/n$ = allowable variation in speed (from $1/50$ to $1/100$)
E = excess and deficiency of energy in foot-pounds
c = factor of energy excess, from the accompanying table
HP = indicated horsepower

Then, if the indicated horsepower is given:

$$W = \frac{387{,}587{,}500 \times cn \times \text{HP}}{D^2N^3} \qquad (1)$$

If the work in foot-pounds is given, then:

$$W = \frac{11{,}745nE}{D^2N^2} \qquad (2)$$

In the second formula, E equals the average work in foot-pounds done by the engine in one revolution, multiplied by the decimal given in the accompanying table, "*Factors for Engine Flywheel Calculations*," which covers both condensing and non-condensing engines:

Factors for Engine Flywheel Calculations

Condensing Engines						
Fraction of stroke at which steam is cut off	$\frac{1}{3}$	$\frac{1}{4}$	$\frac{1}{5}$	$\frac{1}{6}$	$\frac{1}{7}$	$\frac{1}{8}$
Factor of energy excess	0.163	0.173	0.178	0.184	0.189	0.191
Non-condensing Engines						
Steam cut off at			$\frac{1}{2}$	$\frac{1}{3}$	$\frac{1}{4}$	$\frac{1}{5}$
Factor of energy excess			0.160	0.186	0.209	0.232

Example 1: A non-condensing engine of 150 indicated horsepower is to make 200 revolutions per minute, with a speed variation of 2 per cent. The average cut-off is to be at one-quarter stroke, and the flywheel is to have a mean diameter of 6 feet. Find the necessary weight of the rim in pounds.

From the table $c = 0.209$, and from the data given $HP = 150$; $N = 200$; $1/n = 1/50$ or $n = 50$; and, $D = 6$.

Substituting these values in Equation (1):

$$W = \frac{387,587,500 \times 0.209 \times 50 \times 150}{6^2 \times 200^3} = 2110 \text{ pounds, nearly}$$

Example 2: A condensing engine, 24 × 42 inches, cuts off at one-third stroke and has a mean effective pressure of 50 pounds per square inch. The flywheel is to be 18 feet in mean diameter and make 75 revolutions per minute with a variation of 1 per cent. Find the required weight of the rim.

The work done on the piston in one revolution is equal to the pressure on the piston multiplied by the distance traveled or twice the stroke in feet. The area of the piston is 452.4 square inches, and twice the stroke is 7 feet. The work done on the piston in one revolution is, therefore, $452.4 \times 50 \times 7 = 158,340$ foot-pounds. From the table $c = 0.163$, and therefore:

$$E = 158,340 \times 0.163 = 25,810 \text{ foot-pounds}$$

From the data given: $n = 100$; $D = 18$; $N = 75$. Substituting these values in Equation (2):

$$W = \frac{11,745 \times 100 \times 25,810}{18^2 \times 75^2} = 16,650 \text{ pounds, nearly}$$

Spokes or Arms of Flywheels.—Flywheel arms are usually of elliptical cross-section. The major axis of the ellipse is in the plane of rotation to give the arms greater resistance to bending stresses and reduce the air resistance which may be considerable at high velocity. The stresses in the arms may be severe, due to the inertia of a heavy rim when sudden load changes occur. The strength of the arms should equal three-fourths the strength of the shaft in torsion.

If W equals the width of the arm at the hub (length of major axis) and D equals the shaft diameter, then W equals $1.3\,D$ for a wheel having 6 arms; and for an 8-arm wheel W equals $1.2\,D$. The thickness of the arm at the hub (length of minor axis) equals one-half the width. The arms usually taper toward the rim. The cross-sectional area at the rim should not be less than two-thirds the area at the hub.

Critical Speeds

Critical Speeds of Rotating Bodies and Shafts.—If a body or disk mounted upon a shaft rotates about it, the center of gravity of the body or disk must be at the center of the shaft, if a perfect running balance is to be obtained. In most cases, however, the center of gravity of the disk will be slightly removed from the center of the shaft, owing to the difficulty of perfect balancing. Now, if the shaft and disk be rotated, the centrifugal force generated by the heavier side will be greater than that generated by the lighter side geometrically opposite to it, and the shaft will deflect toward the heavier side, causing the center of the disk to rotate in a small circle. A rotating shaft without a body or disk mounted on it can also become dynamically unstable, and the resulting vibrations and deflections can result in damage not only to the shaft but to the machine of which it is a part. These conditions hold true up to a comparatively high speed; but a point is eventually reached (at several thousand revolutions per minute) when momentarily there will be excessive vibration, and then the parts will run quietly again. The speed at which this occurs is called the *critical speed* of the wheel or shaft, and the phenomenon itself for the shaft-mounted disk or body is called the *settling* of the wheel. The explanation of the settling is that at this speed the axis of rotation changes, and the wheel and shaft, instead of rotating about their geometrical center, begin to rotate about an axis through their center of gravity. The shaft itself is then deflected so that for every revolution its geometrical center traces a circle around the center of gravity of the rotating mass.

Critical speeds depend upon the magnitude or location of the load or loads carried by the shaft, the length of the shaft, its diameter and the kind of supporting bearings. The normal operating speed of a machine may or may not be higher than the critical speed. For instance, some steam turbines exceed the critical speed, although they do not run long enough at the critical speed for the vibrations to build up to an excessive amplitude. The practice of the General Electric Co. at Schenectady is to keep below the critical speeds. It is assumed that the maximum speed of a machine may be within 20 per cent high or low of the critical speed without vibration troubles. Thus, in a design of steam turbine sets, critical speed is a factor that determines the size of the shafts for both the generators and turbines. Although a machine may run very close to the critical speed, the alignment and play of the bearings, the balance and construction generally, will require extra care, resulting in a more expensive machine; moreover, while such a machine may run smoothly for a considerable time, any looseness or play that may develop later, causing a slight imbalance, will immediately set up excessive vibrations.

The formulas commonly used to determine critical speeds are sufficiently accurate for general purposes. There are cases, however, where the torque applied to a shaft has an important effect on its critical speed. Investigations have shown that the critical speeds of a uniform shaft are decreased as the applied torque is increased, and that there exist critical torques which will reduce the corresponding critical speed of the shaft to zero. A detailed analysis of the effects of applied torques on critical speeds may be found in a paper, "Critical Speeds of Uniform Shafts under Axial Torque," by Golumb and Rosenberg, presented at the First U.S. National Congress of Applied Mechanics in 1951.

Formulas for Critical Speeds.—The critical speed formulas given in the accompanying table (from the paper on Critical Speed Calculation presented before the ASME by S. H. Weaver) apply to (1) shafts with single concentrated loads and (2) shafts carrying uniformly distributed loads. These formulas also cover different conditions as regards bearings. If the bearings are self-aligning or very short, the shaft is considered supported at the ends; whereas, if the bearings are long and rigid, the shaft is considered fixed. These formulas, for both concentrated and distributed loads, apply to vertical shafts as well as horizontal shafts, the critical speeds having the same value in both cases. The data required for the solution of critical speed problems are the same as for shaft deflection. As the shaft is usually of variable diameter and its stiffness is increased by a long hub, an ideal shaft of uniform diameter and equal stiffness must be assumed.

Critical Speed Formulas

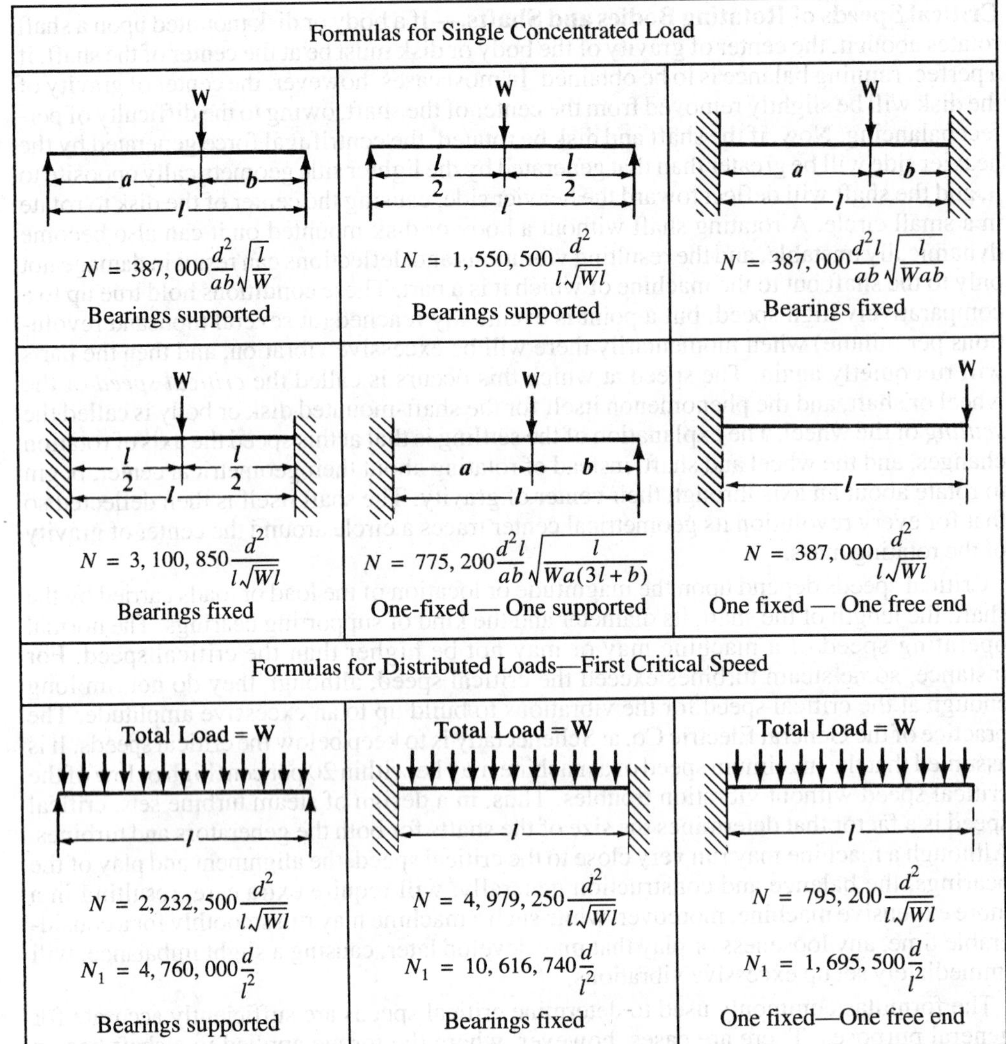

Formulas for Single Concentrated Load		
$N = 387,000\frac{d^2}{ab}\sqrt{\frac{l}{W}}$ Bearings supported	$N = 1,550,500\frac{d^2}{l\sqrt{Wl}}$ Bearings supported	$N = 387,000\frac{d^2 l}{ab}\sqrt{\frac{l}{Wab}}$ Bearings fixed
$N = 3,100,850\frac{d^2}{l\sqrt{Wl}}$ Bearings fixed	$N = 775,200\frac{d^2 l}{ab}\sqrt{\frac{l}{Wa(3l+b)}}$ One-fixed — One supported	$N = 387,000\frac{d^2}{l\sqrt{Wl}}$ One fixed — One free end
Formulas for Distributed Loads—First Critical Speed		
Total Load = W: $N = 2,232,500\frac{d^2}{l\sqrt{Wl}}$ $N_1 = 4,760,000\frac{d}{l^2}$ Bearings supported	Total Load = W: $N = 4,979,250\frac{d^2}{l\sqrt{Wl}}$ $N_1 = 10,616,740\frac{d}{l^2}$ Bearings fixed	Total Load = W: $N = 795,200\frac{d^2}{l\sqrt{Wl}}$ $N_1 = 1,695,500\frac{d}{l^2}$ One fixed—One free end

N = critical speed, RPM

N_1 = critical speed of shaft alone

d = diameter of shaft, in inches

W = load applied to shaft, in pounds

l = distance between centers of bearings, in inches

a and b = distances from bearings to load

In calculating critical speeds, the weight of the shaft is either neglected or, say, one-half to two-thirds of the weight is added to the concentrated load. The formulas apply to steel shafts having a modulus of elasticity $E = 29,000,000$. Although a shaft carrying a number of loads or a distributed load may have an infinite number of critical speeds, ordinarily it is the first critical speed that is of importance in engineering work. The first critical speed is obtained by the formulas given in the distributed loads portion of the table *Critical Speed Formulas*.

Angular Velocity

Angular Velocity of Rotating Bodies.—The angular velocity of a rotating body is the angle through which the body turns in a unit of time. Angular velocity is commonly expressed in terms of revolutions per minute, but in certain engineering applications it is necessary to express it as radians per second. By definition there are 2π radians in 360 degrees, or one revolution, so that one radian = 360 2π = 57.3 degrees. To convert angular velocity in revolutions per minute, n, to angular velocity in radians per second, ω, multiply by π and divide by 30:

$$\omega = \frac{\pi n}{30} \qquad (1)$$

The following table may be used to obtain angular velocity in radians per second for all numbers of revolutions per minute from 1 to 239.

Example: To find the angular velocity in radians per second of a flywheel making 97 revolutions per minute, locate 90 in the left-hand column and 7 at the top of the columns; at the intersection of the two lines, the angular velocity is read off as equal to 10.16 radians per second.

Linear Velocity of Points on a Rotating Body.—The linear velocity, v, of any point on a rotating body expressed in feet per second may be found by multiplying the angular velocity of the body in radians per second, ω, by the radius, r, in feet from the center of rotation to the point:

$$v = \omega r \qquad (2)$$

The metric SI units are v = meters per second; ω = radians per second, r = meters.

Angular Velocity in Revolutions per Minute Converted to Radians per Second

R.P.M.	Angular Velocity in Radians per Second									
	0	1	2	3	4	5	6	7	8	9
0	0.00	0.10	0.21	0.31	0.42	0.52	0.63	0.73	0.84	0.94
10	1.05	1.15	1.26	1.36	1.47	1.57	1.67	1.78	1.88	1.99
20	2.09	2.20	2.30	2.41	2.51	2.62	2.72	2.83	2.93	3.04
30	3.14	3.25	3.35	3.46	3.56	3.66	3.77	3.87	3.98	4.08
40	4.19	4.29	4.40	4.50	4.61	4.71	4.82	4.92	5.03	5.13
50	5.24	5.34	5.44	5.55	5.65	5.76	5.86	5.97	6.07	6.18
60	6.28	6.39	6.49	6.60	6.70	6.81	6.91	7.02	7.12	7.23
70	7.33	7.43	7.54	7.64	7.75	7.85	7.96	8.06	8.17	8.27
80	8.38	8.48	8.59	8.69	8.80	8.90	9.01	9.11	9.21	9.32
90	9.42	9.53	9.63	9.74	9.84	9.95	10.05	10.16	10.26	10.37
100	10.47	10.58	10.68	10.79	10.89	11.00	11.10	11.20	11.31	11.41
110	11.52	11.62	11.73	11.83	11.94	12.04	12.15	12.25	12.36	12.46
120	12.57	12.67	12.78	12.88	12.98	13.09	13.19	13.30	13.40	13.51
130	13.61	13.72	13.82	13.93	14.03	14.14	14.24	14.35	14.45	14.56
140	14.66	14.76	14.87	14.97	15.08	15.18	15.29	15.39	15.50	15.60
150	15.71	15.81	15.92	16.02	16.13	16.23	16.34	16.44	16.55	16.65
160	16.75	16.86	16.96	17.07	17.17	17.28	17.38	17.49	17.59	17.70
170	17.80	17.91	18.01	18.12	18.22	18.33	18.43	18.53	18.64	18.74
180	18.85	18.95	19.06	19.16	19.27	19.37	19.48	19.58	19.69	19.79
190	19.90	20.00	20.11	20.21	20.32	20.42	20.52	20.63	20.73	20.84
200	20.94	21.05	21.15	21.26	21.36	21.47	21.57	21.68	21.78	21.89
210	21.99	22.10	22.20	22.30	22.41	22.51	22.62	22.72	22.83	22.93
220	23.04	23.14	23.25	23.35	23.46	23.56	23.67	23.77	23.88	23.98
230	24.09	24.19	24.29	24.40	24.50	24.61	24.71	24.82	24.92	25.03

PENDULUMS

Types of Pendulums

Types of Pendulums.—A *compound* or *physical* pendulum consists of any rigid body suspended from a fixed horizontal axis about which the body may oscillate in a vertical plane due to the action of gravity.

A *simple* or *mathematical* pendulum is similar to a compound pendulum except that the mass of the body is concentrated at a single point which is suspended from a fixed horizontal axis by a weightless cord. Actually, a simple pendulum cannot be constructed since it is impossible to have either a weightless cord or a body whose mass is entirely concentrated at one point. A good approximation, however, consists of a small, heavy bob suspended by a light, fine wire. If these conditions are not met by the pendulum, it should be considered as a compound pendulum.

A *conical* pendulum is similar to a simple pendulum except that the weight suspended by the cord moves at a uniform speed around the circumference of a circle in a horizontal plane instead of oscillating back and forth in a vertical plane. The principle of the conical pendulum is employed in the Watt fly-ball governor.

A *torsional* pendulum in its simplest form consists of a disk fixed to a slender rod, the other end of which is fastened to a fixed frame. When the disc is twisted through some angle and released, it will then oscillate back and forth about the axis of the rod because of the torque exerted by the rod.

Four Types of Pendulum

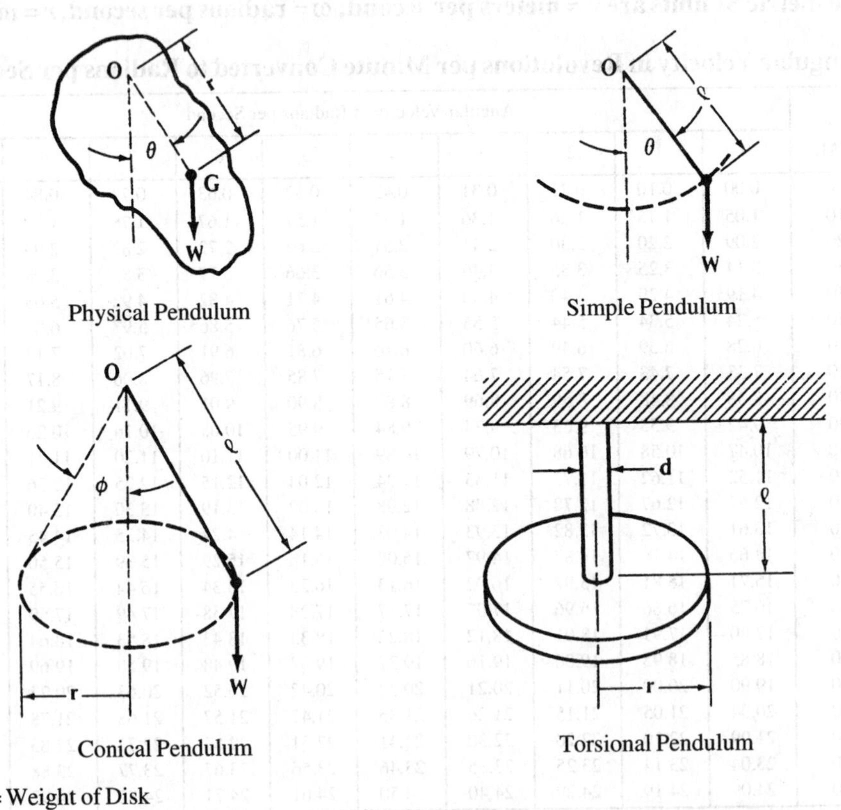

Pendulum Calculations

Pendulum Formulas.—From the formulas that follow, the period of vibration or time required for one complete cycle back and forth may be determined for the types of pendulums shown in the accompanying diagram.

For a *simple* pendulum,

$$T = 2\pi\sqrt{\frac{l}{g}} \qquad (1)$$

where T = period in seconds for one complete cycle; g = acceleration due to gravity = 32.17 feet per second per second (approximately); and l is the length of the pendulum in feet as shown on the accompanying diagram.

For a *physical* or *compound* pendulum,

$$T = 2\pi\sqrt{\frac{k_o^2}{gr}} \qquad (2)$$

where k_0 = radius of gyration of the pendulum about the axis of rotation, in feet, and r is the distance from the axis of rotation to the center of gravity, in feet.

The metric SI units that can be used in the two above formulas are T = seconds; g = approximately 9.81 meters per second squared, which is the value for acceleration due to gravity; l = the length of the pendulum in meters; k_0 = the radius of gyration in meters, and r = the distance from the axis of rotation to the center of gravity, in meters.

Formulas (1) and (2) are accurate when the angle of oscillation θ shown in the diagram is very small. For θ equal to 22 degrees, these formulas give results that are too small by 1 per cent; for θ equal to 32 degrees, by 2 per cent.

For a *conical* pendulum, the time in seconds for one revolution is:

$$T = 2\pi\sqrt{\frac{l\cos\phi}{g}} \quad (3a) \qquad \text{or} \qquad T = 2\pi\sqrt{\frac{r\cot\phi}{g}} \quad (3b)$$

For a *torsional* pendulum consisting of a thin rod and a disk as shown in the figure

$$T = \frac{2}{3}\sqrt{\frac{\pi W r^2 l}{g d^4 G}} \qquad (4)$$

where W = weight of disk in pounds; r = radius of disk in feet; l = length of rod in feet; d = diameter of rod in feet; and G = modulus of elasticity in shear of the rod material in pounds per square inch.

The formula using metric SI units is:

$$T = 8\sqrt{\frac{\pi M r^2 l}{d^4 G}}$$

where T = time in seconds for one complete oscillation; M = mass in kilograms; r = radius in meters; l = length of rod in meters; d = diameter of rod in meters; G = modulus of elasticity in shear of the rod material in pascals (newtons per meter squared). The same formula can be applied using millimeters, providing dimensions are expressed in millimeters throughout, and the modulus of elasticity in megapascals (newtons per millimeter squared).

FRICTION

Properties of Friction

Friction is the resistance to motion that takes place when one body is moved upon another, and is generally defined as "that force which acts between two bodies at their surface of contact, so as to resist their sliding on each other." According to the conditions under which sliding occurs, the force of friction, F, bears a certain relation to the force between the two bodies called the normal force N. The relation between force of friction and normal force is given by the *coefficient of friction*, generally denoted by the Greek letter μ. Thus:

$$F = \mu \times N \qquad \text{and} \qquad \mu = \frac{F}{N}$$

A body weighing 28 pounds rests on a horizontal surface. The force required to keep it in motion along the surface is 7 pounds. Find the coefficient of friction.

$$\mu = \frac{F}{N} = \frac{7}{28} = 0.25$$

If a body is placed on an inclined plane, the friction between the body and the plane will prevent it from sliding down the inclined surface, provided the angle of the plane with the horizontal is not too great. There will be a certain angle, however, at which the body will just barely be able to remain stationary, the frictional resistance being very nearly overcome by the tendency of the body to slide down. This angle is termed the angle of repose, and the tangent of this angle equals the coefficient of friction. The angle of repose is frequently denoted by the Greek letter θ. Thus, $\mu = \tan \theta$.

A greater force is required to start a body moving from a state of rest than to merely keep it in motion, because the *friction of rest* is greater than the *friction of motion.*

Laws of Friction.—The laws of friction for unlubricated or dry surfaces are summarized in the following statements.

1) For low pressures (normal force per unit area) the friction is directly proportional to the normal force between the two surfaces. As the pressure increases, the friction does not rise proportionally; but when the pressure becomes abnormally high, the friction increases at a rapid rate until seizing takes place.

2) The friction both in its total amount and its coefficient is independent of the areas in contact, so long as the normal force remains the same. This is true for moderate pressures only. For high pressures, this law is modified in the same way as in the first case.

3) At very low velocities the friction is independent of the velocity of rubbing. As the velocities increase, the friction decreases.

Lubricated Surfaces: For well lubricated surfaces, the laws of friction are considerably different from those governing dry or poorly lubricated surfaces.

1) The frictional resistance is almost independent of the pressure (normal force per unit area) if the surfaces are flooded with oil.

2) The friction varies directly as the speed, at low pressures; but for high pressures the friction is very great at low velocities, approaching a minimum at about two feet per second linear velocity, and afterwards increasing approximately as the square root of the speed.

3) For well lubricated surfaces the frictional resistance depends, to a very great extent, on the temperature, partly because of the change in the viscosity of the oil and partly because, for a journal bearing, the diameter of the bearing increases with the rise of temperature more rapidly than the diameter of the shaft, thus relieving the bearing of side pressure.

4) If the bearing surfaces are flooded with oil, the friction is almost independent of the nature of the material of the surfaces in contact. As the lubrication becomes less ample, the coefficient of friction becomes more dependent upon the material of the surfaces.

Influence of Friction on the Efficiency of Small Machine Elements.—Friction between machine parts lowers the efficiency of a machine. Average values of the efficiency, in per cent, of the most common machine elements when carefully made are ordinary bearings, 95 to 98; roller bearings, 98; ball bearings, 99; spur gears with cut teeth, including bearings, 99; bevel gears with cut teeth, including bearings, 98; belting, from 96 to 98; high-class silent power transmission chain, 97 to 99; roller chains, 95 to 97.

Coefficients of Friction.—Tables 1 and provide representative values of static friction for various combinations of materials with dry (clean, unlubricated) and lubricated surfaces. The values for static or breakaway friction shown in these tables will generally be higher than the subsequent or sliding friction. Typically, the steel-on-steel static coefficient of 0.8 unlubricated will drop to 0.4 when sliding has been initiated; with oil lubrication, the value will drop from 0.16 to 0.03.

Many factors affect friction, and even slight deviations from normal or test conditions can produce wide variations. Accordingly, when using friction coefficients in design calculations, due allowance or factors of safety should be considered, and in critical applications, specific tests conducted to provide specific coefficients for material, geometry, and/or lubricant combinations.

Rolling Friction.—When a body rolls on a surface, the force resisting the motion is termed *rolling friction* or *rolling resistance*. Let W = total weight of rolling body or load on wheel, in pounds; r = radius of wheel, in inches; f = coefficient of rolling resistance, in inches. Then: resistance to rolling, in pounds = $(W \times f) \div r$.

The coefficient of rolling resistance varies with the conditions. For wood on wood it may be assumed as 0.06 inch; for iron on iron, 0.02 inch; iron on granite, 0.085 inch; iron on asphalt, 0.15 inch; and iron on wood, 0.22 inch.

The coefficient of rolling resistance, f, is in inches and is not the same as the sliding or static coefficient of friction given in Tables 1 and , which is a dimensionless ratio between frictional resistance and normal load. Various investigators are not in close agreement on the true values for these coefficients and the foregoing values should only be used for the approximate calculation of rolling resistance.

Table 1. Coefficients of Static Friction for Steel on Various Materials*

Material	Coefficient of Friction, μ	
	Clean	Lubricated
Steel	0.8	0.16
Copper-lead alloy	0.22	…
Phosphor-bronze	0.35	…
Aluminum-bronze	0.45	…
Brass	0.35	0.19
Cast iron	0.4	0.21
Bronze	…	0.16
Sintered bronze	…	0.13
Hard carbon	0.14	0.11–0.14
Graphite	0.1	0.1
Tungsten carbide	0.4–0.6	0.1–0.2
Plexiglas	0.4–0.5	0.4–0.5
Polystyrene	0.3–0.35	0.3–0.35
Polythene	0.2	0.2
Teflon	0.04	0.04

Tables 1 and used with permission from *The Friction and Lubrication of Solids*, Vol. 1, by Bowden and Tabor, Clarendon Press, Oxford, 1950.

Material Combination	Coefficient of Friction, μ	
	Clean	Lubricated
Aluminum-aluminum	1.35	0.30
Cadmium-cadmium	0.5	0.05
Chromium-chromium	0.41	0.34
Copper-copper	1.0	0.08
Iron-iron	1.0	0.15–0.20
Magnesium-magnesium	0.6	0.08
Nickel-nickel	0.7	0.28
Platinum-platinum	1.2	0.25
Silver-silver	1.4	0.55
Zinc-zinc	0.6	0.04
Glass-glass	0.9–1.0	0.1–0.6
Glass-metal	0.5–0.7	0.2–0.3
Diamond-diamond	0.1	0.05–0.1
Diamond-metal	0.1–0.15	0.1
Sapphire-sapphire	0.2	0.2
Hard carbon on carbon	0.16	0.12–0.14
Graphite-graphite (in vacuum)	0.5–0.8	…
Graphite-graphite	0.1	0.1
Tungsten carbide-tungsten carbide	0.2–0.25	0.12
Plexiglas-plexiglas	0.8	0.8
Polystyrene-polystyrene	0.5	0.5
Teflon-Teflon	0.04	0.04
Nylon-nylon	0.15–0.25	…
Solids on rubber	1–4	…
Wood on wood (clean)	0.25–0.5	…
Wood on wood (wet)	0.2	…
Wood on metals (clean)	0.2–0.6	…
Wood on metals (wet)	0.2	…
Brick on wood	0.6	…
Leather on wood	0.3–0.4	…
Leather on metal (clean)	0.6	…
Leather on metal (wet)	0.4	…
Leather on metal (greasy)	0.2	…
Brake material on cast iron	0.4	…
Brake material on cast iron (wet)	0.2	…

STRENGTH OF MATERIALS

MOMENT OF INERTIA

BEAMS

COLUMNS

PLATES, SHELLS, AND CYLINDERS

SHAFTS

SPRINGS

SPRINGS (cont.)

STRENGTH AND PROPERTIES OF WIRE ROPE

STRENGTH AND PROPERTIES OF WIRE ROPE (cont.)

CRANE CHAIN AND HOOKS

STRENGTH OF MATERIALS

Strength of Materials

Strength of materials deals with the relations between the external forces applied to elastic bodies and the resulting deformations and stresses. In the design of structures and machines, the application of the principles of strength of materials is necessary if satisfactory materials are to be utilized and adequate proportions obtained to resist functional forces.

Forces are produced by the action of gravity, by accelerations and impacts of moving parts, by gasses and fluids under pressure, by the transmission of mechanical power, etc. In order to analyze the stresses and deflections of a body, the magnitudes, directions and points of application of forces acting on the body must be known. Information given in the Mechanics section provides the basis for evaluating force systems.

The time element in the application of a force on a body is an important consideration. Thus a force may be static or change so slowly that its maximum value can be treated as if it were static; it may be suddenly applied, as with an impact; or it may have a repetitive or cyclic behavior.

The environment in which forces act on a machine or part is also important. Such factors as high and low temperatures; the presence of corrosive gases, vapors and liquids; radiation, etc. may have a marked effect on how well parts are able to resist stresses.

Throughout the Strength of Materials section in this Handbook, both English and metric SI data and formulas are given to cover the requirements of working in either system of measurement. Formulas and text relating exclusively to SI units are given in bold-face type.

Mechanical Properties of Materials.—Many mechanical properties of materials are determined from tests, some of which give relationships between stresses and strains as shown by the curves in the accompanying figures.

Stress is force per unit area and is usually expressed in pounds per square inch. If the stress tends to stretch or lengthen the material, it is called *tensile* stress; if to compress or shorten the material, a *compressive* stress; and if to shear the material, a *shearing* stress. Tensile and compressive stresses always act at right-angles to (normal to) the area being considered; shearing stresses are always in the plane of the area (at right-angles to compressive or tensile stresses).

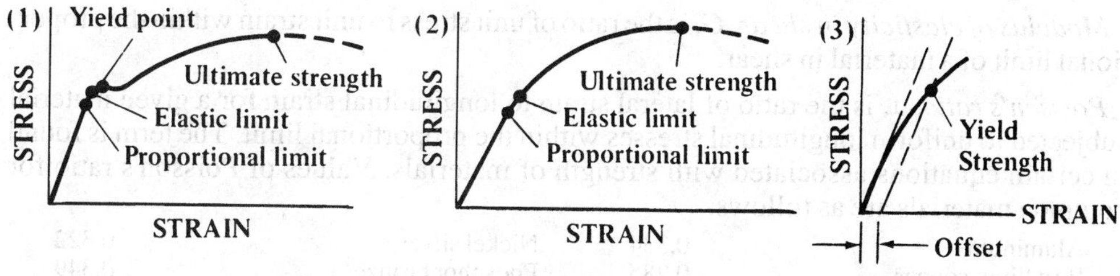

Fig. 1. Stress-strain curves

In the SI, the unit of stress is the pascal (Pa), the newton per meter squared (N/m²). The megapascal (newtons per millimeter squared) is often an appropriate sub-multiple for use in practice.

Unit strain is the amount by which a dimension of a body changes when the body is subjected to a load, divided by the original value of the dimension. The simpler term *strain* is often used instead of unit strain.

Proportional limit is the point on a stress-strain curve at which it begins to deviate from the straight-line relationship between stress and strain.

Elastic limit is the maximum stress to which a test specimen may be subjected and still return to its original length upon release of the load. A material is said to be stressed within the *elastic region* when the working stress does not exceed the elastic limit, and to be stressed in the *plastic region* when the working stress does exceed the elastic limit. The elastic limit for steel is for all practical purposes the same as its proportional limit.

Yield point is a point on the stress-strain curve at which there is a sudden increase in strain without a corresponding increase in stress. Not all materials have a yield point. Some representative values of the yield point (in ksi) are as follows:

Aluminum, wrought, 2014-T6	60	Titanium, pure	55–70
Aluminum, wrought, 6061-T6	35	Titanium, alloy, 5Al, 2.5Sn	110
Beryllium copper	140	Steel for bridges and buildings, ASTM A7-61T, all shapes	33
Brass, naval	25–50		
Cast iron, malleable	32–45	Steel, castings, high strength, for structural purposes, ASTM A148.60 (seven grades)	40–145
Cast iron, nodular	45–65		
Magnesium, AZ80A-T5	38	Steel, stainless (0.08–0.2C, 17Cr, 7Ni) ¼	78

Yield strength, S_y, is the maximum stress that can be applied without permanent deformation of the test specimen. This is the value of the stress at the elastic limit for materials for which there is an elastic limit. Because of the difficulty in determining the elastic limit, and because many materials do not have an elastic region, yield strength is often determined by the offset method as illustrated by the accompanying figure at (3). Yield strength in such a case is the stress value on the stress-strain curve corresponding to a definite amount of permanent set or strain, usually 0.1 or 0.2 per cent of the original dimension.

Ultimate strength, S_u, (also called *tensile strength*) is the maximum stress value obtained on a stress-strain curve.

Modulus of elasticity, E, (also called *Young's modulus*) is the ratio of unit stress to unit strain within the proportional limit of a material in tension or compression. Some representative values of Young's modulus (in 10^6 psi) are as follows:

Aluminum, cast, pure	9	Magnesium, AZ80A-T5	6.5
Aluminum, wrought, 2014-T6	10.6	Titanium, pure	15.5
Beryllium copper	19	Titanium, alloy, 5 Al, 2.5 Sn	17
Brass, naval	15	Steel for bridges and buildings, ASTM A7-61T, all shapes	29
Bronze, phosphor, ASTM B159	15		
Cast iron, malleable	26	Steel, castings, high strength, for structural purposes, ASTM A148-60 (seven grades)	29
Cast iron, nodular	23.5		

Modulus of elasticity in shear, G, is the ratio of unit stress to unit strain within the proportional limit of a material in shear.

Poisson's ratio, μ, is the ratio of lateral strain to longitudinal strain for a given material subjected to uniform longitudinal stresses within the proportional limit. The term is found in certain equations associated with strength of materials. Values of Poisson's ratio for common materials are as follows:

Aluminum	0.334	Nickel silver	0.322
Beryllium copper	0.285	Phosphor bronze	0.349
Brass	0.340	Rubber	0.500
Cast iron, gray	0.211	Steel, cast	0.265
Copper	0.340	high carbon	0.295
Inconel	0.290	mild	0.303
Lead	0.431	nickel	0.291
Magnesium	0.350	Wrought iron	0.278
Monel metal	0.320	Zinc	0.331

Compressive Properties.—From compression tests, *compressive yield strength,* S_{cy}, and *compressive ultimate strength,* S_{cu}, are determined. Ductile materials under compression loading merely swell or buckle without fracture, hence do not have a compressive ultimate strength.

Shear Properties.—The properties of *shear yield strength*, S_{sy}, *shear ultimate strength*, S_{su}, and the *modulus of rigidity*, G, are determined by direct shear and torsional tests. The modulus of rigidity is also known as the modulus of elasticity in shear. It is the ratio of the shear stress, τ, to the shear strain, γ, in radians, within the proportional limit: $G = \tau/\gamma$.

Fatigue Properties.—When a material is subjected to many cycles of stress reversal or fluctuation (variation in magnitude without reversal), failure may occur, even though the maximum stress at any cycle is considerably less than the value at which failure would occur if the stress were constant. Fatigue properties are determined by subjecting test specimens to stress cycles and counting the number of cycles to failure. From a series of such tests in which maximum stress values are progressively reduced, S-N diagrams can be plotted as illustrated by the accompanying figures. The S-N diagram Fig. 2a shows the behavior of a material for which there is an *endurance limit*, S_{en}. Endurance limit is the stress value at which the number of cycles to failure is infinite. Steels have endurance limits that vary according to hardness, composition, and quality; but many non-ferrous metals do not. The S-N diagram Fig. 2b does not have an endurance limit. For a metal that does not have an endurance limit, it is standard practice to specify fatigue strength as the stress value corresponding to a specific number of stress reversals, usually 100,000,000 or 500,000,000.

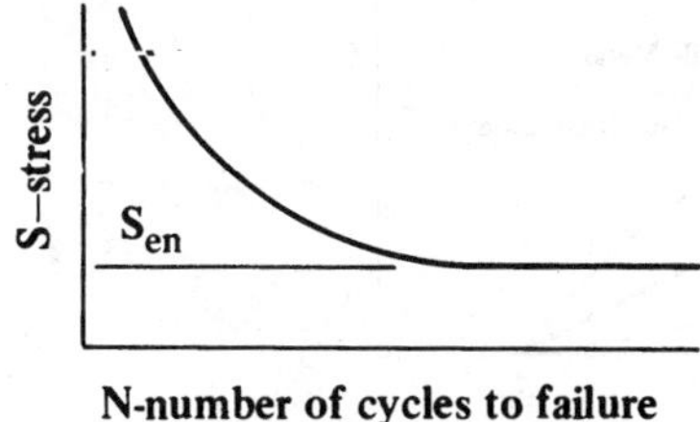

S N

Fig. 2a. S-N endurance limit

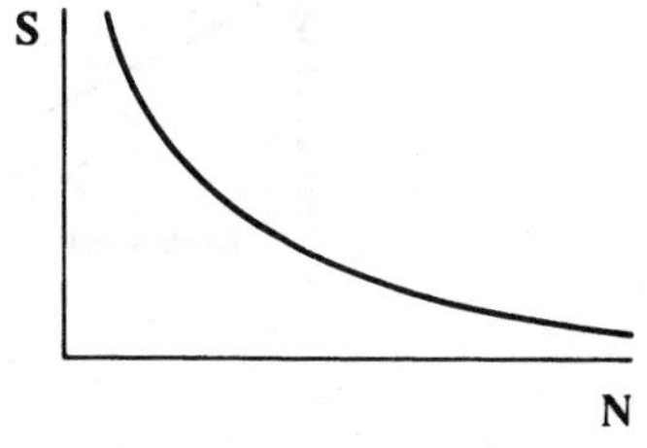

Fig. 2b. S-N no endurance limit

The Influence of Mean Stress on Fatigue.—Most published data on the fatigue properties of metals are for completely reversed alternating stresses, that is, the mean stress of the cycle is equal to zero. However, if a structure is subjected to stresses that fluctuate between different values of tension and compression, then the mean stress is not zero.

When fatigue data for a specified mean stress and design life are not available for a material, the influence of nonzero mean stress can be estimated from empirical relationships that relate failure at a given life, under zero mean stress, to failure at the same life under zero mean cyclic stress. One widely used formula is Goodman's linear relationship, which is

$$S_a = S(1 - S_m/S_u)$$

where S_a is the alternating stress associated with some nonzero mean stress, S_m. S is the alternating fatigue strength at zero mean stress. S_u is the ultimate tensile strength.

Goodman's linear relationship is usually represented graphically on a so-called *Goodman Diagram*, as shown below. The alternating fatigue strength or the alternating stress for a given number of endurance cycles is plotted on the ordinate (y-axis) and the static tensile strength is plotted on the abscissa (x-axis). The straight line joining the alternating fatigue strength, S, and the tensile strength, S_u, is the Goodman line.

The value of an alternating stress S_{ax} at a known value of mean stress S_{mx} is determined as shown by the dashed lines on the diagram.

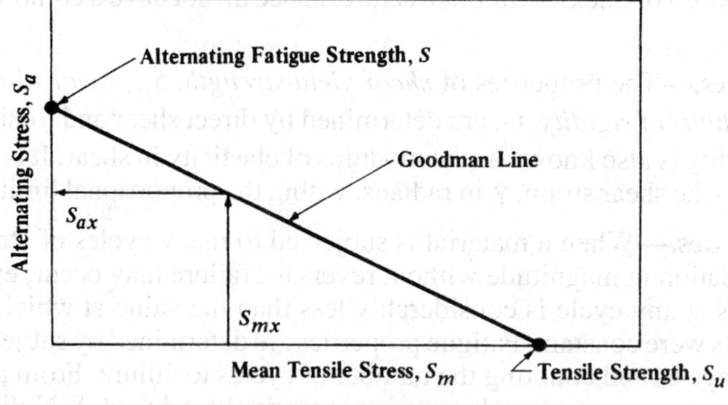

Goodman Diagram

For ductile materials, the Goodman law is usually conservative, since approximately 90 per cent of actual test data for most ferrous and nonferrous alloys fall above the Goodman line, even at low endurance values where the yield strength is exceeded. For many brittle materials, however, actual test values can fall below the Goodman line, as illustrated below:

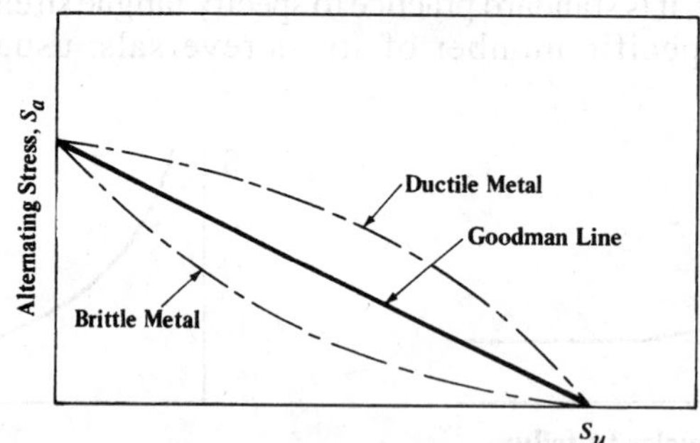

Mean Tensile Stress

As a rule of thumb, materials having an elongation of less than 5 per cent in a tensile test may be regarded as brittle. Those having an elongation of 5 per cent or more may be regarded as ductile.

Cumulative Fatigue Damage.—Most data are determined from tests at a constant stress amplitude. This is easy to do experimentally, and the data can be presented in a straightforward manner. In actual engineering applications, however, the alternating stress amplitude usually changes in some way during service operation. Such changes, referred to as "spectrum loading," make the direct use of standard S-N fatigue curves inappropriate. A problem exists, therefore, in predicting the fatigue life under varying stress amplitude from conventional, constant-amplitude S-N fatigue data.

The assumption in predicting spectrum loading effects is that operation at a given stress amplitude and number of cycles will produce a certain amount of permanent fatigue damage and that subsequent operation at different stress amplitude and number of cycles will produce additional fatigue damage and a sequential accumulation of total damage, which at a critical value will cause fatigue failure. Although the assumption appears simple, the amount of damage incurred at any stress amplitude and number of cycles has proven difficult to determine, and several "cumulative damage" theories have been advanced.

One of the first and simplest methods for evaluating cumulative damage is known as Miner's law or the linear damage rule, where it is assumed that n_1 cycles at a stress of S_1, for which the average number of cycles to failure is N_1, cause an amount of damage n_1/N_1. Failure is predicted to occur when

$$\Sigma n/N = 1$$

The term n/N is known as the "cycle ratio" or the damage fraction.

The greatest advantages of the Miner rule are its simplicity and prediction reliability, which approximates that of more complex theories. For these reasons the rule is widely used. It should be noted, however, that it does not account for all influences, and errors are to be expected in failure prediction ability.

Modes of Fatigue Failure.—Several modes of fatigue failure are:

Low/High-Cycle Fatigue: This fatigue process covers cyclic loading in two significantly different domains, with different physical mechanisms of failure. One domain is characterized by relatively low cyclic loads, strain cycles confined largely to the elastic range, and long lives or a high number of cycles to failure; traditionally, this has been called "high-cycle fatigue." The other domain has cyclic loads that are relatively high, significant amounts of plastic strain induced during each cycle, and short lives or a low number of cycles to failure. This domain has commonly been called "low-cycle fatigue" or cyclic strain-controlled fatigue.

The transition from low- to high-cycle fatigue behavior occurs in the range from approximately 10,000 to 100,000 cycles. Many define low-cycle fatigue as failure that occurs in 50,000 cycles or less.

Thermal Fatigue: Cyclic temperature changes in a machine part will produce cyclic stresses and strains if natural thermal expansions and contractions are either wholly or partially constrained. These cyclic strains produce fatigue failure just as though they were produced by external mechanical loading. When strain cycling is produced by a fluctuating temperature field, the failure process is termed "thermal fatigue."

While thermal fatigue and mechanical fatigue phenomena are very similar, and can be mathematically expressed by the same types of equations, the use of mechanical fatigue results to predict thermal fatigue performance must be done with care. For equal values of plastic strain range, the number of cycles to failure is usually up to 2.5 times lower for thermally cycled than for mechanically cycled samples.

Corrosion Fatigue: Corrosion fatigue is a failure mode where cyclic stresses and a corrosion-producing environment combine to initiate and propagate cracks in fewer stress cycles and at lower stress amplitudes than would be required in a more inert environment. The corrosion process forms pits and surface discontinuities that act as stress raisers to accelerate fatigue cracking. The cyclic loads may also cause cracking and flaking of the corrosion layer, baring fresh metal to the corrosive environment. Each process accelerates the other, making the cumulative result more serious.

Surface or Contact Fatigue: Surface fatigue failure is usually associated with rolling surfaces in contact, and results in pitting, cracking, and spalling of the contacting surfaces from cyclic Hertz contact stresses that cause the maximum values of cyclic shear stresses to be slightly below the surface. The cyclic subsurface shear stresses generate cracks that propagate to the contacting surface, dislodging particles in the process.

Combined Creep and Fatigue: In this failure mode, all of the conditions for both creep failure and fatigue failure exist simultaneously. Each process influences the other in producing failure, but this interaction is not well understood.

Factors of Safety.—There is always a risk that the working stress to which a member is subjected will exceed the strength of its material. The purpose of a factor of safety is to minimize this risk.

Factors of safety can be incorporated into design calculations in many ways. For most calculations the following equation is used:

$$s_w = S_m/f_s \tag{1}$$

where f_s is the factor of safety, S_m is the strength of the material in pounds per square inch, and S_w is the allowable working stress, also in pounds per square inch. Since the factor of safety is greater than 1, the allowable working stress will be less than the strength of the material.

In general, S_m is based on yield strength for ductile materials, ultimate strength for brittle materials, and fatigue strength for parts subjected to cyclic stressing. Most strength values are obtained by testing standard specimens at 68°F. in normal atmospheres. If, however, the character of the stress or environment differs significantly from that used in obtaining standard strength data, then special data must be obtained. If special data are not available, standard data must be suitably modified.

General recommendations for values of factors of safety are given in the following list.

f_s	Application
1.3–1.5	For use with highly reliable materials where loading and environmental conditions are not severe, and where weight is an important consideration.
1.5–2	For applications using reliable materials where loading and environmental conditions are not severe.
2–2.5	For use with ordinary materials where loading and environmental conditions are not severe.
2.5–3	For less tried and for brittle materials where loading and environmental conditions are not severe.
3–4	For applications in which material properties are not reliable and where loading and environmental conditions are not severe, or where reliable materials are to be used under difficult loading and environmental conditions.

Working Stress.—Calculated working stresses are the products of calculated nominal stress values and stress concentration factors. Calculated nominal stress values are based on the assumption of idealized stress distributions. Such nominal stresses may be simple stresses, combined stresses, or cyclic stresses. Depending on the nature of the nominal stress, one of the following equations applies:

$$s_w = K\sigma \quad (2) \qquad s_w = K\sigma' \quad (4) \qquad s_w = K\sigma_{cy} \quad (6)$$
$$s_w = K\tau \quad (3) \qquad s_w = K\tau' \quad (5) \qquad s_w = K\tau_{cy} \quad (7)$$

where K is a stress concentration factor; σ and τ are, respectively, simple normal (tensile or compressive) and shear stresses; σ' and τ' are combined normal and shear stresses; σ_{cy} and τ_{cy} are cyclic normal and shear stresses.

Where there is uneven stress distribution, as illustrated in the table (on page 204) of simple stresses for Cases 3, 4 and 6, the maximum stress is the one to which the stress concentration factor is applied in computing working stresses. The location of the maximum stress in each case is discussed under the section "Simple Stresses" and the formulas for these maximum stresses are given in the on page 204.

Stress Concentration Factors.—Stress concentration is related to type of material, the nature of the stress, environmental conditions, and the geometry of parts. When stress concentration factors that specifically match all of the foregoing conditions are not available, the following equation may be used:

$$K = 1 + q(K_t - 1) \qquad (8)$$

K_t is a theoretical stress concentration factor that is a function only of the geometry of a part and the nature of the stress; q is the *index of sensitivity* of the material. If the geometry is such as to provide no theoretical stress concentration, $K_t = 1$.

Curves for evaluating K_t are on pages 201 through 204. For constant stresses in cast iron and in ductile materials, $q = 0$ (hence $K = 1$). For constant stresses in brittle materials such as hardened steel, q may be taken as 0.15; for very brittle materials such as steels that have been quenched but not drawn, q may be taken as 0.25. When stresses are suddenly applied (impact stresses) q ranges from 0.4 to 0.6 for ductile materials; for cast iron it is taken as 0.5; and, for brittle materials, 1.

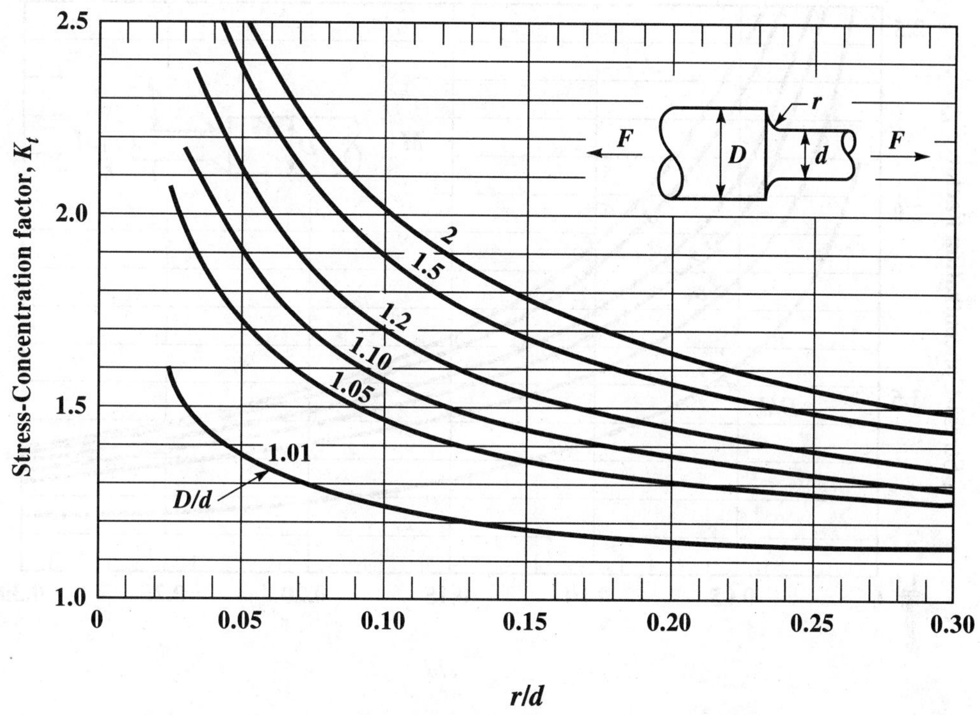

Fig. 3. Stress-concentration factor, K_t, for a filleted shaft in tension

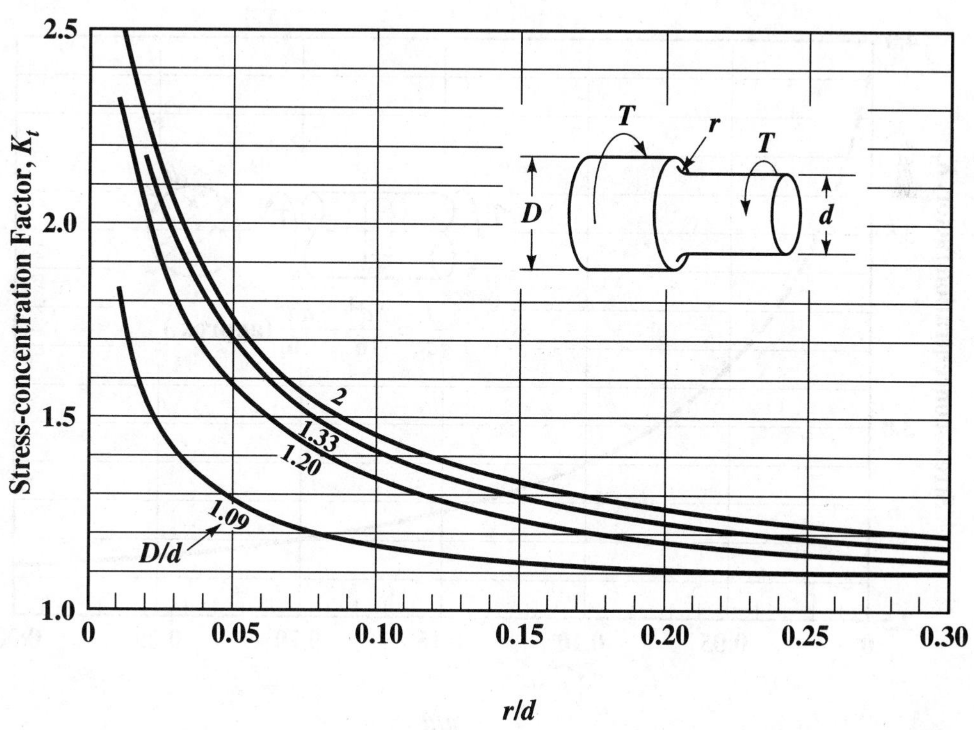

Fig. 4. Stress-concentration factor, K_t, for a filleted shaft in torsion[a]

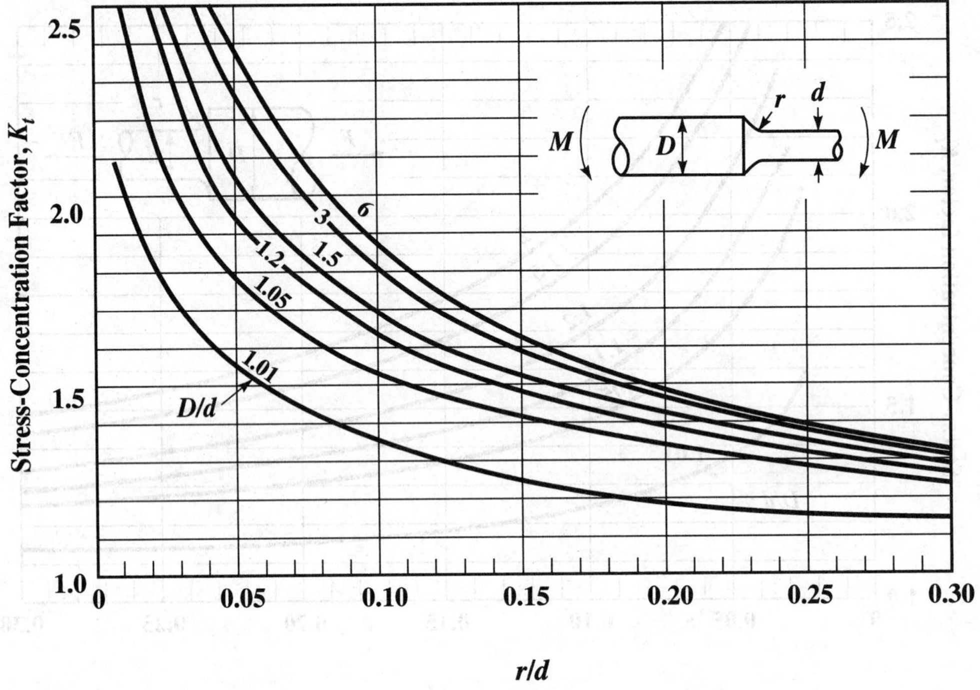

Fig. 5. Stress-concentration factor, K_t, for a shaft with shoulder fillet in bending[a]

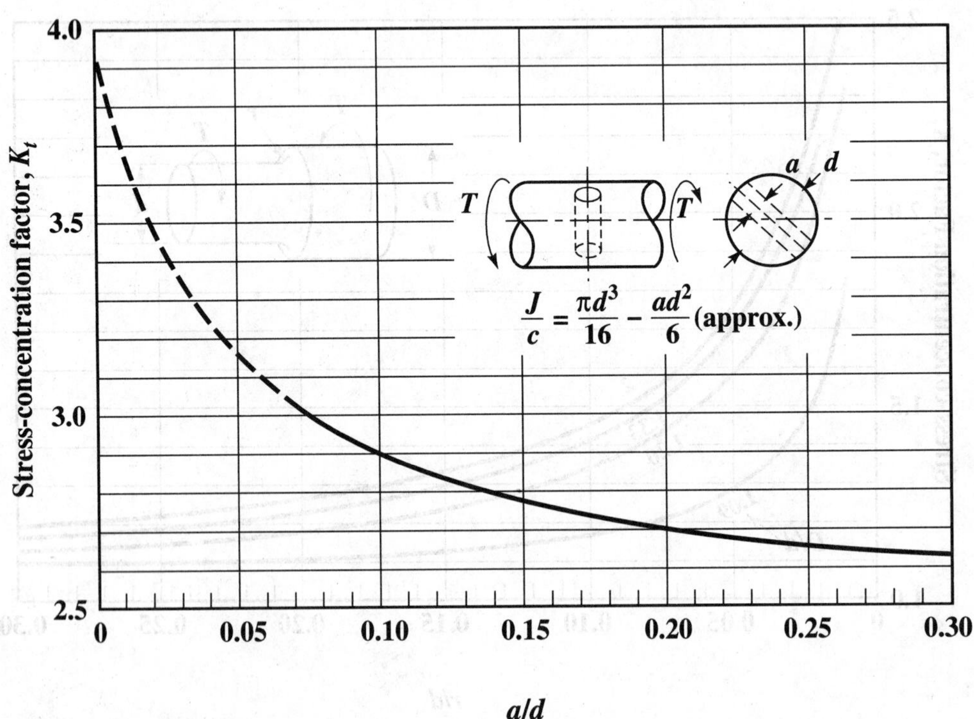

Fig. 6. Stress-concentration factor, K_t, for a shaft, with a transverse hole, in torsion[a]

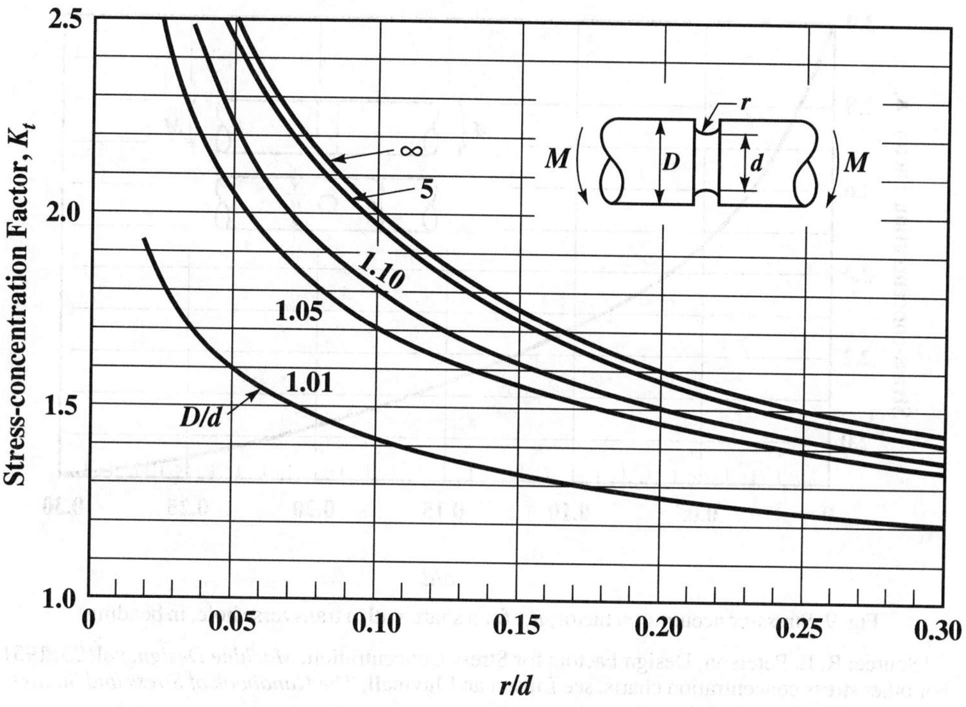

Fig. 7. Stress-concentration factor, K_t, for a grooved shaft in bending[a]

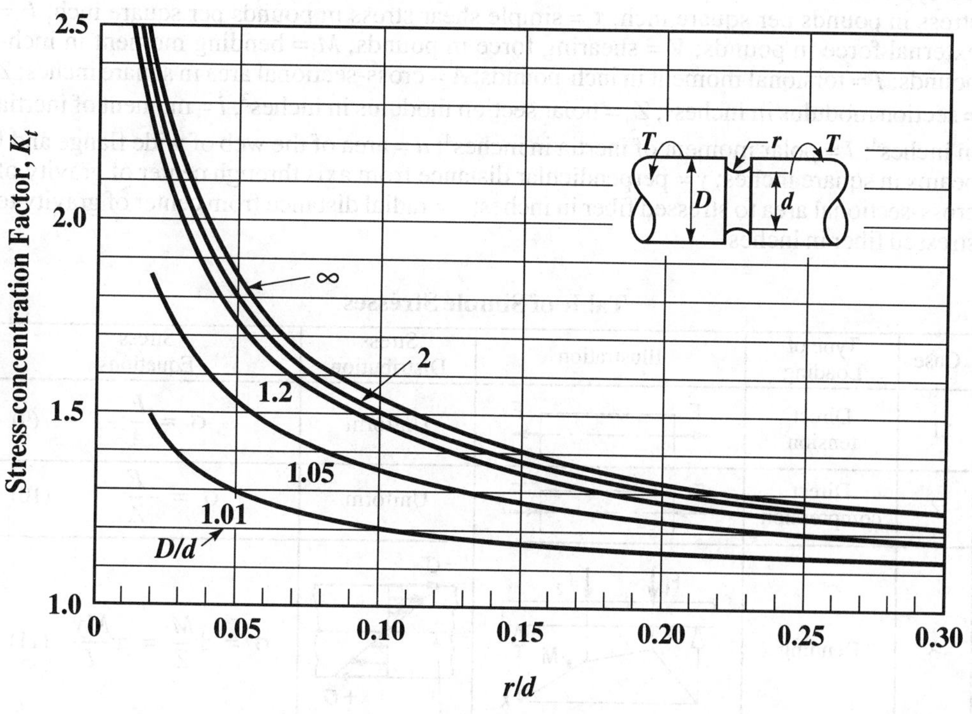

Fig. 8. Stress-concentration factor, K_t, for a grooved shaft in torsion[a]

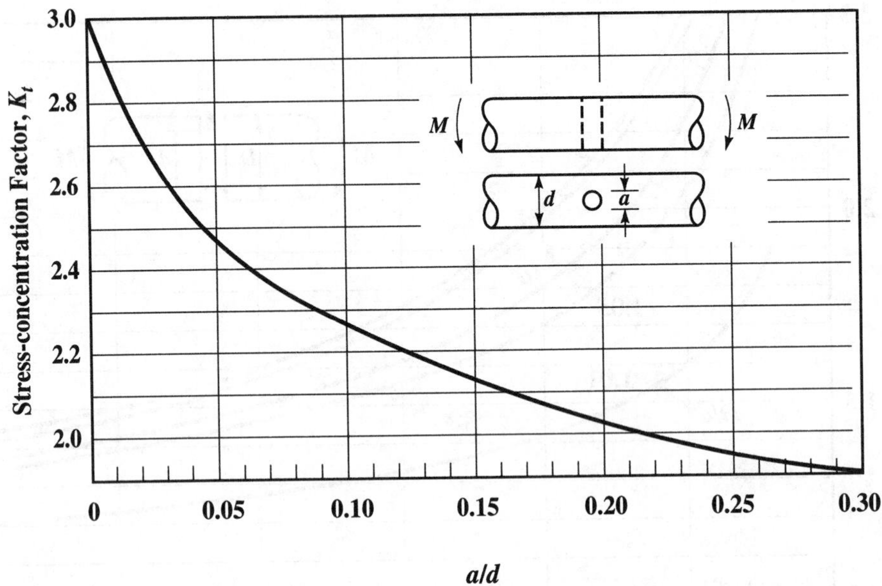

Fig. 9. Stress-concentration factor, K_t, for a shaft, with a transverse hole, in bending[a]

[a] Source: R. E. Peterson, Design Factors for Stress Concentration, *Machine Design*, vol. 23, 1951. For other stress concentration charts, see Lipson and Juvinall, *The Handbook of Stress and Strength*, The Macmillan Co., 1963.

Simple Stresses.—Simple stresses are produced by constant conditions of loading on elements that can be represented as beams, rods, or bars. The table on page 204 summarizes information pertaining to the calculation of simple stresses. Following is an explanation of the symbols used in simple stress formulae: σ = simple normal (tensile or compressive) stress in pounds per square inch; τ = simple shear stress in pounds per square inch; F = external force in pounds; V = shearing force in pounds; M = bending moment in inch-pounds; T = torsional moment in inch-pounds; A = cross-sectional area in square inches; Z = section modulus in inches^3; Z_p = polar section modulus in inches^3; I = moment of inertia in inches^4; J = polar moment of inertia in inches^4; a = area of the web of wide flange and I beams in square inches; y = perpendicular distance from axis through center of gravity of cross-sectional area to stressed fiber in inches; c = radial distance from center of gravity to stressed fiber in inches.

Table of Simple Stresses

Case	Type of Loading	Illustration	Stress Distribution	Stress Equations
1	Direct tension	F F	Uniform	$\sigma = \frac{F}{A}$ (9)
2	Direct compression	F F	Uniform	$\sigma = -\frac{F}{A}$ (10)
3	Bending	F_1 F_2 M x Bending moment diagram	$-\sigma$ $+\sigma$ Neutral plane	$\sigma = \pm\frac{M}{Z} = \pm\frac{My}{I}$ (11)

Table of Simple Stresses *(Continued)*

Case	Type of Loading	Illustration	Stress Distribution	Stress Equations
4	Bending	F_1 F_2 R_1 R_2 R_1 V F_1 x F_2 R_2 Shearing force diagram	τ Neutral plane	For beams of rectangular cross-section: $\tau = \frac{3V}{2A}$ (12) For beams of solid circular cross-section: $\tau = \frac{4V}{3A}$ (13) For wide flange and I beams (approximately): $\tau = \frac{V}{a}$ (14)
5	Direct shear	F F	Uniform	$\tau = \frac{F}{A}$ (15)
6	Torsion	T	τ τ	$\tau = \frac{T}{Zp} = \frac{Tc}{J}$ (16)

SI metric units can be applied in the calculations in place of the English units of measurement without changes to the formulas. The SI units are the newton (N), which is the unit of force; the meter; the meter squared; the pascal (Pa) which is the newton per meter squared (N/M²); and the newton-meter (N · m) for moment of force. Often in design work using the metric system, the millimeter is employed rather than the meter. In such instances, the dimensions can be converted to meters before the stress calculations are begun. Alternatively, the same formulas can be applied using millimeters in place of the meter, providing the treatment is consistent throughout. In such instances, stress and strength properties must be expressed in megapascals (MPa), which is the same as newtons per millimeter squared (N/mm²), and moments in newton-millimeters (N · mm²). ***Note*****: $1\ N/mm^2 = 1\ N/10^{-6}m^2 = 10^6\ N/m^2$ = 1 meganewton/m² = 1 megapascal.**

For direct tension and direct compression loading, Cases 1 and 2 in the on page 204, the force *F* must act along a line through the center of gravity of the section at which the stress is calculated. The equation for direct compression loading applies only to members for which the ratio of length to least radius of gyration is relatively small, approximately 20, otherwise the member must be treated as a column.

The table *Stresses and Deflections in Beams* starting on page 237 give equations for calculating stresses due to bending for common types of beams and conditions of loading. Where these tables are not applicable, stress may be calculated using Equation (11) in the table on page 204. In using this equation it is necessary to determine the value of the bending moment at the point where the stress is to be calculated. For beams of constant cross-section, stress is ordinarily calculated at the point coinciding with the maximum value of bending moment. Bending loading results in the characteristic stress distribution shown in the table for Case 3. It will be noted that the maximum stress values are at the surfaces farthest from the neutral plane. One of the surfaces is stressed in tension and the other in compression. It is for this reason that the ± sign is used in Equation (11). Numerous tables for evaluating section moduli are given in the section starting on page 217.

Shear stresses caused by bending have maximum values at neutral planes and zero values at the surfaces farthest from the neutral axis, as indicated by the stress distribution diagram shown for Case 4 in the . Values for V in Equations (12), (13) and (14) can be determined from shearing force diagrams. The shearing force diagram shown in Case 4 corresponds to the bending moment diagram for Case 3. As shown in this diagram, the value taken for V is represented by the greatest vertical distance from the x axis. The shear stress caused by direct shear loading, Case 5, has a uniform distribution. However, the shear stress caused by torsion loading, Case 6, has a zero value at the axis and a maximum value at the surface farthest from the axis.

Deflections.—For direct tension and direct compression loading on members with uniform cross sections, deflection can be calculated using Equation (17). For direct tension loading, e is an elongation; for direct compression loading, e is a contraction. Deflection is in inches when the load F is in pounds, the length L over which deflection occurs is in inches, the cross-sectional area A is in square inches, and the modulus of elasticity E is in pounds per square inch. The angular deflection of members with uniform circular cross sections subject to torsion loading can be calculated with Equation (18).

$$e = FL/AE \quad (17) \qquad \theta = TL/GJ \quad (18)$$

The angular deflection θ is in radians when the torsional moment T is in inch-pounds, the length L over which the member is twisted is in inches, the modulus of rigidity G is in pounds per square inch, and the polar moment of inertia J is in inches^4.

Metric SI units can be used in Equations (17) and (18), where F = force in newtons (N); L = length over which deflection or twisting occurs in meters; A = cross-sectional area in meters squared; E = the modulus of elasticity in (newtons per meter squared); θ = radians; T = the torsional moment in newton-meters (N·m); G = modulus of rigidity, in pascals; and J = the polar moment of inertia in meters^4. If the load (F) is applied as a weight, it should be noted that the weight of a mass M kilograms is Mg newtons, where g = 9.81 m/s^2. Millimeters can be used in the calculations in place of meters, providing the treatment is consistent throughout.

Combined Stresses.—A member may be loaded in such a way that a combination of simple stresses acts at a point. Three general cases occur, examples of which are shown in the accompanying illustration Fig. 10.

Superposition of stresses: Fig. 10 at (1) illustrates a common situation that results in simple stresses combining by superposition at points **a** and **b**. The equal and opposite forces F_1 will cause a compressive stress $\sigma_1 = -F_1/A$. Force F_2 will cause a bending moment M to exist in the plane of points **a** and **b**. The resulting stress $\sigma_2 = \pm M/Z$. The combined stress at point **a**,

$$\sigma_a' = -\frac{F_1}{A} - \frac{M}{Z} \quad (19) \qquad \text{and at } \mathbf{b}, \qquad \sigma_b' = -\frac{F_1}{A} + \frac{M}{Z} \quad (20)$$

where the minus sign indicates a compressive stress and the plus sign a tensile stress. Thus, the stress at **a** will be compressive and at **b** either tensile or compressive depending on which term in the equation for σ_b' has the greatest value.

Normal stresses at right angles: This is shown in Fig. 10 at (2). This combination of stresses occurs, for example, in tanks subjected to internal or external pressure. The principle normal stresses are $\sigma_x = F_1/A_1$, $\sigma_y = F_2/A_2$, and $\sigma_z = 0$ in this plane stress problem. Determine the values of these three stresses with their signs, order them algebraically, and then calculate the maximum shear stress:

$$\tau = (\sigma_{largest} - \sigma_{smallest})/2 \quad (21)$$

Normal and shear stresses: The example in Fig. 10 at (3) shows a member subjected to a torsional shear stress, $\tau = T/Z_p$, and a direct compressive stress, $\sigma = -F/A$. At some point **a** on the member the principal normal stresses are calculated using the equation,

$$\sigma' = \frac{\sigma}{2} \pm \sqrt{\left(\frac{\sigma}{2}\right)^2 + \tau^2} \qquad (22)$$

The maximum shear stress is calculated by using the equation,

$$\tau' = \sqrt{\left(\frac{\sigma}{2}\right)^2 + \tau^2} \qquad (23)$$

The point **a** should ordinarily be selected where stress is a maximum value. For the example shown in the figure at (3), the point **a** can be anywhere on the cylindrical surface because the combined stress has the same value anywhere on that surface.

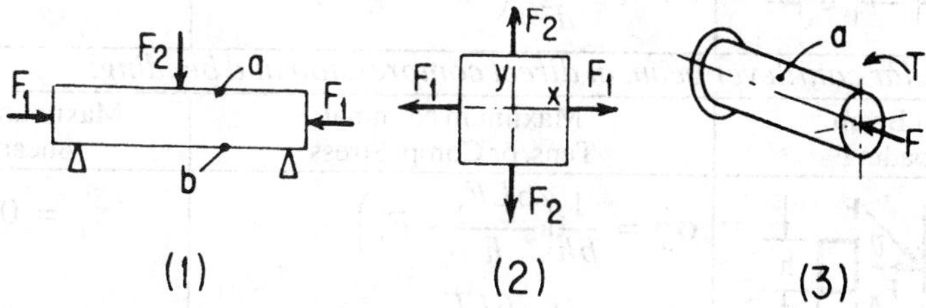

Fig. 10. Types of Combined Loading

Tables of Combined Stresses.—Beginning on page 208, these tables list equations for maximum nominal tensile or compressive (normal) stresses, and maximum nominal shear stresses for common machine elements. These equations were derived using general Equations (19), (20), (22), and (23). The equations apply to the critical points indicated on the figures. Cases 1 through 4 are cantilever beams. These may be loaded with a combination of a vertical and horizontal force, or by a single oblique force. If the single oblique force F and the angle θ are given, then horizontal and vertical forces can be calculated using the equations $F_x = F \cos \theta$ and $F_y = F \sin \theta$. In cases 9 and 10 of the table, the equations for σ_a' can give a tensile and a compressive stress because of the ± sign in front of the radical. Equations involving direct compression are valid only if machine elements have relatively short lengths with respect to their sections, otherwise column equations apply.

Calculation of worst stress condition: Stress failure can occur at any critical point if either the tensile, compressive, or shear stress properties of the material are exceeded by the corresponding working stress. It is necessary to evaluate the factor of safety for each possible failure condition.

The following rules apply to calculations using equations in the , and to calculations based on Equations (19) and (20). *Rule 1:* For every calculated normal stress there is a corresponding induced shear stress; the value of the shear stress is equal to half that of the normal stress. *Rule 2:* For every calculated shear stress there is a corresponding induced normal stress; the value of the normal stress is equal to that of the shear stress. The tables of combined stresses includes equations for calculating both maximum nominal tensile or compressive stresses, and maximum nominal shear stresses.

Formulas for Combined Stresses

(1) Circular cantilever beam in direct compression and bending:

Type of Beam and Loading	Maximum Nominal Tens. or Comp. Stress	Maximum Nominal Shear Stress
	$\sigma_a' = \frac{1.273}{d^2}\left(\frac{8LF_y}{d} - F_x\right)$	$\tau_a' = 0.5\sigma_a'$
	$\sigma_b' = -\frac{1.273}{d^2}\left(\frac{8LF_y}{d} + F_x\right)$	$\tau_b' = 0.5\sigma_b'$

(2) Circular cantilever beam in direct tension and bending:

Type of Beam and Loading	Maximum Nominal Tens. or Comp. Stress	Maximum Nominal Shear Stress
	$\sigma_a' = \frac{1.273}{d^2}\left(F_x + \frac{8LF_y}{d}\right)$	$\tau_a' = 0.5\sigma_a'$
	$\sigma_b' = \frac{1.273}{d^2}\left(F_x - \frac{8LF_y}{d}\right)$	$\tau_b' = 0.5\sigma_b'$

(3) Rectangular cantilever beam in direct compression and bending:

Type of Beam and Loading	Maximum Nominal Tens. or Comp. Stress	Maximum Nominal Shear Stress
	$\sigma_a' = \frac{1}{bh}\left(\frac{6LF_y}{h} - F_x\right)$	$\tau_a' = 0.5\sigma_a'$
	$\sigma_b' = -\frac{1}{bh}\left(\frac{6LF_y}{h} + F_x\right)$	$\tau_b' = 0.5\sigma_b'$

(4) Rectangular cantilever beam in direct tension and bending:

Type of Beam and Loading	Maximum Nominal Tens. or Comp. Stress	Maximum Nominal Shear Stress
	$\sigma_a' = \frac{1}{bh}\left(F_x + \frac{6LF_y}{h}\right)$	$\tau_a' = 0.5\sigma_a'$
	$\sigma_b' = \frac{1}{bh}\left(F_x - \frac{6LF_y}{h}\right)$	$\tau_b' = 0.5\sigma_b'$

(5) Circular beam or shaft in direct compression and bending:

Type of Beam and Loading	Maximum Nominal Tens. or Comp. Stress	Maximum Nominal Shear Stress
	$\sigma_a' = -\frac{1.273}{d^2}\left(\frac{2LF_y}{d} + F_x\right)$	$\tau_a' = 0.5\sigma_a'$
	$\sigma_b' = \frac{1.273}{d^2}\left(\frac{2LF_y}{d} - F_x\right)$	$\tau_b' = 0.5\sigma_b'$

(6) Circular beam or shaft in direct tension and bending:

Type of Beam and Loading	Maximum Nominal Tens. or Comp. Stress	Maximum Nominal Shear Stress
	$\sigma_a' = \frac{1.273}{d^2}\left(F_x - \frac{2LF_y}{d}\right)$	$\tau_a' = 0.5\sigma_a'$
	$\sigma_b' = \frac{1.273}{d^2}\left(F_x + \frac{2LF_y}{d}\right)$	$\tau_b' = 0.5\sigma_b'$

(7) Rectangular beam or shaft in direct compression and bending:

Type of Beam and Loading	Maximum Nominal Tens. or Comp. Stress	Maximum Nominal Shear Stress
F_x, F_y, a, b, h, L/2, L, b	$\sigma_a' = -\frac{1}{bh}\left(\frac{3LF_y}{2h} + F_x\right)$	$\tau_a' = 0.5\sigma_a'$
	$\sigma_b' = \frac{1}{bh}\left(-\frac{3LF_y}{2h} - F_x\right)$	$\tau_b' = 0.5\sigma_b'$

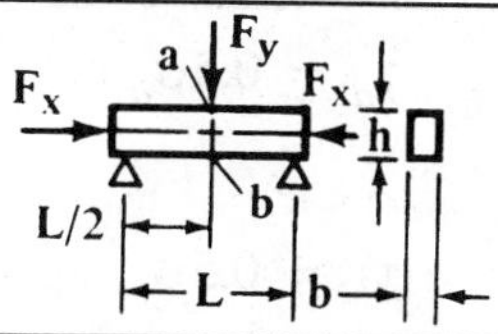

(8) Rectangular beam or shaft in direct tension and bending:

Type of Beam and Loading	Maximum Nominal Tens. or Comp. Stress	Maximum Nominal Shear Stress
F_x, F_y, a, b, h, L/2, L, b	$\sigma_a' = \frac{1}{bh}\left(F_x - \frac{3LF_y}{2h}\right)$	$\tau_a' = 0.5\sigma_a'$
	$\sigma_b' = \frac{1}{bh}\left(F_x + \frac{3LF_y}{2h}\right)$	$\tau_b' = 0.5\sigma_b'$

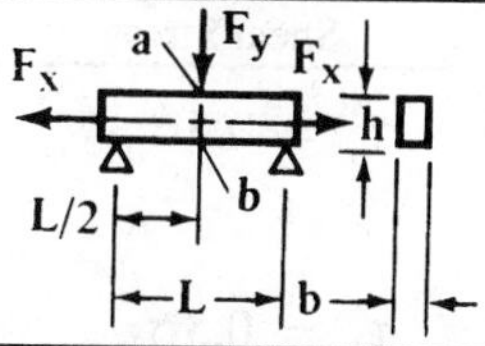

(9) Circular shaft in direct compression and torsion:

Type of Beam and Loading	Maximum Nominal Tens. or Comp. Stress	Maximum Nominal Shear Stress
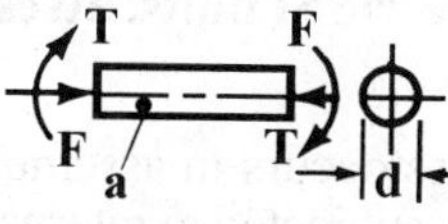 T, F, a, d; a anywhere on surface	$\sigma_a' = -\frac{0.637}{d^2}\left[F \pm \sqrt{F^2 + \left(\frac{8T}{d}\right)^2}\right]$	$\tau_a' = -\frac{0.637}{d^2}\sqrt{F^2 + \left(\frac{8T}{d}\right)^2}$

(10) Circular shaft in direct tension and torsion:

Type of Beam and Loading	Maximum Nominal Tens. or Comp. Stress	Maximum Nominal Shear Stress
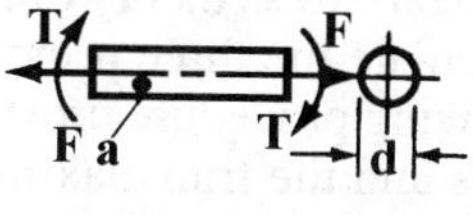 T, F, a, d; a anywhere on surface	$\sigma_a' = -\frac{0.637}{d^2}\left[F \pm \sqrt{F^2 + \left(\frac{8T}{d}\right)^2}\right]$	$\tau_a' = -\frac{0.637}{d^2}\sqrt{F^2 + \left(\frac{8T}{d}\right)^2}$

(11) Offset link, circular cross section, in direct tension:

Type of Beam and Loading	Maximum Nominal Tens. or Comp. Stress	Maximum Nominal Shear Stress
F, a, b, e, d	$\sigma_a' = \frac{1.273F}{d^2}\left(1 - \frac{8e}{d}\right)$	$\tau_a' = 0.5\sigma_a'$
	$\sigma_b' = \frac{1.273F}{d^2}\left(1 + \frac{8e}{d}\right)$	$\tau_b' = 0.5\sigma_b'$

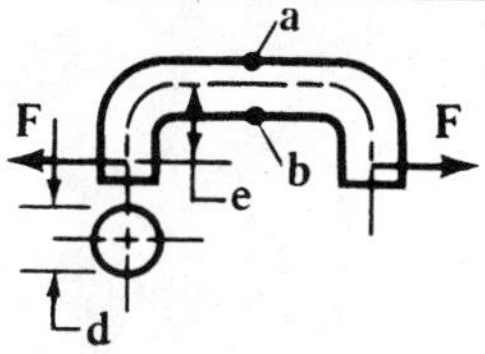

(12) Offset link, circular cross section, in direct compression:

Type of Beam and Loading	Maximum Nominal Tens. or Comp. Stress	Maximum Nominal Shear Stress
F, a, b, e, d	$\sigma_a' = \frac{1.273F}{d^2}\left(\frac{8e}{d} - 1\right)$	$\tau_a' = 0.5\sigma_a'$
	$\sigma_b' = -\frac{1.273F}{d^2}\left(\frac{8e}{d} + 1\right)$	$\tau_b' = 0.5\sigma_b'$

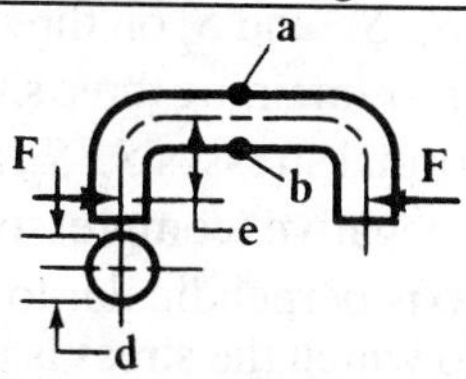

(13) Offset link, rectangular section, in direct tension:

Type of Beam and Loading	Maximum Nominal Tens. or Comp. Stress	Maximum Nominal Shear Stress
e, a, F, F, b, h, b	$\sigma_a' = \frac{F}{bh}\left(1 - \frac{6e}{h}\right)$	$\tau_a' = 0.5\sigma_a'$
	$\sigma_b' = \frac{F}{bh}\left(1 + \frac{6e}{h}\right)$	$\tau_b' = 0.5\sigma_b'$

(14) Offset link, rectangular section, in direct compression:

Type of Beam and Loading	Maximum Nominal Tens. or Comp. Stress	Maximum Nominal Shear Stress
e, a, F, F, b, h, b	$\sigma_a' = \frac{F}{bh}\left(1 - \frac{6e}{h}\right)$	$\tau_a' = 0.5\sigma_a'$
	$\sigma_b' = \frac{F}{bh}\left(1 + \frac{6e}{h}\right)$	$\tau_b' = 0.5\sigma_b'$

Formulas from the simple and combined stress tables, as well as tension and shear factors, can be applied without change in calculations using metric SI units. Stresses are given in newtons per meter squared (N/m²) or in N/mm².

Three-Dimensional Stress.—Three-dimensional or triaxial stress occurs in assemblies such as a shaft press-fitted into a gear bore or in pipes and cylinders subjected to internal or external fluid pressure. Triaxial stress also occurs in two-dimensional stress problems if the loads produce normal stresses that are either both tensile or both compressive. In either case the calculated maximum shear stress, based on the corresponding two-dimensional theory, will be less than the true maximum value because of three-dimensional effects. Therefore, if the stress analysis is to be based on the maximum-shear-stress theory of failure, the triaxial stress cubic equation should first be used to calculate the three principal stresses and from these the true maximum shear stress. The following procedure provides the principal maximum normal tensile and compressive stresses and the true maximum shear stress at any point on a body subjected to any combination of loads.

The basis for the procedure is the stress cubic equation

$$S^3 - AS^2 + BS - C = 0$$

in which:

$$A = S_x + S_y + S_z$$
$$B = S_xS_y + S_yS_z + S_zS_x - S_{xy}^2 - S_{yz}^2 - S_{zx}^2$$
$$C = S_xS_yS_z + 2S_{xy}S_{yz}S_{zx} - S_xS_{yz}^2 - S_yS_{zx}^2 - S_zS_{xy}^2$$

and S_x, S_y, etc., are as shown in Fig. 1.

The coordinate system *XYZ* in Fig. 1 shows the positive directions of the normal and shear stress components on an elementary cube of material. Only six of the nine components shown are needed for the calculations: the normal stresses S_x, S_y, and S_z on three of the faces of the cube; and the three shear stresses S_{xy}, S_{yz}, and S_{zx}. The remaining three shear stresses are known because $S_{yx} = S_{xy}$, $S_{zy} = S_{yz}$, and $S_{xz} = S_{zx}$. The normal stresses S_x, S_y, and S_z are shown as positive (tensile) stresses; the opposite direction is negative (compressive). The first subscript of each shear stress identifies the coordinate axis perpendicular to the plane of the shear stress; the second subscript identifies the axis to which the stress is parallel. Thus, S_{xy}, is the shear stress in the *YZ* plane to which the *X* axis is perpendicular, and the stress is parallel to the *Y* axis.

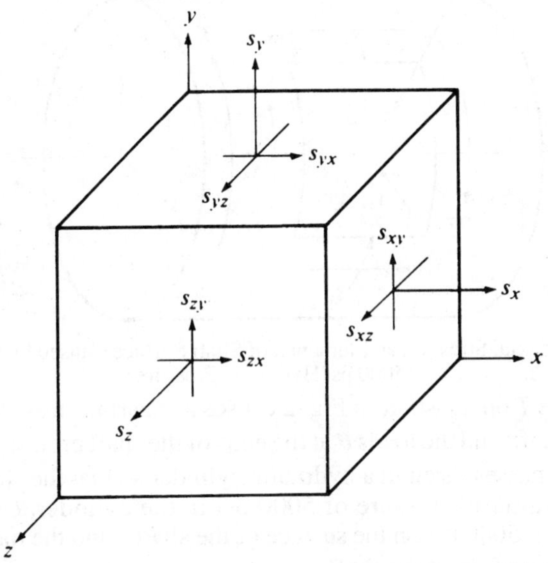

Fig. 1. *XYZ* Coordinate System Showing Positive Directions of Stresses

Step 1. Draw a diagram of the hardware to be analyzed, such as the shaft shown in Fig. 2, and show the applied loads *P*, *T*, and any others.

Step 2. For any point at which the stresses are to be analyzed, draw a coordinate diagram similar to Fig. 1 and show the magnitudes of the stresses resulting from the applied loads (these stresses may be calculated by using standard basic equations from strength of materials, and should include any stress concentration factors).

Step 3. Substitute the values of the six stresses S_x, S_y, S_z, S_{xy}, S_{yz}, and S_{zx}, including zero values, into the formulas for the quantities *A* through *K*. The quantities *I*, *J*, and *K* represent the principal normal stresses at the point analyzed. As a check, if the algebraic sum $I + J + K$ equals *A*, within rounding errors, then the calculations up to this point should be correct.

$$D = A^2/3 - B$$

$$E = A \times B/3 - C - 2 \times A^3/27$$

$$F = \sqrt{(D^3/27)}$$

$$G = \arccos(-E/(2 \times F))$$

$$H = \sqrt{(D/3)}$$

$$I = 2 \times H \times \cos(G/3) + A/3$$

$$J = 2 \times H \times [\cos(G/3 + 120°)] + A/3$$

$$K = 2 \times H \times [\cos(G/3 + 240°)] + A/3$$

Step 4. Calculate the true maximum shear stress, $S_{s(max)}$ using the formula

$$S_{s(max)} = 0.5 \times (S_{large} - S_{small})$$

in which S_{large} is equal to the algebraically largest of the calculated principal stresses *I*, *J*, or *K* and S_{small} is algebraically the smallest.

The maximum principal normal stresses and the maximum true shear stress calculated above may be used with any of the various theories of failure.

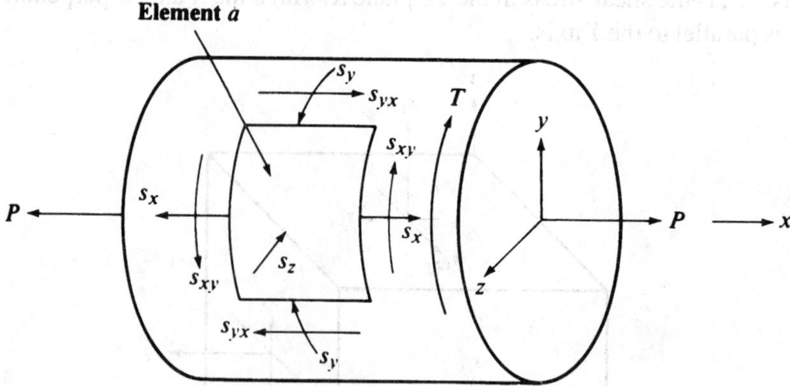

Fig. 2. Example of Triaxial Stress on an Element *a* of Shaft Surface Caused by Load *P*, Torque *T*, and 5000 psi Hydraulic Pressure

Example: A torque *T* on the shaft in Fig. 2 causes a shearing stress S_{xy} of 8000 psi in the outer fibers of the shaft; and the loads *P* at the ends of the shaft produce a tensile stress S_x of 4000 psi. The shaft passes through a hydraulic cylinder so that the shaft circumference is subjected to the hydraulic pressure of 5000 psi in the cylinder, causing compressive stresses S_y and S_z of – 5000 psi on the surface of the shaft. Find the maximum shear stress at any point *A* on the surface of the shaft.

From the statement of the problem $S_x = +4000$ psi, $S_y = -5000$ psi, $S_z = -5000$ psi, $S_{xy} = +8000$ psi, $S_{yz} = 0$ psi, and $S_{zx} = 0$ psi.

$A = 4000 - 5000 - 5000 = -6000$

$B = (4000 \times -5000) + (-5000 \times -5000) + (-5000 \times 4000) - 8000^2 - 0^2 - 0^2 = -7.9 \times 10^7$

$C = (4000 \times -5000 \times -5000) + 2 \times 8000 \times 0 \times 0 - (4000 \times 0^2) - (-5000 \times 0^2) - (-5000 \times 8000^2) = 4.2 \times 10^{11}$

$D = A^2/3 - B = 9.1 \times 10^7$

$E = A \times B/3 - C - 2 \times A^3/27 = -2.46 \times 10^{11}$

$F = \sqrt{(D^3/27)} = 1.6706 \times 10^{11}$

$G = \arccos(-E/(2 \times F)) = 42.586$ degrees, $H = \sqrt{(D/3)} = 5507.57$

$I = 2 \times H \times \cos(G/3 + A/3 = 8678.8$, say, 8680 psi

$J = 2 \times H \times [\cos(G/3 + 120°)] + A/3 = -9678.78$, say, – 9680 psi

$K = 2 \times H [\cos(G/3 + 240°)] + A/3 = -5000$ psi

Check: 8680 + (– 9680) + (– 5000) = – 6000 within rounding error.

$S_{s(\max)} = 0.5 \times (8680 - (-9680)) = 9180$ psi

Sample Calculations.—The following examples illustrate some typical strength of materials calculations, using both English and metric SI units of measurement.

Example 1(a): A round bar made from SAE 1025 low carbon steel is to support a direct tension load of 50,000 pounds. Using a factor of safety of 4, and assuming that the stress concentration factor $K = 1$, a suitable standard diameter is to be determined. Calculations are to be based on a yield strength of 40,000 psi.

Because the factor of safety and strength of the material are known, the allowable working stress s_w may be calculated using Equation (1): 40,000/4 = 10,000 psi. The relationship between working stress s_w and nominal stress σ is given by Equation (2). Since $K = 1$, $\sigma = 10{,}000$ psi. Applying Equation (9) in the , the area of the bar can be solved for: A = 50,000/10,000 or 5 square inches. The next largest standard diameter corresponding to this area is $2\frac{9}{16}$ inches.

Example 1(b): **A similar example to that given in 1(a), using metric SI units is as follows. A round steel bar of 300 meganewtons/meter ^{2} yield strength, is to withstand a direct tension of 200 kilonewtons. Using a safety factor of 4, and assuming that the stress concentration factor $K = 1$, a suitable diameter is to be determined.**

Because the factor of safety and the strength of the material are known, the allowable working stress s_w may be calculated using Equation (1): 300/4 = 75 mega-newtons/meter^2. The relationship between working stress and nominal stress σ is given by Equation (2). Since $K = 1$, $\sigma = 75$ MN/m^2. Applying Equation (9) in the , the area of the bar can be determined from:

$$A = \frac{200\text{ kN}}{75\text{ MN/m}^2} = \frac{200{,}000\text{ N}}{75{,}000{,}000\text{ N/m}^2} = 0.00267\text{m}^2$$

The diameter corresponding to this area is 0.058 meters, or approximately 0.06 m.

Millimeters can be employed in the calculations in place of meters, providing the treatment is consistent throughout. In this instance the diameter would be 60 mm.

Note: **If the tension in the bar is produced by hanging a mass of M kilograms from its end, the value is Mg newtons, where g = approximately 9.81 meters per second^2.**

Example 2(a): What would the total elongation of the bar in Example 1(a) be if its length were 60 inches? Applying Equation (17),

$$e = \frac{50{,}000 \times 60}{5.157 \times 30{,}000{,}000} = 0.019\text{ inch}$$

Example 2(b): **What would be the total elongation of the bar in Example 1(b) if its length were 1.5 meters? The problem is solved by applying Equation (17) in which F = 200 kilonewtons; L = 1.5 meters; $A = \pi 0.06^2/4 = 0.00283$ m^2. Assuming a modulus of elasticity E of 200 giganewtons/meter^2, then the calculation is:**

$$e = \frac{200{,}000 \times 1.5}{0.00283 \times 200{,}000{,}000{,}000} = 0.000530\text{ m}$$

The calculation is less unwieldy if carried out using millimeters in place of meters; then F = 200 kN; L = 1500 mm; A = 2830 mm^2, and E = 200,000 N/mm^2. Thus:

$$e = \frac{200{,}000 \times 1500}{2830 \times 200{,}000} = 0.530\text{ mm}$$

Example 3(a): Determine the size for the section of a square bar which is to be held firmly at one end and is to support a load of 3000 pounds at the outer end. The bar is to be 30 inches long and is to be made from SAE 1045 medium carbon steel with a yield point of 60,000 psi. A factor of safety of 3 and a stress concentration factor of 1.3 are to be used.

From Equation (1) the allowable working stress s_w = 60,000/3 = 20,000 psi. The applicable equation relating working stress and nominal stress is Equation (2); hence, σ = 20,000/1.3 = 15,400 psi. The member must be treated as a cantilever beam subject to a bending moment of 30 × 3000 or 90,000 inch-pounds. Solving Equation (11) in the for section modulus: Z = 90,000/15,400 = 5.85 inch^3. The section modulus for a square section with neutral axis equidistant from either side is $a^3/6$, where a is the dimension of the square, so $a = \sqrt[3]{35.1} = 3.27$ inches. The size of the bar can therefore be $3\frac{5}{16}$ inches.

Example 3(b): **A similar example to that given in Example 3(a), using metric SI units is as follows. Determine the size for the section of a square bar which is to be held firmly at one end and is to support a load of 1600 kilograms at the outer end. The bar is to be 1 meter long, and is to be made from steel with a yield strength of 500 newtons/mm^2. A factor of safety of 3, and a stress concentration factor of 1.3 are to be used. The calculation can be performed using millimeters throughout.**

From Equation (1) the allowable working stress s_w = 500 N/mm²/3 = 167 N/mm². The formula relating working stress and nominal stress is Equation (2); hence σ = 167/1.3 = 128 N/mm². Since a mass of 1600 kg equals a weight of 1600 *g* newtons, where *g* = 9.81 meters/second², the force acting on the bar is 15,700 newtons. The bending moment on the bar, which must be treated as a cantilever beam, is thus 1000 mm × 15,700 N = 15,700,000 N · mm. Solving Equation (11) in the for section modulus: $Z = M/\sigma$ = 15,700,000/128 = 123,000 mm³. Since the section modulus for a square section with neutral axis equidistant from either side is $a^3/6$, where *a* is the dimension of the square,

$$a = \sqrt[3]{6 \times 123{,}000} = 90.4 \text{ mm}$$

Example 4(a): Find the working stress in a 2-inch diameter shaft through which a transverse hole ¼ inch in diameter has been drilled. The shaft is subject to a torsional moment of 80,000 inch-pounds and is made from hardened steel so that the index of sensitivity $q = 0.2$.

The polar section modulus is calculated using the equation shown in the stress concentration curve for a Round Shaft in Torsion with Transverse Hole, page 202.

$$\frac{J}{c} = Z_p = \frac{\pi \times 2^3}{16} - \frac{2^2}{4 \times 6} = 1.4 \text{ inches}^3$$

The nominal shear stress due to the torsion loading is computed using Equation (16) in the :

$$\tau = 80{,}000/1.4 = 57{,}200 \text{ psi}$$

Referring to the previously mentioned stress concentration curve on page 202, K_t is 2.82 since d/D is 0.125. The stress concentration factor may now be calculated by means of Equation (8): $K = 1 + 0.2(2.82 - 1) = 1.36$. Working stress calculated with Equation (3) is $s_w = 1.36 \times 57{,}200 = 77{,}800$ psi.

***Example 4(b):* A similar example to that given in 4(a), using metric SI units is as follows. Find the working stress in a 50 mm diameter shaft through which a transverse hole 6 mm in diameter has been drilled. The shaft is subject to a torsional moment of 8000 newton-meters, and has an index of sensitivity of *q* = 0.2. If the calculation is made in millimeters, the torsional moment is 8,000,000 N · mm.**

The polar section modulus is calculated using the equation shown in the stress concentration curve for a Round Shaft with Transverse Hole, page 202:

$$\frac{J}{c} = Z_p = \frac{\pi \times 50^3}{16} - \frac{6 \times 50^2}{6}$$

$$= 24{,}544 - 2500 = 22{,}044 \text{ mm}^3$$

The nominal shear stress due to torsion loading is computed using Equation (16) in the :

$$\tau = 8{,}000{,}000/22{,}000 = 363 \text{ N/mm}^2 = 363 \text{ megapascals}$$

Referring to the previously mentioned stress concentration curve on page 202, K_t is 2.85, since a/d = 6/50 = 0.12. The stress concentration factor may now be calculated by means of Equation (8): $K = 1 + 0.2(2.85 - 1) = 1.37$. From Equation (3), working stress $s_w = 1.37 \times 363$ = 497 N/mm² = 497 megapascals.

Example 5(a): For Case 3 in the *Tables of Combined Stresses*, calculate the least factor of safety for a 5052-H32 aluminum beam is 10 inches long, one inch wide, and 2 inches high. Yield strengths are 23,000 psi tension; 21,000 psi compression; 13,000 psi shear. The stress concentration factor is 1.5; F_y is 600 lbs; F_x 500 lbs.

From *Tables of Combined Stresses*, Case 3:

$$\sigma_b' = -\frac{1}{1 \times 2}\left(\frac{6 \times 10 \times 600}{2} + 500\right) = -9250 \text{ psi (in compression)}$$

The other formulas for Case 3 give $\sigma_a' = 8750$ psi (in tension); $\tau_a' + 4375$ psi, and $\tau_b' +$ 4625 psi. Using equation (4) for the nominal compressive stress of 9250 psi: $S_w = 1.5 \times 9250 = 13{,}900$ psi. From Equation (1) $f_s = 21{,}000/13{,}900 = 1.51$. Applying Equations (1), (4) and (5) in appropriate fashion to the other calculated nominal stress values for tension and shear will show that the factor of safety of 1.51, governed by the compressive stress at *b* on the beam, is minimum.

***Example 5(b):* What maximum F can be applied in Case 3 if the aluminum beam is 200 mm long; 20 mm wide; 40 mm high; $\theta = 30°$; $f_s = 2$, governing for compression, $K = 1.5$, and $S_m = 144\text{N/mm}^2$ for compression.**

From Equation (1) $S_w = -144\text{N/mm}^2$. Therefore, from Equation (4), $\sigma_b' = -72/1.5 = -48\text{N/mm}^2$. Since $F_x = F\cos 30° = 0.866F$, and $F_y = F \sin 30° = 0.5F$:

$$-48 = -\frac{1}{20 \times 40}\left(0.866F + \frac{6 \times 200 \times 0.5F}{40}\right)$$

$$F = 2420 \text{ N}$$

Stresses and Deflections in a Loaded Ring.—For *thin* rings, that is, rings in which the dimension *d* shown in the accompanying diagram is small compared with *D*, the maximum stress in the ring is due primarily to bending moments produced by the forces *P*. The maximum stress due to bending is:

$$S = \frac{PDd}{4\pi I} \quad (1)$$

For a ring of circular cross section where *d* is the diameter of the bar from which the ring is made,

$$S = \frac{1.621PD}{d^3} \quad \text{or} \quad P = \frac{0.617Sd^3}{D} \quad (2)$$

The increase in the vertical diameter of the ring due to load *P* is:

$$\text{Increase in vertical diameter} = \frac{0.0186PD^3}{EI} \text{ inches} \quad (3)$$

The *decrease* in the horizontal diameter will be about 92% of the increase in the vertical diameter given by Formula (3). In the above formulas, P = load on ring in pounds; D = mean diameter of ring in inches; S = tensile stress in pounds per square inch, I = moment of inertia of section in inches^4; and E = modulus of elasticity of material in pounds per square inch.

Strength of Taper Pins.—The mean diameter of taper pin required to safely transmit a known torque, may be found from the formulas:

$$d = 1.13\sqrt{\frac{T}{DS}} \quad (1) \qquad \text{and} \qquad d = 283\sqrt{\frac{\text{HP}}{NDS}} \quad (2)$$

in which formulas T = torque in inch-pounds; S = safe unit stress in pounds per square inch; HP = horsepower transmitted; N = number of revolutions per minute; and d and D denote dimensions shown in the figure.

Formula (1) can be used with metric SI units where d and D denote dimensions shown in the figure in millimeters; T = torque in newton-millimeters (N · mm); and S = safe unit stress in newtons per millimeter² (N/mm²). Formula (2) is replaced by:

$$\mathbf{d = 110.3\sqrt{\frac{Power}{NDS}}}$$

where d and D denote dimensions shown in the figure in millimeters; S = safe unit stress in N/mm²; N = number of revolutions per minute, and Power = power transmitted in watts.

Examples: A lever secured to a 2-inch round shaft by a steel tapered pin (dimension $d = \frac{3}{8}$ inch) has a pull of 50 pounds at a 30-inch radius from shaft center. Find S, the unit working stress on the pin. By rearranging Formula (1):

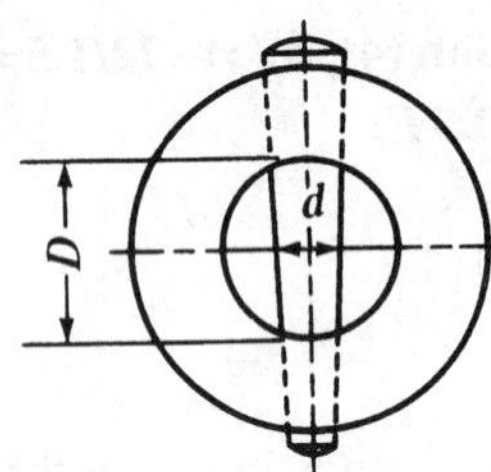

$$S = \frac{1.27T}{Dd^2} = \frac{1.27 \times 50 \times 30}{2 \times \left(\frac{3}{8}\right)^2} = 6770$$

pounds per square inch (nearly), which is a safe unit working stress for machine steel in shear.

Let P = 50 pounds, R = 30 inches, D = 2 inches, and S = 6000 pounds unit working stress. Using Formula (1) to find d:

$$d = 1.13\sqrt{\frac{T}{DS}} = 1.13\sqrt{\frac{50 \times 30}{2 \times 6000}} = 1.13\sqrt{\frac{1}{8}} = 0.4 \text{ inch}$$

A similar example using SI units is as follows: A lever secured to a 50 mm round shaft by a steel tapered pin (d = 10 mm) has a pull of 200 newtons at a radius of 800 mm. Find S, the working stress on the pin. By rearranging Formula (1):

$$\mathbf{S = \frac{1.27T}{Dd^2} = \frac{1.27 \times 200 \times 800}{50 \times 10^2} = 40.6\ N/mm^2 = 40.6\ megapascals}$$

If a shaft of 50 mm diameter is to transmit power of 12 kilowatts at a speed of 500 rpm, find the mean diameter of the pin for a material having a safe unit stress of 40 N/mm². Using the formula:

$$\mathbf{d = 110.3\sqrt{\frac{Power}{NDS}} \qquad then\ d = 110.3\sqrt{\frac{12,000}{500 \times 50 \times 40}}}$$

$$\mathbf{= 110.3 \times 0.1096 = 12.09\ mm}$$

MOMENT OF INERTIA

Calculating Moment of Inertia

Moment of Inertia of Built-up Sections.—The usual method of calculating the moment of inertia of a built-up section involves the calculations of the moment of inertia for each element of the section about its own neutral axis, and the transferring of this moment of inertia to the previously found neutral axis of the whole built-up section. A much simpler method that can be used in the case of any section which can be divided into rectangular elements bounded by lines parallel and perpendicular to the neutral axis is the so-called tabular method based upon the formula: $I = b(h_1^3 - h^3)/3$ in which I = the moment of inertia about axis *DE*, Fig. 1, and b, h and h_1 are dimensions as given in the same illustration.

The method may be illustrated by applying it to the section shown in Fig. 2, and for simplicity of calculation shown "massed" in Fig. 3. The calculation may then be tabulated as shown in the accompanying table. The distance from the axis *DE* to the neutral axis *xx* (which will be designated as d) is found by dividing the sum of the geometrical moments by the area. The moment of inertia about the neutral axis is then found in the usual way by subtracting the area multiplied by d^2 from the moment of inertia about the axis *DE*.

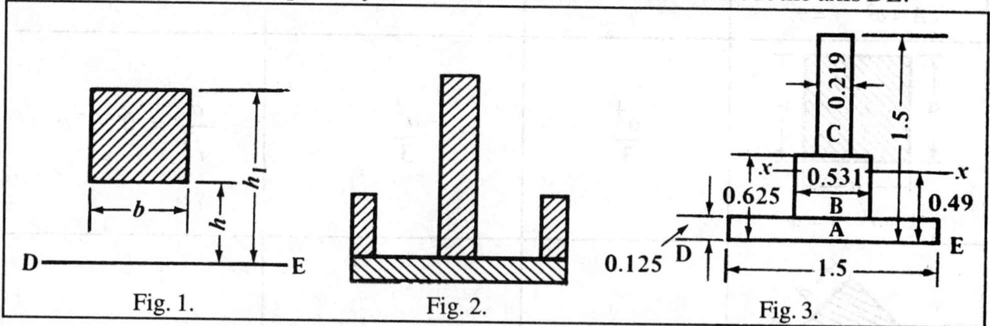

Fig. 1. Fig. 2. Fig. 3.

Tabulated Calculation of Moment of Inertia

Section	Breadth b	Height h_1	Area $b(h_1 - h)$	h_1^2	Moment $\frac{b(h_1^2 - h^2)}{2}$	h_1^3	I about axis DE $\frac{b(h_1^3 - h^3)}{3}$
A	1.500	0.125	0.187	0.016	0.012	0.002	0.001
B	0.531	0.625	0.266	0.391	0.100	0.244	0.043
C	0.219	1.500	0.191	2.250	0.203	3.375	0.228
			$A = 0.644$		$M = 0.315$		$I_{DE} = 0.272$

The distance d from *DE*, the axis through the base of the configuration, to the neutral axis *xx* is:

$$d = \frac{M}{A} = \frac{0.315}{0.644} = 0.49$$

The moment of inertia of the entire section with reference to the neutral axis *xx* is:

$$I_N = I_{DE} - Ad^2$$

$$= 0.272 - 0.644 \times 0.49^2$$

$$= 0.117$$

Formulas for Moments of Inertia, Section Moduli, etc.—On the following pages are given formulas for the moments of inertia and other properties of forty-two different cross-sections. The formulas give the area of the section A, and the distance y from the neutral axis to the extreme fiber, for each example. Where the formulas for the section modulus and radius of gyration are very lengthy, the formula for the section modulus, for example, has been simply given as $I \div y$. The radius of gyration is sometimes given as $\sqrt{I \div A}$ to save space.

Moments of Inertia, Section Moduli, and Radii of Gyration

Section A = area y = distance from axis to extreme fiber	Moment of Inertia I	Section Modulus $Z = \frac{I}{y}$	Radius of Gyration $k = \sqrt{\frac{I}{A}}$
Square and Rectangular Sections			
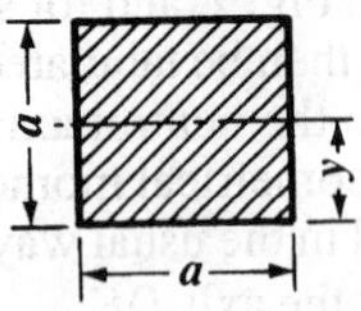 $A = a^2 \quad y = a/2$	$\frac{a^4}{12}$	$\frac{a^3}{6}$	$\frac{a}{\sqrt{12}} = 0.289a$
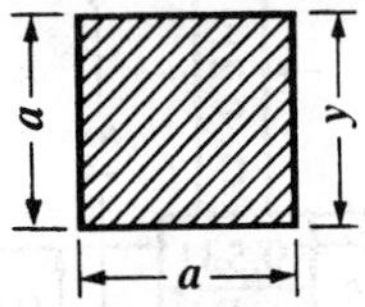 $A = a^2 \quad y = a$	$\frac{a^4}{3}$	$\frac{a^3}{3}$	$\frac{a}{\sqrt{3}} = 0.577a$
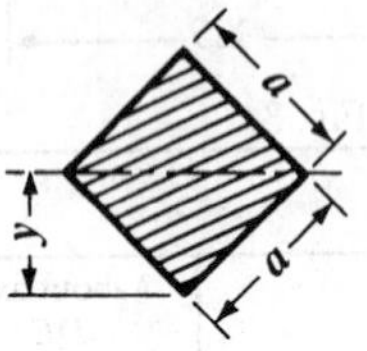 $A = a^2$ $y = \frac{a}{\sqrt{2}} = 0.707a$	$\frac{a^4}{12}$	$\frac{a^3}{6\sqrt{2}} = 0.118a^3$	$\frac{a}{\sqrt{12}} = 0.289a$
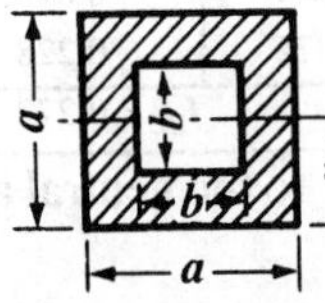 $A = a^2 - b^2 \quad y = a/2$	$\frac{a^4 - b^4}{12}$	$\frac{a^4 - b^4}{6a}$	$\sqrt{\frac{a^2 + b^2}{12}}$ $= 0.289\sqrt{a^2 + b^2}$
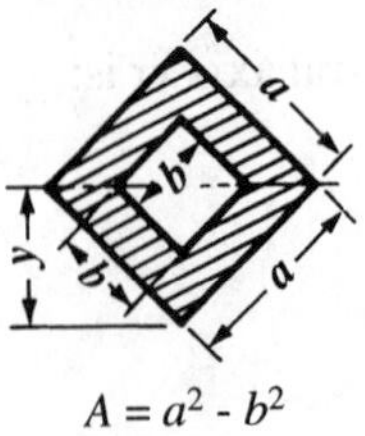 $A = a^2 - b^2$ $y = \frac{a}{\sqrt{2}} = 0.707a$	$\frac{a^4 - b^4}{12}$	$\frac{\sqrt{2}(a^4 - b^4)}{12a}$ $= 0.118\frac{a^4 - b^4}{a}$	$\sqrt{\frac{a^2 + b^2}{12}}$ $= 0.289\sqrt{a^2 + b^2}$

Moments of Inertia, Section Moduli, and Radii of Gyration *(Continued)*

Section A = area y = distance from axis to extreme fiber	Moment of Inertia I	Section Modulus $Z = \frac{I}{y}$	Radius of Gyration $k = \sqrt{\frac{I}{A}}$
Square and Rectangular Sections *(Continued)*			
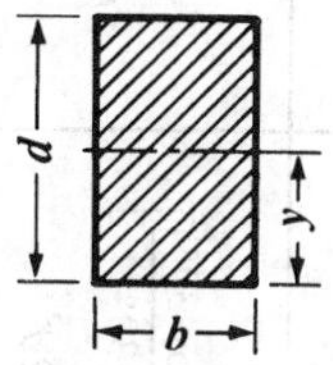 $A = bd \quad y = d/2$	$\frac{bd^3}{12}$	$\frac{bd^2}{6}$	$\frac{d}{\sqrt{12}} = 0.289d$
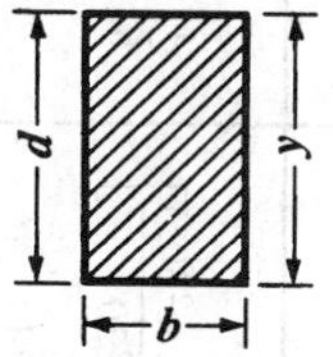 $A = bd \quad y = d$	$\frac{bd^3}{3}$	$\frac{bd^2}{3}$	$\frac{d}{\sqrt{3}} = 0.577d$
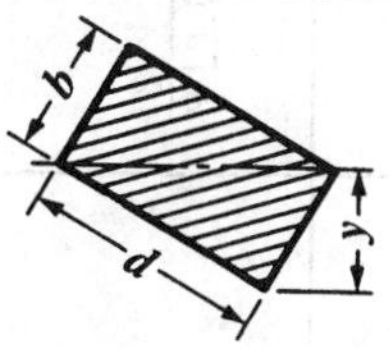 $A = bd$ $y = \frac{bd}{\sqrt{b^2 + d^2}}$	$\frac{b^3 d^3}{6(b^2 + d^2)}$	$\frac{b^2 d^2}{6\sqrt{b^2 + d^2}}$	$\frac{bd}{\sqrt{6(b^2 + d^2)}}$ $= 0.408 \frac{bd}{\sqrt{b^2 + d^2}}$
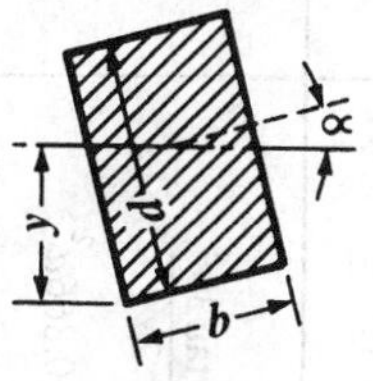 $A = bd$ $y = \frac{1}{2}(d \cos \alpha + b \sin \alpha)$	$\frac{bd}{12}(d^2 \cos^2 \alpha + b^2 \sin^2 \alpha)$	$\frac{bd}{6} \times \left(\frac{d^2 \cos^2 \alpha + b^2 \sin^2 \alpha}{d \cos \alpha + b \sin \alpha}\right)$	$\frac{\sqrt{d^2 \cos^2 \alpha + b^2 \sin^2 \alpha}}{12}$ $= 0.289 \times \sqrt{d^2 \cos^2 \alpha + b^2 \sin^2 \alpha}$
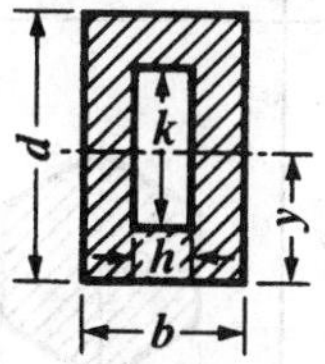 $A = bd - hk$ $y = d/2$	$\frac{bd^3 - hk^3}{12}$	$\frac{bd^3 - hk^3}{6d}$	$\sqrt{\frac{bd^3 - hk^3}{12(bd - hk)}}$ $= 0.289 \sqrt{\frac{bd^3 - hk^3}{bd - hk}}$

Moments of Inertia, Section Moduli, and Radii of Gyration

Section	Area of Section, A	Distance from Neutral Axis to Extreme Fiber, y	Moment of Inertia, I	Section Modulus, $Z = I/y$	Radius of Gyration, $k = \sqrt{I/A}$
Triangular Sections					
	$\frac{1}{2}bd$	$\frac{2}{3}d$	$\frac{bd^3}{36}$	$\frac{bd^2}{24}$	$\frac{d}{\sqrt{18}} = 0.236d$
	$\frac{1}{2}bd$	d	$\frac{bd^3}{12}$	$\frac{bd^2}{12}$	$\frac{d}{\sqrt{6}} = 0.408d$
Polygon Sections					
	$\frac{d(a+b)}{2}$	$\frac{d(a+2b)}{3(a+b)}$	$\frac{d^3(a^2+4ab+b^2)}{36(a+b)}$	$\frac{d^2(a^2+4ab+b^2)}{12(a+2b)}$	$\sqrt{\frac{d^2(a^2+4ab+b^2)}{18(a+b)^2}}$
	$\frac{3d^2\tan 30°}{2} = 0.866d^2$	$\frac{d}{2}$	$\frac{A}{12}\left[\frac{d^2(1+2\cos^2 30°)}{4\cos^2 30°}\right] = 0.06d^4$	$\frac{A}{6}\left[\frac{d(1+2\cos^2 30°)}{4\cos^2 30°}\right] = 0.12d^3$	$\sqrt{\frac{d^2(1+2\cos^2 30°)}{48\cos^2 30°}} = 0.264d$

Moments of Inertia, Section Moduli, and Radii of Gyration *(Continued)*

Section	Area of Section, A	Distance from Neutral Axis to Extreme Fiber, y	Moment of Inertia, I	Section Modulus, $Z = I/y$	Radius of Gyration, $k = \sqrt{I/A}$
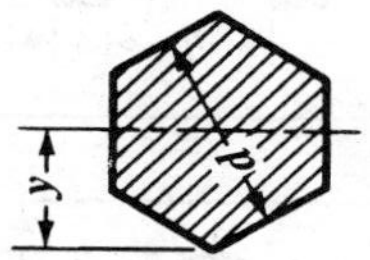	$\frac{3d^2 \tan 30°}{2}$ $= 0.866d^2$	$\frac{d}{2\cos 30°} = 0.577d$	$\frac{A}{12}\left[\frac{d^2(1+2\cos^2 30°)}{4\cos^2 30°}\right]$ $= 0.06d^4$	$\frac{A}{6.9}\left[\frac{d(1+2\cos^2 30°)}{4\cos^2 30°}\right]$ $= 0.104d^3$	$\sqrt{\frac{d^2(1+2\cos^2 30°)}{48\cos^2 30°}}$ $= 0.264d$
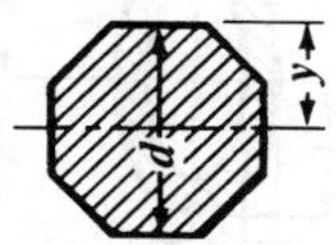	$2d^2 \tan 22\frac{1}{2} = 0.828d^2$	$\frac{d}{2}$	$\frac{A}{12}\left[\frac{d^2(1+2\cos^2 22\frac{1}{2}°)}{4\cos^2 22\frac{1}{2}°}\right]$ $= 0.055d^4$	$\frac{A}{6}\left[\frac{d(1+2\cos^2 22\frac{1}{2}°)}{4\cos^2 22\frac{1}{2}°}\right]$ $= 0.109d^3$	$\sqrt{\frac{d^2(1+2\cos^2 22\frac{1}{2}°)}{48\cos^2 22\frac{1}{2}°}}$ $= 0.257d$
Circular, Elliptical, and Circular Arc Sections					
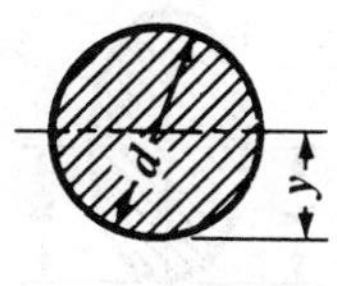	$\frac{\pi d^2}{4} = 0.7854d^2$	$\frac{d}{2}$	$\frac{\pi d^4}{64} = 0.049d^4$	$\frac{\pi d^3}{32} = 0.098d^3$	$\frac{d}{4}$
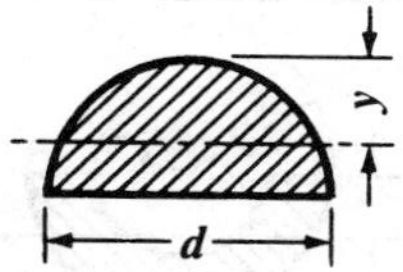	$\frac{\pi d^2}{8} = 0.393d^2$	$\frac{(3\pi - 4)d}{6\pi}$ $= 0.288d$	$\frac{(9\pi^2 - 64)d^4}{1152\pi}$ $= 0.007d^4$	$\frac{(9\pi^2 - 64)d^3}{192(3\pi - 4)}$ $= 0.024d^3$	$\frac{\sqrt{(9\pi^2 - 64)d^2}}{12\pi}$ $= 0.132d$
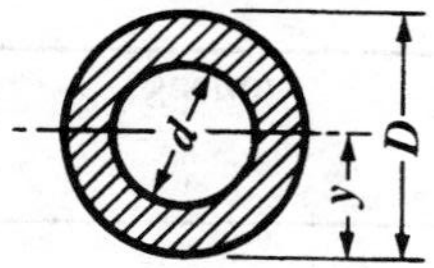	$\frac{\pi(D^2 - d^2)}{4}$ $= 0.7854(D^2 - d^2)$	$\frac{D}{2}$	$\frac{\pi(D^4 - d^4)}{64}$ $= 0.049(D^4 - d^4)$	$\frac{\pi(D^4 - d^4)}{32D}$ $= 0.098\frac{D^4 - d^4}{D}$	$\frac{\sqrt{D^2 + d^2}}{4}$

Moments of Inertia, Section Moduli, and Radii of Gyration *(Continued)*

Section	Area of Section, A	Distance from Neutral Axis to Extreme Fiber, y	Moment of Inertia, I	Section Modulus, $Z = I/y$	Radius of Gyration, $k = \sqrt{I/A}$
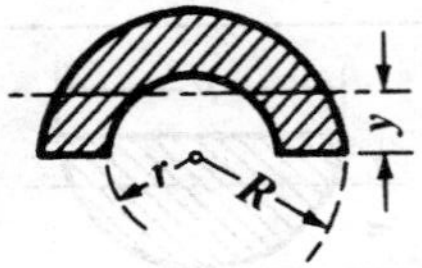	$\frac{\pi(R^2 - r^2)}{2}$ $= 1.5708(R^2 - r^2)$	$\frac{4(R^3 - r^3)}{3\pi(R^2 - r^2)}$ $= 0.424\frac{R^3 - r^3}{R^2 - r^2}$	$0.1098(R^4 - r^4)$ $- \frac{0.283R^2r^2(R - r)}{R + r}$	$\frac{I}{y}$	$\sqrt{\frac{I}{A}}$
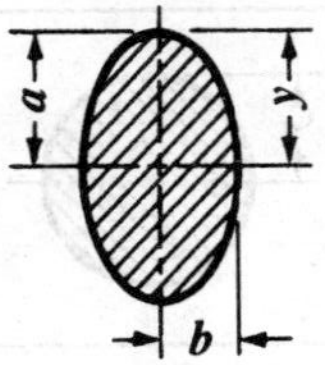	$\pi ab = 3.1416ab$	a	$\frac{\pi a^3 b}{4} = 0.7854a^3 b$	$\frac{\pi a^2 b}{4} = 0.7854a^2 b$	$\frac{a}{2}$
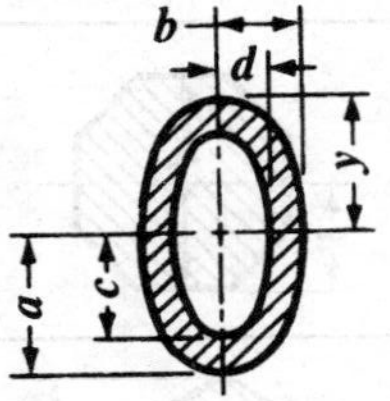	$\pi(ab - cd)$ $= 3.1416(ab - cd)$	a	$\frac{\pi}{4}(a^3b - c^3d)$ $= 0.7854(a^3b - c^3d)$	$\frac{\pi(a^3b - c^3d)}{4a}$ $= 0.7854\frac{a^3b - c^3d}{a}$	$\frac{1}{2}\sqrt{\frac{a^3b - c^3d}{ab - cd}}$
I-Sections					
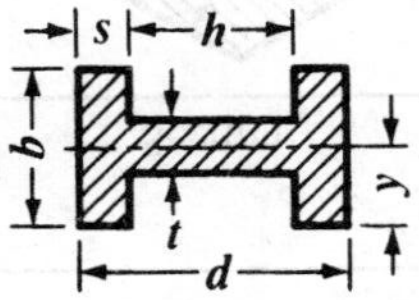	$bd - h(b - t)$	$\frac{b}{2}$	$\frac{2sb^3 + ht^3}{12}$	$\frac{2sb^3 + ht^3}{6b}$	$\sqrt{\frac{2sb^3 + ht^3}{12[bd - h(b - t)]}}$

Moments of Inertia, Section Moduli, and Radii of Gyration *(Continued)*

Section	Area of Section, A	Distance from Neutral Axis to Extreme Fiber, y	Moment of Inertia, I	Section Modulus, $Z = I/y$	Radius of Gyration, $k = \sqrt{I/A}$
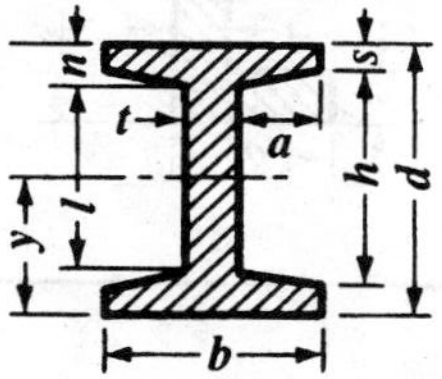	$dt + 2a(s + n)$	$\frac{d}{2}$	$\frac{1}{12}\left[bd^3 - \frac{1}{4g}(h^4 - l^4)\right]$ in which g = slope of flange = $(h - l)/(b - t) = \frac{1}{6}$ for standard I-beams.	$\frac{1}{6d}\left[bd^3 - \frac{1}{4g}(h^4 - l^4)\right]$	$\sqrt{\frac{\frac{1}{12}\left[bd^3 - \frac{1}{4g}(h^4 - l^4)\right]}{dt + 2a(s + n)}}$
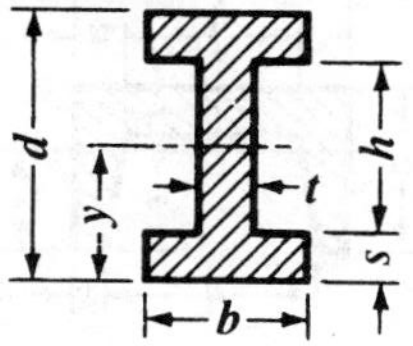	$bd - h(b - t)$	$\frac{d}{2}$	$\frac{bd^3 - h^3(b - t)}{12}$	$\frac{bd^3 - h^3(b - t)}{6d}$	$\sqrt{\frac{bd^3 - h^3(b - t)}{12[bd - h(b - t)]}}$
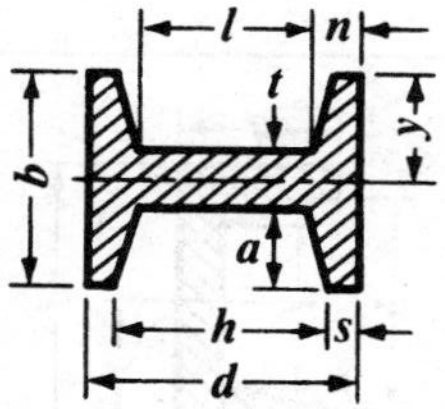	$dt + 2a(s + n)$	$\frac{b}{2}$	$\frac{1}{12}\left[b^3(d - h) + lt^3 + \frac{g}{4}(b^4 - t^4)\right]$ in which g = slope of flange = $(h - l)/(b - t) = \frac{1}{6}$ for standard I-beams.	$\frac{1}{6b}\left[b^3(d - h) + lt^3 + \frac{g}{4}(b^4 - t^4)\right]$	$\sqrt{\frac{I}{A}}$
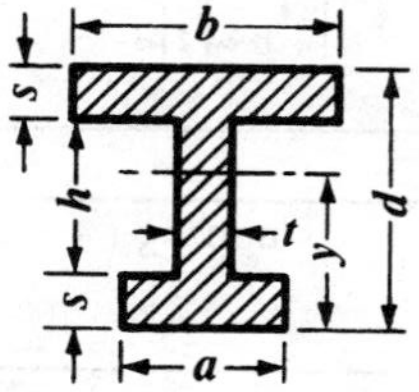	$bs + ht + as$	$d - [td^2 + s^2(b - t) + s(a - t)(2d - s)]\ 2A$	$\frac{1}{3}[b(d - y)^3 + ay^3 - (b - t)(d - y - s)^3 - (a - t)(y - s)^3]$	$\frac{I}{y}$	$\sqrt{\frac{I}{A}}$

Moments of Inertia, Section Moduli, and Radii of Gyration *(Continued)*

Section	Area of Section, A	Distance from Neutral Axis to Extreme Fiber, y	Moment of Inertia, I	Section Modulus, $Z = I/y$	Radius of Gyration, $k = \sqrt{I/A}$
C-Sections					
	$dt + a(s+n)$	$\frac{d}{2}$	$\frac{1}{12}\left[bd^3 - \frac{1}{8g}(h^4 - l^4)\right]$ g = slope of flange $= \frac{h-l}{2(b-t)} = \frac{1}{6}$ for standard channels.	$\frac{1}{6d}\left[bd^3 - \frac{1}{8g}(h^4 - l^4)\right]$	$\sqrt{\frac{\frac{1}{12}\left[bd^3 - \frac{1}{8g}(h^4 - l^4)\right]}{dt + a(s+n)}}$
	$dt + 2a(s+n)$	$b - \left[b^2 s + \frac{ht^2}{2}\right.$ $+ \frac{g}{3}(b-t)^2$ $\left.\times (b+2t)\right] \div A$ g = slope of flange $= \frac{h-l}{2(b-t)}$	$\frac{1}{3}\left[2sb^3 + lt^3 + \frac{g}{2}(b^4 - t^4)\right]$ $-A(b-y)^2$ g = slope of flange $= \frac{h-l}{2(b-t)} = \frac{1}{6}$ for standard channels.	$\frac{I}{y}$	$\sqrt{\frac{I}{A}}$
	$bd - h(b - t)$	$\frac{d}{2}$	$\frac{bd^3 - h^3(b-t)}{12}$	$\frac{bd^3 - h^3(b-t)}{6d}$	$\sqrt{\frac{bd^3 - h^3(b-t)}{12[bd - h(b-t)]}}$

Moments of Inertia, Section Moduli, and Radii of Gyration *(Continued)*

Section	Area of Section, A	Distance from Neutral Axis to Extreme Fiber, y	Moment of Inertia, I	Section Modulus, $Z = I/y$	Radius of Gyration, $k = \sqrt{I/A}$
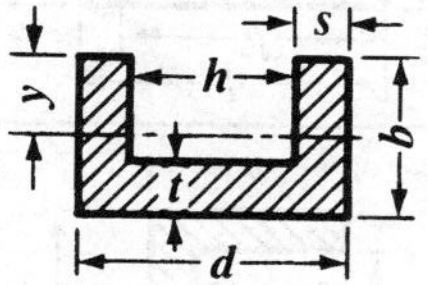	$bd - h(b - t)$	$b - \frac{2b^2s + ht^2}{2bd - 2h(b-t)}$	$\frac{2sb^3 + ht^3}{3} - A(b-y)^2$	$\frac{I}{y}$	$\sqrt{\frac{I}{A}}$
T-Sections					
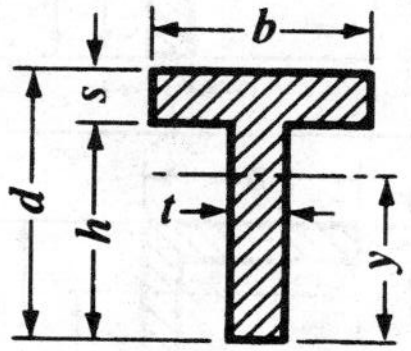	$bs + ht$	$d - \frac{d^2t + s^2(b-t)}{2(bs+ht)}$	$\frac{1}{3}[ty^3 + b(d-y)^3 - (b-t)(d-y-s)^3]$	$\frac{I}{y}$	$\sqrt{\frac{1}{3(bs+ht)}[ty^3 + b(d-y)^3 - (b-t)(d-y-s)^3]}$
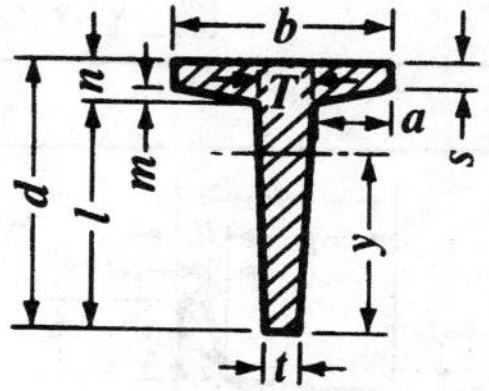	$\frac{l(T+t)}{2} + Tn + a(s+n)$	$d - [3s^2(b-T) + 2am(m+3s) + 3Td^2 - l(T-t)(3d-l)] \div 6A$	$\frac{1}{12}[l^3(T+3t) + 4bn^3 - 2am^3] - A(d-y-n)^2$	$\frac{I}{y}$	$\sqrt{\frac{I}{A}}$
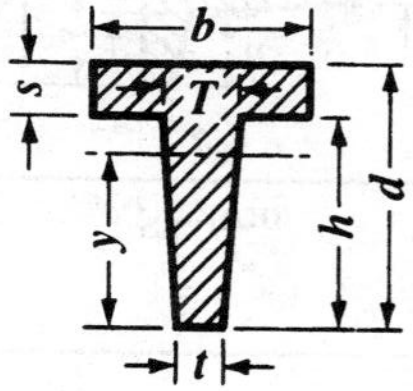	$bs + \frac{h(T+t)}{2}$	$d - [3bs^2 + 3ht(d+s) + h(T-t)(h+3s)]\ 6A$	$\frac{1}{12}[4bs^3 + h^3(3t+T)] - A(d-y-s)^2$	$\frac{I}{y}$	$\sqrt{\frac{I}{A}}$

Moments of Inertia, Section Moduli, and Radii of Gyration *(Continued)*

Section	Area of Section, A	Distance from Neutral Axis to Extreme Fiber, y	Moment of Inertia, I	Section Modulus, $Z = I/y$	Radius of Gyration, $k = \sqrt{I/A}$
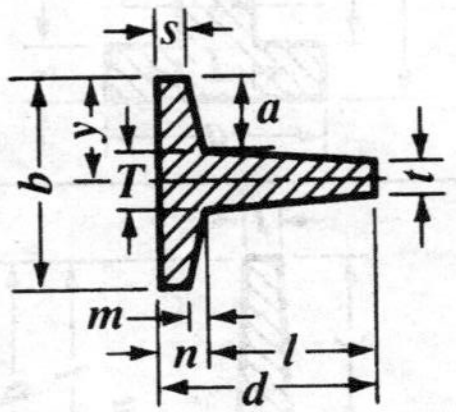	$\frac{l(T+t)}{2} + Tn + a(s+n)$	$\frac{b}{2}$	$\frac{sb^3 + mT^3 + lt^3}{12} + \frac{am[2a^2 + (2a+3T)^2]}{36} + \frac{l(T-t)[(T-t)^2 + 2(T+2t)^2]}{144}$	$\frac{I}{y}$	$\sqrt{\frac{I}{A}}$
L-, Z-, and X-Sections					
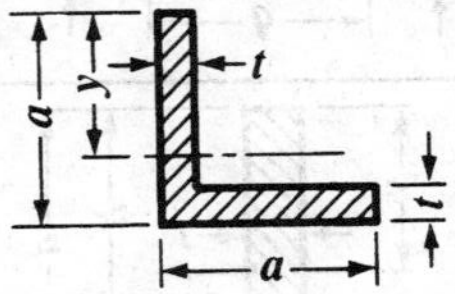	$t(2a - t)$	$a - \frac{a^2 + at - t^2}{2(2a-t)}$	$\frac{1}{3}[ty^3 + a(a-y)^3 - (a-t)(a-y-t)^3]$	$\frac{I}{y}$	$\sqrt{\frac{I}{A}}$
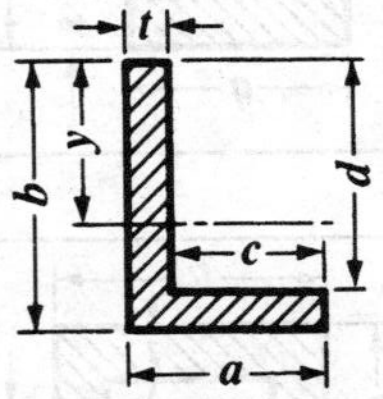	$t(a + b - t)$	$b - \frac{t(2d+a) + d^2}{2(d+a)}$	$\frac{1}{3}[ty^3 + a(b-y)^3 - (a-t)(b-y-t)^3]$	$\frac{I}{y}$	$\sqrt{\frac{1}{3t(a+b-t)}[ty^3 + a(b-y)^3 - (a-t)(b-y-t)^3]}$
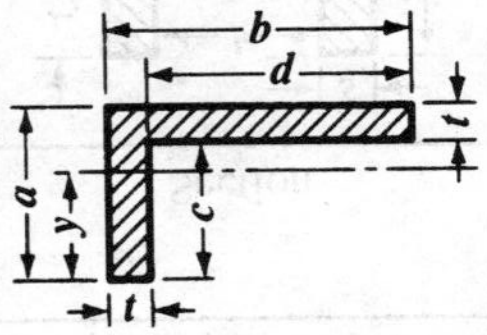	$t(a + b - t)$	$a - \frac{t(2c+b) + c^2}{2(c+b)}$	$\frac{1}{3}[ty^3 + b(a-y)^3 - (b-t)(a-y-t)^3]$	$\frac{I}{y}$	$\sqrt{\frac{1}{3t(a+b-t)}[ty^3 + b(a-y)^3 - (b-t)(a-y-t)^3]}$

Moments of Inertia, Section Moduli, and Radii of Gyration *(Continued)*

Section	Area of Section, A	Distance from Neutral Axis to Extreme Fiber, y	Moment of Inertia, I	Section Modulus, $Z = I/y$	Radius of Gyration, $k = \sqrt{I/A}$
	$t(2a - t)$	$\frac{a^2 + at - t^2}{2(2a - t)\cos 45°}$	$\frac{A}{12}[7(a^2 + b^2) - 12y^2]$ $-2ab^2(a - b)$ in which $b = (a - t)$	$\frac{I}{y}$	$\sqrt{\frac{I}{A}}$
	$t[b + 2(a - t)]$	$\frac{b}{2}$	$\frac{ab^3 - c(b - 2t)^3}{12}$	$\frac{ab^3 - c(b - 2t)^3}{6b}$	$\sqrt{\frac{ab^3 - c(b - 2t)^3}{12t[b + 2(a - t)]}}$
	$t[b + 2(a - t)]$	$\frac{2a - t}{2}$	$\frac{b(a + c)^3 - 2c^3d - 6a^2cd}{12}$	$\frac{b(a + c)^3 - 2c^3d - 6a^2cd}{6(2a - t)}$	$\sqrt{\frac{b(a + c)^3 - 2c^3d - 6a^2cd}{12t[b + 2(a - t)]}}$
	$dt + s(b - t)$	$\frac{d}{2}$	$\frac{td^3 + s^3(b - t)}{12}$	$\frac{td^3 - s^3(b - t)}{6d}$	$\sqrt{\frac{td^3 + s^3(b - t)}{12[td + s(b - t)]}}$

Tabulated Moments of Inertia and Section Moduli for Rectangles and Round Shafts

Moments of Inertia and Section Moduli for Rectangles (Metric Units)

Moments of inertia and section modulus values shown here are for rectangles 1 millimeter wide. To obtain moment of inertia or section modulus for rectangle of given side length, multiply appropriate table value by given width. (See the text starting on page 217 for basic formulas.)

Length of Side (mm)	Moment of Inertia	Section Modulus	Length of Side (mm)	Moment of Inertia	Section Modulus	Length of Side (mm)	Moment of Inertia	Section Modulus
5	10.4167	4.16667	56	14634.7	522.667	107	102087	1908.17
6	18.0000	6.00000	57	15432.8	541.500	108	104976	1944.00
7	28.5833	8.16667	58	16259.3	560.667	109	107919	1980.17
8	42.6667	10.6667	59	17114.9	580.167	110	110917	2016.67
9	60.7500	13.5000	60	18000.0	600.000	111	113969	2053.50
10	83.3333	16.6667	61	18915.1	620.167	112	117077	2090.67
11	110.917	20.1667	62	19860.7	640.667	113	120241	2128.17
12	144.000	24.0000	63	20837.3	661.500	114	123462	2166.00
13	183.083	28.1667	64	21845.3	682.667	115	126740	2204.17
14	228.667	32.6667	65	22885.4	704.167	116	130075	2242.67
15	281.250	37.5000	66	23958.0	726.000	117	133468	2281.50
16	341.333	42.6667	67	25063.6	748.167	118	136919	2320.67
17	409.417	48.1667	68	26202.7	770.667	119	140430	2360.17
18	486.000	54.0000	69	27375.8	793.500	120	144000	2400.00
19	571.583	60.1667	70	28583.3	816.667	121	147630	2440.17
20	666.667	66.6667	71	29825.9	840.167	122	151321	2480.67
21	771.750	73.5000	72	31104.0	864.000	123	155072	2521.50
22	887.333	80.6667	73	32418.1	888.167	124	158885	2562.67
23	1013.92	88.1667	74	33768.7	912.667	125	162760	2604.17
24	1152.00	96.0000	75	35156.3	937.500	126	166698	2646.00
25	1302.08	104.1667	76	36581.3	962.667	127	170699	2688.17
26	1464.67	112.6667	77	38044.4	988.167	128	174763	2730.67
27	1640.25	121.5000	78	39546.0	1014.00	130	183083	2816.67
28	1829.33	130.6667	79	41086.6	1040.17	132	191664	2904.00
29	2032.42	140.167	80	42666.7	1066.67	135	205031	3037.50
30	2250.00	150.000	81	44286.8	1093.50	138	219006	3174.00
31	2482.58	160.167	82	45947.3	1120.67	140	228667	3266.67
32	2730.67	170.667	83	47648.9	1148.17	143	243684	3408.17
33	2994.75	181.500	84	49392.0	1176.00	147	264710	3601.50
34	3275.33	192.667	85	51177.1	1204.17	150	281250	3750.00
35	3572.92	204.167	86	53004.7	1232.67	155	310323	4004.17
36	3888.00	216.000	87	54875.3	1261.50	160	341333	4266.67
37	4221.08	228.167	88	56789.3	1290.67	165	374344	4537.50
38	4572.67	240.667	89	58747.4	1320.17	170	409417	4816.67
39	4943.25	253.500	90	60750.0	1350.00	175	446615	5104.17
40	5333.33	266.667	91	62797.6	1380.17	180	486000	5400.00
41	5743.42	280.167	92	64890.7	1410.67	185	527635	5704.17
42	6174.00	294.000	93	67029.8	1441.50	190	571583	6016.67
43	6625.58	308.167	94	69215.3	1472.67	195	617906	6337.50
44	7098.67	322.667	95	71447.9	1504.17	200	666667	6666.67
45	7593.75	337.500	96	73728.0	1536.00	210	771750	7350.00
46	8111.33	352.667	97	76056.1	1568.17	220	887333	8066.67
47	8651.92	368.167	98	78432.7	1600.67	230	1013917	8816.67
48	9216.00	384.000	99	80858.3	1633.50	240	1152000	9600.00
49	9804.08	400.167	100	83333.3	1666.67	250	1302083	10416.7
50	10416.7	416.667	101	85858.4	1700.17	260	1464667	11266.7
51	11054.3	433.500	102	88434.0	1734.00	270	1640250	12150.0
52	11717.3	450.667	103	91060.6	1768.17	280	1829333	13066.7
53	12406.4	468.167	104	93738.7	1802.67	290	2032417	14016.7
54	13122.0	486.000	105	96468.8	1837.50	300	2250000	15000.0
55	13864.6	504.167	106	99251.3	1872.67	…	…	…

Section Moduli for Rectangles

Length of Side	Section Modulus	Length of Side	Section Modulus	Length of Side	Section Modulus	Length of Side	Section Modulus
⅛	0.0026	2¾	1.26	12	24.00	25	104.2
3/16	0.0059	3	1.50	12½	26.04	26	112.7
¼	0.0104	3¼	1.76	13	28.17	27	121.5
5/16	0.0163	3½	2.04	13½	30.38	28	130.7
⅜	0.0234	3¾	2.34	14	32.67	29	140.2
7/16	0.032	4	2.67	14½	35.04	30	150.0
½	0.042	4½	3.38	15	37.5	32	170.7
⅝	0.065	5	4.17	15½	40.0	34	192.7
¾	0.094	5½	5.04	16	42.7	36	216.0
⅞	0.128	6	6.00	16½	45.4	38	240.7
1	0.167	6½	7.04	17	48.2	40	266.7
1⅛	0.211	7	8.17	17½	51.0	42	294.0
1¼	0.260	7½	9.38	18	54.0	44	322.7
1⅜	0.315	8	10.67	18½	57.0	46	352.7
1½	0.375	8½	12.04	19	60.2	48	384.0
1⅝	0.440	9	13.50	19½	63.4	50	416.7
1¾	0.510	9½	15.04	20	66.7	52	450.7
1⅞	0.586	10	16.67	21	73.5	54	486.0
2	0.67	10½	18.38	22	80.7	56	522.7
2¼	0.84	11	20.17	23	88.2	58	560.7
2½	1.04	11½	22.04	24	96.0	60	600.0

Section modulus values are shown for rectangles 1 inch wide. To obtain section modulus for rectangle of given side length, multiply value in table by given width.

Section Moduli and Moments of Inertia for Round Shafts

Dia.	Section Modulus	Moment of Inertia	Dia.	Section Modulus	Moment of Inertia	Dia.	Section Modulus	Moment of Inertia
⅛	0.00019	0.00001	27/64	0.00737	0.00155	23/32	0.03645	0.01310
9/64	0.00027	0.00002	7/16	0.00822	0.00180	47/64	0.03888	0.01428
5/32	0.00037	0.00003	29/64	0.00913	0.00207	¾	0.04142	0.01553
11/64	0.00050	0.00004	15/32	0.01011	0.00237	49/64	0.04406	0.01687
3/16	0.00065	0.00006	31/64	0.01116	0.00270	25/32	0.04681	0.01829
13/64	0.00082	0.00008	½	0.01227	0.00307	51/64	0.04968	0.01979
7/32	0.00103	0.00011	33/64	0.01346	0.00347	13/16	0.05266	0.02139
15/64	0.00126	0.00015	17/32	0.01472	0.00391	53/64	0.05576	0.02309
¼	0.00153	0.00019	35/64	0.01606	0.00439	27/32	0.05897	0.02488
17/64	0.00184	0.00024	9/16	0.01747	0.00491	55/64	0.06231	0.02677
9/32	0.00218	0.00031	37/64	0.01897	0.00548	⅞	0.06577	0.02877
19/64	0.00257	0.00038	19/32	0.02055	0.00610	57/64	0.06936	0.03089
5/16	0.00300	0.00047	39/64	0.02222	0.00677	29/32	0.07307	0.03311
21/64	0.00347	0.00057	⅝	0.02397	0.00749	59/64	0.07692	0.03545
11/32	0.00399	0.00069	41/64	0.02581	0.00827	15/16	0.08089	0.03792
23/64	0.00456	0.00082	21/32	0.02775	0.00910	61/64	0.08501	0.04051
⅜	0.00518	0.00097	43/64	0.02978	0.01000	31/32	0.08926	0.04323
25/64	0.00585	0.00114	11/16	0.03190	0.01097	63/64	0.09364	0.04609
13/32	0.00658	0.00134	45/64	0.03413	0.01200	…	…	…

In this and succeeding tables, the *Polar Section Modulus* for a shaft of given diameter can be obtained by multiplying its section modulus by 2. Similarly, its *Polar Moment of Inertia* can be obtained by multiplying its moment of inertia by 2.

Section Moduli and Moments of Inertia for Round Shafts (English or Metric Units)

Dia.	Section Modulus	Moment of Inertia	Dia.	Section Modulus	Moment of Inertia	Dia.	Section Modulus	Moment of Inertia
1.00	0.0982	0.0491	1.50	0.3313	0.2485	2.00	0.7854	0.7854
1.01	0.1011	0.0511	1.51	0.3380	0.2552	2.01	0.7972	0.8012
1.02	0.1042	0.0531	1.52	0.3448	0.2620	2.02	0.8092	0.8173
1.03	0.1073	0.0552	1.53	0.3516	0.2690	2.03	0.8213	0.8336
1.04	0.1104	0.0574	1.54	0.3586	0.2761	2.04	0.8335	0.8501
1.05	0.1136	0.0597	1.55	0.3656	0.2833	2.05	0.8458	0.8669
1.06	0.1169	0.0620	1.56	0.3727	0.2907	2.06	0.8582	0.8840
1.07	0.1203	0.0643	1.57	0.3799	0.2982	2.07	0.8708	0.9013
1.08	0.1237	0.0668	1.58	0.3872	0.3059	2.08	0.8835	0.9188
1.09	0.1271	0.0693	1.59	0.3946	0.3137	2.09	0.8963	0.9366
1.10	0.1307	0.0719	1.60	0.4021	0.3217	2.10	0.9092	0.9547
1.11	0.1343	0.0745	1.61	0.4097	0.3298	2.11	0.9222	0.9730
1.12	0.1379	0.0772	1.62	0.4174	0.3381	2.12	0.9354	0.9915
1.13	0.1417	0.0800	1.63	0.4252	0.3465	2.13	0.9487	1.0104
1.14	0.1455	0.0829	1.64	0.4330	0.3551	2.14	0.9621	1.0295
1.15	0.1493	0.0859	1.65	0.4410	0.3638	2.15	0.9757	1.0489
1.16	0.1532	0.0889	1.66	0.4491	0.3727	2.16	0.9894	1.0685
1.17	0.1572	0.0920	1.67	0.4572	0.3818	2.17	1.0032	1.0885
1.18	0.1613	0.0952	1.68	0.4655	0.3910	2.18	1.0171	1.1087
1.19	0.1654	0.0984	1.69	0.4739	0.4004	2.19	1.0312	1.1291
1.20	0.1696	0.1018	1.70	0.4823	0.4100	2.20	1.0454	1.1499
1.21	0.1739	0.1052	1.71	0.4909	0.4197	2.21	1.0597	1.1710
1.22	0.1783	0.1087	1.72	0.4996	0.4296	2.22	1.0741	1.1923
1.23	0.1827	0.1124	1.73	0.5083	0.4397	2.23	1.0887	1.2139
1.24	0.1872	0.1161	1.74	0.5172	0.4500	2.24	1.1034	1.2358
1.25	0.1917	0.1198	1.75	0.5262	0.4604	2.25	1.1183	1.2581
1.26	0.1964	0.1237	1.76	0.5352	0.4710	2.26	1.1332	1.2806
1.27	0.2011	0.1277	1.77	0.5444	0.4818	2.27	1.1484	1.3034
1.28	0.2059	0.1318	1.78	0.5537	0.4928	2.28	1.1636	1.3265
1.29	0.2108	0.1359	1.79	0.5631	0.5039	2.29	1.1790	1.3499
1.30	0.2157	0.1402	1.80	0.5726	0.5153	2.30	1.1945	1.3737
1.31	0.2207	0.1446	1.81	0.5822	0.5268	2.31	1.2101	1.3977
1.32	0.2258	0.1490	1.82	0.5919	0.5386	2.32	1.2259	1.4221
1.33	0.2310	0.1536	1.83	0.6017	0.5505	2.33	1.2418	1.4468
1.34	0.2362	0.1583	1.84	0.6116	0.5627	2.34	1.2579	1.4717
1.35	0.2415	0.1630	1.85	0.6216	0.5750	2.35	1.2741	1.4971
1.36	0.2470	0.1679	1.86	0.6317	0.5875	2.36	1.2904	1.5227
1.37	0.2524	0.1729	1.87	0.6420	0.6003	2.37	1.3069	1.5487
1.38	0.2580	0.1780	1.88	0.6523	0.6132	2.38	1.3235	1.5750
1.39	0.2637	0.1832	1.89	0.6628	0.6264	2.39	1.3403	1.6016
1.40	0.2694	0.1886	1.90	0.6734	0.6397	2.40	1.3572	1.6286
1.41	0.2752	0.1940	1.91	0.6841	0.6533	2.41	1.3742	1.6559
1.42	0.2811	0.1996	1.92	0.6949	0.6671	2.42	1.3914	1.6836
1.43	0.2871	0.2053	1.93	0.7058	0.6811	2.43	1.4087	1.7116
1.44	0.2931	0.2111	1.94	0.7168	0.6953	2.44	1.4262	1.7399
1.45	0.2993	0.2170	1.95	0.7280	0.7098	2.45	1.4438	1.7686
1.46	0.3055	0.2230	1.96	0.7392	0.7244	2.46	1.4615	1.7977
1.47	0.3119	0.2292	1.97	0.7506	0.7393	2.47	1.4794	1.8271
1.48	0.3183	0.2355	1.98	0.7621	0.7545	2.48	1.4975	1.8568
1.49	0.3248	0.2419	1.99	0.7737	0.7698	2.49	1.5156	1.8870

Section Moduli and Moments of Inertia for Round Shafts (English or Metric Units)

Dia.	Section Modulus	Moment of Inertia	Dia.	Section Modulus	Moment of Inertia	Dia.	Section Modulus	Moment of Inertia
2.50	1.5340	1.9175	3.00	2.6507	3.9761	3.50	4.2092	7.3662
2.51	1.5525	1.9483	3.01	2.6773	4.0294	3.51	4.2454	7.4507
2.52	1.5711	1.9796	3.02	2.7041	4.0832	3.52	4.2818	7.5360
2.53	1.5899	2.0112	3.03	2.7310	4.1375	3.53	4.3184	7.6220
2.54	1.6088	2.0432	3.04	2.7582	4.1924	3.54	4.3552	7.7087
2.55	1.6279	2.0755	3.05	2.7855	4.2479	3.55	4.3922	7.7962
2.56	1.6471	2.1083	3.06	2.8130	4.3038	3.56	4.4295	7.8844
2.57	1.6665	2.1414	3.07	2.8406	4.3604	3.57	4.4669	7.9734
2.58	1.6860	2.1749	3.08	2.8685	4.4175	3.58	4.5054	8.0631
2.59	1.7057	2.2089	3.09	2.8965	4.4751	3.59	4.5424	8.1536
2.60	1.7255	2.2432	3.10	2.9247	4.5333	3.60	4.5804	8.2248
2.61	1.7455	2.2779	3.11	2.9531	4.5921	3.61	4.6187	8.3368
2.62	1.7656	2.3130	3.12	2.9817	4.6514	3.62	4.6572	8.4295
2.63	1.7859	2.3485	3.13	3.0105	4.7114	3.63	4.6959	8.5231
2.64	1.8064	2.3844	3.14	3.0394	4.7719	3.64	4.7348	8.6174
2.65	1.8270	2.4208	3.15	3.0685	4.8329	3.65	4.7740	8.7125
2.66	1.8478	2.4575	3.16	3.0979	4.8946	3.66	4.8133	8.8083
2.67	1.8687	2.4947	3.17	3.1274	4.9569	3.67	4.8529	8.9050
2.68	1.8897	2.5323	3.18	3.1570	5.0197	3.68	4.8926	9.0025
2.69	1.9110	2.5703	3.19	3.1869	5.0831	3.69	4.9326	9.1007
2.70	1.9324	2.6087	3.20	3.2170	5.1472	3.70	4.9728	9.1998
2.71	1.9539	2.6476	3.21	3.2472	5.2118	3.71	5.0133	9.2996
2.72	1.9756	2.6869	3.22	3.2777	5.2771	3.72	5.0539	9.4003
2.73	1.9975	2.7266	3.23	3.3083	5.3429	3.73	5.0948	9.5018
2.74	2.0195	2.7668	3.24	3.3391	5.4094	3.74	5.1359	9.6041
2.75	2.0417	2.8074	3.25	3.3702	5.4765	3.75	5.1772	9.7072
2.76	2.0641	2.8484	3.26	3.4014	5.5442	3.76	5.2187	9.8112
2.77	2.0866	2.8899	3.27	3.4328	5.6126	3.77	5.2605	9.9160
2.78	2.1093	2.9319	3.28	3.4643	5.6815	3.78	5.3024	10.0216
2.79	2.1321	2.9743	3.29	3.4961	5.7511	3.79	5.3446	10.1281
2.80	2.1551	3.0172	3.30	3.5281	5.8214	3.80	5.3870	10.2354
2.81	2.1783	3.0605	3.31	3.5603	5.8923	3.81	5.4297	10.3436
2.82	2.2016	3.1043	3.32	3.5926	5.9638	3.82	5.4726	10.4526
2.83	2.2251	3.1486	3.33	3.6252	6.0360	3.83	5.5156	10.5625
2.84	2.2488	3.1933	3.34	3.6580	6.1088	3.84	5.5590	10.6732
2.85	2.2727	3.2385	3.35	3.6909	6.1823	3.85	5.6025	10.7848
2.86	2.2967	3.2842	3.36	3.7241	6.2564	3.86	5.6463	10.8973
2.87	2.3208	3.3304	3.37	3.7574	6.3313	3.87	5.6903	11.0107
2.88	2.3452	3.3771	3.38	3.7910	6.4067	3.88	5.7345	11.1249
2.89	2.3697	3.4242	3.39	3.8247	6.4829	3.89	5.7789	11.2401
2.90	2.3944	3.4719	3.40	3.8587	6.5597	3.90	5.8236	11.3561
2.91	2.4192	3.5200	3.41	3.8928	6.6372	3.91	5.8685	11.4730
2.92	2.4443	3.5686	3.42	3.9272	6.7154	3.92	5.9137	11.5908
2.93	2.4695	3.6178	3.43	3.9617	6.7943	3.93	5.9591	11.7095
2.94	2.4948	3.6674	3.44	3.9965	6.8739	3.94	6.0047	11.8292
2.95	2.5204	3.7176	3.45	4.0314	6.9542	3.95	6.0505	11.9497
2.96	2.5461	3.7682	3.46	4.0666	7.0352	3.96	6.0966	12.0712
2.97	2.5720	3.8194	3.47	4.1019	7.1168	3.97	6.1429	12.1936
2.98	2.5981	3.8711	3.48	4.1375	7.1992	3.98	6.1894	12.3169
2.99	2.6243	3.9233	3.49	4.1733	7.2824	3.99	6.2362	12.4412

Section Moduli and Moments of Inertia for Round Shafts (English or Metric Units)

Dia.	Section Modulus	Moment of Inertia	Dia.	Section Modulus	Moment of Inertia	Dia.	Section Modulus	Moment of Inertia
4.00	6.2832	12.566	4.50	8.946	20.129	5.00	12.272	30.680
4.01	6.3304	12.693	4.51	9.006	20.308	5.01	12.346	30.926
4.02	6.3779	12.820	4.52	9.066	20.489	5.02	12.420	31.173
4.03	6.4256	12.948	4.53	9.126	20.671	5.03	12.494	31.423
4.04	6.4736	13.077	4.54	9.187	20.854	5.04	12.569	31.673
4.05	6.5218	13.207	4.55	9.248	21.039	5.05	12.644	31.925
4.06	6.5702	13.337	4.56	9.309	21.224	5.06	12.719	32.179
4.07	6.6189	13.469	4.57	9.370	21.411	5.07	12.795	32.434
4.08	6.6678	13.602	4.58	9.432	21.599	5.08	12.870	32.691
4.09	6.7169	13.736	4.59	9.494	21.788	5.09	12.947	32.949
4.10	6.7663	13.871	4.60	9.556	21.979	5.10	13.023	33.209
4.11	6.8159	14.007	4.61	9.618	22.170	5.11	13.100	33.470
4.12	6.8658	14.144	4.62	9.681	22.363	5.12	13.177	33.733
4.13	6.9159	14.281	4.63	9.744	22.558	5.13	13.254	33.997
4.14	6.9663	14.420	4.64	9.807	22.753	5.14	13.332	34.263
4.15	7.0169	14.560	4.65	9.871	22.950	5.15	13.410	34.530
4.16	7.0677	14.701	4.66	9.935	23.148	5.16	13.488	34.799
4.17	7.1188	14.843	4.67	9.999	23.347	5.17	13.567	35.070
4.18	7.1702	14.986	4.68	10.063	23.548	5.18	13.645	35.342
4.19	7.2217	15.130	4.69	10.128	23.750	5.19	13.725	35.616
4.20	7.2736	15.275	4.70	10.193	23.953	5.20	13.804	35.891
4.21	7.3257	15.420	4.71	10.258	24.158	5.21	13.884	36.168
4.22	7.3780	15.568	4.72	10.323	24.363	5.22	13.964	36.446
4.23	7.4306	15.716	4.73	10.389	24.571	5.23	14.044	36.726
4.24	7.4834	15.865	4.74	10.455	24.779	5.24	14.125	37.008
4.25	7.5364	16.015	4.75	10.522	24.989	5.25	14.206	37.291
4.26	7.5898	16.166	4.76	10.588	25.200	5.26	14.288	37.576
4.27	7.6433	16.319	4.77	10.655	25.412	5.27	14.369	37.863
4.28	7.6972	16.472	4.78	10.722	25.626	5.28	14.451	38.151
4.29	7.7513	16.626	4.79	10.790	25.841	5.29	14.533	38.441
4.30	7.8056	16.782	4.80	10.857	26.058	5.30	14.616	38.732
4.31	7.8602	16.939	4.81	10.925	26.275	5.31	14.699	39.025
4.32	7.9150	17.096	4.82	10.994	26.495	5.32	14.782	39.320
4.33	7.9701	17.255	4.83	11.062	26.715	5.33	14.866	39.617
4.34	8.0254	17.415	4.84	11.131	26.937	5.34	14.949	39.915
4.35	8.0810	17.576	4.85	11.200	27.160	5.35	15.034	40.215
4.36	8.1369	17.738	4.86	11.270	27.385	5.36	15.118	40.516
4.37	8.1930	17.902	4.87	11.339	27.611	5.37	15.203	40.819
4.38	8.2494	18.066	4.88	11.409	27.839	5.38	15.288	41.124
4.39	8.3060	18.232	4.89	11.480	28.068	5.39	15.373	41.431
4.40	8.3629	18.398	4.90	11.550	28.298	5.40	15.459	41.739
4.41	8.4201	18.566	4.91	11.621	28.530	5.41	15.545	42.049
4.42	8.4775	18.735	4.92	11.692	28.763	5.42	15.631	42.361
4.43	8.5351	18.905	4.93	11.764	28.997	5.43	15.718	42.675
4.44	8.5931	19.077	4.94	11.835	29.233	5.44	15.805	42.990
4.45	8.6513	19.249	4.95	11.907	29.471	5.45	15.892	43.307
4.46	8.7097	19.423	4.96	11.980	29.710	5.46	15.980	43.626
4.47	8.7684	19.597	4.97	12.052	29.950	5.47	16.068	43.946
4.48	8.8274	19.773	4.98	12.125	30.192	5.48	16.156	44.268
4.49	8.8867	19.951	4.99	12.198	30.435	5.49	16.245	44.592

Section Moduli and Moments of Inertia for Round Shafts (English or Metric Units)

Dia.	Section Modulus	Moment of Inertia	Dia.	Section Modulus	Moment of Inertia	Dia.	Section Modulus	Moment of Inertia
5.5	16.3338	44.9180	30	2650.72	39760.8	54.5	15892.4	433068
6	21.2058	63.6173	30.5	2785.48	42478.5	55	16333.8	449180
6.5	26.9612	87.6241	31	2924.72	45333.2	55.5	16783.4	465738
7	33.6739	117.859	31.5	3068.54	48329.5	56	17241.1	482750
7.5	41.4175	155.316	32	3216.99	51471.9	56.5	17707.0	500223
8	50.2655	201.062	32.5	3370.16	54765.0	57	18181.3	518166
8.5	60.2916	256.239	33	3528.11	58213.8	57.5	18663.9	536588
9	71.5694	322.062	33.5	3690.92	61822.9	58	19155.1	555497
9.5	84.1726	399.820	34	3858.66	65597.2	58.5	19654.7	574901
10	98.1748	490.874	34.5	4031.41	69541.9	59	20163.0	594810
10.5	113.650	596.660	35	4209.24	73661.8	59.5	20680.0	615230
11	130.671	718.688	35.5	4392.23	77962.1	60	21205.8	636173
11.5	149.312	858.541	36	4580.44	82448.0	60.5	21740.3	657645
12	169.646	1017.88	36.5	4773.96	87124.7	61	22283.8	679656
12.5	191.748	1198.42	37	4972.85	91997.7	61.5	22836.3	702215
13	215.690	1401.98	37.5	5177.19	97072.2	62	23397.8	725332
13.5	241.547	1630.44	38	5387.05	102354	62.5	23968.4	749014
14	269.392	1885.74	38.5	5602.50	107848	63	24548.3	773272
14.5	299.298	2169.91	39	5823.63	113561	63.5	25137.4	798114
15	331.340	2485.05	39.5	6050.50	119497	64	25735.9	823550
15.5	365.591	2833.33	40	6283.19	125664	64.5	26343.8	849589
16	402.124	3216.99	40.5	6521.76	132066	65	26961.2	876241
16.5	441.013	3638.36	41	6766.30	138709	65.5	27588.2	903514
17	482.333	4099.83	41.5	7016.88	145600	66	28224.9	931420
17.5	526.155	4603.86	42	7273.57	152745	66.5	28871.2	959967
18	572.555	5153.00	42.5	7536.45	160150	67	29527.3	989166
18.5	621.606	5749.85	43	7805.58	167820	67.5	30193.3	1019025
19	673.381	6397.12	43.5	8081.05	175763	68	30869.3	1049556
19.5	727.954	7097.55	44	8362.92	183984	68.5	31555.2	1080767
20	785.398	7853.98	44.5	8651.27	192491	69	32251.3	1112670
20.5	845.788	8669.33	45	8946.18	201289	69.5	32957.5	1145273
21	909.197	9546.56	45.5	9247.71	210385	70	33673.9	1178588
21.5	975.698	10488.8	46	9555.94	219787	70.5	34400.7	1212625
22	1045.36	11499.0	46.5	9870.95	229499	71	35137.8	1247393
22.5	1118.27	12580.6	47	10192.8	239531	71.5	35885.4	1282904
23	1194.49	13736.7	47.5	10521.6	249887	72	36643.5	1319167
23.5	1274.10	14970.7	48	10857.3	260576	72.5	37412.3	1356194
24	1357.17	16286.0	48.5	11200.2	271604	73	38191.7	1393995
24.5	1443.77	17686.2	49	11550.2	282979	73.5	38981.8	1432581
25	1533.98	19174.8	49.5	11907.4	294707	74	39782.8	1471963
25.5	1627.87	20755.4	50	12271.8	306796	74.5	40594.6	1512150
26	1725.52	22431.8	50.5	12643.7	319253	75	41417.5	1553156
26.5	1827.00	24207.7	51	13023.0	332086	75.5	42251.4	1594989
27	1932.37	26087.0	51.5	13409.8	345302	76	43096.4	1637662
27.5	2041.73	28073.8	52	13804.2	358908	76.5	43952.6	1681186
28	2155.13	30171.9	52.5	14206.2	372913	77	44820.0	1725571
28.5	2272.66	32385.4	53	14616.0	387323	77.5	45698.8	1770829
29	2394.38	34718.6	53.5	15033.5	402147	78	46589.0	1816972
29.5	2520.38	37175.6	54	15459.0	417393	78.5	47490.7	1864011

Section Moduli and Moments of Inertia for Round Shafts (English or Metric Units)

Dia.	Section Modulus	Moment of Inertia	Dia.	Section Modulus	Moment of Inertia	Dia.	Section Modulus	Moment of Inertia
79	48404.0	1911958	103.5	108848	5632890	128	205887	13176795
79.5	49328.9	1960823	104	110433	5742530	128.5	208310	13383892
80	50265.5	2010619	104.5	112034	5853762	129	210751	13593420
80.5	51213.9	2061358	105	113650	5966602	129.5	213211	13805399
81	52174.1	2113051	105.5	115281	6081066	130	215690	14019848
81.5	53146.3	2165710	106	116928	6197169	130.5	218188	14236786
82	54130.4	2219347	106.5	118590	6314927	131	220706	14456231
82.5	55126.7	2273975	107	120268	6434355	131.5	223243	14678204
83	56135.1	2329605	107.5	121962	6555469	132	225799	14902723
83.5	57155.7	2386249	108	123672	6678285	132.5	228374	15129808
84	58188.6	2443920	108.5	125398	6802818	133	230970	15359478
84.5	59233.9	2502631	109	127139	6929085	133.5	233584	15591754
85	60291.6	2562392	109.5	128897	7057102	134	236219	15826653
85.5	61361.8	2623218	110	130671	7186884	134.5	238873	16064198
86	62444.7	2685120	110.5	132461	7318448	135	241547	16304406
86.5	63540.1	2748111	111	134267	7451811	135.5	244241	16547298
87	64648.4	2812205	111.5	136089	7586987	136	246954	16792893
87.5	65769.4	2877412	112	137928	7723995	136.5	249688	17041213
88	66903.4	2943748	112.5	139784	7862850	137	252442	17292276
88.5	68050.2	3011223	113	141656	8003569	137.5	255216	17546104
89	69210.2	3079853	113.5	143545	8146168	138	258010	17802715
89.5	70383.2	3149648	114	145450	8290664	138.5	260825	18062131
90	71569.4	3220623	114.5	147372	8437074	139	263660	18324372
90.5	72768.9	3292791	115	149312	8585414	139.5	266516	18589458
91	73981.7	3366166	115.5	151268	8735703	140	269392	18857410
91.5	75207.9	3440759	116	153241	8887955	140.5	272288	19128248
92	76447.5	3516586	116.5	155231	9042189	141	275206	19401993
92.5	77700.7	3593659	117	157238	9198422	141.5	278144	19678666
93	78967.6	3671992	117.5	159262	9356671	142	281103	19958288
93.5	80248.1	3751598	118	161304	9516953	142.5	284083	20240878
94	81542.4	3832492	118.5	163363	9679286	143	287083	20526460
94.5	82850.5	3914688	119	165440	9843686	143.5	290105	20815052
95	84172.6	3998198	119.5	167534	10010172	144	293148	21106677
95.5	85508.6	4083038	120	169646	10178760	144.5	296213	21401356
96	86858.8	4169220	120.5	171775	10349469	145	299298	21699109
96.5	88223.0	4256760	121	173923	10522317	145.5	302405	21999959
97	89601.5	4345671	121.5	176088	10697321	146	305533	22303926
97.5	90994.2	4435968	122	178270	10874498	146.5	308683	22611033
98	92401.3	4527664	122.5	180471	11053867	147	311854	22921300
98.5	93822.8	4620775	123	182690	11235447	147.5	315047	23234749
99	95258.9	4715315	123.5	184927	11419254	148	318262	23551402
99.5	96709.5	4811298	124	187182	11605307	148.5	321499	23871280
100	98174.8	4908739	124.5	189456	11793625	149	324757	24194406
100.5	99654.8	5007652	125	191748	11984225	149.5	328037	24520802
101	101150	5108053	125.5	194058	12177126	150	331340	24850489
101.5	102659	5209956	126	196386	12372347	…	…	…
102	104184	5313376	126.5	198734	12569905	…	…	…
102.5	105723	5418329	127	201100	12769820	…	…	…
103	107278	5524828	127.5	203484	12972110	…	…	…

Strength and Stiffness of Perforated Metals.—It is common practice to use perforated metals in equipment enclosures to provide cooling by the flow of air or fluids. If the perforated material is to serve also as a structural member, then calculations of stiffness and strength must be made that take into account the effect of the perforations on the strength of the panels.

The accompanying table provides equivalent or effective values of the yield strength S^*; modulus of elasticity E^*; and Poisson's ratio ν^* of perforated metals in terms of the values for solid material. The S^*/S and E^*/E ratios, given in the accompanying table for the standard round hole staggered pattern, can be used to determine the safety margins or deflections for perforated metal use as compared to the unperforated metal for any geometry or loading condition.

Perforated material has different strengths depending on the direction of loading; therefore, values of S^*/S in the table are given for the width (strongest) and length (weakest) directions. Also, the effective elastic constants are for plane stress conditions and apply to the in-plane loading of thin perforated sheets; the bending stiffness is greater. However, since most loading conditions involve a combination of bending and stretching, it is more convenient to use the same effective elastic constants for these combined loading conditions. The plane stress effective elastic constants given in the table can be conservatively used for all loading conditions.

Mechanical Properties of Materials Perforated with Round Holes in IPA Standard Staggered Hole Pattern

Length of Sheet
Width of Sheet
STANDARD 60° STAGGERED PATTERN

IPA No.	Perforation Diam. (in.)	Center Distance (in.)	Open Area (%)	S^*/S Width (in.)	S^*/S Length (in.)	E^*/E	ν^*
100	0.020	(625)	20	0.530	0.465	0.565	0.32
106	1/16	1/8	23	0.500	0.435	0.529	0.33
107	5/64	7/64	46	0.286	0.225	0.246	0.38
108	5/64	1/8	36	0.375	0.310	0.362	0.35
109	3/32	5/32	32	0.400	0.334	0.395	0.34
110	3/32	3/16	23	0.500	0.435	0.529	0.33
112	1/10	5/32	36	0.360	0.296	0.342	0.35
113	1/8	3/16	40	0.333	0.270	0.310	0.36
114	1/8	7/32	29	0.428	0.363	0.436	0.33
115	1/8	1/4	23	0.500	0.435	0.529	0.33
116	5/32	7/32	46	0.288	0.225	0.249	0.38
117	5/32	1/4	36	0.375	0.310	0.362	0.35
118	3/16	1/4	51	0.250	0.192	0.205	0.42
119	3/16	5/16	33	0.400	0.334	0.395	0.34
120	1/4	5/16	58	0.200	0.147	0.146	0.47
121	1/4	3/8	40	0.333	0.270	0.310	0.36
122	1/4	7/16	30	0.428	0.363	0.436	0.33
123	1/4	1/2	23	0.500	0.435	0.529	0.33
124	3/8	1/2	51	0.250	0.192	0.205	0.42
125	3/8	9/16	40	0.333	0.270	0.310	0.36
126	3/8	5/8	33	0.400	0.334	0.395	0.34
127	7/16	5/8	45	0.300	0.239	0.265	0.38
128	1/2	11/16	47	0.273	0.214	0.230	0.39
129	9/16	3/4	51	0.250	0.192	0.205	0.42

Value in parentheses specifies holes per square inch instead of center distance. S^*/S = ratio of yield strength of perforated to unperforated material; E^*/E = ratio of modulus of elasticity of perforated to unperforated material; ν^* = Poisson's ratio for given percentage of open area.

IPA is Industrial Perforators Association.

BEAMS

Beam Calculations

Reaction at the Supports.—When a beam is loaded by vertical loads or forces, the sum of the reactions at the supports equals the sum of the loads. In a simple beam, when the loads are symmetrically placed with reference to the supports, or when the load is uniformly distributed, the reaction at each end will equal one-half of the sum of the loads. When the loads are not symmetrically placed, the reaction at each support may be ascertained from the fact that the algebraic sum of the moments must equal zero. In the accompanying illustration, if moments are taken about the support to the left, then: $R_2 \times 40 - 8000 \times 10 - 10{,}000 \times 16 - 20{,}000 \times 20 = 0$; $R_2 = 16{,}000$ pounds. In the same way, moments taken about the support at the right give $R_1 = 22{,}000$ pounds.

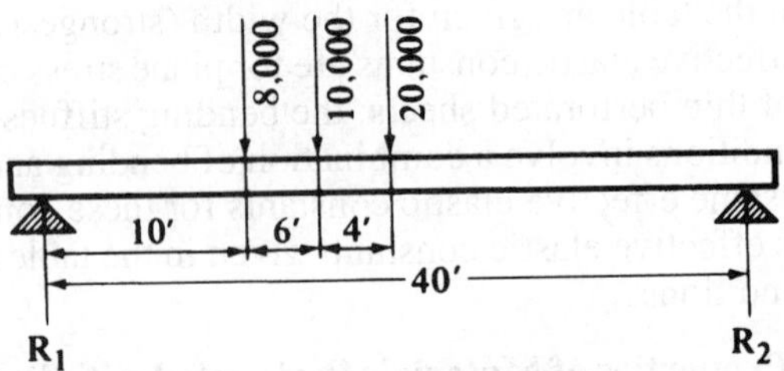

The sum of the reactions equals 38,000 pounds, which is also the sum of the loads. If part of the load is uniformly distributed over the beam, this part is first equally divided between the two supports, or the uniform load may be considered as concentrated at its center of gravity.

If metric SI units are used for the calculations, distances may be expressed in meters or millimeters, providing the treatment is consistent, and loads in newtons. *Note:* If the load is given in kilograms, the value referred to is the mass. A mass of *M* kilograms has a weight (applies a force) of *Mg* newtons, where *g* = approximately 9.81 meters per second².

Stresses and Deflections in Beams.—On the following pages are given an extensive table of formulas for stresses and deflections in beams, shafts, etc. It is assumed that all the dimensions are in inches, all loads in pounds, and all stresses in pounds per square inch. **The formulas are also valid using metric SI units, with all dimensions in millimeters, all loads in newtons, and stresses and moduli in newtons per millimeter² (N/mm²). *Note:* A load due to the weight of a mass of *M* kilograms is *Mg* newtons, where *g* = approximately 9.81 meters per second².** In the tables:

E = modulus of elasticity of the material

I = moment of inertia of the cross-section of the beam

Z = section modulus of the cross-section of the beam = $I \div$ distance from neutral axis to extreme fiber

W = load on beam

s = stress in extreme fiber, or maximum stress in the cross-section considered, due to load W. A positive value of s denotes tension in the upper fibers and compression in the lower ones (as in a cantilever). A negative value of s denotes the reverse (as in a beam supported at the ends). The greatest safe load is that value of W which causes a maximum stress equal to, but not exceeding, the greatest safe value of s

y = deflection measured from the position occupied if the load causing the deflection were removed. A positive value of y denotes deflection below this position; a negative value, deflection upward

u, v, w, x = variable distances along the beam from a given support to any point

Stresses and Deflections in Beams

Type of Beam	Stresses: General Formula for Stress at any Point	Stresses: Stresses at Critical Points	Deflections: General Formula for Deflection at any Point[a]	Deflections: Deflections at Critical Points[a]
Case 1. — Supported at Both Ends, Uniform Load				
TOTAL LOAD W; $\frac{W}{2}$, x, l, $\frac{W}{2}$	$s = -\frac{W}{2Zl}x(l-x)$	Stress at center, $-\frac{Wl}{8Z}$ If cross-section is constant, this is the maximum stress.	$y = \frac{Wx(l-x)}{24EIl}[l^2 + x(l-x)]$	Maximum deflection, at center, $\frac{5}{384}\frac{Wl^3}{EI}$
Case 2. — Supported at Both Ends, Load at Center				
W; $\frac{W}{2}$, x, $l/2$, $l/2$, x, l, $\frac{W}{2}$	Between each support and load, $s = -\frac{Wx}{2Z}$	Stress at center, $-\frac{Wl}{4Z}$ If cross-section is constant, this is the maximum stress.	Between each support and load, $y = \frac{Wx}{48EI}(3l^2 - 4x^2)$	Maximum deflection, at load, $\frac{Wl^3}{48EI}$
Case 3. — Supported at Both Ends, Load at any Point				
W; $\frac{Wb}{l}$, x, a, b, v, l, $\frac{Wa}{l}$; $a+b=l$	For segment of length a, $s = -\frac{Wbx}{Zl}$ For segment of length b, $s = -\frac{Wav}{Zl}$	Stress at load, $-\frac{Wab}{Zl}$ If cross-section is constant, this is the maximum stress.	For segment of length a, $y = \frac{Wbx}{6EIl}(l^2 - x^2 - b^2)$ For segment of length b, $y = \frac{Wav}{6EIl}(l^2 - v^2 - a^2)$	Deflection at load, $\frac{Wa^2b^2}{3EIl}$ Let a be the length of the shorter segment and b of the longer one. The maximum deflection $\frac{Wav_1^3}{3EIl}$ is in the longer segment, at $v = b\sqrt{\frac{1}{3} + \frac{2a}{3b}} = v_1$

Stresses and Deflections in Beams *(Continued)*

Type of Beam	Stresses: General Formula for Stress at any Point	Stresses: Stresses at Critical Points	Deflections: General Formula for Deflection at any Point[a]	Deflections: Deflections at Critical Points[a]
Case 4. — Supported at Both Ends, Two Symmetrical Loads				
W, W, x, v, x, a, a, W, l, W	Between each support and adjacent load, $s = -\frac{Wx}{Z}$ Between loads, $s = -\frac{Wa}{Z}$	Stress at each load, and at all points between, $-\frac{Wa}{Z}$	Between each support and adjacent load, $y = \frac{Wx}{6EI}[3a(l-a) - x^2]$ Between loads, $y = \frac{Wa}{6EI}[3v(l-v) - a^2]$	Maximum deflection at center, $\frac{Wa}{24EI}(3l^2 - 4a^2)$ Deflection at loads $\frac{Wa^2}{6EI}(3l - 4a)$
Case 5. — Both Ends Overhanging Supports Symmetrically, Uniform Load				
TOTAL LOAD W; u, x, u, c, l, c, L; $\frac{W}{2}$, $L = l + 2c$, $\frac{W}{2}$	Between each support and adjacent end, $s = \frac{W}{2Zl}(c-u)^2$ Between supports, $s = \frac{W}{2ZL}(c^2 - x(l-x))$	Stress at each support, $\frac{Wc^2}{2ZL}$ Stress at center, $\frac{W}{2ZL}(c^2 - \frac{1}{4}l^2)$ If cross-section is constant, the greater of these is the maximum stress. If l is greater than $2c$, the stress is zero at points $\sqrt{\frac{1}{4}l^2 - c^2}$ on both sides of the center. If cross-section is constant and if $l = 2.828c$, the stresses at supports and center are equal and opposite, and are $\pm\frac{WL}{46.62Z}$	Between each support and adjacent end, $y = \frac{Wu}{24EIL}[6c^2(l+u) - u^2(4c-u) - l^3]$ Between supports, $y = \frac{Wx(l-x)}{24EIL}[x(l-x) + l^2 - 6c^2]$	Deflection at ends, $\frac{Wc}{24EIL}[3c^2(c+2l) - l^3]$ Deflection at center, $\frac{Wl^2}{384EIL}(5l^2 - 24c^2)$ If l is between $2c$ and $2.449c$, there are maximum upward deflections at points $\sqrt{3(\frac{1}{4}l^2 - c^2)}$ on both sides of the center, which are, $-\frac{W}{96EIL}(6c^2 - l^2)^2$

Stresses and Deflections in Beams *(Continued)*

Type of Beam	Stresses: General Formula for Stress at any Point	Stresses: Stresses at Critical Points	Deflections: General Formula for Deflection at any Point[a]	Deflections: Deflections at Critical Points[a]
Case 6. — Both Ends Overhanging Supports Unsymmetrically, Uniform Load				
TOTAL LOAD W; u, x, c, l, L, w, d; $\frac{W}{2l}(l-d+c)$; $\frac{W}{2l}(l+d-c)$	For overhanging end of length c, $s = \frac{W}{2ZL}(c-u)^2$ Between supports, $s = \frac{W}{2ZL}\left\{c^2\left(\frac{l-x}{l}\right) + d^2\frac{x}{l} - x(l-x)\right\}$ For overhanging end of length d, $s = \frac{W}{2ZL}(d-w)^2$	Stress at support next end of length c, $\frac{Wc^2}{2ZL}$ Critical stress between supports is at $x = \frac{l^2+c^2-d^2}{2l} = x_1$ and is $\frac{W}{2ZL}(c^2 - x_1^2)$ Stress at support next end of length d, $\frac{Wd^2}{2ZL}$ If cross-section is constant, the greatest of these three is the maximum stress. If $x_1 > c$, the stress is zero at points $\sqrt{x_1^2 - c^2}$ on both sides of $x = x_1$.	For overhanging end of length c, $y = \frac{Wu}{24EIL}[2l(d^2+2c^2) + 6c^2u - u^2(4c-u) - l^3]$ Between supports, $y = \frac{Wx(l-x)}{24EIL}\{x(l-x) + l^2 - 2(d^2+c^2) - \frac{2}{l}[d^2x + c^2(l-x)]\}$ For overhanging end of length d, $y = \frac{Ww}{24EIL}[2l(c^2+2d^2) + 6d^2w - w^2(4d-w) - l^3]$	Deflection at end c, $\frac{Wc}{24EIL}[2l(d^2+2c^2) + 3c^3 - l^3]$ Deflection at end d, $\frac{Wd}{24EIL}[2l(c^2+2d^2) + 3d^3 - l^3]$ This case is so complicated that convenient general expressions for the critical deflections between supports cannot be obtained.
Case 7. — Both Ends Overhanging Supports, Load at any Point Between				
W; u, x, c, a, b, v, w, d, l; Wb; Wa	Between supports: For segment of length a, $s = -\frac{Wbx}{Zl}$ For segment of length b, $s = -\frac{Wav}{Zl}$ Beyond supports $s = 0$.	Stress at load, $-\frac{Wab}{Zl}$ If cross-section is constant, this is the maximum stress.	Between supports, same as Case 3. For overhanging end of length c, $y = -\frac{Wabu}{6EIl}(l+b)$ For overhanging end of length d, $y = -\frac{Wabw}{6EIl}(l+a)$	Between supports, same as Case 3. Deflection at end c, $-\frac{Wabc}{6EIl}(l+b)$ Deflection at end d, $-\frac{Wabd}{6EIl}(l+a)$

Stresses and Deflections in Beams *(Continued)*

Type of Beam	Stresses		Deflections	
	General Formula for Stress at any Point	Stresses at Critical Points	General Formula for Deflection at any Point[a]	Deflections at Critical Points[a]
Case 8. — Both Ends Overhanging Supports, Single Overhanging Load				
W, u, c, x, l, w, d	Between load and adjacent support, $s = \frac{W}{Z}(c-u)$ Between supports, $s = \frac{Wc}{Zl}(l-x)$ Between unloaded end and adjacent supports, $s = 0$.	Stress at support adjacent to load, $\frac{Wc}{Z}$ If cross-section is constant, this is the maximum stress. Stress is zero at other support.	Between load and adjacent support, $y = \frac{Wu}{6EI}(3cu - u^2 + 2cl)$ Between supports, $y = -\frac{Wcx}{6EIl}(l-x)(2l-x)$ Between unloaded end and adjacent support, $y = \frac{Wclw}{6EI}$	Deflection at load, $\frac{Wc^2}{3EI}(c+l)$ Maximum upward deflection is at $x = .42265l$, and is $-\frac{Wcl^2}{15.55EI}$ Deflection at unloaded end, $\frac{Wcld}{6EI}$
Case 9. — Both Ends Overhanging Supports, Symmetrical Overhanging Loads				
W, W, u, x, c, l, u, c, W, W	Between each load and adjacent support, $s = \frac{W}{Z}(c-u)$ Between supports, $s = \frac{Wc}{Z}$	Stress at supports and at all points between, $\frac{Wc}{Z}$ If cross-section is constant, this is the maximum stress.	Between each load and adjacent support, $y = \frac{Wu}{6EI}[3c(l+u) - u^2]$ Between supports, $y = -\frac{Wcx}{2EI}(l-x)$	Deflections at loads, $\frac{Wc^2}{6EI}(2c+3l)$ Deflection at center, $-\frac{Wcl^2}{8EI}$
			The above expressions involve the usual approximations of the theory of flexure, and hold only for small deflections. Exact expressions for deflections of any magnitude are as follows: Between supports the curve is a circle of radius $r = \frac{EI}{Wc}$; $y = \sqrt{r^2 - \frac{1}{4}l^2} - \sqrt{r^2 - (\frac{1}{2}l - x)^2}$ Deflection at center, $\sqrt{r^2 - \frac{1}{4}l^2} - r$	

Stresses and Deflections in Beams *(Continued)*

Type of Beam	Stresses: General Formula for Stress at any Point	Stresses: Stresses at Critical Points	Deflections: General Formula for Deflection at any Point[a]	Deflections: Deflections at Critical Points[a]
Case 10. — Fixed at One End, Uniform Load				
TOTAL LOAD W	$s = \frac{W}{2Zl}(l-x)^2$	Stress at support, $\frac{Wl}{2Z}$ If cross-section is constant, this is the maximum stress.	$y = \frac{Wx^2}{24EIl}[2l^2+(2l-x)^2]$	Maximum deflection, at end, $\frac{Wl^3}{8EI}$
Case 11. — Fixed at One End, Load at Other				
	$s = \frac{W}{Z}(l-x)$	Stress at support, $\frac{Wl}{Z}$ If cross-section is constant, this is the maximum stress.	$y = \frac{Wx^2}{6EI}(3l-x)$	Maximum deflection, at end, $\frac{Wl^3}{3EI}$
Case 12. — Fixed at One End, Intermediate Load				
	Between support and load, $s = \frac{W}{Z}(l-x)$ Beyond load, $s = 0$.	Stress at support, $\frac{Wl}{Z}$ If cross-section is constant, this is the maximum stress.	Between support and load, $y = \frac{Wx^2}{6EI}(3l-x)$ Beyond load, $y = \frac{Wl^2}{6EI}(3v-l)$	Deflection at load, $\frac{Wl^3}{3EI}$ Maximum deflection, at end, $\frac{Wl^2}{6EI}(2l+3b)$

Stresses and Deflections in Beams *(Continued)*

Type of Beam	Stresses: General Formula for Stress at any Point	Stresses: Stresses at Critical Points	Deflections: General Formula for Deflection at any Point[a]	Deflections: Deflections at Critical Points[a]
Case 13. — Fixed at One End, Supported at the Other, Load at Center				
$\frac{3}{16}Wl$; W; x; $l/2$; $l/2$; v; l; $\frac{5}{16}W$; $\frac{11}{16}W$	Between point of fixture and load, $s = \frac{W}{16Z}(3l - 11x)$ Between support and load, $s = -\frac{5}{16}\frac{Wv}{Z}$	Maximum stress at point of fixture, $\frac{3}{16}\frac{Wl}{Z}$ Stress is zero at $x = \frac{3}{11}l$ Greatest negative stress at center, $-\frac{5}{32}\frac{Wl}{Z}$	Between point of fixture and load, $y = \frac{Wx^2}{96EI}(9l - 11x)$ Between support and load, $y = \frac{Wv}{96EI}(3l^2 - 5v^2)$	Maximum deflection is at $v = 0.4472l$, and is $\frac{Wl^3}{107.33EI}$ Deflection at load, $\frac{7}{768}\frac{Wl^3}{EI}$
Case 14. — Fixed at One End, Supported at the Other, Load at any Point				
$m = (l + a)(l + b) + al$ $n = al(l + b)$ $\frac{Wab(l+b)}{2l^2}$; W; x; a; b; v; l; $W[1 - \frac{a^2}{2l^3}(3l - a)]$; $\frac{Wa^2(3l-a)}{2l^3}$	Between point of fixture and load, $s = \frac{Wb}{2Zl^3}(n - mx)$ Between support and load, $s = -\frac{Wa^2v}{2Zl^3}(3l - a)$	Greatest positive stress, at point of fixture, $\frac{Wab}{2Zl^2}(l + b)$ Greatest negative stress, at load, $-\frac{Wa^2b}{2Zl^3}(3l - a)$ If $a < 0.5858l$, the first is the maximum stress. If $a = 0.5858l$, the two are equal and are $\pm\frac{Wl}{5.83Z}$. If $a > 0.5858l$, the second is the maximum stress. Stress is zero at $x = \frac{n}{m}$	Between point of fixture and load, $y = \frac{Wx^2b}{12EIl^3}(3n - mx)$ Between support and load, $y = \frac{Wa^2v}{12EIl^3}[3l^2b - v^2(3l - a)]$	Deflection at load, $\frac{Wa^3b^2}{12EIl^3}(3l + b)$ If $a < 0.5858l$, maximum deflection is $\frac{Wa^2b}{6EI}\sqrt{\frac{b}{2l + b}}$ and located between load and support, at $v = l\sqrt{\frac{b}{2l + b}}$ If $a = 0.5858l$, maximum deflection is at load and is $\frac{Wl^3}{101.9EI}$ If $a > 0.5858l$, maximum deflection is $\frac{Wbn^3}{3EIm^2l^3}$ and located between load and point of fixture, at $x = \frac{2n}{m}$

Stresses and Deflections in Beams *(Continued)*

Type of Beam	Stresses: General Formula for Stress at any Point	Stresses: Stresses at Critical Points	Deflections: General Formula for Deflection at any Point[a]	Deflections: Deflections at Critical Points[a]
Case 15. — Fixed at One End, Supported at the Other, Uniform Load				
TOTAL LOAD W; $\frac{Wl}{8}$; x; l; $\frac{3}{8}W$; $\frac{5}{8}W$	$s = \frac{W(l-x)}{2Zl}(\frac{1}{4}l - x)$	Maximum stress at point of fixture, $\frac{Wl}{8Z}$ Stress is zero at $x = \frac{1}{4}l$. Greatest negative stress is at $x = \frac{5}{8}l$ and is $-\frac{9}{128}\frac{Wl}{Z}$	$y = \frac{Wx^2(l-x)}{48EIl}(3l - 2x)$	Maximum deflection is at $x = 0.5785l$, and is $\frac{Wl^3}{185EI}$ Deflection at center, $\frac{Wl^3}{192EI}$ Deflection at point of greatest negative stress, at $x = \frac{5}{8}l$ is $\frac{Wl^3}{187EI}$
Case 16. — Fixed at One End, Free but Guided at the Other, Uniform Load				
TOTAL LOAD W; $\frac{Wl}{3}$; x; l; $\frac{Wl}{6}$; W	$s = \frac{Wl}{Z}\left\{\frac{1}{3} - \frac{x}{l} + \frac{1}{2}\left(\frac{x}{l}\right)^2\right\}$	Maximum stress, at support, $\frac{Wl}{3Z}$ Stress is zero at $x = 0.4227l$ Greatest negative stress, at free end, $-\frac{Wl}{6Z}$	$y = \frac{Wx^2}{24EIl}(2l - x)^2$	Maximum deflection, at free end, $\frac{Wl^3}{24EI}$
Case 17. — Fixed at One End, Free but Guided at the Other, with Load				
$\frac{Wl}{2}$; x; l; W; $\frac{Wl}{2}$; W	$s = \frac{W}{Z}(\frac{1}{2}l - x)$	Stress at support, $\frac{Wl}{2Z}$ Stress at free end $-\frac{Wl}{2Z}$ These are the maximum stresses and are equal and opposite. Stress is zero at center.	$y = \frac{Wx^2}{12EI}(3l - 2x)$	Maximum deflection, at free end, $\frac{Wl^3}{12EI}$

Stresses and Deflections in Beams *(Continued)*

Type of Beam	Stresses: General Formula for Stress at any Point	Stresses: Stresses at Critical Points	Deflections: General Formula for Deflection at any Point[a]	Deflections: Deflections at Critical Points[a]
Case 18. — Fixed at Both Ends, Load at Center				
$\frac{Wl}{8}$, W, $\frac{Wl}{8}$; x, $l/2$, $l/2$, x, l; $\frac{W}{2}$, $\frac{W}{2}$	Between each end and load, $s = \frac{W}{2Z}(\frac{1}{4}l - x)$	Stress at ends $\frac{Wl}{8Z}$ at load $-\frac{Wl}{8Z}$ These are the maximum stresses and are equal and opposite. Stress is zero at $x = \frac{1}{4}l$	$y = \frac{Wx^2}{48EI}(3l - 4x)$	Maximum deflection, at load, $\frac{Wl^3}{192EI}$
Case 19. — Fixed at Both Ends, Load at any Point				
$\frac{Wab^2}{l^2}$, W, $\frac{Wa^2b}{l^2}$; x, v, a, b, l; $\frac{Wb^2}{l^3}(l+2a)$, $\frac{Wa^2}{l^3}(l+2b)$	For segment of length a, $s = \frac{Wb^2}{Zl^3}[al - x(l + 2a)]$ For segment of length b, $\frac{Wl}{8Z}$	Stress at end next segment of length a, $\frac{Wab^2}{Zl^2}$ Stress at end next segment of length b, $\frac{Wa^2b}{Zl^2}$ Maximum stress is at end next shorter segment. Stress is zero at $x = \frac{al}{l + 2a}$ and $v = \frac{bl}{l + 2b}$ Greatest negative stress, at load, $-\frac{2Wa^2b^2}{Zl^3}$	For segment of length a, $y = \frac{Wx^2b^2}{6EIl^3}[2a(l - x) + l(a - x)]$ For segment of length b, $y = \frac{Wv^2a^2}{6EIl^3}[2b(l - v) + l(b - v)]$	Deflection at load, $\frac{Wa^3b^3}{3EIl^3}$ Let b be the length of the longer segment and a of the shorter one. The maximum deflection is in the longer segment, at $v = \frac{2bl}{l + 2b}$ and is $x = \frac{l_1}{W_1}(W_1 - R_1)$

Stresses and Deflections in Beams *(Continued)*

Type of Beam	Stresses: General Formula for Stress at any Point	Stresses: Stresses at Critical Points	Deflections: General Formula for Deflection at any Point[a]	Deflections: Deflections at Critical Points[a]
Case 20. — Fixed at Both Ends, Uniform Load				
TOTAL LOAD W; $\frac{Wl}{12}$ (at each end); x; l; $\frac{W}{2}$ (each support)	$s = \frac{Wl}{2Z}\left\{\frac{1}{6} - \frac{x}{l} + \left(\frac{x}{l}\right)^2\right\}$	Maximum stress, at ends, $x = \frac{2n}{m}$ and is $\frac{Wbn^3}{3EIm^2l^3}$ Stress is zero at $x = 0.7887l$ and at $x = 0.2113l$ Greatest negative stress, at center, $-\frac{Wl}{24Z}$	$y = \frac{Wx^2}{24EIl}(l-x)^2$	Maximum deflection, at center, $\frac{Wl^3}{384EI}$
Case 21. — Continuous Beam, with Two Unequal Spans, Unequal, Uniform Loads				
TOTAL LOAD W_1; TOTAL LOAD W_2; R_1; R; R_2; x; u; l_1; l_2; $\frac{l_1W_1(3l_1+4l_2)-W_2l_2^2}{8l_1(l_1+l_2)}$; $\frac{l_2W_2(3l_2+4l_1)-W_1l_1^2}{8l_2(l_1+l_2)}$; $\left(\frac{W_1+W_2}{2}\right)+\frac{1}{8}\left(\frac{W_1l_1}{l_2}+\frac{W_2l_2}{l_1}\right)$	Between R_1 and R, $s = \frac{l_1-x}{Z}\left\{\frac{(l_1-x)W_1}{2l_1} - R_1\right\}$ Between R_2 and R, $s = \frac{l_2-u}{Z}\left\{\frac{(l_2-u)W_2}{2l_2} - R_2\right\}$	Stress at support R, $\frac{W_1l_1^2 + W_2l_2^2}{8Z(l_1+l_2)}$ Greatest stress in the first span is at $x = \frac{l_1}{W_1}(W_1 - R_1)$ and is $v = \frac{2bl}{l+2b}$ Greatest stress in the second span is at $u = \frac{l_2}{W_2}(W_2 - R_2)$ and is, $-\frac{R_2^2l_2}{2ZW_2}$	Between R_1 and R, $y = \frac{x(l_1-x)}{24EI}\left\{(2l_1-x)(4R_1-W_1) - \frac{W_1(l_1-x)^2}{l_1}\right\}$ Between R_2 an R, $y = \frac{u(l_2-u)}{24EI}\left\{(2l_2-u)(4R_2-W_2) - \frac{W_2(l_2-u)^2}{l_2}\right\}$	This case is so complicated that convenient general expressions for the critical deflections cannot be obtained.

Stresses and Deflections in Beams *(Continued)*

Type of Beam	Stresses: General Formula for Stress at any Point	Stresses: Stresses at Critical Points	Deflections: General Formula for Deflection at any Point[a]	Deflections: Deflections at Critical Points[a]
Case 22. — Continuous Beam, with Two Equal Spans, Uniform Load				
TOTAL LOAD ON EACH SPAN, *W* A; $\frac{3}{8}W$; $\frac{5}{4}W$; $\frac{3}{8}W$; *x*; *x*; *l*; *l*	$s = \frac{W(l-x)}{2Zl}(\frac{1}{4}l - x)$	Maximum stress at point $u = \frac{l_2}{W_2}(W_2 - R_2)$ Stress is zero at $x = \frac{1}{4}l$ Greatest negative stress is at $x = \frac{5}{8}l$ and is, $-\frac{9}{128}\frac{Wl}{Z}$	$y = \frac{Wx^2(l-x)}{48EIl}(3l - 2x)$	Maximum deflection is at $x =$ 0.5785*l*, and is $\frac{Wl^3}{185EI}$ Deflection at center of span, $\frac{Wl^3}{192EI}$ Deflection at point of greatest negative stress, at $x = \frac{5}{8}l$ is $\frac{Wl^3}{187EI}$
Case 23. — Continuous Beam, with Two Equal Spans, Equal Loads at Center of Each				
B; W; A; W; B; *v*; *x*; *x*; *v*; *l*/2; *l*/2; *l*/2; *l*/2; *l*; *l*; $\frac{5}{16}$ W; $\frac{11}{8}$ W; $\frac{5}{16}$ W	Between point *A* and load, $s = \frac{W}{16Z}(3l - 11x)$ Between point *B* and load, $s = -\frac{5}{16}\frac{Wv}{Z}$	Maximum stress at point *A*, $\frac{3}{16}\frac{Wl}{Z}$ Stress is zero at $x = \frac{3}{11}l$ Greatest negative stress at center of span, $-\frac{5}{32}\frac{Wl}{Z}$	Between point *A* and load, $y = \frac{Wx^2}{96EI}(9l - 11x)$ Between point *B* and load, $y = \frac{Wv}{96EI}(3l^2 - 5v^2)$	Maximum deflection is at $v =$ 0.4472*l*, and is $\frac{Wl^3}{107.33EI}$ Deflection at load, $\frac{7}{768}\frac{Wl^3}{EI}$

Stresses and Deflections in Beams *(Continued)*

Type of Beam	Stresses		Deflections	
	General Formula for Stress at any Point	Stresses at Critical Points	General Formula for Deflection at any Point[a]	Deflections at Critical Points[a]
Case 24. — Continuous Beam, with Two Unequal Spans, Unequal Loads at any Point of Each				
$m = \frac{1}{2(l_1 + l_2)}\left(\frac{W_1 a_1 b_1}{l_1}(l_1 + a_1) + \frac{W_2 a_2 b_2}{l_1}(l_2 + a_2)\right)$ R_1, W_1, R, W_2, R_2; w, u, x, v; a_1, b_1, b_2, a_2; l_1, l_2 $\frac{W_1 b_1 - m}{l_1} = r_1$ $\frac{W_1 a_1 + m}{l_1} + \frac{W_2 b_2 - m}{l_2} = r$ $\frac{W_2 b_2 - m}{l_2} = r_2$	Between R_1 and W_1, $s = -\frac{wr_1}{Z}$ Between R and W_1, $s =$ $\frac{1}{l_1 Z}[m(l_1 - u) - W_1 a_1 u]$ Between R and W_2, $s =$ $\frac{1}{l_2 Z}[m(l_2 - x) - W_2 a_2 x]$ Between R_2 and W_2, $s = -\frac{vr_2}{Z}$	Stress at load W_1, $-\frac{a_1 r_1}{Z}$ Stress at support R, $\frac{m}{Z}$ Stress at load W_2, $-\frac{a_2 r_2}{Z}$ The greatest of these is the maximum stress.	Between R_1 and W_1, $y = \frac{w}{6EI}\left\{(l_1 - w)(l_1 + w)r_1 - \frac{W_1 b_1^3}{l_1}\right\}$ Between R and W_1, $y = \frac{u}{6EIl_1}[W_1 a_1 b_1(l_1 + a_1)$ $- W_1 a_1 u^2 - m(2l_1 - u)(l_1 - u)]$ Between R and W_2 $y = \frac{x}{6EIl_2}[W_2 a_2 b_2(l_2 + a_2)$ $- W_2 a_2 x^2 - m(2l_2 - x)(l_2 - x)]$ Between R_2 and W_2, $y = \frac{v}{6EI}\left\{(l_2 - v)(l_2 + v)r_2 - \frac{W_2 b_2^3}{l_2}\right\}$	Deflection at load W_1, $\frac{a_1 b_1}{6EIl_1}[2a_1 b_1 W_1$ $- m(l_1 + a_1)]$ Deflection at load W_2, $\frac{a_2 b_2}{6EIl_2}[2a_2 b_2 W_2$ $- m(l_2 + a_2)]$ This case is so complicated that convenient general expressions for the maximum deflections cannot be obtained.

[a] The deflections apply only to cases where the cross section of the beam is constant for its entire length.

In the diagrammatical illustrations of the beams and their loading, the values indicated near, but below, the supports are the "reactions" or upward forces at the supports. For Cases 1 to 12, inclusive, the reactions, as well as the formulas for the stresses, are the same whether the beam is of constant or variable cross-section. For the other cases, the reactions and the stresses given are for constant cross-section beams only.

The bending moment at any point in inch-pounds is $s \times Z$ and can be found by omitting the divisor Z in the formula for the stress given in the tables. A positive value of the bending moment denotes tension in the upper fibers and compression in the lower ones. A negative value denotes the reverse, The value of W corresponding to a given stress is found by transposition of the formula. For example, in Case 1, the stress at the critical point is $s = -Wl \div 8Z$. From this formula we find $W = -8Zs \div l$. Of course, the negative sign of W may be ignored.

If there are several kinds of loads, as, for instance, a uniform load and a load at any point, or separate loads at different points, the total stress and the total deflection at any point is found by adding together the various stresses or deflections at the point considered due to each load acting by itself. If the stress or deflection due to any one of the loads is negative, it must be subtracted instead of added.

Deflection of Beam Uniformly Loaded for Part of Its Length.—In the following formulas, lengths are in inches, weights in pounds. W = total load; L = total length between supports; E = modulus of elasticity; I = moment of inertia of beam section; a = fraction of length of beam at each end, that is not loaded = $b \div L$; f = deflection.

$$f = \frac{WL^3}{384EI(1-2a)}(5 - 24a^2 + 16a^4)$$

The expression for maximum bending moment is: $M_{max} = \frac{1}{8}WL(1 + 2a)$.

These formulas apply to simple beams resting on supports at the ends.

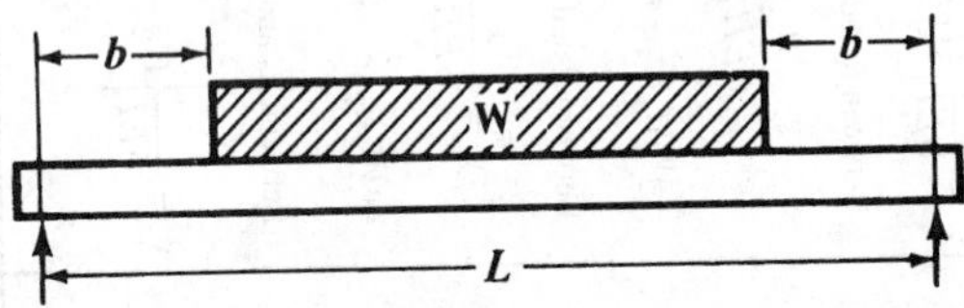

If the formulas are used with metric SI units, W = total load in newtons; L = total length between supports in millimeters; E = modulus of elasticity in newtons per millimeter^2; I = moment of inertia of beam section in millimeters^4; a = fraction of length of beam at each end, that is not loaded = $b \div L$; and f = deflection in millimeters. The bending moment M_{max} is in newton-millimeters (N · mm).

***Note:* A load due to the weight of a mass of *M* kilograms is *Mg* newtons, where *g* = approximately 9.81 meters per second ^{2}.**

Bending Stress Due to an Oblique Transverse Force.—The following illustration shows a beam and a channel being subjected to a transverse force acting at an angle ϕ to the center of gravity. To find the bending stress, the moments of inertia I around axes 3-3 and 4-4 are computed from the following equations: $I_3 = I_x\sin^2\phi + I_y\cos^2\phi$, and $I_4 = I_x\cos^2\phi + I_y\sin^2\phi$.

The computed bending stress f_b is then found from $f_b = M\left(\frac{y}{I_x}\sin\phi + \frac{x}{I_y}\cos\phi\right)$ where M is the bending moment due to force F.

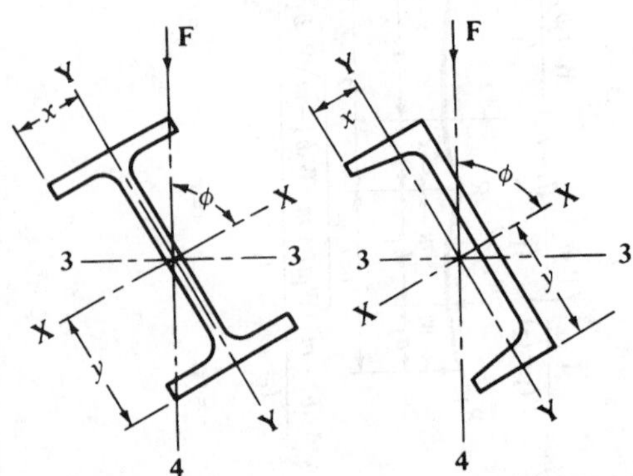

Rectangular Solid Beams

Style of Loading and Support	Diameter of Beam, d inch (mm)	Beam Height, $h$inch (mm)	Stress in Extreme Fibers, f lb/in² (N/mm²)	Beam Length, l inch (mm)	Total Load, W lb (N)
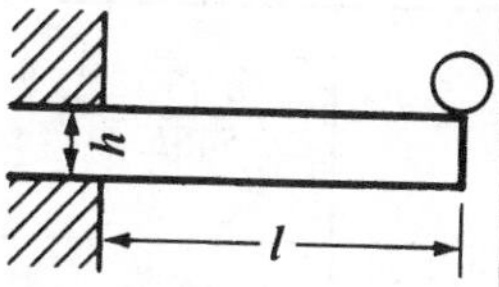	Beam fixed at one end, loaded at the other				
	$\frac{6lW}{fh^2} = b$	$\sqrt{\frac{6lW}{bf}} = h$	$\frac{6lW}{bh^2} = f$	$\frac{bfh^2}{6W} = l$	$\frac{bfh^2}{6l} = W$
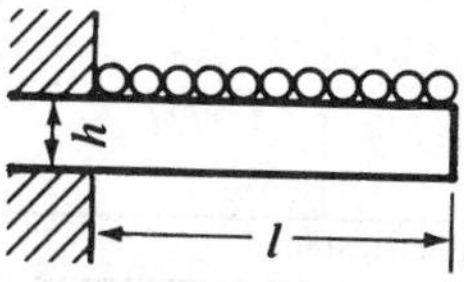	Beam fixed at one end, uniformly loaded				
	$\frac{3lW}{fh^2} = b$	$\sqrt{\frac{3lW}{bf}} = h$	$\frac{3lW}{bh^2} = f$	$\frac{bfh^2}{3W} = l$	$\frac{bfh^2}{3l} = W$
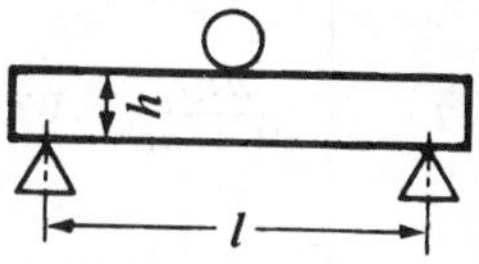	Beam supported at both ends, single load in middle				
	$\frac{3lW}{2fh^2} = b$	$\sqrt{\frac{3lW}{2bf}} = h$	$\frac{3lW}{2bh^2} = f$	$\frac{2bfh^2}{3W} = l$	$\frac{2bfh^2}{3l} = W$
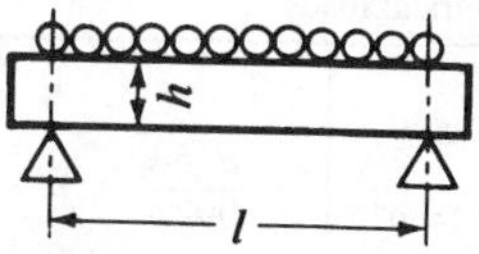	Beam supported at both ends, uniformly loaded				
	$\frac{3lW}{4fh^2} = b$	$\sqrt{\frac{3lW}{4bf}} = h$	$\frac{3lW}{4bh^2} = f$	$\frac{4bfh^2}{3W} = l$	$\frac{4bfh^2}{3l} = W$
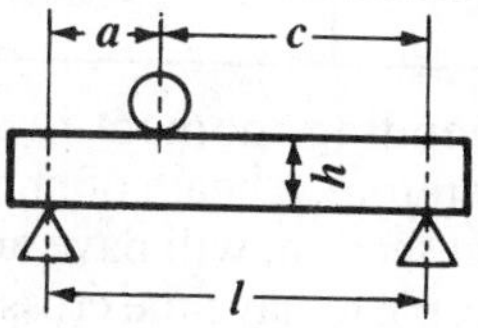	Beam supported at both ends, single unsymmetrical load				
	$\frac{6Wac}{fh^2l} = b$	$\sqrt{\frac{6Wac}{bfl}} = h$	$\frac{6Wac}{bh^2l} = f$	$a + c = l$	$\frac{bh^2fl}{6ac} = W$
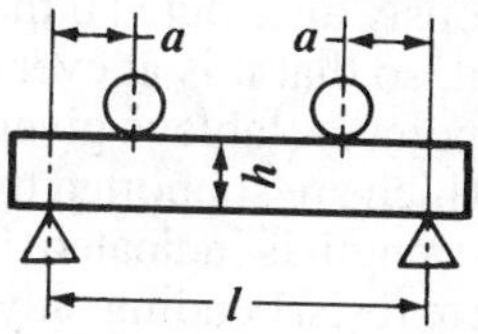	Beam supported at both ends, two symmetrical loads				
	$\frac{3Wa}{fh^2} = b$	$\sqrt{\frac{3Wa}{bf}} = h$	$\frac{3Wa}{bh^2} = f$	l, any length $\frac{bh^2f}{3W} = a$	$\frac{bh^2f}{3a} = W$

Round Solid Beams

Style of Loading and Support	Diameter of Beam, d inch (mm)	Stress in Extreme Fibers, $f$lb/in² (N/mm²)	Beam Length, l inch (mm)	Total Load, Wlb (N)
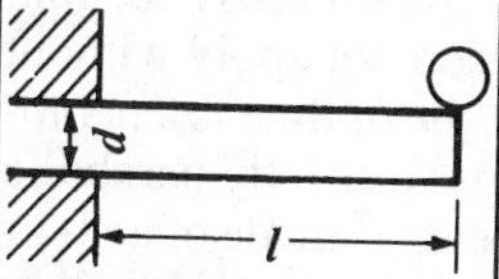	Beam fixed at one end, loaded at the other			
	$\sqrt[3]{\frac{10.18lW}{f}} = d$	$\frac{10.18lW}{d^3} = f$	$\frac{d^3f}{10.18W} = l$	$\frac{d^3f}{10.18l} = W$
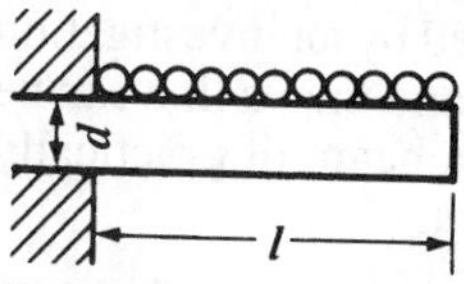	Beam fixed at one end, uniformly loaded			
	$\sqrt[3]{\frac{5.092Wl}{f}} = d$	$\frac{5.092Wl}{d^3} = f$	$\frac{d^3f}{5.092W} = l$	$\frac{d^3f}{5.092l} = W$

Round Solid Beams *(Continued)*

Style of Loading and Support	Diameter of Beam, d inch (mm)	Stress in Extreme Fibers, $f$lb/in² (N/mm²)	Beam Length, l inch (mm)	Total Load, Wlb (N)
	Beam supported at both ends, single load in middle			
	$\sqrt[3]{\frac{2.546Wl}{f}} = d$	$\frac{2.546Wl}{d^3} = f$	$\frac{d^3f}{2.546W} = l$	$\frac{d^3f}{2.546l} = W$
	Beam supported at both ends, uniformly loaded			
	$\sqrt[3]{\frac{1.273Wl}{f}} = d$	$\frac{1.273Wl}{d^3} = f$	$\frac{d^3f}{1.273W} = l$	$\frac{d^3f}{1.273l} = W$
	Beam supported at both ends, single unsymmetrical load			
	$\sqrt[3]{\frac{10.18Wac}{fl}} = d$	$\frac{10.18Wac}{d^3l} = f$	$a + c = l$	$\frac{d^3fl}{10.18ac} = W$
	Beam supported at both ends, two symmetrical loads			
	$\sqrt[3]{\frac{5.092Wa}{f}} = d$	$\frac{5.092Wa}{d^3} = f$	l, any length $\frac{d^3f}{5.092W} = a$	$\frac{d^3f}{5.092a} = W$

Beams of Uniform Strength Throughout Their Length.—The bending moment in a beam is generally not uniform throughout its length, but varies. Therefore, a beam of uniform cross-section which is made strong enough at its most strained section, will have an excess of material at every other section. Sometimes it may be desirable to have the cross-section uniform, but at other times the metal can be more advantageously distributed if the beam is so designed that its cross-section varies from point to point, so that it is at every point just great enough to take care of the bending stresses at that point. A table is given showing beams in which the load is applied in different ways and which are supported by different methods, and the shape of the beam required for uniform strength is indicated. It should be noted that the shape given is the theoretical shape required to resist bending only. It is apparent that sufficient cross-section of beam must also be added either at the points of support (in beams supported at both ends), or at the point of application of the load (in beams loaded at one end), to take care of the vertical shear.

It should be noted that the theoretical shapes of the beams given in the two tables that follow are based on the stated assumptions of uniformity of width or depth of cross-section, and unless these are observed in the design, the theoretical outlines do not apply without modifications. For example, in a cantilever with the load at one end, the outline is a parabola only when the width of the beam is uniform. It is not correct to use a strictly parabolic shape when the thickness is not uniform, as, for instance, when the beam is made of an I- or T-section. In such cases, some modification may be necessary; but it is evident that whatever the shape adopted, the correct depth of the section can be obtained by an investigation of the bending moment and the shearing load at a number of points, and then a line can be drawn through the points thus ascertained, which will provide for a beam of practically uniform strength whether the cross-section be of uniform width or not.

Beams of Uniform Strength Throughout Their Length

Type of Beam	Description	Formula[a]
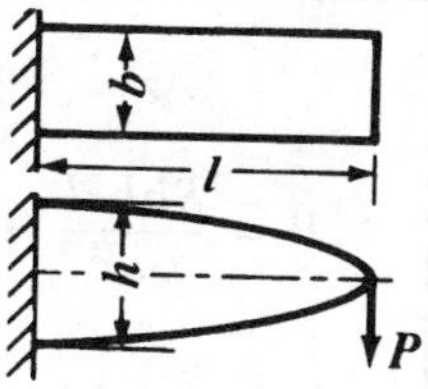	Load at one end. Width of beam uniform. Depth of beam decreasing towards loaded end. Outline of beam-shape, parabola with vertex at loaded end.	$P = \frac{Sbh^2}{6l}$
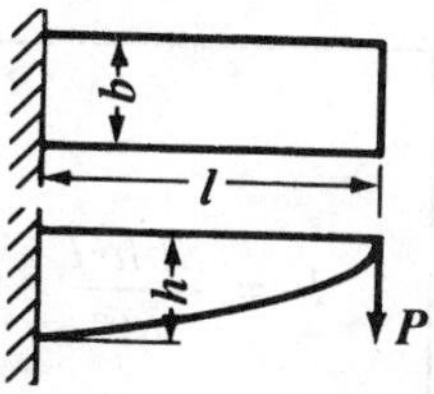	Load at one end. Width of beam uniform. Depth of beam decreasing towards loaded end. Outline of beam, one-half of a parabola with vertex at loaded end. Beam may be reversed so that upper edge is parabolic.	$P = \frac{Sbh^2}{6l}$
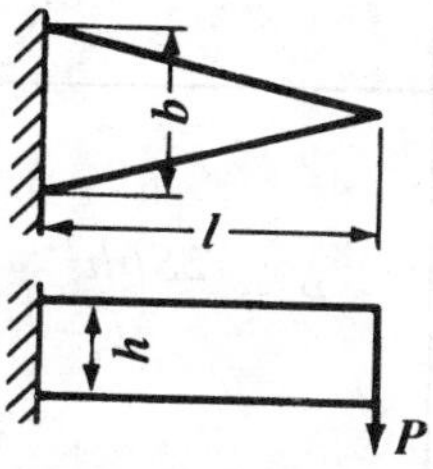	Load at one end. Depth of beam uniform. Width of beam decreasing towards loaded end. Outline of beam triangular, with apex at loaded end.	$P = \frac{Sbh^2}{6l}$
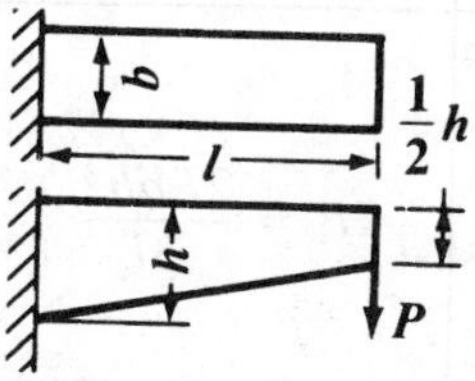	Beam of *approximately* uniform strength. Load at one end. Width of beam uniform. Depth of beam decreasing towards loaded end, but not tapering to a sharp point.	$P = \frac{Sbh^2}{6l}$
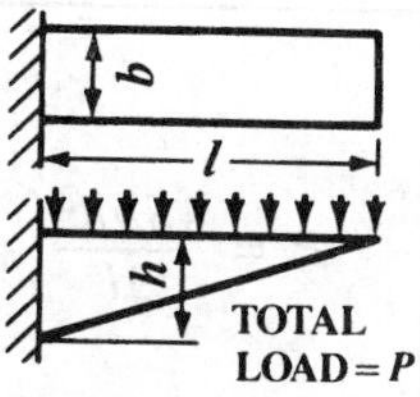 TOTAL LOAD = *P*	Uniformly distributed load. Width of beam uniform. Depth of beam decreasing towards outer end. Outline of beam, right-angled triangle.	$P = \frac{Sbh^2}{3l}$
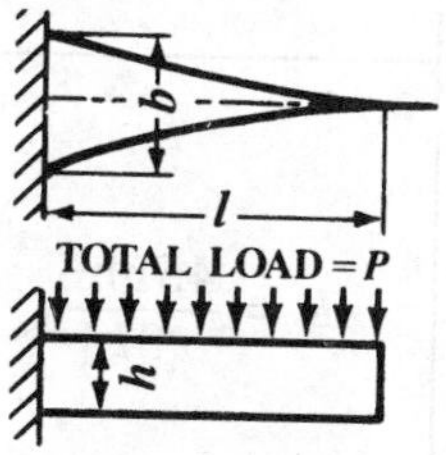 TOTAL LOAD = *P*	Uniformly distributed load. Depth of beam uniform. Width of beam gradually decreasing towards outer end. Outline of beam is formed by two parabolas which tangent each other at their vertexes at the outer end of the beam.	$P = \frac{Sbh^2}{3l}$

[a] In the formulas, P = load in pounds; S = safe stress in pounds per square inch; and a, b, c, h, and l are in inches. **If metric SI units are used, P is in newtons; S = safe stress in N/mm²; and a, b, c, h, and l are in millimeters.**

Beams of Uniform Strength Throughout Their Length

Type of Beam	Description	Formula[a]
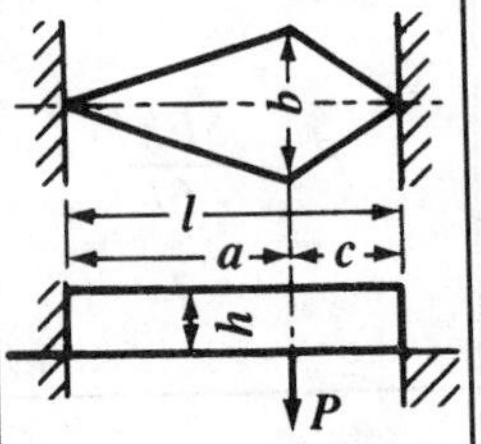	Beam supported at both ends. Load concentrated at any point. Depth of beam uniform. Width of beam maximum at point of loading. Outline of beam, two triangles with apexes at points of support.	$P = \frac{Sbh^2l}{6ac}$
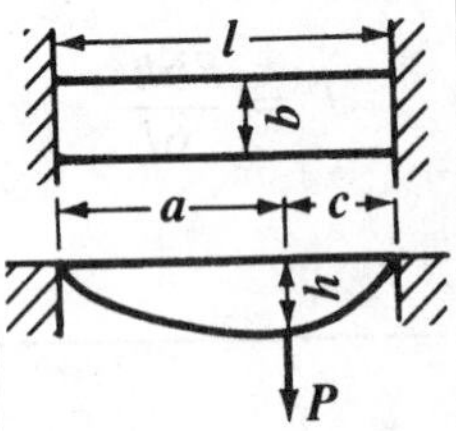	Beam supported at both ends. Load concentrated at any point. Width of beam uniform. Depth of beam maximum at point of loading. Outline of beam is formed by two parabolas with their vertexes at points of support.	$P = \frac{Sbh^2l}{6ac}$
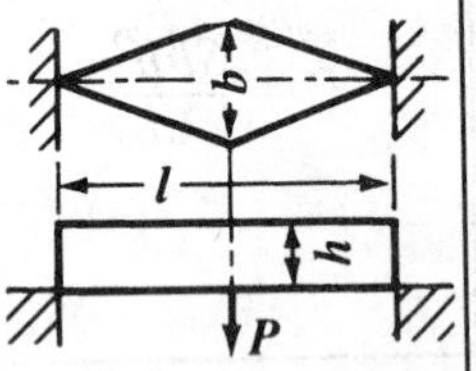	Beam supported at both ends. Load concentrated in the middle. Depth of beam uniform. Width of beam maximum at point of loading. Outline of beam, two triangles with apexes at points of support.	$P = \frac{2Sbh^2}{3l}$
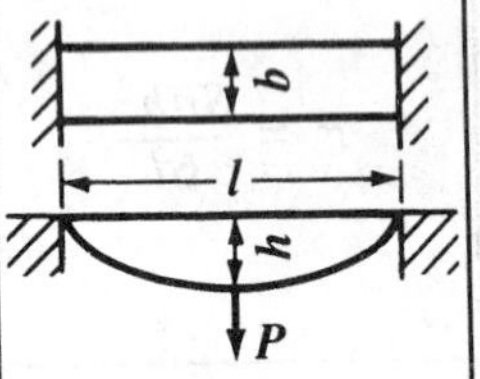	Beam supported at both ends. Load concentrated at center. Width of beam uniform. Depth of beam maximum at point of loading. Outline of beam, two parabolas with vertices at points of support.	$P = \frac{2Sbh^2}{3l}$
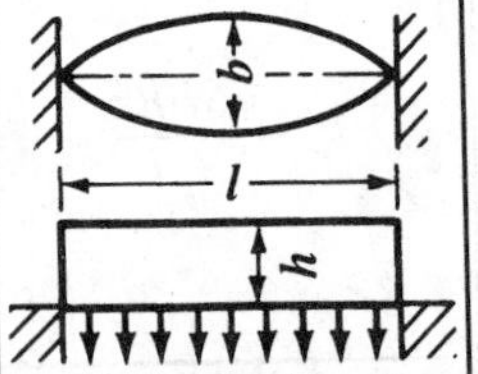	Beam supported at both ends. Load uniformly distributed. Depth of beam uniform. Width of beam maximum at center. Outline of beam, two parabolas with vertexes at middle of beam.	$P = \frac{4Sbh^2}{3l}$
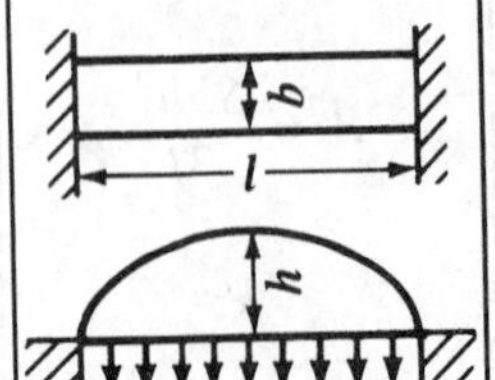	Beam supported at both ends. Load uniformly distributed. Width of beam uniform. Depth of beam maximum at center. Outline of beam one-half of an ellipse.	$P = \frac{4Sbh^2}{3l}$

[a] For details of English and metric SI units used in the formulas, see footnote on page 251.

Deflection as a Limiting Factor in Beam Design.—For some applications, a beam must be stronger than required by the maximum load it is to support, in order to prevent excessive deflection. Maximum allowable deflections vary widely for different classes of service, so a general formula for determining them cannot be given. When exceptionally stiff girders are required, one rule is to limit the deflection to 1 inch per 100 feet of span; hence, if l = length of span in inches, deflection = $l \div 1200$. According to another formula, deflection limit = $l \div 360$ where beams are adjacent to materials like plaster which would be broken by excessive beam deflection. Some machine parts of the beam type must be very rigid to maintain alignment under load. For example, the deflection of a punch press column may be limited to 0.010 inch or less. These examples merely illustrate variations in practice. It is impracticable to give general formulas for determining the allowable deflection in any specific application, because the allowable amount depends on the conditions governing each class of work.

Procedure in Designing for Deflection: Assume that a deflection equal to $l \div 1200$ is to be the limiting factor in selecting a wide-flange (W-shape) beam having a span length of 144 inches. Supports are at both ends and load at center is 15,000 pounds. Deflection y is to be limited to 144 ÷ 1200 = 0.12 inch. According to the formula on page 237 (Case 2), in which W = load on beam in pounds, l = length of span in inches, E = modulus of elasticity of material, I = moment of inertia of cross section:

$$\text{Deflection } y = \frac{Wl^3}{48EI} \quad \text{hence, } I = \frac{Wl^3}{48yE} = \frac{15{,}000 \times 144^3}{48 \times 0.12 \times 29{,}000{,}000} = 268.1$$

A structural wide-flange beam having a depth of 12 inches and weighing 35 pounds per foot has a moment of inertia I of 285 and a section modulus (Z or S) Of 45.6 (see *Steel Wide-Flange Sections—3* on page 2491)). Checking now for maximum stress s (Case 2, page 237):

$$s = \frac{Wl}{4Z} = \frac{15{,}000 \times 144}{4 \times 46.0} = 11{,}842 \text{ lbs. per sq. in.}$$

Although deflection is the limiting factor in this case, the maximum stress is checked to make sure that it is within the allowable limit. As the limiting deflection is decreased, for a given load and length of span, the beam strength and rigidity must be increased, and, consequently, the maximum stress is decreased. Thus, in the preceding example, if the maximum deflection is 0.08 inch instead of 0.12 inch, then the calculated value for the moment of inertia I will be 402; hence a W 12 × 53 beam having an I value of 426 could be used (nearest value above 402). The maximum stress then would be reduced to 7640 pounds per square inch and the calculated deflection is 0.076 inch.

A similar example using metric SI units is as follows. Assume that a deflection equal to $l \div 1000$ millimeters is to be the limiting factor in selecting a W-beam having a span length of 5 meters. Supports are at both ends and the load at the center is 30 kilonewtons. Deflection y is to be limited to 5000 ÷ 1000 = 5 millimeters. The formula on page 237 (Case 2) is applied, and W = load on beam in newtons; l = length of span in mm; E = modulus of elasticity (assume 200,000 N/mm² in this example); and I = moment of inertia of cross-section in millimeters⁴. Thus,

$$\text{Deflection } y = \frac{Wl^3}{48EI}$$

hence

$$I = \frac{Wl^3}{48yE} = \frac{30{,}000 \times 5000^3}{48 \times 5 \times 200{,}000} = 78{,}125{,}000 \text{ mm}^4$$

Although deflection is the limiting factor in this case, the maximum stress is checked to make sure that it is within the allowable limit, using the formula from page 237 (Case 2):

$$s = \frac{Wl}{4Z}$$

The units of s are newtons per square millimeter; W is the load in newtons; l is the length in mm; and Z = section modulus of the cross-section of the beam = $I \div$ distance in mm from neutral axis to extreme fiber.

Curved Beams.—The formula $S = Mc/I$ used to compute stresses due to bending of beams is based on the assumption that the beams are straight before any loads are applied. In beams having initial curvature, however, the stresses may be considerably higher than predicted by the ordinary straight-beam formula because the effect of initial curvature is to shift the neutral axis of a curved member in from the gravity axis toward the center of curvature (the concave side of the beam). This shift in the position of the neutral axis causes an increase in the stress on the concave side of the beam and decreases the stress at the outside fibers.

Hooks, press frames, and other machine members which as a rule have a rather pronounced initial curvature may have a maximum stress at the inside fibers of up to about 3½ times that predicted by the ordinary straight-beam formula.

Stress Correction Factors for Curved Beams: A simple method for determining the maximum fiber stress due to bending of curved members consists of 1) calculating the maximum stress using the straight-beam formula $S = Mc/I$; and; and 2) multiplying the calculated stress by a stress correction factor. The table on page 255 gives stress correction factors for some of the common cross-sections and proportions used in the design of curved members..

An example in the application of the method using English units of measurement is given at the bottom of the table. **A similar example using metric SI units is as follows: The fiber stresses of a curved rectangular beam are calculated as 40 newtons per millimeter², using the straight beam formula, $S = Mc/I$. If the beam is 150 mm deep and its radius of curvature is 300 mm, what are the true stresses? R/c = 300/75 = 4. From the table on page 255, the K factors corresponding to R/c = 4 are 1.20 and 0.85. Thus, the inside fiber stress is $40 \times 1.20 = 48$ N/mm² = 48 megapascals; and the outside fiber stress is $40 \times 0.85 = 34$ N/mm² = 34 megapascals.**

Approximate Formula for Stress Correction Factor: The stress correction factors given in the table on page 255 were determined by Wilson and Quereau and published in the University of Illinois Engineering Experiment Station Circular No. 16, "A Simple Method of Determining Stress in Curved Flexural Members." In this same publication the authors indicate that the following empirical formula may be used to calculate the value of the stress correction factor for the *inside* fibers of sections not covered by the tabular data to within 5 per cent accuracy except in triangular sections where up to 10 per cent deviation may be expected. However, for most engineering calculations, this formula should prove satisfactory for general use in determining the factor for the inside fibers.

$$K = 1.00 + 0.5\frac{I}{bc^2}\left[\frac{1}{R-c} + \frac{1}{R}\right]$$

Values of the Stress Correction Factor *K* for Various Curved Beam Sections

Section	R/c	Factor *K* Inside Fiber	Factor *K* Outside Fiber	y_0[a]
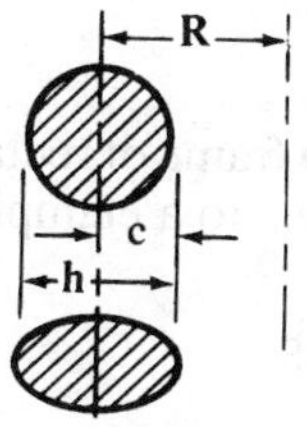	1.2	3.41	.54	.224*R*
	1.4	2.40	.60	.151*R*
	1.6	1.96	.65	.108*R*
	1.8	1.75	.68	.084*R*
	2.0	1.62	.71	.069*R*
	3.0	1.33	.79	.030*R*
	4.0	1.23	.84	.016*R*
	6.0	1.14	.89	.0070*R*
	8.0	1.10	.91	.0039*R*
	10.0	1.08	.93	.0025*R*
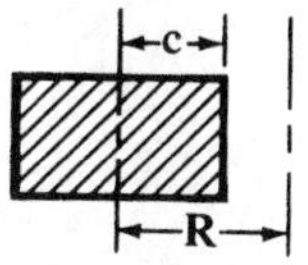	1.2	2.89	.57	.305*R*
	1.4	2.13	.63	.204*R*
	1.6	1.79	.67	.149*R*
	1.8	1.63	.70	.112*R*
	2.0	1.52	.73	.090*R*
	3.0	1.30	.81	.041*R*
	4.0	1.20	.85	.021*R*
	6.0	1.12	.90	.0093*R*
	8.0	1.09	.92	.0052*R*
	10.0	1.07	.94	.0033*R*
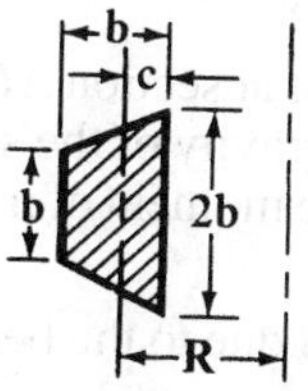	1.2	3.01	.54	.336*R*
	1.4	2.18	.60	.229*R*
	1.6	1.87	.65	.168*R*
	1.8	1.69	.68	.128*R*
	2.0	1.58	.71	.102*R*
	3.0	1.33	.80	.046*R*
	4.0	1.23	.84	.024*R*
	6.0	1.13	.88	.011*R*
	8.0	1.10	.91	.0060*R*
	10.0	1.08	.93	.0039*R*
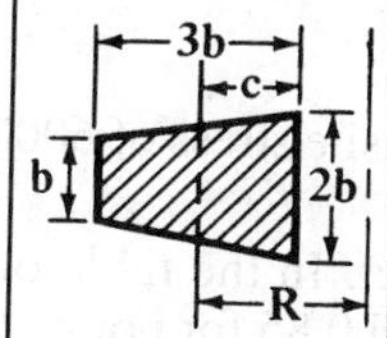	1.2	3.09	.56	.336*R*
	1.4	2.25	.62	.229*R*
	1.6	1.91	.66	.168*R*
	1.8	1.73	.70	.128*R*
	2.0	1.61	.73	.102*R*
	3.0	1.37	.81	.046*R*
	4.0	1.26	.86	.024*R*
	6.0	1.17	.91	.011*R*
	8.0	1.13	.94	.0060*R*
	10.0	1.11	.95	.0039*R*
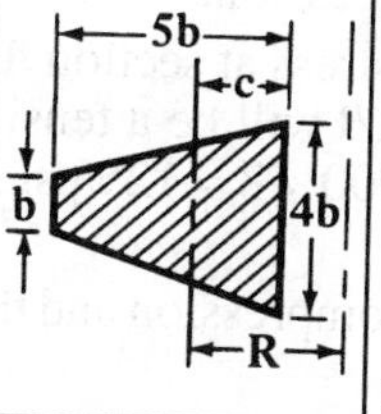	1.2	3.14	.52	.352*R*
	1.4	2.29	.54	.243*R*
	1.6	1.93	.62	.179*R*
	1.8	1.74	.65	.138*R*
	2.0	1.61	.68	.110*R*
	3.0	1.34	.76	.050*R*
	4.0	1.24	.82	.028*R*
	6.0	1.15	.87	.012*R*
	8.0	1.12	.91	.0060*R*
	10.0	1.10	.93	.0039*R*
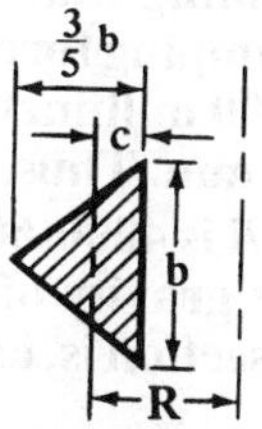	1.2	3.26	.44	.361*R*
	1.4	2.39	.50	.251*R*
	1.6	1.99	.54	.186*R*
	1.8	1.78	.57	.144*R*
	2.0	1.66	.60	.116*R*
	3.0	1.37	.70	.052*R*
	4.0	1.27	.75	.029*R*
	6.0	1.16	.82	.013*R*
	8.0	1.12	.86	.0060*R*
	10.0	1.09	.88	.0039*R*

Section	R/c	Factor *K* Inside Fiber	Factor *K* Outside Fiber	y_0[a]
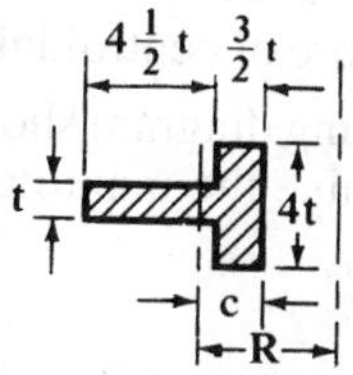	1.2	3.63	.58	.418*R*
	1.4	2.54	.63	.299*R*
	1.6	2.14	.67	.229*R*
	1.8	1.89	.70	.183*R*
	2.0	1.73	.72	.149*R*
	3.0	1.41	.79	.069*R*
	4.0	1.29	.83	.040*R*
	6.0	1.18	.88	.018*R*
	8.0	1.13	.91	.010*R*
	10.0	1.10	.92	.0065*R*
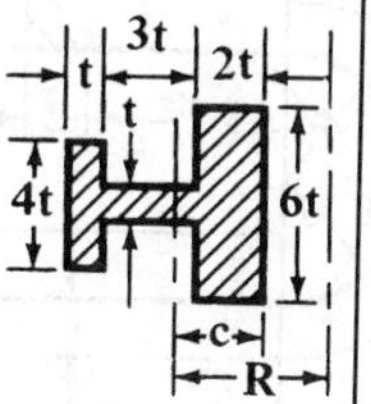	1.2	3.55	.67	.409*R*
	1.4	2.48	.72	.292*R*
	1.6	2.07	.76	.224*R*
	1.8	1.83	.78	.178*R*
	2.0	1.69	.80	.144*R*
	3.0	1.38	.86	.067*R*
	4.0	1.26	.89	.038*R*
	6.0	1.15	.92	.018*R*
	8.0	1.10	.94	.010*R*
	10.0	1.08	.95	.0065*R*
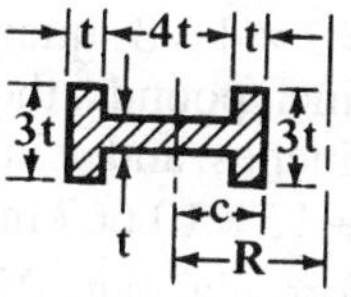	1.2	2.52	.67	.408*R*
	1.4	1.90	.71	.285*R*
	1.6	1.63	.75	.208*R*
	1.8	1.50	.77	.160*R*
	2.0	1.41	.79	.127*R*
	3.0	1.23	.86	.058*R*
	4.0	1.16	.89	.030*R*
	6.0	1.10	.92	.013*R*
	8.0	1.07	.94	.0076*R*
	10.0	1.05	.95	.0048*R*
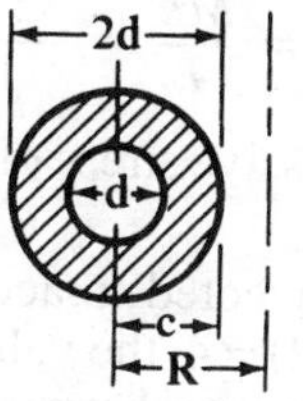	1.2	3.28	.58	.269*R*
	1.4	2.31	.64	.182*R*
	1.6	1.89	.68	.134*R*
	1.8	1.70	.71	.104*R*
	2.0	1.57	.73	.083*R*
	3.0	1.31	.81	.038*R*
	4.0	1.21	.85	.020*R*
	6.0	1.13	.90	.0087*R*
	8.0	1.10	.92	.0049*R*
	10.0	1.07	.93	.0031*R*
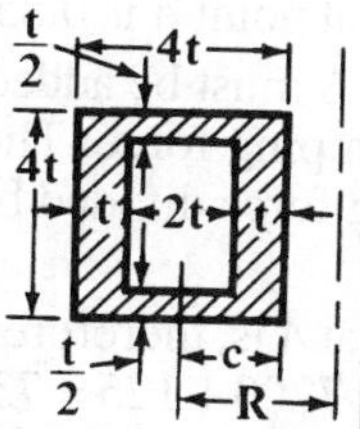	1.2	2.63	.68	.399*R*
	1.4	1.97	.73	.280*R*
	1.6	1.66	.76	.205*R*
	1.8	1.51	.78	.159*R*
	2.0	1.43	.80	.127*R*
	3.0	1.23	.86	.058*R*
	4.0	1.15	.89	.031*R*
	6.0	1.09	.92	.014*R*
	8.0	1.07	.94	.0076*R*
	10.0	1.06	.95	.0048*R*

Example: The fiber stresses of a curved rectangular beam are calculated as 5000 psi using the straight beam formula, $S = Mc/I$. If the beam is 8 inches deep and its radius of curvature is 12 inches, what are the true stresses? $R/c = 12/4 = 3$. The factors in the table corresponding to $R/c = 3$ are 0.81 and 1.30 Outside fiber stress = 5000 × 0.81 = 4050 psi; inside fiber stress = 5000 × 1.30 = 6500 psi.

[a] y_0 is the distance from the centroidal axis to the neutral axis of curved beams subjected to pure bending and is measured from the centroidal axis toward the center of curvature.

(Use 1.05 instead of 0.5 in this formula for circular and elliptical sections.)

I = Moment of inertia of section about centroidal axis

b = maximum width of section

c = distance from centroidal axis to inside fiber, i.e., to the extreme fiber nearest the center of curvature

R = radius of curvature of centroidal axis of beam

Example: The accompanying diagram shows the dimensions of a clamp frame of rectangular cross-section. Determine the maximum stress at points *A* and *B* due to a clamping force of 1000 pounds.

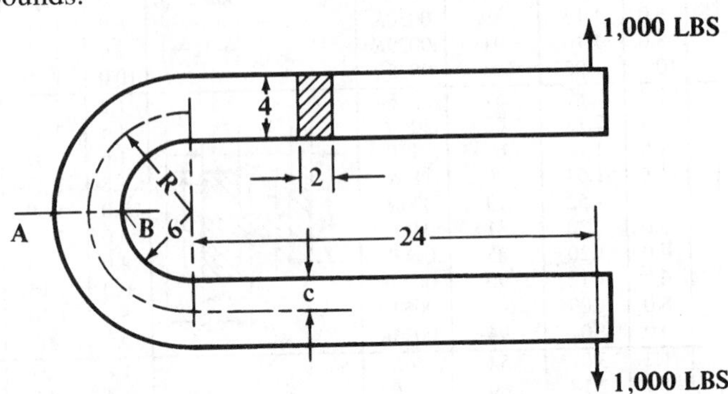

The cross-sectional area = 2 × 4 = 8 square inches; the bending moment at section *AB* is 1000 (24 + 6 + 2) = 32,000 inch pounds; the distance from the center of gravity of the section at *AB* to point *B* is c = 2 inches; and using the formula on page 219, the moment of inertia of the section is $2 \times (4)^3 \div 12 = 10.667$ inches^4.

Using the straight-beam formula, page 254, the stress at points *A* and *B* due to the bending moment is:

$$S = \frac{Mc}{I} = \frac{32{,}000 \times 2}{10.667} = 6000 \text{ psi}$$

The stress at *A* is a compressive stress of 6000 psi and that at *B* is a tensile stress of 6000 psi.

These values must be corrected to account for the curvature effect. In the table on page 255 for $R/c = (6 + 2)/(2) = 4$, the value of K is found to be 1.20 and 0.85 for points *B* and *A* respectively. Thus, the actual stress due to bending at point *B* is 1.20 × 6000 = 7200 psi in tension and the stress at point *A* is 0.85 × 6000 = 5100 psi in compression.

To these stresses at *A* and *B* must be added, algebraically, the direct stress at section *AB* due to the 1000-pound clamping force. The direct stress on section *AB* will be a tensile stress equal to the clamping force divided by the section area. Thus 1000 ÷ 8 = 125 psi in tension.

The maximum unit stress at *A* is, therefore, 5100 − 125 = 4975 psi in compression and the maximum unit stress at *B* is 7200 + 125 = 7325 psi in tension.

The following is a similar calculation using metric SI units, assuming that it is required to determine the maximum stress at points *A* and *B* due to clamping force of 4 kilonewtons acting on the frame. The frame cross-section is 50 by 100 millimeters, the radius R = 200 mm, and the length of the straight portions is 600 mm. Thus, the cross-sectional area = 50 × 100 = 5000 mm^2; the bending moment at *AB* is 4000(600 + 200) = 3,200,000 newton-millimeters; the distance from the center of gravity of the section at *AB* to point *B* is c = 50 mm; and the moment of inertia of the section is, using the formula on page 219, $50 \times (100)^3$ = 4,170,000 mm^4.

Using the straight-beam formula, page 254, the stress at points *A* and *B* due to the bending moment is:

$$s = \frac{Mc}{I} = \frac{3{,}200{,}000 \times 50}{4{,}170{,}000}$$

$$= 38.4 \text{ newtons per millimeter}^2 = 38.4 \text{ megapascals}$$

The stress at *A* is a compressive stress of 38.4 N/mm^2, while that at *B* is a tensile stress of 38.4 N/mm^2. These values must be corrected to account for the curvature effect. From the table on page 255, the *K* factors are 1.20 and 0.85 for points *A* and *B* respectively, derived from R/c = 200/50 = 4. Thus, the actual stress due to bending at point *B* is 1.20 × 38.4 = 46.1 N/mm^2 (46.1 megapascals) in tension; and the stress at point *A* is 0.85 × 38.4 = 32.6 N/mm^2 (32.6 megapascals) in compression.

To these stresses at *A* and *B* must be added, algebraically, the direct stress at section *AB* due to the 4 kN clamping force. The direct stress on section *AB* will be a tensile stress equal to the clamping force divided by the section area. Thus, 4000/5000 = 0.8 N/mm^2. The maximum unit stress at *A* is, therefore, 32.61 − 0.8 = 31.8 N/mm^2 (31.8 megapascals) in compression, and the maximum unit stress at *B* is 46.1 + 0.8 = 46.9 N/mm^2 (46.9 megapascals) in tension.

Stresses Produced by Shocks

Stresses in Beams Produced by Shocks.—Any elastic structure subjected to a shock will deflect until the product of the average resistance, developed by the deflection, and the distance through which it has been overcome, has reached a value equal to the energy of the shock. It follows that for a given shock, the average resisting stresses are inversely proportional to the deflection. If the structure were perfectly rigid, the deflection would be zero, and the stress infinite. The effect of a shock is, therefore, to a great extent dependent upon the elastic property (the springiness) of the structure subjected to the impact.

The energy of a body in motion, such as a falling body, may be spent in each of four ways:

1) In deforming the body struck as a whole.

2) In deforming the falling body as a whole.

3) In partial deformation of both bodies on the surface of contact (most of this energy will be transformed into heat).

4) Part of the energy will be taken up by the supports, if these are not perfectly rigid and inelastic.

How much energy is spent in the last three ways it is usually difficult to determine, and for this reason it is safest to figure as if the whole amount were spent as in Case 1. If a reliable judgment is possible as to what percentage of the energy is spent in other ways than the first, a corresponding fraction of the total energy can be assumed as developing stresses in the body subjected to shocks.

One investigation into the stresses produced by shocks led to the following conclusions: 1) A suddenly applied load will produce the same deflection, and, therefore, the same stress as a static load twice as great; and 2) The unit stress p (see formulas in the table "*Stresses Produced in Beams by Shocks*") for a given load producing a shock, varies directly as the square root of the modulus of elasticity E, and inversely as the square root of the length L of the beam and the area of the section.

Thus, for instance, if the sectional area of a beam is increased by four times, the unit stress will diminish only by half. This result is entirely different from those produced by static loads where the stress would vary inversely with the area, and within certain limits be practically independent of the modulus of elasticity.

In the table, the expression for the approximate value of p, which is applicable whenever the deflection of the beam is small as compared with the total height h through which the body producing the shock is dropped, is always the same for beams supported at both ends and subjected to shock at *any* point between the supports. In the formulas all dimensions are in inches and weights in pounds.

If metric SI units are used, p is in newtons per square millimeter; Q is in newtons; E = modulus of elasticity in N/mm²; I = moment of inertia of section in millimeters⁴; and h, a, and L in millimeters. *Note:* If Q is given in kilograms, the value referred to is mass. The weight Q of a mass M kilograms is Mg newtons, where g = approximately 9.81 meters per second².

Stresses Produced in Beams by Shocks

Method of Support and Point Struck by Falling Body	Fiber (Unit) Stress p produced by Weight Q Dropped Through a Distance h	Approximate Value of p
Supported at both ends; struck in center.	$p = \frac{QaL}{4I}\left(1 + \sqrt{1 + \frac{96hEI}{QL^3}}\right)$	$p = a\sqrt{\frac{6QhE}{LI}}$
Fixed at one end; struck at the other.	$p = \frac{QaL}{I}\left(1 + \sqrt{1 + \frac{6hEI}{QL^3}}\right)$	$p = a\sqrt{\frac{6QhE}{LI}}$
Fixed at both ends; struck in center.	$p = \frac{QaL}{8I}\left(1 + \sqrt{1 + \frac{384hEI}{QL^3}}\right)$	$p = a\sqrt{\frac{6QhE}{LI}}$
I = moment of inertia of section; a = distance of extreme fiber from neutral axis; L = length of beam; E = modulus of elasticity.		

Examples of How Formulas for Stresses Produced by Shocks are Derived: The general formula from which specific formulas for shock stresses in beams, springs, and other machine and structural members are derived is:

$$p = p_s\left(1 + \sqrt{1 + \frac{2h}{y}}\right) \quad (1)$$

In this formula, p = stress in pounds per square inch due to shock caused by impact of a moving load; p_s = stress in pounds per square inch resulting when moving load is applied statically; h = distance in inches that load falls before striking beam, spring, or other member; y = deflection, in inches, resulting from static load.

As an example of how Formula (1) may be used to obtain a formula for a specific application, suppose that the load W shown applied to the beam in Case 2 on page 237 were dropped on the beam from a height of h inches instead of being gradually applied (static loading). The maximum stress p_s due to load W for Case 2 is given as $Wl \div 4Z$ and the maximum deflection y is given as $Wl^3 \div 48EI$. Substituting these values in Formula (1),

$$p = \frac{Wl}{4Z}\left(1 + \sqrt{1 + \frac{2h}{Wl^3 \div 48EI}}\right) = \frac{Wl}{4Z}\left(1 + \sqrt{1 + \frac{96hEI}{Wl^3}}\right) \quad (2)$$

If in Formula (2) the letter Q is used in place of W and if Z, the section modulus, is replaced by its equivalent, $I \div$ distance a from neutral axis to extreme fiber of beam, then Formula (2) becomes the first formula given in the accompanying table *Stresses Produced in Beams by Shocks*

Stresses in Helical Springs Produced by Shocks.—A load suddenly applied on a spring will produce the same deflection, and, therefore, also the same unit stress, as a static load twice as great. When the load drops from a height h, the stresses are as given in the accompanying table. The approximate values are applicable when the deflection is small as compared with the height h. The formulas show that the fiber stress for a given shock will be greater in a spring made from a square bar than in one made from a round bar, if the diameter of coil be the same and the side of the square bar equals the diameter of the round bar. It is, therefore, more economical to use round stock for springs which must withstand shocks, due to the fact that the deflection for the same fiber stress for a square bar spring is smaller than that for a round bar spring, the ratio being as 4 to 5. The round bar spring is therefore capable of storing more energy than a square bar spring for the same stress.

Stresses Produced in Springs by Shocks

Form of Bar from Which Spring is Made	Fiber (Unit) Stress f Produced by Weight Q Dropped a Height h on a Helical Spring	Approximate Value of f
Round	$f = \frac{8QD}{\pi d^3}\left(1 + \sqrt{1 + \frac{Ghd^4}{4QD^3 n}}\right)$	$f = 1.27\sqrt{\frac{QhG}{Dd^2 n}}$
Square	$f = \frac{9QD}{4d^3}\left(1 + \sqrt{1 + \frac{Ghd^4}{0.9\pi(QD)^3 n}}\right)$	$f = 1.34\sqrt{\frac{QhG}{Dd^2 n}}$
G = modulus of elasticity for torsion; d = diameter or side of bar; D = mean diameter of spring; n = number of coils in spring.		

Shocks from Bodies in Motion.—The formulas given can be applied, in general, to shocks from bodies in motion. A body of weight W moving horizontally with the velocity of v feet per second, has a stored-up energy:

$$E_K = \frac{1}{2} \times \frac{Wv^2}{g} \text{ foot-pounds} \qquad \text{or} \qquad \frac{6Wv^2}{g} \text{ inch-po}$$

This expression may be substituted for Qh in the tables in the equations for unit stresses containing this quantity, and the stresses produced by the energy of the moving body thereby determined.

The formulas in the tables give the maximum value of the stresses, providing the designer with some definitive guidance even where there may be justification for assuming that only a part of the energy of the shock is taken up by the member under stress.

The formulas can also be applied using metric SI units. The stored-up energy of a body of mass M kilograms moving horizontally with the velocity of v meters per second is:

$$E_K = \tfrac{1}{2}Mv^2 \text{ newton-meters}$$

This expression may be substituted for Qh in the appropriate equations in the tables. For calculation in millimeters, $Qh = 1000\,E_K$ newton-millimeters.

Size of Rail Necessary to Carry a Given Load.—The following formulas may be employed for determining the size of rail and wheel suitable for carrying a given load. Let, A = the width of the head of the rail in inches; B = width of the tread of the rail in inches; C = the wheel-load in pounds; D = the diameter of the wheel in inches.

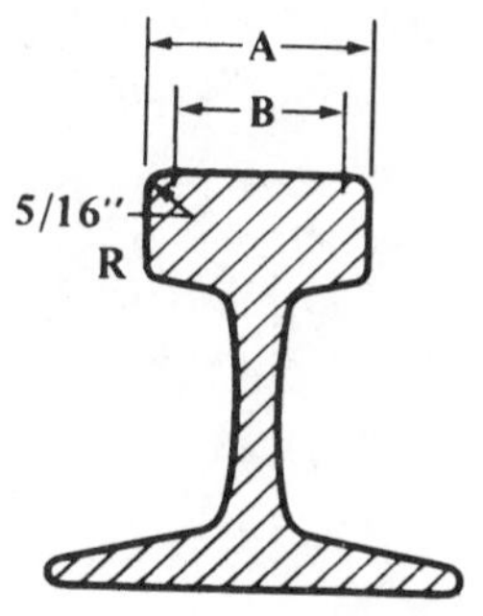

Then the width of the tread of the rail in inches is found from the formula:

$$B = \frac{C}{1250D} \quad (1)$$

The width A of the head equals $B + \frac{5}{8}$ inch. The diameter D of the smallest track wheel that will safely carry the load is found from the formula:

$$D = \frac{C}{A \times K} \quad (2)$$

in which $K = 600$ to 800 for steel castings; $K = 300$ to 400 for cast iron.

As an example, assume that the wheel-load is 10,000 pounds; the diameter of the wheel is 20 inches; and the material is cast steel. Determine the size of rail necessary to carry this load. From Formula (1):

$$B = \frac{10,000}{1250 \times 20} = 0.4 \text{ inch}$$

Hence the width of the rail required equals 0.4 + ⅝ inch = 1.025 inch. Determine also whether a wheel 20 inches in diameter is large enough to safely carry the load. From Formula (2):

$$D = \frac{10,000}{1.025 \times 600} = 16\tfrac{1}{4} \text{ inches}$$

This is the smallest diameter of track wheel that will safely carry the load; hence a 20-inch wheel is ample.

American Railway Engineering Association Formulas.—The American Railway Engineering Association recommends for safe operation of steel cylinders rolling on steel plates that the allowable load p in pounds per inch of length of the cylinder should not exceed the value calculated from the formula

$$p = \frac{\text{y.s.} - 13,000}{20,000} 600\ d \text{ for diameter } d \text{ less than 25 inches}$$

This formula is based on steel having a yield strength, y.s., of 32,000 pounds per square inch. For roller or wheel diameters of up to 25 inches, the Hertz stress (contact stress) resulting from the calculated load p will be approximately 76,000 pounds per square inch.

For a 10-inch diameter roller the safe load per inch of roller length is

$$p = \frac{32,000 - 13,000}{20,000} 600 \times 10 = 5700 \text{ lbs per inch of length}$$

Therefore, to support a 10,000 pound load the roller or wheel would need to be 10,000/5700 = 1.75 inches wide.

COLUMNS

Columns

Strength of Columns or Struts.—Structural members which are subject to compression may be so long in proportion to the diameter or lateral dimensions that failure may be the result 1) of both compression and bending; and 2) of bending or buckling to such a degree that compression stress may be ignored.

In such cases, the *slenderness ratio* is important. This ratio equals the length l of the column in inches divided by the least radius of gyration r of the cross-section. Various formulas have been used for designing columns which are too slender to be designed for compression only.

Rankine or Gordon Formula.—This formula is generally applied when slenderness ratios range between 20 and 100, and sometimes for ratios up to 120. The notation, in English and metric SI units of measurement, is given on page 263.

$$p = \frac{S}{1 + K\left(\frac{l}{r}\right)^2} = \text{ultimate load, lbs. per sq. in.}$$

Factor K may be established by tests with a given material and end condition, and for the probable range of l/r. If determined by calculation, $K = S/C\pi^2E$. Factor C equals 1 for either rounded or pivoted column ends, 4 for fixed ends, and 1 to 4 for square flat ends. The factors 25,000, 12,500, etc., in the Rankine formulas, arranged as on page 263, equal $1/K$, and have been used extensively.

Straight-line Formula.—This general type of formula is often used in designing compression members for buildings, bridges, or similar structural work. It is convenient especially in designing a number of columns that are made of the same material but vary in size, assuming that factor B is known. This factor is determined by tests.

$$p = S_y - B\left(\frac{l}{r}\right) = \text{ultimate load, lbs. per sq. in.}$$

S_y equals yield point, lbs. per square inch, and factor B ranges from 50 to 100. Safe unit stress = $p \div$ factor of safety.

Formulas of American Railway Engineering Association.—The formulas that follow apply to structural steel having an ultimate strength of 60,000 to 72,000 pounds per square inch.

For building columns having l/r ratios not greater than 120, allowable unit stress = $17{,}000 - 0.485\, l^2/r^2$. For columns having l/r ratios greater than 120, allowable unit stress

$$\text{allowable unit stress} = \frac{18{,}000}{1 + l^2/18{,}000r^2}$$

For bridge compression members centrally loaded and with values of l/r not greater than 140:

$$\text{Allowable unit stress, riveted ends} = 15{,}000 - \frac{1}{4}\frac{l^2}{r^2}$$

$$\text{Allowable unit stress, pin ends} = 15{,}000 - \frac{1}{3}\frac{l^2}{r^2}$$

Euler Formula.—This formula is for columns that are so slender that bending or buckling action predominates and compressive stresses are not taken into account.

$$P = \frac{C\pi^2 IE}{l^2} = \text{total ultimate load, in pounds}$$

The notation, in English and metric SI units of measurement, is given in the table *Rankine's and Euler's Formulas for Columns* on page 263. Factors C for different end conditions are included in the Euler formulas at the bottom of the table. According to a series of experiments, Euler formulas should be used if the values of l/r exceed the following ratios: Structural steel and flat ends, 195; hinged ends, 155; round ends, 120; cast iron with flat ends, 120; hinged ends, 100; round ends, 75; oak with flat ends, 130. The *critical slenderness ratio*, which marks the dividing line between the shorter columns and those slender enough to warrant using the Euler formula, depends upon the column material and its end conditions. If the Euler formula is applied when the slenderness ratio is too small, the *calculated* ultimate strength will exceed the yield point of the material and, obviously, will be incorrect.

Eccentrically Loaded Columns.—In the application of the column formulas previously referred to, it is assumed that the action of the load coincides with the axis of the column. If the load is offset relative to the column axis, the column is said to be eccentrically loaded, and its strength is then calculated by using a modification of the Rankine formula, the quantity cz/r^2 being added to the denominator, as shown in the table on the next page. This modified formula is applicable to columns having a slenderness ratio varying from 20 or 30 to about 100.

Machine Elements Subjected to Compressive Loads.—As in structural compression members, an unbraced machine member that is relatively slender (i.e., its length is more than, say, six times the least dimension perpendicular to its longitudinal axis) is usually designed as a column, because failure due to overloading (assuming a compressive load centrally applied in an axial direction) may occur by buckling or a combination of buckling and compression rather than by direct compression alone. In the design of unbraced steel machine "columns" which are to carry compressive loads applied along their longitudinal axes, two formulas are in general use:

(Euler)
$$P_{cr} = \frac{s_y A r^2}{Q} \qquad (1)$$

(J. B. Johnson)
$$P_{cr} = A s_y\left(1 - \frac{Q}{4r^2}\right) \qquad (2) \qquad \text{where} \qquad Q = \frac{s_y l^2}{n\pi^2 E} \qquad (3)$$

In these formulas, P_{cr} = critical load in pounds that would result in failure of the column; A = cross-sectional area, square inches; S_y = yield point of material, pounds per square inch; r = least radius of gyration of cross-section, inches; E = modulus of elasticity, pounds per square inch; l = column length, inches; and n = coefficient for end conditions. For both ends fixed, $n = 4$; for one end fixed, one end free, $n = 0.25$; for one end fixed and the other end free but guided, $n = 2$; for round or pinned ends, free but guided, $n = 1$; and for flat ends, $n = 1$ to 4. It should be noted that these values of n represent ideal conditions that are seldom attained in practice; for example, for both ends fixed, a value of $n = 3$ to 3.5 may be more realistic than $n = 4$.

If metric SI units are used in these formulas, P_{cr} = critical load in newtons that would result in failure of the column; A = cross-sectional area, square millimeters; S_y = yield point of the material, newtons per square mm; r = least radius of gyration of cross-section, mm; E = modulus of elasticity, newtons per square mm; l = column length, mm; and n = a coefficient for end conditions. The coefficients given are valid for calculations in metric units.

Rankine's and Euler's Formulas for Columns

Symbol	Quantity	English Unit	Metric SI Units
p	Ultimate unit load	Lbs./sq. in.	Newtons/sq. mm.
P	Total ultimate load	Pounds	Newtons
S	Ultimate compressive strength of material	Lbs./sq. in.	Newtons/sq. mm.
l	Length of column or strut	Inches	Millimeters
r	Least radius of gyration	Inches	Millimeters
I	Least moment of inertia	Inches⁴	Millimeters⁴
r^2	Moment of inertia/area of section	Inches²	Millimeters²
E	Modulus of elasticity of material	Lbs./sq. in.	Newtons/sq. mm.
c	Distance from neutral axis of cross-section to side under compression	Inches	Millimeters
z	Distance from axis of load to axis coinciding with center of gravity of cross-section	Inches	Millimeters

Rankine's Formulas			
Material	Both Ends of Column Fixed	One End Fixed and One End Rounded	Both Ends Rounded
Steel	$p = \frac{S}{1 + \frac{l^2}{25{,}000r^2}}$	$p = \frac{S}{1 + \frac{l^2}{12{,}500r^2}}$	$p = \frac{S}{1 + \frac{l^2}{6250r^2}}$
Cast Iron	$p = \frac{S}{1 + \frac{l^2}{5000r^2}}$	$p = \frac{S}{1 + \frac{l^2}{2500r^2}}$	$p = \frac{S}{1 + \frac{l^2}{1250r^2}}$
Wrought Iron	$p = \frac{S}{1 + \frac{l^2}{35{,}000r^2}}$	$p = \frac{S}{1 + \frac{l^2}{17{,}500r^2}}$	$p = \frac{S}{1 + \frac{l^2}{8750r^2}}$
Timber	$p = \frac{S}{1 + \frac{l^2}{3000r^2}}$	$p = \frac{S}{1 + \frac{l^2}{1500r^2}}$	$p = \frac{S}{1 + \frac{l^2}{750r^2}}$

Formulas Modified for Eccentrically Loaded Columns			
Material	Both Ends of Column Fixed	One End Fixed and One End Rounded	Both Ends Rounded
Steel	$p = \frac{S}{1 + \frac{l^2}{25{,}000r^2} + \frac{cz}{r^2}}$	$p = \frac{S}{1 + \frac{l^2}{12{,}500r^2} + \frac{cz}{r^2}}$	$p = \frac{S}{1 + \frac{l^2}{6250r^2} + \frac{cz}{r^2}}$
For materials other than steel, such as cast iron, use the Rankine formulas given in the upper table and add to the denominator the quantity cz/r^2			

Euler's Formulas for Slender Columns			
Both Ends of Column Fixed	One End Fixed and One End Rounded	Both Ends Rounded	One End Fixed and One End Free
$P = \frac{4\pi^2 IE}{l^2}$	$P = \frac{2\pi^2 IE}{l^2}$	$P = \frac{\pi^2 IE}{l^2}$	$P = \frac{\pi^2 IE}{4l^2}$

Allowable Working Loads for Columns: To find the total allowable working load for a given section, divide the total ultimate load P (or $p \times$ area), as found by the appropriate formula above, by a suitable factor of safety.

Allowable Concentric Loads for Steel Pipe Columns

STANDARD STEEL PIPE								
Effective Length (KL), Feet[a]	Nominal Diameter of Pipe, Inches							
	12	10	8	6	5	4	3½	3
	Wall Thickness of Pipe, Inch							
	0.375	0.365	0.322	0.280	0.258	0.237	0.226	.216
	Weight per Foot of Pipe, Pounds							
	49.56	40.48	28.55	18.97	14.62	10.79	9.11	7.58
	Allowable Concentric Loads in Thousands of Pounds							
6	303	246	171	110	83	59	48	38
7	301	243	168	108	81	57	46	36
8	299	241	166	106	78	54	44	34
9	296	238	163	103	76	52	41	31
10	293	235	161	101	73	49	38	28
11	291	232	158	98	71	46	35	25
12	288	229	155	95	68	43	32	22
13	285	226	152	92	65	40	29	19
14	282	223	149	89	61	36	25	16
15	278	220	145	86	58	33	22	14
16	275	216	142	82	55	29	19	12
17	272	213	138	79	51	26	17	11
18	268	209	135	75	47	23	15	10
19	265	205	131	71	43	21	14	9
20	261	201	127	67	39	19	12	
22	254	193	119	59	32	15	10	
24	246	185	111	51	27	13		
25	242	180	106	47	25	12		
26	238	176	102	43	23			

EXTRA STRONG STEEL PIPE								
Effective Length (KL), Feet[a]	Nominal Diameter of Pipe, Inches							
	12	10	8	6	5	4	3½	3
	Wall Thickness of Pipe, Inch							
	0.500	0.500	0.500	0.432	0.375	0.337	0.318	.300
	Weight per Foot of Pipe, Pounds							
	65.42	54.74	43.39	28.57	20.78	14.98	12.50	10.25
	Allowable Concentric Loads in Thousands of Pounds							
6	400	332	259	166	118	81	66	52
7	397	328	255	162	114	78	63	48
8	394	325	251	159	111	75	59	45
9	390	321	247	155	107	71	55	41
10	387	318	243	151	103	67	51	37
11	383	314	239	146	99	63	47	33
12	379	309	234	142	95	59	43	28
13	375	305	229	137	91	54	38	24
14	371	301	224	132	86	49	33	21
15	367	296	219	127	81	44	29	18
16	363	291	214	122	76	39	25	16
18	353	281	203	111	65	31	20	12
19	349	276	197	105	59	28	18	11
20	344	271	191	99	54	25	16	
21	337	265	185	92	48	22	14	
22	334	260	179	86	44	21		
24	323	248	166	73	37	17		
26	312	236	152	62	32			
28	301	224	137	54	27			

[a] With respect to radius of gyration. The effective length (KL) is the actual unbraced length, L, in feet, multiplied by the effective length factor (K) which is dependent upon the restraint at the ends of the unbraced length and the means available to resist lateral movements. K may be determined by referring to the last portion of this table.

Allowable Concentric Loads for Steel Pipe Columns *(Continued)*

Effective Length (*KL*), Feet[a]	DOUBLE-EXTRA STRONG STEEL PIPE				
	Nominal Diameter of Pipe, Inches				
	8	6	5	4	3
	Wall Thickness of Pipe, Inch				
	0.875	0.864	0.750	0.674	0.600
	Weight per Foot of Pipe, Pounds				
	72.42	53.16	38.55	27.54	18.58
	Allowable Concentric Loads in Thousands of Pounds				
6	431	306	216	147	91
7	424	299	209	140	84
8	417	292	202	133	77
9	410	284	195	126	69
10	403	275	187	118	60
11	395	266	178	109	51
12	387	257	170	100	43
13	378	247	160	91	37
14	369	237	151	81	32
15	360	227	141	70	28
16	351	216	130	62	24
17	341	205	119	55	22
18	331	193	108	49	
19	321	181	97	44	
20	310	168	87	40	
22	288	142	72	33	
24	264	119	61		
26	240	102	52		
28	213	88	44		

EFFECTIVE LENGTH FACTORS (*K*) FOR VARIOUS COLUMN CONFIGURATIONS						
	(a)	(b)	(c)	(d)	(e)	(f)
Buckled shape of column is shown by dashed line	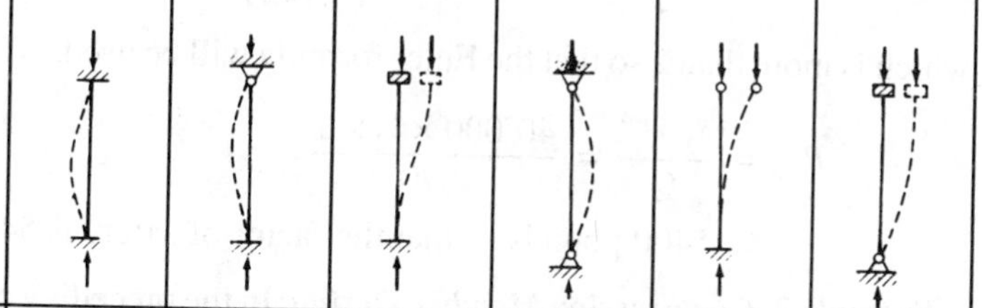					
Theoretical *K* value	0.5	0.7	1.0	1.0	2.0	2.0
Recommended design value when ideal conditions are approximated	0.65	0.80	1.2	1.0	2.10	2.0
End condition code		Rotation fixed and translation fixed Rotation free and translation fixed Rotation fixed and translation free Rotation free and translation free				

Load tables are given for 36 ksi yield stress steel. No load values are given below the heavy horizontal lines, because the *Kl*/*r* ratios (where *l* is the actual unbraced length in inches and *r* is the governing radius of gyration in inches) would exceed 200.

Data from "Manual of Steel Construction," 8th ed., 1980, with permission of the American Institute of Steel Construction.

Factor of Safety for Machine Columns: When the conditions of loading and the physical qualities of the material used are accurately known, a factor of safety as low as 1.25 is sometimes used when minimum weight is important. Usually, however, a factor of safety of 2 to 2.5 is applied for steady loads. The factor of safety represents the ratio of the critical load P_{cr} to the working load.

Application of Euler and Johnson Formulas: To determine whether the Euler or Johnson formula is applicable in any particular case, it is necessary to determine the value of the quantity $Q \div r^2$. If $Q \div r^2$ is greater than 2, then the Euler Formula (1) should be used; if $Q \div r^2$ is less than 2, then the J. B. Johnson formula is applicable. Most compression members in machine design are in the range of proportions covered by the Johnson formula. For this reason a good procedure is to design machine elements on the basis of the Johnson formula and then as a check calculate $Q \div r^2$ to determine whether the Johnson formula applies or the Euler formula should have been used.

Example 1, Compression Member Design: A rectangular machine member 24 inches long and ½× 1 inch in cross-section is to carry a compressive load of 4000 pounds along its axis. What is the factor of safety for this load if the material is machinery steel having a yield point of 40,000 pounds per square inch, the load is steady, and each end of the rod has a ball connection so that $n = 1$?

From Formula (3)

$$Q = \frac{40{,}000 \times 24 \times 24}{1 \times 3.1416 \times 3.1416 \times 30{,}000{,}000} = 0.0778$$

(The values 40,000 and 30,000,000 were obtained from the table *Strength Data for Iron and Steel* on page 476.)

The radius of gyration r for a rectangular section (page 219) is 0.289 × the dimension in the direction of bending. In columns, bending is most apt to occur in the direction in which the section is the weakest, the ½-inch dimension in this example. Hence, least radius of gyration $r = 0.289 \times \frac{1}{2} = 0.145$ inch.

$$\frac{Q}{r^2} = \frac{0.0778}{(0.145)^2} = 3.70$$

which is more than 2 so that the Euler formula will be used.

$$P_{cr} = \frac{s_y A r^2}{Q} = \frac{40{,}000 \times \frac{1}{2} \times 1}{3.70}$$

$$= 5400 \text{ pounds so that the factor of safety is } 5400 \div 4000 = 1.35$$

Example 2, Compression Member Design: In the preceding example, the column formulas were used to check the adequacy of a column of known dimensions. The more usual problem involves determining what the dimensions should be to resist a specified load. For example,:

A 24-inch long bar of rectangular cross-section with width w twice its depth d is to carry a load of 4000 pounds. What must the width and depth be if a factor of safety of 1.35 is to be used?

First determine the critical load P_{cr}:

$$P_{cr} = \text{working load} \times \text{factor of safety}$$

$$= 4000 \times 1.35 = 5400 \text{ pounds}$$

Next determine Q which, as before, will be 0.0778.

Assume Formula (2) applies:

$$P_{cr} = A s_y \left(1 - \frac{Q}{4r^2}\right)$$

$$5400 = w \times d \times 40{,}000\left(1 - \frac{0.0778}{4r^2}\right)$$
$$= 2d^2 \times 40{,}000\left(1 - \frac{0.01945}{r^2}\right)$$
$$\frac{5400}{40{,}000 \times 2} = d^2\left(1 - \frac{0.01945}{r^2}\right)$$

As mentioned in Example 1 the least radius of gyration r of a rectangle is equal to 0.289 times the least dimension, d, in this case. Therefore, substituting for d the value $r \div 0.289$,

$$\frac{5400}{40{,}000 \times 2} = \left(\frac{r}{0.289}\right)^2\left(1 - \frac{0.01945}{r^2}\right)$$
$$\frac{5400 \times 0.289 \times 0.289}{40{,}000 \times 2} = r^2 - 0.01945$$
$$0.005638 = r^2 - 0.01945$$
$$r^2 = 0.0251$$

Checking to determine if $Q \div r^2$ is greater or less than 2,

$$\frac{Q}{r^2} = \frac{0.0778}{0.0251} = 3.1$$

therefore Formula (1) should have been used to determine r and dimensions w and d. Using Formula (1),

$$5400 = \frac{40{,}000 \times 2d^2 \times r^2}{Q} = \frac{40{,}000 \times 2 \times \left(\frac{r}{0.289}\right)^2 r^2}{0.0778}$$
$$r^4 = \frac{5400 \times 0.0778 \times 0.289 \times 0.289}{40{,}000 \times 2}$$
$$d = \frac{0.145}{0.289} = 0.50 \text{ inch}$$

and $w = 2d = 1$ inch as in the previous example.

American Institute of Steel Construction.—For main or secondary compression members with l/r ratios up to 120, safe unit stress = $17{,}000 - 0.485l^2/r^2$. For columns and bracing or other secondary members with l/r ratios above 120,

Safe unit stress, psi $= \dfrac{18{,}000}{1 + l^2/18{,}000r^2}$ for bracing and secondary members. For main members, safe unit stress, psi $= \dfrac{18{,}000}{1 + l^2/18{,}000r^2} \times \left(1.6 - \dfrac{l/r}{200}\right)$

Pipe Columns: Allowable concentric loads for steel pipe columns based on the above formulas are given in the table on page 264.

PLATES, SHELLS, AND CYLINDERS

Flat Stayed Surfaces.—Large flat areas are often held against pressure by stays distributed at regular intervals over the surface. In boiler work, these stays are usually screwed into the plate and the projecting end riveted over to insure steam tightness. The U.S. Board of Supervising Inspectors and the American Boiler Makers Association rules give the following formula for flat stayed surfaces:

$$P = \frac{C \times t^2}{S^2}$$

in which P = pressure in pounds per square inch

C = a constant, which equals 112 for plates $\frac{7}{16}$ inch and under; 120, for plates over $\frac{7}{16}$ inch thick; 140, for plates with stays having a nut and bolt on the inside and outside; and 160, for plates with stays having washers of at least one-half the thickness of the plate, and with a diameter at least one-half of the greatest pitch.

t = thickness of plate in 16ths of an inch (thickness = $\frac{7}{16}$, $t = 7$)

S = greatest pitch of stays in inches

Strength and Deflection of Flat Plates.—Generally, the formulas used to determine stresses and deflections in flat plates are based on certain assumptions that can be closely approximated in practice. These assumptions are:

1) the thickness of the plate is not greater than one-quarter the least width of the plate;

2) the greatest deflection when the plate is loaded is less than one-half the plate thickness;

3) the maximum tensile stress resulting from the load does not exceed the elastic limit of the material; and

4) all loads are perpendicular to the plane of the plate.

Plates of ductile materials fail when the maximum stress resulting from deflection under load exceeds the yield strength; for brittle materials, failure occurs when the maximum stress reaches the ultimate tensile strength of the material involved.

Square and Rectangular Flat Plates.—The formulas that follow give the maximum stress and deflection of flat steel plates supported in various ways and subjected to the loading indicated. These formulas are based upon a modulus of elasticity for steel of 30,000,000 pounds per square inch and a value of Poisson's ratio of 0.3. If the formulas for maximum stress, S, are applied without modification to other materials such as cast iron, aluminum, and brass for which the range of Poisson's ratio is about 0.26 to 0.34, the maximum stress calculations will be in error by not more than about 3 per cent. The deflection formulas may also be applied to materials other than steel by substituting in these formulas the appropriate value for E, the modulus of elasticity of the material (see pages 476 and 477). The deflections thus obtained will not be in error by more than about 3 per cent.

In the stress and deflection formulas that follow,

p = uniformly distributed load acting on plate, pounds per square inch

W = total load on plate, pounds; $W = p \times$ area of plate

L = distance between supports (length of plate), inches. For rectangular plates, L = long side, l = short side

t = thickness of plate, inches

S = maximum tensile stress in plate, pounds per square inch

d = maximum deflection of plate, inches

E = modulus of elasticity in tension. E = 30,000,000 pounds per square inch for steel

If metric SI units are used in the formulas, then,

W = total load on plate, newtons

L = distance between supports (length of plate), millimeters. For rectangular plates, L = long side, l = short side

t = thickness of plate, millimeters

S = maximum tensile stress in plate, newtons per mm squared

d = maximum deflection of plate, mm

E = modulus of elasticity, newtons per mm squared

A) Square flat plate supported at top and bottom of all four edges and a uniformly distributed load over the surface of the plate.

$$S = \frac{0.29W}{t^2} \quad (1) \qquad d = \frac{0.0443WL^2}{Et^3} \quad (2)$$

B) Square flat plate supported at the bottom only of all four edges and a uniformly distributed load over the surface of the plate.

$$S = \frac{0.28W}{t^2} \quad (3) \qquad d = \frac{0.0443WL^2}{Et^3} \quad (4)$$

C) Square flat plate with all edges firmly fixed and a uniformly distributed load over the surface of the plate.

$$S = \frac{0.31W}{t^2} \quad (5) \qquad d = \frac{0.0138WL^2}{Et^3} \quad (6)$$

D) Square flat plate with all edges firmly fixed and a uniform load over small circular area at the center. In Equations (7) and (9), r_0 = radius of area to which load is applied. If $r_0 < 1.7t$, use r_s where $r_s = \sqrt{1.6r_0{}^2 + t^2} - 0.675t$.

$$S = \frac{0.62W}{t^2}\log_e\left(\frac{L}{2r_0}\right) \quad (7) \qquad d = \frac{0.0568WL^2}{Et^3} \quad (8)$$

E) Square flat plate with all edges supported above and below, or below only, and a concentrated load at the center. (See Case 4, above, for definition of r_0).

$$S = \frac{0.62W}{t^2}\left[\log_e\left(\frac{L}{2r_0}\right) + 0.577\right] \quad (9) \qquad d = \frac{0.1266WL^2}{Et^3} \quad (10)$$

F) Rectangular plate with all edges supported at top and bottom and a uniformly distributed load over the surface of the plate.

$$S = \frac{0.75W}{t^2\left(\frac{L}{l} + 1.61\frac{l^2}{L^2}\right)} \quad (11) \qquad d = \frac{0.1422W}{Et^3\left(\frac{L}{l^3} + \frac{2.21}{L^2}\right)} \quad (12)$$

G) Rectangular plate with all edges fixed and a uniformly distributed load over the surface of the plate.

$$S = \frac{0.5W}{t^2\left(\frac{L}{l} + \frac{0.623l^5}{L^5}\right)} \quad (13) \qquad d = \frac{0.0284W}{Et^3\left(\frac{L}{l^3} + \frac{1.056l^2}{L^4}\right)} \quad (14)$$

Circular Flat Plates.—In the following formulas, R = radius of plate to supporting edge in inches; W = total load in pounds; and other symbols are the same as used for square and rectangular plates.

If metric SI units are used, R = radius of plate to supporting edge in millimeters, and the values of other symbols are the same as those used for square and rectangular plates.

A) Edge supported around the circumference and a uniformly distributed load over the surface of the plate.

$$S = \frac{0.39W}{t^2} \quad (1) \qquad d = \frac{0.221WR^2}{Et^3} \quad (2)$$

B) Edge fixed around circumference and a uniformly distributed load over the surface of the plate.

$$S = \frac{0.24W}{t^2} \quad (3) \qquad d = \frac{0.0543WR^2}{Et^3} \quad (4)$$

C) Edge supported around the circumference and a concentrated load at the center.

$$S = \frac{0.48W}{t^2}\left[1 + 1.3\log_e\frac{R}{0.325t} - 0.0185\frac{t^2}{R^2}\right] \quad (5) \qquad d = \frac{0.55WR^2}{Et^3} \quad (6)$$

D) Edge fixed around circumference and a concentrated load at the center.

$$S = \frac{0.62W}{t^2}\left[\log_e\frac{R}{0.325t} + 0.0264\frac{t^2}{R^2}\right] \quad (7) \qquad d = \frac{0.22WR^2}{Et^3} \quad (8)$$

Strength of Cylinders Subjected to Internal Pressure.—In designing a cylinder to withstand internal pressure, the choice of formula to be used depends on 1) the kind of material of which the cylinder is made (whether brittle or ductile); 2) the construction of the cylinder ends (whether open or closed); and 3) whether the cylinder is classed as a thin- or a thick-walled cylinder.

A cylinder is considered to be thin-walled when the ratio of wall thickness to inside diameter is 0.1 or less and thick-walled when this ratio is greater than 0.1. Materials such as cast iron, hard steel, cast aluminum are considered to be brittle materials; low-carbon steel, brass, bronze, etc. are considered to be ductile.

In the formulas that follow, p = internal pressure, pounds per square inch; D = inside diameter of cylinder, inches; t = wall thickness of cylinder, inches; μ = Poisson's ratio, = 0.3 for steel, 0.26 for cast iron, 0.34 for aluminum and brass; and S = allowable tensile stress, pounds per square inch.

Metric SI units can be used in Formulas (1), (3), (4), and (5), where p = internal pressure in newtons per square millimeter; D = inside diameter of cylinder, millimeters; t = wall thickness, mm; μ = Poisson's ratio, = 0.3 for steel, 0.26 for cast iron, and 0.34 for aluminum and brass; and S = allowable tensile stress, N/mm². For the use of metric SI units in Formula (2), see below.

Thin-walled cylinders:

$$t = \frac{Dp}{2S} \quad (1)$$

For low-pressure cylinders of cast iron such as are used for certain engine and press applications, a formula in common use is

$$t = \frac{Dp}{2500} + 0.3 \quad (2)$$

This formula is based on allowable stress of 1250 pounds per square inch and will give a wall thickness 0.3 inch greater than Formula (1) to allow for variations in metal thickness that may result from the casting process.

If metric SI units are used in Formula (2), t = cylinder wall thickness in millimeters; D = inside diameter of cylinder, mm; and the allowable stress is in newtons per square

millimeter. The value of 0.3 inches additional wall thickness is 7.62 mm, and the next highest number in preferred metric basic sizes is 8 mm.

Thick-walled cylinders of brittle material, ends open or closed: Lamé's equation is used when cylinders of this type are subjected to internal pressure.

$$t = \frac{D}{2}\left[\sqrt{\frac{S+p}{S-p}} - 1\right] \qquad (3)$$

The table *Allowable Stress in Metal per Sq. In. of Section* on page 272 is for convenience in calculating the dimensions of cylinders under high internal pressure without the use of Formula (3).

Example, Use of the Table: Assume that a cylinder of 10 inches inside diameter is to withstand a pressure of 2500 pounds per square inch; the material is cast iron and the allowable stress is 6000 pounds per square inch. To solve the problem, locate the allowable stress per square inch in the left-hand column of the table and the working pressure at the top of the columns. Then find the ratio between the outside and inside radii in the body of the table. In this example, the ratio is 1.558, and hence the outside diameter of the cylinder should be 10 × 1.558, or about 15⅝ inches. The thickness of the cylinder wall will therefore be (15.558 − 10)/2 = 2.779 inches.

Unless very high-grade material is used and sound castings assured, cast iron should not be used for pressures exceeding 2000 pounds per square inch. It is well to leave more metal in the bottom of a hydraulic cylinder than is indicated by the results of calculations, because a hole of some size must be cored in the bottom to permit the entrance of a boring bar when finishing the cylinder, and when this hole is subsequently tapped and plugged it often gives trouble if there is too little thickness.

For steady or gradually applied stresses, the maximum allowable fiber stress S may be assumed to be from 3500 to 4000 pounds per square inch for cast iron; from 6000 to 7000 pounds per square inch for brass; and 12,000 pounds per square inch for steel castings. For intermittent stresses, such as in cylinders for steam and hydraulic work, 3000 pounds per square inch for cast iron; 5000 pounds per square inch for brass; and 10,000 pounds per square inch for steel castings, is ordinarily used. These values give ample factors of safety.

***Note:* In metric SI units, 1000 pounds per square inch equals 6.895 newtons per square millimeter.**

Thick-walled cylinders of ductile material, closed ends: Clavarino's equation is used:

$$t = \frac{D}{2}\left[\sqrt{\frac{S+(1-2\mu)p}{S-(1+\mu)p}} - 1\right] \qquad (4)$$

Spherical Shells Subjected to Internal Pressure.—Let:

D = internal diameter of shell in inches
p = internal pressure in pounds per square inch
S = safe tensile stress per square inch
t = thickness of metal in the shell, in inches. Then: $t = \frac{pD}{4S}$

This formula also applies to hemi-spherical shells, such as the hemi-spherical head of a cylindrical container subjected to internal pressure, etc.

If metric SI units are used, then:

D = internal diameter of shell in millimeters
p = internal pressure in newtons per square millimeter
S = safe tensile stress in newtons per square millimeter
t = thickness of metal in the shell in millimeters

Meters can be used in the formula in place of millimeters, providing the treatment is consistent throughout.

Allowable Stress in Metal per Sq. In. of Section	Working Pressure in Cylinder, Pounds per Square Inch						
	1000	2000	3000	4000	5000	6000	7000
2,000	1.732	…	…	…	…	…	…
2,500	1.527	…	…	…	…	…	…
3,000	1.414	2.236	…	…	…	…	…
3,500	1.341	1.915	…	…	…	…	…
4,000	1.291	1.732	2.645	…	…	…	…
4,500	1.253	1.612	2.236	…	…	…	…
5000	1.224	1.527	2.000	3.000	…	…	…
5,500	1.201	1.464	1.844	2.516	…	…	…
6,000	1.183	1.414	1.732	2.236	3.316	…	…
6,500	…	1.374	1.647	2.049	2.768	…	…
7,000	…	1.341	1.581	1.914	2.449	3.605	…
7,500	…	1.314	1.527	1.813	2.236	3.000	…
8,000	…	1.291	1.483	1.732	2.081	2.645	3.872
8,500	…	1.271	1.446	1.666	1.963	2.408	3.214
9,000	…	1.253	1.414	1.612	1.871	2.236	2.828
9,500	…	1.235	1.386	1.566	1.795	2.104	2.569
10,000	…	1.224	1.362	1.527	1.732	2.000	2.380
10,500	…	1.212	1.341	1.493	1.678	1.915	2.236
11,000	…	1.201	1.322	1.464	1.633	1.844	2.121
11,500	…	1.193	1.306	1.437	1.593	1.784	2.027
12,000	…	1.183	1.291	1.414	1.558	1.732	1.949
12,500	…	…	1.277	1.393	1.527	1.687	1.878
13,000	…	…	1.264	1.374	1.500	1.647	1.825
13,500	…	…	1.253	1.357	1.475	1.612	1.775
14,000	…	…	1.243	1.341	1.453	1.581	1.732
14,500	…	…	1.233	1.327	1.432	1.553	1.693
15,000	…	…	1.224	1.314	1.414	1.527	1.658
16,000	…	…	1.209	1.291	1.381	1.483	1.599

Thick-walled cylinders of ductile material; open ends: Birnie's equation is used:

$$t = \frac{D}{2}\left[\sqrt{\frac{S + (1 - \mu)p}{S - (1 + \mu)p}} - 1\right] \tag{5}$$

Example: Find the thickness of metal required in the hemi-spherical end of a cylindrical vessel, 2 feet in diameter, subjected to an internal pressure of 500 pounds per square inch. The material is mild steel and a tensile stress of 10,000 pounds per square inch is allowable.

$$t = \frac{500 \times 2 \times 12}{4 \times 10{,}000} = 0.3 \text{ inch}$$

A similar example using metric SI units is as follows: find the thickness of metal required in the hemi-spherical end of a cylindrical vessel, 750 mm in diameter, subjected to an internal pressure of 3 newtons/mm². The material is mild steel and a tensile stress of 70 newtons/mm² is allowable.

$$t = \frac{3 \times 750}{4 \times 70} = 8.04 \text{ mm}$$

If the radius of curvature of the domed head of a boiler or container subjected to internal pressure is made equal to the diameter of the boiler, the thickness of the cylindrical shell and of the spherical head should be made the same. For example, if a boiler is 3 feet in diameter, the radius of curvature of its head should also be 3 feet, if material of the same thickness is to be used and the stresses are to be equal in both the head and cylindrical portion.

Collapsing Pressure of Cylinders and Tubes Subjected to External Pressures.—The following formulas may be used for finding the collapsing pressures of lap-welded Bessemer steel tubes:

$$P = 86{,}670\frac{t}{D} - 1386 \qquad (1)$$

$$P = 50{,}210{,}000\left(\frac{t}{D}\right)^3 \qquad (2)$$

in which P = collapsing pressure in pounds per square inch; D = outside diameter of tube or cylinder in inches; t = thickness of wall in inches.

Formula (1) is for values of P greater than 580 pounds per square inch, and Formula (2) is for values of P less than 580 pounds per square inch. These formulas are substantially correct for all lengths of pipe greater than six diameters between transverse joints that tend to hold the pipe to a circular form. The pressure P found is the actual collapsing pressure, and a suitable factor of safety must be used. Ordinarily, a factor of safety of 5 is sufficient. In cases where there are repeated fluctuations of the pressure, vibration, shocks and other stresses, a factor of safety of from 6 to 12 should be used.

If metric SI units are used the formulas are:

$$P = 597.6\frac{t}{D} - 9.556 \qquad (3)$$

$$P = 346{,}200\left(\frac{t}{D}\right)^3 \qquad (4)$$

where P = collapsing pressure in newtons per square millimeter; D = outside diameter of tube or cylinder in millimeters; and t = thickness of wall in millimeters. Formula (3) is for values of P greater than 4 N/mm², and Formula (4) is for values of P less than 4 N/mm².

The table "Tubes Subjected to External Pressure" is based upon the requirements of the Steam Boat Inspection Service of the Department of Commerce and Labor and gives the permissible working pressures and corresponding minimum wall thickness for long, plain, lap-welded and seamless steel flues subjected to external pressure only. The table thicknesses have been calculated from the formula:

$$t = \frac{[(F \times p) + 1386]D}{86{,}670}$$

in which D = outside diameter of flue or tube in inches; t = thickness of wall in inches; p = working pressure in pounds per square inch; F = factor of safety. The formula is applicable to working pressures greater than 100 pounds per square inch, to outside diameters from 7 to 18 inches, and to temperatures less than 650°F.

The preceding Formulas (1) and (2) were determined by Prof. R. T. Stewart, Dean of the Mechanical Engineering Department of the University of Pittsburgh, in a series of experiments carried out at the plant of the National Tube Co., McKeesport, Pa.

The apparent fiber stress under which the different tubes failed varied from about 7000 pounds per square inch for the relatively thinnest to 35,000 pounds per square inch for the relatively thickest walls. The average yield point of the material tested was 37,000 pounds and the tensile strength 58,000 pounds per square inch, so it is evident that the strength of a tube subjected to external fluid collapsing pressure is not dependent alone upon the elastic limit or ultimate strength of the material from which it is made.

Tubes Subjected to External Pressure

Outside Diameter of Tube, Inches	Working Pressure in Pounds per Square Inch						
	100	120	140	160	180	200	220
	Thickness of Tube in Inches. Safety Factor, 5						
7	0.152	0.160	0.168	0.177	0.185	0.193	0.201
8	0.174	0.183	0.193	0.202	0.211	0.220	0.229
9	0.196	0.206	0.217	0.227	0.237	0.248	0.258
10	0.218	0.229	0.241	0.252	0.264	0.275	0.287
11	0.239	0.252	0.265	0.277	0.290	0.303	0.316
12	0.261	0.275	0.289	0.303	0.317	0.330	0.344
13	0.283	0.298	0.313	0.328	0.343	0.358	0.373
14	0.301	0.320	0.337	0.353	0.369	0.385	0.402
15	0.323	0.343	0.361	0.378	0.396	0.413	0.430
16	0.344	0.366	0.385	0.404	0.422	0.440	0.459
16	0.366	0.389	0.409	0.429	0.448	0.468	0.488
18	0.387	0.412	0.433	0.454	0.475	0.496	0.516

Dimensions and Maximum Allowable Pressure of Tubes Subjected to External Pressure

Outside Diam., Inches	Thickness of Material, Inches	Maximum Pressure Allowed, psi	Outside Diam., Inches	Thickness of Material, Inches	Maximum Pressure Allowed, psi	Outside Diam., Inches	Thickness of Material, Inches	Maximum Pressure Allowed, psi
2	0.095	427	3	0.109	327	4	0.134	303
2¼	0.095	380	3¼	0.120	332	4½	0.134	238
2½	0.109	392	3½	0.120	308	5	0.148	235
2¾	0.109	356	3¾	0.120	282	6	0.165	199

SHAFTS

Shaft Calculations

Torsional Strength of Shafting.—In the formulas that follow,

α = angular deflection of shaft in degrees

c = distance from center of gravity to extreme fiber

D = diameter of shaft in inches

G = torsional modulus of elasticity = 11,500,000 pounds per square inch for steel

J = polar moment of inertia of shaft cross-section (see table)

l = length of shaft in inches

N = angular velocity of shaft in revolutions per minute

P = power transmitted in horsepower

S_s = allowable torsional shearing stress in pounds per square inch

T = torsional or twisting moment in inch-pounds

Z_p = polar section modulus (see table page 278)

The allowable twisting moment for a shaft of any cross-section such as circular, square, etc., is:

$$T = S_s \times Z_p \qquad (1)$$

For a shaft delivering P horsepower at N revolutions per minute the twisting moment T being transmitted is:

$$T = \frac{63{,}000P}{N} \qquad (2)$$

The twisting moment T as determined by this formula should be less than the value determined by using Formula (1) if the maximum allowable stress S_s is not to be exceeded.

The diameter of a solid circular shaft required to transmit a given torque T is:

$$D = \sqrt[3]{\frac{5.1T}{S_s}} \quad (3a) \qquad \text{or} \qquad D = \sqrt[3]{\frac{321{,}000P}{NS_s}} \quad (3b)$$

The allowable stresses that are generally used in practice are: 4000 pounds per square inch for main power-transmitting shafts; 6000 pounds per square inch for lineshafts carrying pulleys; and 8500 pounds per square inch for small, short shafts, countershafts, etc. Using these allowable stresses, the horsepower P transmitted by a shaft of diameter D, or the diameter D of a shaft to transmit a given horsepower P may be determined from the following formulas:

For main power-transmitting shafts:

$$P = \frac{D^3N}{80} \quad (4a) \qquad \text{or} \qquad D = \sqrt[3]{\frac{80P}{N}} \quad (4b)$$

For lineshafts carrying pulleys:

$$P = \frac{D^3N}{53.5} \quad (5a) \qquad \text{or} \qquad D = \sqrt[3]{\frac{53.5P}{N}} \quad (5b)$$

For small, short shafts:

$$P = \frac{D^3 N}{38} \quad (6a) \qquad \text{or} \qquad D = \sqrt[3]{\frac{38P}{N}} \quad (6b)$$

Shafts that are subjected to shocks, such as sudden starting and stopping, should be given a greater factor of safety resulting in the use of lower allowable stresses than those just mentioned.

Example: What should be the diameter of a lineshaft to transmit 10 horsepower if the shaft is to make 150 revolutions per minute? Using Formula (5b),

$$D = \sqrt[3]{\frac{53.5 \times 10}{150}} = 1.53 \text{ or, say, } 1\tfrac{9}{16} \text{ inches}$$

Example: What horsepower would be transmitted by a short shaft, 2 inches in diameter, carrying two pulleys close to the bearings, if the shaft makes 300 revolutions per minute? Using Formula (6a),

$$P = \frac{2^3 \times 300}{38} = 63 \text{ horsepower}$$

Torsional Strength of Shafting, Calculations in Metric SI Units.—The allowable twisting moment for a shaft of any cross-section such as circular, square, etc., can be calculated from:

$$T = S_s \times Z_p \quad (1)$$

where T = torsional or twisting moment in newton-millimeters; S_s = allowable torsional shearing stress in newtons per square millimeter; and Z_p = polar section modulus in millimeters^3.

For a shaft delivering power of P kilowatts at N revolutions per minute, the twisting moment T being transmitted is:

$$T = \frac{9.55 \times 10^6 P}{N} \quad (2) \qquad \text{or} \qquad T = \frac{10^6 P}{\omega} \quad (2a)$$

where T is in newton-millimeters, and ω = angular velocity in radians per second.

The diameter D of a solid circular shaft required to transmit a given torque T is:

$$D = \sqrt[3]{\frac{5.1T}{S_s}} \quad (3a) \qquad \text{or} \qquad D = \sqrt[3]{\frac{48.7 \times 10^6 P}{NS_s}} \quad (3b)$$

$$\text{or} \qquad D = \sqrt[3]{\frac{5.1 \times 10^6 P}{\omega S_s}} \quad (3c)$$

where D is in millimeters; T is in newton-millimeters; P is power in kilowatts; N = revolutions per minute; S_s = allowable torsional shearing stress in newtons per square millimeter, and ω = angular velocity in radians per second.

If 28 newtons/mm^2 and 59 newtons/mm^2 are taken as the generally allowed stresses for main power-transmitting shafts and small short shafts, respectively, then using these allowable stresses, the power P transmitted by a shaft of diameter D, or the diameter D of a shaft to transmit a given power P may be determined from the following formulas:

For main power-transmitting shafts:

$$P = \frac{D^3N}{1.77 \times 10^6} \quad (4a) \qquad \text{or} \qquad D = \sqrt[3]{\frac{1.77 \times 10^6 P}{N}} \quad (4b)$$

For small, short shafts:

$$P = \frac{D^3N}{0.83 \times 10^6} \quad (5a) \qquad \text{or} \qquad D = \sqrt[3]{\frac{0.83 \times 10^6 P}{N}} \quad (5b)$$

where P is in kilowatts, D is in millimeters, and N = revolutions per minute.

Example: **What should be the diameter of a power-transmitting shaft to transmit 150 kW at 500 rpm?**

$$D = \sqrt[3]{\frac{1.77 \times 10^6 \times 150}{500}} = 81 \text{ millimeters}$$

Example: **What power would a short shaft, 50 millimeters in diameter, transmit at 400 rpm?**

$$P = \frac{50^3 \times 400}{0.83 \times 10^6} = 60 \text{ kilowatts}$$

Polar Moment of Inertia and Section Modulus.—The *polar moment of inertia*, J, of a cross-section with respect to a polar axis, that is, an axis at right angles to the plane of the cross-section, is defined as the moment of inertia of the cross-section with respect to the point of intersection of the axis and the plane. The polar moment of inertia may be found by taking the sum of the moments of inertia about two perpendicular axes lying in the plane of the cross-section and passing through this point. Thus, for example, the polar moment of inertia of a circular or a square area with respect to a polar axis through the center of gravity is equal to two times the moment of inertia with respect to an axis lying in the plane of the cross-section and passing through the center of gravity.

The polar moment of inertia with respect to a polar axis through the center of gravity is required for problems involving the torsional strength of shafts since this axis is usually the axis about which twisting of the shaft takes place.

The *polar section modulus* (also called section modulus of torsion), Z_p, for *circular* sections may be found by dividing the polar moment of inertia, J, by the distance c from the center of gravity to the most remote fiber. This method may be used to find the *approximate* value of the polar section modulus of sections that are *nearly* round. For other than circular cross-sections, however, the polar section modulus *does not* equal the polar moment of inertia divided by the distance c.

The accompanying table gives formulas for the polar section modulus for several different cross-sections. The polar section modulus multiplied by the allowable torsional shearing stress gives the allowable twisting moment to which a shaft may be subjected, see Formula (1).

Torsional Deflection of Circular Shafts.—Shafting must often be proportioned not only to provide the strength required to transmit a given torque, but also to prevent torsional deflection (twisting) through a greater angle than has been found satisfactory for a given type of service.

For a solid circular shaft the torsional deflection in degrees is given by:

$$\alpha = \frac{584Tl}{D^4G} \quad (6)$$

Polar Moment of Inertia and Polar Section Modulus

Section	Polar Moment of Inertia J	Polar Section Modulus Z_p
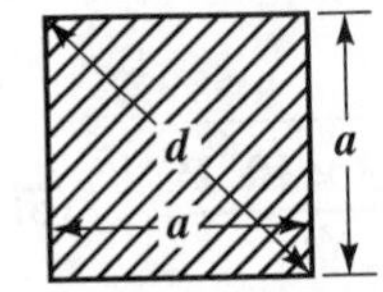 d, a, a	$\frac{a^4}{6} = 0.1667a^4$	$0.208a^3 = 0.074d^3$
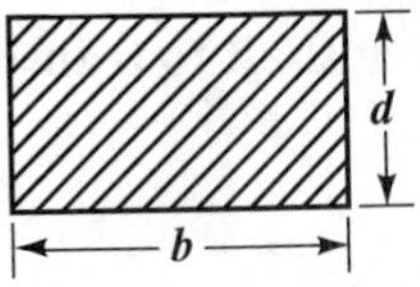 d, b	$\frac{bd(b^2 + d^2)}{12}$	$\frac{bd^2}{3 + 1.8\frac{d}{b}}$ (d is the shorter side)
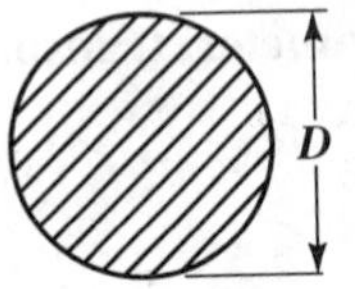 D	$\frac{\pi D^4}{32} = 0.098D^4$ (see also footnote, page 229)	$\frac{\pi D^3}{16} = 0.196D^3$ (see also footnote, page 229)
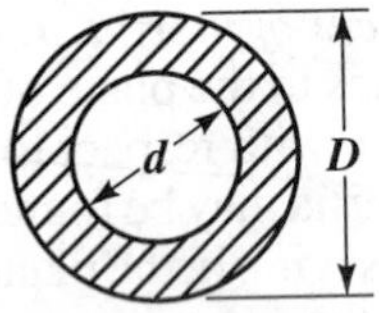 d, D	$\frac{\pi}{32}(D^4 - d^4)$ $= 0.098(D^4 - d^4)$	$\frac{\pi}{16}\left(\frac{D^4 - d^4}{D}\right)$ $= 0.196\left(\frac{D^4 - d^4}{D}\right)$
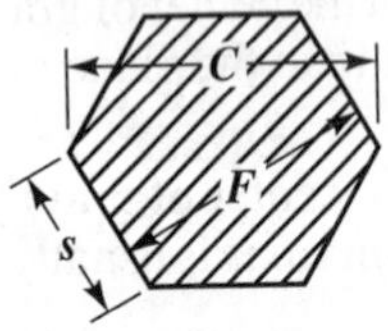 C, F, s	$\frac{5\sqrt{3}}{8}s^4 = 1.0825s^4$ $= 0.12F^4$	$0.20F^3$
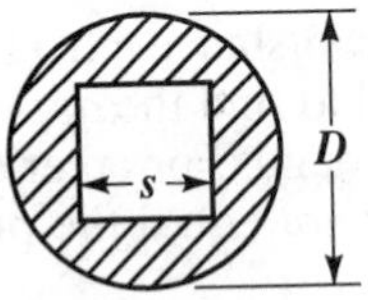 s, D	$\frac{\pi D^4}{32} - \frac{s^4}{6}$ $= 0.098D^4 - 0.167s^4$	$\frac{\pi D^3}{16} - \frac{s^4}{3D}$ $= 0.196D^3 - 0.333\frac{s^4}{D}$
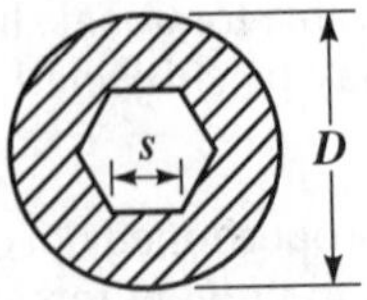 s, D	$\frac{\pi D^4}{32} - \frac{5\sqrt{3}}{8}s^4$ $= 0.098D^4 - 1.0825s^4$	$\frac{\pi D^3}{16} - \frac{5\sqrt{3}}{4D}s^4$ $= 0.196D^3 - 2.165\frac{s^4}{D}$
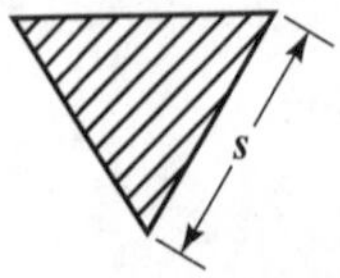 s	$\frac{\sqrt{3}}{48}s^4 = 0.036s^4$	$\frac{s^3}{20} = 0.05s^3$

Example: Find the torsional deflection for a solid steel shaft 4 inches in diameter and 48 inches long, subjected to a twisting moment of 24,000 inch-pounds. By Formula (6),

$$\alpha = \frac{584 \times 24{,}000 \times 48}{4^4 \times 11{,}500{,}000} = 0.23 \text{ degree}$$

Formula (6) can be used with metric SI units, where α = angular deflection of shaft in degrees; T = torsional moment in newton-millimeters; l = length of shaft in millimeters; D = diameter of shaft in millimeters; and G = torsional modulus of elasticity in newtons per square millimeter.

Example: **Find the torsional deflection of a solid steel shaft, 100 mm in diameter and 1300 mm long, subjected to a twisting moment of 3×10^6 newton-millimeters. The torsional modulus of elasticity is 80,000 newtons/mm^2. By Formula (6)**

$$\alpha = \frac{584 \times 3 \times 10^6 \times 1300}{100^4 \times 80{,}000} = 0.285 \text{ degree}$$

The diameter of a shaft that is to have a maximum torsional deflection α is given by:

$$D = 4.9 \times \sqrt[4]{\frac{Tl}{G\alpha}} \quad (7)$$

Formula (7) can be used with metric SI units, where D = diameter of shaft in millimeters; T = torsional moment in newton-millimeters; l = length of shaft in millimeters; G = torsional modulus of elasticity in newtons per square millimeter; and α = angular deflection of shaft in degrees.

According to some authorities, the allowable twist in steel transmission shafting should not exceed 0.08 degree per foot length of the shaft. The diameter D of a shaft that will permit a maximum angular deflection of 0.08 degree per foot of length for a given torque T or for a given horsepower P can be determined from the formulas:

$$D = 0.29\sqrt[4]{T} \quad (8a) \qquad \text{or} \qquad D = 4.6 \times \sqrt[4]{\frac{P}{N}} \quad (8b)$$

Using metric SI units and assuming an allowable twist in steel transmission shafting of 0.26 degree per meter length, Formulas (8a) and (8b) become:

$$D = 2.26\sqrt[4]{T} \qquad \text{or} \qquad D = 125.7 \times \sqrt[4]{\frac{P}{N}}$$

where D = diameter of shaft in millimeters; T = torsional moment in newton-millimeters; P = power in kilowatts; and N = revolutions per minute.

Another rule that has been generally used in mill practice limits the deflection to 1 degree in a length equal to 20 times the shaft diameter. For a given torque or horsepower, the diameter of a shaft having this maximum deflection is given by:

$$D = 0.1\sqrt[3]{T} \quad (9a) \qquad \text{or} \qquad D = 4.0 \times \sqrt[3]{\frac{P}{N}} \quad (9b)$$

Example: Find the diameter of a steel lineshaft to transmit 10 horsepower at 150 revolutions per minute with a torsional deflection not exceeding 0.08 degree per foot of length. By Formula (8b),

$$D = 4.6 \times \sqrt[4]{\frac{10}{150}} = 2.35 \text{ inches}$$

This diameter is larger than that obtained for the same horsepower and rpm in the example given for Formula (5b) in which the diameter was calculated for strength considerations only. The usual procedure in the design of shafting which is to have a specified maximum angular deflection is to compute the diameter first by means of Formulas (7), (8a), (8b), (9a), or (9b) and then by means of Formulas (3a), (3b), (4b), (5b), or (6b), using the larger of the two diameters thus found.

Linear Deflection of Shafting.—For steel lineshafting, it is considered good practice to limit the linear deflection to a maximum of 0.010 inch per foot of length. The maximum distance in feet between bearings, for average conditions, in order to avoid excessive linear deflection, is determined by the formulas:

$$L = 8.95\sqrt[3]{D^2}$$ for shafting subject to no bending action except it's own weight

$$L = 5.2\sqrt[3]{D^2}$$ for shafting subject to bending action of pulleys, etc.

in which D = diameter of shaft in inches and L = maximum distance between bearings in feet. Pulleys should be placed as close to the bearings as possible.

In general, shafting up to three inches in diameter is almost always made from cold-rolled steel. This shafting is true and straight and needs no turning, but if keyways are cut in the shaft, it must usually be straightened afterwards, as the cutting of the keyways relieves the tension on the surface of the shaft produced by the cold-rolling process. Sizes of shafting from three to five inches in diameter may be either cold-rolled or turned, more frequently the latter, and all larger sizes of shafting must be turned because cold-rolled shafting is not available in diameters larger than 5 in.

Diameters, Inches		Minus Tolerances, Inches[a]	Diameters, Inches		Minus Tolerances, Inches[a]	Diameters, Inches		Minus Tolerances, Inches[a]
Transmission Shafting	Machinery Shafting		Transmission Shafting	Machinery Shafting		Transmission Shafting	Machinery Shafting	
	1/2	0.002		1 13/16	0.003		3 3/4	0.004
	9/16	0.002		1 7/8	0.003		3 7/8	0.004
	5/8	0.002	1 15/16	1 15/16	0.003	3 15/16	4	0.004
	11/16	0.002		2	0.003		4 1/4	0.005
	3/4	0.002		2 1/16	0.004	4 7/16	4 1/2	0.005
	13/16	0.002		2 1/8	0.004		4 3/4	0.005
	7/8	0.002	2 3/16	2 3/16	0.004	4 15/16	5	0.005
15/16	15/16	0.002		2 1/4	0.004		5 1/4	0.005
	1	0.002		2 5/16	0.004	5 7/16	5 1/2	0.005
	1 1/16	0.003		2 3/8	0.004		5 3/4	0.005
	1 1/8	0.003	2 7/16	2 7/16	0.004	5 15/16	6	0.005
1 3/16	1 3/16	0.003		2 1/2	0.004		6 1/4	0.006
	1 1/4	0.003		2 5/8	0.004	6 1/2	6 1/2	0.006
	1 5/16	0.003		2 3/4	0.004		6 3/4	0.006
	1 3/8	0.003	2 15/16	2 7/8	0.004	7	7	0.006
1 7/16	1 7/16	0.003		3	0.004		7 1/4	0.006
	1 1/2	0.003		3 1/8	0.004	7 1/2	7 1/2	0.006
	1 9/16	0.003		3 1/4	0.004		7 3/4	0.006
	1 5/8	0.003		3 3/8	0.004	8	8	0.006
1 11/16	1 11/16	0.003	3 7/16	3 1/2	0.004	…	…	…
	1 3/4	0.003		3 5/8	0.004	…	…	…

[a] *Note*:—These tolerances are *negative* or minus and represent the maximum allowable variation *below* the exact nominal size. For instance the maximum diameter of the 1 15/16 inch shaft is 1.938 inch and its minimum allowable diameter is 1.935 inch. Stock lengths of finished transmission shafting shall be: 16, 20 and 24 feet.

Design of Transmission Shafting.—The following guidelines for the design of shafting for transmitting a given amount of power under various conditions of loading are based upon formulas given in the former American Standard ASA B17c Code for the Design of Transmission Shafting. These formulas are based on the *maximum-shear theory* of failure which assumes that the elastic limit of a *ductile* ferrous material in shear is practically one-half its elastic limit in tension. This theory agrees, very nearly, with the results of tests on ductile materials and has gained wide acceptance in practice.

The formulas given apply in all shaft designs including shafts for special machinery. The limitation of these formulas is that they provide only for the strength of shafting and are not concerned with the torsional or lineal deformations which may, in shafts used in machine design, be the controlling factor (see *Torsional Deflection of Circular Shafts* and *Linear Deflection of Shafting* for deflection considerations). In the formulas that follow,

$B = \sqrt[3]{1 \div (1 - K^4)}$ (see Table 3)

D = outside diameter of shaft in inches

D_1 = inside diameter of a hollow shaft in inches

K_m = shock and fatigue factor to be applied in every case to the computed bending moment (see Table 1)

K_t = combined shock and fatigue factor to be applied in every case to the computed torsional moment (see Table 1)

M = maximum bending moment in inch-pounds

N = revolutions per minute

P = maximum power to be transmitted by the shaft in horsepower

p_t = maximum allowable shearing stress under combined loading conditions in pounds per square inch (see Table 2)

S = maximum allowable flexural (bending) stress, in either tension or compression in pounds per square inch (see Table 2)

S_s = maximum allowable torsional shearing stress in pounds per square inch (see Table 2)

T = maximum torsional moment in inch-pounds

V = maximum transverse shearing load in pounds

For shafts subjected to pure torsional loads only,

$$D = B\sqrt[3]{\frac{5.1K_tT}{S_s}} \quad (1a) \qquad \text{or} \qquad D = B\sqrt[3]{\frac{321{,}000K_tP}{S_sN}} \quad (1b)$$

For stationary shafts subjected to bending only,

$$D = B\sqrt[3]{\frac{10.2K_mM}{S}} \quad (2)$$

For shafts subjected to combined torsion and bending,

$$D = B\sqrt[3]{\frac{5.1}{p_t}\sqrt{(K_mM)^2 + (K_tT)^2}} \quad (3a)$$

or

$$D = B\sqrt[3]{\frac{5.1}{p_t}\sqrt{(K_mM)^2 + \left(\frac{63{,}000K_tP}{N}\right)^2}} \quad (3b)$$

Formulas (1a) to (3b) may be used for solid shafts or for hollow shafts. For solid shafts the factor B is equal to 1, whereas for hollow shafts the value of B depends on the value of K which, in turn, depends on the ratio of the inside diameter of the shaft to the outside diameter ($D_1 \div D = K$). Table 3 gives values of B corresponding to various values of K.

For short solid shafts subjected only to heavy transverse shear, the diameter of shaft required is:

$$D = \sqrt{\frac{1.7V}{S_s}} \tag{4}$$

Formulas (1a), (2), (3a) and (4), can be used unchanged with metric SI units. Formula (1b) becomes:

$$D = B\sqrt[3]{\frac{48.7K_tP}{S_sN}}$$ **and Formula (3b) becomes:**

$$D = B\sqrt[3]{\frac{5.1}{p_t}\sqrt{(K_mM)^2 + \left(\frac{9.55K_tP}{N}\right)^2}}$$

Throughout the formulas, D = outside diameter of shaft in millimeters; T = maximum torsional moment in newton-millimeters; S_s = maximum allowable torsional shearing stress in newtons per millimeter squared (see Table 2); P = maximum power to be transmitted in milliwatts; N = revolutions per minute; M = maximum bending moment in newton-millimeters; S = maximum allowable flexural (bending) stress, either in tension or compression in newtons per millimeter squared (see Table 2); p_t = maximum allowable shearing stress under combined loading conditions in newtons per millimeter squared; and V = maximum transverse shearing load in kilograms. The factors K_m, K_t, and B are unchanged, and D_1 = the inside diameter of a hollow shaft in millimeters.

Table 1. Recommended Values of the Combined Shock and Fatigue Factors for Various Types of Load

Type of Load	Stationary Shafts		Rotating Shafts	
	K_m	K_t	K_m	K_t
Gradually applied and steady	1.0	1.0	1.5	1.0
Suddenly applied, minor shocks only	1.5–2.0	1.5–2.0	1.5–2.0	1.0–1.5
Suddenly applied, heavy shocks	…	…	2.0–3.0	1.5–3.0

Table 2. Recommended Maximum Allowable Working Stresses for Shafts Under Various Types of Load

Material	Type of Load		
	Simple Bending	Pure Torsion	Combined Stress
"Commercial Steel" shafting without keyways	$S = 16{,}000$	$S_s = 8000$	$p_t = 8000$
"Commercial Steel" shafting with keyways	$S = 12{,}000$	$S_s = 6000$	$p_t = 6000$
Steel purchased under definite physical specs.	(See note [a])	(See note [b])	(See note [b])

[a] S = 60 per cent of the elastic limit in tension but not more than 36 per cent of the ultimate tensile strength.

[b] S_s and p_t = 30 per cent of the elastic limit in tension but not more than 18 per cent of the ultimate tensile strength.

If the values in the Table are converted to metric SI units, note that 1000 pounds per square inch = 6.895 newtons per square millimeter.

Table 3. Values of the Factor B Corresponding to Various Values of K for Hollow Shafts

$K = \frac{D_1}{D} =$	0.95	0.90	0.85	0.80	0.75	0.70	0.65	0.60	0.55	0.50
$B = \sqrt[3]{1 \div (1 - K^4)}$	1.75	1.43	1.28	1.19	1.14	1.10	1.07	1.05	1.03	1.02

For solid shafts, $B = 1$ since $K = 0$. $[B = \sqrt[3]{1 \div (1 - K^4)} = \sqrt[3]{1 \div (1 - 0)} = 1]$

Effect of Keyways on Shaft Strength.—Keyways cut into a shaft reduce its load carrying ability, particularly when impact loads or stress reversals are involved. To ensure an adequate factor of safety in the design of a shaft with standard keyway (width, one-quarter, and depth, one-eighth of shaft diameter), the former Code for Transmission Shafting tentatively recommended that shafts with keyways be designed on the basis of a solid circular shaft using not more than 75 per cent of the working stress recommended for the solid shaft. See also page 2342.

Formula for Shafts of Brittle Materials.—The preceding formulas are applicable to ductile materials and are based on the maximum-shear theory of failure which assumes that the elastic limit of a *ductile* material in shear is one-half its elastic limit in tension.

Brittle materials are generally stronger in shear than in tension; therefore, the maximum-shear theory is not applicable. The *maximum-normal-stress theory* of failure is now generally accepted for the design of shafts made from brittle materials. A material may be considered to be brittle if its elongation in a 2-inch gage length is less than 5 per cent. Materials such as cast iron, hardened tool steel, hard bronze, etc., conform to this rule. The diameter of a shaft made of a brittle material may be determined from the following formula which is based on the maximum-normal-stress theory of failure:

$$D = B\sqrt[3]{\frac{5.1}{S_t}[(K_m M) + \sqrt{(K_m M)^2 + (K_t T)^2}]}$$

where S_t is the maximum allowable tensile stress in pounds per square inch and the other quantities are as previously defined.

The formula can be used unchanged with metric SI units, where D = outside diameter of shaft in millimeters; S_t = the maximum allowable tensile stress in newtons per millimeter squared; M = maximum bending moment in newton-millimeters; and T = maximum torsional moment in newton-millimeters. The factors K_m, K_t, and B are unchanged.

Critical Speed of Rotating Shafts.—At certain speeds, a rotating shaft will become dynamically unstable and the resulting vibrations and deflections can result in damage not only to the shaft but to the machine of which it is a part. The speeds at which such dynamic instability occurs are called the critical speeds of the shaft. On page 186 are given formulas for the critical speeds of shafts subject to various conditions of loading and support. A shaft may be safely operated either above or below its critical speed, good practice indicating that the operating speed be at least 20 per cent above or below the critical.

The formulas commonly used to determine critical speeds are sufficiently accurate for general purposes. However, the torque applied to a shaft has an important effect on its critical speed. Investigations have shown that the critical speeds of a uniform shaft are decreased as the applied torque is increased, and that there exist critical torques which will reduce the corresponding critical speed of the shaft to zero. A detailed analysis of the effects of applied torques on critical speeds may be found in a paper. "Critical Speeds of Uniform Shafts under Axial Torque," by Golomb and Rosenberg presented at the First U.S. National Congress of Applied Mechanics in 1951.

Comparison of Hollow and Solid Shafting with Same Outside Diameter.—The table that follows gives the per cent decrease in strength and weight of a hollow shaft relative to the strength and weight of a solid shaft of the same diameter. The upper figures in each line give the per cent decrease in strength and the lower figures give the per cent decrease in weight.

Example: A 4-inch shaft, with a 2-inch hole through it, has a weight 25 per cent less than a solid 4-inch shaft, but its strength is decreased only 6.25 per cent.

Comparative Torsional Strengths and Weights of Hollow and Solid Shafting with Same Outside Diameter

Diam. of Solid and Hollow Shaft, Inches	Diameter of Axial Hole in Hollow Shaft, Inches									
	1	1¼	1½	1¾	2	2½	3	3½	4	4½
1½	19.76	48.23	…	…	…	…	…	…	…	…
	44.44	69.44	…	…	…	…	…	…	…	…
1¾	10.67	26.04	53.98	…	…	…	…	…	…	…
	32.66	51.02	73.49	…	…	…	…	…	…	…
2	6.25	15.26	31.65	58.62	…	…	…	…	…	…
	25.00	39.07	56.25	76.54	…	…	…	…	…	…
2¼	3.91	9.53	19.76	36.60	62.43	…	…	…	…	…
	19.75	30.87	44.44	60.49	79.00	…	…	…	…	…
2½	2.56	6.25	12.96	24.01	40.96	…	…	…	…	…
	16.00	25.00	36.00	49.00	64.00	…	…	…	…	…
2¾	1.75	4.28	8.86	16.40	27.98	68.30	…	…	…	…
	13.22	20.66	29.74	40.48	52.89	82.63	…	…	…	…
3	1.24	3.01	6.25	11.58	19.76	48.23	…	…	…	…
	11.11	17.36	25.00	34.01	44.44	69.44	…	…	…	…
3¼	0.87	2.19	4.54	8.41	14.35	35.02	72.61	…	…	…
	9.46	14.80	21.30	29.00	37.87	59.17	85.22	…	…	…
3½	0.67	1.63	3.38	6.25	10.67	26.04	53.98	…	…	…
	8.16	12.76	18.36	25.00	32.66	51.02	73.49	…	…	…
3¾	0.51	1.24	2.56	4.75	8.09	19.76	40.96	75.89	…	…
	7.11	11.11	16.00	21.77	28.45	44.44	64.00	87.10	…	…
4	0.40	0.96	1.98	3.68	6.25	15.26	31.65	58.62	…	…
	6.25	9.77	14.06	19.14	25.00	39.07	56.25	76.56	…	…
4¼	0.31	0.74	1.56	2.89	4.91	11.99	24.83	46.00	78.47	…
	5.54	8.65	12.45	16.95	22.15	34.61	49.85	67.83	88.59	…
4½	0.25	0.70	1.24	2.29	3.91	9.53	19.76	36.60	62.43	…
	4.94	7.72	11.11	15.12	19.75	30.87	44.44	60.49	79.00	…
4¾	0.20	0.50	1.00	1.85	3.15	7.68	15.92	29.48	50.29	80.56
	4.43	6.93	9.97	13.57	17.73	27.70	39.90	54.29	70.91	89.75
5	0.16	0.40	0.81	1.51	2.56	6.25	12.96	24.01	40.96	65.61
	4.00	6.25	8.10	12.25	16.00	25.00	36.00	49.00	64.00	81.00
5½	0.11	0.27	0.55	1.03	1.75	4.27	8.86	16.40	27.98	44.82
	3.30	5.17	7.43	10.12	13.22	20.66	29.76	40.48	52.89	66.94
6	0.09	0.19	0.40	0.73	1.24	3.02	6.25	11.58	19.76	31.65
	2.77	4.34	6.25	8.50	11.11	17.36	25.00	34.02	44.44	56.25
6½	0.06	0.14	0.29	0.59	0.90	2.19	4.54	8.41	14.35	23.98
	2.36	3.70	5.32	7.24	9.47	14.79	21.30	28.99	37.87	47.93
7	0.05	0.11	0.22	0.40	0.67	1.63	3.38	6.25	10.67	17.08
	2.04	3.19	4.59	6.25	8.16	12.76	18.36	25.00	32.66	41.33
7½	0.04	0.08	0.16	0.30	0.51	1.24	2.56	4.75	8.09	12.96
	1.77	2.77	4.00	5.44	7.11	11.11	16.00	21.77	28.45	36.00
8	0.03	0.06	0.13	0.23	0.40	0.96	1.98	3.68	6.25	10.02
	1.56	2.44	3.51	4.78	6.25	9.77	14.06	19.14	25.00	31.64

The upper figures in each line give number of per cent decrease in strength; the lower figures give per cent decrease in weight.

SPRINGS*

Springs

Introduction.—Many advances have been made in the spring industry in recent years. For example: developments in materials permit longer fatigue life at higher stresses; simplified design procedures reduce the complexities of design, and improved methods of manufacture help to speed up some of the complicated fabricating procedures and increase production. New types of testing instruments and revised tolerances also permit higher standards of accuracy. Designers should also consider the possibility of using standard springs now available from stock. They can be obtained from spring manufacturing companies located in different areas, and small shipments usually can be made quickly.

Designers of springs require information in the following order of precedence to simplify design procedures.

1) Spring materials and their applications

2) Allowable spring stresses

3) Spring design data with tables of spring characteristics, tables of formulas, and tolerances.

Only the more commonly used types of springs are covered in detail here. Special types and designs rarely used such as torsion bars, volute springs, Belleville washers, constant force, ring and spiral springs and those made from rectangular wire are only described briefly.

Notation.—The following symbols are used in spring equations:

AC = Active coils
b = Widest width of rectangular wire, inches
CL = Compressed length, inches
D = Mean coil diameter, inches $= OD - d$
d = Diameter of wire or side of square, inches
E = Modulus of elasticity in tension, pounds per square inch
F = Deflection, for N coils, inches
$F°$ = Deflection, for N coils, rotary, degrees
f = Deflection, for one active coil
FL = Free length, unloaded spring, inches
G = Modulus of elasticity in torsion, pounds per square inch
IT = Initial tension, pounds
K = Curvature stress correction factor
L = Active length subject to deflection, inches
N = Number of active coils, total
P = Load, pounds
p = pitch, inches
R = Distance from load to central axis, inches
S or S_t = Stress, torsional, pounds per square inch
S_b = Stress, bending, pounds per square inch
SH = Solid height
S_{it} = Stress, torsional, due to initial tension, pounds per square inch
T = Torque $= P \times R$, pound-inches
TC = Total coils
t = Thickness, inches
U = Number of revolutions $= F°/360°$

* This section was compiled by Harold Carlson, P. E., Consulting Engineer, Lakewood, N.J.

Spring Materials

The spring materials most commonly used include high-carbon spring steels, alloy spring steels, stainless spring steels, copper-base spring alloys, and nickel-base spring alloys.

High-Carbon Spring Steels in Wire Form.—These spring steels are the most commonly used of all spring materials because they are the least expensive, are easily worked, and are readily available. However, they are not satisfactory for springs operating at high or low temperatures or for shock or impact loading. The following wire forms are available:

Music Wire, ASTM A228 (0.80–0.95 per cent carbon): This is the most widely used of all spring materials for small springs operating at temperatures up to about 250 degrees F. It is tough, has a high tensile strength, and can withstand high stresses under repeated loading. The material is readily available in round form in diameters ranging from 0.005 to 0.125 inch and in some larger sizes up to $\frac{3}{16}$ inch. It is not available with high tensile strengths in square or rectangular sections. Music wire can be plated easily and is obtainable pretinned or preplated with cadmium, but plating after spring manufacture is usually preferred for maximum corrosion resistance.

Oil-Tempered MB Grade, ASTM A229 (0.60–0.70 per cent carbon): This general-purpose spring steel is commonly used for many types of coil springs where the cost of music wire is prohibitive and in sizes larger than are available in music wire. It is readily available in diameters ranging from 0.125 to 0.500 inch, but both smaller and larger sizes may be obtained. The material should not be used under shock and impact loading conditions, at temperatures above 350 degrees F., or at temperatures in the sub-zero range. Square and rectangular sections of wire are obtainable in fractional sizes. Annealed stock also can be obtained for hardening and tempering after coiling. This material has a heat-treating scale that must be removed before plating.

Oil-Tempered HB Grade, SAE 1080 (0.75–0.85 per cent carbon): This material is similar to the MB Grade except that it has a higher carbon content and a higher tensile strength. It is obtainable in the same sizes and is used for more accurate requirements than the MB Grade, but is not so readily available. In lieu of using this material it may be better to use an alloy spring steel, particularly if a long fatigue life or high endurance properties are needed. Round and square sections are obtainable in the oil-tempered or annealed conditions.

Hard-Drawn MB Grade, ASTM A227 (0.60–0.70 per cent carbon): This grade is used for general-purpose springs where cost is the most important factor. Although increased use in recent years has resulted in improved quality, it is best not to use it where long life and accuracy of loads and deflections are important. It is available in diameters ranging from 0.031 to 0.500 inch and in some smaller and larger sizes also. The material is available in square sections but at reduced tensile strengths. It is readily plated. Applications should be limited to those in the temperature range of 0 to 250 degrees F.

High-Carbon Spring Steels in Flat Strip Form.—Two types of thin, flat, high-carbon spring steel strip are most widely used although several other types are obtainable for specific applications in watches, clocks, and certain instruments. These two compositions are used for over 95 per cent of all such applications. Thin sections of these materials under 0.015 inch having a carbon content of over 0.85 per cent and a hardness of over 47 on the Rockwell C scale are susceptible to hydrogen-embrittlement even though special plating and heating operations are employed. The two types are described as follows:

Cold-Rolled Spring Steel, Blue-Tempered or Annealed, SAE 1074, also 1064, and 1070 (0.60 to 0.80 per cent carbon): This very popular spring steel is available in thicknesses ranging from 0.005 to 0.062 inch and in some thinner and thicker sections. The material is available in the annealed condition for forming in 4-slide machines and in presses, and can readily be hardened and tempered after forming. It is also available in the heat-treated or blue-tempered condition. The steel is obtainable in several finishes such as straw color, blue color, black, or plain. Hardnesses ranging from 42 to 46 Rockwell C are recommended for spring applications. Uses include spring clips, flat springs, clock springs, and motor, power, and spiral springs.

Cold-Rolled Spring Steel, Blue-Tempered Clock Steel, SAE 1095 (0.90 to 1.05 per cent carbon): This popular type should be used principally in the blue-tempered condition. Although obtainable in the annealed condition, it does not always harden properly during heat-treatment as it is a "shallow" hardening type. It is used principally in clocks and motor springs. End sections of springs made from this steel are annealed for bending or piercing operations. Hardnesses usually range from 47 to 51 Rockwell C.

Other materials available in strip form and used for flat springs are brass, phosphor-bronze, beryllium-copper, stainless steels, and nickel alloys.

Alloy Spring Steels.—These spring steels are used for conditions of high stress, and shock or impact loadings. They can withstand both higher and lower temperatures than the high-carbon steels and are obtainable in either the annealed or pretempered conditions.

Chromium Vanadium, ASTM A231: This very popular spring steel is used under conditions involving higher stresses than those for which the high-carbon spring steels are recommended and is also used where good fatigue strength and endurance are needed. It behaves well under shock and impact loading. The material is available in diameters ranging from 0.031 to 0.500 inch and in some larger sizes also. In square sections it is available in fractional sizes. Both the annealed and pretempered types are available in round, square, and rectangular sections. It is used extensively in aircraft-engine valve springs and for springs operating at temperatures up to 425 degrees F.

Silicon Manganese: This alloy steel is quite popular in Great Britain. It is less expensive than chromium-vanadium steel and is available in round, square, and rectangular sections in both annealed and pretempered conditions in sizes ranging from 0.031 to 0.500 inch. It was formerly used for knee-action springs in automobiles. It is used in flat leaf springs for trucks and as a substitute for more expensive spring steels.

Chromium Silicon, ASTM A401: This alloy is used for highly stressed springs that require long life and are subjected to shock loading. It can be heat-treated to higher hardnesses than other spring steels so that high tensile strengths are obtainable. The most popular sizes range from 0.031 to 0.500 inch in diameter. Very rarely are square, flat, or rectangular sections used. Hardnesses ranging from 50 to 53 Rockwell C are quite common and the alloy may be used at temperatures up to 475 degrees F. This material is usually ordered specially for each job.

Stainless Spring Steels.—The use of stainless spring steels has increased and several compositions are available all of which may be used for temperatures up to 550 degrees F. They are all corrosion resistant. Only the stainless 18-8 compositions should be used at sub-zero temperatures.

Stainless Type 302, ASTM A313 (18 per cent chromium, 8 per cent nickel): This stainless spring steel is very popular because it has the highest tensile strength and quite uniform properties. It is cold-drawn to obtain its mechanical properties and cannot be hardened by heat treatment. This material is nonmagnetic only when fully annealed and becomes slightly magnetic due to the cold-working performed to produce spring properties. It is suitable for use at temperatures up to 550 degrees F. and for sub-zero temperatures. It is very corrosion resistant. The material best exhibits its desirable mechanical properties in diameters ranging from 0.005 to 0.1875 inch although some larger diameters are available. It is also available as hard-rolled flat strip. Square and rectangular sections are available but are infrequently used.

Stainless Type 304, ASTM A313 (18 per cent chromium, 8 per cent nickel): This material is quite similar to Type 302, but has better bending properties and about 5 per cent lower tensile strength. It is a little easier to draw, due to the slightly lower carbon content.

Stainless Type 316, ASTM A313 (18 per cent chromium, 12 per cent nickel, 2 per cent molybdenum): This material is quite similar to Type 302 but is slightly more corrosion resistant because of its higher nickel content. Its tensile strength is 10 to 15 per cent lower than Type 302. It is used for aircraft springs.

Stainless Type 17-7 PH ASTM A313 (17 per cent chromium, 7 per cent nickel): This alloy, which also contains small amounts of aluminum and titanium, is formed in a moderately hard state and then precipitation hardened at relatively low temperatures for several hours to produce tensile strengths nearly comparable to music wire. This material is not readily available in all sizes, and has limited applications due to its high manufacturing cost.

Stainless Type 414, SAE 51414 (12 per cent chromium, 2 per cent nickel): This alloy has tensile strengths about 15 per cent lower than Type 302 and can be hardened by heat-treatment. For best corrosion resistance it should be highly polished or kept clean. It can be obtained hard drawn in diameters up to 0.1875 inch and is commonly used in flat cold-rolled strip for stampings. The material is not satisfactory for use at low temperatures.

Stainless Type 420, SAE 51420 (13 per cent chromium): This is the best stainless steel for use in large diameters above 0.1875 inch and is frequently used in smaller sizes. It is formed in the annealed condition and then hardened and tempered. It does not exhibit its stainless properties until after it is hardened. Clean bright surfaces provide the best corrosion resistance, therefore the heat-treating scale must be removed. Bright hardening methods are preferred.

Stainless Type 431, SAE 51431 (16 per cent chromium, 2 per cent nickel): This spring alloy acquires high tensile properties (nearly the same as music wire) by a combination of heat-treatment to harden the wire plus cold-drawing after heat-treatment. Its corrosion resistance is not equal to Type 302.

Copper-Base Spring Alloys.—Copper-base alloys are important spring materials because of their good electrical properties combined with their good resistance to corrosion. Although these materials are more expensive than the high-carbon and the alloy steels, they nevertheless are frequently used in electrical components and in sub-zero temperatures.

Spring Brass, ASTM B 134 (70 per cent copper, 30 per cent zinc): This material is the least expensive and has the highest electrical conductivity of the copper-base alloys. It has a low tensile strength and poor spring qualities, but is extensively used in flat stampings and where sharp bends are needed. It cannot be hardened by heat-treatment and should not be used at temperatures above 150 degrees F., but is especially good at sub-zero temperatures. Available in round sections and flat strips, this hard-drawn material is usually used in the "spring hard" temper.

Phosphor Bronze, ASTM B 159 (95 per cent copper, 5 per cent tin): This alloy is the most popular of this group because it combines the best qualities of tensile strength, hardness, electrical conductivity, and corrosion resistance with the least cost. It is more expensive than brass, but can withstand stresses 50 per cent higher.The material cannot be hardened by heat-treatment. It can be used at temperatures up to 212 degrees F. and at sub-zero temperatures. It is available in round sections and flat strip, usually in the "extra-hard" or "spring hard" tempers. It is frequently used for contact fingers in switches because of its low arcing properties. An 8 per cent tin composition is used for flat springs and a superfine grain composition called "Duraflex," has good endurance properties.

Beryllium Copper, ASTM B 197 (98 per cent copper, 2 per cent beryllium): This alloy can be formed in the annealed condition and then precipitation hardened after forming at temperatures around 600 degrees F, for 2 to 3 hours. This treatment produces a high hardness combined with a high tensile strength. After hardening, the material becomes quite brittle and can withstand very little or no forming. It is the most expensive alloy in the group and heat-treating is expensive due to the need for holding the parts in fixtures to prevent distortion. The principal use of this alloy is for carrying electric current in switches and in electrical components. Flat strip is frequently used for contact fingers.

Nickel-Base Spring Alloys.—Nickel-base alloys are corrosion resistant, withstand both elevated and sub-zero temperatures, and their non-magnetic characteristic makes them useful for such applications as gyroscopes, chronoscopes, and indicating instruments. These materials have a high electrical resistance and therefore should not be used for conductors of electrical current.

Monel (67 per cent nickel, 30 per cent copper):* This material is the least expensive of the nickel-base alloys. It also has the lowest tensile strength but is useful due to its resistance to the corrosive effects of sea water and because it is nearly non-magnetic. The alloy can be subjected to stresses slightly higher than phosphor bronze and nearly as high as beryllium copper. Its high tensile strength and hardness are obtained as a result of cold-drawing and cold-rolling only, since it can not be hardened by heat-treatment. It can be used at temperatures ranging from −100 to +425 degrees F. at normal operating stresses and is available in round wires up to 3⁄16 inch in diameter with quite high tensile strengths. Larger diameters and flat strip are available with lower tensile strengths.

*"K" Monel * (66 per cent nickel, 29 per cent copper, 3 per cent aluminum):* This material is quite similar to Monel except that the addition of the aluminum makes it a precipitation-hardening alloy. It may be formed in the soft or fairly hard condition and then hardened by a long-time age-hardening heat-treatment to obtain a tensile strength and hardness above Monel and nearly as high as stainless steel. It is used in sizes larger than those usually used with Monel, is non-magnetic and can be used in temperatures ranging from − 100 to + 450 degrees F. at normal working stresses under 45,000 pounds per square inch.

Inconel (78 per cent nickel, 14 per cent chromium, 7 per cent iron):* This is one of the most popular of the non-magnetic nickel-base alloys because of its corrosion resistance and because it can be used at temperatures up to 700 degrees F. It is more expensive than stainless steel but less expensive than beryllium copper. Its hardness and tensile strength is higher than that of "K" Monel and is obtained as a result of cold-drawing and cold-rolling only. It cannot be hardened by heat treatment. Wire diameters up to 1⁄4 inch have the best tensile properties. It is often used in steam valves, regulating valves, and for springs in boilers, compressors, turbines, and jet engines.

Inconel "X" (70 per cent nickel, 16 per cent chromium, 7 per cent iron):* This material is quite similar to Inconel but the small amounts of titanium, columbium and aluminum in its composition make it a precipitation-hardening alloy. It can be formed in the soft or partially hard condition and then hardened by holding it at 1200 degrees F. for 4 hours. It is non-magnetic and is used in larger sections than Inconel. This alloy is used at temperatures up to 850 degrees F. and at stresses up to 55,000 pounds per square inch.

Duranickel ("Z" Nickel) (98 per cent nickel):* This alloy is non-magnetic, corrosion resistant, has a high tensile strength and is hardenable by precipitation hardening at 900 degrees F. for 6 hours. It may be used at the same stresses as Inconel but should not be used at temperatures above 500 degrees F.

Nickel-Base Spring Alloys with Constant Moduli of Elasticity.—Some special nickel alloys have a constant modulus of elasticity over a wide temperature range. These materials are especially useful where springs undergo temperature changes and must exhibit uniform spring characteristics. These materials have a low or zero thermo-elastic coefficient

* Trade name of the International Nickel Company.

and therefore do not undergo variations in spring stiffness because of modulus changes due to temperature differentials. They also have low hysteresis and creep values which makes them preferred for use in food-weighing scales, precision instruments, gyroscopes, measuring devices, recording instruments and computing scales where the temperature ranges from − 50 to + 150 degrees F. These materials are expensive, none being regularly stocked in a wide variety of sizes. They should not be specified without prior discussion with spring manufacturers because some suppliers may not fabricate springs from these alloys due to the special manufacturing processes required. All of these alloys are used in small wire diameters and in thin strip only and are covered by U.S. patents. They are more specifically described as follows:

Elinvar (nickel, iron, chromium):* This alloy, the first constant-modulus alloy used for hairsprings in watches, is an austenitic alloy hardened only by cold-drawing and cold-rolling. Additions of titanium, tungsten, molybdenum and other alloying elements have brought about improved characteristics and precipitation-hardening abilities. These improved alloys are known by the following trade names: Elinvar Extra, Durinval, Modulvar and Nivarox.

Ni-Span C (nickel, iron, chromium, titanium):* This very popular constant-modulus alloy is usually formed in the 50 per cent cold-worked condition and precipitation-hardened at 900 degrees F. for 8 hours, although heating up to 1250 degrees F. for 3 hours produces hardnesses of 40 to 44 Rockwell C, permitting safe torsional stresses of 60,000 to 80,000 pounds per square inch. This material is ferromagnetic up to 400 degrees F; above that temperature it becomes non-magnetic.

Iso-Elastic† (nickel, iron, chromium, molybdenum): This popular alloy is relatively easy to fabricate and is used at safe torsional stresses of 40,000 to 60,000 pounds per square inch and hardnesses of 30 to 36 Rockwell C. It is used principally in dynamometers, instruments, and food-weighing scales.

Elgiloy‡ (nickel, iron, chromium, cobalt): This alloy, also known by the trade names 8J Alloy, Durapower, and Cobenium, is a non-magnetic alloy suitable for sub-zero temperatures and temperatures up to about 1000 degrees F., provided that torsional stresses are kept under 75,000 pounds per square inch. It is precipitation-hardened at 900 degrees F. for 8 hours to produce hardnesses of 48 to 50 Rockwell C. The alloy is used in watch and instrument springs.

*Dynavar** (nickel, iron, chromium, cobalt):* This alloy is a non-magnetic, corrosion-resistant material suitable for sub-zero temperatures and temperatures up to about 750 degrees F., provided that torsional stresses are kept below 75,000 pounds per square inch. It is precipitation-hardened to produce hardnesses of 48 to 50 Rockwell C and is used in watch and instrument springs.

Spring Stresses

Allowable Working Stresses for Springs.—The safe working stress for any particular spring depends to a large extent on the following items:

1) Type of spring — whether compression, extension, torsion, etc.;
2) Size of spring — small or large, long or short;
3) Spring material;
4) Size of spring material;
5) Type of service — light, average, or severe;
6) Stress range — low, average, or high;

* Trade name of Soc. Anon. de Commentry Fourchambault et Decazeville, Paris, France.

† Trade name of John Chatillon & Sons.

‡ Trade name of Elgin National Watch Company.

** Trade name of Hamilton Watch Company.

7) Loading — static, dynamic, or shock;

8) Operating temperature;

9) Design of spring — spring index, sharp bends, hooks.

Consideration should also be given to other factors that affect spring life: corrosion, buckling, friction, and hydrogen embrittlement decrease spring life; manufacturing operations such as high-heat stress-equalizing, presetting, and shot-peening increase spring life.

Item 5, the type of service to which a spring is subjected, is a major factor in determining a safe working stress once consideration has been given to type of spring, kind and size of material, temperature, type of loading, and so on. The types of service are:

Light Service: This includes springs subjected to static loads or small deflections and seldom-used springs such as those in bomb fuses, projectiles, and safety devices. This service is for 1,000 to 10,000 deflections.

Average Service: This includes springs in general use in machine tools, mechanical products, and electrical components. Normal frequency of deflections not exceeding 18,000 per hour permit such springs to withstand 100,000 to 1,000,000 deflections.

Severe Service: This includes springs subjected to rapid deflections over long periods of time and to shock loading such as in pneumatic hammers, hydraulic controls and valves. This service is for 1,000,000 deflections, and above. Lowering the values 10 per cent permits 10,000,000 deflections.

Figs. 1 through 6 show curves that relate the three types of service conditions to allowable working stresses and wire sizes for compression and extension springs, and safe values are provided. Figs. 7 through 10 provide similar information for helical torsion springs. In each chart, the values obtained from the curves may be increased by 20 per cent (but not beyond the top curves on the charts if permanent set is to be avoided) for springs that are baked, and shot-peened, and compression springs that are pressed. Springs stressed slightly above the Light Service curves will take a permanent set.

A curvature correction factor is included in all curves, and is used in spring design calculations (see examples beginning page 300). The curves may be used for materials other than those designated in Figs. 1 through 10, by applying multiplication factors as given in Table 1.

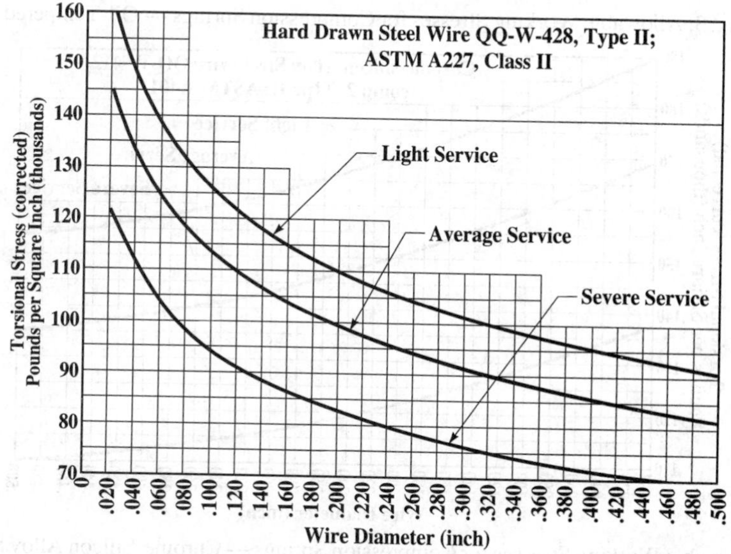

Fig. 1. Allowable Working Stresses for Compression Springs — Hard Drawn Steel Wire[a]

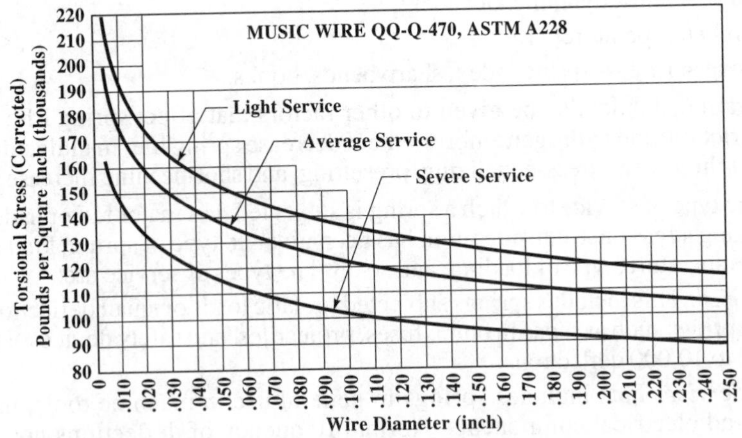

Fig. 2. Allowable Working Stresses for Compression Springs — Music Wire[a]

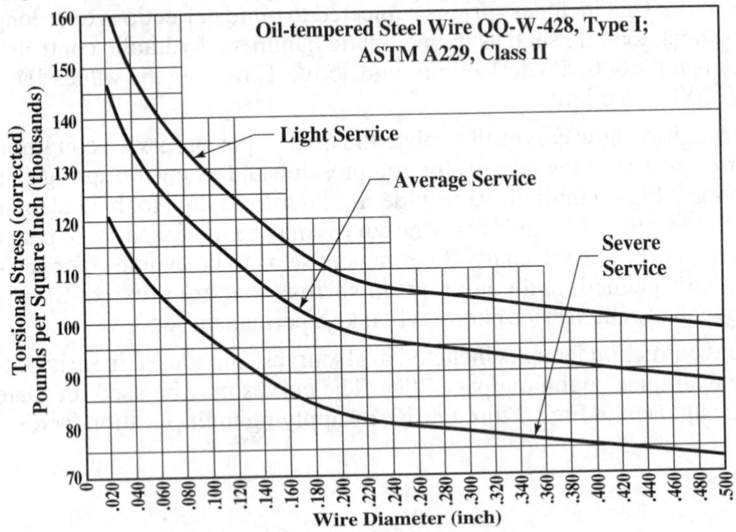

Fig. 3. Allowable Working Stresses for Compression Springs — Oil-Tempered[a]

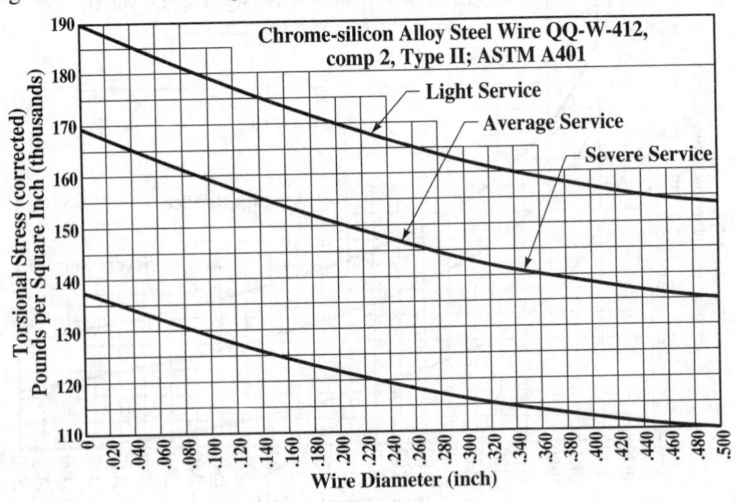

Fig. 4. Allowable Working Stresses for Compression Springs — Chrome-Silicon Alloy Steel Wire[a]

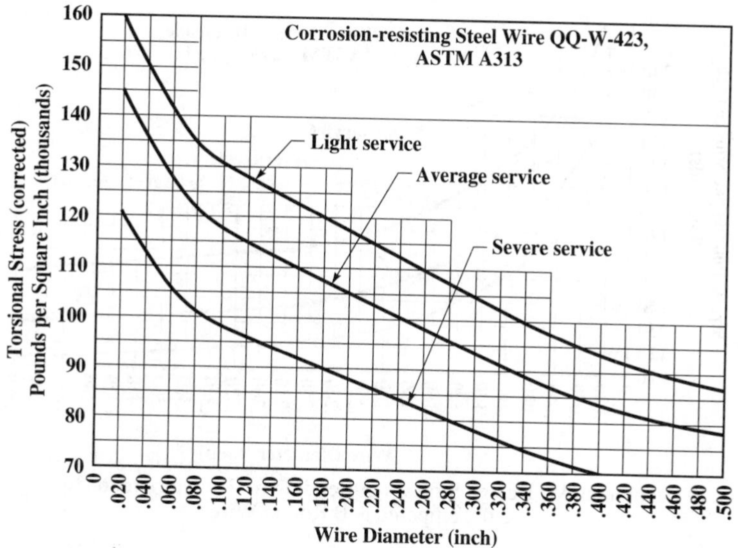

Fig. 5. Allowable Working Stresses for Compression Springs — Corrosion-Resisting Steel Wire[a]

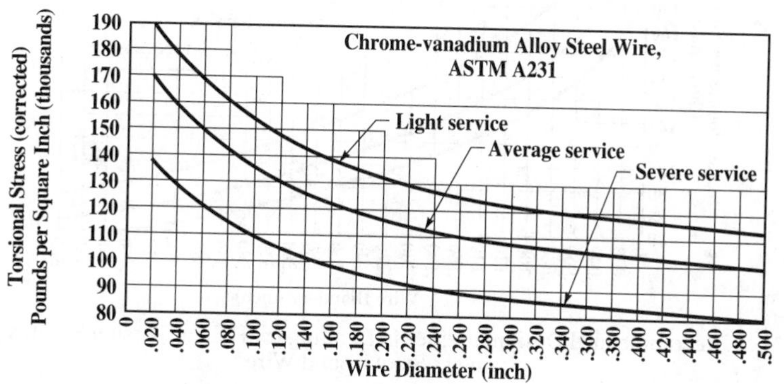

Fig. 6. Allowable Working Stresses for Compression Springs — Chrome-Vanadium Alloy Steel Wire[a]

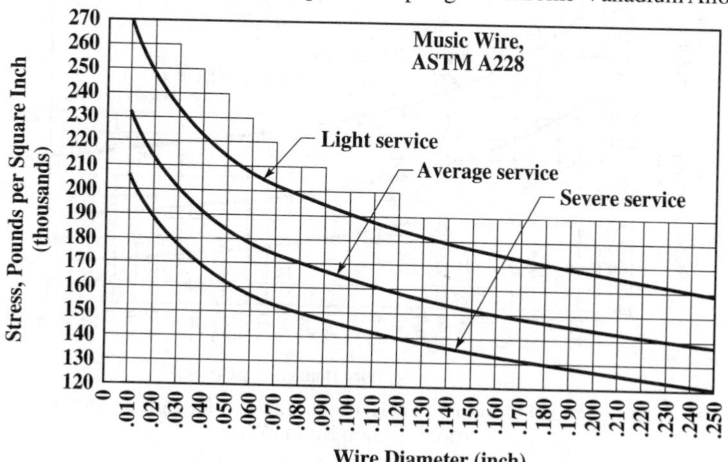

Fig. 7. Recommended Design Stresses in Bending for Helical Torsion Springs — Round Music Wire

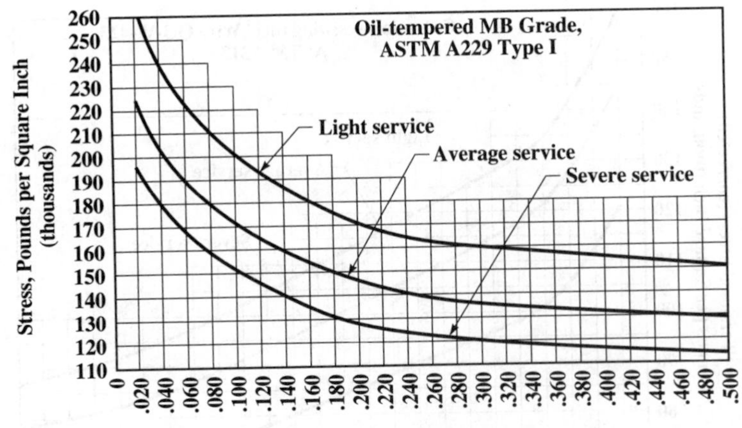

Fig. 8. Recommended Design Stresses in Bending for Helical Torsion Springs — Oil-Tempered MB Round Wire

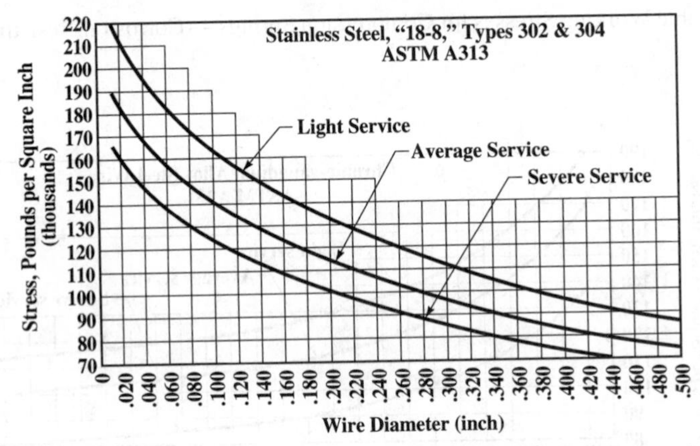

Fig. 9. Recommended Design Stresses in Bending for Helical Torsion Springs — Stainless Steel Round Wire

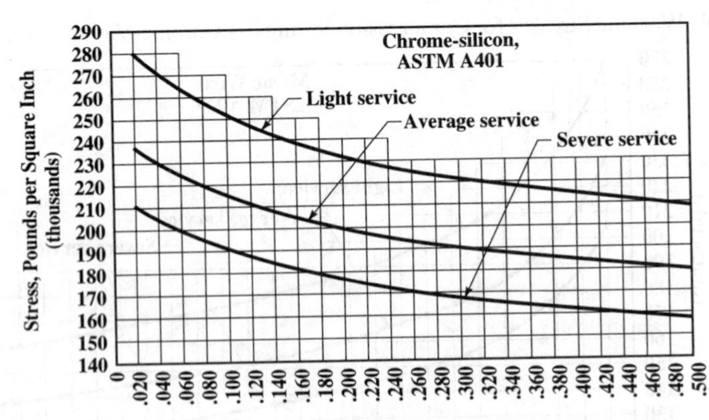

Fig. 10. Recommended Design Stresses in Bending for Helical Torsion Springs — Chrome-Silicon Round Wire

[a] Although Figs. 1 through 6 are for compression springs, they may also be used for extension springs; for extension springs, *reduce* the values obtained from the curves by 10 to 15 per cent.

Table 1. Correction Factors for Other Materials

Compression and Tension Springs			
Material	Factor	Material	Factor
Silicon-manganese	Multiply the values in the chromium-vanadium curves (Fig. 6) by 0.90	Stainless Steel, 316	Multiply the values in the corrosion-resisting steel curves (Fig. 5) by 0.90
Valve-spring quality wire	Use the values in the chromium-vanadium curves (Fig. 6)		
Stainless Steel, 304 and 420	Multiply the values in the corrosion-resisting steel curves (Fig. 5) by 0.95	Stainless Steel, 431 and 17-7PH	Multiply the values in the music wire curves (Fig. 2) by 0.90

Helical Torsion Springs			
Material	Factor[a]	Material	Factor[a]
Hard Drawn MB	0.70	Stainless Steel, 431	
Stainless Steel, 316		Up to $\frac{1}{32}$ inch diameter	0.80
Up to $\frac{1}{32}$ inch diameter	0.75	Over $\frac{1}{32}$ to $\frac{1}{16}$ inch	0.85
Over $\frac{1}{32}$ to $\frac{3}{16}$ inch	0.70	Over $\frac{1}{16}$ to $\frac{1}{8}$ inch	0.95
Over $\frac{3}{16}$ to $\frac{1}{4}$ inch	0.65	Over $\frac{1}{8}$ inch	1.00
Over $\frac{1}{4}$ inch	0.50	Chromium-Vanadium	
Stainless Steel, 17-7 PH		Up to $\frac{1}{16}$ inch diameter	1.05
Up to $\frac{1}{8}$ inch diameter	1.00	Over $\frac{1}{16}$ inch	1.10
Over $\frac{1}{8}$ to $\frac{3}{16}$ inch	1.07	Phosphor Bronze	
Over $\frac{3}{16}$ inch	1.12	Up to $\frac{1}{8}$ inch diameter	0.45
Stainless Steel, 420		Over $\frac{1}{8}$ inch	0.55
Up to $\frac{1}{32}$ inch diameter	0.70	Beryllium Copper[b]	
Over $\frac{1}{32}$ to $\frac{1}{16}$ inch	0.75	Up to $\frac{1}{32}$ inch diameter	0.55
Over $\frac{1}{16}$ to $\frac{1}{8}$ inch	0.80	Over $\frac{1}{32}$ to $\frac{1}{16}$ inch	0.60
Over $\frac{1}{8}$ to $\frac{3}{16}$ inch	0.90	Over $\frac{1}{16}$ to $\frac{1}{8}$ inch	0.70
Over $\frac{3}{16}$ inch	1.00	Over $\frac{1}{8}$ inch	0.80

[a] Multiply the values in the curves for oil-tempered MB grade ASTM A229 Type 1 steel (Fig. 8) by these factors to obtain required values.

[b] Hard drawn and heat treated after coiling.

For use with design stress curves shown in Fig. 2, 5, 6, and 8.

Endurance Limit for Spring Materials.—When a spring is deflected continually it will become "tired" and fail at a stress far below its elastic limit. This type of failure is called *fatigue failure* and usually occurs without warning. *Endurance limit* is the highest stress, or range of stress, in pounds per square inch that can be repeated indefinitely without failure of the spring. Usually ten million cycles of deflection is called "infinite life" and is satisfactory for determining this limit.

For severely worked springs of long life, such as those used in automobile or aircraft engines and in similar applications, it is best to determine the allowable working stresses by referring to the endurance limit curves seen in Fig. 11. These curves are based principally upon the range or difference between the stress caused by the first or initial load and the stress caused by the final load. Experience with springs designed to stresses within the limits of these curves indicates that they should have infinite or unlimited fatigue life. All values include Wahl curvature correction factor. The stress ranges shown may be increased 20 to 30 per cent for springs that have been properly heated, pressed to remove set, and then shot peened, provided that the increased values are lower than the torsional elastic limit by at least 10 per cent.

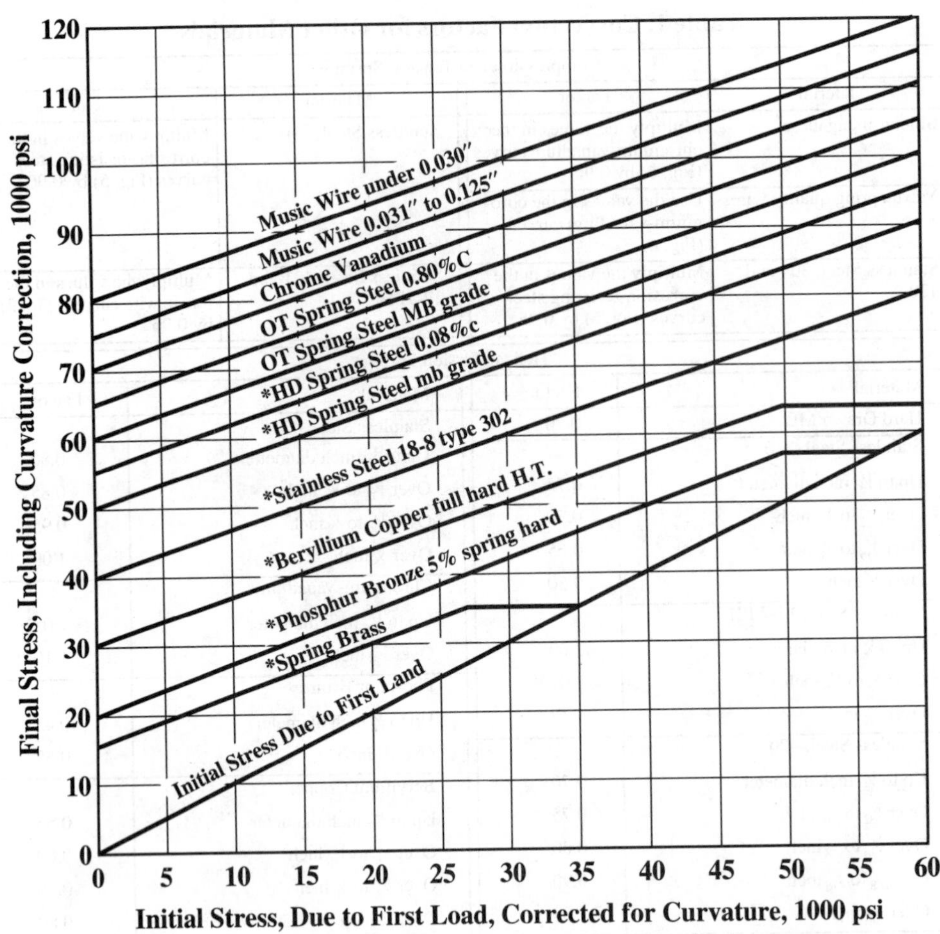

Fig. 11. Endurance Limit Curves for Compression Springs

Notes: For commercial spring materials with wire diameters up to ¼ inch except as noted. Stress ranges may be increased by approximately 30 per cent for properly heated, preset, shot-peened springs.

Materials preceeded by * are not ordinarily recommended for long continued service under severe operating conditions.

Working Stresses at Elevated Temperatures.—Since modulus of elasticity decreases with increase in temperature, springs used at high temperatures exert less load and have larger deflections under load than at room temperature. The torsional modulus of elasticity for steel may be 11,200,000 pounds per square inch at room temperature, but it will drop to 10,600,000 pounds per square inch at 400°F. and will be only 10,000,000 pounds per square inch at 600°F. Also, the elastic limit is reduced, thereby lowering the permissible working stress.

Design stresses should be as low as possible for all springs used at elevated temperatures. In addition, corrosive conditions that usually exist at high temperatures, especially with steam, may require the use of corrosion-resistant material. Table 2 shows the permissible elevated temperatures at which various spring materials may be operated, together with the maximum recommended working stresses at these temperatures. The loss in load at the temperatures shown is less than 5 per cent in 48 hours; however, if the temperatures listed are increased by 20 to 40 degrees, the loss of load may be nearer 10 per cent. Maximum stresses shown in the table are for compression and extension springs and may be increased by 75 per cent for torsion and flat springs. In using the data in Table 2 it should be noted that the values given are for materials in the heat-treated or spring temper condition.

Table 2. Recommended Maximum Working Temperatures and Corresponding Maximum Working Stresses for Springs

Spring Material	Maximum Working Temperature, Degrees, F.	Maximum Working Stress, Pounds per Square Inch
Brass Spring Wire	150	30,000
Phosphor Bronze	225	35,000
Music Wire	250	75,000
Beryllium-Copper	300	40,000
Hard Drawn Steel Wire	325	50,000
Carbon Spring Steels	375	55,000
Alloy Spring Steels	400	65,000
Monel	425	40,000
K-Monel	450	45,000
Permanickel[a]	500	50,000
Stainless Steel 18-8	550	55,000
Stainless Chromium 431	600	50,000
Inconel	700	50,000
High Speed Steel	775	70,000
Inconel X	850	55,000
Chromium-Molybdenum-Vanadium	900	55,000
Cobenium, Elgiloy	1000	75,000

[a] Formerly called Z-Nickel, Type B.

Loss of load at temperatures shown is less than 5 per cent in 48 hours.

Spring Design Data

Spring Characteristics.—This section provides tables of spring characteristics, tables of principal formulas, and other information of a practical nature for designing the more commonly used types of springs.

Standard wire gages for springs: Information on wire gages is given in the section beginning on page 2499, and gages in decimals of an inch are given in the table on page 2500. It should be noted that the range in this table extends from Number 7/0 through Number 80. However, in spring design, the range most commonly used extends only from Gage Number 4/0 through Number 40. When selecting wire use Steel Wire Gage or Washburn and Moen gage for all carbon steels and alloy steels except music wire; use Brown & Sharpe gage for brass and phosphor bronze wire; use Birmingham gage for flat spring steels, and cold rolled strip; and use piano or music wire gage for music wire.

Spring index: The spring index is the ratio of the mean coil diameter of a spring to the wire diameter (D/d). This ratio is one of the most important considerations in spring design because the deflection, stress, number of coils, and selection of either annealed or tempered material depend to a considerable extent on this ratio. The best proportioned springs have an index of 7 through 9. Indexes of 4 through 7, and 9 through 16 are often used. Springs with values larger than 16 require tolerances wider than standard for manufacturing; those with values less than 5 are difficult to coil on automatic coiling machines.

Direction of helix: Unless functional requirements call for a definite hand, the helix of compression and extension springs should be specified as optional. When springs are designed to operate, one inside the other, the helices should be opposite hand to prevent intermeshing. For the same reason, a spring that is to operate freely over a threaded member should have a helix of opposite hand to that of the thread. When a spring is to engage with a screw or bolt, it should, of course, have the same helix as that of the thread.

Helical Compression Spring Design.—After selecting a suitable material and a safe stress value for a given spring, designers should next determine the type of end coil formation best suited for the particular application. Springs with unground ends are less expensive but they do not stand perfectly upright; if this requirement has to be met, closed ground ends are used. Helical compression springs with different types of ends are shown in Fig. 12.

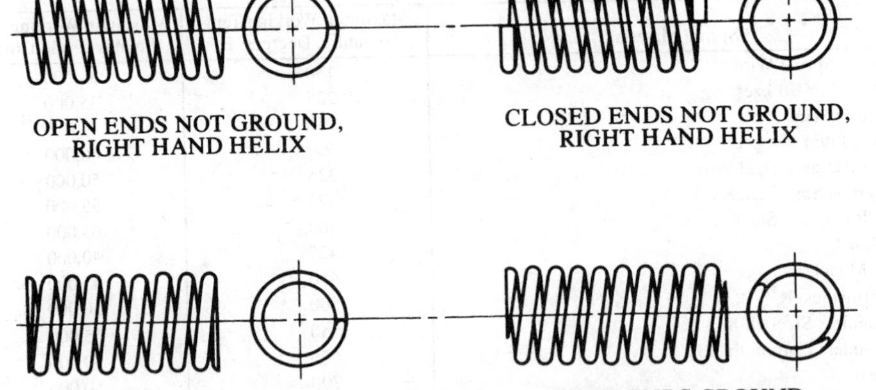

Fig. 12. Types of Helical Compression Spring Ends

Spring design formulas: Table 3 gives formulas for compression spring dimensional characteristics, and Table 4 gives design formulas for compression and extension springs.

Curvature correction: In addition to the stress obtained from the formulas for load or deflection, there is a direct shearing stress and an increased stress on the inside of the section due to curvature. Therefore, the stress obtained by the usual formulas should be multiplied by a factor K taken from the curve in Fig. 13. The corrected stress thus obtained is used only for comparison with the allowable working stress (fatigue strength) curves to determine if it is a safe stress and should not be used in formulas for deflection. The curvature correction factor K is for compression and extension springs made from round wire. For square wire reduce the K value by approximately 4 per cent.

Design procedure: The limiting dimensions of a spring are often determined by the available space in the product or assembly in which it is to be used. The loads and deflections on a spring may also be known or can be estimated, but the wire size and number of coils are usually unknown. Design can be carried out with the aid of the tabular data that appears later in this section (see Table , which is a simple method, or by calculation alone using the formulas in Tables 3 and 4.

Table 3. Formulas for Compression Springs

Feature	Type of End: Open or Plain (not ground)	Open or Plain (with ends ground)	Squared or Closed (not ground)	Closed and Ground
	Formula			
Pitch (p)	$\frac{FL-d}{N}$	$\frac{FL}{TC}$	$\frac{FL-3d}{N}$	$\frac{FL-2d}{N}$
Solid Height (SH)	$(TC+1)d$	$TC \times d$	$(TC+1)d$	$TC \times d$
Number of Active Coils (N)	$N = TC$ $= \frac{FL-d}{p}$	$N = TC-1$ $= \frac{FL}{p} - 1$	$N = TC-2$ $= \frac{FL-3d}{p}$	$N = TC-2$ $= \frac{FL-2d}{p}$
Total Coils (TC)	$\frac{FL-d}{p}$	$\frac{FL}{p}$	$\frac{FL-3d}{p} + 2$	$\frac{FL-2d}{p} + 2$
Free Length (FL)	$(p \times TC) + d$	$p \times TC$	$(p \times N) + 3d$	$(p \times N) + 2d$

The symbol notation is given on page 285.

Table 4. Formulas for Compression and Extension Springs

Feature	Formula[a]: Springs made from round wire	Springs made from square wire
Load, P Pounds	$P = \frac{0.393Sd^3}{D} = \frac{Gd^4F}{8ND^3}$	$P = \frac{0.416Sd^3}{D} = \frac{Gd^4F}{5.58ND^3}$
Stress, Torsional, S Pounds per square inch	$S = \frac{GdF}{\pi ND^2} = \frac{PD}{0.393d^3}$	$S = \frac{GdF}{2.32ND^2} = P\frac{D}{0.416d^3}$
Deflection, F Inch	$F = \frac{8PND^3}{Gd^4} = \frac{\pi SND^2}{Gd}$	$F = \frac{5.58PND^3}{Gd^4} = \frac{2.32SND^2}{Gd}$
Number of Active Coils, N	$N = \frac{Gd^4F}{8PD^3} = \frac{GdF}{\pi SD^2}$	$N = \frac{Gd^4F}{5.58PD^3} = \frac{GdF}{2.32SD^2}$
Wire Diameter, d Inch	$d = \frac{\pi SND^2}{GF} = \sqrt[3]{\frac{2.55PD}{S}}$	$d = \frac{2.32SND^2}{GF} = \sqrt[3]{\frac{PD}{0.416S}}$
Stress due to Initial Tension, S_{it}	$S_{it} = \frac{S}{P} \times IT$	$S_{it} = \frac{S}{P} \times IT$

[a] Two formulas are given for each feature, and designers can use the one found to be appropriate for a given design. The end result from either of any two formulas is the same.

The symbol notation is given on page 285.

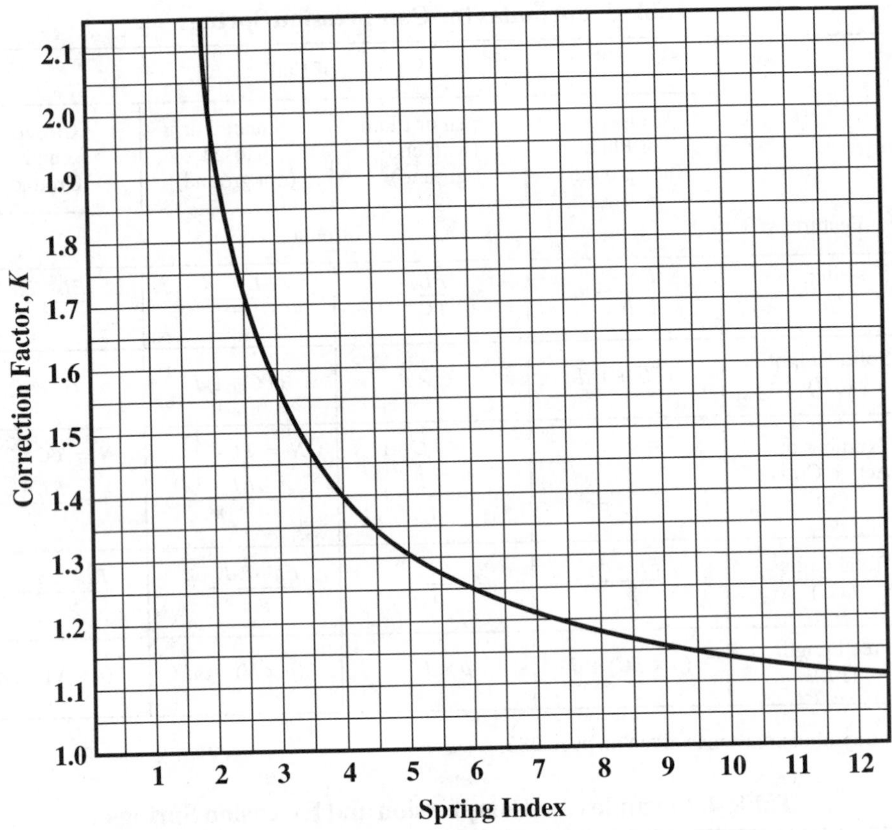

Fig. 13. Compression and Extension Spring-Stress Correction for Curvature*

Example: A compression spring with closed and ground ends is to be made from ASTM A229 high carbon steel wire, as shown in Fig. 14. Determine the wire size and number of coils.

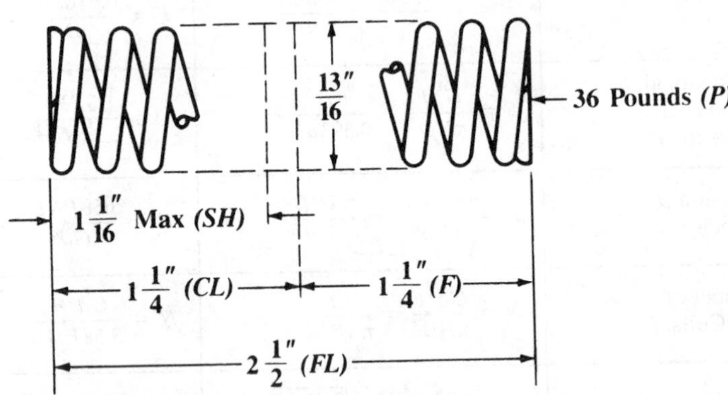

Fig. 14. Compression Spring Design Example

Method 1, using table: Referring to Table , starting on page 302, locate the spring outside diameter (13⁄16 inches, from Fig. 14) in the left-hand column. Note from the drawing that the spring load is 36 pounds. Move to the right in the table to the figure nearest this value, which is 41.7 pounds. This is somewhat above the required value but safe. Immediately above the load value, the deflection *f* is given, which in this instance is 0.1594 inch.

* For springs made from round wire. For springs made from square wire, reduce the *K* factor values by approximately 4 per cent.

This is the deflection of one coil under a load of 41.7 pounds with an uncorrected torsional stress S of 100,000 pounds per square inch (for ASTM A229 oil-tempered MB steel, see table on page 320). Moving vertically in the table from the load entry, the wire diameter is found to be 0.0915 inch.

The remaining spring design calculations are completed as follows:

Step 1: The stress with a load of 36 pounds is obtained by proportion, as follows: The 36 pound load is 86.3 per cent of the 41.7 pound load; therefore, the stress S at 36 pounds = $0.863 \times 100{,}000 = 86{,}300$ pounds per square inch.

Step 2: The 86.3 per cent figure is also used to determine the deflection per coil f at 36 pounds load: $0.863 \times 0.1594 = 0.1375$ inch.

Step 3: The number of active coils $AC = \frac{F}{f} = \frac{1.25}{0.1375} = 9.1$

Step 4: Total Coils $TC = AC + 2$ (Table 3) $= 9 + 2 = 11$
Therefore, a quick answer is: 11 coils of 0.0915 inch diameter wire. However, the design procedure should be completed by carrying out these remaining steps:

Step 5: From Table 3, Solid Height = $SH = TC \times d = 11 \times 0.0915 \cong 1$ inch

Therefore, Total Deflection = $FL - SH = 1.5$ inches

$$Step\ 6:\ \text{Stress Solid} = \frac{86{,}300}{1.25} \times 1.5 = 103{,}500 \text{ pounds per square inch}$$

$$Step\ 7:\ \text{Spring Index} = \frac{O.D.}{d} - 1 = \frac{0.8125}{0.0915} - 1 = 7.9$$

Step 8: From Fig. 13, the curvature correction factor $K = 1.185$

Step 9: Total Stress at 36 pounds load = $S \times K = 86{,}300 \times 1.185 = 102{,}300$ pounds per square inch. This stress is below the 117,000 pounds per square inch permitted for 0.0915 inch wire shown on the middle curve in Fig. 3, so it is a safe working stress.

Step 10: Total Stress at Solid = $103{,}500 \times 1.185 = 122{,}800$ pounds per square inch. This stress is also safe, as it is below the 131,000 pounds per square inch shown on the top curve Fig. 3, and therefore the spring will not set.

Method 2, using formulas: The procedure for design using formulas is as follows (the design example is the same as in Method I, and the spring is shown in Fig. 14):

Step 1: Select a safe stress S below the middle fatigue strength curve Fig. 8 for ASTM A229 steel wire, say 90,000 pounds per square inch. Assume a mean diameter D slightly below the 13⁄16-inch $O.D.$, say 0.7 inch. Note that the value of G is 11,200,000 pounds per square inch (Table).

Step 2: A trial wire diameter d and other values are found by formulas from Table 4 as follows:

$$d = \sqrt[3]{\frac{2.55PD}{S}} = \sqrt[3]{\frac{2.55 \times 36 \times 0.7}{90{,}000}}$$

$$= \sqrt[3]{0.000714} = 0.0894 \text{ inch}$$

Note: Table 21 can be used to avoid solving the cube root.

Step 3: From the table on page 2500, select the nearest wire gauge size, which is 0.0915 inch diameter. Using this value, the mean diameter D = 13⁄16 inch − 0.0915 = 0.721 inch.

Table 5. Compression and Extension Spring Deflections

Outside Diam.		Wire Size or Washburn and Moen Gauge, and Decimal Equivalent[a]																		
																	19	18	17	16
		.010	.012	.014	.016	.018	.020	.022	.024	.026	.028	.030	.032	.034	.036	.038	.041	.0475	.054	.0625
Nom.	Dec.	Deflection f (inch) per coil, at Load P (pounds)[b]																		
7/64	.1094	.0277	.0222	.01824	.01529	.01302	.01121	.00974	.00853	.00751	.00664	.00589	…	…	…	…	…	…	…	…
		.395	.697	1.130	1.722	2.51	3.52	4.79	6.36	8.28	10.59	13.35	…	…	…	…	…	…	…	…
1/8	.125	.0371	.0299	.0247	.0208	.01784	.01548	.01353	.01192	.01058	.00943	.00844	.00758	.00683	.00617	…	…	…	…	…
		.342	.600	.971	1.475	2.14	2.99	4.06	5.37	6.97	8.89	11.16	13.83	16.95	20.6	…	…	…	…	…
9/64	.1406	.0478	.0387	.0321	.0272	.0234	.0204	.01794	.01590	.01417	.01271	.01144	.01034	.00937	.00852	.00777	…	…	…	…
		.301	.528	.852	1.291	1.868	2.61	3.53	4.65	6.02	7.66	9.58	11.84	14.47	17.51	21.0	…	…	…	…
5/32	.1563	.0600	.0487	.0406	.0345	.0298	.0261	.0230	.0205	.01832	0.1649	.01491	.01354	.01234	.01128	.01033	.00909	…	…	…
		.268	.470	.758	1.146	1.656	2.31	3.11	4.10	5.30	6.72	8.39	10.35	12.62	15.23	18.22	23.5	…	…	…
11/64	.1719	.0735	.0598	.0500	.0426	.0369	.0324	.0287	.0256	.0230	.0208	.01883	.01716	.01569	.01439	.01324	.01172	.00914	…	…
		.243	.424	.683	1.031	1.488	2.07	2.79	3.67	4.73	5.99	7.47	9.19	11.19	13.48	16.09	21.8	33.8	…	…
3/16	.1875	.0884	.0720	.0603	.0516	.0448	.0394	.0349	.0313	.0281	.0255	.0232	.0212	.01944	.01788	.01650	.01468	.01157	.00926	…
		.221	.387	.621	.938	1.351	1.876	2.53	3.32	4.27	5.40	6.73	8.27	10.05	12.09	14.41	18.47	30.07	46.3	…
13/64	.2031	.1046	.0854	.0717	.0614	.0534	.0470	.0418	.0375	.0338	.0307	.0280	.0257	.0236	.0218	.0201	.01798	.01430	.01155	…
		.203	.355	.570	.859	1.237	1.716	2.31	3.03	3.90	4.92	6.12	7.52	9.13	10.96	13.05	16.69	27.1	41.5	…
7/32	.2188	…	.1000	.0841	.0721	.0628	.0555	.0494	.0444	.0401	.0365	.0333	.0306	.0282	.0260	.0241	.0216	.01733	.01411	.01096
		…	.328	.526	.793	1.140	1.580	2.13	2.79	3.58	4.52	5.61	6.88	8.35	10.02	11.92	15.22	24.6	37.5	61.3
15/64	.2344	…	.1156	.0974	.0836	.0730	.0645	.0575	.0518	.0469	.0427	.0391	.0359	.0331	.0307	.0285	.0256	.0206	.01690	.01326
		…	.305	.489	.736	1.058	1.465	1.969	2.58	3.21	4.18	5.19	6.35	7.70	9.23	10.97	13.99	22.5	34.3	55.8
1/4	.250	…	…	.1116	.0960	.0839	.0742	.0663	.0597	.0541	.0494	.0453	.0417	.0385	.0357	.0332	.0299	.0242	.01996	.01578
		…	…	.457	.687	.987	1.366	1.834	2.40	3.08	3.88	4.82	5.90	7.14	8.56	10.17	12.95	20.8	31.6	51.1
9/32	.2813	…	…	.1432	.1234	.1080	.0958	.0857	.0774	.0703	.0643	.0591	.0545	.0505	.0469	.0437	.0395	.0323	.0268	.0215
		…	…	.403	.606	.870	1.202	1.613	2.11	2.70	3.40	4.22	5.16	6.24	7.47	8.86	11.26	18.01	27.2	43.8
5/16	.3125	…	…	…	.1541	.1351	.1200	.1076	.0973	.0886	.0811	.0746	.0690	.0640	.0596	.0556	.0504	.0415	.0347	.0281
		…	…	…	.542	.778	1.074	1.440	1.881	2.41	3.03	3.75	4.58	5.54	6.63	7.85	9.97	15.89	23.9	38.3
11/32	.3438	…	…	…	…	.1633	.1470	.1321	.1196	.1090	.0999	.0921	.0852	.0792	.0733	.0690	.0627	.0518	.0436	.0355
		…	…	…	…	.703	.970	1.300	1.697	2.17	2.73	3.38	4.12	4.98	5.95	7.05	8.94	14.21	21.3	34.1
3/8	.375	…	…	…	…	…	.1768	.1589	.1440	.1314	.1206	.1113	.1031	.0960	.0895	.0839	.0764	.0634	.0535	.0438
		…	…	…	…	…	.885	1.185	1.546	1.978	2.48	3.07	3.75	4.53	5.40	6.40	8.10	12.85	19.27	30.7

[a] Round wire. For square wire, multiply f by 0.707, and p, by 1.2

[b] The upper figure is the deflection and the lower figure the load as read against each spring size.

Table 5. *(Continued)* **Compression and Extension Spring Deflections**

Outside Diam.		Wire Size or Washburn and Moen Gauge, and Decimal Equivalent																	
									19	18	17	16	15	14	13	3⁄32	12	11	1⁄8
		.026	.028	.030	.032	.034	.036	.038	.041	.0475	.054	.0625	.072	.080	.0915	.0938	.1055	.1205	.125
Nom.	Dec.	Deflection f (inch) per coil, at Load P (pounds)																	
13⁄32	.4063	.1560	.1434	.1324	.1228	.1143	.1068	.1001	.0913	.0760	.0645	.0531	.0436	.0373	.0304	.0292	.0241	…	…
		1.815	2.28	2.82	3.44	4.15	4.95	5.85	7.41	11.73	17.56	27.9	43.9	61.6	95.6	103.7	153.3	…	…
7⁄16	.4375	.1827	.1680	.1553	.1441	.1343	.1256	.1178	.1075	.0898	.0764	.0631	.0521	.0448	.0367	.0353	.0293	.0234	.0219
		1.678	2.11	2.60	3.17	3.82	4.56	5.39	6.82	10.79	16.13	25.6	40.1	56.3	86.9	94.3	138.9	217.	245.
15⁄32	.4688	.212	.1947	.1800	.1673	.1560	.1459	.1370	.1252	.1048	.0894	.0741	.0614	.0530	.0437	.0420	.0351	.0282	.0265
		1.559	1.956	2.42	2.94	3.55	4.23	5.00	6.33	9.99	14.91	23.6	37.0	51.7	79.7	86.4	126.9	197.3	223.
1⁄2	.500	.243	.223	.207	.1920	.1792	.1678	.1575	.1441	.1209	.1033	.0859	.0714	.0619	.0512	.0494	.0414	.0335	.0316
		1.456	1.826	2.26	2.75	3.31	3.95	4.67	5.90	9.30	13.87	21.9	34.3	47.9	73.6	80.0	116.9	181.1	205.
17⁄32	.5313	.276	.254	.235	.219	.204	.1911	.1796	.1645	.1382	.1183	.0987	.0822	.0714	.0593	.0572	.0482	.0393	.0371
		1.366	1.713	2.12	2.58	3.10	3.70	4.37	5.52	8.70	12.96	20.5	31.9	44.6	68.4	74.1	108.3	167.3	188.8
9⁄16	.5625	…	.286	.265	.247	.230	.216	.203	.1861	.1566	.1343	.1122	.0937	.0816	.0680	.0657	.0555	.0455	.0430
		…	1.613	1.991	2.42	2.92	3.48	4.11	5.19	8.18	12.16	19.17	29.9	41.7	63.9	69.1	100.9	155.5	175.3
19⁄32	.5938	…	…	.297	.277	.259	.242	.228	.209	.1762	.1514	.1267	.1061	.0926	.0774	.0748	.0634	.0522	.0493
		…	…	1.880	2.29	2.76	3.28	3.88	4.90	7.71	11.46	18.04	28.1	39.1	60.0	64.8	94.4	145.2	163.6
5⁄8	.625	…	…	.331	.308	.288	.270	.254	.233	.1969	.1693	.1420	.1191	.1041	.0873	.0844	.0718	.0593	.0561
		…	…	1.782	2.17	2.61	3.11	3.67	4.63	7.29	10.83	17.04	26.5	36.9	56.4	61.0	88.7	136.2	153.4
21⁄32	.6563	…	…	…	.342	.320	.300	.282	.259	.219	.1884	.1582	.1330	.1164	.0978	.0946	.0807	.0668	.0634
		…	…	…	2.06	2.48	2.95	3.49	4.40	6.92	10.27	16.14	25.1	34.9	53.3	57.6	83.7	128.3	144.3
11⁄16	.6875	…	…	…	…	.352	.331	.311	.286	.242	.208	.1753	.1476	.1294	.1089	.1054	.0901	.0748	.0710
		…	…	…	…	2.36	2.81	3.32	4.19	6.58	9.76	15.34	23.8	33.1	50.5	54.6	79.2	121.2	136.3
23⁄32	.7188	…	…	…	…	…	.363	.342	.314	.266	.230	.1933	.1630	.1431	.1206	.1168	.1000	.0833	.0791
		…	…	…	…	…	2.68	3.17	3.99	6.27	9.31	14.61	22.7	31.5	48.0	51.9	75.2	114.9	129.2
3⁄4	.750	…	…	…	…	…	…	.374	.344	.291	.252	.212	.1791	.1574	.1329	.1288	.1105	.0923	.0877
		…	…	…	…	…	…	3.03	3.82	5.99	8.89	13.94	21.6	30.0	45.7	49.4	71.5	109.2	122.7
25⁄32	.7813	…	…	…	…	…	…	…	.375	.318	.275	.232	.1960	.1724	.1459	.1413	.1214	.1017	.0967
		…	…	…	…	…	…	…	3.66	5.74	8.50	13.34	20.7	28.7	43.6	47.1	68.2	104.0	116.9
13⁄16	.8125	…	…	…	…	…	…	…	.407	.346	.299	.253	.214	.1881	.1594	.1545	.1329	.1115	.1061
		…	…	…	…	…	…	…	3.51	5.50	8.15	12.78	19.80	27.5	41.7	45.1	65.2	99.3	111.5

Table 5. *(Continued)* **Compression and Extension Spring Deflections**

Outside Diam.		Wire Size or Washburn and Moen Gauge, and Decimal Equivalent																
		15	14	13	3/32	12	11	1/8	10	9	5/32	8	7	3/16	6	5	7/32	4
		.072	.080	.0915	.0938	.1055	.1205	.125	.135	.1483	.1563	.162	.177	.1875	.192	.207	.2188	.2253
Nom.	Dec.	Deflection *f* (inch) per coil, at Load *P* (pounds)																
7/8	.875	.251	.222	.1882	.1825	.1574	.1325	.1262	.1138	.0999	.0928	.0880	.0772	.0707	.0682	.0605	.0552	.0526
		18.26	25.3	39.4	41.5	59.9	91.1	102.3	130.5	176.3	209.	234.	312.	377.	407.	521.	626.	691.
29/32	.9063	.271	.239	.204	.1974	.1705	.1438	.1370	.1236	.1087	.1010	.0959	.0843	.0772	.0746	.0663	.0606	.0577
		17.57	24.3	36.9	39.9	57.6	87.5	98.2	125.2	169.0	199.9	224.	299.	360.	389.	498.	598.	660.
15/16	.9375	.292	.258	.219	.213	.1841	.1554	.1479	.1338	.1178	.1096	.1041	.0917	.0842	.0812	.0723	.0662	.0632
		16.94	23.5	35.6	38.4	55.4	84.1	94.4	120.4	162.3	191.9	215.	286.	345.	373.	477.	572.	631.
31/32	.9688	.313	.277	.236	.229	.1982	.1675	.1598	.1445	.1273	.1183	.1127	.0994	.0913	.0882	.0786	.0721	.0688
		16.35	22.6	34.3	37.0	53.4	81.0	90.9	115.9	156.1	184.5	207.	275.	332.	358.	457.	548.	604.
1	1.000	.336	.297	.253	.246	.213	.1801	.1718	.1555	.1372	.1278	.1216	.1074	.0986	.0954	.0852	.0783	.0747
		15.80	21.9	33.1	35.8	51.5	78.1	87.6	111.7	150.4	177.6	198.8	264.	319.	344.	439.	526.	580.
1 1/32	1.031	.359	.317	.271	.263	.228	.1931	.1843	.1669	.1474	.1374	.1308	.1157	.1065	.1029	.0921	.0845	.0809
		15.28	21.1	32.0	34.6	49.8	75.5	84.6	107.8	145.1	171.3	191.6	255.	307.	331.	423.	506.	557.
1 1/16	1.063	.382	.338	.289	.281	.244	.207	.1972	.1788	.1580	.1474	.1404	.1243	.1145	.1107	.0993	.0913	.0873
		14.80	20.5	31.0	33.5	48.2	73.0	81.8	104.2	140.1	165.4	185.0	246.	296.	319.	407.	487.	537.
1 1/32	1.094	.407	.360	.308	.299	.260	.221	.211	.1910	.1691	.1578	.1503	.1332	.1229	.1188	.1066	.0982	.0939
		14.34	19.83	30.0	32.4	46.7	70.6	79.2	100.8	135.5	159.9	178.8	238.	286.	308.	393.	470.	517.
1 1/8	1.125	.432	.383	.328	.318	.277	.235	.224	.204	.1804	.1685	.1604	.1424	.1315	.1272	.1142	.1053	.1008
		13.92	19.24	29.1	31.4	45.2	68.4	76.7	97.6	131.2	154.7	173.0	230.	276.	298.	379.	454.	499.
1 3/16	1.188	.485	.431	.368	.358	.311	.265	.254	.231	.204	.1908	.1812	.1620	.1496	.1448	.1303	.1203	.1153
		13.14	18.15	27.5	29.6	42.6	64.4	72.1	91.7	123.3	145.4	162.4	215.	259.	279.	355.	424.	467.
1 1/4	1.250	.541	.480	.412	.400	.349	.297	.284	.258	.230	.215	.205	.1824	.1690	.1635	.1474	.1363	.1308
		12.44	17.19	26.0	28.0	40.3	60.8	68.2	86.6	116.2	137.0	153.1	203.	244.	263.	334.	399.	438.
1 5/16	1.313	.600	.533	.457	.444	.387	.331	.317	.288	.256	.240	.229	.205	.1894	.1836	.1657	.1535	.1472
		11.81	16.31	24.6	26.6	38.2	57.7	64.6	82.0	110.1	129.7	144.7	191.6	230.	248.	315.	376.	413.
1 3/8	1.375	.662	.588	.506	.491	.429	.367	.351	.320	.285	.267	.255	.227	.211	.204	.1848	.1713	.1650
		11.25	15.53	23.4	25.3	36.3	54.8	61.4	77.9	104.4	123.0	137.3	181.7	218.	235.	298.	356.	391
1 7/16	1.438	.727	.647	.556	.540	.472	.404	.387	.353	.314	.295	.282	.252	.234	.227	.205	.1905	.1829
		10.73	14.81	22.3	24.1	34.6	52.2	58.4	74.1	99.4	117.0	130.6	172.6	207.	223.	283.	337.	371.

Table 5. *(Continued)* **Compression and Extension Spring Deflections**

Outside Diam.		Wire Size or Washburn and Moen Gauge, and Decimal Equivalent																	
		11	1/8	10	9	5/32	8	7	3/16	6	5	7/32	4	3	1/4	2	9/32	0	5/16
		.1205	.125	.135	.1483	.1563	.162	.177	.1875	.192	.207	.2188	.2253	.2437	.250	.2625	.2813	.3065	.3125
Nom.	Dec.	Deflection f (inch) per coil, at Load P (pounds)																	
1½	1.500	.443	.424	.387	.350	.324	.310	.277	.258	.250	.227	.210	.202	.1815	.1754	.1612	.1482	.1305	.1267
		49.8	55.8	70.8	94.8	111.5	124.5	164.6	197.1	213.	269.	321.	352.	452.	499.	574.	717.	947.	1008.
1⅝	1.625	.527	.505	.461	.413	.387	.370	.332	.309	.300	.273	.254	.244	.220	.212	.1986	.1801	.1592	.1547
		45.7	51.1	64.8	86.7	102.0	113.9	150.3	180.0	193.9	246.	292.	321.	411.	446.	521.	650.	858.	912.
1¾	1.750	.619	.593	.542	.485	.456	.437	.392	.366	.355	.323	.301	.290	.261	.253	.237	.215	.1908	.1856
		42.2	47.2	59.8	80.0	94.0	104.9	138.5	165.6	178.4	226.	269.	295.	377.	409.	477.	595.	783.	833.
1⅞	1.875	.717	.687	.629	.564	.530	.508	.457	.426	.414	.377	.351	.339	.306	.296	.278	.253	.225	.219
		39.2	43.8	55.5	74.2	87.2	97.3	128.2	153.4	165.1	209.	248.	272.	348.	378.	440.	548.	721.	767.
1 15/16	1.938	.769	.738	.676	.605	.569	.546	.492	.458	.446	.405	.379	.365	.331	.320	.300	.273	.243	.237
		37.8	42.3	53.6	71.6	84.2	93.8	123.6	147.9	159.2	201.	239.	262.	335.	364.	425.	528.	693.	737.
2	2.000	.823	.789	.723	.649	.610	.585	.527	.492	.478	.436	.407	.392	.355	.344	.323	.295	.263	.256
		36.6	40.9	51.8	69.2	81.3	90.6	119.4	142.8	153.7	194.3	231.	253.	324.	351.	409.	509.	668.	710.
2 1/16	2.063	.878	.843	.768	.693	.652	.626	.564	.526	.512	.467	.436	.421	.381	.369	.346	.316	.282	.275
		35.4	39.6	50.1	66.9	78.7	87.6	115.4	138.1	148.5	187.7	223.	245.	312.	339.	395.	491.	644.	685.
2⅛	2.125	.936	.898	.823	.739	.696	.667	.602	.562	.546	.499	.466	.449	.407	.395	.371	.339	.303	.295
		34.3	38.3	48.5	64.8	76.1	84.9	111.8	133.6	143.8	181.6	216.	236.	302.	327.	381.	474.	622.	661.
2 3/16	2.188	.995	.955	.876	.786	.740	.711	.641	.598	.582	.532	.497	.479	.435	.421	.396	.362	.324	.316
		33.3	37.2	47.1	62.8	73.8	82.2	108.3	129.5	139.2	175.8	209.	229.	292.	317.	369.	459.	601.	639.
2¼	2.250	1.056	1.013	.930	.835	.787	.755	.681	.637	.619	.566	.529	.511	.463	.449	.423	.387	.346	.337
		32.3	36.1	45.7	60.9	71.6	79.8	105.7	125.5	135.0	170.5	202.	222.	283.	307.	357.	444.	582.	618.
2 5/16	2.313	1.119	1.074	.986	.886	.834	.801	.723	.676	.657	.601	.562	.542	.493	.478	.449	.411	.368	.359
		31.4	35.1	44.4	59.2	69.5	77.5	101.9	121.8	131.0	165.4	196.3	215.	275.	298.	347.	430.	564.	599.
2⅜	2.375	1.184	1.136	1.043	.938	.884	.848	.763	.716	.696	.637	.596	.576	.523	.507	.477	.437	.392	.382
		30.5	34.1	43.1	57.5	67.6	75.3	99.1	118.3	127.3	160.7	190.7	209.	267.	289.	336.	417.	547.	581.
2 7/16	2.438	…	1.201	1.102	.991	.934	.897	.810	.757	.737	.674	.631	.609	.554	.537	.506	.464	.416	.405
		…	33.2	42.0	56.0	65.7	73.2	96.3	115.1	123.7	156.1	185.3	203.	259.	281.	327.	405.	531.	564.
2½	2.500	…	1.266	1.162	1.046	.986	.946	.855	.800	.778	.713	.667	.644	.586	.568	.536	.491	.441	.430
		…	32.3	40.9	54.5	64.0	71.3	93.7	111.6	120.4	151.9	180.2	197.5	252.	273.	317.	394.	516.	548.

Note: Intermediate values can be obtained within reasonable accuracy by interpolation.

The table is for ASTM A229 oil tempered spring steel with a torsional modulus G of 11,200,000 psi, and an uncorrected torsional stress of 100,000 psi. For other materials use the following factors: stainless steel, multiply f by 1.067; spring brass, multiply f by 2.24; phosphor bronze, multiply f by 1.867; Monel metal, multiply f by 1.244; beryllium copper, multiply f by 1.725; Inconel (non-magnetic), multiply f by 1.045.

Step 4: The stress $S = \dfrac{PD}{0.393d^3} = \dfrac{36 \times 0.721}{0.393 \times 0.0915^3} = 86{,}300 \text{ lb/in}^2$

Step 5: The number of active coils is $N = \dfrac{GdF}{\pi SD^2}$

$$= \frac{11{,}200{,}000 \times 0.0915 \times 1.25}{3.1416 \times 86{,}300 \times 0.721^2} = 9.1 \text{ (say 9)}$$

The answer is the same as before, which is to use 11 total coils of 0.0915-inch diameter wire. The total coils, solid height, etc., are determined in the same manner as in Method 1.

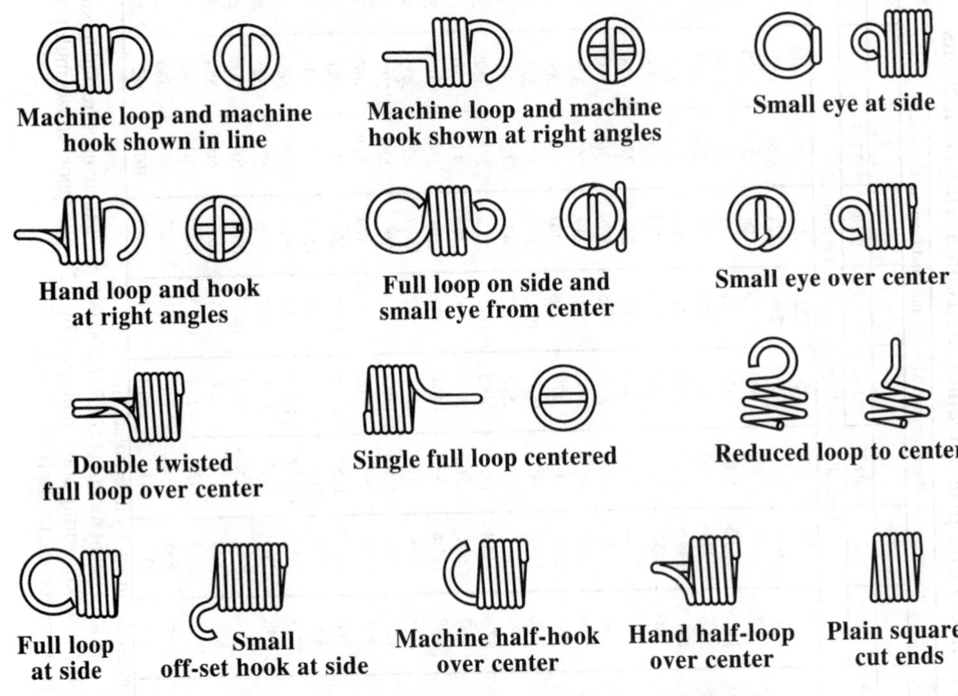

All the Above Ends are Standard Types for Which No Special Tools are Required

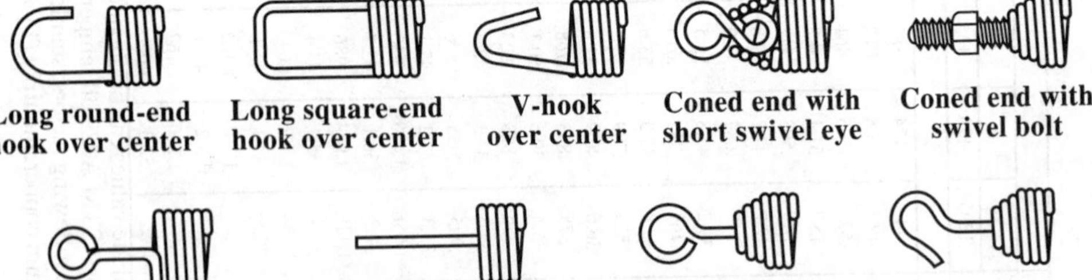

This Group of Special Ends Requires Special Tools

Fig. 15. Types of Helical Extension Spring Ends

Table of Spring Characteristics.—Table 5 gives characteristics for compression and extension springs made from ASTM A229 oil-tempered MB spring steel having a torsional modulus of elasticity *G* of 11,200,000 pounds per square inch, and an uncorrected torsional stress *S* of 100,000 pounds per square inch. The deflection *f* for one coil under a load *P* is shown in the body of the table. The method of using these data is explained in the problems for compression and extension spring design. The table may be used for other materials by applying factors to *f*. The factors are given in a footnote to the table.

Extension Springs.—About 10 per cent of all springs made by many companies are of this type, and they frequently cause trouble because insufficient consideration is given to stress due to initial tension, stress and deflection of hooks, special manufacturing methods, secondary operations and overstretching at assembly. Fig. 15 shows types of ends used on these springs.

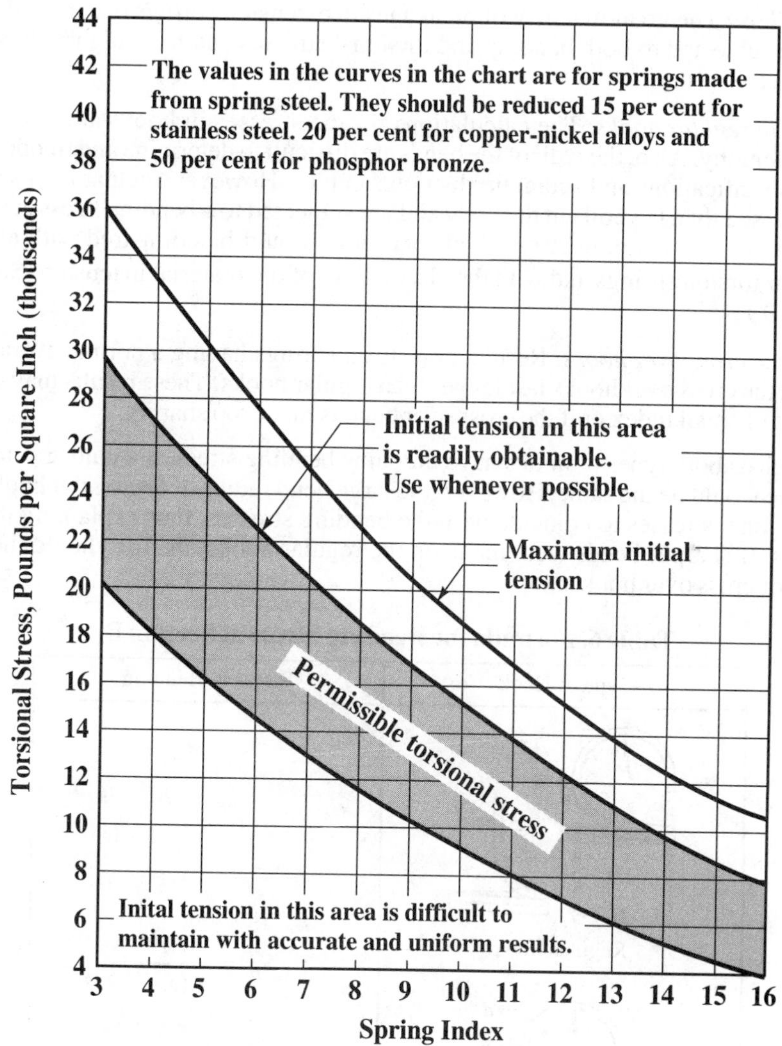

Fig. 16. Permissible Torsional Stress Caused by Initial Tension in Coiled Extension Springs for Different Spring Indexes

Initial tension: In the spring industry, the term "Initial tension" is used to define a force or load, measurable in pounds or ounces, which presses the coils of a close wound extension spring against one another. This force must be overcome before the coils of a spring begin to open up.

Initial tension is wound into extension springs by bending each coil as it is wound away from its normal plane, thereby producing a slight twist in the wire which causes the coil to spring back tightly against the adjacent coil. Initial tension can be wound into cold-coiled extension springs only. Hot-wound springs and springs made from annealed steel are hardened and tempered after coiling, and therefore initial tension cannot be produced. It is possible to make a spring having initial tension only when a high tensile strength, obtained by cold drawing or by heat-treatment, is possessed by the material as it is being wound into springs. Materials that possess the required characteristics for the manufacture of such springs include hard-drawn wire, music wire, pre-tempered wire, 18-8 stainless steel, phosphor-bronze, and many of the hard-drawn copper-nickel, and nonferrous alloys. Permissible torsional stresses resulting from initial tension for different spring indexes are shown in Fig. 16.

Hook failure: The great majority of breakages in extension springs occurs in the hooks. Hooks are subjected to both bending and torsional stresses and have higher stresses than the coils in the spring.

Stresses in regular hooks: The calculations for the stresses in hooks are quite complicated and lengthy. Also, the radii of the bends are difficult to determine and frequently vary between specifications and actual production samples. However, regular hooks are more highly stressed than the coils in the body and are subjected to a bending stress at section B (see Table 6.) The bending stress S_b at section B should be compared with allowable stresses for torsion springs and with the elastic limit of the material in tension (See Figs. 7 through 10.)

Stresses in cross over hooks: Results of tests on springs having a normal average index show that the cross over hooks last longer than regular hooks. These results may not occur on springs of small index or if the cross over bend is made too sharply.

Inasmuch as both types of hooks have the same bending stress, it would appear that the fatigue life would be the same. However, the large bend radius of the regular hooks causes some torsional stresses to coincide with the bending stresses, thus explaining the earlier breakages. If sharper bends were made on the regular hooks, the life should then be the same as for cross over hooks.

Table 6. Formula for Bending Stress at Section B

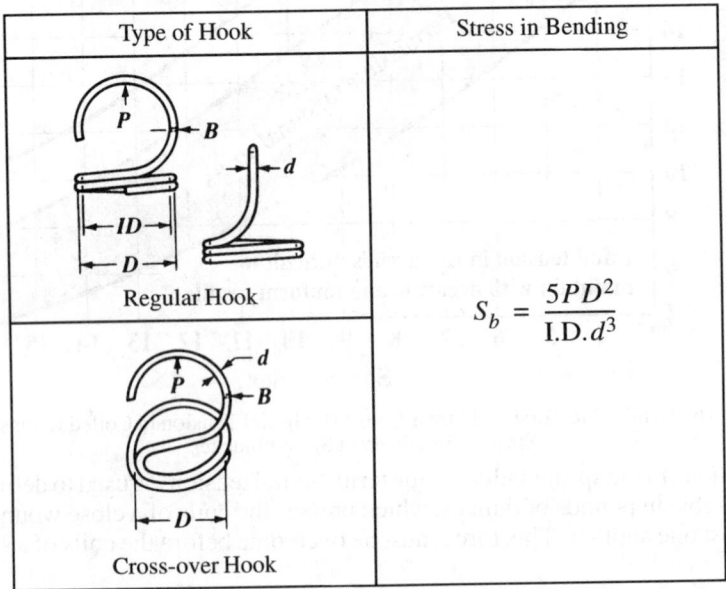

Type of Hook	Stress in Bending
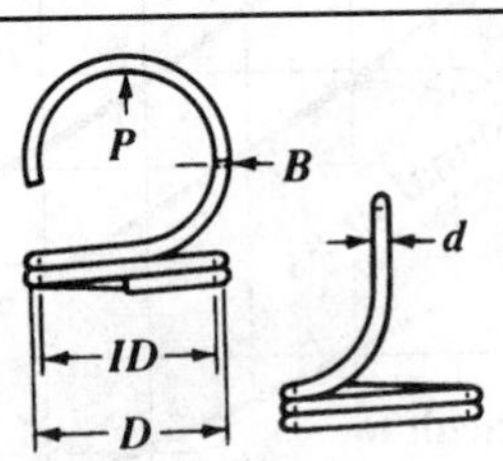 P, B, d, ID, D Regular Hook	$S_b = \dfrac{5PD^2}{I.D.d^3}$
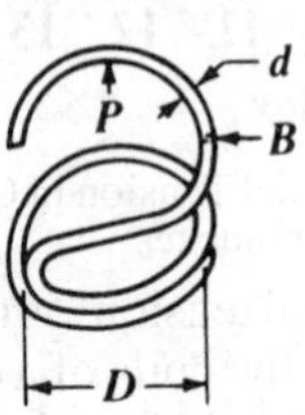 d, P, B, D Cross-over Hook	

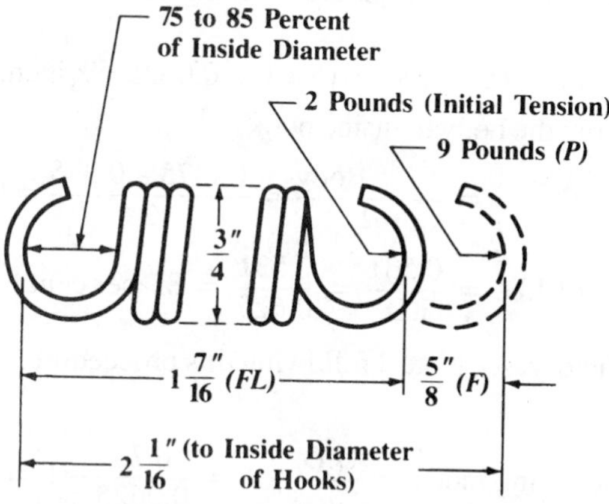

Fig. 17. Extension Spring Design Example

Stresses in half hooks: The formulas for regular hooks can also be used for half hooks, because the smaller bend radius allows for the increase in stress. It will therefore be observed that half hooks have the same stress in bending as regular hooks.

Frequently overlooked facts by many designers are that one full hook deflects an amount equal to one half a coil and each half hook deflects an amount equal to one tenth of a coil. Allowances for these deflections should be made when designing springs. Thus, an extension spring, with regular full hooks and having 10 coils, will have a deflection equal to 11 coils, or 10 per cent more than the calculated deflection.

Extension Spring Design.—The available space in a product or assembly usually determines the limiting dimensions of a spring, but the wire size, number of coils, and initial tension are often unknown.

Example: An extension spring is to be made from spring steel ASTM A229, with regular hooks as shown in Fig. 17. Calculate the wire size, number of coils and initial tension.

Note: Allow about 20 to 25 per cent of the 9 pound load for initial tension, say 2 pounds, and then design for a 7 pound load (not 9 pounds) at ⅝ inch deflection. Also use lower stresses than for a compression spring to allow for overstretching during assembly and to obtain a safe stress on the hooks. Proceed as for compression springs, but locate a load in the tables somewhat higher than the 9 pound load.

Method 1, using table: From Table locate ¾ inch outside diameter in the left column and move to the right to locate a load P of 13.94 pounds. A deflection f of 0.212 inch appears above this figure. Moving vertically from this position to the top of the column a suitable wire diameter of 0.0625 inch is found.

The remaining design calculations are completed as follows:

Step 1: The stress with a load of 7 pounds is obtained as follows:
The 7 pound load is 50.2 per cent of the 13.94 pound load. Therefore, the stress S at 7 pounds = 0.502 per cent × 100,000 = 50,200 pounds per square inch.

Step 2: The 50.2 per cent figure is also used to determine the deflection per coil f: 0.502 per cent × 0.212 = 0.1062 inch.

Step 3: The number of active coils. (say 6)

$$AC = \frac{F}{f} = \frac{0.625}{0.1062} = 5.86$$

This result should be reduced by 1 to allow for deflection of 2 hooks (see notes 1 and 2 that follow these calculations.) Therefore, a quick answer is: 5 coils of 0.0625 inch diameter wire. However, the design procedure should be completed by carrying out the following steps:

Step 4: The body length $= (TC+1)\times d = (5+1)\times 0.0625 = \frac{3}{8}$ inch.

Step 5: The length from the body to inside hook

$$= \frac{FL - \text{Body}}{2} = \frac{1.4375 - 0.375}{2} = 0.531 \text{ inch}$$

$$\text{Percentage of I.D.} = \frac{0.531}{\text{I.D.}} = \frac{0.531}{0.625} = 85 \text{ per cent}$$

This length is satisfactory, see Note 3 following this proceedure.

Step 6:

$$\text{The spring index} = \frac{\text{O.D.}}{d} - 1 = \frac{0.75}{0.0625} - 1 = 11$$

Step 7: The initial tension stress is

$$S_{it} = \frac{S\times IT}{P} = \frac{50{,}200\times 2}{7}$$

$$= 14{,}340 \text{ pounds per square inch}$$

This stress is satisfactory, as checked against curve in Fig. 16.

Step 8: The curvature correction factor $K = 1.12$ (Fig. 13).

Step 9: The total stress $= (50{,}200 + 14{,}340)\times 1.12 = 72.285$ pounds per square inch

This result is less than 106,250 pounds per square inch permitted by the middle curve for 0.0625 inch wire in Fig. 3 and therefore is a safe working stress that permits some additional deflection that is usually necessary for assembly purposes.

Step 10: The large majority of hook breakage is due to high stress in bending and should be checked as follows:

From Table 6, stress on hook in bending is:

$$S_b = \frac{5PD^2}{\text{I.D.}d^3}$$

$$= \frac{5\times 9\times 0.6875^2}{0.625\times 0.0625^3} = 139{,}200 \text{ pounds per square inch}$$

This result is less than the top curve value, Fig. 8, for 0.0625 inch diameter wire, and is therefore safe. Also see Note 5 that follows.

Notes: The following points should be noted when designing extension springs:

1) All coils are active and thus $AC = TC$.

2) Each full hook deflection is approximately equal to ½ coil. Therefore for 2 hooks, reduce the total coils by 1. (Each half hook deflection is nearly equal to $\frac{1}{10}$ of a coil.)

3) The distance from the body to the inside of a regular full hook equals 75 to 85 per cent (90 per cent maximum) of the I.D. For a cross over center hook, this distance equals the I.D.

4) Some initial tension should usually be used to hold the spring together. Try not to exceed the maximum curve shown on Fig. 16. Without initial tension, a long spring with many coils will have a different length in the horizontal position than it will when hung vertically.

5) The hooks are stressed in bending, therefore their stress should be less than the maximum bending stress as used for torsion springs — use top fatigue strength curves Figs. 7 through 10.

Method 2, using formulas: The sequence of steps for designing extension springs by formulas is similar to that for compression springs. The formulas for this method are given in Table 3.

Tolerances for Compression and Extension Springs.—Tolerances for coil diameter, free length, squareness, load, and the angle between loop planes for compression and extension springs are given in Tables 7 through 12. To meet the requirements of load, rate, free length, and solid height, it is necessary to vary the number of coils for compression springs by ± 5 per cent. For extension springs, the tolerances on the numbers of coils are: for 3 to 5 coils, ± 20 per cent; for 6 to 8 coils, ± 30 per cent; for 9 to 12 coils, ± 40 per cent. For each additional coil, a further 1½ per cent tolerance is added to the extension spring values. Closer tolerances on the number of coils for either type of spring lead to the need for trimming after coiling, and manufacturing time and cost are increased. Fig. 18 shows deviations allowed on the ends of extension springs, and variations in end alignments.

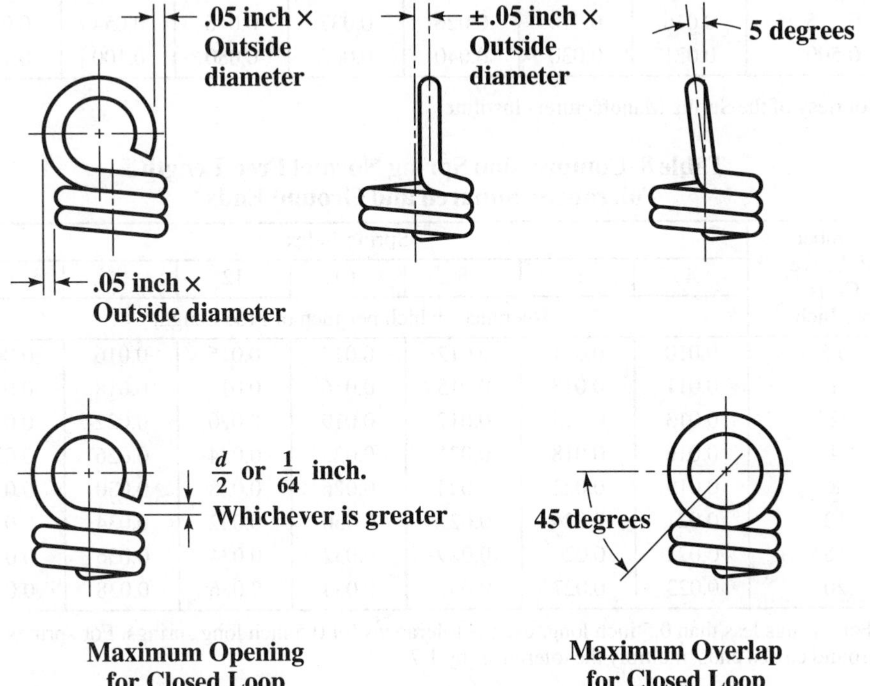

Fig. 18. Maximum Deviations Allowed on Ends and Variation in Alignment of Ends (Loops) for Extension Springs

Table 7. Compression and Extension Spring Coil Diameter Tolerances

Wire Diameter, Inch	Spring Index						
	4	6	8	10	12	14	16
	Tolerance, ± inch						
0.015	0.002	0.002	0.003	0.004	0.005	0.006	0.007
0.023	0.002	0.003	0.004	0.006	0.007	0.008	0.010
0.035	0.002	0.004	0.006	0.007	0.009	0.011	0.013
0.051	0.003	0.005	0.007	0.010	0.012	0.015	0.017
0.076	0.004	0.007	0.010	0.013	0.016	0.019	0.022
0.114	0.006	0.009	0.013	0.018	0.021	0.025	0.029
0.171	0.008	0.012	0.017	0.023	0.028	0.033	0.038
0.250	0.011	0.015	0.021	0.028	0.035	0.042	0.049
0.375	0.016	0.020	0.026	0.037	0.046	0.054	0.064
0.500	0.021	0.030	0.040	0.062	0.080	0.100	0.125

Courtesy of the Spring Manufacturers Institute

Table 8. Compression Spring Normal Free-Length Tolerances, Squared and Ground Ends

Number of Active Coils per Inch	Spring Index						
	4	6	8	10	12	14	16
	Tolerance, ± Inch per Inch of Free Length[a]						
0.5	0.010	0.011	0.012	0.013	0.015	0.016	0.016
1	0.011	0.013	0.015	0.016	0.017	0.018	0.019
2	0.013	0.015	0.017	0.019	0.020	0.022	0.023
4	0.016	0.018	0.021	0.023	0.024	0.026	0.027
8	0.019	0.022	0.024	0.026	0.028	0.030	0.032
12	0.021	0.024	0.027	0.030	0.032	0.034	0.036
16	0.022	0.026	0.029	0.032	0.034	0.036	0.038
20	0.023	0.027	0.031	0.034	0.036	0.038	0.040

[a] For springs less than 0.5 inch long, use the tolerances for 0.5 inch long springs. For springs with unground closed ends, multiply the tolerances by 1.7.

Courtesy of the Spring Manufacturers Institute

Table 9. Extension Spring Normal Free-Length and End Tolerances

Free-Length Tolerances		End Tolerances	
Spring Free-Length (inch)	Tolerance (inch)	Total Number of Coils	Angle Between Loop Planes (degrees)
Up to 0.5	±0.020		
Over 0.5 to 1.0	±0.030	3 to 6	±25
Over 1.0 to 2.0	±0.040	7 to 9	±35
Over 2.0 to 4.0	±0.060	10 to 12	±45
Over 4.0 to 8.0	±0.093	13 to 16	±60
Over 8.0 to 16.0	±0.156	Over 16	Random
Over 16.0 to 24.0	±0.218		

Courtesy of the Spring Manufacturers Institute

Table 10. Compression Spring Squareness Tolerances

Slenderness Ratio *FL/D*[a]	Spring Index						
	4	6	8	10	12	14	16
	Squareness Tolerances (± degrees)						
0.5	3.0	3.0	3.5	3.5	3.5	3.5	4.0
1.0	2.5	3.0	3.0	3.0	3.0	3.5	3.5
1.5	2.5	2.5	2.5	3.0	3.0	3.0	3.0
2.0	2.5	2.5	2.5	2.5	3.0	3.0	3.0
3.0	2.0	2.5	2.5	2.5	2.5	2.5	3.0
4.0	2.0	2.0	2.5	2.5	2.5	2.5	2.5
6.0	2.0	2.0	2.0	2.5	2.5	2.5	2.5
8.0	2.0	2.0	2.0	2.0	2.5	2.5	2.5
10.0	2.0	2.0	2.0	2.0	2.0	2.5	2.5
12.0	2.0	2.0	2.0	2.0	2.0	2.0	2.5

[a] Slenderness Ratio = $FL \div D$

Springs with closed and ground ends, in the free position. Squareness tolerances closer than those shown require special process techniques which increase cost. Springs made from fine wire sizes, and with high spring indices, irregular shapes or long free lengths, require special attention in determining appropriate tolerance and feasibility of grinding ends.

Table 11. Compression Spring Normal Load Tolerances

Length Tolerance, ± inch	Deflection (inch)[a]							
	0.05	0.10	0.15	0.20	0.25	0.30	0.40	0.50
	Tolerance, ± Per Cent of Load							
0.005	12	7	6	5	…	…	…	…
0.010	…	12	8.5	7	6.5	5.5	5	…
0.020	…	22	15.5	12	10	8.5	7	6
0.030	…	…	22	17	14	12	9.5	8
0.040	…	…	…	22	18	15.5	12	10
0.050	…	…	…	…	22	19	14.5	12
0.060	…	…	…	…	25	22	17	14
0.070	…	…	…	…	…	25	19.5	16
0.080	…	…	…	…	…	…	22	18
0.090	…	…	…	…	…	…	25	20
0.100	…	…	…	…	…	…	…	22
0.200	…	…	…	…	…	…	…	…
0.300	…	…	…	…	…	…	…	…
0.400	…	…	…	…	…	…	…	…
0.500	…	…	…	…	…	…	…	…

Length Tolerance, ± inch	Deflection (inch)[a]						
	0.75	1.00	1.50	2.00	3.00	4.00	6.00
	Tolerance, ± Per Cent of Load						
0.005	…	…	…	…	…	…	…
0.010	…	…	…	…	…	…	…
0.020	5	…	…	…	…	…	…
0.030	6	5	…	…	…	…	…
0.040	7.5	6	5	…	…	…	…
0.050	9	7	5.5	…	…	…	…
0.060	10	8	6	5	…	…	…
0.070	11	9	6.5	5.5	…	…	…
0.080	12.5	10	7.5	6	5	…	…
0.090	14	11	8	6	5	…	…
0.100	15.5	12	8.5	7	5.5	…	…
0.200	…	22	15.5	12	8.5	7	5.5
0.300	…	…	22	17	12	9.5	7
0.400	…	…	…	21	15	12	8.5
0.500	…	…	…	25	18.5	14.5	10.5

[a] From free length to loaded position.

Table 12. Extension Spring Normal Load Tolerances

Spring Index	$\frac{FL}{F}$	Wire Diameter (inch)										
		0.015	0.022	0.032	0.044	0.062	0.092	0.125	0.187	0.250	0.375	0.437
		Tolerance, ± Per Cent of Load										
4	12	20.0	18.5	17.6	16.9	16.2	15.5	15.0	14.3	13.8	13.0	12.6
	8	18.5	17.5	16.7	15.8	15.0	14.5	14.0	13.2	12.5	11.5	11.0
	6	16.8	16.1	15.5	14.7	13.8	13.2	12.7	11.8	11.2	9.9	9.4
	4.5	15.0	14.7	14.1	13.5	12.6	12.0	11.5	10.3	9.7	8.4	7.9
	2.5	13.1	12.4	12.1	11.8	10.6	10.0	9.1	8.5	8.0	6.8	6.2
	1.5	10.2	9.9	9.3	8.9	8.0	7.5	7.0	6.5	6.1	5.3	4.8
	0.5	6.2	5.4	4.8	4.6	4.3	4.1	4.0	3.8	3.6	3.3	3.2
6	12	17.0	15.5	14.6	14.1	13.5	13.1	12.7	12.0	11.5	11.2	10.7
	8	16.2	14.7	13.9	13.4	12.6	12.2	11.7	11.0	10.5	10.0	9.5
	6	15.2	14.0	12.9	12.3	11.6	10.9	10.7	10.0	9.4	8.8	8.3
	4.5	13.7	12.4	11.5	11.0	10.5	10.0	9.6	9.0	8.3	7.6	7.1
	2.5	11.9	10.8	10.2	9.8	9.4	9.0	8.5	7.9	7.2	6.2	6.0
	1.5	9.9	9.0	8.3	7.7	7.3	7.0	6.7	6.4	6.0	4.9	4.7
	0.5	6.3	5.5	4.9	4.7	4.5	4.3	4.1	4.0	3.7	3.5	3.4
8	12	15.8	14.3	13.1	13.0	12.1	12.0	11.5	10.8	10.2	10.0	9.5
	8	15.0	13.7	12.5	12.1	11.4	11.0	10.6	10.1	9.4	9.0	8.6
	6	14.2	13.0	11.7	11.2	10.6	10.0	9.7	9.3	8.6	8.1	7.6
	4.5	12.8	11.7	10.7	10.1	9.7	9.0	8.7	8.3	7.8	7.2	6.6
	2.5	11.2	10.2	9.5	8.8	8.3	7.9	7.7	7.4	6.9	6.1	5.6
	1.5	9.5	8.6	7.8	7.1	6.9	6.7	6.5	6.2	5.8	4.9	4.5
	0.5	6.3	5.6	5.0	4.8	4.5	4.4	4.2	4.1	3.9	3.6	3.5
10	12	14.8	13.3	12.0	11.9	11.1	10.9	10.5	9.9	9.3	9.2	8.8
	8	14.2	12.8	11.6	11.2	10.5	10.2	9.7	9.2	8.6	8.3	8.0
	6	13.4	12.1	10.8	10.5	9.8	9.3	8.9	8.6	8.0	7.6	7.2
	4.5	12.3	10.8	10.0	9.5	9.0	8.5	8.1	7.8	7.3	6.8	6.4
	2.5	10.8	9.6	9.0	8.4	8.0	7.7	7.3	7.0	6.5	5.9	5.5
	1.5	9.2	8.3	7.5	6.9	6.7	6.5	6.3	6.0	5.6	5.0	4.6
	0.5	6.4	5.7	5.1	4.9	4.7	4.5	4.3	4.2	4.0	3.8	3.7
12	12	14.0	12.3	11.1	10.8	10.1	9.8	9.5	9.0	8.5	8.2	7.9
	8	13.2	11.8	10.7	10.2	9.6	9.3	8.9	8.4	7.9	7.5	7.2
	6	12.6	11.2	10.2	9.7	9.0	8.5	8.2	7.9	7.4	6.9	6.4
	4.5	11.7	10.2	9.4	9.0	8.4	8.0	7.6	7.2	6.8	6.3	5.8
	2.5	10.5	9.2	8.5	8.0	7.8	7.4	7.0	6.6	6.1	5.6	5.2
	1.5	8.9	8.0	7.2	6.8	6.5	6.3	6.1	5.7	5.4	4.8	4.5
	0.5	6.5	5.8	5.3	5.1	4.9	4.7	4.5	4.3	4.2	4.0	3.3
14	12	13.1	11.3	10.2	9.7	9.1	8.8	8.4	8.1	7.6	7.2	7.0
	8	12.4	10.9	9.8	9.2	8.7	8.3	8.0	7.6	7.2	6.8	6.4
	6	11.8	10.4	9.3	8.8	8.3	7.7	7.5	7.2	6.8	6.3	5.9
	4.5	11.1	9.7	8.7	8.2	7.8	7.2	7.0	6.7	6.3	5.8	5.4
	2.5	10.1	8.8	8.1	7.6	7.1	6.7	6.5	6.2	5.7	5.2	5.0
	1.5	8.6	7.7	7.0	6.7	6.3	6.0	5.8	5.5	5.2	4.7	4.5
	0.5	6.6	5.9	5.4	5.2	5.0	4.8	4.6	4.4	4.3	4.2	4.0
16	12	12.3	10.3	9.2	8.6	8.1	7.7	7.4	7.2	6.8	6.3	6.1
	8	11.7	10.0	8.9	8.3	7.8	7.4	7.2	6.8	6.5	6.0	5.7
	6	11.0	9.6	8.5	8.0	7.5	7.1	6.9	6.5	6.2	5.7	5.4
	4.5	10.5	9.1	8.1	7.5	7.2	6.8	6.5	6.2	5.8	5.3	5.1
	2.5	9.7	8.4	7.6	7.0	6.7	6.3	6.1	5.7	5.4	4.9	4.7
	1.5	8.3	7.4	6.6	6.2	6.0	5.8	5.6	5.3	5.1	4.6	4.4
	0.5	6.7	5.9	5.5	5.3	5.1	5.0	4.8	4.6	4.5	4.3	4.1

FL/F = the ratio of the spring free length FL to the deflection F.

Torsion Spring Design.—Fig. 19 shows the types of ends most commonly used on torsion springs. To produce them requires only limited tooling. The straight torsion end is the least expensive and should be used whenever possible. After determining the spring load or torque required and selecting the end formations, the designer usually estimates suitable space or size limitations. However, the space should be considered approximate until the wire size and number of coils have been determined. The wire size is dependent principally upon the torque. Design data can be devoloped with the aid of the tabular data, which is a simple method, or by calculation alone, as shown in the following sections. Many other factors affecting the design and operation of torsion springs are also covered in the section, *Torsion Spring Design Recommendations* on page page 325. Design formulas are shown in Table .

Curvature correction: In addition to the stress obtained from the formulas for load or deflection, there is a direct shearing stress on the inside of the section due to curvature. Therefore, the stress obtained by the usual formulas should be multiplied by the factor K obtained from the curve in Fig. 20. The corrected stress thus obtained is used only for comparison with the allowable working stress (fatigue strength) curves to determine if it is a safe value, and should not be used in the formulas for deflection.

Torque: Torque is a force applied to a moment arm and tends to produce rotation. Torsion springs exert torque in a circular arc and the arms are rotated about the central axis. It should be noted that the stress produced is in bending, not in torsion. In the spring industry it is customary to specify torque in conjunction with the deflection or with the arms of a spring at a definite position. Formulas for torque are expressed in pound-inches. If ounce-inches are specified, it is necessary to divide this value by 16 in order to use the formulas.

When a load is specified at a distance from a centerline, the torque is, of course, equal to the load multiplied by the distance. The load can be in pounds or ounces with the distances in inches or the load can be in grams or kilograms with the distance in centimeters or millimeters, but to use the design formulas, all values must be converted to pounds and inches. Design formulas for torque are based on the tangent to the arc of rotation and presume that a rod is used to support the spring. The stress in bending caused by the moment $P \times R$ is identical in magnitude to the torque T, provided a rod is used.

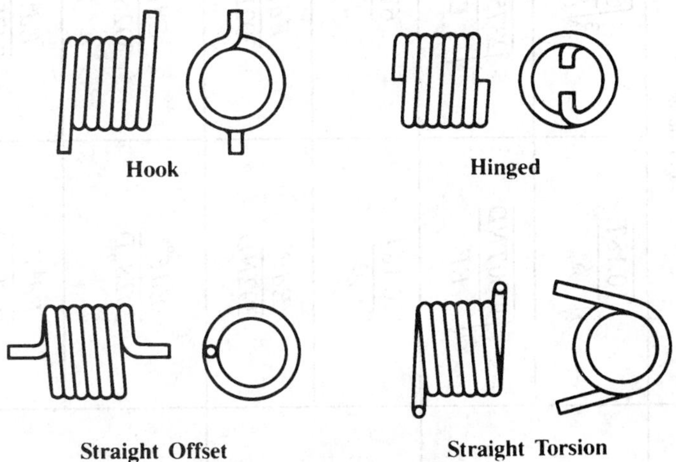

Fig. 19. The Most Commonly Used Types of Ends for Torsion Springs

Theoretically, it makes no difference how or where the load is applied to the arms of torsion springs. Thus, in Fig. 21, the loads shown multiplied by their respective distances produce the same torque; i.e., $20 \times 0.5 = 10$ pound-inches; $10 \times 1 = 10$ pound-inches; and $5 \times 2 = 10$ pound-inches. To further simplify the understanding of torsion spring torque, observe in both Fig. 22 and Fig. 23 that although the turning force is in a circular arc the torque is not

Table 13. Formulas for Torsion Springs

Feature	Springs made from round wire	Springs made from square wire	Feature	Springs made from round wire	Springs made from square wire
	Formula[a]			Formula[a]	
d = Wire diameter, Inches	$\sqrt[3]{\frac{10.18T}{S_b}}$	$\sqrt[3]{\frac{6T}{S_b}}$	$F°$ = Deflection	$\frac{392S_bND}{Ed}$	$\frac{392S_bND}{Ed}$
	$\sqrt[4]{\frac{4000TND}{EF°}}$	$\sqrt[4]{\frac{2375TND}{EF°}}$		$\frac{4000TND}{Ed^4}$	$\frac{2375TND}{Ed^4}$
S_b = Stress, bending pounds per square inch	$\frac{10.18T}{d^3}$	$\frac{6T}{d^3}$	T = Torque Inch lbs. (Also = $P \times R$)	$0.0982S_bd^3$	$0.1666S_bd^3$
	$\frac{EdF°}{392ND}$	$\frac{EdF°}{392ND}$		$\frac{Ed^4F°}{4000ND}$	$\frac{Ed^4F°}{2375ND}$
N = Active Coils	$\frac{EdF°}{392S_bD}$	$\frac{EdF°}{392S_bD}$	ID_1 = Inside Diameter After Deflection, Inches	$\frac{N(ID\ \text{free})}{N + \frac{F°}{360}}$	$\frac{N(ID\ \text{free})}{N + \frac{F°}{360}}$
	$\frac{Ed^4F°}{4000TD}$	$\frac{Ed^4F°}{2375TD}$	The symbol notation is given on page 285.		

[a] Where two formulas are given for one feature, the designer should use the one found to be appropriate for the given design. The end result from either of any two formulas is the same.

equal to P times the radius. The torque in both designs equals $P \times R$ because the spring rests against the support rod at point a.

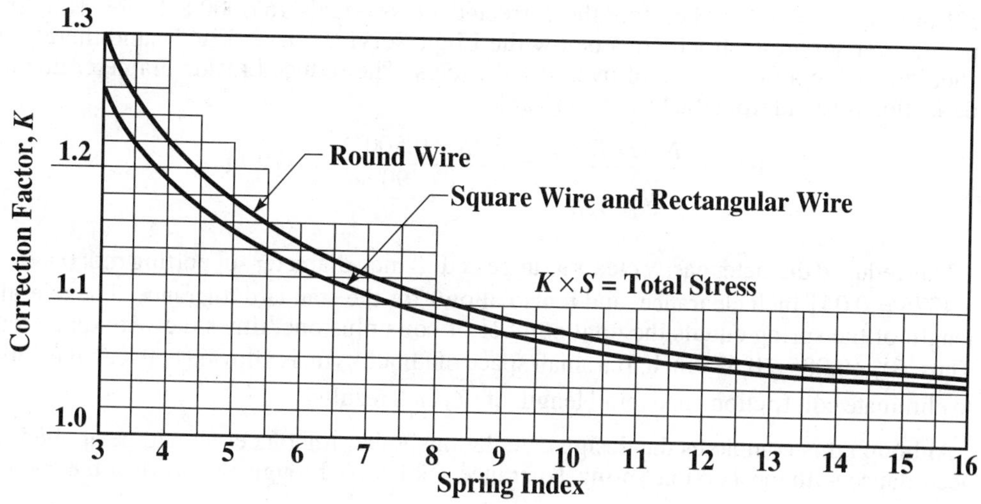

Fig. 20. Torsion Spring Stress Correction for Curvature

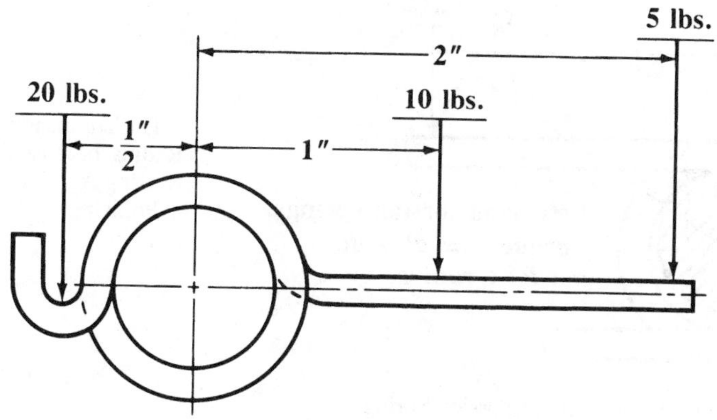

Fig. 21. Right-Hand Torsion Spring

Design Procedure: Torsion spring designs require more effort than other kinds because consideration has to be given to more details such as the proper size of a supporting rod, reduction of the inside diameter, increase in length, deflection of arms, allowance for friction, and method of testing.

Example: What music wire diameter and how many coils are required for the torsion spring shown in Fig. 24, which is to withstand at least 1000 cycles? Determine the corrected stress and the reduced inside diameter after deflection.

Method 1, using table: From Table 15, page 321, locate the ½ inch inside diameter for the spring in the left-hand column. Move to the right and then vertically to locate a torque value nearest to the required 10 pound-inches, which is 10.07 pound-inches. At the top of the same column, the music wire diameter is found, which is Number 31 gauge (0.085 inch). At the bottom of the same column the deflection for one coil is found, which is 15.81 degrees. As a 90-degree deflection is required, the number of coils needed is 90/15.81 = 5.69 (say 5¾ coils).

The spring index $\frac{D}{d} = \frac{0.500 + 0.085}{0.085} = 6.88$ and thus the curvature correction factor K from Fig. 20 = 1.13. Therefore the corrected stress equals 167,000 × 1.13 = 188,700 pounds per square inch which is below the Light Service curve (Fig. 7) and therefore should provide a fatigue life of over 1,000 cycles. The reduced inside diameter due to deflection is found from the formula in Table :

$$ID_1 = \frac{N(ID\ \text{free})}{N + \frac{F}{360}} = \frac{5.75 \times 0.500}{5.75 + \frac{90}{360}} = 0.479 \text{ in.}$$

This reduced diameter easily clears a suggested $^7\!/_{16}$ inch diameter supporting rod: 0.479 – 0.4375 = 0.041 inch clearance, and it also allows for the standard tolerance. The overall length of the spring equals the total number of coils plus one, times the wire diameter. Thus, $6\frac{3}{4} \times 0.085 = 0.574$ inch. If a small space of about $^1\!/_{64}$ in. is allowed between the coils to eliminate coil friction, an overall length of $^{21}\!/_{32}$ inch results.

Although this completes the design calculations, other tolerances should be applied in accordance with the Torsion Spring Tolerance Tables 16 through 18 shown at the end of this section.

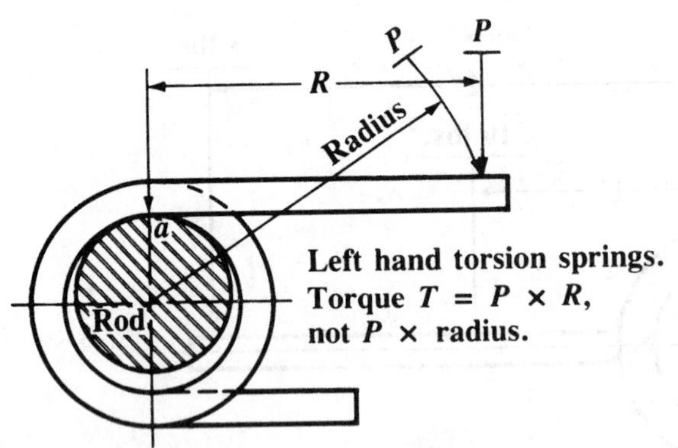

The Torque is $T = P \times R$, Not $P \times$ Radius, because the Spring is Resting Against the Support Rod at Point a

Fig. 22. Left-Hand Torsion Spring

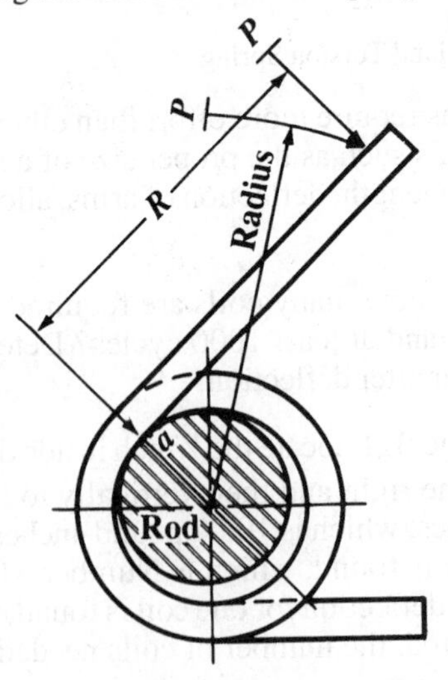

As with the Spring in Fig. 22, the Torque is $T = P \times R$, Not $P \times$ Radius, Because the Support Point Is at a

Fig. 23. Left-Hand Torsion Spring

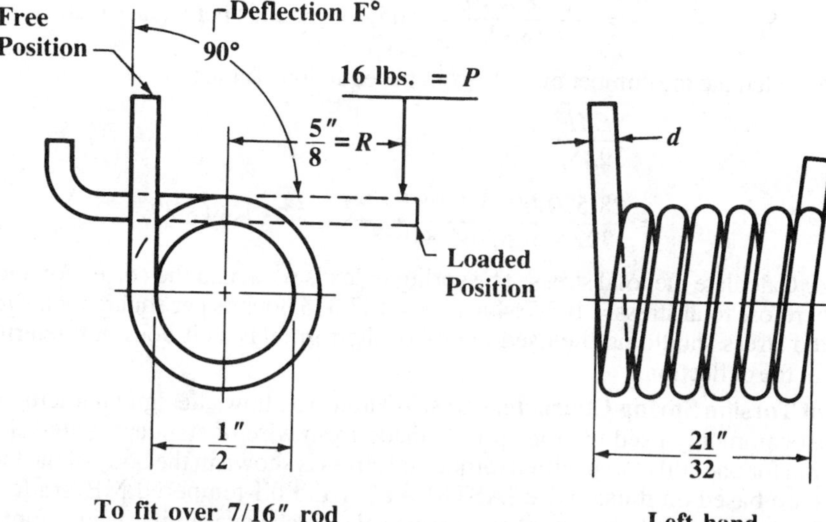

Fig. 24. Torsion Spring Design Example. The Spring Is to be Assembled on a 7/16-Inch Support Rod

Longer fatigue life: If a longer fatigue life is desired, use a slightly larger wire diameter. Usually the next larger gage size is satisfactory. The larger wire will reduce the stress and still exert the same torque, but will require more coils and a longer overall length.

Percentage method for calculating longer life: The spring design can be easily adjusted for longer life as follows:

1) Select the next larger gage size, which is Number 32 (0.090 inch) from Table 15. The torque is 11.88 pound-inches, the design stress is 166,000 pounds per square inch, and the deflection is 14.9 degrees per coil. As a percentage the torque is $10/11.88 \times 100 = 84$ per cent.

2) The new stress is $0.84 \times 166{,}000 = 139{,}440$ pounds per square inch. This value is under the bottom or Severe Service curve, Fig. 7, and thus assures longer life.

3) The new deflection per coil is $0.84 \times 14.97 = 12.57$ degrees. Therefore, the total number of coils required = $90/12.57 = 7.16$ (say 7 ⅛). The new overall length = $8\ ⅛ \times 0.090 = 0.73$ inch (say ¾ inch). A slight increase in the overall length and new arm location are thus necessary.

Method 2, using formulas: When using this method, it is often necessary to solve the formulas several times because assumptions must be made initially either for the stress or for a wire size. The procedure for design using formulas is as follows (the design example is the same as in Method 1, and the spring is shown in Fig. 24):

Step 1: Note from Table , page 315 that the wire diameter formula is:

$$d = \sqrt[3]{\frac{10.18T}{S_b}}$$

Step 2: Referring to Fig. 7, select a trial stress, say 150,000 pounds per square inch.

Step 3: Apply the trial stress, and the 10 pound-inches torque value in the wire diameter formula:

$$d = \sqrt[3]{\frac{10.18T}{S_b}} = \sqrt[3]{\frac{10.18 \times 10}{150{,}000}} = \sqrt[3]{0.000679} = 0.0879 \text{ inch}$$

The nearest gauge sizes are 0.085 and 0.090 inch diameter. *Note*: Table 21, page 330, can be used to avoid solving the cube root.

Step 4: Select 0.085 inch wire diameter and solve the equation for the actual stress:

$$S_b = \frac{10.18T}{d^3} = \frac{10.18 \times 10}{0.085^3} = 165{,}764 \text{ pounds per square inch}$$

Step 5: Calculate the number of coils from the equation, Table :

$$N = \frac{EdF^{\circ}}{392 S_b D}$$

$$= \frac{28{,}500{,}000 \times 0.085 \times 90}{392 \times 165{,}764 \times 0.585} = 5.73 \text{ (say } 5\tfrac{3}{4}\text{)}$$

Step 6: Calculate the total stress. The spring index is 6.88, and the correction factor K is 1.13, therefore total stress = 165,764 × 1.13 = 187,313 pounds per square inch. *Note*: The corrected stress should not be used in any of the formulas as it does not determine the torque or the deflection.

Table of Torsion Spring Characteristics.—Table 15 shows design characteristics for the most commonly used torsion springs made from wire of standard gauge sizes. The deflection for one coil at a specified torque and stress is shown in the body of the table. The figures are based on music wire (ASTM A228) and oil-tempered MB grade (ASTM A229), and can be used for several other materials which have similar values for the modulus of elasticity E. However, the design stress may be too high or too low, and the design stress, torque, and deflection per coil should each be multiplied by the appropriate correction factor in Table 14 when using any of the materials given in that table.

Table 14. Correction Factors for Other Materials

Material	Factor	Material	Factor
Hard Drawn MB	0.75	Stainless 316	
Chrome-Vanadium	1.10	Up to ⅛ inch diameter	0.75
Chrome-Silicon	1.20	Over ⅛ to ¼ inch diameter	0.65
Stainless 302 and 304		Over ¼ inch diameter	0.65
Up to ⅛ inch diameter	0.85	Stainless 17–7 PH	
Over ⅛ to ¼ inch diameter	0.75	Up to ⅛ inch diameter	1.00
Over ¼ inch diameter	0.65	Over ⅛ to 3⁄16 inch diameter	1.07
Stainless 431	0.80	Over 3⁄16 inch diameter	1.12
Stainless 420	0.85	…	…

For use with values in Table 15. *Note*: The figures in Table 15 are for music wire (ASTM A228) and oil-tempered MB grade (ASTM A229) and can be used for several other materials that have a similar modulus of elasticity E. However, the design stress may be too high or too low, and therefore the design stress, torque, and deflection per coil should each be multiplied by the appropriate correction factor when using any of the materials given in this table (Table 14).

Table 15. Torsion Spring Deflections

Inside Diam.		AMW Wire Gauge and Decimal Equivalent[a]															
		1 .010	2 .011	3 .012	4 .013	5 .014	6 .016	7 .018	8 .020	9 .022	10 .024	11 .026	12 .029	13 .031	14 .033	15 .035	16 .037
		Design Stress, pounds per sq. in. (thousands)															
		232	229	226	224	221	217	214	210	207	205	202	199	197	196	194	192
		Torque, pound-inch															
		.0228	.0299	.0383	.0483	.0596	.0873	.1226	.1650	.2164	.2783	.3486	.4766	.5763	.6917	.8168	.9550
Fractional	Decimal	Deflection, degrees per coil															
1/16	0.0625	22.35	20.33	18.64	17.29	16.05	14.15	18.72	11.51	10.56	9.818	9.137	8.343	7.896	…	…	…
5/64	0.078125	27.17	24.66	22.55	20.86	19.32	16.96	15.19	13.69	12.52	11.59	10.75	9.768	9.215	…	…	…
3/32	0.09375	31.98	28.98	26.47	24.44	22.60	19.78	17.65	15.87	14.47	13.36	12.36	11.19	10.53	10.18	9.646	9.171
7/64	0.109375	36.80	33.30	30.38	28.02	25.88	22.60	20.12	18.05	16.43	15.14	13.98	12.62	11.85	11.43	10.82	10.27
1/8	0.125	41.62	37.62	34.29	31.60	29.16	25.41	22.59	20.23	18.38	16.91	15.59	14.04	13.17	12.68	11.99	11.36
9/64	0.140625	46.44	41.94	38.20	35.17	32.43	28.23	25.06	22.41	20.33	18.69	17.20	15.47	14.49	13.94	13.16	12.46
5/32	0.15625	51.25	46.27	42.11	38.75	35.71	31.04	27.53	24.59	22.29	20.46	18.82	16.89	15.81	15.19	14.33	13.56
3/16	0.1875	60.89	54.91	49.93	45.91	42.27	36.67	32.47	28.95	26.19	24.01	22.04	19.74	18.45	17.70	16.67	15.75
7/32	0.21875	70.52	63.56	57.75	53.06	48.82	42.31	37.40	33.31	30.10	27.55	25.27	22.59	21.09	20.21	19.01	17.94
1/4	0.250	80.15	72.20	65.57	60.22	55.38	47.94	42.34	37.67	34.01	31.10	28.49	25.44	23.73	22.72	21.35	20.13

Inside Diam.		AMW Wire Gauge and Decimal Equivalent[a]														
		17 .039	18 .041	19 .043	20 .045	21 .047	22 .049	23 .051	24 .055	25 .059	26 .063	27 .067	28 .071	29 .075	30 .080	31 .085
		Design Stress, pounds per sq. in. (thousands)														
		190	188	187	185	184	183	182	180	178	176	174	173	171	169	167
		Torque, pound-inch														
		1.107	1.272	1.460	1.655	1.876	2.114	2.371	2.941	3.590	4.322	5.139	6.080	7.084	8.497	10.07
Fractional	Decimal	Deflection, degrees per coil														
7/64	0.109375	9.771	9.320	8.957	…	…	…	…	…	…	…	…	…	…	…	…
1/8	0.125	10.80	10.29	9.876	9.447	9.102	8.784	…	…	…	…	…	…	…	…	…
9/64	0.140625	11.83	11.26	10.79	10.32	9.929	9.572	9.244	8.654	8.141	…	…	…	…	…	…
5/32	0.15625	12.86	12.23	11.71	11.18	10.76	10.36	9.997	9.345	8.778	8.279	7.975	…	…	…	…
3/16	0.1875	14.92	14.16	13.55	12.92	12.41	11.94	11.50	10.73	10.05	9.459	9.091	8.663	8.232	7.772	7.364
7/32	0.21875	16.97	16.10	15.39	14.66	14.06	13.52	13.01	12.11	11.33	10.64	10.21	9.711	9.212	8.680	8.208
1/4	0.250	19.03	18.04	17.22	16.39	15.72	15.09	14.52	13.49	12.60	11.82	11.32	10.76	10.19	9.588	9.053

[a] For sizes up to 13 gauge, the table values are for music wire with a modulus E of 29,000,000 psi; and for sizes from 27 to 31 guage, the values are for oil-tempered MB with a modulus of 28,500,000 psi.

Table 15. *(Continued)* Torsion Spring Deflections

Inside Diam.		AMW Wire Gauge and Decimal Equivalent[a]															
		8 .020	9 .022	10 .024	11 .026	12 .029	13 .031	14 .033	15 .035	16 .037	17 .039	18 .041	19 .043	20 .045	21 .047	22 .049	23 .051
		Design Stress, pounds per sq. in. (thousands)															
		210	207	205	202	199	197	196	194	192	190	188	187	185	184	183	182
		Torque, pound-inch															
		.1650	.2164	.2783	.3486	.4766	.5763	.6917	.8168	.9550	1.107	1.272	1.460	1.655	1.876	2.114	2.371
Fractional	Decimal	Deflection, degrees per coil															
9⁄32	0.28125	42.03	37.92	34.65	31.72	28.29	26.37	25.23	23.69	22.32	21.09	19.97	19.06	18.13	17.37	16.67	16.03
5⁄16	0.3125	46.39	41.82	38.19	34.95	31.14	29.01	27.74	26.04	24.51	23.15	21.91	20.90	19.87	19.02	18.25	17.53
11⁄32	0.34375	50.75	45.73	41.74	38.17	33.99	31.65	30.25	28.38	26.71	25.21	23.85	22.73	21.60	20.68	19.83	19.04
3⁄8	0.375	55.11	49.64	45.29	41.40	36.84	34.28	32.76	30.72	28.90	27.26	25.78	24.57	23.34	22.33	21.40	20.55
13⁄32	0.40625	59.47	53.54	48.85	44.63	39.69	36.92	35.26	33.06	31.09	29.32	27.72	26.41	25.08	23.99	22.98	22.06
7⁄16	0.4375	63.83	57.45	52.38	47.85	42.54	39.56	37.77	35.40	33.28	31.38	29.66	28.25	26.81	25.64	24.56	23.56
15⁄32	0.46875	68.19	61.36	55.93	51.00	45.39	42.20	40.28	37.74	35.47	33.44	31.59	30.08	28.55	27.29	26.14	25.07
1⁄2	0.500	72.55	65.27	59.48	54.30	48.24	44.84	42.79	40.08	37.67	35.49	33.53	31.92	30.29	28.95	27.71	26.58

Inside Diam.		AMW Wire Gauge and Decimal Equivalent[a]															
		24 .055	25 .059	26 .063	27 .067	28 .071	29 .075	30 .080	31 .085	32 .090	33 .095	34 .100	35 .106	36 .112	37 .118	1⁄8 125
		Design Stress, pounds per sq. in. (thousands)														
		180	178	176	174	173	171	169	167	166	164	163	161	160	158	156
		Torque, pound-inch														
		2.941	3.590	4.322	5.139	6.080	7.084	8.497	10.07	11.88	13.81	16.00	18.83	22.07	25.49	29.92
Fractional	Decimal	Deflection, degrees per coil														
9⁄32	0.28125	14.88	13.88	13.00	12.44	11.81	11.17	10.50	9.897	9.418	8.934	8.547	8.090	7.727	7.353	6.973
5⁄16	0.3125	16.26	15.15	14.18	13.56	12.85	12.15	11.40	10.74	10.21	9.676	9.248	8.743	8.341	7.929	7.510
11⁄32	0.34375	17.64	16.42	15.36	14.67	13.90	13.13	12.31	11.59	11.00	10.42	9.948	9.396	8.955	8.504	8.046
3⁄8	0.375	19.02	17.70	16.54	15.79	14.95	14.11	13.22	12.43	11.80	11.16	10.65	10.05	9.569	9.080	8.583
13⁄32	0.40625	20.40	18.97	17.72	16.90	15.99	15.09	14.13	13.28	12.59	11.90	11.35	10.70	10.18	9.655	9.119
7⁄16	0.4375	21.79	20.25	18.90	18.02	17.04	16.07	15.04	14.12	13.38	12.64	12.05	11.35	10.80	10.23	9.655
15⁄32	0.46875	23.17	21.52	20.08	19.14	18.09	17.05	15.94	14.96	14.17	13.39	12.75	12.01	11.41	10.81	10.19
1⁄2	0.500	24.55	22.80	21.26	20.25	19.14	18.03	16.85	15.81	14.97	14.13	13.45	12.66	12.03	11.38	10.73

[a] For sizes up to 13 gauge, the table values are for music wire with a modulus *E* of 29,000,000 psi; and for sizes from 27 to 31 guage, the values are for oil-tempered MB with a modulus of 28,500,000 psi.

Table 15. *(Continued)* Torsion Spring Deflections

Inside Diam.		AMW Wire Gauge and Decimal Equivalent[a]														
		16 .037	17 .039	18 .041	19 .043	20 .045	21 .047	22 .049	23 .051	24 .055	25 .059	26 .063	27 .067	28 .071	29 .075	30 .080
		Design Stress, pounds per sq. in. (thousands)														
		192	190	188	187	185	184	183	182	180	178	176	174	173	171	169
		Torque, pound-inch														
Fractional	Decimal	.9550	1.107	1.272	1.460	1.655	1.876	2.114	2.371	2.941	3.590	4.322	5.139	6.080	7.084	8.497
		Deflection, degrees per coil														
17⁄32	0.53125	39.86	37.55	35.47	33.76	32.02	30.60	29.29	28.09	25.93	24.07	22.44	21.37	20.18	19.01	17.76
9⁄16	0.5625	42.05	39.61	37.40	35.59	33.76	32.25	30.87	29.59	27.32	25.35	23.62	22.49	21.23	19.99	18.67
19⁄32	0.59375	44.24	41.67	39.34	37.43	35.50	33.91	32.45	31.10	28.70	26.62	24.80	23.60	22.28	20.97	19.58
5⁄8	0.625	46.43	43.73	41.28	39.27	37.23	35.56	34.02	32.61	30.08	27.89	25.98	24.72	23.33	21.95	20.48
21⁄32	0.65625	48.63	45.78	43.22	41.10	38.97	37.22	35.60	34.12	31.46	29.17	27.16	25.83	24.37	22.93	21.39
11⁄16	0.6875	50.82	47.84	45.15	42.94	40.71	38.87	37.18	35.62	32.85	30.44	28.34	26.95	25.42	23.91	22.30
23⁄32	0.71875	53.01	49.90	47.09	44.78	42.44	40.52	38.76	37.13	34.23	31.72	29.52	28.07	26.47	24.89	23.21
3⁄4	0.750	55.20	51.96	49.03	46.62	44.18	42.18	40.33	38.64	35.61	32.99	30.70	29.18	27.52	25.87	24.12

Inside Diam.		Wire Gauge and Decimal Equivalent[ab]															
		31 .085	32 .090	33 .095	34 .100	35 .106	36 .112	37 .118	1⁄8 .125	10 .135	9 .1483	5⁄32 .1563	8 .162	7 .177	3⁄16 .1875	6 .192	5 .207
		Design Stress, pounds per sq. in. (thousands)															
		167	166	164	163	161	160	158	156	161	158	156	154	150	149	146	143
		Torque, pound-inch															
Fractional	Decimal	10.07	11.88	13.81	16.00	18.83	22.07	25.49	29.92	38.90	50.60	58.44	64.30	81.68	96.45	101.5	124.6
		Deflection, degrees per coil															
17⁄32	0.53125	16.65	15.76	14.87	14.15	13.31	12.64	11.96	11.26	10.93	9.958	9.441	9.064	8.256	7.856	7.565	7.015
9⁄16	0.5625	17.50	16.55	15.61	14.85	13.97	13.25	12.53	11.80	11.44	10.42	9.870	9.473	8.620	8.198	7.891	7.312
19⁄32	0.59375	18.34	17.35	16.35	15.55	14.62	13.87	13.11	12.34	11.95	10.87	10.30	9.882	8.984	8.539	8.218	7.609
5⁄8	0.625	19.19	18.14	17.10	16.25	15.27	14.48	13.68	12.87	12.47	11.33	10.73	10.29	9.348	8.881	8.545	7.906
21⁄32	0.65625	20.03	18.93	17.84	16.95	15.92	15.10	14.26	13.41	12.98	11.79	11.16	10.70	9.713	9.222	8.872	8.202
11⁄16	0.6875	20.88	19.72	18.58	17.65	16.58	15.71	14.83	13.95	13.49	12.25	11.59	11.11	10.08	9.564	9.199	8.499
23⁄32	0.71875	21.72	20.52	19.32	18.36	17.23	16.32	15.41	14.48	14.00	12.71	12.02	11.52	10.44	9.905	9.526	8.796
3⁄4	0.750	22.56	21.31	20.06	19.06	17.88	16.94	15.99	15.02	14.52	13.16	12.44	11.92	10.81	10.25	9.852	9.093

[a] For sizes up to 26 gauge, the table values are for music wire with a modulus E of 29,500,000 psi; for sizes from 27 to 1⁄8 inch diameter the table values are for music wire with a modulus of 28,500,000 psi; for sizes from 10 gauge to 1⁄8 inch diameter, the values are for oil-tempered MB with a modulus of 28,500,000 psi.

[b] Gauges 31 through 37 are AMW gauges. Gauges 10 through 5 are Washburn and Moen.

Table 15. *(Continued)* Torsion Spring Deflections

Inside Diam.		AMW Wire Gauge and Decimal Equivalent[a]														
		24 .055	25 .059	26 .063	27 .067	28 .071	29 .075	30 .080	31 .085	32 .090	33 .095	34 .100	35 .106	36 .112	37 .118	⅛ .125
		Design Stress, pounds per sq. in. (thousands)														
		180	178	176	174	173	171	169	167	166	164	163	161	160	158	156
		Torque, pound-inch														
		2.941	3.590	4.322	5.139	6.080	7.084	8.497	10.07	11.88	13.81	16.00	18.83	22.07	25.49	29.92
Fractional	Decimal	Deflection, degrees per coil														
13⁄16	0.8125	38.38	35.54	33.06	31.42	29.61	27.83	25.93	24.25	22.90	21.55	20.46	19.19	18.17	17.14	16.09
7⁄8	0.875	41.14	38.09	35.42	33.65	31.70	29.79	27.75	25.94	24.58	23.03	21.86	20.49	19.39	18.29	17.17
15⁄16	0.9375	43.91	40.64	37.78	35.88	33.80	31.75	29.56	27.63	26.07	24.52	23.26	21.80	20.62	19.44	18.24
1	1.000	46.67	43.19	40.14	38.11	35.89	33.71	31.38	29.32	27.65	26.00	24.66	23.11	21.85	20.59	19.31
1 1⁄16	1.0625	49.44	45.74	42.50	40.35	37.99	35.67	33.20	31.01	29.24	27.48	26.06	24.41	23.08	21.74	20.38
1⅛	1.125	52.20	48.28	44.86	42.58	40.08	37.63	35.01	32.70	30.82	28.97	27.46	25.72	24.31	22.89	21.46
1 3⁄16	1.1875	54.97	50.83	47.22	44.81	42.18	39.59	36.83	34.39	32.41	30.45	28.86	27.02	25.53	24.04	22.53
1¼	1.250	57.73	53.38	49.58	47.04	44.27	41.55	38.64	36.08	33.99	31.94	30.27	28.33	26.76	25.19	23.60

Inside Diam.		Washburn and Moen Gauge or Size and Decimal Equivalent[a]														
		10 .135	9 .1483	5⁄32 .1563	8 .162	7 .177	3⁄16 .1875	6 .192	5 .207	7⁄32 .2188	4 .2253	3 .2437	¼ .250	9⁄32 .2813	5⁄16 .3125	11⁄32 .3438	⅜ .375
		Design Stress, pounds per sq. in. (thousands)														
		161	158	156	154	150	149	146	143	142	141	140	139	138	137	136	135
		Torque, pound-inch														
		38.90	50.60	58.44	64.30	81.68	96.45	101.5	124.6	146.0	158.3	199.0	213.3	301.5	410.6	542.5	700.0
Fractional	Decimal	Deflection, degrees per coil														
13⁄16	0.8125	15.54	14.08	13.30	12.74	11.53	10.93	10.51	9.687	9.208	8.933	8.346	8.125	7.382	6.784	6.292	5.880
7⁄8	0.875	16.57	15.00	14.16	13.56	12.26	11.61	11.16	10.28	9.766	9.471	8.840	8.603	7.803	7.161	6.632	6.189
15⁄16	0.9375	17.59	15.91	15.02	14.38	12.99	12.30	11.81	10.87	10.32	10.01	9.333	9.081	8.225	7.537	6.972	6.499
1	1.000	18.62	16.83	15.88	15.19	13.72	12.98	12.47	11.47	10.88	10.55	9.827	9.559	8.647	7.914	7.312	6.808
1 1⁄16	1.0625	19.64	17.74	16.74	16.01	14.45	13.66	13.12	12.06	11.44	11.09	10.32	10.04	9.069	8.291	7.652	7.118
1⅛	1.125	20.67	18.66	17.59	16.83	15.18	14.35	13.77	12.66	12.00	11.62	10.81	10.52	9.491	8.668	7.993	7.427
1 3⁄16	1.1875	21.69	19.57	18.45	17.64	15.90	15.03	14.43	13.25	12.56	12.16	11.31	10.99	9.912	9.045	8.333	7.737
1¼	1.250	22.72	20.49	19.31	18.46	16.63	15.71	15.08	13.84	13.11	12.70	11.80	11.47	10.33	9.422	8.673	8.046

[a] For sizes up to 26 gauge, the table values are for music wire with a modulus E of 29,500,000 psi; for sizes from 27 to ⅛ inch diameter the table values are for music wire with a modulus of 28,500,000 psi; for sizes from 10 gauge to ⅛ inch diameter, the values are for oil-tempered MB with a modulus of 28,500,000 psi.

For an example in the use of the table, see the example starting on page 317. Note: Intermediate values may be interpolated within reasonable accuracy.

Torsion Spring Design Recommendations.—The following recommendations should be taken into account when designing torsion springs:

Hand: The hand or direction of coiling should be specified and the spring designed so deflection causes the spring to wind up and to have more coils. This increase in coils and overall length should be allowed for during design. Deflecting the spring in an unwinding direction produces higher stresses and may cause early failure. When a spring is sighted down the longitudinal axis, it is "right hand" when the direction of the wire into the spring takes a clockwise direction or if the angle of the coils follows an angle similar to the threads of a standard bolt or screw, otherwise it is "left hand." A spring must be coiled right-handed to engage the threads of a standard machine screw.

Rods: Torsion springs should be supported by a rod running through the center whenever possible. If unsupported, or if held by clamps or lugs, the spring will buckle and the torque will be reduced or unusual stresses may occur.

Diameter Reduction: The inside diameter reduces during deflection. This reduction should be computed and proper clearance provided over the supporting rod. Also, allowances should be considered for normal spring diameter tolerances.

Winding: The coils of a spring may be closely or loosely wound, but they seldom should be wound with the coils pressed tightly together. Tightly wound springs with initial tension on the coils do not deflect uniformly and are difficult to test accurately. A small space between the coils of about 20 to 25 per cent of the wire thickness is desirable. Square and rectangular wire sections should be avoided whenever possible as they are difficult to wind, expensive, and are not always readily available.

Arm Length: All the wire in a torsion spring is active between the points where the loads are applied. Deflection of long extended arms can be calculated by allowing one third of the arm length, from the point of load contact to the body of the spring, to be converted into coils. However, if the length of arm is equal to or less than one-half the length of one coil, it can be safely neglected in most applications.

Total Coils: Torsion springs having less than three coils frequently buckle and are difficult to test accurately. When thirty or more coils are used, light loads will not deflect all the coils simultaneously due to friction with the supporting rod. To facilitate manufacturing it is usually preferable to specify the total number of coils to the nearest fraction in eighths or quarters such as 5 ⅛, 5 ¼, 5 ½, etc.

Double Torsion: This design consists of one left-hand-wound series of coils and one series of right-hand-wound coils connected at the center. These springs are difficult to manufacture and are expensive, so it often is better to use two separate springs. For torque and stress calculations, each series is calculated separately as individual springs; then the torque values are added together, but the deflections are not added.

Bends: Arms should be kept as straight as possible. Bends are difficult to produce and often are made by secondary operations, so they are therefore expensive. Sharp bends raise stresses that cause early failure. Bend radii should be as large as practicable. Hooks tend to open during deflection; their stresses can be calculated by the same procedure as that for tension springs.

Spring Index: The spring index must be used with caution. In design formulas it is D/d. For shop measurement it is O.D./d. For arbor design it is I.D./d. Conversions are easily performed by either adding or subtracting 1 from D/d.

Proportions: A spring index between 4 and 14 provides the best proportions. Larger ratios may require more than average tolerances. Ratios of 3 or less, often cannot be coiled on automatic spring coiling machines because of arbor breakage. Also, springs with smaller or larger spring indexes often do not give the same results as are obtained using the design formulas.

Torsion Spring Tolerances.—Torsion springs are coiled in a different manner from other types of coiled springs and therefore different tolerances apply. The commercial tolerance on loads is ± 10 per cent and is specified with reference to the angular deflection. For example: 100 pound-inches ± 10 per cent at 45 degrees deflection. One load specified usually suffices. If two loads and two deflections are specified, the manufacturing and testing times are increased. Tolerances smaller than ± 10 per cent require each spring to be individually tested and adjusted, which adds considerably to manufacturing time and cost. Tables 16, 17, and 18 give, respectively, free angle tolerances, coil diameter tolerances, and tolerances on the number of coils.

Table 16. Torsion Spring Tolerances for Angular Relationship of Ends

Number of Coils (*N*)	Spring Index								
	4	6	8	10	12	14	16	18	20
	Free Angle Tolerance, ± degrees								
1	2	3	3.5	4	4.5	5	5.5	5.5	6
2	4	5	6	7	8	8.5	9	9.5	10
3	5.5	7	8	9.5	10.5	11	12	13	14
4	7	9	10	12	14	15	16	16.5	17
5	8	10	12	14	16	18	20	20.5	21
6	9.5	12	14.5	16	19	20.5	21	22.5	24
8	12	15	18	20.5	23	25	27	28	29
10	14	19	21	24	27	29	31.5	32.5	34
15	20	25	28	31	34	36	38	40	42
20	25	30	34	37	41	44	47	49	51
25	29	35	40	44	48	52	56	60	63
30	32	38	44	50	55	60	65	68	70
50	45	55	63	70	77	84	90	95	100

Table 17. Torsion Spring Coil Diameter Tolerances

Wire Diameter, Inch	Spring Index						
	4	6	8	10	12	14	16
	Coil Diameter Tolerance, ± inch						
0.015	0.002	0.002	0.002	0.002	0.003	0.003	0.004
0.023	0.002	0.002	0.002	0.003	0.004	0.005	0.006
0.035	0.002	0.002	0.003	0.004	0.006	0.007	0.009
0.051	0.002	0.003	0.005	0.007	0.008	0.010	0.012
0.076	0.003	0.005	0.007	0.009	0.012	0.015	0.018
0.114	0.004	0.007	0.010	0.013	0.018	0.022	0.028
0.172	0.006	0.010	0.013	0.020	0.027	0.034	0.042
0.250	0.008	0.014	0.022	0.030	0.040	0.050	0.060

Table 18. Torsion Spring Tolerance on Number of Coils

Number of Coils	Tolerance	Number of Coils	Tolerance
up to 5	±5°	over 10 to 20	±15°
over 5 to 10	±10°	over 20 to 40	±30°

Miscellaneous Springs.—This section provides information on various springs, some in common use, some less commonly used.

Conical compression: These springs taper from top to bottom and are useful where an increasing (instead of a constant) load rate is needed, where solid height must be small, and where vibration must be damped. Conical springs with a uniform pitch are easiest to coil. Load and deflection formulas for compression springs can be used – using the average mean coil diameter, and providing the deflection does not cause the largest active coil to lie against the bottom coil. When this happens, each coil must be calculated separately, using the standard formulas for compression springs.

Constant force springs: Those springs are made from flat spring steel and are finding more applications each year. Complicated design procedures can be eliminated by selecting a standard design from thousands now available from several spring manufacturers.

Spiral, clock, and motor springs: Although often used in wind-up type motors for toys and other products, these springs are difficult to design and results cannot be calculated with precise accuracy. However, many useful designs have been developed and are available from spring manufacturing companies.

Flat springs: These springs are often used to overcome operating space limitations in various products such as electric switches and relays. Table 19 lists formulas for designing flat springs. The formulas are based on standard beam formulas where the deflection is small.

Table 19. Formulas for Flat Springs

Feature	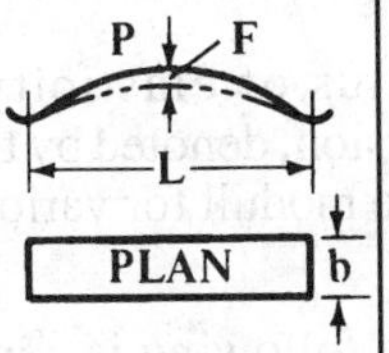	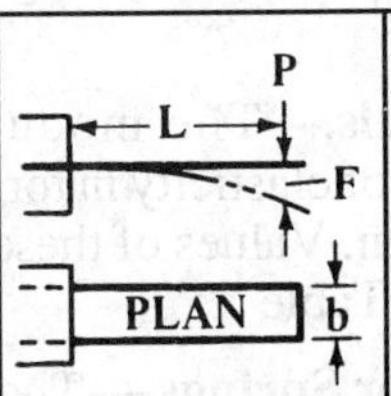	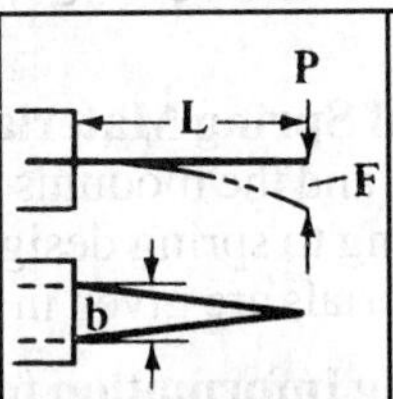	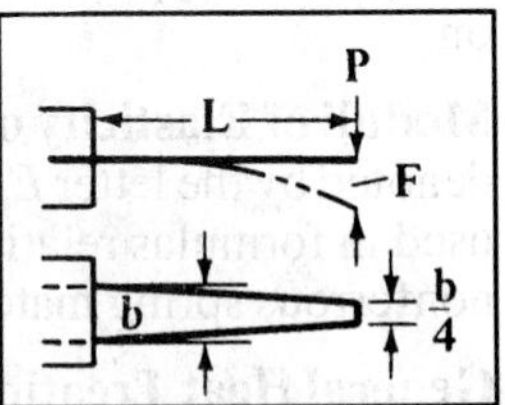
Deflect., f Inches	$f = \frac{PL^3}{4Ebt^3} = \frac{S_bL^2}{6Et}$	$f = \frac{4PL^3}{Ebt^3} = \frac{2S_bL^2}{3Et}$	$f = \frac{6PL^3}{Ebt^3} = \frac{S_bL^2}{Et}$	$f = \frac{5.22PL^3}{Ebt^3} = \frac{0.87S_bL^2}{Et}$
Load, P Pounds	$P = \frac{2S_bbt^2}{3L} = \frac{4Ebt^3F}{L^3}$	$P = \frac{S_bbt^2}{6L} = \frac{Ebt^3F}{4L^3}$	$P = \frac{S_bbt^2}{6L} = \frac{Ebt^3F}{6L^3}$	$P = \frac{S_bbt^2}{6L} = \frac{Ebt^3F}{5.22L^3}$
Stress, S_b Bending Pounds per sq. inch	$S_b = \frac{3PL}{2bt^2} = \frac{6EtF}{L^2}$	$S_b = \frac{6PL}{bt^2} = \frac{3EtF}{2L^2}$	$S_b = \frac{6PL}{bt^2} = \frac{EtF}{L^2}$	$S_b = \frac{6PL}{bt^2} = \frac{EtF}{0.87L^2}$
Thickness, t Inches	$t = \frac{S_bL^2}{6EF} = \sqrt[3]{\frac{PL^3}{4EbF}}$	$t = \frac{2S_bL^2}{3EF} = \sqrt[3]{\frac{4PL^3}{EbF}}$	$t = \frac{S_bL^2}{EF} = \sqrt[3]{\frac{6PL^3}{EbF}}$	$t = \frac{0.87S_bL^2}{EF} = \sqrt[3]{\frac{5.22PL^3}{EbF}}$

Based on standard beam formulas where the deflection is small

See page 285 for notation.

Note: Where two formulas are given for one feature, the designer should use the one found to be appropriate for the given design. The result from either of any two formulas is the same.

Belleville washers: These washer type springs can sustain relatively large loads with small deflections, and the loads and deflections can be increased by stacking the springs as shown in Fig. 25.

Design data is not given here because the wide variations in ratios of O.D. to I.D., height to thickness, and other factors require too many formulas for convenient use and involve constants obtained from more than 24 curves. It is now practicable to select required sizes from the large stocks carried by several of the larger spring manufacturing companies. Most of these companies also stock curved and wave washers.

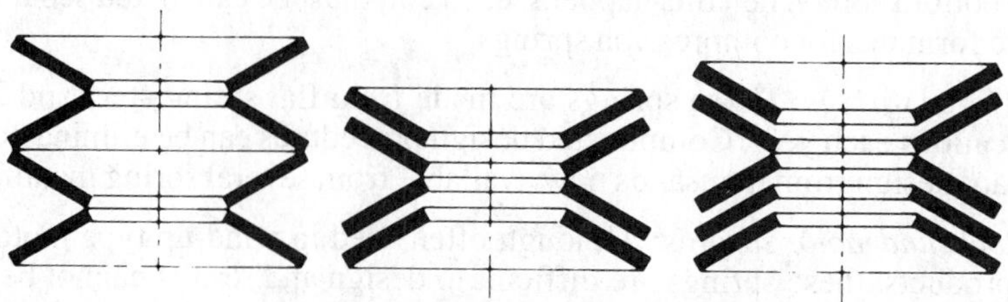

Fig. 25. Examples of Belleville Spring Combinations

Volute springs: These springs are often used on army tanks and heavy field artillery, and seldom find additional uses because of their high cost, long production time, difficulties in manufacture, and unavailability of a wide range of materials and sizes. Small volute springs are often replaced with standard compression springs.

Torsion bars: Although the more simple types are often used on motor cars, the more complicated types with specially forged ends are finding fewer applications as time goes on.

Moduli of Elasticity of Spring Materials.—The modulus of elasticity in tension, denoted by the letter *E*, and the modulus of elasticity in torsion, denoted by the letter *G*, are used in formulas relating to spring design. Values of these moduli for various ferrous and nonferrous spring materials are given in Table .

General Heat Treating Information for Springs.—The following is general information on the heat treatment of springs, and is applicable to pre-tempered or hard-drawn spring materials only.

Compression springs: are baked after coiling (before setting) to relieve residual stresses and thus permit larger deflections before taking a permanent set.

Extension springs: also are baked, but heat removes some of the initial tension. Allowance should be made for this loss. Baking at 500 degrees F for 30 minutes removes approximately 50 per cent of the initial tension. The shrinkage in diameter however, will slightly increase the load and rate.

Outside diameters shrink: when springs of music wire, pretempered MB, and other carbon or alloy steels are baked. Baking also slightly increases the free length and these changes produce a little stronger load and increase the rate.

Outside diameters expand: when springs of stainless steel (18-8) are baked. The free length is also reduced slightly and these changes result in a little lighter load and a decrease the spring rate.

Inconel, Monel, and nickel alloys: do not change much when baked.

Beryllium-copper shrinks and deforms: when heated. Such springs usually are baked in fixtures or supported on arbors or rods during heating.

Brass and phosphor bronze: springs should be given a light heat only. Baking above 450 degrees F will soften the material. Do not heat in salt pots.

Torsion springs: do not require baking because coiling causes residual stresses in a direction that is helpful, but such springs frequently are baked so that jarring or handling will not cause them to lose the position of their ends.

Table 20. Moduli of Elasticity in Torsion and Tension of Spring Materials

Ferrous Materials			Nonferrous Materials		
	Modulus of Elasticity, pounds per square inch			Modulus of Elasticity, pounds per square inch	
Material (Commercial Name)	In Torsion, *G*	In Tension, *E*	Material (Commercial Name)	In Torsion, *G*	In Tension, *E*
Hard Drawn MB			Spring Brass		
Up to 0.032 inch	11,700,000	28,800,000	Type 70–30	5,000,000	15,000,000
0.033 to 0.063 inch	11,600,000	28,700,000	Phosphor Bronze		
0.064 to 0.125 inch	11,500,000	28,600,000	5 per cent tin	6,000,000	15,000,000
0.126 to 0.625 inch	11,400,000	28,500,000	Beryllium-Copper		
Music Wire			Cold Drawn 4 Nos.	7,000,000	17,000,000
Up to 0.032 inch	12,000,000	29,500,000	Pretempered,		
0.033 to 0.063 inch	11,850,000	29,000,000	fully hard	7,250,000	19,000,000
0.064 to 0.125 inch	11,750,000	28,500,000	Inconel[a] 600	10,500,000	31,000,000[b]
0.126 to 0.250 inch	11,600,000	28,000,000	Inconel[a] X 750	10,500,000	31,000,000[b]
Oil-Tempered MB	11,200,000	28,500,000			
Chrome-Vanadium	11,200,000	28,500,000	Monel[a] 400	9,500,000	26,000,000
Chrome-Silicon	11,200,000	29,500,000	Monel[a] K 500	9,500,000	26,000,000
Silicon-Manganese	10,750,000	29,000,000	Duranickel[a] 300	11,000,000	30,000,000
Stainless Steel			Permanickel[a]	11,000,000	30,000,000
Types 302, 304, 316	10,000,000	28,000,000[b]			
Type 17–7 PH	10,500,000	29,500,000	Ni Span C[a] 902	10,000,000	27,500,000
Type 420	11,000,000	29,000,000	Elgiloy[c]	12,000,000	29,500,000
Type 431	11,400,000	29,500,000	Iso-Elastic[d]	9,200,000	26,000,000

[a] Trade name of International Nickel Company.
[b] May be 2,000,000 pounds per square inch less if material is not fully hard.
[c] Trade name of Hamilton Watch Company.
[d] Trade name of John Chatillon & Sons.

Note: Modulus *G* (shear modulus) is used for compression and extension springs; modulus *E* (Young's modulus) is used for torsion, flat, and spiral springs.

Spring brass and phosphor bronze: springs that are not very highly stressed and are not subject to severe operating use may be stress relieved after coiling by immersing them in boiling water for a period of 1 hour.

Positions of loops will: change with heat. Parallel hooks may change as much as 45 degrees during baking. Torsion spring arms will alter position considerably. These changes should be allowed for during looping or forming.

Quick heating: after coiling either in a high-temperature salt pot or by passing a spring through a gas flame is not good practice. Samples heated in this way will not conform with production runs that are properly baked. A small, controlled-temperature oven should be used for samples and for small lot orders.

Plated springs: should always be baked before plating to relieve coiling stresses and again after plating to relieve hydrogen embrittlement.

Hardness: values fall with high heat—but music wire, hard drawn, and stainless steel will increase 2 to 4 points Rockwell C.

Table 21. Squares, Cubes, and Fourth Powers of Wire Diameters

Steel Wire Gage (U.S.)	Music or Piano Wire Gage	Diameter Inch	Section Area	Square	Cube	Fourth Power
7-0	…	0.4900	0.1886	0.24010	0.11765	0.05765
6-0	…	0.4615	0.1673	0.21298	0.09829	0.04536
5-0	…	0.4305	0.1456	0.18533	0.07978	0.03435
4-0	…	0.3938	0.1218	0.15508	0.06107	0.02405
3-0	…	0.3625	0.1032	0.13141	0.04763	0.01727
2-0	…	0.331	0.0860	0.10956	0.03626	0.01200
1-0	…	0.3065	0.0738	0.09394	0.02879	0.008825
1	…	0.283	0.0629	0.08009	0.02267	0.006414
2	…	0.2625	0.0541	0.06891	0.01809	0.004748
3	…	0.2437	0.0466	0.05939	0.01447	0.003527
4	…	0.2253	0.0399	0.05076	0.01144	0.002577
5	…	0.207	0.0337	0.04285	0.00887	0.001836
6	…	0.192	0.0290	0.03686	0.00708	0.001359
…	45	0.180	0.0254	0.03240	0.00583	0.001050
7	…	0.177	0.0246	0.03133	0.00555	0.000982
…	44	0.170	0.0227	0.02890	0.00491	0.000835
8	43	0.162	0.0206	0.02624	0.00425	0.000689
…	42	0.154	0.0186	0.02372	0.00365	0.000563
9	…	0.1483	0.0173	0.02199	0.00326	0.000484
…	41	0.146	0.0167	0.02132	0.00311	0.000455
…	40	0.138	0.0150	0.01904	0.00263	0.000363
10	…	0.135	0.0143	0.01822	0.00246	0.000332
…	39	0.130	0.0133	0.01690	0.00220	0.000286
…	38	0.124	0.0121	0.01538	0.00191	0.000237
11	…	0.1205	0.0114	0.01452	0.00175	0.000211
…	37	0.118	0.0109	0.01392	0.00164	0.000194
…	36	0.112	0.0099	0.01254	0.00140	0.000157
…	35	0.106	0.0088	0.01124	0.00119	0.000126
12	…	0.1055	0.0087	0.01113	0.001174	0.0001239
…	34	0.100	0.0078	0.0100	0.001000	0.0001000
…	33	0.095	0.0071	0.00902	0.000857	0.0000815
13	…	0.0915	0.0066	0.00837	0.000766	0.0000701
…	32	0.090	0.0064	0.00810	0.000729	0.0000656
…	31	0.085	0.0057	0.00722	0.000614	0.0000522
14	30	0.080	0.0050	0.0064	0.000512	0.0000410
…	29	0.075	0.0044	0.00562	0.000422	0.0000316
15	…	0.072	0.0041	0.00518	0.000373	0.0000269
…	28	0.071	0.0040	0.00504	0.000358	0.0000254
…	27	0.067	0.0035	0.00449	0.000301	0.0000202
…	26	0.063	0.0031	0.00397	0.000250	0.0000158
16	…	0.0625	0.0031	0.00391	0.000244	0.0000153
…	25	0.059	0.0027	0.00348	0.000205	0.0000121
…	24	0.055	0.0024	0.00302	0.000166	0.00000915
17	…	0.054	0.0023	0.00292	0.000157	0.00000850
…	23	0.051	0.0020	0.00260	0.000133	0.00000677
…	22	0.049	0.00189	0.00240	0.000118	0.00000576
18	…	0.0475	0.00177	0.00226	0.000107	0.00000509
…	21	0.047	0.00173	0.00221	0.000104	0.00000488
…	20	0.045	0.00159	0.00202	0.000091	0.00000410
…	19	0.043	0.00145	0.00185	0.0000795	0.00000342
19	18	0.041	0.00132	0.00168	0.0000689	0.00000283
…	17	0.039	0.00119	0.00152	0.0000593	0.00000231
…	16	0.037	0.00108	0.00137	0.0000507	0.00000187
…	15	0.035	0.00096	0.00122	0.0000429	0.00000150
20	…	0.0348	0.00095	0.00121	0.0000421	0.00000147
…	14	0.033	0.00086	0.00109	0.0000359	0.00000119
21	…	0.0317	0.00079	0.00100	0.0000319	0.00000101
…	13	0.031	0.00075	0.00096	0.0000298	0.000000924
…	12	0.029	0.00066	0.00084	0.0000244	0.000000707
22	…	0.0286	0.00064	0.00082	0.0000234	0.000000669
…	11	0.026	0.00053	0.00068	0.0000176	0.000000457
23	…	0.0258	0.00052	0.00067	0.0000172	0.000000443
…	10	0.024	0.00045	0.00058	0.0000138	0.000000332
24	…	0.023	0.00042	0.00053	0.0000122	0.000000280
…	9	0.022	0.00038	0.00048	0.0000106	0.000000234

Table 22. Causes of Spring Failure

	Cause	Comments and Recommendations
Group 1	High stress	The majority of spring failures are due to high stresses caused by large deflections and high loads. High stresses should be used only for statically loaded springs. Low stresses lengthen fatigue life.
	Hydrogen embrittlement	Improper electroplating methods and acid cleaning of springs, without proper baking treatment, cause spring steels to become brittle, and are a frequent cause of failure. Nonferrous springs are immune.
	Sharp bends and holes	Sharp bends on extension, torsion, and flat springs, and holes or notches in flat springs, cause high concentrations of stress, resulting in failure. Bend radii should be as large as possible, and tool marks avoided.
	Fatigue	Repeated deflections of springs, especially above 1,000,000 cycles, even with medium stresses, may cause failure. Low stresses should be used if a spring is to be subjected to a very high number of operating cycles.
Group 2	Shock loading	Impact, shock, and rapid loading cause far higher stresses than those computed by the regular spring formulas. High-carbon spring steels do not withstand shock loading as well as do alloy steels.
	Corrosion	Slight rusting or pitting caused by acids, alkalis, galvanic corrosion, stress corrosion cracking, or corrosive atmosphere weakens the material and causes higher stresses in the corroded area.
	Faulty heat treatment	Keeping spring materials at the hardening temperature for longer periods than necessary causes an undesirable growth in grain structure, resulting in brittleness, even though the hardness may be correct.
	Faulty material	Poor material containing inclusions, seams, slivers, and flat material with rough, slit, or torn edges is a cause of early failure. Overdrawn wire, improper hardness, and poor grain structure also cause early failure.
Group 3	High temperature	High operating temperatures reduce spring temper (or hardness) and lower the modulus of elasticity, thereby causing lower loads, reducing the elastic limit, and increasing corrosion. Corrosion-resisting or nickel alloys should be used.
	Low temperature	Temperatures below –40 degrees F reduce the ability of carbon steels to withstand shock loads. Carbon steels become brittle at -70 degrees F. Corrosion-resisting, nickel, or nonferrous alloys should be used.
	Friction	Close fits on rods or in holes result in a wearing away of material and occasional failure. The outside diameters of compression springs expand during deflection but they become smaller on torsion springs.
	Other causes	Enlarged hooks on extension springs increase the stress at the bends. Carrying too much electrical current will cause failure. Welding and soldering frequently destroy the spring temper. Tool marks, nicks, and cuts often raise stresses. Deflecting torsion springs outwardly causes high stresses and winding them tightly causes binding on supporting rods. High speed of deflection, vibration, and surging due to operation near natural periods of vibration or their harmonics cause increased stresses.

Spring failure may be breakage, high permanent set, or loss of load. The causes are listed in groups in this table. Group 1 covers causes that occur most frequently; Group 2 covers causes that are less frequent; and Group 3 lists causes that occur occasionally.

Table 23. Arbor Diameters for Springs Made from Music Wire

Wire Diam. (inch)	Spring Outside Diameter (inch)												
	1/16	3/32	1/8	5/32	3/16	7/32	1/4	9/32	5/16	11/32	3/8	7/16	1/2
	Arbor Diameter (inch)												
0.008	0.039	0.060	0.078	0.093	0.107	0.119	0.129	…	…	…	…	…	…
0.010	0.037	0.060	0.080	0.099	0.115	0.129	0.142	0.154	0.164	…	…	…	…
0.012	0.034	0.059	0.081	0.101	0.119	0.135	0.150	0.163	0.177	0.189	0.200	…	…
0.014	0.031	0.057	0.081	0.102	0.121	0.140	0.156	0.172	0.187	0.200	0.213	0.234	…
0.016	0.028	0.055	0.079	0.102	0.123	0.142	0.161	0.178	0.194	0.209	0.224	0.250	0.271
0.018	…	0.053	0.077	0.101	0.124	0.144	0.161	0.182	0.200	0.215	0.231	0.259	0.284
0.020	…	0.049	0.075	0.096	0.123	0.144	0.165	0.184	0.203	0.220	0.237	0.268	0.296
0.022	…	0.046	0.072	0.097	0.122	0.145	0.165	0.186	0.206	0.224	0.242	0.275	0.305
0.024	…	0.043	0.070	0.095	0.120	0.144	0.166	0.187	0.207	0.226	0.245	0.280	0.312
0.026	…	…	0.067	0.093	0.118	0.143	0.166	0.187	0.208	0.228	0.248	0.285	0.318
0.028	…	…	0.064	0.091	0.115	0.141	0.165	0.187	0.208	0.229	0.250	0.288	0.323
0.030	…	…	0.061	0.088	0.113	0.138	0.163	0.187	0.209	0.229	0.251	0.291	0.328
0.032	…	…	0.057	0.085	0.111	0.136	0.161	0.185	0.209	0.229	0.251	0.292	0.331
0.034	…	…	…	0.082	0.109	0.134	0.159	0.184	0.208	0.229	0.251	0.292	0.333
0.036	…	…	…	0.078	0.106	0.131	0.156	0.182	0.206	0.229	0.250	0.294	0.333
0.038	…	…	…	0.075	0.103	0.129	0.154	0.179	0.205	0.227	0.251	0.293	0.335
0.041	…	…	…	…	0.098	0.125	0.151	0.176	0.201	0.226	0.250	0.294	0.336
0.0475	…	…	…	…	0.087	0.115	0.142	0.168	0.194	0.220	0.244	0.293	0.337
0.054	…	…	…	…	…	0.103	0.132	0.160	0.187	0.212	0.245	0.287	0.336
0.0625	…	…	…	…	…	…	0.108	0.146	0.169	0.201	0.228	0.280	0.330
0.072	…	…	…	…	…	…	…	0.129	0.158	0.186	0.214	0.268	0.319
0.080	…	…	…	…	…	…	…	…	0.144	0.173	0.201	0.256	0.308
0.0915	…	…	…	…	…	…	…	…	…	…	0.181	0.238	0.293
0.1055	…	…	…	…	…	…	…	…	…	…	…	0.215	0.271
0.1205	…	…	…	…	…	…	…	…	…	…	…	…	0.215
0.125	…	…	…	…	…	…	…	…	…	…	…	…	0.239

Wire Diam. (inch)	Spring Outside Diameter (inches)												
	9/16	5/8	11/16	3/4	13/16	7/8	15/16	1	1 1/8	1 1/4	1 3/8	1 1/2	1 3/4	2
	Arbor Diameter (inches)													
0.022	0.332	0.357	0.380	…	…	…	…	…	…	…	…	…	…	…
0.024	0.341	0.367	0.393	0.415	…	…	…	…	…	…	…	…	…	…
0.026	0.350	0.380	0.406	0.430	…	…	…	…	…	…	…	…	…	…
0.028	0.356	0.387	0.416	0.442	0.467	…	…	…	…	…	…	…	…	…
0.030	0.362	0.395	0.426	0.453	0.481	0.506	…	…	…	…	…	…	…	…
0.032	0.367	0.400	0.432	0.462	0.490	0.516	0.540	…	…	…	…	…	…	…
0.034	0.370	0.404	0.437	0.469	0.498	0.526	0.552	0.557	…	…	…	…	…	…
0.036	0.372	0.407	0.442	0.474	0.506	0.536	0.562	0.589	…	…	…	…	…	…
0.038	0.375	0.412	0.448	0.481	0.512	0.543	0.572	0.600	0.650	…	…	…	…	…
0.041	0.378	0.416	0.456	0.489	0.522	0.554	0.586	0.615	0.670	0.718	…	…	…	…
0.0475	0.380	0.422	0.464	0.504	0.541	0.576	0.610	0.643	0.706	0.763	0.812	…	…	…
0.054	0.381	0.425	0.467	0.509	0.550	0.589	0.625	0.661	0.727	0.792	0.850	0.906	…	…
0.0625	0.379	0.426	0.468	0.512	0.556	0.597	0.639	0.678	0.753	0.822	0.889	0.951	1.06	1.17
0.072	0.370	0.418	0.466	0.512	0.555	0.599	0.641	0.682	0.765	0.840	0.911	0.980	1.11	1.22
0.080	0.360	0.411	0.461	0.509	0.554	0.599	0.641	0.685	0.772	0.851	0.930	1.00	1.13	1.26
0.0915	0.347	0.398	0.448	0.500	0.547	0.597	0.640	0.685	0.776	0.860	0.942	1.02	1.16	1.30
0.1055	0.327	0.381	0.433	0.485	0.535	0.586	0.630	0.683	0.775	0.865	0.952	1.04	1.20	1.35
0.1205	0.303	0.358	0.414	0.468	0.520	0.571	0.622	0.673	0.772	0.864	0.955	1.04	1.22	1.38
0.125	0.295	0.351	0.406	0.461	0.515	0.567	0.617	0.671	0.770	0.864	0.955	1.05	1.23	1.39

STRENGTH AND PROPERTIES OF WIRE ROPE

Strength and Properties of Wire Rope

Wire Rope Construction.—Essentially, a wire rope is made up of a number of strands laid helically about a metallic or non-metallic core. Each strand consists of a number of wires also laid helically about a metallic or non-metallic center. Various types of wire rope have been developed to meet a wide range of uses and operating conditions. These types are distinguished by the kind of core; the number of strands; the number, sizes, and arrangement of the wires in each strand; and the way in which the wires and strands are wound or laid about each other. The following descriptive material is based largely on information supplied by the Bethlehem Steel Co.

Rope Wire Materials: Materials used in the manufacture of rope wire are, in order of increasing strength: iron, phosphor bronze, traction steel, plow steel, improved plow steel, and bridge rope steel. Iron wire rope is largely used for low-strength applications such as elevator ropes not used for hoisting, and for stationary guy ropes.

Phosphor bronze wire rope is used occasionally for elevator governor-cable rope and for certain marine applications as life lines, clearing lines, wheel ropes and rigging.

Traction steel wire rope is used primarily as hoist rope for passenger and freight elevators of the traction drive type, an application for which it was specifically designed.

Ropes made of galvanized wire or wire coated with zinc by the electrodeposition process are used in certain applications where additional protection against rusting is required. As will be noted from the tables of wire-rope sizes and strengths, the breaking strength of galvanized wire rope is 10 per cent less than that of ungalvanized (bright) wire rope. Bethanized (zinc-coated) wire rope can be furnished to bright wire rope strength when so specified.

Galvanized carbon steel, tinned carbon steel, and stainless steel are used for small cords and strands ranging in diameter from $\frac{1}{64}$ to $\frac{3}{8}$ inch and larger.

Marline clad wire rope has each strand wrapped with a layer of tarred marline. The cladding provides hand protection for workers and wear protection for the rope.

Rope Cores: Wire-rope cores are made of fiber, cotton, asbestos, polyvinyl plastic, a small wire rope (independent wire-rope core), a multiple-wire strand (wire-strand core) or a cold-drawn wire-wound spring.

Fiber: (manila or sisal) is the type of core most widely used when loads are not too great. It supports the strands in their relative positions and acts as a cushion to prevent nicking of the wires lying next to the core.

Cotton: is used for small ropes such as sash cord and aircraft cord.

Asbestos cores: can be furnished for certain special operations where the rope is used in oven operations.

Polyvinyl plastics cores: are offered for use where exposure to moisture, acids, or caustics is excessive.

A wire-strand core: often referred to as WSC, consists of a multiple-wire strand that may be the same as one of the strands of the rope. It is smoother and more solid than the independent wire rope core and provides a better support for the rope strands.

The *independent wire rope core*, often referred to as IWRC, is a small 6 × 7 wire rope with a wire-strand core and is used to provide greater resistance to crushing and distortion of the wire rope. For certain applications it has the advantage over a wire-strand core in that it stretches at a rate closer to that of the rope itself.

Wire ropes with wire-strand cores are, in general, less flexible than wire ropes with independent wire-rope or non-metallic cores.

Ropes with metallic cores are rated 7½ per cent stronger than those with non-metallic cores.

Wire-Rope Lay: The lay of a wire rope is the direction of the helical path in which the strands are laid and, similarly, the lay of a strand is the direction of the helical path in which the wires are laid. If the wires in the strand or the strands in the rope form a helix similar to the threads of a right-hand screw, i.e., they wind around to the right, the lay is called right hand and, conversely, if they wind around to the left, the lay is called left hand. In the *regular lay*, the wires in the strands are laid in the opposite direction to the lay of the strands in the rope. In right-regular lay, the strands are laid to the right and the wires to the left. In left-regular lay, the strands are laid to the left, the wires to the right. In *Lang lay*, the wires and strands are laid in the same direction, i.e., in right Lang lay, both the wires and strands are laid to the right and in left Lang they are laid to the left.

Alternate lay ropes having alternate right and left laid strands are used to resist distortion and prevent clamp slippage, but because other advantages are missing, have limited use.

The regular lay wire rope is most widely used and right regular lay rope is customarily furnished. Regular lay rope has less tendency to spin or untwist when placed under load and is generally selected where long ropes are employed and the loads handled are frequently removed. Lang lay ropes have greater flexibility than regular lay ropes and are more resistant to abrasion and fatigue.

In preformed wire ropes the wires and strands are preshaped into a helical form so that when laid to form the rope they tend to remain in place. In a non-preformed rope, broken wires tend to "wicker out" or protrude from the rope and strands that are not seized tend to spring apart. Preforming also tends to remove locked-in stresses, lengthen service life, and make the rope easier to handle and to spool.

Strand Construction: Various arrangements of wire are used in the construction of wire rope strands. In the simplest arrangement six wires are grouped around a central wire thus making seven wires, all of the same size. Other types of construction known as "filler-wire," Warrington, Seale, etc. make use of wires of different sizes. Their respective patterns of arrangement are shown diagrammatically in the table of wire weights and strengths.

Specifying Wire Rope.—In specifying wire rope the following information will be required: length, diameter, number of strands, number of wires in each strand, type of rope construction, grade of steel used in rope, whether preformed or not preformed, type of center, and type of lay. The manufacturer should be consulted in selecting the best type of wire rope for a new application.

Properties of Wire Rope.—Important properties of wire rope are strength, wear resistance, flexibility, and resistance to crushing and distortion.

Strength: The strength of wire rope depends upon its size, kind of material of which the wires are made and their number, the type of core, and whether the wire is galvanized or not. Strengths of various types and sizes of wire ropes are given in the accompanying tables together with appropriate factors to apply for ropes with steel cores and for galvanized wire ropes.

Wear Resistance: When wire rope must pass back and forth over surfaces that subject it to unusual wear or abrasion, it must be specially constructed to give satisfactory service.

Such construction may make use of 1) relatively large outer wires; 2) Lang lay in which wires in each strand are laid in the same direction as the strand; and 3) flattened strands.

The object in each type is to provide a greater outside surface area to take the wear or abrasion. From the standpoint of material, improved plow steel has not only the highest tensile strength but also the greatest resistance to abrasion in regularly stocked wire rope.

Flexibility: Wire rope that undergoes repeated and severe bending, such as in passing around small sheaves and drums, must have a high degree of flexibility to prevent premature breakage and failure due to fatigue. Greater flexibility in wire rope is obtained by

1) using small wires in larger numbers; 2) using Lang lay; and 3) preforming, that is, the wires and strands of the rope are shaped during manufacture to fit the position they will assume in the finished rope.

Resistance to Crushing and Distortion: Where wire rope is to be subjected to transverse loads that may crush or distort it, care should be taken to select a type of construction that will stand up under such treatment.

Wire rope designed for such conditions may have 1) large outer wires to spread the load per wire over a greater area; and 2) an independent wire core or a high-carbon cold-drawn wound spring core.

Standard Classes of Wire Rope.—Wire rope is commonly designated by two figures, the first indicating the number of strands and the second, the number of wires per strand, as: 6 × 7, a six-strand rope having seven wires per strand, or 8 × 19, an eight-strand rope having 19 wires per strand. When such numbers are used as designations of standard wire rope classes, the second figure in the designation may be purely nominal in that the number of wires per strand for various ropes in the class may be slightly less or slightly more than the nominal as will be seen from the following brief descriptions. (For ropes with a wire strand core, a second group of two numbers may be used to indicate the construction of the wire core, as 1 × 21, 1 × 43, and so on.)

6 × 7 Class (Standard Coarse Laid Rope): Wire ropes in this class are for use where resistance to wear, as in dragging over the ground or across rollers, is an important requirement. Heavy hauling, rope transmissions, and well drilling are common applications. These wire ropes are furnished in right regular lay and occasionally in Lang lay. The cores may be of fiber, independent wire rope, or wire strand. Since this class is a relatively stiff type of construction, these ropes should be used with large sheaves and drums. Because of the small number of wires, a larger factor of safety may be called for.

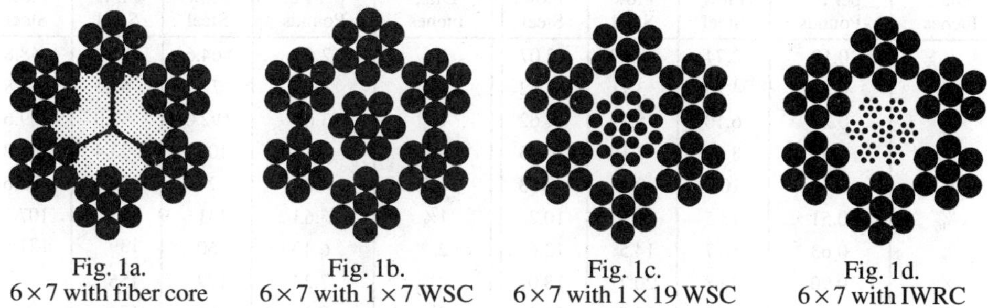

Fig. 1a. 6 × 7 with fiber core | Fig. 1b. 6 × 7 with 1 × 7 WSC | Fig. 1c. 6 × 7 with 1 × 19 WSC | Fig. 1d. 6 × 7 with IWRC

As shown in Figs. 1a through Figs. 1d, this class includes a 6 × 7 construction with fiber core: a 6 × 7 construction with 1 × 7 wire strand core (sometimes called 7 × 7); a 6 × 7 construction with 1 × 19 wire strand core; and a 6 × 7 construction with independent wire rope core. Table 1 provides strength and weight data for this class.

Two special types of wire rope in this class are: aircraft cord, a 6 × 6 or 7 × 7 Bethanized wire rope of high tensile strength and sash cord, a 6 × 7 iron rope used for a variety of purposes where strength is not an important factor.

Table 1. Weights and Strengths of 6 × 7 (Standard Coarse Laid) Wire Ropes, Preformed and Not Preformed

Diam., Inches	Approx. Weight per Ft., Pounds	Breaking Strength, Tons of 2000 Lbs.			Diam., Inches	Approx. Weight per Ft., Pounds	Breaking Strength, Tons of 2000 Lbs.		
		Impr. Plow Steel	Plow Steel	Mild Plow Steel			Impr. Plow Steel	Plow Steel	Mild Plow Steel
1/4	0.094	2.64	2.30	2.00	3/4	0.84	22.7	19.8	17.2
5/16	0.15	4.10	3.56	3.10	7/8	1.15	30.7	26.7	23.2
3/8	0.21	5.86	5.10	4.43	1	1.50	39.7	34.5	30.0
7/16	0.29	7.93	6.90	6.00	1 1/8	1.90	49.8	43.3	37.7
1/2	0.38	10.3	8.96	7.79	1 1/4	2.34	61.0	53.0	46.1
9/16	0.48	13.0	11.3	9.82	1 3/8	2.84	73.1	63.6	55.3
5/8	0.59	15.9	13.9	12.0	1 1/2	3.38	86.2	75.0	65.2

For ropes with steel cores, add 7½ per cent to above strengths.

For galvanized ropes, deduct 10 per cent from above strengths.

Source: Rope diagrams, Bethlehem Steel Co. All data, U.S. Simplified Practice Recommendation 198–50.

6 × 19 Class (Standard Hoisting Rope): This rope is the most popular and widely used class. Ropes in this class are furnished in regular or Lang lay and may be obtained preformed or not preformed. Cores may be of fiber, independent wire rope, or wire strand. As can be seen from Table 2 and Figs. 2a through 2h, there are four common types: 6 × 25 filler wire construction with fiber core (not illustrated), independent wire core, or wire strand core (1 × 25 or 1 × 43); 6 × 19 Warrington construction with fiber core; 6 × 21 filler wire construction with fiber core; and 6 × 19, 6 × 21, and 6 × 17 Seale construction with fiber core.

Table 2. Weights and Strengths of 6 × 19 (Standard Hoisting) Wire Ropes, Preformed and Not Preformed

Dia., Inches	Approx. Weight per Ft., Pounds	Breaking Strength, Tons of 2000 Lbs.			Dia., Inches	Approx. Weight per Ft., Pounds	Breaking Strength, Tons of 2000 Lbs.		
		Impr. Plow Steel	Plow Steel	Mild Plow Steel			Impr. Plow Steel	Plow Steel	Mild Plow Steel
1/4	0.10	2.74	2.39	2.07	1 1/4	2.50	64.6	56.2	48.8
5/16	0.16	4.26	3.71	3.22	1 3/8	3.03	77.7	67.5	58.8
3/8	0.23	6.10	5.31	4.62	1 1/2	3.60	92.0	80.0	69.6
7/16	0.31	8.27	7.19	6.25	1 5/8	4.23	107	93.4	81.2
1/2	0.40	10.7	9.35	8.13	1 3/4	4.90	124	108	93.6
9/16	0.51	13.5	11.8	10.2	1 7/8	5.63	141	123	107
5/8	0.63	16.7	14.5	12.6	2	6.40	160	139	121
3/4	0.90	23.8	20.7	18.0	2 1/8	7.23	179	156	…
7/8	1.23	32.2	28.0	24.3	2 1/4	8.10	200	174	…
1	1.60	41.8	36.4	31.6	2 1/2	10.00	244	212	…
1 1/8	2.03	52.6	45.7	39.8	2 3/4	12.10	292	254	…

The 6 × 25 filler wire with fiber core not illustrated.

For ropes with steel cores, add 7½ per cent to above strengths.

For galvanized ropes, deduct 10 per cent from above strengths.

Source: Rope diagrams, Bethlehem Steel Co. All data, U.S. Simplified Practice Recommendation 198–50.

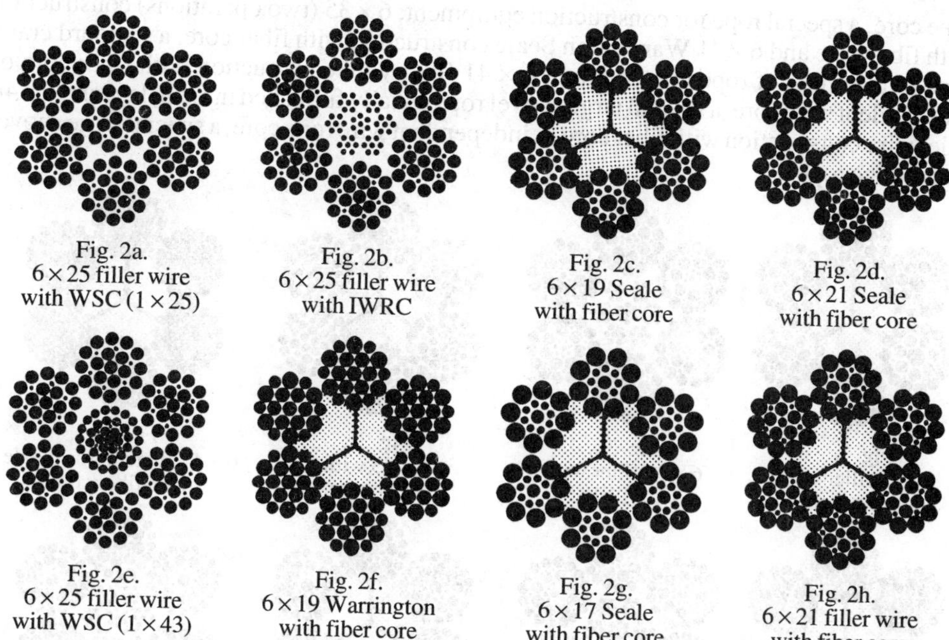

Fig. 2a. 6 × 25 filler wire with WSC (1 × 25)

Fig. 2b. 6 × 25 filler wire with IWRC

Fig. 2c. 6 × 19 Seale with fiber core

Fig. 2d. 6 × 21 Seale with fiber core

Fig. 2e. 6 × 25 filler wire with WSC (1 × 43)

Fig. 2f. 6 × 19 Warrington with fiber core

Fig. 2g. 6 × 17 Seale with fiber core

Fig. 2h. 6 × 21 filler wire with fiber core

6 × 37 Class (Extra Flexible Hoisting Rope): For a given size of rope, the component wires are of smaller diameter than those in the two classes previously described and hence have less resistance to abrasion. Ropes in this class are furnished in regular and Lang lay with fiber core or independent wire rope core, preformed or not preformed.

Table 3. Weights and Strengths of 6 × 37 (Extra Flexible Hoisting) Wire Ropes, Preformed and Not Preformed

Dia., Inches	Approx. Weight per Ft., Pounds	Breaking Strength, Tons of 2000 Lbs. — Impr. Plow Steel	Breaking Strength, Tons of 2000 Lbs. — Plow Steel	Dia., Inches	Approx. Weight per Ft., Pounds	Breaking Strength, Tons of 2000 Lbs. — Impr. Plow Steel	Breaking Strength, Tons of 2000 Lbs. — Plow Steel
1/4	0.10	2.59	2.25	1 1/2	3.49	87.9	76.4
5/16	0.16	4.03	3.50	1 5/8	4.09	103	89.3
3/8	0.22	5.77	5.02	1 3/4	4.75	119	103
7/16	0.30	7.82	6.80	1 7/8	5.45	136	118
1/2	0.39	10.2	8.85	2	6.20	154	134
9/16	0.49	12.9	11.2	2 1/8	7.00	173	150
5/8	0.61	15.8	13.7	2 1/4	7.85	193	168
3/4	0.87	22.6	19.6	2 1/2	9.69	236	205
7/8	1.19	30.6	26.6	2 3/4	11.72	284	247
1	1.55	39.8	34.6	3	14.0	335	291
1 1/8	1.96	50.1	43.5	3 1/4	16.4	390	339
1 1/4	2.42	61.5	53.5	3 1/2	19.0	449	390
1 3/8	2.93	74.1	64.5	…	…	…	…

For ropes with steel cores, add 7½ per cent to above strengths.

For galvanized ropes, deduct 10 per cent from above strengths.

Source: Rope diagrams, Bethlehem Steel Co. All data, U. S. Simplified Practice Recommendation 198-50.

As shown in Table 3 and Figs. 3a through 3h, there are four common types: 6 × 29 filler wire construction with fiber core and 6 × 36 filler wire construction with independent wire rope core, a special rope for construction equipment; 6 × 35 (two operations) construction with fiber core and 6 × 41 Warrington Seale construction with fiber core, a standard crane rope in this class of rope construction; 6 × 41 filler wire construction with fiber core or independent wire core, a special large shovel rope usually furnished in Lang lay; and 6 × 46 filler wire construction with fiber core or independent wire rope core, a special large shovel and dredge rope.

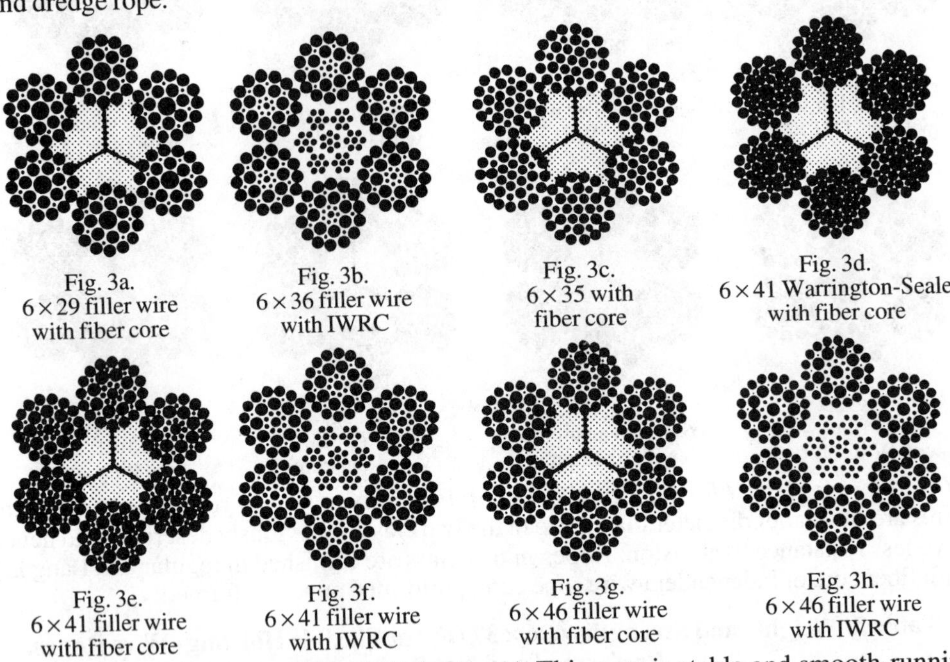

Fig. 3a. 6 × 29 filler wire with fiber core

Fig. 3b. 6 × 36 filler wire with IWRC

Fig. 3c. 6 × 35 with fiber core

Fig. 3d. 6 × 41 Warrington-Seale with fiber core

Fig. 3e. 6 × 41 filler wire with fiber core

Fig. 3f. 6 × 41 filler wire with IWRC

Fig. 3g. 6 × 46 filler wire with fiber core

Fig. 3h. 6 × 46 filler wire with IWRC

8 × 19 Class (Special Flexible Hoisting Rope): This rope is stable and smooth-running, and is especially suitable, because of its flexibility, for high speed operation with reverse bends. Ropes in this class are available in regular lay with fiber core.

As shown in Table 4 and Figs. 4a through 4d, there are four common types: 8 × 25 filler wire construction, the most flexible but the least wear resistant rope of the four types; Warrington type in 8 × 19 construction, less flexible than the 8 × 25; 8 × 21 filler wire construction, less flexible than the Warrington; and Seale type in 8 × 19 construction, which has the greatest wear resistance of the four types but is also the least flexible.

Table 4. Weights and Strengths of 8 × 19 (Special Flexible Hoisting) Wire Ropes, Preformed and Not Preformed

Dia., Inches	Approx. Weight per Ft., Pounds	Breaking Strength, Tons of 2000 Lbs.		Dia., Inches	Approx. Weight per Ft., Pounds	Breaking Strength, Tons of 2000 Lbs.	
		Impr. Plow Steel	Plow Steel			Impr. Plow Steel	Plow Steel
1/4	0.09	2.35	2.04	3/4	0.82	20.5	17.8
5/16	0.14	3.65	3.18	7/8	1.11	27.7	24.1
3/8	0.20	5.24	4.55	1	1.45	36.0	31.3
7/16	0.28	7.09	6.17	1 1/8	1.84	45.3	39.4
1/2	0.36	9.23	8.02	1 1/4	2.27	55.7	48.4
9/16	0.46	11.6	10.1	1 3/8	2.74	67.1	58.3
5/8	0.57	14.3	12.4	1 1/2	3.26	79.4	69.1

For ropes with steel cores, add 7½ per cent to above strengths.

For galvanized ropes, deduct 10 per cent from above strengths.

Source: Rope diagrams, Bethlehem Steel Co. All data, U. S. Simplified Practice Recommendation 198-50.

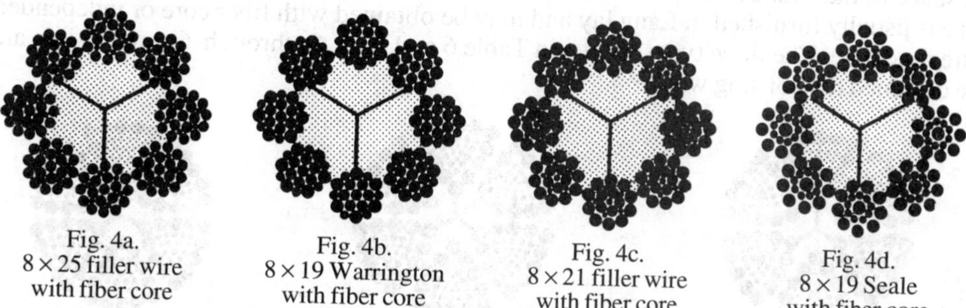

Fig. 4a. 8 × 25 filler wire with fiber core

Fig. 4b. 8 × 19 Warrington with fiber core

Fig. 4c. 8 × 21 filler wire with fiber core

Fig. 4d. 8 × 19 Seale with fiber core

Also in this class, but not shown in Table 4 are elevator ropes made of traction steel and iron.

18 × 7 Non-rotating Wire Rope: This rope is specially designed for use where a minimum of rotating or spinning is called for, especially in the lifting or lowering of free loads with a single-part line. It has an inner layer composed of 6 strands of 7 wires each laid in left Lang lay over a fiber core and an outer layer of 12 strands of 7 wires each laid in right regular lay. The combination of opposing lays tends to prevent rotation when the rope is stretched. However, to avoid any tendency to rotate or spin, loads should be kept to at least one-eighth and preferably one-tenth of the breaking strength of the rope. Weights and strengths are shown in Table 5.

Table 5. Weights and Strengths of Standard 18 × 7 Nonrotating Wire Rope, Preformed and Not Preformed

Fig. 5. 18 × 7 Non-Rotating Rope

Recommended Sheave and Drum Diameters: Single layer on drum … 36 rope diameters. Multiple layers on drum … 48 rope diameters. Mine service … 60 rope diameters.

Dia., Inches	Approx. Weight per Ft., Pounds	Breaking Strength, Tons of 2000 Lbs.		Dia., Inches	Approx. Weight per Ft., Pounds	Breaking Strength, Tons of 2000 Lbs.	
		Impr. Plow Steel	Plow Steel			Impr. Plow Steel	Plow Steel
3/16	0.061	1.42	1.24	7/8	1.32	29.5	25.7
1/4	0.108	2.51	2.18	1	1.73	38.3	33.3
5/16	0.169	3.90	3.39	1 1/8	2.19	48.2	41.9
3/8	0.24	5.59	4.86	1 1/4	2.70	59.2	51.5
7/16	0.33	7.58	6.59	1 3/8	3.27	71.3	62.0
1/2	0.43	9.85	8.57	1 1/2	3.89	84.4	73.4
9/16	0.55	12.4	10.8	1 5/8	4.57	98.4	85.6
5/8	0.68	15.3	13.3	1 3/4	5.30	114	98.8
3/4	0.97	21.8	19.0	…	…	…	…

For galvanized ropes, deduct 10 per cent from above strengths.

Source: Rope diagrams, sheave and drum diameters, and data for 3/16, 1/4 and 5/16-inch sizes, Bethlehem Steel Co. All other data, U. S. Simplified Practice Recommendation 198-50.

Flattened Strand Wire Rope: The wires forming the strands of this type of rope are wound around triangular centers so that a flattened outer surface is provided with a greater area than in the regular round rope to withstand severe conditions of abrasion. The triangular shape of the strands also provides superior resistance to crushing. Flattened strand wire rope is usually furnished in Lang lay and may be obtained with fiber core or independent wire rope core. The three types shown in Table 6 and Figs. 6a through 6c are flexible and are designed for hoisting work.

Fig. 6a.
6 × 25 with fiber core

Fig. 6b.
6 × 30 with fiber core

Fig. 6c.
6 × 27 with fiber core

Table 6. Weights and Strengths of Flattened Strand Wire Rope, Preformed and Not Preformed

Dia., Inches	Approx. Weight per Ft., Pounds	Breaking Strength, Tons of 2000 Lbs.		Dia., Inches	Approx. Weight per Ft., Pounds	Breaking Strength, Tons of 2000 Lbs.	
		Impr. Plow Steel	Mild Plow Steel			Impr. Plow Steel	Mild Plow Steel
3/8[a]	0.25	6.71	…	1 3/8	3.40	85.5	…
1/2[a]	0.45	11.8	8.94	1 1/2	4.05	101	…
9/16[a]	0.57	14.9	11.2	1 5/8	4.75	118	…
5/8	0.70	18.3	13.9	1 3/4	5.51	136	…
3/4	1.01	26.2	19.8	2	7.20	176	…
7/8	1.39	35.4	26.8	2 1/4	9.10	220	…
1	1.80	46.0	34.8	2 1/2	11.2	269	…
1 1/8	2.28	57.9	43.8	2 3/4	13.6	321	…
1 1/4	2.81	71.0	53.7	…	…	…	…

[a] These sizes in Type B only.

Type H is not in U.S. Simplified Practice Recommendation.

Source: Rope diagrams, Bethlehem Steel Co. All other data, U.S. Simplified Practice Recommendation 198-50.

Flat Wire Rope: This type of wire rope is made up of a number of four-strand rope units placed side by side and stitched together with soft steel sewing wire. These four-strand units are alternately right and left lay to resist warping, curling, or rotating in service. Weights and strengths are shown in Table 7.

Simplified Practice Recommendations.—Because the total number of wire rope types is large, manufacturers and users have agreed upon and adopted a U.S. Simplified Practice Recommendation to provide a simplified listing of those kinds and sizes of wire rope which are most commonly used and stocked. These, then, are the types and sizes which are most generally available. Other types and sizes for special or limited uses also may be found in individual manufacturer's catalogs.

Sizes and Strengths of Wire Rope.—The data shown in Tables 1 through 7 have been taken from U.S. Simplified Practice Recommendation 198-50 but do not include those wire ropes shown in that Simplified Practice Recommendation which are intended primarily for marine use.

Wire Rope Diameter: The diameter of a wire rope is the diameter of the circle that will just enclose it, hence when measuring the diameter with calipers, care must be taken to obtain the largest outside dimension, taken across the opposite strands, rather than the smallest dimension across opposite "valleys" or "flats." It is standard practice for the nominal diameter to be the minimum with all tolerances taken on the plus side. Limits for diameter as well as for minimum breaking strength and maximum pitch are given in Federal Specification for Wire Rope, RR-R—571a.

Wire Rope Strengths: The strength figures shown in the accompanying tables have been obtained by a mathematical derivation based on actual breakage tests of wire rope and represent from 80 to 95 per cent of the total strengths of the individual wires, depending upon the type of rope construction.

Table 7. Weights and Strengths of Standard Flat Wire Rope, Not Preformed

Flat Wire Rope

This rope consists of a number of 4-strand rope units placed side by side and stitched together with soft steel sewing wire.

Width and Thickness, Inches	No. of Ropes	Approx. Weight per Ft., Pounds	Breaking Strength, Tons of 2000 Lbs. Plow Steel	Mild Plow-Steel	Width and Thickness, Inches	No. of Ropes	Approx. Weight per Ft., Pounds	Breaking Strength, Tons of 2000 Lbs. Plow Steel	Mild Plow Steel
1/4 × 1 1/2	7	0.69	16.8	14.6	1/2 × 4	9	3.16	81.8	71.2
1/4 × 2	9	0.88	21.7	18.8	1/2 × 4 1/2	10	3.82	90.9	79.1
1/4 × 2 1/2	11	1.15	26.5	23.0	1/2 × 5	12	4.16	109	94.9
1/4 × 3	13	1.34	31.3	27.2	1/2 × 5 1/2	13	4.50	118	103
					1/2 × 6	14	4.85	127	111
5/16 × 1 1/2	5	0.77	18.5	16.0	1/2 × 7	16	5.85	145	126
5/16 × 2	7	1.05	25.8	22.4					
5/16 × 2 1/2	9	1.33	33.2	28.8	5/8 × 3 1/2	6	3.40	85.8	74.6
5/16 × 3	11	1.61	40.5	35.3	5/8 × 4	7	3.95	100	87.1
5/16 × 3 1/2	13	1.89	47.9	41.7	5/8 × 4 1/2	8	4.50	114	99.5
5/16 × 4	15	2.17	55.3	48.1	5/8 × 5	9	5.04	129	112
					5/8 × 5 1/2	10	5.59	143	124
3/8 × 2	6	1.25	31.4	27.3	5/8 × 6	11	6.14	157	137
3/8 × 2 1/2	8	1.64	41.8	36.4	5/8 × 7	13	7.23	186	162
3/8 × 3	9	1.84	47.1	40.9	5/8 × 8	15	8.32	214	186
3/8 × 3 1/2	11	2.23	57.5	50.0					
3/8 × 4	12	2.44	62.7	54.6	3/4 × 5	8	6.50	165	143
3/8 × 4 1/2	14	2.83	73.2	63.7	3/4 × 6	9	7.31	185	161
3/8 × 5	15	3.03	78.4	68.2	3/4 × 7	10	8.13	206	179
3/8 × 5 1/2	17	3.42	88.9	77.3	3/4 × 8	11	9.70	227	197
3/8 × 6	18	3.63	94.1	81.9					
					7/8 × 5	7	7.50	190	165
1/2 × 2 1/2	6	2.13	54.5	47.4	7/8 × 6	8	8.56	217	188
1/2 × 3	7	2.47	63.6	55.4	7/8 × 7	9	9.63	244	212
1/2 × 3 1/2	8	2.82	72.7	63.3	7/8 × 8	10	10.7	271	236

Source: Rope diagram, Bethlehem Steel Co.; all data, U.S. Simplified Practice Recommendation 198–50.

Safe Working Loads and Factors of Safety.—The maximum load for which a wire rope is to be used should take into account such associated factors as friction, load caused by bending around each sheave, acceleration and deceleration, and, if a long length of rope is to be used for hoisting, the weight of the rope at its maximum extension. The condition of the rope — whether new or old, worn or corroded — and type of attachments should also be considered.

Factors of safety for standing rope usually range from 3 to 4; for operating rope, from 5 to 12. Where there is the element of hazard to life or property, higher values are used.

Installing Wire Rope.—The main precaution to be taken in removing and installing wire rope is to avoid kinking which greatly lessens the strength and useful life. Thus, it is preferable when removing wire rope from the reel to have the reel with its axis in a horizontal position and, if possible, mounted so that it will revolve and the wire rope can be taken off straight. If the rope is in a coil, it should be unwound with the coil in a vertical position as by rolling the coil along the ground. Where a drum is to be used, the rope should be run directly onto it from the reel, taking care to see that it is not bent around the drum in a direction opposite to that on the reel, thus causing it to be subject to reverse bending. On flat or smooth-faced drums it is important that the rope be started from the proper end of the drum. A right lay rope that is being overwound on the drum, that is, it passes over the top of the drum as it is wound on, should be started from the right flange of the drum (looking at the drum from the side that the rope is to come) and a left lay rope from the left flange.

When the rope is underwound on the drum, a right lay rope should be started from the left flange and a left lay rope from the right flange, so that the rope will spool evenly and the turns will lie snugly together.

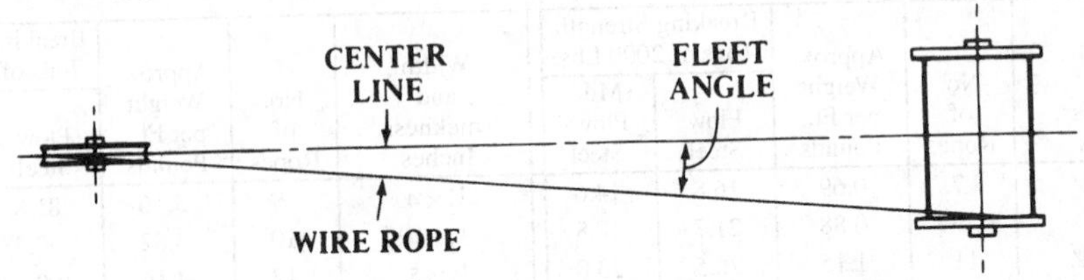

Sheaves and drums should be properly aligned to prevent undue wear. The proper position of the main or lead sheave for the rope as it comes off the drum is governed by what is called the fleet angle or angle between the rope as it stretches from drum to sheave and an imaginary center-line passing through the center of the sheave groove and a point halfway between the ends of the drum. When the rope is at one end of the drum, this angle should not exceed one and a half to two degrees. With the lead sheave mounted with its groove on this center-line, a safe fleet angle is obtained by allowing 30 feet of lead for each two feet of drum width.

Sheave and Drum Dimensions: Sheaves and drums should be as large as possible to obtain maximum rope life. However, factors such as the need for lightweight equipment may call for relatively small sheaves with conse- for easy transport and use at high speeds, quent sacrifice in rope life in the interest of overall economy. No hard and fast rules can be laid down for any particular rope if the utmost in economical performance is to be obtained. Where maximum rope life is of prime importance, the following recommenda- tions of Federal Specification RR-R-571a for minimum sheave or drum diameters D in terms of rope diameter d will be of interest. For 6 × 7 rope (six strands of 7 wires each) $D = 72d$; for 6 × 19 rope, $D = 45d$; for 6 × 25 rope, $D = 45d$; for 6 × 29 rope, $D = 30d$; for 6 × 37 rope, $D = 27d$; and for 8 × 19 rope, $D = 31d$.

Too small a groove for the rope it is to carry will prevent proper seating of the rope in the bottom of the groove and result in uneven distribution of load on the rope. Too large a groove will not give the rope sufficient side support. Federal specification RR-R-571a recommends that sheave groove diameters be larger than the nominal rope diameters by the following minimum amounts: For ropes of $\frac{1}{4}$- to $\frac{5}{16}$-inch diameters, $\frac{1}{64}$ inch larger; for $\frac{3}{8}$- to $\frac{3}{4}$-inch diameter ropes, $\frac{1}{32}$ inch larger; for $\frac{13}{16}$- to $1\frac{1}{8}$-inch diameter ropes, $\frac{3}{64}$ inch larger; for $1\frac{3}{16}$- to $1\frac{1}{2}$-inch ropes, $\frac{1}{16}$ inch larger; for $1\frac{9}{16}$- to $2\frac{1}{4}$-inch ropes, $\frac{3}{32}$ inch larger; and for $2\frac{5}{16}$ and larger diameter ropes, $\frac{1}{8}$ inch larger. For new or regrooved sheaves these values should be doubled; in other words for $\frac{1}{4}$- to $\frac{5}{16}$-inch diameter ropes, the groove diameter should be $\frac{1}{32}$ inch larger, and so on.

Drum or Reel Capacity: The length of wire rope, in feet, that can be spooled onto a drum or reel, is computed by the following formula, where

A = depth of rope space on drum, inches: $A = (H - D - 2Y) \div 2$

B = width between drum flanges, inches

D = diameter of drum barrel, inches

H = diameter of drum flanges, inches

K = factor from Table 8 for size of line selected

Y = depth not filled on drum or reel where winding is to be less than full capacity

L = length of wire rope on drum or reel, feet.

$$L = (A + D) \times A \times B \times K$$

Table 8. Table 8 Factors K Used in Calculating Wire Rope Drum and Reel Capacities

Rope Dia., In.	Factor K	Rope Dia., In.	Factor K	Rope Dia., In.	Factor K
3/32	23.4	1/2	0.925	1 3/8	0.127
1/8	13.6	9/16	0.741	1 1/2	0.107
9/64	10.8	5/8	0.607	1 5/8	0.0886
5/32	8.72	11/16	0.506	1 3/4	0.0770
3/16	6.14	3/4	0.428	1 7/8	0.0675
7/32	4.59	13/16	0.354	2	0.0597
1/4	3.29	7/8	0.308	2 1/8	0.0532
5/16	2.21	1	0.239	2 1/4	0.0476
3/8	1.58	1 1/8	0.191	2 3/8	0.0419
7/16	1.19	1 1/4	0.152	2 1/2	0.0380

Note: The values of "K" allow for normal oversize of ropes, and the fact that it is practically impossible to "thread-wind" ropes of small diameter. However, the formula is based on uniform rope winding and will not give correct figures if rope is wound non-uniformly on the reel. The amount of tension applied when spooling the rope will also affect the length. The formula is based on the same number of wraps of rope in each layer, which is not strictly correct, but which does not result in appreciable error unless the width (B) of the reel is quite small compared with the flange diameter (H).

Example: Find the length in feet of 9/16-inch diameter rope required to fill a drum having the following dimensions: B = 24 inches, D = 18 inches, H = 30 inches,

$$A = (30 - 18 - 0) \div 2 = 6 \text{ inches}$$
$$L = (6 + 18) \times 6 \times 24 \times 0.741 = 2560.0 \text{ or } 2560 \text{ feet}$$

The above formula and factors K allow for normal oversize of ropes but will not give correct figures if rope is wound non-uniformly on the reel.

Load Capacity of Sheave or Drum: To avoid excessive wear and groove corrugation, the radial pressure exerted by the wire rope on the sheave or drum must be kept within certain maximum limits. The radial pressure of the rope is a function of the rope tension, rope diameter, and tread diameter of the sheave and can be determined by the following equation:

$$P = \frac{2T}{D \times d}$$

where P = Radial pressure in pounds per square inch (see Table 9)

T = Rope tension in pounds

D = Tread diameter of sheave or drum in inches

d = Rope diameter in inches

Table 9. Maximum Radial Pressures for Drums and Sheaves

Type of Wire Rope	Drum or Sheave Material		
	Cast Iron	Cast Steel	Manganese Steel[a]
	Recommended Maximum Radial Pressures, Pounds per Square Inch		
6 × 7	300[b]	550[b]	1500[b]
6 × 19	500[b]	900[b]	2500[b]
6 × 37	600	1075	3000
6 × 8 Flattened Strand	450	850	2200
6 × 25 Flattened Strand	800	1450	4000
6 × 30 Flattened Strand	800	1450	4000

[a] 11 to 13 per cent manganese.

[b] These values are for regular lay rope. For Lang lay rope these values may be increased by 15 per cent.

According to the Bethlehem Steel Co. the radial pressures shown in Table 9 are recommended as maximums according to the material of which the sheave or drum is made.

Rope Loads due to Bending: When a wire rope is bent around a sheave, the resulting bending stress s_b in the outer wire, and equivalent bending load P_b (amount that direct tension load on rope is increased by bending) may be computed by the following formulas: $s_b = Ed_w \div D$; $P_b = s_bA$, where $A = d^2Q$. E is the modulus of elasticity of the wire rope (varies with the type and condition of rope from 10,000,000 to 14,000,000. An average value of 12,000,000 is frequently used), d is the diameter of the wire rope, d_w is the diameter of the component wire (for 6 × 7 rope, $d_w = 0.106d$; for 6 × 19 rope, $0.063d$; for 6 × 37 rope, $0.045d$; and for 8 × 19 rope, $d_w = 0.050d$). D is the pitch diameter of the sheave in inches, A is the metal cross-sectional area of the rope, and Q is a constant, values for which are: 6 × 7 (Fiber Core) rope, 0.380; 6 × 7 (IWRC or WSC), 0.437; 6 × 19 (Fiber Core), 0.405; 6 × 19 (IWRC or WSC), 0.475; 6 × 37 (Fiber Core), 0.400; 6 × 37 (IWRC), 0.470; 8 × 19 (Fiber Core), 0.370; and Flattened Strand Rope, 0.440.

Example: Find the bending stress and equivalent bending load due to the bending of a 6 × 19 (Fiber Core) wire rope of ½-inch diameter around a 24-inch pitch diameter sheave.

$$d_w = 0.063 \times 0.5 = 0.0315 \text{ in.} \qquad A = 0.5^2 \times 0.405 = 0.101 \text{ sq. in.}$$

$$s_b = 12{,}000{,}000 \times 0.0315 \div 24 = 15{,}750 \text{ lbs. per sq. in.}$$

$$P_b = 15{,}750 \times 0.101 = 1590 \text{ lbs.}$$

Cutting and Seizing of Wire Rope.—Wire rope can be cut with mechanical wire rope shears, an abrasive wheel, an electric resistance cutter (used for ropes of smaller diameter only), or an acetylene torch. This last method fuses the ends of the wires in the strands. It is important that the rope be seized on either side of where the cut is to be made. Any annealed low carbon steel wire may be used for seizing, the recommended sizes being as follows: For a wire rope of ¼- to 15⁄16-inch diameter, use a seizing wire of 0.054-inch (No. 17 Steel Wire Gage); for a rope of 1- to 1⅝-inch diameter, use a 0.105-inch wire (No. 12); and for rope of 1¾- to 3½-inch diameter, use a 0.135-inch wire (No. 10). Except for preformed wire ropes, a minimum of two seizings on either side of a cut is recommended. Four seizings should be used on either side of a cut for Lang lay rope, a rope with a steel core, or a non-spinning type of rope.

The following method of seizing is given in Federal specification for wire rope, RR-R-571a. Lay one end of the seizing wire in the groove between two strands of wire rope and wrap the other end tightly in a close helix over the portion in the groove. A seizing iron (round bar ½ to ⅝ inch diameter by 18 inches long) should be used to wrap the seizing tightly. This bar is placed at right angles to the rope next to the first turn or two of the seizing wire. The seizing wire is brought around the back of the seizing iron so that it can be wrapped loosely around the wire rope in the opposite direction to that of the seizing coil. As the seizing iron is now rotated around the rope it will carry the seizing wire snugly and tightly into place. When completed, both ends of the seizing should be twisted together tightly.

Maintenance of Wire Rope.—Heavy abrasion, overloading, and bending around sheaves or drums that are too small in diameter are the principal reasons for the rapid deterioration of wire rope. Wire rope in use should be inspected periodically for evidence of wear and damage by corrosion. Such inspections should take place at progressively shorter intervals over the useful life of the rope as wear tends to accelerate with use. Where wear is rapid, the outside of a wire rope will show flattened surfaces in a short time.

If there is any hazard involved in the use of the rope, it may be prudent to estimate the remaining strength and service life. This assessment should be done for the weakest point where the most wear or largest number of broken wires are in evidence. One way to arrive at a conclusion is to set an arbitrary number of broken wires in a given strand as an indication that the rope should be removed from service and an ultimate strength test run on the worn sample. The arbitrary figure can then be revised and rechecked until a practical working formula is arrived at. A piece of waste rubbed along the wire rope will help to reveal broken wires. The effects of corrosion are not easy to detect because the exterior wires may appear to be only slightly rusty, and the damaging effects of corrosion may be confined to the hidden inner wires where it cannot be seen. To prevent damage by corrosion, the rope should be kept well lubricated. Use of zinc coated wire rope may be indicated for some applications.

Periodic cleaning of wire rope by using a stiff brush and kerosene or with compressed air or live steam and relubricating will help to lengthen rope life and reduce abrasion and wear on sheaves and drums. Before storing after use, wire rope should be cleaned and lubricated.

Lubrication of Wire Rope.—Although wire rope is thoroughly lubricated during manufacture to protect it against corrosion and to reduce friction and wear, this lubrication should be supplemented from time to time. Special lubricants are supplied by wire rope manufacturers. These lubricants vary somewhat with the type of rope application and operating condition. Where the preferred lubricant can not be obtained from the wire rope manufacturer, an adhesive type of lubricant similar to that used for open gearing will often be found suitable. At normal temperatures, some wire rope lubricants may be practically solid and will require thinning before application. Thinning may be done by heating to 160 to 200 degrees F. or by diluting with gasoline or some other fluid that will allow the lubricant to penetrate the rope. The lubricant may be painted on the rope or the rope may be passed through a box or tank filled with the lubricant.

Replacement of Wire Rope.—When an old wire rope is to be replaced, all drums and sheaves should be examined for wear. All evidence of scoring or imprinting of grooves from previous use should be removed and sheaves with flat spots, defective bearings, and broken flanges, should be repaired or replaced. It will frequently be found that the area of maximum wear is located relatively near one end of the rope. By cutting off that portion, the remainder of the rope may be salvaged for continued use. Sometimes the life of a rope can be increased by simply changing it end for end at about one-half the estimated normal life. The worn sections will then no longer come at the points that cause the greatest wear.

Wire Rope Slings and Fittings.—A few of the simpler sling arrangements or hitches as they are called, are shown in the accompanying illustration. Normally 6 × 19 Class wire rope is recommended where a diameter in the ¼-inch to 1⅛-inch range is to be used and 6 × 37 Class wire rope where a diameter in the 1¼-inch and larger range is to be used. However, the 6 × 19 Class may be used even in the larger sizes if resistance to abrasion is of primary importance and the 6 × 37 Class in the smaller sizes if greater flexibility is desired.

Wire Rope Slings and Fittings

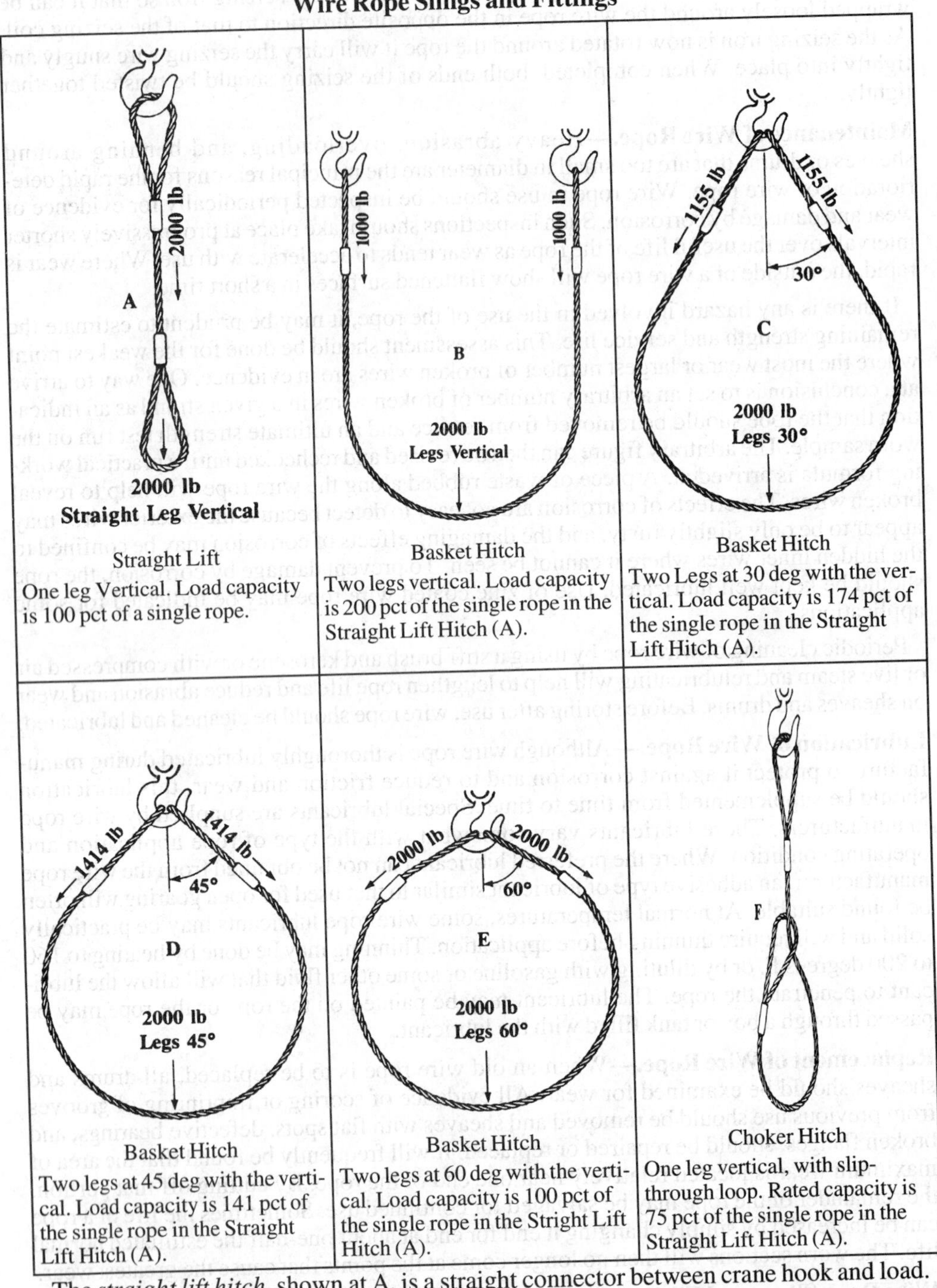
2000 lb
A
2000 lb
Straight Leg Vertical
Straight Lift
One leg Vertical. Load capacity is 100 pct of a single rope.
1000 lb
1000 lb
B
2000 lb
Legs Vertical
Basket Hitch
Two legs vertical. Load capacity is 200 pct of the single rope in the Straight Lift Hitch (A).
1155 lb
1155 lb
30°
C
2000 lb
Legs 30°
Basket Hitch
Two Legs at 30 deg with the vertical. Load capacity is 174 pct of the single rope in the Straight Lift Hitch (A).
1414 lb
1414 lb
45°
D
2000 lb
Legs 45°
Basket Hitch
Two legs at 45 deg with the vertical. Load capacity is 141 pct of the single rope in the Straight Lift Hitch (A).
2000 lb
2000 lb
60°
E
2000 lb
Legs 60°
Basket Hitch
Two legs at 60 deg with the vertical. Load capacity is 100 pct of the single rope in the Stright Lift Hitch (A).
F
Choker Hitch
One leg vertical, with slip-through loop. Rated capacity is 75 pct of the single rope in the Straight Lift Hitch (A).

The *straight lift hitch*, shown at A, is a straight connector between crane hook and load.

The *basket hitch* may be used with two hooks so that the sides are vertical as shown at B or with a single hook with sides at various angles with the vertical as shown at C, D, and E. As the angle with the vertical increases, a greater tension is placed on the rope so that for any given load, a sling of greater lifting capacity must be used.

The *choker hitch*, shown at F, is widely used for lifting bundles of items such as bars, poles, pipe, and similar objects. The choker hitch holds these items firmly but the load must be balanced so that it rides safely. Since additional stress is imposed on the rope due to the choking action, the capacity of this type of hitch is 25 per cent less than that of the comparable straight lift. If two choker hitches are used at an angle, these angles must also be taken into consideration as with the basket hitches.

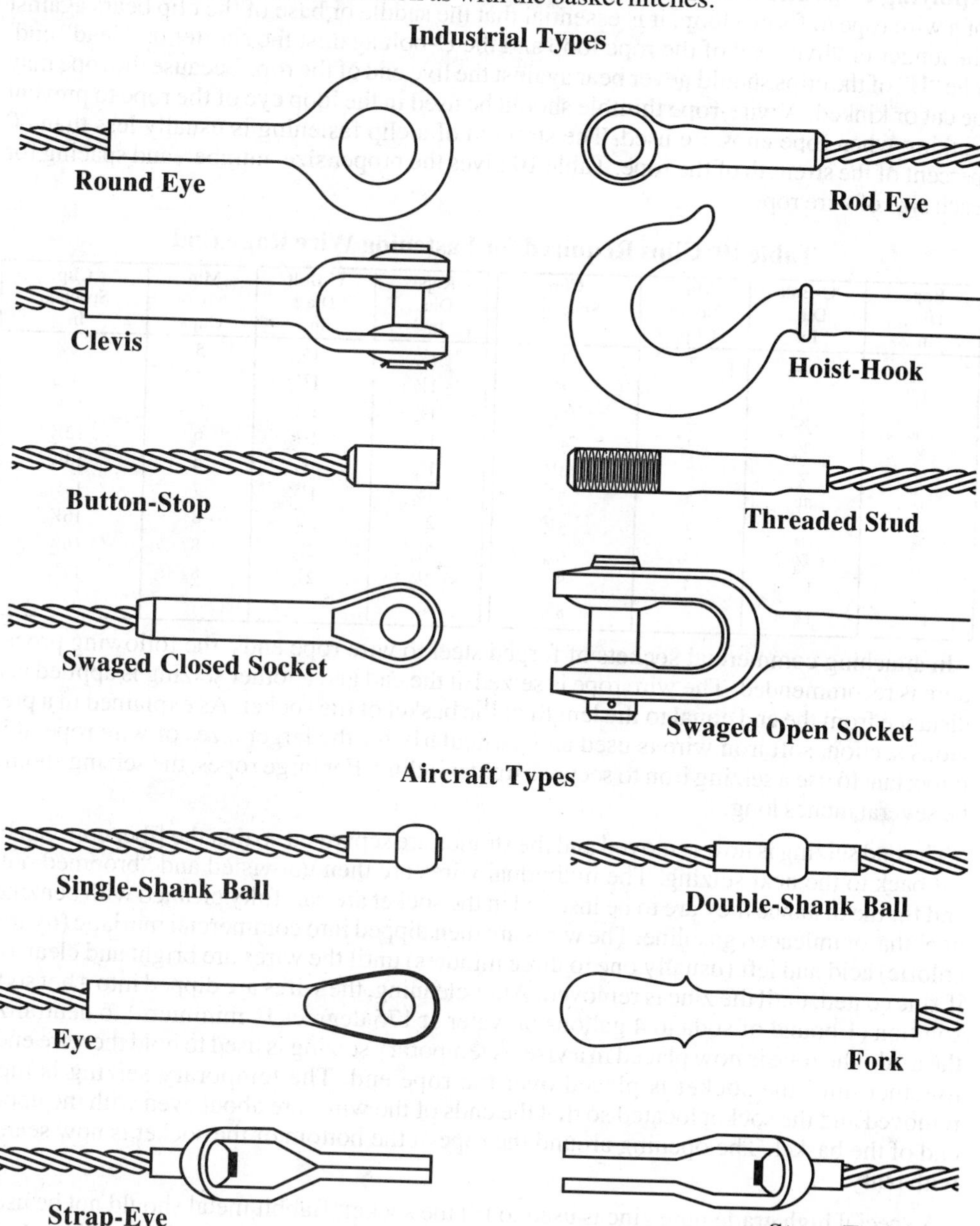

Wire Rope Fittings

Wire Rope Fittings.—Many varieties of swaged fittings are available for use with wire rope and several industrial and aircraft types are shown in the accompanying illustration. Swaged fittings on wire rope have an efficiency (ability to hold the wire rope) of approximately 100 per cent of the catalogue rope strength. These fittings are attached to the end or body of the wire rope by the application of high pressure through special dies that cause the material of the fitting to "flow" around the wires and strands of the rope to form a union that is as strong as the rope itself. The more commonly used types, of swaged fittings range from $\frac{1}{8}$- to $\frac{5}{8}$-inch diameter sizes in industrial types and from the $\frac{1}{16}$- to $\frac{5}{8}$-inch sizes in aircraft types. These fittings are furnished attached to the wire strand, rope, or cable.

Applying Clips and Attaching Sockets.—In attaching U-bolt clips for fastening the end of a wire rope to form a loop, it is essential that the saddle or base of the clip bears against the longer or "live" end of the rope loop and the U-bolt against the shorter or "dead" end. The "U" of the clips should never bear against the live end of the rope because the rope may be cut or kinked. A wire-rope thimble should be used in the loop eye of the rope to prevent kinking when rope clips are used. The strength of a clip fastening is usually less than 80 percent of the strength of the rope. Table 10 gives the proper size, number, and spacing for each size of wire rope.

Table 10. Clips Required for Fastening Wire Rope End

Rope Dia., In.	U-Bolt Dia., In.	Min. No. of Clips	Clip Spacing, In.	Rope Dia., In.	U-Bolt Dia., In.	Min. No. of Clips	Clip Spacing, In.
$\frac{3}{16}$	$\frac{11}{32}$	2	3	$1\frac{1}{8}$	$1\frac{1}{4}$	5	$9\frac{3}{4}$
$\frac{1}{4}$	$\frac{7}{16}$	2	$3\frac{1}{4}$	$1\frac{1}{4}$	$1\frac{7}{16}$	5	$10\frac{3}{4}$
$\frac{5}{16}$	$\frac{1}{2}$	2	$3\frac{1}{4}$	$1\frac{3}{8}$	$1\frac{1}{2}$	6	$11\frac{1}{2}$
$\frac{3}{8}$	$\frac{9}{16}$	2	4	$1\frac{1}{2}$	$1\frac{23}{32}$	6	$12\frac{1}{2}$
$\frac{7}{16}$	$\frac{5}{8}$	2	$4\frac{1}{2}$	$1\frac{5}{8}$	$1\frac{3}{4}$	6	$13\frac{1}{4}$
$\frac{1}{2}$	$\frac{11}{16}$	3	5	$1\frac{3}{4}$	$1\frac{15}{16}$	7	$14\frac{1}{2}$
$\frac{5}{8}$	$\frac{3}{4}$	3	$5\frac{3}{4}$	2	$2\frac{1}{8}$	8	$16\frac{1}{2}$
$\frac{3}{4}$	$\frac{7}{8}$	4	$6\frac{3}{4}$	$2\frac{1}{4}$	$2\frac{5}{8}$	8	$16\frac{1}{2}$
$\frac{7}{8}$	1	4	8	$2\frac{1}{2}$	$2\frac{7}{8}$	8	$17\frac{3}{4}$
1	$1\frac{1}{8}$	4	$8\frac{3}{4}$	…	…	…	…

In attaching commercial sockets of forged steel to wire rope ends, the following procedure is recommended. The wire rope is seized at the end and another seizing is applied at a distance from the end equal to the length of the basket of the socket. As explained in a previous section, soft iron wire is used and particularly for the larger sizes of wire rope, it is important to use a seizing iron to secure a tight winding. For large ropes, the seizing should be several inches long.

The end seizing is now removed and the strands are separated so that the fiber core can be cut back to the next seizing. The individual wires are then untwisted and "broomed out" and for the distance they are to be inserted in the socket are carefully cleaned with benzine, naphtha, or unleaded gasoline. The wires are then dipped into commercial muriatic (hydrochloric) acid and left (usually one to three minutes) until the wires are bright and clean or, if zinc coated, until the zinc is removed. After cleaning, the wires are dipped into a hot soda solution (1 pound of soda to 4 gallons of water at 175 degrees F. minimum) to neutralize the acid. The rope is now placed in a vise. A temporary seizing is used to hold the wire ends together until the socket is placed over the rope end. The temporary seizing is then removed and the socket located so that the ends of the wires are about even with the upper end of the basket. The opening around the rope at the bottom of the socket is now sealed with putty.

A special high grade pure zinc is used to fill the socket. Babbit metal should not be used as it will not hold properly. For proper fluidity and penetration, the zinc is heated to a temperature in the 830- to 900-degree F. range. If a pyrometer is not available to measure the temperature of the molten zinc, a dry soft pine stick dipped into the zinc and quickly withdrawn will show only a slight discoloration and no zinc will adhere to it. If the wood chars, the zinc is too hot. The socket is now permitted to cool and the resulting joint is ready for use. When properly prepared, the strength of the joint should be approximately equal to that of the rope itself.

Rated Capacities for Improved Plow Steel Wire Rope and Wire Rope Slings (in tons of 2,000 lbs)—Independent Wire Rope Core

Rope Diameter (in.)	Vertical			Choker			60° Bridle			45°Bridle			30°Bridle		
	A	B	C	A	B	C	A	B	C	A	B	C	A	B	C
Single Leg, 6 × 19 Wire Rope															
¼	0.59	0.56	0.53	0.44	0.42	0.40	…	…	…	…	…	…	…	…	…
⅜	1.3	1.2	1.1	0.98	0.93	0.86	…	…	…	…	…	…	…	…	…
½	2.3	2.2	2.0	1.7	1.6	1.5	…	…	…	…	…	…	…	…	…
⅝	3.6	3.4	3.0	2.7	2.5	2.2	…	…	…	…	…	…	…	…	…
¾	5.1	4.9	4.2	3.8	3.6	3.1	…	…	…	…	…	…	…	…	…
⅞	6.9	6.6	5.5	5.2	4.9	4.1	…	…	…	…	…	…	…	…	…
1	9.0	8.5	7.2	6.7	6.4	5.4	…	…	…	…	…	…	…	…	…
1⅛	11	10	9.0	8.5	7.8	6.8	…	…	…	…	…	…	…	…	…
Single Leg, 6 × 37 Wire Rope															
1¼	13	12	10	9.9	9.2	7.9	…	…	…	…	…	…	…	…	…
1⅜	16	15	13	12	11	9.6	…	…	…	…	…	…	…	…	…
1½	19	17	15	14	13	11	…	…	…	…	…	…	…	…	…
1¾	26	24	20	19	18	15	…	…	…	…	…	…	…	…	…
2	33	30	26	25	23	20	…	…	…	…	…	…	…	…	…
2¼	41	38	33	31	29	25	…	…	…	…	…	…	…	…	…
Two-Leg Bridle or Basket Hitch, 6 × 19 Wire Rope Sling															
¼	1.2	1.1	1.0	…	…	…	1.0	0.97	0.92	0.83	0.79	0.75	0.59	0.56	0.53
⅜	2.0	2.5	2.3	…	…	…	2.3	2.1	2.0	1.8	1.8	1.8	1.3	1.2	1.1
½	4.0	4.4	3.9	…	…	…	4.0	3.6	3.4	3.2	3.1	2.8	2.3	2.2	2.0
⅝	7.2	6.6	6.0	…	…	…	6.2	5.9	5.2	5.1	4.8	4.2	3.6	3.4	3.0
¾	10	9.7	8.4	…	…	…	8.9	8.4	7.3	7.2	6.9	5.9	5.1	4.9	4.2
⅞	14	13	11	…	…	…	12	11	9.6	9.8	9.3	7.8	6.9	6.6	5.5
1	18	17	14	…	…	…	15	15	12	13	12	10	9.0	8.5	7.2
1⅛	23	21	18	…	…	…	19	18	16	16	15	13	11	10	9.0
Two-Leg Bridle or Basket Hitch, 6 × 37 Wire Rope Sling															
1¼	26	24	21	…	…	…	23	21	18	19	17	15	13	12	10
1⅜	32	29	25	…	…	…	28	25	22	22	21	18	16	15	13
1½	38	35	30	…	…	…	33	30	26	27	25	21	19	17	15
1¾	51	47	41	…	…	…	44	41	35	36	33	29	26	24	20
2	66	61	53	…	…	…	57	53	46	47	43	37	33	30	26
2¼	83	76	66	…	…	…	72	66	67	58	54	47	41	38	33

Rated Capacities for Improved Plow Steel Wire Rope and Wire Rope Slings (in tons of 2,000 lbs)—Fiber Core

Rope Diameter (in.)	Vertical			Choker			60° Bridle			45° Bridle			30° Bridle		
	A	B	C	A	B	C	A	B	C	A	B	C	A	B	C
Single Leg, 6 × 19 Wire Rope															
¼	0.55	0.51	0.49	0.41	0.38	0.37	…	…	…	…	…	…	…	…	…
⅜	1.2	1.1	1.1	0.91	0.85	0.80	…	…	…	…	…	…	…	…	…
½	2.1	2.0	1.8	1.6	1.5	1.4	…	…	…	…	…	…	…	…	…
⅝	3.3	3.1	2.8	2.5	2.3	2.1	…	…	…	…	…	…	…	…	…
¾	4.8	4.4	3.9	3.6	3.3	2.9	…	…	…	…	…	…	…	…	…
⅞	6.4	5.9	5.1	4.8	4.5	3.9	…	…	…	…	…	…	…	…	…
1	8.4	7.7	6.7	6.3	5.8	5.0	…	…	…	…	…	…	…	…	…
1⅛	10	9.5	8.4	7.9	7.1	6.3	…	…	…	…	…	…	…	…	…
Single Leg, 6 × 37 Wire Rope															
1¼	12	11	9.8	9.2	8.3	7.4	…	…	…	…	…	…	…	…	…
1⅜	15	13	12	11	10	8.9	…	…	…	…	…	…	…	…	…
1½	17	16	14	13	12	10	…	…	…	…	…	…	…	…	…
1¾	24	21	19	18	16	14	…	…	…	…	…	…	…	…	…
2	31	28	25	23	21	18	…	…	…	…	…	…	…	…	…
Two-Leg Bridle or Basket Hitch, 6 × 19 Wire Rope Sling															
¼	1.1	1.0	0.99	…	…	…	0.95	0.88	0.85	0.77	0.72	0.70	0.55	0.51	0.49
⅜	2.4	2.2	2.1	…	…	…	2.1	1.9	1.8	1.7	1.6	1.5	1.2	1.1	1.1
½	4.3	3.9	3.7	…	…	…	3.7	3.4	3.2	3.0	2.8	2.6	2.1	2.0	1.8
⅝	6.7	6.2	5.6	…	…	…	6.2	5.3	4.8	4.7	4.4	4.0	3.3	3.1	2.8
¾	9.5	8.8	7.8	…	…	…	8.2	7.6	6.8	6.7	6.2	5.5	4.8	4.4	3.9
⅞	13	12	10	…	…	…	11	10	8.9	9.1	8.4	7.3	6.4	5.9	5.1
1	17	15	13	…	…	…	14	13	11	12	11	9.4	8.4	7.7	6.7
1⅛	21	19	17	…	…	…	18	16	14	15	13	12	10	9.5	8.4
Two-Leg Bridle or Basket Hitch, 6 × 37 Wire Rope Sling															
1¼	25	22	20	…	…	…	21	19	17	17	16	14	12	11	9.8
1⅜	30	27	24	…	…	…	26	23	20	21	19	17	15	13	12
1½	35	32	28	…	…	…	30	27	24	25	22	20	17	16	14
1¾	46	43	39	…	…	…	41	37	33	34	30	27	24	21	19
2	62	55	49	…	…	…	53	43	43	43	39	35	31	26	25

A—socket or swaged terminal attachment; B—mechanical sleeve attachment; C—hand-tucked splice attachment.

Data taken from *Longshoring Industry*, OSHA Safety and Health Standards Digest, OSHA 2232, 1985.

CRANE CHAIN AND HOOKS

Material for Crane Chains.—The best material for crane and hoisting chains is a good grade of wrought iron, in which the percentage of phosphorus, sulfur, silicon, and other impurities is comparatively low. The tensile strength of the best grades of wrought iron does not exceed 46,000 pounds per square inch, whereas mild steel with about 0.15 per cent carbon has a tensile strength nearly double this amount. The ductility and toughness of wrought iron, however, is greater than that of ordinary commercial steel, and for this reason it is preferable for chains subjected to heavy intermittent strains, because wrought iron will always give warning by bending or stretching, before breaking. Another important reason for using wrought iron in preference to steel is that a perfect weld can be effected more easily. Heat-treated alloy steel is also widely used for chains. This steel contains carbon, 0.30 per cent, max; phosphorus, 0.045 per cent, max; and sulfur, 0.045 per cent, max. The selection and amounts of alloying elements are left to the individual manufacturers.

Strength of Chains.—When calculating the strength of chains it should be observed that the strength of a link subjected to tensile stresses is not equal to twice the strength of an iron bar of the same diameter as the link stock, but is a certain amount less, owing to the bending action caused by the manner in which the load is applied to the link. The strength is also reduced somewhat by the weld. The following empirical formula is commonly used for calculating the breaking load, in pounds, of wrought-iron crane chains:

$$W = 54{,}000D^2$$

in which W = breaking load in pounds and D = diameter of bar (in inches) from which links are made. The working load for chains should not exceed one-third the value of W, and, it is often one-fourth or one-fifth of the breaking load. When a chain is wound around a casting and severe bending stresses are introduced, a greater factor of safety should be used.

Care of Hoisting and Crane Chains.—Chains used for hoisting heavy loads are subject to deterioration, both apparent and invisible. The links wear, and repeated loading causes localized deformations to form cracks that spread until the links fail. Chain wear can be reduced by occasional lubrication. The life of a wrought-iron chain can be prolonged by frequent annealing or normalizing unless it has been so highly or frequently stressed that small cracks have formed. If this condition is present, annealing or normalizing will not "heal" the material, and the links will eventually fracture. To anneal a wrought-iron chain, heat it to cherry-red and allow it to cool slowly. Annealing should be done every six months, and oftener if the chain is subjected to unusually severe service.

Maximum Allowable Wear at Any Point of Link

Chain Size (in.)	Maximum Allowable Wear (in.)	Chain Size (in.)	Maximum Allowable Wear (in.)
1⁄4 (9⁄32)	3⁄64	1	3⁄16
3⁄8	5⁄64	1 1⁄8	7⁄32
1⁄2	7⁄64	1 1⁄4	1⁄4
5⁄8	9⁄64	1 3⁄8	9⁄32
3⁄4	5⁄32	1 1⁄2	5⁄16
7⁄8	11⁄64	1 3⁄4	11⁄32

Source:Longshoring Industry, OSHA 2232, 1985.

Chains should be examined periodically for twists, as a twisted chain will wear rapidly. Any links that have worn excessively should be replaced with new ones, so that every link will do its full share of work during the life of the chain, without exceeding the limit of safety. Chains for hoisting purposes should be made with short links, so that they will wrap closely around the sheaves or drums without bending. The diameter of the winding drums should be not less than 25 or 30 times the diameter of the iron used for the links. The accompanying table lists the maximum allowable wear for various sizes of chains.

Safe Loads for Ropes and Chains.—Safe loads recommended for wire rope or chain slings depend not only upon the strength of the sling but also upon the method of applying it to the load, as shown by the accompanying table giving safe loads as prepared by OSHA. The loads recommended in this table are more conservative than those usually specified, in order to provide ample allowance for some unobserved weakness in the sling, or the possibility of excessive strains due to misjudgment or accident.

The working load limit is defined as the maximum load in pounds that should ever be applied to chain, when the chain is new or in "as new" condition, and when the load is uniformly applied in direct tension to a straight length of chain. This limit is also affected by the number of chains used and their configuration. The accompanying table shows the working load limit for various configurations of heat-treated alloy steel chain using a 4 to 1 design factor, which conforms to ISO practice.

Protection from Sharp Corners: When the load to be lifted has sharp corners or edges, as are often encountered with castings, and with structural steel and other similar objects, pads or wooden protective pieces should be applied at the corners, to prevent the slings from being abraded or otherwise damaged where they come in contact with the load. These precautions are especially important when the slings consist of wire cable or fiber rope, although they should also be used even when slings are made of chain. Wooden corner-pieces are often provided for use in hoisting loads with sharp angles. If pads of burlap or other soft material are used, they should be thick and heavy enough to sustain the pressure, and distribute it over a considerable area, instead of allowing it to be concentrated directly at the edges of the part to be lifted.

Strength of Manila Rope

Dia. (in.)	Circumference (in.)	Weight of 100 feet of Rope[a] (lb)	New Rope Tensile Strength[b] (lb)	Working Load[c] (lb)	Dia. (in.)	Circumference (in.)	Weight of 100 feet of Rope[a] (lb)	New Rope Tensile Strength[b] (lb)	Working Load[c] (lb)
3/16	5/8	1.50	406	41	1 5/16	4	47.8	13,500	1930
1/4	3/4	2.00	540	54	1 1/2	4 1/2	60.0	16,700	2380
5/16	1	2.90	900	90	1 5/8	5	74.5	20,200	2880
3/8	1 1/8	4.10	1220	122	1 3/4	5 1/2	89.5	23,800	3400
7/16	1 1/4	5.25	1580	176	2	6	108	28,000	4000
1/2	1 1/2	7.50	2380	264	2 1/8	6 1/2	125	32,400	4620
9/16	1 3/4	10.4	3100	388	2 1/4	7	146	37,000	5300
5/8	2	13.3	3960	496	2 1/2	7 1/2	167	41,800	5950
3/4	2 1/4	16.7	4860	695	2 5/8	8	191	46,800	6700
13/16	2 1/2	19.5	5850	835	2 7/8	8 1/2	215	52,000	7450
7/8	2 3/4	22.4	6950	995	3	9	242	57,500	8200
1	3	27.0	8100	1160	3 1/4	10	298	69,500	9950
1 1/16	3 1/4	31.2	9450	1350	3 1/2	11	366	82,000	11,700
1 1/8	3 1/2	36.0	10,800	1540	4	12	434	94,500	13,500
1 1/4	3 3/4	41.6	12,200	1740	…	…	…	…	…

[a] Average value is shown; maximum is 5 per cent higher.

[b] Based on tests of new and unused rope of standard construction in accordance with Cordage Institute Standard Test Methods.

[c] These values are for rope in good condition with appropriate splices, in noncritical applications, and under normal service conditions. These values should be reduced where life, limb, or valuable propety are involved, or for exceptional service conditions such as shock loads or sustained loads.

Data from Cordage Institute Rope Specifications for three-strand laid and eight-strand plaited manila rope (standard construction).

Strength of Nylon and Double Braided Nylon Rope

Dia. (in.)	Circumference (in.)	Weight of 100 feet of Rope[a] (lb)	New Rope Tensile Strength[b] (lb)	Working Load[c] (lb)	Dia. (in.)	Circumference (in.)	Weight of 100 feet of Rope[a] (lb)	New Rope Tensile Strength[a] (lb)	Working Load[c] (lb)
Nylon Rope									
3⁄16	5⁄8	1.00	900	75	1 5⁄16	4	45.0	38,800	4,320
1⁄4	3⁄4	1.50	1,490	124	1 1⁄2	4 1⁄2	55.0	47,800	5,320
5⁄16	1	2.50	2,300	192	1 5⁄8	5	66.5	58,500	6,500
3⁄8	1 1⁄8	3.50	3,340	278	1 3⁄4	5 1⁄2	83.0	70,000	7,800
7⁄16	1 1⁄4	5.00	4,500	410	2	6	95.0	83,000	9,200
1⁄2	1 1⁄2	6.50	5,750	525	2 1⁄8	6 1⁄2	109	95,500	10,600
9⁄16	1 3⁄4	8.15	7,200	720	2 1⁄4	7	129	113,000	12,600
5⁄8	2	10.5	9,350	935	2 1⁄2	7 1⁄2	149	126,000	14,000
3⁄4	2 1⁄4	14.5	12,800	1,420	2 5⁄8	8	168	146,000	16,200
13⁄16	2 1⁄2	17.0	15,300	1,700	2 7⁄8	8 1⁄2	189	162,000	18,000
7⁄8	2 3⁄4	20.0	18,000	2,000	3	9	210	180,000	20,000
1	3	26.4	22,600	2,520	3 1⁄4	10	264	226,000	25,200
1 1⁄16	3 1⁄4	29.0	26,000	2,880	3 1⁄2	11	312	270,000	30,000
1 1⁄8	3 1⁄2	34.0	29,800	3,320	4	12	380	324,000	36,000
1 1⁄4	3 3⁄4	40.0	33,800	3,760	…	…	…	…	…
Double Braided Nylon Rope (Nylon Cover—Nylon Core)									
1⁄4	3⁄4	1.56	1,650	150	1 5⁄16	4	43.1	44,700	5,590
5⁄16	1	2.44	2,570	234	1 3⁄8	4 1⁄4	47.3	49,000	6,130
3⁄8	1 1⁄8	3.52	3,700	336	1 1⁄2	4 1⁄2	56.3	58,300	7,290
7⁄16	1 5⁄16	4.79	5,020	502	1 5⁄8	5	66.0	68,300	8,540
1⁄2	1 1⁄2	6.25	6,550	655	1 3⁄4	5 1⁄2	76.6	79,200	9,900
9⁄16	1 3⁄4	7.91	8,270	919	2	6	100	103,000	12,900
5⁄8	2	9.77	10,200	1,130	2 1⁄8	6 1⁄2	113	117,000	14,600
3⁄4	2 1⁄4	14.1	14,700	1,840	2 1⁄4	7	127	131,000	18,700
13⁄16	2 1⁄2	16.5	17,200	2,150	2 1⁄2	7 1⁄2	156	161,000	23,000
7⁄8	2 3⁄4	19.1	19,900	2,490	2 5⁄8	8	172	177,000	25,300
1	3	25.0	26,000	3,250	3	9	225	231,000	33,000
1 1⁄16	3 1⁄4	28.2	29,300	3,660	3 1⁄4	10	264	271,000	38,700
1 1⁄8	3 1⁄2	31.6	32,800	4,100	3 1⁄2	11	329	338,000	48,300
1 1⁄4	3 3⁄4	39.1	40,600	5,080	4	12	400	410,000	58,600

[a] Average value is shown. Maximum for nylon rope is 5 per cent higher; tolerance for double braided nylon rope is ± 5 per cent.

[b] Based on tests of new and unused rope of standard construction in accordance with Cordage Institute Standard Test Methods. For double braided nylon rope these values are minimums and are based on a large number of tests by various manufacturers; these values represent results two standard deviations below the mean. The minimum tensile strength is determined by the formula 1057 × (linear density)$^{0.995}$.

[c] These values are for rope in good condition with appropriate splices, in noncritical applications, and under normal service conditions. These values should be reduced where life, limb, or valuable property are involved, or for exceptional service conditions such as shock loads or sustained loads.

Data from Cordage Institute Specifications for nylon rope (three-strand laid and eight-strand plaited, standard construction) and double braided nylon rope.

Safe Working Loads in Pounds for Manila Rope and Chains

Diameter of Rope, or of Rod or Bar for Chain Links, Inch	Rope or Chain Vertical	Sling at 60°	Sling at 45°	Sling at 30°
Manila Rope				
1/4	120	204	170	120
5/16	200	346	282	200
3/8	270	467	380	270
7/16	350	605	493	350
15/32	450	775	635	450
1/2	530	915	798	530
9/16	690	1190	973	690
5/8	880	1520	1240	880
3/4	1080	1870	1520	1080
13/16	1300	2250	1830	1300
7/8	1540	2660	2170	1540
1	1800	3120	2540	1800
1 1/16	2000	3400	2800	2000
1 1/8	2400	4200	3400	2400
1 1/4	2700	4600	3800	2700
1 5/16	3000	5200	4200	3000
1 1/2	3600	6200	5000	3600
1 5/8	4500	7800	6400	4500
1 3/4	5200	9000	7400	5200
2	6200	10,800	8800	6200
2 1/8	7200	12,400	10,200	7200
Crane Chain (Wrought Iron)				
1/4[a]	1060	1835	1500	1060
5/16[a]	1655	2865	2340	1655
3/8	2385	4200	3370	2385
7/16[a]	3250	5600	4600	3250
1/2	4200	7400	6000	4200
9/16[a]	5400	9200	7600	5400
5/8	6600	11,400	9400	6600
3/4	9600	16,600	13,400	9600
7/8	13,000	22,400	18,400	13,000
1	17,000	29,400	24,000	17,000
1 1/8	20,000	34,600	28,400	20,000
1 1/4	24,800	42,600	35,000	24,800
1 3/8	30,000	51,800	42,200	30,000
1 1/2	35,600	61,600	50,400	35,600
1 5/8	41,800	72,400	59,000	41,800
1 3/4	48,400	84,000	68,600	48,400
1 7/8	55,200	95,800	78,200	55,200
2	63,200	109,600	89,600	63,200
Crane Chain (Alloy Steel)				
1/4	3240	5640	4540	3240
3/8	6600	11,400	9300	6600
1/2	11,240	19,500	15,800	11,240
5/8	16,500	28,500	23,300	16,500
3/4	23,000	39,800	32,400	23,000
7/8	28,600	49,800	40,600	28,600
1	38,600	67,000	54,600	38,600
1 1/8	44,400	77,000	63,000	44,400
1 1/4	57,400	99,400	81,000	57,400
1 3/8	67,000	116,000	94,000	67,000
1 1/2	79,400	137,000	112,000	79,400
1 5/8	85,000	147,000	119,000	85,000
1 3/4	95,800	163,000	124,000	95,800

[a] These sizes of wrought chain are no longer manufactured in the United States.

Data from *Longshoring Industry*, OSHA Safety and Health Standards Digest, OSHA 2232, 1985.

Working Load Limit for Heat-Treated Alloy Steel Chain, pounds

Chain Size (in.)	Single Leg	Double Leg			Triple and Quad Leg		
	90°	60°	45°	30°	60°	45°	30°
1/4	3,600	6,200	5,050	3,600	9,300	7,600	5,400
3/8	6,400	11,000	9,000	6,400	16,550	13,500	9,500
1/2	11,400	19,700	16,100	11,400	29,600	24,200	17,100
5/8	17,800	30,800	25,150	17,800	46,250	37,750	26,700
3/4	25,650	44,400	36,250	25,650	66,650	54,400	38,450
7/8	34,900	60,400	49,300	34,900	90,650	74,000	52,350

Source: The Crosby Group.

Loads Lifted by Crane Chains.—To find the approximate weight a chain will lift when rove as a tackle, multiply the safe load given in the table by the number of parts or chains at the movable block, and subtract one-quarter for frictional resistance. To find the size of chain required for lifting a given weight, divide the weight by the number of chains at the movable block, and add one-third for friction; next find in the column headed "Average Safe Working Load" the corresponding load, and then the corresponding size of chain in the column headed "Size." With the heavy chain or where the chain is unusually long, the weight of the chain itself should also be considered.

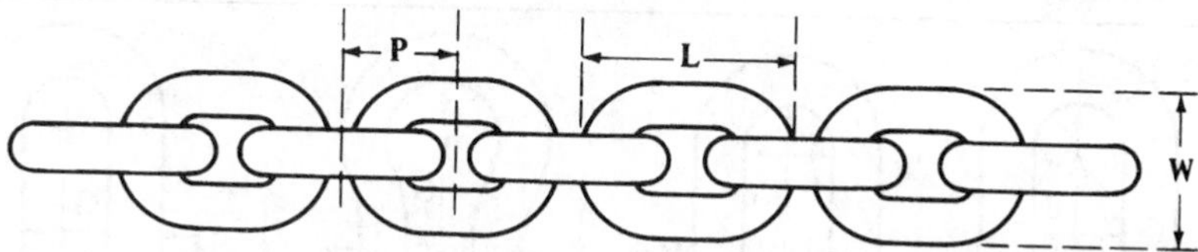

Close-link Hoisting, Sling and Crane Chain

Size	Standard Pitch, P Inches	Average Weight per Foot, Pounds	Outside Length, L Inches	Outside Width, W Inches	Average Safe Working Load, Pounds	Proof Test, Pounds[a]	Approximate Breaking Load, Pounds
1/4	25/32	3/4	1 5/16	7/8	1,200	2,500	5,000
5/16	27/32	1	1 1/2	1 1/16	1,700	3,500	7,000
3/8	31/32	1 1/2	1 3/4	1 1/4	2,500	5,000	10,000
7/16	1 5/32	2	2 1/16	1 3/8	3,500	7,000	14,000
1/2	1 11/32	2 1/2	2 3/8	1 11/16	4,500	9,000	18,000
9/16	1 15/32	3 1/4	2 5/8	1 7/8	5,500	11,000	22,000
5/8	1 23/32	4	3	2 1/16	6,700	14,000	27,000
11/16	1 13/16	5	3 1/4	2 1/4	8,100	17,000	32,500
3/4	1 15/16	6 1/4	3 1/2	2 1/2	10,000	20,000	40,000
13/16	2 1/16	7	3 3/4	2 11/16	10,500	23,000	42,000
7/8	2 3/16	8	4	2 7/8	12,000	26,000	48,000
15/16	2 7/16	9	4 3/8	3 1/16	13,500	29,000	54,000
1	2 1/2	10	4 5/8	3 1/4	15,200	32,000	61,000
1 1/16	2 5/8	12	4 7/8	3 5/16	17,200	35,000	69,000
1 1/8	2 3/4	13	5 1/8	3 3/4	19,500	40,000	78,000
1 3/16	3 1/16	14 1/2	5 9/16	3 7/8	22,000	46,000	88,000
1 1/4	3 1/8	16	5 3/4	4 1/8	23,700	51,000	95,000
1 5/16	3 3/8	17 1/2	6 1/8	4 1/4	26,000	54,000	104,000
1 3/8	3 9/16	19	6 7/16	4 9/16	28,500	58,000	114,000
1 7/16	3 11/16	21 1/2	6 11/16	4 3/4	30,500	62,000	122,000
1 1/2	3 7/8	23	7	5	33,500	67,000	134,000
1 9/16	4	25	7 3/8	5 5/16	35,500	70,500	142,000
1 5/8	4 1/4	28	7 3/4	5 1/2	38,500	77,000	154,000
1 11/16	4 1/2	30	8 1/8	5 11/16	39,500	79,000	158,000
1 3/4	4 3/4	31	8 1/2	5 7/8	41,500	83,000	166,000

Close-link Hoisting, Sling and Crane Chain *(Continued)*

Size	Standard Pitch, *P* Inches	Average Weight per Foot, Pounds	Outside Length, *L* Inches	Outside Width, *W* Inches	Average Safe Working Load, Pounds	Proof Test, Pounds[a]	Approximate Breaking Load, Pounds
1 13/16	5	33	8 7/8	6 1/16	44,500	89,000	178,000
1 7/8	5 1/4	35	9 1/4	6 3/8	47,500	95,000	190,000
1 15/16	5 1/2	38	9 5/8	6 9/16	50,500	101,000	202,000
2	5 3/4	40	10	6 3/4	54,000	108,000	216,000
2 1/16	6	43	10 3/8	6 15/16	57,500	115,000	230,000
2 1/8	6 1/4	47	10 3/4	7 1/8	61,000	122,000	244,000
2 3/16	6 1/2	50	11 1/8	7 5/16	64,500	129,000	258,000
2 1/4	6 3/4	53	11 1/2	7 5/8	68,200	136,500	273,000
2 3/8	6 7/8	58 1/2	11 7/8	8	76,000	152,000	304,000
2 1/2	7	65	12 1/4	8 3/8	84,200	168,500	337,000
2 5/8	7 1/8	70	12 5/8	8 3/4	90,500	181,000	362,000
2 3/4	7 1/4	73	13	9 1/8	96,700	193,500	387,000
2 7/8	7 1/2	76	13 1/2	9 1/2	103,000	206,000	412,000
3	7 3/4	86	14	9 7/8	109,000	218,000	436,000

[a] Chains tested to U.S. Government and American Bureau of Shipping requirements.

Additional Tables

Dimensions of Forged Round Pin, Screw Pin, and Bolt Type Chain Shackles and Bolt Type Anchor Shackles

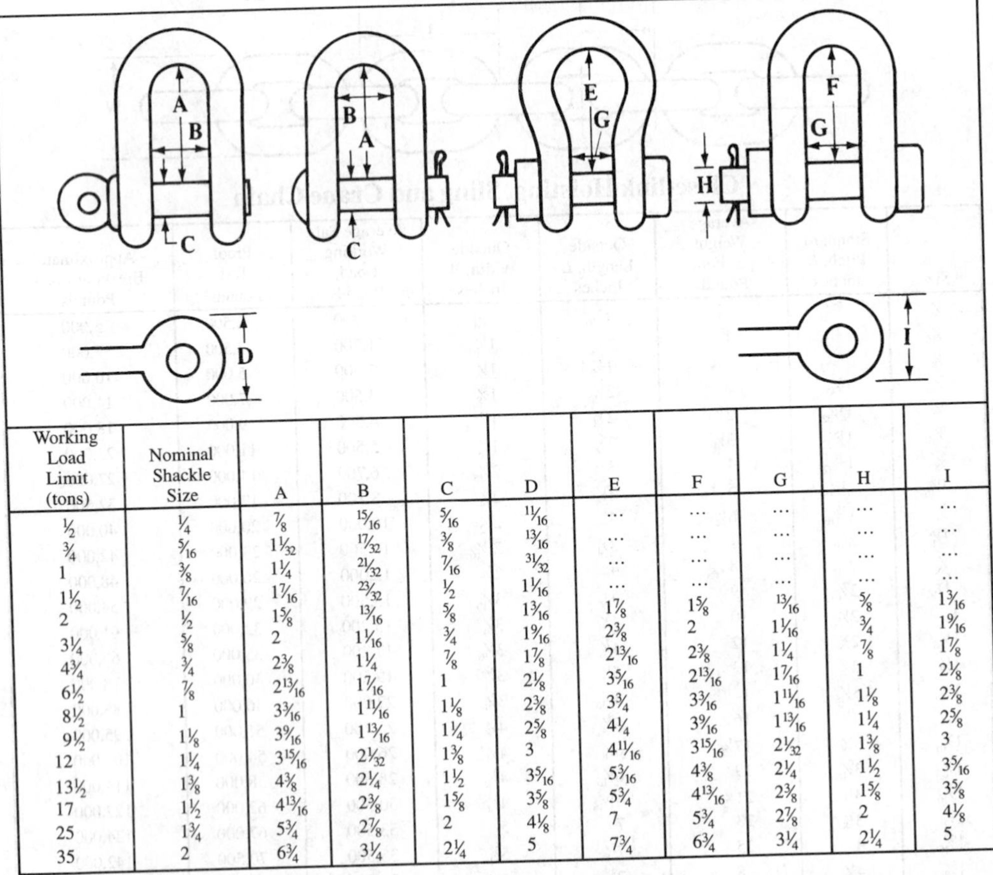

Working Load Limit (tons)	Nominal Shackle Size	A	B	C	D	E	F	G	H	I
1/2	1/4	7/8	15/16	5/16	11/16	…	…	…	…	…
3/4	5/16	1 1/32	17/32	3/8	13/16	…	…	…	…	…
1	3/8	1 1/4	21/32	7/16	31/32	…	…	…	…	…
1 1/2	7/16	1 7/16	23/32	1/2	1 1/16	…	…	…	…	…
2	1/2	1 5/8	13/16	5/8	1 3/16	1 7/8	1 5/8	13/16	5/8	1 3/16
3 1/4	5/8	2	1 1/16	3/4	1 9/16	2 3/8	2	1 1/16	3/4	1 9/16
4 3/4	3/4	2 3/8	1 1/4	7/8	1 7/8	2 13/16	2 3/8	1 1/4	7/8	1 7/8
6 1/2	7/8	2 13/16	1 7/16	1	2 1/8	3 5/16	2 13/16	1 7/16	1	2 1/8
8 1/2	1	3 3/16	1 11/16	1 1/8	2 3/8	3 3/4	3 3/16	1 11/16	1 1/8	2 3/8
9 1/2	1 1/8	3 9/16	1 13/16	1 1/4	2 5/8	4 1/4	3 9/16	1 13/16	1 1/4	2 5/8
12	1 1/4	3 15/16	2 1/32	1 3/8	3	4 11/16	3 15/16	2 1/32	1 3/8	3
13 1/2	1 3/8	4 3/8	2 1/4	1 1/2	3 5/16	5 3/16	4 3/8	2 1/4	1 1/2	3 5/16
17	1 1/2	4 13/16	2 3/8	1 5/8	3 5/8	5 3/4	4 13/16	2 3/8	1 5/8	3 5/8
25	1 3/4	5 3/4	2 7/8	2	4 1/8	7	5 3/4	2 7/8	2	4 1/8
35	2	6 3/4	3 1/4	2 1/4	5	7 3/4	6 3/4	3 1/4	2 1/4	5

All dimensions are in inches. Load limits are in tons of 2000 pounds.

Source: The Crosby Group.

Dimensions of Crane Hooks

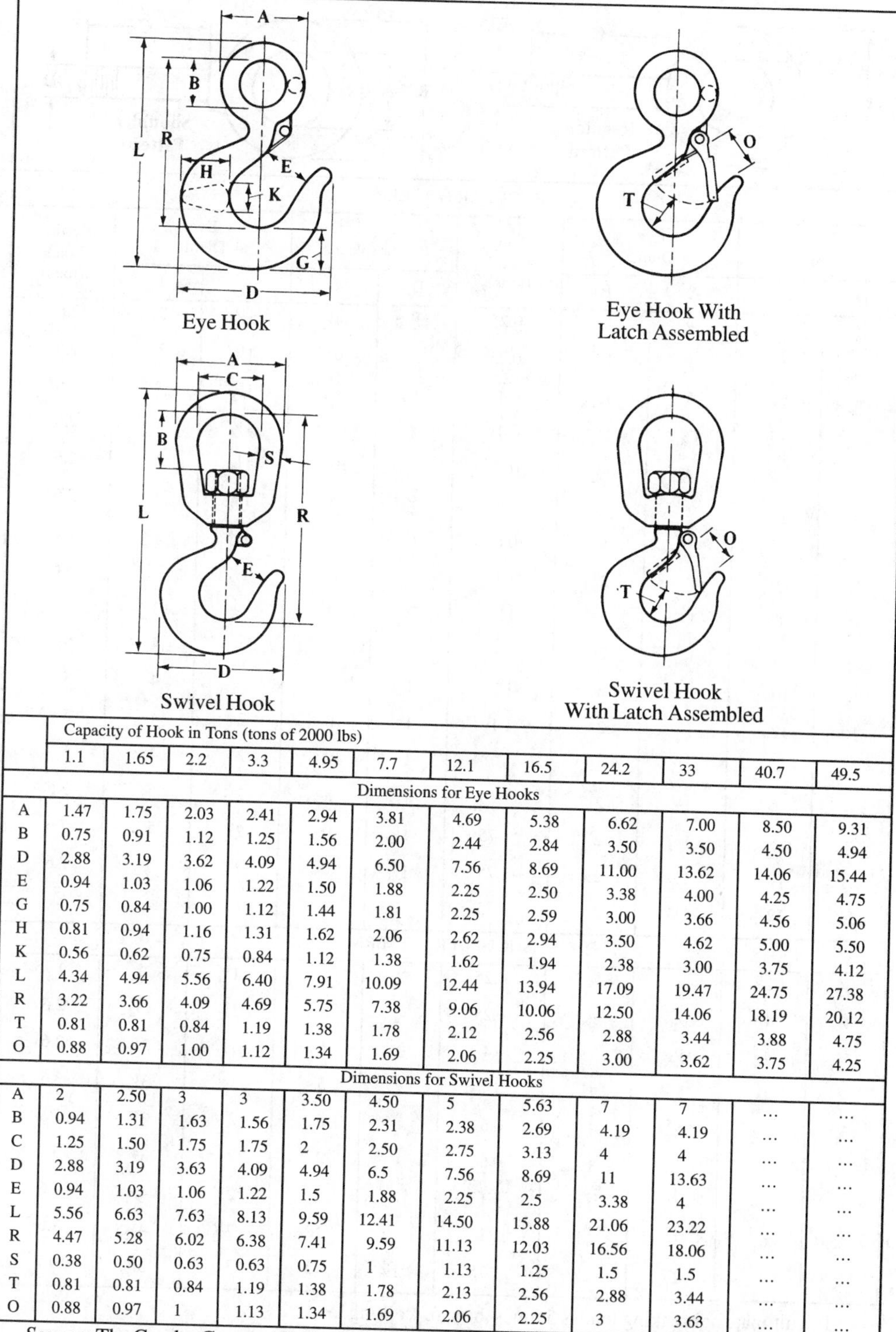

Eye Hook

Eye Hook With Latch Assembled

Swivel Hook

Swivel Hook With Latch Assembled

	Capacity of Hook in Tons (tons of 2000 lbs)											
	1.1	1.65	2.2	3.3	4.95	7.7	12.1	16.5	24.2	33	40.7	49.5
	Dimensions for Eye Hooks											
A	1.47	1.75	2.03	2.41	2.94	3.81	4.69	5.38	6.62	7.00	8.50	9.31
B	0.75	0.91	1.12	1.25	1.56	2.00	2.44	2.84	3.50	3.50	4.50	4.94
D	2.88	3.19	3.62	4.09	4.94	6.50	7.56	8.69	11.00	13.62	14.06	15.44
E	0.94	1.03	1.06	1.22	1.50	1.88	2.25	2.50	3.38	4.00	4.25	4.75
G	0.75	0.84	1.00	1.12	1.44	1.81	2.25	2.59	3.00	3.66	4.56	5.06
H	0.81	0.94	1.16	1.31	1.62	2.06	2.62	2.94	3.50	4.62	5.00	5.50
K	0.56	0.62	0.75	0.84	1.12	1.38	1.62	1.94	2.38	3.00	3.75	4.12
L	4.34	4.94	5.56	6.40	7.91	10.09	12.44	13.94	17.09	19.47	24.75	27.38
R	3.22	3.66	4.09	4.69	5.75	7.38	9.06	10.06	12.50	14.06	18.19	20.12
T	0.81	0.81	0.84	1.19	1.38	1.78	2.12	2.56	2.88	3.44	3.88	4.75
O	0.88	0.97	1.00	1.12	1.34	1.69	2.06	2.25	3.00	3.62	3.75	4.25
	Dimensions for Swivel Hooks											
A	2	2.50	3	3	3.50	4.50	5	5.63	7	7	…	…
B	0.94	1.31	1.63	1.56	1.75	2.31	2.38	2.69	4.19	4.19	…	…
C	1.25	1.50	1.75	1.75	2	2.50	2.75	3.13	4	4	…	…
D	2.88	3.19	3.63	4.09	4.94	6.5	7.56	8.69	11	13.63	…	…
E	0.94	1.03	1.06	1.22	1.5	1.88	2.25	2.5	3.38	4	…	…
L	5.56	6.63	7.63	8.13	9.59	12.41	14.50	15.88	21.06	23.22	…	…
R	4.47	5.28	6.02	6.38	7.41	9.59	11.13	12.03	16.56	18.06	…	…
S	0.38	0.50	0.63	0.63	0.75	1	1.13	1.25	1.5	1.5	…	…
T	0.81	0.81	0.84	1.19	1.38	1.78	2.13	2.56	2.88	3.44	…	…
O	0.88	0.97	1	1.13	1.34	1.69	2.06	2.25	3	3.63	…	…

Source: The Crosby Group. All dimensions are in inches. Hooks are made of alloy steel, quenched and tempered. For swivel hooks, the data are for a bail of carbon steel. The ultimate load is four times the working load limit (capacity). The swivel hook is a positioning device and is not intended to rotate under load; special load swiveling hooks must be used in such applications.

Hot Dip Galvanized, Forged Steel Eye-bolts

Shank		Eye Diam.		Safe Load[a] (tons)	Shank		Eye Diam.		Safe Load[a] (tons)
D	C	A	B		D	C	A	B	
REGULAR PATTERN									
¼	2	½	1	0.25	¾	4½	1½	3	2.6
¼	4	½	1	0.25	¾	6	1½	3	2.6
5/16	2¼	⅝	1¼	0.4	¾	8	1½	3	2.6
5/16	4¼	⅝	1¼	0.4	¾	10	1½	3	2.6
⅜	2½	¾	1½	0.6	¾	10	1½	3	2.6
⅜	4½	¾	1½	0.6	¾	10	1½	3	2.6
⅜	6	¾	1½	0.6	⅞	5	1¾	3½	3.6
½	3¼	1	2	1.1	⅞	8	1¾	3½	3.6
½	6	1	2	1.1	⅞	10	1¾	3½	3.6
½	8	1	2	1.1	1	6	2	4	5
½	10	1	2	1.1	1	9	2	4	5
½	12	1	2	1.1	1	10	2	4	5
⅝	4	1¼	2½	1.75	1	10	2	4	5
⅝	6	1¼	2½	1.75	1¼	8	2½	5	7.6
⅝	8	1¼	2½	1.75	1¼	10	2½	5	7.6
⅝	10	1¼	2½	1.75	1¼	10	2½	5	7.6
⅝	12	1¼	2½	1.75	…	…	…	…	…
SHOULDER PATTERN									
¼	2	½	⅞	0.25	⅝	6	1¼	2¼	1.75
¼	4	½	⅞	0.25	¾	4½	1½	2¾	2.6
5/16	2¼	⅝	1⅛	0.4	¾	6	1½	2¾	2.6
5/16	4¼	⅝	1⅛	0.4	⅞	5	1¾	3¼	3.6
⅜	2½	¾	1⅜	0.6	1	6	2	3¾	5
⅜	4½	¾	1⅜	0.6	1	9	2	3¾	5
½	3¼	1	1¾	1.1	1¼	8	2½	4½	7.6
½	6	1	1¾	1.1	1¼	12	2½	4½	7.6
⅝	4	1¼	2¼	1.75	1½	15	3	5½	10.7

[a] The ultimate or breaking load is 5 times the safe working load.

All dimensions are in inches. Safe loads are in tons of 2000 pounds.
Source: The Crosby Group.

Eye Nuts and Lift Eyes

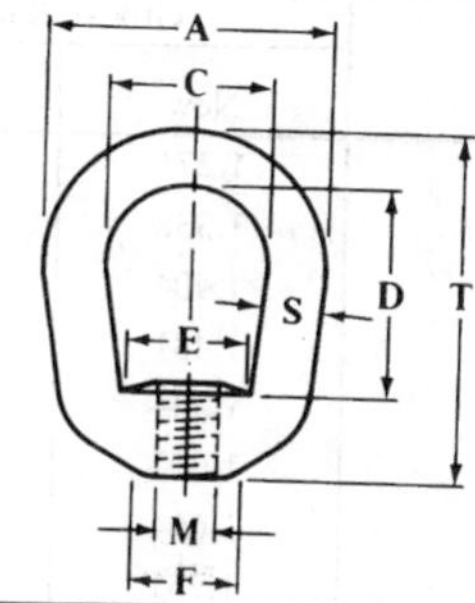

Eye Nuts

The general function of eye nuts is similar to that of eyebolts. Eye nuts are utilized for a variety of applications in either the swivel or tapped design.

M	A	C	D	E	F	S	T	Working Load Limit (lbs)*
1/4	1 1/4	3/4	1 1/16	21/32	1/2	1/4	1 11/16	520
5/16	1 1/4	3/4	1 1/16	21/32	1/2	1/4	1 11/16	850
3/8	1 5/8	1	1 1/4	3/4	9/16	5/16	2 1/16	1,250
7/16	2	1 1/4	1 1/2	1	13/16	3/8	2 1/2	1,700
1/2	2	1 1/4	1 1/2	1	13/16	3/8	2 1/2	2,250
5/8	2 1/2	1 1/2	2	1 3/16	1	1/2	3 3/16	3,600
3/4	3	1 3/4	2 3/8	1 3/8	1 1/8	5/8	3 7/8	5,200
7/8	3 1/2	2	2 5/8	1 5/8	1 5/16	3/4	4 5/16	7,200
1	4	2 1/4	3 1/16	1 7/8	1 9/16	7/8	5	10,000
1 1/8	4	2 1/4	3 1/16	1 7/8	1 9/16	7/8	5	12,300
1 1/4	4 1/2	2 1/2	3 1/2	1 15/16	1 7/8	1	5 3/4	15,500
1 3/8	5	2 3/4	3 3/4	2	2	1 1/8	6 1/4	18,500
1 1/2	5 5/8	3 1/8	4	2 3/8	2 1/4	1 1/4	6 3/4	22,500
2	7	4	6 1/4	4	3 3/8	1 1/2	10	40,000

Lifting Eyes

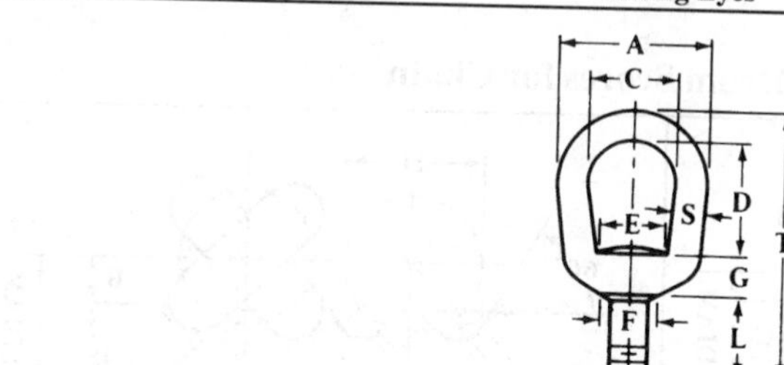

A	C	D	E	F	G	H	L	S	T	Working Load Limit Threaded (lbs)*
1 1/4	3/4	1 1/16	19/32	1/2	3/8	5/16	11/16	1/4	2 3/8	850
1 5/8	1	1 1/4	3/4	9/16	1/2	3/8	15/16	5/16	3	1,250
2	1 1/4	1 1/2	1	13/16	5/8	1/2	1 1/4	3/8	3 3/4	2,250
2 1/2	1 1/2	2	1 3/16	1	11/16	5/8	1 1/2	1/2	4 11/16	3,600
3	1 3/4	2 3/8	1 3/8	1 1/8	7/8	3/4	1 3/4	5/8	5 5/8	5,200
3 1/2	2	2 5/8	1 5/8	1 5/16	15/16	7/8	2	3/4	6 5/16	7,200
4	2 1/4	3 1/16	1 7/8	1 9/16	1 1/16	1	2 1/16	7/8	7 1/16	10,000
4 1/2	2 1/2	3 1/2	1 15/16	1 7/8	1 1/4	1 1/8	2 1/2	1	8 1/4	12,500
5 5/8	3 1/8	4	2 3/8	2 3/8	1 1/2	1 3/8	2 15/16	1 1/4	9 11/16	18,000

All dimensions are in inches. Data for eye nuts are for hot dip galvanized, quenched, and tempered forged steel. Data for lifting eyes are for quenched and tempered forged steel.
Source: The Crosby Group.

Minimum Sheave- and Drum-Groove Dimensions for Wire Rope Applications-I

Nominal Rope Diameter	Groove Radius		Nominal Rope Diameter	Groove Radius	
	New	Worn		New	Worn
1/4	0.135	0.129	2 3/8	1.271	1.199
5/16	0.167	0.160	2 1/2	1.338	1.279
3/8	0.201	0.190	2 5/8	1.404	1.339
7/16	0.234	0.220	2 3/4	1.481	1.409
1/2	0.271	0.256	2 7/8	1.544	1.473
9/16	0.303	0.288	3	1.607	1.538
5/8	0.334	0.320	3 1/8	1.664	1.598
3/4	0.401	0.380	3 1/4	1.731	1.658
7/8	0.468	0.440	3 3/8	1.807	1.730
1	0.543	0.513	3 1/2	1.869	1.794
1 1/8	0.605	0.577	3 3/4	1.997	1.918
1 1/4	0.669	0.639	4	2.139	2.050
1 3/8	0.736	0.699	4 1/4	2.264	2.178
1 1/2	0.803	0.759	4 1/2	2.396	2.298
1 5/8	0.876	0.833	4 3/4	2.534	2.434
1 3/4	0.939	0.897	5	2.663	2.557
1 7/8	1.003	0.959	5 1/4	2.804	2.691
2	1.085	1.025	5 1/2	2.929	2.817
2 1/8	1.137	1.079	5 3/4	3.074	2.947
2 1/4	1.210	1.153	6	3.198	3.075

All dimensions are in inches. Data taken from *Wire Rope Users Manual*, 2nd ed., American Iron and Steel Institute, Washington, D. C. The values given in this table are applicable to grooves in sheaves and drums but are not generally suitable for pitch design, since other factors may be involved.

Winding Drum Scores for Chain

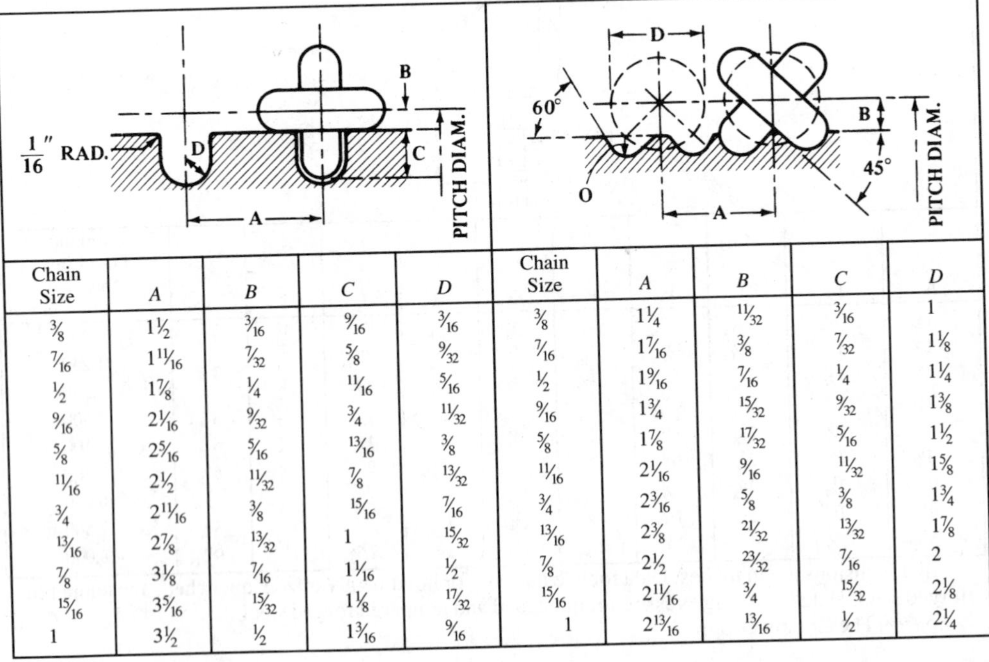

Chain Size	A	B	C	D	Chain Size	A	B	C	D
3/8	1 1/2	3/16	9/16	3/16	3/8	1 1/4	11/32	3/16	1
7/16	1 11/16	7/32	5/8	9/32	7/16	1 7/16	3/8	7/32	1 1/8
1/2	1 7/8	1/4	11/16	5/16	1/2	1 9/16	7/16	1/4	1 1/4
9/16	2 1/16	9/32	3/4	11/32	9/16	1 3/4	15/32	9/32	1 3/8
5/8	2 5/16	5/16	13/16	3/8	5/8	1 7/8	17/32	5/16	1 1/2
11/16	2 1/2	11/32	7/8	13/32	11/16	2 1/16	9/16	11/32	1 5/8
3/4	2 11/16	3/8	15/16	7/16	3/4	2 3/16	5/8	3/8	1 3/4
13/16	2 7/8	13/32	1	15/32	13/16	2 3/8	21/32	13/32	1 7/8
7/8	3 1/8	7/16	1 1/16	1/2	7/8	2 1/2	23/32	7/16	2
15/16	3 5/16	15/32	1 1/8	17/32	15/16	2 11/16	3/4	15/32	2 1/8
1	3 1/2	1/2	1 3/16	9/16	1	2 13/16	13/16	1/2	2 1/4

All dimensions are in inches.

PROPERTIES, TREATMENT, AND TESTING OF MATERIALS

THE ELEMENTS, HEAT, MASS, AND WEIGHT

PROPERTIES OF WOOD, CERAMICS, PLASTICS, METALS, WATER, AND AIR

STANDARD STEELS

STANDARD STEELS (cont.)

TOOL STEELS

PROPERTIES, TREATMENT, AND TESTING OF MATERIALS

HARDENING, TEMPERING, AND ANNEALING

HEAT-TREATING HIGH-SPEED STEELS

NONFERROUS ALLOYS

PLASTICS

THE ELEMENTS, HEAT, MASS, AND WEIGHT

The Elements — Symbols, Atomic Numbers and Weights, Melting Points

Name of Element	Symbol	Atomic Num.	Atomic Weight	Melting Point, °C	Name of Element	Symbol	Atomic Num.	Atomic Weight	Melting Point, °C
Actinium	Ac	89	227.028	1050	Neon	Ne	10	20.1179	−248.67
Aluminum	Al	13	26.9815	660.37	Neptunium	Np	93	237.048	640 ± 1
Americium	Am	95	(243)	994 ± 4	Nickel	Ni	28	58.69	1453
Antimony	Sb	51	121.75	630.74	Niobium	Nb	41	92.9064	2468 ± 10
Argon	A	18	39.948	−189.2	Nitrogen	N	7	14.0067	−209.86
Arsenic	As	33	74.9216	817[a]	Nobelium	No	102	(259)	…
Astatine	At	85	(210)	302	Osmium	Os	76	190.2	3045 ± 30
Barium	Ba	56	137.33	725	Oxygen	O	8	15.9994	−218.4
Berkelium	Bk	97	(247)	…	Palladium	Pd	46	106.42	1554
Beryllium	Be	4	9.01218	1278 ± 5	Phosphorus	P	15	30.9738	44.1
Bismuth	Bi	83	208.980	271.3	Platinum	Pt	78	195.08	1772
Boron	B	5	10.81	2079	Plutonium	Pu	94	(244)	641
Bromine	Br	35	79.904	−7.2	Polonium	Po	84	(209)	254
Cadmium	Cd	48	112.41	320.9	Potassium	K	19	39.0938	63.25
Calcium	Ca	20	40.08	839 ± 2	Praseodymium	Pr	59	140.908	931 ± 4
Californium	Cf	98	(251)	…	Promethium	Pm	61	(145)	1080[b]
Carbon	C	6	12.011	3652[c]	Protactinium	Pa	91	231.0359	1600
Cerium	Ce	58	140.12	798 ± 2	Radium	Ra	88	226.025	700
Cesium	Cs	55	132.9054	28.4 ± 0.01	Radon	Rn	86	(222)	−71
Chlorine	Cl	17	35.453	−100.98	Rhenium	Re	75	186.207	3180
Chromium	Cr	24	51.996	1857 ± 20	Rhodium	Rh	45	102.906	1965 ± 3
Cobalt	Co	27	58.9332	1495	Rubidium	Rb	37	85.4678	38.89
Copper	Cu	29	63.546	1083.4 ± 0.2	Ruthenium	Ru	44	101.07	2310
Curium	Cm	96	(247)	1340 ± 40	Samarium	Sm	62	150.36	1072 ± 5
Dysprosium	Dy	66	162.5	1409	Scandium	Sc	21	44.9559	1539
Einsteinium	Es	99	(252)	…	Selenium	Se	34	78.96	217
Erbium	Er	68	167.26	1522	Silicon	Si	14	28.0855	1410
Europium	Eu	63	151.96	822 ± 5	Silver	Ag	47	107.868	961.93
Fermium	Fm	100	(257)	…	Sodium	Na	11	22.9898	97.81 ± 0.03
Fluorine	F	9	18.9984	−219.62	Strontium	Sr	38	87.62	769
Francium	Fr	87	(223)	27[b]	Sulfur	S	16	32.06	112.8
Gadolinium	Gd	64	157.25	1311 ± 1	Tantalum	Ta	73	180.9479	2996
Gallium	Ga	31	69.72	29.78	Technetium	Tc	43	(98)	2172
Germanium	Ge	32	72.59	937.4	Tellurium	Te	52	127.60	449.5 ± 0.3
Gold	Au	79	196.967	1064.434	Terbium	Tb	65	158.925	1360 ± 4
Hafnium	Hf	72	178.49	2227 ± 20	Thallium	Tl	81	204.383	303.5
Helium	He	2	4.00260	−272.2[d]	Thorium	Th	90	232.038	1750
Holmium	Ho	67	164.930	1470	Thulium	Tm	69	168.934	1545 ± 15
Hydrogen	H	1	1.00794	−259.14	Tin	Sn	50	118.71	231.9681
Indium	In	49	114.82	156.61	Titanium	Ti	22	47.88	1660 ± 10
Iodine	I	53	126.905	113.5	Tungsten	W	74	183.85	3410 ± 20
Iridium	Ir	77	192.22	2410	Unnilhexium	Unh	106	(266)	…
Iron	Fe	26	55.847	1535	Unnilnonium	Unn	109	(266)	…
Krypton	Kr	36	83.80	−156.6	Unniloctium	Uno	108	(265)	…
Lanthanum	La	57	138.906	920 ± 5	Unnilpentium	Unp	105	(262)	…
Lawrencium	Lw	103	(260)	…	Unnilquadium	Unq	104	(261)	…
Lead	Pb	82	207.2	327.502	Unnilseptium	Uns	107	(261)	…
Lithium	Li	3	6.941	180.54	Uranium	U	92	238.029	1132 ± 0.8
Lutetium	Lu	71	174.967	1656 ± 5	Vanadium	V	23	50.9415	1890 ± 10
Magnesium	Mg	12	24.305	648.8 ± 0.5	Xenon	Xe	54	131.29	−111.9
Manganese	Mn	25	54.9380	1244 ± 2	Ytterbium	Yb	70	173.04	824 ± 5
Mendelevium	Md	101	(258)	…	Yttrium	Y	39	88.9059	1523 ± 8
Mercury	Hg	80	200.59	−38.87	Zinc	Zn	30	65.39	419.58
Molybdenum	Mo	42	95.94	2617	Zirconium	Zr	40	91.224	1852 ± 2
Neodymium	Nd	60	144.24	1010					

[a] At 28 atm.

[b] Approximate.

[c] Sublimates.

[d] At 26 atm.

Notes: Values in parentheses are atomic weights of the most stable known isotopes. Melting points at standard pressure except as noted.

Heat and Combustion Related Properties

Latent Heat.—When a body changes from the solid to the liquid state or from the liquid to the gaseous state, a certain amount of heat is used to accomplish this change. This heat does not raise the temperature of the body and is called latent heat. When the body changes again from the gaseous to the liquid, or from the liquid to the solid state, it gives out this quantity of heat. The *latent heat of fusion* is the heat supplied to a solid body at the melting point; this heat is absorbed by the body although its temperature remains nearly stationary during the whole operation of melting. The *latent heat of evaporation* is the heat that must be supplied to a liquid at the boiling point to transform the liquid into a vapor. The latent heat is generally given in British thermal units per pound. When it is said that the latent heat of evaporation of water is 966.6, this means that it takes 966.6 heat units to evaporate 1 pound of water after it has been raised to the boiling point, 212°F.

When a body changes from the solid to the gaseous state without passing through the liquid stage, as solid carbon dioxide does, the process is called *sublimation*.

Latent Heat of Fusion

Substance	Btu per Pound	Substance	Btu per Pound	Substance	Btu per Pound
Bismuth	22.75	Paraffine	63.27	Sulfur	16.86
Beeswax	76.14	Phosphorus	9.06	Tin	25.65
Cast iron, gray	41.40	Lead	10.00	Zinc	50.63
Cast iron, white	59.40	Silver	37.92	Ice	144.00

Latent Heat of Evaporation

Liquid	Btu per Pound	Liquid	Btu per Pound	Liquid	Btu per Pound
Alcohol, ethyl	371.0	Carbon Bisulfide	160.0	Turpentine	133.0
Alcohol, methyl	481.0	Ether	162.8	Water	966.6
Ammonia	529.0	Sulfur dioxide	164.0		

Boiling Points of Various Substances at Atmospheric Pressure

Substance	Boiling Point, °F	Substance	Boiling Point, °F	Substance	Boiling Point, °F
Aniline	363	Chloroform	140	Saturated brine	226
Alcohol	173	Ether	100	Sulfur	833
Ammonia	−28	Linseed oil	597	Sulfuric acid	590
Benzine	176	Mercury	676	Water, pure	212
Bromine	145	Napthaline	428	Water, sea	213.2
Carbon bisulfide	118	Nitric acid	248	Wood alcohol	150
		Oil of turpentine	315		

Specific Heat.—The specific heat of a substance is the ratio of the heat required to raise the temperature of a certain weight of the given substance 1°F to that required to raise the temperature of the same weight of water 1 degree. As the specific heat is not constant at all temperatures, it is generally assumed that it is determined by raising the temperature from 62 to 63°F. For most substances, however, specific heat is practically constant for temperatures up to 212°F.

Average Specific Heats (Btu/lb-°F) of Various Substance

Substance	Specific Heat	Substance	Specific Heat
Alcohol (absolute)	0.700	Kerosene	0.500
Alcohol (density 0.8)	0.622	Lead	0.031
Aluminum	0.214	Limestone	0.217
Antimony	0.051	Magnesia	0.222
Benzine	0.450	Marble	0.210
Brass	0.094	Masonry, brick	0.200
Brickwork	0.200	Mercury	0.033
Cadmium	0.057	Naphtha	0.310
Charcoal	0.200	Nickel	0.109
Chalk	0.215	Oil, machine	0.400
Coal	0.240	Oil, olive	0.350
Coke	0.203	Phosphorus	0.189
Copper, 32° to 212° F	0.094	Platinum	0.032
Copper, 32° to 572° F	0.101	Quartz	0.188
Corundum	0.198	Sand	0.195
Ether	0.503	Silica	0.191
Fusel oil	0.564	Silver	0.056
Glass	0.194	Soda	0.231
Gold	0.031	Steel, high carbon	0.117
Graphite	0.201	Steel, mild	0.116
Ice	0.504	Stone (generally)	0.200
Iron, cast	0.130	Sulfur	0.178
Iron, wrought, 32° to 212° F	0.110	Sulfuric acid	0.330
32° to 392° F	0.115	Tin	0.056
32° to 572° F	0.122	Turpentine	0.472
32° to 662° F	0.126	Water	1.000
Iron, at high temperatures:		Wood, fir	0.650
1382° to 1832° F	0.213	Wood, oak	0.570
1750° to 1840° F	0.218	Wood, pine	0.467
1920° to 2190° F	0.199	Zinc	0.095

Specific Heat of Gases (Btu/lb-°F)

Gas	Constant Pressure	Constant Volume	Gas	Constant Pressure	Constant Volume
Acetic acid	0.412	…	Chloroform	0.157	…
Air	0.238	0.168	Ethylene	0.404	0.332
Alcohol	0.453	0.399	Hydrogen	3.409	2.412
Ammonia	0.508	0.399	Nitrogen	0.244	0.173
Carbonic acid	0.217	0.171	Oxygen	0.217	0.155
Carbonic oxide	0.245	0.176	Steam	0.480	0.346
Chlorine	0.121	…			

Heat Loss from Uncovered Steam Pipes.—The loss of heat from a bare steam or hot-water pipe varies with the difference between the temperature inside the pipe and that of the surrounding air. The loss is 2.15 Btu per hour, per square foot of pipe surface, per degree F of temperature difference when the latter is 100 degrees; for a difference of 200 degrees, the loss is 2.66 Btu; for 300 degrees, 3.26 Btu; for 400 degrees, 4.03 Btu; for 500 degrees, 5.18 Btu. Thus, if the pipe area is 1.18 square feet per foot of length, and the temperature difference 300°F, the loss per hour per foot of length = $1.18 \times 300 \times 3.26 = 1154$ Btu.

Values of Thermal Conductivity (*k*) and of Conductance (*C*) of Common Building and Insulating Materials

Type of Material	Thickness, in.	*k* or *C*[a]	Type of Material	Thick-ness, in.	*k* or *C*[a]	Max. Temp.,° F	Density, lb per cu. ft.	*k*[a]
BUILDING			BUILDING (*Continued*)					
Batt:	…	…	Siding:	…	…	…	…	…
Mineral Fiber	2–2¾	0.14	Metal[b]	Avg.	1.61	…	…	…
Mineral Fiber	3–3½	0.09	Wood, Med. Density	7/16	1.49	…	…	…
Mineral Fiber	3½–6½	0.05	Stone:	…	…	…	…	…
Mineral Fiber	6–7	0.04	Lime or Sand	1	12.50	…	…	…
Mineral Fiber	8½	0.03	Wall Tile:	…	…	…	…	…
Block:	…	…	Hollow Clay, 1-Cell	4	0.9	…	…	…
Cinder	4	0.90	Hollow Clay, 2-Cell	8	0.54	…	…	…
Cinder	8	0.58	Hollow Clay, 3-Cell	12	0.40	…	…	…
Cinder	12	0.53	Hollow Gypsum	Avg.	0.7	…	…	…
Block:	…	…	INSULATING					
Concrete	4	1.40	Blanket, Mineral Fiber:	…	…	…	…	…
Concrete	8	0.90	Felt	…	…	400	3 to 8	0.26
Concrete	12	0.78	Rock or Slag	…	…	1200	6 to 12	0.26[c]
Board:	…	…	Glass	…	…	350	0.65	0.33
Asbestos Cement	¼	16.5	Textile	…	…	350	0.65	0.31
Plaster	½	2.22	Blanket, Hairfelt	…	…	180	10	0.29
Plywood	¾	1.07	Board, Block and Pipe	…	…	…	…	…
Brick:	…	…	Insulation:	…	…	…	…	…
Common	1	5.0	Amosite	…	…	1500	15 to 18	0.32[c]
Face	1	9.0	Asbestos Paper	…	…	700	30	0.40[c]
Concrete (poured)	1	12.0	Glass or Slag (for Pipe)	…	…	350	3 to 4	0.23
Floor:	…	…	Glass or Slag (for Pipe)	…	…	1000	10 to 15	0.33[c]
Wood Subfloor	¾	1.06	Glass, Cellular	…	…	800	9	0.40
Hardwood Finish	¾	1.47	Magnesia (85%)	…	…	600	11 to 12	0.35[c]
Tile	Avg.	20.0	Mineral Fiber	…	…	100	15	0.29
Glass:	…	…	Polystyrene, Beaded	…	…	170	1	0.28
Architectural	…	10.00	Polystyrene, Rigid	…	…	170	1.8	0.25
Mortar:	…	…	Rubber, Rigid Foam	…	…	150	4.5	0.22
Cement	1	5.0	Wood Felt	…	…	180	20	0.31
Plaster:	…	…	Loose Fill:	…	…	…	…	…
Sand	⅜	13.30	Cellulose	…	…	…	2.5 to 3	0.27
Sand and Gypsum	½	11.10	Mineral Fiber	…	…	…	2 to 5	0.28
Stucco	1	5.0	Perlite	…	…	…	5 to 8	0.37
Roofing:	…	…	Silica Aerogel	…	…	…	7.6	0.17
Asphalt Roll	Avg.	6.50	Vermiculite	…	…	…	7 to 8.2	0.47
Shingle, asb. cem.	Avg.	4.76	Mineral Fiber Cement:	…	…	…	…	…
Shingle, asphalt	Avg.	2.27	Clay Binder	…	…	1800	24 to 30	0.49[c]
Shingle, wood	Avg.	1.06	Hydraulic Binder	…	…	1200	30 to 40	0.75[c]

[a] Units are in Btu/hr-ft^2-°F. Where thickness is given as 1 inch, the value given is thermal conductivity (*k*); for other thicknesses the value given is thermal conductance (*C*). All values are for a test mean temperature of 75°F, except those designated with [c], which are for 100°F.

[b] Over hollowback sheathing.

[c] Test mean temperature 100°F, see footnote [a].

Source: American Society of Heating, Refrigerating and Air-Conditioning Engineers, Inc.: *Handbook of Fundamentals.*

Linear Expansion of Various Substances between 32 and 212°F
Expansion of volume = 3 × linear expansion

Substance	Linear Expansion for 1°F	Substance	Linear Expansion for 1°F
Brick	0.0000030	Masonry, brick from	0.0000026
Cement, Portland	0.0000060	to	0.0000050
Concrete	0.0000080	Plaster	0.0000092
Ebonite	0.0000428	Porcelain	0.0000020
Glass, thermometer	0.0000050	Quartz, from	0.0000043
Glass, hard	0.0000040	to	0.0000079
Granite	0.0000044	Slate	0.0000058
Marble, from	0.0000031	Sandstone	0.0000065
to	0.0000079	Wood, pine	0.0000028

Coefficients of Heat Transmission

Metal	Btu per Second	Metal	Btu per Second	Metal	Btu per Second
Aluminum	0.00203	German silver	0.00050	Steel, soft	0.00062
Antimony	0.00022	Iron	0.00089	Silver	0.00610
Brass, yellow	0.00142	Lead	0.00045	Tin	0.00084
Brass, red	0.00157	Mercury	0.00011	Zinc	0.00170
Copper	0.00404	Steel, hard	0.00034	…	…

Heat transmitted, in British thermal units, per second, through metal 1 inch thick, per square inch of surface, for a temperature difference of 1°F

Coefficients of Heat Radiation

Surface	Btu per Hour	Surface	Btu per Hour
Cast-iron, new	0.6480	Sawdust	0.7215
Cast-iron, rusted	0.6868	Sand, fine	0.7400
Copper, polished	0.0327	Silver, polished	0.0266
Glass	0.5948	Tin, polished	0.0439
Iron, ordinary	0.5662	Tinned iron, polished	0.0858
Iron, sheet-, polished	0.0920	Water	1.0853
Oil	1.4800	…	…

Heat radiated, in British thermal units, per square foot of surface per hour, for a temperature difference of 1° F

Freezing Mixtures

Mixture	Temperature Change, °F From	To
Common salt (NaCl), 1 part; snow, 3 parts	32	±0
Common salt (NaCl), 1 part; snow, 1 part	32	−0.4
Calcium chloride ($CaCl_2$), 3 parts; snow, 2 parts	32	−27
Calcium chloride ($CaCl_2$), 2 parts; snow, 1 part	32	−44
Sal ammoniac (NH_4Cl), 5 parts; saltpeter (KNO_3), 5 parts; water, 16 parts	50	+10
Sal ammoniac (NH_4Cl), 1 part; saltpeter (KNO_3), 1 part; water, 1 part	46	−11
Ammonium nitrate (NH_4NO_3), 1 part; water, 1 part	50	+3
Potassium hydrate (KOH), 4 parts; snow, 3 parts	32	−35

Ignition Temperatures.—The following temperatures are required to ignite the different substances specified: Phosphorus, transparent, 120°F; bi sulfide of carbon, 300°F; gun cotton, 430°F; nitro-glycerine, 490°F; phosphorus, amorphous, 500°F; rifle powder, 550°F; charcoal, 660°F; dry pine wood, 800°F; dry oak wood, 900°F.

Typical Thermal Properties of Various Metals

Material and Alloy Designation[a]	Density, ρ lb/in³	Melting Point, °F solidus	Melting Point, °F liquidus	Conductivity, k, Btu/hr-ft-°F	Specific Heat, C, Btu/lb/°F	Coef. of Expansion, α μin./in.-°F
Aluminum Alloys						
2011	0.102	995	1190	82.5	0.23	12.8
2017	0.101	995	1185	99.4	0.22	13.1
2024	0.100	995	1180	109.2	0.22	12.9
3003	0.099	1190	1210	111	0.22	12.9
5052	0.097	1100	1200	80	0.22	13.2
5086	0.096	1085	1185	73	0.23	13.2
6061	0.098	1080	1200	104	0.23	13.0
7075	0.101	890	1180	70	0.23	13.1
Copper-Base Alloys						
Manganese Bronze	0.302	1590	1630	61	0.09	11.8
C11000 (Electrolytic tough pitch)	0.321	1941	1981	226	0.09	9.8
C14500 (Free machining Cu)	0.323	1924	1967	205	0.09	9.9
C17200, C17300 (Beryllium Cu)	0.298	1590	1800	62	0.10	9.9
C18200 (Chromium Cu)	0.321	1958	1967	187	0.09	9.8
C18700 (Leaded Cu)	0.323	1750	1975	218	0.09	9.8
C22000 (Commercial bronze, 90%)	0.318	1870	1910	109	0.09	10.2
C23000 (Red brass, 85%)	0.316	1810	1880	92	0.09	10.4
C26000 (Cartridge brass, 70%)	0.313	1680	1750	70	0.09	11.1
C27000 (Yellow brass)	0.306	1660	1710	67	0.09	11.3
C28000 (Muntz metal, 60%)	0.303	1650	1660	71	0.09	11.6
C33000 (Low-leaded brass tube)	0.310	1660	1720	67	0.09	11.2
C35300 (High-leaded brass)	0.306	1630	1670	67	0.09	11.3
C35600 (Extra-high-leaded brass)	0.307	1630	1660	67	0.09	11.4
C36000 (Free machining brass)	0.307	1630	1650	67	0.09	11.4
C36500 (Leaded Muntz metal)	0.304	1630	1650	71	0.09	11.6
C46400 (Naval brass)	0.304	1630	1650	67	0.09	11.8
C51000 (Phosphor bronze, 5% A)	0.320	1750	1920	40	0.09	9.9
C54400 (Free cutting phos. bronze)	0.321	1700	1830	50	0.09	9.6
C62300 (Aluminum bronze, 9%)	0.276	1905	1915	31.4	0.09	9.0
C62400 (Aluminum bronze, 11%)	0.269	1880	1900	33.9	0.09	9.2
C63000 (Ni-Al bronze)	0.274	1895	1930	21.8	0.09	9.0
Nickel-Silver	0.314	1870	2030	17	0.09	9.0
Nickel-Base Alloys						
Nickel 200, 201, 205	0.321	2615	2635	43.3	0.11	8.5
Hastelloy C-22	0.314	2475	2550	7.5	0.10	6.9
Hastelloy C-276	0.321	2415	2500	7.5	0.10	6.2
Inconel 718	0.296	2300	2437	6.5	0.10	7.2
Monel	0.305	2370	2460	10	0.10	8.7
Monel 400	0.319	2370	2460	12.6	0.10	7.7
Monel K500	0.306	2400	2460	10.1	0.10	7.6
Monel R405	0.319	2370	2460	10.1	0.10	7.6

Typical Thermal Properties of Various Metals *(Continued)*

Material and Alloy Designation[a]	Density, ρ lb/in³	Melting Point, °F solidus	Melting Point, °F liquidus	Conductivity, *k*, Btu/hr-ft-°F	Specific Heat, *C*, Btu/lb/°F	Coef. of Expansion, α µin./in.-°F
Stainless Steels						
S30100	0.290	2550	2590	9.4	0.12	9.4
S30200, S30300, S30323	0.290	2550	2590	9.4	0.12	9.6
S30215	0.290	2500	2550	9.2	0.12	9.0
S30400, S30500	0.290	2550	2650	9.4	0.12	9.6
S30430	0.290	2550	2650	6.5	0.12	9.6
S30800	0.290	2550	2650	8.8	0.12	9.6
S30900, S30908	0.290	2550	2650	9.0	0.12	8.3
S31000, S31008	0.290	2550	2650	8.2	0.12	8.8
S31600, S31700	0.290	2500	2550	9.4	0.12	8.8
S31703	0.290	2500	2550	8.3	0.12	9.2
S32100	0.290	2550	2600	9.3	0.12	9.2
S34700	0.290	2550	2650	9.3	0.12	9.2
S34800	0.290	2550	2650	9.3	0.12	9.3
S38400	0.290	2550	2650	9.4	0.12	9.6
S40300, S41000, S41600, S41623	0.280	2700	2790	14.4	0.11	5.5
S40500	0.280	2700	2790	15.6	0.12	6.0
S41400	0.280	2600	2700	14.4	0.11	5.8
S42000, S42020	0.280	2650	2750	14.4	0.11	5.7
S42200	0.280	2675	2700	13.8	0.11	6.2
S42900	0.280	2650	2750	14.8	0.11	5.7
S43000, S43020, S43023	0.280	2600	2750	15.1	0.11	5.8
S43600	0.280	2600	2750	13.8	0.11	5.2
S44002, S44004	0.280	2500	2700	14.0	0.11	5.7
S44003	0.280	2500	2750	14.0	0.11	5.6
S44600	0.270	2600	2750	12.1	0.12	5.8
S50100, S50200	0.280	2700	2800	21.2	0.11	6.2
Cast Iron and Steel						
Malleable Iron, A220 (50005, 60004, 80002)	0.265	liquidus approximately, 2100 to 2200, depending on composition		29.5	0.12	7.5
Grey Cast Iron	0.25			28.0	0.25	5.8
Ductile Iron, A536 (120–90–02)	0.25				0.16	5.9–6.2
Ductile Iron, A536 (100–70–03)	0.25			20.0	0.16	5.9–6.2
Ductile Iron, A536 (80–55–06)	0.25			18.0	0.15	5.9–6.2
Ductile Iron, A536 (65–45–120)	0.25			20.8	0.15	5.9–6.2
Ductile Iron, A536 (60–40–18)	0.25				0.12	5.9–6.2
Cast Steel, 3%C	0.25	liquidus, 2640		28.0	0.12	7.0
Titanium Alloys						
Commercially Pure	0.163	3000	3040	9.0	0.12	5.1
Ti-5Al-2.5Sn	0.162	2820	3000	4.5	0.13	5.3
Ti-8Mn	0.171	2730	2970	6.3	0.19	6.0

[a] Alloy designations correspond to the AluminumAssociation numbers for aluminum alloys and to the unified numbering system (UNS) for copper and stainless steel alloys. A220 and A536 are ASTM specified irons.

Properties of Mass and Weight

Specific Gravity.—Specific gravity is a number indicating how many times a certain volume of a material is heavier than an equal volume of water. The density of water differs slightly at different temperatures, so the usual custom is to make comparisons on the basis that the water has a temperature of 62°F. The weight of 1 cubic inch of pure water at 62°F is 0.0361 pound. If the specific gravity of any material is known, the weight of a cubic inch of the material, therefore, can be found by multiplying its specific gravity by 0.0361.

Example: The specific gravity of cast iron is 7.2. Find the weight of 5 cubic inches of cast iron.

$$7.2 \times 0.0361 \times 5 = 1.2996 \text{ pounds}$$

To find the weight per cubic foot of a material, multiply the specific gravity by 62.355.

If the weight of a cubic inch of a material is known, the specific gravity is found by dividing the weight per cubic inch by 0.0361.

Example: The weight of a cubic inch of gold is 0.697 pound. Find the specific gravity.

$$0.697 \div 0.0361 = 19.31$$

If the weight per cubic foot of a material is known, the specific gravity is found by multiplying this weight by 0.01604.

Average Specific Gravity of Various Substances

Substance	Specific Gravity	Weight lb/ft³	Substance	Specific Gravity	Weight lb/ft³
ABS	1.05	66	Lead	11.4	711
Acrylic	1.19	74	Limestone	2.6	162
Aluminum bronze	7.8	486	Marble	2.7	168
Aluminum, cast	2.6	160	Masonry	2.4	150
Aluminum, wrought	2.7	167	Mercury	13.56	845.3
Asbestos	2.4	150	Mica	2.8	175
Asphaltum	1.4	87	Mortar	1.5	94
Borax	1.8	112	Nickel, cast	8.3	517
Brick, common	1.8	112	Nickel, rolled	8.7	542
Brick, fire	2.3	143	Nylon 6, Cast	1.16	73
Brick, hard	2.0	125	PTFE	2.19	137
Brick, pressed	2.2	137	Phosphorus	1.8	112
Brickwork, in cement	1.8	112	Plaster of Paris	1.8	112
Brickwork, in mortar	1.6	100	Platinum	21.5	1342
CPVC	1.55	97	Polycarbonate	1.19	74
Cement, Portland (set)	3.1	193	Polyethylene	0.97	60
Chalk	2.3	143	Polypropylene	0.91	57
Charcoal	0.4	25	Polyurethane	1.05	66
Coal, anthracite	1.5	94	Quartz	2.6	162
Coal, bituminous	1.3	81	Salt, common	…	48
Concrete	2.2	137	Sand, dry	…	100
Earth, loose	…	75	Sand, wet	…	125
Earth, rammed	…	100	Sandstone	2.3	143
Emery	4.0	249	Silver	10.5	656
Glass	2.6	162	Slate	2.8	175
Glass, crushed	…	74	Soapstone	2.7	168
Gold, 22 carat fine	17.5	1091	Steel	7.9	491
Gold, pure	19.3	1204	Sulfur	2.0	125
Granite	2.7	168	Tar, bituminous	1.2	75
Gravel	…	109	Tile	1.8	112
Gypsum	2.4	150	Trap rock	3.0	187
Ice	0.9	56	Water at 62°F	1.0	62.355
Iron, cast	7.2	447	White metal	7.3	457
Iron, wrought	7.7	479	Zinc, cast	6.9	429
Iron slag	2.7	168	Zinc, sheet	7.2	450

The weight per cubic foot is calculated on the basis of the specific gravity except for those substances that occur in bulk, heaped, or loose form. In these instances, only the weights per cubic foot are given because the voids present in representative samples make the values of the specific gravities inaccurate.

Specific Gravity of Gases.—The specific gravity of gases is the number that indicates their weight in comparison with that of an equal volume of air. The specific gravity of air is 1, and the comparison is made at 32°F.

Specific Gravity of Gases At 32°F

Gas	Sp. Gr.	Gas	Sp. Gr.	Gas	Sp. Gr.
Air	1.000	Ether vapor	2.586	Marsh gas	0.555
Acetylene	0.920	Ethylene	0.967	Nitrogen	0.971
Alcohol vapor	1.601	Hydrofluoric acid	2.370	Nitric oxide	1.039
Ammonia	0.592	Hydrochloric acid	1.261	Nitrous oxide	1.527
Carbon dioxide	1.520	Hydrogen	0.069	Oxygen	1.106
Carbon monoxide	0.967	Illuminating gas	0.400	Sulfur dioxide	2.250
Chlorine	2.423	Mercury vapor	6.940	Water vapor	0.623

1 cubic foot of air at 32°F and atmospheric pressure weighs 0.0807 pound.

Specific Gravity of Liquids.—The specific gravity of liquids is the number that indicates how much a certain volume of the liquid weighs compared with an equal volume of water, the same as with solid bodies. The density of liquid is often expressed in degrees on the hydrometer, an instrument for determining the density of liquids, provided with graduations made to an arbitrary scale. The hydrometer consists of a glass tube with a bulb at one end containing air, and arranged with a weight at the bottom so as to float in an upright position in the liquid, the density of which is to be measured. The depth to which the hydrometer sinks in the liquid is read off on the graduated scale. The most commonly used hydrometer is the Baumé. The value of the degrees of the Baumé scale differs according to whether the liquid is heavier or lighter than water. The specific gravity for liquids heavier than water equals 145 ÷ (145 – degrees Baumé). For liquids lighter than water, the specific gravity equals 140 ÷ (130 + degrees Baumé).

Specific Gravity of Liquids

Liquid	Sp. Gr.	Liquid	Sp. Gr.	Liquid	Sp. Gr.
Acetic acid	1.06	Fluoric acid	1.50	Petroleum oil	0.82
Alcohol, commercial	0.83	Gasoline	0.70	Phosphoric acid	1.78
Alcohol, pure	0.79	Kerosene	0.80	Rape oil	0.92
Ammonia	0.89	Linseed oil	0.94	sulfuric acid	1.84
Benzine	0.69	Mineral oil	0.92	Tar	1.00
Bromine	2.97	Muriatic acid	1.20	Turpentine oil	0.87
Carbolic acid	0.96	Naphtha	0.76	Vinegar	1.08
Carbon di sulfide	1.26	Nitric acid	1.50	Water	1.00
Cotton-seed oil	0.93	Olive oil	0.92	Water, sea	1.03
Ether, sulfuric	0.72	Palm oil	0.97	Whale oil	0.92

Degrees on Baumé's Hydrometer Converted to Specific Gravity

Deg. Baumé	Specific Gravity		Deg. Baumé	Specific Gravity		Deg. Baumé	Specific Gravity	
	Liquids Heavier than Water	Liquids Lighter than Water		Liquids Heavier than Water	Liquids Lighter than Water		Liquids Heavier than Water	Liquids Lighter than Water
0	1.000	...	27	1.229	0.892	54	1.593	0.761
1	1.007	...	28	1.239	0.886	55	1.611	0.757
2	1.014	...	29	1.250	0.881	56	1.629	0.753
3	1.021	...	30	1.261	0.875	57	1.648	0.749
4	1.028	...	31	1.272	0.870	58	1.667	0.745
5	1.036	...	32	1.283	0.864	59	1.686	0.741
6	1.043	...	33	1.295	0.859	60	1.706	0.737
7	1.051	...	34	1.306	0.854	61	1.726	0.733
8	1.058	...	35	1.318	0.849	62	1.747	0.729
9	1.066	...	36	1.330	0.843	63	1.768	0.725
10	1.074	1.000	37	1.343	0.838	64	1.790	0.721
11	1.082	0.993	38	1.355	0.833	65	1.813	0.718
12	1.090	0.986	39	1.368	0.828	66	1.836	0.714
13	1.099	0.979	40	1.381	0.824	67	1.859	0.710
14	1.107	0.972	41	1.394	0.819	68	1.883	0.707
15	1.115	0.966	42	1.408	0.814	69	1.908	0.704
16	1.124	0.959	43	1.422	0.809	70	1.933	0.700
17	1.133	0.952	44	1.436	0.805	71	1.959	0.696
18	1.142	0.946	45	1.450	0.800	72	1.986	0.693
19	1.151	0.940	46	1.465	0.796	73	2.014	0.689
20	1.160	0.933	47	1.480	0.791	74	2.042	0.686
21	1.169	0.927	48	1.495	0.787	75	2.071	0.683
22	1.179	0.921	49	1.510	0.782	76	2.101	0.679
23	1.189	0.915	50	1.526	0.778	77	2.132	0.676
24	1.198	0.909	51	1.542	0.773	78	2.164	0.673
25	1.208	0.903	52	1.559	0.769	79	2.197	0.669
26	1.219	0.897	53	1.576	0.765	80	2.230	0.666

Average Weights and Volumes of Fuels.—The average weight of a bushel of charcoal is 20 pounds; of a bushel of coke, 40 pounds; of a bushel of anthracite coal, 67 pounds; and of a bushel of bituminous coal, 60 pounds.

Anthracite coal, 1 cubic foot = 55 to 65 pounds.

Anthracite coal, 1 ton (2240 pounds) = 34 to 41 cubic feet.

Bituminous coal, 1 cubic foot = 50 to 55 pounds.

Bituminous coal, 1 ton (2240 pounds) = 41 to 45 cubic feet.

Charcoal, 1 cubic foot = 18 to 18.5 pounds.

Charcoal, 1 ton (2240 pounds) = 120 to 124 cubic feet.

Coke, 1 cubic foot = 28 pounds.

Coke, 1 ton (2240 pounds) = 80 cubic feet.

Weight of Wood.—The weight of seasoned wood per cord is approximately as follows, assuming about 70 cubic feet of *solid wood* per cord: beech, 3300 pounds; chestnut, 2600 pounds; elm, 2900 pounds; maple, 3100 pounds; poplar, 2200 pounds; white pine, 2200 pounds; red oak, 3300 pounds; white oak, 3500 pounds.

Weight per Foot of Wood, Board Measure.—The following is the weight in pounds of various kinds of woods, commercially known as dry timber, per foot board measure: white oak, 4.16; white pine, 1.98; Douglas fir, 2.65; short-leaf yellow pine, 2.65; red pine, 2.60; hemlock, 2.08; spruce, 2.08; cypress, 2.39; cedar, 1.93; chestnut, 3.43; Georgia yellow pine, 3.17; California spruce, 2.08.

How to Estimate the Weight of Natural Piles.—To calculate the upper and lower limits of the weight of a substance piled naturally on a circular plate, so as to form a cone of material, use the equation:

$$W = MD^3 \tag{1}$$

where W = weight, lb; D = diameter of plate, ft. (Fig. 1a); and, M = materials factor, whose upper and lower limits are given in Table 1.

For a rectangular plate, calculate the weight of material piled naturally by means of the following equation:

$$W = MRA^3 \tag{2}$$

where A and B = the length and width in ft., respectively, of the rectangular plate in Fig. 1b, with $B \leq A$; and, R = is a rectangular factor given in Table 2 as a function of the ratio B/A.

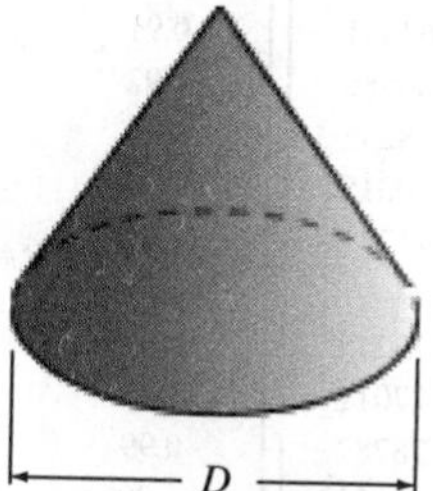

Fig. 1a. Conical Pile

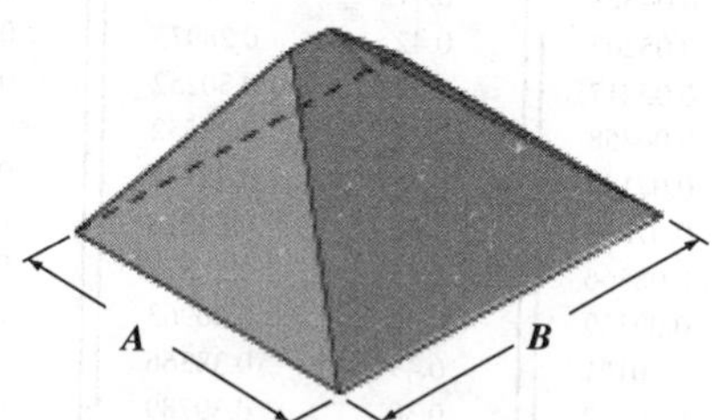

Fig. 1b. Rectangular Pile

Example 1: Find the upper and lower limits of the weight of dry ashes piled naturally on a plate 10 ft. in diameter.

Using Equation (1), $M = 4.58$ from Table 1, the lower limit $W = 4.58 \times 10^3 = 4{,}580$ lb. For $M = 5.89$, the upper limit $W = 5.89 \times 10^3 = 5{,}890$ lb.

Example 2: What weight of dry ashes rests on a rectangular plate 10 ft. by 5 ft.?

For $B/A = 5/10 = 0.5$, $R = 0.39789$ from Table 2. Using Equation (2), for $M = 4.58$, the lower limit $W = 4.58 \times 0.39789 \times 10^3 = 1{,}822$ lb. For $M = 5.89$, the upper limit $W = 5.89 \times 0.39789 \times 10^3 = 2{,}344$lb.

Table 1. Limits of Factor *M* for Various Materials

Material	Factor *M*	Material	Factor *M*	Material	Factor *M*
Aluminum chips	0.92-196	Cast-iron chips	17.02-26.18	Potassium carbonate	3.85-6.68
Aluminum silicate	3.70-6.41	Cinders, coal	3.02-5.24	Potassium sulfate	5.50-6.28
Ammonium chloride	3.93-6.81	Coal, anthracite, chestnut	2.43	Saltpeter	6.05-10.47
Asbestos, shred	2.62-3.27	Coal, ground	2.90	Salt rock, crushed	4.58
Ashes, dry	4.58-5.89	Coke, pulverized	2.21	Sand, very fine	7.36-9.00
Ashes, damp	6.24-7.80	Copper oxide, powdered	20.87	Sawdust, dry	0.95-2.85
Asphalt, crushed	3.4-5.89	Cork, granulated	1.57-1.96	Sodium nitrate	3.96-4.66
Bakelite, powdered	3.93-5.24	Dicalcium phosphate	5.63	Sodium sulfite	10.54
Baking powder	3.10-5.37	Fluorspar	10.73-14.40	Sodium sulfate	6.92
Barium carbonate	9.42	Graphite, flake	3.02-5.24	Sulfur	4.50-6.95
Boric acid	4.16-7.20	Lead silicate, granulated	25.26	Talcum powder	4.37-5.90
Bronze, chips	3.93-6.54	Lead sulfate, pulverized	24.09	Tin oxide, ground	9.17
Calcium lactate	3.40-3.80	Limestone, pulverized	8.84-10.02	Trisodium phosphate	4.53-7.85
Calcium oxide	3.30	Magnesium chloride	4.32	Wood chips, fir	2.49-2.88
Carbon, ground	2.51	Manganese sulfate	5.29-9.16	Zinc sulfate	8.85-11.12
Casein	2.72-4.71	Mica, ground	1.24-1.43		

Table 2. Factor *R* as a function of *B*/*A* ($B \leq A$)

B/*A*	*R*	*B*/*A*	*R*	*B*/*A*	*R*	*B*/*A*	*R*
0.01	0.00019	0.26	0.11792	0.51	0.41231	0.76	0.82367
0.02	0.00076	0.27	0.12670	0.52	0.42691	0.77	0.84172
0.03	0.00170	0.28	0.13576	0.53	0.44170	0.78	0.85985
0.04	0.00302	0.29	0.14509	0.54	0.45667	0.79	0.87807
0.05	0.00470	0.30	0.15470	0.55	0.47182	0.80	0.89636
0.06	0.00674	0.31	0.16457	0.56	0.48713	0.81	0.91473
0.07	0.00914	0.32	0.17471	0.57	0.50262	0.82	0.93318
0.08	0.01190	0.33	0.18511	0.58	0.51826	0.83	0.95169
0.09	0.01501	0.34	0.19576	0.59	0.53407	0.84	0.97027
0.10	0.01846	0.35	0.20666	0.60	0.55004	0.85	0.98891
0.11	0.02226	0.36	0.21782	0.61	0.56616	0.86	1.00761
0.12	0.02640	0.37	0.22921	0.62	0.58243	0.87	1.02636
0.13	0.03088	0.38	0.24085	0.63	0.59884	0.88	1.04516
0.14	0.03569	0.39	0.25273	0.64	0.61539	0.89	1.06400
0.15	0.04082	0.40	0.26483	0.65	0.63208	0.90	1.08289
0.16	0.04628	0.41	0.27717	0.66	0.64891	0.91	1.10182
0.17	0.05207	0.42	0.28973	0.67	0.66586	0.92	1.12078
0.18	0.05817	0.43	0.30252	0.68	0.68295	0.93	1.13977
0.19	0.06458	0.44	0.31552	0.69	0.70015	0.94	1.15879
0.20	0.07130	0.45	0.32873	0.70	0.71747	0.95	1.17783
0.21	0.07833	0.46	0.34216	0.71	0.73491	0.96	1.19689
0.22	0.08566	0.47	0.35579	0.72	0.75245	0.97	1.21596
0.23	0.09329	0.48	0.36963	0.73	0.77011	0.98	1.23505
0.24	0.10121	0.49	0.38366	0.74	0.78787	0.99	1.25414
0.25	0.10942	0.50	0.39789	0.75	0.80572	1.00	1.27324

PROPERTIES OF WOOD, CERAMICS, PLASTICS, METALS, WATER, AND AIR

Properties of Wood

Mechanical Properties of Wood.—Wood is composed of cellulose, lignin, ash-forming minerals, and extractives formed into a cellular structure. (Extractives are substances that can be removed from wood by extraction with such solvents as water, alcohol, acetone, benzene, and ether.) Variations in the characteristics and volumes of the four components and differences in the cellular structure result in some woods being heavy and some light, some stiff and some flexible, and some hard and some soft. For a single species, the properties are relatively constant within limits; therefore, selection of wood by species alone may sometimes be adequate. However, to use wood most effectively in engineering applications, the effects of physical properties or specific characteristics must be considered.

The mechanical properties listed in the accompanying table were obtained from tests on small pieces of wood termed "clear" and "straight grained" because they did not contain such characteristics as knots, cross grain, checks, and splits. However, these test pieces did contain such characteristics as growth rings that occur in consistent patterns within the piece. Since wood products may contain knots, cross grain, etc., these characteristics must be taken into account when assessing actual properties or when estimating actual performance. In addition, the methods of data collection and analysis have changed over the years during which the data in the table have been collected; therefore, the appropriateness of the data should be reviewed when used for critical applications such as stress grades of lumber.

Wood is an orthotropic material; that is, its mechanical properties are unique and independent in three mutually perpendicular directions—longitudinal, radial, and tangential. These directions are illustrated in the following figure.

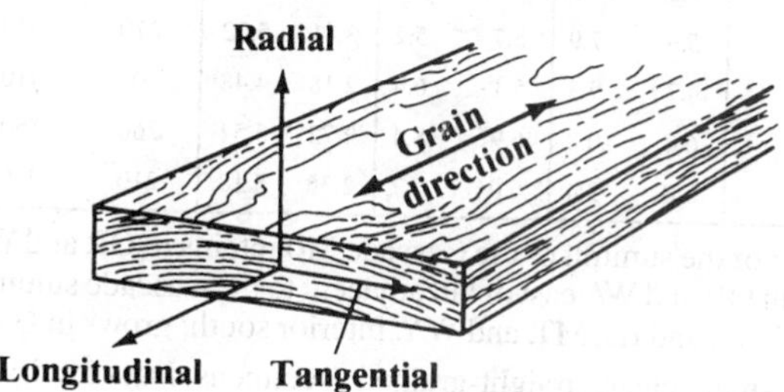

Modulus of Rupture: The modulus of rupture in bending reflects the maximum load-carrying capacity of a member and is proportional to the maximum moment borne by the member. The modulus is an accepted criterion of strength, although it is not a true stress because the formula used to calculate it is valid only to the proportional limit.

Work to Maximum Load in Bending: The work to maximum load in bending represents the ability to absorb shock with some permanent deformation and more or less injury to a specimen; it is a measure of the combined strength and toughness of the wood under bending stress.

Maximum Crushing Strength: The maximum crushing strength is the maximum stress sustained by a compression parallel-to-grain specimen having a ratio of length to least diameter of less than 11.

Compression Perpendicular to Grain: Strength in compression perpendicular to grain is reported as the stress at the proportional limit because there is no clearly defined ultimate stress for this property.

Shear Strength Parallel to Grain: Shear strength is a measure of the ability to resist internal slipping of one part upon another along the grain. The values listed in the table are averages of the radial and tangential shears.

Tensile Strength Perpendicular to Grain: The tensile strength perpendicular to the grain is a measure of the resistance of wood to forces acting across the grain that tend to split the material. Averages of radial and tangential measurements are listed.

Mechanical Properties of Commercially Important U.S. Grown Woods

Use the first number in each column for GREEN wood; use the second number for DRY wood.	Static Bending				Maximum Crushing Strength (10^3 psi)		Compression Strength Perpendicular to Grain (psi)		Shear Strength Parallel to Grain (psi)		Tensile Strength Perp. to Grain (psi)	
	Modulus of Rupture (10^3 psi)		Work to Max Load (in.-lb/in.^{3})									
Basswood, American	5.0	8.7	5.3	7.2	2.22	4.73	170	370	600	990	280	350
Cedar, N. white	4.2	6.5	5.7	4.8	1.90	3.96	230	310	620	850	240	240
Cedar, W. red	5.2	7.5	5.0	5.8	2.77	4.56	240	460	770	990	230	220
Douglas Fir, coast[a]	7.7	12.4	7.6	9.9	3.78	7.23	380	800	900	1,130	300	340
Douglas Fir, interior W.	7.7	12.6	7.2	10.6	3.87	7.43	420	760	940	1,290	290	350
Douglas Fir, interior N.	7.4	13.1	8.1	10.5	3.47	6.90	360	770	950	1,400	340	390
Douglas Fir, interior S.	6.8	11.9	8.0	9.0	3.11	6.23	340	740	950	1,510	250	330
Fir, balsam	5.5	9.2	4.7	5.1	2.63	5.28	190	404	662	944	180	180
Hemlock, Eastern	6.4	8.9	6.7	6.8	3.08	5.41	360	650	850	1,060	230	…
Hemlock, Mountain	6.3	11.5	11.0	10.4	2.88	6.44	370	860	930	1,540	330	…
Hemlock, Western	6.6	11.3	6.9	8.3	3.36	7.20	280	550	860	1,290	290	340
Pine, E. white	4.9	9.9	5.2	8.3	2.44	5.66	220	580	680	1,170	250	420
Pine, Virginia	7.3	13.0	10.9	13.7	3.42	6.71	390	910	890	1,350	400	380
Pine, W. white	4.7	9.7	5.0	8.8	2.43	5.04	190	470	680	1,040	260	…
Redwood, old-growth	7.5	10.0	7.4	6.9	4.20	6.15	420	700	800	940	260	240
Redwood, young-growth	5.9	7.9	5.7	5.2	3.11	5.22	270	520	890	1,110	300	250
Spruce, Engelmann	4.7	9.3	5.1	6.4	2.18	4.48	200	410	640	1,200	240	350
Spruce, red	6.0	10.8	6.9	8.4	2.72	5.54	260	550	750	1,290	220	350
Spruce, white	5.0	9.4	6.0	7.7	2.35	5.18	210	430	640	970	220	360

[a] Coast: grows west of the summit of the Cascade Mountains in OR and WA. Interior west: grows in CA and all counties in OR and WA east of but adjacent to the Cascade summit. Interior north: grows in remainder of OR and WA and ID, MT, and WY. Interior south: grows in UT, CO, AZ, and NM.

Results of tests on small, clear, straight-grained specimens. Data for dry specimens are from tests of seasoned material adjusted to a moisture content of 12%.

Source: U.S. Department of Agriculture: *Wood Handbook.*

Effect of Pressure Treatment on Mechanical Properties of Wood.—The strength of wood preserved with creosote, coal-tar, creosote-coal-tar mixtures, creosote-petroleum mixtures, or pentachlorophenol dissolved in petroleum oil is not reduced. However, waterborne salt preservatives contain chemicals such as copper, arsenic, chromium, and ammonia, which have the potential of affecting mechanical properties of treated wood and causing mechanical fasteners to corrode. Preservative salt-retention levels required for marine protection may reduce bending strength by 10 per cent or more.

Density of Wood.—The following formula can be used to find the density of wood in lb/ft^3 as a function of its moisture content.

$$\rho = 62.4\left(\frac{G}{1 + G \times 0.009 \times M}\right)\left(1 + \frac{M}{100}\right)$$

where ρ is the density, G is the specific gravity of wood, and M is the moisture content expressed in per cent.

Weights of American Woods, in Pounds per Cubic Foot
(United States Department of Agriculture)

Species	Green	Airdry	Species	Green	Airdry
Alder, red	46	28	Hickory, pecan	62	45
Ash, black	52	34	Hickory, true	63	51
Ash, commercial white	48	41	Honeylocust	61	…
Ash, Oregon	46	38	Larch, western	48	36
Aspen	43	26	Locust, black	58	48
Basswood	42	26	Maple, bigleaf	47	34
Beech	54	45	Maple, black	54	40
Birch	57	44	Maple, red	50	38
Birch, paper	50	38	Maple, silver	45	33
Cedar, Alaska	36	31	Maple, sugar	56	44
Cedar, eastern red	37	33	Oak, red	64	44
Cedar, northern white	28	22	Oak, white	63	47
Cedar, southern white	26	23	Pine, lodgepole	39	29
Cedar, western red	27	23	Pine, northern white	36	25
Cherry, black	45	35	Pine, Norway	42	34
Chestnut	55	30	Pine, ponderosa	45	28
Cottonwood, eastern	49	28	Pines, southern yellow:		
Cottonwood, northern black	46	24	Pine, loblolly	53	36
Cypress, southern	51	32	Pine, longleaf	55	41
Douglas fir, coast region	38	34	Pine, shortleaf	52	36
Douglas fir, Rocky Mt. region	35	30	Pine, sugar	52	25
Elm, American	54	35	Pine, western white	35	27
Elm, rock	53	44	Poplar, yellow	38	28
Elm, slippery	56	37	Redwood	50	28
Fir, balsam	45	25	Spruce, eastern	34	28
Fir, commercial white	46	27	Spruce, Engelmann	39	23
Gum, black	45	35	Spruce, Sitka	33	28
Gum, red	50	34	Sycamore	52	34
Hemlock, eastern	50	28	Tamarack	47	37
Hemlock, western	41	29	Walnut, black	58	38

Machinability of Wood.—The ease of working wood with hand tools generally varies directly with the specific gravity of the wood; the lower the specific gravity, the easier the wood is to cut with a sharp tool. A rough idea of the specific gravity of various woods can be obtained from the preceding table by dividing the weight of wood in lb/ft^3 by 62.355.

A wood species that is easy to cut does not necessarily develop a smooth surface when it is machined. Three major factors, other than specific gravity, influence the smoothness of the surface obtained by machining: interlocked and variable grain, hard deposits in the grain, and reaction wood. Interlocked and variable grain is a characteristic of many tropical and some domestic species; this type of grain structure causes difficulty in planing quarter sawn boards unless careful attention is paid to feed rates, cutting angles, and sharpness of the knives. Hard deposits of calcium carbonate, silica, and other minerals in the grain tend to dull cutting edges quickly, especially in wood that has been dried to the usual in service moisture content. Reaction wood results from growth under some physical stress such as occurs in leaning trunks and crooked branches. Generally, reaction wood occurs as tension wood in hardwoods and as compression wood in softwoods. Tension wood is particularly troublesome, often resulting in fibrous and fuzzy surfaces, especially in woods of lower density. Reaction wood may also be responsible for pinching saw blades, resulting in burning and dulling of teeth.

The following table rates the suitability of various domestic hardwoods for machining. The data for each species represent the percentage of pieces machined that successfully met the listed quality requirement for the processes. For example, 62 per cent of the black walnut pieces planed came out perfect, but only 34 per cent of the pieces run on the shaper achieved good to excellent results.

Machinability and Related Properties of Various Domestic Hardwoods

	Planing	Shaping	Turning	Boring	Mortising	Sanding
	Quality Required					
Type of Wood	Perfect	Good to Excellent	Fair to Excellent	Good to Excellent	Fair to Excellent	Good to Excellent
Alder, red	61	20	88	64	52	…
Ash	75	55	79	94	58	75
Aspen	26	7	65	78	60	…
Basswood	64	10	68	76	51	17
Beech	83	24	90	99	92	49
Birch	63	57	80	97	97	34
Birch, paper	47	22	…	…	…	…
Cherry, black	80	80	88	100	100	…
Chestnut	74	28	87	91	70	64
Cottonwood	21	3	70	70	52	19
Elm, soft	33	13	65	94	75	66
Hackberry	74	10	77	99	72	…
Hickory	76	20	84	100	98	80
Magnolia	65	27	79	71	32	37
Maple, bigleaf	52	56	8	100	80	…
Maple, hard	54	72	82	99	95	38
Maple, soft	41	25	76	80	34	37
Oak, red	91	28	84	99	95	81
Oak, white	87	35	85	95	99	83
Pecan	88	40	89	100	98	…
Sweetgum	51	28	86	92	53	23
Sycamore	22	12	85	98	96	21
Tanoak	80	39	81	100	100	…
Tupelo, black	48	32	75	82	24	21
Tupelo, water	55	52	79	62	33	34
Walnut, black	62	34	91	100	98	…
Willow	52	5	58	71	24	24
Yellow-poplar	70	13	81	87	63	19

The data above represent the percentage of pieces attempted that meet the quality requirement listed.

Nominal and Minimum Sizes of Sawn Lumber

Type of Lumber	Thickness (inches)			Face Widths (inches)		
	Nominal, T_n	Dry	Green	Nominal, W_n	Dry	Green
Boards	1	¾	25⁄32	2 to 4	$W_n - ½$	$W_n - 7⁄16$
	1¼	1	1 1⁄32	5 to 7	$W_n - ½$	$W_n - ⅜$
	1½	1¼	1 9⁄32	8 to 16	$W_n - ¾$	$W_n - ½$
Dimension Lumber	2	1½	1 9⁄16	2 to 4	$W_n - ½$	$W_n - 7⁄16$
	2½	2	2 1⁄16	5 to 6	$W_n - ½$	$W_n - ⅜$
	3	2½	2 9⁄16	8 to 16	$W_n - ¾$	$W_n - ½$
	3½	3	3 1⁄16	…	…	…
	4	3½	3 9⁄16	…	…	…
	4½	4	4 1⁄16	…	…	…
Timbers	5 and up	…	$T_n - ½$	5 and up	…	$W_n - ½$

Source: National Forest Products Association: *Design Values for Wood Construction*. Moisture content: dry lumber ≤ 19 percent; green lumber > 19 percent. Dimension lumber refers to lumber 2 to 4 inches thick (nominal) and 2 inches or greater in width. Timbers refers to lumber of approximately square cross-section, 5 × 5 inches or larger, and a width no more than 2 inches greater than the thickness.

Tabulated Properties of Ceramics, Plastics, and Metals

Material		Density[a] (lb/in.^{3})	Dielectric Strength (V/mil)	Coeff. of Expansion[b] (10^{-6} in./in.-°F)	Flexural Strength (10^3 psi)	Mohs's Hardness[c]	Operating Temperature (°F)	Tensile Strength (10^3 psi)	Compressive Strength (10^3 psi)	Thermal Conductivity[d] (Btu-ft-hr-ft^2-°F)
Machinable Glass Ceramic		0.09	1000	4.1–7.0		48 Ra	1472	…	50	0.85
		0.11	400	6	15	5.5	700	…	40	0.24
		0.10	380	5.2	14	5.0	1100	…	32	0.34
Glass-Mica	Machining-Grades{	0.09–0.10	400	10.5–11.2	12.5–13	90 Rh	750	6	40–45	0.24–0.29
		0.10	380	9.4	11	90 Rh	1100	5	32	0.34
	Molding Grades{	0.13–0.17	300–325	11–11.5	9–10	90 Rh	700–750	6–6.5	33–35	0.29–0.31
		0.14	350	10.3	9	90 Rh	1300	6	30	0.3
Aluminum Silicate		0.10	80	2.5	4.5	1–2	1000	…	12	0.92
		0.08	100	2.9	10	6.0	2100	…	25	0.75
Alumina Silicate		0.08	70	…	…	…	2370	…	…	0.38
Silica Foam		0.03	80	0.3	0.4	NA	2000	…	1.4	0.10
TiO_2 (Titania)		0.14	100	4.61	20	8	1800	7.5	100	…
Lava (Grade A)		0.08	80	1.83	9	6	2000	2.5	40	0.92
Zirconium Phosphate		0.11	NA	0.5	7.5	NA	2800	…	30	0.4 (approx.)
ZrO_2		0.21	…	6.1	102	1300 V	…	…	261	1.69
$ZrO_2{\cdot}SiO_2$ (Zircon)		0.11	220	1.94	16	7.5	1825	10	90	…
$2MgO{\cdot}SiO_2$ (Forsterite)		0.11	240	5.56	20	7.5	1825	10	85	4.58
$MgO{\cdot}SiO_2$ (Steatite)		0.09–0.10	210–240	3.83–5.44	18–21	7.5	1825	8.5–10	80–90	3.17–3.42
$2MgO{\cdot}2Al_2O_3{\cdot}5SiO_2$ (Cordierite)		0.06	60	0.33	3.4	6.5	2000	2.5	18.5	1.00
		0.08	100–172	1.22–1.28	8–12	7–7.5	2000	3.5–3.7	30–40	1.00
		0.09	200	1.33	15	8	2000	4	50	1.83
Al_2O_3 (Alumina)	94%	0.13	210	3.33	44	9	2700	20	315	16.00
	96%	0.13–0.14	210	3.5–3.7	48–60	9	2600–2800	25	375	20.3–20.7
	99.5%	0.14	200	3.72	70	9	2700	28	380	21.25
	99.9%	0.14	…	3.75	72	9	2900	…	400	…

[a] Obtain specific gravity by dividing density in lb/in.^{3} by 0.0361; for density in lb/ft^3, multiply lb/in.^{3} by 1728; for g/cm^3, multiply density in lb/in.^{3} by 27.68; for kg/m^3, multiply density in lb/in.^{3} by 27,679.9.

[b] To convert coefficient of expansion to 10^{-6} in./in.-°C, multiply table value by 1.8.

[c] Mohs's Hardness scale is used unless otherwise indicated as follows: Ra and Rh for Rockwell A and H scales, respectively; V for Vickers hardness.

[d] To convert conductivity from Btu-ft/hr-ft^2-°F to cal-cm/sec-cm^2-°C, divide by 241.9.

Typical Properties of Plastics Materials

Material	Density[a] (lb/in^3)	Specific Gravity	Dielectric Strength (V/mil)	Coeff. of Expansion[b] (10^{-6} in/in-°F)	Tensile Modulus (10^5 psi)	Izod Impact (ft-lb/in of notch)	Flexural Modulus (ksi at 73°F)	% Elongation	Hardness[c]	Max. Operating Temp. (°F)
ABS, Extrusion Grade	0.038	1.05	…	53.0	275	7	300	…	105 Rr	200
ABS, High Impact	0.037	1.03	…	…	200	…	330	…	105 Rr	…
Acetal, 20% Glass	0.056	1.55	…	…	1000	0.9	715	…	94 Rm	…
Acetal, Copolymer	0.051	1.41	380	47.0	437	2	400	13	94 Rm	…
Acetyl, Homopolymer	0.051	1.41	…	58.0	310	…	320	…	94 Rm	200
Acrylic	0.043	1.19	500	35.0	400	0.5	400	2.7	94 Rm	180
Azdel	0.043	1.19	500	15.0	750	14	800	2.1	94 Rm	311
CPVC	0.056	1.55	…	34.0	400	3	400	4	…	212
Fiber Glass Sheet	0.067	1.87	…	11.1	…	8	1	…	101 Rm	260
Nylon 6, 30% Glass	0.050	1.39	…	…	1350	2.8	1400	…	119 Rr	…
Nylon 6, Cast	0.042	1.16	295	45.0	380	1.4	450	20	100 Rr	210
Nylon 6/6, Cast	0.047	1.30	…	…	…	…	…	…	…	…
Nylon 6/6, Extruded	0.041	1.14	600	45.0	390	1	…	240	118 Rr	230
Nylon 60L, Cast	0.042	1.16	…	…	…	2.2	…	…	…	…
PET, unfilled	0.049	1.36	1300	39.0	500	0.5	400	70	…	230
PTFE (Teflon)	0.079	2.19	480	50.0	225	3	80	350	…	…
PVC	0.050	1.39	500	29.5	550	0.8	400	31–40	110 Rr	170
PVDF	0.064	1.77	260	60.0	320	3	200	80	100 Rr	180
Phenolics	0.050	1.38	…	11.1	…	2.4	1000	…	100 Rm	248
Polycarbonate	0.043	1.19	380	37.5	345	14	340	110	74 Rm	290
Polyetherimide	0.046	1.27	480	…	430	1.1	480	…	…	…
Polyethylene, HD	0.035	0.97	475	20.0	156	6	160	900	…	180
Polyethylene, UHMW	0.034	0.94	710	19.0	110	No Break	130	450	64 Rr	176
Polymethylpentene	0.030	0.83	…	…	220	2.5	…	…	…	…
Polymid, unfilled	0.051	1.41	560	…	300	1.5	…	…	…	…
Polyphenylene Sulfide	0.047	1.30	380	…	…	0.5	550	…	…	…
Polypropylene	0.033	0.91	600	96.0	155	0.75	200	120	92 Rr	150
Polysulfone	0.045	1.25	425	31.0	360	1.2	390	50	120 Rr	325
Polyurethane	0.038	1.05	…	…	…	…	…	465–520	…	…

[a] To obtain specific gravity, divide density in lb/in^3 by 0.0361; for density in lb/ft^3, multiply lb/in^3 by 1728; for g/cm^3, multiply density in lb/in^3 by 27.68; for kg/m^3, multiply density in lb/in^3 by 27,679.9.

[b] To convert coefficient of expansion to 10^{-6} in/in-°C, multiply table value by 1.8.

[c] Hardness value scales are as follows: Rm for Rockwell M scale; Rr for Rockwell R scale.

Mechanical Properties of Various Investment Casting Alloys

Alloy Designation	Material Condition	Tensile Strength (10^3 psi)	0.2% Yield Strength[a] (10^3 psi)	% Elonga-tion	Hardness
Aluminum					
356	As Cast	32–40	22–30	3–7	...
A356	As Cast	38–40	28–36	3–10	...
A357	As Cast	33–50	27–40	3–9	...
355, C355	As Cast	35–50	28–39	1–8	...
D712 (40E)	As Cast	34–40	25–32	4–8	...
A354	As Cast	47–55	36–45	2–5	...
RR-350	As Cast	32–45	24–38	1.5–5	...
Precedent 71	As Cast	35–55	25–45	2–5	...
KO-1	As Cast	56–60	48–55	3–5	...
Copper-Base Alloys[a]					
Al Bronze C (954)	As Cast	75–85	30–40	10–20	80–85 Rb
	Heat-Treated	90–105	45–55	6–10	91–96 Rb
Al Bronze D (955)	As Cast	90–100	40–50	6–10	91–96 Rb
	Heat-Treated	110–120	60–70	5–8	93–98 Rb
Manganese Bronze, A	...	65–75	25–40	16–24	60–65 Rb
Manganese Bronze, C	...	110–120	60–70	8–16	95–100 Rb
Silicon Bronze	...	45	18	20	...
Tin Bronze	...	40–50	18–30	20–35	40–50 Rb
Lead. Yellow Brass (854)	...	30–50	11–20	15–25	...
Red Brass	...	30–40	14–25	20–30	30–35 Rb
Silicon Brass	...	70	32	24	...
Pure Copper	...	20–30	...	4–50	35–42 Rb
Beryllium Cu 10C (820)	As Cast	45–50	40–45	15–20	50–55 Rb
	Hardened	90–100	90–130	3–8	90–95 Rb
Beryllium Cu 165C (824)	...	70–155	40–140	1–15	60 Rb–38 Rc
Beryllium Cu 20C (825)	As Cast	70–80	50–55	18–23	75–80 Rb
	Hardened	110–160	...	1–4	25–44 Rc
Beryllium Cu 275C (828)	As Cast	80–90	...	15–20	80–85 Rb
Chrome Copper	...	33–50	20–40	20–30	70–78 Rb
Carbon and Low-Alloy Steels and Iron					
IC 1010	Annealed	50–60	30–35	30–35	50–55 Rb
IC 1020	Annealed	60–70	40–45	25–40	80 Rb
IC 1030	Annealed	65–75	45–50	20–30	75 Rb
	Hardened	85–150	60–150	0–15	20–50 Rc
IC 1035	Annealed	70–80	45–55	20–30	80 Rb
	Hardened	90–150	85–150	0–15	25–52 Rc
IC 1045	Annealed	80–90	50–60	20–25	100 Rb
	Hardened	100–180	90–180	0–10	25–57 Rc
IC 1050	Annealed	90–110	50–65	20–25	100 Rb
	Hardened	125–180	100–180	0–10	30–60 Rc
IC 1060	Annealed	100–120	55–70	5–10	25 Rc
	Hardened	120–200	100–180	0–3	30–60 Rc
IC 1090	Annealed	110–150	70–80	12–20	30 Rc
IC 1090	Hardened	130–180	130–180	0–3	37–50 Rc
IC 2345	Hardened	130–200	110–180	5–10	30–58 Rc
IC 4130	Hardened	130–170	100–130	5–20	23–49 Rc
IC 4140	Hardened	130–200	100–155	5–20	29–57 Rc
IC 4150	Hardened	140–200	120–180	5–10	25–58 Rc
IC 4330	Hardened	130–190	100–175	5–20	25–48 Rc
IC 4340	Hardened	130–200	100–180	5–20	20–55 Rc
IC 4620	Hardened	110–150	90–130	10–20	20–32 Rc
IC 6150, IC 8740	Hardened	140–200	120–180	5–10	30–60 Rc
IC 8620	Hardened	100–130	80–110	10–20	20–45 Rc
IC 8630	Hardened	120–170	100–130	7–20	25–50 Rc
IC 8640	Hardened	130–200	100–180	5–20	30–60 Rc

Mechanical Properties of Various Investment Casting Alloys *(Continued)*

Alloy Designation	Material Condition	Tensile Strength (10^3 psi)	0.2% Yield Strength[a] (10^3 psi)	% Elongation	Hardness
Carbon and Low-Alloy Steels and Iron *(Continued)*					
IC 8665	Hardened	170–220	140–200	0–10	…
IC 8730	Hardened	120–170	110–150	7–20	…
IC 52100	Hardened	180–230	140–180	1–7	30–65 Rc
IC 1722AS	Hardened	130–170	100–140	6–12	25–48 Rc
1.2% Si Iron	…	50–60	37–43	30–35	55 Rb
Ductile Iron, Ferritic	Annealed	60–80	40–50	18–24	143–200 Bhn
Ductile Iron, Pearlitic	Normalized	100–120	70–80	3–10	243–303 Bhn
Hardenable Stainless Steel					
CA-15	Hardened	95–200	75–160	5–12	94 Rb–45 Rc
IC 416	Hardened	95–200	75–160	3–8	94 Rb–45 Rc
CA-40	Hardened	200–225	130–210	0–5	30–52 Rc
IC 431	Hardened	110–160	75–105	5–20	20–40 Rc
IC 17–4	Hardened	150–190	140–160	6–20	34–44 Rc
Am-355	Hardened	200–220	150–165	6–12	…
IC 15–5	Hardened	135–170	110–145	5–15	26–38 Rc
CD-4M Cu	Annealed	100–115	75–85	20–30	94–100 Rb
	Hardened	135–145	100–120	10–25	28–32 Rc
Austenitic Stainless Steels					
CF-3, CF-3M, CF-8, CF-8M, IC 316F	Annealed	70–85	40–50	35–50	90 Rb (max)
CF-8C	Annealed	70–85	32–36	30–40	90 Rb (max)
CF-16F	Annealed	65–75	30–35	35–45	90 Rb (max)
CF-20	Annealed	65–75	30–45	35–60	90 Rb (max)
CH-20	Annealed	70–80	30–40	30–45	90 Rb (max)
CN-7M	Annealed	65–75	25–35	35–45	90 Rb (max)
IC 321, CK-20	Annealed	65–75	30–40	35–45	90 Rb (max)
Nickel-Base Alloys					
Alloy B	Annealed	75–85	50–60	8–12	90–100 Rb
Alloy C	As Cast	80–95	45–55	8–12	90–100 Rb
	Annealed	75–95	45–55	8–12	90 Rb–25 Rc
Alloy X[b]	AC to 24°C	63–70	41–45	10–15	85–96 Rb
	AC to 816°C	35–45	…	12–20	…
Invar (Fe–Ni alloy)	As Cast	50–60	25–30	30–40	50–60 Rb
In 600 (Inconel)	As Cast	65–75	35–40	10–20	80–90 Rb
In 625 (Inconel)	Annealed	80–100	40–55	15–30	10–20 Rc
Monel 410	As Cast	65–75	32–38	25–35	65–75 Rb
S Monel	Annealed	100–110	55–65	5–10	20–28 Rc
	Hardened	120–140	85–100	0	32–38 Rc
RH Monel	As Cast	100–110	60–80	10–20	20–30 Rc
Monel E	As Cast	65–80	33–40	25–35	67–78 Rb
M-35 Monel	As Cast	65–80	25–35	25–40	65–85 Rb
Cobalt-Base Alloys					
Cobalt 21	As Cast	95–130	65–95	8–20	24–32 Rc
Cobalt 25	As Cast	90–120	60–75	15–25	20–25 Rc
Cobalt 31	As Cast	105–130	75–90	6–10	20–30 Rc
Cobalt 36	As Cast	90–105	60–70	15–20	30–36 Rc
F75	As Cast	95–110	70–80	8–15	25–34 Rc
N-155	Sol. Anneal	90–100	50–60	15–30	90–100 Rb

[a] For copper alloys, yield strength is determined by 0.5% extension under load or 0.2% offset method. A number in parentheses following a copper alloy indicates the UNS designation of that alloy (for example, Al Bronze C (954) identifies the alloy as UNS C95400).

[b] AC = air cooled to temperature indicated.

Source: Investment Casting Institute. Mechanical properties are average values of separately cast test bars, and are for reference only. Items marked … indicates data are not available. Alloys identified by IC followed by an SAE designation number (IC 1010 steel, for example) are generally similar to the SAE material although properties and chemical composition may be different.

Typical Properties of Compressed and Sintered Powdered Metal Alloys

Alloy Number[a]	Nominal Composition (%)	Density (g/cc)	Hardness	Strength (10^3 psi) Transverse Rupture	Ultimate Tensile	Yield	% Elongation
Copper Base							
…	100Cu	7.7–7.9	81–82 Rh	54–68	24–34	…	10–26
CZP-3002	70Cu, 1.5Pb, Bal. Zn	8	75 Rh	…	33.9	…	24
CNZ-1818	63Cu, 17.5Ni, Bal. Zn	7.9	90 Rh	73	34	20	11
CTG-1004	10Sn, 4.4C, Bal. Cu	7	67 Rh	20	9.4	6.5	6
CTG-1001	10Sn, 1C, Bal. Cu	6.5	45 Rh	25.8	15.1	9.6	9.7
Iron Base (Balance of composition, Fe)							
FC-2015	23.5Cu, 1.5C	6.5	65 Rb	80	52.4	48.5	0
FC-0800	8Cu, 0.4C	6.3–6.8	39–55 Rb	75–100	38–54	32–47	1 or less
FX-2008	20Cu, 1C	7.3	93 Rb	164.2	72.3	57.7	2
FN-0408	4Ni, 1–2Cu, 0.75C	6.3–7	64–84 Rb	70–107	37–63	30–47	1–1.6
F-0000	100Fe	6.5	26 Rf	37.7	15.7	11	5.7
FN-0005	0.45C, 0.50 MnS	6.4–6.8	66–78 Rf	44–61	…	…	…
F-0000	0.02C, 0.45P	6.6–7.2	35–50 Rb	90–125	…	29–38	3.9–5.5
F-0008	0.6–0.9C	6.2–7	50–70 Rb	61–100	35–57	30–40	<0.5 to 1
FC-0508	0.6–0.9C, 4–6Cu	5.9–6.8	60–80 Rb	100–145	58–82	50–70	<0.5 to 1
FN-0405	4Ni, 0.5C	6.6–7.0	73–82 Rb	90–100	47–50	38–40	<1
FN-0208	2Ni, 0.8C	6.6–7.0	50–70 Rb	70–108	47–58	35–51	<1
FN-0205	2Ni, 0.5C	6.6–7.0	51–61 Rb	72–93	35–45	27–31	2.0–2.5
FN-0200	2Ni, 0.25C	6.6	29 Rb	57.5	25.8	19.0	1.3
FC-0208	2Cu, 0.75C	6.5–6.7	68–72 Rb	95–107	56–61	51–54	up to 1
FC-2008	20Cu, 1C	6.2	45 Rb	79.5	47.8	40.0	1.3
…	4Ni, 0.6C, 1.6Cu, 0.55Mo	7.0	92 Rb	190.0	100.0	65.0	2.5
FL-4605	1.8Ni, 0.6C, 1.6Cu, 0.55Mo	7.0	87 Rb	170.0	80.0	55.0	2.5
FL-4605	1.8Ni, 0.6C, 0.55Mo	7.0	80 Rb	150.0	…	…	…
SS-316L	17Cr, 13Ni, 2.2Mo, 0.9Si	6.5	65 Rb	94.0	45.0	30.0	6.0
…	17Cr, 13Ni, 2.2Mo, 0.9Si, 15–20Cu	7.3	66 Rb	108.6	59.2	49.7	4.3
SS-410	13Cr, 0.8Si, 0.8Mn	6.2	15 Rc	85.0	66.7	56.9	0
FL-4608	2Cu, 3.8Ni, 0.9C, 0.75Mo	6.8	24 Rc	107.3	55.8	46.5	1.5
SS-303N1	18Cr, 11Ni, 1Mn	6.4	62 Rb	86.0	39.0	32.0	0.5
SS-304N1	19Cr, 10Ni, 1Mn	6.4	61 Rb	112.0	43.0	38.0	0.5
Tungsten Base							
90W, 6Ni, 4Cu		17.0	24 Rc	…	110	80	6
90W, 7Ni, 3Cu		17.0	25 Rc	…	120	88	10
92.5W, 5.25Ni, 2.25Cu		17.5	26 Rc	…	114	84	7
92.5W, Bal. Ni, Fe, and Mo		17.6	30 Rc	…	120	90	4
93W, Bal. Ni, Fe, and Mo		17.7	32 Rc	…	125	95	4
95W, 3.5Ni, 1.5Cu		18.0	27 Rc	…	110	85	7
95W, 3.5Ni, 1.5Fe		18.0	27 Rc	…	120	90	7
97W, 2.1Ni, 0.9Fe		18.5	28 Rc	…	123	85	5

[a] Copper- and iron-base alloy designations are Metal Powder Industries Federation (MPIF) alloy numbers.

Typical Elastic Properties of Materials

Material	Modulus of Elasticity (10^6 psi)	Shear Modulus (10^6 psi)	Bulk Modulus (10^6 psi)	Poisson's Ratio
Aluminum, var. alloys	9.9–10.3	3.7–3.9	9.9–10.2	0.330–0.334
Aluminum, 6061-T6	10.2	3.8	…	0.35
Aluminum, 2024-T4	10.6	4.0	…	0.32
Beryllium copper	18	7	…	0.29
Brass, 70–30	15.9	6	15.7	0.331
Brass, cast	14.5	5.3	16.8	0.357
Bronze	14.9	6.5	…	0.14
Copper	15.6	5.8	17.9	0.355
Glass	6.7	2.7	…	0.24
Glass ceramic (machinable)	9.7	3.7	…	0.29
Inconel	31	11	…	0.27–0.38
Iron, cast	13.5–21.0	5.2–8.2	8.4–15.5	0.221–0.299
Iron, ductile	23.8–25.2	9.1–9.6	…	0.26–0.31
Iron, grey cast	14.5	6	…	0.211
Iron, malleable	23.6	9.3	17.2	0.271
Lead	5.3	1.9	…	0.43
Magnesium	6.5	2.4	…	0.35
Magnesium alloy	6.3	2.5	4.8	0.281
Molybdenum	48	17	…	0.307
Monel metal	25	9.5	22.5	0.315
Nickel silver	18.5	7	…	0.322
Nickel steel	30	11.5	…	0.291
Phosphor bronze	13.8	5.1	16.3	0.359
Stainless steel 18–8	27.6	10.6	23.6	0.305
Steel, cast	28.5	11.3	20.2	0.265
Steel, cold-rolled	29.5	11.5	23.1	0.287
Steel, all others	28.6–30.0	11.0–11.9	22.6–24.0	0.283–0.292
Titanium (99.0 Ti)	15–16	6.5	…	0.24
Titanium (Ti-8Al-1Mo-1V)	18	6.8	…	0.32
Zinc, cast alloys	10.9–12.4	…	…	0.33
Zinc, wrought alloys	6.2–14	…	…	0.33
Z-nickel	30	11	…	0.36

Data represent typical values, but material properties may vary widely, depending on exact composition, material condition, and processing. Symbol … indicates no data available.

Minimum Tensile Strength of Spring Wire by Diameter

Wire Dia. (in.)	Wire Type						
	Music Wire	Hard-Drawn MB	Oil Temp. MB	Stainless Steel 18-8	Cr–V Alloy	Phosphor Bronze	Chrome Silicon
	Minimum Tensile Strength (10^3 psi)						
0.004	439	…	…	325	…	140	…
0.008	399	…	…	325	…	140	…
0.012	377	…	…	316	…	…	…
0.020	350	283	288	300	…	…	…
0.028	333	271	281	284	…	…	…
0.032	327	265	275	278	281	…	300
0.035	322	261	268	274	276	…	298
0.041	314	255	261	270	270	135	298
0.047	307	248	254	262	263	…	292
0.054	301	243	248	258	257	…	292
0.063	293	237	242	251	251	130	290
0.072	287	232	236	245	245	…	288
0.080	282	227	230	240	240	…	285
0.092	275	220	225	233	235	…	280
0.105	269	216	220	227	229	125	275
0.120	263	210	215	221	222	…	275
0.135	258	206	210	213	219	…	270
0.148	253	203	205	207	215	…	268
0.162	249	200	200	200	212	…	162
0.177	245	195	195	195	210	…	260
0.192	241	192	190	189	206	…	260
0.207	238	190	185	185	204	…	260
0.225	225	186	183	180	200	120	255
0.250	220	182	180	174	196	…	250
0.312	…	174	178	160	189	110	245
0.375	…	167	175	…	187	…	240
0.437	…	165	170	…	186	…	235
0.500	…	156	165	…	185	100	230

For allowable working stresses and recommended design stresses in bending, related to severity of service, refer to Fig. 1 through Fig. 10 on pages 291 through 294, and for endurance limits for compression springs made from these materials refer to Fig. 11 on page 296 in the section on spring stresses.

Pressure and Flow of Water

Water Pressure.—Water is composed of two elements, hydrogen and oxygen, in the ratio of two volumes of hydrogen to one of oxygen. In the common system of measure, water boils under atmospheric pressure at 212 degrees F and freezes at 32 degrees F. Water's greatest density is 62.425 pounds per cubic foot, at 39.1 degrees F. In metric (SI) measure, water boils under atmospheric pressure at 100°C (Celsius) and freezes at 0°C. Its density is equal to 1 kilogram per liter, where 1 liter is 1 cubic decimeter. Also in metric SI, pressure is given in pascals (Pa) or the equivalent newtons per square meter. See page 2523 for additional information on the metric (SI) system of units.

For higher temperatures, the pressure slightly decreases in the proportion indicated by the table *Weight of Water per Cubic Foot at Different Temperatures*. The pressure per square inch is equal in all directions, downwards, upwards, and sideways. Water can be compressed only to a very slight degree, the compressibility being so slight that even at the depth of a mile, a cubic foot of water weighs only about one-half pound more than at the surface.

Pressure in Pounds per Square Inch for Different Heads of Water

Head, ft	0	1	2	3	4	5	6	7	8	9
0	…	0.43	0.87	1.30	1.73	2.16	2.60	3.03	3.46	3.90
10	4.33	4.76	5.20	5.63	6.06	6.49	6.93	7.36	7.79	8.23
20	8.66	9.09	9.53	9.96	10.39	10.82	11.26	11.69	12.12	12.56
30	12.99	13.42	13.86	14.29	14.72	15.15	15.59	16.02	16.45	16.89
40	17.32	17.75	18.19	18.62	19.05	19.48	19.92	20.35	20.78	21.22
50	21.65	22.08	22.52	22.95	23.38	23.81	24.25	24.68	25.11	25.55
60	25.98	26.41	26.85	27.28	27.71	28.14	28.58	29.01	29.44	29.88
70	30.31	30.74	31.18	31.61	32.04	32.47	32.91	33.34	33.77	34.21
80	34.64	35.07	35.51	35.94	36.37	36.80	37.24	37.67	38.10	38.54
90	38.97	39.40	39.84	40.27	40.70	41.13	41.57	42.00	42.43	42.87

Heads of Water in Feet Corresponding to Certain Pressures in Pounds per Square Inch

Pressure, lb/in.^{2}	0	1	2	3	4	5	6	7	8	9
0	…	2.3	4.6	6.9	9.2	11.5	13.9	16.2	18.5	20.8
10	23.1	25.4	27.7	30.0	32.3	34.6	36.9	39.3	41.6	43.9
20	46.2	48.5	50.8	53.1	55.4	57.7	60.0	62.4	64.7	67.0
30	69.3	71.6	73.9	76.2	78.5	80.8	83.1	85.4	87.8	90.1
40	92.4	94.7	97.0	99.3	101.6	103.9	106.2	108.5	110.8	113.2
50	115.5	117.8	120.1	122.4	124.7	127.0	129.3	131.6	133.9	136.3
60	138.6	140.9	143.2	145.5	147.8	150.1	152.4	154.7	157.0	159.3
70	161.7	164.0	166.3	168.6	170.9	173.2	175.5	177.8	180.1	182.4
80	184.8	187.1	189.4	191.7	194.0	196.3	198.6	200.9	203.2	205.5
90	207.9	210.2	212.5	214.8	217.1	219.4	221.7	224.0	226.3	228.6

Comparison of Different Units of Pressure

lb/in.^{2}	oz/in.^{2}	Water[a] Inches of	Water[a] Feet of	Mercury[b] Inches of	Mercury[b] mm of	Pascal (N/m^2)
1.0	16.0	27.7066	2.3089	2.0360	51.7149	6,894.757
0.063	1.0	1.732	0.144	0.127	3.232	430.922
0.036	0.578	1.0	0.083	0.074	1.867	248.640
0.433	6.930	12.000	1.0	0.882	22.398	2,983.680
0.491	7.858	13.608	1.134	1.0	25.400	3,383.493
0.019	0.309	0.536	0.045	0.039	1.0	133.271
0.000145	0.0023	0.00402	0.00033	0.00030	0.00750	1.0
2	32	55.413	4.618	4.072	103.430	13,790
3	48	83.120	6.927	6.108	155.145	20,684
4	64	110.826	9.236	8.144	206.860	27,579
5	80	138.533	11.544	10.180	258.575	34,474
6	96	166.240	13.853	12.216	310.289	41,369
7	112	193.946	16.162	14.252	362.004	48,263
8	128	221.653	18.471	16.288	413.719	55,158
9	144	249.359	20.780	18.324	465.434	62,053
10	160	277.066	23.089	20.360	517.149	68,948
11	176	304.773	25.398	22.396	568.864	75,842
12	192	332.479	27.707	24.432	620.579	82,737
13	208	360.186	30.015	26.468	672.294	89,632
14	224	387.892	32.324	28.504	724.009	96,527
15	240	415.599	34.633	30.540	775.724	103,421
16	256	443.306	36.942	32.576	827.438	110,316
17	272	471.012	39.251	34.612	879.153	117,211
18	288	498.719	41.560	36.648	930.868	124,106
19	304	526.425	43.869	38.684	982.583	131,000
20	320	554.132	46.178	40.720	1034.298	137,895
21	336	581.839	48.487	42.756	1086.013	144,790
22	352	609.545	50.795	44.792	1137.728	151,685
23	368	637.252	53.104	46.828	1189.443	158,579
24	384	664.958	55.413	48.864	1241.158	165,474
25	400	692.665	57.722	50.901	1292.873	172,369

[a] Measured at 60°F
[b] Measured at 32°F (0°C).
To convert pascal to bars, divide by 100,000.

Volumes of Water at Different Temperatures

Degrees F	Volume	Degrees F	Volume	Degrees F	Volume	Degrees F	Volume
39.1	1.00000	86	1.00425	131	1.01423	176	1.02872
50	1.00025	95	1.00586	140	1.01678	185	1.03213
59	1.00083	104	1.00767	149	1.01951	194	1.03570
68	1.00171	113	1.00967	158	1.02241	203	1.03943
77	1.00286	122	1.01186	167	1.02548	212	1.04332

Flow of Water in Pipes.—The quantity of water that will flow through a pipe depends primarily on the head but also on the diameter of the pipe, the character of the interior surface, and the number and shape of the bends. The head may be either the distance between the levels of the surface of water in a reservoir and the point of discharge, or it may be caused by mechanically applied pressure, as by pumping, when the head is calculated as the vertical distance corresponding to the pressure.

One pound per square inch is equal to 2.309 feet head, and a 1-foot head is equal to a pressure of 0.433 pound per square inch.

Weight of Water per Cubic Foot at Different Temperatures

Temp. (°F)	Wt. per Cu Ft (lb/ft³)	Temp. (°F)	Wt. per Cu Ft (lb/ft³)	Temp. (°F)	Wt. per Cu Ft (lb/ft³)	Temp. (°F)	Wt. per Cu Ft (lb/ft³)	Temp. (°F)	Wt. per Cu Ft (lb/ft³)	Temp. (°F)	Wt. per Cu Ft (lb/ft³)
32	62.42	130	61.56	220	59.63	320	56.66	420	52.6	520	47.6
40	62.42	140	61.37	230	59.37	330	56.30	430	52.2	530	47.0
50	62.41	150	61.18	240	59.11	340	55.94	440	51.7	540	46.3
60	62.37	160	60.98	250	58.83	350	55.57	450	51.2	550	45.6
70	62.31	170	60.77	260	58.55	360	55.18	460	50.7	560	44.9
80	62.23	180	60.55	270	58.26	370	54.78	470	50.2	570	44.1
90	62.13	190	60.32	280	57.96	380	54.36	480	49.7	580	43.3
100	62.02	200	60.12	290	57.65	390	53.94	490	49.2	590	42.6
110	61.89	210	59.88	300	57.33	400	53.50	500	48.7	600	41.8
120	61.74	212	59.83	310	57.00	410	53.00	510	48.1	…	…

Table of Horsepower due to Certain Head of Water

Head in Feet	Horse-power	Head in Feet	Horse-power	Head in Feet	Horse-power	Head in Feet	Horse-power	Head in Feet	Horse-power
1	0.0016	170	0.274	340	0.547	520	0.837	1250	2.012
10	0.0161	180	0.290	350	0.563	540	0.869	1300	2.093
20	0.0322	190	0.306	360	0.580	560	0.901	1350	2.173
30	0.0483	200	0.322	370	0.596	580	0.934	1400	2.254
40	0.0644	210	0.338	380	0.612	600	0.966	1450	2.334
50	0.0805	220	0.354	390	0.628	650	1.046	1500	2.415
60	0.0966	230	0.370	400	0.644	700	1.127	1550	2.495
70	0.1127	240	0.386	410	0.660	750	1.207	1600	2.576
80	0.1288	250	0.402	420	0.676	800	1.288	1650	2.656
90	0.1449	260	0.418	430	0.692	850	1.368	1700	2.737
100	0.1610	270	0.435	440	0.708	900	1.449	1750	2.818
110	0.1771	280	0.451	450	0.724	950	1.529	1800	2.898
120	0.1932	290	0.467	460	0.740	1000	1.610	1850	2.978
130	0.2093	300	0.483	470	0.757	1050	1.690	1900	3.059
140	0.2254	310	0.499	480	0.773	1100	1.771	1950	3.139
150	0.2415	320	0.515	490	0.789	1150	1.851	2000	3.220
160	0.2576	330	0.531	500	0.805	1200	1.932	2100	3.381

The table gives the horsepower of 1 cubic foot of water per minute, and is based on an efficiency of 85 per cent.

All formulas for finding the amount of water that will flow through a pipe in a given time are approximate. The formula that follows will give results within 5 or 10 per cent of actual flows, if applied to pipe lines carefully laid and in fair condition.

$$V = C\sqrt{\frac{hD}{L + 54D}}$$

where V = approximate mean velocity in feet per second

C = coefficient from the accompanying table

D = diameter of pipe in feet

h = total head in feet

L = total length of pipe line in feet

Values of Coefficient *C*

Dia. of Pipe			Dia. of Pipe			Dia. of Pipe		
Feet	Inches	*C*	Feet	Inches	*C*	Feet	Inches	*C*
0.1	1.2	23	0.8	9.6	46	3.5	42	64
0.2	2.4	30	0.9	10.8	47	4.0	48	66
0.3	3.6	34	1.0	12.0	48	5.0	60	68
0.4	4.8	37	1.5	18.0	53	6.0	72	70
0.5	6.0	39	2.0	24.0	57	7.0	84	72
0.6	7.2	42	2.5	30.0	60	8.0	96	74
0.7	8.4	44	3.0	36.0	62	10.0	120	77

Example: A pipe line, 1 mile long, 12 inches in diameter, discharges water under a head of 100 feet. Find the velocity and quantity of discharge.

From the table, the coefficient C is found to be 48 for a pipe 1 foot in diameter, hence:

$$V = 48\sqrt{\frac{100 \times 1}{5280 + 54 \times 1}} = 6.57 \text{ feet per second}$$

To find the discharge in cubic feet per second, multiply the velocity found by the area of cross-section of the pipe in square feet:

$$6.57 \times 0.7854 = 5.16 \text{ cubic feet per second}$$

The loss of head due to a bend in the pipe is most frequently given as the equivalent length of straight pipe, which would cause the same loss in head as the bend. Experiments show that a right-angle bend should have a radius of about three times the diameter of the pipe. Assuming this curvature, then, if d is the diameter of the pipe in inches and L is the length of straight pipe in feet that causes the same loss of head as the bend in the pipe, the following formula gives the equivalent length of straight pipe that should be added to simulate a right-angle bend:

$$L = 4d \div 3$$

Thus, the loss of head due to a right-angle bend in a 6-inch pipe would be equal to that in 8 feet of straight pipe. Experiments undertaken to determine the losses due to valves in pipe lines indicate that a fully open gate valve in a pipe causes a loss of head corresponding to the loss in a length of pipe equal to six diameters.

Flow of Water Through Nozzles in Cubic Feet per Second

Head in Feet, at Nozzle	Pressure, lb/in.^{2}	Theoretical Velocity, ft/s	Diameter of Nozzle, Inches							
			1	1.5	2	2.5	3	3.5	4	4.5
5	2.17	17.93	0.10	0.22	0.39	0.61	0.88	1.20	1.56	1.98
10	4.33	25.36	0.14	0.31	0.55	0.86	1.24	1.69	2.21	2.80
20	8.66	35.87	0.20	0.44	0.78	1.22	1.76	2.40	3.13	3.96
30	12.99	43.93	0.24	0.54	0.96	1.50	2.16	2.93	3.83	4.85
40	17.32	50.72	0.28	0.62	1.11	1.73	2.49	3.39	4.43	5.60
50	21.65	56.71	0.31	0.70	1.24	1.93	2.78	3.79	4.95	6.26
60	25.99	62.12	0.34	0.76	1.36	2.12	3.05	4.15	5.42	6.86
70	30.32	67.10	0.37	0.82	1.46	2.29	3.29	4.48	5.86	7.41
80	34.65	71.73	0.39	0.88	1.56	2.45	3.52	4.79	6.26	7.92
90	38.98	76.08	0.41	0.93	1.66	2.59	3.73	5.08	6.64	8.40
100	43.31	80.20	0.44	0.98	1.75	2.73	3.94	5.36	7.00	8.86
120	51.97	87.85	0.48	1.08	1.92	2.99	4.31	5.87	7.67	9.70
140	60.63	94.89	0.52	1.16	2.07	3.23	4.66	6.34	8.28	10.48
160	69.29	101.45	0.55	1.24	2.21	3.46	4.98	6.78	8.85	11.20
180	77.96	107.60	0.59	1.32	2.35	3.67	5.28	7.19	9.39	11.88
200	86.62	113.42	0.62	1.39	2.47	3.87	5.57	7.58	9.90	12.53
250	108.27	126.81	0.69	1.56	2.77	4.32	6.22	8.47	11.07	14.01
300	129.93	138.91	0.76	1.70	3.03	4.74	6.82	9.28	12.12	15.34
350	151.58	150.04	0.82	1.84	3.27	5.11	7.37	10.02	13.09	16.57
400	173.24	160.40	0.87	1.97	3.50	5.47	7.87	10.72	14.00	17.72
450	194.89	170.13	0.93	2.09	3.71	5.80	8.35	11.37	14.85	18.79
500	216.54	179.33	0.98	2.20	3.91	6.11	8.80	11.98	15.65	19.81

Head in Feet, at Nozzle	Pressure, lb/in.^{2}	Theoretical Velocity, ft/s	Diameter of Nozzle, Inches							
			5	6	7	8	9	10	11	12
5	2.17	17.93	2.45	3.52	4.79	6.3	7.9	9.8	11.8	14.1
10	4.33	25.36	3.46	4.98	6.78	8.9	11.2	13.8	16.7	19.9
20	8.66	35.87	4.89	7.04	9.59	12.5	15.8	19.6	23.7	28.2
30	12.99	43.93	5.99	8.63	11.74	15.3	19.4	24.0	29.0	34.5
40	17.32	50.72	6.92	9.96	13.56	17.7	22.4	27.7	33.5	39.8
50	21.65	56.71	7.73	11.13	15.16	19.8	25.1	30.9	37.4	44.5
60	25.99	62.12	8.47	12.20	16.60	21.7	27.4	33.9	41.0	48.8
70	30.32	67.10	9.15	13.18	17.93	23.4	29.6	36.6	44.3	52.7
80	34.65	71.73	9.78	14.08	19.17	25.0	31.7	39.1	47.3	56.3
90	38.98	76.08	10.37	14.94	20.33	26.6	33.6	41.5	50.2	59.8
100	43.31	80.20	10.94	15.75	21.43	28.0	35.4	43.7	52.9	63.0
120	51.97	87.85	11.98	17.25	23.48	30.7	38.8	47.9	58.0	69.0
140	60.63	94.89	12.94	18.63	25.36	33.1	41.9	51.8	62.6	74.5
160	69.29	101.45	13.83	19.92	27.11	35.4	44.8	55.3	66.9	79.7
180	77.96	107.60	14.67	21.13	28.76	37.6	47.5	58.7	71.0	84.5
200	86.62	113.42	15.47	22.27	30.31	39.6	50.1	61.9	74.9	89.1
250	108.27	126.81	17.29	24.90	33.89	44.3	56.0	69.2	83.7	99.6
300	129.93	138.91	18.94	27.27	37.12	48.5	61.4	75.8	91.7	109.1
350	151.58	150.04	20.46	29.46	40.10	52.4	66.3	81.8	99.0	117.8
400	173.24	160.40	21.87	31.49	42.87	56.0	70.9	87.5	105.9	126.0
450	194.89	170.13	23.20	33.40	45.47	59.4	75.2	92.8	112.3	133.6
500	216.54	179.33	24.45	35.21	47.93	62.6	79.2	97.8	118.4	140.8

Theoretical Velocity of Water Due to Head in Feet

Head in Feet	Theoretical Velocity		Head in Feet	Theoretical Velocity		Head in Feet	Theoretical Velocity	
	ft/s	ft/min		ft/s	ft/min		ft/s	ft/min
1	8.01	481	48	55.56	3334	95	78.16	4690
2	11.34	681	49	56.13	3368	96	78.57	4715
3	13.89	833	50	56.70	3403	97	78.98	4739
4	16.04	962	51	57.27	3436	98	79.39	4764
5	17.93	1076	52	57.83	3470	99	79.79	4788
6	19.64	1179	53	58.38	3503	100	80.19	4812
7	21.21	1273	54	58.93	3536	105	82.18	4931
8	22.68	1361	55	59.47	3569	110	84.11	5047
9	24.05	1444	56	60.01	3601	115	86.00	5160
10	25.36	1522	57	60.54	3633	120	87.85	5271
11	26.59	1596	58	61.07	3665	125	89.66	5380
12	27.78	1667	59	61.60	3696	130	91.44	5487
13	28.91	1735	60	62.12	3727	135	93.18	5591
14	30.00	1800	61	62.63	3758	140	94.89	5694
15	31.06	1864	62	63.14	3789	145	96.57	5794
16	32.07	1925	63	63.65	3819	150	98.22	5893
17	33.06	1984	64	64.15	3850	155	99.84	5991
18	34.02	2042	65	64.65	3880	160	101.44	6087
19	34.95	2097	66	65.15	3909	165	103.01	6181
20	35.86	2152	67	65.64	3939	170	104.56	6274
21	36.75	2205	68	66.13	3968	175	106.09	6366
22	37.61	2257	69	66.61	3997	180	107.59	6456
23	38.46	2308	70	67.09	4026	185	109.08	6545
24	39.28	2357	71	67.57	4055	190	110.54	6633
25	40.09	2406	72	68.05	4083	195	111.99	6720
26	40.89	2454	73	68.52	4111	200	113.42	6805
27	41.67	2500	74	68.99	4139	205	114.82	6890
28	42.43	2546	75	69.45	4167	210	116.22	6973
29	43.18	2591	76	69.91	4195	215	117.59	7056
30	43.92	2636	77	70.37	4222	220	118.95	7137
31	44.65	2679	78	70.83	4250	225	120.30	7218
32	45.36	2722	79	71.28	4277	230	121.62	7298
33	46.07	2764	80	71.73	4304	235	122.94	7377
34	46.76	2806	81	72.17	4331	240	124.24	7455
35	47.44	2847	82	72.62	4357	245	125.53	7532
36	48.11	2887	83	73.06	4384	250	126.80	7608
37	48.78	2927	84	73.50	4410	255	128.06	7684
38	49.43	2966	85	73.94	4436	260	129.31	7759
39	50.08	3005	86	74.37	4462	270	131.78	7907
40	50.72	3043	87	74.80	4488	280	134.20	8052
41	51.35	3081	88	75.23	4514	290	136.57	8195
42	51.97	3119	89	75.66	4540	300	138.91	8335
43	52.59	3155	90	76.08	4565	310	141.20	8472
44	53.19	3192	91	76.50	4590	320	143.46	8608
45	53.79	3228	92	76.92	4615	330	145.69	8741
46	54.39	3264	93	77.34	4641	340	147.88	8873
47	54.98	3299	94	77.75	4665	350	150.04	9002

Gallons of Water per Foot of Pipe

Nominal Pipe Size (in.)	Iron or Steel		Copper		
	Sched. 40	Sched. 80	Type K	Type L	Type M
⅛	0.0030	0.0019	0.0014	0.0016	0.0016
¼	0.0054	0.0037	0.0039	0.0040	0.0043
⅜	0.0099	0.0073	0.0066	0.0075	0.0083
½	0.0158	0.0122	0.0113	0.0121	0.0132
⅝	…	…	0.0173	0.0181	0.0194
¾	0.0277	0.0225	0.0226	0.0251	0.0268
1	0.0449	0.0374	0.0404	0.0429	0.0454

Multiply the length of pipe in feet by the factor from the table to find the volume contained in gallons.

Friction Loss in Fittings—Equivalent Length of Pipe in Feet

Nominal Pipe Size (in.)	Elbows						Standard Tee	
	90° Std.	45° Std.	90° Long Radius	90° Street	45° Street	Square Corner	Flow thru Run	Flow thru Branch
¼	0.9	0.5	0.6	1.5	0.8	1.7	0.6	1.8
½	1.6	0.8	1.0	2.6	1.3	3.0	1.0	4.0
¾	2.1	1.1	1.4	3.4	1.8	3.9	1.4	5.1
1	2.6	1.4	1.7	4.4	2.3	5.0	1.7	6.0
1¼	3.5	1.8	2.3	5.8	3.0	6.5	2.3	6.9
1½	4.0	2.1	2.7	6.7	3.5	7.6	2.7	8.1
2	5.5	2.8	4.3	8.6	4.5	9.8	4.3	12.0
2½	6.2	3.3	5.1	10.3	5.4	11.7	5.1	14.3
3	7.7	4.1	6.3	12.8	6.6	14.6	6.3	16.3
4	10.1	5.4	8.3	16.8	8.7	19.1	8.3	22.1
6	15.2	8.1	12.5	25.3	13.1	28.8	12.5	32.2
8	20.0	10.6	16.5	33.3	17.3	37.9	16.5	39.9
10	25.1	13.4	20.7	41.8	21.7	47.6	20.7	50.1
12	29.8	15.9	24.7	49.7	25.9	56.7	24.7	59.7

Pipe Expansion Due to Temperature Changes.—The expansion for any length of pipe caused by a given temperature change can be determined from the following table. Find the expansion factor corresponding to the expected difference in the minimum and maximum pipe temperatures and divide by 100 to obtain the increase in length per foot of pipe. Multiply the increase per foot result by the length of the pipe run to get the total change in pipe length.

Linear Expansion and Contraction Factors per 100 Feet of Pipe

Temperature Change, °F	Pipe Material				
	Steel	Copper	PVC	FRP	PP & PVDF
0	0	0	0	0	0
20	0.15	0.25	0.62	0.26	2.00
40	0.30	0.45	1.30	0.52	4.00
60	0.46	0.65	2.20	0.78	6.00
80	0.61	0.87	2.80	1.05	8.00
100	0.77	1.10	3.50	1.31	10.00
120	0.92	1.35	4.25	1.57	12.00
140	1.08	1.57	4.80	1.83	14.00
160	1.24	1.77	5.50	2.09	16.00
180	1.40	2.00	6.30	2.35	18.00
200	1.57	2.25	7.12	2.62	20.00

Multiply the length of pipe by the table factor and divide by 100 for the increase or decrease in length.

Properties, Compression, and Flow of Air

Properties of Air.—Air is a mechanical mixture composed of 78 per cent of nitrogen, 21 per cent of oxygen, and 1 per cent of argon, by volume. The density of dry air at 32 degrees F and atmospheric pressure (29.92 inches of mercury or 14.70 pounds per square inch) is 0.08073 pound per cubic foot. The density of air at any other temperature or pressure is

$$\rho = \frac{1.325 \times B}{T}$$

in which ρ = density in pounds per cubic foot; B = height of barometric pressure in inches of mercury; T = absolute temperature in degrees Rankine. (When using pounds as a unit, here and elsewhere, care must be exercised to differentiate between pounds mass and pounds force. See *Acceleration of Gravity g Used in Mechanics Formulas* on page 114 and *The Use of the Metric SI System in Mechanics Calculations* on page 116.)

Volumes and Weights of Air at Different Temperatures, at Atmospheric Pressure

Temperature, °F	Volume of 1 lb of Air in Cubic Feet	Density, Pounds per Cubic Foot	Temperature, °F	Volume of 1 lb of Air in Cubic Feet	Density, Pounds per Cubic Foot	Temperature, °F	Volume of 1 lb of Air in Cubic Feet	Density, Pounds per Cubic Foot
0	11.57	0.0864	172	15.92	0.0628	800	31.75	0.0315
12	11.88	0.0842	182	16.18	0.0618	900	34.25	0.0292
22	12.14	0.0824	192	16.42	0.0609	1000	37.31	0.0268
32	12.39	0.0807	202	16.67	0.0600	1100	39.37	0.0254
42	12.64	0.0791	212	16.92	0.0591	1200	41.84	0.0239
52	12.89	0.0776	230	17.39	0.0575	1300	44.44	0.0225
62	13.14	0.0761	250	17.89	0.0559	1400	46.95	0.0213
72	13.39	0.0747	275	18.52	0.0540	1500	49.51	0.0202
82	13.64	0.0733	300	19.16	0.0522	1600	52.08	0.0192
92	13.89	0.0720	325	19.76	0.0506	1700	54.64	0.0183
102	14.14	0.0707	350	20.41	0.0490	1800	57.14	0.0175
112	14.41	0.0694	375	20.96	0.0477	2000	62.11	0.0161
122	14.66	0.0682	400	21.69	0.0461	2200	67.11	0.0149
132	14.90	0.0671	450	22.94	0.0436	2400	72.46	0.0138
142	15.17	0.0659	500	24.21	0.0413	2600	76.92	0.0130
152	15.41	0.0649	600	26.60	0.0376	2800	82.64	0.0121
162	15.67	0.0638	700	29.59	0.0338	3000	87.72	0.0114

The absolute zero from which all temperatures must be counted when dealing with the weight and volume of gases is assumed to be −459.7 degrees F. Hence, to obtain the absolute temperature T used in preceding formula, add the value 459.7 to the temperature observed on a regular Fahrenheit thermometer.

In obtaining the value of B, 1 inch of mercury at 32 degrees F may be taken as equal to a pressure of 0.491 pound per square inch.

Example: What would be the weight of a cubic foot of air at atmospheric pressure (29.92 inches of mercury) at 100 degrees F? The weight, W, is given by $W = \rho V$.

$$W = \rho V = \frac{1.325 \times 29.92}{100 + 459.7} \times 1 = 0.0708 \text{ pound}$$

Density of Air at Different Pressures and Temperatures

Temp. of Air, °F	Gage Pressure, Pounds														
	0	5	10	20	30	40	50	60	80	100	120	150	200	250	300
	Density in Pounds per Cubic Foot														
-20	0.0900	0.1205	0.1515	0.2125	0.274	0.336	0.397	0.458	0.580	0.702	0.825	1.010	1.318	1.625	1.930
-10	0.0882	0.1184	0.1485	0.2090	0.268	0.328	0.388	0.448	0.567	0.687	0.807	0.989	1.288	1.588	1.890
0	0.0864	0.1160	0.1455	0.2040	0.263	0.321	0.380	0.438	0.555	0.672	0.790	0.968	1.260	1.553	1.850
10	0.0846	0.1136	0.1425	0.1995	0.257	0.314	0.372	0.429	0.543	0.658	0.774	0.947	1.233	1.520	1.810
20	0.0828	0.1112	0.1395	0.1955	0.252	0.307	0.364	0.420	0.533	0.645	0.757	0.927	1.208	1.489	1.770
30	0.0811	0.1088	0.1366	0.1916	0.246	0.301	0.357	0.412	0.522	0.632	0.742	0.908	1.184	1.460	1.735
40	0.0795	0.1067	0.1338	0.1876	0.241	0.295	0.350	0.404	0.511	0.619	0.727	0.890	1.161	1.431	1.701
50	0.0780	0.1045	0.1310	0.1839	0.237	0.290	0.343	0.396	0.501	0.607	0.713	0.873	1.139	1.403	1.668
60	0.0764	0.1025	0.1283	0.1803	0.232	0.284	0.336	0.388	0.493	0.596	0.700	0.856	1.116	1.376	1.636
80	0.0736	0.0988	0.1239	0.1738	0.224	0.274	0.324	0.374	0.473	0.572	0.673	0.824	1.074	1.325	1.573
100	0.0710	0.0954	0.1197	0.1676	0.215	0.264	0.312	0.360	0.455	0.551	0.648	0.794	1.035	1.276	1.517
120	0.0680	0.0921	0.1155	0.1618	0.208	0.255	0.302	0.348	0.440	0.533	0.626	0.767	1.001	1.234	1.465
140	0.0663	0.0889	0.1115	0.1565	0.201	0.246	0.291	0.336	0.426	0.516	0.606	0.742	0.968	1.194	1.416
150	0.0652	0.0874	0.1096	0.1541	0.198	0.242	0.286	0.331	0.419	0.508	0.596	0.730	0.953	1.175	1.392
175	0.0626	0.0840	0.1054	0.1482	0.191	0.233	0.275	0.318	0.403	0.488	0.573	0.701	0.914	1.128	1.337
200	0.0603	0.0809	0.1014	0.1427	0.184	0.225	0.265	0.305	0.388	0.470	0.552	0.674	0.879	1.084	1.287
225	0.0581	0.0779	0.0976	0.1373	0.177	0.216	0.255	0.295	0.374	0.452	0.531	0.649	0.846	1.043	1.240
250	0.0560	0.0751	0.0941	0.1323	0.170	0.208	0.247	0.284	0.360	0.436	0.513	0.627	0.817	1.007	1.197
275	0.0541	0.0726	0.0910	0.1278	0.164	0.201	0.238	0.274	0.348	0.421	0.494	0.605	0.789	0.972	1.155
300	0.0523	0.0707	0.0881	0.1237	0.159	0.194	0.230	0.265	0.336	0.407	0.478	0.585	0.762	0.940	1.118
350	0.0491	0.0658	0.0825	0.1160	0.149	0.183	0.216	0.249	0.316	0.382	0.449	0.549	0.715	0.883	1.048
400	0.0463	0.0621	0.0779	0.1090	0.140	0.172	0.203	0.235	0.297	0.360	0.423	0.517	0.674	0.831	0.987
450	0.0437	0.0586	0.0735	0.1033	0.133	0.163	0.192	0.222	0.281	0.340	0.399	0.488	0.637	0.786	0.934
500	0.0414	0.0555	0.0696	0.978	0.126	0.154	0.182	0.210	0.266	0.322	0.379	0.463	0.604	0.746	0.885
550	0.0394	0.0528	0.0661	0.930	0.120	0.146	0.173	0.200	0.253	0.306	0.359	0.440	0.573	0.749	0.841
600	0.0376	0.0504	0.0631	0.885	0.114	0.139	0.165	0.190	0.241	0.292	0.343	0.419	0.547	0.675	0.801

Relation Between Pressure, Temperature, and Volume of Air.—This relationship is expressed by the formula:

$$\frac{P \times V}{T} = 53.3$$

in which P = absolute pressure in pounds per square foot; V = volume in cubic feet of one pound of air at the given pressure and temperature; T = absolute temperature in degrees R.

Example: What is the volume of one pound of air at a pressure of 24.7 pounds per square inch and at a temperature of 210 degrees F?

$$\frac{24.7 \times 144 \times V}{210 + 459.7} = 53.3 \text{ or } V = \frac{53.3 \times 669.7}{24.7 \times 144} = 10.04 \text{ cubic feet}$$

Relation Between Barometric Pressure, and Pressures in Pounds per Square Inch and Square Foot

Barometer, Inches	Pressure in Pounds per Square Inch	Pressure in Pounds per Square Foot	Barometer, Inches	Pressure in Pounds per Square Inch	Pressure in Pounds per Square Foot	Barometer, Inches	Pressure in Pounds per Square Inch	Pressure in Pounds per Square Foot
28.00	13.75	1980	29.25	14.36	2068	30.50	14.98	2156
28.25	13.87	1997	29.50	14.48	2086	30.75	15.10	2174
28.50	13.99	2015	29.75	14.61	2103	31.00	15.22	2192
28.75	14.12	2033	30.00	14.73	2121	31.25	15.34	2210
29.00	14.24	2050	30.25	14.85	2139	…	…	…

Expansion and Compression of Air.—The formula for the relationship between pressure, temperature, and volume of air just given indicates that when the pressure remains constant the volume is directly proportional to the absolute temperature. If the temperature remains constant, the volume is inversely proportional to the absolute pressure. Theoretically, air (as well as other gases) can be expanded or compressed according to different laws. *Adiabatic* expansion or compression takes place when the air is expanded or compressed without transmission of heat to or from it, as, for example, if the air could be expanded or compressed in a cylinder of an absolutely nonconducting material.

Let: P_1 = initial absolute pressure in pounds per square foot
V_1 = initial volume in cubic feet
T_1 = initial absolute temperature in degrees R
P_2 = absolute pressure in pounds per square foot, after compression
V_2 = volume in cubic feet, after compression
T_2 = absolute temperature in degrees R, after compression

Then:

$$\frac{V_2}{V_1} = \left(\frac{P_1}{P_2}\right)^{0.71} \qquad \frac{P_2}{P_1} = \left(\frac{V_1}{V_2}\right)^{1.41} \qquad \frac{T_2}{T_1} = \left(\frac{V_1}{V_2}\right)^{0.41}$$

$$\frac{V_2}{V_1} = \left(\frac{T_1}{T_2}\right)^{2.46} \qquad \frac{P_2}{P_1} = \left(\frac{T_2}{T_1}\right)^{3.46} \qquad \frac{T_2}{T_1} = \left(\frac{P_2}{P_1}\right)^{0.29}$$

These formulas are also applicable if all pressures are in pounds per square inch; if all volumes are in cubic inches; or if any other consistent set of units is used for pressure or volume.

Isothermal: expansion or compression takes place when a gas is expanded or compressed with an addition or transmission of sufficient heat to maintain a constant temperature.

Let: P_1 = initial absolute pressure in pounds per square foot
V_1 = initial volume in cubic feet
P_2 = absolute pressure in pounds per square foot, after compression
V_2 = volume in cubic feet, after compression
R = 53.3
T = temperature in degrees Rankine maintained during isothermal expansion or contraction

Then:

$$P_1 \times V_1 = P_2 \times V_2 = RT$$

Example: A volume of 165 cubic feet of air, at a pressure of 15 pounds per square inch, is compressed adiabatically to a pressure of 80 pounds per square inch. What will be the volume at this pressure?

$$V_2 = V_1\left(\frac{P_1}{P_2}\right)^{0.71} = 165\left(\frac{15}{80}\right)^{0.71} = 50 \text{ cubic feet, approx.}$$

Example: The same volume of air is compressed isothermally from 15 to 80 pounds per square inch. What will be the volume after compression?

$$V_2 = \frac{P_1 \times V_1}{P_2} = \frac{15 \times 165}{80} = 31 \text{ cubic feet}$$

Foot-pounds of Work Required in Compression of Air
Initial Pressure = 1 atmosphere = 14.7 pounds per square inch

Gage Pressure in Pounds per Square Inch	Isothermal Compression	Adiabatic Compression	Actual Compression	Gage Pressure in Pounds per Square Inch	Isothermal Compression	Adiabatic Compression	Actual Compression
	Foot-pounds Required per Cubic Foot of Air at Initial Pressure				Foot-pounds Required per Cubic Foot of Air at Initial Pressure		
5	619.6	649.5	637.5	55	3393.7	4188.9	3870.8
10	1098.2	1192.0	1154.6	60	3440.4	4422.8	4029.8
15	1488.3	1661.2	1592.0	65	3577.6	4645.4	4218.2
20	1817.7	2074.0	1971.4	70	3706.3	4859.6	4398.1
25	2102.6	2451.6	2312.0	75	3828.0	5063.9	4569.5
30	2353.6	2794.0	2617.8	80	3942.9	5259.7	4732.9
35	2578.0	3111.0	2897.8	85	4051.5	5450.0	4890.1
40	2780.8	3405.5	3155.6	90	4155.7	5633.1	5042.1
45	2966.0	3681.7	3395.4	95	4254.3	5819.3	5187.3
50	3136.2	3942.3	3619.8	100	4348.1	5981.2	5327.9

Work Required in Compression of Air.—The total work required for compression and expulsion of air, adiabatically compressed, is:

$$\text{Total work in foot-pounds} = 3.46 P_1 V_1\left[\left(\frac{P_2}{P_1}\right)^{0.29} - 1\right]$$

where P_1 = initial absolute pressure in pounds per square foot
P_2 = absolute pressure in pounds per square foot, after compression
V_1 = initial volume in cubic feet

The total work required for isothermal compression is:

$$\text{Total work in foot-pounds} = P_1 V_1 \log_e \frac{V_1}{V_2}$$

in which P_1, P_2, and V_1 denote the same quantities as in the previous equation, and V_2 = volume of air in cubic feet, after compression.

The work required to compress air isothermally, that is, when the heat of compression is removed as rapidly as produced, is considerably less than the work required for compressing air adiabatically, or when all the heat is retained. In practice, neither of these two theoretical extremes is obtainable, but the power required for air compression is about the median between the powers that would be required for each. The accompanying table gives the average number of foot-pounds of work required to compress air.

Horsepower Required to Compress Air.—In the accompanying tables is given the horsepower required to compress one cubic foot of free air per minute (isothermally and adiabatically) from atmospheric pressure (14.7 pounds per square inch) to various gage pressures, for one-, two-, and three-stage compression. The formula for calculating the horsepower required to compress, adiabatically, a given volume of free air to a given pressure is:

$$\text{HP} = \frac{144NPVn}{33{,}000(n-1)}\left[\left(\frac{P_2}{P}\right)^{\frac{n-1}{Nn}} - 1\right]$$

where N = number of stages in which compression is accomplished

P = atmospheric pressure in pounds per square inch

P_2 = absolute terminal pressure in pounds per square inch

V = volume of air, in cubic feet, compressed per minute, at atmospheric pressure

n = exponent of the compression curve = 1.41 for adiabatic compression

For different methods of compression and for one cubic foot of air per minute, this formula may be simplified as follows:

For one-stage compression: $\text{HP} = 0.015P(R^{0.29} - 1)$

For two-stage compression: $\text{HP} = 0.030\,P(R^{0.145} - 1)$

For three-stage compression: $\text{HP} = 0.045\,P(R^{0.0975} - 1)$

For four-stage compression: $\text{HP} = 0.060\,P(R^{0.0725} - 1)$

In these latter formulas $R = \frac{P^2}{P}$ = number of atmospheres to be compressed

The formula for calculating the horsepower required to compress isothermally a given volume of free air to a given pressure is:

$$\text{HP} = \frac{144PV}{33000}\left(\log_e \frac{P_2}{P}\right)$$

Natural logarithms are obtained by multiplying common logarithms by 2.30259 or by using a handheld calculator.

Horsepower Required to Compress Air

Horsepower Required To Compress One Cubic Foot of Free Air per Minute (Isothermally and Adiabatically) from Atmospheric Pressure (14.7 pounds per square inch) to Various Gage Pressures. — Single-Stage Compression (Initial Temperature of Air, 60°F — Jacket cooling not considered)								
Gage Pressure, Pounds	Absolute Pressure, Pounds	Number of Atmospheres	Isothermal Compression		Adiabatic Compression			
			Mean Effective Pressure[a]	Horsepower	Mean Effective Pressure,[a] Theoretical	Mean Eff. Pressure plus 15 per cent Friction	Horsepower, Theoretical	Horsepower plus 15 per cent Friction
5	19.7	1.34	4.13	0.018	4.46	5.12	0.019	0.022
10	24.7	1.68	7.57	0.033	8.21	9.44	0.036	0.041
15	29.7	2.02	11.02	0.048	11.46	13.17	0.050	0.057
20	34.7	2.36	12.62	0.055	14.30	16.44	0.062	0.071
25	39.7	2.70	14.68	0.064	16.94	19.47	0.074	0.085
30	44.7	3.04	16.30	0.071	19.32	22.21	0.084	0.096
35	49.7	3.38	17.90	0.078	21.50	24.72	0.094	0.108
40	54.7	3.72	19.28	0.084	25.53	27.05	0.103	0.118
45	59.7	4.06	20.65	0.090	25.40	29.21	0.111	0.127
50	64.7	4.40	21.80	0.095	27.23	31.31	0.119	0.136
55	69.7	4.74	22.95	0.100	28.90	33.23	0.126	0.145
60	74.7	5.08	23.90	0.104	30.53	35.10	0.133	0.153
65	79.7	5.42	24.80	0.108	32.10	36.91	0.140	0.161
70	84.7	5.76	25.70	0.112	33.57	38.59	0.146	0.168
75	89.7	6.10	26.62	0.116	35.00	40.25	0.153	0.175
80	94.7	6.44	27.52	0.120	36.36	41.80	0.159	0.182
85	99.7	6.78	28.21	0.123	37.63	43.27	0.164	0.189
90	104.7	7.12	28.93	0.126	38.89	44.71	0.169	0.195
95	109.7	7.46	29.60	0.129	40.11	46.12	0.175	0.201
100	114.7	7.80	30.30	0.132	41.28	47.46	0.180	0.207
110	124.7	8.48	31.42	0.137	43.56	50.09	0.190	0.218
120	134.7	9.16	32.60	0.142	45.69	52.53	0.199	0.229
130	144.7	9.84	33.75	0.147	47.72	54.87	0.208	0.239
140	154.7	10.52	34.67	0.151	49.64	57.08	0.216	0.249
150	164.7	11.20	35.59	0.155	51.47	59.18	0.224	0.258
160	174.7	11.88	36.30	0.158	53.70	61.80	0.234	0.269
170	184.7	12.56	37.20	0.162	55.60	64.00	0.242	0.278
180	194.7	13.24	38.10	0.166	57.20	65.80	0.249	0.286
190	204.7	13.92	38.80	0.169	58.80	67.70	0.256	0.294
200	214.7	14.60	39.50	0.172	60.40	69.50	0.263	0.303

[a] Mean Effective Pressure (MEP) is defined as that single pressure rise, above atmospheric, which would require the same horsepower as the actual varying pressures during compression.

Horsepower Required to Compress Air

Horsepower Required to Compress One Cubic Foot of Free Air per Minute (Isothermally and Adiabatically) from Atmospheric Pressure (14.7 pounds per square inch) to Various Gage Pressures.—Two-Stage Compression (Initial Temperature of Air, 60°F —Jacket cooling not considered)

Gage Pressure, Pounds	Absolute Pressure, Pounds	Number of Atmospheres	Correct Ratio of Cylinder Volumes	Intercooler Gage Pressure	Isothermal Compression: Mean Effective Pressure[a]	Isothermal Compression: Horsepower	Adiabatic Compression: Mean Eff. Pressure, Theoretical[a]	Adiabatic Compression: Mean Eff. Pressure plus 15 per cent Friction	Adiabatic Compression: Horsepower, Theoretical	Adiabatic Compression: HP plus 15 per cent Friction	Percentage of Saving over One-stage Compression
50	64.7	4.40	2.10	16.2	21.80	0.095	24.30	27.90	0.106	0.123	10.9
60	74.7	5.08	2.25	18.4	23.90	0.104	27.20	31.30	0.118	0.136	11.3
70	84.7	5.76	2.40	20.6	25.70	0.112	29.31	33.71	0.128	0.147	12.3
80	94.7	6.44	2.54	22.7	27.52	0.120	31.44	36.15	0.137	0.158	13.8
90	104.7	7.12	2.67	24.5	28.93	0.126	33.37	38.36	0.145	0.167	14.2
100	114.7	7.80	2.79	26.3	30.30	0.132	35.20	40.48	0.153	0.176	15.0
110	124.7	8.48	2.91	28.1	31.42	0.137	36.82	42.34	0.161	0.185	15.2
120	134.7	9.16	3.03	29.8	32.60	0.142	38.44	44.20	0.168	0.193	15.6
130	144.7	9.84	3.14	31.5	33.75	0.147	39.86	45.83	0.174	0.200	16.3
140	154.7	10.52	3.24	32.9	34.67	0.151	41.28	47.47	0.180	0.207	16.7
150	164.7	11.20	3.35	34.5	35.59	0.155	42.60	48.99	0.186	0.214	16.9
160	174.7	11.88	3.45	36.1	36.30	0.158	43.82	50.39	0.191	0.219	18.4
170	184.7	12.56	3.54	37.3	37.20	0.162	44.93	51.66	0.196	0.225	19.0
180	194.7	13.24	3.64	38.8	38.10	0.166	46.05	52.95	0.201	0.231	19.3
190	204.7	13.92	3.73	40.1	38.80	0.169	47.16	54.22	0.206	0.236	19.5
200	214.7	14.60	3.82	41.4	39.50	0.172	48.18	55.39	0.210	0.241	20.1
210	224.7	15.28	3.91	42.8	40.10	0.174	49.35	56.70	0.216	0.247	…
220	234.7	15.96	3.99	44.0	40.70	0.177	50.30	57.70	0.220	0.252	…
230	244.7	16.64	4.08	45.3	41.30	0.180	51.30	59.10	0.224	0.257	…
240	254.7	17.32	4.17	46.6	41.90	0.183	52.25	60.10	0.228	0.262	…
250	264.7	18.00	4.24	47.6	42.70	0.186	52.84	60.76	0.230	0.264	…
260	274.7	18.68	4.32	48.8	43.00	0.188	53.85	62.05	0.235	0.270	…
270	284.7	19.36	4.40	50.0	43.50	0.190	54.60	62.90	0.238	0.274	…
280	294.7	20.04	4.48	51.1	44.00	0.192	55.50	63.85	0.242	0.278	…
290	304.7	20.72	4.55	52.2	44.50	0.194	56.20	64.75	0.246	0.282	…
300	314.7	21.40	4.63	53.4	45.80	0.197	56.70	65.20	0.247	0.283	…
350	364.7	24.80	4.98	58.5	47.30	0.206	60.15	69.16	0.262	0.301	…
400	414.7	28.20	5.31	63.3	49.20	0.214	63.19	72.65	0.276	0.317	…
450	464.7	31.60	5.61	67.8	51.20	0.223	65.93	75.81	0.287	0.329	…
500	514.7	35.01	5.91	72.1	52.70	0.229	68.46	78.72	0.298	0.342	…

[a] Mean Effective Pressure (MEP) is defined as that single pressure rise, above atmospheric, which would require the same horsepower as the actual varying pressures during compression.

Horsepower Required to Compress Air

Horsepower Required for Compressing One Cubic Foot of Free Air per Minute (Isothermally and Adiabatically) from Atmospheric Pressure (14.7 pounds per square inch) to Various Gage Pressures.—Three-stage Compression (Initial Temperature of Air, 60°F.—Jacket-cooling not considered)												
Gage Pressure, Pounds	Absolute Pressure, Pounds	Number of Atmospheres	Correct Ratio of Cylinder Volumes	Intercooler Gage Pressure, First and Second Stages	Isothermal Compression		Adiabatic Compression				Percentage of Saving over Two-stage Compression
					Mean Effective Pressure[a]	Horsepower	Mean Eff. Pressure,[a] Theoretical	Mean Eff. Pressure plus 15 per cent Friction	Horsepower Theoretical	HP plus 15 per cent Friction	
100	114.7	7.8	1.98	14.4– 42.9	30.30	0.132	33.30	38.30	0.145	0.167	5.23
150	164.7	11.2	2.24	18.2– 59.0	35.59	0.155	40.30	46.50	0.175	0.202	5.92
200	214.7	14.6	2.44	21.2– 73.0	39.50	0.172	45.20	52.00	0.196	0.226	6.67
250	264.7	18.0	2.62	23.8– 86.1	42.70	0.186	49.20	56.60	0.214	0.246	6.96
300	314.7	21.4	2.78	26.1– 98.7	45.30	0.197	52.70	60.70	0.229	0.264	7.28
350	364.7	24.8	2.92	28.2–110.5	47.30	0.206	55.45	63.80	0.242	0.277	7.64
400	414.7	28.2	3.04	30.0–121.0	49.20	0.214	58.25	66.90	0.253	0.292	8.33
450	464.7	31.6	3.16	31.8–132.3	51.20	0.223	60.40	69.40	0.263	0.302	8.36
500	514.7	35.0	3.27	33.4–142.4	52.70	0.229	62.30	71.70	0.273	0.314	8.38
550	564.7	38.4	3.38	35.0–153.1	53.75	0.234	65.00	74.75	0.283	0.326	8.80
600	614.7	41.8	3.47	36.3–162.3	54.85	0.239	66.85	76.90	0.291	0.334	8.86
650	664.7	45.2	3.56	37.6–171.5	56.00	0.244	67.90	78.15	0.296	0.340	9.02
700	714.7	48.6	3.65	38.9–180.8	57.15	0.249	69.40	79.85	0.303	0.348	9.18
750	764.7	52.0	3.73	40.1–189.8	58.10	0.253	70.75	81.40	0.309	0.355	…
800	814.7	55.4	3.82	41.4–199.5	59.00	0.257	72.45	83.25	0.315	0.362	…
850	864.7	58.8	3.89	42.5–207.8	60.20	0.262	73.75	84.90	0.321	0.369	…
900	914.7	62.2	3.95	43.4–214.6	60.80	0.265	74.80	86.00	0.326	0.375	…
950	964.7	65.6	4.03	44.6–224.5	61.72	0.269	76.10	87.50	0.331	0.381	…
1000	1014.7	69.0	4.11	45.7–233.3	62.40	0.272	77.20	88.80	0.336	0.383	…
1050	1064.7	72.4	4.15	46.3–238.3	63.10	0.275	78.10	90.10	0.340	0.391	…
1100	1114.7	75.8	4.23	47.5–248.3	63.80	0.278	79.10	91.10	0.344	0.396	…
1150	1164.7	79.2	4.30	48.5–256.8	64.40	0.281	80.15	92.20	0.349	0.401	…
1200	1214.7	82.6	4.33	49.0–261.3	65.00	0.283	81.00	93.15	0.353	0.405	…
1250	1264.7	86.0	4.42	50.3–272.3	65.60	0.286	82.00	94.30	0.357	0.411	…
1300	1314.7	89.4	4.48	51.3–280.8	66.30	0.289	82.90	95.30	0.362	0.416	…
1350	1364.7	92.8	4.53	52.0–287.3	66.70	0.291	84.00	96.60	0.366	0.421	…
1400	1414.7	96.2	4.58	52.6–293.5	67.00	0.292	84.60	97.30	0.368	0.423	…
1450	1464.7	99.6	4.64	53.5–301.5	67.70	0.295	85.30	98.20	0.371	0.426	…
1500	1514.7	103.0	4.69	54.3–309.3	68.30	0.298	85.80	98.80	0.374	0.430	…
1550	1564.7	106.4	4.74	55.0–317.3	68.80	0.300	86.80	99.85	0.378	0.434	…
1600	1614.7	109.8	4.79	55.8–323.3	69.10	0.302	87.60	100.80	0.382	0.438	…

[a] Mean Effective Pressure (MEP) is defined as that single pressure rise, above atmospheric, which would require the same horsepower as the actual varying pressures during compression.

Flow of Air in Pipes.—The following formulas are used:

$$v = \sqrt{\frac{25,000\ dp}{L}} \qquad p = \frac{Lv^2}{25,000\ d}$$

where v = velocity of air in feet per second
p = loss of pressure due to flow through the pipes in ounces per square inch
d = inside diameter of pipe in inches
L = length of pipe in feet

The quantity of air discharged in cubic feet per second is the product of the velocity as obtained from the preceding formula and the area of the pipe in square feet. The horsepower required to drive air through a pipe equals the volume of air in cubic feet per second multiplied by the pressure in pounds per square foot, and this product divided by 550.

Volume of Air Transmitted Through Pipes, in Cubic Feet per Minute

Velocity of Air in Feet per Second	Actual Inside Diameter of Pipe, Inches									
	1	2	3	4	6	8	10	12	16	24
1	0.33	1.31	2.95	5.2	11.8	20.9	32.7	47.1	83.8	188
2	0.65	2.62	5.89	10.5	23.6	41.9	65.4	94.2	167.5	377
3	0.98	3.93	8.84	15.7	35.3	62.8	98.2	141.4	251.3	565
4	1.31	5.24	11.78	20.9	47.1	83.8	131.0	188.0	335.0	754
5	1.64	6.55	14.7	26.2	59.0	104.0	163.0	235.0	419.0	942
6	1.96	7.85	17.7	31.4	70.7	125.0	196.0	283.0	502.0	1131
7	2.29	9.16	20.6	36.6	82.4	146.0	229.0	330.0	586.0	1319
8	2.62	10.50	23.5	41.9	94.0	167.0	262.0	377.0	670.0	1508
9	2.95	11.78	26.5	47.0	106.0	188.0	294.0	424.0	754.0	1696
10	3.27	13.1	29.4	52.0	118.0	209.0	327.0	471.0	838.0	1885
12	3.93	15.7	35.3	63.0	141.0	251.0	393.0	565.0	1005.0	2262
15	4.91	19.6	44.2	78.0	177.0	314.0	491.0	707.0	1256.0	2827
18	5.89	23.5	53.0	94.0	212.0	377.0	589.0	848.0	1508.0	3393
20	6.55	26.2	59.0	105.0	235.0	419.0	654.0	942.0	1675.0	3770
24	7.86	31.4	71.0	125.0	283.0	502.0	785.0	1131.0	2010.0	4524
25	8.18	32.7	73.0	131.0	294.0	523.0	818.0	1178.0	2094.0	4712
28	9.16	36.6	82.0	146.0	330.0	586.0	916.0	1319.0	2346.0	5278
30	9.80	39.3	88.0	157.0	353.0	628.0	982.0	1414.0	2513.0	5655

Flow of Compressed Air in Pipes.—When there is a comparatively small difference of pressure at the two ends of the pipe, the volume of flow in cubic feet per minute is found by the formula:

$$V = 58\sqrt{\frac{pd^5}{WL}}$$

where V = volume of air in cubic feet per minute
p = difference in pressure at the two ends of the pipe in pounds per square inch
d = inside diameter of pipe in inches
W = weight in pounds of one cubic foot of entering air
L = length of pipe in feet

Velocity of Escaping Compressed Air.—If air, or gas, flows from one chamber to another, as from a chamber or tank through an orifice or nozzle into the open air, large changes in velocity may take place owing to the difference in pressures. Since the change takes place almost instantly, little heat can escape from the fluid and the flow may be assumed to be adiabatic.

For a large container with a small orifice or hole from which the air escapes, the velocity of escape (theoretical) may be calculated from the formula:

$$v_2 = \sqrt{2g \cdot \frac{k}{k-1} \cdot 53.3(459.7 + F)\left[1 - \left(\frac{p_2}{p_1}\right)^{\frac{k-1}{k}}\right]}$$

In this formula, v_2 = velocity of escaping air in feet per second; g = acceleration due to gravity, 32.16 feet per second squared; k = 1.41 for adiabatic expansion or compression of air; F = temperature, degrees F; p_2 = atmospheric pressure = 14.7 pounds per square inch; and p_1 = pressure of air in container, pounds per square inch. In applying the preceding formula, when the ratio p_2/p_1 approximately equals 0.53, under normal temperature conditions at sea level, the escape velocity v_2 will be equal to the velocity of sound. Increasing the pressure p_1 will not increase the velocity of escaping air beyond this limiting velocity unless a special converging diverging nozzle design is used rather than an orifice.

The accompanying table provides velocity of escaping air for various values of p_1. These values were calculated from the preceding formula simplified by substituting the appropriate constants:

$$v_2 = 108.58\sqrt{(459.7 + F)\left[1 - \left(\frac{14.7}{p_1}\right)^{0.29}\right]}$$

Velocity of Escaping Air at 70-Degrees F

Pressure Above Atmospheric Pressure			Theoretical Velocity, Feet per Second	Pressure Above Atmospheric Pressure			Theoretical Velocity, Feet per Second
In Atmo-spheres	In Inches Mercury	In lbs per sq. in.		In Atmo-spheres	In Inches Mercury	In lbs per sq. in.	
0.010	0.30	0.147	134	0.408	12.24	6.00	769
0.068	2.04	1.00	344	0.500	15.00	7.35	833
0.100	3.00	1.47	413	0.544	16.33	8.00	861
0.136	4.08	2.00	477	0.612	18.37	9.00	900
0.204	6.12	3.00	573	0.680	20.41	10.0	935
0.272	8.16	4.00	650	0.816	24.49	12.0	997
0.340	10.20	5.00	714	0.884	26.53	13.0	1025

The theoretical velocities in the preceding table must be reduced by multiplying by a "coefficient of discharge," which varies with the orifice and the pressure. The following coefficients are used for orifices in thin plates and short tubes.

Type of Orifice	Pressures in Atmospheres Above Atmospheric Pressure			
	0.01	0.1	0.5	1
Orifice in thin plate	0.65	0.64	0.57	0.54
Orifice in short tube	0.83	0.82	0.71	0.67

STANDARD STEELS

Standard Steels—Compositions, Applications, and Heat Treatments

Steel is the generic term for a large family of iron–carbon alloys, which are malleable, within some temperature range, immediately after solidification from the molten state. The principal raw materials used in steelmaking are iron ore, coal, and limestone. These materials are converted in a blast furnace into a product known as "pig iron," which contains considerable amounts of carbon, manganese, sulfur, phosphorus, and silicon. Pig iron is hard, brittle, and unsuitable for direct processing into wrought forms. Steelmaking is the process of refining pig iron as well as iron and steel scrap by removing undesirable elements from the melt and then adding desirable elements in predetermined amounts. A primary reaction in most steelmaking is the combination of carbon with oxygen to form a gas. If dissolved oxygen is not removed from the melt prior to or during pouring, the gaseous products continue to evolve during solidification. If the steel is strongly deoxidized by the addition of deoxidizing elements, no gas is evolved, and the steel is called "killed" because it lies quietly in the molds. Increasing degrees of gas evolution (decreased deoxidation) characterize steels called "semikilled", "capped," or "rimmed." The degree of deoxidation affects some of the properties of the steel. In addition to oxygen, liquid steel contains measurable amounts of dissolved hydrogen and nitrogen. For some critical steel applications, special deoxidation practices as well as vacuum treatments may be used to reduce and control dissolved gases.

The carbon content of common steel grades ranges from a few hundredths of a per cent to about 1 per cent. All steels also contain varying amounts of other elements, principally manganese, which acts as a deoxidizer and facilitates hot working. Silicon, phosphorus, and sulfur are also always present, if only in trace amounts. Other elements may be present, either as residuals that are not intentionally added, but result from the raw materials or steelmaking practice, or as alloying elements added to effect changes in the properties of the steel.

Steels can be cast to shape, or the cast ingot or strand can be reheated and hot worked by rolling, forging, extrusion, or other processes into a wrought mill shape. Wrought steels are the most widely used of engineering materials, offering a multitude of forms, finishes, strengths, and usable temperature ranges. No other material offers comparable versatility for product design.

Numbering Systems for Metals and Alloys.—Several different numbering systems have been developed for metals and alloys by various trade associations, professional engineering societies, standards organizations, and by private industries for their own use. The numerical code used to identify the metal or alloy may or may not be related to a specification, which is a statement of the technical and commercial requirements that the product must meet. Numbering systems in use include those developed by the American Iron and Steel Institute (AISI), Society of Automotive Engineers (SAE), American Society for Testing and Materials (ASTM), American National Standards Institute (ANSI), Steel Founders Society of America, American Society of Mechanical Engineers (ASME), American Welding Society (AWS), Aluminum Association, Copper Development Association, U.S. Department of Defense (Military Specifications), and the General Accounting Office (Federal Specifications).

The Unified Numbering System (UNS) was developed through a joint effort of the ASTM and the SAE to provide a means of correlating the different numbering systems for metals and alloys that have a commercial standing. This system avoids the confusion caused when more than one identification number is used to specify the same material, or when the same number is assigned to two entirely different materials. It is important to understand that a UNS number is not a specification; it is an identification number for metals and alloys for which detailed specifications are provided elsewhere. UNS numbers are shown in Table 1; each number consists of a letter prefix followed by five digits. In some, the letter is suggestive of the family of metals identified by the series, such as A for aluminum and C for copper. Whenever possible, the numbers in the UNS groups contain numbering sequences taken directly from other systems to facilitate identification of the material; e.g., the corresponding UNS number for AISI 1020 steel is G10200. The UNS numbers corresponding to the commonly used AISI-SAE numbers that are used to identify plain carbon alloy and tool steels are given in Table 1.

Table 1. Unified Numbering System (UNS) for Metals and Alloys

UNS Series	Metal
A00001 to A99999	Aluminum and aluminum alloys
C00001 to C99999	Copper and copper alloys
D00001 to D99999	Specified mechanical property steels
E00001 to E99999	Rare earth and rare earthlike metals and alloys
F00001 to F99999	Cast irons
G00001 to G99999	AISI and SAE carbon and alloy steels (except tool steels)
H00001 to H99999	AISI and SAE H-steels
J00001 to J99999	Cast steels (except tool steels)
K00001 to K99999	Miscellaneous steels and ferrous alloys
L00001 to L99999	Low-melting metals and alloys
M00001 to M99999	Miscellaneous nonferrous metals and alloys
N00001 to N99999	Nickel and nickel alloys
P00001 to P99999	Precious metals and alloys
R00001 to R99999	Reactive and refractory metals and alloys
S00001 to S99999	Heat and corrosion resistant (stainless) steels
T00001 to T99999	Tool steels, wrought and cast
W00001 to W99999	Welding filler metals
Z00001 to Z99999	Zinc and zinc alloys

Table 1. AISI and SAE Numbers and Their Corresponding UNS Numbers for Plain Carbon, Alloy, and Tool Steels

AISI-SAE Numbers	UNS Numbers	AISI-SAE Numbers	UNS Numbers	AISI-SAE Numbers	UNS Numbers	AISI-SAE Numbers	UNS Numbers
Plain Carbon Steels							
1005	G10050	1030	G10300	1070	G10700	1566	G15660
1006	G10060	1035	G10350	1078	G10780	1110	G11100
1008	G10080	1037	G10370	1080	G10800	1117	G11170
1010	G10100	1038	G10380	1084	G10840	1118	G11180
1012	G10120	1039	G10390	1086	G10860	1137	G11370
1015	G10150	1040	G10400	1090	G10900	1139	G11390
1016	G10160	1042	G10420	1095	G10950	1140	G11400
1017	G10170	1043	G10430	1513	G15130	1141	G11410
1018	G10180	1044	G10440	1522	G15220	1144	G11440
1019	G10190	1045	G10450	1524	G15240	1146	G11460
1020	G10200	1046	G10460	1526	G15260	1151	G11510
1021	G10210	1049	G10490	1527	G15270	1211	G12110
1022	G10220	1050	G10500	1541	G15410	1212	G12120
1023	G10230	1053	G10530	1548	G15480	1213	G12130
1025	G10250	1055	G10550	1551	G15510	1215	G12150
1026	G10260	1059	G10590	1552	G15520	12L14	G12144
1029	G10290	1060	G10600	1561	G15610	…	…
Alloy Steels							
1330	G13300	4150	G41500	5140	G51400	8642	G86420
1335	G13350	4161	G41610	5150	G51500	8645	G86450
1340	G13400	4320	G43200	5155	G51550	8655	G86550
1345	G13450	4340	G43400	5160	G51600	8720	G87200
4023	G40230	E4340	G43406	E51100	G51986	8740	G87400
4024	G40240	4615	G46150	E52100	G52986	8822	G88220
4027	G40270	4620	G46200	6118	G61180	9260	G92600
4028	G40280	4626	G46260	6150	G61500	50B44	G50441
4037	G40370	4720	G47200	8615	G86150	50B46	G50461
4047	G40470	4815	G48150	8617	G86170	50B50	G50501
4118	G41180	4817	G48170	8620	G86200	50B60	G50601
4130	G41300	4820	G48200	8622	G86220	51B60	G51601
4137	G41370	5117	G51170	8625	G86250	81B45	G81451
4140	G41400	5120	G51200	8627	G86270	94B17	G94171
4142	G41420	5130	G51300	8630	G86300	94B30	G94301
4145	G41450	5132	G51320	8637	G86370	…	…
4147	G41470	5135	G51350	8640	G86400	…	…
Tool Steels (AISI and UNS Only)							
M1	T11301	T6	T12006	A6	T30106	P4	T51604
M2	T11302	T8	T12008	A7	T30107	P5	T51605
M4	T11304	T15	T12015	A8	T30108	P6	T51606
M6	T11306	H10	T20810	A9	T30109	P20	T51620
M7	T11307	H11	T20811	A10	T30110	P21	T51621
M10	T11310	H12	T20812	D2	T30402	F1	T60601
M3-1	T11313	H13	T20813	D3	T30403	F2	T60602
M3-2	T11323	H14	T20814	D4	T30404	L2	T61202
M30	T11330	H19	T20819	D5	T30405	L3	T61203
M33	T11333	H21	T20821	D7	T30407	L6	T61206
M34	T11334	H22	T20822	O1	T31501	W1	T72301
M36	T11336	H23	T20823	O2	T31502	W2	T72302
M41	T11341	H24	T20824	O6	T31506	W5	T72305
M42	T11342	H25	T20825	O7	T31507	CA2	T90102
M43	T11343	H26	T20826	S1	T41901	CD2	T90402
M44	T11344	H41	T20841	S2	T41902	CD5	T90405
M46	T11346	H42	T20842	S4	T41904	CH12	T90812
M47	T11347	H43	T20843	S5	T41905	CH13	T90813
T1	T12001	A2	T30102	S6	T41906	CO1	T91501
T2	T12002	A3	T30103	S7	T41907	CS5	T91905
T4	T12004	A4	T30104	P2	T51602	…	…
T5	T12005	A5	T30105	P3	T51603	…	…

Standard Steel Classification.—Wrought steels may be classified systematically into groups based on some common characteristic, such as chemical composition, deoxidation practice, finishing method, or product form. Chemical composition is the most often used basis for identifying and assigning standard designations to wrought steels. Although carbon is the principal hardening and strengthening element in steel, no single element controls the steel's characteristics. The combined effect of several elements influences response to heat treatment, hardness, strength, microstructure, corrosion resistance, and formability. The standard steels can be divided broadly into three main groups: carbon steels, alloy steels, and stainless steels.

Carbon Steels: A steel qualifies as a carbon steel when its manganese content is limited to 1.65 per cent (max), silicon to 0.60 per cent (max), and copper to 0.60 per cent (max). With the exception of deoxidizers and boron when specified, no other alloying elements are added intentionally, but they may be present as residuals. If any of these incidental elements are considered detrimental for special applications, maximum acceptable limits may be specified. In contrast to most alloy steels, carbon steels are most often used without a final heat treatment; however, they may be annealed, normalized, case hardened, or quenched and tempered to enhance fabrication or mechanical properties. Carbon steels may be killed, semikilled, capped, or rimmed, and, when necessary, the method of deoxidation may be specified.

Alloy Steels: Alloy steels comprise not only those grades that exceed the element content limits for carbon steel, but also any grade to which different elements than used for carbon steel are added, within specific ranges or specific minimums, to enhance mechanical properties, fabricating characteristics, or any other attribute of the steel. By this definition, alloy steels encompass all steels other than carbon steels; however, by convention, steels containing over 3.99 per cent chromium are considered "special types" of alloy steel, which include the stainless steels and many of the tool steels.

In a technical sense, the term alloy steel is reserved for those steels that contain a modest amount of alloying elements (about 1–4 per cent) and generally depend on thermal treatments to develop specific mechanical properties. Alloy steels are always killed, but special deoxidation or melting practices, including vacuum, may be specified for special critical applications. Alloy steels generally require additional care throughout their manufacture, because they are more sensitive to thermal and mechanical operations.

Stainless Steels: Stainless steels are high-alloy steels and have superior corrosion resistance to the carbon and conventional low-alloy steels because they contain relatively large amounts of chromium. Although other elements may also increase corrosion resistance, their usefulness in this respect is limited.

Stainless steels generally contain at least 10 per cent chromium, with or without other elements. It has been customary in the United States, however, to include in the stainless steel classification those steels that contain as little as 4 per cent chromium. Together, these steels form a family known as the stainless and heat-resisting steels, some of which possess very high strength and oxidation resistance. Few, however, contain more than 30 per cent chromium or less than 50 per cent iron.

In the broadest sense, the standard stainless steels can be divided into three groups based on their structures: austenitic, ferritic, and martensitic. In each of the three groups, there is one composition that represents the basic, general-purpose alloy. All other compositions are derived from the basic alloy, with specific variations in composition being made to obtain very specific properties.

The *austenitic grades* are nonmagnetic in the annealed condition, although some may become slightly magnetic after cold working. They can be hardened only by cold working, and not by heat treatment, and combine outstanding corrosion and heat resistance with good mechanical properties over a wide temperature range. The austenitic grades are further classified into two subgroups: the chromium–nickel types and the less frequently used

Table 2. Composition of AISI-SAE Standard Carbon Steels

AISI-SAE No.	UNS No.	Composition(%)[a]			
		C	Mn	P(max)[b]	S(max)[b]
Nonresulfurized Grades — 1 per cent Mn (max)					
1005[c]	G10050	0.06 max	0.35 max	0.040	0.050
1006[c]	G10060	0.08 max	0.25–0.40	0.040	0.050
1008	G10080	0.10 max	0.30–0.50	0.040	0.050
1010	G10100	0.08–0.13	0.30–0.60	0.040	0.050
1012	G10120	0.10–0.15	0.30–0.60	0.040	0.050
1015	G10150	0.13–0.18	0.30–0.60	0.040	0.050
1016	G10160	0.13–0.18	0.60–0.90	0.040	0.050
1017	G10170	0.15–0.20	0.30–0.60	0.040	0.050
1018	G10180	0.15–0.20	0.60–0.90	0.040	0.050
1019	G10190	0.15–0.20	0.70–1.00	0.040	0.050
1020	G10200	0.18–0.23	0.30–0.60	0.040	0.050
1021	G10210	0.18–0.23	0.60–0.90	0.040	0.050
1022	G10220	0.18–0.23	0.70–1.00	0.040	0.050
1023	G10230	0.20–0.25	0.30–0.60	0.040	0.050
1025	G10250	0.22–0.28	0.30–0.60	0.040	0.050
1026	G10260	0.22–0.28	0.60–0.90	0.040	0.050
1029	G10290	0.25–0.31	0.60–0.90	0.040	0.050
1030	G10300	0.28–0.34	0.60–0.90	0.040	0.050
1035	G10350	0.32–0.38	0.60–0.90	0.040	0.050
1037	G10370	0.32–0.38	0.70–1.00	0.040	0.050
1038	G10380	0.35–0.42	0.60–0.90	0.040	0.050
1039	G10390	0.37–0.44	0.70–1.00	0.040	0.050
1040	G10400	0.37–0.44	0.60–0.90	0.040	0.050
1042	G10420	0.40–0.47	0.60–0.90	0.040	0.050
1043	G10430	0.40–0.47	0.70–1.00	0.040	0.050
1044	G10440	0.43–0.50	0.30–0.60	0.040	0.050
1045	G10450	0.43–0.50	0.60–0.90	0.040	0.050
1046	G10460	0.43–0.50	0.70–1.00	0.040	0.050
1049	G10490	0.46–0.53	0.60–0.90	0.040	0.050
1050	G10500	0.48–0.55	0.60–0.90	0.040	0.050
1053	G10530	0.48–0.55	0.70–1.00	0.040	0.050
1055	G10550	0.50–0.60	0.60–0.90	0.040	0.050
1059[c]	G10590	0.55–0.65	0.50–0.80	0.040	0.050
1060	G10600	0.55–0.65	0.60–0.90	0.040	0.050
1064[c]	G10640	0.60–0.70	0.50–0.80	0.040	0.050
1065[c]	G10650	0.60–0.70	0.60–0.90	0.040	0.050
1069[c]	G10690	0.65–0.75	0.40–0.70	0.040	0.050
1070	G10700	0.65–0.75	0.60–0.90	0.040	0.050
1078	G10780	0.72–0.85	0.30–0.60	0.040	0.050
1080	G10800	0.75–0.88	0.60–0.90	0.040	0.050
1084	G10840	0.80–0.93	0.60–0.90	0.040	0.050
1086[c]	G10860	0.80–0.93	0.30–0.50	0.040	0.050
1090	G10900	0.85–0.98	0.60–0.90	0.040	0.050
1095	G10950	0.90–1.03	0.30–0.50	0.040	0.050

Table 2. *(Continued)* **Composition of AISI-SAE Standard Carbon Steels**

AISI-SAE No.	UNS No.	Composition(%)[a]			
		C	Mn	P(max)[b]	S(max)[b]
Nonresulfurized Grades — Over 1 per cent Mn					
1513	G15130	0.10–0.16	1.10–1.40	0.040	0.050
1522	G15220	0.18–0.24	1.10–1.40	0.040	0.050
1524	G15240	0.19–0.25	1.35–1.65	0.040	0.050
1526	G15260	0.22–0.29	1.10–1.40	0.040	0.050
1527	G15270	0.22–0.29	1.20–1.50	0.040	0.050
1541	G15410	0.36–0.44	1.35–1.65	0.040	0.050
1548	G15480	0.44–0.52	1.10–1.40	0.040	0.050
1551	G15510	0.45–0.56	0.85–1.15	0.040	0.050
1552	G15520	0.47–0.55	1.20–1.50	0.040	0.050
1561	G15610	0.55–0.65	0.75–1.05	0.040	0.050
1566	G15660	0.60–0.71	0.85–1.15	0.040	0.050
Free-Machining Grades — Resulfurized					
1110	G11100	0.08–0.13	0.30–0.60	0.040	0.08–0.13
1117	G11170	0.14–0.20	1.00–1.30	0.040	0.08–0.13
1118	G11180	0.14–0.20	1.30–1.60	0.040	0.08–0.13
1137	G11370	0.32–0.39	1.35–1.65	0.040	0.08–0.13
1139	G11390	0.35–0.43	1.35–1.65	0.040	0.13–0.20
1140	G11400	0.37–0.44	0.70–1.00	0.040	0.08–0.13
1141	G11410	0.37–0.45	1.35–1.65	0.040	0.08–0.13
1144	G11440	0.40–0.48	1.35–1.65	0.040	0.24–0.33
1146	G11460	0.42–0.49	0.70–1.00	0.040	0.08–0.13
1151	G11510	0.48–0.55	0.70–1.00	0.040	0.08–0.13
Free-Machining Grades — Resulfurized and Rephosphorized					
1211	G12110	0.13 max	0.60–0.90	0.07–0.12	0.10–0.15
1212	G12120	0.13 max	0.70–1.00	0.07–0.12	0.16–0.23
1213	G12130	0.13 max	0.70–1.00	0.07–0.12	0.24–0.33
1215	G12150	0.09 max	0.75–1.05	0.04–0.09	0.26–0.35
12L14[d]	G12144	0.15 max	0.85–1.15	0.04–0.09	0.26–0.35

[a] The following notes refer to boron, copper, lead, and silicon additions: Boron: Standard killed carbon steels, which are generally fine grain, may be produced with a boron treatment addition to improve hardenability. Such steels are produced to a range of 0.0005–0.003 per cent B. These steels are identified by inserting the letter "B" between the second and third numerals of the AISI or SAE number, e.g., 10B46. Copper: When copper is required, 0.20 per cent (min) is generally specified. Lead: Standard carbon steels can be produced with a lead range of 0.15–0.35 per cent to improve machinability. Such steels are identified by inserting the letter "L" between the second and third numerals of the AISI or SAE number, e.g., 12L15 and 10L45. Silicon: It is not common practice to produce the 12XX series of resulfurized and rephosphorized steels to specified limits for silicon because of its adverse effect on machinability. When silicon ranges or limits are required for resulfurized or nonresulfurized steels, however, these values apply: a range of 0.08 per cent Si for Si max up to 0.15 per cent inclusive, a range of 0.10 per cent Si for Si max over 0.15 to 0.20 per cent inclusive, a range of 0.15 per cent Si for Si max over 0.20 to 0.30 per cent inclusive, and a range of 0.20 per cent Si for Si max over 0.30 to 0.60 per cent inclusive. Example: Si max is 0.25 per cent, range is 0.10–0.25 per cent.

[b] Values given are maximum percentages, except where a range of values is given.

[c] Standard grades for wire rods and wire only.

[d] 0.15–0.35 per cent Pb.

chromium–manganese–low–nickel types. The basic composition in the chromium–nickel group is widely known as 18–8 (Cr–Ni) and is the general-purpose austenitic grade. This grade is the basis for over 20 modifications that can be characterized as follows: the chromium–nickel ratio has been modified to change the forming characteristics; the carbon content has been decreased to prevent intergranular corrosion; the elements niobium or titanium have been added to stabilize the structure; or molybdenum has been added or the chromium and nickel contents have been increased to improve corrosion or oxidation resistance.

The standard *ferritic grades* are always magnetic and contain chromium but no nickel. They can be hardened to some extent by cold working, but not by heat treatment, and they combine corrosion and heat resistance with moderate mechanical properties and decorative appeal. The ferritic grades generally are restricted to a narrower range of corrosive conditions than the austenitic grades. The basic ferritic grade contains 17 per cent chromium. In this series, there are free-machining modifications and grades with increased chromium content to improve scaling resistance. Also in this ferritic group is a 12 per cent chromium steel (the basic composition of the martensitic group) with other elements, such as aluminum or titanium, added to prevent hardening.

The standard *martensitic grades* are magnetic and can be hardened by quenching and tempering. They contain chromium and, with two exceptions, no nickel. The basic martensitic grade normally contains 12 per cent chromium. There are more than 10 standard compositions in the martensitic series; some are modified to improve machinability and others have small additions of nickel or other elements to improve the mechanical properties or their response to heat treatment. Still others have greatly increased carbon content, in the tool steel range, and are hardenable to the highest levels of all the stainless steels. The martensitic grades are excellent for service in mild environments such as the atmosphere, freshwater, steam, and weak acids, but are not resistant to severely corrosive solutions.

Standard Steel Numbering System.—The most widely used systems for identifying wrought carbon, low–alloy, and stainless steels are based on chemical composition, and are those of the American Iron and Steel Institute (AISI) and the Society of Automotive Engineers (SAE). These systems are almost identical, but they are carefully coordinated. The standard steels so designated have been developed cooperatively by producers and users and have been found through long experience to cover most of the wrought ferrous metals used in automotive vehicles and related equipment. These designations, however, are not specifications, and should not be used for purchasing unless accompanied by supplementary information necessary to describe commercially the product desired. Engineering societies, associations, and institutes whose members make, specify, or purchase steel products publish standard specifications, many of which have become well known and respected. The most comprehensive and widely used specifications are those published by the American Society for Testing and Materials (ASTM). The U.S. government and various companies also publish their own specification for steel products to serve their own special procurement needs. The Unified Numbering System (UNS) for metals and alloys is also used to designate steels (see pages 403 and 405).

The numerical designation system used by both AISI and SAE for wrought carbon, alloy, and stainless steels is summarized in Table . Table 2 lists the compositions of the standard carbon steels; Table 4 lists the standard low–alloy steel compositions; and Table 5 includes the typical compositions of the standard stainless steels.

Table 3. AISI-SAE System of Designating Carbon and Alloy Steels

AISI-SAE Designation[a]		Type of Steel and Nominal Alloy Content (%)
		Carbon Steels
10xx		Plain Carbon (Mn 1.00% max.)
11xx		Resulfurized
12xx		Resulfurized and Rephosphorized
15xx		Plain Carbon (Max. Mn range 1.00 to 1.65%)
		Manganese Steels
13xx		Mn 1.75
		Nickel Steels
23xx		Ni 3.50
25xx		Ni 5.00
		Nickel–Chromium Steels
31xx		Ni 1.25; Cr 0.65 and 0.80
32xx		Ni 1.75; Cr 1.07
33xx		Ni 3.50; Cr 1.50 and 1.57
34xx		Ni 3.00; Cr 0.77
		Molybdenum Steels
40xx		Mo 0.20 and 0.25
44xx		Mo 0.40 and 0.52
		Chromium–Molybdenum Steels
41xx		Cr 0.50, 0.80, and 0.95; Mo 0.12, 0.20, 0.25, and 0.30
		Nickel–Chromium–Molybdenum Steels
43xx		Ni 1.82; Cr 0.50 and 0.80; Mo 0.25
43BVxx		Ni 1.82; Cr 0.50; Mo 0.12 and 0.35; V 0.03 min.
47xx		Ni 1.05; Cr 0.45; Mo 0.20 and 0.35
81xx		Ni 0.30; Cr 0.40; Mo 0.12
86xx		Ni 0.55; Cr 0.50; Mo 0.20
87xx		Ni 0.55; Cr 0.50; Mo 0.25
88xx		Ni 0.55; Cr 0.50; Mo 0.35
93xx		Ni 3.25; Cr 1.20; Mo 0.12
94xx		Ni 0.45; Cr 0.40; Mo 0.12
97xx		Ni 0.55; Cr 0.20; Mo 0.20
98xx		Ni 1.00; Cr 0.80; Mo 0.25
		Nickel–Molybdenum Steels
46xx		Ni 0.85 and 1.82; Mo 0.20 and 0.25
48xx		Ni 3.50; Mo 0.25
		Chromium Steels
50xx		Cr 0.27, 0.40, 0.50, and 0.65
51xx		Cr 0.80, 0.87, 0.92, 0.95, 1.00, and 1.05
50xxx		Cr 0.50; C 1.00 min.
51xxx		Cr 1.02; C 1.00 min.
52xxx		Cr 1.45; C 1.00 min.
		Chromium–Vanadium Steels
61xx		Cr 0.60, 0.80, and 0.95; V 0.10 and 0.15 min
		Tungsten–Chromium Steels
72xx		W 1.75; Cr 0.75
		Silicon–Manganese Steels
92xx		Si 1.40 and 2.00; Mn 0.65, 0.82, and 0.85; Cr 0.00 and 0.65
		High-Strength Low-Alloy Steels
9xx		Various SAE grades
xxBxx		B denotes boron steels
xxLxx		L denotes leaded steels
AISI	SAE	Stainless Steels
2xx	302xx	Chromium–Manganese–Nickel Steels
3xx	303xx	Chromium–Nickel Steels
4xx	514xx	Chromium Steels
5xx	515xx	Chromium Steels

[a] xx in the last two digits of the carbon and low–alloy designations (but not the stainless steels) indicates that the carbon content (in hundredths of a per cent) is to be inserted.

Table 4. Compositions of AISI-SAE Standard Alloy Steels

AISI-SAE No.	UNS No.	Composition (%)[a,b]							
		C	Mn	P (max)	S (max)	Si	Ni	Cr	Mo
1330	G13300	0.28–0.33	1.60–1.90	0.035	0.040	0.15–0.35	…	…	…
1335	G13350	0.33–0.38	1.60–1.90	0.035	0.040	0.15–0.35	…	…	…
1340	G13400	0.38–0.43	1.60–1.90	0.035	0.040	0.15–0.35	…	…	…
1345	G13450	0.43–0.48	1.60–1.90	0.035	0.040	0.15–0.35	…	…	…
4023	G40230	0.20–0.25	0.70–0.90	0.035	0.040	0.15–0.35	…	…	0.20–0.30
4024	G40240	0.20–0.25	0.70–0.90	0.035	0.035–0.050	0.15–0.35	…	…	0.20–0.30
4027	G40270	0.25–0.30	0.70–0.90	0.035	0.040	0.15–0.35	…	…	0.20–0.30
4028	G40280	0.25–0.30	0.70–0.90	0.035	0.035–0.050	0.15–0.35	…	…	0.20–0.30
4037	G40370	0.35–0.40	0.70–0.90	0.035	0.040	0.15–0.35	…	…	0.20–0.30
4047	G40470	0.45–0.50	0.70–0.90	0.035	0.040	0.15–0.35	…	…	0.20–0.30
4118	G41180	0.18–0.23	0.70–0.90	0.035	0.040	0.15–0.35	…	0.40–0.60	0.08–0.15
4130	G41300	0.28–0.33	0.40–0.60	0.035	0.040	0.15–0.35	…	0.80–1.10	0.15–0.25
4137	G41370	0.35–0.40	0.70–0.90	0.035	0.040	0.15–0.35	…	0.80–1.10	0.15–0.25
4140	G41400	0.38–0.43	0.75–1.00	0.035	0.040	0.15–0.35	…	0.80–1.10	0.15–0.25
4142	G41420	0.40–0.45	0.75–1.00	0.035	0.040	0.15–0.35	…	0.80–1.10	0.15–0.25
4145	G41450	0.43–0.48	0.75–1.00	0.035	0.040	0.15–0.35	…	0.80–1.10	0.15–0.25
4147	G41470	0.45–0.50	0.75–1.00	0.035	0.040	0.15–0.35	…	0.80–1.10	0.15–0.25
4150	G41500	0.48–0.53	0.75–1.00	0.035	0.040	0.15–0.35	…	0.80–1.10	0.15–0.25
4161	G41610	0.56–0.64	0.75–1.00	0.035	0.040	0.15–0.35	…	0.70–0.90	0.25–0.35
4320	G43200	0.17–0.22	0.45–0.65	0.035	0.040	0.15–0.35	1.65–2.00	0.40–0.60	0.20–0.30
4340	G43400	0.38–0.43	0.60–0.80	0.035	0.040	0.15–0.35	1.65–2.00	0.70–0.90	0.20–0.30
E4340[c]	G43406	0.38–0.43	0.65–0.85	0.025	0.025	0.15–0.35	1.65–2.00	0.70–0.90	0.20–0.30
4615	G46150	0.13–0.18	0.45–0.65	0.035	0.040	0.15–0.35	1.65–2.00	…	0.20–0.30
4620	G46200	0.17–0.22	0.45–0.65	0.035	0.040	0.15–0.35	1.65–2.00	…	0.20–0.30
4626	G46260	0.24–0.29	0.45–0.65	0.035	0.040	0.15–0.35	0.70–1.00	…	0.15–0.25
4720	G47200	0.17–0.22	0.50–0.70	0.035	0.040	0.15–0.35	0.90–1.20	0.35–0.55	0.15–0.25
4815	G48150	0.13–0.18	0.40–0.60	0.035	0.040	0.15–0.35	3.25–3.75	…	0.20–0.30
4817	G48170	0.15–0.20	0.40–0.60	0.035	0.040	0.15–0.35	3.25–3.75	…	0.20–0.30
4820	G48200	0.18–0.23	0.50–0.70	0.035	0.040	0.15–0.35	3.25–3.75	…	0.20–0.30
5117	G51170	0.15–0.20	0.70–0.90	0.035	0.040	0.15–0.35	…	070–0.90	…
5120	G51200	0.17–0.22	0.70–0.90	0.035	0.040	0.15–0.35	…	0.70–0.90	…
5130	G51300	0.28–0.33	0.70–0.90	0.035	0.040	0.15–0.35	…	0.80–1.10	…
5132	G51320	0.30–0.35	0.60–0.80	0.035	0.040	0.15–0.35	…	0.75–1.00	…
5135	G51350	0.33–0.38	0.60–0.80	0.035	0.040	0.15–0.35	…	0.80–1.05	…
5140	G51400	0.38–0.43	0.70–0.90	0.035	0.040	0.15–0.35	…	0.70–0.90	…
5150	G51500	0.48–0.53	0.70–0.90	0.035	0.040	0.15–0.35	…	0.70–0.90	…
5155	G51550	0.51–0.59	0.70–0.90	0.035	0.040	0.15–0.35	…	0.70–0.90	…
5160	G51600	0.56–0.64	0.75–1.00	0.035	0.040	0.15–0.35	…	0.70–0.90	…

Table 4. *(Continued)* **Compositions of AISI-SAE Standard Alloy Steels**

AISI-SAE No.	UNS No.	Composition (%)[a,b]							
		C	Mn	P (max)	S (max)	Si	Ni	Cr	Mo
E51100[c]	G51986	0.98–1.10	0.25–0.45	0.025	0.025	0.15–0.35	…	0.90–1.15	…
E52100[c]	G52986	0.98–1.10	0.25–0.45	0.025	0.025	0.15–0.35	…	1.30–1.60	…
6118	G61180	0.16–0.21	0.50–0.70	0.035	0.040	0.15–0.35	…	0.50–0.70	0.10–0.15 V
6150	G61500	0.48–0.53	0.70–0.90	0.035	0.040	0.15–0.35	…	0.80–1.10	0.15 V min
8615	G86150	0.13–0.18	0.70–0.90	0.035	0.040	0.15–0.35	0.40–0.70	0.40–0.60	0.15–0.25
8617	G86170	0.15–0.20	0.70–0.90	0.035	0.040	0.15–0.35	0.40–0.70	0.40–0.60	0.15–0.25
8620	G86200	0.18–0.23	0.70–0.90	0.035	0.040	0.15–0.35	0.40–0.70	0.40–0.60	0.15–0.25
8622	G86220	0.20–0.25	0.70–0.90	0.035	0.040	0.15–0.35	0.40–0.70	0.40–0.60	0.15–0.25
8625	G86250	0.23–0.28	0.70–0.90	0.035	0.040	0.15–0.35	0.40–0.70	0.40–0.60	0.15–0.25
8627	G86270	0.25–0.30	0.70–0.90	0.035	0.040	0.15–0.35	0.40–0.70	0.40–0.60	0.15–0.25
8630	G86300	0.28–0.33	0.70–0.90	0.035	0.040	0.15–0.35	0.40–0.70	0.40–0.60	0.15–0.25
8637	G86370	0.35–0.40	0.75–1.00	0.035	0.040	0.15–0.35	0.40–0.70	0.40–0.60	0.15–0.25
8640	G86400	0.38–0.43	0.75–1.00	0.035	0.040	0.15–0.35	0.40–0.70	0.40–0.60	0.15–0.25
8642	G86420	0.40–0.45	0.75–1.00	0.035	0.040	0.15–0.35	0.40–0.70	0.40–0.60	0.15–0.25
8645	G86450	0.43–0.48	0.75–1.00	0.035	0.040	0.15–0.35	0.40–0.70	0.40–0.60	0.15–0.25
8655	G86550	0.51–0.59	0.75–1.00	0.035	0.040	0.15–0.35	0.40–0.70	0.40–0.60	0.15–0.25
8720	G87200	0.18–0.23	0.70–0.90	0.035	0.040	0.15–0.35	0.40–0.70	0.40–0.60	0.20–0.30
8740	G87400	0.38–0.43	0.75–1.00	0.035	0.040	0.15–0.35	0.40–0.70	0.40–0.60	0.20–0.30
8822	G88220	0.20–0.25	0.75–1.00	0.035	0.040	0.15–0.35	0.40–0.70	0.40–0.60	0.30–0.40
9260	G92600	0.56–0.64	0.75–1.00	0.035	0.040	1.80–2.20	…	…	…
Standard Boron Grades[d]									
50B44	G50441	0.43–0.48	0.75–1.00	0.035	0.040	0.15–0.35	…	0.40–0.60	…
50B46	G50461	0.44–0.49	0.75–1.00	0.035	0.040	0.15–0.35	…	0.20–0.35	…
50B50	G50501	0.48–0.53	0.75–1.00	0.035	0.040	0.15–0.35	…	0.40–0.60	…
50B60	G50601	0.56–0.64	0.75–1.00	0.035	0.040	0.15–0.35	…	0.40–0.60	…
51B60	G51601	0.56–0.64	0.75–1.00	0.035	0.040	0.15–0.35	…	0.70–0.90	…
81B45	G81451	0.43–0.48	0.75–1.00	0.035	0.040	0.15–0.35	0.20–0.40	0.35–0.55	0.08–0.15
94B17	G94171	0.15–0.20	0.75–1.00	0.035	0.040	0.15–0.35	0.30–0.60	0.30–0.50	0.08–0.15
94B30	G94301	0.28–0.33	0.75–1.00	0.035	0.040	0.15–0.35	0.30–0.60	0.30–0.50	0.08–0.15

[a] Small quantities of certain elements are present that are not specified or required. These incidental elements may be present to the following maximum amounts: Cu, 0.35 per cent; Ni, 0.25 per cent; Cr, 0.20 per cent; and Mo, 0.06 per cent.

[b] Standard alloy steels can also be produced with a lead range of 0.15–0.35 per cent. Such steels are identified by inserting the letter "L" between the second and third numerals of the AISI or SAE number, e.g., 41L40.

[c] Electric furnace steel.

[d] 0.0005–0.003 per cent B.

Source: American Iron and Steel Institute: *Steel Products Manual.*

Table 5. Standard Stainless Steels — Typical Compositions

AISI Type (UNS)	Typical Composition (%)	AISI Type (UNS)	Typical Composition (%)
Austenitic			
201 (S20100)	16–18 Cr, 3.5–5.5 Ni, 0.15 C, 5.5–7.5 Mn, 0.75 Si, 0.060 P, 0.030 S, 0.25 N	310 (S31000)	24–26 Cr, 19–22 Ni, 0.25 C, 2.0 Mn, 1.5 Si, 0.045 P, 0.030 S
202 (S20200)	17–19 Cr, 4–6 Ni, 0.15 C, 7.5–10.0 Mn, 0.75 Si, 0.060 P, 0.030 S, 0.25 N	310S (S31008)	24–26 Cr, 19–22 Ni, 0.08 C, 2.0 Mn, 1.5 Si, 0.045 P, 0.30 S
205 (S20500)	16.5–18 Cr, 1–1.75 Ni, 0.12–0.25 C, 14–15.5 Mn, 0.75 Si, 0.060 P, 0.030 S, 0.32–0.40 N	314 (S31400)	23–26 Cr, 19–22 Ni, 0.25 C, 2.0 Mn, 1.5–3.0 Si, 0.045 P, 0.030 S
301 (S30100)	16–18 Cr, 6–8 Ni, 0.15 C, 2.0 Mn, 0.75 Si, 0.045 P, 0.030 S	316 (S31600)	16–18 Cr, 10–14 Ni, 0.08 C, 2.0 Mn, 0.75 Si, 0.045 P, 0.030 S, 2.0–3.0 Mo, 0.10 N
302 (S30200)	17–19 Cr, 8–10 Ni, 0.15 C, 2.0 Mn, 0.75 Si, 0.045 P, 0.030 S, 0.10 N	316L (S31603)	16–18 Cr, 10–14 Ni, 0.03 C, 2.0 Mn, 0.75 Si, 0.045 P, 0.030 S, 2.0–3.0 Mo, 0.10 N
302B (S30215)	17–19 Cr, 8–10 Ni, 0.15 C, 2.0 Mn, 2.0–3.0 Si, 0.045 P, 0.030 S	316F (S31620)	16–18 Cr, 10–14 Ni, 0.08 C, 2.0 Mn, 1.0 Si, 0.20 P, 0.10 S min, 1.75–2.50 Mo
303 (S30300)	17–19 Cr, 8–10 Ni, 0.15 C, 2.0 Mn, 1.0 Si, 0.20 P, 0.015 S min, 0.60 Mo (optional)	316N (S31651)	16–18 Cr, 10–14 Ni, 0.08 C, 2.0 Mn, 0.75 Si, 0.045 P, 0.030 S, 2–3 Mo, 0.10–0.16 N
303Se (S30323)	17–19 Cr, 8–10 Ni, 0.15 C, 2.0 Mn, 1.0 Si, 0.20 P, 0.060 S, 0.15 Se min	317 (S31700)	18–20 Cr, 11–15 Ni, 0.08 C, 2.0 Mn, 0.75 Si, 0.045 P, 0.030 S, 3.0–4.0 Mo, 0.10 N max
304 (S30400)	18–20 Cr, 8–10.50 Ni, 0.08 C, 2.0 Mn, 0.75 Si, 0.045 P, 0.030 S, 0.10 N	317L (S31703)	18–20 Cr, 11–15 Ni, 0.03 C, 2.0 Mn, 0.75 Si, 0.045 P, 0.030 S, 3–4 Mo, 0.10 N max
304L (S30403)	18–20 Cr, 8–12 Ni, 0.03 C, 2.0 Mn, 0.75 Si, 0.045 P, 0.030 S, 0.10 N	321 (S32100)	17–19 Cr, 9–12 Ni, 0.08 C, 2.0 Mn, 0.75 Si, 0.045 P, 0.030 S [Ti, 5(C + N) min, 0.70 max], 0.10 max
304 Cu (S30430)	17–19 Cr, 8–10 Ni, 0.08 C, 2.0 Mn, 0.75 Si, 0.045 P, 0.030 S, 3–4 Cu	329 (S32900)	23–28 Cr, 2.5–5 Ni, 0.08 C, 2.0 Mn, 0.75 Si, 0.040 P, 0.030 S, 1–2 Mo
304N (S30451)	18–20 Cr, 8–10.5 Ni, 0.08 C, 2.0 Mn, 0.75 Si, 0.045 P, 0.030 S, 0.10–0.16 N	330 (N08330)	17–20 Cr, 34–37 Ni, 0.08 C, 2.0 Mn, 0.75–1.50 Si, 0.040 P, 0.030 S
305 (S30500)	17–19 Cr, 10.50–13 Ni, 0.12 C, 2.0 Mn, 0.75 Si, 0.045 P, 0.030 S	347 (S34700)	17–19 Cr, 9–13 Ni, 0.08 C, 2.0 Mn, 0.75 Si, 0.045 P, 0.030 S (Nb + Ta, 10 × C min, 1 max)
308 (S30800)	19–21 Cr, 10–12 Ni, 0.08 C, 2.0 Mn, 1.0 Si, 0.045 P, 0.030 S	348 (S34800)	17–19 Cr, 9–13 Ni, 0.08 C, 2.0 Mn, 0.75 Si, 0.045 P, 0.030 S (Nb + Ta, 10 × C min, 1 max, but 0.10 Ta max), 0.20 Ca
309 (S30900)	22–24 Cr, 12–15 Ni, 0.20 C, 2.0 Mn, 1.0 Si, 0.045 P, 0.030 S	384 (S38400)	15–17 Cr, 17–19 Ni, 0.08 C, 2.0 Mn, 1.0 Si, 0.045 P, 0.030 S
309S (S30908)	22–24 Cr, 12–15 Ni, 0.08 C, 2.0 Mn, 1.0 Si, 0.045 P, 0.030 S	…	…
Ferritic			
405 (S40500)	11.5–14.5 Cr, 0.08 C, 1.0 Mn, 1.0 Si, 0.040 P, 0.030 S, 0.1–0.3 Al, 0.60 max	430FSe (S43023)	16–18 Cr, 0.12 C, 1.25 Mn, 1.0 Si, 0.060 P, 0.060 S, 0.15 Se min
409 (S40900)	10.5–11.75 Cr, 0.08 C, 1.0 Mn, 1.0 Si, 0.045 P, 0.030 S, 0.05 Ni (Ti 6 × C, but with 0.75 max)	434 (S43400)	16–18 Cr, 0.12 C, 1.0 Mn, 1.0 Si, 0.040 P, 0.030 S, 0.75–1.25 Mo
429 (S42900)	14–16 Cr, 0.12 C, 1.0 Mn, 1.0 Si, 0.040 P, 030 S, 0.75 Ni	436 (S43600)	16–18 Cr, 0.12 C, 1.0 Mn, 1.0 Si, 0.040 P, 0.030 S, 0.75–1.25 Mo (Nb + Ta 5 × C min, 0.70 max)
430 (S43000)	16–18 Cr, 0.12 C, 1.0 Mn, 1.0 Si, 0.040 P, 030 S, 0.75 Ni	442 (S44200)	18–23 Cr, 0.20 C, 1.0 Mn, 1.0 Si, 0.040 P, 0.030 S
430F (S43020)	16–18 Cr, 0.12 C, 1.25 Mn, 1.0 Si, 0.060 P, 0.15 S min, 0.60 Mo (optional)	446 (S44600)	23–27 Cr, 0.20 C, 1.5 Mn, 1.0 Si, 0.040 P, 0.030 S, 0.025 N

Table 5. *(Continued)* **Standard Stainless Steels — Typical Compositions**

AISI Type (UNS)	Typical Composition (%)	AISI Type (UNS)	Typical Composition (%)
Martensitic			
403 (S40300)	11.5–13.0 Cr, 1.15 C, 1.0 Mn, 0.5 Si, 0.040 P, 0.030 S, 0.60 Ni	420F (S42020)	12–14 Cr, over 0.15 C, 1.25 Mn, 1.0 Si, 0.060 P, 0.15 S min, 0.60 Mo max (optional)
410 (S41000)	11.5–13.5 Cr, 0.15 C, 1.0 Mn, 1.0 Si, 0.040 P, 0.030 S, 0.75 Ni	422 (S42200)	11–12.50 Cr, 0.50–1.0 Ni, 0.20– 0.25 C, 0.50–1.0 Mn, 0.50 Si, 0.025 P, 0.025 S, 0.90–1.25 Mo, 0.20–0.30 V, 0.90–1.25 W
414 (S41400)	11.5–13.5 Cr, 1.25–2.50 Ni, 0.15 C, 1.0 Mn, 1.0 Si, 0.040 P, 0.030 S, 1.25–2.50 Ni	431 (S41623)	15–17 Cr, 1.25–2.50 Ni, 0.20 C, 1.0 Mn, 1.0 Si, 0.040 P, 0.030 S
416 (S41600)	12–14 Cr, 0.15 C, 1.25 Mn, 1.0 Si, 0.060 P, 0.15 S min, 0.060 Mo (optional)	440A (S44002)	16–18 Cr, 0.60–0.75 C, 1.0 Mn, 1.0 Si, 0.040 P, 0.030 S, 0.75 Mo
416Se (S41623)	12–14 Cr, 0.15 C, 1.25 Mn, 1.0 Si, 0.060 P, 0.060 S, 0.15 Se min	440B (S44003)	16–18 Cr, 0.75–0.95 C, 1.0 Mn, 1.0 Si, 0.040 P, 0.030 S, 0.75 Mo
420 (S42000)	12–14 Cr, 0.15 C min, 1.0 Mn, 1.0 Si, 0.040 P, 0.030 S	440C (S44004)	16–18 Cr, 0.95–1.20 C, 1.0 Mn, 1.0 Si, 0040 P, 0.030 S, 0.75 Mo
Heat-Resisting			
501 (S50100)	4–6 Cr, 0.10 C min, 1.0 Mn, 1.0 Si, 0.040 P, 0.030 S, 0.40–0.65 Mo	502 (S50200)	4–6 Cr, 0.10 C. 1.0 Mn, 1.0 Si, 0.040 P, 0.030 S, 0.40–0.65 Mo

Thermal Treatment of Steel.—Steel's versatility is due to its response to thermal treatment. Although most steel products are used in the as-rolled or un-heat-treated condition, thermal treatment greatly increases the number of properties that can be obtained, because at certain "critical temperatures" iron changes from one type of crystal structure to another. This structural change, known as an allotropic transformation, is spontaneous and reversible and can be made to occur by simply changing the temperature of the metal.

In steel, the transformation in crystal structure occurs over a range of temperatures, bounded by lower and upper critical points. When heated, most carbon and low–alloy steels have a critical temperature range between 1300 and 1600 degrees F. Steel above this temperature, but below the melting range, has a crystalline structure known as austenite, in which the carbon and alloying elements are dissolved in a solid solution. Below this critical range, the crystal structure changes to a phase known as ferrite, which is capable of maintaining only a very small percentage of carbon in solid solution. The remaining carbon exists in the form of carbides, which are compounds of carbon and iron and certain of the other alloying elements. Depending primarily on cooling rate, the carbides may be present as thin plates alternating with the ferrite (pearlite); as spheroidal globular particles at ferrite grain boundaries or dispersed throughout the ferrite; or as a uniform distribution of extremely fine particles throughout a "ferritelike" phase, which has an acicular (needle-like) appearance, named martensite. In some of the highly alloyed stainless steels the addtion of certain elements stabilizes the austenite structure so that it persists even at very low temperatures (austenitic grades). Other alloying elements can prevent the formation of austenite entirely up to the melting point (ferritic grades).

Fundamentally, all steel heat treatments are intended to either harden or soften the metal. They involve one or a series of operations in which the solid metal is heated and cooled under specified conditions to develop a required structure and properties. In general, there are five major forms of heat treatment for the standard steels that modify properties to suit either fabrication or end use.

Quenching and Tempering: The primary hardening treatment for steel, quenching and tempering, usually consists of three successive operations: heating the steel above the critical range and holding it at these temperatures for a sufficient time to approach a uniform solid solution (austenitizing); cooling the steel rapidly by quenching in oil, water, brine, salt or air to form a hard, usually brittle, metastable structure known as untempered or white martensite; tempering the steel by reheating it to a temperature below the critical range in order to obtain the required combination of hardness, strength, ductility, toughness, and structural stability (tempered martensite).

Two well-known modifications of conventional quenching and tempering are "austempering" and "martempering." They involve interrupted quenching techniques (two or more quenching media) that can be utilized for some steels to obtain desired structures and properties while minimizing distortion and cracking problems that may occur in conventional hardening.

Normalizing: The steel is heated to a temperature above the critical range, after which it is cooled in still air to produce a generally fine pearlite structure. The purpose is to promote uniformity of structure and properties after a hot-working operation such as forging or extrusion. Steels may be placed in service in the normalized condition, or they may be subjected to additional thermal treatment after subsequent machining or other operations.

Annealing: The steel is heated to a temperature above or within the critical range, then cooled at a predetermined slow rate (usually in a furnace) to produce a coarse pearlite structure. This treatment is used to soften the steel for improved machinability; to improve or restore ductility for subsequent forming operations; or to eliminate the residual stresses and microstructural effects of cold working.

Spheroidize Annealing: This is a special form of annealing that requires prolonged heating at an appropriate temperature followed by slow cooling in order to produce globular carbides, a structure desirable for machining, cold forming, or cold drawing, or for the effect it will have on subsequent heat treatment.

Stress Relieving: This process reduces internal stresses, caused by machining, cold working, or welding, by heating the steel to a temperature below the critical range and holding it there long enough to equalize the temperature throughout the piece.

See the sections *HEAT TREATMENT OF STANDARD STEELS* on page 477 and *HEAT-TREATING HIGH-SPEED STEELS* on page 508 for more information about the heat treatment of steels.

Hardness and Hardenability.—Hardenability is the property of steel that determines the *depth and distribution of hardness* induced by quenching from the austenitizing temperature. Hardenability should not be confused with hardness as such or with maximum hardness. Hardness is a measure of the ability of a metal to resist penetration as determined by any one of a number of standard tests (Brinell, Rockwell, Vickers, etc). The maximum attainable hardness of any steel depends solely on carbon content and is not significantly affected by alloy content. Maximum hardness is realized only when the cooling rate in quenching is rapid enough to ensure full transformation to martensite.

The as-quenched surface hardness of a steel part is dependent on carbon content and cooling rate, but the *depth* to which a certain hardness level is maintained with given quenching conditions is a function of its hardenability. Hardenability is largely determined by the percentage of alloying elements in the steel; however, austenite grain size, time and temperature during austenitizing, and prior microstructure also significantly affect the hardness depth. The hardenability required for a particular part depends on size, design, and service stresses. For highly stressed parts, the best combination of strength and toughness is obtained by through hardening to a martensitic structure followed by adequate tempering. There are applications, however, where through hardening is not necessary or even desirable. For parts that are stressed principally at or near the surface, or in which wear resistance or resistance to shock loading is anticipated, a shallow hardening steel with a moderately soft core may be appropriate.

For through hardening of thin sections, carbon steels may be adequate; but as section size increases, alloy steels of increasing hardenability are required. The usual practice is to select the most economical grade that can meet the desired properties consistently. It is not good practice to utilize a higher alloy grade than necessary, because excessive use of alloying elements adds little to the properties and can sometimes induce susceptibility to quenching cracks.

Quenching Media: The choice of quenching media is often a critical factor in the selection of steel of the proper hardenability for a particular application. Quenching severity can be varied by selection of quenching medium, agitation control, and additives that improve the cooling capability of the quenchant. Increasing the quenching severity permits the use of less expensive steels of lower hardenability; however, consideration must also be given to the amount of distortion that can be tolerated and the susceptibility to quench cracking. In general, the more severe the quenchant and the less symmetrical the part being quenched, the greater are the size and shape changes that result from quenching and the greater is the risk of quench cracking. Consequently, although water quenching is less costly than oil quenching, and water quenching steels are less expensive than those requiring oil quenching, it is important to know that the parts being hardened can withstand the resulting distortion and the possibility of cracking.

Oil, salt, and synthetic water-polymer quenchants are also used, but they often require steels of higher alloy content and hardenability. A general rule for the selection of steel and quenchant for a particular part is that the steel should have a hardenability not exceeding that required by the severity of the quenchant selected. The carbon content of the steel should also not exceed that required to meet specified hardness and strength, because quench cracking susceptibility increases with carbon content.

The choice of quenching media is important in hardening, but another factor is agitation of the quenching bath. The more rapidly the bath is agitated, the more rapidly heat is removed from the steel and the more effective is the quench.

Hardenability Test Methods: The most commonly used method for determining hardenability is the end-quench test developed by Jominy and Boegehold, and described in detail in both SAE J06 and ASTM A255. In this test a normalized 1-inch-round, approximately 4-inch-long specimen of the steel to be evaluated is heated uniformly to its austenitizing temperature. The specimen is then removed from the furnace, placed in a jig, and immediately end quenched by a jet of room-temperature water. The water is played on the end face of the specimen, without touching the sides, until the entire specimen has cooled. Longitudinal flat surfaces are ground on opposite sides of the piece and Rockwell C scale hardness readings are taken at $\frac{1}{16}$-inch intervals from the quenched end. The resulting data are plotted on graph paper with the hardness values as ordinates (y-axis) and distances from the quenched end as abscissas (x-axis). Representative data have been accumulated for a variety of standard steel grades and are published by SAE and AISI as "H-bands." These data show graphically and in tabular form the high and low limits applicable to each grade. The suffix H following the standard AISI/SAE numerical designation indicates that the steel has been produced to specific hardenability limits.

Experiments have confirmed that the cooling rate at a given point along the Jominy bar corresponds closely to the cooling rate at various locations in round bars of various sizes. In general, when end-quench curves for different steels coincide approximately, similar treatments will produce similar properties in sections of the same size. On occasion it is necessary to predict the end-quench hardenability of a steel not available for testing, and reasonably accurate means of calculating hardness for any Jominy location on a section of steel of known analysis and grain size have been developed.

Tempering: As-quenched steels are in a highly stressed condition and are seldom used without tempering. Tempering imparts plasticity or toughness to the steel, and is inevitably accompanied by a loss in hardness and strength. The loss in strength, however, is only incidental to the very important increase in toughness, which is due to the relief of residual stresses induced during quenching and to precipitation, coalescence, and spheroidization of iron and alloy carbides resulting in a microstructure of greater plasticity.

Alloying slows the tempering rate, so that alloy steel requires a higher tempering temperature to obtain a given hardness than carbon steel of the same carbon content. The higher tempering temperature for a given hardness permits a greater relaxation of residual stress and thereby improves the steel's mechanical properties. Tempering is done in furnaces or in oil or salt baths at temperatures varying from 300 to 1200 degrees F. With most grades of alloy steel, the range between 500 and 700 degrees F is avoided because of a phenomenon known as "blue brittleness," which reduces impact properties. Tempering the martensitic stainless steels in the range of 800-1100 degrees F is not recommended because of the low and erratic impact properties and reduced corrosion resistance that result. Maximum toughness is achieved at higher temperatures. It is important to temper parts as soon as possible after quenching, because any delay greatly increases the risk of cracking resulting from the high-stress condition in the as-quenched part.

Surface Hardening Treatment (Case Hardening).—Many applications require high hardness or strength primarily at the surface, and complex service stresses frequently require not only a hard, wear–resistant surface, but also core strength and toughness to withstand impact stress.

To achieve these different properties, two general processes are used: 1) The chemical composition of the surface is altered, prior to or after quenching and tempering; the processes used include carburizing, nitriding, cyaniding, and carbonitriding; and 2) Only the surface layer is hardened by the heating and quenching process; the most common processes used for surface hardening are flame hardening and induction hardening.

Carburizing: Carbon is diffused into the part's surface to a controlled depth by heating the part in a carbonaceous medium. The resulting depth of carburization, commonly referred to as case depth, depends on the carbon potential of the medium used and the time and temperature of the carburizing treatment. The steels most suitable for carburizing to enhance toughness are those with sufficiently low carbon contents, usually below 0.03 per cent. Carburizing temperatures range from 1550 to 1750 degrees F, with the temperature and time at temperature adjusted to obtain various case depths. Steel selection, hardenability, and type of quench are determined by section size, desired core hardness, and service requirements.

Three types of carburizing are most often used: 1) *Liquid carburizing* involves heating the steel in molten barium cyanide or sodium cyanide. The case absorbs some nitrogen in addition to carbon, thus enhancing surface hardness; 2) *Gas carburizing* involves heating the steel in a gas of controlled carbon content. When used, the carbon level in the case can be closely controlled; and 3) *Pack carburizing*, which involves sealing both the steel and solid carbonaceous material in a gas-tight container, then heating this combination.

With any of these methods, the part may be either quenched after the carburizing cycle without reheating or air cooled followed by reheating to the austenitizing temperature prior to quenching. The case depth may be varied to suit the conditions of loading in service. However, service characteristics frequently require that only selective areas of a part have to be case hardened. Covering the areas not to be cased, with copper plating or a layer of commercial paste, allows the carbon to penetrate only the exposed areas. Another method involves carburizing the entire part, then removing the case in selected areas by machining, prior to quench hardening.

Nitriding: The steel part is heated to a temperature of 900–1150 degrees F in an atmosphere of ammonia gas and dissociated ammonia for an extended period of time that depends on the case depth desired. A thin, very hard case results from the formation of nitrides. Strong nitride-forming elements (chromium and molybdenum) are required to be present in the steel, and often special nonstandard grades containing aluminum (a strong nitride former) are used. The major advantage of this process is that parts can be quenched and tempered, then machined, prior to nitriding, because only a little distortion occurs during nitriding.

Cyaniding: This process involves heating the part in a bath of sodium cyanide to a temperature slightly above the transformation range, followed by quenching, to obtain a thin case of high hardness.

Carbonitriding: This process is similar to cyaniding except that the absorption of carbon and nitrogen is accomplished by heating the part in a gaseous atmosphere containing hydrocarbons and ammonia. Temperatures of 1425–1625 degrees F are used for parts to be quenched, and lower temperatures, 1200–1450 degrees F, may be used where a liquid quench is not required.

Flame Hardening: This process involves rapid heating with a direct high-temperature gas flame, such that the surface layer of the part is heated above the transformation range, followed by cooling at a rate that causes the desired hardening. Steels for flame hardening are usually in the range of 0.30–0.60 per cent carbon, with hardenability appropriate for the case depth desired and the quenchant used. The quenchant is usually sprayed on the surface a short distance behind the heating flame. Immediate tempering is required and may be done in a conventional furnace or by a flame-tempering process, depending on part size and costs.

Induction Hardening: This process is similar in many respects to flame hardening except that the heating is caused by a high-frequency electric current sent through a coil or inductor surrounding the part. The depth of heating depends on the frequency, the rate of heat conduction from the surface, and the length of the heating cycle. Quenching is usually accomplished with a water spray introduced at the proper time through jets in or near the inductor block or coil. In some instances, however, parts are oil-quenched by immersing them in a bath of oil after they reach the hardening temperature.

Applications.—Many factors enter into the selection of a steel for a particular application. These factors include the mechanical and physical properties needed to satisfy the design requirements and service environment; the cost and availability of the material; the cost of processing (machining, heat treatment, welding, etc.); and the suitability of available processing equipment or the cost of any new equipment required.

These steel selection considerations require input from designers, metallurgists, manufacturing engineers, service engineers, and procurement specialists, and can be considered proper or optimum when the part is made from the lowest cost material consistent with satisfying engineering and service requirements. The factors in selection can vary widely among different organizations, so that several different steels may be used successfully for similar applications. The best choice of a steel for any application most often results from a balance or trade-offs among the various selection considerations.

The AISI/SAE designated "standard steels" provide a convenient way for engineers and metallurgists to state briefly but clearly the chemical composition and, in some instances, some of the properties desired, and they are widely recognized and used in the United States and in many other countries. There are, however, numerous nonstandard carbon, alloy, and stainless steel grades that are widely used for special applications.

The following sections and tables illustrate the general characteristics and typical applications of most of the standard carbon, alloy, and stainless steel grades.

General Application of SAE Steels: These applications are intended as a general guide only since the selection may depend on the exact character of the service, cost of material, machinability when machining is required, or other factors. When more than one steel is recommended for a given application, information on the characteristics of each steel listed will be found in the section beginning on page 420.

Adapters, 1145

Agricultural steel, 1070, 1080

Aircraft forgings, 4140

Axles front or rear, 1040, 4140

Axle shafts, 1045, 2340, 2345, 3135, 3140, 3141, 4063, 4340

Ball-bearing races, 52100

Balls for ball bearings, 52100

Body stock for cars, rimmed*

Bolts and screws, 1035
Bolts
anchor, 1040
cold-headed, 4042
connecting-rod, 3130
heat-treated, 2330
heavy-duty, 4815, 4820
steering-arm, 3130
Brake levers, 1030, 1040
Bumper bars, 1085
Cams free-wheeling, 4615, 4620
Camshafts, 1020, 1040
Carburized parts, 1020, 1022, 1024, 1117, 1118, 1320, 2317, 2515, 3310, 3115, 3120, 4023, 4032
Chain pins transmission, 4320, 4815, 4820
Chains transmission, 3135, 3140
Clutch disks, 1060, 1070, 1085
Clutch springs, 1060
Coil springs, 4063
Cold-headed bolts, 4042
Cold-heading
steel, 30905, 1070
wire or rod, rimmed*, 1035
Cold-rolled steel, 1070
Connecting-rods, 1040, 3141
Connecting-rod bolts, 3130
Corrosion resisting, 51710, 30805
Covers transmission, rimmed*
Crankshafts, 1045, 1145, 3135, 3140, 3141
Crankshafts Diesel engine, 4340
Cushion springs, 1060
Cutlery stainless, 51335
Cylinder studs, 3130
Deep-drawing steel, rimmed*, 30905
Differential gears, 4023
Disks clutch, 1070, 1060
Ductile steel, 30905
Fan blades, 1020
Fatigue resisting 4340, 4640
Fender stock for cars, rimmed*
Forgings
aircraft, 4140
carbon steel, 1040, 1045
heat-treated, 3240, 5140, 6150
high-duty, 6150
small or medium, 1035
large, 1036
Free-cutting steel
carbon, 1111, 1113
chromium-nickel steel, 30615
manganese steel, 1132, 1137
Gears
carburized, 1320, 2317, 3115, 3120, 3310, 4119, 4125, 4320, 4615, 4620, 4815, 4820
heat-treated, 2345
car and truck, 4027, 4032
cyanide-hardening, 5140
differential, 4023
high duty, 4640, 6150
oil-hardening, 3145, 3150, 4340, 5150
ring, 1045, 3115, 3120, 4119
transmission, 3115, 3120, 4119
truck and bus, 3310, 4320
Gear shift levers, 1030
Harrow disks, 1080
Hay-rake teeth, 1095
Key stock, 1030, 2330, 3130
Leaf springs, 1085, 9260
Levers
brake, 1030, 1040
gear shift, 1030
heat-treated, 2330
Lock washers, 1060
Mower knives, 1085
Mower sections, 1070
Music wire, 1085
Nuts, 3130
heat-treated, 2330
Oil pans automobile, rimmed*
Pinions carburized, 3115, 3120, 4320
Piston pins, 3115, 3120
Plow
beams, 1070
disks, 1080
shares, 1080
Propeller shafts, 2340, 2345, 4140
Races ball-bearing, 52100
Ring gears, 3115, 3120, 4119
Rings snap, 1060, 1070, 1090
Rivets, rimmed*
Rod and wire, killed*
Rod cold-heading, 1035
Roller bearings, 4815
Rollers for bearings, 52100
Screws and bolts, 1035
Screw stock
Bessemer, 1111, 1112, 1113
open-hearth, 1115

* The "rimmed" and "killed" steels listed are in the SAE 1008, 1010, and 1015 group. See general description of these steels.

Screws heat-treated, 2330
Seat springs, 10956
Shafts
axle, 1045
cyanide-hardening, 5140
heavy-duty, 4340, 6150, 4615, 4620
oil-hardening, 5150
propeller, 2340, 2345, 4140
transmission, 4140
Sheets and strips, rimmed*
Snap rings, 1060, 1070, 1090
Spline shafts, 1045, 1320, 2340, 2345, 3115, 3120, 3135, 3140, 4023
Spring clips, 1060
Springs
coil, 1095, 4063, 6150
clutch, 1060
cushion, 1060
hard-drawn coiled, 1066
leaf, 1085, 1095, 4063, 4068, 9260, 6150
oil-hardening, 5150
oil-tempered wire, 1066
seat, 1095
valve, 1060
Spring wire, 1045
hard-drawn, 1055
oil-tempered, 1055
Stainless irons, 51210, 51710
Steel
cold-rolled, 1070
cold-heading, 30905
free-cutting carbon, 11111, 1113
free-cutting chrome-nickel, 30615
free-cutting manganese, 1132
minimum distortion, 4615, 4620, 4640
soft ductile, 30905
Steering arms, 4042
Steering-arm bolts, 3130
Steering knuckles, 3141
Steering-knuckle pins, 4815, 4820
Tacks, rimmed*
Thrust washers, 1060
oil-hardened, 5150
Transmission shafts, 4140
Tubing, 1040
front axle, 4140
seamless, 1030
welded, 1020
Universal joints, 1145
Valve springs, 1060
Washers lock, 1060
Welded structures, 30705
Wire and rod, killed*
Wire
cold-heading, rimmed*
hard-drawn spring, 1045, 1055
music, 1085
oil-tempered spring, 1055
Wrist-pins automobile, 1020
Yokes, 1145

Carbon Steels.—*SAE Steels 1006, 1008, 1010, 1015:* These steels are the lowest carbon steels of the plain carbon type, and are selected where cold formability is the primary requisite of the user. They are produced both as rimmed and killed steels. Rimmed steel is used for sheet, strip, rod, and wire where excellent surface finish or good drawing qualities are required, such as body and fender stock, hoods, lamps, oil pans, and other deep-drawn and -formed products. This steel is also used for cold-heading wire for tacks, and rivets and low carbon wire products. Killed steel (usually aluminum killed or special killed) is used for difficult stampings, or where nonaging properties are needed. Killed steels (usually silicon killed) should be used in preference to rimmed steel for forging or heat-treating applications.

These steels have relatively low tensile values and should not be selected where much strength is desired. Within the carbon range of the group, strength and hardness will rise with increases in carbon and/or with cold work, but such increases in strength are at the sacrifice of ductility or the ability to withstand cold deformation. Where cold rolled strip is used, the proper temper designation should be specified to obtain the desired properties.

With less than 0.15 carbon, the steels are susceptible to serious grain growth, causing brittleness, which may occur as the result of a combination of critical strain (from cold work) followed by heating to certain elevated temperatures. If cold-worked parts formed from these steels are to be later heated to temperatures in excess of 1100 degrees F, the user should exercise care to avoid or reduce cold working. When this condition develops, it can be overcome by heating the parts to a temperature well in excess of the upper critical point, or at least 1750 degrees F.

Steels in this group, being nearly pure iron or ferritic in structure, do not machine freely and should be avoided for cut screws and operations requiring broaching or smooth finish on turning. The machinability of bar, rod, and wire products is improved by cold drawing. Steels in this group are readily welded.

SAE 1016, 1017, 1018, 1019, 1020, 1021, 1022, 1023, 1024, 1025, 1026, 1027, 1030: Steels in this group, due to the carbon range covered, have increased strength and hardness, and reduced cold formability compared to the lowest carbon group. For heat-treating purposes, they are known as carburizing or case hardening grades. When uniform response to heat treatment is required, or for forgings, killed steel is preferred; for other uses, semi-killed or rimmed steel may be indicated, depending on the combination of properties desired. Rimmed steels can ordinarily be supplied up to 0.25 carbon.

Selection of one of these steels for carburizing applications depends on the nature of the part, the properties desired, and the processing practice preferred. Increases in carbon give greater core hardness with a given quench, or permit the use of thicker sections. Increases in manganese improve the hardenability of both the core and case; in carbon steels this is the only change in composition that will increase case hardenability. The higher manganese variants also machine much better. For carburizing applications, SAE 1016, 1018, and 1019 are widely used for thin sections or water-quenched parts. SAE 1022 and 1024 are used for heavier sections or where oil quenching is desired, and SAE 1024 is sometimes used for such parts as transmission and rear axle gears. SAE 1027 is used for parts given a light case to obtain satisfactory core properties without drastic quenching. SAE 1025 and 1030, although not usually regarded as carburizing types, are sometimes used in this manner for larger sections or where greater core hardness is needed.

For cold-formed or -headed parts, the lowest manganese grades (SAE 1017, 1020, and 1025) offer the best formability at their carbon level. SAE 1020 is used for fan blades and some frame members, and SAE 1020 and 1025 are widely used for low-strength bolts. The next higher manganese types (SAE 1018, 1021, and 1026) provide increased strength.

All steels listed may be readily welded or brazed by the common commercial methods. SAE 1020 is frequently used for welded tubing. These steels are used for numerous forged parts, the lower-carbon grades where high strength is not essential. Forgings from the lower-carbon steels usually machine better in the as-forged condition without annealing, or after normalizing.

SAE 1030, 1033, 1034, 1035, 1036, 1038, 1039, 1040, 1041, 1042, 1043, 1045, 1046, 1049, 1050, 1052: These steels, of the medium-carbon type, are selected for uses where higher mechanical properties are needed and are frequently further hardened and strengthened by heat treatment or by cold work. These grades are ordinarily produced as killed steels.

Steels in this group are suitable for a wide variety of automotive-type applications. The particular carbon and manganese level selected is affected by a number of factors. Increases in the mechanical properties required in section thickness, or in depth of hardening, ordinarily indicate either higher carbon or manganese or both. The heat-treating practice preferred, particularly the quenching medium, has a great effect on the steel selected. In general, any of the grades over 0.30 carbon may be selectively hardened by induction or flame methods.

The lower-carbon and manganese steels in this group find usage for certain types of cold-formed parts. SAE 1030 is used for shift and brake levers. SAE 1034 and 1035 are used in the form of wire and rod for cold upsetting such as bolts, and SAE 1038 for bolts and studs. The parts cold-formed from these steels are usually heat-treated prior to use. Stampings are generally limited to flat parts or simple bends. The higher-carbon SAE 1038, 1040, and 1042 are frequently cold drawn to specified physical properties for use without heat treatment for some applications such as cylinder head studs.

Any of this group of steels may be used for forgings, the selection being governed by the section size and the physical properties desired after heat treatment. Thus, SAE 1030 and 1035 are used for shifter forks and many small forgings where moderate properties are desired, but the deeper-hardening SAE 1036 is used for more critical parts where a higher strength level and more uniformity are essential, such as some front suspension parts. Forgings such as connecting rods, steering arms, truck front axles, axle shafts, and tractor wheels are commonly made from the SAE 1038 to 1045 group. Larger forgings at similar strength levels need more carbon and perhaps more manganese. Examples are crankshafts made from SAE 1046 and 1052. These steels are also used for small forgings where high hardness after oil quenching is desired. Suitable heat treatment is necessary on forgings from this group to provide machinability. These steels are also widely used for parts machined from bar stock, the selection following an identical pattern to that described for forgings. They are used both with and without heat treatment, depending on the application and the level of properties needed. As a class, they are considered good for normal machining operations. It is also possible to weld these steels by most commercial methods, but precautions should be taken to avoid cracking from too rapid cooling.

SAE 1055, 1060, 1062, 1064, 1065, 1066, 1070, 1074, 1078, 1080, 1085, 1086, 1090, 1095: Steels in this group are of the high-carbon type, having more carbon than is required to achieve maximum as quenched hardness. They are used for applications where the higher carbon is needed to improve wear characteristics for cutting edges, to make springs, and for special purposes. Selection of a particular grade is affected by the nature of the part, its end use, and the manufacturing methods available.

In general, cold-forming methods are not practical on this group of steels, being limited to flat stampings and springs coiled from small-diameter wire. Practically all parts from these steels are heat treated before use, with some variations in heat-treating methods to obtain optimum properties for the particular use to which the steel is to be put.

Uses in the spring industry include SAE 1065 for pretempered wire and SAE 1066 for cushion springs of hard-drawn wire, SAE 1064 may be used for small washers and thin stamped parts, SAE 1074 for light flat springs formed from annealed stock, and SAE 1080 and 1085 for thicker flat springs. SAE 1085 is also used for heavier coil springs. Valve spring wire and music wire are special products.

Due to good wear properties when properly heat-treated, the high-carbon steels find wide usage in the farm implement industry. SAE 1070 has been used for plow beams, SAE 1074 for plow shares, and SAE 1078 for such parts as rake teeth, scrapers, cultivator shovels, and plow shares. SAE 1085 has been used for scraper blades, disks, and for spring tooth harrows. SAE 1086 and 1090 find use as mower and binder sections, twine holders, and knotter disks.

Free Cutting Steels.—*SAE 1111, 1112, 1113:* This class of steels is intended for those uses where easy machining is the primary requirement. They are characterized by a higher sulfur content than comparable carbon steels. This composition results in some sacrifice of cold-forming properties, weldability, and forging characteristics. In general, the uses are similar to those for carbon steels of similar carbon and manganese content.

These steels are commonly known as Bessemer screw stock, and are considered the best machining steels available, machinability improving within the group as sulfur increases. They are used for a wide variety of machined parts. Although of excellent strength in the cold-drawn condition, they have an unfavorable property of cold shortness and are not commonly used for vital parts. These steels may be cyanided or carburized, but when uniform response to heat-treating is necessary, open-hearth steels are recommended.

SAE 1109, 1114, 1115, 1116, 1117, 1118, 1119, 1120, 1126: Steels in this group are used where a combination of good machinability and more uniform response to heat treatment is needed. The lower-carbon varieties are used for small parts that are to be cyanided or carbonitrided. SAE 1116, 1117, 1118, and 1119 carry more manganese for better hardenability, permitting oil quenching after case-hardening heat treatments in many instances. The higher-carbon SAE 1120 and 1126 provide more core hardness when this is needed.

SAE 1132, 1137, 1138, 1140, 1141, 1144, 1145, 1146, 1151: This group of steels has characteristics comparable to carbon steels of the same carbon level, except for changes due to higher sulfur as noted previously. They are widely used for parts where large amounts of machining are necessary, or where threads, splines, or other contours present special problems with tooling. SAE 1137, for example, is widely used for nuts and bolts and studs with machined threads. The higher-manganese SAE 1132, 1137, 1141, and 1144 offer greater hardenability, the higher-carbon types being suitable for oil quenching for many parts. All these steels may be selectively hardened by induction or flame heating if desired.

Carburizing Grades of Alloy Steels.—*Properties of the Case:* The properties of carburized and hardened cases (surface layers) depend on the carbon and alloy content, the structure of the case, and the degree and distribution of residual stresses. The carbon content of the case depends on the details of the carburizing process, and the response of iron and the alloying elements present, to carburization. The original carbon content of the steel has little or no effect on the carbon content produced in the case. The hardenability of the case, therefore, depends on the alloy content of the steel and the final carbon content produced by carburizing, but not on the initial carbon content of the steel.

With complete carbide solution, the effect of alloying elements on the hardenability of the case is about the same as the effect of these elements on the hardenability of the core. As an exception to this statement, any element that inhibits carburizing may reduce the hardenability of the case. Some elements that raise the hardenability of the core may tend to produce more retained austenite and consequently somewhat lower hardness in the case.

Alloy steels are frequently used for case hardening because the required surface hardness can be obtained by moderate speeds of quenching. Slower quenching may mean less distortion than would be encountered with water quenching. It is usually desirable to select a steel that will attain a minimum surface hardness of 58 or 60 Rockwell C after carburizing and oil quenching. Where section sizes are large, a high-hardenability alloy steel may be necessary, whereas for medium and light sections, low-hardenability steels will suffice.

In general, the case-hardening alloy steels may be divided into two classes as far as the hardenability of the case is concerned. Only the general type of steel (SAE 3300–4100, etc.) is discussed. The original carbon content of the steel has no effect on the carbon content of the case, so the last two digits in the specification numbers are not meaningful as far as the case is concerned.

A) High-Hardenability Case: SAE 2500, 3300, 4300, 4800, 9300

As these are high-alloy steels, both the case and the core have high hardenability. They are used particularly for carburized parts having thick sections, such as bevel drive pinions and heavy gears. Good case properties can be obtained by oil quenching. These steels are likely to have retained austenite in the case after carburizing and quenching; consequently, special precautions or treatments, such as refrigeration, may be required.

B) Medium-Hardenability Case: SAE 1300, 2300, 4000, 4100, 4600, 5100, 8600, 8700

Carburized cases of these steels have medium hardenability, which means that their hardenability is intermediate between that of plain carbon steel and the higher-alloy carburizing steels discussed earlier. In general, these steels can be used for average-size case-hardened automotive parts such as gears, pinions, piston pins, ball studs, universal joint crosses, crankshafts, etc. Satisfactory case hardness is usually produced by oil quenching.

Core Properties: The core properties of case-hardened steels depend on both carbon and alloy content of the steel. Each of the general types of alloy case-hardening steel is usually made with two or more carbon contents to permit different hardenability in the core.

The most desirable hardness for the core depends on the design and functioning of the individual part. In general, where high compressive loads are encountered, relatively high core hardness is beneficial in supporting the case. Low core hardnesses may be desirable where great toughness is essential.

The case-hardening steels may be divided into three general classes, depending on hardenability of the core.

A) Low-Hardenability Core: SAE 4017, 4023, 4024, 4027,* 4028,* 4608, 4615, 4617,* 8615,* 8617*

B) Medium-Hardenability Core: SAE 1320, 2317, 2512, 2515,* 3115, 3120, 4032, 4119, 4317, 4620, 4621, 4812, 4815,* 5115, 5120, 8620, 8622, 8720, 9420

C) High-Hardenability Core: SAE 2517, 3310, 3316, 4320, 4817, 4820, 9310, 9315, 9317

Heat Treatments: In general, all the alloy carburizing steels are made with fine grain and most are suitable for direct quenching from the carburizing temperature. Several other types of heat treatment involving single and double quenching are also used for most of these steels. (See on page 504 and on page 505)

Directly Hardenable Grades of Alloy Steels.—These steels may be considered in five groups on the basis of approximate mean carbon content of the SAE specification. In general, the last two figures of the specification agree with the mean carbon content. Consequently the heading "0.30–0.37 Mean Carbon Content of SAE Specification" includes steels such as SAE 1330, 3135, and 4137.

It is necessary to deviate from the above plan in the classification of the carbon molybdenum steels. When carbon molybdenum steels are used, it is customary to specify higher carbon content for any given application than would be specified for other alloy steels, due to the low alloy content of these steels. For example, SAE 4063 is used for the same applications as SAE 4140, 4145, and 5150. Consequently, in the following discussion, the carbon molybdenum steels have been shown in the groups where they belong on the basis of applications rather than carbon content.

Mean Carbon Content of SAE Specification	Common Applications
(a) 0.30–0.37 per cent	Heat-treated parts requiring moderate strength and great toughness.
(b) 0.40–0.42 per cent	Heat-treated parts requiring higher strength and good toughness.
(c) 0.45–0.50 per cent	Heat-treated parts requiring fairly high hardness and strength with moderate toughness.
(d) 0.50–0.62 per cent	Springs and hand tools.
(e) 1.02 per cent	Ball and roller bearings.

For the present discussion, steels of each carbon content are divided into two or three groups on the basis of hardenability. Transformation ranges and consequently heat-treating practices vary somewhat with different alloying elements even though the hardenability is not changed.

0.30–0.37 Mean Carbon Content of SAE Specification: These steels are frequently used for water-quenched parts of moderate section size and for oil-quenched parts of small section size. Typical applications of these steels are connecting rods, steering arms and steering knuckles, axle shafts, bolts, studs, screws, and other parts requiring strength and toughness where section size is small enough to permit the desired physical properties to be obtained with the customary heat treatment.

Steels falling in this classification may be subdivided into two groups on the basis of hardenability:

* Borderline classifications might be considered in the next higher hardenability group.

A) Low Hardenability: SAE 1330, 1335, 4037, 4042, 4130, 5130, 5132, 8630

B) Medium Hardenability: SAE 2330, 3130, 3135, 4137, 5135, 8632, 8635, 8637, 8735, 9437

0.40–0.42 Mean Carbon Content of SAE Specification: In general, these steels are used for medium and large size parts requiring high degree of strength and toughness. The choice of the proper steel depends on the section size and the mechanical properties that must be produced. The low and medium hardenabilty steels are used for average size automotive parts such as steering knuckles, axle shafts, propeller shafts, etc. The high hardenability steels are used particularly for large axles and shafts for large aircraft parts.

These steels are usually considered as oil quenching steels, although some large parts made of the low and medium hardenability classifications may be quenched in water under properly controlled conditions.

These steels may be divided into three groups on the basis of hardenability:

A) Low Hardenability: SAE 1340, 4047, 5140, 9440

B) Medium Hardenability: SAE 2340, 3140, 3141, 4053, 4063, 4140, 4640, 8640, 8641, 8642, 8740, 8742, 9442

C) High Hardenability: SAE 4340, 9840

0.45–0.50 Mean Carbon Content of SAE Specification: These steels are used primarily for gears and other parts requiring fairly high hardness as well as strength and toughness. Such parts are usually oil-quenched and a minimum of 90 per cent martensite in the as-quenched condition is desirable.

A) Low Hardenability: SAE 5045, 5046, 5145, 9747, 9763

B) Medium Hardenability: SAE 2345, 3145, 3150, 4145, 5147, 5150, 8645, 8647, 8650, 8745, 8747, 8750, 9445, 9845

C) High Hardenability: SAE 4150, 9850

0.50–0.63 Mean Carbon Content of SAE Specification: These steels are used primarily for springs and hand tools. The hardenability necessary depends on the thickness of the material and the quenching practice.

A) Medium hardenability: SAE 4068, 5150, 5152, 6150, 8650, 9254, 9255, 9260, 9261

B) High Hardenability: SAE 8653, 8655, 8660, 9262

1.02 Mean Carbon Content of SAE Specification—SAE 50100, 51100, 52100: These straight chromium electric furnace steels are used primarily for the races and balls or rollers of antifriction bearings. They are also used for other parts requiring high hardness and wear resistance. The compositions of the three steels are identical, except for a variation in chromium, with a corresponding variation in hardenability.

A) Low Hardenability: SAE 50100

B) Medium Hardenability: SAE 51100, 52100

Resulfurized Steel: Some of the alloy steels, SAE 4024, 4028, and 8641, are made resulfurized so as to give better machinability at a relatively high hardness. In general, increased sulfur results in decreased transverse ductility, notched impact toughness, and weldability.

Characteristics and Typical Applications of Standard Stainless Steels.—Typical applications of various stainless steel alloys are given in the following. The first number given is the AISI designation followed by the UNS number in parenthesis. (See also *Numbering Systems for Metals and Alloys* on page 403)

201 (S20100): High work-hardening rate; low-nickel equivalent of type 301. Flatware; automobile wheel covers, trim.

202 (S20200): General-purpose low-nickel equivalent of type 302. Kitchen equipment; hub caps; milk handling.

205 (S20500): Lower work-hardening rate than type 202; used for spinning and special drawing operations. Nonmagnetic and cryogenic parts.

301 (S30100): High work-hardening rate; used for structural applications where high strength plus high ductility are required. Railroad cars; trailer bodies; aircraft structurals; fasteners; automobile wheel covers, trim; pole line hardware.

302 (S30200): General-purpose austenitic stainless steel. Trim; food-handling equipment; aircraft cowlings; antennas; springs; cookware; building exteriors; tanks; hospital, household appliances; jewelry; oil refining equipment; signs.

302B (S30215): More resistant to scale than type 302. Furnace parts; still liners; heating elements; annealing covers; burner sections.

303 (S30300): Free-machining modification of type 302, for heavier cuts. Screw machine products; shafts; valves; bolts; bushings; nuts.

303Se (S30323): Free-machining modification of type 302, for lighter cuts; used where hot working or cold heading may be involved. Aircraft fittings; bolts; nuts; rivets; screws; studs.

304 (S30400): Low-carbon modification of type 302 for restriction of carbide precipitation during welding. Chemical and food processing equipment; brewing equipment; cryogenic vessels; gutters; downspouts; flashings.

304L (S30403): Extra-low-carbon modification of type 304 for further restriction of carbide precipitation during welding. Coal hopper linings; tanks for liquid fertilizer and tomato paste.

304Cu (S30430): Lower work-hardening rate than type 304. Severe cold-heading applications.

304N (S30451): Higher nitrogen than type 304 to increase strength with minimum effect on ductility and corrosion resistance, more resistant to increased magnetic permeability. Type 304 applications requiring higher strength.

305 (S30500): Low work-hardening rate; used for spin forming, severe drawing, cold heading, and forming. Coffee urn tops; mixing bowls; reflectors.

308 (S30800): Higher-alloy steel having high corrosion and heat resistance. Welding filler metals to compensate for alloy loss in welding; industrial furnaces.

309 (S30900): High-temperature strength and scale resistance. Aircraft heaters; heat-treating equipment; annealing covers; furnace parts; heat exchangers; heat-treating trays; oven linings; pump parts.

309S (S30908): Low-carbon modification of type 309. Welded constructions; assemblies subject to moist corrosion conditions.

310 (S31000): Higher elevated temperature strength and scale resistance than type 309. Heat exchangers; furnace parts; combustion chambers; welding filler metals; gas-turbine parts; incinerators; recuperators; rolls for roller hearth furnaces.

310S (S31008): Low-carbon modification of type 310. Welded constructions; jet engine rings.

314 (S31400): More resistant to scale than type 310. Severe cold-heading or -forming applications. Annealing and carburizing boxes; heat-treating fixtures; radiant tubes.

316 (S31600): Higher corrosion resistance than types 302 and 304; high creep strength. Chemical and pulp handling equipment; photographic equipment; brandy vats; fertilizer parts; ketchup cooking kettles; yeast tubs.

316L (S31603): Extra-low-carbon modification of type 316. Welded construction where intergranular carbide precipitation must be avoided. Type 316 applications requiring extensive welding.

316F (S31620): Higher phosphorus and sulfur than type 316 to improve machining and nonseizing characteristics. Automatic screw machine parts.

316N (S31651): Higher nitrogen than type 316 to increase strength with minimum effect on ductility and corrosion resistance. Type 316 applications requiring extra strength.

317 (S31700): Higher corrosion and creep resistance than type 316. Dyeing and ink manufacturing equipment.

317L (S31703): Extra-low-carbon modification of type 317 for restriction of carbide precipitation during welding. Welded assemblies.

321 (S32100): Stabilized for weldments subject to severe corrosive conditions, and for service from 800 to 1650°F Aircraft exhaust manifolds; boiler shells; process equipment; expansion joints; cabin heaters; fire walls; flexible couplings; pressure vessels.

329 (S32900): Austenitic-ferritic type with general corrosion resistance similar to type 316 but with better resistance to stress-corrosion cracking; capable of age hardening. Valves; valve fittings; piping; pump parts.

330 (N08330): Good resistance to carburization and oxidation and to thermal shock. Heat-treating fixtures.

347 (S34700): Similar to type 321 with higher creep strength. Airplane exhaust stacks; welded tank cars for chemicals; jet engine parts.

348 (S34800): Similar to type 321; low retentivity. Tubes and pipes for radioactive systems; nuclear energy uses.

384 (S38400): Suitable for severe cold heading or cold forming; lower cold-work-hardening rate than type 305. Bolts; rivets; screws; instrument parts.

403 (S40300): "Turbine quality" grade. Steam turbine blading and other highly stressed parts including jet engine rings.

405 (S40500): Nonhardenable grade for assemblies where air-hardening types such as 410 or 403 are objectionable. Annealing boxes; quenching racks; oxidation-resistant partitions.

409 (S40900): General-purpose construction stainless. Automotive exhaust systems; transformer and capacitor cases; dry fertilizer spreaders; tanks for agricultural sprays.

410 (S41000): General-purpose heat-treatable type. Machine parts; pump shafts; bolts; bushings; coal chutes; cutlery; hardware; jet engine parts; mining machinery; rifle barrels; screws; valves.

414 (41400): High hardenability steel. Springs; tempered rules; machine parts, bolts; mining machinery; scissors; ships' bells; spindles; valve seats.

416 (S41600): Free-machining modification of type 410, for heavier cuts. Aircraft fittings; bolts; nuts; fire extinguisher inserts; rivets; screws.

416Se (S41623): Free-machining modification of type 410, for lighter cuts. Machined parts requiring hot working or cold heading.

420 (S42000): Highercarbon modification of type 410. Cutlery; surgical instruments; valves; wear-resisting parts; glass molds; hand tools; vegetable choppers.

420F (S42020): Free-machining modification of type 420. Applications similar to those for type 420 requiring better machinability.

422 (S42200): High strength and toughness at service temperatures up to 1200 degrees F. Steam turbine blades; fasteners.

429 (S42900): Improved weldability as compared to type 430. Nitric acid and nitrogen-fixation equipment.

430 (S43000): General-purpose nonhardenable chromium type. Decorative trim; nitric acid tanks; annealing baskets; combustion chambers; dishwashers; heaters; mufflers; range hoods; recuperators; restaurant equipment.

430F (S43020): Free-machining modification of type 430, for heavier cuts. Screw machine parts.

430FSe (S43023): Free-machining modification of type 430, for lighter cuts. Machined parts requiring light cold heading or forming.

431 (S43100): Special-purpose hardenable steel used where particularly high mechanical properties are required. Aircraft fittings; beater bars; paper machinery; bolts.

434 (S43400): Modification of type 430 designed to resist atmospheric corrosion in the presence of winter road conditioning and dust-laying compounds. Automotive trim and fasteners.

436 (S43600): Similar to types 430 and 434. Used where low "roping" or "ridging" required. General corrosion and heat-resistant applications such as automobile trim.

440A (S44002): Hardenable to higher hardness than type 420 with good corrosion resistance. Cutlery; bearings; surgical tools.

440B (S44003): Cutlery grade. Cutlery, valve parts; instrument bearings.

440C (S44004): Yields highest hardnesses of hardenable stainless steels. Balls; bearings; races; nozzles; balls and seats for oil well pumps; valve parts.

442 (S44200): High-chromium steel, principally for parts that must resist high service temperatures without scaling. Furnace parts; nozzles; combustion chambers.

446 (S44600): High-resistance to corrosion and scaling at high temperatures, especially for intermittent service; often used in sulfur-bearing atmosphere. Annealing boxes; combustion chambers; glass molds; heaters; pyrometer tubes; recuperators; stirring rods; valves.

501 (S50100): Heat resistance; good mechanical properties at moderately elevated temperatures. Heat exchangers; petroleum refining equipment.

502 (S50200): More ductility and less strength than type 501. Heat exchangers; petroleum refining equipment; gaskets.

Chromium Nickel Austenitic Steels (Not capable of heat treatment).—*SAE 30201:*

This steel is an austenitic chromium–nickel–manganese stainless steel usually required in flat products. In the annealed condition, it exhibits higher strength values than the corresponding chromium–nickel stainless steel (SAE 30301). It is nonmagnetic in the annealed condition, but may be magnetic when cold-worked. SAE 30201 is used to obtain high strength by work-hardening and is well suited for corrosion-resistant structural members requiring high strength with low weight. It has excellent resistance to a wide variety of corrosive media, showing behavior comparable to stainless grade SAE 30301. It has high ductility and excellent forming properties. Owing to this steel's work-hardening rate and yield strength, tools for forming must be designed to allow for a higher springback or recovery rate. It is used for automotive trim, automotive wheel covers, railroad passenger car bodies and structural members, and truck trailer bodies.

SAE 30202: Like its corresponding chromium–nickel stainless steel SAE 30302, this is a general-purpose stainless steel. It has excellent corrosion resistance and deep drawing qualities. It is nonhardenable by thermal treatments, but may be cold worked to high tensile strengths. In the annealed condition, it is nonmagnetic but slightly magnetic when cold-worked. Applications for this stainless steel are hub cap, railcar and truck trailer bodies, and spring wire.

SAE 30301: Capable of attaining high tensile strength and ductility by moderate or severe cold working. It is used largely in the cold-rolled or cold-drawn condition in the form of sheet, strip, and wire. Its corrosion resistance is good but not equal to SAE 30302.

SAE 30302: The most widely used of the general-purpose austenitic chromium–nickel stainless steels. It is used for deep drawing largely in the annealed condition. It can be worked to high tensile strengths but with slightly lower ductility than SAE 30301.

SAE 30303F: A free-machining steel recommended for the manufacture of parts produced on automatic screw machines. Caution must be used in forging this steel.

SAE 30304: Similar to SAE 30302 but somewhat superior in corrosion resistance and having superior welding properties for certain types of equipment.

SAE 30305: Similar to SAE 30304 but capable of lower hardness. Has greater ductility with slower work-hardening tendency.

SAE 30309: A steel with high heat-resisting qualities which is resistant to oxidation at temperatures up to about 1800 degrees F.

SAE 30310: This teel has the highest heat-resisting properties of any of the chromium nickel steels listed here and will resist oxidation at temperatures up to about 1900 degrees F.

SAE 30316: Recommended for use in parts where unusual resistance to chemical or salt water corrosion is necessary. It has superior creep strength at elevated temperatures.

SAE 30317: Similar to SAE 30316 but has the highest corrosion resistance of all these alloys in many environments.

SAE 30321: Recommended for use in the manufacture of welded structures where heat treatment after welding is not feasible. It is also recommended for use where temperatures up to 1600 degrees F are encountered in service.

SAE 30325: Used for such parts as heat control shafts.

SAE 30347: This steel is similar to SAE 30321. This niobium alloy is sometimes preferred to titanium because niobium is less likely to be lost in welding operations.

Stainless Chromium Irons and Steels.—*SAE 51409:* An 11 per cent chromium alloy developed, especially for automotive mufflers and tailpipes. Resistance to corrosion and oxidation is very similar to SAE 51410. It is nonhardenable and has good forming and welding characteristics. This alloy is recommended for mildly corrosive applications where surface appearance is not critical.

SAE 51410: A general-purpose stainless steel capable of heat treatment to show good physical properties. It is used for general stainless applications, both in the heat-treated and annealed condition but is not as resistant to corrosion as SAE 51430 in either the annealed or heat-treated condition.

SAE 51414: A corrosion and heat-resisting nickel-bearing chromium steel with somewhat better corrosion resistance than SAE 51410. It will attain slightly higher mechanical properties when heat-treated than SAE 51410. It is used in the form of tempered strip or wire, and in bars and forgings for heat-treated parts.

SAE 51416F: A free-machining grade for the manufacture of parts produced in automatic screw machines.

SAE 51420: This steel is capable of being heat-treated to a relatively high hardness. It will harden to a maximum of approximately 500 Brinell. Maximum corrosion resisting qualities exist only in the fully hardened condition. It is used for cutlery, hardened pump shafts, etc.

SAE 51420F: This is similar to SAE 51420 except for its free-machining properties.

SAE 51430: This high-chromium steel is not capable of heat treatment and is recommended for use in shallow parts requiring moderate draw. Corrosion and heat resistance are superior to SAE 51410.

SAE 51430F: This steel is similar to SAE 51430 except for its free-machining properties.

SAE 51431: This nickel-bearing chromium steel is designed for heat treatment to high mechanical properties. Its corrosion resistance is superior to other hardenable steels.

SAE 51440A: A hardenable chromium steel with greater quenched hardness than SAE 51420 and greater toughness than SAE 51440B and 51440C. Maximum corrosion resistance is obtained in the fully hardened and polished condition.

SAE 51440B: A hardenable chromium steel with greater quenched hardness than SAE 51440A. Maximum corrosion resistance is obtained in the fully hardened and polished condition. Capable of hardening to 50–60 Rockwell C depending on carbon content.

SAE 51440C: This steel has the greatest quenched hardness and wear resistance on heat treatment of any corrosion- or heat-resistant steel.

SAE 51440F: The same as SAE 51440C, except for its free-machining characteristics.

SAE 51442: A corrosion- and heat-resisting chromium steel with corrosion-resisting properties slightly better than SAE 51430 and with good scale resistance up to 1600 degrees F.

SAE 51446: A corrosion- and heat-resisting steel with maximum amount of chromium consistent with commercial malleability. Used principally for parts that must resist high temperatures in service without scaling. Resists oxidation up to 2000 degrees F.

SAE 51501: Used for its heat and corrosion resistance and good mechanical properties at temperatures up to approximately 1000 degrees F.

Typical Mechanical Properties of Steel.—Tables 6 through 8 provide expected minimum and/or typical mechanical properties of selected standard carbon and alloy steels and stainless steels.

Table 6. Expected Minimum Mechanical Properties of Cold-Drawn Carbon-Steel Rounds, Squares, and Hexagons

Size, in.	As Cold-Drawn					Cold-Drawn Followed by Low-Temperature Stress Relief					Cold-Drawn Followed by High-Temperature Stress Relief				
	Strength		Elongation in 2 in., Per cent	Reduction in Area, Per cent	Hardness, Bhn	Strength		Elongation in 2 in., Per cent	Reduction in Area, Per cent	Hardness, Bhn	Strength		Elongation in 2 in., Per cent	Reduction in Area, Per cent	Hardness, Bhn
	Tensile	Yield				Tensile	Yield				Tensile	Yield			
	1000 lb/in.^{2}					1000 lb/in^2					1000 lb/in^2				
AISI 1018 and 1025 Steels															
⅝–⅞	70	60	18	40	143	....	....	....	....	...	65	45	20	45	131
Over ⅞–1¼	65	55	16	40	131	....	....	....	....	...	60	45	20	45	121
Over 1¼–2	60	50	15	35	121	....	....	....	....	...	55	45	16	40	111
Over 2–3	55	45	15	35	111	....	....	....	....	...	50	40	15	40	101
AISI 1117 and 1118 Steels															
⅝–⅞	75	65	15	40	149	80	70	15	40	163	70	50	18	45	143
Over ⅞–¼	70	60	15	40	143	75	65	15	40	149	65	50	16	45	131
Over 1¼–2	65	55	13	35	131	70	60	13	35	143	60	50	15	40	121
Over 2–3	60	50	12	30	121	65	55	12	35	131	55	45	15	40	111
AISI 1035 Steel															
⅝–⅞	85	75	13	35	170	90	80	13	35	179	80	60	16	45	163
Over ⅞–1¼	80	70	12	35	163	85	75	12	35	170	75	60	15	45	149
Over 1¼–2	75	65	12	35	149	80	70	12	35	163	70	60	15	40	143
Over 2–3	70	60	10	30	143	75	65	10	30	149	65	55	12	35	131
AISI 1040 and 1140 Steels															
⅝–⅞	90	80	12	35	179	95	85	12	35	187	85	65	15	45	170
Over ⅞–1¼	85	75	12	35	170	90	80	12	35	179	80	65	15	45	163
Over 1¼–2	80	70	10	30	163	85	75	10	30	170	75	60	15	40	149
Over 2–3	75	65	10	30	149	80	70	10	30	163	70	55	12	35	143

Table 6. *(Continued)* **Expected Minimum Mechanical Properties of Cold-Drawn Carbon-Steel Rounds, Squares, and Hexagons**

Size, in.	As Cold-Drawn					Cold-Drawn Followed by Low-Temperature Stress Relief					Cold-Drawn Followed by High-Temperature Stress Relief				
	Strength		Elonga-tion in 2 in., Per cent	Reduc-tion in Area, Per cent	Hard-ness, Bhn	Strength		Elonga-tion in 2 in., Per cent	Reduc-tion in Area, Per cent	Hard-ness, Bhn	Strength		Elonga-tion in 2 in., Per cent	Reduc-tion in Area, Per cent	Hard-ness, Bhn
	Tensile	Yield				Tensile	Yield				Tensile	Yield			
	1000 lb/in.^{2}					1000 lb/in^2					1000 lb/in^2				
AISI 1045, 1145, and 1146 Steels															
$\frac{5}{8}$–$\frac{7}{8}$	95	85	12	35	187	100	90	12	35	197	90	70	15	45	179
Over $\frac{7}{8}$–$1\frac{1}{4}$	90	80	11	30	179	95	85	11	30	187	85	70	15	45	170
Over $1\frac{1}{4}$–2	85	75	10	30	170	90	80	10	30	179	80	65	15	40	163
Over 2–3	80	70	10	30	163	85	75	10	25	170	75	60	12	35	149
AISI 1050, 1137, and 1151 Steels															
$\frac{5}{8}$–$\frac{7}{8}$	100	90	11	35	197	105	95	11	35	212	95	75	15	45	187
Over $\frac{7}{8}$–$1\frac{1}{4}$	95	85	11	30	187	100	90	11	30	197	90	75	15	40	179
Over $1\frac{1}{4}$–2	90	80	10	30	179	95	85	10	30	187	85	70	15	40	170
Over 2–3	85	75	10	30	170	90	80	10	25	179	80	65	12	35	163
AISI 1141 Steel															
$\frac{5}{8}$–$\frac{7}{8}$	105	95	11	30	212	110	100	11	30	223	100	80	15	40	197
Over $\frac{7}{8}$–$1\frac{1}{4}$	100	90	10	30	197	105	95	10	30	212	95	80	15	40	187
Over $1\frac{1}{4}$–2	95	85	10	30	187	100	90	10	25	197	90	75	15	40	179
Over 2–3	90	80	10	20	179	95	85	10	20	187	85	70	12	30	170
AISI 1144 Steel															
$\frac{5}{8}$–$\frac{7}{8}$	110	100	10	30	223	115	105	10	30	229	105	85	15	40	212
Over $\frac{7}{8}$–$1\frac{1}{4}$	105	95	10	30	212	110	100	10	30	223	100	85	15	40	197
Over $1\frac{1}{4}$–2	100	90	10	25	197	105	95	10	25	212	95	80	15	35	187
Over 2–3	95	85	10	20	187	100	90	10	20	197	90	75	12	30	179

Source: AISI Committee of Hot-Rolled and Cold-Finished Bar Producers and published in 1974 DATABOOK issue of the American Society for Metals' *METAL PROGRESS* magazine and used with its permission.

Table 7a. Charactertics and Typical Applications of Standard Stainless Steels

AISI No.[a]	Treatment	Strength Tensile	Strength Yield	Elongation, Per cent	Reduction in Area, Per cent	Hardness, Bhn	Impact Strength (Izod), ft-lb
		lb/in.²					
1015	As-rolled	61,000	45,500	39.0	61.0	126	81.5
	Normalized (1700 F)	61,500	47,000	37.0	69.6	121	85.2
	Annealed (1600 F)	56,000	41,250	37.0	69.7	111	84.8
1020	As-rolled	65,000	48,000	36.0	59.0	143	64.0
	Normalized (1600 F)	64,000	50,250	35.8	67.9	131	86.8
	Annealed (1600 F)	57,250	42,750	36.5	66.0	111	91.0
1022	As-rolled	73,000	52,000	35.0	67.0	149	60.0
	Normalized (1700 F)	70,000	52,000	34.0	67.5	143	86.5
	Annealed (1600 F)	65,250	46,000	35.0	63.6	137	89.0
1030	As-rolled	80,000	50,000	32.0	57.0	179	55.0
	Normalized (1700 F)	75,000	50,000	32.0	60.8	149	69.0
	Annealed (1550 F)	67,250	49,500	31.2	57.9	126	51.2
1040	As-rolled	90,000	60,000	25.0	50.0	201	36.0
	Normalized (1650 F)	85,500	54,250	28.0	54.9	170	48.0
	Annealed (1450 F)	75,250	51,250	30.2	57.2	149	32.7
1050	As-rolled	105,000	60,000	20.0	40.0	229	23.0
	Normalized (1650 F)	108,500	62,000	20.0	39.4	217	20.0
	Annealed (1450 F)	92,250	53,000	23.7	39.9	187	12.5
1060	As-rolled	118,000	70,000	17.0	34.0	241	13.0
	Normalized (1650 F)	112,500	61,000	18.0	37.2	229	9.7
	Annealed (1450 F)	90,750	54,000	22.5	38.2	179	8.3
1080	As-rolled	140,000	85,000	12.0	17.0	293	5.0
	Normalized (1650 F)	146,500	76,000	11.0	20.6	293	5.0
	Annealed (1450 F)	89,250	54,500	24.7	45.0	174	4.5
1095	As-rolled	140,000	83,000	9.0	18.0	293	3.0
	Normalized (1650 F)	147,000	72,500	9.5	13.5	293	4.0
	Annealed (1450 F)	95,250	55,000	13.0	20.6	192	2.0
1117	As-rolled	70,600	44,300	33.0	63.0	143	60.0
	Normalized (1650 F)	67,750	44,000	33.5	63.8	137	62.8
	Annealed (1575 F)	62,250	40,500	32.8	58.0	121	69.0
1118	As-rolled	75,600	45,900	32.0	70.0	149	80.0
	Normalized (1700 F)	69,250	46,250	33.5	65.9	143	76.3
	Annealed (1450 F)	65,250	41,250	34.5	66.8	131	78.5
1137	As-rolled	91,000	55,00	28.0	61.0	192	61.0
	Normalized (1650 F)	97,000	57,500	22.5	48.5	197	47.0
	Annealed (1450 F)	84,750	50,000	26.8	53.9	174	36.8
1141	As-rolled	98,000	52,000	22.0	38.0	192	8.2
	Normalized (1650 F)	102,500	58,750	22.7	55.5	201	38.8
	Annealed (1500 F)	86,800	51,200	25.5	49.3	163	25.3
1144	As-rolled	102,000	61,000	21.0	41.0	212	39.0
	Normalized (1650 F)	96,750	58,000	21.0	40.4	197	32.0
	Annealed (1450 F)	84,750	50,250	24.8	41.3	167	48.0

Table 7a. Charactertics and Typical Applications of Standard Stainless Steels

AISI No.[a]	Treatment	Strength Tensile	Strength Yield	Elongation, Per cent	Reduction in Area, Per cent	Hard-ness, Bhn	Impact Strength (Izod), ft-lb
		lb/in.²					
1340	Normalized (1600 F)	121,250	81,000	22.0	62.9	248	68.2
	Annealed (1475 F)	102,000	63,250	25.5	57.3	207	52.0
3140	Normalized (1600 F)	129,250	87,000	19.7	57.3	262	39.5
	Annealed (1500 F)	100,000	61,250	24.5	50.8	197	34.2
4130	Normalized (1600 F)	97,000	63,250	25.5	59.5	197	63.7
	Annealed (1585 F)	81,250	52,250	28.2	55.6	156	45.5
4140	Normalized (1600 F)	148,000	95,000	17.7	46.8	302	16.7
	Annealed (1500 F)	95,000	60,500	25.7	56.9	197	40.2
4150	Normalized (1600 F)	167,500	106,500	11.7	30.8	321	8.5
	Annealed (1500 F)	105,750	55,000	20.2	40.2	197	18.2
4320	Normalized (1640 F)	115,000	67,250	20.8	50.7	235	53.8
	Annealed (1560 F)	84,000	61,625	29.0	58.4	163	81.0
4340	Normalized (1600 F)	185,500	125,000	12.2	36.3	363	11.7
	Annealed (1490 F)	108,000	68,500	22.0	49.9	217	37.7
4620	Normalized (1650 F)	83,250	53,125	29.0	66.7	174	98.0
	Annealed (1575 F)	74,250	54,000	31.3	60.3	149	69.0
4820	Normalized (1580 F)	109,500	70,250	24.0	59.2	229	81.0
	Annealed (1500 F)	98,750	67,250	22.3	58.8	197	68.5
5140	Normalized (1600 F)	115,000	68,500	22.7	59.2	229	28.0
	Annealed (1525 F)	83,000	42,500	28.6	57.3	167	30.0
5150	Normalized (1600 F)	126,250	76,750	20.7	58.7	255	23.2
	Annealed (1520 F)	98,000	51,750	22.0	43.7	197	18.5
5160	Normalized (1575 F)	138,750	77,000	17.5	44.8	269	8.0
	Annealed (1495 F)	104,750	40,000	17.2	30.6	197	7.4
6150	Normalized (1600 F)	136,250	89,250	21.8	61.0	269	26.2
	Annealed (1500 F)	96,750	59,750	23.0	48.4	197	20.2
8620	Normalized (1675 F)	91,750	51,750	26.3	59.7	183	73.5
	Annealed (1600 F)	77,750	55,875	31.3	62.1	149	82.8
8630	Normalized (1600 F)	94,250	62,250	23.5	53.5	187	69.8
	Annealed (1550 F)	81,750	54,000	29.0	58.9	156	70.2
8650	Normalized (1600 F)	148,500	99,750	14.0	40.4	302	10.0
	Annealed (1465 F)	103,750	56,000	22.5	46.4	212	21.7
8740	Normalized (1600 F)	134,750	88,000	16.0	47.9	269	13.0
	Annealed (1500 F)	100,750	60,250	22.2	46.4	201	29.5
9255	Normalized (1650 F)	135,250	84,000	19.7	43.4	269	10.0
	Annealed (1550 F)	112,250	70,500	21.7	41.1	229	6.5
9310	Normalized (1630 F)	131,500	82,750	18.8	58.1	269	88.0
	Annealed (1550 F)	119,000	63,750	17.3	42.1	241	58.0

[a] All grades are fine-grained except those in the 1100 series that are coarse-grained. Austenitizing temperatures are given in parentheses. Heat-treated specimens were oil-quenched unless otherwise indicated.

Source: Bethlehem Steel Corp. and Republic Steel Corp. as published in 1974 DATABOOK issue of the American Society for Metals' METAL PROGRESS magazine and used with its permission.

Table 7b. Typical Mechanical Properties of Selected Carbon and Alloy Steels (Quenched and Tempered)

AISI No.[a]	Tempering Temperature, °F	Strength: Tensile (1000 lb/in.^{2})	Strength: Yield (1000 lb/in.^{2})	Elongation, Per cent	Reduction in Area, Per cent	Hardness, Bhn
1030[b]	400	123	94	17	47	495
	600	116	90	19	53	401
	800	106	84	23	60	302
	1000	97	75	28	65	255
	1200	85	64	32	70	207
1040[b]	400	130	96	16	45	514
	600	129	94	18	52	444
	800	122	92	21	57	352
	1000	113	86	23	61	269
	1200	97	72	28	68	201
1040	400	113	86	19	48	262
	600	113	86	20	53	255
	800	110	80	21	54	241
	1000	104	71	26	57	212
	1200	92	63	29	65	192
1050[b]	400	163	117	9	27	514
	600	158	115	13	36	444
	800	145	110	19	48	375
	1000	125	95	23	58	293
	1200	104	78	28	65	235
1050	400	…	…	…	…	…
	600	142	105	14	47	321
	800	136	95	20	50	277
	1000	127	84	23	53	262
	1200	107	68	29	60	223
1060	400	160	113	13	40	321
	600	160	113	13	40	321
	800	156	111	14	41	311
	1000	140	97	17	45	277
	1200	116	76	23	54	229
1080	400	190	142	12	35	388
	600	189	142	12	35	388
	800	187	138	13	36	375
	1000	164	117	16	40	321
	1200	129	87	21	50	255
1095[b]	400	216	152	10	31	601
	600	212	150	11	33	534
	800	199	139	13	35	388
	1000	165	110	15	40	293
	1200	122	85	20	47	235
1095	400	187	120	10	30	401
	600	183	118	10	30	375
	800	176	112	12	32	363
	1000	158	98	15	37	321
	1200	130	80	21	47	269
1137	400	157	136	5	22	352
	600	143	122	10	33	285
	800	127	106	15	48	262
	1000	110	88	24	62	229
	1200	95	70	28	69	197

Table 7b. *(Continued)* **Typical Mechanical Properties of Selected Carbon and Alloy Steels** (Quenched and Tempered)

AISI No.[a]	Tempering Temperature, °F	Strength: Tensile (1000 lb/in.^{2})	Strength: Yield (1000 lb/in.^{2})	Elongation, Per cent	Reduction in Area, Per cent	Hardness, Bhn
1137[b]	400	217	169	5	17	415
	600	199	163	9	25	375
	800	160	143	14	40	311
	1000	120	105	19	60	262
	1200	94	77	25	69	187
1141	400	237	176	6	17	461
	600	212	186	9	32	415
	800	169	150	12	47	331
	1000	130	111	18	57	262
	1200	103	86	23	62	217
1144	400	127	91	17	36	277
	600	126	90	17	40	262
	800	123	88	18	42	248
	1000	117	83	20	46	235
	1200	105	73	23	55	217
1330[b]	400	232	211	9	39	459
	600	207	186	9	44	402
	800	168	150	15	53	335
	1000	127	112	18	60	263
	1200	106	83	23	63	216
1340	400	262	231	11	35	505
	600	230	206	12	43	453
	800	183	167	14	51	375
	1000	140	120	17	58	295
	1200	116	90	22	66	252
4037	400	149	110	6	38	310
	600	138	111	14	53	295
	800	127	106	20	60	270
	1000	115	95	23	63	247
	1200	101	61	29	60	220
4042	400	261	241	12	37	516
	600	234	211	13	42	455
	800	187	170	15	51	380
	1000	143	128	20	59	300
	1200	115	100	28	66	238
4130[b]	400	236	212	10	41	467
	600	217	200	11	43	435
	800	186	173	13	49	380
	1000	150	132	17	57	315
	1200	118	102	22	64	245
4140	400	257	238	8	38	510
	600	225	208	9	43	445
	800	181	165	13	49	370
	1000	138	121	18	58	285
	1200	110	95	22	63	230
4150	400	280	250	10	39	530
	600	256	231	10	40	495
	800	220	200	12	45	440
	1000	175	160	15	52	370
	1200	139	122	19	60	290
4340	400	272	243	10	38	520
	600	250	230	10	40	486
	800	213	198	10	44	430
	1000	170	156	13	51	360
	1200	140	124	19	60	280

Table 7b. *(Continued)* **Typical Mechanical Properties of Selected Carbon and Alloy Steels** (Quenched and Tempered)

AISI No.[a]	Tempering Temperature, °F	Strength		Elongation, Per cent	Reduction in Area, Per cent	Hardness, Bhn
		Tensile	Yield			
		1000 lb/in.^{2}				
5046	400	253	204	9	25	482
	600	205	168	10	37	401
	800	165	135	13	50	336
	1000	136	111	18	61	282
	1200	114	95	24	66	235
50B46	400	…	…	…	…	560
	600	258	235	10	37	505
	800	202	181	13	47	405
	1000	157	142	17	51	322
	1200	128	115	22	60	273
50B60	400	…	…	…	…	600
	600	273	257	8	32	525
	800	219	201	11	34	435
	1000	163	145	15	38	350
	1200	130	113	19	50	290
5130	400	234	220	10	40	475
	600	217	204	10	46	440
	800	185	175	12	51	379
	1000	150	136	15	56	305
	1200	115	100	20	63	245
5140	400	260	238	9	38	490
	600	229	210	10	43	450
	800	190	170	13	50	365
	1000	145	125	17	58	280
	1200	110	96	25	66	235
5150	400	282	251	5	37	525
	600	252	230	6	40	475
	800	210	190	9	47	410
	1000	163	150	15	54	340
	1200	117	118	20	60	270
5160	400	322	260	4	10	627
	600	290	257	9	30	555
	800	233	212	10	37	461
	1000	169	151	12	47	341
	1200	130	116	20	56	269
51B60	400	…	…	…	…	600
	600	…	…	…	…	540
	800	237	216	11	36	460
	1000	175	160	15	44	355
	1200	140	126	20	47	290
6150	400	280	245	8	38	538
	600	250	228	8	39	483
	800	208	193	10	43	420
	1000	168	155	13	50	345
	1200	137	122	17	58	282
81B45	400	295	250	10	33	550
	600	256	228	8	42	475
	800	204	190	11	48	405
	1000	160	149	16	53	338
	1200	130	115	20	55	280

Table 7b. *(Continued)* **Typical Mechanical Properties of Selected Carbon and Alloy Steels** (Quenched and Tempered)

AISI No.[a]	Tempering Temperature, °F	Strength: Tensile (1000 lb/in.^{2})	Strength: Yield (1000 lb/in.^{2})	Elongation, Per cent	Reduction in Area, Per cent	Hardness, Bhn
8630	400	238	218	9	38	465
	600	215	202	10	42	430
	800	185	170	13	47	375
	1000	150	130	17	54	310
	1200	112	100	23	63	240
8640	400	270	242	10	40	505
	600	240	220	10	41	460
	800	200	188	12	45	400
	1000	160	150	16	54	340
	1200	130	116	20	62	280
86B45	400	287	238	9	31	525
	600	246	225	9	40	475
	800	200	191	11	41	395
	1000	160	150	15	49	335
	1200	131	127	19	58	280
8650	400	281	243	10	38	525
	600	250	225	10	40	490
	800	210	192	12	45	420
	1000	170	153	15	51	340
	1200	140	120	20	58	280
8660	400	…	…	…	…	580
	600	…	…	…	…	535
	800	237	225	13	37	460
	1000	190	176	17	46	370
	1200	155	138	20	53	315
8740	400	290	240	10	41	578
	600	249	225	11	46	495
	800	208	197	13	50	415
	1000	175	165	15	55	363
	1200	143	131	20	60	302
9255	400	305	297	1	3	601
	600	281	260	4	10	578
	800	233	216	8	22	477
	1000	182	160	15	32	352
	1200	144	118	20	42	285
9260	400	…	…	…	…	600
	600	…	…	…	…	540
	800	255	218	8	24	470
	1000	192	164	12	30	390
	1200	142	118	20	43	295
94B30	400	250	225	12	46	475
	600	232	206	12	49	445
	800	195	175	13	57	382
	1000	145	135	16	65	307
	1200	120	105	21	69	250

[a] All grades are fine-grained except those in the 1100 series that are coarse-grained. Austenitizing temperatures are given in parentheses. Heat-treated specimens were oil-quenched unless otherwise indicated.

[b] Water quenched.

Source: Bethlehem Steel Corp. and Republic Steel Corp. as published in 1974 DATABOOK issue of the American Society for Metals' *METAL PROGRESS* magazine and used with its permission.

Table 8. Nominal Mechanical Properties of Selected Standard Stainless Steels

Grade	Condition	Tensile Strength (psi)	0.2 Per Cent Yield Strength (psi)	Elongation in 2 in. (%)	Reduction of Area (%)	Hardness Rockwell	Hardness Bhn
Austenitic Steels							
201	Annealed	115,000	55,000	55	…	B90	…
	¼-hard	125,000[a]	75,000[a]	20[a]	…	C25	…
	½-hard	150,000[a]	110,000[a]	10[a]	…	C32	…
	¾-hard	175,000[a]	135,000[a]	5[a]	…	C37	…
	Full-hard	185,000[a]	140,000[a]	4[a]	…	C41	…
202	Annealed	105,000	55,000	55	…	B90	…
	¼-hard	125,000[a]	75,000[a]	12[a]	…	C27	…
301	Annealed	110,000	40,000	60	…	B85	165
	¼-hard	125,000[a]	75,000[a]	25[a]	…	C25	…
	½-hard	150,000[a]	110,000[a]	15[a]	…	C32	…
	¾-hard	175,000[a]	135,000[a]	12[a]	…	C37	…
	Full-hard	185,000	140,000[a]	8[a]	…	C41	…
302	Annealed	90,000	37,000	55	65	B82	155
	1/4-hard (sheet, strip)	125,000[a]	75,000[a]	12[a]	…	C25	…
	Cold-drawn (bar, wire)[b]	To 350,000	…	…	…	…	…
302B	Annealed	95,000	40,000	50	65	B85	165
303, 303 (Se)	Annealed	90,000	35,000	50	55	B84	160
304	Annealed	85,000	35,000	55	65	B80	150
304L	Annealed	80,000	30,000	55	65	B76	140
305	Annealed	85,000	37,000	55	70	B82	156
308	Annealed	85,000	35,000	55	65	B80	150
309, 309S	Annealed	90,000	40,000	45	65	B85	165
310, 310S	Annealed	95,000	40,000	45	65	B87	170
314	Annealed	100,000	50,000	45	60	B87	170
316	Annealed	85,000	35,000	55	70	B80	150
	Cold-drawn (bar, wire)[b]	To 300,000	…	…	…	…	…
316L	Annealed	78,000	30,000	55	65	B76	145
317	Annealed	90,000	40,000	50	55	B85	160
321	Annealed	87,000	35,000	55	65	B80	150
347, 348	Annealed	92,000	35,000	50	65	B84	160
Martensitic Steels							
403, 410, 416, 416 (Se)	Annealed	75,000	40,000	30	65	B82	155
	Hardened[c]	…	…	…	…	C43	410
	Tempered at						
	400°F	190,000	145,000	15	55	C41	390
	600°F	180,000	140,000	15	55	C39	375
	800°F	195,000	150,000	17	55	C41	390
	1000°F	145,000	115,000	20	65	C31	300
	1200°F	110,000	85,000	23	65	B97	225
	1400°F	90,000	60,000	30	70	B89	180

Table 8. *(Continued)* **Nominal Mechanical Properties of Selected Standard Stainless**

Grade	Condition	Tensile Strength (psi)	0.2 Per Cent Yield Strength (psi)	Elongation in 2 in. (%)	Reduction of Area (%)	Hardness Rockwell	Hardness Bhn
414	Annealed	120,000	95,000	17	55	C22	235
	Hardened[c]	…	…	…	…	C44	426
	Tempered at						
	400°F	200,000	150,000	15	55	C43	415
	600°F	190,000	145,000	15	55	C41	400
	800°F	200,000	150,000	16	58	C43	415
	1000°F	145,000	120,000	20	60	C34	325
	1200°F	120,000	105,000	20	65	C24	260
420, 420F	Annealed	95,000	50,000	25	55	B92	195
	Hardened[d]	…	…	…	…	C54	540
	Tempered at						
	600°F	230,000	195,000	8	25	C50	500
431	Annealed	125,000	95,000	20	60	C24	260
	Hardened[d]	…	…	…	…	C45	440
	Tempered at						
	400°F	205,000	155,000	15	55	C43	415
	600°F	195,000	150,000	15	55	C41	400
	800°F	205,000	155,000	15	60	C43	415
	1000°F	150,000	130,000	18	60	C34	325
	1200°F	125,000	95,000	20	60	C24	260
440A	Annealed	105,000	60,000	20	45	B95	215
	Hardened[d]	…	…	…	…	C56	570
	Tempered						
	600°F	260,000	240,000	5	20	C51	510
440B	Annealed	107,000	62,000	18	35	B96	220
	Hardened[d]	…	…	…	…	C58	590
	Tempered						
	600°F	280,000	270,000	3	15	C55	555
440C, 440F	Annealed	110,000	65,000	13	25	B97	230
	Hardened[d]	…	…	…	…	C60	610
	Tempered						
	600°F	285,000	275,000	2	10	C57	580
501	Annealed	70,000	30,000	28	65	…	160
502	Annealed	70,000	30,000	30	75	B80	150
			Ferritic Steels				
405	Annealed	70,000	40,000	30	60	B80	150
430	Annealed	75,000	45,000	30	60	B82	155
430F, 430F (Se)	Annealed	80,000	55,000	25	60	B86	170
446	Annealed	80,000	50,000	23	50	B86	170

[a] Minimum.
[b] Depending on size and amount of cold reduction.
[c] Hardening temperature 1800 degrees F, 1-in.-diam. bars.
[d] Hardening temperature 1900 degrees F, 1-in.-diam. bars.
Source: Metals Handbook, 8th edition, Volume 1.

High-Strength, Low-Alloy Steels.—High-strength, low-alloy (HSLA) steel represents a specific group of steels in which enhanced mechanical properties and, sometimes, resistance to atmospheric corrosion are obtained by the addition of moderate amounts of one or more alloying elements other than carbon. Different types are available, some of which are carbon–manganese steels and others contain further alloy additions, governed by special requirements for weldability, formability, toughness, strength, and economics. These steels may be obtained in the form of sheet, strip, plates, structural shapes, bars, and bar size sections.

HSLA steels are especially characterized by their mechanical properties, obtained in the as-rolled condition. They are not intended for quenching and tempering. For certain applications, they are sometimes annealed, normalized, or stress relieved with some influence on mechanical properties.

Where these steels are used for fabrication by welding, care must be exercised in selection of grade and in the details of the welding process. Certain grades may be welded without preheat or postheat.

Because of their high strength-to-weight ratio, abrasion resistance, and, in certain compositions, improved atmospheric corrosion resistance, these steels are adapted particularly for use in mobile equipment and other structures where substantial weight savings are generally desirable. Typical applications are truck bodies, frames, structural members, scrapers, truck wheels, cranes, shovels, booms, chutes, and conveyors.

Grade 942X: A niobium- or vanadium-treated carbon–manganese high-strength steel similar to 945X and 945C except for somewhat improved welding and forming properties.

Grade 945A: A HSLA steel with excellent welding characteristics, both arc and resistance, and the best formability, weldability, and low-temperature notch toughness of the high-strength steels. It is generally used in sheets, strip, and light plate thicknesses.

Grade 945C: A carbon–manganese high-strength steel with satisfactory arc welding properties if adequate precautions are observed. It is similar to grade 950C, except that lower carbon and manganese improve arc welding characteristics, formability, and low-temperature notch toughness at some sacrifice in strength.

Grade 945X: A niobium- or vanadium-treated carbon–manganese high-strength steel similar to 945C, except for somewhat improved welding and forming properties.

Grade 950A: A HSLA steel with good weldability, both arc and resistance, with good low-temperature notch toughness, and good formability. It is generally used in sheet, strip, and light plate thicknesses.

Grade 950B: A HSLA steel with satisfactory arc welding properties and fairly good low-temperature notch toughness and formability.

Grade 950C: A carbon–manganese high-strength steel that can be arc welded with special precautions, but is unsuitable for resistance welding. The formability and toughness are fair.

Grade 950D: A HSLA steel with good weldability, both arc and resistance, and fairly good formability. Where low-temperature properties are important, the effect of phosphorus in conjunction with other elements present should be considered.

Grade 950X: A niobium- or vanadium-treated carbon–manganese high-strength steel similar to 950C, except for somewhat improved welding and forming properties.

Grades 955X, 960X, 965X, 970X, 980X: These are steels similar to 945X and 950X with higher strength obtained by increased amounts of strengthening elements, such as carbon or manganese, or by the addition of nitrogen up to about 0.015 per cent. This increased strength involves reduced formability and usually decreased weldability. Toughness will vary considerably with composition and mill practice.

The composition, minimum mechanical properties, and formability of the HSLA steel grades are shown in Tables 9 through 13.

Table 9. Chemical Composition Ladle Analysis (max. per cent)

Grade	C	Mn	P
942X	0.21	1.35	0.04
945A	0.15	1.00	0.04
945C	0.23	1.40	0.04
945X	0.22	1.35	0.04
950A	0.15	1.30	0.04
950B	0.22	1.30	0.04
950C	0.25	1.60	0.04
950D	0.15	1.00	0.15
950X	0.23	1.35	0.04
955X	0.25	1.35	0.04
960X	0.26	1.45	0.04
965X	0.26	1.45	0.04
970X	0.26	1.65	0.04
980X	0.26	1.65	0.04

Sulfur, 0.05 per cent max; silicon, 0.90 per cent max.

Source: SAE Handbook, 1990. Reprinted with permission.

Table 10. Minimum Mechanical Properties

Grade	Form	Yield Strength[a] (psi min)	Tensile Strength (psi min)	Elongation (% min) 2 in.	Elongation (% min) 8 in.
942X	Plates, shapes, bars to 4 in. incl.	42,000	60,000	24	20
945A, C	Sheet and strip	45,000	60,000	22	…
	Plates, shapes, bars				
	To ½ in. incl.	45,000	65,000	22	18
	½–1½ in. incl.	42,000	62,000	24	19
	1½–3 in. incl.	40,000	62,000	24	19
945X	Sheet and strip	45,000	60,000	25	…
	Plates, shapes, bars				
	To 1½ in. incl.	45,000	60,000	22	19
950A, B, C, D	Sheet and strip	50,000	70,000	22	…
	Plates, shapes, bars				
	To ½ in. incl.	50,000	70,000	22	18
	½–1½ in. incl.	45,000	67,000	24	19
	1½–3 in. incl.	42,000	63,000	24	19
950X	Sheet and strip	50,000	65,000	22	…
	Plates, shapes, bars				
	To 1½ in. incl.	50,000	65,000	…	18
955X	Sheet and strip	55,000	70,000	20	…
	Plates, shapes, bars				
	To 1½ in. incl.	55,000	70,000	…	17
960X	Sheet and strip	60,000	75,000	18	…
	Plates, shapes, bars				
	To 1½ in. incl.	60,000	75,000	…	16
965X	Sheet and strip	65,000	80,000	16	…
	Plates, shapes, bars				
	To ¾ in. incl.	65,000	80,000	…	15
970X	Sheet and strip	70,000	85,000	14	…
	Plates, shapes, bars				
	To ¾ in. incl.	70,000	85,000	…	14
980X	Sheet and strip	80,000	95,000	12	…
	Plates to ⅜ in. incl.	80,000	95,000	…	10

[a] Yield strength to be measured at 0.2 per cent offset.

Mechanical properties to be determined in accordance with ASTM A 370.

Source: SAE Handbook, 1990. Reprinted with permission.

Table 11. Steel Grades in Approximate Order of Increasing Excellence

Weldability	Formability	Toughness
980X	980X	980X
970X	970X	970X
965X	965X	965X
960X	960X	960X
955X, 950C, 942X	955X	955X
945C	950C	945C, 950C, 942X
950B, 950X	950D	945X, 950X
945X	950B, 950X, 942X	950D
950D	945C, 945X	950B
950A	950A	950A
945A	945A	945A

Source: SAE Handbook, 1990. Reprinted with permission.

Table 12. Typical SAE heat treatments for Grades of Chromium–Nickel Austenitic Steels Not Hardenable by Thermal Treatment

SAE Steels	AISI No.	Annealing[a] Temperature (degrees F)	Annealing Temperature (deg. C)	Quenching Medium
30201	201	1850–2050	1010–1120	Air
30202	202	1850–2050	1010–1120	Air
30301	301	1850–2050	1010–1120	Air
30302	302	1850–2050	1010–1120	Air
30303	303	1850–2050	1010–1120	Air
30304	304	1850–2050	1010–1120	Air
30305	305	1850–2050	1010–1120	Air
30309	309	1900–2050	1040–1120	Air
30310	310	1900–2100	1040–1150	Air
30316	316	1850–2050	1010–1120	Air
30317	317	1850–2050	1010–1120	Air
30321	321	1750–2050	955–1120	Air
30325	325	1800–2100	980–1150	Air
30330	…	1950–2150	1065–1175	Air
30347	347	1850–2050	1010–1120	Air

[a] Quench to produce full austenitic structure in accordance with the thickness of the section. Annealing temperatures given cover process and full annealing as already established and used by industry, the lower end of the range being used for process annealing.

Source: SAE Handbook, 1990. Reprinted with permission.

Table 13. Typical SAE Heat Treatments for Stainless Chromium Steels

SAE Steels	AISI No.	Normalizing Temperature (degrees F)	Subcritical Annealing Temperature (degrees F)	Full Annealing[a] Temperature (degrees F)	Hardening Temperature (degrees F)	Quenching Medium	Temper
51409	…	…	…	1625	…	Air	…
51410	410	…	1300–1350[b]	1500–1650	… }		
		…	…	…	1700–1850 }	Oil or air	To desired hardness
51414	414	…	1200–1250[b]	…	… }		
		…	…	…	1800–1900 }	Oil or air	To desired hardness
51416	416	…	1300–1350[c]	1500–1650	… }		
		…	…	…	1700–1850 }	Oil or air	To desired hardness
51420	420	…	1350–1450[b]	1550–1650	… }		
		…	…	…	1800–1900 }	Oil or air	To desired hardness
51420F[d]	…	…	1350–1450[b]	1550–1650	… }		
		…	…	…	1800–1900 }	Oil or air	To desired hardness
51430	430	…	1400–1500[c]	…	…	…	…
51430F[d]	…	…	1250–1400[c]	…	…	…	…
51431	431	…	1150–1225[b]	…	1800–1950	Oil or air	To desired hardness
51434	… }						
51436	… }	…	1400–1600[c]	…	…	…	…
51440A[d]	440A[d] }						
51440B[d]	440B[d]						
51440C[d]	440C[d]	…	1350–1440[b]	1550–1650	1850–1950	Oil or air	To desired hardness
51440F[d]	…						
51442	442	…	1350–1500[c]	…	…	…	…
51446	446	…	1450–1600[b]	…	…	…	…
51501	501	…	1325–1375[c]	1525–1600	1600–1700	Oil or air	To desired hardness

[a] Cool slowly in furnace.
[b] Usually air cooled but may be furnace cooled.
[c] Cool rapidly in air.
[d] Suffixes A, B, and C denote three types of steel differing only in carbon content. Suffix F denotes a free-machining steel.

Source: SAE Handbook, 1990. Reprinted with permission.

TOOL STEELS

Tool Steels

As the designation implies, tool steels serve primarily for making tools used in manufacturing and in the trades for the working and forming of metals, wood, plastics, and other industrial materials. Tools must withstand high specific loads, often concentrated at exposed areas, may have to operate at elevated or rapidly changing temperatures and in continual contact with abrasive types of work materials, and are often subjected to shocks, or may have to perform under other varieties of adverse conditions. Nevertheless, when employed under circumstances that are regarded as normal operating conditions, the tool should not suffer major damage, untimely wear resulting in the dulling of the edges, or be susceptible to detrimental metallurgical changes.

Tools for less demanding uses, such as ordinary handtools, including hammers, chisels, files, mining bits, etc., are often made of standard AISI steels that are not considered as belonging to any of the tool steel categories.

The steel for most types of tools must be used in a heat-treated state, generally hardened and tempered, to provide the properties needed for the particular application. The adaptability to heat treatment with a minimurn of harmful effects, which dependably results in the intended beneficial changes in material properties, is still another requirement that tool steels must satisfy.

To meet such varied requirements, steel types of different chemical composition, often produced by special metallurgical processes, have been developed. Due to the large number of tool steel types produced by the steel mills, which generally are made available with proprietary designations, it is rather difficult for the user to select those types that are most suitable for any specific application, unless the recommendations of a particular steel producer or producers are obtained.

Substantial clarification has resulted from the development of a classification system that is now widely accepted throughout the industry, on the part of both the producers and the users of tool steels. That system is used in the following as a base for providing concise information on tool steel types, their properties, and methods of tool steel selection.

The tool steel classification system establishes seven basic categories of tool and die steels. These categories are associated with the predominant applicational characteristics of the tool steel types they comprise. A few of these categories are composed of several groups to distinguish between families of steel types that, while serving the same general purpose, differ with regard to one or more dominant characteristics.

To provide an easily applicable guide for the selection of tool steel types best suited for a particular application, the subsequent discussions and tables are based on the previously mentioned application-related categories. As an introduction to the detailed surveys, a concise discussion is presented of the principal tool steel characteristics that govern the suitability for varying service purposes and operational conditions. A brief review of the major steel alloying elements and of the effect of these constituents on the significant characteristics of tool steels is also given in the following sections.

The Properties of Tool Steels.—Tool steels must possess certain properties to a higher than ordinary degree to make them adaptable for uses that require the ability to sustain heavy loads and perform dependably even under adverse conditions.

The extent and the types of loads, the characteristics of the operating conditions, and the expected performance with regard to both the duration and the level of consistency are the principal considerations, in combination with the aspects of cost, that govern the selection of tool steels for specific applications.

Although it is not possible to define and apply exact parameters for measuring significant tool steel characteristics, certain properties can be determined that may greatly assist in appraising the suitability of various types of tool steels for specific uses.

Because tool steels are generally heat-treated to make them adaptable to the intended use by enhancing the desirable properties, *the behavior of the steel during heat treatment* is of prime importance. The behavior of the steel comprises, in this respect, both the resistance to harmful effects and the attainment of the desirable properties. The following are considered the major properties related to heat treatment:

Safety in Hardening: This designation expresses the ability of the steel to withstand the harmful effects of exposure to very high heat and particularly to the sudden temperature changes during quenching, without harmful effects. One way of obtaining this property is by adding alloying elements that reduce the critical speed at which the quenching must be carried out, thus permitting the use of milder quenching media such as oil, salt, or just still air.

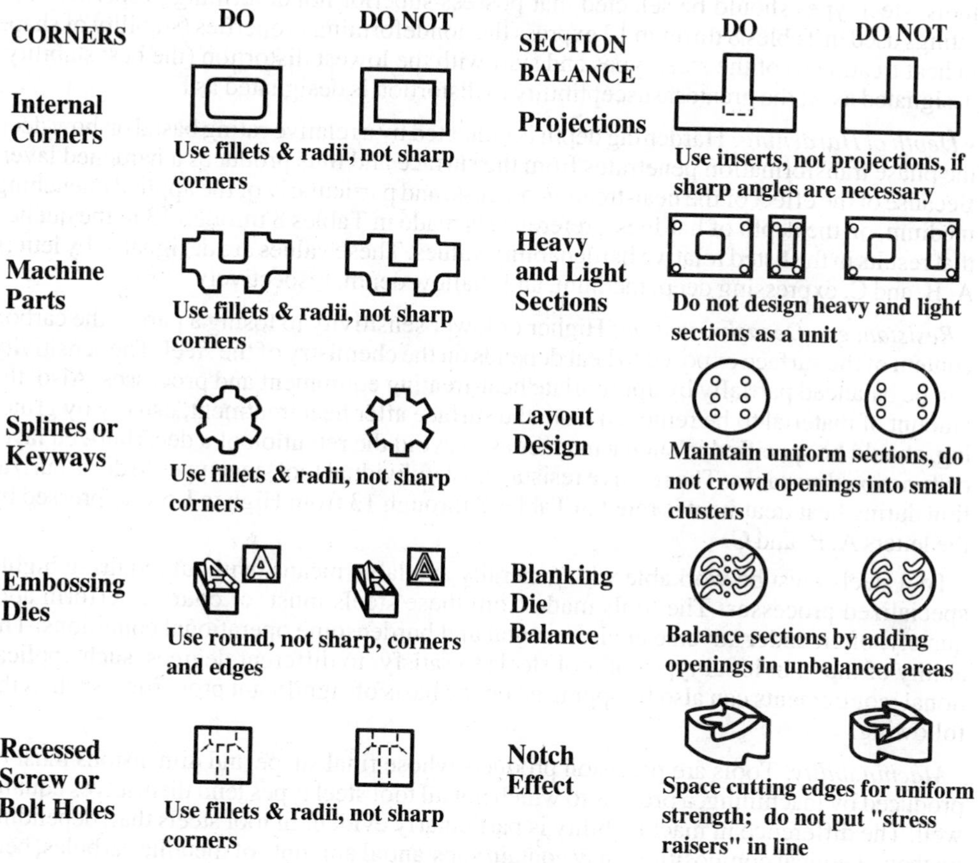

Fig. 1. Tool and die design tips to reduce breakage in heat treatment.
Courtesy of Society of Automotive Engineers, Inc.

The most common harm parts made of tool steel suffer from during heat treatment is the development of cracks. In addition to the composition of the steel and the applied heat-treating process, the configuration of the part can also affect the sensitivity to cracking. The preceding figure illustrates a few design characteristics related to cracking and warpage in heat treatment; the observation of these design tips, which call for generous filleting, avoidance of sharp angles, and major changes without transition in the cross-section, is particularly advisable when using tool steel types with a low index value for safety in hardening.

In current practice, the previously mentioned property of tool steels is rated in the order of decreasing safety (i.e., increasing sensitivity) as Highest, Very High, High, Medium, and Low safety, expressed in Tables 8 through 13 by the letters A, B, C, D, and E.

Distortions in Heat Treating: In parts made from tool steels, distortions are often a consequence of inadequate design (See Fig. 1.) or improper heat treatment (e.g., lack of stress relieving). However, certain types of tool steels display different degrees of sensitivity to distortion. Steels that are less stable require safer design of the parts for which they are used, more careful heat treatment, including the proper support for long and slender parts, or thin sections, and possibly greater grinding allowance to permit subsequent correction of the distorted shape. Some parts made of a type of steel generally sensitive to distortions can be heat-treated with very little damage when the requirements of the part call for a relatively shallow hardened layer over a soft core. However, for intricate shapes and large tools, steel types should be selected that possess superior nondeforming properties. The ratings used in Tables 8 through 13 express the nondeforming properties (stability of shape in heat treatment) of the steel types and start with the lowest distortion (the best stability) designated as A; the greatest susceptibility to distortion is designated as E.

Depth of Hardening: Hardening depth is indicated by a relative rating based on how deep the phase transformation penetrates from the surface and thus produces a hardened layer. Because of the effect of the heat-treating process, and particularly of the applied quenching medium, on the depth of hardness, reference is made in Tables 8 through 13 to the quench that results in the listed relative hardenability values. These values are designated by letters A, B, and C, expressing deep, medium, and shallow depth, respectively.

Resistance to Decarburization: Higher or lower sensitivity to losing a part of the carbon content of the surface exposed to heat depends on the chemistry of the steel. The sensitivity can be balanced partially by appropriate heat-treating equipment and processes. Also, the amount of material to be removed from the surface after heat treatment, usually by grinding, should be specified in such a manner as to avoid the retention of a decarburized layer on functional surfaces. The relative resistance of individual tool steel types to decarburization during heat treatment is rated in Tables 8 through 13 from High to Low, expressed by the letters A, B, and C.

Tool steels must be workable with generally available means, without requiring highly specialized processes. The tools made from these steels must, of course, perform adequately, often under adverse environmental and burdensome operational conditions. The ability of the individual types of tool steels to satisfy, to different degrees, such applicational requirements can also be appraised on the basis of significant properties, such as the following.

Machinability: Tools are precision products whose final shape and dimensions must be produced by machining, a process to which not all tool steel types lend themselves equally well. The difference in machinability is particularly evident in tool steels that, depending on their chemical composition, may contain substantial amounts of metallic carbides, beneficial to increased wear resistance, yet detrimental to the service life of tools with which the steel has to be worked. The microstructure of the steel type can also affect the ease of machining and, in some types, certain phase conditions, such as those due to low carbon content, may cause difficulties in achieving a fine surface finish. Certain types of tool steels have their machinability improved by the addition of small amounts of sulfur or lead. Machinability affects the cost of making the tool, particularly for intricate tool shapes, and must be considered in selection of the steel to be used. The ratings in Tables 8 through 13, starting with A for the greatest ease of machining to E for the lowest machinability, refer to working of the steel in an unhardened condition. Machinability is not necessarily identical with grindability, which expresses how well the steel is adapted to grinding after heat treating. The ease of grinding, however, may become an important consideration in tool steel selection, particularly for cutting tools and dies, which require regular sharpening involving extensive grinding. AVCO Bay State Abrasives Company compiled information on the relative grindability of frequently used types of tool steels. A simplified version of that information is presented in Table 1, which assigns the listed tool steel types to one of the following grindability grades: High (A), Medium (B), Low (C), and Very Low (D), expressing decreasing ratios of volume of metal removed to wheel wear.

Table 1. Relative Grindability of Selected Types of Frequently Used Tool Steels

AISI Tool Steel Type	H41	H42	H43	Other H	D2	D3	D5	D7	A Types	O Types	L Types	F Types
Relative Grind-ability Index	B	B	B	A	B	B	B	C	A	A	A	B

AISI High-Speed Tool Steel Type	M1	M2	M3 (1)	M3 (2)	M4	M7	M8	M10	M15	M36	M43	T1	T2	T3	T5	T6	T15
Relative Grindability Index	A	B	C	C	D	B	A	B	D	B	B	A	B	C	B	B	D

Hot Hardness: This property designates the steel's resistance to the softening effect of elevated temperature. This characteristic is related to the tempering temperature of the type of steel, which is controlled by various alloying elements such as tungsten, molybdenum, vanadium, cobalt, and chromium.

Hot hardness is a necessary property of tools used for hot work, like forging, casting, and hot extrusion. Hot hardness is also important in cutting tools operated at high-speed, which generate sufficient heat to raise their temperature well above the level where ordinary steels lose their hardness; hence the designation *high-speed steels*, which refers to a family of tool steels developed for use at high cutting speeds. Frequently it is the degree of the tool steel's resistance to softening at elevated temperature that governs important process data, such as the applicable cutting speed. In the ratings of Tables 8 through 13, tool steel types having the highest hot hardness are marked with A, subsequent letters expressing gradually decreasing capacity to endure elevated temperature without losing hardness.

Wear Resistance: The gradual erosion of the tool's operating surface, most conspicuously occurring at the exposed edges, is known as wear. Resistance to wear prolongs the useful life of the tool by delaying the degradation of its surface through abrasive contact with the work at regular operating temperatures; these temperatures vary according to the type of process. Wear resistance is observable experimentally and measurable by comparison. Certain types of metallic carbides embedded into the steel matrix are considered to be the prime contributing factors to wear resistance, besides the hardness of the heat-treated steel material. The ratings of Tables 8 through 13, starting with A for the best to E for poor, are based on conditions thought to be normal in operations for which various types of tool materials are primarily used.

Toughness: In tool steels, this property expresses ability to sustain shocks, suddenly applied and relieved loads, or major impacts, without breaking. Steels used for making tools must also be able to absorb such forces with only a minimum of elastic deformation and without permanent deformation to any extent that would interfere with the proper functioning of the tool. Certain types of tool steels, particularly those with high carbon content and without the presence of beneficial alloying constituents, tend to be the most sensitive to shocks, although they can also be made to act tougher when used for tools that permit a hardened case to be supported by a soft core. Tempering improves toughness, while generally reducing hardness. The rating indexes in Tables 8 through 13, A for the highest toughness through E for the types most sensitive to shocks, apply to tools heat treated to hardness values normally used for the particular type of tool steel.

Common Tool Faults and Failures.—The proper selection of the steel grade used for any particular type of tool is of great importance, but it should be recognized that many of the failures experienced in common practice originate from causes other than those related to the tool material.

To permit a better appraisal of the actual causes of failure and possible corrective action, a general, although not complete, list of common tool faults, resulting failures, and corrective actions is shown in Tables 2 through Tables 5. In this list, the potential failure causes are grouped into four categories. The possibility of more than a single cause being responsible for the experienced failure should not be excluded.

Finally, it must be remembered that the proper usage of tools is indispensable for obtaining satisfactory performance and tool life. Using the tools properly involves, for example, the avoidance of damage to the tool; overloading; excessive speeds and feeds; the application of adequate coolant when called for; a rigid setup; proper alignment; and firm tool and work holding.

Table 2. Common Tool Faults, Failures, and Cures
Improper Tool Design

Fault Description	Probable Failure	Possible Cure
Drastic section changes—widely different thicknesses of adjacent wall sections or protruding elements	In liquid quenching, the thin section will cool and then harden more rapidly than the adjacent thicker section, setting up stresses that may exceed the strength of the steel.	Make such parts of two pieces or use an air-hardening tool steel that avoids the harsh action of a liquid quench.
Sharp corners on shoulders or in square holes	Cracking can occur, particularly in liquid quenching, due to stress concentrations.	Apply fillets to the corners and/or use an air-hardening tool steel.
Sharp cornered keyways	Failure may arise during service, and is usually considered to be caused by fatigue.	The use of round keyways should be preferred when the general configuration of the part makes it prone to failure due to square keyways.
Abrupt section changes in battering tools	Due to impact in service, pneumatic tools are particularly sensitive to stress concentrations that lead to fatigue failures.	Use taper transitions, which are better than even generous fillets.
Functional inadequacy of tool design—e.g., insufficient guidance for a punch	Excessive wear or breakage in service may occur.	Assure solid support, avoid unnecessary play, adapt travel length to operational conditions (e.g., punch to penetrate to four-fifths of thickness in hard work material).
Improper tool clearance, such as in blanking and punching tools	Deformed and burred parts may be produced, excessive tool wear or breakage can result.	Adapt clearances to material conditions and dimensions to reduce tool load and to obtain clean sheared surfaces.

Table 3. Common Tool Faults, Failures, and Cures
Faulty Condition or Inadequate Grade of Tool Steel

Fault Description	Probable Failure	Possible Cure
Improper tool steel grade selection	Typical failures: Chipping—insufficient toughness. Wear—poor abrasion resistance. Softening—inadequate "red hardness."	Choose the tool steel grade by following recommendations and improve selection when needed, guided by property ratings.
Material defects—voids, streaks, tears, flakes, surface cooling cracks, etc.	When not recognized during material inspection, tools made of defective steel often prove to be useless.	Obtain tool steels from reliable sources and inspect tool material for detectable defects.

Table 3. *(Continued)* **Common Tool Faults, Failures, and Cures**
Faulty Condition or Inadequate Grade of Tool Steel

Fault Description	Probable Failure	Possible Cure
Decarburized surface layer ("bark") in rolled tool steel bars	Cracking may originate from the decarburized layer or it will not harden ("soft skin").	Provide allowance for stock to be removed from all surfaces of hot-rolled tool steel. Recommended amounts are listed in tool steel catalogs and vary according to section size, generally about 10 per cent for smaller and 5 per cent for larger diameters.
Brittleness caused by poor carbide distribution in high-alloy tool steels	Excessive brittleness can cause chipping or breakage during service.	Bars with large diameter (above about 4 inches) tend to be prone to nonuniform carbide distribution. Choose upset forged discs instead of large-diameter bars.
Unfavorable grain flow	Improper grain flow of the steel used for milling cutters and similar tools can cause teeth to break out.	Upset forged discs made with an upset ratio of about 2 to 1 (starting to upset thickness) display radial grain flow. Highly stressed tools, such as gear-shaper cutters, may require the cross forging of blanks.

Table 4. Common Tool Faults, Failures, and Cures
Heat-Treatment Faults

Fault Description	Probable Failure	Possible Cure
Improper preparation for heat treatment. Certain tools may require stress relieving or annealing, and often preheating, too	Tools highly stressed during machining or forming, unless stress relieved, may aggravate the thermal stresses of heat treatment, thus causing cracks. Excessive temperature gradients developed in nonpreheated tools with different section thicknesses can cause warpage.	Stress relieve, when needed, before hardening. Anneal prior to heavy machining or cold forming (e.g., hobbing). Preheat tools (a) having substantial section thickness variations or (b) requiring high quenching temperatures, as those made of high-speed tool steels.
Overheating during hardening; quenching from too high a temperature	Causes grain coarsening and a sensitivity to cracking that is more pronounced in tools with drastic section changes.	Overheated tools have a characteristic microstructure that aids recognition of the cause of failure and indicates the need for improved temperature control.
Low hardening temperature	The tool may not harden at all, or in its outer portion only, thereby setting up stresses that can lead to cracks.	Controlling both the temperature of the furnace and the time of holding the tool at quenching temperature will prevent this not too frequent deficiency.
Inadequate composition or condition of the quenching media	Water-hardening tool steels are particularly sensitive to inadequate quenching media, which can cause soft spots or even violent cracking.	For water-hardening tool steels, use water free of dissolved air and contaminants, also assure sufficient quantity and proper agitation of the quench.
Improper handling during and after quenching	Cracking, particularly of tools with sharp corners, during the heat treatment can result from holding the part too long in the quench or incorrectly applied tempering.	Following the steel producer's specifications is a safe way to assure proper heat-treatment handling. In general, the tool should be left in the quench until it reaches a temperature of 150 to 200°F, and should then be transferred promptly into a warm tempering furnace.
Insufficient tempering	Omission of double tempering for steel types that require it may cause early failure by heat checking in hot-work steels or make the tool abnormally sensitive to grinding checks.	Double temper highly alloyed tool steel of the high-speed, hot-work, and high-chromium categories, to remove stresses caused by martensite formed during the first tempering phase. Second temper also increases hardness of most high-speed steels.

Table 4. *(Continued)* **Common Tool Faults, Failures, and Cures**

Heat-Treatment Faults

Fault Description	Probable Failure	Possible Cure
Decarburization and carburization	Unless hardened in a neutral atmosphere the original carbon content of the tool surface may be changed: Reduced carbon (decarburization) causes a soft layer that wears rapidly. Increased carbon (carburization) when excessive may cause brittleness.	Heating in neutral atmosphere or well-maintained salt bath and controlling the furnace temperature and the time during which the tool is subjected to heating can usually keep the carbon imbalance within acceptable limits.

Table 5. Common Tool Faults, Failures, and Cures

Grinding Damages

Fault Description	Probable Failure	Possible Cure
	Grinding Damages	
Recessive stock removal rate causing heating of the part surface beyond the applied tempering temperature	Scorched tool surface displaying temper colors varying from yellow to purple, depending on the degree of heat, causes softening of the ground surface. When coolant is used, a local rehardening can take place, often resulting in cracks.	Prevention: by reducing speed and feed, or using coarser, softer, more open-structured grinding wheel, with ample coolant. Correction: eliminate the discolored layer by subsequent light stock removal. Not always a cure, because the effects of abusive grinding may not be corrected.
Improper grinding wheel specifications; grain too fine or bond too hard	Intense localized heating during grinding may set up surface stresses causing grinding cracks. These cracks are either parallel but at right angles to the direction of grinding or, when more advanced, form a network. May need cold etch or magnetic particle testing to become recognizable.	Prevention: by correcting the grinding wheel specifications. Correction: in shallow (0.002- to 0.004-inch) cracks, by removing the damaged layer, when permitted by the design of the tool, using very light grinding passes.
Incorrectly dressed or loaded grinding wheel	Heating of the work surface can cause scorching or cracking. Incorrect dressing can also cause a poor finish of the ground work surface.	Dress wheel with sharper diamond and faster diamond advance to produce coarser wheel surface. Alternate dressing methods, like crush-dressing, can improve wheel surface conditions. Dress wheel regularly to avoid loading or glazing of the wheel surface.
Inadequate coolant, with regard to composition, amount, distribution, and cleanliness	Introducing into the tool surface heat that is not adequately dissipated or absorbed by the coolant can cause softening, or even the development of cracks.	Improve coolant supply and quality, or reduce stock removal rate to reduce generation of heat in grinding.
Damage caused by abusive abrasive cutoff	The intensive heat developed during this process can cause a hardening of the steel surface, or may even result in cracks.	Reduce rate of advance; adopt wheel specifications better suited for the job. Use ample coolant or, when harmful effect not eliminated, replace abrasive cutoff by some cooler-acting stock separation method (e.g., sawing or lathe cutoff) unless damaged surface is being removed by subsequent machining.

Note: Illustrated examples of tool failures from causes such as listed above may be found in "The Tool Steel Trouble Shooter" handbook, published by Bethlehem Steel Corporation.

The Effect of Alloying Elements on Tool Steel Properties.—*Carbon (C):* The presence of carbon, usually in excess of 0.60 per cent for nonalloyed types, is essential for raising the hardenability of steels to the levels needed for tools. Raising the carbon content by different amounts up to a maximum of about 1.3 per cent increases the hardness slightly and the wear resistance considerably. The amount of carbon in tool steels is designed to attain certain properties (such as in the water-hardening category where higher carbon content may be chosen to improve wear resistance, although to the detriment of toughness) or, in the alloyed types of tool steels, in conformance with the other constituents to produce well-balanced metallurgical and performance properties.

Manganese (Mn): In small amounts, to about 0.60 per cent, manganese is added to reduce brittleness and to improve forgeability. Larger amounts of manganese improve hardenability, permitting oil quenching for nonalloyed carbon steels, thus reducing deformation, although with regard to several other properties, manganese is not an equivalent replacement for the regular alloying elements.

Silicon (Si): In itself, silicon may not be considered an alloying element of tool steels, but it is needed as a deoxidizer and improves the hot-forming properties of the steel. In combination with certain alloying elements, the silicon content is sometimes raised to about 2 per cent to increase the strength and toughness of steels used for tools that have to sustain shock loads.

Tungsten (W): Tungsten is one of the important alloying elements of tool steels, particularly because of two valuable properties: it improves "hot hardness," that is, the resistance of the steel to the softening effect of elevated temperature, and it forms hard, abrasion-resistant carbides, thus improving the wear properties of tool steels.

Vanadium (V): Vanadium contributes to the refinement of the carbide structure and thus improves the forgeability of alloy tool steels. Vanadium has a very strong tendency to form a hard carbide, which improves both the hardness and the wear properties of tool steels. However, a large amount of vanadium carbide makes the grinding of the tool very difficult (causing low grindability).

Molybdenum (Mo): In small amounts, molybdenum improves certain metallurgical properties of alloy steels such as deep hardening and toughness. It is used often in larger amounts in certain high-speed tool steels to replace tungsten, primarily for economic reasons, often with nearly equivalent results.

Cobalt (Co): As an alloying element of tool steels, cobalt increases hot hardness and is used in applications where that property is needed. Substantial addition of cobalt, however, raises the critical quenching temperature of the steel with a tendency to increase the decarburization of the surface, and reduces toughness.

Chromium (Cr): This element is added in amounts of several per cent to high-alloy tool steels, and up to 12 per cent to types in which chromium is the major alloying element. Chromium improves hardenability and, together with high carbon, provides both wear resistance and toughness, a combination valuable in certain tool applications. However, high chromium raises the hardening temperature of the tool steel, and thus can make it prone to hardening deformations. A high percentage of chromium also affects the grindability of the tool steel.

Nickel (Ni): Generally in combination with other alloying elements, particularly chromium, nickel is used to improve the toughness and, to some extent, the wear resistance of tool steels.

The addition of more than one element to a steel often produces what is called a synergistic effect. Thus, the combined effects of two or more alloy elements may be greater than the sum of the individual effects of each element.

Classification of Tool Steels.—Steels for tools must satisfy a number of different, often conflicting requirements. The need for specific steel properties arising from widely varying applications has led to the development of many compositions of tool steels, each intended to meet a particular combination of applicational requirements. The resultant diversity of tool steels, their number being continually expanded by the addition of new developments, makes it extremely difficult for the user to select the type best suited to his needs, or to find equivalent alternatives for specific types available from particular sources.

As a cooperative industrial effort under the sponsorship of AISI and SAE, a tool classification system has been developed in which the commonly used tool steels are grouped into seven major categories. These categories, several of which contain more than a single group, are listed in the following with the letter symbols used for their identification. The individual types of tool steels within each category are identified by suffix numbers following the letter symbols.

Category Designation	Letter Symbol	Group Designation
High-Speed Tool Steels	M	Molybdenum types
	T	Tungsten types
Hot-Work Tool Steels	H1–H19	Chromium types
	H20–H39	Tungsten types
	H40–H59	Molybdenum types
Cold-Work Tool Steels	D	High-carbon, high-chromium types
	A	Medium-alloy, air-hardening types
	O	Oil-hardening types
Shock-Resisting Tool Steels	S	…
Mold Steels	P	…
Special-Purpose Tool Steels	L	Low-alloy types
	F	Carbon–tungsten types
Water-Hardening Tool Steels	W	…

The following detailed discussion of tool steels will be in agreement with these categories, showing for each type the percentages of the major alloying elements. However, these values are for identification only; elements in tool steels of different producers in the mean analysis of the individual types may deviate from the listed percentages.

The Selection of Tool Steels for Particular Applications.—Although the advice of the specialized steel producer is often sought as a reliable source of information, the engineer is still faced with the task of selecting the tool steel. It must be realized that frequently the designation of the tool or of the process will not define the particular tool steel type best suited for the job. For that reason, tool steel selection tables naming a single type for each listed application cannot take into consideration such often conflicting work factors as ease of tool fabrication and maintenance (resharpening), productivity, product quality, and tooling cost.

When data related to past experience with tool steels for identical or similar applications are not available, a tool steel selection procedure may be followed, based on information in this Handbook section as follows:

1) Identify the AISI category that contains the sought type of steel by consulting the Quick Reference Table on starting on page 455.

Within the defined category

a) find from the listed applications of the most frequently used types of tool steels the particular type that corresponds to the job on hand; or

b) evaluate from the table of property ratings the best compromise between any conflicting properties (e.g., compromising on wear resistance to obtain better toughness).

For those willing to refine even further the first choice or to improve on it when there is not entirely satisfactory experience in one or more meaningful respects, the identifying analyses of the different types of tool steels within each general category may provide additional guidance. In this procedure, the general discussion of the effects of different alloying elements on the properties of tool steels, in a previous section, will probably be found useful.

The following two examples illustrate the procedure for refining an original choice with the purpose of adopting a tool steel grade best suited to a particular set of conditions:

Table 6. Classification, Approximate Compositions, and Properties Affecting Selection of Tool andDie Steels
(From SAE Recommended Practice)

Type of Tool Steel	Chemical Composition[a]								Non-warping Prop.	Safety in Harden-ing	Tough-ness	Depth of Hardening	Wear Resis-tance
	C	Mn	Si	Cr	V	W	Mo	Co					
Water Hardening													
0.80 Carbon	70–0.85	[b]	[b]	[b]	…	…	…	…	Poor	Fair	Good[c]	Shallow	Fair
0.90 Carbon	0.85–0.95	[b]	[b]	[b]	…	…	…	…	Poor	Fair	Good[c]	Shallow	Fair
1.00 Carbon	0.95–1.10	[b]	[b]	[b]	…	…	…	…	Poor	Fair	Good[c]	Shallow	Good
1.20 Carbon	1.10–1.30	[b]	[b]	[b]	…	…	…	…	Poor	Fair	Good[c]	Shallow	Good
0.90 Carbon–V	0.85–0.95	[b]	[b]	[b]	0.15–0.35	…	…	…	Poor	Fair	Good	Shallow	Fair
1.00 Carbon–V	0.95–1.10	[b]	[b]	[b]	0.15–0.35	…	…	…	Poor	Fair	Good	Shallow	Good
1.00 Carbon–VV	0.90–1.10	[b]	[b]	[b]	0.35–0.50	…	…	…	Poor	Fair	Good	Shallow	Good
Oil Hardening													
Low Manganese	0.90	1.20	0.25	0.50	0.20[d]	0.50	…	…	Good	Good	Fair	Deep	Good
High Manganese	0.90	1.60	0.25	0.35[d]	0.20[d]	…	0.30[d]	…	Good	Good	Fair	Deep	Good
High-Carbon, High-Chromium[e]	2.15	0.35	0.35	12.00	0.80[d]	0.75[d]	0.80[d]	…	Good	Good	Poor	Through	Best
Chromium	1.00	0.35	0.25	1.40	…	…	0.40	…	Fair	Good	Fair	Deep	Good
Molybdenum Graphitic	1.45	0.75	1.00	…	…	…	0.25	…	Fair	Good	Fair	Deep	Good
Nickel–Chromium[f]	0.75	0.70	0.25	0.85	0.25[d]	…	0.50[d]	…	Fair	Good	Fair	Deep	Fair
Air Hardening													
High-Carbon, High-Chromium	1.50	0.40	0.40	12.00	0.80[d]	…	0.90	0.60[d]	Best	Best	Fair	Through	Best
5 Per Cent Chromium	1.00	0.60	0.25	5.25	0.40[d]	…	1.10	…	Best	Best	Fair	Through	Good
High-Carbon, High-Chromium–Cobalt	1.50	0.40	0.40	12.00	0.80[d]	…	0.90	3.10	Best	Best	Fair	Through	Best
Shock-Resisting													
Chromium–Tungsten	0.50	0.25	0.35	1.40	0.20	2.25	0.40[d]	…	Fair	Good	Good	Deep	Fair
Silicon–Molybdenum	0.50	0.40	1.00	…	0.25[d]	…	0.50	…	Poor[g]	Poor[h]	Best	Deep	Fair
Silicon–Manganese	0.55	0.80	2.00	0.30[d]	0.25[d]	…	0.40[d]	…	Poor[g]	Poor[h]	Best	Deep	Fair
Hot Work													
Chromium–Molybdenum–Tungsten	0.35	0.30	1.00	5.00	0.25[d]	1.25	1.50	…	Good	Good	Good	Through	Fair
Chromium–Molybdenum–V	0.35	0.30	1.00	5.00	0.40	…	1.50	…	Good	Good	Good	Through	Fair
Chromium–Molybdenum–VV	0.35	0.30	1.00	5.00	0.90	…	1.50	…	Good	Good	Good	Through	Fair
Tungsten	0.32	0.30	0.20	3.25	0.40	9.00	…	…	Good	Good	Good	Through	Fair

Table 6. *(Continued)* **Classification, Approximate Compositions, and Properties Affecting Selection of Tool andDie Steels**
(From SAE Recommended Practice)

Type of Tool Steel	Chemical Composition[a]								Non-warping Prop.	Safety in Hardening	Toughness	Depth of Hardening	Wear Resistance
	C	Mn	Si	Cr	V	W	Mo	Co					
High Speed													
Tungsten, 18–4–1	0.70	0.30	0.30	4.10	1.10	18.00	…	…	Good	Good	Poor	Through	Good
Tungsten, 18–4–2	0.80	0.30	0.30	4.10	2.10	18.50	0.80	…	Good	Good	Poor	Through	Good
Tungsten, 18–4–3	1.05	0.30	0.30	4.10	3.25	18.50	0.70	…	Good	Good	Poor	Through	Best
Cobalt–Tungsten, 14–4–2–5	0.80	0.30	0.30	4.10	2.00	14.00	0.80	5.00	Good	Fair	Poor	Through	Good
Cobalt–Tungsten, 18–4–1–5	0.75	0.30	0.30	4.10	1.00	18.00	0.80	5.00	Good	Fair	Poor	Through	Good
Cobalt–Tungsten, 18–4–2–8	0.80	0.30	0.30	4.10	1.75	18.50	0.80	8.00	Good	Fair	Poor	Through	Good
Cobalt–Tungsten, 18–4–2–12	0.80	0.30	0.30	4.10	1.75	20.00	0.80	12.00	Good	Fair	Poor	Through	Good
Molybdenum, 8–2–1	0.80	0.30	0.30	4.00	1.15	1.50	8.50	…	Good	Fair	Poor	Through	Good
Molybdenum–Tungsten, 6–6–2	0.83	0.30	0.30	4.10	1.90	6.25	5.00	…	Good	Fair	Poor	Through	Good
Molybdenum–Tungsten, 6–6–3	1.15	0.30	0.30	4.10	3.25	5.75	5.25	…	Good	Fair	Poor	Through	Best
Molybdenum–Tungsten, 6–6–4	1.30	0.30	0.30	4.25	4.25	5.75	5.25	…	Good	Fair	Poor	Through	Best
Cobalt–Molybdenum–Tungsten, 6–6–2–8	0.85	0.30	0.30	4.10	2.00	6.00	5.00	8.00	Good	Fair	Poor	Through	Good

[a] C = carbon; Mn = manganese; Si = silicon; Cr = chromium; V = vanadium; W = tungsten; Mo = molybdenum; Co = cobalt.

[b] Carbon tool steels are usually available in four grades or qualities: *Special (Grade 1)*—The highest quality water-hardening carbon tool steel, controlled for hardenability, chemistry held to closest limits, and subject to rigid tests to ensure maximum uniformity in performance; *Extra (Grade 2)*—A high-quality water-hardening carbon tool steel, controlled for hardenability, subject to tests to ensure good service; *Standard (Grade 3)*—A good-quality water-hardening carbon tool steel, not controlled for hardenability, recommended for application where some latitude with respect to uniformity is permissible; *Commercial (Grade 4)*—A commercial-quality water-hardening carbon tool steel, not controlled for hardenability, not subject to special tests. On *special* and *extra* grades, limits on manganese, silicon, and chromium are not generally required if Shepherd hardenability limits are specified. For *standard* and *commercial* grades, limits are 0.35 max. each for Mn and Si; 0.15 max. Cr for standard; 0.20 max. Cr for commercial.

[c] Toughness decreases somewhat when increasing depth of hardening.

[d] Optional element. Steels have found satisfactory application either with or without the element present. In silicon–manganese steel listed under Shock-Resisting Steels, if chromium, vanadium, and molybdenum are not present, then hardenability will be affected.

[e] This steel may have 0.50 per cent nickel as an optional element. The steel has been found to give satisfactory application either with or without the element present.

[f] Approximate nickel content of this steel is 1.50 per cent.

[g] Poor when water quenched, fair when oil quenched.

[h] Poor when water quenched, good when oil quenched.

Table 7. Quick Reference Guide for Tool Steel Selection

Application Areas	Tool Steel Categories and AISI Letter Symbol						
	High-Speed Tool Steels, M and T	Hot-Work Tool Steels, H	Cold-Work Tool Steels, D, A, and O	Shock-Resisting Tool Steels, S	Mold Steels, P	Special-Purpose Tool Steels, L and F	Water-Hardening Tool Steels, W
	Examples of Typical Applications						
Cutting Tools Single-point types (lathe, planer, boring) Milling cutters Drills Reamers Taps Threading dies Form cutters	General-purpose production tools: M2, T1 For increased abrasion resistance: M3, M4, and M10 Heavy-duty work calling for high hot hardness: T5, T15 Heavy-duty work calling for high abrasion resistance: M42, M44		Tools with keen edges (knives, razors) Tools for operations where no high-speed is involved, yet stability in heat treatment and substantial abrasion resistance are needed	Pipe cutter wheels			Uses that do not require hot hardness or high abrasion resistance. Examples with carbon content of applicable group: Taps (1.05/1.10% C) Reamers (1.10/1.15% C) Twist drills (1.20/1.25% C) Files (1.35/1.40% C)
Hot Forging Tools and Dies Dies and inserts Forging machine plungers and pierces	For combining hot hardness with high abrasion resistance: M2, T1	Dies for presses and hammers: H20, H21 For severe conditions over extended service periods: H22 to H26, also H43	Hot trimming dies: D2	Hot trimming dies Blacksmith tools Hot swaging dies			Smith's tools (1.65/0.70% C) Hot chisels (0.70/0.75% C) Drop forging dies (0.90/1.00% C) Applications limited to short-run production
Hot Extrusion Tools and Dies Extrusion dies and mandrels, Dummy blocks Valve extrusion tools	Brass extrusion dies: T1	Extrusion dies and dummy blocks: H20 to H26 For tools that are exposed to less heat: H10 to H19		Compression molding: S1			

Table 7. *(Continued)* **Quick Reference Guide for Tool Steel Selection**

Application Areas	Tool Steel Categories and AISI Letter Symbol						
	High-Speed Tool Steels, M and T	Hot-Work Tool Steels, H	Cold-Work Tool Steels, D, A, and O	Shock-Resisting Tool Steels, S	Mold Steels, P	Special-Purpose Tool Steels, L and F	Water-Hardening Tool Steels, W
	Examples of Typical Applications						
Cold-Forming Dies Bending, forming, drawing, and deep drawing dies and punches	Burnishing tools: M1, T1	Cold heading: die casting dies: H13	Drawing dies: 01 Coining tools: O1, D2 Forming and bending dies: A2 Thread rolling dies: D2	Hobbing and short-run applications: S1, S7 Rivet sets and rivet busters		Blanking, forming, and trimmer dies when toughness has precedence over abrasion resistance: L6	Cold-heading dies: W1 or W2 (C ≅ 1.00%) Bending dies: W1 (C ≅ 1.00%)
Shearing Tools Dies for piercing, punching, and trimming Shear blades	Special dies for cold and hot work: T1 For work requiring high abrasion resistance: M2, M3	For shearing knives: H11, H12 For severe hot shearing applications: M21, M25	Dies for medium runs: A2, A6 also O1 and O4 Dies for long runs: D2, D3 Trimming dies (also for hot trimming): A2	Cold and hot shear blades Hot punching and piercing tools Boilermaker's tools		Knives for work requiring high toughness: L6	Trimming dies (0.90/0.95% C) Cold blanking and punching dies (1.00% C)
Die Casting Dies and Plastics Molds		For zinc and lead: H11 For aluminum: H13 For brass: H21	A2 and A6 O1		Plastics molds: P2 to P4, and P20		
Structural Parts for Severe Service Conditions	Roller bearings for high-temperature environment: T1 Lathe centers: M2 and T1	For aircraft components (landing gear, arrester hooks, rocket cases): H11	Lathe centers: D2, D3 Arbors: O1 Bushings: A4 Gages: D2	Pawls Clutch parts		Spindles, clutch parts (where high toughness is needed): L6	Spring steel (1.10/1.15% C)
Battering Tools for Hand and Power Tool Use				Pneumatic chisels for cold work: S5 For higher performance: S7			For intermittent use: W1 (0.80% C)

Example 1, Workpiece—Trimming Dies: For the manufacture of a type of trimming die, the first choice was grade A2, because for the planned medium rate of production, the lower material cost was considered an advantage.

A subsequent rise in the production rate indicated the use of a higher-alloy tool steel, such as D2, whose increased abrasion resistance would permit longer runs between regrinds.

A still further increase in the abrasion-resistant properties was then sought, which led to the use of D7, the high carbon and high chromium content of which provided excellent edge retainment, although at the cost of greatly reduced grindability. Finally, it became a matter of economic appraisal, whether the somewhat shorter tool regrind intervals (for D2) or the more expensive tool sharpening (for D7) constituted the lesser burden.

Example 2, Workpiece—Circular form cutter made of high-speed tool steel for use on multiple-spindle automatic turning machines: The first choice from the Quick Reference Guide may be the classical tungsten-base high-speed tool steel T1, because of its good performance and ease of heat treatment, or its alternate in the molybdenum high-speed tool steel category, the type M2.

In practice, neither of these grades provided a tool that could hold its edge and profile over the economical tool change time, because of the abrasive properties of the work material and the high cutting speeds applied in the cycle. An overrating of the problem resulted in reaching for the top of the scale, making the tool from T15, a high-alloy high-speed tool steel (high vanadium and high cobalt).

Although the performance of the tools made of T15 was excellent, the cost of this steel type was rather high, and the grinding of the tool, both for making it and in the regularly needed resharpening, proved to be very time-consuming and expensive. Therefore, an intermediate tool steel type was tried, the M3 that provided added abrasion resistance (due to increased carbon and vanadium content), and was less expensive and much easier to grind than the T15.

High-Speed Tool Steels

The primary application of high-speed steels is to tools used for the working of metals at high cutting speeds. Cutting metal at high speed generates heat, the penetration of the cutting tool edge into the work material requires great hardness and strength, and the continued frictional contact of the tool with both the parent material and the detached chips can only be sustained by an abrasion-resistant tool edge.

Accordingly, the dominant properties of high-speed steel are B) resistance to the softening effect of elevated temperature; C) great hardness penetrating to substantial depth from the surface; and D) excellent abrasion resistance.

High-speed tool steels are listed in the AISI specifications in two groups: molybdenum types and tungsten types, these designations expressing the dominant alloying element of the respective group.

Molybdenum-Type High-Speed Tool Steels.—Unlike the traditional tungsten-base high-speed steels, the tool steels listed in this category are considered to have molybdenum as the principal alloying constituent, this element also being used in the designation of the group. Other significant elements like tungsten and cobalt might be present in equal, or even greater amounts in several types listed in this category. The available range of types also includes high-speed tool steels with higher than usual carbon and vanadium content. Amounts of these alloying elements have been increased to obtain better abrasion resistance although such a change in composition may adversely affect the machinability and the grindability of the steel. The series in whose AISI identification numbers the number 4 is the first digit was developed to attain exceptionally high hardness in heat treatment that, for these types, usually requires triple tempering rather than the double tempering generally applied for high-speed tool steels.

Properties and Applications of Frequently Used Molybdenum Types: AISI M1: This alloy was developed as a substitute for the classical T1 to save on the alloying element tungsten by replacing most of it with molybdenum. In most uses, this steel is an acceptable substitute, although it requires greater care or more advanced equipment for its heat treatment than the tungsten alloyed type it replaces. The steel is often selected for cutting tools like drills, taps, milling cutters, reamers, lathe tools used for lighter cuts, and for shearing dies.

AISI M2: Similar to M1, yet with substantial tungsten content replacing a part of the molybdenum. This is one of the general-purpose high-speed tool steels, combining the economic advantages of the molybdenum-type steels with greater ease of hardening, excellent wear resistance, and improved toughness. It is a preferred steel type for the manufacture of general-purpose lathe tools; of most categories of multiple-edge cutting tools, like milling cutters, taps, dies, reamers, and for form tools in lathe operations.

AISI M3: A high-speed tool steel with increased vanadium content for improved wear resistance, yet still below the level where vanadium would interfere with the ease of grinding. This steel is preferred for cutting tools requiring improved wear resistance, like broaches, form tools, milling cutters, chasers, and reamers.

AISI M7: The chemical composition of this type is similar to that of M1, except for the higher carbon and vanadium content that raises the cutting efficiency without materially reducing the toughness. Because of sensitivity to decarburization, heat treatment in a salt bath or a controlled atmosphere is advisable. Used for blanking and trimming dies, shear blades, lathe tools, and thread rolling dies.

AISI M10: Although the relatively high vanadium content assures excellent wear and cutting properties, the only slightly increased carbon does not cause brittleness to an extent that is harmful in many applications. Form cutters and single-point lathe tools, broaches, planer tools, punches, blanking dies, and shear blades are examples of typical uses.

AISI M42: In applications where high hardness both at regular and at elevated temperatures is needed, this type of high-speed steel with high cobalt content can provide excellent service. Typical applications are tool bits, form tools, shaving tools, fly cutters, roll turning tools, and thread rolling dies. Important uses are found for M42, and for other types of the "M40" group in the working of "difficult-to-machine" alloys.

Tungsten-Type High-Speed Tool Steels.—For several decades following their introduction, the tungsten-base high-speed steels were the only types available for cutting operations involving the generation of substantial heat, and are still preferred by users who do not have the kind of advanced heat-treating equipment that efficient hardening of the molybdenum-type high-speed tool steels requires. Most tungsten high-speed steels display excellent resistance to decarburization and can be brought to good hardness by simple heat treatment. However, even with tungsten-type high-speed steels, heat treatment using modern methods and furnaces can appreciably improve the metallurgical qualities of the hardened material and the performance of the cutting tools made from these steels.

Properties and Applications of Frequently Used Tungsten Types: AISI T1: Also mentioned as the 18–4–1 type with reference to the nominal percentage of its principal alloying elements (W–Cr–V), it is considered to be the classical type of high-speed tool steel. The chemical composition of T1 was developed in the early 1900s, and has changed very little since. T1 is still considered to be perhaps the best general-purpose high-speed tool steel because of the comparative ease of its machining and heat treatment. It combines a high degree of cutting ability with relative toughness. T1 steel is used for all types of multiple-edge cutting tools like drills, reamers, milling cutters, threading taps and dies, light- and medium-duty lathe tools, and is also used for punches, dies, and machine knives, as well as for structural parts that are subjected to elevated temperatures, like lathe centers, and certain types of antifriction bearings.

AISI T2: Similar to T1 except for somewhat higher carbon content and twice the vanadium contained in the former grade. Its handling ease, both in machining and heat treating,

Table 8. Molybdenum High-Speed Steels

Identifying Chemical Composition and Typical Heat-Treatment Data																			
	AISI Type	M1	M2	M3 Cl. 1	M3 Cl. 2	M4	M6	M7	M10	M30	M33	M34	M36	M41	M42	M43	M44	M46	M47
Identifying Chemical Elements in Per Cent	C	0.80	0.85; 1.00	1.05	1.20	1.30	0.80	1.00	0.85; 1.00	0.80	0.90	0.90	0.80	1.10	1.10	1.20	1.15	1.25	1.10
	W	1.50	6.00	6.00	6.00	5.50	4.00	1.75	…	2.00	1.50	2.00	6.00	6.75	1.50	2.75	5.25	2.00	1.50
	Mo	8.00	5.00	5.00	5.00	4.50	5.00	8.75	8.00	8.00	9.50	8.00	5.00	3.75	9.50	8.00	6.25	8.25	9.50
	Cr	4.00	4.00	4.00	4.00	4.00	4.00	4.00	4.00	4.00	4.00	4.00	4.00	4.25	3.75	3.75	4.25	4.00	3.75
	V	1.00	2.00	2.40	3.00	4.00	1.50	2.00	2.00	1.25	1.15	2.00	2.00	2.00	1.15	1.60	2.25	3.20	1.25
	Co	…	…	…	…	…	12.00	…	…	5.00	8.00	8.00	8.00	5.00	8.00	8.25	12.00	8.25	5.00
Heat-Treat. Data	Hardening Temperature Range, °F	2150–2225	2175–2225	2200–2250	2200–2250	2200–2250	2150–2200	2150–2225	2150–2225	2200–2250	2200–2250	2200–2250	2225–2275	2175–2220	2175–2210	2175–2220	2190–2240	2175–2225	2150–2200
	Tempering Temperature Range, °F	1000–1100	1000–1160	1000–1100	1000–1100	1000–1100	1000–1100	1000–1100	1000–1100	1000–1100	1000–1100	1000–1100	1000–1100	1000–1100	950–1100	950–1100	1000–1160	975–1050	975–1100
	Approx. Tempered Hardness, Rc	65–60	65–60	66–61	66–61	66–61	66–61	66–61	65–60	65–60	65–60	65–60	65–60	70–65	70–65	70–65	70–62	69–67	70–65
Relative Ratings of Properties (A = greatest to E = least)																			
Characteristics in Heat Treatment	Safety in Hardening	D	D	D	D	D	D	D	D	D	D	D	D	D	D	D	D	D	D
	Depth of Hardening	A	A	A	A	A	A	A	A	A	A	A	A	A	A	A	A	A	A
	Resistance to Decarburization	C	B	B	B	B	C	C	C	C	C	C	C	C	C	C	C	C	C
	Stability of Shape in Heat Treatment — Quenching Medium: Air or Salt	C	C	C	C	C	C	C	C	C	C	C	C	C	C	C	C	C	C
	Stability of Shape in Heat Treatment — Quenching Medium: Oil	D	D	D	D	D	D	D	D	D	D	D	D	D	D	D	D	D	D
Service Properties	Machinability	D	D	D	D/E	D	D	D	D	D	D	D	D	D	D	D	D	D	D
	Hot Hardness	B	B	B	B	B	A	B	B	A	A	A	A	A	A	A	A	A	A
	Wear Resistance	B	B	B	B	A	B	B	B	B	B	B	B	B	B	B	B	B	B
	Toughness	E	E	E	E	E	E	E	E	E	E	E	E	E	E	E	E	E	E

is comparable to that of T1, although it should be held at the quenching temperature slightly longer, particularly when the heating is carried out in a controlled atmosphere furnace. The applications are similar to that of T1, however, because of its increased wear resistance T2 is preferred for tools required for finer cuts, and where the form or size retention of the tool is particularly important, such as for form and finishing tools.

AISI T5: The essential characteristic of this type of high-speed steel, its superior red hardness, stems from its substantial cobalt content that, combined with the relatively high amount of vanadium, provides this steel with excellent wear resistance. In heat treatment, the tendency for decarburization must be considered, and heating in a controlled, slightly reducing atmosphere is recommended. This type of high-speed tool steel is mainly used for single-point tools and inserts; it is well adapted for working at high-speeds and feeds, for cutting hard materials and those that produce discontinuous chips, also for nonferrous metals and, in general, for all kinds of tools needed for hogging (removing great bulks of material).

AISI T15: The performance qualities of this high-alloy tool steel surpass most of those found in other grades of high-speed tool steels. The high vanadium content, supported by uncommonly high carbon assures superior cutting ability and wear resistance. The addition of high cobalt increases the "hot hardness," and therefore tools made of T15 can sustain cutting speeds in excess of those commonly applicable to tools made of steel. The machining and heat treatment of T15 does not cause extraordinary problems, although for best results, heating to high temperature is often applied in its heat treatment, and double or even triple tempering is recommended. On the other hand, T15 is rather difficult to grind because of the presence of large amounts of very hard metallic carbides; therefore, it is considered to have a very low "grindability" index. The main uses are in the field of high-speed cutting and the working of hard metallic materials, T15 being often considered to represent in its application a transition from the regular high-speed tool steels to cemented carbides. Lathe tool bits, form cutters, and solid and inserted blade milling cutters are examples of uses of this steel type for cutting tools; excellent results may also be obtained with such tools as cold-work dies, punches, blanking, and forming dies, etc. The low toughness rating of the T15 steel excludes its application for operations that involve shock or sudden variations in load.

Hot-Work Tool Steels

A family of special tool steels has been developed for tools that in their regular service are in contact with hot metals over a shorter or longer period of time, with or without cooling being applied, and are known as hot-work steels. The essential property of these steels is their capability to sustain elevated temperature without seriously affecting the usefulness of the tools made from them. Depending on the purpose of the tools for which they were developed, the particular types of hot-work tool steels have different dominant properties and are assigned to one of three groups, based primarily on their principal alloying elements.

Hot-Work Tool Steels, Chromium Types.—As referred to in the group designation, the chromium content is considered the characteristic element of these tool steels. Their predominant properties are high hardenability, excellent toughness, and great ductility, even at the cost of wear resistance. Some members of this family are made with the addition of tungsten, and in one type, cobalt as well. These alloying elements improve the resistance to the softening effect of elevated temperatures, but reduce ductility.

Properties and Applications of Frequently Used Chromium Types: AISI H11: This hot-work tool steel of the Chromium–molybdenum–vanadium type has excellent ductility, can be machined easily, and retains its strength at temperatures up to 1000 degrees F.

Table 9. Tungsten High-Speed Tool Steels

Identifying Chemical Composition and Typical Heat-Treatment Data									
	AISI Type		T1	T2	T4	T5	T6	T8	T15
Identifying Chemical Elements in Per Cent	C		0.75	0.80	0.75	0.80	0.80	0.75	1.50
	W		18.00	18.00	18.00	18.00	20.00	14.00	12.00
	Cr		4.00	4.00	4.00	4.00	4.50	4.00	4.00
	V		1.00	2.00	1.00	2.00	1.50	2.00	5.00
	Co		…	…	5.00	…	…	5.00	5.00
Heat-Treat. Data	Hardening Temperature Range, °F		2300–2375	2300–2375	2300–2375	2325–2375	2325–2375	2300–2375	2200–2300
	Tempering Temperature Range, °F		1000–1100	1000–1100	1000–1100	1000–1100	1000–1100	1000–1100	1000–1200
	Approx. Tempered Hardness, R_c		65–60	66–61	66–62	65–60	65–60	65–60	68–63
Relative Ratings of Properties (A = greatest to E = least)									
Characteristics in Heat Treatment	Safety in Hardening		C	C	D	D	D	D	D
	Depth of Hardening		A	A	A	A	A	A	A
	Resistance to Decarburization		A	A	B	C	C	B	B
	Stability of Shape in Heat Treatment — Quenching Medium	Air or Salt	C	C	C	C	C	C	C
		Oil	D	D	D	D	D	D	D
Service Properties	Machinability		D	D	D	D	D/E	D	D/E
	Hot Hardness		B	B	A	A	A	A	A
	Wear Resistance		B	B	B	B	B	B	A
	Toughness		E	E	E	E	E	E	E

These properties, combined with relatively good abrasion and shock resistance, account for the varied fields of application of H11, which include the following typical uses:

E) structural applications where high strength is needed at elevated operating temperatures, as for gas turbine engine components; and F) hot-work tools, particularly of the kind whose service involves shocks and drastic cooling of the tool, such as in extrusion tools, pierce and draw punches, bolt header dies, etc.

AISI H12: The properties of this type of steel are comparable to those of H11, with increased abrasion resistance and hot hardness, resulting from the addition of tungsten, yet in an amount that does not affect the good toughness of this steel type. The applications, based on these properties, are hot-work tools that often have to withstand severe impact, such as various punches, bolt header dies, trimmer dies, and hot shear blades. H12 is also used to make aluminum extrusion dies and die-casting dies.

AISI H13: This type of tool steel differs from the preceding ones particularly in properties related to the addition of about 1 per cent vanadium, which contributes to increased hot hardness, abrasion resistance, and reduced sensitivity to heat checking. Such properties are needed in die casting, particularly of aluminum, where the tools are subjected to drastic heating and cooling at high operating temperatures. Besides die-casting dies, H13 is also widely used for extrusion dies, trimmer dies, hot gripper and header dies, and hot shear blades.

AISI H19: This high-alloyed hot-work tool steel, containing chromium, tungsten, cobalt, and vanadium, has excellent resistance to abrasion and shocks at elevated temperatures. It is particularly well adapted to severe hot-work uses where the tool, to retain its size and shape, must withstand wear and the washing-out effect of molten work material. Typical applications include brass extrusion dies and dummy blocks, inserts for forging and valve extrusion dies, press forging dies, and hot punches.

Hot-Work Tool Steels, Tungsten Types.—Substantial amounts of tungsten, yet very low-carbon content characterize the hot-work tool steels of this group. These tool steels have been developed for applications where the tool is in contact with the hot-work material over extended periods of time; therefore, the resistance of the steel to the softening effect of elevated temperatures is of prime importance. even to the extent of accepting a lower degree of toughness.

Properties and Applications of Frequently Used Tungsten Types: AISI H21: This medium-tungsten alloyed hot-work tool steel has substantially increased abrasion resistance over the chromium alloyed types, yet possesses a degree of toughness that represents a transition between the chromium and the higher-alloyed tungsten-steel types. The principal applications are for tools subjected to continued abrasion, yet to only a limited amount of shock loads, like tools for the extrusion of brass, both dies and dummy blocks, pierces for forging machines, inserts for forging tools, and hot nut tools. Another typical application is dies for the hot extrusion of automobile valves.

AISI H24: The comparatively high tungsten content (about 14 per cent) of this steel results in good hardness, great compression strength, and excellent abrasion resistance, but makes it sensitive to shock loads. By taking these properties into account, the principal applications include extrusion dies for brass in long-run operations, hot-forming and gripper dies with shallow impressions, punches that are subjected to great wear yet only to moderate shocks, and hot shear blades.

AISI H20: The composition of this high-alloyed tungsten-type hot-work steel resembles the tungsten-type high-speed steel AISI T1, except for the somewhat lower carbon content for improved toughness. The high amount of tungsten provides the maximum resistance to the softening effect of elevated temperature and assures excellent wear-resistant properties, including withstanding the washing-out effect of certain processes. However, this steel is less resistant to thermal shocks than the chromium hot-work steels. Typical applications comprise extrusion dies for long production runs, extrusion mandrels operated

Table 10. Hot-Work Tool Steels

Identifying Chemical Composition and Typical Heat-Treatment Data																
AISI	Group	Chromium Types						Tungsten Types						Molybdenum Types		
	Type	H10	H11	H12	H13	H14	H19	H21	H22	H23	H24	H25	H26	H41	H42	H43
Identifying Chemical Elements in Per Cent	C	0.40	0.35	0.35	0.35	0.40	0.40	0.35	0.35	0.35	0.45	0.25	0.50	0.65	0.60	0.55
	W	…	…	1.50	…	5.00	4.25	9.00	11.00	12.00	15.00	15.00	18.00	1.50	6.00	…
	Mo	2.50	1.50	1.50	1.50	…	…	…	…	…	…	…	…	8.00	5.00	8.00
	Cr	3.25	5.00	5.00	5.00	5.00	4.25	3.50	2.00	12.00	3.00	4.00	4.00	4.00	4.00	4.00
	V	0.40	0.40	0.40	1.00	…	2.00	…	…	…	…	…	1.00	1.00	2.00	2.00
	Co	…	…	…	…	…	4.25	…	…	…	…	…	…	…	…	…
Heat-Treat. Data	Hardening Temperature Range, °F	1850–1900	1825–1875	1825–1875	1825–1900	1850–1950	2000–2200	2000–2200	2000–2200	2000–2300	2000–2250	2100–2300	2150–2300	2000–2175	2050–2225	2000–2175
	Tempering Temperature Range, °F	1000–1200	1000–1200	1000–1200	1000–1200	1100–1200	1000–1300	1100–1250	1100–1250	1200–1500	1050–1200	1050–1250	1050–1250	1050–1200	1050–1200	1050–1200
	Approx. Tempered Hardness, Rc	56–39	54–38	55–38	53–38	47–40	59–40	54–36	52–39	47–30	55–45	44–35	58–43	60–50	60–50	58–45
Relative Ratings of Properties (A = greatest to D = least)																
Characteristics in Heat Treatment	Safety in Hardening	A	A	A	A	A	B	B	B	B	B	B	B	C	C	C
	Depth of Hardening	A	A	A	A	A	A	A	A	A	A	A	A	A	A	A
	Resistance to Decarburization	B	B	B	B	B	B	B	B	B	B	B	B	C	B	C
	Stability of Shape in Heat Treatment — Quenching Medium: Air or Salt	B	B	B	B	C	C	C	C	…	C	C	C	C	C	C
	Stability of Shape in Heat Treatment — Quenching Medium: Oil	…	…	…	…	…	D	D	D	D	D	D	D	D	D	D
Service Properties	Machinability	C/D	C/D	C/D	C/D	D	D	D	D	D	D	D	D	D	D	D
	Hot Hardness	C	C	C	C	C	C	C	C	B	B	B	B	B	B	B
	Wear Resistance	D	D	D	D	D	C/D	C/D	C/D	C/D	C	D	C	C	C	C
	Toughness	C	B	B	B	C	C	C	C	D	D	C	D	D	D	D

without cooling, hot piercing punches, hot forging dies and inserts. It is also used as special structural steel for springs operating at elevated temperatures.

Hot-Work Tool Steels, Molybdenum Types.—These steels are closely related to certain types of molybdenum high-speed steels and possess excellent resistance to the softening effect of elevated temperature but their ductility is rather low. These steel types are generally available on special orders only.

Properties and Applications of Frequently Used Molybdenum Types: AISI H43: The principal constituents of this hot-work steel, chromium, molybdenum, and vanadium, provide excellent abrasion- and wear-resistant properties at elevated temperatures. H43 has a good resistance to the development of heat checks and a toughness adequate for many different purposes. Applications include tools and operations that tend to cause surface wear in high-temperature work, like hot headers, punch and die inserts, hot heading and hot nut dies, as well as different kinds of punches operating at high temperature in service involving considerable wear.

Cold-Work Tool Steels

Tool steels of the cold-working category are primarily intended for die work, although their use is by no means restricted to that general field. Cold-work tool steels are extensively used for tools whose regular service does not involve elevated temperatures. They are available in chemical compositions adjusted to the varying requirements of a wide range of different applications. According to their predominant properties, characterized either by the chemical composition or by the quenching medium in heat treatment, the cold-work tool steels are assigned to three different groups, as discussed in what follows.

Cold-Work Tool Steels, High-Carbon, High-Chromium Types.—The chemical composition of tool steels of this family is characterized by the very high chromium content, to the order of 12 to 13 per cent, and the uncommonly high carbon content, in the range of about 1.50 to 2.30 per cent. Additional alloying elements that are present in different amounts in some of the steel types of this group are vanadium, molybdenum, and cobalt, each of which contributes desirable properties.

The predominant properties of the whole group are: 1) excellent dimensional stability in heat treatment, where, with one exception, air quench is used; 2) great wear resistance, particularly in the types with the highest carbon content; and 3) rather good machinability.

Properties and Applications of Frequently Used High-Carbon, High-Chromium Types:

AISI D2: An air-hardening die steel with high-carbon, high-chromium content having several desirable tool steel properties, such as abrasion resistance. high hardness, and nondeforming characteristics. The carbon content of this type, although relatively high, is not particularly detrimental to its machining. The ease of working can be further improved by selecting the same basic type with the addition of sulfur. Several steel producers supply the sulfurized version of D2, in which the uniformly distributed sulfide particles substantially improve the machinability and the resulting surface finish. The applications comprise primarily cold-working press tools for shearing (blanking and stamping dies, punches, shear blades), for forming (bending, seaming), also for thread rolling dies, solid gages, and wear-resistant structural parts. Dies for hot trimming of forgings are also made of D2 which is then heated treated to a lower hardness for the purpose of increasing toughness.

AISI D3: The high carbon content of this high-chromium tool steel type results in excellent resistance to wear and abrasion and provides superior compressive strength as long as the pressure is applied gradually, without exerting sudden shocks. In hardening, an oil quench is used, without affecting the excellent nondeforming properties of this type. Its deep-hardening properties make it particularly suitable for tools that require repeated regrinding during their service life, such as different types of dies and punches. The more important applications comprise blanking, stamping, and trimming dies and punches for

Table 11. Cold-Work Tool Steels

AISI		Identifying Chemical Composition and Typical Heat-Treatment Data																
	Group	High-Carbon, High-Chromium Types					Medium-Alloy, Air-Hardening Types								Oil-Hardening Types			
	Types	D2	D3	D4	D5	D7	A2	A3	A4	A6	A7	A8	A9	A10	O1	O2	O6	O7
Identifying Chemical Elements in Per Cent	C	1.50	2.25	2.25	1.50	2.35	1.00	1.25	1.00	0.70	2.25	0.55	0.50	1.35	0.90	0.90	1.45	1.20
	Mn	…	…	…	…	…	…	…	2.00	2.00	…	…	…	1.80	1.00	1.60	…	…
	Si	…	…	…	…	…	…	…	…	…	…	…	…	1.25	…	…	1.00	…
	W	…	…	…	…	…	…	…	…	…	1.00	1.25	…	…	0.50	…	…	1.75
	Mo	1.00	…	1.00	1.00	1.00	1.00	1.00	1.00	1.25	1.00	1.25	1.40	1.50	…	…	0.25	…
	Cr	12.00	12.00	12.00	12.00	12.00	5.00	5.00	1.00	1.00	5.25	5.00	5.00	…	0.50	…	…	0.75
	V	1.00	…	…	…	4.00	…	1.00	…	…	4.75	…	1.00	…	…	…	…	…
	Co	…	…	…	3.00	…	…	…	…	…	…	…	…	…	…	…	…	…
	Ni	…	…	…	…	…	…	…	…	…	…	…	1.50	1.80	…	…	…	…
Heat-Treatment Data	Hardening Temperature Range, °F	1800–1875	1700–1800	1775–1850	1800–1875	1850–1950	1700–1800	1750–1850	1500–1600	1525–1600	1750–1800	1800–1850	1800–1875	1450–1500	1450–1500	1400–1475	1450–1500	1550–1525
	Quenching Medium	Air	Oil	Air	Air	Air	Air	Air	Air	Air	Air	Air	Air	Air	Oil	Oil	Oil	Oil
	Tempering Temperature Range, °F	400–1000	400–1000	400–1000	400–1000	300–1000	350–1000	350–1000	350–800	300–800	300–1000	350–1100	950–1150	350–800	350–500	350–500	350–600	350–550
	Approx. Tempered Hardness, Rc	61–54	61–54	61–54	61–54	65–58	62–57	65–57	62–54	60–54	67–57	60–50	56–35	62–55	62–57	62–57	63–58	64–58
	Relative Ratings of Properties (A = greatest to E = least)																	
Characteristics in Heat Treatment	Safety in Hardening	A	C	A	A	A	A	A	A	A	A	A	A	A	B	B	B	B
	Depth of Hardening	A	A	A	A	A	A	A	A	A	A	A	A	A	B	B	B	B
	Resistance to Decarburization	B	B	B	B	B	B	B	A/B	A/B	B	B	B	A/B	A	A	A	A
	Stability of Shape in Heat Treatment	A	B	A	A	A	A	A	A	A	A	A	A	A	B	B	B	B
Service Properties	Machinability	E	E	E	E	E	D	D	D/E	D/E	E	D	D	C/D	C	C	B	C
	Hot Hardness	C	C	C	C	C	C	C	D	D	C	C	C	D	E	E	E	E
	Water Resistance	B/C	B	B	B/C	A	C	B	C/D	C/D	A	C/D	C/D	C	D	D	D	D
	Toughness	E	E	E	E	E	D	D	D	D	E	C	C	D	D	D	D	C

long production runs; forming, bending and drawing tools; and structural elements like plug and ring gages, and lathe centers, in applications where high wear resistance is important.

Cold-Work Tool Steels, Oil-Hardening Types.—With a relatively low percentage of alloying elements, yet with a substantial amount of manganese, these less expensive types of tool steels attain good depth of hardness in an oil quench, although at the cost of reduced resistance to deformation. Their good machinability supports general-purpose applications, yet because of relatively low wear resistance, they are mostly selected for comparatively short-run work.

Properties and Applications of Frequently Used Oil-Hardening Types: AISI O1: A low-alloy tool steel that is hardened in oil and exhibits only a low tendency to shrinking or warping. It is used for cutting tools, the operation of which does not generate high heat, such as taps and threading dies, reamers, and broaches, and for press tools like blanking, trimming, and forming dies in short- or medium-run operations.

AISI O2: Manganese is the dominant alloying element in this type of oil-hardening tool steel that has good nondeforming properties, can be machined easily, and performs satisfactorily in low-volume production. The low hardening temperature results in good safety in hardening, both with regard to form stability and freedom from cracking. The combination of handling ease, including free-machining properties, with good wear resistance, makes this type of tool steel adaptable to a wide range of common applications such as cutting tools for low- and medium-speed operations; forming tools including thread rolling dies; structural parts such as bushings and fixed gages, and for plastics molds.

AISI O6: This oil-hardening type of tool steel belongs to a group often designated as graphitic because of the presence of small particles of graphitic carbon that are uniformly dispersed throughout the steel. Usually, about one-third of the total carbon is present as free graphite in nodular form, which contributes to the uncommon ease of machining. In the service of parts made of this type of steel, the free graphite acts like a lubricant, reducing wear and galling. The ease of hardening is also excellent, requiring only a comparatively low quenching temperature. Deep hardness penetration is produced and the oil quench causes very little dimensional change. The principal applications of the O6 tool steel are in the field of structural parts, like arbors, bushings, bodies for inserted tool cutters, and shanks for cutting tools, jigs, and machine parts, and fixed gages like plugs, rings, and snap gages. It is also used for blanking, forming, and trimming dies and punches, in applications where the stability of the tool material is more important than high wear resistance.

Cold-Work Tool Steels, Medium-Alloy, Air-Hardening Types.—The desirable nondeforming properties of the high-chromium types are approached by the members of this family, with substantially lower alloy content that, however, is sufficient to permit hardening by air quenching. The machinability is good, and the comparatively low wear resistance is balanced by relatively high toughness, a property that, in certain applications, may be considered of prime importance.

Properties and Applications of Frequently Used Medium-Alloy, Air-Hardening Types:

AISI A2: The lower chromium content, about 5 per cent, makes this air-hardening tool steel less expensive than the high-chromium types, without affecting its nondeforming properties. The somewhat reduced wear resistance is balanced by greater toughness, making this type suitable for press work where the process calls for tough tool materials. The machinability is improved by the addition of about 0.12 percent sulfur, offered as a variety of the basic composition by several steel producers. The prime uses of this tool steel type are punches for blanking and forming, cold and hot trimming dies (the latter heat treated to a lower hardness), thread rolling dies, and plastics molds.

AISI A6: The composition of this type of tool steel makes it adaptable to air hardening from a relatively low temperature, comparable to that of oil-hardening types, yet offering improved stability in heat treating. Its reduced tendency to heat-treatment distortions makes this tool steel type well adapted for die work, forming tools, and gages, which do not require the highest degree of wear resistance.

Shock-Resisting, Mold, and Special-Purpose Tool Steels

There are fields of tool application in which specific properties of the tool steels have dominant significance, determining to a great extent the performance and the service life of tools made of these materials. To meet these requirements, special types of tool steels have been developed. These individual types grew into families with members that, while similar in their major characteristics, provide related properties to different degrees. Originally developed for a specific use, the resulting particular properties of some of these tool steels made them desirable for other uses as well. In the tool steel classification system, they are shown in three groups, as discussed in what follows.

Shock-Resisting Tool Steels.—These steels are made with low-carbon content for increased toughness, even at the expense of wear resistance, which is generally low. Each member of this group also contains alloying elements, different in composition and amount, selected to provide properties particularly adjusted to specific applications. Such varying properties are the degree of toughness (generally, high in all members), hot hardness, abrasion resistance, and machinability.

Properties and Applications of Frequently Used Shock-Resisting Types: AISI S1: This Chromium–tungsten alloyed tool steel combines, in its hardened state, great toughness with high hardness and strength. Although it has a low-carbon content for reasons of good toughness, the carbon-forming alloys contribute to deep hardenability and abrasion resistance. When high wear resistance is also required, this property can be improved by carburizing the surface of the tool while still retaining its shock-resistant characteristics. Primary uses are for battering tools, including hand and pneumatic chisels. The chemical composition, particularly the silicon and tungsten content, provides good hot hardness, too, up to operating temperatures of about 1050 °F, making this tool steel type also adaptable for such hot-work tool applications involving shock loads, as headers, pierces, forming tools, drop forge die inserts, and heavy shear blades.

AISI S2: This steel type serves primarily for hand chisels and pneumatic tools, although it also has limited applications for hot work. Although its wear-resistance properties are only moderate, S2 is sometimes used for forming and thread rolling applications, when the resistance to rupturing is more important than extended service life. For hot-work applications, this steel requires heat treatment in a neutral atmosphere to avoid either carburization or decarburization of the surface. Such conditions make this tool steel type particularly susceptible to failure in hot-work uses.

AISI S5: This composition is essentially a Silicon–manganese type tool steel with small additions of chromium, molybdenum, and vanadium for the purpose of improved deep hardening and refinement of the grain structure. The most important properties of this steel are its high elastic limit and good ductility, resulting in excellent shock-resisting characteristics, when used at atmospheric temperatures. Its recommended quenching medium is oil, although a water quench may also be applied as long as the design of the tools avoids sharp corners or drastic sectional changes. Typical applications include pneumatic tools in severe service, like chipping chisels, also shear blades, heavy-duty punches, and bending rolls. Occasionally, this steel is also used for structural applications, like shanks for carbide tools and machine parts subject to shocks.

Table 12. Shock-Resisting, Mold, and Special-Purpose Tool Steels

AISI	Category	Shock-Resisting Tool Steels				Mold Steels							Special-Purpose Tool Steels				
		Identifying Chemical Composition and Typical Heat-Treatment Data															
	Types	S1	S2	S5	S7	P2	P3	P4	P5	P6	P20	P21[a]	L2[b]	L3[b]	L6	F1	F2
Identifying Elements in Per Cent	C	0.50	0.50	0.55	0.50	0.07	0.10	0.07	0.10	0.10	0.35	0.20	0.50/ 1.10	1.00	0.70	1.00	1.25
	Mn	…	…	0.80	…	…	…	…	…	…	…	…	…	…	…	…	…
	Si	…	1.00	2.00	…	…	…	…	…	…	…	…	…	…	…	…	…
	W	2.50	…	…	…	…	…	…	…	…	…	…	…	…	…	1.25	3.50
	Mo	…	0.50	0.40	1.40	0.20	…	0.75	…	…	0.40	…	…	…	0.25	…	…
	Cr	1.50	…	…	3.25	2.00	0.60	5.00	2.25	1.50	1.25	…	1.00	1.50	0.75	…	…
	V	…	…	…	…	…	…	…	…	…	…	…	0.20	0.20	…	…	…
	Ni	…	…	…	…	0.50	1.25	…	…	3.50	…	4.00	…	…	1.50	…	…
Heat-Treat. Data	Hardening Temperature, °F	1650–1750	1550–1650	1600–1700	1700–1750	1525–1550[c]	1475–1525[c]	1775–1825[c]	1550–1600[c]	1450–1500[c]	1500–1600[c]	Soln. treat.	1550–1700	1500–1600	1450–1550	1450–1600	1450–1600
	Tempering Temp. Range, °F	400–1200	350–800	350–800	400–1150	350–500	350–500	350–900	350–500	350–450	900–1100	Aged	350–1000	350–600	350–1000	350–500	350–500
	Approx. Tempered Hardness, Rc	58–40	60–50	60–50	57–45	64–58 [d]	64–58[d]	64–58[d]	64–58[d]	61–58[d]	37–28[d]	40–30	63–45	63–56	62–45	64–60	65–62
	Relative Ratings of Properties (A = greatest to E = least)																
Characteristics in Heat Treatment	Safety in Hardening	C	E	C	B/C	C	C	C	C	C	C	A	D	D	C	E	E
	Depth of Hardening	B	B	B	A	B[e]	B[e]	B[e]	B[e]	A[e]	B	A	B	B	B	C	C
	Resist. to Decarb.	B	C	C	B	A	A	A	A	A	A	A	A	A	A	A	A
	Stability of Shape in Heat Treatment — Quench. Med.: Air	…	…	…	A	…	…	B	…	B	C	A	…	…	…	…	…
	Stability of Shape in Heat Treatment — Quench. Med.: Oil	D	…	D	C	C	C	…	C	C	…	A	D	D	C	…	…
	Stability of Shape in Heat Treatment — Quench. Med.: Water[f]	…	E	…	…	…	…	…	E	…	…	…	E	E	…	E	E
Service Properties	Machinability	D	C/D	C/D	D	C/D	D	D/E	D	D	C/D	D	C	C	D	C	D
	Hot Hardness	D	E	E	C	E	E	D	E	E	E	D	E	E	E	E	E
	Wear Resistance	D/E	D/E	D/E	D/E	D	D	C	D	D	D/E	D	D/E	D	D	D	B/C
	Toughness	B	A	A	B	C	C	C	C	C	C	D	B	D	B	E	E

[a] Contains also about 1.20 per cent A1. Solution treated in hardening.
[b] Quenched in oil.
[c] After carburizing.
[d] Carburized case.
[e] Core hardenability.
[f] Sometimes brine is used.

Mold Steels.—These materials differ from all other types of tool steels by their very low-carbon content, generally requiring carburizing to obtain a hard operating surface. A special property of most steel types in this group is the adaptability to shaping by impression (hobbing) instead of by conventional machining. They also have high resistance to decarburization in heat treatment and dimensional stability, characteristics that obviate the need for grinding following heat treatment. Molding dies for plastics materials require an excellent surface finish, even to the degree of high luster; the generally high-chromium content of these types of tool steels greatly aids in meeting this requirement.

Properties and Applications of Frequently Used Mold Steel Types: AISI P3 and P4: Essentially, both types of tool steels were developed for the same special purpose, that is, the making of plastics molds. The application conditions of plastics molds require high core strength, good wear resistance at elevated temperature, and excellent surface finish. Both types are carburizing steels that possess good dimensional stability. Because hobbing, that is, sinking the cavity by pressing a punch representing the inverse replica of the cavity into the tool material, is the process by which many plastics mold cavities are produced, good "hobbability" of the tool steels used for this purpose is an important requirement. The different chemistry of these two types of mold steels is responsible for the high core hardness of the P4, which makes it better suited for applications requiring high strength at elevated temperature.

AISI P6: This nickel–chromium-type plastics mold steel has exceptional core strength and develops a deep carburized case. Due to the high nickel–chromium content, the cavities of molds made of this steel type are produced by machining rather than by hobbing. An outstanding characteristic of this steel type is the high luster that is produced by polishing of the hard case surface.

AISI P20: This general-type mold steel is adaptable to both through hardening and carburized case hardening. In through hardening, an oil quench is used and a relatively lower, yet deeply penetrating hardness is obtained, such as is needed for zinc die-casting dies and injection molds for plastics. After the direct quenching and tempering, carburizing produces a very hard case and comparatively high core hardness. When thus heat treated, this steel is particularly well adapted for making compression, transfer, and plunger-type plastics molds.

Special-Purpose Tool Steels.—These steels include several low-alloy types of tool steels that were developed to provide transitional types between the more commonly used basic types of tool steels, and thereby contribute to the balancing of certain conflicting properties such as wear resistance and toughness; to offer intermediate depth of hardening; and to be less expensive than the higher-alloyed types of tool steels.

Properties and Applications of Frequently Used Special-Purpose Types: AISI L6: This material is a low-alloy-type special-purpose tool steel. The comparatively safe hardening and the fair nondeforming properties, combined with the service advantage of good toughness in comparison to most other oil-hardening types, explains the acceptance of this steel with a rather special chemical composition. The uses of L6 are for tools whose toughness requirements prevail over abrasion-resistant properties, such as forming rolls and forming and trimmer dies in applications where combinations of moderate shock- and wear-resistant properties are sought. The areas of use also include structural parts, like clutch members, pawls, and knuckle pins, that must withstand shock loads and still display good wear properties.

AISI F2: This carbon–tungsten type is one of the most abrasion-resistant of all water-hardening tool steels. However, it is sensitive to thermal changes, such as are involved in heat treatment and it is also susceptible to distortions. Consequently, its use is limited to tools of simple shape in order to avoid cracking in hardening. The shallow hardening characteristics of F2 result in a tough core and are desirable properties for certain tool types that, at the same time, require excellent wear-resistant properties.

Water-Hardening Tool Steels.—Steel types in this category are made without, or with only a minimum amount of alloying elements and, their heat treatment needs the harsh quenching action of water or brine, hence the general designation of the category.

Water-hardening steels are usually available with different percentages of carbon, to provide properties required for different applications; the classification system lists a carbon range of 0.60 to 1.40 per cent. In practice, however, the steel mills produce these steels in a few varieties of differing carbon content, often giving proprietary designations to each particular group. Typical carbon content limits of frequently used water-hardening tool steels are 0.70–0.90, 0.90–1.10, 1.05–1.20, and 1.20–1.30 per cent. The appropriate group should be chosen according to the intended use, as indicated in the steel selection guide for this category, keeping in mind that whereas higher carbon content results in deeper hardness penetration, it also reduces toughness.

The general system distinguishes the following four grades: 1) special; 2) extra; 3) standard; and 4) commercial.

listed in the order of decreasing quality. The differences between these grades, which are not offered by all steel mills, are defined in principle only. The distinguishing characteristics are purity and consistency, resulting from different degrees of process refinement and inspection steps applied in making the steel. Higher qualities are selected for assuring dependable uniformity and performance of the tools made from the steel.

The groups with higher carbon content are more sensitive to heat-treatment defects and are generally used for the more demanding applications, so the better grades are usually chosen for the high-carbon types and the lower grades for applications where steels with lower carbon content only are needed.

Water-hardening tool steels, although the least expensive, have several drawbacks, but these are quite acceptable in many types of applications. Some limiting properties are the tendency to deformation in heat treatment due to harsh effects of the applied quenching medium, the sensitivity to heat during the use of the tools made of these steels, the only fair degree of toughness, and the shallow penetration of hardness. However, this last-mentioned property may prove a desirable characteristic in certain applications, such as cold-heading dies, because the relatively shallow hard case is supported by the tough, although softer core.

The AISI designation for water-hardening tool steels is W, followed by a numeral indicating the type, primarily defined by the steel's chemical composition, as shown in the Table 13.

Recommended Applications of Water-Hardening Type W1 (Plain Carbon) Tool Steels:

Group I (C-0.70 to 0.90%): This group is relatively tough and therefore preferred for tools that are subjected to shocks or abusive treatment. Used for such applications as: hand tools, chisels, screwdriver blades, cold punches, and nail sets, and fixture elements, vise jaws, anvil faces, and chuck jaws.

Group II (C-0.90 to 1.10%): This group combines greater hardness with fair toughness, resulting in improved cutting capacity and moderate ability to sustain shock loads. Used for such applications as: hand tools, knives, center punches, pneumatic chisels, cutting tools, reamers, hand taps, and threading dies, wood augers; die parts, drawing and heading dies, shear knives, cutting and forming dies; and fixture elements, drill bushings, lathe centers, collets, and fixed gages.

Group III (C-1.05 to 1.20%): The higher carbon content of this group increases the depth of hardness penetrations, yet reduces toughness, thus the resistance to shock loads. Preferred for applications where wear resistance and cutting ability are the prime considerations. Used for such applications as: hand tools, woodworking chisels, paper knives, cutting tools (for low-speed applications), milling cutters, reamers, planer tools, thread chasers, center drills, die parts, cold blanking, coining, bending dies.

Group IV (C-1.20 to 1–30%): The high carbon content of this group produces a hard case of considerable depth with improved wear resistance yet sensitive to shock and concentrated stresses. Selected for applications where the capacity to withstand abrasive wear is needed, and where the retention of a keen edge or the original shape of the tool is important. Used for such applications as: cutting tools for finishing work, like cutters and reamers, and for cutting chilled cast iron and forming tools, for ferrous and nonferrous metals, and burnishing tools.

By adding small amounts of alloying elements to W-steel types 2 and 5, certain characteristics that are desirable for specific applications are improved. The vanadium in type 2 contributes to retaining a greater degree of fine-grain structure after heat treating. Chromium in type 5 improves the deep-hardening characteristics of the steel, a property needed for large sections, and assists in maintaining the keen cutting edge that is desirable in cutting tools like broaches, reamers, threading taps, and dies.

Table 13. Water-Hardening Tool Steels—Identifying Chemical Composition and Heat-Treatment Data

Chemical Composition in Per Cent	AISI Types		
	W1	W2	W5
C	0.60–1.40	0.60–1.40	1.10
	Varying carbon content may be available		
V	…	0.25	…
Cr	These elements are adjusted to satisfy the hardening requirements		0.50
Mn			
Si			

Heat-Treatment Data		
Hardening Temperature Ranges, °F Varying with Carbon Content	0.60–0.80%	1450–1500
	0.85–1.05%	1425–1550
	1.10–1.40%	1400–1525
Quenching Medium		Brine or Water
Tempering Temperature Range, °F		350–650
Approx. Tempered Hardness, Rc		64–50

Relative Ratings of Properties (A = greatest to E = least)							
Characteristics in Heat Treatment				Service Properties			
Safety in Hardening	Depth of Hardening	Resistance to Decarburization	Stability of Shape in Heat Treatment	Machinability	Hot Hardness	Wear Resistance	Toughness
D	C	A	E	A	E	D/E	C/D

Mill Production Forms of Tool Steels

Tool steels are produced in many different forms, although not all those listed in the following are always readily available; certain forms and shapes are made for special orders only.

Hot-Finished Bars and Cold-Finished Bars.—These bars are the most commonly produced forms of tool steels. Bars can be furnished in many different cross-sections, the round shape being the most common. Sizes can vary over a wide range, with a more limited number of standard stock sizes. Various conditions may also be available, however, technological limitations prevent all conditions applying to every size, shape, or type of steel. Tool steel bars may be supplied in one of the following conditions and surface finishes:

Conditions: Hot-rolled or forged (natural); hot-rolled or forged and annealed; hot-rolled or forged and heat-treated; cold- or hot-drawn (as drawn); and cold- or hot-drawn and annealed.

Finishes: Hot-rolled finish (scale not removed); pickled or blast-cleaned; cold-drawn; turned or machined; rough ground; centerless ground or precision flat ground; and polished (rounds only).

Other forms in which tool steels are supplied are the following:

Rolled or Forged Special Shapes: These shapes are usually produced on special orders only, for the purpose of reducing material loss and machining time in the large-volume manufacture of certain frequently used types of tools.

Forgings: All types of tool steels may be supplied in the form of forgings, which are usually specified for special shapes and for dimensions that are beyond the range covered by bars.

Wires: Tool steel wires are produced either by hot or cold drawing and are specified when special shapes, controlled dimensional accuracy, improved surface finish, or special mechanical properties are required. Round wire is commonly produced within an approximate size range of 0.015 to 0.500 inch, and these dimensions also indicate the limits within which other shapes of tool steel wires, like oval, square, or rectangular, may be produced.

Drill Rods: Rods are produced in round, rectangular, square, hexagonal, and octagonal shapes, usually with tight dimensional tolerances to eliminate subsequent machining, thereby offering manufacturing economies for the users.

Hot-Rolled Plates and Sheets, and Cold-Rolled Strips: Such forms of tool steel are generally specified for the high-volume production of specific tool types.

Tool Bits: These pieces are semifinished tools and are used by clamping in a tool holder or shank in a manner permitting ready replacement. Tool bits are commonly made of high-speed types of tool steels, mostly in square, but also in round, rectangular, andother shapes. Tool bits are made of hot rolled bars and are commonly, yet not exclusively, supplied in hardened and ground form, ready for use after the appropriate cutting edges are ground, usually in the user's plant.

Hollow Bars: These bars are generally produced by trepanning, boring, or drilling of solid round rods and are used for making tools or structural parts of annular shapes, like rolls, ring gages, bushings, etc.

Tolerances of Dimensions.—Such tolerances have been developed and published by the American Iron and Steel Institute (AISI) as a compilation of available industry experience that, however, does not exclude the establishment of closer tolerances, particularly for hot rolled products manufactured in large quantities. The tolerances differ for various categories of production processes (e.g., forged, hot-rolled, cold-drawn, centerless ground) and of general shapes.

Allowances for Machining.—These allowances provide freedom from soft spots and defects of the tool surface, thereby preventing failures in heat treatment or in service. After a layer of specific thickness, known as the allowance, has been removed, the bar or other form of tool steel material should have a surface without decarburization and other surface defects, such as scale marks or seams. The industry wide accepted machining allowance values for tool steels in different conditions, shapes, and size ranges are spelled out in AISI specifications and are generally also listed in the tool steel catalogs of the producer companies.

Decarburization Limits.—Heating of steel for production operation causes the oxidation of the exposed surfaces resulting in the loss of carbon. That condition, called decarburization, penetrates to a certain depth from the surface, depending on the applied process, the shape and the dimensions of the product. Values of tolerance for decarburization must be considered as one of the factors for defining the machining allowances, which must also compensate for expected variations of size and shape, the dimensional effects of heat treatment, and so forth. Decarburization can be present not only in hot-rolled and forged, but also in rough turned and cold-drawn conditions.

Advances in Tool Steel Making Technology.—Significant advances in processes for tool steel production have been made that offer more homogeneous materials of greater density and higher purity for applications where such extremely high quality is required. Two of these methods of tool steel production are of particular interest.

Vacuum-melted tool steels: These steels are produced by the consumable electrode method, which involves remelting of the steel originally produced by conventional processes. Inside a vacuum-tight shell that has been evacuated, the electrode cast of tool steel of the desired chemical analysis is lowered into a water-cooled copper mold where it strikes a low-voltage, high-amperage arc causing the electrode to be consumed by gradual melting. The undesirable gases and volatiles are drawn off by the vacuum, and the inclusions float on the surface of the pool, accumulating on the top of the produced ingot, to be removed later by cropping. In the field of tool steels, the consumable-electrode vacuum-melting (CVM) process is applied primarily to the production of special grades of hot-work and high-speed tool steels.

High-speed tool steels produced by powder metallurgy: The steel produced by conventional methods is reduced to a fine powder by a gas atomization process. The powder is compacted by a hot isostatic method with pressures in the range of 15,000 to 17,000 psi. The compacted billets are hot-rolled to the final bar size, yielding a tool-steel material which has 100 per cent theoretical density. High-speed tool steels produced by the P/M method offer a tool material providing increased tool wear life and high impact strength, of particular advantage in interrupted cuts.

Physical Properties

Physical Properties of Heat-Treated Steels.—Steels that have been "fully hardened" to the same hardness when quenched will have about the same tensile and yield strengths regardless of composition and alloying elements. When the hardness of such a steel is known, it is also possible to predict its reduction of area and tempering temperature. The accompanying figures illustrating these relationships have been prepared by the Society of Automotive Engineers.

Fig. 1 gives the range of Brinell hardnesses that could be expected for any particular tensile strength or it may be used to determine the range of tensile strengths that would correspond to any particular hardness. Fig. 2 shows the relationship between the tensile strength or hardness and the yield point. The solid line is the normal-expectancy curve. The dotted-line curves give the range of the variation of scatter of the plotted data. Fig. 3 shows the relationship that exists between the tensile strength (or hardness) and the reduction of area. The curve to the left represents the alloy steels and that on the right the carbon steels. Both are normal-expectancy curves and the extremities of the perpendicular lines that intersect them represent the variations from the normal-expectancy curves that may be caused by quality differences and by the magnitude of parasitic stresses induced by quenching. Fig. 4 shows the relationship between the hardness (or approximately equivalent tensile strength) and the tempering temperature. Three curves are given, one for fully hardened steels with a carbon content between 0.40 and 0.55 per cent, one for fully hardened steels with a carbon content between 0.30 and 0.40 per cent, and one for steels that are not fully hardened.

From Fig. 1, it can be seen that for a tensile strength of, say, 200,000 pounds per square inch, the Brinell hardness could range between 375 and 425. By taking 400 as the mean hardness value and using Fig. 4, it can be seen that the tempering temperature of fully hardened steels of 0.40 to 0.55 per cent carbon content would be 990 degrees F and that of fully hardened steels of 0.30 to 0.40 per cent carbon would be 870 degrees F. This chart also shows that the tempering temperature for a steel not fully hardened would approach 520 degrees F. A yield point of 0.9 × 200,000, or 180,000, pounds per square inch is indicated (Fig. 2) for the fully hardened steel with a tensile strength of 200,000 pounds per square inch. Most alloy steels of 200,000 pounds per square inch tensile strength would probably have a reduction in area of close to 44 per cent (Fig. 3) but some would have values in the range of 35 to 53 per cent. Carbon steels of the same tensile strength would probably have a reduction in area of close to 24 per cent but could possibly range from 17 to 31 per cent.

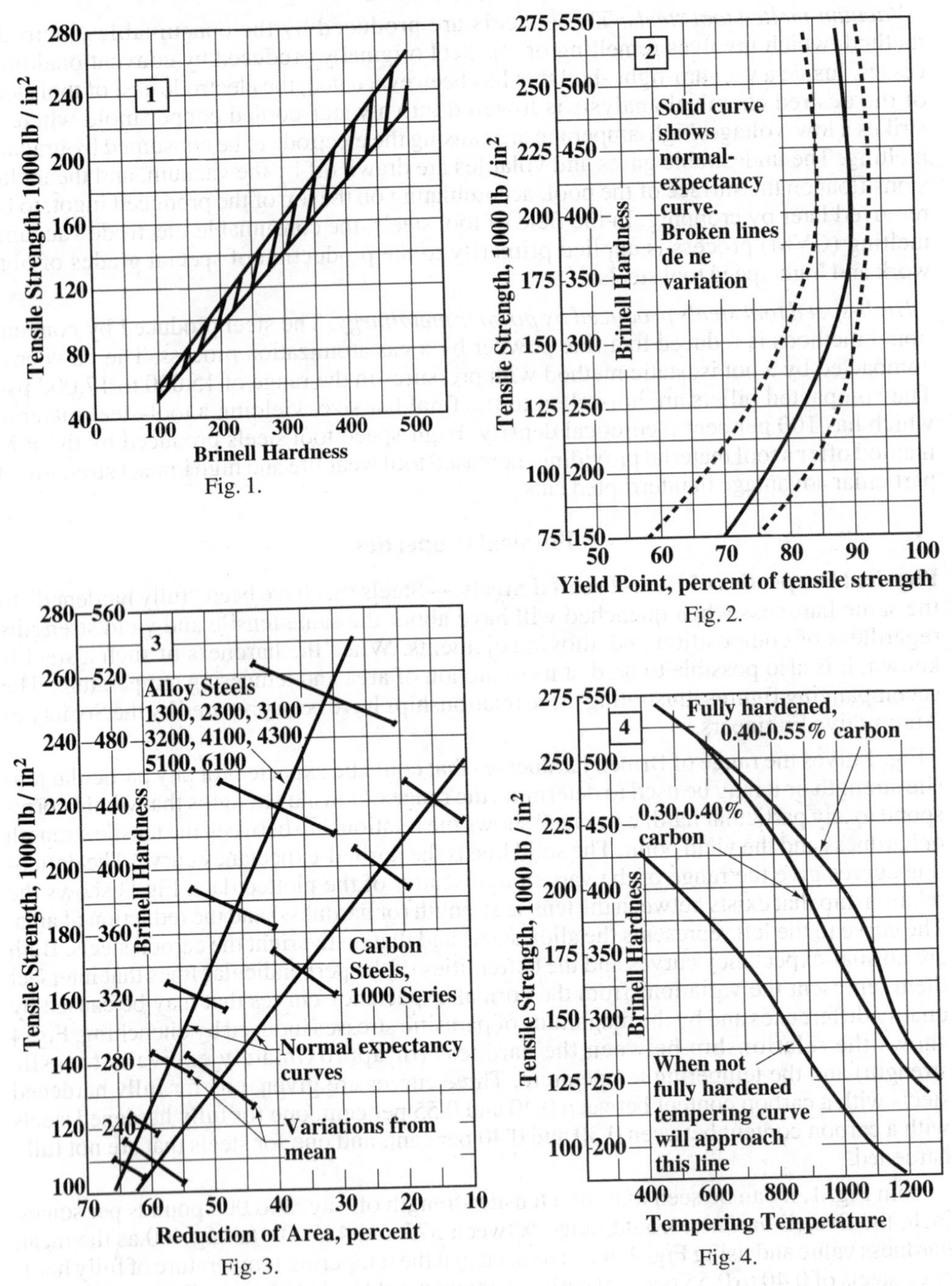

Fig. 1. Fig. 2. Fig. 3. Fig. 4.

Figs. 2 and 3 represent steel in the quenched and tempered condition and Fig. 1 represents steel in the hardened and tempered, as-rolled, annealed, and normalized conditions.

These charts give a good general indication of mechanical properties; however, more exact information when required should be obtained from tests on samples of the individual heats of steel under consideration.

Strength Data for Ferrous Metals.—The accompanying Table 1 gives ultimate strengths, yield points, and moduli of elasticity for various ferrous metals. Values are given as ranges, minimum values, and average values. Ranges of values are due to differences in size and shape of sections, heat treatments undergone, and composition, where several slightly different materials are listed under one general classification. The values in the table are meant to serve as a guide in the selection of ferrous materials and should not be used to write specifications. More specific data should be obtained from the supplier.

Strength Data for Nonferrous Metals.—The ultimate tensile, shear, and yield strengths and moduli of elasticity of many nonferrous metals are given in Table 2. Values for the most part are given in ranges rather than as single values because of differences in composition, forms, sizes, and shapes for the aluminum alloys plus differences in heat treatments undergone for the other nonferrous metals. The values in the table are meant to serve as a guide, not as specifications. More specific data should be obtained from the supplier.

Effect of Temperature on Strength and Elasticity of Metals.—Most ferrous metals have a maximum strength at approximately 400 degrees F, whereas the strength of nonferrous alloys is a maximum at about room temperature. The table on page 478 gives general data for variation in metal strength with temperature.

The modulus of elasticity of metals decreases regularly with increasing temperatures above room temperature until at some elevated temperature it falls off rapidly and reaches zero at the melting point.

Table 1. Strength Data for Iron and Steel

Material	Ultimate Strength: Tension, Thousands of Pounds per Square Inch, *T*	Ultimate Strength: Compression, in terms of *T*	Ultimate Strength: Shear, in terms of *T*	Yield Point, Thousands of Pounds per Square Inch	Modulus of Elasticity: Tension, Millions of psi, *E*	Modulus of Elasticity: Shear,[a] in terms of *E*
Cast iron, gray, class 20	20[b]	3.6*T* to 4.4*T*	1.6*T*	…	11.6	0.40*E*
class 25	25[b]	3.6*T* to 4.4*T*	1.4*T*	…	14.2	0.40*E*
class 30	30[b]	3.7*T*	1.4*T*	…	14.5	0.40*E*
class 35	35[b]	3.2*T* to 3.9*T*	1.4*T*	…	16.0	0.40*E*
class 40	40[b]	3.1*T* to 3.4*T*	1.3*T*	…	17	0.40*E*
class 50	50[b]	3.0*T* to 3.4*T*	1.3*T*	…	18	0.40*E*
class 60	60[b]	2.8*T*	1.0*T*	…	19.9	0.40*E*
malleable	40 to 100[c]	…	…	30 to 80[c]	25	0.43*E*
nodular (ductile iron)	60 to 120[d]	…	…	40 to 90[d]	23	…
Cast steel, carbon	60 to 100	*T*	0.75*T*	30 to 70	30	0.38*E*
low-alloy	70 to 200	*T*	0.75*T*	45 to 170	30	0.38*E*
Steel, SAE 950 (low-alloy)	65 to 70	*T*	0.75*T*	45 to 50	30	0.38*E*
1025 (low-carbon)	60 to 103	*T*	0.75*T*	40 to 90	30	0.38*E*
1045 (medium-carbon)	80 to 182	*T*	0.75*T*	50 to 162	30	0.38*E*
1095 (high-carbon)	90 to 213	*T*	0.75*T*	20 to 150	30	0.39*E*
1112 (free-cuttting)*	60 to 100	*T*	0.75*T*	30 to 95	30	0.38*E*
1212 (free-cuttting)	57 to 80	*T*	0.75*T*	25 to 72	30	0.38*E*
1330 (alloy)	90 to 162	*T*	0.75*T*	27 to 149	30	0.38*E*
2517 (alloy)[e]	88 to 190	*T*	0.75*T*	60 to 155	30	0.38*E*
3140 (alloy)	93 to 188	*T*	0.75*T*	62 to 162	30	0.38*E*
3310 (alloy)[e]	104 to 172	*T*	0.75*T*	56 to 142	30	0.38*E*
4023 (alloy)[e]	105 to 170	*T*	0.75*T*	60 to 114	30	0.38*E*
4130 (alloy)	81 to 179	*T*	0.75*T*	46 to 161	30	0.38*E*
4340 (alloy)	109 to 220	*T*	0.75*T*	68 to 200	30	0.38*E*
4640 (alloy)	98 to 192	*T*	0.75*T*	62 to 169	30	0.38*E*
4820 (alloy)[e]	98 to 209	*T*	0.75*T*	68 to 184	30	0.38*E*
5150 (alloy)	98 to 210	*T*	0.75*T*	51 to 190	30	0.38*E*
52100 (alloy)	100 to 238	*T*	0.75*T*	81 to 228	30	0.38*E*
6150 (alloy)	96 to 228	*T*	0.75*T*	59 to 210	30	0.38*E*
8650 (alloy)	110 to 228	*T*	0.75*T*	69 to 206	30	0.38*E*
8740 (alloy)	100 to 179	*T*	0.75*T*	60 to 165	30	0.38*E*
9310 (alloy)[e]	117 to 187	*T*	0.75*T*	63 to 162	30	0.38*E*
9840 (alloy)	120 to 285	*T*	0.75*T*	45 to 50	30	0.38*E*
Steel, stainless, SAE						
30302[f]	85 to 125	*T*	…	35 to 95	28	0.45*E*
30321[f]	85 to 95	*T*	…	30 to 60	28	…
30347[f]	90 to 100	*T*	…	35 to 65	28	…
51420[g]	95 to 230	*T*	…	50 to 195	29	0.40*E*
51430[h]	75 to 85	*T*	…	40 to 70	29	…
51446[h]	80 to 85	*T*	…	50 to 70	29	…
51501[g]	70 to 175	*T*	…	30 to 135	29	…
Steel, structural						
common	60 to 75	*T*	0.75*T*	33[b]	29	0.41*E*
rivet	52 to 62	*T*	0.75*T*	28[b]	29	…
rivet, high-strength	68 to 82	*T*	0.75*T*	38[b]	29	…
Wrought iron	34 to 54	*T*	0.83*T*	23 to 32	28	…

[a] Synonymous in other literature to the modulus of elasticity in torsion and the modulus of rigidity, *G*.

[b] Minimum specified value of the American Society for Testing and Materials. The specifications for the various materials are as follows: Cast iron, ASTM A48; structural steel for bridges and structures, ASTM A7; structural rivet steel, ASTM A141; high-strength structural rivet steel, ASTM A195.

[c] Range of minimum specified values of the ASTM (ASTM A47, A197, and A220).

[d] Range of minimum specified values of the ASTM (ASTM A339) and the Munitions Board Standards Agency (MIL-I-17166A and MIL-I-11466).

[e] Carburizing grades of steel.

[f] Nonhardenable nickel–chromium and Chromium–nickel–manganese steel (austenitic).

[g] Hardenable chromium steel (martensitic).

[h] Nonhardenable chromium steel (ferritic).

Table 2. Strength Data for Nonferrous Metals

Material	Ultimate Strength, Thousands of Pounds per Square Inch		Yield Strength (0.2 per cent offset), Thousands of Pounds per Square Inch	Modulus of Elasticity, Millions of Pounds per Square Inch	
	in Tension	in Shear		in Tension, E	in Shear, G
Aluminum alloys, cast,					
sand-cast	19 to 35	14 to 26	8 to 25	10.3	…
heat-treated	20 to 48	20 to 34	16 to 40	10.3	…
permanent-mold-cast,	23 to 35	16 to 27	9 to 24	10.3	…
heat-treated	23 to 48	15 to 36	8.5 to 43	10.3	…
die-cast	30 to 46	19 to 29	16 to 27	10.3	…
Aluminum alloys, wrought,					
annealed	10 to 42	7 to 26	4 to 22	10.0 to 10.6	…
cold-worked	12 to 63	8 to 34	11 to 59	10.0 to 10.3	…
heat-treated	22 to 83	14 to 48	13 to 73	10.0 to 11.4	…
Aluminum bronze, cast,	62 to 90	…	25 to 37	15 to 18	…
heat-treated	80 to 110	…	32 to 65	15 to 18	…
Aluminum bronze, wrought,					
annealed	55 to 80	…	20 to 40	16 to 19	…
cold-worked	71 to 110	…	62 to 66	16 to 19	…
heat-treated	101 to 151	…	48 to 94	16 to 19	…
Brasses, leaded, cast	32 to 40	29 to 31	12 to 15	12 to 14	…
flat products, wrought	46 to 85	31 to 45	14 to 62	14 to 17	5.3 to 6.4
wire, wrought	50 to 88	34 to 46	…	15	5.6
Brasses, nonleaded,					
flat products, wrought	34 to 99	28 to 48	10 to 65	15 to 17	5.6 to 6.4
wire, wrought	40 to 130	29 to 60	…	15 to 17	5.6 to 6.4
Copper, wrought,					
flat products	32 to 57	22 to 29	10 to 53	17	6.4
wire	35 to 66	24 to 33	…	17	6.4
Inconel, cast	70 to 95	…	30 to 45	23	…
flat products, wrought	80 to 170	…	30 to 160	31	11
wire, wrought	80 to 185	…	25 to 175	31	11
Lead	2.2 to 4.9	…	…	0.8 to 2.0	…
Magnesium, cast,					
sand & permanent mold	22 to 40	17 to 22	12 to 23	6.5	2.4
die-cast	33	20	22	6.5	2.4
Magnesium, wrought,					
sheet and plate	35 to 42	21 to 23	20 to 32	6.5	2.4
bars, rods, and shapes	37 to 55	19 to 27	26 to 44	6.5	2.4
Monel, cast	65 to 90	…	32 to 40	19	…
flat products, wrought	70 to 140	…	25 to 130	26	9.5
wire, wrought	70 to 170	…	25 to 160	26	9.5
Nickel, cast,	45 to 60	…	20 to 30	21.5	…
flat products, wrought	55 to 130	…	15 to 115	30	11
wire, wrought	50 to 165	…	10 to 155	30	11
Nickel silver, cast	40 to 50	…	24 to 25	…	…
flat products, wrought	49 to 115	41 to 59	18 to 90	17.5 to 18	6.6 to 6.8
wire, wrought	50 to 145	…	25 to 90	17.5 to 18	6.6 to 6.8
Phosphor bronze, wrought,					
flat products	40 to 128	…	14 to 80	15 to 17	5.6 to 6.4
wire	50 to 147	…	20 to 80	16 to 17	6 to 6.4
Silicon bronze, wrought,					
flat products	56 to 110	42 to 63	21 to 62	15	5.6
wire	50 to 145	36 to 70	25 to 70	15 to 17	5.6 to 6.4
Tin bronze, leaded, cast	21 to 38	23 to 43	15 to 18	10 to 14.5	…
Titanium	50 to 135	…	40 to 120	15.0 to 16.5	…
Zinc, commercial rolled	19.5 to 31	…	…	…	…
Zirconium	22 to 83	…	…	9 to 14.5	4.8

Consult the index for data on metals not listed and for more data on metals listed.

Average Ultimate Strength of Common Materials other than Metals (pounds per square inch)

Material	Compression	Tension
Bricks, best hard	12,000	400
Bricks, light red	1,000	40
Brickwork, common	1,000	50
Brickwork, best	2,000	300
Cement, Portland, 1 month old	2,000	400
Cement, Portland, 1 year old	3,000	500
Concrete, Portland	1,000	200
Concrete, Portland, 1 year old	2,000	400
Granite	19,000	700
Limestone and sandstone	9,000	300
Trap rock	20,000	800
Slate	14,000	500
Vulcanized fiber	39,000	13,000

Influence of Temperature on the Strength of Metals

Material	Degrees Fahrenheit							
	210	400	570	750	930	1100	1300	1475
	Strength in Per Cent of Strength at 70 Degrees F							
Wrought iron	104	112	116	96	76	42	25	15
Cast iron	…	100	99	92	76	42	…	…
Steel castings	109	125	121	97	57	…	…	…
Structural steel	103	132	122	86	49	28	…	…
Copper	95	85	73	59	42	…	…	…
Bronze	101	94	57	26	18	…	…	…

Strength of Copper–Zinc–Tin Alloys (U.S. Government Tests)

Percentage of			Tensile Strength, lb/in²	Percentage of			Tensile Strength, lb/in²	Percentage of			Tensile Strength, lb/in²
Copper	Zinc	Tin		Copper	Zinc	Tin		Copper	Zinc	Tin	
45	50	5	15,000	60	20	20	10,000	75	20	5	45,000
50	45	5	50,000	65	30	5	50,000	75	15	10	45,000
50	40	10	15,000	65	25	10	42,000	75	10	15	43,000
55	43	2	65,000	65	20	15	30,000	75	5	20	41,000
55	40	5	62,000	65	15	20	18,000	80	15	5	45,000
55	35	10	32,500	65	10	25	12,000	80	10	10	45,000
55	30	15	15,000	70	25	5	45,000	80	5	15	47,500
60	37	3	60,000	70	20	10	44,000	85	10	5	43,500
60	35	5	52,500	70	15	15	37,000	85	5	10	46,500
60	30	10	40,000	70	10	20	30,000	90	5	5	42,000

HARDENING, TEMPERING, AND ANNEALING

Heat Treatment Of Standard Steels

Heat-Treating Definitions.—This glossary of heat-treating terms has been adopted by the American Foundrymen's Association, the American Society for Metals, the American Society for Testing and Materials, and the Society of Automotive Engineers. Since it is not intended to be a specification but is strictly a set of definitions, temperatures have purposely been omitted.

Aging: Describes a time–temperature-dependent change in the properties of certain alloys. Except for strain aging and age softening, it is the result of precipitation from a solid solution of one or more compounds whose solubility decreases with decreasing temperature. For each alloy susceptible to aging, there is a unique range of time–temperature combinations to which it will respond.

Annealing: A term denoting a treatment, consisting of heating to and holding at a suitable temperature followed by cooling at a suitable rate, used primarily to soften but also to simultaneously produce desired changes in other properties or in microstructure. The purpose of such changes may be, but is not confined to, improvement of machinability; facilitation of cold working; improvement of mechanical or electrical properties; or increase in stability of dimensions. The time–temperature cycles used vary widely both in maximum temperature attained and in cooling rate employed, depending on the composition of the material, its condition, and the results desired. When applicable, the following more specific process names should be used: Black Annealing, Blue Annealing, Box Annealing, Bright Annealing, Cycle Annealing, Flame Annealing, Full Annealing, Graphitizing, Intermediate Annealing, Isothermal Annealing, Process Annealing, Quench Annealing, and Spheroidizing. When the term is used without qualification, full annealing is implied. When applied only for the relief of stress, the process is properly called stress relieving.

Black Annealing: Box annealing or pot annealing, used mainly for sheet, strip, or wire.

Blue Annealing: Heating hot-rolled sheet in an open furnace to a temperature within the transformation range and then cooling in air, to soften the metal. The formation of a bluish oxide on the surface is incidental.

Box Annealing: Annealing in a sealed container under conditions that minimize oxidation. In box annealing, the charge is usually heated slowly to a temperature below the transformation range, but sometimes above or within it, and is then cooled slowly; this process is also called "close annealing" or "pot annealing."

Bright Annealing: Annealing in a protective medium to prevent discoloration of the bright surface.

Cycle Annealing: An annealing process employing a predetermined and closely controlled time–temperature cycle to produce specific properties or microstructure.

Flame Annealing: Annealing in which the heat is applied directly by a flame.

Full Annealing: Austenitizing and then cooling at a rate such that the hardness of the product approaches a minimum.

Graphitizing: Annealing in such a way that some or all of the carbon is precipitated as graphite.

Intermediate Annealing: Annealing at one or more stages during manufacture and before final thermal treatment.

Isothermal Annealing: Austenitizing and then cooling to and holding at a temperature at which austenite transforms to a relatively soft ferrite-carbide aggregate.

Process Annealing: An imprecise term used to denote various treatments that improve workability. For the term to be meaningful, the condition of the material and the time–temperature cycle used must be stated.

Quench Annealing: Annealing an austenitic alloy by *Solution Heat Treatment*.

Spheroidizing: Heating and cooling in a cycle designed to produce a spheroidal or globular form of carbide.

Austempering: Quenching from a temperature above the transformation range, in a medium having a rate of heat abstraction high enough to prevent the formation of high-temperature transformation products, and then holding the alloy, until transformation is complete, at a temperature below that of pearlite formation and above that of martensite formation.

Austenitizing: Forming austenite by heating into the transformation range (partial austenitizing) or above the transformation range (complete austenitizing). When used without qualification, the term implies complete austenitizing.

Baking: Heating to a low temperature in order to remove entrained gases.

Bluing: A treatment of the surface of iron-base alloys, usually in the form of sheet or strip, on which, by the action of air or steam at a suitable temperature, a thin blue oxide film is formed on the initially scale-free surface, as a means of improving appearance and resistance to corrosion. This term is also used to denote a heat treatment of springs after fabrication, to reduce the internal stress created by coiling and forming.

Carbon Potential: A measure of the ability of an environment containing active carbon to alter or maintain, under prescribed conditions, the carbon content of the steel exposed to it. In any particular environment, the carbon level attained will depend on such factors as temperature, time, and steel composition.

Carbon Restoration: Replacing the carbon lost in the surface layer from previous processing by carburizing this layer to substantially the original carbon level.

Carbonitriding: A case-hardening process in which a suitable ferrous material is heated above the lower transformation temperature in a gaseous atmosphere of such composition as to cause simultaneous absorption of carbon and nitrogen by the surface and, by diffusion, create a concentration gradient. The process is completed by cooling at a rate that produces the desired properties in the workpiece.

Carburizing: A process in which carbon is introduced into a solid iron-base alloy by heating above the transformation temperature range while in contact with a carbonaceous material that may be a solid, liquid, or gas. Carburizing is frequently followed by quenching to produce a hardened case.

Case: 1) The surface layer of an iron-base alloy that has been suitably altered in composition and can be made substantially harder than the interior or core by a process of case hardening; and 2) the term case is also used to designate the hardened surface layer of a piece of steel that is large enough to have a distinctly softer core or center.

Cementation: The process of introducing elements into the outer layer of metal objects by means of high-temperature diffusion.

Cold Treatment: Exposing to suitable subzero temperatures for the purpose of obtaining desired conditions or properties, such as dimensional or microstructural stability. When the treatment involves the transformation of retained austenite, it is usually followed by a tempering treatment.

Conditioning Heat Treatment: A preliminary heat treatment used to prepare a material for a desired reaction to a subsequent heat treatment. For the term to be meaningful, the treatment used must be specified.

Controlled Cooling: A term used to describe a process by which a steel object is cooled from an elevated temperature, usually from the final hot-forming operation in a predetermined manner of cooling to avoid hardening, cracking, or internal damage.

Core: 1) The interior portion of an iron-base alloy that after case hardening is substantially softer than the surface layer or case; and 2) the term core is also used to designate the relatively soft central portion of certain hardened tool steels.

Critical Range or Critical Temperature Range: Synonymous with *Transformation Range*, which is preferred.

Cyaniding: A process of case hardening an iron-base alloy by the simultaneous absorption of carbon and nitrogen by heating in a cyanide salt. Cyaniding is usually followed by quenching to produce a hard case.

Decarburization: The loss of carbon from the surface of an iron-base alloy as the result of heating in a medium that reacts with the carbon.

Drawing: Drawing, or drawing the temper, is synonymous with *Tempering*, which is preferable.

Eutectic Alloy: The alloy composition that freezes at constant temperature similar to a pure metal. The lowest melting (or freezing) combination of two or more metals. The alloy structure (homogeneous) of two or more solid phases formed from the liquid eutectically.

Hardenability: In a ferrous alloy, the property that determines the depth and distribution of hardness induced by quenching.

Hardening: Any process of increasing hardness of metal by suitable treatment, usually involving heating and cooling. See also *Aging*.

Hardening, Case: A process of surface hardening involving a change in the composition of the outer layer of an iron-base alloy followed by appropriate thermal treatment. Typical case-hardening processes are *Carburizing*, *Cyaniding*, *Carbonitriding*, and *Nitriding*.

Hardening, Flame: A process of heating the surface layer of an iron-base alloy above the transformation temperature range by means of a high-temperature flame, followed by quenching.

Hardening, Precipitation: A process of hardening an alloy in which a constituent precipitates from a supersaturated solid solution. See also *Aging*.

Hardening, Secondary: An increase in hardness following the normal softening that occurs during the tempering of certain alloy steels.

Heating, Differential: A heating process by which the temperature is made to vary throughout the object being heated so that on cooling, different portions may have such different physical properties as may be desired.

Heating, Induction: A process of local heating by electrical induction.

Heat Treatment: A combination of heating and cooling operations applied to a metal or alloy in the solid state to obtain desired conditions or properties. Heating for the sole purpose of hot working is excluded from the meaning of this definition.

Heat Treatment, Solution: A treatment in which an alloy is heated to a suitable temperature and held at this temperature for a sufficient length of time to allow a desired constituent to enter into solid solution, followed by rapid cooling to hold the constituent in solution. The material is then in a supersaturated, unstable state, and may subsequently exhibit *Age Hardening*.

Homogenizing: A high-temperature heat-treatment process intended to eliminate or to decrease chemical segregation by diffusion.

Isothermal Transformation: A change in phase at constant temperature.

Malleablizing: A process of annealing white cast iron in which the combined carbon is wholly or in part transformed to graphitic or free carbon and, in some cases, part of the carbon is removed completely. See *Temper Carbon*.

Maraging: A precipitation hardening treatment applied to a special group of iron-base alloys to precipitate one or more intermetallic compounds in a matrix of essentially carbon-free martensite.

Martempering: A hardening procedure in which an austenitized ferrous workpiece is quenched into an appropriate medium whose temperature is maintained substantially at the M_s of the workpiece, held in the medium until its temperature is uniform throughout but not long enough to permit bainite to form, and then cooled in air. The treatment is followed by tempering.

Nitriding: A process of case hardening in which an iron-base alloy of special composition is heated in an atmosphere of ammonia or in contact with nitrogenous material. Surface hardening is produced by the absorption of nitrogen without quenching.

Normalizing: A process in which an iron-base alloy is heated to a temperature above the transformation range and subsequently cooled in still air at room temperature.

Overheated: A metal is said to have been overheated if, after exposure to an unduly high temperature, it develops an undesirably coarse grain structure but is not permanently damaged. The structure damaged by overheating can be corrected by suitable heat treatment or by mechanical work or by a combination of the two. In this respect it differs from a Burnt structure.

Patenting: A process of heat treatment applied to medium- or high-carbon steel in wire making prior to the wire drawing or between drafts. It consists in heating to a temperature above the transformation range, followed by cooling to a temperature below that range in air or in a bath of molten lead or salt maintained at a temperature appropriate to the carbon content of the steel and the properties required of the finished product.

Preheating: Heating to an appropriate temperature immediately prior to austenitizing when hardening high-hardenability constructional steels, many of the tool steels, and heavy sections.

Quenching: Rapid cooling. When applicable, the following more specific terms should be used: Direct Quenching, Fog Quenching, Hot Quenching, Interrupted Quenching, Selective Quenching, Slack Quenching, Spray Quenching, and Time Quenching.

Direct Quenching: Quenching carburized parts directly from the carburizing operation.

Fog Quenching: Quenching in a mist.

Hot Quenching: An imprecise term used to cover a variety of quenching procedures in which a quenching medium is maintained at a prescribed temperature above 160 degrees F (71 degrees C).

Interrupted Quenching: A quenching procedure in which the workpiece is removed from the first quench at a temperature substantially higher than that of the quenchant and is then subjected to a second quenching system having a different cooling rate than the first.

Selective Quenching: Quenching only certain portions of a workpiece.

Slack Quenching: The incomplete hardening of steel due to quenching from the austenitizing temperature at a rate slower than the critical cooling rate for the particular steel, resulting in the formation of one or more transformation products in addition to martensite.

Spray Quenching: Quenching in a spray of liquid.

Time Quenching: Interrupted quenching in which the duration of holding in the quenching medium is controlled.

Soaking: Prolonged heating of a metal at a selected temperature.

Stabilizing Treatment: A treatment applied to stabilize the dimensions of a workpiece or the structure of a material such as 1) before finishing to final dimensions, heating a workpiece to or somewhat beyond its operating temperature and then cooling to room temperature a sufficient number of times to ensure stability of dimensions in service; 2) transforming retained austenite in those materials that retain substantial amounts when quench hardened (see cold treatment); and 3) heating a solution-treated austenitic stainless steel that contains controlled amounts of titanium or niobium plus tantalum to a temperature below the solution heat-treating temperature to cause precipitation of finely divided, uniformly distributed carbides of those elements, thereby substantially reducing the amount of carbon available for the formation of chromium carbides in the grain boundaries on subsequent exposure to temperatures in the sensitizing range.

Stress Relieving: A process to reduce internal residual stresses in a metal object by heating the object to a suitable temperature and holding for a proper time at that temperature. This treatment may be applied to relieve stresses induced by casting, quenching, normalizing, machining, cold working, or welding.

Temper Carbon: The free or graphitic carbon that comes out of solution usually in the form of rounded nodules in the structure during *Graphitizing* or *Malleablizing*.

Tempering: Heating a quench-hardened or normalized ferrous alloy to a temperature below the transformation range to produce desired changes in properties.

Double Tempering: A treatment in which quench hardened steel is given two complete tempering cycles at substantially the same temperature for the purpose of ensuring completion of the tempering reaction and promoting stability of the resulting microstructure.

Snap Temper: A precautionary interim stress-relieving treatment applied to high hardenability steels immediately after quenching to prevent cracking because of delay in tempering them at the prescribed higher temperature.

Temper Brittleness: Brittleness that results when certain steels are held within, or are cooled slowly through, a certain range of temperatures below the transformation range. The brittleness is revealed by notched-bar impact tests at or below room temperature.

Transformation Ranges or Transformation Temperature Ranges: Those ranges of temperature within which austenite forms during heating and transforms during cooling. The two ranges are distinct, sometimes overlapping but never coinciding. The limiting temperatures of the ranges depend on the composition of the alloy and on the rate of change of temperature, particularly during cooling.

Transformation Temperature: The temperature at which a change in phase occurs. The term is sometimes used to denote the limiting temperature of a transformation range. The following symbols are used for iron and steels:

Ac_{cm} = In hypereutectoid steel, the temperature at which the solution of cementite in austenite is completed during heating

Ac_1 = The temperature at which austenite begins to form during heating

Ac_3 = The temperature at which transformation of ferrite to austenite is completed during heating

Ac_4 = The temperature at which austenite transforms to delta ferrite during heating

$Ae_1, Ae_3, Ae_{cm}, Ae_4$ = The temperatures of phase changes at equilibrium

Ar_{cm} = In hypereutectoid steel, the temperature at which precipitation of cementite starts during cooling

Ar_1 = The temperature at which transformation of austenite to ferrite or to ferrite plus cementite is completed during cooling

Ar_3 = The temperature at which austenite begins to transform to ferrite during cooling

Ar_4 = The temperature at which delta ferrite transforms to austenite during cooling

M_s = The temperature at which transformation of austenite to martensite starts during cooling

M_f = The temperature, during cooling, at which transformation of austenite to martensite is substantially completed

All these changes except the formation of martensite occur at lower temperatures during cooling than during heating, and depend on the rate of change of temperature.

Structure of Fully Annealed Carbon Steel.—In carbon steel that has been fully annealed, there are normally present, apart from such impurities as phosphorus and sulfur, two constituents: the element iron in a form metallurgically known as *ferrite* and the chemical compound iron carbide in the form metallurgically known as *cementite*. This latter constituent consists of 6.67 per cent carbon and 93.33 per cent iron. A certain proportion of these two constituents will be present as a mechanical mixture. This mechanical mixture, the amount of which depends on the carbon content of the steel, consists of alternate bands or layers of ferrite and cementite. Under the microscope, the matrix frequently has the appearance of mother-of-pearl and hence has been named *pearlite*. Pearlite contains about 0.85 per cent carbon and 99.15 per cent iron, neglecting impurities. A fully annealed steel containing 0.85 per cent carbon would consist entirely of pearlite. Such a steel is known as *eutectoid* steel and has a laminated structure characteristic of a eutectic alloy. Steel that has less than 0.85 per cent carbon (*hypoeutectoid* steel) has an excess of ferrite above that required to mix with the cementite present to form pearlite; hence, both ferrite and pearlite are present in the fully annealed state. Steel having a carbon content greater than 0.85 per cent (*hypereutectoid* steel) has an excess of cementite over that required to mix with the ferrite to form pearlite; hence, both cementite and pearlite are present in the fully annealed state. The structural constitution of carbon steel in terms of ferrite, cementite, pearlite and austenite for different carbon contents and at different temperatures is shown by the accompanying figure, *Phase Diagram of Carbon Steel.*

Effect of Heating Fully Annealed Carbon Steel.—When carbon steel in the fully annealed state is heated above the lower critical point, which is some temperature in the range of 1335 to 1355 degrees F (depending on the carbon content), the alternate bands or layers of ferrite and cementite that make up the pearlite begin to merge into each other. This process continues until the pearlite is thoroughly "dissolved," forming what is known as *austenite*. If the temperature of the steel continues to rise and there is present, in addition to the pearlite, any excess ferrite or cementite, this also will begin to dissolve into the austenite until finally only austenite will be present. The temperature at which the excess ferrite or cementite is completely dissolved in the austenite is called the *upper critical point*. This temperature varies with the carbon content of the steel much more widely than the lower critical point (see Fig. 1).

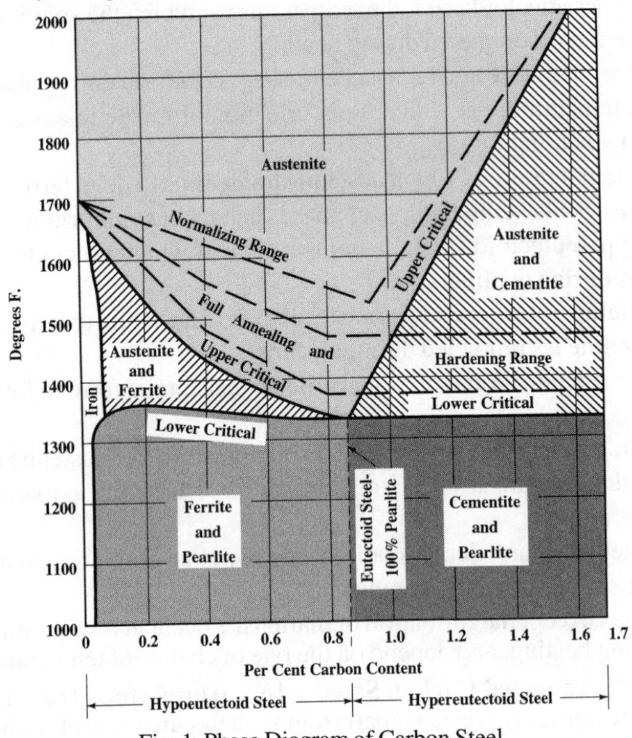

Fig. 1. Phase Diagram of Carbon Steel

Effect of Slow Cooling on Carbon Steel.—If carbon steel that has been heated to the point where it consists entirely of austenite is slowly cooled, the process of transformation that took place during the heating will be reversed, but the upper and lower critical points will occur at somewhat lower temperatures than they do on heating. Assuming that the steel was originally fully annealed, its structure on returning to atmospheric temperature after slow cooling will be the same as before in terms of the proportions of ferrite or cementite and pearlite present. The austenite will have entirely disappeared.

Effect of Rapid Cooling or Quenching on Carbon Steel.—Observations have shown that as the rate at which carbon steel is cooled from an austenitic state is increased, the temperature at which the austenite begins to change into pearlite drops more and more below the slow cooling transformation temperature of about 1300 degrees F. (For example, a 0.80 per cent carbon steel that is cooled at such a rate that the temperature drops 500 degrees in one second will show transformation of austenite beginning at 930 degrees F.) As the cooling rate is increased, the laminations of the pearlite formed by the transformation of the austenite become finer and finer up to the point where they cannot be detected under a high-power microscope, while the steel itself increases in hardness and tensile strength. As the rate of cooling is still further increased, this transformation temperature suddenly drops to around 500 degrees F or lower, depending on the carbon content of the steel. The cooling rate at which this sudden drop in transformation temperature takes place is called the *critical cooling rate*. When a piece of carbon steel is quenched at this rate or faster, a new structure is formed. The austenite is transformed into *martensite*, which is characterized by an angular needlelike structure and a very high hardness.

If carbon steel is subjected to a severe quench or to extremely rapid cooling, a small percentage of the austenite, instead of being transformed into martensite during the quenching operation, may be retained. Over a period of time, however, this remaining austenite tends to be gradually transformed into martensite even though the steel is not subjected to further heating or cooling. Martensite has a lower density than austenite, and such a change, or "aging" as it is called, often results in an appreciable increase in volume or "growth" and the setting up of new internal stresses in the steel.

Steel Heat-Treating Furnaces.—Various types of furnaces heated by gas, oil, or electricity are used for the heat treatment of steel. These furnaces include the oven or box type in various modifications for "in-and-out" or for continuous loading and unloading; the retort type; the pit type; the pot type; and the salt-bath electrode type.

Oven or Box Furnaces: This type of furnace has a box or oven-shaped heating chamber. The "in-and-out" oven furnaces are loaded by hand or by a track-mounted car that, when rolled into the furnace, forms the bottom of the heating chamber. The car type is used where heavy or bulky pieces must be handled. Some oven-type furnaces are provided with a full muffle or a semimuffle, which is an enclosed refractory chamber into which the parts to be heated are placed. The full-muffle, being fully enclosed, prevents any flames or burning gases from coming in contact with the work and permits a special atmosphere to be used to protect or condition the work. The semimuffle, which is open at the top, protects the work from direct impingement of the flame although it does not shut off the work from the hot gases. In the direct-heat-type oven furnace, the work is open to the flame. In the electric oven furnace, a retort is provided when gas atmospheres are to be employed to confine the gas and prevent it from attacking the heating elements. Where muffles are used, they must be replaced periodically, and a greater amount of fuel is required than in a direct-heat type of oven furnace.

For continuous loading and unloading, there are several types of furnaces such as rotary hearth car; roller-, furnace belt-, walking-beam, or pusher-conveyor; and a continuous-kiln-type through which track-mounted cars are run. In the continuous type of furnace, the work may pass through several zones that are maintained at different temperatures for preheating, heating, soaking, and cooling.

Retort Furnace: This is a vertical type of furnace provided with a cylindrical metal retort into which the parts to be heat-treated are suspended either individually, if large enough, or in a container of some sort. The use of a retort permits special gas atmospheres to be employed for carburizing, nitriding, etc.

Pit-Type Furnace: This is a vertical furnace arranged for the loading of parts in a metal basket. The parts within the basket are heated by convection, and when the basket is lowered into place, it fits into the furnace chamber in such a way as to provide a dead-air space to prevent direct heating.

Pot-Type Furnace: This furnace is used for the immersion method of heat treating small parts. A cast-alloy pot is employed to hold a bath of molten lead or salt in which the parts are placed for heating.

Salt Bath Electrode Furnace: In this type of electric furnace, heating is accomplished by means of electrodes suspended directly in the salt bath. The patented grouping and design of electrodes provide an electromagnetic action that results in an automatic stirring action. This stirring tends to produce an even temperature throughout the bath.

Vacuum Furnace: Vacuum heat treatment is a relatively new development in metallurgical processing, with a vacuum substituting for the more commonly used protective gas atmospheres. The most often used furnace is the "cold wall" type, consisting of a water-cooled vessel that is maintained near ambient temperature during operation. During quenching, the chamber is backfilled up to or above atmospheric pressure with an inert gas, which is circulated by an internal fan. When even faster cooling rates are needed, furnaces are available with capability for liquid quenching, performed in an isolated chamber.

Fluidized-Bed Furnace: Fluidized-bed techniques are not new; however, new furnace designs have extended the technology into the temperature ranges required for most common heat treatments. In fluidization, a bed of dry, finely divided particles, typically aluminum oxide, is made to behave like a liquid by feeding gas upward through the bed. An important characteristic of the bed is high-efficiency heat transfer. Applications include continuous or batch-type units for all general heat treatments.

Hardening

Basic Steps in Hardening.—The operation of hardening steel consists fundamentally of two steps. The first step is to heat the steel to some temperature above (usually at least 100 degrees F above) its transformation point so that it becomes entirely austenitic in structure. The second step is to quench the steel at some rate faster than the critical rate (which depends on the carbon content, the amounts of alloying elements present other than carbon, and the grain size of the austenite) to produce a martensitic structure. The hardness of a martensitic steel depends on its carbon content and ranges from about 460 Brinell at 0.20 per cent carbon to about 710 Brinell above 0.50 carbon. In comparison, ferrite has a hardness of about 90 Brinell, pearlite about 240 Brinell, and cementite around 550 Brinell.

Critical Points of Decalescence and Recalescence.—The critical or transformation point at which pearlite is transformed into austenite as it is being heated is also called the *decalescence point*. If the temperature of the steel was observed as it passed through the decalescence point, it would be noted that it would continue to absorb heat without appreciably rising in temperature, although the immediate surroundings were hotter than the steel. Similarly, the critical or transformation point at which austenite is transformed back into pearlite on cooling is called the *recalescence point*. When this point is reached, the steel will give out heat so that its temperature instead of continuing to fall, will momentarily increase.

The recalescence point is lower than the decalescence point by anywhere from 85 to 215 degrees F, and the lower of these points does not manifest itself unless the higher one has first been fully passed. These critical points have a direct relation to the hardening of steel. Unless a temperature sufficient to reach the decalescence point is obtained, so that the pearlite is changed into austenite, no hardening action can take place; and unless the steel is cooled suddenly before it reaches the recalescence point, thus preventing the changing back again from austenite to pearlite, no hardening can take place. The critical points vary for different kinds of steel and must be determined by tests. This variation in the critical points makes it necessary to heat different steels to different temperatures when hardening.

Hardening Temperatures.—The maximum temperature to which a steel is heated before quenching to harden it is called the hardening temperature. Hardening temperatures vary for different steels and different classes of service, although, in general, it may be said that the hardening temperature for any given steel is above the lower critical point of that steel.

Just how far above this point the hardening temperature lies for any particular steel depends on three factors: 1) the chemical composition of the steel; 2) the amount of excess ferrite (if the steel has less than 0.85 per cent carbon content) or the amount of excess cementite (if the steel has more than 0.85 per cent carbon content) that is to be dissolved in the austenite; and 3) the maximum grain size permitted, if desired.

The general range of full-hardening temperatures for carbon steels is shown by the diagram. This range is merely indicative of general practice and is not intended to represent absolute hardening temperature limits. It can be seen that for steels of less than 0.85 per cent carbon content, the hardening range is above the upper critical point — that is, above the temperature at which all the excess ferrite has been dissolved in the austenite. On the other hand, for steels of more than 0.85 per cent carbon content, the hardening range lies somewhat below the upper critical point. This indicates that in this hardening range, some of the excess cementite still remains undissolved in the austenite. If steel of more than 0.85 per cent carbon content were heated above the upper critical point and then quenched, the resulting grain size would be excessively large.

At one time, it was considered desirable to heat steel only to the minimum temperature at which it would fully harden, one of the reasons being to avoid grain growth that takes place at higher temperature. It is now realized that no such rule as this can be applied generally since there are factors other than hardness that must be taken into consideration. For example, in many cases, toughness can be impaired by too low a temperature just as much as by too high a temperature. It is true, however, that too high hardening temperatures result in warpage, distortion, increased scale, and decarburization.

Hardening Temperatures for Carbon Tool Steels.—The best hardening temperatures for any given tool steel are dependent on the type of tool and the intended class of service. Wherever possible, the specific recommendations of the tool steel manufacturer should be followed. General recommendations for hardening temperatures of carbon tool steels based on carbon content are as follows: For steel of 0.65 to 0.80 per cent carbon content, 1450 to 1550 degrees F; for steel of 0.80 to 0.95 per cent carbon content, 1410 to 1460 degrees F; for steel of 0.95 to 1.10 per cent carbon content, 1390 to 1430 degrees F; and for steels of 1.10 per cent and over carbon content, 1380 to 1420 degrees F. For a given hardening temperature range, the higher temperatures tend to produce deeper hardness penetration and increased compressional strength, whereas the lower temperatures tend to result in shallower hardness penetration but increased resistance to splitting or bursting stresses.

Determining Hardening Temperatures.—A hardening temperature can be specified directly or it may be specified indirectly as a certain temperature rise above the lower critical point of the steel. Where the temperature is specified directly, a pyrometer of the type that indicates the furnace temperature or a pyrometer of the type that indicates the work temperature may be employed. If the pyrometer shows furnace temperature, care must be taken to allow sufficient time for the work to reach the furnace temperature after the pyrometer indicates that the required hardening temperature has been attained. If the pyrometer indicates work temperature, then, where the workpiece is large, time must be allowed for the interior of the work to reach the temperature of the surface, which is the temperature indicated by the pyrometer.

Where the hardening temperature is specified as a given temperature rise above the critical point of the steel, a pyrometer that indicates the temperature of the work should be used. The critical point, as well as the given temperature rise, can be more accurately determined with this type of pyrometer. As the work is heated, its temperature, as indicated by the pyrometer, rises steadily until the lower critical or decalescence point of the steel is reached. At this point, the temperature of the work ceases to rise and the pyrometer indicating or recording pointer remains stationary or fluctuates slightly. After a certain elapsed period, depending on the heat input rate, the internal changes in structure of the steel that take place at the lower critical point are completed and the temperature of the work again begins to rise. A small fluctuations in temperature may occur in the interval during which structural changes are taking place, so for uniform practice, the critical point may be considered as the temperature at which the pointer first becomes stationary.

Heating Steel in Liquid Baths.—The liquid bath commonly used for heating steel tools preparatory to hardening are molten lead, sodium cyanide, barium chloride, a mixture of barium and potassium chloride, and other metallic salts. The molten substance is retained in a crucible or pot and the heat required may be obtained from gas, oil, or electricity. The principal advantages of heating baths are as follows: No part of the work can be heated to a temperature above that of the bath; the temperature can be easily maintained at whatever degree has proved, in practice, to give the best results; the submerged steel can be heated uniformly, and the finished surfaces are protected against oxidation.

Salt Baths.—Molten baths of various salt mixtures or compounds are used extensively for heat-treating operations such as hardening and tempering; they are also utilized for annealing ferrous and nonferrous metals. Commercial salt-bath mixtures are available that meet a wide range of temperature and other metallurgical requirements. For example, there are neutral baths for heating tool and die steels without carburizing the surfaces; baths for carburizing the surfaces of low-carbon steel parts; baths adapted for the usual tempering temperatures of, say, 300 to 1100 degrees F; and baths that may be heated to temperatures up to approximately 2400 degrees F for hardening high-speed steels. Salt baths are also adapted for local or selective hardening, the type of bath being selected to suit the requirements. For example, a neutral bath may be used for annealing the ends of tubing or other parts, or an activated cyanide bath for carburizing the ends of shafts or other parts. Surfaces that are not to be carburized are protected by copper plating. When the work is immersed, [illegible]plated surfaces are subjected to the carburizing action.

Baths may consist of a mixture of sodium, potassium, barium, and calcium chlorides or nitrates of sodium, potassium, barium, and calcium in varying proportions, to which sodium carbonate and sodium cyanide are sometimes added to prevent decarburization. Various proportions of these salts provide baths of different properties. Potassium cyanide is seldom used as sodium cyanide costs less. The specific gravity of a salt bath is not as high as that of a lead bath; consequently, the work may be suspended in a salt bath and does not have to be held below the surface as in a lead bath.

The Lead Bath.—The lead bath is extensively used, but is not adapted to the high temperatures required for hardening high-speed steel, as it begins to vaporize at about 1190 degrees F. As the temperature increases, the lead volatilizes and gives off poisonous vapors; hence, lead furnaces should be equipped with hoods to carry away the fumes. Lead baths are generally used for temperatures below 1500 or 1600 degrees F. They are often employed for heating small pieces that must be hardened in quantities. It is important to use pure lead that is free from sulfur. The work should be preheated before plunging it into the molten lead.

Defects in Hardening.—Uneven heating is the cause of most of the defects in hardening. Cracks of a circular form, from the corners or edges of a tool, indicate uneven heating in hardening. Cracks of a vertical nature and dark-colored fissures indicate that the steel has been burned and should be put on the scrap heap. Tools that have hard and soft places have been either unevenly heated, unevenly cooled, or "soaked," a term used to indicate prolonged heating. A tool not thoroughly moved about in the hardening fluid will show hard and soft places, and have a tendency to crack. Tools that are hardened by dropping them to the bottom of the tank sometimes have soft places, owing to contact with the floor or sides.

Scale on Hardened Steel.—The formation of scale on the surface of hardened steel is due to the contact of oxygen with the heated steel; hence, to prevent scale, the heated steel must not be exposed to the action of the air. When using an oven heating furnace, the flame should be so regulated that it is not visible in the heating chamber. The heated steel should be exposed to the air as little as possible, when transferring it from the furnace to the quenching bath. An old method of preventing scale and retaining a fine finish on dies used in jewelry manufacture, small taps, etc., is as follows: Fill the die impression with powdered boracic acid and place near the fire until the acid melts; then add a little more acid to ensure covering all the surfaces. The die is then hardened in the usual way. If the boracic acid does not come off entirely in the quenching bath, immerse the work in boiling water. Dies hardened by this method are said to be as durable as those heated without the acid.

Hardening or Quenching Baths.—The purpose of a quenching bath is to remove heat from the steel being hardened at a rate that is faster than the critical cooling rate. Generally speaking, the more rapid the rate of heat extraction above the cooling rate, the higher will be the resulting hardness. To obtain the different rates of cooling required by different classes of work, baths of various kinds are used. These include plain or fresh water, brine, caustic soda solutions, oils of various classes, oil–water emulsions, baths of molten salt or lead for high-speed steels, and air cooling for some high-speed steel tools when a slow rate of cooling is required. To minimize distortion and cracking where such tendencies are present, without sacrificing depth-of-hardness penetration, a quenching medium should be selected that will cool rapidly at the higher temperatures and more slowly at the lower temperatures, that is below 750 degrees F. Oil quenches in general meet this requirement.

Oil Quenching Baths: Oil is used very extensively as a quenching medium as it results in a good proportion of hardness, toughness, and freedom from warpage when used with standard steels. Oil baths are used extensively for alloy steels. Various kinds of oils are employed, such as prepared mineral oils and vegetable, animal, and fish oils, either singly or in combination. Prepared mineral quenching oils are widely used because they have good quenching characteristics, are chemically stable, do not have an objectionable odor, and are relatively inexpensive. Special compounded oils of the soluble type are used in many plants instead of such oils as fish oil, linseed oil, cottonseed oil, etc. The soluble properties enable the oil to form an emulsion with water.

Oil cools steel at a slower rate than water, but the rate is fast enough for alloy steel. Oils have different cooling rates, however, and this rate may vary through the initial and final stages of the quenching operation. Faster cooling in the initial stage and slower cooling at lower temperatures are preferable because there is less danger of cracking the steel. The temperature of quenching oil baths should range ordinarily between 90 and 130 degrees F. A fairly constant temperature may be maintained either by circulating the oil through cooling coils or by using a tank provided with a cold-water jacket.

A good quenching oil should possess a flash and fire point sufficiently high to be safe under the conditions used and 350 degrees F should be about the minimum point. The specific heat of the oil regulates the hardness and toughness of the quenched steel; and the greater the specific heat, the higher will be the hardness produced. Specific heats of quenching oils vary from 0.20 to 0.75, the specific heats of fish, animal, and vegetable oils usually being from 0.2 to 0.4, and of soluble and mineral oils from 0.5 to 0.7. The efficient temperature range for quenching oil is from 90 to 140 degrees F.

Quenching in Water.—Many carbon tool steels are hardened by immersing them in a bath of fresh water, but water is not an ideal quenching medium. Contact between the water and work and the cooling of the hot steel are impaired by the formation of gas bubbles or an insulating vapor film especially in holes, cavities, or pockets. The result is uneven cooling and sometimes excessive strains which may cause the tool to crack; in fact, there is greater danger of cracking in a fresh-water bath than in one containing salt water or brine.

In order to secure more even cooling and reduce danger of cracking, either rock salt (8 or 9 per cent) or caustic soda (3 to 5 per cent) may be added to the bath to eliminate or prevent the formation of a vapor film or gas pockets, thus promoting rapid early cooling. Brine is commonly used and ¾ pound of rock salt per gallon of water is equivalent to about 8 per cent of salt. Brine is not inherently a more severe or drastic quenching medium than plain water, although it may seem to be because the brine makes better contact with the heated steel and, consequently, cooling is more effective. In still-bath quenching, a slow up-and-down movement of the tool is preferable to a violent swishing around.

The temperature of water-base quenching baths should preferably be kept around 70 degrees F, but 70 to 90 or 100 degrees F is a safe range. The temperature of the hardening bath has a great deal to do with the hardness obtained. The higher the temperature of the quenching water, the more nearly does its effect approach that of oil; and if boiling water is used for quenching, it will have an effect even more gentle than that of oil — in fact, it would leave the steel nearly soft. Parts of irregular shape are sometimes quenched in a water bath that has been warmed somewhat to prevent sudden cooling and cracking.

When water is used, it should be "soft" because unsatisfactory results will be obtained with "hard" water. Any contamination of water-base quenching liquids by soap tends to decrease their rate of cooling. A water bath having 1 or 2 inches of oil on the top is sometimes employed to advantage for quenching tools made of high-carbon steel as the oil through which the work first passes reduces the sudden quenching action of the water.

The bath should be amply large to dissipate the heat rapidly and the temperature should be kept about constant so that successive pieces will be cooled at the same rate. Irregularly shaped parts should be immersed so that the heaviest or thickest section enters the bath first. After immersion, the part to be hardened should be agitated in the bath; the agitation reduces the tendency of the formation of a vapor coating on certain surfaces, and a more uniform rate of cooling is obtained. The work should never be dropped to the bottom of the bath until quite cool.

Flush or Local Quenching by Pressure-Spraying: When dies for cold heading, drawing, extruding, etc., or other tools, require a hard working surface and a relatively soft but tough body, the quenching may be done by spraying water under pressure against the interior or other surfaces to be hardened. Special spraying fixtures are used to hold the tool and apply the spray where the hardening is required. The pressure spray prevents the formation of gas pockets previously referred to in connection with the fresh-water quenching bath; hence, fresh water is effective for flush quenching and there is no advantage in using brine.

Quenching in Molten Salt Bath.—A molten salt bath may be used in preference to oil for quenching high-speed steel. The object in using a liquid salt bath for quenching (instead of an oil bath) is to obtain maximum hardness with minimum cooling stresses and distortion that might result in cracking expensive tools, especially if there are irregular sections. The temperature of the quenching bath may be around 1100 or 1200 degrees F. Quenching is followed by cooling to room temperature and then the tool is tempered or drawn in a bath having a temperature range of 950 to 1100 degrees F. In many cases, the tempering temperature is about 1050 degrees F.

Tanks for Quenching Baths.—The main point to be considered in a quenching bath is to keep it at a uniform temperature, so that successive pieces quenched will be subjected to the same heat treatment. The next consideration is to keep the bath agitated, so that it will not be of different temperatures in different places; if thoroughly agitated and kept in motion, as the case with the bath shown in Fig. 1, it is not even necessary to keep the pieces in motion in the bath, as steam will not be likely to form around the pieces quenched. Experience has proved that if a piece is held still in a thoroughly agitated bath, it will come out much straighter than if it has been moved around in an unagitated bath, an important consideration, especially when hardening long pieces. It is, besides, no easy matter to keep heavy and long pieces in motion unless it be done by mechanical means.

In Fig. 1 is shown a water or brine tank for quenching baths. Water is forced by a pump or other means through the supply pipe into the intermediate space between the outer and inner tank. From the intermediate space, it is forced into the inner tank through holes as indicated. The water returns to the storage tank by overflowing from the inner tank into the outer one and then through the overflow pipe as indicated. In Fig. 3 is shown another water or brine tank of a more common type. In this case, the water or brine is pumped from the storage tank and continuously returned to it. If the storage tank contains a large volume of water, there is no need for a special means for cooling. Otherwise, arrangements must be made for cooling the water after it has passed through the tank. The bath is agitated by the force with which the water is pumped into it. The holes at A are drilled at an angle, so as to throw the water toward the center of the tank. In Fig. 2 is shown an oil-quenching tank in which water is circulated in an outer surrounding tank to keep the oil bath cool. Air is forced into the oil bath to keep it agitated. Fig. 4 shows the ordinary type of quenching tank cooled by water forced through a coil of pipe. This arrangement can be used for oil, water, or brine. Fig. 5 shows a similar type of quenching tank, but with two coils of pipe. Water flows through one of these and steam through the other. By these means, it is possible to keep the bath at a constant temperature.

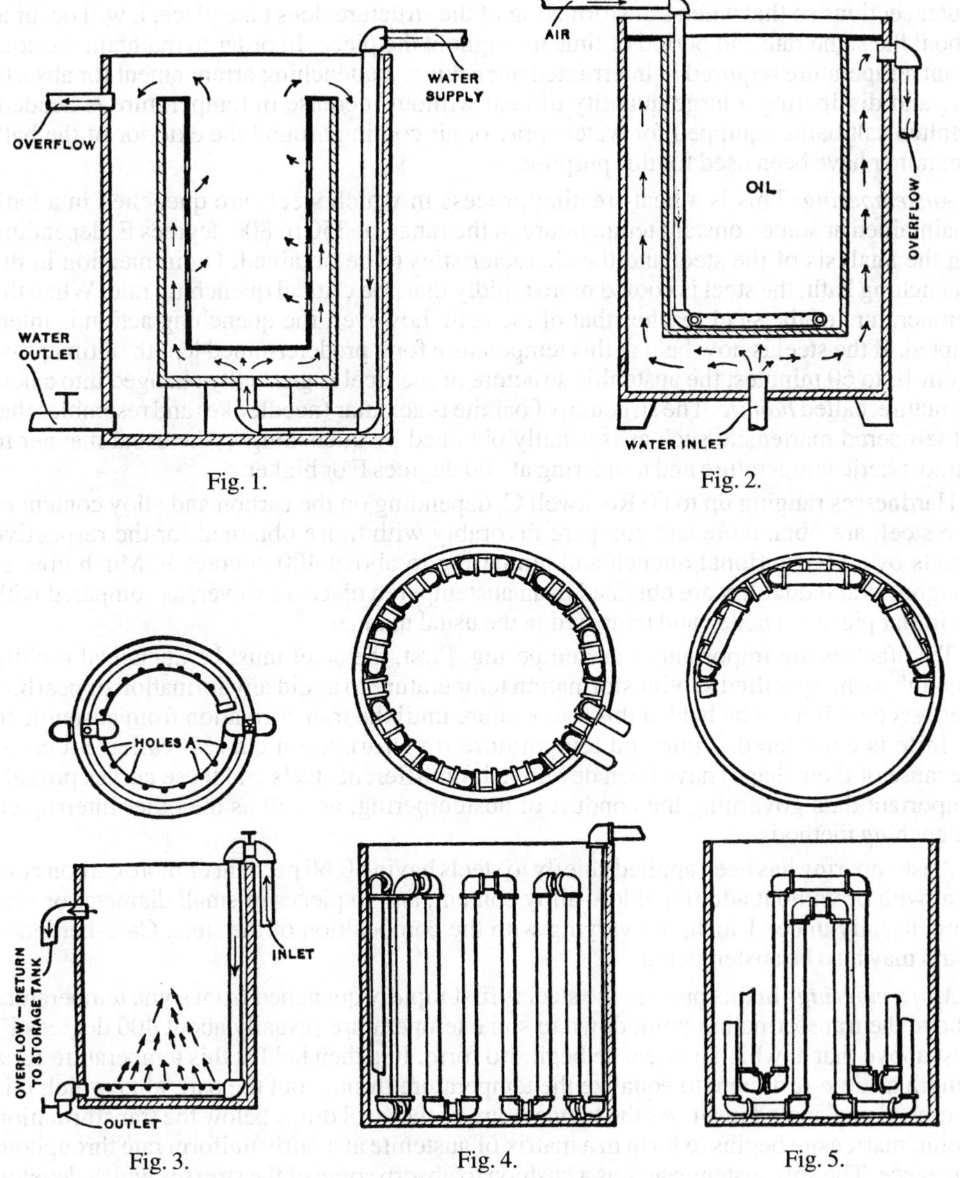

Fig. 1. Fig. 2.

Fig. 3. Fig. 4. Fig. 5.

Interrupted Quenching.—*Austempering, martempering, and isothermal quenching* are three methods of interrupted quenching that have been developed to obtain greater toughness and ductility for given hardnesses and to avoid the difficulties of quench cracks, internal stresses, and warpage, frequently experienced when the conventional method of quenching steel directly and rapidly from above the transformation point to atmospheric temperature is employed. In each of these three methods, quenching is begun when the work has reached some temperature above the transformation point and is conducted at a rate faster than the critical rate. The rapid cooling of the steel is interrupted, however, at some temperature above that at which martensite begins to form. The three methods differ in the temperature range at which interruption of the rapid quench takes place, the length of time that the steel is held at this temperature, and whether the subsequent cooling to atmospheric temperature is rapid or slow, and is or is not preceded by a tempering operation.

One of the reasons for maintaining the steel at a constant temperature for a definite period of time is to permit the inside sections of the piece to reach the same temperature as the outer sections so that when transformation of the structure does take place, it will occur at about the same rate and period of time throughout the piece. In order to maintain the constant temperature required in interrupted quenching, a quenching arrangement for absorbing and dissipating a large quantity of heat without increase in temperature is needed. Molten salt baths equipped for water spray or air cooling around the exterior of the bath container have been used for this purpose.

Austempering: This is a heat-treating process in which steels are quenched in a bath maintained at some constant temperature in the range of 350 to 800 degrees F, depending on the analysis of the steel and the characteristics to be obtained. On immersion in the quenching bath, the steel is cooled more rapidly than the critical quenching rate. When the temperature of the steel reaches that of the bath, however, the quenching action is interrupted. If the steel is now held at this temperature for a predetermined length of time, say, from 10 to 60 minutes, the austenitic structure of the steel is gradually changed into a new structure, called *bainite*. The structure of bainite is acicular (needlelike) and resembles that of tempered martensite such as is usually obtained by quenching in the usual manner to atmospheric temperature and tempering at 400 degrees F or higher.

Hardnesses ranging up to 60 Rockwell C, depending on the carbon and alloy content of the steel, are obtainable and compare favorably with those obtained for the respective steels by a conventional quench and tempering to above 400 degrees F. Much greater toughness and ductility are obtained in an austempered piece, however, as compared with a similar piece quenched and tempered in the usual manner.

Two factors are important in austempering. First, the steel must be quenched rapidly enough to the specified subtransformation temperature to avoid any formation of pearlite, and, second, it must be held at this temperature until the transformation from austenite to bainite is completed. Time and temperature transformation curves (called S-curves because of their shape) have been developed for different steels and these curves provide important data governing the conduct of austempering, as well as the other interrupted quenching methods.

Austempering has been applied chiefly to steels having 0.60 per cent or more carbon content with or without additional low-alloy content, and to pieces of small diameter or section, usually under 1 inch, but varying with the composition of the steel. Case-hardened parts may also be austempered.

Martempering: In this process the steel is first rapidly quenched from some temperature above the transformation point down to some temperature (usually about 400 degrees F) just above that at which martensite begins to form. It is then held at this temperature for a length of time sufficient to equalize the temperature throughout the part, after which it is removed and cooled in air. As the temperature of the steel drops below the transformation point, martensite begins to form in a matrix of austenite at a fairly uniform rate throughout the piece. The soft austenite acts as a cushion to absorb some of the stresses which develop as the martensite is formed. The difficulties presented by quench cracks, internal stresses, and dimensional changes are largely avoided, thus a structure of high hardness can be obtained. If greater toughness and ductility are required, conventional tempering may follow. In general, heavier sections can be hardened more easily by the martempering process than by the austempering process. The martempering process is especially suited to the higher-alloyed steels.

Isothermal Quenching: This process resembles austempering in that the steel is first rapidly quenched from above the transformation point down to a temperature that is above that at which martensite begins to form and is held at this temperature until the austenite is completely transformed into bainite. The constant temperature to which the piece is quenched and then maintained is usually 450 degrees F or above. The process differs from austempering in that after transformation to a bainite structure has been completed, the steel is immersed in another bath and is brought up to some higher temperature, depending on the characteristics desired, and is maintained at this temperature for a definite period of time, followed by cooling in air. Thus, tempering to obtain the desired toughness or ductility takes place immediately after the structure of the steel has changed to bainite and before it is cooled to atmospheric temperature.

Tempering

The object of *tempering* or *drawing* is to reduce the brittleness in hardened steel and to remove the internal strains caused by the sudden cooling in the quenching bath. The tempering process consists in heating the steel by various means to a certain temperature and then cooling it. When steel is in a fully hardened condition, its structure consists largely of *martensite*. On reheating to a temperature of from about 300 to 750 degrees F, a softer and tougher structure known as *troostite* is formed. If the steel is reheated to a temperature of from 750 to 1290 degrees F, a structure known as *sorbite* is formed that has somewhat less strength than troostite but much greater ductility.

Tempering Temperatures.—If steel is heated in an oxidizing atmosphere, a film of oxide forms on the surface that changes color as the temperature increases. These oxide colors (see table) have been used extensively in the past as a means of gaging the correct amount of temper; but since these colors are affected to some extent by the composition of the metal, the method is not dependable.

The availability of reliable pyrometers in combination with tempering baths of oil, salt, or lead make it possible to heat the work uniformly and to a given temperature within close limits.

Suggested temperatures for tempering various tools are given in the accompanying table.

Tempering in Oil.—Oil baths are extensively used for tempering tools (especially in quantity), the work being immersed in oil heated to the required temperature, which is indicated by a thermometer. It is important that the oil have a uniform temperature throughout and that the work be immersed long enough to acquire this temperature. Cold steel should not be plunged into a bath heated for tempering, owing to the danger of cracking. The steel should either be preheated to about 300 degrees F, before placing it in the bath, or the latter should be at a comparatively low temperature before immersing the steel, and then be heated to the required degree. A temperature of from 650 to 700 degrees F can be obtained with heavy tempering oils; for higher temperatures, either a bath of nitrate salts or a lead bath may be used.

In tempering, the best method is to immerse the pieces to be tempered before starting to heat the oil, so that they are heated with the oil. After the pieces tempered are taken out of the oil bath, they should be immediately dipped in a tank of caustic soda, and after that in a tank of hot water. This will remove all oil that might adhere to the tools. The following tempering oil has given satisfactory results: mineral oil, 94 per cent; saponifiable oil, 6 per cent; specific gravity, 0.920; flash point, 550 degrees F; fire test, 625 degrees F.

Tempering in Salt Baths.—Molten salt baths may be used for tempering or drawing operations. Nitrate baths are particularly adapted for the usual drawing temperature range of, say, 300 to 1100 degrees F. Tempering in an oil bath usually is limited to temperatures of 500 to 600 degrees F, and some heat-treating specialists recommend the use of a salt bath for temperatures above 350 or 400 degrees F, as it is considered more efficient and economical. Tempering in a bath (salt or oil) has several advantages, such as ease in controlling the temperature range and maintenance of a uniform temperature. The work is also heated much more rapidly in a molten bath. A gas- or oil-fired muffle or semimuffle furnace may be used for tempering, but a salt bath or oil bath is preferable. A salt bath is recommended for tempering high-speed steel, although furnaces may also be used. The bath or furnace temperature should be increased gradually, say, from 300 to 400 degrees F up to the tempering temperature, which may range from 1050 to 1150 degrees F for high-speed steel.

Tempering in a Lead Bath.—The lead bath is commonly used for heating steel in connection with tempering, as well as for hardening. The bath is first heated to the temperature at which the steel should be tempered; the preheated work is then placed in the bath long enough to acquire this temperature, after which it is removed and cooled. As the melting temperature of pure lead is about 620 degrees F, tin is commonly added to it to lower the temperature sufficiently for tempering. Reductions in temperature can be obtained by varying the proportions of lead and tin, as shown by the table, "Temperatures of Lead Bath Alloys."

Temperatures of Lead Bath Alloys

Parts Lead	Parts Tin	Melting Temp., Deg. F	Parts Lead	Parts Tin	Melting Temp., Deg. F	Parts Lead	Parts Tin	Melting Temp., Deg. F
200	8	560	39	8	510	19	8	460
100	8	550	33	8	500	17	8	450
75	8	540	28	8	490	16	8	440
60	8	530	24	8	480	15	8	430
48	8	520	21	8	470	14	8	420

To Prevent Lead from Sticking to Steel.—To prevent hot lead from sticking to parts heated in it, mix common whiting with wood alcohol, and paint the part that is to be heated. Water can be used instead of alcohol, but in that case, the paint must be thoroughly dry, as otherwise the moisture will cause the lead to "fly." Another method is to make a thick paste according to the following formula: Pulverized charred leather, 1 pound; fine wheat flour, 1½ pounds; fine table salt, 2 pounds. Coat the tool with this paste and heat slowly until dry, then proceed to harden. Still another method is to heat the work to a blue color, or about 600 degrees F, and then dip it in a strong solution of salt water, prior to heating in the lead bath. The lead is sometimes removed from parts having fine projections or teeth, by using a stiff brush just before immersing in the cooling bath. Removal of lead is necessary to prevent the formation of soft spots.

Tempering in Sand.—The sand bath is used for tempering certain classes of work. One method is to deposit the sand on an iron plate or in a shallow box that has burners beneath it. With this method of tempering, tools such as boiler punches, etc., can be given a varying temper by placing them endwise in the sand. As the temperature of the sand bath is higher toward the bottom, a tool can be so placed that the color of the lower end will become a deep dark blue when the middle portion is a very dark straw, and the working end or top a light straw color, the hardness gradually increasing from the bottom up.

Double Tempering.—In tempering high-speed steel tools, it is common practice to repeat the tempering operation or "double temper" the steel. Double tempering is done by heating the steel to the tempering temperature (say, 1050 degrees F) and holding it at that temperature for two hours. It is then cooled to room temperature, reheated to 1050 degrees F for another two-hour period, and again cooled to room temperature. After the first tempering operation, some untempered martensite remains in the steel. This martensite is not only tempered by a second tempering operation but is relieved of internal stresses, thus improving the steel for service conditions. The hardening temperature for the higher-alloy steels may affect the hardness after tempering. For example, molybdenum high-speed steel when heated to 2100 degrees F had a hardness of 61 Rockwell C after tempering, whereas a temperature of 2250 degrees F resulted in a hardness of 64.5 Rockwell C after tempering.

Temperatures as Indicated by the Color of Plain Carbon Steel

Degrees Centigrade	Degrees Fahrenheit	Color of Steel	Degrees Centigrade	Degrees Fahrenheit	Color of Steel
221.1	430	Very pale yellow	265.6	510	Spotted red-brown
226.7	440	Light yellow	271.1	520	Brown-purple
232.2	450	Pale straw-yellow	276.7	530	Light purple
237.8	460	Straw-yellow	282.2	540	Full purple
243.3	470	Deep straw-yellow	287.8	550	Dark purple
248.9	480	Dark yellow	293.3	560	Full blue
254.4	490	Yellow-brown	298.9	570	Dark blue
260.0	500	Brown-yellow	337.8	640	Light blue

Tempering Temperatures for Various Plain Carbon Steel Tools

Degrees F	Class of Tool
495 to 500	Taps ½ inch or over, for use on automatic screw machines
495 to 500	Nut taps ½ inch and under
515 to 520	Taps ¼ inch and under, for use on automatic screw machines
525 to 530	Thread dies to cut thread close to shoulder
500 to 510	Thread dies for general work
495	Thread dies for tool steel or steel tube
525 to 540	Dies for bolt threader threading to shoulder
460 to 470	Thread rolling dies
430 to 435	Hollow mills (solid type) for roughing on automatic screw machines
485	Knurls
450	Twist drills for hard service
450	Centering tools for automatic screw machines
430	Forming tools for automatic screw machines
430 to 435	Cut-off tools for automatic screw machines
440 to 450	Profile cutters for milling machines
430	Formed milling cutters
435 to 440	Milling cutters
430 to 440	Reamers
460	Counterbores and countersinks
480	Cutters for tube- or pipe-cutting machines
460 to 520	Snaps for pneumatic hammers — harden full length, temper to 460 degrees, then bring point to 520 degrees

Annealing, Spheroidizing, and Normalizing

Annealing of steel is a heat-treating process in which the steel is heated to some elevated temperature, usually in or near the critical range, is held at this temperature for some period of time, and is then cooled, usually at a slow rate. Spheroidizing and normalizing may be considered as special cases of annealing.

The *full annealing* of carbon steel consists in heating it slightly above the *upper* critical point for hypoeutectoid steels (steels of less than 0.85 per cent carbon content) and slightly above the *lower* critical point for hypereutectoid steels (steels of more than 0.85 per cent carbon content), holding it at this temperature until it is uniformly heated and then slowly cooling it to 1000 degrees F or below. The resulting structure is layerlike, or lamellar, in character due to the pearlite that is formed during the slow cooling.

Anealing is employed 1) to soften steel for machining, cutting, stamping, etc., or for some particular service; 2) to alter ductility, toughness, electrical or magnetic characteristics or other physical properties; 3) to refine the crystal structure; 4) to produce grain reorientation; and 5) to relieve stresses and hardness resulting from cold working.

The *spheroidizing* of steel, according to the American Society of Metals, is "any process of heating and cooling that produces a rounded or globular form of carbide." High-carbon steels are spheroidized to improve their machinability especially in continuous cutting operations such as are performed by lathes and screw machines. In low-carbon steels, spheroidizing may be employed to meet certain strength requirements before subsequent heat treatment. Spheroidizing also tends to increase resistance to abrasion.

The *normalizing* of steel consists in heating it to some temperature above that used for annealing, usually about 100 degrees F above the upper critical range, and then cooling it in still air at room temperature. Normalizing is intended to put the steel into a uniform, unstressed condition of proper grain size and refinement so that it will properly respond to further heat treatments. It is particularly important in the case of forgings that are to be later heat treated. Normalizing may or may not (depending on the composition) leave steel in a sufficiently soft state for machining with available tools. Annealing for machinability is often preceded by normalizing and the combined treatment — frequently called a *double anneal* — produces a better result than a simple anneal.

Annealing Practice.—For carbon steels, the following annealing temperatures are recommended by the American Society for Testing and Materials: Steels of less than 0.12 per cent carbon content, 1600 to 1700 degrees F; steels of 0.12 to 0.29 per cent carbon content, 1550 to 1600 degrees F, steels of 0.30 to 0.49 per cent carbon content, 1500 to 1550 degrees F; and for 0.50 to 1.00 per cent carbon steels, from 1450 to 1500 degrees F. Slightly lower temperatures are satisfactory for steels having more than 0.75 per cent manganese content. Heating should be uniform to avoid the formation of additional stresses. In the case of large workpieces, the heating should be slow enough so that the temperature of the interior does not lag too far behind that of the surface.

It has been found that in annealing steel, the higher the temperature to which it is heated to produce an austenitic structure, the greater the tendency of the structure to become lamellar (pearlitic) in cooling. On the other hand, the closer the austenitizing temperature to the critical temperature, the greater is the tendency of the annealed steel to become spheroidal.

Rate of Cooling: After heating the steel to some temperature within the annealing range, it should be cooled slowly enough to permit the development of the desired softness and ductility. In general, the slower the cooling rate, the greater the resulting softness and ductility. Steel of a high-carbon content should be cooled more slowly than steel of a low-carbon content; and the higher the alloy content, the slower is the cooling rate usually required. Where extreme softness and ductility are not required. the steel may be cooled in the annealing furnace to some temperature well below the critical point, say, to about 1000 degrees F and then removed and cooled in air.

Annealing by Constant-Temperature Transformation.—It has been found that steel that has been heated above the critical point so that it has an austenitic structure can be transformed into a lamellar (pearlitic) or a spheroidal structure by holding it for a definite period of time at some constant subcritical temperature. In other words, it is feasible to anneal steel by means of a constant-temperature transformation as well as by the conventional continuous cooling method. When the constant-temperature transformation method is employed, the steel, after being heated to some temperature above the critical and held at this temperature until it is austenitized, is cooled as rapidly as feasible to some relatively high subcritical transformation temperature. The selection of this temperature is governed by the desired microstructure and hardness required and is taken from a transformation time and temperature curve (often called a TTT curve). As drawn for a particular steel, such a curve shows the length of time required to transform that steel from an austenitic state at various subcritical temperatures. After being held at the selected sub-critical temperature for the required length of time, the steel is cooled to room temperature — again, as rapidly as feasible. This rapid cooling down to the selected transformation temperature and then down to room temperature has a negligible effect on the structure of the steel and often produces a considerable saving in time over the conventional slow cooling method of annealing.

The softest condition in steel can be developed by heating it to a temperature usually less than 100 degrees F above the lower critical point and then cooling it to some temperature, usually less than 100 degrees, below the critical point, where it is held until the transformation is completed. Certain steels require a very lengthy period of time for transformation of the austenite when held at a constant temperature within this range. For such steels, a practical procedure is to allow most of the transformation to take place in this temperature range where a soft product is formed and then to finish the transformation at a lower temperature where the time for the completion of the transformation is short.

Spheroidizing Practice.—A common method of spheroidizing steel consists in heating it to or slightly below the lower critical point, holding it at this temperature for a period of time, and then cooling it slowly to about 1000 degrees F or below. The length of time for which the steel is held at the spheroidizing temperature largely governs the degree of spheroidization. High-carbon steel may be spheroidized by subjecting it to a temperature that alternately rises and falls between a point within and a point without the critical range. Tool steel may be spheroidized by heating to a temperature slightly above the critical range and then, after being held at this temperature for a period of time, cooling without removal from the furnace.

Normalizing Practice.—When using the lower-carbon steels, simple normalizing is often sufficient to place the steel in its best condition for machining and will lessen distortion in carburizing or hardening. In the medium- and higher-carbon steels, combined normalizing and annealing constitutes the best practice. For unimportant parts, the normalizing may be omitted entirely or annealing may be practiced only when the steel is otherwise difficult to machine. Both processes are recommended in the following heat treatments (for SAE steels) as representing the best metallurgical practice. The temperatures recommended for normalizing and annealing have been made indefinite in many instances because of the many different types of furnaces used in various plants and the difference in results desired.

Case Hardening

In order to harden low-carbon steel, it is necessary to increase the carbon content of the surface of the steel so that a thin outer "case" can be hardened by heating the steel to the hardening temperature and then quenching it. The process, therefore, involves two separate operations. The first is the *carburizing* operation for impregnating the outer surface with sufficient carbon, and the second operation is that of heat treating the carburized parts so as to obtain a hard outer case and, at the same time, give the "core" the required physical properties. The term "case hardening" is ordinarily used to indicate the complete process of carburizing and hardening.

Carburization.—Carburization is the result of heating iron or steel to a temperature below its melting point in the presence of a solid, liquid, or gaseous material that decomposes so as to liberate carbon when heated to the temperature used. In this way, it is possible to obtain by the gradual penetration, diffusion, or absorption of the carbon by the steel, a "zone" or "case" of higher-carbon content at the outer surfaces than that of the original object. When a carburized object is rapidly cooled or quenched in water, oil, brine, etc., from the proper temperature, this case becomes hard, leaving the inside of the piece soft, but of great toughness.

Use of Carbonaceous Mixtures.—When carburizing materials of the solid class are used, the case-hardening process consists in packing steel articles in metal boxes or pots, with a carbonaceous compound surrounding the steel objects. The boxes or pots are sealed and placed in a carburizing oven or furnace maintained usually at a temperature of from about 1650 to 1700 degrees F for a length of time depending on the extent of the carburizing action desired. The carbon from the carburizing compound will then be absorbed by the steel on the surfaces desired, and the low-carbon steel is converted into high-carbon steel at these portions. The internal sections and the insulated parts of the object retain practically their original low-carbon content. The result is a steel of a dual structure, a high-carbon and a low-carbon steel in the same piece. The carburized steel may now be heat treated by heating and quenching, in much the same way as high-carbon steel is hardened, in order to develop the properties of hardness and toughness; but as the steel is, in reality, two steels in one, one high-carbon and one low-carbon, the correct heat treatment after carburizing includes two distinct processes, one suitable for the high-carbon portion or the "case," as it is generally called, and one suitable for the low-carbon portion or core. The method of heat treatment varies according to the kind of steel used. Usually, an initial heating and slow cooling is followed by reheating to 1400–1450 degrees F, quenching in oil or water, and a final tempering. More definite information is given in the following section on S.A.E. steels.

Carburizers: There are many commercial carburizers on the market in which the materials used as the generator may be hard and soft wood charcoal, animal charcoal, coke, coal, beans and nuts, bone and leather, or various combinations of these. The energizers may be barium, cyanogen, and ammonium compounds, various salts, soda ash, or lime and oil hydrocarbons.

Pack-Hardening.—When cutting tools, gages, and other parts made from high-carbon steels are heated for hardening while packed in some carbonaceous material in order to protect delicate edges, corners, or finished surfaces, the process usually is known as pack-hardening. Thus, the purpose is to protect the work, prevent scale formation, ensure uniform heating, and minimize the danger of cracking and warpage. The work is packed, as in carburizing, and in the same type of receptacle. Common hardwood charcoal often is used, especially if it has had an initial heating to eliminate shrinkage and discharge its more impure gases. The lowest temperature required for hardening should be employed for pack-hardening — usually 1400 to 1450 degrees F for carbon steels. Pack-hardening has also been applied to high-speed steels, but modern developments in heat-treating salts have made it possible to harden high-speed steel without decarburization, injury to sharp edges, or marring the finished surfaces. See paragraph on Salt Baths.

Cyanide Hardening.—When low-carbon steel requires a very hard outer surface but does not need high shock-resisting qualities, the cyanide-hardening process may be employed to produce what is known as superficial hardness. This superficial hardening is the result of carburizing a very thin outer skin (which may be only a few thousandths inch thick) by immersing the steel in a bath containing sodium cyanide. The temperatures usually vary from 1450 to 1650 degrees F and the percentage of sodium cyanide in the bath extends over a wide range, depending on the steel used and properties required.

Nitriding Process.—Nitriding is a process for surface hardening certain alloy steels by heating the steel in an atmosphere of nitrogen (ammonia gas) at approximately 950 degrees F. The steel is then cooled slowly. Finish machined surfaces hardened by nitriding are subject to minimum distortion. The physical properties, such as toughness, high impact strength, etc., can be imparted to the core by previous heat treatments and are unaffected by drawing temperatures up to 950 degrees F. The "Nitralloy" steels suitable for this process may be readily machined in the heat-treated as well as in the annealed state, and they forge as easily as alloy steels of the same carbon content. Certain heat treatments must be applied prior to nitriding, the first being annealing to relieve rolling, forging, or machining strains. Parts or sections not requiring heat treating should be machined or ground to the exact dimensions required. Close tolerances must be maintained in finish machining, but allowances for growth due to adsorption of nitrogen should be made, and this usually amounts to about 0.0005 inch for a case depth of 0.02 inch. Parts requiring heat treatment for definite physical properties are forged or cut from annealed stock, heat treated for the desired physical properties, rough machined, normalized, and finish machined. If quenched and drawn parts are normalized afterwards, the drawing and normalizing temperatures should be alike. The normalizing temperature may be below but should never be above the drawing temperature.

Ion Nitriding.—Ion nitriding, also referred to as glow discharge nitriding, is a process for case hardening of steel parts such as tool spindles, cutting tools, extrusion equipment, forging dies, gears, and crankshafts. An electrical potential ionizes low-pressure nitrogen gas, and the ions produced are accelerated to and impinge on the workpiece, heating it to the appropriate temperature for diffusion to take place. Therefore, there is no requirement for a supplemental heat source. The inward diffusion of the nitrogen ions forms the iron and alloy nitrides in the case. White layer formation, familiar in conventional gas nitriding, is readily controlled by this process.

Liquid Carburizing.—Activated liquid salt baths are now used extensively for carburizing. Sodium cyanide and other salt baths are used. The salt bath is heated by electrodes immersed in it, the bath itself acting as the conductor and resistor. One or more groups of electrodes, with two or more electrodes per group, may be used. The heating is accompanied by a stirring action to ensure uniform temperature and carburizing activity throughout the bath. The temperature may be controlled by a thermocouple immersed in the bath and connecting with a pyrometer designed to provide automatic regulation. The advantages of liquid baths include rapid action; uniform carburization; minimum distortion; and elimination of the packing and unpacking required when carbonaceous mixtures are used. In selective carburizing, the portions of the work that are not to be carburized are copper-plated and the entire piece is then immersed in an activated cyanide bath. The copper inhibits any carburizing action on the plated parts, and this method offers a practical solution for selectively carburizing any portion of a steel part.

Gas Carburizing.—When carburizing gases are used, the mixture varies with the type of case and quality of product desired. The gaseous hydrocarbons most widely used are methane (natural gas), propane, and butane. These carbon-bearing gases are mixed with air, with manufactured gases of several types, with flue gas, or with other specially prepared "diluent" gases. It is necessary to maintain a continuous fresh stream of carburizing gases to the carburizing retort or muffle, as well as to remove the spent gases from the muffle continuously, in order to obtain the correct mixture of gases inside the muffle. A slight pressure is maintained on the muffle to exclude unwanted gases.

The horizontal rotary type of gas carburizing furnace has a retort or muffle that revolves slowly. This type of furnace is adapted to small parts such as ball and roller bearings, chain links, small axles, bolts, etc. With this type of furnace, very large pieces such as gears, for example, may be injured by successive shocks due to tumbling within the rotor.

The vertical pit type of gas carburizer has a stationary workholder that is placed vertically in a pit. The work, instead of circulating in the gases as with the rotary type, is stationary and the gases circulate around it. This type is applicable to long large shafts or other parts or shapes that cannot be rolled in a rotary type of furnace.

There are three types of continuous gas furnaces that may be designated as 1) direct quench and manually operated; 2) direct quench and mechanically operated; and 3) cooling-zone type.

Where production does not warrant using a large continuous-type furnace, a horizontal muffle furnace of the batch type may be used, especially if the quantities of work are varied and the production not continuous.

Vacuum Carburizing.—Vacuum carburizing is a high-temperature gas carburizing process that is performed at pressures below atmospheric. The furnace atmosphere usually consists solely of an enriching gas, such as natural gas, pure methane, or propane; nitrogen is sometimes used as a carrier gas. Vacuum carburizing offers several advantages such as combining of processing operations and reduced total processing time.

Carburizing Steels.—A low-carbon steel containing, say, from 0.10 to 0.20 per cent of carbon is suitable for carburized case hardening. In addition to straight-carbon steels, the low-carbon alloy steels are employed. The alloys add to case-hardened parts the same advantageous properties that they give to other classes of steel. Various steels suitable for case hardening will be found in the section on SAE steels.

To Clean Work after Case Hardening.—To clean work, especially if knurled, where dirt is likely to stick into crevices after case hardening, wash it in caustic soda (1 part soda to 10 parts water). In making this solution, the soda should be put into hot water gradually, and the mixture stirred until the soda is thoroughly dissolved. A still more effective method of cleaning is to dip the work into a mixture of 1 part sulfuric acid and 2 parts water. Leave the pieces in this mixture about three minutes; then wash them off immediately in a soda solution.

Flame Hardening.—This method of hardening is especially applicable to the selective hardening of large steel forgings or castings that must be finish-machined prior to heat-treatment, or that because of size or shape cannot be heat treated by using a furnace or bath. An oxyacetylene torch is used to heat quickly the surface to be hardened; this surface is then quenched to secure a hardened layer that may vary in depth from a mere skin to ¼ inch and with hardness ranging from 400 to 700 Brinell. A multiflame torchhead may be equipped with quenching holes or a spray nozzle back of the flame. This is not a carburizing or a case-hardening process as the torch is only a heating medium. Most authorities recommend tempering or drawing of the hardened surface at temperatures between 200 and 350 degrees F. This treatment may be done in a standard furnace, an oil bath, or with a gas flame. It should follow the hardening process as closely as possible. Medium-carbon and many low-alloy steels are suitable for flame hardening. Plain carbon steels ranging from 0.35 to 0.60 per cent carbon will give hardnesses of from 400 to 700 Brinell. Steels in the 0.40 to 0.45 per cent carbon range are preferred, as they have excellent core properties and produce hardnesses of from 400 to 500 Brinell without checking or cracking. Higher-carbon steels will give greater hardnesses, but extreme care must be taken to prevent cracking. Careful control of the quenching operation is required.

Spinning Method of Flame Hardening: This method is employed on circular objects that can be rotated or spun past a stationary flame. It may be subdivided according to the speed of rotation, as where the part is rotated slowly in front of a stationary flame and the quench is applied immediately after the flame. This method is used on large circular pieces such as track wheels and bearing surfaces. There will be a narrow band of material with lower hardness between adjacent torches if more than one path of the flame is required to harden the surface. There will also be an area of lower hardness where the flame is extinguished. A second method is applicable to small rollers or pinions. The work is spun at a speed of 50 to 150 rpm in front of the flame until the entire piece has reached the proper temperature; then it is quenched as a unit by a cooling spray or by ejecting it into a cooling bath.

The Progressive Method: With this method the torch travels along the face of the work while the work remains stationary. It is used to harden lathe ways, gear teeth, and track rails.

The Stationary or Spot-hardening Method: When this method is employed, the work and torch are both stationary. When the spot to be hardened reaches the quenching temperature, the flame is removed and the quench applied.

The Combination Method: This approach is a combination of the spinning and progressive methods, and is used for long bearing surfaces. The work rotates slowly past the torch as the torch travels longitudinally across the face of the work at the rate of the torch width per revolution of the work.

Equipment for the stationary method of flame hardening consists merely of an acetylene torch, an oxyacetylene supply, and a suitable means of quenching; but when the other methods are employed, work-handling tools are essential and specialty designed torches are desirable. A lathe is ideally suited for the spinning or combination hardening method, whereas a planer is easily adapted for progressive hardening. Production jobs, such as the hardening of gears, require specially designed machines. These machines reduce handling and hardening time, as well as assuring consistent results.

Induction Hardening.—The hardening of steel by means of induction heating and subsequent quenching in either liquid or air is particularly applicable to parts that require localized hardening or controlled depth of hardening and to irregularly shaped parts, such as cams that require uniform surface hardening around their contour.

Advantages offered by induction hardening are: 1) a short heating cycle that may range from a fraction of a second to several seconds (heat energy can be induced in a piece of steel at the rate of 100 to 250 Btu per square inch per minute by induction heating, as compared with a rate of 3 Btu per square inch per minute for the same material at room temperature when placed in a furnace with a wall temperature of 2000 degrees F); 2) absence of tendency to produce oxidation or decarburization; 3) exact control of depth and area of hardening; 4) close regulation of degree of hardness obtained by automatic timing of heating and quenching cycles; 5) minimum amount of warpage or distortion; and 6) possibility of substituting carbon steels for higher-cost alloy steels.

The principal advantage of induction hardening to the designer lies in its application to localized zones. Thus, specific areas in a given part can be heat treated separately to the respective hardnesses required. Parts can be designed so that the stresses at any given point in the finished piece can be relieved by local heating. Parts can be designed in which welded or brazed assemblies are built up prior to heat treating with only internal surfaces or projections requiring hardening.

Types of Induction Heating Equipment.—Induction heating is secured by placing the metal part inside or close to an "applicator" coil of one or more turns, through which alternating current is passed. The coil, formed to suit the general class of workto be heated, is usually made of copper tubing through which water is passed to prevent overheating of the coil itself. The workpiece is held either in a fixed position or is rotated slowly within or close to the applicator coil. Where the length of work is too great to permit heating in a fixed position, progressive heating may be employed. Thus, a rod or tube of steel may be fed through an applicator coil of one or more turns so that the heating zone travels progressively along the entire length of the workpiece.

The frequency of the alternating current used and the type of generator employed to supply this current to the applicator coil depend on the character of the work to be done.

There are three types of equipment used commercially to produce high-frequency current for induction heating: 1) motor generator sets that deliver current at frequencies of approximately 1000, 2000, 3000, and 10,000 cycles; 2) spark gap oscillator units that produce frequencies ranging from 80,000 to 300,000 cycles; and 3) vacuum tube oscillator sets, which produce currents at frequencies ranging from 350,000 to 15,000,000 cycles or more.

Depth of Heat Penetration.—Generally speaking, the higher the frequency used, the shallower the depth of heat penetration. For heating clear through, for deep hardening and for large workpieces, low power concentrations and low frequencies are usually employed. For very shallow and closely controlled depths of heating, as in surface hardening, and in localized heat treating of small workpieces, currents at high frequencies are employed.

For example, a ½-inch round bar of hardenable steel will be heated through its entire structure quite rapidly by an induced current of 2000 cycles. After quenching, the bar would show through hardness with a decrease in hardness from surface to center. The same piece of steel could be readily heated and surface hardened to a depth of 0.100 inch with current at 9600 cycles, and to an even shallower depth with current at 100,000 cycles. A ¼-inch bar, however, would not reach a sufficiently high temperature at 2000 cycles to permit hardening, but at 9600 cycles through hardening would be accomplished. Current at over 100,000 cycles would be needed for surface hardening such a bar.

Types of Steel for Induction Hardening.—Most of the standard types of steels can be hardened by induction heating, providing the carbon content is sufficient to produce the desired degree of hardness by quenching. Thus, low-carbon steels with a carburized case, medium- and high-carbon steels (both plain and alloy), and cast iron with a portion of the carbon in combined form, may be used for this purpose. Induction heating of alloy steels should be limited primarily to the shallow hardening type, that is those of low alloy content, otherwise the severe quench usually required may result in a highly stressed surface with consequent reduced load-carrying capacity and danger of cracking.

Through Hardening, Annealing, and Normalizing by Induction.—For through hardening, annealing, and normalizing by induction, low power concentrations are desirable to prevent too great a temperature differential between the surface and the interior of the work. A satisfactory rate of heating is obtained when the total power input to the work is slightly greater than the radiation losses at the desired temperature. If possible, as low a frequency should be used as is consistent with good electrical coupling. A number of applicator coils may be connected in a series so that several workpieces can be heated simultaneously, thus reducing the power input to each. Widening the spacing between work and applicator coil also will reduce the amount of power delivered to the work.

Induction Surface Hardening.—As indicated earlier in "Depth of Heat Penetration," currents at much higher frequencies are required in induction surface hardening than in through hardening by induction. In general, the smaller the workpiece, the thinner the section, or the shallower the depth to be hardened, the higher will be the frequency required. High power concentrations are also needed to make possible a short heating period so that an undue amount of heat will not be conducted to adjacent or interior areas, where a change in hardness is not desired. Generators of large capacity and applicator coils of but a few turns, or even a single turn, provide the necessary concentration of power in the localized area to be hardened.

Induction heating of internal surfaces, such as the interior of a hollow cylindrical part or the inside of a hole, can be accomplished readily with applicator coils shaped to match the cross-section of the opening, which may be round, square, elliptical or other form. If the internal surface is of short length, a multiturn applicator coil extending along its entire length may be employed. Where the power available is insufficient to heat the entire internal surface at once, progressive heating is used. For this purpose, an applicator coil of few turns — often but a single turn — is employed, and either coil or work is moved so that the heated zone passes progressively from one end of the hole or opening to the other. For bores of small diameter, a hairpin-shaped applicator, extending the entire length of the hole, may be employed and the work rotated about the axis of the hole to ensure even heating.

Quenching After Induction Heating.—After induction heating, quenching may be by immersion in a liquid bath (usually oil), by liquid spray (usually water), or by self-quenching. (The term "self-quenching" is used when there is no quenching medium and hardening of the heated section is due chiefly to rapid absorption of heat by the mass of cool metal adjacent to it.) Quenching by immersion offers the advantage of even cooling and is particularly satisfactory for through heated parts. Spray quenching may be arranged so that the quenching ring and applicator coil are in the same or adjacent units, permitting the quenching cycle to follow the heating cycle immediately without removal of the work from the holding fixture. Automatic timing to a fraction of a second may also be employed for both heating and quenching with this arrangement to secure the exact degree of hardness desired. Self-quenching is applicable only in thin-surface hardening where the mass of adjacent cool metal in the part is great enough to conduct the heat rapidly out of the surface layer that is being hardened. It has been recommended that for adequate self-quenching, the mass of the unheated section should be at least ten times that of the heated shell. It has been found difficult to use the self-quenching technique to produce hardened shells of much more than about 0.060 inch thickness. Close to this limit, self-quenching can only be accomplished with the easily hardenable steels. By using a combination of self-quench and liquid quench, however, it is possible to produced hardened shells on work too thin to self-quench completely. In general, self-quenching is confined chiefly to relatively small parts and simple shapes.

Induction Hardening of Gear Teeth.—Several advantages are claimed for the induction hardening of gear teeth. One advantage is that the gear teeth can be completely machined, including shaving, when in the soft-annealed or normalized condition, and then hardened, because when induction heating is used, distortion is held to a minimum. Another advantage claimed is that bushings and inserts can be assembled in the gears before hardening. A wide latitude in choice of built-up webs and easily machined hubs is afforded because the hardness of neither web nor hub is affected by the induction-hardening operation although slight dimensional changes may occur in certain designs. Regular carbon steels can be used in place of alloy steels for a wide variety of gears, and a steel with a higher carbon content can frequently be substituted for a carburizing steel so that the carburizing operation can be eliminated. Another saving in time is the elimination of cleaning after hardening.

In heating spur gear teeth by induction, the gear is usually placed inside a circular unit that combines the applicator coil and quenching ring. An automatic timing device controls both the heating and quenching cycles. During the heating cycle, the gear is rotated at 25 to 35 rpm to ensure uniform heating.

In hardening bevel gears, the applicator coil is wound to conform to the face angle of the gear. In some spiral-bevel gears, there is a tendency to obtain more heat on one side of the tooth than on the other. In some sizes of spiral-bevel gears, this tendency can be overcome by applying slightly more heat to ensure hardening of the concave side. In some forms of spiral-bevel gears, it has been the practice to carburize that part of the gear surface which is to be hardened, after the teeth have been rough-cut. Carburizing is followed by the finish-cutting operation, after which the teeth can be induction heated, using a long enough period to heat the entire tooth. When the gear is quenched, only the carburized surface will become hardened.

Table 1a. Typical Heat Treatments for SAE Carbon Steels (Carburizing Grades)

SAE No.	Normalize, Deg. F	Carburize, Deg. F	Cool[a]	Reheat, Deg. F	Cool[a]	2nd Reheat, Deg. F	Cool[a]	Temper,[b] Deg. F
	…	1650–1700	A	…	…	…	…	250–400
	…	1650–1700	B	1400–1450	A	…	…	250–400
1010	…	1650–1700	C	1400–1450	A	…	…	250–400
to	…	1650–1700	C	1650–1700	B	1400–1450	A	250–400
1022	…	1500–1650[c,d]	B	…	…	…	…	Optional
	…	1350–1575[e,d]	D	…	…	…	…	Optional
	1650–1750[f]	1650–1700	E	…	…	…	…	250–400
1024		1350–1575[e,d]	D	…	…	…	…	Optional
1025	…	1650–1700	A	…	…	…	…	250–400
1026	…	1500–1650[c,d]	B	…	…	…	…	Optional
1027	…	1350–1575[e,d]	D	…	…	…	…	Optional
1030	…	1500–1650[c,d]	B	…	…	…	…	Optional
	…	1350–1575[e,d]	D	…	…	…	…	Optional
1111	…	1500–1650[c,d]	B	…	…	…	…	Optional
1112	…	1350–1575[e,d]	D	…	…	…	…	Optional
1113	…	1650–1700	A	…	…	…	…	250–400
	…	1650–1700	B	1400–1450	A	…	…	250–400
1109	…	1650–1700	C	1400–1450	A	…	…	250–400
to	…	1650–1700	C	1650–1700	B	1400–1450	A	250–400
1120	…	1500–1650[c,d]	B	…	…	…	…	Optional
	…	1350–1575[e,d]	D	…	…	…	…	Optional
	…	1500–1650[c,d]	B	…	…	…	…	Optional
1126	…	1350–1575[e,d]	D	…	…	…	…	Optional

[a] Symbols: A = water or brine; B = water or oil; C = cool slowly; D = air or oil; E = oil; F = water, brine, or oil.

[b] Even where tempering temperatures are shown, tempering is not mandatory in many applications. Tempering is usually employed for partial stress relief and improves resistance to grinding cracks.

[c] Activated or cyanide baths.

[d] May be given refining heat as in other processes.

[e] Carbonitriding atmospheres

[f] Normalizing temperatures at least 50 deg. F above the carburizing temperature are sometimes recommended where minimum heat-treatment distortion is of vital importance.

Table 1b. Typical Heat Treatments for SAE Carbon Steels (Heat-Treating Grades)

SAE Number	Normalize, Deg. F		Anneal, Deg. F	Harden, Deg. F	Quench[a]	Temper, Deg. F
1025 & 1030	…		…	1575–1650	A	
1033 to 1035	…		…	1525–1575	B	
	1600–1700		…	1525–1575	B	
1036	…		…	1525–1575	B	
	1600–1700		…	1525–1575	B	
1038 to 1040	…		…	1525–1575	B	
1041	1600–1700	and/or	1400–1500	1475–1550	E	
1042 to 1050	1600–1700		…	1475–1550	B	
1052 & 1055	1550–1650	and/or	1400–1500	1475–1550	E	
1060 to 1074	1550–1650	and/or	1400–1500	1475–1550	E	To Desired Hardness
1078	…		1400–1500[a]	1450–1500	A	
1080 to 1090	1550–1650	and/or	1400–1500[a]	1450–1500	E[b]	
	…		1400–1500[a]	1450–1500	F	
1095	…		1400–1500[a]	1500–1600	E	
1132 & 1137	1600–1700	and/or	1400–1500	1525–1575	B	
	…		…	1500–1550	B	
1138 & 1140	1600–1700		…	1500–1550	B	
	…		1400–1500	1475–1550	E	
1141 & 1144	1600–1700		1400–1500	1475–1550	E	
	…		…	1475–1550	B	
1145 to 1151	1600–1700		…	1475–1550	B	

[a] Slow cooling produces a spheroidal structure in these high-carbon steels that is sometimes required for machining purposes.

[b] May be water- or brine-quenched by special techniques such as partial immersion or time quenched; otherwise they are subject to quench cracking.

Table 2a. Typical Heat Treatments for SAE Alloy Steels (Carburizing Grades)

SAE No.	Normal-ize[a]	Cycle Anneal[b]	Carburized, Deg. F	Cool[c]	Reheat, Deg. F	Cool[c]	Temper,[d] Deg. F
1320	yes	…	1650–1700	E	1400–1450[e]	E	250–350
	yes	…	1650–1700	E	1475–1525[f]	E	250–350
	yes	…	1650–1700	C	1400–1450[e]	E	250–350
	yes	…	1650–1700	C	1500–1550[f]	E	250–350
	yes	…	1650–1700	E[g]	…	…	250–350
	yes	…	1500–1650[h]	E	…	…	250–350
2317	yes	yes	1650–1700	E	1375–1425[e]	E	250–350
	yes	yes	1650–1700	E	1450–1500[f]	E	250–350
	yes	yes	1650–1700	C	1375–1425[e]	E	250–350
	yes	yes	1650–1700	C	1475–1525[f]	E	250–350
	yes	yes	1650–1700	E[g]	…	…	250–350
	yes	yes	1450–1650[h]	E	…	…	250–350
2512 to 2517	yes[i]	…	1650–1700	C	1325–1375[e]	E	250–350
	yes[i]	…	1650–1700	C	1425–1475[f]	E	250–350
3115 & 3120	yes	…	1650–1700	E	1400–1450[e]	E	250–350
	yes	…	1650–1700	E	1475–1525[f]	E	250–350
	yes	…	1650–1700	C	1400–1450[e]	E	250–350
	yes	…	1650–1700	C	1500–1550[f]	E	250–350
	yes	…	1650–1700	E[g]	…	…	250–350
	yes	……	1500–1650[h]	E	…	…	250–350
3310 & 3316	yes[i]	…	1650–1700	E	1400–1450[e]	E	250–350
	yes[i]	…	1650–1700	C	1475–1500[f]	E	250–350
4017 to 4032	yes	yes	1650–1700	E[g]	…	…	250–350
4119 & 4125	yes	…	1650–1700	E[g]	…	…	250–350
4317 & 4320 4608 to 4621	yes	yes	1650–1700	E	1425–1475[e]	E	250–350
	yes	yes	1650–1700	E	1475–1527[f]	E	250–350
	yes	yes	1650–1700	C	1425–1475[e]	E	250–350
	yes	yes	1650–1700	C	1475–1525[f]	E	250–350
	yes	yes	1650–1700	E[g]	…	…	250–350
	yes	yes	1650–1700	E[g]	…	…	250–350
	yes	…	1500–1650[h]	E	…	…	250–350
4812 to 4820	yes[i]	yes	1650–1700	E	1375–1425[e]	E	250–350
	yes[i]	yes	1650–1700	E	1450–1500[f]	E	250–350
	yes[i]	yes	1650–1700	C	1375–1425[e]	E	250–350
	yes[i]	yes	1650–1700	C	1450–1500[f]	E	250–350
	…	…	1650–1700	E[g]	…	…	250–350
5115 & 5120	yes	…	1650–1700	E	1425–1475[e]	E	250–350
	yes	…	1650–1700	E	1500–1550[f]	E	250–350
	yes	…	1650–1700	C	1425–1475[e]	E	250–350
	yes	…	1650–1700	C	1500–1550[f]	E	250–350
	yes	…	1500–1650[h]	E	…	…	250–350
8615 to 8625 8720	yes	yes	1650–1700	E	1475–1525[e]	E	250–350
	yes	yes	1650–1700	E	1525–1575[f]	E	250–350
	yes	yes	1650–1700	C	1475–1525[e]	E	250–350
	yes	yes	1650–1700	C	1525–1575[f]	E	250–350
	yes	yes	1650–1700	E[g]	…	…	250–350
	yes	yes	1500–1650[h]	E	…	…	250–350
9310 to 9317	yes[i]	…	1650–1700	E	1400–1450[e]	E	250–350
	yes[i]	…	1650–1700	C	1500–1525	E	250–350

[a] Normalizing temperatures should be not less than 50 deg. F higher than the carburizing temperature. Follow by air cooling.

[b] For cycle annealing, heat to normalizing temperature—hold for uniformity—cool rapidly to 1000–1250 deg. F; hold 1 to 3 hours, then air or furnace cool to obtain a structure suitable for machining and finishing.

[c] Symbols: C = cool slowly; E = oil.

[d] Tempering treatment is optional and is generally employed for partial stress relief and improved resistance to cracking from grinding operations.

[e] For use when case hardness only is paramount.

[f] For use when higher core hardness is desired.

[g] Treatment is for fine-grained steels only, when a second reheat is often unnecessary.

[h] Treatment is for activated or cyanide baths. Parts may be given refining heats as indicated for other heat-treating processes.

[i] After normalizing, reheat to temperatures of 1000–1200 deg. F and hold approximately 4 hours.

Table 3a. Typical Heat Treatments for SAE Alloy Steels (Directly Hardenable Grades)

SAE No.	Normalize, Deg. F		Anneal, Deg. F	Harden, Deg. F	Quench[a]	Temper, Deg. F
1330	…		…	1525–1575	B	To desired hardness
	1600–1700	and/or	1500–1600	1525–1575	B	To desired hardness
1335 & 1340	…		…	1500–1550	E	To desired hardness
	1600–1700	and/or	1500–1600	1525–1575	E	To desired hardness
2330	…		…	1450–1500	E	To desired hardness
	1600–1700	and/or	1400–1500	1450–1500	E	To desired hardness
2340 & 2345	…		…	1425–1475	E	To desired hardness
	1600–1700	and/or	1400–1500	1425–1475	E	To desired hardness
3130	1600–1700		…	1500–1550	B	To desired hardness
3135 to 3141	…		…	1500–1550	E	To desired hardness
	1600–1700	and/or	1450–1550	1500–1550	E	To desired hardness
3145 & 3150	…		…	1500–1550	E	To desired hardness
	1600–1700	and/or	1400–1500	1500–1550	E	To desired hardness
4037 & 4042	…		1525–1575	1500–1575	E	Gears, 350–450 To desired hardness
4047 & 4053	…		1450–1550	1500–1575	E	To desired hardness
4063 & 4068	…		1450–1550	1475–1550	E	To desired hardness
4130	1600–1700	and/or	1450–1550	1600–1650	B	To desired hardness
4137 & 4140	1600–1700	and/or	1450–1550	1550–1600	E	To desired hardness
4145 & 4150	1600–1700	and/or	1450–1550	1500–1600	E	To desired hardness
4340	1600–1700	and draw	1100–1225	1475–1525	E	To desired hardness
4640	1600–1700	and/or	1450–1550	1450–1500	E	To desired hardness
	1600–1700	and/or	1450–1500	1450–1500	E	Gears, 350–450
5045 & 5046	1600–1700	and/or	1450–1550	1475–1500	E	250–300
5130 & 5132	1650–1750	and/or	1450–1550	1500–1550	G	To desired hardness
5135 to 5145	1650–1750	and/or	1450–1550	1500–1550	E	To desired hardness Gears, 350–400
5147 to 5152	1650–1750	and/or	1450–1550	1475–1550	E	To desired hardness Gears, 350–400
50100 51100 52100	…		1350–1450	1425–1475	H	To desired hardness
	…		1350–1450	1500–1600	E	To desired hardness
6150	1650–1750	and/or	1550–1650	1600–1650	E	To desired hardness
9254 to 9262	…		…	1500–1650	E	To desired hardness
8627 to 8632	1600–1700	and/or	1450–1550	1550–1650	B	To desired hardness
8635 to 8641	1600–1700	and/or	1450–1550	1525–1575	E	To desired hardness
8642 to 8653	1600–1700	and/or	1450–1550	1500–1550	E	To desired hardness
8655 & 8660	1650–1750	and/or	1450–1550	1475–1550	E	To desired hardness
8735 & 8740	1600–1700	and/or	1450–1550	1525–1575	E	To desired hardness
8745 & 8750	1600–1700	and/or	1450–1500	1500–1550	E	To desired hardness
9437 & 9440	1600–1700	and/or	1450–1550	1550–1600	E	To desired hardness
9442 to 9747	1600–1700	and/or	1450–1550	1500–1600	E	To desired hardness
9840	1600–1700	and/or	1450–1550	1500–1550	E	To desired hardness
9845 & 9850	1600–1700	and/or	1450–1550	1500–1550	E	To desired hardness

[a] Symbols: B = water or oil; E = oil; G = water, caustic solution, or oil; H = water.

Table 4a. Typical Heat Treatments for SAE Alloy Steels
(Heat-Treating Grades—Chromium–Nickel Austenitic Steels)

SAE No.	Normalize	Anneal,[a] Deg. F	Harden, Deg. F	Quenching Medium	Temper
30301 to 30347	…	1800–2100	…	Water or Air	…

[a] Quench to produce full austenitic structure using water or air in accordance with thickness of section. Annealing temperatures given cover process and full annealing as used by industry, the lower end of the range being used for process annealing.

Table 5a. Typical Heat Treatments for SAE Alloy Steels
(Heat-Treating Grades — Stainless Chromium Irons and Steels)

SAE No.[a]	Normalize	Sub-critical Anneal, Deg. F	Full Anneal Deg. F	Harden Deg. F	Quenching Medium	Temper Deg. F
51410	… …	1300–1350[b] …	1550–1650[c] …	… 1750–1850	Oil or air	To desired hardness
51414	… …	1200–1250[b] …	… …	… 1750–1850	Oil or air	To desired hardness
51416	… …	1300–1350[b] …	1550–1650[c] …	… 1750–1850	Oil or air	To desired hardness
51420 51420F	… …	1350–1450[b] …	1550–1650[c] …	… 1800–1850	Oil or air	To desired hardness
51430	…	1400–1500[d]	…	…	…	…
51430F	…	1250–1500[d]	…	…	…	…
51431	…	1150–1225[b]	…	1800–1900	Oil or air	To desired hardness
51440A 51440B 51440C 51440F	…	1350–1440[b]	1550–1650[c]	1850–1950	Oil or air	To desired hardness
51442	…	1400–1500[d]	…	…	…	…
51446	…	1500–1650[d]	…	…	…	…
51501	…	1325–1375[b]	1525–1600[c]	1600–1700	Oil or air	To desired hardness

[a] Suffixes A, B, and C denote three types of steel differing in carbon content only. Suffix F denotes a free-machining steel.

[b] Usually air cooled, but may be furnace cooled.

[c] Cool slowly in furnace.

[d] Cool rapidly in air.

Laser and Electron-Beam Surface Hardening.—Industrial lasers and electron-beam equipment are now available for surface hardening of steels. The laser and electron beams can generate very intense energy fluxes and steep temperature profiles in the workpiece, so that external quench media are not needed. This self-quenching is due to a cold interior with sufficient mass acting as a large heat sink to rapidly cool the hot surface by conducting heat to the interior of a part. The laser beam is a beam of light and does not require a vacuum for operation. The electron beam is a stream of electrons and processing usually takes place in a vacuum chamber or envelope. Both processes may normally be applied to finished machined or ground surfaces, because little distortion results.

HEAT-TREATING HIGH-SPEED STEELS

Heat Treating High-Speed Steels

Cobaltcrom Steel.—A tungstenless alloy steel or high-speed steel that contains approximately 1.5 per cent carbon, 12.5 per cent chromium, and 3.5 per cent cobalt. Tools such as dies and milling cutters, made from cobaltcrom steel can be cast to shape in suitable molds, the teeth of cutters being formed so that it is necessary only to grind them.

Before the blanks can be machined, they must be annealed; this operation is performed by pack annealing at the temperature of 1800 degrees F, for a period of from three to six hours, according to the size of the castings being annealed. The following directions are given for the hardening of blanking and trimming dies, milling cutters, and similar tools made from cobaltcrom steel: Heat slowly in a hardening furnace to about 1830 degrees F, and hold at this temperature until the tools are thoroughly soaked. Reduce the temperature about 50 degrees, withdraw the tools from the furnace, and allow them to cool in the atmosphere. As soon as the red color disappears from the cooling tool, place it in quenching oil until cold. The slight drop of 50 degrees in temperature while the tool is still in the hardening furnace is highly important to obtain proper results. The steel will be injured if the tool is heated above 1860 degrees F. In cooling milling cutters or other rotary tools, it is suggested that they be suspended on a wire to ensure a uniform rate of cooling.

Tools that are to be subjected to shocks or vibration, such as pneumatic rivet sets, shear blades, etc., should be heated slowly to 1650 degrees F, after which the temperature should be reduced to about 1610 degrees F, at which point the tool should be removed from the furnace and permitted to cool in the atmosphere. No appreciable scaling occurs in the hardening of cobaltcrom steel tools.

Preheating Tungsten High-Speed Steel.—Tungsten high-speed steel must be hardened at a very high temperature; consequently, tools made from such steel are seldom hardened without at least one preheating stage to avoid internal strain. This requirement applies especially to milling cutters, taps, and other tools having thin teeth and thick bodies and to forming tools of irregular shape and section. The tools should be heated slowly and carefully to a temperature somewhat below the critical point of the steel, usually in the range of 1500 to 1600 degrees F. Limiting the preheating temperature prevents the operation from being unduly sensitive, and the tool may be safely left in the furnace until it reaches a uniform temperature throughout its length and cross-section.

A single stage of preheating is customary for tools of simple form that are not more than from 1 to 1½ inches in thickness. For large, intricate tools, two stages of preheating are frequently used. The first brings the tool up to a temperature of about 1100 to 1200 degrees F, and the second raises its temperature to 1550 to 1600 degrees F. A preheating time of 5 minutes for each ¼ inch in tool thickness has been recommended for a furnace temperature of 1600 degrees F. This is where a single stage of preheating is used and the furnace capacity should be sufficient to maintain practically constant temperature when the tools are changed. To prevent undue chilling, it is common practice to insert a single tool or a small lot in the hardening furnace whenever a tool or lot is removed, rather than to insert a full charge of cold metal at one time.

Preheating is usually done in a simple type of oven furnace heated by gas, electricity, or oil. Atmospheric control is seldom used, although for 18–4–1 steel a slightly reducing atmosphere (2 to 6 per cent carbon monoxide) has been found to produce the least amount of scale and will result in a better surface after final hardening.

Hardening of Tungsten High-Speed Steel.—All tungsten high-speed steels must be heated to a temperature close to their fusion point to develop their maximum efficiency as metal-cutting tools. Hardening temperatures ranging from 2200 to 2500 deg. F may be needed. The effects of changes in the hardening temperature on the cutting efficiency of several of the more common high-speed steels are shown in Table 1. The figures given are ratios, the value 1.00 for each steel being assigned to the highest observed cutting speed for that steel. The figures for different steels, therefore, cannot be directly compared with each other, except to note changes in the point of maximum cutting efficiency.

Table 1. Relation of Hardening Temperature to Cutting Efficiency

Hardening Temperature, Deg. F	Typical Analyses of High-Speed Steels			
	18 – 4 – 1	14 – 4 – 2	18 – 4 –1 Cobalt	14 – 4 – 2 Cobalt
2200	0.86	0.83	0.84	0.85
2250	0.88	0.88	0.86	0.88
2300	0.90	0.93	0.90	0.91
2350	0.95	0.98	0.94	0.94
2400	0.99	0.98	0.98	0.98
2450	1.00	…	0.99	1.00
2500	0.98	…	1.00	0.97

The figures in the table refer to tools heated in an oven-type furnace in which a neutral atmosphere is maintained. The available data indicate that a steel reaches its best cutting qualities at a temperature approximately 50 deg. F lower than the figures in the table if it is hardened in a bath-type furnace. It is, however, desirable to use a hardening temperature approximately 50 deg. F lower than that giving maximum cutting qualities, to avoid the possibility of overheating the tool.

Length of Time for Heating: The cutting efficiency of a tool is affected by the time that it is kept at the hardening temperature, almost as much as by the hardening temperature itself. It has been common practice to heat a tool for hardening until a "sweat" appeared on its surface. This sweat is presumably a melting of the oxide film on the surface of a tool heated in an oxidizing atmosphere. It does not appear when the tool is heated in an inert atmosphere. This method of determining the proper heating time is at best an approximation and indicates only the temperature on the outside of the tool rather than the condition of the interior. As such, it cannot be relied upon to give consistent results.

The only safe method is to heat the tool for a definite predetermined time, based on the size and the thickness of metal that the heat must penetrate to reach the interior. The values given in Table 2 are based on a series of experiments to determine the relative cutting efficiency of a group of tools hardened in an identical manner, except for variations in the time the tools were kept at the hardening temperature. The time given is based on that required to harden throughout a tool resting on a conducting hearth; the tool receives heat freely from three sides, on its large top surface and its smaller side surfaces. (The table does not apply to a disk lying flat on the hearth.) For a tool having a projecting cutting edge, such as a tap, the thickness or depth of the projecting portion on which the cutting edge is formed should be used when referring to the table.

Table 2. Length of Heating Time for Through Hardening

High-Speed Steel Tool Thickness, in Inches	Time in Furnace at High Heat, in Minutes	High-Speed Steel Tool Thickness, in Inches	Time in Furnace at High Heat, in Minutes	High-Speed Steel Tool Thickness, in Inches	Time in Furnace at High Heat, in Minutes
¼	2	1½	7	5	18
½	3	2	8	6	20
¾	4	3	12	8	25
1	5	4	15	10	30

The time periods given in Table 2 are based on complete penetration of the hardening effect. For very thick tools, the practical procedure is to harden to a depth sufficient to produce an adequate cutting edge, leaving the interior of the tool relatively soft.

Where atmosphere control is not provided, it often will be found impracticable to use both the temperature for maximum cutting efficiency, given in Table 1, and the heating time, given in Table 2, because abnormal scaling, grain growth, and surface decarburization of the tool will result. The principal value of an accurate control of the furnace atmosphere appears to lie in the fact that its use makes possible the particular heat treatment that produces the best structure in the tool without destruction of the tool surface or grain.

Quenching Tungsten High-Speed Steel.—High-speed steel is usually quenched in oil. The oil bath offers a convenient quench; it calls for no unusual care in handling and brings about a uniform and satisfactory rate of cooling, which does not vary appreciably with the temperature of the oil. Some authorities believe it desirable to withdraw the tool from the oil bath for a few seconds after it has reached a dull red. It is also believed desirable to move the tool around in the quenching oil, particularly immediately after it has been placed in it, to prevent the formation of a gas film on the tool. Such a film is usually a poor conductor of heat and slows the rate of cooling.

Salt Bath: Quenching in a lead or salt bath at from 1000 to 1200 deg. F has the advantage that cooling of the tool from hardening to room temperature is accomplished in two stages, thus reducing the possibility of setting up internal strains that may tend to crack the tool. The quenching temperature is sufficiently below the lower critical point for a tool so quenched to be allowed to cool to room temperature in still air. This type of quench is particularly advantageous for tools of complicated section that would easily develop hardening cracks. The salt quench has the advantage that the tool sinks and requires only a support, whereas the same tool will float in the lead bath and must be held under the surface. It is believed that the lead quench gives a somewhat higher matrix hardness, and is of advantage for tools that tend to fail by nose abrasion. Tools treated as described are brittle unless given a regular tempering treatment, because the 1000-deg. F quenching temperature is not a substitute for later tempering at the same temperature, after the tool has cooled to room temperature.

Air Cooling: Many high-speed steel tools are quenched in air, either in a stream of dry compressed air or in still air. Small sections harden satisfactorily in still air, but heavier sections should be subjected to air under pressure. One advantage of air cooling is that the tool can be kept straight and free from distortion, although it is likely that there will be more scale on a tool thus quenched than when oil, lead, or salt is used. Cooling between steel plates may help to keep thin flat tools straight and flat.

Straightening High-Speed Tools when Quenching.—The final straightness required in a tool must be considered when it is quenched. When several similar tools are to be hardened, a jig can be used to advantage for holding the tools while quenching. When long slender tools are quenched without holders, they frequently warp and must be straightened later. The best time for this straightening is during the first few minutes after the tools have been quenched, as the steel is then quite pliable and may be straightened without difficulty. The straightening must be done at once, as the tools become hard in a few minutes.

Anneal Before Rehardening.—Tools that are too soft after hardening must be annealed before rehardening. A quick anneal, such as previously described, is all that is required to put such a tool into the proper condition for rehardening. This treatment is absolutely essential. For milling cutters and forming tools of irregular section, a full anneal should be used.

Tempering or Drawing Tungsten High-Speed Steel.—The tempering or drawing temperature for high-speed steel tools usually varies from 900 to 1200 deg. F. This temperature is higher for turning and planing tools than for such tools as milling cutters, forming tools, etc. If the temperature is below 800 deg. F, the tool is likely to be too brittle. The general idea is to temper tools at the highest temperature likely to occur in service. Because this temperature ordinarily would not be known, the general practice is to temper at whatever temperature experience with that particular steel and tool has proved to be the best.

The furnace used for tempering usually is kept at a temperature of from 1000 to 1100 deg. F for ordinary high-speed steels and from 1200 to 1300 deg. F for steels of the cobalt type. These furnace temperatures apply to tools of the class used on lathes and planers. Such tools, in service, frequently heat to the point of visible redness. Milling cutters, forming tools, or any other tools for lighter duty may be tempered as low as 850 or 900 deg. F. When the tool has reached the temperature of the furnace, it should be held at this temperature for from one to several hours until it has been heated evenly throughout. It should then be allowed to cool gradually in the air and in a place that is dry and free from air drafts. In tempering, the tool should not be quenched, because quenching tends to produce strains that may result later in cracks.

Annealing Tungsten High-Speed Steel.—The following method of annealing high-speed steel has been used extensively. Use an iron box or pipe of sufficient size to allow at least ½ inch of packing between the pieces of steel to be annealed and the sides of the box or pipe. It is not necessary that each piece of steel be kept separate from every other piece, but only that the steel be prevented from touching the sides of the annealing pipe or box. Pack carefully with powdered charcoal, fine dry lime, or mica (preferably charcoal), and cover with an airtight cap or lute with fire clay; heat slowly to 1600 to 1650 deg. F and keep at this heat from 2 to 8 hours, depending on the size of the pieces to be annealed. A piece measuring 2 by 1 by 8 inches requires about 3 hours. Cool as slowly as possible, and do not expose to the air until cold, because cooling in air is likely to cause partial hardening. A good method is to allow the box or pipe to remain in the furnace until cold.

Hardening Molybdenum High-Speed Steels.—Table 3 gives the compositions of several molybdenum high-speed steels that are widely used for general commercial tool applications. The general method of hardening molybdenum high-speed steels resembles that used for 18–4–1 tungsten high-speed steel except that the hardening temperatures are lower and more precautions must be taken to avoid decarburization, especially on tools made from Type I or Type II steels, when the surface is not ground after hardening. Either salt baths or atmosphere-controlled furnaces are recommended for hardening molybdenum high-speed steels.

Table 3. Compositions of Molybdenum High-Speed Steels

	Molybdenum–Tungsten		Molybdenum–Vanadium	Tungsten–Molybdenum
Element	Type Ia (Per Cent)	Type Ib[a] (Per Cent)	Type II (Per Cent)	Type III (Per Cent)
Carbon	0.70–0.85	0.76–0.82	0.70–0.90	0.75–0.90
Tungsten	1.25–2.00	1.60–2.30	…	5.00–6.00
Chromium	3.00–5.00	3.70–4.20	3.00–5.00	3.50–5.00
Vanadium	0.90–1.50	1.05–1.35	1.50–2.25	1.25–1.75
Molybdenum	8.00–9.50	8.00–9.00	7.50–9.50	3.50–5.50
Cobalt	See footnote	4.50–5.50	See footnote	See footnote

[a] Cobalt may be used in any of these steels in varying amounts up to 9 per cent, and the vanadium content may be as high as 2.25 percent. When cobalt is used in Type III steel, the vanadium content may be as high as 2.25 per cent. When cobalt is used in Type III steel, this steel becomes susceptible to decarburization. As an illustration of the use of cobalt, Type Ib steel is included. This is steel T10 in the U.S. Navy Specification 46S37, dated November 1, 1939.

The usual method is to preheat uniformly in a separate furnace to 1250 to 1550 deg. F then transfer to a high-heat furnace maintained within the hardening temperature range given in Table 4. Single-point cutting tools, in general, should be hardened at the upper end of the temperature range indicated by Table 4. Slight grain coarsening is not objectionable in such tools when they are properly supported in service and are not subjected to chattering; however, when these tools are used for intermittent cuts, it is better to use the middle of the temperature range. All other cutting tools, such as drills, countersinks, taps, milling cutters, reamers, broaches, and form tools, should be hardened in the middle of the range shown. For certain tools, such as slender taps, cold punches, and blanking and trimming dies, where greater toughness to resist shocks is required, the lower end of the hardening temperature range should be used.

Table 4. Heat Treatment of Molybdenum High-Speed Steels

Heat-Treating Operation	Molybdenum–Tungsten Types Ia and Ib[a] (Temp., in Deg. F)	Molybdenum–Vanadium Type II (Temp., in Deg. F)	Tungsten–Molybdenum Type III (Temp., in Deg. F)
Forging	1850–2000	1850–2000	1900–2050
Not below	1600	1600	1600
Annealing	1450–1550	1450–1550	1450–1550
Strain relief	1150–1350	1150–1350	1150–1350
Preheating	1250–1500	1250–1500	1250–1550
Hardening[b]	2150–2250[a]	2150–2250	2175–2275
Salt	2150–2225	2150–2225	2150–2250
Tempering	950–1100	950–1100	950–1100

[a] For similar working conditions, Type Ib steel requires a slightly higher hardening heat than Type Ia.

[b] The higher side of the hardening range should be used for large sections, and the lower side for small sections.

Molybdenum high-speed steels can be pack-hardened following the same practice as is used for tungsten high-speed steels, but keeping on the lower side of the hardening range (approximately 1850 degrees F). Special surface treatments such as nitriding by immersion in molten cyanide that are used for tungsten high-speed steels are also applicable to molybdenum high-speed tools.

When heated in an open fire or in furnaces without atmosphere control, these steels do not sweat like 18–4–1 steels; consequently, determining the proper time in the high-heat chamber is a matter of experience. This time approximates that used with 18–4–1 steels, although it may be slightly longer when the lower part of the hardening range is used. Much can be learned by preliminary hardening of test pieces and checking on the hardness fracture and structure. It is difficult to give the exact heating time, because it is affected by temperature, type of furnace, size and shape, and furnace atmosphere. Rate of heat transfer is most rapid in salt baths, and slowest in controlled-atmosphere furnaces with high carbon monoxide content.

Quenching and Tempering of Molybdenum High-Speed Tools.—Quenching may be done in oil, air, or molten bath. To reduce the possibility of breakage and undue distortion of intricately shaped tools, it is advisable to quench in a molten bath at approximately 1100 degrees F. The tool also may be quenched in oil and removed while still red, or at approximately 1100 degrees F. The tool is then cooled in air to room temperature, and tempered immediately to avoid cracking.

When straightening is necessary, it should be done after quenching and before cooling to room temperature prior to tempering.

To temper, the tools should be reheated slowly and uniformly to 950 to 1100 degrees F. For general work, 1050 degrees F is most common. The tools should be held at this temperature at least 1 hour. Two hours is a safer minimum, and 4 hours is maximum. The time and temperature depend on the hardness and toughness required. Where tools are subjected to more or less shock, multiple temperings are suggested.

Protective Coatings for Molybdenum Steels.—To protect the surface from oxidation during heat treatment, borax may be applied by sprinkling it lightly over the steel when the latter is heated in a furnace to a low temperature (1200 to 1400 deg. F). Small tools may be rolled in a box of borax before heating. Another method more suitable for finished tools is to apply the borax or boric acid in the form of a supersaturated water solution. The tools are then immersed in the solution at 180 to 212 deg. F, or the solution may be applied with a brush or spray. Pieces so treated are heated as usual, taking care in handling to ensure good adherence of the coating. Special protective coatings or paints, when properly applied, have been found extremely useful. These materials do not fuse or run at the temperatures used, and therefore do not affect the furnace hearth. When applying these coatings, it is necessary to have a surface free from scale or grease to ensure good adherence. Coatings may be sprayed or brushed on, and usually one thin coat is sufficient. Heavy coats tend to pit the surface of the tool and are difficult to remove. Tools covered with these coatings should be allowed to dry before they are charged into the preheat furnace. After hardening and tempering, the coating can be easily removed by light blasting with sand or steel shot. When tools are lightly ground, these coatings come off immediately. Protection may also be obtained by wrapping pieces in stainless steel foil.

Nitriding High-Speed Steel Tools.—Nitriding is applied to high-speed steel for the purpose of increasing tool life by producing a very hard skin or case, the thickness of which ordinarily is from 0.001 to 0.002 inch. Nitriding is done after the tool has been fully heat treated and finish-ground. (The process differs entirely from that which is applied to surface harden certain alloy steels by heating in an atmosphere of nitrogen or ammonia gas.) The temperature of the high-speed steel nitriding bath, which is a mixture of sodium and potassium cyanides, is equal to or slightly lower than the tempering temperature. For ordinary tools, this temperature usually varies from about 1025 to 1050 deg. F; but if the tools are exceptionally fragile, the range may be reduced to 950 or 1000 deg. F. Accurate temperature control is essential to prevent exceeding the final tempering temperature. The nitriding time may vary from 10 or 15 minutes to 30 minutes or longer, and should be determined by experiment. The shorter periods are applied to tools for iron or steel, or any shock-resisting tools, and the longer periods are for tools used in machining nonferrous metals and plastics. This nitriding process is applied to tools such as hobs, reamers, taps, box tools, form tools, and milling cutters. Nitriding may increase tool life 50 to 200 per cent, or more, but it should always be preceded by correct heat treatment.

Nitriding Bath Mixtures and Temperatures: A mixture of 60 per cent sodium cyanide and 40 per cent potassium cyanide is commonly used for nitriding. This mixture has a melting point of 925 deg. F, which is gradually reduced to 800 deg. F as the cyanate content of the bath increases. A more economical mixture of 70 per cent sodium cyanide and 30 per cent potassium cyanide may be used if the operating temperature of the bath is only 1050 deg. F. Nitriding bath temperatures should not exceed 1100 deg. F because higher temperatures accelerate the formation of carbonate at the expense of the essential cyanide. A third mixture suitable for nitriding consists of 55 per cent sodium cyanide, 25 per cent potassium chloride, and 20 per cent sodium carbonate. This mixture melts at 930 deg. F.

Equipment for Hardening High-Speed Steel.—Equipment for hardening high-speed steel consists of a hardening furnace capable of maintaining a temperature of 2350 to 2450 deg. F; a preheating furnace capable of maintaining a temperature of 1700 to 1800 deg. F, and of sufficient size to hold a number of pieces of the work; a tempering (drawing) furnace capable of maintaining a temperature of 1000 to 1200 deg. F as a general rule; and a water-cooled tank of quenching oil.

High-speed steels usually are heated for hardening either in some type of electric furnace or in a gas-fired furnace of the muffle type. The small furnaces used for high-speed steel seldom are oil-fired. It is desirable to use automatic temperature control and, where an oven type of furnace is employed, a controlled atmosphere is advisable because of the variations in cutting qualities caused by hardening under uncontrolled conditions. Some furnaces of both electric and fuel-fired types are equipped with a salt bath suitable for high-speed steel hardening temperatures. Salt baths have the advantage of providing protection against the atmosphere during the heating period. A type of salt developed for commercial use is water-soluble, so that all deposits from the hardening bath may be removed by immersion in water after quenching in oil or salt, or after air cooling. One type of electric furnace heats the salt bath internally by electrodes immersed in it. The same type of furnace is also applied to various heat-treating operations, such as cyanide hardening, liquid carburizing, tempering, and annealing.

An open-forge fire has many disadvantages, especially in hardening cutters or other tools that cannot be ground all over after hardening. The air blast decarburizes the steel and lack of temperature control makes it impossible to obtain uniform results. Electric and gas furnaces provide continuous uniform heat, and the temperature may be regulated accurately, especially when pyrometers are used. In shops equipped with only one furnace for carbon steel and one for high-speed steel, the tempering can be done in the furnace used for hardening carbon steel after the preheating is finished and the steel has been removed for hardening.

Heating High-Speed Steel for Forging.—Care should be taken not to heat high-speed steel for forging too abruptly. In winter, the steel may be extremely cold when brought into the forge shop. If the steel is put directly into the hot forge fire, it is likely to develop cracks that will show up later in the finished tool. The steel, therefore, should be warmed gradually before heating for forging.

Subzero Treatment of Steel

Subzero treatment consists of subjecting the steel, after hardening and either before or after tempering, to a subzero temperature (that usually ranges from −100 to −120 deg. F) and for a period of time varying with the size or volume of the tool, gage, or other part. Commercial equipment is available for obtaining these low temperatures.

The subzero treatment is employed by most gage manufacturers to stabilize precision gages and prevent subsequent changes in size or form. Subzero treatment is also applied to some high-speed steel cutting tools. The object here is to increase the durability or life of the tools; however, up to the present time, the results of tests by metallurgists and tool engineers often differ considerably and in some instances are contradictory. Methods of procedure also vary, especially with regard to the order and number of operations in the complete heat-treating and cooling cycle.

Changes Resulting From Subzero Treatment.—When steel is at the hardening temperature it contains a solid solution of carbon and iron known as *austenite*. When the steel is hardened by sudden cooling, most of the austenite, which is relatively soft, tough, and ductile even at room temperatures, is transformed into martensite, a hard and strong constituent. If all the austenite were changed to martensite upon reaching room temperature, this process would be an ideal hardening operation, but many steels retain some austenite. In general, the higher the carbon and alloy contents and the higher the hardening temperature, the greater the tendency to retain austenite. When steel is cooled to subzero temperatures, the stability of the retained austenite is reduced so that it is more readily transformed. To obtain more complete transformation, the subzero treatment may be repeated. The ultimate transformation of austenite to martensite may take place in carbon steel without the aid of subzero treatment, but this natural transformation might require 6 months or longer, whereas by refrigeration this change occurs in a few hours.

The thorough, uniform heating that is always recommended in heat-treating operations should be accompanied by thorough, uniform cooling when the subzero treatment is applied. To ensure uniform cooling, the subzero cooling period should be increased for the larger tools and it may range from 2 to 6 hours. The tool or other part is sometimes surrounded by one or more layers of heavy wrapping or asbestos paper to delay the cooling somewhat and ensure uniformity. After the cooling cycle is started, it should continue without interruption.

Subzero treatment may sometimes cause cracking. Normally, the austenite in steel provides a cushioning effect that may prevent cracking or breakage resulting from treatments involving temperature and dimensional changes; but if this cushioning effect is removed, particularly at very low temperatures as in subzero treatments, there may be danger of cracking, especially with tools having large or irregular sections and sharp corners offering relatively low resistance to stresses. This effect is one reason why subzero treatments may differ in regard to the cooling and tempering cycle.

Stabilizing Dimensions of Gages or Precision Parts by Subzero Cooling.—Transformation of austenite into martensite is accompanied by an increase in volume; consequently, the transformation of austenite, that may occur naturally over a period of months or years, tends to change the dimensions and form of steel parts, and such changes may be serious in the case of precision gages, close-fitting machine parts, etc. To prevent such changes, the subzero treatment has proved effective. Gage-blocks, for example, may be stabilized by hardening followed by repeated cycles of chilling and tempering, to transform a large percentage of the austenite into martensite.

Order of Operations for Stabilizing Precision Gages: If precision gages and sine-bars, are heat-treated in the ordinary manner and then are finished without some stabilizing treatment, dimensional changes and warpage are liable to occur. Sub-zero cooling provides a practical and fairly rapid method of obtaining the necessary stabilization by transforming the austenite into martensite. In stabilization treatments of this kind, tempering is the final operation. One series of treatments that has been recommended after hardening and rough-grinding is as follows:

A) Cool to −120 degrees F. (This cooling period may require from one to six hours, depending on the size and form of the gage.)

B) Place gage in boiling water for two hours (oil or salt bath may also be used).

Note: Steps (a) and (b) may be repeated from two to six times, depending on the size and form of the gage. These repeated cycles will eventually transform practically all the austenite into martensite. Two or three cooling and drawing operations usually are sufficient for such work as thread gages and gage-blocks.

C) Follow with regular tempering or drawing operation and finish gage by lapping.

Series of Stabilizing Treatments for Chromium Steel: The following series of treatments has proved successful in stabilizing precision gage-blocks made from SAE 52100 chromium steel.

A) Preheat to 600 degrees F and then heat to 1575 degrees F for a period of four minutes.

B) Quench in oil at 85 degrees F. (Uniform quenching is essential.)

C) Temper at 275 degrees F for one hour.

D) Cool in tempering furnace to room temperature.

E) Continue cooling in atmosphere of industrial refrigerator for six hours with temperature of atmosphere at −120 degrees F.

F) Allow gage-blocks to return to room temperature and again temper.

Note: The complete treatment consists of six subzero cooling periods, each followed by a tempering operation. The transformation to martensite is believed to be complete even after the fifth cooling period. The hardness is about 66 Rockwell C. Transformation is checked by magnetic tests based upon the magnetism of martensite and the nonmagnetic qualities of austenite.

Stabilizing Dimensions of Close-Fitting Machine Parts.—Subzero treatment will always cause an increase in size. Machine parts subjected to repeated and perhaps drastic changes in temperature, as in aircraft, may eventually cause trouble due to growth or warpage as the austenite gradually changes to martensite. In some instances, the sizes of close-fitting moving parts have increased sufficiently to cause seizure. Such treatment, for example, may be applied to precision bearings made from SAE 52100 or alloy carburizing steels for stabilizing or aging them. *Time* aging of 52100 steels after hardening has been found to cause changes as large as 0.0025 inch in medium size sections. A practical remedy is to apply the subzero treatment before the final grinding or other machining operation.

Subzero Treatment of Carburized Parts to Improve Physical Properties.—The subzero treatment has been applied to carburized machine parts. For example, the amount of retained austenite in carburized gears may be sufficient to reduce the life of the gears. In one component, the Rockwell hardness was increased from 55 C to 65 C without loss of impact resistance qualities; in fact, impact and fatigue resistance may be increased in some examples.

Application of Subzero Treatments to High-Speed Steel.—The subzero treatment has been applied to such tools as milling cutters, hobs, taps, broaches, and drills. It is applicable to different classes of high-speed steels, such as the 18–4–1 tungsten, 18–4–14 cobalt, and the molybdenum high-speed steels. This *cold* treatment is applied preferably in conjunction with the heat treatment, both being combined in a continuous cycle of operations. The general procedure is either to harden the steel, cool it to a subzero temperature, and then temper; or, especially if there is more than one tempering operation, the first one may *precede* subzero cooling. The cooling and tempering cycle may be repeated two or more times. The number and order of the operations, or the complete cycle, may be varied to suit the class of work and, to minimize the danger of cracking, particularly if the tool has large or irregular sections, sharp corners or edges, or a high cobalt content. A subzero treatment of some kind with a final tempering operation for stress relief, is intended to increase strength and toughness without much loss in hardness; consequently, if there is greater strength at a given hardness, tools subjected to subzero treatment can operate with a higher degree of hardness than those heat treated in the ordinary manner, or, if greater toughness is preferred, it can be obtained by tempering to the original degree of hardness.

Order of Cooling and Tempering Periods for High-speed Steel.—The order or cycle for the cooling and tempering periods has not been standardized. The methods that follow have been applied to high-speed steel tools. They are given as examples of procedure and are subject to possible changes due to subsequent developments. The usual ranges of preheating and hardening temperatures are given; but for a particular steel, the recommended temperatures should be obtained from the manufacturer.

1) *Double Subzero Treatment:* (For rugged simple tool forms without irregular sections, sharp corners or edges where cracks might develop during the subzero treatment).

a) Preheat between 1400 and 1600 degrees F (double preheating is preferable, the first preheating ranging from 700 to 1000 degrees F).

b) Heat to the hardening temperature. (*Note:* Tests indicate that the effect of subzero treatment on high-speed steel may be influenced decidedly by the hardening temperature. If this temperature is near the lower part of the range, the results are unsatisfactory. Effective temperatures for ordinary high-speed steels appear to range from 2300 to 2350 degrees F).

c) Quench in oil, salt, lead, or air, down to a workpiece temperature of 150–200 degrees F. (*Note:* One method is to quench in oil; a second method is to quench in oil to about 200–225 degrees F and then air cool; a third method is to quench in salt bath at 1050–1100 degrees F and then air cool.)

d) Cool in refrigerating unit to temperature of −100 to −120 degrees F *right after quenching*. (*Note:* Tests have shown that a delay of one hour has a detrimental effect, and in ten hours the efficiency of the subzero treatment is reduced 50 per cent. This is because the austenite becomes more and more stabilized when the subzero treatment is delayed; consequently, the austenite is more difficult to transform into martensite.) The refrigerating period usually varies from two to six hours, depending on the size of the tool. Remove the tool from the refrigerating unit and allow it to return to room temperature.

e) Temper to required hardness for a period of two and one-half to three hours. The tempering temperature usually varies from a minimum of 1000 to 1100 degrees F for ordinary high-speed steels. Tests indicate that if this first tempering is less than two and one-half hours at 1050 degrees F, there will not be sufficient precipitation of carbides at the tempering temperature to allow complete transformation of the retained austenite on cooling, whereas more than three hours causes some loss in room temperature hardness, hot hardness, strength, and toughness.

f) Repeat subzero treatment, step (d).

g) Repeat the tempering operation, step (e). (*Note:* The time for the second tempering operation is sometimes reduced to about one-half the time required for the first tempering.)

2) *Single Subzero Treatment:* This treatment is the same as procedure No. 1 except that a second subzero cooling is omitted; hence, the cycle consists of hardening, subzero cooling, and double tempering. Procedure No. 3, which follows, also has one subzero cooling period in the cycle, but this *follows* the first tempering operation.

3) *Tempering Followed by Subzero Treatment:*(This treatment is for tools having irregular sections, sharp corners, or edges where cracks might develop if the hardening operation were followed immediately by subzero cooling.)

a) Preheat and heat for hardening.

b) Preheat and heat for hardening.

c) Quench as described under Procedure No. I.

d) Temper to required hardness.

e) Cool to subzero temperature −100 to −120 degrees F and then allow the tool to return to room temperature.

f) Repeat tempering operation.

Testing the Hardness of Metals

Brinell Hardness Test.—The Brinell test for determining the hardness of metallic materials consists in applying a known load to the surface of the material to be tested through a hardened steel ball of known diameter. The diameter of the resulting permanent impression in the metal is measured and the Brinell Hardness Number (BHN) is then calculated from the following formula in which D = diameter of ball in millimeters, d = measured diameter at the rim of the impression in millimeters, and P = applied load in kilograms.

$$\text{BHN} = \frac{\text{load on indenting tool in kilograms}}{\text{surface area of indentation in sq. mm.}} = \frac{P}{\frac{\pi D}{2}(D - \sqrt{D^2 - d^2})}$$

If the steel ball were not deformed under the applied load and if the impression were truly spherical, then the preceding formula would be a general one, and any combination of applied load and size of ball could be used. The impression, however, is not quite a spherical surface because there must always be some deformation of the steel ball and some recovery of form of the metal in the impression; hence, for a standard Brinell test, the size and characteristics of the ball and the magnitude of the applied load must be standardized. In the standard Brinell test, a ball 10 millimeters in diameter and a load of 3000, 1500, or 500 kilograms is used. It is desirable, although not mandatory, that the test load be of such magnitude that the diameter of the impression be in the range of 2.50 to 4.75 millimeters. The following test loads and approximate Brinell numbers for this range of impression diameters are: 3000 kg, 160 to 600 BHN; 1500 kg, 80 to 300 BHN; 500 kg, 26 to 100 BHN. In making a Brinell test, the load should be applied steadily and without a jerk for at least 15 seconds for iron and steel, and at least 30 seconds in testing other metals. A minimum period of 2 minutes, for example, has been recommended for magnesium and magnesium alloys. (For the softer metals, loads of 250, 125, or 100 kg are sometimes used.)

According to the American Society for Testing and Materials Standard E10-66, a steel ball may be used on material having a BHN not over 450, a Hultgren ball on material not over 500, or a carbide ball on material not over 630. The Brinell hardness test is not recommended for material having a BHN over 630.

Rockwell Hardness Test.—The Rockwell hardness tester is essentially a machine that measures hardness by determining the depth of penetration of a penetrator into the specimen under certain fixed conditions of test. The penetrator may be either a steel ball or a diamond spheroconical penetrator. The hardness number is related to the depth of indentation and the number is higher the harder the material. A minor load of 10 kg is first applied, causing an initial penetration; the dial is set at zero on the black-figure scale, and the major load is applied. This major load is customarily 60 or 100 kg when a steel ball is used as a penetrator, but other loads may be used when necessary. The ball penetrator is $\frac{1}{16}$ inch in diameter normally, but other penetrators of larger diameter, such as $\frac{1}{8}$ inch, may be employed for soft metals. When a diamond spheroconical penetrator is employed, the load usually is 150 kg. Experience decides the best combination of load and penetrator for use. After the major load is applied and removed, according to standard procedure, the reading is taken while the minor load is still applied.

The Rockwell Hardness Scales.—The various Rockwell scales and their applications are shown in the following table. The type of penetrator and load used with each are shown in Tables and , which give comparative hardness values for different hardness scales.

Scale	Testing Application
A	For tungsten carbide and other extremely hard materials. Also for thin, hard sheets.
B	For materials of medium hardness such as low- and medium-carbon steels in the annealed condition.
C	For materials harder than Rockwell B-100.
D	Where a somewhat lighter load is desired than on the C scale, as on case-hardened pieces.
E	For very soft materials such as bearing metals.
F	Same as the E scale but using a $\frac{1}{16}$-inch ball.
G	For metals harder than tested on the B scale.
H & K	For softer metals.
15–N; 30–N; 45–N	Where a shallow impression or a small area is desired. For hardened steel and hard alloys.
15–T; 30–T; 45-T	Where a shallow impression or a small area is desired for materials softer than hardened steel.

Shore's Scleroscope.—The scleroscope is an instrument that measures the hardness of the work in terms of elasticity. A diamond-tipped hammer is allowed to drop from a known height on the metal to be tested. As this hammer strikes the metal, it rebounds, and the harder the metal, the greater the rebound. The extreme height of the rebound is recorded, and an average of a number of readings taken on a single piece will give a good indication of the hardness of the work. The surface smoothness of the work affects the reading of the instrument. The readings are also affected by the contour and mass of the work and the depth of the case, in carburized work, the soft core of light-depth carburizing, pack-hardening, or cyanide hardening, absorbing the force of the hammer fall and decreasing the rebound. The hammer weighs about 40 grains, the height of the rebound of hardened steel is in the neighborhood of 100 on the scale, or about 6¼ inches, and the total fall is about 10 inches or 255 millimeters.

Vickers Hardness Test.—The Vickers test is similar in principle to the Brinell test. The standard Vickers penetrator is a square-based diamond pyramid having an included point angle of 136 degrees. The numerical value of the hardness number equals the applied load in kilograms divided by the area of the pyramidal impression: A smooth, firmly supported, flat surface is required. The load, which usually is applied for 30 seconds, may be 5, 10, 20, 30, 50, or 120 kilograms. The 50-kilogram load is the most usual. The hardness number is based upon the diagonal length of the square impression. The Vickers test is considered to be very accurate, and may be applied to thin sheets as well as to larger sections with proper load regulation.

Knoop Hardness Numbers.—The Knoop hardness test is applicable to extremely thin metal, plated surfaces, exceptionally hard and brittle materials, very shallow carburized or nitrided surfaces, or whenever the applied load must be kept below 3600 grams. The Knoop indentor is a diamond ground to an elongated pyramidal form and it produces an indentation having long and short diagonals with a ratio of approximately 7 to 1. The longitudinal angle of the indentor is 172 degrees, 30 minutes, and the transverse angle 130 degrees. The Tukon Tester in which the Knoop indentor is used is fully automatic under electronic control. The Knoop hardness number equals the load in kilograms divided by the projected area of indentation in square millimeters. The indentation number corresponding to the long diagonal and for a given load may be determined from a table computed for a theoretically perfect indentor. The load, which may be varied from 25 to 3600 grams, is applied for a definite period and always normal to the surface tested. Lapped plane surfaces free from scratches are required.

Monotron Hardness Indicator.—With this instrument, a diamond-ball impressor point $\frac{3}{4}$ mm in diameter is forced into the material to a depth of $\frac{9}{5000}$ inch and the pressure required to produce this constant impression indicates the hardness. One of two dials shows the pressure in kilograms and pounds, and the other shows the depth of the impression in millimeters and inches. Readings in Brinell numbers may be obtained by means of a scale designated as *M*–1.

Keep's Test.—With this apparatus, a standard steel drill is caused to make a definite number of revolutions while it is pressed with standard force against the specimen to be tested. The hardness is automatically recorded on a diagram on which a dead soft material gives a horizontal line, and a material as hard as the drill itself gives a vertical line, intermediate hardness being represented by the corresponding angle between 0 and 90 degrees.

Comparison of Hardness Scales.—Tables and show comparisons of various hardness scales. All such tables are based on the assumption that the metal tested is homogeneous to a depth several times that of the indentation. To the extent that the metal being tested is not homogeneous, errors are introduced because different loads and different shapes of penetrators meet the resistance of metal of varying hardness, depending on the depth of indentation. Another source of error is introduced in comparing the hardness of different materials as measured on different hardness scales. This error arises from the fact that in any hardness test, metal that is severely cold-worked actually supports the penetrator, and different metals, different alloys, and different analyses of the same type of alloy have different cold-working properties. In spite of the possible inaccuracies introduced by such factors, it is of considerable value to be able to compare hardness values in a general way.

The data shown in Table are for hardness measurements of unhardened steel, steel of soft temper, grey and malleable cast iron, and most nonferrous metals. Again these hardness comparisons are not as accurate for annealed metals of high Rockwell B hardness such as austenitic stainless steel, nickel and high nickel alloys, and cold-worked metals of low B-scale hardness such as aluminum and the softer alloys.

The data shown in Table are based on extensive tests on carbon and alloy steels mostly in the heat-treated condition, but have been found to be reliable on constructional alloy steels and tool steels in the as-forged, annealed, normalized, quenched, and tempered conditions, providing they are homogeneous. These hardness comparisons are not as accurate for special alloys such as high manganese steel, 18–8 stainless steel and other austenitic steels, nickel-base alloys, constructional alloy steels, and nickel-base alloys in the cold-worked condition.

Table 1. Comparative Hardness Scales for Steel

Rockwell C-Scale Hardness Number	Diamond Pyramid Hardness Number Vickers	Brinell Hardness Number 10-mm Ball, 3000-kgf Load			Rockwell Hardness Number		Rockwell Superficial Hardness Number Superficial Diam. Penetrator			Shore Scleroscope Hardness Number
		Standard Ball	Hultgren Ball	Tungsten Carbide Ball	A-Scale 60-kgf Load Diam. Penetrator	D-Scale 100-kgf Load Diam. Penetrator	15-N Scale 15-kgf Load	30-N Scale 30-kgf Load	45-N Scale 45-kgf Load	
68	**940**	…	…	…	**85.6**	**76.9**	**93.2**	**84.4**	**75.4**	97
67	**900**	…	…	…	**85.0**	**76.1**	**92.9**	**83.6**	**74.2**	95
66	**865**	…	…	…	**84.5**	**75.4**	**92.5**	**82.8**	**73.3**	92
65	**832**	…	…	**739**	**83.9**	**74.5**	**92.2**	**81.9**	**72.0**	91
64	**800**	…	…	**722**	**83.4**	**73.8**	**91.8**	**81.1**	**71.0**	88
63	**772**	…	…	**705**	**82.8**	**73.0**	**91.4**	**80.1**	**69.9**	87
62	**746**	…	…	**688**	**82.3**	**72.2**	**91.1**	**79.3**	**68.8**	85
61	**720**	…	…	**670**	**81.8**	**71.5**	**90.7**	**78.4**	**67.7**	83
60	**697**	…	**613**	**654**	**81.2**	**70.7**	**90.2**	**77.5**	**66.6**	81
59	**674**	…	**599**	**634**	**80.7**	**69.9**	**89.8**	**76 6**	**65.5**	80
58	**653**	…	**587**	**615**	**80.1**	**69.2**	**89.3**	**75.7**	**64.3**	78
57	**633**	…	**575**	**595**	**79.6**	**68.5**	**88.9**	**74.8**	**63.2**	76
56	**613**	…	**561**	**577**	**79.0**	**67.7**	**88.3**	**73.9**	**62.0**	75
55	**595**	…	**546**	**560**	**78.5**	**66.9**	**87.9**	**73.0**	**60.9**	74
54	**577**	…	**534**	**543**	**78.0**	**66.1**	**87.4**	**72.0**	**59.8**	72
53	**560**	…	**519**	**525**	**77.4**	**65.4**	**86.9**	**71.2**	**58.6**	71
52	**544**	**500**	**508**	**512**	**76.8**	**64.6**	**86.4**	**70.2**	**57.4**	69
51	**528**	**487**	**494**	**496**	**76.3**	**63.8**	**85.9**	**69.4**	**56.1**	68
50	**513**	**475**	**481**	**481**	**75.9**	**63.1**	**85.5**	**68.5**	**55.0**	67
49	**498**	**464**	**469**	**469**	**75.2**	**62.1**	**85.0**	**67.6**	**53.8**	66
48	**484**	**451**	**455**	**455**	**74.7**	**61.4**	**84.5**	**66.7**	**52.5**	64
47	**471**	**442**	**443**	**443**	**74.1**	**60.8**	**83.9**	**65.8**	**51.4**	63
46	**458**	**432**	**432**	**432**	**73.6**	**60.0**	**83.5**	**64.8**	**50.3**	62
45	**446**	**421**	**421**	**421**	**73.1**	**59.2**	**83.0**	**64.0**	**49.0**	60
44	**434**	**409**	**409**	**409**	**72.5**	**58.5**	**82.5**	**63.1**	**47.8**	58
43	**423**	**400**	**400**	**400**	**72.0**	**57.7**	**82.0**	**62.2**	**46.7**	57
42	**412**	**390**	**390**	**390**	**71.5**	**56.9**	**81.5**	**61.3**	**45.5**	56
41	**402**	**381**	**381**	**381**	**70.9**	**56.2**	**80.9**	**60.4**	**44.3**	55
40	**392**	**371**	**371**	**371**	**70.4**	**55.4**	**80.4**	**59.5**	**43.1**	54
39	**382**	**362**	**362**	**362**	**69.9**	**54.6**	**79.9**	**58.6**	**41.9**	52
38	**372**	**353**	**353**	**353**	**69.4**	**53.8**	**79.4**	**57.7**	**40.8**	51
37	**363**	**344**	**344**	**344**	**68.9**	**53.1**	**78.8**	**56.8**	**39.6**	50
36	**354**	**336**	**336**	**336**	**68.4**	**52.3**	**78.3**	**55.9**	**38.4**	49
35	**345**	**327**	**327**	**327**	**67.9**	**51.5**	**77.7**	**55.0**	**37.2**	48
34	**336**	**319**	**319**	**319**	**67.4**	**50.8**	**77.2**	**54.2**	**36.1**	47
33	**327**	**311**	**311**	**311**	**66.8**	**50.0**	**76.6**	**53.3**	**34.9**	46
32	**318**	**301**	**301**	**301**	**66.3**	**49.2**	**76.1**	**52.1**	**33.7**	44
31	**310**	**294**	**294**	**294**	**65.8**	**48.4**	**75.6**	**51.3**	**32.5**	43
30	**302**	**286**	**286**	**286**	**65.3**	**47.7**	**75.0**	**50.4**	**31.3**	42
29	**294**	**279**	**279**	**279**	**64.7**	**47.0**	**74.5**	**49.5**	**30.1**	41
28	**286**	**271**	**271**	**271**	**64.3**	**46.1**	**73.9**	**48.6**	**28.9**	41
27	**279**	**264**	**264**	**264**	**63.8**	**45.2**	**73.3**	**47.7**	**27.8**	40
26	**272**	**258**	**258**	**258**	**63.3**	**44.6**	**72.8**	**46.8**	**26.7**	38
25	**266**	**253**	**253**	**253**	**62.8**	**43.8**	**72.2**	**45.9**	**25.5**	38
24	**260**	**247**	**247**	**247**	**62.4**	**43.1**	**71.6**	**45.0**	**24.3**	37
23	**254**	**243**	**243**	**243**	**62.0**	**42.1**	**71.0**	**44.0**	**23.1**	36
22	**248**	**237**	**237**	**237**	**61.5**	**41.6**	**70.5**	**43.2**	**22.0**	35

Table 1. *(Continued)* **Comparative Hardness Scales for Steel**

Rockwell C-Scale Hardness Number	Diamond Pyramid Hardness Number Vickers	Brinell Hardness Number 10-mm Ball, 3000-kgf Load			Rockwell Hardness Number		Rockwell Superficial Hardness Number Superficial Diam. Penetrator			Shore Scleroscope Hardness Number
		Standard Ball	Hultgren Ball	Tungsten Carbide Ball	A-Scale 60-kgf Load Diam. Penetrator	D-Scale 100-kgf Load Diam. Penetrator	15-N Scale 15-kgf Load	30-N Scale 30-kgf Load	45-N Scale 45-kgf Load	
21	**243**	**231**	**231**	**231**	**61.0**	**40.9**	**69.9**	**42.3**	**20.7**	35
20	**238**	**226**	**226**	**226**	**60.5**	**40.1**	**69.4**	**41.5**	**19.6**	34
(18)	230	219	219	219	…	…	…	…	…	33
(16)	222	212	212	212	…	…	…	…	…	32
(14)	213	203	203	203	…	…	…	…	…	31
(12)	204	194	194	194	…	…	…	…	…	29
(10)	196	187	187	187	…	…	…	…	…	28
(8)	188	179	179	179	…	…	…	…	…	27
(6)	180	171	171	171	…	…	…	…	…	26
(4)	173	165	165	165	…	…	…	…	…	25
(2)	166	158	158	158	…	…	…	…	…	24
(0)	160	152	152	152	…	…	…	…	…	24

Note: The values in this table shown in **boldface** type correspond to those shown in American Society for Testing and Materials Specification E140-67.

Values in ( ) are beyond the normal range and are given for information only.

Turner's Sclerometer.—In making this test a weighted diamond point is drawn, once forward and once backward, over the smooth surface of the material to be tested. The hardness number is the weight in grams required to produce a standard scratch.

Mohs's Hardness Scale.—Hardness, in general, is determined by what is known as Mohs's scale, a standard for hardness that is applied mainly to nonmetallic elements and minerals. In this hardness scale, there are ten degrees or steps, each designated by a mineral, the difference in hardness of the different steps being determined by the fact that any member in the series will scratch any of the preceding members.

This scale is as follows: 1) talc; 2) gypsum; 3) calcite; 4) fluor spar; 5) apatite; 6) orthoclase; 7) quartz; 8) topaz; 9) sapphire or corundum; and 10) diamond.

These minerals, arbitrarily selected as standards, are successively harder, from talc, the softest of all minerals, to diamond, the hardest. This scale, which is now universally used for nonmetallic minerals, is not applied to metals.

Relation Between Hardness and Tensile Strength.—The approximate relationship between the hardness and tensile strength is shown by the following formula:

Tensile strength = $Bhn \times 515$ (for Brinell numbers up to 175).

Tensile strength = $Bhn \times 490$ (for Brinell numbers larger than 175).

The above formulas give the tensile strength in pounds per square inch for steels. These approximate relationships between hardness and tensile strength do not apply to nonferrous metals with the possible exception of certain aluminum alloys.

Durometer Tests.—The durometer is a portable hardness tester for measuring hardness of rubber, plastics, and some soft metals. The instrument is designed to apply pressure to the specimen and the hardness is read from a scale while the pressure is maintained. Various scales can be used by changing the indentor and the load applied.

Table 2. Comparative Hardness Scales for Unhardened Steel, Soft-Temper Steel, Grey and Malleable Cast Iron, and Nonferrous Alloys

Rockwell Hardness Number			Rockwell Superficial Hardness Number			Rockwell Hardness Number			Brinell Hardness Number	
Rockwell B scale 1/16″ Ball Penetrator 100-kg Load	Rockwell F scale 1/16″ Ball Penetrator 60-kg Load	Rockwell G scale 1/16″ Ball Penetrator 150-kg Load	Rockwell Superficial 15-T scale 1/16″ Ball Penetrator	Rockwell Superficial 30-T scale 1/16″ Ball Penetrator	Rockwell Superficial 45-T scale 1/16″ Ball Penetrator	Rockwell E scale 1/8″ Ball Penetrator 100-kg Load	Rockwell K scale 1/8″ Ball Penetrator 150-kg Load	Rockwell A scale "Brale" Penetrator 60-kg Load	Brinell Scale 10-mm Standard Ball 500-kg Load	Brinell Scale 10-mm Standard Ball 3000-kg Load
100	…	82.5	93.0	82.0	72.0	…	…	61.5	201	240
99	…	81.0	92.5	81.5	71.0	…	…	61.0	195	234
98	…	79.0	…	81.0	70.0	…	…	60.0	189	228
97	…	77.5	92.0	80.5	69.0	…	…	59.5	184	222
96	…	76.0	…	80.0	68.0	…	…	59.0	179	216
95	…	74.0	91.5	79.0	67.0	…	…	58.0	175	210
94	…	72.5	…	78.5	66.0	…	…	57.5	171	205
93	…	71.0	91.0	78.0	65.5	…	…	57.0	167	200
92	…	69.0	90.5	77.5	64.5	…	100	56.5	163	195
91	…	67.5	…	77.0	63.5	…	99.5	56.0	160	190
90	…	66.0	90.0	76.0	62.5	…	98.5	55.5	157	185
89	…	64.0	89.5	75.5	61.5	…	98.0	55.0	154	180
88	…	62.5	…	75.0	60.5	…	97.0	54.0	151	176
87	…	61.0	89.0	74.5	59.5	…	96.5	53.5	148	172
86	…	59.0	88.5	74.0	58.5	…	95.5	53.0	145	169
85	…	57.5	…	73.5	58.0	…	94.5	52.5	142	165
84	…	56.0	88.0	73.0	57.0	…	94.0	52.0	140	162
83	…	54.0	87.5	72.0	56.0	…	93.0	51.0	137	159
82	…	52.5	…	71.5	55.0	…	92.0	50.5	135	156
81	…	51.0	87.0	71.0	54.0	…	91.0	50.0	133	153
80	…	49.0	86.5	70.0	53.0	…	90.5	49.5	130	150
79	…	47.5	…	69.5	52.0	…	89.5	49.0	128	147
78	…	46.0	86.0	69.0	51.0	…	88.5	48.5	126	144
77	…	44.0	85.5	68.0	50.0	…	88.0	48.0	124	141
76	…	42.5	…	67.5	49.0	…	87.0	47.0	122	139
75	99.5	41.0	85.0	67.0	48.5	…	86.0	46.5	120	137
74	99.0	39.0	…	66.0	47.5	…	85.0	46.0	118	135
73	98.5	37.5	84.5	65.5	46.5	…	84.5	45.5	116	132
72	98.0	36.0	84.0	65.0	45.5	…	83.5	45.0	114	130
71	97.5	34.5	…	64.0	44.5	100	82.5	44.5	112	127
70	97.0	32.5	83.5	63.5	43.5	99.5	81.5	44.0	110	125
69	96.0	31.0	83.0	62.5	42.5	99.0	81.0	43.5	109	123
68	95.5	29.5	…	62.0	41.5	98.0	80.0	43.0	107	121
67	95.0	28.0	82.5	61.5	40.5	97.5	79.0	42.5	106	119
66	94.5	26.5	82.0	60.5	39.5	97.0	78.0	42.0	104	117
65	94.0	25.0	…	60.0	38.5	96.0	77.5	…	102	116
64	93.5	23.5	81.5	59.5	37.5	95.5	76.5	41.5	101	114
63	93.0	22.0	81.0	58.5	36.5	95.0	75.5	41.0	99	112
62	92.0	20.5	…	58.0	35.5	94.5	74.5	40.5	98	110
61	91.5	19.0	80.5	57.0	34.5	93.5	74.0	40.0	96	108
60	91.0	17.5	…	56.5	33.5	93.0	73.0	39.5	95	107
59	90.5	16.0	80.0	56.0	32.0	92.5	72.0	39.0	94	106
58	90.0	14.5	79.5	55.0	31.0	92.0	…	71.0	38.5	92
57	89.5	13.0	…	54.5	30.0	91.0	…	70.5	38.0	91
56	89.0	11.5	79.0	54.0	29.0	90.5	…	69.5	…	90
55	88.0	10.0	78.5	53.0	28.0	90.0	…	68.5	37.5	89
54	87.5	8.5	…	52.5	27.0	89.5	…	68.0	37.0	87
53	87.0	7.0	78.0	51.5	26.0	89.0	…	67.0	36.5	86
52	86.5	5.5	77.5	51.0	25.0	88.0	…	66.0	36.0	85
51	86.0	4.0	…	50.5	24.0	87.5	…	65.0	35.5	84
50	85.5	2.5	77.0	49.5	23.0	87.0	…	64.5	35.0	83
50	85.5	2.5	77.0	49.5	23.0	87.0	…	64.5	35.0	83
49	85.0	1.0	76.5	49.0	22.0	86.5	…	63.5	…	82

Table 2. *(Continued)* **Comparative Hardness Scales for Unhardened Steel, Soft-Temper Steel, Grey and Malleable Cast Iron, and Nonferrous Alloys**

Rockwell Hardness Number			Rockwell Superficial Hardness Number			Rockwell Hardness Number			Brinell Hardness Number	
Rockwell B scale $\frac{1}{16}$″ Ball Penetrator 100-kg Load	Rockwell F scale $\frac{1}{16}$″ Ball Penetrator 60-kg Load	Rockwell G scale $\frac{1}{16}$″ Ball Penetrator 150-kg Load	Rockwell Superficial 15-T scale $\frac{1}{16}$″ Ball Penetrator	Rockwell Superficial 30-T scale $\frac{1}{16}$″ Ball Penetrator	Rockwell Superficial 45-T scale $\frac{1}{16}$″ Ball Penetrator	Rockwell E scale $\frac{1}{8}$″ Ball Penetrator 100-kg Load	Rockwell K scale $\frac{1}{8}$″ Ball Penetrator 150-kg Load	Rockwell A scale "Brale" Penetrator 60-kg Load	Brinell Scale 10-mm Standard Ball 500-kg Load	Brinell Scale 10-mm Standard Ball 3000-kg Load
48	84.5	…	…	48.5	20.5	85.5	…	62.5	34.5	81
47	84.0	…	76.0	47.5	19.5	85.0	…	61.5	34.0	80
46	83.0	…	75.5	47.0	18.5	84.5	…	61.0	33.5	…
45	82.5	…	…	46.0	17.5	84.0	…	60.0	33.0	79
44	82.0	…	75.0	45.5	16.5	83.5	…	59.0	32.5	78
43	81.5	…	74.5	45.0	15.5	82.5	…	58.0	32.0	77
42	81.0	…	…	44.0	14.5	82.0	…	57.5	31.5	76
41	80.5	…	74.0	43.5	13.5	81.5	…	56.5	31.0	75
40	79.5	…	73.5	43.0	12.5	81.0	…	55.5	…	…
39	79.0	…	…	42.0	11.0	80.0	…	54.5	30.5	74
38	78.5	…	73.0	41.5	10.0	79.5	…	54.0	30.0	73
37	78.0	…	72.5	40.5	9.0	79.0	…	53.0	29.5	72
36	77.5	…	…	40.0	8.0	78.5	100	52.0	29.0	…
35	77.0	…	72.0	39.5	7.0	78.0	99.5	51.5	28.5	71
34	76.5	…	71.5	38.5	6.0	77.0	99.0	50.5	28.0	70
33	75.5	…	…	38.0	5.0	76.5	…	49.5	…	69
32	75.0	…	71.0	37.5	4.0	76.0	98.5	48.5	27.5	…
31	74.5	…	…	36.5	3.0	75.5	98.0	48.0	27.0	68
30	74.0	…	70.5	36.0	2.0	75.0	…	47.0	26.5	67
29	73.5	…	70.0	35.5	1.0	74.0	97.5	46.0	26.0	…
28	73.0	…	…	34.5	…	73.5	97.0	45.0	25.5	66
27	72.5	…	69.5	34.0	…	73.0	96.5	44.5	25.0	…
26	72.0	…	69.0	33.0	…	72.5	…	43.5	24.5	65
25	71.0	…	…	32.5	…	72.0	96.0	42.5	…	64
24	70.5	…	68.5	32.0	…	71.0	95.5	41.5	24.0	…
23	70.0	…	68.0	31.0	…	70.5	…	41.0	23.5	63
22	69.5	…	…	30.5	…	70.0	95.0	40.0	23.0	…
21	69.0	…	67.5	29.5	…	69.5	94.5	39.0	22.5	62
20	68.5	…	…	29.0	…	68.5	…	38.0	22.0	…
19	68.0	…	67.0	28.5	…	68.0	94.0	37.5	21.5	61
18	67.0	…	66.5	27.5	…	67.5	93.5	36.5	…	…
17	66.5	…	…	27.0	…	67.0	93.0	35.5	21.0	60
16	66.0	…	66.0	26.0	…	66.5	…	35.0	20.5	…
15	65.5	…	65.5	25.5	…	65.5	92.5	34.0	20.0	59
14	65.0	…	…	25.0	…	65.0	92.0	33.0	…	…
13	64.5	…	65.0	24.0	…	64.5	…	32.0	…	58
12	64.0	…	64.5	23.5	…	64.0	91.5	31.5	…	…
11	63.5	…	…	23.0	…	63.5	91.0	30.5	…	…
10	63.0	…	64.0	22.0	…	62.5	90.5	29.5	…	57
9	62.0	…	…	21.5	…	62.0	…	29.0	…	…
8	61.5	…	63.5	20.5	…	61.5	90.0	28.0	…	…
7	61.0	…	63.0	20.0	…	61.0	89.5	27.0	…	56
6	60.5	…	…	19.5	…	60.5	…	26.0	…	…
5	60.0	…	62.5	18.5	…	60.0	89.0	25.5	…	55
4	59.5	…	62.0	18.0	…	59.0	88.5	24.5	…	…
3	59.0	…	…	17.0	…	58.5	88.0	23.5	…	…
2	58.0	…	61.5	16.5	…	58.0	…	23.0	…	54
1	57.5	…	61.0	16.0	…	57.5	87.5	22.0	…	…
0	57.0	…	…	15.0	…	57.0	87.0	21.0	…	53

Not applicable to annealed metals of high B-scale hardness such as austenitic stainless steels, nickel and high-nickel alloys nor to cold-worked metals of low B-scale hardness such as aluminum and the softer alloys.

(Compiled by Wilson Mechanical Instrument Co.)

Creep.—Continuing changes in dimensions of a stressed material over time is called creep, and it varies with different materials and periods under stress, also with temperature. Creep tests may take some time as it is necessary to apply a constant tensile load to a specimen under a selected temperature. Measurements are taken to record the resulting elongation at time periods sufficiently long for a relationship to be established. The data are then plotted as elongation against time. The load is applied to the specimen only after it has reached the testing temperature, and causes an initial elastic elongation that includes some plastic deformation if the load is above the proportional limit for the material.

Some combinations of stress and temperature may cause failure of the specimen. Others show initial high rates of deformation, followed by decreasing, then constant, rates over long periods. Generally testing times to arrive at the constant rate of deformation are over 1000 hours.

Creep Rupture.—Tests for creep rupture are similar to creep tests but are prolonged until the specimen fails. Further data to be obtained from these tests include time to rupture, amount of elongation, and reduction of area. Stress-rupture tests are performed without measuring the elongation, so that no strain data are recorded, time to failure, elongation and reduction of area being sufficient. Sometimes, a V-notch is cut in the specimen to allow measurement of notch sensitivity under the testing conditions.

Stress Analysis.—Stresses, deflections, strains, and loads may be determined by application of strain gages or lacquers to the surface of a part, then applying loads simulating those to be encountered in service. Strain gages are commercially available in a variety of configurations and are usually cemented to the part surface. The strain gages are then calibrated by application of a known moment, load, torque, or pressure. The electrical characteristics of the strain gages change in proportion to the amount of strain, and the magnitude of changes in these characteristics under loads to be applied in service indicate changes caused by stress in the shape of the components being tested.

Lacquers are compounded especially for stress analysis and are applied to the entire part surface. When the part is loaded, and the lacquer is viewed under light of specific wavelength, stresses are indicated by color shading in the lacquer. The presence and intensity of the strains can then be identified and measured on the part(s) or on photographs of the setup. From such images, it is possible to determine the need for thicker walls, strengthening ribs and other modifications to component design that will enable the part to withstand stresses in service.

Most of these tests have been standardized by the American Society for Testing and Materials (ASTM), and are published in their *Book of Standards* in separate sections for metals, plastics, rubber, and wood. Many of the test methods are also adopted by the American National Standards Institute (ANSI).

Identifying Metals.—When it is necessary to sort materials, several rough methods may be used without elaborate chemical analysis. The most obvious of these is by using a magnet to pick out those materials that contain magnetic elements. To differentiate various levels of carbon and other elements in a steel bar, hold the bar in contact with a grinding wheel and observe the sparks. With high levels of carbon, for instance, sparks are produced that appear to split into several bright tracers. Patterns produced by several other elements, including small amounts of aluminum and titanium, for instance, can be identified with the aid of Data Sheet 13, issued by the American Society for Metals (ASM), Metals Park, OH.

NONFERROUS ALLOYS

Copper and Copper Alloys

Pure copper is a reddish, highly malleable metal, and was one of the first to be found and utilized. Copper and its alloys are widely used because of their excellent electrical and thermal conductivities, outstanding resistance to corrosion, ease of fabrication, and broad ranges of obtainable strengths and special properties. Almost 400 commercial copper and copper-alloy compositions are available from mills as wrought products (rod, plate, sheet, strip, tube, pipe, extrusions, foil, forgings, and wire) and from foundries as castings.

Copper alloys are grouped into several general categories according to composition:

coppers and high-copper alloys; brasses; bronzes; copper nickels; copper–nickel–zinc alloys (nickel silvers); leaded coppers; and special alloys.

The designation system originally developed by the U.S. copper and brass industry for identifying copper alloys used a three-digit number preceded by the letters CA. These designations have now been made part of the Unified Numbering System (UNS) simply by expanding the numbers to five digits preceded by the letter C. Because the old numbers are embedded in the new UNS numbers, no confusion results. UNS C10000 to C79999 are assigned to wrought compositions, and UNS C80000 to C99999 are assigned to castings. The designation system is not a specification, but a method for identifying the composition of mill and foundry products. The precise technical and quality assurance requirements to be satisfied are defined in relevant standard specifications issued by the federal government, the military, and the ASTM.

Classification of Copper and Copper Alloys

Family	Principal Alloying Element	UNS Numbers[a]
Coppers, high-copper alloys		C1xxxx
Brasses	Zn	C2xxxx, C3xxxx, C4xxxx, C66400 to C69800
Phosphor bronzes	Sn	C5xxxx
Aluminum bronzes	Al	C60600 to C64200
Silicon bronzes	Si	C64700 to C66100
Copper nickels, nickel silvers	Ni	C7xxxx

[a] Wrought alloys.

Cast Copper Alloys.—Generally, casting permits greater latitude in the use of alloying elements than in the fabrication of wrought products, which requires either hot or cold working. The cast compositions of coppers and high-copper alloys have a designated minimum copper content and may include other elements to impart special properties. The cast brasses comprise copper–zinc–tin alloys (red, semired, and yellow brasses); manganese bronze alloys (high-strength yellow brasses); leaded manganese bronze alloys (leaded high-strength yellow brasses); and copper–zinc–silicon alloys (silicon brasses and bronzes).

The cast bronze alloys have four main families: copper–tin alloys (tin bronzes); copper–tin–lead alloys (leaded and high leaded tin bronzes); copper–tin–nickel alloys (nickel–tin bronzes); and copper–aluminum alloys (aluminum bronzes).

The cast copper–nickel alloys contain nickel as the principal alloying element. The leaded coppers are cast alloys containing 20 per cent or more lead.

Table lists the properties and applications of common cast copper alloys.

Table 1. Properties and Applications of Cast Coppers and Copper Alloys

UNS Designation	Nominal Composition (%)	Typical Mechanical Properties, as Cast or Heat Treated[a]				Typical Applications
		Tensile Strength (ksi)	Yield Strength (ksi)	Elongation in 2 in. (%)	Machinability Rating[b]	
Copper Alloys						
C80100	99.95 Cu + Ag min, 0.05 others max	25	9	40	10	Electrical and thermal conductors; corrosion and oxidation-resisant applications.
C80300	99.95 Cu + Ag min, 0.034 Ag min, 0.05 others max	25	9	40	10	Electrical and thermal conductors; corrosion and oxidation-resistant applications.
C80500	99.75 Cu + Ag min, 0.034 Ag min, 0.02 B max, 0.23 others max	25	9	40	10	Electrical and thermal conductors; corrosion and oxidation-resistant applications.
C80700	99.75 Cu + Ag min, 0.02 B max, 0.23 others max	25	9	40	10	Electrical and thermal conductors; corrosion and oxidation-resistant applications.
C80900	99.70 Cu + Ag min, 0.034 Ag min, 0.30 others max	25	9	40	10	Electrical and thermal conductors; corrosion and oxidation-resistant applications.
C81100	99.70 Cu + Ag min, 0.30 others max	25	9	40	10	Electrical and thermal conductors; corrosion and oxidation resstant applications.
High-Copper Alloys						
C81300	98.5 Cu min, 0.06 Be, 0.80 Co, 0.40 others max	(53)	(36)	(11)	20	Higher hardness electrical and thermal conductors.
C81400	98.5 Cu min, 0.06 Be, 0.80 Cr, 0.40 others max	(53)	(36)	(11)	20	Higher hardness electrical and thermal conductors.
C81500	98.0 Cu min, 1.0 Cr, 0.50 others max	(51)	(40)	(17)	20	Electrical and/or thermal conductors used as structural members where strength and hardness greater than that of C80100–81100 are required.

Table 1. *(Continued)* **Properties and Applications of Cast Coppers and Copper Alloys**

UNS Designation	Nominal Composition (%)	Typical Mechanical Properties, as Cast or Heat Treated[a]				Typical Applications
		Tensile Strength (ksi)	Yield Strength (ksi)	Elongation in 2 in. (%)	Machinability Rating[b]	
C81700	94.2 Cu min, 1.0 Ag, 0.4 Be, 0.9 Co, 0.9 Ni	(92)	(68)	(8)	30	Electrical and/or thermal conductors used as structural members where strength and hardness greater than that of C80100–81100 are required. Also used in place of C81500 where electrical and/or thermal conductivities can be sacrificed for hardness and strength.
C81800	95.6 Cu min,1.0 Ag, 0.4 Be,1.6 Co	50 (102)	25 (75)	20 (8)	20	Resistance-welding electrodes, dies.
C82000	96.8 Cu, 0.6 Be, 2.6 Co	50 (100)	20 (75)	20 (8)	20	Current-carrying parts, contact and switch blades, bushings and bearings, and soldering iron and resistance-welding tips.
C82100	97.7 Cu, 0.5 Be, 0.9 Co, 0.9 Ni	(92)	(68)	(8)	30	Electrical and/or thermal conductors used as structural members where strength and hardness greater than that of C80100–81100 are required. Also used in place of C81500 where electrical and/or thermal conductivities can be sacrificed for hardness and strength.
C82200	96.5 Cu min, 0.6 Be, 1.5 Ni	57 (95)	30 (75)	20 (8)	20	Clutch rings, brake drums, seam-welder electrodes, projection welding dies, spot-welding tips, beam-welder shapes, bushings, water-cooled holders.
C82400	96.4 Cu min, 1.70 Be, 0.25 Co	72 (150)	37 (140)	20 (1)	20	Safety tools, molds for plastic parts, cams, bushings, bearings, valves, pump parts, gears.
C82500	97.2 Cu, 2.0 Be, 0.5 Co, 0.25 Si	80 (160)	45	20 (1)	20	Safety tools, molds for plastic parts, cams, bushings, bearings, valves, pump parts.
C82600	95.2 Cu min,2.3 Be, 0.5 Co, 0.25 Si	82 (165)	47 (155)	20 (1)	20	Bearings and molds for plastic parts.
C82700	96.3 Cu, 2.45 Be,1.25 Ni	(155)	(130)	(0)	20	Bearings and molds for plastic parts.
C82800	96.6 Cu, 2.6 Be, 0.5 Co, 0.25 Si	97 (165)	55 (145)	20 (1)	10	Molds for plastic parts, cams, bushings, bearings, valves, pump parts, sleeves.
Red Brasses and Leaded Red Brasses						
C83300	93 Cu, 1.5 Sn,1.5 Pb, 4 Zn	32	10	35	35	Terminal ends for electrical cables.
C83400	90 Cu, 10 Zn	35	10	30	60	Moderate strength, moderate conductivity castings; rotating bands.

Table 1. *(Continued)* **Properties and Applications of Cast Coppers and Copper Alloys**

UNS Designation	Nominal Composition (%)	Typical Mechanical Properties, as Cast or Heat Treated[a]				Typical Applications
		Tensile Strength (ksi)	Yield Strength (ksi)	Elongation in 2 in. (%)	Machinability Rating[b]	
C83600	85 Cu, 5 Sn, 5 Pb, 5 Zn	37	17	30	84	Valves, flanges, pipe fittings, plumbing goods, pump castings, water pump impellers and housings, ornamental fixtures, small gears.
C83800	83 Cu, 4 Sn, 6 Pb,7 Zn	35	16	25	90	Low-pressure valves and fittings, plumbing supplies and fittings, general hardware, air–gas–water fittings, pump components, railroad catenary fittings.
Semired Brasses and Leaded Semired Brasses						
C84200	80 Cu, 5 Sn,2.5 Pb, 12.5 Zn	35	14	27	80	Pipe fittings, elbows, T's, couplings, bushings, locknuts, plugs, unions.
C84400	81 Cu, 3 Sn, 7 Pb,9 Zn	34	15	26	90	General hardware, ornamental castings, plumbing supplies and fixtures, low-pressure valves and fittings.
C84500	78 Cu, 3 Sn, 7 Pb,12 Zn	35	14	28	90	Plumbing fixtures, cocks, faucets, stops, waste, air and gas fittings, low-pressure valve fittings.
C84800	76 Cu, 3 Sn, 6 Pb,15 Zn	36	14	30	90	Plumbing fixtures, cocks, faucets, stops, waste, air, and gas, general hardware, and low-pressure valve fittings.
Yellow Brasses and Leaded Yellow Brasses						
C85200	72 Cu, 1 Sn, 3 Pb,24 Zn	38	13	35	80	Plumbing fittings and fixtures, ferrules, valves, hardware, ornamental brass, chandeliers, and-irons.
C85400	67 Cu, 1 Sn, 3 Pb,29 Zn	34	12	35	80	General-purpose yellow casting alloy not subject to high internal pressure. Furniture hardware, ornamental castings, radiator fittings, ship trimmings, battery clamps, valves, and fittings.
C85500	61 Cu, 0.8 Al, bal Zn	60	23	40	80	Ornamental castings.
C85700	63 Cu, 1 Sn, 1 Pb, 34.7 Zn, 0.3 Al	50	18	40	80	Bushings, hardware fittings, ornamental castings.
C85800	58 Cu, 1 Sn, 1 Pb, 40 Zn	55	30	15	80	General-purpose die-casting alloy having moderate strength.
Manganese and Leaded Manganese Bronze Alloys						
C86100	67 Cu, 21 Zn, 3 Fe, 5 Al, 4 Mn	95	50	20	30	Marine castings, gears, gun mounts, bushings and bearings, marine racing propellers.
C86200	64 Cu, 26 Zn, 3 Fe, 4 Al, 3 Mn	95	48	20	30	Marine castings, gears, gun mounts, bushings and bearings.

Table 1. *(Continued)* **Properties and Applications of Cast Coppers and Copper Alloys**

UNS Designation	Nominal Composition (%)	Typical Mechanical Properties, as Cast or Heat Treated[a]				Typical Applications
		Tensile Strength (ksi)	Yield Strength (ksi)	Elongation in 2 in. (%)	Machinability Rating[b]	
C86300	63 Cu, 25 Zn, 3 Fe, 6 Al, 3 Mn	115	83	15	8	Extra-heavy duty, high-strength alloy. Large valve stems, gears, cams, slow-speed heavy-load bearings, screwdown nuts, hydraulic cylinder-parts.
C86400	59 Cu, 1 Pb, 40 Zn	65	25	20	65	Free-machining manganese bronze. Valve stems, marine fittings, lever arms, brackets, light-duty gears.
C86500	58 Cu, 0.5 Sn, 39.5 Zn, 1 Fe, 1 Al	71	28	30	26	Machinery parts requiring strength and toughness, lever arms, valve stems, gears.
C86700	58 Cu, 1 Pb, 41 Zn	85	42	20	55	High strength, free-machining manganese bronze. Valve stems.
C86800	55 Cu, 37 Zn, 3 Ni, 2 Fe, 3 Mn	82	38	22	30	Marine fittings, marine propellers.
Silicon Bronzes and Silicon Brasses						
C87200	89 Cu min, 4 Si	55	25	30	40	Bearings, bells, impellers, pump and valve components, marine fittings, corrosion-resistant castings.
C87400	83 Cu, 14 Zn, 3 Si	55	24	30	50	Bearings, gears, impellers, rocker arms, valve stems, clamps.
C87500	82 Cu, 14 Zn, 4 Si	67	30	21	50	Bearings, gears, impellers, rocker arms, valve stems, small boat proellers.
C87600	90 Cu, 5.5 Zn,4.5 Si	66	32	20	40	Valve stems.
C87800	82 Cu, 14 Zn, 4 Si	85	50	25	40	High-strength, thin-wall die castings; brush holders, lever arms, brackets, clamps, hexagonal nuts.
C87900	65 Cu, 34 Zn, 1 Si	70	35	25	80	General-purpose die-casting alloy having moderate strength.
Tin Bronzes						
C90200	93 Cu, 7 Sn	38	16	30	20	Bearings and bushings.
C90300	88 Cu, 8 Sn, 4 Zn	45	21	30	30	Bearings, bushings, pump impellers, piston rings, valve components, seal rings, steam fittings, gears.
C90500	88 Cu, 10 Sn, 2 Zn	45	22	25	30	Bearings, bushings, pump impellers, piston rings, valve components, steam fittings, gears.
C90700	89 Cu, 11 Sn	44 (55)	22 (30	20 (16)	20	Gears, bearings, bushings.

Table 1. *(Continued)* **Properties and Applications of Cast Coppers and Copper Alloys**

UNS Designation	Nominal Composition (%)	Typical Mechanical Properties, as Cast or Heat Treated[a]				Typical Applications
		Tensile Strength (ksi)	Yield Strength (ksi)	Elongation in 2 in. (%)	Machinability Rating[b]	
C90900	87 Cu, 13 Sn	40	20	15	20	Bearings and bushings.
C91000	85 Cu, 14 Sn, 1 Zn	32	25	2	20	Piston rings and bearings.
C91100	84 Cu, 16 Sn	35	25	2	10	Piston rings, bearings, bushings, bridge plates.
C91300	81 Cu, 19 Sn	35	30	0.5	10	Piston rings, bearings, bushings, bridge plates, bells.
C91600	88 Cu, 10.5 Sn,1.5 Ni	44 (60)	22 (32)	16 (16)	20	Gears.
C91700	86.5 Cu, 12 Sn,1.5 Ni	44 (60)	22 (32)	16 (16)	20	Gears.
Leaded Tin Bronzes						
C92200	88 Cu, 6 Sn,1.5 Pb, 4.5 Zn	40	20	30	42	Valves, fittings, and pressure-containing parts for use up to 550°F.
C92300	87 Cu, 8 Sn, 4 Zn	40	20	25	42	Valves, pipe fittings, and high-pressure steam castings. Superior machinability to C90300.
C92500	87 Cu, 11 Sn,1 Pb, 1 Ni	44	20	20	30	Gears, automotive synchronizer rings.
C92600	87 Cu, 10 Sn,1 Pb, 2 Zn	44	20	30	40	Bearings, bushings, pump impellers, piston rings, valve components, steam fittings, and gears. Superior machinability to C90500.
C92700	88 Cu, 10 Sn,2 Pb	42	21	20	45	Bearings, bushings, pump impellers, piston rings, and gears. Superior machinability to C90500.
C92800	79 Cu, 16 Sn, 5 Pb	40	30	1	70	Piston rings.
C92900	82 Cu min, 9 Sn min, 2 Pb min,2.8 Ni min	47 (47)	26 (26)	20 (20)	40	Gears, wear plates, guides, cams, parts requiring machinability superior to that of C91600 or 91700.
High-Leaded Tin Bronzes						
C93200	83 Cu, 6.3 Sn min,7 Pb, 3 Zn	35	18	20	70	General-utility bearings and bushings.
C93400	84 Cu, 8 Sn, 8 Pb	32	16	20	70	Bearings and bushings.
C93500	85 Cu, 5 Sn, 9 Pb	32	16	20	70	Small bearings and bushings, bronze backing for babbit-lined automotive bearings.
C93700	80 Cu, 10 Sn,10 Pb	35	18	20	80	Bearings for high speed and heavy pressures, pumps, impellers, corrosion-resistant applications, pressure tight castings.

Table 1. *(Continued)* **Properties and Applications of Cast Coppers and Copper Alloys**

UNS Designation	Nominal Composition (%)	Typical Mechanical Properties, as Cast or Heat Treated[a]				Typical Applications
		Tensile Strength (ksi)	Yield Strength (ksi)	Elongation in 2 in. (%)	Machinability Rating[b]	
C93800	78 Cu, 7 Sn, 15 Pb	30	16	18	80	Bearings for general service and moderate pressure, pump impellers, and bodies for use in acid mine water.
C93900	79 Cu, 6 Sn, 15 Pb	32	22	7	80	Continuous castings only. Bearings for general service, pump bodies, and impellers for mine waters.
C94300	70 Cu, 5 Sn, 25 Pb	27	13	15	80	High-speed bearings for light loads.
C94400	81 Cu, 8 Sn, 11 Pb,0.35 P	32	16	18	80	General-utility alloy for bushings and bearings.
C94500	73 Cu, 7 Sn, 20 Pb	25	12	12	80	Locomotive wearing parts; high-low, low-speed bearings.
Nickel–Tin Bronzes						
C94700	88 Cu, 5 Sn, 2 Zn,5 Ni	50 (85)	23 (60)	35 (10)	30 (20)	Valve stems and bodies, bearings, wear guides, shift forks, feeding mechanisms, circuit breaker parts, gears, piston cylinders, nozzles.
C94800	87 Cu, 5 Sn, 5 Ni	45 (60)	23 (30)	35 (8)	50 (40)	Structural castings, gear components, motion-translation devices, machinery parts, bearings.
Aluminum Bronzes						
C95200	88 Cu, 3 Fe, 9 Al	80	27	35	50	Acid-resisting pumps, bearing, gears, valve seats, guides, plungers, pump rods, bushings.
C95300	89 Cu, 1 Fe, 10 Al	75 (85)	27 (42)	25 (15)	55	Pickling baskets, nuts, gears, steel mill slippers, marine equipment, welding jaws.
C95400	85 Cu, 4 Fe, 11 Al	85 (105)	35 (54)	18 (8)	60	Bearings, gears, worms, bushings, valve seats and guides, pickling hooks.
C95500	81 Cu, 4 Ni, 4 Fe, 11 Al	100 (120)	44 (68)	12 (10)	50	Valve guides and seats in aircraft engines, corrosion-resistant parts, bushings, gears, worms, pickling hooks and baskets, agitators.
C95600	91 Cu, 7 Al, 2 Si	75	34	18	60	Cable connectors, terminals, valve stems, marine hardware, gears, worms, pole-line hardware.
C95700	75 Cu, 2 Ni, 3 Fe, 8 Al, 12 Mn	95	45	26	50	Propellers, impellers, stator clamp segments, safety tools, welding rods, valves, pump casings.
C95800	81 Cu, 5 Ni, 4 Fe, 9 Al, 1 Mn	95	38	25	50	Propeller hubs, blades, and other parts in contact with salt water.

Table 1. *(Continued)* **Properties and Applications of Cast Coppers and Copper Alloys**

UNS Designation	Nominal Composition (%)	Typical Mechanical Properties, as Cast or Heat Treated[a]				Typical Applications
		Tensile Strength (ksi)	Yield Strength (ksi)	Elongation in 2 in. (%)	Machinability Rating[b]	
			Copper–Nickels			
C96200	88.6 Cu, 10 Ni,1.4 Fe	45 min	25 min	20 min	10	Components of items being used for seawater corrosion resistance.
C96300	79.3 Cu, 20 Ni,0.7 Fe	75 min	55 min	10 min	15	Centrifugally cast tailshaft sleeves.
C96400	69.1 Cu, 30 Ni,0.9 Fe	68	37	28	20	Valves, pump bodies, flanges, elbows used for seawater corrosion resistance.
C96600	68.5 Cu, 30 Ni, 1 Fe, 0.5 Be	(110)	(70)	(7)	20	High-strength constructional parts for seawater corrosion resistance.
			Nickel Silvers			
C97300	56 Cu, 2 Sn,10 Pb, 12 Ni,20 Zn	35	17	20	70	Hardware fittings, valves and valve trim, statuary, ornamental castings.
C97400	59 Cu, 3 Sn, 5 Pb,17 Ni, 16 Zn	38	17	20	60	Valves, hardware, fittings, ornamental castings.
C97600	64 Cu, 4 Sn, 4 Pb, 20 Ni, 8 Zn	45	24	20	70	Marine castings, sanitary fittings, ornamental hardware, valves,pumps.
C97800	66 Cu, 5 Sn, 2 Pb, 25 Ni, 2 Zn	55	30	15	60	Ornamental and sanitary castings, valves and valve seats, musical instrument components.
			Special Alloys			
C99300	71.8 Cu,15 Ni,0.7 Fe, 11 Al,1.5 Co	95	55	2	20	Glass-making molds, plate glass rolls, marine hardware.
C99400	90.4 Cu, 2.2 Ni, 2.0 Fe, 1.2 Al,1.2 Si, 3.0 Zn	66 (79)	34 (54)	25	50	Valve stems, marine and other uses requiring resistance to dezincification and dealuminification, propeller wheels, electrical parts, mining equipment gears.
C99500	87.9 Cu, 4.5 Ni, 4.0 Fe, 1.2 Al,1.2 Si, 1.2 Zn	70 min	40 min	12 min	50	Same as C99400, but where higher yield strength is required.
C99700	56.5 Cu, Al,1.5 Pb, 12 Mn,5 Ni, 24 Zn	55	25	25	80	…
C99750	58 Cu,1 Al,1 Pb, 20 Mn, 20 Zn	65 (75)	32 (40)	30 (20)	…	…

[a] Values in parentheses are for heat-treated condition.

[b] Free cutting brass = 100.

Source: Copper Development Association, New York.

Wrought Copper Alloys.—Wrought copper alloys can be utilized in the annealed, cold-worked, stress-relieved, or hardened-by-heat-treatment conditions, depending on composition and end use. The "temper designation" for copper alloys is defined in ASTM Standard Recommended Practice B601, which is applicable to all product forms.

Wrought copper and high-copper alloys, like cast alloys, have a designated minimum copper content and may include other elements to impart special properties. Wrought brasses have zinc as the principal alloying element and may have other designated elements. They comprise the copper–zinc alloys; copper–zinc–lead alloys (leaded brasses); and copper–zinc–tin alloys (tin brasses).

Wrought bronzes comprise four main groups:; copper–tin–phosphorus alloys (phosphor bronze); copper–tin–lead–phosphorus alloys (leaded phosphor bronze); copper–aluminum alloys (aluminum bronzes); and copper–silicon alloys (silicon bronze).

Wrought copper–nickel alloys, like the cast alloys, have nickel as the principal alloying element. The wrought copper–nickel–zinc alloys are known as "nickel silvers" because of their color.

Table lists the nominal composition, properties, and applications of common wrought copper alloys.

Table 2. Properties and Applications of Wrought Coppers and Copper Alloys

Name and Number	Nominal Composition (%)	Strength (ksi) Tensile	Strength (ksi) Yield	Elongation in 2 in. (%)	Machinability Rating[a]	Fabricating Characteristics and Typical Applications
C10100 Oxygen-free electronic	99.99 Cu	32 –66	10 – 53	55	20	Excellent hot and cold workability; good forgeability. Fabricated by blanking, coining, coppersmithing, drawing and upsetting, hot forging and pressing, spinning, swaging, stamping. Uses: busbars, bus conductors, waveguides, hollow conductors, lead-in wires and anodes for vacuum tubes, vacuum seals, transistor components, glass to metal seals, coaxial cables and tubes, klystrons, microwave tubes, rectifiers.
C10200 Oxygen-free copper	99.95 Cu	32–66	10–53	55	20	Fabricating characteristics same as C10100. Uses: busbars, waveguides.
C10300 Oxygen-free, extra-low phosphorus	99.95 Cu, 0.003 P	32–55	10–50	50	20	Fabricating characteristics same as C10100. Uses: busbars, electrical conductors, tubular bus, and applications requiring good conductivity and welding or brazing properties.
C10400, C10500, C10700 Oxygen-free, silver-bearing	99.95 Cu	32–66	10–53	55	20	Fabricating characteristics same as C10100. Uses: auto gaskets, radiators, busbars, conductivity wire, contacts, radio parts, winding, switches, terminals, commutator segments; chemical process equipment, printing rolls, clad metals, printed-circuit foil.
C10800 Oxygen-free, low phosphorus	99.95 Cu, 0.009 P	32–55	10–50	50	20	Fabricating characteristics same as C10100. Uses: refrigerators, air conditioners, gas and heater lines, oil burner tubes, plumbing pipe and tube, brewery tubes, condenser and heat-exchanger tubes, dairy and distiller tubes, pulp and paper lines, tanks; air, gasoline, and hydraulic lines.
C11000 Electrolytic tough pitch copper	99.90 Cu, 0.04 O	32–66	10–53	55	20	Fabricating characteristics same as C10100. Uses: downspouts, gutters, roofing, gaskets, auto radiators, busbars, nails, printing rolls, rivets, radio parts.
C11000 Electrolytic tough pitch, anneal-resistant	99.90 Cu, 0.04 O, 0.01 Cd	66	…	…	20	Fabricated by drawing and stranding, stamping. Uses: electrical power transmission where resistance to softening under overloads is desired.

Table 2. *(Continued)* **Properties and Applications of Wrought Coppers and Copper**

Name and Number	Nominal Composi-tion (%)	Strength (ksi) Ten-sile	Yield	Elon-gation in 2 in. (%)	Machin ability Rating[a]	Fabricating Characteristics and Typical Applications
C11300, C11400, C11500, C11600 Silver- bearing tough pitch copper	99.90 Cu, 0.04 O, Ag	32–66	10–53	55	20	Fabricating characteristics same as C10100. Uses: gaskets, radiators, busbars, windings, switches, chemical process equipment, clad metals, printed-circuit foil.
C12000, C12100 Phosphorus deoxidized, low residual phosphorus	99.9 Cu	32–57	10–53	55	20	Fabricating characteristics same as C10100. Uses: busbars, electrical conductors, tubular bus, and applications requiring welding or brazing.
C12200, C12210 Phosphorus deoxidized copper, high residual phosphorus	99.90 Cu, 0.02 P	32–55	10–53	55	20	Fabricating characteristics same as C10100. Uses: gas and heater lines; oil burner tubing; plumbing pipe and tubing; condenser, evaporator, heat exchanger, dairy, and distiller tubing; steam and water lines; air, gasoline, and hydraulic lines.
C12500,C12700, C12800, C12900, C13000 Fire-refined tough pitch with silver	99.88 Cu	32–66	10–53	55	20	Fabricating characteristics same as C10100. Uses: same as C11000, Electrolytic tough pitch copper.
C14200 Phosphorus deoxidized, arsenical	99.68 Cu, 0.3 As, 0.02 P	32–55	10–50	45	20	Fabricating characteristics same as C10100. Uses: staybolts, heat-exchanger and condenser tubes.
C14300, C14310 Cadmium copper, deoxidized	99.9 Cu, 0.1 Cd	32–58	11–56	42	20	Fabricating characteristics same as C10100. Uses: anneal-resistant electrical applications requiring thermal softening and embrittlement resistance, lead frames, contacts, terminals, solder-coated and solder-fabricated parts, furnace-brazed assemblies and welded components, cable wrap.
C14500, C14510, C14520 Tellurium bearing	99.5 Cu, 0.50 Te, 0.008 P	32–56	10–51	50	85	Fabricating characteristics same as C10100. Uses: Forgings and screw-machine products, and parts requiring high conductivity, extensive machining, corrosion resistance, copper color, or a combination of these; electrical connectors, motor and switch parts, plumbing fittings, soldering coppers, welding torch tips, transistor bases, and furnace-brazed articles.
C14700, C14710, C14720 Sulfur bearing	99.6 Cu, 0.40 S	32–57	10–55	52	85	Fabricating characteristics same as C10100. Uses: screw-machine products and parts requiring high conductivity, extensive machining, corrosion resistance, copper color, or a combination of these; electrical connectors, motor and switch components, plumbing fittings, cold-headed and machined parts, cold forgings, furnace-brazed articles, screws, soldering coppers, rivets and welding torch tips.
C15000 Zirconium copper	99.8 Cu, 0.15 Zr	29–76	6–72	54	20	Fabricating characteristics same as C10100. Uses: switches, high-temperature circuit breakers, commutators, stud bases for power transmitters, rectifiers, soldering welding tips.
C15500	99.75 Cu, 0.06 P, 0.11 Mg, Ag	40–80	18–72	40	20	Fabricating characteristics same as C10100. Uses: high-conductivity light-duty springs, electrical contacts, fittings, clamps, connectors, diahragms, electronic components, resistance-welding electrodes.
C15715	99.6 Cu, 0.13 Al_2O_3	52–88	44–84	27	20	Excellent cold workability. Fabricated by extrusion, drawing, rolling, heading, swaging, machining, blanking, roll threading. Uses: integrated-circuit lead frames, diode leads; vacuum, microwave, and x-ray tube components; electrical components; brush springs; commutators, electric generator and motor components.

Table 2. *(Continued)* **Properties and Applications of Wrought Coppers and Copper**

Name and Number	Nominal Composition (%)	Strength (ksi) Tensile	Yield	Elongation in 2 in. (%)	Machinability Rating[a]	Fabricating Characteristics and Typical Applications
C15720	99.5 Cu, 0.18 Al_2O_3	64–98	54–96	25	…	Excellent cold workability, Fabricated by extrusion, drawing, rolling, heading, swaging, machining, blanking. Uses: relay and switch springs, lead frames, contact supports, heat sinks, circuit breaker parts, rotor bars, resistance-welding electrodes and wheels, connectors, soldering gun tips.
C15760	98.8 Cu, 0.58 Al_2O_3	70–90	65–87	22	…	Excellent cold workability. Fabricated by extrusion and drawing. Uses: resistance-welding electrodes, soldering gun tips, MIG welding contact tips, continuous-casting molds.
C16200, C16210 Cadmium copper	99.0 Cu, 1.0 Cd	35–100	7–69	57	20	Excellent cold workability; good hot formability. Uses: trolley wires, heating pads, electric-blanket elements, spring contacts, railbands, high-strength transmission lines, connectors, cable wrap, switch-gear components, and waveguide cavities.
C16500	98.6 Cu, 0.8 Cd, 0.6 Sn	40–95	14–71	53	20	Fabricating characteristics same as C16200. Uses: electrical springs and contacts, trolley wire, clips, flat cable, resistance-welding electrodes.
C17000 Beryllium copper	98.3 Cu, 1.7 Be, 0.20 Co	70–190	32–170	45	20	Fabricating characteristics same as C16200. Commonly fabricated by blanking, forming and bending, turning, drilling, tapping. Uses: bellows, Bourdon tubing, diaphragms, fuse clips, fasteners, lock-washers, springs, switch parts, roll pins, valves, welding equipment.
C17200 Beryllium copper	98.1 Cu, 1.9 Be, 0.20 Co	68–212	25–195	48	20	Similar to C17000, particularly for its non-sparking characteristics.
C17300 Beryllium copper	98.1 Cu, 1.9 Be, 0.40 Pb	68–212	25–195	48	50	Combines superior machinability with good fabricating characteristics of C17200.
C17500, C17510 Beryllium copper	96.9 Cu, 2.5 Co, 0.6 Be	45–115	25–110	28	…	Fabricating characteristics same as C16200. Uses: fuse clips, fasteners, springs, switch and relay parts, electrical conductors, welding equipment.
C18200, C18400, C18500 Chromium copper	99.2 Cu	34–86	14–77	40	20	Excellent cold workability, good hot workability. Uses: resistance-welding electrodes, seam-welding wheels, switch gear, electrode holder jaws, cable connectors, current-carrying arms and shafts, circuit-breaker parts, molds, spot-welding tips, flash-welding electrodes, electrical and thermal conductors requiring strength, switch contacts.
C18700 Leaded copper	99.0 Cu, 1.0 Pb	32–55	10–50	45	85	Good cold workability; poor hot formability. Uses: connectors, motor and switch parts, screw-machine parts requiring high conductivity.
C18900	98.7 Cu, 0.8 Sn, 0.3 Si, 0.20 Mn	38–95	9–52	48	20	Fabricating characteristics same as C10100. Uses: welding rod and wire for inert gas tungsten arc and metal arc welding and oxyacetylene welding of copper.
C19000 Copper–nickel–phosphorus alloy	98.6 Cu, 1.1 Ni, 0.3 P	38–115	20–81	50	30	Fabricating characteristics same as C10100. Uses: springs, clips, electrical connectors, power tube and electron tube components, high-strength electrical conductors, bolts, nails, screws, cotter pins, and parts requiring some combination of high strength, high electrical or thermal conductivity, high resistance to fatigue and creep, and good workability.

Table 2. *(Continued)* **Properties and Applications of Wrought Coppers and Copper**

Name and Number	Nominal Composition (%)	Strength (ksi)		Elongation in 2 in. (%)	Machinability Rating[a]	Fabricating Characteristics and Typical Applications
		Tensile	Yield			
C19100 Copper–nickel–phosphorus–tellurium alloy	98.2 Cu, 1.1 Ni, 0.5 Te, 0.2 P	36–104	10–92	27	75	Good hot and cold workability. Uses: forgings and screw-machine parts requiring high strength, hardenability, extensive machining, corrosion resistance, copper color, good conductivity, or a combination of these; bolts, bushings, electrical connectors, gears, marine hardware, nuts, pinions, tie rods, turnbuckle barrels, welding torch tips.
C19200	99 Cu, 1.0 Fe, 0.03 P	37–77	11–74	40	20	Excellent hot and cold workability. Uses: automotive hydraulic brake lines, flexible hose, electrical terminals, fuse clips, gaskets, gift hollow ware, applications requiring resistance to softening and stress corrosion, air-conditioning and heat-exchanger tubing.
C19400	97.4 Cu, 2.4 Fe, 0.13 Zn, 0.04 P	45–76	24–73	32	20	Excellent hot and cold workability. Uses: circuit-breaker components, contact springs, electrical clamps, electrical springs, electrical terminals, flexible hose, fuse clips, gaskets, gift hollow ware, plug contacts, rivets, and welded condenser tubes.
C19500	97.0 Cu, 1.5 Fe, 0.6 Sn, 0.10 P, 0.80 Co	80–97	65–95	15	20	Excellent hot and cold workability. Uses: electrical springs, sockets, terminals, connectors, clips, and other current-carrying parts requiring strength.
C21000 Gilding, 95%	95.0 Cu, 5.0 Zn	34–64	10–58	45	20	Excellent cold workability, good hot workability for blanking, coining, drawing, piercing and punching, shearing, spinning, squeezing and swaging, stamping. Uses: coins, medals, bullet jackets, fuse caps, primers, plaques, jewelry base for gold plate.
C22000 Commercial bronze, 90%	90.0 Cu, 10.0 Zn	37–72	10–62	50	20	Fabricating characteristics same as C21000, plus heading and up-setting, roll threading and knurling, hot forging and pressing. Uses: etching bronze, grillwork, screen cloth, weatherstripping, lipstick cases, compacts, marine hardware, screws, rivets.
C22600 Jewelry bronze, 87.5%	87.5 Cu, 12.5 Zn	39–97	11–62	46	30	Fabricating characteristics same as C21000, plus heading and up-setting, roll threading and knurling. Uses: angles, channels, chain, fasteners, costume jewelry, lipstick cases, powder compacts, base for gold plate.
C23000 Red brass, 85%	85.0 Cu, 15.0 Zn	39–105	10–63	55	30	Excellent cold workability; good hot formability. Uses: weather-stripping, conduit, sockets, fasteners, fire extinguishers, condenser and heat-exchanger tubing, plumbing pipe, radiator cores.
C24000 Low brass, 80%	80.0 Cu, 20.0 Zn	42–125	12–65	55	30	Excellent cold workability. Fabricating characteristics same as C23000. Uses: battery caps, bellows, musical instruments, clock dials, pump lines, flexible hose.
C26000, C26100, C26130, C26200 Cartridge brass, 70%	70.0 Cu, 30.0 Zn	44–130	11–65	66	…	Excellent cold workability. Uses: radiator cores and tanks, flashlight shells, lamp fixtures, fasteners, screws, springs, grillwork, stencils, plumbing accessories, plumbing brass goods, locks, hinges, ammunition components, plumbing accessories, pins, rivets.
C26800, C27000 Yellow brass	65.0 Cu, 35.0 Zn	46–128	14–62	65	30	Excellent cold workability. Fabricating characteristics same as C23000. Uses: same as C26000 except not used for ammunition.

Table 2. *(Continued)* **Properties and Applications of Wrought Coppers and Copper**

Name and Number	Nominal Composition (%)	Strength (ksi)		Elongation in 2 in. (%)	Machinability Rating[a]	Fabricating Characteristics and Typical Applications
		Tensile	Yield			
C28000 Muntz metal, 60%	60.0 Cu, 40.0 Zn	54–74	21–55	52	40	Excellent hot formability and forgeability for blanking, forming and bending, hot forging and pressing, hot heading and upsetting, shearing. Uses: architectural, large nuts and bolts, brazing rod, condenser plates, heat-exchanger and condenser tubing, hot forgings.
C31400 Leaded commercial bronze	89.0 Cu, 1.9 Pb, 0.1 Zn	37–60	12–55	45	80	Excellent machinability. Uses: screws, machine parts, pickling crates.
C31600 Leaded commercial bronze, nickel-bearing	89.0 Cu, 1.9 Pb, 1.0 Ni, 8.1 Zn	37–67	12–59	45	80	Good cold workability; poor hot formability. Uses: electrical connectors, fasteners, hardware, nuts, screws, screw-machine parts.
C33000 Low-leaded brass tube	66.0 Cu, 0.5 Pb, 33.5 Zn	47–75	15–60	60	60	Combines good machinability and excellent cold workability. Fabricated by forming and bending, machining, piercing and punching. Uses: pump and power cylinders and liners, ammunition primers, plumbing accessories.
C33200 High-leaded brass tube	66.0 Cu, 2.0 Pb, 32.0 Zn	47–75	15–60	50	80	Excellent machinability. Fabricated by piercing, punching, and machining. Uses: general-purpose screw-machine parts.
C33500 Low-leaded brass	63.5 Cu, 0.5 Pb, 36 Zn	46–74	14–60	65	60	Similar to C33200. Commonly fabricated by blanking, drawing, machining, piercing and punching, stamping. Uses: butts, hinges, watch backs.
C34000 Medium-leaded brass	63.5 Cu, 1.0 Pb, 35.5 Zn	47–88	15–60	60	70	Similar to C33200. Fabricated by blanking, heading and upsetting, machining, piercing and punching, roll threading and knurling, stamping. Uses: butts, gears, nuts, rivets, screws, dials, engravings, instrument plates.
C34200 High-leaded brass	63.5 Cu, 2.0 Pb, 34.5 Zn	49–85	17–62	52	90	Combines excellent machinability with moderate cold workability. Uses: clock plates and nuts, clock and watch backs, gears, wheels and channel plate.
C35000 Medium-leaded brass	62.5 Cu, 1.1 Pb, 36.4 Zn	45–95	13–70	66	70	Fair cold workability; poor hot formability. Uses: bearing cages, book dies, clock plates, gears, hinges, hose couplings, keys, lock parts, lock tumblers, meter parts, nuts, sink strainers, strike plates, templates, type characters, washers, wear plates.
C35300 High-leaded brass	61.5 Cu, 2.8 Pb, 36.5 Zn	49–85	17–62	52	90	Similar to C34200.
C35600 Extra-high-leaded brass	61.5 Cu, 2.5 Pb, 36 Zn	47–97	17–87	60	100	Excellent machinability. Fabricated by blanking, machining, piercing and punching, stamping. Uses: clock plates and nuts, clock and watch backs, gears, wheels, and channel plate.
C36000 Free-cutting brass	61.5 Cu, 3.1 Pb, 35.4 Zn	49–68	18–45	53	100	Excellent machinability. Fabricated by machining, roll threading, and knurling. Uses: gears, pinions, automatic high-speed screw-machine parts.
C36500 to C36800 Leaded Muntz metal	59.5 Cu, 0.5 Pb, 40.0 Zn	54 (As hot rolled)	20	45	60	Combines good machinability with excellent hot formability. Uses: condenser-tube plates.
C37000 Free-cutting Muntz metal	60.0 Cu, 1.0 Pb, 39.0 Zn	54–80	20–60	40	70	Fabricating characteristics similar to C36500 to 36800. Uses: automatic screw-machine parts.
C37700 Forging brass	59.5 Cu, 2.0 Pb, 38.0 Zn	52 (As extruded)	20	45	80	Excellent hot workability. Fabricated by heading and upsetting, hot forging and pressing, hot heading and upsetting, machining. Uses: forgings and pressings of all kinds.

Table 2. *(Continued)* **Properties and Applications of Wrought Coppers and Copper**

Name and Number	Nominal Composi-tion (%)	Strength (ksi)		Elon-gation in 2 in. (%)	Machin ability Rating[a]	Fabricating Characteristics and Typical Applications
		Ten-sile	Yield			
C38500 Architectural bronze	57.0 Cu, 3.0 Pb, 40.0 Zn	60 (As extrud ed)	20	30	90	Excellent machinability and hot workability. Fabricated by hot forging and pressing, forming, bending, and machining. Uses: architectural extrusions, store fronts, thresholds, trim, butts, hinges, lock bodies, and forgings.
C40500	95 Cu, 1 Sn, 4 Zn	39–78	12–70	49	20	Excellent cold workability. Fabricated by blanking, forming, and drawing. Uses: meter clips, terminals, fuse clips, contact and relay springs, washers.
C40800	95 Cu, 2 Sn, 3 Zn	42–79	13–75	43	20	Excellent cold workability. Fabricated by blanking, stamping, and shearing. Uses: electrical connectors.
C41100	91 Cu, 0.5 Sn, 8.5 Zn	39–106	11–72	43	20	Excellent cold workability, good hot formability. Fabricated by blanking, forming and bending, drawing, piercing and punching, shearing, spinning, and stamping. Uses: bushings, bearing sleeves, thrust washers, flexible metal hose.
C41300	90.0 Cu, 1.0 Sn, 9.0 Zn	41–105	12–82	45	20	Excellent cold workability; good hot formability. Uses: plater bar for jewelry products, flat springs for electrical switchgear.
C41500	91 Cu, 1.8 Sn, 7.2 Zn	46–81	17–75	44	30	Excellent cold workability. Fabricated by blanking, drawing, bending, forming, shearing, and stamping. Uses: spring applications for electrical switches.
C42200	87.5 Cu, 1.1 Sn, 11.4 Zn	43–88	15–75	46	30	Excellent cold workability; good hot formability. Fabricated by blanking, piercing, forming, and drawing. Uses: sash chains, fuse clips, terminals, spring washers, contact springs, electrical connectors.
C42500	88.5 Cu, 2.0 Sn, 9.5 Zn	45–92	18–76	49	30	Excellent cold workability. Fabricated by blanking, piercing, forming, and drawing. Uses: electrical switches, springs, terminals, connectors, fuse clips, pen clips, weather stripping.
C43000	87.0 Cu, 2.2 Sn, 10.8 Zn	46–94	18–73	55	30	Excellent cold workability; good hot formability. Fabricated by blanking, coining, drawing, forming, bending, heading, and upsetting. Uses: same as C42500.
C43400	85.0 Cu, 0.7 Sn, 14.3 Zn	45–90	15–75	49	30	Excellent cold workability. Fabricated by blanking, drawing, bonding, forming, stamping, and shearing. Uses: electrical switch parts, blades, relay springs, contacts.
C43500	81.0 Cu, 0.9 Sn, 18.1 Zn	46–80	16–68	46	30	Excellent cold workability for fabrication by forming and bending. Uses: Bourdon tubing and musical instruments.
C44300, C44400, C44500 Inhibited admiralty	71.0 Cu, 28.0 Zn, 1.0 Sn	48–55	18–22	65	30	Excellent cold workability for forming and bending. Uses: condenser, evaporator and heat-exchanger tubing, condenser tubing plates, distiller tubing, ferrules.
C46400 to C46700 Naval brass	60.0 Cu, 39.2 Zn, 0.8 Sn	55–88	25–66	50	30	Excellent hot workability and hot forgeability. Fabricated by blanking, drawing, bending, heading and upsetting, hot forging, pressing. Uses: aircraft turnbuckle barrels, balls, bolts, marine hardware, nuts, propeller shafts, rivets, valve stems, condenser plates, welding rod.
C48200 Naval brass, medium-leaded	60.5 Cu, 0.7 Pb, 0.8 Sn, 38.0 Zn	56–75	25–53	43	50	Good hot workability for hot forging, pressing, and machining operations. Uses: marine hardware, screw-machine products, valve stems.
C48500 Leaded naval brass	60.0 Cu, 1.8 Pb, 37.5 Zn, 0.7 Sn	57–75	25–53	40	70	Combines good hot forgeability and machinability. Fabricated by hot forging and pressing, machining. Uses: marine hardware, screw-machine parts, valve stems.

Table 2. *(Continued)* **Properties and Applications of Wrought Coppers and Copper**

Name and Number	Nominal Composition (%)	Strength (ksi)		Elongation in 2 in. (%)	Machinability Rating[a]	Fabricating Characteristics and Typical Applications
		Tensile	Yield			
C50500 Phosphor bronze, 1.25% E	98.7 Cu, 1.3 Sn, trace P	40–79	14–50	48	20	Excellent cold workability; good hot formability. Fabricated by blanking, bending, heading and upsetting, shearing and swaging. Uses: electrical contacts, flexible hose, pole-line hardware.
C51000 Phosphor bronze, 5% A	94.8 Cu, 5.0 Sn, trace P	47–140	19–80	64	20	Excellent cold workability. Fabricated by blanking, drawing, bending, heading and upsetting, roll threading and knurling, shearing, stamping. Uses: bellows, Bourdon tubing, clutch discs, cotter pins, diaphragms, fasteners, lock washers, wire brushes, chemical hardware, textile machinery, welding rod.
C51100	95.6 Cu, 4.2 Sn, 0.2 P	46–103	50–80	48	20	Excellent cold workability. Uses: bridge bearing plates, locator bars, fuse clips, sleeve bushings, springs, switch parts, truss wire, wire brushes, chemical hardware, perforated sheets, textile machinery.
C52100 Phosphor bronze, 8% C	92.0 Cu, 8.0 Sn, trace P	55–140	24–80	70	20	Good cold workability for blanking, drawing, forming and bending, shearing, stamping. Uses: generally for more severe service conditions than C51000.
C52400 Phosphor bronze, 10% D	90.0 Cu, 10.0 Sn, trace P	66–147	28	70	20	Good cold workability for blanking, forming and bending, shearing. Uses: heavy bars and plates for severe compression, bridge and expansion plates and fittings, articles requiring good spring qualities, resilience, fatigue resistance, good wear and corrosion resistance.
C54400	88.0 Cu, 4.0 Pb, 4.0 Zn, 4.0 Sn	44–75	19–63	50	80	Excellent machinability; good cold workability. Fabricated by blanking, drawing, bending, machining, shearing, stamping. Uses: bearings, bushings, gears, pinions, shafts, thrust washers, valve parts.
C60800	95.0 Cu, 5.0 Al	60	27	55	20	Good cold workability; fair hot formability. Uses: condenser, evaporator and heat-exchanger tubes, distiller tubes, ferrules.
C61000	92.0 Cu, 8.0 Al	52–60	17–27	45	20	Good hot and cold workability. Uses: bolts, pump parts, shafts, tie rods, overlay on steel for wearing surfaces.
C61300	90.3 Cu, 0.35 Sn, 6.8 Al, 0.35 Sn	70–85	30–58	42	30	Good hot and cold formability. Uses: nuts, bolts, corrosion resistant vessels and tanks, structural components, machine parts, condenser tube and piping systems, marine protective sheathing and fasteners, munitions mixing troughs and blending chambers.
C61400 Aluminum bronze, D	91.0 Cu, 7.0 Al, 2.0 Fe	76–89	33–60	45	20	Similar to C61300.
C61500	90.0 Cu, 8.0 Al, 2.0 Ni	70–145	22–140	55	30	Good hot and cold workability. Fabricating characteristics similar to C52100. Uses: hardware, decorative metal trim, interior furnishings and other articles requiring high tarnish resistance.
C61800	89.0 Cu, 1.0 Fe, 10.0 Al	80–85	39–42.5	28	40	Fabricated by hot forging and hot pressing. Uses: bushings, bearings, corrosion-resistant applications, welding rods.
C61900	86.5 Cu, 4.0 Fe, 9.5 Al	92–152	49–145	30	…	Excellent hot formability for fabricating by blanking, forming, bending, shearing, and stamping. Uses: springs, contacts, and switch components.
C62300	87.0 Cu, 3.0 Fe, 10.0 Al	75–98	35–52	35	50	Good hot and cold formability. Fabricated by bending, hot forging, hot pressing, forming, and welding. Uses: bearings, bushings, valve guides, gears, valve seats, nuts, bolts, pump rods, worm gears, and cams.

Table 2. *(Continued)* **Properties and Applications of Wrought Coppers and Copper**

Name and Number	Nominal Composition (%)	Strength (ksi) Tensile	Strength (ksi) Yield	Elongation in 2 in. (%)	Machinability Rating[a]	Fabricating Characteristics and Typical Applications
C62400	86.0 Cu, 3.0 Fe, 11.0 Al	90–105	40–52	18	50	Excellent hot formability for fabrication by hot forging and hot bending. Uses: bushings, gears, cams, wear strips, nuts, drift pins, tie rods.
C62500	82.7 Cu, 4.3 Fe, 13.0 Al	100 (As extruded)	55	1	20	Excellent hot formability for fabrication by hot forging and machining. Uses: guide bushings, wear strips, cams, dies, forming rolls.
C63000	82.0 Cu, 3.0 Fe, 10.0 Al, 5.0 Ni	90–118	50–75	20	30	Good hot formability. Fabricated by hot forming and forging. Uses: nuts, bolts, valve seats, plunger tips, marine shafts, valve guides, aircraft parts, pump shafts, structural members.
C63200	82.0 Cu, 4.0 Fe, 9.0 Al, 5.0 Ni	90–105	45–53	25	30	Good hot formability. Fabricated by hot forming and welding. Uses: nuts, bolts, structural pump parts, shafting requiring corrosion resistance.
C63600	95.5 Cu, 3.5 Al, 1.0 Si	60–84	…	64	40	Excellent cold workability; fair hot formability. Fabricated by cold heading. Uses: components for pole-line hardware, cold-headed nuts for wire and cable connectors, bolts and screw products.
C63800	95.0 Cu, 2.8 Al, 1.8 Si, 0.40 Co	82–130	54–114	36	…	Excellent cold workability and hot formability. Uses: springs, switch parts, contacts, relay springs, glass sealing, and porcelain enameling.
C64200	91.2 Cu, 7.0 Al, 1.8 Si	75–102	35–68	32	60	Excellent hot formability. Fabricated by hot forming, forging, machining. Uses: valve stems, gears, marine hardware, pole-line hardware, bolts, nuts, valve bodies, and components.
C65100 Low-silicon bronze, B	98.5 Cu, 1.5 Si	40–105	15–71	55	30	Excellent hot and cold workability. Fabricated by forming and bending, heading and upsetting, hot forging and pressing, roll threading and knurling, squeezing and swaging. Uses: hydraulic pressure lines, anchor screws, bolts, cable clamps, cap screws, machine screws, marine hardware, nuts, pole-line hardware, rivets, U-bolts, electrical conduits, heat-exchanger tubing, welding rod.
C65500 High-silicon bronze, A	97.0 Cu, 3.0 Si	56–145	21–71	63	30	Excellent hot and cold workability. Fabricated by blanking, drawing, forming and bending, heading and upsetting, hot forging and pressing, roll threading and knurling, shearing, squeezing and swaging. Uses: similar to C65100 including propeller shafts.
C66700 Manganese brass	70.0 Cu, 28.8 Zn, 1.2 Mn	45.8–100	12–92.5	60	30	Excellent cold formability. Fabricated by blanking, bending, forming, stamping, welding. Uses: brass products resistance welded by spot, seam, and butt welding.
C67400	58.5 Cu, 36.5 Zn, 1.2 Al, 2.8 Mn, 1.0 Sn	70–92	34–55	28	25	Excellent hot formability. Fabricated by hot forging and pressing, machining. Uses: bushings, gears, connecting rods, shafts, wear plates.
C67500 Manganese bronze, A	58.5 Cu, 1.4 Fe, 39.0 Zn, 1.0 Sn, 0.1 Mn	65–84	30–60	33	30	Excellent hot workability. Fabricated by hot forging and pressing, hot heading and upsetting. Uses: clutch discs, pump rods, shafting, balls, valve stems and bodies.
C68700 Aluminum brass, arsenical	77.5 Cu, 20.5 Zn, 2.0 Al, trace As	60	27	55	30	Excellent cold workability for forming and bending. Uses: condenser, evaporator- and heat-exchanger tubing, condenser tubing plates, distiller tubing, ferrules.

Table 2. *(Continued)* **Properties and Applications of Wrought Coppers and Copper**

Name and Number	Nominal Composition (%)	Strength (ksi) Tensile	Yield	Elongation in 2 in. (%)	Machinability Rating[a]	Fabricating Characteristics and Typical Applications
C68800	73.5 Cu, 22.7 Zn, 3.4 Al, 0.40 Co	82–129	55–114	36	…	Excellent hot and cold formability. Fabricated by blanking, drawing, forming and bending, shearing and stamping. Uses: springs, switches, contacts, relays, drawn parts.
C69000	73.3 Cu, 3.4 Al, 0.6 Ni, 22.7 Zn	82–130	52–117	35	…	Fabricating characteristics same as C68800. Uses: contacts, relays, switches, springs, drawn parts.
C69400 Silicon red brass	81.5 Cu, 14.5 Zn, 4.0 Si	80–100	40–57	25	30	Excellent hot formability for fabrication by forging, screw-machine operations. Uses: valve stems where corrosion resistance and high strength are critical.
C70400 Copper nickel, 5%	92.4 Cu, 1.5 Fe, 5.5 Ni, 0.6 Mn	38–77	40–76	46	20	Excellent cold workability; good hot formability. Fabricated by forming, bending, and welding. Uses: condensers, evaporators, heat exchangers, ferrules, salt water piping, lithium bromide absorption tubing, shipboard condenser intake systems.
C70600 Copper nickel, 10%	88.6 Cu, 1.4 Fe, 10.0 Ni	44–60	16–57	42	20	Good hot and cold workability. Fabricated by forming and bending, welding. Uses: condensers, condenser plates, distiller tubing, evaporator and heat-exchanger tubing, ferrules.
C71000 Copper nickel, 20%	79.0 Cu, 21.0 Ni	49–95	13–85	40	20	Good hot and cold formability. Fabricated by blanking, forming and bending, welding. Uses: communication relays, condensers, condenser plates, electrical springs, evaporator and heat-exchanger tubes, ferrules, resistors.
C71500 Copper nickel, 30%	69.5 Cu, 30.0 Ni, 0.5 Fe	54–75	20–70	45	20	Similar to C70600.
C72200	82.2 Cu, 16.5 Ni, 0.8 Fe, 0.5 Cr	46–70	18–66	46	…	Good hot and cold formability. Fabricated by forming, bending, and welding. Uses: condenser tubing, heat-exchanger tubing, salt water piping.
C72500	88.2 Cu, 9.5 Ni, 2.3 Sn	55–120	22–108	35	20	Excellent cold and hot formability. Fabricated by blanking, brazing, coining, drawing, etching, forming and bending, heading and upsetting, roll threading and knurling, shearing, spinning, squeezing, stamping, and swaging. Uses: relay and switch springs, connectors, brazing alloy, lead frames, control and sensing bellows.
C73500	72.0 Cu, 10.0 Zn, 18.0 Ni	50–100	15–84	37	20	Fabricating characteristics same as C74500. Uses: hollow ware, medallions, jewelry, base for silver plate, cosmetic cases, musical instruments, name plates, contacts.
C74500 Nickel silver, 65–10	65.0 Cu, 25.0 Zn, 10.0 Ni	49–130	18–76	50	20	Excellent cold workability. Fabricated by blanking, drawing, etching, forming and bending, heading and upsetting, roll threading and knurling, shearing, spinning, squeezing, and swaging. Uses: rivets, screws, slide fasteners, optical parts, etching stock, hollow ware, nameplates, platers' bars.
C75200 Nickel silver, 65–18	65.0 Cu, 17.0 Zn, 18.0 Ni	56–103	25–90	45	20	Fabricating characteristics similar to C74500. Uses: rivets, screws, table flatware, truss wire, zippers, bows, camera parts, core bars, temples, base for silver plate, costume jewelry, etching stock, hollow ware, nameplates, radio dials.
C75400 Nickel silver, 65–15	65.0 Cu, 20.0 Zn, 15.0 Ni	53–92	18–79	43	20	Fabricating characteristics similar to C74500. Uses: camera parts, optical equipment, etching stock, jewelry.

Table 2. *(Continued)* **Properties and Applications of Wrought Coppers and Copper**

Name and Number	Nominal Composition (%)	Strength (ksi) Tensile	Yield	Elongation in 2 in. (%)	Machinability Rating[a]	Fabricating Characteristics and Typical Applications
C75700 Nickel silver, 65–12	65.0 Cu, 23.0 Zn, 12.0 Ni	52–93	18–79	48	20	Fabricating characteristics similar to C74500. Uses: slide fasteners, camera parts, optical parts, etching stock, name plates.
C76390	61 Cu, 13 Zn, 24.5 Ni, 1 Pb, 0.5 Sn	90	85	6	40	Fabricated by machining, roll threading, and knurling. Uses: hardware, fasteners, connectors for electronic applications.
C77000 Nickel silver, 55–18	55.0 Cu, 27.0 Zn, 18.0 Ni	60–145	27–90	40	30	Good cold workability. Fabricated by blanking, forming and bending, and shearing. Uses: optical goods, springs, and resistance wire.
C78200	65.0 Cu, 2.0 Pb, 25.0 Zn, 8.0 Ni	53–91	23–76	40	60	Good cold formability. Fabricated by blanking, milling, and drilling. Uses: key blanks, watch plates, watch parts.

[a] Free-cutting brass = 100.

Source: Copper Development Association, New York.

Aluminum and Aluminum Alloys

Pure aluminum is a silver-white metal characterized by a slightly bluish cast. It has a specific gravity of 2.70, resists the corrosive effects of many chemicals, and has a malleability approaching that of gold. When alloyed with other metals, numerous properties are obtained that make these alloys useful over a wide range of applications.

Aluminum alloys are light in weight compared with steel, brass, nickel, or copper; can be fabricated by all common processes; are available in a wide range of sizes, shapes, and forms; resist corrosion; readily accept a wide range of surface finishes; have good electrical and thermal conductivities; and are highly reflective to both heat and light.

Characteristics of Aluminum and Aluminum Alloys.—Aluminum and its alloys lose part of their strength at elevated temperatures, although some alloys retain good strength at temperatures from 400 to 500 degrees F. At subzero temperatures, however, their strength increases without loss of ductility so that aluminum is a particularly useful metal for low-temperature applications.

When aluminum surfaces are exposed to the atmosphere, a thin invisible oxide skin forms immediately that protects the metal from further oxidation. This self-protecting characteristic gives aluminum its high resistance to corrosion. Unless exposed to some substance or condition that destroys this protective oxide coating, the metal remains protected against corrosion. Aluminum is highly resistant to weathering, even in industrial atmospheres. It is also corrosion resistant to many acids. Alkalis are among the few substances that attack the oxide skin and therefore are corrosive to aluminum. Although the metal can safely be used in the presence of certain mild alkalis with the aid of inhibitors, in general, direct contact with alkaline substances should be avoided. Direct contact with certain other metals should be avoided in the presence of an electrolyte; otherwise, galvanic corrosion of the aluminum may take place in the contact area. Where other metals must be fastened to aluminum, the use of a bituminous paint coating or insulating tape is recommended.

Aluminum is one of the two common metals having an electrical conductivity high enough for use as an electric conductor. The conductivity of electric-conductor (EC) grade is about 62 per cent that of the International Annealed Copper Standard. Because aluminum has less than one-third the specific gravity of copper, however, a pound of aluminum will go almost twice as far as a pound of copper when used as a conductor. Alloying lowers the conductivity somewhat so that wherever possible the EC grade is used in electric conductor applications. However, aluminum takes a set, which often results in loosening of screwed connectors, leading to arcing and fires. Special clamping designs are therefore required when aluminum is used for electrical wiring, especially in buildings.

Aluminum has nonsparking and nonmagnetic characteristics that make the metal useful for electrical shielding purposes such as in bus bar housings or enclosures for other electrical equipment and for use around inflammable or explosive substances.

Aluminum can be cast by any method known. It can be rolled to any desired thickness down to foil thinner than paper and in sheet form can be stamped, drawn, spun, or roll-formed. The metal also may be hammered or forged. Aluminum wire, drawn from rolled rod, may be stranded into cable of any desired size and type. The metal may be extruded into a variety of shapes. It may be turned, milled, bored, or otherwise machined in equipment often operating at their maximum speeds. Aluminum rod and bar may readily be employed in the high-speed manufacture of parts made on automatic screw-machine.

Almost any method of joining is applicable to aluminum—riveting, welding, or brazing. A wide variety of mechanical aluminum fasteners simplifies the assembly of many products. Resin bonding of aluminum parts has been successfully employed, particularly in aircraft components.

For the majority of applications, aluminum needs no protective coating. Mechanical finishes such as polishing, sandblasting, or wire brushing meet the majority of needs. When additional protection is desired, chemical, electrochemical, and paint finishes are all used. Vitreous enamels have been developed for aluminum, and the metal may also be electroplated.

Temper Designations for Aluminum Alloys.—The temper designation system adopted by the Aluminum Association and used in industry pertains to all forms of wrought and cast aluminum and aluminum alloys except ingot. It is based on the sequences of basic treatments used to produce the various tempers. The temper designation follows the alloy designation, being separated by a dash.

Basic temper designations consist of letters. Subdivisions of the basic tempers, where required, are indicated by one or more digits following the letter. These digits designate specific sequences of basic treatments, but only operations recognized as significantly influencing the characteristics of the product are indicated. Should some other variation of the same sequence of basic operations be applied to the same alloy, resulting in different characteristics, then additional digits are added.

The basic temper designations and subdivisions are as follows:

–F, as fabricated: Applies to products that acquire some temper from shaping processes not having special control over the amount of strain-hardening or thermal treatment. For wrought products, there are no mechanical property limits.

–O, annealed, recrystallized (wrought products only): Applies to the softest temper of wrought products.

–H, strain-hardened (wrought products only): Applies to products that have their strength increased by strain-hardening with or without supplementary thermal treatments to produce partial softening.

The –H is always followed by two or more digits. The first digit indicates the specific combination of basic operations, as follows:

–H1, strain-hardened only: Applies to products that are strain-hardened to obtain the desired mechanical properties without supplementary thermal treatment. The number following this designation indicates the degree of strain-hardening.

–H2, strain-hardened and then partially annealed: Applies to products that are strain-hardened more than the desired final amount and then reduced in strength to the desired level by partial annealing. For alloys that age-soften at room temperature, the –H2 tempers have approximately the same ultimate strength as the corresponding –H3 tempers. For other alloys, the –H2 tempers have approximately the same ultimate strengths as the corresponding –H1 tempers and slightly higher elongations.The number following this desig-

nation indicates the degree of strain-hardening remaining after the product has been partially annealed.

–H3, strain-hardened and then stabililized: Applies to products which are strain-hardened and then stabilized by a low-temperature heating to slightly lower their strength and increase ductility. This designation applies only to the magnesium-containing alloys that, unless stabilized, gradually age-soften at room temperature.The number following this designation indicates the degree of strain-hardening remaining after the product has been strain-hardened a specific amount and then stabilized.

The second digit following the designations –H1, –H2, and –H3 indicates the final degree of strain-hardening. Numeral 8 has been assigned to indicate tempers having a final degree of strain-hardening equivalent to that resulting from approximately 75 per cent reduction of area. Tempers between –O (annealed) and 8 (full hard) are designated by numerals 1 through 7. Material having an ultimate strength about midway between that of the –O temper and that of the 8 temper is designated by the numeral 4 (half hard); between –O and 4 by the numeral 2 (quarter hard); and between 4 and 8 by the numeral 6 (three-quarter hard). (*Note:* For two-digit –H tempers whose second figure is odd, the standard limits for ultimate strength are exactly midway between those for the adjacent two-digit –H tempers whose second figures are even.) Numeral 9 designates extra-hard tempers.

The third digit, when used, indicates a variation of a two-digit –H temper, and is used when the degree of control of temper or the mechanical properties are different from but close to those for the two-digit –H temper designation to which it is added. (*Note:* The minimum ultimate strength of a three-digit –H temper is at least as close to that of the corresponding two-digit –H temper as it is to the adjacent two-digit –H tempers.) Numerals 1 through 9 may be arbitrarily assigned and registered with the Aluminum Association for an alloy and product to indicate a specific degree of control of temper or specific mechanical property limits. Zero has been assigned to indicate degrees of control of temper or mechanical property limits negotiated between the manufacturer and purchaser that are not used widely enough to justify registration with the Aluminum Association.

The following three-digit –H temper designations have been assigned for wrought products in all alloys:

–H111: Applies to products that are strain-hardened less than the amount required for a controlled H11 temper.

–H112: Applies to products that acquire some temper from shaping processes not having special control over the amount of strain-hardening or thermal treatment, but for which there are mechanical property limits, or mechanical property testing is required.

The following three-digit H temper designations have been assigned for wrought products in alloys containing more than a normal 4 per cent magnesium.

–H311: Applies to products that are strain-hardened less than the amount required for a controlled H31 temper.

–H321: Applies to products that are strain-hardened less than the amount required for a controlled H32 temper.

–H323: Applies to products that are specially fabricated to have acceptable resistance to stress-corrosion cracking.

–H343: Applies to products that are specially fabricated to have acceptable resistance to stress-corrosion cracking.

The following three-digit –H temper designations have been assigned for

Patterned or Embossed Sheet	Fabricated Form
–H114	–O temper
–H124, –H224, –H324	–H11, –H21, –H31 temper, respectively
–H134, –H234, –H334	–H12, –H22, –H32 temper, respectively
–H144, –H244, –H344	–H13, –H23, –H33 temper, respectively
–H154, –H254, –H354	–H14, –H24, –H34 temper, respectively
–H164, –H264, –H364	–H15, –H25, –H35 temper, respectively

Patterned or Embossed Sheet	Fabricated Form
–H174, –H274, –H374	–H16, –H26, –H36 temper, respectively
–H184, –H284, –H384	–H17, –H27, –H37 temper, respectively
–H194, –H294, –H394	–H18, –H28, –H38 temper, respectively
–H195, –H395	–H19, –H39 temper, respectively

–W, solution heat-treated: An unstable temper applicable only to alloys that spontaneously age at room temperature after solution heat treatment. This designation is specific only when the period of natural aging is indicated.

–T, thermally treated to produce stable tempers other than –F, –O, or –H: Applies to products that are thermally treated, with or without supplementary strain-hardening, to produce stable tempers.The –T is always followed by one or more digits. Numerals 2 through 10 have been assigned to indicate specific sequences of basic treatments, as follows:

–T1, naturally aged to a substantially stable condition: Applies to products for which the rate of cooling from an elevated temperature-shaping process, such as casting or extrusion, is such that their strength is increased by room-temperature aging.

–T2, annealed (cast products only): Designates a type of annealing treatment used to improve ductility and increase dimensional stability of castings.

–T3, solution heat-treated and then cold-worked: Applies to products that are cold-worked to improve strength, or in which the effect of cold work in flattening or straightening is recognized in applicable specifications.

–T4, solution heat-treated and naturally aged to a substantially stable condition: Applies to products that are not cold-worked after solution heat treatment, or in which the effect of cold work in flattening or straightening may not be recognized in applicable specifications.

–T5, artificially aged only: Applies to products that are artificially aged after an elevated-temperature rapid-cool fabrication process, such as casting or extrusion, to improve mechanical properties or dimensional stability, or both.

–T6, solution heat-treated and then artificially aged: Applies to products that are not cold-worked after solution heat-treatment, or in which the effect of cold work in flattening or straightening may not be recognized in applicable specifications.

–T7, solution heat-treated and then stabilized: Applies to products that are stabilized to carry them beyond the point of maximum hardness, providing control of growth or residual stress or both.

–T8, solution heat-treated, cold-worked, and then artificially aged: Applies to products that are cold-worked to improve strength, or in which the effect of cold work in flattening or straightening is recognized in applicable specifications.

–T9, solution heat-treated, artificially aged, and then cold-worked: Applies to products that are cold-worked to improve strength.

–T10, artificially aged and then cold-worked: Applies to products that are artificially aged after an elevated-temperature rapid-cool fabrication process, such as casting or extrusion, and then cold-worked to improve strength.

Additional digits may be added to designations –T1 through –T10 to indicate a variation in treatment that significantly alters the characteristics of the product. These may be arbitrarily assigned and registered with The Aluminum Association for an alloy and product to indicate a specific treatment or specific mechanical property limits.

These additional digits have been assigned for wrought products in all alloys:

–T__51, stress-relieved by stretching: Applies to products that are stress-relieved by stretching the following amounts after solution heat-treatment:

Plate	1½ to 3 per cent permanent set
Rod, Bar and Shapes	1 to 3 per cent permanent set
Drawn tube	0.5 to 3 per cent permanent set

Applies directly to plate and rolled or cold-finished rod and bar.

These products receive no further straightening after stretching.

Applies to extruded rod and bar shapes and tube when designated as follows:

–T__510 : Products that receive no further straightening after stretching.

–T__511: Products that receive minor straightening after stretching to comply with standard tolerances.

–T__52, stress-relieved by compressing: Applies to products that are stress-relieved by compressing after solution heat-treatment, to produce a nominal permanent set of $2^1/_2$ per cent.

–T__54, stress-relieved by combined stretching and compressing: applies to die forgings that are stress relieved by restriking cold in the finish die.

The following two-digit –T temper designations have been assigned for wrought products in all alloys:

–T42: Applies to products solution heat-treated and naturally aged that attain mechanical properties different from those of the –T4 temper.

–T62: Applies to products solution heat-treated and artificially aged that attain mechanical properties different from those of the –T6 temper.

Aluminum Alloy Designation Systems.—Aluminum casting alloys are listed in many specifications of various standardizing agencies. The numbering systems used by each differ and are not always correlatable. Casting alloys are available from producers who use a commercial numbering system and this numbering system is the one used in the tables of aluminum casting alloys given further along in this section.

A system of four-digit numerical designations for wrought aluminum and wrought aluminum alloys was adopted by the Aluminum Association in 1954. This system is used by the commercial producers and is similar to the one used by the SAE; the difference being the addition of two prefix letters.

The first digit of the designation identifies the alloy type: 1) indicating an aluminum of 99.00 per cent or greater purity; 2) copper; 3) manganese; 4) silicon; 5) magnesium; 6) magnesium and silicon; 7) zinc; 8) some element other than those aforementioned; and 9) unused (not assigned at present).

If the second digit in the designation is zero, it indicates that there is no special control on individual impurities; integers 1 through 9 indicate special control on one or more individual impurities.

In the 1000 series group for aluminum of 99.00 per cent or greater purity, the last two of the four digits indicate to the nearest hundredth the amount of aluminum above 99.00 per cent. Thus designation 1030 indicates 99.30 per cent minimum aluminum. In the 2000 to 8000 series groups the last two of the four digits have no significance but are used to identify different alloys in the group. At the time of adoption of this designation system most of the existing commercial designation numbers were used for these last two digits, as for example, 14S became 2014, 3S became 3003, and 75S became 7075. When new alloys are developed and are commercially used these last two digits are assigned consecutively beginning with –01, skipping any numbers previously assigned at the time of initial adoption.

Experimental alloys are also designated in accordance with this system but they are indicated by the prefix X. The prefix is dropped upon standardization.

Table 3 lists the nominal composition of commonly used aluminum casting alloys, and Tables 4a and 4b list the typical tensile properties of separately cast bars. Table shows the product forms and nominal compositions of common wrought aluminum alloys, and Table lists their typical mechanical properties.

Heat-treatability of Wrought Aluminum Alloys.—In high-purity form, aluminum is soft and ductile. Most commercial uses, however, require greater strength than pure aluminum affords. This extra strength is achieved in aluminum first by the addition of other elements to produce various alloys, which singly or in combination impart strength to the metal. Further strengthening is possible by means that classify the alloys roughly into two categories, non-heat-treatable and heat-treatable.

Non-heat-treatable alloys: The initial strength of alloys in this group depends upon the hardening effect of elements such as manganese, silicon, iron and magnesium, singly or in various combinations. The non-heat-treatable alloys are usually designated, therefore, in the 1000, 3000, 4000, or 5000 series. These alloys are work-hardenable, so further strengthening is made possible by various degrees of cold working, denoted by the "H" series of tempers. Alloys containing appreciable amounts of magnesium when supplied in strain-hardened tempers are usually given a final elevated-temperature treatment called *stabilizing* for property stability.

Heat-treatable alloys: The initial strength of alloys in this group is enhanced by the addition of alloying elements such as copper, magnesium, zinc, and silicon. These elements singly or in various combinations show increasing solid solubility in aluminum with increasing temperature, so it is possible to subject them to thermal treatments that will impart pronounced strengthening.

The first step, called *heat-treatment* or *solution heat-treatment,* is an elevated-temperature process designed to put the soluble element in solid solution. This step is followed by rapid quenching, usually in water, which momentarily "freezes" the structure and for a short time renders the alloy very workable. Some fabricators retain this more workable structure by storing the alloys at below freezing temperatures until they can be formed. At room or elevated temperatures the alloys are not stable after quenching, however, and precipitation of the constituents from the supersaturated solution begins. After a period of several days at room temperature, termed *aging* or *room-temperature precipitation,* the alloy is considerably stronger. Many alloys approach a stable condition at room temperature, but some alloys, particularly those containing magnesium and silicon or magnesium and zinc, continue to age-harden for long periods of time at room temperature.

Heating for a controlled time at slightly elevated temperatures provides even further strengthening and properties are stabilized. This process is called *artificial aging* or *precipitation hardening.* By application of the proper combination of solution heat-treatment, quenching, cold working and artificial aging, the highest strengths are obtained.

Clad Aluminum Alloys.—The heat-treatable alloys in which copper or zinc are major alloying constituents are less resistant to corrosive attack than the majority of non-heat-treatable alloys. To increase the corrosion resistance of these alloys in sheet and plate form they are often clad with high-purity aluminum, a low magnesium-silicon alloy, or an alloy containing 1 per cent zinc. The cladding, usually from 2½ to 5 per cent of the total thickness on each side, not only protects the composite due to its own inherently excellent corrosion resistance but also exerts a galvanic effect that further protects the core material.

Special composites may be obtained such as clad non-heat-treatable alloys for extra corrosion protection, for brazing purposes, or for special surface finishes. Some alloys in wire and tubular form are clad for similar reasons and on an experimental basis extrusions also have been clad.

Table 3. Nominal Compositions (in per cent) of Common Aluminum Casting Alloys (AA/ANSI)

Alloy Designation)	Product	Si	Fe	Cu	Mn	Mg	Cr	Ni	Zn	Ti	Others	
											Each	Total
201.0	S	0.10	0.15	4.0–5.2	0.20–0.50	0.15–0.55	...	...	...	0.15–0.35	0.05[a]	0.10
204.0	S&P	0.20	0.35	4.2–5.0	0.10	0.15–0.35	...	0.05	0.10	0.15–0.30	0.05[b]	0.15
208.0	S&P	2.5–3.5	1.2	3.5–4.5	0.50	0.10	...	0.35	1.0	0.25	...	0.50
222.0	S&P	2.0	1.5	9.2–10.7	0.50	0.15–0.35	...	0.50	0.8	0.25	...	0.35
242.0	S&P	0.7	1.0	3.5–4.5	0.35	1.2–1.8	0.25	1.7–2.3	0.35	0.25	0.05	0.15
295.0	S	0.7–1.5	1.0	4.0–5.0	0.35	0.03	...	...	0.35	0.25	0.05	0.15
308.0	P	5.0–6.0	1.0	4.0–5.0	0.50	0.10	...	...	1.0	0.25	...	0.50
319.0	S&P	5.5–6.5	1.0	3.0–4.0	0.50	0.10	...	0.35	1.0	0.25	...	0.50
328.0	S	7.5–8.5	1.0	1.0–2.0	0.20–0.6	0.20–0.6	0.35	0.25	1.5	0.25	...	0.50
332.0	P	8.5–10.5	1.2	2.0–4.0	0.50	0.50–1.5	...	0.50	1.0	0.25	...	0.50
333.0	P	8.0–10.0	1.0	3.0–4.0	0.50	0.05–0.50	...	0.50	1.0	0.25	...	0.50
336.0	P	11.0–13.0	1.2	0.50–1.5	0.35	0.7–1.3	...	2.0–3.0	0.35	0.25	0.05	...
355.0	S&P	4.5–5.5	0.6[c]	1.0–1.5	0.50[c]	0.40–0.6	0.25	...	0.35	0.25	0.05	0.15
C355.0	S&P	4.5–5.5	0.20	1.0–1.5	0.10	0.40–0.6	...	...	0.10	0.20	0.05	0.15
356.0	S&P	6.5–7.5	0.6[c]	0.25	0.35[c]	0.20–0.45	...	...	0.35	0.25	0.05	0.15
356.0	S&P	6.5–7.5	0.20	0.20	0.10	0.25–0.45	...	...	0.10	0.20	0.05	0.15
357.0	S&P	6.5–7.5	0.15	0.05	0.03	0.45–0.6	...	...	0.05	0.20	0.05	0.15
A357.0	S&P	6.5–7.5	0.20	0.20	0.10	0.40–0.7	...	...	0.10	0.04–0.20	0.05[d]	0.15
443.0	S&P	4.5–6.0	0.8	0.6	0.50	0.05	0.25	...	0.50	0.25	...	0.35
B443.0	S&P	4.5–6.0	0.8	0.15	0.35	0.05	...	...	0.35	0.25	0.05	0.15
A444.0	P	6.5–7.5	0.20	0.10	0.10	0.05	...	...	0.10	0.20	0.05	0.15
512.0	S	1.4–2.2	0.6	0.35	0.8	3.5–4.5	0.25	...	0.35	0.25	0.05	0.15
513.0	P	0.30	0.40	0.10	0.30	3.5–4.5	...	...	1.4–2.2	0.20	0.05	0.15
514.0	S	0.35	0.50	0.15	0.35	3.5–4.5	...	...	0.15	0.25	0.05	0.15
520.0	S	0.25	0.30	0.25	0.15	9.5–10.6	...	...	0.15	0.25	0.05	0.15
705.0	S&P	0.20	0.8	0.20	0.40–0.6	1.4–1.8	0.20–0.40	...	2.7–3.3	0.25	0.05	0.15
707.0	S&P	0.20	0.8	0.20	0.40–0.6	1.8–2.4	0.20–0.40	...	4.0–4.5	0.25	0.05	0.15
710.0	S	0.15	0.50	0.35–0.65	0.05	0.6–0.8	...	...	6.0–7.0	0.25	0.05	0.15
711.0	P	0.30	0.7–1.4	0.35–0.65	0.05	0.25–0.45	...	...	6.0–7.0	0.20	0.05	0.15
712.0	S	0.30	0.50	0.25	0.10	0.50–0.65	0.40–0.6	...	5.0–6.5	0.15–0.25	0.05	0.20
850.0	S&P	0.7	0.7	0.7–1.3	0.10	0.10	...	0.7–1.3	...	0.20	—[e]	0.30
851.0	S&P	2.0–3.0	0.7	0.7–1.3	0.10	0.10	...	0.30–0.7	...	0.20	—[e]	0.30

[a] Also contains 0.40–1.0 per cent silver.

[b] Also contains 0.05 max. per cent tin.

[c] If iron exceeds 0.45 per cent, manganese content should not be less than one-half the iron content.

[d] Also contains 0.04-0.07 per cent beryllium.

[e] Also contains 5.5–7.0 per cent tin.

S = sand cast; P = permanent mold cast. The sum of those "Others" metallic elements 0.010 per cent or more each, expressed to the second decimal before determining the sum. *Source:* Standards for Aluminum Sand and Permanent Mold Castings. Courtesy of the Aluminum Association.

Characteristics of Principal Aluminum Alloy Series Groups.— *1000 series:* These alloys are characterized by high corrosion resistance, high thermal and electrical conductivity, low mechanical properties and good workability. Moderate increases in strength may be obtained by strain-hardening. Iron and silicon are the major impurities.

2000 series: Copper is the principal alloying element in this group. These alloys require solution heat-treatment to obtain optimum properties; in the heat-treated condition mechanical properties are similar to, and sometimes exceed, those of mild steel. In some instances artificial aging is employed to further increase the mechanical properties. This treatment materially increases yield strength, with attendant loss in elongation; its effect on tensile (ultimate) strength is not as great. The alloys in the 2000 series do not have as good corrosion resistance as most other aluminum alloys and under certain conditions they may be subject to intergranular corrosion. Therefore, these alloys in the form of sheet are usually clad with a high-purity alloy or a magnesium-silicon alloy of the 6000 series which provides galvanic protection to the core material and thus greatly increases resistance to corrosion. Alloy 2024 is perhaps the best known and most widely used aircraft alloy.

3000 series: Manganese is the major alloying element of alloys in this group, which are generally non-heat-treatable. Because only a limited percentage of manganese, up to about 1.5 per cent, can be effectively added to aluminum, it is used as a major element in only a few instances. One of these, however, is the popular 3003, used for moderate-strength applications requiring good workability.

4000 series: The major alloying element of this group is silicon, which can be added in sufficient quantities to cause substantial lowering of the melting point without producing brittleness in the resulting alloys. For these reasons aluminum-silicon alloys are used in welding wire and as brazing alloys where a lower melting point than that of the parent metal is required. Most alloys in this series are non-heat-treatable, but when used in welding heat-treatable alloys they will pick up some of the alloying constituents of the latter and so respond to heat-treatment to a limited extent. The alloys containing appreciable amounts of silicon become dark gray when anodic oxide finishes are applied, and hence are in demand for architectural applications.

5000 series: Magnesium is one of the most effective and widely used alloying elements for aluminum. When it is used as the major alloying element or with manganese, the result is a moderate to high strength non-heat-treatable alloy. Magnesium is considerably more effective than manganese as a hardener, about 0.8 per cent magnesium being equal to 1.25 per cent manganese, and it can be added in considerably higher quantities. Alloys in this series possess good welding characteristics and good resistance to corrosion in marine atmospheres. However, certain limitations should be placed on the amount of cold work and the safe operating temperatures permissible for the higher magnesium content alloys (over about 3½ per cent for operating temperatures over about 150 deg. F) to avoid susceptibility to stress corrosion.

6000 series: Alloys in this group contain silicon and magnesium in approximate proportions to form magnesium silicide, thus making them capable of being heat-treated. The major alloy in this series is 6061, one of the most versatile of the heat-treatable alloys. Though less strong than most of the 2000 or 7000 alloys, the magnesium-silicon (or magnesium-silicide) alloys possess good formability and corrosion resistance, with medium strength. Alloys in this heat-treatable group may be formed in the –T4 temper (solution heat-treated but not artificially aged) and then reach full –T6 properties by artificial aging.

7000 series: Zinc is the major alloying element in this group, and when coupled with a smaller percentage of magnesium, results in heat-treatable alloys of very high strength. Other elements such as copper and chromium are ussually added in small quantities. A notable member of this group is 7075, which is among the highest strength aluminum alloys available and is used in air-frame structures and for highly stressed parts.

Table 4a. Mechanical Property Limits for Commonly Used Aluminum Sand Casting Alloys

Alloy	Temper[a]	Minimum Properties			Typical Brinell Hardness (500 kgf load, 10-mm ball)
		Tensile Strength (ksi)		Elongation In 2 inches (%)	
		Ultimate	Yield		
201.0	T7	60.0	50.0	3.0	110–140
204.0	T4	45.0	28.0	6.0	…
208.0	F	19.0	12.0	1.5	40–70
222.0	O	23.0	…	…	65–95
222.0	T61	30.0	…	…	100–130
242.0	O	23.0	…	…	55–85
242.0	T571	29.0	…	…	70–100
242.0	T61	32.0	20.0	…	90–120
242.0	T77	24.0	13.0	1.0	60–90
295.0	T4	29.0	13.0	6.0	45–75
295.0	T6	32.0	20.0	3.0	60–90
295.0	T62	36.0	28.0	…	80–110
295.0	T7	29.0	16.0	3.0	55–85
319.0	F	23.0	13.0	1.5	55–85
319.0	T5	25.0	…	…	65–95
319.0	T6	31.0	20.0	1.5	65–95
328.0	F	25.0	14.0	1.0	45–75
328.0	T6	34.0	21.0	1.0	65–95
354.0	[b]	…	…	…	…
355.0	T51	25.0	18.0	…	50–80
355.0	T6	32.0	20.0	2.0	70–105
355.0	T7	35.0	…	…	70–100
355.0	T71	30.0	22.0	…	60–95
C355.0	T6	36.0	25.0	2.5	75–105
356.0	F	19.0	…	2.0	40–70
356.0	T51	23.0	16.0	…	45–75
356.0	T6	30.0	20.0	3.0	55–90
356.0	T7	31.0	29.0	…	60–90
356.0	T71	25.0	18.0	3.0	45–75
A356.0	T6	34.0	24.0	3.5	70–105
443.0	F	17.0	7.0	3.0	25–55
B443.0	F	17.0	6.0	3.0	25–55
512.0	F	17.0	10.0	…	35–65
514.0	F	22.0	9.0	6.0	35–65
520.0	T4[c]	42.0	22.0	12.0	60–90
535.0	F or T5	35.0	18.0	9.0	60–90
705.0	F or T5	30.0	17.0	5.0	50–80
707.0	T5	33.0	22.0	2.0	70–100
707.0	T7	37.0	30.0	1.0	65–95
710.0	F or T5	32.0	20.0	2.0	60–90
712.0	F or T5	34.0	25.0	4.0	60–90
713.0	F or T5	32.0	22.0	3.0	60–90
771.0	T5	42.0	38.0	1.5	85–115
771.0	T51	32.0	27.0	3.0	70–100
771.0	T52	36.0	30.0	1.5	70–100
771.0	T53	36.0	27.0	1.5	…
771.0	T6	42.0	35.0	5.0	75–105
771.0	T71	48.0	45.0	2.0	105–135
850.0	T5	16.0	…	5.0	30–60
851.0	T5	17.0	…	3.0	30–60
852.0	T5	24.0	18.0	…	45–75

[a] F indicates "as cast" condition.

[b] Mechanical properties for these alloys depend on the casting process. For further information consult the individual foundries.

[c] The T4 temper of Alloy 520.0 is unstable; significant room temperature aging occurs within life expectancy of most castings. Elongation may decrease by as much as 80 percent.

For separately cast test bars.

Source: Standards for Aluminum Sand and Permanent Mold Castings. Courtesy of the Aluminum Association.

Table 4b. Mechanical Property Limits for Commonly Used Aluminum Permanent Mold Casting Alloys

Alloy	Temper[a]	Minimum Properties			Typical Brinell Hardness (500 kgf load, 10-mm ball)
		Tensile Strength (ksi)		Elongation In 2 inches (%)	
		Ultimate	Yield		
204.0	T4	48.0	29.0	8.0	…
208.0	T4	33.0	15.0	4.5	60–90
208.0	T6	35.0	22.0	2.0	75–105
208.0	T7	33.0	16.0	3.0	65–95
222.0	T551	30.0	…	…	100–130
222.0	T65	40.0	…	…	125–155
242.0	T571	34.0	…	…	90–120
242.0	T61	40.0	…	…	95–125
296.0	T6	35.0	…	2.0	75–105
308.0	F	24.0	…	…	55–85
319.0	F	28.0	14.0	1.5	70–100
319.0	T6	34.0	…	2.0	75–105
332.0	T5	31.0	…	…	90–120
333.0	F	28.0	…	…	65–100
333.0	T5	30.0	…	…	70–105
333.0	T6	35.0	…	…	85–115
333.0	T7	31.0	…	…	75–105
336.0	T551	31.0	…	…	90–120
336.0	T65	40.0	…	…	110–140
354.0	T61	48.0	37.0	3.0	…
354.0	T62	52.0	42.0	2.0	…
355.0	T51	27.0	…	…	60–90
355.0	T6	37.0	…	1.5	75–105
355.0	T62	42.0	…	…	90–120
355.0	T7	36.0	…	…	70–100
355.0	T71	34.0	27.0	…	65–95
C355.0	T61	40.0	30.0	3.0	75–105
356.0	F	21.0	…	3.0	40–70
356.0	T51	25.0	…	…	55–85
356.0	T6	33.0	22.0	3.0	65–95
356.0	T7	25.0	…	3.0	60–90
356.0	T71	25.0	…	3.0	60–90
A356.0	T61	37.0	26.0	5.0	70–100
357.0	T6	45.0	…	3.0	75–105
A357.0	T61	45.0	36.0	3.0	85–115
359.0	T61	45.0	34.0	4.0	75–105
359.0	T62	47.0	38.0	3.0	85–115
443.0	F	21.0	7.0	2.0	30–60
B443.0	F	21.0	6.0	2.5	30–60
A444.0	T4	20.0	…	20.0	…
513.0	F	22.0	12.0	2.5	45–75
535.0	F	35.0	18.0	8.0	60–90
705.0	T5	37.0	17.0	10.0	55–85
707.0	T7	45.0	35.0	3.0	80–110
711.0	T1	28.0	18.0	7.0	55–85
713.0	T5	32.0	22.0	4.0	60–90
850.0	T5	18.0	…	8.0	30–60
851.0	T5	17.0	…	3.0	30–60
851.0	T6	18.0	…	8.0	…
852.0	T5	27.0	…	3.0	55–85

[a] F indicates "as cast" condition.

For separately cast test bars.

Source: Standards for Aluminum Sand and Permanent Mold Castings. Courtesy of the Aluminum Association.

Table 5. Typical Mechanical Properties of Wrought Aluminum Alloys

Alloy and Temper	Tension				Brinell Hardness Number (500 kg load, 10-mm ball)	Ultimate Shearing Strength (ksi)	Endurance Limit[a] (ksi)
	Strength (ksi)		Elongation in 2 inches (%)				
	Ultimate	Yield	1/16-inch Thick Specimen	1/2-inch Diameter Specimen			
1060-O	10	4	43	…	19	7	3
1060-H12	12	11	16	…	23	8	4
1060-H14	14	13	12	…	26	9	5
1060-H16	16	15	8	…	30	10	6.5
1060-H18	19	18	6	…	35	11	6.5
1100-O	13	5	35	45	23	9	5
1100-H12	16	15	12	25	28	10	6
1100-H14	18	17	9	20	32	11	7
1100-H16	21	20	6	17	38	12	9
1100-H18	24	22	5	15	44	13	9
1350-O	12	4	…	…[b]	…	8	…
1350-H12	14	12	…	…	…	9	…
1350-H14	16	14	…	…	…	10	…
1350-H16	18	16	…	…	…	11	…
1350-H19	27	24	…	…[c]	…	15	7
2011-T3	55	43	…	15	95	32	18
2011-T8	59	45	…	12	100	35	18
2014-O	27	14	…	18	45	18	13
2014-T4, T451	62	42	…	20	105	38	20
2014-T6, T651	70	60	…	13	135	42	18
Alclad 2014-O	25	10	21	…	…	18	…
Alclad 2014-T3	63	40	20	…	…	37	…
Alclad 2014-T4, T451	61	37	22	…	…	37	…
Alclad 2014-T6, T651	68	60	10	…	…	41	…
2017-O	26	10	…	22	45	18	13
2017-T4, T451	62	40	…	22	105	38	18
2018-T61	61	46	…	12	120	39	17
2024-O	27	11	20	22	47	18	13
2024-T3	70	50	18	…	120	41	20
2024-T4, T351	68	47	20	19	120	41	20
2024-T361[d]	72	57	13	…	130	42	18
Alclad 2024-O	26	11	20	…	…	18	…
Alclad 2024-T3	65	45	18	…	…	40	…
Alclad 2024-T4, T351	64	42	19	…	…	40	…
Alclad 202024-T361[d]	67	53	11	…	…	41	…
Alclad 2024-T81, T851	65	60	6	…	…	40	…
Alclad 2024-T861[d]	70	66	6	…	…	42	…
2025-T6	58	37	…	19	110	35	18
2036-T4	49	28	24	…	…	…	18[e]
2117-T4	43	24	…	27	70	28	14
2218-T72	48	37	…	11	95	30	…
2219-O	25	11	18	…	…	…	…
2219-T42	52	27	20	…	…	…	…
2219-T31, T351	52	36	17	…	…	…	…
2219-T37	57	46	11	…	…	…	…
2219-T62	60	42	10	…	…	…	15
2219-T81, T851	66	51	10	…	…	…	15

Table 5. *(Continued)* **Typical Mechanical Properties of Wrought Aluminum Alloys**

Alloy and Temper	Tension				Brinell Hardness Number (500 kg load, 10-mm ball)	Ultimate Shearing Strength (ksi)	Endurance Limit[a] (ksi)
	Strength (ksi)		Elongation in 2 inches (%)				
	Ultimate	Yield	1/16-inch Thick Specimen	1/2-inch Diameter Specimen			
2219-T87	69	57	10	…	…	…	15
3003-O	16	6	30	40	28	11	7
3003-H12	19	18	10	20	35	12	8
3003-H14	22	21	8	16	40	14	9
3003-H16	26	25	5	14	47	15	10
3003-H18	29	27	4	10	55	16	10
Alclad 3003-O	16	6	30	40	…	11	…
Alclad 3003-H12	19	18	10	20	…	12	…
Alclad 3003-H14	22	21	8	16	…	14	…
Alclad 3003-H16	26	25	5	14	…	15	…
Alclad 3003-H18	29	27	4	10	…	16	…
3004-O	26	10	20	25	45	16	14
3004-H32	31	25	10	17	52	17	15
3004-H34	35	29	9	12	63	18	15
3004-H36	38	33	5	9	70	20	16
3004-H38	41	36	5	6	77	21	16
Alclad 3004-O	26	10	20	25	…	16	…
Alclad 3004-H32	31	25	10	17	…	17	…
Alclad 3004-H34	35	29	9	12	…	18	…
Alclad 3004-H36	38	33	5	9	…	20	…
Alclad 3004-H38	41	36	5	6	…	21	…
3105-O	17	8	24	…	…	12	…
3105-H12	22	19	7	…	…	14	…
3105-H14	25	22	5	…	…	15	…
3105-H16	28	25	4	…	…	16	…
3105-H18	31	28	3	…	…	17	…
3105-H25	26	23	8	…	…	15	…
4032-T6	55	46	…	9	120	38	16
5005-O	18	6	25	…	28	11	…
5005-H12	20	19	10	…	…	14	…
5005-H14	23	22	6	…	…	14	…
5005-H16	26	25	5	…	…	15	…
5005-H18	29	28	4	…	…	16	…
5005-H32	20	17	11	…	36	14	…
5005-H34	23	20	8	…	41	14	…
5005-H36	26	24	6	…	46	15	…
5005-H38	29	27	5	…	51	16	…
5050-O	21	8	24	…	36	15	12
5050-H32	25	21	9	…	46	17	13
5050-H34	28	24	8	…	53	18	13
5050-H36	30	26	7	…	58	19	14
5050-H38	32	29	6	…	63	20	14
5052-O	28	13	25	30	47	18	16
5052-H32	33	28	12	18	60	20	17
5052-H34	38	31	10	14	68	21	18
5052-H36	40	35	8	10	73	23	19
5052-H38	42	37	7	8	77	24	20
5056-O	42	22	…	35	65	26	20

Table 5. *(Continued)* **Typical Mechanical Properties of Wrought Aluminum Alloys**

Alloy and Temper	Tension				Brinell Hardness Number (500 kg load, 10-mm ball)	Ultimate Shearing Strength (ksi)	Endurance Limit[a] (ksi)
	Strength (ksi)		Elongation in 2 inches (%)				
	Ultimate	Yield	1/16-inch Thick Specimen	1/2-inch Diameter Specimen			
5056-H18	63	59	…	10	105	34	22
5056-H38	60	50	…	15	100	32	22
5083-O	42	21	…	22	…	25	…
5083-H321, H116	46	33	…	16	…	…	23
5086-O	38	17	22	…	…	23	…
5086-H32, H116	42	30	12	…	…	…	…
5086-H34	47	37	10	…	…	27	…
5086-H112	39	19	14	…	…	…	…
5154-O	35	17	27	…	58	22	17
5154-H32	39	30	15	…	67	22	18
5154-H34	42	33	13	…	73	24	19
5154-H36	45	36	12	…	78	26	20
5154-H38	48	39	10	…	80	28	21
5154-H112	35	17	25	…	63	…	17
5252-H25	34	25	11	…	68	21	…
5252-H38, H28	41	35	5	…	75	23	…
5254-O	35	17	27	…	58	22	17
5254-H32	39	30	15	…	67	22	18
5254-H34	42	33	13	…	73	24	19
5254-H36	45	36	12	…	78	26	20
5254-H38	48	39	10	…	80	28	21
5254-H112	35	17	25	…	63	…	17
5454-O	36	17	22	…	62	23	…
5454-H32	40	30	10	…	73	24	…
5454-H34	44	35	10	…	81	26	…
5454-H111	38	26	14	…	70	23	…
5454-H112	36	18	18	…	62	23	…
5456-O	45	23	…	24	…	…	…
5456-H112	45	24	…	22	…	…	…
5456-H321, H116	51	37	…	16	90	30	…
5457-O	19	7	22	…	32	12	…
5457-H25	26	23	12	…	48	16	…
5457-H38, H28	30	27	6	…	55	18	…
5652-O	28	13	25	30	47	18	16
5652-H32	33	28	12	18	60	20	17
5652-H34	38	31	10	14	68	21	18
5652-H36	40	35	8	10	73	23	19
5652-H38	42	37	7	8	77	24	20
5657-H25	23	20	12	…	40	14	…
5657-H38, H28	28	24	7	…	50	15	…
6061-O	18	8	25	30	30	12	9
6061-T4, T451	35	21	22	25	65	24	14
6061-T6, T651	45	40	12	17	95	30	14
Alclad 6061-O	17	7	25	…	…	11	…
Alclad 6061-T4, T451	33	19	22	…	…	22	…
Alclad 6061-T6, T651	42	37	12	…	…	27	…
6063-O	13	7	…	…	25	10	8
6063-T1	22	13	20	…	42	14	9

Table 5. *(Continued)* **Typical Mechanical Properties of Wrought Aluminum Alloys**

Alloy and Temper	Tension				Brinell Hardness Number (500 kg load, 10-mm ball)	Ultimate Shearing Strength (ksi)	Endurance Limit[a] (ksi)
	Strength (ksi)		Elongation in 2 inches (%)				
	Ultimate	Yield	1/16-inch Thick Specimen	1/2-inch Diameter Specimen			
6063-T4	25	13	22	...	...	...	...
6063-T5	27	21	12	...	60	17	10
6063-T6	35	31	12	...	73	22	10
6063-T83	37	35	9	...	82	22	...
6063-T831	30	27	10	...	70	18	...
6063-T832	42	39	12	...	95	27	...
6066-O	22	12	...	18	43	14	...
6066-T4, T451	52	30	...	18	90	29	...
6066-T6, T651	57	52	...	12	120	34	16
6070-T6	55	51	10	...	...	34	14
6101-H111	14	11	...	...	...	...	...
6101-T6	32	28	15	...	71	20	...
6262-T9	58	55	...	10	120	35	13
6351-T4	36	22	20	...	...	...	...
6351-T6	45	41	14	...	95	29	13
6463-T1	22	13	20	...	42	14	10
6463-T5	27	21	12	...	60	17	10
6463-T6	35	31	12	...	74	22	10
7049-T73	75	65	...	12	135	44	...
7049-T7352	75	63	...	11	135	43	...
7050-T73510, T73511	72	63	...	12	...	...	...
7050-T7451[f]	76	68	...	11	...	44	...
7050-T7651	80	71	...	11	...	47	...
7075-O	33	15	17	16	60	22	...
7075-T6, T651	83	73	11	11	150	48	23
Alclad 7075-O	32	14	17	...	...	22	...
Alclad 7075-T6, T651	76	67	11	...	...	46	...
7178-O	33	15	15	16	...	...	...
7178-T6, T651	88	78	10	11	...	...	...
7178-T76, T7651	83	73	...	11	...	...	...
Alclad 7178-O	32	14	16	...	...	...	...
Alclad 7178-T6, T651	81	71	10	...	...	...	...
8176-H24	17	14	15	...	...	10	...

[a] Based on 500,000,000 cycles of completely reversed stress using the R. R. Moore type of machine and specimen.

[b] 1350-O wire should have an elongation of approximately 23 per cent in 10 inches.

[c] 1350-H19 wire should have an elongation of approximately 1.5 per cent in 10 inches.

[d] Tempers T361 and T861 were formerly designated T36 and T86, respectively.

[e] Based on 10^7 cycles using flexural type testing of sheet specimens.

[f] T7451, although not previously registered, has appeared in the literature and in some specifications as T73651.

The data given in this table are intended only as a basis for comparing alloys and tempers and should not be specified as engineering requirements or used for design purposes. The indicated typical mechanical properties for all except O temper material are higher than the specified minimum properties. For O temper products, typical ultimate and yield values are slightly lower than specified (maximum) values.

Source: Aluminum Standards and Data. Courtesy of the Aluminum Association.

Table 6. Nominal Compositions of Common Wrought Aluminum Alloys

Alloy	Alloying Elements — Aluminum and Normal Impurities Constitute Remainder							
	Si	Cu	Mn	Mg	Cr	Ni	Zn	Ti
1050	…	…	99.50 per cent minimum aluminum			…	…	…
1060	…	…	99.60 per cent minimum aluminum			…	…	…
1100	…	0.12	99.00 per cent minimum aluminum			…	…	…
1145	…	…	99.45 per cent minimum aluminum			…	…	…
1175	…	…	99.75 per cent minimum aluminum			…	…	…
1200	…	…	99.00 per cent minimum aluminum			…	…	…
1230	…	…	99.30 per cent minimum aluminum			…	…	…
1235	…	…	99.35 per cent minimum aluminum			…	…	…
1345	…	…	99.45 per cent minimum aluminum			…	…	…
1350[a]	…	…	99.50 per cent minimum aluminum			…	…	…
2011[b]	…	5.5	…	…	…	…	…	…
2014	0.8	4.4	0.8	0.50	…	…	…	…
2017	0.50	4.0	0.7	0.6	…	…	…	…
2018	…	4.0	…	0.7	…	2.0	…	…
2024	…	4.4	0.6	1.5	…	…	…	…
2025	0.8	4.4	0.8	…	…	…	…	…
2036	…	2.6	0.25	0.45	…	…	…	…
2117	…	2.6	…	0.35	…	…	…	…
2124	…	4.4	0.6	1.5	…	…	…	…
2218	…	4.0	…	1.5	…	2.0	…	…
2219[c]	…	6.3	0.30	…	…	…	…	0.06
2319[c]	…	6.3	0.30	…	…	…	…	0.15
2618[d]	0.18	2.3	…	1.6	…	1.0	…	0.07
3003	…	0.12	1.2	…	…	…	…	…
3004	…	…	1.2	1.0	…	…	…	…
3005	…	…	1.2	0.40	…	…	…	…
4032	12.2	0.9	…	1.0	…	0.9	…	…
4043	5.2	…	…	…	…	…	…	…
4045	10.0	…	…	…	…	…	…	…
4047	12.0	…	…	…	…	…	…	…
4145	10.0	4.0	…	…	…	…	…	…
5005	…	…	…	0.8	…	…	…	…
5050	…	…	…	1.4	…	…	…	…
5052	…	…	…	2.5	0.25	…	…	…
5056	…	…	0.12	5.0	0.12	…	…	…
5083	…	…	0.7	4.4	0.15	…	…	…
5086	…	…	0.45	4.0	0.15	…	…	…
5183	…	…	0.8	4.8	0.15	…	…	…
5252	…	…	…	2.5	…	…	…	…
5254	…	…	…	3.5	0.25	…	…	…
5356	…	…	0.12	5.0	0.12	…	…	0.13
5456	…	…	0.8	5.1	0.12	…	…	…
5457	…	…	0.30	1.0	…	…	…	…
5554	…	…	0.8	2.7	0.12	…	…	0.12

Table 6. *(Continued)* **Nominal Compositions of Common Wrought Aluminum Alloys**

Alloy	Alloying Elements — Aluminum and Normal Impurities Constitute Remainder							
	Si	Cu	Mn	Mg	Cr	Ni	Zn	Ti
5556	…	…	0.8	5.1	0.12	…	…	0.12
5652	…	…	…	2.5	0.25	…	…	…
5654	…	…	…	3.5	0.25	…	…	0.10
6003	0.7	…	…	1.2	…	…	…	…
6005	0.8	…	…	0.50	…	…	…	…
6053	0.7	…	…	1.2	0.25	…	…	…
6061	0.6	0.28	…	1.0	0.20	…	…	…
6066	1.4	1.0	0.8	1.1	…	…	…	…
6070	1.4	0.28	0.7	0.8	…	…	…	…
6101	0.50	…	…	0.6	…	…	…	…
6105	0.8	…	…	0.62	…	…	…	…
6151	0.9	…	…	0.6	0.25	…	…	…
6201	0.7	…	…	0.8	…	…	…	…
6253	0.7	…	…	1.2	0.25	…	2.0	…
6262[e]	0.6	0.28	…	1.0	0.09	…	…	…
6351	1.0	…	0.6	0.6	…	…	…	…
6463	0.40	…	…	0.7	…	…	…	…
7005[f]	…	…	0.45	1.4	0.13	…	4.5	0.04
7008	…	…	…	1.0	0.18	…	5.0	…
7049	…	1.6	…	2.4	0.16	…	7.7	…
7050[g]	…	2.3	…	2.2	6.2	…	…	…
7072	…	…	…	…	…	…	1.0	…
7075	…	1.6	…	2.5	0.23	…	5.6	…
7108[h]	…	…	…	1.0	…	…	5.0	…
7178	…	2.0	…	2.8	0.23	…	6.8	…
8017[i]	…	0.15	…	0.03	…	…	…	…
8030[j]	…	0.22	…	…	…	…	…	…
8177[k]	…	…	…	0.08	…	…	…	…

[a] Formerly designated EC.
[b] Lead and bismuth, 0.40 per cent each.
[c] Vanadium 0.10 per cent; zirconium 0.18 per cent.
[d] Iron 1.1 per cent.
[e] Lead and bismuth, 0.6 per cent each.
[f] Zirconium 0.14 per cent.
[g] Zirconium 0.12 per cent.
[h] Zirconium 0.18 per cent.
[i] Iron 0.7 per cent.
[j] Boron 0.02 per cent.
[k] Iron 0.35 per cent.
Source: Aluminum Standards and Data. Courtesy of the Aluminum Association.

Magnesium Alloys

Magnesium Alloys.—Magnesium is the lightest of all structural metals. Silver-white in color, pure magnesium is relatively soft, so is rarely used for structural purposes in the pure state. Principal metallurgical uses for pure magnesium are as an alloying element for aluminum and other metals; as a reducing agent in the extraction of such metals as titanium, zirconium, hafnium, and uranium; as a nodularizing agent in the manufacture of ductile iron; and as a sulfur removal agent in steel manufacture. Magnesium alloys are made by alloying up to about 10 per cent of other metals and have low density and an excellent combination of mechanical properties, as shown in Table 7a, resulting in high strength-to-weight ratios.

Magnesium alloys are the easiest of all the structural metals to machine, and these alloys have very high weld efficiencies. Magnesium is readily processed by all the standard casting and fabrication techniques used in metalworking, especially by pressure die casting. Because the metal work hardens rapidly, cold forming is limited to mild deformation, but magnesium alloys have excellent working characteristics at temperatures between 300 and 500 degrees F.

These alloys have relatively low elastic moduli, so they will absorb energy with good resistance to dents and high damping capacities. Fatigue strength also is good, particularly in the low-stress, high-cycle range. The alloys can be precipitation hardened, so mechanical properties can be improved by solution heat treatment and aging. Corrosion resistance was greatly improved recently, when methods were found to limit heavy metal impurities to “parts per million.”

Applications of Magnesium Alloys.—Magnesium alloys are used in a wide variety of structural applications including industrial, materials handling, automotive, consumer-durable, and aerospace equipment. In industrial machinery, the alloys are used for parts that operate at high speeds, which must have light weight to allow rapid acceleration and minimize inertial forces. Materials handling equipment applications include hand trucks, dockboards, grain shovels, and gravity conveyors. Automotive applications include wheels, gearboxes, clutch housings, valve covers, and brake pedal and other brackets. Consumer durables include luggage, softball bats, tennis rackets, and housings for cameras and projectors. Their high strength-to-weight ratio suits magnesium alloys to use in a variety of aircraft structures, particularly helicopters. Very intricate shapes that are uneconomical to produce in other materials are often cast in magnesium, sometimes without draft. Wrought magnesium alloys are made in the form of bars, forgings, extrusions, wire, sheet, and plate.

Alloy and Temper Designation.—Magnesium alloys are designated by a standard four-part system established by the ASTM, and now also used by the SAE, that indicates both chemical composition and temper. Designations begin with two letters representing the two alloying elements that are specified in the greatest amount; these letters are arranged in order of decreasing percentage of alloying elements or alphabetically if they are present in equal amounts. The letters are followed by digits representing the respective composition percentages, rounded off to whole numbers, and then by a serial letter indicating some variation in composition of minor constituents. The final part, separated by a hyphen, consists of a letter followed by a number, indicating the temper condition. The letters that designate the more common alloying elements are A, aluminum; E, rare earths; H, thorium; K, zirconium; M, manganese; Q, silver; S, silicon; T, tin; Z, zinc.

The letters and numbers that indicate the temper designation are: F, as fabricated; O, annealed; H10, H11, strain hardened; H23, H24, H26, strain hardened and annealed; T4, solution heat treated; T5, artificially aged; T6, solution heat treated and artificially aged;

and T8, solution heat treated, cold-worked, and artificially aged.

The nominal composition and typical properties of magnesium alloys are listed in Table 7a.

Table 7a. Nominal Compositions of Magnesium Alloys

Alloy	Al	Zn	Mn[a]	Si	Zr	Ag	Th	Y	Rare Earth
Sand and Permanent Mold (Gravity Die) Castings									
AM100A-T61	10.0	…	0.10	…	…	…	…	…	…
AZ63A-T6	6.0	3.0	0.15	…	…	…	…	…	…
AZ81A-T4	7.6	0.7	0.13	…	…	…	…	…	…
AZ91C-T6	8.7	0.7	0.13	…	…	…	…	…	…
AZ91E-T6[b]	8.7	0.7	0.17	…	…	…	…	…	…
AZ92A-T6	9.0	2.0	0.10	…	…	…	…	…	…
EZ33A-T5	…	2.6	…	…	0.8	…	…	…	3.3
HK31A-T6	…	0.3	…	…	0.7	…	3.3	…	…
HZ32A-T6	…	2.1	…	…	0.8	…	3.3	…	0.1
K1A-F	…	…	…	…	0.7	…	…	…	…
QE22A-T6	…	…	…	…	0.7	2.5	…	…	2.2
QH21A-T6	…	0.2	…	…	0.7	…	1.1	…	1.1
ZE41A-T5	…	4.3	0.15	…	0.7	…	…	…	1.3
ZE63A-T6	…	5.8	…	…	0.7	…	…	…	2.6
ZH62A-T5	…	5.7	…	…	0.8	…	1.8	…	…
ZK51A-T5	…	4.6	…	…	0.8	…	…	…	…
ZK61A-T6	…	6.0	…	…	0.8	…	…	…	…
WE54A-F	…	…	…	…	0.5	…	…	5.3	3.5
Pressure Die Castings									
AZ91A-F	9.0	0.7	0.13	…	…	…	…	…	…
AZ91B-F[c]	9.0	0.7	0.13	…	…	…	…	…	…
AZ91D-F[b]	9.0	0.7	0.15	…	…	…	…	…	…
AM60A-F	6.0	…	0.13	…	…	…	…	…	…
AM60B-F[b]	6.0	…	0.25	…	…	…	…	…	…
AS41A-F[d]	4.3	…	0.35	1.0	…	…	…	…	…
Extruded Bars and Shapes									
AZ10A-F	1.3	0.4	0.20	…	…	…	…	…	…
AZ31B-F	3.0	1.0	0.20	…	…	…	…	…	…
AZ31C-F	3.0	1.0	0.15	…	…	…	…	…	…
AZ61A-F	6.5	1.0	0.15	…	…	…	…	…	…
AZ80A-T5	8.5	0.5	0.12	…	…	…	…	…	…
HM31A-F	…	…	1.20	…	…	…	3.0	…	…
M1A-F	…	…	1.20	…	…	…	…	…	…
ZK40A-T5	…	4.0	…	…	0.45	…	…	…	…
ZK60A-F	…	5.5	…	…	0.45	…	…	…	…
Sheet and Plate									
AZ31B-H24	3.0	1.0	0.20	…	…	…	…	…	…
AZ31C-H24	3.0	1.0	0.15	…	…	…	…	…	…
HK31A-H24	…	…	…	…	0.7	…	3.3	…	…
HM21A-T8	…	…	0.80	…	…	…	2.0	…	…

[a] All manganese vales are minium.

[b] High-purity alloy, Ni, Fe, and Cu severely restricted.

[c] 0.30 per cent maximum residual copper is allowed.

[d] For battery applications.

Source: Metals Handbook, 9th edition, Vol. 2, American Society for Metals.

Table 7b. Typical Room-Temperature Mechanical Properties of Magnesium Alloys

Alloy	Tensile Strength (ksi)	Yield Strength			Elongation in 2 in. (%)	Shear Strength (ksi)	Hardness Rockwell B[a]
		Tensile (ksi)	Compressive (ksi)	Bearing (ksi)			
Sand and Permanent Mold (Gravity Die) Castings							
AM100A-T61	40	22	22	68	1	…	69
AZ63A-T6	40	14	14	44	12	18	55
AZ81A-T4	40	12	12	35	15	21	55
AZ91C-T6	40	21	21	52	6	21	70
AZ91E-T6[b]	40	21	21	52	6	21	70
AZ92A-T6	40	22	22	65	3	21	81
EZ33A-T5	23	16	16	40	3	20	50
HK31A-T6	32	15	15	40	8	21	55
HZ32A-T6	27	13	13	37	4	20	55
K1A-F	26	8	8	18	19	8	…
QE22A-T6	38	28	28	…	3	…	80
QH21A-T6	40	30	30	…	4	22	…
ZE41A-T5	30	20	20	51	4	23	62
ZE63A-T6	44	28	28	…	10	…	60–85
ZH62A-T5	35	22	22	49	4	23	70
ZK51A-T5	30	20	20	51	4	22	62
ZK61A-T6	45	28	28	…	10	…	…
WE54A-F	40	29	29	…	4	…	…
Pressure Die Castings							
AZ91A-F	34	23	23	…	3	20	63
AZ91B-FAZ91B-F[c]	34	23	23	…	3	20	63
AZ91D-F[b]	34	23	23	…	3	20	63
AM60A-F	32	19	19	…	8	…	…
AM60B-F[b]	32	19	19	…	8	…	…
AS41A-F[d]	31	20	20	…	6	…	…
Extruded Bars and Shapes							
AZ10A-F	35	21	10	…	10	…	…
AZ31B-F	38	29	14	33	15	19	49
AZ31C-F	38	29	14	33	15	19	49
AZ61A-F	45	33	19	41	16	20	60
AZ80A-T5	55	40	35	…	7	24	82
HM31A-F	42	33	27	50	10	22	…
M1A-F	37	26	12	28	12	18	44
ZK40A-T5	40	37	20	…	4	…	…
ZK60A-F	51	41	36	59	11	26	88
Sheet and Plate							
AZ31B-H24	42	32	26	47	15	23	73
AZ31C-H24	42	32	26	47	15	23	73
HK31A-H24	38	30	23	41	9	…	68
HM21A-T8	34	25	19	39	11	18	…

[a] 500 kg load, 10-mm ball.
[b] High-purity alloy, Ni, Fe, and Cu severely restricted.
[c] 0.30 per cent maximum residual copper is allowed.
[d] For battery applications.

Source: Metals Handbook, 9th edition, Vol. 2, American Society for Metals.

Nickel and Nickel Alloys

Nickel is a white metal, similar in some respects to iron but with good oxidation and corrosion resistances. Nickel and its alloys are used in a variety of applications, usually requiring specific corrosion resistance or high strength at high temperature. Some nickel alloys exhibit very high toughness; others have very high strength, high proportional limits, and high moduli compared with steel. Commercially, pure nickel has good electrical, magnetic, and magnetostrictive properties. Nickel alloys are strong, tough, and ductile at cryogenic temperatures, and several of the so-called nickel-based superalloys have good strength at temperatures up to 2000 degrees F.

Most wrought nickel alloys can be hot and cold-worked, machined, and welded successfully; an exception is the most highly alloyed nickel compound—forged nickel-based superalloys—in which these operations are more difficult. The casting alloys can be machined or ground, and many can be welded and brazed.

There are five categories into which the common nickel-based metals and alloys can be separated: the pure nickel and high nickel (over 94 per cent Ni) alloys; the nickel–molybdenum and nickel–molybdenum–chromium superalloys, which are specifically for corrosive or high-temperature, high-strength service; the nickel–molybdenum–chromium–copper alloys, which are also specified for corrosion applications; the nickel–copper (Monel) alloys, which are used in actively corrosive environments; and the nickel–chromium and nickel–chromium–iron superalloys, which are noted for their strength and corrosion resistance at high temperatures.

Descriptions and compositions of some commonly used nickel and high nickel alloys are shown in Table .

Titanium and Titanium Alloys

Titanium is a gray, light metal with a better strength-to-weight ratio than any other metal at room temperature, and is used in corrosive environments or in applications that take advantage of its light weight, good strength, and nonmagnetic properties. Titanium is available commercially in many alloys, but multiple requirements can be met by a single grade of the commercially pure metal. The alloys of titanium are of three metallurgical types: alpha, alpha–beta, and beta, with these designations referring to the predominant phases present in the microstructure.

Titanium has a strong affinity for hydrogen, oxygen, and nitrogen gases, which tend to embrittle the material; carbon is another embrittling agent. Titanium is outstanding in its resistance to strongly oxidizing acids, aqueous chloride solutions, moist chlorine gas, sodium hypochlorite, and seawater and brine solutions. Nearly all nonaircraft applications take advantage of this corrosion resistance. Its uses in aircraft engine compressors and in airframe structures are based on both its high corrosion resistance and high strength-to-weight ratio.

Procedures for forming titanium are similar to those for forming stainless steel. Titanium and its alloys can be machined and abrasive ground; however, sharp tools and continuous feed are required to prevent work hardening. Tapping is difficult because the metal galls.

Titanium castings can be produced by investment or graphite mold methods; however, because of the highly reactive nature of the metal in the presence of oxygen, casting must be done in a vacuum.

Generally, titanium is welded by gas-tungsten arc or plasma arc techniques, and the key to successful welding lies in proper cleaning and shielding. The alpha–beta titanium alloys can be heat treated for higher strength, but they are not easily welded. Beta and alpha–beta alloys are designed for formability; they are formed in the soft state, and then heat treated for high strength.

The properties of some wrought titanium alloys are shown in Table 9.

Table 8. Common Cast and Wrought Nickel and High Nickel Alloys — Designations, Compositions, Typical Properties, and Uses

UNS Designation	Description and Common Name	Nominal Composition (Weight %)	Typical Room-Temperature Properties			Form	Typical Uses
			Tensile (ksi)	0.2% Yield (ksi)	Elong. (%)		
N02200	Commercially pure Ni (Nickel 200)	99.5 Ni	67	22	47	Wrought	Food processing and chemical equipment.
N04400	Nickel–copper alloy (Monel 400)	65 Ni, 32 Cu, 2 Fe	79	30	48	Wrought	Valves, pumps, shafts, marine fixtures and fasteners, electrical and petroleum refining equipment.
N05500	Age-hardened Ni–Cu alloy (Monel K 500)	65 Ni, 30 Cu, 2 Fe, 3 Al + Ti	160	111	24	Wrought	Pump shafts, impellers, springs, fasteners, and electronic and oil well components.
N06002	Ni–Cr Alloy (Hastelloy X)	60 Ni, 22 Cr, 19 Fe, 9 Mo, 0.6 W	114	52	43	Wrought	Turbine and furnace parts, petrochemical equipment.
N06003	Ni–Cr alloy (Nichrome V)	80 Ni, 20 Cr	100	60	30	Wrought	Heating elements, resistors, electronic parts.
N06333	Ni–Cr alloy (RA 333)	48 Ni, 25 Cr, 18 Fe, 3 Mo, 3 W, 3 Co	100	50	50	Wrought	Turbine and furnace parts.
N06600	Ni–Cr alloy (Inconel 600)	75 Ni, 15 Cr, 10 Fe	90	36	47	Wrought	Chemical, electronic, food processing and heat treating equipment; nuclear steam generator tubing.
N06625	Ni–Cr alloy (Inconel 625)	61 Ni, 21 Cr, 2 Fe, 9 Mo, 4 Nb	142	86	42	Wrought	Turbine parts, marine and chemical equipment.
N07001	Age-hardened Ni–Cr alloy (Waspalloy)	58 Ni, 20 Cr, 14 Co, 4 Mo, 3 Al, 1.3 Ti, B, Zr	185	115	25	Wrought	Turbine parts.
N07500	Age-hardened Ni–Cr alloy (Udimet 500)	52 Ni, 18 Cr, 19 Co, 4 Mo, 3 Al, 3 Ti, B, Zr	176	110	16	Wrought & Cast	Turbine parts.

Table 8. *(Continued)* **Common Cast and Wrought Nickel and High Nickel Alloys — Designations, Compositions, Typical Properties, and**

UNS Designation	Description and Common Name	Nominal Composition (Weight %)	Typical Room-Temperature Properties Tensile (ksi)	0.2% Yield (ksi)	Elong. (%)	Form	Typical Uses
N07750	Age-hardened Ni–Cr alloy (Inconel X-750)	73 Ni, 16 Cr, 7 Fe, 2.5 Ti, 1 Al, 1 Nb	185	130	20	Wrought	Turbine parts, nuclear reactor springs, bolts, extrusion dies, forming tools.
N08800	Ni–Cr–Fe alloy (Incoloy 800)	32 Ni, 21 Cr, 46 Fe, 0.4 Ti, 0.4 Al	87	42	44	Wrought	Heat exchangers, furnace parts, chemical and power plant piping.
N08825	Ni–Cr–Fe alloy (Incoloy 825)	42 Ni, 22 Cr, 30 Fe, 3 Mo, 2 Cu, 1 Ti, Al	91	35	50	Wrought	Heat treating and chemical handling equipment.
N09901	Age-hardened Ni–Cr–Fe alloy (Incoloy 901)	43 Ni, 12 Cr, 36 Fe, 6 Mo, 3 Ti + Al, B	175	130	14	Wrought	Turbine parts.
N10001	Ni–Mo alloy (Hastelloy B)	67 Ni, 28 Mo, 5 Fe	121	57	63	Wrought	Chemical handling equipment.
N10004	Ni–Cr–Mo alloy (Hastelloy W)	59 Ni, 5 Cr, 25 Mo, 5 Fe, 0.6 V	123	53	55	Wrought	Weld wire for joining dissimilar metals, engine repair and maintenance.
N10276	Ni–Cr–Mo alloy (Hastelloy C-276)	57 Ni, 15 Cr, 16 Mo, 5 Fe, 4 W, 2 Co	116	52	60	Wrought	Chemical handling equipment.
N13100	Ni–Co alloy (IN 100)	60 Ni, 10 Cr, 15 Co, 3 Mo, 5.5 Al, 5 Ti, 1 V. B, Zr	147	123	9	Cast	Turbine parts.

Table 9. Mechanical Properties of Wrought Titanium Alloys

Nominal Composition (%)	Condition	Room Temperature			Reduction in Area (%)
		Tensile Strength (ksi)	Yield Strength (ksi)	Elongation (%)	
Commercially Pure					
99.5 Ti	Annealed	48	35	30	55
99.2 Ti	Annealed	63	50	28	50
99.1 Ti	Annealed	75	65	25	45
99.0 Ti	Annealed	96	85	20	40
99.2 Ti[a]	Annealed	63	50	28	50
98.9[b]	Annealed	75	65	25	42
Alpha Alloys					
5 Al, 2.5 Sn	Annealed	125	117	16	40
5 Al, 2.5 Sn (low O_2)	Annealed	117	108	16	…
Near Alpha Alloys					
8 Al, 1 Mo, 1 V	Duplex annealed	145	138	15	28
11 Sn, 1 Mo, 2.25 Al, 5.0 Zr, 1 Mo, 0.2 Si	Duplex annealed	160	144	15	35
6 Al, 2 Sn, 4 Zr, 2 Mo	Duplex annealed	142	130	15	35
5 Al, 5 Sn, 2 Zr, 2 Mo, 0.25 Si	975°C (1785°F) (½ h), AC 595°C (1100°F)(2 h), AC	152	140	13	…
6 Al, 2 Nb, 1 Ta, 1 Mo	As rolled 2.5 cm (1 in.) plate	124	110	13	34
6 Al, 2 Sn, 1.5 Zr, 1 Mo, 0.35 Bi, 0.1 Si	Beta forge + duplex anneal	147	137	11	…
Alpha–Beta Alloys					
8 Mn	Annealed	137	125	15	32
3 Al, 2.5 V	Annealed	100	85	20	…
6 Al, 4 V	Annealed	144	134	14	30
	Solution + age	170	160	10	25
6 Al, 4 V (low O_2)	Annealed	130	120	15	35
6 Al, 6 V, 2 Sn	Annealed	155	145	14	30
	Solution + age	185	170	10	20
7 Al, 4 Mo	Solution + age	160	150	16	22
6 Al, 2 Sn, 4 Zr, 6 Mo	Solution + age	184	170	10	23
6 Al, 2 Sn, 2 Zr, 2 Mo, 2 Cr, 0.25 Si	Solution + age	185	165	11	33
10 V, 2 Fe, 3 Al	Solution + age	185	174	10	19
Beta Alloys					
13 V, 11 Cr, 3 Al	Solution + age	177	170	8	…
	Solution + age	185	175	8	…
8 Mo, 8 V, 2 Fe, 3 Al	Solution + age	190	180	8	…
3 Al, 8 V, 6 Cr, 4 Mo, 4 Zr	Solution + age	210	200	7	…
	Annealed	128	121	15	…
11.5 Mo, 6 Zr, 4.5 Sn	Solution + age	201	191	11	…

[a] Also contains 0.2 Pd.

[b] Also contains 0.8 Ni and 0.3 Mo.

Source: Titanium Metals Corp. of America and RMI Co.

Copper–Silicon and Copper–Beryllium Alloys

Everdur.—This copper–silicon alloy is available in five slightly different nominal compositions for applications that require high strength, good fabricating and fusing qualities, immunity to rust, free-machining and a corrosion resistance equivalent to copper. The following table gives the nominal compositions and tensile strengths, yield strengths, and per cent elongations for various tempers and forms.

Table 10. Nominal Composition and Properties of Everdur

Desig. No.	Nominal Composition					Temper[a]	Strength		Elongation (%)
	Cu	Si	Mn	Pb	Al		Tensile (ksi)	Yield (ksi)	
655	95.80	3.10	1.10	…	…	A	52	15	35[b]
						HRA	50	18	40
						CRA	52	18	35
						CRHH	71	40	10
						CRH	87	60	3
						H	70 to 85	38 to 50	17 to 8[b]
651	98.25	1.50	0.25	…	…	AP	38	10	35
						HP	50	40	7
						XHB	75 to 85	45 to 55	8 to 6[b]
661	95.60	3.00	1.00	0.40	…	A	52	15	35[b]
						H	85	50	13 to 8[b]
6552	94.90	4.00	1.10	…	…	AC	45	…	15
637	90.75	2.00	…	…	7.25	A	75 to 90	37.5 to 45	12 to 9[b]

[a] Symbols used are: HRA for hot-rolled and annealed tank plates; CRA for cold-rolled sheets and strips; CRHH for cold-rolled half hard strips; abd CRH for cold-rolled hard strips. For round, square, hexagonal, and octagonal rods: A for anealed; H for hard; and XHB for extra-hard bolt temper (in coils for cold-heading). For pipe and tube: AP for annealed; and HP for hard. For castings: AC for as cast.

[b] Per cent elongation in 4 times the diameter or thickness of the specimen. All other values are per cent elongation in 2 inches.

Designation numbers are those of the American Brass Co.

The values given for the tensile srength, yield strength, and elongation are all minimum values. Where ranges are shown, the first values given are for the largest diameter or largest size specimens. Yield strength values were determined at 0.50 per cent elongation under load.

Copper–Beryllium Alloys.—Alloys of copper and beryllium present health hazards. Particles produced by machining may be absorbed into the body through the skin, the mouth, the nose, or an open wound, resulting in a condition requiring immediate medical attention. Working of these alloys requires protective clothing or other shielding in a monitored environment. Copper–beryllium alloys involved in a fire give off profuse toxic fumes that must not be inhaled.

These alloys contain copper, beryllium, cobalt, and silver, and fall into two groups. One group whose beryllium content is greater than one per cent is characterized by its high strength and hardness and the other, whose beryllium content is less than one per cent, by its high electrical and thermal conductivity. The alloys have many applications in the electrical and aircraft industries or wherever strength, corrosion resistance, conductivity, nonmagnetic and nonsparking properties are essential. Beryllium copper is obtainable in the form of strips, rods and bars, wire, platers, bars, billets, tubes, and casting ingots.

Composition and Properties: Table 11 lists some of the more common wrought alloys and gives some of their mechanical properties.

Table 11. Wrought Copper–Beryllium Properties

Alloy[a]	Form	Temper[b]	Heat Treatment	Tensile Strength (ksi)	Yield Strength 0.2% Offset (ksi)	Elongation in 2 in. (%)
25	Rod, Bar, and Plate	A	…	60–85	20–30	35–60
		½ H or H	…	85–130	75–105	10–20
		AT	3 hr at 600°F or mill heat treated	165–190	145–175	3–10
		½ HT or HT	2 hr at 600°F or mill heat treated	175–215	150–200	2–5
	Wire	A	…	58–78	20–35	35–55
		¼ H	…	90–115	70–95	10–35
		½ H	…	110–135	90–110	5–10
		¾ H	…	130–155	110–135	2–8
		AT	3 hr at 600°F	165–190	145–175	3–8
		¼ HT	2 hr at 600°F	175–205	160–190	2–5
		½ HT	2 hr at 600°F	190–215	175–200	1–3
		¾ HT	2 hr at 600°F	195–220	180–205	1–3
		XHT	Mill heat treated	115–165	95–145	2–8
165	Rod, Bar, and Plate	A	…	60–85	20–30	35–60
		½ H or H	…	85–130	75–105	10–20
		AT	3 hr at 650°F or mill heat treated	150–180	125–155	4–10
		½ HT or HT	2 hr at 650°F or mill heat treated	165–200	135–165	2–5
10	Rod, Bar, and Plate	A	…	35–55	20–30	20–35
		½ H or H	…	65–80	55–75	10–15
		AT	3 hr at 900°F or mill heat treated	100–120	80–100	10–25
		½ HT or HT	2 hr at 900°F or mill heat treated	110–130	100–120	8–20
50	Rod, Bar, and Plate	A	…	35–55	20–30	20–35
		½ H or H	…	65–80	55–75	10–15
		AT	3 hr at 900°F or mill heat treated	100–120	80–100	10–25
		½ HT or HT	2 hr at 900°F or mill heat treated	110–130	100–120	8–20
35	Rod, Bar, and Plate	A	…	35–55	20–30	20–35
		½ H or H	…	65–80	55–75	10–15
		AT	3 hr at 900°F or mill heat treated	100–120	80–100	10–25
		½ HT or HT	2 hr at 900°F or mill heat treated	110–130	100–120	8–20

[a] Composition (in per cent) of alloys is asfollows: alloy 25: 1.80–2.05 Be, 0.20–0.35 Co, balance Cu; alloy 165: 1.6–1.8 Be, 0.20–0.35 Co, balance Cu; alloy 10: 0.4–0.7 Be, 2.35–2.70 Co, balance Cu; alloy 50, 0.25–0.50 Be, 1.4–1.7 Co, 0.9–1.1 Ag, balance Cu; alloy 35, 0.25–0.50 Be, 1.4–1.6 Ni, balance Cu.

[b] Temper symbol designations: A, solution annealed; H, hard; HT, heat-treated from hard; At, heat-treated from solution annealed.

PLASTICS

Properties of Plastics

Characteristics of Important Plastics Families

ABS (acrylonitrile-butadiene-styrene)	Rigid, low-cost thermoplastic, easily machined and thermo-formed.
Acetal	Engineering thermoplastic with good strength, wear resistance, and dimensional stability. More dimensionally stable than nylon under wet and humid conditions.
Acrylic	Clear, transparent, strong, break-resistant thermoplastic with excellent chemical resistance and weatherability.
CPVC (chlorinated PVC)	Thermoplastic with properties similar to PVC, but operates to a 40-60°F higher temperature.
Fiberglass	Thermosetting composite with high strength-to-weight ratio, excellent dielectric properties, and unaffected by corrosion.
Nylon	Thermoplastic with excellent impact resistance, ideal for wear applications such as bearings and gears, self-lubricating under some circumstances.
PEEK (polyetherether-ketone)	Engineering thermoplastic, excellent temperature resistance, suitable for continuous use above 500°F, excellent flexural and tensile properties.
PET (polyethylene-terephthalate)	Dimensionally stable thermoplastic with superior machining characteristics compared to acetal.
Phenolic	Thermosetting family of plastics with minimal thermal expansion, high compressive strength, excellent wear and abrasion resistance, and a low coefficient of friction. Used for bearing applications and molded parts.
Polycarbonate	Transparent tough thermoplastic with high impact strength, excellent chemical resistance and electrical properties, and good dimensional stability.
Polypropylene	Good chemical resistance combined with low moisture absorption and excellent electrical properties, retains strength up to 250°F.
Polysulfone	Durable thermoplastic, good electrical properties, operates at temperatures in excess of 300°F.
Polyurethane	Thermoplastic, excellent impact and abrasion resistance, resists sunlight and weathering.
PTFE (polytetrafluoro-ethylene)	Thermoplastic, low coefficient of friction, withstands up to 500°F, inert to chemicals and solvents, self-lubricating with a low thermal-expansion rate.
PVC (polyvinyl chloride)	Thermoplastic, resists corrosive solutions and gases both acid and alkaline, good stiffness.
PVDF (polyvinylidene-fluoride)	Thermoplastic, outstanding chemical resistance, excellent substitute for PVC or polypropylene. Good mechanical strength and dielectric properties.

Plastics Materials.—Plastics materials, often called resins, are made up of many repeating groups of atoms or molecules linked in long chains (called polymers) that combine such elements as oxygen, hydrogen, nitrogen, carbon, silicon, fluorine, and sulfur. Both the lengths of the chains and the mechanisms that bond the links of the chains together are related directly to the mechanical and physical properties of the materials. There are two main groups: thermoplastics and thermosets.

Thermoplastic materials become soft and moldable when heated, and change back to solids when allowed to cool. Examples of thermoplastics are acetal, acrylic, cellulose acetate, nylon, polyethylene, polystyrene, vinyl, and nylon. Thermoplastic materials that are flexible even when cool are known as thermoplastic elastomers or TPEs. When thermoplastic materials are heated, the linked chains of molecules can move relative to each other, allowing the mass to flow into a different shape. Cooling prevents further flow. Although the heating/cooling cycle can be repeated, recycling reduces mechanical properties and appearance.

Thermoset plastics such as amino, epoxy, phenolic, and unsaturated polyesters, are so named because they are changed chemically during processing and become hard solids. Although the structures of thermoset materials are similar to those of thermoplastic materials, processing develops cross-links between adjacent molecules, forming complex networks that prevent relative movement between the chains at any temperature. Many rubbers that are processed by vulcanizing, such as butyl, latex, neoprene, nitrile, polyurethane, and silicone, also are classified as thermosets. Heating a thermoset degrades the material so that it cannot be reprocessed satisfactorily.

Elastomers are flexible materials that can be stretched up to about double their length at room temperature and can return to their original length when released. Thermoplastic elastomers are often used in place of rubber, and may also be used as additives to improve the impact strength of rigid thermoplastics.

Structures.—Thermoplastics can be classified by their structures into categories such as amorphous (noncrystalline), crystalline, and liquid crystalline polymers (LCP). Amorphous thermoplastics include polycarbonate, polystyrene, ABS (acrylonitrile-butadiene-styrene), SAN (styrene-acrylonitrile), and PVC (polyvinylchloride). Crystalline thermoplastics have polymer chains that are packed together in an organized way, as distinct from the unorganized structures of amorphous plastics, and include acetal, nylon, polyethylene, polypropylene, and polyester. The organized regions in crystalline thermoplastics are joined by noncrystalline (amorphous) zones, and the structures are such that the materials are stronger and stiffer, though less resistant to impact, than completely noncrystalline materials. Crystalline thermoplastics have higher melting temperatures and higher shrinkage and warpage factors than amorphous plastics. Liquid crystalline plastics are polymers with highly ordered rod-like structures and have high mechanical property values, good dimensional stability and chemical resistance, and are easy to process, with melting temperatures similar to those of crystalline plastics. Unlike amorphous and crystalline plastics, liquid crystalline plastics retain significant order in the melt phase. As a result, they have the lowest shrinkage and warpage of the three types of thermoplastics.

Mixtures.—Characteristics of plastics materials can be changed by mixing or combining different types of polymers and by adding nonplastics materials. Particulate fillers such as wood flour, silica, sand, ceramic and carbon powder, tiny glass balls, and powdered metal are added to increase modulus and electrical conductivity, to improve resistance to heat or ultraviolet light, and to reduce cost, for example. Plasticizers may be added to decrease modulus and increase flexibility. Other additives may be used to increase resistance to effects of ultraviolet light and heat or to prevent oxidation, and for a variety of other purposes.

Reinforcing fibers of glass, carbon, or Aramid (aromatic polyamide fibers having high tensile strength, a range of moduli, good toughness, and stress–strain behavior similar to that of metals) are added to improve mechanical properties. Careful design and process selection must be used to position the fibers so that they will provide the required strength where it is needed. Continuous fiber may be positioned carefully in either a thermoplastics or thermoset matrix to produce basic parts generally called composites, which have the highest mechanical properties and cost of the reinforced plastics.

Copolymers embody two or more different polymers and may have properties that are completely different from those of the individual polymers (homopolymers) from which they are made. An approach known as alloying consists of pure mechanical blending of two or more different polymers, often with special additives to make them compatible. These "alloys" are compounded so as to retain the most desirable characteristics of each constituent, especially in impact strength and flame resistance. However, properties usually are intermediate between those of the constituent materials.

Physical Properties.—Almost all proposed uses of plastics require some knowledge of the physical properties of the materials, and this information is generally readily available from manufacturers. Properties such as density, ductility, elasticity and plasticity, homogeneity, uniformity of composition, shrinkage during cooling from the molding temperature, transmittal of light, toughness (resistance to impact), brittleness, notch sensitivity, isotropy (properties that are the same when measured in any direction) and anisotropy (properties that vary when measured in different directions), and lubricity (load-bearing characteristics under relative motion) may all need consideration when a material suitable for a specific application is to be specified.

Most of the terms used to describe the physical characteristics of metals, such as density, ductility, brittleness, elasticity, notch sensitivity, specific gravity, and toughness, have similar meanings when they are applied to plastics, but different measures are often used with plastics. Like cast metals, many plastics are isotropic so that their characteristics are the same measured in any direction. Properties of rolled metals and extruded plastics vary when measured in the longitudinal and transverse directions, so these materials are anisotropic.

Density is a measure of the mass per unit volume, usually expressed in lb/in.^{3} or g/cm^3 at a temperature of 73.4 degrees F (23 degrees C). Density information is used mainly to calculate the amount of material required to make a part of a given volume, the volume being calculated from drawing dimensions.

Specific gravity is the ratio of the mass of a given volume of a material to the mass of the same volume of water, both measured at 73.4 degrees F (23 degrees C). The ratio is dimensionless so is useful for comparing different materials, and is used in cost estimating and quality control.

Shrinkage is the ratio of the dimension of the plastics molding to the corresponding dimension of the mold, expressed in in./in. or cm/cm, both at room temperature. As with a die casting die, the moldmaker uses this ratio to determine mold cavity measurements that will produce a part of the required dimensions. Shrinkage in a given material can vary with wall thickness, direction of flow of the plastics in the mold, and molding conditions. Amorphous and liquid crystalline thermoplastics have lower shrinkage ratios than crystalline thermoplastics. Glass-reinforced and filled materials have lower shrinkage than unfilled materials.

Water absorption is the amount of increase in weight of a material due to absorption of water, expressed as a percentage of the original weight. Standard test specimens are first dried for 24 hr, then weighed before and after immersion in water at 73.4 degrees F (23 degrees C) for various lengths of time. Water absorption affects both mechanical and electrical properties and part dimensions. Parts made from materials with low water absorption rates tend to have greater dimensional stability.

Opacity (or transparency) is a measure of the amount of light transmitted through a given material under specific conditions. Measures are expressed in terms of haze and luminous transmittance. Haze measurements indicate the percentage of light transmitted through a test specimen that is scattered more than 2.5 degrees from the incident beam. Luminous transmittance is the ratio of transmitted light to incident light.

Elasticity is the ability of a material to return to its original size and shape after being deformed. Most plastics have limited elasticity, although rubber and materials classified as thermoplastic elastomers (TPEs) have excellent elasticity.

Plasticity is the inverse of elasticity, and a material that tends to stay in the shape or size to which it has been deformed has high plasticity. Some plastics can be formed cold by being stressed beyond the yield point and such plastics then exhibit plasticity. When thermoplastics are heated to their softening temperature, they have almost perfect plasticity.

Ductility is the ability of a material to be stretched, pulled, or rolled into shape without destroying the integrity of the material.

Toughness is a measure of the ability of a material to absorb mechanical energy without cracking or breaking. Tough material can absorb mechanical energy with either elastic or plastic deformation. High-impact unfilled plastics generally have excellent toughness, and low- or moderate-impact materials may also be tough if their ultimate strength is high enough (see *Typical Stress–Strain Curves* on page 573). The area under the stress–strain curve is often used as the measure of toughness for a particular plastics material.

Brittleness is the lack of toughness. Brittle plastics frequently have low impact and high stiffness properties. Many glass-reinforced and mineral-filled materials are brittle.

Notch sensitivity is a measure of the ease with which a crack progresses through a material from an existing notch, crack, or sharp corner.

Lubricity describes the load-bearing characteristics of a material under relative motion. Plastics with good lubricity have low coefficients of friction with other materials (or sometimes with themselves) and no tendency to gall.

Homogeneous means uniform. The degree of homogeneity indicates the uniformity of composition of a material throughout its mass. In a completely homogeneous body, the smallest sample has the same physical properties as the body. An unfilled thermoplastics is a reasonably homogeneous material.

Heterogeneous means varying. In a heterogeneous body, for example, a glass-reinforced material, the composition varies from point to point. Many heterogeneous materials are treated as homogeneous for design purposes because a small sample of the material has the same properties as the body.

Isotropy means that the properties at any point in a body are the same, regardless of the direction in which they are measured.

Anisotropy means that the physical properties of a material depend on the direction of measurement. Various degrees of anisotropy exist, depending on the amount of symmetry of the material or component shape. For example, cast metals and plastics tend to be isotropic so that samples cut in any direction within a cast body tend to have the same physical properties. However, rolled metals tend to develop crystal orientation in the direction of rolling so that they have different mechanical properties in the rolling and transverse-to-rolling directions.

Extruded plastics film also may have different properties in the extruding and transverse directions so that these materials are oriented biaxially and are anisotropic. Composite materials that have fiber reinforcements carefully oriented in the direction of applied loads, surrounded by a plastics matrix, have a high degree of property orientation with direction at various points in the structure and are anisotropic.

As another example, wood page 375 is an anisotropic material with distinct properties in three directions and is very stiff and strong in the direction of growth. Fair properties are also found in one direction perpendicular to the growth direction, but in a third direction at right angles to the other two directions, the mechanical properties are much lower.

The preceding examples involve mechanical properties, but anisotropy is also used in referring to the way a material shrinks in the mold. Anisotropic shrinkage is important in molding crystalline and glass-fiber-reinforced materials for which shrinkage values are usually listed for the flow direction and the cross-flow direction. These values are of most concern to the tool designer and molder, but the existence of anisotropy and its severity must be considered when a material is chosen for a part having tight tolerances.

Significance of Elasticity, Homogeneity, and Isotropy: Structural analysis during design of components uses two independent constants, Young's modulus (E) and Poisson's ratio (ν), but two constants are sufficient only for elastic, isotropic materials that respond linearly to loads (when load is proportional to deformation). Designers often use the same values for these constants everywhere in the structure, which is correct only if the structure is homogeneous.

Assumptions of linear elasticity, isotropy, and homogeneity are reasonable for many analyses and are a good starting point, but use of these assumptions can lead to significant design errors with plastics, particularly with glass-reinforced and liquid crystalline polymers, which are highly anisotropic. In the following, plastics are assumed to be linearly elastic, homogeneous, and isotropic to allow a simpler presentation of mechanical properties in line with the data provided in plastics manufacturers' marketing data sheets. The standard equations of structural analysis (bending, torsion, pressure in a pipe, etc.) also require these assumptions.

As the degree of anisotropy increases, the number of constants or moduli required to describe the material also increases, up to a maximum of 21. Uncertainty about material properties and the questionable applicability of the simple analysis techniques employed point to the need for extensive end-use testing of plastics parts before approval of a particular application. A partial solution to this problem lies in the use of finite-element-analysis (FEA) methods. The applicability of FEA methods requires good understanding of the anisotropic nature of plastics materials.

Mechanical Properties.— Almost all end-use applications involve some degree of loading, so mechanical properties are of prime importance in designing with plastics. Material selection is usually based on manufacturers' marketing data sheets listing tensile strength, modulus of elasticity (E), elongation, impact strength, stress and strain behavior, and shear strength. Suppliers' data often are generated under standard test conditions so may not be directly transferable to the components produced. Because of the somewhat lower modulus of elasticity of plastics materials (10^5 for plastics compared with 10^6 lb_f/in^2 for metals), different units of measure are used to express the results.

Determination of the true meaning of mechanical properties and their relation to end-use requirements is of vital importance in design. In practical applications, materials are seldom subjected to steady deformation without the influence of other factors such as environment and temperature. A thorough understanding of mechanical properties and tests used to determine such properties, and the effects of adverse or beneficial conditions on mechanical properties over long time periods, is extremely important. Some manufacturers offer design and technical advice to customers who do not possess this understanding.

Stress: A three-dimensional body having a balanced system of external forces F_1 through F_5 acting on it, such that the body is at rest, is shown in Fig. 1. Such a body develops internal forces to transfer and distribute the external loads. If the body is cut at an arbitrary cross-section and one part is removed, as shown at the right in Fig. 1, a new system of forces acting on the cut surface is developed to balance the remaining external forces. Similar forces (stresses) exist within the uncut body.

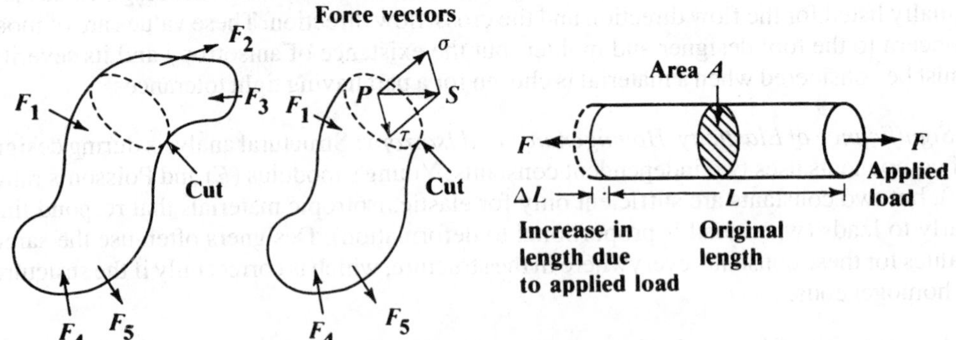

Fig. 1. Internal Forces and Stresses in a Body

Fig. 2. Simple Tension Load

Stresses must be defined with both magnitude and direction. The stress *S* acting in the direction shown in Fig. 1, on the point *P* of the cut surface, has two stress components. One of these components, σ, acts perpendicular to the surface and is called a normal or direct stress. The other stress, τ, acts parallel to the surface and is called a shear stress.

Normal stress is illustrated by the simple tension test shown in Fig. 2, where the direct stress is the ratio of applied load to the original cross-sectional area in lb_f/in^2. In the Système International (SI or metric system, see pages 2520, starting on page 2532, and 2548) the stress, σ, is expressed in newtons/meter² (N/m^2).

$$\text{Stress} = \frac{\text{Load}}{\text{Area}} \text{ or } \sigma = \frac{F}{A} \tag{1}$$

If the load is applied as shown in Fig. 2, the test piece is in tension, and if reversed, it is in compression.

Normal strain is also illustrated by the diagram in Fig. 2, where the load or stress applied to the test piece causes it to change its length. If the bar has an original length *L*, and changes its length by Δ*L*, the strain, ∈, is defined as

$$\text{Strain} = \frac{\text{Change of Length}}{\text{Original Length}} \text{ or } \in = \frac{\Delta L}{L} \tag{2}$$

Strain is the ratio between the amount of deformation of the material and its original length and is a dimensionless quantity. Extensions of most materials under load are generally very small. Strain (μ∈ or microstrain in most metals) is measured and expressed in microinches (millionths of an inch) per inch, or 10^{-6} in./in. (10^{-6} cm/cm). Alternatively, strain is expressed as a percentage. The three methods compare as follows:

$$1000\mu\in = 0.001 = 0.1 \text{ per cent strain}$$

$$10000\mu\in = 0.010 = 1 \text{ per cent strain}$$

Modulus of Elasticity: Most metals and plastics have deformations that are proportional to the imposed loads over a range of loads. Stress is proportional to load and strain is proportional to deformation, so stress is proportional to strain and is expressed by Hooke's law:

$$\frac{\text{Stress}}{\text{Strain}} = \text{Constant} = E \tag{3}$$

The constant *E* is called the modulus of elasticity, Young's modulus, or, in the plastics industry, tensile modulus. Referring to Fig. 2, tensile modulus is given by the formula:

$$E = \frac{\sigma}{\in} = \frac{F/A}{\Delta L/L} = \frac{FL}{A\Delta L} \tag{4}$$

Thus, the modulus is the slope of the initial portion of the stress–strain curve. An elastic material does not necessarily obey Hooke's law, since it is possible for a material to return to its original shape without the stress being proportional to the strain. If a material does obey Hooke's law, however, it is elastic.

The straight portion of the stress–strain curve for many plastics is difficult to locate, and it is necessary to construct a straight line tangent to the initial portion of the curve to use as a modulus. The shape of a line so obtained is called the initial modulus. In some plastics, the initial modulus can be misleading, owing to the nonlinear elasticity of the material. Some suppliers therefore provide the so-called 1 per cent secant modulus, which is the ratio of stress to strain at 1 per cent strain on the stress–strain curve. In the illustration of typical stress–strain curves in Fig. 3, the secant modulus at the point *E* is the slope of the line *OE*.

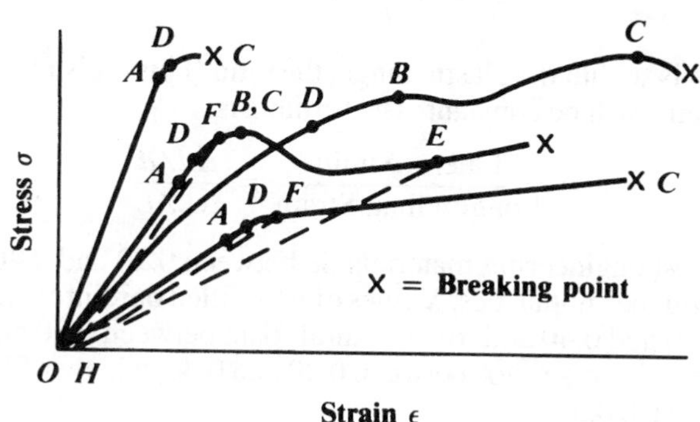

Fig. 3. Typical Stress–Strain Curves

For metals, Young's modulus is expressed in terms of 10^6 $lb_f/in.^2$, N/m^2, or Pa, as convenient (see starting on page 2525). For plastics, tensile modulus is expressed as 10^5 lb_f/in^2 or in GPa (1 GPa = 145,000 $lb_f/in.^2$).

Secant modulus is the ratio of stress to corresponding strain at any point on the stress–strain curve (see *Modulus of Elasticity*).

Proportional limit is the greatest stress at which a material is capable of sustaining the applied load without losing the proportionality of stress to strain. This limit is the point on the stress–strain curve where the slope begins to change, as shown at A on each of the curves in Fig. 3. Proportional limit is expressed in $lb_f/in.^2$ (MPa or GPa).

Yield point is the first point on the stress–strain curve where an increase in strain occurs without an increase in stress, and is indicated by *B* on some of the curves in Fig. 3. The slope of the curve is zero at this point; however, some materials do not have a yield point.

Ultimate strength is the maximum stress a material withstands when subjected to a load, and is indicated by *C* in Fig. 3. Ultimate strength is expressed in $lb_f/in.^2$ (MPa or GPa).

Elastic limit is indicated by the point *D* on the stress–strain curve in Fig. 3, and is the level beyond which the material is permanently deformed when the load is removed.Although many materials can be loaded beyond their proportional limit and still return to zero strain when the load is removed, some plastics have no proportional limit in that no region exists where the stress is proportional to strain (i.e., where the material obeys Hooke's law).

Yield strength is the stress at which a material shows a specified deviation from stress to strain proportionality. Some materials do not show a yield strength clearly, and it may be desirable to choose an arbitrary stress level beyond the elastic limit, especially with plastics that have a very high strain at the yield point, to establish a realistic yield strength. Such a point is seen at F on some of the curves in Fig. 3, and is defined by constructing a line parallel to OA at a specified offset strain, H. The stress at the intersection of the line with the stress–strain curve at F would be the yield strength at H offset. If H were at 2 per cent strain, F would be described as the yield strength at a 2 per cent strain offset.

Poisson's ratio is defined on page 196. Under a tensile load, a rectangular bar of length L, with sides of widths b and d, lengthens by an amount ΔL, producing a longitudinal strain of

$$\epsilon = \frac{\Delta L}{L} \tag{5}$$

The bar is reduced in its lateral dimensions and the associated lateral strains will be opposite in sign, resulting in

$$\epsilon = -\frac{\Delta b}{b} = -\frac{\Delta d}{d} \tag{6}$$

If the deformation is within the elastic range, the ratio (Poisson's ratio ν) of the lateral to the longitudinal strains will be constant. The formula is:

$$\nu = \frac{\text{Lateral Strain}}{\text{Longitudinal Strain}} = \frac{\Delta d/d}{\Delta L/L} \tag{7}$$

Values of ν for most engineering materials lie between 0.20 and 0.40, and these values hold for unfilled rigid thermoplastics. Values of ν for filled or reinforced rigid thermoplastics fall between 0.10 and 0.40 and for structural foam between 0.30 and 0.40. Rigid thermoset plastics have Poisson's ratios between 0.20 and 0.40, whether filled or unfilled, and elastomers can approach 0.5.

Shear stress is treated on page 207. Any block of material is subject to a set of equal and opposite shearing forces Q. If the block is envisaged as an infinite number of infinitesimally thin layers as shown diagrammatically in Fig. 4, it is easy to imagine a tendency for one layer subject to a force to slide over the next layer, producing a shear form of deformation or failure. The shear stress τ is defined as

$$\tau = \frac{\text{Shear Load}}{\text{Area Resisting Load}} = \frac{Q}{A} \tag{8}$$

Shear stress is always tangential to the area on which it acts. Shearing strain is the angle of deformation γ and is measured in radians.

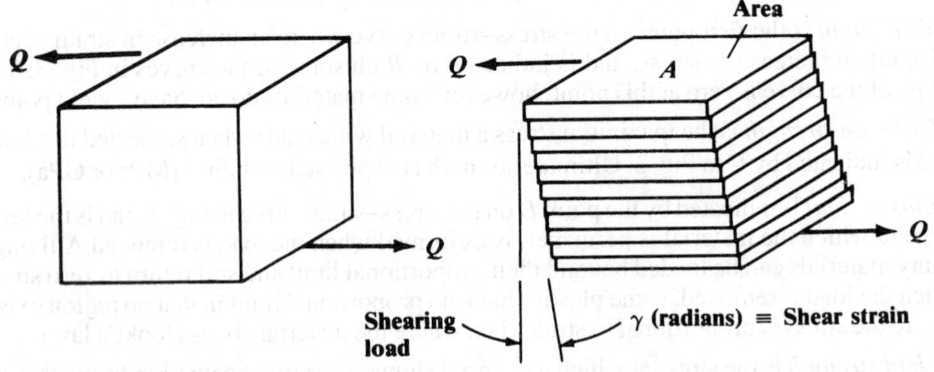

Fig. 4. Shear Stress is Visualized as a Force Q Causing Infinitely Thin Layers of a Component to Slide Past Each Other, Producing a Shear Form of Failure

Shear modulus is a constant G, otherwise called the modulus of rigidity, and, for materials that behave according to Hooke's law, is directly comparable to the modulus of elasticity used in direct stress calculations. The constant is derived from

$$G = \frac{\text{Shear Stress}}{\text{Shear Strain}} = \frac{\tau}{\gamma} \qquad (9)$$

Relating Material Constants: Although only two material constants are required to characterize a material that is linearly elastic, homogeneous, and isotropic, three such constants have been introduced here. These three constants are tensile modulus E, Poisson's ratio ν, and shear modulus G, and they are related by the following equation, based on elasticity principles:

$$\frac{E}{G} = 2(1 + \nu) \qquad (10)$$

This relationship holds for most metals and is generally applicable to injection-moldable thermoplastics. It must be remembered, however, that most plastics, and particularly fiber-reinforced and liquid crystalline materials, are inherently either nonlinear, or anisotropic, or both.

Direct shear refers to a shear strength test much used in the plastics industry with a setup similar to that shown in Fig. 5, and the results of such tests are often described in manufacturers' marketing data sheets as the shear strength of the material. The shear strength reported from such a test is not a pure shear strength because a considerable part of the load is transferred by bending or compressing, or both, rather than by pure shear, and results can be affected by the susceptibility of the material to the sharpness of the load faces in the test apparatus. Thus, the test cannot be used to develop shear stress–strain curves or to determine the shear modulus.

When analyzing plastics in a pure shear situation or when the maximum shear stress is calculated in a complex stress environment, designers often use a shear strength value of about half the tensile strength, or the direct shear strength obtained from the test referred to above, whichever is least.

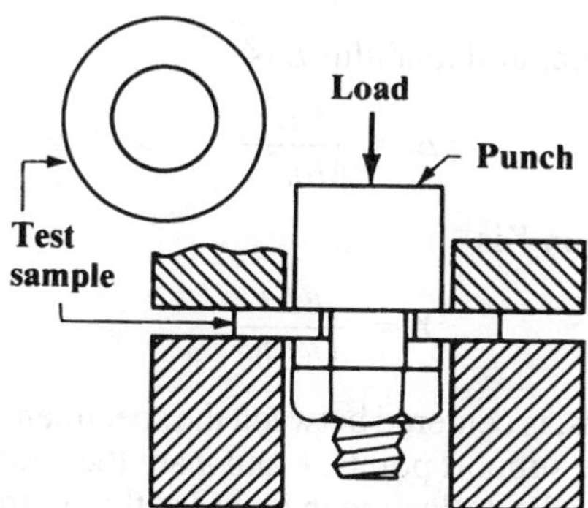

Fig. 5. Direct Shear Test Used in the Plastics Industry

True stress and true strain are terms not in frequent use. In Fig. 2, the stress, sometimes called the engineering stress, is calculated from an increasing load F, acting over a constant area A. Because the cross-sectional area is reduced with most materials, use of that smaller cross-sectional area in the calculation yields what is called the "true stress." In addition, the direct strain referred to earlier, that is, the total change in length divided by the original length, is often called the "engineering strain." The true strain would be the instantaneous deformation divided by the instantaneous length. Therefore, the shape of such a stress–strain curve would not be the same as a simple stress–strain curve. Modulus values and stress–strain curves are almost universally based on engineering stress and strain.

Other Measures of Strength and Modulus.—Tensile and compression properties of many engineering materials, which are treated as linearly elastic, homogeneous, and isotropic, are often considered to be identical, so as to eliminate the need to measure properties in compression. Also, if tension and compression properties are identical, under standard beam bending theory, there is no need to measure the properties in bending. In a concession to the nonlinear, anisotropic nature of most plastics, these properties, particularly flexural properties, are often reported on manufacturers' marketing data sheets.

Compression Strength and Modulus: Because of the relative simplicity of testing in tension the elastic modulus of a material is usually measured and reported as a tension value. For design purposes it often is necessary to know the stress–strain relationship for compression loading. With most elastic materials at low stress levels, the tensile and compressive stress–strain curves are nearly equivalent. At higher stress levels, the compressive strain is less than the tensile strain. Unlike tensile loading, which usually results in a clear-cut failure, stressing in compression produces a slow and indefinite yielding that seldom leads to failure. Because of this phenomenon, compressive strength is customarily expressed as the stress in $lb_f/in.^2$ (Pa) required to deform a standard plastics test specimen to a certain strain. Compression modulus is not always reported, because defining a stress at a given strain is equivalent to reporting a secant modulus. If a compression modulus is given, it is usually an initial modulus.

Bending Strength and Modulus: When material of rectangular cross-section is bent, it is apparent that one surface is stretched in tension and the other side is compressed. Within the material is a line or plane of zero stress called the neutral axis. Simple beam bending theory makes the following assumptions: that the beam is initially straight, unstressed, and symmetric; the material is linearly elastic, homogeneous, and isotropic; the proportional limit is not exceeded; Young's modulus for the material is the same in tension and compression; and all deflections are small so that planar cross-sections remain planar before and after bending.

In these conditions, the formula for bending stress σ is

$$\sigma = \frac{3FL}{2bh^2} \tag{11}$$

the formula for bending or flexural modulus E is

$$E = \frac{FL^3}{4bh^3Y} \tag{12}$$

and the formula for deflection Y is

$$Y = \frac{FL^3}{4Ebh^3} \tag{13}$$

where F, the force in pounds, is centered between the specimen support points, L is the distance in inches between the support points, b and h are the width and thickness of the test specimen in inches, and Y is the deflection in inches at the central load point.

Using the preceding relationships, the flexural strength and flexural modulus (of elasticity) for any material can be determined in the laboratory. The flexural modulus reported is usually the initial modulus from the load deflection curve. Most plastics parts must be analyzed in bending, so use of flexural values should give more accurate results than corresponding tensile values.

Rate Dependence of Mechanical Properties: Tensile and flexural data in manufacturers' literature are measured at specific displacement rates. These rates are usually not consistent with the loading environment encountered in use of the product. The same plastics material, under differing rates or in other environmental conditions, can produce different stress–strain curves. Designers should be aware of the loading rates in specific applications and request the appropriate data. End-use testing must always be considered, but particularly when adequate data are not available.

Time-Related Mechanical Properties.—Mechanical properties discussed previously were related to loads applied gradually and applied for short periods. Long-term and very short-term loading may give somewhat different results. With high-performance thermoplastics it is important to consider creep, impact, fatigue, and related issues. Even the best laboratory test methods do not always predict structural response of production parts accurately, and other factors may also affect results.

Creep is defined as increasing strain over time in the presence of a constant stress when deformation continues without increases in load or stress. The rate of creep for a given material depends on applied stress, temperature, and time.

Creep behavior of a material is important, and a crucial issue with plastics, where parts are to be subjected to loads for extended periods and where the maximum deflection is critical. To determine the creep behavior, test samples may be loaded in tension, compression, or flexure in a constant-temperature environment. Under constant loads, deflection is recorded at regular intervals over suitable periods. Results are generally obtained for four or more stress levels and recorded as creep curves of strain versus time on a logarithmic scale. In general, crystalline materials have lower creep rates than amorphous plastics. Glass reinforcement generally improves the creep resistance.

Apparent or Creep Modulus: If the deflection of a part subjected to continuous loading is calculated by using the modulus of elasticity E, results are likely to be inaccurate because the effects of creep have not been considered. If the stress level and temperature are known and creep curves are available for the temperature in question, an apparent or creep modulus E_{app} can be calculated from the creep curves by the formula: $E_{app} = \sigma/\epsilon_c$, where σ is the calculated stress level and ϵ_c is the strain from the creep curve at the expected time and temperature.

This value E_{app} can be used instead of E in predicting the maximum deflection, using the methods described subsequently (page 577).

Manufacturers' data often include curves of creep modulus (or log creep modulus) versus log time at either constant stress or constant strain, derived from creep data. This information may also be provided as tables of values at constant stress and temperature for various time periods. Some manufacturers provide creep data in the form of creep modulus figures rather than curves.

Creep rupture data are obtained in the same manner as creep data except that higher stresses are used and time is measured to failure. Such failures may be brittle or ductile with some degree of necking. Results are generally plotted as log stress versus log time to failure.

Stress relaxation occurs when plastics parts are assembled into a permanent deflected condition, as in a press fit, a bolted assembly, or some plastics springs. Under constant strain over a period of time, the stress level decreases due to the same internal molecular movement that produces creep. Stress relaxation is important with such applications as bolt preloading and springs, where loading must be maintained.The relaxation can be assessed by applying a fixed strain to a sample and measuring the load over time. A relaxation modulus similar to the creep modulus can be derived from the relaxation data. Relaxation data are not as readily available as creep data, but the decrease in load due to stress relaxation can be approximated by using the creep modulus E_{app} calculated from the creep curves.

Plastics parts often fail due to imposition of excessive fixed strains over extended periods of time, for example, a plastics tube that is a press fit over a steel shaft. No relaxation rupture equivalent to creep rupture exists, so for initial design purposes a strain limit of 20 per cent of the strain at the yield point or yield strength is suggested for high-elongation plastics. For low-elongation brittle plastics that have no yield point, 20 per cent of the elongation at break is also recommended. These figures should be regarded only as guidelines for development of initial design concepts; prototype parts should be thoroughly tested under end-use conditions to confirm the suitability of the design. Higher or lower property limits may also be indicated in manufacturers' data on specific materials.

Extrapolating creep and relaxation data must be done with caution. When creep and relaxation data are plotted as log property against log time, the curves are generally less pronounced, facilitating extrapolation. This procedure is common practice, particularly with creep modulus and creep rupture data. Extrapolation should not exceed one unit of log time, and the strain limit of 20 per cent of the yield or ultimate strength mentioned above should not be exceeded.

Impact loading describes a situation in which a load is imposed rapidly. Any moving body has kinetic energy and when the motion is stopped by a collision, the energy is dissipated. Ability of a plastics part to absorb energy is determined by the shape, size, thickness, and type of material. Impact testing methods now available do not provide designers with information that can be used analytically. The tests can be used for comparing relative notch sensitivity or relative impact resistance, so can be useful in choosing a series of materials to be evaluated for an application or in grading materials within a series.

Impact testing by the Izod and Charpy methods, in which a pendulum arm is swung from a certain height to impact a notched test specimen, is the most widely used for measuring impact strength. Impact with the test specimen reduces the energy remaining in the arm, and this energy loss is recorded in ft-lb (J). The value of such tests is that they permit comparison of the relative notch toughness of two or more materials under specific conditions.

Tensile impact tests mount the test specimen on the swinging arm. Attached to the test specimen is a cross piece that is arrested by a notched anvil as the bar swings down, allowing the energy stored in the arm to break the specimen under tension as it passes through the notch. Another impact test used for plastics allows a weighted, round-ended cylindrical "dart" to fall on a flat disk of the plastics to be tested. This test is good for ranking materials because it represents conditions that are encountered by actual parts in certain applications.

Fatigue tests are designed to measure the relative ability of plastics materials to withstand repeated stresses or other cyclic phenomena. For example, a snap-action, or snap-fit latch that is continually opened and closed, a gear tooth, a bearing, a structural component subject to vibration or to repeated impacts. Cyclic loading can cause mechanical deterioration and progressive fracture, leading to failure in service. Typical fatigue tests are carried out on machines designed to subject a cantilever test piece to reversing flexural loading cycles at different stress levels. Numbers of cycles before failure are recorded for each stress level. Data are normally presented in plots of log stress versus log cycles called *S-N* curves for specific cycle rates and environmental temperatures. With thermoplastics materials there is the added complication that heat built up by the frequency of the cyclic stress may contribute to failure. Significantly different *S-N* curves can be produced for the same materials by testing at different frequencies, mean stresses, waveforms, and methods, such as testing in tension rather than in bending. Testing usually cannot reproduce the conditions under which components will work. Only tests on the end product can determine whether the design is suitable for the purpose to be served.

Thermal Properties.—Melting temperatures of crystalline thermoplastics are sharp and clearly defined, but amorphous and liquid crystalline materials soften and become more fluid over wider temperature ranges. Melting points have greater significance in molding and assembly operations than in product design, which usually deals with the product's temperatures.

Glass transition temperature is a level at which a plastics material undergoes a significant change in properties. Below this temperature T_g, the material has a stiff, glassy, brittle response to loads. Above T_g the material has a more ductile, rubbery response.

Vicat softening point is the temperature at which a small, circular, lightly gravity-loaded, heated probe penetrates a specific distance into a thermoplastics test specimen. This test measures the ability of a thermoplastics material to withstand a short-term contact with a heated surface, and is most useful for crystalline plastics. Amorphous thermoplastics materials tend to creep during the test, which reduces its usefulness for such materials.

Deflection temperature under load (DTUL) is the temperature at which a test bar of 0.5 in. thickness, loaded to a specified bending stress, will deflect by 0.010 in. This test is run at bending stresses of 66 $lb_f/in.^2$ or 264 $lb_f/in.^2$ or both. The value obtained is sometimes referred to as the heat distortion temperature (HDT), and is an indication of the ability of the material to perform at elevated temperatures under load. Both stress and deflection for a specific design of test bar are given so the test may be regarded as establishing the temperature at which the flexural modulus is reduced to particular values, 35,200 $lb_f/in.^2$ at 66 $lb_f/in.^2$ stress, and 140,000 $lb_f/in.^2$ at 264 $lb_f/in.^2$ stress.

Table 1. Typical Values of Coefficient of Linear Thermal Expansion for Thermoplastics and Other Commonly Used Materials

Material[a]	in./in./deg F $\times 10^{-5}$	cm/cm/deg C $\times 10^{-5}$	Material[b]	in./in./deg F $\times 10^{-5}$	cm/cm/deg C $\times 10^{-5}$
Liquid Crystal—GR	0.3	0.6	ABS—GR	1.7	3.1
Glass	0.4	0.7	Polypropylene—GR	1.8	3.2
Steel	0.6	1.1	Epoxy—GR	2.0	3.6
Concrete	0.8	1.4	Polyphenylene sulfide—GR	2.0	3.6
Copper	0.9	1.6	Acetal—GR	2.2	4.0
Bronze	1.0	1.8	Epoxy	3.0	5.4
Brass	1.0	1.8	Polycarbonate	3.6	6.5
Aluminum	1.2	2.2	Acrylic	3.8	6.8
Polycarbonate—GR	1.2	2.2	ABS	4.0	7.2
Nylon—GR	1.3	2.3	Nylon	4.5	8.1
TP polyester—GR	1.4	2.5	Acetal	4.8	8.5
Magnesium	1.4	2.5	Polypropylene	4.8	8.6
Zinc	1.7	3.1	TP Polyester	6.9	12.4
ABS—GR	1.7	3.1	Polyethylene	7.2	13.0

[a] GR = Typical glass fiber-reinforced material. Other plastics materials shown are unfilled.
[b] GR = Typical glass fiber-reinforced material. Other plastics materials shown are unfilled.

Linear thermal expansion: Like metals, thermoplastic materials expand when heated and contract when cooled. For a given temperature range, most plastics change dimensions much more than metals. The coefficient of linear thermal expansion (CLTE) is the ratio of the change in a linear dimension to the original dimension for a unit change of temperature and is expressed as in./in./degree F, or cm/cm/degree C. Typical average values for common materials are shown in Table 1. These values do not take account of grades, molding conditions, wall thickness, or direction of flow in molding.

Thermal conductivity is the rate at which a material conducts heat energy along its length or through its thickness.

Aging at elevated temperatures may affect physical, mechanical, electrical, or thermal properties of plastics materials. Data from tests on specimens stored at specific temperatures for suitable periods are presented as plots of properties versus aging time at various temperatures, and may be used as an indication of thermal stability of the material.

Temperature index is a rating by Underwriters Laboratories (UL) of electrical and mechanical properties (with and without effects of impacts) of plastics materials used in electrical equipment for certain continuous operating conditions.

Flammability ratings also are produced by Underwriters Laboratories. UL tests measure the ability to continue burning after a flame is removed, and the percentage of oxygen needed for the material to continue burning. Other tests measure combustibility, ignition temperatures, and smoke generation.

Effect of Temperature on Mechanical Properties.—The inverse relationship between strain rate and temperature must be kept in mind when designing with plastics materials. Stress/strain curves for tests performed with one strain rate at several temperatures are similar to those for tests with one temperature and several strain rates. Therefore, very high strain rates and very low temperatures produce similar responses in materials. Conversely, the effects of very low strain rates, that is, creep effects, can be determined more quickly by testing at elevated temperatures. Testing at temperatures near or above the highest values expected in everyday use of a product helps the designer estimate long-term performance of components.

Strength, modulus, and elongation behavior are similar for tensile, compressive, flexural, and shear properties. Generally, strength and modulus decrease with increasing temperature. The effect of temperature increases is shown by the curves in Fig. 6 for crystalline and amorphous materials, where a gradual drop in modulus is seen as the glass transition temperature T_g is approached. Above the glass transition temperature, amorphous materials have a rapid loss of modulus, and even with glass-fiber reinforcement they display a rapid drop in modulus above the glass transition temperature. Crystalline materials maintain a significant usable modulus at temperatures approaching the crystalline melting point, and glass-fiber reinforcement can significantly improve the modulus of crystalline materials between the glass transition and melting temperatures. Generally, strength versus temperature curves are similar to modulus curves and elongation increases with rising temperatures.

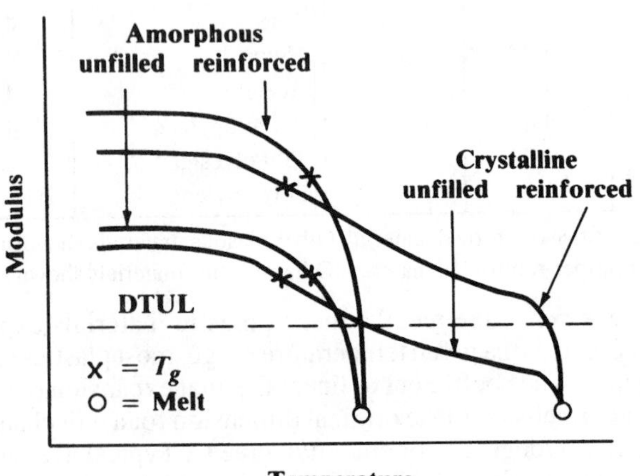

Fig. 6. Modulus Behavior of Crystalline and Amorphous Plastics Showing T_g and Melt Temperatures and the Effect of Reinforcement on Deflection Temperature Under Load (DTUL)

Isochronous stress and isometric stress curves are taken from measurements made at fixed temperatures, although sometimes curves are available for other-than-ambient temperatures. Creep rupture and apparent modulus curves also are often plotted against the log of time, with temperature as a parameter. Refer to sections on creep, creep modulus, and creep rupture beginning on page 577.

As temperatures drop significantly below ambient, most plastics materials lose much of their room temperature impact strength, although a few materials show only a gradual decrease. Plastics reinforced with long glass fibers have relatively high Izod impact values at room temperature, and retain these values at −40 degrees F (− 40 degrees C).

Electrical Properties.—The most notable electrical property of plastics is that they are good insulators, but there are many other electrical properties that must be considered in plastics part design.

Conductivity in solids depends on the availability and mobility of movable charge carriers within the material. Metals are good conductors because the metal atom has a loosely held, outermost electron, and the close proximity of the atoms allows these outer electrons to break free and move within the lattice structure. These free electrons give metals the ability to conduct large currents, even at low voltages. Outer electrons in materials such as glass, porcelain, and plastics are tightly bound to the atoms or molecules so there are no free electrons. Electrical current cannot be conducted and the materials act as insulators.

Volume resistivity is the electrical resistance of a material when a current is applied to it. The resistance is measured in ohm-cm. Materials having values above 10^8 ohm-cm are considered to be insulators, and materials with values between 10^8 and 10^3 ohm-cm are considered to be partial conductors. Most plastics have volume resistivity in the range of 10^{12} to 10^{18} ohm-cm.

Surface resistivity is a measure of the susceptibility of a material to surface contamination, particularly moisture. The tests use electrodes that are placed on the same side of the material.

Dielectric strength is a measure of the voltage required to cause an insulator to break down and allow an electric current to pass and is expressed in volts per 0.001 in. of thickness. Variables that may affect test results include temperature, sample thickness and condition, rate of voltage increase and duration of test, sample contamination, and internal voids.

Dielectric constant or permittivity is a dimensionless constant that indicates how easily a material can become polarized by imposition of an electrical field on an insulator. Reversal of the direction of flow of the current results in reversal of the polarization. The dielectric constant is the ratio of the permittivity of the material in normal ambient conditions to the permittivity of a vacuum. Permittivity is important when plastics are used as insulating materials in high-frequency electrical apparatus. Changes in temperature, moisture levels, electrical frequency, and part thickness may affect the dielectric constant.

Heat dissipation factor is a measure of heat energy dissipated by rapidly repeated reversals of polarization, as with an alternating current. The dissipation factor may also be thought of as the ratio of heat energy lost compared to that transmitted at a given frequency, often 1 MHz (10^6 cycles/sec). Some dielectric constants and heat dissipation factor values are shown in Table 2.

Arc resistance is the length of time required for an electric arc imposed on the surface of an insulating material to develop a conductive path. Materials that resist such a development are preferred for parts of switchgear and other high-voltage apparatus. Tests are used mostly for thermosetting materials because conductive paths can be formed on such materials from the decomposition products resulting from heating by an electric arc.

Table 2. Typical Values of Dielectric Constants and Heat Dissipation Factors for Various Thermoplastics at Room Temperature

Material	Dielectric Constant	Heat Dissipation Factor	Material	Dielectric Constant	Heat Dissipation Factor
Acetal	3.7–3.9	0.001–0.007	Polypropylene	2.3–2.9	0.003–0.014
Acrylic	2.1–3.9	0.001–0.060	Polysulfone	2.7–3.8	0.0008–0.009
ABS	2.9–3.4	0.006–0.021	Modified PPO	2.4–3.1	0.0002–0.005
Nylon 6⁄6	3.1–8.3	0.006–0.190	Polyphenylene sulfide	2.9–4.5	0.001–0.002
Polycarbonate	2.9–3.8	0.0006–0.026	Polyarylate	2.6–3.1	0.001–0.022
TP Polyester	3.0–4.5	0.0012–0.022	Liquid crystal	3.7–10	0.010–0.060

Comparative tracking index (CTI) is another UL test that is similar to the arc resistance test except that the surface to be tested is precoated with an ammonium chloride electrolyte. The test measures the voltage required to cause a conductive path to form between the electrodes, and indicates the arc resistance of a contaminated surface, often found in electrical and electronic equipment.

End-Use Environmental Considerations.—The environment that will be encountered by the product is a prime consideration at the design stage. Problems with cracking, crazing, discoloration, loss of properties, melting, or dissolving can be encountered in the presence of high or low temperatures, chemical substances, energy sources, and radiation. Plastics components also are often subjected to processing, assembly, finishing, and cleaning operations before reaching their ultimate environment.

The stress level in the plastics product greatly affects performance. Generally, increased stress levels resulting from injection molding, forming, assembly work, and end-use forces reduce resistance to environmental factors. Although many plastics are hygroscopic and absorption of water results in dimensional and property changes, plastics are widely accepted because of their relative compatibility with the environment compared with metals. Some chemicals attack the polymer chain directly by reaction, resulting in a progressive lowering of the molecular weight of the polymer and changes in the short-term mechanical properties. Others dissolve the material, although high-molecular-weight plastics dissolve very slowly. Swelling, changes in weight and dimensions, and loss of properties are evidence of solvation.

Plasticization may result if the chemical is miscible with the polymer, resulting in loss of strength, stiffness, and creep resistance, and increased impact resistance. The material may swell and warp due to relaxation of molded-in stresses. Environmental stress cracking may cause catastrophic failure when plastics are stressed, even when the product appears to be unaffected by exposure to a chemical.

Chemical compatibility data are obtained from standard test bars exposed to or placed in the chemical of study and tested as previously described for such properties as tensile strength, flexural modulus, dimensional change, weight, and discoloration. Chemical resistance from some commonly used thermoplastics materials are shown in Table 3, but are only general guidelines and cannot substitute for tests on the end product. More extensive tests expose samples to a chemical in the presence of fixed stress or fixed strain distribution along its length, followed by examination for the stress or strain location at which damage begins.

The preceding tests may provide data about chemical compatibility but do not generate reliable information on performance properties for design purposes. The only test that provides such information is the creep rupture test, conducted at appropriate temperatures in the environment that will be encountered by the product, preferably on prototype parts. Plastics are degraded to varying degrees by ultraviolet light, which causes fading, chalking, and embrittlement. Plastics that will resist the action of ultraviolet rays are available on the market.

Table 3. Chemical Resistance of Various Materials by Chemical Classes

	Acetal Copolymer	Acetal Homopolymer	Nylon 6/6	Thermoplastic Polyester (PBT)	Thermoplastic Polyester (PET)	Polyester Elastomer	Liquid Crystal Polymer*	Polyphenylene Sulfide*	Polyarylate	Polycarbonate	Polysulfone*	Modified Polyphenylene Oxide	Polypropylene	ABS	316 Stainless Steel	Carbon Steel	Aluminum	
ACIDS AND BASES																		
Acids, weak	A	B	C	A	A	A	A	A	A	A	B	A	A	A	A	A	C	Dilute mineral acids
Acids, strong	C	C	C	B	—	C	B	A	—	C	C	—	A	A	B	C	C	Concentrated mineral acids
Bases, weak	A	C	A	B	B	A	B	A	—	C	A	A	A	A	A	B	C	Dilute sodium hydroxide
Bases, strong	A	C	C	—	—	B	C	A	—	C	A	—	A	A	B	B	C	Concentrated sodium hydroxide
Acids, organic, weak	A	B	C	A	A	A	A	A	A	A	B	A	A	A	A	C	C	Acetic acid, vinegar
Acids, organic, strong	C	C	C	B	—	C	B	A	—	C	C	A	A	A	B	C	C	Trichloroacetic acid
AUTOMOTIVE																		
Automotive, fuel	A	A	A	A	A	A	A	A	C	C	A	C	C	A	A	A	A	
Automotive, lubricants	A	A	A	A	A	A	A	A	C	C	A	A	A	A	A	B	A	
Automotive, hydraulic	A	A	—	A	A	—	A	A	C	C	C	A	A	A	—	—	—	
SOLVENTS																		
Aliphatic hydrocarbons	A	A	A	A	A	A	A	A	A	A	A	B	C	A	A	A	A	Heptane, hexane
Aliphatic hydrocarbons, halogenated	A	B	C	B	B	A	A	A	C	C	C	—	—	—	B	B	B	Ethylene chloride, chloroform
Alcohols	A	A	B	A	A	A	A	A	A	A	A	A	A	A	A	A	B	Ethanol, cyclohexanol
Aldehydes	A	A	A	A	B	B	A	A	—	C	B	—	A	—	A	B	A	Acetaldehyde, formaldehyde
Amines	—	—	—	—	—	—	C	B	—	C	C	—	A	—	A	B	B	Aniline, triethanolamine
Aromatic hydrocarbons	A	B	A	A	B	B	A	A	C	C	C	C	C	C	A	A	A	Toluene, xylene, naphtha
Aromatic hydrocarbons, halogenated	—	—	—	—	—	C	—	A	C	C	C	—	—	—	A	A	A	Chlorobenzene
Aromatic, hydroxy	C	C	C	C	—	C	A	A	—	C	C	—	A	—	B	C	A	Phenol
Esters	B	B	A	B	B	B	A	A	C	C	C	—	C	—	B	B	B	Ethyl acetate, dioctyl phthalate
Ethers	B	—	A	A	—	—	—	A	—	A	B	—	C	—	A	A	A	Butyl ether, diethyl ether
Ketones	B	B	A	B	B	B	A	A	C	C	C	—	B	C	A	A	A	Methyl ethyl ketone, acetone
MISCELLANEOUS																		
Detergents	A	—	A	—	B	—	—	A	A	A	—	B	A	—	A	A	B	Laundry and dishwashing detergents, soaps
Inorganic salts	B	B	B	—	A	—	—	A	—	A	—	—	A	A	B	B	B	Zinc chloride, cupric sulfate
Oxidizing agents, strong	C	C	C	—	C	—	B	B	—	C	—	—	A	—	C	C	C	30% hydrogen peroxide, bromine (wet)
Oxidizing agents, weak	C	C	C	A	—	A	A	A	—	A	—	A	A	A	B	C	A	Sodium hypochlorite solution
Water, ambient	A	A	B	A	A	A	A	A	A	A	A	A	A	—	A	C	B	
Water, hot	B	C	B	C	C	B	A	A	—	C	—	A	C	—	A	C	B	
Water, steam	C	C	C	C	C	C	B	A	—	C	—	—	C	—	A	C	—	

This information is presented for instructional purposes and is not intended for design. The data were extracted from numerous sources making consistent rating assignments difficult. Furthermore, the response of any given material to specific chemicals in any one class can vary significantly. Indeed, during the preparation of the table, the effect on one plastics of various chemicals in the same category ranged from essentially no effect to total dissolution. Therefore, an "A" rating for a particular plastics exposed to a particular class of chemicals should not be interpreted as applying to all chemicals in that class. The rating simply means that for the chemicals in that class found in the literature reviewed, the rating was generally an "A." There may be other chemicals in the same class for which the rating would be "C." Finally, the typical chemicals listed do not necessarily correspond to the ones on which the individual ratings are based.

A—minimal effect; B—some effect; C—generally not recommended.

Room temperature except for hot water, steam, and materials marked with a * ≡ 200°. Generally, extended exposure (more than a week) data were used.

DESIGN ANALYSIS

Structural Analysis.—Even the simplest plastics parts may be subjected to stresses caused by assembly, handling, temperature variations, and other environmental effects. Simple analysis using information in *Calculating Moment of Inertia* starting on page 217 and *Beam Calculations* on page 236 can be used to make sure that newly designed parts can withstand these stresses. These methods may also be used for product improvement, cost reduction, and failure analysis of existing parts.

Safety Factors: In setting safety factors for plastics parts there are no hard and fast rules. The most important consideration is the consequence of failure. For example, a little extra deflection in an outside wall or a crack in one of six internal screw bosses may not cause much concern, but the failure of a pressure vessel or water valve might have serious safety or product liability implications. Tests should be run on actual parts at the most extreme operating conditions that could possibly be encountered before any product is marketed. For example, maximum working load should be applied at the maximum temperature and in the presence of any chemicals that might be encountered in service. Loads, temperatures, and chemicals to which a product may be exposed prior to its end use also should be investigated. Impact loading tests should be performed at the lowest temperature expected, including during assembly and shipping. Effects of variations in resin lots and molding conditions must also be considered.

Failures in testing of preproduction lots often can be corrected by increasing the wall thickness, using ribs or gussets, and eliminating stress concentrations. Changing the material to another grade of the same resin or to a different plastics with more suitable mechanical properties is another possible solution. Reviews of product data and discussions with experienced engineers suggest the design stresses shown in Table 4 are suitable for use with the structural analysis information indicated above and the equations presented here, for preliminary design analysis and evaluating general product dimensions. Products designed under these guidelines must be thoroughly tested before being marketed.

Table 4. Design Stresses for Preliminary Part Designs Expressed as a Percentage of Manufacturers' Data Sheet Strength Values

	Failure Not Critical	Failure Critical
Intermittent (Nonfatigue) loading	25–50	10–25
Continuous loading	10–25	5–10

Failure Criteria: Setting of failure criteria is beyond the scope of this section, which is intended to give only basic general information on plastics. Designers who wish to rationalize complex stress states and analyses might investigate the maximum shear theory of failure (otherwise known as Coulomb or Tresca theory). It is further suggested that the shear strength be taken as the manufacturer's published shear strength, or half the tensile strength, whichever is lower. Better still, use half the stress at the elastic limit, if known.

Pressure Vessels: The most common plastics pressure vessel takes the form of a tube with internal pressure. In selecting a wall thickness for the tube, it is convenient to use the thin-wall hoop stress equation:

$$\text{hoop stress } \sigma = \frac{Pd}{2t} \qquad (14)$$

where P = the uniform internal pressure in the tube, d = inside diameter of the tube, and t = the tube wall thickness. This equation is reasonably accurate for tubes where the wall thickness is less than 0.1 of the inside diameter of the tube. As the wall thickness increases, the error becomes quite large.

For thick-walled tubes the maximum hoop stress on the wall surface inside the tube can be calculated from

$$\text{hoop stress } \sigma = P\frac{1+R}{1-R} \tag{15}$$

where $R = (d_i / d_o)^2$, and d_i and d_o are the inside and outside diameters of the tube, respectively.

Press Fits: Press fits are used widely in assembly work for speed and convenience, although they sometimes are unsatisfactory with thermoplastics parts. Common applications are to a plastics hub or boss accepting a plastics or metal shaft or pin. Forcing the pin into the hole expands the hub, creating a tensile or hoop stress.

If the interference is too great, very high strain and stress develop and the plastics part will J) fail immediately by developing a crack parallel to the hub axis to relieve the stress, a typical hoop stress failure; K) survive assembly but fail prematurely due to creep rupture caused by the high induced-stress levels; and L) undergo stress relaxation sufficient to reduce the stress to a level that can be sustained.

For a typical press fit, the allowable design stress depends on the particular plastics material, temperature, and other environmental considerations. Hoop stress equations for such a design make use of a geometry factor γ:

$$\gamma = \frac{1+(d_s/d_o)^2}{1-(d_s/d_o)^2} \tag{16}$$

where d_s = diameter of the pin to be inserted and d_o = outside diameter of the boss.

When both the shaft and the hub are of the same, or essentially the same, materials, the hoop stress σ, given the diametral interference, $i = d_s - d_i$, is

$$\sigma = \frac{i}{d_s}E_p\frac{\gamma}{\gamma+1} \tag{17}$$

and the allowable interference i_a, given the permissible design stress σ_a, is

$$i_a = d_s\frac{\sigma_a}{E_p}\frac{\gamma+1}{\gamma} \tag{18}$$

When the shaft is metal and the hub is plastics, the hoop stress, given i, is obtained from

$$\sigma = \frac{i}{d_s}E_p\frac{\gamma}{\gamma+\nu_p} \tag{19}$$

and the allowable interference i_a, given the permissible design stress for plastics σ_a, is

$$i_a = d_s\frac{\sigma_a}{E_p}\frac{\gamma+\nu_p}{\gamma} \tag{20}$$

where E_p = modulus of elasticity of plastics and ν_p = Poisson's ratio for plastics.

Pipe Threads: Pipe threads on plastics pipes and other parts used in plastics plumbing and pneumatic assemblies require only hand tight assembly to effect a good seat, especially if a compatible sealant tape or compound is used. Assembling a tapered male pipe thread into a mating female thread in a plastics part is analogous to driving a cone into a round hole and may result in a split boss. Sometimes straight threads and an O-ring seal can avoid the need for pipe threads. When pipe threads must be used, torque control is essential.

When mating metal to plastics pipe threads, the threaded plastics component should be the male member, so that the plastics are in compression. If torque can be controlled during assembly, use fluoroplastics tape on female plastics pipe threads. If torque cannot be controlled, consider using an external hoop ring, either pressed on or molded in. Do not design flats into plastics parts for assembly purposes, because they will encourage overtightening. If some provision for improved gripping must be made, use wings or a textured surface. An approximate formula for the hoop stress σ produced in a plastics boss with internal pipe threads is

$$\sigma = \frac{3T}{tdL} \tag{21}$$

where T = torque in in.-lb, t = wall thickness of the plastics boss in in., d = pipe outside diameter in in., and L = length of thread engagement in in.

This equation assumes certain geometric relationships and a coefficient of friction of 0.15. If compatible thread lubricants are used during assembly, the torque must be reduced. To ensure safety and reliability, all threaded assemblies must be subjected to long-term testing under operating pressures, temperatures, and stresses caused by installation procedures exceeding those likely to be encountered in service.

Thermal Stresses.— When materials with different coefficients of thermal expansion are bolted, riveted, bonded, crimped, pressed, welded, or fastened by any method that prevents relative movement between the parts, there is potential for thermal stress to exist. Typical examples are joining of nonreinforced thermoplastics parts with materials such as metals, glass, or ceramics that usually have much lower coefficients of thermal expansion. The basic relationship for thermal expansion is

$$\Delta L = \alpha L \Delta T \tag{22}$$

where ΔL = change in length, α = coefficient of thermal expansion (see Table 4), L = linear dimension under consideration (including hole diameters), and ΔT = temperature change.

If the plastics component is constrained so that it cannot expand or contract, the strain $\in_T$, induced by a temperature change, is calculated by

$$\in_T = \frac{\Delta L}{L} = \alpha \Delta T \tag{23}$$

The stress can then be calculated by multiplying the strain $\in_T$ by the tensile modulus of the material at the temperature involved. A typical example is of a plastics part to be mounted to a metal part, such as a window in a housing. Both components expand with changes in temperature. The plastics imposes insignificant load to the metal but considerable stress is generated in the plastics. For such an example, the approximate thermal stress σ_T in the plastics is given by

$$\sigma_T = (\alpha_m - \alpha_p) E_p \Delta T \tag{24}$$

where α_m = coefficient of thermal expansion of the metal, α_p = coefficient of thermal expansion of the plastics, and E_p = tensile modulus of the plastics at the temperature involved.

Other equations for thermal expansion in various situations are shown in Fig. 7.

Most plastics expand more than metals with temperature increases and their modulus drops. The result is a compressive load in the plastics that often results in buckling. Conversely, as the temperature drops, the plastics shrinks more than the metal and develops an increased tensile modulus. These conditions can cause tensile rupture of the plastics part. Clearances around fasteners, warpage, creep, or failure, or yield of adhesives tend to relieve the thermal stress. Allowances must be made for temperature changes, especially with large parts subjected to wide variations. Provision is often made for relative motion ΔL_{rel}, between two materials, as illustrated in Fig. 7:

$$\Delta L_{rel} = (\alpha_p - \alpha_m) L \Delta T \tag{25}$$

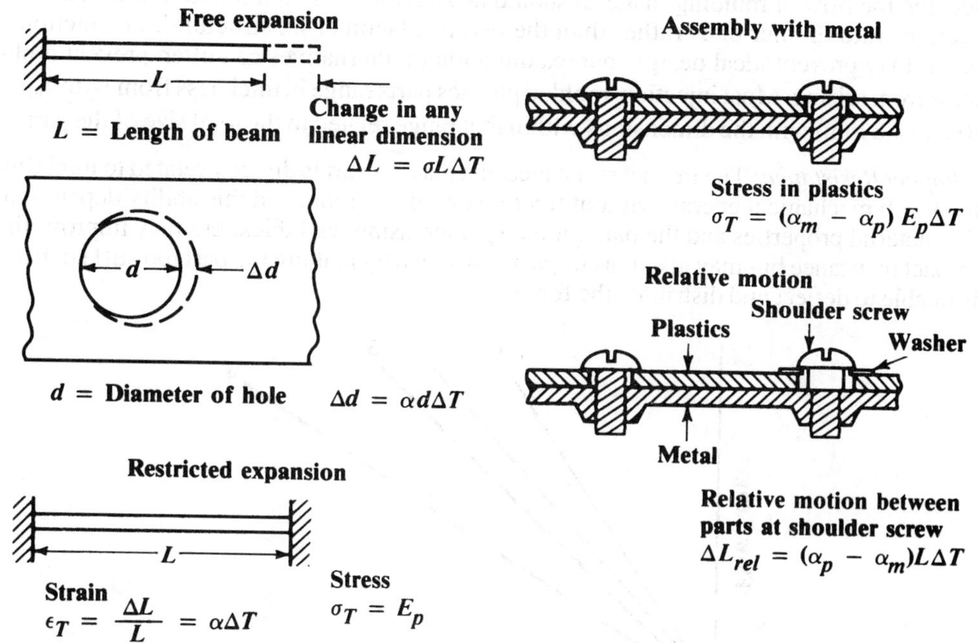

Fig. 7. Thermal Expansion Equations for Various Combinations of Materials and Situations

Design for Injection Moldings.—Injection molding uses equipment similar to that for die casting, in that a precision steel mold is clamped shut, and melted material (here, plasticized plastics) is forced into the cavity between the mold components. The pelletized plastics material is fed into a heated chamber, or barrel, by a large, slowly rotating screw, and is melted. When a sufficient quantity to fill the cavity has been prepared, the screw is moved axially under high pressure to force the material into the cavity. The mold has channels through which coolant is circulated to remove heat and to chill the plastics. When the plastics has cooled sufficiently, the mold is unclamped and opened, and the molding is forced out by strategically located ejectors. During cooling and removal, material for the next part is plasticized within the barrel, ready for the cycle to be repeated.

Product analysis provides a good approach to design for plastics molding. A basic principle of plastics molding design, when moldings are being substituted for parts made by other means, is to incorporate as many functions into the molding as possible, especially those requiring nuts, bolts, and washers, for instance. Material should be selected that fulfills the maximum requirements, such as the functions mentioned, as well as insulation from the passage of heat or electricity, and allows use of the minimum amount of material.

Important material selection criteria include ability to withstand the heat of assembly, finishing, shipping, operating, and heat from internal sources. Effects of chemicals in the environment and approvals of government and other agencies also should be checked. Many such approvals specify wall thicknesses, color additives, fillers, and operating temperatures. Plans for assembly by bonding may dictate use of certain materials, and the question of painting, plating, or other surface coatings must be considered. Cost of candidate materials compared with the alternatives must be weighed, using the formula for cost per in.^{3} = 0.0361 × specific gravity × material cost per lb. Material cost required for a part is obtained by multiplying the cost per in.^{3} by the part volume. A rough estimate of likely part cost is double the cost of material for the part.

Wall Thicknesses: The thickness of material used in a plastics molding is of the greatest importance, and should be settled before the mold is made, since modifications are costly. In general, wall thicknesses should be kept as thin as practical and as uniform as possible.

Ideally, the flow of molding material should be so arranged that it moves through thicker sections into thinner ones rather than the reverse. Geometric, structural, or functional needs may prevent ideal design, but examination of alternatives can often prevent problems from arising. Most injection-molded plastics parts range in thickness from 1⁄32 to 3⁄16 in. (0.8 to 4.8 mm) with the dimensions within that range related to the total size of the part.

Impact Resistance: The impact resistance of a plastics part is directly related to its ability to absorb mechanical energy without fracture or deformation, and this ability depends on the material properties and the part geometry. Increasing wall thickness may improve the impact resistance but may also hurt impact resistance by making the part too stiff so that it is unable to deflect and distribute the force.

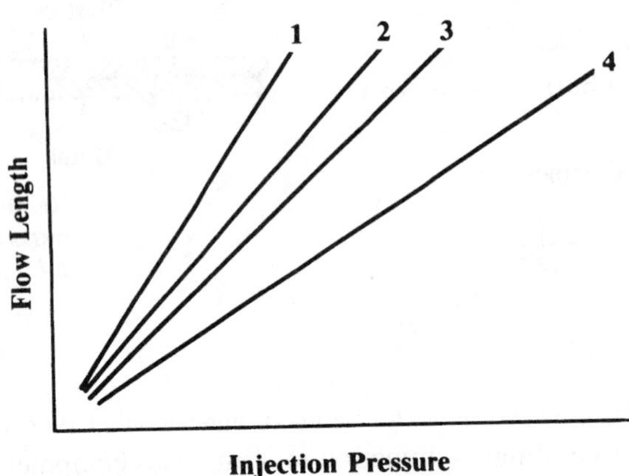

Fig. 8. Typical Spiral-Flow Curves for (1) Nylon 6/6, (2) Polyester Thermoplastics PBT, Liquid-Crystal-Glass-Reinforced, and Polyphenylene-Sulfide-Glass-Reinforced, (3) Acetal Copolymer, and (4) PBT-Glass-Reinforced Plastics Materials

Design engineers must also have some knowledge of mold design and, in determining wall thickness, should consider the ability of plastics to flow into the narrow mold channels. This flowability depends on temperature and pressure to some extent, but varies for different materials, as shown in Fig. 8.

Table 5 shows some typical nominal wall thicknesses for various types of thermoplastics.

Table 5. Typical Nominal Wall Thicknesses for Various Classes of Thermoplastics

Thermoplastics Group	Typical Working Range (in.)
Acrylonitrile-butadiene-styrene (ABS)	0.045–0.140
Acetal	0.030–0.120
Acrylic	0.025–0.150
Liquid-crystal polymer	0.008–0.120
Long-fiber-reinforced plastics	0.075–1.000
Modified polyphenylene ether	0.045–0.140
Nylon	0.010–0.115
Polyarylate	0.045–0.150
Polycarbonate	0.040–0.150
Polyester	0.025–0.125
Polyester elastomer	0.025–0.125
Polyethylene	0.030–0.200
Polyphenylene sulfide	0.020–0.180
Polypropylene	0.025–0.150
Polystyrene	0.035–0.150
Polysulfone	0.050–0.150
Polyurethane	0.080–0.750
Polyvinyl chloride (PVC)	0.040–0.150
Styrene-acrylonitrile (SAN)	0.035–0.150

If the plastics part is to carry loads, load-bearing areas should be analyzed for stress and deflection. When stress or deflection is too high, solutions are to use ribs or contours to increase section modulus; to use a higher-strength, higher-modulus (fiber-reinforced) material; or to increase the wall thickness if it is not already too thick. Where space allows, adding or thickening ribs can increase structural integrity without thickening walls.

Equations (11), (12), and (13) can be related to formulas using the section modulus and moment of inertia on page 237, where Case 2, for (i), stress at the beam center is given by $\sigma = -Wl/4Z$.

On page 236, note that $Z = I \div$ distance from neutral axis to extreme fiber ($h \div 2$ in the plastics example). The rectangular beam section diagrammed on page 219 gives the equivalent of $I = bh^3/12$ for the rectangular section in the plastics example. Therefore,

$$Z = \frac{I}{h/2} = \frac{bh^3}{12} \times \frac{2}{h} = \frac{bh^2}{6}$$

In $\sigma = -Wl/4Z$, the (–) sign indicates that the beam is supported at the ends, so that the upper fibers are in compression and the lower fibers are in tension. Also, $W = F$ and $l = L$ in the respective equations, so that stress, $\sigma = FL/4(bh^2)/6$, and $\sigma = 3FL/2bh^2$.

To calculate (ii) maximum deflection Y at load, use $Y = Wl^3/48EI$ from page 237, where $W = F$, $l = L$, $E = E$, from Equation (12) and $I = bh^3/12$. Therefore,

$$Y = \frac{FL^3}{48E(bh^2/12)} = \frac{FL^3}{4Ebh^3}$$

As an example, assume that a beam as described in connection with Equations (11), (12), and (13) is 0.75 in. wide, with a constant wall thickness of 0.080 in., so that the cross-sectional area is 0.060 in.^{2}, and there is a central load W of 5 lb. Based on a bending or flexural modulus of 300,000 lb/in.^{2}, the maximum stress is calculated at 6250 lb/in.^{2} and the maximum deflection at 0.694 in. Both the stress and the deflection are too high, so a decision is made to add a rib measuring 0.040 in. thick by 0.400 in. deep, with a small draft of ½ degree per side, to reinforce the structure.

The equations on page 237, the drawing page representing the ribbed section (neglecting radii), and the accompanying formulas, permit calculation of the maximum stress and deflection for the ribbed section.

With the new cross-sectional area only slightly larger at 0.0746 in.^{2}, the calculated stress is reduced to 2270 lb/in.^{2}, and the deflection goes down to 0.026 in., which is acceptable for both the material and the application. To achieve the same result from a heavier beam would require a thickness of 0.239 in., tripling the weight of the beam and increasing molding difficulties. The rib adds only 25 percent to the total section weight.

Use of ribs allows the structural characteristics of a part to be tailored to suit its function, but ribs can cause warping and appearance problems, so are best avoided if they are not structurally necessary. If the first parts produced require strengthening, ribs can be added or thickened without high cost after the tool is finished because the work consists only of removing steel from the mold. In general, ribs should have a base thickness of about half the thickness of the adjacent wall, and be kept as thin as possible where they are positioned near faces that need to have a good appearance.

Where structural strength is more important than appearance, or when using materials that have low shrinkage, ribs can be made 75 or 100 per cent of the wall thickness. However, where the rib base joins the main molding there is an increase in thickness forming a heavy mass of material. Shrinkage of this mass can produce a cavity or void, a hollow area or sink, or can distort the molding. If the mass is very large, cooling time may be prolonged, leading to low output from the machine. Large masses of material in other parts of a molding are also best avoided. These problems can usually be addressed by good mold design.

Ribs need not be of constant height or width, and are often varied in proportions to suit the stress distribution in the part. All ribs should have a minimum of ½ degree of draft per side for ease of removal from the mold, and a minimum radius of 0.005 in. at the base to avoid stress-raising corners. Draft and thickness requirements will usually limit the height of the rib, which can be from 1.5 to 5 times the base thickness, and several evenly spaced ribs are generally preferred to a single large one. Smooth transitions should be made to other structural features such as bosses, walls, and pads.

Other ways to improve section properties include use of top-hat and corrugated sections, crowning or doming of some areas, and reinforcement with metal or other inserts that are placed in the mold before it is closed. To keep molded parts uniform in wall thickness, cores or projections may be provided in the mold to prevent a space being filled with molding material. Blind holes can be cored by pins that are supported on only one side of the mold and through holes by pins that pass through both sides. The length to width ratio should be kept as low as possible to prevent bending or breakage under the high pressures used in the injection molding process.

Agency approvals for resistance to flammability or heat, electrical properties, or other characteristics are usually based on specific wall thicknesses. These restrictions sometimes necessitate thicker walls than are required for structural strength purposes.

Table 6. Dimensional Changes for Various Combinations of Draft Angles and Draw Depths (Values to Nearest 0.001 in.)

Parallel draft
Draft angle
Depth of draw
Dimensional difference

Draw Depth (in)	Draft Angle (degrees)							
	⅛	¼	½	1	2	3	4	5
1	0.002	0.004	0.009	0.017	0.035	0.052	0.070	0.080
2	0.004	0.009	0.017	0.035	0.070	0.105	0.140	0.175
3	0.007	0.013	0.026	0.053	0.105	0.157	0.210	0.263
4	0.009	0.018	0.035	0.070	0.140	0.210	0.280	0.350
5	0.011	0.022	0.044	0.088	0.175	0.262	0.350	0.437
6	0.013	0.026	0.052	0.105	0.209	0.314	0.420	0.525
7	0.015	0.031	0.061	0.123	0.244	0.367	0.490	0.612
8	0.018	0.035	0.070	0.140	0.279	0.419	0.559	0.700
9	0.020	0.040	0.078	0.158	0.314	0.472	0.629	0.787
10	0.022	0.044	0.087	0.175	0.349	0.524	0.699	0.875

Draft: Most molded parts have features that must be cut into the mold perpendicular to the parting line. Removal of these parts from the mold is easier if they are tapered in the direction of mold opening. This taper is called draft in the line of draw or mold movement, and it allows the part to break free of the mold by creating a clearance as soon as the mold starts to open. Plastics materials shrink as they cool, so they grip mold projections very tightly and ejection can be difficult without sufficient draft. A draft of ½ degree on each side of a projection on the part is generally considered as a minimum, although up to 3 degrees per side is often used. Draft angles in degrees for various draw depths, and the resulting dimensional changes per side in inches (rounded to three decimal places), between the dimensions at the base and at the top of a projection are shown in Table 6. A rule of thumb is that 1 degree of draft yields 0.017 in. of difference in dimension per inch of draw length. Where a minimum of variation in wall thickness is needed to produce walls that are perpendicular to the direction of draw, the mold sometimes can be designed to produce parallel draft, as seen at the left in Table 6. The amount of draft required also depends on the surface finish of the mold walls. Any surface texture will increase the draft requirement by at least 1 degree per side for every 0.001 in. of texture depth.

Fillets, Radii, and Undercuts: Sharp corners are always to be avoided in injection-molded part designs because they represent points of stress concentration. Sharp corners in metal parts often are less important because the stresses are low compared with the strength of the material or because local yielding redistributes the loads. Sharp inside corners are particularly to be avoided in moldings because severe molded-in stresses are generated as the material shrinks onto the mold corner. Sharp corners also cause poor material flow patterns, reduced mechanical properties, and increased tool wear. Therefore, inside corner radii should be made equal to half the nominal wall thickness, with a minimum of 0.020 in. for parts subject to stress and 0.005 in. radius for stress-free parts. Outside corners should have a radius equal to the inside corner radius plus the wall thickness.

With an inside radius of half the wall thickness, a stress concentration of 1.5 is a reasonable assumption, and for radii down to 0.1 times the wall thickness, a stress concentration of 3 is likely. More information on stress concentrations is found in *Working Stress* on page 200, *Stress Concentration Factors* on page 200, and in the charts pages 201 through 204. A suitable value for q in Equation (8) on page 200, for plastics materials, is 1. Most plastics parts are so designed that they can be ejected parallel with the direction of mold parting. Complex parts with undercuts may require mold designs with cavity-forming projections that must move at an angle to the direction of opening. Between these two extremes lie such items as "windows," or simple openings in the side of a molding, which can be produced by the normal interaction of the two main parts of the mold.

Design for Assembly.—An advantage of the flexibility of plastics parts is that they can often be designed for assembly by means of molded-in snap-fit, press-fit, pop-on, and thread fasteners, so that no additional fasteners, adhesives, solvents, or special equipment is required. Improper assembly can be minimized, but tooling is often made more complex and disassembly may be difficult with these methods.

Chemical bonding involves fixtures, substances, and safety equipment, is suited to applications that must be leak-tight, and does not create stresses. However, adhesives and solvents can be dangerous and preparation and cure times can be prolonged.

Thermal welding methods include ultrasonic, hot-plate, spin, induction, and radio-frequency energy and require special equipment. Thermal methods are also used for staking, swaging, and other heat deformation procedures. Materials must be compatible and have similar melting temperatures.

Mechanical fasteners designed for metals are generally usable with plastics, and there are many other fasteners designed specifically for plastics. Typical are bolts, self-tapping and thread-forming screws, rivets, threaded inserts, and spring clips. Care must be taken to avoid overstressing the parts. Creep can result in loss of preload in poorly designed systems.

Snap-fit designs are widely used, a typical application being to battery compartment covers. All snap-fit designs have a molded part that must flex like a spring, usually past a designed-in interference, then return to its unflexed position to hold the parts together. There must be sufficient holding power without exceeding the elastic or fatigue limits of the material. With the typical snap-fit designs in Fig. 9, beam equations can be used to calculate the maximum strain during assembly. If the stress is kept below the yield point of the material, the flexing finger returns to its original position.

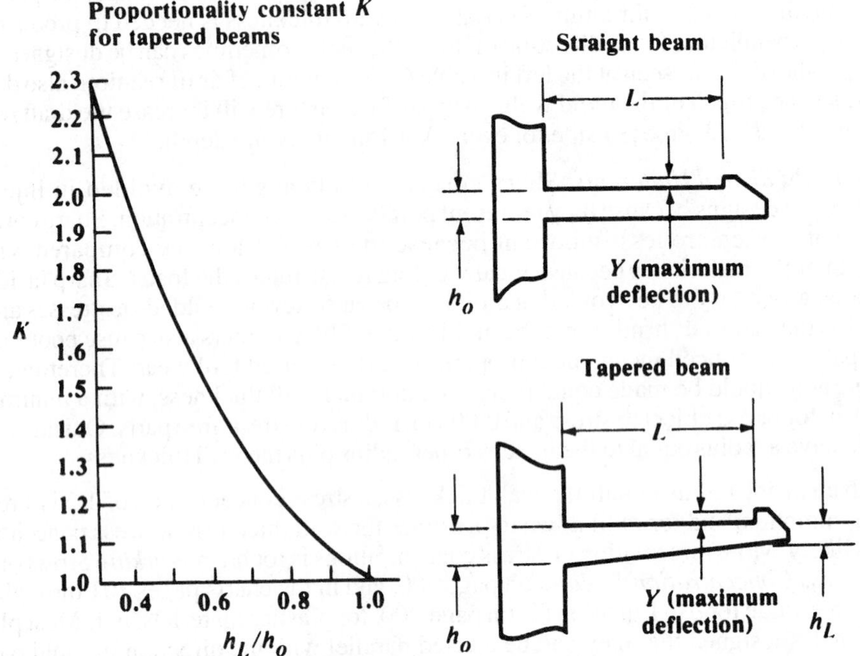

Fig. 9. Snap-fit Designs for Cantilever Beams with Rectangular Cross Sections

With some materials the calculated bending stress can exceed the yield point stress considerably if the movement is done rapidly. In other words, the flexing finger passes through its maximum deflection or strain and the material does not respond as it should if the yield stress has been greatly exceeded. It is common to evaluate snap-ins by calculating strain instead of stress.

Dynamic strain ∈, for the straight beam, is calculated from

$$\in = \frac{3Yh_o}{2L^2} \tag{26}$$

and for the tapered beam, from

$$\in = \frac{3Yh_o}{2L^2K} \tag{27}$$

The derived values should be compared with the permissible dynamic strain limits for the material in question, if known. A tapered finger provides more-uniform stress distribution and is recommended where possible. Sharp corners or structural discontinuities that will cause stress concentrations on fingers such as those shown must be avoided.

Snap-in arrangements usually require undercuts produced by a sliding core in the mold as shown in Fig. 10a. Sometimes the snap finger can be simply popped off when the mold is opened. An alternative to the sliding core is shown in Fig. 10b, which requires an opening in the molding at the base of the flexing finger. Other snap-in assembly techniques that take advantage of the flexibility of plastics are shown in Fig. 11.

Molded-in threads in holes usually are formed by cores that require some type of unscrewing or collapsing mechanism leading to tooling complications. External threads can often be molded by positioning them across the parting plane of the mold. Molding of threads finer than 28 to the inch is generally not practical.

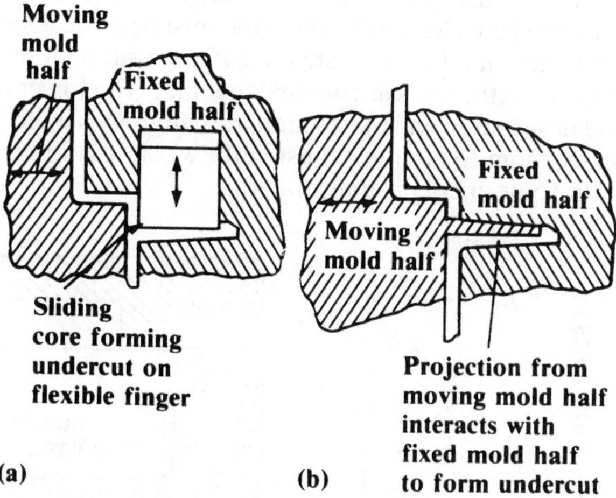

Fig. 10. (left) (a) Arrangement for Molding an Undercut on the End of a Flexible Finger Using a Sliding Core; (b) With the Undercut Formed by a Mold Projection, the Sliding Core is Eliminated.

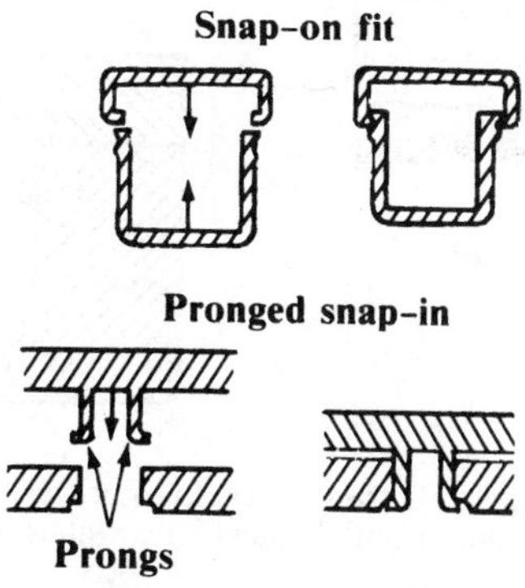

Fig. 11. (right) Examples of Snap-In and Snap-On Arrangements

Chemical bonding may use solvents or adhesives. Use of solvents is limited to compatible materials that can be dissolved by the same solvent. Chemical resistance of many plastics, especially crystalline materials, limits the use of this method. Safety precautions must also be considered in handling the solvents to protect workers and for solvent recovery. With adhesive bonding, a third adhesive substance is introduced at the interface between the parts to be joined. Adhesives can join plastics, metals, ceramics, glass, wood, or other bondable substances.

Typical adhesives used for thermoplastics are epoxies, acrylics, polyurethanes, phenolics, rubbers, polyesters, and vinyls. Cyanoacrylates are often used because of their rapid adhesion to many materials. Manufacturers' recommendations should be sought because many adhesives contain solvents that partially dissolve the plastics surfaces, giving improved adhesion. However, some adhesives can attack certain plastics, leading to deterioration and failure. The main disadvantages of adhesives are that they are slow, use long clamp times, require fixtures, and may involve special ovens or curing conditions. Surface preparation also may be difficult because the presence of grease, oil, mold-release material, or even a fingerprint can spoil a bond. Some materials may need surface preparation such as chemical etching or mechanical roughening to improve joint strength.

Ultrasonic welding is frequently used for joining parts of similar material, of small and medium size, is rapid, and can be automated. High-frequency (20–40 kHz) vibrational energy is directed to the interfaces to be joined, creating localized molecular excitation that causes the plastics to melt. With proper joint designs, welds can be made in only 2 seconds that are as strong as the base materials. When the energy is switched off, the plastics solidifies immediately. Parts to be welded ultrasonically must be so designed that the energy is concentrated in an initially small contact area, creating rapid melting and melt flow that progresses along the joint as the parts are pressed together. The lower part of the assembly is supported in a rigid nest fixture and the upper part is aligned, usually by the joint design. This upper part has freedom to couple acoustically when it is in contact with the horn through which the ultrasonic energy is transmitted.

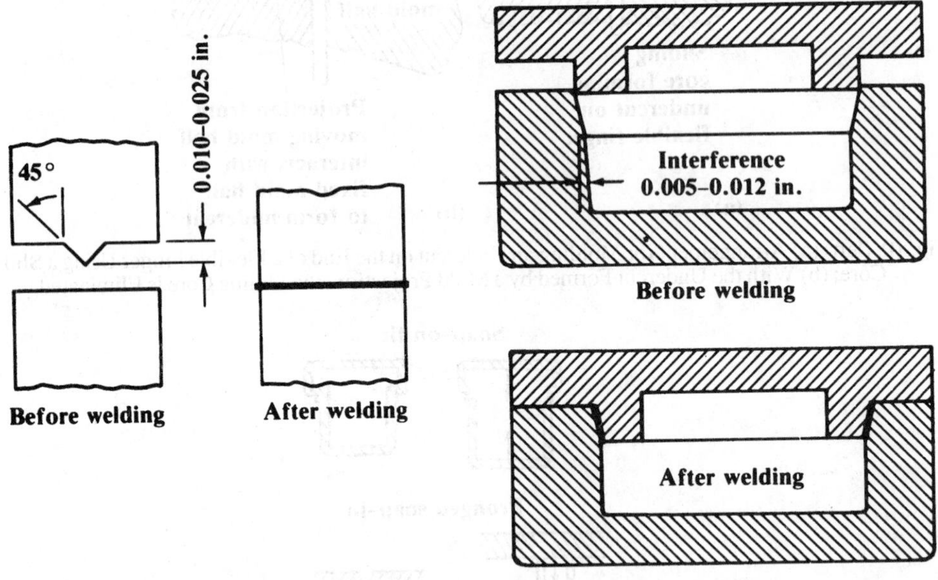

Fig. 12. (left) Energy Director Types of Ultrasonic Weld Joint Designs for Assembly of Plastics Moldings, and (right) Typical Shear Interference

Typical joint designs are shown in Fig. 12, where the example at the left is of a simple butt-type energy director design that works well with amorphous materials. The inverted V-projection, known as an energy director, concentrates the energy in a small area on both sides of the joint. This area melts quickly and the material flows as the parts are pressed together. A basic shear interference joint is seen at the right in Fig. 12. Melting of both components starts at the small initial contact area and flow continues along the near-vertical wall as the parts are pressed together, creating a continuous, leakproof joint with a strength that often exceeds that of the parts joined.

This design is preferred with liquid crystal polymers and crystalline materials such as nylon, acetal, and thermoplastics polyester, and for any application of these materials where high strength and a hermetic seal are required. Many variations of these basic designs are possible and manufacturers of materials and ultrasonic equipment offer literature and design assistance. Ultrasonic vibrations may also be used for staking, swaging, and spot-welding in assembly of plastics parts.

With hygroscopic materials, welding should be completed as soon as possible after molding because moisture can cause weaker bonds. Drying may be advisable immediately before welding. Drawbacks with ultrasonic welding are that design, quality control, equipment maintenance, and settings are of critical importance for consistent, high-strength welds; the equipment is costly; the process uses large amounts of electric power especially with large parts; and parts to be joined must be of the same or similar materials. Filled and reinforced materials also present difficulties with compatibility.

Operating frequencies used in ultrasonic welding are in the range of 20–40 kHz, above the range detectable by the ear. However, discomforting sounds may be generated when plastics parts vibrate at lower frequencies, and may make sound-proofing necessary.

Vibration welding resembles ultrasonic welding except that the parts to be joined are rubbed together to produce heat to melt the joint faces by friction. The energy is transferred in the form of high-amplitude, low-frequency, reciprocating motion. When the vibration stops, the weld area cools and the parts remain joined in the alignment provided by the welding fixture. Typical frequencies used are 120–240 Hz and amplitudes range between 0.10 and 0. 20 in. of linear displacement. When the geometry or assembly design prevents linear movement, vibration-welding equipment can be designed to produce angular displacement of parts.

Like ultrasonic welding, vibration welding produces high-strength joints and is better suited to large parts and irregular joint faces. Moisture in hygroscopic materials such as nylon has less effect on the joint strength than it does with ultrasonic methods.

Spin welding is a rapid and economical method of joining parts that have circular joint interfaces. The process usually is completed in about 3 seconds and can be automated easily. Frictional heat for welding is generated by rotating one part against the other (usually fixed) with a controlled pressure. When the rotation is stopped, pressure is maintained during cooling and solidification of the melted material. Simple equipment such as a drill press is often sufficient for this process.

Radio-frequency welding, often called heat sealing, is widely used with flexible thermoplastics films and sheets of materials such as vinyl (plasticized PVC) and polyurethane, and for joining injection-molded parts, usually to film. Heat for welding is generated by a strong radio-frequency field to the joint region through a metal die formed to suit the joint shape. The die also applies the pressure required to complete the weld. Some plastics are transparent to radio frequency, so cannot be welded by this method.

Electromagnetic or induction welding uses inductive heating to generate fusion temperatures in thermoplastics materials as shown at the top in Fig. 13. Fine, magnetizable particles embedded in a gasket, preform, filament, ribbon, adhesive, coextruded film, or molded part are excited by the radio frequency and are thus heated to welding temperatures. The heated parts are pressed together, and as the temperature rises, the material of the particle carrier flows under pressure through the joint interface, filling voids and cavities and becoming an integral part of the weld. Ideally, the melted material should be contained and subjected to an internal pressure by the surrounding component surfaces. Proper joint design is essential to successful welding and some basic designs are also shown in Fig. 13.

Requirements of the preform often add cost to this welding method but the cost is offset by low reject rates resulting from good reliability of the welds. Structural, hermetic welds can be produced in most thermoplastics materials and automation can be used for large-volume production. The process also offers great latitude in joint size, configuration, tolerance requirements, and ability to bond some dissimilar materials. A disadvantage is that no metal can be near the joint line during energization of the inductor coil. All components of an assembly to be induction-welded must therefore be nonmetallic, or metallic components must be placed where they will not be subjected to the radio-frequency field from the inductor.

Assembly with Fasteners.—Metal fasteners of high strength can overstress plastics parts, so torque-controlled tightening or special design provisions are needed. Examples of poor and preferred designs are shown in Fig. 14. Where torque cannot be controlled, even with a shoulder screw, various types of washers can be used to spread the compression force over wider areas.

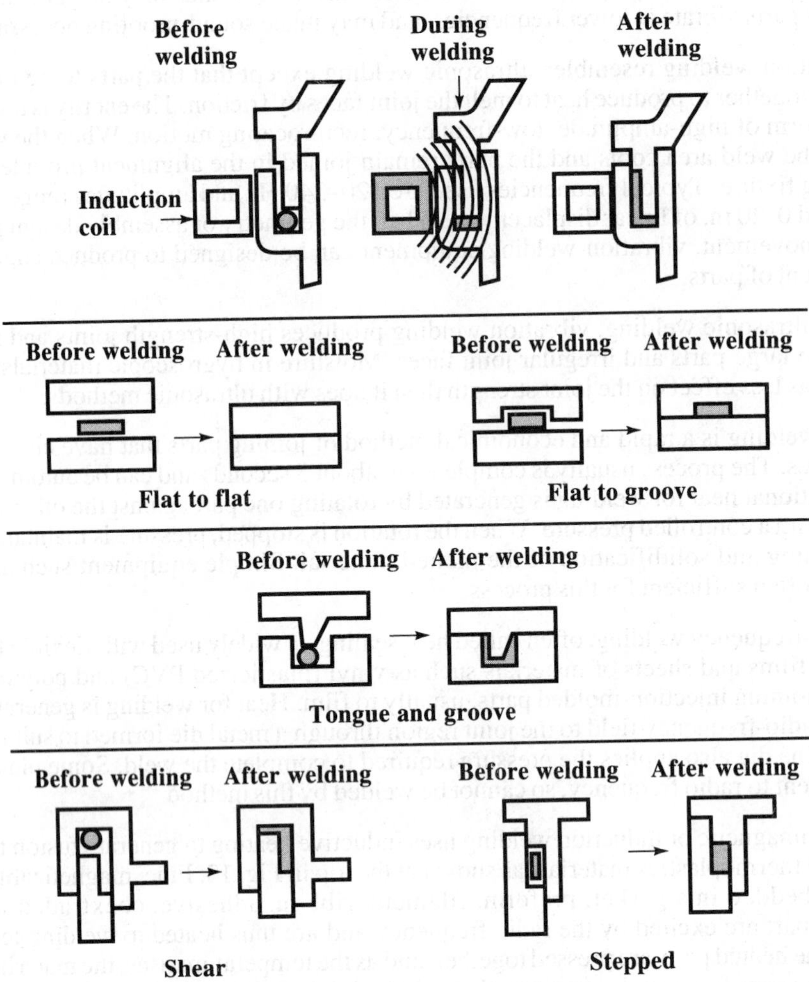

Fig. 13. Typical Joint Designs Used in Induction Welding of Plastics Materials

Metal inserts are available in a wide range of shapes and sizes for permanent installation of metal threads or bushings in plastics parts. Inserts are typically installed in molded bosses, designed with holes to suit the insert to be used. Some inserts are pressed into place and others are installed by methods designed to limit stress and increase strength. Generally, the outside of the insert is provided with projections of various configurations that penetrate the plastics and prevent movement under normal forces exerted during assembly. Inserts can also be installed with equipment similar to that used for ultrasonic welding, the plastics being melted to enhance contact with the metal and reduce insertion stresses.

Thread-cutting and -forming screws are widely used with plastics parts. Information on standard self-threading screws is found in *SELF-THREADING SCREWS* starting on page 1620. Thread-forming screws must be used carefully with high-modulus, low-creep materials, as high hoop stresses can be generated during insertion. Screws with multiple lobes and screws with alternating low and high threads have excellent holding power in plastics. Molded holes must have sufficient depth to prevent bottoming, and boss walls must be thick enough to resist stresses. A rule of thumb is that the outside diameter of the boss should be double the major diameter of the screw.

Hollow aluminum or other metal rivets are often used in plastics assembly, as are stamped sheet metal components, especially push-on or -in designs. Molded plastics fasteners also are frequently used.

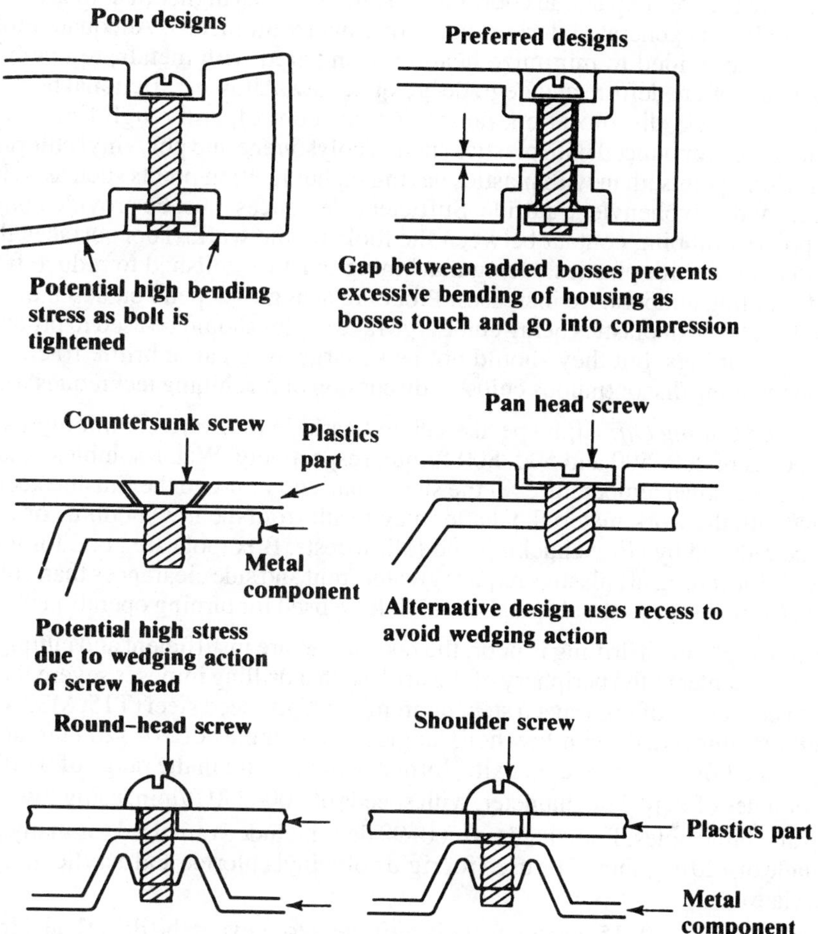

Fig. 14. Examples of Poor and Good Designs Used for Assembly of Plastics Parts with Metal Fasteners

Machining Plastics.—Plastics can be molded into complex shapes so do not usually need to be machined. However, machining is sometimes more cost-effective than making a complex tool, especially when requirements are for prototype development, low-volume production, undercuts, angular holes, or other openings that are difficult to produce in a mold. Specialized methods for development of prototypes are discussed later. All machining of plastics requires dust control, adequate ventilation, safety guards, and eye protection.

Like some metals, plastics may need to be annealed before machining to avoid warpage. Specific annealing instructions can be obtained from plastics suppliers. The modulus of elasticity of plastics is 10–60 times smaller than for metals, and this resilience permits much greater deflection of the work material during cutting. Thermoplastics materials must be held and supported firmly to prevent distortion, and sharp tools are essential to keep cutting forces to a minimum.

Plastics recover elastically during and after machining so that drilled or tapped holes often end up tapered or of smaller diameter than the tool. Turned diameters also can end up larger than the dimensions measured immediately after the finishing cut. The low thermal conductivity of plastics causes most of the heat generated in cutting to be absorbed by the tool. Heat in the plastics tends to stay at the surface. The heat must be removed by an air blast or a liquid coolant for good results in machining.

Plastics have thermal expansion coefficients some 10 times higher than those of metals, so that more heat is generated during machining than with metals. Adequate tool clearances must be provided to minimize heating. Compared with metals, temperatures at which plastics soften, deform and degrade are quite low. Allowing frictional heat to build up causes gumming, discoloration, poor tolerance control, and rough finishes. These effects are more pronounced with plastics such as polystyrene and polyvinyl chloride, having low melting points, than with plastics having higher melting points such as nylon, fluoroplastics, and polyphenylene sulfide. Sufficient clearances must be provided on cutting tools to prevent rubbing contact between the tool and the work. Tool surfaces that will come into contact with plastics during machining must be polished to reduce frictional drag and resulting temperature increases. Proper rake angles depend on depth of cut, cutting speed, and type of plastics being cut. Large rake angles should be used to produce continuous-type cuttings, but they should not be so large as to cause brittle fracture of the work, and resulting discontinuous chips. A discussion of machining techniques follows.

Turning and Cutting Off: High speed steel and carbide tools are commonly used with cutting speeds of 200–500 and 500–800 ft/min, respectively. Water-soluble coolants can be used to keep down temperatures at the shear zone and improve the finish, except when they react with the work material. Chatter may result from the low modulus of elasticity and can be reduced by close chucking and follow rests. Box tools are good for long, thin parts. Tools for cutting off plastics require greater front and side clearances than are needed for metal. Cutting speeds should be about half those used for turning operations.

Drilling: Chip flow in drilling is poor, the rake angles are insufficient and cutting speeds vary from the center to the periphery of the drill, so that drilling imposes severe loading on the workpiece. Drills of high speed steel or premium high speed steel (T15, M33, or M41–M47) are recommended, with low helix angles, point angles of 70–120 deg, and wide, highly polished flutes to ease chip exit. Normal feed rates are in the range of 0.001–0.012 in./rev for holes of $\frac{1}{16}$ to 2 in. diameter, with speeds of 100–250 ft/min, using lower speeds for deep and blind holes. Point angles of 60–90 deg (included) are used for many plastics, but an angle of 120 deg should be used for rigid polyvinyl chloride and acrylic (polymethyl methacrylate).

Clearance angles of 9–15 deg are usually sufficient to prevent the drill flanks from rubbing in the bottom of the hole, but acrylic materials require angles of 12–20 deg. Tests may be needed to determine the drill diameter for accurately sized holes, taking thermal expansion and elastic recovery into account. Reaming may be used to size holes accurately, but diameters produced may also be affected by thermal expansion of the plastics. Close-fitting bushings in drill jigs may increase friction on the drill and cause chips to plug up the drill flutes. For positioning accuracy, removable templates may be used to spot the hole position, then removed for the drilling to be completed. Pilot holes are not necessary, except when the hole is to be reamed or counterbored. Peck feeds to remove chips and compressed air cooling may be needed, especially for deep holes.

Drilling and reaming speed and feed recommendations for various materials are shown in Table 7.

These speeds and feeds can be increased where there is no melting, burning, discoloration, or poor surface finish. Drilling is best done with commercially available drills designed for plastics, usually having large helix angles, narrow lands, and highly polished or chromium-plated flutes to expel chips rapidly and minimize frictional heating. Circle cutters are often preferred for holes in thin materials. Deep holes may require peck feeds. Drills must be kept sharp and cool, and carbide tools may be needed in high production, especially with glass-reinforced materials. Cool with clean compressed air to avoid contamination. Use aqueous solutions for deep drilling because metalcutting fluids and oils may degrade or attack the plastics and may cause a cleaning problem. Hold plastics parts firmly during drilling to counter the tendency for the tooling to grab and spin the work.

Table 7. Speeds and Feeds for Drilling Holes of 0.25 to 0.375 in. Diameter in Various Thermoplastics

Material	Speed (rpm)	Feed[a]	Comments
Polyethylene	1,000–2,000	H	Easy to machine
Polyvinyl chloride	1,000–2,000	M	Tends to become gummy
Acrylic	500–1,500	M–H	Easy to drill with lubricant
Polystyrene	500–1,500	H	Must have coolant
ABS	500–1,000	M–H	
Polytetrafluoroethylene	1,000	L–M	Easy to drill
Nylon 6/6	1,000	H	Easy to drill
Polycarbonate	500–1,500	M–H	Easy to drill, some gumming
Acetal	1,000–2,000	H	Easy to drill
Polypropylene	1,000–2,000	H	Easy to drill
Polyester	1,000–1,500	H	Easy to drill

[a] H = high; M = medium; L = low.

Tapping and Threading of Plastics: Many different threaded fasteners can be used with plastics, including thread-tapping and -forming screws, threaded metal inserts, and molded-in threads, but threads must sometimes be machined after molding. For tapping of through-holes in thin cast, molded, or extruded thermoplastics and thermosets, a speed of 50 ft/min is appropriate. Tapping of filled materials is done at 25 ft/min. These speeds should be reduced for deep or blind holes, and when the percentage of thread is greater than 65–75 per cent. Taps should be of M10, M7, or M1, molybdenum high-speed steel, with finish-ground and -polished flutes. Two-flute taps are recommended for holes up to 0.125 in. diameter. Oversize taps may be required to make up for elastic recovery of the plastics. The danger of retapping on the return stroke can be reduced by blunting the withdrawal edges of the tool.

Sawing Thermoset Cast or Molded Plastics: Circular or band saws may be used for sawing. Circular saws provide smoother cut faces than band saws, but band saws run cooler so are often preferred even for straight cuts. Projection of the circular saw above the table should be minimized. Saws should have skip teeth or buttress teeth with zero front rake and a raker set. Precision-tooth saw blades should be used for thicknesses up to 1 in., and saws with buttress teeth are recommended for thicknesses above 1 in. Dull edges to the teeth cause chipping of the plastics and may cause breakage of the saw. Sawing speeds and other recommendations for using blades of high-carbon steel are shown in the accompanying table.

Speeds and Numbers of Teeth for Sawing Plastics Materials with High-Carbon Steel Saw Blades

Material Thickness (in.)	Number of Teeth on Blade	Peripheral Speed (ft/min)	
		Thermoset Cast or Molded Plastics	Thermoplastics (and Epoxy, Melamine, Phenolic and Allyl Thermosets)
0–0.5	8–14	2000–3000	4000–5000
0.5–1	6–8	1800–2200	3500–4300
1–3	3	1500–2200	3000–3500
>3	>3	1200–1800	2500–3000

Milling of Plastics: Peripheral cutting with end mills is used for edge preparation, slotting and similar milling operations, and end cutting can also be used for facing operations. Speeds for milling range from 800 to 1400 ft/min for peripheral end milling of many thermoplastics and from 400 to 800 ft/min for many thermosets. However, slower speeds are generally used for other milling operations, with some thermoplastics being machined at 300–500 ft/min, and some thermosets at 150–300 ft/min. Adequate support and suitable feed rates are very important. A table feed that is too low will generate excessive heat and cause surface cracks, loss of dimensional accuracy, and poor surface finish. Too high a feed rate will produce a rough surface. High-speed steel tools (M2, M3, M7, or T15) are generally used, but for glass-reinforced nylon, silicone, polyimide, and allyl, carbide (C2) is recommended.

New Techniques: Lasers can be used for machining plastics, especially sheet laminates, although their use may generate internal stresses. Ultrasonic machining has no thermal, chemical, or electrical reaction with the workpiece and can produce holes down to 0.003 in. diameter, tight tolerances (0.0005 in.), and very smooth finishes (0.15 μin. with No. 600 boron carbide abrasive powder). Water-jet cutting using pressures up to 60,000 lb/in.^{2} is widely used for plastics and does not introduce stresses into the material. Tolerances of ± 0.004 in. can be held, depending on the equipment available. Process variables, pressures, feed rates, and the nozzle diameter depend on the material being cut. This method does not work with hollow parts unless they can be filled with a solid core.

Development of Prototypes.—Prototypes are made for testing of properties such as stress and fatigue resistance, to find ways to improve quality and reliability, to improve tooling, and to reduce time to market. Prototyping may answer questions about finish, sink marks that result from contraction, witness lines from mold joints, ejector pin marks, knit or weld lines, texturing, moldability, shrinkage, mechanical strength, pull-out resistance of inserts, electrical properties, and problems of mating with other parts.

Prototypes of moldings are made in five major steps including design; refining the design; making a model (physical or computer); making a mold; and producing parts. The model may be made from wood, plaster, plastics (by machining), or a metal. Some 90 per cent of prototypes are made by modern CAD/CAM methods that allow holding of dimensional tolerances of 2–3 per cent of drawing specifications.

Prototypes can also be made by a process called stereo lithography that uses a tank of photosensitive liquid polymer, an *x-y* scanning, ultraviolet laser with a beam diameter of 0.010 in., a *z*-axis elevator platform, and a controlling computer. The platform height is adjusted so that a suitable thickness of liquid polymer covers its surface. The laser beam is focused on the liquid surface and hardens the polymer at this point by heating.

The CAD representation of the prototype is described by a model in which thin (0.005–0.020 in.) cross sections can be isolated. Data representing the lowest level of the prototype are used to move the platform so that a layer of the polymer corresponding to the lowest "slice" is hardened. The platform is then lowered, the liquid polymer flows over the hardened layer, and the platform is again raised, less an amount equal to the next "slice." The process is repeated for successive "slices" of the prototype, which is thus built up gradually to form a hollow, three-dimensional shape corresponding to the model in the CAD program. The part thus produced is fairly brittle but can be used for visual examination, design verification, and marketing evaluation, and can be replicated from other materials such as plastics or metals by casting or other methods.

Finishing and decorating methods used for plastics parts include spray painting, vacuum metallizing, hot stamping, silk screening, and plating. Conductive coatings may be applied to inside surfaces, usually by flame- or arc-spraying, to dissipate static electricity and provide electromagnetic shielding. Thorough cleaning is essential. Materials such as polyethylene, polypropylene, and acetal have waxlike surfaces that may not be painted easily or may need pretreatment or special primers. Many amorphous plastics are easy to paint.

Suitable coatings include polyurethane-, epoxy-, acrylic-, alkyd-, and vinyl-based paints. Oven curing may distort parts made from non-heat-resistant materials.

Vacuum metallizing and sputter-plating require application of a special base coat and a protective clear top coat before and after treatment. Resistance heating or an electron beam can be used to melt the metallizing materials such as aluminum, silver, copper, and gold, which usually are pure elements. Sputter plating uses a plasma to produce the metallic vapor and can use brass as well as the metals mentioned. Chromium plating requires etched surfaces to ensure good adhesion.

Plastics may be polished by buffing methods similar to those used on metals, but experiments to determine the effect of frictional heat are recommended. Surfaces can be heated to 300–400 deg. F by buffing, and some plastics soften and melt at these temperatures. Heating sometimes causes plastics to give off toxic gases, so masks should be worn to filter out such gases and dust. Parting lines, imperfections, scratches, saw lines, imperfections and scars resulting from fabrication can be treated with abrasives prior to buffing. Wet or dry abrasives such as silicon carbide or aluminum oxide are generally used, in grain sizes of 60 to as fine as 320. Some buffing compounds are ineffective on plastics. Scratch lines should be presented at a slight angle to the buff surface for best results. Light, tallow-free grease will help keep the abrasive surface free from buildup, and speeds of 5,000 to 6,000 surface feet per minute are recommended.

For low-melting point plastics, soft cotton buffs are best, with surface speeds of 4,000 to 5,000 feet per minute, using a wet or greasy tripoli or silica compound. For finishing, only rouge may be needed for a satisfactory finish. If a cleaning solvent is used it should be checked to see that it does not dissolve the plastics, and it should be used only in a well-ventilated area. Acrylics such as Acrylite or Plexiglass may also be 'flame polished,' under advice from the materials supplier.

Plastics Gearing.—Plastics gears may be cut from blanks, as with metal gears, or molded to shape in an injection-molding machine, for lower production costs, though tooling may cost more. Cut plastics gears may be of similar design to their metal counterparts, but molded gears are usually of modified form to suit the material characteristics. Plastics materials also may be preferred for gears because of superior sliding properties with reduced noise and need for lubrication, chemical or electrical properties, or resistance to wear. However, plastics gear teeth slide more smoothly and easily against metal teeth than do plastics against plastics, and wear is less. For power transmission, plastics gear teeth are usually of involute form. See also *Non-metallic Gearing* on page 2119.

Most plastics gears are made from nylons and acetals, although acrylonitrile-butadiene-styrenes (ABS), polycarbonates, polysulfones, phenylene oxides, poly-urethanes, and thermoplastic polyesters can also be used. Additives used in plastics gears include glass fiber for added strength, and fibers, beads, and powders for reduced thermal expansion and improved dimensional stability. Other materials, such as molybdenum disulfide, tetrafluoroethylene (TFE), and silicones, may be added as lubricants to improve wear resistance.

Choice of plastics gear material depends on requirements for size and nature of loads to be transmitted, speeds, required life, working environment, type of cooling, lubrication, and operating precision. Because of cost, plastics gears are sometimes not enclosed in sealed housings, so are often given only a single coating of lubricant grease. Overloading of lubricated plastics gear teeth will usually cause tooth fracture, and unlubricated teeth often suffer excessive wear. Thermoplastics strength varies with temperature, with higher temperatures reducing root stress and permitting tooth deformation. In calculating power to be transmitted by spur, helical, and straight bevel gearing, the following formulas should be used with the factors given in Tables 2, 3 and 1.

For internal and external spur gears,

$$HP = \frac{S_s FYV}{55(600 + V)PC_s}$$

For internal and external helical gears,

$$HP = \frac{S_s FYV}{423(78 + \sqrt{V})P_n C_s}$$

For straight bevel gears,

$$HP = \frac{S_s FYV(C - F)}{55(600 + V)PCC_s}$$

where S_s = safe stress in bending (from Table 2); F = face width in inches; Y = tooth form factor (from Table 1); C = pitch cone distance in inches; C_s = service factor (from Table 3); P = diametral pitch; P_n = normal diametral pitch; and V = velocity at pitch circle diameter in ft/min.

Table 1. Tooth Form Factors *Y* for Plastics Gears

Number of Teeth	14½-deg Involute or Cycloidal	20-deg Full Depth Involute	20-deg Stub Tooth Involute	20-deg Internal Full Depth	
				Pinion	Gear
12	0.210	0.245	0.311	0.327	…
13	0.220	0.261	0.324	0.327	…
14	0.226	0.276	0.339	0.330	…
15	0.236	0.289	0.348	0.330	…
16	0.242	0.259	0.361	0.333	…
17	0.251	0.302	0.367	0.342	…
18	0.261	0.308	0.377	0.349	…
19	0.273	0.314	0.386	0.358	…
20	0.283	0.320	0.393	0.364	…
21	0.289	0.327	0.399	0.371	…
22	0.292	0.330	0.405	0.374	…
24	0.298	0.336	0.415	0.383	…
26	0.307	0.346	0.424	0.393	…
28	0.314	0.352	0.430	0.399	0.691
30	0.320	0.358	0.437	0.405	0.679
34	0.327	0.371	0.446	0.415	0.660
38	0.336	0.383	0.456	0.424	0.644
43	0.346	0.396	0.462	0.430	0.628
50	0.352	0.480	0.474	0.437	0.613
60	0.358	0.421	0.484	0.446	0.597
75	0.364	0.434	0.496	0.452	0.581
100	0.371	0.446	0.506	0.462	0.565
150	0.377	0.459	0.518	0.468	0.550
300	0.383	0.471	0.534	0.478	0.534
Rack	0.390	0.484	0.550	…	…

These values assume a moderate temperature increase and some initial lubrication. With bevel gearing, divide the number of teeth by the cosine of the pitch angle and use the data in the table. For example, if a 20-deg PA bevel gear has 40 teeth and a pitch angle of 58 deg, 40 divided by the cosine of 58 deg = 40 ÷ 0.529919 ~ 75, and Y = 0.434.

Table 2. Safe Bending Stress (lb/in²) Values for Plastics Gears

Plastics Type	Safe Stress	
	Unfilled	Glass-filled
ABS	3,000	6,000
Acetal	5,000	7,000
Nylon	6,000	12,000
Polycarbonate	6,000	9,000
Polyester	3,500	8,000
Polyurethane	2,500	…

Table 3. Service Factors for Plastics Gears

Type of Load	8–10 Hr/Day	24 Hr/Day	Intermittent, 3 Hr/Day	Occasional, ½ Hr/Day
Steady	1.00	1.25	0.80	0.50
Light shock	1.25	1.5	1.00	0.80
Medium shock	1.5	1.75	1.25	1.00
Heavy shock	1.75	2.00	1.5	1.25

As an example, assume that a material is to be selected for a spur gear that must transmit ⅛ hp at 350 rpm, for 8 hrs/day under a steady load. The gear is to have 75 teeth, 32 diametral pitch, 20 deg pressure angle, 0.375 in. face width, and a pitch diameter of 2.3438 in. Using the first formula above,

$$HP = \frac{S_s FYV}{55(600+V)PC_s} \quad \text{or} \quad S_s = \frac{55(600+V)PC_s HP}{FYV}$$

$$\text{hp} = 0.125, \quad Y = 0.434 \quad \text{and}$$

$$V = \frac{rpm \times \pi \times D}{12} = \frac{350 \times 3.1416 \times 2.3438}{12} = 215 \text{ ft/min}$$

therefore,

$$S_s = \frac{55(600+215)32 \times 1.00 \times 0.125}{0.375 \times 0.434 \times 215} = 5{,}124 \text{ lb/in.}^2$$

From Table 2 it is apparent that the gear could be molded from several materials. Available physical and chemical characteristics must now be considered in relation to the operating environment for the gear. Strengths of plastics materials decrease with increasing temperatures and not all plastics resist the effects of some liquids, including some lubricants. Some plastics deteriorate when in sunlight for long periods; some are more dimensionally stable than others; and wear resistance varies from one to another. Manufacturers' data sheets will answer some of these questions.

Backlash: Plastics gears should be so dimensioned that they will provide sufficient backlash at the highest temperatures likely to be encountered in service. Dimensional allowances must also be made for gears made of hygroscopic plastics that may be exposed to damp service conditions. Teeth of heavily loaded gears usually have tip relief to reduce effects of deflection, and have full fillet radii to reduce stress concentrations. Such modifications to tooth form are also desirable in plastics gears. If the pinion in a pair of gears has a small number of teeth, undercutting may result. Undercutting weakens teeth, causes undue wear, and may affect continuity of action. The undercutting can be reduced by using the long-short addendum system, which involves increasing the addendum of the pinion teeth and reducing that of the gear teeth. The modified addendum method will also reduce the amount of initial wear that takes place during the initial stages of contact between the teeth.

Accuracy: The Gear Handbook, AGMA 390-03a-1980, Part 2, Gear Classification, provides a system whereby results of gear accuracy measurements are expressed in terms of maximum tooth-to-tooth and composite tolerances. This system uses AGMA quality numbers related to maximum tolerances, by pitch and diameter, and is equally applicable to plastics gears as to metal gears. AGMA quality numbers must be chosen for a pair of mating gears early in the design process, and the finished gears must be inspected by being run in close mesh with a master gear in a center-distance measuring instrument to make sure that the errors do not exceed the specified tolerances.

To prevent failure from fatigue and wear caused by excessive flexing of the teeth, plastics gears must be made to the same standards of acccuracy as metal gears. Solidification shrinkage of plastics requires that dimensions of molds for gears be larger than the dimensions of the parts to be produced from them. The amount of the shrinkage is usually added to the mold dimension (with the mold at operating temperature). However, this procedure cannot be followed for the tooth profile as it would introduce large errors in the pressure angle. Increases in pressure angle cause gear teeth to become wider at the root and more pointed. Sliding conditions are improved and the teeth are stronger, so that higher loading values can be used.

Shrinkage allowances have the greatest effect on the accuracy of the molded gears, so tooth profiles must be calculated extremely carefully in terms of mold profile. If a tooth is merely made larger by using a standard hobbing cutter to cut the tool whereby the teeth in the mold are electroeroded, differential shrinkage caused by the molded tooth being thicker at the root than at the tip will distort the shape of the molded tooth, making it thinner at the tip and thicker at the root. With two mating gears, these faulty shapes will affect the pressure angle resulting in binding, wear, and general malfunction. If the tooth thickness limits for a molded gear are to be held to +0.000 in., −0.001 in., the outside diameter must be permitted to vary up to 0.0027 in. for 20-deg, and 0.0039 in. for 14½-deg pressure angle gears. All high-accuracy gears should be specified with AGMA quality numbers and inspected with center-distance measuring machines if the required accuracy is to be achieved.

DIMENSIONING, GAGING, AND MEASURING

DRAFTING PRACTICES

ALLOWANCES AND TOLERANCES FOR FITS

ALLOWANCES AND TOLERANCES FOR FITS (cont.)

MEASURING INSTRUMENTS AND INSPECTION METHODS

SURFACE TEXTURE

DRAFTING PRACTICES

American National Standard Drafting Practices

Several American National Standards for use in preparing engineering drawings and related documents are referred to for use.

Sizes of Drawing Sheets.—Recommended trimmed sheet sizes, based on ANSI Y14.1-1980 (R1987), are shown in the following table.

Size, inches				Metric Size, mm			
A	8½×11	*D*	22×34	*A0*	841×1189	*A3*	297×420
B	11×17	*E*	34×44	*A1*	594×841	*A4*	210×297
C	17×22	*F*	28×40	*A2*	420×594		

The standard sizes shown by the left-hand section of the table are based on the dimensions of the commercial letter head, 8½× 11 inches, in general use in the United States. The use of the basic sheet size 8½× 11 inches and its multiples permits filing of small tracings and folded blueprints in commercial standard letter files with or without correspondence. These sheet sizes also cut without unnecessary waste from the present 36-inch rolls of paper and cloth.

For drawings made in the metric system of units or for foreign correspondence, it is recommended that the metric standard trimmed sheet sizes be used. (Right-hand section of table.) These sizes are based on the width-to-length ratio of 1 to $\sqrt{2}$.

Line Conventions and Drawings.—American National Standard Y14.2M-1979 (R1987) establishes line and lettering practices for engineering drawings. The line conventions and the symbols for section lining are as shown on pages 607 and 608.

Approximate width of THICK lines for metric drawings are 0.6 mm, and for inch drawings, 0.032 inch. Approximate width of THIN lines for metric drawings are 0.3 mm, and for inch drawings, 0.016 inch. These approximate line widths are intended to differentiate between THICK and THIN lines and are not values for control of acceptance or rejection of the drawings.

Surface-Texture Symbols.—A detailed explanation of the use of surface-texture symbols from American National Standard Y14.36-1978 (R1987) begins on page 709.

Geometric Dimensioning and Tolerancing.—ANSI/ASME Y14.5M-1994, "Dimensioning and Tolerancing," covers dimensioning, tolerancing, and similar practices for engineering drawings and related documentation. The mathematical definitions of dimensioning and tolerancing principles are given in the standard ANSI/ASME Y14.5.1M-1994. ISO standards ISO 8015 and ISO 26921 contain a detailed explanation of ISO geometric dimensioning and tolerancing practices.

Geometric dimensioning and tolerancing provides a comprehensive system for symbolically defining the geometrical tolerance zone within which features must be contained. It provides an accurate transmission of design specifications among the three primary users of engineering drawings; design, manufacturing and quality assurance.

Some techniques introduced in ANSI/ASME Y14.5M-1994 have been accepted by ISO. These techniques include projected tolerance zone, three-plane datum concept, total runout tolerance, multiple datums, and datum targets. Although this Standard follows ISO practice closely, there are still differences between ISO and U.S. practice. (A comparison of the symbols used in ISO standards and Y14.5M is given on page 609.)

American National Standard for Engineering Drawings
ANSI/ASME Y14.2M-1992

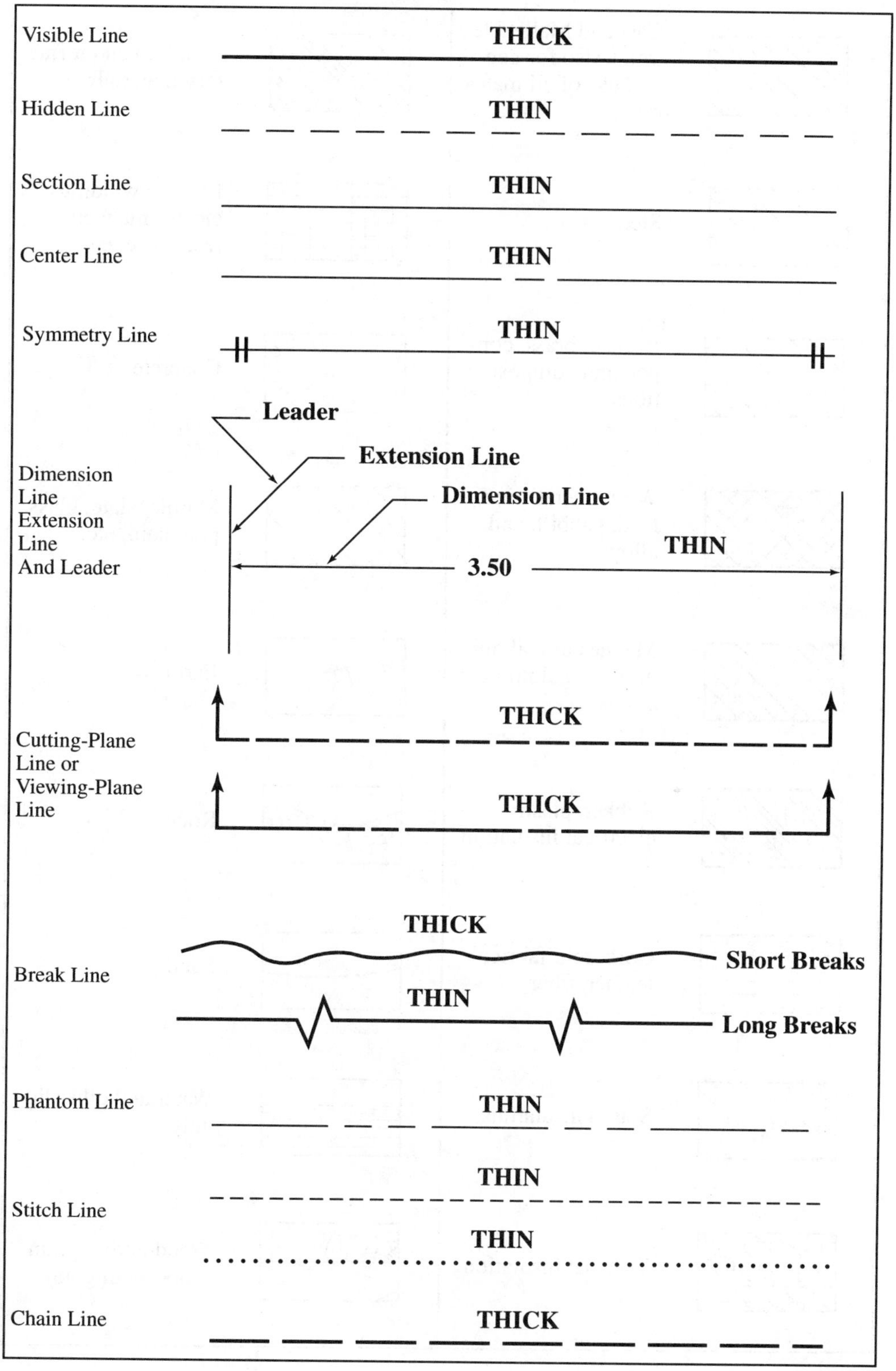

Visible Line — THICK

Hidden Line — THIN

Section Line — THIN

Center Line — THIN

Symmetry Line — THIN

Dimension Line Extension Line And Leader — THIN

Leader

Extension Line

Dimension Line

3.50

Cutting-Plane Line or Viewing-Plane Line — THICK

THICK

Break Line — THICK — Short Breaks

THIN — Long Breaks

Phantom Line — THIN

Stitch Line — THIN

THIN

Chain Line — THICK

American National Standard Symbols for Section Lining
ANSI Y14.2M-1979 (R1987)

Material	Material
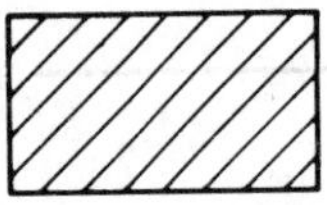 Cast and Malleable iron (Also for general use of all materials)	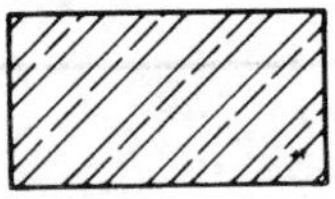 Titanium and refractory material
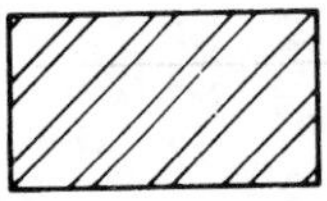 Steel	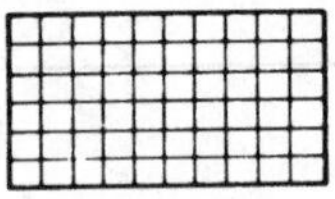 Electric windings, electro magnets, resistance, etc.
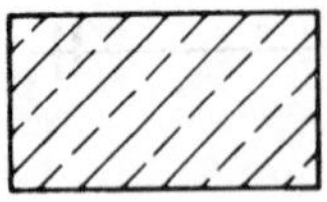 Bronze, brass, copper, and compositions	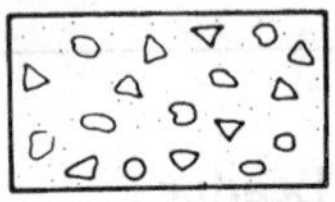 Concrete
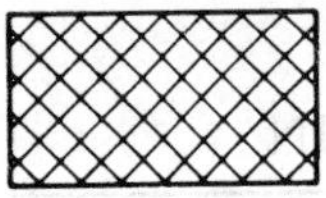 White metal, zinc, lead, babbitt, and alloys	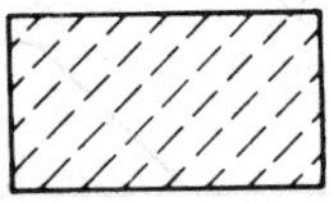 Marble, slate, glass, porcelain, etc.
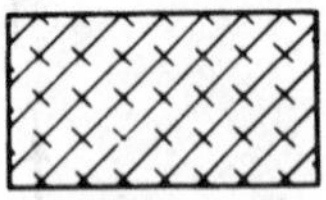 Magnesium, aluminum, and aluminum alloys	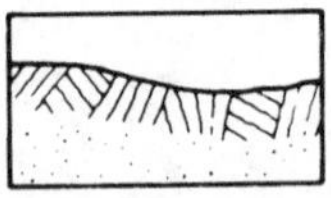 Earth
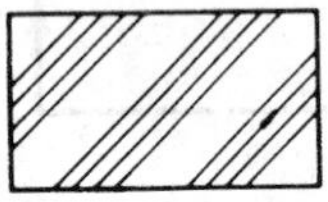 Rubber, plastic electrical insulation	 Rock
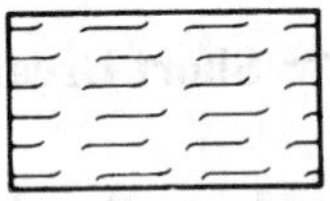 Cork, felt, fabric, leather, fiber	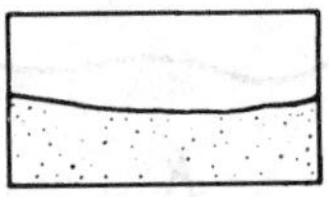 Sand
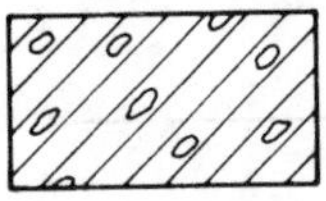 Sound insulation	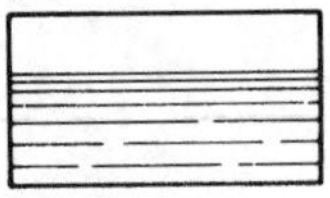 Water and other liquids
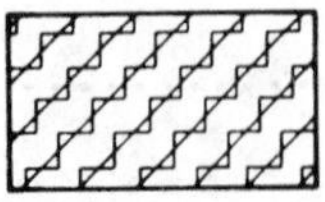 Thermal insulation	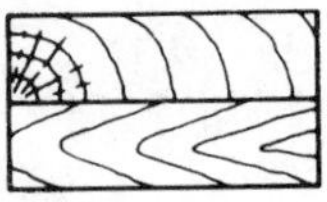 Wood-across grain Wood-with grain

Comparison of ANSI and ISO Geometric Symbols *ASME Y14.5M-1994*

Symbol for	ANSI Y14.5	ISO	Symbol for	ANSI Y14.5	ISO	Symbol for	ANSI Y14.5	ISO
Straightness	⏤	⏤	Circular Runout[a]	↗	↗	Feature Control Frame	⌖ Ø 0.5 Ⓜ A B C	⌖ Ø 0.5 Ⓜ A B C
Flatness	⏥	⏥	Total Runout[a]	⌰	⌰	Datum Feature[a]	-A-	⏚ OR ⏚ A
Circularity	○	○	At Maximum Material Condition	Ⓜ	Ⓜ	All Around - Profile	⌀→	⌀→ (proposed)
Cylindricity	⌭	⌭	At Least Material Condition	Ⓛ	Ⓛ	Conical Taper	⌲	⌲
Profile of a Line	⌒	⌒	Regardless of Feature Size	NONE	NONE	Slope	⌳	⌳
Profile of a Surface	⌓	⌓	Projected Tolerance Zone	Ⓟ	Ⓟ	Counterbore/Spotface	⌴	⌴ (proposed)
Angularity	∠	∠	Diameter	Ø	Ø	Countersink	⌵	⌵ (proposed)
Perpendicularity	⊥	⊥	Basic Dimension	50 (boxed)	50 (boxed)	Depth/Deep	↧	↧ (proposed)
Parallelism	//	//	Reference Dimension	(50)	(50)	Square (Shape)	□	□
Position	⌖	⌖	Datum Target	Ø6/A1 (circled)	Ø6/A1 (circled)	Dimension Not to Scale	15	15
Concentricity/Coaxiality	◎	◎	Target Point	×	×	Number of Times/Places	8X	8X
Symmetry	⌯	⌯	Dimension Origin	⌀→	⌀→	Arc Length	⌒105	⌒105
Radius	R	R	Spherical Radius	SR	SR	Sperical Diameter	SØ	SØ

[a] Arrowheads may be filled in.

One major area of disagreement is the ISO "principle of independency" versus the "Taylor principle." Y14.5M and standard U.S. practice both follow the Taylor principle, in which a geometric tolerancing zone may not extend beyond the boundary (or envelope) of perfect form at MMC (maximum material condition). This boundary is prescribed to control variations as well as the size of individual features. The U.S. definition of independency further defines features of size as being independent and not required to maintain a perfect relationship with other features. The envelope principle is optional in treatment of these principles. A summary of the application of ANSI/ASME geometric control symbols and their use with basic dimensions and modifiers is given in Table 1.

Table 1. Application of Geometric Control Symbols

<table>
<tr><th>Type</th><th>Geometric Characteristics</th><th>Pertains To</th><th>Basic Dimensions</th><th>Feature Modifier</th><th>Datum Modifier</th></tr>
<tr><td rowspan="4">Form</td><td>Straightness</td><td rowspan="4">ONLY individual feature</td><td rowspan="4"></td><td rowspan="6">Modifier not applicable</td><td rowspan="4">NO datum</td></tr>
<tr><td>Circularity</td></tr>
<tr><td>Flatness</td></tr>
<tr><td>Cylindricity</td></tr>
<tr><td rowspan="2">Profile</td><td>Profile (Line)</td><td rowspan="2">Individual or related</td><td rowspan="2">Yes if related</td><td rowspan="6">RFS implied unless MMC or LMC is stated</td></tr>
<tr><td>Profile (Surface)</td></tr>
<tr><td rowspan="3">Orientation</td><td>Angularity</td><td rowspan="6">ALWAYS related feature(s)</td><td>Yes</td><td rowspan="4">RFS implied unless MMC or LMC is stated</td></tr>
<tr><td>Perpendicularity</td><td rowspan="2"></td></tr>
<tr><td>Parallelism</td></tr>
<tr><td rowspan="3">Location</td><td>Position</td><td>Yes</td></tr>
<tr><td>Concentricity</td><td rowspan="4"></td><td rowspan="4">Only RFS</td><td rowspan="4">Only RFS</td></tr>
<tr><td>Symmetry</td></tr>
<tr><td rowspan="2">Runout</td><td>Circular Runout</td><td rowspan="2"></td></tr>
<tr><td>Total Runout</td></tr>
</table>

Five types of geometric control, when datums are indicated, when basic dimensions are required, and when MMC and LMC modifiers may be used.

ANSI/ASME Y14.5M features metric SI units (the International System of Units), but customary units may be used without violating any principles. On drawings where all dimensions are either in millimeters or in inches, individual identification of linear units is not required. However, the drawing should contain a note stating UNLESS OTHERWISE SPECIFIED, ALL DIMENSIONS ARE IN MILLIMETERS (or IN INCHES, as applicable). According to this Standard, all dimensions are applicable at a temperature of 20 C (68 F) unless otherwise specified. Compensation may be made for measurements taken at other temperatures.

Angular units are expressed in degrees and decimals of a degree (35.4) or in degrees (°), minutes (′), and seconds (″), as in 35° 25′ 10″.A 90-degree angle is implied where center lines and depicting features are shown on a drawing at right angles and no angle is specified. A 90-degree BASIC angle applies where center lines of features in a pattern or surface shown at right angles on a drawing are located or defined by basic dimensions and no angle is specified.

Definitions.—The following terms are defined as their use applies to ANSI/ASME Y14.5M.

Datum Feature: The feature of a part that is used to establish a datum.

Datum Identifier: The graphic symbol on a drawing used to indicate the datum feature.

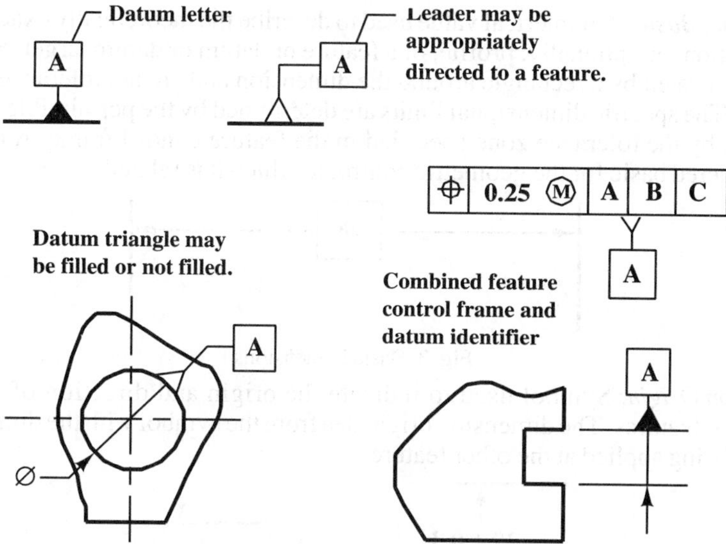

Fig. 1. Datum Feature Symbol

Datum Plane: The individual theoretical planes of the reference frame derived from a specified datum feature. A datum is the origin from which the location or other geometric characteristics of features of a part are established.

Datum Reference Frame: Sufficient features on a part are chosen to position the part in relationship to three planes. The three planes are mutually perpendicular and together called the datum reference frame. The planes follow an order of precedence and allow the part to be immobilized. This immobilization in turn creates measurable relationships among features.

Datum Simulator: Formed by the datum feature contacting a precision surface such as a surface plate, gage surface or by a mandrel contacting the datum. Thus, the plane formed by contact restricts motion and constitutes the specific reference surface from which measurements are taken and dimensions verified. The datum simulator is the practical embodiment of the datum feature during manufacturing and quality assurance.

Datum Target: A specified point, line, or area on a part, used to establish a datum.

Degrees of Freedom: The six directions of movement or translation are called degrees of freedom in a three-dimensional environment. They are up-down, left-right, fore-aft, roll, pitch and yaw.

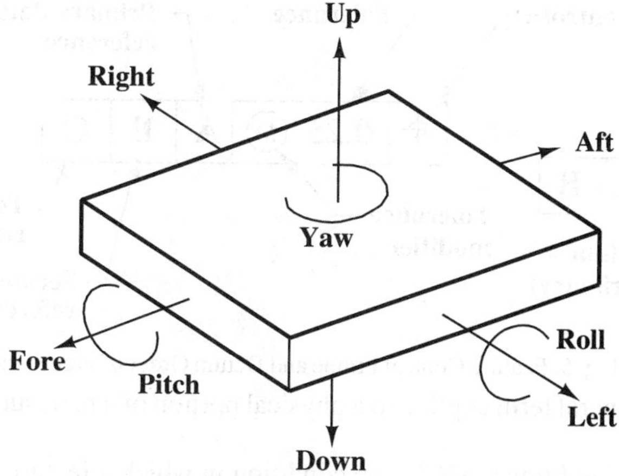

Fig. 2. Degrees of Freedom (Movement) That Must be Controlled, Depending on the Design Requirements.

Dimension, Basic: A numerical value used to describe the theoretically exact size, orientation, location, or optionally, profile, of a feature or datum or datum target. Basic dimensions are indicated by a rectangle around the dimension and are not toleranced directly or by default. The specific dimensional limits are determined by the permissible variations as established by the tolerance zone specified in the feature control frame. A dimension is only considered basic for the geometric control to which it is related.

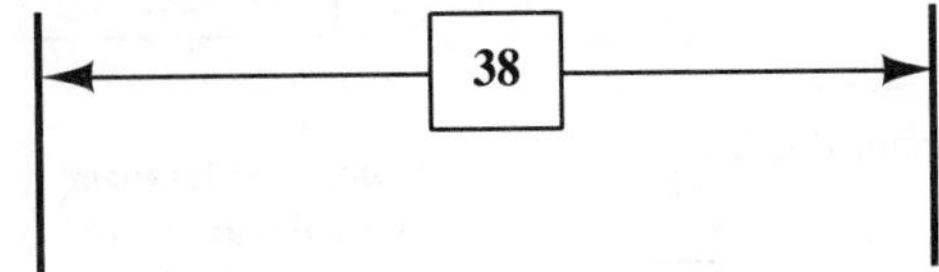

Fig. 3. Basic Dimensions

Dimension Origin: Symbol used to indicate the origin and direction of a dimension between two features. The dimension originates from the symbol with the dimension tolerance zone being applied at the other feature.

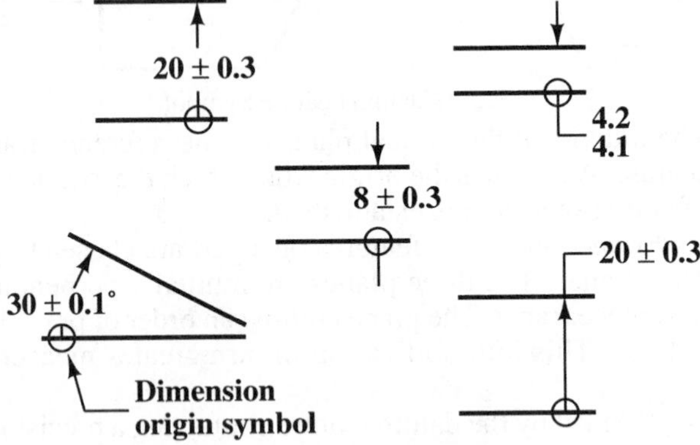

Fig. 4. Dimension Origin Symbol

Dimension, Reference: A dimension, usually without tolerance, used for information purposes only. Considered to be auxiliary information and not governing production or inspection operations. A reference dimension is a repeat of a dimension or is derived from a calculation or combination of other values shown on the drawing or on related drawings.

Feature Control Frame: Specification on a drawing that indicates the type of geometric control for the feature, the tolerance for the control, and the related datums, if applicable.

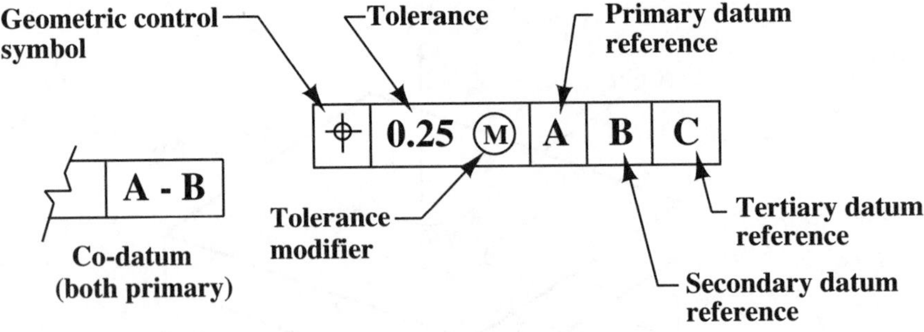

Fig. 5. Feature Control Frame and Datum Order of Precedence

Feature: The general term applied to a physical portion of a part, such as a surface, hole, pin, tab, or slot.

Least Material Condition (LMC): The condition in which a feature of size contains the least amount of material within the stated limits of size, for example, upper limit or maximum hole diameter and lower limit or minimum shaft diameter.

Limits, Upper and Lower (UL and LL): The arithmetic values representing the maximum and minimum size allowable for a dimension or tolerance. The upper limit represents the maximum size allowable. The lower limit represents the minimum size allowable.

Maximum Material Condition (MMC): The condition in which a feature of size contains the maximum amount of material within the stated limits of size. For example, the lower limit of a hole is the minimum hole diameter. The upper limit of a shaft is the maximum shaft diameter.

Position: Formerly called true position, position is the theoretically exact location of a feature established by basic dimensions.

Regardless of Feature Size (RFS): The term used to indicate that a geometric tolerance or datum reference applies at any increment of size of the feature within its tolerance limits. RFS is the default condition unless MMC or LMC is specified. The concept is now the default in ANSI/ASME Y14.5M-1994, unless specifically stated otherwise. Thus the symbol for RFS is no longer supported in ANSI/ASME Y14.5M-1994.

Size, Actual: The term indicating the size of a feature as produced.

Size, Feature of: A feature that can be described dimensionally. May include a cylindrical or spherical surface, or a set of two opposed parallel surfaces associated with a size dimension.

Tolerance Zone Symmetry: In geometric tolerancing, the tolerance value stated in the feature control frame is always a single value. Unless otherwise specified, it is assumed that the boundaries created by the stated tolerance are bilateral and equidistant about the perfect form control specified. However, if desired, the tolerance may be specified as unilateral or unequally bilateral. (See Figs. 6 through 8)

Tolerance, Bilateral: A tolerance where variation is permitted in both directions from the specified dimension. Bilateral tolerances may be equal or unequal.

Tolerance, Geometric: The general term applied to the category of tolerances used to control form, profile, orientation, location, and runout.

Tolerance, Unilateral: A tolerance where variation is permitted in only one direction from the specified dimension.

True Geometric Counterpart: The theoretically perfect plane of a specified datum feature.

Virtual Condition: A constant boundary generated by the collective effects of the feature size, its specified MMC or LMC material condition, and the geometric tolerance for that condition.

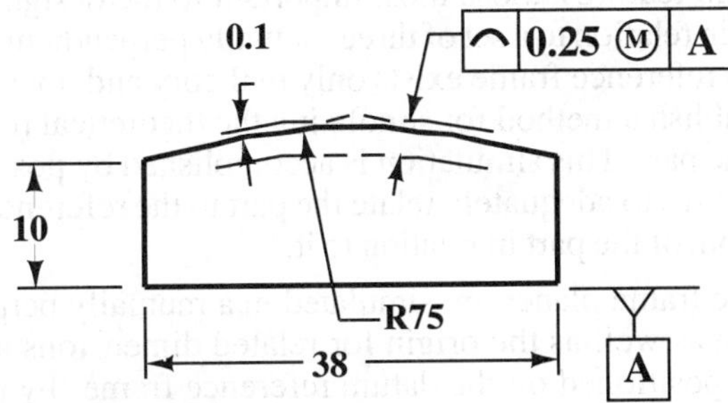

Bilateral zone with 0.1 of the 0.25 tolerance outside perfect form.

Fig. 6. Application of a bilateral geometric tolerance

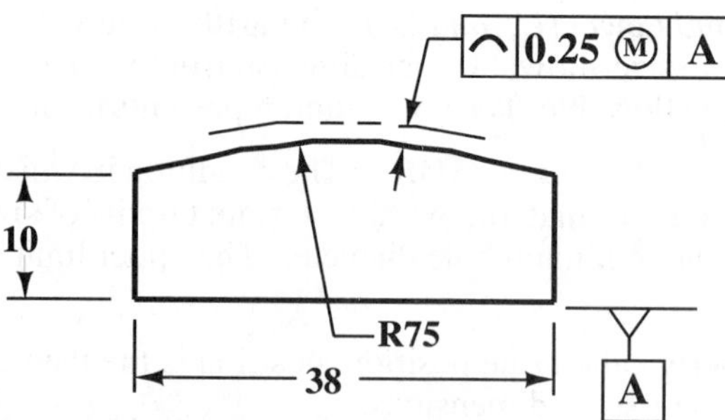

Fig. 7. Application of a unilateral geometric tolerance zone outside perfect form

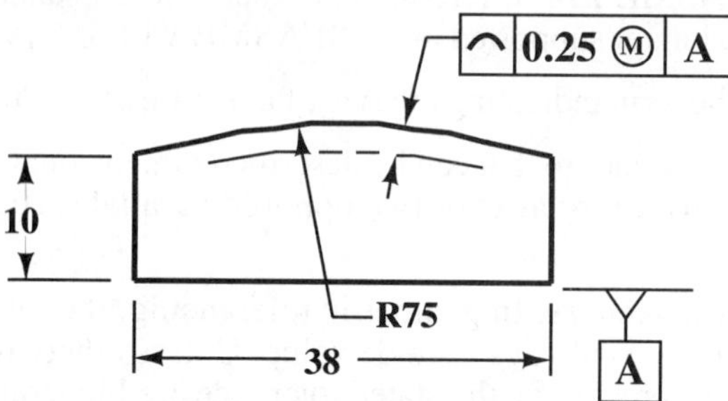

Fig. 8. Application of a unilateral geometric tolerance zone inside a perfect form

Datum Referencing.—A datum indicates the origin of a dimensional relationship between a toleranced feature and a designated feature or features on a part. The designated feature serves as a datum feature, whereas its true geometric counterpart establishes the datum plane. Because measurements cannot be made from a true geometric counterpart, which is theoretical, a datum is assumed to exist in, and be simulated by the associated processing equipment.

For example, machine tables and surface plates, although not true planes, are of such quality that they are used to simulate the datums from which measurements are taken and dimensions are verified. When magnified, flat surfaces of manufactured parts are seen to have irregularities, so that contact is made with a datum plane formed at a number of surface extremities or high points.

Sufficient datum features, those most important to the design of the part, are chosen to position the part in relation to a set of three mutually perpendicular planes, the datum reference frame. This reference frame exists only in theory and not on the part. Therefore, it is necessary to establish a method for simulating the theoretical reference frame from existing features of the part. This simulation is accomplished by positioning the part on appropriate datum features to adequately relate the part to the reference frame and to restrict the degrees of freedom of the part in relation to it.

These reference frame planes are simulated in a mutually perpendicular relationship to provide direction as well as the origin for related dimensions and measurements. Thus, when the part is positioned on the datum reference frame (by physical contact between each datum feature and its counterpart in the associated processing equipment), dimensions related to the datum reference frame by a feature control frame are thereby mutually perpendicular. This theoretical reference frame constitutes the three-plane dimensioning system used for datum referencing.

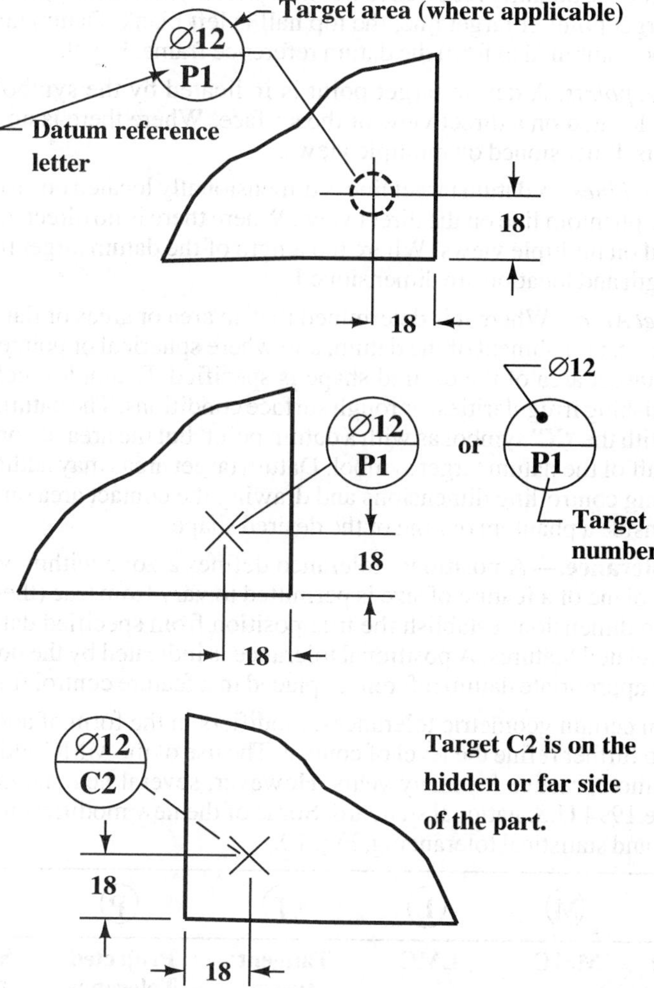

Fig. 9. Datum target symbols

Depending on the degrees of freedom that must be controlled, a simple reference frame may suffice. At other times, additional datum reference frames may be necessary where physical separation occurs or the functional relationship. Depending on the degrees of freedom that must be controlled, a single datum of features require that datum reference frames be applied at specific locations on the part. Each feature control frame must contain the datum feature references that are applicable.

Datum Targets: Datum targets are used to establish a datum plane. They may be points, lines or surface areas. Datum targets are used when the datum feature contains irregularities, the surface is blocked by other features or the entire surface cannot be used. Examples where datum targets may be indicated include uneven surfaces, forgings and castings, weldments, non-planar surfaces or surfaces subject to warping or distortion. The datum target symbol is located outside the part outline with a leader directed to the target point, area or line. The targets are dimensionally located on the part using basic or toleranced dimensions. If basic dimensions are used, established tooling or gaging tolerances apply. A solid leader line from the symbol to the target is used for visible or near side locations with a dashed leader line used for hidden or far side locations. The datum target symbol is divided horizontally into two halves. The top half contains the target point area if applicable; the bottom half contains a datum feature identifying letter and target number. Target numbers indicate the quantity required to define a primary, secondary, or tertiary datum. If indicating a target point or target line, the top half is left blank. Datum targets and datum features may be combined to form the datum reference frame, Fig. 9.

Datum Target points: A datum target point is indicated by the symbol "X," which is dimensionally located on a direct view of the surface. Where there is no direct view, the point location is dimensioned on multiple views.

Datum Target Lines: A datum target line is dimensionally located on an edge view of the surface using a phantom line on the direct view. Where there is no direct view, the location is dimensioned on multiple views. Where the length of the datum target line must be controlled, its length and location are dimensioned.

Datum Target Areas: Where it is determined that an area or areas of flat contact are necessary to ensure establishment of the datum, and where spherical or pointed pins would be inadequate, a target area of the desired shape is specified. Examples include the need to span holes, finishing irregularities, or rough surface conditions. The datum target area may be indicated with the "X" symbol as with a datum point, but the area of contact is specified in the upper half of the datum target symbol. Datum target areas may additionally be specified by defining controlling dimensions and drawing the contact area on the feature with section lines inside a phantom outline of the desired shape.

Positional Tolerance.—A positional tolerance defines a zone within which the center, axis, or center plane of a feature of size is permitted to vary from true (theoretically exact) position. Basic dimensions establish the true position from specified datum features and between interrelated features. A positional tolerance is indicated by the position symbol, a tolerance, and appropriate datum references placed in a feature control frame.

Modifiers: In certain geometric tolerances, modifiers in the form of additional symbols may be used to further refine the level of control. The use of the MMC and LMC modifiers has been common practice for many years. However, several new modifiers were introduced with the 1994 U.S. national standard. Some of the new modifiers include free state, tangent plane and statistical tolerancing, Fig. 10.

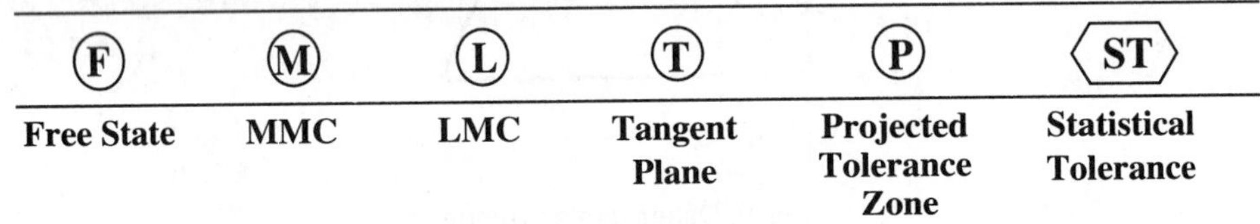

F
M
L
T
P
ST
Free State
MMC
LMC
Tangent Plane
Projected Tolerance Zone
Statistical Tolerance

Fig. 10. Tolerance modifiers

Projected Tolerance Zone: Application of this concept is recommended where any variation in perpendicularity of the threaded or press-fit holes could cause fasteners such as screws, studs, or pins to interfere with mating parts. An interference with subsequent parts can occur even though the hole axes are inclined within allowable limits. This interference occurs because, without a projected tolerance zone, a positional tolerance is applied only to the depth of threaded or press-fit holes. Unlike the floating fastener application involving clearance holes only, the attitude of a fixed fastener is restrained by the inclination of the produced hole into which it assembles.

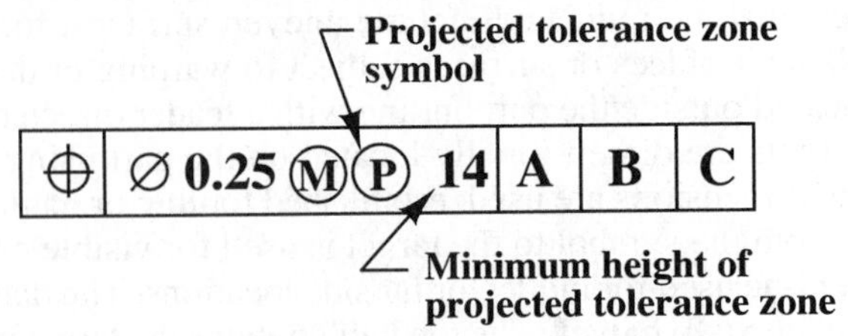

Fig. 11. Projected tolerance zone callout

With a projected tolerance zone equal to the thickness of the mating part, the inclinational error is accounted for in both parts. The minimum extent and direction of the projected tolerance zone is shown as a value in the feature control frame. The zone may be shown in a drawing view as a dimensioned value with a heavy chain line drawn closely adjacent to an extension of the center line of the hole.

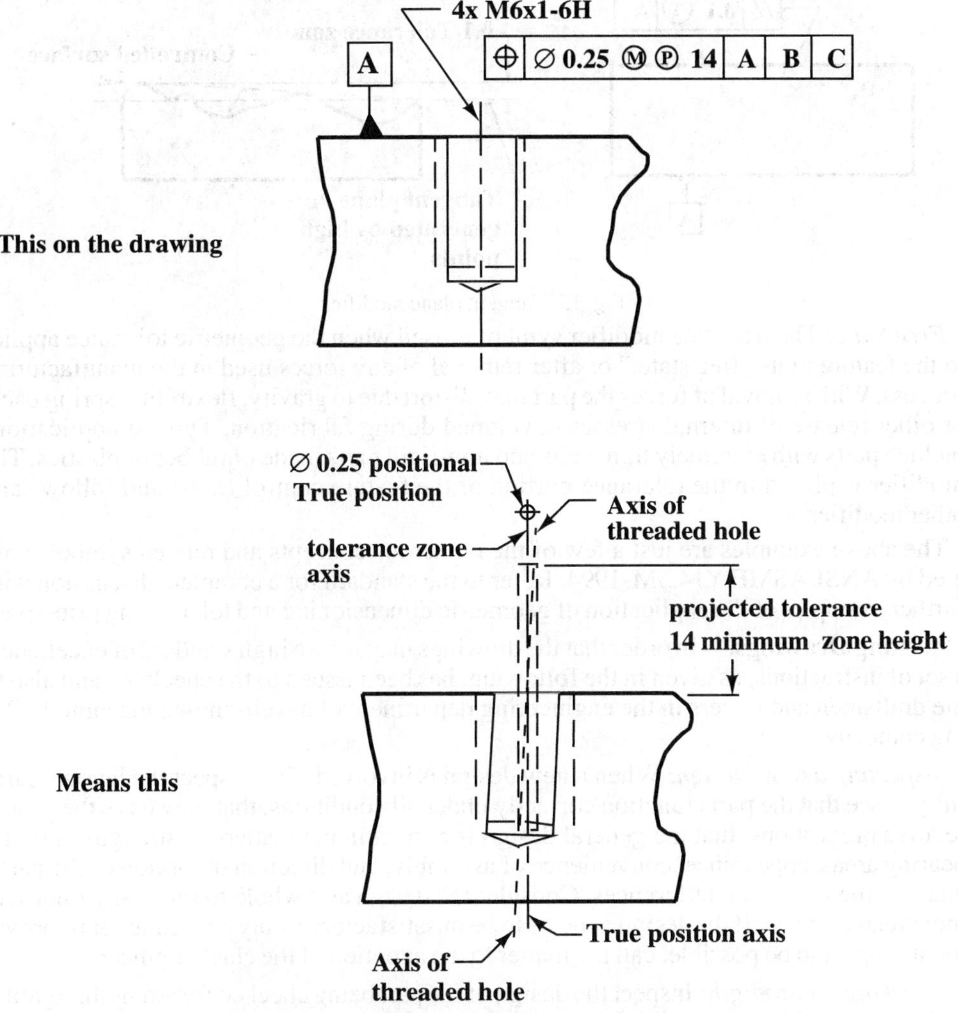

Fig. 12. Projected tolerance zone application

Statistical Tolerance: The statistical tolerancing symbol is a modifier that may be used to indicate that a tolerance is controlled statistically as opposed to being controlled arithmetically. With arithmetic control, assembly tolerances are typically divided arithmetically among the individual components of the assembly. This division results in the assumption that assemblies based on "worst case" conditions would be guaranteed to fit because the worst case set of parts fit — so that anything better would fit as well.

When this technique is restrictive, statistical tolerancing, via the symbol, may be specified in the feature control frame as a method of increasing tolerances for individual parts. This procedure may reduce manufacturing costs because its use changes the assumption that statistical process control may make a statistically significant quantity of parts fit, but not absolutely all. The technique should only be used when sound statistical methods are employed.

Tangent Plane: When it is desirable to control the surface of a feature by the contacting or high points of the surface, a tangent plane symbol is added as a modifier to the tolerance in the feature control frame, Fig. 13.

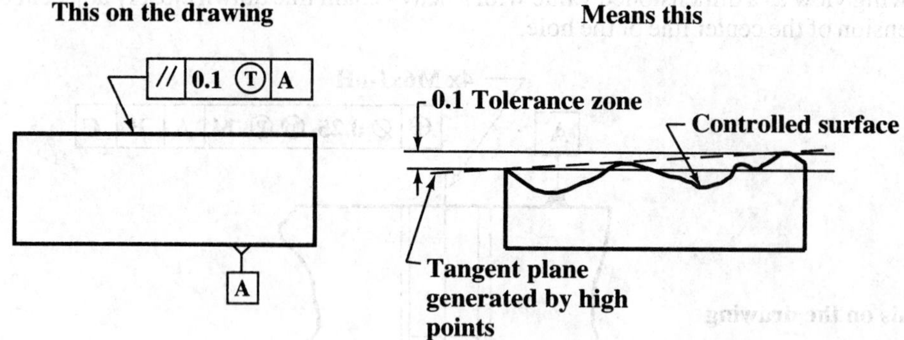

Fig. 13. Tangent plane modifier

Free State: The free state modifier symbol is used when the geometric tolerance applies to the feature in its "free state," or after removal of any forces used in the manufacturing process. With removal of forces the part may distort due to gravity, flexibility, spring back, or other release of internal stresses developed during fabrication. Typical applications include parts with extremely thin walls and non-rigid parts made of rubber or plastics. The modifier is placed in the tolerance portion of the feature control frame and follows any other modifier.

The above examples are just a few of the numerous concepts and related symbols covered by ANSI/ASME Y14.5M-1994. Refer to the standard for a complete discussion with further examples of the application of geometric dimensioning and tolerancing principles.

Checking Drawings.—In order that the drawings may have a high standard of excellence, a set of instructions, as given in the following, has been issued to the checkers, and also to the draftsmen and tracers in the engineering department of a well-known machine-building company.

Inspecting a New Design: When a new design is involved, first inspect the layouts carefully to see that the parts function correctly under all conditions, that they have the proper relative proportions, that the general design is correct in the matters of strength, rigidity, bearing areas, appearance, convenience of assembly, and direction of motion of the parts, and that there are no interferences. Consider the design as a whole to see if any improvements can be made. If the design appears to be unsatisfactory in any particular, or improvements appear to be possible, call the matter to the attention of the chief engineer.

Checking for Strength: Inspect the design of the part being checked for strength, rigidity, and appearance by comparing it with other parts for similar service whenever possible, giving preference to the later designs in such comparison, unless the later designs are known to be unsatisfactory. If there is any question regarding the matter, compute the stresses and deformations or find out whether the chief engineer has approved the stresses or deformations that will result from the forces applied to the part in service. In checking parts that are to go on a machine of increased size, be sure that standard parts used in similar machines and proposed for use on the larger machine, have ample strength and rigidity under the new and more severe service to which they will be put.

Materials Specified: Consider the kind of material required for the part and the various possibilities of molding, forging, welding, or otherwise forming the rough part from this material. Then consider the machining operations to see whether changes in form or design will reduce the number of operations or the cost of machining.

See that parts are designed with reference to the economical use of material, and whenever possible, utilize standard sizes of stock and material readily obtainable from local dealers. In the case of alloy steel, special bronze, and similar materials, be sure that the material can be obtained in the size required.

Method of Making Drawing: Inspect the drawing to see that the projections and sections are made in such a way as to show most clearly the form of the piece and the work to be done on it. Make sure that any worker looking at the drawing will understand what the shape of the piece is and how it is to be molded or machined. Make sure that the delineation is correct in every particular, and that the information conveyed by the drawing as to the form of the piece is complete.

Checking Dimensions: Check all dimensions to see that they are correct. Scale all dimensions and see that the drawing is to scale. See that the dimensions on the drawing agree with the dimensions scaled from the lay-out. Wherever any dimension is out of scale, see that the dimension is so marked. Investigate any case where the dimension, the scale of the drawing, and the scale of the lay-out do not agree. All dimensions not to scale must be underlined on the tracing. In checking dimensions, note particularly the following points:

See that all figures are correctly formed and that they will print clearly, so that the workers can easily read them correctly.

See that the overall dimensions are given.

See that all witness lines go to the correct part of the drawing.

See that all arrow points go to the correct witness lines.

See that proper allowance is made for all fits.

See that the tolerances are correctly given where necessary.

See that all dimensions given agree with the corresponding dimensions of adjacent parts.

Be sure that the dimensions given on a drawing are those that the machinist will use, and that the worker will not be obliged to do addition or subtraction to obtain the necessary measurements for machining or checking his work.

Avoid strings of dimensions where errors can accumulate. It is generally better to give a number of dimensions from the same reference surface or center line.

When holes are to be located by boring on a horizontal spindle boring machine or other similar machine, give dimensions to centers of bored holes in rectangular coordinates and from the center lines of the first hole to be bored, so that the operator will not be obliged to add measurements or transfer gages.

Checking Assembly: See that the part can readily be assembled with the adjacent parts. If necessary, provide tapped holes for eyebolts and cored holes for tongs, lugs, or other methods of handling.

Make sure that, in being assembled, the piece will not interfere with other pieces already in place and that the assembly can be taken apart without difficulty.

Check the sum of a number of tolerances; this sum must not be great enough to permit two pieces that should not be in contact to come together.

Checking Castings: In checking castings, study the form of the pattern, the methods of molding, the method of supporting and venting the cores, and the effect of draft and rough molding on clearances.

Avoid undue metal thickness, and especially avoid thick and thin sections in the same casting.

Indicate all metal thicknesses, so that the molder will know what chaplets to use for supporting the cores.

See that ample fillets are provided, and that they are properly dimensioned.

See that the cores can be assembled in the mold without crushing or interference.

See that swelling, shrinkage, or misalignment of cores will not make trouble in machining.

See that the amount of extra material allowed for finishing is indicated.

See that there is sufficient extra material for finishing on large castings to permit them to be "cleaned up," even though they warp. In such castings, make sure that the metal thickness will be sufficient after finishing, even though the castings do warp.

Make sure that sufficient sections are shown so that the pattern makers and molders will not be compelled to make assumptions about the form of any part of the casting. These details are particularly important when a number of sections of the casting are similar in form, while others differ slightly.

Checking Machined Parts: Study the sequences of operations in machining and see that all finish marks are indicated.

See that the finish marks are placed on the lines to which dimensions are given.

See that methods of machining are indicated where necessary.

Give all drill, reamer, tap, and rose bit sizes.

See that jig and gage numbers are indicated at the proper places.

See that all necessary bosses, lugs, and openings are provided for lifting, handling, clamping, and machining the piece.

See that adequate wrench room is provided for all nuts and bolt heads.

Avoid special tools, such as taps, drills, reamers, etc., unless such tools are specifically authorized.

Where parts are right- and left-hand, be sure that the hand is correctly designated. When possible, mark parts as symmetrical, so as to avoid having them right- and left-hand, but do not sacrifice correct design or satisfactory operation on this account.

When heat-treatment is required, the heat-treatment should be specified.

Check the title, size of machine, the scale, and the drawing number on both the drawing and the drawing record card.

ALLOWANCES AND TOLERANCES FOR FITS

Limits and Fits.—Fits between cylindrical parts, i.e., cylindrical fits, govern the proper assembly and performance of many mechanisms. Clearance fits permit relative freedom of motion between a shaft and a hole—axially, radially, or both. Interference fits secure a certain amount of tightness between parts, whether these are meant to remain permanently assembled or to be taken apart from time to time. Or again, two parts may be required to fit together snugly—without apparent tightness or looseness. The designer's problem is to specify these different types of fits in such a way that the shop can produce them. Establishing the specifications requires the adoption of two manufacturing limits for the hole and two for the shaft, and, hence, the adoption of a manufacturing tolerance on each part.

In selecting and specifying limits and fits for various applications, it is essential in the interests of interchangeable manufacturing that 1) standard definitions of terms relating to limits and fits be used; 2) preferred basic sizes be selected wherever possible to reduce material and tooling costs; 3) limits be based upon a series of preferred tolerances and allowances; and 4) a uniform system of applying tolerances (preferably unilateral) be used. These principles have been incorporated in both the American and British standards for limits and fits. Information about these standards is given beginning on page 627.

Basic Dimensions.—The basic size of a screw thread or machine part is the theoretical or nominal standard size from which variations are made. For example, a shaft may have a *basic* diameter of 2 inches, but a maximum variation of minus 0.010 inch may be permitted. The minimum hole should be of basic size wherever the use of standard tools represents the greatest economy. The maximum shaft should be of basic size wherever the use of standard purchased material, without further machining, represents the greatest economy, even though special tools are required to machine the mating part.

Tolerances.—Tolerance is the amount of variation permitted on dimensions or surfaces of machine parts. The tolerance is equal to the difference between the maximum and minimum limits of any specified dimension. For example, if the maximum limit for the diameter of a shaft is 2.000 inches and its minimum limit 1.990 inches, the tolerance for this diameter is 0.010 inch. The extent of these tolerances is established by determining the maximum and minimum clearances required on operating surfaces. As applied to the fitting of machine parts, the word tolerance means the amount that duplicate parts are allowed to vary in size in connection with manufacturing operations, owing to unavoidable imperfections of workmanship. Tolerance may also be defined as the amount that duplicate parts are permitted to vary in size to secure sufficient accuracy without unnecessary refinement. The terms "tolerance" and "allowance" are often used interchangeably, but, according to common usage, *allowance* is a difference in dimensions prescribed to secure various classes of fits between different parts.

Unilateral and Bilateral Tolerances.—The term "unilateral tolerance" means that the total tolerance, as related to a basic dimension, is in *one* direction only. For example, if the basic dimension were 1 inch and the tolerance were expressed as 1.000 – 0.002, or as 1.000 + 0.002, these would be unilateral tolerances because the total tolerance in each is in one direction. On the contrary, if the tolerance were divided, so as to be partly plus and partly minus, it would be classed as "bilateral."

Thus,

$$1.000^{+0.001}_{-0.001}$$

is an example of bilateral tolerance, because the total tolerance of 0.002 is given in two directions—plus and minus.

When unilateral tolerances are used, one of the three following methods should be used to express them:

1) Specify, limiting dimensions only as
Diameter of hole: 2.250, 2.252
Diameter of shaft: 2.249, 2.247
2) One limiting size may be specified with its tolerances as
Diameter of hole: 2.250 + 0.002, −0.000
Diameter of shaft: 2.249 + 0.000, −0.002
3) The nominal size may be specified for both parts, with a notation showing both allowance and tolerance, as
Diameter of hole: 2¼ + 0.002, −0.000
Diameter of shaft: 2¼ − 0.001, −0.003

Bilateral tolerances should be specified as such, usually with plus and minus tolerances of equal amount. An example of the expression of bilateral tolerances is

$$2 \pm 0.001 \quad \text{or} \quad 2\begin{matrix}+0.001\\-0.001\end{matrix}$$

Application of Tolerances.—According to common practice, tolerances are applied in such a way as to show the permissible amount of dimensional variation in the direction that is less dangerous. When a variation in either direction is equally dangerous, a bilateral tolerance should be given. When a variation in one direction is more dangerous than a variation in another, a unilateral tolerance should be given in the less dangerous direction.

For nonmating surfaces, or atmospheric fits, the tolerances may be bilateral, or unilateral, depending entirely upon the nature of the variations that develop in manufacture. On mating surfaces, with few exceptions, the tolerances should be unilateral.

Where tolerances are required on the distances between holes, usually they should be bilateral, as variation in either direction is normally equally dangerous. The variation in the distance between shafts carrying gears, however, should always be unilateral and plus; otherwise, the gears might run too tight. A slight increase in the backlash between gears is seldom of much importance.

One exception to the use of unilateral tolerances on mating surfaces occurs when tapers are involved; either bilateral or unilateral tolerances may then prove advisable, depending upon conditions. These tolerances should be determined in the same manner as the tolerances on the distances between holes. When a variation either in or out of the position of the mating taper surfaces is equally dangerous, the tolerances should be bilateral. When a variation in one direction is of less danger than a variation in the opposite direction, the tolerance should be unilateral and in the less dangerous direction.

Locating Tolerance Dimensions.—Only one dimension in the same straight line can be controlled within fixed limits. That dimension is the distance between the cutting surface of the tool and the locating or registering surface of the part being machined. Therefore, it is incorrect to locate any point or surface with tolerances from more than one point in the same straight line.

Every part of a mechanism must be located in each plane. Every operating part must be located with proper operating allowances. After such requirements of location are met, all other surfaces should have liberal clearances. Dimensions should be given between those points or surfaces that it is essential to hold in a specific relation to each other. This restriction applies particularly to those surfaces in each plane that control the location of other component parts. Many dimensions are relatively unimportant in this respect. It is good practice to establish a common locating point in each plane and give, as far as possible, all such dimensions from these common locating points. The locating points on the drawing, the locatingor registering points used for machining the surfaces and the locating points for measuring should all be identical.

The initial dimensions placed on component drawings should be the exact dimensions that would be used if it were possible to work without tolerances. Tolerances should be given in that direction in which variations will cause the least harm or danger. When a variation in either direction is equally dangerous, the tolerances should be of equal amount in both directions, or bilateral. The initial clearance, or allowance, between operating parts should be as small as the operation of the mechanism will permit. The maximum clearance should be as great as the proper functioning of the mechanism will permit.

Direction of Tolerances on Gages.—The extreme sizes for all plain limit gages shall not exceed the extreme limits of the part to be gaged. All variations in the gages, whatever their cause or purpose, shall bring these gages within these extreme limits.

The data for gage tolerances on page 656 cover gages to inspect workpieces held to tolerances in the American National Standard ANSI B4.4M-1981.

Allowance for Forced Fits.—The allowance per inch of diameter usually ranges from 0.001 inch to 0.0025 inch, 0.0015 being a fair average. Ordinarily the allowance per inch decreases as the diameter increases; thus the total allowance for a diameter of 2 inches might be 0.004 inch, whereas for a diameter of 8 inches the total allowance might not be over 0.009 or 0.010 inch. The parts to be assembled by forced fits are usually made cylindrical, although sometimes they are slightly tapered. The advantages of the taper form are that the possibility of abrasion of the fitted surfaces is reduced; that less pressure is required in assembling; and that the parts are more readily separated when renewal is required. On the other hand, the taper fit is less reliable, because if it loosens, the entire fit is free with but little axial movement. Some lubricant, such as white lead and lard oil mixed to the consistency of paint, should be applied to the pin and bore before assembling, to reduce the tendency toward abrasion.

Pressure for Forced Fits.—The pressure required for assembling cylindrical parts depends not only upon the allowance for the fit, but also upon the area of the fitted surfaces, the pressure increasing in proportion to the distance that the inner member is forced in. The approximate ultimate pressure in tons can be determined by the use of the following formula in conjunction with the accompanying table of "Pressure Factors." Assuming that A = area of surface in contact in "fit"; a = total allowance in inches; P = ultimate pressure required, in tons; F = pressure factor based upon assumption that the diameter of the hub is twice the diameter of the bore, that the shaft is of machine steel, and that the hub is of cast iron:

$$P = \frac{A \times a \times F}{2}$$

Pressure Factors

Diameter, Inches	Pressure Factor	Diameter, Inches	Pressure Factor	Diameter, Inches	Pressure Factor	Diameter, Inches	Pressure Factor	Diameter, Inches	Pressure Factor
1	500	3½	132	6	75	9	48.7	14	30.5
1¼	395	3¾	123	6¼	72	9½	46.0	14½	29.4
1½	325	4	115	6½	69	10	43.5	15	28.3
1¾	276	4¼	108	6¾	66	10½	41.3	15½	27.4
2	240	4½	101	7	64	11	39.3	16	26.5
2¼	212	4¾	96	7¼	61	11½	37.5	16½	25.6
2½	189	5	91	7½	59	12	35.9	17	24.8
2¾	171	5¼	86	7¾	57	12½	34.4	17½	24.1
3	156	5½	82	8	55	13	33.0	18	23.4
3¼	143	5¾	78	8½	52	13½	31.7	…	…

Allowance for Given Pressure.—By transposing the preceding formula, the approximate allowance for a required ultimate tonnage can be determined. Thus, $a = \frac{2P}{AF}$. The average ultimate pressure in tons commonly used ranges from 7 to 10 times the diameter in inches.

Expansion Fits.—In assembling certain classes of work requiring a very tight fit, the inner member is contracted by sub-zero cooling to permit insertion into the outer member and a tight fit is obtained as the temperature rises and the inner part expands. To obtain the sub-zero temperature, solid carbon dioxide or "dry ice" has been used but its temperature of about 109 degrees F. below zero will not contract some parts sufficiently to permit insertion in holes or recesses. Greater contraction may be obtained by using high purity liquid nitrogen which has a temperature of about 320 degrees F. below zero. During a temperature reduction from 75 degrees F. to −321 degrees F., the shrinkage per inch of diameter varies from about 0.002 to 0.003 inch for steel; 0.0042 inch for aluminum alloys; 0.0046 inch for magnesium alloys; 0.0033 inch for copper alloys; 0.0023 inch for monel metal; and 0.0017 inch for cast iron (not alloyed). The cooling equipment may vary from an insulated bucket to a special automatic unit, depending upon the kind and quantity of work. One type of unit is so arranged that parts are precooled by vapors from the liquid nitrogen before immersion. With another type, cooling is entirely by the vapor method.

Shrinkage Fits.—General practice seems to favor a smaller allowance for shrinkage fits than for forced fits, although in many shops the allowances are practically the same for each, and for some classes of work, shrinkage allowances exceed those for forced fits. The shrinkage allowance also varies to a great extent with the form and construction of the part that has to be shrunk into place. The thickness or amount of metal around the hole is the most important factor. The way in which the metal is distributed also has an influence on the results. Shrinkage allowances for locomotive driving wheel tires adopted by the American Railway Master Mechanics Association are as follows:

Center diameter, inches	38	44	50	56	62	66
Allowances, inches	0.040	0.047	0.053	0.060	0.066	0.070

Whether parts are to be assembled by forced or shrinkage fits depends upon conditions. For example, to press a tire over its wheel center, without heating, would ordinarily be a rather awkward and difficult job. On the other hand, pins, etc., are easily and quickly forced into place with a hydraulic press and there is the additional advantage of knowing the exact pressure required in assembling, whereas there is more or less uncertainty connected with a shrinkage fit, unless the stresses are calculated. Tests to determine the difference in the quality of shrinkage and forced fits showed that the resistance of a shrinkage fit to slippage for an axial pull was 3.66 times greater than that of a forced fit, and in rotation or torsion, 3.2 times greater. In each comparative test, the dimensions and allowances were the same.

Allowances for Shrinkage Fits.—The most important point to consider when calculating shrinkage fits is the stress in the hub at the bore, which depends chiefly upon the shrinkage allowance. If the allowance is excessive, the elastic limit of the material will be exceeded and permanent set will occur, or, in extreme conditions, the ultimate strength of the metal will be exceeded and the hub will burst. The intensity of the grip of the fit and the resistance to slippage depends mainly upon the thickness of the hub; the greater the thickness, the stronger the grip, and *vice versa.* Assuming the modulus of elasticity for steel to be 30,000,000, and for cast iron, 15,000,000, the shrinkage allowance per inch of nominal diameter can be determined by the following formula, in which A = allowance per inch of diameter; T = true tangential tensile stress at inner surface of outer member; C = factor taken from one of the accompanying tables, *Factors for Calculating Shrinkage Fit Allowances.*

For a cast-iron hub and steel shaft:

$$A = \frac{T(2 + C)}{30{,}000{,}000} \tag{1}$$

When both hub and shaft are of steel:

$$A = \frac{T(1 + C)}{30{,}000{,}000} \tag{2}$$

If the shaft is solid, the factor C is taken from Table 1; if it is hollow and the hub is of steel, factor C is taken from Table 2; if it is hollow and the hub is of cast iron, the factor is taken from Table 3.

Table 1. Factors for Calculating Shrinkage Fit Allowances

Ratio of Diameters $\frac{D_2}{D_1}$	Steel Hub	Cast-iron Hub	Ratio of Diameters $\frac{D_2}{D_1}$	Steel Hub	Cast-iron Hub
1.5	0.227	0.234	2.8	0.410	0.432
1.6	0.255	0.263	3.0	0.421	0.444
1.8	0.299	0.311	3.2	0.430	0.455
2.0	0.333	0.348	3.4	0.438	0.463
2.2	0.359	0.377	3.6	0.444	0.471
2.4	0.380	0.399	3.8	0.450	0.477
2.6	0.397	0.417	4.0	0.455	0.482

Values of factor C for solid steel shafts of nominal diameter D_1, and hubs of steel or cast iron of nominal external and internal diameters D_2 and D_1, respectively.

Example 1: A steel crank web 15 inches outside diameter is to be shrunk on a 10-inch solid steel shaft. Required the allowance per inch of shaft diameter to produce a maximum tensile stress in the crank of 25,000 pounds per square inch, assuming the stresses in the crank to be equivalent to those in a ring of the diameter given.

The ratio of the external to the internal diameters equals $15 \div 10 = 1.5$; $T = 25{,}000$ pounds; from Table 1, $C = 0.227$. Substituting in Formula (2):

$$A = \frac{25{,}000 \times (1 + 0.227)}{30{,}000{,}000} = 0.001 \text{ inch}$$

Example 2: Find the allowance per inch of diameter for a 10-inch shaft having a 5-inch axial through hole, other conditions being the same as in Example 1.

The ratio of external to internal diameters of the hub equals $15 \div 10 = 1.5$, as before, and the ratio of external to internal diameters of the shaft equals $10 \div 5 = 2$. From Table 2, we find that factor $C = 0.455$; $T = 25{,}000$ pounds. Substituting these values in Formula (2):

$$A = \frac{25{,}000(1 + 0.455)}{30{,}000{,}000} = 0.0012 \text{ inch}$$

The allowance is increased, as compared with Example 1, because the hollow shaft is more compressible.

Table 2. Factors for Calculating Shrinkage Fit Allowances

$\frac{D_2}{D_1}$	$\frac{D_1}{D_0}$	C	$\frac{D_2}{D_1}$	$\frac{D_1}{D_0}$	C	$\frac{D_2}{D_1}$	$\frac{D_1}{D_0}$	C
1.5	2.0	0.468	2.4	2.0	0.798	3.4	2.0	0.926
	2.5	0.368		2.5	0.628		2.5	0.728
	3.0	0.322		3.0	0.549		3.0	0.637
	3.5	0.296		3.5	0.506		3.5	0.587
1.6	2.0	0.527	2.6	2.0	0.834	3.6	2.0	0.941
	2.5	0.414		2.5	0.656		2.5	0.740
	3.0	0.362		3.0	0.574		3.0	0.647
	3.5	0.333		3.5	0.528		3.5	0.596
1.8	2.0	0.621	2.8	2.0	0.864	3.8	2.0	0.953
	2.5	0.488		2.5	0.679		2.5	0.749
	3.0	0.427		3.0	0.594		3.0	0.656
	3.5	0.393		3.5	0.547		3.5	0.603
2.0	2.0	0.696	3.0	2.0	0.888	4.0	2.0	0.964
	2.5	0.547		2.5	0.698		2.5	0.758
	3.0	0.479		3.0	0.611		3.0	0.663
	3.5	0.441		3.5	0.562		3.5	0.610
2.2	2.0	0.753	3.2	2.0	0.909	…	…	…
	2.5	0.592		2.5	0.715		…	…
	3.0	0.518		3.0	0.625		…	…
	3.5	0.477		3.5	0.576		…	…

Values of factor C for hollow steel shafts and cast-iron hubs.
Notation as in Table 1.

Table 3. Factors for Calculating Shrinkage Fit Allowances

$\frac{D_2}{D_1}$	$\frac{D_1}{D_0}$	C	$\frac{D_2}{D_1}$	$\frac{D_1}{D_0}$	C	$\frac{D_2}{D_1}$	$\frac{D_1}{D_0}$	C
1.5	2.0	0.455	2.4	2.0	0.760	3.4	2.0	0.876
	2.5	0.357		2.5	0.597		2.5	0.689
	3.0	0.313		3.0	0.523		3.0	0.602
	3.5	0.288		3.5	0.481		3.5	0.555
1.6	2.0	0.509	2.6	2.0	0.793	3.6	2.0	0.888
	2.5	0.400		2.5	0.624		2.5	0.698
	3.0	0.350		3.0	0.546		3.0	0.611
	3.5	0.322		3.5	0.502		3.5	0.562
1.8	2.0	0.599	2.8	2.0	0.820	3.8	2.0	0.900
	2.5	0.471		2.5	0.645		2.5	0.707
	3.0	0.412		3.0	0.564		3.0	0.619
	3.5	0.379		3.5	0.519		3.5	0.570
2.0	2.0	0.667	3.0	2.0	0.842	4.0	2.0	0.909
	2.5	0.524		2.5	0.662		2.5	0.715
	3.0	0.459		3.0	0.580		3.0	0.625
	3.5	0.422		3.5	0.533		3.5	0.576
2.2	2.0	0.718	3.2	2.0	0.860	…	…	…
	2.5	0.565		2.5	0.676		…	…
	3.0	0.494		3.0	0.591		…	…
	3.5	0.455		3.5	0.544		…	…

Values of factor C for hollow steel shafts of external and internal diameters D_1 and D_0, respectively, and steel hubs of nominal external diameter D_2.

Example 3: If the crank web in Example 1 is of cast iron and 4000 pounds per square inch is the maximum tensile stress in the hub, what is the allowance per inch of diameter?

$$\frac{D_2}{D_1} = 1.5 \qquad T = 4000$$

In Table 1, we find that $C = 0.234$. Substituting in Formula (1), for cast-iron hubs, $A =$ 0.0003 inch, which, owing to the lower tensile strength of cast iron, is endout one-third the shrinkage allowance in Example 1, although the stress is two-thirds of the elastic limit.

Temperatures for Shrinkage Fits.—The temperature to which the outer member in a shrinkage fit should be heated for clearance in assembling the parts depends on the total expansion required and on the coefficient α of linear expansion of the metal (i.e., the increase in length of any section of the metal in any direction for an increase in temperature of 1 degree F). The total expansion in diameter that is required consists of the total allowance for shrinkage and an added amount for clearance. The value of the coefficient α is, for nickel-steel, 0.000007; for steel in general, 0.0000065; for cast iron, 0.0000062. As an example, take an outer member of steel to be expanded 0.005 inch per inch of internal diameter, 0.001 being the shrinkage allowance and the remainder for clearance. Then

$$\alpha \times t^{\circ} = 0.005$$

$$t = \frac{0.005}{0.0000065} = 769 \text{ degrees F}$$

The value t is the number of degrees F that the temperature of the member must be raised above that of the room temperature.

ANSI Standard Limits and Fits (ANSI B4.1-1967 (R1994)).—This American National Standard for Preferred Limits and Fits for Cylindrical Parts presents definitions of terms applying to fits between plain (non threaded) cylindrical parts and makes recommendations on preferred sizes, allowances, tolerances, and fits for use wherever they are applicable. This standard is in accord with the recommendations of American-British-Canadian (ABC) conferences up to a diameter of 20 inches. Experimental work is being carried on with the objective of reaching agreement in the range above 20 inches. The recommendations in the standard are presented for guidance and for use where they might serve to improve and simplify products, practices, and facilities. They should have application for a wide range of products.

As revised in 1967, and reaffirmed in 1979, the definitions in ANSI B4.1 have been expanded and some of the limits in certain classes have been changed.

Factors Affecting Selection of Fits.—Many factors, such as length of engagement, bearing load, speed, lubrication, temperature, humidity, and materials must be taken into consideration in the selection of fits for a particular application, and modifications in the ANSI recommendations may be required to satisfy extreme conditions. Subsequent adjustments may also be found desirable as a result of experience in a particular application to suit critical functional requirements or to permit optimum manufacturing economy.

Definitions.—The following terms are defined in this standard:

Nominal Size: The nominal size is the designation used for the purpose of general identification.

Dimension: A dimension is a geometrical characteristic such as diameter, length, angle, or center distance.

Size: Size is a designation of magnitude. When a value is assigned to a dimension, it is referred to as the size of that dimension. (It is recognized that the words "dimension" and "size" are both used at times to convey the meaning of magnitude.)

Allowance: An allowance is a prescribed difference between the maximum material limits of mating parts. (See definition of *Fit*). It is a minimum clearance (positive allowance) or maximum interference (negative allowance) between such parts.

Tolerance: A tolerance is the total permissible variation of a size. The tolerance is the difference between the limits of size.

Basic Size: The basic size is that size from which the limits of size are derived by the application of allowances and tolerances.

Design Size: The design size is the basic size with allowance applied, from which the limits of size are derived by the application of tolerances. Where there is no allowance, the design size is the same as the basic size.

Actual Size: An actual size is a measured size.

Limits of Size: The limits of size are the applicable maximum and minimum sizes.

Maximum Material Limit: A maximum material limit is that limit of size that provides the maximum amount of material for the part. Normally it is the maximum limit of size of an external dimension or the minimum limit of size of an internal dimension.*

Minimum Material Limit: A minimum material limit is that limit of size that provides the minimum amount of material for the part. Normally it is the minimum limit of size of an external dimension or the maximum limit of size of an internal dimension.*

Tolerance Limit: A tolerance limit is the variation, positive or negative, by which a size is permitted to depart from the design size.

Unilateral Tolerance: A unilateral tolerance is a tolerance in which variation is permitted in only one direction from the design size.

Bilateral Tolerance: A bilateral tolerance is a tolerance in which variation is permitted in both directions from the design size.

Unilateral Tolerance System: A design plan that uses only unilateral tolerances is known as a Unilateral Tolerance System.

Bilateral Tolerance System: A design plan that uses only bilateral tolerances is known as a Bilateral Tolerance System.

Fit: Fit is the general term used to signify the range of tightness that may result from the application of a specific combination of allowances and tolerances in the design of mating parts.

Actual Fit: The actual fit between two mating parts is the relation existing between them with respect to the amount of clearance or interference that is present when they are assembled. (Fits are of three general types: clearance, transition, and interference.)

Clearance Fit: A clearance fit is one having limits of size so specified that a clearance always results when mating parts are assembled.

Interference Fit: An interference fit is one having limits of size so specified that an interference always results when mating parts are assembled.

Transition Fit: A transition fit is one having limits of size so specified that either a clearance or an interference may result when mating parts are assembled.

Basic Hole System: A basic hole system is a system of fits in which the design size of the hole is the basic size and the allowance, if any, is applied to the shaft.

Basic Shaft System: A basic shaft system is a system of fits in which the design size of the shaft is the basic size and the allowance, if any, is applied to the hole.

* An example of exceptions: an exterior corner radius where the maximum radius is the minimum material limit and the minimum radius is the maximum material limit.

Preferred Basic Sizes.—In specifying fits, the basic size of mating parts may be chosen from the decimal series or the fractional series in the following table.

Table 1. Preferred Basic Sizes

Decimal			Fractional					
0.010	2.00	8.50	1/64	0.015625	2¼	2.2500	9½	9.5000
0.012	2.20	9.00	1/32	0.03125	2½	2.5000	10	10.0000
0.016	2.40	9.50	1/16	0.0625	2¾	2.7500	10½	10.5000
0.020	2.60	10.00	3/32	0.09375	3	3.0000	11	11.0000
0.025	2.80	10.50	⅛	0.1250	3¼	3.2500	11½	11.5000
0.032	3.00	11.00	5/32	0.15625	3½	3.5000	12	12.0000
0.040	3.20	11.50	3/16	0.1875	3¾	3.7500	12½	12.5000
0.05	3.40	12.00	¼	0.2500	4	4.0000	13	13.0000
0.06	3.60	12.50	5/16	0.3125	4¼	4.2500	13½	13.5000
0.08	3.80	13.00	⅜	0.3750	4½	4.5000	14	14.0000
0.10	4.00	13.50	7/16	0.4375	4¾	4.7500	14½	14.5000
0.12	4.20	14.00	½	0.5000	5	5.0000	15	15.0000
0.16	4.40	14.50	9/16	0.5625	5¼	5.2500	15½	15.5000
0.20	4.60	15.00	⅝	0.6250	5½	5.5000	16	16.0000
0.24	4.80	15.50	11/16	0.6875	5¾	5.7500	16½	16.5000
0.30	5.00	16.00	¾	0.7500	6	6.0000	17	17.0000
0.40	5.20	16.50	⅞	0.8750	6½	6.5000	17½	17.5000
0.50	5.40	17.00	1	1.0000	7	7.0000	18	18.0000
0.60	5.60	17.50	1¼	1.2500	7½	7.5000	18½	18.5000
0.80	5.80	18.00	1½	1.5000	8	8.0000	19	19.0000
1.00	6.00	18.50	1¾	1.7500	8½	8.5000	19½	19.5000
1.20	6.50	19.00	2	2.0000	9	9.0000	20	20.0000
1.40	7.00	19.50	…	…	…	…	…	…
1.60	7.50	20.00	…	…	…	…	…	…
1.80	8.00	…	…	…	…	…	…	…

All dimensions are in inches.

Preferred Series of Tolerances and Allowances (In thousandths of an inch)

0.1	1	10	100	0.3	3	30	…
…	1.2	12	125	…	3.5	35	…
0.15	1.4	14	…	0.4	4	40	…
…	1.6	16	160	…	4.5	45	…
…	1.8	18	…	0.5	5	50	…
0.2	2	20	200	0.6	6	60	…
…	2.2	22	…	0.7	7	70	…
0.25	2.5	25	250	0.8	8	80	…
…	2.8	28	…	0.9	9	…	…

Standard Tolerances.—The series of standard tolerances shown in Table 1 are so arranged that for any one grade they represent approximately similar production difficulties throughout the range of sizes. This table provides a suitable range from which appropriate tolerances for holes and shafts can be selected and enables standard gages to be used. The tolerances shown in Table 1 have been used in the succeeding tables for different classes of fits.

Table 1. ANSI Standard Tolerances *ANSI B4.1-1967 (R1987)*

Nominal Size, Inches		Grade									
		4	5	6	7	8	9	10	11	12	13
Over	To	Tolerances in thousandths of an inch[a]									
0	0.12	0.12	0.15	0.25	0.4	0.6	1.0	1.6	2.5	4	6
0.12	0.24	0.15	0.20	0.3	0.5	0.7	1.2	1.8	3.0	5	7
0.24	0.40	0.15	0.25	0.4	0.6	0.9	1.4	2.2	3.5	6	9
0.40	0.71	0.2	0.3	0.4	0.7	1.0	1.6	2.8	4.0	7	10
0.71	1.19	0.25	0.4	0.5	0.8	1.2	2.0	3.5	5.0	8	12
1.19	1.97	0.3	0.4	0.6	1.0	1.6	2.5	4.0	6	10	16
1.97	3.15	0.3	0.5	0.7	1.2	1.8	3.0	4.5	7	12	18
3.15	4.73	0.4	0.6	0.9	1.4	2.2	3.5	5	9	14	22
4.73	7.09	0.5	0.7	1.0	1.6	2.5	4.0	6	10	16	25
7.09	9.85	0.6	0.8	1.2	1.8	2.8	4.5	7	12	18	28
9.85	12.41	0.6	0.9	1.2	2.0	3.0	5.0	8	12	20	30
12.41	15.75	0.7	1.0	1.4	2.2	3.5	6	9	14	22	35
15.75	19.69	0.8	1.0	1.6	2.5	4	6	10	16	25	40
19.69	30.09	0.9	1.2	2.0	3	5	8	12	20	30	50
30.09	41.49	1.0	1.6	2.5	4	6	10	16	25	40	60
41.49	56.19	1.2	2.0	3	5	8	12	20	30	50	80
56.19	76.39	1.6	2.5	4	6	10	16	25	40	60	100
76.39	100.9	2.0	3	5	8	12	20	30	50	80	125
100.9	131.9	2.5	4	6	10	16	25	40	60	100	160
131.9	171.9	3	5	8	12	20	30	50	80	125	200
171.9	200	4	6	10	16	25	40	60	100	160	250

[a] All tolerances above heavy line are in accordance with American-British-Canadian (ABC) agreements.

Table 2. Relation of Machining Processes to Tolerance Grades

This chart may be used as a general guide to determine the machining processes that will under normal conditions, produce work withen the tolerance grades indicated.

(See also *Relation of Surface Roughness to Tolerances* starting on page 702.)

MACHINING OPERATION	TOLERANCE GRADES									
	4	5	6	7	8	9	10	11	12	13
Lapping & Honing	X	X								
Cylindrical Grinding		X	X	X						
Surface Grinding		X	X	X	X					
Diamond Turning		X	X	X						
Diamond Boring		X	X	X						
Broaching		X	X	X	X					
Reaming			X	X	X	X	X			
Turning				X	X	X	X	X	X	X
Boring					X	X	X	X	X	X
Milling							X	X	X	X
Planing & Shaping							X	X	X	X
Drilling							X	X	X	X

ANSI Standard Fits.—Tables 3 through 9 inclusive show a series of standard types and classes of fits on a unilateral hole basis, such that the fit produced by mating parts in any one class will produce approximately similar performance throughout the range of sizes. These tables prescribe the fit for any given size, or type of fit; they also prescribe the standard limits for the mating parts that will produce the fit. The fits listed in these tables contain all those that appear in the approved American-British-Canadian proposal.

Selection of Fits: In selecting limits of size for any application, the type of fit is determined first, based on the use or service required from the equipment being designed; then the limits of size of the mating parts are established, to insure that the desired fit will be produced.

Theoretically, an infinite number of fits could be chosen, but the number of standard fits shown in the accompanying tables should cover most applications.

Designation of Standard Fits: Standard fits are designated by means of the following symbols which, facilitate reference to classes of fit for educational purposes. The symbols are not intended to be shown on manufacturing drawings; instead, sizes should be specified on drawings.

The letter symbols used are as follows:

RC = Running or Sliding Clearance Fit

LC = Locational Clearance Fit

LT = Transition Clearance or Interference Fit

LN = Locational Interference Fit

FN = Force or Shrink Fit

These letter symbols are used in conjunction with numbers representing the class of fit; thus FN 4 represents a Class 4, force fit.

Each of these symbols (two letters and a number) represents a complete fit for which the minimum and maximum clearance or interference and the limits of size for the mating parts are given directly in the tables.

Description of Fits.—The classes of fits are arranged in three general groups: running and sliding fits, locational fits, and force fits.

Running and Sliding Fits (RC): Running and sliding fits, for which limits of clearance are given in Table 2, are intended to provide a similar running performance, with suitable lubrication allowance, throughout the range of sizes. The clearances for the first two classes, used chiefly as slide fits, increase more slowly with the diameter than for the other classes, so that accurate location is maintained even at the expense of free relative motion.

These fits may be described as follows:

RC 1 *Close sliding fits* are intended for the accurate location of parts that must assemble without perceptible play.

RC 2 *Sliding fits* are intended for accurate location, but with greater maximum clearance than class RC 1. Parts made to this fit move and turn easily but are not intended to run freely, and in the larger sizes may seize with small temperature changes.

RC 3 *Precision running fits* are about the closest fits that can be expected to run freely, and are intended for precision work at slow speeds and light journal pressures, but are not suitable where appreciable temperature differences are likely to be encountered.

RC 4 *Close running fits* are intended chiefly for running fits on accurate machinery with moderate surface speeds and journal pressures, where accurate location and minimum play are desired.

RC 5 and RC 6 *Medium running fits* are intended for higher running speeds, or heavy journal pressures, or both.

RC 7 *Free running fits* are intended for use where accuracy is not essential, or where large temperature variations are likely to be encountered, or under both these conditions.

RC 8 and RC 9 *Loose running fits* are intended for use where wide commercial tolerances may be necessary, together with an allowance, on the external member.

Locational Fits (LC, LT, and LN): Locational fits are fits intended to determine only the location of the mating parts; they may provide rigid or accurate location, as with interference fits, or provide some freedom of location, as with clearance fits. Accordingly, they are divided into three groups: clearance fits (LC), transition fits (LT), and interference fits (LN).

These are described as follows:

LC *Locational clearance fits* are intended for parts which are normally stationary, but that can be freely assembled or disassembled. They range from snug fits for parts requiring accuracy of location, through the medium clearance fits for parts such as spigots, to the looser fastener fits where freedom of assembly is of prime importance.

LT *Locational transition fits* are a compromise between clearance and interference fits, for applications where accuracy of location is important, but either a small amount of clearance or interference is permissible.

LN *Locational interference fits* are used where accuracy of location is of prime importance, and for parts requiring rigidity and alignment with no special requirements for bore pressure. Such fits are not intended for parts designed to transmit frictional loads from one part to another by virtue of the tightness of fit. These conditions are covered by force fits.

Force Fits: (FN): Force or shrink fits constitute a special type of interference fit, normally characterized by maintenance of constant bore pressures throughout the range of sizes. The interference therefore varies almost directly with diameter, and the difference between its minimum and maximum value is small, to maintain the resulting pressures within reasonable limits.

These fits are described as follows:

FN 1 *Light drive fits* are those requiring light assembly pressures, and produce more or less permanent assemblies. They are suitable for thin sections or long fits, or in cast-iron external members.

FN 2 *Medium drive fits* are suitable for ordinary steel parts, or for shrink fits on light sections. They are about the tightest fits that can be used with high-grade cast-iron external members.

FN 3 *Heavy drive fits* are suitable for heavier steel parts or for shrink fits in medium sections.

FN 4 and FN 5 *Force fits* are suitable for parts that can be highly stressed, or for shrink fits where the heavy pressing forces required are impractical.

Graphical Representation of Limits and Fits.—A visual comparison of the hole and shaft tolerances and the clearances or interferences provided by the various types and classes of fits can be obtained from the diagrams on page 633. These diagrams have been drawn to scale for a nominal diameter of 1 inch.

Use of Standard Fit Tables.—*Example 1:* A Class RC 1 fit is to be used in assembling a mating hole and shaft of 2-inch nominal diameter. This class of fit was selected because the application required accurate location of the parts with no perceptible play (see *Description of Fits*, RC 1 close sliding fits). From the data in Table 2, establish the limts of size and clearance of the hole and shaft.

Maximum hole = 2 + 0.005 = 2.00005; minimum hole = 2 inches

Maximum shaft = 2 − 0.0004 = 1.9996; minimum shaft = 2 − 0.0007 = 1.9993 inches

Minimum clearance = 0.0004; maximum clearance = 0.0012 inch

Graphical Representation of ANSI Standard Limits and Fits

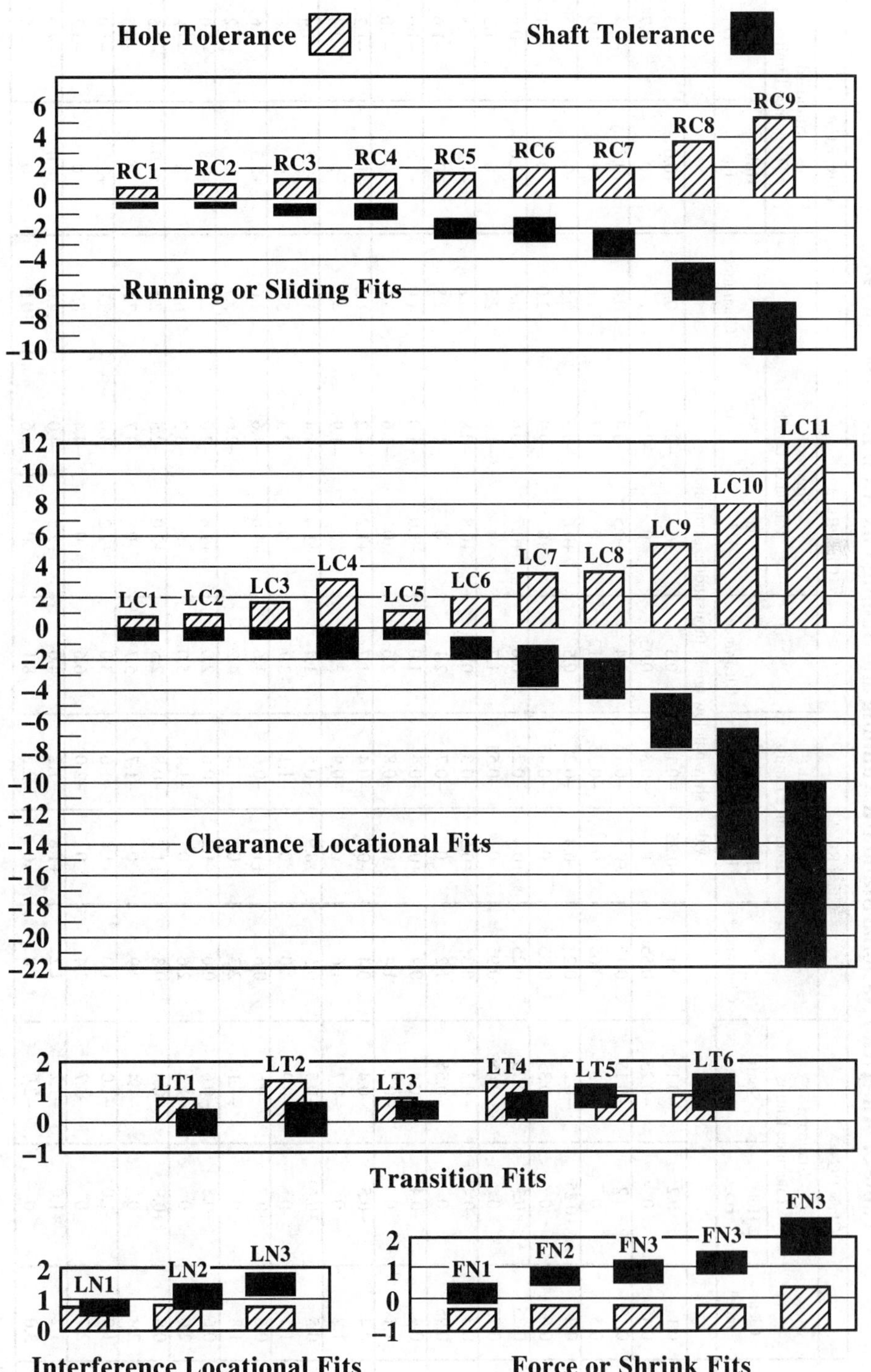

Diagrams show disposition of hole and shaft tolerances (in thousandths of an inch) with respect to basic size (0) for a diameter of 1 inch.

Table 3. American National Standard Running and Sliding Fits *ANSI B4.1-1967 (R1987)*

Nominal Size Range, Inches	Class RC 1			Class RC 2			Class RC 3			Class RC 4		
	Clearance[a]	Standard Tolerance Limits		Clearance[a]	Standard Tolerance Limits		Clearance[a]	Standard Tolerance Limits		Clearance[a]	Standard Tolerance Limits	
Over – To		Hole H5	Shaft g4		Hole H6	Shaft g5		Hole H7	Shaft f6		Hole H8	Shaft f7
	Values shown below are in thousandths of an inch											
0 – 0.12	0.1	+0.2	−0.1	0.1	+0.25	−0.1	0.3	+0.4	−0.3	0.3	+0.6	−0.3
	0.45	0	−0.25	0.55	0	−0.3	0.95	0	−0.55	1.3	0	−0.7
0.12 – 0.24	0.15	+0.2	−0.15	0.15	+0.3	−0.15	0.4	+0.5	−0.4	0.4	+0.7	−0.4
	0.5	0	−0.3	0.65	0	−0.35	1.12	0	−0.7	1.6	0	−0.9
0.24 – 0.40	0.2	+0.25	−0.2	0.2	+0.4	−0.2	0.5	+0.6	−0.5	0.5	+0.9	−0.5
	0.6	0	−0.35	0.85	0	−0.45	1.5	0	−0.9	2.0	0	−1.1
0.40 – 0.71	0.25	+0.3	−0.25	0.25	+0.4	−0.25	0.6	+0.7	−0.6	0.6	+1.0	−0.6
	0.75	0	−0.45	0.95	0	−0.55	1.7	0	−1.0	2.3	0	−1.3
0.71 – 1.19	0.3	+0.4	−0.3	0.3	+0.5	−0.3	0.8	+0.8	−0.8	0.8	+1.2	−0.8
	0.95	0	−0.55	1.2	0	−0.7	2.1	0	−1.3	2.8	0	−1.6
1.19 – 1.97	0.4	+0.4	−0.4	0.4	+0.6	−0.4	1.0	+1.0	−1.0	1.0	+1.6	−1.0
	1.1	0	−0.7	1.4	0	−0.8	2.6	0	−1.6	3.6	0	−2.0
1.97 – 3.15	0.4	+0.5	−0.4	0.4	+0.7	−0.4	1.2	+1.2	−1.2	1.2	+1.8	−1.2
	1.2	0	−0.7	1.6	0	−0.9	3.1	0	−1.9	4.2	0	−2.4
3.15 – 4.73	0.5	+0.6	−0.5	0.5	+0.9	−0.5	1.4	+1.4	−1.4	1.4	+2.2	−1.4
	1.5	0	−0.9	2.0	0	−1.1	3.7	0	−2.3	5.0	0	−2.8
4.73 – 7.09	0.6	+0.7	−0.6	0.6	+1.0	−0.6	1.6	+1.6	−1.6	1.6	+2.5	−1.6
	1.8	0	−1.1	2.3	0	−1.3	4.2	0	−2.6	5.7	0	−3.2
7.09 – 9.85	0.6	+0.8	−0.6	0.6	+1.2	−0.6	2.0	+1.8	−2.0	2.0	+2.8	−2.0
	2.0	0	−1.2	2.6	0	−1.4	5.0	0	−3.2	6.6	0	−3.8
9.85 – 12.41	0.8	+0.9	−0.8	0.8	+1.2	−0.8	2.5	+2.0	−2.5	2.5	+3.0	−2.5
	2.3	0	−1.4	2.9	0	−1.7	5.7	0	−3.7	7.5	0	−4.5
12.41 – 15.75	1.0	+1.0	−1.0	1.0	+1.4	−1.0	3.0	+2.2	−3.0	3.0	+3.5	−3.0
	2.7	0	−1.7	3.4	0	−2.0	6.6	0	−4.4	8.7	0	−5.2
15.75 – 19.69	1.2	+1.0	−1.2	1.2	+1.6	−1.2	4.0	+2.5	−4.0	4.0	+4.0	−4.0
	3.0	0	−2.0	3.8	0	−2.2	8.1	0	−5.6	10.5	0	−6.5

[a] Pairs of values shown represent minimum and maximum amounts of clearance resulting from application of standard tolerance limits.

Table 4. American National Standard Running and Sliding Fits *ANSI B4.1-1967 (R1987)*

Nominal Size Range, Inches		Class RC 5			Class RC 6			Class RC 7			Class RC 8			Class RC 9		
			Standard Tolerance Limits			Standard Tolerance Limits			Standard Tolerance Limits			Standard Tolerance Limits			Standard Tolerance Limits	
Over	To	Clearance[a]	Hole H8	Shaft e7	Clearance[a]	Hole H9	Shaft e8	Clearance[a]	Hole H9	Shaft d8	Clearance[a]	Hole H10	Shaft c9	Clearance[a]	Hole H11	Shaft
		Values shown below are in thousandths of an inch														
0 –	0.12	0.6	+0.6	− 0.6	0.6	+1.0	− 0.6	1.0	+1.0	− 1.0	2.5	+1.6	− 2.5	4.0	+2.5	− 4.0
		1.6	0	− 1.0	2.2	0	− 1.2	2.6	0	− 1.6	5.1	0	− 3.5	8.1	0	− 5.6
0.12 –	0.24	0.8	+0.7	− 0.8	0.8	+1.2	− 0.8	1.2	+1.2	− 1.2	2.8	+1.8	− 2.8	4.5	+3.0	− 4.5
		2.0	0	− 1.3	2.7	0	− 1.5	3.1	0	− 1.9	5.8	0	− 4.0	9.0	0	− 6.0
0.24 –	0.40	1.0	+0.9	− 1.0	1.0	+1.4	− 1.0	1.6	+1.4	− 1.6	3.0	+2.2	− 3.0	5.0	+3.5	− 5.0
		2.5	0	− 1.6	3.3	0	− 1.9	3.9	0	− 2.5	6.6	0	− 4.4	10.7	0	− 7.2
0.40 –	0.71	1.2	+1.0	− 1.2	1.2	+1.6	− 1.2	2.0	+1.6	− 2.0	3.5	+2.8	− 3.5	6.0	+4.0	− 6.0
		2.9	0	− 1.9	3.8	0	− 2.2	4.6	0	− 3.0	7.9	0	− 5.1	12.8	0	− 8.8
0.71 –	1.19	1.6	+1.2	− 1.6	1.6	+2.0	− 1.6	2.5	+2.0	− 2.5	4.5	+3.5	− 4.5	7.0	+5.0	− 7.0
		3.6	0	− 2.4	4.8	0	− 2.8	5.7	0	− 3.7	10.0	0	− 6.5	15.5	0	−10.5
1.19 –	1.97	2.0	+1.6	− 2.0	2.0	+2.5	− 2.0	3.0	+2.5	− 3.0	5.0	+4.0	− 5.0	8.0	+6.0	− 8.0
		4.6	0	− 3.0	6.1	0	− 3.6	7.1	0	− 4.6	11.5	0	− 7.5	18.0	0	−12.0
1.97 –	3.15	2.5	+1.8	− 2.5	2.5	+3.0	− 2.5	4.0	+3.0	− 4.0	6.0	+4.5	− 6.0	9.0	+7.0	− 9.0
		5.5	0	− 3.7	7.3	0	− 4.3	8.8	0	− 5.8	13.5	0	− 9.0	20.5	0	−13.5
3.15 –	4.73	3.0	+2.2	− 3.0	3.0	+3.5	− 3.0	5.0	+3.5	− 5.0	7.0	+5.0	− 7.0	10.0	+9.0	−10.0
		6.6	0	− 4.4	8.7	0	− 5.2	10.7	0	− 7.2	15.5	0	−10.5	24.0	0	−15.0
4.73 –	7.09	3.5	+2.5	− 3.5	3.5	+4.0	− 3.5	6.0	+4.0	− 6.0	8.0	+6.0	− 8.0	12.0	+10.0	−12.0
		7.6	0	− 5.1	10.0	0	− 6.0	12.5	0	− 8.5	18.0	0	−12.0	28.0	0	−18.0
7.09 –	9.85	4.0	+2.8	− 4.0	4.0	+4.5	− 4.0	7.0	+4.5	− 7.0	10.0	+7.0	−10.0	15.0	+12.0	−15.0
		8.6	0	− 5.8	11.3	0	− 6.8	14.3	0	− 9.8	21.5	0	−14.5	34.0	0	−22.0
9.85 –	12.41	5.0	+3.0	− 5.0	5.0	+5.0	− 5.0	8.0	+5.0	− 8.0	12.0	+8.0	−12.0	18.0	+12.0	−18.0
		10.0	0	− 7.0	13.0	0	− 8.0	16.0	0	−11.0	25.0	0	−17.0	38.0	0	−26.0
12.41 –	15.75	6.0	+3.5	− 6.0	6.0	+6.0	− 6.0	10.0	+6.0	−10.0	14.0	+9.0	−14.0	22.0	+14.0	−22.0
		11.7	0	− 8.2	15.5	0	− 9.5	19.5	0	−13.5	29.0	0	−20.0	45.0	0	−31.0
15.75 –	19.69	8.0	+4.0	− 8.0	8.0	+6.0	− 8.0	12.0	+6.0	−12.0	16.0	+10.0	−16.0	25.0	+16.0	−25.0
		14.5	0	−10.5	18.0	0	−12.0	22.0	0	−16.0	32.0	0	−22.0	51.0	0	−35.0

[a] Tolerance limits given in body of table are added to or subtracted from basic size (as indicated by + or − sign) to obtain maximum and minimum sizes of mating parts.

All data above heavy lines are in accord with ABC agreements. Symbols H5, g4, etc. are hole and shaft designations in ABC system. Limits for sizes above 19.69 inches are also given in the ANSI Standard.

Table 5. American National Standard Clearance Locational Fits *ANSI B4.1-1967 (R1987)*

Nominal Size Range, Inches	Class LC 1			Class LC 2			Class LC 3			Class LC 4			Class LC 5		
		Standard Tolerance Limits			Standard Tolerance Limits			Standard Tolerance Limits			Standard Tolerance Limits			Standard Tolerance Limits	
Over To	Clearance[a]	Hole H6	Shaft h5	Clearance[a]	Hole H7	Shaft h6	Clearance[a]	Hole H8	Shaft h7	Clearance[a]	Hole H10	Shaft h9	Clearance[a]	Hole H7	Shaft g6
	Values shown below are in thousandths of an inch														
0– 0.12	0 0.45	+0.25 0	0 −0.2	0 0.65	+0.4 0	0 −0.25	0 1	+0.6 0	0 −0.4	0 2.6	+1.6 0	0 −1.0	0.1 0.75	+0.4 0	−0.1 −0.35
0.12– 0.24	0 0.5	+0.3 0	0 −0.2	0 0.8	+0.5 0	0 −0.3	0 1.2	+0.7 0	0 −0.5	0 3.0	+1.8 0	0 −1.2	0.15 0.95	+0.5 0	−0.15 −0.45
0.24– 0.40	0 0.65	+0.4 0	0 −0.25	0 1.0	+0.6 0	0 −0.4	0 1.5	+0.9 0	0 −0.6	0 3.6	+2.2 0	0 −1.4	0.2 1.2	+0.6 0	−0.2 −0.6
0.40– 0.71	0 0.7	+0.4 0	0 −0.3	0 1.1	+0.7 0	0 −0.4	0 1.7	+1.0 0	0 −0.7	0 4.4	+2.8 0	0 −1.6	0.25 1.35	+0.7 0	−0.25 −0.65
0.71– 1.19	0 0.9	+0.5 0	0 −0.4	0 1.3	+0.8 0	0 −0.5	0 2	+1.2 0	0 −0.8	0 5.5	+3.5 0	0 −2.0	0.3 1.6	+0.8 0	−0.3 −0.8
1.19– 1.97	0 1.0	+0.6 0	0 −0.4	0 1.6	+1.0 0	0 −0.6	0 2.6	+1.6 0	0 −1	0 6.5	+4.0 0	0 −2.5	0.4 2.0	+1.0 0	−0.4 −1.0
1.97– 3.15	0 1.2	+0.7 0	0 −0.5	0 1.9	+1.2 0	0 −0.7	0 3	+1.8 0	0 −1.2	0 7.5	+4.5 0	0 −3	0.4 2.3	+1.2 0	−0.4 −1.1
3.15– 4.73	0 1.5	+0.9 0	0 −0.6	0 2.3	+1.4 0	0 −0.9	0 3.6	+2.2 0	0 −1.4	0 8.5	+5.0 0	0 −3.5	0.5 2.8	+1.4 0	−0.5 −1.4
4.73– 7.09	0 1.7	+1.0 0	0 −0.7	0 2.6	+1.6 0	0 −1.0	0 4.1	+2.5 0	0 −1.6	0 10.0	+6.0 0	0 −4	0.6 3.2	+1.6 0	−0.6 −1.6
7.09– 9.85	0 2.0	+1.2 0	0 −0.8	0 3.0	+1.8 0	0 −1.2	0 4.6	+2.8 0	0 −1.8	0 11.5	+7.0 0	0 −4.5	0.6 3.6	+1.8 0	−0.6 −1.8
9.85– 12.41	0 2.1	+1.2 0	0 −0.9	0 3.2	+2.0 0	0 −1.2	0 5	+3.0 0	0 −2.0	0 13.0	+8.0 0	0 −5	0.7 3.9	+2.0 0	−0.7 −1.9
12.41– 15.75	0 2.4	+1.4 0	0 −1.0	0 3.6	+2.2 0	0 −1.4	0 5.7	+3.5 0	0 −2.2	0 15.0	+9.0 0	0 −6	0.7 4.3	+2.2 0	−0.7 −2.1
15.75– 19.69	0 2.6	+1.6 0	0 −1.0	0 4.1	+2.5 0	0 −1.6	0 6.5	+4 0	0 −2.5	0 16.0	+10.0 0	0 −6	0.8 4.9	+2.5 0	−0.8 −2.4

[a] Pairs of values shown represent minimum and maximum amounts of interference resulting from application of standard tolerance limits.

Table 6. American National Standard Clearance Locational Fits *ANSI B4.1-1967 (R1987)*

Nominal Size Range, Inches	Class LC 6			Class LC 7			Class LC 8			Class LC 9			Class LC 10			Class LC 11		
		Std. Tolerance Limits			Std. Tolerance Limits			Std. Tolerance Limits			Std. Tolerance Limits			Std. Tolerance Limits			Std. Tolerance Limits	
Over To	Clearance[a]	Hole H9	Shaft f8	Clearance[a]	Hole H10	Shaft e9	Clearance[a]	Hole H10	Shaft d9	Clearance[a]	Hole H11	Shaft c10	Clearance[a]	Hole H12	Shaft	Clearance[a]	Hole H13	Shaft
	Values shown below are in thousandths of an inch																	
0 – 0.12	0.3	+1.0	−0.3	0.6	+1.6	− 0.6	1.0	+1.6	− 1.0	2.5	+2.5	− 2.5	4	+4	− 4	5	+6	− 5
	1.9	0	−0.9	3.2	0	− 1.6	2.0	0	− 2.0	6.6	0	− 4.1	12	0	− 8	17	0	− 11
0.12 – 0.24	0.4	+1.2	−0.4	0.8	+1.8	− 0.8	1.2	+1.8	− 1.2	2.8	+3.0	− 2.8	4.5	+5	− 4.5	6	+7	− 6
	2.3	0	−1.1	3.8	0	− 2.0	4.2	0	− 2.4	7.6	0	− 4.6	14.5	0	− 9.5	20	0	−13
0.24 – 0.40	0.5	+1.4	−0.5	1.0	+2.2	− 1.0	1.6	+2.2	− 1.6	3.0	+3.5	− 3.0	5	+6	− 5	7	+9	− 7
	2.8	0	−1.4	4.6	0	− 2.4	5.2	0	− 3.0	8.7	0	− 5.2	17	0	−11	25	0	−16
0.40 – 0.71	0.6	+1.6	−0.6	1.2	+2.8	− 1.2	2.0	+2.8	− 2.0	3.5	+4.0	− 3.5	6	+7	− 6	8	+10	− 8
	3.2	0	−1.6	5.6	0	− 2.8	6.4	0	− 3.6	10.3	0	− 6.3	20	0	−13	28	0	−18
0.71 – 1.19	0.8	+2.0	−0.8	1.6	+3.5	− 1.6	2.5	+3.5	− 2.5	4.5	+5.0	− 4.5	7	+8	− 7	10	+12	−10
	4.0	0	−2.0	7.1	0	− 3.6	8.0	0	− 4.5	13.0	0	− 8.0	23	0	−15	34	0	−22
1.19 – 1.97	1.0	+2.5	−1.0	2.0	+4.0	− 2.0	3.6	+4.0	− 3.0	5.0	+6	− 5.0	8	+10	− 8	12	+16	−12
	5.1	0	−2.6	8.5	0	− 4.5	9.5	0	− 5.5	15.0	0	− 9.0	28	0	−18	44	0	−28
1.97 – 3.15	1.2	+3.0	−1.0	2.5	+4.5	− 2.5	4.0	+4.5	− 4.0	6.0	+7	− 6.0	10	+12	−10	14	+18	−14
	6.0	0	−3.0	10.0	0	− 5.5	11.5	0	− 7.0	17.5	0	−10.5	34	0	−22	50	0	−32
3.15 – 4.73	1.4	+3.5	−1.4	3.0	+5.0	− 3.0	5.0	+5.0	− 5.0	7	+9	− 7	11	+14	−11	16	+22	−16
	7.1	0	−3.6	11.5	0	− 6.5	13.5	0	− 8.5	21	0	−12	39	0	−25	60	0	−38
4.73 – 7.09	1.6	+4.0	−1.6	3.5	+6.0	− 3.5	6	+6	− 6	8	+10	− 8	12	+16	−12	18	+25	−18
	8.1	0	−4.1	13.5	0	− 7.5	16	0	−10	24	0	−14	44	0	−28	68	0	−43
7.09 – 9.85	2.0	+4.5	−2.0	4.0	+7.0	− 4.0	7	+7	− 7	10	+12	−10	16	+18	−16	22	+28	−22
	9.3	0	−4.8	15.5	0	− 8.5	18.5	0	−11.5	29	0	−17	52	0	−34	78	0	−50
9.85 – 12.41	2.2	+5.0	−2.2	4.5	+8.0	− 4.5	7	+8	− 7	12	+12	−12	20	+20	−20	28	+30	−28
	10.2	0	−5.2	17.5	0	− 9.5	20	0	−12	32	0	−20	60	0	−40	88	0	−58
12.41 – 15.75	2.5	+6.0	−2.5	5.0	+9.0	− 5	8	+9	− 8	14	+14	−14	22	+22	−22	30	+35	−30
	12.0	0	−6.0	20.0	0	−11	23	0	−14	37	0	−23	66	0	−44	100	0	−65
15.75 – 19.69	2.8	+6.0	−2.8	5.0	+10.0	− 5	9	+10	− 9	16	+16	−16	25	+25	−25	35	+40	−35
	12.8	0	−6.8	21.0	0	−11	25	0	−15	42	0	−26	75	0	−50	115	0	−75

[a] Tolerance limits given in body of table are added or subtracted to basic size (as indicated by + or − sign) to obtain maximum and minimum sizes of mating parts.

All data above heavy lines are in accordance with American-British-Canadian (ABC) agreements. Symbols H6, H7, s6, etc. are hole and shaft designations in ABC system. Limits for sizes above 19.69 inches are not covered by ABC agreements but are given in the ANSI Standard.

Table 7. ANSI Standard Transition Locational Fits *ANSI B4.1-1967 (R1987)*

Nominal Size Range, Inches	Class LT 1			Class LT 2			Class LT 3			Class LT 4			Class LT 5			Class LT 6		
		Std. Tolerance Limits			Std. Tolerance Limits			Std. Tolerance Limits			Std. Tolerance Limits			Std. Tolerance Limits			Std. Tolerance Limits	
Over – To	Fit[a]	Hole H7	Shaft js6	Fit[a]	Hole H8	Shaft js7	Fit[a]	Hole H7	Shaft k6	Fit[a]	Hole H8	Shaft k7	Fit[a]	Hole H7	Shaft n6	Fit[a]	Hole H7	Shaft n7
	Values shown below are in thousandths of an inch																	
0 – 0.12	−0.12 +0.52	+0.4 0	+0.12 −0.12	−0.2 +0.8	+0.6 0	+0.2 −0.2							−0.5 +0.15	+0.4 0	+0.5 +0.25	−0.65 +0.15	+0.4 0	+0.65 +0.25
0.12 – 0.24	−0.15 +0.65	+0.5 0	+0.15 −0.15	−0.25 +0.95	+0.7 0	+0.25 −0.25							−0.6 +0.2	+0.5 0	+0.6 +0.3	−0.8 +0.2	+0.5 0	+0.8 +0.3
0.24 – 0.40	−0.2 +0.8	+0.6 0	+0.2 −0.2	−0.3 +1.2	+0.9 0	+0.3 −0.3	−0.5 +0.5	+0.6 0	+0.5 +0.1	−0.7 +0.8	+0.9 0	+0.7 +0.1	−0.8 +0.2	+0.6 0	+0.8 +0.4	−1.0 +0.2	+0.6 0	+1.0 +0.4
0.40 – 0.71	−0.2 +0.9	+0.7 0	+0.2 −0.2	−0.35 +1.35	+1.0 0	+0.35 −0.35	−0.5 +0.6	+0.7 0	+0.5 +0.1	−0.8 +0.9	+1.0 0	+0.8 +0.1	−0.9 +0.2	+0.7 0	+0.9 +0.5	−1.2 +0.2	+0.7 0	+1.2 +0.5
0.71 – 1.19	−0.25 +1.05	+0.8 0	+0.25 −0.25	−0.4 +1.6	+1.2 0	+0.4 −0.4	−0.6 +0.7	+0.8 0	+0.6 +0.1	−0.9 +1.1	+1.2 0	+0.9 +0.1	−1.1 +0.2	+0.8 0	+1.1 +0.6	−1.4 +0.2	+0.8 0	+1.4 +0.6
1.19 – 1.97	−0.3 +1.3	+1.0 0	+0.3 −0.3	−0.5 +2.1	+1.6 0	+0.5 −0.5	−0.7 +0.9	+1.0 0	+0.7 +0.1	−1.1 +1.5	+1.6 0	+1.1 +0.1	−1.3 +0.3	+1.0 0	+1.3 +0.7	−1.7 +0.3	+1.0 0	+1.7 +0.7
1.97 – 3.15	−0.3 +1.5	+1.2 0	+0.3 −0.3	−0.6 +2.4	+1.8 0	+0.6 −0.6	−0.8 +1.1	+1.2 0	+0.8 +0.1	−1.3 +1.7	+1.8 0	+1.3 +0.1	−1.5 +0.4	+1.2 0	+1.5 +0.8	−2.0 +0.4	+1.2 0	+2.0 +0.8
3.15 – 4.73	−0.4 +1.8	+1.4 0	+0.4 −0.4	−0.7 +2.9	+2.2 0	+0.7 −0.7	−1.0 +1.3	+1.4 0	+1.0 +0.1	−1.5 +2.1	+2.2 0	+1.5 +0.1	−1.9 +0.4	+1.4 0	+1.9 +1.0	−2.4 +0.4	+1.4 0	+2.4 +1.0
4.73 – 7.09	−0.5 +2.1	+1.6 0	+0.5 −0.5	−0.8 +3.3	+2.5 0	+0.8 −0.8	−1.1 +1.5	+1.6 0	+1.1 +0.1	−1.7 +2.4	+2.5 0	+1.7 +0.1	−2.2 +0.4	+1.6 0	+2.2 +1.2	−2.8 +0.4	+1.6 0	+2.8 +1.2
7.09 – 9.85	−0.6 +2.4	+1.8 0	+0.6 −0.6	−0.9 +3.7	+2.8 0	+0.9 −0.9	−1.4 +1.6	+1.8 0	+1.4 +0.2	−2.0 +2.6	+2.8 0	+2.0 +0.2	−2.6 +0.4	+1.8 0	+2.6 +1.4	−3.2 +0.4	+1.8 0	+3.2 +1.4
9.85 – 12.41	−0.6 +2.6	+2.0 0	+0.6 −6.6	−1.0 +4.0	+3.0 0	+1.0 −1.0	−1.4 +1.8	+2.0 0	+1.4 +0.2	−2.2 +2.8	+3.0 0	+2.2 +0.2	−2.6 +0.6	+2.0 0	+2.6 +1.4	−3.4 +0.6	+2.0 0	+3.4 +1.4
12.41 – 15.75	−0.7 +2.9	+2.2 0	+0.7 −0.7	−1.0 +4.5	+3.5 0	+1.0 −1.0	−1.6 +2.0	+2.2 0	+1.6 +0.2	−2.4 +3.3	+3.5 0	+2.4 +0.2	−3.0 +0.6	+2.2 0	+3.0 +1.6	−3.8 +0.6	+2.2 0	+3.8 +1.6
15.75 – 19.69	−0.8 +3.3	+2.5 0	+0.8 −0.8	−1.2 +5.2	+4.0 0	+1.2 −1.2	−1.8 +2.3	+2.5 0	+1.8 +0.2	−2.7 +3.8	+4.0 0	+2.7 +0.2	−3.4 +0.7	+2.5 0	+3.4 +1.8	−4.3 +0.7	+2.5 0	+4.3 +1.8

[a] Pairs of values shown represent maximum amount of interference (−) and maximum amount of clearance (+) resulting from application of standard tolerance limits.

All data above heavy lines are in accord with ABC agreements. Symbols H7, js6, etc., are hole and shaft designations in the ABC system.

Table 8. ANSI Standard Interference Location Fits *ANSI B4.1-1967 (R1987)*

Nominal Size Range, Inches	Class LN 1			Class LN 2			Class LN 3		
	Limits of Interference	Standard Limits		Limits of Interference	Standard Limits		Limits of Interference	Standard Limits	
Over To		Hole H6	Shaft n5		Hole H7	Shaft p6		Hole H7	Shaft r6
	Values shown below are given in thousandths of an inch								
0– 0.12	0 0.45	+0.25 0	+0.45 +0.25	0 0.65	+0.4 0	+0.65 +0.4	0.1 0.75	+0.4 0	+0.75 +0.5
0.12– 0.24	0 0.5	+0.3 0	+0.5 +0.3	0 0.8	+0.5 0	+0.8 +0.5	0.1 0.9	+0.5 0	+0.9 +0.6
0.24– 0.40	0 0.65	+0.4 0	+0.65 +0.4	0 1.0	+0.6 0	+1.0 +0.6	0.2 1.2	+0.6 0	+1.2 +0.8
0.40– 0.71	0 0.8	+0.4 0	+0.8 +0.4	0 1.1	+0.7 0	+1.1 +0.7	0.3 1.4	+0.7 0	+1.4 +1.0
0.71– 1.19	0 1.0	+0.5 0	+1.0 +0.5	0 1.3	+0.8 0	+1.3 +0.8	0.4 1.7	+0.8 0	+1.7 +1.2
1.19– 1.97	0 1.1	+0.6 0	+1.1 +0.6	0 1.6	+1.0 0	+1.6 +1.0	0.4 2.0	+1.0 0	+2.0 +1.4
1.97– 3.15	0.1 1.3	+0.7 0	+1.3 +0.8	0.2 2.1	+1.2 0	+2.1 +1.4	0.4 2.3	+1.2 0	+2.3 +1.6
3.15– 4.73	0.1 1.6	+0.9 0	+1.6 +1.0	0.2 2.5	+1.4 0	+2.5 +1.6	0.6 2.9	+1.4 0	+2.9 +2.0
4.73– 7.09	0.2 1.9	+1.0 0	+1.9 +1.2	0.2 2.8	+1.6 0	+2.8 +1.8	0.9 3.5	+1.6 0	+3.5 +2.5
7.09– 9.85	0.2 2.2	+1.2 0	+2.2 +1.4	0.2 3.2	+1.8 0	+3.2 +2.0	1.2 4.2	+1.8 0	+4.2 +3.0
9.85– 12.41	0.2 2.3	+1.2 0	+2.3 +1.4	0.2 3.4	+2.0 0	+3.4 +2.2	1.5 4.7	+2.0 0	+4.7 +3.5
12.41– 15.75	0.2 2.6	+1.4 0	+2.6 +1.6	0.3 3.9	+2.2 0	+3.9 +2.5	2.3 5.9	+2.2 0	+5.9 +4.5
15.75– 19.69	0.2 2.8	+1.6 0	+2.8 +1.8	0.3 4.4	+2.5 0	+4.4 +2.8	2.5 6.6	+2.5 0	+6.6 +5.0

All data in this table are in accordance with American-British-Canadian (ABC) agreements.

Limits for sizes above 19.69 inches are not covered by ABC agreements but are given in the ANSI Standard.

Symbols H7, p6, etc., are hole and shaft designations in the ABC system.

Tolerance limits given in body of table are added or subtracted to basic size (as indicated by + or – sign) to obtain maximum and minimum sizes of mating parts.

Table 9. ANSI Standard Force and Shrink Fits *ANSI B4.1-1967 (R1987)*

Nominal Size Range, Inches	Class FN 1			Class FN 2			Class FN 3			Class FN 4			Class FN 5		
	Interference[a]	Standard Tolerance Limits		Interference[a]	Standard Tolerance Limits		Interference[a]	Standard Tolerance Limits		Interference[a]	Standard Tolerance Limits		Interference[a]	Standard Tolerance Limits	
Over To		Hole H6	Shaft		Hole H7	Shaft s6		Hole H7	Shaft t6		Hole H7	Shaft u6		Hole H8	Shaft x7
	Values shown below are in thousandths of an inch														
0– 0.12	0.05	+0.25	+0.5	0.2	+0.4	+0.85				0.3	+0.4	+0.95	0.3	+0.6	+1.3
	0.5	0	+0.3	0.85	0	+0.6				0.95	0	+0.7	1.3	0	+0.9
0.12– 0.24	0.1	+0.3	+0.6	0.2	+0.5	+1.0				0.4	+0.5	+1.2	0.5	+0.7	+1.7
	0.6	0	+0.4	1.0	0	+0.7				1.2	0	+0.9	1.7	0	+1.2
0.24– 0.40	0.1	+0.4	+0.75	0.4	+0.6	+1.4				0.6	+0.6	+1.6	0.5	+0.9	+2.0
	0.75	0	+0.5	1.4	0	+1.0				1.6	0	+1.2	2.0	0	+1.4
0.40– 0.56	0.1	+0.4	+0.8	0.5	+0.7	+1.6				0.7	+0.7	+1.8	0.6	+1.0	+2.3
	0.8	0	+0.5	1.6	0	+1.2				1.8	0	+1.4	2.3	0	+1.6
0.56– 0.71	0.2	+0.4	+0.9	0.5	+0.7	+1.6				0.7	+0.7	+1.8	0.8	+1.0	+2.5
	0.9	0	+0.6	1.6	0	+1.2				1.8	0	+1.4	2.5	0	+1.8
0.71– 0.95	0.2	+0.5	+1.1	0.6	+0.8	+1.9				0.8	+0.8	+2.1	1.0	+1.2	+3.0
	1.1	0	+0.7	1.9	0	+1.4				2.1	0	+1.6	3.0	0	+2.2
0.95– 1.19	0.3	+0.5	+1.2	0.6	+0.8	+1.9	0.8	+0.8	+2.1	+1.0	+0.8	+2.3	1.3	+1.2	+3.3
	1.2	0	+0.8	1.9	0	+1.4	2.1	0	+1.6	2.3	0	+1.8	3.3	0	+2.5
1.19– 1.58	0.3	+0.6	+1.3	0.8	+1.0	+2.4	1.0	+1.0	+2.6	1.5	+1.0	+3.1	1.4	+1.6	+4.0
	1.3	0	+0.9	2.4	0	+1.8	2.6	0	+2.0	3.1	0	+2.5	4.0	0	+3.0
1.58– 1.97	0.4	+0.6	+1.4	0.8	+1.0	+2.4	1.2	+1.0	+2.8	1.8	+1.0	+3.4	2.4	+1.6	+5.0
	1.4	0	+1.0	2.4	0	+1.8	2.8	0	+2.2	3.4	0	+2.8	5.0	0	+4.0
1.97– 2.56	0.6	+0.7	+1.8	0.8	+1.2	+2.7	1.3	+1.2	+3.2	2.3	+1.2	+4.2	3.2	+1.8	+6.2
	1.8	0	+1.3	2.7	0	+2.0	3.2	0	+2.5	4.2	0	+3.5	6.2	0	+5.0
2.56– 3.15	0.7	+0.7	+1.9	1.0	+1.2	+2.9	1.8	+1.2	+3.7	2.8	+1.2	+4.7	4.2	+1.8	+7.2
	1.9	0	+1.4	2.9	0	+2.2	3.7	0	+3.0	4.7	0	+4.0	7.2	0	+6.0
3.15– 3.94	0.9	+0.9	+2.4	1.4	+1.4	+3.7	2.1	+1.4	+4.4	3.6	+1.4	+5.9	4.8	+2.2	+8.4
	2.4	0	+1.8	3.7	0	+2.8	4.4	0	+3.5	5.9	0	+5.0	8.4	0	+7.0
3.94– 4.73	1.1	+0.9	+2.6	1.6	+1.4	+3.9	2.6	+1.4	+4.9	4.6	+1.4	+6.9	5.8	+2.2	+9.4
	2.6	0	+2.0	3.9	0	+3.0	4.9	0	+4.0	6.9	0	+6.0	9.4	0	+8.0

Table 9. *(Continued)* **ANSI Standard Force and Shrink Fits** *ANSI B4.1-1967 (R1987)*

Nominal Size Range, Inches		Class FN 1			Class FN 2			Class FN 3			Class FN 4			Class FN 5		
		Interference[a]	Standard Tolerance Limits		Interference[a]	Standard Tolerance Limits		Interference[a]	Standard Tolerance Limits		Interference[a]	Standard Tolerance Limits		Interference[a]	Standard Tolerance Limits	
Over	To		Hole H6	Shaft		Hole H7	Shaft s6		Hole H7	Shaft t6		Hole H7	Shaft u6		Hole H8	Shaft x7
		Values shown below are in thousandths of an inch														
4.73–	5.52	1.2	+1.0	+2.9	1.9	+1.6	+4.5	3.4	+1.6	+6.0	5.4	+1.6	+8.0	7.5	+2.5	+11.6
		2.9	0	+2.2	4.5	0	+3.5	6.0	0	+5.0	8.0	0	+7.0	11.6	0	+10.0
5.52–	6.30	1.5	+1.0	+3.2	2.4	+1.6	+5.0	3.4	+1.6	+6.0	5.4	+1.6	+8.0	9.5	+2.5	+13.6
		3.2	0	+2.5	5.0	0	+4.0	6.0	0	+5.0	8.0	0	+7.0	13.6	0	+12.0
6.30–	7.09	1.8	+1.0	+3.5	2.9	+1.6	+5.5	4.4	+1.6	+7.0	6.4	+1.6	+9.0	9.5	+2.5	+13.6
		3.5	0	+2.8	5.5	0	+4.5	7.0	0	+6.0	9.0	0	+8.0	13.6	0	+12.0
7.09–	7.88	1.8	+1.2	+3.8	3.2	+1.8	+6.2	5.2	+1.8	+8.2	7.2	+1.8	+10.2	11.2	+2.8	+15.8
		3.8	0	+3.0	6.2	0	+5.0	8.2	0	+7.0	10.2	0	+9.0	15.8	0	+14.0
7.88–	8.86	2.3	+1.2	+4.3	3.2	+1.8	+6.2	5.2	+1.8	+8.2	8.2	+1.8	+11.2	13.2	+2.8	+17.8
		4.3	0	+3.5	6.2	0	+5.0	8.2	0	+7.0	11.2	0	+10.0	17.8	0	+16.0
8.86–	9.85	2.3	+1.2	+4.3	4.2	+1.8	+7.2	6.2	+1.8	+9.2	10.2	+1.8	+13.2	13.2	+2.8	+17.8
		4.3	0	+3.5	7.2	0	+6.0	9.2	0	+8.0	13.2	0	+12.0	17.8	0	+16.0
9.85–	11.03	2.8	+1.2	+4.9	4.0	+2.0	+7.2	7.0	+2.0	+10.2	10.0	+2.0	+13.2	15.0	+3.0	+20.0
		4.9	0	+4.0	7.2	0	+6.0	10.2	0	+9.0	13.2	0	+12.0	20.0	0	+18.0
11.03–	12.41	2.8	+1.2	+4.9	5.0	+2.0	+8.2	7.0	+2.0	+10.2	12.0	+2.0	+15.2	17.0	+3.0	+22.0
		4.9	0	+4.0	8.2	0	+7.0	10.2	0	+9.0	15.2	0	+14.0	22.0	0	+20.0
12.41–	13.98	3.1	+1.4	+5.5	5.8	+2.2	+9.4	7.8	+2.2	+11.4	13.8	+2.2	+17.4	18.5	+3.5	+24.2
		5.5	0	+4.5	9.4	0	+8.0	11.4	0	+10.0	17.4	0	+16.0	24.2	0	+22.0
13.98–	15.75	3.6	+1.4	+6.1	5.8	+2.2	+9.4	9.8	+2.2	+13.4	15.8	+2.2	+19.4	21.5	+3.5	+27.2
		6.1	0	+5.0	9.4	0	+8.0	13.4	0	+12.0	19.4	0	+18.0	27.2	0	+25.0
15.75–	17.72	4.4	+1.6	+7.0	6.5	+2.5	+10.6	+9.5	+2.5	+13.6	17.5	+2.5	+21.6	24.0	+4.0	+30.5
		7.0	0	+6.0	10.6	0	+9.0	13.6	0	+12.0	21.6	0	+20.0	30.5	0	+28.0
17.72–	19.69	4.4	+1.6	+7.0	7.5	+2.5	+11.6	11.5	+2.5	+15.6	19.5	+2.5	+23.6	26.0	+4.0	+32.5
		7.0	0	+6.0	11.6	0	+10.0	15.6	0	+14.0	23.6	0	+22.0	32.5	0	+30.0

[a] Pairs of values shown represent minimum and maximum amounts of interference resulting from application of standard tolerance limits.

All data above heavy lines are in accordance with American-British-Canadian (ABC) agreements. Symbols H6, H7, s6, etc., are hole and shaft designations in the ABC system. Limits for sizes above 19.69 inches are not covered by ABC agreements but are given in the ANSI standard.

Modified Standard Fits.—Fits having the same limits of clearance or interference as those shown in Tables 2 to 6 may sometimes have to be produced by using holes or shafts having limits of size other than those shown in these tables. These modifications may be accomplished by using either a *Bilateral Hole* (*System B*) or a *Basic Shaft System* (*Symbol S*). Both methods will result in nonstandard holes and shafts.

Bilateral Hole Fits: (*Symbol B*): The common situation is where holes are produced with fixed tools such as drills or reamers; to provide a longer wear life for such tools, a bilateral tolerance is desired.

The symbols used for these fits are identical with those used for standard fits except that they are followed by the letter B. Thus, LC 4B is a clearance locational fit, Class 4, except that it is produced with a bilateral hole.

The limits of clearance or interference are identical with those shown in Tables 2 to 6 for the corresponding fits.

The hole tolerance, however, is changed so that the plus limit is that for one grade finer than the value shown in the tables and the minus limit equals the amount by which the plus limit was lowered. The shaft limits are both lowered by the same amount as the lower limit of size of the hole. The finer grade of tolerance required to make these modifications may be obtained from Table 1. For example, an LC 4B fit for a 6-inch diameter hole would have tolerance limits of +4.0, −2.0 (+0.0040 inch, −0.0020 inch); the shaft would have tolerance limits of −2.0, −6.0 (−0.0020 inch, −0.0060 inch).

Basic Shaft Fits: (*Symbol S*): For these fits, the maximum size of the shaft is basic. The limits of clearance or interference are identical with those shown in Tables 2 to 6 for the corresponding fits and the symbols used for these fits are identical with those used for standard fits except that they are followed by the letter S. Thus, LC 4S is a clearance locational fit, Class 4, except that it is produced on a basic shaft basis.

The limits for hole and shaft as given in Tables 2 to 6 are increased for clearance fits (*decreased* for transition or interference fits) by the value of the upper shaft limit; that is, by the amount required to change the maximum shaft to the basic size.

American National Standard Preferred Metric Limits and Fits.—This standard ANSI B4.2-1978 (R1994) describes the ISO system of metric limits and fits for mating parts as approved for general engineering usage in the United States.

It establishes: 1) the designation symbols used to define dimensional limits on drawings, material stock, related tools, gages, etc.; 2) the preferred basic sizes (first and second choices); 3) the preferred tolerance zones (first, second, and third choices); 4) the preferred limits and fits for sizes (first choice only) up to and including 500 millimeters; and 5) the definitions of related terms.

The general terms "hole" and "shaft" can also be taken to refer to the space containing or contained by two parallel faces of any part, such as the width of a slot, or the thickness of a key.

Definitions.—The most important terms relating to limits and fits are shown in Fig. 1 and are defined as follows:

Basic Size: The size to which limits of deviation are assigned. The basic size is the same for both members of a fit. For example, it is designated by the numbers 40 in 40H7.

Deviation: The algebraic difference between a size and the corresponding basic size.

Upper Deviation: The algebraic difference between the maximum limit of size and the corresponding basic size.

Lower Deviation: The algebraic difference between the minimum limit of size and the corresponding basic size.

Fundamental Deviation: That one of the two deviations closest to the basic size. For example, it is designated by the letter H in 40H7.

Tolerance: The difference between the maximum and minimum size limits on a part.

Tolerance Zone: A zone representing the tolerance and its position in relation to the basic size.

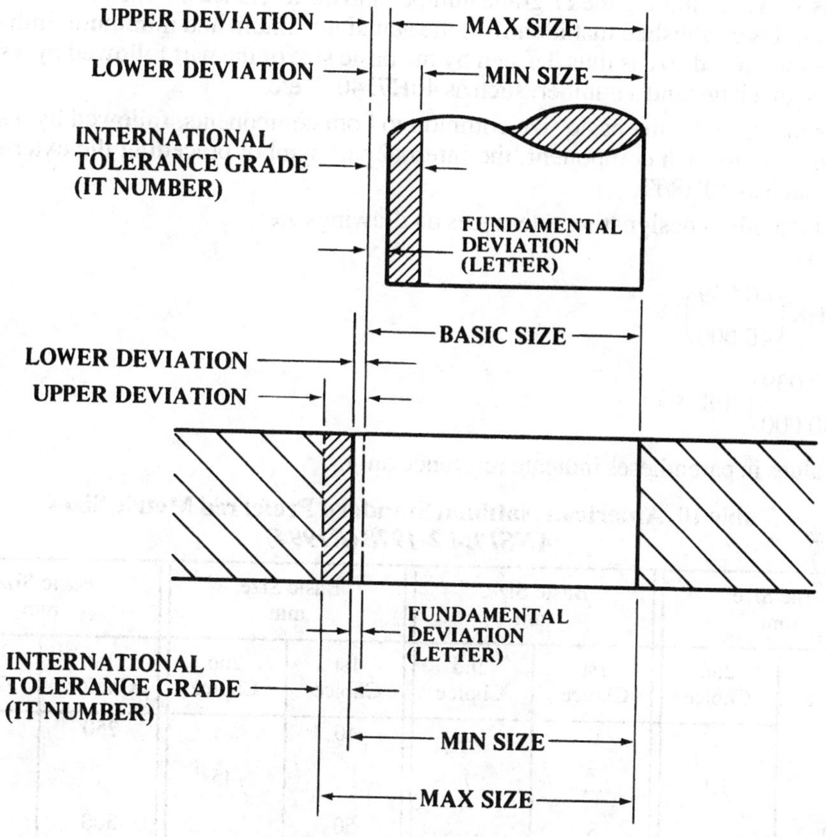

Fig. 1. Illustration of Definitions

International Tolerance Grade: (*IT*): A group of tolerances that vary depending on the basic size, but that provide the same relative level of accuracy within a given grade. For example, it is designated by the number 7 in 40H7 or as IT7.

Hole Basis: The system of fits where the minimum hole size is basic. The fundamental deviation for a hole basis system is H.

Shaft Basis: The system of fits where the maximum shaft size is basic. The fundamental deviation for a shaft basis system is h.

Clearance Fit: The relationship between assembled parts when clearance occurs under all tolerance conditions.

Interference Fit: The relationship between assembled parts when interference occurs under all tolerance conditions.

Transition Fit: The relationship between assembled parts when either a clearance or an interference fit can result, depending on the tolerance conditions of the mating parts.

Tolerances Designation.—An "International Tolerance grade" establishes the magnitude of the tolerance zone or the amount of part size variation allowed for external and internal dimensions alike (see Fig. 1). Tolerances are expressed in grade numbers that are consistent with International Tolerance grades identified by the prefix IT, such as IT6, IT11, etc. A smaller grade number provides a smaller tolerance zone.

A fundamental deviation establishes the position of the tolerance zone with respect to the basic size (see Fig. 1). Fundamental deviations are expressed by tolerance position letters.

Capital letters are used for internal dimensions and lowercase or small letters for external dimensions.

Symbols.—By combining the IT grade number and the tolerance position letter, the tolerance symbol is established that identifies the actual maximum and minimum limits of the part. The toleranced size is thus defined by the basic size of the part followed by a symbol composed of a letter and a number, such as 40H7, 40f7, etc.

A fit is indicated by the basic size common to both components, followed by a symbol corresponding to each component, the internal part symbol preceding the external part symbol, such as 40H8/f7.

Some methods of designating tolerances on drawings are:

A) 40H8

B) 40H8 $\begin{pmatrix} 40.039 \\ 40.000 \end{pmatrix}$

C) $\begin{pmatrix} 40.039 \\ 40.000 \end{pmatrix}$ 40H8

The values in parentheses indicate reference only.

Table 10. American National Standard Preferred Metric Sizes
ANSI B4.2-1978 (R1994)

Basic Size, mm		Basic Size, mm		Basic Size, mm		Basic Size, mm	
1st Choice	2nd Choice	1st Choice	2nd Choice	1st Choice	2nd Choice	1st Choice	2nd Choice
1	…	6	…	40	…	250	…
…	1.1	…	7	…	45	…	280
1.2	…	8	…	50	…	300	…
…	1.4	…	9	…	55	…	350
1.6	…	10	…	60	…	400	…
…	1.8	…	11	…	70	…	450
2	…	12	…	80	…	500	…
…	2.2	…	14	…	90	…	550
2.5	…	16	…	100	…	600	…
…	2.8	…	18	…	110	…	700
3	…	20	…	120	…	800	…
…	3.5	…	22	…	140	…	900
4	…	25	…	160	…	1000	…
…	4.5	…	28	…	180	…	…
5	…	30	…	200	…	…	…
…	5.5	…	35	…	220	…	…

Preferred Metric Sizes.—American National Standard ANSI B32.4M-1980 (R1994), presents series of preferred metric sizes for round, square, rectangular, and hexagonal metal products. Table 10 gives preferred metric diameters from 1 to 320 millimeters for round metal products. Wherever possible, sizes should be selected from the Preferred Series shown in the table. A Second Preference series is also shown. A Third Preference Series not shown in the table is: 1.3, 2.1, 2.4, 2.6, 3.2, 3.8, 4.2, 4.8, 7.5, 8.5, 9.5, 36, 85, and 95.

Most of the Preferred Series of sizes are derived from the American National Standard "10 series" of preferred numbers (see *American National Standard for Preferred Numbers* on page 20). Most of the Second Preference Series are derived from the "20 series" of preferred numbers. Third Preference sizes are generally from the "40 series" of preferred numbers.

For preferred metric diameters less than 1 millimeter, preferred across flat metric sizes of square and hexagon metal products, preferred across flat metric sizes of rectangular metal products, and preferred metric lengths of metal products, reference should be made to the Standard.

Preferred Fits.—First-choice tolerance zones are used to establish preferred fits in the Standard for Preferred Metric Limits and Fits, ANSI B4.2, as shown in Figs. 2 and 3. A complete listing of first-, second-, and third- choice tolerance zones is given in the Standard.

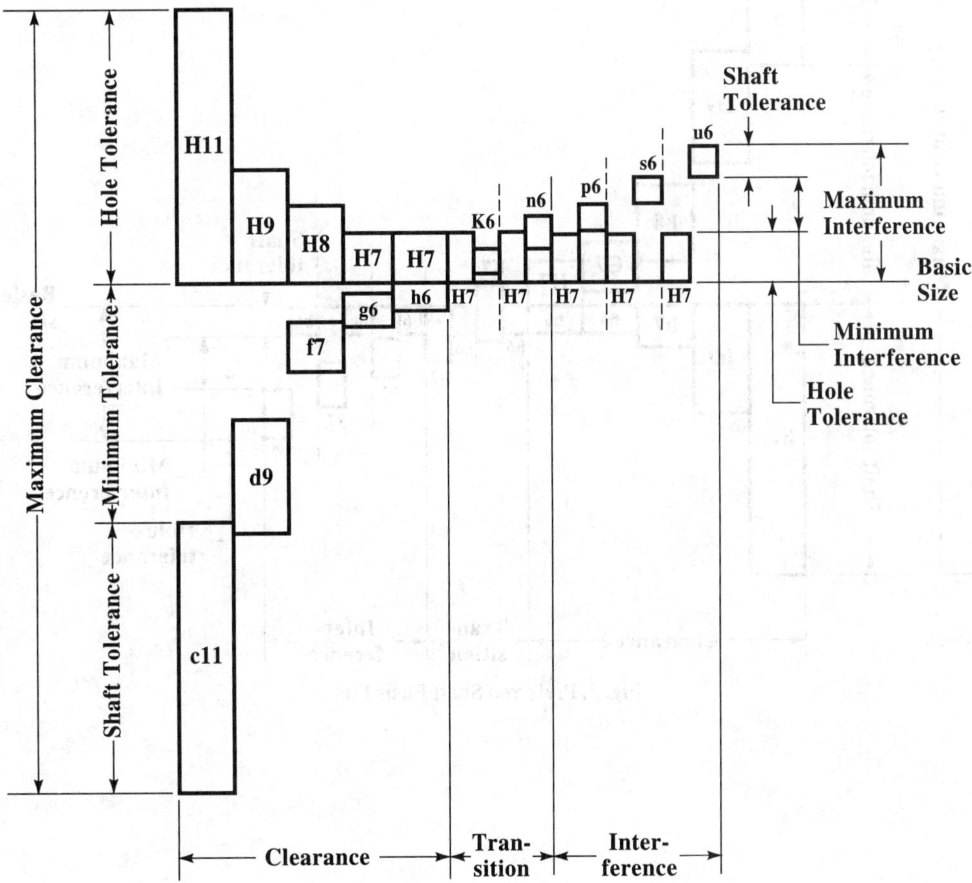

Fig. 2. Preferred Hole Basis Fits

Hole basis fits have a fundamental deviation of H on the hole, and shaft basis fits have a fundamental deviation of h on the shaft and are shown in Fig. 2 for hole basis and Fig. 3 for shaft basis fits. A description of both types of fits, that have the same relative fit condition, is given in Table 11. Normally, the hole basis system is preferred; however, when a common shaft mates with several holes, the shaft basis system should be used.

The hole basis and shaft basis fits shown in Table 11 are combined with the first-choice sizes shown in Table 10 to form Tables 12, 13, 14, and 15, where specific limits as well as the resultant fits are tabulated.

If the required size is not tabulated in Tables 12 through 15 then the preferred fit can be calculated from numerical values given in an appendix of ANSI B4.2-1978 (R1984). It is anticipated that other fit conditions may be necessary to meet special requirements, and a preferred fit can be loosened or tightened simply by selecting a standard tolerance zone as given in the Standard. Information on how to calculate limit dimensions, clearances, and interferences, for nonpreferred fits and sizes can also be found in an appendix of this Standard.

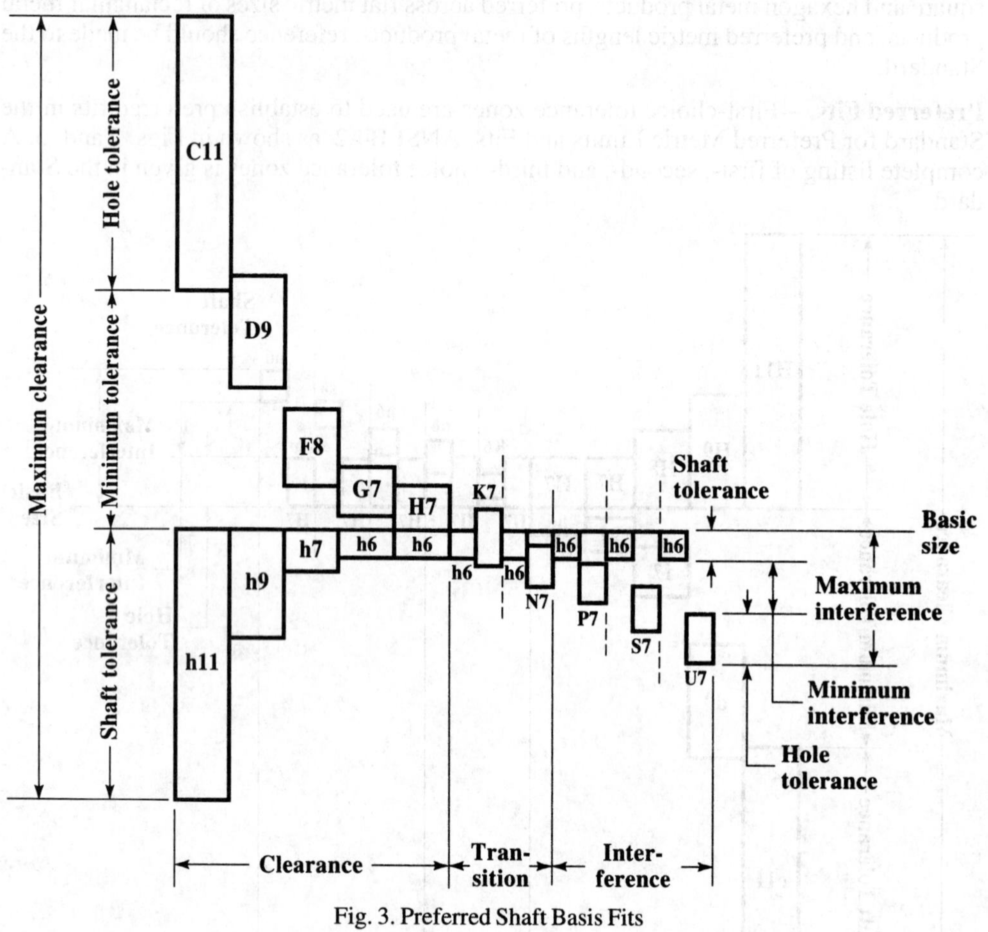

Fig. 3. Preferred Shaft Basis Fits

Table 11. Description of preferred fits

	ISO SYMBOL		DESCRIPTION	
	Hole Basis	Shaft Basis		
Clearance Fits	H11/c11	C11/h11	*Loose running* fit for wide commercial tolerances or allowances on external members.	↑ More Clearance
	H9/d9	D9/h9	*Free running* fit not for use where accuracy is essential, but good for large temperature variations, high running speeds, or heavy journal pressures.	
	H8/f7	F8/h7	*Close Running* fit for running on accuarate machines and for accurate moderate speeds and journal pressures.	
	H7/g6	G7/h6	*Sliding fit* not intended to run freely, but to move and turn freely and locate accurately.	
	H7/h6	H7/h6	*Locational clearance* fit provides snug fit for locating stationary parts; but can be freely assembled and disassembled.	
Transition Fits	H7/k6	K7/h6	*Locational transition* fit for accurate location, a compromise between clearance and interferance.	
	H7/n6	N7/h6	*Locational transition* fit for more accuarate location where greater interferance is permissible.	
Interference Fits	H7/p6[a]	P7/h6	*Locational interference* fit for parts requiring and alignment with prime accuracy of location but without special bore pressure requirements.	More Interferance ↓
	H7/s6	S7/h6	*Medium drive* fit for ordinary steel parts or shrink fits on light sections, the tightest fit usable with cast iron.	
	H7/u6	U7/h6	*Force* fit suitable for parts which can be highly stressed or for shrink fits where the heavy pressing forces required are impractical.	

[a] Transition fit for basic sizes in range from 0 through 3 mm.

Table 12. American National Standard Preferred Hole Basis Metric Clearance Fits *ANSI B4.2-1978 (R1994)*

Basic Size[a]		Loose Running			Free Running			Close Running			Sliding			Locational Clearance		
		Hole H11	Shaft C11	Fit[b]	Hole H9	Shaft d9	Fit[b]	Hole H8	Shaft f7	Fit[b]	Hole H7	Shaft g6	Fit[b]	Hole H7	Shaft h6	Fit[b]
1	Max	1.060	0.940	0.180	1.025	0.980	0.070	1.014	0.994	0.030	1.010	0.998	0.018	1.010	1.000	0.016
	Min	1.000	0.880	0.060	1.000	0.995	0.020	1.000	0.984	0.006	1.000	0.992	0.002	1.000	0.994	0.000
1.2	Max	1.260	1.140	0.180	1.225	1.180	0.070	1.214	1.194	0.030	1.210	1.198	0.018	1.210	1.200	0.016
	Min	1.200	1.080	0.060	1.200	1.155	0.020	1.200	1.184	0.006	1.200	1.192	0.002	1.200	1.194	0.000
1.6	Max	1.660	1.540	0.180	1.625	1.580	0.070	1.614	1.594	0.030	1.610	1.598	0.018	1.610	1.600	0.016
	Min	1.600	1.480	0.060	1.600	1.555	0.020	1.600	1.584	0.006	1.600	1.592	0.002	1.600	1.594	0.000
2	Max	2.060	1.940	0.180	2.025	1.980	0.070	2.014	1.994	0.030	2.010	1.998	0.018	2.010	2.000	0.016
	Min	2.000	1.880	0.060	2.000	1.955	0.020	2.000	1.984	0.006	2.000	1.992	0.002	2.000	1.994	0.000
2.5	Max	2.560	2.440	0.180	2.525	2.480	0.070	2.514	2.494	0.030	2.510	2.498	0.018	2.510	2.500	0.016
	Min	2.500	2.380	0.060	2.500	2.455	0.020	2.500	2.484	0.006	2.500	2.492	0.002	2.500	2.494	0.000
3	Max	3.060	2.940	0.180	3.025	2.980	0.070	3.014	2.994	0.030	3.010	2.998	0.018	3.010	3.000	0.016
	Min	3.000	2.880	0.060	3.000	2.955	0.020	3.000	2.984	0.006	3.000	2.992	0.002	3.000	2.994	0.000
4	Max	4.075	3.930	0.220	4.030	3.970	0.090	4.018	3.990	0.040	4.012	3.996	0.024	4.012	4.000	0.020
	Min	4.000	3.855	0.070	4.000	3.940	0.030	4.000	3.978	0.010	4.000	3.988	0.004	4.000	3.992	0.000
5	Max	5.075	4.930	0.220	5.030	4.970	0.090	5.018	4.990	0.040	5.012	4.996	0.024	5.012	5.000	0.020
	Min	5.000	4.855	0.070	5.000	4.940	0.030	5.000	4.978	0.010	5.000	4.988	0.004	5.000	4.992	0.000
6	Max	6.075	5.930	0.220	6.030	5.970	0.090	6.018	5.990	0.040	6.012	5.996	0.024	6.012	6.000	0.020
	Min	6.000	5.855	0.070	6.000	5.940	0.030	6.000	5.978	0.010	6.000	5.988	0.004	6.000	5.992	0.000
8	Max	8.090	7.920	0.260	8.036	7.960	0.112	8.022	7.987	0.050	8.015	7.995	0.029	8.015	8.000	0.024
	Min	8.000	7.830	0.080	8.000	7.924	0.040	8.000	7.972	0.013	8.000	7.986	0.005	8.000	7.991	0.000
10	Max	10.090	9.920	0.260	10.036	9.960	0.112	10.022	9.987	0.050	10.015	9.995	0.029	10.015	10.000	0.024
	Min	10.000	9.830	0.080	10.000	9.924	0.040	10.000	9.972	0.013	10.000	9.986	0.005	10.000	9.991	0.000
12	Max	12.110	11.905	0.315	12.043	11.956	0.136	12.027	11.984	0.061	12.018	11.994	0.035	12.018	12.000	0.029
	Min	12.000	11.795	0.095	12.000	11.907	0.050	12.000	11.966	0.016	12.000	11.983	0.006	12.000	11.989	0.000
16	Max	16.110	15.905	0.315	16.043	15.950	0.136	16.027	15.984	0.061	16.018	15.994	0.035	16.018	16.000	0.029
	Min	16.000	15.795	0.095	16.000	15.907	0.050	16.000	15.966	0.016	16.000	15.983	0.006	16.000	15.989	0.000
20	Max	20.130	19.890	0.370	20.052	19.935	0.169	20.033	19.980	0.074	20.021	19.993	0.041	20.021	20.000	0.034
	Min	20.000	19.760	0.110	20.000	19.883	0.065	20.000	19.959	0.020	20.000	19.980	0.007	20.000	19.987	0.000
25	Max	25.130	24.890	0.370	25.052	24.935	0.169	25.033	24.980	0.074	25.021	24.993	0.041	25.021	25.000	0.034
	Min	25.000	24.760	0.110	25.000	24.883	0.065	25.000	24.959	0.020	25.000	24.980	0.007	25.000	24.987	0.000

Table 12. *(Continued)* **American National Standard Preferred Hole Basis Metric Clearance Fits** *ANSI B4.2-1978 (R1994)*

Basic Size[a]		Loose Running			Free Running			Close Running			Sliding			Locational Clearance		
		Hole H11	Shaft C11	Fit[b]	Hole H9	Shaft d9	Fit[b]	Hole H8	Shaft f7	Fit[b]	Hole H7	Shaft g6	Fit[b]	Hole H7	Shaft h6	Fit[b]
30	Max	30.130	29.890	0.370	30.052	29.935	0.169	30.033	29.980	0.074	30.021	29.993	0.041	30.021	30.000	0.034
	Min	30.000	29.760	0.110	30.000	29.883	0.065	30.000	29.959	0.020	30.000	29.980	0.007	30.000	29.987	0.000
40	Max	40.160	39.880	0.440	40.062	39.920	0.204	40.039	39.975	0.089	40.025	39.991	0.050	40.025	40.000	0.041
	Min	40.000	39.720	0.120	40.000	39.858	0.080	40.000	39.950	0.025	40.000	39.975	0.009	40.000	39.984	0.000
50	Max	50.160	49.870	0.450	50.062	49.920	0.204	50.039	49.975	0.089	50.025	49.991	0.050	50.025	50.000	0.041
	Min	50.000	49.710	0.130	50.000	49.858	0.080	50.000	49.950	0.025	50.000	49.975	0.009	50.000	49.984	0.000
60	Max	60.190	59.860	0.520	60.074	59.900	0.248	60.046	59.970	0.106	60.030	59.990	0.059	60.030	60.000	0.049
	Min	60.000	59.670	0.140	60.000	59.826	0.100	60.000	59.940	0.030	60.000	59.971	0.010	60.000	59.981	0.000
80	Max	80.190	79.850	0.530	80.074	79.900	0.248	80.046	79.970	0.106	80.030	79.990	0.059	80.030	80.000	0.049
	Min	80.000	79.660	0.150	80.000	79.826	0.100	80.000	79.940	0.030	80.000	79.971	0.010	80.000	79.981	0.000
100	Max	100.220	99.830	0.610	100.087	99.880	0.294	100.054	99.964	0.125	100.035	99.988	0.069	100.035	100.000	0.057
	Min	100.000	99.610	0.170	100.000	99.793	0.120	100.000	99.929	0.036	100.000	99.966	0.012	100.000	99.978	0.000
120	Max	120.220	119.820	0.620	120.087	119.880	0.294	120.054	119.964	0.125	120.035	119.988	0.069	120.035	120.000	0.057
	Min	120.000	119.600	0.180	120.000	119.793	0.120	120.000	119.929	0.036	120.000	119.966	0.012	120.000	119.978	0.000
160	Max	160.250	159.790	0.710	160.100	159.855	0.345	160.063	159.957	0.146	160.040	159.986	0.079	160.040	160.000	0.065
	Min	160.000	159.540	0.210	160.000	159.755	0.145	160.000	159.917	0.043	160.000	159.961	0.014	160.000	159.975	0.000
200	Max	200.290	199.760	0.820	200.115	199.830	0.400	200.072	199.950	0.168	200.046	199.985	0.090	200.046	200.000	0.075
	Min	200.000	199.470	0.240	200.000	199.715	0.170	200.000	199.904	0.050	200.000	199.956	0.015	200.000	199.971	0.000
250	Max	250.290	249.720	0.860	250.115	249.830	0.400	250.072	249.950	0.168	250.046	249.985	0.090	250.046	250.000	0.075
	Min	250.000	249.430	0.280	250.000	249.715	0.170	250.000	249.904	0.050	250.000	249.956	0.015	250.000	249.971	0.000
300	Max	300.320	299.670	0.970	300.130	299.810	0.450	300.081	299.944	0.189	300.052	299.983	0.101	300.052	300.000	0.084
	Min	300.000	299.350	0.330	300.000	299.680	0.190	300.000	299.892	0.056	300.000	299.951	0.017	300.000	299.968	0.000
400	Max	400.360	399.600	1.120	400.140	399.790	0.490	400.089	399.938	0.208	400.057	399.982	0.111	400.057	400.000	0.093
	Min	400.000	399.240	0.400	400.000	399.650	0.210	400.000	399.881	0.062	400.000	399.946	0.018	400.000	399.964	0.000
500	Max	500.400	499.520	1.280	500.155	499.770	0.540	500.097	499.932	0.228	500.063	499.980	0.123	500.063	500.000	0.103
	Min	500.000	499.120	0.480	500.000	499.615	0.230	500.000	499.869	0.068	500.000	499.940	0.020	500.000	499.960	0.000

[a] The sizes shown are first-choice basic sizes (see Table 10). Preferred fits for other sizes can be calculated from data given in ANSI B4.2-1978 (R1984).

[b] All fits shown in this table have clearance.

All dimensions are in millimeters.

Table 13. American National Standard Preferred Hole Basis Metric Transition and Interference Fits *ANSI B4.2-1978 (R1994)*

Basic Size[a]		Locational Transition			Locational Transition			Locational Interference			Medium Drive			Force		
		Hole H7	Shaft k6	Fit[b]	Hole H7	Shaft n6	Fit[b]	Hole H7	Shaft p6	Fit[b]	Hole H7	Shaft s6	Fit[b]	Hole H7	Shaft u6	Fit[b]
1	Max	1.010	1.006	+0.010	1.010	1.010	+0.006	1.010	1.012	+0.004	1.010	1.020	−0.004	1.010	1.024	−0.008
	Min	1.000	1.000	−0.006	1.000	1.004	−0.010	1.000	1.006	−0.012	1.000	1.014	−0.020	1.000	1.018	−0.024
1.2	Max	1.210	1.206	+0.010	1.210	1.210	+0.006	1.210	1.212	+0.004	1.210	1.220	−0.004	1.210	1.224	−0.008
	Min	1.200	1.200	−0.006	1.200	1.204	−0.010	1.200	1.206	−0.012	1.200	1.214	−0.020	1.200	1.218	−0.024
1.6	Max	1.610	1.606	+0.010	1.610	1.610	+0.006	1.610	1.612	+0.004	1.610	1.620	−0.004	1.610	1.624	−0.008
	Min	1.600	1.600	−0.006	1.600	1.604	−0.010	1.600	1.606	−0.012	1.600	1.614	−0.020	1.600	1.618	−0.024
2	Max	2.010	2.006	+0.010	2.010	2.010	+0.006	2.010	2.012	+0.004	2.010	2.020	−0.004	2.010	2.024	−0.008
	Min	2.000	2.000	−0.006	2.000	2.004	−0.010	2.000	2.006	−0.012	2.000	2.014	−0.020	2.000	2.018	−0.024
2.5	Max	2.510	2.506	+0.010	2.510	2.510	+0.006	2.510	2.512	+0.004	2.510	2.520	−0.004	2.510	2.524	−0.008
	Min	2.500	2.500	−0.006	2.500	2.504	−0.010	2.500	2.506	−0.012	2.500	2.514	−0.020	2.500	2.518	−0.024
3	Max	3.010	3.006	+0.010	3.010	3.010	+0.006	3.010	3.012	+0.004	3.010	3.020	−0.004	3.010	3.024	−0.008
	Min	3.000	3.000	−0.006	3.000	3.004	−0.010	3.000	3.006	−0.012	3.000	3.014	−0.020	3.000	3.018	−0.024
4	Max	4.012	4.009	+0.011	4.012	4.016	+0.004	4.012	4.020	0.000	4.012	4.027	−0.007	4.012	4.031	−0.011
	Min	4.000	4.001	−0.009	4.000	4.008	−0.016	4.000	4.012	−0.020	4.000	4.019	−0.027	4.000	4.023	−0.031
5	Max	5.012	5.009	+0.011	5.012	5.016	+0.004	5.012	5.020	0.000	5.012	5.027	−0.007	5.012	5.031	−0.011
	Min	5.000	5.001	−0.009	5.000	5.008	−0.016	5.000	5.012	−0.020	5.000	5.019	−0.027	5.000	5.023	−0.031
6	Max	6.012	6.009	+0.011	6.012	6.016	+0.004	6.012	6.020	0.000	6.012	6.027	−0.007	6.012	6.031	−0.011
	Min	6.000	6.001	−0.009	6.000	6.008	−0.016	6.000	6.012	−0.020	6.000	6.019	−0.027	6.000	6.023	−0.031
8	Max	8.015	8.010	+0.014	8.015	8.019	+0.005	8.015	8.024	0.000	8.015	8.032	−0.008	8.015	8.037	−0.013
	Min	8.000	8.001	−0.010	8.000	8.010	−0.019	8.000	8.015	−0.024	8.000	8.023	−0.032	8.000	8.028	−0.037
10	Max	10.015	10.010	+0.014	10.015	10.019	+0.005	10.015	10.024	0.000	10.015	10.032	−0.008	10.015	10.034	−0.013
	Min	10.000	10.001	−0.010	10.000	10.010	−0.019	10.000	10.015	−0.024	10.000	10.023	−0.032	10.000	10.028	−0.037
12	Max	12.018	12.012	+0.017	12.018	12.023	+0.006	12.018	12.029	0.000	12.018	12.039	−0.010	12.018	12.044	−0.015
	Min	12.000	12.001	−0.012	12.000	12.012	−0.023	12.000	12.018	−0.029	12.000	12.028	−0.039	12.000	12.033	−0.044
16	Max	16.018	16.012	+0.017	16.018	16.023	+0.006	16.018	16.029	0.000	16.018	16.039	−0.010	16.018	16.044	−0.015
	Min	16.000	16.001	−0.012	16.000	16.012	−0.023	16.000	16.018	−0.029	16.000	16.028	−0.039	16.000	16.033	−0.044
20	Max	20.021	20.015	+0.019	20.021	20.028	+0.006	20.021	20.035	−0.001	20.021	20.048	−0.014	20.021	20.054	−0.020
	Min	20.000	20.002	−0.015	20.000	20.015	−0.028	20.000	20.022	−0.035	20.000	20.035	−0.048	20.000	20.041	−0.054
25	Max	25.021	25.015	+0.019	25.021	25.028	+0.006	25.021	25.035	−0.001	25.021	25.048	−0.014	25.021	25.061	−0.027
	Min	25.000	25.002	−0.015	25.000	25.015	−0.028	25.000	25.022	−0.035	25.000	25.035	−0.048	25.000	25.048	−0.061

Table 13. *(Continued)* **American National Standard Preferred Hole Basis Metric Transition and Interference Fits** *ANSI B4.2-1978 (R1994)*

Basic Size[a]		Locational Transition			Locational Transition			Locational Interference			Medium Drive			Force		
		Hole H7	Shaft k6	Fit[b]	Hole H7	Shaft n6	Fit[b]	Hole H7	Shaft p6	Fit[b]	Hole H7	Shaft s6	Fit[b]	Hole H7	Shaft u6	Fit[b]
30	Max	30.021	30.015	+0.019	30.021	30.028	+0.006	30.021	30.035	−0.001	30.021	30.048	−0.014	30.021	30.061	−0.027
	Min	30.000	30.002	−0.015	30.000	30.015	−0.028	30.000	30.022	−0.035	30.000	30.035	−0.048	30.000	30.048	−0.061
40	Max	40.025	40.018	+0.023	40.025	40.033	+0.008	40.025	40.042	−0.001	40.025	40.059	−0.018	40.025	40.076	−0.035
	Min	40.000	40.002	−0.018	40.000	40.017	−0.033	40.000	40.026	−0.042	40.000	40.043	−0.059	40.000	40.060	−0.076
50	Max	50.025	50.018	+0.023	50.025	50.033	+0.008	50.025	50.042	−0.001	50.025	50.059	−0.018	50.025	50.086	−0.045
	Min	50.000	50.002	−0.018	50.000	50.017	−0.033	50.000	50.026	−0.042	50.000	50.043	−0.059	50.000	50.070	−0.086
60	Max	60.030	60.021	+0.028	60.030	60.039	+0.010	60.030	60.051	−0.002	60.030	60.072	−0.023	60.030	60.106	−0.057
	Min	60.000	60.002	−0.021	60.000	60.020	−0.039	60.000	60.032	−0.051	60.000	60.053	−0.072	60.000	60.087	−0.106
80	Max	80.030	80.021	+0.028	80.030	80.039	+0.010	80.030	80.051	−0.002	80.030	80.078	−0.029	80.030	80.121	−0.072
	Min	80.000	80.002	−0.021	80.000	80.020	−0.039	80.000	80.032	−0.051	80.000	80.059	−0.078	80.000	80.102	−0.121
100	Max	100.035	100.025	+0.032	100.035	100.045	+0.012	100.035	100.059	−0.002	100.035	100.093	−0.036	100.035	100.146	−0.089
	Min	100.000	100.003	−0.025	100.000	100.023	−0.045	100.000	100.037	−0.059	100.000	100.071	−0.093	100.000	100.124	−0.146
120	Max	120.035	120.025	+0.032	120.035	120.045	+0.012	120.035	120.059	−0.002	120.035	120.101	−0.044	120.035	120.166	−0.109
	Min	120.000	120.003	−0.025	120.000	120.023	−0.045	120.000	120.037	−0.059	120.000	120.079	−0.101	120.000	120.144	−0.166
160	Max	160.040	160.028	+0.037	160.040	160.052	+0.013	160.040	160.068	−0.003	160.040	160.125	−0.060	160.040	160.215	−0.150
	Min	160.000	160.003	−0.028	160.000	160.027	−0.052	160.000	160.043	−0.068	160.000	160.100	−0.125	160.000	160.190	−0.215
200	Max	200.046	200.033	+0.042	200.046	200.060	+0.015	200.046	200.079	−0.004	200.046	200.151	−0.076	200.046	200.265	−0.190
	Min	200.000	200.004	−0.033	200.000	200.031	−0.060	200.000	200.050	−0.079	200.000	200.122	−0.151	200.000	200.236	−0.265
250	Max	250.046	250.033	+0.042	250.046	250.060	+0.015	250.046	250.079	−0.004	250.046	250.169	−0.094	250.046	250.313	−0.238
	Min	250.000	250.004	−0.033	250.000	250.031	−0.060	250.000	250.050	−0.079	250.000	250.140	−0.169	250.000	250.284	−0.313
300	Max	300.052	300.036	+0.048	300.052	300.066	+0.018	300.052	300.088	−0.004	300.052	300.202	−0.118	300.052	300.382	−0.298
	Min	300.000	300.004	−0.036	300.000	300.034	−0.066	300.000	300.056	−0.088	300.000	300.170	−0.202	300.000	300.350	−0.382
400	Max	400.057	400.040	+0.053	400.057	400.073	+0.020	400.057	400.098	−0.005	400.057	400.244	−0.151	400.057	400.471	−0.378
	Min	400.000	400.004	−0.040	400.000	400.037	−0.073	400.000	400.062	−0.098	400.000	400.208	−0.244	400.000	400.435	−0.471
500	Max	500.063	500.045	+0.058	500.063	500.080	+0.023	500.063	500.108	−0.005	500.063	500.292	−0.189	500.063	500.580	−0.477
	Min	500.000	500.005	−0.045	500.000	500.040	−0.080	500.000	500.068	−0.108	500.000	500.252	−0.292	500.000	500.540	−0.580

[a] The sizes shown are first-choice basic sizes (see Table 10). Preferred fits for other sizes can be calculated from data given in ANSI B4.2-1978 (R1984).

[b] A plus sign indicates clearance; a minus sign indicates interference.

All dimensions are in millimeters.

Table 14. American National Standard Preferred Shaft Basis Metric Clearance Fits *ANSI B4.2-1978 (R1994)*

Basic Size[a]		Loose Running			Free Running			Close Running			Sliding			Locational Clearance		
		Hole C11	Shaft h11	Fit[b]	Hole D9	Shaft h9	Fit[b]	Hole F8	Shaft h7	Fit[b]	Hole G7	Shaft h6	Fit[b]	Hole H7	Shaft h6	Fit[b]
1	Max	1.120	1.000	0.180	1.045	1.000	0.070	1.020	1.000	0.030	1.012	1.000	0.018	1.010	1.000	0.016
	Min	1.060	0.940	0.060	1.020	0.975	0.020	1.006	0.990	0.006	1.002	0.994	0.002	1.000	0.994	0.000
1.2	Max	1.320	1.200	0.180	1.245	1.200	0.070	1.220	1.200	0.030	1.212	1.200	0.018	1.210	1.200	0.016
	Min	1.260	1.140	0.060	1.220	1.175	0.020	1.206	1.190	0.006	1.202	1.194	0.002	1.200	1.194	0.000
1.6	Max	1.720	1.600	0.180	1.645	1.600	0.070	1.620	1.600	0.030	1.612	1.600	0.018	1.610	1.600	0.016
	Min	1.660	1.540	0.060	1.620	1.575	0.020	1.606	1.590	0.006	1.602	1.594	0.002	1.600	1.594	0.000
2	Max	2.120	2.000	0.180	2.045	2.000	0.070	2.020	2.000	0.030	2.012	2.000	0.018	2.010	2.000	0.016
	Min	2.060	1.940	0.060	2.020	1.975	0.020	2.006	1.990	0.006	2.002	1.994	0.002	2.000	1.994	0.000
2.5	Max	2.620	2.500	0.180	2.545	2.500	0.070	2.520	2.500	0.030	2.512	2.500	0.018	2.510	2.500	0.016
	Min	2.560	2.440	0.060	2.520	2.475	0.020	2.506	2.490	0.006	2.502	2.494	0.002	2.500	2.494	0.000
3	Max	3.120	3.000	0.180	3.045	3.000	0.070	3.020	3.000	0.030	3.012	3.000	0.018	3.010	3.000	0.016
	Min	3.060	2.940	0.060	3.020	2.975	0.020	3.006	2.990	0.006	3.002	2.994	0.002	3.000	2.994	0.000
4	Max	4.145	4.000	0.220	4.060	4.000	0.090	4.028	4.000	0.040	4.016	4.000	0.024	4.012	4.000	0.020
	Min	4.070	3.925	0.070	4.030	3.970	0.030	4.010	3.988	0.010	4.004	3.992	0.004	4.000	3.992	0.000
5	Max	5.145	5.000	0.220	5.060	5.000	0.090	5.028	5.000	0.040	5.016	5.000	0.024	5.012	5.000	0.020
	Min	5.070	4.925	0.070	5.030	4.970	0.030	5.010	4.988	0.010	5.004	4.992	0.004	5.000	4.992	0.000
6	Max	6.145	6.000	0.220	6.060	6.000	0.090	6.028	6.000	0.040	6.016	6.000	0.024	6.012	6.000	0.020
	Min	6.070	5.925	0.070	6.030	5.970	0.030	6.010	5.988	0.010	6.004	5.992	0.004	6.000	5.992	0.000
8	Max	8.170	8.000	0.260	8.076	8.000	0.112	8.035	8.000	0.050	8.020	8.000	0.029	8.015	8.000	0.024
	Min	8.080	7.910	0.080	8.040	7.964	0.040	8.013	7.985	0.013	8.005	7.991	0.005	8.000	7.991	0.000
10	Max	10.170	10.000	0.260	10.076	10.000	0.112	10.035	10.000	0.050	10.020	10.000	0.029	10.015	10.000	0.024
	Min	10.080	9.910	0.080	10.040	9.964	0.040	10.013	9.985	0.013	10.005	9.991	0.005	10.000	9.991	0.000
12	Max	12.205	12.000	0.315	12.093	12.000	0.136	12.043	12.000	0.061	12.024	12.000	0.035	12.018	12.000	0.029
	Min	12.095	11.890	0.095	12.050	11.957	0.050	12.016	11.982	0.016	12.006	11.989	0.006	12.000	11.989	0.000
16	Max	16.205	16.000	0.315	16.093	16.000	0.136	16.043	16.000	0.061	16.024	16.000	0.035	16.018	16.000	0.029
	Min	16.095	15.890	0.095	16.050	15.957	0.050	16.016	15.982	0.016	16.006	15.989	0.006	16.000	15.989	0.000
20	Max	20.240	20.000	0.370	20.117	20.000	0.169	20.053	20.000	0.074	20.028	20.000	0.041	20.021	20.000	0.034
	Min	20.110	19.870	0.110	20.065	19.948	0.065	20.020	19.979	0.020	20.007	19.987	0.007	20.000	19.987	0.000
25	Max	25.240	25.000	0.370	25.117	25.000	0.169	25.053	25.000	0.074	25.028	25.000	0.041	25.021	25.000	0.034
	Min	25.110	24.870	0.110	25.065	24.948	0.065	25.020	24.979	0.020	25.007	24.987	0.007	25.000	24.987	0.000

Table 14. *(Continued)* **American National Standard Preferred Shaft Basis Metric Clearance Fits** *ANSI B4.2-1978 (R1994)*

Basic Size[a]		Loose Running			Free Running			Close Running			Sliding			Locational Clearance		
		Hole C11	Shaft h11	Fit[b]	Hole D9	Shaft h9	Fit[b]	Hole F8	Shaft h7	Fit[b]	Hole G7	Shaft h6	Fit[b]	Hole H7	Shaft h6	Fit[b]
30	Max	30.240	30.000	0.370	30.117	30.000	0.169	30.053	30.000	0.074	30.028	30.000	0.041	30.021	30.000	0.034
	Min	30.110	29.870	0.110	30.065	29.948	0.065	30.020	29.979	0.020	30.007	29.987	0.007	30.000	29.987	0.000
40	Max	40.280	40.000	0.440	40.142	40.000	0.204	40.064	40.000	0.089	40.034	40.000	0.050	40.025	40.000	0.041
	Min	40.120	39.840	0.120	40.080	39.938	0.080	40.025	39.975	0.025	40.009	39.984	0.009	40.000	39.984	0.000
50	Max	50.290	50.000	0.450	50.142	50.000	0.204	50.064	50.000	0.089	50.034	50.000	0.050	50.025	50.000	0.041
	Min	50.130	49.840	0.130	50.080	49.938	0.080	50.025	49.975	0.025	50.009	49.984	0.009	50.000	49.984	0.000
60	Max	60.330	60.000	0.520	60.174	60.000	0.248	60.076	60.000	0.106	60.040	60.000	0.059	60.030	60.000	0.049
	Min	60.140	59.810	0.140	60.100	59.926	0.100	60.030	59.970	0.030	60.010	59.981	0.010	60.000	59.981	0.000
80	Max	80.340	80.000	0.530	80.174	80.000	0.248	80.076	80.000	0.106	80.040	80.000	0.059	80.030	80.000	0.049
	Min	80.150	79.810	0.150	80.100	79.926	0.100	80.030	79.970	0.030	80.010	79.981	0.010	80.000	79.981	0.000
100	Max	100.390	100.000	0.610	100.207	100.000	0.294	100.090	100.000	0.125	100.047	100.000	0.069	100.035	100.000	0.057
	Min	100.170	99.780	0.170	100.120	99.913	0.120	100.036	99.965	0.036	100.012	99.978	0.012	100.000	99.978	0.000
120	Max	120.400	120.000	0.620	120.207	120.000	0.294	120.090	120.000	0.125	120.047	120.000	0.069	120.035	120.000	0.057
	Min	120.180	119.780	0.180	120.120	119.913	0.120	120.036	119.965	0.036	120.012	119.978	0.012	120.000	119.978	0.000
160	Max	160.460	160.000	0.710	160.245	160.000	0.345	160.106	160.000	0.146	160.054	160.000	0.079	160.040	160.000	0.065
	Min	160.210	159.750	0.210	160.145	159.900	0.145	160.043	159.960	0.043	160.014	159.975	0.014	160.000	159.975	0.000
200	Max	200.530	200.000	0.820	200.285	200.000	0.400	200.122	200.000	0.168	200.061	200.000	0.090	200.046	200.000	0.075
	Min	200.240	199.710	0.240	200.170	199.885	0.170	200.050	199.954	0.050	200.015	199.971	0.015	200.000	199.971	0.000
250	Max	250.570	250.000	0.860	250.285	250.000	0.400	250.122	250.000	0.168	250.061	250.000	0.090	250.046	250.000	0.075
	Min	250.280	249.710	0.280	250.170	249.885	0.170	250.050	249.954	0.050	250.015	249.971	0.015	250.000	249.971	0.000
300	Max	300.650	300.000	0.970	300.320	300.000	0.450	300.137	300.000	0.189	300.069	300.000	0.101	300.052	300.000	0.084
	Min	300.330	299.680	0.330	300.190	299.870	0.190	300.056	299.948	0.056	300.017	299.968	0.017	300.000	299.968	0.000
400	Max	400.760	400.000	1.120	400.350	400.000	0.490	400.151	400.000	0.208	400.075	400.000	0.111	400.057	400.000	0.093
	Min	400.400	399.640	0.400	400.210	399.860	0.210	400.062	399.943	0.062	400.018	399.964	0.018	400.000	399.964	0.000
500	Max	500.880	500.000	1.280	500.385	500.000	0.540	500.165	500.000	0.228	500.083	500.000	0.123	500.063	500.000	0.103
	Min	500.480	499.600	0.480	500.230	499.845	0.230	500.068	499.937	0.068	500.020	499.960	0.020	500.000	499.960	0.000

[a] The sizes shown are first-choice basic sizes (see Table 10). Preferred fits for other sizes can be calculated from data given in ANSI B4.2-1978 (R1984).

[b] All fits shown in this table have clearance.

All dimensions are in millimeters.

Table 15. American National Standard Preferred Shaft Basis Metric Transition and Interference Fits *ANSI B4.2-1978 (R1994)*

Basic Size[a]		Locational Transition			Locational Transition			Locational Interference			Medium Drive			Force		
		Hole K7	Shaft h6	Fit[b]	Hole N7	Shaft h6	Fit[b]	Hole P7	Shaft h6	Fit[b]	Hole S7	Shaft h6	Fit[b]	Hole U7	Shaft h6	Fit[b]
1	Max	1.000	1.000	+0.006	0.996	1.000	+0.002	0.994	1.000	0.000	0.986	1.000	−0.008	0.982	1.000	−0.012
	Min	0.990	0.994	−0.010	0.986	0.954	−0.014	0.984	0.994	−0.016	0.976	0.994	−0.024	0.972	0.994	−0.028
1.2	Max	1.200	1.200	+0.006	1.196	1.200	+0.002	1.194	1.200	0.000	1.186	1.200	−0.008	1.182	1.200	−0.012
	Min	1.190	1.194	−0.010	1.186	1.194	−0.014	1.184	1.194	−0.016	1.176	1.194	−0.024	1.172	1.194	−0.028
1.6	Max	1.600	1.600	+0.006	1.596	1.600	+0.002	1.594	1.600	0.000	1.586	1.600	−0.008	1.582	1.600	−0.012
	Min	1.590	1.594	−0.010	1.586	1.594	−0.014	1.584	1.594	−0.016	1.576	1.594	−0.024	1.572	1.594	−0.028
2	Max	2.000	2.000	+0.006	1.996	2.000	+0.002	1.994	2.000	0.000	1.986	2.000	−0.008	1.982	2.000	−0.012
	Min	1.990	1.994	−0.010	1.986	1.994	−0.014	1.984	1.994	−0.016	1.976	1.994	−0.024	1.972	1.994	−0.028
2.5	Max	2.500	2.500	+0.006	2.496	2.500	+0.002	2.494	2.500	0.000	2.486	2.500	−0.008	2.482	2.500	−0.012
	Min	2.490	2.494	−0.010	2.486	2.494	−0.014	2.484	2.494	−0.016	2.476	2.494	−0.024	2.472	2.494	−0.028
3	Max	3.000	3.000	+0.006	2.996	3.000	+0.002	2.994	3.000	0.000	2.986	3.000	−0.008	2.982	3.000	−0.012
	Min	2.990	2.994	−0.010	2.986	2.994	−0.014	2.984	2.994	−0.016	2.976	2.994	−0.024	2.972	2.994	−0.028
4	Max	4.003	4.000	+0.011	3.996	4.000	+0.004	3.992	4.000	0.000	3.985	4.000	−0.007	3.981	4.000	−0.011
	Min	3.991	3.992	−0.009	3.984	3.992	−0.016	3.980	3.992	−0.020	3.973	3.992	−0.027	3.969	3.992	−0.031
5	Max	5.003	5.000	+0.011	4.996	5.000	+0.004	4.992	5.000	0.000	4.985	5.000	−0.007	4.981	5.000	−0.011
	Min	4.991	4.992	−0.009	4.984	4.992	−0.016	4.980	4.992	−0.020	4.973	4.992	−0.027	4.969	4.992	−0.031
6	Max	6.003	6.000	+0.011	5.996	6.000	+0.004	5.992	6.000	0.000	5.985	6.000	−0.007	5.981	6.000	−0.011
	Min	5.991	5.992	−0.009	5.984	5.992	−0.016	5.980	5.992	−0.020	5.973	5.992	−0.027	5.969	5.992	−0.031
8	Max	8.005	8.000	+0.014	7.996	8.000	+0.005	7.991	8.000	0.000	7.983	8.000	−0.008	7.978	8.000	−0.013
	Min	7.990	7.991	−0.010	7.981	7.991	−0.019	7.976	7.991	−0.024	7.968	7.991	−0.032	7.963	7.991	−0.037
10	Max	10.005	10.000	+0.014	9.996	10.000	+0.005	9.991	10.000	0.000	9.983	10.000	−0.008	9.978	10.000	−0.013
	Min	9.990	9.991	−0.010	9.981	9.991	−0.019	9.976	9.991	−0.024	9.968	9.991	−0.032	9.963	9.991	−0.037
12	Max	12.006	12.000	+0.017	11.995	12.000	+0.006	11.989	12.000	0.000	11.979	12.000	−0.010	11.974	12.000	−0.015
	Min	11.988	11.989	−0.012	11.977	11.989	−0.023	11.971	11.989	−0.029	11.961	11.989	−0.039	11.956	11.989	−0.044
16	Max	16.006	16.000	+0.017	15.995	16.000	+0.006	15.989	16.000	0.000	15.979	16.000	−0.010	15.974	16.000	−0.015
	Min	15.988	15.989	−0.012	15.977	15.989	−0.023	15.971	15.989	−0.029	15.961	15.989	−0.039	15.956	15.989	−0.044
20	Max	20.006	20.000	+0.019	19.993	20.000	+0.006	19.986	20.000	−0.001	19.973	20.000	−0.014	19.967	20.000	−0.020
	Min	19.985	19.987	−0.015	19.972	19.987	−0.028	19.965	19.987	−0.035	19.952	19.987	−0.048	19.946	19.987	−0.054
25	Max	25.006	25.000	+0.019	24.993	25.000	+0.006	24.986	25.000	−0.001	24.973	25.000	−0.014	24.960	25.000	−0.027
	Min	24.985	24.987	−0.015	24.972	24.987	−0.028	24.965	24.987	−0.035	24.952	24.987	−0.048	24.939	24.987	−0.061

Table 15. *(Continued)* **American National Standard Preferred Shaft Basis Metric Transition and Interference Fits** *ANSI B4.2-1978 (R1994)*

Basic Size[a]		Locational Transition			Locational Transition			Locational Interference			Medium Drive			Force		
		Hole K7	Shaft h6	Fit[b]	Hole N7	Shaft h6	Fit[b]	Hole P7	Shaft h6	Fit[b]	Hole S7	Shaft h6	Fit[b]	Hole U7	Shaft h6	Fit[b]
30	Max	30.006	30.000	+0.019	29.993	30.000	+0.006	29.986	30.000	−0.001	29.973	30.000	−0.014	29.960	30.000	−0.027
	Min	29.985	29.987	−0.015	29.972	29.987	−0.028	29.965	29.987	−0.035	29.952	29.987	−0.048	29.939	29.987	−0.061
40	Max	40.007	40.000	+0.023	39.992	40.000	+0.008	39.983	40.000	−0.001	39.966	40.000	−0.018	39.949	40.000	−0.035
	Min	39.982	39.984	−0.018	39.967	39.984	−0.033	39.958	39.984	−0.042	39.941	39.984	−0.059	39.924	39.984	−0.076
50	Max	50.007	50.000	+0.023	49.992	50.000	+0.008	49.983	50.000	−0.001	49.966	50.000	−0.018	49.939	50.000	−0.045
	Min	49.982	49.984	−0.018	49.967	49.984	−0.033	49.958	49.984	−0.042	49.941	49.984	−0.059	49.914	49.984	−0.086
60	Max	60.009	60.000	+0.028	59.991	60.000	+0.010	59.979	60.000	−0.002	59.958	60.000	−0.023	59.924	60.000	−0.087
	Min	59.979	59.981	−0.021	59.961	59.981	−0.039	59.949	59.981	−0.051	59.928	59.981	−0.072	59.894	59.981	−0.106
80	Max	80.009	80.000	+0.028	79.991	80.000	+0.010	79.979	80.000	−0.002	79.952	80.000	−0.029	79.909	80.000	−0.072
	Min	79.979	79.981	−0.021	79.961	79.981	−0.039	79.949	79.981	−0.051	79.922	79.981	−0.078	79.879	79.981	−0.121
100	Max	100.010	100.000	+0.032	99.990	100.000	+0.012	99.976	100.000	−0.002	99.942	100.000	−0.036	99.889	100.000	−0.089
	Min	99.975	99.978	−0.025	99.955	99.978	−0.045	99.941	99.978	−0.059	99.907	99.978	−0.093	99.854	99.978	−0.146
120	Max	120.010	120.000	+0.032	119.990	120.000	+0.012	119.976	120.000	−0.002	119.934	120.000	−0.044	119.869	120.000	−0.109
	Min	119.975	119.978	−0.025	119.955	119.978	−0.045	119.941	119.978	−0.059	119.899	119.978	−0.101	119.834	119.978	−0.166
160	Max	160.012	160.000	+0.037	159.988	160.000	+0.013	159.972	160.000	−0.003	159.915	160.000	−0.060	159.825	160.000	−0.150
	Min	159.972	159.975	−0.028	159.948	159.975	−0.052	159.932	159.975	−0.068	159.875	159.975	−0.125	159.785	159.975	−0.215
200	Max	200.013	200.00	+0.042	199.986	200.000	+0.015	199.967	200.000	−0.004	199.895	200.000	−0.076	199.781	200.000	−0.190
	Min	199.967	199.971	−0.033	199.940	199.971	−0.060	199.921	199.971	−0.079	199.849	199.971	−0.151	199.735	199.971	−0.265
250	Max	250.013	250.000	+0.042	249.986	250.000	+0.015	249.967	250.000	−0.004	249.877	250.000	−0.094	249.733	250.000	−0.238
	Min	249.967	249.971	−0.033	249.940	249.971	−0.060	249.921	249.971	−0.079	249.831	249.971	−0.169	249.687	249.971	−0.313
300	Max	300.016	300.000	+0.048	299.986	300.000	+0.018	299.964	300.000	−0.004	299.850	300.000	−0.118	299.670	300.000	−0.298
	Min	299.964	299.968	−0.036	299.934	299.968	−0.066	299.912	299.968	−0.088	299.798	299.968	−0.202	299.618	299.968	−0.382
400	Max	400.017	400.000	+0.053	399.984	400.000	+0.020	399.959	400.000	−0.005	399.813	400.000	−0.151	399.586	400.000	−0.378
	Min	399.960	399.964	−0.040	399.927	399.964	−0.073	399.902	399.964	−0.098	399.756	399.964	−0.244	399.529	399.964	−0.471
500	Max	500.018	500.000	+0.058	499.983	500.000	+0.023	499.955	500.000	−0.005	499.771	500.000	−0.189	499.483	500.000	−0.477
	Min	499.955	499.960	−0.045	499.920	499.960	−0.080	499.892	499.960	−0.108	499.708	499.960	−0.292	499.420	499.960	−0.580

[a] The sizes shown are first-choice basic sizes (see Table 10). Preferred fits for other sizes can be calculated from data given in ANSI B4.2-1978 (R1984).

[b] A plus sign indicates clearance; a minus sign indicates interference.

All dimensions are in millimeters.

Table 16. American National Standard Gagemakers Tolerances *ANSI B4.4M-1981 (R1987)*

	Gagemakers Tolerance		Workpiece Tolerance	
	Class	ISO Symbol[a]	IT Grade	Recommended Gage Usage
Rejection of Good Parts Increase ↑	ZM	0.05 IT11	IT11	Low-precision gages recommended to be used to inspect workpieces held to internal (hole) tolerances C11 and H11 and to external (shaft) tolerances c11 and h11.
	YM	0.05 IT9	IT9	Gages recommended to be used to inspect workpieces held to internal (hole) tolerances D9 and H9 and to external (shaft) tolerances d9 and h9.
	XM	0.05 IT8	IT8	Precision gages recommended to be used to inspect workpieces held to internal (hole) tolerances F8 and H8.
Gage Cost Increase ↓	XXM	0.05 IT7	IT7	Recommended to be used for gages to inspect workpieces held to internal (hole) tolerances G7, H7, K7, N7, P7, S7, and U7, and to external (shaft) tolerances f7 and h7.
	XXX M	0.05 IT6	IT6	High-precision gages recommended to be used to inspect workpieces held to external (shaft) tolerances g6, h6, k6, n6, p6, s6, and u6.

[a] Gagemakers tolerance is equal to 5 per cent of workpiece tolerance or 5 per cent of applicable IT grade value. See table *American National Standard Gagemakers Tolerances ANSI B4.4M-1981 (R1987).*

For workpiece tolerance class values, see previous Tables 12 through 15, incl.

Table 17. American National Standard Gagemakers Tolerances *ANSI B4.4M-1981 (R1987)*

Basic Size		Class ZM	Class YM	Class XM	Class XXM	Clas XXXM
Over	To	(0.05 IT11)	(0.05 IT9)	(0.05 IT8)	(0.05 IT7)	(0.05 IT6)
0	3	0.0030	0.0012	0.0007	0.0005	0.0003
3	6	0.0037	0.0015	0.0009	0.0006	0.0004
6	10	0.0045	0.0018	0.0011	0.0007	0.0005
10	18	0.0055	0.0021	0.0013	0.0009	0.0006
18	30	0.0065	0.0026	0.0016	0.0010	0.0007
30	50	0.0080	0.0031	0.0019	0.0012	0.0008
50	80	0.0095	0.0037	0.0023	0.0015	0.0010
80	120	0.0110	0.0043	0.0027	0.0017	0.0011
120	180	0.0125	0.0050	0.0031	0.0020	0.0013
180	250	0.0145	0.0057	0.0036	0.0023	0.0015
250	315	0.0160	0.0065	0.0040	0.0026	0.0016
315	400	0.0180	0.0070	0.0044	0.0028	0.0018
400	500	0.0200	0.0077	0.0048	0.0031	0.0020

All dimensions are in millimeters. For closer gagemakers tolerance classes than Class XXXM, specify 5 per cent of IT5, IT4, or IT3 and use the designation 0.05 IT5, 0.05 IT4, etc.

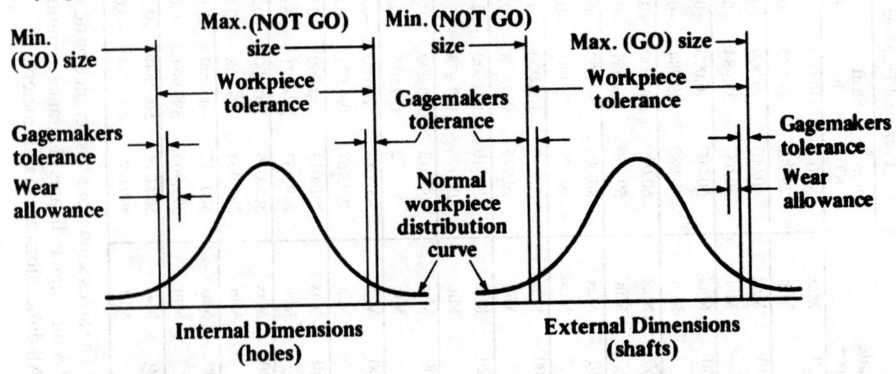

Fig. 4. Relationship between Gagemakers Tolerance, Wear Allowance and Workpiece Tolerance

Applications.—Many factors such as length of engagement, bearing load, speed, lubrication, operating temperatures, humidity, surface texture, and materials must be taken into account in fit selections for a particular application.

Choice of other than the preferred fits might be considered necessary to satisfy extreme conditions. Subsequent adjustments might also be desired as the result of experience in a particular application to suit critical functional requirements or to permit optimum manufacturing economy. Selection of a departure from these recommendations will depend upon consideration of the engineering and economic factors that might be involved; however, the benefits to be derived from the use of preferred fits should not be overlooked.

A general guide to machining processes that may normally be expected to produce work within the tolerances indicated by the IT grades given in ANSI B4.2-1978 (R1994) is shown in the chart in Table 18.

Table 18. Relation of Machining Processes to IT Tolerance Grades

	IT Grades							
	4	5	6	7	8	9	10	11
Lapping & Honing	■	■						
Cylindrical Grinding		■	■	■				
Surface Grinding		■	■	■	■			
Diamond Turning		■	■	■				
Diamond Boring		■	■	■				
Broaching		■	■	■	■			
Powder Metal sizes			■	■	■			
Reaming			■	■	■	■	■	
Turning				■	■	■	■	■
Powder Metal sintered				■	■	■	■	
Boring				■	■	■	■	■
Milling							■	■
Planing & Shaping							■	■
Drilling							■	■
Punching							■	■
Die Casting								■

British Standard for Metric ISO Limits and Fits.—Based on ISO Recommendation R286, this British Standard BS 4500:1969 is intended to provide a comprehensive range of metric limits and fits for engineering purposes, and meets the requirements of metrication in the United Kingdom. Sizes up to 3,150 mm are covered by the Standard, but the condensed information presented here embraces dimensions up to 500 mm only. The system is based on a series of tolerances graded to suit all classes of work from the finest to the most coarse, and the different types of fits that can be obtained range from coarse clearance to heavy interference. In the Standard, only cylindrical parts, designated holes and shafts are referred to explicitly, but it is emphasized that the recommendations apply equally well to other sections, and the general term *hole* or *shaft* can be taken to mean the space contained by or containing two parallel faces or tangent planes of any part, such as the width of a slot, or the thickness of a key. It is also strongly emphasized that the grades series of tolerances are intended for the most general application, and should be used wherever possible whether the features of the component involved are members of a fit or not.

Definitions.—The definitions given in the Standard include the following:

Limits of Size: The maximum and minimum sizes permitted for a feature.

Basic Size: The reference size to which the limits of size are fixed. The basic size is the same for both members of a fit.

Upper Deviation: The algebraical difference between the maximum limit of size and the corresponding basic size. It is designated as ES for a hole, and as es for a shaft, which stands for the French term *écart supérieur.*

Lower Deviation: The algebraical difference between the minimum limit of size and the corresponding basic size. It is designated as EI for a hole, and as ei for a shaft, which stands for the French term *écart inférieur.*

Zero Line: In a graphical representation of limits and fits, the straight line to which the deviations are referred. The zero line is the line of zero deviation and represents the basic size.

Tolerance: The difference between the maximum limit of size and the minimum limit of size. It is an absolute value without sign.

Tolerance Zone: In a graphical representation of tolerances, the zone comprised between the two lines representing the limits of tolerance and defined by its magnitude (tolerance) and by its position in relation to the zero line.

Fundamental Deviation: That one of the two deviations, being the one nearest to the zero line, which is conventionally chosen to define the position of the tolerance zone in relation to the zero line.

Shaft-Basis System of Fits: A system of fits in which the different clearances and interferences are obtained by associating various holes with a single shaft. In the ISO system, the basic shaft is the shaft the upper deviation of which is zero.

Hole-Basis System of Fits: A system of fits in which the different clearances and interferences are obtained by associating various shafts with a single hole. In the ISO system, the basic hole is the hole the lower deviation of which is zero.

Selected Limits of Tolerance, and Fits.—The number of fit combinations that can be built up with the ISO system is very large. However, experience shows that the majority of fits required for usual engineering products can be provided by a limited selection of tolerances. Limits of tolerance for selected holes are shown in Table 19, and for shafts, in Table 20. Selected fits, based on combinations of the selected hole and shaft tolerances, are given in Table 21.

Tolerances and Fundamental Deviations.—There are 18 tolerance grades intended to meet the requirements of different classes of work, and they are designated IT 01, IT 02, and IT 1 to IT 16. (IT stands for ISO series of tolerances.) Table 22 shows the standardized numerical values for the 18 tolerance grades, which are known as standard tolerances. The system provides 27 fundamental deviations for sizes up to and including 500 mm, and Tables 15 and 25 contain the values for shafts and holes, respectively. Uppercase (capital) letters designate hole deviations, and the same letters in lower case designate shaft deviations. The deviation j_s (J_s for holes) is provided to meet the need for symmetrical bilateral tolerances. In this instance, there is no fundamental deviation, and the tolerance zone, of whatever magnitude, is equally disposed about the zero line.

Calculated Limits of Tolerance.—The deviations and fundamental tolerances provided by the ISO system can be combined in any way that appears necessary to give a required fit. Thus, for example, the deviations H (basic hole) and f (clearance shaft) could be associated, and with each of these deviations any one of the tolerance grades IT 01 to IT 16 could be used. All the limits of tolerance that the system is capable of providing for sizes up to and including 500 mm can be calculated from the standard tolerances given in Table 22, and the fundamental deviations given in Tables 15 and 25. The range includes limits of tolerance for shafts and holes used in small high-precision work and horology.

The system provides for the use of either hole-basis or shaft-basis fits, and the Standard includes details of procedures for converting from one type of fit to the other.

The limits of tolerance for a shaft or hole are designated by the appropriate letter indicating the fundamental deviation, followed by a suffix number denoting the tolerance grade. This suffix number is the numerical part of the tolerance grade designation. Thus, a hole tolerance with deviation H and tolerance grade IT7 is designated H7. Likewise, a shaft with deviation p and tolerance grade IT 6 is designated p6. The limits of size of a component feature are defined by the basic size, say, 45 mm, followed by the appropriate tolerance designation, for example, 45 H7 or 45 p6. A fit is indicated by combining the basic size common to both features with the designation appropriate to each of them, for example, 45 H7-p6 or 45 H7/p6.

When calculating the limits of size for a shaft, the upper deviation es, or the lower deviation ei, is first obtained from Table 15, depending on the particular letter designation, and nominal dimension. If an upper deviation has been determined, the lower deviation ei = es – IT. The IT value is obtained from Table 22 for the particular tolerance grade being applied. If a lower deviation has been obtained from Table 15, the upper deviation es = ei + IT. When the upper deviation ES has been determined for a hole from Table 25, the lower deviation EI = ES – IT. If a lower deviation EI has been obtained from Table 25, then the upper deviation ES = EI + IT.

The upper deviations for holes K, M, and N with tolerance grades up to and including IT8, and for holes P to ZC with tolerance grades up to and including IT7 must be calculated by adding the delta (Δ) values given in Table 25 as indicated.

Example of Calculations: The limits of size for a part of 133 mm basic size with a tolerance designation g9 are derived as follows:

From Table 15, the upper deviation (es) is – 0.014 mm. From Table 22, the tolerance grade (ITg) is 0.100 mm. The lower deviation (ei) = es – IT = 0.114 mm, and the limits of size are thus 132.986 and 132.886 mm.

The limits of size for a part 20 mm in size, with tolerance designation D3, are derived as follows: From Table 25, the lower deviation (EI) is + 0.065 mm. From Table 22, the tolerance grade (IT9) is 0.004 mm. The upper deviation (ES) = EI + IT = 0.069 mm, and thus the limits of size for the part are 20.069 and 20.065 mm.

The limits of size for a part 32 mm in size, with tolerance designation M5, which involves a delta value, are obtained as follows: From Table 25, the upper deviation ES is – 0.009 mm + Δ = –0.005 mm. (The delta value given at the end of this table for this size and grade IT 5 is 0.004 mm.) From Table 22, the tolerance grade (IT5) is 0.011 mm. The lower deviation (EI) = ES – IT = – 0.016 mm, and thus the limits of size for the part are 31.995 and 31.984 mm.

Where the designations h and H or j_s and J_s are used, it is only necessary to refer to Table 22. For h and H, the fundamental deviation is always zero, and the disposition of the tolerance is always negative (–) for a shaft, and positive (+) for a hole. Thus, the limits for a part 40 mm in size, designated h8 are derived as follows:

From Table 22, the tolerance grade (IT 8) is 0.039 mm, and the limits are therefore 40.000 and 39.961 mm.

The limits for a part 60 mm in size, designated j_s7 or J_s7 are derived as follows:

From Table 1, the tolerance grade (IT 7) is 0.030 mm, and this value is divided equally about the basic size to give limits of 60.015 and 59.985 mm.

Table 19. British Standard Limits of Tolerance for Selected Holes (Upper and Lower Deviations) *BS 4500:1969*

Nominal Sizes, mm		H7		H8		H9		H11	
Over	Up to and Including	ES +	EI	ES +	EI	ES +	EI	ES +	EI
…	3	10	0	14	0	25	0	60	0
3	6	12	0	18	0	30	0	75	0
6	10	15	0	22	0	36	0	90	0
10	18	18	0	27	0	43	0	110	0
18	30	21	0	33	0	52	0	130	0
30	50	25	0	39	0	62	0	160	0
50	80	30	0	46	0	74	0	190	0
80	120	35	0	54	0	87	0	220	0
120	180	40	0	63	0	100	0	250	0
180	250	46	0	72	0	115	0	290	0
250	315	52	0	81	0	130	0	320	0
315	400	57	0	89	0	140	0	360	0
400	500	63	0	97	0	155	0	400	0

ES = Upper deviation.
EI = Lower deviation.
The dimensions are given in 0.001 mm, except for the nominal sizes, which are in millimeters.

Table 20. British Standard Limits of Tolerance for Selected Shafts (Upper and Lower Deviations) *BS 4500:1969*

Nominal Sizes, mm		c11		d10		e9		f7		g6		h6		k6		n6		p6		s6	
Over	Up to and Incl.	es −	ei −	es −	ei −	es −	ei −	es −	ei −	es −	ei −	es −	ei −	es +	ei +	es +	ei +	es +	ei +	es +	ei +
…	3	60	120	20	60	14	39	6	16	2	8	0	6	6	0	10	4	12	6	20	14
3	6	70	145	30	78	20	50	10	22	4	12	0	8	9	1	16	8	20	12	27	19
6	10	80	170	40	98	25	61	13	28	5	14	0	9	10	1	19	10	24	15	32	23
10	18	95	205	50	120	32	75	16	34	6	17	0	11	12	1	23	12	29	18	39	28
18	30	110	240	65	149	40	92	20	41	7	20	0	13	15	2	28	15	35	22	48	35
30	40	120	280	80	180	50	112	25	50	9	25	0	16	18	2	33	17	42	26	59	43
40	50	130	290	80	180	50	112	25	50	9	25	0	16	18	2	33	17	42	26	59	43
50	65	140	330	100	220	60	134	30	60	10	29	0	19	21	2	39	20	51	32	72	53
65	80	150	340	100	220	60	134	30	60	10	29	0	19	21	2	39	20	51	32	78	59
80	100	170	390	120	260	72	159	36	71	12	34	0	22	25	3	45	23	59	37	93	71
100	120	180	400	120	260	72	159	36	71	12	34	0	22	25	3	45	23	59	37	101	79
120	140	200	450	145	305	85	185	43	83	14	39	0	25	28	3	52	27	68	43	117	92
140	160	210	460	145	305	85	185	43	83	14	39	0	25	28	3	52	27	68	43	125	100
160	180	230	480	145	305	85	185	43	83	14	39	0	25	28	3	52	27	68	43	133	108
180	200	240	530	170	355	100	215	50	96	15	44	0	29	33	4	60	31	79	50	151	122
200	225	260	550	170	355	100	215	50	96	15	44	0	29	33	4	60	31	79	50	159	130
225	250	280	570	170	355	100	215	50	96	15	44	0	29	33	4	60	31	79	50	169	140
250	280	300	620	190	400	110	240	56	108	17	49	0	32	36	4	66	34	88	56	190	158
280	315	330	650	190	400	110	240	56	108	17	49	0	32	36	4	66	34	88	56	202	170
315	355	360	720	210	440	125	265	62	119	18	54	0	36	40	4	73	37	98	62	226	190
355	400	400	760	210	440	125	265	62	119	18	54	0	36	40	4	73	37	98	62	244	208
400	450	440	840	230	480	135	290	68	131	20	60	0	40	45	5	80	40	108	68	272	232
450	500	480	880	230	480	135	290	68	131	20	60	0	40	45	5	80	40	108	68	292	252

es = Upper deviation.
ei = Lower deviation.
The dimensions are given in 0.001 mm, except for the nominal sizes, which are in millimeters.

Table 21. British Standard Selected Fits. Minimum and Maximum Clearances ***BS 4500:1969***

Nominal Sizes, mm		H11—c11		H9—d10		H9—e9		H8—f7		H7—g6		H7—h6		H7—k6		H7—n6		H7—p6		H7—s6	
Over	Up to and Incl.	Min	Max	Min	Max	Min	Max	Min	Max	Min	Max	Min	Max	Min	Max	Min	Max	Min	Max	Min	Max
…	3	60	180	20	85	14	64	6	30	2	18	0	16	−6	+10	−10	+6	−12	+4	−20	−4
3	6	70	220	30	108	20	80	10	40	4	24	0	20	−9	+11	−16	+4	−20	0	−27	−7
6	10	80	260	40	134	25	97	13	50	5	29	0	24	−10	+14	−19	+5	−24	0	−32	−8
10	18	95	315	50	163	32	118	16	61	6	35	0	29	−12	+17	−23	+6	−29	0	−39	−10
18	30	110	370	65	201	40	144	20	74	7	41	0	34	−15	+19	−28	+6	−35	−1	−48	−14
30	40	120	440	80	242	50	174	25	89	9	50	0	41	−18	+23	−33	+8	−42	−1	−59	−18
40	50	130	450	80	242	50	174	25	89	9	50	0	41	−18	+23	−33	+8	−42	−1	−59	−18
50	65	140	520	100	294	60	208	30	106	10	59	0	49	−21	+28	−39	+10	−51	−2	−72	−23
65	80	150	530	100	294	60	208	30	106	10	59	0	49	−21	+28	−39	+10	−51	−2	−78	−29
80	100	170	610	120	347	72	246	36	125	12	69	0	57	−25	+32	−45	+12	−59	−2	−93	−36
100	120	180	620	120	347	72	246	36	125	12	69	0	57	−25	+32	−45	+12	−59	−2	−101	−44
120	140	200	700	145	405	85	285	43	146	14	79	0	65	−28	+37	−52	+13	−68	−3	−117	−52
140	160	210	710	145	405	85	285	43	146	14	79	0	65	−28	+37	−52	+13	−68	−3	−125	−60
160	180	230	730	145	405	85	285	43	146	14	79	0	65	−28	+37	−52	+13	−68	−3	−133	−68
180	200	240	820	170	470	100	330	50	168	15	90	0	75	−33	+42	−60	+15	−79	−4	−151	−76
200	225	260	840	170	470	100	330	50	168	15	90	0	75	−33	+42	−60	+15	−79	−4	−159	−84
225	250	280	860	170	470	100	330	50	168	15	90	0	75	−33	+42	−60	+15	−79	−4	−169	−94
250	280	300	940	190	530	110	370	56	189	17	101	0	84	−36	+48	−66	+18	−88	−4	−190	−126
280	315	330	970	190	530	110	370	56	189	17	101	0	84	−36	+48	−66	+18	−88	−4	−202	−112
315	355	360	1080	210	580	125	405	62	208	18	111	0	93	−40	−53	−73	+20	−98	−5	−226	−133
355	400	400	1120	210	580	125	405	62	208	18	111	0	93	−40	−53	−73	+20	−98	−5	−244	−151
400	450	440	1240	230	635	135	445	68	228	20	123	0	103	−45	+58	−80	+23	−108	−5	−272	−169
450	500	480	1280	230	635	135	445	68	228	20	123	0	103	−45	+58	−80	+23	−108	−5	−292	−189

The dimensions are given in 0.001 mm, except for the nominal sizes, which are in millimeters.
Minus (−) sign indicates negative clearance, i.e., interference.

Table 22. British Standard Limits and Fits *BS 4500:1969*

Nominal Sizes, mm		Tolerance Grades																
Over	To	IT 01	IT 0	IT 1	IT 2	IT 3	IT 4	IT 5	IT 6	IT 7	IT 8	IT 9	IT 10	IT 11	IT 12	IT 13	IT 14[a]	IT 15[a]	IT 16[a]
…	3	0.3	0.5	0.8	1.2	2	3	4	6	10	14	25	40	60	100	140	250	400	600
3	6	0.4	0.6	1	1.5	2.5	4	5	8	12	18	30	48	75	120	180	300	480	750
6	10	0.4	0.6	1	1.5	2.5	4	6	9	15	22	36	58	90	150	220	360	580	900
10	18	0.5	0.8	1.2	2	3	5	8	11	18	27	43	70	110	180	270	430	700	1100
18	30	0.6	1	1.5	2.5	4	6	9	13	21	33	52	84	130	210	330	520	840	1300
30	50	0.6	1	1.5	2.5	4	7	11	16	25	39	62	100	160	250	390	620	1000	1600
50	80	0.8	1.2	2	3	5	8	13	19	30	46	74	120	190	300	460	740	1200	1900
80	120	1	1.5	2.5	4	6	10	15	22	35	54	87	140	220	350	540	870	1400	2200
120	180	1.2	2	3.5	5	8	12	18	25	40	63	100	160	250	400	630	1000	1600	2500
180	250	2	3	4.5	7	10	14	20	29	46	72	115	185	290	460	720	1150	1850	2900
250	315	2.5	4	6	8	12	16	23	32	52	81	130	210	320	520	810	1300	2100	3200
315	400	3	5	7	9	13	18	25	36	57	89	140	230	360	570	890	1400	2300	3600
400	500	4	6	8	10	15	20	27	40	63	97	155	250	400	630	970	1550	2500	4000

[a] Not applicable to sizes below 1 mm.

The dimensions are given in 0.001 mm, except for the nominal sizes which are in millimeters.

Table 23. British Standard Fundamental Deviations for Shafts *BS 4500:1969*

Nominal Sizes, mm		Grade																
		01 to 16												5–6	7	8	4–7	≤3 >7
		Fundamental (Upper) Deviation es												Fundamental (Lower) Deviation ei				
Over	To	a[a]	b[a]	c	cd	d	e	ef	f	fg	g	h	js[b]	j			k	
…	3	−270	−140	−60	−34	−20	−14	−10	−6	−4	−2	0		−2	−4	−6	0	0
3	6	−270	−140	−70	−46	−30	−20	−14	−10	−6	−4	0		−2	−4	…	+1	0
6	10	−280	−150	−80	−56	−40	−25	−18	−13	−8	−5	0		−2	−5	…	+1	0
10	14	−290	−150	−95	…	−50	−32	…	−16	…	−6	0		−3	−6	…	+1	0
14	18	−290	−150	−95	…	−50	−32	…	−16	…	−6	0		−3	−6	…	+1	0
18	24	−300	−160	−110	…	−65	−40	…	−20	…	−7	0		−4	−8	…	+2	0
24	30	−300	−160	−110	…	−65	−40	…	−20	…	−7	0		−4	−8	…	+2	0
30	40	−310	−170	−120	…	−80	−50	…	−25	…	−9	0		−5	−10	…	+2	0
40	50	−320	−180	−130	…	−80	−50	…	−25	…	−9	0		−5	−10	…	+2	0
50	65	−340	−190	−140	…	−100	−60	…	−30	…	−10	0		−7	−12	…	+2	0
65	80	−360	−200	−150	…	−100	−60	…	−30	…	−10	0		−7	−12	…	+2	0
80	100	−380	−220	−170	…	−120	−72	…	−36	…	−12	0		−9	−15	…	+3	0
100	120	−410	−240	−180	…	−120	−72	…	−36	…	−12	0		−9	−15	…	+3	0
120	140	−460	−260	−200	…	−145	−85	…	−43	…	−14	0	±IT/2	−11	−18	…	+3	0
140	160	−520	−280	−210	…	−145	−85	…	−43	…	−14	0		−11	−18	…	+3	0
160	180	−580	−310	−230	…	−145	−85	…	−43	…	−14	0		−11	−18	…	+3	0
180	200	−660	−340	−240	…	−170	−100	…	−50	…	−15	0		−13	−21	…	+4	0
200	225	−740	−380	−260	…	−170	−100	…	−50	…	−15	0		−13	−21	…	+4	0
225	250	−820	−420	−280	…	−170	−100	…	−50	…	−15	0		−13	−21	…	+4	0
250	280	−920	−480	−300	…	−190	−110	…	−56	…	−17	0		−16	−26	…	+4	0
280	315	−1050	−540	−330	…	−190	−110	…	−56	…	−17	0		−16	−26	…	+4	0
315	355	−1200	−600	−360	…	−210	−125	…	−62	…	−18	0		−18	−28	…	+4	0
355	400	−1350	−680	−400	…	−210	−125	…	−62	…	−18	0		−18	−28	…	+4	0
400	450	−1500	−760	−440	…	−230	−135	…	−68	…	−20	0		−20	−32	…	+5	0
450	500	−1650	−840	−480	…	−230	−135	…	−68	…	−20	0		−20	−32	…	+5	0

[a] Not applicable to sizes up to 1 mm.

[b] In grades 7 to 11, the two symmetrical deviations ±IT/2 should be rounded if the IT value in micrometers is an odd value by replacing it with the even value immediately below. For example, if IT = 175, replace it by 174.

Table 24. British Standard Fundamental Deviations for Shafts *BS 4500:1969*

Nominal Sizes, mm		Grade 01 to 16 — Fundamental (Lower) Deviation ei													
Over	To	m	n	p	r	s	t	u	v	x	y	z	za	zb	zc
…	3	+2	+4	+6	+10	+14	…	+18	…	+20	…	+26	+32	+40	+60
3	6	+4	+8	+12	+15	+19	…	+23	…	+28	…	+35	+42	+50	+80
6	10	+6	+10	+15	+19	+23	…	+28	…	+34	…	+42	+52	+67	+97
10	14	+7	+12	+18	+23	+28	…	+33	…	+40	…	+50	+64	+90	+130
14	18	+7	+12	+18	+23	+28	…	+33	+39	+45	…	+60	+77	+108	+150
18	24	+8	+15	+22	+28	+35	…	+41	+47	+54	+63	+73	+98	+136	+188
24	30	+8	+15	+22	+28	+35	+41	+48	+55	+64	+75	+88	+118	+160	+218
30	40	+9	+17	+26	+34	+43	+48	+60	+68	+80	+94	+112	+148	+200	+274
40	50	+9	+17	+26	+34	+43	+54	+70	+81	+97	+114	+136	+180	+242	+325
50	65	+11	+20	+32	+41	+53	+66	+87	+102	+122	+144	+172	+226	+300	+405
65	80	+11	+20	+32	+43	+59	+75	+102	+120	+146	+174	+210	+274	+360	+480
80	100	+13	+23	+37	+51	+71	+91	+124	+146	+178	+214	+258	+335	+445	+585
100	120	+13	+23	+37	+54	+79	+104	+144	+172	+210	+254	+310	+400	+525	+690
120	140	+15	+27	+43	+63	+92	+122	+170	+202	+248	+300	+365	+470	+620	+800
140	160	+15	+27	+43	+65	+100	+134	+190	+228	+280	+340	+415	+535	+700	+900
160	180	+15	+27	+43	+68	+108	+146	+210	+252	+310	+380	+465	+600	+780	+1000
180	200	+17	+31	+50	+77	+122	+166	+236	+284	+350	+425	+520	+670	+880	+1150
200	225	+17	+31	+50	+80	+130	+180	+258	+310	+385	+470	+575	+740	+960	+1250
225	250	+17	+31	+50	+84	+140	+196	+284	+340	+425	+520	+640	+820	+1050	+1350
250	280	+20	+34	+56	+94	+158	+218	+315	+385	+475	+580	+710	+920	+1200	+1550
280	315	+20	+34	+56	+98	+170	+240	+350	+425	+525	+650	+790	+1000	+1300	+1700
315	355	+21	+37	+62	+108	+190	+268	+390	+475	+590	+730	+900	+1150	+1500	+1900
355	400	+21	+37	+62	+114	+208	+294	+435	+530	+660	+820	+1000	+1300	+1650	+2100
400	450	+23	+40	+68	+126	+232	+330	+490	+595	+740	+920	+1100	+1450	+1850	+2400
450	500	+23	+40	+68	+132	+252	+360	+540	+660	+820	+1000	+1250	+1600	+2100	+2600

The dimensions are in 0.001 mm, except the nominal sizes, which are in millimeters.

Table 25. British Standard Fundamental Deviations for Holes *BS 4500:1969*

Nominal Sizes, mm		Grade																				
		01 to 16												6	7	8	≤8	>8	≤8[a]	>8	≤8	>8[b]
		Fundamental (Lower) Deviation EI												Fundamental (Upper) Deviation ES								
Over	To	A[b]	B[b]	C	CD	D	E	EF	F	FG	G	H	Js[c]	J			K[d]		M[d]		N[d]	
…	3	+270	+140	+60	+34	+20	+14	+10	+6	+4	+2	0		+2	+4	+6	0	0	−2	−2	−4	−4
3	6	+270	+140	+70	+46	+30	+20	+14	+10	+6	+4	0		+5	+6	+10	−1+Δ	…	−4+Δ	−4	−8+Δ	0
6	10	+280	+150	+80	+56	+40	+25	+18	+13	+8	+5	0		+5	+8	+12	−1+Δ	…	−6+Δ	−6	−10+Δ	0
10	14	+290	+150	+95	…	+50	+32	…	+16	…	+6	0		+6	+10	+15	−1+Δ	…	−7+Δ	−7	−12+Δ	0
14	18	+290	+150	+95	…	+50	+32	…	+16	…	+6	0		+6	+10	+15	−1+Δ	…	−7+Δ	−7	−12+Δ	0
18	24	+300	+160	+110	…	+65	+40	…	+20	…	+7	0		+8	+12	+20	−2+Δ	…	−8+Δ	−8	−15+Δ	0
24	30	+300	+160	+110	…	+65	+40	…	+20	…	+7	0		+8	+12	+20	−2+Δ	…	−8+Δ	−8	−15+Δ	0
30	40	+310	+170	+120	…	+80	+50	…	+25	…	+9	0		+10	+14	+24	−2+Δ	…	−9+Δ	−9	−17+Δ	0
40	50	+320	+180	+130	…	+80	+50	…	+25	…	+9	0		+10	+14	+24	−2+Δ	…	−9+Δ	−9	−17+Δ	0
50	65	+340	+190	+140	…	+100	+60	…	+30	…	+10	0		+13	+18	+28	−2+Δ	…	−11+Δ	−11	−20+Δ	0
65	80	+360	+200	+150	…	+100	+60	…	+30	…	+10	0		+13	+18	+28	−2+Δ	…	−11+Δ	−11	−20+Δ	0
80	100	+380	+220	+170	…	+120	+72	…	+36	…	+12	0		+16	+22	+34	−3+Δ	…	−13+Δ	−13	−23+Δ	0
100	120	+410	+240	+180	…	+120	+72	…	+36	…	+12	0		+16	+22	+34	−3+Δ	…	−13+Δ	−13	−23+Δ	0
120	140	+460	+260	+200	…	+145	+85	…	+43	…	+14	0	±IT/2	+18	+26	+41	−3+Δ	…	−15+Δ	−15	−27+Δ	0
140	160	+520	+280	+210	…	+145	+85	…	+43	…	+14	0		+18	+26	+41	−3+Δ	…	−15+Δ	−15	−27+Δ	0
160	180	+580	+310	+230	…	+145	+85	…	+43	…	+14	0		+18	+26	+41	−3+Δ	…	−15+Δ	−15	−27+Δ	0
180	200	+660	+340	+240	…	+170	+100	…	+50	…	+15	0		+22	+30	+47	−4+Δ	…	−17+Δ	−17	−31+Δ	0
200	225	+740	+380	+260	…	+170	+100	…	+50	…	+15	0		+22	+30	+47	−4+Δ	…	−17−Δ	−17	−31+Δ	0
225	250	+820	+420	+280	…	+170	+100	…	+50	…	+15	0		+22	+30	+47	−4+Δ	…	−17+Δ	−17	−31+Δ	0
250	280	+920	+480	+300	…	+190	+110	…	+56	…	+17	0		+25	+36	+55	−4+Δ	…	−20+Δ	−20	−34+Δ	0
280	315	+1050	+540	+330	…	+190	+110	…	+56	…	+17	0		+25	+36	+55	−4+Δ	…	−20+Δ	−20	−34+Δ	0
315	355	+1200	+600	+360	…	+210	+125	…	+62	…	+18	0		+29	+39	+60	−4+Δ	…	−21+Δ	−21	−37+Δ	0
355	400	+1350	+680	+400	…	+210	+125	…	+62	…	+18	0		+29	+39	+60	−4+Δ	…	−21+Δ	−21	−37+Δ	0
400	450	+1500	+760	+440	…	+230	+135	…	+68	…	+20	0		+33	+43	+66	−5+4	…	−23+Δ	−23	−40+Δ	0
450	500	+1650	+840	+480	…	+230	+135	…	+68	…	+20	0		+33	+43	+66	−5+4	…	−23+Δ	−23	−40+Δ	0

[a] Special case: for M6, ES = −9 for sizes from 250 to 315 mm, instead of −11.

[b] Not applicable to sizes up to 1 mm.

[c] In grades 7 to 11, the two symmetrical deviations ±IT/2 should be rounded if the IT value in micrometers is an odd value, by replacing it with the even value below. For example, if IT = 175, replace it by 174.

[d] When calculating deviations for holes K, M, and N with tolerance grades up to and including IT 8, and holes F to ZC with tolerance grades up to and including IT 7, the delta (Δ) values are added to the upper deviation ES. For example, for 25 P7, ES = −0.022 + 0.008 = −0.014 mm.

Table 26. British Standard Fundamental Deviations for Holes *BS 4500:1969*

Nominal Sizes, mm		Grade ≤7	Grade >7: Fundamental (Upper) Deviation ES												Values for delta (Δ)[d]: Grade					
Over	To	P to ZC	P	R	S	T	U	V	X	Y	Z	ZA	ZB	ZC	3	4	5	6	7	8
…	3	Same deviation as for grades above 7 increased by Δ	−6	−10	−14	…	−18	…	−20	…	−26	−32	−40	−60	0	0	0	0	0	0
3	6		−12	−15	−19	…	−23	…	−28	…	−35	−42	−50	−80	1	1.5	1	3	4	6
6	10		−15	−19	−23	…	−28	…	−34	…	−42	−52	−67	−97	1	1.5	2	3	6	7
10	14		−18	−23	−28	…	−33	…	−40	…	−50	−64	−90	−130	1	2	3	3	7	9
14	18		−18	−23	−28	…	−33	−39	−45	…	−60	−77	−108	−150	1	2	3	3	7	9
18	24		−22	−28	−35	…	−41	−47	−54	−63	−73	−98	−136	−188	1.5	2	3	4	8	12
24	30		−22	−28	−35	−41	−48	−55	−64	−75	−88	−118	−160	−218	1.5	2	3	4	8	12
30	40		−26	−34	−43	−48	−60	−68	−80	−94	−112	−148	−200	−274	1.5	3	4	5	9	14
40	50		−26	−34	−43	−54	−70	−81	−97	−114	−136	−180	−242	−325	1.5	3	4	5	9	14
50	65		−32	−41	−53	−66	−87	−102	−122	−144	−172	−226	−300	−405	2	3	5	6	11	16
65	80		−32	−43	−59	−75	−102	−120	−146	−174	−210	−274	−360	−480	2	3	5	6	11	16
80	100		−37	−51	−71	−91	−124	−146	−178	−214	−258	−335	−445	−585	2	4	5	7	13	19
100	120		−37	−54	−79	−104	−144	−172	−210	−254	−310	−400	−525	−690	2	4	5	7	13	19
120	140		−43	−63	−92	−122	−170	−202	−248	−300	−365	−470	−620	−800	3	4	6	7	15	23
140	160		−43	−65	−100	−134	−190	−228	−280	−340	−415	−535	−700	−900	3	4	6	7	15	23
160	180		−43	−68	−108	−146	−210	−252	−310	−380	−465	−600	−780	−1000	3	4	6	7	15	23
180	200		−50	−77	−122	−166	−226	−284	−350	−425	−520	−670	−880	−1150	3	4	6	9	17	26
200	225		−50	−80	−130	−180	−258	−310	−385	−470	−575	−740	−960	−1250	3	4	6	9	17	26
225	250		−50	−84	−140	−196	−284	−340	−425	−520	−640	−820	−1050	−1350	3	4	6	9	17	26
250	280		−56	−94	−158	−218	−315	−385	−475	−580	−710	−920	−1200	−1550	4	4	7	9	20	29
280	315		−56	−98	−170	−240	−350	−425	−525	−650	−790	−1000	−1300	−1700	4	4	7	9	20	29
315	355		−62	−108	−190	−268	−390	−475	−590	−730	−900	−1150	−1500	−1800	4	5	7	11	21	32
355	400		−62	−114	−208	−294	−435	−530	−660	−820	−1000	−1300	−1650	−2100	4	5	7	11	21	32
400	450		−68	−126	−232	−330	−490	−595	−740	−920	−1100	−1450	−1850	−2400	5	5	7	13	23	34
450	500		−68	−132	−252	−360	−540	−660	−820	−1000	−1250	−1600	−2100	−2600	5	5	7	13	23	34

The dimensions are given in 0.001 mm, except the nominal sizes, which are in millimeters.

British Standard Preferred Numbers and Preferred Sizes.—This British Standard, PD 6481:1977 1983, gives recommendations for the use of preferred numbers and preferred sizes for functional characteristics and dimensions of various products.

The preferred number system is internationally standardized in ISO 3. It is also referred to as the Renard, or R, series (see *American National Standard for Preferred Numbers*, on page 20).

The series in the preferred number system are geometric series, that is, there is a constant ratio between each figure and the succeeding one, within a decimal framework. Thus, the R5 series has five steps between 1 and 10, the R10 series has 10 steps between 1 and 10, the R20 series, 20 steps, and the R40 series, 40 steps, giving increases between steps of approximately 60, 25, 12, and 6 per cent, respectively.

The preferred size series have been developed from the preferred number series by rounding off the inconvenient numbers in the basic series and adjusting for linear measurement in millimeters. These series are shown in the following table.

After taking all normal considerations into account, it is recommended that (a) for ranges of values of the primary *functional* characteristics (outputs and capacities) of a series of products, the preferred number series R5 to R40 (see page 20) should be used, and (b) whenever linear sizes are concerned, the preferred sizes as given in the following table should be used. The presentation of preferred sizes gives designers and users a logical selection and the benefits of rational variety reduction.

The second-choice size given should only be used when it is not possible to use the first choice, and the third choice should be applied only if a size from the second choice cannot be selected. With this procedure, common usage will tend to be concentrated on a limited range of sizes, and a contribution is thus made to variety reduction. However, the decision to use a particular size cannot be taken on the basis that one is first choice and the other not. Account must be taken of the effect on the design, the availability of tools, and other relevant factors.

British Standard Preferred Sizes, *PD 6481: 1977 (1983)*

Choice			Choice			Choice			Choice			Choice			Choice		
1st	2nd	3rd	1st	2nd	3rd	1st	2nd	3rd	1st	2nd	3rd	1st	2nd	3rd	1st	2nd	3rd
1					5.2			23	65					122			188
	1.1			5.5				24			66		125		190		
1.2					5.8	25				68				128			192
		1.3	6					26	70			130				195	
	1.4				6.2		28			72				132			198
		1.5		6.5		30					74		135		200		
1.6					6.8		32		75					138			205
		1.7		7				34			76	140				210	
	1.8				7.5	35				78				142			215
		1.9	8					36	80				145		220		
2					8.5		38				82			148			225
		2.1		9		40				85		150				230	
	2.2				9.5		42				88			152			235
		2.4	10					44	90				155		240		
2.5				11		45					92			158			245
		2.6	12					46		95		160				250	
2.8					13		48				98			162			255
3				14		50			100				165		260		
		3.2			15		52				102			168			265
	3.5		16					54	105			170				270	
		3.8			17	55					108			172			275
4				18				56	110				175		280		
		4.2			19		58				112			178			285
	4.5		20			60				115		180				290	
		4.8			21		62				118			182			295
5				22				64	120				185		300		

For dimensions above 300, each series continues in a similar manner, i.e., the intervals between each series number are the same as between 200 and 300.

Length Differences Due to Temperature Changes.—The following table gives changes in length for variations from the standard reference temperature of 68 deg. F (20 deg. C) for materials of known coefficients of expansion. Coefficients of expansion are given in tables on pages 367 and 368.

In the table below, for coefficients between those listed, add appropriate listed values. For example, a length change for a coefficient of 7 is the sum of values in the 5 and 2 columns. Fractional interpolation also is possible. Thus, in a steel bar with a coefficient of thermal expansion of 6.3×10^{-6} [= 0.0000063 in./in. (µin./in.) of length/deg. F], the increase in length at 73 deg. F is 25 + 5 + 1.5 = 31.5 µin./in. of length. For a steel with the same coefficient of expansion, the change in length, measured in deg. C, is expressed in microns (micrometers)/meter (µm/m) of length.

Table Showing Differences in Length in Inches/Inch (Microns/Meter) for Changes from the Standard Temperature of 68 Deg. F (20 Deg. C)

Temperature Deg.		Coefficient of Thermal Expansion of Material per Degree F (C) × 10^4									
		1	2	3	4	5	10	15	20	25	30
		Total Change in Length from Standard for F Deg. Microinches/Inch (µin./in.) and for C deg. (K) microns/meter (µm/m) of length									
F	C										
48	0	−20	−40	−60	−80	−100	−200	−300	−400	−500	−600
49	1	−19	−38	−57	−76	−95	−190	−285	−380	−475	−570
50	2	−18	−36	−54	−72	−90	−180	−270	−360	−450	−540
51	3	−17	−34	−51	−68	−85	−170	−255	−340	−425	−510
52	4	−16	−32	−48	−64	−80	−160	−240	−320	−400	−480
53	5	−15	−30	−45	−60	−75	−150	−225	−300	−375	−450
54	6	−14	−28	−42	−56	−70	−140	−210	−280	−350	−420
55	7	−13	−26	−39	−52	−65	−130	−195	−260	−325	−390
56	8	−12	−24	−36	−48	−60	−120	−180	−240	−300	−360
57	9	−11	−22	−33	−44	−55	−110	−165	−220	−275	−330
58	10	−10	−20	−30	−40	−50	−100	−150	−200	−250	−300
59	11	−9	−18	−27	−36	−45	−90	−135	−180	−225	−270
60	12	−8	−16	−24	−32	−40	−80	−120	−160	−200	−240
61	13	−7	−14	−21	−28	−35	−70	−105	−140	−175	−210
62	14	−6	−12	−18	−24	−30	−60	−90	−120	−150	−180
63	15	−5	−10	−15	−20	−25	−50	−75	−100	−125	−150
64	16	−4	−8	−12	−16	−20	−40	−60	−80	−100	−120
65	17	−3	−6	−9	−12	−15	−30	−45	−60	−75	−90
66	18	−2	−4	−6	−8	−10	−20	−30	−40	−50	−60
67	19	−1	−2	−3	−4	−5	−10	−15	−20	−25	−30
68	20	0	0	0	0	0	0	0	0	0	0
69	21	1	2	3	4	5	10	15	20	25	30
70	22	2	4	6	8	10	20	30	40	50	60
71	23	3	6	9	12	15	30	45	60	75	90
72	24	4	8	12	16	20	40	60	80	100	120
73	25	5	10	15	20	25	50	75	100	125	150
74	26	6	12	18	24	30	60	90	120	150	180
75	27	7	14	21	28	35	70	105	140	175	210
76	28	8	16	24	32	40	80	120	160	200	240
77	29	9	18	27	36	45	90	135	180	225	270
78	30	10	20	30	40	50	100	150	200	250	300
79	31	11	22	33	44	55	110	165	220	275	330
80	32	12	24	36	48	60	120	180	240	300	360
81	33	13	26	39	52	65	130	195	260	325	390
82	34	14	28	42	56	70	140	210	280	350	420
83	35	15	30	45	60	75	150	225	300	375	450
84	36	16	32	48	64	80	160	240	320	400	480
85	37	17	34	51	68	85	170	255	340	425	510
86	38	18	36	54	72	90	180	270	360	450	540
87	39	19	38	57	76	95	190	285	380	475	570
88	40	20	40	60	80	100	200	300	400	500	600

MEASURING INSTRUMENTS AND INSPECTION METHODS

Verniers and Micrometers

Reading a Vernier.— A general rule for taking readings with a vernier scale is as follows: Note the number of inches and sub-divisions of an inch that the zero mark of the vernier scale has moved along the true scale, and then add to this reading as many thousandths, or hundredths, or whatever fractional part of an inch the vernier reads to, as there are spaces between the vernier zero and that line on the vernier which coincides with one on the true scale. For example, if the zero line of a vernier which reads to thousandths is slightly beyond the 0.5 inch division on the main or true scale, as shown in Fig. 1, and graduation line 10 on the vernier exactly coincides with one on the true scale, the reading is 0.5 + 0.010 or 0.510 inch. In order to determine the reading or fractional part of an inch that can be obtained by a vernier, multiply the denominator of the finest sub-division given on the true scale by the total number of divisions on the vernier. For example, if one inch on the true scale is divided into 40 parts or fortieths (as in Fig. 1), and the vernier into twenty-five parts, the vernier will read to thousandths of an inch, as $25 \times 40 = 1000$. Similarly, if there are sixteen divisions to the inch on the true scale and a total of eight on the vernier, the latter will enable readings to be taken within one-hundred-twenty-eighths of an inch, as $8 \times 16 = 128$.

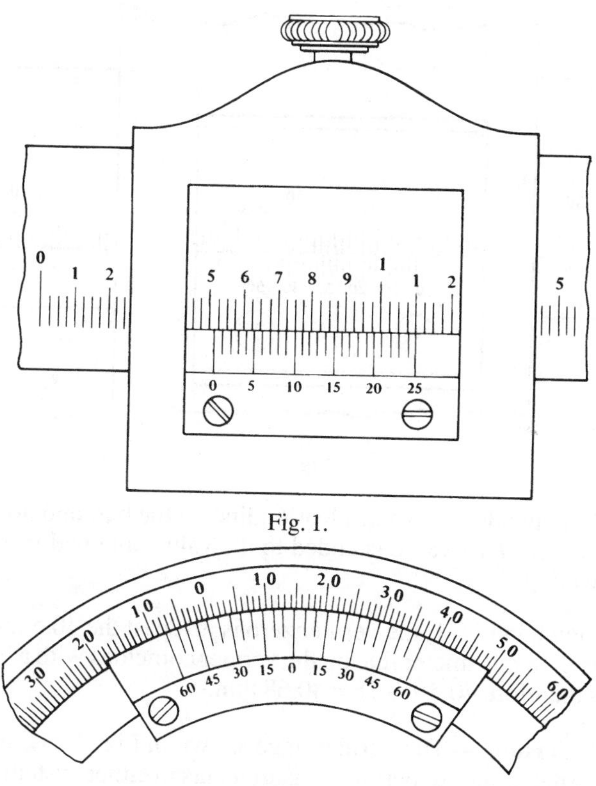

Fig. 1.

Fig. 2.

If the vernier is on a protractor, note the whole number of degrees passed by the vernier zero mark and then count the spaces between the vernier zero and that line which coincides with a graduation on the protractor scale. If the vernier indicates angles within five minutes or one-twelfth degree (as in Fig. 2), the number of spaces multiplied by 5 will, of course, give the number of minutes to be added to the whole number of degrees. The reading of the protractor set as illustrated would be 14 whole degrees (the number passed by the zero mark on the vernier) plus 30 minutes, as the graduation 30 on the vernier is the only one to the right of the vernier zero which exactly coincides with a line on the protractor scale. It will be noted that there are duplicate scales on the vernier, one being to the right and the other to the left of zero. The left-hand scale is used when the vernier zero is moved to the left of the zero of the protractor scale, whereas the right-hand graduations are used when the movement is to the right.

Reading a Metric Vernier.—The smallest graduation on the bar (true or main scale) of the metric vernier gage shown in Fig. 1, is 0.5 millimeter. The scale is numbered at each twentieth division, and thus increments of 10, 20, 30, 40 millimeters, etc., are indicated. There are 25 divisions on the vernier scale, occupying the same length as 24 divisions on the bar, which is 12 millimeters. Therefore, one division on the vernier scale equals one twenty-fifth of 12 millimeters $= 0.04 \times 12 = 0.48$ millimeter. Thus, the difference between one bar division (0.50 mm) and one vernier division (2.48 mm) is $0.50 - 0.48 = 0.02$ millimeter, which is the minimum measuring increment that the gage provides. To permit direct readings, the vernier scale has graduations to represent tenths of a millimeter (0.1 mm) and fiftieths of a millimeter (0.02 mm).

To read a vernier gage, first note how many millimeters the zero line on the vernier is from the zero line on the bar. Next, find the graduation on the vernier

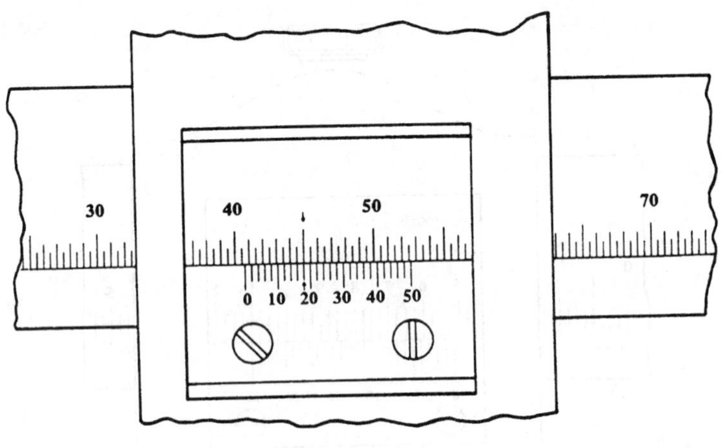

Fig. 1.

scale which exactly coincides with a graduation line on the bar, and note the value of the vernier scale graduation. This value is added to the value obtained from the bar, and the result is the total reading.

In the example shown in Fig. 1, the vernier zero is just past the 40.5 millimeters graduation on the bar. The 0.18 millimeter line on the vernier coincides with a line on the bar, and the total reading is therefore $40.5 + 0.18 = 40.68$ mm.

Dual Metric-Inch Vernier.—The vernier gage shown in Fig. 2 has separate metric and inch 50-division vernier scales to permit measurements in either system.

A 50-division vernier has more widely spaced graduations than the 25-division vernier shown on the previous pages, and is thus easier to read. On the bar, the smallest metric graduation is 1 millimeter, and the 50 divisions of the vernier occupy the same length as 49 divisions on the bar, which is 49 mm. Therefore, one division on the vernier scale equals one-fiftieth of 49 millimeters $= 0.02 \times 49 = 0.98$ mm. Thus, the difference between one bar division (1.0 mm) and one vernier division (0.98 mm) is 0.02 mm, which is the minimum measuring increment the gage provides.

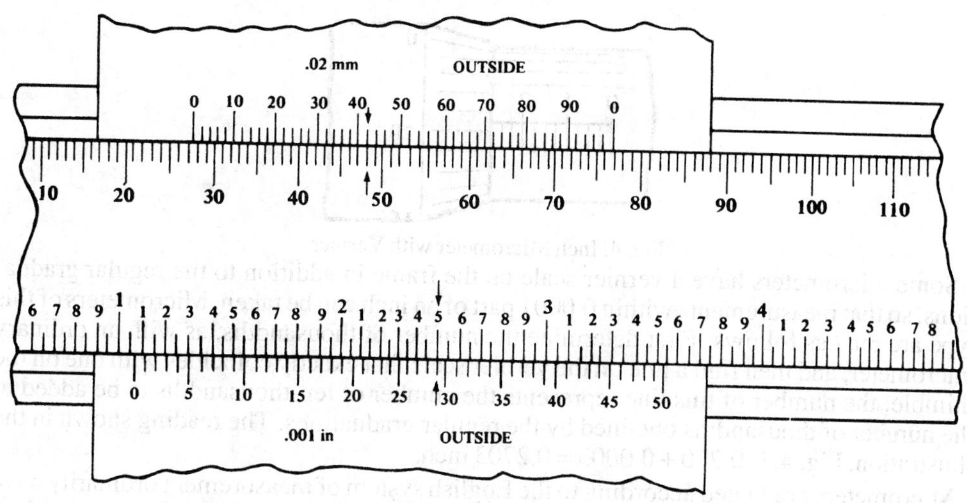

Fig. 2.

The vernier scale is graduated for direct reading to 0.02 mm. In the figure, the vernier zero is just past the 27 mm graduation on the bar, and the 0.42 mm graduation on the vernier coincides with a line on the bar. The total reading is therefore 27.42 mm.

The smallest inch graduation on the bar is 0.05 inch, and the 50 vernier divisions occupy the same length as 49 bar divisions, which is 2.45 inches. Therefore, one vernier division equals one-fiftieth of 2.45 inches = 0.02 × 2.45 = 0.049 inch. Thus, the difference between the length of a bar division and a vernier division is 0.050-0.049 = 0.001 inch. The vernier scale is graduated for direct reading to 0.001 inch. In the example, the vernier zero is past the 1.05 graduation on the bar, and the 0.029 graduation on the vernier coincides with a line on the bar. Thus, the total reading is 1.079 inches.

Reading a Micrometer.—The spindle of an inch-system micrometer has 40 threads per inch, so that one turn moves the spindle axially 0.025 inch (1 ÷ 40 = 0.025), equal to the distance between two graduations on the frame. The 25 graduations on the thimble allow the 0.025 inch to be further divided, so that turning the thimble through one division moves the spindle axially 0.001 inch (0.025 ÷ 25 = 0.001). To read a micrometer, count the number of whole divisions that are visible on the scale of the frame, multiply this number by 25 (the number of thousandths of an inch that each division represents) and add to the product the number of that division on the thimble which coincides with the axial zero line on the frame. The result will be the diameter expressed in thousandths of an inch. As the numbers 1, 2, 3, etc., opposite every fourth sub-division on the frame, indicate hundreds of thousandths, the reading can easily be taken mentally. Suppose the thimble were screwed out so that graduation 2, and three additional sub-divisions, were visible (as shown in Fig. 3), and that graduation 10 on the thimble coincided with the axial line on the frame. The reading then would be 0.200 + 0.075 + 0.010, or 0.285 inch.

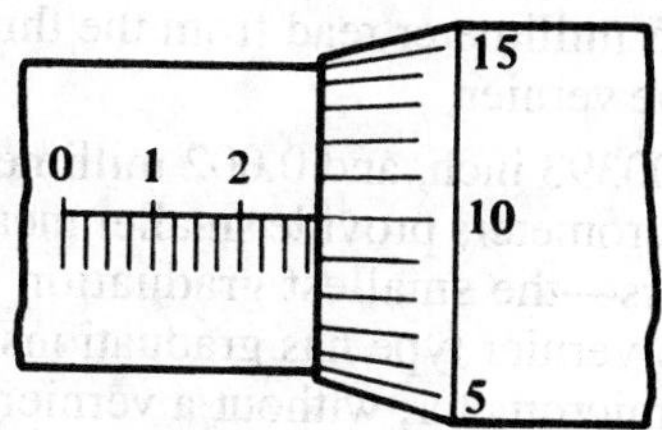

Fig. 3. Inch Micrometer

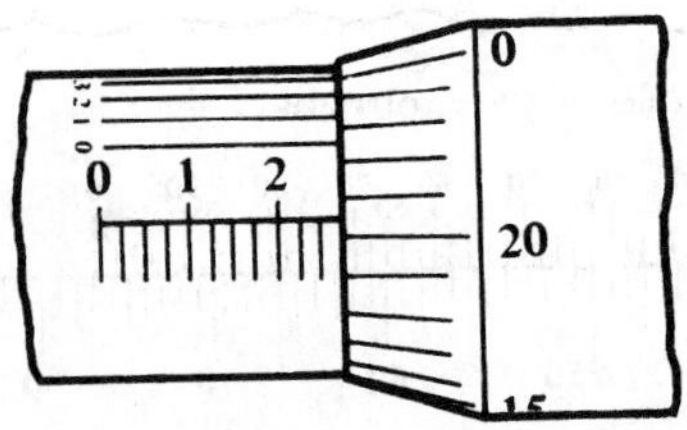

Fig. 4. Inch Micrometer with Vernier

Some micrometers have a vernier scale on the frame in addition to the regular graduations, so that measurements within 0.0001 part of an inch can be taken. Micrometers of this type are read as follows: First determine the number of thousandths, as with an ordinary micrometer, and then find a line on the vernier scale that exactly coincides with one on the thimble; the number of this line represents the number of ten-thousandths to be added to the number of thousandths obtained by the regular graduations. The reading shown in the illustration, Fig. 4, is 0.270 + 0.0003 = 0.2703 inch.

Micrometers graduated according to the English system of measurement ordinarily have a table of decimal equivalents stamped on the sides of the frame, so that fractions such as sixty-fourths, thirty-seconds, etc., can readily be converted into decimals.

Reading a Metric Micrometer.—The spindle of an ordinary metric micrometer has 2 threads per millimeter, and thus one complete revolution moves the spindle through a distance of 0.5 millimeter. The longitudinal line on the frame is graduated with 1 millimeter divisions and 0.5 millimeter sub-divisions. The thimble has 50 graduations, each being 0.01 millimeter (one-hundredth of a millimeter).

To read a metric micrometer, note the number of millimeter divisions visible on the scale of the sleeve, and add the total to the particular division on the thimble which coincides with the axial line on the sleeve. Suppose that the thimble were screwed out so that graduation 5, and one additional 0.5 sub-division were visible (as shown in Fig. 5), and that graduation 28 on the thimble coincided with the axial line on the sleeve. The reading then would be 5.00 + 0.5 + 0.28 = 5.78 mm.

Some micrometers are provided with a vernier scale on the sleeve in addition to the regular graduations to permit measurements within 0.002 millimeter to be made. Micrometers of this type are read as follows: First determine the number of whole millimeters (if any) and the number of hundredths of a millimeter, as with an ordinary micrometer, and then find a line on the sleeve vernier scale which exactly coincides

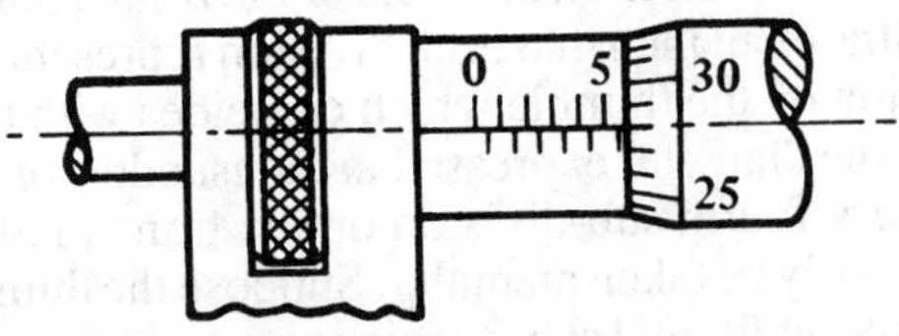

Fig. 5. Metric Micrometer

with one on the thimble. The number of this coinciding vernier line represents the number of two-thousandths of a millimeter to be added to the reading already obtained. Thus, for example, a measurement of 2.958 millimeters would be obtained by reading 2.5 millimeters on the sleeve, adding 0.45 millimeter read from the thimble, and then adding 0.008 millimeter as determined by the vernier.

Note: 0.01 millimeter = 0.000393 inch, and 0.002 millimeter = 0.000078 inch (78 millionths). Therefore, metric micrometers provide smaller measuring increments than comparable inch unit micrometers—the smallest graduation of an ordinary inch reading micrometer is 0.001 inch; the vernier type has graduations down to 0.0001 inch. When using either a metric or inch micrometer, without a vernier, smaller readings than those graduated may of course be obtained by visual interpolation between graduations.

Sine-bar

The sine-bar is used either for very accurate angular measurements or for locating work at a given angle as, for example, in surface grinding templets, gages, etc. The sine-bar is especially useful in measuring or checking angles when the limit of accuracy is 5 minutes or less. Some bevel protractors are equipped with verniers which read to 5 minutes but the setting depends upon the alignment of graduations whereas a sine-bar usually is located by positive contact with precision gage-blocks selected for whatever dimension is required for obtaining a given angle.

Types of Sine-bars.—A sine-bar consists of a hardened, ground and lapped steel bar with very accurate cylindrical plugs of equal diameter attached to or near each end. The form illustrated by Fig. 1 has notched ends for receiving the cylindrical plugs so that they are held firmly against both faces of the notch. The standard center-to-center distance C between the plugs is either 5 or 10 inches. The upper and lower sides of sine-bars are parallel to the center line of the plugs within very close limits. The body of the sine-bar ordinarily has several through holes to reduce the weight. In the making of the sine-bar shown in Fig. 2, if too much material is removed from one locating notch, regrinding the shoulder at the opposite end would make it possible to obtain the correct center distance. That is the reason for this change in form. The type of sine-bar illustrated by Fig. 3 has the cylindrical disks or plugs attached to one side. These differences in form or arrangement do not, of course, affect the principle governing the use of the sine-bar. An accurate surface plate or master flat is always used in conjunction with a sine-bar in order to form the base from which the vertical measurements are made

.

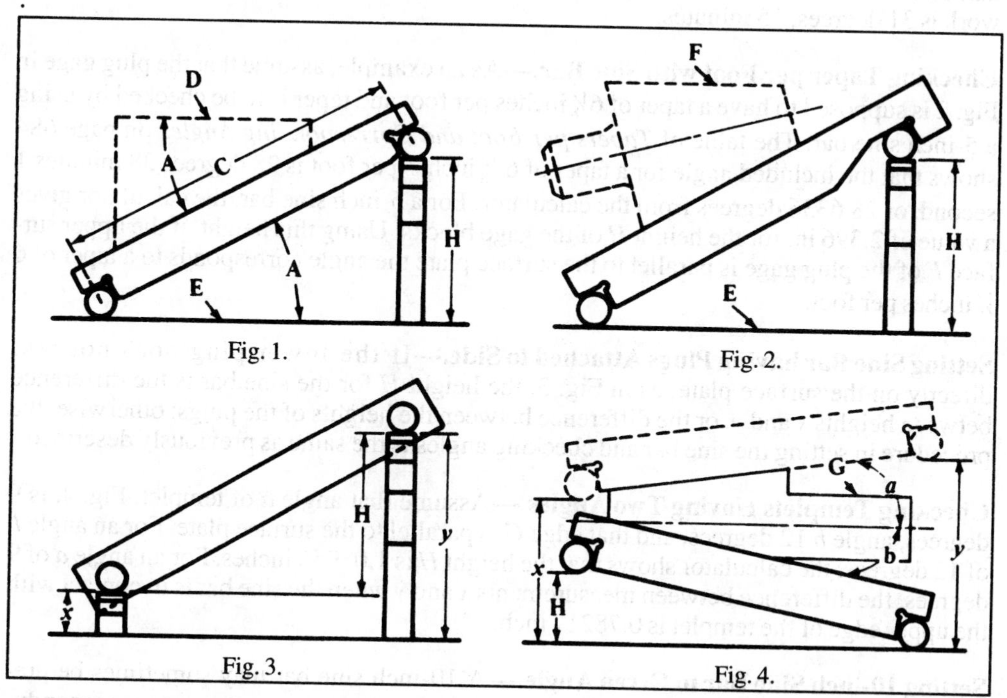

Fig. 1. Fig. 2. Fig. 3. Fig. 4.

Setting a Sine Bar to a Given Angle.—To find the vertical distance *H*, for setting a sine bar to the required angle, convert the angle to decimal form on a pocket calculator, take the sine of that angle, and multiply by the distance between the cylinders. For example, if an angle of 31 degrees, 30 minutes is required, the equivalent angle is 31 degrees plus $^{30}/_{60}$ = 31 + 0.5, or 31.5 degrees. (For conversions from minutes and seconds to decimals of degrees and vice versa, see page 90). The sine of 31.5 degrees is 0.5225 and multiplying this value by the sine bar length gives 2.613 in. for the height *H*, Fig. 1 and 3, of the gage blocks.

Finding Angle when Height *H* of Sine Bar is Known.—To find the angle equivalent to a given height *H*, reverse the above procedure. Thus, if the height *H* is 1.4061 in., dividing by 5 gives a sine of 0.28122, which corresponds to an angle of 16.333 degrees, or 16 degrees 20 minutes.

Checking Angle of Templet or Gage by Using Sine Bar.—Place templet or gage on sine bar as indicated by dotted lines, Fig. 1. Clamps may be used to hold work in place. Place upper end of sine bar on gage blocks having total height *H* corresponding to the required angle. If upper edge *D* of work is parallel with surface plate *E*, then angle *A* of work equals angle *A* to which sine bar is set. Parallelism between edge *D* and surface plate may be tested by checking the height at each end with a dial gage or some type of indicating comparator.

Measuring Angle of Templet or Gage with Sine Bar.—To measure such an angle, adjust height of gage blocks and sine bar until edge *D*, Fig. 1, is parallel with surface plate *E*; then find angle corresponding to height *H*, of gage blocks. For example, if height *H* is 2.5939 inches when *D* and *E* are parallel, the calculator will show that the angle *A* of the work is 31 degrees, 15 minutes.

Checking Taper per Foot with Sine Bar.—As an example, assume that the plug gage in Fig. 2 is supposed to have a taper of 6⅛ inches per foot and taper is to be checked by using a 5-inch sine bar. The table of *Tapers per Foot and Corresponding Angles* on page 684 shows that the included angle for a taper of 6 ⅛ inches per foot is 28 degrees 38 minutes 1 second, or 28.6336 degrees from the calculator. For a 5-inch sine bar, the calculator gives a value of 2.396 in. for the height *H* of the gage blocks. Using this height, if the upper surface *F* of the plug gage is parallel to the surface plate the angle corresponds to a taper of 6 ⅛ inches per foot.

Setting Sine Bar having Plugs Attached to Side.—If the lower plug does not rest directly on the surface plate, as in Fig. 3, the height *H* for the sine bar is the difference between heights *x* and *y*, or the difference between the heights of the plugs; otherwise, the procedure in setting the sine bar and checking angles is the same as previously described.

Checking Templets Having Two Angles.—Assume that angle *a* of templet, Fig. 4, is 9 degrees, angle *b* 12 degrees, and that edge *G* is parallel to the surface plate. For an angle *b* of 12 degrees, the calculator shows that the height *H* is 1.03956 inches. For an angle *a* of 9 degrees, the difference between measurements *x* and *y* when the sine bar is in contact with the upper edge of the templet is 0.78217 inch.

Setting 10-inch Sine Bar to Given Angle.—A 10-inch sine bar may sometimes be preferred because of its longer working surface or because the longer center distance is conducive to greater precision. To obtain the vertical distances *H* for setting a 10-inch sine bar, multiply the sine of the angle by 10, by shifting the decimal point one place to the right.

For example, the sine of 39 degrees is 0.62932, hence the vertical height *H* for setting a 10-inch sine bar is 6.2932 inches.

Constants for Setting a 5-inch Sine-Bar for 1° to 7°

Min.	0°	1°	2°	3°	4°	5°	6°	7°
0	0.00000	0.08726	0.17450	0.26168	0.34878	0.43578	0.52264	0.60935
1	0.00145	0.08872	0.17595	0.26313	0.35023	0.43723	0.52409	0.61079
2	0.00291	0.09017	0.17740	0.26458	0.35168	0.43868	0.52554	0.61223
3	0.00436	0.09162	0.17886	0.26604	0.35313	0.44013	0.52698	0.61368
4	0.00582	0.09308	0.18031	0.26749	0.35459	0.44157	0.52843	0.61512
5	0.00727	0.09453	0.18177	0.26894	0.35604	0.44302	0.52987	0.61656
6	0.00873	0.09599	0.18322	0.27039	0.35749	0.44447	0.53132	0.61801
7	0.01018	0.09744	0.18467	0.27185	0.35894	0.44592	0.53277	0.61945
8	0.01164	0.09890	0.18613	0.27330	0.36039	0.44737	0.53421	0.62089
9	0.01309	0.10035	0.18758	0.27475	0.36184	0.44882	0.53566	0.62234
10	0.01454	0.10180	0.18903	0.27620	0.36329	0.45027	0.53710	0.62378
11	0.01600	0.10326	0.19049	0.27766	0.36474	0.45171	0.53855	0.62522
12	0.01745	0.10471	0.19194	0.27911	0.36619	0.45316	0.54000	0.62667
13	0.01891	0.10617	0.19339	0.28056	0.36764	0.45461	0.54144	0.62811
14	0.02036	0.10762	0.19485	0.28201	0.36909	0.45606	0.54289	0.62955
15	0.02182	0.10907	0.19630	0.28346	0.37054	0.45751	0.54433	0.63099
16	0.02327	0.11053	0.19775	0.28492	0.37199	0.45896	0.54578	0.63244
17	0.02473	0.11198	0.19921	0.28637	0.37344	0.46040	0.54723	0.63388
18	0.02618	0.11344	0.20066	0.28782	0.37489	0.46185	0.54867	0.63532
19	0.02763	0.11489	0.20211	0.28927	0.37634	0.46330	0.55012	0.63677
20	0.02909	0.11634	0.20357	0.29072	0.37779	0.46475	0.55156	0.63821
21	0.03054	0.11780	0.20502	0.29218	0.37924	0.46620	0.55301	0.63965
22	0.03200	0.11925	0.20647	0.29363	0.38069	0.46765	0.55445	0.64109
23	0.03345	0.12071	0.20793	0.29508	0.38214	0.46909	0.55590	0.64254
24	0.03491	0.12216	0.20938	0.29653	0.38360	0.47054	0.55734	0.64398
25	0.03636	0.12361	0.21083	0.29798	0.38505	0.47199	0.55879	0.64542
26	0.03782	0.12507	0.21228	0.29944	0.38650	0.47344	0.56024	0.64686
27	0.03927	0.12652	0.21374	0.30089	0.38795	0.47489	0.56168	0.64830
28	0.04072	0.12798	0.21519	0.30234	0.38940	0.47633	0.56313	0.64975
29	0.04218	0.12943	0.21664	0.30379	0.39085	0.47778	0.56457	0.65119
30	0.04363	0.13088	0.21810	0.30524	0.39230	0.47923	0.56602	0.65263
31	0.04509	0.13234	0.21955	0.30669	0.39375	0.48068	0.56746	0.65407
32	0.04654	0.13379	0.22100	0.30815	0.39520	0.48212	0.56891	0.65551
33	0.04800	0.13525	0.22246	0.30960	0.39665	0.48357	0.57035	0.65696
34	0.04945	0.13670	0.22391	0.31105	0.39810	0.48502	0.57180	0.65840
35	0.05090	0.13815	0.22536	0.31250	0.39954	0.48647	0.57324	0.65984
36	0.05236	0.13961	0.22681	0.31395	0.40099	0.48791	0.57469	0.66128
37	0.05381	0.14106	0.22827	0.31540	0.40244	0.48936	0.57613	0.66272
38	0.05527	0.14252	0.22972	0.31686	0.40389	0.49081	0.57758	0.66417
39	0.05672	0.14397	0.23117	0.31831	0.40534	0.49226	0.57902	0.66561
40	0.05818	0.14542	0.23263	0.31976	0.40679	0.49370	0.58046	0.66705
41	0.05963	0.14688	0.23408	0.32121	0.40824	0.49515	0.58191	0.66849
42	0.06109	0.14833	0.23553	0.32266	0.40969	0.49660	0.58335	0.66993
43	0.06254	0.14979	0.23699	0.32411	0.41114	0.49805	0.58480	0.67137
44	0.06399	0.15124	0.23844	0.32556	0.41259	0.49949	0.58624	0.67281
45	0.06545	0.15269	0.23989	0.32702	0.41404	0.50094	0.58769	0.67425
46	0.06690	0.15415	0.24134	0.32847	0.41549	0.50239	0.58913	0.67570
47	0.06836	0.15560	0.24280	0.32992	0.41694	0.50383	0.59058	0.67714
48	0.06981	0.15705	0.24425	0.33137	0.41839	0.50528	0.59202	0.67858
49	0.07127	0.15851	0.24570	0.33282	0.41984	0.50673	0.59346	0.68002
50	0.07272	0.15996	0.24715	0.33427	0.42129	0.50818	0.59491	0.68146
51	0.07417	0.16141	0.24861	0.33572	0.42274	0.50962	0.59635	0.68290
52	0.07563	0.16287	0.25006	0.33717	0.42419	0.51107	0.59780	0.68434
53	0.07708	0.16432	0.25151	0.33863	0.42564	0.51252	0.59924	0.68578
54	0.07854	0.16578	0.25296	0.34008	0.42708	0.51396	0.60068	0.68722
55	0.07999	0.16723	0.25442	0.34153	0.42853	0.51541	0.60213	0.68866
56	0.08145	0.16868	0.25587	0.34298	0.42998	0.51686	0.60357	0.69010
57	0.08290	0.17014	0.25732	0.34443	0.43143	0.51830	0.60502	0.69154
58	0.08435	0.17159	0.25877	0.34588	0.43288	0.51975	0.60646	0.69298
59	0.08581	0.17304	0.26023	0.34733	0.43433	0.52120	0.60790	0.69443
60	0.08726	0.17450	0.26168	0.34878	0.43578	0.52264	0.60935	0.69587

Constants for Setting a 5-inch Sine-Bar for 8° to 15°

Min.	8°	9°	10°	11°	12°	13°	14°	15°
0	0.69587	0.78217	0.86824	0.95404	1.03956	1.12476	1.20961	1.29410
1	0.69731	0.78361	0.86967	0.95547	1.04098	1.12617	1.21102	1.29550
2	0.69875	0.78505	0.87111	0.95690	1.04240	1.12759	1.21243	1.29690
3	0.70019	0.78648	0.87254	0.95833	1.04383	1.12901	1.21384	1.29831
4	0.70163	0.78792	0.87397	0.95976	1.04525	1.13042	1.21525	1.29971
5	0.70307	0.78935	0.87540	0.96118	1.04667	1.13184	1.21666	1.30112
6	0.70451	0.79079	0.87683	0.96261	1.04809	1.13326	1.21808	1.30252
7	0.70595	0.79223	0.87827	0.96404	1.04951	1.13467	1.21949	1.30393
8	0.70739	0.79366	0.87970	0.96546	1.05094	1.13609	1.22090	1.30533
9	0.70883	0.79510	0.88113	0.96689	1.05236	1.13751	1.22231	1.30673
10	0.71027	0.79653	0.88256	0.96832	1.05378	1.13892	1.22372	1.30814
11	0.71171	0.79797	0.88399	0.96974	1.05520	1.14034	1.22513	1.30954
12	0.71314	0.79941	0.88542	0.97117	1.05662	1.14175	1.22654	1.31095
13	0.71458	0.80084	0.88686	0.97260	1.05805	1.14317	1.22795	1.31235
14	0.71602	0.80228	0.88829	0.97403	1.05947	1.14459	1.22936	1.31375
15	0.71746	0.80371	0.88972	0.97545	1.06089	1.14600	1.23077	1.31516
16	0.71890	0.80515	0.89115	0.97688	1.06231	1.14742	1.23218	1.31656
17	0.72034	0.80658	0.89258	0.97830	1.06373	1.14883	1.23359	1.31796
18	0.72178	0.80802	0.89401	0.97973	1.06515	1.15025	1.23500	1.31937
19	0.72322	0.80945	0.89544	0.98116	1.06657	1.15166	1.23640	1.32077
20	0.72466	0.81089	0.89687	0.98258	1.06799	1.15308	1.23781	1.32217
21	0.72610	0.81232	0.89830	0.98401	1.06941	1.15449	1.23922	1.32357
22	0.72754	0.81376	0.89973	0.98544	1.07084	1.15591	1.24063	1.32498
23	0.72898	0.81519	0.90117	0.98686	1.07226	1.15732	1.24204	1.32638
24	0.73042	0.81663	0.90260	0.98829	1.07368	1.15874	1.24345	1.32778
25	0.73185	0.81806	0.90403	0.98971	1.07510	1.16015	1.24486	1.32918
26	0.73329	0.81950	0.90546	0.99114	1.07652	1.16157	1.24627	1.33058
27	0.73473	0.82093	0.90689	0.99256	1.07794	1.16298	1.24768	1.33199
28	0.73617	0.82237	0.90832	0.99399	1.07936	1.16440	1.24908	1.33339
29	0.73761	0.82380	0.90975	0.99541	1.08078	1.16581	1.25049	1.33479
30	0.73905	0.82524	0.91118	0.99684	1.08220	1.16723	1.25190	1.33619
31	0.74049	0.82667	0.91261	0.99826	1.08362	1.16864	1.25331	1.33759
32	0.74192	0.82811	0.91404	0.99969	1.08504	1.17006	1.25472	1.33899
33	0.74336	0.82954	0.91547	1.00112	1.08646	1.17147	1.25612	1.34040
34	0.74480	0.83098	0.91690	1.00254	1.08788	1.17288	1.25753	1.34180
35	0.74624	0.83241	0.91833	1.00396	1.08930	1.17430	1.25894	1.34320
36	0.74768	0.83384	0.91976	1.00539	1.09072	1.17571	1.26035	1.34460
37	0.74911	0.83528	0.92119	1.00681	1.09214	1.17712	1.26175	1.34600
38	0.75055	0.83671	0.92262	1.00824	1.09355	1.17854	1.26316	1.34740
39	0.75199	0.83815	0.92405	1.00966	1.09497	1.17995	1.26457	1.34880
40	0.75343	0.83958	0.92547	1.01109	1.09639	1.18136	1.26598	1.35020
41	0.75487	0.84101	0.92690	1.01251	1.09781	1.18278	1.26738	1.35160
42	0.75630	0.84245	0.92833	1.01394	1.09923	1.18419	1.26879	1.35300
43	0.75774	0.84388	0.92976	1.01536	1.10065	1.18560	1.27020	1.35440
44	0.75918	0.84531	0.93119	1.01678	1.10207	1.18702	1.27160	1.35580
45	0.76062	0.84675	0.93262	1.01821	1.10349	1.18843	1.27301	1.35720
46	0.76205	0.84818	0.93405	1.01963	1.10491	1.18984	1.27442	1.35860
47	0.76349	0.84961	0.93548	1.02106	1.10632	1.19125	1.27582	1.36000
48	0.76493	0.85105	0.93691	1.02248	1.10774	1.19267	1.27723	1.36140
49	0.76637	0.85248	0.93834	1.02390	1.10916	1.19408	1.27863	1.36280
50	0.76780	0.85391	0.93976	1.02533	1.11058	1.19549	1.28004	1.36420
51	0.76924	0.85535	0.94119	1.02675	1.11200	1.19690	1.28145	1.36560
52	0.77068	0.85678	0.94262	1.02817	1.11342	1.19832	1.28285	1.36700
53	0.77211	0.85821	0.94405	1.02960	1.11483	1.19973	1.28426	1.36840
54	0.77355	0.85965	0.94548	1.03102	1.11625	1.20114	1.28566	1.36980
55	0.77499	0.86108	0.94691	1.03244	1.11767	1.20255	1.28707	1.37119
56	0.77643	0.86251	0.94833	1.03387	1.11909	1.20396	1.28847	1.37259
57	0.77786	0.86394	0.94976	1.03529	1.12050	1.20538	1.28988	1.37399
58	0.77930	0.86538	0.95119	1.03671	1.12192	1.20679	1.29129	1.37539
59	0.78074	0.86681	0.95262	1.03814	1.12334	1.20820	1.29269	1.37679
60	0.78217	0.86824	0.95404	1.03956	1.12476	1.20961	1.29410	1.37819

Constants for Setting a 5-inch Sine-Bar for 16° to 23°

Min.	16°	17°	18°	19°	20°	21°	22°	23°
0	1.37819	1.46186	1.54509	1.62784	1.71010	1.79184	1.87303	1.95366
1	1.37958	1.46325	1.54647	1.62922	1.71147	1.79320	1.87438	1.95499
2	1.38098	1.46464	1.54785	1.63059	1.71283	1.79456	1.87573	1.95633
3	1.38238	1.46603	1.54923	1.63197	1.71420	1.79591	1.87708	1.95767
4	1.38378	1.46742	1.55062	1.63334	1.71557	1.79727	1.87843	1.95901
5	1.38518	1.46881	1.55200	1.63472	1.71693	1.79863	1.87977	1.96035
6	1.38657	1.47020	1.55338	1.63609	1.71830	1.79998	1.88112	1.96169
7	1.38797	1.47159	1.55476	1.63746	1.71966	1.80134	1.88247	1.96302
8	1.38937	1.47298	1.55615	1.63884	1.72103	1.80270	1.88382	1.96436
9	1.39076	1.47437	1.55753	1.64021	1.72240	1.80405	1.88516	1.96570
10	1.39216	1.47576	1.55891	1.64159	1.72376	1.80541	1.88651	1.96704
11	1.39356	1.47715	1.56029	1.64296	1.72513	1.80677	1.88786	1.96837
12	1.39496	1.47854	1.56167	1.64433	1.72649	1.80812	1.88920	1.96971
13	1.39635	1.47993	1.56306	1.64571	1.72786	1.80948	1.89055	1.97105
14	1.39775	1.48132	1.56444	1.64708	1.72922	1.81083	1.89190	1.97238
15	1.39915	1.48271	1.56582	1.64845	1.73059	1.81219	1.89324	1.97372
16	1.40054	1.48410	1.56720	1.64983	1.73195	1.81355	1.89459	1.97506
17	1.40194	1.48549	1.56858	1.65120	1.73331	1.81490	1.89594	1.97639
18	1.40333	1.48687	1.56996	1.65257	1.73468	1.81626	1.89728	1.97773
19	1.40473	1.48826	1.57134	1.65394	1.73604	1.81761	1.89863	1.97906
20	1.40613	1.48965	1.57272	1.65532	1.73741	1.81897	1.89997	1.98040
21	1.40752	1.49104	1.57410	1.65669	1.73877	1.82032	1.90132	1.98173
22	1.40892	1.49243	1.57548	1.65806	1.74013	1.82168	1.90266	1.98307
23	1.41031	1.49382	1.57687	1.65943	1.74150	1.82303	1.90401	1.98440
24	1.41171	1.49520	1.57825	1.66081	1.74286	1.82438	1.90535	1.98574
25	1.41310	1.49659	1.57963	1.66218	1.74422	1.82574	1.90670	1.98707
26	1.41450	1.49798	1.58101	1.66355	1.74559	1.82709	1.90804	1.98841
27	1.41589	1.49937	1.58238	1.66492	1.74695	1.82845	1.90939	1.98974
28	1.41729	1.50075	1.58376	1.66629	1.74831	1.82980	1.91073	1.99108
29	1.41868	1.50214	1.58514	1.66766	1.74967	1.83115	1.91207	1.99241
30	1.42008	1.50353	1.58652	1.66903	1.75104	1.83251	1.91342	1.99375
31	1.42147	1.50492	1.58790	1.67041	1.75240	1.83386	1.91476	1.99508
32	1.42287	1.50630	1.58928	1.67178	1.75376	1.83521	1.91610	1.99641
33	1.42426	1.50769	1.59066	1.67315	1.75512	1.83657	1.91745	1.99775
34	1.42565	1.50908	1.59204	1.67452	1.75649	1.83792	1.91879	1.99908
35	1.42705	1.51046	1.59342	1.67589	1.75785	1.83927	1.92013	2.00041
36	1.42844	1.51185	1.59480	1.67726	1.75921	1.84062	1.92148	2.00175
37	1.42984	1.51324	1.59617	1.67863	1.76057	1.84198	1.92282	2.00308
38	1.43123	1.51462	1.59755	1.68000	1.76193	1.84333	1.92416	2.00441
39	1.43262	1.51601	1.59893	1.68137	1.76329	1.84468	1.92550	2.00574
40	1.43402	1.51739	1.60031	1.68274	1.76465	1.84603	1.92685	2.00708
41	1.43541	1.51878	1.60169	1.68411	1.76601	1.84738	1.92819	2.00841
42	1.43680	1.52017	1.60307	1.68548	1.76737	1.84873	1.92953	2.00974
43	1.43820	1.52155	1.60444	1.68685	1.76873	1.85009	1.93087	2.01107
44	1.43959	1.52294	1.60582	1.68821	1.77010	1.85144	1.93221	2.01240
45	1.44098	1.52432	1.60720	1.68958	1.77146	1.85279	1.93355	2.01373
46	1.44237	1.52571	1.60857	1.69095	1.77282	1.85414	1.93490	2.01506
47	1.44377	1.52709	1.60995	1.69232	1.77418	1.85549	1.93624	2.01640
48	1.44516	1.52848	1.61133	1.69369	1.77553	1.85684	1.93758	2.01773
49	1.44655	1.52986	1.61271	1.69506	1.77689	1.85819	1.93892	2.01906
50	1.44794	1.53125	1.61408	1.69643	1.77825	1.85954	1.94026	2.02039
51	1.44934	1.53263	1.61546	1.69779	1.77961	1.86089	1.94160	2.02172
52	1.45073	1.53401	1.61683	1.69916	1.78097	1.86224	1.94294	2.02305
53	1.45212	1.53540	1.61821	1.70053	1.78233	1.86359	1.94428	2.02438
54	1.45351	1.53678	1.61959	1.70190	1.78369	1.86494	1.94562	2.02571
55	1.45490	1.53817	1.62096	1.70327	1.78505	1.86629	1.94696	2.02704
56	1.45629	1.53955	1.62234	1.70463	1.78641	1.86764	1.94830	2.02837
57	1.45769	1.54093	1.62371	1.70600	1.78777	1.86899	1.94964	2.02970
58	1.45908	1.54232	1.62509	1.70737	1.78912	1.87034	1.95098	2.03103
59	1.46047	1.54370	1.62647	1.70873	1.79048	1.87168	1.95232	2.03235
60	1.46186	1.54509	1.62784	1.71010	1.79184	1.87303	1.95366	2.03368

Constants for Setting a 5-inch Sine-Bar for 24° to 31°

Min.	24°	25°	26°	27°	28°	29°	30°	31°
0	2.03368	2.11309	2.19186	2.26995	2.34736	2.42405	2.50000	2.57519
1	2.03501	2.11441	2.19316	2.27125	2.34864	2.42532	2.50126	2.57644
2	2.03634	2.11573	2.19447	2.27254	2.34993	2.42659	2.50252	2.57768
3	2.03767	2.11704	2.19578	2.27384	2.35121	2.42786	2.50378	2.57893
4	2.03900	2.11836	2.19708	2.27513	2.35249	2.42913	2.50504	2.58018
5	2.04032	2.11968	2.19839	2.27643	2.35378	2.43041	2.50630	2.58142
6	2.04165	2.12100	2.19970	2.27772	2.35506	2.43168	2.50755	2.58267
7	2.04298	2.12231	2.20100	2.27902	2.35634	2.43295	2.50881	2.58391
8	2.04431	2.12363	2.20231	2.28031	2.35763	2.43422	2.51007	2.58516
9	2.04563	2.12495	2.20361	2.28161	2.35891	2.43549	2.51133	2.58640
10	2.04696	2.12626	2.20492	2.28290	2.36019	2.43676	2.51259	2.58765
11	2.04829	2.12758	2.20622	2.28420	2.36147	2.43803	2.51384	2.58889
12	2.04962	2.12890	2.20753	2.28549	2.36275	2.43930	2.51510	2.59014
13	2.05094	2.13021	2.20883	2.28678	2.36404	2.44057	2.51636	2.59138
14	2.05227	2.13153	2.21014	2.28808	2.36532	2.44184	2.51761	2.59262
15	2.05359	2.13284	2.21144	2.28937	2.36660	2.44311	2.51887	2.59387
16	2.05492	2.13416	2.21275	2.29066	2.36788	2.44438	2.52013	2.59511
17	2.05625	2.13547	2.21405	2.29196	2.36916	2.44564	2.52138	2.59635
18	2.05757	2.13679	2.21536	2.29325	2.37044	2.44691	2.52264	2.59760
19	2.05890	2.13810	2.21666	2.29454	2.37172	2.44818	2.52389	2.59884
20	2.06022	2.13942	2.21796	2.29583	2.37300	2.44945	2.52515	2.60008
21	2.06155	2.14073	2.21927	2.29712	2.37428	2.45072	2.52640	2.60132
22	2.06287	2.14205	2.22057	2.29842	2.37556	2.45198	2.52766	2.60256
23	2.06420	2.14336	2.22187	2.29971	2.37684	2.45325	2.52891	2.60381
24	2.06552	2.14468	2.22318	2.30100	2.37812	2.45452	2.53017	2.60505
25	2.06685	2.14599	2.22448	2.30229	2.37940	2.45579	2.53142	2.60629
26	2.06817	2.14730	2.22578	2.30358	2.38068	2.45705	2.53268	2.60753
27	2.06950	2.14862	2.22708	2.30487	2.38196	2.45832	2.53393	2.60877
28	2.07082	2.14993	2.22839	2.30616	2.38324	2.45959	2.53519	2.61001
29	2.07214	2.15124	2.22969	2.30745	2.38452	2.46085	2.53644	2.61125
30	2.07347	2.15256	2.23099	2.30874	2.38579	2.46212	2.53769	2.61249
31	2.07479	2.15387	2.23229	2.31003	2.38707	2.46338	2.53894	2.61373
32	2.07611	2.15518	2.23359	2.31132	2.38835	2.46465	2.54020	2.61497
33	2.07744	2.15649	2.23489	2.31261	2.38963	2.46591	2.54145	2.61621
34	2.07876	2.15781	2.23619	2.31390	2.39091	2.46718	2.54270	2.61745
35	2.08008	2.15912	2.23749	2.31519	2.39218	2.46844	2.54396	2.61869
36	2.08140	2.16043	2.23880	2.31648	2.39346	2.46971	2.54521	2.61993
37	2.08273	2.16174	2.24010	2.31777	2.39474	2.47097	2.54646	2.62117
38	2.08405	2.16305	2.24140	2.31906	2.39601	2.47224	2.54771	2.62241
39	2.08537	2.16436	2.24270	2.32035	2.39729	2.47350	2.54896	2.62364
40	2.08669	2.16567	2.24400	2.32163	2.39857	2.47477	2.55021	2.62488
41	2.08801	2.16698	2.24530	2.32292	2.39984	2.47603	2.55146	2.62612
42	2.08934	2.16830	2.24660	2.32421	2.40112	2.47729	2.55271	2.62736
43	2.09066	2.16961	2.24789	2.32550	2.40239	2.47856	2.55397	2.62860
44	2.09198	2.17092	2.24919	2.32679	2.40367	2.47982	2.55522	2.62983
45	2.09330	2.17223	2.25049	2.32807	2.40494	2.48108	2.55647	2.63107
46	2.09462	2.17354	2.25179	2.32936	2.40622	2.48235	2.55772	2.63231
47	2.09594	2.17485	2.25309	2.33065	2.40749	2.48361	2.55896	2.63354
48	2.09726	2.17616	2.25439	2.33193	2.40877	2.48487	2.56021	2.63478
49	2.09858	2.17746	2.25569	2.33322	2.41004	2.48613	2.56146	2.63602
50	2.09990	2.17877	2.25698	2.33451	2.41132	2.48739	2.56271	2.63725
51	2.10122	2.18008	2.25828	2.33579	2.41259	2.48866	2.56396	2.63849
52	2.10254	2.18139	2.25958	2.33708	2.41386	2.48992	2.56521	2.63972
53	2.10386	2.18270	2.26088	2.33836	2.41514	2.49118	2.56646	2.64096
54	2.10518	2.18401	2.26217	2.33965	2.41641	2.49244	2.56771	2.64219
55	2.10650	2.18532	2.26347	2.34093	2.41769	2.49370	2.56895	2.64343
56	2.10782	2.18663	2.26477	2.34222	2.41896	2.49496	2.57020	2.64466
57	2.10914	2.18793	2.26606	2.34350	2.42023	2.49622	2.57145	2.64590
58	2.11045	2.18924	2.26736	2.34479	2.42150	2.49748	2.57270	2.64713
59	2.11177	2.19055	2.26866	2.34607	2.42278	2.49874	2.57394	2.64836
60	2.11309	2.19186	2.26995	2.34736	2.42405	2.50000	2.57519	2.64960

Constants for Setting a 5-inch Sine-Bar for 32° to 39°

Min.	32°	33°	34°	35°	36°	37°	38°	39°
0	2.64960	2.72320	2.79596	2.86788	2.93893	3.00908	3.07831	3.14660
1	2.65083	2.72441	2.79717	2.86907	2.94010	3.01024	3.07945	3.14773
2	2.65206	2.72563	2.79838	2.87026	2.94128	3.01140	3.08060	3.14886
3	2.65330	2.72685	2.79958	2.87146	2.94246	3.01256	3.08174	3.14999
4	2.65453	2.72807	2.80079	2.87265	2.94363	3.01372	3.08289	3.15112
5	2.65576	2.72929	2.80199	2.87384	2.94481	3.01488	3.08403	3.15225
6	2.65699	2.73051	2.80319	2.87503	2.94598	3.01604	3.08518	3.15338
7	2.65822	2.73173	2.80440	2.87622	2.94716	3.01720	3.08632	3.15451
8	2.65946	2.73295	2.80560	2.87741	2.94833	3.01836	3.08747	3.15564
9	2.66069	2.73416	2.80681	2.87860	2.94951	3.01952	3.08861	3.15676
10	2.66192	2.73538	2.80801	2.87978	2.95068	3.02068	3.08976	3.15789
11	2.66315	2.73660	2.80921	2.88097	2.95185	3.02184	3.09090	3.15902
12	2.66438	2.73782	2.81042	2.88216	2.95303	3.02300	3.09204	3.16015
13	2.66561	2.73903	2.81162	2.88335	2.95420	3.02415	3.09318	3.16127
14	2.66684	2.74025	2.81282	2.88454	2.95538	3.02531	3.09433	3.16240
15	2.66807	2.74147	2.81402	2.88573	2.95655	3.02647	3.09547	3.16353
16	2.66930	2.74268	2.81523	2.88691	2.95772	3.02763	3.09661	3.16465
17	2.67053	2.74390	2.81643	2.88810	2.95889	3.02878	3.09775	3.16578
18	2.67176	2.74511	2.81763	2.88929	2.96007	3.02994	3.09890	3.16690
19	2.67299	2.74633	2.81883	2.89048	2.96124	3.03110	3.10004	3.16803
20	2.67422	2.74754	2.82003	2.89166	2.96241	3.03226	3.10118	3.16915
21	2.67545	2.74876	2.82123	2.89285	2.96358	3.03341	3.10232	3.17028
22	2.67668	2.74997	2.82243	2.89403	2.96475	3.03457	3.10346	3.17140
23	2.67791	2.75119	2.82364	2.89522	2.96592	3.03572	3.10460	3.17253
24	2.67913	2.75240	2.82484	2.89641	2.96709	3.03688	3.10574	3.17365
25	2.68036	2.75362	2.82604	2.89759	2.96827	3.03803	3.10688	3.17478
26	2.68159	2.75483	2.82723	2.89878	2.96944	3.03919	3.10802	3.17590
27	2.68282	2.75605	2.82843	2.89996	2.97061	3.04034	3.10916	3.17702
28	2.68404	2.75726	2.82963	2.90115	2.97178	3.04150	3.11030	3.17815
29	2.68527	2.75847	2.83083	2.90233	2.97294	3.04265	3.11143	3.17927
30	2.68650	2.75969	2.83203	2.90351	2.97411	3.04381	3.11257	3.18039
31	2.68772	2.76090	2.83323	2.90470	2.97528	3.04496	3.11371	3.18151
32	2.68895	2.76211	2.83443	2.90588	2.97645	3.04611	3.11485	3.18264
33	2.69018	2.76332	2.83563	2.90707	2.97762	3.04727	3.11599	3.18376
34	2.69140	2.76453	2.83682	2.90825	2.97879	3.04842	3.11712	3.18488
35	2.69263	2.76575	2.83802	2.90943	2.97996	3.04957	3.11826	3.18600
36	2.69385	2.76696	2.83922	2.91061	2.98112	3.05073	3.11940	3.18712
37	2.69508	2.76817	2.84042	2.91180	2.98229	3.05188	3.12053	3.18824
38	2.69630	2.76938	2.84161	2.91298	2.98346	3.05303	3.12167	3.18936
39	2.69753	2.77059	2.84281	2.91416	2.98463	3.05418	3.12281	3.19048
40	2.69875	2.77180	2.84401	2.91534	2.98579	3.05533	3.12394	3.19160
41	2.69998	2.77301	2.84520	2.91652	2.98696	3.05648	3.12508	3.19272
42	2.70120	2.77422	2.84640	2.91771	2.98813	3.05764	3.12621	3.19384
43	2.70243	2.77543	2.84759	2.91889	2.98929	3.05879	3.12735	3.19496
44	2.70365	2.77664	2.84879	2.92007	2.99046	3.05994	3.12848	3.19608
45	2.70487	2.77785	2.84998	2.92125	2.99162	3.06109	3.12962	3.19720
46	2.70610	2.77906	2.85118	2.92243	2.99279	3.06224	3.13075	3.19831
47	2.70732	2.78027	2.85237	2.92361	2.99395	3.06339	3.13189	3.19943
48	2.70854	2.78148	2.85357	2.92479	2.99512	3.06454	3.13302	3.20055
49	2.70976	2.78269	2.85476	2.92597	2.99628	3.06568	3.13415	3.20167
50	2.71099	2.78389	2.85596	2.92715	2.99745	3.06683	3.13529	3.20278
51	2.71221	2.78510	2.85715	2.92833	2.99861	3.06798	3.13642	3.20390
52	2.71343	2.78631	2.85834	2.92950	2.99977	3.06913	3.13755	3.20502
53	2.71465	2.78752	2.85954	2.93068	3.00094	3.07028	3.13868	3.20613
54	2.71587	2.78873	2.86073	2.93186	3.00210	3.07143	3.13982	3.20725
55	2.71709	2.78993	2.86192	2.93304	3.00326	3.07257	3.14095	3.20836
56	2.71831	2.79114	2.86311	2.93422	3.00443	3.07372	3.14208	3.20948
57	2.71953	2.79235	2.86431	2.93540	3.00559	3.07487	3.14321	3.21059
58	2.72076	2.79355	2.86550	2.93657	3.00675	3.07601	3.14434	3.21171
59	2.72198	2.79476	2.86669	2.93775	3.00791	3.07716	3.14547	3.21282
60	2.72320	2.79596	2.86788	2.93893	3.00908	3.07831	3.14660	3.21394

Constants for Setting a 5-inch Sine-Bar for 40° to 47°

Min.	40°	41°	42°	43°	44°	45°	46°	47°
0	3.21394	3.28030	3.34565	3.40999	3.47329	3.53553	3.59670	3.65677
1	3.21505	3.28139	3.34673	3.41106	3.47434	3.53656	3.59771	3.65776
2	3.21617	3.28249	3.34781	3.41212	3.47538	3.53759	3.59872	3.65875
3	3.21728	3.28359	3.34889	3.41318	3.47643	3.53862	3.59973	3.65974
4	3.21839	3.28468	3.34997	3.41424	3.47747	3.53965	3.60074	3.66073
5	3.21951	3.28578	3.35105	3.41531	3.47852	3.54067	3.60175	3.66172
6	3.22062	3.28688	3.35213	3.41637	3.47956	3.54170	3.60276	3.66271
7	3.22173	3.28797	3.35321	3.41743	3.48061	3.54273	3.60376	3.66370
8	3.22284	3.28907	3.35429	3.41849	3.48165	3.54375	3.60477	3.66469
9	3.22395	3.29016	3.35537	3.41955	3.48270	3.54478	3.60578	3.66568
10	3.22507	3.29126	3.35645	3.42061	3.48374	3.54580	3.60679	3.66667
11	3.22618	3.29235	3.35753	3.42168	3.48478	3.54683	3.60779	3.66766
12	3.22729	3.29345	3.35860	3.42274	3.48583	3.54785	3.60880	3.66865
13	3.22840	3.29454	3.35968	3.42380	3.48687	3.54888	3.60981	3.66964
14	3.22951	3.29564	3.36076	3.42486	3.48791	3.54990	3.61081	3.67063
15	3.23062	3.29673	3.36183	3.42592	3.48895	3.55093	3.61182	3.67161
16	3.23173	3.29782	3.36291	3.42697	3.48999	3.55195	3.61283	3.67260
17	3.23284	3.29892	3.36399	3.42803	3.49104	3.55297	3.61383	3.67359
18	3.23395	3.30001	3.36506	3.42909	3.49208	3.55400	3.61484	3.67457
19	3.23506	3.30110	3.36614	3.43015	3.49312	3.55502	3.61584	3.67556
20	3.23617	3.30219	3.36721	3.43121	3.49416	3.55604	3.61684	3.67655
21	3.23728	3.30329	3.36829	3.43227	3.49520	3.55707	3.61785	3.67753
22	3.23838	3.30438	3.36936	3.43332	3.49624	3.55809	3.61885	3.67852
23	3.23949	3.30547	3.37044	3.43438	3.49728	3.55911	3.61986	3.67950
24	3.24060	3.30656	3.37151	3.43544	3.49832	3.56013	3.62086	3.68049
25	3.24171	3.30765	3.37259	3.43649	3.49936	3.56115	3.62186	3.68147
26	3.24281	3.30874	3.37366	3.43755	3.50039	3.56217	3.62286	3.68245
27	3.24392	3.30983	3.37473	3.43861	3.50143	3.56319	3.62387	3.68344
28	3.24503	3.31092	3.37581	3.43966	3.50247	3.56421	3.62487	3.68442
29	3.24613	3.31201	3.37688	3.44072	3.50351	3.56523	3.62587	3.68540
30	3.24724	3.31310	3.37795	3.44177	3.50455	3.56625	3.62687	3.68639
31	3.24835	3.31419	3.37902	3.44283	3.50558	3.56727	3.62787	3.68737
32	3.24945	3.31528	3.38010	3.44388	3.50662	3.56829	3.62887	3.68835
33	3.25056	3.31637	3.38117	3.44494	3.50766	3.56931	3.62987	3.68933
34	3.25166	3.31746	3.38224	3.44599	3.50869	3.57033	3.63087	3.69031
35	3.25277	3.31854	3.38331	3.44704	3.50973	3.57135	3.63187	3.69130
36	3.25387	3.31963	3.38438	3.44810	3.51077	3.57236	3.63287	3.69228
37	3.25498	3.32072	3.38545	3.44915	3.51180	3.57338	3.63387	3.69326
38	3.25608	3.32181	3.38652	3.45020	3.51284	3.57440	3.63487	3.69424
39	3.25718	3.32289	3.38759	3.45126	3.51387	3.57542	3.63587	3.69522
40	3.25829	3.32398	3.38866	3.45231	3.51491	3.57643	3.63687	3.69620
41	3.25939	3.32507	3.38973	3.45336	3.51594	3.57745	3.63787	3.69718
42	3.26049	3.32615	3.39080	3.45441	3.51697	3.57846	3.63886	3.69816
43	3.26159	3.32724	3.39187	3.45546	3.51801	3.57948	3.63986	3.69913
44	3.26270	3.32832	3.39294	3.45651	3.51904	3.58049	3.64086	3.70011
45	3.26380	3.32941	3.39400	3.45757	3.52007	3.58151	3.64186	3.70109
46	3.26490	3.33049	3.39507	3.45862	3.52111	3.58252	3.64285	3.70207
47	3.26600	3.33158	3.39614	3.45967	3.52214	3.58354	3.64385	3.70305
48	3.26710	3.33266	3.39721	3.46072	3.52317	3.58455	3.64484	3.70402
49	3.26820	3.33375	3.39827	3.46177	3.52420	3.58557	3.64584	3.70500
50	3.26930	3.33483	3.39934	3.46281	3.52523	3.58658	3.64683	3.70598
51	3.27040	3.33591	3.40041	3.46386	3.52627	3.58759	3.64783	3.70695
52	3.27150	3.33700	3.40147	3.46491	3.52730	3.58861	3.64882	3.70793
53	3.27260	3.33808	3.40254	3.46596	3.52833	3.58962	3.64982	3.70890
54	3.27370	3.33916	3.40360	3.46701	3.52936	3.59063	3.65081	3.70988
55	3.27480	3.34025	3.40467	3.46806	3.53039	3.59164	3.65181	3.71085
56	3.27590	3.34133	3.40573	3.46910	3.53142	3.59266	3.65280	3.71183
57	3.27700	3.34241	3.40680	3.47015	3.53245	3.59367	3.65379	3.71280
58	3.27810	3.34349	3.40786	3.47120	3.53348	3.59468	3.65478	3.71378
59	3.27920	3.34457	3.40893	3.47225	3.53451	3.59569	3.65578	3.71475
60	3.28030	3.34565	3.40999	3.47329	3.53553	3.59670	3.65677	3.71572

Constants for Setting a 5-inch Sine-Bar for 48° to 55°

Min.	48°	49°	50°	51°	52°	53°	54°	55°
0	3.71572	3.77355	3.83022	3.88573	3.94005	3.99318	4.04508	4.09576
1	3.71670	3.77450	3.83116	3.88665	3.94095	3.99405	4.04594	4.09659
2	3.71767	3.77546	3.83209	3.88756	3.94184	3.99493	4.04679	4.09743
3	3.71864	3.77641	3.83303	3.88847	3.94274	3.99580	4.04765	4.09826
4	3.71961	3.77736	3.83396	3.88939	3.94363	3.99668	4.04850	4.09909
5	3.72059	3.77831	3.83489	3.89030	3.94453	3.99755	4.04936	4.09993
6	3.72156	3.77927	3.83583	3.89122	3.94542	3.99842	4.05021	4.10076
7	3.72253	3.78022	3.83676	3.89213	3.94631	3.99930	4.05106	4.10159
8	3.72350	3.78117	3.83769	3.89304	3.94721	4.00017	4.05191	4.10242
9	3.72447	3.78212	3.83862	3.89395	3.94810	4.00104	4.05277	4.10325
10	3.72544	3.78307	3.83956	3.89487	3.94899	4.00191	4.05362	4.10409
11	3.72641	3.78402	3.84049	3.89578	3.94988	4.00279	4.05447	4.10492
12	3.72738	3.78498	3.84142	3.89669	3.95078	4.00366	4.05532	4.10575
13	3.72835	3.78593	3.84235	3.89760	3.95167	4.00453	4.05617	4.10658
14	3.72932	3.78688	3.84328	3.89851	3.95256	4.00540	4.05702	4.10741
15	3.73029	3.78783	3.84421	3.89942	3.95345	4.00627	4.05787	4.10823
16	3.73126	3.78877	3.84514	3.90033	3.95434	4.00714	4.05872	4.10906
17	3.73222	3.78972	3.84607	3.90124	3.95523	4.00801	4.05957	4.10989
18	3.73319	3.79067	3.84700	3.90215	3.95612	4.00888	4.06042	4.11072
19	3.73416	3.79162	3.84793	3.90306	3.95701	4.00975	4.06127	4.11155
20	3.73513	3.79257	3.84886	3.90397	3.95790	4.01062	4.06211	4.11238
21	3.73609	3.79352	3.84978	3.90488	3.95878	4.01148	4.06296	4.11320
22	3.73706	3.79446	3.85071	3.90579	3.95967	4.01235	4.06381	4.11403
23	3.73802	3.79541	3.85164	3.90669	3.96056	4.01322	4.06466	4.11486
24	3.73899	3.79636	3.85257	3.90760	3.96145	4.01409	4.06550	4.11568
25	3.73996	3.79730	3.85349	3.90851	3.96234	4.01495	4.06635	4.11651
26	3.74092	3.79825	3.85442	3.90942	3.96322	4.01582	4.06720	4.11733
27	3.74189	3.79919	3.85535	3.91032	3.96411	4.01669	4.06804	4.11816
28	3.74285	3.80014	3.85627	3.91123	3.96500	4.01755	4.06889	4.11898
29	3.74381	3.80109	3.85720	3.91214	3.96588	4.01842	4.06973	4.11981
30	3.74478	3.80203	3.85812	3.91304	3.96677	4.01928	4.07058	4.12063
31	3.74574	3.80297	3.85905	3.91395	3.96765	4.02015	4.07142	4.12145
32	3.74671	3.80392	3.85997	3.91485	3.96854	4.02101	4.07227	4.12228
33	3.74767	3.80486	3.86090	3.91576	3.96942	4.02188	4.07311	4.12310
34	3.74863	3.80581	3.86182	3.91666	3.97031	4.02274	4.07395	4.12392
35	3.74959	3.80675	3.86274	3.91756	3.97119	4.02361	4.07480	4.12475
36	3.75056	3.80769	3.86367	3.91847	3.97207	4.02447	4.07564	4.12557
37	3.75152	3.80863	3.86459	3.91937	3.97296	4.02533	4.07648	4.12639
38	3.75248	3.80958	3.86551	3.92027	3.97384	4.02619	4.07732	4.12721
39	3.75344	3.81052	3.86644	3.92118	3.97472	4.02706	4.07817	4.12803
40	3.75440	3.81146	3.86736	3.92208	3.97560	4.02792	4.07901	4.12885
41	3.75536	3.81240	3.86828	3.92298	3.97649	4.02878	4.07985	4.12967
42	3.75632	3.81334	3.86920	3.92388	3.97737	4.02964	4.08069	4.13049
43	3.75728	3.81428	3.87012	3.92478	3.97825	4.03050	4.08153	4.13131
44	3.75824	3.81522	3.87104	3.92568	3.97913	4.03136	4.08237	4.13213
45	3.75920	3.81616	3.87196	3.92658	3.98001	4.03222	4.08321	4.13295
46	3.76016	3.81710	3.87288	3.92748	3.98089	4.03308	4.08405	4.13377
47	3.76112	3.81804	3.87380	3.92839	3.98177	4.03394	4.08489	4.13459
48	3.76207	3.81898	3.87472	3.92928	3.98265	4.03480	4.08572	4.13540
49	3.76303	3.81992	3.87564	3.93018	3.98353	4.03566	4.08656	4.13622
50	3.76399	3.82086	3.87656	3.93108	3.98441	4.03652	4.08740	4.13704
51	3.76495	3.82179	3.87748	3.93198	3.98529	4.03738	4.08824	4.13785
52	3.76590	3.82273	3.87840	3.93288	3.98616	4.03823	4.08908	4.13867
53	3.76686	3.82367	3.87931	3.93378	3.98704	4.03909	4.08991	4.13949
54	3.76782	3.82461	3.88023	3.93468	3.98792	4.03995	4.09075	4.14030
55	3.76877	3.82554	3.88115	3.93557	3.98880	4.04081	4.09158	4.14112
56	3.76973	3.82648	3.88207	3.93647	3.98967	4.04166	4.09242	4.14193
57	3.77068	3.82742	3.88298	3.93737	3.99055	4.04252	4.09326	4.14275
58	3.77164	3.82835	3.88390	3.93826	3.99143	4.04337	4.09409	4.14356
59	3.77259	3.82929	3.88481	3.93916	3.99230	4.04423	4.09493	4.14437
60	3.77355	3.83022	3.88573	3.94005	3.99318	4.04508	4.09576	4.14519

Measuring Tapers with Vee-Block and Sine-Bar.—The taper on a conical part may be checked or found by placing the part in a vee-block which rests on the surface of a sine-plate or sine-bar as shown in the accompanying diagram. The advantage of this method is that the axis of the vee-block may be aligned with the sides of the sine-bar. Thus when the tapered part is placed in the vee-block it will be aligned perpendicular to the transverse axis of the sine-bar.

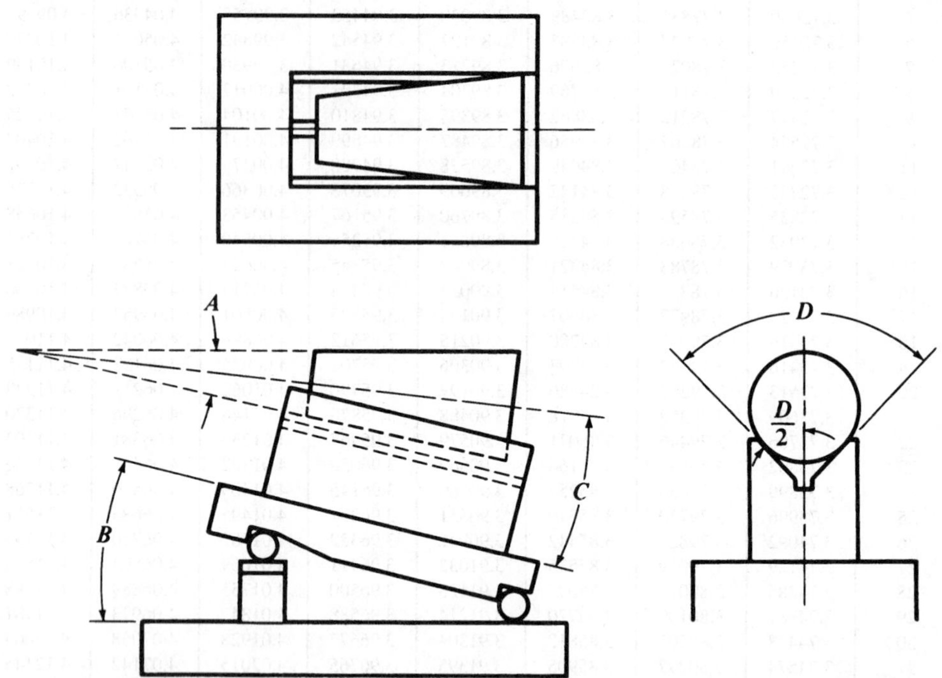

The sine-bar is set to angle $B = (C + A/2)$ where $A/2$ is one-half the included angle of the tapered part. If D is the included angle of the precision vee-block, the angle C is calculated from the formula:

$$\sin C = \frac{\sin(A/2}{\sin(D/2}$$

If dial indicator readings show no change across all points along the top of the taper surface, then this checks that the angle A of the taper is correct.

If the indicator readings vary, proceed as follows to find the actual angle of taper: 1) Adjust the angle of the sine-bar until the indicator reading is constant. Then find the new angleB' as explained in the paragraph *Measuring Angle of Templet or Gage with Sine Bar* on page 674; and 2) Using the angle B' calculate the actual half-angle $A'/2$ of the taper from the formula:.

$$\tan\frac{A'}{2} = \frac{\sin B'}{\csc\frac{D}{2} + \cos B'}$$

The taper per foot corresponding to certain half-angles of taper may be found in the table on page 684.

Measuring Dovetail Slides.—Dovetail slides that must be machined accurately to a given width are commonly gaged by using pieces of cylindrical rod or wire and measuring as indicated by the dimensions x and y of the accompanying illustrations.

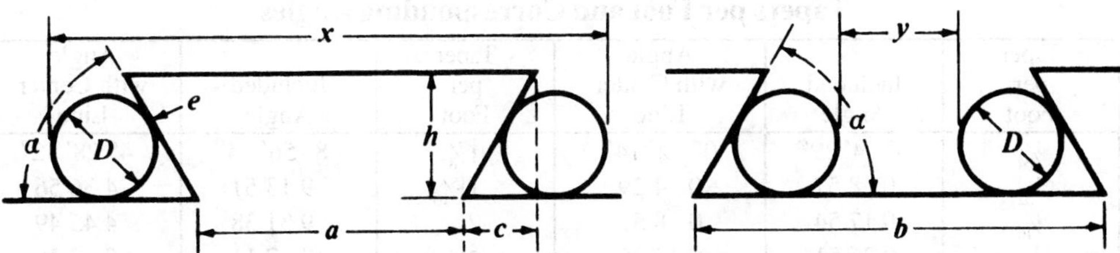

To obtain dimension x for measuring male dovetails, add I to the cotangent of one-half the dovetail angle α, multiply by diameter D of the rods used, and add the product to dimension α. To obtain dimension y for measuring a female dovetail, add 1 to the cotangent of one-half the dovetail angle α, multiply by diameter D of the rod used, and subtract the result from dimension b. Expressing these rules as formulas:

$$x = D(1 + \cot \tfrac{1}{2}\alpha) + a$$
$$y = b - D(1 + \cot \tfrac{1}{2}\alpha)$$
$$c = h \times \cot\alpha$$

The rod or wire used should be small enough so that the point of contact e is somewhat below the corner or edge of the dovetail.

Accurate Measurement of Angles and Tapers

When great accuracy is required in the measurement of angles, or when originating tapers, disks are commonly used. The principle of the disk method of taper measurement is that if two disks of unequal diameters are placed either in contact or a certain distance apart, lines tangent to their peripheries will represent an angle or taper, the degree of which depends upon the diameters of the two disks and the distance between them.

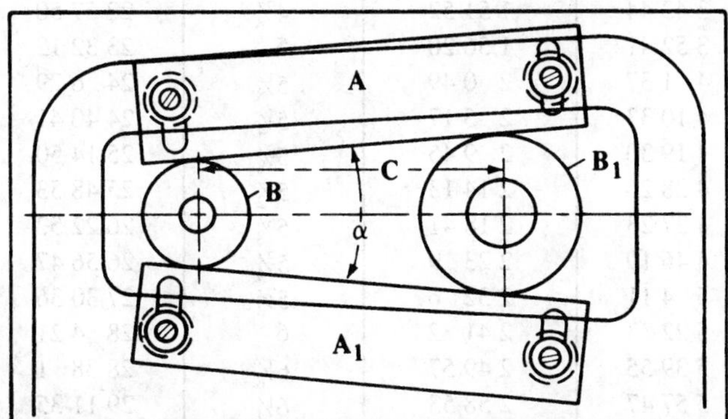

The gage shown in the accompanying illustration, which is a form commonly used for originating tapers or measuring angles accurately, is set by means of disks. This gage consists of two adjustable straight edges A and A_1, which are in contact with disks B and B_1. The angle α or the taper between the straight edges depends, of course, upon the diameters of the disks and the center distance C, and as these three dimensions can be measured accurately, it is possible to set the gage to a given angle within very close limits. Moreover, if a record of the three dimensions is kept, the exact setting of the gage can be reproduced quickly at any time. The following rules may be used for adjusting a gage of this type, and cover all problems likely to arise in practice. Disks are also occasionally used for the setting of parts in angular positions when they are to be machined accurately to a given angle: the rules are applicable to these conditions also.

Tapers per Foot and Corresponding Angles

Taper per Foot	Included Angle	Angle with Center Line	Taper per Foot	Included Angle	Angle with Center Line
1/64	0° 4′ 29″	0° 2′ 14″	1 7/8	8° 56′ 4″	4° 28′ 2″
1/32	0 8 57	0 4 29	1 15/16	9 13 51	4 36 56
1/16	0 17 54	0 8 57	2	9 31 38	4 45 49
3/32	0 26 51	0 13 26	2 1/8	10 7 11	5 3 36
1/8	0 35 49	0 17 54	2 1/4	10 42 42	5 21 21
5/32	0 44 46	0 22 23	2 3/8	11 18 11	5 39 5
3/16	0 53 43	0 26 51	2 1/2	11 53 37	5 56 49
7/32	1 2 40	0 31 20	2 5/8	12 29 2	6 14 31
1/4	1 11 37	0 35 49	2 3/4	13 4 24	6 32 12
9/32	1 20 34	0 40 17	2 7/8	13 39 43	6 49 52
5/16	1 29 31	0 44 46	3	14 15 0	7 7 30
11/32	1 38 28	0 49 14	3 1/8	14 50 14	7 25 7
3/8	1 47 25	0 53 43	3 1/4	15 25 26	7 42 43
13/32	1 56 22	0 58 11	3 3/8	16 0 34	8 0 17
7/16	2 5 19	1 2 40	3 1/2	16 35 39	8 17 50
15/32	2 14 16	1 7 8	3 5/8	17 10 42	8 35 21
1/2	2 23 13	1 11 37	3 3/4	17 45 41	8 52 50
17/32	2 32 10	1 16 5	3 7/8	18 20 36	9 10 18
9/16	2 41 7	1 20 33	4	18 55 29	9 27 44
19/32	2 50 4	1 25 2	4 1/8	19 30 17	9 45 9
5/8	2 59 1	1 29 30	4 1/4	20 5 3	10 2 31
21/32	3 7 57	1 33 59	4 3/8	20 39 44	10 19 52
11/16	3 16 54	1 38 27	4 1/2	21 14 22	10 37 11
23/32	3 25 51	1 42 55	4 5/8	21 48 55	10 54 28
3/4	3 34 47	1 47 24	4 3/4	22 23 25	11 11 42
25/32	3 43 44	1 51 52	4 7/8	22 57 50	11 28 55
13/16	3 52 41	1 56 20	5	23 32 12	11 46 6
27/32	4 1 37	2 0 49	5 1/8	24 6 29	12 3 14
7/8	4 10 33	2 5 17	5 1/4	24 40 41	12 20 21
29/32	4 19 30	2 9 45	5 3/8	25 14 50	12 37 25
15/16	4 28 26	2 14 13	5 1/2	25 48 53	12 54 27
31/32	4 37 23	2 18 41	5 5/8	26 22 52	13 11 26
1	4 46 19	2 23 9	5 3/4	26 56 47	13 28 23
1 1/16	5 4 11	2 32 6	5 7/8	27 30 36	13 45 18
1 1/8	5 22 3	2 41 2	6	28 4 21	14 2 10
1 3/16	5 39 55	2 49 57	6 1/8	28 38 1	14 19 0
1 1/4	5 57 47	2 58 53	6 1/4	29 11 35	14 35 48
1 5/16	6 15 38	3 7 49	6 3/8	29 45 5	14 52 32
1 3/8	6 33 29	3 16 44	6 1/2	30 18 29	15 9 15
1 7/16	6 51 19	3 25 40	6 5/8	30 51 48	15 25 54
1 1/2	7 9 10	3 34 35	6 3/4	31 25 2	15 42 31
1 9/16	7 27 0	3 43 30	6 7/8	31 58 11	15 59 5
1 5/8	7 44 49	3 52 25	7	32 31 13	16 15 37
1 11/16	8 2 38	4 1 19	7 1/8	33 4 11	16 32 5
1 3/4	8 20 27	4 10 14	7 1/4	33 37 3	16 48 31
1 13/16	8 38 16	4 19 8	7 3/8	34 9 49	17 4 54

For conversions into decimal degrees and radians see *Conversion Tables of Angular Measure* on page 90.

Rules for Figuring Tapers

Given	To Find	Rule
The taper per foot.	The taper per inch.	Divide the taper per foot by 12.
The taper per inch.	The taper per foot.	Multiply the taper per inch by 12.
End diameters and length of taper in inches.	The taper per foot.	Subtract small diameter from large; divide by length of taper; and multiply quotient by 12.
Large diameter and length of taper in inches, and taper per foot.	Diameter at small end in inches	Divide taper per foot by 12; multiply by length of taper; and subtract result from large diameter.
Small diameter and length of taper in inches, and taper per foot.	Diameter at large end in inches.	Divide taper per foot by 12; multiply by length of taper; and add result to small diameter.
The taper per foot and two diameters in inches.	Distance between two given diameters in inches.	Subtract small diameter from large; divide remainder by taper per foot; and multiply quotient by 12.
The taper per foot.	Amount of taper in a certain length in inches.	Divide taper per foot by 12; multiply by given length of tapered part.

To find angle α for given taper *T* in inches per foot.—

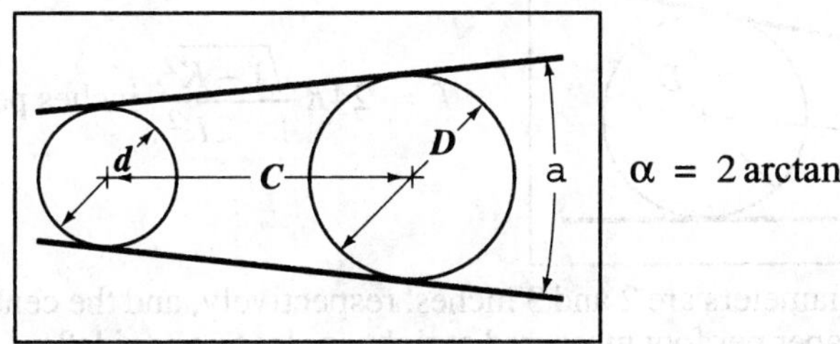

$\alpha = 2\arctan(T/24)$.

Example: What angle α is equivalent to a taper of 1.5 inches per foot?

$$\alpha = 2 \times \arctan(1.5/24) = 7.153°$$

To find taper per foot *T* given angle α in degrees.—

$$T = 24 \tan(\alpha/2) \text{ inches per foot}$$

Example: What taper *T* is equivalent to an angle of 7.153°?

$$T = 24\tan(7.153/2) = 1.5 \text{ inches per foot}$$

To find angle α given dimensions *D*, *d*, and *C*.— Let *K* be the difference in the disk diameters divided by twice the center distance. $K = (D - d)/(2C)$, then $\alpha = 2\arcsin K$

Example: If the disk diameters *d* and *D* are 1 and 1.5 inches, respectively, and the center distance *C* is 5 inches, find the included angle α.

$$K = (1.5 - 1)/(2 \times 5) = 0.05 \qquad \alpha = 2 \times \arcsin 0.05 = 5.732°$$

To find taper *T* measured at right angles to a line through the disk centers given dimensions *D*, *d*, and distance *C*.— Find *K* using the formula in the previous example, then $T = 24K/\sqrt{1 - K^2}$ inches per foot

Example: If disk diameters *d* and *D* are 1 and 1.5 inches, respectively, and the center distance *C* is 5 inches, find the taper per foot.

$$K = (1.5 - 1)/(2 \times 5) = 0.05 \qquad T = \frac{24 \times 0.05}{\sqrt{1 - (0.05)^2}} = 1.2015 \text{ inches per foot}$$

To find center distance C for a given taper T in inches per foot.—

$$C = \frac{D-d}{2} \times \frac{\sqrt{1+(T/24)^2}}{T/24} \text{ inches}$$

Example: Gage is to be set to ¾ inch per foot, and disk diameters are 1.25 and 1.5 inches, respectively. Find the required center distance for the disks.

$$C = \frac{1.5-1.25}{2} \times \frac{\sqrt{1+(0.75/24)^2}}{0.75/24} = 4.002 \text{ inches}$$

To find center distance C for a given angle α and dimensions D and d.—

$$C = (D-d)/2\sin(\alpha/2) \text{ inches}$$

Example: If an angle α of 20° is required, and the disks are 1 and 3 inches in diameter, respectively, find the required center distance C.

$$C = (3-1)/(2 \times \sin 10°) = 5.759 \text{ inches}$$

To find taper T measured at right angles to one side .—When one side is taken as a base line and the taper is measured at right angles to that side, calculate K as explained above and use the following formula for determining the taper T:

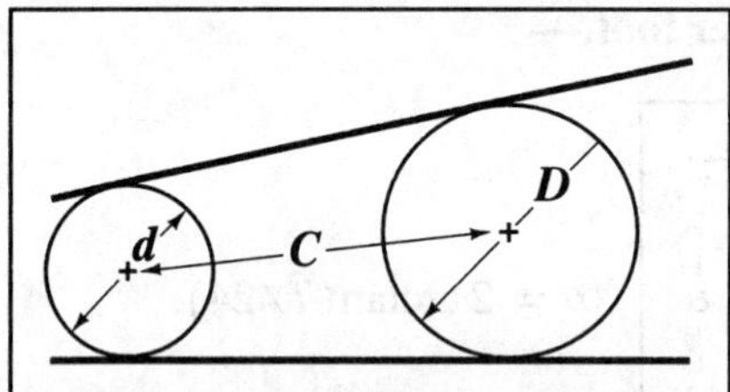

$$T = 24K\frac{\sqrt{1-K^2}}{1-2K^2} \text{ inches per foot}$$

Example: If the disk diameters are 2 and 3 inches, respectively, and the center I distance is 5 inches, what is the taper per foot measured at right angles to one side?

$$K = \frac{3-2}{2 \times 5} = 0.1 \qquad T = 24 \times 0.1 \times \frac{\sqrt{1-(0.1)^2}}{1-[2 \times (0.1)^2]} = 2.4367 \text{ in. per ft.}$$

To find center distance C when taper T is measured from one side.—

$$C = \frac{D-d}{\sqrt{2-2/\sqrt{1+(T/12)^2}}} \text{ inches}$$

Example: If the taper measured at right angles to one side is 6.9 inches per foot, and the disks are 2 and 5 inches in diameter, respectively, what is center distance C?

$$C = \frac{5-2}{\sqrt{2-2/\sqrt{1+(6.9/12)^2}}} = 5.815 \text{ inches.}$$

To find diameter D of a large disk in contact with a small disk of diameter d given angle α.—

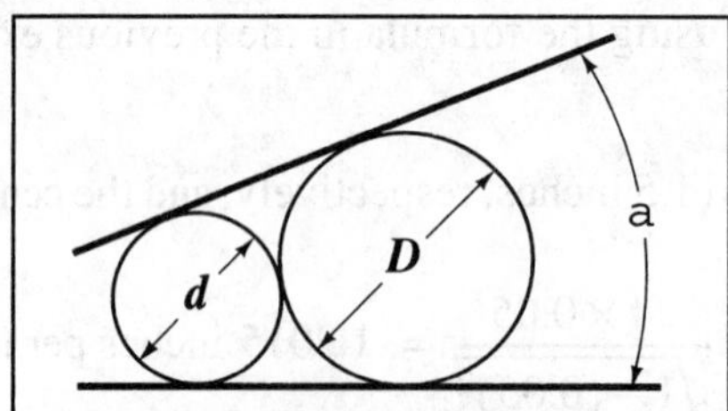

$$D = d \times \frac{1+\sin(\alpha/2)}{1-\sin(\alpha/2)} \text{ inches}$$

Example: The required angle α is 15°. Find diameter D of a large disk that is in contact with a standard 1-inch reference disk.

$$D = 1 \times \frac{1 + \sin 7.5°}{1 - \sin 7.5°} = 1.3002 \text{ inches}$$

Measurement over Pins.—When the distance across a bolt circle is too large to measure using ordinary measuring tools, then the required distance may be found from the distance across adacent or alternate holes using one of the methods that follow:

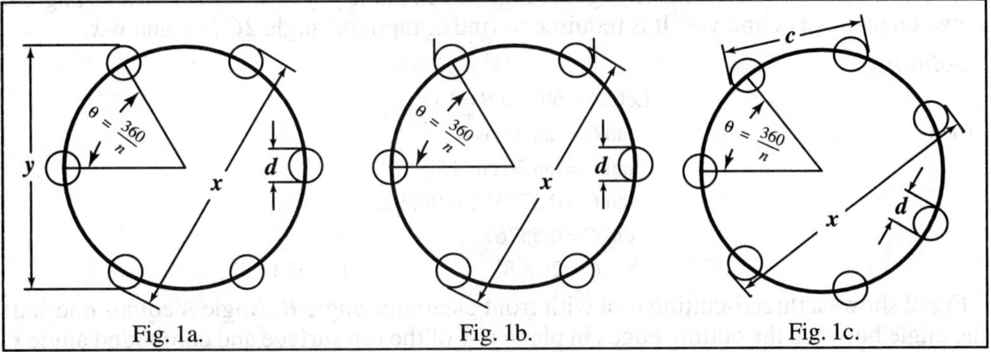

Fig. 1a. Fig. 1b. Fig. 1c.

Even Number of Holes in Circle: To measure the unknown distance x over opposite plugs in a bolt circle of n holes (n is even and greater than 4), as shown in Fig. 1a, where y is the distance over alternate plugs, d is the diameter of the holes, and $\theta = 360°/n$ is the angle between adjacent holes, use the following general equation for obtaining x:

$$x = \frac{y - d}{\sin\theta} + d$$

Example: In a die that has six 3/4-inch diameter holes equally spaced on a circle, where the distance y over alternate holes is 4½ inches, and the angle θ between adjacent holes is 60°, then

$$x = \frac{4.500 - 0.7500}{\sin 60°} + 0.7500 = 5.0801$$

In a similar problem, the distance c over adjacent plugs is given, as shown in Fig. 1b. If the number of holes is even and greater than 4, the distance x over opposite plugs is given in the following formula:

$$x = 2(c - d)\left(\frac{\sin\left(\frac{180 - \theta}{2}\right)}{\sin\theta}\right) + d$$

where d and θ are as defined above.

Odd Number of Holes in Circle: In a circle as shown in Fig. 1c, where the number of holes n is odd and greater than 3, and the distance c over adjacent holes is given, then θ equals $360/n$ and the distance x across the most widely spaced holes is given by:

$$x = \frac{\frac{c - d}{2}}{\sin\frac{\theta}{4}} + d$$

Compound Angles

Three types of compound angles are illustrated by Figs. 1 through 6. The first type is shown in Figs. 1, 2, and 3; the second in Fig. 4; and the third in Figs. 5 and 6.

In Fig. 1 is shown what might be considered as a thread-cutting tool without front clearance. A is a known angle in plane $y–y$ of the top surface. C is the corresponding angle in plane $x–x$ that is at some given angle B with plane $y–y$. Thus, angles A and B are components of the compound angle C.

Example Problem Referring to Fig. 1: Angle $2A$ in plane $y–y$ is known, as is also angle B between planes $x–x$ and $y–y$. It is required to find compound angle $2C$ in plane $x–x$.

Solution:

$$\text{Let } 2A = 60 \text{ and } B = 15$$

Then

$$\tan C = \tan A \cos B$$

$$\tan C = \tan 30 \cos 15$$

tan C = 0.57735 ¥ 0.96592

$$\tan C = 0.55767$$

$$C = 29\ 8.8' \qquad 2C = 58\ 17.6'$$

Fig. 2 shows a thread-cutting tool with front clearance angle B. Angle A equals one-half the angle between the cutting edges in plane $y–y$ of the top surface and compound angle C is one-half the angle between the cutting edges in a plane $x–x$ at right angles to the inclined front edge of the tool. The angle between planes $y–y$ and $x–x$ is, therefore, equal to clearance angle B.

Example Problem Referring to Fig. 2: Find the angle $2C$ between the front faces of a thread-cutting tool having a known clearance angle B, which will permit the grinding of these faces so that their top edges will form the desired angle $2A$ for cutting the thread.

Solution:

$$\text{Let } 2A = 60 \text{ and } B = 15$$

Then

$$\tan C = \frac{\tan A}{\cos B} = \frac{\tan 30°}{\cos 15°} = \frac{0.57735}{0.96592}$$

$$\tan C = 0.59772$$

$$C = 30\ 52' \qquad 2C = 61\ 44'$$

In Fig. 3 is shown a form-cutting tool in which the angle A is one-half the angle between the cutting edges in plane $y–y$ of the top surface; B is the front clearance angle; and C is one-half the angle between the cutting edges in plane $x–x$ at right angles to the front edges of the tool. The formula for finding angle C when angles A and B are known is the same as that for Fig. 2.

Example Problem Referring to Fig. 3: Find the angle $2C$ between the front faces of a form-cutting tool having a known clearance angle B that will permit the grinding of these faces so that their top edges will form the desired angle $2A$ for form cutting.

Solution:

$$\text{Let } 2A = 46 \text{ and } B = 12$$

Then

$$\tan C = \frac{\tan A}{\cos B} = \frac{\tan 23°}{\cos 12°} = \frac{0.42447}{0.97815}$$

$$\tan C = 0.43395$$

$$C = 23\ 27.5' \qquad 2C = 46\ 55'$$

In Fig. 4 is shown a wedge-shaped block, the top surface of which is inclined at compound angle C with the base in a plane at right angles with the base and at angle R with the front edge. Angle A in the vertical plane of the front of the plate and angle B in the vertical plane of one side that is at right angles to the front are components of angle C.

Formulas for Compound Angles

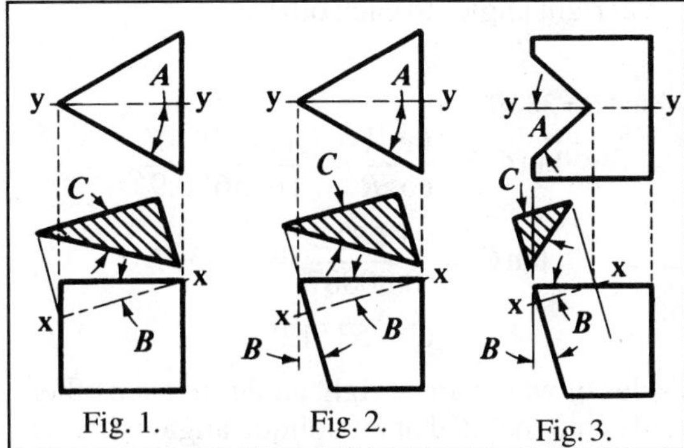 Fig. 1. Fig. 2. Fig. 3.	For given angles A and B, find the resultant angle C in plane x–x. Angle B is measured in vertical plane y–y of midsection. (Fig. 1) $\tan C = \tan A \times \cos B$ (Fig. 2) $\tan C = \dfrac{\tan A}{\cos B}$ (Fig. 3) (Same formula as for Fig. 2)
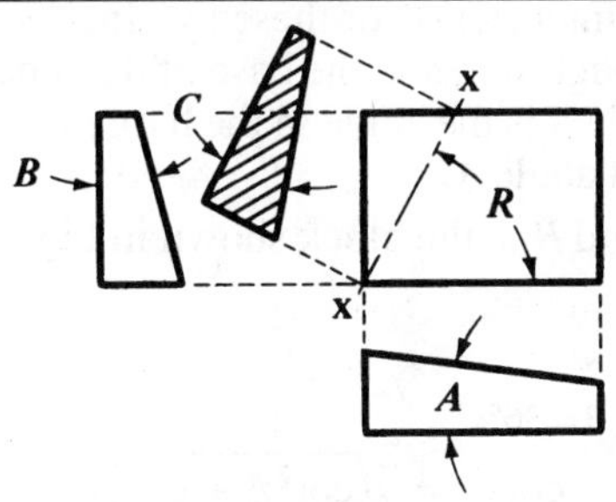 Fig. 4.	Fig. 4. In machining plate to angles A and B, it is held at angle C in plane x–x. Angle of rotation R in plane parallel to base (or complement of R) is for locating plate so that plane x–x is perpendicular to axis of pivot on angle-plate or work-holding vise. $\tan R = \dfrac{\tan B}{\tan A}$; $\tan C = \dfrac{\tan A}{\cos R}$
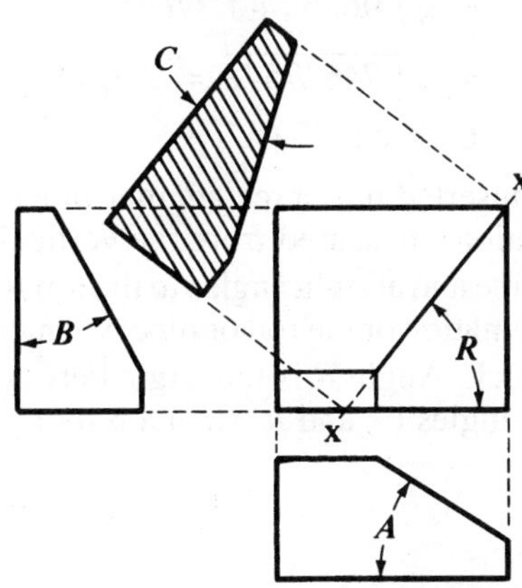	Fig. 5. Angle R in horizontal plane parallel to base is angle from plane x–x to side having angle A. $\tan R = \dfrac{\tan A}{\tan B}$ $\tan C = \tan A \cos R = \tan B \sin R$ Compound angle C is angle in plane x–x from base to corner formed by intersection of planes inclined to angles A and B. This formula for C may be used to find cot of complement of C_1, Fig. 6.
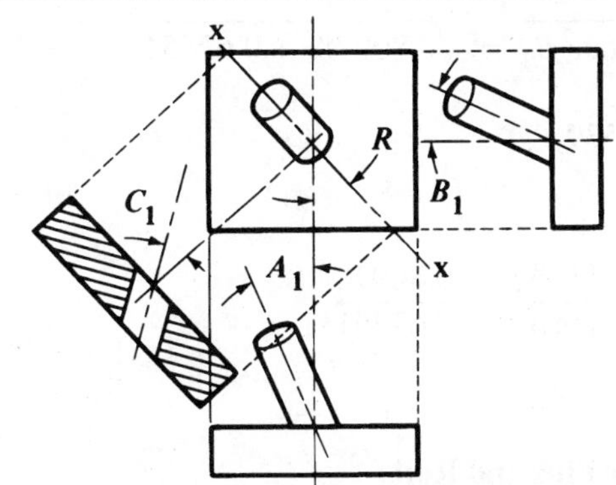	Fig. 6. Angles A_1 and B_1 are measured in vertical planes of front and side elevations. Plane x–x is located by angle R from center-line or from plane of angle B_1. $\tan R = \dfrac{\tan A_1}{\tan B_1}$ $\tan C_1 = \dfrac{\tan A_1}{\sin R} = \dfrac{\tan B_1}{\cos R}$ The resultant angle C_1 would be required in drilling hole for pin.

C = compound angle in plane x–x and is the resultant of angles A and B

Problem Referring to Fig. 4: Find the compound angle C of a wedge-shaped block having known component angles A and B in sides at right angles to each other.

Solution:

Let $A = 47\ 14'$ and $B = 38\ 10'$

$$\tan R = \frac{\tan B}{\tan A} = \frac{\tan 38°10'}{\tan 47°14'} \qquad \tan C = \frac{\tan A}{\cos R} = \frac{\tan 47°14'}{\cos 36°0.9'}$$

$$\tan R = \frac{0.78598}{1.0812} = 0.72695 \qquad \tan C = \frac{1.0812}{0.80887} = 1.3367$$

$$R = 36°09' \qquad C = 53°12'$$

In Fig. 5 is shown a four-sided block, two sides of which are at right angles to each other and to the base of the block. The other two sides are inclined at an oblique angle with the base. Angle C is a compound angle formed by the intersection of these two inclined sides and the intersection of a vertical plane passing through x–x, and the base of the block. The components of angle C are angles A and B and angle R is the angle in the base plane of the block between the plane of angle C and the plane of angle A.

Problem Referring to Fig. 5**:** Find the angles C and R in the block shown in Fig. 5 when angles A and B are known.

Solution:

Let angle $A = 27°$ and $B = 36°$

$$\tan R = \frac{\cot B}{\cot A} = \frac{\cot 36°}{\cot 27°} = \frac{1.3764}{1.9626} \qquad \cot C = \sqrt{\cot^2 A + \cot^2 B}$$

$$\tan R = 0.70131 \qquad R = 35°2.5' \qquad = \sqrt{1.9626^2 + 1.3764^2}$$

$$= \sqrt{5.74627572} = 2.3971$$

$$C = 22°38.6'$$

Problem Referring to Fig. 6**:** A rod or pipe is inserted into a rectangular block at an angle. Angle C_1 is the compound angle of inclination (measured from the vertical) in a plane passing through the center line of the rod or pipe and at right angles to the top surface of the block. Angles A_1 and B_1 are the angles of inclination of the rod or pipe when viewed respectively in the front and side planes of the block. Angle R is the angle between the plane of angle C_1 and the plane of angle B_1. Find angles C_1 and R when a rod or pipe is inclined at known angles A_1 and B_1.

Solution:

Let $A_1 = 39$ and $B_1 = 34$

Then

$$\tan C_1 = \sqrt{\tan^2 A_1 + \tan^2 B_1} = \sqrt{0.80978^2 + 0.67451^2}$$

$$\tan C_1 = \sqrt{1.1107074} = 1.0539$$

$$C_1 = 46\ 30.2'$$

$$\tan R = \frac{\tan A_1}{\tan B_1} = \frac{0.80978}{0.67451}$$

$$\tan R = 1.2005 \qquad R = 50\ 12.4'$$

Measurement over Pins and Rolls

Checking a V-shaped Groove by Measurement Over Pins.—In checking a groove of the shape shown in Fig. 7, it is necessary to measure the dimension X over the pins of radius

R. If values for the radius *R*, dimension *Z*, and the angles α and β are known, the problem is to determine the distance *Y*, to arrive at the required overall dimension for *X*. If a line *AC* is drawn from the bottom of the V to the center of the pin at the left in Fig. 7, and a line *CB* from the center of this pin to its point of tangency with the side of the V, a right-angled triangle is formed in which one side, *CB*, is known and one angle *CAB*, can be determined. A line drawn from the center of a circle to the point of intersection of two tangents to the circle bisects the angle made by the tangent lines, and angle *CAB* therefore equals ½(α + β). The length *AC* and the angle *DAC* can now be found, and with *AC* known in the right-angled triangle *ADC*, *AD*, which is equal to *Y*. can be found.

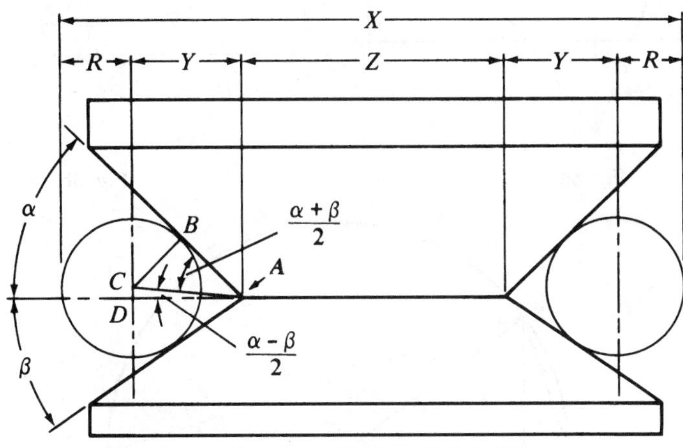

Fig. 7.

The value for *X* can be obtained from the formula

$$X = Z + 2R\left(\csc\frac{\alpha+\beta}{2}\cos\frac{\alpha-\beta}{2} + 1\right)$$

For example, if $R = 0.500$, $Z = 1.824$, $\alpha = 45$ degrees, and $\beta = 35$ degrees,

$$X = 1.824 + (2 \times 0.5)\left(\csc\frac{45° + 35°}{2}\cos\frac{45° - 35°}{2} + 1\right)$$

$$X = 1.824 + \csc 40° \cos 5° + 1$$

$$X = 1.824 + 1.5557 \times 0.99619 + 1$$

$$X = 1.824 + 1.550 + 1 = 4.374$$

Checking Radius of Arc by Measurement Over Rolls.—The radius *R* of large-radius concave and convex gages of the type shown in Figs. 8a, 8b and 8c can be checked by measurement *L* over two rolls with the gage resting on the rolls as shown. If the diameter of the rolls *D*, the length *L*, and the height *H* of the top of the arc above the surface plate (for the concave gage, Fig. 8a) are known or can be measured, the radius *R* of the workpiece to be checked can be calculated trigonometrically, as follows.

Referring to Fig. 8a for the concave gage, if *L* and *D* are known, *cb* can be found, and if *H* and *D* are known, *ce* can be found. With *cb* and *ce* known, *ab* can be found by means of a diagram as shown in Fig. 8c.

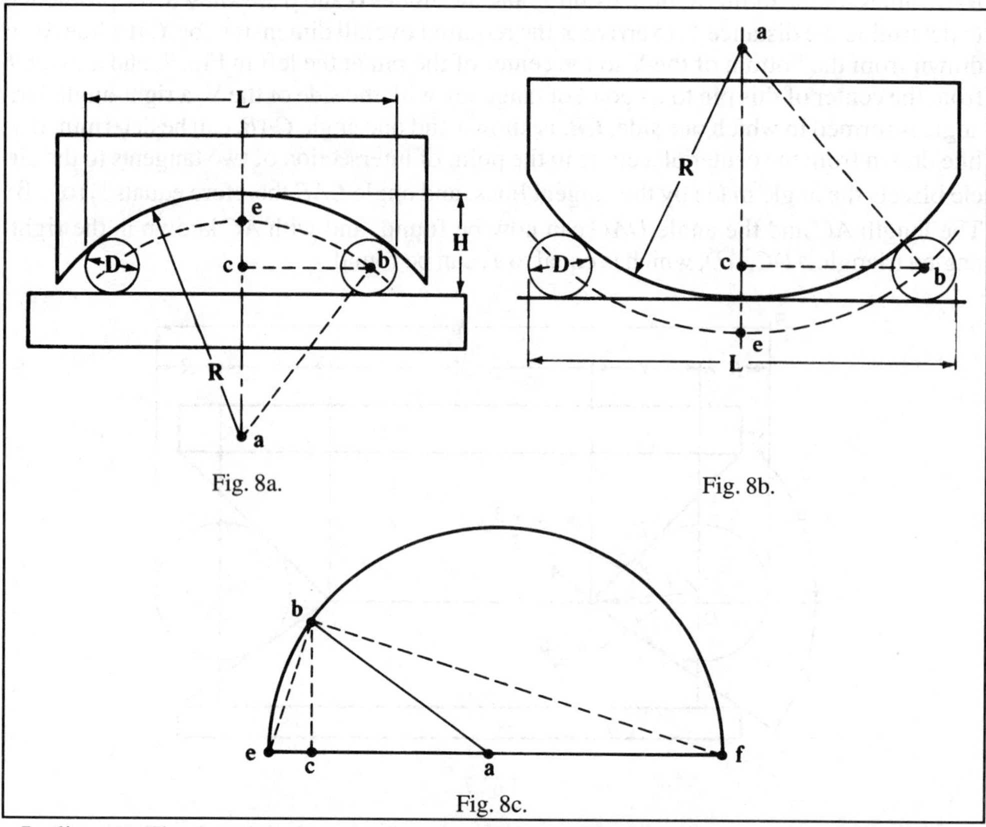

Fig. 8a.

Fig. 8b.

Fig. 8c.

In diagram Fig. 8c, *cb* and *ce* are shown at right angles as in Fig. 8a. A line is drawn connecting points *b* and *e* and line *ce* is extended to the right. A line is now drawn from point *b* perpendicular to *be* and intersecting the extension of *ce* at point *f*. A semicircle can now be drawn through points *b*, *e*, and *f* with point *a* as the center. Triangles *bce* and *bcf* are similar and have a common side. Thus *ce*:*bc*::*bc*:*cf*. With *ce* and *bc* known, *cf* can be found from this proportion and hence *ef* which is the diameter of the semicircle and radius *ab*. Then $R = ab + D/2$.

The procedure for the convex gage is similar. The distances *cb* and *ce* are readily found and from these two distances *ab* is computed on the basis of similar triangles as before. Radius *R* is then readily found.

The derived formulas for concave and convex gages are as follows:

Formulas:

$$R = \frac{(L-D)^2}{8(H-D)} + \frac{H}{2} \qquad \text{(Concave gage Fig. 8a)}$$

$$R = \frac{(L-D)^2}{8D} \qquad \text{(Convex gage Fig. 8b)}$$

For example: For Fig. 8a, let $L = 17.8$, $D = 3.20$, and $H = 5.72$, then

$$R = \frac{(17.8 - 3.20)^2}{8(5.72 - 3.20)} + \frac{5.72}{2} = \frac{(14.60)^2}{8 \times 2.52} + 2.86$$

$$R = \frac{213.16}{20.16} + 2.86 = 13.43$$

For Fig. 8b, let $L = 22.28$ and $D = 3.40$, then

$$R = \frac{(22.28 - 3.40)^2}{8 \times 3.40} = \frac{356.45}{27.20} = 13.1$$

Checking Shaft Conditions

Checking for Various Shaft Conditions.—An indicating height gage, together with V-blocks can be used to check shafts for ovality, taper, straightness (bending or curving), and concentricity of features (as shown exaggerated in Fig. 9). If a shaft on which work has been completed shows lack of concentricity. it may be due to the shaft having become bent or bowed because of mishandling or oval or tapered due to poor machine conditions. In checking for concentricity, the first step is to check for ovality, or out-of-roundness, as in Fig. 9a. The shaft is supported in a suitable V-block on a surface table and the dial indicator plunger is placed over the workpiece, which is then rotated beneath the plunger to obtain readings of the amount of eccentricity.

This procedure (sometimes called clocking, owing to the resemblance of the dial indicator to a clock face) is repeated for other shaft diameters as necessary, and, in addition to making a written record of the measurements, the positions of extreme conditions should be marked on the workpiece for later reference.

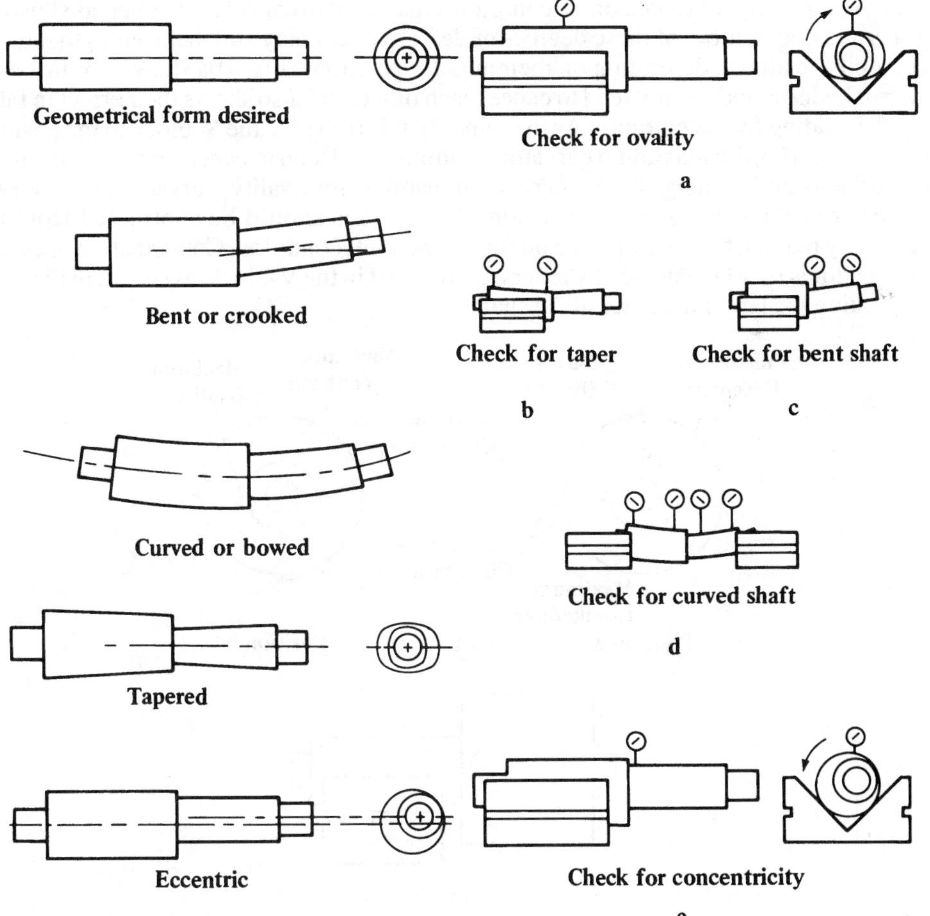

Fig. 9.

To check for taper, the shaft is supported in the V-block and the dial indicator is used to measure the maximum height over the shaft at various positions along its length, as shown in Fig. 9b, without turning the workpiece. Again, the shaft should be marked with the reading positions and values, also the direction of the taper, and a written record should be made of the amount and direction of any taper discovered.

Checking for a bent shaft requires that the shaft be clocked at the shoulder and at the farther end, as shown in Fig. 9c. For a second check the shaft is rotated only 90° or a quarter turn. When the recorded readings are compared with those from the ovality and taper checks, the three conditions can be distinguished.

To detect a curved or bowed condition, the shaft should be suspended in two V-blocks with only about ⅛ inch of each end in each vee. Alternatively, the shaft can be placed between centers. The shaft is then clocked at several points, as shown in Fig. 9d, but preferably not at those locations used for the ovality, taper, or crookedness checks. If the single element due to curvature is to be distinguished from the effects of ovality, taper, and crookedness, and its value assessed, great care must be taken to differentiate between the conditions detected by the measurements.

Finally, the amount of eccentricity between one shaft diameter and another may be tested by the setup shown in Fig. 9e. With the indicator plunger in contact with the smaller diameter, close to the shoulder, the shaft is rotated in the V-block and the indicator needle position is monitored to find the maximum and minimum readings.

Curvature, ovality, or crookedness conditions may tend to cancel each other, as shown in Fig. 10, and one or more of these degrees of defectiveness may add themselves to the true eccentricity readings, depending on their angular positions. Fig. 10a shows, for instance, how crookedness and ovality tend to cancel each other, and also shows their effect in falsifying the reading for eccentricity. As the same shaft is turned in the V-block to the position shown in Fig. 10b, the maximum curvature reading could tend to cancel or reduce the maximum eccentricity reading. Where maximum readings for ovality, curvature, or crookedness occur at the same angular position, their values should be subtracted from the eccentricity reading to arrive at a true picture of the shaft condition. Confirmation of eccentricity readings may be obtained by reversing the shaft in the V-block, as shown in Fig. 10c, and clocking the larger diameter of the shaft.

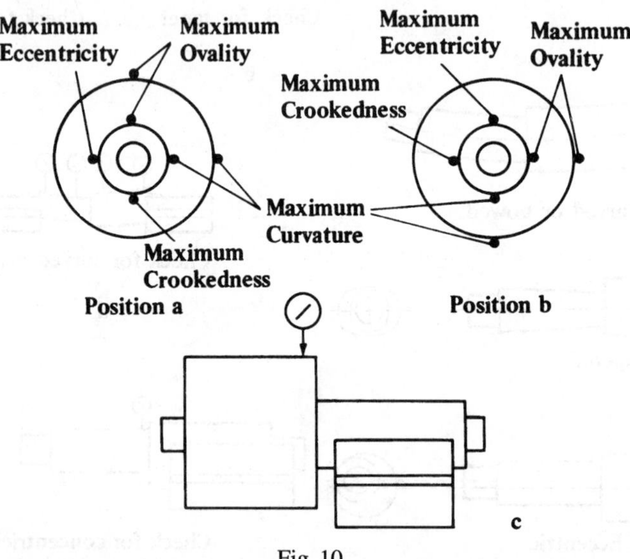

Fig. 10.

Out-of-Roundness—Lobing.—With the imposition of finer tolerances and the development of improved measurement methods, it has become apparent that no hole,' cylinder, or sphere can be produced with a perfectly symmetrical round shape. Some of the conditions are diagrammed in Fig. 11, where Fig. 11a shows simple ovality and Fig. 11b shows ovality occurring in two directions. From the observation of such conditions have come the terms lobe and lobing. Fig. 11c shows the three-lobed shape common with centerless-ground components, and Fig. 11d is typical of multi-lobed shapes. In Fig. 11e are shown surface waviness, surface roughness, and out-of-roundness, which often are combined with lobing.

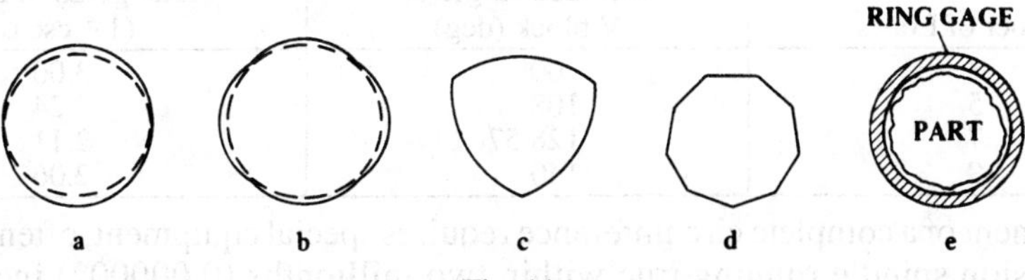

Fig. 11.

In Figs. 11a through 11d the cylinder (or hole) diameters are shown at full size but the lobes are magnified some 10,000 times to make them visible. In precision parts, the deviation from the round condition is usually only in the range of millionths of an inch, although it occasionally can be 0.0001 inch, 0.0002 inch, or more. For instance, a 3-inch-diameter part may have a lobing condition amounting to an inaccuracy of only 30 millionths (0.000030 inch). Even if the distortion (ovality, waviness, roughness) is small, it may cause hum, vibration, heat buildup, and wear, possibly leading to eventual failure of the component or assembly.

Plain elliptical out-of-roundness (two lobes), or any even number of lobes, can be detected by rotating the part on a surface plate under a dial indicator of adequate resolution, or by using an indicating caliper or snap gage. However, supporting such a part in a V-block during measurement will tend to conceal roundness errors. Ovality in a hole can be detected by a dial-type bore gage or internal measuring machine. Parts with odd numbers of lobes require an instrument that can measure the envelope or complete circumference. Plug and ring gages will tell whether a shaft can be assembled into a bearing, but not whether there will be a good fit, as illustrated in Fig. 11e.

A standard, 90-degree included-angle V-block can be used to detect and count the number of lobes, but to measure the exact amount of lobing indicated by *R-r* in Fig. 12 requires a V-block with an angle α, which is related to the number of lobes. This angle α can be calculated from the formula $2\alpha = 180° - 360°/N$, where *N* is the number of lobes. Thus, for a three-lobe form, α becomes 30 degrees, and the V-block used should have a 60-degree included angle. The distance *M*, which is obtained by rotating the part under the comparator plunger, is converted to a value for the radial variation in cylinder contour by the formula $M = (R-r)(1 + \csc\alpha)$.

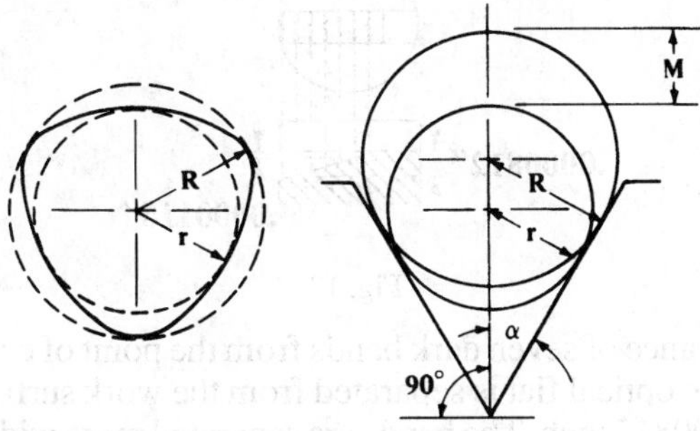

Fig. 12.

Using a V-block (even of appropriate angle) for parts with odd numbers of lobes will give exaggerated readings when the distance *R* - *r* (Fig. 12) is used as the measure of the amount of out-of-roundness. The accompanying table shows the appropriate V-block angles for various odd numbers of lobes, and the factors (1 + csc α) by which the readings are increased over the actual out-of-roundness values.

Table of Lobes, V-block Angles and Exaggeration Factors in Measuring Out-of-round Conditions in Shafts

Number of Lobes	Included Angle of V-block (deg)	Exaggeration Factor (1 + csc α)
3	60	3.00
5	108	2.24
7	128.57	2.11
9	140	2.06

Measurement of a complete circumference requires special equipment, often incorporating a precision spindle running true within two millionths (0.000002) inch. A stylus attached to the spindle is caused to traverse the internal or external cylinder being inspected, and its divergences are processed electronically to produce a polar chart similar to the wavy outline in Fig. 11e. The electronic circuits provide for the variations due to surface effects to be separated from those of lobing and other departures from the "true" cylinder traced out by the spindle.

Measurements Using Light

Measuring by Light-wave Interference Bands.—Surface variations as small as two millionths (0.000002:) inch can be detected by light-wave interference methods, using an optical flat. An optical flat is a transparent block, usually of plate glass, clear fused quartz, or borosilicate glass, the faces of which are finished to extremely fine limits (of the order of 1 to 8 millionths [0.000001 to 0.000008] inch, depending on the application) for flatness. When an optical flat is placed on a "flat" surface, as shown in Fig. 13, any small departure from flatness will result in formation of a wedge-shaped layer of air between the work surface and the underside of the flat.

Light rays reflected from the work surface and the underside of the flat either interfere with or reinforce each other. Interference of two reflections results when the air gap measures exactly half the wavelength of the light used, and produces a dark band across the work surface when viewed perpendicularly, under monochromatic helium light. A light band is produced halfway between the dark bands when the rays reinforce each other. With the 0.0000232-inch-wavelength helium light used, the dark bands occur where the optical flat and the work surface are separated by 11.6 millionths (0.0000116) inch, or multiples thereof.

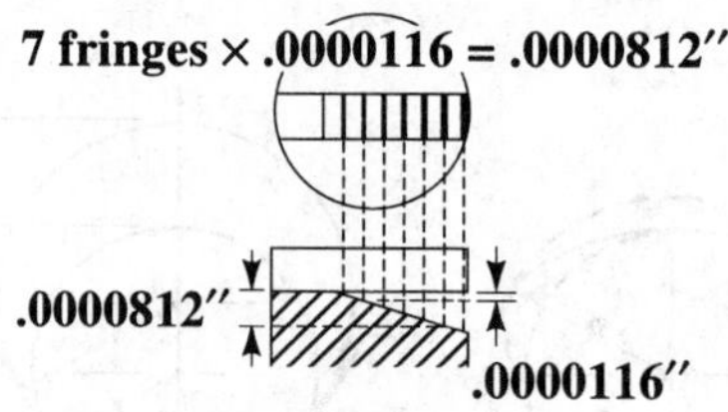

Fig. 13.

For instance, at a distance of seven dark bands from the point of contact, as shown in Fig. 13, the underface of the optical flat is separated from the work surface by a distance of 7 × 0.000116 inch or 0.0000812 inch. The bands are separated more widely and the indications become increasingly distorted as the viewing angle departs from the perpendicular. If the bands appear straight, equally spaced and parallel with each other, the work surface is flat. Convex or concave surfaces cause the bands to curve correspondingly, and a cylindrical tendency in the work surface will produce unevenly spaced, straight bands.

SURFACE TEXTURE

American National Standard Surface Texture (Surface Roughness, Waviness, and Lay).—American National Standard ANSI/ASME B46.1-1995 is concerned with the geometric irregularities of surfaces of solid materials, physical specimens for gaging roughness, and the characteristics of stylus instrumentation for measuring roughness. The standard defines surface texture and its constituents: roughness, waviness, lay, and flaws. A set of symbols for drawings, specifications, and reports is established. To ensure a uniform basis for measurements the standard also provides specifications for Precision Reference Specimens, and Roughness Comparison Specimens, and establishes requirements for stylus-type instruments. The standard is not concerned with luster, appearance, color, corrosion resistance, wear resistance, hardness, subsurface microstructure, surface integrity, and many other characteristics that may be governing considerations in specific applications.

The standard is expressed in SI metric units but U.S. customary units may be used without prejudice. The standard does not define the degrees of surface roughness and waviness or type of lay suitable for specific purposes, nor does it specify the means by which any degree of such irregularities may be obtained or produced. However, criteria for selection of surface qualities and information on instrument techniques and methods of producing, controlling and inspecting surfaces are included in Appendixes attached to the standard. The Appendix sections are not considered a part of the standard: they are included for clarification or information purposes only.

Surfaces, in general, are very complex in character. The standard deals only with the height, width, and direction of surface irregularities because these characteristics are of practical importance in specific applications. Surface texture designations as delineated in this standard may not be a sufficient index to performance. Other part characteristics such as dimensional and geometrical relationships, material, metallurgy, and stress must also be controlled.

Definitions of Terms Relating to the Surfaces of Solid Materials.—The terms and ratings in the standard relate to surfaces produced by such means as abrading, casting, coating, cutting, etching, plastic deformation, sintering, wear, and erosion.

Error of form is considered to be that deviation from the nominal surface caused by errors in machine tool ways, guides, insecure clamping or incorrect alignment of the workpiece or wear, all of which are not included in surface texture. Out-of-roundness and out-of-flatness are examples of errors of form. See ANSI/ASME B46.3.1-1988 for measurement of out-of-roundness.

Flaws are unintentional, unexpected, and unwanted interruptions in the topography typical of a part surface and are defined as such only when agreed upon by buyer and seller. If flaws are defined, the surface should be inspected specifically to determine whether flaws are present, and rejected or accepted prior to performing final surface roughness measurements. If defined flaws are not present, or if flaws are not defined, then interruptions in the part surface may be included in roughness measurements.

Lay is the direction of the predominant surface pattern, ordinarily determined by the production method used.

Roughness consists of the finer irregularities of the surface texture, usually including those irregularities that result from the inherent action of the production process. These irregularities are considered to include traverse feed marks and other irregularities within the limits of the roughness sampling length.

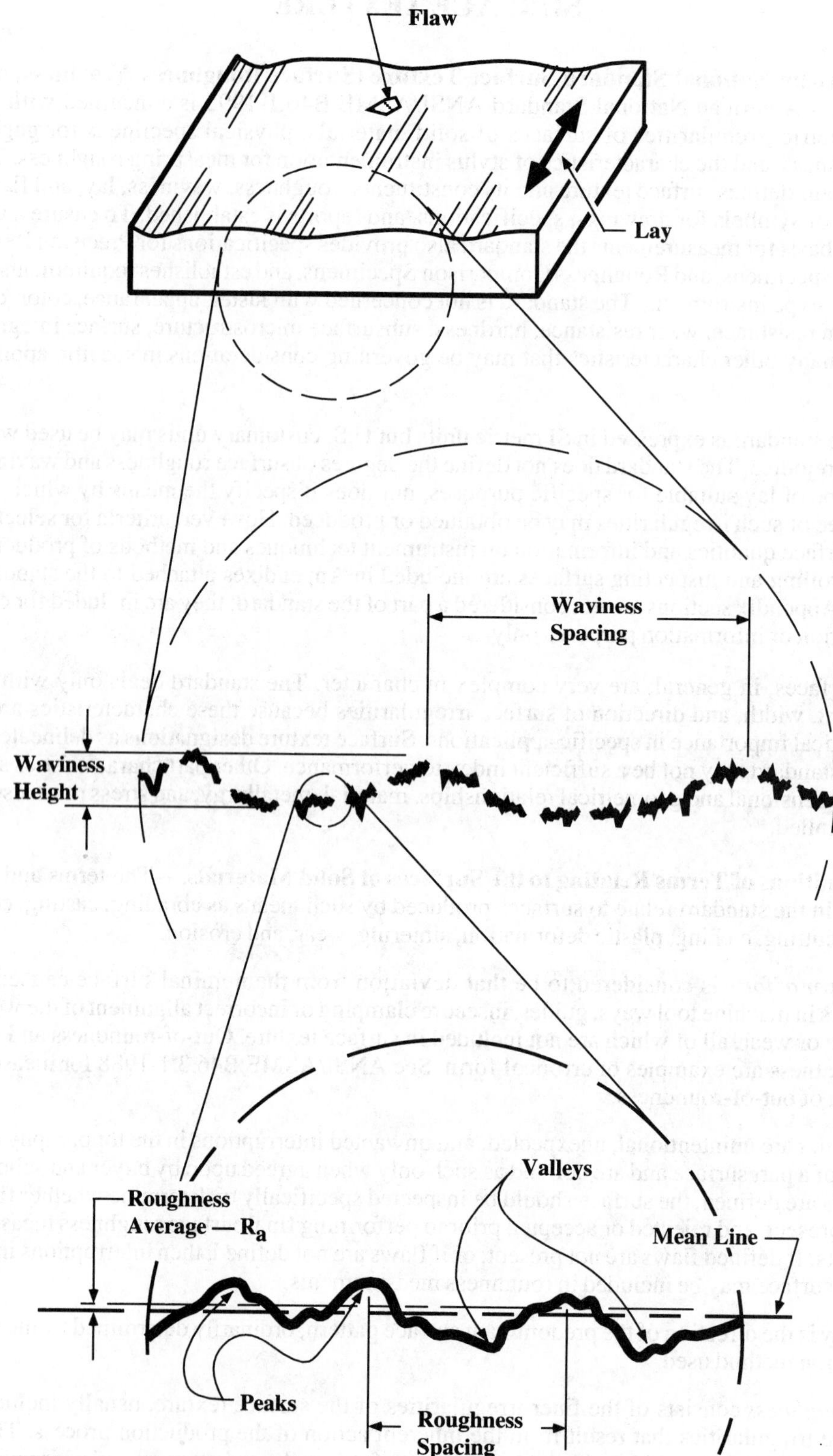

Fig. 1. Pictorial Display of Surface Characteristics

Surface is the boundary of an object that separates that object from another object, substance or space.

Surface, measured is the real surface obtained by instrumental or other means.

Surface, nominal is the intended surface contour (exclusive of any intended surface roughness), the shape and extent of which is usually shown and dimensioned on a drawing or descriptive specification.

Surface, real is the actual boundary of the object. Manufacturing processes determine its deviation from the nominal surface.

Surface texture is repetitive or random deviations from the real surface that forms the three-dimensional topography of the surface. Surface texture includes roughness, waviness, lay and flaws. Fig. 1 is an example of a unidirectional lay surface. Roughness and waviness parallel to the lay are not represented in the expanded views.

Waviness is the more widely spaced component of surface texture. Unless otherwise noted, waviness includes all irregularities whose spacing is greater than the roughness sampling length and less than the waviness sampling length. Waviness may result from such factors as machine or work deflections, vibration, chatter, heat-treatment or warping strains. Roughness may be considered as being superposed on a 'wavy' surface.

Definitions of Terms Relating to the Measurement of Surface Texture.—Terms regarding surface texture pertain to the geometric irregularities of surfaces and include roughness, waviness and lay.

Profile is the contour of the surface in a plane measured normal, or perpendicular, to the surface, unless another other angle is specified.

Graphical centerline. See Mean Line.

Height (z) is considered to be those measurements of the profile in a direction normal, or perpendicular, to the nominal profile. For digital instruments, the profile Z(x) is approximated by a set of digitized values. Height parameters are expressed in micrometers (μm).

Height range (z) is the maximum peak-to-valley surface height that can be detected accurately with the instrument. It is measurement normal, or perpendicular, to the nominal profile and is another key specification.

Mean line (M) is the line about which deviations are measured and is a line parallel to the general direction of the profile within the limits of the sampling length. See Fig. 2. The mean line may be determined in one of two ways. The filtered mean line is the centerline established by the selected cutoff and its associated circuitry in an electronic roughness average measuring instrument. The least squares mean line is formed by the nominal profile but by dividing into selected lengths the sum of the squares of the deviations minimizes the deviation from the nominal form. The form of the nominal profile could be a curve or a straight line.

Peak is the point of maximum height on that portion of a profile that lies above the mean line and between two intersections of the profile with the mean line.

Profile measured is a representation of the real profile obtained by instrumental or other means. When the measured profile is a graphical representation, it will usually be distorted through the use of different vertical and horizontal magnifications but shall otherwise be as faithful to the profile as technically possible.

Profile, modified is the measured profile where filter mechanisms (including the instrument datum) are used to minimize certain surface texture characteristics and emphasize others. Instrument users apply profile modifications typically to differentiate surface roughness from surface waviness.

Profile, nominal is the profile of the nominal surface; it is the intended profile (exclusive of any intended roughness profile). Profile is usually drawn in an x-z coordinate system. See Fig. 2.

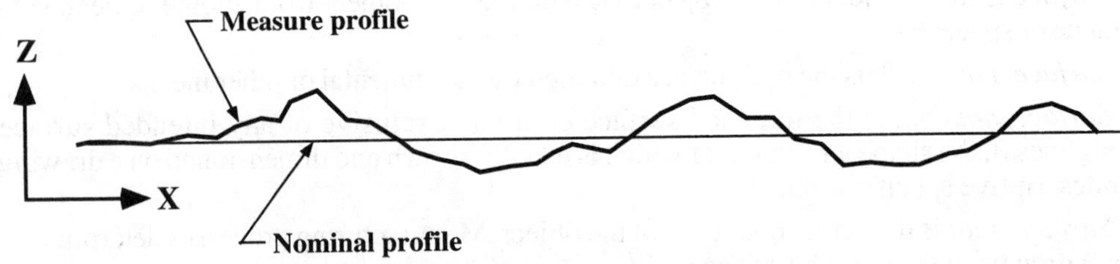

Fig. 2. Nominal and Measured Profiles

Profile, real is the profile of the real surface.

Profile, total is the measured profile where the heights and spacing may be amplified differently but otherwise no filtering takes place.

Roughness profile is obtained by filtering out the longer wavelengths characteristic of waviness.

Roughness spacing is the average spacing between adjacent peaks of the measured profile within the roughness sampling length.

Roughness topography is the modified topography obtained by filtering out the longer wavelengths of waviness and form error.

Sampling length is the nominal spacing within which a surface characteristic is determined. The range of sampling lengths is a key specification of a measuring instrument.

Spacing is the distance between specified points on the profile measured parallel to the nominal profile.

Spatial (x) resolution is the smallest wavelength which can be resolved to 50% of the actual amplitude. This also is a key specification of a measuring instrument.

System height resolution is the minimum height that can be distinguished from background noise of the measurement instrument. Background noise values can be determined by measuring approximate rms roughness of a sample surface where actual roughness is significantly less than the background noise of the measuring instrument. It is a key instrumentation specification.

Topography is the three-dimensional representation of geometric surface irregularities.

Topography, measured is the three-dimensional representation of geometric surface irregularities obtained by measurement.

Topography, modified is the three-dimensional representation of geometric surface irregularities obtained by measurement but filtered to minimize certain surface characteristics and accentuate others.

Valley is the point of maximum depth on that portion of a profile that lies below the mean line and between two intersections of the profile with the mean line.

Waviness, evaluation length (L), is the length within which waviness parameters are determined.

Waviness, long-wavelength cutoff (lcw) the spatial wavelength above which the undulations of waviness profile are removed to identify form parameters. A digital Gaussian filter can be used to separate form error from waviness but its use must be specified.

Waviness profile is obtained by filtering out the shorter roughness wavelengths characteristic of roughness and the longer wavelengths associated with the part form parameters.

Waviness sampling length is a concept no longer used. See waviness long-wavelength cutoff and waviness evaluation length.

Waviness short-wavelength cutoff (lsw) is the spatial wavelength below which roughness parameters are removed by electrical or digital filters.

Waviness topography is the modified topography obtained by filtering out the shorter wavelengths of roughness and the longer wavelengths associated with form error.

Waviness spacing is the average spacing between adjacent peaks of the measured profile within the waviness sampling length.

Sampling Lengths.—Sampling length is the normal interval for a single value of a surface parameter. Generally it is the longest spatial wavelength to be included in the profile measurement. Range of sampling lengths is an important specification for a measuring instrument.

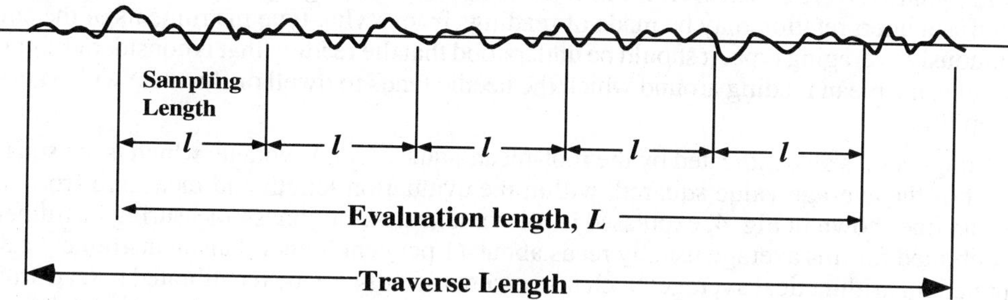

Fig. 3. Traverse Length

Roughness sampling length (l) is the sampling length within which the roughness average is determined. This length is chosen to separate the profile irregularities which are designated as roughness from those irregularities designated as waviness. It is different from evaluation length (L) and the traversing length. See Fig. 3.

Evaluation length (L) is the length the surface characteristics are evaluated. The evaluation length is a key specification of a measuring instrument.

Traversing length is profile length traversed to establish a representative evaluation length. It is always longer than the evaluation length. See Section 4.4.4 of ANSI/ASME B46.1-1995 for values which should be used for different type measurements.

Cutoff is the electrical response characteristic of the measuring instrument which is selected to limit the spacing of the surface irregularities to be included in the assessment of surface texture. Cutoff is rated in millimeters. In most electrical averaging instruments, the cutoff can be user selected and is a characteristic of the instrument rather than of the surface being measured. In specifying the cutoff, care must be taken to choose a value which will include all the surface irregularities to be assessed.

Waviness sampling length (l) is a concept no longer used. See waviness long-wavelength cutoff and waviness evaluation length.

Roughness Parameters.—Roughness is the fine irregularities of the surface texture resulting from the production process or material condition.

Roughness average (Ra), also known as arithmetic average (AA) is the arithmetic average of the absolute values of the measured profile height deviations divided by the evaluation length, L. This is shown as the shaded area of Fig. 4 and generally includes sampling lengths or cutoffs. For graphical determinations of roughness average, the height deviations are measured normal, or perpendicular, to the chart center line.

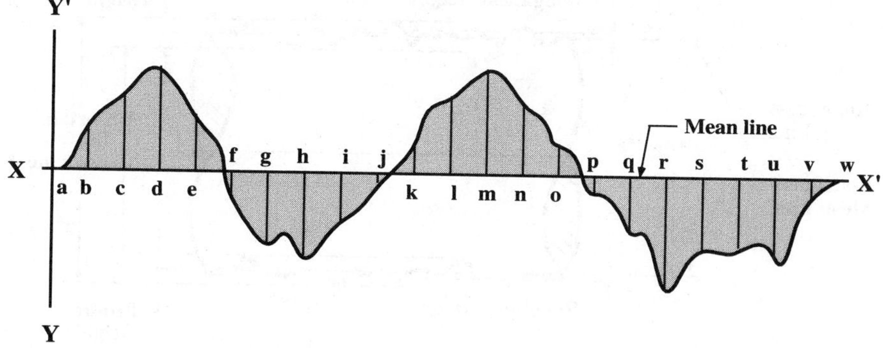

Fig. 4.

Roughness average is expressed in micrometers (μm). A micrometer is one millionth of a meter (0.000001 meter). A microinch (μin) is one millionth of an inch (0.000001 inch). One microinch equals 0.0254 micrometer (1 μin. = 0.0254 μm).

Roughness Average Value (Ra) From Continuously Averaging Meter Reading. So that uniform interpretation may be made of readings from stylus-type instruments of the continuously averaging type, it should be understood that the reading that is considered significant is the mean reading around which the needle tends to dwell or fluctuate with a small amplitude.

Roughness is also indicated by the root-mean-square (rms) average, which is the square root of the average value squared, within the evaluation length and measured from the mean line shown in Fig. 4, expressed in micrometers. A roughness-measuring instrument calibrated for rms average usually reads about 11 per cent higher than an instrument calibrated for arithmetical average. Such instruments usually can be recalibrated to read arithmetical average. Some manufacturers consider the difference between rms and AA to be small enough that rms on a drawing may be read as AA for many purposes.

Roughness evaluation length (L), for statistical purposes should, whenever possible, consist of five sampling lengths (l). Use of other than five sampling lengths must be clearly indicated.

Waviness Parameters.—Waviness is the more widely spaced component of surface texture. Roughness may be thought of as superimposed on waviness.

Waviness height (Wt) is the peak-to-valley height of the modified profile with roughness and part form errors removed by filtering, smoothing or other means. This value is typically three or more times the roughness average. The measurement is taken normal, or perpendicular, to the nominal profile within the limits of the waviness sampling length.

Waviness evaluation length (Lw) is the evaluation length required to determine waviness parameters. For waviness, the sampling length concept is no longer used. Rather, only waviness evaluation length (*Lw*) and waviness long-wavelength cutoff (*lew*) are defined. For better statistics, the waviness evaluation length should be several times the waviness long-wavelength cutoff.

Relation of Surface Roughness to Tolerances.—Because the measurement of surface roughness involves the determination of the average linear deviation of the measured surface from the nominal surface, there is a direct relationship between the dimensional tolerance on a part and the permissible surface roughness. It is evident that a requirement for the accurate measurement of a dimension is that the variations introduced by surface roughness should not exceed the dimensional tolerances. If this is not the case, the measurement of the dimension will be subject to an uncertainty greater than the required tolerance, as illustrated in Fig. 5.

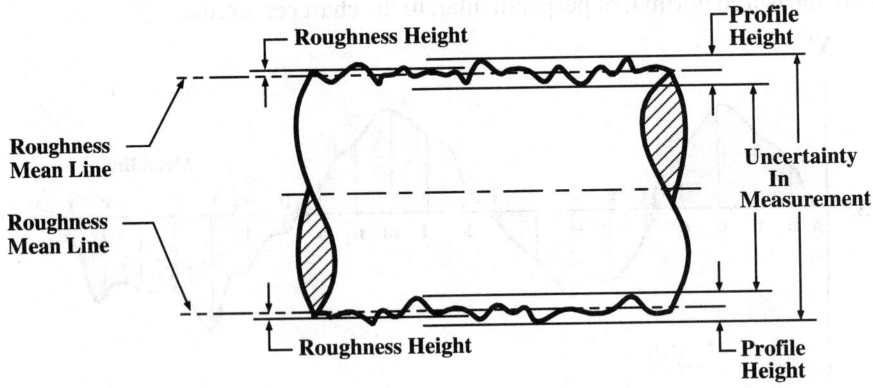

Fig. 5.

The standard method of measuring surface roughness involves the determination of the average deviation from the mean surface. On most surfaces the total profile height of the surface roughness (peak-to-valley height) will be approximately four times (4×) the measured average surface roughness. This factor will vary somewhat with the character of the surface under consideration, but the value of four may be used to establish approximate profile heights.

From these considerations it follows that if the arithmetical average value of surface roughness specified on a part exceeds one eighth of the dimensional tolerance, the whole tolerance will be taken up by the roughness height. In most cases, a smaller roughness specification than this will be found; but on parts where very small dimensional tolerances are given, it is necessary to specify a suitably small surface roughness so useful dimensional measurements can be made. The tables on pages pages 630 and 657 show the relations between machining processes and working tolerances.

Values for surface roughness produced by common processing methods are shown in Table 1. The ability of a processing operation to produce a specific surface roughness depends on many factors. For example, in surface grinding, the final surface depends on the peripheral speed of the wheel, the speed of the traverse, the rate of feed, the grit size, bonding material and state of dress of the wheel, the amount and type of lubrication at the point of cutting, and the mechanical properties of the piece being ground. A small change in any of the above factors can have a marked effect on the surface produced.

Table 1. Surface Roughness Produced by Common Production Methods

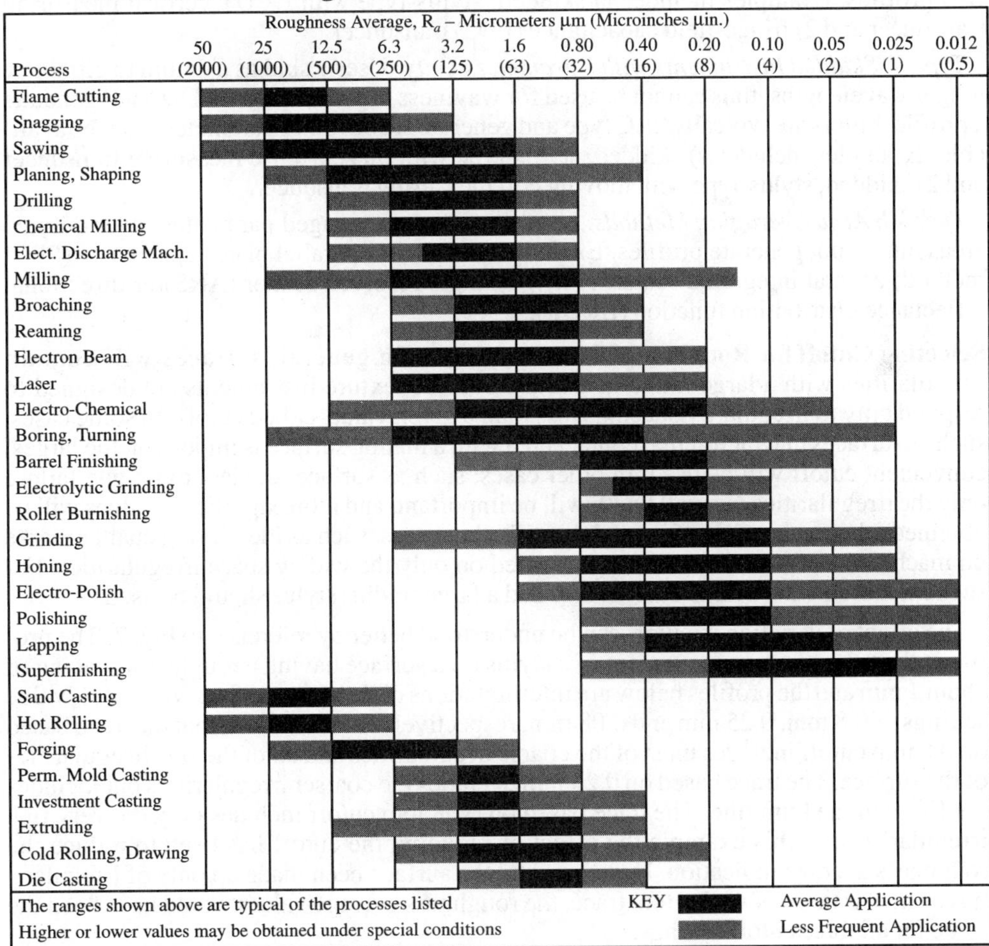

Instrumentation for Surface Texture Measurement.—Instrumentation used for measurement of surface texture, including roughness and waviness generally falls into six types. These include:

Type I, Profiling Contact Skidless Instruments: Used for very smooth to very rough surfaces. Used for roughness and may measure waviness. Can generate filtered or unfiltered profiles and may have a selection of filters and parameters for data analysis. Examples include: 1) skidless stylus-type with LVDT (linear variable differential transformer) vertical transducers; 2) skidless-type using an interferometric transducer; 3)skidless stylus-type using capacitance transducer.

Type II, Profiling Non-contact Instruments: Capable of full profiling or topographical analysis. Non-contact operation may be advantageous for softness but may vary with sample type and reflectivity. Can generate filtered or unfiltered profiles but may have difficulty with steeply inclined surfaces. Examples include: 1) interferometric microscope; 2) optical focus sending; 3) Nomarski differential profiling; 4) laser triangulation; 5) scanning electron microscope (SEM) stereoscopy; 6) confocal optical microscope.

Type III, Scanned Probe Microscope: Feature high spatial resolution (at or near the atomic scale) but area of measurement may be limited. Examples include: 1) scanning tunneling microscope (STM) and 2) atomic force microscope (AFM).

Type IV, Profiling Contact Skidded Instruments: Uses a skid as a datum to eliminate longer wavelengths; thus cannot be used for waviness or errors of form. May have a selection of filters and parameters and generates an output recording of filtered and skid-modified profiles. Examples include: 1) skidded, stylus-type with LVDT vertical measuring transducer and 2) fringe-field capacitance (FFC) transducer.

Type V, Skidded Instruments with Parameters Only: Uses a skid as a datum to eliminate longer wavelengths; thus cannot be used for waviness or errors of form. Does not generate a profile. Filters are typically 2RC type and generate Ra but other parameters may be available. Examples include: 1) skidded, stylus-type with piezoelectric measuring transducer and 2) skidded, stylus-type with moving coil measuring transducer.

Type VI, Area Averaging Methods: Used to measure averaged parameters over defined areas but do not generate profiles. Examples include: 1) parallel plate capacitance (PPC) method; 2) total integrated scatter (TIS); 3) angle resolved scatter (ARS)/bi-directional reflectance distribution function (BRDF).

Selecting Cutoff for Roughness Measurements.—In general, surfaces will contain irregularities with a large range of widths. Surface texture instruments are designed to respond only to irregularity spacings less than a given value, called cutoff. In some cases, such as surfaces in which actual contact area with a mating surface is important, the largest convenient cutoff will be used. In other cases, such as surfaces subject to fatigue failure only the irregularities of small width will be important, and more significant values will be obtained when a short cutoff is used. In still other cases, such as identifying chatter marks on machined surfaces, information is needed on only the widely space irregularities. For such measurements, a large cutoff value and a larger radius stylus should be used.

The effect of variation in cutoff can be understood better by reference to Fig. 7. The profile at the top is the true movement of a stylus on a surface having a roughness spacing of about 1 mm and the profiles below are interpretations of the same surface with cutoff value settings of 0.8 mm, 0.25 mm and 0.08 mm, respectively. It can be seen that the trace based on 0.8 mm cutoff includes most of the coarse irregularities and all of the fine irregularities of the surface. The trace based on 0.25 mm excludes the coarser irregularities but includes the fine and medium fine. The trace based on 0.08 mm cutoff includes only the very fine irregularities. In this example the effect of reducing the cutoff has been to reduce the roughness average indication. However, had the surface been made up only of irregularities as fine as those of the bottom trace, the roughness average values would have been the same for all three cutoff settings.

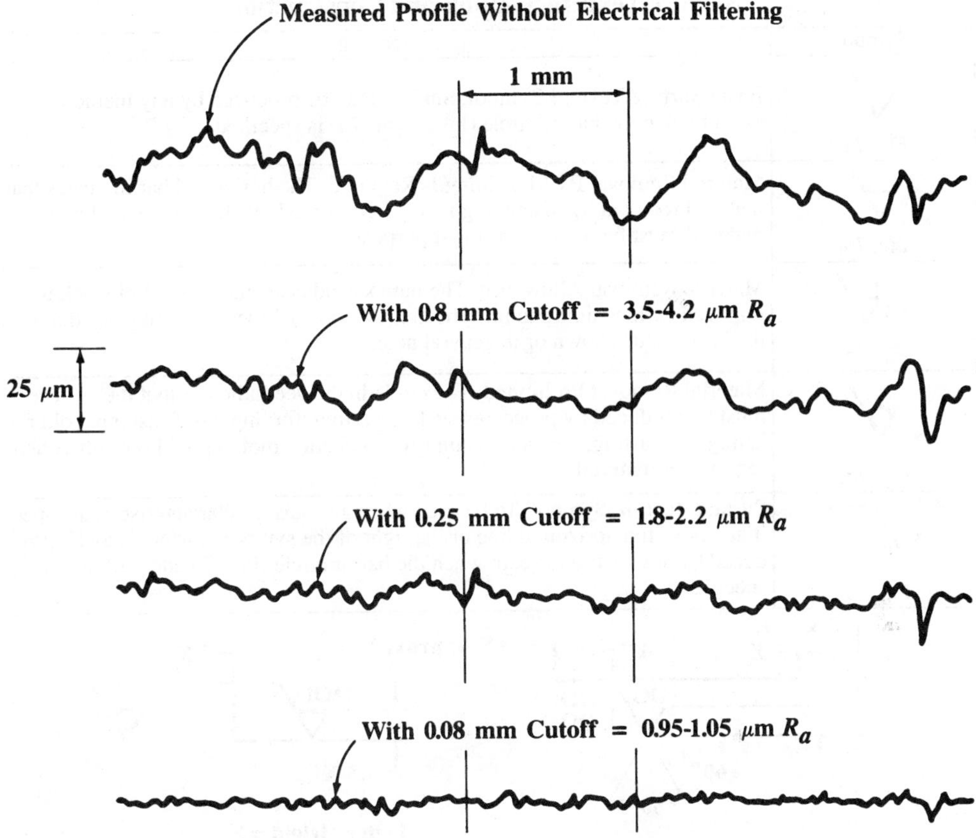

Fig. 6. Effects of Various Cutoff Values

In other words, all irregularities having a spacing less than the value of the cutoff used are included in a measurement. Obviously, if the cutoff value is too small to include coarser irregularities of a surface, the measurements will not agree with those taken with a larger cutoff. For this reason, care must be taken to choose a cutoff value which will include all of the surface irregularities it is desired to assess.

To become proficient in the use of continuously averaging stylus-type instruments the inspector or machine operator must realize that for uniform interpretation, the reading which is considered significant is the mean reading around which the needle tends to dwell or fluctuate under small amplitude.

Drawing Practices for Surface Texture Symbols.—American National Standard ANSI/ASME Y14.36M-1996 establishes the method to designate symbolic controls for surface texture of solid materials. It includes methods for controlling roughness, waviness, and lay, and provides a set of symbols for use on drawings, specifications, or other documents. The standard is expressed in SI metric units but U.S. customary units may be used without prejudice. Units used (metric or non-metric) should be consistent with the other units used on the drawing or documents. Approximate non-metric equivalents are shown for reference.

Surface Texture Symbol.—The symbol used to designate control of surface irregularities is shown in Fig. 7b and Fig. 7d. Where surface texture values other than roughness average are specified, the symbol must be drawn with the horizontal extension as shown in Fig. 7f.

Surface Texture Symbols and Construction

Symbol	Meaning
 Fig. 7a.	Basic Surface Texture Symbol. Surface may be produced by any method except when the bar or circle (Fig. 7b or 7d) is specified.
 Fig. 7b.	Material Removal By Machining Is Required. The horizontal bar indicates that material removal by machining is required to produce the surface and that material must be provided for that purpose.
 3.5 Fig. 7c.	Material Removal Allowance. The number indicates the amount of stock to be removed by machining in millimeters (or inches). Tolerances may be added to the basic value shown or in general note.
 Fig. 7d.	Material Removal Prohibited. The circle in the vee indicates that the surface must be produced by processes such as casting, forging, hot finishing, cold finishing, die casting, powder metallurgy or injection molding without subsequent removal of material.
 Fig. 7e.	Surface Texture Symbol. To be used when any surface characteristics are specified above the horizontal line or the right of the symbol. Surface may be produced by any method except when the bar or circle (Fig. 7b and 7d) is specified.

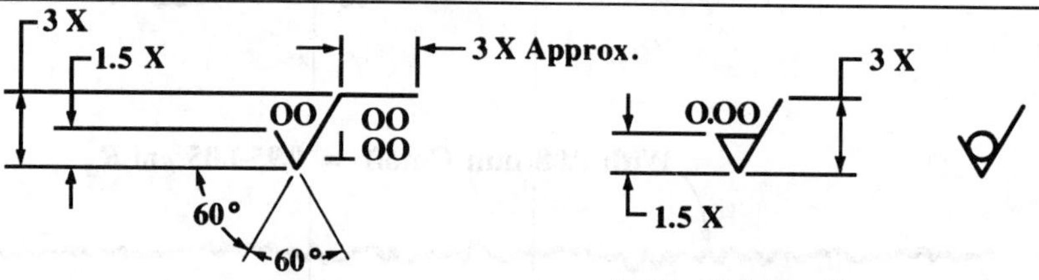

Letter Height = X

Fig. 7f.

Use of Surface Texture Symbols: When required from a functional standpoint, the desired surface characteristics should be specified. Where no surface texture control is specified, the surface produced by normal manufacturing methods is satisfactory provided it is within the limits of size (and form) specified in accordance with ANSI/ASME Y14.5M-1994, Dimensioning and Tolerancing. It is considered good practice to always specify some maximum value, either specifically or by default (for example, in the manner of the note shown in Fig. 2).

Material Removal Required or Prohibited: The surface texture symbol is modified when necessary to require or prohibit removal of material. When it is necessary to indicate that a surface must be produced by removal of material by machining, specify the symbol shown in Fig. 7b. When required, the amount of material to be removed is specified as shown in Fig. 7c, in millimeters for metric drawings and in inches for non-metric drawings. Tolerance for material removal may be added to the basic value shown or specified in a general note. When it is necessary to indicate that a surface must be produced without material removal, specify the machining prohibited symbol as shown in Fig. 7d.

Proportions of Surface Texture Symbols: The recommended proportions for drawing the surface texture symbol are shown in Fig. 7f. The letter height and line width should be the same as that for dimensions and dimension lines.

Applying Surface Texture Symbols.—The point of the symbol should be on a line representing the surface, an extension line of the surface, or a leader line directed to the surface, or to an extension line. The symbol may be specified following a diameter dimension. Although ANSI/ASME Y14.5M-1994, "Dimensioning and Tolerancing" specifies that normally all textual dimensions and notes should be read from the bottom of the drawing, the surface texture symbol itself with its textual values may be rotated as required. Regardless, the long leg (and extension) must be to the right as the symbol is read. For parts requiring extensive and uniform surface roughness control, a general note may be added to the drawing which applies to each surface texture symbol specified without values as shown in Fig. 8.

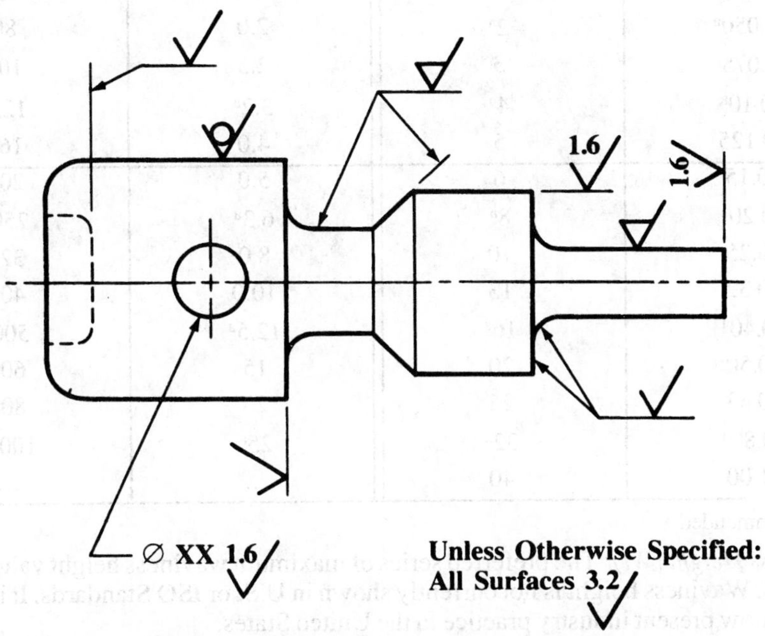

Fig. 8. Application of Surface Texture Symbols

When the symbol is used with a dimension, it affects the entire surface defined by the dimension. Areas of transition, such as chamfers and fillets, shall conform with the roughest adjacent finished area unless otherwise indicated.

Surface texture values, unless otherwise specified, apply to the complete surface. Drawings or specifications for plated or coated parts shall indicate whether the surface texture values apply before plating, after plating, or both before and after plating.

Only those values required to specify and verify the required texture characteristics should be included in the symbol. Values should be in metric units for metric drawing and non-metric units for non-metric drawings. Minority units on dual dimensioned drawings are enclosed in brackets.

Roughness and waviness measurements, unless otherwise specified, apply in a direction which gives the maximum reading; generally across the lay.

Cutoff or Roughness Sampling Length, (l): Standard values are listed in Table 2. When no value is specified, the value 0.8 mm (0.030 in.) applies.

Table 2. Standard Roughness Sampling Length (Cutoff) Values

mm	in.	mm	in.
0.08	0.003	2.5	0.1
0.25	0.010	8.0	0.3
0.80	0.030	25.0	1.0

Roughness Average (Ra): The preferred series of specified roughness average values is given in Table 3.

Table 3. Preferred Series Roughness Average Values (R_a)

μm	μin	μm	μin
0.012	0.5	1.25	50
0.025[a]	1[a]	1.60[a]	63[a]
0.050[a]	2[a]	2.0	80
0.075[a]	3	2.5	100
0.10[a]	4[a]	3.2[a]	125[a]
0.125	5	4.0	160
0.15	6	5.0	200
0.20[a]	8[a]	6.3[a]	250[a]
0.25	10	8.0	320
0.32	13	10.0	400
0.40[a]	16[a]	12.5[a]	500[a]
0.50	20	15	600
0.63	25	20	800
0.80[a]	32[a]	25[a]	1000[a]
1.00	40	…	…

[a] Recommended

Waviness Height (Wt): The preferred series of maximum waviness height values is listed in Table 3. Waviness height is not currently shown in U.S. or ISO Standards. It is included here to follow present industry practice in the United States.

Table 4. Preferred Series Maximum Waviness Height Values

mm	in.	mm	in.	mm	in.
0.0005	0.00002	0.008	0.0003	0.12	0.005
0.0008	0.00003	0.012	0.0005	0.20	0.008
0.0012	0.00005	0.020	0.0008	0.25	0.010
0.0020	0.00008	0.025	0.001	0.38	0.015
0.0025	0.0001	0.05	0.002	0.50	0.020
0.005	0.0002	0.08	0.003	0.80	0.030

Lay: Symbols for designating the direction of lay are shown and interpreted in Table 5.

Example Designations.—Table 6 illustrates examples of designations of roughness, waviness, and lay by insertion of values in appropriate positions relative to the symbol.

Where surface roughness control of several operations is required within a given area, or on a given surface, surface qualities may be designated, as in Fig. 9a. If a surface must be produced by one particular process or a series of processes, they should be specified as shown in Fig. 9b. Where special requirements are needed on a designated surface, a note should be added at the symbol giving the requirements and the area involved. An example is illustrated in Fig. 9c.

Surface Texture of Castings.—Surface characteristics should not be controlled on a drawing or specification unless such control is essential to functional performance or appearance of the product. Imposition of such restrictions when unnecessary may increase production costs and in any event will serve to lessen the emphasis on the control specified for important surfaces. Surface characteristics of castings should never be considered on

Table 5. Lay Symbols

Lay Symbol	Meaning	Example Showing Direction of Tool Marks
=	Lay approximately parallel to the line representing the surface to which the symbol is applied.	=
⊥	Lay approximately perpendicular to the line representing the surface to which the symbol is applied.	⊥
X	Lay angular in both directions to line representing the surface to which the symbol is applied.	X
M	Lay multidirectional	M
C	Lay approximately circular relative to the center of the surface to which the symbol is applied.	C
R	Lay approximately radial relative to the center of the surface to which the symbol is applied.	R
P	Lay particulate, non-directional, or protuberant	P

Table 6. Application of Surface Texture Values to Symbol

Symbol	Description	Symbol	Description
1.6 √	Roughness average rating is placed at the left of the long leg. The specification of only one rating shall indicate the maximum value and any lesser value shall be acceptable. Specify in micrometers (microinch).	1.6 / 3.5 √	Material removal by machining is required to produce the surface. The basic amount of stock provided forf material removal is specified at the left of the short leg of the symbol. Specify in millimeters (inch).
1.6 / 0.8 √	The specification of maximum and minimum roughness average values indicates permissible range of roughness. Specify in micrometers (microinch).	1.6 √	Removal of material is prohibited.
0.8 √ 0.005-5	Maximum waviness height rating is the first rating place above the horizontal extension. Any lesser rating shall be acceptable. Specify in millimeters (inch). Maximum waviness spacing rating is the second rating placed above the horizontal extension and to the right of the waviness height rating. Any lesser rating shall be acceptable. Specify in millimeters (inch).	0.8 √ ⊥	Lay designation is indicated by the lay symbol placed at the right of the long leg.
		0.8 √ 2.5	Roughness sampling length or cutoff rating is placed below the horizontal extension. When no value is shown, 0.80 mm (0.030 inch) applies. Specify in millimeters (inch).
		0.8 √ ⊥0.5	Where required maximum roughness spacing shall be placed at the right of the lay symbol. Any lesser rating shall be acceptable. Specify in millimeters (inch).

the same basis as machined surfaces. Castings are characterized by random distribution of non-directional deviations from the nominal surface.

Surfaces of castings rarely need control beyond that provided by the production method necessary to meet dimensional requirements. Comparison specimens are frequently used for evaluating surfaces having specific functional requirements. Surface texture control should not be specified unless required for appearance or function of the surface. Specification of such requirements may increase cost to the user.

Engineers should recognize that different areas of the same castings may have different surface textures. It is recommended that specifications of the surface be limited to defined areas of the casting. Practicality of and methods of determining that a casting's surface texture meets the specification shall be coordinated with the producer. The Society of Automotive Engineers standard J435 "Automotive Steel Castings" describes methods of evaluating steel casting surface texture used in the automotive and related industries.

Metric Dimensions on Drawings.—The length units of the metric system that are most generally used in connection with any work relating to mechanical engineering are the meter (39.37 inches) and the millimeter (0.03937 inch). One meter equals 1000 millimeters. On mechanical drawings, all dimensions are generally given in millimeters, no matter how large the dimensions may be. In fact, dimensions of such machines as locomotives and large electrical apparatus are given exclusively in millimeters. This practice is adopted to avoid mistakes due to misplacing decimal points, or misreading dimensions as when other units are used as well. When dimensions are given in millimeters, many of them can

Table 7. Examples of Special Designations

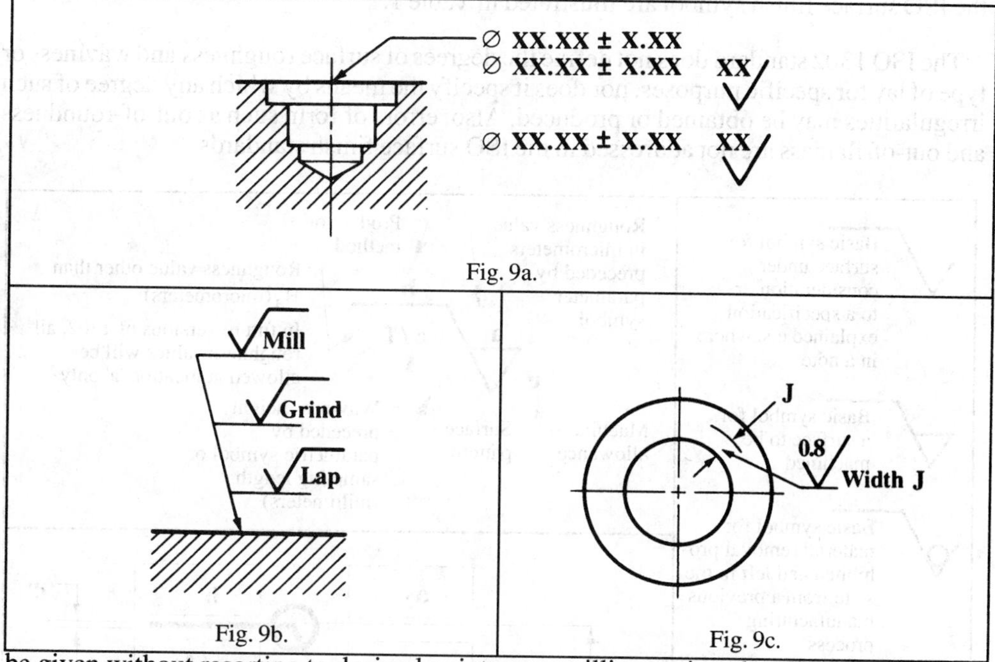

Fig. 9a. Fig. 9b. Fig. 9c.

be given without resorting to decimal points, as a millimeter is only a little more than $\frac{1}{32}$ inch. Only dimensions of precision need be given in decimals of a millimeter; such dimensions are generally given in hundredths of a millimeter—for example, 0.02 millimeter, which is equal to 0.0008 inch. As 0.01 millimeter is equal to 0.0004 inch, dimensions are seldom given with greater accuracy than to hundredths of a millimeter.

Scales of Metric Drawings: Drawings made to the metric system are not made to scales of $\frac{1}{2}$, $\frac{1}{4}$, $\frac{1}{8}$, etc., as with drawings made to the English system. If the object cannot be drawn full size, it may be drawn $\frac{1}{2}$, $\frac{1}{5}$, $\frac{1}{10}$, $\frac{1}{20}$, $\frac{1}{50}$, $\frac{1}{100}$, $\frac{1}{200}$, $\frac{1}{500}$, or $\frac{1}{1000}$ size. If the object is too small and has to be drawn larger, it is drawn 2, 5, or 10 times its actual size.

ISO Surface Finish

Differences Between ISO and ANSI Surface Finish Symbology.—ISO surface finish standards are comprised of numerous individual standards that taken as a whole form a set of standards roughly comparable in scope to American National Standard ANSI/ASME Y14.36M.

The primary standard dealing with surface finish, ISO 1302:1992, is concerned with the methods of specifying surface texture symbology and additional indications on engineering drawings. The parameters in ISO surface finish standards relate to surfaces produced by abrading, casting, coating, cutting, etching, plastic deformation, sintering, wear, erosion, and some other methods.

ISO 1302 defines how surface texture and its constituents, roughness, waviness, and lay, are specified on the symbology. Surface defects are specifically excluded from consideration during inspection of surface texture, but definitions of flaws and imperfections are discussed in ISO 8785.

As with American National Standard ASME Y14.36, ISO 1302 is not concerned with luster, appearance, color, corrosion resistance, wear resistance, hardness, sub-surface microstructure, surface integrity, and many other characteristics that may govern considerations in specific applications. Visually, the ISO surface finish symbol is similar to the ANSI symbol, but the proportions of the symbol in relationship to text height differs from

ANSI, as do some of the parameters as described in Fig. 1. Examples of the application of the ISO surface finish symbol are illustrated in Table 1.

The ISO 1302 standard does not define the degrees of surface roughness and waviness or type of lay for specific purposes, nor does it specify the means by which any degree of such irregularities may be obtained or produced. Also, errors of form such as out-of-roundness and out-of-flatness are not addressed in the ISO surface finish standards.

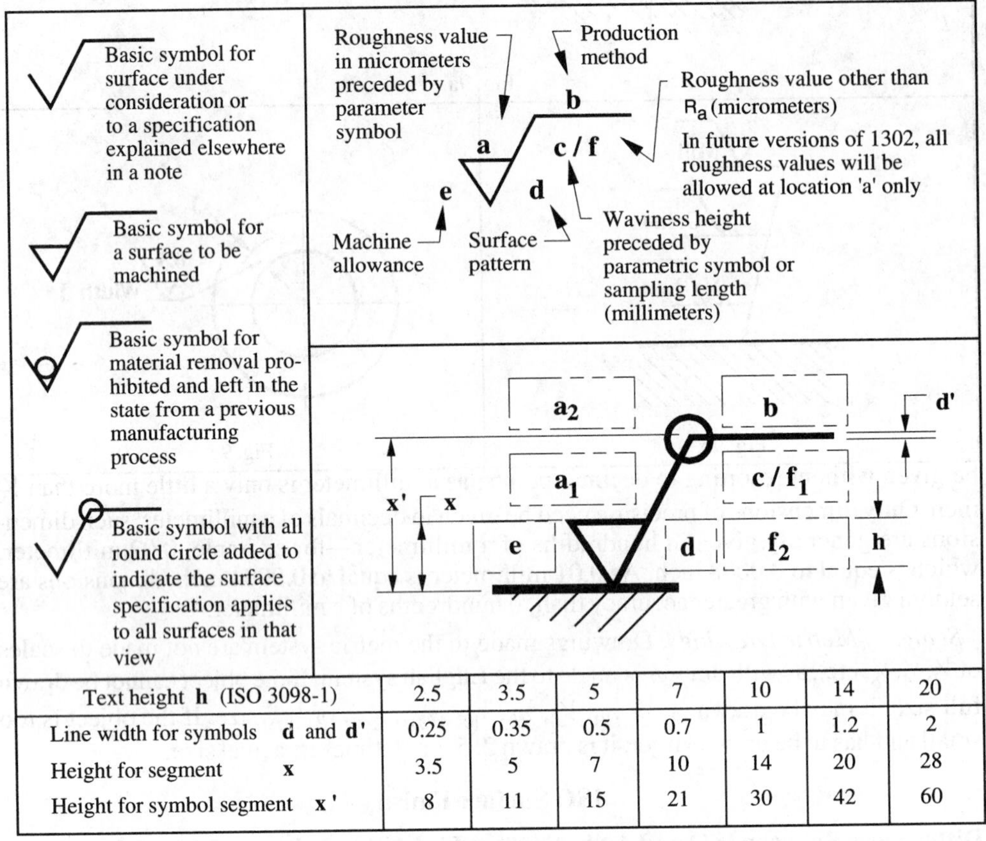

Text height **h** (ISO 3098-1)	2.5	3.5	5	7	10	14	20
Line width for symbols **d** and **d'**	0.25	0.35	0.5	0.7	1	1.2	2
Height for segment **x**	3.5	5	7	10	14	20	28
Height for symbol segment **x'**	8	11	15	21	30	42	60

Fig. 1. ISO Surface Finish Symbol

Other Iso Standards Related To Surface Finish

ISO 468:1982	"Surface roughness — parameters. Their values and general rules for specifying requirements."
ISO 4287:1997	"Surface texture: Profile method — Terms, definitions and surface texture parameters."
ISO 4288:1996	"Surface texture: Profile method — Rules and procedures for the assessment of surface texture." Includes specifications for precision reference specimens, and roughness comparison specimens, and establishes requirements for stylus-type instruments."
ISO 8785:1998	"Surface imperfections — Terms, definitions and parameters."
ISO 10135-1:CD	"Representation of parts produced by shaping processes — Part 1: Molded parts."

Table 1. Examples of ISO Applications of Surface Texture Symbology

Interpretation	Example
Surface roughness is produced by milling and between upper limit of Ra = 50 μm and Ra = 6.3 μm; direction of lay is crossed in oblique directions relative to plane of projection; sampling length is 5 mm.	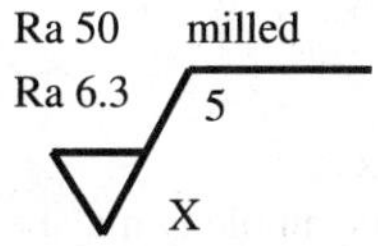 Ra 50 milled Ra 6.3 5 X
Surface roughness of Rz = 6.3 μm is the default for all surfaces as indicated by the Rz = 6.3 specification, plus basic symbol within parentheses. Any deviating specification is called out with local notes such as the Ra = 0.8 μm specification.	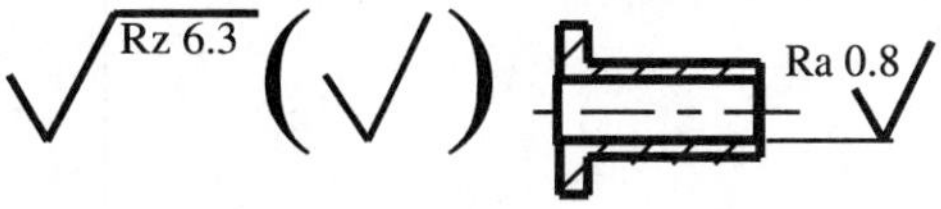 Rz 6.3 Ra 0.8
Surface roughness is produced by grinding to Ra = 1.2 μm and limited to Ry = 6.3 μm max; direction of lay is perpendicular relative to the plane of projection; sampling length is 2.4 mm.	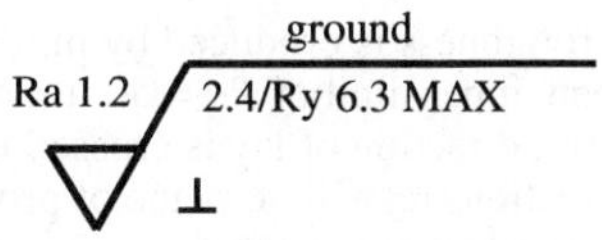 ground Ra 1.2 2.4/Ry 6.3 MAX ⊥
Surface treatment without any machining; nickel-chrome plated to Rz = 1 μm on all surfaces.	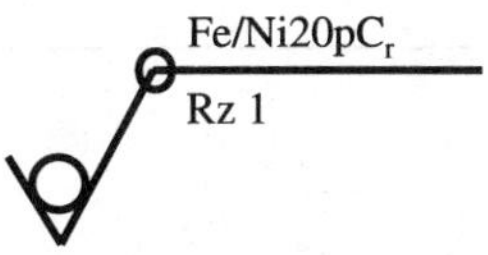 $Fe/Ni20pC_r$ Rz 1
Surface is nickel-chrome plated to roughness of Ra = 3.2 μm with a sampling length of 0.8 mm; limited to Rz = 16 μm to Rz = 6.3 μm with a sampling length of 2.5 mm.	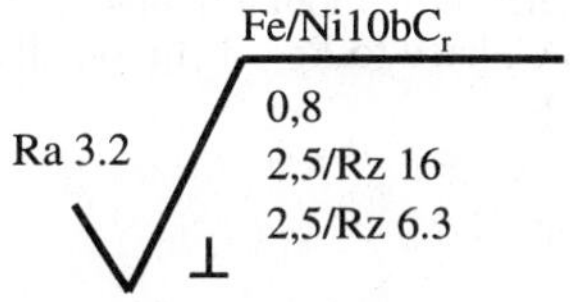 $Fe/Ni10bC_r$ 0,8 Ra 3.2 2,5/Rz 16 2,5/Rz 6.3 ⊥
Surface roughness of Rz = 6.3 μm is the default for all surfaces except the inside diameter which is Ra = 0.8 mm.	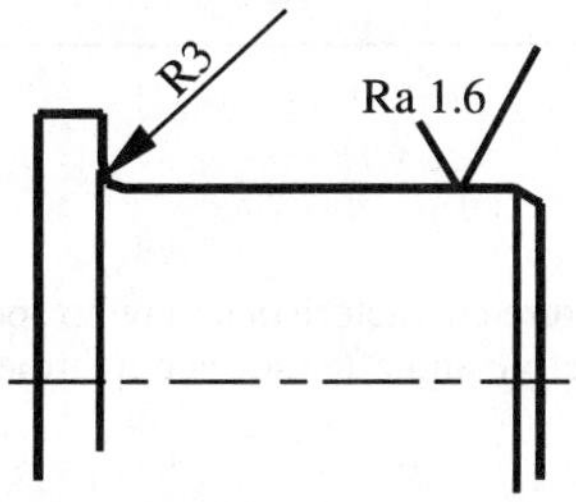 R3 Ra 1.6
Surface texture symbology may be combined with dimension leaders and witness (extension) lines.	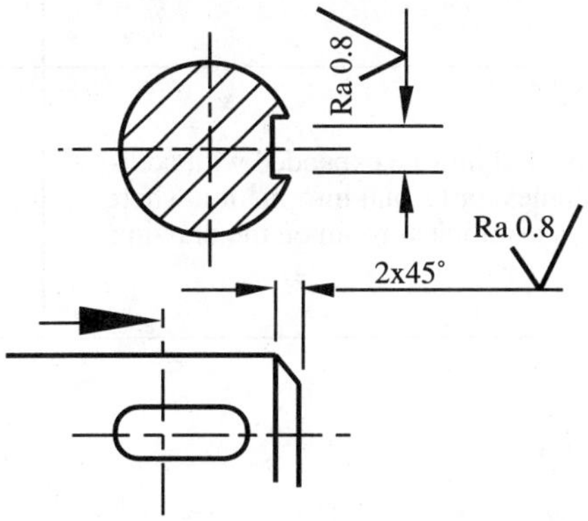 Ra 0.8 Ra 0.8 2x45°

Table 1. *(Continued)* **Examples of ISO Applications of Surface Texture Symbology**

Interpretation	Example
Surface texture symbology may be applied to extended extension lines or on extended projection lines.	Ra 1.6 Ra 0.8 Rz 4.0 R3 43 Ø45
Surface roughness is produced by milling and between upper limit of *Ra* = 50 μm and *Ra* = 6.3 μm; direction of lay is crossed in oblique directions relative to plane of projection; sampling length is 5 mm.	Rz 40 3x Ø5
Surface treatment without any machining; nickel-chrome plated to *Rz* = 1 μm on all surfaces.	Fe/Cr 50 Ry 6.2 Ry 1.6 Ø45 30
Surface texture characteristics may be specified both before and after surface treatment.	Chromium plated a_2 a_1 Ø
The symbol may be expanded with additional lines for textual information where there is insufficient room on the drawing.	Built-up surface Ground

ISO Surface Parameter Symbols

Rp = max height profile	*R*δ*c* = profile section height difference
Rv = max profile valley depth	*lp* = sampling length – primary profile
*Rz** = max height of the profile	*lw* = sampling length – waviness profile
Rc = mean height of profile	*lr* = sampling length – roughness profile
Rt = total height of the profile	*ln* = evaluation length
Ra = arithmetic mean deviation of the profile	*Z(x)* = ordinate value
Rq = root mean square deviation of the profile	*dZ/dX* = local slope
Rsk = skewness of the profile	*Zp* = profile peak height
Rku = kurtosis of the profile	*Zv* = profile valley depth
RSm = mean width of the profile	*Zt* = profile element height
*R*Δ*q* = root mean square slope of the profile	*Xs* = profile element width
Rmr = material ration of the profile	*Ml* = material length of profile

Rules for Comparing Measured Values to Specified Limits

Max rule: When a maximum requirement is specified for a surface finish parameter on a drawing (e.g. *Rz*1.5max), none of the inspected values may extend beyond the upper limit over the entire surface. MAX must be added to the parametric symbol in the surface finish symbology on the drawing.

16% rule: When upper and lower limits are specified, no more than 16% of all measured values of the selected parameter within the evaluation length may exceed the upper limit. No more than 16% of all measured values of the selected parameter within the evaluation length may be less than the lower limit.

Exceptions to the 16% rule: Where the measured values of roughness profiles being inspected follow a normal distribution, the 16% rule may be overridden. This is allowed when greater than 16% of the measured values exceed the upper limit, but the total roughness profile conforms with the sum of the arithmetic mean and standard deviation ($\mu + \sigma$). Effectively this means that the greater the value of σ, the further μ must be from the upper limit (see Fig. 2).

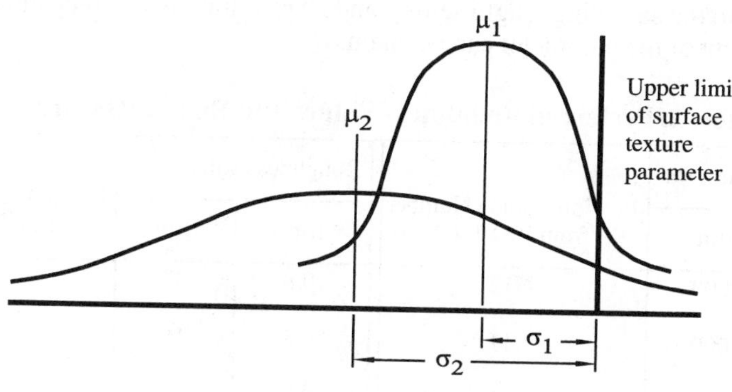

Fig. 2.

Basic rules for determining cut-off wavelength: When the sampling length is specified on the drawing or in documentation, the cut-off wavelength λc is equal to the sample length. When no sampling length is specified, the cut-off wavelength is estimated using Table .

Curves for Non-periodic Profiles such as Ground Surfaces		Curves for Periodic and Non-periodic Profiles	Sampling length, lr (mm)	Eevaluation length, ln (mm)
For Ra, Rq, Rsk, Rku, $R\Delta q$	For Rz, Rv, Rp, Rc, Rt	For R-parameters and RSm		
Ra, μm	Rz, $Rz1_{max}$, μm	RSm, μm		
$(0.006) < Ra \le 0.02$	$(0.025) < Rz, Rz1_{max} \le 0.1$	$0.013 < RSm \le 0.04$	0.08	0.4
$0.02 < Ra \le 0.1$	$0.1 < Rz, Rz1_{max} \le 0.5$	$0.04 < RSm \le 0.13$	0.25	1.25
$0.1 < Ra \le 2$	$0.5 < Rz, Rz1_{max} \le 10$	$0.13 < RSm \le 0.4$	0.8	4
$2 < Ra \le 10$	$10 < Rz, Rz1_{max} \le 50$	$0.4 < RSm \le 1.3$	2.5	12.5
$10 < Ra \le 80$	$50 < Rz, Rz1_{max} \le 200$	$1.3 < RSm \le 4$	8	40

Basic rules for measurement of roughness parameters: For non-periodic roughness the parameter Ra, Rz, $Rz1_{max}$ or RSm are first estimated using visual inspection, comparison to specimens, graphic analysis, etc. The sampling length is then selected from Table , based on the use of Ra, Rz, $Rz1_{max}$ or RSm. Then with instrumentation, a representative sample is taken using the sampling length chosen above.

The measured values are then compared to the ranges of values in Table for the particular parameter. If the value is outside the range of values for the estimated sampling length, the measuring instrument is adjusted for the next higher or lower sampling length and the measurement repeated. If the final setting corresponds to Table , then both the sampling length setting and Ra, Rz, $Rz1_{max}$ or RSm values are correct and a representative measurement of the parameter can be taken.

For periodic roughness, the parameter RSm is estimated graphically and the recommended cut-off values selected using Table . If the value is outside the range of values for the estimated sampling length, the measuring instrument is adjusted for the next higher or lower sampling length and the measurement repeated. If the final setting corresponds to Table , then both the sampling length setting and RSm values are correct and a representative measurement of the parameter can be taken.

Table 2. Preferred Roughness Values and Roughness Grades

Roughness values, Ra		Previous Grade Number from ISO 1302	Roughness values, Ra		Previous Grade Number from ISO 1302
μm	μin		μm	μin	
50	2000	N12	0.8	32	N6
25	1000	N11	0.4	16	N5
12.5	500	N10	0.2	8	N4
6.3	250	N9	0.1	4	N3
3.2	125	N8	0.05	2	N2
1.6	63	N7	0.025	1	N1

Gage Blocks

Precision Gage Blocks.—Precision gage blocks are usually purchased in sets comprising a specific number of blocks of different sizes. The nominal gage lengths of individual blocks in a set are determined mathematically so that particular desired lengths can be obtained by combining selected blocks. They are made to several different tolerance grades which categorize them as master blocks, calibration blocks, inspection blocks, and workshop blocks. *Master blocks* are employed as basic reference standards; *calibration blocks* are used for high precision gaging work and calibrating inspection blocks; *inspection blocks* are used as toolroom standards and for checking and setting limit and comparator gages, for example. The *workshop blocks* are working gages used as shop standards for a variety of direct precision measurements and gaging applications, including sine bar settings.

Federal Specification GGG-G-15C, Gage Blocks (see below), lists typical sets, and gives details of materials, design, and manufacturing requirements, and tolerance grades. When there is in a set no single block of the exact size that is wanted, two or more blocks are combined by "wringing" them together. Wringing is achieved by first placing one block crosswise on the other and applying some pressure. Then a swiveling motion is used to twist the blocks to a parallel position, causing them to adhere firmly to one another.

When combining blocks for a given dimension, the object is to use as few blocks as possible to obtain the dimension. The procedure for selecting blocks is based on successively eliminating the right-hand figure of the desired dimension.

Example: Referring to gage block set number 1 in Table 1, determine the blocks required to obtain 3.6742 inches. *Step 1:* Eliminate 0.0002 by selecting a 0.1002 block. Subtract 0.1002 from 3.6743 = 3.5740. *Step 2:* Eliminate 0.004 by selecting a 0.124 block. Subtract 0.124 from 3.5740 = 3.450. *Step 3:* Eliminate 0.450 with a block this size. Subtract 0.450 from 3.450 = 3.000. *Step 4:* Select a 3.000 inch block. The combined blocks are 0.1002 + 0.124 + 0.450 + 3.000 = 3.6742 inches.

Federal Specification for Gage Blocks, Inch and Metric Sizes.—This Specification, GGG-G-15C, March 20, 1975, which supersedes GGG-G-15B, November 6, 1970, covers design, manufacturing, and purchasing details for precision gage blocks in inch and metric sizes up to and including 20 inches and 500 millimeters gage lengths. The shapes of blocks are designated Style 1, which is rectangular; Style 2, which is square with a center accessory hole, and Style 3, which defines other shapes as may be specified by the purchaser. Blocks may be made from steel, chromium-plated steel, chromium carbide, or tungsten carbide. There are four tolerance grades, which are designated Grade 0.5 (formerly Grade AAA in the GGG-G-15A issue of the Specification); Grade 1 (formerly Grade AA); Grade 2 (formerly Grade A +); and Grade 3 (a compromise between former Grades A and B). Grade 0.5 blocks are special reference gages used for extremely high precision gaging work, and are not recommended for general use. Grade 1 blocks are laboratory reference standards used for calibrating inspection gage blocks and high precision gaging work. Grade 2 blocks are used as inspection and toolroom standards, and Grade 3 blocks are used as shop standards.

Inch and metric sizes of blocks in specific sets are given in Tables 1 and 2, which is not a complete list of available sizes. It should be noted that some gage blocks must be ordered as specials, some may not be available in all materials, and some may not be available from all manufacturers. Gage block set number 4 (88 blocks), listed in the Specification, is not given in Table 1. It is the same as set number 1 (81 blocks) but contains seven additional blocks measuring 0.0625, 0.078125, 0.093750, 0.100025, 0.100050, 0.100075, and 0.109375 inch. In Table 2, gage block set number 3M (112 blocks) is not given. It is similar to set number 2M (88 blocks), and the chief difference is the inclusion of a larger number of blocks in the 0.5 millimeter increment series up to 24.5 mm. Set numbers 5M (88 blocks), 6M (112 blocks), and 7M (17 blocks) also are not listed.

Table 1. Gage Block Sets—Inch Sizes *Federal Specification GGG-G-15C*

Set Number 1 (81 Blocks)									
First Series: 0.0001 Inch Increments (9 Blocks)									
.1001	.1002	.1003	.1004	.1005	.1006	.1007	.1008	.1009	
Second Series: 0.001 Inch Increments (49 Blocks)									
.101	.102	.103	.104	.105	.106	.107	.108	.109	.110
.111	.112	.113	.114	.115	.116	.117	.118	.119	.120
.121	.122	.123	.124	.125	.126	.127	.128	.129	.130
.131	.132	.133	.134	.135	.136	.137	.138	.139	.140
.141	.142	.143	.144	.145	.146	.147	.148	.149	
Third Series: 0.050 Inch Increments (19 Blocks)									
.050	.100	.150	.200	.250	.300	.350	.400	.450	.500
.550	.600	.650	.700	.750	.800	.850	.900	.950	
Fourth Series: 1.000 Inch Increments (4 Blocks)									
1.000	2.000	3.000	4.000						
Set Number 5 (21 Blocks)									
First Series: 0.0001 Inch Increments (9 Blocks)									
.0101	.0102	.0103	.0104	.0105	.0106	.0107	.0108	.0109	
Second Series: 0.001 Inch Increments (11 Blocks)									
.010	.011	.012	.013	.014	.015	.016	.017	.018	.019
.020									
One Block 0.01005 Inch									
Set Number 6 (28 Blocks)									
First Series: 0.0001 Inch Increments (9 Blocks)									
.0201	.0202	.0203	.0204	.0205	.0206	.0207	.0208	.0209	
Second Series: 0.001 Inch Increments (9 Blocks)									
.021	.022	.023	.024	.025	.026	.027	.028	.029	
Third Series: 0.010 Inch Increments (9 Blocks)									
.010	.020	.030	.040	.050	.060	.070	.080	.090	
One Block 0.02005 Inch									
Long Gage Block Set Number 7 (8 Blocks)									
Whole Inch Series (8 Blocks)									
5	6	7	8	10	12	16	20		
Set Number 8 (36 Blocks)									
First Series: 0.0001 Inch Increments (9 Blocks)									
.1001	.1002	.1003	.1004	.1005	.1006	.1007	.1008	.1009	
Second Series: 0.001 Inch Increments (11 Blocks)									
.100	.101	.102	.103	.104	.105	.106	.107	.108	.109
.110									
Third Series: 0.010 Inch Increments (8 Blocks)									
.120	.130	.140	.150	.160	.170	.180	.190		
Fourth Series: 0.100 Inch Increments (4 Blocks)									
.200	.300	.400	.500						
Whole Inch Series (3 Blocks)									
1	2	4							
One Block 0.050 Inch									
Set Number 9 (20 Blocks)									
First Series: 0.0001 Inch Increments (9 Blocks)									
.0501	.0502	.0503	.0504	.0505	.0506	.0507	.0508	.0509	
Second Series: 0.001 Inch Increments (10 Blocks)									
.050	.051	.052	.053	.054	.055	.056	.057	.058	.059
One Block 0.05005 Inch									

Set number 4 is not shown, and the Specification does not list a set 2 or 3.

Arranged here in incremental series for convenience of use.

Table 2. Gage Block Sets—Metric Sizes ***Federal Specification GGG-G-15C***

Set Number 1M (45 Blocks)									
First Series: 0.001 Millimeter Increments (9 Blocks)									
1.001	1.002	1.003	1.004	1.005	1.006	1.007	1.008	1.009	
Second Series: 0.01 Millimeter Increments (9 Blocks)									
1.01	1.02	1.03	1.04	1.05	1.06	1.07	1.08	1.09	
Third Series: 0.10 Millimeter Increments (9 Blocks)									
1.10	1.20	1.30	1.40	1.50	1.60	1.70	1.80	1.90	
Fourth Series: 1.0 Millimeter Increments (9 Blocks)									
1.0	2.0	3.0	4.0	5.0	6.0	7.0	8.0	9.0	
Fifth Series: 10 Millimeter Increments (9 Blocks)									
10	20	30	40	50	60	70	80	90	
Set Number 2M (88 Blocks)									
First Series: 0.001 Millimeter Increments (9 Blocks)									
1.001	1.002	1.003	1.004	1.005	1.006	1.007	1.008	1.009	
Second Series: 0.01 Millimeter Increments (49 Blocks)									
1.01	1.02	1.03	1.04	1.05	1.06	1.07	1.08	1.09	1.10
1.11	1.12	1.13	1.14	1.15	1.16	1.17	1.18	1.19	1.20
1.21	1.22	1.23	1.24	1.25	1.26	1.27	1.28	1.29	1.30
1.31	1.32	1.33	1.34	1.35	1.36	1.37	1.38	1.39	1.40
1.41	1.42	1.43	1.44	1.45	1.46	1.47	1.48	1.49	
Third Series: 0.50 Millimeter Increments (19 Blocks)									
0.5	1.0	1.5	2.0	2.5	3.0	3.5	4.0	4.5	5.0
5.5	6.0	6.5	7.0	7.5	8.0	8.5	9.0	9.5	
Fourth Series: 10 Millimeter Increments (10 Blocks)									
10	20	30	40	50	60	70	80	90	100
One Block 1.0005 mm									
Set Number 4M (45 Blocks)									
First Series: 0.001 Millimeter Increments (9 Blocks)									
2.001	2.002	2.003	2.004	2.005	2.006	2.007	2.008	2.009	
Second Series: 0.01 Millimeter Increments (9 Blocks)									
2.01	2.02	2.03	2.04	2.05	2.06	2.07	2.08	2.09	
Third Series: 0.10 Millimeter Increments (9 Blocks)									
2.1	2.2	2.3	2.4	2.5	2.6	2.7	2.8	2.9	
Fourth Series: 1 Millimeter Increments (9 Blocks)									
1.0	2.0	3.0	4.0	5.0	6.0	7.0	8.0	9.0	
Fifth Series: 10 Millimeter Increments (9 Blocks)									
10	20	30	40	50	60	70	80	90	
Long Gage Block Set Number 8M (8 Blocks)									
Whole Millimeter Series (8 Blocks)									
125	150	175	200	250	300	400	500		

Set numbers 3M, 5M, 6M, and 7M are not listed.

Arranged here in incremental series for convenience of use.

Note: Gage blocks measuring 1.09 millimeters and under in set number 1M, blocks measuring 1.5 millimeters and under in set number 2M, and block measuring 1.0 millimeter in set number 4M are not available in tolerance grade 0.5.

TOOLING AND TOOLMAKING

CUTTING TOOLS

CEMENTED CARBIDES

FORMING TOOLS

MILLING CUTTERS

REAMERS

TWIST DRILLS AND COUNTERBORES

TWIST DRILLS AND COUNTERBORES (cont.)

TAPS AND THREADING DIES

STANDARD TAPERS

STANDARD TAPERS (cont.)

BROACHES AND BROACHING

BROACHES AND BROACHING (cont.)

JIGS AND FIXTURES

CUTTING TOOLS

Tool Contour.— Tools for turning, planing, etc., are made in straight, bent, offset, and other forms to place the cutting edges in convenient positions for operating on differently located surfaces. The contour or shape of the cutting edge may also be varied to suit different classes of work. Tool shapes, however, are not only related to the kind of operation, but, in roughing tools particularly, the contour may have a decided effect upon the cutting efficiency of the tool. To illustrate, an increase in the side cutting-edge angle of a roughing tool, or in the nose radius, tends to permit higher cutting speeds because the chip will be thinner for a given feed rate. Such changes, however, may result in chattering or vibrations unless the work and the machine are rigid; hence, the most desirable contour may be a compromise between the ideal form and one that is needed to meet practical requirements.

Terms and Definitions.— The terms and definitions relating to single-point tools vary somewhat in different plants, but the following are in general use.

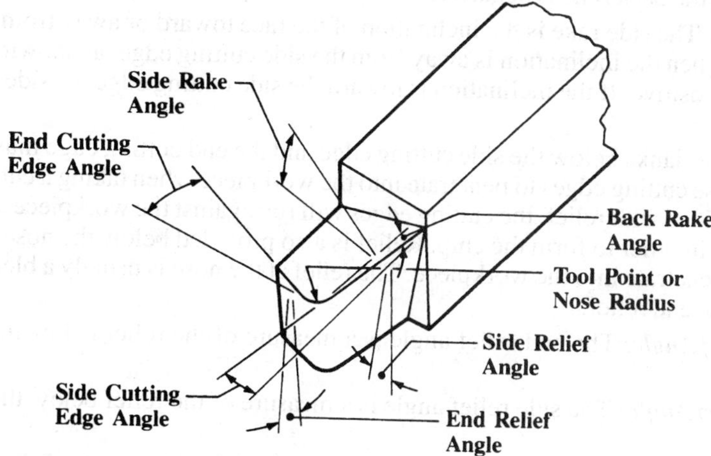

Fig. 1. Terms Applied to Single-point Turning Tools

Single-point Tool: This term is applied to tools for turning, planing, boring, etc., which have a cutting edge at one end. This cutting edge may be formed on one end of a solid piece of steel, or the cutting part of the tool may consist of an insert or tip which is held to the body of the tool by brazing, welding, or mechanical means.

Shank: The shank is the main body of the tool. If the tool is an inserted cutter type, the shank supports the cutter or bit. (See diagram, Fig. 1.)

Nose: A general term sometimes used to designate the cutting end but usually relating more particularly to the rounded tip of the cutting end.

Face: The surface against which the chips bear, as they are severed in turning or planing operations, is called the face.

Flank: The flank is that end surface adjacent to the cutting edge and below it when the tool is in a horizontal position as for turning.

Base: The base is the surface of the tool shank that bears against the supporting toolholder or block.

Side Cutting Edge: The side cutting edge is the cutting edge on the side of the tool. Tools such as shown in Fig. 1 do the bulk of the cutting with this cutting edge and are, therefore, sometimes called side cutting edge tools.

End Cutting Edge: The end cutting edge is the cutting edge at the end of the tool.

On side cutting edge tools, the end cutting edge can be used for light plunging and facing cuts. Cutoff tools and similar tools have only one cutting edge located on the end. These tools and other tools that are intended to cut primarily with the end cutting edge are sometimes called end cutting edge tools.

Rake: A metal-cutting tool is said to have rake when the tool face or surface against which the chips bear as they are being severed, is inclined for the purpose of either increasing or diminishing the keenness or bluntness of the edge. The magnitude of the rake is most conveniently measured by two angles called the back rake angle and the side rake angle. The tool shown in Fig. 1 has rake. If the face of the tool did not incline but was parallel to the base, there would be no rake; the rake angles would be zero.

Positive Rake: If the inclination of the tool face is such as to make the cutting edge keener or more acute than when the rake angle is zero, the rake angle is defined as positive.

Negative Rake: If the inclination of the tool face makes the cutting edge less keen or more blunt than when the rake angle is zero, the rake is defined as negative.

Back Rake: The back rake is the inclination of the face toward or away from the end or the end cutting edge of the tool. When the inclination is away from the end cutting edge, as shown in Fig. 1, the back rake is positive. If the inclination is downward toward the end cutting edge the back rake is negative.

Side Rake: The side rake is the inclination of the face toward or away from the side cutting edge. When the inclination is away from the side cutting edge, as shown in Fig. 1, the side rake is positive. If the inclination is toward the side cutting edge the side rake is negative.

Relief: The flanks below the side cutting edge and the end cutting edge must be relieved to allow these cutting edges to penetrate into the workpiece when taking a cut. If the flanks are not provided with relief, the cutting edges will rub against the workpiece and be unable to penetrate in order to form the chip. Relief is also provided below the nose of the tool to allow it to penetrate into the workpiece. The relief at the nose is usually a blend of the side relief and the end relief.

End Relief Angle: The end relief angle is a measure of the relief below the end cutting edge.

Side Relief Angle: The side relief angle is a measure of the relief below the side cutting edge.

Back Rake Angle: The back rake angle is a measure of the back rake. It is measured in a plane that passes through the side cutting edge and is perpendicular to the base. Thus, the back rake angle can be defined by measuring the inclination of the side cutting edge with respect to a line or plane that is parallel to the base. The back rake angle may be positive, negative, or zero depending upon the magnitude and direction of the back rake.

Side Rake Angle: The side rake angle is a measure of the side rake. This angle is always measured in a plane that is perpendicular to the side cutting edge and perpendicular to the base. Thus, the side rake angle is the angle of inclination of the face perpendicular to the side cutting edge with reference to a line or a plane that is parallel to the base.

End Cutting Edge Angle: The end cutting edge angle is the angle made by the end cutting edge with respect to a plane perpendicular to the axis of the tool shank. It is provided to allow the end cutting edge to clear the finish machined surface on the workpiece.

Side Cutting Edge Angle: The side cutting edge angle is the angle made by the side cutting edge and a plane that is parallel to the side of the shank.

Nose Radius: The nose radius is the radius of the nose of the tool. The performance of the tool, in part, is influenced by nose radius so that it must be carefully controlled.

Lead Angle: The lead angle, shown in Fig. 2, is not ground on the tool. It is a tool setting angle which has a great influence on the performance of the tool. The lead angle is bounded by the side cutting edge and a plane perpendicular to the workpiece surface when the tool is in position to cut; or, more exactly, the lead angle is the angle between the side cutting edge and a plane perpendicular to the direction of the feed travel.

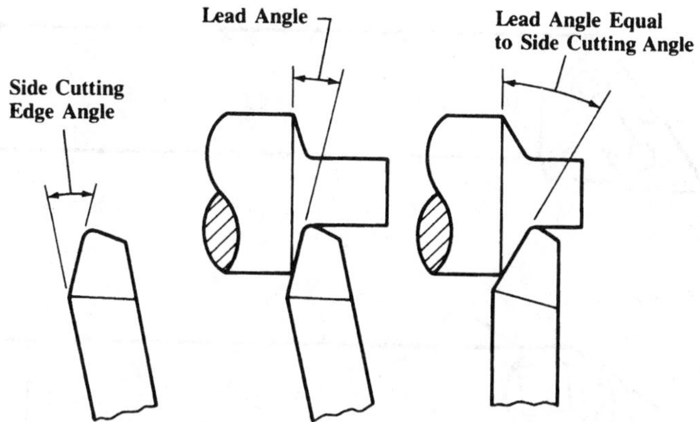

Fig. 2. Lead Angle on Single-point Turning Tool

Solid Tool: A solid tool is a cutting tool made from one piece of tool material.

Brazed Tool: A brazed tool is a cutting tool having a blank of cutting-tool material permanently brazed to a steel shank.

Blank: A blank is an unground piece of cutting-tool material from which a brazed tool is made.

Tool Bit: A tool bit is a relatively small cutting tool that is clamped in a holder in such a way that it can readily be removed and replaced. It is intended primarily to be reground when dull and not indexed.

Tool-bit Blank: The tool-bit blank is an unground piece of cutting-tool material from which a tool bit can be made by grinding. It is available in standard sizes and shapes.

Tool-bit Holder: Usually made from forged steel, the tool-bit holder is used to hold the tool bit, to act as an extended shank for the tool bit, and to provide a means for clamping in the tool post.

Straight-shank Tool-bit Holder: A straight-shank tool-bit holder has a straight shank when viewed from the top. The axis of the tool bit is held parallel to the axis of the shank.

Offset-shank Tool-bit Holder: An offset-shank tool-bit holder has the shank bent to the right or left, as seen in Fig. 3. The axis of the tool bit is held at an angle with respect to the axis of the shank.

Side cutting Tool: A side cutting tool has its major cutting edge on the side of the cutting part of the tool. The major cutting edge may be parallel or at an angle with respect to the axis of the tool.

Indexable Inserts: An indexable insert is a relatively small piece of cutting-tool material that is geometrically shaped to have two or several cutting edges that are used until dull. The insert is then indexed on the holder to apply a sharp cutting edge. When all the cutting edges have been dulled, the insert is discarded. The insert is held in a pocket or against other locating surfaces on an indexable insert holder by means of a mechanical clamping device that can be tightened or loosened easily.

Indexable Insert Holder: Made of steel, an indexable insert holder is used to hold indexable inserts. It is equipped with a mechanical clamping device that holds the inserts firmly in a pocket or against other seating surfaces.

Straight-shank Indexable Insert Holder: A straight-shank indexable insert tool-holder is essentially straight when viewed from the top, although the cutting edge of the insert may be oriented parallel, or at an angle to, the axis of the holder.

Offset-shank Indexable Insert Holder: An offset-shank indexable insert holder has the head end, or the end containing the insert pocket, offset to the right or left, as shown in Fig. 3.

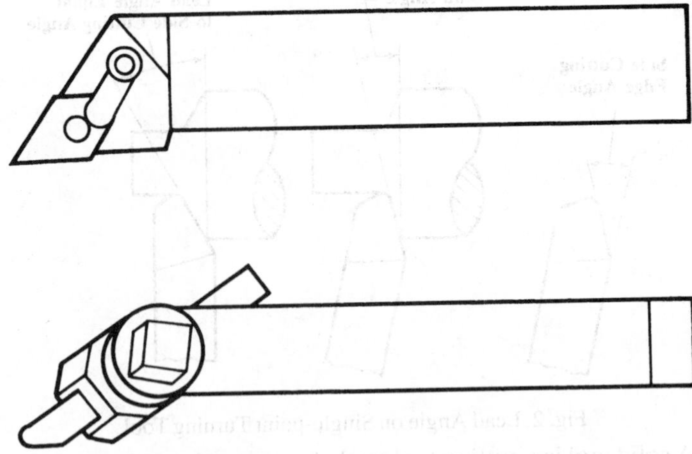

Fig. 3. Top: Right-hand Offset-shank, Indexable Insert Holder
Bottom: Right-hand Offset-shank Tool-bit Holder

End cutting Tool: An end cutting tool has its major cutting edge on the end of the cutting part of the tool. The major cutting edge may be perpendicular or at an angle, with respect to the axis of the tool.

Curved Cutting-edge Tool: A curved cutting-edge tool has a continuously variable side cutting edge angle. The cutting edge is usually in the form of a smooth, continuous curve along its entire length, or along a large portion of its length.

Right-hand Tool: A right-hand tool has the major, or working, cutting edge on the right-hand side when viewed from the cutting end with the face up. As used in a lathe, such a tool is usually fed into the work from right to left, when viewed from the shank end.

Left-hand Tool: A left-hand tool has the major or working cutting edge on the left-hand side when viewed from the cutting end with the face up. As used in a lathe, the tool is usually fed into the work from left to right, when viewed from the shank end.

Neutral-hand Tool: A neutral-hand tool is a tool to cut either left to right or right to left; or the cut may be parallel to the axis of the shank as when plunge cutting.

Chipbreaker: A groove formed in or on a shoulder on the face of a turning tool back of the cutting edge to break up the chips and prevent the formation of long,continuous chips which would be dangerous to the operator and also bulky and cumbersome to handle. A chipbreaker of the shoulder type may be formed directly on the tool face or it may consist of a separate piece that is held either by brazing or by clamping.

Relief Angles.—The end relief angle and the side relief angle on single-point cutting tools are usually, though not invariably, made equal to each other. The relief angle under the nose of the tool is a blend of the side and end relief angles.

The size of the relief angles has a pronounced effect on the performance of the cutting tool. If the relief angles are too large, the cutting edge will be weakened and in danger of breaking when a heavy cutting load is placed on it by a hard and tough material. On finish cuts, rapid wear of the cutting edge may cause problems with size control on the part. Relief angles that are too small will cause the rate of wear on the flank of the tool below the cutting edge to increase, thereby significantly reducing the tool life. In general, when cutting hard and tough materials, the relief angles should be 6 to 8 degrees for high-speed steel tools and 5 to 7 degrees for carbide tools. For medium steels, mild steels, cast iron, and other average work the recommended values of the relief angles are 8 to 12 degrees for high-speed steel tools and 5 to 10 degrees for carbides. Ductile materials having a relatively low modulus of elasticity should be cut using larger relief angles. For example, the relief angles recommended for turning copper, brass, bronze, aluminum, ferritic malleable iron, and similar metals are 12 to 16 degrees for high-speed steel tools and 8 to 14 degrees for carbides.

Larger relief angles generally tend to produce a better finish on the finish machined surface because less surface of the worn flank of the tool rubs against the workpiece. For this reason, single-point thread-cutting tools should be provided with relief angles that are as large as circumstances will permit. Problems encountered when machining stainless steel may be overcome by increasing the size of the relief angle. The relief angles used should never be smaller than necessary.

Rake Angles.—Machinability tests have confirmed that when the rake angle along which the chip slides, called the true rake angle, is made larger in the positive direction, the cutting force and the cutting temperature will decrease. Also, the tool life for a given cutting speed will increase with increases in the true rake angle up to an optimum value, after which it will decrease again. For turning tools which cut primarily with the side cutting edge, the true rake angle corresponds rather closely with the side rake angle except when taking shallow cuts. Increasing the side rake angle in the positive direction lowers the cutting force and the cutting temperature, while at the same time it results in a longer tool life or a higher permissible cutting speed up to an optimum value of the side rake angle. After the optimum value is exceeded, the cutting force and the cutting temperature will continue to drop; however, the tool life and the permissible cutting speed will decrease.

As an approximation, the magnitude of the cutting force will decrease about one per cent per degree increase in the side rake angle. While not exact, this rule of thumb does correspond approximately to test results and can be used to make rough estimates. Of course, the cutting force also increases about one per cent per degree decrease in the side rake angle. The limiting value of the side rake angle for optimum tool life or cutting speed depends upon the work material and the cutting tool material. In general, lower values can be used for hard and tough work materials. Cemented carbides are harder and more brittle than high-speed steel; therefore, the rake angles usually used for cemented carbides are less positive than for high-speed steel.

Negative rake angles cause the face of the tool to slope in the opposite direction from positive rake angles and, as might be expected, they have an opposite effect. For side cutting edge tools, increasing the side rake angle in a negative direction will result in an increase in the cutting force and an increase in the cutting temperature of approximately one per cent per degree change in rake angle. For example, if the side rake angle is changed from 5 degrees positive to 5 degrees negative, the cutting force will be about 10 per cent larger. Usually the tool life will also decrease when negative side rake angles are used, although the tool life will sometimes increase when the negative rake angle is not too large and when a fast cutting speed is used.

Negative side rake angles are usually used in combination with negative back rake angles on single-point cutting tools. The negative rake angles strengthen the cutting edges enabling them to sustain heavier cutting loads and shock loads. They are recommended for turning very hard materials and for heavy interrupted cuts. There is also an economic advantage in favor of using negative rake indexable inserts and tool holders inasmuch as the cutting edges provided on both the top and bottom of the insert can be used.

On turning tools that cut primarily with the side cutting edge, the effect of the back rake angle alone is much less than the effect of the side rake angle although the direction of the change in cutting force, cutting temperature, and tool life is the same. The effect that the back rake angle has can be ignored unless, of course, extremely large changes in this angle are made. A positive back rake angle does improve the performance of the nose of the tool somewhat and is helpful in taking light finishing cuts. A negative back rake angle strengthens the nose of the tool and is helpful when interrupted cuts are taken. The back rake angle has a very significant effect on the performance of end cutting edge tools, such as cut-off tools. For these tools, the effect of the back rake angle is very similar to the effect of the side rake angle on side cutting edge tools.

Side Cutting Edge and Lead Angles.—These angles are considered together because the side cutting edge angle is usually designed to provide the desired lead angle when the tool is being used. The side cutting edge angle and the lead angle will be equal when the shank of the cutting tool is positioned perpendicular to the workpiece, or, more correctly, perpendicular to the direction of the feed. When the shank is not perpendicular, the lead angle is determined by the side cutting edge and an imaginary line perpendicular to the feed direction.

The flow of the chips over the face of the tool is approximately perpendicular to the side cutting edge except when shallow cuts are taken. The thickness of the undeformed chip is measured perpendicular to the side cutting edge. As the lead angle is increased, the length of chip in contact with the side cutting edge is increased, and the chip will become longer and thinner. This effect is the same as increasing the depth of cut and decreasing the feed, although the actual depth of cut and feed remain the same and the same amount of metal is removed. The effect of lengthening and thinning the chip by increasing the lead angle is very beneficial as it increases the tool life for a given cutting speed or that speed can be increased. Increasing the cutting speed while the feed and the tool life remain the same leads to faster production.

However, an adverse effect must be considered. Chatter can be caused by a cutting edge that is oriented at a high lead angle when turning and sometimes, when turning long and slender shafts, even a small lead angle can cause chatter. In fact, an unsuitable lead angle of the side cutting edge is one of the principal causes of chatter. When chatter occurs, often simply reducing the lead angle will cure it. Sometimes, very long and slender shafts can be turned successfully with a tool having a zero degree lead angle (and having a small nose radius). Boring bars, being usually somewhat long and slender, are also susceptible to chatter if a large lead angle is used. The lead angle for boring bars should be kept small, and for very long and slender boring bars a zero degree lead angle is recommended. It is impossible to provide a rule that will determine when chatter caused by a lead angle will occur and when it will not. In making a judgment, the first consideration is the length to diameter ratio of the part to be turned, or of the boring bar. Then the method of holding the workpiece must be considered — a part that is firmly held is less apt to chatter. Finally, the overall condition and rigidity of the machine must be considered because they may be the real cause of chatter.

Although chatter can be a problem, the advantages gained from high lead angles are such that the lead angle should be as large as possible at all times.

End Cutting Edge Angle.—The size of the end cutting edge angle is important when tool wear by cratering occurs. Frequently, the crater will enlarge until it breaks through the end cutting edge just behind the nose, and tool failure follows shortly. Reducing the size of the end cutting edge angle tends to delay the time of crater breakthrough. When cratering takes place, the recommended end cutting edge angle is 8 to 15 degrees. If there is no cratering, the angle can be made larger. Larger end cutting edge angles may be required to enable profile turning tools to plunge into the work without interference from the end cutting edge.

Nose Radius.—The tool nose is a very critical part of the cutting edge since it cuts the finished surface on the workpiece. If the nose is made to a sharp point, the finish machined surface will usually be unacceptable and the life of the tool will be short. Thus, a nose radius is required to obtain an acceptable surface finish and tool life. The surface finish obtained is determined by the feed rate and by the nose radius if other factors such as the work material, the cutting speed, and cutting fluids are not considered. A large nose radius will give a better surface finish and will permit a faster feed rate to be used.

Machinability tests have demonstrated that increasing the nose radius will also improve the tool life or allow a faster cutting speed to be used. For example, high-speed steel tools were used to turn an alloy steel in one series of tests where complete or catastrophic tool failure was used as a criterion for the end of tool life. The cutting speed for a 60-minute tool life was found to be 125 fpm when the nose radius was $\frac{1}{16}$ inch and 160 fpm when the nose radius was $\frac{1}{4}$ inch.

A very large nose radius can often be used but a limit is sometimes imposed because the tendency for chatter to occur is increased as the nose radius is made larger. A nose radius that is too large can cause chatter and when it does, a smaller nose radius must be used on the tool. It is always good practice to make the nose radius as large as is compatible with the operation being performed.

Chipbreakers.—Many steel turning tools are equipped with chipbreaking devices to prevent the formation of long continuous chips in connection with the turning of steel at the high speeds made possible by high-speed steel and especially cemented carbide tools. Long steel chips are dangerous to the operator, and cumbersome to handle, and they may twist around the tool and cause damage. Broken chips not only occupy less space, but permit a better flow of coolant to the cutting edge. Several different forms of chipbreakers are illustrated in Fig. 4.

Angular Shoulder Type: The angular shoulder type shown at *A* is one of the commonly used forms. As the enlarged sectional view shows, the chipbreaking shoulder is located back of the cutting edge. The angle *a* between the shoulder and cutting edge may vary from 6 to 15 degrees or more, 8 degrees being a fair average. The ideal angle, width *W* and depth *G*, depend upon the speed and feed, the depth of cut, and the material. As a general rule, width *W*, at the end of the tool, varies from $\frac{3}{32}$ to $\frac{7}{32}$ inch, and the depth *G* may range from $\frac{1}{64}$ to $\frac{1}{16}$ inch. The shoulder radius equals depth *G*. If the tool has a large nose radius, the corner of the shoulder at the nose end may be beveled off, as illustrated at *B*, to prevent it from coming into contact with the work. The width *K* for type *B* should equal approximately 1.5 times the nose radius.

Parallel Shoulder Type: Diagram C shows a design with a chipbreaking shoulder that is parallel with the cutting edge. With this form, the chips are likely to come off in short curled sections. The parallel form may also be applied to straight tools which do not have a side cutting-edge angle. The tendency with this parallel shoulder form is to force the chips against the work and damage it.

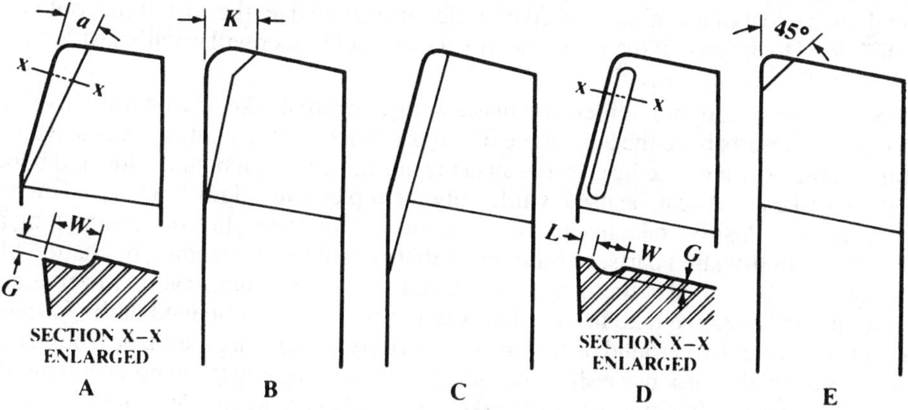

Fig. 4. Different Forms of Chipbreakers for Turning Tools

Groove Type: This type (diagram D) has a groove in the face of the tool produced by grinding. Between the groove and the cutting edge, there is a land *L*. Under ideal conditions, this width *L*, the groove width *W*, and the groove depth *G*, would be varied to suit the feed, depth of cut and material. For average use, *L* is about $\frac{1}{32}$ inch; *G*, $\frac{1}{32}$ inch; and *W*, $\frac{1}{16}$ inch. There are differences of opinion concerning the relative merits of the groove type and the shoulder type. Both types have proved satisfactory when properly proportioned for a given class of work.

Chipbreaker for Light Cuts: Diagram E illustrates a form of chipbreaker that is sometimes used on tools for finishing cuts having a maximum depth of about $\frac{1}{32}$ inch. This chipbreaker is a shoulder type having an angle of 45 degrees and a maximum width of about $\frac{1}{16}$ inch. It is important in grinding all chipbreakers to give the chip-bearing surfaces a fine finish, such as would be obtained by honing. This finish greatly increases the life of the tool.

Planing Tools.—Many of the principles which govern the shape of turning tools also apply in the grinding of tools for planing. The amount of rake depends upon the hardness of the material, and the direction of the rake should be away from the *working part* of the cutting edge. The angle of clearance should be about 4 or 5 degrees for planer tools, which is less than for lathe tools. This small clearance is allowable because a planer tool is held about square with the platen, whereas a lathe tool, the height and inclination of which can be varied, may not always be clamped in the same position.

Carbide Tools: Carbide tools for planing usually have negative rake. Round-nose and square-nose end-cutting tools should have a "negative back rake" (or front rake) of 2 or 3 degrees. Side cutting tools may have a negative back rake of 10 degrees, a negative side rake of 5 degrees, and a side cutting-edge angle of 8 degrees.

Indexable Inserts.—A large proportion of the cemented carbide, single-point cutting tools are indexable inserts and indexable insert tool holders. Dimensional specifications for solid sintered carbide indexable inserts are given inAmerican National Standard ANSI B212.12-1991. Samples of the many insert shapes are shown in Table 3. Most modern, cemented carbide, face milling cutters are of the indexable insert type. Larger size end milling cutters, side milling or slotting cutters, boring tools, and a wide variety of special tools are made to use indexable inserts. These inserts are primarily made from cemented carbide, although most of the cemented oxide cutting tools are also indexable inserts.

The objective of this type of tooling is to provide an insert with several cutting edges. When an edge is worn, the insert is indexed in the tool holder until all the cutting edges are used up, after which it is discarded. The insert is not intended to be reground. The advantages are that the cutting edges on the tool can be rapidly changed without removing the tool holder from the machine, tool-grinding costs are eliminated, and the cost of the insert is less than the cost of a similar, brazed carbide tool. Of course, the cost of the tool holder must be added to the cost of the insert; however, one tool holder will usually last for a long time before it, too, must be replaced.

Indexable inserts and tool holders are made with a negative rake or with a positive rake. Negative rake inserts have the advantage of having twice as many cutting edges available as comparable positive rake inserts, because the cutting edges on both the top and bottom of negative rake inserts can be used, while only the top cutting edges can be used on positive rake inserts. Positive rake inserts have a distinct advantage when machining long and slender parts, thin-walled parts, or other parts that are subject to bending or chatter when the cutting load is applied to them, because the cutting force is significantly lower as compared to that for negative rake inserts. Indexable inserts can be obtained in the following forms: utility ground, or ground on top and bottom only; precision ground, or ground on all surfaces; prehoned to produce a slight rounding of the cutting edge; and precision molded, which are unground. Positive-negative rake inserts also are available. These inserts are held on a negative-rake tool holder and have a chipbreaker groove that is formed to produce an effective positive-rake angle while cutting. Cutting edges may be available on the top surface only, or on both top and bottom surfaces. The positive-rake chipbreaker surface may be ground or precision molded on the insert.

Many materials, such as gray cast iron, form a discontinuous chip. For these materials an insert that has plain faces without chipbreaker grooves should always be used. Steels and other ductile materials form a continuous chip that must be broken into small segments when machined on lathes and planers having single-point, cemented-carbide and cemented-oxide cutting tools; otherwise, the chips can cause injury to the operator. In this case a chipbreaker must be used. Some inserts are made with chipbreaker grooves molded or ground directly on the insert. When inserts with plain faces are used, a cemented-carbide plate-type chipbreaker is clamped on top of the insert.

Identification System for Indexable Inserts.—The size of indexable inserts is determined by the diameter of an inscribed circle (I.C.), except for rectangular and parallelogram inserts where the length and width dimensions are used. To describe an insert in its entirety, a standard ANSI B212.4-1986 identification system is used where each position number designates a feature of the insert. The ANSI Standard includes items now commonly used and facilitates identification of items not in common use. Identification consists of up to ten positions; each position defines a characteristic of the insert as shown below:

1	2	3	4	5	6	7	8[a]	9[a]	10[a]
T	**N**	**M**	**G**	**5**	**4**	**3**			**A**

[a] Eighth, Ninth, and Tenth Positions are used only when required.

1) *Shape:* The shape of an insert is designated by a letter: **R** for round; **S,** square; **T,** triangle; **A,** 85° parallelogram; **B,** 82° parallelogram; **C,** 80° diamond; **D,** 55° diamond; **E,** 75° diamond; **H,** hexagon; **K,** 55° parallelogram; **L,** rectangle; **M,** 86° diamond; **O,** octagon; **P,** pentagon; **V,** 35° diamond; and **W,** 80° trigon.

2) *Relief Angle (Clearances):* The second position is a letter denoting the relief angles; **N** for 0°; **A,** 3°; **B,** 5°; **C,** 7°; **P,** 11°; **D,** 15°; **E,** 20°; **F,** 25°; **G,** 30°; **H,** 0° & 11°*; **J,** 0° & 14°*; **K,** 0° & 17°*; **L,** 0° & 20°*; **M,** 11° & 14°*; **R,** 11° & 17°*; **S,** 11° & 20°*. When mounted on a holder, the actual relief angle may be different from that on the insert.

3) *Tolerances:* The third position is a letter and indicates the tolerances which control the indexability of the insert. Tolerances specified do not imply the method of manufacture.

Symbol	Tolerance (± from nominal) Inscribed Circle, Inch	Tolerance (± from nominal) Thicknes, Inch	Symbol	Tolerance (± from nominal) Inscribed Circle, Inch	Tolerance (± from nominal) Thickness, Inch
A	0.001	0.001	**H**	0.0005	0.001
B	0.001	0.005	**J**	0.002–0.005	0.001
C	0.001	0.001	**K**	0.002–0.005	0.001
D	0.001	0.005	**L**	0.002–0.005	0.001
E	0.001	0.001	**M**	0.002–0.004[a]	0.005
F	0.0005	0.001	**U**	0.005–0.010[a]	0.005
G	0.001	0.005	**N**	0.002–0.004[a]	0.001

[a] Exact tolerance is determined by size of insert. See ANSI B94.25.

4) *Type:* The type of insert is designated by a letter. **A,** with hole; **B,** with hole and countersink; **C,** with hole and two countersinks; **F,** chip grooves both surfaces, no hole; **G,** same as **F** but with hole; **H,** with hole, one countersink, and chip groove on one rake surface; **J,** with hole, two countersinks and chip grooves on two rake surfaces; **M,** with hole and chip groove on one rake surface; **N,** without hole; **Q,** with hole and two countersinks; **R,** without hole but with chip groove on one rake surface; **T,** with hole, one countersink, and chip groove on one rake face; **U,** with hole, two countersinks, and chip grooves on two rake faces; and **W,** with hole and one countersink. *Note:* a dash may be used after position 4 to

* Second angle is secondary facet angle, which may vary by ± 1°.

separate the shape-describing portion from the following dimensional description of the insert and is not to be considered a position in the standard description.

5) *Size:* The size of the insert is designated by a one- or a two-digit number. For regular polygons and diamonds, it is the number of eighths of an inch in the nominal size of the inscribed circle, and will be a one- or two-digit number when the number of eighths is a whole number. It will be a two-digit number, including one decimal place, when it is not a whole number. Rectangular and parallelogram inserts require two digits: the first digit indicates the number of eighths of an inch width and the second digit, the number of quarters of an inch length.

6) *Thickness:* The thickness is designated by a one- or two-digit number, which indicates the number of sixteenths of an inch in the thickness of the insert. It is a one-digit number when the number of sixteenths is a whole number; it is a two-digit number carried to one decimal place when the number of sixteenths of an inch is not a whole number.

7) *Cutting Point Configuration:* The cutting point, or nose radius, is designated by a number representing $\frac{1}{64}$ths of an inch; a flat at the cutting point or nose, is designated by a letter: **0** for sharp corner; **1,** $\frac{1}{64}$ inch radius; **2,** $\frac{1}{32}$ inch radius; **3,** $\frac{3}{64}$inch radius; **4,** $\frac{1}{16}$ inch radius; **5,** $\frac{5}{64}$ inch radius; **6,** $\frac{3}{32}$ inch radius; **7,** $\frac{7}{64}$ inch radius; **8,** $\frac{1}{8}$ inch radius; **A,** square insert with 45° chamfer; **D,** square insert with 30° chamfer; **E,** square insert with 15° chamfer; **F,** square insert with 3° chamfer; **K,** square insert with 30° double chamfer; **L,** square insert with 15° double chamfer; **M,** square insert with 3° double chamfer; **N,** truncated triangle insert; and **P,** flatted corner triangle insert.

8) *Special Cutting Point Definition:* The eighth position, if it follows a letter in the 7th position, is a number indicating the number of $\frac{1}{64}$ths of an inch measured parallel to the edge of the facet.

9) *Hand:* **R,** right; **L,** left; to be used when required in ninth position.

10) *Other Conditions:* The tenth position defines special conditions (such as edge treatment, surface finish) as follows: **A,** honed, 0.0005 inch to less than 0.003 inch; **B,** honed, 0.003 inch to less than 0.005 inch; **C,** honed, 0.005 inch to less than 0.007 inch; **J,** polished, 4 microinch arithmetic average (AA) on rake surfaces only; **T,** chamfered, manufacturer's standard negative land, rake face only.

Indexable Insert Tool Holders.—Indexable insert tool holders are made from a good grade of steel which is heat treated to a hardness of 44 to 48 Rc for most normal applications. Accurate pockets that serve to locate the insert in position and to provide surfaces against which the insert can be clamped are machined in the ends of tool holders. A cemented carbide seat usually is provided, and is held in the bottom of the pocket by a screw or by the clamping pin, if one is used. The seat is necessary to provide a flat bearing surface upon which the insert can rest and, in so doing, it adds materially to the ability of the insert to withstand the cutting load. The seating surface of the holder may provide a positive-, negative-, or a neutral-rake orientation to the insert when it is in position on the holder. Holders, therefore, are classified as positive, negative, or neutral rake.

Four basic methods are used to clamp the insert on the holder: 1) Clamping, usually top clamping; 2) Pin-lock clamping; 3) Multiple clamping using a clamp, usually a top clamp, and a pin lock; and 4) Clamping the insert with a machine screw.

All top clamps are actuated by a screw that forces the clamp directly against the insert. When required, a cemented-carbide, plate-type chipbreaker is placed between the clamp and the insert. Pin-lock clamps require an insert having a hole: the pin acts against the walls of the hole to clamp the insert firmly against the seating surfaces of the holder. Multiple or combination clamping, simultaneously using both a pin-lock and a top clamp, is recommended when taking heavier or interrupted cuts. Holders are available on which all the above-mentioned methods of clamping may be used. Other holders are made with only a top clamp or a pin lock. Screw-on type holders use a machine screw to hold the insert in the pocket. Most standard indexable insert holders are either straight-shank or offset-shank, although special holders are made having a wide variety of configurations.

The common shank sizes of indexable insert tool holders are shown in Table 1. Not all styles are available in every shank size. Positive- and negative-rake tools are also not available in every style or shank size. Some manufacturers provide additional shank sizes for certain tool holder styles. For more complete details the manufacturers' catalogs must be consulted.

Table 1. Standard Shank Sizes for Indexable Insert Holders

Basic Shank Size	Shank Dimensions for Indexable Insert Holders					
	A		B		C[a]	
	In.	mm	In.	mm	In.	mm
½×½×4½	0.500	12.70	0.500	12.70	4.500	114.30
⅝×⅝×4½	0.625	15.87	0.625	15.87	4.500	114.30
⅝×1¼×6	0.625	15.87	1.250	31.75	6.000	152.40
¾×¾×4½	0.750	19.05	0.750	19.05	4.500	114.30
¾×1×6	0.750	19.05	1.000	25.40	6.000	152.40
¾×1¼×6	0.750	19.05	1.250	31.75	6.000	152.40
1×1×6	1.000	25.40	1.000	25.40	6.000	152.40
1×1¼×6	1.000	25.40	1.250	31.75	6.000	152.40
1×1½×6	1.000	25.40	1.500	38.10	6.000	152.40
1¼×1¼×7	1.250	31.75	1.250	31.75	7.000	177.80
1¼×1½×8	1.250	31.75	1.500	38.10	8.000	203.20
1⅜×2¹⁄₁₆×6⅜	1.375	34.92	2.062	52.37	6.380	162.05
1½×1½×7	1.500	38.10	1.500	38.10	7.000	177.80
1¾×1¾×9½	1.750	44.45	1.750	44.45	9.500	241.30
2×2×8	2.000	50.80	2.000	50.80	8.000	203.20

[a] Holder length; may vary by manufacturer. Actual shank length depends on holder style.

Identification System for Indexable Insert Holders.—The following identification system conforms to the American National Standard, ANSI B212.5-1986, Metric Holders for Indexable Inserts.

Each position in the system designates a feature of the holder in the following sequence:

1	2	3	4	5	—	6	—	7	—	8[a]	—	9	—	10[a]
C	**T**	**N**	**A**	**R**	**—**	**85**	**—**	**25**	**—**	**D**	**—**	**16**	**—**	**Q**

1) *Method of Holding Horizontally Mounted Insert:* The method of holding or clamping is designated by a letter: **C,** top clamping, insert without hole; **M,** top and hole clamping, insert with hole; **P,** hole clamping, insert with hole; **S,** screw clamping through hole, insert with hole; **W,** wedge clamping.

2) *Insert Shape:* The insert shape is identified by a letter: **H,** hexagonal; **O,** octagonal; **P,** pentagonal; **S,** square; **T,** triangular; **C,** rhombic, 80° included angle; **D,** rhombic, 55° included angle; **E,** rhombic, 75° included angle; **M,** rhombic, 86° included angle; **V,** rhombic, 35° included angle; **W,** hexagonal, 80° included angle; **L,** rectangular; **A,** parallelogram, 85° included angle; **B,** parallelogram, 82° included angle; **K,** parallelogram, 55° included angle; **R,** round. The included angle is always the smaller angle.

3) *Holder Style:* The holder style designates the shank style and the side cutting edge angle, or end cutting edge angle, or the purpose for which the holder is used. It is desig-

nated by a letter: **A,** for straight shank with 0° side cutting edge angle; **B,** straight shank with 15° side cutting edge angle; **C,** straight-shank end cutting tool with 0° end cutting edge angle; **D,** straight shank with 45° side cutting edge angle; **E,** straight shank with 30° side cutting edge angle; **F,** offset shank with 0° end cutting edge angle; **G,** offset shank with 0° side cutting edge angle; **J,** offset shank with negative 3° side cutting edge angle; **K,** offset shank with 15° end cutting edge angle; **L,** offset shank with negative 5° side cutting edge angle and 5° end cutting edge angle; **M,** straight shank with 40° side cutting edge angle; **N,** straight shank with 27° side cutting edge angle; **R,** offset shank with 15° side cutting edge angle; **S,** offset shank with 45° side cutting edge angle; **T,** offset shank with 30° side cutting edge angle; **U,** offset shank with negative 3° end cutting edge angle; **V,** straight shank with 17½° side cutting edge angle; **W,** offset shank with 30° end cutting edge angle; **Y,** offset shank with 5° end cutting edge angle.

4) *Normal Clearances:* The normal clearances of inserts are identified by letters: **A,** 3°; **B,** 5°; **C,** 7°; **D,** 15°; **E,** 20°; **F,** 25°; **G,** 30°; **N,** 0°; **P,** 11°.

5) *Hand of tool:* The hand of the tool is designated by a letter: **R** for right-hand; **L,** left-hand; and **N,** neutral, or either hand.

6) *Tool Height for Rectangular Shank Cross Sections:* The tool height for tool holders with a rectangular shank cross section and the height of cutting edge equal to shank height is given as a two-digit number representing this value in millimeters. For example, a height of 32 mm would be encoded as 32; 8 mm would be encoded as 08, where the one-digit value is preceded by a zero.

7) *Tool Width for Rectangular Shank Cross Sections:* The tool width for tool holders with a rectangular shank cross section is given as a two-digit number representing this value in millimeters. For example, a width of 25 mm would be encoded as 25; 8 mm would be encoded as 08, where the one-digit value is preceded by a zero.

8) *Tool Length:* The tool length is designated by a letter: **A,** 32 mm; **B,** 40 mm; **C,** 50 mm; **D,** 60 mm; **E,** 70 mm; **F,** 80 mm; **G,** 90 mm; **H,** 100 mm; **J,** 110 mm; **K,** 125 mm; **L,** 140 mm; **M,** 150 mm; **N,** 160 mm; **P,** 170 mm; **Q,** 180 mm; **R,** 200 mm; **S,** 250 mm; **T,** 300 mm; **U,** 350 mm; **V,** 400 mm; **W,** 450 mm; **X,** special length to be specified; **Y,** 500 mm.

9) *Indexable Insert Size:* The size of indexable inserts is encoded as follows: For insert shapes **C, D, E, H. M, O, P, R, S, T, V,** the side length (the diameter for **R** inserts) in millimeters is used as a two-digit number, with decimals being disregarded. For example, the symbol for a side length of 16.5 mm is 16. For insert shapes **A, B, K, L,** the length of the main cutting edge or of the longer cutting edge in millimeters is encoded as a two-digit number, disregarding decimals. If the symbol obtained has only one digit, then it should be preceded by a zero. For example, the symbol for a main cutting edge of 19.5 mm is 19; for an edge of 9.5 mm, the symbol is 09.

10) *Special Tolerances:* Special tolerances are indicated by a letter: **Q,** back and end qualified tool; **F,** front and end qualified tool; **B,** back, front, and end qualified tool. A qualified tool is one that has tolerances of ±0.08 mm for dimensions *F*, *G*, and *C*. (See Table 2.)

Table 2. Letter Symbols for Qualification of Tool Holders — Position 10
ANSI B212.5-1986

	Letter Symbol		
	Q	F	B
Qualification of Tool Holder	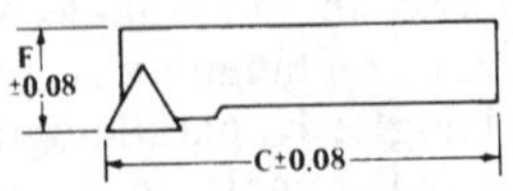 F ±0.08; C±0.08 Back and end qualified tool	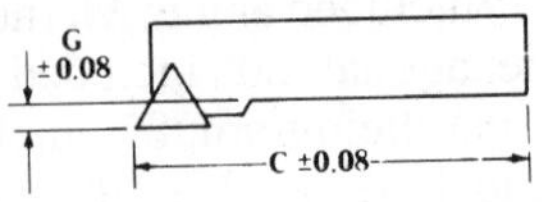 G ±0.08; C ±0.08 Front and end qualified tool	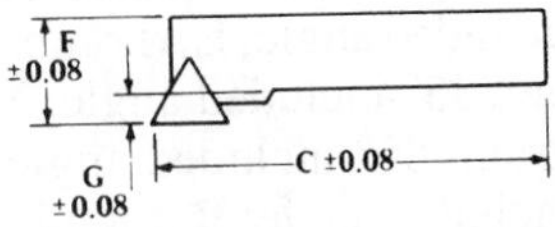 F ±0.08; G ±0.08; C ±0.08 Back, front, and end qualified tool

Selecting Indexable Insert Holders.—A guide for selecting indexable insert holders is provided by Table 3b. Some operations such as deep grooving, cut-off, and threading are not given in this table. However, tool holders designed specifically for these operations are available. The boring operations listed in Table 3b refer primarily to larger holes, into which the holders will fit. Smaller holes are bored using boring bars. An examination of this table shows that several tool-holder styles can be used and frequently are used for each operation. Selection of the best holder for a given job depends largely on the job and there are certain basic facts that should be considered in making the selection.

Rake Angle: A negative-rake insert has twice as many cutting edges available as a comparable positive-rake insert. Sometimes the tool life obtained when using the second face may be less than that obtained on the first face because the tool wear on the cutting edges of the first face may reduce the insert strength. Nevertheless, the advantage of negative-rake inserts and holders is such that they should be considered first in making any choice. Positive-rake holders should be used where lower cutting forces are required, as when machining slender or small-diameter parts, when chatter may occur, and for machining some materials, such as aluminum, copper, and certain grades of stainless steel, when positive-negative rake inserts can sometimes be used to advantage. These inserts are held on negative-rake holders that have their rake surfaces ground or molded to form a positive-rake angle.

Insert Shape: The configuration of the workpiece, the operation to be performed, and the lead angle required often determine the insert shape. When these factors need not be considered, the insert shape should be selected on the basis of insert strength and the maximum number of cutting edges available. Thus, a round insert is the strongest and has a maximum number of available cutting edges. It can be used with heavier feeds while producing a good surface finish. Round inserts are limited by their tendency to cause chatter, which may preclude their use. The square insert is the next most effective shape, providing good corner strength and more cutting edges than all other inserts except the round insert. The only limitation of this insert shape is that it must be used with a lead angle. Therefore, the square insert cannot be used for turning square shoulders or for back-facing. Triangle inserts are the most versatile and can be used to perform more operations than any other insert shape. The 80-degree diamond insert is designed primarily for heavy turning and facing operations, using the 100-degree corners, and for turning and back-facing square shoulders using the 80-degree corners. The 55- and 35-degree diamond inserts are intended primarily for tracing.

Lead Angle: Tool holders should be selected to provide the largest possible lead angle, although limitations are sometimes imposed by the nature of the job. For example, when tuning and back-facing a shoulder, a negative lead angle must be used. Slender or small-diameter parts may deflect, causing difficulties in holding size, or chatter when the lead angle is too large.

End Cutting Edge Angle: When tracing or contour turning, the plunge angle is determined by the end cutting edge angle. A 2-deg minimum clearance angle should be provided between the workpiece surface and the end cutting edge of the insert. Table 3a provides the maximum plunge angle for holders commonly used to plunge when tracing where insert shape identifiers are S = square; T = triangle; D = 55-deg diamond, V = 35-deg diamond. When severe cratering cannot be avoided, an insert having a small, end cutting edge angle is desirable to delay the crater breakthrough behind the nose. For very heavy cuts a small, end cutting edge angle will strengthen the corner of the tool. Tool holders for numerical control machines are discussed in the NC section, beginning page 1280.

Table 3a. Maximum Plunge Angle for Tracing or Contour Turning

Tool Holder Style	Insert Shape	Maximum Plunge Angle	Tool Holder Style	Insert Shape	Maximum Plunge Angle
E	T	58°	J	D	30°
D and S	S	43°	J	V	50°
H	D	71°	N	T	55°
J	T	25°	N	D	58°–60°

Table 3b. Indexable Insert Holder Application Guide

Tool	Tool Holder Style	Insert Shape	N-Negative P-Positive Rake	Application								
				Turn	Face	Turn and Face	Turn and Backface	Trace	Groove	Chamfer	Bore	Plane
0°	A	T	N	●	●						●	
			P	●	●						●	
0°	A	T	N	●	●			●				
			P	●	●			●				
	A	R	N	●	●	●						●
	A	R	N	●	●	●		●				●
15°	B	T	N	●	●						●	
			P	●	●						●	
15°	B	T	N	●	●			●			●	
			P	●	●			●			●	
15°	B	S	N	●	●						●	
			P	●	●						●	
5° 15°	B	C	N	●	●	●					●	●

Table 3b. *(Continued)* **Indexable Insert Holder Application Guide**

Tool	Tool Holder Style	Insert Shape	Rake N-Negative P-Positive	Application								
				Turn	Face	Turn and Face	Turn and Backface	Trace	Groove	Chamfer	Bore	Plane
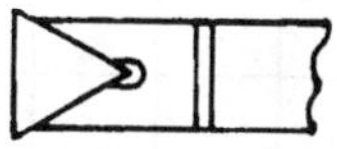	C	T	N	●	●				●	●		
			P	●	●				●	●		
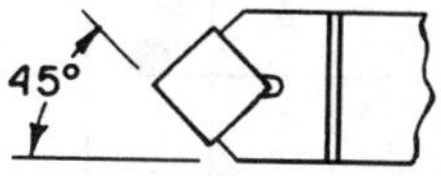	D	S	N	●	●	●		●		●	●	●
			P	●	●	●		●		●	●	●
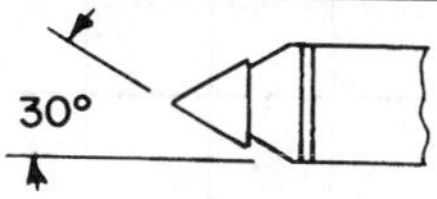	E	T	N	●	●			●	●	●		
			P	●	●			●	●	●		
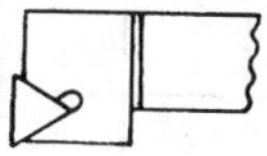	F	T	N	●	●						●	
			P	●	●						●	
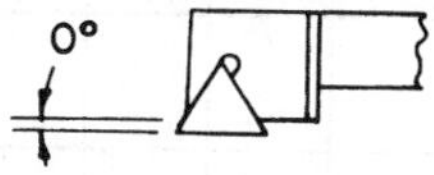	G	T	N	●	●						●	
			P	●	●						●	
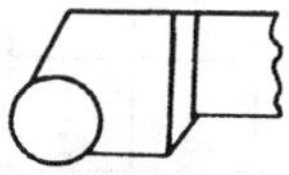	G	R	N	●	●	●						
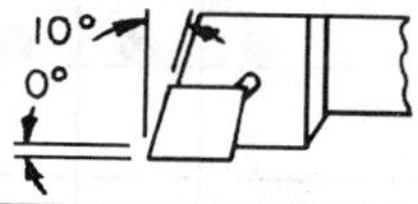	G	C	N	●	●	●						
			P	●	●	●						
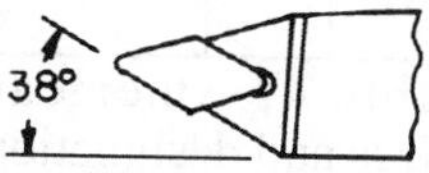	H	D	N	●	●			●				
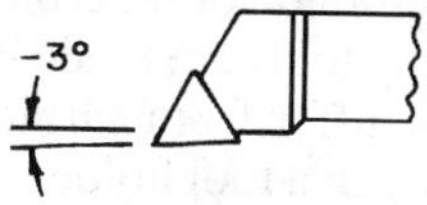	J	T	N				●	●				
			P				●	●				
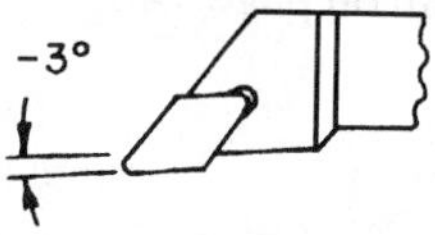	J	D	N				●	●				
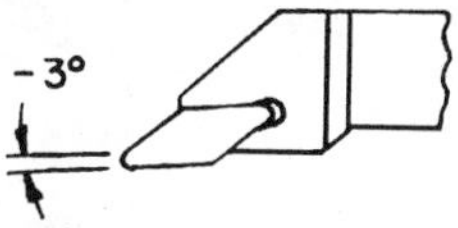	J	V	N				●	●				

Table 3b. *(Continued)* **Indexable Insert Holder Application Guide**

Tool	Tool Holder Style	Insert Shape	Rake N-Negative P-Positive	Application								
				Turn	Face	Turn and Face	Turn and Backface	Trace	Groove	Chamfer	Bore	Plane
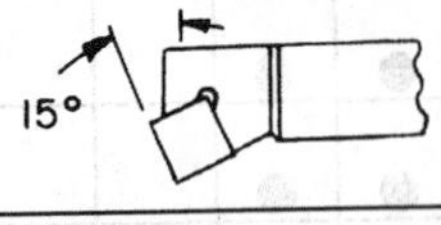 15°	K	S	N	●	●						●	
			P	●	●						●	
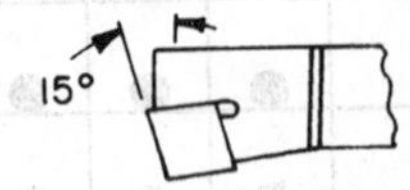 15°	K	C	N	●	●						●	
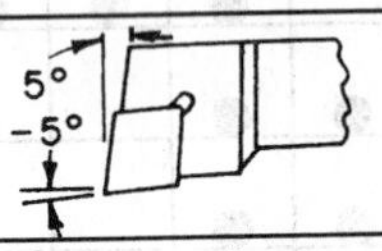 5° −5°	L	C	N			●	●					
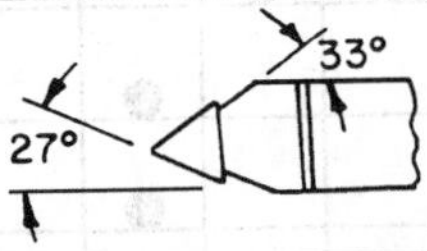 27° 33°	N	T	N	●	●			●				
			P	●	●			●				
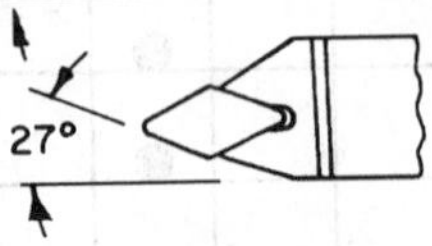 27°	N	D	N	●	●			●				
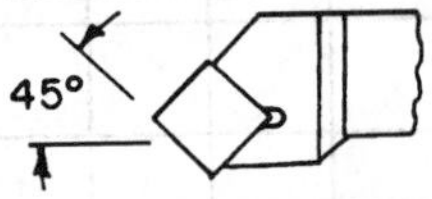 45°	S	S	N	●	●	●		●		●	●	●
			P	●	●	●		●		●	●	●
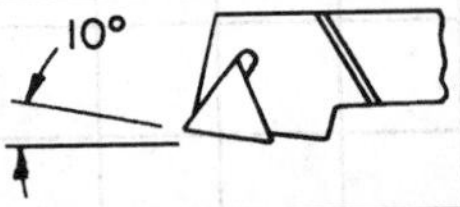 10°	W	S	N	●	●							

Sintered Carbide Blanks and Cutting Tools.—As shown in Table 4, American National Standard ANSI B212.1-1984 (R1997) provides standard sizes and designations for eight styles of sintered carbide blanks. These blanks are the unground solid carbide from which either solid or tipped cutting tools are made. Tipped cutting tools are made by brazing a blank onto a shank to produce the cutting tool; these tools differ from carbide *insert* cutting tools which consist of a carbide insert held mechanically in a tool holder. A typical single-point carbide-tipped cutting tool is shown in the diagram on page 740.

Table 4. American National Standard Sizes and Designations for Carbide Blanks
ANSI B212.1-1984 (R1997)

T	W	L	Style 1000 Blank Designation	Style 2000 Blank Designation
1/16	1/8	5/8	1010	2010
1/16	5/32	1/4	1015	2015
1/16	3/16	1/4	1020	2020
1/16	1/4	1/4	1025	2025
1/16	1/4	5/16	1030	2030
3/32	1/8	3/4	1035	2035
3/32	3/16	5/16	1040	2040
3/32	3/16	1/2	1050	2050
3/32	1/4	3/8	1060	2060
3/32	1/4	1/2	1070	2070
3/32	5/16	3/8	1080	2080
3/32	3/8	3/8	1090	2090
3/32	3/8	1/2	1100	2100
3/32	7/16	1/2	1105	2105
1/8	3/16	3/4	1110	2110
1/8	1/4	1/2	1120	2120
1/8	1/4	5/8	1130	2130
1/8	1/4	3/4	1140	2140
1/8	5/16	7/16	1150	2150
1/8	5/16	1/2	1160	2160
1/8	5/16	5/8	1170	2170
1/8	3/8	1/2	1180	2180
1/8	3/8	3/4	1190	2190
1/8	1/2	1/2	1200	2200
1/8	1/2	3/4	1210	2210
1/8	3/4	3/4	1215	2215
5/32	3/8	9/16	1220	2220
5/32	3/8	3/4	1230	2230
5/32	5/8	5/8	1240	2240
3/16	5/16	7/16	1250	2250
3/16	5/16	5/8	1260	2260
3/16	3/8	1/2	1270	2270
3/16	3/8	5/8	1280	2280
3/16	3/8	3/4	1290	2290
3/16	7/16	5/8	1300	2300
3/16	7/16	13/16	1310	2310
3/16	1/2	1/2	1320	2320
3/16	1/2	3/4	1330	2330
3/16	3/4	3/4	1340	2340

T	W	L	Style 0000 Blank Designation	Style 1000 Blank Designation	Style 3000 Blank Designation	Style 4000 Blank Designation
1/4	3/8	9/16	0350	1350	3350	4350
1/4	3/8	3/4	0360	1360	3360	4360
1/4	7/16	5/8	0370	1370	3370	4370
1/4	1/2	3/4	0380	1380	3380	4380
1/4	9/16	1	0390	1390	3390	4390
1/4	5/8	5/8	0400	1400	3400	4400
1/4	3/4	3/4	0405	1405	3405	4405
1/4	3/4	1	0410	1410	3410	4410
1/4	1	1	0415	1415	3415	4415
5/16	7/16	5/8	0420	1420	3420	4420
5/16	7/16	15/16	0430	1430	3430	4430
5/16	1/2	3/4	0440	1440	3440	4440
5/16	1/2	1	0450	1450	3450	4450
5/16	5/8	1	0460	1460	3460	4460
5/16	3/4	3/4	0470	1470	3470	4470
5/16	3/4	1	0475	1475	3475	4475
5/16	3/4	1 1/4	0480	1480	3480	4480
3/8	1/2	3/4	0490	1490	3490	4490
3/8	1/2	1	0500	1500	3500	4500
3/8	5/8	1	0510	1510	3510	4510
3/8	5/8	1 1/4	0515	1515	3515	4515
3/8	3/4	1 1/4	0520	1520	3520	4520
3/8	3/4	1 1/2	0525	1525	3525	4525
1/2	3/4	1	0530	1530	3530	4530
1/2	3/4	1 1/4	0540	1540	3540	4540
1/2	3/4	1 1/2	0550	1550	3550	4550

T	W	L	F	Style 5000	Style 6000	Style 7000
1/16	1/4	5/16	…	5030	…	…
3/32	1/4	3/8	1/16	…	…	7060
3/32	5/16	3/8	…	5080	6080	…
3/32	3/8	1/2	…	5100	6100	…
3/32	7/16	1/2	…	5105	…	…
1/8	5/16	5/8	3/32	…	…	7170
1/8	1/2	1/2	…	5200	6200	…
5/32	3/8	3/4	1/8	…	…	7230
5/32	5/8	5/8	…	5240	6240	…
3/16	3/4	3/4	…	5340	6340	…
1/4	1	3/4	…	5410	…	…

All dimensions are in inches.

See diagram on page 740.

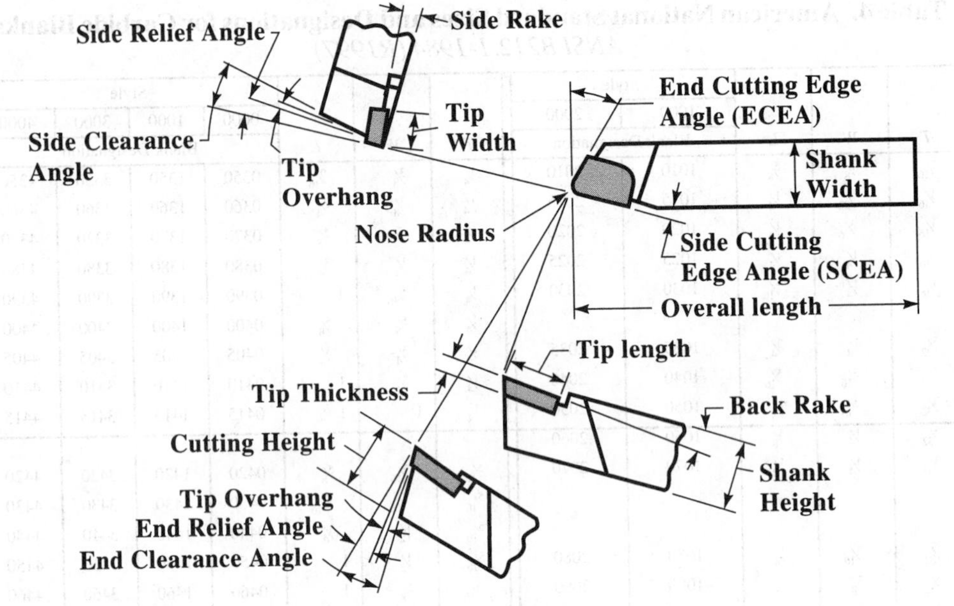

A typical single-point carbide tipped cutting tool. The side rake, side relief, and the clearance angles are normal to the side-cutting edge, rather than the shank, to facilitate its being ground on a tilting-table grinder. The end-relief and clearance angles are *normal* to the end-cutting edge. The back-rake angle is parallel to the side-cutting edge. The tip of the brazed carbide blank overhangs the shank of the tool by either ⅓₂ or ¹⁄₁₆ inch, depending on the size of the tool. For tools in Tables 5, 6, 7, 8, 11 and 12, the maximum overhang is $\frac{1}{32}$ inch for shank sizes 4, 5, 6, 7, 8, 10, 12 and 44; for other shank sizes in these tables, the maximum overhang is $\frac{1}{16}$ inch. In Tables 9 and 10 all tools have maximum overhang of $\frac{1}{32}$ inch.

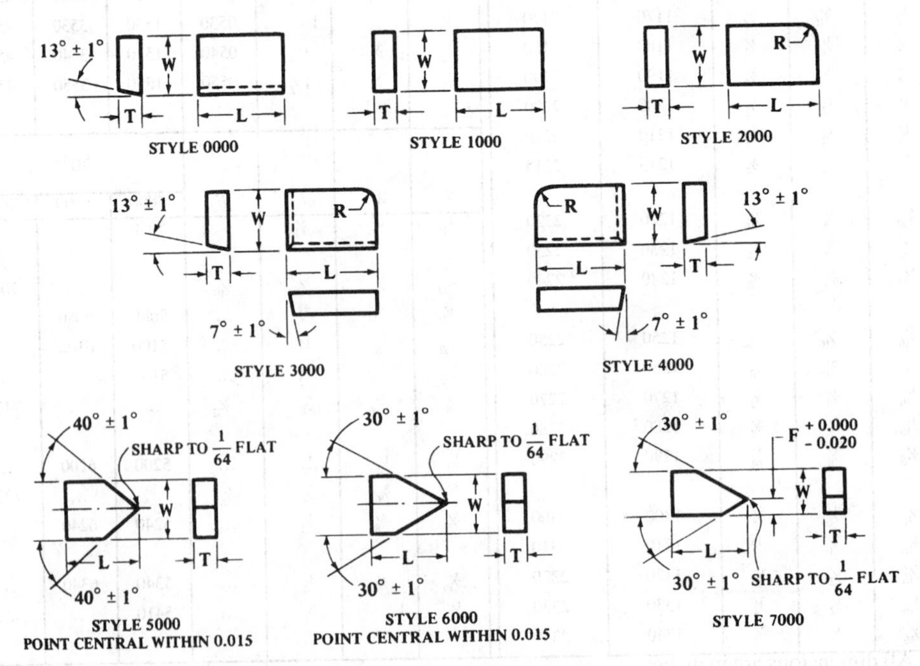

Eight styles of sintered carbide blanks. Standard dimensions for these blanks are given in Table 4.

Table 5. American National Standard Style A Carbide Tipped Tools
ANSI B212.1-1984 (R1997)

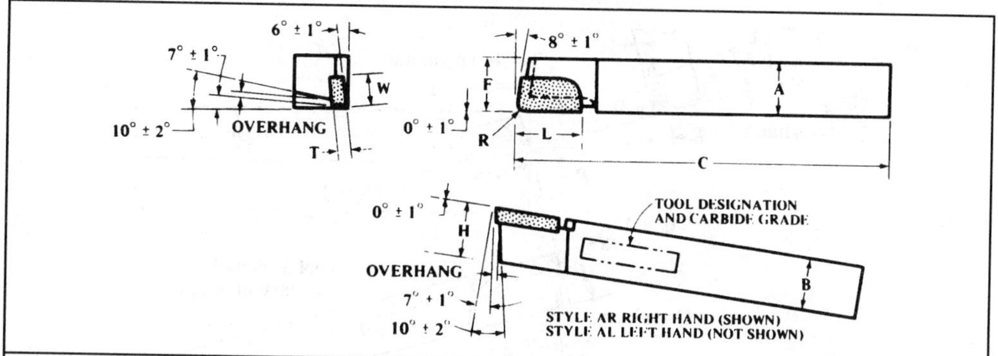

Designation		Shank Dimensions			Tip Designation[a]	Tip Dimensions		
Style AR[a]	Style AL[a]	Width *A*	Height *B*	Length *C*		Thickness *T*	Width *W*	Length *L*
Square Shank								
AR 4	AL 4	1/4	1/4	2	2040	3/32	3/16	5/16
AR 5	AL 5	5/16	5/16	2 1/4	2070	3/32	1/4	1/2
AR 6	AL 6	3/8	3/8	2 1/2	2070	3/32	1/4	1/2
AR 7	AL 7	7/16	7/16	3	2070	3/32	1/4	1/2
AR 8	AL 8	1/2	1/2	3 1/2	2170	1/8	5/16	5/8
AR 10	AL 10	5/8	5/8	4	2230	5/32	3/8	3/4
AR 12	AL 12	3/4	3/4	4 1/2	2310	3/16	7/16	13/16
AR 16	AL 16	1	1	6	{ P3390 P4390	1/4	9/16	1
AR 20	AL 20	1 1/4	1 1/4	7	{ P3460 P4460	5/16	5/8	1
AR 24	AL 24	1 1/2	1 1/2	8	{ P3510 P4510	3/8	5/8	1
Rectangular Shank								
AR 44	AL 44	1/2	1	6	P2260	3/16	5/16	5/8
AR 54	AL 54	5/8	1	6	{ P3360 P4360	1/4	3/8	3/4
AR 55	AL 55	5/8	1 1/4	7	{ P3360 P4360	1/4	3/8	3/4
AR 64	AL 64	3/4	1	6	{ P3380 P4380	1/4	1/2	3/4
AR 66	AL 66	3/4	1 1/2	8	{ P3430 P4430	5/16	7/16	15/16
AR 85	AL 85	1	1 1/4	7	{ P3460 P4460	5/16	5/8	1
AR 86	AL 86	1	1 1/2	8	{ P3510 P4510	3/8	5/8	1
AR 88	AL 88	1	2	10	{ P3510 P4510	3/8	5/8	1
AR 90	AL 90	1 1/2	2	10	{ P3540 P4540	1/2	3/4	1 1/4

[a] "A" is straight shank, 0 deg., SCEA (side-cutting-edge angle). "R" is right-cut. "L" is left-cut. Where a pair of tip numbers is shown, the upper number applies to AR tools, the lower to AL tools. All dimensions are in inches.

Single-Point, Sintered-Carbide-Tipped Tools.—American National Standard ANSI B212.1-1984 (R1997) covers eight different styles of single-point, carbide-tipped general purpose tools. These styles are designated by the letters A to G inclusive. Styles A, B, F, G, and E with offset point are either right- or left-hand cutting as indicated by the letters R or L. Dimensions of tips and shanks are given in Tables 5 to 11. For dimensions and tolerances not shown, and for the identification system, dimensions, and tolerances of sintered carbide boring tools, see the Standard.

Table 6. American National Standard Style B Carbide Tipped Tools with 15-degree Side-cutting-edge Angle *ANSI B212.1-1984 (R1997)*

Designation		Shank Dimensions				Tip Dimensions		
Style BR	Style BL	Width A	Height B	Length C	Tip Designation[a]	Thickness T	Width W	Length L
Square Shank								
BR 4	BL 4	1/4	1/4	2	2015	1/16	5/32	1/4
BR 5	BL 5	5/16	5/16	2 1/4	2040	3/32	3/16	5/16
BR 6	BL 6	3/8	3/8	2 1/2	2070	3/32	1/4	1/2
BR 7	BL 7	7/16	7/16	3	2070	3/32	1/4	1/2
BR 8	BL 8	1/2	1/2	3 1/2	2170	1/8	5/16	5/8
BR 10	BL 10	5/8	5/8	4	2230	5/32	3/8	3/4
BR 12	BL 12	3/4	3/4	4 1/2	2310	3/16	7/16	13/16
BR 16	BL 16	1	1	6	3390 4390	1/4	9/16	1
BR 20	BL 20	1 1/4	1 1/4	7	3460 4460	5/16	5/8	1
BR 24	BL 24	1 1/2	1 1/2	8	3510 4510	3/8	5/8	1
Rectangular Shank								
BR 44	BL 44	1/2	1	6	2260	3/16	5/16	5/8
BR 54	BL 54	5/8	1	6	3360 4360	1/4	3/8	3/4
BR 55	BL 55	5/8	1 1/4	7	3360 4360	1/4	3/8	3/4
BR 64	BL 64	3/4	1	6	3380 4380	1/4	1/2	3/4
BR 66	BL 66	3/4	1 1/2	8	3430 4430	5/16	7/16	15/16
BR 85	BL 85	1	1 1/4	7	3460 4460	5/16	5/8	1
BR 86	BL 86	1	1 1/2	8	3510 4510	3/8	5/8	1
BR 88	BL 88	1	2	10	3510 4510	3/8	5/8	1
BR 90	BL 90	1 1/2	2	10	3540 4540	1/2	3/4	1 1/4

[a] Where a pair of tip numbers is shown, the upper number applies to BR tools, the lower to BL tools. All dimensions are in inches.

A number follows the letters of the tool style and hand designation and for square shank tools, represents the number of sixteenths of an inch of width, *W*, and height, *H*. With rectangular shanks, the first digit of the number indicates the number of eighths of an inch in the shank width, *W*, and the second digit the number of quarters of an inch in the shank height, *H*. One exception is the 1½ × 2-inch size which has been arbitrarily assigned the number 90.

Table 7. American National Standard Style C Carbide Tipped Tools *ANSI B212.1-1984 (R1997)*

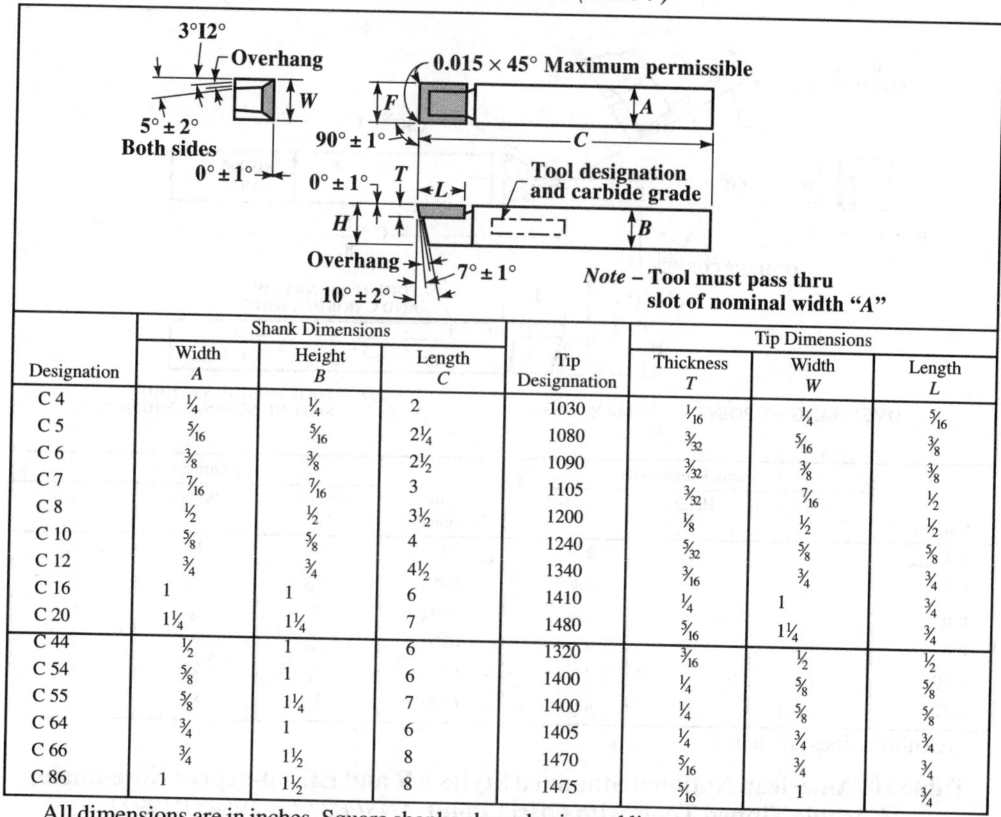

Designation	Shank Dimensions Width *A*	Height *B*	Length *C*	Tip Designnation	Tip Dimensions Thickness *T*	Width *W*	Length *L*
C 4	1/4	1/4	2	1030	1/16	1/4	5/16
C 5	5/16	5/16	2 1/4	1080	3/32	5/16	3/8
C 6	3/8	3/8	2 1/2	1090	3/32	3/8	3/8
C 7	7/16	7/16	3	1105	3/32	7/16	1/2
C 8	1/2	1/2	3 1/2	1200	1/8	1/2	1/2
C 10	5/8	5/8	4	1240	5/32	5/8	5/8
C 12	3/4	3/4	4 1/2	1340	3/16	3/4	3/4
C 16	1	1	6	1410	1/4	1	3/4
C 20	1 1/4	1 1/4	7	1480	5/16	1 1/4	3/4
C 44	1/2	1	6	1320	3/16	1/2	1/2
C 54	5/8	1	6	1400	1/4	5/8	5/8
C 55	5/8	1 1/4	7	1400	1/4	5/8	5/8
C 64	3/4	1	6	1405	1/4	3/4	3/4
C 66	3/4	1 1/2	8	1470	5/16	3/4	3/4
C 86	1	1 1/2	8	1475	5/16	1	3/4

All dimensions are in inches. Square shanks above horizontal line; rectangular below.

Table 8. American National Standard Style D, 80-degree Nose-angle Carbide Tipped Tools *ANSI B212.1-1984 (R1997)*

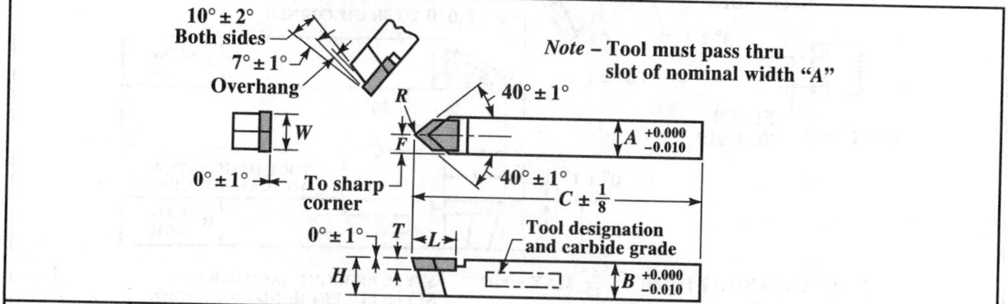

Designation	Shank Dimensions Width *A*	Height *B*	Length *C*	Tip Designation	Tip Dimensions Thickness *T*	Width *W*	Length *L*
D 4	1/4	1/4	2	5030	1/16	1/4	5/16
D 5	5/16	5/16	2 1/4	5080	3/32	5/16	3/8
D 6	3/8	3/8	2 1/2	5100	3/32	3/8	1/2
D 7	7/16	7/16	3	5105	3/32	7/16	1/2
D 8	1/2	1/2	3 1/2	5200	1/8	1/2	1/2
D 10	5/8	5/8	4	5240	5/32	5/8	5/8
D 12	3/4	3/4	4 1/2	5340	3/16	3/4	3/4
D 16	1	1	6	5410	1/4	1	3/4

All dimensions are in inches.

Table 9. American National Standard Style E, 60-degree Nose-angle, Carbide Tipped Tools *ANSI B212.1-1984 (R1997)*

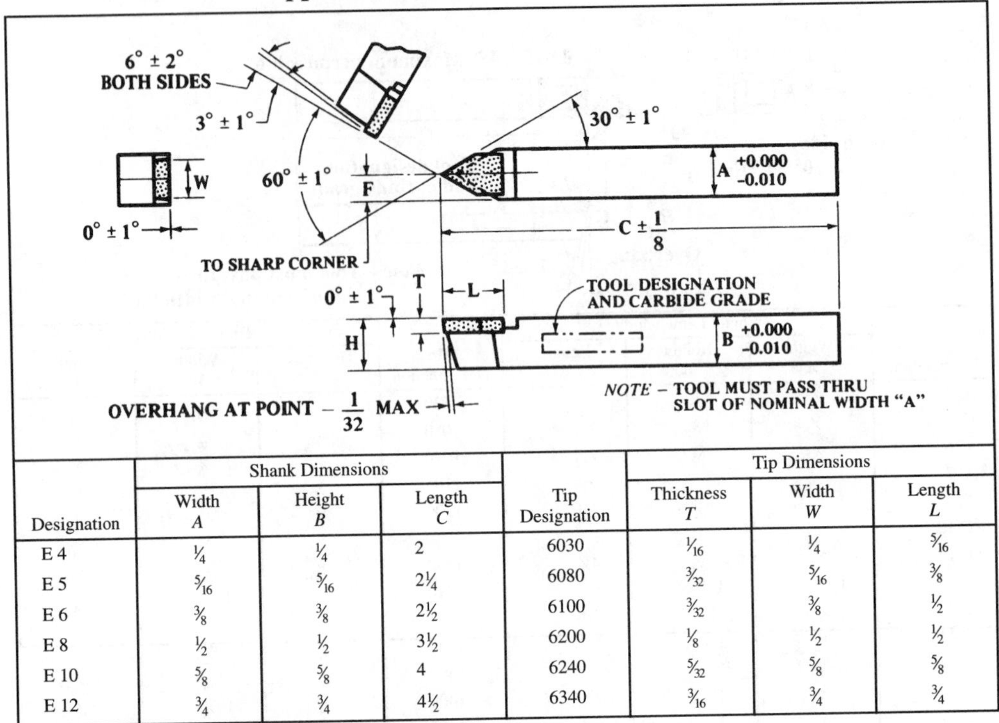

Designation	Shank Dimensions			Tip Designation	Tip Dimensions		
	Width *A*	Height *B*	Length *C*		Thickness *T*	Width *W*	Length *L*
E 4	¼	¼	2	6030	1/16	¼	5/16
E 5	5/16	5/16	2¼	6080	3/32	5/16	⅜
E 6	⅜	⅜	2½	6100	3/32	⅜	½
E 8	½	½	3½	6200	⅛	½	½
E 10	⅝	⅝	4	6240	5/32	⅝	⅝
E 12	¾	¾	4½	6340	3/16	¾	¾

All dimensions are in inches.

Table 10. American National Standard Styles ER and EL, 60-degree Nose-angle, Carbide Tipped Tools with Offset Point *ANSI B212.1-1984 (R1997)*

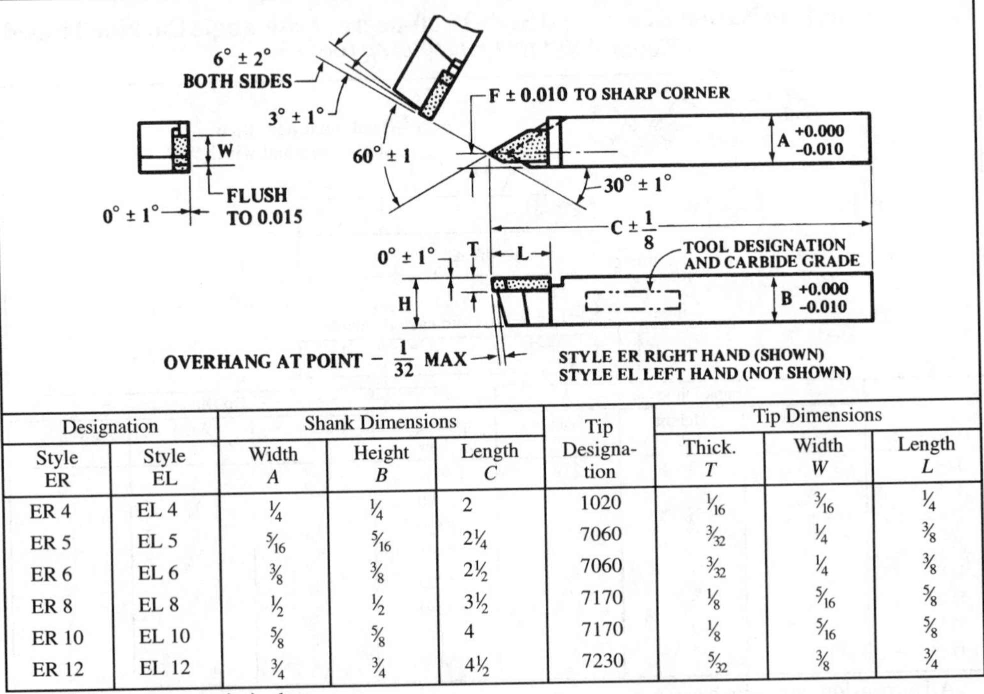

Designation		Shank Dimensions			Tip Designation	Tip Dimensions		
Style ER	Style EL	Width *A*	Height *B*	Length *C*		Thick. *T*	Width *W*	Length *L*
ER 4	EL 4	¼	¼	2	1020	1/16	3/16	¼
ER 5	EL 5	5/16	5/16	2¼	7060	3/32	¼	⅜
ER 6	EL 6	⅜	⅜	2½	7060	3/32	¼	⅜
ER 8	EL 8	½	½	3½	7170	⅛	5/16	⅝
ER 10	EL 10	⅝	⅝	4	7170	⅛	5/16	⅝
ER 12	EL 12	¾	¾	4½	7230	5/32	⅜	¾

All dimensions are in inches.

Table 11. American National Standard Style F, Offset, End-cutting Carbide Tipped Tools *ANSI B212.1-1984 (R1997)*

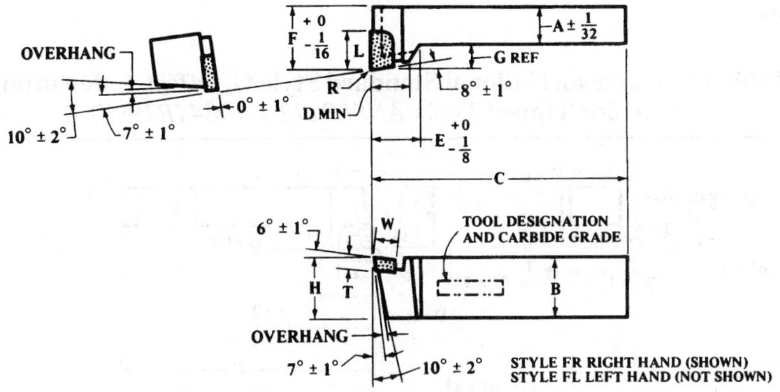

Designation		Shank Dimensions						Tip Dimensions		
Style FR	Style FL	Width *A*	Height *B*	Length *C*	Offset *G*	Length of Offset *E*	Tip Designation	Thickness *T*	Width *W*	Length *L*
Square Shank										
FR 8	FL 8	½	½	3½	¼	¾	P4170 P3170	⅛	5/16	⅝
FR 10	FL 10	⅝	⅝	4	⅜	1	P1230 P3230	5/32	⅜	¾
FR 12	FL 12	¾	¾	4½	⅝	1⅛	P4310 P3310	3/16	7/16	13/16
FR 16	FL 16	1	1	6	¾	1⅜	P4390 P3390	¼	9/16	1
FR 20	FL 20	1¼	1¼	7	¾	1½	P4460 P3460	5/16	⅝	1
FR 24	FL 24	1½	1½	8	¾	1½	P4510 P3510	⅜	⅝	1
Rectangular Shank										
FR 44	FL 44	½	1	6	½	⅞	P4260 P1260	3/16	5/16	⅝
FR 55	FL 55	⅝	1¼	7	⅝	1⅛	P4360 P3360	¼	⅜	¾
FR 64	FL 64	¾	1	6	⅝	1 3/16	P4380 P3380	¼	½	¾
FR 66	FL 66	¾	1½	8	¾	1¼	P4430 P3430	5/16	7/16	15/16
FR 85	FL 85	1	1¼	7	¾	1½	P4460 P3460	5/16	⅝	1
FR 86	FL 86	1	1½	8	¾	1½	P4510 P3510	⅜	⅝	1
FR 90	FL 90	1½	2	10	¾	1⅝	P4540 P3540	½	¾	1¼

All dimensions are in inches. Where a pair of tip numbers is shown, the upper number applies to FR tools, the lower number to FL tools.

Single-point Tool Nose Radii.—The tool nose radii recommended in the American National Standard are as follows: For square-shank tools up to and including ⅜-inch square tools, 1/64 inch; for those over ⅜-inch square through 1¼-inches square, 1/32 inch; and for those above 1¼-inches square, 1/16 inch. For rectangular-shank tools with shank section of ½ × 1 inch through 1 × 1½ inches, the nose radii are 1/32 inch, and for 1 × 2 and 1½ × 2 inch shanks, the nose radius is 1/16 inch.

Single-point Tool Angle Tolerances.—The tool angles shown on the diagrams in the Tables 5 through 11 are general recommendations. Tolerances applicable to these angles are ± 1 degree on all angles except end and side clearance angles; for these the tolerance is ± 2 degrees.

Table 12. American National Standard Style G, Offset, Side-cutting, Carbide Tipped Tools *ANSI B212.1-1984 (R1997)*

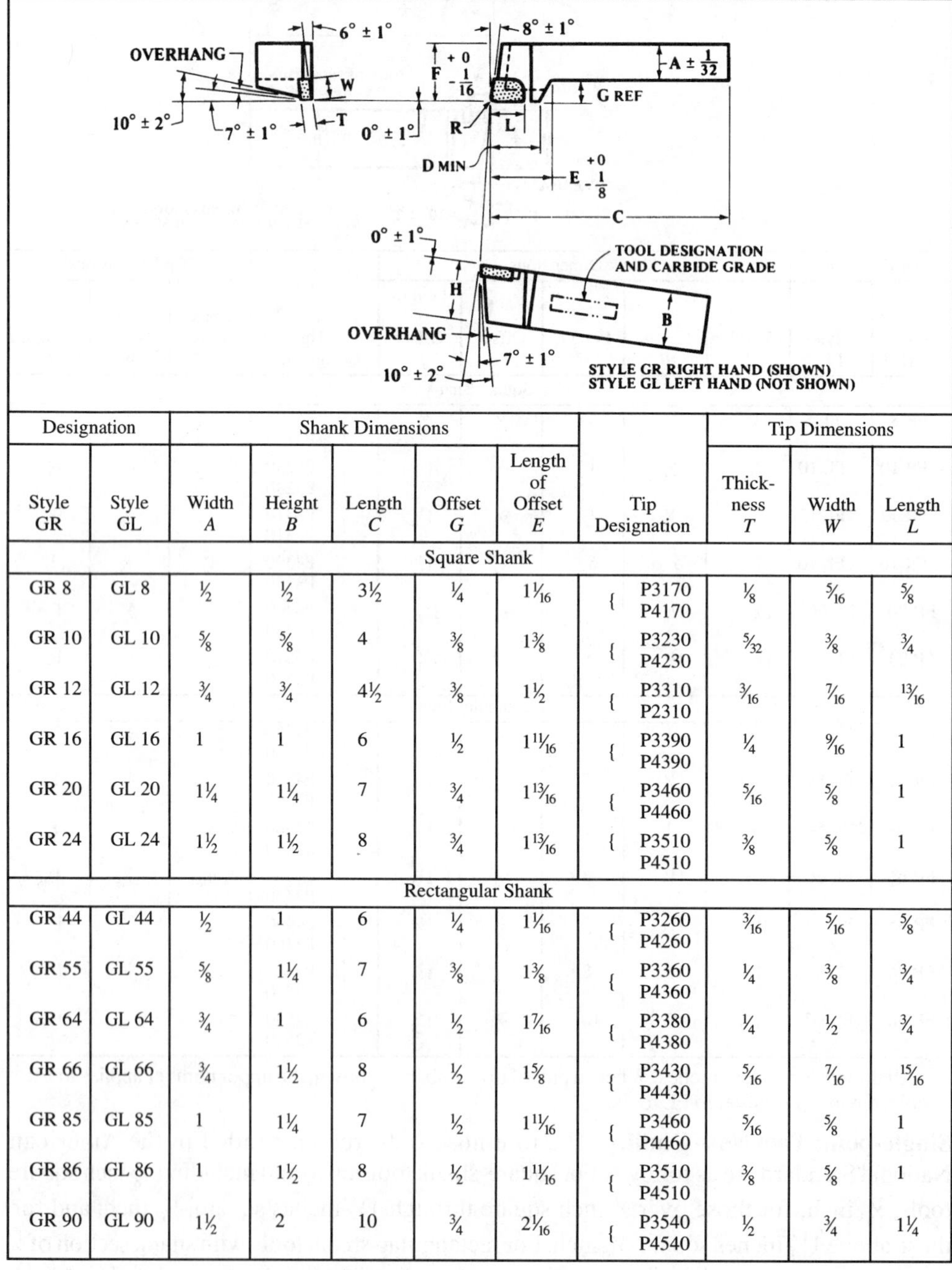

Designation		Shank Dimensions						Tip Dimensions		
Style GR	Style GL	Width *A*	Height *B*	Length *C*	Offset *G*	Length of Offset *E*	Tip Designation	Thickness *T*	Width *W*	Length *L*
Square Shank										
GR 8	GL 8	1/2	1/2	3 1/2	1/4	1 1/16	{ P3170 P4170	1/8	5/16	5/8
GR 10	GL 10	5/8	5/8	4	3/8	1 3/8	{ P3230 P4230	5/32	3/8	3/4
GR 12	GL 12	3/4	3/4	4 1/2	3/8	1 1/2	{ P3310 P2310	3/16	7/16	13/16
GR 16	GL 16	1	1	6	1/2	1 11/16	{ P3390 P4390	1/4	9/16	1
GR 20	GL 20	1 1/4	1 1/4	7	3/4	1 13/16	{ P3460 P4460	5/16	5/8	1
GR 24	GL 24	1 1/2	1 1/2	8	3/4	1 13/16	{ P3510 P4510	3/8	5/8	1
Rectangular Shank										
GR 44	GL 44	1/2	1	6	1/4	1 1/16	{ P3260 P4260	3/16	5/16	5/8
GR 55	GL 55	5/8	1 1/4	7	3/8	1 3/8	{ P3360 P4360	1/4	3/8	3/4
GR 64	GL 64	3/4	1	6	1/2	1 7/16	{ P3380 P4380	1/4	1/2	3/4
GR 66	GL 66	3/4	1 1/2	8	1/2	1 5/8	{ P3430 P4430	5/16	7/16	15/16
GR 85	GL 85	1	1 1/4	7	1/2	1 11/16	{ P3460 P4460	5/16	5/8	1
GR 86	GL 86	1	1 1/2	8	1/2	1 11/16	{ P3510 P4510	3/8	5/8	1
GR 90	GL 90	1 1/2	2	10	3/4	2 1/16	{ P3540 P4540	1/2	3/4	1 1/4

All dimensions are in inches. Where a pair of tip numbers is shown, the upper number applies to GR tools, the lower number to GL tools.

CEMENTED CARBIDES

Cemented Carbides and Other Hard Materials

Carbides and Carbonitrides.—Though high-speed steel retains its importance for such applications as drilling and broaching, most metal cutting is carried out with carbide tools. For materials that are very difficult to machine, carbide is now being replaced by carbonitrides, ceramics, and superhard materials. Cemented (or sintered) carbides and carbonitrides, known collectively in most parts of the world as hard metals, are a range of very hard, refractory, wear-resistant alloys made by powder metallurgy techniques. The minute carbide or nitride particles are "cemented" by a binder metal that is liquid at the sintering temperature. Compositions and properties of individual hardmetals can be as different as those of brass and high-speed steel.

All hardmetals are *cermets*, combining *cer*amic particles with a *met*allic binder. It is unfortunate that (owing to a mistranslation) the term *cermet* has come to mean either all hardmetals with a titanium carbide (TiC) base or simply cemented titanium carbonitrides. Although no single element other than carbon is present in all hard-metals, it is no accident that the generic term is "tungsten carbide." The earliest successful grades were based on carbon, as are the majority of those made today, as listed in Table 1.

The outstanding machining capabilities of high-speed steel are due to the presence of very hard carbide particles, notably tungsten carbide, in the iron-rich matrix. Modern methods of making cutting tools from pure tungsten carbide were based on this knowledge. Early pieces of cemented carbide were much too brittle for industrial use, but it was soon found that mixing tungsten carbide powder with up to 10 per cent of metals such as iron, nickel, or cobalt, allowed pressed compacts to be sintered at about 1500°C to give a product with low porosity, very high hardness, and considerable strength. This combination of properties made the materials ideally suitable for use as tools for cutting metal.

Cemented carbides for cutting tools were introduced commercially in 1927, and although the key discoveries were made in Germany, many of the later developments have taken place in the United States, Austria, Sweden, and other countries. Recent years have seen two "revolutions" in carbide cutting tools, one led by the United States and the other by Europe. These were the change from brazed to clamped carbide inserts and the rapid development of coating technology.

When indexable tips were first introduced, it was found that so little carbide was worn away before they were discarded that a minor industry began to develop, regrinding the so-called "throwaway" tips and selling them for reuse in adapted toolholders. Hardmetal consumption, which had grown dramatically when indexable inserts were introduced, leveled off and began to decline. This situation was changed by the advent and rapid acceptance of carbide, nitride, and oxide coatings. Application of an even harder, more wear-resistant surface to a tougher, more shock-resistant substrate allowed production of new generations of longer-lasting inserts. Regrinding destroyed the enhanced properties of the coatings, so was abandoned for coated tooling.

Brazed tools have the advantage that they can be reground over and over again, until almost no carbide is left, but the tools must always be reset after grinding to maintain machining accuracy. However, all brazed tools suffer to some extent from the stresses left by the brazing process, which in unskilled hands or with poor design can shatter the carbide even before it has been used to cut metal. In present conditions it is cheaper to use indexable inserts, which are tool tips of precise size, clamped in similarly precise holders, needing no time-consuming and costly resetting but usable only until each cutting edge or corner has lost its initial sharpness (see *Indexable Inserts* and related topics starting on page 730 and *Indexable Insert Holders for NC* on page 1280. The absence of brazing stresses and the "one-use" concept also means that harder, longer-lasting grades can be used.

Table 1. Typical Properties of Tungsten-Carbide-Based Cutting-Tool Hardmetals

ISO Application Code	Composition (%)				Density	Hardness	Transverse Rupture Strength (N/mm^2)
	WC	TiC	TaC	Co			
P01	50	35	7	6	8.5	1900	1100
P05	78	16		6	11.4	1820	1300
P10	69	15	8	8	11.5	1740	1400
P15	78	12	3	7	11.7	1660	1500
P20	79	8	5	8	12.1	1580	1600
P25	82	6	4	8	12.9	1530	1700
P30	84	5	2	9	13.3	1490	1850
P40	85	5		10	13.4	1420	1950
P50	78	3	3	16	13.1	1250	2300
M10	85	5	4	6	13.4	1590	1800
M20	82	5	5	8	13.3	1540	1900
M30	86	4		10	13.6	1440	2000
M40	84	4	2	10	14.0	1380	2100
K01	97			3	15.2	1850	1450
K05	95		1	4	15.0	1790	1550
K10	92		2	6	14.9	1730	1700
K20	94			6	14.8	1650	1950
K30	91			9	14.4	1400	2250
K40	89			11	14.1	1320	2500

A complementary development was the introduction of ever-more complex chip-breakers, derived from computer-aided design and pressed and sintered to precise shapes and dimensions. Another advance was the application of hot isostatic pressing (HIP), which has moved hardmetals into applications that were formerly uneconomic. This method allows virtually all residual porosity to be squeezed out of the carbide by means of inert gas at high pressure, applied at about the sintering temperature. Toughness, rupture strength, and shock resistance can be doubled or tripled by this method, and the reject rates of very large sintered components are reduced to a fraction of their previous levels.

Further research has produced a substantial number of excellent cutting-tool materials based on titanium carbonitride. Generally called "cermets," as noted previously, carbonitride-based cutting inserts offer excellent performance and considerable prospects for the future.

Compositions and Structures: Properties of hardmetals are profoundly influenced by microstructure. The microstructure in turn depends on many factors including basic chemical composition of the carbide and matrix phases; size, shape, and distribution of carbide particles; relative proportions of carbide and matrix phases; degree of intersolubility of carbides; excess or deficiency of carbon; variations in composition and structure caused by diffusion or segregation; production methods generally, but especially milling, carburizing, and sintering methods, and the types of raw materials; post sintering treatments such as hot isostatic pressing; and coatings or diffusion layers applied after initial sintering.

Tungsten Carbide/Cobalt (WC/Co): The first commercially available cemented carbides consisted of fine angular particles of tungsten carbide bonded with metallic cobalt. Intended initially for wire-drawing dies, this composition type is still considered to have the greatest resistance to simple abrasive wear and therefore to have many applications in machining.

For maximum hardness to be obtained from closeness of packing, the tungsten carbide grains should be as small as possible, preferably below 1 μm swaging 0.00004 in.) and considerably less for special purposes. Hardness and abrasion resistance increase as the cobalt content is lowered, provided that a minimum of cobalt is present (2 per cent can be enough, although 3 per cent is the realistic minimum) to ensure complete sintering. In general, as carbide grain size or cobalt content or both are increased—frequently in unison—tougher and less hard grades are obtained. No porosity should be visible, even under the highest optical magnification.

WC/Co compositions used for cutting tools range from about 2 to 13 per cent cobalt, and from less than 0.5 to more than 5 μm (0.00002–0.0002 in.) in grain size. For stamping tools, swaying dies, and other wear applications for parts subjected to moderate or severe shock, cobalt content can be as much as 30 per cent, and grain size a maximum of about 10 μm (0.0004 in.). In recent years, "micrograin" carbides, combining submicron (less than 0.00004 in.) carbide grains with relatively high cobalt content have found increasing use for machining at low speeds and high feed rates. An early use was in high-speed woodworking cutters such as are used for planing.

For optimum properties, porosity should be at a minimum, carbide grain size as regular as possible, and carbon content of the tungsten carbide phase close to the theoretical (stoichiometric) value. Many tungsten carbide/cobalt compositions are modified by small but important additions—from 0.5 to perhaps 3 per cent of tantalum, niobium, chromium, vanadium, titanium, hafnium, or other carbides. The basic purpose of these additions is generally inhibition of grain growth, so that a consistently fine structure is maintained.

Tungsten – Titanium Carbide/Cobalt (WC/TiC/Co): These grades are used for tools to cut steels and other ferrous alloys, the purpose of the TiC content being to resist the high-temperature diffusive attack that causes chemical breakdown and cratering. Tungsten carbide diffuses readily into the chip surface, but titanium carbide is extremely resistant to such diffusion. A solid solution or "mixed crystal" of WC in TiC retains the anticratering property to a great extent.

Unfortunately, titanium carbide and TiC-based solid solutions are considerably more brittle and less abrasion resistant than tungsten carbide. TiC content, therefore, is kept as low as possible, only sufficient TiC being provided to avoid severe cratering wear. Even 2 or 3 per cent of titanium carbide has a noticeable effect, and as the relative content is substantially increased, the cratering tendency becomes more severe.

In the limiting formulation the carbide is tungsten-free and based entirely on TiC, but generally TiC content extends to no more than about 18 per cent. Above this figure the carbide becomes excessively brittle and is very difficult to braze, although this drawback is not a problem with throwaway inserts.

WC/TiC/Co grades generally have two distinct carbide phases, angular crystals of almost pure WC and rounded TiC/WC mixed crystals. Among progressive manufacturers, although WC/TiC/Co hardmetals are very widely used, in certain important respects they are obsolescent, having been superseded by the WC/TiC/Ta(Nb)C/Co series in the many applications where higher strength combined with crater resistance is an advantage. TiC, TiN, and other coatings on tough substrates have also diminished the attractions of high-TiC grades for high-speed machining of steels and ferrous alloys.

Tungsten-Titanium-Tantalum (-Niobium) Carbide/Cobalt: Except for coated carbides, tungsten-titanium-tantalum (-niobium) grades could be the most popular class of hardmetals. Used mainly for cutting steel, they combine and improve upon most of the best features of the longer-established WC/TiC/Co compositions. These carbides compete directly with carbonitrides and silicon nitride ceramics, and the best cemented carbides of this class can undertake very heavy cuts at high speeds on all types of steels, including austenitic stainless varieties. These tools also operate well on ductile cast irons and nickel-base superalloys, where great heat and high pressures are generated at the cutting edge. However, they do not have the resistance to abrasive wear possessed by micrograin straight tungsten carbide grades nor the good resistance to cratering of coated grades and titanium carbide-based cermets.

Titanium Carbide/Molybdenum/Nickel (TiC/Mo/Ni): The extreme indentation hardness and crater resistance of titanium carbide, allied to the cheapness and availability of its main raw material (titanium dioxide, TiO_2), provide a strong inducement to use grades based on this carbide alone. Although developed early in the history of hardmetals, these carbides were difficult to braze satisfactorily and consequently were little used until the advent of clamped, throwaway inserts. Moreover, the carbides were notoriously brittle and could take only fine cuts in minimal-shock conditions.

Titanium-carbide-based grades again came into prominence about 1960, when nickel-molybdenum began to be used as a binder instead of nickel. The new grades were able to perform a wider range of tasks including interrupted cutting and cutting under shock conditions.

The very high indentation hardness values recorded for titanium carbide grades are not accompanied by correspondingly greater resistance to abrasive wear, the apparently less hard tungsten carbide being considerably superior in this property. Moreover, carbonitrides, advanced tantalum-containing multicarbides, and coated variants generally provide better all-round cutting performances.

Titanium-Base Carbonitrides: Development of titanium-carbonitride-based cutting-tool materials predates the use of coatings of this type on more conventional hardmetals by many years. Appreciable, though uncontrolled, amounts of carbonitride were often present, if only by accident, when cracked ammonia was used as a less expensive substitute for hydrogen in some stages of the production process in the 1950's and perhaps for two decades earlier.

Much of the recent, more scientific development of this class of materials has taken place in the United States, particularly by Teledyne Firth Sterling with its SD_3 grade and in Japan by several companies. Many of the compositions currently in use are extremely complex, and their structures—even with apparently similar compositions—can vary enormously. For instance, Mitsubishi characterizes its Himet NX series of cermets as TiC/WC/Ta(Nb)C/Mo_2C/TiN/Ni/Co/Al, with a structure comprising both large and medium-size carbide particles (mainly TiC according to the quoted density) in a superalloy-type matrix containing an aluminum-bearing intermetallic compound.

Steel- and Alloy-Bonded Titanium Carbide: The class of material exemplified by Ferro-Tic, as it is known, consists primarily of titanium carbide bonded with heat-treatable steel, but some grades also contain tungsten carbide or are bonded with nickel- or copper-base alloys. These cemented carbides are characterized by high binder contents (typically 50–60 per cent by volume) and lower hardnesses, compared with the more usual hardmetals, and by the great variation in properties obtained by heat treatment.

In the annealed condition, steel-bonded carbides have a relatively soft matrix and can be machined with little difficulty, especially by CBN (superhard cubic boron nitride) tools. After heat treatment, the degree of hardness and wear resistance achieved is considerably greater than that of normal tool steels, although understandably much less than that of traditional sintered carbides. Microstructures are extremely varied, being composed of 40–50 per cent TiC by volume and a matrix appropriate to the alloy composition and the stage of heat treatment. Applications include stamping, blanking and drawing dies, machine components, and similar items where the ability to machine before hardening reduces production costs substantially.

Coating: As a final stage in carbide manufacture, coatings of various kinds are applied mainly to cutting tools, where for cutting steel in particular it is advantageous to give the rank and clearance surfaces characteristics that are quite different from those of the body of the insert. Coatings of titanium carbide, nitride, or carbonitride; of aluminum oxide; and of other refractory compounds are applied to a variety of hardmetal substrates by chemical or physical vapor deposition (CVD or PVD) or by newer plasma methods.

The most recent types of coatings include hafnium, tantalum, and zirconium carbides and nitrides; alumina/titanium oxide; and multiple carbide/carbonitride/nitride/oxide, oxynitride or oxycarbonitride combinations. Greatly improved properties have been claimed for variants with as many as 13 distinct CVD coatings. A markedly sharper cutting edge compared with other CVD-coated hardmetals is claimed, permitting finer cuts and the successful machining of soft but abrasive alloys.

The keenest edges on coated carbides are achieved by the techniques of physical vapor deposition. In this process, ions are deposited directionally from the electrodes, rather than evenly on all surfaces, so the sharpness of cutting edges is maintained and may even be enhanced. PVD coatings currently available include titanium nitride and carbonitride, their distinctive gold color having become familiar throughout the world on high-speed steel tooling. The high temperatures required for normal CVD tends to soften heat-treated high-speed steel. PVD-coated hardmetals have been produced commercially for several years, especially for precision milling inserts.

Recent developments in extremely hard coatings, generally involving exotic techniques, include boron carbide, cubic boron nitride, and pure diamond. Almost the ultimate in wear resistance, the commercial applications of thin plasma-generated diamond surfaces at present are mainly in manufacture of semiconductors, where other special properties are important.

For cutting tools the substrate is of equal importance to the coating in many respects, its critical properties including fracture toughness (resistance to crack propagation), elastic modulus, resistance to heat and abrasion, and expansion coefficient. Some manufacturers are now producing inserts with graded composition, so that structures and properties are optimized at both surface and interior, and coatings are less likely to crack or break away.

Specifications: Compared with other standardized materials, the world of sintered hardmetals is peculiar. For instance, an engineer who seeks a carbide grade for the finish-machining of a steel component may be told to use *ISO Standard Grade P10 or Industry Code C7.* If the composition and nominal properties of the designated tool material are then requested, the surprising answer is that, in basic composition alone, the tungsten carbide content of P10 (or of the now superseded C7) can vary from zero to about 75, titanium carbide from 8 to 80, cobalt 0 to 10, and nickel 0 to 15 per cent. There are other possible constituents, also, in this so-called standard alloy, and many basic properties can vary as much as the composition. All that these dissimilar materials have in common, and all that the so-called standards mean, is that their suppliers—and sometimes their suppliers alone—consider them suitable for one particular and ill-defined machining application (which for P10 or C7 is the finish machining of steel).

This peculiar situation arose because the production of cemented carbides in occupied Europe during World War II was controlled by the German Hartmetallzentrale, and no factory other than Krupp was permitted to produce more than one grade. By the end of the war, all German-controlled producers were equipped to make the G, S, H, and F series to German standards. In the postwar years, this series of carbides formed the basis of unofficial European standardization. With the advent of the newer multicarbides, the previous identities of grades were gradually lost. The applications relating to the old grades were retained, however, as a new German DIN standard, eventually being adopted, in somewhat modified form, by the International Standards Organization (ISO) and by ANSI in the United States.

The American cemented carbides industry developed under diverse ownership and solid competition. The major companies actively and independently developed new varieties of hardmetals, and there was little or no standardization, although there were many attempts to compile equivalent charts as a substitute for true standardization. Around 1942, the Buick division of GMC produced a simple classification code that arranged nearly 100 grades derived from 10 manufacturers under only 14 symbols (TC-1 to TC-14). In spite of serious deficiencies, this system remained in use for many years as an American industry standard; that is, Buick TC-1 was equivalent to industry code C1. Buick itself went much further, using the tremendous influence, research facilities, and purchasing potential of its parent company to standardize the products of each carbide manufacturer by properties that could be tested, rather than by the indeterminate recommended applications. Many large-scale carbide users have developed similar systems in attempts to exert some degree of in-house standardization and quality control. Small and medium-sized users, however, still suffer from so-called industry standards, which only provide a starting point for grade selection.

ISO standard 513, summarized in Table , divides all machining grades into three color-coded groups: straight tungsten carbide grades (letter K, color red) for cutting gray cast iron, nonferrous metals, and nonmetallics; highly alloyed grades (letter, P. color blue) for machining steel; and less alloyed grades (letter M, color yellow, generally with less TiC than the corresponding P series), which are multipurpose and may be used on steels, nickel-base superalloys, ductile cast irons, and so on. Each grade within a group is also given a number to represent its position in a range from maximum hardness to maximum toughness (shock resistance). Typical applications are described for grades at more or less regular numerical intervals. Although coated grades scarcely existed when the ISO standard was prepared, it is easy to classify coated as uncoated carbides—or carbonitrides, ceramics, and superhard materials—according to this system.

In this situation, it is easy to see how one plant will prefer one manufacturer's carbide and a second plant will prefer that of another. Each has found the carbide most nearly ideal for the particular conditions involved. In these circumstances it pays each manufacturer to make grades that differ in hardness, toughness, and crater resistance, so that they can provide a product that is near the optimum for a specific customer's application.

Although not classified as a hard metal, new particle or powder metallurgical methods of manufacture, coupled with new coating technology have led in recent years to something of an upsurge in the use of high speed steel. Lower cost is a big factor, and the development of such coatings as titanium nitride, cubic boron nitride, and pure diamond, has enabled some high speed steel tools to rival tools made from tungsten and other carbides in their ability to maintain cutting accuracy and prolong tool life. Multiple layers may be used to produce optimum properties in the coating, with adhesive strength where there is contact with the substrate, combined with hardness at the cutting surface to resist abrasion. Total thickness of such coating, even with multiple layers, is seldom more than 15 microns (0.000060 in.).

Importance of Correct Grades: A great diversity of hardmetal types is required to cope with all possible combinations of metals and alloys, machining operations, and working conditions. Tough, shock-resistant grades are needed for slow speeds and interrupted cutting, harder grades for high-speed finishing, heat-resisting alloyed grades for machining superalloys, and crater-resistant compositions, including most of the many coated varieties, for machining steels and ductile iron.

Ceramics.—Moving up the hardness scale, ceramics provide increasing competition for cemented carbides, both in performance and in cost-effectiveness, though not yet in reliability. Hardmetals themselves consist of ceramics—nonmetallic refractory compounds, usually carbides or carbonitrides—with a metallic binder of much lower melting point. In such systems, densification generally takes place by liquid-phase sintering. Pure ceramics have no metallic binder, but may contain lower-melting-point compounds or ceramic mixtures that permit liquid-phase sintering to take place. Where this condition is not possible, hot pressing or hot isostatic pressing can often be used to make a strong, relatively pore-

Table 2. ISO Classifications of Hardmetals (Cemented Carbides and Carbonitrides) by Application

Main Types of Chip Removal		Groups of Applications			Direction of Decrease in Characteristic	
Symbol and Color	Broad Categories of Materials to be Machined	Designation (Grade)	Specific Material to be Machined	Use and Working Conditions	of cut	of carbide
P Blue	Ferrous with long chips	P01	Steel, steel castings	Finish turning and boring; high cutting speeds, small chip sections, accurate dimensions, fine finish, vibration-free operations	↑ speed	↑ wear
		P10	Steel, steel casting	Turning, copying, threading, milling; high cutting speeds; small or medium chip sections		
		P20	Steel, steel castings, ductile cast iron with long chips	Turning, copying, milling; medium cutting speeds and chip sections, planing with small chip sections		
		P30	Steel, steel castings, ductile cast iron with long chips	Turning, milling, planing; medium or large chip sections, unfavorable machining conditions		
		P40	Steel, steel castings with sand inclusions and cavities	Turning, planing, slotting; low cutting speeds, large chip sections, with possible large cutting angles, unfavorable cutting conditions, and work on automatic machines		
		P50	Steel, steel castings of medium or low tensile strength, with sand inclusions and cavities	Operations demanding very tough carbides; turning, planing, slotting; low cutting speeds, large chip sections, with possible large cutting angles, unfavorable conditions and work on automatic machines		
M Yellow	Ferrous metals with long or short chips, and non ferrous metals	M10	Steel, steel castings, manganese steel, gray cast iron, alloy cast iron	Turning; medium or high cutting speeds, small or medium chip sections		
		M20	Steel, steel castings, austenitic or manganese steel, gray cast iron	Turning, milling; medium cutting speeds and chip sections		
		M30	Steel, steel castings, austenitic steel, gray cast iron, high-temperature-resistant alloys	Turning, milling, planing; medium cutting speeds, medium or large chip sections		
		M40	Mild, free-cutting steel, low-tensile steel, nonferrous metals and light alloys	Turning, parting off; particularly on automatic machines		
K Red	Ferrous metals with short chips, non-ferrous metals and non-metallic materials	K01	Very hard gray cast iron, chilled castings over 85 Shore, high-silicon aluminum alloys, hardened steel, highly abrasive plastics, hard cardboard, ceramics	Turning, finish turning, boring, milling, scraping		
		K10	Gray cast iron over 220 Brinell, malleable cast iron with short chips, hardened steel, silicon-aluminum and copper alloys, plastics, glass, hard rubber, hard cardboard, porcelain, stone	Turning, milling, drilling, boring, broaching, scraping		
		K20	Gray cast iron up to 220 Brinell, nonferrous metals, copper, brass, aluminum	Turning, milling, planing, boring, broaching, demanding very tough carbide		
		K30	Low-hardness gray cast iron, low-tensile steel, compressed wood	Turning, milling, planing, slotting, unfavorable conditions, and possibility of large cutting angles	feed ↓	toughness ↓
		K40	Softwood or hard wood, nonferrous metals	Turning, milling, planing, slotting, unfavorable conditions, and possibility of large cutting angles		

free component or cutting insert. This section is restricted to those ceramics that compete directly with hardmetals, mainly in the cutting-tool category as shown in Table 3.

Ceramics are hard, completely nonmetallic substances that resist heat and abrasive wear. Increasingly used as clamped indexable tool inserts, ceramics differ significantly from tool steels, which are completely metallic. Ceramics also differ from cermets such as cemented carbides and carbonitrides, which comprise minute ceramic particles held together by metallic binders.

Table 3. Typical Properties of Cutting Tool Ceramics

Group	Alumina	Alumina/TiC	Silicon Nitride	PCD	PCBN
Typical composition types	Al_2O_3 or Al_2O_3/ZrO_2	70/30 Al_2O_3/TiC	Si_3N_4/Y_2O_3 plus		
Density (g/cm^3)	4.0	4.25	3.27	3.4	3.1
Transverse rupture strength (N/mm^2)	700	750	800		800
Compressive strength (kN/mm^2)	4.0	4.5	4.0	4.7	3.8
Hardness (HV)	1750	1800	1600		
Hardness HK (kN/mm^2)				50	28
Young's modulus (kN/mm^2)	380	370	300	925	680
Modulus of rigidity (kN/mm^2)	150	160	150	430	280
Poisson's ratio	0.24	0.22	0.20	0.09	0.22
Thermal expansion coefficient ($10^{-6}/K$)	8.5	7.8	3.2	3.8	4.9
Thermal conductivity (W/m K)	23	17	22	120	100
Fracture toughness($K_{1c}MN/m^{3/2}$)	2.3	3.3	5.0	7.9	10

Alumina-based ceramics were introduced as cutting inserts during World War II, and were for many years considered too brittle for regular machine-shop use. Improved machine tools and finer-grain, tougher compositions incorporating zirconia or silicon carbide "whiskers" now permit their use in a wide range of applications. Silicon nitride, often combined with alumina (aluminum oxide), yttria (yttrium oxide), and other oxides and nitrides, is used for much of the high-speed machining of superalloys, and newer grades have been formulated specifically for cast iron—potentially a far larger market.

In addition to improvements in toolholders, great advances have been made in machine tools, many of which now feature the higher powers and speeds required for the efficient use of ceramic tooling. Brittleness at the cutting edge is no longer a disadvantage, with the improvements made to the ceramics themselves, mainly in toughness, but also in other critical properties.

Although very large numbers of useful ceramic materials are now available, only a few combinations have been found to combine such properties as minimum porosity, hardness, wear resistance, chemical stability, and resistance to shock to the extent necessary for cutting-tool inserts. Most ceramics used for machining are still based on high-purity, fine-grained alumina (aluminum oxide), but embody property-enhancing additions of other ceramics such as zirconia (zirconium oxide), titania (titanium oxide), titanium carbide, tungsten carbide, and titanium nitride. For commercial purposes, those more commonly used are often termed "white" (alumina with or without zirconia) or "black" (roughly 70/30 alumina/titanium carbide). More recent developments are the distinctively green alumina ceramics strengthened with silicon carbide whiskers and the brown-tinged silicon nitride types.

Ceramics benefit from hot isostatic pressing, used to remove the last vestiges of porosity and raise substantially the material's shock resistance, even more than carbide-based hardmetals. Significant improvements are derived by even small parts such as tool inserts, although, in principle, they should not need such treatment if raw materials and manufacturing methods are properly controlled.

Oxide Ceramics: Alumina cutting tips have extreme hardness—more than HV 2000 or HRA 94—and give excellent service in their limited but important range of uses such as the machining of chilled iron rolls and brake drums. A substantial family of alumina-based materials has been developed, and fine-grained alumina-based composites now have sufficient strength for milling cast iron at speeds up to 2500 ft/min (800 m/min). Resistance to cratering when machining steel is exceptional.

Oxide/Carbide Ceramics: A second important class of alumina-based cutting ceramics combines aluminum oxide or alumina-zirconia with a refractory carbide or carbides, nearly always 30 per cent TiC. The compound is black and normally is hot pressed or hot isostatically pressed (HIPed). As shown in Table 3, the physical and mechanical properties of this material are generally similar to those of the pure alumina ceramics, but strength and shock resistance are generally higher, being comparable with those of higher-toughness simple alumina-zirconia grades. Current commercial grades are even more complex, combining alumina, zirconia, and titanium carbide with the further addition of titanium nitride.

Silicon Nitride Base: One of the most effective ceramic cutting-tool materials developed in the UK is Syalon (from SiAlON or silicon-aluminum-oxynitride) though it incorporates a substantial amount of yttria for efficient liquid-phase sintering). The material combines high strength with hot hardness, shock resistance, and other vital properties. Syalon cutting inserts are made by Kennametal and Sandvik and sold as Kyon 2000 and CC680, respectively. The brown Kyon 200 is suitable for machining high-nickel alloys and cast iron, but a later development, Kyon 3000 has good potential for machining cast iron.

Resistance to thermal stress and thermal shock of Kyon 2000 are comparable to those of sintered carbides. Toughness is substantially less than that of carbides, but roughly twice that of oxide-based cutting-tool materials at temperatures up to 850°C. Syon 200 can cut at high edge temperatures and is harder than carbide and some other ceramics at over 700°C, although softer than most at room temperature.

Whisker-Reinforced Ceramics: To improve toughness, Greenleaf Corp. has reinforced alumina ceramics with silicon carbide single-crystal "whiskers" that impart a distinctive green color to the material, marketed as WG300. Typically as thin as human hairs, the immensely strong whiskers improve tool life under arduous conditions. Whisker-reinforced ceramics and perhaps hardmetals are likely to become increasingly important as cutting and wear-resistant materials. Their only drawback seems to be the carcinogenic nature of the included fibers, which requires stringent precautions during manufacture.

Superhard Materials.—Polycrystalline synthetic diamond (PCD) and cubic boron nitride (PCBN), in the two columns at the right in Table 3, are almost the only cutting-insert materials in the "superhard" category. Both PCD and PCBN are usually made with the highest practicable concentration of the hard constituent, although ceramic or metallic binders can be almost equally important in providing overall strength and optimizing other properties. Variations in grain size are another critical factor in determining cutting characteristics and edge stability. Some manufacturers treat CBN in similar fashion to tungsten carbide, varying the composition and amount of binder within exceptionally wide limits to influence the physical and mechanical properties of the sintered compact.

In comparing these materials, users should note that some inserts comprise solid polycrystalline diamond or CBN and are double-sized to provide twice the number of cutting edges. Others consist of a layer, from 0.020 to 0.040 in. (0.5 to 1 mm) thick, on a tough carbide backing. A third type is produced with a solid superhard material almost surrounded by sintered carbide. A fourth type, used mainly for cutting inserts, comprises solid hard metal with a tiny superhard insert at one or more (usually only one) cutting corners or edges. Superhard cutting inserts are expensive—up to 30 times the cost of equivalent shapes or sizes in ceramic or cemented carbide—but their outstanding properties, exceptional performance and extremely long life can make them by far the most cost-effective for certain applications.

Diamond: Diamond is the hardest material found or made. As harder, more abrasive ceramics and other materials came into widespread use, diamond began to be used for grinding-wheel grits. Cemented carbide tools virtually demanded diamond grinding wheels for fine edge finishing. Solid single-crystal diamond tools were and are used to a small extent for special purposes, such as microtomes, for machining of hard materials, and for exceptionally fine finishes. These diamonds are made from comparatively large, high-quality gem-type diamonds, have isotropic properties, and are very expensive. By comparison, diamond abrasive grits cost only a few dollars a carat.

Synthetic diamonds are produced from graphite using high temperatures and extremely high pressures. The fine diamond particles produced are sintered together in the presence of a metal "catalyst" to produce high-efficiency anisotropic cutting tool inserts. These tools comprise either a solid diamond compact or a layer of sintered diamond on a carbide backing, and are made under conditions similar to, though less severe than, those used in diamond synthesis. Both natural and synthetic diamond can be sintered in this way, although the latter method is the most frequently used.

Polycrystalline diamond (PCD) compacts are immensely hard and can be used to machine many substances, from highly abrasive hardwoods and glass fiber to nonferrous metals, hardmetals, and tough ceramics. Important classes of tools that are also available with cubic boron nitride inserts include brazed-tip drills, single-point turning tools, and face-milling cutters.

Boron Nitride: Polycrystalline diamond has one big limitation: it cannot be used to machine steel or any other ferrous material without rapid chemical breakdown. Boron nitride does not have this limitation. Normally soft and slippery like graphite, the soft hexagonal crystals (HBN) become cubic boron nitride (CBN) when subjected to ultrahigh pressures and temperatures, with a structure similar to and hardness second only to diamond. As a solid insert of polycrystalline cubic boron nitride (PCBN), the compound machines even the hardest steel with relative immunity from chemical breakdown or cratering.

Backed by sintered carbide, inserts of PCBN can readily be brazed, increasing the usefulness of the material and the range of tooling in which it can be used. With great hardness and abrasion resistance, coupled with extreme chemical stability when in contact with ferrous alloys at high temperatures, PCBN has the ability to machine both steels and cast irons at high speeds for long operating cycles. Only its currently high cost in relation to hardmetals prevents its wider use in mass-production machining.

Similar in general properties to PCBN, the recently developed "Wurbon" consists of a mixture of ultrafine (0.02 μm grain size) hexagonal and cubic boron nitride with a "wurtzite" structure, and is produced from soft hexagonal boron nitride in a microsecond by an explosive shock-wave.

Basic Machining Data: Most mass-production metalcutting operations are carried out with carbide-tipped tools but their correct application is not simple. Even apparently similar batches of the same material vary greatly in their machining characteristics and may require different tool settings to attain optimum performance. Depth of cut, feed, surface speed, cutting rate, desired surface finish, and target tool life often need to be modified to suit the requirements of a particular component.

For the same downtime, the life of an insert between indexings can be less than that of an equivalent brazed tool between regrinds, so a much higher rate of metal removal is possible with the indexable or throwaway insert. It is commonplace for the claims for a new coating to include increases in surface-speed rates of 200–300 per cent, and for a new insert design to offer similar improvements. Many operations are run at metal removal rates that are far from optimum for tool life because the rates used maximize productivity and cost-effectiveness.

Thus any recommendations for cutting speeds and feeds must be oversimplified or extremely complex, and must be hedged with many provisos, dependent on the technical and economic conditions in the manufacturing plant concerned. A preliminary grade selection should be made from the ISO-based tables and manufacturers' literature consulted for recommendations on the chosen grades and tool designs. If tool life is much greater than that desired under the suggested conditions, speeds, feeds, or depths of cut may be increased. If tools fail by edge breakage, a tougher (more shock-resistant) grade should be selected, with a numerically higher ISO code.

Alternatively, increasing the surface speed and decreasing the feed may be tried. If tools fail prematurely from what appears to be abrasive wear, a harder grade with numerically lower ISO designation should be tried. If cratering is severe, use a grade with higher titanium carbide content; that is, switch from an ISO K to M or M to P grade, use a P grade with lower numerical value, change to a coated grade, or use a coated grade with a (claimed) more-resistant surface layer.

Built-Up Edge and Cratering: The big problem in cutting steel with carbide tools is associated with the built-up edge and the familar phenomenon called cratering. Research has shown that the built-up edge is continuous with the chip itself during normal cutting. Additions of titanium, tantalum, and niobium to the basic carbide mixture have a remarkable effect on the nature and degree of cratering, which is related to adhesion between the tool and the chip.

Hardmetal Tooling for Wood and Nonmetallics.—Carbide-tipped circular saws are now conventional for cutting wood, wood products such as chipboard, and plastics, and tipped bandsaws of large size are also gaining in popularity. Tipped handsaws and mechanical equivalents are seldom needed for wood, but they are extremely useful for cutting abrasive building boards, glass-reinforced plastics, and similar material. Like the hardmetal tips used on most other woodworking tools, saw tips generally make use of straight (unalloyed) tungsten carbide/cobalt grades. However, where excessive heat is generated as with the cutting of high-silica hardwoods and particularly abrasive chipboards, the very hard but tough tungsten-titanium-tantalum-niobium carbide solid-solution grades, normally reserved for steel finishing, may be preferred. Saw tips are usually brazed and reground a number of times during service, so coated grades appear to have little immediate potential in this field.

Cutting Blades and Plane Irons: These tools comprise long, thin, comparatively wide slabs of carbide on a minimal-thickness steel backing. Compositions are straight tungsten carbide, preferably micrograin (to maintain a keen cutting edge with an included angle of 30° or less), but with relatively high amounts of cobalt, 11–13 per cent, for toughness. Considerable expertise is necessary to braze and grind these cutters without inducing or failing to relieve the excessive stresses that cause distortion or cracking.

Other Woodworking Cutters: Routers and other cutters are generally similar to those used on metals and include many indexable-insert designs. The main difference with wood is that rotational and surface speeds can be the maximum available on the machine. High-speed routing of aluminum and magnesium alloys was developed largely from machines and techniques originally designed for work on wood.

Cutting Other Materials: The machining of plastics, fiber-reinforced plastics, graphite, asbestos, and other hard and abrasive constructional materials mainly requires abrasion resistance. Cutting pressures and power requirements are generally low. With thermoplastics and some other materials, particular attention must be given to cooling because of softening or degradation of the work material that might be caused by the heat generated in cutting. An important application of cemented carbides is the drilling and routing of printed circuit boards. Solid tungsten carbide drills of extremely small sizes are used for this work.

FORMING TOOLS

When curved surfaces or those of stepped, angular or irregular shape are required in connection with turning operations, especially on turret lathes and "automatics," forming tools are used. These tools are so made that the contour of the cutting edge corresponds to the shape required and usually they may be ground repeatedly without changing the shape of the cutting edge. There are two general classes of forming tools—the straight type and the circular type. The circular forming tool is generally used on small narrow forms, whereas the straight type is more suitable for wide forming operations. Some straight forming tools are clamped in a horizontal position upon the cut-off slide, whereas the others are held in a vertical position in a special holder. A common form of holder for these vertical tools is one having a dovetail slot in which the forming tool is clamped; hence they are often called "dovetail forming tools." In many cases, two forming tools are used, especially when a very smooth surface is required, one being employed for roughing and the other for finishing.

There was an American standard for forming tool blanks which covered both straight or dovetailed, and circular forms. The formed part of the finished blanks must be shaped to suit whatever job the tool is to be used for. This former standard includes the important dimensions of holders for both straight and circular forms.

Dimensions of Steps on Straight or Dovetail Forming Tools.—The diagrams at the top of the accompanying table illustrate a straight or "dovetail" forming tool. The upper or cutting face lies in the same plane as the center of the work and there is no rake. (Many forming tools have rake to increase the cutting efficiency, and this type will be referred to later.) In making a forming tool, the various steps measured perpendicular to the front face (as at *d*) must be proportioned so as to obtain the required radial dimensions on the work. For example, if *D* equals the difference between two radial dimensions on the work, then:

$$\text{Step} d = D \times \text{cosine front clearance angle}$$

Angles on Straight Forming Tools.—In making forming tools to the required shape or contour, any angular surfaces (like the steps referred to in the previous paragraph) are affected by the clearance angle. For example, assume that angle *A* on the work (see diagram at top of accompanying table) is 20 degrees. The angle on the tool in plane *x-x*, in that case, will be slightly less than 20 degrees. In making the tool, this modified or reduced angle is required because of the convenience in machining and measuring the angle square to the front face of the tool or in the plane *x–x*.

If the angle on the work is measured from a line parallel to the axis (as at *A* in diagram), then the reduced angle on the tool as measured square to the front face (or in plane *x–x*) is found as follows:

$$\text{tan reduced angle on tool} = \tan A \times \cos \text{ front clearance angle}$$

If angle *A* on the work is larger than, say, 45 degrees, it may be given on the drawing as indicated at *B*. In this case, the angle is measured from a plane perpendicular to the axis of the work. When the angle is so specified, the angle on the tool in plane *x–x* may be found as follows:

$$\text{tan reduced angle on tool} = \frac{\tan B}{\cos \text{ clearance angle}}$$

Table Giving Step Dimensions and Angles on Straight or Dovetailed Forming Tools.—The accompanying table "Dimensions of Steps and Angles on Straight Forming Tools" gives the required dimensions and angles within its range, direct or without calculation.

Dimensions of Steps and Angles on Straight Forming Tools

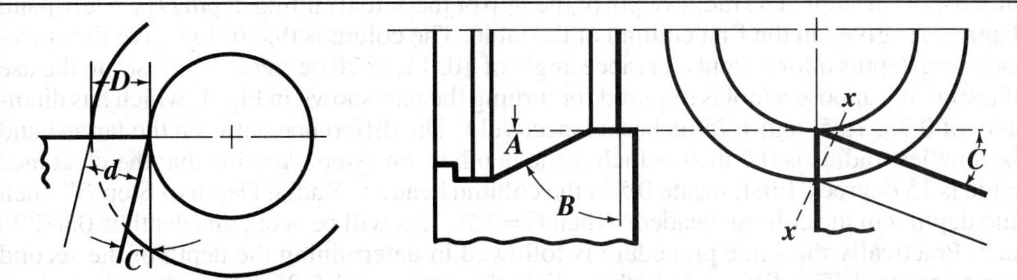

Upper section of table gives depth *d* of step on forming tool for a given dimension *D* that equals the actual depth of the step on the work, measured radially and along the cutting face of the tool (see diagram at left). First, locate depth *D* required on work; then find depth *d* on tool under tool clearance angle *C*. Depth *d* is measured perpendicular to front face of tool.

Radial Depth of Step *D*	Depth *d* of step on tool When *C* = 10°	When *C* = 15°	When *C* = 20°	Radial Depth of Step *D*	Depth *d* of step on tool When *C* = 10°	When *C* = 15°	When *C* = 20°
0.001	0.00098	0.00096	0.00094	0.040	0.03939	0.03863	0.03758
0.002	0.00197	0.00193	0.00187	0.050	0.04924	0.04829	0.04698
0.003	0.00295	0.00289	0.00281	0.060	0.05908	0.05795	0.05638
0.004	0.00393	0.00386	0.00375	0.070	0.06893	0.06761	0.06577
0.005	0.00492	0.00483	0.00469	0.080	0.07878	0.07727	0.07517
0.006	0.00590	0.00579	0.00563	0.090	0.08863	0.08693	0.08457
0.007	0.00689	0.00676	0.00657	0.100	0.09848	0.09659	0.09396
0.008	0.00787	0.00772	0.00751	0.200	0.19696	0.19318	0.18793
0.009	0.00886	0.00869	0.00845	0.300	0.29544	0.28977	0.28190
0.010	0.00984	0.00965	0.00939	0.400	0.39392	0.38637	0.37587
0.020	0.01969	0.01931	0.01879	0.500	0.49240	0.48296	0.46984
0.030	0.02954	0.02897	0.02819	…	…	…	…

Section of table below gives angles as measured in plane *x–x* perpendicular to front face of forming tool (see diagram on right). Find in first column the angle *A* required on work; then find reduced angle in plane *x–x* under given clearance angle *C*.

Angle *A* in Plane of Tool Cutting Face	Angle on tool in plane *x–x* When *C* = 10°		When *C* = 15°		When *C* = 20°		Angle *A* in Plane of Tool Cutting Face	Angle on tool in plane *x–x* When *C* = 10°		When *C* = 15°		When *C* = 20°	
5°	4°	55′	4°	50′	4°	42′	50°	49°	34′	49°	1′	48°	14′
10	9	51	9	40	9	24	55	54	35	54	4	53	18
15	14	47	14	31	14	8	60	59	37	59	8	58	26
20	19	43	19	22	18	53	65	64	40	64	14	63	36
25	24	40	24	15	23	40	70	69	43	69	21	68	50
30	29	37	29	9	28	29	75	74	47	74	30	74	5
35	34	35	34	4	33	20	80	79	51	79	39	79	22
40	39	34	39	1	38	15	85	84	55	84	49	84	41
45	44	34	44	0	43	13	…	…		…		…	

To Find Dimensions of Steps: The upper section of the table is used in determining the dimensions of steps. The radial depth of the step or the actual cutting depth *D* (see left-hand diagram) is given in the first column of the table. The columns that follow give the corresponding depths *d* for a front clearance angle of 10, 15, or 20 degrees. To illustrate the use of the table, suppose a tool is required for turning the part shown in Fig. 1, which has diameters of 0.75, 1.25, and 1.75 inches, respectively. The difference between the largest and the smallest radius is 0.5 inch, which is the depth of one step. Assume that the clearance angle is 15 degrees. First, locate 0.5 in the column headed "Radial Depth of Step *D*"; then find depth *d* in the column headed "when C = 15°." As will be seen, this depth is 0.48296 inch. Practically the same procedure is followed in determining the depth of the second step on the tool. The difference in the radii in this case equals0.25. This value is not given directly in the table, so first find the depth equivalent to 0.200 and add to it the depth equivalent to 0.050. Thus, we have 0.19318 + 0.04829 = 0.24147. In using this table, it is assumed that the top face of the tool is set at the height of the work axis.

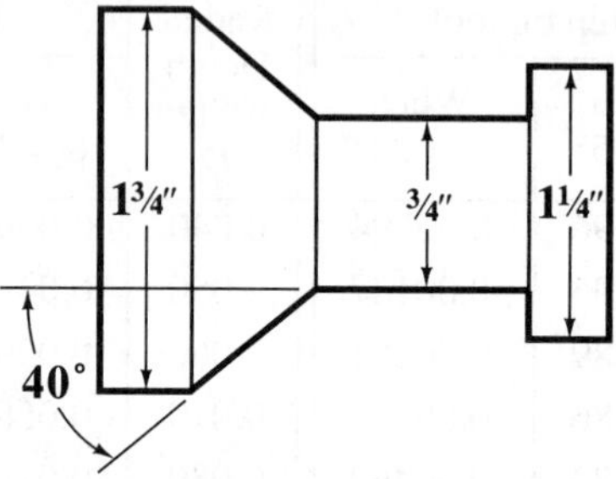

Fig. 1.

To Find Angle: The lower section of the table applies to angles when they are measured relative to the axis of the work. The application of the table will again be illustrated by using the part shown in Fig. 1. The angle used here is 40 degrees (which is also the angle in the plane of the cutting face of the tool). If the clearance angle is 15 degrees, the angle measured in plane *x–x* square to the face of the tool is shown by the table to be 39° 1′- a reduction of practically 1 degree.

If a straight forming tool has rake, the depth *x* of each step (see Fig. 2), measured perpendicular to the front or clearance face, is affected not only by the clearance angle, but by the rake angle *F* and the radii *R* and *r* of the steps on the work. First, it is necessary to find three angles, designated *A*, *B*, and *C*, that are not shown on the drawing.

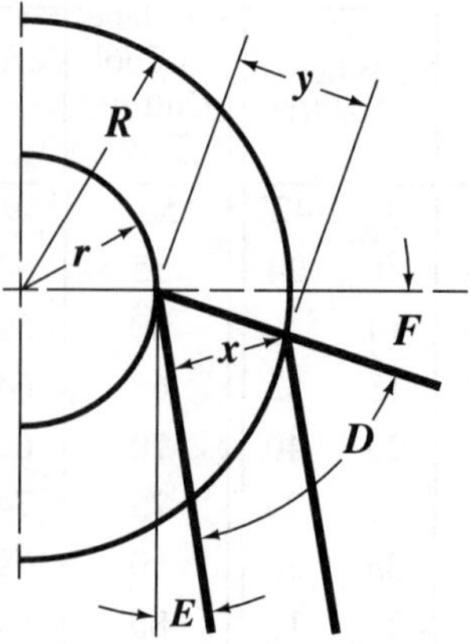

Fig. 2.

$$\text{Angle } A = 180° - \text{ rake angle } F$$

$$\sin B = \frac{r \sin A}{R}$$

$$\text{Angle } C = 180° - (A + B)$$

$$y = \frac{R \sin C}{\sin A}$$

$$\text{Angle } D \text{ of tool} = 90° - (E + F)$$

$$\text{Depth } x = y \sin D$$

If the work has two or more shoulders, the depth x for other steps on the tool may be determined for each radius r. If the work has curved or angular forms, it is more practical to use a tool without rake because its profile, in the plane of the cutting face, duplicates that of the work.

Example: Assume that radius R equals 0.625 inch and radius r equals 0.375 inch, so that the step on the work has a radial depth of 0.25 inch. The tool has a rake angle F of 10 degrees and a clearance angle E of 15 degrees. Then angle $A = 180 - 10 = 170$ degrees.

$$\sin B = \frac{0.375 \times 0.17365}{0.625} = 0.10419$$

$$\text{Angle } B = 5°59' \text{ nearly.} \qquad \text{Angle } C = 180 - (170° + 5°59') = 4°1'$$

$$\text{Dimension } y = \frac{0.625 \times 0.07005}{0.17365} = 0.25212$$

$$\text{Angle } D = 90° - (15 + 10) = 65 \text{ degrees}$$

$$\text{Depth } x \text{ of step} = 0.25212 \times 0.90631 = 0.2285 \text{ inch}$$

Circular Forming Tools.—To provide sufficient peripheral clearance on circular forming tools, the cutting face is offset with relation to the center of the tool a distance C, as shown in Fig. 3. Whenever a circular tool has two or more diameters, the difference in the radii of the steps on the tool will not correspond exactly to the difference in the steps on the work. The form produced with the tool also changes, although the change is very slight, unless the amount of offset C is considerable. Assume that a circular tool is required to produce the piece A having two diameters as shown.

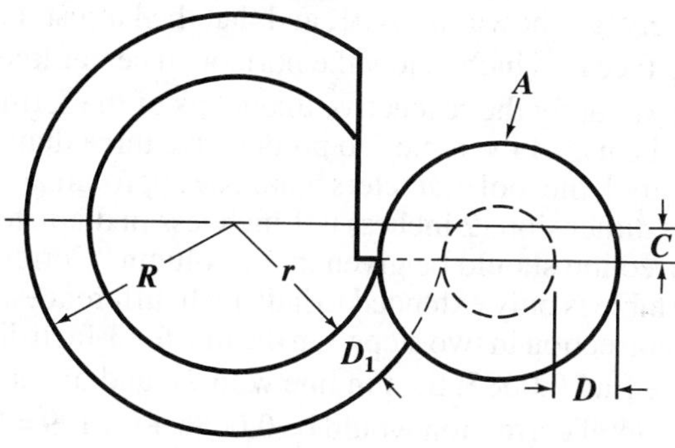

Fig. 3.

If the difference D_1 between the large and small radii of the tool were made equal to dimension D required on the work, D would be a certain amount oversize, depending upon the offset C of the cutting edge. The following formulas can be used to determine the radii of circular forming tools for turning parts to different diameters:

Let R = largest radius of tool in inches; D = difference in radii of steps on work; C = amount cutting edge is offset from center of tool; r = required radius in inches; then

$$r = \sqrt{(\sqrt{R^2 - C^2} - D)^2 + C^2} \tag{1}$$

If the small radius r is given and the large radius R is required, then

$$R = \sqrt{(\sqrt{r^2 - C^2} + D)^2 + C^2} \tag{2}$$

To illustrate, if D (Fig. 3) is to be ⅛ inch, the large radius R is 1⅛ inches, and C is 5⁄32 inch, what radius r would be required to compensate for the offset C of the cutting edge? Inserting these values in Formula (1):

$$r = \sqrt{\sqrt{(1\tfrac{1}{8})^2 - (\tfrac{5}{32})^2} - (\tfrac{1}{8})^2 + (\tfrac{5}{32})^2} = 1.0014 \text{ inches}$$

The value of r is thus found to be 1.0014 inches; hence, the diameter = 2 × 1.0014 = 2.0028 inches instead of 2 inches, as it would have been if the cutting edge had been exactly on the center line. Formulas for circular tools used on different makes of screw machines can be simplified when the values R and C are constant for each size of machine. The accompanying table, "Formulas for Circular Forming Tools," gives the standard values of R and C for circular tools used on different automatics. The formulas for determining the radius r (see column at right-hand side of table) contain a constant that represents the value of the expression $\sqrt{R^2 - C^2}$ in Formula (1).

The table "Constants for Determining Diameters of Circular Forming Tools" has been compiled to facilitate proportioning tools of this type and gives constants for computing the various diameters of forming tools, when the cutting face of the tool is ⅛, 3⁄16, ¼, or 5⁄16 inch below the horizontal center line. As there is no standard distance for the location of the cutting face, the table has been prepared to correspond with distances commonly used. As an example, suppose the tool is required for a part having three diameters of 1.75, 0.75, and 1.25 inches, respectively, as shown in Fig. 1, and that the largest diameter of the tool is 3 inches and the cutting face is ¼ inch below the horizontal center line. The first step would be to determine approximately the respective diameters of the forming tool and then correct the diameters by the use of the table. To produce the three diameters shown in Fig. 1, with a 3-inch forming tool, the tool diameters would be approximately 2, 3, and 2.5 inches, respectively. The first dimension (2 inches) is 1 inch less in diameter than that of the tool, and the necessary correction should be given in the column "Correction for Difference in Diameter"; but as the table is only extended to half-inch differences, it will be necessary to obtain this particular correction in two steps. On the line for 3-inch diameter and under corrections for ½ inch, we find 0.0085; then in line with 2½ and under the same heading, we find 0.0129, hence the total correction would be 0.0085 + 0.0129 = 0.0214 inch. This correction is added to the approximate diameter, making the exact diameter of the first step 2 + 0.0214 = 2.0214 inches. The next step would be computed in the same way, by noting on the 3-inch line the correction for ½ inch and adding it to the approximate diameter of the second step, giving an exact diameter of 2.5 + 0.0085 + 2.5085 inches. Therefore, to produce the part shown in Fig. 1, the tool should have three steps of 3, 2.0214, and 2.5085 inches, respectively, provided the cutting face is ¼ inch below the center. All diameters are computed in this way, from the largest diameter of the tool.

Formulas for Circular Forming Tools (For notation, see Fig. 3)

Make of Machine	Size of Machine	Radius R, Inches	Offset C, Inches	Radius r, Inches
Brown & Sharpe	No. 00	0.875	0.125	$r = \sqrt{(0.8660 - D)^2 + 0.0156}$
	No. 0	1.125	0.15625	$r = \sqrt{(1.1141 - D)^2 + 0.0244}$
	No. 2	1.50	0.250	$r = \sqrt{(1.4790 - D)^2 + 0.0625}$
	No. 6	2.00	0.3125	$r = \sqrt{(1.975 - D)^2 + 0.0976}$
Acme	No. 51	0.75	0.09375	$r = \sqrt{(1.7441 - D)^2 + 0.0088}$
	No. 515	0.75	0.09375	$r = \sqrt{(0.7441 - D)^2 + 0.0088}$
	No. 52	1.0	0.09375	$r = \sqrt{(0.9956 - D)^2 + 0.0088}$
	No. 53	1.1875	0.125	$r = \sqrt{(1.1809 - D)^2 + 0.0156}$
	No. 54	1.250	0.15625	$r = \sqrt{(1.2402 - D)^2 + 0.0244}$
	No. 55	1.250	0.15625	$r = \sqrt{(1.2402 - D)^2 + 0.0244}$
	No. 56	1.50	0.1875	$r = \sqrt{(1.4882 - D)^2 + 0.0352}$
Cleveland	¼″	0.625	0.03125	$r = \sqrt{(0.6242 - D)^2 + 0.0010}$
	⅜″	0.084375	0.0625	$r = \sqrt{(0.8414 - D)^2 + 0.0039}$
	⅝″	1.15625	0.0625	$r = \sqrt{(1.1546 - D)^2 + 0.0039}$
	⅞″	1.1875	0.0625	$r = \sqrt{(1.1859 - D)^2 + 0.0039}$
	1¼″	1.375	0.0625	$r = \sqrt{(1.3736 - D)^2 + 0.0039}$
	2″	1.375	0.0625	$r = \sqrt{(1.3736 - D)^2 + 0.0039}$
	2¼″	1.625	0.125	$r = \sqrt{(1.6202 - D)^2 + 0.0156}$
	2¾″	1.875	0.15625	$r = \sqrt{(1.8685 - D)^2 + 0.0244}$
	3¼″	1.875	0.15625	$r = \sqrt{(1.8685 - D)^2 + 0.0244}$
	4¼″	2.50	0.250	$r = \sqrt{(2.4875 - D)^2 + 0.0625}$
	6″	2.625	0.250	$r = \sqrt{(2.6131 - D)^2 + 0.0625}$

The tables "Corrected Diameters of Circular Forming Tools" are especially applicable to tools used on Brown & Sharpe automatic screw machines. Directions for using these tables are given at the end of Table 4.

Circular Tools Having Top Rake.—Circular forming tools without top rake are satisfactory for brass, but tools for steel or other tough metals cut better when there is a rake angle of 10 or 12 degrees. For such tools, the small radius *r* (see Fig. 3) for an outside radius *R* may be found by the formula

$$r = \sqrt{P^2 + R^2 - 2PR\cos\theta}$$

To find the value of *P*, proceed as follows: $\sin\phi$ = small radius on work × sin rake angle ÷ large radius on work. Angle β = rake angle − ϕ. *P* = large radius on work × $\sin\beta$ ÷ sin rake angle. Angle θ = rake angle + δ. $\sin\delta$ = vertical height *C* from center of tool to center of work ÷ *R*. It is assumed that the tool point is to be set at the same height as the work center.

Dimensions for Circular Cut-Off Tools

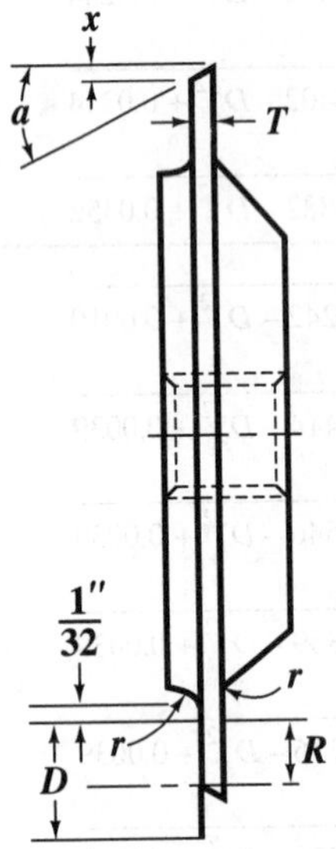

Dia. of Stock	Soft Brass, Copper		Norway Iron, Machine Steel		Drill Rod, Tool Steel	
	a = 23 Deg.		*a* = 15 Deg.		*a* = 12 Deg.	
	T	*x*	*T*	*x*	*T*	*x*
1⁄16	0.031	0.013	0.039	0.010	0.043	0.009
1⁄8	0.044	0.019	0.055	0.015	0.062	0.013
3⁄16	0.052	0.022	0.068	0.018	0.076	0.016
1⁄4	0.062	0.026	0.078	0.021	0.088	0.019
5⁄16	0.069	0.029	0.087	0.023	0.098	0.021
3⁄8	0.076	0.032	0.095	0.025	0.107	0.023
7⁄16	0.082	0.035	0.103	0.028	0.116	0.025
1⁄2	0.088	0.037	0.110	0.029	0.124	0.026
9⁄16	0.093	0.039	0.117	0.031	0.131	0.028
5⁄8	0.098	0.042	0.123	0.033	0.137	0.029
11⁄16	0.103	0.044	0.129	0.035	0.145	0.031
3⁄4	0.107	0.045	0.134	0.036	0.152	0.032
13⁄16	0.112	0.047	0.141	0.038	0.158	0.033
7⁄8	0.116	0.049	0.146	0.039	0.164	0.035
15⁄16	0.120	0.051	0.151	0.040	0.170	0.036
1	0.124	0.053	0.156	0.042	0.175	0.037

The length of the blade equals radius of stock *R* + *x* + *r* + 1⁄32 inch (for notation, see illustration above); *r* = 1⁄16 inch for 3⁄8- to 3⁄4-inch stock, and 3⁄32 inch for 3⁄4- to 1-inch stock.

Constant for Determining Diameters of Circular Forming Tools

Dia. of Tool	Radius of Tool	Cutting Face 1/8 Inch Below Center			Cutting Face 3/16 Inch Below Center			Cutting Face 1/4 Inch Below Center			Cutting Face 5/16 Inch Below Center		
		Correction for Difference in Diameter			Correction for Difference in Diameter			Correction for Difference in Diameter			Correction for Difference in Diameter		
		1/8 Inch	1/4 Inch	1/2 Inch	1/8 Inch	1/4 Inch	1/2 Inch	1/8 Inch	1/4 Inch	1/2 Inch	1/8 Inch	1/4 Inch	1/2 Inch
1	0.500	…	…	…	…	…	…	…	…	…	…	…	…
1 1/8	0.5625	0.0036	…	…	0.0086	…	…	0.0167	…	…	0.0298	…	…
1 1/4	0.625	0.0028	0.0065	…	0.0067	0.0154	…	0.0128	0.0296	…	0.0221	0.0519	…
1 3/8	0.6875	0.0023	…	…	0.0054	…	…	0.0102	…	…	0.0172	…	…
1 1/2	0.750	0.0019	0.0042	0.0107	0.0045	0.0099	0.0253	0.0083	0.0185	0.0481	0.0138	0.0310	0.0829
1 5/8	0.8125	0.0016	…	…	0.0037	…	…	0.0069	…	…	0.0114	…	…
1 3/4	0.875	0.0014	0.0030	…	0.0032	0.0069	…	0.0058	0.0128	…	0.0095	0.0210	…
1 7/8	0.9375	0.0012	…	…	0.0027	…	…	0.0050	…	…	0.0081	…	…
2	1.000	0.0010	0.0022	0.0052	0.0024	0.0051	0.0121	0.0044	0.0094	0.0223	0.0070	0.0152	0.0362
2 1/8	1.0625	0.0009	…	…	0.0021	…	…	0.0038	…	…	0.0061	…	…
2 1/4	1.125	0.0008	0.0017	…	0.0018	0.0040	…	0.0034	0.0072	…	0.0054	0.0116	…
2 3/8	1.1875	0.0007	…	…	0.0016	…	…	0.0029	…	…	0.0048	…	…
2 1/2	1.250	0.0006	0.0014	0.0031	0.0015	0.0031	0.0071	0.0027	0.0057	0.0129	0.0043	0.0092	0.0208
2 5/8	1.3125	0.0006	…	…	0.0013	…	…	0.0024	…	…	0.0038	…	…
2 3/4	1.375	0.0005	0.0011	…	0.0012	0.0026	…	0.0022	0.0046	…	0.0035	0.0073	…
2 7/8	1.4375	0.0005	…	…	0.0011	…	…	0.0020	…	…	0.0032	…	…
3	1.500	0.0004	0.0009	0.0021	0.0010	0.0021	0.0047	0.0018	0.0038	0.0085	0.0029	0.0061	0.0135
3 1/8	1.5625	0.00004	…	…	0.0009	…	…	0.0017	…	…	0.0027	…	…
3 1/4	1.625	0.0003	0.0008	…	0.0008	0.0018	…	0.0015	0.0032	…	0.0024	0.0051	…
3 3/8	1.6875	0.0003	…	…	0.0008	…	…	0.0014	…	…	0.0023	…	…
3 1/2	1.750	0.0003	0.0007	0.0015	0.0007	0.0015	0.0033	0.0013	0.0028	0.0060	0.0021	0.0044	0.0095
3 5/8	1.8125	0.0003	…	…	0.0007	…	…	0.0012	…	…	0.0019	…	…
3 3/4	1.875	0.0002	0.0006	…	0.0.0006	0.0013	…	0.0011	0.0024	…	0.0018	0.0038	…

Corrected Diameters of Circular Forming Tools

Length *c* on Tool	Number of B. & S. Automatic Screw Machine			Length *c* on Tool	Number of B. & S. Automatic Screw Machine		
	No. 00	No. 0	No. 2		No. 00	No. 0	No. 2
0.001	1.7480	2.2480	2.9980	0.058	1.6353	2.1352	2.8857
0.002	1.7460	2.2460	2.9961	0.059	1.6333	2.1332	2.8837
0.003	1.7441	2.2441	2.9941	0.060	1.6313	2.1312	2.8818
0.004	1.7421	2.2421	2.9921	0.061	1.6294	2.1293	2.8798
0.005	1.7401	2.2401	2.9901	0.062	1.6274	2.1273	2.8778
0.006	1.7381	2.2381	2.9882	$\frac{1}{16}$	1.6264	2.1263	2.8768
0.007	1.7362	2.2361	2.9862	0.063	1.6254	2.1253	2.8759
0.008	1.7342	2.2341	2.9842	0.064	1.6234	2.1233	2.8739
0.009	1.7322	2.2321	2.9823	0.065	1.6215	2.1213	2.8719
0.010	1.7302	2.2302	2.9803	0.066	1.6195	2.1194	2.8699
0.011	1.7282	2.2282	2.9783	0.067	1.6175	2.1174	2.8680
0.012	1.7263	2.2262	2.9763	0.068	1.6155	2.1154	2.8660
0.013	1.7243	2.2243	2.9744	0.069	1.6136	2.1134	2.8640
0.014	1.7223	2.2222	2.9724	0.070	1.6116	2.1115	2.8621
0.015	1.7203	2.2203	2.9704	0.071	1.6096	2.1095	2.8601
$\frac{1}{64}$	1.7191	2.2191	2.9692	0.072	1.6076	2.1075	2.8581
0.016	1.7184	2.2183	2.9685	0.073	1.6057	2.1055	2.8561
0.017	1.7164	2.2163	2.9665	0.074	1.6037	2.1035	2.8542
0.018	1.7144	2.2143	2.9645	0.075	1.6017	2.1016	2.8522
0.019	1.7124	2.2123	2.9625	0.076	1.5997	2.0996	2.8503
0.020	1.7104	2.2104	2.9606	0.077	1.5978	2.0976	2.8483
0.021	1.7085	2.2084	2.9586	0.078	1.5958	2.0956	2.8463
0.022	1.7065	2.2064	2.9566	$\frac{5}{64}$	1.5955	2.0954	2.8461
0.023	1.7045	2.2045	2.9547	0.079	1.5938	2.0937	2.8443
0.024	1.7025	2.2025	2.9527	0.080	1.5918	2.0917	2.8424
0.025	1.7005	2.2005	2.9507	0.081	1.5899	2.0897	2.8404
0.026	1.6986	2.1985	2.9488	0.082	1.5879	2.0877	2.8384
0.027	1.6966	2.1965	2.9468	0.083	1.5859	2.0857	2.8365
0.028	1.6946	2.1945	2.9448	0.084	1.5839	2.0838	2.8345
0.029	1.6926	2.1925	2.9428	0.085	1.5820	2.0818	2.8325
0.030	1.6907	2.1906	2.9409	0.086	1.5800	2.0798	2.8306
0.031	1.6887	2.1886	2.9389	0.087	1.5780	2.0778	2.8286
$\frac{1}{32}$	1.6882	2.1881	2.9384	0.088	1.5760	2.0759	2.8266
0.032	1.6867	2.1866	2.9369	0.089	1.5740	2.0739	2.8247
0.033	1.6847	2.1847	2.9350	0.090	1.5721	2.0719	2.8227
0.034	1.6827	2.1827	2.9330	0.091	1.5701	2.0699	2.8207
0.035	1.6808	2.1807	2.9310	0.092	1.5681	2.0679	2.8187
0.036	1.6788	2.1787	2.9290	0.093	1.5661	2.0660	2.8168
0.037	1.6768	2.1767	2.9271	$\frac{3}{32}$	1.5647	2.0645	2.8153
0.038	1.6748	2.1747	2.9251	0.094	1.5642	2.0640	2.8148
0.039	1.6729	2.1727	2.9231	0.095	1.5622	2.0620	2.8128
0.040	1.6709	2.1708	2.9211	0.096	1.5602	2.0600	2.8109
0.041	1.6689	2.1688	2.9192	0.097	1.5582	2.0581	2.8089
0.042	1.6669	2.1668	2.9172	0.098	1.5563	2.0561	2.8069
0.043	1.6649	2.1649	2.9152	0.099	1.5543	2.0541	2.8050
0.044	1.6630	2.1629	2.9133	0.100	1.5523	2.0521	2.8030
0.045	1.6610	2.1609	2.9113	0.101	1.5503	2.0502	2.8010
0.046	1.6590	2.1589	2.9093	0.102	1.5484	2.0482	2.7991
$\frac{3}{64}$	1.6573	2.1572	2.9076	0.103	1.5464	2.0462	2.7971
0.047	1.6570	2.1569	2.9073	0.104	1.5444	2.0442	2.7951
0.048	1.6550	2.1549	2.9054	0.105	1.5425	2.0422	2.7932
0.049	1.6531	2.1529	2.9034	0.106	1.5405	2.0403	2.7912
0.050	1.6511	2.1510	2.9014	0.107	1.5385	2.0383	2.7892
0.051	1.6491	2.1490	2.8995	0.108	1.5365	2.0363	2.7873
0.052	1.6471	2.1470	2.8975	0.109	1.5346	2.0343	2.7853
0.053	1.6452	2.1451	2.8955	$\frac{7}{64}$	1.5338	2.0336	2.7846
0.054	1.6432	2.1431	2.8936	0.110	1.5326	2.0324	2.7833
0.055	1.6412	2.1411	2.8916	0.111	1.5306	2.0304	2.7814
0.056	1.6392	2.1391	2.8896	0.112	1.5287	2.0284	2.7794

Corrected Diameters of Circular Forming Tools *(Continued)*

Length c on Tool	Number of B. & S. Automatic Screw Machine			Length c on Tool	Number of B. & S. Automatic Screw Machine		
	No. 00	No. 0	No. 2		No. 00	No. 0	No. 2
0.057	1.6373	2.1372	2.8877	0.113	1.5267	2.0264	2.7774
0.113	1.5267	2.0264	2.7774	0.171	1.4124	1.9119	2.6634
0.114	1.5247	2.0245	2.7755	11/64	1.4107	1.9103	2.6617
0.115	1.5227	2.0225	2.7735	0.172	1.4104	1.9099	2.6614
0.116	1.5208	2.0205	2.7715	0.173	1.4084	1.9080	2.6595
0.117	1.5188	2.0185	2.7696	0.174	1.4065	1.9060	2.6575
0.118	1.5168	2.0166	2.7676	0.175	1.4045	1.9040	2.6556
0.119	1.5148	2.0146	2.7656	0.176	1.4025	1.9021	2.6536
0.120	1.5129	2.0126	2.7637	0.177	1.4006	1.9001	2.6516
0.121	1.5109	2.0106	2.7617	0.178	1.3986	1.8981	2.6497
0.122	1.5089	2.0087	2.7597	0.179	1.3966	1.8961	2.6477
0.123	1.5070	2.0067	2.7578	0.180	1.3947	1.8942	2.6457
0.124	1.5050	2.0047	2.7558	0.181	1.3927	1.8922	2.6438
0.125	1.5030	2.0027	2.7538	0.182	1.3907	1.8902	2.6418
0.126	1.5010	2.0008	2.7519	0.183	1.3888	1.8882	2.6398
0.127	1.4991	1.9988	2.7499	0.184	1.3868	1.8863	2.6379
0.128	1.4971	1.9968	2.7479	0.185	1.3848	1.8843	2.6359
0.129	1.4951	1.9948	2.7460	0.186	1.3829	1.8823	2.6339
0.130	1.4932	1.9929	2.7440	0.187	1.3809	1.8804	2.6320
0.131	1.4912	1.9909	2.7420	3/16	1.3799	1.8794	2.6310
0.132	1.4892	1.9889	2.7401	0.188	1.3789	1.8784	2.6300
0.133	1.4872	1.9869	2.7381	0.189	1.3770	1.8764	2.6281
0.134	1.4853	1.9850	2.7361	0.190	1.3750	1.8744	2.6261
0.135	1.4833	1.9830	2.7342	0.191	1.3730	1.8725	2.6241
0.136	1.4813	1.9810	2.7322	0.192	1.3711	1.8705	2.6222
0.137	1.4794	1.9790	2.7302	0.193	1.3691	1.8685	2.6202
0.138	1.4774	1.9771	2.7282	0.194	1.3671	1.8665	2.6182
0.139	1.4754	1.9751	2.7263	0.195	1.3652	1.8646	2.6163
0.140	1.4734	1.9731	2.7243	0.196	1.3632	1.8626	2.6143
9/64	1.4722	1.9719	2.7231	0.197	1.3612	1.8606	2.6123
0.141	1.4715	1.9711	2.7224	0.198	1.3592	1.8587	2.6104
0.142	1.4695	1.9692	2.7204	0.199	1.3573	1.8567	2.6084
0.143	1.4675	1.9672	2.7184	0.200	1.3553	1.8547	2.6064
0.144	1.4655	1.9652	2.7165	0.201	…	1.8527	2.6045
0.145	1.4636	1.9632	2.7145	0.202	…	1.8508	2.6025
0.146	1.4616	1.9613	2.7125	0.203	…	1.8488	2.6006
0.147	1.4596	1.9593	2.7106	13/64	…	1.8486	2.6003
0.148	1.4577	1.9573	2.7086	0.204	…	1.8468	2.5986
0.149	1.4557	1.9553	2.7066	0.205	…	1.8449	2.5966
0.150	1.4537	1.9534	2.7047	0.206	…	1.8429	2.5947
0.151	1.4517	1.9514	2.7027	0.207	…	1.8409	2.5927
0.152	1.4498	1.9494	2.7007	0.208	…	1.8390	2.5908
0.153	1.4478	1.9474	2.6988	0.209	…	1.8370	2.5888
0.154	1.4458	1.9455	2.6968	0.210	…	1.8350	2.5868
0.155	1.4439	1.9435	2.6948	0.211	…	1.8330	2.5849
0.156	1.4419	1.9415	2.6929	0.212	…	1.8311	2.5829
5/32	1.4414	1.9410	2.6924	0.213	…	1.8291	2.5809
0.157	1.4399	1.9395	2.6909	0.214	…	1.8271	2.5790
0.158	1.4380	1.9376	2.6889	0.215	…	1.8252	2.5770
0.159	1.4360	1.9356	2.6870	0.216	…	1.8232	2.5751
0.160	1.4340	1.9336	2.6850	0.217	…	1.8212	2.5731
0.161	1.4321	1.9317	2.6830	0.218	…	1.8193	2.5711
0.162	1.4301	1.9297	2.6811	7/32	…	1.8178	2.5697
0.163	1.4281	1.9277	2.6791	0.219	…	1.8173	2.5692
0.164	1.4262	1.9257	2.6772	0.220	…	1.8153	2.5672
0.165	1.4242	1.9238	2.6752	0.221	…	1.8133	2.5653
0.166	1.4222	1.9218	2.6732	0.222	…	1.8114	2.5633
0.167	1.4203	1.9198	2.6713	0.223	…	1.8094	2.5613
0.168	1.4183	1.9178	2.6693	0.224	…	1.8074	2.5594
0.169	1.4163	1.9159	2.6673	0.225	…	1.8055	2.5574

Corrected Diameters of Circular Forming Tools *(Continued)*

Length *c* on Tool	Number of B. & S. Automatic Screw Machine			Length *c* on Tool	Number of B. & S. Automatic Screw Machine		
	No. 00	No. 0	No. 2		No. 00	No. 0	No. 2
0.170	1.4144	1.9139	2.6654	0.226	....	1.8035	2.5555
0.227	1.8015	2.5535	0.284	1.6894	2.4418	0.341	2.3303
0.228	1.7996	2.5515	0.285	1.6874	2.4398	0.342	2.3284
0.229	1.7976	2.5496	0.286	1.6854	2.4378	0.343	2.3264
0.230	1.7956	2.5476	0.287	1.6835	2.4359	11/32	2.3250
0.231	1.7936	2.5456	0.288	1.6815	2.4340	0.344	2.3245
0.232	1.7917	2.5437	0.289	1.6795	2.4320	0.345	2.3225
0.233	1.7897	2.5417	0.290	1.6776	2.4300	0.346	2.3206
0.234	1.7877	2.5398	0.291	1.6756	2.4281	0.347	2.3186
15/64	1.7870	2.5390	0.292	1.6736	2.4261	0.348	2.3166
0.235	1.7858	2.5378	0.293	1.6717	2.4242	0.349	2.3147
0.236	1.7838	2.5358	0.294	1.6697	2.4222	0.350	2.3127
0.237	1.7818	2.5339	0.295	1.6677	2.4203	0.351	2.3108
0.238	1.7799	2.5319	0.296	1.6658	2.4183	0.352	2.3088
0.239	1.7779	2.5300	19/64	1.6641	2.4166	0.353	2.3069
0.240	1.7759	2.5280	0.297	1.6638	2.4163	0.354	2.3049
0.241	1.7739	2.5260	0.298	1.6618	2.4144	0.355	2.3030
0.242	1.7720	2.5241	0.299	1.6599	2.4124	0.356	2.3010
0.243	1.7700	2.5221	0.300	1.6579	2.4105	0.357	2.2991
0.244	1.7680	2.5201	0.301	…	2.4085	0.358	2.2971
0.245	1.7661	2.5182	0.302	…	2.4066	0.359	2.2952
0.246	1.7641	2.5162	0.303	…	2.4046	23/64	2.2945
0.247	1.7621	2.5143	0.304	…	2.4026	0.360	2.2932
0.248	1.7602	2.5123	0.305	…	2.4007	0.361	2.2913
0.249	1.7582	2.5104	0.306	…	2.3987	0.362	2.2893
0.250	1.7562	2.5084	0.307	…	2.3968	0.363	2.2874
0.251	1.7543	2.5064	0.308	…	2.3948	0.364	2.2854
0.252	1.7523	2.5045	0.309	…	2.3929	0.365	2.2835
0.253	1.7503	2.5025	0.310	…	2.3909	0.366	2.2815
0.254	1.7484	2.5005	0.311	…	2.3890	0.367	2.2796
0.255	1.7464	2.4986	0.312	…	2.3870	0.368	2.2776
0.256	1.7444	2.4966	5/16	…	2.3860	0.369	2.2757
0.257	1.7425	2.4947	0.313	…	2.3851	0.370	2.2737
0.258	1.7405	2.4927	0.314	…	2.3831	0.371	2.2718
0.259	1.7385	2.4908	0.315	…	2.3811	0.372	2.2698
0.260	1.7366	2.4888	0.316	…	2.3792	0.373	2.2679
0.261	1.7346	2.4868	0.317	…	2.3772	0.374	2.2659
0.262	1.7326	2.4849	0.318	…	2.3753	0.375	2.2640
0.263	1.7306	2.4829	0.319	…	2.3733	0.376	2.2620
0.264	1.7287	2.4810	0.320	…	2.3714	0.377	2.2601
0.265	1.7267	2.4790	0.321	…	2.3694	0.378	2.2581
17/64	1.7255	2.4778	0.322	…	2.3675	0.379	2.2562
0.266	1.7248	2.4770	0.323	…	2.3655	0.380	2.2542
0.267	1.7228	2.4751	0.324	…	2.3636	0.381	2.2523
0.268	1.7208	2.4731	0.325	…	2.3616	0.382	2.2503
0.269	1.7189	2.4712	0.326	…	2.3596	0.383	2.2484
0.270	1.7169	2.4692	0.327	…	2.3577	0.384	2.2464
0.271	1.7149	2.4673	0.328	…	2.3557	0.385	2.2445
0.272	1.7130	2.4653	21/64	…	2.3555	0.386	2.2425
0.273	1.7110	2.4633	0.329	…	2.3538	0.387	2.2406
0.274	1.7090	2.4614	0.330	…	2.3518	0.388	2.2386
0.275	1.7071	2.4594	0.331	…	2.3499	0.389	2.2367
0.276	1.7051	2.4575	0.332	…	2.3479	0.390	2.2347
0.277	1.7031	2.4555	0.333	…	2.3460	25/64	2.2335
0.278	1.7012	2.4535	0.334	…	2.3440	0.391	2.2328
0.279	1.6992	2.4516	0.335	…	2.3421	0.392	2.2308
0.280	1.6972	2.4496	0.336	…	2.3401	0.393	2.2289
0.281	1.6953	2.4477	0.337	…	2.3381	0.394	2.2269
9/32	1.6948	2.4472	0.338	…	2.3362	0.395	2.2250
0.282	1.6933	2.4457	0.339	…	2.3342	0.396	2.2230

Corrected Diameters of Circular Forming Tools

Length c on Tool	Number of B. & S. Automatic Screw Machine			Length c on Tool	Number of B. & S. Automatic Screw Machine		
	No. 00	No. 0	No. 2		No. 00	No. 0	No. 2
0.283	1.6913	2.4438	0.340	…	2.3323	0.397	2.2211
0.398	2.2191	0.423	2.1704	0.449	2.1199	0.474	2.0713
0.399	2.2172	0.424	2.1685	0.450	2.1179	0.475	2.0694
0.400	2.2152	0.425	2.1666	0.451	2.1160	0.476	2.0674
0.401	2.2133	0.426	2.1646	0.452	2.1140	0.477	2.0655
0.402	2.2113	0.427	2.1627	0.453	2.1121	0.478	2.0636
0.403	2.2094	0.428	2.1607	29⁄64	2.1118	0.479	2.0616
0.404	2.2074	0.429	2.1588	0.454	2.1101	0.480	2.0597
0.405	2.2055	0.430	2.1568	0.455	2.1082	0.481	2.0577
0.406	2.2035	0.431	2.1549	0.456	2.1063	0.482	2.0558
13⁄32	2.2030	0.432	2.1529	0.457	2.1043	0.483	2.0538
0.407	2.2016	0.433	2.1510	0.458	2.1024	0.484	2.0519
0.408	2.1996	0.434	2.1490	0.459	2.1004	0.485	2.0500
0.409	2.1977	0.435	2.1471	0.460	2.0985	0.486	2.0480
0.410	2.1957	0.436	2.1452	0.461	2.0966	0.487	2.0461
0.411	2.1938	0.437	2.1432	0.462	2.0946	0.488	2.0441
0.412	2.1919	7⁄16	2.1422	0.463	2.0927	0.489	2.0422
0.413	2.1899	0.438	2.1413	0.464	2.0907	0.490	2.0403
0.414	2.1880	0.439	2.1393	0.465	2.0888	0.491	2.0383
0.415	2.1860	0.440	2.1374	0.466	2.0868	0.492	2.0364
0.416	2.1841	0.441	2.1354	0.467	2.0849	0.493	2.0344
0.417	2.1821	0.442	2.1335	0.468	2.0830	0.494	2.0325
0.418	2.1802	0.443	2.1315	15⁄32	2.0815	0.495	2.0306
0.419	2.1782	0.444	2.1296	0.469	2.0810	0.496	2.0286
0.420	2.1763	0.445	2.1276	0.470	2.0791	0.497	2.0267
0.421	2.1743	0.446	2.1257	0.471	2.0771	0.498	2.0247
27⁄64	2.1726	0.447	2.1237	0.472	2.0752	0.499	2.0228
0.422	2.1724	0.448	2.1218	0.473	2.0733	0.500	2.0209

Method of Using Tables for "Corrected Diameters of Circular Forming Tools".— These tables are especially applicable to Brown & Sharpe automatic screw machines. The maximum diameter D of forming tools for these machines should be as follows: For No. 00 machine, 1¾ inches; for No. 0 machine, 2¼ inches; for No. 2 machine, 3 inches. To find the other diameters of the tool for any piece to be formed, proceed as follows: Subtract the smallest diameter of the work from the diameter of the work that is to be formed by the required tool diameter; divide the remainder by 2; locate the quotient obtained in the column headed "Length c on Tool," and opposite the figure thus located and in the column headed by the number of the machine used, read off directly the diameter to which the tool is to be made. The quotient obtained, which is located in the column headed "Length c on Tool," is the length c, as shown in the following table.

Dimensions of Forming Tools for B. & S. Automatic Screw Machines

No. of Machine	Max. Dia., D	h	T	W
00	1¾	⅛	⅜–16	¼
0	2¼	5⁄32	½–14	5⁄16
2	3	¼	⅝–12	⅜
6	4	5⁄16	¾–12	⅜

Example: A piece of work is to be formed on a No. 0 machine to two diameters, one being ¼ inch and one 0.550 inch; find the diameters of the tool. The maximum tool diameter is 2¼ inches, or the diameter that will cut the ¼-inch diameter of the work. To find the other diameter, proceed according to the rule given: 0.550 − ¼ = 0.300; 0.300 ÷ 2 = 0.150. In Table 2, opposite 0.150, we find that the required tool diameter is 1.9534 inches. These tables are for tools without rakes.

Arrangement of Circular Tools.—When applying circular tools to automatic screw machines, their arrangement has an important bearing on the results obtained. The various ways of arranging the circular tools, with relation to the rotation of the spindle, are shown at A, B, C, and D in the illustration. These diagrams represent the view obtained when looking toward the chuck. The arrangement shown at A gives good results on long forming operations on brass and steel because the pressure of the cut on the front tool is downward; the support is more rigid than when the forming tool is turned upside down on the front slide, as shown at B; here the stock, turning up toward the tool, has a tendency to lift the cross-slide, causing chattering; therefore, the arrangement shown at A is recommended when a high-quality finish is desired. The arrangement at B works satisfactorily for short steel pieces that do not require a high finish; it allows the chips to drop clear of the work, and is especially advantageous when making screws, when the forming and cut-off tools operate after the die, as no time is lost in reversing the spindle. The arrangement at C is recommended for heavy cutting on large work, when both tools are used for forming the piece; a rigid support is then necessary for both tools and a good supply of oil is also required. The arrangement at D is objectionable and should be avoided; it is used only when a left-hand thread is cut on the piece and when the cut-off tool is used on the front slide, leaving the heavy cutting to be performed from the rear slide. In all "cross-forming" work, it is essential that the spindle bearings be kept in good condition, and that the collet or chuck has a parallel contact upon the bar that is being formed.

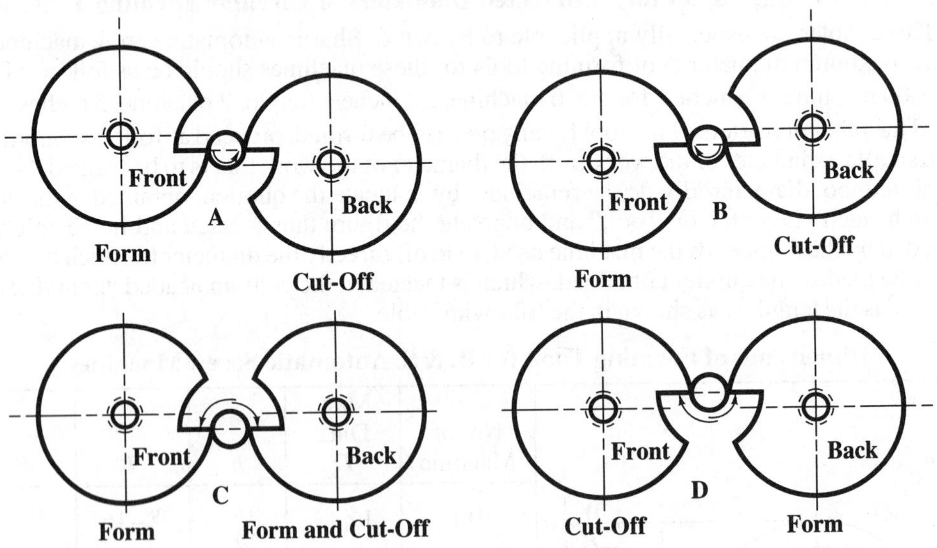

Feeds and Speeds for Forming Tools.—Approximate feeds and speeds for forming tools are given in the table beginning on page 1095. The feeds and speeds are average values, and if the job at hand has any features out of the ordinary, the figures given should be altered accordingly.

MILLING CUTTERS

Selection of Milling Cutters.—The most suitable type of milling cutter for a particular milling operation depends on such factors as the kind of cut to be made, the material to be cut, the number of parts to be machined, and the type of milling machine available. Solid cutters of small size will usually cost less, initially, than inserted blade types; for long-run production, inserted-blade cutters will probably have a lower overall cost. Depending on either the material to be cut or the amount of production involved, the use of carbide-tipped cutters in preference to high-speed steel or other cutting tool materials may be justified.

Rake angles depend on both the cutter material and the work material. Carbide and cast alloy cutting tool materials generally have smaller rake angles than high-speed steel tool materials because of their lower edge strength and greater abrasion resistance. Soft work materials permit higher radial rake angles than hard materials; thin cutters permit zero or practically zero axial rake angles; and wide cutters operate smoother with high axial rake angles. See *Rake Angles for Milling Cutters* on page 801.

Cutting edge relief or clearance angles are usually from 3 to 6 degrees for hard or tough materials, 4 to 7 degrees for average materials, and 6 to 12 degrees for easily machined materials. See *Clearance Angles for Milling Cutter Teeth* on page 800.

The number of teeth in the milling cutter is also a factor that should be given consideration, as explained in the next paragraph.

Number of Teeth in Milling Cutters.—In determining the number of teeth a milling cutter should have for optimum performance, there is no universal rule.

There are, however, two factors that should be considered in making a choice: 1) The number of teeth should never be so great as to reduce the chip space between the teeth to a point where a free flow of chips is prevented; and 2) The chip space should be smooth and without sharp corners that would cause clogging of the chips in the space.

For milling ductile materials that produce a continuous and curled chip, a cutter with large chip spaces is preferable. Such coarse tooth cutters permit an easier flow of the chips through the chip space than would be obtained with fine tooth cutters, and help to eliminate cutter "chatter." For cutting operations in thin materials, fine tooth cutters reduce cutter and workpiece vibration and the tendency for the cutter teeth to "straddle" the workpiece and dig in. For slitting copper and other soft nonferrous materials, teeth that are either chamfered or alternately flat and V-shaped are best.

As a general rule, to give satisfactory performance the number of teeth in milling cutters should be such that *no more than two teeth at a time are engaged in the cut.* Based on this rule, the following formulas are recommended:

For face milling cutters,

$$T = \frac{6.3D}{W} \tag{1}$$

For peripheral milling cutters,

$$T = \frac{12.6D\cos A}{D + 4d} \tag{2}$$

where T = number of teeth in cutter; D = cutter diameter in inches; W = width of cut in inches; d = depth of cut in inches; and A = helix angle of cutter.

To find the number of teeth that a cutter should have when other than two teeth in the cut at the same time is desired, Formulas (1) and (2) should be divided by 2 and the result multiplied by the number of teeth desired in the cut.

Example: Determine the required number of teeth in a face mill where $D = 6$ inches and $W = 4$ inches. Using Formula (1),

$$T = \frac{6.3 \times 6}{4} = 10 \text{ teeth, approximately}$$

Example: Determine the required number of teeth in a plain milling cutter where $D = 4$ inches and $d = \frac{1}{4}$ inch. Using Formula (2),

$$T = \frac{12.6 \times 4 \times \cos 0^\circ}{4 + (4 \times \frac{1}{4})} = 10 \text{ teeth, approximately}$$

In *high speed milling* with sintered carbide, high-speed steel, and cast non-ferrous cutting tool materials, a formula that permits full use of the power available at the cutter but prevents overloading of the motor driving the milling machine is:

$$T = \frac{K \times H}{F \times N \times d \times W} \qquad (3)$$

where T = number of cutter teeth; H = horsepower available at the cutter; F = feed per tooth in inches; N = revolutions per minute of cutter; d = depth of cut in inches; W = width of cut in inches; and K = a constant which may be taken as 0.65 for average steel, 1.5 for cast iron, and 2.5 for aluminum. These values are conservative and take into account dulling of the cutter in service.

Example: Determine the required number of teeth in a sintered carbide tipped face mill for high speed milling of 200 Brinell hardness alloy steel if $H = 10$ horsepower; $F = 0.008$ inch; $N = 272$ rpm; $d = 0.125$ inch; $W = 6$ inches; and K for alloy steel is 0.65. Using Formula (3),

$$T = \frac{0.65 \times 10}{0.008 \times 272 \times 0.125 \times 6} = 4 \text{ teeth, approximately}$$

American National Standard Milling Cutters.—According to American National Standard ANSI/ASME B94.19-1997 milling cutters may be classified in two general ways, which are given as follows:

By Type of Relief on Cutting Edges: Milling cutters may be described on the basis of one of two methods of providing relief for the cutting edges. *Profile sharpened* cutters are those on which relief is obtained and which are resharpened by grinding a narrow land back of the cutting edges. Profile sharpened cutters may produce flat, curved, or irregular surfaces. *Form relieved* cutters are those which are so relieved that by grinding only the faces of the teeth the original form is maintained throughout the life of the cutters. Form relieved cutters may produce flat, curved or irregular surfaces.

By Method of Mounting: Milling cutters may be described by one of two methods used to mount the cutter. *Arbor type* cutters are those which have a hole for mounting on an arbor and usually have a keyway to receive a driving key. These are sometimes called *Shell type. Shank type* cutters are those which have a straight or tapered shank to fit the machine tool spindle or adapter.

Explanation of the "Hand" of Milling Cutters.—In the ANSI Standard the terms "right hand" and "left hand" are used to describe hand of rotation, hand of cutter and hand of flute helix.

Hand of Rotation or Hand of Cut: is described as either "right hand" if the cutter revolves counterclockwise as it cuts when viewed from a position in front of a horizontal milling machine and facing the spindle or "left hand" if the cutter revolves clockwise as it cuts when viewed from the same position.

American National Standard Plain Milling Cutters ***ANSI/ASME B94.19-1997***

Cutter Diameter			Range of Face Widths Nom.[a]	Hole Diameter		
Nom.	Max.	Min.		Nom.	Max.	Min.
Light-duty Cutters[b]						
2½	2.515	2.485	3⁄16, ¼, 5⁄16, ⅜, ½, ⅝, ¾, 1, 1½, 2 and 3	1	1.00075	1.0000
3	3.015	2.985	3⁄16, ¼, 5⁄16, ⅜, ⅝, ¾, and 1½	1	1.00075	1.0000
3	3.015	2.985	½, ⅝, ¾, 1, 1¼, 1½, 2 and 3	1¼	1.2510	1.2500
4	4.015	3.985	¼, 5⁄16 and ⅜	1	1.00075	1.0000
4	4.015	3.985	⅜, ½, ⅝, ¾, 1, 1½, 2, 3 and 4	1¼	1.2510	1.2500
Heavy-duty Cutters[c]						
2½	2.515	2.485	2	1	1.00075	1.0000
2½	2.515	2.485	4	1	1.0010	1.0000
3	3.015	2.985	2, 2½, 3, 4 and 6	1¼	1.2510	1.2500
4	4.015	3.985	2, 3, 4 and 6	1½	1.5010	1.5000
High-helix Cutters[d]						
3	3.015	2.985	4 and 6	1¼	1.2510	1.2500
4	4.015	3.985	8	1½	1.5010	1.5000

[a] *Tolerances on Face Widths:* Up to 1 inch, inclusive, ± 0.001 inch; over 1 to 2 inches, inclusive, +0.010, −0.000 inch; over 2 inches, +0.020, −0.000 inch.

[b] Light-duty plain milling cutters with face widths under ¾ inch have straight teeth. Cutters with ¾-inch face and wider have helix angles of not less than 15 degrees nor greater than 25 degrees.

[c] Heavy-duty plain milling cutters have a helix angle of not less than 25 degrees nor greater than 45 degrees.

[d] High-helix plain milling cutters have a helix angle of not less than 45 degrees nor greater than 52 degrees.

All dimensions are in inches. All cutters are high-speed steel. Plain milling cutters are of cylindrical shape, having teeth on the peripheral surface only.

Hand of Cutter: Some types of cutters require special consideration when referring to their hand. These are principally cutters with unsymmetrical forms, face type cutters, or cutters with threaded holes. *Symmetrical* cutters may be reversed on the arbor in the same axial position and rotated in the cutting direction without altering the contour produced on the work-piece, and may be considered as either right or left hand. *Unsymmetrical* cutters reverse the contour produced on the work-piece when reversed on the arbor in the same axial position and rotated in the cutting direction. A *single-angle* cutter is considered to be a right-hand cutter if it revolves counterclockwise, or a left-hand cutter if it revolves clockwise, when cutting as viewed from the side of the larger diameter. The *hand of rotation* of a single angle milling cutter need not necessarily be the same as its *hand of cutter.* A *single corner rounding* cutter is considered to be a right-hand cutter if it revolves counterclockwise, or a left-hand cutter if it revolves clockwise, when cutting as viewed from the side of the smaller diameter.

American National Standard Side Milling Cutters *ANSI/ASME B94.19-1997*

Cutter Diameter			Range of Face Widths Nom.[a]	Hole Diameter		
Nom.	Max.	Min.		Nom.	Max.	Min.
			Side Cutters[b]			
2	2.015	1.985	3/16, 1/4, 3/8	5/8	0.62575	0.6250
2 1/2	2.515	2.485	1/4, 3/8, 1/2	7/8	0.87575	0.8750
3	3.015	2.985	1/4, 5/16, 3/8, 7/16, 1/2	1	1.00075	1.0000
4	4.015	3.985	1/4, 3/8, 1/2, 5/8, 3/4, 7/8	1	1.00075	1.0000
4	4.015	3.985	1/2, 5/8, 3/4	1 1/4	1.2510	1.2500
5	5.015	4.985	1/2, 5/8, 3/4	1	1.00075	1.0000
5	5.015	4.985	1/2, 5/8, 3/4, 1	1 1/4	1.2510	1.2500
6	6.015	5.985	1/2	1	1.00075	1.0000
6	6.015	5.985	1/2, 5/8, 3/4, 1	1 1/4	1.2510	1.2500
7	7.015	6.985	3/4	1 1/4	1.2510	1.2500
7	7.015	6.985	3/4	1 1/2	1.5010	1.5000
8	8.015	7.985	3/4, 1	1 1/4	1.2510	1.2500
8	8.015	7.985	3/4, 1	1 1/2	1.5010	1.5000
			Staggered-tooth Side Cutters[c]			
2 1/2	2.515	2.485	1/4, 5/16, 3/8, 1/2	7/8	0.87575	0.8750
3	3.015	2.985	3/16, 1/4, 5/16, 3/8	1	1.00075	1.0000
3	3.015	2.985	1/2, 5/8, 3/4	1 1/4	1.2510	1.2500
4	4.015	3.985	1/4, 5/16, 3/8, 7/16, 1/2, 5/8, 3/4 and 7/8	1 1/4	1.2510	1.2500
5	5.015	4.985	1/2, 5/8, 3/4	1 1/4	1.2510	1.2500
6	6.015	5.985	3/8, 1/2, 5/8, 3/4, 7/8, 1	1 1/4	1.2510	1.2500
8	8.015	7.985	3/8, 1/2, 5/8, 3/4, 1	1 1/2	1.5010	1.5000
			Half Side Cutters[d]			
4	4.015	3.985	3/4	1 1/4	1.2510	1.2500
5	5.015	4.985	3/4	1 1/4	1.2510	1.2500
6	6.015	5.985	3/4	1 1/4	1.2510	1.2500

[a] *Tolerances on Face Widths:* For side cutters, +0.002, –0.001 inch; for staggered-tooth side cutters up to 3/4 inch face width, inclusive, +0.000 –0.0005 inch, and over 3/4 to 1 inch, inclusive, +0.000 –0.0010 inch; and for half side cutters, +0.015, –0.000 inch.

[b] Side milling cutters have straight peripheral teeth and side teeth on both sides.

[c] Staggered-tooth side milling cutters have peripheral teeth of alternate right- and left-hand helix and alternate side teeth.

[d] Half side milling cutters have side teeth on one side only. The peripheral teeth are helical of the same hand as the cut. Made either with right-hand or left-hand cut.

All dimensions are in inches. All cutters are high-speed steel. Side milling cutters are of cylindrical shape, having teeth on the periphery and on one or both sides.

Hand of Flute Helix: Milling cutters may have *straight flutes* which means that their cutting edges are in planes parallel to the cutter axis. Milling cutters with flute helix in one direction only are described as having a right-hand helix if the flutes twist away from the observer in a clockwise direction when viewed from either end of the cutter or as having a left-hand helix if the flutes twist away from the observer in a counterclockwise direction when viewed from either end of the cutter. *Staggered tooth cutters* are milling cutters with every other flute of opposite (right and left hand) helix.

An illustration describing the various milling cutter elements of both a profile cutter and a form-relieved cutter is given on page 776.

American National Standard Staggered Teeth, T-Slot Milling Cutters with Brown & Sharpe Taper and Weldon Shanks *ANSI/ASME B94.19-1997*

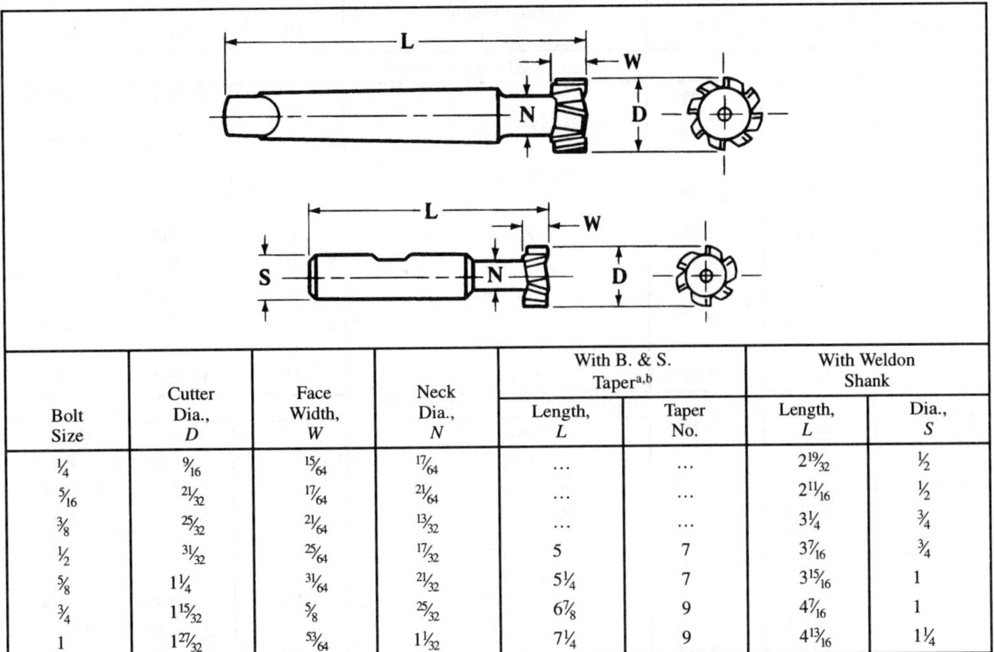

Bolt Size	Cutter Dia., *D*	Face Width, *W*	Neck Dia., *N*	With B. & S. Taper[a,b] Length, *L*	With B. & S. Taper[a,b] Taper No.	With Weldon Shank Length, *L*	With Weldon Shank Dia., *S*
1/4	9/16	15/64	17/64	…	…	2 19/32	1/2
5/16	21/32	17/64	21/64	…	…	2 11/16	1/2
3/8	25/32	21/64	13/32	…	…	3 1/4	3/4
1/2	31/32	25/64	17/32	5	7	3 7/16	3/4
5/8	1 1/4	31/64	21/32	5 1/4	7	3 15/16	1
3/4	1 15/32	5/8	25/32	6 7/8	9	4 7/16	1
1	1 27/32	53/64	1 1/32	7 1/4	9	4 13/16	1 1/4

[a] For dimensions of Brown & Sharpe taper shanks, see information given on page 916.

[b] Brown & Sharpe taper shanks have been removed from ANSI/ASME B94.19 they are included for reference only.

All dimensions are in inches. All cutters are high-speed steel and only right-hand cutters are standard.

Tolerances: On *D*, +0.000, −0.010 inch; on *W*, +0.000, −0.005 inch; on *N*, +0.000, −0.005 inch; on *L*, ± 1/16 inch; on *S*, −00001 to −0.0005 inch.

American National Standard Form Relieved Corner Rounding Cutters with Weldon Shanks *ANSI/ASME B94.19-1997*

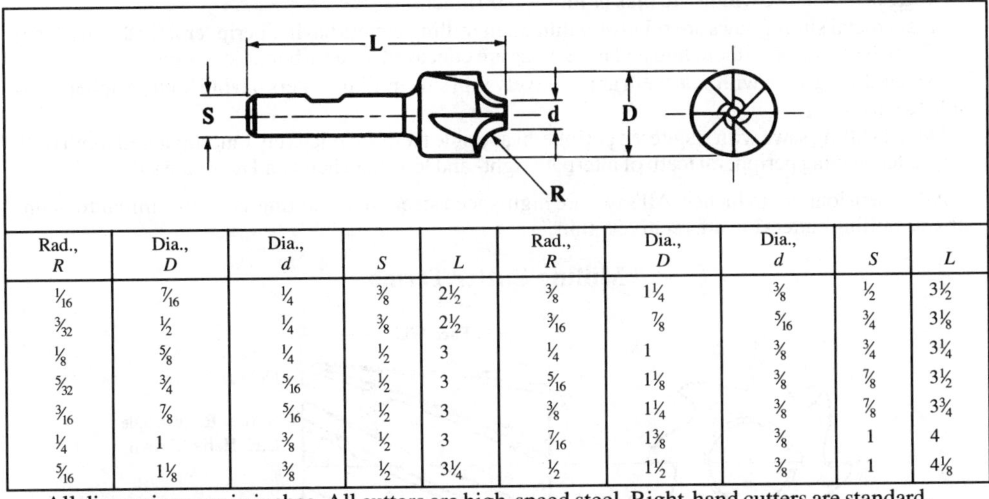

Rad., *R*	Dia., *D*	Dia., *d*	*S*	*L*	Rad., *R*	Dia., *D*	Dia., *d*	*S*	*L*
1/16	7/16	1/4	3/8	2 1/2	3/8	1 1/4	3/8	1/2	3 1/2
3/32	1/2	1/4	3/8	2 1/2	3/16	7/8	5/16	3/4	3 1/8
1/8	5/8	1/4	1/2	3	1/4	1	3/8	3/4	3 1/4
5/32	3/4	5/16	1/2	3	5/16	1 1/8	3/8	7/8	3 1/2
3/16	7/8	5/16	1/2	3	3/8	1 1/4	3/8	7/8	3 3/4
1/4	1	3/8	1/2	3	7/16	1 3/8	3/8	1	4
5/16	1 1/8	3/8	1/2	3 1/4	1/2	1 1/2	3/8	1	4 1/8

All dimensions are in inches. All cutters are high-speed steel. Right-hand cutters are standard.

Tolerances: On *D*, ±0.010 inch; on diameter of circle, 2*R*, ±0.001 inch for cutters up to and including 1/8-inch radius, +0.002, −0.001 inch for cutters over 1/8-inch radius; on *S*, −0.0001 to −0.0005 inch; and on *L*, ± 1/16 inch.

American National Standard Metal Slitting Saws *ANSI/ASME B94.19-1997*

Cutter Diameter			Range of Face Widths Nom.[a]	Hole Diameter		
Nom.	Max.	Min.		Nom.	Max.	Min.
Plain Metal Slitting Saws[b]						
2½	2.515	2.485	1/32, 3/64, 1/16, 3/32, 1/8	7/8	0.87575	0.8750
3	3.015	2.985	1/32, 3/64, 1/16, 3/32, 1/8 and 5/32	1	1.00075	1.0000
4	4.015	3.985	1/32, 3/64, 1/16, 3/32, 1/8, 5/32 and 3/16	1	1.00075	1.0000
5	5.015	4.985	1/16, 3/32, 1/8	1	1.00075	1.0000
5	5.015	4.985	1/8	1¼	1.2510	1.2500
6	6.015	5.985	1/16, 3/32, 1/8	1	1.00075	1.0000
6	6.015	5.985	1/8, 3/16	1¼	1.2510	1.2500
8	8.015	7.985	1/8	1	1.00075	1.0000
8	8.015	7.985	1/8	1¼	1.2510	1.2500
Metal Slitting Saws with Side Teeth[c]						
2½	2.515	2.485	1/16, 3/32, 1/8	7/8	0.87575	0.8750
3	3.015	2.985	1/16, 3/32, 1/8, 5/32	1	1.00075	1.0000
4	4.015	3.985	1/16, 3/32, 1/8, 5/32, 3/16	1	1.00075	1.0000
5	5.015	4.985	1/16, 3/32, 1/8, 5/32, 3/16	1	1.00075	1.0000
5	5.015	4.985	1/8	1¼	1.2510	1.2500
6	6.015	5.985	1/16, 3/32, 1/8, 3/16	1	1.00075	1.0000
6	6.015	5.985	1/8, 3/16	1¼	1.2510	1.2500
8	8.015	7.985	1/8	1	1.00075	1.0000
8	8.015	7.985	1/8, 3/16	1¼	1.2510	1.2500
Metal Slitting Saws with Staggered Peripheral and Side Teeth[d]						
3	3.015	2.985	3/16	1	1.00075	1.0000
4	4.015	3.985	3/16	1	1.00075	1.0000
5	5.015	4.985	3/16, 1/4	1	1.00075	1.0000
6	6.015	5.985	3/16, 1/4	1	1.00075	1.0000
6	6.015	5.985	3/16, 1/4	1¼	1.2510	1.2500
8	8.015	7.985	3/16, 1/4	1¼	1.2510	1.2500
10	10.015	9.985	3/16, 1/4	1¼	1.2510	1.2500
12	12.015	11.985	1/4, 5/16	1½	1.5010	1.5000

[a] Tolerances on face widths are plus or minus 0.001 inch.

[b] Plain metal slitting saws are relatively thin plain milling cutters having peripheral teeth only. They are furnished with or without hub and their sides are concaved to the arbor hole or hub.

[c] Metal slitting saws with side teeth are relatively thin side milling cutters having both peripheral and side teeth.

[d] Metal slitting saws with staggered peripheral and side teeth are relatively thin staggered tooth milling cutters having peripheral teeth of alternate right- and left-hand helix and alternate side teeth.

All dimensions are in inches. All saws are high-speed steel. Metal slitting saws are similar to plain or side milling cutters but are relatively thin.

Milling Cutter Terms

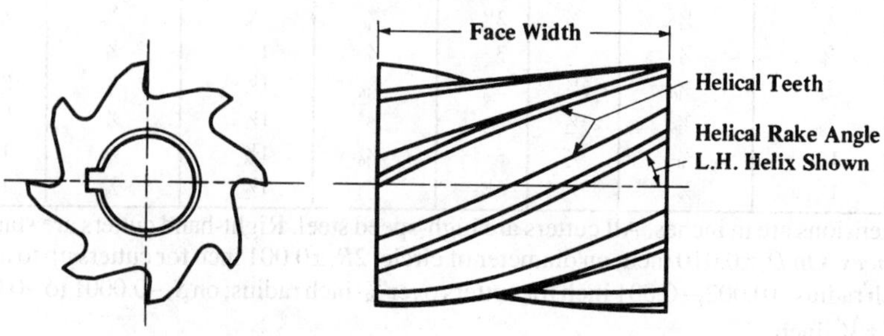

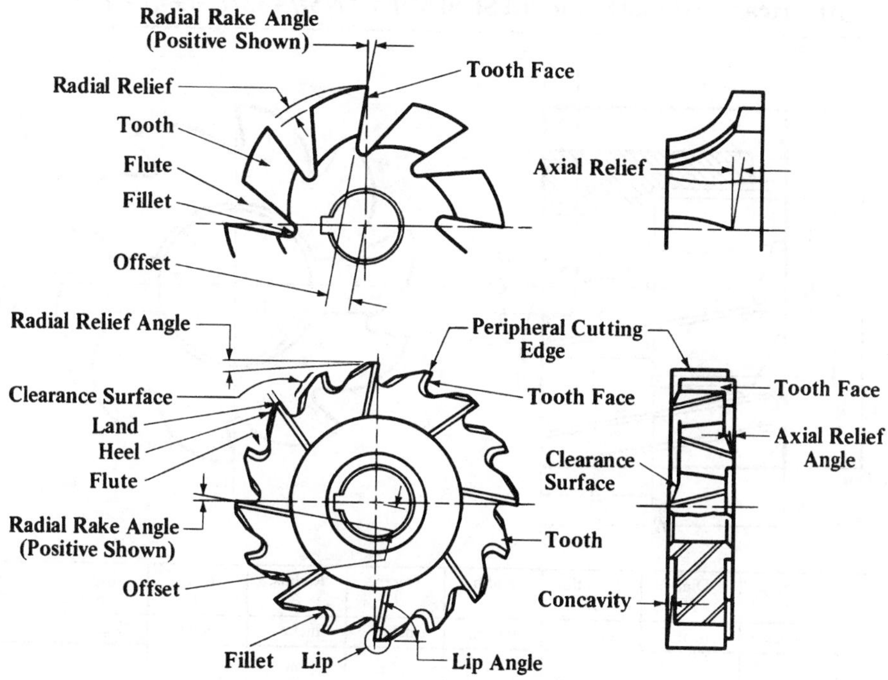

American National Standard Single- and Double-Angle Milling Cutters *ANSI/ASME B94.19-1997*

Cutter Diameter			Nominal Face Width[a]	Hole Diameter		
Nom.	Max.	Min.		Nom.	Max.	Min.
Single-angle Cutters[b]						
[c]1¼	1.265	1.235	7/16	⅜-24 UNF-2B RH ⅜-24 UNF-2B LH		
[c]1⅝	1.640	1.610	9/16	½-20 UNF-2B RH		
2¾	2.765	2.735	½	1	1.00075	1.0000
3	3.015	2.985	½	1¼	1.2510	1.2500
Double-angle Cutters[d]						
2¾	2.765	2.735	½	1	1.00075	1.0000

[a] Face width tolerances are plus or minus 0.015 inch.

[b] Single-angle milling cutters have peripheral teeth, one cutting edge of which lies in a conical surface and the other in the plane perpendicular to the cutter axis. There are two types: one has a plain keywayed hole and has an included tooth angle of either 45 or 60 degrees plus or minus 10 minutes; the other has a threaded hole and has an included tooth angle of 60 degrees plus or minus 10 minutes. Cutters with a right-hand threaded hole have a right-hand hand of rotation and a right-hand hand of cutter. Cutters with a left-hand threaded hole have a left-hand hand of rotation and a left-hand hand of cutter. Cutters with plain keywayed holes are standard as either right-hand or left-hand cutters.

[c] These cutters have threaded holes, the sizes of which are given under "Hole Diameter."

[d] Double-angle milling cutters have symmetrical peripheral teeth both sides of which lie in conical surfaces. They are designated by the included angle, which may be 45, 60 or 90 degrees. Tolerances are plus or minus 10 minutes for the half angle on each side of the center.

All dimensions are in inches. All cutters are high-speed steel.

American National Standard Shell Mills *ANSI/ASME B94.19-1997*

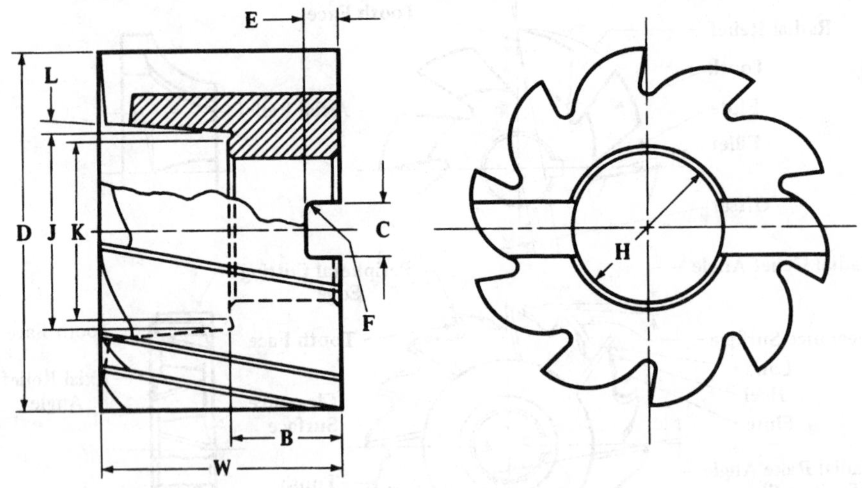

Dia., *D*	Width, *W*	Dia., *H*	Length, *B*	Width, *C*	Depth, *E*	Radius, *F*	Dia., *J*	Dia., *K*	Angle, *L*
inches	inches	inches	inches	inches	inches	inches	inches	degrees	inches
1 1/4	1	1/2	5/8	1/4	5/32	1/64	11/16	5/8	0
1 1/2	1 1/8	1/2	5/8	1/4	5/32	1/64	11/16	5/8	0
1 3/4	1 1/4	3/4	3/4	5/16	3/16	1/32	15/16	7/8	0
2	1 3/8	3/4	3/4	5/16	3/16	1/32	15/16	7/8	0
2 1/4	1 1/2	1	3/4	3/8	7/32	1/32	1 1/4	1 3/16	0
2 1/2	1 5/8	1	3/4	3/8	7/32	1/32	1 3/8	1 3/16	0
2 3/4	1 5/8	1	3/4	3/8	7/32	1/32	1 1/2	1 3/16	5
3	1 3/4	1 1/4	3/4	1/2	9/32	1/32	1 21/32	1 1/2	5
3 1/2	1 7/8	1 1/4	3/4	1/2	9/32	1/32	1 11/16	1 1/2	5
4	2 1/4	1 1/2	1	5/8	3/8	1/16	2 1/32	1 7/8	5
4 1/2	2 1/4	1 1/2	1	5/8	3/8	1/16	2 1/16	1 7/8	10
5	2 1/4	1 1/2	1	5/8	3/8	1/16	2 9/16	1 7/8	10
6	2 1/4	2	1	3/4	7/16	1/16	2 13/16	2 1/2	15

All cutters are high-speed steel. Right-hand cutters with right-hand helix and square corners are standard.

Tolerances: On *D*, +1/64 inch; on *W*, ±1/64 inch; on *H*, +0.0005 inch; on *B*, +1/64 inch; on *C*, at least +0.008 but not more than +0.012 inch; on *E*, +1/64 inch; on *J*, ±1/64 inch; on *K*, ±1/64 inch.

End Mill Terms

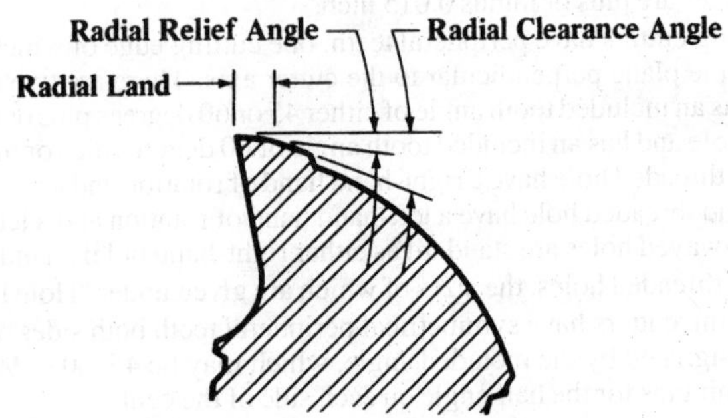

Enlarged Section of End Mill Tooth

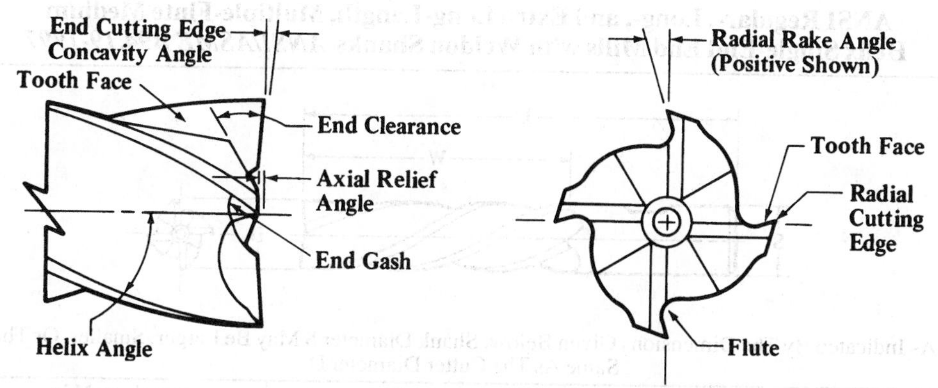

Enlarged Section of End Mill

American National Standard Multiple- and Two-Flute Single-End Helical End Mills with Plain Straight and Weldon Shanks *ANSI/ASME B94.19-1997*

Cutter Diameter, *D*			Shank Diameter, *S*		Length of Cut, *W*	Length Overall, *L*
Nom.	Max.	Min.	Max.	Min.		
Multiple-flute with Plain Straight Shanks						
1/8	.130	.125	.125	.1245	5/16	1 1/4
3/16	.1925	.1875	.1875	.1870	1/2	1 3/8
1/4	.255	.250	.250	.2495	5/8	1 11/16
3/8	.380	.375	.375	.3745	3/4	1 13/16
1/2	.505	.500	.500	.4995	15/16	2 1/4
3/4	.755	.750	.750	.7495	1 1/4	2 5/8
Two-flute for Keyway Cutting with Weldon Shanks						
1/8	.125	.1235	.375	.3745	3/8	2 5/16
3/16	.1875	.1860	.375	.3745	7/16	2 5/16
1/4	.250	.2485	.375	.3745	1/2	2 5/16
5/16	.3125	.3110	.375	.3745	9/16	2 5/16
3/8	.375	.3735	.375	.3745	9/16	2 5/16
1/2	.500	.4985	.500	.4995	1	3
5/8	.625	.6235	.625	.6245	1 5/16	3 7/16
3/4	.750	.7485	.750	.7495	1 5/16	3 9/16
7/8	.875	.8735	.875	.8745	1 1/2	3 3/4
1	1.000	.9985	1.000	.9995	1 5/8	4 1/8
1 1/4	1.250	1.2485	1.250	1.2495	1 5/8	4 1/8
1 1/2	1.500	1.4985	1.250	1.2495	1 5/8	4 1/8

All dimensions are in inches. All cutters are high-speed steel. Right-hand cutters with right-hand helix are standard.

The helix angle is not less than 10 degrees for multiple-flute cutters with plain straight shanks; the helix angle is optional with the manufacturer for two-flute cutters with Weldon shanks.

Tolerances: On *W*, ±1/32 inch; on *L*, ±1/16 inch.

ANSI Regular-, Long-, and Extra Long-Length, Multiple-Flute Medium Helix Single-End End Mills with Weldon Shanks *ANSI/ASME B94.19-1997*

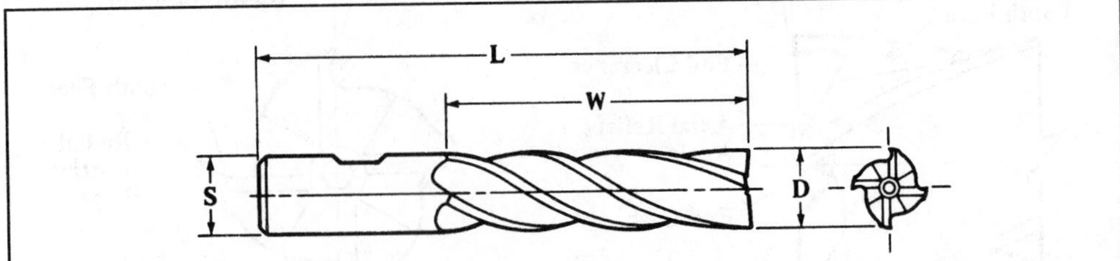

As Indicated By The Dimensions Given Below, Shank Diameter S May Be Larger, Smaller, Or The Same As The Cutter Diameter D

Cutter Dia., D	Regular Mills				Long Mills				Extra Long Mills			
	S	*W*	*L*	*N*[a]	*S*	*W*	*L*	*N*[a]	*S*	*W*	*L*	*N*[a]
1/8 [b]	3/8	3/8	2 5/16	4	…	…	…	…	…	…	…	…
3/16 [b]	3/8	1/2	2 3/8	4	…	…	…	…	…	…	…	…
1/4 [b]	3/8	5/8	2 7/16	4	3/8	1 1/4	3 1/16	4	3/8	1 3/4	3 9/16	4
5/16 [b]	3/8	3/4	2 1/2	4	3/8	1 3/8	3 1/8	4	3/8	2	3 3/4	4
3/8 [b]	3/8	3/4	2 1/2	4	3/8	1 1/2	3 1/4	4	3/8	2 1/2	4 1/4	4
7/16	3/8	1	2 11/16	4	1/2	1 3/4	3 3/4	4	…	…	…	…
1/2	3/8	1	2 11/16	4	1/2	2	4	4	1/2	3	5	4
1/2 [b]	1/2	1 1/4	3 1/4	4	…	…	…	…	…	…	…	…
9/16	1/2	1 3/8	3 3/8	4	…	…	…	…	…	…	…	…
5/8	1/2	1 3/8	3 3/8	4	5/8	2 1/2	4 5/8	4	5/8	4	6 1/8	4
11/16	1/2	1 5/8	3 5/8	4	…	…	…	…	…	…	…	…
3/4	1/2	1 5/8	3 5/8	4	3/4	3	5 1/4	4	3/4	4	6 1/4	4
5/8 [b]	5/8	1 5/8	3 3/4	4	…	…	…	…	…	…	…	…
11/16	5/8	1 5/8	3 3/4	4	…	…	…	…	…	…	…	…
3/4 [b]	5/8	1 5/8	3 3/4	4	…	…	…	…	…	…	…	…
13/16	5/8	1 7/8	4	6	…	…	…	…	…	…	…	…
7/8	5/8	1 7/8	4	6	7/8	3 1/2	5 3/4	4	7/8	5	7 1/4	4
1	5/8	1 7/8	4	6	1	4	6 1/2	4	1	6	8 1/2	4
7/8	7/8	1 7/8	4 1/8	4	…	…	…	…	…	…	…	…
1	7/8	1 7/8	4 1/8	4	…	…	…	…	…	…	…	…
1 1/8	7/8	2	4 1/4	6	1	4	6 1/2	6	…	…	…	…
1 1/4	7/8	2	4 1/4	6	1	4	6 1/2	6	1 1/4	6	8 1/2	6
1	1	2	4 1/2	4	…	…	…	…	…	…	…	…
1 1/8	1	2	4 1/2	6	…	…	…	…	…	…	…	…
1 1/4	1	2	4 1/2	6	…	…	…	…	…	…	…	…
1 3/8	1	2	4 1/2	6	…	…	…	…	…	…	…	…
1 1/2	1	2	4 1/2	6	1	4	6 1/2	6	…	…	…	…
1 1/4	1 1/4	2	4 1/2	6	1 1/4	4	6 1/2	6	…	…	…	…
1 1/2	1 1/4	2	4 1/2	6	1 1/4	4	6 1/2	6	1 1/4	8	10 1/2	6
1 3/4	1 1/4	2	4 1/2	6	1 1/4	4	6 1/2	6	…	…	…	…
2	1 1/4	2	4 1/2	8	1 1/4	4	6 1/2	8	…	…	…	…

[a] *N* = Number of flutes.

[b] In this size of regular mill a left-hand cutter with left-hand helix is also standard.

All dimensions are in inches. All cutters are high-speed steel. Helix angle is greater than 19 degrees but not more than 39 degrees. Right-hand cutters with right-hand helix are standard.

Tolerances: On *D*, +0.003 inch; on *S*, −0.0001 to −0.0005 inch; on *W*, ±1/32 inch; on *L*, ±1/16 inch.

ANSI Two-Flute, High Helix, Regular-, Long-, and Extra Long-Length, Single-End End Mills with Weldon Shanks *ANSI/ASME B94.19-1997*

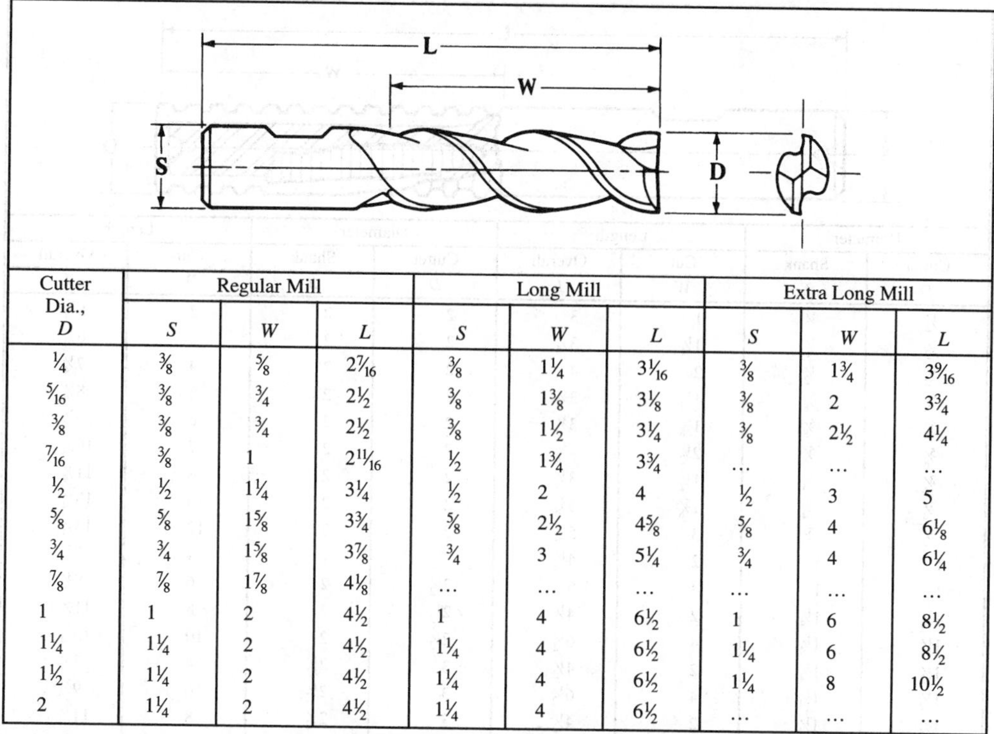

Cutter Dia., *D*	Regular Mill			Long Mill			Extra Long Mill		
	S	*W*	*L*	*S*	*W*	*L*	*S*	*W*	*L*
1/4	3/8	5/8	2 7/16	3/8	1 1/4	3 1/16	3/8	1 3/4	3 9/16
5/16	3/8	3/4	2 1/2	3/8	1 3/8	3 1/8	3/8	2	3 3/4
3/8	3/8	3/4	2 1/2	3/8	1 1/2	3 1/4	3/8	2 1/2	4 1/4
7/16	3/8	1	2 11/16	1/2	1 3/4	3 3/4	…	…	…
1/2	1/2	1 1/4	3 1/4	1/2	2	4	1/2	3	5
5/8	5/8	1 5/8	3 3/4	5/8	2 1/2	4 5/8	5/8	4	6 1/8
3/4	3/4	1 5/8	3 7/8	3/4	3	5 1/4	3/4	4	6 1/4
7/8	7/8	1 7/8	4 1/8	…	…	…	…	…	…
1	1	2	4 1/2	1	4	6 1/2	1	6	8 1/2
1 1/4	1 1/4	2	4 1/2	1 1/4	4	6 1/2	1 1/4	6	8 1/2
1 1/2	1 1/4	2	4 1/2	1 1/4	4	6 1/2	1 1/4	8	10 1/2
2	1 1/4	2	4 1/2	1 1/4	4	6 1/2	…	…	…

All dimensions are in inches. All cutters are high-speed steel. Right-hand cutters with right-hand helix are standard. Helix angle is greater than 39 degrees.

Tolerances: On *D*, +0.003 inch; on *S*, −0.0001 to −0.0005 inch; on *W*, ±1/32 inch; and on *L*, ±1/16 inch.

Combination Shanks for End Mills *ANSI/ASME B94.19-1997*

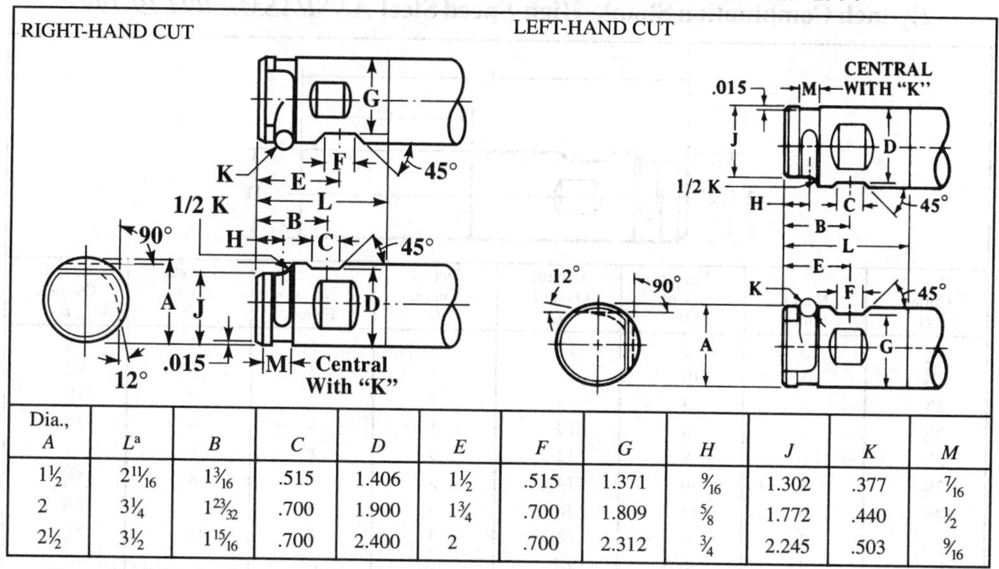

Dia., *A*	*L*[a]	*B*	*C*	*D*	*E*	*F*	*G*	*H*	*J*	*K*	*M*
1 1/2	2 11/16	1 3/16	.515	1.406	1 1/2	.515	1.371	9/16	1.302	.377	7/16
2	3 1/4	1 23/32	.700	1.900	1 3/4	.700	1.809	5/8	1.772	.440	1/2
2 1/2	3 1/2	1 15/16	.700	2.400	2	.700	2.312	3/4	2.245	.503	9/16

[a] Length of shank.

All dimensions are in inches.

Modified for use as Weldon or Pin Drive shank.

ANSI Roughing, Single-End End Mills with Weldon Shanks, High-Speed Steel *ANSI/ASME B94.19-1997*

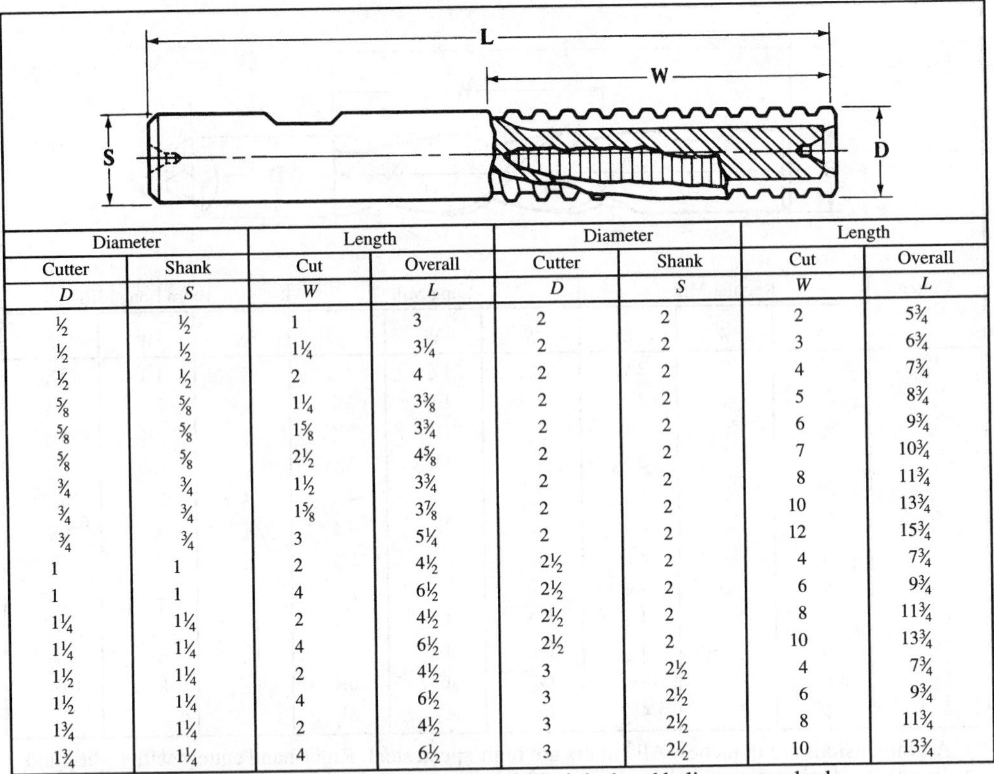

Diameter		Length		Diameter		Length	
Cutter	Shank	Cut	Overall	Cutter	Shank	Cut	Overall
D	*S*	*W*	*L*	*D*	*S*	*W*	*L*
½	½	1	3	2	2	2	5¾
½	½	1¼	3¼	2	2	3	6¾
½	½	2	4	2	2	4	7¾
⅝	⅝	1¼	3⅜	2	2	5	8¾
⅝	⅝	1⅝	3¾	2	2	6	9¾
⅝	⅝	2½	4⅜	2	2	7	10¾
¾	¾	1½	3¾	2	2	8	11¾
¾	¾	1⅝	3⅞	2	2	10	13¾
¾	¾	3	5¼	2	2	12	15¾
1	1	2	4½	2½	2	4	7¾
1	1	4	6½	2½	2	6	9¾
1¼	1¼	2	4½	2½	2	8	11¾
1¼	1¼	4	6½	2½	2	10	13¾
1½	1¼	2	4½	3	2½	4	7¾
1½	1¼	4	6½	3	2½	6	9¾
1¾	1¼	2	4½	3	2½	8	11¾
1¾	1¼	4	6½	3	2½	10	13¾

All dimensions are in inches. Right-hand cutters with right-hand helix are standard.

Tolerances: Outside diameter, +0.025, −0.005 inch; length of cut, +⅛, −1/32 inch.

American National Standard Heavy Duty, Medium Helix Single-End End Mills, 2½-inch Combination Shank, High-Speed Steel *ANSI/ASME B94.19-1997*

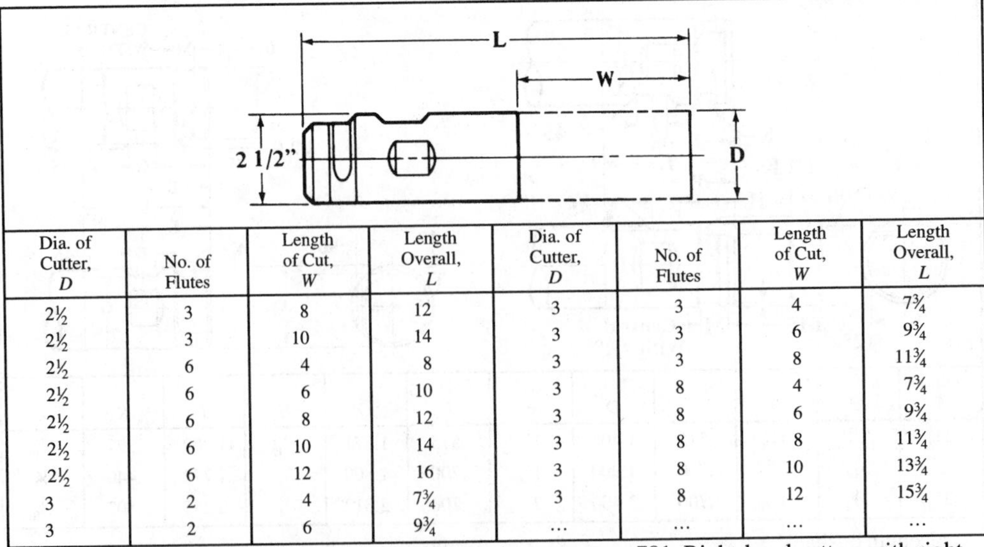

Dia. of Cutter, *D*	No. of Flutes	Length of Cut, *W*	Length Overall, *L*	Dia. of Cutter, *D*	No. of Flutes	Length of Cut, *W*	Length Overall, *L*
2½	3	8	12	3	3	4	7¾
2½	3	10	14	3	3	6	9¾
2½	6	4	8	3	3	8	11¾
2½	6	6	10	3	8	4	7¾
2½	6	8	12	3	8	6	9¾
2½	6	10	14	3	8	8	11¾
2½	6	12	16	3	8	10	13¾
3	2	4	7¾	3	8	12	15¾
3	2	6	9¾	…	…	…	…

All dimensions are in inches. For shank dimensions see page 781. Right-hand cutters with right-hand helix are standard. Helix angle is greater than 19 degrees but not more than 39 degrees.

Tolerances: On *D*, +0.005 inch; on *W*, ±1/32 inch; on *L*, ±1/16 inch.

ANSI Stub-, Regular-, and Long-Length, Four-Flute, Medium Helix, Plain-End, Double-End Miniature End Mills with 3/16-Inch Diameter Straight Shanks *ANSI/ASME B94.19-1997*

Dia. *D*	Stub Length *W*	Stub Length *L*	Regular Length *W*	Regular Length *L*
1/16	3/32	2	3/16	2 1/4
3/32	9/64	2	9/32	2 1/4
1/8	3/16	2	3/8	2 1/4
5/32	15/64	2	7/16	2 1/4
3/16	9/32	2	1/2	2 1/4

Dia. *D*	Long Length *B*	Long Length *W*	Long Length *L*
1/16	3/8	7/32	2 1/2
3/32	1/2	9/32	2 5/8
1/8	3/4	3/4	3 1/8
5/32	7/8	7/8	3 1/4
3/16	1	1	3 3/8

All dimensions are in inches. All cutters are high-speed steel. Right-hand cutters with right-hand helix are standard. Helix angle is greater than 19 degrees but not more than 39 degrees.

Tolerances: On *D*, + 0.003 inch (if the shank is the same diameter as the cutting portion, however, then the tolerance on the cutting diameter is − 0.0025 inch.); on *W*, + 1/32, − 1/64 inch; and on *L*, ±1/16 inch.

American National Standard 60-Degree Single-Angle Milling Cutters with Weldon Shanks *ANSI/ASME B94.19-1997*

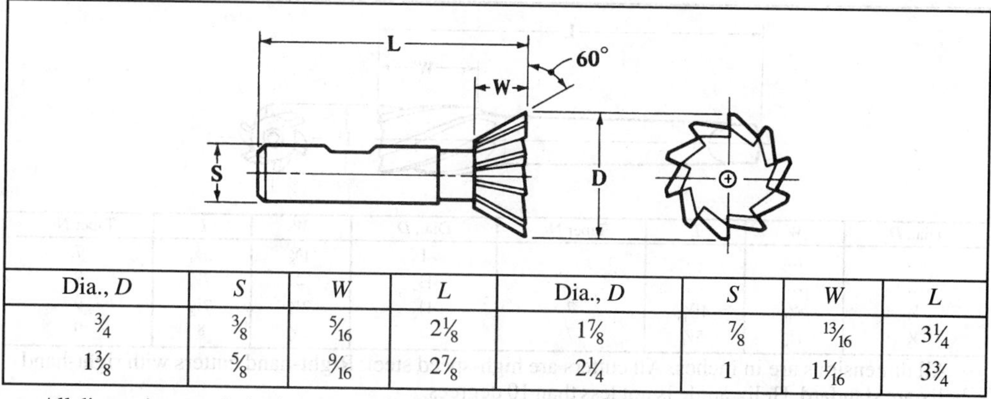

Dia., *D*	*S*	*W*	*L*	Dia., *D*	*S*	*W*	*L*
3/4	3/8	5/16	2 1/8	1 7/8	7/8	13/16	3 1/4
1 3/8	5/8	9/16	2 7/8	2 1/4	1	1 1/16	3 3/4

All dimensions are in inches. All cutters are high-speed steel. Right-hand cutters are standard.

Tolerances: On *D*, ± 0.015 inch; on *S*, − 0.0001 to − 0.0005 inch; on *W*, ± 0.015 inch; and on *L*, ±1/16 inch.

American National Standard Stub-, Regular-, and Long-Length, Two-Flute, Medium Helix, Plain- and Ball-End, Double-End Miniature End Mills with 3/16-Inch Diameter Straight Shanks *ANSI/ASME B94.19-1997*

Dia., *C* and *D*	Stub Length				Regular Length			
	Plain End		Ball End		Plain End		Ball End	
	W	*L*	*W*	*L*	*W*	*L*	*W*	*L*
1/32	3/64	2	…	…	3/32	2 1/4	…	…
3/64	1/16	2	…	…	9/64	2 1/4	…	…
1/16	3/32	2	3/32	2	3/16	2 1/4	3/16	2 1/4
5/64	1/8	2	…	…	15/64	2 1/4	…	…
3/32	9/64	2	9/64	2	9/32	2 1/4	9/32	2 1/4
7/64	5/32	2	…	…	21/64	2 1/4	…	…
1/8	3/16	2	3/16	2	3/8	2 1/4	3/8	2 1/4
9/64	7/32	2	…	…	13/32	2 1/4	…	…
5/32	15/64	2	15/64	2	7/16	2 1/4	7/16	2 1/4
11/64	1/4	2	…	…	1/2	2 1/4	…	…
3/16	9/32	2	9/32	2	1/2	2 1/4	1/2	2 1/4

Dia., *D*	Long Length		
	Plain End		
	B [a]	*W*	*L*
1/16	3/8	7/32	2 1/2
3/32	1/2	9/32	2 5/8
1/8	3/4	3/4	3 1/8
5/32	7/8	7/8	3 1/4
3/16	1	1	3 3/8

[a] *B* is the length below the shank.

All dimensions are in inches. All cutters are high-speed steel. Right-hand cutters with right-hand helix are standard. Helix angle is greater than 19 degrees but not more than 39 degrees.

Tolerances: On *C* and *D*, − 0.0015 inch for stub and regular length; + 0.003 inch for long length (if the shank is the same diameter as the cutting portion, however, then the tolerance on the cutting diameter is − 0.0025 inch.); on *W*, + 1/32, − 1/64 inch; and on *L*, ± 1/16 inch.

American National Standard Multiple Flute, Helical Series End Mills with Brown & Sharpe Taper Shanks

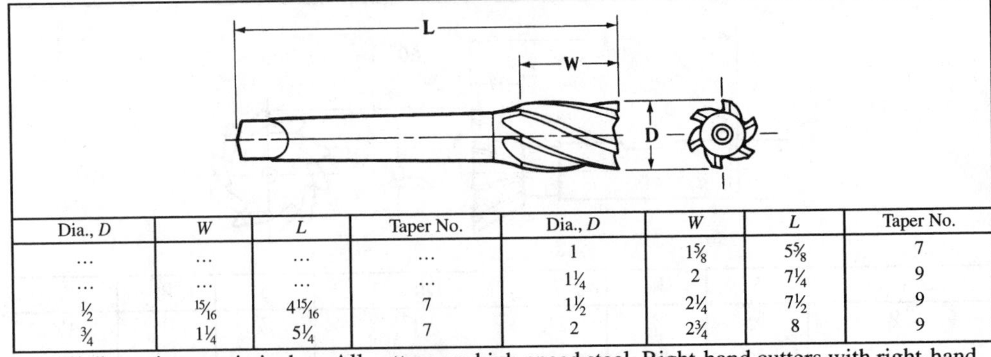

Dia., *D*	*W*	*L*	Taper No.	Dia., *D*	*W*	*L*	Taper No.
…	…	…	…	1	1 5/8	5 5/8	7
…	…	…	…	1 1/4	2	7 1/4	9
1/2	15/16	4 15/16	7	1 1/2	2 1/4	7 1/2	9
3/4	1 1/4	5 1/4	7	2	2 3/4	8	9

All dimensions are in inches. All cutters are high-speed steel. Right-hand cutters with right-hand helix are standard. Helix angle is not less than 10 degrees.

No. 5 taper is standard without tang; Nos. 7 and 9 are standard with tang only.

Tolerances: On *D*, +0.005 inch; on *W*, ±1/32 inch; and on *L* ±1/16 inch.

For dimensions of B & S taper shanks, see information given on page 916.

American National Standard Stub- and Regular-Length, Two-Flute, Medium Helix, Plain- and Ball-End, Single-End End Mills with Weldon Shanks *ANSI/ASME B94.19-1997*

Regular Length — Plain End

Dia., *D*	*S*	*W*	*L*
1/8	3/8	3/8	2 5/16
3/16	3/8	7/16	2 5/16
1/4	3/8	1/2	2 5/16
5/16	3/8	9/16	2 5/16
3/8	3/8	9/16	2 5/16
7/16	3/8	13/16	2 1/2
1/2	3/8	13/16	2 1/2
1/2	1/2	1	3
9/16	1/2	1 1/8	3 1/8
5/8	1/2	1 1/8	3 1/8
11/16	1/2	1 5/16	3 5/16
3/4	1/2	1 5/16	3 5/16
5/8	5/8	1 5/16	3 7/16
11/16	5/8	1 5/16	3 7/16
3/4	5/8	1 5/16	3 7/16
13/16	5/8	1 1/2	3 5/8
7/8	5/8	1 1/2	3 5/8
1	5/8	1 1/2	3 5/8
7/8	7/8	1 1/2	3 3/4
1	7/8	1 1/2	3 3/4
1 1/8	7/8	1 5/8	3 7/8
1 1/4	7/8	1 5/8	3 7/8
1	1	1 5/8	4 1/8
1 1/8	1	1 5/8	4 1/8
1 1/4	1	1 5/8	4 1/8
1 3/8	1	1 5/8	4 1/8
1 1/2	1	1 5/8	4 1/8
1 1/4	1 1/4	1 5/8	4 1/8
1 1/2	1 1/4	1 5/8	4 1/8
1 3/4	1 1/4	1 5/8	4 1/8
2	1 1/4	1 5/8	4 1/8

Stub Length — Plain End

Cutter Dia., *D*	Shank Dia., *S*	Length of Cut. *W*	Length Overall. *L*
1/8	3/8	3/16	2 1/8
3/16	3/8	9/32	2 3/16
1/4	3/8	3/8	2 1/4

Regular Length — Ball End

Dia., *C* and *D*	Shank Dia., *S*	Length of Cut. *W*	Length Overall. *L*
1/8	3/8	3/8	2 5/16
3/16	3/8	1/2	2 3/8
1/4	3/8	5/8	2 7/16
5/16	3/8	3/4	2 1/2
3/8	3/8	3/4	2 1/2
7/16	1/2	1	3
1/2	1/2	1	3
9/16	1/2	1 1/8	3 1/8
5/8	1/2	1 1/8	3 1/8
5/8	5/8	1 3/8	3 1/2
3/4	1/2	1 5/16	3 5/16
3/4	3/4	1 5/8	3 7/8
7/8	7/8	2	4 1/4
1	1	2 1/4	4 3/4
1 1/8	1	2 1/4	4 3/4
1 1/4	1 1/4	2 1/2	5
1 1/2	1 1/4	2 1/2	5

All dimensions are in inches. All cutters are high-speed steel. Right-hand cutters with right-hand helix are standard. Helix angle is greater than 19 degrees but not more than 39 degrees.

Tolerances: On *C* and *D*, −0.0015 inch for stub-length mills, + 0.003 inch for regular-length mills; on *S*, −0.0001 to −0.0005 inch; on *W*, ± 1/32 inch; and on *L*, ± 1/16 inch.

The following single-end end mills are available in premium high speed steel: ball end, two flute, with *D* ranging from 1/8 to 1 1/2 inches; ball end, multiple flute, with *D* ranging from 1/8 to 1 inch; and plain end, two flute, with *D* ranging from 1/8 to 1 1/2 inches.

American National Standard Long-Length Single-End and Stub-, and Regular Length, Double-End, Plain- and Ball-End, Medium Helix, Two-Flute End Mills with Weldon Shanks *ANSI/ASME B94.19-1997*

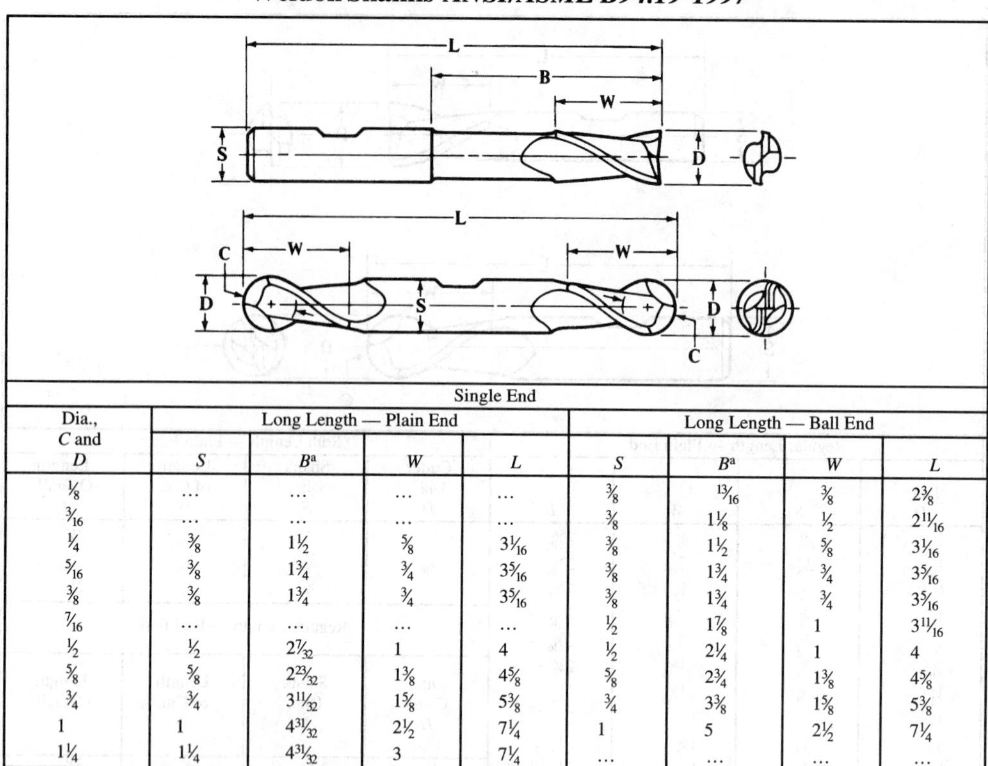

Dia., C and D	Single End: Long Length — Plain End S	B^a	W	L	Long Length — Ball End S	B^a	W	L
1/8	…	…	…	…	3/8	13/16	3/8	2 3/8
3/16	…	…	…	…	3/8	1 1/8	1/2	2 11/16
1/4	3/8	1 1/2	5/8	3 1/16	3/8	1 1/2	5/8	3 1/16
5/16	3/8	1 3/4	3/4	3 5/16	3/8	1 3/4	3/4	3 5/16
3/8	3/8	1 3/4	3/4	3 5/16	3/8	1 3/4	3/4	3 5/16
7/16	…	…	…	…	1/2	1 7/8	1	3 11/16
1/2	1/2	2 7/32	1	4	1/2	2 1/4	1	4
5/8	5/8	2 23/32	1 3/8	4 5/8	5/8	2 3/4	1 3/8	4 5/8
3/4	3/4	3 11/32	1 5/8	5 3/8	3/4	3 3/8	1 5/8	5 3/8
1	1	4 31/32	2 1/2	7 1/4	1	5	2 1/2	7 1/4
1 1/4	1 1/4	4 31/32	3	7 1/4	…	…	…	…

[a] *B* is the length below the shank.

Dia., C and D	Double End: Stub Length — Plain End S	W	L	Regular Length — Plain End S	W	L	Regular Length — Ball End S	W	L
1/8	3/8	3/16	2 3/4	3/8	3/8	3 1/16	3/8	3/8	3 1/16
5/32	3/8	15/64	2 3/4	3/8	7/16	3 1/8	…	…	…
3/16	3/8	9/32	2 3/4	3/8	7/16	3 1/8	3/8	7/16	3 1/8
7/32	3/8	21/64	2 7/8	3/8	1/2	3 1/8	…	…	…
1/4	3/8	3/8	2 7/8	3/8	1/2	3 1/8	3/8	1/2	3 1/8
9/32	…	…	…	3/8	9/16	3 1/8	…	…	…
5/16	…	…	…	3/8	9/16	3 1/8	3/8	9/16	3 1/8
11/32	…	…	…	3/8	9/16	3 1/8	…	…	…
3/8	…	…	…	3/8	9/16	3 1/8	3/8	9/16	3 1/8
13/32	…	…	…	1/2	13/16	3 3/4	…	…	…
7/16	…	…	…	1/2	13/16	3 3/4	1/2	13/16	3 3/4
15/32	…	…	…	1/2	13/16	3 3/4	…	…	…
1/2	…	…	…	1/2	13/16	3 3/4	1/2	13/16	3 3/4
9/16	…	…	…	5/8	1 1/8	4 1/2	…	…	…
5/8	…	…	…	5/8	1 1/8	4 1/2	5/8	1 1/8	4 1/2
11/16	…	…	…	3/4	1 5/16	5	…	…	…
3/4	…	…	…	3/4	1 5/16	5	3/4	1 5/16	5
7/8	…	…	…	7/8	1 9/16	5 1/2	…	…	…
1	…	…	…	1	1 5/8	5 7/8	1	1 5/8	5 7/8

All dimensions are in inches. All cutters are high-speed steel. Right-hand cutters with right-hand helix are standard. Helix angle is greater than 19 degrees but not more than 39 degrees.

Tolerances: On *C* and *D*, +0.003 inch for single-end mills, −0.0015 inch for double-end mills; on *S*, −0.0001 to −0.0005 inch; on *W*, ±1/32 inch; and on *L*, ±1/16 inch.

American National Standard Regular-, Long-, and Extra Long-Length, Three-and Four-Flute, Medium Helix, Center Cutting, Single-End End Mills with Weldon Shanks *ANSI/ASME B94.19-1997*

Four Flute

Dia., *D*	Regular Length			Long Length			Extra Long Length		
	S	*W*	*L*	*S*	*W*	*L*	*S*	*W*	*L*
1/8	3/8	3/8	2 5/16	…	…	…	…	…	…
3/16	3/8	1/2	2 3/8	…	…	…	…	…	…
1/4	3/8	5/8	2 7/16	3/8	1 1/4	3 3/16	3/8	1 3/4	3 9/16
5/16	3/8	3/4	2 1/2	3/8	1 3/8	3 1/8	3/8	2	3 3/4
3/8	3/8	3/4	2 1/2	3/8	1 1/2	3 1/4	3/8	2 1/2	4 1/4
1/2	1/2	1 1/4	3 1/4	1/2	2	4	1/2	3	5
5/8	5/8	1 3/8	3 3/4	5/8	2 1/2	4 5/8	5/8	4	6 1/8
11/16	5/8	1 5/8	3 3/4	…	…	…	…	…	…
3/4	3/4	1 5/8	3 7/8	3/4	3	5 1/4	3/4	4	6 1/4
7/8	7/8	1 7/8	4 1/8	7/8	3 1/2	5 3/4	7/8	5	7 1/4
1	1	2	4 1/2	1	4	6 1/2	1	6	8 1/2
1 1/8	1	2	4 1/2	…	…	…	…	…	…
1 1/4	1 1/4	2	4 1/2	1 1/4	4	6 1/2	1 1/4	6	8 1/2
1 1/2	1 1/4	2	4 1/2	…	…	…	…	…	…

Three Flute

Dia., *D*	*S*	*W*	*L*	Dia., *D*	*S*	*W*	*L*
Regular Length				Regular Length (*cont.*)			
1/8	3/8	3/8	2 5/16	1 1/8	1	2	4 1/2
3/16	3/8	1/2	2 3/8	1 1/4	1	2	4 1/2
1/4	3/8	5/8	2 7/16	1 1/2	1	2	4 1/2
5/16	3/8	3/4	2 1/2	1 1/4	1 1/4	2	4 1/2
3/8	3/8	3/4	2 1/2	1 1/2	1 1/4	2	4 1/2
7/16	3/8	1	2 11/16	1 3/4	1 1/4	2	4 1/2
1/2	3/8	1	2 11/16	2	1 1/4	2	4 1/2
1/2	1/2	1 1/4	3 1/4	Long Length			
9/16	1/2	1 3/8	3 3/8	1/4	3/8	1 1/4	3 11/16
9/16	1/2	1 3/8	3 3/8	5/16	3/8	1 3/8	3 1/8
5/8	1/2	1 3/8	3 3/8	3/8	3/8	1 1/2	3 1/4
3/4	1/2	1 5/8	3 5/8	7/16	1/2	1 3/4	3 3/4
5/8	5/8	1 5/8	3 3/4	1/2	1/2	2	4
3/4	5/8	1 5/8	3 3/4	5/8	5/8	2 1/2	4 5/8
7/8	5/8	1 7/8	4	3/4	3/4	3	5 1/4
1	5/8	1 7/8	4	1	1	4	6 1/2
3/4	3/4	1 5/8	3 7/8	1 1/4	1 1/4	4	6 1/2
7/8	3/4	1 7/8	4 1/8	1 1/2	1 1/4	4	6 1/2
1	3/4	1 7/8	4 1/8	1 3/4	1 1/4	4	6 1/2
1	7/8	1 7/8	4 1/8	2	1 1/4	4	6 1/2
1	1	2	4 1/2				

All dimensions are in inches. All cutters are high-speed steel. Right-hand cutters with right-hand helix are standard. Helix angle is greater than 19 degrees but not more than 39 degrees.

Tolerances: On *D*, +0.003 inch; on *S*, −0.0001 to −0.0005 inch; on *W*, ±1/32 inch; and on *L*, ±1/16 inch.

The following center-cutting, single-end end mills are available in premium high speed steel: regular length, multiple flute, with *D* ranging from 1/8 to 1 1/2 inches; long length, multiple flute, with *D* ranging from 3/8 to 1 1/4 inches; and extra long-length, multiple flute, with *D* ranging from 3/8 to 1 1/4 inches.

American National Standard Stub- and Regular-length, Four-flute, Medium Helix, Double-end End Mills with Weldon Shanks *ANSI/ASME B94.19-1997*

Dia., D	S	W	L	Dia., D	S	W	L	Dia., D	S	W	L
Stub Length											
1/8	3/8	3/16	2 3/4	3/16	3/8	9/32	2 3/4	1/4	3/8	3/8	2 7/8
5/32	3/8	15/64	2 3/4	7/32	3/8	21/64	2 7/8	…	…	…	…
Regular Length											
1/8 [a]	3/8	3/8	3 1/16	11/32	3/8	3/4	3 1/2	5/8 [a]	5/8	1 3/8	5
5/32 [a]	3/8	7/16	3 1/8	3/8 [a]	3/8	3/4	3 1/2	11/16	3/4	1 5/8	5 5/8
3/16 [a]	3/8	1/2	3 1/4	13/32	1/2	1	4 1/8	3/4 [a]	3/4	1 5/8	5 5/8
7/32	3/8	9/16	3 1/4	7/16	1/2	1	4 1/8	13/16	7/8	1 7/8	6 1/8
1/4 [a]	3/8	5/8	3 3/8	15/32	1/2	1	4 1/8	7/8	7/8	1 7/8	6 1/8
9/32	3/8	11/16	3 3/8	1/2 [a]	1/2	1	4 1/8	1	1	1 7/8	6 3/8
5/16 [a]	3/8	3/4	3 1/2	9/16	5/8	1 3/8	5	…	…	…	…

[a] In this size of regular mill a left-hand cutter with a left-hand helix is also standard.

All dimensions are in inches. All cutters are high-speed steel. Right-hand cutters with right-hand helix are standard. Helix angle is greater than 19 degrees but not more than 39 degrees.

Tolerances: On *D*, +0.003 inch (if the shank is the same diameter as the cutting portion, however, then the tolerance on the cutting diameter is −0.0025 inch); on *S*, −0.0001 to −0.0005 inch; on *W*, ±1/32 inch; and on *L*, ±1/16 inch.

American National Standard Stub- and Regular-Length, Four-Flute, Medium Helix, Double-End End Mills with Weldon Shanks *ANSI/ASME B94.19-1997*

Dia., D	S	W	L	Dia., D	S	W	L
Three Flute				Four Flute			
1/8	3/8	3/8	3 1/16	1/8	3/8	3/8	3 1/16
3/16	3/8	1/2	3 1/4	3/16	3/8	1/2	3 1/4
1/4	3/8	5/8	3 3/8	1/4	3/8	5/8	3 3/8
5/16	3/8	3/4	3 1/2	5/16	3/8	3/4	3 1/2
3/8	3/8	3/4	3 1/2	3/8	3/8	3/4	3 1/2
7/16	1/2	1	4 1/8	1/2	1/2	1	4 1/8
1/2	1/2	1	4 1/8	5/8	5/8	1 3/8	5
9/16	5/8	1 3/8	5	3/4	3/4	1 5/8	5 5/8
5/8	5/8	1 3/8	5	7/8	7/8	1 7/8	6 1/8
3/4	3/4	1 5/8	5 5/8	1	1	1 7/8	6 3/8
1	1	1 7/8	6 3/8	…	…	…	…

All dimensions are in inches. All cutters are high-speed steel. Right-hand cutters with right-hand helix are standard. Helix angle is greater than 19 degrees but not more than 39 degrees.

Tolerances: On *D*, +0.0015 inch; on *S*, −0.0001 to −0.0005 inch; on *W*, ±1/32 inch; and on *L*, ±1/16 inch.

American National Standard Plain- and Ball-End, Heavy Duty, Medium Helix, Single-End End Mills with 2-Inch Diameter Shanks *ANSI/ASME B94.19-1997*

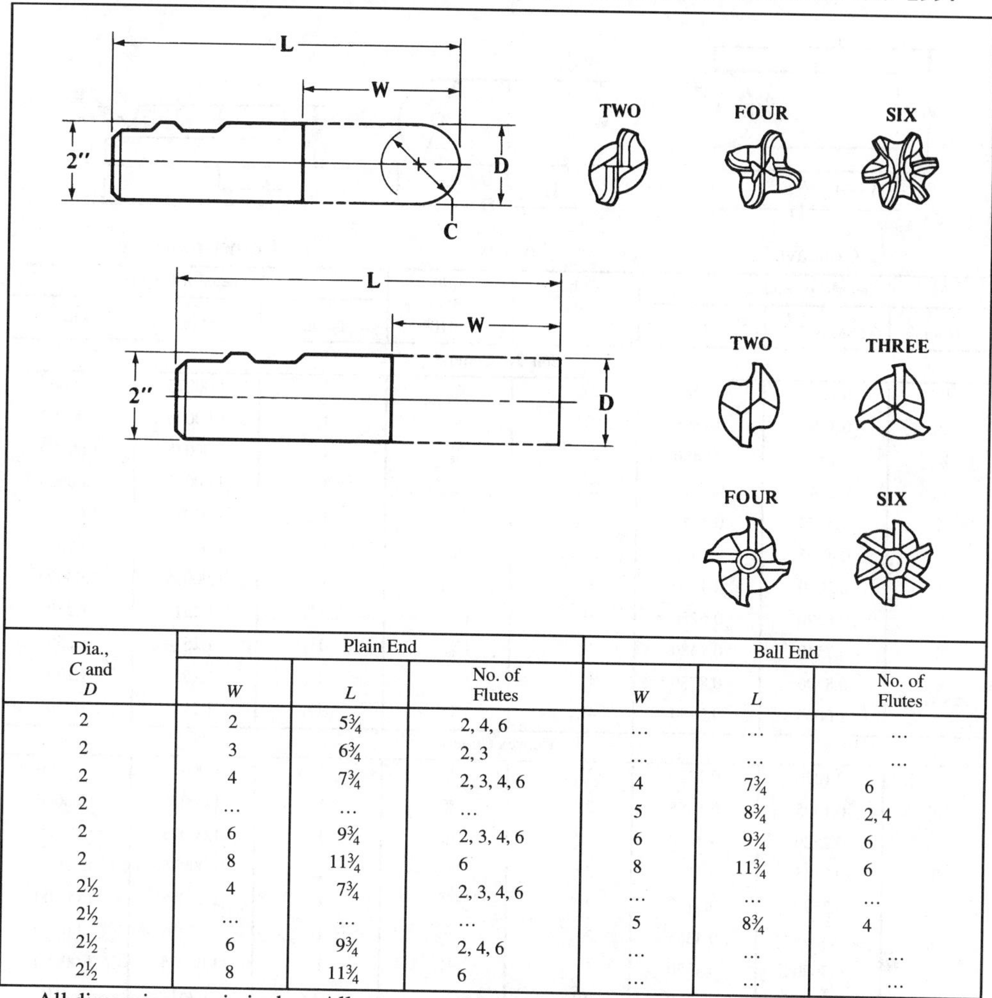

Dia., *C* and *D*	Plain End			Ball End		
	W	*L*	No. of Flutes	*W*	*L*	No. of Flutes
2	2	5¾	2, 4, 6	…	…	…
2	3	6¾	2, 3	…	…	…
2	4	7¾	2, 3, 4, 6	4	7¾	6
2	…	…	…	5	8¾	2, 4
2	6	9¾	2, 3, 4, 6	6	9¾	6
2	8	11¾	6	8	11¾	6
2½	4	7¾	2, 3, 4, 6	…	…	…
2½	…	…	…	5	8¾	4
2½	6	9¾	2, 4, 6	…	…	…
2½	8	11¾	6	…	…	…

All dimensions are in inches. All cutters are high-speed steel. Right-hand cutters with right-hand helix are standard. Helix angle is greater than 19 degrees but not more than 39 degrees.

Tolerances: On *C* and *D*, + 0.005 inch for 2, 3, 4 and 6 flutes: on *W*, ± 1⁄16 inch; and on *L*, ± 1⁄16 inch.

Dimensions of American National Standard Weldon Shanks *ANSI/ASME B94.19-1997*

Shank		Flat		Shank		Flat	
Dia.	Length	*X*[a]	Length[b]	Dia.	Length	*X*[a]	Length[b]
⅜	1 9⁄16	0.325	0.280	1	2 9⁄32	0.925	0.515
½	1 25⁄32	0.440	0.330	1¼	2 9⁄32	1.156	0.515
⅝	1 29⁄32	0.560	0.400	1½	2 11⁄16	1.406	0.515
¾	2 1⁄32	0.675	0.455	2	3¼	1.900	0.700
⅞	2 1⁄32	0.810	0.455	2½	3½	2.400	0.700

[a] *X* is distance from bottom of flat to opposite side of shank.

[b] Minimum.

All dimensions are in inches.

Centerline of flat is at half-length of shank except for 1½-, 2- and 2½-inch shanks where it is 1 3⁄16, 1 27⁄32 and 1 15⁄16 from shank end, respectively.

Tolerance on shank diameter, – 0.0001 to – 0.0005 inch.

American National Standard Form Relieved, Concave, Convex, and Corner-Rounding Arbor-Type Cutters *ANSI/ASME B94.19-1997*

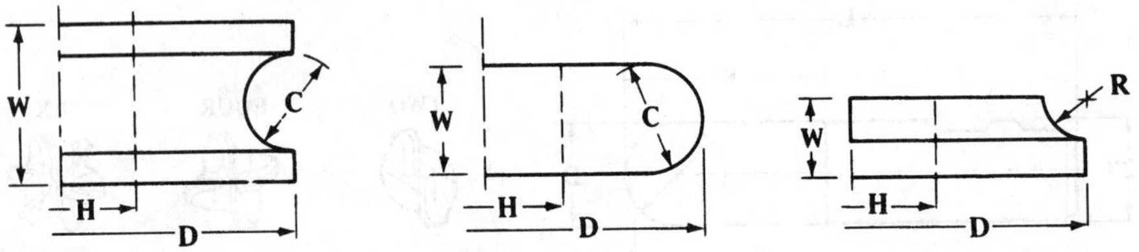

Concave Convex Corner-rounding

Diameter C or Radius R			Cutter Dia. D[a]	Width W ± .010[b]	Diameter of Hole H		
Nom.	Max.	Min.			Nom.	Max.	Min.
Concave Cutters[c]							
1/8	0.1270	0.1240	2 1/4	1/4	1	1.00075	1.00000
3/16	0.1895	0.1865	2 1/4	3/8	1	1.00075	1.00000
1/4	0.2520	0.2490	2 1/2	7/16	1	1.00075	1.00000
5/16	0.3145	0.3115	2 3/4	9/16	1	1.00075	1.00000
3/8	0.3770	0.3740	2 3/4	5/8	1	1.00075	1.00000
7/16	0.4395	0.4365	3	3/4	1	1.00075	1.00000
1/2	0.5040	0.4980	3	13/16	1	1.00075	1.00000
5/8	0.6290	0.6230	3 1/2	1	1 1/4	1.251	1.250
3/4	0.7540	0.7480	3 3/4	1 3/16	1 1/4	1.251	1.250
7/8	0.8790	0.8730	4	1 3/8	1 1/4	1.251	1.250
1	1.0040	0.9980	4 1/4	1 9/16	1 1/4	1.251	1.250
Convex Cutters[c]							
1/8	0.1270	0.1230	2 1/4	1/8	1	1.00075	1.00000
3/16	0.1895	0.1855	2 1/4	3/16	1	1.00075	1.00000
1/4	0.2520	0.2480	2 1/2	1/4	1	1.00075	1.00000
5/16	0.3145	0.3105	2 3/4	5/16	1	1.00075	1.00000
3/8	0.3770	0.3730	2 3/4	3/8	1	1.00075	1.00000
7/16	0.4395	0.4355	3	7/16	1	1.00075	1.00000
1/2	0.5020	0.4980	3	1/2	1	1.00075	1.00000
5/8	0.6270	0.6230	3 1/2	5/8	1 1/4	1.251	1.250
3/4	0.7520	0.7480	3 3/4	3/4	1 1/4	1.251	1.250
7/8	0.8770	0.8730	4	7/8	1 1/4	1.251	1.250
1	1.0020	0.9980	4 1/4	1	1 1/4	1.251	1.250
Corner-rounding Cutters[d]							
1/8	0.1260	0.1240	2 1/2	1/4	1	1.00075	1.00000
1/4	0.2520	0.2490	3	13/32	1	1.00075	1.00000
3/8	0.3770	0.3740	3 3/4	9/16	1 1/4	1.251	1.250
1/2	0.5020	0.4990	4 1/4	3/4	1 1/4	1.251	1.250
5/8	0.6270	0.6240	4 1/4	15/16	1 1/4	1.251	1.250

[a] Tolerances on cutter diameter are + 1/16, − 1/16 inch for all sizes.

[b] Tolerance does not apply to convex cutters.

[c] Size of cutter is designated by specifying diameter *C* of circular form.

[d] Size of cutter is designated by specifying radius *R* of circular form.

All dimensions in inches. All cutters are high-speed steel and are form relieved.

Right-hand corner rounding cutters are standard, but left-hand cutter for 1/4-inch size is also standard.

For key and keyway dimensions for these cutters, see page 794.

American National Standard Roughing and Finishing Gear Milling Cutters for Gears with 14½-Degree Pressure Angles *ANSI/ASME B94.19-1997*

ROUGHING FINISHING

Diametral Pitch	Dia. of Cutter, *D*	Dia. of Hole, *H*	Diametral Pitch	Dia. of Cutter, *D*	Dia. of Hole, *H*	Diametral Pitch	Dia. of Cutter, *D*	Dia. of Hole, *H*
Roughing Gear Milling Cutters								
1	8½	2	3	5¼	1½	5	3⅝	1
1¼	7¾	2	3	4¾	1¼	6	3⅞	1½
1½	7	1¾	4	4¾	1¾	6	3½	1¼
1¾	6½	1¾	4	4½	1½	6	3⅛	1
2	6½	1¾	4	4¼	1¼	7	3⅜	1¼
2	5¾	1½	4	3⅝	1	7	2⅞	1
2½	6⅛	1¾	5	4⅜	1¾	8	3¼	1¼
2½	5¾	1½	5	4¼	1½	8	2⅞	1
3	5⅝	1¾	5	3¾	1¼	…	…	…
Finishing Gear Milling Cutters								
1	8½	2	6	3⅞	1½	14	2⅛	⅞
1¼	7¾	2	6	3½	1¼	16	2½	1
1½	7	1¾	6	3⅛	1	16	2⅛	⅞
1¾	6½	1¾	7	3⅝	1½	18	2⅜	1
2	6½	1¾	7	3⅜	1¼	18	2	⅞
2	5¾	1½	7	2⅞	1	20	2⅜	1
2½	6⅛	1¾	8	3½	1½	20	2	⅞
2½	5¾	1½	8	3¼	1¼	22	2¼	1
3	5⅝	1¾	8	2⅞	1	22	2	⅞
3	5¼	1½	9	3⅛	1¼	24	2¼	1
3	4¾	1¼	9	2¾	1	24	1¾	⅞
4	4¾	1¾	10	3	1¼	26	1¾	⅞
4	4½	1½	10	2¾	1	28	1¾	⅞
4	4¼	1¼	10	2⅜	⅞	30	1¾	⅞
4	3⅝	1	11	2⅝	1	32	1¾	⅞
5	4⅜	1¾	11	2⅜	⅞	36	1¾	⅞
5	4¼	1½	12	2⅞	1¼	40	1¾	⅞
5	3¾	1¼	12	2⅝	1	48	1¾	⅞
5	3⅜	1	12	2¼	⅞	…	…	…
6	4¼	1¾	14	2½	1	…	…	…

All dimensions are in inches.

All gear milling cutters are high-speed steel and are form relieved.

For keyway dimensions see page 794.

Tolerances: On outside diameter, $+\frac{1}{16}, -\frac{1}{16}$ inch; on hole diameter, through 1-inch hole diameter, +0.00075 inch, over 1-inch and through 2-inch hole diameter, +0.0010 inch.

For cutter number relative to numbers of gear teeth, see page 2021. Roughing cutters are made with No. 1 cutter form only.

American National Standard Gear Milling Cutters for Mitre and Bevel Gears with 14½-Degree Pressure Angles *ANSI/ASME B94.19-1997*

Diametral Pitch	Diameter of Cutter, *D*	Diameter of Hole, *H*	Diametral Pitch	Diameter of Cutter, *D*	Diameter of Hole, *H*
3	4	1 1/4	10	2 3/8	7/8
4	3 5/8	1 1/4	12	2 1/4	7/8
5	3 3/8	1 1/4	14	2 1/8	7/8
6	3 1/8	1	16	2 1/8	7/8
7	2 7/8	1	20	2	7/8
8	2 7/8	1	24	1 3/4	7/8

All dimensions are in inches.

All cutters are high-speed steel and are form relieved.

For keyway dimensions see page 794. For cutter selection see page 2060.

Tolerances: On outside diameter, +1/16, −1/16 inch; on hole diameter, through 1-inch hole diameter, +0.00075 inch, for 1 1/4-inch hole diameter, +0.0010 inch.

To select the cutter number for bevel gears with the axis at any angle, double the back cone radius and multiply the result by the diametral pitch. This procedure gives the number of equivalent spur gear teeth and is the basis for selecting the cutter number from the table on page 2023.

American National Standard Roller Chain Sprocket Milling Cutters *ANSI/ASME B94.19-1997*

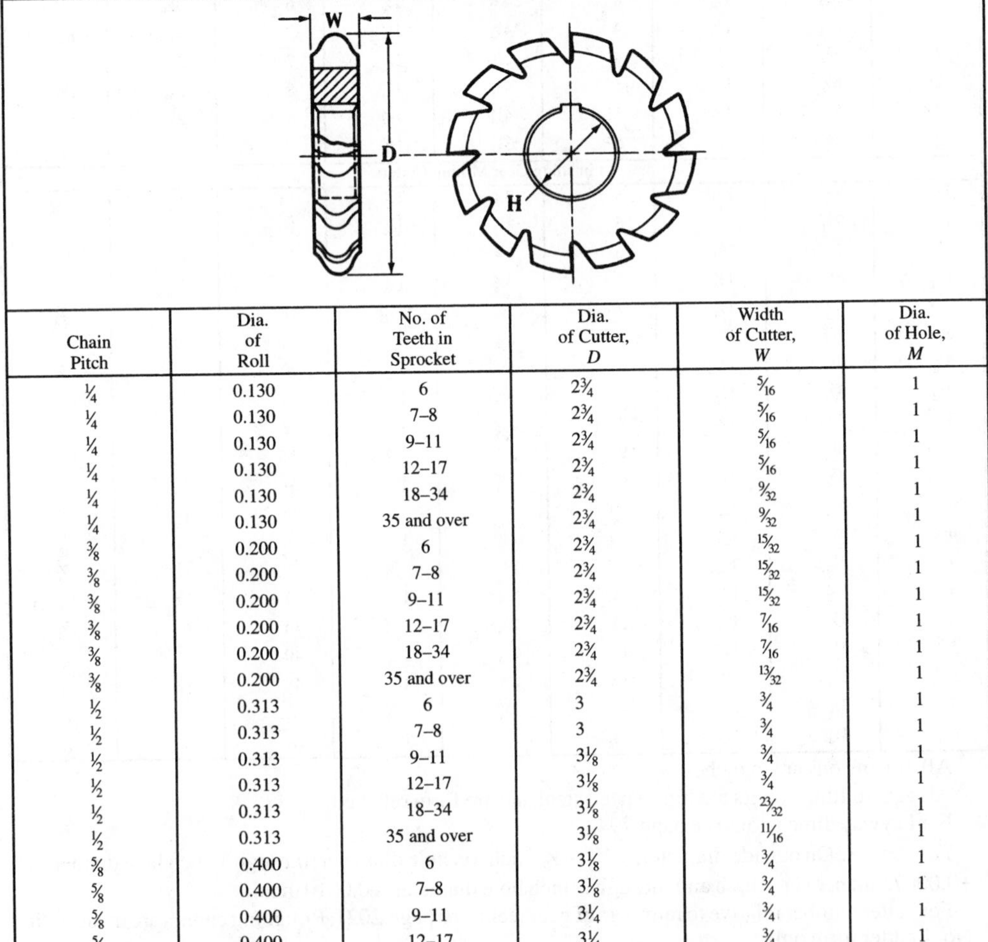

Chain Pitch	Dia. of Roll	No. of Teeth in Sprocket	Dia. of Cutter, *D*	Width of Cutter, *W*	Dia. of Hole, *M*
1/4	0.130	6	2 3/4	5/16	1
1/4	0.130	7–8	2 3/4	5/16	1
1/4	0.130	9–11	2 3/4	5/16	1
1/4	0.130	12–17	2 3/4	5/16	1
1/4	0.130	18–34	2 3/4	9/32	1
1/4	0.130	35 and over	2 3/4	9/32	1
3/8	0.200	6	2 3/4	15/32	1
3/8	0.200	7–8	2 3/4	15/32	1
3/8	0.200	9–11	2 3/4	15/32	1
3/8	0.200	12–17	2 3/4	7/16	1
3/8	0.200	18–34	2 3/4	7/16	1
3/8	0.200	35 and over	2 3/4	13/32	1
1/2	0.313	6	3	3/4	1
1/2	0.313	7–8	3	3/4	1
1/2	0.313	9–11	3 1/8	3/4	1
1/2	0.313	12–17	3 1/8	3/4	1
1/2	0.313	18–34	3 1/8	23/32	1
1/2	0.313	35 and over	3 1/8	11/16	1
5/8	0.400	6	3 1/8	3/4	1
5/8	0.400	7–8	3 1/8	3/4	1
5/8	0.400	9–11	3 1/4	3/4	1
5/8	0.400	12–17	3 1/4	3/4	1

American National Standard Roller Chain Sprocket Milling Cutters *ANSI/ASME B94.19-1997*

Chordal Pitch	Dia. of Roll	No. of Teeth in Sprocket	Dia. of Cutter, *D*	Width of Cutter, *D*	Dia.of Hole, *H*
$\frac{5}{8}$	0.400	18–34	$3\frac{1}{4}$	$\frac{23}{32}$	1
$\frac{5}{8}$	0.400	35 and over	$3\frac{1}{4}$	$\frac{11}{16}$	1
$\frac{3}{4}$	0.469	6	$3\frac{1}{4}$	$\frac{29}{32}$	1
$\frac{3}{4}$	0.469	7–8	$3\frac{1}{4}$	$\frac{29}{32}$	1
$\frac{3}{4}$	0.469	9–11	$3\frac{3}{8}$	$\frac{29}{32}$	1
$\frac{3}{4}$	0.469	12–17	$3\frac{3}{8}$	$\frac{7}{8}$	1
$\frac{3}{4}$	0.469	18–34	$3\frac{3}{8}$	$\frac{27}{32}$	1
$\frac{3}{4}$	0.469	35 and over	$3\frac{3}{8}$	$\frac{13}{16}$	1
1	0.625	6	$3\frac{7}{8}$	$1\frac{1}{2}$	$1\frac{1}{4}$
1	0.625	7–8	4	$1\frac{1}{2}$	$1\frac{1}{4}$
1	0.625	9–11	$4\frac{1}{8}$	$1\frac{15}{32}$	$1\frac{1}{4}$
1	0.625	18–34	$4\frac{1}{4}$	$1\frac{13}{32}$	$1\frac{1}{4}$
1	0.625	35 and over	$4\frac{1}{4}$	$1\frac{11}{32}$	$1\frac{1}{4}$
$1\frac{1}{4}$	0.750	6	$4\frac{1}{4}$	$1\frac{13}{16}$	$1\frac{1}{4}$
$1\frac{1}{4}$	0.750	7–8	$4\frac{3}{8}$	$1\frac{13}{16}$	$1\frac{1}{4}$
$1\frac{1}{4}$	0.750	9–11	$4\frac{1}{2}$	$1\frac{25}{32}$	$1\frac{1}{4}$
$1\frac{1}{4}$	0.750	18–34	$4\frac{5}{8}$	$1\frac{11}{16}$	$1\frac{1}{4}$
$1\frac{1}{4}$	0.750	35 and over	$4\frac{5}{8}$	$1\frac{5}{8}$	$1\frac{1}{4}$
$1\frac{1}{2}$	0.875	6	$4\frac{3}{8}$	$1\frac{13}{16}$	$1\frac{1}{4}$
$1\frac{1}{2}$	0.875	7–8	$4\frac{1}{2}$	$1\frac{13}{16}$	$1\frac{1}{4}$
$1\frac{1}{2}$	0.875	9–11	$4\frac{5}{8}$	$1\frac{25}{32}$	$1\frac{1}{4}$
$1\frac{1}{2}$	0.875	12–17	$4\frac{5}{8}$	$1\frac{3}{4}$	$1\frac{1}{4}$
$1\frac{1}{2}$	0.875	18–34	$4\frac{3}{4}$	$1\frac{11}{16}$	$1\frac{1}{4}$
$1\frac{1}{2}$	0.875	35 and over	$4\frac{3}{4}$	$1\frac{5}{8}$	$1\frac{1}{4}$
$1\frac{3}{4}$	1.000	6	5	$2\frac{3}{32}$	$1\frac{1}{2}$
$1\frac{3}{4}$	1.000	7–8	$5\frac{1}{8}$	$2\frac{3}{32}$	$1\frac{1}{2}$
$1\frac{3}{4}$	1.000	9–11	$5\frac{1}{4}$	$2\frac{1}{16}$	$1\frac{1}{2}$
$1\frac{3}{4}$	1.000	12–17	$5\frac{3}{8}$	$2\frac{1}{32}$	$1\frac{1}{2}$
$1\frac{3}{4}$	1.000	18–34	$5\frac{1}{2}$	$1\frac{31}{32}$	$1\frac{1}{2}$
$1\frac{3}{4}$	1.000	35 and over	$5\frac{1}{2}$	$1\frac{7}{8}$	$1\frac{1}{2}$
2	1.125	6	$5\frac{3}{8}$	$2\frac{13}{32}$	$1\frac{1}{2}$
2	1.125	7–8	$5\frac{1}{2}$	$2\frac{13}{32}$	$1\frac{1}{2}$
2	1.125	9–11	$5\frac{5}{8}$	$2\frac{3}{8}$	$1\frac{1}{2}$
2	1.125	12–17	$5\frac{3}{4}$	$2\frac{5}{16}$	$1\frac{1}{2}$
2	1.125	18–34	$5\frac{7}{8}$	$2\frac{1}{4}$	$1\frac{1}{2}$
2	1.125	35 and over	$5\frac{7}{8}$	$2\frac{5}{32}$	$1\frac{1}{2}$
$2\frac{1}{4}$	1.406	6	$5\frac{7}{8}$	$2\frac{11}{16}$	$1\frac{1}{2}$
$2\frac{1}{4}$	1.406	7–8	6	$2\frac{11}{16}$	$1\frac{1}{2}$
$2\frac{1}{4}$	1.406	9–11	$6\frac{1}{4}$	$2\frac{21}{32}$	$1\frac{1}{2}$
$2\frac{1}{4}$	1.406	12–17	$6\frac{3}{8}$	$2\frac{19}{32}$	$1\frac{1}{2}$
$2\frac{1}{4}$	1.406	18–34	$6\frac{1}{2}$	$2\frac{15}{32}$	$1\frac{1}{2}$
$2\frac{1}{4}$	1.406	35 and over	$6\frac{1}{2}$	$2\frac{13}{32}$	$1\frac{1}{2}$
$2\frac{1}{2}$	1.563	6	$6\frac{3}{8}$	3	$1\frac{3}{4}$
$2\frac{1}{2}$	1.563	7–8	$6\frac{5}{8}$	3	$1\frac{3}{4}$
$2\frac{1}{2}$	1.563	9–11	$6\frac{3}{4}$	$2\frac{15}{16}$	$1\frac{3}{4}$
$2\frac{1}{2}$	1.563	12–17	$6\frac{7}{8}$	$2\frac{29}{32}$	$1\frac{3}{4}$
$2\frac{1}{2}$	1.563	18–34	7	$2\frac{3}{4}$	$1\frac{3}{4}$
$2\frac{1}{2}$	1.563	35 and over	$7\frac{1}{8}$	$2\frac{11}{16}$	$1\frac{3}{4}$
3	1.875	6	$7\frac{1}{2}$	$3\frac{19}{32}$	2
3	1.875	7–8	$7\frac{3}{4}$	$3\frac{19}{32}$	2
3	1.875	9–11	$7\frac{7}{8}$	$3\frac{17}{32}$	2
3	1.875	12–17	8	$3\frac{15}{32}$	2
3	1.875	18–34	8	$3\frac{11}{32}$	2
3	1.875	35 and over	$8\frac{1}{4}$	$3\frac{7}{32}$	2

All dimensions are in inches.

All cutters are high-speed steel and are form relieved.

For keyway dimensions see page 794.

Tolerances: Outside diameter, $+\frac{1}{16}$, $-\frac{1}{16}$ inch; hole diameter, through 1-inch diameter, + 0.00075 inch, above 1-inch diameter and through 2-inch diameter, + 0.0010 inch.

For tooth form, see ANSI sprocket tooth form table on page 2438.

American National Standard Keys and Keyways for Milling Cutters and Arbors *ANSI/ASME B94.19-1997*

Nom. Arbor and Cutter Hole Dia.	Nom. Size Key (Square)	Arbor and Keyseat				Hole and Keyway					Arbor and Key			
		A Max.	*A* Min.	*B* Max.	*B* Min.	*C* Max.	*C* Min.	*D*[a] Min.	*H* Nom.	Corner Radius	*E* Max.	*E* Min.	*F* Max.	*F* Min.
½	3/32	0.0947	0.0937	0.4531	0.4481	0.106	0.099	0.5578	3/64	0.020	0.0932	0.0927	0.5468	0.5408
⅝	⅛	0.1260	0.1250	0.5625	0.5575	0.137	0.130	0.6985	1/16	1/32	0.1245	0.1240	0.6875	0.6815
¾	⅛	0.1260	0.1250	0.6875	0.6825	0.137	0.130	0.8225	1/16	1/32	0.1245	0.1240	0.8125	0.8065
⅞	⅛	0.1260	0.1250	0.8125	0.8075	0.137	0.130	0.9475	1/16	1/32	0.1245	0.1240	0.9375	0.9315
1	¼	0.2510	0.2500	0.8438	0.8388	0.262	0.255	1.1040	3/32	3/64	0.2495	0.2490	1.0940	1.0880
1¼	5/16	0.3135	0.3125	1.0630	1.0580	0.343	0.318	1.3850	⅛	1/16	0.3120	0.3115	1.3750	1.3690
1½	⅜	0.3760	0.3750	1.2810	1.2760	0.410	0.385	1.6660	5/32	1/16	0.3745	0.3740	1.6560	1.6500
1¾	7/16	0.4385	0.4375	1.5000	1.4950	0.473	0.448	1.9480	3/16	1/16	0.4370	0.4365	1.9380	1.9320
2	½	0.5010	0.5000	1.6870	1.6820	0.535	0.510	2.1980	3/16	1/16	0.4995	0.4990	2.1880	2.1820
2½	⅝	0.6260	0.6250	2.0940	2.0890	0.660	0.635	2.7330	7/32	1/16	0.6245	0.6240	2.7180	2.7120
3	¾	0.7510	0.7500	2.5000	2.4950	0.785	0.760	3.2650	¼	3/32	0.7495	0.7490	3.2500	3.2440
3½	⅞	0.8760	0.8750	3.0000	2.9950	0.910	0.885	3.8900	⅜	3/32	0.8745	0.8740	3.8750	3.8690
4	1	1.0010	1.0000	3.3750	3.3700	1.035	1.010	4.3900	⅜	3/32	0.9995	0.9990	4.3750	4.3690
4½	1⅛	1.1260	1.1250	3.8130	3.8080	1.160	1.135	4.9530	7/16	⅛	1.1245	1.1240	4.9380	4.9320
5	1¼	1.2510	1.2500	4.2500	4.2450	1.285	1.260	5.5150	½	⅛	1.2495	1.2490	5.5000	5.4940

[a] *D* max. is 0.010 inch larger than *D* min.

All dimensions given in inches.

American National Standard Woodruff Keyseat Cutters—Shank-Type Straight-Teeth and Arbor-Type Staggered-Teeth *ANSI/ASME B94.19-1997*

Shank-type Cutters											
Cutter Number	Nom. Dia.of Cutter, *D*	Width of Face, *W*	Length Over-all, *L*	Cutter Number	Nom. Dia. of Cutter, *D*	Width of Face, *W*	Length Over-all, *L*	Cutter Number	Nom. Dia.of Cutter, *D*	Width of Face, *W*	Length Over-all, *L*
202	$\frac{1}{4}$	$\frac{1}{16}$	$2\frac{1}{16}$	506	$\frac{3}{4}$	$\frac{5}{32}$	$2\frac{5}{32}$	809	$1\frac{1}{8}$	$\frac{1}{4}$	$2\frac{1}{4}$
202 $\frac{1}{2}$	$\frac{5}{16}$	$\frac{1}{16}$	$2\frac{1}{16}$	606	$\frac{3}{4}$	$\frac{3}{16}$	$2\frac{3}{16}$	1009	$1\frac{1}{8}$	$\frac{5}{16}$	$2\frac{5}{16}$
302 $\frac{1}{2}$	$\frac{5}{16}$	$\frac{3}{32}$	$2\frac{3}{32}$	806	$\frac{3}{4}$	$\frac{1}{4}$	$2\frac{1}{4}$	610	$1\frac{1}{4}$	$\frac{3}{16}$	$2\frac{3}{16}$
203	$\frac{3}{8}$	$\frac{1}{16}$	$2\frac{1}{16}$	507	$\frac{7}{8}$	$\frac{5}{32}$	$2\frac{5}{32}$	710	$1\frac{1}{4}$	$\frac{7}{32}$	$2\frac{7}{32}$
303	$\frac{3}{8}$	$\frac{3}{32}$	$2\frac{3}{32}$	607	$\frac{7}{8}$	$\frac{3}{16}$	$2\frac{3}{16}$	810	$1\frac{1}{4}$	$\frac{1}{4}$	$2\frac{1}{4}$
403	$\frac{3}{8}$	$\frac{1}{8}$	$2\frac{1}{8}$	707	$\frac{7}{8}$	$\frac{7}{32}$	$2\frac{7}{32}$	1010	$1\frac{1}{4}$	$\frac{5}{16}$	$2\frac{5}{16}$
204	$\frac{1}{2}$	$\frac{1}{16}$	$2\frac{1}{16}$	807	$\frac{7}{8}$	$\frac{1}{4}$	$2\frac{1}{4}$	1210	$1\frac{1}{4}$	$\frac{3}{8}$	$2\frac{3}{8}$
304	$\frac{1}{2}$	$\frac{3}{32}$	$2\frac{3}{32}$	608	1	$\frac{3}{16}$	$2\frac{3}{16}$	811	$1\frac{3}{8}$	$\frac{1}{4}$	$2\frac{1}{4}$
404	$\frac{1}{2}$	$\frac{1}{8}$	$2\frac{1}{8}$	708	1	$\frac{7}{32}$	$2\frac{7}{32}$	1011	$1\frac{3}{8}$	$\frac{5}{16}$	$2\frac{5}{16}$
305	$\frac{5}{8}$	$\frac{3}{32}$	$2\frac{3}{32}$	808	1	$\frac{1}{4}$	$2\frac{1}{4}$	1211	$1\frac{3}{8}$	$\frac{3}{8}$	$2\frac{3}{8}$
405	$\frac{5}{8}$	$\frac{1}{8}$	$2\frac{1}{8}$	1008	1	$\frac{5}{16}$	$2\frac{5}{16}$	812	$1\frac{1}{2}$	$\frac{1}{4}$	$2\frac{1}{4}$
505	$\frac{5}{8}$	$\frac{5}{32}$	$2\frac{5}{32}$	1208	1	$\frac{3}{8}$	$2\frac{3}{8}$	1012	$1\frac{1}{2}$	$\frac{5}{16}$	$2\frac{5}{16}$
605	$\frac{5}{8}$	$\frac{3}{16}$	$2\frac{3}{16}$	609	$1\frac{1}{8}$	$\frac{3}{16}$	$2\frac{3}{16}$	1212	$1\frac{1}{2}$	$\frac{3}{8}$	$2\frac{3}{8}$
406	$\frac{3}{4}$	$\frac{1}{8}$	$2\frac{1}{8}$	709	$1\frac{1}{8}$	$\frac{7}{32}$	$2\frac{7}{32}$	…	…	…	…

Arbor-type Cutters											
Cutter Number	Nom. Dia.of Cutter, *D*	Width of Face, *W*	Dia. of Hole, *H*	Cutter Number	Nom. Dia.of Cutter, *D*	Width of Face, *W*	Dia. of Hole, *H*	Cutter Number	Nom. Dia.of Cutter, *D*	Width of Face, *W*	Dia. of Hole, *H*
617	$2\frac{1}{8}$	$\frac{3}{16}$	$\frac{3}{4}$	1022	$2\frac{3}{4}$	$\frac{5}{16}$	1	1628	$3\frac{1}{2}$	$\frac{1}{2}$	1
817	$2\frac{1}{8}$	$\frac{1}{4}$	$\frac{3}{4}$	1222	$2\frac{3}{4}$	$\frac{3}{8}$	1	1828	$3\frac{1}{2}$	$\frac{9}{16}$	1
1017	$2\frac{1}{8}$	$\frac{5}{16}$	$\frac{3}{4}$	1422	$2\frac{3}{4}$	$\frac{7}{16}$	1	2028	$3\frac{1}{2}$	$\frac{5}{8}$	1
1217	$2\frac{1}{8}$	$\frac{3}{8}$	$\frac{3}{4}$	1622	$2\frac{3}{4}$	$\frac{1}{2}$	1	2428	$3\frac{1}{2}$	$\frac{3}{4}$	1
822	$2\frac{3}{4}$	$\frac{1}{4}$	1	1228	$3\frac{1}{2}$	$\frac{3}{8}$	1	…	…	…	…

All dimensions are given in inches. All cutters are high-speed steel.

Shank type cutters are standard with right-hand cut and straight teeth. All sizes have $\frac{1}{2}$-inch diameter straight shank.

Arbor type cutters have staggered teeth.

For Woodruff key and key-slot dimensions, see pages 2348 through 2350.

Tolerances: Face with *W* for shank type cutters: $\frac{1}{16}$- to $\frac{5}{32}$-inch face, + 0.0000, −0.0005; $\frac{3}{16}$ to $\frac{7}{32}$, − 0.0002, − 0.0007; $\frac{1}{4}$, −0.0003, −0.0008; $\frac{5}{16}$, −0.0004, −0.0009; $\frac{3}{8}$, − 0.0005, −0.0010 inch. Face width *W* for arbor type cutters; $\frac{3}{16}$ inch face, −0.0002, −0.0007; $\frac{1}{4}$, −0.0003, −0.0008; $\frac{5}{16}$, −0.0004, −0.0009; $\frac{3}{8}$ and over, −0.0005, −0.0010 inch. Hole size *H:* +0.00075, −0.0000 inch. Diameter *D* for shank type cutters: $\frac{1}{4}$- through $\frac{3}{4}$-inch diameter, +0.010, +0.015, $\frac{7}{8}$ through $1\frac{1}{8}$, +0.012, +0.017; $1\frac{1}{4}$ through $1\frac{1}{2}$, +0.015, +0.020 inch. These tolerances include an allowance for sharpening. For arbor type cutters diameter *D* is furnished $\frac{1}{32}$ inch larger than listed and a tolerance of ±0.002 inch applies to the oversize diameter.

Setting Angles for Milling Straight Teeth of Uniform Land Width in End Mills, Angular Cutters, and Taper Reamers.—The accompanying tables give setting angles for the dividing head when straight teeth, having a land of uniform width throughout their length, are to be milled using single-angle fluting cutters. These setting angles depend upon three factors: the number of teeth to be cut; the angle of the blank in which the teeth are to be cut; and the angle of the fluting cutter. Setting angles for various combinations of these three factors are given in the tables. For example, assume that 12 teeth are to be cut on the end of an end mill using a 60-degree cutter. By following the horizontal line from 12 teeth, read in the column under 60 degrees that the dividing head should be set to an angle of 70 degrees and 32 minutes.

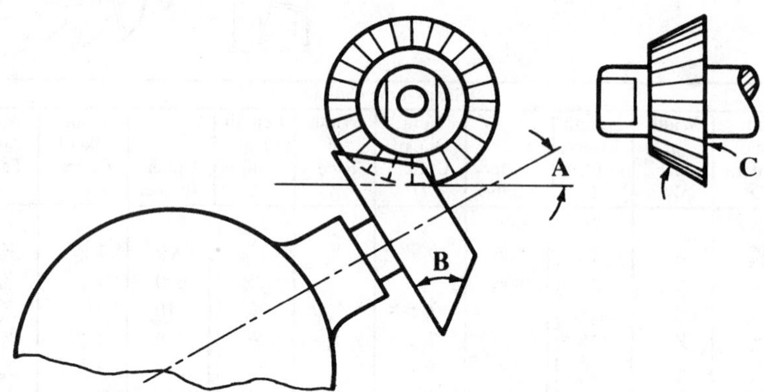

The following formulas, which were used to compile these tables, may be used to calculate the setting-angles for combinations of number of teeth, blank angle, and cutter angle not covered by the tables. In these formulas, A = setting-angle for dividing head, B = angle of blank in which teeth are to be cut, C = angle of fluting cutter, N = number of teeth to be cut, and D and E are angles not shown on the accompanying diagram and which are used only to simplify calculations.

$$\tan D = \cos(360°/N) \times \cot B \tag{1}$$

$$\sin E = \tan(360°/N) \times \cot C \times \sin D \tag{2}$$

$$\text{Setting-angle } A = D - E \tag{3}$$

Example: Suppose 9 teeth are to be cut in a 35-degree blank using a 55-degree single-angle fluting cutter. Then, $N = 9$, $B = 35°$, and $C = 55°$.

$$\tan D = \cos(360°/9) \times \cot 35° = 0.76604 \times 1.4281 = 1.0940;\text{ and } D = 47°34'$$

$$\sin E = \tan(360°/9) \times \cot 55° \times \sin 47°34' = 0.83910 \times 0.70021 \times 0.73806$$
$$= 0.43365;\text{ and } E = 25°42'$$

$$\text{Setting angle } A = 47°34' - 25°42' = 21°52'$$

For end mills and side mills the angle of the blank B is 0 degrees and the following simplified formula may be used to find the setting angle A

$$\cos A = \tan(360°/N) \times \cot C \tag{4}$$

Example: If in the previous example the blank angle was 0 degrees,

$\cos A = \tan(360°/9) \times \cot 55°$

$= 0.83910 \times 0.70021 = 0.58755$; and setting-angle $A = 54°1'$

Angles of Elevation for Milling Straight Teeth in 0-, 5-, 10-, 15-, 20-, 25-, 30-, and 35-degree Blanks Using Single-Angle Fluting Cutters

No. of Teeth	Angle of Fluting Cutter									
	90°	80°	70°	60°	50°	90°	80°	70°	60°	50°
	0° Blank (End Mill)					5° Blank				
6	…	72° 13′	50° 55′	…	…	80° 4′	62° 34′	41° 41′	…	…
8	…	79 51	68 39	54° 44′	32° 57′	82 57	72 52	61 47	48° 0′	25° 40′
10	…	82 38	74 40	65 12	52 26	83 50	76 31	68 35	59 11	46 4
12	…	84 9	77 52	70 32	61 2	84 14	78 25	72 10	64 52	55 5
14	…	85 8	79 54	73 51	66 10	84 27	79 36	74 24	68 23	60 28
16	…	85 49	81 20	76 10	69 40	84 35	80 25	75 57	70 49	64 7
18	…	86 19	82 23	77 52	72 13	84 41	81 1	77 6	72 36	66 47
20	…	86 43	83 13	79 11	74 11	84 45	81 29	77 59	73 59	68 50
22	…	87 2	83 52	80 14	75 44	84 47	81 50	78 40	75 4	70 26
24	…	87 18	84 24	81 6	77 0	84 49	82 7	79 15	75 57	71 44
	10° Blank					15° Blank				
6	70° 34′	53° 50′	34° 5′	…	…	61° 49′	46° 12′	28° 4′	…	…
8	76 0	66 9	55 19	41° 56′	20° 39′	69 15	59 46	49 21	36° 34′	17° 34′
10	77 42	70 31	62 44	53 30	40 42	71 40	64 41	57 8	48 12	36 18
12	78 30	72 46	66 37	59 26	49 50	72 48	67 13	61 13	54 14	45 13
14	78 56	74 9	69 2	63 6	55 19	73 26	68 46	63 46	57 59	50 38
16	79 12	75 5	70 41	65 37	59 1	73 50	69 49	65 30	60 33	54 20
18	79 22	75 45	71 53	67 27	61 43	74 5	70 33	66 46	62 26	57 0
20	79 30	76 16	72 44	68 52	63 47	74 16	71 6	67 44	63 52	59 3
22	79 35	76 40	73 33	69 59	65 25	74 24	71 32	68 29	65 0	60 40
24	79 39	76 59	74 9	70 54	66 44	74 30	71 53	69 6	65 56	61 59
	20° Blank					25° Blank				
6	53° 57′	39° 39′	23° 18′	…	…	47° 0′	34° 6′	19° 33′	…	…
8	62 46	53 45	43 53	31° 53′	14° 31′	56 36	48 8	38 55	27° 47′	11° 33′
10	65 47	59 4	51 50	43 18	32 1	60 2	53 40	46 47	38 43	27 47
12	67 12	61 49	56 2	49 18	40 40	61 42	56 33	51 2	44 38	36 10
14	68 0	63 29	58 39	53 4	46 0	62 38	58 19	53 41	48 20	41 22
16	68 30	64 36	60 26	55 39	49 38	63 13	59 29	55 29	50 53	44 57
18	68 50	65 24	61 44	57 32	52 17	63 37	60 19	56 48	52 46	47 34
20	69 3	65 59	62 43	58 58	54 18	63 53	60 56	57 47	54 11	49 33
22	69 14	66 28	63 30	60 7	55 55	64 5	61 25	58 34	55 19	51 9
24	69 21	66 49	64 7	61 2	57 12	64 14	61 47	59 12	56 13	52 26
	30° Blank					35° Blank				
6	40° 54′	29° 22′	16° 32′	…	…	35° 32′	25° 19′	14° 3′	…	…
8	50 46	42 55	34 24	24° 12′	10° 14′	45 17	38 5	30 18	21° 4′	8° 41′
10	54 29	48 30	42 3	34 31	24 44	49 7	43 33	37 35	30 38	21 40
12	56 18	51 26	46 14	40 12	32 32	51 3	46 30	41 39	36 2	28 55
14	57 21	53 15	48 52	43 49	37 27	52 9	48 19	44 12	39 28	33 33
16	58 0	54 27	50 39	46 19	40 52	52 50	49 20	45 56	41 51	36 45
18	58 26	55 18	51 57	48 7	43 20	53 18	50 21	47 12	43 36	39 8
20	58 44	55 55	52 56	49 30	45 15	53 38	50 59	48 10	44 57	40 57
22	58 57	56 24	53 42	50 36	46 46	53 53	51 29	48 56	46 1	42 24
24	59 8	56 48	54 20	51 30	48 0	54 4	51 53	49 32	46 52	43 35

Angles of Elevation for Milling Straight Teeth in 40-, 45-, 50-, 55-, 60-, 65-, 70-, and 75-degree Blanks Using Single-Angle Fluting Cutters

No. of Teeth	Angle of Fluting Cutter									
	90°	80°	70°	60°	50°	90°	80°	70°	60°	50°
	40° Blank					45° Blank				
6	30° 48′	21° 48′	11° 58′	…	…	26° 34′	18° 43′	10° 11′	…	…
8	40 7	33 36	26 33	18° 16′	7° 23′	35 16	29 25	23 8	15° 48′	5° 58′
10	43 57	38 51	33 32	27 3	18 55	38 58	34 21	29 24	23 40	16 10
12	45 54	41 43	37 14	32 3	25 33	40 54	37 5	33 0	28 18	22 13
14	47 3	43 29	39 41	35 19	29 51	42 1	38 46	35 17	31 18	26 9
16	47 45	44 39	41 21	37 33	32 50	42 44	39 54	36 52	33 24	28 57
18	48 14	45 29	42 34	39 13	35 5	43 13	40 42	38 1	34 56	30 1
20	48 35	46 7	43 30	40 30	36 47	43 34	41 18	38 53	36 8	32 37
22	48 50	46 36	44 13	41 30	38 8	43 49	41 46	39 34	37 5	34 53
24	49 1	46 58	44 48	42 19	39 15	44 0	42 7	40 7	37 50	35 55
	50° Blank					55° Blank				
6	22° 45′	15° 58′	8° 38′	…	…	19° 17′	13° 30′	7° 15′	…	…
8	30 41	25 31	19 59	13° 33′	5° 20′	26 21	21 52	17 3	11° 30′	4° 17′
10	34 10	30 2	25 39	20 32	14 9	29 32	25 55	22 3	17 36	11 52
12	36 0	32 34	28 53	24 42	19 27	31 14	28 12	24 59	21 17	16 32
14	37 5	34 9	31 1	27 26	22 58	32 15	29 39	26 53	23 43	19 40
16	37 47	35 13	32 29	29 22	25 30	32 54	30 38	28 12	25 26	21 54
18	38 15	35 58	33 33	30 46	27 21	33 21	31 20	29 10	26 43	23 35
20	38 35	36 32	34 21	31 52	28 47	33 40	31 51	29 54	27 42	24 53
22	38 50	36 58	34 59	32 44	29 57	33 54	32 15	30 29	28 28	25 55
24	39 1	37 19	35 30	33 25	30 52	34 5	32 34	30 57	29 7	26 46
	60° Blank					65° Blank				
6	16° 6′	11° 12′	6° 2′	…	…	13° 7′	9° 8′	4° 53′	…	…
8	22 13	18 24	14 19	9° 37′	3° 44′	18 15	15 6	11 42	7° 50′	3° 1′
10	25 2	21 56	18 37	14 49	10 5	20 40	18 4	15 19	12 9	8 15
12	26 34	23 57	21 10	17 59	14 13	21 59	19 48	17 28	14 49	11 32
14	27 29	25 14	22 51	20 6	16 44	22 48	20 55	18 54	16 37	13 48
16	28 5	26 7	24 1	21 37	18 40	23 18	21 39	19 53	17 53	15 24
18	28 29	26 44	24 52	22 44	20 6	23 40	22 11	20 37	18 50	16 37
20	28 46	27 11	25 30	23 35	21 14	23 55	22 35	21 10	19 33	17 34
22	29 0	27 34	26 2	24 17	22 8	24 6	22 53	21 36	20 8	18 20
24	29 9	27 50	26 26	24 50	22 52	24 15	23 8	21 57	20 36	18 57
	70° Blank					75° Blank				
6	10° 18′	7° 9′	3° 48′	…	…	7° 38′	5° 19′	2° 50′	…	…
8	14 26	11 55	9 14	6° 9′	2° 21′	10 44	8 51	6 51	4° 34′	1° 45′
10	16 25	14 21	12 8	9 37	6 30	12 14	10 40	9 1	7 8	4 49
12	17 30	15 45	13 53	11 45	9 8	13 4	11 45	10 21	8 45	6 47
14	18 9	16 38	15 1	13 11	10 55	13 34	12 26	11 13	9 50	8 7
16	18 35	17 15	15 50	14 13	12 13	13 54	12 54	11 50	10 37	9 7
18	18 53	17 42	16 26	14 59	13 13	14 8	13 14	12 17	11 12	9 51
20	19 6	18 1	16 53	15 35	13 59	14 18	13 29	12 38	11 39	10 27
22	19 15	18 16	17 15	16 3	14 35	14 25	13 41	12 53	12 0	10 54
24	19 22	18 29	17 33	16 25	15 5	14 31	13 50	13 7	12 18	11 18

Angles of Elevation for Milling Straight Teeth in 80- and 85-degree Blanks UsingSingle-Angle Fluting Cutters

No.of Teeth	Angle of Fluting Cutter									
	90°	80°	70°	60°	50°	90°	80°	70°	60°	50°
	80° Blank					85° Blank				
6	5° 2′	3° 30′	1° 52′	…	…	2° 30′	1° 44′	0° 55′	…	…
8	7 6	5 51	4 31	3° 2′	1° 8′	3 32	2 55	2 15	1° 29′	0° 34′
10	8 7	7 5	5 59	4 44	3 11	4 3	3 32	2 59	2 21	1 35
12	8 41	7 48	6 52	5 48	4 29	4 20	3 53	3 25	2 53	2 15
14	9 2	8 16	7 28	6 32	5 24	4 30	4 7	3 43	3 15	2 42
16	9 15	8 35	7 51	7 3	6 3	4 37	4 17	3 56	3 30	3 1
18	9 24	8 48	8 10	7 26	6 33	4 42	4 24	4 5	3 43	3 16
20	9 31	8 58	8 24	7 44	6 56	4 46	4 29	4 12	3 52	3 28
22	9 36	9 6	8 35	7 59	7 15	4 48	4 33	4 18	3 59	3 37
24	9 40	9 13	8 43	8 11	7 30	4 50	4 36	4 22	4 5	3 45

Spline-Shaft Milling Cutter.—The most efficient method of forming splines on shafts is by hobbing, but special milling cutters may also be used. Since the cutter forms the space between adjacent splines, it must be made to suit the number of splines and the root diameter of the shaft. The cutter angle *B* equals 360 degrees divided by the number of splines. The following formulas are for determining the chordal width *C* at the root of the splines or the chordal width across the concave edge of the cutter. In these formulas, A = angle between center line of spline and a radial line passing through the intersection of the root circle and one side of the spline; W = width of spline; d = root diameter of splined shaft; C = chordal width at root circle between adjacent splines; N = number of splines.

$$\sin A = \frac{W}{d} \qquad C = d \times \sin\left(\frac{180}{N} - A\right)$$

Splines of involute form are often used in preference to the straight-sided type. Dimensions of the American Standard involute splines and hobs are given in the section on splines.

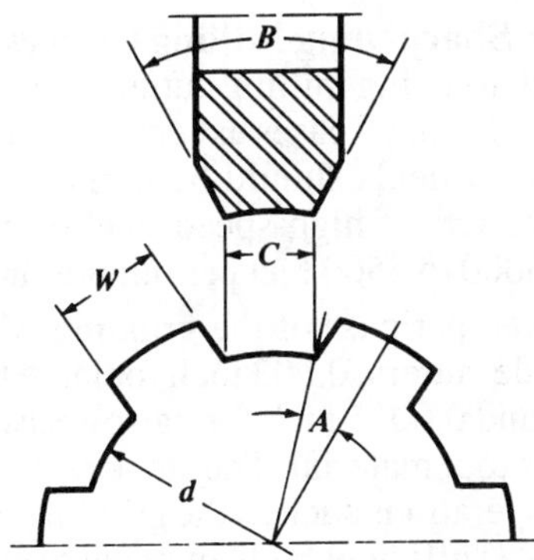

Cutter Grinding

Wheels for Sharpening Milling Cutters.—Milling cutters may be sharpened either by using the periphery of a disk wheel or the face of a cup wheel. The latter grinds the lands of the teeth flat, whereas the periphery of a disk wheel leaves the teeth slightly concave back of the cutting edges. The concavity produced by disk wheels reduces the effective clearance angle on the teeth, the effect being more pronounced for wheels of small diameter than for wheels of large diameter. For this reason, large diameter wheels are preferred when sharpening milling cutters with disk type wheels. Irrespective of what type of wheel is used to sharpen a milling cutter, any burrs resulting from grinding should be carefully removed by a hand stoning operation. Stoning also helps to reduce the roughness of grinding marks and improves the quality of the finish produced on the surface being machined. Unless done very carefully, hand stoning may dull the cutting edge. Stoning may be avoided and a sharper cutting edge produced if the wheel rotates toward the cutting edge, which requires that the operator maintain contact between the tool and the rest while the wheel rotation is trying to move the tool away from the rest. Though slightly more difficult, this method will eliminate the burr.

Specifications of Grinding Wheels for Sharpening Milling Cutters

Cutter Material	Operation	Grinding Wheel			
		Abrasive Material	Grain Size	Grade	Bond
Carbon Tool Steel	Roughing	Aluminum Oxide	46–60	K	Vitrified
	Finishing	Aluminum Oxide	100	H	Vitrified
High-speed Steel:					
18-4-1	Roughing	Aluminum Oxide	60	K,H	Vitrified
	Finishing	Aluminum Oxide	100	H	Vitrified
18-4-2	Roughing	Aluminum Oxide	80	F,G,H	Vitrified
	Finishing	Aluminum Oxide	100	H	Vitrified
Cast Non-Ferrous Tool Material	Roughing	Aluminum Oxide	46	H,K,L,N	Vitrified
	Finishing	Aluminum Oxide	100–120	H	Vitrified
Sintered Carbide	Roughing after Brazing	Silicon Carbide	60	G	Vitrified
	Roughing	Diamond	100	[a]	Resinoid
	Finishing	Diamond	Up to 500	[a]	Resinoid
Carbon Tool Steel and High-Speed Steel[b]	Roughing	Cubic Boron Nitride	80–100	R,P	Resinoid
	Finishing	Cubic Boron Nitride	100–120	S,T	Resinoid

[a] Not indicated in diamond wheel markings.

[b] For hardnesses above Rockwell C 56.

Wheel Speeds and Feeds for Sharpening Milling Cutters.—Relatively low cutting speeds should be used when sharpening milling cutters to avoid tempering and heat checking. Dry grinding is recommended in all cases except when diamond wheels are employed. The surface speed of grinding wheels should be in the range of 4500 to 6500 feet per minute for grinding milling cutters of high-speed steel or cast non-ferrous tool material. For sintered carbide cutters, 5000 to 5500 feet per minute should be used.

The maximum stock removed per pass of the grinding wheel should not exceed about 0.0004 inch for sintered carbide cutters; 0.003 inch for large high-speed steel and cast non-ferrous tool material cutters; and 0.0015 inch for narrow saws and slotting cutters of high-speed steel or cast non-ferrous tool material. The stock removed per pass of the wheel may be increased for backing-off operations such as the grinding of secondary clearance behind the teeth since there is usually a sufficient body of metal to carry off the heat.

Clearance Angles for Milling Cutter Teeth.—The clearance angle provided on the cutting edges of milling cutters has an important bearing on cutter performance, cutting efficiency, and cutter life between sharpenings. It is desirable in all cases to use a clearance angle as small as possible so as to leave more metal back of the cutting edges for better heat dissipation and to provide maximum support. Excessive clearance angles not only weaken the cutting edges, but also increase the likelihood of "chatter" which will result in poor finish on the machined surface and reduce the life of the cutter. According to The Cincinnati Milling Machine Co., milling cutters used for general purpose work and having diameters from $\frac{1}{8}$ to 3 inches should have clearance angles from 13 to 5 degrees, respectively, decreasing proportionately as the diameter increases. General purpose cutters over 3 inches in diameter should be provided with a clearance angle of 4 to 5 degrees. The land width is usually $\frac{1}{64}$, $\frac{1}{32}$, and $\frac{1}{16}$ inch, respectively, for small, medium, and large cutters.

The primary clearance or relief angle for best results varies according to the material being milled about as follows: low carbon, high carbon, and alloy steels, 3 to 5 degrees; cast iron and medium and hard bronze, 4 to 7 degrees; brass, soft bronze, aluminum, magnesium, plastics, etc., 10 to 12 degrees. When milling cutters are resharpened, it is customary to grind a secondary clearance angle of 3 to 5 degrees behind the primary clearance angle to reduce the land width to its original value and thus avoid interference with the surface to be milled. A general formula for plain milling cutters, face mills, and form relieved cutters which gives the clearance angle C, in degrees, necessitated by the feed per revolution F, in inches, the width of land L, in inches, the depth of cut d, in inches, the cutter diameter D, in inches, and the Brinell hardness number B of the work being cut is:

$$C = \frac{45860}{DB}\left(1.5L + \frac{F}{\pi D}\sqrt{d(D-d)}\right)$$

Rake Angles for Milling Cutters.—In peripheral milling cutters, the rake angle is generally defined as the angle in degrees that the tooth face deviates from a radial line to the cutting edge. In face milling cutters, the teeth are inclined with respect to both the radial and axial lines. These angles are called *radial* and *axial* rake, respectively. The radial and axial rake angles may be positive, zero, or negative.

Positive rake angles should be used whenever possible for all types of high-speed steel milling cutters. For sintered carbide tipped cutters, zero and negative rake angles are frequently employed to provide more material back of the cutting edge to resist shock loads.

Rake Angles for High-speed Steel Cutters: Positive rake angles of 10 to 15 degrees are satisfactory for milling steels of various compositions with plain milling cutters. For softer materials such as magnesium and aluminum alloys, the rake angle may be 25 degrees or more. Metal slitting saws for cutting alloy steel usually have rake angles from 5 to 10 degrees, whereas zero and sometimes negative rake angles are used for saws to cut copper and other soft non-ferrous metals to reduce the tendency to "hog in." Form relieved cutters usually have rake angles of 0, 5, or 10 degrees. Commercial face milling cutters usually have 10 degrees positive radial and axial rake angles for general use in milling cast iron, forged and alloy steel, brass, and bronze; for milling castings and forgings of magnesium and free-cutting aluminum and their alloys, the rake angles may be increased to 25 degrees positive or more, depending on the operating conditions; a smaller rake angle is used for abrasive or difficult to machine aluminum alloys.

Cast Non-ferrous Tool Material Milling Cutters: Positive rake angles are generally provided on milling cutters using cast non-ferrous tool materials although negative rake angles may be used advantageously for some operations such as those where shock loads are encountered or where it is necessary to eliminate vibration when milling thin sections.

Sintered Carbide Milling Cutters: Peripheral milling cutters such as slab mills, slotting cutters, saws, etc., tipped with sintered carbide, generally have negative radial rake angles of 5 degrees for soft low carbon steel and 10 degrees or more for alloy steels. Positive axial rake angles of 5 and 10 degrees, respectively, may be provided, and for slotting saws and cutters, 0 degree axial rake may be used. On soft materials such as free-cutting aluminum alloys, positive rake angles of 10 to so degrees are used. For milling abrasive or difficult to machine aluminum alloys, small positive or even negative rake angles are used.

Eccentric Type Radial Relief.—When the radial relief angles on peripheral teeth of milling cutters are ground with a disc type grinding wheel in the conventional manner the ground surfaces on the lands are slightly concave, conforming approximately to the radius of the wheel. A flat land is produced when the radial relief angle is ground with a cup wheel. Another entirely different method of grinding the radial angle is by the eccentric method, which produces a slightly convex surface on the land. If the radial relief angle at the cutting edge is equal for all of the three types of land mentioned, it will be found that the land with the eccentric relief will drop away from the cutting edge a somewhat greater distance for a given distance around the land than will the others. This is evident from a study of Table entitled, "Indicator Drops for Checking Radial Relief Angles on Peripheral Teeth." This feature is an advantage of the eccentric type relief which also produces an excellent finish.

Indicator Drops for Checking the Radial Relief Angle on Peripheral Teeth

Cutter Diameter, Inch	Recom. Range of Radial Relief Angles, Degrees	Checking Distance, Inch	Indicator Drops, Inches				Recom. Max. Primary Land Width, Inch
			For Flat and Concave Relief		For Eccentric Relief		
			Min.	Max.	Min.	Max.	
1/16	20–25	.005	.0014	.0019	.0020	.0026	.007
3/32	16–20	.005	.0012	.0015	.0015	.0019	.007
1/8	15–19	.010	.0018	.0026	.0028	.0037	.015
5/32	13–17	.010	.0017	.0024	.0024	.0032	.015
3/16	12–16	.010	.0016	.0023	.0022	.0030	.015
7/32	11–15	.010	.0015	.0022	.0020	.0028	.015
1/4	10–14	.015	.0017	.0028	.0027	.0039	.020
9/32	10–14	.015	.0018	.0029	.0027	.0039	.020
5/16	10–13	.015	.0019	.0027	.0027	.0035	.020
11/32	10–13	.015	.0020	.0028	.0027	.0035	.020
3/8	10–13	.015	.0020	.0029	.0027	.0035	.020
13/32	9–12	.020	.0022	.0032	.0032	.0044	.025
7/16	9–12	.020	.0022	.0033	.0032	.0043	.025
15/32	9–12	.020	.0023	.0034	.0032	.0043	.025
1/2	9–12	.020	.0024	.0034	.0032	.0043	.025
9/16	9–12	.020	.0024	.0035	.0032	.0043	.025
5/8	8–11	.020	.0022	.0032	.0028	.0039	.025
11/16	8–11	.030	.0029	.0045	.0043	.0059	.035
3/4	8–11	.030	.0030	.0046	.0043	.0059	.035
13/16	8–11	.030	.0031	.0047	.0043	.0059	.035
7/8	8–11	.030	.0032	.0048	.0043	.0059	.035
15/16	7–10	.030	.0027	.0043	.0037	.0054	.035
1	7–10	.030	.0028	.0044	.0037	.0054	.035
1 1/8	7–10	.030	.0029	.0045	.0037	.0053	.035
1 1/4	6–9	.030	.0024	.0040	.0032	.0048	.035
1 3/8	6–9	.030	.0025	.0041	.0032	.0048	.035
1 1/2	6–9	.030	.0026	.0041	.0032	.0048	.035
1 5/8	6–9	.030	.0026	.0042	.0032	.0048	.035
1 3/4	6–9	.030	.0026	.0042	.0032	.0048	.035
1 7/8	6–9	.030	.0027	.0043	.0032	.0048	.035
2	6–9	.030	.0027	.0043	.0032	.0048	.035
2 1/4	5–8	.030	.0022	.0038	.0026	.0042	.040
2 1/2	5–8	.030	.0023	.0039	.0026	.0042	.040
2 3/4	5–8	.030	.0023	.0039	.0026	.0042	.040
3	5–8	.030	.0023	.0039	.0026	.0042	.040
3 1/2	5–8	.030	.0024	.0040	.0026	.0042	.047
4	5–8	.030	.0024	.0040	.0026	.0042	.047
5	4–7	.030	.0019	.0035	.0021	.0037	.047
6	4–7	.030	.0019	.0035	.0021	.0037	.047
7	4–7	.030	.0020	.0036	.0021	.0037	.060
8	4–7	.030	.0020	.0036	.0021	.0037	.060
10	4–7	.030	.0020	.0036	.0021	.0037	.060
12	4–7	.030	.0020	.0036	.0021	.0037	.060

The setup for grinding an eccentric relief is shown in Fig. 2. In this setup the point of contact between the cutter and the tooth rest must be in the same plane as the centers, or axes, of the grinding wheel and the cutter. A wide face is used on the grinding wheel, which is trued and dressed at an angle with respect to the axis of the cutter. An alternate method is to tilt the wheel at this angle. Then as the cutter is traversed and rotated past the grinding wheel while in contact with the tooth rest, an eccentric relief will be generated by the angular face of the wheel. This type of relief can only be ground on the peripheral teeth on milling cutters having helical flutes because the combination of the angular wheel face and the twisting motion of the cutter is required to generate the eccentric relief. Therefore, an eccentric relief cannot be ground on the peripheral teeth of straight fluted cutters.

Table 4 is a table of wheel angles for grinding an eccentric relief for different combinations of relief angles and helix angles. When angles are required that cannot be found in this table, the wheel angle, *W*, can be calculated by using the following formula, in which *R* is the radial relief angle and *H* is the helix angle of the flutes on the cutter.

$$\tan W = \tan R \times \tan H$$

Table 4. Grinding Wheel Angles for Grinding Eccentric Type Radial Relief Angle

Radial Relief Angle, *R*, Degrees	Helix Angle of Cutter Flutes, *H*, Degrees							
	12	18	20	30	40	45	50	52
	Wheel Angle, *W*, Degrees							
1	0°13′	0°19′	0°22′	0°35′	0°50′	1°00′	1°12′	1°17′
2	0°26′	0°39′	0°44′	1°09′	1°41′	2°00′	2°23′	2°34′
3	0°38′	0°59′	1°06′	1°44′	2°31′	3°00′	3°34′	3°50′
4	0°51′	1°18′	1°27′	2°19′	3°21′	4°00′	4°46′	5°07′
5	1°04′	1°38′	1°49′	2°53′	4°12′	5°00′	5°57′	6°23′
6	1°17′	1°57′	2°11′	3°28′	5°02′	6°00′	7°08′	7°40′
7	1°30′	2°17′	2°34′	4°03′	5°53′	7°00′	8°19′	8°56′
8	1°43′	2°37′	2°56′	4°38′	6°44′	8°00′	9°30′	10°12′
9	1°56′	2°57′	3°18′	5°13′	7°34′	9°00′	10°41′	11°28′
10	2°09′	3°17′	3°40′	5°49′	8°25′	10°00′	11°52′	12°43′
11	2°22′	3°37′	4°03′	6°24′	9°16′	11°00′	13°03′	13°58′
12	2°35′	3°57′	4°25′	7°00′	10°07′	12°00′	14°13′	15°13′
13	2°49′	4°17′	4°48′	7°36′	10°58′	13°00′	15°23′	16°28′
14	3°02′	4°38′	5°11′	8°11′	11°49′	14°00′	16°33′	17°42′
15	3°16′	4°59′	5°34′	8°48′	12°40′	15°00′	17°43′	18°56′
16	3°29′	5°19′	5°57′	9°24′	13°32′	16°00′	18°52′	20°09′
17	3°43′	5°40′	6°21′	10°01′	14°23′	17°00′	20°01′	21°22′
18	3°57′	6°02′	6°45′	10°37′	15°15′	18°00′	21°10′	22°35′
19	4°11′	6°23′	7°09′	11°15′	16°07′	19°00′	22°19′	23°47′
20	4°25′	6°45′	7°33′	11°52′	16°59′	20°00′	23°27′	24°59′
21	4°40′	7°07′	7°57′	12°30′	17°51′	21°00′	24°35′	26°10′
22	4°55′	7°29′	8°22′	13°08′	18°44′	22°00′	25°43′	27°21′
23	5°09′	7°51′	8°47′	13°46′	19°36′	23°00′	26°50′	28°31′
24	5°24′	8°14′	9°12′	14°25′	20°29′	24°00′	27°57′	29°41′
25	5°40′	8°37′	9°38′	15°04′	21°22′	25°00′	29°04′	30°50′

Indicator Drop Method of Checking Relief and Rake Angles.—The most convenient and inexpensive method of checking the relief and rake angles on milling cutters is by the indicator drop method. Three tables, Tables , 5 and 6, of indicator drops are provided in this section, for checking radial relief angles on the peripheral teeth, relief angles on side and end teeth, and rake angles on the tooth faces.

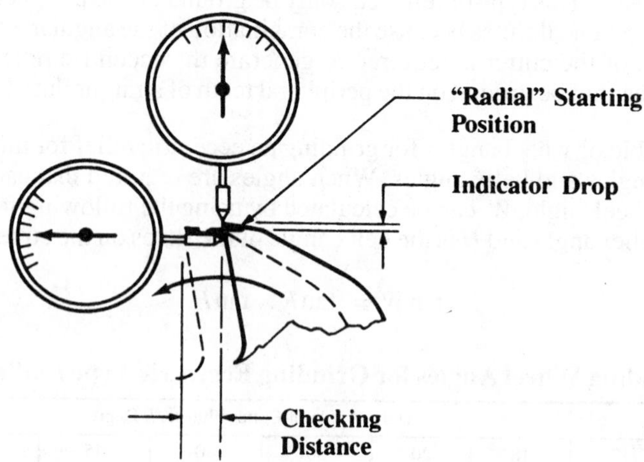

Fig. 1. Setup for Checking the Radial Relief Angle by Indicator Drop Method

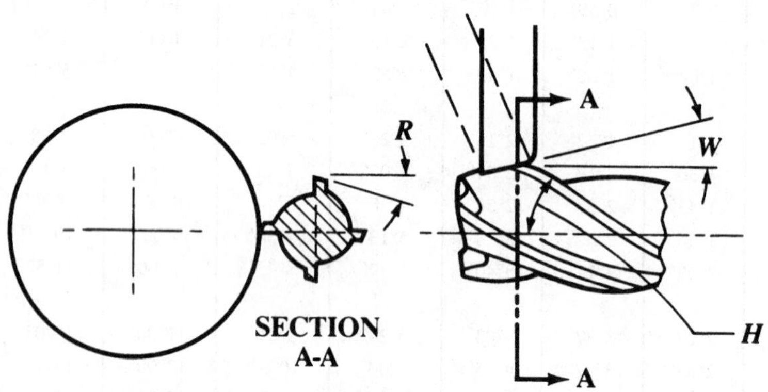

Fig. 2. Setup for Grinding Eccentric Type Radial Relief Angle

Table 5. Indicator Drops for Checking Relief Angles on Side Teeth and End Teeth

Checking Distance, Inch	Given Relief Angle								
	1°	2°	3°	4°	5°	6°	7°	8°	9°
	Indicator Drop, inch								
.005	.00009	.00017	.00026	.00035	.0004	.0005	.0006	.0007	.0008
.010	.00017	.00035	.00052	.0007	.0009	.0011	.0012	.0014	.0016
.015	.00026	.0005	.00079	.0010	.0013	.0016	.0018	.0021	.0024
.031	.00054	.0011	.0016	.0022	.0027	.0033	.0038	.0044	.0049
.047	.00082	.0016	.0025	.0033	.0041	.0049	.0058	.0066	.0074
.062	.00108	.0022	.0032	.0043	.0054	.0065	.0076	.0087	.0098

Table 6. Indicator Drops for Checking Rake Angles on Milling Cutter Face

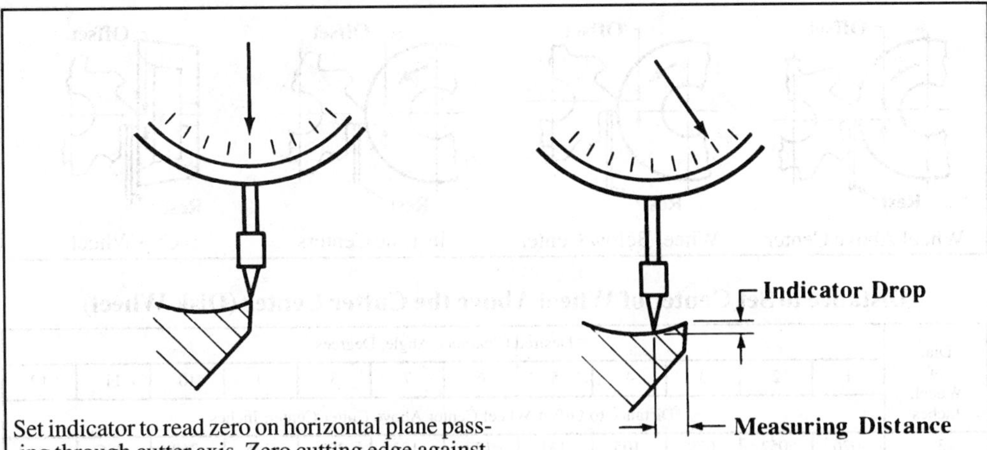

Set indicator to read zero on horizontal plane passing through cutter axis. Zero cutting edge against indicator.

Move cutter or indicator measuring distance.

Rate Angle, Deg.	Measuring Distance, inch				Rate Angle, Deg.	Measuring Distance, inch			
	.031	.062	.094	.125		.031	.062	.094	.125
	Indicator Drop, inch					Indicator Drop, inch			
1	.0005	.0011	.0016	.0022	11	.0060	.0121	.0183	.0243
2	.0011	.0022	.0033	.0044	12	.0066	.0132	.0200	.0266
3	.0016	.0032	.0049	.0066	13	.0072	.0143	.0217	.0289
4	.0022	.0043	.0066	.0087	14	.0077	.0155	.0234	.0312
5	.0027	.0054	.0082	.0109	15	.0083	.0166	.0252	.0335
6	.0033	.0065	.0099	.0131	16	.0089	.0178	.0270	.0358
7	.0038	.0076	.0115	.0153	17	.0095	.0190	.0287	.0382
8	.0044	.0087	.0132	.0176	18	.0101	.0201	.0305	.0406
9	.0049	.0098	.0149	.0198	19	.0107	.0213	.0324	.0430
10	.0055	.0109	.0166	.0220	20	.0113	.0226	.0342	.0455

The setup for checking the radial relief angle is illustrated in Fig. 1. Two dial test indicators are required, one of which should have a sharp pointed contact point. This indicator is positioned so that the axis of its spindle is vertical, passing through the axis of the cutter. The cutter may be held by its shank in the spindle of a tool and cutter grinder workhead, or between centers while mounted on a mandrel. The cutter is rotated to the position where the vertical indicator contacts a cutting edge. The second indicator is positioned with its spindle axis horizontal and with the contact point touching the tool face just below the cutting edge. With both indicators adjusted to read zero, the cutter is rotated a distance equal to the checking distance, as determined by the reading on the second indicator. Then the indicator drop is read on the vertical indicator and checked against the values in the tables. The indicator drops for radial relief angles ground by a disc type grinding wheel and those ground with a cup wheel are so nearly equal that the values are listed together; values for the eccentric type relief are listed separately, since they are larger. A similar procedure is used to check the relief angles on the side and end teeth of milling cutters; however, only one indicator is used. Also, instead of rotating the cutter, the indicator or the cutter must be moved a distance equal to the checking distance in a straight line.

Various Set-ups Used in Grinding the Clearance Angle on Milling Cutter Teeth

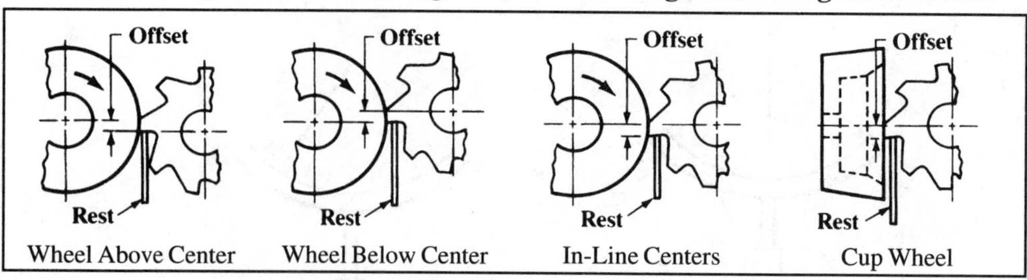

Wheel Above Center | Wheel Below Center | In-Line Centers | Cup Wheel

Distance to Set Center of Wheel Above the Cutter Center (Disk Wheel)

Dia. of Wheel, Inches	Desired Clearance Angle, Degrees											
	1	2	3	4	5	6	7	8	9	10	11	12
	[a]Distance to Offset Wheel Center Above Cutter Center, Inches											
3	.026	.052	.079	.105	.131	.157	.183	.209	.235	.260	.286	.312
4	.035	.070	.105	.140	.174	.209	.244	.278	.313	.347	.382	.416
5	.044	.087	.131	.174	.218	.261	.305	.348	.391	.434	.477	.520
6	.052	.105	.157	.209	.261	.314	.366	.417	.469	.521	.572	.624
7	.061	.122	.183	.244	.305	.366	.427	.487	.547	.608	.668	.728
8	.070	.140	.209	.279	.349	.418	.488	.557	.626	.695	.763	.832
9	.079	.157	.236	.314	.392	.470	.548	.626	.704	.781	.859	.936
10	.087	.175	.262	.349	.436	.523	.609	.696	.782	.868	.954	1.040

[a] Calculated from the formula: Offset = Cutter Diameter × ½ × Sine of Clearance Angle.

Distance to Set Center of Wheel Below the Cutter Center (Disk Wheel)

Dia. of Cutter, Inches	Desired Clearance Angle, Degrees											
	1	2	3	4	5	6	7	8	9	10	11	12
	[a]Distance to Offset Wheel Center Below Cutter Center, Inches											
2	.017	.035	.052	.070	.087	.105	.122	.139	.156	.174	.191	.208
3	.026	.052	.079	.105	.131	.157	.183	.209	.235	.260	.286	.312
4	.035	.070	.105	.140	.174	.209	.244	.278	.313	.347	.382	.416
5	.044	.087	.131	.174	.218	.261	.305	.348	.391	.434	.477	.520
6	.052	.105	.157	.209	.261	.314	.366	.417	.469	.521	.572	.624
7	.061	.122	.183	.244	.305	.366	.427	.487	.547	.608	.668	.728
8	.070	.140	.209	.279	.349	.418	.488	.557	.626	.695	.763	.832
9	.079	.157	.236	.314	.392	.470	.548	.626	.704	.781	.859	.936
10	.087	.175	.262	.349	.436	.523	.609	.696	.782	.868	.954	1.040

Distance to Set Tooth Rest Below Center Line of Wheel and Cutter.—When the clearance angle is ground with a disk type wheel by keeping the center line of the wheel in line with the center line of the cutter, the tooth rest should be lowered by an amount given by the following formula:

$$\text{Offset} = \frac{\text{Wheel Diam.} \times \text{Cutter Dia.} \times \text{Sine of One-half the Clearance Angle}}{\text{Wheel Dia.} + \text{Cutter Dia.}}$$

Distance to Set Tooth Rest Below Cutter Center When Cup Wheel is Used.—When the clearance is ground with a cup wheel, the tooth rest is set below the center of the cutter the same amount as given in the table for "Distance to Set Center of Wheel Below the Cutter Center (Disk Wheel)."

REAMERS

Hand Reamers.—Hand reamers are made with both straight and helical flutes. Helical flutes provide a shearing cut and are especially useful in reaming holes having keyways or grooves, as these are bridged over by the helical flutes, thus preventing binding or chattering. Hand reamers are made in both solid and expansion forms. The American standard dimensions for solid forms are given in the accompanying table. The expansion type is useful whenever, in connection with repair or other work, it is necessary to enlarge a reamed hole by a few thousandths of an inch. The expansion form is split through the fluted section and a slight amount of expansion is obtained by screwing in a tapering plug. The diameter increase may vary from 0.005 to 0.008 inch for reamers up to about 1 inch diameter and from 0.010 to 0.012 inch for diameters between 1 and 2 inches. Hand reamers are tapered slightly on the end to facilitate starting them properly. The actual diameter of the shanks of commercial reamers may be from 0.002 to 0.005 inch under the reamer size. That part of the shank that is squared should be turned smaller in diameter than the shank itself, so that, when applying a wrench, no burr may be raised that may mar the reamed hole if the reamer is passed clear through it.

When fluting reamers, the cutter is so set with relation to the center of the reamer blank that the tooth gets a slight negative rake; that is, the cutter should be set *ahead* of the center, as shown in the illustration accompanying the table giving the amount to set the cutter ahead of the radial line. The amount is so selected that a tangent to the circumference of the reamer at the cutting point makes an angle of approximately 95 degrees with the front face of the cutting edge.

Amount to Set Cutter Ahead of Radial Line to Obtain Negative Front Rake

Size of Reamer	Dimension *a*, Inches	Size of Reamer	Dimension *a*, Inches	Size of Reamer	Dimension *a*, Inches
1/4	0.011	7/8	0.038	2	0.087
3/8	0.016	1	0.044	2 1/4	0.098
1/2	0.022	1 1/4	0.055	2 1/2	0.109
5/8	0.027	1 1/2	0.066	2 3/4	0.120
3/4	0.033	1 3/4	0.076	3	0.131

When fluting reamers, it is necessary to "break up the flutes"; that is, to space the cutting edges unevenly around the reamer. The difference in spacing should be very slight and need not exceed two degrees one way or the other. The manner in which the breaking up of the flutes is usually done is to move the index head to which the reamer is fixed a certain amount more or less than it would be moved if the spacing were regular. A table is given showing the amount of this additional movement of the index crank for reamers with different numbers of flutes. When a reamer is provided with helical flutes, the angle of spiral should be such that the cutting edges make an angle of about 10 or at most 15 degrees with the axis of the reamer.

The relief of the cutting edges should be comparatively slight. An eccentric relief, that is, one where the land back of the cutting edge is convex, rather than flat, is used by one or two manufacturers, and is preferable for finishing reamers, as the reamer will hold its size longer. When hand reamers are used merely for removing stock, or simply for enlarging holes, the flat relief is better, because the reamer has a keener cutting edge. The width of the land of the cutting edges should be about 1/32 inch for a 1/4-inch, 1/16 inch for a 1-inch, and 3/32 inch for a 3-inch reamer.

Irregular Spacing of Teeth in Reamers

Number of flutes in reamer	4	6	8	10	12	14	16
Index circle to use	39	39	39	39	39	49	20
Before cutting	Move Spindle the Number of Holes below More or Less than for Regular Spacing						
2d flute	8 less	4 less	3 less	2 less	4 less	3 less	2 less
3d flute	4 more	5 more	5 more	3 more	4 more	2 more	2 more
4th flute	6 less	7 less	2 less	5 less	1 less	2 less	1 less
5th flute	…	6 more	4 more	2 more	3 more	4 more	2 more
6th flute	…	5 less	6 less	2 less	4 less	1 less	2 less
7th flute	…	…	2 more	3 more	4 more	3 more	1 more
8th flute	…	…	3 less	2 less	3 less	2 less	2 less
9th flute	…	…	…	5 more	2 more	1 more	2 more
10th flute	…	…	…	1 less	2 less	3 less	2 less
11th flute	…	…	…	…	3 more	3 more	1 more
12th flute	…	…	…	…	4 less	2 less	2 less
13th flute	…	…	…	…	…	2 more	2 more
14th flute	…	…	…	…	…	3 less	1 less
15th flute	…	…	…	…	…	…	2 more
16th flute	…	…	…	…	…	…	2 less

Threaded-end Hand Reamers.—Hand reamers are sometimes provided with a thread at the extreme point in order to give them a uniform feed when reaming. The diameter on the top of this thread at the point of the reamer is slightly smaller than the reamer itself, and the thread tapers upward until it reaches a dimension of from 0.003 to 0.008 inch, according to size, below the size of the reamer; at this point, the thread stops and a short neck about 1/16-inch wide separates the threaded portion from the actual reamer, which is provided with a short taper from 3/16 to 7/16 inch long up to where the standard diameter is reached. The length of the threaded portion and the number of threads per inch for reamers of this kind are given in the accompanying table. The thread employed is a sharp V-thread.

Dimensions for Threaded-End Hand Reamers

Sizes of Reamers	Length of Threaded Part	No. of Threads per Inch	Dia. of Thread at Point of Reamer	Sizes of Reamers	Length of Threaded Part	No. of Threads per Inch	Dia. of Thread at Point of Reamer
			Full diameter				Full diameter
1/8–5/16	3/8	32	−0.006	1 1/32–1 1/2	9/16	18	−0.010
11/32–1/2	7/16	28	−0.006	1 17/32–2	9/16	18	−0.012
17/32–3/4	1/2	24	−0.008	2 1/32–2 1/2	9/16	18	−0.015
25/32–1	9/16	18	−0.008	2 17/32–3	9/16	18	−0.020

Fluted Chucking Reamers.—Reamers of this type are used in turret lathes, screw machines, etc., for enlarging holes and finishing them smooth and to the required size. The best results are obtained with a floating type of holder that permits a reamer to align itself with the hole being reamed. These reamers are intended for removing a small amount of metal, 0.005 to 0.010 inch being common allowances. Fluted chucking reamers are provided either with a straight shank or a standard taper shank. (See table for standard dimensions.)

Rose Chucking Reamers.—The rose type of reamer is used for enlarging cored or other holes. The cutting edges at the end are ground to a 45-degree bevel. This type of reamer will remove considerable metal in one cut. The cylindrical part of the reamer has no cutting edges, but merely grooves cut for the full length of the reamer body, providing a way for the chips to escape and a channel for lubricant to reach the cutting edges. There is no relief on the cylindrical surface of the body part, but it is slightly back-tapered so that the diameter at the point with the beveled cutting edges is slightly larger than the diameter farther back. The back-taper should not exceed 0.001 inch per inch. This form of reamer usually produces holes slightly larger than its size and it is, therefore, always made from 0.005 to 0.010 inch smaller than its nominal size, so that it may be followed by a fluted reamer for finishing. The grooves on the cylindrical portion are cut by a convex cutter having a width equal to from one-fifth to one-fourth the diameter of the rose reamer itself. The depth of the groove should be from one-eighth to one-sixth the diameter of the reamer. The teeth at the end of the reamer are milled with a 75-degree angular cutter; the width of the land of the cutting edge should be about one-fifth the distance from tooth to tooth. If an angular cutter is preferred to a convex cutter for milling the grooves on the cylindrical portion, because of the higher cutting speed possible when milling, an 80-degree angular cutter slightly rounded at the point may be used.

Fluting Cutters for Reamers

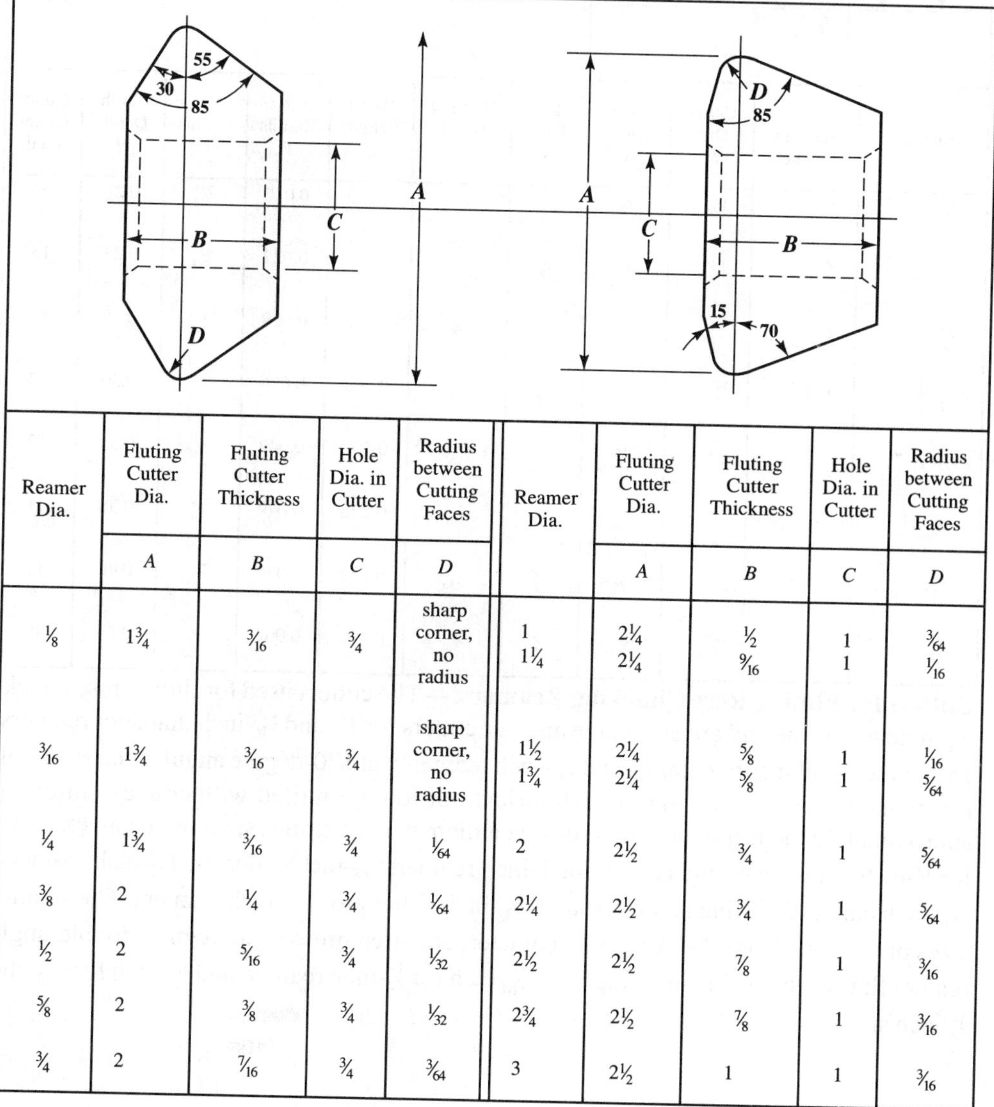

Reamer Dia.	Fluting Cutter Dia. A	Fluting Cutter Thickness B	Hole Dia. in Cutter C	Radius between Cutting Faces D	Reamer Dia.	Fluting Cutter Dia. A	Fluting Cutter Thickness B	Hole Dia. in Cutter C	Radius between Cutting Faces D
1⁄8	1 3⁄4	3⁄16	3⁄4	sharp corner, no radius	1 1 1⁄4	2 1⁄4 2 1⁄4	1⁄2 9⁄16	1 1	3⁄64 1⁄16
3⁄16	1 3⁄4	3⁄16	3⁄4	sharp corner, no radius	1 1⁄2 1 3⁄4	2 1⁄4 2 1⁄4	5⁄8 5⁄8	1 1	1⁄16 5⁄64
1⁄4	1 3⁄4	3⁄16	3⁄4	1⁄64	2	2 1⁄2	3⁄4	1	5⁄64
3⁄8	2	1⁄4	3⁄4	1⁄64	2 1⁄4	2 1⁄2	3⁄4	1	5⁄64
1⁄2	2	5⁄16	3⁄4	1⁄32	2 1⁄2	2 1⁄2	7⁄8	1	3⁄16
5⁄8	2	3⁄8	3⁄4	1⁄32	2 3⁄4	2 1⁄2	7⁄8	1	3⁄16
3⁄4	2	7⁄16	3⁄4	3⁄64	3	2 1⁄2	1	1	3⁄16

Dimensions of Formed Reamer Fluting Cutters

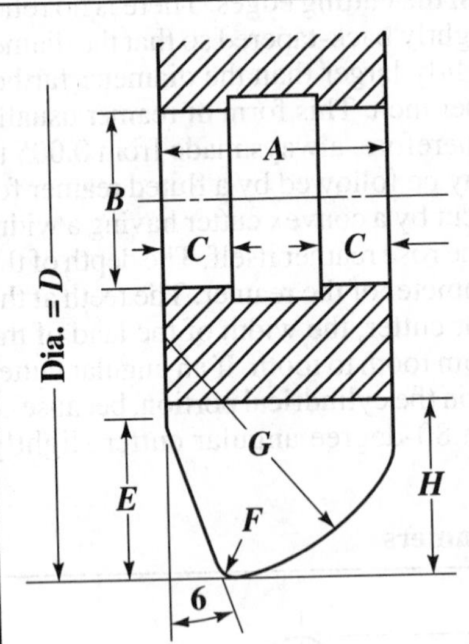

The making and maintenance of cutters of the formed type involves greater expense than the use of angular cutters of which dimensions are given on the previous page; but the form of flute produced by the formed type of cutter is preferred by many reamer users. The claims made for the formed type of flute are that the chips can be more readily removed from the reamer, and that the reamer has greater strength and is less likely to crack or spring out of shape in hardening.

Reamer Size	No. of Teeth in Reamer	Cutter Dia. *D*	Cutter Width *A*	Hole Dia. *B*	Bearing Width *C*	Bevel Length *E*	Radius *F*	Radius *F*	Tooth Depth *H*	No. of Cutter Teeth
1⁄8–3⁄16	6	1¾	3⁄16	7⁄8	…	0.125	0.016	7⁄32	0.21	14
1⁄4–5⁄16	6	1¾	1⁄4	7⁄8	…	0.152	0.022	9⁄32	0.25	13
3⁄8–7⁄16	6	1⅞	3⁄8	7⁄8	1⁄8	0.178	0.029	1⁄2	0.28	12
1⁄2–11⁄16	6–8	2	7⁄16	7⁄8	1⁄8	0.205	0.036	9⁄16	0.30	12
3⁄4–1	8	2⅛	1⁄2	7⁄8	5⁄32	0.232	0.042	11⁄16	0.32	12
1 1⁄16–1½	10	2¼	9⁄16	7⁄8	5⁄32	0.258	0.049	3⁄4	0.38	11
1 9⁄16–2⅛	12	2⅜	5⁄8	7⁄8	3⁄16	0.285	0.056	27⁄32	0.40	11
2¼–3	14	2⅝	11⁄16	7⁄8	3⁄16	0.312	0.062	7⁄8	0.44	10

Cutters for Fluting Rose Chucking Reamers.—The cutters used for fluting rose chucking reamers on the end are 80-degree angular cutters for 1⁄4- and 5⁄16-inch diameter reamers; 75-degree angular cutters for 3⁄8- and 7⁄16-inch reamers; and 70-degree angular cutters for all larger sizes. The grooves on the cylindrical portion are milled with convex cutters of approximately the following sizes for given diameters of reamers: 5⁄32-inch convex cutter for 1⁄2-inch reamers; 5⁄16-inch cutter for 1-inch reamers; 3⁄8-inch cutter for 1½-inch reamers; 13⁄32-inch cutters for 2-inch reamers; and 15⁄32-inch cutters for 2½-inch reamers. The smaller sizes of reamers, from 1⁄4 to 3⁄8 inch in diameter, are often milled with regular double-angle reamer fluting cutters having a radius of 1⁄64 inch for 1⁄4-inch reamer, and 1⁄32 inch for 5⁄16- and 3⁄8-inch sizes.

Vertical Adjustment of Tooth-rest for Grinding Clearance on Reamers

Size of Reamer	Hand Reamer for Steel. Cutting Clearance Land 0.006 inch Wide		Hand Reamer for Cast Iron and Bronze. Cutting Clearance Land 0.025 inch Wide		Chucking Reamer for Cast Iron and Bronze. Cutting Clearance Land 0.025 inch Wide		Rose Chucking Reamers for Steel
	For Cutting Clearance	For Second Clearance	For Cutting Clearance	For Second Clearance	For Cutting Clearance	For Second Clearance	For Cutting Clearance on Angular Edge at End
½	0.012	0.052	0.032	0.072	0.040	0.080	0.080
⅝	0.012	0.062	0.032	0.072	0.040	0.090	0.090
¾	0.012	0.072	0.035	0.095	0.040	0.100	0.100
⅞	0.012	0.082	0.040	0.120	0.045	0.125	0.125
1	0.012	0.092	0.040	0.120	0.045	0.125	0.125
1⅛	0.012	0.102	0.040	0.120	0.045	0.125	0.125
1¼	0.012	0.112	0.045	0.145	0.050	0.160	0.160
1⅜	0.012	0.122	0.045	0.145	0.050	0.160	0.175
1½	0.012	0.132	0.048	0.168	0.055	0.175	0.175
1⅝	0.012	0.142	0.050	0.170	0.060	0.200	0.200
1¾	0.012	0.152	0.052	0.192	0.060	0.200	0.200
1⅞	0.012	0.162	0.056	0.196	0.060	0.200	0.200
2	0.012	0.172	0.056	0.216	0.064	0.224	0.225
2⅛	0.012	0.172	0.059	0.219	0.064	0.224	0.225
2¼	0.012	0.172	0.063	0.223	0.064	0.224	0.225
2⅜	0.012	0.172	0.063	0.223	0.068	0.228	0.230
2½	0.012	0.172	0.065	0.225	0.072	0.232	0.230
2⅝	0.012	0.172	0.065	0.225	0.075	0.235	0.235
2¾	0.012	0.172	0.065	0.225	0.077	0.237	0.240
2⅞	0.012	0.172	0.070	0.230	0.080	0.240	0.240
3	0.012	0.172	0.072	0.232	0.080	0.240	0.240
3⅛	0.012	0.172	0.075	0.235	0.083	0.240	0.240
3¼	0.012	0.172	0.078	0.238	0.083	0.243	0.245
3⅜	0.012	0.172	0.081	0.241	0.087	0.247	0.245
3½	0.012	0.172	0.084	0.244	0.090	0.250	0.250
3⅝	0.012	0.172	0.087	0.247	0.093	0.253	0.250
3¾	0.012	0.172	0.090	0.250	0.097	0.257	0.255
3⅞	0.012	0.172	0.093	0.253	0.100	0.260	0.255
4	0.012	0.172	0.096	0.256	0.104	0.264	0.260
4⅛	0.012	0.172	0.096	0.256	0.104	0.264	0.260
4¼	0.012	0.172	0.096	0.256	0.106	0.266	0.265
4⅜	0.012	0.172	0.096	0.256	0.108	0.268	0.265
4½	0.012	0.172	0.100	0.260	0.108	0.268	0.265
4⅝	0.012	0.172	0.100	0.260	0.110	0.270	0.270
4¾	0.012	0.172	0.104	0.264	0.114	0.274	0.275
4⅞	0.012	0.172	0.106	0.266	0.116	0.276	0.275
5	0.012	0.172	0.110	0.270	0.118	0.278	0.275

Reamer Difficulties.—Certain frequently occurring problems in reaming require remedial measures. These difficulties include the production of oversize holes, bellmouth holes, and holes with a poor finish. The following is taken from suggestions for correction of these difficulties by the National Twist Drill and Tool Co. and Winter Brothers Co.*

Oversize Holes: The cutting of a hole oversize from the start of the reaming operations usually indicates a mechanical defect in the setup or reamer. Thus, the wrong reamer for the workpiece material may have been used or there may be inadequate workpiece support, inadequate or worn guide bushings, or misalignment of the spindles, bushings, or workpiece or runout of the spindle or reamer holder. The reamer itself may be defective due to chamfer runout or runout of the cutting end due to a bent or nonconcentric shank.

When reamers gradually start to cut oversize, it is due to pickup or galling, principally on the reamer margins. This condition is partly due to the workpiece material. Mild steels, certain cast irons, and some aluminum alloys are particularly troublesome in this respect.

Corrective measures include reducing the reamer margin widths to about 0.005 to 0.010 inch, use of hard case surface treatments on high-speed-steel reamers, either alone or in combination with black oxide treatments, and the use of a high-grade finish on the reamer faces, margins, and chamfer relief surfaces.

Bellmouth Holes: The cutting of a hole that becomes oversize at the entry end with the oversize decreasing gradually along its length always reflects misalignment of the cutting portion of the reamer with respect to the hole. The obvious solution is to provide improved guiding of the reamer by the use of accurate bushings and pilot surfaces. If this solution is not feasible, and the reamer is cutting in a vertical position, a flexible element may be employed to hold the reamer in such a way that it has both radial and axial float, with the hope that the reamer will follow the original hole and prevent the bellmouth condition.

In horizontal setups where the reamer is held fixed and the workpiece rotated, any misalignment exerts a sideways force on the reamer as it is fed to depth, resulting in the formation of a tapered hole. This type of bellmouthing can frequently be reduced by shortening the bearing length of the cutting portion of the reamer. One way to do this is to reduce the reamer diameter by 0.010 to 0.030 inch, depending on size and length, behind a short full-diameter section, $\frac{1}{8}$ to $\frac{1}{2}$ inch long according to length and size, following the chamfer. The second method is to grind a high back taper, 0.008 to 0.015 inch per inch, behind the short full-diameter section. Either of these modifications reduces the length of the reamer tooth that can cause the bellmouth condition.

Poor Finish: The most obvious step toward producing a good finish is to reduce the reamer feed per revolution. Feeds as low as 0.0002 to 0.0005 inch per tooth have been used successfully. However, reamer life will be better if the maximum feasible feed is used.

The minimum practical amount of reaming stock allowance will often improve finish by reducing the volume of chips and the resulting heat generated on the cutting portion of the chamfer. Too little reamer stock, however, can be troublesome in that the reamer teeth may not cut freely but will deflect or push the work material out of the way. When this happens, excessive heat, poor finish, and rapid reamer wear can occur.

Because of their superior abrasion resistance, carbide reamers are often used when fine finishes are required. When properly conditioned, carbide reamers can produce a large number of good-quality holes. Careful honing of the carbide reamer edges is very important.

* "Some Aspects of Reamer Design and Operation," *Metal Cuttings,* April 1963.

Dimensions of Centers for Reamers and Arbors

Arbor. Dia. *A*	Large Center Dia. *B*	Drill No. *C*	Hole Depth *D*	Arbor Dia. *A*	Large Center Dia. *B*	Drill No. *C*	Hole Depth *D*	Arbor Dia. *A*	Large Center Dia. *B*	Drill No. *C*	Hole Depth *D*
				3/4	3/8	25	7/16	2 1/2	11/16	**J**	27/32
				13/16	13/32	20	1/2	2 5/8	45/64	**K**	7/8
				7/8	7/16	17	17/32	2 3/4	23/32	**L**	29/32
				15/16	15/32	12	9/16	2 7/8	47/64	**M**	29/32
				1	1/2	8	19/32	3	3/4	**N**	15/16
				1 1/8	33/64	5	5/8	3 1/8	49/64	**N**	31/32
				1 1/4	17/32	3	21/32	3 1/4	25/32	**O**	31/32
				1 3/8	35/64	2	21/32	3 3/8	51/64	**O**	1
				1 1/2	9/16	1	11/16	3 1/2	13/16	**P**	1
1/4	1/8	55	5/32	…	…	Letter	…	3 5/8	53/64	**Q**	1 1/16
5/16	5/32	52	3/16	1 5/8	37/64	**A**	23/32	3 3/4	27/32	**R**	1 1/16
3/8	3/16	48	7/32	1 3/4	19/32	B	23/32	3 7/8	55/64	R	1 1/16
7/16	7/32	43	1/4	1 7/8	39/64	**C**	3/4	4	7/8	**S**	1 1/8
1/2	1/4	39	5/16	2	5/8	**E**	3/4	4 1/4	29/32	**T**	1 1/8
9/16	9/32	33	11/32	2 1/8	41/64	**F**	25/32	4 1/2	15/16	**V**	1 3/16
5/8	5/16	30	3/8	2 1/4	21/32	**G**	13/16	4 3/4	31/32	**W**	1 1/4
11/16	11/32	29	13/32	2 3/8	43/64	**H**	27/32	5	1	**X**	1 1/4

Straight Shank Center Reamers and Machine Countersinks

ANSI B94.2-1983, R1988

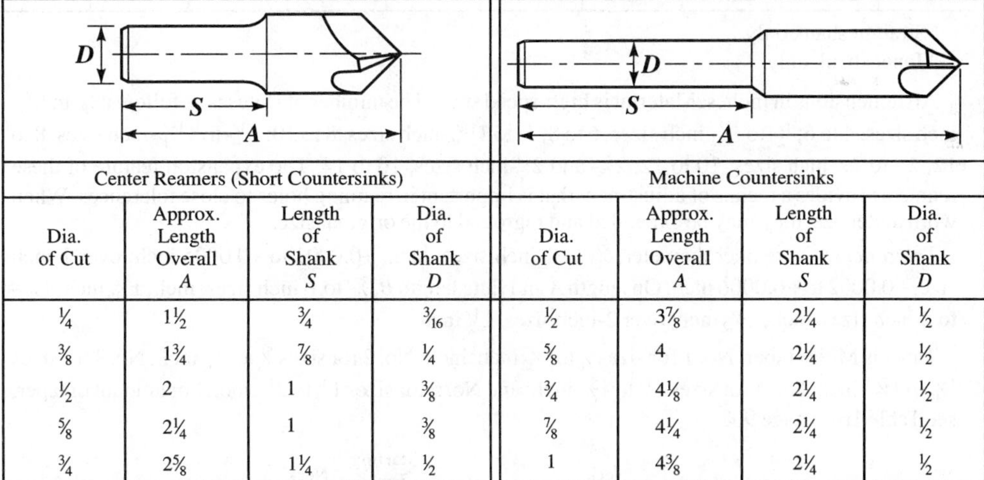

Center Reamers (Short Countersinks)				Machine Countersinks			
Dia. of Cut	Approx. Length Overall *A*	Length of Shank *S*	Dia. of Shank *D*	Dia. of Cut	Approx. Length Overall *A*	Length of Shank *S*	Dia. of Shank *D*
1/4	1 1/2	3/4	3/16	1/2	3 7/8	2 1/4	1/2
3/8	1 3/4	7/8	1/4	5/8	4	2 1/4	1/2
1/2	2	1	3/8	3/4	4 1/8	2 1/4	1/2
5/8	2 1/4	1	3/8	7/8	4 1/4	2 1/4	1/2
3/4	2 5/8	1 1/4	1/2	1	4 3/8	2 1/4	1/2

All dimensions are given in inches. Material is high-speed steel. Reamers and countersinks have 3 or 4 flutes. Center reamers are standard with 60, 82, 90, or 100 degrees included angle. Machine countersinks are standard with either 60 or 82 degrees included angle.

Tolerances: On overall length *A*, the tolerance is ±1/8 inch for center reamers in a size range of from 1/4 to 3/8 inch, incl., and machine countersinks in a size range of from 1/2 to 5/8 inch. incl.; ± 3/16 inch for center reamers, 1/2 to 3/4 inch, incl.; and machine countersinks, 3/4 to 1 inch, incl. On shank diameter *D*, the tolerance is –0.0005 to –0.002 inch. On shank length *S*, the tolerance is ±1/16 inch.

Expansion Chucking Reamers—Straight and Taper Shanks

ANSI B94.2-1983, R1988

Dia of Reamer	Length, *A*	Flute Length, *B*	Shank Dia., *D*		Dia. of Reamer	Length, *A*	Flute Length, *B*	Shank Dia., *D*	
			Max.	Min.				Max.	Min.
3/8	7	3/4	0.3105	0.3095	1 3/32	10 1/2	1 5/8	0.8745	0.8730
13/32	7	3/4	0.3105	0.3095	1 1/8	11	1 3/4	0.8745	0.8730
7/16	7	7/8	0.3730	0.3720	1 5/32	11	1 3/4	0.8745	0.8730
15/32	7	7/8	0.3730	0.3720	1 3/16	11	1 3/4	0.9995	0.9980
1/2	8	1	0.4355	0.4345	1 7/32	11	1 3/4	0.9995	0.9980
17/32	8	1	0.4355	0.4345	1 1/4	11 1/2	1 7/8	0.9995	0.9980
9/16	8	1 1/8	0.4355	0.4345	1 5/16	11 1/2	1 7/8	0.9995	0.9980
19/32	8	1 1/8	0.4355	0.4345	1 3/8	12	2	0.9995	0.9980
5/8	9	1 1/4	0.5620	0.5605	1 7/16	12	2	1.2495	1.2480
21/32	9	1 1/4	0.5620	0.5605	1 1/2	12 1/2	2 1/8	1.2495	1.2480
11/16	9	1 1/4	0.5620	0.5605	1 9/16[a]	12 1/2	2 1/8	1.2495	1.2480
23/32	9	1 1/4	0.5620	0.5605	1 5/8	13	2 1/4	1.2495	1.2480
3/4	9 1/2	1 3/8	0.6245	0.6230	1 11/16[a]	13	2 1/4	1.2495	1.2480
25/32	9 1/2	1 3/8	0.6245	0.6230	1 3/4	13 1/2	2 3/8	1.2495	1.2480
13/16	9 1/2	1 3/8	0.6245	0.6230	1 13/16[a]	13 1/2	2 3/8	1.4995	1.4980
27/32	9 1/2	1 3/8	0.6245	0.6230	1 7/8	14	2 1/2	1.4995	1.4980
7/8	10	1 1/2	0.7495	0.7480	1 15/16[a]	14	2 1/2	1.4995	1.4980
29/32	10	1 1/2	0.7495	0.7480	2	14	2 1/2	1.4995	1.4980
15/16	10	1 1/2	0.7495	0.7480	2 1/8[b]	14 1/2	2 3/4	…	…
31/32	10	1 1/2	0.7495	0.7480	2 1/4[b]	14 1/2	2 3/4	…	…
1	10 1/2	1 5/8	0.8745	0.8730	2 3/8[b]	15	3	…	…
1 1/32	10 1/2	1 5/8	0.8745	0.8730	2 1/2[b]	15	3	…	…
1 1/16	10 1/2	1 5/8	0.8745	0.8730	…	…	…	…	…

[a] Straight shank only.

[b] Taper shank only.

All dimensions in inches. Material is high-speed steel. The number of flutes is as follows: 3/8- to 15/32-inch sizes, 4 to 6; 1/2- to 31/32-inch sizes, 6 to 8; 1- to 1 11/16-inch sizes, 8 to 10; 1 3/4- to 1 15/16-inch sizes, 8 to 12; 2 - to 2 1/4-inch sizes, 10 to 12; 2 3/8- and 2 1/2-inch sizes, 10 to 14. The expansion feature of these reamers provides a means of adjustment that is important in reaming holes to close tolerances. When worn undersize, they may be expanded and reground to the original size.

Tolerances: On reamer diameter, 3/8- to 1-inch sizes, incl., +0.0001 to +0.0005 inch; over 1-inch size, +0.0002 to +0.0006 inch. On length *A* and flute length *B*, 3/8- to 1-inch sizes, incl., ±1/16 inch; 1 1/32- to 2-inch sizes, incl., ±3/32 inch; over 2-inch sizes, ±1/8 inch.

Taper is Morse taper: No. 1 for sizes 3/8 to 19/32 inch, incl.; No. 2 for sizes 5/8 to 29/32 incl.; No. 3 for sizes 15/16 to 1 7/32, incl.; No. 4 for sizes 1 1/4 to 1 5/8, incl.; and No. 5 for sizes 1 3/4 to 2 1/2, incl. For amount of taper, see Table 1b on page 908.

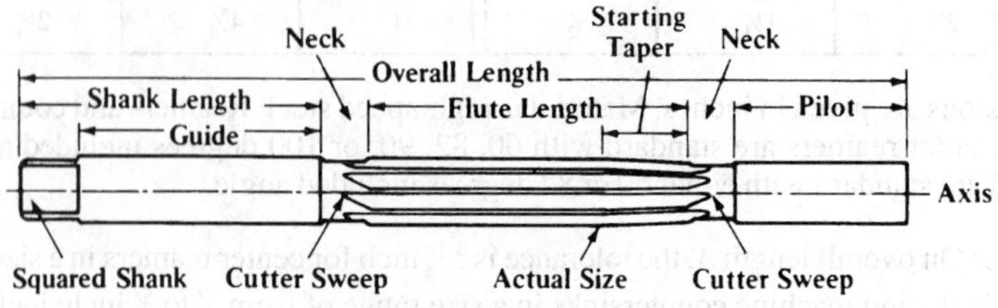

Hand Reamer, Pilot and Guide

Illustration of Terms Applying to Reamers—1

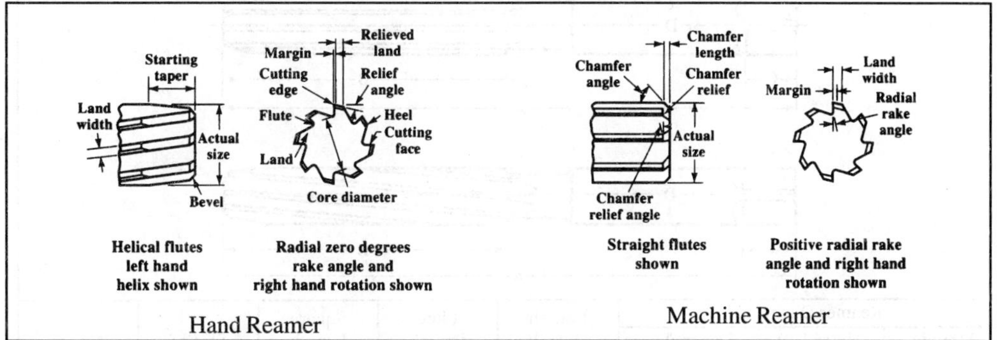

Illustration of Terms Applying to Reamers—2

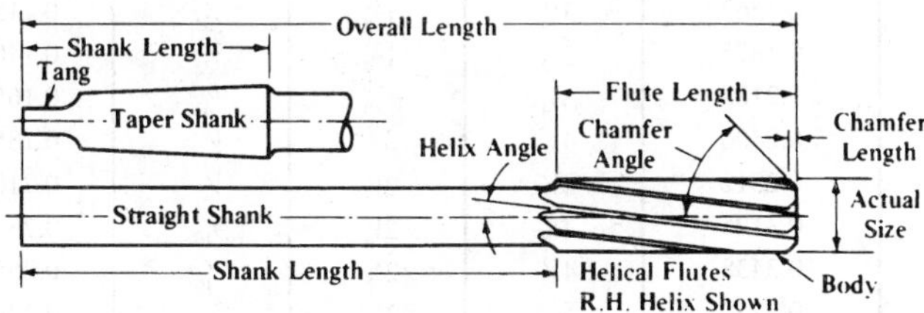

Chucking Reamer, Straight and Taper Shank

American National Standard Fluted Taper Shank Chucking Reamers—Straight and Helical Flutes, Fractional Sizes *ANSI B94.2-1983, R1988*

Reamer Dia.	Length Overall *A*	Flute Length *B*	No. of Morse Taper Shank[a]	No. of Flutes	Reamer Dia.	Length Overall *A*	Flute Length *B*	No. of Morse Taper Shank[a]	No. of Flutes
1/4	6	1 1/2	1	4 to 6	27/32	9 1/2	2 1/2	2	8 to 10
5/16	6	1 1/2	1	4 to 6	7/8	10	2 5/8	2	8 to 10
3/8	7	1 3/4	1	4 to 6	29/32	10	2 5/8	2	8 to 10
7/16	7	1 3/4	1	6 to 8	15/16	10	2 5/8	3	8 to 10
1/2	8	2	1	6 to 8	31/32	10	2 5/8	3	8 to 10
17/32	8	2	1	6 to 8	1	10 1/2	2 3/4	3	8 to 12
9/16	8	2	1	6 to 8	1 1/16	10 1/2	2 3/4	3	8 to 12
19/32	8	2	1	6 to 8	1 1/8	11	2 7/8	3	8 to 12
5/8	9	2 1/4	2	6 to 8	1 3/16	11	2 7/8	3	8 to 12
21/32	9	2 1/4	2	6 to 8	1 1/4	11 1/2	3	4	8 to 12
11/16	9	2 1/4	2	6 to 8	1 5/16	11 1/2	3	4	8 to 12
23/32	9	2 1/4	2	6 to 8	1 3/8	12	3 1/4	4	10 to 12
3/4	9 1/2	2 1/2	2	6 to 8	1 7/16	12	3 1/4	4	10 to 12
25/32	9 1/2	2 1/2	2	8 to 10	1 1/2	12 1/2	3 1/2	4	10 to 12
13/16	9 1/2	2 1/2	2	8 to 10	…	…	…	…	…

[a] American National Standard self-holding tapers (see Table on page 913.)

All dimensions are given in inches. Material is high-speed steel.

Helical flute reamers with right-hand helical flutes are standard.

Tolerances: On reamer diameter, 1/4-inch size, +.0001 to +.0004 inch; over 1/4- to 1-inch size, +.0001 to +.0005 inch; over 1-inch size, +.0002 to +.0006 inch. On length overall *A* and flute length *B*, 1/4- to 1-inch size, incl., ±1/16 inch; 1 1/16-to 1 1/2-inch size, incl., 3/32 inch.

Hand Reamers—Straight and Helical Flutes *ANSI B94.2-1983, R1988*

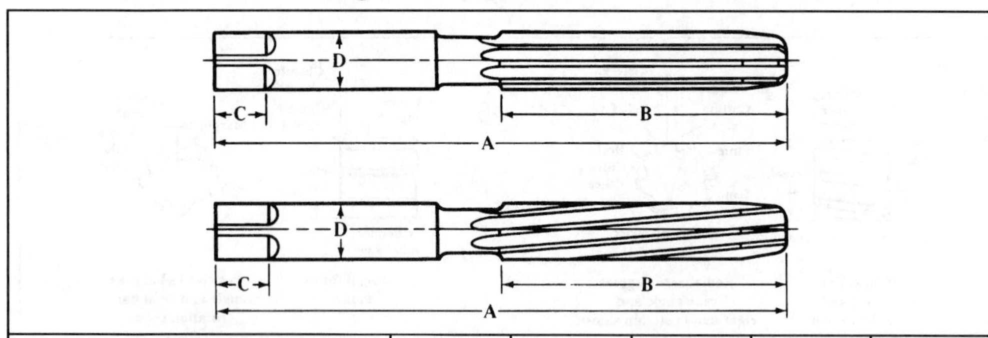

Reamer Diameter			Length Overall	Flute Length	Square Length	Size of Square	No. of Flutes
Straight Flutes	Helical Flutes	Decimal Equivalent	*A*	*B*	*C*		
1/8	…	0.1250	3	1 1/2	5/32	0.095	4 to 6
5/32	…	0.1562	3 1/4	1 5/8	7/32	0.115	4 to 6
3/16	…	0.1875	3 1/2	1 3/4	7/32	0.140	4 to 6
7/32	…	0.2188	3 3/4	1 7/8	1/4	0.165	4 to 6
1/4	1/4	0.2500	4	2	1/4	0.185	4 to 6
9/32	…	0.2812	4 1/4	2 1/8	1/4	0.210	4 to 6
5/16	5/16	0.3125	4 1/2	2 1/4	5/16	0.235	4 to 6
11/32	…	0.3438	4 3/4	2 3/8	5/16	0.255	4 to 6
3/8	3/8	0.3750	5	2 1/2	3/8	0.280	4 to 6
13/32	…	0.4062	5 1/4	2 5/8	3/8	0.305	6 to 8
7/16	7/16	0.4375	5 1/2	2 3/4	7/16	0.330	6 to 8
15/32	…	0.4688	5 3/4	2 7/8	7/16	0.350	6 to 8
1/2	1/2	0.5000	6	3	1/2	0.375	6 to 8
17/32	…	0.5312	6 1/4	3 1/8	1/2	0.400	6 to 8
9/16	9/16	0.5625	6 1/2	3 1/4	9/16	0.420	6 to 8
19/32	…	0.5938	6 3/4	3 3/8	9/16	0.445	6 to 8
5/8	5/8	0.6250	7	3 1/2	5/8	0.470	6 to 8
21/32	…	0.6562	7 3/8	3 11/16	5/8	0.490	6 to 8
11/16	11/16	0.6875	7 3/4	3 7/8	11/16	0.515	6 to 8
23/32	…	0.7188	8 1/8	4 1/16	11/16	0.540	6 to 8
3/4	3/4	0.7500	8 3/8	4 3/16	3/4	0.560	6 to 8
…	13/16	0.8125	9 1/8	4 9/16	13/16	0.610	8 to 10
7/8	7/8	0.8750	9 3/4	4 7/8	7/8	0.655	8 to 10
…	15/16	0.9375	10 1/4	5 1/8	15/16	0.705	8 to 10
1	1	1.0000	10 7/8	5 7/16	1	0.750	8 to 10
1 1/8	1 1/8	1.1250	11 5/8	5 13/16	1	0.845	8 to 10
1 1/4	1 1/4	1.2500	12 1/4	6 1/8	1	0.935	8 to 12
1 3/8	1 3/8	1.3750	12 5/8	6 5/16	1	1.030	10 to 12
1 1/2	1 1/2	1.5000	13	6 1/2	1 1/8	1.125	10 to 14

All dimensions in inches. Material is high-speed steel. The nominal shank diameter *D* is the same as the reamer diameter. Helical-flute hand reamers with left-hand helical flutes are standard. Reamers are tapered slightly on the end to facilitate proper starting.

Tolerances: On diameter of reamer, up to 1/4-inch size, incl., + .0001 to + .0004 inch; over 1/4-to 1-inch size, incl., +.0001 to + .0005 inch; over 1-inch size, +.0002 to +.0006 inch. On length overall *A* and flute length *B*, 1/8- to 1-inch size, incl., ± 1/16 inch; 1 1/8- to 1 1/2-inch size, incl., ±3/32 inch. On length of square *C*, 1/8- to 1 inch size, incl., ±1/32 inch; 1 1/8-to 1 1/2-inch size, incl., ±1/16 inch. On shank diameter *D*, 1/8- to 1-inch size, incl., –.001 to –.005 inch; 1 1/8- to 1 1/2-inch size, incl., –.0015 to – .006 inch. On size of square, 1/8- to 1/2-inch size, incl., –.004 inch; 17/32- to 1-inch size, incl., –.006 inch; 1 1/8- to 1 1/2-inch size, incl., –.008 inch.

American National Standard Expansion Hand Reamers—Straight and Helical Flutes, Squared Shank *ANSI B94.2-1983, R1988*

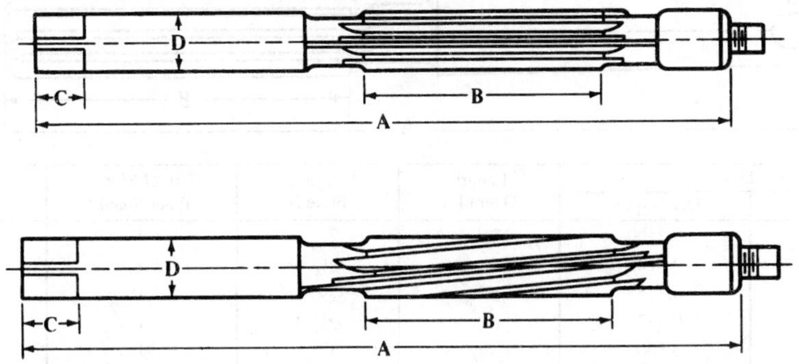

Reamer Dia.	Length Overall A		Flute Length B		Length of Square C	Shank Dia. D	Size of Square	Number of Flutes
	Max	Min	Max	Min				
Straight Flutes								
1/4	4 3/8	3 3/4	1 3/4	1 1/2	1/4	1/4	0.185	6 to 8
5/16	4 3/8	4	1 7/8	1 1/2	5/16	5/16	0.235	6 to 8
3/8	5 3/8	4 1/4	2	1 3/4	3/8	3/8	0.280	6 to 9
7/16	5 3/8	4 1/2	2	1 3/4	7/16	7/16	0.330	6 to 9
1/2	6 1/2	5	2 1/2	1 3/4	1/2	1/2	0.375	6 to 9
9/16	6 1/2	5 3/8	2 1/2	1 7/8	9/16	9/16	0.420	6 to 9
5/8	7	5 3/4	3	2 1/4	5/8	5/8	0.470	6 to 9
11/16	7 5/8	6 1/4	3	2 1/2	11/16	11/16	0.515	6 to 10
3/4	8	6 1/2	3 1/2	2 5/8	3/4	3/4	0.560	6 to 10
7/8	9	7 1/2	4	3 1/8	7/8	7/8	0.655	8 to 10
1	10	8 3/8	4 1/2	3 1/8	1	1	0.750	8 to 10
1 1/8	10 1/2	9	4 3/4	3 1/2	1	1 1/8	0.845	8 to 12
1 1/4	11	9 3/4	5	4 1/4	1	1 1/4	0.935	8 to 12
Helical Flutes								
1/4	4 3/8	3 7/8	1 3/4	1 1/2	1/4	1/4	0.185	6 to 8
5/16	4 3/8	4	1 3/4	1 1/2	5/16	5/16	0.235	6 to 8
3/8	6 1/8	4 1/4	2	1 3/4	3/8	3/8	0.280	6 to 9
7/16	6 1/4	4 1/2	2	1 3/4	7/16	7/16	0.330	6 to 9
1/2	6 1/2	5	2 1/2	1 3/4	1/2	1/2	0.375	6 to 9
5/8	8	6	3	2 1/4	5/8	5/8	0.470	6 to 9
3/4	8 5/8	6 1/2	3 1/2	2 5/8	3/4	3/4	0.560	6 to 10
7/8	9 3/8	7 1/2	4	3 1/8	7/8	7/8	0.655	6 to 10
1	10 1/4	8 3/8	4 1/2	3 1/8	1	1	0.750	6 to 10
1 1/4	11 3/8	9 3/4	5	4 1/4	1	1 1/4	0.935	8 to 12

All dimensions are given in inches. Material is carbon steel. Reamers with helical flutes that are left hand are standard. Expansion hand reamers are primarily designed for work where it is necessary to enlarge reamed holes by a few thousandths. The pilots and guides on these reamers are ground undersize for clearance. The maximum expansion on these reamers is as follows: .006 inch for the 1/4- to 7/16-inch sizes. .010 inch for the 1/2- to 7/8-inch sizes and .012 inch for the 1- to 1 1/4-inch sizes.

Tolerances: On length overall *A* and flute length *B*, ±1/16 inch for 1/4- to 1-inch sizes, ± 3/32 inch for 1 1/8- to 1 1/4-inch sizes; on length of square *C*, ±1/32 inch for 1/4- to 1-inch sizes, ± 1/16 inch for 1 1/8-to 1 1/4-inch sizes; on shank diameter *D* −.001 to −.005 inch for 1/4- to 1-inch sizes, −.0015 to −.006 inch for 1 1/8- to 1 1/4-inch sizes; on size of square, −.004 inch for 1/4- to 1/2-inch sizes. −.006 inch for 9/16- to 1-inch sizes, and −.008 inch for 1 1/8- to 1 1/4-inch sizes.

Taper Shank Jobbers Reamers—Straight Flutes *ANSI B94.2-1983, R1988*

Reamer Diameter		Length Overall *A*	Length of Flute *B*	No. of Morse Taper Shank[a]	No. of Flutes
Fractional	Dec. Equiv.				
1/4	0.2500	5 3/16	2	1	6 to 8
5/16	0.3125	5 1/2	2 1/4	1	6 to 8
3/8	0.3750	5 13/16	2 1/2	1	6 to 8
7/16	0.4375	6 1/8	2 3/4	1	6 to 8
1/2	0.5000	6 7/16	3	1	6 to 8
9/16	0.5625	6 3/4	3 1/4	1	6 to 8
5/8	0.6250	7 9/16	3 1/2	2	6 to 8
11/16	0.6875	8	3 7/8	2	8 to 10
3/4	0.7500	8 3/8	4 3/16	2	8 to 10
13/16	0.8125	8 13/16	4 9/16	2	8 to 10
7/8	0.8750	9 3/16	4 7/8	2	8 to 10
15/16	0.9375	10	5 1/8	3	8 to 10
1	1.0000	10 3/8	5 7/16	3	8 to 10
1 1/16	1.0625	10 5/8	5 5/8	3	8 to 10
1 1/8	1.1250	10 7/8	5 13/16	3	8 to 10
1 3/16	1.1875	11 1/8	6	3	8 to 12
1 1/4	1.2500	12 9/16	6 1/8	4	8 to 12
1 3/8	1.3750	12 13/16	6 5/16	4	10 to 12
1 1/2	1.5000	13 1/8	6 1/2	4	10 to 12

[a] American National Standard self-holding tapers (Table on page 913.)

All dimensions in inches. Material is high-speed steel.

Tolerances: On reamer diameter, 1/4-inch size, +.0001 to +.0004 inch; over 1/4- to 1-inch size, incl., +.0001 to +.0005 inch; over 1-inch size, +.0002 to +.0006 inch. On overall length *A* and length of flute *B*, 1/4- to 1-inch size, incl., ±1/16 inch; and 1 1/16- to 1 1/2-inch size, incl., ±3/32 inch.

American National Standard Driving Slots and Lugs for Shell Reamers or Shell Reamer Arbors *ANSI B94.2-1983, R1988*

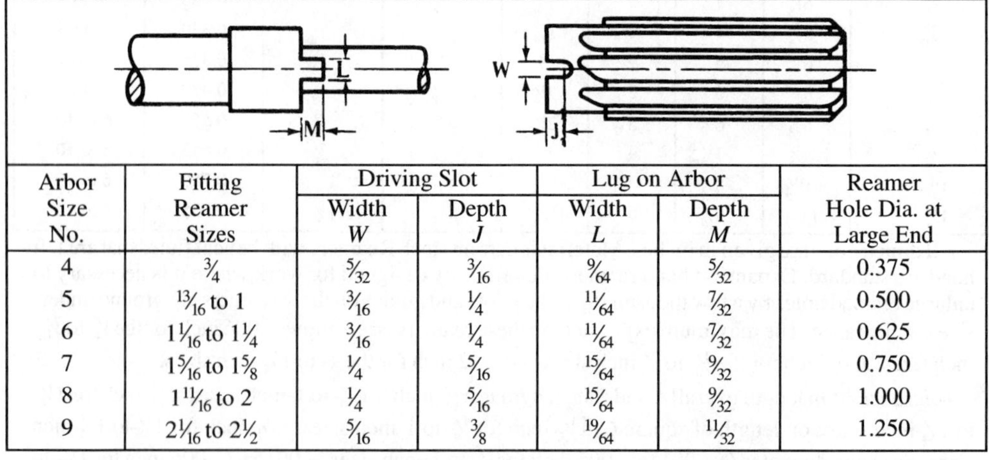

Arbor Size No.	Fitting Reamer Sizes	Driving Slot		Lug on Arbor		Reamer Hole Dia. at Large End
		Width *W*	Depth *J*	Width *L*	Depth *M*	
4	3/4	5/32	3/16	9/64	5/32	0.375
5	13/16 to 1	3/16	1/4	11/64	7/32	0.500
6	1 1/16 to 1 1/4	3/16	1/4	11/64	7/32	0.625
7	1 5/16 to 1 5/8	1/4	5/16	15/64	9/32	0.750
8	1 11/16 to 2	1/4	5/16	15/64	9/32	1.000
9	2 1/16 to 2 1/2	5/16	3/8	19/64	11/32	1.250

All dimension are given in inches. The hole in shell reamers has a taper of 1/8 inch per foot, with arbors tapered to correspond. Shell reamer arbor tapers are made to permit a driving fit with the reamer.

Straight Shank Chucking Reamers—Straight Flutes, Wire Gage Sizes

ANSI B94.2-1983, R1988

Reamer Diameter		Lgth. Over-all *A*	Lgth. of Flute *B*	Shank Dia. *D*		No. of Flutes	Reamer Diameter		Lgth. Over-all *A*	Lgth. of Flute *B*	Shank Dia. *D*		No. of Flutes
Wire Gage	Inch			Max	Min		Wire Gage	Inch			Max	Min	
60	.0400	2½	½	.0390	.0380	4	49	.0730	3	¾	.0660	.0650	4
59	.0410	2½	½	.0390	.0380	4	48	.0760	3	¾	.0720	.0710	4
58	.0420	2½	½	.0390	.0380	4	47	.0785	3	¾	.0720	.0710	4
57	.0430	2½	½	.0390	.0380	4	46	.0810	3	¾	.0771	.0701	4
56	.0465	2½	½	.0455	.0445	4	45	.0820	3	¾	.0771	.0761	4
55	.0520	2½	½	.0510	.0500	4	44	.0860	3	¾	.0810	.0800	4
54	.0550	2½	½	.0510	.0500	4	43	.0890	3	¾	.0810	.0800	4
53	.0595	2½	½	.0585	.0575	4	42	.0935	3	¾	.0880	.0870	4
52	.0635	2½	½	.0585	.0575	4	41	.0960	3½	⅞	.0928	.0918	4 to 6
51	.0670	3	¾	.0660	.0650	4	40	.0980	3½	⅞	.0928	.0918	4 to 6
50	.0700	3	¾	.0660	.0650	4	39	.0995	3½	⅞	.0928	.0918	4 to 6
38	.1015	3½	⅞	.0950	.0940	4 to 6	19	.1660	4½	1⅛	.1595	.1585	4 to 6
37	.1040	3½	⅞	.0950	.0940	4 to 6	18	.1695	4½	1⅛	.1595	.1585	4 to 6
36	.1065	3½	⅞	.1030	.1020	4 to 6	17	.1730	4½	1⅛	.1645	.1635	4 to 6
35	.1100	3½	⅞	.1030	.1020	4 to 6	16	.1770	4½	1⅛	.1704	.1694	4 to 6
34	.1110	3½	⅞	.1055	.1045	4 to 6	15	.1800	4½	1⅛	.1755	.1745	4 to 6
33	.1130	3½	⅞	.1055	.1045	4 to 6	14	.1820	4½	1⅛	.1755	.1745	4 to 6
32	.1160	3½	⅞	.1120	.1110	4 to 6	13	.1850	4½	1⅛	.1805	.1795	4 to 6
31	.1200	3½	⅞	.1120	.1110	4 to 6	12	.1890	4½	1⅛	.1805	.1795	4 to 6
30	.1285	3½	⅞	.1190	.1180	4 to 6	11	.1910	5	1¼	.1860	.1850	4 to 6
29	.1360	4	1	.1275	.1265	4 to 6	10	.1935	5	1¼	.1860	.1850	4 to 6
28	.1405	4	1	.1350	.1340	4 to 6	9	.1960	5	1¼	.1895	.1885	4 to 6
27	.1440	4	1	.1350	.1340	4 to 6	8	.1990	5	1¼	.1895	.1885	4 to 6
26	.1470	4	1	.1430	.1420	4 to 6	7	.2010	5	1¼	.1945	.1935	4 to 6
25	.1495	4	1	.1430	.1420	4 to 6	6	.2040	5	1¼	.1945	.1935	4 to 6
24	.1520	4	1	.1460	.1450	4 to 6	5	.2055	5	1¼	.2016	.2006	4 to 6
23	.1540	4	1	.1460	.1450	4 to 6	4	.2090	5	1¼	.2016	.2006	4 to 6
22	.1570	4	1	.1510	.1500	4 to 6	3	.2130	5	1¼	.2075	.2065	4 to 6
21	.1590	4½	1⅛	.1530	.1520	4 to 6	2	2210	6	1½	.2173	.2163	4 to 6
20	.1610	4½	1⅛	.1530	.1520	4 to 6	1	.2280	6	1½	.2173	.2163	4 to 6

All dimensions in inches. Material is high-speed steel.

Tolerances: On diameter of reamer, plus .0001 to plus .0004 inch. On overall length *A*, plus or minus 1⁄16 inch. On length of flute *B*, plus or minus 1⁄16 inch.

Straight Shank Chucking Reamers—Straight Flutes, Letter Sizes

ANSI B94.2-1983, R1988

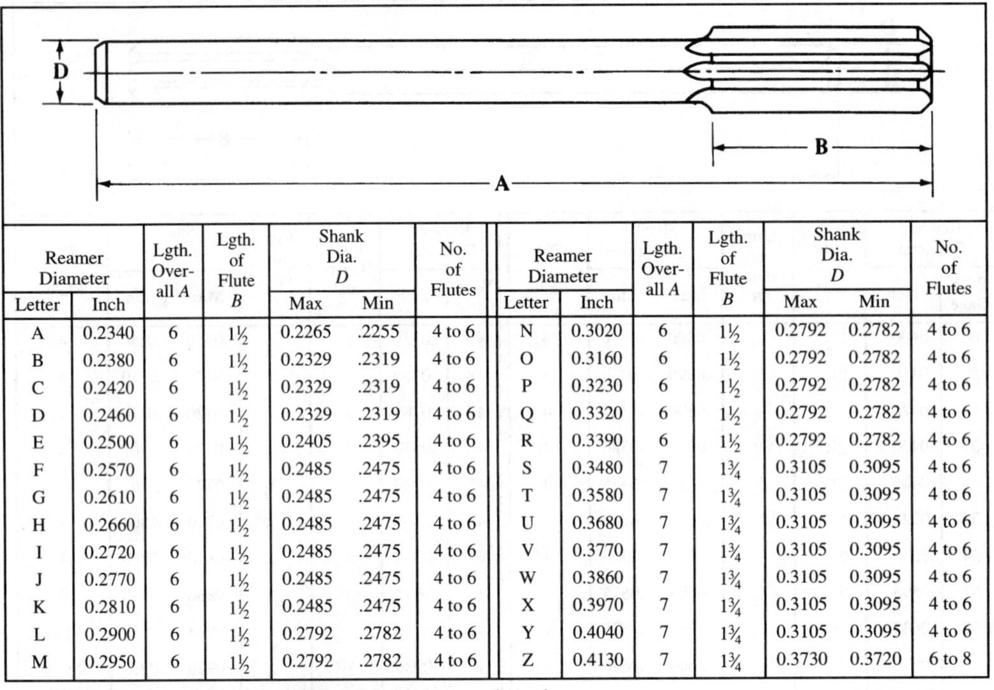

Reamer Diameter		Lgth. Over-all *A*	Lgth. of Flute *B*	Shank Dia. *D*		No. of Flutes	Reamer Diameter		Lgth. Over-all *A*	Lgth. of Flute *B*	Shank Dia. *D*		No. of Flutes
Letter	Inch			Max	Min		Letter	Inch			Max	Min	
A	0.2340	6	1½	0.2265	.2255	4 to 6	N	0.3020	6	1½	0.2792	0.2782	4 to 6
B	0.2380	6	1½	0.2329	.2319	4 to 6	O	0.3160	6	1½	0.2792	0.2782	4 to 6
C	0.2420	6	1½	0.2329	.2319	4 to 6	P	0.3230	6	1½	0.2792	0.2782	4 to 6
D	0.2460	6	1½	0.2329	.2319	4 to 6	Q	0.3320	6	1½	0.2792	0.2782	4 to 6
E	0.2500	6	1½	0.2405	.2395	4 to 6	R	0.3390	6	1½	0.2792	0.2782	4 to 6
F	0.2570	6	1½	0.2485	.2475	4 to 6	S	0.3480	7	1¾	0.3105	0.3095	4 to 6
G	0.2610	6	1½	0.2485	.2475	4 to 6	T	0.3580	7	1¾	0.3105	0.3095	4 to 6
H	0.2660	6	1½	0.2485	.2475	4 to 6	U	0.3680	7	1¾	0.3105	0.3095	4 to 6
I	0.2720	6	1½	0.2485	.2475	4 to 6	V	0.3770	7	1¾	0.3105	0.3095	4 to 6
J	0.2770	6	1½	0.2485	.2475	4 to 6	W	0.3860	7	1¾	0.3105	0.3095	4 to 6
K	0.2810	6	1½	0.2485	.2475	4 to 6	X	0.3970	7	1¾	0.3105	0.3095	4 to 6
L	0.2900	6	1½	0.2792	.2782	4 to 6	Y	0.4040	7	1¾	0.3105	0.3095	4 to 6
M	0.2950	6	1½	0.2792	.2782	4 to 6	Z	0.4130	7	1¾	0.3730	0.3720	6 to 8

All dimensions in inches. Material is high-speed steel.

Tolerances: On diameter of reamer, for sizes A to E, incl., plus .0001 to plus .0004 inch and for sizes F to Z, incl., plus .0001 to plus .0005 inch. On overall length *A*, plus or minus $1/16$ inch. On length of flute *B*, plus or minus $1/16$ inch.

Straight Shank Chucking Reamers— Straight Flutes, Decimal Sizes

ANSI B94.2-1983, R1988

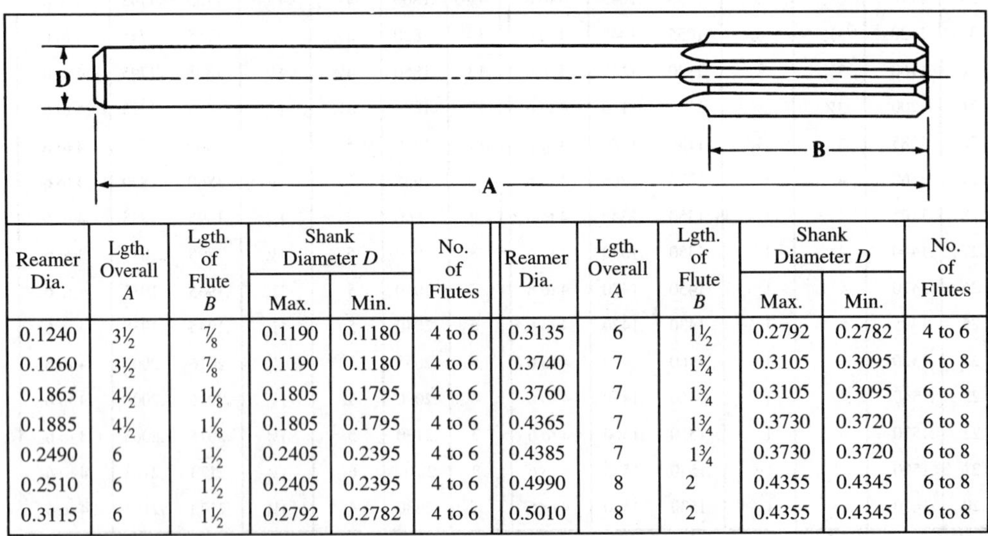

Reamer Dia.	Lgth. Overall *A*	Lgth. of Flute *B*	Shank Diameter *D*		No. of Flutes	Reamer Dia.	Lgth. Overall *A*	Lgth. of Flute *B*	Shank Diameter *D*		No. of Flutes
			Max.	Min.					Max.	Min.	
0.1240	3½	⅞	0.1190	0.1180	4 to 6	0.3135	6	1½	0.2792	0.2782	4 to 6
0.1260	3½	⅞	0.1190	0.1180	4 to 6	0.3740	7	1¾	0.3105	0.3095	6 to 8
0.1865	4½	1⅛	0.1805	0.1795	4 to 6	0.3760	7	1¾	0.3105	0.3095	6 to 8
0.1885	4½	1⅛	0.1805	0.1795	4 to 6	0.4365	7	1¾	0.3730	0.3720	6 to 8
0.2490	6	1½	0.2405	0.2395	4 to 6	0.4385	7	1¾	0.3730	0.3720	6 to 8
0.2510	6	1½	0.2405	0.2395	4 to 6	0.4990	8	2	0.4355	0.4345	6 to 8
0.3115	6	1½	0.2792	0.2782	4 to 6	0.5010	8	2	0.4355	0.4345	6 to 8

All dimensions in inches. Material is high-speed steel.

Tolerances: On diameter of reamer, for 0.124 to 0.249-inch sizes, plus .0001 to plus .0004 inch and for 0.251 to 0.501-inch sizes, plus .0001 to plus .0005 inch. On overall length *A*, plus or minus $1/16$ inch. On length of flute *B*, plus or minus $1/16$ inch.

American National Standard Straight Shank Rose Chucking and Chucking Reamers—Straight and Helical Flutes, Fractional Sizes *ANSI B94.2-1983 (R1988)*

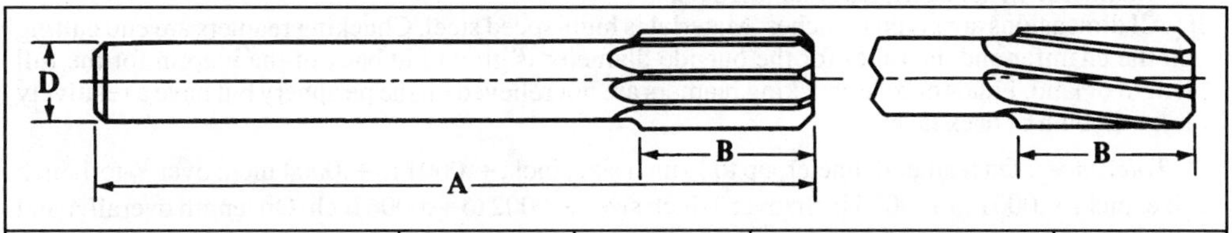

Reamer Diameter		Length Overall *A*	Flute Length *B*	Shank Dia. *D*		No. of Flutes
Chucking	Rose Chucking			Max	Min	
3/64[a]	…	2½	½	0.0455	0.0445	4
1/16	…	2½	½	0.0585	0.0575	4
5/64	…	3	¾	0.0720	0.0710	4
3/32	…	3	¾	0.0880	0.0870	4
7/64	…	3½	7/8	0.1030	0.1020	4 to 6
1/8	1/8[a]	3½	7/8	0.1190	0.1180	4 to 6
9/64	…	4	1	0.1350	0.1340	4 to 6
5/32	…	4	1	0.1510	0.1500	4 to 6
11/64	…	4½	1 1/8	0.1645	0.1635	4 to 6
3/16	3/16[a]	4½	1 1/8	0.1805	0.1795	4 to 6
13/64	…	5	1¼	0.1945	0.1935	4 to 6
7/32	…	5	1¼	0.2075	0.2065	4 to 6
15/64	…	6	1½	0.2265	0.2255	4 to 6
¼	¼[a]	6	1½	0.2405	0.2395	4 to 6
17/64	…	6	1½	0.2485	0.2475	4 to 6
9/32	…	6	1½	0.2485	0.2475	4 to 6
19/64	…	6	1½	0.2792	0.2782	4 to 6
5/16	5/16[a]	6	1½	0.2792	0.2782	4 to 6
21/64	…	6	1½	0.2792	0.2782	4 to 6
11/32	…	6	1½	0.2792	0.2782	4 to 6
23/64	…	7	1¾	0.3105	0.3095	4 to 6
3/8	3/8[a]	7	1¾	0.3105	0.3095	4 to 6
25/64	…	7	1¾	0.3105	0.3095	4 to 6
13/32	…	7	1¾	0.3105	0.3095	4 to 6
27/64	…	7	1¾	0.3730	0.3720	6 to 8
7/16	7/16[a]	7	1¾	0.3730	0.3720	6 to 8
29/64	…	7	1¾	0.3730	0.3720	6 to 8
15/32	…	7	1¾	0.3730	0.3720	6 to 8
31/64	…	8	2	0.4355	0.4345	6 to 8
½	½[a]	8	2	0.4355	0.4345	6 to 8
17/32	…	8	2	0.4355	0.4345	6 to 8
9/16	…	8	2	0.4355	0.4345	6 to 8
19/32	…	8	2	0.4355	0.4345	6 to 8
5/8	…	9	2¼	0.5620	0.5605	6 to 8
21/32	…	9	2¼	0.5620	0.5605	6 to 8
11/16	…	9	2¼	0.5620	0.5605	6 to 8
23/32	…	9	2¼	0.5620	0.5605	6 to 8
¾	…	9½	2½	0.6245	0.6230	6 to 8
25/32	…	9½	2½	0.6245	0.6230	8 to 10
13/16	…	9½	2½	0.6245	0.6230	8 to 10
27/32	…	9½	2½	0.6245	0.6230	8 to 10
7/8	…	10	2 5/8	0.7495	0.7480	8 to 10
29/32	…	10	2 5/8	0.7495	0.7480	8 to 10
15/16	…	10	2 5/8	0.7495	0.7480	8 to 10
31/32	…	10	2 5/8	0.7495	0.7480	8 to 10
1	…	10½	2¾	0.8745	0.8730	8 to 12
1 1/16	…	10½	2¾	0.8745	0.8730	8 to 12
1 1/8	…	11	2 7/8	0.8745	0.8730	8 to 12
1 3/16	…	11	2 7/8	0.9995	0.9980	8 to 12
1¼	…	11½	3	0.9995	0.9980	8 to 12
1 5/16[b]	…	11½	3	0.9995	0.9980	10 to 12
1 3/8	…	12	3¼	0.9995	0.9980	10 to 12
1 7/16[b]	…	12	3¼	1.2495	1.2480	10 to 12
1½	…	12½	3½	1.2495	1.2480	10 to 12

[a] Reamer with straight flutes is standard only.
[b] Reamer with helical flutes is standard only.

All dimensions are given in inches. Material is high-speed steel. Chucking reamers are end cutting on the chamfer and the relief for the outside diameter is ground in back of the margin for the full length of land. Lands of rose chucking reamers are not relieved on the periphery but have a relatively large amount of back taper.

Tolerances: On reamer diameter, up to 1⁄4-inch size, incl., +.0001 to +.0004 inch; over 1⁄4-to 1-inch size, incl., +.0001 to +.0005 inch; over 1-inch size, +.0002 to +.0006 inch. On length overall *A* and flute length *B*, up to 1-inch size, incl., ±1⁄16 inch; 1 1⁄16- to 1 1⁄2-inch size, incl., ±3⁄32 inch.

Helical flutes are right- or left-hand helix, right-hand cut, except sizes 1 1⁄16 through 1 1⁄2 inches, which are right-hand helix only.

Shell Reamers—Straight and Helical Flutes *ANSI B94.2-1983, R1988*

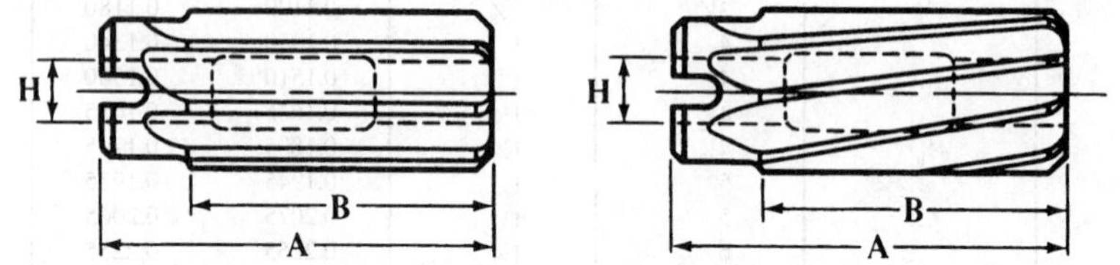

Diameter of Reamer	Length Overall *A*	Flute Length *B*	Hole Diameter Large End *H*	Fitting Arbor No.	Number of Flutes
3⁄4	2 1⁄4	1 1⁄2	0.375	4	8 to 10
7⁄8	2 1⁄2	1 3⁄4	0.500	5	8 to 10
15⁄16 [a]	2 1⁄2	1 3⁄4	0.500	5	8 to 10
1	2 1⁄2	1 3⁄4	0.500	5	8 to 10
1 1⁄16	2 3⁄4	2	0.625	6	8 to 12
1 1⁄8	2 3⁄4	2	0.625	6	8 to 12
1 3⁄16	2 3⁄4	2	0.625	6	8 to 12
1 1⁄4	2 3⁄4	2	0.625	6	8 to 12
1 5⁄16	3	2 1⁄4	0.750	7	8 to 12
1 3⁄8	3	2 1⁄4	0.750	7	8 to 12
1 7⁄16	3	2 1⁄4	0.750	7	8 to 12
1 1⁄2	3	2 1⁄4	0.750	7	10 to 14
1 9⁄16	3	2 1⁄4	0.750	7	10 to 14
1 5⁄8	3	2 1⁄4	0.750	7	10 to 14
1 11⁄16	3 1⁄2	2 1⁄2	1.000	8	10 to 14
1 3⁄4	3 1⁄2	2 1⁄2	1.000	8	12 to 14
1 13⁄16	3 1⁄2	2 1⁄2	1.000	8	12 to 14
1 7⁄8	3 1⁄2	2 1⁄2	1.000	8	12 to 14
1 15⁄16	3 1⁄2	2 1⁄2	1.000	8	12 to 14
2	3 1⁄2	2 1⁄2	1.000	8	12 to 14
2 1⁄16 [a]	3 3⁄4	2 3⁄4	1.250	9	12 to 16
2 1⁄8	3 3⁄4	2 3⁄4	1.250	9	12 to 16
2 3⁄16 [a]	3 3⁄4	2 3⁄4	1.250	9	12 to 16
2 1⁄4	3 3⁄4	2 3⁄4	1.250	9	12 to 16
2 3⁄8 [a]	3 3⁄4	2 3⁄4	1.250	9	14 to 16
2 1⁄2 [a]	3 3⁄4	2 3⁄4	1.250	9	14 to 16

[a] Helical flutes only.

All dimensions are given in inches. Material is high-speed steel. Helical flute shell reamers with left-hand helical flutes are standard. Shell reamers are designed as a sizing or finishing reamer and are held on an arbor provided with driving lugs. The holes in these reamers are ground with a taper of 1⁄8 inch per foot.

Tolerances: On diameter of reamer, 3⁄4- to 1-inch size, incl., + .0001 to + .0005 inch; over 1-inch size, +.0002 to +.0006 inch. On length overall *A* and flute length *B*, 3⁄4- to 1-inch size, incl., ± 1⁄16 inch; 1 1⁄16- to 2-inch size, incl., ± 3⁄32 inch; 2 1⁄16- to 2 1⁄2-inch size, incl., ± 1⁄8 inch.

American National Standard Arbors for Shell Reamers—Straight and Taper Shanks *ANSI B94.2-1983, R1988*

Arbor Size No.	Overall Length *A*	Approximate Length of Taper *L*	Reamer Size	Taper Shank No.[a]	Straight Shank Dia. *D*
4	9	2 1/4	3/4	2	1/2
5	9 1/2	2 1/2	13/16 to 1	2	5/8
6	10	2 3/4	1 1/16 to 1 1/4	3	3/4
7	11	3	1 5/16 to 1 5/8	3	7/8
8	12	3 1/2	1 11/16 to 2	4	1 1/8
9	13	3 3/4	2 1/16 to 2 1/2	4	1 3/8

[a] American National Standard self-holding tapers (see Table on page 913.)

All dimensions are given in inches. These arbors are designed to fit standard shell reamers (see table). End which fits reamer has taper of 1/8 inch per foot.

Stub Screw Machine Reamers—Helical Flutes *ANSI B94.2-1983, R1988*

Series No.	Diameter Range	Length Over-all *A*	Length of Flute *B*	Dia. of Shank *D*	Size of Hole *H*	Flute No.	Series No.	Diameter Range	Length Over-all *A*	Length of Flute *B*	Dia. of Shank *D*	Size of Hole *H*	Flute No.
00	.0600-.066	1 3/4	1/2	1/8	1/16	4	12	.3761-.407	2 1/2	1 1/4	1/2	3/16	6
0	.0661-.074	1 3/4	1/2	1/8	1/16	4	13	.4071-.439	2 1/2	1 1/4	1/2	3/16	6
1	.0741-.084	1 3/4	1/2	1/8	1/16	4	14	.4391-.470	2 1/2	1 1/4	1/2	3/16	6
2	.0841-.096	1 3/4	1/2	1/8	1/16	4	15	.4701-.505	2 1/2	1 1/4	1/2	3/16	6
3	.0961-.126	2	3/4	1/8	1/16	4	16	.5051-.567	3	1 1/2	5/8	1/4	6
4	.1261-.158	2 1/4	1	1/4	3/32	4	17	.5671-.630	3	1 1/2	5/8	1/4	6
5	.1581-.188	2 1/4	1	1/4	3/32	4	18	.6301-.692	3	1 1/2	5/8	1/4	6
6	.1881-.219	2 1/4	1	1/4	3/32	6	19	.6921-.755	3	1 1/2	3/4	5/16	8
7	.2191-.251	2 1/4	1	1/4	3/32	6	20	.7551-.817	3	1 1/2	3/4	5/16	8
8	.2511-.282	2 1/4	1	3/8	1/8	6	21	.8171-.880	3	1 1/2	3/4	5/16	8
9	.2821-.313	2 1/4	1	3/8	1/8	6	22	.8801-.942	3	1 1/2	3/4	5/16	8
10	.3131-.344	2 1/2	1 1/4	3/8	1/8	6	23	.9421-1.010	3	1 1/2	3/4	5/16	8
11	.3441-.376	2 1/2	1 1/4	3/8	1/8	6	…	…	…	…	…	…	…

All dimensions in inches. Material is high-speed steel.

These reamers are standard with right-hand cut and left-hand helical flutes within the size ranges shown.

Tolerances: On diameter of reamer, for sizes 00 to 7, incl., plus .0001 to plus .0004 inch and for sizes 8 to 23, incl., plus .0001 to plus .0005 inch. On overall length *A*, plus or minus 1/16 inch. On length of flute *B*, plus or minus 1/16 inch. On diameter of shank *D*, minus .0005 to minus .002 inch.

American National Standard Morse Taper Finishing Reamers
ANSI B94.2-1983, R1988

Straight Flutes and Squared Shank

Taper No.[a]	Small End Dia. (Ref.)	Large End Dia. (Ref.)	Length Overall *A*	Flute Length *B*	Square Length *C*	Shank Dia. *D*	Square Size
0	0.2503	0.3674	3¾	2¼	5⁄16	5⁄16	0.235
1	0.3674	0.5170	5	3	7⁄16	7⁄16	0.330
2	0.5696	0.7444	6	3½	5⁄8	5⁄8	0.470
3	0.7748	0.9881	7¼	4¼	7⁄8	7⁄8	0.655
4	1.0167	1.2893	8½	5¼	1	1⅛	0.845
5	1.4717	1.8005	9¾	6¼	1⅛	1½	1.125

Straight and Spiral Flutes and Taper Shank						Squared and Taper Shank
Taper No.[a]	Small End Dia. (Ref.)	Large End Dia. (Ref.)	Length Overall *A*	Flute Length *B*	Taper Shank No.[a]	Number of Flutes
0	0.2503	0.3674	5 11⁄32	2¼	0	4 to 6 incl.
1	0.3674	0.5170	6 5⁄16	3	1	6 to 8 incl.
2	0.5696	0.7444	7⅜	3½	2	6 to 8 incl.
3	0.7748	0.9881	8⅞	4¼	3	8 to 10 incl.
4	1.0167	1.2893	10⅞	5¼	4	8 to 10 incl.
5	1.4717	1.8005	13⅛	6¼	5	10 to 12 incl.

[a] Morse. For amount of taper see Table 1b on page 908.

All dimension are given in inches. Material is high-speed steel. The chamfer on the cutting end of the reamer is optional. Squared shank reamers are standard with straight flutes. Tapered shank reamers are standard with straight or spiral flutes. Spiral flute reamers are standard with left-had spiral flutes.

Tolerances: On overall length *A* and flute length *B*, in taper numbers 0 to 3, incl., ±1⁄16 inch, in taper numbers 4 and 5, ±3⁄32 inch. On length of square *C*, in taper numbers 0 to 3, incl., ±1⁄32 inch; in taper numbers 4 and 5, ±1⁄16 inch. On shank diameter *D*, −.0005 to −.002 inch. On size of square, in taper numbers 0 and 1, −.004 inch; in taper numbers 2 and 3, −.006 inch; in taper numbers 4 and 5, −.008 inch.

Taper Pipe Reamers—Spiral Flutes *ANSI B94.2-1983, R1988*

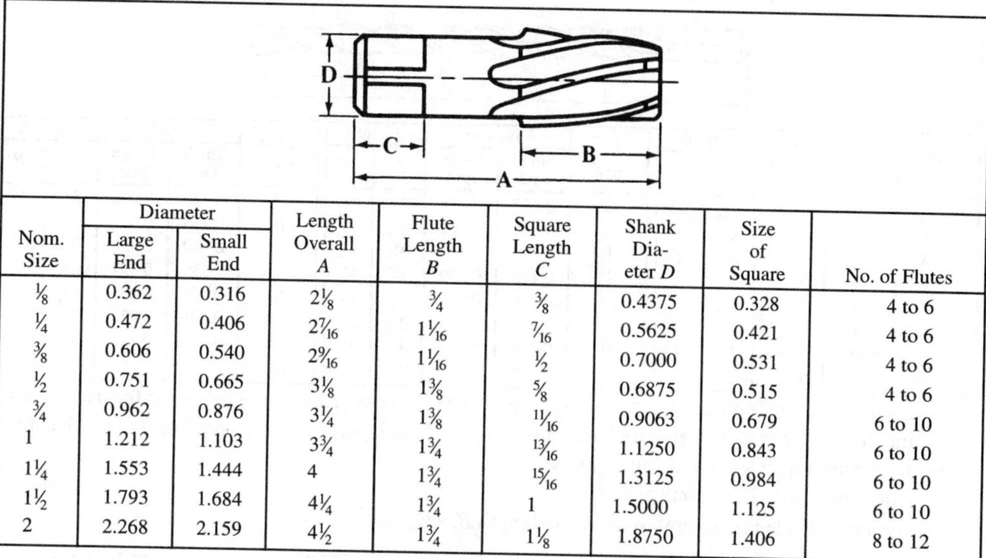

Nom. Size	Diameter Large End	Diameter Small End	Length Overall *A*	Flute Length *B*	Square Length *C*	Shank Diameter *D*	Size of Square	No. of Flutes
1/8	0.362	0.316	2 1/8	3/4	3/8	0.4375	0.328	4 to 6
1/4	0.472	0.406	2 7/16	1 1/16	7/16	0.5625	0.421	4 to 6
3/8	0.606	0.540	2 9/16	1 1/16	1/2	0.7000	0.531	4 to 6
1/2	0.751	0.665	3 1/8	1 3/8	5/8	0.6875	0.515	4 to 6
3/4	0.962	0.876	3 1/4	1 3/8	11/16	0.9063	0.679	6 to 10
1	1.212	1.103	3 3/4	1 3/4	13/16	1.1250	0.843	6 to 10
1 1/4	1.553	1.444	4	1 3/4	15/16	1.3125	0.984	6 to 10
1 1/2	1.793	1.684	4 1/4	1 3/4	1	1.5000	1.125	6 to 10
2	2.268	2.159	4 1/2	1 3/4	1 1/8	1.8750	1.406	8 to 12

All dimensions are given in inches. These reamers are tapered 3/4 inch per foot and are intended for reaming holes to be tapped with American National Standard Taper Pipe Thread taps. Material is high-speed steel. Reamers are standard with left-hand spiral flutes.

Tolerances: On length overall *A* and flute length *B*, 1/8- to 3/4-inch size, incl., ±1/16 inch; 1- to 1 1/2-inch size, incl., ±3/32 inch; 2-inch size, ±1/8 inch. On length of square *C*, 1/8- to 3/4-inch size, incl., ±1/32 inch; 1- to 2-inch size, incl., ±1/16 inch. On shank diameter *D*, 1/8-inch size, – .0015 inch; 1/4- to 1-inch size, incl., – .002 inch; 1 1/4- to 2-inch size, incl., – .003 inch. On size of square, 1/8-inch size, – .004 inch; 1/4- to 3/4-inch size, incl., – .006 inch; 1- to 2-inch size, incl., – .008 inch.

B & S Taper Reamers—Straight and Spiral Flutes, Squared Shank

Taper No.[a]	Dia., Small End	Dia., Large End	Overall Length	Square Length	Flute Length	Dia. of Shank	Size of Square	No. of Flutes
1	0.1974	0.3176	4 3/4	1/4	2 7/8	9/32	0.210	4 to 6
2	0.2474	0.3781	5 1/8	5/16	3 1/8	11/32	0.255	4 to 6
3	0.3099	0.4510	5 1/2	3/8	3 3/8	13/32	0.305	4 to 6
4	0.3474	0.5017	5 7/8	7/16	3 11/16	7/16	0.330	4 to 6
5	0.4474	0.6145	6 3/8	1/2	4	9/16	0.420	4 to 6
6	0.4974	0.6808	6 7/8	5/8	4 3/8	5/8	0.470	4 to 6
7	0.5974	0.8011	7 1/2	3/4	4 7/8	3/4	0.560	6 to 8
8	0.7474	0.9770	8 1/8	13/16	5 1/2	13/16	0.610	6 to 8
9	0.8974	1.1530	8 7/8	7/8	6 1/8	1	0.750	6 to 8
10	1.0420	1.3376	9 3/4	1	6 7/8	1 1/8	0.845	6 to 8

[a] For taper per foot, see Table 10 on page 916.

These reamers are no longer ANSI Standard.

All dimensions are given in inches. Material is high-speed steel. The chamfer on the cutting end of the reamer is optional. All reamers are finishing reamers. Spiral flute reamers are standard with left-hand spiral flutes. (Tapered reamers, especially those with left-hand spirals, should not have circular lands because cutting must take place on the outer diameter of the tool.) B & S taper reamers are designed for use in reaming out Brown & Sharpe standard taper sockets.

Tolerances: On length overall *A* and flute length *B*, taper nos. 1 to 7, incl., ±1/16 inch; taper nos. 8 to 10, incl., ±3/32 inch. On length of square *C*, taper nos. 1 to 9, incl., ±1/32 inch; taper no. 10, ±1/16 inch. On shank diameter *D*, – .0005 to – .002 inch. On size of square, taper nos. 1 to 3, incl., – .004 inch; taper nos. 4 to 9, incl., – .006 inch; taper no. 10, – .008 inch.

American National Standard Die-Maker's Reamers *ANSI B94.2-1983, R1988*

Letter Size	Diameter Small End	Diameter Large End	Length A	Length B	Letter Size	Diameter Small End	Diameter Large End	Length A	Length B	Letter Size	Diameter Small End	Diameter Large End	Length A	Length B
AAA	0.055	0.070	2 1/4	1 1/8	G	0.135	0.158	3	1 3/4	O	0.250	0.296	5	3 1/2
AA	0.065	0.080	2 1/4	1 1/8	H	0.145	0.169	3 1/4	1 7/8	P	0.275	0.327	5 1/2	4
A	0.075	0.090	2 1/4	1 1/8	I	0.160	0.184	3 1/4	1 7/8	Q	0.300	0.358	6	4 1/2
B	0.085	0.103	2 3/8	1 3/8	J	0.175	0.199	3 1/4	1 7/8	R	0.335	0.397	6 1/2	4 3/4
C	0.095	0.113	2 1/2	1 3/8	K	0.190	0.219	3 1/2	2 1/4	S	0.370	0.435	6 3/4	5
D	0.105	0.126	2 5/8	1 5/8	L	0.205	0.234	3 1/2	2 1/4	T	0.405	0.473	7	5 1/4
E	0.115	0.136	2 3/4	1 5/8	M	0.220	0.252	4	2 1/2	U	0.440	0.511	7 1/4	5 1/2
F	0.125	0.148	3	1 3/4	N	0.235	0.274	4 1/2	3	…	…	…	…	…

All dimensions in inches. Material is high-speed steel. These reamers are designed for use in die-making, have a taper of $\frac{3}{4}$ degree included angle or 0.013 inch per inch, and have 2 or 3 flutes. Reamers are standard with left-hand spiral flutes.

Tip of reamer may have conical end.

Tolerances: On length overall *A* and flute length *B*, $\pm\frac{1}{16}$ inch.

Taper Pin Reamers — Straight and Left-Hand Spiral Flutes, Squared Shank; and Left-Hand High-Spiral Flutes, Round Shank *ANSI B94.2-1983, R1988*

No. of Taper Pin Reamer	Diameter at Large End of Reamer (Ref.)	Diameter at Small End of Reamer (Ref.)	Overall Lengthof Reamer A	Length of Flute B	Length of Square C[a]	Diameter of Shank D	Size of Square[a]
8/0[b]	0.0514	0.0351	1 5/8	25/32	…	1/16	…
7/0	0.0666	0.0497	1 13/16	13/16	5/32	5/64	0.060
6/0	0.0806	0.0611	1 15/16	15/16	5/32	3/32	0.070
5/0	0.0966	0.0719	2 3/16	1 3/16	5/32	7/64	0.080
4/0	0.1142	0.0869	2 5/16	1 5/16	5/32	1/8	0.095
3/0	0.1302	0.1029	2 5/16	1 5/16	5/32	9/64	0.105
2/0	0.1462	0.1137	2 9/16	1 9/16	7/32	5/32	0.115
0	0.1638	0.1287	2 15/16	1 11/16	7/32	11/64	0.130
1	0.1798	0.1447	2 15/16	1 11/16	7/32	3/16	0.140
2	0.2008	0.1605	3 3/16	1 15/16	1/4	13/64	0.150
3	0.2294	0.1813	3 11/16	2 5/16	1/4	15/64	0.175
4	0.2604	0.2071	4 1/16	2 9/16	1/4	17/64	0.200
5	0.2994	0.2409	4 5/16	2 13/16	5/16	5/16	0.235
6	0.3540	0.2773	5 7/16	3 11/16	3/8	23/64	0.270
7	0.4220	0.3297	6 5/16	4 7/16	3/8	13/32	0.305
8	0.5050	0.3971	7 3/16	5 3/16	7/16	7/16	0.330
9	0.6066	0.4805	8 5/16	6 1/16	9/16	9/16	0.420
10	0.7216	0.5799	9 5/16	6 13/16	5/8	5/8	0.470

[a] Not applicable to high-spiral flute reamers.

[b] Not applicable to straight and left-hand spiral fluted, squared shank reamers.

All dimensions in inches. Reamers have a taper of $\frac{1}{4}$ inch per foot and are made of high-speed steel. Straight flute reamers of carbon steel are also standard. The number of flutes is as follows; 3 or 4, for 7/0 to 4/0 sizes; 4 to 6, for 3/0 to 0 sizes; 5 or 6, for 1 to 5 sizes; 6 to 8, for 6 to 9 sizes; 7 or 8, for the 10 size in the case of straight- and spiral-flute reamers; and 2 or 3, for 8/0 to 8 sizes; 2 to 4, for the 9 and 10 sizes in the case of high-spiral flute reamers.

Tolerances: On length overall *A* and flute length *B*, $\pm\frac{1}{16}$ inch. On length of square *C*, $\pm\frac{1}{32}$ inch. On shank diameter *D*, −.001 to −.005 inch for straight- and spiral-flute reamers and −.0005 to −.002 inch for high-spiral flute reamers. On size of square, −.004 inch for 7/0 to 7 sizes and −.006 inch for 8 to 10 sizes.

TWIST DRILLS AND COUNTERBORES

Twist drills are rotary end-cutting tools having one or more cutting lips and one or more straight or helical flutes for the passage of chips and cutting fluids. Twist drills are made with straight or tapered shanks, but most have straight shanks. All but the smaller sizes are ground with "back taper," reducing the diameter from the point toward the shank, to prevent binding in the hole when the drill is worn.

Straight Shank Drills: Straight shank drills have cylindrical shanks which may be of the same or of a different diameter than the body diameter of the drill and may be made with or without driving flats, tang, or grooves.

Taper Shank Drills: Taper shank drills are preferable to the straight shank type for drilling medium and large size holes. The taper on the shank conforms to one of the tapers in the American Standard (Morse) Series.

American National Standard.—American National Standard B94.11M-1993 covers nomenclature, definitions, sizes and tolerances for High Speed Steel Straight and Taper Shank Drills and Combined Drills and Countersinks, Plain and Bell types. It covers both inch and metric sizes. Dimensional tables from the Standard will be found on the following pages.

Definitions of Twist Drill Terms.—The following definitions are included in the Standard.

Axis: The imaginary straight line which forms the longitudinal center of the drill.

Back Taper: A slight decrease in diameter from point to back in the body of the drill.

Body: The portion of the drill extending from the shank or neck to the outer corners of the cutting lips.

Body Diameter Clearance: That portion of the land that has been cut away so it will not rub against the wall of the hole.

Chisel Edge: The edge at the ends of the web that connects the cutting lips.

Chisel Edge Angle: The angle included between the chisel edge and the cutting lip as viewed from the end of the drill.

Clearance Diameter: The diameter over the cutaway portion of the drill lands.

Drill Diameter: The diameter over the margins of the drill measured at the point.

Flutes: Helical or straight grooves cut or formed in the body of the drill to provide cutting lips, to permit removal of chips, and to allow cutting fluid to reach the cutting lips.

Helix Angle: The angle made by the leading edge of the land with a plane containing the axis of the drill.

Land: The peripheral portion of the drill body between adjacent flutes.

Land Width: The distance between the leading edge and the heel of the land measured at a right angle to the leading edge.

Lips—Two Flute Drill: The cutting edges extending from the chisel edge to the periphery.

Lips—Three or Four Flute Drill (Core Drill): The cutting edges extending from the bottom of the chamfer to the periphery.

Lip Relief: The axial relief on the drill point.

Lip Relief Angle: The axial relief angle at the outer corner of the lip. It is measured by projection into a plane tangent to the periphery at the outer corner of the lip. (Lip relief angle is usually measured across the margin of the twist drill.)

Margin: The cylindrical portion of the land which is not cut away to provide clearance.

Neck: The section of reduced diameter between the body and the shank of a drill.

Overall Length: The length from the extreme end of the shank to the outer corners of the cutting lips. It does not include the conical shank end often used on straight shank drills, nor does it include the conical cutting point used on both straight and taper shank drills. (For core drills with an external center on the cutting end it is the same as for two-flute drills. For core drills with an internal center on the cutting end, the overall length is to the extreme ends of the tool.)

Point: The cutting end of a drill made up of the ends of the lands, the web, and the lips. In form, it resembles a cone, but departs from a true cone to furnish clearance behind the cutting lips.

Point Angle: The angle included between the lips projected upon a plane parallel to the drill axis and parallel to the cutting lips.

Shank: The part of the drill by which it is held and driven.

Tang: The flattened end of a taper shank, intended to fit into a driving slot in the socket.

Tang Drive: Two opposite parallel driving flats on the end of a straight shank.

Web: The central portion of the body that joins the end of the lands. The end of the web forms the chisel edge on a two-flute drill.

Web Thickness: The thickness of the web at the point unless another specific location is indicated.

Web Thinning: The operation of reducing the web thickness at the point to reduce drilling thrust.

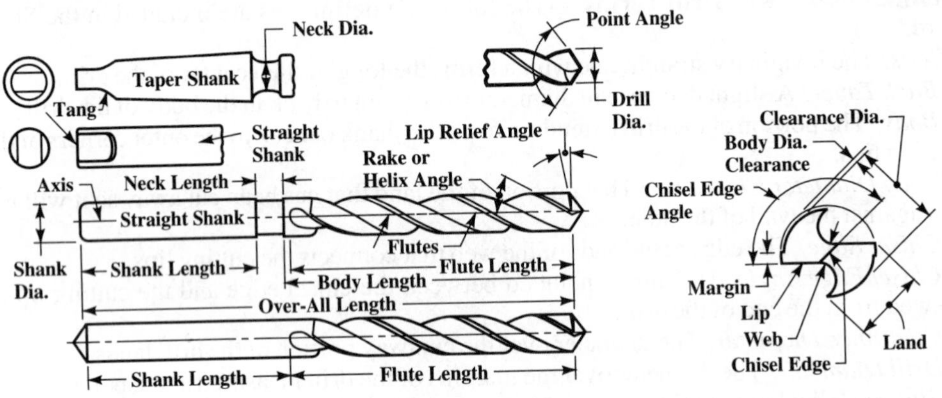

ANSI Standard Twist Drill Nomenclature

Types of Drill.—Drills may be classified based on the type of shank, number of flutes or hand of cut.

Straight Shank Drills: Those having cylindrical shanks which may be the same or different diameter than the body of the drill. The shank may be with or without driving flats, tang, grooves, or threads.

Taper Shank Drills: Those having conical shanks suitable for direct fitting into tapered holes in machine spindles, driving sleeves, or sockets. Tapered shanks generally have a driving tang.

Two-Flute Drills: The conventional type of drill used for originating holes.

Three-Flute Drills (Core Drills): Drill commonly used for enlarging and finishing drilled, cast or punched holes. They will not produce original holes.

Four-Flute Drills (Core Drills): Used interchangeably with three-flute drills. They are of similar construction except for the number of flutes.

Right-Hand Cut: When viewed from the cutting point, the counterclockwise rotation of a drill in order to cut.

Left-Hand Cut: When viewed from the cutting point, the clockwise rotation of a drill in order to cut.

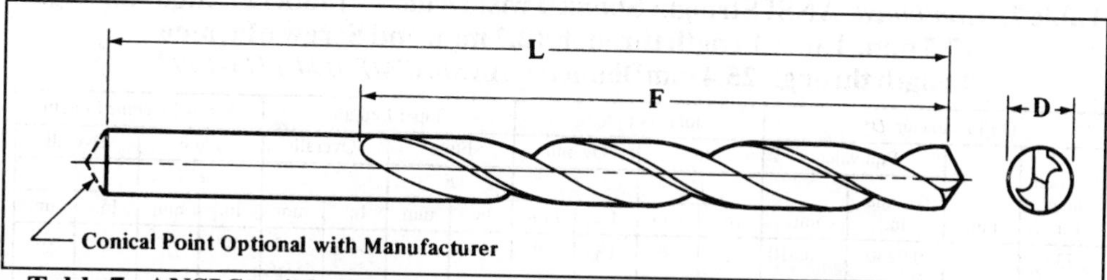

Table 7. ANSI Straight Shank Twist Drills — Jobbers Length through 17.5 mm, Taper Length through 12.7 mm, and Screw Machine Length through 25.4 mmDiameter *ANSI/ASME B94.11M-1993*

Drill Diameter, *D*[a]				Jobbers Length				Taper Length				Screw Machine Length			
Fraction No. or Ltr.	mm	Equivalent		Flute		Overall		Flute		Overall		Flute		Overall	
		Decimal In.	mm	*F*		*L*		*F*		*L*		*F*		*L*	
				In.	mm	In.	mm	In.	mm	In.	mm	In.	mm	In.	mm
97	0.15	0.0059	0.150	1/16	1.6	3/4	19	…	…	…	…	…	…	…	…
96	0.16	0.0063	0.160	1/16	1.6	3/4	19	…	…	…	…	…	…	…	…
95	0.17	0.0067	0.170	1/16	1.6	3/4	19	…	…	…	…	…	…	…	…
94	0.18	0.0071	0.180	1/16	1.6	3/4	19	…	…	…	…	…	…	…	…
93	0.19	0.0075	0.190	1/16	1.6	3/4	19	…	…	…	…	…	…	…	…
92	0.20	0.0079	0.200	1/16	1.6	3/4	19	…	…	…	…	…	…	…	…
91		0.0083	0.211	5/64	2.0	3/4	19	…	…	…	…	…	…	…	…
90	0.22	0.0087	0.221	5/64	2.0	3/4	19	…	…	…	…	…	…	…	…
89		0.0091	0.231	5/64	2.0	3/4	19	…	…	…	…	…	…	…	…
88		0.0095	0.241	5/64	2.0	3/4	19	…	…	…	…	…	…	…	…
	0.25	0.0098	0.250	5/64	2.0	3/4	19	…	…	…	…	…	…	…	…
87		0.0100	0.254	5/64	2.0	3/4	19	…	…	…	…	…	…	…	…
86		0.0105	0.267	3/32	2.4	3/4	19	…	…	…	…	…	…	…	…
85	0.28	0.0110	0.280	3/32	2.4	3/4	19	…	…	…	…	…	…	…	…
84		0.0115	0.292	3/32	2.4	3/4	19	…	…	…	…	…	…	…	…
	0.30	0.0118	0.300	3/32	2.4	3/4	19	…	…	…	…	…	…	…	…
83		0.0120	0.305	3/32	2.4	3/4	19	…	…	…	…	…	…	…	…
82		0.0125	0.318	3/32	2.4	3/4	19	…	…	…	…	…	…	…	…
	0.32	0.0126	0.320	3/32	2.4	3/4	19	…	…	…	…	…	…	…	…
81		0.0130	0.330	3/32	2.4	3/4	19	…	…	…	…	…	…	…	…
80		0.0135	0.343	1/8	3	3/4	19	…	…	…	…	…	…	…	…
	0.35	0.0138	0.350	1/8	3	3/4	19	…	…	…	…	…	…	…	…
79		0.0145	0.368	1/8	3	3/4	19	…	…	…	…	…	…	…	…
	0.38	0.0150	0.380	3/16	5	3/4	19	…	…	…	…	…	…	…	…
1/64		0.0156	0.396	3/16	5	3/4	19	…	…	…	…	…	…	…	…
	0.40	0.0157	0.400	3/16	5	3/4	19	…	…	…	…	…	…	…	…
78		0.0160	0.406	3/16	5	7/8	22	…	…	…	…	…	…	…	…
	0.42	0.0165	0.420	3/16	5	7/8	22	…	…	…	…	…	…	…	…
	0.45	0.0177	0.450	3/16	5	7/8	22	…	…	…	…	…	…	…	…
77		0.0180	0.457	3/16	5	7/8	22	…	…	…	…	…	…	…	…
	0.48	0.0189	0.480	3/16	5	7/8	22	…	…	…	…	…	…	…	…
	0.50	0.0197	0.500	3/16	5	7/8	22	…	…	…	…	…	…	…	…
76		0.0200	0.508	3/16	5	7/8	22	…	…	…	…	…	…	…	…
75		0.0210	0.533	1/4	6	1	25	…	…	…	…	…	…	…	…
	0.55	0.0217	0.550	1/4	6	1	25	…	…	…	…	…	…	…	…
74		0.0225	0.572	1/4	6	1	25	…	…	…	…	…	…	…	…
	0.60	0.0236	0.600	5/16	8	1 1/8	29	…	…	…	…	…	…	…	…

Table 7. *(Continued)* **ANSI Straight Shank Twist Drills — Jobbers Length through 17.5 mm, Taper Length through 12.7 mm, and Screw Machine Length through 25.4 mmDiameter** *ANSI/ASME B94.11M-1993*

Drill Diameter, D^a				Jobbers Length				Taper Length				Screw Machine Length			
		Equivalent		Flute		Overall		Flute		Overall		Flute		Overall	
Fraction No. or Ltr.	mm	Decimal In.	mm	*F*		*L*		*F*		*L*		*F*		*L*	
				In.	mm	In.	mm	In.	mm	In.	mm	In.	mm	In.	mm
73		0.0240	0.610	5/16	8	1 1/8	29	…	…	…	…	…	…	…	…
72		0.0250	0.635	5/16	8	1 1/8	29	…	…	…	…	…	…	…	…
	0.65	0.0256	0.650	3/8	10	1 1/4	32	…	…	…	…	…	…	…	…
71		0.0260	0.660	3/8	10	1 1/4	32	…	…	…	…	…	…	…	…
	0.70	0.0276	0.700	3/8	10	1 1/4	32	…	…	…	…	…	…	…	…
70		0.0280	0.711	3/8	10	1 1/4	32	…	…	…	…	…	…	…	…
69		0.0292	0.742	1/2	13	1 3/8	35	…	…	…	…	…	…	…	…
	0.75	0.0295	0.750	1/2	13	1 3/8	35	…	…	…	…	…	…	…	…
68		0.0310	0.787	1/2	13	1 3/8	35	…	…	…	…	…	…	…	…
1/32		0.0312	0.792	1/2	13	1 3/8	35	…	…	…	…	…	…	…	…
	0.80	0.0315	0.800	1/2	13	1 3/8	35	…	…	…	…	…	…	…	…
67		0.0320	0.813	1/2	13	1 3/8	35	…	…	…	…	…	…	…	…
66		0.0330	0.838	1/2	13	1 3/8	35	…	…	…	…	…	…	…	…
	0.85	0.0335	0.850	5/8	16	1 1/2	38	…	…	…	…	…	…	…	…
65		0.0350	0.889	5/8	16	1 1/2	38	…	…	…	…	…	…	…	…
	0.90	0.0354	0.899	5/8	16	1 1/2	38	…	…	…	…	…	…	…	…
64		0.0360	0.914	5/8	16	1 1/2	38	…	…	…	…	…	…	…	…
63		0.0370	0.940	5/8	16	1 1/2	38	…	…	…	…	…	…	…	…
	0.95	0.0374	0.950	5/8	16	1 1/2	38	…	…	…	…	…	…	…	…
62		0.0380	0.965	5/8	16	1 1/2	38	…	…	…	…	…	…	…	…
61		0.0390	0.991	11/16	17	1 5/8	41	…	…	…	…	…	…	…	…
	1.00	0.0394	1.000	11/16	17	1 5/8	41	1 1/8	29	2 1/4	57	1/2	13	1 3/8	35
60		0.0400	1.016	11/16	17	1 5/8	41	1 1/8	29	2 1/4	57	1/2	13	1 3/8	35
59		0.0410	1.041	11/16	17	1 5/8	41	1 1/8	29	2 1/4	57	1/2	13	1 3/8	35
	1.05	0.0413	1.050	11/16	17	1 5/8	41	1 1/8	29	2 1/4	57	1/2	13	1 3/8	35
58		0.0420	1.067	11/16	17	1 5/8	41	1 1/8	29	2 1/4	57	1/2	13	1 3/8	35
57		0.0430	1.092	3/4	19	1 3/4	44	1 1/8	29	2 1/4	57	1/2	13	1 3/8	35
	1.10	0.0433	1.100	3/4	19	1 3/4	44	1 1/8	29	2 1/4	57	1/2	13	1 3/8	35
	1.15	0.0453	1.150	3/4	19	1 3/4	44	1 1/8	29	2 1/4	57	1/2	13	1 3/8	35
56		0.0465	1.181	3/4	19	1 3/4	44	1 1/8	29	2 1/4	57	1/2	13	1 3/8	35
3/64		0.0469	1.191	3/4	19	1 3/4	44	1 1/8	29	2 1/4	57	1/2	13	1 3/8	35
	1.20	0.0472	1.200	7/8	22	1 7/8	48	1 3/4	44	3	76	5/8	16	1 5/8	41
	1.25	0.0492	1.250	7/8	22	1 7/8	48	1 3/4	44	3	76	5/8	16	1 5/8	41
	1.30	0.0512	1.300	7/8	22	1 7/8	48	1 3/4	44	3	76	5/8	16	1 5/8	41
55		0.0520	1.321	7/8	22	1 7/8	48	1 3/4	44	3	76	5/8	16	1 5/8	41
	1.35	0.0531	1.350	7/8	22	1 7/8	48	1 3/4	44	3	76	5/8	16	1 5/8	41
54		0.0550	1.397	7/8	22	1 7/8	48	1 3/4	44	3	76	5/8	16	1 5/8	41
	1.40	0.0551	1.400	7/8	22	1 7/8	48	1 3/4	44	3	76	5/8	16	1 5/8	41
	1.45	0.0571	1.450	7/8	22	1 7/8	48	1 3/4	44	3	76	5/8	16	1 5/8	41
	1.50	0.0591	1.500	7/8	22	1 7/8	48	1 3/4	44	3	76	5/8	16	1 5/8	41
53		0.0595	1.511	7/8	22	1 7/8	48	1 3/4	44	3	76	5/8	16	1 5/8	41
	1.55	0.0610	1.550	7/8	22	1 7/8	48	1 3/4	44	3	76	5/8	16	1 5/8	41
1/16		0.0625	1.588	7/8	22	1 7/8	48	1 3/4	44	3	76	5/8	16	1 5/8	41
	1.60	0.0630	1.600	7/8	22	1 7/8	48	2	51	3 3/4	95	11/16	17	1 11/16	43
52		0.0635	1.613	7/8	22	1 7/8	48	2	51	3 3/4	95	11/16	17	1 11/16	43
	1.65	0.0650	1.650	1	25	2	51	2	51	3 3/4	95	11/16	17	1 11/16	43

Table 7. *(Continued)* **ANSI Straight Shank Twist Drills — Jobbers Length through 17.5 mm, Taper Length through 12.7 mm, and Screw Machine Length through 25.4 mmDiameter** *ANSI/ASME B94.11M-1993*

Drill Diameter, D[a]				Jobbers Length				Taper Length				Screw Machine Length			
Fraction No. or Ltr.	mm	Equivalent		Flute		Overall		Flute		Overall		Flute		Overall	
		Decimal In.	mm	F		L		F		L		F		L	
				In.	mm	In.	mm	In.	mm	In.	mm	In.	mm	In.	mm
	1.70	0.0669	1.700	1	25	2	51	2	51	3 3/4	95	11/16	17	1 11/16	43
51		0.0670	1.702	1	25	2	51	2	51	3 3/4	95	11/16	17	1 11/16	43
	1.75	0.0689	1.750	1	25	2	51	2	51	3 3/4	95	11/16	17	1 11/16	43
50		0.0700	1.778	1	25	2	51	2	51	3 3/4	95	11/16	17	1 11/16	43
	1.80	0.0709	1.800	1	25	2	51	2	51	3 3/4	95	11/16	17	1 11/16	43
	1.85	0.0728	1.850	1	25	2	51	2	51	3 3/4	95	11/16	17	1 11/16	43
49		0.0730	1.854	1	25	2	51	2	51	3 3/4	95	11/16	17	1 11/16	43
	1.90	0.0748	1.900	1	25	2	51	2	51	3 3/4	95	11/16	17	1 11/16	43
48		0.0760	1.930	1	25	2	51	2	51	3 3/4	95	11/16	17	1 11/16	43
	1.95	0.0768	1.950	1	25	2	51	2	51	3 3/4	95	11/16	17	1 11/16	43
5/64		0.0781	1.984	1	25	2	51	2	51	3 3/4	95	11/16	17	1 11/16	43
47		0.0785	1.994	1	25	2	51	2 1/4	57	4 1/4	108	11/16	17	1 11/16	43
	2.00	0.0787	2.000	1	25	2	51	2 1/4	57	4 1/4	108	11/16	17	1 11/16	43
	2.05	0.0807	2.050	1 1/8	29	2 1/8	54	2 1/4	57	4 1/4	108	3/4	19	1 3/4	44
46		0.0810	2.057	1 1/8	29	2 1/8	54	2 1/4	57	4 1/4	108	3/4	19	1 3/4	44
45		0.0820	2.083	1 1/8	29	2 1/8	54	2 1/4	57	4 1/4	108	3/4	19	1 3/4	44
	2.10	0.0827	2.100	1 1/8	29	2 1/8	54	2 1/4	57	4 1/4	108	3/4	19	1 3/4	44
	2.15	0.0846	2.150	1 1/8	29	2 1/8	54	2 1/4	57	4 1/4	108	3/4	19	1 3/4	44
44		0.0860	2.184	1 1/8	29	2 1/8	54	2 1/4	57	4 1/4	108	3/4	19	1 3/4	44
	2.20	0.0866	2.200	1 1/4	32	2 1/4	57	2 1/4	57	4 1/4	108	3/4	19	1 3/4	44
	2.25	0.0886	2.250	1 1/4	32	2 1/4	57	2 1/4	57	4 1/4	108	3/4	19	1 3/4	44
43		0.0890	2.261	1 1/4	32	2 1/4	57	2 1/4	57	4 1/4	108	3/4	19	1 3/4	44
	2.30	0.0906	2.300	1 1/4	32	2 1/4	57	2 1/4	57	4 1/4	108	3/4	19	1 3/4	44
	2.35	0.0925	2.350	1 1/4	32	2 1/4	57	2 1/4	57	4 1/4	108	3/4	19	1 3/4	44
42		0.0935	2.375	1 1/4	32	2 1/4	57	2 1/4	57	4 1/4	108	3/4	19	1 3/4	44
3/32		0.0938	2.383	1 1/4	32	2 1/4	57	2 1/4	57	4 1/4	108	3/4	19	1 3/4	44
	2.40	0.0945	2.400	1 3/8	35	2 3/8	60	2 1/2	64	4 5/8	117	13/16	21	1 13/16	46
41		0.0960	2.438	1 3/8	35	2 3/8	60	2 1/2	64	4 5/8	117	13/16	21	1 13/16	46
	2.46	0.0965	2.450	1 3/8	35	2 3/8	60	2 1/2	64	4 5/8	117	13/16	21	1 13/16	46
40		0.0980	2.489	1 3/8	35	2 3/8	60	2 1/2	64	4 5/8	117	13/16	21	1 13/16	46
	2.50	0.0984	2.500	1 3/8	35	2 3/8	60	2 1/2	64	4 5/8	117	13/16	21	1 13/16	46
39		0.0995	2.527	1 3/8	35	2 3/8	60	2 1/2	64	4 5/8	117	13/16	21	1 13/16	46
38		0.1015	2.578	1 7/16	37	2 1/2	64	2 1/2	64	4 5/8	117	13/16	21	1 13/16	46
	2.60	0.1024	2.600	1 7/16	37	2 1/2	64	2 1/2	64	4 5/8	117	13/16	21	1 13/16	46
37		0.1040	2.642	1 7/16	37	2 1/2	64	2 1/2	64	4 5/8	117	13/16	21	1 13/16	46
	2.70	0.1063	2.700	1 7/16	37	2 1/2	64	2 1/2	64	4 5/8	117	13/16	21	1 13/16	46
36		0.1065	2.705	1 7/16	37	2 1/2	64	2 1/2	64	4 5/8	117	13/16	21	1 13/16	46
7/64		0.1094	2.779	1 1/2	38	2 5/8	67	2 1/2	64	4 5/8	117	13/16	21	1 13/16	46
35		0.1100	2.794	1 1/2	38	2 5/8	67	2 3/4	70	5 1/8	130	7/8	22	1 7/8	48
	2.80	0.1102	2.800	1 1/2	38	2 5/8	67	2 3/4	70	5 1/8	130	7/8	22	1 7/8	48
34		0.1110	2.819	1 1/2	38	2 5/8	67	2 3/4	70	5 1/8	130	7/8	22	1 7/8	48
33		0.1130	2.870	1 1/2	38	2 5/8	67	2 3/4	70	5 1/8	130	7/8	22	1 7/8	48
	2.90	0.1142	2.900	1 5/8	41	2 3/4	70	2 3/4	70	5 1/8	130	7/8	22	1 7/8	48
32		0.1160	2.946	1 5/8	41	2 3/4	70	2 3/4	70	5 1/8	130	7/8	22	1 7/8	48
	3.00	0.1181	3.000	1 5/8	41	2 3/4	70	2 3/4	70	5 1/8	130	7/8	22	1 7/8	48
31		0.1200	3.048	1 5/8	41	2 3/4	70	2 3/4	70	5 1/8	130	7/8	22	1 7/8	48

Table 7. *(Continued)* **ANSI Straight Shank Twist Drills — Jobbers Length through 17.5 mm, Taper Length through 12.7 mm, and Screw Machine Length through 25.4 mmDiameter** *ANSI/ASME B94.11M-1993*

Drill Diameter, D[a]				Jobbers Length				Taper Length				Screw Machine Length			
Fraction No. or Ltr.	mm	Equivalent		Flute F		Overall L		Flute F		Overall L		Flute F		Overall L	
		Decimal In.	mm	In.	mm	In.	mm	In.	mm	In.	mm	In.	mm	In.	mm
	3.10	0.1220	3.100	1 5/8	41	2 3/4	70	2 3/4	70	5 1/8	130	7/8	22	1 7/8	48
1/8		0.1250	3.175	1 5/8	41	2 3/4	70	2 3/4	70	5 1/8	130	7/8	22	1 7/8	48
	3.20	0.1260	3.200	1 5/8	41	2 3/4	70	3	76	5 3/8	137	15/16	24	1 15/16	49
30		0.1285	3.264	1 5/8	41	2 3/4	70	3	76	5 3/8	137	15/16	24	1 15/16	49
	3.30	0.1299	3.300	1 3/4	44	2 7/8	73	3	76	5 3/8	137	15/16	24	1 15/16	49
	3.40	0.1339	3.400	1 3/4	44	2 7/8	73	3	76	5 3/8	137	15/16	24	1 15/16	49
29		0.1360	3.454	1 3/4	44	2 7/8	73	3	76	5 3/8	137	15/16	24	1 15/16	49
	3.50	0.1378	3.500	1 3/4	44	2 7/8	73	3	76	5 3/8	137	15/16	24	1 15/16	49
28		0.1405	3.569	1 3/4	44	2 7/8	73	3	76	5 3/8	137	15/16	24	1 15/16	49
9/64		0.1406	3.571	1 3/4	44	2 7/8	73	3	76	5 3/8	137	15/16	24	1 15/16	49
	3.60	0.1417	3.600	1 7/8	48	3	76	3	76	5 3/8	137	1	25	2 1/16	52
27		0.1440	3.658	1 7/8	48	3	76	3	76	5 3/8	137	1	25	2 1/16	52
	3.70	0.1457	3.700	1 7/8	48	3	76	3	76	5 3/8	137	1	25	2 1/16	52
26		0.1470	3.734	1 7/8	48	3	76	3	76	5 3/8	137	1	25	2 1/16	52
25		0.1495	3.797	1 7/8	48	3	76	3	76	5 3/8	137	1	25	2 1/16	52
	3.80	0.1496	3.800	1 7/8	48	3	76	3	76	5 3/8	137	1	25	2 1/16	52
24		0.1520	3.861	2	51	3 1/8	79	3	76	5 3/8	137	1	25	2 1/16	52
	3.90	0.1535	3.900	2	51	3 1/8	79	3	76	5 3/8	137	1	25	2 1/16	52
23		0.1540	3.912	2	51	3 1/8	79	3	76	5 3/8	137	1	25	2 1/16	52
5/32		0.1562	3.967	2	51	3 1/8	79	3	76	5 3/8	137	1	25	2 1/16	52
22		0.1570	3.988	2	51	3 1/8	79	3 3/8	86	5 3/4	146	1 1/16	27	2 1/8	54
	4.00	0.1575	4.000	2 1/8	54	3 1/4	83	3 3/8	86	5 3/4	146	1 1/16	27	2 1/8	54
21		0.1590	4.039	2 1/8	54	3 1/4	83	3 3/8	86	5 3/4	146	1 1/16	27	2 1/8	54
20		0.1610	4.089	2 1/8	54	3 1/4	83	3 3/8	86	5 3/4	146	1 1/16	27	2 1/8	54
	4.10	0.1614	4.100	2 1/8	54	3 1/4	83	3 3/8	86	5 3/4	146	1 1/16	27	2 1/8	54
	4.20	0.1654	4.200	2 1/8	54	3 1/4	83	3 3/8	86	5 3/4	146	1 1/16	27	2 1/8	54
19		0.1660	4.216	2 1/8	54	3 1/4	83	3 3/8	86	5 3/4	146	1 1/16	27	2 1/8	54
	4.30	0.1693	4.300	2 1/8	54	3 1/4	83	3 3/8	86	5 3/4	146	1 1/16	27	2 1/8	54
18		0.1695	4.305	2 1/8	54	3 1/4	83	3 3/8	86	5 3/4	146	1 1/16	27	2 1/8	54
11/64		0.1719	4.366	2 1/8	54	3 1/4	83	3 3/8	86	5 3/4	146	1 1/16	27	2 1/8	54
17		0.1730	4.394	2 3/16	56	3 3/8	86	3 3/8	86	5 3/4	146	1 1/8	29	2 3/16	56
	4.40	0.1732	4.400	2 3/16	56	3 3/8	86	3 3/8	86	5 3/4	146	1 1/8	29	2 3/16	56
16		0.1770	4.496	2 3/16	56	3 3/8	86	3 3/8	86	5 3/4	146	1 1/8	29	2 3/16	56
	4.50	0.1772	4.500	2 3/16	56	3 3/8	86	3 3/8	86	5 3/4	146	1 1/8	29	2 3/16	56
15		0.1800	4.572	2 3/16	56	3 3/8	86	3 3/8	86	5 3/4	146	1 1/8	29	2 3/16	56
	4.60	0.1811	4.600	2 3/16	56	3 3/8	86	3 3/8	86	5 3/4	146	1 1/8	29	2 3/16	56
14		0.1820	4.623	2 3/16	56	3 3/8	86	3 3/8	86	5 3/4	146	1 1/8	29	2 3/16	56
13	4.70	0.1850	4.700	2 5/16	59	3 1/2	89	3 3/8	86	5 3/4	146	1 1/8	29	2 3/16	56
3/16		0.1875	4.762	2 5/16	59	3 1/2	89	3 3/8	86	5 3/4	146	1 1/8	29	2 3/16	56
12	4.80	0.1890	4.800	2 5/16	59	3 1/2	89	3 5/8	92	6	152	1 3/16	30	2 1/4	57
11		0.1910	4.851	2 5/16	59	3 1/2	89	3 5/8	92	6	152	1 3/16	30	2 1/4	57
	4.90	0.1929	4.900	2 7/16	62	3 5/8	92	3 5/8	92	6	152	1 3/16	30	2 1/4	57
10		0.1935	4.915	2 7/16	62	3 5/8	92	3 5/8	92	6	152	1 3/16	30	2 1/4	57
9		0.1960	4.978	2 7/16	62	3 5/8	92	3 5/8	92	6	152	1 3/16	30	2 1/4	57
	5.00	0.1969	5.000	2 7/16	62	3 5/8	92	3 5/8	92	6	152	1 3/16	30	2 1/4	57
8		0.1990	5.054	2 7/16	62	3 5/8	92	3 5/8	92	6	152	1 3/16	30	2 1/4	57

Table 7. *(Continued)* **ANSI Straight Shank Twist Drills — Jobbers Length through 17.5 mm, Taper Length through 12.7 mm, and Screw Machine Length through 25.4 mmDiameter** *ANSI/ASME B94.11M-1993*

Drill Diameter, D[a]				Jobbers Length				Taper Length				Screw Machine Length			
Fraction No. or Ltr.	mm	Equivalent		Flute		Overall		Flute		Overall		Flute		Overall	
		Decimal In.	mm	F In.	F mm	L In.	L mm	F In.	F mm	L In.	L mm	F In.	F mm	L In.	L mm
	5.10	0.2008	5.100	2 7/16	62	3 5/8	92	3 5/8	92	6	152	1 3/16	30	2 1/4	57
7		0.2010	5.105	2 7/16	62	3 5/8	92	3 5/8	92	6	152	1 3/16	30	2 1/4	57
13/64		0.2031	5.159	2 7/16	62	3 5/8	92	3 5/8	92	6	152	1 3/16	30	2 1/4	57
6		0.2040	5.182	2 1/2	64	3 3/4	95	3 5/8	92	6	152	1 1/4	32	2 3/8	60
	5.20	0.2047	5.200	2 1/2	64	3 3/4	95	3 5/8	92	6	152	1 1/4	32	2 3/8	60
5		0.2055	5.220	2 1/2	64	3 3/4	95	3 5/8	92	6	152	1 1/4	32	2 3/8	60
	5.30	0.2087	5.300	2 1/2	64	3 3/4	95	3 5/8	92	6	152	1 1/4	32	2 3/8	60
4		0.2090	5.309	2 1/2	64	3 3/4	95	3 5/8	92	6	152	1 1/4	32	2 3/8	60
	5.40	0.2126	5.400	2 1/2	64	3 3/4	95	3 5/8	92	6	152	1 1/4	32	2 3/8	60
3		0.2130	5.410	2 1/2	64	3 3/4	95	3 5/8	92	6	152	1 1/4	32	2 3/8	60
	5.50	0.2165	5.500	2 1/2	64	3 3/4	95	3 5/8	92	6	152	1 1/4	32	2 3/8	60
7/32		0.2188	5.558	2 1/2	64	3 3/4	95	3 5/8	92	6	152	1 1/4	32	2 3/8	60
	5.60	0.2205	5.600	2 5/8	67	3 7/8	98	3 3/4	95	6 1/8	156	1 5/16	33	2 7/16	62
2		0.2210	5.613	2 5/8	67	3 7/8	98	3 3/4	95	6 1/8	156	1 5/16	33	2 7/16	62
	5.70	0.2244	5.700	2 5/8	67	3 7/8	98	3 3/4	95	6 1/8	156	1 5/16	33	2 7/16	62
1		0.2280	5.791	2 5/8	67	3 7/8	98	3 3/4	95	6 1/8	156	1 5/16	33	2 7/16	62
	5.80	0.2283	5.800	2 5/8	67	3 7/8	98	3 3/4	95	6 1/8	156	1 5/16	33	2 7/16	62
	5.90	0.2323	5.900	2 5/8	67	3 7/8	98	3 3/4	95	6 1/8	156	1 5/16	33	2 7/16	62
A		0.2340	5.944	2 5/8	67	3 7/8	98	…	…	…	…	1 5/16	33	2 7/16	62
15/64		0.2344	5.954	2 5/8	67	3 7/8	98	3 3/4	95	6 1/8	156	1 5/16	33	2 7/16	62
	6.00	0.2362	6.000	2 3/4	70	4	102	3 3/4	95	6 1/8	156	1 3/8	35	2 1/2	64
B		0.2380	6.045	2 3/4	70	4	102	…	…	…	…	1 3/8	35	2 1/2	64
	6.10	0.2402	6.100	2 3/4	70	4	102	3 3/4	95	6 1/8	156	1 3/8	35	2 1/2	64
C		0.2420	6.147	2 3/4	70	4	102	…	…	…	…	1 3/8	35	2 1/2	64
	6.20	0.2441	6.200	2 3/4	70	4	102	3 3/4	95	6 1/8	156	1 3/8	35	2 1/2	64
D		0.2460	6.248	2 3/4	70	4	102	…	…	…	…	1 3/8	35	2 1/2	64
	6.30	0.2480	6.300	2 3/4	70	4	102	3 3/4	95	6 1/8	156	1 3/8	35	2 1/2	64
E, 1/4		0.2500	6.350	2 3/4	70	4	102	3 3/4	95	6 1/8	156	1 3/8	35	2 1/2	64
	6.40	0.2520	6.400	2 7/8	73	4 1/8	105	3 7/8	98	6 1/4	159	1 7/16	37	2 5/8	67
	6.50	0.2559	6.500	2 7/8	73	4 1/8	105	3 7/8	98	6 1/4	159	1 7/16	37	2 5/8	67
F		0.2570	6.528	2 7/8	73	4 1/8	105	…	…	…	…	1 7/16	37	2 5/8	67
	6.60	0.2598	6.600	2 7/8	73	4 1/8	105	…	…	…	…	1 7/16	37	2 5/8	67
G		0.2610	6.629	2 7/8	73	4 1/8	105	…	…	…	…	1 7/16	37	2 5/8	67
	6.70	0.2638	6.700	2 7/8	73	4 1/8	105	…	…	…	…	1 7/16	37	2 5/8	67
17/64		0.2656	6.746	2 7/8	73	4 1/8	105	3 7/8	98	6 1/4	159	1 7/16	37	2 5/8	67
H		0.2660	6.756	2 7/8	73	4 1/8	105	…	…	…	…	1 1/2	38	2 11/16	68
	6.80	0.2677	6.800	2 7/8	73	4 1/8	105	3 7/8	98	6 1/4	159	1 1/2	38	2 11/16	68
	6.90	0.2717	6.900	2 7/8	73	4 1/8	105	…	…	…	…	1 1/2	38	2 11/16	68
I		0.2720	6.909	2 7/8	73	4 1/8	105	…	…	…	…	1 1/2	38	2 11/16	68
	7.00	0.2756	7.000	2 7/8	73	4 1/8	105	3 7/8	98	6 1/4	159	1 1/2	38	2 11/16	68
J		0.2770	7.036	2 7/8	73	4 1/8	105	…	…	…	…	1 1/2	38	2 11/16	68
	7.10	0.2795	7.100	2 15/16	75	4 1/4	108	…	…	…	…	1 1/2	38	2 11/16	68
K		0.2810	7.137	2 15/16	75	4 1/4	108	…	…	…	…	1 1/2	38	2 11/16	68
9/32		0.2812	7.142	2 15/16	75	4 1/4	108	3 7/8	98	6 1/4	159	1 1/2	38	2 11/16	68
	7.20	0.2835	7.200	2 15/16	75	4 1/4	108	4	102	6 3/8	162	1 9/16	40	2 3/4	70
	7.30	0.2874	7.300	2 15/16	75	4 1/4	108	…	…	…	…	1 9/16	40	2 3/4	70

Table 7. *(Continued)* **ANSI Straight Shank Twist Drills — Jobbers Length through 17.5 mm, Taper Length through 12.7 mm, and Screw Machine Length through 25.4 mmDiameter** *ANSI/ASME B94.11M-1993*

Drill Diameter, *D*[a]				Jobbers Length				Taper Length				Screw Machine Length			
Fraction No. or Ltr.	mm	Equivalent Decimal In.	Equivalent mm	Flute *F* In.	Flute *F* mm	Overall *L* In.	Overall *L* mm	Flute *F* In.	Flute *F* mm	Overall *L* In.	Overall *L* mm	Flute *F* In.	Flute *F* mm	Overall *L* In.	Overall *L* mm
L		0.2900	7.366	2 15/16	75	4 1/4	108	…	…	…	…	1 9/16	40	2 3/4	70
	7.40	0.2913	7.400	3 1/16	78	4 3/8	111	…	…	…	…	1 9/16	40	2 3/4	70
M		0.2950	7.493	3 1/16	78	4 3/8	111	…	…	…	…	1 9/16	40	2 3/4	70
	7.50	0.2953	7.500	3 1/16	78	4 3/8	111	4	102	6 3/8	162	1 9/16	40	2 3/4	70
19/64		0.2969	7.541	3 1/16	78	4 3/8	111	4	102	6 3/8	162	1 9/16	40	2 3/4	70
	7.60	0.2992	7.600	3 1/16	78	4 3/8	111	…	…	…	…	1 5/8	41	2 13/16	71
N		0.3020	7.671	3 1/16	78	4 3/8	111	…	…	…	…	1 5/8	41	2 13/16	71
	7.70	0.3031	7.700	3 3/16	81	4 1/2	114	…	…	…	…	1 5/8	41	2 13/16	71
	7.80	0.3071	7.800	3 3/16	81	4 1/2	114	4	102	6 3/8	162	1 5/8	41	2 13/16	71
	7.90	0.3110	7.900	3 3/16	81	4 1/2	114	…	…	…	…	1 5/8	41	2 13/16	71
5/16		0.3125	7.938	3 3/16	81	4 1/2	114	4	102	6 3/8	162	1 5/8	41	2 13/16	71
	8.00	0.3150	8.000	3 3/16	81	4 1/2	114	4 1/8	105	6 1/2	165	1 11/16	43	2 15/16	75
O		0.3160	8.026	3 3/16	81	4 1/2	114	…	…	…	…	1 11/16	43	2 15/16	75
	8.10	0.3189	8.100	3 5/16	84	4 5/8	117	…	…	…	…	1 11/16	43	2 15/16	75
	8.20	0.3228	8.200	3 5/16	84	4 5/8	117	4 1/8	105	6 1/2	165	1 11/16	43	2 15/16	75
P		0.3230	8.204	3 5/16	84	4 5/8	117	…	…	…	…	1 11/16	43	2 15/16	75
	8.30	0.3268	8.300	3 5/16	84	4 5/8	117	…	…	…	…	1 11/16	43	2 15/16	75
21/64		0.3281	8.334	3 5/16	84	4 5/8	117	4 1/8	105	6 1/2	165	1 11/16	43	2 15/16	75
	8.40	0.3307	8.400	3 7/16	87	4 3/4	121	…	…	…	…	1 11/16	43	3	76
Q		0.3320	8.433	3 7/16	87	4 3/4	121	…	…	…	…	1 11/16	43	3	76
	8.50	0.3346	8.500	3 7/16	87	4 3/4	121	4 1/8	105	6 1/2	165	1 11/16	43	3	76
	8.60	0.3386	8.600	3 7/16	87	4 3/4	121	…	…	…	…	1 11/16	43	3	76
R		0.3390	8.611	3 7/16	87	4 3/4	121	…	…	…	…	1 11/16	43	3	76
	8.70	0.3425	8.700	3 7/16	87	4 3/4	121	…	…	…	…	1 11/16	43	3	76
11/32		0.3438	8.733	3 7/16	87	4 3/4	121	4 1/8	105	6 1/2	165	1 11/16	43	3	76
	8.80	0.3465	8.800	3 1/2	89	4 7/8	124	4 1/4	108	6 3/4	171	1 3/4	44	3 1/16	78
S		0.3480	8.839	3 1/2	89	4 7/8	124	…	…	…	…	1 3/4	44	3 1/16	78
	8.90	0.3504	8.900	3 1/2	89	4 7/8	124	…	…	…	…	1 3/4	44	3 1/16	78
	9.00	0.3543	9.000	3 1/2	89	4 7/8	124	4 1/4	108	6 3/4	171	1 3/4	44	3 1/16	78
T		0.3580	9.093	3 1/2	89	4 7/8	124	…	…	…	…	1 3/4	44	3 1/16	78
	9.10	0.3583	9.100	3 1/2	89	4 7/8	124	…	…	…	…	1 3/4	44	3 1/16	78
23/64		0.3594	9.129	3 1/2	89	4 7/8	124	4 1/4	108	6 3/4	171	1 3/4	44	3 1/16	78
	9.20	0.3622	9.200	3 5/8	92	5	127	4 1/4	108	6 3/4	171	1 13/16	46	3 1/8	79
	9.30	0.3661	9.300	3 5/8	92	5	127	…	…	…	…	1 13/16	46	3 1/8	79
U		0.3680	9.347	3 5/8	92	5	127	…	…	…	…	1 13/16	46	3 1/8	79
	9.40	0.3701	9.400	3 5/8	92	5	127	…	…	…	…	1 13/16	46	3 1/8	79
	9.50	0.3740	9.500	3 5/8	92	5	127	4 1/4	108	6 3/4	171	1 13/16	46	3 1/8	79
3/8		0.3750	9.525	3 5/8	92	5	127	4 1/4	108	6 3/4	171	1 13/16	46	3 1/8	79
V		0.3770	9.576	3 5/8	92	5	127	…	…	…	…	1 7/8	48	3 1/4	83
	9.60	0.3780	9.600	3 3/4	95	5 1/8	130	…	…	…	…	1 7/8	48	3 1/4	83
	9.70	0.3819	9.700	3 3/4	95	5 1/8	130	…	…	…	…	1 7/8	48	3 1/4	83
	9.80	0.3858	9.800	3 3/4	95	5 1/8	130	4 3/8	111	7	178	1 7/8	48	3 1/4	83
W		0.3860	9.804	3 3/4	95	5 1/8	130	…	…	…	…	1 7/8	48	3 1/4	83
	9.90	0.3898	9.900	3 3/4	95	5 1/8	130	…	…	…	…	1 7/8	48	3 1/4	83
25/64		0.3906	9.921	3 3/4	95	5 1/8	130	4 3/8	111	7	178	1 7/8	48	3 1/4	83
	10.00	0.3937	10.000	3 3/4	95	5 1/8	130	4 3/8	111	7	178	1 15/16	49	3 5/16	84

Table 7. *(Continued)* **ANSI Straight Shank Twist Drills — Jobbers Length through 17.5 mm, Taper Length through 12.7 mm, and Screw Machine Length through 25.4 mmDiameter** *ANSI/ASME B94.11M-1993*

Drill Diameter, D^a				Jobbers Length				Taper Length				Screw Machine Length			
Fraction No. or Ltr.	mm	Equivalent		Flute		Overall		Flute		Overall		Flute		Overall	
		Decimal In.	mm	*F* In.	*F* mm	*L* In.	*L* mm	*F* In.	*F* mm	*L* In.	*L* mm	*F* In.	*F* mm	*L* In.	*L* mm
X		0.3970	10.084	3 3/4	95	5 1/8	130	…	…	…	…	1 15/16	49	3 5/16	84
	10.20	0.4016	10.200	3 7/8	98	5 1/4	133	4 3/8	111	7	178	1 15/16	49	3 5/16	84
Y		0.4040	10.262	3 7/8	98	5 1/4	133	…	…	…	…	1 15/16	49	3 5/16	84
13/32		0.4062	10.317	3 7/8	98	5 1/4	133	4 3/8	111	7	178	1 15/16	49	3 5/16	84
Z		0.4130	10.490	3 7/8	98	5 1/4	133	…	…	…	…	2	51	3 3/8	86
	10.50	0.4134	10.500	3 7/8	98	5 1/4	133	4 5/8	117	7 1/4	184	2	51	3 3/8	86
27/64		0.4219	10.716	3 15/16	100	5 3/8	137	4 5/8	117	7 1/4	184	2	51	3 3/8	86
	10.80	0.4252	10.800	4 1/16	103	5 1/2	140	4 5/8	117	7 1/4	184	2 1/16	52	3 7/16	87
	11.00	0.4331	11.000	4 1/16	103	5 1/2	140	4 5/8	117	7 1/4	184	2 1/16	52	3 7/16	87
7/16		0.4375	11.112	4 1/16	103	5 1/2	140	4 5/8	117	7 1/4	184	2 1/16	52	3 7/16	87
	11.20	0.4409	11.200	4 3/16	106	5 5/8	143	4 3/4	121	7 1/2	190	2 1/8	54	3 9/16	90
	11.50	0.4528	11.500	4 3/16	106	5 5/8	143	4 3/4	121	7 1/2	190	2 1/8	54	3 9/16	90
29/64		0.4531	11.509	4 3/16	106	5 5/8	143	4 3/4	121	7 1/2	190	2 1/8	54	3 9/16	90
	11.80	0.4646	11.800	4 5/16	110	5 3/4	146	4 3/4	121	7 1/2	190	2 1/8	54	3 5/8	92
15/32		0.4688	11.908	4 5/16	110	5 3/4	146	4 3/4	121	7 1/2	190	2 1/8	54	3 5/8	92
	12.00	0.4724	12.000	4 3/8	111	5 7/8	149	4 3/4	121	7 3/4	197	2 3/16	56	3 11/16	94
	12.20	0.4803	12.200	4 3/8	111	5 7/8	149	4 3/4	121	7 3/4	197	2 3/16	56	3 11/16	94
31/64		0.4844	12.304	4 3/8	111	5 7/8	149	4 3/4	121	7 3/4	197	2 3/16	56	3 11/16	94
	12.50	0.4921	12.500	4 1/2	114	6	152	4 3/4	121	7 3/4	197	2 1/4	57	3 3/4	95
1/2		0.5000	12.700	4 1/2	114	6	152	4 3/4	121	7 3/4	197	2 1/4	57	3 3/4	95
	12.80	0.5039	12.800	4 1/2	114	6	152	…	…	…	…	2 3/8	60	3 7/8	98
	13.00	0.5118	13.000	4 1/2	114	6	152	…	…	…	…	2 3/8	60	3 7/8	98
33/64		0.5156	13.096	4 13/16	122	6 5/8	168	…	…	…	…	2 3/8	60	3 7/8	98
	13.20	0.5197	13.200	4 13/16	122	6 5/8	168	…	…	…	…	2 3/8	60	3 7/8	98
17/32		0.5312	13.492	4 13/16	122	6 5/8	168	…	…	…	…	2 3/8	60	3 7/8	98
	13.50	0.5315	13.500	4 13/16	122	6 5/8	168	…	…	…	…	2 3/8	60	3 7/8	98
	13.80	0.5433	13.800	4 13/16	122	6 5/8	168	…	…	…	…	2 1/2	64	4	102
35/64		0.5469	13.891	4 13/16	122	6 5/8	168	…	…	…	…	2 1/2	64	4	102
	14.00	0.5512	14.000	4 13/16	122	6 5/8	168	…	…	…	…	2 1/2	64	4	102
	14.25	0.5610	14.250	4 13/16	122	6 5/8	168	…	…	…	…	2 1/2	64	4	102
9/16		0.5625	14.288	4 13/16	122	6 5/8	168	…	…	…	…	2 1/2	64	4	102
	14.50	0.5709	14.500	4 13/16	122	6 5/8	168	…	…	…	…	2 5/8	67	4 1/8	105
37/64		0.5781	14.684	4 13/16	122	6 5/8	168	…	…	…	…	2 5/8	67	4 1/8	105
	14.75	0.5807	14.750	5 3/16	132	7 1/8	181	…	…	…	…	2 5/8	67	4 1/8	105
	15.00	0.5906	15.000	5 3/16	132	7 1/8	181	…	…	…	…	2 5/8	67	4 1/8	105
19/32		0.5938	15.083	5 3/16	132	7 1/8	181	…	…	…	…	2 5/8	67	4 1/8	105
	15.25	0.6004	15.250	5 3/16	132	7 1/8	181	…	…	…	…	2 3/4	70	4 1/4	108
39/64		0.6094	15.479	5 3/16	132	7 1/8	181	…	…	…	…	2 3/4	70	4 1/4	108
	15.50	0.6102	15.500	5 3/16	132	7 1/8	181	…	…	…	…	2 3/4	70	4 1/4	108
	15.75	0.6201	15.750	5 3/16	132	7 1/8	181	…	…	…	…	2 3/4	70	4 1/4	108
5/8		0.6250	15.875	5 3/16	132	7 1/8	181	…	…	…	…	2 3/4	70	4 1/4	108
	16.00	0.6299	16.000	5 3/16	132	7 1/8	181	…	…	…	…	2 7/8	73	4 1/2	114
	16.25	0.6398	16.250	5 3/16	132	7 1/8	181	…	…	…	…	2 7/8	73	4 1/2	114
41/64		0.6406	16.271	5 3/16	132	7 1/8	181	…	…	…	…	2 7/8	73	4 1/2	144
	16.50	0.6496	16.500	5 3/16	132	7 1/8	181	…	…	…	…	2 7/8	73	4 1/2	114
21/32		0.6562	16.669	5 3/16	132	7 1/8	181	…	…	…	…	2 7/8	73	4 1/2	114

Table 7. *(Continued)* **ANSI Straight Shank Twist Drills — Jobbers Length through 17.5 mm, Taper Length through 12.7 mm, and Screw Machine Length through 25.4 mmDiameter** *ANSI/ASME B94.11M-1993*

Drill Diameter, D[a]				Jobbers Length				Taper Length				Screw Machine Length			
		Equivalent		Flute		Overall		Flute		Overall		Flute		Overall	
Fraction No. or Ltr.	mm	Decimal In.	mm	F In.	F mm	L In.	L mm	F In.	F mm	L In.	L mm	F In.	F mm	L In.	L mm
	16.75	0.6594	16.750	5⅝	143	7⅝	194	…	…	…	…	2⅞	73	4½	114
	17.00	0.6693	17.000	5⅝	143	7⅝	194	…	…	…	…	2⅞	73	4½	114
43/64		0.6719	17.066	5⅝	143	7⅝	194	…	…	…	…	2⅞	73	4½	114
	17.25	0.6791	17.250	5⅝	143	7⅝	194	…	…	…	…	2⅞	73	4½	114
11/16		0.6875	17.462	5⅝	143	7⅝	194	…	…	…	…	2⅞	73	4½	114
	17.50	0.6890	17.500	5⅝	143	7⅝	194	…	…	…	…	3	76	4¾	121
45/64		0.7031	17.859	…	…	…	…	…	…	…	…	3	76	4¾	121
	18.00	0.7087	18.000	…	…	…	…	…	…	…	…	3	76	4¾	121
23/32		0.7188	18.258	…	…	…	…	…	…	…	…	3	76	4¾	121
	18.50	0.7283	18.500	…	…	…	…	…	…	…	…	3⅛	79	5	127
47/64		0.7344	18.654	…	…	…	…	…	…	…	…	3⅛	79	5	127
	19.00	0.7480	19.000	…	…	…	…	…	…	…	…	3⅛	79	5	127
3/4		0.7500	19.050	…	…	…	…	…	…	…	…	3⅛	79	5	127
49/64		0.7656	19.446	…	…	…	…	…	…	…	…	3¼	83	5⅛	130
	19.50	0.7677	19.500	…	…	…	…	…	…	…	…	3¼	83	5⅛	130
25/32		0.7812	19.845	…	…	…	…	…	…	…	…	3¼	83	5⅛	130
	20.00	0.7879	20.000	…	…	…	…	…	…	…	…	3⅜	86	5¼	133
51/64		0.7969	20.241	…	…	…	…	…	…	…	…	3⅜	86	5¼	133
	20.50	0.8071	20.500	…	…	…	…	…	…	…	…	3⅜	86	5¼	133
13/16		0.8125	20.638	…	…	…	…	…	…	…	…	3⅜	86	5¼	133
	21.00	0.8268	21.000	…	…	…	…	…	…	…	…	3½	89	5⅜	137
53/64		0.8281	21.034	…	…	…	…	…	…	…	…	3½	89	5⅜	137
27/32		0.8438	21.433	…	…	…	…	…	…	…	…	3½	89	5⅜	137
	21.50	0.8465	21.500	…	…	…	…	…	…	…	…	3½	89	5⅜	137
55/64		0.8594	21.829	…	…	…	…	…	…	…	…	3½	89	5⅜	137
	22.00	0.8661	22.000	…	…	…	…	…	…	…	…	3½	89	5⅜	137
7/8		0.8750	22.225	…	…	…	…	…	…	…	…	3½	89	5⅜	137
	22.50	0.8858	22.500	…	…	…	…	…	…	…	…	3⅝	92	5⅝	143
57/64		0.8906	22.621	…	…	…	…	…	…	…	…	3⅝	92	5⅝	143
	23.00	0.9055	23.000	…	…	…	…	…	…	…	…	3⅝	92	5⅝	143
29/32		0.9062	23.017	…	…	…	…	…	…	…	…	3⅝	92	5⅝	143
59/64		0.9219	23.416	…	…	…	…	…	…	…	…	3¾	95	5¾	146
	23.50	0.9252	23.500	…	…	…	…	…	…	…	…	3¾	95	5¾	146
15/16		0.9375	23.812	…	…	…	…	…	…	…	…	3¾	95	5¾	146
	24.00	0.9449	24.000	…	…	…	…	…	…	…	…	3⅞	98	5⅞	149
61/64		0.9531	24.209	…	…	…	…	…	…	…	…	3⅞	98	5⅞	149
	24.50	0.9646	24.500	…	…	…	…	…	…	…	…	3⅞	98	5⅞	149
31/32		0.9688	24.608	…	…	…	…	…	…	…	…	3⅞	98	5⅞	149
	25.00	0.9843	25.000	…	…	…	…	…	…	…	…	4	102	6	152
63/64		0.9844	25.004	…	…	…	…	…	…	…	…	4	102	6	152
1		1.0000	25.400	…	…	…	…	…	…	…	…	4	102	6	152

[a] Fractional inch, number, letter, and metric sizes.

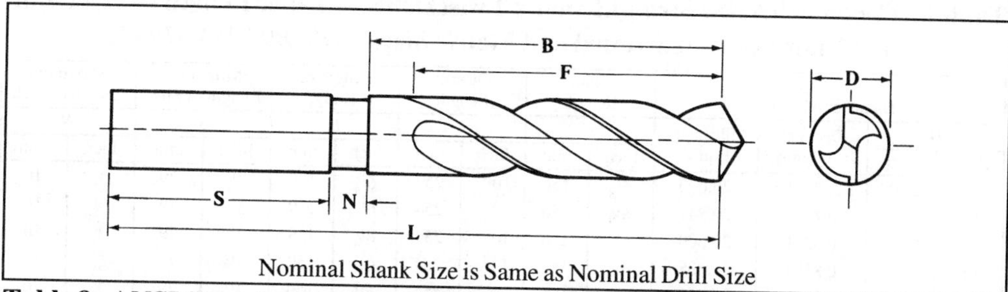

Nominal Shank Size is Same as Nominal Drill Size

Table 8. ANSI Straight Shank Twist Drills — Taper Length — Over ½ in. (12.7 mm) Dia., Fractional and Metric Sizes *ANSI B94.11M-1993*

Diameter of Drill				Flute Length		Overall Length		Length of Body		Minimum Length of Shk.		Maximum Length of Neck	
D		Decimal Inch Equiv.	Millimeter Equiv.	*F*		*L*		*B*		*S*		*N*	
Frac.	mm			Inch	mm	Inch	mm	Inch	mm	Inch	mm	Inch	mm
	12.80	0.5039	12.800	4¾	121	8	203	4⅞	124	2⅝	66	½	13
	13.00	0.5117	13.000	4¾	121	8	203	4⅞	124	2⅝	66	½	13
33/64		0.5156	13.096	4¾	121	8	203	4⅞	124	2⅝	66	½	13
	13.20	0.5197	13.200	4¾	121	8	203	4⅞	124	2⅝	66	½	13
17/32		0.5312	13.492	4¾	121	8	203	4⅞	124	2⅝	66	½	13
	13.50	0.5315	13.500	4¾	121	8	203	4⅞	124	2⅝	66	½	13
	13.80	0.5433	13.800	4⅞	124	8¼	210	5	127	2¾	70	½	13
35/64		0.5419	13.891	4⅞	124	8¼	210	5	127	2¾	70	½	13
	14.00	0.5512	14.000	4⅞	124	8¼	210	5	127	2¾	70	½	13
	14.25	0.5610	14.250	4⅞	124	8¼	210	5	127	2¾	70	½	13
9/16		0.5625	14.288	4⅞	124	8¼	210	5	127	2¾	70	½	13
	14.50	0.5709	14.500	4⅞	124	8¾	222	5	127	3⅛	79	⅝	16
37/64		0.5781	14.684	4⅞	124	8¾	222	5	127	3⅛	79	⅝	16
	14.75	0.5807	14.750	4⅞	124	8¾	222	5	127	3⅛	79	⅝	16
	15.00	0.5906	15.000	4⅞	124	8¾	222	5	127	3⅛	79	⅝	16
19/32		0.5938	15.083	4⅞	124	8¾	222	5	127	3⅛	79	⅝	16
	15.25	0.6004	15.250	4⅞	124	8¾	222	5	127	3⅛	79	⅝	16
39/64		0.6094	15.479	4⅞	124	8¾	222	5	127	3⅛	79	⅝	16
	15.50	0.6102	15.500	4⅞	124	8¾	222	5	127	3⅛	79	⅝	16
	15.75	0.6201	15.750	4⅞	124	8¾	222	5	127	3⅛	79	⅝	16
⅝		0.6250	15.875	4⅞	124	8¾	222	5	127	3⅛	79	⅝	16
	16.00	0.6299	16.000	5⅛	130	9	228	5¼	133	3⅛	79	⅝	16
	16.25	0.6398	16.250	5⅛	130	9	228	5¼	133	3⅛	79	⅝	16
41/64		0.6406	16.271	5⅛	130	9	228	5¼	133	3⅛	79	⅝	16
	16.50	0.6496	16.500	5⅛	130	9	228	5¼	133	3⅛	79	⅝	16
21/32		0.6562	16.667	5⅛	130	9	228	5¼	133	3⅛	79	⅝	16
	16.75	0.6594	16.750	5⅜	137	9¼	235	5½	140	3⅛	79	⅝	16
	17.00	0.6693	17.000	5⅜	137	9¼	235	5½	140	3⅛	79	⅝	16
43/64		0.6719	17.066	5⅜	137	9¼	235	5½	140	3⅛	79	⅝	16
	17.25	0.6791	17.250	5⅜	137	9¼	235	5½	140	3⅛	79	⅝	16
11/16		0.6875	17.462	5⅜	137	9¼	235	5½	140	3⅛	79	⅝	16
	17.50	0.6890	17.500	5⅝	143	9½	241	5¾	146	3⅛	79	⅝	16
45/64		0.7031	17.859	5⅝	143	9½	241	5¾	146	3⅛	79	⅝	16
	18.00	0.7087	18.000	5⅝	143	9½	241	5¾	146	3⅛	79	⅝	16
23/32		0.7188	18.258	5⅝	143	9½	241	5¾	146	3⅛	79	⅝	16
	18.50	0.7283	18.500	5⅞	149	9¾	247	6	152	3⅛	79	⅝	16
47/64		0.7344	18.654	5⅞	149	9¾	247	6	152	3⅛	79	⅝	16
	19.00	0.7480	19.000	5⅞	149	9¾	247	6	152	3⅛	79	⅝	16
¾		0.7500	19.050	5⅞	149	9¾	247	6	152	3⅛	79	⅝	16
49/64		0.7656	19.446	6	152	9⅞	251	6⅛	156	3⅛	79	⅝	16
	19.50	0.7677	19.500	6	152	9⅞	251	6⅛	156	3⅛	79	⅝	16
25/32		0.7812	19.842	6	152	9⅞	251	6⅛	156	3⅛	79	⅝	16

Table 8. *(Continued)* **ANSI Straight Shank Twist Drills — Taper Length — Over ½ in. (12.7 mm) Dia., Fractional and Metric Sizes** *ANSI B94.11M-1993*

Diameter of Drill				Flute Length		Overall Length		Length of Body		Minimum Length of Shk.		Maximum Length ofNeck	
D		Decimal Inch Equiv.	Millimeter Equiv.	*F*		*L*		*B*		*S*		*N*	
Frac.	mm			Inch	mm	Inch	mm	Inch	mm	Inch	mm	Inch	mm
	20.00	0.7874	20.000	6⅛	156	10	254	6¼	159	3⅛	79	⅝	16
51/64		0.7969	20.241	6⅛	156	10	254	6¼	159	3⅛	79	⅝	16
	20.50	0.8071	20.500	6⅛	156	10	254	6¼	159	3⅛	79	⅝	16
13/16		0.8125	20.638	6⅛	156	10	254	6¼	159	3⅛	79	⅝	16
	21.00	0.8268	21.000	6⅛	156	10	254	6¼	159	3⅛	79	⅝	16
53/64		0.8281	21.034	6⅛	156	10	254	6¼	159	3⅛	79	⅝	16
27/32		0.8438	21.433	6⅛	156	10	254	6¼	159	3⅛	79	⅝	16
	21.50	0.8465	21.500	6⅛	156	10	254	6¼	159	3⅛	79	⅝	16
55/64		0.8594	21.829	6⅛	156	10	254	6¼	159	3⅛	79	⅝	16
	22.00	0.8661	22.000	6⅛	156	10	254	6¼	159	3⅛	79	⅝	16
7/8		0.8750	22.225	6⅛	156	10	254	6¼	159	3⅛	79	⅝	16
	22.50	0.8858	22.500	6⅛	156	10	254	6¼	159	3⅛	79	⅝	16
57/64		0.8906	22.621	6⅛	156	10	254	6¼	159	3⅛	79	⅝	16
	23.00	0.9055	23.000	6⅛	156	10	254	6¼	159	3⅛	79	⅝	16
29/32		0.9062	23.017	6⅛	156	10	254	6¼	159	3⅛	79	⅝	16
59/64		0.9219	23.416	6⅛	156	10¾	273	6¼	159	3⅞	98	⅝	16
	23.50	0.9252	23.500	6⅛	156	10¾	273	6¼	159	3⅞	98	⅝	16
15/16		0.9375	23.812	6⅛	156	10¾	273	6¼	159	3⅞	98	⅝	16
	24.00	0.9449	24.000	6⅜	162	11	279	6½	165	3⅞	98	⅝	16
61/64		0.9531	24.209	6⅜	162	11	279	6½	165	3⅞	98	⅝	16
	24.50	0.9646	24.500	6⅜	162	11	279	6½	165	3⅞	98	⅝	16
31/32		0.9688	24.608	6⅜	162	11	279	6½	165	3⅞	98	⅝	16
	25.00	0.9843	25.000	6⅜	162	11	279	6½	165	3⅞	98	⅝	16
63/64		0.9844	25.004	6⅜	162	11	279	6½	165	3⅞	98	⅝	16
1		1.0000	25.400	6⅜	162	11	279	6½	165	3⅞	98	⅝	16
	25.50	1.0039	25.500	6½	165	11⅛	282	6⅝	168	3⅞	98	⅝	16
1 1/64		1.0156	25.796	6½	165	11⅛	282	6⅝	168	3⅞	98	⅝	16
	26.00	1.0236	26.000	6½	165	11⅛	282	6⅝	168	3⅞	98	⅝	16
1 1/32		1.0312	26.192	6½	165	11⅛	282	6⅝	168	3⅞	98	⅝	16
	26.50	1.0433	26.560	6⅝	168	11¼	286	6¾	172	3⅞	98	⅝	16
1 3/64		1.0469	26.591	6⅝	168	11¼	286	6¾	172	3⅞	98	⅝	16
1 1/16		1.0625	26.988	6⅝	168	11¼	286	6¾	172	3⅞	98	⅝	16
	27.00	1.0630	27.000	6⅝	168	11¼	286	6¾	172	3⅞	98	⅝	16
1 5/64		1.0781	27.384	6⅞	175	11½	292	7	178	3⅞	98	⅝	16
	27.50	1.0827	27.500	6⅞	175	11½	292	7	178	3⅞	98	⅝	16
1 3/32		1.0938	27.783	6⅞	175	11½	292	7	178	3⅞	98	⅝	16
	28.00	1.1024	28.000	7⅛	181	11¾	298	7¼	184	3⅞	98	⅝	16
1 7/64		1.1094	28.179	7⅛	181	11¾	298	7¼	184	3⅞	98	⅝	16
	28.50	1.1220	28.500	7⅛	181	11¾	298	7¼	184	3⅞	98	⅝	16
1⅛		1.1250	28.575	7⅛	181	11¾	298	7¼	184	3⅞	98	⅝	16
1 9/64		1.1406	28.971	7¼	184	11⅞	301	7⅜	187	3⅞	98	⅝	16
	29.00	1.1417	29.000	7¼	184	11⅞	301	7⅜	187	3⅞	98	⅝	16
1 5/32		1.1562	29.367	7¼	184	11⅞	301	7⅜	187	3⅞	98	⅝	16
	29.50	1.1614	29.500	7⅜	187	12	305	7½	191	3⅞	98	⅝	16
1 11/64		1.1719	29.766	7⅜	187	12	305	7½	191	3⅞	98	⅝	16
	30.00	1.1811	30.000	7⅜	187	12	305	7½	191	3⅞	98	⅝	16
1 3/16		1.1875	30.162	7⅜	187	12	305	7½	191	3⅞	98	⅝	16
	30.50	1.2008	30.500	7½	190	12⅛	308	7⅝	194	3⅞	98	⅝	16
1 13/64		1.2031	30.559	7½	190	12⅛	308	7⅝	194	3⅞	98	⅝	16
1 7/32		1.2188	30.958	7½	190	12⅛	308	7⅝	194	3⅞	98	⅝	16
	31.00	1.2205	31.000	7⅞	200	12½	317	8	203	3⅞	98	⅝	16
1 15/64		1.2344	31.354	7⅞	200	12½	317	8	203	3⅞	98	⅝	16
	31.50	1.2402	31.500	7⅞	200	12½	317	8	203	3⅞	98	⅝	16

Table 8. *(Continued)* **ANSI Straight Shank Twist Drills — Taper Length — Over ½ in. (12.7 mm) Dia., Fractional and Metric Sizes** *ANSI B94.11M-1993*

Diameter of Drill				Flute Length		Overall Length		Length of Body		Minimum Length of Shk.		Maximum Length of Neck	
D		Decimal Inch Equiv.	Millimeter Equiv.	*F*		*L*		*B*		*S*		*N*	
Frac.	mm			Inch	mm	Inch	mm	Inch	mm	Inch	mm	Inch	mm
1¼		1.2500	31.750	7⅞	200	12½	317	8	203	3⅞	98	⅝	16
	32.00	1.2598	32.000	8½	216	14⅛	359	8⅝	219	4⅞	124	⅝	16
	32.50	1.2795	32.500	8½	216	14⅛	359	8⅝	219	4⅞	124	⅝	16
1 9/32		1.2812	32.542	8½	216	14⅛	359	8⅝	219	4⅞	124	⅝	16
	33.00	1.2992	33.000	8⅝	219	14¼	362	8¾	222	4⅞	124	⅝	16
1 5/16		1.3125	33.338	8⅝	219	14¼	362	8¾	222	4⅞	124	⅝	16
	33.50	1.3189	33.500	8¾	222	14⅜	365	8⅞	225	4⅞	124	⅝	16
	34.00	1.3386	34.000	8¾	222	14⅜	365	8⅞	225	4⅞	124	⅝	16
1 11/32		1.3438	34.133	8¾	222	14⅜	365	8⅞	225	4⅞	124	⅝	16
	34.50	1.3583	34.500	8⅞	225	14½	368	9	229	4⅞	124	⅝	16
1⅜		1.3750	34.925	8⅞	225	14½	368	9	229	4⅞	124	⅝	16
	35.00	1.3780	35.000	9	229	14⅝	372	9⅛	232	4⅞	124	⅝	16
	35.50	1.3976	35.500	9	229	14⅝	372	9⅛	232	4⅞	124	⅝	16
1 13/32		1.4062	35.717	9	229	14⅝	372	9⅛	232	4⅞	124	⅝	16
	36.00	1.4173	36.000	9⅛	232	14¾	375	9¼	235	4⅞	124	⅝	16
	36.50	1.4370	36.500	9⅛	232	14¾	375	9¼	235	4⅞	124	⅝	16
1 7/16		1.4375	36.512	9⅛	232	14¾	375	9¼	235	4⅞	124	⅝	16
	37.00	1.4567	37.000	9¼	235	14⅞	378	9⅜	238	4⅞	124	⅝	16
1 15/32		1.4688	37.308	9¼	235	14⅞	378	9⅜	238	4⅞	124	⅝	16
	37.50	1.4764	37.500	9⅜	238	15	381	9½	241	4⅞	124	⅝	16
	38.00	1.4961	38.000	9⅜	238	15	381	9½	241	4⅞	124	⅝	16
1½		1.5000	38.100	9⅜	238	15	381	9½	241	4⅞	124	⅝	16
1 9/16		1.5625	39.688	9⅝	244	15¼	387	9¾	247	4⅞	124	⅝	16
1⅝		1.6250	41.275	9⅞	251	15⅝	397	10	254	4⅞	124	¾	19
1¾		1.7500	44.450	10½	267	16¼	413	10⅝	270	4⅞	124	¾	19

Table 9. American National Standard Tangs for Straight Shank Drills *ANSI/ASME B94.11M-1993*

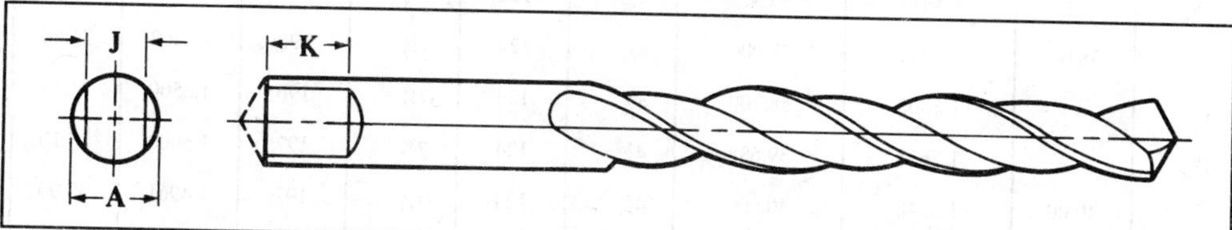

Nominal Diameter of Drill Shank		Thickness of Tang				Length of Tang	
A		*J*				*K*	
Inches	Millimeters	Inches		Millimeters		Inches	Millimeters
		Max.	Min.	Max.	Min.		
⅛ thru 3/16	3.18 thru 4.76	0.094	0.090	2.39	2.29	9/32	7.0
over 3/16 thru ¼	over 4.76 thru 6.35	0.122	0.118	3.10	3.00	5/16	8.0
over ¼ thru 5/16	over 6.35 thru 7.94	0.162	0.158	4.11	4.01	11/32	8.5
over 5/16 thru ⅜	over 7.94 thru 9.53	0.203	0.199	5.16	5.06	⅜	9.5
over ⅜ thru 15/32	over 9.53 thru 11.91	0.243	0.239	6.17	6.07	7/16	11.0
over 15/32 thru 9/16	over 11.91 thru 14.29	0.303	0.297	7.70	7.55	½	12.5
over 9/16 thru 21/32	over 14.29 thru 16.67	0.373	0.367	9.47	9.32	9/16	14.5
over 21/32 thru ¾	over 16.67 thru 19.05	0.443	0.437	11.25	11.10	⅝	16.0
over ¾ thru ⅞	over 19.05 thru 22.23	0.514	0.508	13.05	12.90	11/16	17.5
over ⅞ thru 1	over 22.23 thru 25.40	0.609	0.601	15.47	15.27	¾	19.0
over 1 thru 1 3/16	over 25.40 thru 30.16	0.700	0.692	17.78	17.58	13/16	20.5
over 1 3/16 thru 1⅜	over 30.16 thru 34.93	0.817	0.809	20.75	20.55	⅞	22.0

To fit split sleeve collet type drill drivers. See page 850.

Table 10. American National Standard Straight Shank Twist Drills — Screw Machine Length — Over 1 in. (25.4 mm) Dia. *ANSI/ASME B94.11M-1993*

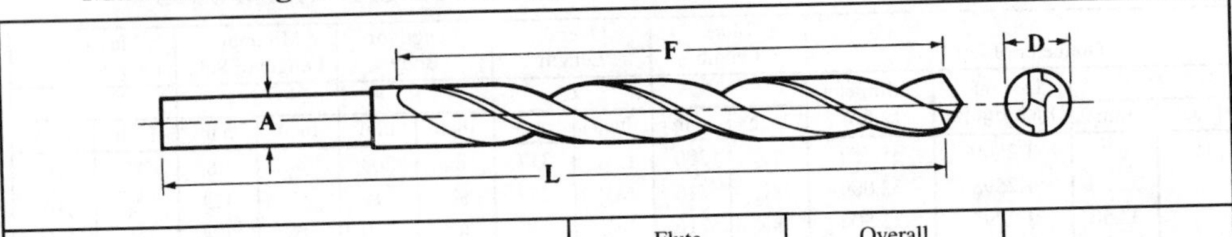

Diameter of Drill				Flute Length		Overall Length		Shank Diameter	
D		Decimal Inch Equivalent	Millimeter Equivalent	*F*		*L*		*A*	
Frac.	mm			Inch	mm	Inch	mm	Inch	mm
	25.50	1.0039	25.500	4	102	6	152	0.9843	25.00
	26.00	1.0236	26.000	4	102	6	152	0.9843	25.00
1 1/16		1.0625	26.988	4	102	6	152	1.0000	25.40
	28.00	1.1024	28.000	4	102	6	152	0.9843	25.00
1 1/8		1.1250	28.575	4	102	6	152	1.0000	25.40
	30.00	1.1811	30.000	4 1/4	108	6 5/8	168	0.9843	25.00
1 3/16		1.1875	30.162	4 1/4	108	6 5/8	168	1.0000	25.40
1 1/4		1.2500	31.750	4 3/8	111	6 3/4	171	1.0000	25.40
	32.00	1.2598	32.000	4 3/8	111	7	178	1.2402	31.50
1 5/16		1.3125	33.338	4 3/8	111	7	178	1.2500	31.75
	34.00	1.3386	34.000	4 1/2	114	7 1/8	181	1.2402	31.50
1 3/8		1.3750	34.925	4 1/2	114	7 1/8	181	1.2500	31.75
	36.00	1.4173	36.000	4 3/4	121	7 3/8	187	1.2402	31.50
1 7/16		1.4375	36.512	4 3/4	121	7 3/8	187	1.2500	31.75
	38.00	1.4961	38.000	4 7/8	124	7 1/2	190	1.2402	31.50
1 1/2		1.5000	38.100	4 7/8	124	7 1/2	190	1.2500	31.75
1 9/16		1.5625	39.688	4 7/8	124	7 3/4	197	1.5000	38.10
	40.00	1.5748	40.000	4 7/8	124	7 3/4	197	1.4961	38.00
1 5/8		1.6250	41.275	4 7/8	124	7 3/4	197	1.5000	38.10
	42.00	1.6535	42.000	5 1/8	130	8	203	1.4961	38.00
1 11/16		1.6875	42.862	5 1/8	130	8	203	1.5000	38.10
	44.00	1.7323	44.000	5 1/8	130	8	203	1.4961	38.00
1 3/4		1.7500	44.450	5 1/8	130	8	203	1.5000	38.10
	46.00	1.8110	46.000	5 3/8	137	8 1/4	210	1.4961	38.00
1 13/16		1.8125	46.038	5 3/8	137	8 1/4	210	1.5000	38.10
1 7/8		1.8750	47.625	5 3/8	137	8 1/4	210	1.5000	38.10
	48.00	1.8898	48.000	5 5/8	143	8 1/2	216	1.4961	38.00
1 15/16		1.9375	49.212	5 5/8	143	8 1/2	216	1.5000	38.10
	50.00	1.9685	50.000	5 5/8	143	8 1/2	216	1.4961	38.00
2		2.0000	50.800	5 5/8	143	8 1/2	216	1.5000	38.10

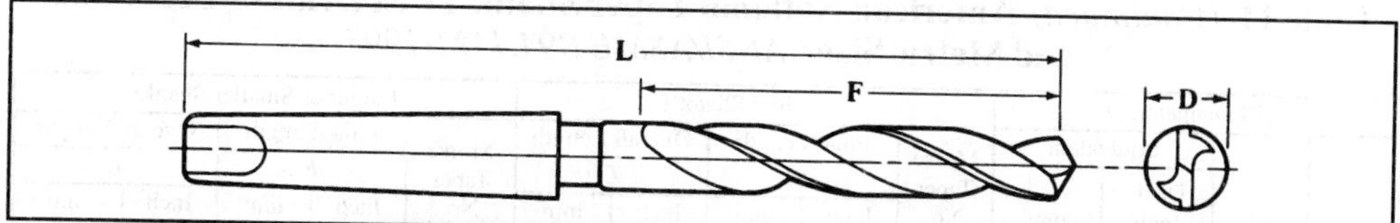

Table 11. American National Taper Shank Twist Drills — Fractional and Metric Sizes *ANSI/ASME B94.11M-1993*

Drill Diameter, D				Regular Shank					Larger or Smaller Shank[a]				
		Equivalent		Morse Taper No.	Flute Length F		Overall Length L		Morse Taper No.	Flute Length F		Overall Length L	
Fraction	mm	Deci. Inch	mm		Inch	mm	Inch	mm		Inch	mm	Inch	mm
	3.00	0.1181	3.000	1	1⅞	48	5⅛	130	…	…	…	…	…
⅛		0.1250	3.175	1	1⅞	48	5⅛	130	…	…	…	…	…
	3.20	0.1260	3.200	1	2⅛	54	5⅜	137	…	…	…	…	…
	3.50	0.1378	3.500	1	2⅛	54	5⅜	137	…	…	…	…	…
9/64		0.1406	3.571	1	2⅛	54	5⅜	137	…	…	…	…	…
	3.80	0.1496	3.800	1	2⅛	54	5⅜	137	…	…	…	…	…
5/32		0.1562	3.967	1	2⅛	54	5⅜	137	…	…	…	…	…
	4.00	0.1575	4.000	1	2½	64	5¾	146	…	…	…	…	…
	4.20	0.1654	4.200	1	2½	64	5¾	146	…	…	…	…	…
11/64		0.1719	4.366	1	2½	64	5¾	146	…	…	…	…	…
	4.50	0.1772	4.500	1	2½	64	5¾	146	…	…	…	…	…
3/16		0.1875	4.762	1	2½	64	5¾	146	…	…	…	…	…
	4.80	0.1890	4.800	1	2¾	70	6	152	…	…	…	…	…
	5.00	0.1969	5.000	1	2¾	70	6	152	…	…	…	…	…
13/64		0.2031	5.159	1	2¾	70	6	152	…	…	…	…	…
	5.20	0.2047	5.200	1	2¾	70	6	152	…	…	…	…	…
	5.50	0.2165	5.500	1	2¾	70	6	152	…	…	…	…	…
7/32		0.2183	5.558	1	2¾	70	6	152	…	…	…	…	…
	5.80	0.2223	5.800	1	2⅞	73	6⅛	156	…	…	…	…	…
15/64		0.2344	5.954	1	2⅞	73	6⅛	156	…	…	…	…	…
	6.00	0.2362	6.000	1	2⅞	73	6⅛	156	…	…	…	…	…
	6.20	0.2441	6.200	1	2⅞	73	6⅛	156	…	…	…	…	…
¼		0.2500	6.350	1	2⅞	73	6⅛	156	…	…	…	…	…
	6.50	0.2559	6.500	1	3	76	6¼	159	…	…	…	…	…
17/64		0.2656	6.746	1	3	76	6¼	159	…	…	…	…	…
	6.80	0.2677	6.800	1	3	76	6¼	159	…	…	…	…	…
	7.00	0.2756	7.000	1	3	76	6¼	159	…	…	…	…	…
9/32		0.2812	7.142	1	3	76	6¼	159	…	…	…	…	…
	7.20	0.2835	7.200	1	3⅛	79	6⅜	162	…	…	…	…	…
	7.50	0.2953	7.500	1	3⅛	79	6⅜	162	…	…	…	…	…
19/64		0.2969	7.541	1	3⅛	79	6⅜	162	…	…	…	…	…
	7.80	0.3071	7.800	1	3⅛	79	6⅜	162	…	…	…	…	…
5/16		0.3125	7.938	1	3⅛	79	6⅜	162	…	…	…	…	…
	8.00	0.3150	8.000	1	3¼	83	6½	165	…	…	…	…	…
	8.20	0.3228	8.200	1	3¼	83	6½	165	…	…	…	…	…
21/64		0.3281	8.334	1	3¼	83	6½	165	…	…	…	…	…
	8.50	0.3346	8.500	1	3¼	83	6½	165	…	…	…	…	…
11/32		0.3438	8.733	1	3¼	83	6½	165	…	…	…	…	…
	8.80	0.3465	8.800	1	3½	89	6¾	171	…	…	…	…	…
	9.00	0.3543	9.000	1	3½	89	6¾	171	…	…	…	…	…
23/64		0.3594	9.129	1	3½	89	6¾	171	…	…	…	…	…
	9.20	0.3622	9.200	1	3½	89	6¾	171	…	…	…	…	…
	9.50	0.3740	9.500	1	3½	89	6¾	171	…	…	…	…	…
⅜		0.3750	9.525	1	3½	89	6¾	171	2	3½	89	7⅜	187
	9.80	0.3858	9.800	1	3⅝	92	7	178	…	…	…	…	…
25/64		0.3906	9.921	1	3⅝	92	7	178	2	3⅝	92	7½	190
	10.00	0.3937	10.000	1	3⅝	92	7	178	…	…	…	…	…

Table 11. *(Continued)* **American National Taper Shank Twist Drills — Fractional and Metric Sizes** *ANSI/ASME B94.11M-1993*

Drill Diameter, D				Regular Shank					Larger or Smaller Shank[a]				
		Equivalent		Morse Taper No.	Flute Length F		Overall Length L		Morse Taper No.	Flute Length F		Overall Length L	
Fraction	mm	Deci. Inch	mm		Inch	mm	Inch	mm		Inch	mm	Inch	mm
	10.20	0.4016	10.200	1	3⅝	92	7	178	…	…	…	…	…
13/32		0.4062	10.320	1	3⅝	92	7	178	2	3⅝	92	7½	190
	10.50	0.4134	10.500	1	3⅞	98	7¼	184	…	…	…	…	…
27/64		0.4219	10.716	1	3⅞	98	7¼	184	2	3⅞	98	7¾	197
	10.80	0.4252	10.800	1	3⅞	98	7¼	184	…	…	…	…	…
	11.00	0.4331	11.000	1	3⅞	98	7¼	184	…	…	…	…	…
7/16		0.4375	11.112	1	3⅞	98	7¼	184	2	3⅞	98	7¾	197
	11.20	0.4409	11.200	1	4⅛	105	7½	190	…	…	…	…	…
	11.50	0.4528	11.500	1	4⅛	105	7½	190	…	…	…	…	…
29/64		0.4531	11.509	1	4⅛	105	7½	190	2	4⅛	105	8	203
	11.80	0.4646	11.800	1	4⅛	105	7½	190	…	…	…	…	…
15/32		0.4688	11.906	1	4⅛	105	7½	190	2	4⅛	105	8	203
	12.00	0.4724	12.000	2	4⅜	111	8¼	210	1	4⅜	111	7¾	197
	12.20	0.4803	12.200	2	4⅜	111	8¼	210	1	4⅜	111	7¾	197
31/64		0.4844	12.304	2	4⅜	111	8¼	210	1	4⅜	111	7¾	197
	12.50	0.4921	12.500	2	4⅜	111	8¼	210	1	4⅜	111	7¾	197
1/2		0.5000	12.700	2	4⅜	111	8¼	210	1	4⅜	111	7¾	197
	12.80	0.5034	12.800	2	4⅝	117	8½	216	1	4⅝	117	8	203
	13.00	0.5118	13.000	2	4⅝	117	8½	216	1	4⅝	117	8	203
33/64		0.5156	13.096	2	4⅝	117	8½	216	1	4⅝	117	8	203
	13.20	0.5197	13.200	2	4⅝	117	8½	216	1	4⅝	117	8	203
17/32		0.5312	13.492	2	4⅝	117	8½	216	1	4⅝	117	8	203
	13.50	0.5315	13.500	2	4⅝	117	8½	216	1	4⅝	117	8	203
	13.80	0.5433	13.800	2	4⅞	124	8¾	222	1	4⅞	124	8¼	210
35/64		0.5469	13.891	2	4⅞	124	8¾	222	1	4⅞	124	8¼	210
	14.00	0.5572	14.000	2	4⅞	124	8¾	222	1	4⅞	124	8¼	210
	14.25	0.5610	14.250	2	4⅞	124	8¾	222	1	4⅞	124	8¼	210
9/16		0.5625	14.288	2	4⅞	124	8¾	222	1	4⅞	124	8¼	210
	14.50	0.5709	14.500	2	4⅞	124	8¾	222	…	…	…	…	…
37/64		0.5781	14.684	2	4⅞	124	8¾	222	…	…	…	…	…
	14.75	0.5807	14.750	2	4⅞	124	8¾	222	…	…	…	…	…
	15.00	0.5906	15.000	2	4⅞	124	8¾	222	…	…	…	…	…
19/32		0.5938	15.083	2	4⅞	124	8¾	222	…	…	…	…	…
	15.25	0.6004	15.250	2	4⅞	124	8¾	222	…	…	…	…	…
39/64		0.6094	15.479	2	4⅞	124	8¾	222	…	…	…	…	…
	15.50	0.6102	15.500	2	4⅞	124	8¾	222	…	…	…	…	…
	15.75	0.6201	15.750	2	4⅞	124	8¾	222	…	…	…	…	…
5/8		0.6250	15.875	2	4⅞	124	8¾	222	…	…	…	…	…
	16.00	0.6299	16.000	2	5⅛	130	9	229	…	…	…	…	…
	16.25	0.6398	16.250	2	5⅛	130	9	229	…	…	…	…	…
41/64		0.6406	16.271	2	5⅛	130	9	229	3	5⅛	130	9¾	248
	16.50	0.6496	16.500	2	5⅛	130	9	229	…	…	…	…	…
21/32		0.6562	16.667	2	5⅛	130	9	229	3	5⅛	130	9¾	248
	16.75	0.6594	16.750	2	5⅜	137	9¼	235	…	…	…	…	…
	17.00	0.6693	17.000	2	5⅜	137	9¼	235	…	…	…	…	…
43/64		0.6719	17.066	2	5⅜	137	9¼	235	3	5⅜	137	10	254
	17.25	0.6791	17.250	2	5⅜	137	9¼	235	…	…	…	…	…
11/16		0.6875	17.462	2	5⅜	137	9¼	235	3	5⅜	137	10	254
	17.50	0.6880	17.500	2	5⅝	143	9½	241	…	…	…	…	…
45/64		0.7031	17.859	2	5⅝	143	9½	241	3	5⅝	143	10¼	260
	18.00	0.7087	18.000	2	5⅝	143	9½	241	…	…	…	…	…
23/32		0.7188	18.258	2	5⅝	143	9½	241	3	5⅝	143	10¼	260
	18.50	0.7283	18.500	2	5⅞	149	9¾	248	…	…	…	…	…
47/64		0.7344	18.654	2	5⅞	149	9¾	248	3	5⅞	149	10½	267

Table 11. *(Continued)* **American National Taper Shank Twist Drills — Fractional and Metric Sizes** *ANSI/ASME B94.11M-1993*

Drill Diameter, *D*				Regular Shank					Larger or Smaller Shank[a]				
		Equivalent		Morse Taper No.	Flute Length *F*		Overall Length *L*		Morse Taper No.	Flute Length *F*		Overall Length *L*	
Fraction	mm	Deci. Inch	mm		Inch	mm	Inch	mm		Inch	mm	Inch	mm
	19.00	0.7480	19.000	2	5⅞	149	9¾	248	…	…	…	…	…
¾		0.7500	19.050	2	5⅞	149	9¾	248	3	5⅞	149	10½	267
49/64		0.7656	19.446	2	6	152	9⅞	251	3	6	152	10⅝	270
	19.50	0.7677	19.500	2	6	152	9⅞	251	…	…	…	…	…
25/32		0.7812	19.843	2	6	152	9⅞	251	3	6	152	10⅝	270
	20.00	0.7821	20.000	3	6⅛	156	10¾	273	2	6⅛	156	10	254
51/64		0.7969	20.241	3	6⅛	156	10¾	273	2	6⅛	156	10	254
	20.50	0.8071	20.500	3	6⅛	156	10¾	273	2	6⅛	156	10	254
13/16		0.8125	20.638	3	6⅛	156	10¾	273	2	6⅛	156	10	254
	21.00	0.8268	21.000	3	6⅛	156	10¾	273	2	6⅛	156	10	254
53/64		0.8281	21.034	3	6⅛	156	10¾	273	2	6⅛	156	10	254
27/32		0.8438	21.433	3	6⅛	156	10¾	273	2	6⅛	156	10	254
	21.50	0.8465	21.500	3	6⅛	156	10¾	273	2	6⅛	156	10	254
55/64		0.8594	21.829	3	6⅛	156	10¾	273	2	6⅛	156	10	254
	22.00	0.8661	22.000	3	6⅛	156	10¾	273	2	6⅛	156	10	254
⅞		0.8750	22.225	3	6⅛	156	10¾	273	2	6⅛	156	10	254
	22.50	0.8858	22.500	3	6⅛	156	10¾	273	2	6⅛	156	10	254
57/64		0.8906	22.621	3	6⅛	156	10¾	273	2	6⅛	156	10	254
	23.00	0.9055	23.000	3	6⅛	156	10¾	273	2	6⅛	156	10	254
29/32		0.9062	23.017	3	6⅛	156	10¾	273	2	6⅛	156	10	254
59/64		0.9219	23.416	3	6⅛	156	10¾	273	…	…	…	…	…
	23.50	0.9252	23.500	3	6⅛	156	10¾	273	…	…	…	…	…
15/16		0.9375	23.813	3	6⅛	156	10¾	273	…	…	…	…	…
	24.00	0.9449	24.000	3	6⅜	162	11	279	…	…	…	…	…
61/64		0.9531	24.209	3	6⅜	162	11	279	…	…	…	…	…
	24.50	0.9646	24.500	3	6⅜	162	11	279	…	…	…	…	…
31/32		0.9688	24.608	3	6⅜	162	11	279	…	…	…	…	…
	25.00	0.9843	25.000	3	6⅜	162	11	279	…	…	…	…	…
63/64		0.9844	25.004	3	6⅜	162	11	279	…	…	…	…	…
1		1.0000	25.400	3	6⅜	162	11	279	4	6⅜	162	12	305
	25.50	1.0039	25.500	3	6½	165	11⅛	283	…	…	…	…	…
1 1/64		1.0156	25.796	3	6½	165	11⅛	283	…	…	…	…	…
	26.00	1.0236	26.000	3	6½	165	11⅛	283	…	…	…	…	…
1 1/32		1.0312	26.192	3	6½	165	11⅛	283	4	6½	165	12⅛	308
	26.50	1.0433	26.500	3	6⅝	168	11¼	286	…	…	…	…	…
1 3/64		1.0469	26.591	3	6⅝	168	11¼	286	…	…	…	…	…
1 1/16		1.0625	26.988	3	6⅝	168	11¼	286	4	6⅝	168	12¼	311
	27.00	1.0630	27.000	3	6⅝	168	11¼	286	…	…	…	…	…
1 5/64		1.0781	27.384	4	6⅞	175	12½	318	3	6⅞	175	11½	292
	27.50	1.0827	27.500	4	6⅞	175	12½	318	3	6⅞	175	11½	292
1 3/32		1.0938	27.783	4	6⅞	175	12½	318	3	6⅞	175	11½	292
	28.00	1.1024	28.000	4	7⅛	181	12¾	324	3	7⅛	181	11¾	298
1 7/64		1.1094	28.179	4	7⅛	181	12¾	324	3	7⅛	181	11¾	298
	28.50	1.1220	28.500	4	7⅛	181	12¾	324	3	7⅛	181	11¾	298
1⅛		1.1250	28.575	4	7⅛	181	12¾	324	3	7⅛	181	11¾	298
1 9/64		1.1406	28.971	4	7¼	184	12⅞	327	3	7¼	184	11⅞	302
	29.00	1.1417	29.000	4	7¼	184	12⅞	327	3	7¼	184	11⅞	302
1 5/32		1.1562	29.367	4	7¼	184	12⅞	327	3	7¼	184	11⅞	302
	29.50	1.1614	29.500	4	7⅜	187	13	330	3	7⅜	187	12	305
1 11/64		1.1719	29.797	4	7⅜	187	13	330	3	7⅜	187	12	305
	30.00	1.1811	30.000	4	7⅜	187	13	330	3	7⅜	187	12	305
1 3/16		1.1875	30.162	4	7⅜	187	13	330	3	7⅜	187	12	305
	30.50	1.2008	30.500	4	7½	190	13⅛	333	3	7½	190	12⅛	308
1 13/64		1.2031	30.559	4	7½	190	13⅛	333	3	7½	190	12⅛	308

Table 11. *(Continued)* **American National Taper Shank Twist Drills — Fractional and Metric Sizes** *ANSI/ASME B94.11M-1993*

Drill Diameter, *D*				Regular Shank					Larger or Smaller Shank[a]				
		Equivalent		Morse Taper No.	Flute Length *F*		Overall Length *L*		Morse Taper No.	Flute Length *F*		Overall Length *L*	
Fraction	mm	Deci. Inch	mm		Inch	mm	Inch	mm		Inch	mm	Inch	mm
1 7/32		1.2188	30.958	4	7½	190	13⅛	333	3	7½	190	12⅛	308
	31.00	1.2205	31.000	4	7⅞	200	13½	343	3	7⅞	200	12½	318
1 15/64		1.2344	31.354	4	7⅞	200	13½	343	3	7⅞	200	12½	318
	31.50	1.2402	31.500	4	7⅞	200	13½	343	3	7⅞	200	12½	318
1¼		1.2500	31.750	4	7⅞	200	13½	343	3	7⅞	200	12½	318
	32.00	1.2598	32.000	4	8½	216	14⅛	359	…	…	…	…	…
1 17/64		1.2656	32.146	4	8½	216	14⅛	359	…	…	…	…	…
	32.50	1.2795	32.500	4	8½	216	14⅛	359	…	…	…	…	…
1 9/32		1.2812	32.542	4	8½	216	14⅛	359	…	…	…	…	…
1 19/64		1.2969	32.941	4	8⅝	219	14¼	362	…	…	…	…	…
	33.00	1.2992	33.000	4	8⅝	219	14¼	362	…	…	…	…	…
1 5/16		1.3125	33.338	4	8⅝	219	14¼	362	…	…	…	…	…
	33.50	1.3189	33.500	4	8¾	222	14⅜	365	…	…	…	…	…
1 21/64		1.3281	33.734	4	8¾	222	14⅜	365	…	…	…	…	…
	34.00	1.3386	34.000	4	8¾	222	14⅜	365	…	…	…	…	…
1 11/32		1.3438	34.133	4	8¾	222	14⅜	365	…	…	…	…	…
	34.50	1.3583	34.500	4	8⅞	225	14½	368	…	…	…	…	…
1 23/64		1.3594	34.529	4	8⅞	225	14½	368	…	…	…	…	…
1⅜		1.3750	34.925	4	8⅞	225	14½	368	…	…	…	…	…
	35.00	1.3780	35.000	4	9	229	14⅝	371	…	…	…	…	…
1 25/64		1.3906	35.321	4	9	229	14⅝	371	…	…	…	…	…
	35.50	1.3976	35.500	4	9	229	14⅝	371	…	…	…	…	…
1 13/32		1.4062	35.717	4	9	229	14⅝	371	…	…	…	…	…
	36.00	1.4173	36.000	4	9⅛	232	14¾	375	…	…	…	…	…
1 27/64		1.4219	36.116	4	9⅛	232	14¾	375	…	…	…	…	…
	36.50	1.4370	36.500	4	9⅛	232	14¾	375	…	…	…	…	…
1 7/16		1.4375	36.512	4	9⅛	232	14¾	375	…	…	…	…	…
1 29/64		1.4531	36.909	4	9¼	235	14⅞	378	…	…	…	…	…
	37.00	1.4567	37.000	4	9¼	235	14⅞	378	…	…	…	…	…
1 15/32		1.4688	37.308	4	9¼	235	14⅞	378	…	…	…	…	…
	37.50	1.4764	37.500	4	9⅜	238	15	381	…	…	…	…	…
1 31/64		1.4844	37.704	4	9⅜	238	15	381	…	…	…	…	…
	38.00	1.4961	38.000	4	9⅜	238	15	381	…	…	…	…	…
1½		1.5000	38.100	4	9⅜	238	15	381	…	…	…	…	…
1 33/64		1.5156	38.496	…	…	…	…	…	4	9¾	238	15	381
1 17/32		1.5312	38.892	5	9⅜	238	16⅜	416	4	9⅜	238	15	381
	39.00	1.5354	39.000	5	9⅝	244	16⅝	422	4	9⅝	244	15¼	387
1 35/64		1.5469	39.291	…	…	…	…	…	4	9⅝	244	15¼	387
1 9/16		1.5625	39.688	5	9⅝	244	16⅝	422	4	9⅝	244	15¼	387
	40.00	1.5748	40.000	5	9⅞	251	16⅞	429	4	9⅞	251	15½	394
1 37/64		1.5781	40.084	…	…	…	…	…	4	9⅞	251	15½	394
1 19/32		1.5938	40.483	5	9⅞	251	16⅞	429	4	9⅞	251	15½	394
1 39/64		1.6094	40.879	…	…	…	…	…	4	10	254	15⅝	397
	41.00	1.6142	41.000	5	10	254	17	432	4	10	254	15⅝	397
1⅝		1.6250	41.275	5	10	254	17	432	4	10	254	15⅝	397
1 41/64		1.6406	41.671	…	…	…	…	…	4	10⅛	257	15¾	400
	42.00	1.6535	42.000	5	10⅛	257	17⅛	435	4	10⅛	257	15¾	400
1 21/32		1.6562	42.067	5	10⅛	257	17⅛	435	4	10⅛	257	15¾	400
1 43/64		1.6719	42.466	…	…	…	…	…	4	10⅛	257	15¾	400
1 11/16		1.6875	42.862	5	10⅛	257	17⅛	435	4	10⅛	257	15¾	400
	43.00	1.6929	43.000	5	10⅛	257	17⅛	435	4	10⅛	257	15¾	400
1 45/64		1.7031	43.259	…	…	…	…	…	4	10⅛	257	15¾	400
1 23/32		1.7188	43.658	5	10⅛	257	17⅛	435	4	10⅛	257	15¾	400
	44.00	1.7323	44.000	5	10⅛	257	17⅛	435	4	10⅜	264	16¼	413

Table 11. *(Continued)* **American National Taper Shank Twist Drills — Fractional and Metric Sizes** *ANSI/ASME B94.11M-1993*

Drill Diameter, D				Regular Shank					Larger or Smaller Shank[a]				
		Equivalent		Morse Taper No.	Flute Length F		Overall Length L		Morse Taper No.	Flute Length F		Overall Length L	
Fraction	mm	Deci. Inch	mm		Inch	mm	Inch	mm		Inch	mm	Inch	mm
1 47/64		1.7344	44.054	…	…	…	…	…	4	10 3/8	264	16 1/4	413
1 3/4		1.7500	44.450	5	10 1/8	257	17 1/8	435	4	10 3/4	264	16 1/4	413
	45.00	1.7717	45.000	5	10 1/8	257	17 1/8	435	4	10 3/8	264	16 1/4	413
1 25/32		1.7812	45.242	5	10 1/8	257	17 1/8	435	4	10 3/8	264	16 1/4	413
	46.00	1.8110	46.000	5	10 1/8	257	17 1/8	435	4	10 3/8	264	16 1/4	413
1 13/16		1.8125	46.038	5	10 1/8	257	17 1/8	435	4	10 3/8	264	16 1/4	413
1 27/32		1.8438	46.833	5	10 1/8	257	17 1/8	435	4	10 3/8	264	16 1/4	413
	47.00	1.8504	47.000	5	10 3/8	264	17 3/8	441	4	10 1/2	267	16 1/2	419
1 7/8		1.8750	47.625	5	10 3/8	264	17 3/8	441	4	10 1/2	267	16 1/2	419
	48.00	1.8898	48.000	5	10 3/8	264	17 3/8	441	4	10 1/2	267	16 1/2	419
1 29/32		1.9062	48.417	5	10 3/8	264	17 3/8	441	4	10 1/2	267	16 1/2	419
	49.00	1.9291	49.000	5	10 3/8	264	17 3/8	441	4	10 5/8	270	16 5/8	422
1 15/16		1.9375	49.212	5	10 3/8	264	17 3/8	441	4	10 5/8	270	16 5/8	422
	50.00	1.9625	50.000	5	10 3/8	264	17 3/8	441	4	10 5/8	270	16 5/8	422
1 31/32		1.9688	50.008	5	10 3/8	264	17 3/8	441	4	10 5/8	270	16 5/8	422
2		2.0000	50.800	5	10 3/8	264	17 3/8	441	4	10 5/8	270	16 5/8	422
	51.00	2.0079	51.000	5	10 3/8	264	17 3/8	441	…	…	…	…	…
2 1/32		2.0312	51.592	5	10 3/8	264	17 3/8	441	…	…	…	…	…
	52.00	2.0472	52.000	5	10 1/4	260	17 3/8	441	…	…	…	…	…
2 1/16		2.0625	52.388	5	10 1/4	260	17 3/8	441	…	…	…	…	…
	53.00	2.0866	53.000	5	10 1/4	260	17 3/8	441	…	…	…	…	…
2 3/32		2.0938	53.183	5	10 1/4	260	17 3/8	441	…	…	…	…	…
2 1/8		2.1250	53.975	5	10 1/4	260	17 3/8	441	…	…	…	…	…
	54.00	2.1260	54.000	5	10 1/4	260	17 3/8	441	…	…	…	…	…
2 5/32		2.1562	54.767	5	10 1/4	260	17 3/8	441	…	…	…	…	…
	55.00	2.1654	55.000	5	10 1/4	260	17 3/8	441	…	…	…	…	…
2 3/16		2.1875	55.563	5	10 1/4	260	17 3/4	441	…	…	…	…	…
	56.00	2.2000	56.000	5	10 1/8	257	17 3/8	441	…	…	…	…	…
2 7/32		2.2188	56.358	5	10 1/8	257	17 3/8	441	…	…	…	…	…
	57.00	2.2441	57.000	5	10 1/8	257	17 3/8	441	…	…	…	…	…
2 1/4		2.2500	57.150	5	10 1/8	257	17 3/8	441	…	…	…	…	…
	58.00	2.2835	58.000	5	10 1/8	257	17 3/8	441	…	…	…	…	…
2 5/16		2.3125	58.738	5	10 1/8	257	17 3/8	441	…	…	…	…	…
	59.00	2.3228	59.000	5	10 1/8	257	17 3/8	441	…	…	…	…	…
	60.00	2.3622	60.000	5	10 1/8	257	17 3/8	441	…	…	…	…	…
2 3/8		2.3750	60.325	5	10 1/8	257	17 3/8	441	…	…	…	…	…
	61.00	2.4016	61.000	5	11 1/4	286	18 3/4	476	…	…	…	…	…
2 7/16		2.4375	61.912	5	11 1/4	286	18 3/4	476	…	…	…	…	…
	62.00	2.4409	62.000	5	11 1/4	286	18 3/4	476	…	…	…	…	…
	63.00	2.4803	63.000	5	11 1/4	286	18 3/4	476	…	…	…	…	…
2 1/2		2.5000	63.500	5	11 1/4	286	18 3/4	476	…	…	…	…	…
	64.00	2.5197	64.000	5	11 7/8	302	19 1/2	495	…	…	…	…	…
	65.00	2.5591	65.000	5	11 7/8	302	19 1/2	495	…	…	…	…	…
2 9/16		2.5625	65.088	5	11 7/8	302	19 1/2	495	…	…	…	…	…
	66.00	2.5984	66.000	5	11 7/8	302	19 1/2	495	…	…	…	…	…
2 5/8		2.6250	66.675	5	11 7/8	302	19 1/2	495	…	…	…	…	…
	67.00	2.6378	67.000	5	12 3/4	324	20 3/8	518	…	…	…	…	…
	68.00	2.6772	68.000	5	12 3/4	324	20 3/8	518	…	…	…	…	…
2 11/16		2.6875	68.262	5	12 3/4	324	20 3/8	518	…	…	…	…	…
	69.00	2.7165	69.000	5	12 3/4	324	20 3/8	518	…	…	…	…	…
2 3/4		2.7500	69.850	5	12 3/4	324	20 3/8	518	…	…	…	…	…
	70.00	2.7559	70.000	5	13 3/8	340	21 1/8	537	…	…	…	…	…
	71.00	2.7953	71.000	5	13 3/8	340	21 1/8	537	…	…	…	…	…
2 13/16		2.8125	71.438	5	13 3/8	340	21 1/8	537	…	…	…	…	…

Table 11. *(Continued)* **American National Taper Shank Twist Drills — Fractional and Metric Sizes** *ANSI/ASME B94.11M-1993*

Drill Diameter, D				Regular Shank					Larger or Smaller Shank[a]				
		Equivalent		Morse Taper No.	Flute Length F		Overall Length L		Morse Taper No.	Flute Length F		Overall Length L	
Fraction	mm	Deci. Inch	mm		Inch	mm	Inch	mm		Inch	mm	Inch	mm
	72.00	2.8346	72.000	5	13 3/8	340	21 1/8	537	…	…	…	…	…
	73.00	2.8740	73.000	5	13 3/8	340	21 1/8	537	…	…	…	…	…
2 7/8		2.8750	73.025	5	13 3/8	340	21 1/8	537	…	…	…	…	…
	74.00	2.9134	74.000	5	14	356	21 3/4	552	…	…	…	…	…
2 15/16		2.9375	74.612	5	14	356	21 3/4	552	…	…	…	…	…
	75.00	2.9528	75.000	5	14	356	21 3/4	552	…	…	…	…	…
	76.00	2.9921	76.000	5	14	356	21 3/4	552	…	…	…	…	…
3		3.0000	76.200	5	14	356	21 3/4	552	…	…	…	…	…
	77.00	3.0315	77.000	6	14 5/8	371	24 1/2	622	5	14 1/4	362	22	559
	78.00	3.0709	78.000	6	14 5/8	371	24 1/2	622	5	14 1/4	362	22	559
3 1/8		3.1250	79.375	6	14 5/8	371	24 1/2	622	5	14 1/4	362	22	559
3 1/4		3.2500	82.550	6	15 1/2	394	25 1/2	648	5	15 1/4	387	23	584
3 1/2		3.5000	88.900	…	…	…	…	…	5	16 1/4	413	24	610

[a] Larger or smaller than regular shank.

Table 12. American National Standard Combined Drills and Countersinks — Plain and Bell Types *ANSI/ASME B94.11M-1993*

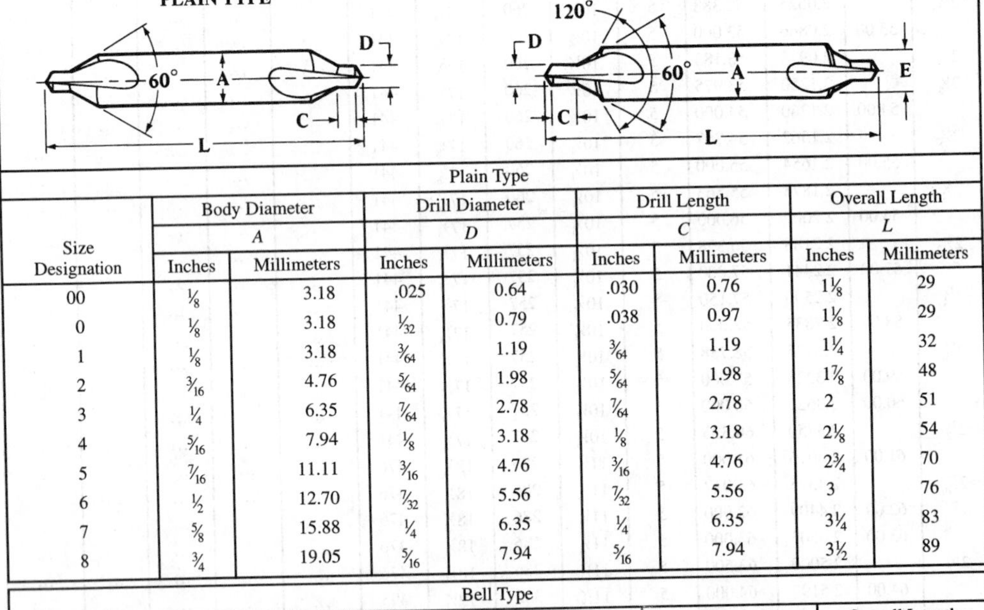

Plain Type								
Size Designation	Body Diameter A		Drill Diameter D		Drill Length C		Overall Length L	
	Inches	Millimeters	Inches	Millimeters	Inches	Millimeters	Inches	Millimeters
00	1/8	3.18	.025	0.64	.030	0.76	1 1/8	29
0	1/8	3.18	1/32	0.79	.038	0.97	1 1/8	29
1	1/8	3.18	3/64	1.19	3/64	1.19	1 1/4	32
2	3/16	4.76	5/64	1.98	5/64	1.98	1 7/8	48
3	1/4	6.35	7/64	2.78	7/64	2.78	2	51
4	5/16	7.94	1/8	3.18	1/8	3.18	2 1/8	54
5	7/16	11.11	3/16	4.76	3/16	4.76	2 3/4	70
6	1/2	12.70	7/32	5.56	7/32	5.56	3	76
7	5/8	15.88	1/4	6.35	1/4	6.35	3 1/4	83
8	3/4	19.05	5/16	7.94	5/16	7.94	3 1/2	89

Bell Type										
Size Designation	Body Diameter A		Drill Diameter D		Bell Diameter E		Drill Length C		Overall Length L	
	Inches	mm	Inches	mm	Inches	mm	Inches	mm	Inches	mm
11	1/8	3.18	3/64	1.19	0.10	2.5	3/64	1.19	1 1/4	32
12	3/16	4.76	1/16	1.59	0.15	3.8	1/16	1.59	1 7/8	48
13	1/4	6.35	3/32	2.38	0.20	5.1	3/32	2.38	2	51
14	5/16	7.94	7/64	2.78	0.25	6.4	7/64	2.78	2 1/8	54
15	7/16	11.11	5/32	3.97	0.35	8.9	5/32	3.97	2 3/4	70
16	1/2	12.70	3/16	4.76	0.40	10.2	3/16	4.76	3	76
17	5/8	15.88	7/32	5.56	0.50	12.7	7/32	5.56	3 1/4	83
18	3/4	19.05	1/4	6.35	0.60	15.2	1/4	6.35	3 1/2	89

Table 13. American National Standard Three- and Four-Flute Taper Shank Core Drills — Fractional Sizes Only *ANSI/ASME B94.11M-1993*

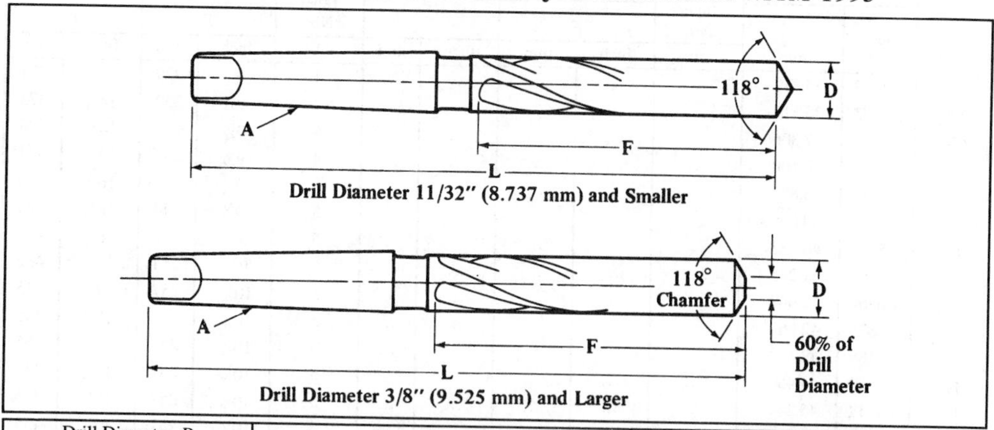

Drill Diameter, *D*			Three-Flute Drills					Four-Flute Drills				
	Equivalent		Morse Taper No.	Flute Length		Overall Length		Morse Taper No.	Flute Length		Overall Length	
				F		*L*			*F*		*L*	
Inch	Deci. Inch	mm	A	Inch	mm	Inch	mm	A	Inch	mm	Inch	mm
1/4	0.2500	6.350	1	2 7/8	73	6 1/8	156	…	…	…	…	…
9/32	0.2812	7.142	1	3	76	6 1/4	159	…	…	…	…	…
5/16	0.3175	7.938	1	3 1/8	79	6 3/8	162	…	…	…	…	…
11/32	0.3438	8.733	1	3 1/4	83	6 1/2	165	…	…	…	…	…
3/8	0.3750	9.525	1	3 1/2	89	6 3/4	171	…	…	…	…	…
13/32	0.4062	10.319	1	3 5/8	92	7	178	…	…	…	…	…
7/16	0.4375	11.112	1	3 7/8	98	7 1/4	184	…	…	…	…	…
15/32	0.4688	11.908	1	4 1/8	105	7 1/2	190	…	…	…	…	…
1/2	0.5000	12.700	2	4 3/8	111	8 1/4	210	2	4 3/8	111	8 1/4	210
17/32	0.5312	13.492	2	4 5/8	117	8 1/2	216	2	4 5/8	117	8 1/2	216
9/16	0.5625	14.288	2	4 7/8	124	8 3/4	222	2	4 7/8	124	8 3/4	222
19/32	0.5938	15.083	2	4 7/8	124	8 3/4	222	2	4 7/8	124	8 3/4	222
5/8	0.6250	15.815	2	4 7/8	124	8 3/4	222	2	4 7/8	124	8 3/4	222
21/32	0.6562	16.668	2	5 1/8	130	9	229	2	5 1/8	130	9	229
11/16	0.6875	17.462	2	5 3/8	137	9 1/4	235	2	5 3/8	137	9 1/4	235
23/32	0.7188	18.258	2	5 5/8	143	9 1/2	241	2	5 5/8	143	9 1/2	241
3/4	0.7500	19.050	2	5 7/8	149	9 3/4	248	2	5 7/8	149	9 3/4	248
25/32	0.7812	19.842	2	6	152	9 7/8	251	2	6	152	9 7/8	251
13/16	0.8125	20.638	3	6 1/8	156	10 3/4	273	3	6 1/8	156	10 3/4	273
27/32	0.8438	21.433	3	6 1/8	156	10 3/4	273	3	6 1/8	156	10 3/4	273
7/8	0.8750	22.225	3	6 1/8	156	10 3/4	273	3	6 1/8	156	10 3/4	273
29/32	0.9062	23.019	3	6 1/8	156	10 3/4	273	3	6 1/8	156	10 3/4	273
15/16	0.9375	23.812	3	6 1/8	156	10 3/4	273	3	6 1/8	156	10 3/4	273
31/32	0.9688	24.608	3	6 3/8	162	11	279	3	6 3/8	162	11	279
1	1.0000	25.400	3	6 3/8	162	11	279	3	6 3/8	162	11	279
1 1/32	1.0312	26.192	3	6 1/2	165	11 1/8	283	3	6 1/2	165	11 1/8	283
1 1/16	1.0625	26.988	3	6 5/8	168	11 1/4	286	3	6 5/8	168	11 1/4	286
1 3/32	1.0938	27.783	4	6 7/8	175	12 1/2	318	4	6 7/8	175	12 1/2	318
1 1/8	1.1250	28.575	4	7 1/8	181	12 3/4	324	4	7 1/8	181	12 3/4	324
1 5/32	1.1562	29.367	4	7 1/4	184	12 7/8	327	4	7 1/4	184	12 7/8	327
1 3/16	1.1875	30.162	4	7 3/8	187	13	330	4	7 3/8	187	13	330
1 7/32	1.2188	30.958	4	7 1/2	190	13 1/8	333	4	7 1/2	190	13 1/8	333
1 1/4	1.2500	31.750	4	7 7/8	200	13 1/2	343	4	7 7/8	200	13 1/2	343
1 9/32	1.2812	32.542	…	…	…	…	…	4	8 1/2	216	14 1/8	359
1 5/16	1.3125	33.338	…	…	…	…	…	4	8 5/8	219	14 1/4	362
1 11/32	1.3438	34.133	…	…	…	…	…	4	8 3/4	222	14 3/8	365
1 3/8	1.3750	34.925	…	…	…	…	…	4	8 7/8	225	14 1/2	368

Drill Diameter, D			Three-Flute Drills					Four-Flute Drills				
	Equivalent		Morse Taper No.	Flute Length F		Overall Length L		Morse Taper No.	Flute Length F		Overall Length L	
Inch	Deci. Inch	mm	A	Inch	mm	Inch	mm	A	Inch	mm	Inch	mm
$1\frac{13}{32}$	1.4062	35.717	…	…	…	…	…	4	9	229	$14\frac{5}{8}$	371
$1\frac{7}{16}$	1.4375	36.512	…	…	…	…	…	4	$9\frac{1}{8}$	232	$14\frac{3}{4}$	375
$1\frac{15}{32}$	1.4688	37.306	…	…	…	…	…	4	$9\frac{1}{4}$	235	$14\frac{7}{8}$	378
$1\frac{1}{2}$	1.5000	38.100	…	…	…	…	…	4	$9\frac{3}{8}$	238	15	381
$1\frac{17}{32}$	1.5312	38.892	…	…	…	…	…	5	$9\frac{3}{8}$	238	$16\frac{3}{8}$	416
$1\frac{9}{16}$	1.5675	39.688	…	…	…	…	…	5	$9\frac{5}{8}$	244	$16\frac{5}{8}$	422
$1\frac{19}{32}$	1.5938	40.483	…	…	…	…	…	5	$9\frac{7}{8}$	251	$16\frac{7}{8}$	429
$1\frac{5}{8}$	1.6250	41.275	…	…	…	…	…	5	10	254	17	432
$1\frac{21}{32}$	1.6562	42.067	…	…	…	…	…	5	$10\frac{1}{8}$	257	$17\frac{1}{8}$	435
$1\frac{11}{16}$	1.6875	42.862	…	…	…	…	…	5	$10\frac{1}{8}$	257	$17\frac{1}{8}$	435
$1\frac{23}{32}$	1.7188	43.658	…	…	…	…	…	5	$10\frac{1}{8}$	257	$17\frac{1}{8}$	435
$1\frac{3}{4}$	1.7500	44.450	…	…	…	…	…	5	$10\frac{1}{8}$	257	$17\frac{1}{8}$	435
$1\frac{25}{32}$	1.7812	45.244	…	…	…	…	…	5	$10\frac{1}{8}$	257	$17\frac{1}{8}$	435
$1\frac{13}{16}$	1.8125	46.038	…	…	…	…	…	5	$10\frac{1}{8}$	257	$17\frac{1}{8}$	435
$1\frac{27}{32}$	1.8438	46.833	…	…	…	…	…	5	$10\frac{1}{8}$	257	$17\frac{1}{8}$	435
$1\frac{7}{8}$	1.8750	47.625	…	…	…	…	…	5	$10\frac{3}{8}$	264	$17\frac{3}{8}$	441
$1\frac{29}{32}$	1.9062	48.417	…	…	…	…	…	5	$10\frac{3}{8}$	264	$17\frac{3}{8}$	441
$1\frac{15}{16}$	1.9375	49.212	…	…	…	…	…	5	$10\frac{3}{8}$	264	$17\frac{3}{8}$	441
$1\frac{31}{32}$	1.9688	50.008	…	…	…	…	…	5	$10\frac{3}{8}$	264	$17\frac{3}{8}$	441
2	2.0000	50.800	…	…	…	…	…	5	$10\frac{3}{8}$	264	$17\frac{3}{8}$	441
$2\frac{1}{8}$	2.1250	53.975	…	…	…	…	…	5	$10\frac{1}{4}$	260	$17\frac{3}{8}$	441
$2\frac{1}{4}$	2.2500	57.150	…	…	…	…	…	5	$10\frac{1}{8}$	257	$17\frac{3}{8}$	441
$2\frac{3}{8}$	2.3750	60.325	…	…	…	…	…	5	$10\frac{1}{8}$	257	$17\frac{3}{8}$	441
$2\frac{1}{2}$	2.5000	63.500	…	…	…	…	…	5	$11\frac{1}{4}$	286	$18\frac{3}{4}$	476

British Standard Combined Drills and Countersinks (Center Drills).—BS 328: Part 2: 1972 (1990) provides dimensions of combined drills and countersinks for center holes. Three types of drill and countersink combinations are shown in this standard but are not given here. These three types will produce center holes without protecting chamfers, with protecting chamfers, and with protecting chamfers of radius form.

American National Standard Drill Drivers — Split-Sleeve, Collet Type

ANSI B94.35-1972 (R1995)

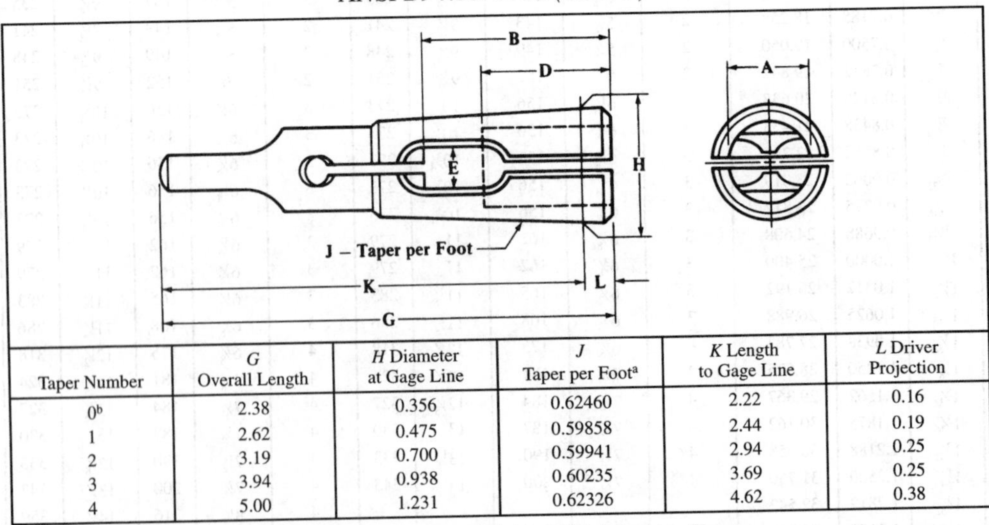

Taper Number	*G* Overall Length	*H* Diameter at Gage Line	*J* Taper per Foot[a]	*K* Length to Gage Line	*L* Driver Projection
0[b]	2.38	0.356	0.62460	2.22	0.16
1	2.62	0.475	0.59858	2.44	0.19
2	3.19	0.700	0.59941	2.94	0.25
3	3.94	0.938	0.60235	3.69	0.25
4	5.00	1.231	0.62326	4.62	0.38

[a] Taper rate in accordance with ANSI/ASME B5.10-1994, Machine Tapers.

[b] Size 0 is not an American National Standard but is included here to meet special needs.

All dimensions are in inches.

Table 14. ANSI Three- and Four-Flute Straight Shank Core Drills — Fractional Sizes Only *ANSI/ASME B94.11M-1993*

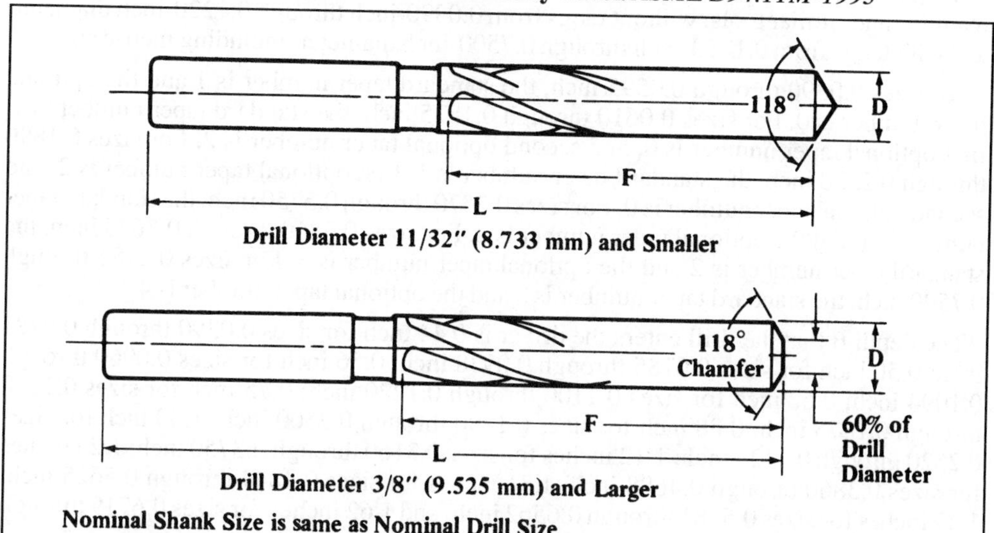

Drill Diameter 11/32″ (8.733 mm) and Smaller

Drill Diameter 3/8″ (9.525 mm) and Larger

Nominal Shank Size is same as Nominal Drill Size

Drill Diameter, *D*			Three-Flute Drills				Four-Flute Drills			
	Equivalent		Flute Length *F*		Overall Length *L*		Flute Length *F*		Overall Length *L*	
Inch	Deci. Inch	mm	Inch	mm	Inch	mm	Inch	mm	Inch	mm
1/4	0.2500	6.350	3 3/4	95	6 1/8	156	…	…	…	…
9/32	0.2812	7.142	3 7/8	98	6 1/4	159	…	…	…	…
5/16	0.3125	7.938	4	102	6 3/8	162	…	…	…	…
11/32	0.3438	8.733	4 1/8	105	6 1/2	165	…	…	…	…
3/8	0.3750	9.525	4 1/8	105	6 3/4	171	…	…	…	…
13/32	0.4062	10.317	4 3/8	111	7	178	…	…	…	…
7/16	0.4375	11.112	4 5/8	117	7 1/4	184	…	…	…	…
15/32	0.4688	11.908	4 3/4	121	7 1/2	190	…	…	…	…
1/2	0.5000	12.700	4 3/4	121	7 3/4	197	4 3/4	121	7 3/4	197
17/32	0.5312	13.492	4 3/4	121	8	203	4 3/4	121	8	203
9/16	0.5625	14.288	4 7/8	124	8 1/4	210	4 7/8	124	8 1/4	210
19/32	0.5938	15.083	4 7/8	124	8 3/4	222	4 7/8	124	8 3/4	222
5/8	0.6250	15.875	4 7/8	124	8 3/4	222	4 7/8	124	8 3/4	222
21/32	0.6562	16.667	5 1/8	130	9	229	5 1/8	130	9	229
11/16	0.6875	17.462	5 3/8	137	9 1/4	235	5 3/8	137	9 1/4	235
23/32	0.7188	18.258	…	…	…	…	5 5/8	143	9 1/2	241
3/4	0.7500	19.050	5 7/8	149	9 3/4	248	5 7/8	149	9 3/4	248
25/32	0.7812	19.842	…	…	…	…	6	152	9 7/8	251
13/16	0.8125	20.638	…	…	…	…	6 1/8	156	10	254
27/32	0.8438	21.433	…	…	…	…	6 1/8	156	10	254
7/8	0.8750	22.225	…	…	…	…	6 1/8	156	10	254
29/32	0.9062	23.017	…	…	…	…	6 1/8	156	10	254
15/16	0.9375	23.812	…	…	…	…	6 1/8	156	10 3/4	273
31/32	0.9688	24.608	…	…	…	…	6 3/8	162	11	279
1	1.0000	25.400	…	…	…	…	6 3/8	162	11	279
1 1/32	1.0312	26.192	…	…	…	…	6 1/2	165	11 1/8	283
1 1/16	1.0625	26.988	…	…	…	…	6 5/8	168	11 1/4	286
1 3/32	1.0938	27.783	…	…	…	…	6 7/8	175	11 1/2	292
1 1/8	1.1250	28.575	…	…	…	…	7 1/8	181	11 3/4	298
1 1/4	1.2500	31.750	…	…	…	…	7 7/8	200	12 1/2	318

Drill Drivers—Split-Sleeve, Collet Type.—American National Standard ANSI B94.35-1972 (R1995) covers split-sleeve, collet-type drivers for driving straight shank drills, reamers, and similar tools, without tangs from 0.0390-inch through 0.1220-inch diameter, and with tangs from 0.1250-inch through 0.7500-inch diameter, including metric sizes.

For sizes 0.0390 through 0.0595 inch, the standard taper number is 1 and the optional taper number is 0. For sizes 0.0610 through 0.1875 inch, the standard taper number is 1, first optional taper number is 0, and second optional taper number is 2. For sizes 0.1890 through 0.2520 inch, the standard taper number is 1, first optional taper number is 2, and second optional taper number is 0. For sizes 0.2570 through 0.3750 inch, the standard taper number is 1 and the optional taper number is 2. For sizes 0.3860 through 0.5625 inch, the standard taper number is 2 and the optional taper number is 3. For sizes 0.5781 through 0.7500 inch, the standard taper number is 3 and the optional taper number is 4.

The depth *B* that the drill enters the driver is 0.44 inch for sizes 0.0390 through 0.0781 inch; 0.50 inch for sizes 0.0785 through 0.0938 inch; 0.56 inch for sizes 0.0960 through 0.1094 inch; 0.62 inch for sizes 0.1100 through 0.1220 inch; 0.75 inch for sizes 0.1250 through 0.1875 inch; 0.88 inch for sizes 0.1890 through 0.2500 inch; 1.00 inch for sizes 0.2520 through 0.3125 inch; 1.12 inches for sizes 0.3160 through 0.3750 inch; 1.25 inches for sizes 0.3860 through 0.4688 inch; 1.31 inches for sizes 0.4844 through 0.5625 inch; 1.47 inches for sizes 0.5781 through 0.6562 inch; and 1.62 inches for sizes 0.6719 through 0.7500 inch.

British Standard Metric Twist Drills.—BS 328: Part I: 1959 (incorporating amendments issued March 1960 and March 1964) covers twist drills made to inch and metric dimensions that are intended for general engineering purposes. ISO recommendations are taken into account. The accompanying tables give the standard metric sizes of Morse taper shank twist drills and core drills, parallel shank jobbing and long series drills, and stub drills.

All drills are right-hand cutting unless otherwise specified, and normal, slow, or quick helix angles may be provided. A "back-taper" is ground on the diameter from point to shank to provide longitudinal clearance. Core drills may have three or four flutes, and are intended for opening up cast holes or enlarging machined holes, for example. The parallel shank jobber, and long series drills, and stub drills are made without driving tenons.

Morse taper shank drills with oversize dimensions are also listed, and Table 15 shows metric drill sizes superseding gage and letter size drills, which are now obsolete in Britain. To meet special requirements, the Standard lists nonstandard sizes for the various types of drills.

The limits of tolerance on cutting diameters, as measured across the lands at the outer corners of a drill, shall be h8, in accordance with BS 1916, Limits and Fits for Engineering (Part I, Limits and Tolerances), and Table 3 shows the values common to the different types of drills mentioned before.

The drills shall be permanently and legibly marked whenever possible, preferably by rolling, showing the size, and the manufacturer's name or trademark. If they are made from high-speed steel, they shall be marked with the letters H.S. where practicable.

Drill Elements: The following definitions of drill elements are given.

Axis: The longitudinal center line.

Body: That portion of the drill extending from the extreme cutting end to the commencement of the shank.

Shank: That portion of the drill by which it is held and driven.

Flutes: The grooves in the body of the drill that provide lips and permit the removal of chips and allow cutting fluid to reach the lips.

Web (Core): The central portion of the drill situated between the roots of the flutes and extending from the point end toward the shank; the point end of the web or core forms the chisel edge.

Lands: The cylindrical-ground surfaces on the leading edges of the drill flutes. The width of the land is measured at right angles to the flute helix.

Body Clearance: The portion of the body surface that is reduced in diameter to provide diametral clearance.

Heel: The edge formed by the intersection of the flute surface and the body clearance.

Point: The sharpened end of the drill, consisting of all that part of the drill that is shaped to produce lips, faces, flanks, and chisel edge.

Face: That portion of the flute surface adjacent to the lip on which the chip impinges as it is cut from the work.

Flank: The surface on a drill point that extends behind the lip to the following flute.

Lip (Cutting Edge): The edge formed by the intersection of the flank and face.

Relative Lip Height: The relative position of the lips measured at the outer corners in a direction parallel to the drill axis.

Outer Corner: The corner formed by the intersection of the lip and the leading edge of the land.

Chisel Edge: The edge formed by the intersection of the flanks.

Chisel Edge Corner: The corner formed by the intersection of a lip and the chisel edge.

Table 15. British Standard Drills — Metric Sizes Superseding Gauge and Letter Sizes *BS 328: Part 1: 1959 Appendix B*

Obsolete Drill Size	Recommended MetricSize (mm)	Obsolete Drill Size	Recommended Metric Size (mm)	Obsolete Drill Size	Recommended Metric Size (mm)	Obsolete Drill Size	Recommended Metric Size (mm)	Obsolete Drill Size	Recommended Metric Size (mm)
80	0.35	58	1.05	36	2.70	14	4.60	I	6.90
79	0.38	57	1.10			13	4.70	J	7.00
78	0.40	56	$^{3}/_{64}$ in.	35	2.80	12	4.80		
77	0.45			34	2.80	11	4.90		
76	0.50			33	2.85			K	$^{9}/_{32}$ in.
		55	1.30	33	2.85			L	7.40
		54	1.40	32	2.95	10	4.90	M	7.50
75	0.52	53	1.50	31	3.00	9	5.00	N	7.70
74	0.58	52	1.60			8	5.10	O	8.00
73	0.60	51	1.70	30	3.30	7	5.10		
72	0.65			29	3.50	6	5.20		
71	0.65			28	$^{9}/_{64}$ in.				
		50	1.80	27	3.70			P	8.20
		49	1.85	26	3.70	5	5.20	Q	8.40
70	0.70	48	1.95			4	5.30	R	8.60
69	0.75	47	2.00	25	3.80	3	5.40	S	8.80
68	$^{1}/_{32}$ in.	46	2.05	24	3.90	2	5.60	T	9.10
67	0.82			23	3.90	1	5.80		
66	0.85	45	2.10	22	4.00			U	9.30
		44	2.20	21	4.00	A	$^{15}/_{64}$ in.	V	$^{3}/_{8}$ in.
65	0.90	43	2.25			B	6.00	W	9.80
64	0.92	42	$^{3}/_{32}$ in.	20	4.10	C	6.10	X	10.10
63	0.95	41	2.45	19	4.20	D	6.20	Y	10.30
62	0.98			18	4.30	E	$^{1}/_{4}$ in.	Z	10.50
61	1.00	40	2.50	17	4.40			…	…
		39	2.55	16	4.50	F	6.50	…	…
60	1.00	38	2.60			G	6.60	…	…
59	1.05	37	2.65	15	4.60	H	$^{17}/_{64}$ in.	…	…

Gauge and letter size drills are now obsolete in the United Kingdom and should not be used in the production of new designs. The table is given to assist users in changing over to the recommended standard sizes.

Table 1. British Standard Morse Taper Shank Twist Drills and Core Drills — Standard Metric Sizes *BS 328: Part 1: 1959*

Diameter	Flute Length	Overall Length
3.00	33	114
3.20	36	117
3.50	39	120
3.80	43	123
4.00		
4.20		
4.50	47	128
4.80	52	133
5.00		
5.20		
5.50	57	138
5.80		
6.00		
6.20	63	144
6.50		
6.80	69	150
7.00		
7.20		
7.50		
7.80	75	156
8.00		
8.20		
8.50		
8.80	81	162
9.00		
9.20		
9.50		
9.80	87	168
10.00		
10.20		
10.50		
10.80	94	175
11.00		
11.20		
11.50		
11.80		
12.00	101	182
12.20		
12.50		
12.80		
13.00		
13.20		
13.50	108	189
13.80		
14.00		
14.25	114	212
14.50		
14.75		
15.00		
15.25	120	218
15.50		
15.75		
16.00		

Diameter	Flute Length	Overall Length
16.25	125	223
16.50		
16.75		
17.00		
17.25	130	228
17.50		
17.75		
18.00		
18.25	135	233
18.50		
18.75		
19.00		
19.25	140	238
19.50		
19.75		
20.00		
20.25	145	243
20.50		
20.75		
21.00		
21.25	150	248
21.50		
21.75		
22.00		
22.25		
22.50	155	253
22.75		
23.00		
23.25	155	276
23.50		
23.75	160	281
24.00		
24.25		
24.50		
24.75		
25.00		
25.25	165	286
25.50		
25.75		
26.00		
26.25		
26.50		
26.75	170	291
27.00		
27.25		
27.50		
27.75		
28.00		
28.25	175	296
28.50		
28.75		
29.00		
29.25		

Diameter	Flute Length	Overall Length
29.50	175	296
29.75		
30.00		
30.25	180	301
30.50		
30.75		
31.00		
31.25		
31.50		
31.75	185	306
32.00	185	334
32.50		
33.00		
33.50		
34.00	190	339
34.50		
35.00		
35.50		
36.00	195	344
36.50		
37.00		
37.50		
38.00	200	349
38.50		
39.00		
39.50		
40.00		
40.50	205	354
41.00		
41.50		
42.00		
42.50		
43.00	210	359
43.50		
44.00		
44.50		
45.00		
45.50	215	364
46.00		
46.50		
47.00		
47.50		
48.00	220	369
48.50		
49.00		
49.50		
50.00		
50.50	225	374
51.00	225	412
52.00		
53.00		
54.00	230	417
55.00		

Table 1. British Standard Morse Taper Shank Twist Drills and Core Drills — Standard Metric Sizes *BS 328: Part 1: 1959*

Diameter	Flute Length	Overall Length	Diameter	Flute Length	Overall Length	Diameter	Flute Length	Overall Length
56.00	230	417	71.00	250	437	86.00	270	524
57.00	235	422	72.00	255	442	87.00		
58.00			73.00			88.00		
59.00			74.00			89.00		
60.00			75.00			90.00		
61.00	240	427	76.00	260	477	91.00	275	529
62.00			77.00	260	514	92.00		
63.00			78.00			93.00		
64.00	245	432	79.00			94.00		
65.00			80.00			95.00		
66.00			81.00	265	519	96.00	280	534
67.00			82.00			97.00		
68.00	250	437	83.00			98.00		
69.00			84.00			99.00		
70.00			85.00			100.00		

All dimensions are in millimeters. Tolerances on diameters are given in the table below.

Table 2, shows twist drills that may be supplied with the shank and length oversize, but they should be regarded as nonpreferred.

The Morse taper shanks of these twist and core drills are as follows: 3.00 to 14.00 mm diameter, M.T. No. 1; 14.25 to 23.00 mm diameter, M.T. No. 2; 23.25 to 31.50 mm diameter, M.T. No. 3; 31.75 to 50.50 mm diameter, M.T. No. 4; 51.00 to 76.00 mm diameter, M.T. No. 5; 77.00 to 100.00 mm diameter, M.T. No. 6.

Table 2. British Standard Morse Taper Shank Twist Drills — Metric Oversize Shank and Length Series *BS 328: Part 1: 1959*

Dia. Range	Overall Length	M. T. No.	Dia. Range	Overall Length	M. T. No.	Dia. Range	Overall Length	M. T. No.
12.00 to 13.20	199	2	22.50 to 23.00	276	3	45.50 to 47.50	402	5
13.50 to 14.00	206	2	26.75 to 28.00	319	4	48.00 to 50.00	407	5
18.25 to 19.00	256	3	29.00 to 30.00	324	4	50.50	412	5
19.25 to 20.00	251	3	30.25 to 31.50	329	4	64.00 to 67.00	499	6
20.25 to 21.00	266	3	40.50 to 42.50	392	5	68.00 to 71.00	504	6
21.25 to 22.25	271	3	43.00 to 45.00	397	5	72.00 to 75.00	509	6

Diameters and lengths are given in millimeters. For the individual sizes within the diameter ranges given, see Table 1.

This series of drills should be regarded as non-preferred.

Table 3. British Standard Limits of Tolerance on Diameter for Twist Drills and Core Drills — Metric Series *BS 328: Part 1: 1959*

Drill Size (Diameter measured across lands at outer corners)	Tolerance (h8)
0 to 1 inclusive	Plus 0.000 to Minus 0.014
Over 1 to 3 inclusive	Plus 0.000 to Minus 0.014
Over 3 to 6 inclusive	Plus 0.000 to Minus 0.018
Over 6 to 10 inclusive	Plus 0.000 to Minus 0.022
Over 10 to 18 inclusive	Plus 0.000 to Minus 0.027
Over 18 to 30 inclusive	Plus 0.000 to Minus 0.033
Over 30 to 50 inclusive	Plus 0.000 to Minus 0.039
Over 50 to 80 inclusive	Plus 0.000 to Minus 0.046
Over 80 to 120 inclusive	Plus 0.000 to Minus 0.054

All dimensions are given in millimeters.

Table 4. British Standard Parallel Shank Jobber Series Twist Drills — Standard Metric Sizes *BS 328: Part 1: 1959*

Diameter	Flute Length	Overall Length
0.20	2.5	19
0.22		
0.25	3.0	19
0.28		
0.30	4.0	19
0.32	4	19
0.35		
0.38		
0.40	5	20
0.42		
0.45		
0.48		
0.50	6	22
0.52		
0.55	7	24
0.58		
0.60		
0.62	8	26
0.65		
0.68	9	28
0.70		
0.72		
0.75		
0.78	10	30
0.80		
0.82		
0.85		
0.88	11	32
0.90		
0.92		
0.95		
0.98	12	34
1.00		
1.05		
1.10	14	36
1.15		
1.20	16	38
1.25		
1.30		
1.35	18	40
1.40		
1.45		
1.50		
1.55	20	43
1.60		
1.65		
1.70		

Diameter	Flute Length	Overall Length
1.75	22	46
1.80		
1.85		
1.90		
1.95	24	49
2.00		
2.05		
2.10		
2.15	27	53
2.20		
2.25		
2.30		
2.35		
2.40	30	57
2.45		
2.50		
2.55		
2.60		
2.65		
2.70	33	61
2.75		
2.80		
2.85		
2.90		
2.95		
3.00		
3.10	36	65
3.20		
3.30		
3.40	39	70
3.50	39	70
3.60		
3.70		
3.80	43	75
3.90		
4.00		
4.10		
4.20		
4.30	47	80
4.40		
4.50		
4.60		
4.70		
4.80	52	86
4.90		
5.00		
5.10		
5.20		
5.30		

Diameter	Flute Length	Overall Length
5.40	57	93
5.50		
5.60		
5.70		
5.80		
5.90		
6.00		
6.10	63	101
6.20		
6.30		
6.40		
6.50		
6.60		
6.70		
6.80	69	109
6.90		
7.00		
7.10		
7.20		
7.30		
7.40		
7.50		
7.60	75	117
7.70		
7.80		
7.90		
8.00		
8.10		
8.20		
8.30		
8.40		
8.50		
8.60	81	125
8.70		
8.80		
8.90		
9.00		
9.10		
9.20		
9.30		
9.40		
9.50		
9.60	87	133
9.70		
9.80		
9.90		
10.00		
10.10		

Diameter	Flute Length	Overall Length
10.20	87	133
10.30		
10.40		
10.50		
10.60		
10.70	94	142
10.80		
10.90		
11.00		
11.10		
11.20		
11.30		
11.40		
11.50		
11.60		
11.70		
11.80		
11.90	101	151
12.00		
12.10		
12.20		
12.30		
12.40		
12.50		
12.60		
12.70		
12.80		
12.90		
13.00		
13.10		
13.20		
13.30	108	160
13.40		
13.50		
13.60		
13.70		
13.80		
13.90		
14.00		
14.25	114	169
14.50		
14.75		
15.00		
15.25	120	178
15.50		
15.75		
16.00		

All dimensions are in millimeters. Tolerances on diameters are given in Table 3.

Table 5. British Standard Parallel Shank Long Series Twist Drills — Standard Metric Sizes *BS 328: Part 1: 1959*

Diameter	Flute Length	Overall Length
2.00 2.05 2.10	56	85
2.15 2.20 2.25 2.30 2.35	59	90
2.40 2.45 2.50 2.55 2.60 2.65	62	95
2.70 2.75 2.80 2.85 2.90 2.95 3.00	66	100
3.10 3.20 3.30	69	106
3.40 3.50 3.60 3.70	73	112
3.80 3.90 4.00 4.10 4.20	78	119
4.30 4.40 4.50 4.60 4.70	82	126
4.80 4.90 5.00 5.10 5.20 5.30	87	132
5.40 5.50 5.60 5.70 5.80 5.90 6.00	91	139
6.10 6.20 6.30 6.40 6.50 6.60 6.70	97	148

Diameter	Flute Length	Overall Length
6.80 6.90 7.00 7.10 7.20 7.30 7.40 7.50	102	156
7.60 7.70 7.80 7.90 8.00 8.10 8.20 8.30 8.40 8.50	109	165
8.60 8.70 8.80 8.90 9.00 9.10 9.20 9.30 9.40 9.50	115	175
9.60 9.70 9.80 9.90 10.00 10.10 10.20 10.30 10.40 10.50 10.60	121	184
10.70 10.80 10.90 11.00 11.10 11.20 11.30 11.40 11.50 11.60 11.70 11.80	128	195
11.90 12.00 12.10 12.20 12.30 12.40 12.50 12.60	134	205

Diameter	Flute Length	Overall Length
12.70 12.80 12.90 13.00 13.10 13.20	134	205
13.30 13.40 13.50 13.60 13.70 13.80 13.90 14.00	140	214
14.25 14.50 14.75 15.00	144	220
15.25 15.50 15.75 16.00	149	227
16.25 16.50 16.75 17.00	154	235
17.25 17.50 17.75 18.00	158	241
18.25 18.50 18.75 19.00	162	247
19.25 19.50 19.75 20.00	166	254
20.25 20.50 20.75 21.00	171	261
21.25 21.50 21.75 22.00 22.25	176	268
22.50 22.75 23.00 23.25 23.50	180	275
23.75 24.00 24.25 24.50 24.75 25.00	185	282

All dimensions are in millimeters. Tolerances on diameters are given in Table 3.

Table 6. British Standard Stub Drills — Metric Sizes *BS 328: Part 1: 1959*

Diameter	Flute Length	Overall Length
0.50	3	20
0.80	5	24
1.00	6	26
1.20	8	30
1.50	9	32
1.80	11	36
2.00	12	38
2.20	13	40
2.50	14	43
2.80	16	46
3.00		
3.20	18	49
3.50	20	52
3.80	22	55
4.00		
4.20		
4.50	24	58
4.80	26	62

Diameter	Flute Length	Overall Length
5.00	26	62
5.20		
5.50	28	66
5.80		
6.00		
6.20	31	70
6.50		
6.80	34	74
7.00		
7.20		
7.50		
7.80	37	79
8.00		
8.20		
8.50		
8.80	40	84
9.00		
9.20		

Diameter	Flute Length	Overall Length
9.50	40	84
9.80	43	89
10.00		
10.20		
10.50		
10.80	47	95
11.00		
11.20		
11.50		
11.80		
12.00	51	102
12.20		
12.50		
12.80		
13.00		
13.20		
13.50	54	107
13.80		

Diameter	Flute Length	Overall Length
14.00	54	107
14.50	56	111
15.00		
15.50	58	115
16.00		
16.50	60	119
17.00		
17.50	62	123
18.00		
18.50	64	127
19.00		
19.50	66	131
20.00		
21.00	68	136
22.00	70	141
23.00	72	146
24.00	75	151
25.00		

All dimensions are given in millimeters. Tolerances on diameters are given in Table 3.

Steels for Twist Drills.—Twist drill steels need good toughness, abrasion resistance, and ability to resist softening due to heat generated by cutting. The amount of heat generated indicates the type of steel that should be used.

Carbon Tool Steel: may be used where little heat is generated during drilling.

High-Speed Steel: is preferred because of its combination of red hardness and wear resistance, which permit higher operating speeds and increased productivity. Optimum properties can be obtained by selection of alloy analysis and heat treatment.

Cobalt High-Speed Steel: alloys have higher red hardness than standard high-speed steels, permitting drilling of materials such as heat-resistant alloys and materials with hardness greater than Rockwell 38 C. These high-speed drills can withstand cutting speeds beyond the range of conventional high-speed-steel drills and have superior resistance to abrasion but are not equal to tungsten-carbide tipped tools.

Accuracy of Drilled Holes.—Normally the diameter of drilled holes is not given a tolerance; the size of the hole is expected to be as close to the drill size as can be obtained.

The accuracy of holes drilled with a two-fluted twist drill is influenced by many factors, which include: the accuracy of the drill point; the size of the drill; length and shape of the chisel edge; whether or not a bushing is used to guide the drill; the work material; length of the drill; runout of the spindle and the chuck; rigidity of the machine tool, workpiece, and the setup; and also the cutting fluid used, if any.

The diameter of the drilled holes will be oversize in most materials. The table following provides the results of tests reported by The United States Cutting Tool Institute in which the diameters of over 2800 holes drilled in steel and cast iron were measured. The values in this table indicate what might be expected under average shop conditions; however, when the drill point is accurately ground and the other machining conditions are correct, the resulting hole size is more likely to be between the mean and average minimum values given in this table. If the drill is ground and used incorrectly, holes that are even larger than the average maximum values can result.

Oversize Diameters in Drilling

Drill Dia., Inch	Amount Oversize, Inch			Drill Dia., Inch	Amount Oversize, Inch		
	Average Max.	Mean	Average Min.		Average Max.	Mean	Average Min.
1⁄16	0.002	0.0015	0.001	1⁄2	0.008	0.005	0.003
1⁄8	0.0045	0.003	0.001	3⁄4	0.008	0.005	0.003
1⁄4	0.0065	0.004	0.0025	1	0.009	0.007	0.004

Courtesy of The United States Cutting Tool Institute

Some conditions will cause the drilled hole to be undersize. For example, holes drilled in light metals and in other materials having a high coefficient of thermal expansion such as plastics, may contract to a size that is smaller than the diameter of the drill as the material surrounding the hole is cooled after having been heated by the drilling. The elastic action of the material surrounding the hole may also cause the drilled hole to be undersize when drilling high strength materials with a drill that is dull at its outer corner.

The accuracy of the drill point has a great effect on the accuracy of the drilled hole. An inaccurately ground twist drill will produce holes that are excessively over-size. The drill point must be symmetrical; i.e., the point angles must be equal, as well as the lip lengths and the axial height of the lips. Any alterations to the lips or to the chisel edge, such as thinning the web, must be done carefully to preserve the symmetry of the drill point. Adequate relief should be provided behind the chisel edge to prevent heel drag. On conventionally ground drill points this relief can be estimated by the chisel edge angle.

When drilling a hole, as the drill point starts to enter the workpiece, the drill will be unstable and will tend to wander. Then as the body of the drill enters the hole the drill will tend to stabilize. The result of this action is a tendency to drill a bellmouth shape in the hole at the entrance and perhaps beyond. Factors contributing to bellmouthing are: an unsymmetrically ground drill point; a large chisel edge length; inadequate relief behind the chisel edge; runout of the spindle and the chuck; using a slender drill that will bend easily; and lack of rigidity of the machine tool, workpiece, or the setup. Correcting these conditions as required will reduce the tendency for bellmouthing to occur and improve the accuracy of the hole diameter and its straightness. Starting the hole with a short stiff drill, such as a center drill, will quickly stabilize the drill that follows and reduce or eliminate bellmouthing; this procedure should always be used when drilling in a lathe, where the work is rotating. Bellmouthing can also be eliminated almost entirely and the accuracy of the hole improved by using a close fitting drill jig bushing placed close to the workpiece. Although specific recommendations cannot be made, many cutting fluids will help to increase the accuracy of the diameters of drilled holes. Double margin twist drills, available in the smaller sizes, will drill a more accurate hole than conventional twist drills having only a single margin at the leading edge of the land. The second land, located on the trailing edge of each land, provides greater stability in the drill bushing and in the hole. These drills are especially useful in drilling intersecting off-center holes. Single and double margin step drills, also available in the smaller sizes, will produce very accurate drilled holes, which are usually less than 0.002 inch larger than the drill size.

Counterboring.—Counterboring (called spot-facing if the depth is shallow)is the enlargement of a previously formed hole. Counterbores for screw holes are generally made in sets. Each set contains three counterbores: one with the body of the size of the screw head and the pilot the size of the hole to admit the body of the screw; one with the body the size of the head of the screw and the pilot the size of the tap drill; and the third with the body the size of the body of the screw and the pilot the size of the tap drill. Counterbores are usually provided with helical flutes to provide positive effective rake on the cutting edges. The four flutes are so positioned that the end teeth cut ahead of center to provide a shearing action and eliminate chatter in the cut. Three designs are most common: solid, two-piece, and three-piece. Solid designs have the body, cutter, and pilot all in one piece. Two-piece designs have an integral shank and counterbore cutter, with an interchangeable pilot, and provide true concentricity of the cutter diameter with the shank, but allowing use of various pilot diameters. Three-piece counterbores have separate holder, counterbore cutter, and pilot, so that a holder will take any size of counterbore cutter. Each counterbore cutter, in turn, can be fitted with any suitable size diameter of pilot. Counterbores for brass are fluted straight.

Counterbores with Interchangeable Cutters and Guides

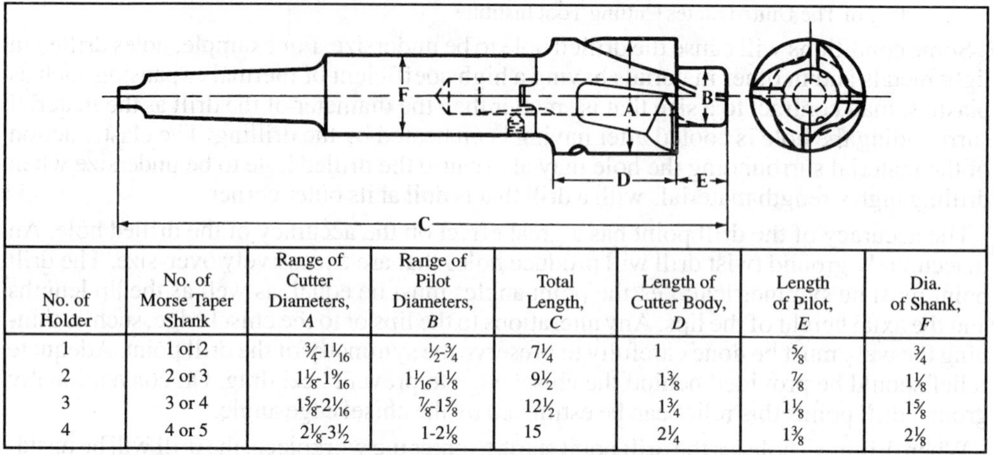

No. of Holder	No. of Morse Taper Shank	Range of Cutter Diameters, *A*	Range of Pilot Diameters, *B*	Total Length, *C*	Length of Cutter Body, *D*	Length of Pilot, *E*	Dia. of Shank, *F*
1	1 or 2	3/4-1 1/16	1/2-3/4	7 1/4	1	5/8	3/4
2	2 or 3	1 1/8-1 9/16	1 1/16-1 1/8	9 1/2	1 3/8	7/8	1 1/8
3	3 or 4	1 5/8-2 1/16	7/8-1 5/8	12 1/2	1 3/4	1 1/8	1 5/8
4	4 or 5	2 1/8-3 1/2	1-2 1/8	15	2 1/4	1 3/8	2 1/8

Solid Counterbores with Integral Pilot

Counterbore Diameters	Pilot Diameters			Straight Shank Diameter	Overall Length	
	Nominal	+1/64	+1/32		Short	Long
13/32	1/4	17/64	9/32	3/8	3 1/2	5 1/2
1/2	5/16	21/64	11/32	3/8	3 1/2	5 1/2
19/32	3/8	25/64	13/32	1/2	4	6
11/16	7/16	29/64	15/32	1/2	4	6
25/32	1/2	33/64	17/32	1/2	5	7
0.110	0.060	0.076	…	7/64	2 1/2	…
0.133	0.073	0.089	…	1/8	2 1/2	…
0.155	0.086	0.102	…	5/32	2 1/2	…
0.176	0.099	0.115	…	11/64	2 1/2	…
0.198	0.112	0.128	…	3/16	2 1/2	…
0.220	0.125	0.141	…	3/16	2 1/2	…
0.241	0.138	0.154	…	7/32	2 1/2	…
0.285	0.164	0.180	…	1/4	2 1/2	…
0.327	0.190	0.206	…	9/32	2 3/4	…
0.372	0.216	0.232	…	5/16	2 3/4	…

All dimensions are in inches.

Small counterbores are often made with three flutes, but should then have the size plainly stamped on them before fluting, as they cannot afterwards be conveniently measured. The flutes should be deep enough to come below the surface of the pilot. The counterbore should be relieved on the end of the body only, and not on the cylindrical surface. To facilitate the relieving process, a small neck is turned between the guide and the body for clearance. The amount of clearance on the cutting edges is, for general work, from 4 to 5 degrees. The accompanying table gives dimensions for straight shank counterbores.

Three Piece Counterbores.—Data shown for the first two styles of counterbores are for straight shank designs. These tools are also available with taper shanks in most sizes. Sizes of taper shanks for cutter diameters of 1/4 to 9/16 in. are No. 1, for 19/32 to 7/8 in., No. 2; for 15/16 to 1 3/8 in., No. 3; for 1 1/2 to 2 in., No. 4; and for 2 1/8 to 2 1/2 in., No. 5.

Table 1. American National Standard Sintered Carbide Boring Tools — Style Designations *ANSI B212.1-1984 (R1997)*

Side Cutting Edge Angle *E*		Boring Tool Styles			
Degrees	Designation	Solid Square (SS)	Tipped Square (TS)	Solid Round (SR)	Tipped Round (TR)
0	A		TSA		
10	B		TSB		
30	C	SSC	TSC	SRC	TRC
40	D		TSD		
45	E	SSE	TSE	SRE	TRE
55	F		TSF		
90 (0° Rake)	G				TRG
90 (10° Rake)	H				TRH

Table 2. American National Standard Solid Carbide Square Boring Tools—Style SSC for 60° Boring Bar and Style SSE for 45° Boring Bar *ANSI B212.1-1984 (R1997)*

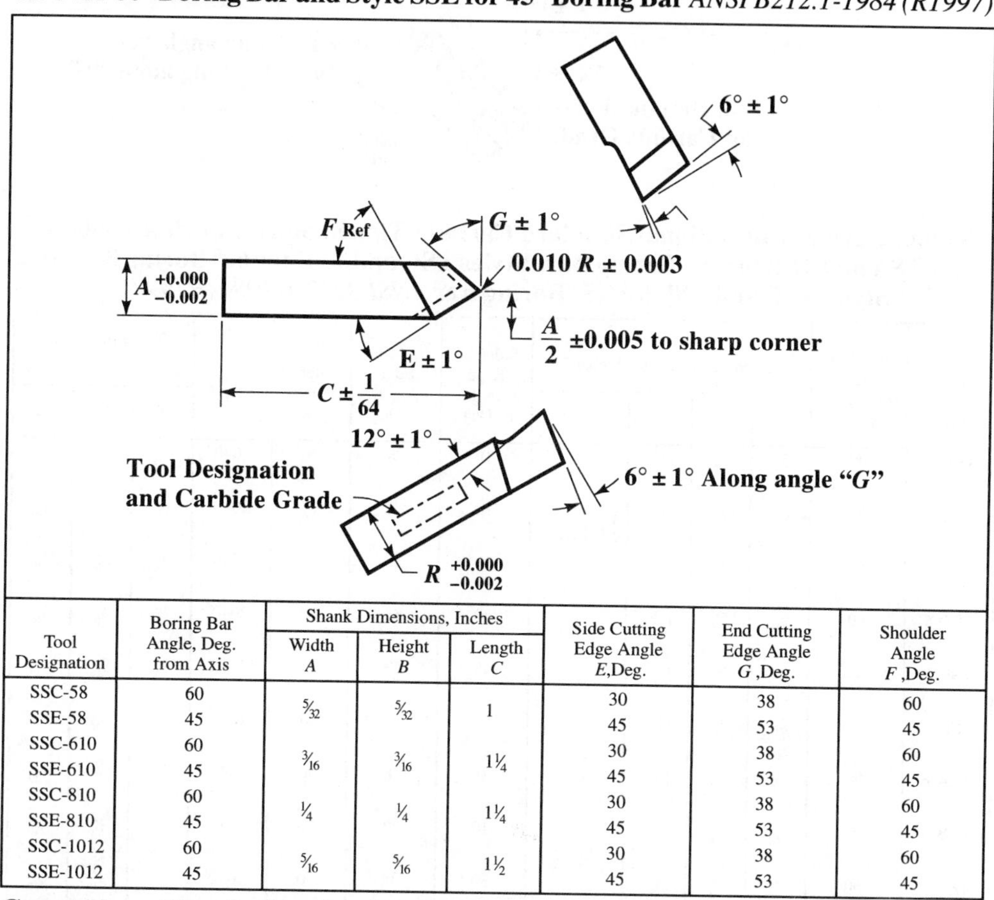

Tool Designation	Boring Bar Angle, Deg. from Axis	Shank Dimensions, Inches Width *A*	Height *B*	Length *C*	Side Cutting Edge Angle *E*, Deg.	End Cutting Edge Angle *G*, Deg.	Shoulder Angle *F*, Deg.
SSC-58	60	5⁄32	5⁄32	1	30	38	60
SSE-58	45				45	53	45
SSC-610	60	3⁄16	3⁄16	1¼	30	38	60
SSE-610	45				45	53	45
SSC-810	60	¼	¼	1¼	30	38	60
SSE-810	45				45	53	45
SSC-1012	60	5⁄16	5⁄16	1½	30	38	60
SSE-1012	45				45	53	45

Counterbore Sizes for Hex-head Bolts and Nuts.—Table 2, page 1511, shows the maximum socket wrench dimensions for standard ¼-, ½- and ¾-inch drive socket sets. For a given socket size (nominal size equals the maximum width across the flats of nut or bolt head), the dimension *K* given in the table is the minimum counterbore diameter required to provide socket wrench clearance for access to the bolt or nut.

Sintered Carbide Boring Tools.—Industrial experience has shown that the shapes of tools used for boring operations need to be different from those of single-point tools ordinarily used for general applications such as lathe work. Accordingly, Section 5 of American National Standard ANSI B212.1-1984 (R1997) gives standard sizes, styles and designations for four basic types of sintered carbide boring tools, namely: solid carbide square; carbide-tipped square; solid carbide round; and carbide-tipped round boring tools. In addition to these ready-to-use standard boring tools, solid carbide round and square unsharpened boring tool bits are provided.

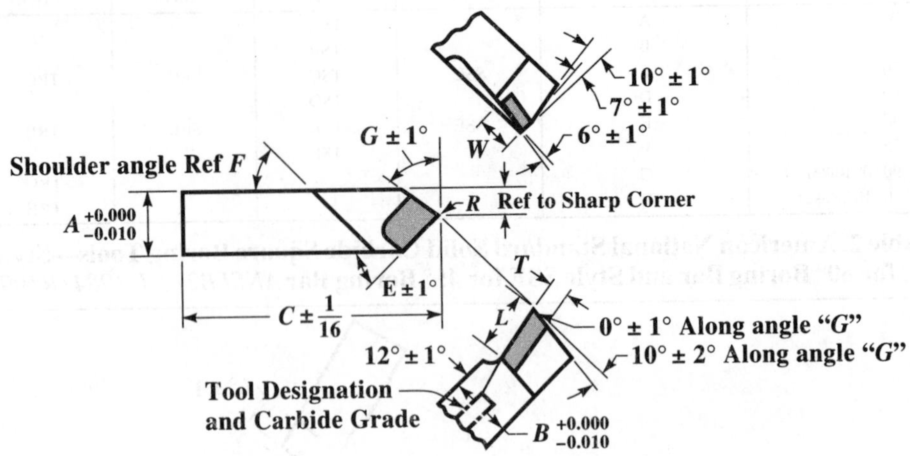

Table 3. American National Standard Carbide-Tipped Square Boring Tools — Styles TSA and TSB for 90° Boring Bar, Styles TSC and TSD for 60° Boring Bar, and Styles TSE and TSF for 45° Boring Bar *ANSI B212.1-1984 (R1997)*

Tool Designation	Bor. Bar Angle-from Axis, Deg.	Shank Dimensions, Inches				SideCut. Edge Angle E, Deg.	End Cut. Edge Angle G, Deg.	Shoulder Angle F, Deg.	Tip No.	Tip Dimensions, Inches		
		A	B	C	R					T	W	L
TSA-5	90	5⁄16	5⁄16	1½	1⁄64 ± 0.005	0	8	90	2040	3⁄32	3⁄16	5⁄16
TSB-5	90	5⁄16	5⁄16	1½		10	8	90	2040	3⁄32	3⁄16	5⁄16
TSC-5	60	5⁄16	5⁄16	1½		30	38	60	2040	3⁄32	3⁄16	5⁄16
TSD-5	60	5⁄16	5⁄16	1½		40	38	60	2040	3⁄32	3⁄16	5⁄16
TSE-5	45	5⁄16	5⁄16	1½		45	53	45	2040	3⁄32	3⁄16	5⁄16
TSF-5	45	5⁄16	5⁄16	1½		55	53	45	2040	3⁄32	3⁄16	5⁄16
TSA-6	90	3⁄8	3⁄8	1¾	1⁄64 ± 0.005	0	8	90	2040	3⁄32	3⁄16	5⁄16
TSB-6	90	3⁄8	3⁄8	1¾		10	8	90	2040	3⁄32	3⁄16	5⁄16
TSC-6	60	3⁄8	3⁄8	1¾		30	38	60	2040	3⁄32	3⁄16	5⁄16
TSD-6	60	3⁄8	3⁄8	1¾		40	38	60	2040	3⁄32	3⁄16	5⁄16
TSE-6	45	3⁄8	3⁄8	1¾		45	53	45	2040	3⁄32	3⁄16	5⁄16
TSF-6	45	3⁄8	3⁄8	1¾		55	53	45	2040	3⁄32	3⁄16	5⁄16

Table 3. *(Continued)* **American National Standard Carbide-Tipped Square Boring Tools — Styles TSA and TSB for 90° Boring Bar, Styles TSC and TSD for 60° Boring Bar, and Styles TSE and TSF for 45° Boring Bar** *ANSI B212.1-1984 (R1997)*

Tool Designation	Bor. Bar Angle from Axis, Deg.	Shank Dimensions, Inches				SideCut. Edge Angle *E*, Deg.	End Cut. Edge Angle *G*, Deg.	Shoulder Angle *F*, Deg.	Tip No.	Tip Dimensions, Inches		
		A	*B*	*C*	*R*					*T*	*W*	*L*
TSA-7	90	7/16	7/16	2 1/2	1/32 ± 0.010	0	8	90	2060	3/32	1/4	3/8
TSB-7	90	7/16	7/16	2 1/2		10	8	90	2060	3/32	1/4	3/8
TSC-7	60	7/16	7/16	2 1/2		30	38	60	2060	3/32	1/4	3/8
TSD-7	60	7/16	7/16	2 1/2		40	38	60	2060	3/32	1/4	3/8
TSE-7	45	7/16	7/16	2 1/2		45	53	45	2060	3/32	1/4	3/8
TSF-7	45	7/16	7/16	2 1/2		55	53	45	2060	3/32	1/4	3/8
TSA-8	90	1/2	1/2	2 1/2	1/32 ± 0.010	0	8	90	2150	1/8	5/16	7/16
TSB-8	90	1/2	1/2	2 1/2		10	8	90	2150	1/8	5/16	7/16
TSC-8	60	1/2	1/2	2 1/2		30	38	60	2150	1/8	5/16	7/16
TSD-8	60	1/2	1/2	2 1/2		40	38	60	2150	1/8	5/16	7/16
TSE-8	45	1/2	1/2	2 1/2		45	53	45	2150	1/8	5/16	7/16
TSF-8	45	1/2	1/2	2 1/2		55	53	45	2150	1/8	5/16	7/16
TSA-10	90	5/8	5/8	3	1/32 ± 0.010	0	8	90	2220	5/32	3/8	9/16
TSB-10	90	5/8	5/8	3		10	8	90	2220	5/32	3/8	9/16
TSC-10	60	5/8	5/8	3		30	38	60	2220	5/32	3/8	9/16
TSD-10	60	5/8	5/8	3		40	38	60	2220	5/32	3/8	9/16
TSE-10	45	5/8	5/8	3		45	53	45	2220	5/32	3/8	9/16
TSF-10	45	5/8	5/8	3		55	53	45	2220	5/32	3/8	9/16
TSA-12	90	3/4	3/4	3 1/2	1/32 ± 0.010	0	8	90	2300	3/16	7/16	5/8
TSB-12	90	3/4	3/4	3 1/2		10	8	90	2300	3/16	7/16	5/8
TSC-12	60	3/4	3/4	3 1/2		30	38	60	2300	3/16	7/16	5/8
TSD-12	60	3/4	3/4	3 1/2		40	38	60	2300	3/16	7/16	5/8
TSE-12	45	3/4	3/4	3 1/2		45	53	45	2300	3/16	7/16	5/8
TSF-12	45	3/4	3/4	3 1/2		55	53	45	2300	3/16	7/16	5/8

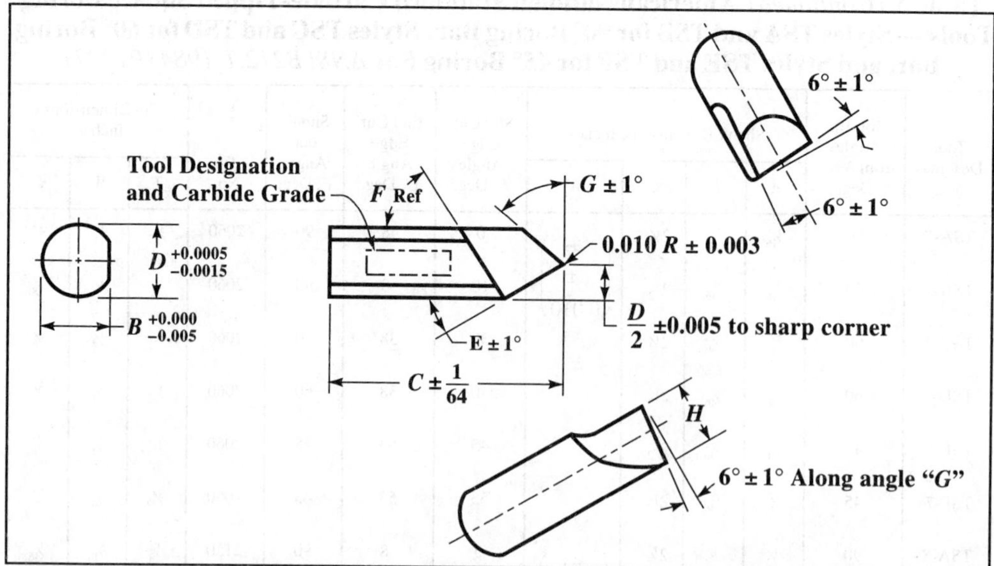

Table 4. American National Standard Solid Carbide Round Boring Tools — Style SRC for 60° Boring Bar and Style SRE for 45° Boring Bar *ANSI B212.1-1984 (R1997)*

Tool Designation	Bor. Bar Angle from Axis, Deg.	Shank Dimensions, Inches					Side Cut. Edge Angle *E*, Deg.	End Cut. Edge Angle *G*, Deg.	Shoulder Angle *F*, Deg.
		Dia. *D*	Length *C*	Dim. Over Flat *B*	Nose Height *H*				
SRC-33	60	3/32	3/8	0.088	0.070	[+0.000 −0.005]	30	38	60
SRE-33	45	3/32	3/8	0.088	0.070		45	53	45
SRC-44	60	1/8	1/2	0.118	0.094	[+0.000 −0.005]	30	38	60
SRE-44	45	1/8	1/2	0.118	0.094		45	53	45
SRC-55	60	5/32	5/8	0.149	0.117	±0.005	30	38	60
SRE-55	45	5/32	5/8	0.149	0.117	±0.005	45	53	45
SRC-66	60	3/16	3/4	0.177	0.140	±0.005	30	38	60
SRE-66	45	3/16	3/4	0.177	0.140	±0.005	45	53	45
SRC-88	60	1/4	1	0.240	0.187	±0.005	30	38	60
SRE-88	45	1/4	1	0.240	0.187	±0.005	45	53	45
SRC-1010	60	5/16	1 1/4	0.300	0.235	±0.005	30	38	60
SRE-1010	45	5/16	1 1/4	0.300	0.235	±0.005	45	53	45

Style Designations for Carbide Boring Tools: Table 1 shows designations used to specify the styles of American Standard sintered carbide boring tools. The first letter denotes solid (S) or tipped (T). The second letter denotes square (S) or round (R). The side cutting edge angle is denoted by a third letter (A through H) to complete the style designation. Solid square and round bits with the mounting surfaces ground but the cutting edges unsharpened (Table 7) are designated using the same system except that the third letter indicating the side cutting edge angle is omitted.

Size Designation of Carbide Boring Tools: Specific sizes of boring tools are identified by the addition of numbers after the style designation. The first number denotes the diameter or square size in number of 1/32nds for types SS and SR and in number of 1/16ths for types

TS and TR. The second number denotes length in number of 1/8ths for types SS and SR. For styles TRG and TRH, a letter "U" after the number denotes a semi-finished tool (cutting edges unsharpened). Complete designations for the various standard sizes of carbide boring tools are given in Tables 2 through 7. In the diagrams in the tables, angles shown without tolerance are ± 1°.

Table 5. American National Standard Carbide-Tipped Round Boring Tools — Style TRC for 60° Boring Bar and Style TRE for 45° Boring Bar
ANSI B212.1-1984 (R1997)

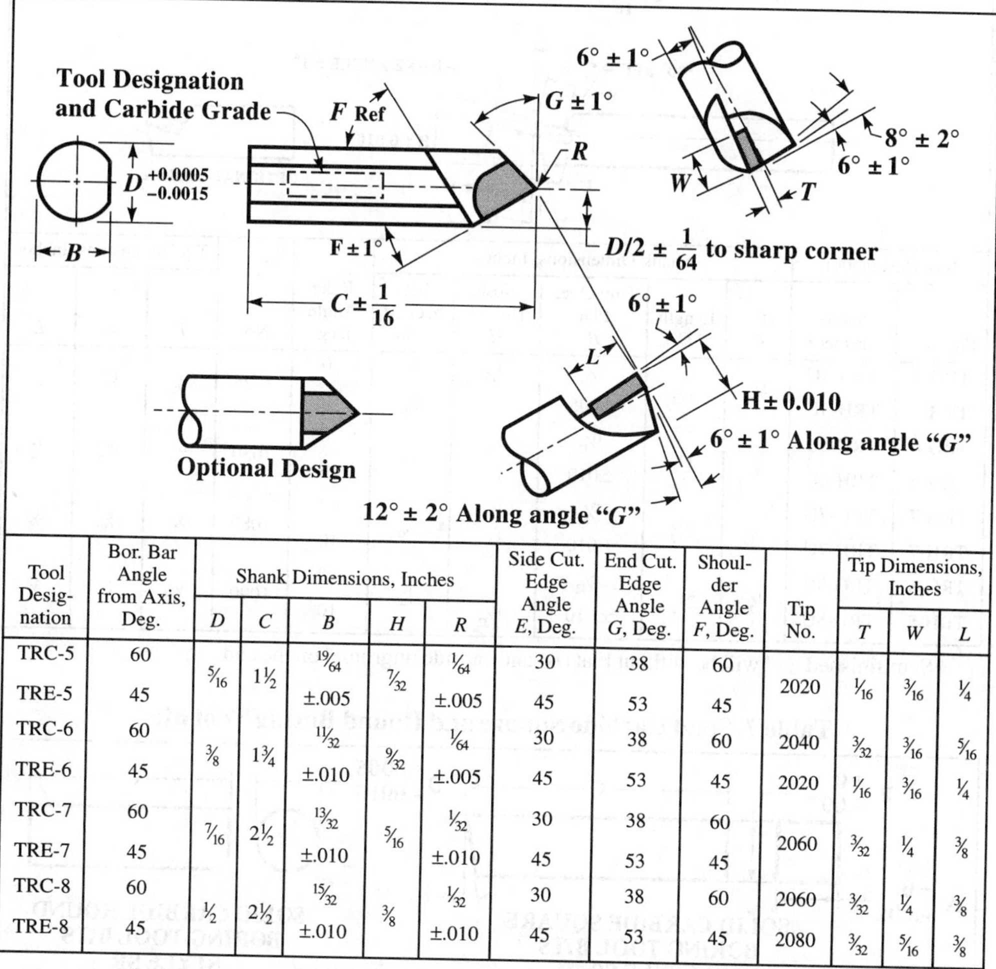

Tool Designation	Bor. Bar Angle from Axis, Deg.	Shank Dimensions, Inches D	C	B	H	R	Side Cut. Edge Angle E, Deg.	End Cut. Edge Angle G, Deg.	Shoulder Angle F, Deg.	Tip No.	Tip Dimensions, Inches T	W	L
TRC-5	60	5/16	1½	19/64	7/32	1/64	30	38	60	2020	1/16	3/16	¼
TRE-5	45			±.005		±.005	45	53	45				
TRC-6	60	3/8	1¾	11/32	9/32	1/64	30	38	60	2040	3/32	3/16	5/16
TRE-6	45			±.010		±.005	45	53	45	2020	1/16	3/16	¼
TRC-7	60	7/16	2½	13/32	5/16	1/32	30	38	60	2060	3/32	¼	3/8
TRE-7	45			±.010		±.010	45	53	45				
TRC-8	60	½	2½	15/32	3/8	1/32	30	38	60	2060	3/32	¼	3/8
TRE-8	45			±.010		±.010	45	53	45	2080	3/32	5/16	3/8

Examples of Tool Designation: The designation TSC-8 indicates: a carbide-tipped tool (T); square cross-section (S); 30-degree side cutting edge angle (C); and 8/16 or ½ inch square size (8).

The designation SRE-66 indicates: a solid carbide tool (S); round cross-section (R); 45 degree side cutting edge angle (E); 6/32 or 3/16 inch diameter (6); and 6/8 or ¾ inch long (6).

The designation SS-610 indicates: a solid carbide tool (S); square cross-section (S); 6/32 or 3/16 inch square size (6); 10/8 or 1¼ inches long (10).

It should be noted in this last example that the absence of a third letter (from A to H) indicates that the tool has its mounting surfaces ground but that the cutting edges are unsharpened.

Table 6. American National Standard Carbide-Tipped Round General-Purpose Square-End Boring Tools — Style TRG with 0° Rake and Style TRH with 10° Rake
ANSI B212.1-1984 (R1997)

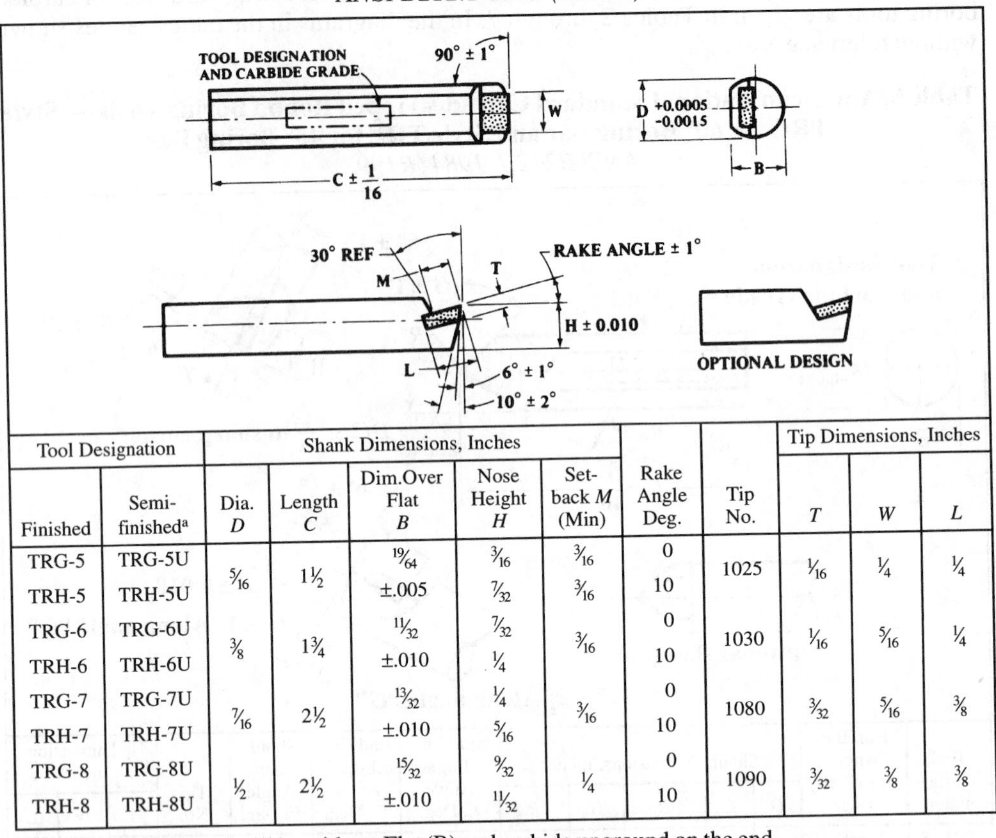

Tool Designation		Shank Dimensions, Inches							Tip Dimensions, Inches		
Finished	Semi-finished[a]	Dia. *D*	Length *C*	Dim.Over Flat *B*	Nose Height *H*	Set-back *M* (Min)	Rake Angle Deg.	Tip No.	*T*	*W*	*L*
TRG-5	TRG-5U	5/16	1 1/2	19/64	3/16	3/16	0	1025	1/16	1/4	1/4
TRH-5	TRH-5U			±.005	7/32	3/16	10				
TRG-6	TRG-6U	3/8	1 3/4	11/32	7/32	3/16	0	1030	1/16	5/16	1/4
TRH-6	TRH-6U			±.010	1/4		10				
TRG-7	TRG-7U	7/16	2 1/2	13/32	1/4	3/16	0	1080	3/32	5/16	3/8
TRH-7	TRH-7U			±.010	5/16		10				
TRG-8	TRG-8U	1/2	2 1/2	15/32	9/32	1/4	0	1090	3/32	3/8	3/8
TRH-8	TRH-8U			±.010	11/32		10				

[a] Semifinished tool will be without Flat (B) and carbide unground on the end.

Table 7. Solid Carbide Square and Round Boring Tool Bits

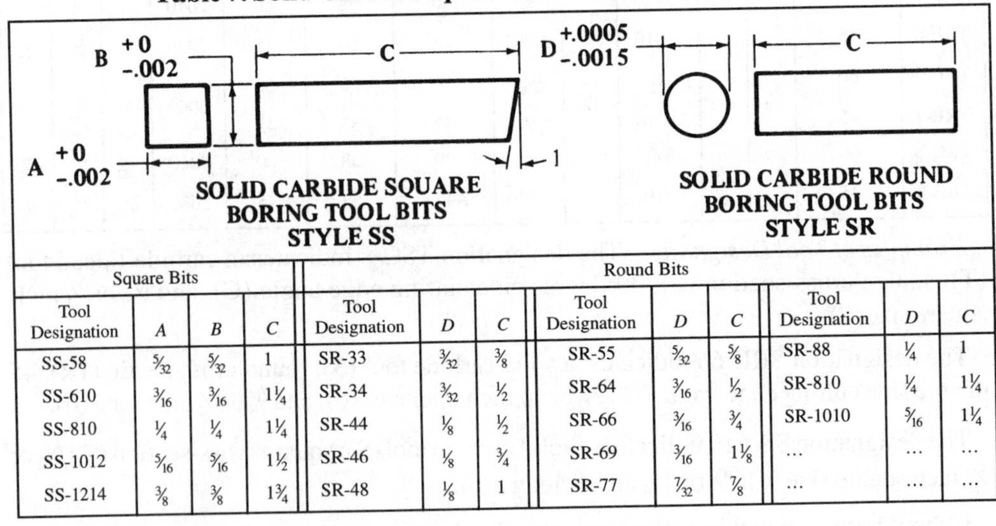

Square Bits				Round Bits								
Tool Designation	*A*	*B*	*C*	Tool Designation	*D*	*C*	Tool Designation	*D*	*C*	Tool Designation	*D*	*C*
SS-58	5/32	5/32	1	SR-33	3/32	3/8	SR-55	5/32	5/8	SR-88	1/4	1
SS-610	3/16	3/16	1 1/4	SR-34	3/32	1/2	SR-64	3/16	1/2	SR-810	1/4	1 1/4
SS-810	1/4	1/4	1 1/4	SR-44	1/8	1/2	SR-66	3/16	3/4	SR-1010	5/16	1 1/4
SS-1012	5/16	5/16	1 1/2	SR-46	1/8	3/4	SR-69	3/16	1 1/8	…	…	…
SS-1214	3/8	3/8	1 3/4	SR-48	1/8	1	SR-77	7/32	7/8	…	…	…

All dimensions are in inches.

Tolerance on Length: Through 1 inch, + 1/32, – 0; over 1 inch, +1/16, –0.

Spade Drills and Drilling

Spade drills are used to produce holes ranging in size from about 1 inch to 6 inches diameter, and even larger. Very deep holes can be drilled and blades are available for core drilling, counterboring, and for bottoming to a flat or contoured shape. There are two principal parts to a spade drill, the blade and the holder. The holder has a slot into which the blade fits; a wide slot at the back of the blade engages with a tongue in the holder slot to locate the blade accurately. A retaining screw holds the two parts together. The blade is usually made from high-speed steel, although cast nonferrous metal and cemented carbide-tipped blades are also available. Spade drill holders are classified by a letter symbol designating the range of blade sizes that can be held and by their length. Standard stub, short, long, and extra long holders are available; for very deep holes, special holders having wear strips to support and guide the drill are often used. Long, extra long, and many short length holders have coolant holes to direct cutting fluid, under pressure, to the cutting edges. In addition to its function in cooling and lubricating the tool, the cutting fluid also flushes the chips out of the hole. The shank of the holder may be straight or tapered; special automotive shanks are also used. A holder and different shank designs are shown in Fig. 1; Figs. 2a through Fig. 2f show some typical blades.

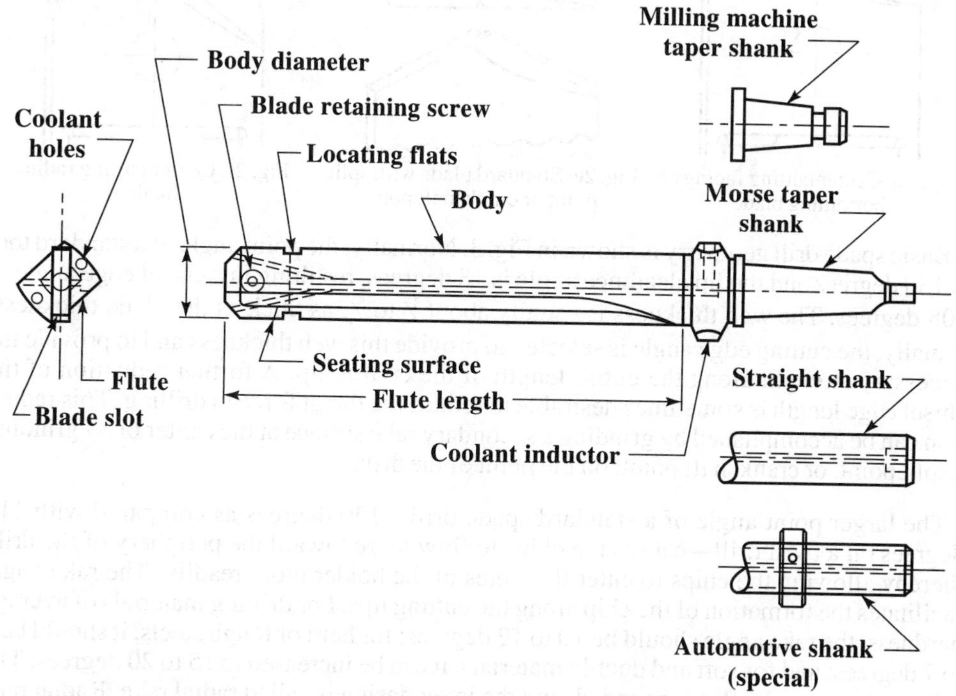

Fig. 1. Spade Drill Blade Holder

Spade Drill Geometry.—Metal separation from the work is accomplished in a like manner by both twist drills and spade drills, and the same mechanisms are involved for each. The two cutting lips separate the metal by a shearing action that is identical to that of chip formation by a single-point cutting tool. At the chisel edge, a much more complex condition exists. Here the metal is extruded sideways and at the same time is sheared by the rotation of the blunt wedge-formed chisel edge. This combination accounts for the very high thrust force required to penetrate the work. The chisel edge of a twist drill is slightly rounded, but on spade drills, it is a straight edge. Thus, it is likely that it is more difficult for the extruded metal to escape from the region of the chisel edge with spade drills. However, the chisel edge is shorter in length than on twist drills and the thrust for spade drilling is less.

Typical Spade Drill Blades

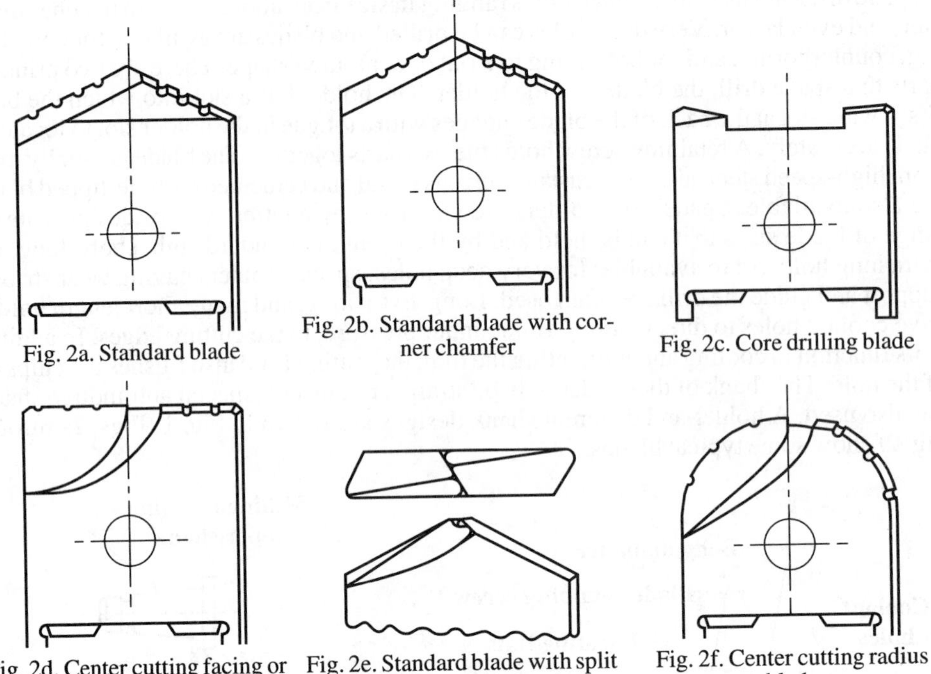

Fig. 2a. Standard blade

Fig. 2b. Standard blade with corner chamfer

Fig. 2c. Core drilling blade

Fig. 2d. Center cutting facing or bottoming blade

Fig. 2e. Standard blade with split point or crankshaft point

Fig. 2f. Center cutting radius blade

Basic spade drill geometry is shown in Fig. 3. Normally, the point angle of a standard tool is 130 degrees and the lip clearance angle is 18 degrees, resulting in a chisel edge angle of 108 degrees. The web thickness is usually about ¼ to 5⁄16 as thick as the blade thickness. Usually, the cutting edge angle is selected to provide this web thickness and to provide the necessary strength along the entire length of the cutting lip. A further reduction of the chisel edge length is sometimes desirable to reduce the thrust force in drilling. This reduction can be accomplished by grinding a secondary rake surface at the center or by grinding a split point, or crankshaft point, on the point of the drill.

The larger point angle of a standard spade drill—130 degrees as compared with 118 degrees on a twist drill—causes the chips to flow more toward the periphery of the drill, thereby allowing the chips to enter the flutes of the holder more readily. The rake angle facilitates the formation of the chip along the cutting lips. For drilling materials of average hardness, the rake angle should be 10 to 12 degrees; for hard or tough steels, it should be 5 to 7 degrees; and for soft and ductile materials, it can be increased to 15 to 20 degrees. The rake surface may be flat or rounded, and the latter design is called radial rake. Radial rake is usually ground so that the rake angle is maximum at the periphery and decreases uniformly toward the center to provide greater cutting edge strength at the center. A flat rake surface is recommended for drilling hard and tough materials in order to reduce the tendency to chipping and to reduce heat damage.

A most important feature of the cutting edge is the chip splitters, which are also called chip breaker grooves. Functionally, these grooves are chip dividers; instead of forming a single wide chip along the entire length of the cutting edge, these grooves cause formation of several chips that can be readily disposed of through the flutes of the holder. Chip splitters must be carefully ground to prevent the chips from packing in the grooves, which greatly reduces their effectiveness. Splitters should be ground perpendicular to the cutting lip and parallel to the surface formed by the clearance angle. The grooves on the two cutting lips must not overlap when measured radially along the cutting lip. Fig. 4 and the accompanying table show the groove form and dimensions.

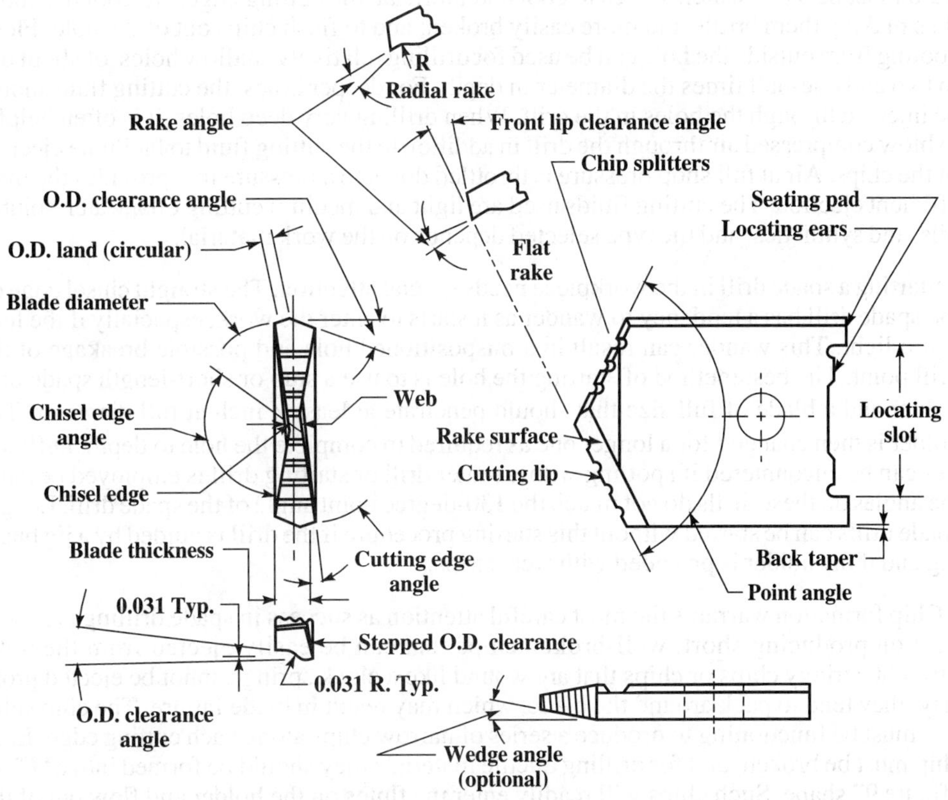

Fig. 3. Spade Drill Blade

On spade drills, the front lip clearance angle provides the relief. It may be ground on a drill grinding machine but usually it is ground flat. The normal front lip clearance angle is 8 degrees; in some instances, a secondary relief angle of about 14 degrees is ground below the primary clearance. The wedge angle on the blade is optional. It is generally ground on thicker blades having a larger diameter to prevent heel dragging below the cutting lip and to reduce the chisel edge length. The outside-diameter land is circular, serving to support and guide the blade in the hole. Usually it is ground to have a back taper of 0.001 to 0.002 inch per inch per side. The width of the land is approximately 20 to 25 per cent of the blade thickness. Normally, the outside-diameter clearance angle behind the land is 7 to 10 degrees. On many spade drill blades, the outside-diameter clearance surface is stepped about 0.030 inch below the land.

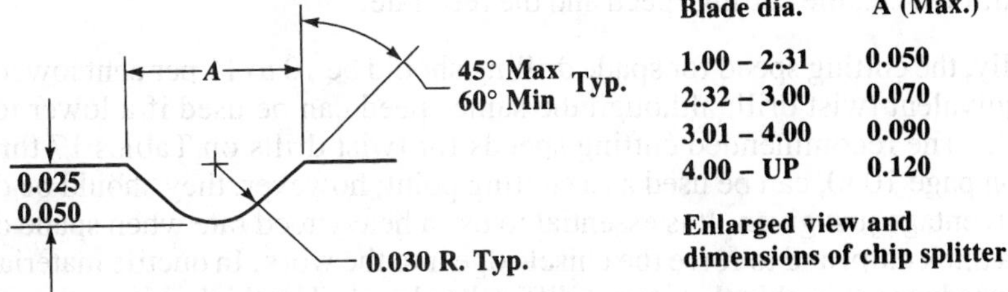

Blade dia.	A (Max.)
1.00 – 2.31	0.050
2.32 – 3.00	0.070
3.01 – 4.00	0.090
4.00 – UP	0.120

Enlarged view and dimensions of chip splitter

Fig. 4. Spade Drill Chip Splitter Dimensions

Spade Drilling.—Spade drills are used on drilling machines and other machine tools where the cutting tool rotates; they are also used on turning machines where the work rotates and the tool is stationary. Although there are some slight operational differences, the methods of using spade drills are basically the same. An adequate supply of cutting fluid must be used, which serves to cool and lubricate the cutting edges; to cool the chips, thus making them brittle and more easily broken; and to flush chips out of the hole. Flood cooling from outside the hole can be used for drilling relatively shallow holes, of about one to two and one-half times the diameter in depth. For deeper holes, the cutting fluid should be injected through the holes in the drill. When drilling very deep holes, it is often helpful to blow compressed air through the drill in addition to the cutting fluid to facilitate ejection of the chips. Air at full shop pressure is throttled down to a pressure that provides the most efficient ejection. The cutting fluids used are light and medium cutting oils, water-soluble oils, and synthetics, and the type selected depends on the work material.

Starting a spade drill in the workpiece needs special attention. The straight chisel edge on the spade drill has a tendency to wander as it starts to enter the work, especially if the feed is too light. This wander can result in a mispositioned hole and possible breakage of the drill point. The best method of starting the hole is to use a stub or short-length spade drill holder and a blade of full size that should penetrate at least ⅛ inch at full diameter. The holder is then changed for a longer one as required to complete the hole to depth. Difficulties can be encountered if spotting with a center drill or starting drill is employed because the angles on these drills do not match the 130-degree point angle of the spade drill. Longer spade drills can be started without this starting procedure if the drill is guided by a jig bushing and if the holder is provided with wear strips.

Chip formation warrants the most careful attention as success in spade drilling is dependent on producing short, well-broken chips that can be easily ejected from the hole. Straight, stringy chips or chips that are wound like a clock spring cannot be ejected properly; they tend to pack around the blade, which may result in blade failure. The chip splitters must be functioning to produce a series of narrow chips along each cutting edge. Each chip must be broken, and for drilling ductile materials they should be formed into a "C" or "figure 9" shape. Such chips will readily enter the flutes on the holder and flow out of the hole.

Proper chip formation is dependent on the work material, the spade drill geometry, and the cutting conditions. Brittle materials such as gray cast iron seldom pose a problem because they produce a discontinuous chip, but austenitic stainless steels and very soft and ductile materials require much attention to obtain satisfactory chip control. Thinning the web or grinding a split point on the blade will sometimes be helpful in obtaining better chip control, as these modifications allow use of a heavier feed. Reducing the rake angle to obtain a tighter curl on the chip and grinding a corner chamfer on the tool will sometimes help to produce more manageable chips.

In most instances, it is not necessary to experiment with the spade drill blade geometry to obtain satisfactory chip control. Control usually can be accomplished by adjusting the cutting conditions; i.e., the cutting speed and the feed rate.

Normally, the cutting speed for spade drilling should be 10 to 15 per cent lower than that for an equivalent twist drill, although the same speed can be used if a lower tool life is acceptable. The recommended cutting speeds for twist drills on Tables 17 through 23, starting on page 1030, can be used as a starting point; however, they should be decreased by the percentage just given. It is essential to use a heavy feed rate when spade drilling to produce a thick chip. and to force the chisel edge into the work. In ductile materials, a light feed will produce a thin chip that is very difficult to break. The thick chip on the other hand, which often contains many rupture planes, will curl and break readily. Table 1 gives suggested feed rates for different spade drill sizes and materials. These rates should be used as a starting point and some adjustments may be necessary as experience is gained.

Table 1. Feed Rates for Spade Drilling

Material	Hardness, Bhn	Feed—Inches per Revolution					
		Spade Drill Diameter—Inches					
		1–1¼	1¼–2	2–3	3–4	4–5	5–8
Free Machining Steel	100–240	0.014	0.016	0.018	0.022	0.025	0.030
	240–325	0.010	0.014	0.016	0.020	0.022	0.025
Plain Carbon Steels	100–225	0.012	0.015	0.018	0.022	0.025	0.030
	225–275	0.010	0.013	0.015	0.018	0.020	0.025
	275–325	0.008	0.010	0.013	0.015	0.018	0.020
Free Machining Alloy Steels	150–250	0.014	0.016	0.018	0.022	0.025	0.030
	250–325	0.012	0.014	0.016	0.018	0.020	0.025
	325–375	0.010	0.010	0.014	0.016	0.018	0.020
Alloy Steels	125–180	0.012	0.015	0.018	0.022	0.025	0.030
	180–225	0.010	0.012	0.016	0.018	0.022	0.025
	225–325	0.009	0.010	0.013	0.015	0.018	0.020
	325–400	0.006	0.008	0.010	0.012	0.014	0.016
Tool Steels							
Water Hardening	150–250	0.012	0.014	0.016	0.018	0.020	0.022
Shock Resisting	175–225	0.012	0.014	0.015	0.016	0.017	0.018
Cold Work	200–250	0.007	0.008	0.009	0.010	0.011	0.012
Hot Work	150–250	0.012	0.013	0.015	0.016	0.018	0.020
Mold	150–200	0.010	0.012	0.014	0.016	0.018	0.018
Special-Purpose	150–225	0.010	0.012	0.014	0.016	0.016	0.018
High-Speed	200–240	0.010	0.012	0.013	0.015	0.017	0.018
Gray Cast Iron	110–160	0.020	0.022	0.026	0.028	0.030	0.034
	160–190	0.015	0.018	0.020	0.024	0.026	0.028
	190–240	0.012	0.014	0.016	0.018	0.020	0.022
	240–320	0.010	0.012	0.016	0.018	0.018	0.018
Ductile or Nodular Iron	140–190	0.014	0.016	0.018	0.020	0.022	0.024
	190–250	0.012	0.014	0.016	0.018	0.018	0.020
	250–300	0.010	0.012	0.016	0.018	0.018	0.018
Malleable Iron							
Ferritic	110–160	0.014	0.016	0.018	0.020	0.022	0.024
Pearlitic	160–220	0.012	0.014	0.016	0.018	0.020	0.020
	220–280	0.010	0.012	0.014	0.016	0.018	0.018
Free Machining Stainless Steel							
Ferritic	…	0.016	0.018	0.020	0.024	0.026	0.028
Austenitic	…	0.016	0.018	0.020	0.022	0.024	0.026
Martensitic	…	0.012	0.014	0.016	0.016	0.018	0.020
Stainless Steel							
Ferritic	…	0.012	0.014	0.018	0.020	0.020	0.022
Austenitic	…	0.012	0.014	0.016	0.018	0.020	0.020
Martensitic	…	0.010	0.012	0.012	0.014	0.016	0.018
Aluminum Alloys	…	0.020	0.022	0.024	0.028	0.030	0.040
Copper Alloys	(Soft)	0.016	0.018	0.020	0.026	0.028	0.030
	(Hard)	0.010	0.012	0.014	0.016	0.018	0.018
Titanium Alloys	…	0.008	0.010	0.012	0.014	0.014	0.016
High-Temperature Alloys	…	0.008	0.010	0.012	0.012	0.014	0.014

Power Consumption and Thrust for Spade Drilling.—In each individual setup, there are factors and conditions influencing power consumption that cannot be accounted for in a simple equation; however, those given below will enable the user to estimate power consumption and thrust accurately enough for most practical purposes. They are based on experimentally derived values of unit horsepower, as given in Table 2. As a word of caution, these values are for sharp tools. In spade drilling, it is reasonable to estimate that a dull tool will increase the power consumption and the thrust by 25 to 50 per cent. The unit horsepower values in the table are for the power consumed at the cutting edge, to which must be added the power required to drive the machine tool itself, in order to obtain the horsepower required by the machine tool motor. An allowance for power to drive the machine is provided by dividing the horsepower at the cutter by a mechanical efficiency factor, e_m. This factor can be estimated to be 0.90 for a direct spindle drive with a belt, 0.75 for a back gear drive, and 0.70 to 0.80 for geared head drives. Thus, for spade drilling the formulas are

$$hp_c = uhp\left(\frac{\pi D^2}{4}\right)fN$$

$$B_s = 148{,}500\ uhp fD$$

$$hp_m = \frac{hp_c}{e_m}$$

$$f = \frac{f_m}{N}$$

where hp_c = horsepower at the cutter
hp_m = horsepower at the motor
B_s = thrust for spade drilling in pounds
uhp = unit horsepower
D = drill diameter in inches
f = feed in inches per revolution
f_m = feed in inches per minute
N = spindle speed in revolutions per minute
e_m = mechanical efficiency factor

Table 2. Unit Horsepower for Spade Drilling

Material	Hardness	uhp	Material	Hardness	uhp
Plain Carbon and Alloy Steel	85–200 Bhn	0.79	Titanium Alloys	250–375 Bhn	0.72
	200–275	0.94	High-Temp Alloys	200–360 Bhn	1.44
	275–375	1.00	Aluminum Alloys	…	0.22
	375–425	1.15	Magnesium Alloys	…	0.16
	45–52 Rc	1.44	Copper Alloys	20–80 Rb	0.43
Cast Irons	110–200 Bhn	0.5		80–100 Rb	0.72
	200–300	1.08			
Stainless Steels	135–275 Bhn	0.94			
	30–45 Rc	1.08			

Example: Estimate the horsepower and thrust required to drive a 2-inch diameter spade drill in AISI 1045 steel that is quenched and tempered to a hardness of 275 Bhn. From Table 17 on page 1030, the cutting speed, *V*, for drilling this material with a twist drill is 50 feet per minute. This value is reduced by 10 per cent for spade drilling and the speed selected is thus 0.9 × 50 = 45 feet per minute. The feed rate (from Table 1, page 869) is 0.015 in/rev. and the unit horsepower from Table 2 above is 0.94. The machine efficiency factor is estimated to be 0.80 and it will be assumed that a 50 per cent increase in the unit horsepower must be allowed for dull tools.

Step 1. Calculate the spindle speed from the following formula:

$$N = \frac{12V}{\pi D}$$

where: N = spindle speed in revolutions per minute

V = cutting speed in feet per minute

D = drill diameter in inches

$$\text{Thus:}\quad N = \frac{12 \times 45}{\pi \times 2} = 86 \text{ revolutions per minute}$$

Step 2. Calculate the horsepower at the cutter:

$$\text{hp}_c = \text{uhp}\left(\frac{\pi D^2}{4}\right)fN = 0.94\left(\frac{\pi \times 2^2}{4}\right)0.015 \times 86 = 3.8$$

Step 3. Calculate the horsepower at the motor and provide for a 50 per cent power increase for the dull tool:

$$\text{hp}_m = \frac{\text{hp}_c}{e_m} = \frac{3.8}{0.80} = 4.75 \text{ horsepower}$$

$$\text{hp}_m \text{ (with dull tool)} = 1.5 \times 4.75 = 7.125 \text{ horsepower}$$

Step 4. Estimate the spade drill thrust:

$$B_s = 148{,}500 \times \text{uhp} \times fD = 148{,}500 \times 0.94 \times 0.015 \times 2$$

$$= 4188 \text{ lb (for sharp tool)}$$

$$B_s = 1.5 \times 4188$$

$$= 6282 \text{ lb (for dull tool)}$$

Trepanning.—Cutting a groove in the form of a circle or boring or cutting a hole by removing the center or core in one piece is called trepanning. Shallow trepanning, also called face grooving, can be performed on a lathe using a single-point tool that is similar to a grooving tool but has a curved blade. Generally, the minimum outside diameter that can be cut by this method is about 3 inches and the maximum groove depth is about 2 inches. Trepanning is probably the most economical method of producing deep holes that are 2 inches, and larger, in diameter. Fast production rates can be achieved. The tool consists of a hollow bar, or stem, and a hollow cylindrical head to which a carbide or high-speed steel, single-point cutting tool is attached. Usually, only one cutting tool is used although for some applications a multiple cutter head must be used; e.g., heads used to start the hole have multiple tools. In operation, the cutting tool produces a circular groove and a residue core that enters the hollow stem after passing through the head. On outside-diameter exhaust trepanning tools, the cutting fluid is applied through the stem and the chips are flushed around the outside of the tool; inside-diameter exhaust tools flush the chips out through the stem with the cutting fluid applied from the outside. For starting the cut, a tool that cuts a starting groove in the work must be used, or the trepanning tool must be guided by a bushing. For holes less than about five diameters deep, a machine that rotates the trepanning tool can be used. Often, an ordinary drill press is satisfactory; deeper holes should be machined on a lathe with the work rotating. A hole diameter tolerance of ±0.010 inch can be obtained easily by trepanning and a tolerance of ±0.001 inch has sometimes been held. Hole runout can be held to ±0.003 inch per foot and, at times, to ±0.001 inch per foot. On heat-treated metal, a surface finish of 125 to 150 μm AA can be obtained and on annealed metals 100 to 250 μm AA is common.

TAPS AND THREADING DIES

General dimensions and tap markings given in the ASME/ANSI Standard B94.9-1987 for straight fluted taps, spiral pointed taps, spiral pointed only taps, spiral fluted taps, fast spiral fluted taps, thread forming taps, pulley taps, nut taps, and pipe taps are shown in the tables on the pages that follow. This Standard also gives the thread limits for taps with cut threads and ground threads. The thread limits for cut thread and ground thread taps for screw threads are given in Tables 3 through 7 and Tables 8a and 8b; thread limits for cut thread and ground thread taps for pipe threads are given in Tables 9a through 10c. Taps recommended for various classes of Unified screw threads are given in Tables 11a through 14 in numbered sizes and Table 12 for nuts in fractional sizes.

Types of Taps.—Taps included in ASME/ANSI B94.9-1987 are categorized either by the style of fluting or by the specific application for which the taps are designed. The following types 1 through 6 are generally short in length, and were originally called "Hand Taps" but this design is generally used in machine applications. The remaining types have special lengths, which are detailed in the tables.

The thread size specifications for these types may be fractional or machine screw inch sizes, or metric sizes. The thread form may be ground or cut (unground) as further defined in each table. Additionally, the cutting chamfer on the thread may be Bottoming (B), Plug (P), or Taper (T).

(1) Straight Flute Taps: These taps have straight flutes of a number specified as either standard or optional, and are for general purpose applications.

(2) Spiral Pointed Taps: These taps have straight flutes and the cutting face of the first few threads is ground at an angle to force the chips ahead and prevent clogging in the flutes.

(3) Spiral Pointed Only Taps: These taps are made with the spiral point feature only without longitudinal flutes. These taps are especially suitable for tapping thin materials.

(4) Spiral Fluted Taps: These taps have right-hand helical flutes with a helix angle of 25 to 35 deg. These features are designed to help draw chips from the hole or to bridge a keyway.

(5) Fast Spiral Fluted Taps: These taps are similar to spiral fluted taps, except the helix angle is from 45 to 60 deg.

(6) Thread Forming Taps: These taps are fluteless except as optionally designed with one or more lubricating grooves. The thread form on the tap is lobed, so that there are a finite number of points contacting the work thread form. The tap does not cut, but forms the thread by extrusion.

(7) Pulley Taps: These taps have shanks that are extended in length by a standard amount for use where added reach is required. The shank is the same nominal diameter as the thread.

(8) Nut Taps: These taps are designed for tapping nuts on a low-production basis. Approximately one-half to three-quarters of the threaded portion has a chamfered section, which distributes the cutting over many teeth and facilitates entering the hole to be tapped. The length overall, the length of the thread, and the length of the shank are appreciably longer than on a regular straight fluted tap.

(9) Pipe Taps: These taps are used to produce standard straight or tapered pipe threads.

Definitions of Tap Terms.—The definitions that follow are taken from ANSI/ASME B94.9 but include only the more important terms. Some tap terms are the same as screw thread terms; therefore, see *Definitions of Screw Threads* starting on page 1707.

Back Taper: A gradual decrease in the diameter of the thread form on a tap from the chamfered end of the land toward the back, which creates a slight radial relief in the threads.

Base of Thread: Coincides with the cylindrical or conical surface from which the thread projects.

Chamfer: Tapering of the threads at the front end of each land or chaser of a tap by cutting away and relieving the crest of the first few teeth to distribute the cutting action over several teeth.

Chamfer Angle: Angle formed between the chamfer and the axis of the tap measured in an axial plane at the cutting edge.

Chamfer Relief Angle: Complement of the angle formed between a tangent to the relieved surface at the cutting edge and a radial line to the same point on the cutting edge.

Core Diameter: Diameter of a circle which is tangent to the bottom of the flutes at a given point on the axis.

First Full Thread: First full thread on the cutting edge back of the chamfer. It is at this point that rake, hook, and thread elements are measured.

Crest Clearance: Radial distance between the root of the internal thread and the crest of the external thread of the coaxially assembled design forms of mating threads.

Class of Thread: Designation of the class that determines the specification of the size, allowance, and tolerance to which a given threaded product is to be manufactured. It is not applicable to the tools used for threading.

Tap Terms

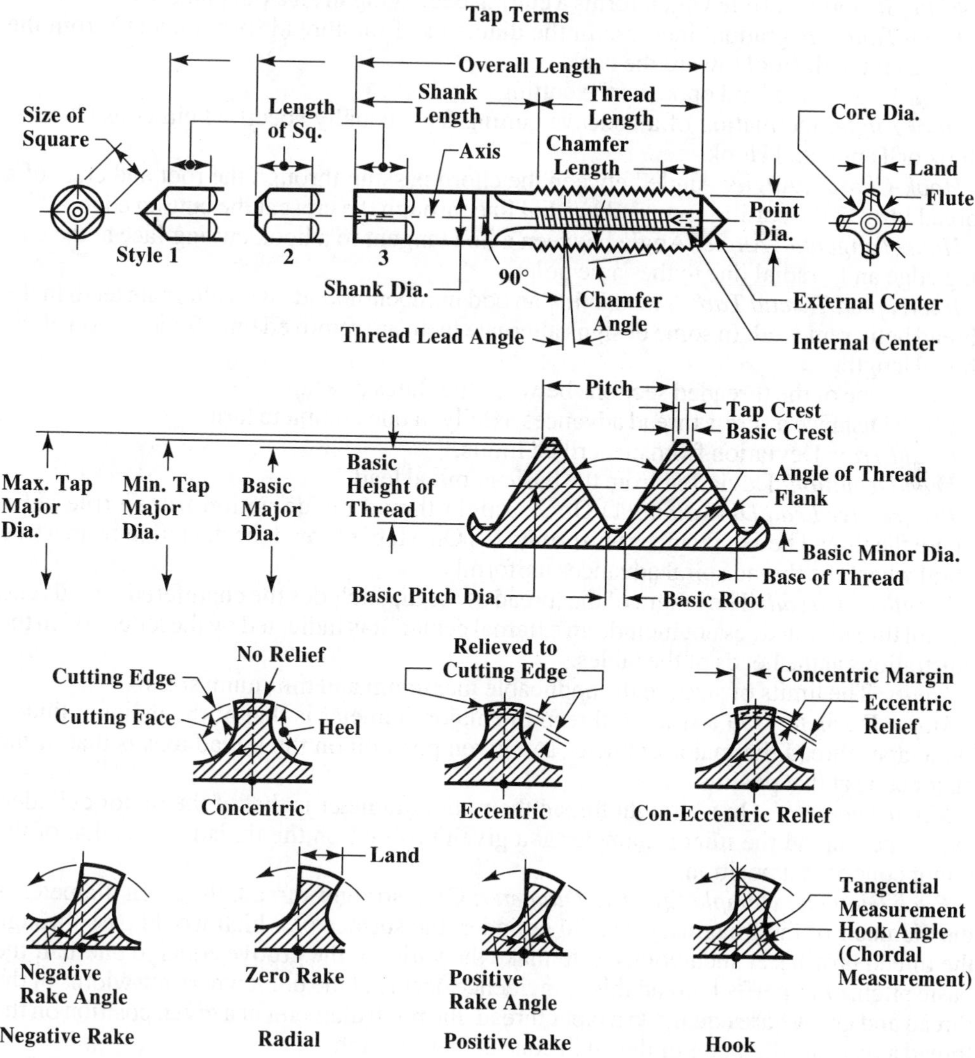

Flank Angle: Angle between the individual flank and the perpendicular to the axis of the thread, measured in an axial plane. A flank angle of a symmetrical thread is commonly termed the "half angle of thread."

Flank—Leading: 1) Flank of a thread facing toward the chamfered end of a threading tool; and 2) The leading flank of a thread is the one which, when the thread is about to be assembled with a mating thread, faces the mating thread.

Flank—Trailing: The trailing flank of a thread is the one opposite the leading flank.

Flutes: Longitudinal channels formed in a tap to create cutting edges on the thread profile and to provide chip spaces and cutting fluid passages. On a parallel or straight thread tap they may be straight, angular or helical; on a taper thread tap they may be straight, angular or spiral.

Flute-Angular: A flute lying in a plane intersecting the tool axis at an angle.

Flute-Helical: A flute with uniform axial lead and constant helix in a helical path around the axis of a cylindrical tap.

Flute-Spiral: A flute with uniform axial lead in a spiral path around the axis of a conical tap.

Flute Lead Angle: Angle at which a helical or spiral cutting edge at a given point makes with an axial plane through the same point.

Flute-Straight: A flute which forms a cutting edge lying in an axial plane.

Front Taper: A gradual increase in the diameter of the thread form on a tap from the leading end of the tool toward the back.

Heel: Edge of the land opposite the cutting edge.

Hook Angle: Inclination of a concave cutting face, usually specified either as Chordal Hook or Tangential Hook.

Hook-Chordal Angle: Angle between the chord passing through the root and crest of a thread form at the cutting face, and a radial line through the crest at the cutting edge.

Hook-Tangential Angle: Angle between a line tangent to a hook cutting face at the cutting edge and a radial line to the same point.

Interrupted Thread Tap: A tap having an odd number of lands with alternate teeth in the thread helix removed. In some designs alternate teeth are removed only for a portion of the thread length.

Land: One of the threaded sections between the flutes of a tap.

Lead: Distance a screw thread advances axially in one complete turn.

Lead Error: Deviation from prescribed limits.

Lead Deviation: Deviation from the basic nominal lead.

Progressive Lead Deviation: (1) On a straight thread the deviation from a true helix where the thread helix advances uniformly. (2) On a taper thread the deviation from a true spiral where the thread spiral advances uniformly.

Length of Thread: The length of the thread of the tap includes the chamfered threads and the full threads but does not include an external center. It is indicated by the letter "B" in the illustrations at the heads of the tables.

Limits: The limits of size are the applicable maximum and minimum sizes.

Major Diameter: On a straight thread the major diameter is that of the major cylinder. On a taper thread the major diameter at a given position on the thread axis is that of the major cone at that position.

Minor Diameter: On a straight thread the minor diameter is that of the minor cylinder. On a taper thread the minor diameter at a given position on the thread axis is that of the minor cone at that position.

Pitch Diameter (Simple Effective Diameter: On a straight thread, the pitch diameter is the diameter of the imaginary coaxial cylinder, the surface of which would pass through the thread profiles at such points as to make the width of the groove equal to one-half the basic pitch. On a perfect thread this coincidence occurs at the point where the widths of the thread and groove are equal. On a taper thread, the pitch diameter at a given position on the thread axis is the diameter of the pitch cone at that position.

Point Diameter: Diameter at the cutting edge of the leading end of the chamfered section.

Rake: Angular relationship of the straight cutting face of a tooth with respect to a radial line through the crest of the tooth at the cutting edge. Positive rake means that the crest of the cutting face is angularly ahead of the balance of the cutting face of the tooth. Negative rake means that the crest of the cutting face is angularly behind the balance of the cutting face of the tooth. Zero rake means that the cutting face is directly on a radial line.

Relief: Removal of metal behind the cutting edge to provide clearance between the part being threaded and the threaded land.

Relief-Center: Clearance produced on a portion of the tap land by reducing the diameter of the entire thread form between cutting edge and heel.

Relief-Chamfer: Gradual decrease in land height from cutting edge to heel on the chamfered portion of the land on a tap to provide radial clearance for the cutting edge.

Relief-Con-eccentric Thread: Radial relief in the thread form starting back of a concentric margin.

Relief-Double Eccentric Thread: Combination of a slight radial relief in the thread form starting at the cutting edge and continuing for a portion of the land width, and a greater radial relief for the balance of the land.

Relief-Eccentric Thread: Radial relief in the thread form starting at the cutting edge and continuing to the heel.

Relief-Flatted Land: Clearance produced on a portion of the tap land by truncating the thread between cutting edge and heel.

Relief-Grooved Land: Clearance produced on a tap land by forming a longitudinal groove in the center of the land.

Relief-Radial: Clearance produced by removal of metal from behind the cutting edge. Taps should have the chamfer relieved and should have back taper, but may or may not have relief in the angle and on the major diameter of the threads. When the thread angle is relieved, starting at the cutting edge and continuing to the heel, the tap is said to have "eccentric" relief. If the thread angle is relieved back of a concentric margin (usually one-third of land width), the tap is said to have "con-eccentric" relief.

Size-Actual: Measured size of an element on an individual part.

Size-Basic: That size from which the limits of size are derived by the application of allowances and tolerances.

Size-Functional: The functional diameter of an external or internal thread is the pitch diameter of the enveloping thread of perfect pitch, lead and flank angles, having full depth of engagement but clear at crests and roots, and of a specified length of engagement. It may be derived by adding to the pitch diameter in an external thread, or subtracting from the pitch diameter in an internal thread, the cumulative effects of deviations from specified profile, including variations in lead and flank angle over a specified length of engagement. The effects of taper, out-of-roundness,and surface defects may be positive or negative on either external or internal threads.

Size-Nominal: Designation used for the purpose of general identification.

Spiral Flute: See *Flutes.*

Spiral Point: Angular fluting in the cutting face of the land at the chamfered end. It is formed at an angle with respect to the tap axis of opposite hand to that of rotation. Its length is usually greater than the chamfer length and its angle with respect to the tap axis is usually made great enough to direct the chips ahead of the tap. The tap may or may not have longitudinal flutes.

Thread Lead Angle: On a straight thread, the lead angle is the angle made by the helix of the thread at the pitch line with a plane perpendicular to the axis. On a taper thread, the lead angle at a given axial position is the angle made by the conical spiral of the thread, with the plane perpendicular to the axis, at the pitch line.

Table 3. ANSI Standard Fraction-Size Taps — Cut Thread Limits
ASME/ANSI B94.9-1987

Tap Size	Threads per Inch			Major Diameter			Pitch Diameter		
	NC UNC	NF UNF	NS UNS	Basic	Min.	Max.	Basic	Min.	Max.
1/8	…	…	40	0.1250	0.1266	0.1286	0.1088	0.1090	0.1105
5/32	…	…	32	0.1563	0.1585	0.1605	0.1360	0.1365	0.1380
3/16	…	…	24	0.1875	0.1903	0.1923	0.1604	0.1609	0.1624
3/16	…	…	32	0.1875	0.1897	0.1917	0.1672	0.1677	0.1692
1/4	20	…	…	0.2500	0.2532	0.2557	0.2175	0.2180	0.2200
1/4	…	28	…	0.2500	0.2524	0.2549	0.2268	0.2273	0.2288
5/16	18	…	…	0.3125	0.3160	0.3185	0.2764	0.2769	0.2789
5/16	…	24	…	0.3125	0.3153	0.3178	0.2854	0.2859	0.2874
3/8	16	…	…	0.3750	0.3789	0.3814	0.3344	0.3349	0.3369
3/8	…	24	…	0.3750	0.3778	0.3803	0.3479	0.3484	0.3499
7/16	14	…	…	0.4375	0.4419	0.4449	0.3911	0.3916	0.3941
7/16	…	20	…	0.4375	0.4407	0.4437	0.4050	0.4055	0.4075
1/2	13	…	…	0.5000	0.5047	0.5077	0.4500	0.4505	0.4530
1/2	…	20	…	0.5000	0.5032	0.5062	0.4675	0.4680	0.4700
9/16	12	…	…	0.5625	0.5675	0.5705	0.5084	0.5089	0.5114
9/16	…	18	…	0.5625	0.5660	0.5690	0.5264	0.5269	0.5289
5/8	11	…	…	0.6250	0.6304	0.6334	0.5660	0.5665	0.5690
5/8	…	18	…	0.6250	0.6285	0.6315	0.5889	0.5894	0.5914
3/4	10	…	…	0.7500	0.7559	0.7599	0.6850	0.6855	0.6885
3/4	…	16	…	0.7500	0.7539	0.7579	0.7094	0.7099	0.7124
7/8	9	…	…	0.8750	0.8820	0.8860	0.8028	0.8038	0.8068
7/8	…	14	…	0.8750	0.8799	0.8839	0.8286	0.8296	0.8321
1	8	…	…	1.0000	1.0078	1.0118	0.9188	0.9198	0.9228
1	…	12	…	1.0000	1.0055	1.0095	0.9459	0.9469	0.9494
1	…	…	14	1.0000	1.0049	1.0089	0.9536	0.9546	0.9571
1 1/8	7	…	…	1.1250	1.1337	1.1382	1.0322	1.0332	1.0367
1 1/8	…	12	…	1.1250	1.1305	1.1350	1.0709	1.0719	1.0749
1 1/4	7	…	…	1.2500	1.2587	1.2632	1.1572	1.1582	1.1617
1 1/4	…	12	…	1.2500	1.2555	1.2600	1.1959	1.1969	1.1999
1 3/8	6	…	…	1.3750	1.3850	1.3895	1.2667	1.2677	1.2712
1 3/8	…	12	…	1.3750	1.3805	1.3850	1.3209	1.3219	1.3249
1 1/2	6	…	…	1.5000	1.5100	1.5145	1.3917	1.3927	1.3962
1 1/2	…	12	…	1.5000	1.5055	1.5100	1.4459	1.4469	1.4499
1 3/4	5	…	…	1.7500	1.7602	1.7657	1.6201	1.6216	1.6256
2	4 1/2	…	…	2.0000	2.0111	2.0166	1.8557	1.8572	1.8612

All dimensions are given in inches.

Lead Tolerance: Plus or minus 0.003 inch max. per inch of thread.

Angle Tolerance: Plus or minus 35 min. in half angle or 53 min. in full angle for 4½ to 5½ thds. per in.; 40 min. half angle and 60 min. full angle for 6 to 9 thds.; 45 min. half angle and 68 min. full angle for 10 to 28 thds.; 60 min. half angle and 90 min. full angle for 30 to 64 thds. per in.

Table 4. ANSI Standard Fractional-Size Taps — Ground Thread Limits *ASME/ANSI B94.9-1987*

Size	Threads per Inch			Major Diameter			Pitch Diameter Limits								
	NC UNC	NF UNF	NS UNS	Basic	Min.	Max.	Basic Pitch Dia.	H1 Limit Min.	H1 Limit Max.	H2 Limit Min.	H2 Limit Max.	H3 & H4[a] Limits Min.	H3 & H4[a] Limits Max.	H4,[a] H5[b] & H6[c] Limits Min.	H4,[a] H5[b] & H6[c] Limits Max.
1/4	20	…	…	0.2500	0.2533	0.2565	0.2175	0.2175	0.2180	0.2180	0.2185	0.2185	0.2190	0.2195[b]	0.2200[b]
1/4	…	28	…	0.2500	0.2523	0.2546	0.2268	0.2268	0.2273	0.2273	0.2278	0.2278	0.2283	0.2283[a]	0.2288[a]
5/16	18	…	…	0.3125	0.3161	0.3197	0.2764	0.2764	0.2769	0.2769	0.2774	0.2774	0.2779	0.2784[b]	0.2789[b]
5/16	…	24	…	0.3125	0.3152	0.3179	0.2854	0.2854	0.2859	0.2859	0.2864	0.2864	0.2869	0.2869[a]	0.2874[a]
3/8	16	…	…	0.3750	0.3790	0.3831	0.3344	0.3344	0.3349	0.3349	0.3354	0.3354	0.3359	0.3364[b]	0.3369[b]
3/8	…	24	…	0.3750	0.3777	0.3804	0.3479	0.3479	0.3484	0.3484	0.3489	0.3489	0.3494	0.3494[a]	0.3499[a]
7/16	14	…	…	0.4375	0.4422	0.4468	0.3911	…	…	0.3916	0.3921	0.3921	0.3926	0.3931[b]	0.3936[b]
7/16	…	20	…	0.4375	0.4408	0.4440	0.4050	…	…	…	…	0.4060	0.4065	0.4070[b]	0.4075[b]
1/2	13	…	…	0.5000	0.5050	0.5100	0.4500	0.4500	0.4505	0.4505	0.4510	0.4510	0.4515	0.4520[b]	0.4525[b]
1/2	…	20	…	0.5000	0.5033	0.5065	0.4675	0.4675	0.4680	0.4680	0.4685	0.4685	0.4690	0.4695[b]	0.4700[b]
9/16	12	…	…	0.5625	0.5679	0.5733	0.5084	…	…	…	…	0.5094	0.5099	0.5104[b]	0.5109[b]
9/16	…	18	…	0.5625	0.5661	0.5697	0.5264	…	…	0.5269	0.5274	0.5274	0.5279	0.5284[b]	0.5289[b]
5/8	11	…	…	0.6250	0.6309	0.6368	0.5660	…	…	0.5665	0.5670	0.5670	0.5675	0.5680[b]	0.5685[b]
5/8	…	18	…	0.6250	0.6286	0.6322	0.5889	…	…	0.5894	0.5899	0.5899	0.5904	0.5909[b]	0.5914[b]
11/16	…	…	11	0.6875	0.6934	0.6993	0.6285	…	…	…	…	0.6295	0.6300	…	…
11/16	…	…	16	0.6875	0.6915	0.6956	0.6469	…	…	…	…	0.6479	0.6484	…	…
3/4	10	…	…	0.7500	0.7565	0.7630	0.6850	…	…	0.6855	0.6860	0.6860	0.6865	0.6870[b]	0.6875[b]
3/4	…	16	…	0.7500	0.7540	0.7581	0.7094	0.7094	0.7099	0.7099	0.7104	0.7104	0.7109	0.7114[b]	0.7119[b]
7/8	9	…	…	0.8750	0.8822	0.8894	0.8028	…	…	…	…	0.8043[a]	0.8048[a]	0.8053[c]	0.8058[c]
7/8	…	14	…	0.8750	0.8797	0.8843	0.8286	…	…	0.8291	0.8296	0.8301[a]	0.8306[a]	…	…
1	8	…	…	1.0000	1.0081	1.0162	0.9188	…	…	…	…	0.9203[a]	0.9208[a]	0.9213[c]	0.9218[c]
1	…	12	…	1.0000	1.0054	1.0108	0.9459	…	…	…	…	0.9474[a]	0.9479[a]	…	…
1	…	…	14	1.0000	1.0047	1.0093	0.9536	…	…	…	…	0.9551[a]	0.9556[a]	…	…

[a] H4 limit value.
[b] H5 limit value.
[c] H6 li.

Table 5. ANSI Standard Fractional –Size Taps—Ground Thread Limits (ASME/ANSI B94.9–1987)

Size	Threads per Inch			Major Diameter			Pitch Diameter Limits		
	NC UNC	NF UNF	NS UNS	Basic	Min.	Max.	Basic Pitch Dia.	H4 Limit Min.	H4 Limit Max.
1⅛	7	…	…	1.1250	1.1343	1.1436	1.0322	1.0332	1.0342
1⅛	…	12	…	1.1250	1.1304	1.1358	1.0709	1.0719	1.0729
1¼	7	…	…	1.2500	1.2593	1.2686	1.1572	1.1582	1.1592
1¼	…	12	…	1.2500	1.2554	1.2608	1.1959	1.1969	1.1979
1⅜	6	…	…	1.3750	1.3859	1.3967	1.2667	1.2677	1.2687
1⅜	…	12	…	1.3750	1.3804	1.3858	1.3209	1.3219	1.3229
1½	6	…	…	1.5000	1.5109	1.5217	1.3917	1.3927	1.3937
1½	…	12	…	1.5000	1.5054	1.5108	1.4459	1.4469	1.4479

All dimensions are given in inches.

Lead Tolerance: Plus or minus 0.0005 inch within any two threads not farther apart than one inch.

Angle Tolerance: Plus or minus 25 min. in half angle for 6 to 9 threads per inch; plus or minus 30 min. in half angle for 10 to 28 threads per inch.

For an explanation of the significance of the H4 limit value range see *Standard System Tap Thread Limits and Identification for Unified Inch Screw Threads, Ground Thread* starting on page 896.

Table 6. ANSI Standard Machine Screw Taps — Ground Thread Limits *ASME/ANSI B94.9-1987*

Size	Threads per Inch			Major Diameter			Pitch Diameter Limits[a]						
	NC UNC	NF UNF	NS UNS	Basic	Min.	Max.	Basic Pitch Dia.	H1 Limit Min.	H1 Limit Max.	H2 Limit Min.	H2 Limit Max.	H3 Limit Min.	H3 Limit Max.
0	…	80	…	0.0600	0.0605	0.0616	0.0519	0.0519	0.0524	0.0524	0.0529	…	…
1	64	…	…	0.0730	0.0736	0.0750	0.0629	0.0629	0.0634	0.0634	0.0639	…	…
1	…	72	…	0.0730	0.0736	0.0748	0.0640	0.0640	0.0645	0.0645	0.0650	…	…
2	56	…	…	0.0860	0.0867	0.0883	0.0744	0.0744	0.0749	0.0749	0.0754	…	…
2	…	64	…	0.0860	0.0866	0.0880	0.0759	…	…	0.0764	0.0769	…	…
3	48	…	…	0.0990	0.0999	0.1017	0.0855	0.0855	0.0860	0.0860	0.0865	…	…
3	…	56	…	0.0990	0.0997	0.1013	0.0874	0.0874	0.0879	0.0879	0.0884	…	…
4	…	…	36	0.1120	0.1135	0.1156	0.0940	…	…	0.0945	0.0950	…	…
4	40	…	…	0.1120	0.1133	0.1152	0.0958	0.0958	0.0963	0.0963	0.0968	…	…
4	…	48	…	0.1120	0.1129	0.1147	0.0985	0.0985	0.0990	0.0990	0.0995	…	…
5	40	…	…	0.1250	0.1263	0.1282	0.1088	0.1088	0.1093	0.1093	0.1098	…	…
5	…	44	…	0.1250	0.1263	0.1280	0.1102	…	…	0.1107	0.1112	…	…
6	32	…	…	0.1380	0.1401	0.1421	0.1177	0.1177	0.1182	0.1182	0.1187	0.1187	0.1192
6	…	40	…	0.1380	0.1393	0.1412	0.1218	0.1218	0.1223	0.1223	0.1228	…	…
8	32	…	…	0.1640	0.1661	0.1681	0.1437	0.1437	0.1442	0.1442	0.1447	0.1447	0.1452
8	…	36	…	0.1640	0.1655	0.1676	0.1460	…	…	0.1465	0.1470	…	…
10	24	…	…	0.1900	0.1927	0.1954	0.1629	0.1629	0.1634	0.1634	0.1639	0.1639	0.1644
10	…	32	…	0.1900	0.1921	0.1941	0.1697	0.1697	0.1702	0.1702	0.1707	0.1707	0.1712
12	24	…	…	0.2160	0.2187	0.2214	0.1889	…	…	…	…	0.1899	0.1904
12	…	28	…	0.2160	0.2183	0.2206	0.1928	…	…	…	…	0.1938	0.1943

[a] H7 limits (formerly designated as G) apply to same threads as H3 limits with the exception of the 12–24 and 12–28 threads. H7 limits have minimum and maximum major diameters 0.0020 inch larger than shown and minimum and maximum pitch diameters 0.0020 inch larger than shown for H3 limits.

All dimensions are given in inches.

Lead Tolerance: Plus or minus 0.0005 inch within any two threads not farther apart than one inch.

Angle Tolerance: Plus or minus 30 min. in half angle for 20 to 80 threads per inch.

For an explanation of the significance of the limit value ranges see *Standard System Tap Thread Limits and Identification for Unified Inch Screw Threads, Ground Thread* starting on page 896.

Table 7. ANSI Standard Machine Screw Taps — Cut Threads Limits
ASME/ANSI B94.9-1987

Size	Threads per Inch NC UNC	NF UNF	NS UNS	Major Diameter Basic	Min.	Max.	Pitch Diameter Basic	Min.	Max.
0	…	80	…	0.0600	0.0609	0.0624	0.0519	0.0521	0.0531
1	64	…	…	0.0730	0.0740	0.0755	0.0629	0.0631	0.0641
1	…	72	…	0.0730	0.0740	0.0755	0.0640	0.0642	0.0652
2	56	…	…	0.0860	0.0872	0.0887	0.0744	0.0746	0.0756
2	…	64	…	0.0860	0.0870	0.0885	0.0759	0.0761	0.0771
3	48	…	…	0.0990	0.1003	0.1018	0.0855	0.0857	0.0867
3	…	56	…	0.0990	0.1002	0.1017	0.0874	0.0876	0.0886
4	…	…	36	0.1120	0.1137	0.1157	0.0940	0.0942	0.0957
4	40	…	…	0.1120	0.1136	0.1156	0.0958	0.0960	0.0975
4	…	48	…	0.1120	0.1133	0.1153	0.0985	0.0987	0.1002
5	40	…	…	0.1250	0.1266	0.1286	0.1088	0.1090	0.1105
6	32	…	…	0.1380	0.1402	0.1422	0.1177	0.1182	0.1197
6	…	…	36	0.1380	0.1397	0.1417	0.1200	0.1202	0.1217
6	…	40	…	0.1380	0.1396	0.1416	0.1218	0.1220	0.1235
8	32	…	…	0.1640	0.1662	0.1682	0.1437	0.1442	0.1457
8	…	36	…	0.1640	0.1657	0.1677	0.1460	0.1462	0.1477
8	…	…	40	0.1640	0.1656	0.1676	0.1478	0.1480	0.1495
10	24	…	…	0.1900	0.1928	0.1948	0.1629	0.1634	0.1649
10	…	32	…	0.1900	0.1922	0.1942	0.1697	0.1702	0.1717
12	24	…	…	0.2160	0.2188	0.2208	0.1889	0.1894	0.1909
12	…	28	…	0.2160	0.2184	0.2204	0.1928	0.1933	0.1948
14	…	…	24	0.2420	0.2448	0.2473	0.2149	0.2154	0.2174

All dimensions are given in inches.

Lead Tolerance: Plus or minus 0.003 inch per inch of thread. *Angle Tolerance:* Plus or minus 45 min. in half angle and 68 min. in full angle for 20 to 28 threads per inch; plus or minus 60 min. in half angle and 90 min. in full angle for 30 or more threads per inch.

Table 8a. ANSI Standard Metric Tap Ground Thread Limits in Inches — M Profile
ASME/ANSI B94.9-1987

Nominal Diam, mm	Pitch, mm	Major Diameter (Inches) Basic	Min	Max	Pitch Diameter (Inches) Basic	Min	Max
1.6	0.35	0.06299	0.06409	0.06508	0.05406	0.05500	0.05559
2	0.4	0.07874	0.08000	0.08098	0.06850	0.06945	0.07004
2.5	0.45	0.09843	0.09984	0.10083	0.08693	0.08787	0.08846
3	0.5	0.11811	0.11969	0.12067	0.10531	0.10626	0.10685
3.5	0.6	0.13780	0.13969	0.14067	0.12244	0.12370	0.12449
4	0.7	0.15748	0.15969	0.16130	0.13957	0.14083	0.14161
4.5	0.75	0.17717	0.17953	0.18114	0.15799	0.15925	0.16004
5	0.8	0.19685	0.19937	0.20098	0.17638	0.17764	0.17843
6	1	0.23622	0.23937	0.24098	0.21063	0.21220	0.21319
7	1	0.27559	0.27874	0.28035	0.25000	0.25157	0.25256
8	1.25	0.31496	0.31890	0.32142	0.28299	0.28433	0.28555
10	1.5	0.39370	0.39843	0.40094	0.35535	0.35720	0.35843
12	1.75	0.47244	0.47795	0.48047	0.42768	0.42953	0.43075
14	2	0.55118	0.55748	0.56000	0.50004	0.50201	0.50362
16	2	0.62992	0.63622	0.63874	0.57878	0.58075	0.58236
20	2.5	0.78740	0.79538	0.79780	0.72346	0.72543	0.72705
24	3	0.94488	0.95433	0.95827	0.86815	0.87063	0.87224
30	3.5	1.18110	1.19213	1.19606	1.09161	1.09417	1.09622
36	4	1.41732	1.42992	1.43386	1.31504	1.31760	1.31965

Basic pitch diameter is the same as minimum pitch diameter of internal thread, Class 6H as shown in table starting on page 1769.

Pitch diameter limits are designated in the Standard as D3 for 1.6 to 3 mm diameter sizes, incl.: D4 for 3.5 to 5 mm sizes, incl.; D5 for 6 and 8 mm sizes; D6 for 10 and 12 mm sizes; D7 for 14 to 20 mm sizes, incl.; D8 for 24 mm size; and D9 for 30 and 36 mm sizes.

Angle tolerances are plus or minus 30 minutes in half angle for pitches ranging from 0.35 through 2.5 mm, incl. and plus or minus 25 minutes in half angle for pitches ranging from 3 to 4 mm, incl.

A maximum deviation of plus or minus 0.0005 inch within any two threads not farther apart than one inch is permitted.

Table 8b. ANSI Standard Metric Tap Ground Thread Limits in Millimeters— M Profile *ASME/ANSI B94.9-1987*

Nominal Diam, mm	Pitch, mm	Major Diameter (mm)			Pitch Diameter (mm)		
		Basic	Min	Max	Basic	Min	Max
1.6	0.35	1.600	1.628	1.653	1.373	1.397	1.412
2	0.4	2.000	2.032	2.057	1.740	1.764	1.779
2.5	0.45	2.500	2.536	2.561	2.208	2.232	2.247
3	0.5	3.000	3.040	3.065	2.675	2.699	2.714
3.5	0.6	3.500	3.548	3.573	3.110	3.142	3.162
4	0.7	4.000	4.056	4.097	3.545	3.577	3.597
4.5	0.75	4.500	4.560	4.601	4.013	4.045	4.065
5	0.8	5.000	5.064	5.105	4.480	4.512	4.532
6	1	6.000	6.080	6.121	5.350	5.390	5.415
7	1	7.000	7.080	7.121	6.350	6.390	6.415
8	1.25	8.000	8.100	8.164	7.188	7.222	7.253
10	1.5	10.000	10.120	10.184	9.026	9.073	9.104
12	1.75	12.000	12.140	12.204	10.863	10.910	10.941
14	2	14.000	14.160	14.224	12.701	12.751	12.792
16	2	16.000	16.160	16.224	14.701	14.751	14.792
20	2.5	20.000	20.200	20.264	18.376	18.426	18.467
24	3	24.000	24.240	24.340	22.051	22.114	22.155
30	3.5	30.000	30.280	30.380	27.727	27.792	27.844
36	4	36.000	36.320	36.420	33.402	33.467	33.519

Basic pitch diameter is the same as minimum pitch diameter of internal thread, Class 6H as shown in table starting on page 1769.

Pitch diameter limits are designated in the Standard as D3 for 1.6 to 3 mm diameter sizes, incl.: D4 for 3.5 to 5 mm sizes, incl.; D5 for 6 and 8 mm sizes; D6 for 10 and 12 mm sizes; D7 for 14 to 20 mm sizes, incl.; D8 for 24 mm size; and D9 for 30 and 36 mm sizes.

Angle tolerances are plus or minus 30 minutes in half angle for pitches ranging from 0.35 through 2.5 mm, incl. and plus or minus 25 minutes in half angle for pitches ranging from 3 to 4 mm, incl.

A maximum lead deviation of plus or minus 0.013 mm within any two threads not farther apart than 25 mm is permitted.

Table 9a. ANSI Standard Taper Pipe Taps — Cut Thread Tolerances for NPT and Ground Thread Tolerances for NPT, NPTF, and ANPT *ASME/ANSI B94.9-1987*

Nominal Size	Threads per Inch NPT, NPTF, or ANPT	Gage Measurement[a]			Taper per Foot, Inches			
			Tolerance Plus or Minus		Cut Thread		Ground Thread	
		Projection Inches	Cut Thread	Ground Thread	Min.	Max.	Min.	Max.
1/16	27	0.312	1/16	1/16	23/32	27/32	23/32	25/32
1/8	27	0.312	1/16	1/16	23/32	27/32	23/32	25/32
1/4	18	0.459	1/16	1/16	23/32	27/32	23/32	25/32
3/8	18	0.454	1/16	1/16	23/32	27/32	23/32	25/32
1/2	14	0.579	1/16	1/16	23/32	13/16	23/32	25/32
3/4	14	0.565	1/16	1/16	23/32	13/16	23/32	25/32
1	11½	0.678	3/32	3/32	23/32	13/16	23/32	25/32
1¼	11½	0.686	3/32	3/32	23/32	13/16	23/32	25/32
1½	11½	0.699	3/32	3/32	23/32	13/16	23/32	25/32
2	11½	0.667	3/32	3/32	23/32	13/16	23/32	25/32
2½	8	0.925	3/32	3/32	47/64	51/64	47/64	25/32
3	8	0.925	3/32	3/32	47/64	51/64	47/64	25/32
3½	8	0.938	1/8	1/8	47/64	51/64	47/64	25/32
4	8	0.950	1/8	1/8	47/64	51/64	47/64	25/32

[a] Distance that small end of tap projects through L1 taper ring gage (see ANSI B1.20.3).

All dimensions are given in inches.

Lead Tolerance: Plus or minus 0.003 inch per inch of cut thread and plus or minus 0.0005 inch per inch of ground thread.

Angle Tolerance: Plus or minus 40 min. in half angle and 60 min. in full angle for 8 cut threads per inch; plus or minus 45 min. in half angle and 68 min. in full angle for 11½ to 27 cut threads per inch; plus or minus 25 min. in half angle for 8 ground threads per inch; and plus and minus 30 min. in half angle for 11½ to 27 ground threads per inch.

Table 9b. ANSI Standard Taper Pipe Thread — Widths of Flats at Tap Crests and Roots for Cut Thread NPT and Ground Thread NPT, ANPT, and NPTF
ASME/ANSI B94.9-1987

Threads per Inch	Tap Flat Width at	Column I NPT—Cut and Ground Thread ANPT—Ground Thread		Column II NPTF—Cut and Ground Thread	
		Minimum[a]	Maximum	Minimum[a]	Maximum
27	{ Major Diameter	0.0014	0.0041	0.0040	0.0055
	{ Minor Diameter	…	0.0041	…	0.0040
18	{ Major Diameter	0.0021	0.0057	0.0050	0.0065
	{ Minor Diameter	…	0.0057	…	0.0050
14	{ Major Diameter	0.0027	0.0064	0.0050	0.0065
	{ Minor Diameter	…	0.0064	…	0.0050
11½	{ Major Diameter	0.0033	0.0073	0.0060	0.0083
	{ Minor Diameter	…	0.0073	…	0.0060
8	{ Major Diameter	0.0048	0.0090	0.0080	0.0103
	{ Minor Diameter	…	0.0090	…	0.0080

[a] Minimum minor diameter falts are not specified. May be sharp as practicable.

All dimensions are given in inches.

Note: Cut Thread taps made to Column I are marked NPT but are not recommended for ANPT applications. Ground Thread taps made to Column I are marked NPT and may be used for NPT and ANPT applications. Ground Thread taps made to Column II are marked NPTF and used for Dryseal application.

Table 10a. ANSI Standard Straight Pipe Taps (NPSF—Dryseal)—Ground Thread Limits *ASME/ANSI B94.9-1987*

Nominal Size, Inches	Threads per Inch	Major Diameter		Pitch Diameter			
		Min. *G*	Max. *H*	Plug at Gaging Notch *E*	Min. *K*	Max. *L*	Minor[a] Dia. Flat, Max.
1/16	27	0.3008	0.3018	0.2812	0.2772	0.2777	0.004
1/8	27	0.3932	0.3942	0.3736	0.3696	0.3701	0.004
1/4	18	0.5239	0.5249	0.4916	0.4859	0.4864	0.005
3/8	18	0.6593	0.6603	0.6270	0.6213	0.6218	0.005
1/2	14	0.8230	0.8240	0.7784	0.7712	0.7717	0.005
3/4	14	1.0335	1.0345	0.9889	0.9817	0.9822	0.005

[a] As specified or sharper.

Formulas For American Dryseal (NPSF) Ground Thread Taps					
Nominal Size, Inches	Major Diameter		Pitch Diameter		Max. Minor Dia.
	Min. *G*	Max. *H*	Min. *K*	Max. *L*	
1/16	$H - 0.0010$	$K + Q - 0.0005$	$L - 0.0005$	$E - F$	$M - Q$
1/8	$H - 0.0010$	$K + Q - 0.0005$	$L - 0.0005$	$E - F$	$M - Q$
1/4	$H - 0.0010$	$K + Q - 0.0005$	$L - 0.0005$	$E - F$	$M - Q$
3/8	$H - 0.0010$	$K + Q - 0.0005$	$L - 0.0005$	$E - F$	$M - Q$
1/2	$H - 0.0010$	$K + Q - 0.0005$	$L - 0.0005$	$E - F$	$M - Q$
3/4	$H - 0.0010$	$K + Q - 0.0005$	$L - 0.0005$	$E - F$	$M - Q$

Values to Use in Formulas				
Threads per Inch	*E*	*F*	*M*	*Q*
27	Pitch diameter of plug at gaging notch	0.0035	Actual measured pitch diameter	0.0251
18		0.0052		0.0395
14		0.0067		0.0533

All dimensions are given in inches.

Lead Tolerance: Plus or minus 0.0005 inch within any two threads not farther apart than one inch.

Angle Tolerance: Plus or minus 30 min. in half angle for 14 to 27 threads per inch.

Table 10b. ANSI Standard Straight Pipe Taps (NPS)—Cut Thread Limits
ASME/ANSI B94.9-1987

Nominal Size	Threads per Inch, NPS, NPSC	Size at Gaging Notch	Pitch Diameter		Values to Use in Formulas		
			Min.	Max.	*A*	*B*	*C*
⅛	27	0.3736	0.3721	0.3751	0.0267	0.0296	0.0257
¼	18	0.4916	0.4908	0.4938	} 0.0408	0.0444	0.0401
⅜	18	0.6270	0.6257	0.6292			
½	14	0.7784	0.7776	0.7811	} 0.0535	0.0571	0.0525
¾	14	0.9889	0.9876	0.9916			
1	11½	1.2386	1.2372	1.2412	0.0658	0.0696	0.0647

The following are approximate formulas, in which M = measured pitch diameter in inches:
Major dia., min. = $M + A$
Major dia., max. = $M + B$ Minor dia., max. = $M - C$

All dimensions are given in inches.

Lead Tolerance: Plus or minus 0.003 inch per inch of thread.

Angle Tolerance: All pitches, plus or minus 45 min. in half angle and 68 min. in full angle. Taps made to these specifications are to be marked NPS and used for NPSC thread form.

Table 10c. ANSI Standard Straight Pipe Taps (NPS)—Ground Thread Limits
ASME/ANSI B94.9-1987

Nominal Size, Inches	Threads per Inch, NPS, NPSC, NPSM	Major Diameter			Pitch Diameter		
		Plug at Gaging Notch	Min. *G*	Max. *H*	Plug at Gaging Notch *E*	Min. *K*	Max. *L*
⅛	27	0.3983	0.4022	0.4032	0.3736	0.3746	0.3751
¼	18	0.5286	0.5347	0.5357	0.4916	0.4933	0.4938
⅜	18	0.6640	0.6701	0.6711	0.6270	0.6287	0.6292
½	14	0.8260	0.8347	0.8357	0.7784	0.7806	0.7811
¾	14	1.0364	1.0447	1.0457	0.9889	0.9906	0.9916
1	11½	1.2966	1.3062	1.3077	1.2386	1.2402	1.2412

Formulas for NPS Ground Thread Taps[a]

Nominal Size	Major Diameter		Minor Dia.	Threads per Inch	*A*	*B*
	Min. *G*	Max. *H*	Max.	27	0.0296	0.0257
⅛	$H - 0.0010$	$(K + A) - 0.0010$	$M - B$	18	0.0444	0.0401
¼ to ¾	$H - 0.0010$	$(K + A) - 0.0020$	$M - B$	14	0.0571	0.0525
1	$H - 0.0015$	$(K + A) - 0.0021$	$M - B$	11½	0.0696	0.0647

The maximum Pitch Diameter of tap is based upon an allowance deducted from the maximum product pitch diameter of NPSC or NPSM, whichever is smaller.

The minimum Pitch Diameter of tap is derived by subtracting the ground thread pitch diameter tolerance for actual equivalent size.

[a] In the formulas, M equals the actual measured pitch diameter.

All dimensions are given in inches.

Lead tolerance: Plus or minus 0.0005 inch within any two threads not farther apart than one inch.

Angle Tolerance: All pitches, plus or minus 30 min. in half angle. Taps made to these specifications are to be marked NPS and used for NPSC and NPSM.

Table 11a. ANSI Standard Ground Thread Straight Fluted Taps—Machine Screw Sizes *ASME/ANSI B94.9-1987*

STANDARD NUMBER OF FLUTES

OPTIONAL NUMBER OF FLUTES

Size	Basic Major Diameter	Threads per Inch			No. of Flutes	Pitch Dia. Limits and Chamfers[a]				Length Overall *A*	Length of Thread	Length of Square	Diameter of Shank	*E*
		NC UNC	NF UNF	NS UNS		H1	H2	H3	H7					
0	0.060	…	80	…	2	TPB	PB	…	…	$1\frac{5}{8}$	$\frac{5}{16}$	$\frac{3}{16}$	0.141	0.110
1	0.073	64	…	…	2	TPB	P	…	…	$1\frac{11}{16}$	$\frac{3}{16}$	$\frac{3}{16}$	0.141	0.110
1	0.073	…	72	…	2	TPB	PB	…	…	$1\frac{11}{16}$	$\frac{3}{8}$	$\frac{3}{16}$	0.141	0.110
2	0.086	56	…	…	2[b]	…	PB	…	…	$1\frac{3}{4}$	$\frac{7}{16}$	$\frac{3}{16}$	0.141	0.110
2	0.086	56	…	…	3	TPB	TPB	…	…	$1\frac{3}{4}$	$\frac{7}{16}$	$\frac{3}{16}$	0.141	0.110
2	0.086	…	64	…	3	…	TPB	…	…	$1\frac{3}{4}$	$\frac{7}{16}$	$\frac{3}{16}$	0.141	0.110
3	0.099	48	…	…	2[b]	…	PB	…	…	$1\frac{13}{16}$	$\frac{1}{2}$	$\frac{3}{16}$	0.141	0.110
3	0.099	48	…	…	3	P	TPB	…	…	$1\frac{13}{16}$	$\frac{1}{2}$	$\frac{3}{16}$	0.141	0.110
3	0.099	…	56	…	3	…	TPB	…	…	$1\frac{13}{16}$	$\frac{1}{2}$	$\frac{3}{16}$	0.141	0.110
4	0.112	…	…	36	3	…	TPB	…	…	$1\frac{7}{8}$	$\frac{9}{16}$	$\frac{3}{16}$	0.141	0.110
4	0.112	40	…	…	2[b]	P	PB	…	…	$1\frac{7}{8}$	$\frac{9}{16}$	$\frac{3}{16}$	0.141	0.110
4	0.112	40	…	…	3	…	TPB	…	…	$1\frac{7}{8}$	$\frac{9}{16}$	$\frac{3}{16}$	0.141	0.110
4	0.112	…	48	…	3	…	TPB	…	…	$1\frac{7}{8}$	$\frac{9}{16}$	$\frac{3}{16}$	0.141	0.110
5	0.125	40	…	…	2[b]	…	PB	…	…	$1\frac{15}{16}$	$\frac{5}{8}$	$\frac{3}{16}$	0.141	0.110
5	0.125	40	…	…	3	P	TPB	…	…	$1\frac{15}{16}$	$\frac{5}{8}$	$\frac{3}{16}$	0.141	0.110
5	0.125	…	44	…	3	…	TPB	…	…	$1\frac{15}{16}$	$\frac{5}{8}$	$\frac{3}{16}$	0.141	0.110
6	0.138	32	…	…	2[b]	P	PB	PB	…	2	$\frac{11}{16}$	$\frac{3}{16}$	0.141	0.110
6	0.138	32	…	…	3	TPB	TPB	TPB	PB	2	$\frac{11}{16}$	$\frac{3}{16}$	0.141	0.110
6	0.138	…	40	…	2[b]	…	P	…	…	2	$\frac{11}{16}$	$\frac{3}{16}$	0.141	0.110
6	0.138	…	40	…	3	P	TPB	…	…	2	$\frac{11}{16}$	$\frac{3}{16}$	0.141	0.110
8	0.164	32	…	…	2[b]	P	PB	PB	…	$2\frac{1}{8}$	$\frac{3}{4}$	$\frac{1}{4}$	0.168	0.131
8	0.164	32	…	…	3[b]	…	PB	PB	PB	$2\frac{1}{8}$	$\frac{3}{4}$	$\frac{1}{4}$	0.168	0.131
8	0.164	32	…	…	4	TPB	TPB	TPB	PB	$2\frac{1}{8}$	$\frac{3}{4}$	$\frac{1}{4}$	0.168	0.131
8	0.164	…	36	…	4	…	TPB	…	…	$2\frac{1}{8}$	$\frac{3}{4}$	$\frac{1}{4}$	0.168	0.131
10	0.190	24	…	…	2*	…	PB	PB	…	$2\frac{3}{8}$	$\frac{7}{8}$	$\frac{1}{4}$	0.194	0.152
10	0.190	24	…	…	3[b]	…	P	PB	…	$2\frac{3}{8}$	$\frac{7}{8}$	$\frac{1}{4}$	0.194	0.152
10	0.190	…	32	…	2[b]	P	PB	PB	…	$2\frac{3}{8}$	$\frac{7}{8}$	$\frac{1}{4}$	0.194	0.152
10	0.190	…	32	…	3[b]	…	PB	PB	PB	$2\frac{3}{8}$	$\frac{7}{8}$	$\frac{1}{4}$	0.194	0.152
10	0.190	24	32	…	4	TPB	TPB	TPB	PB	$2\frac{3}{8}$	$\frac{7}{8}$	$\frac{1}{4}$	0.194	0.152
12	0.216	24	…	…	4	…	…	TPB	…	$2\frac{3}{8}$	$\frac{15}{16}$	$\frac{9}{32}$	0.220	0.165
12	0.216	…	28	…	4	…	…	TPB	…	$2\frac{3}{8}$	$\frac{15}{16}$	$\frac{9}{32}$	0.220	0.165

[a] Chamfer designations are: T = taper, P = plug, and B = bottoming.

[b] Optional number of flutes.

All dimensions are given in inches.

These taps are standard as high-speed steel taps with ground threads, with standard and optional number of flutes and pitch diameter limits and chamfers as given in the table.

These are style 1 taps and have external centers on thread and shank ends (may be removed on thread end of bottoming taps).

For standard thread limits see Table 6. For eccentricity tolerances see Table 25.

Tolerances: Numbers 0 to 12 size range — $A, \pm\frac{1}{32}$; $B, \pm\frac{3}{64}$; $C, \pm\frac{1}{32}$; $D, -0.0015$; $E, -0.004$.

Table 11b. ANSI Standard Cut Thread Straight Fluted Taps — Machine Screw Sizes *ASME/ANSI B94.9-1987*

Size	Basic Major Diameter	Threads per Inch: Carbon Steel NC UNC	Carbon Steel NF UNF	Carbon Steel NS UNS	HS Steel NC UNC	HS Steel NF UNF	Number of Flutes	Length Overall, *A*	Length of Thread, *B*	Length of Square, *C*	Diameter of Shank, *D*	Size of Square, *E*
0	0.060	…	80[a]	…	…	…	2	$1\frac{5}{8}$	$\frac{5}{16}$	$\frac{3}{16}$	0.141	0.110
1	0.073	64[a]	72[a]	…	…	…	2	$1\frac{11}{16}$	$\frac{3}{8}$	$\frac{3}{16}$	0.141	0.110
2	0.086	56	64[a]	…	…	…	3	$1\frac{3}{4}$	$\frac{7}{16}$	$\frac{3}{16}$	0.141	0.110
3	0.099	48[a]	56[a]	…	…	…	3	$1\frac{13}{16}$	$\frac{1}{2}$	$\frac{3}{16}$	0.141	0.110
4	0.112	40	48[a]	36[a]	40[a]	…	3	$1\frac{7}{8}$	$\frac{9}{16}$	$\frac{3}{16}$	0.141	0.110
5	0.125	40	…	…	40[a]	…	3	$1\frac{15}{16}$	$\frac{5}{8}$	$\frac{3}{16}$	0.141	0.110
6	0.138	32	40[a]	36[a]	32	…	3	2	$\frac{11}{16}$	$\frac{3}{16}$	0.141	0.110
8	0.164	32	36[a]	40[a]	32	…	4	$2\frac{1}{8}$	$\frac{3}{4}$	$\frac{1}{4}$	0.168	0.131
10	0.190	24	32	…	24	32	4	$2\frac{3}{8}$	$\frac{7}{8}$	$\frac{1}{4}$	0.194	0.152
12	0.216	24	28[a]	…	24	…	4	$2\frac{3}{8}$	$\frac{15}{16}$	$\frac{9}{32}$	0.220	0.165
14	0.242	…	…	24[a]	…	…	4	$2\frac{1}{2}$	1	$\frac{5}{16}$	0.255	0.191

[a] These taps are standard with plug chamfer only. All others are standard with taper, plug or bottoming chamfer.

Tolerances for General Dimensions					
Element	Range	Tolerance	Element	Range	Tolerance
Length Overall, *A*	0 to 14 incl	$\pm\frac{1}{32}$	Diameter of Shank, *D*	0 to 12 incl	−0.004
Length of Thread, *B*	0 to 12 incl	$\pm\frac{3}{64}$		14	−0.005
	14	$\pm\frac{1}{16}$	Size of Square, *E*	0 to 14 incl	−0.004
Length of Square, *C*	0 to 14 incl	$\pm\frac{1}{32}$			

All dimensions are given in inches.

Styles 1 and 2 cut thread taps have optional style centers on thread and shank ends.

For standard thread limits see Table 7. For eccentricity tolerances see Table 25.

Table 12. ANSI Standard Nut Taps *ASME/ANSI B94.9-1987*

Dia. of Tap	Threads per Inch NC,UNC	Number of Flutes	Length Overall, *A*	Length of Thread, *B*	Length of Square, *C*	Diameter of Shank, *D*	Size of Square, *E*
$\frac{1}{4}$	20	4	5	$1\frac{5}{8}$	$\frac{9}{16}$	0.185	0.139
$\frac{5}{16}$	18	4	$5\frac{1}{2}$	$1\frac{13}{16}$	$\frac{5}{8}$	0.240	0.180
$\frac{3}{8}$	16	4	6	2	$\frac{11}{16}$	0.294	0.220
$\frac{1}{2}$	13	4	7	$2\frac{1}{2}$	$\frac{7}{8}$	0.400	0.300

Tolerances for General Dimensions					
Element	Diameter Range	Tolerance	Element	Diameter Range	Tolerance
Overall Length, *A*	$\frac{1}{4}$ to $\frac{1}{2}$	$\pm\frac{1}{16}$	Shank Diameter,D	$\frac{1}{4}$ to $\frac{1}{2}$	−0.005
Thread Length, *B*	$\frac{1}{4}$ to $\frac{1}{2}$	$\pm\frac{1}{16}$	Size of Square,E	$\frac{1}{4}$ to $\frac{1}{2}$	−0.004
Square Length, *C*	$\frac{1}{4}$ to $\frac{1}{2}$	$\pm\frac{1}{32}$			

All dimensions are given in inches. These ground thread high-speed steel taps are standard in H3 limit only. All taps have an internal center in thread end. For standard limits see Table 4.

Chamfer *J* is made $\frac{1}{2}$ ro $\frac{3}{4}$ the thread length of *B*.

Table 13. ANSI Standard Spiral-Pointed Taps—Machine Screw Sizes

ASME/ANSI B94.9-1987

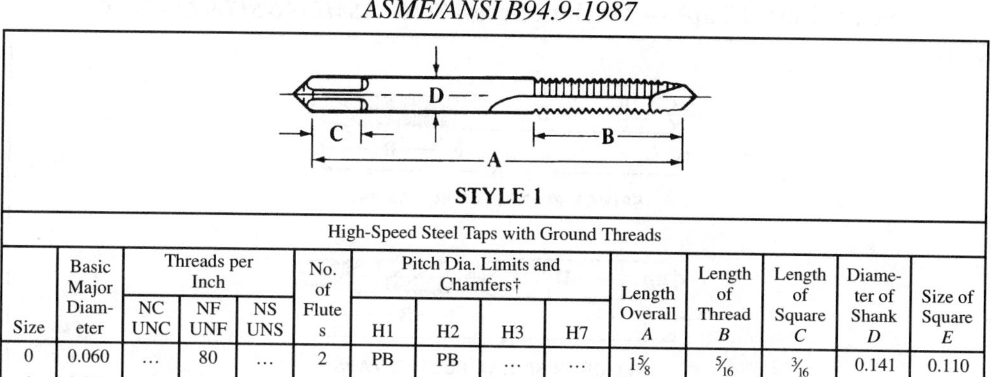

High-Speed Steel Taps with Ground Threads

Size	Basic Major Diameter	Threads per Inch NC UNC	NF UNF	NS UNS	No. of Flutes	Pitch Dia. Limits and Chamfers† H1	H2	H3	H7	Length Overall *A*	Length of Thread *B*	Length of Square *C*	Diameter of Shank *D*	Size of Square *E*
0	0.060	…	80	…	2	PB	PB	…	…	$1\frac{5}{8}$	$\frac{5}{16}$	$\frac{3}{16}$	0.141	0.110
1	0.073	64	72	…	2	P	P	…	…	$1\frac{11}{16}$	$\frac{3}{8}$	$\frac{3}{16}$	0.141	0.110
2	0.086	56	…	…	2	PB	PB	…	…	$1\frac{3}{4}$	$\frac{7}{16}$	$\frac{3}{16}$	0.141	0.110
2	0.086	…	64	…	2	…	P	…	…	$1\frac{3}{4}$	$\frac{7}{16}$	$\frac{3}{16}$	0.141	0.110
3	0.099	48	…	…	2	…	PB	…	…	$1\frac{13}{16}$	$\frac{1}{2}$	$\frac{3}{16}$	0.141	0.110
3	0.099	…	56	…	2	P	P	…	…	$1\frac{13}{16}$	$\frac{1}{2}$	$\frac{3}{16}$	0.141	0.110
4	0.112	…	…	36	2	…	P	…	…	$1\frac{7}{8}$	$\frac{9}{16}$	$\frac{3}{16}$	0.141	0.110
4	0.112	40	…	…	2	P	PB	…	…	$1\frac{7}{8}$	$\frac{9}{16}$	$\frac{3}{16}$	0.141	0.110
4	0.112	…	48	…	2	P	PB	…	…	$1\frac{7}{8}$	$\frac{9}{16}$	$\frac{3}{16}$	0.141	0.110
5	0.125	40	…	…	2	P	PB	…	…	$1\frac{15}{16}$	$\frac{5}{8}$	$\frac{3}{16}$	0.141	0.110
5	0.125	…	44	…	2	…	P	…	…	$1\frac{15}{16}$	$\frac{5}{8}$	$\frac{3}{16}$	0.141	0.110
6	0.138	32	…	…	2	P	PB	PB	PB	2	$\frac{11}{16}$	$\frac{3}{16}$	0.141	0.110
6	0.138	…	40	…	2	…	PB	…	…	2	$\frac{11}{16}$	$\frac{3}{16}$	0.141	0.110
8	0.164	32	…	…	2	P	PB	PB	PB	$2\frac{1}{8}$	$\frac{3}{4}$	$\frac{1}{4}$	0.168	0.131
8	0.164	…	36	…	2	…	P	…	…	$2\frac{1}{8}$	$\frac{3}{4}$	$\frac{1}{4}$	0.168	0.131
10	0.190	24	…	…	2	P	PB	PB	P	$2\frac{3}{8}$	$\frac{7}{8}$	$\frac{1}{4}$	0.194	0.152
10	0.190	…	32	…	2	PB	PB	PB	P	$2\frac{3}{8}$	$\frac{7}{8}$	$\frac{1}{4}$	0.194	0.152
12	0.216	24	…	…	2	…	…	PB	…	$2\frac{3}{8}$	$\frac{15}{16}$	$\frac{9}{32}$	0.220	0.165
12	0.216	…	28	…	2	…	…	P	…	$2\frac{3}{8}$	$\frac{15}{16}$	$\frac{9}{32}$	0.220	0.165

High-Speed and Carbon Steel Taps with Cut Threads

Size	Basic Major Diameter	Threads per Inch, Carbon Steel NC UNC	Carbon Steel NF UNF	HS Steel NC UNC	HS Steel NF UNF	No. of Flutes	Length Overall, *A*	Length of Thread, *B*	Length of Square, *C*	Diameter of Shank, *D*	Size of Square, *E*
4	0.112	…	…	40	…	2	$1\frac{7}{8}$	$\frac{9}{16}$	$\frac{3}{16}$	0.141	0.110
5	0.125	…	…	40	…	2	$1\frac{15}{16}$	$\frac{5}{8}$	$\frac{3}{16}$	0.141	0.110
6	0.138	32	…	32	…	2	2	$\frac{11}{16}$	$\frac{3}{16}$	0.141	0.110
8	0.164	32	…	32	…	2	$2\frac{1}{8}$	$\frac{3}{4}$	$\frac{1}{4}$	0.168	0.131
10	0.190	24	32	24	32	2	$2\frac{3}{8}$	$\frac{7}{8}$	$\frac{1}{4}$	0.194	0.152
12	0.216	…	…	24	…	2	$2\frac{3}{8}$	$\frac{15}{16}$	$\frac{9}{32}$	0.220	0.165

Tolerances for General Dimensions

Element	Size Range	Tolerance Ground Thread	Tolerance Cut Thread	Element	Size Range	Tolerance Ground Thread	Tolerance Cut Thread
Overall Length, *A*	0 to 12	$\pm\frac{1}{32}$	$\pm\frac{1}{32}$	Shank Diameter, *D*	0 to 12	−0.0015	−0.004
Thread Length, *B*	0 to 12	$\pm\frac{3}{64}$	$\pm\frac{3}{64}$	Size of Square, *E*	0 to 12	−0.004	−0.004
Square Length, *C*	0 to 12	$\pm\frac{1}{32}$	$\pm\frac{1}{32}$				

All dimensions are in inches. Chamfer designations are: P = plug and B = bottoming. Cut thread taps are standard with plug chamfer only. Style 1 ground thread taps have external centers on thread and shank ends (may be removed on thread end of bottoming taps). Style 1 cut thread taps have optional style centers on thread and shank ends. Standard thread limits for ground threads are given in Table 6 and for cut threads in Table 7. For eccentricity tolerances see Table 25.

Table 14. ANSI Standard Spiral Pointed Only and Regular and Fast Spiral-Fluted Taps — Machine Screw Sizes *ASME/ANSI B94.9-1987*

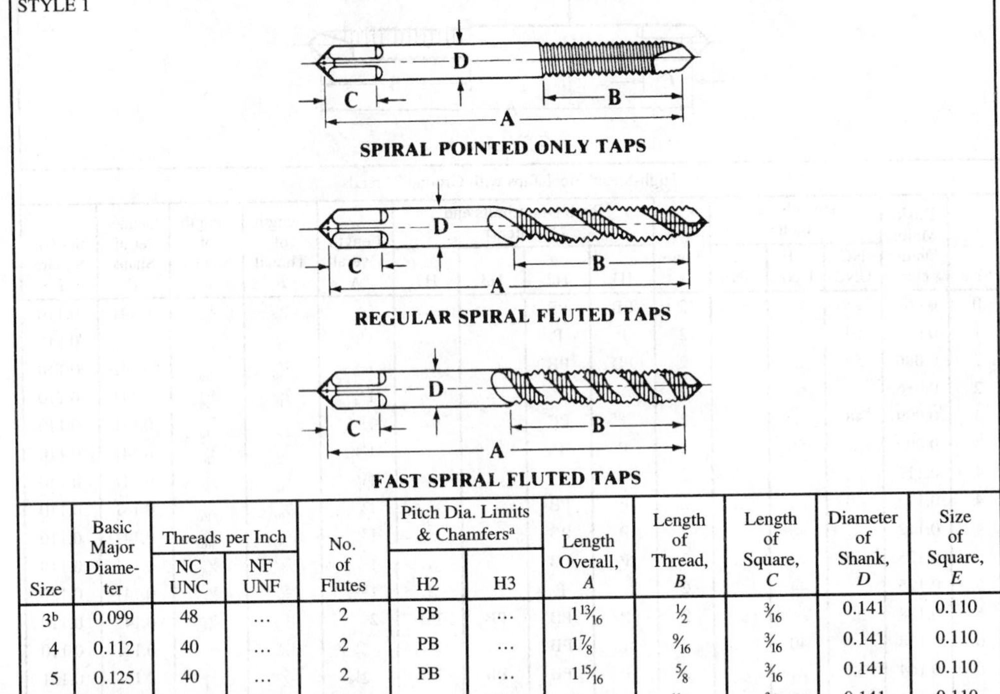

Size	Basic Major Diameter	Threads per Inch		No. of Flutes	Pitch Dia. Limits & Chamfers[a]		Length Overall, *A*	Length of Thread, *B*	Length of Square, *C*	Diameter of Shank, *D*	Size of Square, *E*
		NC UNC	NF UNF		H2	H3					
3[b]	0.099	48	…	2	PB	…	$1\frac{13}{16}$	$\frac{1}{2}$	$\frac{3}{16}$	0.141	0.110
4	0.112	40	…	2	PB	…	$1\frac{7}{8}$	$\frac{9}{16}$	$\frac{3}{16}$	0.141	0.110
5	0.125	40	…	2	PB	…	$1\frac{15}{16}$	$\frac{5}{8}$	$\frac{3}{16}$	0.141	0.110
6	0.138	32	…	2	…	PB	2	$\frac{11}{16}$	$\frac{3}{16}$	0.141	0.110
8	0.164	32	…	2[c], 3[b]	…	PB	$2\frac{1}{8}$	$\frac{3}{4}$	$\frac{1}{4}$	0.168	0.131
10	0.190	24	32	2[c], 3[b]	…	PB	$2\frac{3}{8}$	$\frac{7}{8}$	$\frac{1}{4}$	0.194	0.152
12[d]	0.216	24	…	2[c], 3[b]	…	PB	$2\frac{3}{8}$	$\frac{15}{16}$	$\frac{9}{32}$	0.220	0.165

[a] Bottom chamfer applies only to regular and fast spiral-fluted machine screw taps.
[b] Applies only to fast spiral-fluted machine screw taps.
[c] Does not apply to fast spiral-fluted machine screw taps.
[d] Does not apply to regular spiral-fluted machine screw taps.

Tolerances for General Dimensions					
Element	Size Range	Tolerance	Element	Size Range	Tolerance
Overall Length, *A*	3 to 12	$\pm\frac{1}{32}$	Shank Diameter, *D*	3 to 12	−0.0015
Thread Length, *B*	3 to 12	$\pm\frac{3}{64}$	Size of Square, *E*	3 to 12	−0.004
Square Length, *C*	3 to 12	$\pm\frac{1}{32}$			

All dimensions are given in inches. These standard taps are made of high-speed steel with ground threads. For standard thread limits see Table 6. For eccentricity tolerances see Table 25.

Spiral Pointed Only Taps: These taps are standard with plug chamfer only. They are provided with a spiral point only; the balance of the threaded section is left unfluted. These Style 1 taps have external centers on thread and shank ends.

Regular Spiral Fluted Taps: These taps have right-hand spiral flutes with a helix angle of from 25 to 35 degrees.

Fast Spiral Fluted Taps: These taps have right-hand spiral flutes with a helix angle of from 45 to 60 degrees.

Both regular and fast spiral-fluted Style 1 taps have external centers on thread and shank ends (may be removed on thread end of bottoming taps).

Chamfer designations: P = plug and B = bottoming.

Table 15a. ANSI Standard Ground Thread Straight Fluted Taps— Fractional Sizes *ASME/ANSI B94.9-1987*

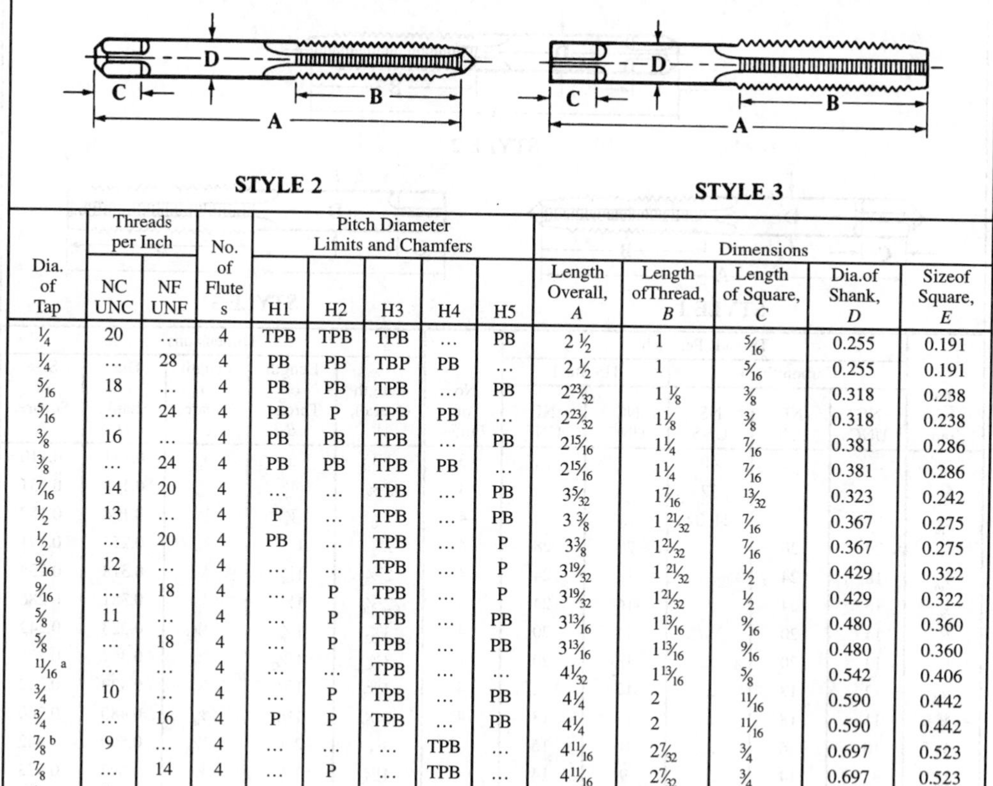

Dia. of Tap	Threads per Inch		No. of Flutes	Pitch Diameter Limits and Chamfers					Dimensions				
	NC UNC	NF UNF		H1	H2	H3	H4	H5	Length Overall, *A*	Length of Thread, *B*	Length of Square, *C*	Dia. of Shank, *D*	Size of Square, *E*
1/4	20	…	4	TPB	TPB	TPB	…	PB	2 1/2	1	5/16	0.255	0.191
1/4	…	28	4	PB	PB	TBP	PB	…	2 1/2	1	5/16	0.255	0.191
5/16	18	…	4	PB	PB	TPB	…	PB	2 23/32	1 1/8	3/8	0.318	0.238
5/16	…	24	4	PB	P	TPB	PB	…	2 23/32	1 1/8	3/8	0.318	0.238
3/8	16	…	4	PB	PB	TPB	…	PB	2 15/16	1 1/4	7/16	0.381	0.286
3/8	…	24	4	PB	PB	TPB	PB	…	2 15/16	1 1/4	7/16	0.381	0.286
7/16	14	20	4	…	…	TPB	…	PB	3 5/32	1 7/16	13/32	0.323	0.242
1/2	13	…	4	P	…	TPB	…	PB	3 3/8	1 21/32	7/16	0.367	0.275
1/2	…	20	4	PB	…	TPB	…	P	3 3/8	1 21/32	7/16	0.367	0.275
9/16	12	…	4	…	…	TPB	…	P	3 19/32	1 21/32	1/2	0.429	0.322
9/16	…	18	4	…	P	TPB	…	P	3 19/32	1 21/32	1/2	0.429	0.322
5/8	11	…	4	…	P	TPB	…	PB	3 13/16	1 13/16	9/16	0.480	0.360
5/8	…	18	4	…	P	TPB	…	PB	3 13/16	1 13/16	9/16	0.480	0.360
11/16 [a]	…	…	4	…	…	TPB	…	…	4 1/32	1 13/16	5/8	0.542	0.406
3/4	10	…	4	…	P	TPB	…	PB	4 1/4	2	11/16	0.590	0.442
3/4	…	16	4	P	P	TPB	…	PB	4 1/4	2	11/16	0.590	0.442
7/8 [b]	9	…	4	…	…	…	TPB	…	4 11/16	2 7/32	3/4	0.697	0.523
7/8	…	14	4	…	P	…	TPB	…	4 11/16	2 7/32	3/4	0.697	0.523
1 [b]	8	…	4	…	…	…	TPB	…	5 1/8	2 1/2	13/16	0.800	0.600
1	…	12	4	…	…	…	TPB	…	5 1/8	2 1/2	13/16	0.800	0.600
1 [c]	…	…	4	…	…	…	TPB	…	5 1/8	2 1/2	13/16	0.800	0.600
1 1/8	7	12	4	…	…	…	TPB	…	5 7/16	2 9/16	7/8	0.896	0.672
1 1/4	7	12 [d]	4	…	…	…	TPB	…	5 3/4	2 9/16	1	1.021	0.766
1 3/8	6	12 [d]	4	…	…	…	TPB	…	6 1/16	3	1 1/16	1.108	0.831
1 1/2	6	12 [d]	4	…	…	…	TPB	…	6 3/8	3	1 1/8	1.233	0.925

[a] This size has 11 or 16 threads per inch NS-UNS.

[b] These sizes are also available with plug chamfer in H6 pitch diameter limits.

[c] This size has 14 threads per inch NS-UNS.

[d] In these sizes NF-UNF thread taps have six flutes.

Tolerances for General Dimensions					
Element	Diameter Range	Tolerance	Element	Diameter Range	Tolerance
Length Overall, *A*	1/4 to 1 incl	±1/32	Diameter of Shank, *D*	1/4 to 5/8 incl	−0.0015
	1 1/8 to 1 1/2 incl	±1/16		11/16 to 1 1/2 incl	−0.002
Length of Thread, *B*	1/4 to 1/2 incl	±1/16	Size of Square, *E*	1/4 to 1/2 incl	−0.004
	9/16 to 1 1/2 incl	±3/32		9/16 to 1 incl	−0.006
Length of Square, *C*	1/4 to 1 incl	±1/32		1 1/8 to 1 1/2 incl	−0.008
	1 1/8 to 1 1/2 incl	±1/16			

All dimensions are given in inches.

These taps are standard in high-speed steel.

Chamfer designations are: T = taper, P = plug, and B = bottoming.

Style 2 taps, 3/8 inch and smaller, have external center on thread end (may be removed on bottoming taps) and external partial cone center on shank end with length of come approximately one-quarter of diameter of shank.

Style 3 taps, larger than 3/8 inch, have internal center in thread and shank ends.

For standared thread limits see Table 4. For eccentricity tolerances see Table 25.

Table 15b. ANSI Standard Cut Thread Straight Fluted Taps— Fractional Sizes *ASME/ANSI B94.9-1987*

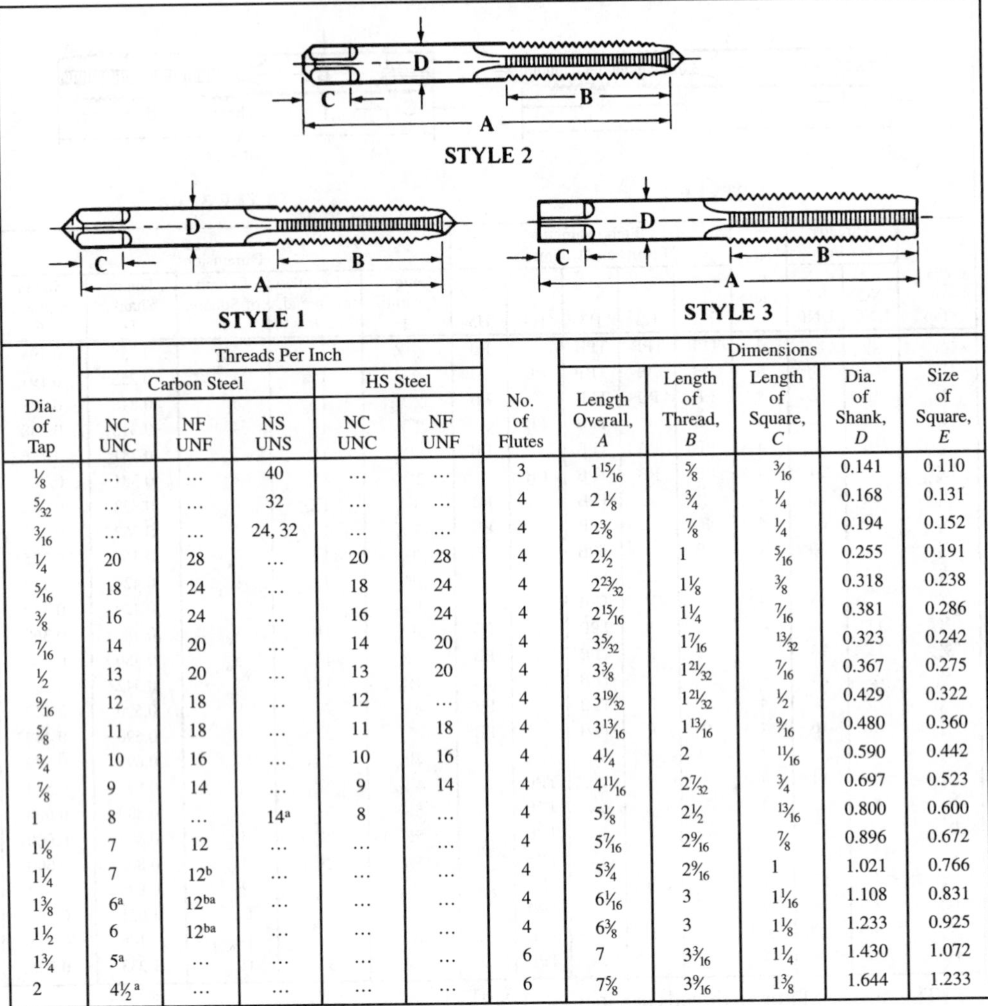

Dia. of Tap	Threads Per Inch					No. of Flutes	Dimensions				
	Carbon Steel			HS Steel			Length Overall, *A*	Length of Thread, *B*	Length of Square, *C*	Dia. of Shank, *D*	Size of Square, *E*
	NC UNC	NF UNF	NS UNS	NC UNC	NF UNF						
1/8	…	…	40	…	…	3	1 15/16	5/8	3/16	0.141	0.110
5/32	…	…	32	…	…	4	2 1/8	3/4	1/4	0.168	0.131
3/16	…	…	24, 32	…	…	4	2 3/8	7/8	1/4	0.194	0.152
1/4	20	28	…	20	28	4	2 1/2	1	5/16	0.255	0.191
5/16	18	24	…	18	24	4	2 23/32	1 1/8	3/8	0.318	0.238
3/8	16	24	…	16	24	4	2 15/16	1 1/4	7/16	0.381	0.286
7/16	14	20	…	14	20	4	3 5/32	1 7/16	13/32	0.323	0.242
1/2	13	20	…	13	20	4	3 3/8	1 21/32	7/16	0.367	0.275
9/16	12	18	…	12	…	4	3 19/32	1 21/32	1/2	0.429	0.322
5/8	11	18	…	11	18	4	3 13/16	1 13/16	9/16	0.480	0.360
3/4	10	16	…	10	16	4	4 1/4	2	11/16	0.590	0.442
7/8	9	14	…	9	14	4	4 11/16	2 7/32	3/4	0.697	0.523
1	8	…	14[a]	8	…	4	5 1/8	2 1/2	13/16	0.800	0.600
1 1/8	7	12	…	…	…	4	5 7/16	2 9/16	7/8	0.896	0.672
1 1/4	7	12[b]	…	…	…	4	5 3/4	2 9/16	1	1.021	0.766
1 3/8	6[a]	12[ba]	…	…	…	4	6 1/16	3	1 1/16	1.108	0.831
1 1/2	6	12[ba]	…	…	…	4	6 3/8	3	1 1/8	1.233	0.925
1 3/4	5[a]	…	…	…	…	6	7	3 3/16	1 1/4	1.430	1.072
2	4 1/2[a]	…	…	…	…	6	7 5/8	3 9/16	1 3/8	1.644	1.233

[a] Standard in plug chamfer only.

[b] In these sizes NF-UNF thread taps have six flutes.

Tolerances for General Dimensions					
Elements	Range	Tolerance	Elements	Range	Tolerance
Length Overall, *A*	1/16 to 1	±1/32	Diameter of Shank, *D*	1/16 to 3/16	−0.004
	1 1/8 to 2	±1/16		1/4 to 1	−0.005
Length of Thread, *B*	1/16 to 3/16	±3/64		1 1/8 to 2	−0.007
	1/4 to 1/2	±1/16	Size of Square, *E*	1/16 to 1/2	−0.004
	9/16 to 1 1/2	±3/32		9/16 to 1	−0.006
	1 5/8 to 2	±1/8		1/8 to 2	−0.008
Length of Square, *C*	1/16 to 1	±1/32			
	1 1/8 to 2	±1/16			

All dimensions are given in inches.

These taps are standard in carbon steel and high-speed steel.

Except where indicated, these taps are standard with taper, plug, or bottoming chamfer.

Cut thread taps, sizes 3/8 inch and smaller have optional style center on thread and shank ends; sizes larger than 3/8 inch have internal centers in thread and shank ends.

For standard thread limits see Table 3. For eccentricity tolerances see Table 25.

Table 16. ANSI Standard Straight Fluted (Optional Number of Flutes) and Spiral Pointed Taps—Fractional Sizes *ASME/ANSI B94.9-1987*

STYLE 2 STYLE 3

STRAIGHT FLUTED TAPS (OPTIONAL NUMBER OF FLUTES)

SPIRAL POINTED TAPS

Dia. of Tap	Threads per-Inch		No. of Flutes	Pitch Diameter Limits and Chamfers[a,b]					Length Overall, *A*	Length of Thread, *B*	Length of Square, *C*	Dia. of Shank, *D*	Size of Square, *E*
	NC UNC	NF UNF		H1	H2	H3	H4	H5					
Ground Thread High-Speed-Steel Straight Fluted Taps													
1/4	20	…	2	…	…	PB	…	…	2 1/2	1	5/16	0.255	0.191
1/4	20	…	3	P	P	PB	…	P	2 1/2	1	5/16	0.255	0.191
1/4	…	28	2, 3	…	…	PB	…	…	2 1/2	1	5/16	0.255	0.191
5/16	18	…	2	…	…	PB	…	…	2 23/32	1 1/8	3/8	0.318	0.238
5/16	18	…	3	…	…	PB	…	…	2 23/32	1 1/8	3/8	0.318	0.238
5/16	…	24	3	…	…	PB	…	…	2 23/32	1 1/8	3/8	0.318	0.238
3/8	16	…	3	…	…	PB	…	…	2 15/16	1 1/4	7/16	0.381	0.286
3/8	…	24	3	…	…	PB	…	…	2 15/16	1 1/4	7/16	0.381	0.286
7/16	14	…	3	…	…	P	…	…	3 5/32	1 7/16	13/32	0.323	0.242
7/16	…	20	3	…	…	P	…	…	3 5/32	1 7/16	13/32	0.323	0.242
1/2	13	…	3	…	…	PB	…	…	3 3/8	1 21/32	7/16	0.367	0.275
1/2	…	20	3	…	…	P	…	…	3 3/8	1 21/32	7/16	0.367	0.275
Ground Thread High-Speed-Steel and Cut Thread High-Speed-Steel Spiral Pointed Taps													
1/4	20	…	2	P	P	PB	…	P	2 1/2	1	5/16	0.255	0.191
1/4 [a]	20	…	3	…	…	P	…	P	2 1/2	1	5/16	0.255	0.191
1/4	…	28	2	P	P	PB	P	…	2 1/2	1	5/16	0.255	0.191
1/4 [a]	…	28	3	…	P	…	P	…	2 1/2	1	5/16	0.255	0.191
5/16	18	…	2	P	P	PB	…	P	2 23/32	1 1/8	3/8	0.318	0.238
5/16 [a]	18	…	3	…	…	P	…	P	2 23/32	1 1/8	3/8	0.318	0.238
5/16	…	24	2	P	P	PB	P	…	2 23/32	1 1/8	3/8	0.318	0.238
5/16 [a]	…	24	3	…	P	P	P	…	2 23/32	1 1/8	3/8	0.318	0.238
3/8	16	…	3	P	P	P	…	P	2 15/16	1 1/4	7/16	0.381	0.286
3/8	…	24	3	P	P	P	P	…	2 15/16	1 1/4	7/16	0.381	0.286
7/16 [a]	14	20	3	…	P[c]	P	…	P	3 5/32	1 7/16	13/32	0.323	0.242
1/2	13	20[a]	3	P	P	P	…	P	3 3/8	1 21/32	7/16	0.367	0.275
5/8 [a]	11	18	3	…	…	P	…	P[d]	3 13/16	1 13/16	9/16	0.480	0.360
3/4 [a]	10	16	3	…	…	P	…	P[e]	4 1/4	2	11/16	0.590	0.442

[a] Applies only to ground thread high-speed-steel taps.

[b] Cut thread high-speed-steel taps are standard with plug chamfer only.

[c] Applies only to 7/16-14 tap.

[d] Applies only to 5/8-11 tap.

[e] Applies ony to 3/4-10 tap. For eccentricity tolerances see Table 25.

Tolerances for General Dimensions							
Element	Diameter Range	Tolerance		Element	Diameter Range	Tolerance	
		Ground Thread	Cut Thread			Ground Thread	CutThread
Overall-Length, *A*	1/4 to 3/4	±1/32	±1/32	ShankDiameter,D	1/4 to 5/8	−0.0015	−0.005
					3/4	−0.0020	…
Thread-Length, *B*	1/4 to 1/2	±1/16	±1/16	Size ofSquare,*E*	1/4 to 1/2	−0.0040	−0.004
	5/8 to 3/4	±1/32			5/8 to 3/4	−0.0060	…
				Square Length, *C*	1/4 to 3/4	±1/32	

All dimensions are given in inches. P = plug and B = bottoming. Ground thread taps — Style 2, 3/8 inch and smaller, have external center on thread end (may be removed on bottoming taps) and external partial cone center on shank end, with length of cone approximately 1/4 of shank diameter. Ground thread taps—Style 3, larger than 3/8 inch, have internal center in thread and shank ends. Cut thread-taps, 3/8 inch and smaller have optional style center on thread and shank ends; sizes larger than 3/8 inch have internal centers in thread and shank ends. For standard thread limits see Tables 3 and 4.

Table 17. Other Types of ANSI Standard Taps *ASME/ANSI B94.9-198*

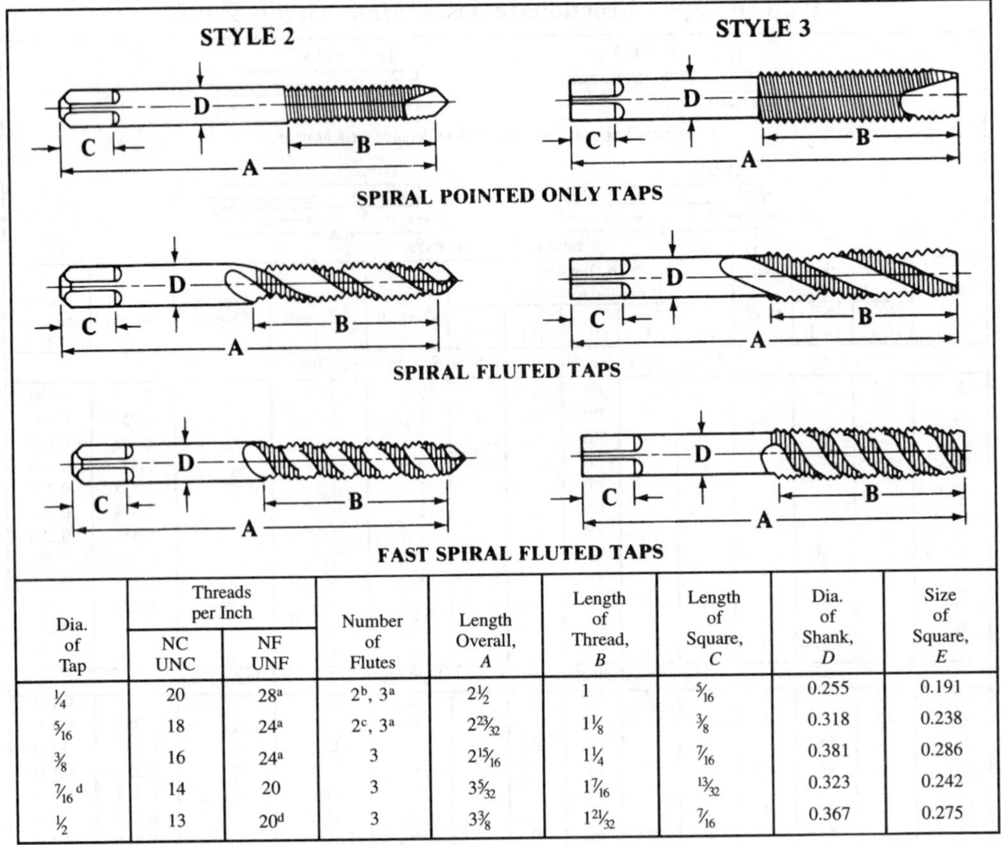

Dia. of Tap	Threads per Inch NC UNC	Threads per Inch NF UNF	Number of Flutes	Length Overall, *A*	Length of Thread, *B*	Length of Square, *C*	Dia. of Shank, *D*	Size of Square, *E*
1/4	20	28[a]	2[b], 3[a]	2 1/2	1	5/16	0.255	0.191
5/16	18	24[a]	2[c], 3[a]	2 23/32	1 1/8	3/8	0.318	0.238
3/8	16	24[a]	3	2 15/16	1 1/4	7/16	0.381	0.286
7/16 [d]	14	20	3	3 5/32	1 7/16	13/32	0.323	0.242
1/2	13	20[d]	3	3 3/8	1 21/32	7/16	0.367	0.275

[a] Does not apply to spiral pointed only taps.
[b] Does not apply to spiral fluted taps or to spiral fluted taps with 28 threads per inch.
[c] Applies only to spiral pointed only taps.
[d] Applies only to fast spiral fluted taps.

Tolerances for General Dimensions					
Element	Diameter Range	Tolerance	Element	Diameter Range	Tolerance
Overall Length, *A*	1/4 to 1/2	±1/32	Shank Diameter, *D*	1/4 to 1/2	−0.0015
Thread Length, *B*	1/4 to 1/2	±1/16			
Square Length, *C*	1/4 to 1/2	±1/32	Size of Square, *E*	1/4 to 1/2	−0.004

All dimensions are given in inches. These standard taps are made of high-speed steel with ground threads. For standard thread limits see Table 4.

Spiral Pointed Only Taps: These taps are standard with plulg chamfer only in H3 limit. They are provided with spiral point only. The balance of the threaded section is left unfluted.

Spiral Fluted Taps: These taps are standard with plug or bottoming chamfer in H3 limit and have right-hand spiral flutes with a helix angle of from 25 to 35 degrees.

Fast Spiral Fluted Taps: These taps are standard with plug or bottoming chamfer in H3 limit and have right-hand spiral flutes with a helix angle of from 45 to 60 degrees.

Style 2 taps, 3/8 inch and smaller, have external center on thread end (may be removed on bottoming taps) and external partial cone center on shank end with cone length approximately 1/4 shank diameter.

Style 3 taps larger than 3/8 inch have internal center in thread and shank ends.

For standard thread limits see Table 4. For eccentricity tolerances see Table 25.

Table 18. ANSI Standard Pulley Taps *ASME/ANSI B94.9-1987*

Dia. of Tap	Threads per Inch NC UNC	Number of Flutes	Length Overall, *A*	Length of Thread, *B*	Length of Square, *C*	Dia. of Shank, *D*	Length of Close Toler-ance, *T*[a]	Size of Square, *E*	Length of Neck, *K*[b]
1/4	20	4	6,8	1	5/16	0.255	1 1/2	0.191	3/8
5/16	18	4	6,8	1 1/8	3/8	0.318	1 9/16	0.238	3/8
3/8	16	4	6,8,10	1 1/4	7/16	0.381	1 5/8	0.286	3/8
7/16	14	4	6	1 7/16	1/2	0.444	1 11/16	0.333	7/16
1/2	13	4	6,8,10,12	1 21/32	9/16	0.507	1 11/16	0.380	1/2
5/8	11	4	6,8,10	1 13/16	11/16	0.633	2	0.475	5/8
3/4	10	4	10	2	3/4	0.759	2 1/4	0.569	3/4

[a] *T* is minimum length of shank which is held to eccentricity tolerances.
[b] *K* neck optional with manufacturer.

Tolerances for General Dimensions					
Element	Diameter Range	Tolerance	Element	Diameter Range	Tolerance
Overall Length, *A*	1/4 to 3/4	±1/16	Shank Diameter, *D*	1/4 to 3/4	−0.005
Thread Length, *B*	1/4 to 3/4	±1/16	Size of Square, *E*	1/4 to 1/2 5/8 to 3/4	−0.004 −0.006
Square Length, *C*	1/4 to 3/4	±1/32			

All dimensions are given in inches. These ground thread high-speed steel taps are standard with plug chamfer in H3 limit only. All taps have an internal center in thread end. For standard thread limits see Table 4. For eccentricity tolerances see Table 25.

Table 19. ANSI Standard Ground Thread Spark Plug Taps—Metric Sizes
ASME/ANSI B94.9-1987

Tap Diameter, mm	Pitch, mm	Number of Flutes	Overall Length, In. *A*	Thread Length, In. *B*	Square Length, In. *C*	Shank Dia., In. *D*	Square Size, In. *E*
14	1.25	4	3 19/32	1 21/32	1/2	0.429	0.322
18	1.50	4	4 1/32	1 13/16	5/8	0.542	0.406

These are high-speed steel Style 3 taps and have internal center in thread and shank ends. They are standard with plug chamfer only, right-hand threads with 60-degree form of thread.

Tolerances: Overall length, ± 1/32 inch; thread length, ± 3/32 inch; square length, ± 1/32 inch; shank diameter, 14 mm, −0.0015 inch, 18 mm, −0.0020 inch; and size of square, −0.0040 inch.

Table 20a. ANSI Standard Ground Thread Straight Fluted Taps — M Profile — Metric Sizes *ASME/ANSI B94.9-1987*

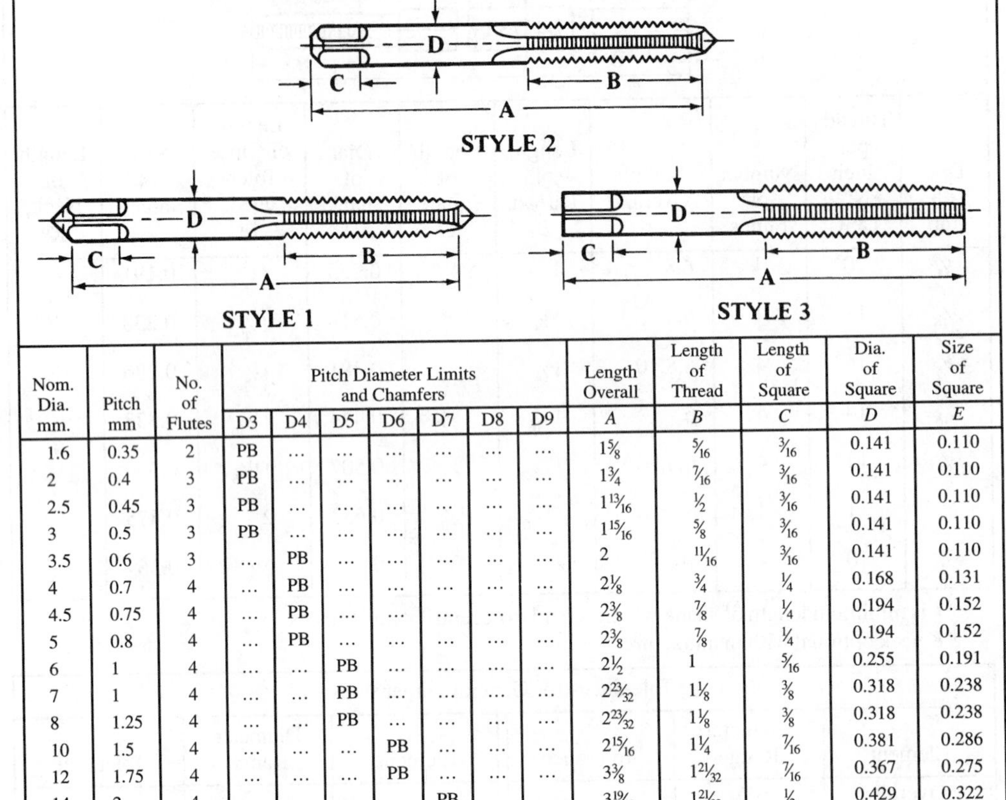

Nom. Dia. mm.	Pitch mm	No. of Flutes	Pitch Diameter Limits and Chamfers							Length Overall	Length of Thread	Length of Square	Dia. of Square	Size of Square
			D3	D4	D5	D6	D7	D8	D9	*A*	*B*	*C*	*D*	*E*
1.6	0.35	2	PB	…	…	…	…	…	…	1 5/8	5/16	3/16	0.141	0.110
2	0.4	3	PB	…	…	…	…	…	…	1 3/4	7/16	3/16	0.141	0.110
2.5	0.45	3	PB	…	…	…	…	…	…	1 13/16	1/2	3/16	0.141	0.110
3	0.5	3	PB	…	…	…	…	…	…	1 15/16	5/8	3/16	0.141	0.110
3.5	0.6	3	…	PB	…	…	…	…	…	2	11/16	3/16	0.141	0.110
4	0.7	4	…	PB	…	…	…	…	…	2 1/8	3/4	1/4	0.168	0.131
4.5	0.75	4	…	PB	…	…	…	…	…	2 3/8	7/8	1/4	0.194	0.152
5	0.8	4	…	PB	…	…	…	…	…	2 3/8	7/8	1/4	0.194	0.152
6	1	4	…	…	PB	…	…	…	…	2 1/2	1	5/16	0.255	0.191
7	1	4	…	…	PB	…	…	…	…	2 23/32	1 1/8	3/8	0.318	0.238
8	1.25	4	…	…	PB	…	…	…	…	2 23/32	1 1/8	3/8	0.318	0.238
10	1.5	4	…	…	…	PB	…	…	…	2 15/16	1 1/4	7/16	0.381	0.286
12	1.75	4	…	…	…	PB	…	…	…	3 3/8	1 21/32	7/16	0.367	0.275
14	2	4	…	…	…	…	PB	…	…	3 19/32	1 21/32	1/2	0.429	0.322
16	2	4	…	…	…	…	PB	…	…	3 13/16	1 13/16	9/16	0.480	0.360
20	2.5	4	…	…	…	…	PB	…	…	4 15/32	2	11/16	0.652	0.489
24	3	4	…	…	…	…	…	PB	…	4 29/32	2 7/32	3/4	0.760	0.570
30	3.5	4	…	…	…	…	…	…	PB	5 7/16	2 9/16	1	1.021	0.766
36	4	4	…	…	…	…	…	…	PB	6 1/16	3	1 1/8	1.233	0.925

Tolerances					
Element	Nom. Dia. Range, mm	Toler., Inch	Element	Nom. Dia. Range, mm	Toler., Inch
Overall Length, *A*	M1.6 to M24, incl.	±1/32	Shank Diameter, *D*	M1.6 to M14, incl.	−0.0015
	M30 and M36	±1/16		M16 to M36	−0.002
Thread Length, *B*	M1.6 to M5, incl.	±3/64			
	M6 to M12 incl.	±1/16			
	M14 to M36	±3/32	Size of Square, *E*	M1.6 to M12, incl.	−0.004
Square Length, *C*	M1.6 to M24, incl.	±1/32		M14 to M24, incl.	−0.006
	M30 and M36	±1/16		M30 and M36	−0.008

All dimensions are in inches except where otherwise stated.

Chamfer Designation: P — Plug, B — Bottoming. These taps are high-speed steel.

Style 1 taps, sizes M1.6 through M5, have external center on thread and shank ends (may be removed on thread end of bottoming taps).

Style 2 taps, sizes M6, M7, M8, and M10, have external center on thread end (may be removed on bottoming taps) and external partial cone center on shank end with length of cone approximately 1/4 of diameter of shank.

Style 3 taps, sizes larger than M10 have external center on thread and shank ends.

For standard thread limits see Tables 8a and 8b.

For eccentricity tolerances of tap elements see Table 25.

Table 20b. ANSI Standard Spiral Pointed Ground Thread Taps — M Profile — Metric Sizes *ASME/ANSI B94.9-1987*

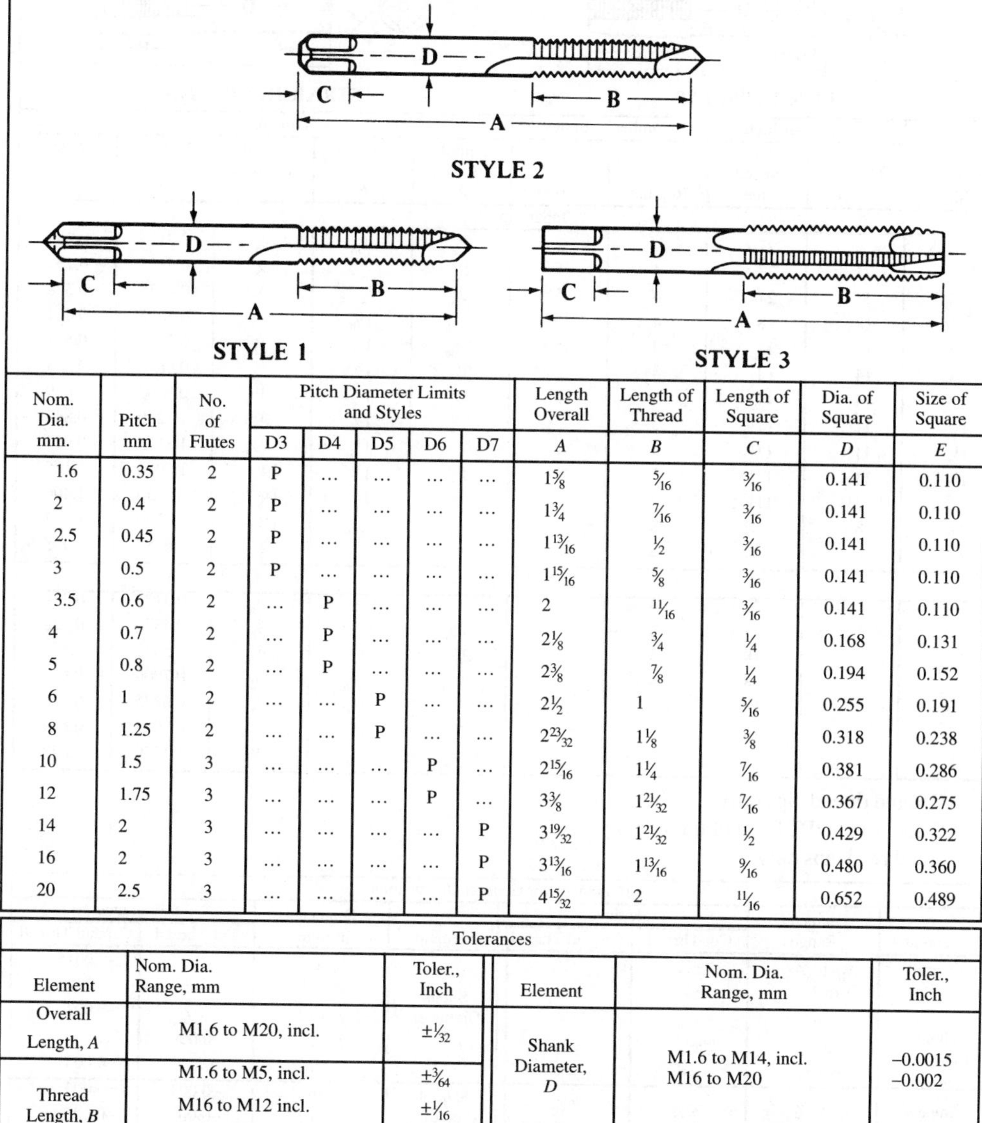

Nom. Dia. mm.	Pitch mm	No. of Flutes	Pitch Diameter Limits and Styles					Length Overall	Length of Thread	Length of Square	Dia. of Square	Size of Square
			D3	D4	D5	D6	D7	*A*	*B*	*C*	*D*	*E*
1.6	0.35	2	P	…	…	…	…	$1\frac{5}{8}$	$\frac{5}{16}$	$\frac{3}{16}$	0.141	0.110
2	0.4	2	P	…	…	…	…	$1\frac{3}{4}$	$\frac{7}{16}$	$\frac{3}{16}$	0.141	0.110
2.5	0.45	2	P	…	…	…	…	$1\frac{13}{16}$	$\frac{1}{2}$	$\frac{3}{16}$	0.141	0.110
3	0.5	2	P	…	…	…	…	$1\frac{15}{16}$	$\frac{5}{8}$	$\frac{3}{16}$	0.141	0.110
3.5	0.6	2	…	P	…	…	…	2	$\frac{11}{16}$	$\frac{3}{16}$	0.141	0.110
4	0.7	2	…	P	…	…	…	$2\frac{1}{8}$	$\frac{3}{4}$	$\frac{1}{4}$	0.168	0.131
5	0.8	2	…	P	…	…	…	$2\frac{3}{8}$	$\frac{7}{8}$	$\frac{1}{4}$	0.194	0.152
6	1	2	…	…	P	…	…	$2\frac{1}{2}$	1	$\frac{5}{16}$	0.255	0.191
8	1.25	2	…	…	P	…	…	$2\frac{23}{32}$	$1\frac{1}{8}$	$\frac{3}{8}$	0.318	0.238
10	1.5	3	…	…	…	P	…	$2\frac{15}{16}$	$1\frac{1}{4}$	$\frac{7}{16}$	0.381	0.286
12	1.75	3	…	…	…	P	…	$3\frac{3}{8}$	$1\frac{21}{32}$	$\frac{7}{16}$	0.367	0.275
14	2	3	…	…	…	…	P	$3\frac{19}{32}$	$1\frac{21}{32}$	$\frac{1}{2}$	0.429	0.322
16	2	3	…	…	…	…	P	$3\frac{13}{16}$	$1\frac{13}{16}$	$\frac{9}{16}$	0.480	0.360
20	2.5	3	…	…	…	…	P	$4\frac{15}{32}$	2	$\frac{11}{16}$	0.652	0.489

Tolerances					
Element	Nom. Dia. Range, mm	Toler., Inch	Element	Nom. Dia. Range, mm	Toler., Inch
Overall Length, *A*	M1.6 to M20, incl.	$\pm\frac{1}{32}$	Shank Diameter, *D*	M1.6 to M14, incl. M16 to M20	−0.0015 −0.002
Thread Length, *B*	M1.6 to M5, incl. M16 to M12 incl. M14 to M20	$\pm\frac{3}{64}$ $\pm\frac{1}{16}$ $\pm\frac{3}{32}$			
Square Length, *C*	M1.6 to M20	$\pm\frac{1}{32}$	Size of Square, *E*	M1.6 to M12, incl. M14 to M20, incl.	−0.004 −0.006

All dimensions are in inches except where otherwise stated.

Chamfer Designation: P — Plug. These taps are high-speed steel.

Style 1 taps, sizes M1.6 through M5, have external center on thread and shank ends.

Style 2 taps, sizes M6, M8 and M10, have external center on thread end and external partial cone center on shank end with length of cone approximately $\frac{1}{4}$ of diameter of shank.

Style 3 taps, sizes larger than M10 have external center on thread and shank ends.

For standards thread limits see Table 8a and 8b.

For eccentricity tolerances of tap elements see Table 25.

Table 21. ANSI Standard Taper and Straight Pipe Taps *ASME/ANSI B94.9-1987*

TAPER PIPE TAP STRAIGHT PIPE TAP

Nominal Size	Threads per Inch: Carbon Steel	Threads per Inch: High-Speed Steel	Number of Flutes: Regular	Number of Flutes: Interrupted	Dimensions: Length Overall, *A*	Length of Thread, *B*	Length of Square, *C*	Diameter of Shank, *D*	Size of Square, *E*
Taper Pipe Taps									
1/16 [a]	…	27	4	…	2 1/8	11/16	3/8	0.3125	0.234
1/8	27	27	4	5	2 1/8	3/4	3/8	0.3125	0.234
1/8	27	27	4	5	2 1/8	3/4	3/8	0.4375	0.328
1/4	18	18	4	5	2 7/16	1 1/16	7/16	0.5625	0.421
3/8	18	18	4	5	2 9/16	1 1/16	1/2	0.7000	0.531
1/2	14	14	4	5	3 1/8	1 3/8	5/8	0.6875	0.515
3/4	14	14	5	5	3 1/4	1 3/8	11/16	0.9063	0.679
1	11 1/2	11 1/2	5	5	3 3/4	1 3/4	13/16	1.1250	0.843
1 1/4	11 1/2	11 1/2	5	5	4	1 3/4	15/16	1.3125	0.984
1 1/2	11 1/2	11 1/2	7	7[ba]	4 1/4	1 3/4	1	1.5000	1.125
2	11 1/2	11 1/2	7	7[ba]	4 1/2	1 3/4	1 1/8	1.8750	1.406
2 1/2 [c]	8	…	8	…	5 1/2	2 9/16	1 1/4	2.2500	1.687
3 [c]	8	…	8	…	6	2 5/8	1 3/8	2.6250	1.968
Straight Pipe Taps									
1/8 [a]	…	27	4	…	2 1/8	3/4	3/8	0.3125	0.234
1/8	…	27	4	…	2 1/8	3/4	3/8	0.4375	0.328
1/4	…	18	4	…	2 7/16	1 1/16	7/16	0.5625	0.421
3/8	…	18	4	…	2 9/16	1 1/16	1/2	0.7000	0.531
1/2	…	14	4	…	3 1/8	1 3/8	5/8	0.6875	0.515
3/4	…	14	5	…	3 1/4	1 3/8	11/16	0.9063	0.679
1	…	11 1/2	5	…	3 3/4	1 3/4	13/16	1.1250	0.843

[a] Ground thread taps only.
[b] Standard in NPT form of thread only.
[c] Cut thread taps only.

Tolerances for General Dimensions

Element	Diameter Range	Tolerance: Cut Thread	Tolerance: Ground Thread
Overall Length, *A*	1/16 to 3/4	±1/32	±1/32
	1 to 3	±1/16	±1/16
Thread Length, *B*	1/16 to 3/4	±1/16	±1/16
	1 to 1 1/4	±3/32	±3/32
	1 1/2 to 3	±1/8	±1/8
Square Length, *C*	1/16 to 3/4	±1/32	±1/32
	1 to 3	±1/16	±1/16

Element	Diameter Range	Tolerance: Cut Thread	Tolerance: Ground Thread
Shank Diameter, *D*	1/16 to 1/8	…	−0.0015
	1/8 to 1/2	−0.007	…
	1/4 to 1	…	−0.002
	3/4 to 3	−0.009	…
	1 1/4 to 2	…	−0.003
Size of Square, *E*	1/16 to 1/8	−0.004	−0.004
	1/4 to 3/4	−0.006	−0.006
	1 to 3	−0.008	−0.008

All dimensions are given in inches. These taps have an internal center in the thread end. *Taper Pipe Threads*: The 1/8-inch pipe tap is furnished with large size shank unless the small shank is specified. These taps have 2 to 3 1/2 threads chamfer. The first few threads on interrupted thread pipe taps are left full. The following styles and sizes are standard: 1/16 to 2 inches regular ground thread, NPT, NPTF, and ANPT: 1/8 to 2 inches interrupted ground thread, NPT, NPTF and ANPT: 1/8 to 3 inches carbon steel regular cut thread, NPT; 1/8 to 2 inches high-speed steel, regular cut thread, NPT; 1/8 to 1 1/4 inches high-speed steel interrupted cut thread, NPT. For standard thread limits see Tables 9a and 9b. *Straight Pipe Threads:* The 1/8-inch pipe tap is furnished with large size shank unless the small size is specified. These taps are standard with plug chamfer only. The following styles and sizes are standard: ground threads — 1/8 to 1 inch, NPSC and NPSM; 1/8 to 3/4 inch, NPSF; cut threads — 1/8 to 1 inch, NPSC and NPSM. For standard thread limits see Tables 10a, 10b, and 10c. For eccentricity tolerances see Table 25.

Table 22. Taps Recommended for Classes 2B and 3B Unified Screw Threads — Numbered and Fractional Sizes *ASME/ANSI B94.9-1987*

Size	Threads per Inch		Recommended Tap For Class of Thread		Pitch Diameter Limits For Class of Thread		
	NC UNC	NF UNF	Class 2B[a]	Class 3B	Min All Classes (Basic)	Max Class 2B	Max Class 3B
Machine Screw Numbered Size Taps							
0	…	80	G H2	G H1	0.0519	0.0542	0.0536
1	64	…	G H2	G H1	0.0629	0.0655	0.0648
1	…	72	G H2	G H1	0.0640	0.0665	0.0659
2	56	…	G H2	G H1	0.0744	0.0772	0.0765
2	…	64	G H2	G H1	0.0759	0.0786	0.0779
3	48	…	G H2	G H1	0.0855	0.0885	0.0877
3	…	56	G H2	G H1	0.0874	0.0902	0.0895
4	40	…	G H2	G H2	0.0958	0.0991	0.0982
4	…	48	G H2	G H1	0.0985	0.1016	0.1008
5	40	…	G H2	G H2	0.1088	0.1121	0.1113
5	…	44	G H2	G H1	0.1102	0.1134	0.1126
6	32	…	G H3	G H2	0.1177	0.1214	0.1204
6	…	40	G H2	G H2	0.1218	0.1252	0.1243
8	32	…	G H3	G H2	0.1437	0.1475	0.1465
8	…	36	G H2	G H2	0.1460	0.1496	0.1487
10	24	…	G H3	G H3	0.1629	0.1672	0.1661
10	…	32	G H3	G H2	0.1697	0.1736	0.1726
12	24	…	G H3	G H3	0.1889	0.1933	0.1922
12	…	28	G H3	G H3	0.1928	0.1970	0.1959
Fractional Size Taps							
1/4	20	…	G H5	G H3	0.2175	0.2224	0.2211
1/4	…	28	G H4	G H3	0.2268	0.2311	0.2300
5/16	18	…	G H5	G H3	0.2764	0.2817	0.2803
5/16	…	24	G H4	G H3	0.2854	0.2902	0.2890
3/8	16	…	G H5	G H3	0.3344	0.3401	0.3387
3/8	…	24	G H4	G H3	0.3479	0.3528	0.3516
7/16	14	…	G H5	G H3	0.3911	0.3972	0.3957
7/16	…	20	G H5	G H3	0.4050	0.4104	0.4091
1/2	13	…	G H5	G H3	0.4500	0.4565	0.4548
1/2	…	20	G H5	G H3	0.4675	0.4731	0.4717
9/16	12	…	G H5	G H3	0.5084	0.5152	0.5135
9/16	…	18	G H5	G H3	0.5264	0.5323	0.5308
5/8	11	…	G H5	G H3	0.5660	0.5732	0.5714
5/8	…	18	G H5	G H3	0.5889	0.5949	0.5934
3/4	10	…	G H5	G H5	0.6850	0.6927	0.6907
3/4	…	16	G H5	G H3	0.7094	0.7159	0.7143
7/8	9	…	G H6[b]	G H4	0.8028	0.8110	0.8089
7/8	…	14	G H6[b]	G H4	0.8286	0.8356	0.8339
1	8	…	G H6[b]	G H4	0.9188	0.9276	0.9254
1	…	12	G H6[b]	G H4	0.9459	0.9535	0.9516
1	14NS		G H6[b]	G H4	0.9536	0.9609	0.9590
1 1/8	7	…	G H8[b]	G H4	1.0322	1.0416	1.0393
1 1/8	…	12	G H6[b]	G H4	1.0709	1.0787	1.0768
1 1/4	7	…	G H8[b]	G H4	1.1572	1.1668	1.1644
1 1/4	…	12	G H6[b]	G H4	1.1959	1.2039	1.2019
1 3/8	6	…	G H8[b]	G H4	1.2667	1.2771	1.2745
1 3/8	…	12	G H6[b]	G H4	1.3209	1.3291	1.3270
1 1/2	6	…	G H8[b]	G H4	1.3917	1.4022	1.3996
1 1/2	…	12	G H6[b]	G H4	1.4459	1.4542	1.4522

[a] Cut thread taps in all fractional sizes and in numbered sizes 3 to 12 NC and NF may be used under normal conditions and in average materials to produce tapped holes in this classification.

[b] Standard G H4 taps are also suitable for this class of thread.

All dimensions are given in inches.

The above recommended taps normally produce the class of thread indicated in average materials when used with reasonable care. However, if the tap specified does not give a satisfactory gage fit in the work, a choice of some other limit tap will be necessary.

Standard System of Tap Marking.—Ground thread taps, inch screw threads, are marked with the nominal size, number of threads per inch, the proper symbol to identify the thread form, "HS" for high-speed steel, "G" for ground thread, and designators for tap pitch diameter and special features, such as left-hand and multi-start threads.

Cut thread taps, inch screw threads, are marked with the nominal size, number of threads per inch, and the proper symbol to identify the thread form. High-speed steel taps are marked "HS," but carbon steel taps need not be marked.

Ground thread taps made with metric screw threads, M profile, are marked with "M," followed by the nominal size and pitch in millimeters, separated by "x." Marking also includes "HS" for high-speed steel, "G" for ground thread, designators for tap pitch diameter and special features, such as left-hand and multi-start threads.

Thread symbol designators are listed in the accompanying table. Tap pitch diameter designators, systems of limits, special features, and examples for ground threads are given in the following section.

Standard System Tap Thread Limits and Identification for Unified Inch Screw Threads, Ground Thread.—*H or L Limits:* For Unified inch screw threads, when the maximum tap pitch diameter is over basic pitch diameter by an even multiple of 0.0005 in. or the minimum tap pitch diameter limit is under basic pitch diameter by an even multiple of 0.0005 in., the taps are marked "H" or "L," respectively, followed by a limit number, determined as follows:

H limit number =Amount maximum tap PD limit is over basic PD divided by 0.0005

L limit number =Amount minimum tap PD limit is under basic PD divided by 0.0005

Table 23. Thread Series Designations

Standard Tap Marking	Product Thread Designation	Third Series
M	M	Metric Screw Threads—M Profile, with basic ISO 68 profile
M	MJ	Metric Screw Threads—M Profile, with rounded root of radius 0.15011P to 0.18042P
		Class 5 interference-fit thread
NC	NC_5IF	Entire ferrous material range
NC	NC_5INF	Entire nonferrous material range
NPS	NPSC	American Standard straight pipe threads in pipe couplings
NPSF	NPSF	Dry seal American Standard fuel internal straight pipe threads
NPSH	NPSH	American Standard straight hose coupling threads for joining to American Standard taper pipe threads
NPSI	NPSI	Dryseal American Standard intermediate internal straight pipe threads
NPSL	NPSL	American Standard straight pipe threads for loose-fitting mechanical joints with locknuts
NPS	NPSM	American Standard straight pipe threads for free-fitting mechanical joints for fixtures
NPT	NPT	American Standard taper pipe threads for general use
NPTF	NPTF	Dryseal American Standard taper pipe threads
NPTR	NPTR	American Standard taper pipe threads for railing joints
Unified Inch Screw Thread		
N	UN	Constant-pitch series
NC	UNC	Coarse pitch series
NF	UNF	Fine pitch series
NEF	UNEF	Extra-fine pitch series
N	UNJ	Constant-pitch series, with rounded root of radius 0.15011P to 0.18042P (ext. thd. only)
NC	UNJC	Coarse pitch series, with rounded root of radius 0.15011P to 0.18042 P (ext. thd. only)
NF	UNJF	Fine pitch series, with rounded root of radius 0.15011P to 0.18042P (ext. thd. only)
NEF	UNJEF	Extra-fine pitch series, with rounded root of radius 0.15011P to 0.18042P (ext. thd. only)
N	UNR	Constant-pitch series, with rounded root of radius not less than 0.108P (ext. thd. only)
NC	UNRC	Coarse thread series, with rounded root of radius not less than 0.108P (ext. thd. only)
NF	UNRF	Fine pitch series, with rounded root of radius not less than 0.108P (ext. thd. only)
NEF	UNREF	Extra-fine pitch series, with rounded root of radius not less than 0.108P (ext. thd. only)
NS	UNS	Special diameter pitch, or length of engagement

The PD limits for various H limit numbers are given in Table 4. The PD limits for L limit numbers are determined as follows. The minimum tap PD equals the basic PD minus the number of half-thousandths (0.0005 in.) represented by the limit number. The maximum tap PD equals the minimum PD plus the PD tolerance given in Table 24.

Table 24. PD Tolerance for Unified Inch Screw Threads— Ground Thread *ASME/ANSI B94.9-1987*

Threads per Inch	To 1 in., incl.	Over 1 in. to 1½ in., incl.	Over 1½ to 2½ in., incl.	Over 2 ½ in.
80-28	0.0005	0.0010	0.0010	0.0015
24-18	0.0005	0.0010	0.0015	0.0015
16-18	0.0005	0.0010	0.0015	0.0020
7-6	0.0010	0.0010	0.0020	0.0025
5½-4	0.0010	0.0015	0.0020	0.0025

Examples: ⅜-16 NC HS H1

Max. tap PD = 0.3349

Min. tap PD = 0.3344

⅜-16 NC HS G L2

Min. tap PD = Basic PD − 0.0010 in. = 0.3344 − 0.0010 = 0.3334

Max. tap PD = Min. Tap PD + 0.0005 = 0.3334 + 0.0005 = 0.3339

Oversize or Undersize: When the maximum tap PD over basic PD or the minimum tap PD under basic PD is not an even multiple of 0.0005, the tap PD is usually designated as an amount oversize or undersize. The amount oversize is added to the basic PD to establish the *minimum* tap PD. The amount undersize is subtracted from the basic PD to establish the *minimum* tap PD. The PD tolerance in Table 24 is added to the minimum tap PD to establish the maximum tap PD for both.

Example : 7⁄16-14 NC plus 0.0017 HS G

Min. tap PD = Basic PD + 0.0017 in.

Max. tap PD = Min. tap PD + 0.0005 in.

Whenever possible for oversize or other special tap PD requirements, the maximum and minimum tap PD requirements should be specified.

Special Tap Pitch Diameter: Taps not made to H or L limit numbers, to Table 25, or to the formula for oversize or undersize taps, may be marked with the letter "S" enclosed by a circle or by some other special identifier. *Example:* ½-16 NC HS G .

Table 25. ANSI Standard Eccentricity Tolerances of Tap Elements When Tested on Dead Centers *ASME/ANSI B94.9-1987*

Element	Range Sizes are Inclusive			Cut Thread		Ground Thread	
	Hand, Mch. Screw	Metric	Pipe	Eccentricity	tiv[a]	Eccentricity	tiv[a]
Square (at central point)	#0–½″	M1.6–M12	1⁄16–⅛″	0.0030	0.0060	0.0030	0.0060
	17⁄32–4″	M14–M100	¼–4″	0.0040	0.0080	0.0040	0.0080
Shank	#0–5⁄16″	M1.6–M8	1⁄16″	0.0030	0.0060	0.0005	0.0010
	11⁄32–4″	M10–M100	⅛–4″	0.0040	0.0080	0.0008	0.0016
Major Diameter	#0–5⁄16″	M1.6–M8	1⁄16″	0.0025	0.0050	0.0005	0.0010
	11⁄32–4″	M10–M100	⅛–4″	0.0040	0.0080	0.0008	0.0016
Pitch Diameter (at first full thread)	#0–5⁄16″	M1.6–M8	1⁄16″	0.0025	0.0050	0.0005	0.0010
	11⁄32–4″	M10–M100	⅛–4″	0.0040	0.0080	0.0008	0.0016
Chamfer[b]	#0–½″	M1.6–M12	1⁄16–⅛″	0.0020	0.0040	0.0010	0.0020
	17⁄32–4″	M14–M100	¼–4″	0.0030	0.0060	0.0015	0.0030

[a] tiv = total indicator variation. Figures are given for both eccentricity and total indicator variation to avoid misunderstanding.

[b] Chamfer should preferably be inspected by light projection to avoid errors due to indicator contact points dropping into the thread groove.

All dimensions are given in inches.

Left-Hand Taps: Taps with left-hand threads are marked "LEFT HAND" or "LH." *Example:* 3⁄8-16 NC LH HS G H3.

Multiple-Start Threads: Taps with multiple-start threads are marked with the lead designated as a fraction, also "Double," "Triple," etc. The Unified Screw Thread form symbol is always designated as "NS" for multiple-start threads. *Example:* 3⁄8-16 NS Double 1⁄8 Lead HS G H5.

Standard System of Ground Thread Tap Limits and Identification for Metric Screw Threads — M Profile.—All calculations for metric taps use millimeter values. When U.S. customary values are needed, they are translated from the three-place millimeter tap diameters only after the calculations are completed.

Table 26. PD Tolerance for Metric Screw Threads— M Profile—Ground Threads *ASME/ANSI B94.9-1987*

Pitch, *P* (mm)	M1.6 to M6.3, incl.	Over M6.3 to M25, incl.	Over M25 to M90, incl.	Over M90
0.3	0.015	0.015	0.020	0.020
0.35	0.015	0.015	0.020	0.020
0.4	0.015	0.015	0.020	0.025
0.45	0.015	0.020	0.020	0.025
0.5	0.015	0.020	0.025	0.025
0.6	0.020	0.020	0.025	0.025
0.7	0.020	0.020	0.025	0.025
0.75	0.020	0.025	0.025	0.031
0.8	0.020	0.025	0.025	0.031
0.9	0.020	0.025	0.025	0.031
1	0.025	0.025	0.031	0.031
1.25	0.025	0.031	0.031	0.041
1.5	0.025	0.031	0.031	0.041
1.75	…	0.031	0.041	0.041
2	…	0.041	0.041	0.041
2.5	…	0.041	0.041	0.052
3	…	0.041	0.052	0.052
3.5	…	0.041	0.052	0.052
4	…	0.052	0.052	0.064
4.5	…	0.052	0.052	0.064
5	…	…	0.064	0.064
5.5	…	…	0.064	0.064
6	…	…	0.064	0.064

D or DU Limits: When the maximum tap pitch diameter is over basic pitch diameter by an even multiple of 0.013 mm (0.000512 in. reference), or the minimum tap pitch diameter limit is under basic pitch diameter by an even multiple of 0.013 mm, the taps are marked with the letters "D" or "DU," respectively, followed by a limit number. The limit number is determined as follows:

D limit number = Amount maximum tap PD limit is over basic PD divided by 0.013

DU limit number = Amount minimum tap PD limit is under basic PD divided by 0.013

The PD limits for various D limit numbers are given in Table 8b. The PD limits for DU limit numbers are determined as follows. The minimum tap PD equals the basic PD minus the number of millimeters represented by the limit number (multiples of 0.013 mm). The maximum tap PD equals the minimum tap PD plus the PD tolerance given in Table 26. E

Table 27. Dimensions of Acme Threads Taps in Sets of Three Taps

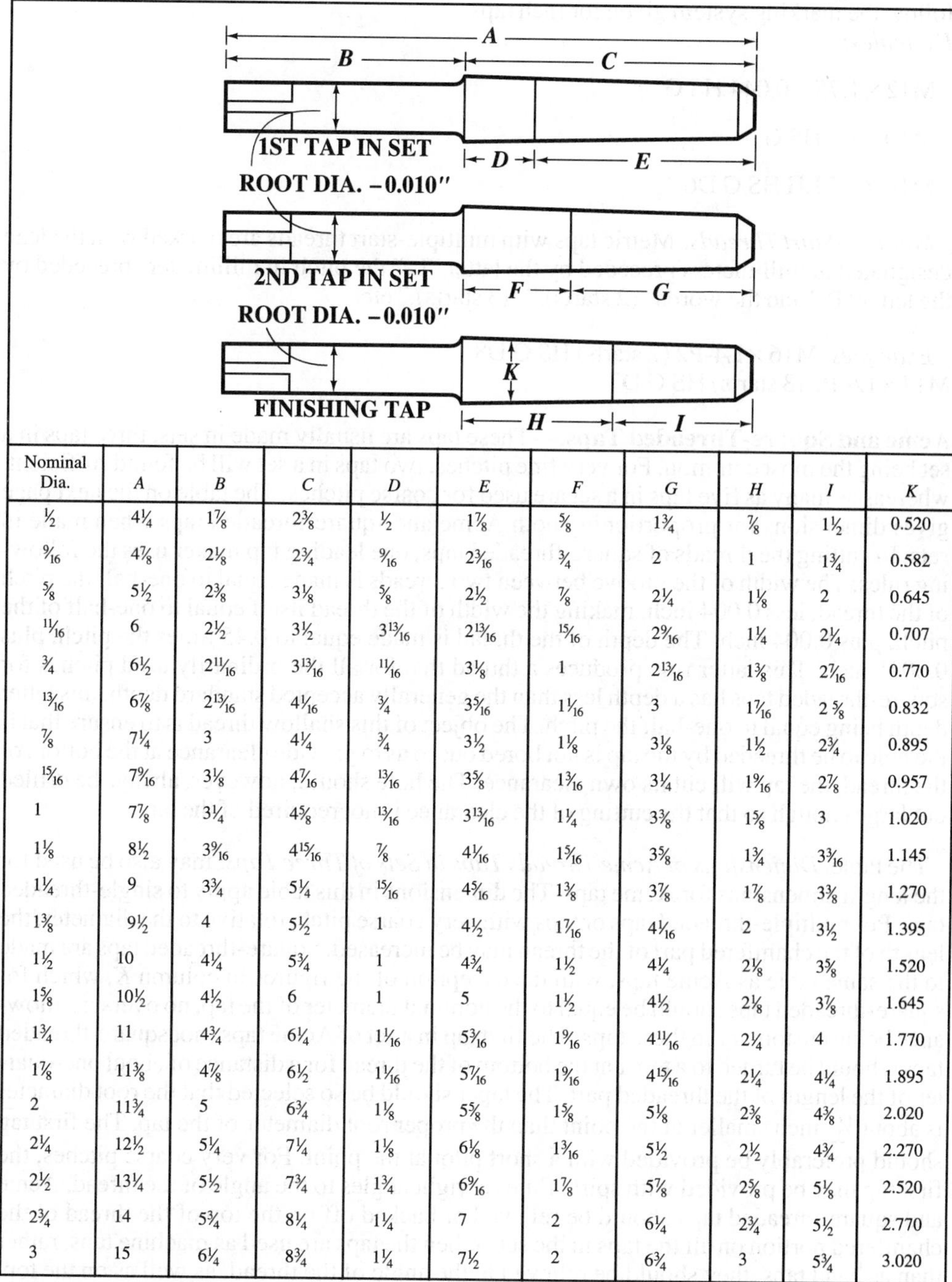

Nominal Dia.	A	B	C	D	E	F	G	H	I	K
1/2	4 1/4	1 7/8	2 3/8	1/2	1 7/8	5/8	1 3/4	7/8	1 1/2	0.520
9/16	4 7/8	2 1/8	2 3/4	9/16	2 3/16	3/4	2	1	1 3/4	0.582
5/8	5 1/2	2 3/8	3 1/8	5/8	2 1/2	7/8	2 1/4	1 1/8	2	0.645
11/16	6	2 1/2	3 1/2	3 13/16	2 13/16	15/16	2 9/16	1 1/4	2 1/4	0.707
3/4	6 1/2	2 11/16	3 13/16	11/16	3 1/8	1	2 13/16	1 3/8	2 7/16	0.770
13/16	6 7/8	2 13/16	4 1/16	3/4	3 5/16	1 1/16	3	1 7/16	2 5/8	0.832
7/8	7 1/4	3	4 1/4	3/4	3 1/2	1 1/8	3 1/8	1 1/2	2 3/4	0.895
15/16	7 9/16	3 1/8	4 7/16	13/16	3 5/8	1 3/16	3 1/4	1 9/16	2 7/8	0.957
1	7 7/8	3 1/4	4 5/8	13/16	3 13/16	1 1/4	3 3/8	1 5/8	3	1.020
1 1/8	8 1/2	3 9/16	4 15/16	7/8	4 1/16	1 5/16	3 5/8	1 3/4	3 3/16	1.145
1 1/4	9	3 3/4	5 1/4	15/16	4 5/16	1 3/8	3 7/8	1 7/8	3 3/8	1.270
1 3/8	9 1/2	4	5 1/2	1	4 1/2	1 7/16	4 1/16	2	3 1/2	1.395
1 1/2	10	4 1/4	5 3/4	1	4 3/4	1 1/2	4 1/4	2 1/8	3 5/8	1.520
1 5/8	10 1/2	4 1/2	6	1	5	1 1/2	4 1/2	2 1/8	3 7/8	1.645
1 3/4	11	4 3/4	6 1/4	1 1/16	5 3/16	1 9/16	4 11/16	2 1/4	4	1.770
1 7/8	11 3/8	4 7/8	6 1/2	1 1/16	5 7/16	1 9/16	4 15/16	2 1/4	4 1/4	1.895
2	11 3/4	5	6 3/4	1 1/8	5 5/8	1 5/8	5 1/8	2 3/8	4 3/8	2.020
2 1/4	12 1/2	5 1/4	7 1/4	1 1/8	6 1/8	1 3/16	5 1/2	2 1/2	4 3/4	2.270
2 1/2	13 1/4	5 1/2	7 3/4	1 3/4	6 9/16	1 7/8	5 7/8	2 5/8	5 1/8	2.520
2 3/4	14	5 3/4	8 1/4	1 1/4	7	2	6 1/4	2 3/4	5 1/2	2.770
3	15	6 1/4	8 3/4	1 1/4	7 1/2	2	6 3/4	3	5 3/4	3.020

Examples:

M1.6 × 0.35 HS G D3

Max. tap PD = 1.412

Min. tap PD = 1.397

M6 × 1 HS G DU4

Min. tap PD = Basic PD − 0.052 mm = 5.350 − 0.052 = 5.298

Max. tap PD = Min. tap PD + 0.025 mm = 5.323

Metric oversize or undersize taps, taps with special pitch diameters, and left-hand taps follow the marking system given for inch taps.
Examples:

M12 × 1.75 + 0.044 HS G

M10 × 1.5 HS G

M10 × 1.5 LH HS G D6

Multiple-Start Threads: Metric taps with multiple-start threads are marked with the lead designated in millimeters preceded by the letter "L," the pitch in millimeters preceded by the letter "P," and the words "(2 starts)," "(3 starts)," etc.

Examples: M16 × L4-P2 (2 starts) HS G D8
M14 × L6-P2 (3 starts) HS G D7

Acme and Square-Threaded Taps.—These taps are usually made in sets, three taps in a set being the most common. For very fine pitches, two taps in a set will be found sufficient, whereas as many as five taps in a set are used for coarse pitches. The table on the next page gives dimensions for proportioning both Acme and square-threaded taps when made in sets. In cutting the threads of square-threaded taps, one leading tap maker uses the following rules: The width of the groove between two threads is made equal to one-half the pitch of the thread, less 0.004 inch, making the width of the thread itself equal to one-half of the pitch, plus 0.004 inch. The depth of the thread is made equal to 0.45 times the pitch, plus 0.0025 inch. This latter rule produces a thread that for all the ordinarily used pitches for square-threaded taps has a depth less than the generally accepted standard depth, this latter depth being equal to one-half the pitch. The object of this shallow thread is to ensure that if the hole to be threaded by the tap is not bored out so as to provide clearance at the bottom of the thread, the tap will cut its own clearance. The hole should, however, always be drilled out large enough so that the cutting of the clearance is not required of the tap.

The table, *Dimensions of Acme Threads Taps in Sets of Three Taps*, may also be used for the length dimensions for Acme taps. The dimensions in this table apply to single-threaded taps. For multiple-threaded taps or taps with very coarse pitch, relative to the diameter, the length of the chamfered part of the thread may be increased. Square-threaded taps are made to the same table as Acme taps, with the exception of the figures in column *K*, which for square-threaded taps should be equal to the nominal diameter of the tap, no oversize allowance being customary in these taps. The first tap in a set of Acme taps (not square-threaded taps) should be turned to a taper at the bottom of the thread for a distance of about one-quarter of the length of the threaded part. The taper should be so selected that the root diameter is about $\frac{1}{32}$ inch smaller at the point than the proper root diameter of the tap. The first tap should preferably be provided with a short pilot at the point. For very coarse pitches, the first tap may be provided with spiral flutes at right angles to the angle of the thread. Acme and square-threaded taps should be relieved or backed off on the top of the thread of the chamfered portion on all the taps in the set. When the taps are used as machine taps, rather than as hand taps, they should be relieved in the angle of the thread, as well as on the top, for the whole length of the chamfered portion. Acme taps should also always be relieved on the front side of the thread to within $\frac{1}{32}$ inch of the cutting edge.

Adjustable Taps: Many adjustable taps are now used, especially for accurate work. Some taps of this class are made of a solid piece of tool steel that is split and provided with means of expanding sufficiently to compensate for wear. Most of the larger adjustable taps have inserted blades or chasers that are held rigidly, but are capable of radial adjustment. The use of taps of this general class enables standard sizes to be maintained readily.

Table 28. Proportions of Acme and Square-Threaded Taps Made in Sets

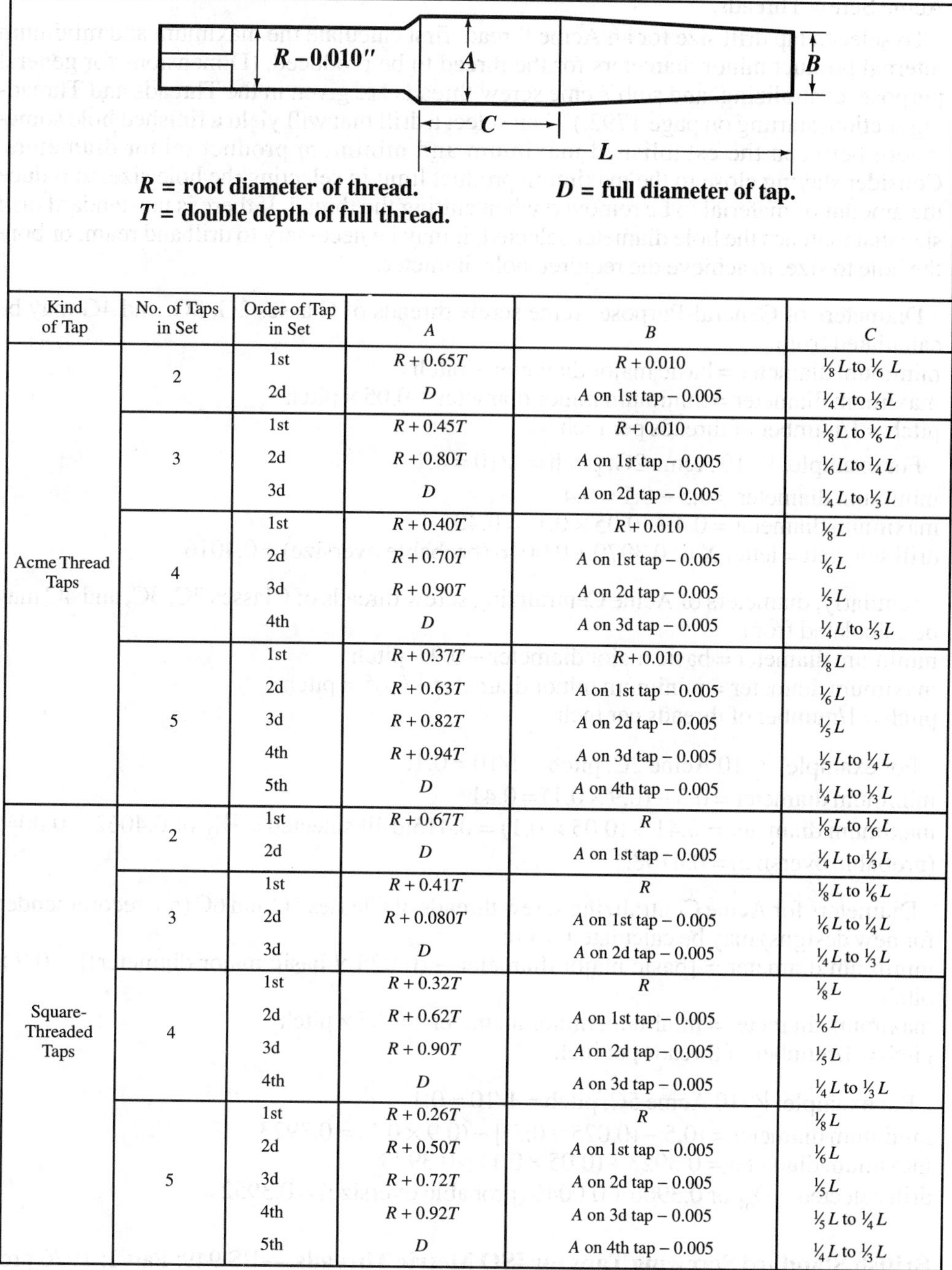

R = root diameter of thread. *D* = full diameter of tap.
T = double depth of full thread.

Kind of Tap	No. of Taps in Set	Order of Tap in Set	A	B	C
Acme Thread Taps	2	1st	$R + 0.65T$	$R + 0.010$	$\frac{1}{8}L$ to $\frac{1}{6}L$
		2d	D	A on 1st tap − 0.005	$\frac{1}{4}L$ to $\frac{1}{3}L$
	3	1st	$R + 0.45T$	$R + 0.010$	$\frac{1}{8}L$ to $\frac{1}{6}L$
		2d	$R + 0.80T$	A on 1st tap − 0.005	$\frac{1}{6}L$ to $\frac{1}{4}L$
		3d	D	A on 2d tap − 0.005	$\frac{1}{4}L$ to $\frac{1}{3}L$
	4	1st	$R + 0.40T$	$R + 0.010$	$\frac{1}{8}L$
		2d	$R + 0.70T$	A on 1st tap − 0.005	$\frac{1}{6}L$
		3d	$R + 0.90T$	A on 2d tap − 0.005	$\frac{1}{5}L$
		4th	D	A on 3d tap − 0.005	$\frac{1}{4}L$ to $\frac{1}{3}L$
	5	1st	$R + 0.37T$	$R + 0.010$	$\frac{1}{8}L$
		2d	$R + 0.63T$	A on 1st tap − 0.005	$\frac{1}{6}L$
		3d	$R + 0.82T$	A on 2d tap − 0.005	$\frac{1}{5}L$
		4th	$R + 0.94T$	A on 3d tap − 0.005	$\frac{1}{5}L$ to $\frac{1}{4}L$
		5th	D	A on 4th tap − 0.005	$\frac{1}{4}L$ to $\frac{1}{3}L$
Square-Threaded Taps	2	1st	$R + 0.67T$	R	$\frac{1}{8}L$ to $\frac{1}{6}L$
		2d	D	A on 1st tap − 0.005	$\frac{1}{4}L$ to $\frac{1}{3}L$
	3	1st	$R + 0.41T$	R	$\frac{1}{8}L$ to $\frac{1}{6}L$
		2d	$R + 0.080T$	A on 1st tap − 0.005	$\frac{1}{6}L$ to $\frac{1}{4}L$
		3d	D	A on 2d tap − 0.005	$\frac{1}{4}L$ to $\frac{1}{3}L$
	4	1st	$R + 0.32T$	R	$\frac{1}{8}L$
		2d	$R + 0.62T$	A on 1st tap − 0.005	$\frac{1}{6}L$
		3d	$R + 0.90T$	A on 2d tap − 0.005	$\frac{1}{5}L$
		4th	D	A on 3d tap − 0.005	$\frac{1}{4}L$ to $\frac{1}{3}L$
	5	1st	$R + 0.26T$	R	$\frac{1}{8}L$
		2d	$R + 0.50T$	A on 1st tap − 0.005	$\frac{1}{6}L$
		3d	$R + 0.72T$	A on 2d tap − 0.005	$\frac{1}{5}L$
		4th	$R + 0.92T$	A on 3d tap − 0.005	$\frac{1}{5}L$ to $\frac{1}{4}L$
		5th	D	A on 4th tap − 0.005	$\frac{1}{4}L$ to $\frac{1}{3}L$

Drill Hole Sizes for Acme Threads

Many tap and die manufacturers and vendors make available to their customers computer programs designed to calculate drill hole sizes for all the Acme threads in their ranges from the basic dimensions. The large variety and combination of dimensions for such tools prevent inclusion of a complete set of tables of tap drills for Acme taps in this Handbook. The following formulas (dimensions in inches) for calculating drill hole sizes for Acme threads are derived from the American National Standard, ASME/ANSI B1.5-1988, Acme Screw Threads.

To select a tap drill size for an Acme thread, first calculate the maximum and minimum internal product minor diameters for the thread to be produced. (Dimensions for general purpose, centralizing, and stub Acme screw threads are given in the Threads and Threading section, starting on page 1792.) Then select a drill that will yield a finished hole somewhere between the established maximum and minimum product minor diameters. Consider staying close to the maximum product limit in selecting the hole size, to reduce the amount of material to be removed when cutting the thread. If there is no standard drill size that matches the hole diameter selected, it may be necessary to drill and ream, or bore the hole to size, to achieve the required hole diameter.

Diameters of General-Purpose Acme screw threads of Classes 2G, 3G, and 4G may be calculated from:
minimum diameter = basic major diameter − pitch
maximum diameter = minimum minor diameter + 0.05 × pitch
pitch = 1/number of threads per inch

For example, ½-10 Acme 2G, pitch = 1/10 = 0.1
minimum diameter = 0.5 − 0.1 = 0.4
maximum diameter = 0.4 + (0.05 × 0.1) = 0.405
drill selected = letter X or 0.3970 + 0.0046 (probable oversize) = 0.4016

Similarly, diameters of Acme Centralizing screw threads of Classes 2C, 3C, and 4C may be calculated from:
minimum diameter = basic major diameter − 0.9 × pitch
maximum diameter = minimum minor diameter + 0.05 × pitch
pitch = 1/number of threads per inch

For example, ½-10 Acme 2C, pitch = 1/10 = 0.1:
minimum diameter = 0.5 − (0.9 × 0.1) = 0.41
maximum diameter = 0.41 + (0.05 × 0.1) = 0.415.drill selected = 13⁄32 or 0.4062 + 0.0046 (probable oversize) = 0.4108.

Diameters for Acme Centralizing screw threads of Classes 5C and 6C (not recommended for new designs) may be calculated from:
minimum diameter = [basic major diameter − (0.025 √ basic major diameter)] − 0.9 × pitch;
maximum diameter = minimum minor diameter + 0.05 × pitch
pitch = 1/number of threads per inch.

For example, ½-10 Acme 5C, pitch = 1/10 = 0.1
minimum diameter = [0.5 − (0.025 √ 0.5)] − (0.9 × 0.1) = 0.3923
maximum diameter = 0.3923 + (0.05 × 0.1) = 0.3973
drill selected = 25⁄64 or 0.3906 + 0.0046 (probable oversize) = 0.3952

British Standard Screwing Taps for ISO Metric Threads.—BS 949: Part 1: 1976 provides dimensions and tolerances for screwing taps for ISO metric coarse-pitch series threads in accordance with BS 3643: Part 2; and for metric fine-pitch series threads in accordance with BS 3643: Part 3.

Table 1 provides dimensional data for the cutting portion of cut-thread taps for coarse-series threads of ISO metric sizes. The sizes shown were selected from the first-choice combinations of diameter and pitch listed in BS 3643:Part 1:1981 (1998). Table 16 provides similar data for ground-thread taps for both coarse- and fine-pitch series threads of ISO metric sizes.

Table 1. British Standard Screwing Taps for ISO Metric Threads; Dimensional Limits for the Threaded Portion of Cut Taps—Coarse Pitch Series *BS 949: Part 1: 1976*

Designation	Pitch	Major Diameter	Pitch Diameter			Tolerance on Thread Angle, Degrees
		Minimum[a]	Basic	Max.	Min.	
M1	0.25	1.030	0.838	0.875	0.848	4.0
M1.2	0.25	1.230	1.038	1.077	1.048	4.0
M1.6	0.35	1.636	1.373	1.417	1.385	3.4
M2	0.40	2.036	1.740	1.786	1.752	3.2
M2.5	0.45	2.539	2.208	2.259	2.221	3.0
M3	0.50	3.042	2.675	2.730	2.689	2.9
M4	0.70	4.051	3.545	3.608	3.562	2.4
M5	0.80	5.054	4.480	4.547	4.498	2.3
M6	1.00	6.060	5.350	5.424	5.370	2.0
M8	1.25	8.066	7.188	7.270	7.210	1.8
M10	1.50	10.072	9.026	9.116	9.050	1.6
M12	1.75	12.078	10.863	10.961	10.889	1.5
M16	2.00	16.084	14.701	14.811	14.729	1.4
M20	2.50	20.093	18.376	18.497	18.407	1.3
M24	3.00	24.102	22.051	22.183	22.085	1.2
M30	3.50	30.111	27.727	27.874	27.764	1.1
M36	4.00	36.117	33.402	33.563	33.441	1.0

[a] See notes under Table 2.

Table 2. British Standard Screwing Taps for ISO Metric Threads; Dimensional Limits for the Threaded Portion of Ground Taps—Coarse-and Fine-Pitch *BS 949: Part 1: 1976*

Thread			All Classes of Taps		Class 1 Taps		Class 2 Taps		Class 3 Taps		Tolerance on ½ Thd Angle
					Pitch Diameter						
Designation	Nominal Major Dia. (basic) d	Pitch p	Min. Major Dia. d_{min}[a]	Basic Pitch Dia. d_2	$d_2$min	$d_2$max	$d_2$min	$d_2$max	$d_2$min	$d_2$max	
COARSE-PITCH THREAD SERIES											
M1	1	0.25	1.022	0.838	0.844	0.855	…	…	…	…	±60′
M1.2	1.2	0.25	1.222	1.038	1.044	1.055	…	…	…	…	±60′
M1.6	1.6	0.35	1.627	1.373	1.380	1.393	1.393	1.407	…	…	±50′
M2	2	0.40	2.028	1.740	1.747	1.761	1.761	1.776	…	…	±40′
M2.5	2.5	0.45	2.530	2.208	2.216	2.231	2.231	2.246	…	…	±38′
M3	3	0.50	3.032	2.675	2.683	2.699	2.699	2.715	2.715	2.731	±36′
M4	4	0.70	4.038	3.545	3.555	3.574	3.574	3.593	3.593	3.612	±30′
M5	5	0.80	5.040	4.480	4.490	4.510	4.510	4.530	4.530	4.550	±26′
M6	6	1.00	6.047	5.350	5.362	5.385	5.385	5.409	5.409	5.433	±24′
M8	8	1.25	8.050	7.188	7.201	7.226	7.226	7.251	7.251	7.276	±22′
M10	10	1.50	10.056	9.026	9.040	9.068	9.068	9.096	9.096	9.124	±20′
M12	12	1.75	12.064	10.863	10.879	10.911	10.911	10.943	10.943	10.975	±19′
M16	16	2.00	16.068	14.701	14.718	14.752	14.752	14.786	14.786	14.820	±18′
M20	20	2.50	20.072	18.376	18.394	18.430	18.430	18.466	18.466	18.502	±16′
M24	24	3.00	24.085	22.051	22.072	22.115	22.115	22.157	22.157	22.199	±14′

Table 2. *(Continued)* **British Standard Screwing Taps for ISO Metric Threads; Dimensional Limits for the Threaded Portion of Ground Taps— Coarse-and Fine-Pitch** *BS 949: Part 1: 1976*

Thread			All Classes of Taps		Class 1 Taps		Class 2 Taps		Class 3 Taps		
					Pitch Diameter						Tolerance on ½ Thd Angle
Designation	Nominal Major Dia. (basic) d	Pitch p	Min. Major Dia. d_{min}[a]	Basic Pitch Dia. d_2	$d_2$min	$d_2$max	$d_2$min	$d_2$max	$d_2$min	$d_2$max	
M30	30	3.50	30.090	27.727	27.749	27.794	27.794	27.839	27.839	27.884	±13′
M36	36	4.00	36.094	33.402	33.426	33.473	33.473	33.520	33.520	33.567	±12′
FINE-PITCH THREAD SIZES											
M1 × 0.2	1	0.20	1.020	0.870	0.875	0.885	…	…	…	…	±70′
M1.2 × 0.2	1.2	0.20	1.220	1.070	1.075	1.085	…	…	…	…	±70′
M1.6 × 0.2	1.6	0.20	1.621	1.470	1.475	1.485	…	…	…	…	±70′
M2 × 0.25	2	0.25	2.024	1.838	1.844	1.856	…	…	…	…	±60′
M2.5 × 0.35	2.5	0.35	2.527	2.273	2.280	2.293	2.293	2.307	…	…	±50′
M3 × 0.35	3	0.35	3.028	2.773	2.780	2.794	2.794	2.809	…	…	±50′
M4 × 0.5	4	0.50	4.032	3.675	3.683	3.699	3.699	3.715	3.715	3.731	±36′
M5 × 0.5	5	0.50	5.032	4.675	4.683	4.699	4.699	4.715	4.715	4.731	±36′
M6 × 0.75	6	0.75	6.042	5.513	5.524	5.545	5.545	5.566	5.566	5.587	±28′
M8 × 1	8	1.00	8.047	7.350	7.362	7.385	7.385	7.409	7.409	7.433	±24′
M10 × 1.25	10	1.25	10.050	9.188	9.201	9.226	9.226	9.251	9.251	9.276	±22′
M12 × 1.25	12	1.25	12.056	11.188	11.202	11.230	11.230	11.258	11.258	11.286	±22′
M16 × 1.5	16	1.50	16.060	15.026	15.041	15.071	15.071	15.101	15.101	15.131	±20′
M20 × 1.5	20	1.50	20.060	19.026	19.041	19.071	19.071	19.101	19.101	19.131	±20′
M24 × 2	24	2.00	24.072	22.701	22.719	22.755	22.755	22.791	22.791	22.827	±18′
M30 × 2	30	2.00	30.072	28.701	28.719	28.755	28.755	28.791	28.791	28.827	±18′

[a] The maximum tap major diameter, d max, is not specified and is left to the manufacturer's discretion.

All dimension are in millimeters. The thread sizes in the table have been selected from the preferred series shown in BS 3643:Part 1:1981 (1998). For other sizes, and for second and third choice combinations of diameters and pitches, see the Standard.

Tolerance Classes of Taps: Three tolerance classes (class 1, class 2, and class 3) are used for the designation of taps used for the production of nuts of the following classes:

nut classes 4H, 5H, 6H, 7H, and 8H, all having zero minimum clearance;

nut classes 4G, 5G, and 6G, all having positive minimum clearance.

The tolerances for the three classes of taps are stated in terms of a tolerance unit t, the value of which is equal to the pitch diameter tolerance, T_{D2}, grade 5, of the nut. Thus, $t = T_{D2}$, grade 5, of the nut. Taps of the different classes vary in the limits of size of the tap pitch diameter. The tolerance on the tap pitch diameter, T_{d2}, is the same for all three classes of taps (20 percent of t), but the position of the tolerance zone with respect to the basic pitch diameter depends upon the lower deviation value Em which is: for tap class 1, $Em = +0.1t$; for tap class 2, $Em = +0.3t$; and for tap class 3, $Em = +0.5t$.

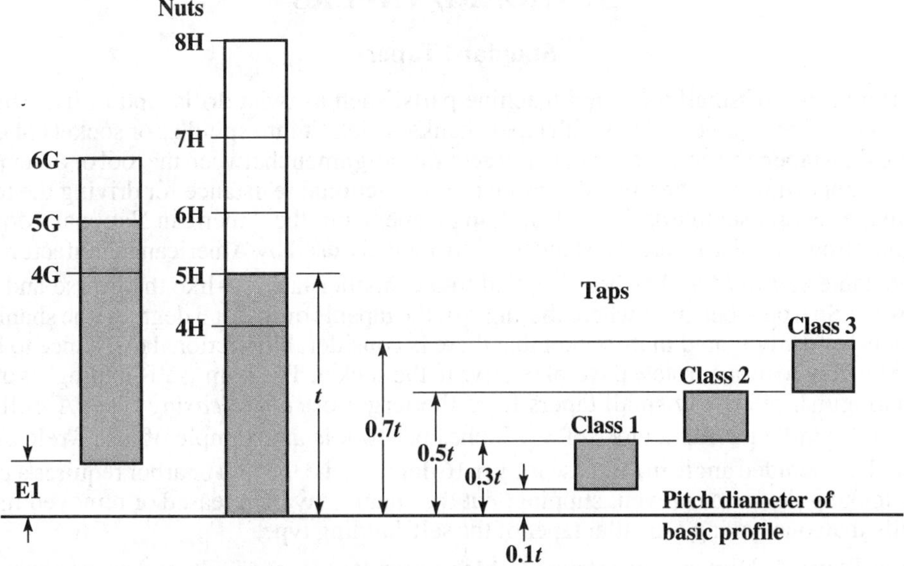

The disposition of the tolerances described is shown in the accompanying illustration of nut class tolerances compared against tap class tolerances. The distance *EI* shown in this illustration is the minumum clearance, which is zero for *H* classes and positive for *G* classes of nuts.

Choice of Tap Tolerance Class: Unless otherwise specified, class 1 taps are used for nuts of classes 4H and 5H; class 2 taps for nuts of classes 6H, 4G, and 5G; and class 3 taps for nuts of classes 7H, 8H, and 6G. This relationship of tap and nut classes is a general one, since the accuracy of tapping varies with a number of factors such as the material being tapped, the condition of the machine tool used, the tapping attachment used, the tapping speed, and the lubricant.

Tap Major Diameter: Except when a screwed connection has to be tight against gaseous or liquid pressure, it is undesirable for the mating threads to bear on the roots and crests. By avoiding contact in these regions of the threads, the opposite flanks of the two threads are allowed to make proper load bearing contact when the connection is tightened. In general, the desired clearance between crests and roots of mating threads is obtained by increasing the major and minor diameters of the internal thread. Such an increase in the minor diameter is already provided on threads such as the ISO metric thread, in which there is a basic clearance between the crests of minimum size nuts and the roots of maximum size bolts. For this reason, and the fact that taps are susceptible to wear on the crests of their threads, a minimum size is specified for the major diameter of new taps which provides a reasonable margin for the wear of their crests and at the same time provides the desired clearance at the major diameter of the hole. These minimum major diameters for taps are shown in Tables 1 and 16. The maximum tap major diameter is not specified and is left to the manufacturer to take advantage of this concession to produce taps with as liberal a margin possible for wear on the major diameter.

Tapping Square Threads.—If it is necessary to tap square threads, this should be done by using a set of taps that will form the thread by a progressive cutting action, the taps varying in size in order to distribute the work, especially for threads of comparatively coarse pitch. From three to five taps may be required in a set, depending upon the pitch. Each tap should have a pilot to steady it. The pilot of the first tap has a smooth cylindrical end from 0.003 to 0.005 inch smaller than the hole, and the pilots of following taps should have teeth.

STANDARD TAPERS

Standard Tapers

Certain types of small tools and machine parts, such as twist drills, end mills, arbors, lathe centers, etc., are provided with taper shanks which fit into spindles or sockets of corresponding taper, thus providing not only accurate alignment between the tool or other part and its supporting member, but also more or less frictional resistance for driving the tool. There are several standards for "self-holding" tapers, but the American National, Morse, and the Brown & Sharpe are the standards most widely used by American manufacturers.

The name *self-holding* has been applied to the smaller tapers—like the Morse and the Brown & Sharpe—because, where the angle of the taper is only 2 or 3 degrees, the shank of a tool is so firmly seated in its socket that there is considerable frictional resistance to any force tending to turn or rotate the tool relative to the socket. The term "self-holding" is used to distinguish relatively small tapers from the larger or *self-releasing* type. A milling machine spindle having a taper of 3½ inches per foot is an example of a self-releasing taper. The included angle in this case is over 16 degrees and the tool or arbor requires a positive locking device to prevent slipping, but the shank may be released or removed more readily than one having a smaller taper of the self-holding type.

Morse Taper.—Dimensions relating to Morse standard taper shanks and sockets may be found in an accompanying table. The taper for different numbers of Morse tapers is slightly different, but it is approximately ⅝ inch per foot in most cases. The table gives the actual tapers, accurate to five decimal places. Morse taper shanks are used on a variety of tools, and exclusively on the shanks of twist drills. Dimensions for Morse Stub Taper Shanks are given in Table 1a.

Brown & Sharpe Taper.—This standard taper is used for taper shanks on tools such as end mills and reamers, the taper being approximately ½ inch per foot for all sizes except for taper No. 10, where the taper is 0.5161 inch per foot. Brown & Sharpe taper sockets are used for many arbors, collets, and machine tool spindles, especially milling machines and grinding machines. In many cases there are a number of different lengths of sockets corresponding to the same number of taper; all these tapers, however, are of the same diameter at the small end.

Jarno Taper.—The Jarno taper was originally proposed by Oscar J. Beale of the Brown & Sharpe Mfg. Co. This taper is based on such simple formulas that practically no calculations are required when the number of taper is known. The taper per foot of all Jarno taper sizes is 0.600 inch on the diameter. The diameter at the large end is as many eighths, the diameter at the small end is as many tenths, and the length as many half inches as are indicated by the number of the taper. For example, a No. 7 Jarno taper is ⅞ inch in diameter at the large end; $\frac{7}{10}$, or 0.700 inch at the small end; and $\frac{7}{2}$, or 3½ inches long; hence, diameter at large end = No. of taper ÷ 8; diameter at small end = No. of taper ÷ 10; length of taper = No. of taper ÷ 2. The Jarno taper is used on various machine tools, especially profiling machines and die-sinking machines. It has also been used for the headstock and tailstock spindles of some lathes.

American National Standard Machine Tapers: This standard includes a self-holding series (Table , 2, 3, 4 and 5) and a steep taper series, Table 6. The self-holding taper series consists of 22 sizes which are listed in Table . The reference gage for the self-holding tapers is a plug gage. Table 7 gives the dimensions and tolerances for both plug and ring gages applying to this series. Tables 2 through 5 inclusive give the dimensions for self-holding taper shanks and sockets which are classified as to (1) means of transmitting torque from spindle to the tool shank, and (2) means of retaining the shank in the socket. The steep machine tapers consist of a preferred series (bold-face type, Table 6) and an intermediate series (light-face type). A self-holding taper is defined as "a taper with an angle small enough to hold a shank in place ordinarily by friction without holding means. (Sometimes referred to as slow taper.)" A steep taper is defined as "a taper having an angle sufficiently large to insure the easy or self-releasing feature." The term "gage line" indicates the basic diameter at or near the large end of the taper.

Table 1a. Morse Stub Taper Shanks

No. of Taper	Taper per Foot[a]	Taper per Inch[b]	Small End of Plug,[b] *D*	Dia. End of Socket,[a] *A*	Shank		Tang	
					Total Length, *B*	Depth, *C*	Thickness, *E*	Length, *F*
1	0.59858	0.049882	0.4314	0.475	1 5/16	1 1/8	13/64	5/16
2	0.59941	0.049951	0.6469	0.700	1 11/16	1 7/16	19/64	7/16
3	0.60235	0.050196	0.8753	0.938	2	1 3/4	25/64	9/16
4	0.62326	0.051938	1.1563	1.231	2 3/8	2 1/16	33/64	11/16
5	0.63151	0.052626	1.6526	1.748	3	2 11/16	3/4	15/16

No. of Taper	Tang		Socket				Tang Slot	
	Radius of Mill, *G*	Diameter, *H*	Plug Depth, *P*	Min. Depth of Tapered Hole: Drilled *X*	Min. Depth of Tapered Hole: Reamed *Y*	Socket End to Tang Slot, *M*	Width, *N*	Length, *O*
1	3/16	13/32	7/8	5/16	29/32	25/32	7/32	23/32
2	7/32	39/64	1 1/16	1 5/32	1 7/64	15/16	5/16	15/16
3	9/32	13/16	1 1/4	1 3/8	1 5/16	1 1/16	13/32	1 1/8
4	3/8	1 3/32	1 7/16	1 9/16	1 1/2	1 3/16	17/32	1 3/8
5	9/16	1 19/32	1 13/16	1 15/16	1 7/8	1 7/16	25/32	1 3/4

[a] These are basic dimensions.

[b] These dimensions are calculated for reference only.

All dimensions in inches.

Radius *J* is 3/64, 1/16, 5/64, 3/32, and 1/8 inch respectively for Nos. 1, 2, 3, 4, and 5 tapers.

Table 1b. Morse Standard Taper Shanks

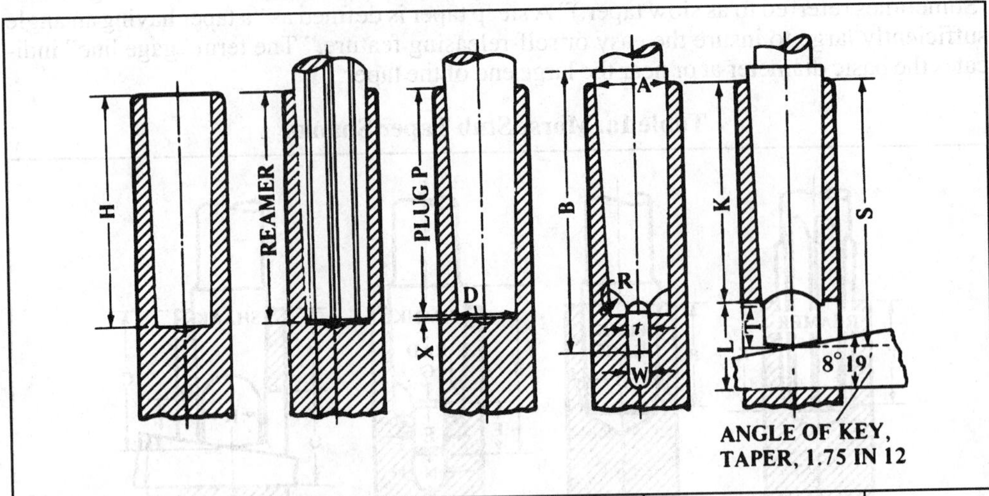

No. of Taper	Taper per Foot	Taper per Inch	Small End of Plug D	Diameter End of Socket A	Shank Length B	Shank Depth S	Depth of Hole H
0	0.62460	0.05205	0.252	0.3561	2 11⁄32	2 7⁄32	2 1⁄32
1	0.59858	0.04988	0.369	0.475	2 9⁄16	2 7⁄16	2 5⁄32
2	0.59941	0.04995	0.572	0.700	3 1⁄8	2 15⁄16	2 39⁄64
3	0.60235	0.05019	0.778	0.938	3 7⁄8	3 11⁄16	3 1⁄4
4	0.62326	0.05193	1.020	1.231	4 7⁄8	4 5⁄8	4 1⁄8
5	0.63151	0.05262	1.475	1.748	6 1⁄8	5 7⁄8	5 1⁄4
6	0.62565	0.05213	2.116	2.494	8 9⁄16	8 1⁄4	7 21⁄64
7	0.62400	0.05200	2.750	3.270	11 5⁄8	11 1⁄4	10 5⁄64

Plug Depth P	Tang or Tongue Thickness t	Tang or Tongue Length T	Tang or Tongue Radius R	Tang or Tongue Dia.	Keyway Width W	Keyway Length L	Keyway to End K
2	0.1562	1⁄4	5⁄32	0.235	11⁄64	9⁄16	1 15⁄16
2 1⁄8	0.2031	3⁄8	3⁄16	0.343	0.218	3⁄4	2 1⁄16
2 9⁄16	0.2500	7⁄16	1⁄4	17⁄32	0.266	7⁄8	2 1⁄2
3 3⁄16	0.3125	9⁄16	9⁄32	23⁄32	0.328	1 3⁄16	3 1⁄16
4 1⁄16	0.4687	5⁄8	5⁄16	31⁄32	0.484	1 1⁄4	3 7⁄8
5 3⁄16	0.6250	3⁄4	3⁄8	1 13⁄32	0.656	1 1⁄2	4 15⁄16
7 1⁄4	0.7500	1 1⁄8	1⁄2	2	0.781	1 3⁄4	7
10	1.1250	1 3⁄8	3⁄4	2 5⁄8	1.156	2 5⁄8	9 1⁄2

Table 2. American National Standard Taper Drive with Tang, Self-Holding Tapers *ANSI/ASME B5.10-1994*

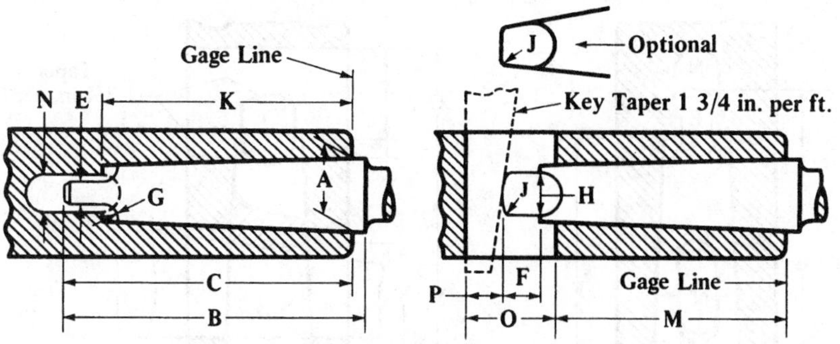

No. of Taper	Diameter at Gage Line (1) *A*	Shank: Total Length of Shank *B*	Shank: Gage Line to End of Shank *C*	Tang: Thickness *E*	Tang: Length *F*	Tang: Radius of Mill *G*	Tang: Diameter *H*
0.239	0.23922	1.28	1.19	0.125	0.19	0.19	0.18
0.299	0.29968	1.59	1.50	0.156	0.25	0.19	0.22
0.375	0.37525	1.97	1.88	0.188	0.31	0.19	0.28
1	0.47500	2.56	2.44	0.203	0.38	0.19	0.34
2	0.70000	3.13	2.94	0.250	0.44	0.25	0.53
3	0.93800	3.88	3.69	0.312	0.56	0.22	0.72
4	1.23100	4.88	4.63	0.469	0.63	0.31	0.97
4½	1.50000	5.38	5.13	0.562	0.69	0.38	1.20
5	1.74800	6.12	5.88	0.625	0.75	0.38	1.41
6	2.49400	8.25	8.25	0.750	1.13	0.50	2.00

No. of Taper	Radius *J*	Socket: Min. Depth of Hole *K*: Drilled	Socket: Min. Depth of Hole *K*: Reamed	Socket: Gage Line to Tang Slot *M*	Tang Slot: Width *N*	Tang Slot: Length *O*	Tang Slot: Shank End to Back of Tang Slot *P*
0.239	0.03	1.06	1.00	0.94	0.141	0.38	0.13
0.299	0.03	1.31	1.25	1.17	0.172	0.50	0.17
0.375	0.05	1.63	1.56	1.47	0.203	0.63	0.22
1	0.05	2.19	2.16	2.06	0.218	0.75	0.38
2	0.06	2.66	2.61	2.50	0.266	0.88	0.44
3	0.08	3.31	3.25	3.06	0.328	1.19	0.56
4	0.09	4.19	4.13	3.88	0.484	1.25	0.50
4½	0.13	4.62	4.56	4.31	0.578	1.38	0.56
5	0.13	5.31	5.25	4.94	0.656	1.50	0.56
6	0.16	7.41	7.33	7.00	0.781	1.75	0.50

All dimensions are in inches. (1) See Table 7 for plug and ring gage dimensions.

Tolerances: For shank diameter *A* at gage line, + 0.002 – 0.000; for hole diameter *A*, + 0.000 – 0.002. For tang thickness *E* up to No. 5 inclusive, + 0.000 – 0.006; No. 6, + 0.000 – 0.008. For width *N* of tang slot up to No. 5 inclusive, + 0.006; – 0.000; No. 6, + 0.008 – 0.000. For centrality of tang *E* with center line of taper, 0.0025 (0.005 total indicator variation). These centrality tolerances also apply to the tang slot *N*. On rate of taper, all sizes 0.002 per foot. This tolerance may be applied on *shanks* only in the direction which *increases* the rate of taper and on *sockets* only in the direction which *decreases* the rate of taper. Tolerances for two-decimal dimensions are plus or minus 0.010, unless otherwise specified.

Table 3. American National Standard Taper Drive with Keeper Key Slot, Self-Holding Tapers *ANSI/ASME B5.10-1994*

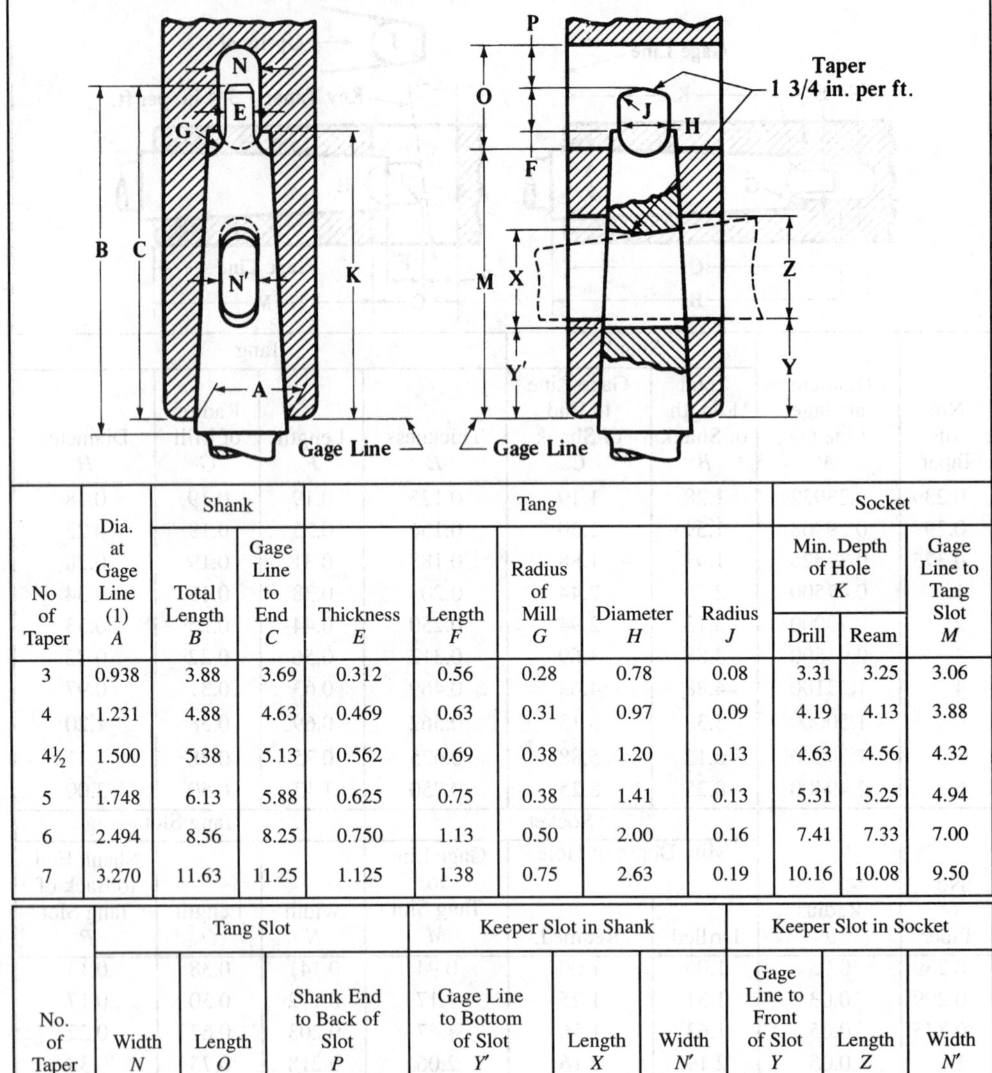

No of Taper	Dia. at Gage Line (1) *A*	Shank: Total Length *B*	Shank: Gage Line to End *C*	Tang: Thickness *E*	Tang: Length *F*	Tang: Radius of Mill *G*	Tang: Diameter *H*	Tang: Radius *J*	Socket: Min. Depth of Hole *K*, Drill	Socket: Min. Depth of Hole *K*, Ream	Socket: Gage Line to Tang Slot *M*
3	0.938	3.88	3.69	0.312	0.56	0.28	0.78	0.08	3.31	3.25	3.06
4	1.231	4.88	4.63	0.469	0.63	0.31	0.97	0.09	4.19	4.13	3.88
4½	1.500	5.38	5.13	0.562	0.69	0.38	1.20	0.13	4.63	4.56	4.32
5	1.748	6.13	5.88	0.625	0.75	0.38	1.41	0.13	5.31	5.25	4.94
6	2.494	8.56	8.25	0.750	1.13	0.50	2.00	0.16	7.41	7.33	7.00
7	3.270	11.63	11.25	1.125	1.38	0.75	2.63	0.19	10.16	10.08	9.50

No. of Taper	Tang Slot: Width *N*	Tang Slot: Length *O*	Tang Slot: Shank End to Back of Slot *P*	Keeper Slot in Shank: Gage Line to Bottom of Slot *Y′*	Keeper Slot in Shank: Length *X*	Keeper Slot in Shank: Width *N′*	Keeper Slot in Socket: Gage Line to Front of Slot *Y*	Keeper Slot in Socket: Length *Z*	Keeper Slot in Socket: Width *N′*
3	0.328	1.19	0.56	1.03	1.13	0.266	1.13	1.19	0.266
4	0.484	1.25	0.50	1.41	1.19	0.391	1.50	1.25	0.391
4½	0.578	1.38	0.56	1.72	1.25	0.453	1.81	1.38	0.453
5	0.656	1.50	0.56	2.00	1.38	0.516	2.13	1.50	0.516
6	0.781	1.75	0.50	2.13	1.63	0.641	2.25	1.75	0.641
7	1.156	2.63	0.88	2.50	1.69	0.766	2.63	1.81	0.766

All dimensions are in inches. (1) See Table 7 for plug and ring gage dimensions.

Tolerances: For shank diameter *A* at gage line, +0.002, −0; for hole diameter *A*, +0, −0.002. For tang thickness *E* up to No. 5 inclusive, +0, −0.006; larger than No. 5, +0, −0.008. For width of slots *N* and *N′* up to No. 5 inclusive, +0.006, −0; larger than No. 5, +0.008, −0. For centrality of tang *E* with center line of taper 0.0025 (0.005 total indicator variation). These centrality tolerances also apply to slots *N* and *N′*. On rate of taper, see footnote in Table 2. Tolerances for two-decimal dimensions are ±0.010 unless otherwise specified.

Table 4. American National Standard Nose Key Drive with Keeper Key Slot, Self-Holding Tapers *ANSI/ASME B5.10-1994*

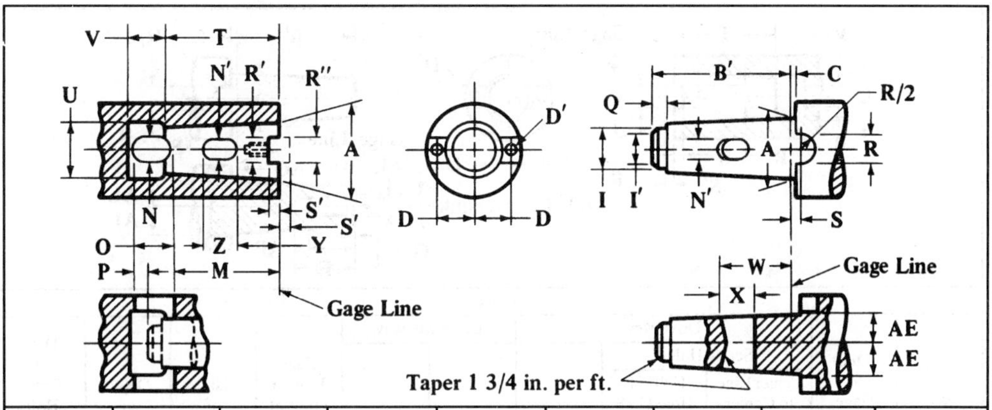

Taper 1 3/4 in. per ft.

Taper	A(1)	B′	C	Q	I′	I	R	S
200	2.000	5.13	Min 0.003 Max 0.035 for all sizes	0.25	1.38	1.63	1.010	0.562
250	2.500	5.88		0.25	1.38	2.06	1.010	0.562
300	3.000	6.63		0.25	1.63	2.50	2.010	0.562
350	3.500	7.44		0.31	2.00	2.94	2.010	0.562
400	4.000	8.19		0.31	2.13	3.31	2.010	0.562
450	4.500	9.00		0.38	2.38	3.81	3.010	0.812
500	5.000	9.75		0.38	2.50	4.25	3.010	0.812
600	6.000	11.31		0.44	3.00	5.19	3.010	0.812
800	8.000	14.38		0.50	3.50	7.00	4.010	1.062
1000	10.000	17.44		0.63	4.50	8.75	4.010	1.062
1200	12.000	20.50		0.75	5.38	10.50	4.010	1.062
Taper	**D**	**D′[a]**	**W**	**X**	**N′**	**R′**	**S′**	**T**
200	1.41	0.375	3.44	1.56	0.656	1.000	0.50	4.75
250	1.66	0.375	3.69	1.56	0.781	1.000	0.50	5.50
300	2.25	0.375	4.06	1.56	1.031	2.000	0.50	6.25
350	2.50	0.375	4.88	2.00	1.031	2.000	0.50	6.94
400	2.75	0.375	5.31	2.25	1.031	2.000	0.50	7.69
450	3.00	0.500	5.88	2.44	1.031	3.000	0.75	8.38
500	3.25	0.500	6.44	2.63	1.031	3.000	0.75	9.13
600	3.75	0.500	7.44	3.00	1.281	3.000	0.75	10.56
800	4.75	0.500	9.56	4.00	1.781	4.000	1.00	13.50
1000	…	…	11.50	4.75	2.031	4.000	1.00	16.31
1200	…	…	13.75	5.75	2.031	4.000	1.00	19.00
Taper	**U**	**V**	**M**	**N**	**O**	**P**	**Y**	**Z**
200	1.81	1.00	4.50	0.656	1.56	0.94	2.00	1.69
250	2.25	1.00	5.19	0.781	1.94	1.25	2.25	1.69
300	2.75	1.00	5.94	1.031	2.19	1.50	2.63	1.69
350	3.19	1.25	6.75	1.031	2.19	1.50	3.00	2.13
400	3.63	1.25	7.50	1.031	2.19	1.50	3.25	2.38
450	4.19	1.50	8.00	1.031	2.75	1.75	3.63	2.56
500	4.63	1.50	8.75	1.031	2.75	1.75	4.00	2.75
600	5.50	1.75	10.13	1.281	3.25	2.06	4.63	3.25
800	7.38	2.00	12.88	1.781	4.25	2.75	5.75	4.25
1000	9.19	2.50	15.75	2.031	5.00	3.31	7.00	5.00
1200	11.00	3.00	18.50	2.531	6.00	4.00	8.25	6.00

[a] Thread is UNF-2B for hole; UNF-2A for screw. (1) See Table 7 for plug and ring gage dimensions.

All dimensions are in inches. *AE* is 0.005 greater than one-half of *A*.

Width of drive key *R″* is 0.001 less than width *R″* of keyway.

Tolerances: For diameter *A* of hole at gage line, +0, −0.002; for diameter *A* of shank at gage line, +0.002, −0; for width of slots *N* and *N′*, +0.008, −0; for width of drive keyway *R′* in socket, +0, −0.001; for width of drive keyway *R* in shank, 0.010, −0; for centrality of slots *N* and *N′* with center line of spindle, 0.007; for centrality of keyway with spindle center line: for *R*, 0.004 and for *R′*, 0.002 T.I.V. On rate of taper, see footnote in Table 2. Two-decimal dimensions, ±0.010 unless otherwise specified.

Table 5. American National Standard Nose Key Drive with Drawbolt, Self-Holding Tapers *ANSI/ASME B5.10-1994*

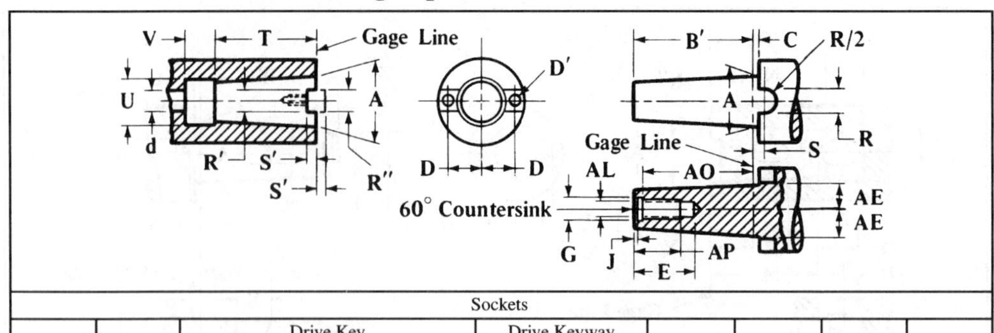

Sockets										
		Drive Key			Drive Keyway					
		Screw Holes								
No. of Taper	Dia. at Gage Line A^a	Center Line to Center of Screw *D*	UNF 2B Hole UNF 2A Screw *D′*	Width *R″*	Width *R′*	Depth *S′*	Gage Line to Front of Relief *T*	Dia. of Relief *U*	Depth of Relief *V*	Dia. of Draw Bolt Hole *d*
200	2.000	1.41	0.38	0.999	1.000	0.50	4.75	1.81	1.00	1.00
250	2.500	1.66	0.38	0.999	1.000	0.50	5.50	2.25	1.00	1.00
300	3.000	2.25	0.38	1.999	2.000	0.50	6.25	2.75	1.00	1.13
350	3.500	2.50	0.38	1.999	2.000	0.50	6.94	3.19	1.25	1.13
400	4.000	2.75	0.38	1.999	2.000	0.50	7.69	3.63	1.25	1.63
450	4.500	3.00	0.50	2.999	3.000	0.75	8.38	4.19	1.50	1.63
500	5.000	3.25	0.50	2.999	3.000	0.75	9.13	4.63	1.50	1.63
600	6.000	3.75	0.50	2.999	3.000	0.75	10.56	5.50	1.75	2.25
800	8.000	4.75	0.50	3.999	4.000	1.00	13.50	7.38	2.00	2.25
1000	10.000	…	…	3.999	4.000	1.00	16.31	9.19	2.50	2.25
1200	12.000	…	…	3.999	4.000	1.00	19.00	11.00	3.00	2.25

[a] See Table 7 for plug and ring gage dimensions.

Shanks										
		Drawbar Hole						Drive Keyway		
No. of Taper	Length from Gage Line B′	Dia. UNC-2B *AL*	Depth of Drilled Hole *E*	Depth of Thread *AP*	Dia. of Counter Bore *G*	Gage Line to First Thread *AO*	Depth of 60° Chamfer *J*	Width *R*	Depth *S*	Center Line to Bottom of Keyway *AE*
200	5.13	⅞–9	2.44	1.75	0.91	4.78	0.13	1.010	0.562	1.005
250	5.88	⅞–9	2.44	1.75	0.91	5.53	0.13	1.010	0.562	1.255
300	6.63	1–8	2.75	2.00	1.03	6.19	0.19	2.010	0.562	1.505
350	7.44	1–8	2.75	2.00	1.03	7.00	0.19	2.010	0.562	1.755
400	8.19	1½–6	4.00	3.00	1.53	7.50	0.31	2.010	0.562	2.005
450	9.00	1½–6	4.00	3.00	1.53	8.31	0.31	3.010	0.812	2.255
500	9.75	1½–6	4.00	3.00	1.53	9.06	0.31	3.010	0.812	2.505
600	11.31	2–4½	5.31	4.00	2.03	10.38	0.50	3.010	0.812	3.005
800	14.38	2–4½	5.31	4.00	2.03	13.44	0.50	4.010	1.062	4.005
1000	17.44	2–4½	5.31	4.00	2.03	16.50	0.50	4.010	1.062	5.005
1200	20.50	2–4½	5.31	4.00	2.03	19.56	0.50	4.010	1.062	6.005

All dimensions in inches.

Exposed length *C* is 0.003 minimum and 0.035 maximum for all sizes.

Drive Key *D′* screw sizes are ⅜–24 UNF-2A up to taper No. 400 inclusive and ½–20 UNF-2A for larger tapers.

Tolerances: For diameter *A* of hole at gage line, +0.000, –0.002 for all sizes; for diameter *A* of shank at gage line, +0.002, –0.000; for all sizes; for width of drive keyway *R′* in socket, +0.000, –0.001; for width of drive keyway *R* in shank, +0.010, –0.000; for centrality of drive keyway *R′*, with center line of shank, 0.004 total indicator variation, and for drive keyway *R′*, with center line of spindle, 0.002. On rate of taper, see footnote in Table 2. Tolerances for two-decimal dimensions are ±0.010 unless otherwise specified.

Table 6. ANSI Standard Steep Machine Tapers *ANSI/ASME B5.10-1994*

No. of Taper	Taper per Foot[a]	Dia. at Gage Line[b]	Length Along Axis	No. of Taper	Taper per Foot[a]	Dia.at Gage Line[b]	Length Along Axis
5	3.500	0.500	0.6875	35	3.500	1.500	2.2500
10	**3.500**	**0.625**	**0.8750**	**40**	**3.500**	**1.750**	**2.5625**
15	3.500	0.750	1.0625	45	3.500	2.250	3.3125
20	**3.500**	**0.875**	**1.3125**	**50**	**3.500**	**2.750**	**4.0000**
25	3.500	1.000	1.5625	55	3.500	3.500	5.1875
30	**3.500**	**1.250**	**1.8750**	**60**	**3.500**	**4.250**	**6.3750**

[a] This taper corresponds to an included angle of 16°, 35′, 39.4″.
[b] The basic diameter at gage line is at large end of taper.

All dimensions given in inches.

The tapers numbered 10, 20, 30, 40, 50, and 60 that are printed in heavy-faced type are designated as the "Preferred Series." The tapers numbered 5, 15, 25, 35, 45, and 55 that are printed in light-faced type are designated as the "Intermediate Series."

Table 7. American National Standard Self-holding Tapers — Basic Dimensions *ANSI/ASME B5.10-1994*

No. of Taper	Taper per Foot	Dia.at Gage Line[a] *A*	Means of Driving and Holding[a]	Origin of Series
.239	0.50200	0.23922	Tang Drive With Shank Held in by Friction (See Table 2)	Brown & Sharpe Taper Series
.299	0.50200	0.29968		
.375	0.50200	0.37525		
1	0.59858	0.47500	Tang Drive With Shank Held in by Key (See Table 3)	Morse Taper Series
2	0.59941	0.70000		
3	0.60235	0.93800		
4	0.62326	1.23100		
4½	0.62400	1.50000		
5	0.63151	1.74800		
6	0.62565	2.49400		
7	0.62400	3.27000		
200	0.750	2.000	Key Drive With Shank Held in by Key (See Table 4)	¾ Inch per Foot Taper Series
250	0.750	2.500		
300	0.750	3.000		
350	0.750	3.500		
400	0.750	4.000	Key Drive With Shank Heldin by Draw-bolt (See Table 5)	
450	0.750	4.500		
500	0.750	5.000		
600	0.750	6.000		
800	0.750	8.000		
1000	0.750	10.000		
1200	0.750	12.000		

[a] See illustrations above Tables 2 through 5.

All dimensions given in inches.

Table 7. American National Standard Plug and Ring Gages for the Self-Holding Taper Series *ANSI/ASME B5.10-1994*

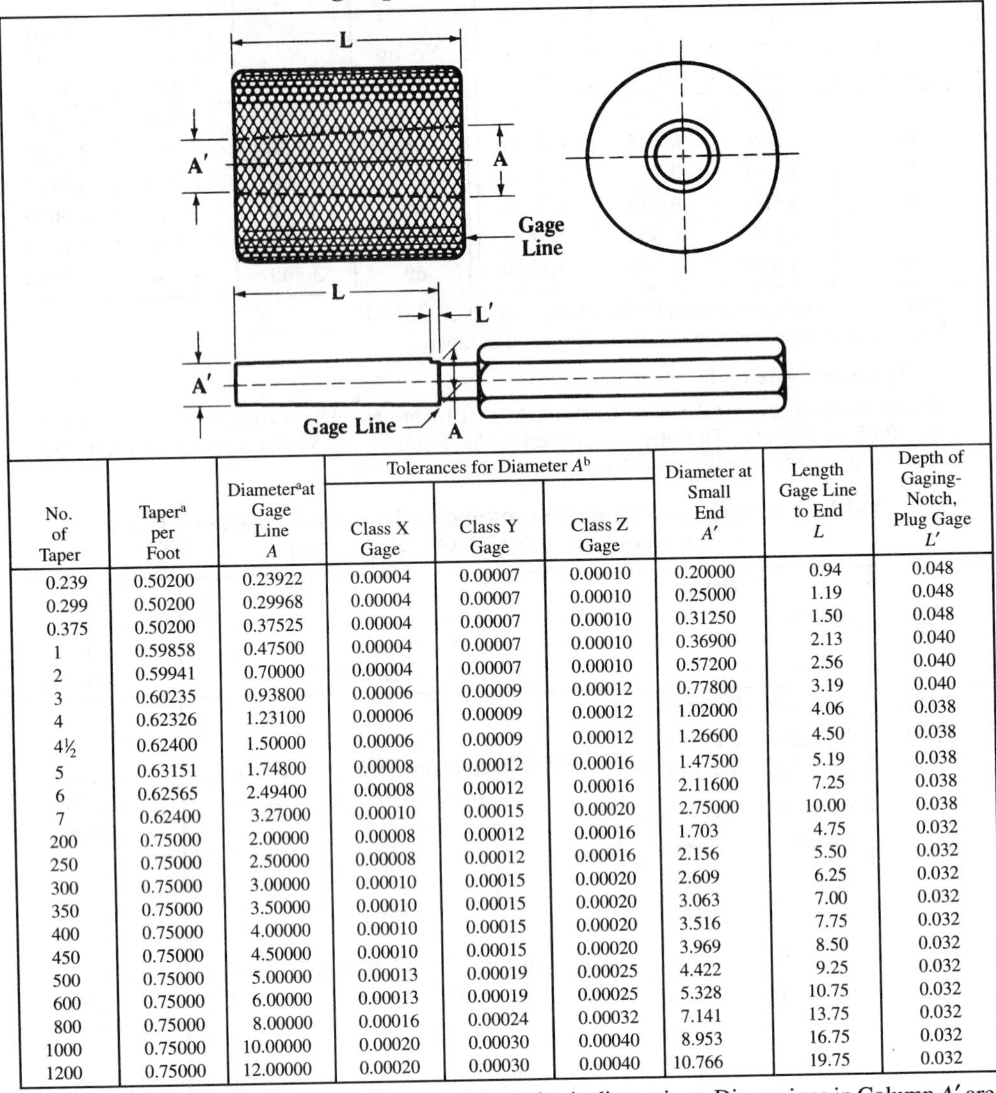

No. of Taper	Taper[a] per Foot	Diameter[a] at Gage Line *A*	Tolerances for Diameter *A*[b] Class X Gage	Class Y Gage	Class Z Gage	Diameter at Small End *A′*	Length Gage Line to End *L*	Depth of Gaging-Notch, Plug Gage *L′*
0.239	0.50200	0.23922	0.00004	0.00007	0.00010	0.20000	0.94	0.048
0.299	0.50200	0.29968	0.00004	0.00007	0.00010	0.25000	1.19	0.048
0.375	0.50200	0.37525	0.00004	0.00007	0.00010	0.31250	1.50	0.048
1	0.59858	0.47500	0.00004	0.00007	0.00010	0.36900	2.13	0.040
2	0.59941	0.70000	0.00004	0.00007	0.00010	0.57200	2.56	0.040
3	0.60235	0.93800	0.00006	0.00009	0.00012	0.77800	3.19	0.040
4	0.62326	1.23100	0.00006	0.00009	0.00012	1.02000	4.06	0.038
4½	0.62400	1.50000	0.00006	0.00009	0.00012	1.26600	4.50	0.038
5	0.63151	1.74800	0.00008	0.00012	0.00016	1.47500	5.19	0.038
6	0.62565	2.49400	0.00008	0.00012	0.00016	2.11600	7.25	0.038
7	0.62400	3.27000	0.00010	0.00015	0.00020	2.75000	10.00	0.038
200	0.75000	2.00000	0.00008	0.00012	0.00016	1.703	4.75	0.032
250	0.75000	2.50000	0.00008	0.00012	0.00016	2.156	5.50	0.032
300	0.75000	3.00000	0.00010	0.00015	0.00020	2.609	6.25	0.032
350	0.75000	3.50000	0.00010	0.00015	0.00020	3.063	7.00	0.032
400	0.75000	4.00000	0.00010	0.00015	0.00020	3.516	7.75	0.032
450	0.75000	4.50000	0.00010	0.00015	0.00020	3.969	8.50	0.032
500	0.75000	5.00000	0.00013	0.00019	0.00025	4.422	9.25	0.032
600	0.75000	6.00000	0.00013	0.00019	0.00025	5.328	10.75	0.032
800	0.75000	8.00000	0.00016	0.00024	0.00032	7.141	13.75	0.032
1000	0.75000	10.00000	0.00020	0.00030	0.00040	8.953	16.75	0.032
1200	0.75000	12.00000	0.00020	0.00030	0.00040	10.766	19.75	0.032

[a] The taper per foot and diameter *A* at gage line are basic dimensions. Dimensions in Column *A′* are calculated for reference only.

[b] Tolerances for diameter *A* are plus for plug gages and minus for ring gages.

All dimensions are in inches.

The amount of taper deviation for Class X, Class Y, and Class Z gages are the same, respectively, as the amounts shown for tolerances on diameter *A*. Taper deviation is the permissible allowance from true taper at any point of diameter in the length of the gage. On taper *plug* gages, this deviation may be applied only in the direction which *decreases* the rate of taper. On taper *ring* gages, this deviation may be applied only in the direction which *increases* the rate of taper. Tolerances on two-decimal dimensions are ±0.010.

British Standard Tapers.—British Standard 1660: 1972, "Machine Tapers, Reduction Sleeves, and Extension Sockets," contains dimensions for self-holding and self-releasing tapers, reduction sleeves, extension sockets, and turret sockets for tools having Morse and metric 5 per cent taper shanks. Adapters for use with $\frac{7}{24}$ tapers and dimensions for spindle noses and tool shanks with self-release tapers and cotter slots are included in this Standard.

Table 8. Dimensions of Morse Taper Sleeves

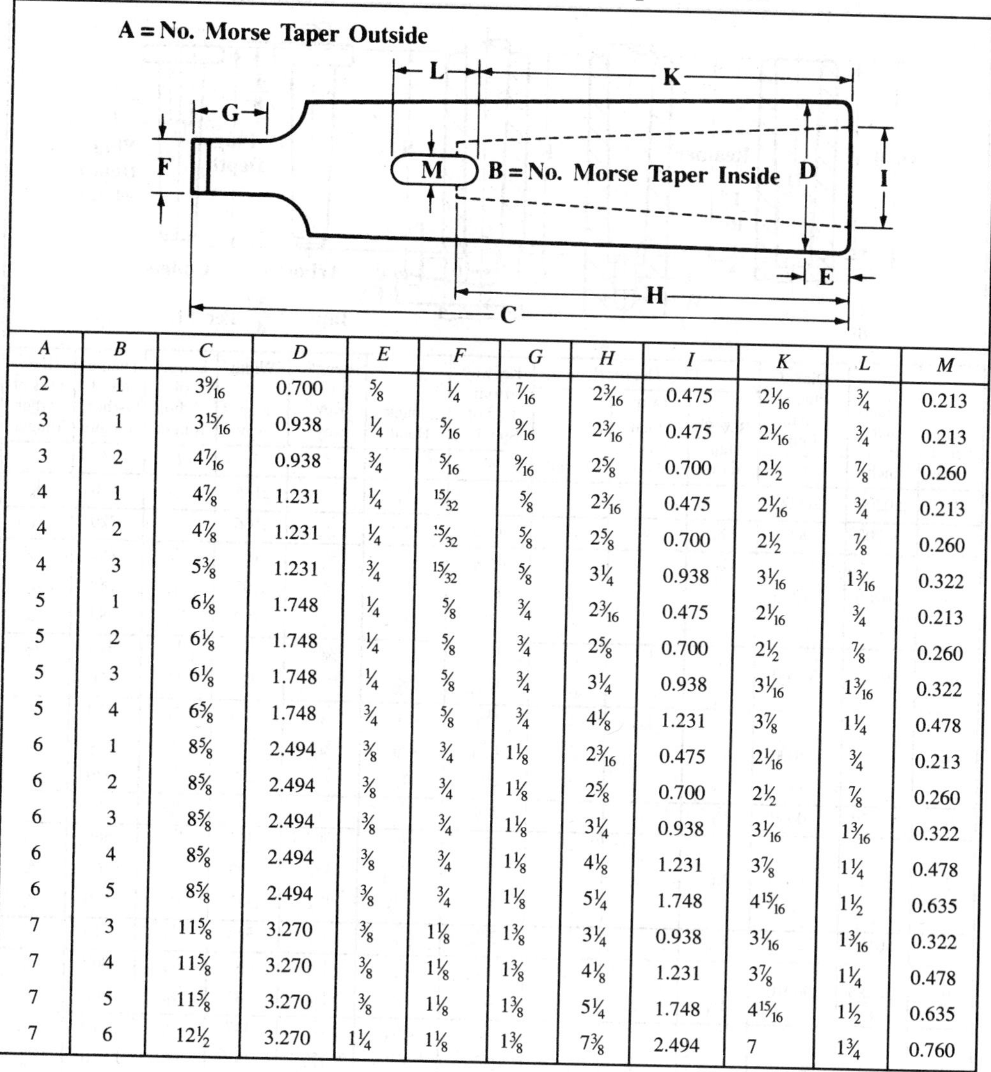

A	*B*	*C*	*D*	*E*	*F*	*G*	*H*	*I*	*K*	*L*	*M*
2	1	3 9/16	0.700	5/8	1/4	7/16	2 3/16	0.475	2 1/16	3/4	0.213
3	1	3 15/16	0.938	1/4	5/16	9/16	2 3/16	0.475	2 1/16	3/4	0.213
3	2	4 7/16	0.938	3/4	5/16	9/16	2 5/8	0.700	2 1/2	7/8	0.260
4	1	4 7/8	1.231	1/4	15/32	5/8	2 3/16	0.475	2 1/16	3/4	0.213
4	2	4 7/8	1.231	1/4	15/32	5/8	2 5/8	0.700	2 1/2	7/8	0.260
4	3	5 3/8	1.231	3/4	15/32	5/8	3 1/4	0.938	3 1/16	1 3/16	0.322
5	1	6 1/8	1.748	1/4	5/8	3/4	2 3/16	0.475	2 1/16	3/4	0.213
5	2	6 1/8	1.748	1/4	5/8	3/4	2 5/8	0.700	2 1/2	7/8	0.260
5	3	6 1/8	1.748	1/4	5/8	3/4	3 1/4	0.938	3 1/16	1 3/16	0.322
5	4	6 5/8	1.748	3/4	5/8	3/4	4 1/8	1.231	3 7/8	1 1/4	0.478
6	1	8 5/8	2.494	3/8	3/4	1 1/8	2 3/16	0.475	2 1/16	3/4	0.213
6	2	8 5/8	2.494	3/8	3/4	1 1/8	2 5/8	0.700	2 1/2	7/8	0.260
6	3	8 5/8	2.494	3/8	3/4	1 1/8	3 1/4	0.938	3 1/16	1 3/16	0.322
6	4	8 5/8	2.494	3/8	3/4	1 1/8	4 1/8	1.231	3 7/8	1 1/4	0.478
6	5	8 5/8	2.494	3/8	3/4	1 1/8	5 1/4	1.748	4 15/16	1 1/2	0.635
7	3	11 5/8	3.270	3/8	1 1/8	1 3/8	3 1/4	0.938	3 1/16	1 3/16	0.322
7	4	11 5/8	3.270	3/8	1 1/8	1 3/8	4 1/8	1.231	3 7/8	1 1/4	0.478
7	5	11 5/8	3.270	3/8	1 1/8	1 3/8	5 1/4	1.748	4 15/16	1 1/2	0.635
7	6	12 1/2	3.270	1 1/4	1 1/8	1 3/8	7 3/8	2.494	7	1 3/4	0.760

Table 9. Morse Taper Sockets — Hole and Shank Sizes

Size	Morse Taper		Size	Morse Taper		Size	Morse Taper	
	Hole	Shank		Hole	Shank		Hole	Shank
1 by 2	No. 1	No. 2	2 by 5	No. 2	No. 5	4 by 4	No. 4	No. 4
1 by 3	No. 1	No. 3	3 by 2	No. 3	No. 2	4 by 5	No. 4	No. 5
1 by 4	No. 1	No. 4	3 by 3	No. 3	No. 3	4 by 6	No. 4	No. 6
1 by 5	No. 1	No. 5	3 by 4	No. 3	No. 4	5 by 4	No. 5	No. 4
2 by 3	No. 2	No. 3	3 by 5	No. 3	No. 5	5 by 5	No. 5	No. 5
2 by 4	No. 2	No. 4	4 by 3	No. 4	No. 3	5 by 6	No. 5	No. 6

Table 10. Brown & Sharpe Taper Shanks

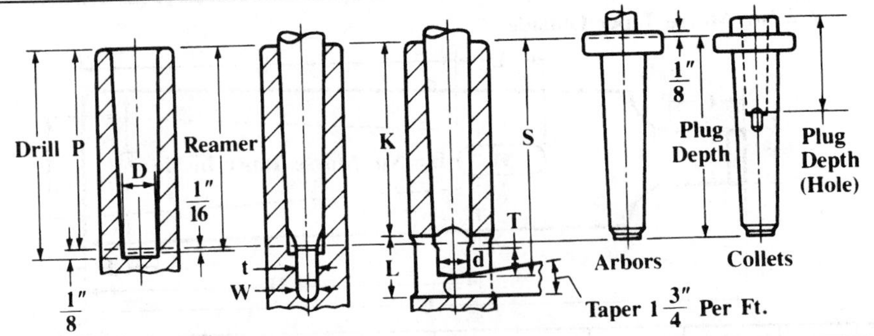

Number of Taper	Taper per Foot (inch)	Dia. of Plug at Small End	Plug Depth, P			Keyway from End of Spindle	Shank Depth	Length of Keyway[a]	Width of Keyway	Length of Arbor Tongue	Diameter of Arbor Tongue	Thickness of Arbor Tongue
		D	B & S[b] Standard	Mill. Mach. Standard	Miscell.	K	S	L	W	T	d	t
1[c]	.50200	.20000	15/16	…	…	15/16	1 3/16	3/8	.135	3/16	.170	1/8
2[c]	.50200	.25000	1 3/16	…	…	1 11/64	1 1/2	1/2	.166	1/4	.220	5/32
3[c]	.50200	.31250	1 1/2	…	…	1 15/32	1 7/8	5/8	.197	5/16	.282	3/16
			…	…	1 3/4	1 23/32	2 1/8	5/8	.197	5/16	.282	3/16
			…	…	2	1 31/32	2 3/8	5/8	.197	5/16	.282	3/16
4	.50240	.35000	…	1 1/4	…	1 13/64	1 21/32	11/16	.228	11/32	.320	7/32
			1 11/16	…	…	1 41/64	2 3/32	11/16	.228	11/32	.320	7/32
5	.50160	.45000	…	1 3/4	…	1 11/16	2 3/16	3/4	.260	3/8	.420	1/4
			…	…	2	1 15/16	2 7/16	3/4	.260	3/8	.420	1/4
			2 1/8	…	…	2 1/16	2 9/16	3/4	.260	3/8	.420	1/4
6	.50329	.50000	2 3/8	…	…	2 19/64	2 7/8	7/8	.291	7/16	.460	9/32
7	.50147	.60000	…	…	2 1/2	2 13/32	3 1/32	15/16	.322	15/32	.560	5/16
			2 7/8	…	…	2 25/32	3 13/32	15/16	.322	15/32	.560	5/16
			…	3	…	2 29/32	3 17/32	15/16	.322	15/32	.560	5/16
8	.50100	.75000	3 9/16	…	…	3 29/64	4 1/8	1	.353	1/2	.710	11/32
9	.50085	.90010	…	4	…	3 7/8	4 5/8	1 1/8	.385	9/16	.860	3/8
			4 1/4	…	…	4 1/8	4 7/8	1 1/8	.385	9/16	.860	3/8
10	.51612	1.04465	5	…	…	4 27/32	5 23/32	1 5/16	.447	21/32	1.010	7/16
			…	5 11/16	…	5 17/32	6 13/32	1 5/16	.447	21/32	1.010	7/16
			…	…	6 7/32	6 1/16	6 15/16	1 5/16	.447	21/32	1.010	7/16
11	.50100	1.24995	5 15/16	…	…	5 25/32	6 21/32	1 5/16	.447	21/32	1.210	7/16
			…	6 3/4	…	6 19/32	7 15/32	1 5/16	.447	21/32	1.210	7/16
12	.49973	1.50010	7 1/8	7 1/8	…	6 15/16	7 15/16	1 1/2	.510	3/4	1.460	1/2
			…	…	6 1/4	…	…	…	…	…	…	…
13	.50020	1.75005	7 3/4	…	…	7 9/16	8 9/16	1 1/2	.510	3/4	1.710	1/2
14	.50000	2.00000	8 1/4	8 1/4	…	8 1/32	9 5/32	1 11/16	.572	27/32	1.960	9/16
15	.5000	2.25000	8 3/4	…	…	8 17/32	9 21/32	1 11/16	.572	27/32	2.210	9/16
16	.50000	2.50000	9 1/4	…	…	9	10 1/4	1 7/8	.635	15/16	2.450	5/8
17	.50000	2.75000	9 3/4	…	…	…	…	…	…	…	…	…
18	.50000	3.00000	10 1/4	…	…	…	…	…	…	…	…	…

[a] Special lengths of keyway are used instead of standard lengths in some places. Standard lengths need not be used when keyway is for driving only and not for admitting key to force out tool.

[b] "B & S Standard" Plug Depths are not used in all cases.

[c] Adopted by American Standards Association.

Table 11. Jarno Taper Shanks

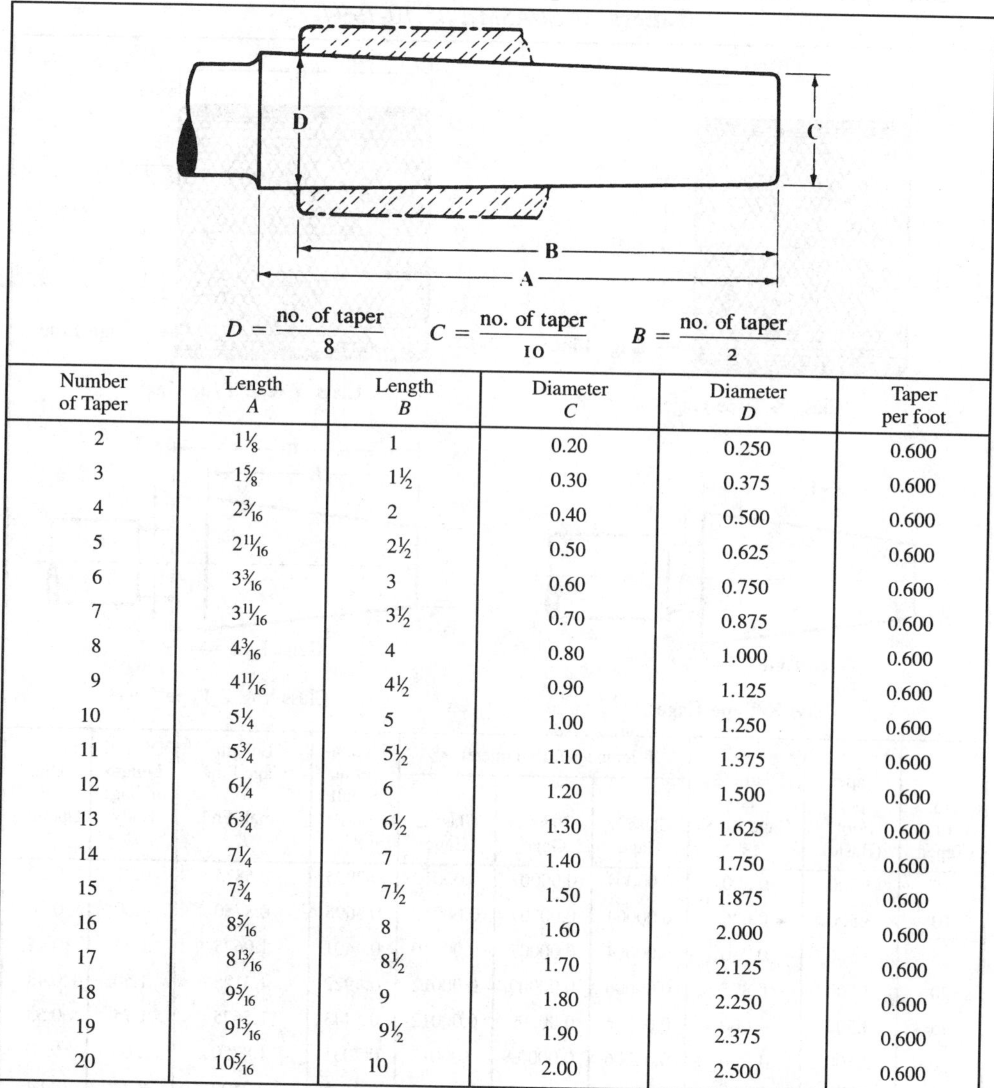

$$D = \frac{\text{no. of taper}}{8} \qquad C = \frac{\text{no. of taper}}{10} \qquad B = \frac{\text{no. of taper}}{2}$$

Number of Taper	Length A	Length B	Diameter C	Diameter D	Taper per foot
2	$1\frac{1}{8}$	1	0.20	0.250	0.600
3	$1\frac{5}{8}$	$1\frac{1}{2}$	0.30	0.375	0.600
4	$2\frac{3}{16}$	2	0.40	0.500	0.600
5	$2\frac{11}{16}$	$2\frac{1}{2}$	0.50	0.625	0.600
6	$3\frac{3}{16}$	3	0.60	0.750	0.600
7	$3\frac{11}{16}$	$3\frac{1}{2}$	0.70	0.875	0.600
8	$4\frac{3}{16}$	4	0.80	1.000	0.600
9	$4\frac{11}{16}$	$4\frac{1}{2}$	0.90	1.125	0.600
10	$5\frac{1}{4}$	5	1.00	1.250	0.600
11	$5\frac{3}{4}$	$5\frac{1}{2}$	1.10	1.375	0.600
12	$6\frac{1}{4}$	6	1.20	1.500	0.600
13	$6\frac{3}{4}$	$6\frac{1}{2}$	1.30	1.625	0.600
14	$7\frac{1}{4}$	7	1.40	1.750	0.600
15	$7\frac{3}{4}$	$7\frac{1}{2}$	1.50	1.875	0.600
16	$8\frac{5}{16}$	8	1.60	2.000	0.600
17	$8\frac{13}{16}$	$8\frac{1}{2}$	1.70	2.125	0.600
18	$9\frac{5}{16}$	9	1.80	2.250	0.600
19	$9\frac{13}{16}$	$9\frac{1}{2}$	1.90	2.375	0.600
20	$10\frac{5}{16}$	10	2.00	2.500	0.600

Tapers for Machine Tool Spindles.—Most lathe spindles have Morse tapers, most milling machine spindles have American Standard tapers, almost all smaller milling machine spindles have R8 tapers, and large vertical milling machine spindles have American Standard tapers. The spindles of drilling machines and the taper shanks of twist drills are made to fit the Morse taper. For lathes, the Morse taper is generally used, but lathes may have the Jarno, Brown & Sharpe, or a special taper. Of 33 lathe manufacturers, 20 use the Morse taper; 5, the Jarno; 3 use special tapers of their own; 2 use modified Morse (longer than the standard but the same taper); 2 use Reed (which is a short Jarno); 1 uses the Brown & Sharpe standard. For grinding machine centers, Jarno, Morse, and Brown & Sharpe tapers are used. Of ten grinding machine manufacturers, 3 use Brown & Sharpe; 3 use Morse; and 4 use Jarno. The Brown & Sharpe taper is used extensively for milling machine and dividing head spindles. The standard milling machine spindle adopted in 1927 by the milling machine manufacturers of the National Machine Tool Builders' Association (now The Association for Manufacturing Technology [AMT]), has a taper of $3\frac{1}{2}$ inches per foot. This comparatively steep taper was adopted to ensure easy release of arbors.

Table 12. American National Standard Plug and Ring Gages for Steep Machine Tapers *ANSI/ASME B5.10-1994*

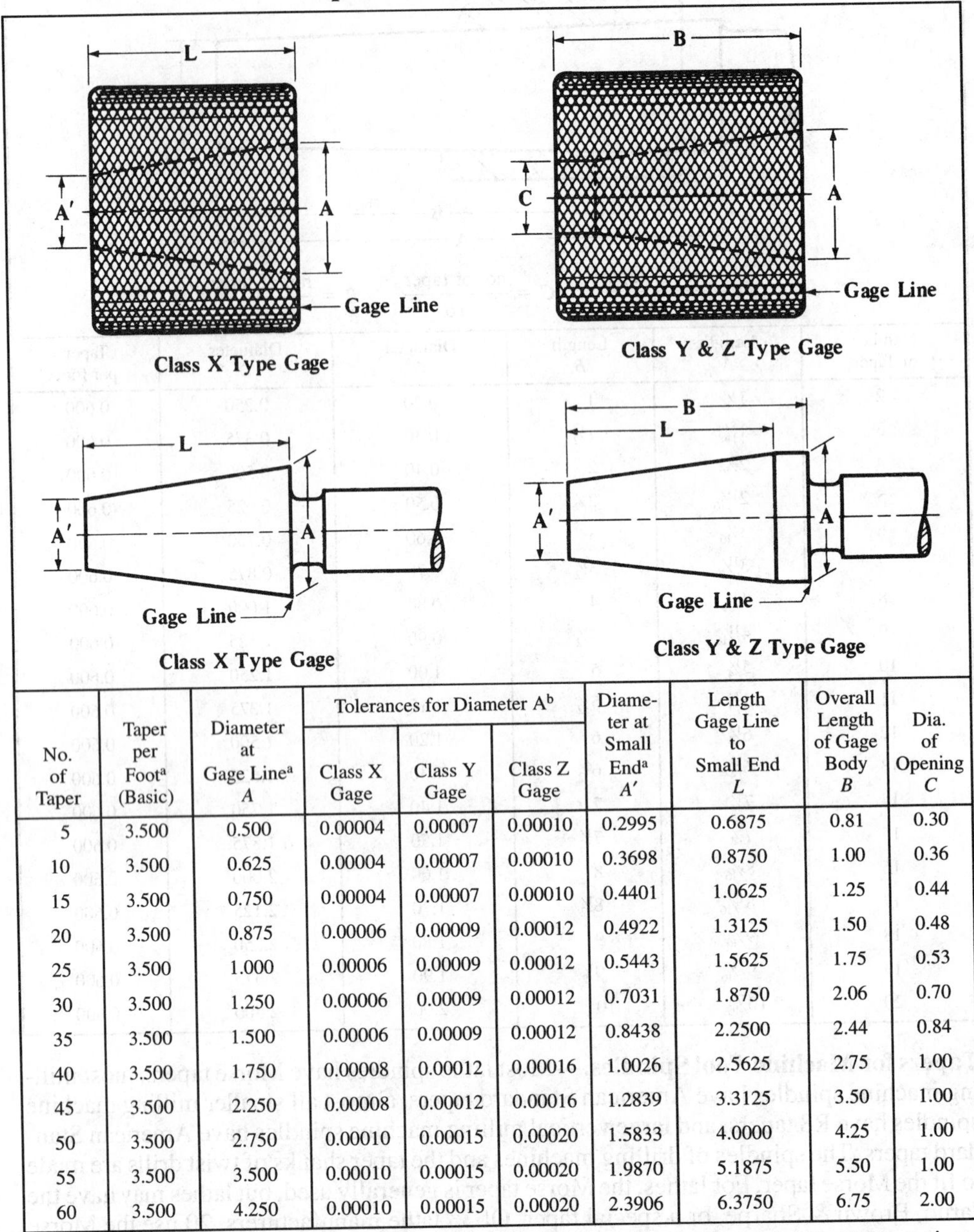

No. of Taper	Taper per Foot[a] (Basic)	Diameter at Gage Line[a] *A*	Tolerances for Diameter *A*[b]			Diameter at Small End[a] *A′*	Length Gage Line to Small End *L*	Overall Length of Gage Body *B*	Dia. of Opening *C*
			Class X Gage	Class Y Gage	Class Z Gage				
5	3.500	0.500	0.00004	0.00007	0.00010	0.2995	0.6875	0.81	0.30
10	3.500	0.625	0.00004	0.00007	0.00010	0.3698	0.8750	1.00	0.36
15	3.500	0.750	0.00004	0.00007	0.00010	0.4401	1.0625	1.25	0.44
20	3.500	0.875	0.00006	0.00009	0.00012	0.4922	1.3125	1.50	0.48
25	3.500	1.000	0.00006	0.00009	0.00012	0.5443	1.5625	1.75	0.53
30	3.500	1.250	0.00006	0.00009	0.00012	0.7031	1.8750	2.06	0.70
35	3.500	1.500	0.00006	0.00009	0.00012	0.8438	2.2500	2.44	0.84
40	3.500	1.750	0.00008	0.00012	0.00016	1.0026	2.5625	2.75	1.00
45	3.500	2.250	0.00008	0.00012	0.00016	1.2839	3.3125	3.50	1.00
50	3.500	2.750	0.00010	0.00015	0.00020	1.5833	4.0000	4.25	1.00
55	3.500	3.500	0.00010	0.00015	0.00020	1.9870	5.1875	5.50	1.00
60	3.500	4.250	0.00010	0.00015	0.00020	2.3906	6.3750	6.75	2.00

[a] The taper per foot and diameter *A* at gage line are basic dimensions. Dimensions in Column *A′* are calculated for reference only.

[b] Tolerances for diameter *A* are plus for plug gages and minus for ring gages.

All dimensions are in inches.

The amounts of taper deviation for Class X, Class Y, and Class Z gages are the same, respectively, as the amounts shown for tolerances on diameter *A*. Taper deviation is the permissible allowance from true taper at any point of diameter in the length of the gage. On taper *plug* gages, this deviation may be applied only in the direction which *decreases* the rate of taper. On taper *ring* gages, this deviation may be applied only in the direction which *increases* the rate of taper. Tolerances on two-decimal dimensions are ±0.010.

Table 13. Jacobs Tapers and Threads for Drill Chucks and Spindles

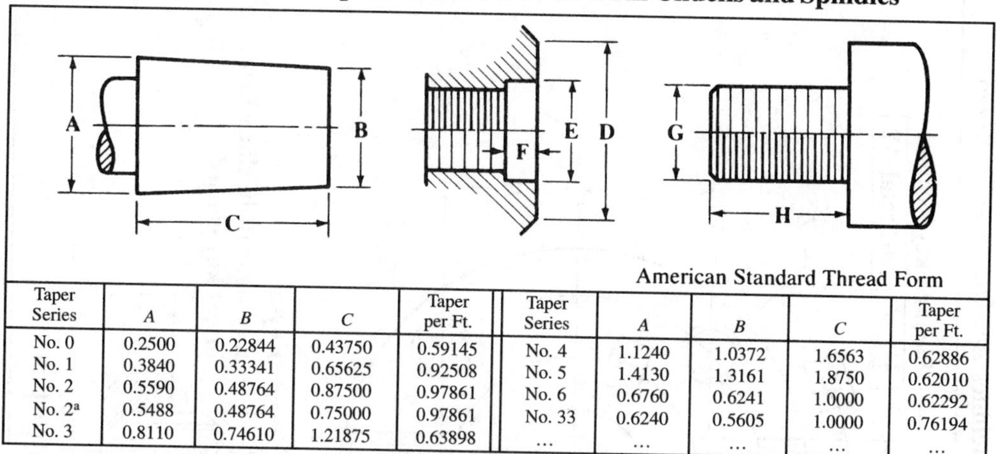

Taper Series	A	B	C	Taper per Ft.	Taper Series	A	B	C	Taper per Ft.
No. 0	0.2500	0.22844	0.43750	0.59145	No. 4	1.1240	1.0372	1.6563	0.62886
No. 1	0.3840	0.33341	0.65625	0.92508	No. 5	1.4130	1.3161	1.8750	0.62010
No. 2	0.5590	0.48764	0.87500	0.97861	No. 6	0.6760	0.6241	1.0000	0.62292
No. 2[a]	0.5488	0.48764	0.75000	0.97861	No. 33	0.6240	0.5605	1.0000	0.76194
No. 3	0.8110	0.74610	1.21875	0.63898	…	…	…	…	…

[a] These dimensions are for the No. 2 "short" taper.

Thread Size	Diameter D		Diameter E		Dimension F	
	Max.	Min.	Max.	Min.	Max.	Min.
5/16–24	0.531	0.516	0.3245	0.3195	0.135	0.115
5/16–24	0.633	0.618	0.3245	0.3195	0.135	0.115
3/8–24	0.633	0.618	0.385	0.380	0.135	0.115
1/2–20	0.860	0.845	0.510	0.505	0.135	0.115
5/8–11	1.125	1.110	0.635	0.630	0.166	0.146
5/8–16	1.125	1.110	0.635	0.630	0.166	0.146
45/64–16	1.250	1.235	0.713	0.708	0.166	0.146
3/4–16	1.250	1.235	0.760	0.755	0.166	0.146
1–8	1.437	1.422	1.036	1.026	0.281	0.250
1–10	1.437	1.422	1.036	1.026	0.281	0.250
1 1/2–8	1.871	1.851	1.536	1.526	0.343	0.312

Thread[a] Size	G		H[b]	Plug Gage Pitch Dia.		Ring Gage Pitch Dia.	
	Max	Min		Go	Not Go	Go	Not Go
5/16–24	0.3114	0.3042	0.437[c]	0.2854	0.2902	0.2843	0.2806
3/8–24	0.3739	0.3667	0.562[d]	0.3479	0.3528	0.3468	0.3430
1/2–20	0.4987	0.4906	0.562	0.4675	0.4731	0.4662	0.4619
5/8–11	0.6234	0.6113	0.687	0.5660	0.5732	0.5644	0.5589
5/8–16	0.6236	0.6142	0.687	0.5844	0.5906	0.5830	0.5782
45/64–16	0.7016	0.6922	0.687	0.6625	0.6687	0.6610	0.6561
3/4–16	0.7485	0.7391	0.687	0.7094	0.7159	0.7079	0.7029
1–8	1.000	0.9848	1.000	0.9188	0.9242	0.9188	0.9134
1–10	1.000	0.9872	1.000	0.9350	0.9395	0.9350	0.9305
1 1/2–8	1.500	1.4848	1.000	1.4188	1.4242	1.4188	1.4134

[a] Except for 1–8, 1–10, 1 1/2–8 all threads are now manufactured to the American National Standard Unified Screw Thread System, Internal Class 2B, External Class 2A. Effective date 1976.

[b] Tolerances for dimension *H* are as follows: 0.030 inch for thread sizes 5/16 –24 to 3/4–16, inclusive and 0.125 inch for thread sizes 1–8 to 1 1/2–8, inclusive.

[c] Length for Jacobs 0B5/16 chuck is 0.375 inch, length for 1B5/16 chuck is 0.437 inch.

[d] Length for Jacobs No. 1BS chuck is 0.437 inch.

Usual Chuck Capacities for Different Taper Series Numbers: No. 0 taper, drill diameters, 0–5/32 inch; No. 1, 0–1/4 inch; No. 2, 0–1/2 inch; No. 2 "Short," 0–5/16 inch; No. 3, 0–1/2, 1/8–5/8, 3/16–3/4, or 1/4–13/16 inch; No. 4, 1/8–3/4 inch; No. 5, 3/8–1; No. 6, 0–1/2 inch; No. 33, 0–1/2 inch.

Usual Chuck Capacities for Different Thread Sizes: Size 5/16–24, drill diameters 0–1/4 inch; size 3/8–24, drill diameters 0–3/8, 1/16–3/8, or 5/64–1/2 inch; size 1/2–20, drill diameters 0–1/2, 1/16–3/8, or 5/64–1/2 inch; size 5/8–11, drill diameters 0–1/2 inch; size 5/8–16, drill diameters 0–1/2, 1/8– –5/8, or 3/16–3/4 inch; size 45/64 –16, drill diameters 0–1/2 inch; size 3/4–16, drill diameters 0–1/2 or 3/16–3/4.

Table 1. Essential Dimensions of American National Standard Spindle Noses for Milling Machines *ANSI B5.18-1972 (R1998)*

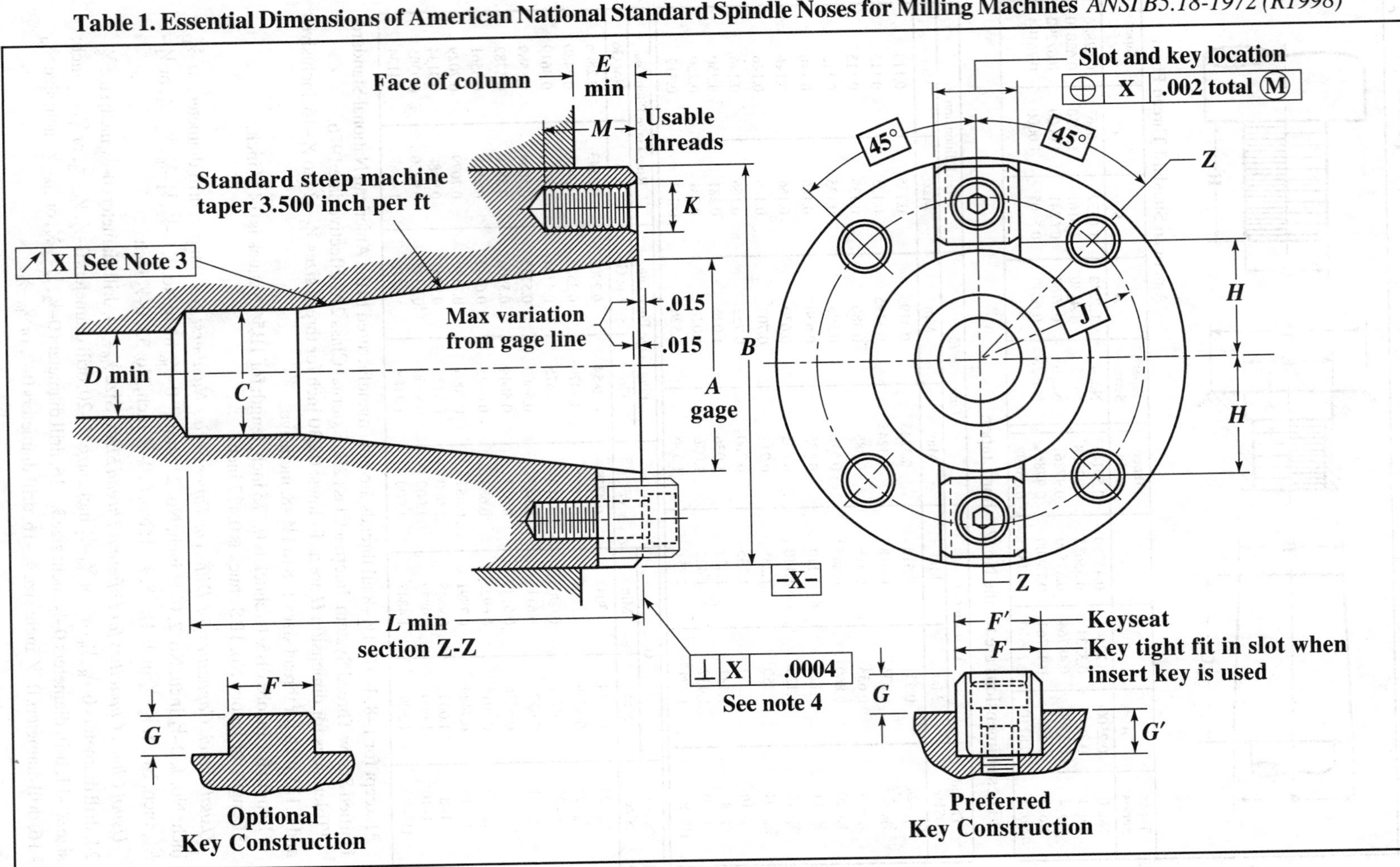

Table 1. *(Continued)* **Essential Dimensions of American National Standard Spindle Noses for Milling Machines** *ANSI B5.18-1972 (R1998)*

Size No.	Gage Dia.of Taper *A*	Dia.of Spindle *B*	Pilot Dia. *C*	Clearance Hole for Draw-in Bolt Min. *D*	Minimum Dimension Spindle End to Column *E*	Width of Driving Key *F*	Width of Keyseat *F′*	Maximum Height of Driving Key *G*	Minimum Depth of Keyseat *G′*	Distance fromCenter to Driving Keys *H*	Radius of Bolt Hole Circle *J*	Size of Threads for Bolt Holes UNC-2B *K*	Full Depth of Arbor Hole in Spindle Min. *L*	Depth of Usable Thread for Bolt Hole *M*
30	1.250	2.7493 2.7488	0.692 0.685	0.66	0.50	0.6255 0.6252	0.624 0.625	0.31	0.31	0.660 0.654	1.0625 (Note 1)	0.375–16	2.88	0.62
40	1.750	3.4993 3.4988	1.005 0.997	0.66	0.62	0.6255 0.6252	0.624 0.625	0.31	0.31	0.910 0.904	1.3125 (Note 1)	0.500–13	3.88	0.81
45	2.250	3.9993 3.9988	1.286 1.278	0.78	0.62	0.7505 0.7502	0.749 0.750	0.38	0.38	1.160 1.154	1.500 (Note 1)	0.500–13	4.75	0.81
50	2.750	5.0618 5.0613	1.568 1.559	1.06	0.75	1.0006 1.0002	0.999 1.000	0.50	0.50	1.410 1.404	2.000(Note 2)	0.625–11	5.50	1.00
60	4.250	8.7180 8.7175	2.381 2.371	1.38	1.50	1.0006 1.0002	0.999 1.000	0.50	0.50	2.420 2.414	3.500 (Note 2)	0.750–10	8.62	1.25

All dimensions are given in inches.

Tolerances:

Two-digit decimal dimensions ± 0.010 unless otherwise specified.

A—Taper: Tolerance on rate of taper to be 0.001 inch per foot applied only in direction which decreases rate of taper.

F′—Centrality of keyway with axis of taper 0.002 total at maximum material condition. (0.002 Total indicator variation)

F—Centrality of solid key with axis of taper 0.002 total at maximum material condition. (0.002 Total indicator variation)

Note 1: Holes spaced as shown and located within 0.006 inch diameter of true position.

Note 2: Holes spaced as shown and located within 0.010 inch diameter of true position.

Note 3: Maximum turnout on test plug:
0.0004 at 1 inch projection from gage line.
0.0010 at 12 inch projection from gage line.

Note 4: Squareness of mounting face measured near mounting bolt hole circle.

Table 2. Essential Dimensions of American Nationalndard Tool Shanks for Milling Machines *ANSI B5.18--1972, R1991*

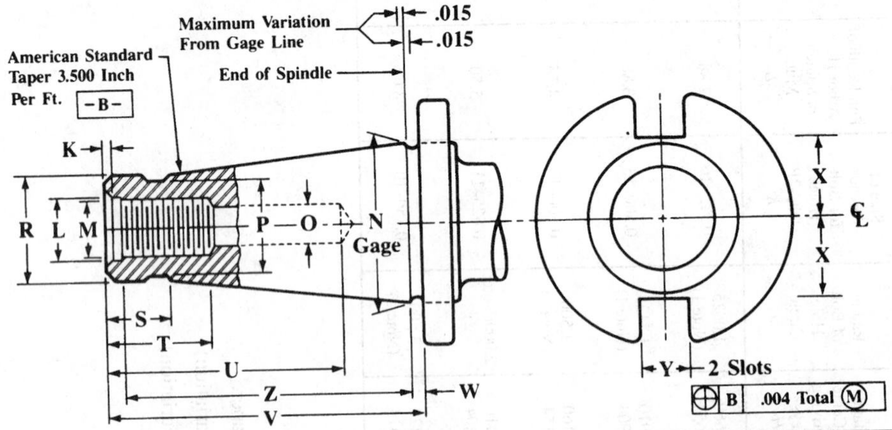

Size No.	Gage Dia.of Taper *N*	Tap Drill Size for Draw-in Thread *O*	Dia.of Neck *P*	Size of Thread for Draw-in Bolt UNC-2B *M*	Pilot Dia. *R*	Length of Pilot *S*	Minimum Length of Usable Thread *T*	Minimum Depth of Clearance Hole *U*
30	1.250	0.422 0.432	0.66 0.65	0.500–13	0.675 0.670	0.81	1.00	2.00
40	1.750	0.531 0.541	0.94 0.93	0.625–11	0.987 0.980	1.00	1.12	2.25
45	2.250	0.656 0.666	1.19 1.18	0.750–10	1.268 1.260	1.00	1.50	2.75
50	2.750	0.875 0.885	1.50 1.49	1.000–8	1.550 1.540	1.00	1.75	3.50
60	4.250	1.109 1.119	2.28 2.27	1.250–7	2.360 2.350	1.75	2.25	4.25

Size. No.	Distance from Rear of Flange to End of Arbor *V*	Clearance of Flange from Gage Diameter *W*	Tool Shank Centerline to Driving Slot *X*	Width of Driving Slot *Y*	Distance from Gage Line to Bottom of C'bore *Z*	Depth of 60° Center *K*	Diameter of C'bore *L*
30	2.75	0.045 0.075	0.640 0.625	0.635 0.645	2.50	0.05 0.07	0.525 0.530
40	3.75	0.045 0.075	0.890 0.875	0.635 0.645	3.50	0.05 0.07	0.650 0.655
45	4.38	0.105 0.135	1.140 1.125	0.760 0.770	4.06	0.05 0.07	0.775 0.780
50	5.12	0.105 0.135	1.390 1.375	1.010 1.020	4.75	0.05 0.12	1.025 1.030
60	8.25	0.105 0.135	2.400 2.385	1.010 1.020	7.81	0.05 0.12	1.307 1.312

All dimensions are given in inches.

Tolerances: Two digit decimal dimensions ± 0.010 inch unless otherwise specified.

M—Permissible for Class 2B "NoGo" gage to enter five threads before interference.

N—Taper tolerance on rate of taper to be 0.001 inch per foot applied only in direction which increases rate of taper.

Y—Centrality of drive slot with axis of taper shank 0.004 inch at maximum material condition. (0.004 inch total indicator variation)

Table 3. American National Standard Draw-in Bolt Ends
ANSI B5.18–1972, R1991

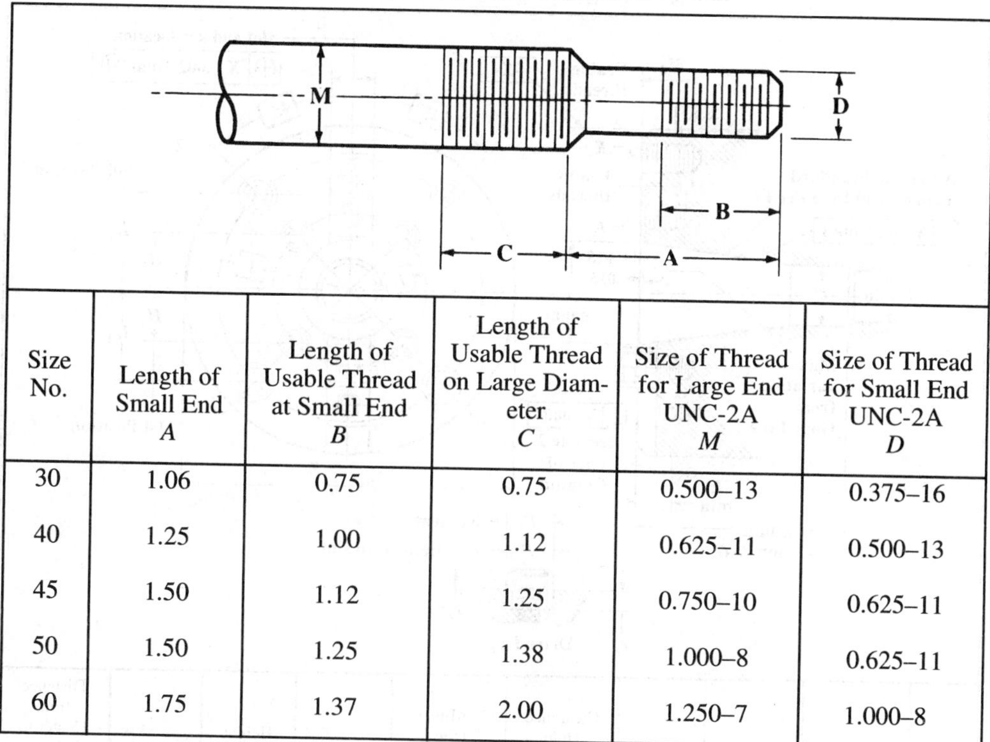

Size No.	Length of Small End *A*	Length of Usable Thread at Small End *B*	Length of Usable Thread on Large Diameter *C*	Size of Thread for Large End UNC-2A *M*	Size of Thread for Small End UNC-2A *D*
30	1.06	0.75	0.75	0.500–13	0.375–16
40	1.25	1.00	1.12	0.625–11	0.500–13
45	1.50	1.12	1.25	0.750–10	0.625–11
50	1.50	1.25	1.38	1.000–8	0.625–11
60	1.75	1.37	2.00	1.250–7	1.000–8

All dimensions are given in inches.

Table 4. American National Standard Pilot Lead on Centering Plugs for Flatback Milling Cutters *ANSI B5.18-1972 (R1998)*

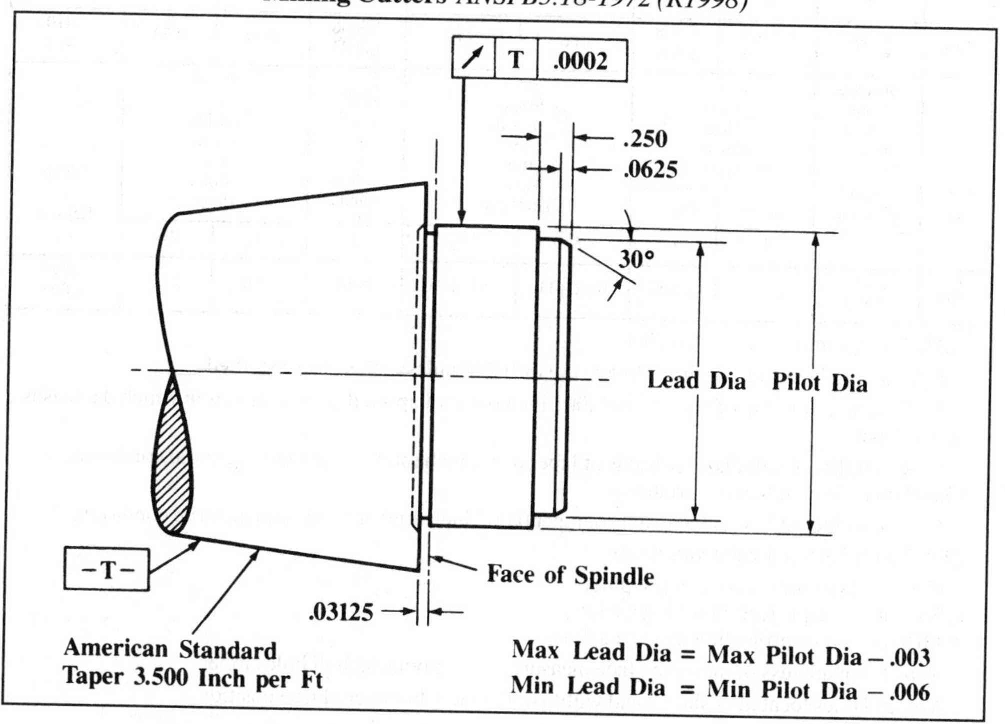

Table 5. Essential Dimensions for American National Standard Spindle Nose with Large Flange *ANSI B5.18-1972 (R1998)*

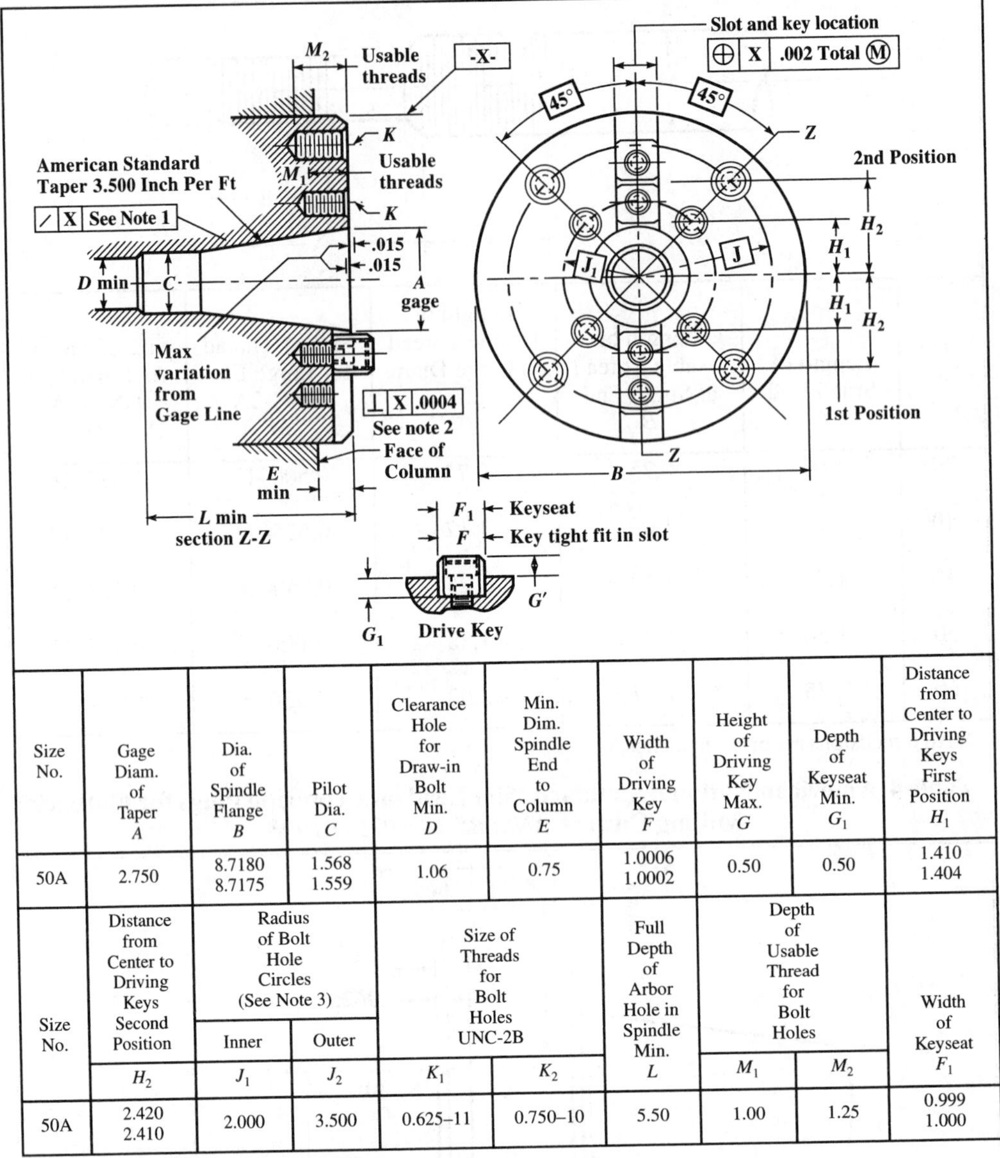

Size No.	Gage Diam. of Taper *A*	Dia. of Spindle Flange *B*	Pilot Dia. *C*	Clearance Hole for Draw-in Bolt Min. *D*	Min. Dim. Spindle End to Column *E*	Width of Driving Key *F*	Height of Driving Key Max. *G*	Depth of Keyseat Min. G_1	Distance from Center to Driving Keys First Position H_1
50A	2.750	8.7180 8.7175	1.568 1.559	1.06	0.75	1.0006 1.0002	0.50	0.50	1.410 1.404

Size No.	Distance from Center to Driving Keys Second Position	Radius of Bolt Hole Circles (See Note 3)		Size of Threads for Bolt Holes UNC-2B		Full Depth of Arbor Hole in Spindle Min.	Depth of Usable Thread for Bolt Holes		Width of Keyseat
		Inner	Outer						
	H_2	J_1	J_2	K_1	K_2	*L*	M_1	M_2	F_1
50A	2.420 2.410	2.000	3.500	0.625–11	0.750–10	5.50	1.00	1.25	0.999 1.000

All dimensions are given in inches.

Tolerances: Two-digit decimal dimensions ± 0.010 unless otherwise specified.

A—Tolerance on rate of taper to be 0.001 inch per foot applied only in direction which decreases rate of taper.

F—Centrality of solid key with axis of taper 0.002 inch total at maximum material condition. (0.002 inch Total indicator variation)

F_1—Centrality of keyseat with axis of taper 0.002 inch total at maximum material condition. (0.002 inch Total indicator variation)

Note 1: Maximum runout on test plug:
0.0004 at 1 inch projection from gage line.
0.0010 at 12 inch projection from gage line.

Note 2: Squareness of mounting face measured near mounting bolt hole circle.

Note 3: Holes located as shown and within 0.010 inch diameter of true position.

Length of Point on Twist Drills and Centering Tools

Size of Drill	Decimal Equivalent	Length of Point when Included Angle =90°	Length of Point when Included Angle =118°	Size of Drill	Decimal Equivalent	Length of Point when Included Angle =90°	Length of Point when Included Angle =118°	Size or Dia. of Drill	Decimal Equivalent	Length of Point when Included Angle =90°	Length of Point when Included Angle =118°	Dia. of Drill	Decimal Equivalent	Length of Point when Included Angle =90°	Length of Point when Included Angle =118°
60	0.0400	0.020	0.012	37	0.1040	0.052	0.031	14	0.1820	0.091	0.055	3/8	0.3750	0.188	0.113
59	0.0410	0.021	0.012	36	0.1065	0.054	0.032	13	0.1850	0.093	0.056	25/64	0.3906	0.195	0.117
58	0.0420	0.021	0.013	35	0.1100	0.055	0.033	12	0.1890	0.095	0.057	13/32	0.4063	0.203	0.122
57	0.0430	0.022	0.013	34	0.1110	0.056	0.033	11	0.1910	0.096	0.057	27/64	0.4219	0.211	0.127
56	0.0465	0.023	0.014	33	0.1130	0.057	0.034	10	0.1935	0.097	0.058	7/16	0.4375	0.219	0.131
55	0.0520	0.026	0.016	32	0.1160	0.058	0.035	9	0.1960	0.098	0.059	29/64	0.4531	0.227	0.136
54	0.0550	0.028	0.017	31	0.1200	0.060	0.036	8	0.1990	0.100	0.060	15/32	0.4688	0.234	0.141
53	0.0595	0.030	0.018	30	0.1285	0.065	0.039	7	0.2010	0.101	0.060	31/64	0.4844	0.242	0.145
52	0.0635	0.032	0.019	29	0.1360	0.068	0.041	6	0.2040	0.102	0.061	1/2	0.5000	0.250	0.150
51	0.0670	0.034	0.020	28	0.1405	0.070	0.042	5	0.2055	0.103	0.062	33/64	0.5156	0.258	0.155
50	0.0700	0.035	0.021	27	0.1440	0.072	0.043	4	0.2090	0.105	0.063	17/32	0.5313	0.266	0.159
49	0.0730	0.037	0.022	26	0.1470	0.074	0.044	3	0.2130	0.107	0.064	35/64	0.5469	0.273	0.164
48	0.0760	0.038	0.023	25	0.1495	0.075	0.045	2	0.2210	0.111	0.067	9/16	0.5625	0.281	0.169
47	0.0785	0.040	0.024	24	0.1520	0.076	0.046	1	0.2280	0.114	0.068	37/64	0.5781	0.289	0.173
46	0.0810	0.041	0.024	23	0.1540	0.077	0.046	15/64	0.2344	0.117	0.070	19/32	0.5938	0.297	0.178
45	0.0820	0.041	0.025	22	0.1570	0.079	0.047	1/4	0.2500	0.125	0.075	39/64	0.6094	0.305	0.183
44	0.0860	0.043	0.026	21	0.1590	0.080	0.048	17/64	0.2656	0.133	0.080	5/8	0.6250	0.313	0.188
43	0.0890	0.045	0.027	20	0.1610	0.081	0.048	9/32	0.2813	0.141	0.084	41/64	0.6406	0.320	0.192
42	0.0935	0.047	0.028	19	0.1660	0.083	0.050	19/64	0.2969	0.148	0.089	21/32	0.6563	0.328	0.197
41	0.0960	0.048	0.029	18	0.1695	0.085	0.051	5/16	0.3125	0.156	0.094	43/64	0.6719	0.336	0.202
40	0.0980	0.049	0.029	17	0.1730	0.087	0.052	21/64	0.3281	0.164	0.098	11/16	0.6875	0.344	0.206
39	0.0995	0.050	0.030	16	0.1770	0.089	0.053	11/32	0.3438	0.171	0.103	23/32	0.7188	0.359	0.216
38	0.1015	0.051	0.030	15	0.1800	0.090	0.054	23/64	0.3594	0.180	0.108	3/4	0.7500	0.375	0.225

BROACHES AND BROACHING

The Broaching Process.—The broaching process may be applied in machining holes or other internal surfaces and also to many flat or other external surfaces. Internal broaching is applied in forming either symmetrical or irregular holes, grooves, or slots in machine parts, especially when the size or shape of the opening, or its length in proportion to diameter or width, make other machining processes impracticable. Broaching originally was utilized for such work as cutting keyways, machining round holes into square, hexagonal, or other shapes, forming splined holes, and for a large variety of other internal operations. The development of broaching machines and broaches finally resulted in extensive application of the process to external, flat, and other surfaces. Most external or surface broaching is done on machines of vertical design, but horizontal machines are also used for some classes of work. The broaching process is very rapid, accurate, and it leaves a finish of good quality. It is employed extensively in automotive and other plants where duplicate parts must be produced in large quantities and frequently to given dimensions within small tolerances.

Types of Broaches.—A number of typical broaches and the operations for which they are intended are shown by the diagrams, Fig. 1. Broach A produces a round-cornered, square hole. Prior to broaching square holes, it is usually the practice to drill a round hole having a diameter *d* somewhat larger than the width of the square. Hence, the sides are not completely finished, but this unfinished part is not objectionable in most cases. In fact, this clearance space is an advantage during the broaching operation in that it serves as a channel for the broaching lubricant; moreover, the broach has less metal to remove. Broach *B* is for finishing round holes. Broaching is superior to reaming for some classes of work, because the broach will hold its size for a much longer period, thus insuring greater accuracy. Broaches *C* and *D* are for cutting single and double keyways, respectively. Broach *C* is of rectangular section and, when in use, slides through a guiding bushing which is inserted in the hole. Broach *E* is for forming four integral splines in a hub. The broach at *F* is for producing hexagonal holes. Rectangular holes are finished by broach *G*. The teeth on the sides of this broach are inclined in opposite directions, which has the following advantages: The broach is stronger than it would be if the teeth were opposite and parallel to each other; thin work cannot drop between the inclined teeth, as it tends to do when the teeth are at right angles, because at least two teeth are always cutting; the inclination in opposite directions neutralizes the lateral thrust. The teeth on the edges are staggered, the teeth on one side being midway between the teeth on the other edge, as shown by the dotted line. A double cut broach is shown at *H*. This type is for finishing, simultaneously, both sides *f* of a slot, and for similar work. Broach *I* is the style used for forming the teeth in internal gears. It is practically a series of gear-shaped cutters, the outside diameters of which gradually increase toward the finishing end of the broach, Broach *J* is for round holes but differs from style *B* in that it has a continuous helical cutting edge. Some prefer this form because it gives a shearing cut. Broach *K* is for cutting a series of helical grooves in a hub or bushing. In helical broaching, either the work or the broach is rotated to form the helical grooves as the broach is pulled through.

In addition to the typical broaches shown in Fig. 1, many special designs are now in use for performing more complex operations. Two surfaces on opposite sides of a casting or forging are sometimes machined simultaneously by twin broaches and, in other cases, three or four broaches are drawn through a part at the same time, for finishing as many duplicate holes or surfaces. Notable developments have been made in the design of broaches for external or "surface" broaching.

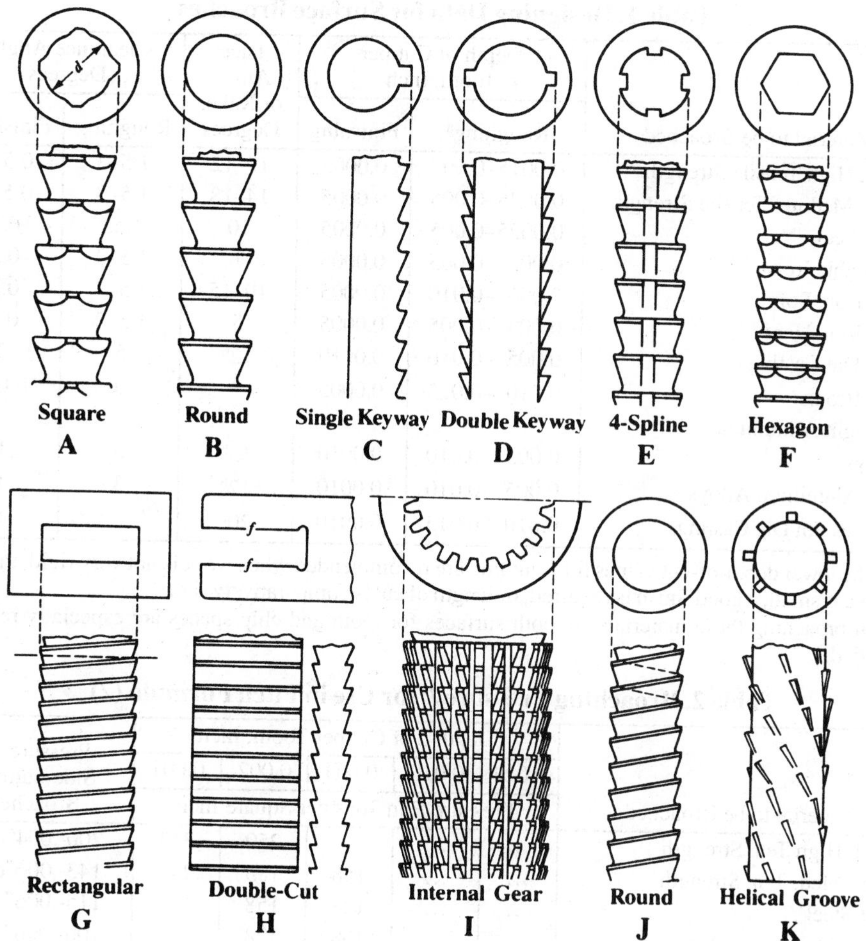

Fig. 1. Types of Broaches

Pitch of Broach Teeth.—The pitch of broach teeth depends upon the depth of cut or chip thickness, length of cut, the cutting force required and power of the broaching machine. In the pitch formulas which follow

L = length, in inches, of layer to be removed by broaching

d = depth of cut per tooth as shown by Table 1 (For internal broaches, d = depth of cut as measured on one side of broach or one-half difference in diameters of successive teeth in case of a round broach)

F = a factor. (For brittle types of material, F = 3 or 4 for roughing teeth, and 6 for finishing teeth. For ductile types of material, F = 4 to 7 for roughing teeth and 8 for finishing teeth.)

b = width of inches, of layer to be removed by broaching

P = pressure required in tons per square inch, of an area equal to depth of cut times width of cut, in inches (Table 2)

T = usable capacity, in tons, of broaching machine = 70 per cent of maximum tonnage

Table 1. Designing Data for Surface Broaches

Material to be Broached	Depth of Cut per Tooth, Inch		Face Angle or Rake, Degrees	Clearance Angle, Degrees	
	Roughing[a]	Finishing		Roughing	Finishing
Steel, High Tensile Strength	0.0015–0.002	0.0005	10–12	1.5–3	0.5–1
Steel, Medium Tensile Strength	0.0025–0.005	0.0005	14–18	1.5–3	0.5–1
Cast Steel	0.0025–0.005	0.0005	10	1.53	0.5
Malleable Iron	0.0025–0.005	0.0005	7	1.5–3	0.5
Cast Iron, Soft	0.006 –0.010	0.0005	10–15	1.5–3	0.5
Cast Iron, Hard	0.003 –0.005	0.0005	5	1.5–3	0.5
Zinc Die Castings	0.005 –0.010	0.0010	12[b]	5	2
Cast Bronze	0.010 –0.025	0.0005	8	0	0
Wrought Aluminum Alloys	0.005 –0.010	0.0010	15[b]	3	1
Cast Aluminum Alloys	0.005 –0.010	0.0010	12[b]	3	1
Magnesium Die Castings	0.010 –0.015	0.0010	20[b]	3	1

[a] The lower depth-of-cut values for roughing are recommended when work is not very rigid, the tolerance is small, a good finish is required, or length of cut is comparatively short.

[b] In broaching these materials, smooth surfaces for tooth and chip spaces are especially recommended.

Table 2. Broaching Pressure *P* for Use in Pitch Formula (2)

Material to be Broached	Depth *d* of Cut per Tooth, Inch					Pressure *P*, Side-cutting Broaches
	0.024	0.010	0.004	0.002	0.001	
	Pressure *P* in Tons per Square Inch					
Steel, High Ten. Strength	…	…	…	250	312	200-.004″cut
Steel, Med. Ten. Strength	…	…	158	185	243	143-.006″cut
Cast Steel	…	…	128	158	…	115-.006″ cut
Malleable Iron	…	…	108	128	…	100-.006″ cut
Cast Iron	…	115	115	143	…	115-.020″ cut
Cast Brass	…	50	50	…	…	…………
Brass, Hot Pressed	…	85	85	…	…	…………
Zinc Die Castings	…	70	70	…	…	…………
Cast Bronze	35	35	…	…	…	…………
Wrought Aluminum	…	70	70	…	…	…………
Cast Aluminum	…	85	85	…	…	…………
Magnesium Alloy	35	35	…	…	…	…………

The minimum pitch shown by Formula (1) is based upon the receiving capacity of the chip space. The minimum, however, should not be less than 0.2 inch unless a smaller pitch is required for exceptionally short cuts to provide at least two teeth in contact simultaneously, with the part being broached. A reduction below 0.2 inch is seldom required in surface broaching but it may be necessary in connection with internal broaching.

$$\text{Minimum pitch} = 3\sqrt{LdF} \qquad (1)$$

Whether the minimum pitch may be used or not depends upon the power of the available machine. The factor *F* in the formula provides for the increase in volume as the material is broached into chips. If a broach has adjustable inserts for the finishing teeth, the pitch of the finishing teeth may be smaller than the pitch of the roughing teeth because of the smaller depth *d* of the cut. The higher value of *F* for finishing teeth prevents the pitch from becoming too small, so that the spirally curled chips will not be crowded into too small a space.

The pitch of the roughing and finishing teeth should be equal for broaches without separate inserts (notwithstanding the different values of d and F) so that some of the finishing teeth may be ground into roughing teeth after wear makes this necessary.

$$\text{Allowable pitch} = \frac{dLbP}{T} \qquad (2)$$

If the pitch obtained by Formula (2) is larger than the minimum obtained by Formula (1), this larger value should be used because it is based upon the usable power of the machine. As the notation indicates, 70 per cent of the maximum tonnage T is taken as the usable capacity. The 30 per cent reduction is to provide a margin for the increase in broaching load resulting from the gradual dulling of the cutting edges. The procedure in calculating both minimum and allowable pitches will be illustrated by an example.

Example: Determine pitch of broach for cast iron when $L = 9$ inches; $d = 0.004$; and $F = 4$.

$$\text{Minimum pitch} = 3\sqrt{9 \times 0.004 \times 4} = 1.14$$

Next, apply Formula (2). Assume that $b = 3$ and $T = 10$; for cast iron and depth d of 0.004, $P = 115$ (Table 2). Then,

$$\text{Allowable pitch} = \frac{0.004 \times 9 \times 3 \times 115}{10} = 1.24$$

This pitch is safely above the minimum. If in this case the usable tonnage of an available machine were, say, 8 tons instead of 10 tons, the pitch as shown by Formula (2) might be increased to about 1.5 inches, thus reducing the number of teeth cutting simultaneously and, consequently, the load on the machine; or the cut per tooth might be reduced instead of increasing the pitch, especially if only a few teeth are in cutting contact, as might be the case with a short length of cut. If the usable tonnage in the preceding example were, say, 15, then a pitch of 0.84 would be obtained by Formula (2); hence the pitch in this case should not be less than the minimum of approximately 1.14 inches.

Depth of Cut per Tooth.—The term "depth of cut" as applied to surface or external broaches means the difference in the heights of successive teeth. This term, as applied to internal broaches for round, hexagonal or other holes, may indicate the total increase in the diameter of successive teeth; however, to avoid confusion, the term as here used means in all cases and regardless of the type of broach, the depth of cut as measured on one side.

In broaching free cutting steel, the Broaching Tool Institute recommends 0.003 to 0.006 inch depth of cut for surface broaching; 0.002 to 0.003 inch for multispline broaching; and 0.0007 to 0.0015 inch for round hole broaching. The accompanying table contains data from a German source and applies specifically to surface broaches. All data relating to depth of cut are intended as a general guide only. While depth of cut is based primarily upon the machinability of the material, some reduction from the depth thus established may be required particularly when the work supporting fixture in surface broaching is not sufficiently rigid to resist the thrust from the broaching operation. In some cases, the pitch and cutting length may be increased to reduce the thrust force. Another possible remedy in surface broaching certain classes of work is to use a side-cutting broach instead of the ordinary depth cutting type. A broach designed for side cutting takes relatively deep narrow cuts which extend nearly to the full depth required. The side cutting section is followed by teeth arranged for depth cutting to obtain the required size and surface finish on the work. In general, small tolerances in surface broaching require a reduced cut per tooth to minimize work deflection resulting from the pressure of the cut. See *Cutting Speed for Broaching* starting on page 1043 for broaching speeds.

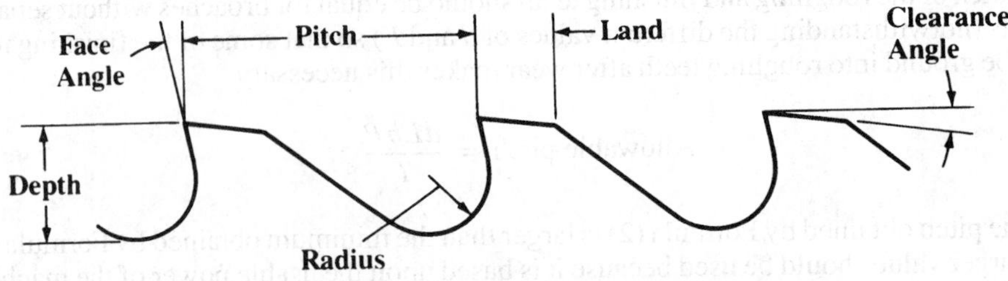

Terms Commonly Used in Broach Design

Face Angle or Rake.—The face angle (see diagram) of broach teeth affects the chip flow and varies considerably for different materials. While there are some variations in practice, even for the same material, the angles given in the accompanying table are believed to represent commonly used values. Some broach designers increase the rake angle for finishing teeth in order to improve the finish on the work.

Clearance Angle.—The clearance angle (see illustration) for roughing steel varies from 1.5 to 3 degrees and for finishing steel from 0.5 to 1 degree. Some recommend the same clearance angles for cast iron and others, larger clearance angles varying from 2 to 4 or 5 degrees. Additional data will be found in Table 1.

Land Width.—The width of the land usually is about 0.25 × pitch. It varies, however, from about one-fourth to one-third of the pitch. The land width is selected so as to obtain the proper balance between tooth strength and chip space.

Depth of Broach Teeth.—The tooth depth as established experimentally and on the basis of experience, usually varies from about 0.37 to 0.40 of the pitch. This depth is measured radially from the cutting edge to the bottom of the tooth fillet.

Radius of Tooth Fillet.—The "gullet" or bottom of the chip space between the teeth should have a rounded fillet to strengthen the broach, facilitate curling of the chips, and safeguard against cracking in connection with the hardening operation. One rule is to make the radius equal to one-fourth the pitch. Another is to make it equal 0.4 to 0.6 the tooth depth. A third method preferred by some broach designers is to make the radius equal one-third of the sum obtained by adding together the land width, one-half the tooth depth, and one-fourth of the pitch.

Total Length of Broach.—After the depth of cut per tooth has been determined, the total amount of material to be removed by a broach is divided by this decimal to ascertain the number of cutting teeth required. This number of teeth multiplied by the pitch gives the length of the active portion of the broach. By adding to this dimension the distance over three or four straight teeth, the length of a pilot to be provided at the finishing end of the broach, and the length of a shank which must project through the work and the faceplate of the machine to the draw-head, the overall length of the broach is found. This calculated length is often greater than the stroke of the machine, or greater than is practical for a broach of the diameter required. In such cases, a set of broaches must be used.

Chip Breakers.—The teeth of broaches frequently have rounded chip-breaking grooves located at intervals along the cutting edges. These grooves break up wide curling chips and prevent them from clogging the chip spaces, thus reducing the cutting pressure and strain on the broach. These chip-breaking grooves are on the roughing teeth only. They are staggered and applied to both round and flat or surface broaches. The grooves are formed by a round edged grinding wheel and usually vary in width from about $\frac{1}{32}$ to $\frac{3}{32}$ inch depending upon the size of broach. The more ductile the material, the wider the chip breaker grooves should be and the smaller the distance between them. Narrow slotting broaches may have the right- and left-hand corners of alternate teeth beveled to obtain chip-breaking action.

Shear Angle.—The teeth of surface broaches ordinarily are inclined so they are not at right angles to the broaching movement. The object of this inclination is to obtain a shearing cut which results in smoother cutting action and an improvement in surface finish. The shearing cut also tends to eliminate troublesome vibration. Shear angles for surface broaches are not suitable for broaching slots or any profiles that resist the outward movement of the chips. When the teeth are inclined, the fixture should be designed to resist the resulting thrusts unless it is practicable to incline the teeth of right- and left-hand sections in opposite directions to neutralize the thrust. The shear angle usually varies from 10 to 25 degrees.

Types of Broaching Machines.—Broaching machines may be divided into horizontal and vertical designs, and they may be classified further according to the method of operation, as, for example, whether a broach in a vertical machine is pulled up or pulled down in forcing it through the work. Horizontal machines usually pull the broach through the work in internal broaching but short rigid broaches may be pushed through. External surface broaching is also done on some machines of horizontal design, but usually vertical machines are employed for flat or other external broaching. Although parts usually are broached by traversing the broach itself, some machines are designed to hold the broach or broaches stationary during the actual broaching operation. This principle has been applied both to internal and surface broaching.

Vertical Duplex Type: The vertical duplex type of surface broaching machine has two slides or rams which move in opposite directions and operate alternately. While the broach connected to one slide is moving downward on the cutting stroke, the other broach and slide is returning to the starting position, and this returning time is utilized for reloading the fixture on that side; consequently, the broaching operation is practically continuous. Each ram or slide may be equipped to perform a separate operation on the same part when two operations are required.

Pull-up Type: Vertical hydraulically operated machines which pull the broach or broaches up through the work are used for internal broaching of holes of various shapes, for broaching bushings, splined holes, small internal gears, etc. A typical machine of this kind is so designed that all broach handling is done automatically.

Pull-down Type: The various movements in the operating cycle of a hydraulic pull-down type of machine equipped with an automatic broach-handling slide, are the reverse of the pull-up type. The broaches for a pull-down type of machine have shanks on each end, there being an upper one for the broach-handling slide and a lower one for pulling through the work.

Hydraulic Operation: Modern broaching machines, as a general rule, are operated hydraulically rather than by mechanical means. Hydraulic operation is efficient, flexible in the matter of speed adjustments, low in maintenance cost, and the "smooth" action required for fine precision finishing may be obtained. The hydraulic pressures required, which frequently are 800 to 1000 pounds per square inch, are obtained from a motor-driven pump forming part of the machine. The cutting speeds of broaching machines frequently are between 20 and 30 feet per minute, and the return speeds often are double the cutting speed, or higher, to reduce the idle period.

Broaching Difficulties.—The accompanying table has been compiled from information supplied by the National Broach and Machine Co. and presents some of the common broaching difficulties, their causes and means of correction.

Causes of Broaching Difficulties

Broaching Difficulty	Possible Causes
Stuck broach	Insufficient machine capacity; dulled teeth; clogged chip gullets; failure of power during cutting stroke. To remove a stuck broach, workpiece and broach are removed from the machine as a unit; never try to back out broach by reversing machine. If broach does not loosen by tapping workpiece lightly and trying to slide it off its starting end, mount workpiece and broach in a lathe and turn down workpiece to the tool surface. Workpiece may be sawed longitudinally into several sections in order to free the broach. Check broach design, perhaps tooth relief (back off) angle is too small or depth of cut per tooth is too great.
Galling and pickup	Lack of homogeneity of material being broached—uneven hardness, porosity; improper or insufficient coolant; poor broach design, mutilated broach; dull broach; improperly sharpened broach; improperly designed or outworn fixtures. Good broach design will do away with possible chip build-up on tooth faces and excessive heating. Grinding of teeth should be accurate so that the correct gullet contour is maintained. Contour should be fair and smooth.
Broach breakage	Overloading; broach dullness; improper sharpening; interrupted cutting stroke; backing up broach with workpiece in fixture; allowing broach to pass entirely through guide hole; ill fitting and/or sharp edged key; crooked holes; untrue locating surface; excessive hardness of workpiece; insufficient clearance angle; sharp corners on pull end of broach. When grinding bevels on pull end of broach use wheel that is not too pointed.
Chatter	Too few teeth in cutting contact simultaneously; excessive hardness of material being broached; loose or poorly constructed tooling; surging of ram due to load variations. Chatter can be alleviated by changing the broaching speed, by using shear cutting teeth instead of right angle teeth, and by changing the coolant and the face and relief angles of the teeth.
Drifting or misalignment of tool during cutting stroke	Lack of proper alignment when broach is sharpened in grinding machine, which may be caused by dirt in the female center of the broach; inadequate support of broach during the cutting stroke, on a horizontal machine especially; body diameter too small; cutting resistance variable around I.D. of hole due to lack of symmetry of surfaces to be cut; variations in hardness around I.D. of hole; too few teeth in cutting contact.
Streaks in broached surface	Lands too wide; presence of forging, casting or annealing scale; metal pickup; presence of grinding burrs and grinding and cleaning abrasives.
Rings in the broached hole	Due to surging resulting from uniform pitch of teeth; presence of sharpening burrs on broach; tooth clearance angle too large; locating face not smooth or square; broach not supported for all cutting teeth passing through the work. The use of differential tooth spacing or shear cutting teeth helps in preventing surging. Sharpening burrs on a broach may be removed with a wood block.

Tool Wear

Metal cutting tools wear constantly when they are being used. A normal amount of wear should not be a cause for concern until the size of the worn region has reached the point where the tool should be replaced. Normal wear cannot be avoided and should be differentiated from abnormal tool breakage or excessively fast wear. Tool breakage and an excessive rate of wear indicate that the tool is not operating correctly and steps should be taken to correct this situation.

There are several basic mechanisms that cause tool wear. It is generally understood that tools wear as a result of abrasion which is caused by hard particles of work material plowing over the surface of the tool. Wear is also caused by diffusion or alloying between the work material and the tool material. In regions where the conditions of contact are favorable, the work material reacts with the tool material causing an attrition of the tool material. The rate of this attrition is dependent upon the temperature in the region of contact and the reactivity of the tool and the work materials with each other. Diffusion or alloying also occurs where particles of the work material are welded to the surface of the tool. These welded deposits are often quite visible in the form of a built-up edge, as particles or a layer of work material inside a crater or as small mounds attached to the face of the tool. The diffusion or alloying occurring between these deposits and the tool weakens the tool material below the weld. Frequently these deposits are again rejoined to the chip by welding or they are simply broken away by the force of collision with the passing chip. When this happens, a small amount of the tool material may remain attached to the deposit and be plucked from the surface of the tool, to be carried away with the chip. This mechanism can cause chips to be broken from the cutting edge and the formation of small craters on the tool face called pull-outs. It can also contribute to the enlargement of the larger crater that sometimes forms behind the cutting edge. Among the other mechanisms that can cause tool wear are severe thermal gradients and thermal shocks, which cause cracks to form near the cutting edge, ultimately leading to tool failure. This condition can be caused by improper tool grinding procedures, heavy interrupted cuts, or by the improper application of cutting fluids when machining at high cutting speeds. Chemical reactions between the active constituents in some cutting fluids sometimes accelerate the rate of tool wear. Oxidation of the heated metal near the cutting edge also contributes to tool wear, particularly when fast cutting speeds and high cutting temperatures are encountered. Breakage of the cutting edge caused by overloading, heavy shock loads, or improper tool design is not normal wear and should be corrected.

The wear mechanisms described bring about visible manifestations of wear on the tool which should be understood so that the proper corrective measures can be taken, when required. These visible signs of wear are described in the following paragraphs and the corrective measures that might be required are given in the accompanying Tool Trouble-Shooting Check List. The best procedure when trouble shooting is to try to correct only one condition at a time. When a correction has been made it should be checked. After one condition has been corrected, work can then start to correct the next condition.

Flank Wear: Tool wear occurring on the flank of the tool below the cutting edge is called flank wear. Flank wear always takes place and cannot be avoided. It should not give rise to concern unless the rate of flank wear is too fast or the flank wear land becomes too large in size. The size of the flank wear can be measured as the distance between the top of the cutting edge and the bottom of the flank wear land. In practice, a visual estimate is usually made instead of a precise measurement, although in many instances flank wear is ignored and the tool wear is "measured" by the loss of size on the part. The best measure of tool wear, however, is flank wear. When it becomes too large, the rubbing action of the wear land against the workpiece increases and the cutting edge must be replaced. Because conditions vary, it is not possible to give an exact amount of flank wear at which the tool should be replaced. Although there are many exceptions, as a rough estimate, high-speed steel tools should be replaced when the width of the flank wear land reaches 0.005 to 0.010 inch for finish turning and 0.030 to 0.060 inch for rough turning; and for cemented carbides 0.005 to 0.010 inch for finish turning and 0.020 to 0.040 inch for rough turning.

Under ideal conditions which, surprisingly, occur quite frequently, the width of the flank wear land will be very uniform along its entire length. When the depth of cut is uneven, such as when turning out-of-round stock, the bottom edge of the wear land may become somewhat slanted, the wear land being wider toward the nose. A jagged-appearing wear land usually is evidence of chipping at the cutting edge. Sometimes, only one or two sharp depressions of the lower edge of the wear land will appear, to indicate that the cutting edge has chipped above these depressions. A deep notch will sometimes occur at the "depth of cut line," or that part of the cutting opposite the original surface of the work. This can be caused by a hard surface scale on the work, by a work-hardened surface layer on the work, or when machining high-temperature alloys. Often the size of the wear land is enlarged at the nose of the tool. This can be a sign of crater breakthrough near the nose or of chipping in this region. Under certain conditions, when machining with carbides, it can be an indication of deformation of the cutting edge in the region of the nose.

When a sharp tool is first used, the initial amount of flank wear is quite large in relation to the subsequent total amount. Under normal operating conditions, the width of the flank wear land will increase at a uniform rate until it reaches a critical size after which the cutting edge breaks down completely. This is called catastrophic failure and the cutting edge should be replaced before this occurs. When cutting at slow speeds with high-speed steel tools, there may be long periods when no increase in the flank wear can be observed. For a given work material and tool material, the rate of flank wear is primarily dependent on the cutting speed and then the feed rate.

Cratering: A deep crater will sometimes form on the face of the tool which is easily recognizable. The crater forms at a short distance behind the side cutting edge leaving a small shelf between the cutting edge and the edge of the crater. This shelf is sometimes covered with the built-up edge and at other times it is uncovered. Often the bottom of the crater is obscured with work material that is welded to the tool in this region. Under normal operating conditions, the crater will gradually enlarge until it breaks through a part of the cutting edge. Usually this occurs on the end cutting edge just behind the nose. When this takes place, the flank wear at the nose increases rapidly and complete tool failure follows shortly. Sometimes cratering cannot be avoided and a slow increase in the size of the crater is considered normal. However, if the rate of crater growth is rapid, leading to a short tool life, corrective measures must be taken.

Cutting Edge Chipping: Small chips are sometimes broken from the cutting edge which accelerates tool wear but does not necessarily cause immediate tool failure. Chipping can be recognized by the appearance of the cutting edge and the flank wear land. A sharp depression in the lower edge of the wear land is a sign of chipping and if this edge of the wear land has a jagged appearance it indicates that a large amount of chipping has taken place. Often the vacancy or cleft in the cutting edge that results from chipping is filled up with work material that is tightly welded in place. This occurs very rapidly when chipping is caused by a built-up edge on the face of the tool. In this manner the damage to the cutting edge is healed; however, the width of the wear land below the chip is usually increased and the tool life is shortened.

Deformation: Deformation occurs on carbide cutting tools when taking a very heavy cut using a slow cutting speed and a high feed rate. A large section of the cutting edge then becomes very hot and the heavy cutting pressure compresses the nose of the cutting edge, thereby lowering the face of the tool in the area of the nose. This reduces the relief under the nose, increases the width of the wear land in this region, and shortens the tool life.

Surface Finish: The finish on the machined surface does not necessarily indicate poor cutting tool performance unless there is a rapid deterioration. A good surface finish is, however, sometimes a requirement. The principal cause of a poor surface finish is the built-up edge which forms along the edge of the cutting tool. The elimination of the built-up edge will always result in an improvement of the surface finish. The most effective way to eliminate the built-up edge is to increase the cutting speed. When the cutting speed is increased beyond a certain critical cutting speed, there will be a rather sudden and large improvement in the surface finish. Cemented carbide tools can operate successfully at higher cutting speeds, where the built-up edge does not occur and where a good surface finish is obtained. Whenever possible, cemented carbide tools should be operated at cutting speeds where a good surface finish will result. There are times when such speeds are not possible. Also, high-speed tools cannot be operated at the speed where the built-up edge does not form. In these conditions the most effective method of obtaining a good surface finish is to employ a cutting fluid that has active sulphur or chlorine additives.

Cutting tool materials that do not alloy readily with the work material are also effective in obtaining an improved surface finish. Straight titanium carbide and diamond are the two principal tool materials that fall into this category.

The presence of feed marks can mar an otherwise good surface finish and attention must be paid to the feed rate and the nose radius of the tool if a good surface finish is desired. Changes in the tool geometry can also be helpful. A small "flat," or secondary cutting edge, ground on the end cutting edge behind the nose will sometimes provide the desired surface finish. When the tool is in operation, the flank wear should not be allowed to become too large, particularly in the region of the nose where the finished surface is produced.

Sharpening Twist Drills.—Twist drills are cutting tools designed to perform concurrently several functions, such as penetrating directly into solid material, ejecting the removed chips outside the cutting area, maintaining the essentially straight direction of the advance movement and controlling the size of the drilled hole. The geometry needed for these multiple functions is incorporated into the design of the twist drill in such a manner that it can be retained even after repeated sharpening operations. Twist drills are resharpened many times during their service life, with the practically complete restitution of their original operational characteristics. However, in order to assure all the benefits which the design of the twist drill is capable of providing, the surfaces generated in the sharpening process must agree with the original form of the tool's operating surfaces, unless a change of shape is required for use on a different work material.

The principal elements of the tool geometry which are essential for the adequate cutting performance of twist drills are shown in Fig. 1. The generally used values for these dimensions are the following:

Point angle: Commonly 118°, except for high strength steels, 118° to 135°; aluminum alloys, 90° to 140°; and magnesium alloys, 70° to 118°.

Helix angle: Commonly 24° to 32°, except for magnesium and copper alloys, 10° to 30°.

Lip relief angle: Commonly 10° to 15°, except for high strength or tough steels, 7° to 12°. The lower values of these angle ranges are used for drills of larger diameter, the higher values for the smaller diameters. For drills of diameters less than ¼ inch, the lip relief angles are increased beyond the listed maximum values up to 24°. For soft and free machining materials, 12° to 18° except for diameters less than ¼ inch, 20° to 26°.

Relief Grinding of the Tool Flanks.—In sharpening twist drills the tool flanks containing the two cutting edges are ground. Each flank consists of a curved surface which provides the relief needed for the easy penetration and free cutting of the tool edges. In grinding the flanks, Fig. 2, the drill is swung around the axis *A* of an imaginary cone while resting in a support which holds the drill at one-half the point angle *B* with respect to the face of the grinding wheel. Feed *f* for stock removal is in the direction of the drill axis. The relief angle is usually measured at the periphery of the twist drill and is also specified by that value. It is not a constant but should increase toward the center of the drill.

The relief grinding of the flank surfaces will generate the chisel angle on the web of the twist drill. The value of that angle, typically 55°, which can be measured, for example, with the protractor of an optical projector, is indicative of the correctness of the relief grinding.

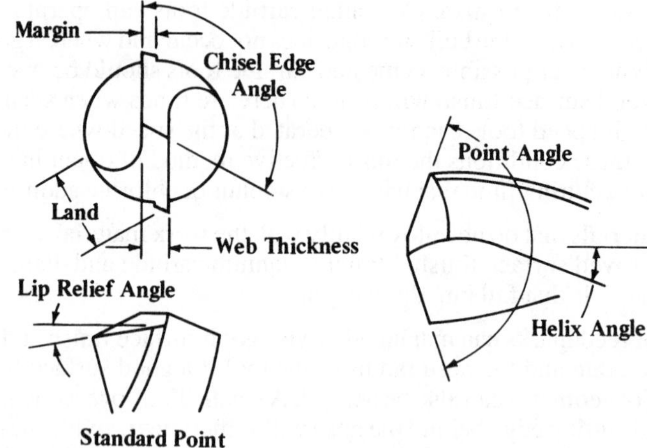

Fig. 1. The principal elements of tool geometry on twist drills.

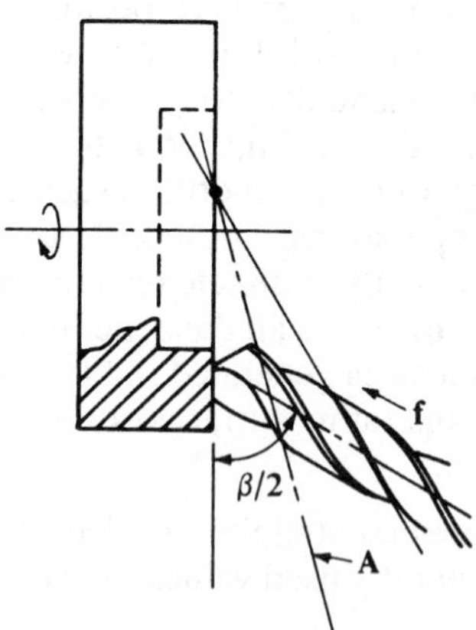

Fig. 2. In grinding the face of the twist drill the tool is swung around the axis *A* of an imaginary cone, while resting in a support tilted by half of the point angle β with respect to the face of the grinding wheel. Feed *f* for stock removal is in the direction of the drill axis.

Fig. 3. The chisel edge *C* after thinning the web by grinding off area *T*.

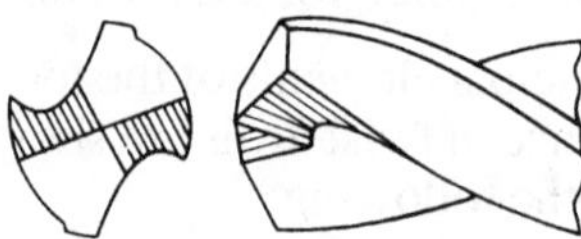

Fig. 4. Split point or "crankshaft" type web thinning.

Drill Point Thinning.—The chisel edge is the least efficient operating surface element of the twist drill because it does not cut, but actually squeezes or extrudes the work material. To improve the inefficient cutting conditions caused by the chisel edge, the point width is often reduced in a drill-point thinning operation, resulting in a condition such as that shown in Fig. 3. Point thinning is particularly desirable on larger size drills and also on those which become shorter in usage, because the thickness of the web increases toward the shaft of the twist drill, thereby adding to the length of the chisel edge. The extent of point thinning is limited by the minimum strength of the web needed to avoid splitting of the drill point under the influence of cutting forces.

Both sharpening operations—the relieved face grinding and the point thinning—should be carried out in special drill grinding machines or with twist drill grinding fixtures mounted on general-purpose tool grinding machines, designed to assure the essential accuracy of the required tool geometry. Off-hand grinding may be used for the important web thinning when a special machine is not available; however, such operation requires skill and experience.

Improperly sharpened twist drills, e.g. those with unequal edge length or asymmetrical point angle, will tend to produce holes with poor diameter and directional control.

For deep holes and also drilling into stainless steel, titanium alloys, high temperature alloys, nickel alloys, very high strength materials and in some cases tool steels, split point grinding, resulting in a "crankshaft" type drill point, is recommended. In this type of pointing, see Fig. 4, the chisel edge is entirely eliminated, extending the positive rake cutting edges to the center of the drill, thereby greatly reducing the required thrust in drilling. Points on modified-point drills must be restored after sharpening to maintain their increased drilling efficiency.

Sharpening Carbide Tools.—Cemented carbide indexable inserts are usually not resharpened but sometimes they require a special grind in order to form a contour on the cutting edge to suit a special purpose. Brazed type carbide cutting tools are resharpened after the cutting edge has become worn. On brazed carbide tools the cutting-edge wear should not be allowed to become excessive before the tool is re-sharpened. One method of determining when brazed carbide tools need resharpening is by periodic inspection of the flank wear and the condition of the face. Another method is to determine the amount of production which is normally obtained before excessive wear has taken place, or to determine the equivalent period of time. One disadvantage of this method is that slight variations in the work material will often cause the wear rate not to be uniform and the number of parts machined before regrinding will not be the same each time. Usually, sharpening should not require the removal of more than 0.005 to 0.010 inch of carbide.

General Procedure in Carbide Tool Grinding: The general procedure depends upon the kind of grinding operation required. If the operation is to resharpen a dull tool, a diamond wheel of 100 to 120 grain size is recommended although a finer wheel—up to 150 grain size—is sometimes used to obtain a better finish. If the tool is new or is a "standard" design and changes in shape are necessary, a 100-grit diamond wheel is recommended for roughing and a finer grit diamond wheel can be used for finishing. Some shops prefer to rough grind the carbide with a vitrified silicon carbide wheel, the finish grinding being done with a diamond wheel. A final operation commonly designated as lapping may or may not be employed for obtaining an extra-fine finish.

Wheel Speeds: The speed of silicon carbide wheels usually is about 5000 feet per minute. The speeds of diamond wheels generally range from 5000 to 6000 feet per minute; yet lower speeds (550 to 3000 fpm) can be effective.

Offhand Grinding: In grinding single-point tools (excepting chip breakers) the common practice is to hold the tool by hand, press it against the wheel face and traverse it continuously across the wheel face while the tool is supported on the machine rest or table which is adjusted to the required angle. This is known as "offhand grinding" to distinguish it from the machine grinding of cutters as in regular cutter grinding practice. The selection of wheels adapted to carbide tool grinding is very important.

Silicon Carbide Wheels.—The green colored silicon carbide wheels generally are preferred to the dark gray or gray-black variety, although the latter are sometimes used.

Grain or Grit Sizes: For roughing, a grain size of 60 is very generally used. For finish grinding with silicon carbide wheels, a finer grain size of 100 or 120 is common. A silicon carbide wheel such as C60-I-7V may be used for grinding both the steel shank and carbide tip. However, for under-cutting steel shanks up to the carbide tip, it may be advantageous to use an aluminum oxide wheel suitable for grinding softer, carbon steel.

Grade: According to the standard system of marking, different grades from soft to hard are indicated by letters from A to Z. For carbide tool grinding fairly soft grades such as G, H, I, and J are used. The usual grades for roughing are I or J and for finishing H, I, and J. The grade should be such that a sharp free-cutting wheel will be maintained without excessive grinding pressure. Harder grades than those indicated tend to overheat and crack the carbide.

Structure: The common structure numbers for carbide tool grinding are 7 and 8. The larger cup-wheels (10 to 14 inches) may be of the porous type and be designated as 12P. The standard structure numbers range from 1 to 15 with progressively higher numbers indicating less density and more open wheel structure.

Diamond Wheels.—Wheels with diamond-impregnated grinding faces are fast and cool cutting and have a very low rate of wear. They are used extensively both for resharpening and for finish grinding of carbide tools when preliminary roughing is required. Diamond wheels are also adapted for sharpening multi-tooth cutters such as milling cutters, reamers, etc., which are ground in a cutter grinding machine.

Resinoid bonded: wheels are commonly used for grinding chip breakers, milling cutters, reamers or other multi-tooth cutters. They are also applicable to precision grinding of carbide dies, gages, and various external, internal and surface grinding operations. Fast, cool cutting action is characteristic of these wheels.

Metal bonded: wheels are often used for offhand grinding of single-point tools especially when durability or long life and resistance to grooving of the cutting face, are considered more important than the rate of cutting. *Vitrified bonded* wheels are used both for roughing of chipped or very dull tools and for ordinary resharpening and finishing. They provide rigidity for precision grinding, a porous structure for fast cool cutting, sharp cutting action and durability.

Diamond Wheel Grit Sizes.—For roughing with diamond wheels a grit size of 100 is the most common both for offhand and machine grinding.

Grit sizes of 120 and 150 are frequently used in offhand grinding of single point tools 1) for resharpening; 2) for a combination roughing and finishing wheel; and 3) for chip-breaker grinding.

Grit sizes of 220 or 240 are used for ordinary finish grinding all types of tools (offhand and machine) and also for cylindrical, internal and surface finish grinding. Grits of 320 and 400 are used for "lapping" to obtain very fine finishes, and for hand hones. A grit of 500 is for lapping to a mirror finish on such work as carbide gages and boring or other tools for exceptionally fine finishes.

Diamond Wheel Grades.—Diamond wheels are made in several different grades to better adapt them to different classes of work. The grades vary for different types and shapes of wheels. Standard Norton grades are H, J, and L, for resinoid bonded wheels, grade N for metal bonded wheels and grades J, L, N, and P, for vitrified wheels. Harder and softer grades than standard may at times be used to advantage.

Diamond Concentration.—The relative amount (by carat weight) of diamond in the diamond section of the wheel is known as the "diamond concentration." Concentrations of 100 (high), 50 (medium) and 25 (low) ordinarily are supplied. A concentration of 50 represents one-half the diamond content of 100 (if the depth of the diamond is the same in each case) and 25 equals one-fourth the content of 100 or one-half the content of 50 concentration.

100 Concentration: Generally interpreted to mean 72 carats of diamond/in.^{3} of abrasive section. (A 75 concentration indicates 54 carats/in.^{3}.) Recommended (especially in grit sizes up to about 220) for general machine grinding of carbides, and for grinding cutters and chip breakers. Vitrified and metal bonded wheels usually have 100 concentration.

50 Concentration: In the finer grit sizes of 220, 240, 320, 400, and 500, a 50 concentration is recommended for offhand grinding with resinoid bonded cup-wheels.

25 Concentration: A low concentration of 25 is recommended for offhand grinding with resinoid bonded cup-wheels with grit sizes of 100, 120 and 150.

Depth of Diamond Section: The radial depth of the diamond section usually varies from $\frac{1}{16}$ to $\frac{1}{4}$ inch. The depth varies somewhat according to the wheel size and type of bond.

Dry Versus Wet Grinding of Carbide Tools.—In using silicon carbide wheels, grinding should be done either absolutely dry or with enough coolant to flood the wheel and tool. Satisfactory results may be obtained either by the wet or dry method. However, dry grinding is the most prevalent usually because, in wet grinding, operators tend to use an inadequate supply of coolant to obtain better visibility of the grinding operation and avoid getting wet; hence checking or cracking in many cases is more likely to occur in wet grinding than in dry grinding.

Wet Grinding with Silicon Carbide Wheels: One advantage commonly cited in connection with wet grinding is that an ample supply of coolant permits using wheels about one grade harder than in dry grinding thus increasing the wheel life. Plenty of coolant also prevents thermal stresses and the resulting cracks, and there is less tendency for the wheel to load. A dust exhaust system also is unnecessary.

Wet Grinding with Diamond Wheels: In grinding with diamond wheels the general practice is to use a coolant to keep the wheel face clean and promote free cutting. The amount of coolant may vary from a small stream to a coating applied to the wheel face by a felt pad.

Coolants for Carbide Tool Grinding.—In grinding either with silicon carbide or diamond wheels a coolant that is used extensively consists of water plus a small amount either of soluble oil, sal soda, or soda ash to prevent corrosion. One prominent manufacturer recommends for silicon carbide wheels about 1 ounce of soda ash per gallon of water and for diamond wheels kerosene. The use of kerosene is quite general for diamond wheels and usually it is applied to the wheel face by a felt pad. Another coolant recommended for diamond wheels consists of 80 per cent water and 20 per cent soluble oil.

Peripheral Versus Flat Side Grinding.—In grinding single point carbide tools with silicon carbide wheels, the roughing preparatory to finishing with diamond wheels may be done either by using the flat face of a cup-shaped wheel (side grinding) or the periphery of a "straight" or disk-shaped wheel. Even where side grinding is preferred, the periphery of a straight wheel may be used for heavy roughing as in grinding back chipped or broken tools (see left-hand diagram). Reasons for preferring peripheral grinding include faster cutting with less danger of localized heating and checking especially in grinding broad surfaces. The advantages usually claimed for side grinding are that proper rake or relief angles are easier to obtain and the relief or land is ground flat. The diamond wheels used for tool sharpening are designed for side grinding. (See right-hand diagram.)

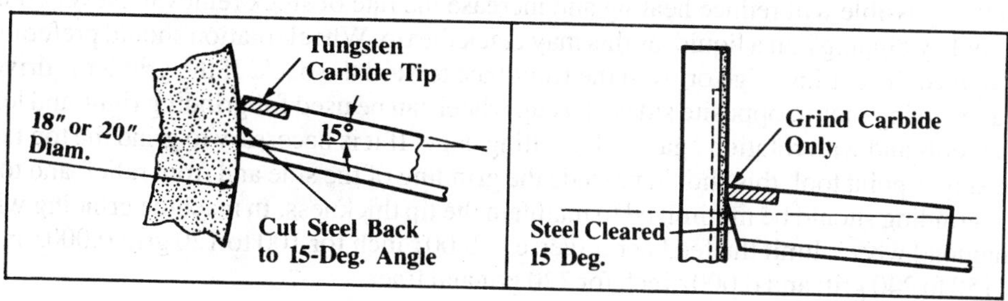

Lapping Carbide Tools.—Carbide tools may be finished by lapping, especially if an exceptionally fine finish is required on the work as, for example, tools used for precision boring or turning non-ferrous metals. If the finishing is done by using a diamond wheel of very fine grit (such as 240, 320, or 400), the operation is often called "lapping." A second lapping method is by means of a power-driven lapping disk charged with diamond dust, Norbide powder, or silicon carbide finishing compound. A third method is by using a hand lap or hone usually of 320 or 400 grit. In many plants the finishes obtained with carbide tools meet requirements without a special lapping operation. In all cases any feather edge which may be left on tools should be removed and it is good practice to bevel the edges of roughing tools at 45 degrees to leave a chamfer 0.005 to 0.010 inch wide. This is done by hand honing and the object is to prevent crumbling or flaking off at the edges when hard scale or heavy chip pressure is encountered.

Hand Honing: The cutting edge of carbide tools, and tools made from other tool materials, is sometimes hand honed before it is used in order to strengthen the cutting edge. When interrupted cuts or heavy roughing cuts are to be taken, or when the grade of carbide is slightly too hard, hand honing is beneficial because it will prevent chipping, or even possibly, breakage of the cutting edge. Whenever chipping is encountered, hand honing the cutting edge before use will be helpful. It is important, however, to hone the edge lightly and only when necessary. Heavy honing will always cause a reduction in tool life. Normally, removing 0.002 to 0.004 inch from the cutting edge is sufficient. When indexable inserts are used, the use of pre-honed inserts is preferred to hand honing although sometimes an additional amount of honing is required. Hand honing of carbide tools in between cuts is sometimes done to defer grinding or to increase the life of a cutting edge on an indexable insert. If correctly done, so as not to change the relief angle, this procedure is sometimes helpful. If improperly done, it can result in a reduction in tool life.

Chip Breaker Grinding.—For this operation a straight diamond wheel is used on a universal tool and cutter grinder, a small surface grinder, or a special chipbreaker grinder. A resinoid bonded wheel of the grade J or N commonly is used and the tool is held rigidly in an adjustable holder or vise. The width of the diamond wheel usually varies from 1⁄8 to 1⁄4 inch. A vitrified bond may be used for wheels as thick as 1⁄4 inch, and a resinoid bond for relatively narrow wheels.

Summary of Miscellaneous Points.—In grinding a single-point carbide tool, traverse it across the wheel face continuously to avoid localized heating. This traverse movement should be quite rapid in using silicon carbide wheels and comparatively slow with diamond wheels. A hand traversing and feeding movement, whenever practicable, is generally recommended because of greater sensitivity. In grinding, maintain a constant, moderate pressure. Manipulating the tool so as to keep the contact area with the wheel as small as possible will reduce heating and increase the rate of stock removal. Never cool a hot tool by dipping it in a liquid, as this may crack the tip. Wheel rotation should preferably be *against* the cutting edge or from the front face toward the back. If the grinder is driven by a reversing motor, opposite sides of a cup wheel can be used for grinding right-and left-hand tools and with rotation against the cutting edge. If it is necessary to grind the top face of a single-point tool, this should precede the grinding of the side and front relief, and top-face grinding should be minimized to maintain the tip thickness. In machine grinding with a diamond wheel, limit the feed per traverse to 0.001 inch for 100 to 120 grit; 0.0005 inch for 150 to 240 grit; and 0.0002 inch for 320 grit and finer.

JIGS AND FIXTURES

Material for Jig Bushings.—Bushings are generally made of a good grade of tool steel to ensure hardening at a fairly low temperature and to lessen the danger of fire cracking. They can also be made from machine steel, which will answer all practical purposes, provided the bushings are properly casehardened to a depth of about $\frac{1}{16}$ inch. Sometimes, bushings for guiding tools may be made of cast iron, but only when the cutting tool is of such a design that no cutting edges come within the bushing itself. For example, bushings used simply to support the smooth surface of a boring-bar or the shank of a reamer might, in some instances, be made of cast iron, but hardened steel bushings should always be used for guiding drills, reamers, taps, etc., when the cutting edges come in direct contact with the guiding surfaces. If the outside diameter of the bushing is very large, as compared with the diameter of the cutting tool, the cost of the bushing can sometimes be reduced by using an outer cast-iron body and inserting a hardened tool steel bushing.

When tool steel bushings are made and hardened, it is recommended that A-2 steel be used. The furnace should be set to 1750°F and the bushing placed in the furnace and held there approximately 20 minutes after the furnace reaches temperature. Remove the bushing and cool in still air. After the part cools to 100–150°F, immediately place in a tempering furnace that has been heated to 300°F. Remove the bushing after one hour and cool in still air. If an atmospherically controlled furnace is unavailable, the part should be wrapped in stainless foil to prevent scaling and oxidation at the 1750°F temperature.

American National Standard Jig Bushings.—Specifications for the following types of jig bushings are given in American National Standard B94.33-1974 (R1986). Head Type Press Fit Wearing Bushings, Type H (Fig. 1 and Tables 1 and 3); Headless Type Press Fit Wearing Bushings, Type P (Fig. 2 and Tables 1 and 3); Slip Type Renewable Wearing Bushings, Type S (Fig. 3 and Tables 4 and 5); Fixed Type Renewable Wearing Bushings, Type F (Fig. 4 and Tables 5 and 6); Headless Type Liner Bushings, Type L (Fig. 5 and Table 7); and Head Type Liner Bushings, Type HL (Fig. 6 and Table 8). Specifications for locking mechanisms are also given in Table 9.

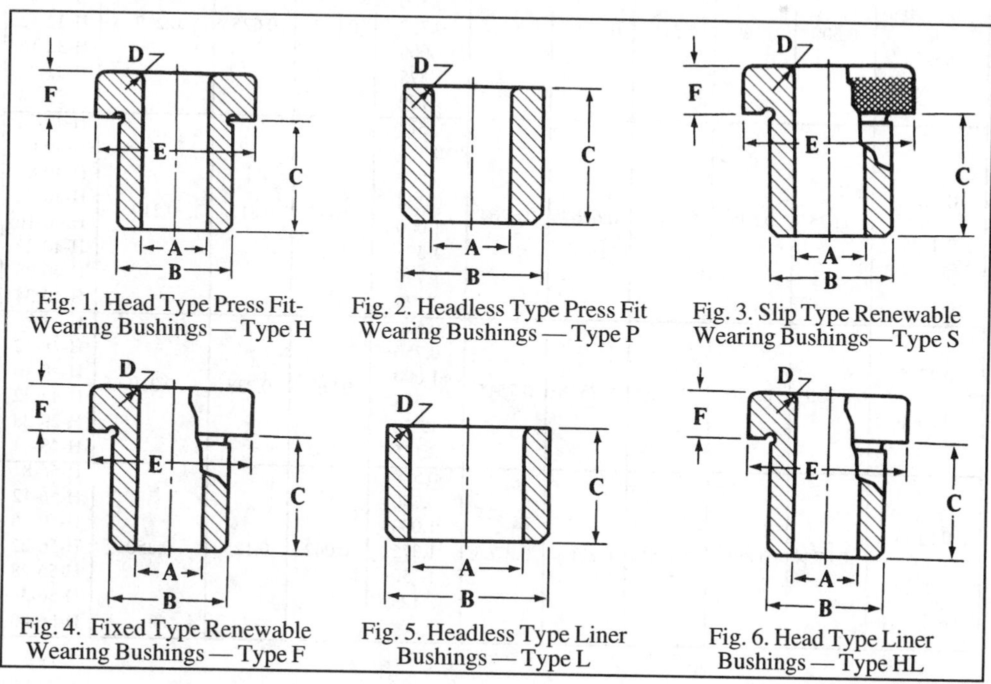

Fig. 1. Head Type Press Fit-Wearing Bushings — Type H

Fig. 2. Headless Type Press Fit Wearing Bushings — Type P

Fig. 3. Slip Type Renewable Wearing Bushings—Type S

Fig. 4. Fixed Type Renewable Wearing Bushings — Type F

Fig. 5. Headless Type Liner Bushings — Type L

Fig. 6. Head Type Liner Bushings — Type HL

Table 1. American National Standard Head Type Press Fit Wearing Bushings — Type H *ANSI B94.33-1974, R1986*

Range of Hole Sizes *A*	Body Diameter *B*					Body Length *C*	Radius *D*	Head Diam. *E* Max	Head Thickness *F* Max	Number
	Nom	Unfinished Max	Unfinished Min	Finished Max	Finished Min					
0.0135 up to and including 0.0625	0.156	0.166	0.161	0.1578	0.1575	0.250 0.312 0.375 0.500	0.016	0.250	0.094	H-10-4 H-10-5 H-10-6 H-10-8
0.0630 to 0.0995	0.203	0.213	0.208	0.2046	0.2043	0.250 0.312 0.375 0.500 0.750	0.016	0.312	0.094	H-13-4 H-13-5 H-13-6 H-13-8 H-13-12
0.1015 to 0.1405	0.250	0.260	0.255	0.2516	0.2513	0.250 0.312 0.375 0.500 0.750	0.016	0.375	0.094	H-16-4 H-16-5 H-16-6 H-16-8 H-16-12
0.1406 to 0.1875	0.312	0.327	0.322	0.3141	0.3138	0.250 0.312 0.375 0.500 0.750 1.000	0.031	0.438	0.125	H-20-4 H-20-5 H-20-6 H-20-8 H-20-12 H-20-16
0.189 to 0.2500	0.406	0.421	0.416	0.4078	0.4075	0.250 0.312 0.375 0.500 0.750 1.000 1.375 1.750	0.031	0.531	0.156	H-26-4 H-26-5 H-26-6 H-26-8 H-26-12 H-26-16 H-26-22 H-26-28
0.2570 to 0.3125	0.500	0.520	0.515	0.5017	0.5014	0.312 0.375 0.500 0.750 1.000 1.375 1.750	0.047	0.625	0.219	H-32-5 H-32-6 H-32-8 H-32-12 H-32-16 H-32-22 H-32-28
0.3160 to 0.4219	0.625	0.645	0.640	0.6267	0.6264	0.312 0.375 0.500 0.750 1.000 1.375 1.750 2.125	0.047	0.812	0.219	H-40-5 H-40-6 H-40-8 H-40-12 H-40-16 H-40-22 H-40-28 H-40-34
0.4375 to 0.5000	0.750	0.770	0.765	0.7518	0.7515	0.500 0.750 1.000 1.375 1.750 2.125	0.062	0.938	0.219	H-48-8 H-48-12 H-48-16 H-48-22 H-29-28 H-48-34
0.5156 to 0.6250	0.875	0.895	0.890	0.8768	0.8765	0.500 0.750 1.000 1.375 1.750 2.125 2.500	0.062	0.125	0.250	H-56-8 H-56-12 H-56-16 H-56-22 H-56-28 H-56-34 H-56-40

Table 1. *(Continued)* **American National Standard Head Type Press Fit Wearing Bushings — Type H** *ANSI B94.33-1974, R1986*

Range of Hole Sizes *A*	Body Diameter *B*					Body Length *C*	Radius *D*	Head Diam. *E* Max	Head Thickness *F* Max	Number
	Nom	Unfinished		Finished						
		Max	Min	Max	Min					
0.6406 to 0.7500	1.000	1.020	1.015	1.0018	1.0015	0.500	0.094	1.250	0.312	H-64-8
						0.750				H-64-12
						1.000				H-64-16
						1.375				H-64-22
						1.750				H-64-28
						2.125				H-64-34
						2.500				H-64-40
0.7656 to 1.0000	1.375	1.395	1.390	1.3772	1.3768	0.750	0.094	1.625	0.375	H-88-12
						1.000				H-88-16
						1.375				H-88-22
						1.750				H-88-28
						2.125				H-88-34
						2.500				H-88-40
1.0156 to 1.3750	1.750	1.770	1.765	1.7523	1.7519	1.000	0.094	2.000	0.375	H-112-16
						1.375				H-112-22
						1.750				H-112-28
						2.125				H-112-34
						2.500				H-112-40
						3.000				H-112-48
1.3906 to 1.7500	2.250	2.270	2.265	2.2525	2.2521	1.000	0.094	2.500	0.375	H-144-16
						1.375				H-144-22
						1.750				H-144-28
						2.125				H-144-34
						2.500				H-144-40
						3.000				H-144-48

All dimensions are in inches.
See also Table 3 for additional specifications.

Table 2. American National Standard Headless Type Press Fit Wearing Bushings — Type P *ANSI B94.33-1974, R1986*

Range of Hole Sizes *A*	Body Diameter *B*					Body Length *C*	Radius *D*	Number
	Nom	Unfinished		Finished				
		Max	Min	Max	Min			
0.0135 up to and including 0.0625	0.156	0.166	0.161	0.1578	0.1575	0.250	0.016	P-10-4
						0.312		P-10-5
						0.375		P-10-6
						0.500		P-10-8
0.0630 to 0.0995	0.203	0.213	0.208	0.2046	0.2043	0.250	0.016	P-13-4
						0.312		P-13-5
						0.375		P-13-6
						0.500		P-13-8
						0.750		P-13-12
0.1015 to 0.1405	0.250	0.260	0.255	0.2516	0.2513	0.250	0.016	P-16-4
						0.312		P-16-5
						0.375		P-16-6
						0.500		P-16-8
						0.750		P-16-12
0.1406 to 0.1875	0.312	0.327	0.322	0.3141	0.3138	0.250	0.031	P-20-4
						0.312		P-20-5
						0.375		P-20-6
						0.500		P-20-8
						0.750		P-20-12
						1.000		P-20-16

Table 2. American National Standard Headless Type Press Fit Wearing Bushings — Type P *ANSI B94.33-1974, R1986*

Range of Hole Sizes *A*	Body Diameter *B*					Body Length *C*	Radius *D*	Number
		Unfinished		Finished				
	Nom	Max	Min	Max	Min			
0.1890 to 0.2500	0.406	0.421	0.416	0.4078	0.4075	0.250	0.031	P-26-4
						0.312		P-26-5
						0.375		P-26-6
						0.500		P-26-8
						0.750		P-26-12
						1.000		P-26-16
						1.375		P-26-22
						1.750		P-26-28
0.2570 to 0.3125	0.500	0.520	0.515	0.5017	0.5014	0.312	0.047	P-32-5
						0.375		P-32-6
						0.500		P-32-8
						0.750		P-32-12
						1.000		P-32-16
						1.375		P-32-22
						1.750		P-32-28
0.3160 to 0.4219	0.625	0.645	0.640	0.6267	0.6264	0.312	0.047	P-40-5
						0.375		P-40-6
						0.500		P-40-8
						0.750		P-40-12
						1.000		P-40-16
						1.375		P-40-22
						1.750		P-40-28
						2.125		P-40-34
0.4375 to 0.5000	0.750	0.770	0.765	0.7518	0.7515	0.500	0.062	P-48-8
						0.750		P-48-12
						1.000		P-48-16
						1.375		P-48-22
						1.750		P-48-28
						2.125		P-48-34
0.5156 to 0.6250	0.875	0.895	0.890	0.8768	0.8765	0.500	0.062	P-56-8
						0.750		P-56-12
						1.000		P-56-16
						1.375		P-56-22
						1.750		P-56-28
						2.125		P-56-34
						2.500		P-56-40
0.6406 to 0.7500	1.000	1.020	1.015	1.0018	1.0015	0.500	0.062	P-64-8
						0.750		P-64-12
						1.000		P-64-16
						1.375		P-64-22
						1.750		P-64-28
						2.125		P-64-34
						2.500		P-64-40
0.7656 to 1.0000	1.375	1.395	1.390	1.3772	1.3768	0.750	0.094	P-88-12
						1.000		P-88-16
						1.375		P-88-22
						1.750		P-88-28
						2.125		P-88-34
						2.500		P-88-40
1.0156 to 1.3750	1.750	1.770	1.765	1.7523	1.7519	1.000	0.094	P-112-16
						1.375		P-112-22
						1.750		P-112-28
						2.125		P-112-34
						2.500		P-112-40
						3.000		P-112-48
1.3906 to 1.7500	2.250	2.270	2.265	2.2525	2.2521	1.000	0.094	P-144-16
						1.375		P-144-22
						1.750		P-144-28
						2.125		P-144-34
						2.500		P-144-40
						3.000		P-144-48

All dimensions are in inches. See Table 3 for additional specifications.

Table 3. Specifications for Head Type H and Headless Type P Press Fit Wearing Bushings *ANSI B94.33-1974, R1986*

All dimensions given in inches. Tolerance on dimensions where not otherwise specified shall be ±0.010 inch.

Size and type of chamfer on lead end to be manufacturer's option.

The length, *C*, is the overall length for the headless type and length underhead for the head type.

The head design shall be in accordance with the manufacturer's practice.

Diameter *A* must be concentric to diameter *B* within 0.0005 T.I.V. on finish ground bushings.

The body diameter, *B*, for unfinished bushings is larger than the nominal diameter in order to provide grinding stock for fitting to jig plate holes. The grinding allowance is:

0.005 to 0.010 in. for sizes 0.156, 0.203 and 0.250 in.

0.010 to 0.015 in. for sizes 0.312 and 0.406 in.

0.015 to 0.020 in. for sizes 0.500 in. and up.

Hole sizes are in accordance with American National Standard Twist Drill Sizes.

The maximum and minimum values of the hole size, *A*, shall be as follows:

Nominal Size of Hole	Maximum	Minimum
Above 0.0135 to 0.2500 in., incl.	Nominal + 0.0004 in.	Nominal + 0.0001 in.
Above 0.2500 to 0.7500 in., incl.	Nominal + 0.0005 in.	Nominal + 0.0001 in.
Above 0.7500 to 1.5000 in., incl.	Nominal + 0.0006 in.	Nominal + 0.0002 in.
Above 1.5000 in.	Nominal + 0.0007 in.	Nominal + 0.0003 in.

Bushings in the size range from 0.0135 through 0.3125 will be counterbored to provide for lubrication and chip clearance.

Bushings without counterbore are optional and will be furnished upon request.

The size of the counterbore shall be inside diameter of the bushing + 0.031 inch.

The included angle at the bottom of the counterbore shall be 118 deg, ± 2 deg.

The depth of the counterbore shall be in accordance with the table below to provide adequate drill bearing.

Body Length	Drill Bushing Hole Size											
	0.0135 to 0.0625		0.0630 to 0.0995		0.1015 to 0.1405		0.1406 to 0.1875		0.1890 to 0.2500		0.2570 to 0.3125	
	P	H	P	H	P	H	P	H	P	H	P	H
	Minimum Drill Bearing Length—Inch											
0.250	X	0.250	X	X	X	X	X	X	X	X	X	X
0.312	X	0.250	X	X	X	X	X	X	X	X	X	X
0.375	0.250	0.250	X	X	X	X	X	X	X	X	X	X
0.500	0.250	0.250	X	0.312	X	0.312	X	0.375	X	X	X	X
0.750	+	+	0.375	0.375	0.375	0.375	X	0.375	X	X	X	X
1.000	+	+	+	+	+	+	0.625	0.625	0.625	0.625	0.625	0.625
1.375	+	+	+	+	+	+	+	+	0.625	0.625	0.625	0.625
1.750	+	+	+	+	+	+	+	+	0.625	0.625	0.625	0.625

All dimensions are in inches.

X indicates no counterbore.

+ indicates not American National Standard

Table 4. American National Standard Slip Type Renewable Wearing Bushings—Type S *ANSI B94.33-1974, R1986*

Range of Hole Sizes *A*	Body Diameter *B* Nom	Max	Min	Length Under-Head *C*	Radius *D*	Head Diam. *E* Max	Head Thickness *F* Max	Number
0.0135 up to and including 0.0469	0.188	0.1875	0.1873	0.250 0.312 0.375 0.500	0.031	0.312	0.188	S-12-4 S-12-5 S-12-6 S-12-8
0.0492 to 0.1562	0.312	0.3125	0.3123	0.312 0.500 0.750 1.000	0.047	0.562	0.375	S-20-5 S-20-8 S-20-12 S-20-16
0.1570 to 0.3125	0.500	0.5000	0.4998	0.312 0.500 0.750 1.000 1.375 1.750	0.047	0.812	0.438	S-32-5 S-32-8 S-32-12 S-32-16 S-32-22 S-32-28
0.3160 to 0.5000	0.750	0.7500	0.7498	0.500 0.750 1.000 1.375 1.750 2.125	0.094	1.062	0.438	S-48-8 S-48-12 S-48-16 S-48-22 S-48-28 S-48-34

Table 4. *(Continued)* American National Standard Slip Type Renewable Wearing Bushings—Type S *ANSI B94.33-1974, R1986*

Range of Hole Sizes *A*	Body Diameter *B*			Length Under-Head *C*	Radius *D*	Head Diam. *E* Max	Head Thickness *F* Max	Number
	Nom	Max	Min					
0.5156 to 0.7500	1.000	1.0000	0.9998	0.500	0.094	1.438	0.438	S-64-8
				0.750				S-64-12
				1.000				S-64-16
				1.375				S-64-22
				1.750				S-64-28
				2.125				S-64-34
				2.500				S-64-40
0.7656 to 1.0000	1.375	1.3750	1.3747	0.750	0.094	1.812	0.438	S-88-12
				1.000				S-88-16
				1.375				S-88-22
				1.750				S-88-28
				2.125				S-88-34
				2.500				S-88-40
1.0156 to 1.3750	1.750	1.7500	1.7497	1.000	0.125	2.312	0.625	S-112-16
				1.375				S-112-22
				1.750				S-112-28
				2.125				S-112-34
				2.500				S-112-40
				3.000				S-112-48
1.3906 to 1.7500	2.250	2.2500	2.2496	1.000	0.125	2.812	0.625	S-144-16
				1.375				S-144-22
				1.750				S-144-28
				2.125				S-144-34
				2.500				S-144-40
				3.000				S-144-48

All dimensions are in inches. See also Table 5 for additional specifications.

Table 5. Specifications for Slip Type S and Fixed Type F Renewable Wearing Bushings *ANSI B94.33-1974, R1986*

Tolerance on dimensions where not otherwise specified shall be plus or minus 0.010 inch.
Hole sizes are in accordance with the American Standard Twist Drill Sizes.
The maximum and minimum values of hole size, *A*, shall be as follows:

Nominal Size of Hole	Maximum	Minimum
Above 0.0135 to 0.2500 in. incl.	Nominal + 0.0004 in.	Nominal + 0.0001 in.
Above 0.2500 to 0.7500 in. incl.	Nominal + 0.0005 in.	Nominal + 0.0001 in.
Above 0.7500 to 1.5000 in. incl.	Nominal + 0.0006 in.	Nominal + 0.0002 in.
Above 1.5000	Nominal + 0.0007 in.	Nominal + 0.0003 in.

The head design shall be in accordance with the manufacturer's practice.
Head of slip type is usually knurled.
When renewable wearing bushings are used with liner bushings of the head type, the length under the head will still be equal to the thickness of the jig plate, because the head of the liner bushing will be countersunk into the jig plate.
Diameter *A* must be concentric to diameter *B* within 0.0005 T.I.R. on finish ground bushings.
Size and type of chamfer on lead end to be manufacturer's option.
Bushings in the size range from 0.0135 through 0.3125 will be counterbored to provide for lubrication and chip clearance.
Bushings without counterbore are optional and will be furnished upon request.
The size of the counterbore shall be inside diameter of the bushings plus 0.031 inch.
The included angle at the bottom of the counterbore shall be 118 deg., plus or minus 2 deg.
The depth of the counterbore shall be in accordance with the table below to provide adequate drill bearing.

Body Length	Drill Bearing Hole Size											
	0.0135 to 0.0625		0.0630 to 0.0995		0.1015 to 0.1405		0.1406 to 0.1875		0.1890 to 0.2500		0.2500 to 0.3125	
	S	F	S	F	S	F	S	F	S	F	S	F
	Minimum Drill Bearing Length											
0.250	0.250	0.250	0.375	0.375	X	X	X	X	X	X	X	X
0.312	0.250	0.250	0.375	0.375	0.375	0.375	0.375	0.375	0.375	0.375	X	X
0.375	0.250	0.250	0.375	0.375	0.375	0.375	0.375	0.375	0.375	0.375	X	X
0.500	0.250	0.250	0.375	0.375	0.375	0.375	0.375	0.375	0.375	0.375	X	X
0.750	0.250	0.250	0.375	0.375	0.375	0.375	0.375	0.375	0.625	0.625	0.625	0.625
1.000	0.312	0.312	0.375	0.375	0.375	0.375	0.625	0.625	0.625	0.625	0.625	0.625
1.375	+	+	+	+	+	+	0.625	0.625	0.625	0.625	0.625	0.625
1.750	+	+	+	+	+	+	0.625	0.625	0.625	0.625	0.625	0.625

All dimensions are in inches.
X indicates no counterbore.
+ indicates not American National Standard length.

Table 6. American National Standard Fixed Type Renewable Wearing Bushings — Type F *ANSI B94.33-1974, R1986*

Range of Hole Sizes *A*	Body Diameter *B*			Length Under Head *C*	Radius *D*	Head Diam. *E* Max	Head Thickness *F* Max	Number
	Nom	Max	Min					
0.0135 up to and including 0.0469	0.188	0.1875	0.1873	0.250	0.031	0.312	0.188	F-12-4
				0.312				F-12-5
				0.375				F-12-6
				0.500				F-12-8
0.0492 to 0.1562	0.312	0.3125	0.3123	0.312	0.047	0.562	0.250	F-20-5
				0.500				F-20-8
				0.750				F-20-12
				1.000				F-20-16
0.1570 to 0.3125	0.500	0.5000	0.4998	0.312	0.047	0.812	0.250	F-32-5
				0.500				F-32-8
				0.750				F-32-12
				1.000				F-32-16
				1.375				F-32-22
				1.750				F-32-28
0.3160 to 0.5000	0.750	0.7500	0.7498	0.500	0.094	1.062	0.250	F-48-8
				0.750				F-48-12
				1.000				F-48-16
				1.375				F-48-22
				1.750				F-48-28
				2.125				F-48-34
0.5156 to 0.7500	1.000	1.0000	0.9998	0.500	0.094	1.438	0.375	F-64-8
				0.750				F-64-12
				1.000				F-64-16
				1.375				F-64-22
				1.750				F-64-28
				2.125				F-64-34
				2.500				F-64-40
0.7656 to 1.0000	1.375	1.3750	1.3747	0.750	0.094	1.812	0.375	F-88-12
				1.000				F-88-16
				1.375				F-88-22
				1.750				F-88-28
				2.125				F-88-34
				2.500				F-88-40
1.0156 to 1.3750	1.750	1.7500	1.7497	1.000	0.125	2.312	0.375	F-112-16
				1.375				F-112-22
				1.750				F-112-28
				2.125				F-112-34
				2.500				F-112-40
				3.000				F-112-48
1.3906 to 1.7500	2.250	2.2500	2.2496	1.000	0.125	2.812	0.375	F-144-16
				1.375				F-144-22
				1.750				F-144-28
				2.125				F-144-34
				2.500				F-144-40
				3.000				F-144-48

All dimensions are in inches. See also Table 5 for additional specifications.

Table 7. American National Standard Headless Type Liner Bushings — Type L
ANSI B94.33-1974, R1986

Range of Hole Sizes in Renewable Bushings	Inside Diameter *A*			Body Diameter *B*					Over-all Length *C*	Radius *D*	Number
					Unfinished		Finished				
	Nom	Max	Min	Nom	Max	Min	Max	Min			
0.0135 up to and including 0.0469	0.188	0.1879	0.1876	0.312	0.3341	0.3288	0.3141	0.3138	0.250 0.312 0.375 0.500	0.031	L-20-4 L-20-5 L-20-6 L-20-8
0.0492 to 0.1562	0.312	0.3129	0.3126	0.500	0.520	0.515	0.5017	0.5014	0.312 0.500 0.750 1.000	0.047	L-32-5 L-32-8 L-32-12 L-32-16
0.1570 to 0.3125	0.500	0.5005	0.5002	0.750	0.770	0.765	0.7518	0.7515	0.312 0.500 0.750 1.000 1.375 1.750	0.062	L-48-5 L-48-8 L-48-12 L-48-16 L-48-22 L-48-28
0.3160 to 0.5000	0.750	0.7506	0.7503	1.000	1.020	1.015	1.0018	1.0015	0.500 0.750 1.000 1.375 1.750 2.125	0.062	L-64-8 L-64-12 L-64-16 L-64-22 L-64-28 L-64-34
0.5156 to 0.7500	1.000	1.0007	1.0004	1.375	1.395	1.390	1.3772	1.3768	0.500 1.750 1.000 1.375 1.750 2.125 2.500	0.094	L-88-8 L-88-12 L-88-16 L-88-22 L-88-28 L-88-34 L-88-40
0.7656 to 1.0000	1.375	1.3760	1.3756	1.750	1.770	1.765	1.7523	1.7519	0.750 1.000 1.375 1.750 2.125 2.500	0.094	L-112-12 L-112-16 L-112-22 L-112-28 L-112-34 L-112-40
1.0156 to 1.3750	1.750	1.7512	1.7508	2.250	2.270	2.265	2.2525	2.2521	1.000 1.375 1.750 2.125 2.500 3.000	0.094	L-144-16 L-144-22 L-144-28 L-144-34 L-144-40 L-144-48
1.3906 to 1.7500	2.250	2.2515	2.2510	2.750	2.770	2.765	2.7526	2.7522	1.000 1.375 1.750 2.125 2.500 3.000	0.125	L-176-16 L-176-22 L-176-28 L-176-34 L-176-40 L-176-48

All dimensions are in inches.

Tolerances on dimensions where otherwise not specified are ± 0.010 in.

The body diameter, *B*, for unfinished bushings is 0.015 to 0.020 in. larger than the nominal diameter in order to provide grinding stock for fitting to jig plate holes.

Diameter *A* must be concentric to diameter *B* within 0.0005 T.I.R. on finish ground bushings.

Table 8. American National Standard Head Type Liner Bushing — Type HL *ANSI B94.33-1974, R1986*

Range of Hole Sizes in Renewable Bushings	Inside Diameter *A*			Body Diameter *B*					Overall Length *C*	Radius *D*	Head Dia. *E*	Head Thickness *F* Max	Number
					Unfinished		Finished						
	Nom	Max	Min	Nom	Max	Min	Max	Min					
0.0135 to 0.1562	0.312	0.3129	0.3126	0.500	0.520	0.515	0.5017	0.5014	0.312	0.047	0.625	0.094	HL-32-5
									0.500				HL-32-8
									0.750				HL-32-12
									1.000				HL-32-16
0.1570 to 0.3125	0.500	0.5005	0.5002	0.750	0.770	0.765	0.7518	0.7515	0.312	0.062	0.875	0.094	HL-48-5
									0.500				HL-48-8
									0.750				HL-48-12
									1.000				HL-48-16
									1.375				HL-48-22
									1.750				HL-48-28
0.3160 to 0.5000	0.750	0.7506	0.7503	1.000	1.020	1.015	1.0018	1.0015	0.500	0.062	1.125	0.125	HL-64-8
									0.750				HL-64-12
									1.000				HL-64-16
									1.375				HL-64-22
									1.750				HL-64-28
									2.125				HL-64-34
0.5156 to 0.7500	1.000	1.0007	1.0004	1.375	1.395	1.390	1.3772	1.3768	0.500	0.094	1.500	0.125	HL-88-8
									0.750				HL-88-12
									1.000				HL-88-16
									1.375				HL-88-22
									1.750				HL-88-28
									2.125				HL-88-34
									2.500				HL-88-40
0.7656 to 1.0000	1.375	1.3760	1.3756	1.750	1.770	1.765	1.7523	1.7519	0.750	0.094	1.875	0.188	HL-112-12
									1.000				HL-112-16
									1.375				HL-112-22
									1.750				HL-112-28
									2.125				HL-112-34
									2.500				HL-112-40
1.0156 to 1.3750	1.750	1.7512	1.7508	2.250	2.27	2.265	2.2525	2.2521	1.000	0.094	2.375	0.188	HL-144-16
									1.375				HL-144-22
									1.750				HL-144-28
									2.125				HL-144-34
									2.500				HL-144-40
									3.000				HL-144-48
1.3906 to 1.7500	2.250	2.2515	2.2510	2.750	2.770	2.765	2.7526	2.7522	1.000	0.125	2.875	0.188	HL-176-16
									1.375				HL-176-22
									1.750				HL-176-28
									2.125				HL-176-34
									2.500				HL-176-40
									3.000				HL-176-48

All dimensions are in inches.

See also footnotes to Table 7.

Table 9. American National Standard Locking Mechanisms for Jig Bushings *ANSI B94.33-1974, R1986*

Lock Screw for Use with Slip or Fixed Renewable Bushings

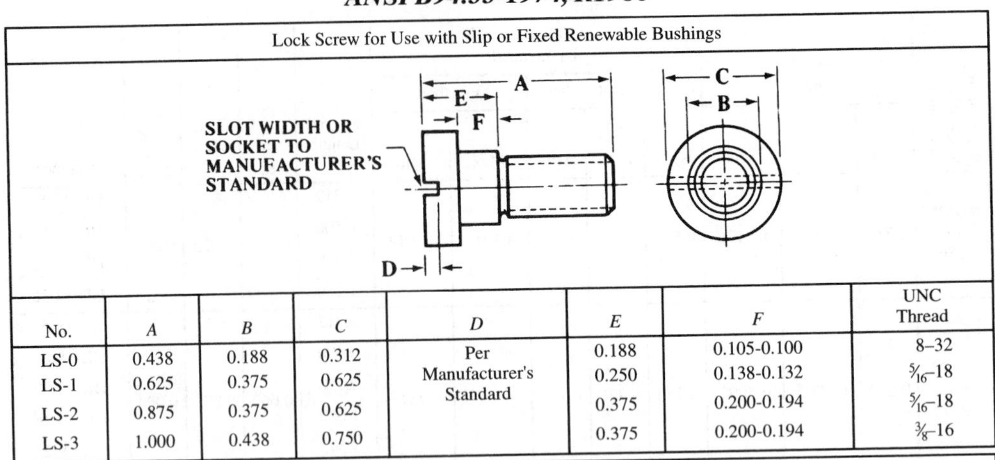

No.	*A*	*B*	*C*	*D*	*E*	*F*	UNC Thread
LS-0	0.438	0.188	0.312	Per Manufacturer's Standard	0.188	0.105-0.100	8–32
LS-1	0.625	0.375	0.625		0.250	0.138-0.132	$\frac{5}{16}$–18
LS-2	0.875	0.375	0.625		0.375	0.200-0.194	$\frac{5}{16}$–18
LS-3	1.000	0.438	0.750		0.375	0.200-0.194	$\frac{3}{8}$–16

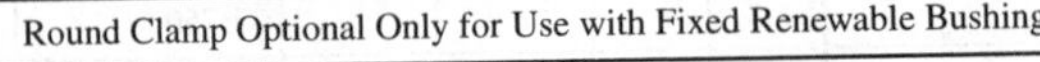
Round Clamp Optional Only for Use with Fixed Renewable Bushing

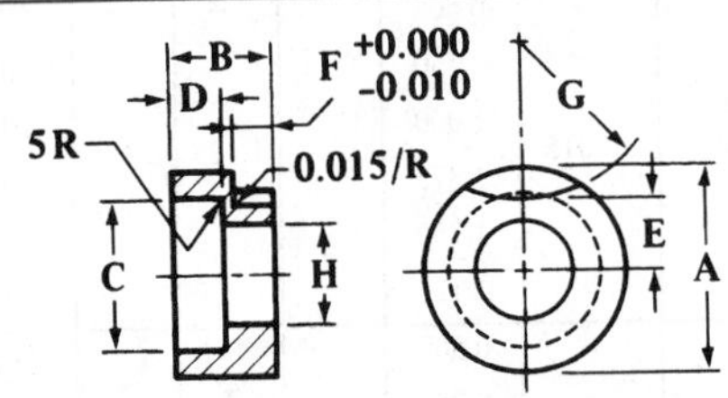

NOTE: F DIMENSION ALLOWS FOR CLAMPING. MATERIAL AND HARDNESS TO MANUFACTURER'S STANDARD. TO CHANGE TO THE ROUND CLAMP IN OLD FIXTURES, REMOVE THE CONVENTIONAL SCREW AND USE THE SAME TAPPED HOLE TO SECURE THE NEW CLAMP WITH STANDARD SOCKET HEAD SCREW.

Number	*A*	*B*	*C*	*D*	*E*	*F*	*G*	*H*	Use With Socket Head Screw
RC-1	0.625	0.312	0.484	0.150	0.203	0.125	0.531	0.328	$\frac{5}{16}$–18
RC-2	0.625	0.438	0.484	0.219	0.187	0.188	0.906	0.328	$\frac{5}{16}$–18
RC-3	0.750	0.500	0.578	0.281	0.219	0.188	1.406	0.391	$\frac{3}{8}$–16

Locking Mechanism Dimensions of Slip and Fixed Renewable Bushings

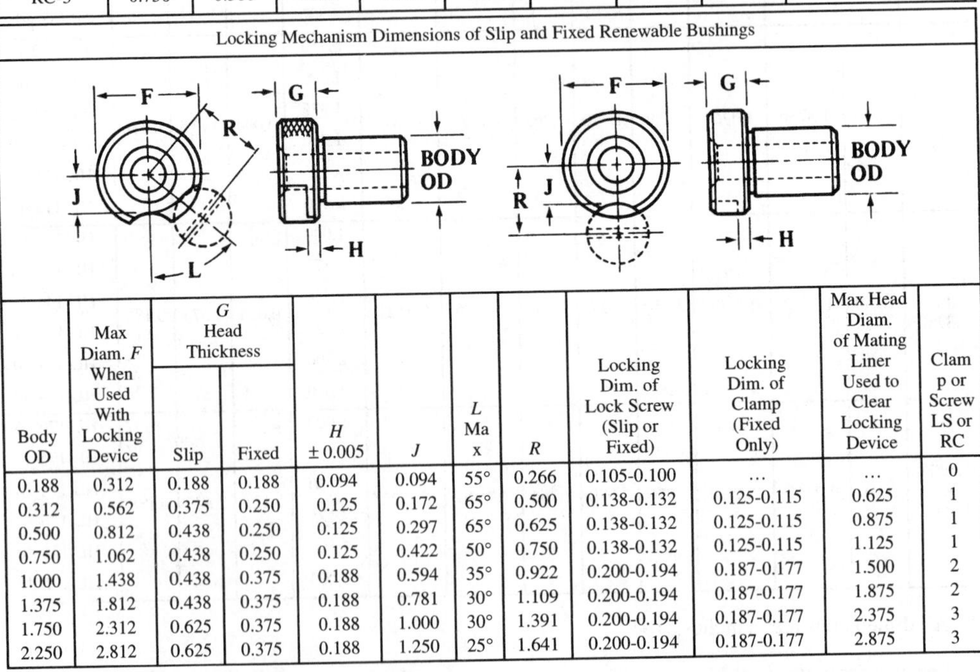

Body OD	Max Diam. *F* When Used With Locking Device	*G* Head Thickness		*H* ± 0.005	*J*	*L* Max	*R*	Locking Dim. of Lock Screw (Slip or Fixed)	Locking Dim. of Clamp (Fixed Only)	Max Head Diam. of Mating Liner Used to Clear Locking Device	Clamp or Screw LS or RC
		Slip	Fixed								
0.188	0.312	0.188	0.188	0.094	0.094	55°	0.266	0.105-0.100	…	…	0
0.312	0.562	0.375	0.250	0.125	0.172	65°	0.500	0.138-0.132	0.125-0.115	0.625	1
0.500	0.812	0.438	0.250	0.125	0.297	65°	0.625	0.138-0.132	0.125-0.115	0.875	1
0.750	1.062	0.438	0.250	0.125	0.422	50°	0.750	0.138-0.132	0.125-0.115	1.125	1
1.000	1.438	0.438	0.375	0.188	0.594	35°	0.922	0.200-0.194	0.187-0.177	1.500	2
1.375	1.812	0.438	0.375	0.188	0.781	30°	1.109	0.200-0.194	0.187-0.177	1.875	2
1.750	2.312	0.625	0.375	0.188	1.000	30°	1.391	0.200-0.194	0.187-0.177	2.375	3
2.250	2.812	0.625	0.375	0.188	1.250	25°	1.641	0.200-0.194	0.187-0.177	2.875	3

All dimensions are in inches.

Jig Bushing Definitions.— *Renewable Bushings:* Renewable wearing bushings to guide the tool are for use in liners which in turn are installed in the jig. They are used where the bushing will wear out or become obsolete before the jig or where several bushings are to be interchangeable in one hole. Renewable wearing bushings are divided into two classes, "Fixed" and "Slip." Fixed renewable bushings are installed in the liner with the intention of leaving them in place until worn out. Slip renewable bushings are interchangeable in a given size of liner and, to facilitate removal, they are usually made with a knurled head. They are most frequently used where two or more operations requiring different inside diameters are performed in a single jig, such as where drilling is followed by reaming, tapping, spot facing, counterboring, or some other secondary operation.

Press Fit Bushings: Press fit wearing bushings to guide the tool are for installation directly in the jig without the use of a liner and are employed principally where the bushings are used for short production runs and will not require replacement. They are intended also for short center distances.

Liner Bushings: Liner bushings are provided with and without heads and are permanently installed in a jig to receive the renewable wearing bushings. They are sometimes called master bushings.

Jig Plate Thickness.—The standard length of the press fit portion of jig bushings as established are based on standardized uniform jig plate thicknesses of 5⁄16, 3⁄8, 1⁄2, 3⁄4, 1, 13⁄8, 13⁄4, 21⁄8, 21⁄2, and 3 inches.

Jig Bushing Designation System.—*Inside Diameter:* The inside diameter of the hole is specified by a decimal dimension.

Type Bushing: The type of bushing is specified by a letter: S for Slip Renewable, F for Fixed Renewable, L for Headless Liner, HL for Head Liner, P for Headless Press Fit, and H for Head Press Fit.

Body Diameter: The body diameter is specified in multiples of 0.0156 inch. For example, a 0.500-inch body diameter = 0.500/0.0156 = 32.

Body Length: The effective or body length is specified in multiples of 0.0625 inch. For example, a 0.500-inch length = 0.500/0.0625 = 8.

Unfinished Bushings: All bushings with grinding stock on the body diameter are designated by the letter U following the number.

Example: A slip renewable bushing having a hole diameter of 0.5000 inch, a body diameter of 0.750 inch, and a body length of 1.000 inch would be designated as .5000-S-48-16.

Definition of Jig and Fixture.—The distinction between a jig and fixture is not easy to define, but, as a general rule, it is as follows: A jig either holds or is held on the work, and, at the same time, contains guides for the various cutting tools, whereas a fixture holds the work while the cutting tools are in operation, but does not contain any special arrangements for guiding the tools. A fixture, therefore, must be securely held or fixed to the machine on which the operation is performed—hence the name. A fixture is sometimes provided with a number of gages and stops, but not with bushings or other devices for guiding and supporting the cutting tools.

Jig Borers.—Jig borers are used for precision hole-location work. For this reason, the coordinate measuring systems on these machines are designed to provide longitudinal and transverse movements that are accurate to 0.0001 in. One widely used method of obtaining this accuracy utilizes ultraprecision lead screws. Another measuring system employs precision end measuring rods and a micrometer head that are placed in a trough which is parallel to the table movement. However, the purpose of all coordinate measuring systems used is the same: to provide a method of aligning the spindle at the precise location where a hole is to be produced. Since the work table of a jig borer moves in two directions, the coordinate system of dimensioning is used, where dimensions are given from two perpendicular reference axes, usually the sides of the workpiece, frequently its upper left-hand corner. See Fig. 1C.

Jig-Boring Practice.—The four basic steps to follow to locate and machine a hole on a jig borer are: 1) align and clamp the workpiece on the jig-borer table; 2) locate the two reference axes of the workpiece with respect to the jig-borer spindle; 3) locate the hole to be machined; and 4) drill and bore the hole to size.

Align and Clamp the Workpiece: The first consideration in placing the workpiece on the jig-borer table should be the relation of the coordinate measuring system of the jig borer to the coordinate dimensions on the drawing. Therefore, the coordinate measuring system is designed so that the readings of the coordinate measurements are direct when the table is moved toward the left and when it is moved toward the column of the jig borer. The result would be the same if the spindle were moved toward the right and away from the column, with the workpiece situated in such a position that one reference axis is located at the left and the other axis at the back, toward the column.

If the holes to be bored are to pass through the bottom of the workpiece, then the workpiece must be placed on precision parallel bars. In order to prevent the force exerted by the clamps from bending the workpiece the parallel bars are placed directly under the clamps, which hold the workpiece on the table. The reference axes of the workpiece must also be aligned with respect to the transverse and longitudinal table movements before it is firmly clamped. This alignment can be done with a dial-test indicator held in the spindle of the jig borer and bearing against the longitudinal reference edge. As the table is traversed in the longitudinal direction, the workpiece is adjusted until the dial-test indicator readings are the same for all positions.

Locate the Two Reference Axes of the Workpiece with Respect to the Spindle: The jig-borer table is now moved to position the workpiece in a precise and known location from where it can be moved again to the location of the holes to be machined. Since all the holes are dimensioned from the two reference axes, the most convenient position to start from is where the axis of the jig-borer spindle and the intersection of the two workpiece reference axes are aligned. This is called the starting position, which is similar to a zero reference position. When so positioned, the longitudinal and transverse measuring systems of the jig borer are set to read zero. Occasionally, the reference axes are located outside the body of the workpiece: a convenient edge or hole on the workpiece is picked up as the starting position, and the dimensions from this point to the reference axes are set on the positioning measuring system.

Locate the Hole: Precise coordinate table movements are used to position the workpiece so that the spindle axis is located exactly where the hole is to be machined. When the measuring system has been set to zero at the starting position, the coordinate readings at the hole location will be the same as the coordinate dimensions of the hole center.

The movements to each hole must be made in one direction for both the transverse and longitudinal directions, to eliminate the effect of any backlash in the lead screw. The usual table movements are toward the left and toward the column.

The most convenient sequence on machines using micrometer dials as position indicators (machines with lead screws) is to machine the hole closest to the starting position first and then the next closest, and so on. On jig borers using end measuring rods, the opposite sequence is followed: The farthest hole is machined first and then the next farthest, and so on, since it is easier to remove end rods and replace them with shorter rods.

Drill and Bore Hole to Size: The sequence of operations used to produce a hole on a jig borer is as follows: 1) a short, stiff drill, such as a center drill, that will not deflect when cutting should be used to spot a hole when the work and the axis of the machine tool spindle are located at the exact position where the hole is wanted; 2) the initial hole is made by a twist drill; and 3) a single-point boring tool that is set to rotate about the axis of the machine tool spindle is then used to generate a cut surface that is concentric to the axis of rotation.

Heat will be generated by the drilling operation, so it is good practice to drill all the holes first, and then allow the workpiece to cool before the holes are bored to size.

Transfer of Tolerances.—All of the dimensions that must be accurately held on precision machines and engine parts are usually given a tolerance. And when such dimensions are changed from the conventional to the coordinate system of dimensioning, the tolerances must also be included. Because of their importance, the transfer of the tolerances must be done with great care, keeping in mind that the sum of the tolerances of any pair of dimensions in the coordinate system must not be larger than the tolerance of the dimension that they replaced in the conventional system. An example is given in Fig. 1.

The first step in the procedure is to change the tolerances given in Fig. 1A to equal, bilateral tolerances given in Fig. 1B. For example, the dimension $2.125^{+.003}_{-.001}$ has a total tolerance of 0.004. The equal, bilateral tolerance would be plus or minus one-half of this value, or ±.002. Then to keep the limiting dimensions the same, the basic dimension must be changed to 2.126, in order to give the required values of 2.128 and 2.124. When changing to equal, bilateral tolerances, if the upper tolerance is decreased (as in this example), the basic dimension must be increased by a like amount. The upper tolerance was decreased by $0.003 - 0.002 = 0.001$; therefore, the basic dimension was increased by 0.001 to 2.126. Conversely, if the upper tolerance is increased, the basic dimension is decreased.

The next step is to transfer the revised basic dimension to the coordinate dimensioning system. To transfer the 2.126 dimension, the distance of the applicable holes from the left reference axis must be determined. The first holes to the right are 0.8750 from the reference axis. The second hole is 2.126 to the right of the first holes. Therefore, the second hole is $0.8750 + 2.126 = 3.0010$ to the right of the reference axis. This value is then the coordinate dimension for the second hole, while the 0.8750 value is the coordinate dimension of the first two, vertically aligned holes. This procedure is followed for all the holes to find their distances from the two reference axes. These values are given in Fig. 1C.

The final step is to transfer the tolerances. The 2.126 value in Fig. 1B has been replaced by the 0.8750 and 3.0010 values in Fig. 1C. The 2.126 value has an available tolerance of ±0.002. Dividing this amount equally between the two replacement values gives 0.8750 ± 0.001 and 3.0010 ± 0.001. The sum of these tolerances is .002, and as required, does not exceed the tolerance that was replaced. Next transfer the tolerance of the 0.502 dimension. Divide the available tolerance, ±0.002, equally between the two replacement values to yield 3.0010 ±0.001 and 3.5030 ±0.001. The sum of these two tolerances equals the replaced tolerance, as required. However, the 1.125 value of the last hole to the right (coordinate dimension 4.6280 in.) has a tolerance of only ±0.001. Therefore, the sum of the tolerances on the 3.5030 and 4.6280 values cannot be larger than 0.001. Dividing this tolerance equally would give 3.5030 ± .0005 and 4.6280 ±0.0005. This new, smaller tolerance replaces the ± 0.001 tolerance on the 3.5030 value in order to satisfy all tolerance sum requirements. This example shows how the tolerance of a coordinate value is affected by more than one other dimensional requirement.

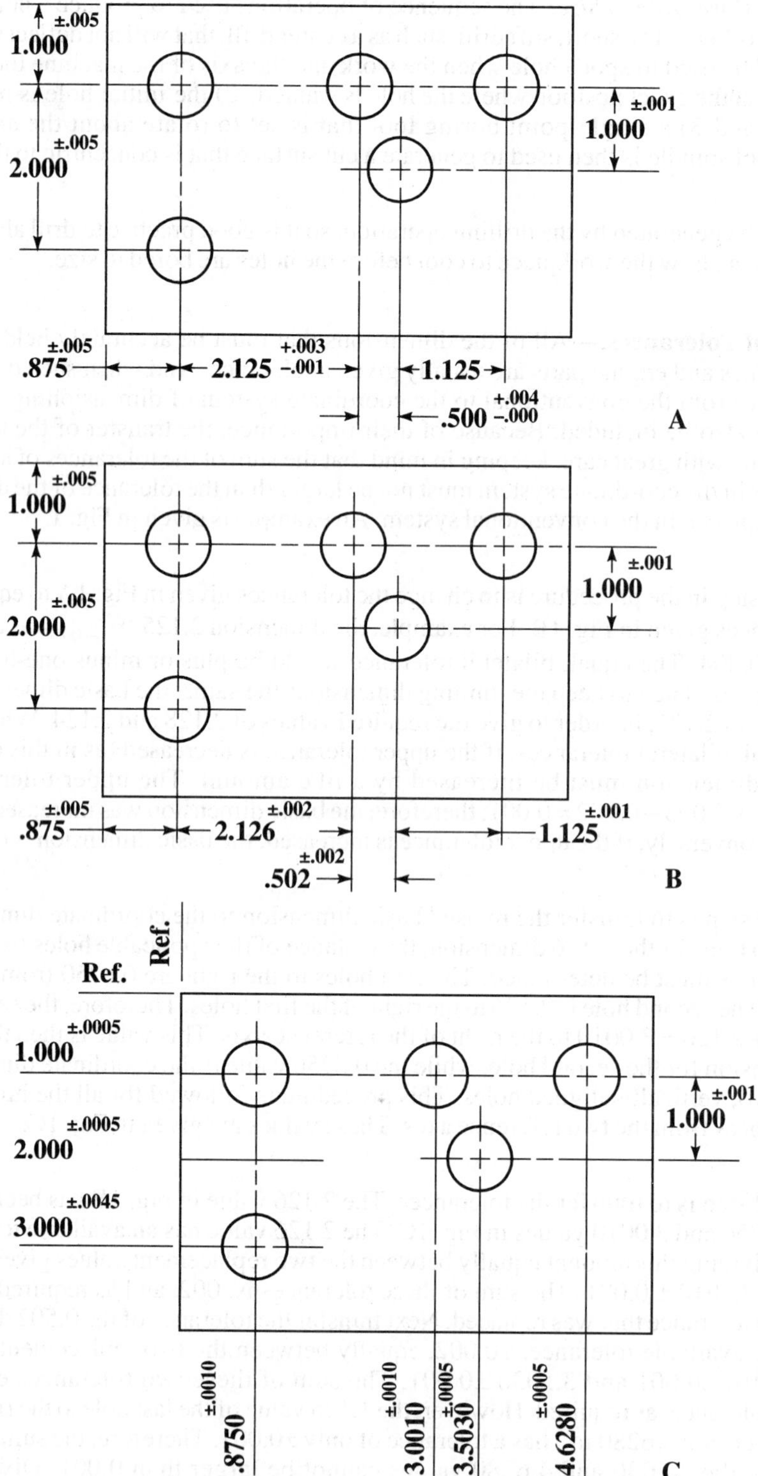

Fig. 1. (A) Conventional Dimensions, Mixed Tolerances; (B) Conventional Dimensions, All Equal, Bilateral Tolerances; and (C) Coordinate Dimensions

The following discussion will summarize the various tolerances listed in Fig. 1C. For the 0.8750 ± 0.0010 dimension, the ± 0.0010 tolerance together with the ± 0.0010 tolerance on the 3.0010 dimension is required to maintain the ± 0.002 tolerance of the 2.126 dimension. The ± .0005 tolerances on the 3.5030 and 4.2680 dimensions are required to maintain the ± 0.001 tolerance of the 1.125 dimension, at the same time as the sum of the ± .0005 tolerance on the 3.5030 dimension and the ± 0.001 tolerance on the 3.0010 dimension does not exceed the ± 0.002 tolerance on the replaced 0.503 dimension. The ± 0.0005 tolerances on the 1.0000 and 2.0000 values maintain the ± 0.001 tolerance on the 1.0000 value given at the right in Fig. 1A. The ± 0.0045 tolerance on the 3.0000 dimension together with the ± 0.0005 tolerance on the 1.0000 value maintains the ± .005 tolerance on the 2.0000 dimension of Fig. 1A. It should be noted that the 2.000 ± .005 dimension in Fig. 1A was replaced by the 1.0000 and 3.0000 dimensions in Fig. 1C. Each of these values could have had a tolerance of ± 0.0025, except that the tolerance on the 1.0000 dimension on the left in Fig. 1A is also bound by the ± 0.001 tolerance on the 1.0000 dimension on the right, thus the ± 0.0005 tolerance value is used. This procedure requires the tolerance on the 3.0000 value to be increased to ± 0.0045.

Lengths of Chords for Spacing Off the Circumferences of Circles

On the following pages are given tables of the lengths of chords for spacing off the circumferences of circles. The object of these tables is to make possible the division of the periphery into a number of equal parts without trials with the dividers. The first table is calculated for circles having a diameter equal to 1. For circles of other diameters, the length of chord given in the table should be multiplied by the diameter of the circle. This first table may be used by toolmakers when setting "buttons" in circular formation. Assume that it is required to divide the periphery of a circle of 20 inches diameter into thirty-two equal parts. From the table the length of the chord is found to be 0.098017 inch, if the diameter of the circle were 1 inch. With a diameter of 20 inches the length of the chord for one division would be $20 \times 0.098017 = 1.9603$ inches. Another example in metric units: For a 100 millimeter diameter requiring 5 equal divisions, the length of the chord for one division would be $100 \times 0.587785 = 58.7785$ millimeters.

The two following pages give an additional table for the spacing off of circles, the table, in this case, being worked out for diameters from $\frac{1}{16}$ inch to 14 inches. As an example, assume that it is required to divide a circle having a diameter of $6\frac{1}{2}$ inches into seven equal parts. Find first, in the column headed "6" and in line with 7 divisions, the length of the chord for a 6-inch circle, which is 2.603 inches. Then find the length of the chord for a $\frac{1}{2}$-inch diameter circle, 7 divisions, which is 0.217. The sum of these two values, $2.603 + 0.217 = 2.820$ inches, is the length of the chord required for spacing off the circumference of a $6\frac{1}{2}$-inch circle into seven equal divisions.

As another example, assume that it is required to divide a circle having a diameter of $9\frac{23}{32}$ inches into 15 equal divisions. First find the length of the chord for a 9-inch circle, which is 1.871 inch. The length of the chord for a $\frac{23}{32}$-inch circle can easily be estimated from the table by taking the value that is exactly between those given for $\frac{11}{16}$ and $\frac{3}{4}$ inch. The value for $\frac{11}{16}$ inch is 0.143, and for $\frac{3}{4}$ inch, 0.156. For $\frac{23}{32}$ the value would be 0.150. Then, $1.871 + 0.150 = 2.021$ inches.

Lengths of Chords for Spacing Off the Circumferences of Circles with a Diameter Equal to 1 (English or metric units)

No. of Spaces	Length of Chord	No. of Spaces	Length of Chord	No. of Spaces	Length of Chord	No. of Spaces	Length of Chord
3	0.866025	22	0.142315	41	0.076549	60	0.052336
4	0.707107	23	0.136167	42	0.074730	61	0.051479
5	0.587785	24	0.130526	43	0.072995	62	0.050649
6	0.500000	25	0.125333	44	0.071339	63	0.049846
7	0.433884	26	0.120537	45	0.069756	64	0.049068
8	0.382683	27	0.116093	46	0.068242	65	0.048313
9	0.342020	28	0.111964	47	0.066793	66	0.047582
10	0.309017	29	0.108119	48	0.065403	67	0.046872
11	0.281733	30	0.104528	49	0.064070	68	0.046183
12	0.258819	31	0.101168	50	0.062791	69	0.045515
13	0.239316	32	0.098017	51	0.061561	70	0.044865
14	0.222521	33	0.095056	52	0.060378	71	0.044233
15	0.207912	34	0.092268	53	0.059241	72	0.043619
16	0.195090	35	0.089639	54	0.058145	73	0.043022
17	0.183750	36	0.087156	55	0.057089	74	0.042441
18	0.173648	37	0.084806	56	0.056070	75	0.041876
19	0.164595	38	0.082579	57	0.055088	76	0.041325
20	0.156434	39	0.080467	58	0.054139	77	0.040789
21	0.149042	40	0.078459	59	0.053222	78	0.040266

For circles of other diameters, multiply length given in table by diameter of circle.

Hole Coordinate Dimension Factors for Jig Boring.—Tables of hole coordinate dimension factors for use in jig boring are given in Tables 1 through 4 starting on page 959. The coordinate axes shown in the figure accompanying each table are used to reference the tool path; the values listed in each table are for the end points of the tool path. In this machine coordinate system, a positive *Y* value indicates that the effective motion of the tool with reference to the work is toward the front of the jig borer (the actual motion of the jig borer table is toward the column). Similarly, a positive *X* value indicates that the effective motion of the tool with respect to the work is toward the right (the actual motion of the jig borer table is toward the left). When entering data into most computer-controlled jig borers, current practice is to use the more familiar Cartesian coordinate axis system in which the positive *Y* direction is "up" (i.e., pointing toward the column of the jig borer). The computer will automatically change the signs of the entered *Y* values to the signs that they would have in the machine coordinate system. Therefore, before applying the coordinate dimension factors given in the tables, it is important to determine the coordinate system to be used. If a Cartesian coordinate system is to be used for the tool path, then the sign of the *Y* values in the tables must be changed, from positive to negative and from negative to positive. For example, when programming for a three-hole type *A* circle using Cartesian coordinates, the *Y* values from Table 3 would be $y1 = +0.50000$, $y2 = -0.25000$, and $y3 = -0.25000$.

Table 10. Table for Spacing Off the Circumferences of Circles

No. of Divisions	Degrees in Arc	Diameter of Circle to be Spaced Off														
		1/16	1/8	3/16	1/4	5/16	3/8	7/16	1/2	9/16	5/8	11/16	3/4	13/16	7/8	15/16
		Length of Chord														
3	120	0.054	0.108	0.162	0.217	0.271	0.325	0.379	0.433	0.487	0.541	0.595	0.650	0.704	0.758	0.812
4	90	0.044	0.088	0.133	0.177	0.221	0.265	0.309	0.354	0.398	0.442	0.486	0.530	0.575	0.619	0.663
5	72	0.037	0.073	0.110	0.147	0.184	0.220	0.257	0.294	0.331	0.367	0.404	0.441	0.478	0.514	0.551
6	60	0.031	0.063	0.094	0.125	0.156	0.188	0.219	0.250	0.281	0.313	0.344	0.375	0.406	0.438	0.469
7	51 3/7	0.027	0.054	0.081	0.108	0.136	0.163	0.190	0.217	0.244	0.271	0.298	0.325	0.353	0.380	0.407
8	45	0.024	0.048	0.072	0.096	0.120	0.144	0.167	0.191	0.215	0.239	0.263	0.287	0.311	0.335	0.359
9	40	0.021	0.043	0.064	0.086	0.107	0.128	0.150	0.171	0.192	0.214	0.235	0.257	0.278	0.299	0.321
10	36	0.019	0.039	0.058	0.077	0.097	0.116	0.135	0.155	0.174	0.193	0.212	0.232	0.251	0.270	0.290
11	32 8/11	0.018	0.035	0.053	0.070	0.088	0.106	0.123	0.141	0.158	0.176	0.194	0.211	0.229	0.247	0.264
12	30	0.016	0.032	0.049	0.065	0.081	0.097	0.113	0.129	0.146	0.162	0.178	0.194	0.210	0.226	0.243
13	27 9/13	0.015	0.030	0.045	0.060	0.075	0.090	0.105	0.120	0.135	0.150	0.165	0.179	0.194	0.209	0.224
14	25 5/7	0.014	0.028	0.042	0.056	0.069	0.083	0.097	0.111	0.125	0.139	0.153	0.167	0.181	0.195	0.209
15	24	0.013	0.026	0.039	0.052	0.065	0.078	0.091	0.104	0.117	0.130	0.143	0.156	0.169	0.182	0.195
16	22 1/2	0.012	0.024	0.037	0.049	0.061	0.073	0.085	0.098	0.110	0.122	0.134	0.146	0.159	0.171	0.183
17	21 3/17	0.011	0.023	0.034	0.046	0.057	0.069	0.080	0.092	0.103	0.115	0.126	0.138	0.149	0.161	0.172
18	20	0.011	0.022	0.033	0.043	0.054	0.065	0.076	0.087	0.098	0.109	0.119	0.130	0.141	0.152	0.163
19	18 18/19	0.010	0.021	0.031	0.041	0.051	0.062	0.072	0.082	0.093	0.103	0.113	0.123	0.134	0.144	0.154
20	18	0.010	0.020	0.029	0.039	0.049	0.059	0.068	0.078	0.088	0.098	0.108	0.117	0.127	0.137	0.147
21	17 1/7	0.009	0.019	0.028	0.037	0.047	0.056	0.065	0.075	0.084	0.093	0.102	0.112	0.121	0.130	0.140
22	16 4/11	0.009	0.018	0.027	0.036	0.044	0.053	0.062	0.071	0.080	0.089	0.098	0.107	0.116	0.125	0.133
23	15 15/23	0.009	0.017	0.026	0.034	0.043	0.051	0.060	0.068	0.077	0.085	0.094	0.102	0.111	0.119	0.128
24	15	0.008	0.016	0.024	0.033	0.041	0.049	0.057	0.065	0.073	0.082	0.090	0.098	0.106	0.114	0.122
25	14 2/5	0.008	0.016	0.023	0.031	0.039	0.047	0.055	0.063	0.070	0.078	0.086	0.094	0.102	0.110	0.117
26	13 11/13	0.008	0.015	0.023	0.030	0.038	0.045	0.053	0.060	0.068	0.075	0.083	0.090	0.098	0.105	0.113
28	12 6/7	0.007	0.014	0.021	0.028	0.035	0.042	0.049	0.056	0.063	0.070	0.077	0.084	0.091	0.098	0.105
30	12	0.007	0.013	0.020	0.026	0.033	0.039	0.046	0.052	0.059	0.065	0.072	0.078	0.085	0.091	0.098
32	11 1/4	0.006	0.012	0.018	0.025	0.031	0.037	0.043	0.049	0.055	0.061	0.067	0.074	0.080	0.086	0.092

See *Lengths of Chords for Spacing Off the Circumferences of Circles* on page 955 for explanatory matter.

Table for Spacing Off the Circumferences of Circles

No. of Divisions	Degrees in Arc	Diameter of Circle to be Spaced Off													
		1	2	3	4	5	6	7	8	9	10	11	12	13	14
		Length of Chord													
3	120	0.866	1.732	2.598	3.464	4.330	5.196	6.062	6.928	7.794	8.660	9.526	10.392	11.258	12.124
4	90	0.707	1.414	2.121	2.828	3.536	4.243	4.950	5.657	6.364	7.071	7.778	8.485	9.192	9.899
5	72	0.588	1.176	1.763	2.351	2.939	3.527	4.114	4.702	5.290	5.878	6.466	7.053	7.641	8.229
6	60	0.500	1.000	1.500	2.000	2.500	3.000	3.500	4.000	4.500	5.000	5.500	6.000	6.500	7.000
7	$51\frac{3}{7}$	0.434	0.868	1.302	1.736	2.169	2.603	3.037	3.471	3.905	4.339	4.773	5.207	5.640	6.074
8	45	0.383	0.765	1.148	1.531	1.913	2.296	2.679	3.061	3.444	3.827	4.210	4.592	4.975	5.358
9	40	0.342	0.684	1.026	1.368	1.710	2.052	2.394	2.736	3.078	3.420	3.762	4.104	4.446	4.788
10	36	0.309	0.618	0.927	1.236	1.545	1.854	2.163	2.472	2.781	3.090	3.399	3.708	4.017	4.326
11	$32\frac{8}{11}$	0.282	0.563	0.845	1.127	1.409	1.690	1.972	2.254	2.536	2.817	3.099	3.381	3.663	3.944
12	30	0.259	0.518	0.776	1.035	1.294	1.553	1.812	2.071	2.329	2.588	2.847	3.106	3.365	3.623
13	$27\frac{9}{13}$	0.239	0.479	0.718	0.957	1.197	1.436	1.675	1.915	2.154	2.393	2.632	2.872	3.111	3.350
14	$25\frac{5}{7}$	0.223	0.445	0.668	0.890	1.113	1.335	1.558	1.780	2.003	2.225	2.448	2.670	2.893	3.115
15	24	0.208	0.416	0.624	0.832	1.040	1.247	1.455	1.663	1.871	2.079	2.287	2.495	2.703	2.911
16	$22\frac{1}{2}$	0.195	0.390	0.585	0.780	0.975	1.171	1.366	1.561	1.756	1.951	2.146	2.341	2.536	2.731
17	$21\frac{3}{17}$	0.184	0.367	0.551	0.735	0.919	1.102	1.286	1.470	1.654	1.837	2.021	2.205	2.389	2.572
18	20	0.174	0.347	0.521	0.695	0.868	1.042	1.216	1.389	1.563	1.736	1.910	2.084	2.257	2.431
19	$18\frac{18}{19}$	0.165	0.329	0.494	0.658	0.823	0.988	1.152	1.317	1.481	1.646	1.811	1.975	2.140	2.304
20	18	0.156	0.313	0.469	0.626	0.782	0.939	1.095	1.251	1.408	1.564	1.721	1.877	2.034	2.190
21	$17\frac{1}{7}$	0.149	0.298	0.447	0.596	0.745	0.894	1.043	1.192	1.341	1.490	1.639	1.789	1.938	2.087
22	$16\frac{4}{11}$	0.142	0.285	0.427	0.569	0.712	0.854	0.996	1.139	1.281	1.423	1.565	1.708	1.850	1.992
23	$15\frac{15}{23}$	0.136	0.272	0.408	0.545	0.681	0.817	0.953	1.089	1.225	1.362	1.498	1.634	1.770	1.906
24	15	0.131	0.261	0.392	0.522	0.653	0.783	0.914	1.044	1.175	1.305	1.436	1.566	1.697	1.827
25	$14\frac{2}{5}$	0.125	0.251	0.376	0.501	0.627	0.752	0.877	1.003	1.128	1.253	1.379	1.504	1.629	1.755
26	$13\frac{11}{13}$	0.121	0.241	0.362	0.482	0.603	0.723	0.844	0.964	1.085	1.205	1.326	1.446	1.567	1.688
28	$12\frac{6}{7}$	0.112	0.224	0.336	0.448	0.560	0.672	0.784	0.896	1.008	1.120	1.232	1.344	1.456	1.568
30	12	0.105	0.209	0.314	0.418	0.523	0.627	0.732	0.836	0.941	1.045	1.150	1.254	1.359	1.463
32	$11\frac{1}{4}$	0.098	0.196	0.294	0.392	0.490	0.588	0.686	0.784	0.882	0.980	1.078	1.176	1.274	1.372

Table 1. Hole Coordinate Dimension Factors for Jig Boring — Type "A" Hole Circles (English or Metric Units)

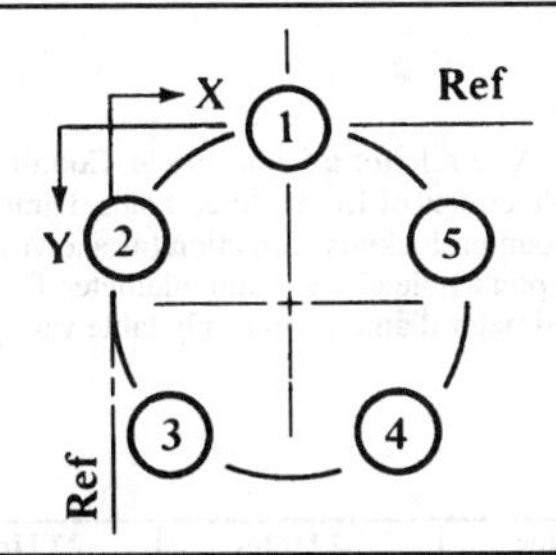

The diagram shows a type "A" circle for a 5-hole circle. Coordinates *x*, *y* are given in the table for hole circles of from 3 to 28 holes. Dimensions are for holes numbered in a counterclockwise direction (as shown). Dimensions given are based upon a hole circle of unit diameter. For a hole circle of, say, 3-inch or 3-centimeter diameter, multiply table values by 3.

3 Holes		4 Holes		5 Holes		6 Holes		7 Holes		8 Holes		9 Holes	
x1	0.50000	x1	0.50000	x1	0.50000	x1	0.50000	x1	0.50000	x1	0.50000	x1	0.50000
y1	0.00000	y1	0.00000	y1	0.00000	y1	0.00000	y1	0.00000	y1	0.00000	y1	0.00000
x2	0.06699	x2	0.00000	x2	0.02447	x2	0.06699	x2	0.10908	x2	0.14645	x2	0.17861
y2	0.75000	y2	0.50000	y2	0.34549	y2	0.25000	y2	0.18826	y2	0.14645	y2	0.11698
x3	0.93301	x3	0.50000	x3	0.20611	x3	0.06699	x3	0.01254	x3	0.00000	x3	0.00760
y3	0.75000	y3	1.00000	y3	0.90451	y3	0.75000	y3	0.61126	y3	0.50000	y3	0.41318
		x4	1.00000	x4	0.79389	x4	0.50000	x4	0.28306	x4	0.14645	x4	0.06699
		y4	0.50000	y4	0.90451	y4	1.00000	y4	0.95048	y4	0.85355	y4	0.75000
				x5	0.97553	x5	0.93301	x5	0.71694	x5	0.50000	x5	0.32899
				y5	0.34549	y5	0.75000	y5	0.95048	y5	1.00000	y5	0.96985
						x6	0.93301	x6	0.98746	x6	0.85355	x6	0.67101
						y6	0.25000	y6	0.61126	y6	0.85355	y6	0.96985
								x7	0.89092	x7	1.00000	x7	0.93301
								y7	0.18826	y7	0.50000	y7	0.75000
										x8	0.85355	x8	0.99240
										y8	0.14645	y8	0.41318
												x9	0.82139
												y9	0.11698

10 Holes		11 Holes		12 Holes		13 Holes		14 Holes		15 Holes		16 Holes	
x1	0.50000	x1	0.50000	x1	0.50000	x1	0.50000	x1	0.50000	x1	0.50000	x1	0.50000
y1	0.00000	y1	0.00000	y1	0.00000	y1	0.00000	y1	0.00000	y1	0.00000	y1	0.00000
x2	0.20611	x2	0.22968	x2	0.25000	x2	0.26764	x2	0.28306	x2	0.29663	x2	0.30866
y2	0.09549	y2	0.07937	y2	0.06699	y2	0.05727	y2	0.04952	y2	0.04323	y2	0.03806
x3	0.02447	x3	0.04518	x3	0.06699	x3	0.08851	x3	0.10908	x3	0.12843	x3	0.14645
y3	0.34549	y3	0.29229	y3	0.25000	y3	0.21597	y3	0.18826	y3	0.16543	y3	0.14645
x4	0.02447	x4	0.00509	x4	0.00000	x4	0.00365	x4	0.01254	x4	0.02447	x4	0.03806
y4	0.65451	y4	0.57116	y4	0.50000	y4	0.43973	y4	0.38874	y4	0.34549	y4	0.30866
x5	0.20611	x5	0.12213	x5	0.06699	x5	0.03249	x5	0.01254	x5	0.00274	x5	0.00000
y5	0.90451	y5	0.82743	y5	0.75000	y5	0.67730	y5	0.61126	y5	0.55226	y5	0.50000
x6	0.50000	x6	0.35913	x6	0.25000	x6	0.16844	x6	0.10908	x6	0.06699	x6	0.03806
y6	1.00000	y6	0.97975	y6	0.93301	y6	0.87426	y6	0.81174	y6	0.75000	y6	0.69134
x7	0.79389	x7	0.64087	x7	0.50000	x7	0.38034	x7	0.28306	x7	0.20611	x7	0.14645
y7	0.90451	y7	0.97975	y7	1.00000	y7	0.98547	y7	0.95048	y7	0.90451	y7	0.85355
x8	0.97553	x8	0.87787	x8	0.75000	x8	0.61966	x8	0.50000	x8	0.39604	x8	0.30866
y8	0.65451	y8	0.82743	y8	0.93301	y8	0.98547	y8	1.00000	y8	0.98907	y8	0.96194
x9	0.97553	x9	0.99491	x9	0.93301	x9	0.83156	x9	0.71694	x9	0.60396	x9	0.50000
y9	0.34549	y9	0.57116	y9	0.75000	y9	0.87426	y9	0.95048	y9	0.98907	y9	1.00000
x10	0.79389	x10	0.95482	x10	1.00000	x10	0.96751	x10	0.89092	x10	0.79389	x10	0.69134
y10	0.09549	y10	0.29229	y10	0.50000	y10	0.67730	y10	0.81174	y10	0.90451	y10	0.96194
		x11	0.77032	x11	0.93801	x11	0.99635	x11	0.98746	x11	0.93301	x11	0.85355
		y11	0.07937	y11	0.25000	y11	0.43973	y11	0.61126	y11	0.75000	y11	0.85355
				x12	0.75000	x12	0.91149	x12	0.98746	x12	0.99726	x12	0.96194
				y12	0.06699	y12	0.21597	y12	0.38874	y12	0.55226	y12	0.69134
						x13	0.73236	x13	0.89092	x13	0.97553	x13	1.00000
						y13	0.05727	y13	0.18826	y13	0.34549	y13	0.50000
								x14	0.71694	x14	0.87157	x14	0.96194
								y14	0.04952	y14	0.16543	y14	0.30866
										x15	0.70337	x15	0.85355
										y15	0.04323	y15	0.14645
												x16	0.69134
												y16	0.03806

Table 1. *(Continued)* **Hole Coordinate Dimension Factors for Jig Boring — Type "A" Hole Circles** (English or Metric Units)

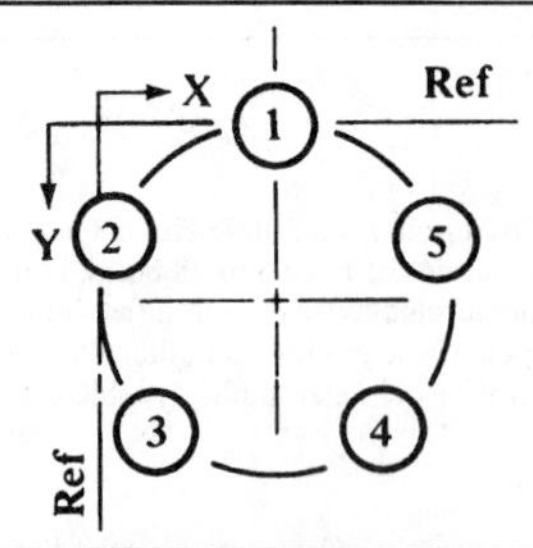

The diagram shows a type "A" circle for a 5-hole circle. Coordinates *x*, *y* are given in the table for hole circles of from 3 to 28 holes. Dimensions are for holes numbered in a counterclockwise direction (as shown). Dimensions given are based upon a hole circle of unit diameter. For a hole circle of, say, 3-inch or 3-centimeter diameter, multiply table values by 3.

17 Holes		18 Holes		19 Holes		20 Holes		21 Holes		22 Holes		23 Holes	
x1	0.50000	x1	0.50000	x1	0.50000	x1	0.50000	x1	0.50000	x1	0.50000	x1	0.50000
y1	0.00000	y1	0.00000	y1	0.00000	y1	0.00000	y1	0.00000	y1	0.00000	y1	0.00000
x2	0.31938	x2	0.32899	x2	0.33765	x2	0.34549	x2	0.35262	x2	0.35913	x2	0.36510
y2	0.03376	y2	0.03015	y2	0.02709	y2	0.02447	y2	0.02221	y2	0.02025	y2	0.01854
x3	0.16315	x3	0.17861	x3	0.19289	x3	0.20611	x3	0.21834	x3	0.22968	x3	0.24021
y3	0.13050	y3	0.11698	y3	0.10543	y3	0.09549	y3	0.08688	y3	0.07937	y3	0.07279
x4	0.05242	x4	0.06699	x4	0.08142	x4	0.09549	x4	0.10908	x4	0.12213	x4	0.13458
y4	0.27713	y4	0.25000	y4	0.22653	y4	0.20611	y4	0.18826	y4	0.17257	y4	0.15872
x5	0.00213	x5	0.00760	x5	0.01530	x5	0.02447	x5	0.03456	x5	0.04518	x5	0.05606
y5	0.45387	y5	0.41318	y5	0.37726	y5	0.34549	y5	0.31733	y5	0.29229	y5	0.26997
x6	0.01909	x6	0.00760	x6	0.00171	x6	0.00000	x6	0.00140	x6	0.00509	x6	0.01046
y6	0.63683	y6	0.58682	y6	0.54129	y6	0.50000	y6	0.46263	y6	0.42884	y6	0.39827
x7	0.10099	x7	0.06699	x7	0.04211	x7	0.02447	x7	0.01254	x7	0.00509	x7	0.00117
y7	0.80132	y7	0.75000	y7	0.70085	y7	0.65451	y7	0.61126	y7	0.57116	y7	0.53412
x8	0.23678	x8	0.17861	x8	0.13214	x8	0.09549	x8	0.06699	x8	0.04518	x8	0.02887
y8	0.92511	y8	0.88302	y8	0.83864	y8	0.79389	y8	0.75000	y8	0.70771	y8	0.66744
x9	0.40813	x9	0.32899	x9	0.26203	x9	0.20611	x9	0.15991	x9	0.12213	x9	0.09152
y9	0.99149	y9	0.96985	y9	0.93974	y9	0.90451	y9	0.86653	y9	0.82743	y9	0.78834
x10	0.59187	x10	0.50000	x10	0.41770	x10	0.34549	x10	0.28306	x10	0.22968	x10	0.18446
y10	0.99149	y10	1.00000	y10	0.99318	y10	0.97553	y10	0.95048	y10	0.92063	y10	0.88786
x11	0.76322	x11	0.67101	x11	0.58230	x11	0.50000	x11	0.42548	x11	0.35913	x11	0.30080
y11	0.92511	y11	0.96985	y11	0.99318	y11	1.00000	y11	0.99442	y11	0.97975	y11	0.95861
x12	0.89901	x12	0.82139	x12	0.73797	x12	0.65451	x12	0.57452	x12	0.50000	x12	0.43192
y12	0.80132	y12	0.88302	y12	0.93974	y12	0.97553	y12	0.99442	y12	1.00000	y12	0.99534
x13	0.98091	x13	0.93301	x13	0.86786	x13	0.79389	x13	0.71694	x13	0.64087	x13	0.56808
y13	0.63683	y13	0.75000	y13	0.83864	y13	0.90451	y13	0.95048	y13	0.97975	y13	0.99534
x14	0.99787	x14	0.99240	x14	0.95789	x14	0.90451	x14	0.84009	x14	0.77032	x14	0.69920
y14	0.45387	y14	0.58682	y14	0.70085	y14	0.79389	y14	0.86653	y14	0.92063	y14	0.95861
x15	0.94758	x15	0.99240	x15	0.99829	x15	0.97553	x15	0.93301	x15	0.87787	x15	0.81554
y15	0.27713	y15	0.41318	y15	0.54129	y15	0.65451	y15	0.75000	y15	0.82743	y15	0.88786
x16	0.83685	x16	0.93301	x16	0.98470	x16	1.00000	x16	0.98746	x16	0.95482	x16	0.90848
y16	0.13050	y16	0.25000	y16	0.37726	y16	0.50000	y16	0.61126	y16	0.70771	y16	0.78834
x17	0.68062	x17	0.82139	x17	0.91858	x17	0.97553	x17	0.99860	x17	0.99491	x17	0.97113
y17	0.03376	y17	0.11698	y17	0.22658	y17	0.34549	y17	0.46263	y17	0.57116	y17	0.66744
		x18	0.67101	x18	0.80711	x18	0.90451	x18	0.96544	x18	0.99491	x18	0.99883
		y18	0.03015	y18	0.10543	y18	0.20611	y18	0.31733	y18	0.42884	y18	0.53412
				x19	0.66235	x19	0.79389	x19	0.89092	x19	0.95482	x19	0.98954
				y19	0.02709	y19	0.09549	y19	0.18826	y19	0.29229	y19	0.39827
						x20	0.65451	x20	0.78166	x20	0.87787	x20	0.94394
						y20	0.02447	y20	0.08688	y20	0.17257	y20	0.26997
								x21	0.64738	x21	0.77032	x21	0.86542
								y21	0.02221	y21	0.07937	y21	0.15872
										x22	0.64087	x22	0.75979
										y22	0.02025	y22	0.07279
												x23	0.63490
												y23	0.01854

24Holes		25 Holes		26 Holes		27 Holes		28 Holes	
x1	0.50000	x1	0.50000	x1	0.50000	x1	0.50000	x1	0.50000
y1	0.00000	y1	0.00000	y1	0.00000	y1	0.00000	y1	0.00000
x2	0.37059	x2	0.37566	x2	0.38034	x2	0.38469	x2	0.38874
y2	0.01704	y2	0.01571	y2	0.01453	y2	0.01348	y2	0.01254
x3	0.25000	x3	0.25912	x3	0.26764	x3	0.27560	x3	0.28306

Table 1. *(Continued)* Hole Coordinate Dimension Factors for Jig Boring — Type "A" Hole Circles (English or Metric Units)

The diagram shows a type "A" circle for a 5-hole circle. Coordinates *x*, *y* are given in the table for hole circles of from 3 to 28 holes. Dimensions are for holes numbered in a counterclockwise direction (as shown). Dimensions given are based upon a hole circle of unit diameter. For a hole circle of, say, 3-inch or 3-centimeter diameter, multiply table values by 3.

y3	0.06699	y3	0.06185	y3	0.05727	y3	0.05318	y3	0.04952
x4	0.14645	x4	0.15773	x4	0.16844	x4	0.17861	x4	0.18826
y4	0.14645	y4	0.13552	y4	0.12574	y4	0.11698	y4	0.10908
x5	0.06699	x5	0.07784	x5	0.08851	x5	0.09894	x5	0.10908
y5	0.25000	y5	0.23209	y5	0.21597	y5	0.20142	y5	0.18826
x6	0.01704	x6	0.02447	x6	0.03249	x6	0.04089	x6	0.04952
y6	0.37059	y6	0.34549	y6	0.32270	y6	0.30196	y6	0.28306
x7	0.00000	x7	0.00099	x7	0.00365	x7	0.00760	x7	0.01254
y7	0.50000	y7	0.46860	y7	0.43973	y7	0.41318	y7	0.38874
x8	0.01704	x8	0.00886	x8	0.00365	x8	0.00085	x8	0.00000
y8	0.62941	y8	0.59369	y8	0.56027	y8	0.52907	y8	0.50000
x9	0.06699	x9	0.04759	x9	0.03249	x9	0.02101	x9	0.01254
y9	0.75000	y9	0.71289	y9	0.67730	y9	0.64340	y9	0.61126
x10	0.14645	x10	0.11474	x10	0.08851	x10	0.06699	x10	0.04952
y10	0.85355	y10	0.81871	y10	0.78403	y10	0.75000	y10	0.71694
x11	0.25000	x11	0.20611	x11	0.16844	x11	0.13631	x11	0.10908
y11	0.93301	y11	0.90451	y11	0.87426	y11	0.84312	y11	0.81174
x12	0.37059	x12	0.31594	x12	0.26764	x12	0.22525	x12	0.18826
y12	0.98296	y12	0.96489	y12	0.94273	y12	0.91774	y12	0.89092
x13	0.50000	x13	0.43733	x13	0.38034	x13	0.32899	x13	0.28306
y13	1.00000	y13	0.99606	y13	0.98547	y13	0.96985	y13	0.95048
x14	0.62941	x14	0.56267	x14	0.50000	x14	0.44195	x14	0.38874
y14	0.98296	y14	0.99606	y14	1.00000	y14	0.99662	y14	0.98746
x15	0.75000	x15	0.68406	x15	0.61966	x15	0.55805	x15	0.50000
y15	0.93301	y15	0.96489	y15	0.98547	y15	0.99662	y15	1.00000
x16	0.85355	x16	0.79389	x16	0.73236	x16	0.67101	x16	0.61126
y16	0.85355	y16	0.90451	y16	0.94273	y16	0.96985	y16	0.98746
x17	0.93301	x17	0.88526	x17	0.83156	x17	0.77475	x17	0.71694
y17	0.75000	y17	0.81871	y17	0.87426	y17	0.91774	y17	0.95048
x18	0.98296	x18	0.95241	x18	0.91149	x18	0.86369	x18	0.81174
y18	0.62941	y18	0.71289	y18	0.78403	y18	0.84312	y18	0.89092
x19	1.00000	x19	0.99114	x19	0.96751	x19	0.93301	x19	0.89092
y19	0.50000	y19	0.59369	y19	0.67730	y19	0.75000	y19	0.81174
x20	0.98296	x20	0.99901	x20	0.99635	x20	0.97899	x20	0.95048
y20	0.37059	y20	0.46860	y20	0.56027	y20	0.64340	y20	0.71694
x21	0.93301	x21	0.97553	x21	0.99635	x21	0.99915	x21	0.98746
y21	0.25000	y21	0.34549	y21	0.43973	y21	0.52907	y21	0.61126
x22	0.85355	x22	0.92216	x22	0.96751	x22	0.99240	x22	1.00000
y22	0.14645	y22	0.23209	y22	0.32270	y22	0.41318	y22	0.50000
x23	0.75000	x23	0.84227	x23	0.91149	x23	0.95911	x23	0.98746
y23	0.6699	y23	0.13552	y23	0.21597	y23	0.30196	y23	0.38874
x24	0.62941	x24	0.74088	x24	0.83156	x24	0.90106	x24	0.95048
y24	0.01704	y24	0.06185	y24	0.12574	y24	0.20142	y24	0.28306
		x25	0.62434	x25	0.73236	x25	0.82139	x25	0.89092
		y25	0.01571	y25	0.05727	y25	0.11698	y25	0.18826
				x26	0.61966	x26	0.72440	x26	0.81174
				y26	0.01453	y26	0.05318	y26	0.10908
						x27	0.61531	x27	0.71694
						y27	0.01348	y27	0.04952
								x28	0.61126
								y28	0.01254

Table 2. Hole Coordinate Dimension Factors for Jig Boring — Type "B" Hole Circles (English or Metric Units)

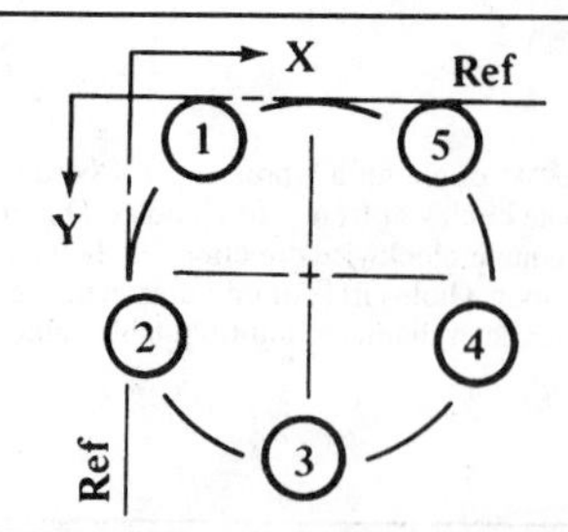

The diagram shows a type "B" circle for a 5-hole circle. Coordinates *x*, *y* are given in the table for hole circles of from 3 to 28 holes. Dimensions are for holes numbered in a counterclockwise direction (as shown). Dimensions given are based upon a hole circle of unit diameter. For a hole circle of, say, 3-inch or 3-centimeter diameter, multiply table values by 3.

3 Holes		4 Holes		5 Holes		6 Holes		7 Holes		8 Holes		9 Holes	
x1	0.06699	x1	0.14645	x1	0.20611	x1	0.25000	x1	0.28306	x1	0.30866	x1	0.32899
y1	0.25000	y1	0.14645	y1	0.09549	y1	0.06699	y1	0.04952	y1	0.03806	y1	0.03015
x2	0.50000	x2	0.14645	x2	0.02447	x2	0.00000	x2	0.01254	x2	0.03806	x2	0.06699
y2	1.00000	y2	0.85355	y2	0.65451	y2	0.50000	y2	0.38874	y2	0.30866	y2	0.25000
x3	0.93301	x3	0.85355	x3	0.50000	x3	0.25000	x3	0.10908	x3	0.03806	x3	0.00760
y3	0.25000	y3	0.85355	y3	1.00000	y3	0.93301	y3	0.81174	y3	0.69134	y3	0.58682
		x4	0.85355	x4	0.97553	x4	0.75000	x4	0.50000	x4	0.30866	x4	0.17861
		y4	0.14645	y4	0.65451	y4	0.93301	y4	1.00000	y4	0.96194	y4	0.88302
				x5	0.79389	x5	1.00000	x5	0.89092	x5	0.69134	x5	0.50000
				y5	0.09549	y5	0.50000	y5	0.81174	y5	0.96194	y5	1.00000
						x6	0.75000	x6	0.98746	x6	0.96194	x6	0.82139
						y6	0.06699	y6	0.38874	y6	0.69134	y6	0.88302
								x7	0.71694	x7	0.96194	x7	0.99240
								y7	0.04952	y7	0.30866	y7	0.58682
										x8	0.69134	x8	0.93301
										y8	0.03806	y8	0.25000
												x9	0.67101
												y9	0.03015

10 Holes		11 Holes		12 Holes		13 Holes		14 Holes		15 Holes		16 Holes	
x1	0.34549	x1	0.35913	x1	0.37059	x1	0.38034	x1	0.38874	x1	0.39604	x1	0.40245
y1	0.02447	y1	0.02025	y1	0.01704	y1	0.01453	y1	0.01254	y1	0.01093	y1	0.00961
x2	0.09549	x2	0.12213	x2	0.14645	x2	0.16844	x2	0.18826	x2	0.20611	x2	0.22221
y2	0.20611	y2	0.17257	y2	0.14645	y2	0.12574	y2	0.10908	y2	0.09549	y2	0.08427
x3	0.00000	x3	0.00509	x3	0.01704	x3	0.03249	x3	0.04952	x3	0.06699	x3	0.08427
y3	0.50000	y3	0.42884	y3	0.37059	y3	0.32270	y3	0.28306	y3	0.25000	y3	0.22221
x4	0.09549	x4	0.04518	x4	0.01704	x4	0.00365	x4	0.00000	x4	0.00274	x4	0.00961
y4	0.79389	y4	0.70771	y4	0.62941	y4	0.56027	y4	0.50000	y4	0.44774	y4	0.40245
x5	0.34549	x5	0.22968	x5	0.14645	x5	0.08851	x5	0.04952	x5	0.02447	x5	0.00961
y5	0.97553	y5	0.92063	y5	0.85355	y5	0.78403	y5	0.71694	y5	0.65451	y5	0.59755
x6	0.65451	x6	0.50000	x6	0.37059	x6	0.26764	x6	0.18826	x6	0.12843	x6	0.08427
y6	0.97553	y6	1.00000	y6	0.98296	y6	0.94273	y6	0.89092	y6	0.83457	y6	0.77779
x7	0.90451	x7	0.77032	x7	0.62941	x7	0.50000	x7	0.38874	x7	0.29663	x7	0.22221
y7	0.79389	y7	0.92063	y7	0.98296	y7	1.00000	y7	0.98746	y7	0.95677	y7	0.91573
x8	1.00000	x8	0.95482	x8	0.85355	x8	0.73236	x8	0.61126	x8	0.50000	x8	0.40245
y8	0.50000	y8	0.70771	y8	0.85355	y8	0.94273	y8	0.98746	y8	1.00000	y8	0.99039
x9	0.90451	x9	0.99491	x9	0.98296	x9	0.91149	x9	0.81174	x9	0.70337	x9	0.59755
y9	0.20611	y9	0.42884	y9	0.62941	y9	0.78403	y9	0.89092	y9	0.95677	y9	0.99039
x10	0.65451	x10	0.87787	x10	0.98296	x10	0.99635	x10	0.95048	x10	0.87157	x10	0.77779
y10	0.02447	y10	0.17257	y10	0.37059	y10	0.56027	y10	0.71694	y10	0.83457	y10	0.91573
		x11	0.64087	x11	0.85355	x11	0.96751	x11	1.00000	x11	0.97553	x11	0.91573
		y11	0.02025	y11	0.14645	y11	0.32270	y11	0.50000	y11	0.65451	y11	0.77779
				x12	0.62941	x12	0.83156	x12	0.95048	x12	0.99726	x12	0.99039
				y12	0.01704	y12	0.12574	y12	0.28306	y12	0.44774	y12	0.59755
						x13	0.61966	x13	0.81174	x13	0.93301	x13	0.99039
						y13	0.01453	y13	0.10908	y13	0.25000	y13	0.40245
								x14	0.61126	x14	0.79389	x14	0.91573
								y14	0.01254	y14	0.09549	y14	0.22221
										x15	0.60396	x15	0.77779
										y15	0.01093	y15	0.08427
												x16	0.59755
												y16	0.00961

Table 2. *(Continued)* **Hole Coordinate Dimension Factors for Jig Boring — Type "B" Hole Circles** (English or Metric Units)

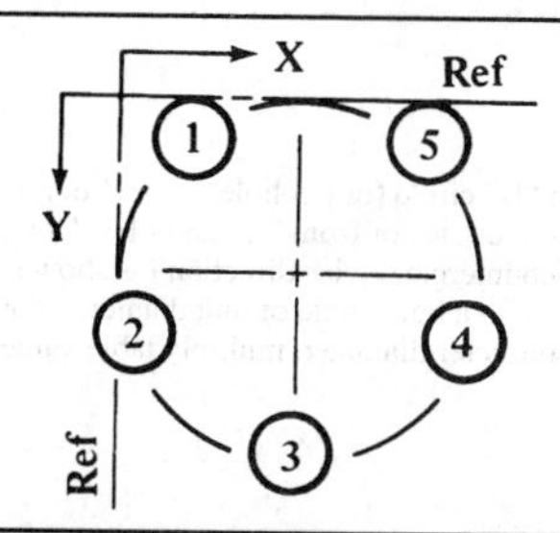

The diagram shows a type "B" circle for a 5-hole circle. Coordinates *x*, *y* are given in the table for hole circles of from 3 to 28 holes. Dimensions are for holes numbered in a counterclockwise direction (as shown). Dimensions given are based upon a hole circle of unit diameter. For a hole circle of, say, 3-inch or 3-centimeter diameter, multiply table values by 3.

17 Holes		18 Holes		19 Holes		20 Holes		21 Holes		22 Holes		23 Holes	
*x*1	0.40813	*x*1	0.41318	*x*1	0.41770	*x*1	0.42178	*x*1	0.42548	*x*1	0.42884	*x*1	0.43192
*y*1	0.00851	*y*1	0.00760	*y*1	0.00682	*y*1	0.00616	*y*1	0.00558	*y*1	0.00509	*y*1	0.00466
*x*2	0.23678	*x*2	0.25000	*x*2	0.26203	*x*2	0.27300	*x*2	0.28306	*x*2	0.29229	*x*2	0.30080
*y*2	0.07489	*y*2	0.06699	*y*2	0.06026	*y*2	0.05450	*y*2	0.04952	*y*2	0.04518	*y*2	0.04139
*x*3	0.10099	*x*3	0.11698	*x*3	0.13214	*x*3	0.14645	*x*3	0.15991	*x*3	0.17257	*x*3	0.18446
*y*3	0.19868	*y*3	0.17861	*y*3	0.16136	*y*3	0.14645	*y*3	0.13347	*y*3	0.12213	*y*3	0.11214
*x*4	0.01909	*x*4	0.03015	*x*4	0.04211	*x*4	0.05450	*x*4	0.06699	*x*4	0.07937	*x*4	0.09152
*y*4	0.36317	*y*4	0.32899	*y*4	0.29915	*y*4	0.27300	*y*4	0.25000	*y*4	0.22968	*y*4	0.21166
*x*5	0.00213	*x*5	0.00000	*x*5	0.00171	*x*5	0.00616	*x*5	0.01254	*x*5	0.02025	*x*5	0.02887
*y*5	0.54613	*y*5	0.50000	*y*5	0.45871	*y*5	0.42178	*y*5	0.38874	*y*5	0.35913	*y*5	0.33256
*x*6	0.05242	*x*6	0.03015	*x*6	0.01530	*x*6	0.00616	*x*6	0.00140	*x*6	0.00000	*x*6	0.00117
*y*6	0.72287	*y*6	0.67101	*y*6	0.62274	*y*6	0.57822	*y*6	0.53737	*y*6	0.50000	*y*6	0.46588
*x*7	0.16315	*x*7	0.11698	*x*7	0.08142	*x*7	0.05450	*x*7	0.03456	*x*7	0.02025	*x*7	0.01046
*y*7	0.86950	*y*7	0.82139	*y*7	0.77347	*y*7	0.72700	*y*7	0.68267	*y*7	0.64087	*y*7	0.60173
*x*8	0.31938	*x*8	0.25000	*x*8	0.19289	*x*8	0.14645	*x*8	0.10908	*x*8	0.07937	*x*8	0.05606
*y*8	0.96624	*y*8	0.93301	*y*8	0.89457	*y*8	0.85355	*y*8	0.81174	*y*8	0.77032	*y*8	0.73003
*x*9	0.50000	*x*9	0.41318	*x*9	0.33765	*x*9	0.27300	*x*9	0.21834	*x*9	0.17257	*x*9	0.13458
*y*9	1.00000	*y*9	0.99240	*y*9	0.97291	*y*9	0.94550	*y*9	0.91312	*y*9	0.87787	*y*9	0.84128
*x*10	0.68062	*x*10	0.58682	*x*10	0.50000	*x*10	0.42178	*x*10	0.35262	*x*10	0.29229	*x*10	0.24021
*y*10	0.96624	*y*10	0.99240	*y*10	1.00000	*y*10	0.99384	*y*10	0.97779	*y*10	0.95482	*y*10	0.92721
*x*11	0.83685	*x*11	0.75000	*x*11	0.66235	*x*11	0.57822	*x*11	0.50000	*x*11	0.42884	*x*11	0.36510
*y*11	0.86950	*y*11	0.93301	*y*11	0.97291	*y*11	0.99384	*y*11	1.00000	*y*11	0.99491	*y*11	0.98146
*x*12	0.94758	*x*12	0.88302	*x*12	0.80711	*x*12	0.72700	*x*12	0.64738	*x*12	0.57116	*x*12	0.50000
*y*12	0.72287	*y*12	0.82139	*y*12	0.89457	*y*12	0.94550	*y*12	0.97779	*y*12	0.99491	*y*12	1.00000
*x*13	0.99787	*x*13	0.96985	*x*13	0.91858	*x*13	0.85355	*x*13	0.78166	*x*13	0.70771	*x*13	0.63490
*y*13	0.54613	*y*13	0.67101	*y*13	0.77347	*y*13	0.85355	*y*13	0.91312	*y*13	0.95482	*y*13	0.98146
*x*14	0.98091	*x*14	1.00000	*x*14	0.98470	*x*14	0.94550	*x*14	0.89092	*x*14	0.82743	*x*14	0.75979
*y*14	0.36317	*y*14	0.50000	*y*14	0.62274	*y*14	0.72700	*y*14	0.81174	*y*14	0.87787	*y*14	0.92721
*x*15	0.89901	*x*15	0.96985	*x*15	0.99829	*x*15	0.99384	*x*15	0.96544	*x*15	0.92063	*x*15	0.86542
*y*15	0.19868	*y*15	0.32899	*y*15	0.45871	*y*15	0.57822	*y*15	0.68267	*y*15	0.77032	*y*15	0.84128
*x*16	0.76322	*x*16	0.88302	*x*16	0.95789	*x*16	0.99384	*x*16	0.99860	*x*16	0.97975	*x*16	0.94394
*y*16	0.07489	*y*16	0.17861	*y*16	0.29915	*y*16	0.42178	*y*16	0.53737	*y*16	0.64087	*y*16	0.73003
*x*17	0.59187	*x*17	0.75000	*x*17	0.86786	*x*17	0.94550	*x*17	0.98746	*x*17	1.00000	*x*17	0.98954
*y*17	0.00851	*y*17	0.06699	*y*17	0.16136	*y*17	0.27300	*y*17	0.38874	*y*17	0.50000	*y*17	0.60173
		*x*18	0.58682	*x*18	0.73797	*x*18	0.85355	*x*18	0.93301	*x*18	0.97975	*x*18	0.99883
		*y*18	0.00760	*y*18	0.06026	*y*18	0.14645	*y*18	0.25000	*y*18	0.35913	*y*18	0.46588
				*x*19	0.58230	*x*19	0.72700	*x*19	0.84009	*x*19	0.92063	*x*19	0.97113
				*y*19	0.00682	*y*19	0.05450	*y*19	0.13347	*y*19	0.22968	*y*19	0.33256
						*x*20	0.57822	*x*20	0.71694	*x*20	0.82743	*x*20	0.90848
						*y*20	0.00616	*y*20	0.04952	*y*20	0.12213	*y*20	0.21166
								*x*21	0.57452	*x*21	0.70771	*x*21	0.81554
								*y*21	0.00558	*y*21	0.04518	*y*21	0.11214
										*x*22	0.57116	*x*22	0.69920
										*y*22	0.00509	*y*22	0.04139
												*x*23	0.56808
												*y*23	0.00466

24 Holes		25 Holes		26 Holes		27 Holes		28 Holes	
*x*1	0.43474	*x*1	0.43733	*x*1	0.43973	*x*1	0.44195	*x*1	0.44402
*y*1	0.00428	*y*1	0.00394	*y*1	0.00365	*y*1	0.00338	*y*1	0.00314
*x*2	0.30866	*x*2	0.31594	*x*2	0.32270	*x*2	0.32899	*x*2	0.33486
*y*2	0.03806	*y*2	0.03511	*y*2	0.03249	*y*2	0.03015	*y*2	0.02806
*x*3	0.19562	*x*3	0.20611	*x*3	0.21597	*x*3	0.22525	*x*3	0.23398

Table 2. *(Continued)* **Hole Coordinate Dimension Factors for Jig Boring — Type "B" Hole Circles** (English or Metric Units)

The diagram shows a type "B" circle for a 5-hole circle. Coordinates *x*, *y* are given in the table for hole circles of from 3 to 28 holes. Dimensions are for holes numbered in a counterclockwise direction (as shown). Dimensions given are based upon a hole circle of unit diameter. For a hole circle of, say, 3-inch or 3-centimeter diameter, multiply table values by 3.

y3	0.10332	y3	0.09549	y3	0.08851	y3	0.08226	y3	0.07664
x4	0.10332	x4	0.11474	x4	0.12574	x4	0.13631	x4	0.14645
y4	0.19562	y4	0.18129	y4	0.16844	y4	0.15688	y4	0.14645
x5	0.03806	x5	0.04759	x5	0.05727	x5	0.06699	x5	0.07664
y5	0.30866	y5	0.28711	y5	0.26764	y5	0.25000	y5	0.23398
x6	0.00428	x6	0.00886	x6	0.01453	x6	0.02101	x6	0.02806
y6	0.43474	y6	0.40631	y6	0.38034	y6	0.35660	y6	0.33486
x7	0.00428	x7	0.00099	x7	0.00000	x7	0.00085	x7	0.00314
y7	0.56526	y7	0.53140	y7	0.50000	y7	0.47093	y7	0.44402
x8	0.03806	x8	0.02447	x8	0.01453	x8	0.00760	x8	0.00314
y8	0.69134	y8	0.65451	y8	0.61966	y8	0.58682	y8	0.55598
x9	0.10332	x9	0.07784	x9	0.05727	x9	0.04089	x9	0.02806
y9	0.80438	y9	0.76791	y9	0.73236	y9	0.69804	y9	0.66514
x10	0.19562	x10	0.15773	x10	0.12574	x10	0.09894	x10	0.07664
y10	0.89668	y10	0.86448	y10	0.83156	y10	0.79858	y10	0.76602
x11	0.30866	x11	0.25912	x11	0.21597	x11	0.17861	x11	0.14645
y11	0.96194	y11	0.93815	y11	0.91149	y11	0.88302	y11	0.85355
x12	0.43474	x12	0.37566	x12	0.32270	x12	0.27560	x12	0.23398
y12	0.99572	y12	0.98429	y12	0.96751	y12	0.94682	y12	0.92336
x13	0.56526	x13	0.50000	x13	0.43973	x13	0.38469	x13	0.33486
y13	0.99572	y13	1.00000	y13	0.99635	y13	0.98652	y13	0.97194
x14	0.69134	x14	0.62434	x14	0.56027	x14	0.50000	x14	0.44402
y14	0.96194	y14	0.98429	y14	0.99635	y14	1.00000	y14	0.99686
x15	0.80438	x15	0.74088	x15	0.67730	x15	0.61531	x15	0.55598
y15	0.89668	y15	0.93815	y15	0.96751	y15	0.98652	y15	0.99686
x16	0.89668	x16	0.84227	x16	0.78403	x16	0.72440	x16	0.66514
y16	0.80438	y16	0.86448	y16	0.91149	y16	0.94682	y16	0.97194
x17	0.96194	x17	0.92216	x17	0.87426	x17	0.82139	x17	0.76602
y17	0.69134	y17	0.76791	y17	0.83156	y17	0.88302	y17	0.92336
x18	0.99572	x18	0.97553	x18	0.94273	x18	0.90106	x18	0.85355
y18	0.56526	y18	0.65451	y18	0.73236	y18	0.79858	y18	0.85355
x19	0.99572	x19	0.99901	x19	0.98547	x19	0.95911	x19	0.92336
y19	0.43474	y19	0.53140	y19	0.61966	y19	0.69804	y19	0.76602
x20	0.96194	x20	0.99114	x20	1.00000	x20	0.99240	x20	0.97194
y20	0.30866	y20	0.40631	y20	0.50000	y20	0.58682	y20	0.66514
x21	0.89668	x21	0.95241	x21	0.98547	x21	0.99915	x21	0.99686
y21	0.19562	y21	0.28711	y21	0.38034	y21	0.47093	y21	0.55598
x22	0.80438	x22	0.88526	x22	0.94273	x22	0.97899	x22	0.99686
y22	0.10332	y22	0.18129	y22	0.26764	y22	0.35660	y22	0.44402
x23	0.69134	x23	0.79389	x23	0.87426	x23	0.93301	x23	0.97194
y23	0.03806	y23	0.09549	y23	0.16844	y23	0.25000	y23	0.33486
x24	0.56526	x24	0.68406	x24	0.78403	x24	0.86369	x24	0.92336
y24	0.00428	y24	0.03511	y24	0.08851	y24	0.15688	y24	0.23398
		x25	0.56267	x25	0.67730	x25	0.77475	x25	0.85355
		y25	0.00394	y25	0.03249	y25	0.08226	y25	0.14645
				x26	0.56027	x26	0.67101	x26	0.76602
				y26	0.00365	y26	0.03015	y26	0.07664
						x27	0.55805	x27	0.66514
						y27	0.00338	y27	0.02806
								x28	0.55598
								y28	0.00314

Table 3. Hole Coordinate Dimension Factors for Jig Boring — Type "A" Hole Circles, Central Coordinates (English or Metric Units)

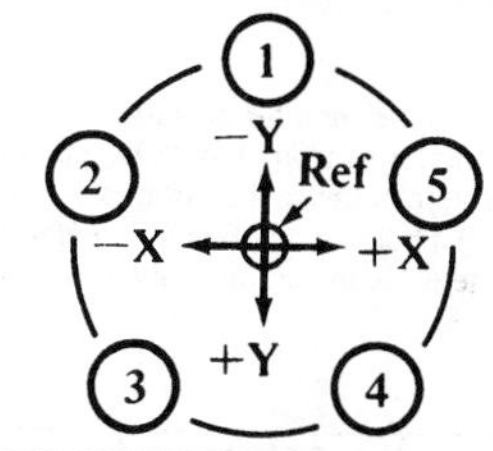

The diagram shows a type "A" circle for a 5-hole circle. Coordinates *x*, *y* are given in the table for hole circles of from 3 to 28 holes. Dimensions are for holes numbered in a counterclockwise direction (as shown). Dimensions given are based upon a hole circle of unit diameter. For a hole circle of, say, 3-inch or 3-centimeter diameter, multiply table values by 3.

	3 Holes	4 Holes	5 Holes	6 Holes	7 Holes	8 Holes	9 Holes
*x*1	0.00000	0.00000	0.00000	0.00000	0.00000	0.00000	0.00000
*y*1	−0.50000	−0.50000	−0.50000	−0.50000	−0.50000	−0.50000	−0.50000
*x*2	−0.43301	−0.50000	−0.47553	−0.43301	−0.39092	−0.35355	−0.32139
*y*2	+0.25000	0.00000	−0.15451	−0.25000	−0.31174	−0.35355	−0.38302
*x*3	+0.43301	0.00000	−0.29389	−0.43301	−0.48746	−0.50000	−0.49240
*y*3	+0.25000	+0.50000	+0.40451	+0.25000	+0.11126	0.00000	−0.08682
*x*4		+0.50000	+0.29389	0.00000	−0.21694	−0.35355	−0.43301
*y*4		0.00000	+0.40451	+0.50000	+0.45048	+0.35355	+0.25000
*x*5			+0.47553	+0.43301	+0.21694	0.00000	−0.17101
*y*5			−0.15451	+0.25000	+0.45048	+0.50000	+0.46985
*x*6				+0.43301	+0.48746	+0.35355	+0.17101
*y*6				−0.25000	+0.11126	+0.35355	+0.46985
*x*7					+0.39092	+0.50000	+0.43301
*y*7					−0.31174	0.00000	+0.25000
*x*8						+0.35355	+0.49240
*y*8						−0.35355	−0.08682
*x*9							+0.32139
*y*9							−0.38302

	10 Holes	11 Holes	12 Holes	13 Holes	14 Holes	15 Holes	16 Holes
*x*1	0.00000	0.00000	0.00000	0.00000	0.00000	0.00000	0.00000
*y*1	−0.50000	−0.5000	−0.50000	−0.50000	−0.50000	−0.50000	−0.50000
*x*2	−0.29389	−0.27032	−0.25000	−0.23236	−0.21694	−0.20337	−0.19134
*y*2	−0.40451	−0.42063	−0.43301	−0.44273	−0.45048	−0.45677	−0.46194
*x*3	−0.47553	−0.45482	−0.43301	−0.41149	−0.39092	−0.37157	−0.35355
*y*3	−0.15451	−0.20771	−0.25000	−2.28403	−0.31174	−0.33457	−0.35355
*x*4	−0.47553	−0.49491	−0.50000	−0.49635	−0.48746	−0.47553	−0.46194
*y*4	+0.15451	+0.07116	0.00000	−0.06027	−0.11126	−0.15451	−0.19134
*x*5	−0.29389	−0.37787	−0.43301	−0.46751	−0.48746	−0.49726	−0.50000
*y*5	+0.40451	+0.32743	+0.25000	+0.17730	+0.11126	+0.05226	0.00000
*x*6	0.00000	−0.14087	−0.25000	− 0.33156	−0.39092	−0.43301	−0.46194
*y*6	+0.50000	+0.47975	+0.43301	+0.37426	+0.31174	+0.25000	+0.19134
*x*7	+0.29389	+0.14087	0.00000	−0.11966	−0.21694	−0.29389	−0.35355
*y*7	+0.40451	+0.47975	+0.50000	+0.48547	+0.45048	+0.40451	+0.35355
*x*8	+0.47553	+0.37787	+0.25000	+0.11966	0.00000	−0.10396	−0.19134
*y*8	+0.15451	+0.32743	+0.43301	+0.48547	+0.50000	+0.48907	+0.46194
*x*9	+0.47553	+0.49491	+0.43301	+0.33156	+0.21694	+0.10396	0.00000
*y*9	−0.15451	+0.07116	+0.25000	+0.37426	+0.45048	+0.48907	+0.50000
*x*10	+0.29389	+0.45482	+0.50000	+0.46751	+0.39092	+0.29389	+0.19134
*y*10	−0.40451	−0.20771	0.00000	+0.17730	+0.31174	+0.40451	+0.46194
*x*11		+0.27032	+0.43301	+0.49635	+0.48746	+0.43301	+0.35355
*y*11		−0.42063	−0.25000	−0.06027	+0.11126	+0.25000	+0.35355
*x*12			+0.25000	+0.41149	+0.48746	+0.49726	+0.46194
*y*12			−0.43301	−0.28403	−0.11126	+0.05226	+0.19134
*x*13				+0.23236	+0.39092	+0.47553	+0.50000
*y*13				−0.44273	−0.31174	−0.15451	0.00000
*x*14					+0.21694	+0.37157	+0.46194
*y*14					−0.45048	−0.33457	−0.19134
*x*15						+0.20337	+0.35355
*y*15						−0.45677	−0.35355
*x*16							+0.19134
*y*16							−0.46194

Table 3. *(Continued)* **Hole Coordinate Dimension Factors for Jig Boring — Type "A" Hole Circles, Central Coordinates** (English or Metric Units)

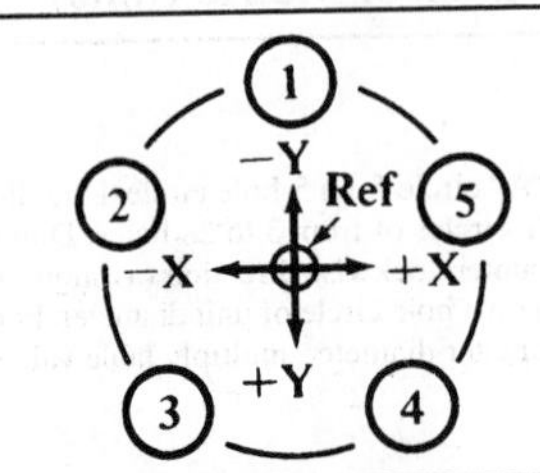

The diagram shows a type "A" circle for a 5-hole circle. Coordinates *x*, *y* are given in the table for hole circles of from 3 to 28 holes. Dimensions are for holes numbered in a counterclockwise direction (as shown). Dimensions given are based upon a hole circle of unit diameter. For a hole circle of, say, 3-inch or 3-centimeter diameter, multiply table values by 3.

17 Holes		18 Holes		19 Holes		20 Holes		21 Holes		22 Holes		23 Holes	
*x*1	0.00000	*x*1	0.00000	*x*1	0.00000	*x*1	0.000000	*x*1	0.00000	*x*1	0.00000	*x*1	0.00000
*y*1	−0.50000	*y*1	−0.50000	*y*1	−0.50000	*y*1	−0.50000	*y*1	−0.50000	*y*1	−0.50000	*y*1	−0.50000
*x*2	−0.18062	*x*2	−0.17101	*x*2	−0.16235	*x*2	−0.15451	*x*2	−0.14738	*x*2	−0.14087	*x*2	−0.13490
*y*2	−0.46624	*y*2	−0.46985	*y*2	−0.47291	*y*2	−0.47553	*y*2	−0.47779	*y*2	−0.47975	*y*2	−0.48146
*x*3	−0.33685	*x*3	+0.32139	*x*3	−0.30711	*x*3	−0.29389	*x*3	−0.28166	*x*3	−0.27032	*x*3	−0.25979
*y*3	−0.36950	*y*3	−0.38302	*y*3	−0.39457	*y*3	−0.40451	*y*3	−0.41312	*y*3	−0.42063	*y*3	−0.42721
*x*4	−0.44758	*x*4	−0.43301	*x*4	−0.41858	*x*4	−0.40451	*x*4	−0.39092	*x*4	−0.37787	*x*4	−0.36542
*y*4	−0.22287	*y*4	−0.25000	*y*4	−0.27347	*y*4	−0.29389	*y*4	−0.31174	*y*4	−0.32743	*y*4	−0.34128
*x*5	−0.49787	*x*5	−0.49240	*x*5	−0.48470	*x*5	−0.47553	*x*5	−.046544	*x*5	−0.45482	*x*5	−0.44394
*y*5	−0.04613	*y*5	−0.08682	*y*5	−0.12274	*y*5	−0.15451	*y*5	−0.18267	*y*5	−0.20771	*y*5	−0.23003
*x*6	−0.48091	*x*6	−0.49420	*x*6	−0.49829	*x*6	−0.50000	*x*6	−0.49860	*x*6	−0.49491	*x*6	−0.48954
*y*6	+0.13683	*y*6	+0.08682	*y*6	+0.04129	*y*6	0.00000	*y*6	−0.03737	*y*6	−0.07116	*y*6	−0.10173
*x*7	−0.39901	*x*7	−0.43301	*x*7	−0.45789	*x*7	−0.47553	*x*7	−0.48746	*x*7	−0.49491	*x*7	−0.49883
*y*7	+0.30132	*y*7	+0.25000	*y*7	+0.20085	*y*7	+0.15451	*y*7	+0.11126	*y*7	+0.07116	*y*7	+0.03412
*x*8	−0.26322	*x*8	−0.32139	*x*8	−0.36786	*x*8	−0.40451	*x*8	−0.43301	*x*8	−0.45482	*x*8	−0.47113
*y*8	+0.42511	*y*8	+0.38302	*y*8	+0.33864	*y*8	+0.29389	*y*8	+0.25000	*y*8	+0.20771	*y*8	+0.16744
*x*9	−0.09187	*x*9	−0.17101	*x*9	−0.23797	*x*9	−0.29389	*x*9	−0.34009	*x*9	−0.37787	*x*9	−0.40848
*y*9	+0.49149	*y*9	+0.46985	*y*9	+0.43974	*y*9	+0.40451	*y*9	+0.36653	*y*9	+0.32743	*y*9	+0.28834
*x*10	+0.09187	*x*10	0.00000	*x*10	−0.08230	*x*10	−0.15451	*x*10	−0.21694	*x*10	−0.27032	*x*10	−0.31554
*y*10	+0.49149	*y*10	+0.50000	*y*10	+0.49318	*y*10	+0.47553	*y*10	+0.45048	*y*10	+0.42063	*y*10	+0.38786
*x*11	+0.26322	*x*11	+0.17101	*x*11	+0.08230	*x*11	0.00000	*x*11	−0.07452	*x*11	−0.14087	*x*11	−0.19920
*y*11	+0.42511	*y*11	+0.46985	*y*11	+0.49318	*y*11	+0.50000	*y*11	+0.49442	*y*11	+0.47975	*y*11	+0.45861
*x*12	+0.39901	*x*12	+0.32139	*x*12	+0.23797	*x*12	+0.15451	*x*12	+0.07452	*x*12	0.00000	*x*12	−0.06808
*y*12	+0.30132	*y*12	+0.38302	*y*12	+0.43974	*y*12	+0.47553	*y*12	+0.49442	*y*12	+0.50000	*y*12	+0.49534
*x*13	+0.48091	*x*13	+0.43301	*x*13	+0.36786	*x*13	+0.29389	*x*13	+0.21694	*x*13	+0.14087	*x*13	+0.06808
*y*13	+0.13683	*y*13	+0.25000	*y*13	+0.33864	*y*13	+0.40451	*y*13	+0.45048	*y*13	+0.47975	*y*13	+0.49534
*x*14	+0.49787	*x*14	+0.49240	*x*14	+0.45789	*x*14	+0.40451	*x*14	+0.34009	*x*14	+0.27032	*x*14	+0.19920
*y*14	−0.04613	*y*14	+0.08682	*y*14	+0.20085	*y*14	+0.29389	*y*14	+0.36653	*y*14	+0.42063	*y*14	+0.45861
*x*15	+0.44758	*x*15	+0.49240	*x*15	+0.49829	*x*15	+0.47553	*x*15	+0.43301	*x*15	+0.37787	*x*15	+0.31554
*y*15	−0.22287	*y*15	−0.08682	*y*15	+0.04129	*y*15	+0.15451	*y*15	+0.25000	*y*15	+0.32743	*y*15	+0.38786
*x*16	+0.33685	*x*16	+0.43301	*x*16	+0.48470	*x*16	+0.50000	*x*16	+0.48746	*x*16	+0.45482	*x*16	+0.40848
*y*16	−0.36950	*y*16	−0.25000	*y*16	−0.12274	*y*16	0.00000	*y*16	+0.11126	*y*16	+0.20771	*y*16	+0.28834
*x*17	+0.18062	*x*17	+0.32139	*x*17	+0.41858	*x*17	+0.47553	*x*17	+0.49860	*x*17	+0.49491	*x*17	+0.47113
*y*17	−0.46624	*y*17	−0.38302	*y*17	−0.27347	*y*17	−0.15451	*y*17	−0.03737	*y*17	+0.07116	*y*17	+0.16744
		*x*18	+0.17101	*x*18	+0.30711	*x*18	+0.40451	*x*18	+0.46544	*x*18	+0.49491	*x*18	+0.49883
		*y*18	−0.46985	*y*18	−0.39457	*y*18	−0.29389	*y*18	−0.18267	*y*18	−0.07116	*y*18	+0.03412
				*x*19	+ 0.16235	*x*19	+0.29389	*x*19	+0.39092	*x*19	+0.45482	*x*19	+0.48954
				*y*19	−0.47291	*y*19	−0.40451	*y*19	−0.31174	*y*19	−0.20771	*y*19	−0.10173
						*x*20	+0.15451	*x*20	+0.28166	*x*20	+0.37787	*x*20	+0.44394
						*y*20	−0.47553	*y*20	−0.41312	*y*20	−0.32743	*y*20	−0.23003
								*x*21	+0.14738	*x*21	+0.27032	*x*21	+0.36542
								*y*21	−0.47779	*y*21	−0.42063	*y*21	−0.34128
										*x*22	+0.14087	*x*22	+0.25979
										*y*22	−0.47975	*y*22	−0.42721
												*x*23	+0.13490
												*y*23	−0.48146

24 Holes		25 Holes		26 Holes		27 Holes		28 Holes	
*x*1	0.00000	*x*1	0.00000	*x*1	0.00000	*x*1	0.00000	*x*1	0.00000
*y*1	−0.50000	*y*1	−0.50000	*y*1	−0.50000	*y*1	−0.50000	*y*1	−0.50000
*x*2	−0.12941	*x*2	−0.12434	*x*2	−0.11966	*x*2	−0.11531	*x*2	−0.11126
*y*2	−0.48296	*y*2	−0.48429	*y*2	−0.48547	*y*2	−0.48652	*y*2	−0.48746
*x*3	−0.25000	*x*3	−0.24088	*x*3	−0.23236	*x*3	−0.22440	*x*3	−0.21694

Table 3. *(Continued)* **Hole Coordinate Dimension Factors for Jig Boring — Type "A" Hole Circles, Central Coordinates** (English or Metric Units)

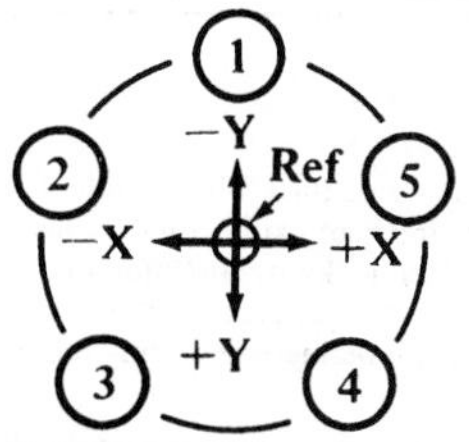

The diagram shows a type "A" circle for a 5-hole circle. Coordinates *x*, *y* are given in the table for hole circles of from 3 to 28 holes. Dimensions are for holes numbered in a counterclockwise direction (as shown). Dimensions given are based upon a hole circle of unit diameter. For a hole circle of, say, 3-inch or 3-centimeter diameter, multiply table values by 3.

*y*3	−0.43301	*y*3	−0.43815	*y*3	−0.44273	*y*3	−0.44682	*y*3	−0.45048
*x*4	−0.35355	*x*4	−0.34227	*x*4	−0.33156	*x*4	−0.32139	*x*4	−0.31174
*y*4	−0.35355	*y*4	−0.36448	*y*4	−0.37426	*y*4	−0.38302	*y*4	−0.39092
*x*5	−0.43301	*x*5	−0.42216	*x*5	−0.41149	*x*5	−0.40106	*x*5	−0.39092
*y*5	−0.25000	*y*5	−0.26791	*y*5	−0.28403	*y*5	−0.29858	*y*5	−0.31174
*x*6	−0.48296	*x*6	−0.47553	*x*6	−0.46751	*x*6	−0.45911	*x*6	−0.45048
*y*6	−0.12941	*y*6	−0.15451	*y*6	−0.17730	*y*6	−0.19804	*y*6	−0.21694
*x*7	−0.50000	*x*7	−0.49901	*x*7	−0.49635	*x*7	−0.49240	*x*7	−0.48746
*y*7	0.00000	*y*7	−0.03140	*y*7	−0.06027	*y*7	−0.08682	*y*7	−0.11126
*x*8	−0.48296	*x*8	−0.49114	*x*8	−0.49635	*x*8	−0.49915	*x*8	−0.50000
*y*8	+0.12941	*y*8	+0.09369	*y*8	+0.06027	*y*8	+0.02907	*y*8	0.00000
*x*9	−0.43301	*x*9	−0.45241	*x*9	−0.46751	*x*9	−0.47899	*x*9	−0.48746
*y*9	+0.25000	*y*9	+0.21289	*y*9	+0.17730	*y*9	+0.14340	*y*9	+0.11126
*x*10	−0.35355	*x*10	−0.38526	*x*10	−0.41149	*x*10	−0.43301	*x*10	−0.45048
*y*10	+0.35355	*y*10	+0.31871	*y*10	+0.28403	*y*10	+0.25000	*y*10	+0.21694
*x*11	−0.25000	*x*11	−0.29389	*x*11	−0.33156	*x*11	−0.36369	*x*11	−0.39092
*y*11	+0.43301	*y*11	+0.40451	*y*11	+0.37426	*y*11	+0.34312	*y*11	+0.31174
*x*12	−0.12941	*x*12	−0.18406	*x*12	−0.23236	*x*12	−0.27475	*x*12	−0.31174
*y*12	+0.48296	*y*12	+0.46489	*y*12	+0.44273	*y*12	+0.41774	*y*12	+0.39092
*x*13	0.00000	*x*13	−0.06267	*x*13	−0.11966	*x*13	−0.17101	*x*13	−0.21694
*y*13	+0.50000	*y*13	+0.49606	*y*13	+0.48547	*y*13	+0.46985	*y*13	+0.45048
*x*14	+0.12941	*x*14	+0.06267	*x*14	0.00000	*x*14	−0.05805	*x*14	−0.11126
*y*14	+0.48296	*y*14	+0.49606	*y*14	+0.50000	*y*14	+0.49662	*y*14	+0.48746
*x*15	+0.25000	*x*15	+0.18406	*x*15	+0.11966	*x*15	+0.05805	*x*15	0.00000
*y*15	+0.43301	*y*15	+0.46489	*y*15	+0.48547	*y*15	+0.49662	*y*15	+0.50000
*x*16	+0.35355	*x*16	+0.29389	*x*16	+0.23236	*x*16	+0.17101	*x*16	+0.11126
*y*16	+0.35355	*y*16	+0.40451	*y*16	+0.44273	*y*16	+0.46985	*y*16	+0.48746
*x*17	+0.43301	*x*17	+0.38526	*x*17	+0.33156	*x*17	+0.27475	*x*17	+0.21694
*y*17	+0.25000	*y*17	+0.31871	*y*17	+0.37426	*y*17	+0.41774	*y*17	+0.45048
*x*18	+0.48296	*x*18	+0.45241	*x*18	+0.41149	*x*18	+0.36369	*x*18	+0.31174
*y*18	+0.12941	*y*18	+0.21289	*y*18	+0.28403	*y*18	+0.34312	*y*18	+0.39092
*x*19	+0.50000	*x*19	+0.49114	*x*19	+0.46751	*x*19	+0.43301	*x*19	+0.39092
*y*19	0.00000	*y*19	+0.09369	*y*19	+0.17730	*y*19	+0.25000	*y*19	+0.31174
*x*20	+0.48296	*x*20	+0.49901	*x*20	+0.49635	*x*20	+0.47899	*x*20	+0.45048
*y*20	−0.12941	*y*20	−0.03140	*y*20	+0.06027	*y*20	+0.14340	*y*20	+0.21694
*x*21	+0.43301	*x*21	+0.47553	*x*21	+0.49635	*x*21	+0.49915	*x*21	+0.48746
*y*21	−0.25000	*y*21	−0.15451	*y*21	−0.06027	*y*21	+0.02907	*y*21	+0.11126
*x*22	+0.35355	*x*22	+0.42216	*x*22	+0.46751	*x*22	+0.49240	*x*22	+0.50000
*y*22	−0.35355	*y*22	−0.26791	*y*22	−0.17730	*y*22	−0.08682	*y*22	0.00000
*x*23	+0.25000	*x*23	+0.34227	*x*23	+0.41149	*x*23	+0.45911	*x*23	+0.48746
*y*23	−0.43301	*y*23	−0.36448	*y*23	−0.28403	*y*23	−0.19804	*y*23	−0.11126
*x*24	+0.12941	*x*24	+0.24088	*x*24	+0.33156	*x*24	+0.40106	*x*24	+0.45048
*y*24	−0.48296	*y*24	−0.43815	*y*24	−0.37426	*y*24	−0.29858	*y*24	−0.21694
		*x*25	+0.12434	*x*25	+0.23236	*x*25	+0.32139	*x*25	+0.39092
		*y*25	−0.48429	*y*25	−0.44273	*y*25	−0.38302	*y*25	−0.31174
				*x*26	+0.11966	*x*26	+0.22440	*x*26	+0.31174
				*y*26	−0.48547	*y*26	−0.44682	*y*26	−0.39092
						*x*27	+0.11531	*x*27	+0.21694
						*y*27	−0.48652	*y*27	−0.45048
								*x*28	+0.11126
								*y*28	−0.48746

Table 4. Hole Coordinate Dimension Factors for Jig Boring —Type "B" Hole Circles Central Coordinates (English or Metric units)

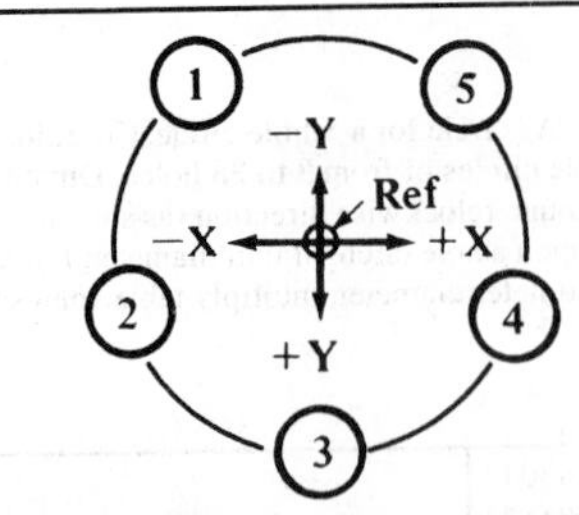

The diagram shows a type "B" circle for a 5-hole circle. Coordinates *x*, *y* are given in the table for hole circles of from 3 to 28 holes. Dimensions are for holes numbered in a counterclockwise direction (as shown). Dimensions given are based upon a hole circle of unit diameter. For a hole circle of, say, 3-inch or 3-centimeter diameter, multiply table values by 3.

3 Holes	4 Holes	5 Holes	6 Holes	7 Holes	8 Holes	9 Holes
x1 −0.43301	x1 −0.35355	x1 −0.29389	x1 −0.25000	x1 −0.21694	x1 −0.19134	x1 −0.17101
y1 −0.25000	y1 −0.35355	y1 −0.40451	y1 −0.43301	y1 −0.45048	y1 −0.46194	y1 −0.46985
x2 0.00000	x2 −0.35355	x2 −0.47553	x2 −0.50000	x2 −0.48746	x2 −0.46194	x2 −0.43301
y2 +0.50000	y2 +0.35355	y2 +0.15451	y2 0.00000	y2 −0.11126	y2 −0.19134	y2 −0.25000
x3 +0.43301	x3 +0.35355	x3 0.00000	x3 −0.25000	x3 −0.39092	x3 −0.46194	x3 −0.49240
y3 −0.25000	y3 +0.35355	y3 +0.50000	y3 +0.43301	y3 +0.31174	y3 +0.19134	y3 +0.08682
	x4 +0.35355	x4 +0.47553	x4 +0.25000	x4 0.00000	x4 −0.19134	x4 −0.32139
	y4 −0.35355	y4 +0.15451	y4 +0.43301	y4 +0.50000	y4 +0.46194	y4 +0.38302
		x5 +0.29389	x5 +0.50000	x5 +0.39092	x5 +0.19134	x5 0.00000
		y5 −0.40451	y5 0.00000	y5 +0.31174	y5 +0.46194	y5 +0.50000
			x6 +0.25000	x6 +0.48746	x6 +0.46194	x6 +0.32139
			y6 −0.43301	y6 −0.11126	y6 +0.19134	y6 +0.38302
				x7 +0.21694	x7 +0.46194	x7 +0.49240
				y7 −0.45048	y7 −0.19134	y7 +0.08682
					x8 +0.19134	x8 +0.43301
					y8 −0.46194	y8 −0.25000
						x9 +0.17101
						y9 −0.46985

10 Holes	11 Holes	12 Holes	13 Holes	14 Holes	15 Holes	16 Holes
x1 −0.15451	x1 −0.14087	x1 −0.12941	x1 −0.11966	x1 −0.11126	x1 −0.10396	x1 −0.09755
y1 −0.47553	y1 −0.47975	y1 −0.48296	y1 −0.48547	y1 −0.48746	y1 −0.48907	y1 −0.49039
x2 −0.40451	x2 −0.37787	x2 −0.35355	x2 −0.33156	x2 −0.31174	x2 −0.29389	x2 −0.27779
y2 −0.29389	y2 −0.32743	y2 −0.35355	y2 −0.37426	y2 −0.39092	y2 −0.40451	y2 −0.41573
x3 −0.50000	x3 −0.49491	x3 −0.48296	x3 −0.46751	x3 −0.45048	x3 −0.43301	x3 −0.41573
y3 0.00000	y3 −0.07116	y3 −0.12941	y3 −0.17730	y3 −0.21694	y3 −0.25000	y3 −0.27779
x4 −0.40451	x4 −0.45482	x4 −0.48296	x4 −0.49635	x4 −0.50000	x4 −0.49726	x4 −0.49039
y4 +0.29389	y4 +0.20771	y4 +0.12941	y4 +0.06027	y4 0.00000	y4 −0.05226	y4 −0.09755
x5 −0.15451	x5 −0.27032	x5 −0.35355	x5 −0.41149	x5 −0.45048	x5 −0.47553	x5 −0.49039
y5 +0.47553	y5 +0.42063	y5 +0.35355	y5 +0.28403	y5 +0.21694	y5 +0.15451	y5 +0.09755
x6 +0.15451	x6 0.00000	x6 −0.12941	x6 −0.23236	x6 −0.31174	x6 −0.37157	x6 −0.41573
y6 +0.47553	y6 +0.50000	y6 +0.48296	y6 +0.44273	y6 +0.39092	y6 +0.33457	y6 +0.27779
x7 +0.40451	x7 +0.27032	x7 +0.12941	x7 0.00000	x7 −0.11126	x7 −0.20337	x7 −0.27779
y7 +0.29389	y7 +0.42063	y7 +0.48296	y7 + 0.50000	y7 +0.48746	y7 +0.45677	y7 +0.41573
x8 +0.50000	x8 +0.45482	x8 +0.35355	x8 +0.23236	x8 +0.11126	x8 0.00000	x8 −0.09755
y8 0.00000	y8 +0.20771	y8 +0.35355	y8 +0.44273	y8 +0.48746	y8 +0.50000	y8 +0.49039
x9 +0.40451	x9 +0.49491	x9 +0.48296	x9 +0.41149	x9 +0.31174	x9 +0.20337	x9 +0.09755
y9 −0.29389	y9 −0.07116	y9 +0.12941	y9 +0.28403	y9 +0.39092	y9 +0.45677	y9 +0.49039
x10 +0.15451	x10 +0.37787	x10 +0.48296	x10 +0.49635	x10 +0.45048	x10 +0.37157	x10 +0.27779
y10 −0.47553	y10 −0.32743	y10 −0.12941	y10 +0.06027	y10 +0.21694	y10 +0.33457	y10 +0.41573
	x11 +0.14087	x11 +0.35355	x11 +0.46751	x11 +0.50000	x11 +0.47553	x11 +0.41573
	y11 −0.47975	y11 −0.35355	y11 −0.17730	y11 0.00000	y11 +0.15451	y11 +0.27779
		x12 +0.12941	x12 +0.33156	x12 +0.45048	x12 +0.49726	x12 +0.49039
		y12 −0.48296	y12 −0.37426	y12 −0.21694	y12 −0.05226	y12 +0.09755
			x13 +0.11966	x13 +0.31174	x13 +0.43301	x13 +0.49039
			y13 −0.48547	y13 −0.39092	y13 −0.25000	y13 −0.09755
				x14 +0.11126	x14 +0.29389	x14 +0.41573
				y14 – 0.48746	y14 −0.40451	y14 −0.27779
					x15 +0.10396	x15 +0.27779
					y15 −0.48907	y15 −0.41573
						x16 +0.09755
						y16 −0.49039

Table 4. *(Continued)* **Hole Coordinate Dimension Factors for Jig Boring —Type "B" Hole Circles Central Coordinates** (English or Metric units)

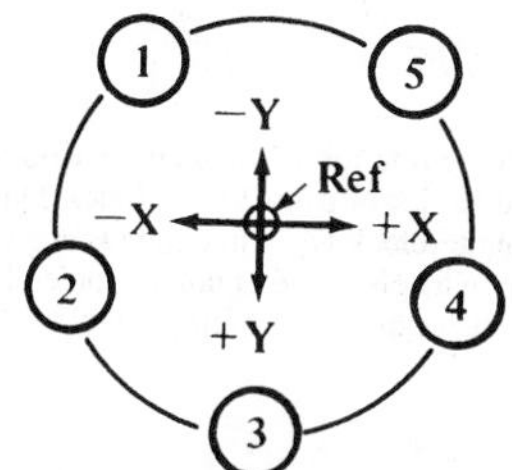

The diagram shows a type "B" circle for a 5-hole circle. Coordinates *x*, *y* are given in the table for hole circles of from 3 to 28 holes. Dimensions are for holes numbered in a counterclockwise direction (as shown). Dimensions given are based upon a hole circle of unit diameter. For a hole circle of, say, 3-inch or 3-centimeter diameter, multiply table values by 3.

	17 Holes	18 Holes	19 Holes	20 Holes	21 Holes	22 Holes	23 Holes
*x*1	−0.09187	−0.08682	−0.08230	−0.07822	−0.07452	−0.07116	−0.06808
*y*1	− 0.49149	−0.49240	−0.49318	−0.49384	−0.49442	−0.49491	−0.49534
*x*2	−0.26322	−0.25000	−0.23797	−0.22700	−0.21694	−0.20771	−0.19920
*y*2	−0.42511	−0.43301	−0.43974	−0.44550	−0.45048	−0.45482	−0.45861
*x*3	− 0.39901	−0.38302	−0.36786	−0.35355	−0.34009	−0.32743	−0.31554
*y*3	−0.30132	−0.32139	−0.33864	−0.35355	−0.36653	−0.37787	−0.38786
*x*4	−0.48091	−0.46985	−0.45789	−0.44550	−0.43301	−0.42063	−0.40848
*y*4	−0.13683	−0.17101	−0.20085	−0.22700	−0.25000	−0.27032	−0.28834
*x*5	−0.49787	−0.50000	−0.49829	−0.49384	−0.48746	−0.47975	−0.47113
*y*5	+0.04613	0.00000	−0.04129	−0.07822	−0.11126	−0.14087	−0.16744
*x*6	−0.44758	−0.46985	−0.48470	−0.49384	−0.49860	−0.50000	−0.49883
*y*6	+0.22287	+0.17101	+0.12274	+0.07822	+0.03737	0.00000	−0.03412
*x*7	−0.33685	−0.38302	−0.41858	−0.44550	−0.46544	−0.47975	−0.48954
*y*7	+0.36950	+0.32139	+0.27347	+0.22700	+0.18267	+0.14087	+0.10173
*x*8	−0.18062	−0.25000	−0.30711	−0.35355	−0.39092	−0.42063	−0.44394
*y*8	+0.46624	+0.43301	+0.39457	+0.35355	+0.31174	+0.27032	+0.23003
*x*9	0.00000	−0.08682	−0.16235	−0.22700	−0.28166	−0.32743	−0.36542
*y*9	+0.50000	+0.49240	+0.47291	+0.44550	+0.41312	+0.37787	+0.34128
*x*10	+0.18062	+0.08682	0.00000	−0.07822	−0.14738	−0.20771	−0.25979
*y*10	+0.46624	+0.49240	+0.50000	+0.49384	+0.47779	+0.45482	+0.42721
*x*11	+0.33685	+0.25000	+0.16235	+0.07822	0.00000	−0.07116	−0.13490
*y*11	+0.36950	+0.43301	+0.47291	+0.49384	+0.50000	+0.49491	+0.48146
*x*12	+0.44758	+0.38302	+0.30711	+0.22700	+0.14738	+ 0.07116	0.00000
*y*12	+0.22287	+0.32139	+0.39457	+0.44550	+0.47779	+0.49491	+0.50000
*x*13	+0.49787	+0.46985	+0.41858	+0.35355	+0.28166	+0.20771	+0.13490
*y*13	+0.04613	+0.17101	+0.27347	+0.35355	+0.41312	+0.45482	+0.48146
*x*14	+0.48091	+0.50000	+0.48470	+0.44550	+0.39092	+0.32743	+0.25979
*y*14	−0.13683	0.00000	+0.12274	+0.22700	+0.31174	+0.37787	+0.42721
*x*15	+0.39901	+0.46985	+0.49829	+0.49384	+0.46544	+0.42063	+0.36542
*y*15	−0.30132	−0.17101	−0.04129	+0.07822	+0.18267	+0.27032	+0.34128
*x*16	+0.26322	+0.38302	+0.45789	+0.49384	+0.49860	+0.47975	+0.44394
*y*16	−0.42511	−0.32139	−0.20085	−0.07822	+0.03737	+0.14087	+0.23003
*x*17	+0.09187	+0.25000	+0.36786	+0.44550	+0.48746	+0.50000	+0.48954
*y*17	− 0.49149	−0.43301	−0.33864	−0.22700	−0.11126	0.00000	+0.10173
*x*18		+0.08682	+0.23797	+0.35355	+0.43301	+0.47975	+0.49883
*y*18		−0.49240	−0.43974	−0.35355	−0.25000	−0.14087	−0.03412
*x*19			+0.08230	+0.22700	+0.34009	+0.42063	+0.47113
*y*19			−0.49318	−0.44550	−0.36653	−0.27032	−0.16744
*x*20				+0.07822	+0.21694	+0.32743	+0.40848
*y*20				−0.49384	−0.45048	−0.37787	−0.28834
*x*21					+0.07452	+0.20771	+0.31554
*y*21					−0.49442	−0.45482	−0.38786
*x*22						+0.07116	+0.19920
*y*22						−0.49491	−0.45861
*x*23							+0.06808
*y*23							−0.49534

	24 Holes	25 Holes	26 Holes	27 Holes	28 Holes
*x*1	−0.06526	−0.06267	−0.06027	−0.05805	−0.05598
*y*1	−0.49572	−0.49606	−0.49635	−0.49662	−0.49686
*x*2	−0.19134	−0.18406	−0.17730	−0.17101	−0.16514
*y*2	−0.46194	−0.46489	−0.46751	−0.46985	−0.47194
*x*3	−0.30438	−0.29389	−0.28403	−0.27475	−0.26602

Table 4. *(Continued)* **Hole Coordinate Dimension Factors for Jig Boring —Type "B" Hole Circles Central Coordinates** (English or Metric units)

The diagram shows a type "B" circle for a 5-hole circle. Coordinates *x*, *y* are given in the table for hole circles of from 3 to 28 holes. Dimensions are for holes numbered in a counterclockwise direction (as shown). Dimensions given are based upon a hole circle of unit diameter. For a hole circle of, say, 3-inch or 3-centimeter diameter, multiply table values by 3.

y3 −0.39668	y3 −0.40451	y3 −0.41149	y3 −0.41774	y3 −0.42336
x4 −0.39668	x4 −0.38526	x4 −0.37426	x4 −0.36369	x4 −0.35355
y4 −0.30438	y4 −0.31871	y4 −0.33156	y4 −0.34312	y4 −0.35355
x5 −0.46194	x5 −0.45241	x5 −0.44273	x5 −0.43301	x5 −0.42336
y5 −0.19134	y5 −0.21289	y5 −0.23236	y5 − 0.25000	y5 −0.26602
x6 −0.49572	x6 −0.49114	x6 −0.48547	x6 −0.47899	x6 −0.47194
y6 −0.06526	y6 −0.09369	y6 −0.11966	y6 −0.14340	y6 −0.16514
x7 −0.49572	x7 −0.49901	x7 −0.50000	x7 −0.49915	x7 −0.49686
y7 +0.06526	y7 +0.03140	y7 0.00000	y7 − 0.02907	y7 −0.05598
x8 −0.46194	x8 −0.47553	x8 −0.48547	x8 −0.49240	x8 −0.49686
y8 +0.19134	y8 +0.15451	y8 +0.11966	y8 +0.08682	y8 +0.05598
x9 −0.39668	x9 −0.42216	x9 −0.44273	x9 −0.45911	x9 −0.47194
y9 +0.30438	y9 +0.26791	y9 +0.23236	y9 +0.19804	y9 +0.16514
x10 −0.30438	x10 −0.34227	x10 −0.37426	x10 −0.40106	x10 −0.42336
y10 +0.39668	y10 + 0.36448	y10 +0.33156	y10 +0.29858	y10 +0.26602
x11 −0.19134	x11 −0.24088	x11 −0.28403	x11 −0.32139	x11 −0.35355
y11 +0.46194	y11 +0.43815	y11 +0.41149	y11 +0.38302	y11 +0.35355
x12 −0.06526	x12 −0.12434	x12 −0.17730	x12 −0.22440	x12 −0.26602
y12 +0.49572	y12 +0.48429	y12 +0.46751	y12 +0.44682	y12 +0.42336
x13 +0.06526	x13 0.00000	x13 −0.06027	x13 −0.11531	x13 −0.16514
y13 +0.49572	y13 +0.50000	y13 +0.49635	y13 +0.48652	y13 +0.47194
x14 +0.19134	x14 +0.12434	x14 +0.06027	x14 0.00000	x14 −0.05598
y14 +0.46194	y14 +0.48429	y14 +0.49635	y14 +0.50000	y14 +0.49686
x15 +0.30438	x15 +0.24088	x15 +0.17730	x15 +0.11531	x15 +0.05598
y15 +0.39668	y15 +0.43815	y15 +0.46751	y15 +0.48652	y15 +0.49686
x16 +0.39668	x16 +0.34227	x16 +0.28403	x16 +0.22440	x16 +0.16514
y16 +0.30438	y16 +0.36448	y16 +0.41149	y16 +0.44682	y16 +0.47194
x17 +0.46194	x17 +0.42216	x17 +0.37426	x17 +0.32139	x17 +0.26602
y17 +0.19134	y17 +0.26791	y17 +0.33156	y17 +0.38302	y17 +0.42336
x18 +0.49572	x18 +0.47553	x18 +0.44273	x18 +0.40106	x18 +0.35355
y18 +0.06526	y18 +0.15451	y18 +0.23236	y18 + 0.29858	y18 +0.35355
x19 +0.49572	x19 +0.49901	x19 +0.48547	x19 +0.45911	x19 +0.42336
y19 −0.06526	y19 +0.03140	y19 +0.11966	y19 +0.19804	y19 +0.26602
x20 +0.46194	x20 +0.49114	x20 +0.50000	x20 +0.49240	x20 +0.47194
y20 −0.19134	y20 −0.09369	y20 0.00000	y20 +0.08682	y20 +0.16514
x21 +0.39668	x21 +0.45241	x21 + 0.48547	x21 +0.49915	x21 +0.49686
y21 −0.30438	y21 −0.21289	y21 −0.11966	y21 −0.02907	y21 +0.05598
x22 +0.30438	x22 +0.38526	x22 +0.44273	x22 +0.47899	x22 +0.49686
y22 −0.39668	y22 −0.31871	y22 −0.23236	y22 − 0.14340	y22 −0.05598
x23 +0.19134	x23 +0.29389	x23 +0.37426	x23 +0.43301	x23 +0.47194
y23 −0.46194	y23 −0.40451	y23 −0.33156	y23 −0.25000	y23 −0.16514
x24 +0.06526	x24 +0.18406	x24 +0.28403	x24 +0.36369	x24 +0.42336
y24 −0.49572	y24 −0.46489	y24 −0.41149	y24 −0.34312	y24 −0.26602
	x25 +0.06267	x25 +0.17730	x25 +0.27475	x25 +0.35355
	y25 −0.49606	y25 −0.46751	y25 −0.41774	y25 −0.35355
		x26 +0.06027	x26 +0.17101	x26 +0.26602
		y26 −0.49635	y26 −0.46985	y26 −0.42336
			x27 +0.05805	x27 +0.16514
			y27 −0.49662	y27 −0.47194
				x28 +0.05598
				y28 −0.49686

Collets

Collets for Lathes, Mills, Grinders, and Fixtures

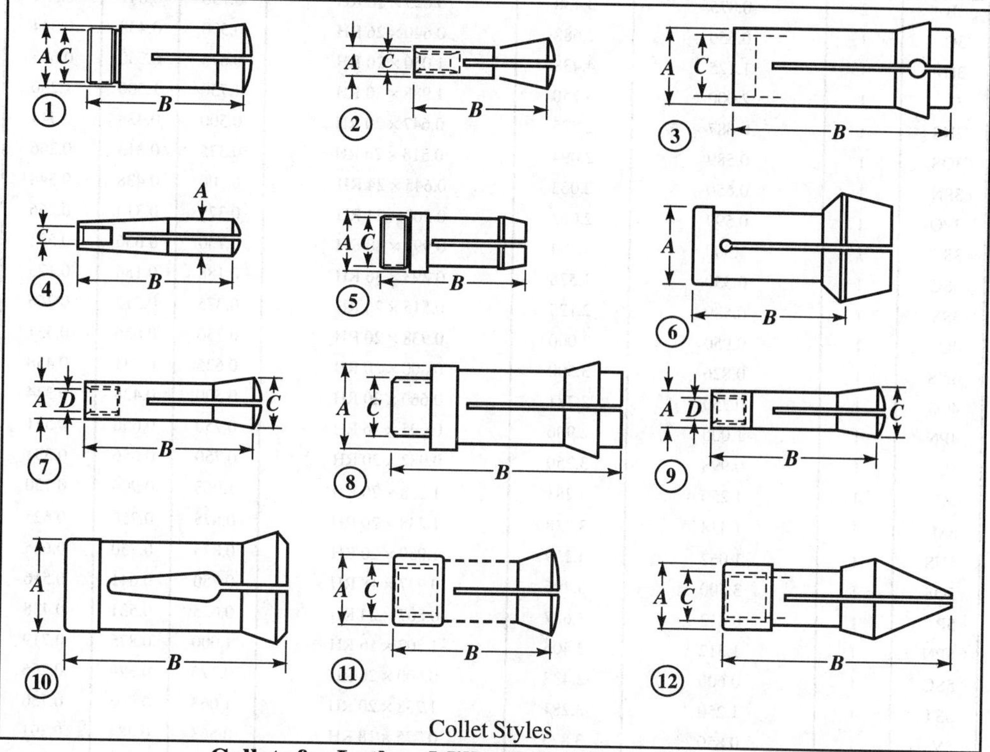

Collet Styles

Collets for Lathes, Mills, Grinders, and Fixtures

Collet	Style	Dimensions			Max. Capacity (inches)		
		Bearing Diam., A	Length, B	Thread, C	Round	Hex	Square
1A	1	0.650	2.563	0.640 × 26 RH	0.500	0.438	0.344
1AM	1	1.125	3.906	1.118 × 24 RH	1.000	0.875	0.719
1B	2	0.437	1.750	0.312 × 30 RH	0.313	0.219	0.188
1C	1	0.335	1.438	0.322 × 40 RH	0.250	0.219	0.172
1J	1	1.250	3.000	1.238 × 20 RH	1.063	0.875	0.750
1K	3	1.250	2.813	None	1.000	0.875	0.719
2A	1	0.860	3.313	0.850 × 20 RH	0.688	0.594	0.469
2AB	2	0.750	2.563	0.500 × 20 RH	0.625	0.484	0.391
2AM	1	0.629	3.188	0.622 × 24 RH	0.500	0.438	0.344
2B	2	0.590	2.031	0.437 × 26 RH	0.500	0.438	0.344
2C	1	0.450	1.812	0.442 × 30 RH	0.344	0.594	0.234
2H	1	0.826	4.250	0.799 × 20 RH	0.625	0.531	1.000
2J	1	1.625	3.250	1.611 × 18 RH	1.375	1.188	0.438
2L	1	0.950	3.000	0.938 × 20 RH	0.750	0.656	1.000
2M	4	2 Morse	2.875	0.375 × 16 RH	0.500	0.438	0.344
2NS	1	0.324	1.562	0.318 × 40 RH	0.250	0.203	0.172
2OS	1	0.299	1.250	0.263 × 40 RH	0.188	0.156	0.125
2S	1	0.750	3.234	0.745 × 18 RH	0.563	0.484	0.391
2VB	2	0.595	2.438	0.437 × 26 RH	0.500	0.438	0.344
3AM	1	0.750	3.188	0.742 × 24 RH	0.625	0.531	0.438
3AT	1	0.687	2.313	0.637 × 26 RH	0.500	0.438	0.344

Collets for Lathes, Mills, Grinders, and Fixtures *(Continued)*

Collet	Style	Dimensions			Max. Capacity (inches)		
		Bearing Diam., A	Length, B	Thread, C	Round	Hex	Square
3B	2	0.875	3.438	0.625 × 16 RH	0.750	0.641	0.531
3C	1	0.650	2.688	0.640 × 26 RH	0.500	0.438	0.344
3H	1	1.125	4.438	1.050 × 20 RH	0.875	0.750	0.625
3J	1	2.000	3.750	1.988 × 20 RH	1.750	1.500	1.250
3NS	1	0.687	2.875	0.647 × 20 RH	0.500	0.438	0.344
3OS	1	0.589	2.094	0.518 × 26 RH	0.375	0.313	0.266
3PN	1	0.650	2.063	0.645 × 24 RH	0.500	0.438	0.344
3PO	1	0.599	2.063	0.500 × 24 RH	0.375	0.313	0.266
3S	1	1.000	4.594	0.995 × 20 RH	0.750	0.656	0.531
3SC	1	0.350	1.578	0.293 × 36 RH	0.188	0.156	0.125
3SS	1	0.589	2.125	0.515 × 26 RH	0.375	0.313	0.266
4C	1	0.950	3.000	0.938 × 20 RH	0.750	0.656	0.531
4NS	1	0.826	3.500	0.800 × 20 RH	0.625	0.531	0.438
4OS	1	0.750	2.781	0.660 × 20 RH	0.500	0.438	0.344
4PN	1	1.000	2.906	0.995 × 16 RH	0.750	0.656	0.531
4S	1	0.998	3.250	0.982 × 20 RH	0.750	0.656	0.531
5C	1	1.250	3.281	1.238 × 20 RH[a]	1.063	0.906	0.750
5M	5	1.438	3.438	1.238 × 20 RH	0.875	0.750	0.625
5NS	1	1.062	4.219	1.050 × 20 RH	0.875	0.750	0.625
5OS	1	3.500	3.406	0.937 × 18 RH	0.750	0.641	0.516
5P	1	0.812	3.687	0.807 × 24 RH	0.625	0.531	0.438
5PN	1	1.312	3.406	1.307 × 16 RH	1.000	0.875	0.719
5SC	1	0.600	2.438	0.500 × 26 RH	0.375	0.328	0.266
5ST	1	1.250	3.281	1.238 × 20 RH	1.063	0.906	0.750
5V	1	0.850	3.875	0.775 × 18 RH	0.563	0.484	0.391
6H	1	1.375	4.750	1.300 × 10 RH	1.125	0.969	0.797
6K	1	0.842	3.000	0.762 × 26 RH	0.625	0.531	0.438
6L	1	1.250	4.438	1.178 × 20 RH	1.000	0.875	0.719
6NS	1	1.312	5.906	1.234 × 14 RH	1.000	0.859	0.703
6R	1	1.375	4.938	1.300 × 20 RH	1.125	0.969	0.781
7B	4	7 B&S	3.125	0.375 × 16 RH	0.500	0.406	0.344
7 B&S	4	7 B&S	2.875	0.375 × 16 RH	0.500	0.406	0.344
7P	1	1.125	4.750	1.120 × 20 RH	0.875	0.750	0.625
7R	6	1.062	3.500	None	0.875	0.750	0.625
8H	1	1.500	4.750	1.425 × 20 RH	1.250	1.063	0.875
8ST	1	2.375	5.906	2.354 × 12 RH	2.125	1.844	1.500
8WN	1	1.250	3.875	1.245 × 16 RH	1.000	0.875	0.719
9B	4	9 B&S	4.125	0.500 × 13 RH	0.750	0.641	0.531
10L	1	1.562	5.500	1.490 × 18 RH	1.250	1.063	0.875
10P	1	1.500	4.750	1.495 × 20 RH	1.250	1.063	0.875
16C	1	1.889	4.516	1.875 × 1.75 mm RH[b]	1.625	1.406	1.141
20W	1	0.787	2.719	0.775 × 6–1 cm	0.563	0.484	0.391
22J	1	2.562	4.000	2.550 × 18 RH	2.250	1.938	1.563
32S	1	0.703	2.563	0.690 × 24 RH	0.500	0.438	0.344
35J	1	3.875	5.000	3.861 × 18 RH	3.500	3.000	2.438
42S	1	1.250	3.688	1.236 × 20 RH	1.000	0.875	0.719
50V	8	1.250	4.000	1.125 × 24 RH	0.938	0.813	0.656
52SC	1	0.800	3.688	0.795 × 20 RH	0.625	0.531	0.438
115	1	1.344	3.500	1.307 × 20 LH	1.125	0.969	0.797

Collets for Lathes, Mills, Grinders, and Fixtures *(Continued)*

Collet	Style	Dimensions			Max. Capacity (inches)		
		Bearing Diam., A	Length, B	Thread, C	Round	Hex	Square
215	1	2.030	4.750	1.990 × 18 LH	1.750	1.500	1.219
315	1	3.687	5.500	3.622 × 16 LH	3.250	2.813	2.250
B3	7	0.650	3.031	0.437 × 20 RH	0.500	0.438	0.344
D5	7	0.780	3.031	0.500 × 20 RH	0.625	0.531	0.438
GTM	7	0.625	2.437	0.437 × 20 RH	0.500	0.438	0.344
J&L	9	0.999	4.375	None	0.750	0.641	0.516
JC	8	1.360	4.000	None	1.188	1.000	0.813
LB	10	0.687	2.000	None	0.500	0.438	0.344
RO	11	1.250	2.938	0.875 × 16 RH	1.125	0.969	0.781
RO	12	1.250	4.437	0.875 × 16 RH	0.800	0.688	0.563
RO	12	1.250	4.437	0.875 × 16 RH	1.125	0.969	0.781
RO	11	1.250	2.938	0.875 × 16 RH	0.800	0.688	0.563
R8	7	0.950	4.000	0.437 × 20 RH	0.750	0.641	0.531

[a] Internal stop thread is 1.041 × 24 RH.
[b] Internal stop thread is 1.687 × 20 RH.
Dimensions in inches unless otherwise noted. Courtesy of Hardinge Brothers, Inc.

DIN 6388, Type B, and DIN 6499, ER Type Collets

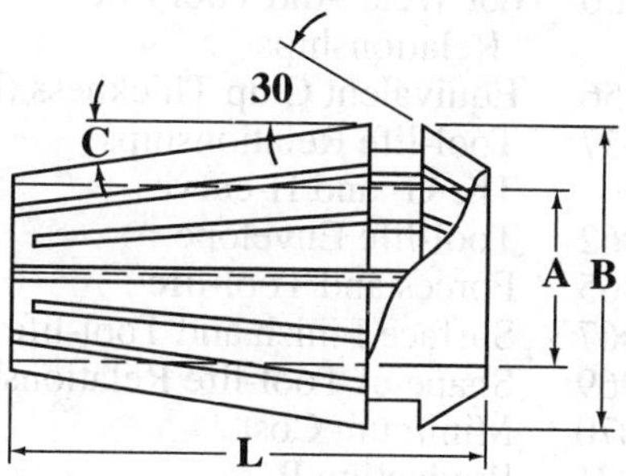

ER Type

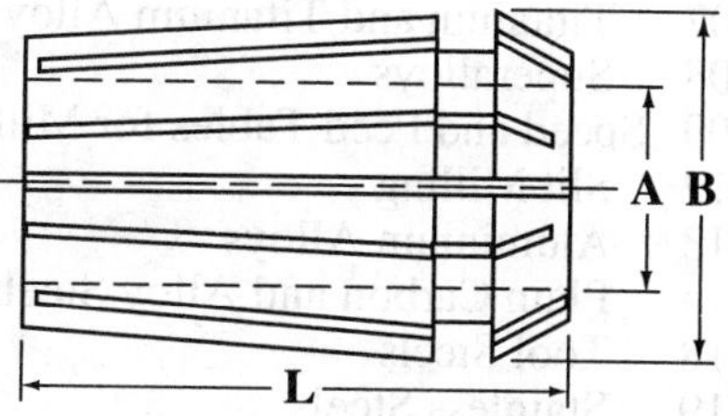

Type B

Collet Standard	Type	Dimensions			
		B (mm)	*L* (mm)	*A* (mm)	*C*
Type B, DIN 6388	16	25.50	40	4.5–16	…
	20	29.80	45	5.5–20	…
	25	35.05	52	5.5–25	…
	32	43.70	60	9.5–32	…
ER Type, DIN 6499	ERA8	8.50	13.5	0.5–5	8°
	ERA11	11.50	18	0.5–7	8°
	ERA16	17	27	0.5–10	8°
	ERA20	21	31	0.5–13	8°
	ERA25	26	35	0.5–16	8°
	ERA32	33	40	2–20	8°
	ERA40	41	46	3–26	8°
		41	39	26–30	8°
	ERA50	52	60	5–34	8°

MACHINING OPERATIONS

CUTTING SPEEDS AND FEEDS

SPEED AND FEED TABLES

ESTIMATING SPEEDS AND MACHINING POWER

MACHINING ECONOMETRICS

OTHER FEEDS AND SPEEDS

CUTTING FLUIDS

MACHINING NONFERROUS METALS

GRINDING FEEDS AND SPEEDS

GRINDING AND OTHER ABRASIVE PROCESSES

GRINDING AND OTHER ABRASIVE PROCESSES (cont.)

ARBORS, CHUCKS, AND SPINDLES

KNURLS AND KNURLING

MACHINE TOOL ACCURACY

NUMERICAL CONTROL

NUMERICAL CONTROL PROGRAMMING

CAD/CAM

CUTTING SPEEDS AND FEEDS

Work Materials.—The large number of work materials that are commonly machined vary greatly in their basic structure and the ease with which they can be machined. Yet it is possible to group together certain materials having similar machining characteristics, for the purpose of recommending the cutting speed at which they can be cut. Most materials that are machined are metals and it has been found that the most important single factor influencing the ease with which a metal can be cut is its microstructure, followed by any cold work that may have been done to the metal, which increases its hardness. Metals that have a similar, but not necessarily the same microstructure, will tend to have similar machining characteristics. Thus, the grouping of the metals in the accompanying tables has been done on the basis of their microstructure.

With the exception of a few soft and gummy metals, experience has shown that harder metals are more difficult to cut than softer metals. Furthermore, any given metal is more difficult to cut when it is in a harder form than when it is softer. It is more difficult to penetrate the harder metal and more power is required to cut it. These factors in turn will generate a higher cutting temperature at any given cutting speed, thereby making it necessary to use a slower speed, for the cutting temperature must always be kept within the limits that can be sustained by the cutting tool without failure. Hardness, then, is an important property that must be considered when machining a given metal. Hardness alone, however, cannot be used as a measure of cutting speed. For example, if pieces of AISI 11L17 and AISI 1117 steel both have a hardness of 150 Bhn, their recommended cutting speeds for high-speed steel tools will be 140 fpm and 130 fpm, respectively. In some metals, two entirely different microstructures can produce the same hardness. As an example, a fine pearlite microstructure and a tempered martensite microstructure can result in the same hardness in a steel. These microstructures will not machine alike. For practical purposes, however, information on hardness is usually easier to obtain than information on microstructure; thus, hardness alone is usually used to differentiate between different cutting speeds for machining a metal. In some situations, the hardness of a metal to be machined is not known. When the hardness is not known, the material condition can be used as a guide.

The surface of ferrous metal castings has a scale that is more difficult to machine than the metal below. Some scale is more difficult to machine than others, depending on the foundry sand used, the casting process, the method of cleaning the casting, and the type of metal cast. Special electrochemical treatments sometimes can be used that almost entirely eliminate the effect of the scale on machining, although castings so treated are not frequently encountered. Usually, when casting scale is encountered, the cutting speed is reduced approximately 5 or 10 per cent. Difficult-to-machine surface scale can also be encountered when machining hot-rolled or forged steel bars.

Metallurgical differences that affect machining characteristics are often found within a single piece of metal. The occurrence of hard spots in castings is an example. Different microstructures and hardness levels may occur within a casting as a result of variations in the cooling rate in different parts of the casting. Such variations are less severe in castings that have been heat treated. Steel bar stock is usually harder toward the outside than toward the center of the bar. Sometimes there are slight metallurgical differences along the length of a bar that can affect its cutting characteristics.

Cutting Tool Materials.—The recommended cutting feeds and speeds in the accompanying tables are given for high-speed steel, coated and uncoated carbides, ceramics, cermets, and polycrystalline diamonds. More data are available for HSS and carbides because these materials are the most commonly used. Other materials that are used to make cutting tools are cemented oxides or ceramics, cermets, cast nonferrous alloys (Stellite), single-crystal diamonds, polycrystalline diamonds, and cubic boron nitride.

Carbon Tool Steel: It is used primarily to make the less expensive drills, taps, and reamers. It is seldom used to make single-point cutting tools. Hardening in carbon steels is very shallow, although some have a small amount of vanadium and chromium added to improve their hardening quality. The cutting speed to use for plain carbon tool steel should be approximately one-half of the recommended speed for high-speed steel.

High-Speed Steel: This designates a number of steels having several properties that enhance their value as cutting tool material. They can be hardened to a high initial or room-temperature hardness ranging from 63 Rc to 65 Rc for ordinary high-speed steels and up to 70 Rc for the so-called superhigh-speed steels. They can retain sufficient hardness at temperatures up to 1,000 to 1,100°F to enable them to cut at cutting speeds that will generate these tool temperatures, and they will return to their original hardness when cooled to room temperature. They harden very deeply, enabling high-speed steels to be ground to the tool shape from solid stock and to be reground many times without sacrificing hardness at the cutting edge. High-speed steels can be made soft by annealing so that they can be machined into complex cutting tools such as drills, reamers, and milling cutters and then hardened.

The principal alloying elements of high-speed steels are tungsten (W), molybdenum (Mo), chromium (Cr), vanadium (V), together with carbon (C). There are a number of grades of high-speed steel that are divided into two types: tungsten high-speed steels and molybdenum high-speed steels. Tungsten high-speed steels are designated by the prefix T before the number that designates the grade. Molybdenum high-speed steels are designated by the prefix letter M. There is little performance difference between comparable grades of tungsten or molybdenum high-speed steel.

The addition of 5 to 12 per cent cobalt to high-speed steel increases its hardness at the temperatures encountered in cutting, thereby improving its wear resistance and cutting efficiency. Cobalt slightly increases the brittleness of high-speed steel, making it susceptible to chipping at the cutting edge. For this reason, cobalt high-speed steels are primarily made into single-point cutting tools that are used to take heavy roughing cuts in abrasive materials and through rough abrasive surface scales.

The M40 series and T15 are a group of high-hardness or so-called super high-speed steels that can be hardened to 70 Rc; however, they tend to be brittle and difficult to grind. For cutting applications, they are usually heat treated to 67–68 Rc to reduce their brittleness and tendency to chip. The M40 series is appreciably easier to grind than T15. They are recommended for machining tough die steels and other difficult-to-cut materials; they are not recommended for applications where conventional high-speed steels perform well. High-speed steels made by the powder-metallurgy process are tougher and have an improved grindability when compared with similar grades made by the customary process. Tools made of these steels can be hardened about 1 Rc higher than comparable high-speed steels made by the customary process without a sacrifice in toughness. They are particularly useful in applications involving intermittent cutting and where tool life is limited by chipping. All these steels augment rather than replace the conventional high-speed steels.

Cemented Carbides: They are also called sintered carbides or simply carbides. They are harder than high-speed steels and have excellent wear resistance. Information on cemented carbides and other hard metal tools is included in the section *CEMENTED CARBIDES* starting on page 747.

Cemented carbides retain a very high degree of hardness at temperatures up to 1400°F and even higher; therefore, very fast cutting speeds can be used. When used at fast cutting speeds, they produce good surface finishes on the workpiece. Carbides are more brittle than high-speed steel and, therefore, must be used with more care.

Hundreds of grades of carbides are available and attempts to classify these grades by area of application have not been entirely successful.

There are four distinct types of carbides: 1) straight tungsten carbides; 2) crater-resistant carbides; 3) titanium carbides; and 4) coated carbides.

Straight Tungsten Carbide: This is the most abrasion-resistant cemented carbide and is used to machine gray cast iron, most nonferrous metals, and nonmetallic materials, where abrasion resistance is the primary criterion. Straight tungsten carbide will rapidly form a crater on the tool face when used to machine steel, which reduces the life of the tool. Titanium carbide is added to tungsten carbide in order to counteract the rapid formation of the crater. In addition, tantalum carbide is usually added to prevent the cutting edge from deforming when subjected to the intense heat and pressure generated in taking heavy cuts.

Crater-Resistant Carbides: These carbides, containing titanium and tantalum carbides in addition to tungsten carbide, are used to cut steels, alloy cast irons, and other materials that have a strong tendency to form a crater.

Titanium Carbides: These carbides are made entirely from titanium carbide and small amounts of nickel and molybdenum. They have an excellent resistance to cratering and to heat. Their high hot hardness enables them to operate at higher cutting speeds, but they are more brittle and less resistant to mechanical and thermal shock. Therefore, they are not recommended for taking heavy or interrupted cuts. Titanium carbides are less abrasion-resistant and not recommended for cutting through scale or oxide films on steel. Although the resistance to cratering of titanium carbides is excellent, failure caused by crater formation can sometimes occur because the chip tends to curl very close to the cutting edge, thereby forming a small crater in this region that may break through.

Coated Carbides: These are available only as indexable inserts because the coating would be removed by grinding. The principal coating materials are titanium carbide (TiC), titanium nitride (TiN), and aluminum oxide (Al_2O_3). A very thin layer (approximately 0.0002 in.) of coating material is deposited over a cemented carbide insert; the material below the coating is called the substrate. The overall performance of the coated carbide is limited by the substrate, which provides the required toughness and resistance to deformation and thermal shock. With an equal tool life, coated carbides can operate at higher cutting speeds than uncoated carbides. The increase may be 20 to 30 per cent and sometimes up to 50 per cent faster. Titanium carbide and titanium nitride coated carbides usually operate in the medium (200–800 fpm) cutting speed range, and aluminum oxide coated carbides are used in the higher (800–1600 fpm) cutting speed range.

Carbide Grade Selection: The selection of the best grade of carbide for a particular application is very important. An improper grade of carbide will result in a poor performance—it may even cause the cutting edge to fail before any significant amount of cutting has been done. Because of the many grades and the many variables that are involved, the carbide producers should be consulted to obtain recommendations for the application of their grades of carbide. A few general guidelines can be given that are useful to form an orientation. Metal cutting carbides usually range in hardness from about 89.5 Ra (Rockwell A Scale) to 93.0 Ra with the exception of titanium carbide, which has a hardness range of 90.5 Ra to 93.5 Ra. Generally, the harder carbides are more wear-resistant and more brittle, whereas the softer carbides are less wear-resistant but tougher. A choice of hardness must be made to suit the given application. The very hard carbides are generally used for taking light finishing cuts. For other applications, select the carbide that has the highest hardness with sufficient strength to prevent chipping or breaking. Straight tungsten carbide grades should always be used unless cratering is encountered. Straight tungsten carbides are used to machine gray cast iron, ferritic malleable iron, austenitic stainless steel, high-temperature alloys, copper, brass, bronze, aluminum alloys, zinc alloy die castings, and plastics. Crater-resistant carbides should be used to machine plain carbon steel, alloy steel, tool steel, pearlitic malleable iron, nodular iron, other highly alloyed cast irons, ferritic stainless steel, martensitic stainless steel, and certain high-temperature alloys. Titanium carbides are recommended for taking high-speed finishing and semifinishing cuts on steel, especially the low-carbon, low-alloy steels, which are less abrasive and have a strong tendency to form a crater. They are also used to take light cuts on alloy cast iron and on some high-nickel alloys. Nonferrous materials, such as some aluminum alloys and brass, that are essentially nonabrasive may also be machined with titanium carbides. Abrasive materials and others that should not be machined with titanium carbides include gray cast iron, titanium alloys, cobalt- and nickel-base superalloys, stainless steel, bronze, many aluminum alloys, fiberglass, plastics, and graphite. The feed used should not exceed about 0.020 inch per revolution.

Coated carbides can be used to take cuts ranging from light finishing to heavy roughing on most materials that can be cut with these carbides. The coated carbides are recommended for machining all free-machining steels, all plain carbon and alloy steels, tool steels, martensitic and ferritic stainless steels, precipitation-hardening stainless steels, alloy cast iron, pearlitic and martensitic malleable iron, and nodular iron. They are also recommended for taking light finishing and roughing cuts on austenitic stainless steels. Coated carbides should not be used to machine nickel- and cobalt-base superalloys, titanium and titanium alloys, brass, bronze, aluminum alloys, pure metals, refractory metals, and nonmetals such as fiberglass, graphite, and plastics.

Ceramic Cutting Tool Materials: These are made from finely powdered aluminum oxide particles sintered into a hard dense structure without a binder material. Aluminum oxide is also combined with titanium carbide to form a composite, which is called a cermet. These materials have a very high hot hardness enabling very high cutting speeds to be used. For example, ceramic cutting tools have been used to cut AISI 1040 steel at a cutting speed of 18,000 fpm with a satisfactory tool life. However, much lower cutting speeds, in the range of 1000 to 4000 fpm and lower, are more common because of limitations placed by the machine tool, cutters, and chucks. Although most applications of ceramic and cermet cutting tool materials are for turning, they have also been used successfully for milling. Ceramics and cermets are relatively brittle and a special cutting edge preparation is required to prevent chipping or edge breakage. This preparation consists of honing or grinding a narrow flat land, 0.002 to 0.006 inch wide, on the cutting edge that is made about 30 degrees with respect to the tool face. For some heavy-duty applications, a wider land is used. The setup should be as rigid as possible and the feed rate should not normally exceed 0.020 inch, although 0.030 inch has been used successfully. Ceramics and cermets are recommended for roughing and finishing operations on all cast irons, plain carbon and alloy steels, and stainless steels. Materials up to a hardness of 60 Rockwell C Scale can be cut with ceramic and cermet cutting tools. These tools should not be used to machine aluminum and aluminum alloys, magnesium alloys, titanium, and titanium alloys.

Cast Nonferrous Alloy: Cutting tools of this alloy are made from tungsten, tantalum, chromium, and cobalt plus carbon. Other alloying elements are also used to produce materials with high temperature and wear resistance. These alloys cannot be softened by heat treatment and must be cast and ground to shape. The room-temperature hardness of cast nonferrous alloys is lower than for high-speed steel, but the hardness and wear resistance is retained to a higher temperature. The alloys are generally marketed under trade names such as Stellite, Crobalt, and Tantung. The initial cutting speed for cast nonferrous tools can be 20 to 50 per cent greater than the recommended cutting speed for high-speed steel as given in the accompanying tables.

Diamond Cutting Tools: These are available in three forms: single-crystal natural diamonds shaped to a cutting edge and mounted on a tool holder on a boring bar; polycrystalline diamond indexable inserts made from synthetic or natural diamond powders that have been compacted and sintered into a solid mass, and chemically vapor-deposited diamond. Single-crystal and polycrystalline diamond cutting tools are very wear-resistant, and are recommended for machining abrasive materials that cause other cutting tool materials to wear rapidly. Typical of the abrasive materials machined with single-crystal and polycrystalline diamond tools and cutting speeds used are the following: fiberglass, 300 to 1000 fpm; fused silica, 900 to 950 fpm; reinforced melamine plastics, 350 to 1000 fpm; reinforced phenolic plastics, 350 to 1000 fpm; thermosetting plastics, 300 to 2000 fpm; Teflon, 600 fpm; nylon, 200 to 300 fpm; mica, 300 to 1000 fpm; graphite, 200 to 2000 fpm; babbitt bearing metal, 700 fpm; and aluminum-silicon alloys, 1000 to 2000 fpm. Another important application of diamond cutting tools is to produce fine surface finishes on soft nonferrous metals that are difficult to finish by other methods. Surface finishes of 1 to 2 microinches can be readily obtained with single-crystal diamond tools, and finishes down to 10 microinches can be obtained with polycrystalline diamond tools. In addition to babbitt and the aluminum-silicon alloys, other metals finished with diamond tools include: soft aluminum, 1000 to 2000 fpm; all wrought and cast aluminum alloys, 600 to 1500 fpm; copper, 1000 fpm; brass, 500 to 1000 fpm; bronze, 300 to 600 fpm; oilite bearing metal, 500 fpm; silver, gold, and platinum, 300 to 2500 fpm; and zinc, 1000 fpm. Ferrous alloys, such as cast iron and steel, should not be machined with diamond cutting tools because the high cutting temperatures generated will cause the diamond to transform into carbon.

Chemically Vapor-Deposited (CVD) Diamond: This is a new tool material offering performance characteristics well suited to highly abrasive or corrosive materials, and hard-to-machine composites. CVD diamond is available in two forms: thick-film tools, which are fabricated by brazing CVD diamond tips, approximately 0.020 inch (0.5 mm) thick, to carbide substrates; and thin-film tools, having a pure diamond coating over the rake and flank surfaces of a ceramic or carbide substrate.

CVD is pure diamond, made at low temperatures and pressures, with no metallic binder phase. This diamond purity gives CVD diamond tools extreme hardness, high abrasion resistance, low friction, high thermal conductivity, and chemical inertness. CVD tools are generally used as direct replacements for PCD (polycrystalline diamond) tools, primarily in finishing, semifinishing, and continuous turning applications of extremely wear-intensive materials. The small grain size of CVD diamond (ranging from less than 1 μm to 50 μm) yields superior surface finishes compared with PCD, and the higher thermal conductivity and better thermal and chemical stability of pure diamond allow CVD tools to operate at faster speeds without generating harmful levels of heat. The extreme hardness of CVD tools may also result in significantly longer tool life.

CVD diamond cutting tools are recommended for the following materials: aluminum and other ductile; nonferrous alloys such as copper, brass, and bronze; and highly abrasive composite materials such as graphite, carbon-carbon, carbon-filled phenolic, fiberglass, and honeycomb materials.

Cubic Boron Nitride (CBN): Next to diamond, CBN is the hardest known material. It will retain its hardness at a temperature of 1800°F and higher, making it an ideal cutting tool material for machining very hard and tough materials at cutting speeds beyond those possible with other cutting tool materials. Indexable inserts and cutting tool blanks made from this material consist of a layer, approximately 0.020 inch thick, of polycrystalline cubic boron nitride firmly bonded to the top of a cemented carbide substrate. Cubic boron nitride is recommended for rough and finish turning hardened plain carbon and alloy steels, hardened tool steels, hard cast irons, all hardness grades of gray cast iron, and superalloys. As a class, the superalloys are not as hard as hardened steel; however, their combination of high strength and tendency to deform plastically under the pressure of the cut, or gumminess, places them in the class of hard-to-machine materials. Conventional materials that can be readily machined with other cutting tool materials should not be machined with cubic boron nitride. Round indexable CBN inserts are recommended when taking severe cuts in order to provide maximum strength to the insert. When using square or triangular inserts, a large lead angle should be used, normally 15°, and whenever possible, 45°. A negative rake angle should always be used, which for most applications is negative 5°. The relief angle should be 5° to 9°. Although cubic boron nitride cutting tools can be used without a coolant, flooding the tool with a water-soluble type coolant is recommended.

Cutting Speed, Feed, Depth of Cut, Tool Wear, and Tool Life.—The cutting conditions that determine the rate of metal removal are the cutting speed, the feed rate, and the depth of cut. These cutting conditions and the nature of the material to be cut determine the power required to take the cut. The cutting conditions must be adjusted to stay within the power available on the machine tool to be used. Power requirements are discussed in Estimating Machining Power later in this section.

The cutting conditions must also be considered in relation to the tool life. Tool life is defined as the cutting time to reach a predetermined amount of wear, usually flank wear. Tool life is determined by assessing the time—the tool life—at which a given predetermined flank wear is reached (0.01, 0.015, 0.025, 0.03 inch, for example). This amount of wear is called the tool wear criterion, and its size depends on the tool grade used. Usually, a tougher grade can be used with a bigger flank wear, but for finishing operations, where close tolerances are required, the wear criterion is relatively small. Other wear criteria are a predetermined value of the machined surface roughness and the depth of the crater that develops on the rake face of the tool.

The ANSI standard, Specification For Tool Life Testing With Single-Point Tools (ANSI B94.55M-1985), defines the end of tool life as a given amount of wear on the flank of a tool. This standard is followed when making scientific machinability tests with single-point cutting tools in order to achieve uniformity in testing procedures so that results from different machinability laboratories can be readily compared. It is not practicable or necessary to follow this standard in the shop; however, it should be understood that the cutting conditions and tool life are related.

Tool life is influenced most by cutting speed, then by the feed rate, and least by the depth of cut. When the depth of cut is increased to about 10 times greater than the feed, a further increase in the depth of cut will have no significant effect on the tool life. This characteristic of the cutting tool performance is very important in determining the operating or cutting conditions for machining metals. Conversely, if the cutting speed or the feed is decreased, the increase in the tool life will be proportionately greater than the decrease in the cutting speed or the feed.

Tool life is reduced when either feed or cutting speed is increased. For example, the cutting speed and the feed may be increased if a shorter tool life is accepted; furthermore, the reduction in the tool life will be proportionately greater than the increase in the cutting speed or the feed. However, it is less well understood that a higher feed rate (feed/rev × speed) may result in a longer tool life if a higher feed/rev is used in combination with a lower cutting speed. This principle is well illustrated in the speed tables of this section, where two sets of feed and speed data are given (labeled *optimum* and *average*) that result in the same tool life. The *optimum* set results in a greater feed rate (i.e., increased productivity) although the feed/rev is higher and cutting speed lower than the *average* set. Complete instructions for using the speed tables and for estimating tool life are given in *How to Use the Feeds and Speeds Tables* starting on page 991.

Selecting Cutting Conditions.—The first step in establishing the cutting conditions is to select the depth of cut. The depth of cut will be limited by the amount of metal that is to be machined from the workpiece, by the power available on the machine tool, by the rigidity of the workpiece and the cutting tool, and by the rigidity of the setup. The depth of cut has the least effect upon the tool life, so the heaviest possible depth of cut should always be used.

The second step is to select the feed (feed/rev for turning, drilling, and reaming, or feed/tooth for milling). The available power must be sufficient to make the required depth of cut at the selected feed. The maximum feed possible that will produce an acceptable surface finish should be selected.

The third step is to select the cutting speed. Although the accompanying tables provide recommended cutting speeds and feeds for many materials, experience in machining a certain material may form the best basis for adjusting the given cutting speeds to a particular job. However, in general, the depth of cut should be selected first, followed by the feed, and last the cutting speed.

Table 1. Tool Troubleshooting Check List

Problem	Tool Material	Remedy
Excessive flank wear—Tool life too short	Carbide	1. Change to harder, more wear-resistant grade 2. Reduce the cutting speed 3. Reduce the cutting speed and increase the feed to maintain production 4. Reduce the feed 5. For work-hardenable materials—increase the feed 6. Increase the lead angle 7. Increase the relief angles
	HSS	1. Use a coolant 2. Reduce the cutting speed 3. Reduce the cutting speed and increase the feed to maintain production 4. Reduce the feed 5. For work-hardenable materials—increase the feed 6. Increase the lead angle 7. Increase the relief angle
Excessive cratering	Carbide	1. Use a crater-resistant grade 2. Use a harder, more wear-resistant grade 3. Reduce the cutting speed 4. Reduce the feed 5. Widen the chip breaker groove
	HSS	1. Use a coolant 2. Reduce the cutting speed 3. Reduce the feed 4. Widen the chip breaker groove
Cutting edge chipping	Carbide	1. Increase the cutting speed 2. Lightly hone the cutting edge 3. Change to a tougher grade 4. Use negative-rake tools 5. Increase the lead angle 6. Reduce the feed 7. Reduce the depth of cut 8. Reduce the relief angles 9. If low cutting speed must be used, use a high-additive EP cutting fluid
	HSS	1. Use a high additive EP cutting fluid 2. Lightly hone the cutting edge before using 3. Increase the lead angle 4. Reduce the feed 5. Reduce the depth of cut 6. Use a negative rake angle 7. Reduce the relief angles
	Carbide and HSS	1. Check the setup for cause if chatter occurs 2. Check the grinding procedure for tool overheating 3. Reduce the tool overhang
Cutting edge deformation	Carbide	1. Change to a grade containing more tantalum 2. Reduce the cutting speed 3. Reduce the feed
Poor surface finish	Carbide	1. Increase the cutting speed 2. If low cutting speed must be used, use a high additive EP cutting fluid 4. For light cuts, use straight titanium carbide grade 5. Increase the nose radius 6. Reduce the feed 7. Increase the relief angles 8. Use positive rake tools

Table 1. *(Continued)* **Tool Troubleshooting Check List**

Problem	Tool Material	Remedy
Poor surface finish *(Continued)*	HSS	1. Use a high additive EP cutting fluid 2. Increase the nose radius 3. Reduce the feed 4. Increase the relief angles 5. Increase the rake angles
	Diamond	1. Use diamond tool for soft materials
Notching at the depth of cut line	Carbide and HSS	1. Increase the lead angle 2. Reduce the feed

Cutting Speed Formulas.—Most machining operations are conducted on machine tools having a rotating spindle. Cutting speeds are usually given in feet or meters per minute and these speeds must be converted to spindle speeds, in revolutions per minute, to operate the machine. Conversion is accomplished by use of the following formulas:

For U.S. units:

$$N = \frac{12V}{\pi D} = 3.82\frac{V}{D} \text{ rpm}$$

For metric units:

$$N = \frac{1000V}{\pi D} = 318.3\frac{V}{D} \text{ rpm}$$

where N is the spindle speed in revolutions per minute (rpm); V is the cutting speed in feet per minute (fpm) for U.S. units and meters per minute (m/min) for metric units. In turning, D is the diameter of the workpiece; in milling, drilling, reaming, and other operations that use a rotating tool, D is the cutter diameter in inches for U.S. units and in millimeters for metric units. $\pi = 3.1417$.

Example: The cutting speed for turning a 4-inch (102-mm) diameter bar has been found to be 575 fpm (175.3 m/min). Using both the inch and metric formulas, calculate the lathe spindle speed.

$$N = \frac{12V}{\pi D} = \frac{12 \times 575}{3.1417 \times 4} = 549 \text{ rpm} \qquad N = \frac{1000V}{\pi D} = \frac{1000 \times 175.3}{3.1417 \times 102} = 547 \text{ rpm}$$

The small difference in the answers is due to rounding off the numbers and to the lack of precision of the inch–metric conversion.

When the cutting tool or workpiece diameter and the spindle speed in rpm are known, it is often necessary to calculate the cutting speed in feet or meters per minute. In this event, the following formulas are used.

For U.S. units:

$$V = \frac{\pi DN}{12} \text{ fpm}$$

For metric units:

$$V = \frac{\pi DN}{1000} \text{ m/min}$$

As in the previous formulas, N is the rpm and D is the diameter in inches for the U.S. unit formula and in millimeters for the metric formula.

Example: Calculate the cutting speed in feet per minute and in meters per minute if the spindle speed of a ¾-inch (19.05-mm) drill is 400 rpm.

$$V = \frac{\pi DN}{12} = \frac{\pi \times 0.75 \times 400}{12} = 78.5 \text{ fpm}$$

$$V = \frac{\pi DN}{1000} = \frac{\pi \times 19.05 \times 400}{1000} = 24.9 \text{ m/min}$$

Cutting Speeds and Equivalent RPM for Drills of Number and Letter Sizes

Size No.	Cutting Speed, Feet per Minute										
	30′	40′	50′	60′	70′	80′	90′	100′	110′	130′	150′
	Revolutions per Minute for Number Sizes										
1	503	670	838	1005	1173	1340	1508	1675	1843	2179	2513
2	518	691	864	1037	1210	1382	1555	1728	1901	2247	2593
4	548	731	914	1097	1280	1462	1645	1828	2010	2376	2741
6	562	749	936	1123	1310	1498	1685	1872	2060	2434	2809
8	576	768	960	1151	1343	1535	1727	1919	2111	2495	2879
10	592	790	987	1184	1382	1579	1777	1974	2171	2566	2961
12	606	808	1010	1213	1415	1617	1819	2021	2223	2627	3032
14	630	840	1050	1259	1469	1679	1889	2099	2309	2728	3148
16	647	863	1079	1295	1511	1726	1942	2158	2374	2806	3237
18	678	904	1130	1356	1582	1808	2034	2260	2479	2930	3380
20	712	949	1186	1423	1660	1898	2135	2372	2610	3084	3559
22	730	973	1217	1460	1703	1946	2190	2433	2676	3164	3649
24	754	1005	1257	1508	1759	2010	2262	2513	2764	3267	3769
26	779	1039	1299	1559	1819	2078	2338	2598	2858	3378	3898
28	816	1088	1360	1631	1903	2175	2447	2719	2990	3534	4078
30	892	1189	1487	1784	2081	2378	2676	2973	3270	3864	4459
32	988	1317	1647	1976	2305	2634	2964	3293	3622	4281	4939
34	1032	1376	1721	2065	2409	2753	3097	3442	3785	4474	5162
36	1076	1435	1794	2152	2511	2870	3228	3587	3945	4663	5380
38	1129	1505	1882	2258	2634	3010	3387	3763	4140	4892	5645
40	1169	1559	1949	2339	2729	3118	3508	3898	4287	5067	5846
42	1226	1634	2043	2451	2860	3268	3677	4085	4494	5311	6128
44	1333	1777	2221	2665	3109	3554	3999	4442	4886	5774	6662
46	1415	1886	2358	2830	3301	3773	4244	4716	5187	6130	7074
48	1508	2010	2513	3016	3518	4021	4523	5026	5528	6534	7539
50	1637	2183	2729	3274	3820	4366	4911	5457	6002	7094	8185
52	1805	2406	3008	3609	4211	4812	5414	6015	6619	7820	9023
54	2084	2778	3473	4167	4862	5556	6251	6945	7639	9028	10417
Size	Revolutions per Minute for Letter Sizes										
A	491	654	818	982	1145	1309	1472	1636	1796	2122	2448
B	482	642	803	963	1124	1284	1445	1605	1765	2086	2407
C	473	631	789	947	1105	1262	1420	1578	1736	2052	2368
D	467	622	778	934	1089	1245	1400	1556	1708	2018	2329
E	458	611	764	917	1070	1222	1375	1528	1681	1968	2292
F	446	594	743	892	1040	1189	1337	1486	1635	1932	2229
G	440	585	732	878	1024	1170	1317	1463	1610	1903	2195
H	430	574	718	862	1005	1149	1292	1436	1580	1867	2154
I	421	562	702	842	983	1123	1264	1404	1545	1826	2106
J	414	552	690	827	965	1103	1241	1379	1517	1793	2068
K	408	544	680	815	951	1087	1223	1359	1495	1767	2039
L	395	527	659	790	922	1054	1185	1317	1449	1712	1976
M	389	518	648	777	907	1036	1166	1295	1424	1683	1942
N	380	506	633	759	886	1012	1139	1265	1391	1644	1897
O	363	484	605	725	846	967	1088	1209	1330	1571	1813
P	355	473	592	710	828	946	1065	1183	1301	1537	1774
Q	345	460	575	690	805	920	1035	1150	1266	1496	1726
R	338	451	564	676	789	902	1014	1127	1239	1465	1690
S	329	439	549	659	769	878	988	1098	1207	1427	1646
T	320	426	533	640	746	853	959	1066	1173	1387	1600
U	311	415	519	623	727	830	934	1038	1142	1349	1557
V	304	405	507	608	709	810	912	1013	1114	1317	1520
W	297	396	495	594	693	792	891	989	1088	1286	1484
X	289	385	481	576	672	769	865	962	1058	1251	1443
Y	284	378	473	567	662	756	851	945	1040	1229	1418
Z	277	370	462	555	647	740	832	925	1017	1202	1387

For fractional drill sizes, use the following table.

Revolutions per Minute for Various Cutting Speeds and Diameters

Dia., Inches	Cutting Speed, Feet per Minute											
	40	50	60	70	80	90	100	120	140	160	180	200
	Revolutions per Minute											
1/4	611	764	917	1070	1222	1376	1528	1834	2139	2445	2750	3056
5/16	489	611	733	856	978	1100	1222	1466	1711	1955	2200	2444
3/8	408	509	611	713	815	916	1018	1222	1425	1629	1832	2036
7/16	349	437	524	611	699	786	874	1049	1224	1398	1573	1748
1/2	306	382	459	535	611	688	764	917	1070	1222	1375	1528
9/16	272	340	407	475	543	611	679	813	951	1086	1222	1358
5/8	245	306	367	428	489	552	612	736	857	979	1102	1224
11/16	222	273	333	389	444	500	555	666	770	888	999	1101
3/4	203	254	306	357	408	458	508	610	711	813	914	1016
13/16	190	237	284	332	379	427	474	569	664	758	853	948
7/8	175	219	262	306	349	392	438	526	613	701	788	876
15/16	163	204	244	285	326	366	407	488	570	651	733	814
1	153	191	229	267	306	344	382	458	535	611	688	764
1 1/16	144	180	215	251	287	323	359	431	503	575	646	718
1 1/8	136	170	204	238	272	306	340	408	476	544	612	680
1 3/16	129	161	193	225	258	290	322	386	451	515	580	644
1 1/4	123	153	183	214	245	274	306	367	428	490	551	612
1 5/16	116	146	175	204	233	262	291	349	407	466	524	582
1 3/8	111	139	167	195	222	250	278	334	389	445	500	556
1 7/16	106	133	159	186	212	239	265	318	371	424	477	530
1 1/2	102	127	153	178	204	230	254	305	356	406	457	508
1 9/16	97.6	122	146	171	195	220	244	293	342	390	439	488
1 5/8	93.9	117	141	165	188	212	234	281	328	374	421	468
1 11/16	90.4	113	136	158	181	203	226	271	316	362	407	452
1 3/4	87.3	109	131	153	175	196	218	262	305	349	392	436
1 7/8	81.5	102	122	143	163	184	204	244	286	326	367	408
2	76.4	95.5	115	134	153	172	191	229	267	306	344	382
2 1/8	72.0	90.0	108	126	144	162	180	216	252	288	324	360
2 1/4	68.0	85.5	102	119	136	153	170	204	238	272	306	340
2 3/8	64.4	80.5	96.6	113	129	145	161	193	225	258	290	322
2 1/2	61.2	76.3	91.7	107	122	138	153	184	213	245	275	306
2 5/8	58.0	72.5	87.0	102	116	131	145	174	203	232	261	290
2 3/4	55.6	69.5	83.4	97.2	111	125	139	167	195	222	250	278
2 7/8	52.8	66.0	79.2	92.4	106	119	132	158	185	211	238	264
3	51.0	63.7	76.4	89.1	102	114	127	152	178	203	228	254
3 1/8	48.8	61.0	73.2	85.4	97.6	110	122	146	171	195	219	244
3 1/4	46.8	58.5	70.2	81.9	93.6	105	117	140	164	188	211	234
3 3/8	45.2	56.5	67.8	79.1	90.4	102	113	136	158	181	203	226
3 1/2	43.6	54.5	65.5	76.4	87.4	98.1	109	131	153	174	196	218
3 5/8	42.0	52.5	63.0	73.5	84.0	94.5	105	126	147	168	189	210
3 3/4	40.8	51.0	61.2	71.4	81.6	91.8	102	122	143	163	184	205
3 7/8	39.4	49.3	59.1	69.0	78.8	88.6	98.5	118	138	158	177	197
4	38.2	47.8	57.3	66.9	76.4	86.0	95.6	115	134	153	172	191
4 1/4	35.9	44.9	53.9	62.9	71.8	80.8	89.8	108	126	144	162	180
4 1/2	34.0	42.4	51.0	59.4	67.9	76.3	84.8	102	119	136	153	170
4 3/4	32.2	40.2	48.2	56.3	64.3	72.4	80.4	96.9	113	129	145	161
5	30.6	38.2	45.9	53.5	61.1	68.8	76.4	91.7	107	122	138	153
5 1/4	29.1	36.4	43.6	50.9	58.2	65.4	72.7	87.2	102	116	131	145
5 1/2	27.8	34.7	41.7	48.6	55.6	62.5	69.4	83.3	97.2	111	125	139
5 3/4	26.6	33.2	39.8	46.5	53.1	59.8	66.4	80.0	93.0	106	120	133
6	25.5	31.8	38.2	44.6	51.0	57.2	63.6	76.3	89.0	102	114	127
6 1/4	24.4	30.6	36.7	42.8	48.9	55.0	61.1	73.3	85.5	97.7	110	122
6 1/2	23.5	29.4	35.2	41.1	47.0	52.8	58.7	70.4	82.2	93.9	106	117
6 3/4	22.6	28.3	34.0	39.6	45.3	50.9	56.6	67.9	79.2	90.6	102	113
7	21.8	27.3	32.7	38.2	43.7	49.1	54.6	65.5	76.4	87.4	98.3	109
7 1/4	21.1	26.4	31.6	36.9	42.2	47.4	52.7	63.2	73.8	84.3	94.9	105
7 1/2	20.4	25.4	30.5	35.6	40.7	45.8	50.9	61.1	71.0	81.4	91.6	102
7 3/4	19.7	24.6	29.5	34.4	39.4	44.3	49.2	59.0	68.9	78.7	88.6	98.4
8	19.1	23.9	28.7	33.4	38.2	43.0	47.8	57.4	66.9	76.5	86.0	95.6

Revolutions per Minute for Various Cutting Speeds and Diameters

Dia., Inches	Cutting Speed, Feet per Minute											
	225	250	275	300	325	350	375	400	425	450	500	550
	Revolutions per Minute											
1/4	3438	3820	4202	4584	4966	5348	5730	6112	6493	6875	7639	8403
5/16	2750	3056	3362	3667	3973	4278	4584	4889	5195	5501	6112	6723
3/8	2292	2546	2801	3056	3310	3565	3820	4074	4329	4584	5093	5602
7/16	1964	2182	2401	2619	2837	3056	3274	3492	3710	3929	4365	4802
1/2	1719	1910	2101	2292	2483	2675	2866	3057	3248	3439	3821	4203
9/16	1528	1698	1868	2037	2207	2377	2547	2717	2887	3056	3396	3736
5/8	1375	1528	1681	1834	1987	2139	2292	2445	2598	2751	3057	3362
11/16	1250	1389	1528	1667	1806	1941	2084	2223	2362	2501	2779	3056
3/4	1146	1273	1401	1528	1655	1783	1910	2038	2165	2292	2547	2802
13/16	1058	1175	1293	1410	1528	1646	1763	1881	1998	2116	2351	2586
7/8	982	1091	1200	1310	1419	1528	1637	1746	1855	1965	2183	2401
15/16	917	1019	1120	1222	1324	1426	1528	1630	1732	1834	2038	2241
1	859	955	1050	1146	1241	1337	1432	1528	1623	1719	1910	2101
1 1/16	809	899	988	1078	1168	1258	1348	1438	1528	1618	1798	1977
1 1/8	764	849	933	1018	1103	1188	1273	1358	1443	1528	1698	1867
1 3/16	724	804	884	965	1045	1126	1206	1287	1367	1448	1609	1769
1 1/4	687	764	840	917	993	1069	1146	1222	1299	1375	1528	1681
1 5/16	654	727	800	873	946	1018	1091	1164	1237	1309	1455	1601
1 3/8	625	694	764	833	903	972	1042	1111	1181	1250	1389	1528
1 7/16	598	664	730	797	863	930	996	1063	1129	1196	1329	1461
1 1/2	573	636	700	764	827	891	955	1018	1082	1146	1273	1400
1 9/16	550	611	672	733	794	855	916	978	1039	1100	1222	1344
1 5/8	528	587	646	705	764	822	881	940	999	1057	1175	1293
1 11/16	509	566	622	679	735	792	849	905	962	1018	1132	1245
1 3/4	491	545	600	654	709	764	818	873	927	982	1091	1200
1 13/16	474	527	579	632	685	737	790	843	895	948	1054	1159
1 7/8	458	509	560	611	662	713	764	815	866	917	1019	1120
1 15/16	443	493	542	591	640	690	739	788	838	887	986	1084
2	429	477	525	573	620	668	716	764	811	859	955	1050
2 1/8	404	449	494	539	584	629	674	719	764	809	899	988
2 1/4	382	424	468	509	551	594	636	679	721	764	849	933
2 3/8	362	402	442	482	522	563	603	643	683	724	804	884
2 1/2	343	382	420	458	496	534	573	611	649	687	764	840
2 5/8	327	363	400	436	472	509	545	582	618	654	727	800
2 3/4	312	347	381	416	451	486	520	555	590	625	694	763
2 7/8	299	332	365	398	431	465	498	531	564	598	664	730
3	286	318	350	381	413	445	477	509	541	572	636	700
3 1/8	274	305	336	366	397	427	458	488	519	549	611	672
3 1/4	264	293	323	352	381	411	440	470	499	528	587	646
3 3/8	254	283	311	339	367	396	424	452	481	509	566	622
3 1/2	245	272	300	327	354	381	409	436	463	490	545	600
3 5/8	237	263	289	316	342	368	395	421	447	474	527	579
3 3/4	229	254	280	305	331	356	382	407	433	458	509	560
3 7/8	221	246	271	295	320	345	369	394	419	443	493	542
4	214	238	262	286	310	334	358	382	405	429	477	525
4 1/4	202	224	247	269	292	314	337	359	383	404	449	494
4 1/2	191	212	233	254	275	297	318	339	360	382	424	466
4 3/4	180	201	221	241	261	281	301	321	341	361	402	442
5	171	191	210	229	248	267	286	305	324	343	382	420
5 1/4	163	181	199	218	236	254	272	290	308	327	363	399
5 1/2	156	173	190	208	225	242	260	277	294	312	347	381
5 3/4	149	166	182	199	215	232	249	265	282	298	332	365
6	143	159	174	190	206	222	238	254	270	286	318	349
6 1/4	137	152	168	183	198	213	229	244	259	274	305	336
6 1/2	132	146	161	176	190	205	220	234	249	264	293	322
6 3/4	127	141	155	169	183	198	212	226	240	254	283	311
7	122	136	149	163	177	190	204	218	231	245	272	299
7 1/4	118	131	144	158	171	184	197	210	223	237	263	289
7 1/2	114	127	139	152	165	178	190	203	216	229	254	279
7 3/4	111	123	135	148	160	172	185	197	209	222	246	271
8	107	119	131	143	155	167	179	191	203	215	238	262

Revolutions per Minute for Various Cutting Speeds and Diameters (Metric Units)

Dia., mm	Cutting Speed, Meters per Minute											
	5	6	8	10	12	16	20	25	30	35	40	45
	Revolutions per Minute											
5	318	382	509	637	764	1019	1273	1592	1910	2228	2546	2865
6	265	318	424	530	637	849	1061	1326	1592	1857	2122	2387
8	199	239	318	398	477	637	796	995	1194	1393	1592	1790
10	159	191	255	318	382	509	637	796	955	1114	1273	1432
12	133	159	212	265	318	424	531	663	796	928	1061	1194
16	99.5	119	159	199	239	318	398	497	597	696	796	895
20	79.6	95.5	127	159	191	255	318	398	477	557	637	716
25	63.7	76.4	102	127	153	204	255	318	382	446	509	573
30	53.1	63.7	84.9	106	127	170	212	265	318	371	424	477
35	45.5	54.6	72.8	90.9	109	145	182	227	273	318	364	409
40	39.8	47.7	63.7	79.6	95.5	127	159	199	239	279	318	358
45	35.4	42.4	56.6	70.7	84.9	113	141	177	212	248	283	318
50	31.8	38.2	51	63.7	76.4	102	127	159	191	223	255	286
55	28.9	34.7	46.3	57.9	69.4	92.6	116	145	174	203	231	260
60	26.6	31.8	42.4	53.1	63.7	84.9	106	133	159	186	212	239
65	24.5	29.4	39.2	49	58.8	78.4	98	122	147	171	196	220
70	22.7	27.3	36.4	45.5	54.6	72.8	90.9	114	136	159	182	205
75	21.2	25.5	34	42.4	51	68	84.9	106	127	149	170	191
80	19.9	23.9	31.8	39.8	47.7	63.7	79.6	99.5	119	139	159	179
90	17.7	21.2	28.3	35.4	42.4	56.6	70.7	88.4	106	124	141	159
100	15.9	19.1	25.5	31.8	38.2	51	63.7	79.6	95.5	111	127	143
110	14.5	17.4	23.1	28.9	34.7	46.2	57.9	72.3	86.8	101	116	130
120	13.3	15.9	21.2	26.5	31.8	42.4	53.1	66.3	79.6	92.8	106	119
130	12.2	14.7	19.6	24.5	29.4	39.2	49	61.2	73.4	85.7	97.9	110
140	11.4	13.6	18.2	22.7	27.3	36.4	45.5	56.8	68.2	79.6	90.9	102
150	10.6	12.7	17	21.2	25.5	34	42.4	53.1	63.7	74.3	84.9	95.5
160	9.9	11.9	15.9	19.9	23.9	31.8	39.8	49.7	59.7	69.6	79.6	89.5
170	9.4	11.2	15	18.7	22.5	30	37.4	46.8	56.2	65.5	74.9	84.2
180	8.8	10.6	14.1	17.7	21.2	28.3	35.4	44.2	53.1	61.9	70.7	79.6
190	8.3	10	13.4	16.8	20.1	26.8	33.5	41.9	50.3	58.6	67	75.4
200	8	39.5	12.7	15.9	19.1	25.5	31.8	39.8	47.7	55.7	63.7	71.6
220	7.2	8.7	11.6	14.5	17.4	23.1	28.9	36.2	43.4	50.6	57.9	65.1
240	6.6	8	10.6	13.3	15.9	21.2	26.5	33.2	39.8	46.4	53.1	59.7
260	6.1	7.3	9.8	12.2	14.7	19.6	24.5	30.6	36.7	42.8	49	55.1
280	5.7	6.8	9.1	11.4	13.6	18.2	22.7	28.4	34.1	39.8	45.5	51.1
300	5.3	6.4	8.5	10.6	12.7	17	21.2	26.5	31.8	37.1	42.4	47.7
350	4.5	5.4	7.3	9.1	10.9	14.6	18.2	22.7	27.3	31.8	36.4	40.9
400	4	4.8	6.4	8	9.5	12.7	15.9	19.9	23.9	27.9	31.8	35.8
450	3.5	4.2	5.7	7.1	8.5	11.3	14.1	17.7	21.2	24.8	28.3	31.8
500	3.2	3.8	5.1	6.4	7.6	10.2	12.7	15.9	19.1	22.3	25.5	28.6

Revolutions per Minute for Various Cutting Speeds and Diameters (Metric Units)

Dia., mm	Cutting Speed, Meters per Minute											
	50	55	60	65	70	75	80	85	90	95	100	200
	Revolutions per Minute											
5	3183	3501	3820	4138	4456	4775	5093	5411	5730	6048	6366	12,732
6	2653	2918	3183	3448	3714	3979	4244	4509	4775	5039	5305	10,610
8	1989	2188	2387	2586	2785	2984	3183	3382	3581	3780	3979	7958
10	1592	1751	1910	2069	2228	2387	2546	2706	2865	3024	3183	6366
12	1326	1459	1592	1724	1857	1989	2122	2255	2387	2520	2653	5305
16	995	1094	1194	1293	1393	1492	1591	1691	1790	1890	1989	3979
20	796	875	955	1034	1114	1194	1273	1353	1432	1512	1592	3183
25	637	700	764	828	891	955	1019	1082	1146	1210	1273	2546
30	530	584	637	690	743	796	849	902	955	1008	1061	2122
35	455	500	546	591	637	682	728	773	819	864	909	1818
40	398	438	477	517	557	597	637	676	716	756	796	1592
45	354	389	424	460	495	531	566	601	637	672	707	1415
50	318	350	382	414	446	477	509	541	573	605	637	1273
55	289	318	347	376	405	434	463	492	521	550	579	1157
60	265	292	318	345	371	398	424	451	477	504	530	1061
65	245	269	294	318	343	367	392	416	441	465	490	979
70	227	250	273	296	318	341	364	387	409	432	455	909
75	212	233	255	276	297	318	340	361	382	403	424	849
80	199	219	239	259	279	298	318	338	358	378	398	796
90	177	195	212	230	248	265	283	301	318	336	354	707
100	159	175	191	207	223	239	255	271	286	302	318	637
110	145	159	174	188	203	217	231	246	260	275	289	579
120	133	146	159	172	186	199	212	225	239	252	265	530
130	122	135	147	159	171	184	196	208	220	233	245	490
140	114	125	136	148	159	171	182	193	205	216	227	455
150	106	117	127	138	149	159	170	180	191	202	212	424
160	99.5	109	119	129	139	149	159	169	179	189	199	398
170	93.6	103	112	122	131	140	150	159	169	178	187	374
180	88.4	97.3	106	115	124	133	141	150	159	168	177	354
190	83.8	92.1	101	109	117	126	134	142	151	159	167	335
200	79.6	87.5	95.5	103	111	119	127	135	143	151	159	318
220	72.3	79.6	86.8	94	101	109	116	123	130	137	145	289
240	66.3	72.9	79.6	86.2	92.8	99.5	106	113	119	126	132	265
260	61.2	67.3	73.4	79.6	85.7	91.8	97.9	104	110	116	122	245
280	56.8	62.5	68.2	73.9	79.6	85.3	90.9	96.6	102	108	114	227
300	53.1	58.3	63.7	69	74.3	79.6	84.9	90.2	95.5	101	106	212
350	45.5	50	54.6	59.1	63.7	68.2	72.8	77.3	81.8	99.1	91	182
400	39.8	43.8	47.7	51.7	55.7	59.7	63.7	67.6	71.6	75.6	79.6	159
450	35.4	38.9	42.4	46	49.5	53.1	56.6	60.1	63.6	67.2	70.7	141
500	31.8	35	38.2	41.4	44.6	47.7	50.9	54.1	57.3	60.5	63.6	127

SPEED AND FEED TABLES

How to Use the Feeds and Speeds Tables

Introduction to the Feed and Speed Tables.—The principal tables of feed and speed values are listed in the table below. In this section, Tables 1 through 9 give data for turning, Tables 10 through 15e give data for milling, and Tables 17 through 23 give data for reaming, drilling, threading.

The materials in these tables are categorized by description, and Brinell hardness number (Bhn) range or material condition. So far as possible, work materials are grouped by similar machining characteristics. The types of cutting tools (HSS end mill, for example) are identified in one or more rows across the tops of the tables. Other important details concerning the use of the tables are contained in the footnotes to Tables 1, 10 and 17. Information concerning specific cutting tool grades is given in notes at the end of each table.

Principal Feeds and Speeds Tables

Feeds and Speeds for Turning
Table 1. Cutting Feeds and Speeds for Turning Plain Carbon and Alloy Steels
Table 2. Cutting Feeds and Speeds for Turning Tool Steels
Table 3. Cutting Feeds and Speeds for Turning Stainless Steels
Table 4a. Cutting Feeds and Speeds for Turning Ferrous Cast Metals
Table 4b. Cutting Feeds and Speeds for Turning Ferrous Cast Metals
Table 5c. Cutting-Speed Adjustment Factors for Turning with HSS Tools
Table 5a. Turning-Speed Adjustment Factors for Feed, Depth of Cut, and Lead Angle
Table 5b. Tool Life Factors for Turning with Carbides, Ceramics, Cermets, CBN, and Polycrystalline Diamond
Table 6. Cutting Feeds and Speeds for Turning Copper Alloys
Table 7. Cutting Feeds and Speeds for Turning Titanium and Titanium Alloys
Table 8. Cutting Feeds and Speeds for Turning Light Metals
Table 9. Cutting Feeds and Speeds for Turning Superalloys
Feeds and Speeds for Milling
Table 10. Cutting Feeds and Speeds for Milling Aluminum Alloys
Table 11. Cutting Feeds and Speeds for Milling Plain Carbon and Alloy Steels
Table 12. Cutting Feeds and Speeds for Milling Tool Steels
Table 13. Cutting Feeds and Speeds for Milling Stainless Steels
Table 14. Cutting Feeds and Speeds for Milling Ferrous Cast Metals
Table 15a. Recommended Feed in Inches per Tooth (ft) for Milling with High Speed Steel Cutters
Table 15b. End Milling (Full Slot) Speed Adjustment Factors for Feed, Depth of Cut, and Lead Angle
Table 15c. End, Slit, and Side Milling Speed Adjustment Factors for Radial Depth of Cut
Table 15d. Face Milling Speed Adjustment Factors for Feed, Depth of Cut, and Lead Angle
Table 15e. Tool Life Adjustment Factors for Face Milling, End Milling, Drilling, and Reaming
Table 16. Cutting Tool Grade Descriptions and Common Vendor Equivalents
Feeds and Speeds for Drilling, Reaming, and Threading
Table 17. Feeds and Speeds for Drilling, Reaming, and Threading Plain Carbon and Alloy Steels
Table 18. Feeds and Speeds for Drilling, Reaming, and Threading Tool Steels
Table 19. Feeds and Speeds for Drilling, Reaming, and Threading Stainless Steels
Table 20. Feeds and Speeds for Drilling, Reaming, and Threading Ferrous Cast Metals
Table 21. Feeds and Speeds for Drilling, Reaming, and Threading Light Metals
Table 22. Feed and Diameter Speed Adjustment Factors for HSS Twist Drills and Reamers
Table 23. Feeds and Speeds for Drilling and Reaming Copper Alloys

Each of the cutting speed tables in this section contains two distinct types of cutting speed data. The speed columns at the left of each table contain traditional Handbook cutting speeds for use with high-speed steel (HSS) tools. For many years, this extensive collection of cutting data has been used successfully as starting speed values for turning, milling, drilling, and reaming operations. Instructions and adjustment factors for use with these speeds are given in Table 5c (feed and depth-of-cut factors) for turning, and in Table 15a (feed, depth of cut, and cutter diameter) for milling. Feeds for drilling and reaming are discussed in Using the Feed and Speed Tables for Drilling, Reaming, and Threading. With traditional speeds and feeds, tool life may vary greatly from material to material, making it very difficult to plan efficient cutting operations, in particular for setting up unattended jobs on CNC equipment where the tool life must exceed cutting time, or at least be predictable so that tool changes can be scheduled. This limitation is reduced by using the combined feed/speed data contained in the remaining columns of the speed tables.

The combined feed/speed portion of the speed tables gives two sets of feed and speed data for each material represented. These feed/speed pairs are the *optimum* and *average* data (identified by *Opt.* and *Avg.*); the *optimum* set is always on the left side of the column and the *average* set is on the right. The *optimum* feed/speed data are approximate values of feed and speed that achieve minimum-cost machining by combining a high productivity rate with low tooling cost at a fixed tool life. The *average* feed/speed data are expected to achieve approximately the same tool life and tooling costs, but productivity is usually lower, so machining costs are higher. The data in this portion of the tables are given in the form of two numbers, of which the first is the feed in thousandths of an inch per revolution (or per tooth, for milling) and the second is the cutting speed in feet per minute. For example, the feed/speed set 15/215 represents a feed of 0.015 in./rev at a speed of 215 fpm. Blank cells in the data tables indicate that feed/speed data for these materials were not available at the time of publication.

Generally, the feed given in the *optimum* set should be interpreted as the maximum safe feed for the given work material and cutting tool grade, and the use of a greater feed may result in premature tool wear or tool failure before the end of the expected tool life. The primary exception to this rule occurs in milling, where the feed may be greater than the *optimum* feed if the radial depth of cut is less than the value established in the table footnote; this topic is covered later in the milling examples. Thus, except for milling, the speed and tool life adjustment tables, to be discussed later, do not permit feeds that are greater than the *optimum* feed. On the other hand, the speed and tool life adjustment factors often result in cutting speeds that are well outside the given *optimum* to *average* speed range.

The combined feed/speed data in this section were contributed by Dr. Colding of Colding International Corp., Ann Arbor, MI. The speed, feed, and tool life calculations were made by means of a special computer program and a large database of cutting speed and tool life testing data. The COMP computer program uses tool life equations that are extensions of the F. W. Taylor tool life equation, first proposed in the early 1900s. The Colding tool life equations use a concept called equivalent chip thickness (*ECT*), which simplifies cutting speed and tool life predictions, and the calculation of cutting forces, torque, and power requirements. *ECT* is a basic metal cutting parameter that combines the four basic turning variables (depth of cut, lead angle, nose radius, and feed per revolution) into one basic parameter. For other metal cutting operations (milling, drilling, and grinding, for example), *ECT* also includes additional variables such as the number of teeth, width of cut, and cutter diameter. The *ECT* concept was first presented in 1931 by Prof. R. Woxen, who showed that equivalent chip thickness is a basic metal cutting parameter for high-speed cutting tools. Dr. Colding later extended the theory to include other tool materials and metal cutting operations, including grinding.

The equivalent chip thickness is defined by $ECT = A/CEL$, where A is the cross-sectional area of the cut (approximately equal to the feed times the depth of cut), and CEL is the cutting edge length or tool contact rubbing length. *ECT* and several other terms related to tool geometry are illustrated in Figs. 1 and 2. Many combinations of feed, lead angle, nose radius and cutter diameter, axial and radial depth of cut, and numbers of teeth can give the same value of *ECT*. However, for a constant cutting speed, no matter how the depth of cut, feed, or lead angle, etc., are varied, if a constant value of *ECT* is maintained, the tool life will also remain constant. A constant value of *ECT* means that a constant cutting speed gives a constant tool life and an increase in speed results in a reduced tool life. Likewise, if *ECT* were increased and cutting speed were held constant, as illustrated in the generalized cutting speed vs. *ECT* graph that follows, tool life would be reduced.

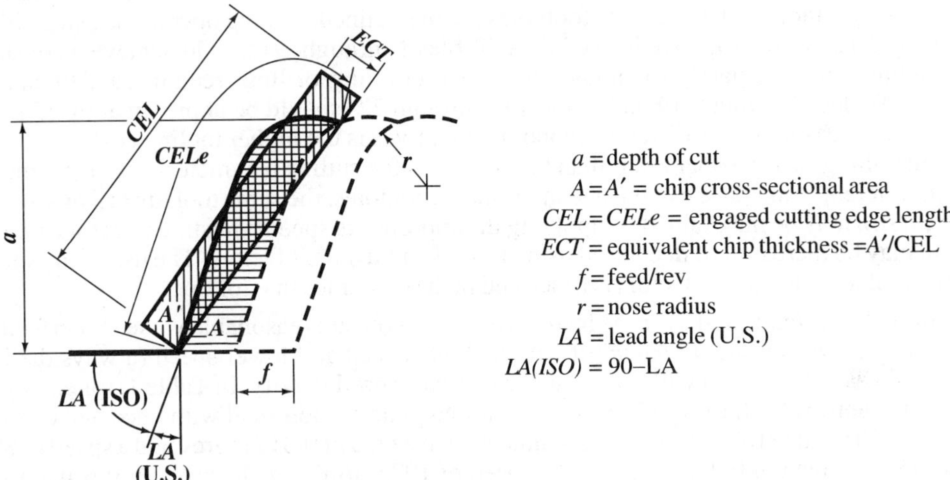

Fig. 1. Cutting Geometry, Equivalent Chip Thickness, and Cutting Edge Length

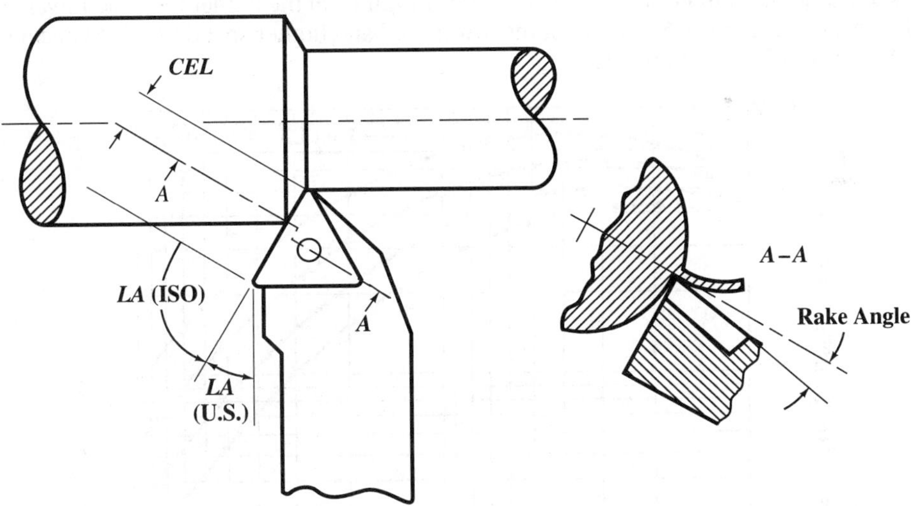

Fig. 2. Cutting Geometry for Turning

In the tables, the *optimum* feed/speed data have been calculated by COMP to achieve a fixed tool life based on the maximum *ECT* that will result in successful cutting, without premature tool wear or early tool failure. The same tool life is used to calculate the *average* feed/speed data, but these values are based on one-half of the maximum *ECT*. Because the data are not linear except over a small range of values, both *optimum* and *average* sets are required to adjust speeds for feed, lead angle, depth of cut, and other factors.

Tool life is the most important factor in a machining system, so feeds and speeds cannot be selected as simple numbers, but must be considered with respect to the many parameters that influence tool life. The accuracy of the combined feed/speed data presented is believed to be very high. However, machining is a variable and complicated process and use of the feed and speed tables requires the user to follow the instructions carefully to achieve good predictability. The results achieved, therefore, may vary due to material condition, tool material, machine setup, and other factors, and cannot be guaranteed.

The feed values given in the tables are valid for the standard tool geometries and fixed depths of cut that are identified in the table footnotes. If the cutting parameters and tool geometry established in the table footnotes are maintained, turning operations using either the *optimum* or *average* feed/speed data (Tables 1 through 9) should achieve a constant tool life of approximately 15 minutes; tool life for milling, drilling, reaming, and threading data (Tables 10 through 14 and Tables 17 through 22) should be approximately 45 minutes. The reason for the different economic tool lives is the higher tooling cost associated with milling-drilling operations than for turning. If the cutting parameters or tool geometry are different from those established in the table footnotes, the same tool life (15 or 45 minutes) still may be maintained by applying the appropriate speed adjustment factors, or tool life may be increased or decreased using tool life adjustment factors. The use of the speed and tool life adjustment factors is described in the examples that follow.

Both the *optimum* and *average* feed/speed data given are reasonable values for effective cutting. However, the *optimum* set with its higher feed and lower speed (always the left entry in each table cell) will usually achieve greater productivity. In Table 1, for example, the two entries for turning 1212 free-machining plain carbon steel with uncoated carbide are 17/805 and 8/1075. These values indicate that a feed of 0.017 in./rev and a speed of 805 ft/min, or a feed of 0.008 in./rev and a speed of 1075 ft/min can be used for this material. The tool life, in each case, will be approximately 15 minutes. If one of these feed and speed pairs is assigned an arbitrary cutting time of 1 minute, then the relative cutting time of the second pair to the first is equal to the ratio of their respective feed × speed products. Here, the same amount of material that can be cut in 1 minute, at the higher feed and lower speed (17/805), will require 1.6 minutes at the lower feed and higher speed (8/1075) because 17 × 805/(8 × 1075) = 1.6 minutes.

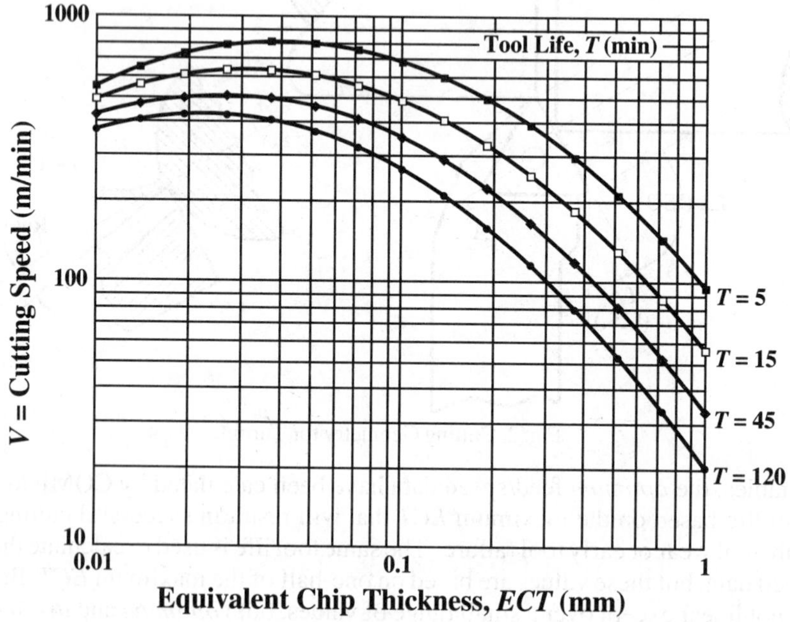

Cutting Speed versus Equivalent Chip Thickness with Tool Life as a Parameter

Speed and Feed Tables for Turning.—Speeds for HSS (high-speed steel) tools are based on a feed of 0.012 inch/rev and a depth of cut of 0.125 inch; use Table 5c to adjust the given speeds for other feeds and depths of cut. The combined feed/speed data in the remaining columns are based on a depth of cut of 0.1 inch, lead angle of 15 degrees, and nose radius of 3⁄64 inch. Use Table 5a to adjust given speeds for other feeds, depths of cut, and lead angles; use Table 5b to adjust given speeds for increased tool life up to 180 minutes. Examples are given in the text.

Examples Using the Feed and Speed Tables for Turning: The examples that follow give instructions for determining cutting speeds for turning. In general, the same methods are also used to find cutting speeds for milling, drilling, reaming, and threading, so reading through these examples may bring some additional insight to those other metalworking processes as well. The first step in determining cutting speeds is to locate the work material in the left column of the appropriate table for turning, milling, or drilling, reaming, and threading.

Example 1, Turning: Find the cutting speed for turning SAE 1074 plain carbon steel of 225 to 275 Brinell hardness, using an uncoated carbide insert, a feed of 0.015 in./rev, and a depth of cut of 0.1 inch.

In Table 1, feed and speed data for two types of uncoated carbide tools are given, one for hard tool grades, the other for tough tool grades. In general, use the speed data from the tool category that most closely matches the tool to be used because there are often significant differences in the speeds and feeds for different tool grades. From the uncoated carbide hard grade values, the *optimum* and *average* feed/speed data given in Table 1 are 17/615 and 8/815, or 0.017 in./rev at 615 ft/min and 0.008 in./rev at 815 ft/min. Because the selected feed (0.015 in./rev) is different from either of the feeds given in the table, the cutting speed must be adjusted to match the feed. The other cutting parameters to be used must also be compared with the general tool and cutting parameters given in the speed tables to determine if adjustments need to be made for these parameters as well. The general tool and cutting parameters for turning, given in the footnote to Table 1, are depth of cut = 0.1 inch, lead angle = 15°, and tool nose radius = 3⁄64 inch.

Table 5a is used to adjust the cutting speeds for turning (from Tables 1 through 9) for changes in feed, depth of cut, and lead angle. The new cutting speed V is found from $V = V_{opt} \times F_f \times F_d$, where V_{opt} is the *optimum* speed from the table (always the lower of the two speeds given), and F_f and F_d are the adjustment factors from Table 5a for feed and depth of cut, respectively.

To determine the two factors F_f and F_d, calculate the ratio of the selected feed to the *optimum* feed, 0.015/0.017 = 0.9, and the ratio of the two given speeds V_{avg} and V_{opt}, 815/615 = 1.35 (approximately). The feed factor $F_d = 1.07$ is found in Table 5a at the intersection of the feed ratio row and the speed ratio column. The depth-of-cut factor $F_d = 1.0$ is found in the same row as the feed factor in the column for depth of cut = 0.1 inch and lead angle = 15°, or for a tool with a 45° lead angle, $F_d = 1.18$. The final cutting speed for a 15° lead angle is $V = V_{opt} \times F_f \times F_d = 615 \times 1.07 \times 1.0 = 658$ fpm. Notice that increasing the lead angle tends to permit higher cutting speeds; such an increase is also the general effect of increasing the tool nose radius, although nose radius correction factors are not included in this table. Increasing lead angle also increases the radial pressure exerted by the cutting tool on the workpiece, which may cause unfavorable results on long, slender workpieces.

Example 2, Turning: For the same material and feed as the previous example, what is the cutting speed for a 0.4-inch depth of cut and a 45° lead angle?

As before, the feed is 0.015 in./rev, so F_f is 1.07, but $F_d = 1.03$ for depth of cut equal to 0.4 inch and a 45° lead angle. Therefore, $V = 615 \times 1.07 \times 1.03 = 676$ fpm. Increasing the lead angle from 15° to 45° permits a much greater (four times) depth of cut, at the same feed and nearly constant speed. Tool life remains constant at 15 minutes. *(Continued on page 1005)*

Table 1. Cutting Feeds and Speeds for Turning Plain Carbon and Alloy Steels

Material AISI/SAE Designation	Brinell Hardness	HSS Speed (fpm)		Uncoated Carbide Hard Opt.	Uncoated Carbide Hard Avg.	Uncoated Carbide Tough Opt.	Uncoated Carbide Tough Avg.	Coated Carbide Hard Opt.	Coated Carbide Hard Avg.	Coated Carbide Tough Opt.	Coated Carbide Tough Avg.	Ceramic Hard Opt.	Ceramic Hard Avg.	Ceramic Tough Opt.	Ceramic Tough Avg.	Cermet Opt.	Cermet Avg.
				Tool Material: f = feed (0.001 in./rev), s = speed (ft/min)													
Free-machining plain carbon steels (resulfurized): 1212, 1213, 1215	100–150	150	f	17	8	36	17	17	8	28	13	15	8	15	8	7	3
			s	805	1075	405	555	1165	1295	850	1200	3340	4985	1670	2500	1610	2055
	150–200	160	f	17	8	36	17	28	13	28	13	15	8	15	8	7	3
			s	745	935	345	470	915	1130	785	1110	1795	2680	1485	2215	1490	1815
1108, 1109, 1115, 1117, 1118, 1120, 1126, 1211	100–150	130															
	150–200	120	f	17	8	36	17	17	8	28	13	15	8	15	8	7	3
			s	730	990	300	430	1090	1410	780	1105	1610	2780	1345	2005	1355	1695
1132, 1137, 1139, 1140, 1144, 1146, 1151	175–225	120	f	17	8	36	17	17	8	28	13	13	7	13	7		
			s	615	815	300	405	865	960	755	960	1400	1965	1170	1640		
	275–325	75															
	325–375	50	f	17	8	36	17	17	8	28	13	10	5	10	5		
			s	515	685	235	340	720	805	650	810	1430	1745	1070	1305		
	375–425	40															
(Leaded): 11L17, 11L18, 12L13, 12L14	100–150	140															
	150–200	145	f	17	8	36	17	28	13	28	13	15	8	15	8	7	3
			s	745	935	345	470	915	1130	785	1110	1795	2680	1485	2215	1490	1815
	200–250	110	f	17	8	36	17	17	8	28	13	13	7	13	7		
			s	615	815	300	405	865	960	755	960	1400	1965	1170	1640		
Plain carbon steels: 1006, 1008, 1009, 1010, 1012, 1015, 1016, 1017, 1018, 1019, 1020, 1021, 1022, 1023, 1024, 1025, 1026, 1513, 1514	100–125	120	f	17	8	36	17	17	8	28	13	15	8	15	8	7	3
			s	805	1075	405	555	1165	1295	850	1200	3340	4985	1670	2500	1610	2055
	125–175	110	f	17	8	36	17	28	13	28	13	15	8	15	8	7	3
			s	745	935	345	470	915	1130	785	1110	1795	2680	1485	2215	1490	1815
	175–225	90	f	17	8	36	17	17	8	28	13	13	7	13	7		
			s	615	815	300	405	865	960	755	960	1400	1965	1170	1640		
	225–275	70															

Table 1. *(Continued)* **Cutting Feeds and Speeds for Turning Plain Carbon and Alloy Steels**

Material AISI/SAE Designation	Brinell Hardness	HSS Speed (fpm)		Uncoated Carbide Hard Opt.	Uncoated Carbide Hard Avg.	Uncoated Carbide Tough Opt.	Uncoated Carbide Tough Avg.	Coated Carbide Hard Opt.	Coated Carbide Hard Avg.	Coated Carbide Tough Opt.	Coated Carbide Tough Avg.	Ceramic Hard Opt.	Ceramic Hard Avg.	Ceramic Tough Opt.	Ceramic Tough Avg.	Cermet Opt.	Cermet Avg.
				f = feed (0.001 in./rev), s = speed (ft/min)													
Plain carbon steels *(continued)*: 1027, 1030, 1033, 1035, 1036, 1037, 1038, 1039, 1040, 1041, 1042, 1043, 1045, 1046, 1048, 1049, 1050, 1052, 1524, 1526, 1527, 1541	125–175	100	f	17	8	36	17	28	13	28	13	15	8	15	8	7	3
			s	745	935	345	470	915	1130	785	1110	1795	2680	1485	2215	1490	1815
	175–225	85	f	17	8	36	17	17	8	28	13	13	7	13	7		
	225–275	70	s	615	815	300	405	865	960	755	960	1400	1965	1170	1640		
	275–325	60															
	325–375	40	f	17	8	36	17	17	8	28	13	10	5	10	5		
			s	515	685	235	340	720	805	650	810	1430	1745	1070	1305		
	375–425	30															
Plain carbon steels *(continued)*: 1055, 1060, 1064, 1065, 1070, 1074, 1078, 1080, 1084, 1086, 1090, 1095, 1548, 1551, 1552, 1561, 1566	125–175	100	f	17	8	36	17	17	8	28	13	15	8	15	8	7	3
	175–225	80	s	730	990	300	430	1090	1410	780	1105	1610	2780	1345	2005	1355	1695
	225–275	65	f	17	8	36	17	17	8	28	13	13	7	13	7	7	3
			s	615	815	300	405	865	960	755	960	1400	1965	1170	1640	1365	1695
	275–325	50															
	325–375	35	f	17	8	36	17	17	8	28	13	10	5	10	5		
			s	515	685	235	340	720	805	650	810	1430	1745	1070	1305		
	375–425	30															
Free-machining alloy steels, (resulfurized): 4140, 4150	175–200	110	f	17	8	36	17	17	8	28	13	15	8	15	8	7	3
	200–250	90	s	525	705	235	320	505	525	685	960	1490	2220	1190	1780	1040	1310
	250–300	65	f	17	8	36	17	17	8	28	13	10	5	10	5	7	3
			s	355	445	140	200	630	850	455	650	1230	1510	990	1210	715	915
	300–375	50	f	17	8	36	17	17	8	28	13	8	4	8	4	7	3
	375–425	40	s	330	440	125	175	585	790	125	220	1200	1320	960	1060	575	740

Table 1. *(Continued)* **Cutting Feeds and Speeds for Turning Plain Carbon and Alloy Steels**

Material AISI/SAE Designation	Brinell Hardness	HSS Speed (fpm)		Uncoated Carbide Hard Opt.	Uncoated Carbide Hard Avg.	Uncoated Carbide Tough Opt.	Uncoated Carbide Tough Avg.	Coated Carbide Hard Opt.	Coated Carbide Hard Avg.	Coated Carbide Tough Opt.	Coated Carbide Tough Avg.	Ceramic Hard Opt.	Ceramic Hard Avg.	Ceramic Tough Opt.	Ceramic Tough Avg.	Cermet Opt.	Cermet Avg.
				f = feed (0.001 in./rev), s = speed (ft/min)													
Free-machining alloy steels: (leaded): 41L30, 41L40, 41L47, 41L50, 43L47, 51L32, 52L100, 86L20, 86L40	150–200	120	f	17	8	36	17	17	8	28	13	15	8	15	8	7	3
			s	730	990	300	430	1090	1410	780	1105	1610	2780	1345	2005	1355	1695
	200–250	100	f	17	8	36	17	17	8	28	13	13	7	13	7	7	3
			s	615	815	300	405	865	960	755	960	1400	1965	1170	1640	1355	1695
	250–300	75															
	300–375	55	f	17	8	36	17	17	8	28	13	10	5	10	5		
			s	515	685	235	340	720	805	650	810	1430	1745	1070	1305		
	375–425	50															
Alloy steels: 4012, 4023, 4024, 4028, 4118, 4320, 4419, 4422, 4427, 4615, 4620, 4621, 4626, 4718, 4720, 4815, 4817, 4820, 5015, 5117, 5120, 6118, 8115, 8615, 8617, 8620, 8622, 8625, 8627, 8720, 8822, 94B17	125–175	100															
	175–225	90	f	17	8	36	17	17	8	28	13	15	8	15	8	7	3
			s	525	705	235	320	505	525	685	960	1490	2220	1190	1780	1040	1310
	225–275	70	f	17	8	36	1	17	8	28	13	10	5	10	5	7	3
			s	355	445	140	200	630	850	455	650	1230	1510	990	1210	715	915
	275–325	60	f	17	8	36	17	17	8	28	13	9	5	8	5	7	3
			s	330	440	135	190	585	790	240	350	1230	1430	990	1150	655	840
	325–35	50															
	375–425	30 (20)	f	17	8	36	17	17	8	28	13	8	4	8	4	7	3
			s	330	440	125	175	585	790	125	220	1200	1320	960	1060	575	740
Alloy steels: 1330, 1335, 1340, 1345, 4032, 4037, 4042, 4047, 4130, 4135, 4137, 4140, 4142, 4145, 4147, 4150, 4161, 4337, 4340, 50B44, 50B46, 50B50, 50B60, 5130, 5132, 5140, 5145, 5147, 5150, 5160, 51B60, 6150, 81B45, 8630, 8635, 8637, 8640, 8642, 8645, 8650, 8655, 8660, 8740, 9254, 9255, 9260, 9262, 94B30 E51100, E52100 use (HSS Speeds)	175–225	85 (70)	f	17	8	36	17	17	8	28	13	15	8	15	8	7	3
			s	525	705	235	320	505	525	685	960	1490	2220	1190	1780	1020	1310
	225–275	70 (65)	f	17	8	36	17	17	8	28	13	10	5	10	5	7	3
			s	355	445	140	200	630	850	455	650	1230	1510	990	1210	715	915
	275–325	60 (50)	f	17	8	36	17	17	8	28	13	9	5	8	5	7	3
			s	330	440	135	190	585	790	240	350	1230	1430	990	1150	655	840
	325–375	40 (30)															
	375–425	30 (20)	f	17	8	36	17	17	8	28	13	8	4	8	4	7	3
			s	330	440	125	175	585	790	125	220	1200	1320	960	1060	575	740

Table 1. *(Continued)* **Cutting Feeds and Speeds for Turning Plain Carbon and Alloy Steels**

Material AISI/SAE Designation	Brinell Hardness	HSS Speed (fpm)		Uncoated Carbide Hard Opt.	Uncoated Carbide Hard Avg.	Uncoated Carbide Tough Opt.	Uncoated Carbide Tough Avg.	Coated Carbide Hard Opt.	Coated Carbide Hard Avg.	Coated Carbide Tough Opt.	Coated Carbide Tough Avg.	Ceramic Hard Opt.	Ceramic Hard Avg.	Ceramic Tough Opt.	Ceramic Tough Avg.	Cermet Opt.	Cermet Avg.
				f = feed (0.001 in./rev), s = speed (ft/min)													
	220–300	65															
	300–350	50	f	17	8	36	17	20	10	28	13			10	5	7	3
			s	220	295	100	150	355	525	600	865			660	810	570	740
Ultra-high-strength steels (not ASI): AMS alloys 6421 (98B37 Mod.), 6422 (98BV40), 6424, 6427, 6428, 6430, 6432, 6433, 6434, 6436, and 6442; 300M and D6ac	350–400	35	f	17	8	36	17	17	8	28	13			8	4	7	3
			s	165	185	55	105	325	350	175	260			660	730	445	560
	43–48 Rc	25															
	48–52 Rc	10	f			17	8					7	3	10	5		
			s			55†	90					385	645	270	500		
Maraging steels (not AISI): 18% Ni, Grades 200, 250, 300, and 350	250–325	60	f	17	8	36	17	20	10	28	13			10	5	7	3
			s	220	295	100	150	355	525	600	865	660	810	570	740		
	50–52 Rc	10	f			17	8					7	3	10	5		
			s			55†	90					385‡	645	270	500		
Nitriding steels (not AISI): Nitralloy 125, 135, 135 Mod., 225, and 230, Nitralloy N, Nitralloy EZ, Nitrex 1	200–250	70	f	17	8	36	17	17	8	28	13	15	8	15	8	7	3
			s	525	705	235	320	505	525	685	960	1490	2220	1190	1780	1040	1310
	300–350	30	f	17	8	36	17	17	8	28	13	8	4	8	4	7	3
			s	330	440	125	175	585	790	125	220	1200	1320	960	1060	575	740

Speeds for HSS (high-speed steel) tools are based on a feed of 0.012 inch/rev and a depth of cut of 0.125 inch; use Table 5c to adjust the given speeds for other feeds and depths of cut. The combined feed/speed data in the remaining columns are based on a depth of cut of 0.1 inch, lead angle of 15 degrees, and nose radius of $\frac{3}{64}$ inch. Use Table 5a to adjust given speeds for other feeds, depths of cut, and lead angles; use Table 5b to adjust given speeds for increased tool life up to 180 minutes. Examples are given in the text.

The combined feed/speed data in this table are based on tool grades (identified in Table 16) as follows: uncoated carbides, hard = 17, tough = 19, † = 15; coated carbides, hard = 11, tough = 14; ceramics, hard = 2, tough = 3, ‡ = 4; cermet = 7 .

Table 2. Cutting Feeds and Speeds for Turning Tool Steels

Material AISI Designation	Brinell Hardness	Uncoated HSS Speed (fpm)		Uncoated Carbide Hard Opt.	Uncoated Carbide Hard Avg.	Uncoated Carbide Tough Opt.	Uncoated Carbide Tough Avg.	Coated Carbide Hard Opt.	Coated Carbide Hard Avg.	Coated Carbide Tough Opt.	Coated Carbide Tough Avg.	Ceramic Hard Opt.	Ceramic Hard Avg.	Ceramic Tough Opt.	Ceramic Tough Avg.	Cermet Opt.	Cermet Avg.
				f = feed (0.001 in./rev), s = speed (ft/min)													
Water hardening: W1, W2, W5	150–200	100															
Shock resisting: S1, S2, S5, S6, S7	175–225	70	f	17	8	36	17	17	8	28	13	13	7	13	7	7	3
Cold work, oil hardening: O1, O2, O6, O7	175–225	70	s	455	610	210	270	830	1110	575	805	935	1310	790	1110	915	1150
Cold work, high carbon, high chromium: D2, D3, D4, D5, D7	200–250	45															
Cold work, air hardening: A2, A3, A8, A9, A10	200–250	70															
A4, A6	200–250	55	f	17	8	36	17	17	8	28	13	13	7	13	7	7	3
A7	225–275	45	s	445	490	170	235	705	940	515	770	660	925	750	1210	1150	1510
Hot work, chromium type: H10, H11, H12, H13, H14, H19	150–200	80															
	200–250	65															
	325–375	50	f	17	8	36	17	17	8	28	13			8	4	7	3
			s	165	185	55	105	325	350	175	260			660	730	445	560
	48–50 Rc	20	f			17	8					7	3	10	5		
	50–52 Rc	10	s			55†	90					385‡	645	270	500		
	52–56 Rc	—															
Hot work, tungsten type: H21, H22, H23, H24, H25, H26	150–200	60															
	200–250	50	f	17	8	36	17	17	8	28	13	13	7	13	7	7	3
Hot work, molybdenum type: H41, H42, H43	150–200	55	s	445	490	170	235	705	940	515	770	660	925	750	1210	1150	1510
	200–250	45															
Special purpose, low alloy: L2, L3, L6	150–200	75	f	17	8	36	17	17	8	28	13	13	7	13	7	7	3
			s	445	610	210	270	830	1110	575	805	935	1310	790	1110	915	1150
Mold: P2, P3, P4, P5, P6, P26, P21	100–150	90	f	17	8	36	17	17	8	28	13	13	7	13	7	7	3
	150–200	80	s	445	610	210	270	830	1110	575	805	935	1310	790	1110	915	1150
High-speed steel: M1, M2, M6, M10, T1, T2,T6	200–250	65															
M3-1, M4 M7, M30, M33, M34, M36, M41, M42, M43, M44, M46, M47, T5, T8	225–275	55	f	17	8	36	17	17	8	28	13	13	7	13	7	7	3
			s	445	490	170	235	705	940	515	770	660	925	750	1210	1150	1510
T15, M3-2	225–275	45															

Speeds for HSS (high-speed steel) tools are based on a feed of 0.012 inch/rev and a depth of cut of 0.125 inch; use Table 5c to adjust the given speeds for other feeds and depths of cut. The combined feed/speed data in the remaining columns are based on a depth of cut of 0.1 inch, lead angle of 15 degrees, and nose radius of $^3/_{64}$ inch. Use Table 5a to adjust given speeds for other feeds, depths of cut, and lead angles; use Table 5b to adjust given speeds for increased tool life up to 180 minutes. Examples are given in the text. The combined feed/speed data in this table are based on tool grades (identified in Table 16) as follows: uncoated carbides, hard = 17, tough = 19, † = 15; coated carbides, hard = 11, tough = 14; ceramics, hard = 2, tough = 3, ‡ = 4; cermet = 7.

Table 3. Cutting Feeds and Speeds for Turning Stainless Steels

Material	Brinell Hardness	Tool Material: Uncoated HSS Speed (fpm)		Uncoated Carbide Hard Opt.	Uncoated Carbide Hard Avg.	Uncoated Carbide Tough Opt.	Uncoated Carbide Tough Avg.	Coated Carbide Hard Opt.	Coated Carbide Hard Avg.	Coated Carbide Tough Opt.	Coated Carbide Tough Avg.	Cermet Opt.	Cermet Avg.
				f = feed (0.001 in./rev), s = speed (ft/min)									
Free-machining stainless steel (Ferritic): 430F, 430FSe	135–185	110	f	20	10	36	17	17	8	28	13	7	3
			s	480	660	370	395	755	945	640	810	790	995
(Austenitic): 203EZ, 303, 303Se, 303MA, 303Pb, 303Cu, 303 Plus X	135–185	100	f	13	7	36	17			28	13	7	3
	225–275	80	s	520	640	310	345			625	815	695	875
(Martensitic): 416, 416Se, 416 Plus X, 420F, 420FSe, 440F, 440FSe	135–185	110	f	13	7	36				28	13	7	3
	185–240	100	s	520	640	310				625	815	695	875
	275–325	60	f	13	7	36	17			28	13		
	375–425	30	s	210	260	85	135			130	165		
Stainless steels (Ferritic): 405, 409 429, 430, 434, 436, 442, 446, 502	135–185	90	f	20	10	36	17	17	8	28	13	7	3
			s	480	660	370	395	755	945	640	810	790	995
(Austenitic): 201, 202, 301, 302, 304, 304L, 305, 308, 321, 347, 348	135–185	75											
	225–275	65											
(Austenitic): 302B, 309, 309S, 310, 310S, 314, 316, 316L, 317, 330	135–185	70	f	13	7	36	17			28	13	7	3
(Martensitic): 403, 410, 420, 501	135–175	95	s	520	640	310	345			625	165	695	875
	175–225	85											
	275–325	55											
	375–425	35											
(Martensitic): 414, 431, Greek Ascoloy, 440A, 440B, 440C	225–275	55–60	f	13	7	36	17			28	13	13	7
	275–325	45–50	s	210	260	85	135			130	165	200†	230
	375–425	30											
(Precipitation hardening):15 -5PH, 17-4PH, 17-7PH, AF-71, 17-14CuMo, AFC-77, AM-350, AM-355, AM-362, Custom 455, HNM, PH13-8, PH14-8Mo, PH15-7Mo, Stainless W	150–200	60	f	13	7	36	17			28	13	13	7
	275–325	50	s	520	640	310	345			625	815	695	875
	325–375	40	f	13	7	36	17						
	375–450	25	s	195	240	85	155						

See footnote to Table 1 for more information. The combined feed/speed data in this table are based on tool grades (identified in Table Table 16) as follows: uncoated carbides, hard = 17, tough = 19; coated carbides, hard = 11, tough = 14; cermet = 7, † = 18.

Table 4a. Cutting Feeds and Speeds for Turning Ferrous Cast Metals

Material	Brinell Hardness	HSS Speed (fpm)		Uncoated Carbide Tough Opt.	Uncoated Carbide Tough Avg.	Coated Carbide Hard Opt.	Coated Carbide Hard Avg.	Coated Carbide Tough Opt.	Coated Carbide Tough Avg.	Ceramic Hard Opt.	Ceramic Hard Avg.	Ceramic Tough Opt.	Ceramic Tough Avg.	Cermet Opt.	Cermet Avg.	CBN Opt.	CBN Avg.
				f = feed (0.001 in./rev), s = speed (ft/min)													
Gray Cast Iron																	
ASTM Class 20	120–150	120															
ASTM Class 25	160–200	90	f	28	13	28	13	28	13	15	8	15	8	8	4	24	11
			s	240	365	665	1040	585	945	1490	2220	1180	1880	395	510	8490	36380
ASTM Class 30, 35, and 40	190–220	80															
ASTM Class 45 and 50	220–260	60	f	28	13	28	13	28	13	11	6	11	6	8	4	24	11
ASTM Class 55 and 60	250–320	35	s	160	245	400	630	360	580	1440	1880	1200	1570	335	420	1590	2200
ASTM Type 1, 1b, 5 (Ni resist)	100–215	70															
ASTM Type 2, 3, 6 (Ni resist)	120–175	65	f	28	13			28	13	15	8	15	8	8	4		
			s	110	175			410	575	1060	1590	885	1320	260	325		
ASTM Type 2b, 4 (Ni resist)	150–250	50															
Malleable Iron																	
(Ferritic): 32510, 35018	110–160	130	f	28	13	28	13	28	13	15	8	15	8				
			s	180	280	730	940	660	885	1640	2450	1410	2110				
(Pearlitic): 40010, 43010, 45006, 45008, 48005, 50005	160–200	95	f	28	13	28	13	28	13	13	7	13	7				
	200–240	75	s	125	200	335	505	340	510	1640	2310	1400	1970				
(Martensitic): 53004, 60003, 60004	200–255	70															
(Martensitic): 70002, 70003	220–260	60	f	28	13			28	13	11	6	11	6				
(Martensitic): 80002	240–280	50	s	100	120			205	250	1720	2240	1460	1910				
(Martensitic): 90001	250–320	30															

Speeds for HSS (high-speed steel) tools are based on a feed of 0.012 inch/rev and a depth of cut of 0.125 inch; use Table 5c to adjust the given speeds for other feeds and depths of cut. The combined feed/speed data in the remaining columns are based on a depth of cut of 0.1 inch, lead angle of 15 degrees, and nose radius of $\frac{3}{64}$ inch. Use Table 5a to adjust the given speeds for other feeds, depths of cut, and lead angles; use Table 5b to adjust given speeds for increased tool life up to 180 minutes. Examples are given in the text.

The combined feed/speed data in this table are based on tool grades (identified in Table 16) as follows: uncoated carbides, tough = 15; Coated carbides, hard = 11, tough = 14; ceramics, hard = 2, tough = 3; cermet = 7; CBN = 1.

Table 4b. Cutting Feeds and Speeds for Turning Ferrous Cast Metals

Material	Brinell Hardness	Uncoated HSS		Tool Material														
				Uncoated Carbide				Coated Carbide				Ceramic				Cermet	
				Hard		Tough		Hard		Tough		Hard		Tough			
		Speed (fpm)		f = feed (0.001 in./rev), s = speed (ft/min)													
				Opt.	*Avg.*	*Opt.*	*Avg.*	*Opt.*	*Avg.*	*Opt.*	*Avg.*	*Opt.*	*Avg.*	*Opt.*	*Avg.*	*Opt.*	*Avg.*
Nodular (Ductile) Iron																	
(Ferritic): 60-40-18, 65-45-12	140–190	100	f			28	13	28	13	28	13	15	8	15	8	8	4
			s			200	325	490	700	435	665	970	1450	845	1260	365	480
(Ferritic-Pearlitic): 80-55-06	190–225	80	f			28	13	28	13	28	13	11	6	11	6	8	4
	225–260	65	s			130	210	355	510	310	460	765	995	1260	1640	355	445
(Pearlitic-Martensitic): 100-70-03	240–300	45															
(Martensitic): 120-90-02	270–330	30	f			28	13			28	13	10	5	10	5	8	4
	300–400	15	s			40	65			145	175	615	750	500	615	120	145
Cast Steels																	
(Low-carbon): 1010, 1020	100–150	110	f	17	8	36	17	17	8	28	13	15	8			7	3
	125–175	100	s	370	490	230	285	665	815	495	675	2090	3120			625	790
(Medium-carbon): 1030, 1040, 1050	175–225	90	f	17	8	36	17	17	8	28	13	15	8			7	3
	225–300	70	s	370	490	150	200	595	815	410	590	1460	2170			625	790
(Low-carbon alloy): 1320, 2315, 2320, 4110, 4120, 4320, 8020, 8620	150–200	90															
	200–250	80															
	250–300	60															
(Medium-carbon alloy): 1330, 1340, 2325, 2330, 4125, 4130, 4140, 4330, 4340, 8030, 80B30, 8040, 8430, 8440, 8630, 8640, 9525, 9530, 9535	175–225	80	f	17	8	36	17	17	8			15	8				
	225–250	70	s	310	415	115	150	555	760			830	1240				
	250–300	55	f			28	13					15	8				
	300–350	45	s			70†	145					445	665				
	350–400	30	f			28	13			28	13			15	8		
			s			115†	355			335	345			955	1430		

The combined feed/speed data in this table are based on tool grades (identified in Table 16) as shown: uncoated carbides, hard = 17; tough = 19, † = 15; coated carbides, hard = 11; tough = 14; ceramics, hard = 2; tough = 3; cermet = 7. Also, see footnote to Table 4a.

Table 5a. Turning-Speed Adjustment Factors for Feed, Depth of Cut, and Lead Angle

Ratio of Chosen Feed to *Optimum* Feed	Ratio of the two cutting speeds given in the tables V_{avg}/V_{opt}							Depth of Cut and Lead Angle									
								1 in. (25.4 mm)		0.4 in. (10.2 mm)		0.2 in. (5.1 mm)		0.1 in. (2.5 mm)		0.04 in. (1.0 mm)	
	1.00	1.10	1.25	1.35	1.50	1.75	2.00	15°	45°	15°	45°	15°	45°	15°	45°	15°	45°
	Feed Factor, F_f							Depth of Cut and Lead Angle Factor, F_d									
1.00	1.0	1.0	1.0	1.0	1.0	1.0	1.0	0.74	1.0	0.79	1.03	0.85	1.08	1.0	1.18	1.29	1.35
0.90	1.00	1.02	1.05	1.07	1.09	1.10	1.12	0.75	1.0	0.80	1.03	0.86	1.08	1.0	1.17	1.27	1.34
0.80	1.00	1.03	1.09	1.10	1.15	1.20	1.25	0.77	1.0	0.81	1.03	0.87	1.07	1.0	1.15	1.25	1.31
0.70	1.00	1.05	1.13	1.22	1.22	1.32	1.43	0.77	1.0	0.82	1.03	0.87	1.08	1.0	1.15	1.24	1.30
0.60	1.00	1.08	1.20	1.25	1.35	1.50	1.66	0.78	1.0	0.82	1.03	0.88	1.07	1.0	1.14	1.23	1.29
0.50	1.00	1.10	1.25	1.35	1.50	1.75	2.00	0.78	1.0	0.82	1.03	0.88	1.07	1.0	1.14	1.23	1.28
0.40	1.00	1.09	1.28	1.44	1.66	2.03	2.43	0.78	1.0	0.84	1.03	0.89	1.06	1.0	1.13	1.21	1.26
0.30	1.00	1.06	1.32	1.52	1.85	2.42	3.05	0.81	1.0	0.85	1.02	0.90	1.06	1.0	1.12	1.18	1.23
0.20	1.00	1.00	1.34	1.60	2.07	2.96	4.03	0.84	1.0	0.89	1.02	0.91	1.05	1.0	1.10	1.15	1.19
0.10	1.00	0.80	1.20	1.55	2.24	3.74	5.84	0.88	1.0	0.91	1.01	0.92	1.03	1.0	1.06	1.10	1.12

Use with Tables 1 through 9. Not for HSS tools. Tables 1 through 9 data, except for HSS tools, are based on depth of cut = 0.1 inch, lead angle = 15 degrees, and tool life = 15 minutes. For other depths of cut, lead angles, or feeds, use the two feed/speed pairs from the tables and calculate the ratio of desired (new) feed to *optimum* feed (largest of the two feeds given in the tables), and the ratio of the two cutting speeds (V_{avg}/V_{opt}). Use the value of these ratios to find the feed factor F_f at the intersection of the feed ratio row and the speed ratio column in the left half of the table. The depth-of-cut factor F_d is found in the same row as the feed factor in the right half of the table under the column corresponding to the depth of cut and lead angle. The adjusted cutting speed can be calculated from $V = V_{opt} \times F_f \times F_d$, where V_{opt} is the smaller (*optimum*) of the two speeds from the speed table (from the left side of the column containing the two feed/speed pairs). See the text for examples.

Table 5b. Tool Life Factors for Turning with Carbides, Ceramics, Cermets, CBN, and Polycrystalline Diamond

Tool Life, T (minutes)	Turning with Carbides: Workpiece < 300 Bhn			Turning with Carbides: Workpiece > 300 Bhn; Turning with Ceramics: Any Hardness			Turning with Mixed Ceramics: Any Workpiece Hardness		
	f_s	f_m	f_l	f_s	f_m	f_l	f_s	f_m	f_l
15	1.0	1.0	1.0	1.0	1.0	1.0	1.0	1.0	1.0
45	0.86	0.81	0.76	0.80	0.75	0.70	0.89	0.87	0.84
90	0.78	0.71	0.64	0.70	0.63	0.56	0.82	0.79	0.75
180	0.71	0.63	0.54	0.61	0.53	0.45	0.76	0.72	0.67

Except for HSS speed tools, feeds and speeds given in Tables 1 through 9 are based on 15-minute tool life. To adjust speeds for another tool life, multiply the cutting speed for 15-minute tool life V_{15} by the tool life factor from this table according to the following rules: for small feeds where feed $\leq \frac{1}{2} f_{opt}$, the cutting speed for desired tool life is $V_T = f_s \times V_{15}$; for medium feeds where $\frac{1}{2} f_{opt} < \text{feed} < \frac{3}{4} f_{opt}$, $V_T = f_m \times V_{15}$; and for larger feeds where $\frac{3}{4} f_{opt} \leq \text{feed} \leq f_{opt}$, $V_T = f_l \times V_{15}$. Here, f_{opt} is the largest (*optimum*) feed of the two feed/speed values given in the speed tables.

Table 5c. Cutting-Speed Adjustment Factors for Turning with HSS Tools

Feed		Feed Factor	Depth of Cut		Depth-of-Cut Factor
in.	mm	F_f	in.	mm	F_d
0.002	0.05	1.50	0.005	0.13	1.50
0.003	0.08	1.50	0.010	0.25	1.42
0.004	0.10	1.50	0.016	0.41	1.33
0.005	0.13	1.44	0.031	0.79	1.21
0.006	0.15	1.34	0.047	1.19	1.15
0.007	0.18	1.25	0.062	1.57	1.10
0.008	0.20	1.18	0.078	1.98	1.07
0.009	0.23	1.12	0.094	2.39	1.04
0.010	0.25	1.08	0.100	2.54	1.03
0.011	0.28	1.04	0.125	3.18	1.00
0.012	0.30	1.00	0.150	3.81	0.97
0.013	0.33	0.97	0.188	4.78	0.94
0.014	0.36	0.94	0.200	5.08	0.93
0.015	0.38	0.91	0.250	6.35	0.91
0.016	0.41	0.88	0.312	7.92	0.88
0.018	0.46	0.84	0.375	9.53	0.86
0.020	0.51	0.80	0.438	11.13	0.84
0.022	0.56	0.77	0.500	12.70	0.82
0.025	0.64	0.73	0.625	15.88	0.80
0.028	0.71	0.70	0.688	17.48	0.78
0.030	0.76	0.68	0.750	19.05	0.77
0.032	0.81	0.66	0.812	20.62	0.76
0.035	0.89	0.64	0.938	23.83	0.75
0.040	1.02	0.60	1.000	25.40	0.74
0.045	1.14	0.57	1.250	31.75	0.73
0.050	1.27	0.55	1.250	31.75	0.72
0.060	1.52	0.50	1.375	34.93	0.71

For use with HSS tool data only from Tables 1 through 9. Adjusted cutting speed $V = V_{HSS} \times F_f \times F_d$, where V_{HSS} is the tabular speed for turning with high-speed tools.

Example 3, Turning: Determine the cutting speed for turning 1055 steel of 175 to 225 Brinell hardness using a hard ceramic insert, a 15° lead angle, a 0.04-inch depth of cut and 0.0075 in./rev feed.

The two feed/speed combinations given in Table 5a for 1055 steel are 15/1610 and 8/2780, corresponding to 0.015 in./rev at 1610 fpm and 0.008 in./rev at 2780 fpm, respectively. In Table 5a, the feed factor $F_f = 1.75$ is found at the intersection of the row corresponding to feed/$f_{opt} = 7.5/15 = 0.5$ and the column corresponding to $V_{avg}/V_{opt} = 2780/1610 = 1.75$ (approximately). The depth-of-cut factor $F_d = 1.23$ is found in the same row, under the column heading for a depth of cut = 0.04 inch and lead angle = 15°. The adjusted cutting speed is $V = 1610 \times 1.75 \times 1.23 = 3466$ fpm.

Example 4, Turning: The cutting speed for 1055 steel calculated in Example 3 represents the speed required to obtain a 15-minute tool life. Estimate the cutting speed needed to obtain a tool life of 45, 90, and 180 minutes using the results of Example 3.

To estimate the cutting speed corresponding to another tool life, multiply the cutting speed for 15-minute tool life V_{15} by the adjustment factor from the Table 5b, Tool Life Factors for Turning. This table gives three factors for adjusting tool life based on the feed used, f_s for feeds less than or equal to $\frac{1}{2}f_{opt}$, $\frac{3}{4}f_m$ for midrange feeds between $\frac{1}{2}$ and $\frac{3}{4}f_{opt}$ and f_l for large feeds greater than or equal to $\frac{3}{4}f_{opt}$ and less than f_{opt}. In Example 3, f_{opt} is 0.015 in./rev and the selected feed is 0.0075 in./rev = $\frac{1}{2}f_{opt}$. The new cutting speeds for the various tool lives are obtained by multiplying the cutting speed for 15-minute tool life V_{15} by the factor for small feeds f_s from the column for turning with ceramics in Table 5b. These calculations, using the cutting speed obtained in Example 3, follow.

Tool Life	*Cutting Speed*
15 min	$V_{15} = 3466$ fpm
45 min	$V_{45} = V_{15} \times 0.80 = 2773$ fpm
90 min	$V_{90} = V_{15} \times 0.70 = 2426$ fpm
180 min	$V_{180} = V_{15} \times 0.61 = 2114$ fpm

Depth of cut, feed, and lead angle remain the same as in Example 3. Notice, increasing the tool life from 15 to 180 minutes, a factor of 12, reduces the cutting speed by only about one-third of the V_{15} speed.

Table 6. Cutting Feeds and Speeds for Turning Copper Alloys

Group 1
Architectural bronze (C38500); Extra-high-headed brass (C35600); Forging brass (C37700); Free-cutting phosphor bronze, B2 (C54400); Free-cutting brass (C36000); Free-cutting Muntz metal (C37000); High-leaded brass (C33200; C34200); High-leaded brass tube (C35300); Leaded commercial bronze (C31400); Leaded naval brass (C48500); Medium-leaded brass (C34000)
Group 2
Aluminum brass, arsenical (C68700); Cartridge brass, 70% (C26000); High-silicon bronze, B (C65500); Admiralty brass (inhibited) (C44300, C44500); Jewelry bronze, 87.5% (C22600); Leaded Muntz metal (C36500, C36800); Leaded nickel silver (C79600); Low brass, 80% (C24000); Low-leaded brass (C33500); Low-silicon bronze, B (C65100); Manganese bronze, A (C67500); Muntz metal, 60% (C28000); Nickel silver, 55-18 (C77000); Red brass, 85% (C23000); Yellow brass (C26800)
Group 3
Aluminum bronze, D (C61400); Beryllium copper (C17000, C17200, C17500); Commercial-bronze, 90% (C22000); Copper nickel, 10% (C70600); Copper nickel, 30% (C71500); Electrolytic tough pitch copper (C11000); Guilding, 95% (C21000); Nickel silver, 65-10 (C74500); Nickel silver, 65-12 (C75700); Nickel silver, 65-15 (C75400); Nickel silver, 65-18 (C75200); Oxygen-free copper (C10200) ; Phosphor bronze, 1.25% (C50200); Phosphor bronze, 10% D (C52400) Phosphor bronze, 5% A (C51000); Phosphor bronze, 8% C (C52100); Phosphorus deoxidized copper (C12200)

Wrought Alloys Description and UNS Alloy Numbers	Material Condition	HSS		Uncoated Carbide		Polycrystalline Diamond	
		Speed (fpm)		f = feed (0.001 in./rev), s = speed (ft/min)			
				Opt.	*Avg.*	*Opt.*	*Avg.*
Group 1	A	300	f	28	13		
	CD	350	s	1170	1680		
Group 2	A	200	f	28	13		
	CD	250	s	715	900		
Group 3	A	100	f	28	13	7	13
	CD	110	s	440	610	1780	2080

Abbreviations designate: A, annealed; CD, cold drawn.

The combined feed/speed data in this table are based on tool grades (identified in Table 16) as follows: uncoated carbide, 15; diamond, 9. See the footnote to Table 7.

Table 7. Cutting Feeds and Speeds for Turning Titanium and Titanium Alloys

Material	Brinell Hardness	Tool Material			
		HSS	Uncoated Carbide (Tough)		
			f = feed (0.001 in./rev), s = speed (ft/min)		
		Speed (fpm)		*Opt.*	*Avg.*
Commercially Pure and Low Alloyed					
99.5Ti, 99.5Ti-0.15Pd	110–150	100–105	f s	28 55	13 190
99.1Ti, 99.2Ti, 99.2Ti-0.15Pd, 98.9Ti-0.8Ni-0.3Mo	180–240	85–90	f s	28 50	13 170
99.0 Ti	250–275	70	f s	20 75	10 210
Alpha Alloys and Alpha-Beta Alloys					
5Al-2.5Sn, 8Mn, 2Al-11Sn-5Zr-1Mo, 4Al-3Mo-1V, 5Al-6Sn-2Zr-1Mo, 6Al-2Sn-4Zr-2Mo, 6Al-2Sn-4Zr-6Mo, 6Al-2Sn-4Zr-2Mo-0.25Si	300–350	50	f s	17 95	8 250
6Al-4V	310–350	40			
6Al-6V-2Sn, Al-4Mo,	320–370	30			
8V-5Fe-IAl	320–380	20			
6Al-4V, 6Al-2Sn-4Zr-2Mo, 6Al-2Sn-4Zr-6Mo, 6Al-2Sn-4Zr-2Mo-0.25Si	320–380	40			
4Al-3Mo-1V, 6Al-6V-2Sn, 7Al-4Mo	375–420	20			
I Al-8V-5Fe	375–440	20			
Beta Alloys					
13V-11Cr-3Al, 8Mo-8V-2Fe-3Al, 3Al-8V-6Cr-4Mo-4Zr, 11.5Mo-6ZR-4.5Sn {	275–350 375–440	25 20	f s	17 55	8 150

The speed recommendations for turning with HSS (high-speed steel) tools may be used as starting speeds for milling titanium alloys, using Table 15a to estimate the feed required. Speeds for HSS (high-speed steel) tools are based on a feed of 0.012 inch/rev and a depth of cut of 0.125 inch; use Table 5c to adjust the given speeds for other feeds and depths of cut. The combined feed/speed data in the remaining columns are based on a depth of cut of 0.1 inch, lead angle of 15 degrees, and nose radius of $\frac{3}{64}$ inch. Use Table 5a to adjust given speeds for other feeds, depths of cut, and lead angles; use Table 5b to adjust given speeds for increased tool life up to 180 minutes. Examples are given in the text. The combined feed/speed data in this table are based on tool grades (identified in Table 16) as follows: uncoated carbide, 15.

Table 8. Cutting Feeds and Speeds for Turning Light Metals

Material Description	Material Condition	Tool Material					
		HSS		Uncoated Carbide (Tough)		Polycrystalline Diamond	
		Speed (fpm)	f = feed (0.001 in./rev), s = speed (ft/min)				
				Opt.	*Avg.*	*Opt.*	*Avg.*
All wrought and cast magnesium alloys	A, CD, ST, and A	800					
All wrought aluminum alloys, including 6061-T651, 5000, 6000, and 7000 series	CD ST and A	600 500	f s	36 2820	17 4570		
All aluminum sand and permanent mold casting alloys	AC ST and A	750 600					
Aluminum Die-Casting Alloys							
Alloys 308.0 and 319.0	—	—	f s	36 865	17 1280	11 5890[a]	8 8270
Alloys 390.0 and 392.0	AC ST and A	80 60	f s	24 2010	11 2760	8 4765	4 5755
Alloy 413	—	—	f s	32 430	15 720	10 5085	5 6570
All other aluminum die-casting alloys including alloys 360.0 and 380.0	ST and A	100					
	AC	125	f s	36 630	17 1060	11 7560	6 9930

[a] The feeds and speeds for turning Al alloys 308.0 and 319.0 with (polycrystalline) diamond tooling represent an expected tool life $T = 960$ minutes = 16 hours; corresponding feeds and speeds for 15-minute tool life are 11/28600 and 6/37500.

Abbreviations for material condition: A, annealed; AC, as cast; CD, cold drawn; and ST and A, solution treated and aged, respectively. Speeds for HSS (high-speed steel) tools are based on a feed of 0.012 inch/rev and a depth of cut of 0.125 inch; use Table 5c to adjust the HSS speeds for other feeds and depths of cut. The combined feed/speed data are based on a depth of cut of 0.1 inch, lead angle of 15 degrees, and nose radius of $^3/_{64}$ inch. Use Table 5a to adjust given speeds for other feeds, depths of cut, and lead angles; use Table 5b to adjust given speeds for increased tool life up to 180 minutes. The data are based on tool grades (identified in Table 16) as follows: uncoated carbide, 15; diamond, 9.

Table 9. Cutting Feeds and Speeds for Turning Superalloys

	Tool Material										
	HSS Turning			Uncoated Carbide		Ceramic				CBN	
	Rough	Finish		Tough		Hard		Tough			
				f = feed (0.001 in./rev), s = speed (ft/min)							
Material Description	Speed (fpm)			*Opt.*	*Avg.*	*Opt.*	*Avg.*	*Opt.*	*Avg.*	*Opt.*	*Avg.*
T-D Nickel	70–80	80–100									
Discalloy	15–35	35–40									
19-9DL, W-545	25–35	30–40	f	24	11					20	10
16-25-6, A-286, Incoloy 800, 801, and 802, V-57	30–35	35–40	s	90	170					365	630
Refractaloy 26	15–20	20–25	f	20	10					20	10
J1300	15–25	20–30	s	75	135					245	420
Inconel 700 and 702, Nimonic 90 and 95	10–12	12–15									
S-816, V-36	10–15	15–20									
S-590	10–20	15–30									
Udimet 630		20–25									
N-155		15–25	f	20	10	11	6	11	6	20	10
Air Resist 213; Hastelloy B, C, G and X (wrought); Haynes 25 and 188; J1570; M252 (wrought); Mar-M905 and M918; Nimonic 75 and 80	15–20	20–25	s	75	125	1170	2590	405	900	230	400
CW-12M; Hastelloy B and C (cast); N-12M	8–12	10–15									
Rene 95 (Hot Isostatic Pressed)	—	—									
HS 6, 21, 2, 31 (X 40), 36, and 151; Haynes 36 and 151; Mar-M302, M322, and M509, WI-52	10–12	10–15									
Rene 41	10–15	12–20									
Incoloy 901	10–20	20–35									
Waspaloy	10–30	25–35	f	28	13	11	6	10	5	20	10
Inconel 625, 702, 706, 718 (wrought), 721, 722, X750, 751, 901, 600, and 604	15–20	20–35	s	20	40	895	2230	345	815	185	315
AF2-1DA, Unitemp 1753	8–10	10–15									
Colmonoy, Inconel 600, 718, K-Monel, Stellite	—	—									
Air Resist 13 and 215, FSH-H14, Nasa C-W-Re, X-45	10–12	10–15									
Udimet 500, 700, and 710	10–15	12–20									
Astroloy	5–10	5–15	f	28	13	11	6	10	5	20	10
Mar-M200, M246, M421, and Rene 77, 80, and 95 (forged)		10–12 10–15	s	15	15	615	1720	290	700	165	280
B-1900, GMR-235 and 235D, IN 100 and 738, Inconel 713C and 718 (cast), M252 (cast)	8–10	8–10									

The speed recommendations for rough turning may be used as starting values for milling and drilling with HSS tools. The combined feed/speed data in this table are based on tool grades (identified in Table 16) as follows: uncoated carbide = 15; ceramic, hard = 4, tough = 3; CBN = 1.

Speeds for HSS (high-speed steel) tools are based on a feed of 0.012 inch/rev and a depth of cut of 0.125 inch; use Table 5c to adjust the given speeds for other feeds and depths of cut. The combined feed/speed data in the remaining columns are based on a depth of cut of 0.1 inch, lead angle of 15 degrees, and nose radius of 3⁄64 inch. Use Table 5a to adjust given speeds for other feeds, depths of cut, and lead angles; use Table 5b to adjust given speeds for increased tool life up to 180 minutes. Examples are given in the text.

Speed and Feed Tables for Milling.—Tables 10 through 14 give feeds and speeds for milling. The data in the first speed column can be used with high-speed steel tools using the feeds given in Table 15a; these are the same speeds contained in previous editions of the Handbook. The remaining data in Tables 10 through 14 are combined feeds and speeds for end, face, and slit, slot, and side milling that use the speed adjustment factors given in Tables 15b, 15c, and 15d. Tool life for the combined feed/speed data can also be adjusted using the factors in Table 15e. Table 16 lists cutting tool grades and vendor equivalents.

End Milling: Table data for end milling are based on a 3-tooth, 20-degree helix angle tool with a diameter of 1.0 inch, an axial depth of cut of 0.2 inch, and a radial depth of cut of 1 inch (full slot). Use Table 15b to adjust speeds for other feeds and axial depths of cut, and Table 15c to adjust speeds if the radial depth of cut is less than the tool diameter. Speeds are valid for all tool diameters.

Face Milling: Table data for face milling are based on a 10-tooth, 8-inch diameter face mill, operating with a 15-degree lead angle, 3⁄64-inch nose radius, axial depth of cut = 0.1 inch, and radial depth (width) of cut = 6 inches (i.e., width of cut to cutter diameter ratio = 3⁄4). These speeds are valid if the cutter axis is above or close to the center line of the workpiece (eccentricity is small). Under these conditions, use Table 15d to adjust speeds for other feeds and axial and radial depths of cut. For larger eccentricity (i.e., when the cutter axis to workpiece center line offset is one half the cutter diameter or more), use the end and side milling adjustment factors (Tables 15b and 15c) instead of the face milling factors.

Slit and Slot Milling: Table data for slit milling are based on an 8-tooth, 10-degree helix angle tool with a cutter width of 0.4 inch, diameter D of 4.0 inch, and a depth of cut of 0.6 inch. Speeds are valid for all tool diameters and widths. See the examples in the text for adjustments to the given speeds for other feeds and depths of cut.

Tool life for all tabulated values is approximately 45 minutes; use Table 15e to adjust tool life from 15 to 180 minutes.

Using the Feed and Speed Tables for Milling: The basic feed for milling cutters is the feed per tooth (f), which is expressed in inches per tooth. There are many factors to consider in selecting the feed per tooth and no formula is available to resolve these factors. Among the factors to consider are the cutting tool material; the work material and its hardness; the width and the depth of the cut to be taken; the type of milling cutter to be used and its size; the surface finish to be produced; the power available on the milling machine; and the rigidity of the milling machine, the workpiece, the workpiece setup, the milling cutter, and the cutter mounting.

The cardinal principle is to always use the maximum feed that conditions will permit. Avoid, if possible, using a feed that is less than 0.001 inch per tooth because such low feeds reduce the tool life of the cutter. When milling hard materials with small-diameter end mills, such small feeds may be necessary, but otherwise use as much feed as possible. Harder materials in general will require lower feeds than softer materials. The width and the depth of cut also affect the feeds. Wider and deeper cuts must be fed somewhat more slowly than narrow and shallow cuts. A slower feed rate will result in a better surface finish; however, always use the heaviest feed that will produce the surface finish desired. Fine chips produced by fine feeds are dangerous when milling magnesium because spontaneous combustion can occur. Thus, when milling magnesium, a fast feed that will produce a relatively thick chip should be used. Cutting stainless steel produces a work-hardened layer on the surface that has been cut. Thus, when milling this material, the feed should be large enough to allow each cutting edge on the cutter to penetrate below the work-hardened layer produced by the previous cutting edge. The heavy feeds recommended for face milling cutters are to be used primarily with larger cutters on milling machines having an adequate amount of power. For smaller face milling cutters, start with smaller feeds and increase as indicated by the performance of the cutter and the machine.

When planning a milling operation that requires a high cutting speed and a fast feed, always check to determine if the power required to take the cut is within the capacity of the milling machine. Excessive power requirements are often encountered when milling with cemented carbide cutters. The large metal removal rates that can be attained require a high horsepower output. An example of this type of calculation is given in the section on Machining Power that follows this section. If the size of the cut must be reduced in order to stay within the power capacity of the machine, start by reducing the cutting speed rather than the feed in inches per tooth.

The formula for calculating the table feed rate, when the feed in inches per tooth is known, is as follows:

$$f_m = f_t n_t N$$

where f_m = milling machine table feed rate in inches per minute (ipm)

f_t = feed in inch per tooth (ipt)

n_t = number of teeth in the milling cutter

N = spindle speed of the milling machine in revolutions per minute (rpm)

Example: Calculate the feed rate for milling a piece of AISI 1040 steel having a hardness of 180 Bhn. The cutter is a 3-inch diameter high-speed steel plain or slab milling cutter with 8 teeth. The width of the cut is 2 inches, the depth of cut is 0.062 inch, and the cutting speed from Table 11 is 85 fpm. From Table 15a, the feed rate selected is 0.008 inch per tooth.

$$N = \frac{12V}{\pi D} = \frac{12 \times 85}{3.14 \times 3} = 108 \text{ rpm}$$

$$f_m = f_t n_t N = 0.008 \times 8 \times 108$$

$$= 7 \text{ ipm (approximately)}$$

Example 1, Face Milling: Determine the cutting speed and machine operating speed for face milling an aluminum die casting (alloy 413) using a 4-inch polycrystalline diamond cutter, a 3-inch width of cut, a 0.10-inch depth of cut, and a feed of 0.006 inch/tooth.

Table 10 gives the feeds and speeds for milling aluminum alloys. The feed/speed pairs for face milling die cast alloy 413 with polycrystalline diamond (PCD) are 8/2320 (0.008 in./tooth feed at 2320 fpm) and 4/4755 (0.004 in./tooth feed at 4755 fpm). These speeds are based on an axial depth of cut of 0.10 inch, an 8-inch cutter diameter *D*, a 6-inch radial depth (width) of cut *ar*, with the cutter approximately centered above the workpiece, i.e., eccentricity is low, as shown in Fig. 3. If the preceding conditions apply, the given feeds and speeds can be used without adjustment for a 45-minute tool life. The given speeds are valid for all cutter diameters if a radial depth of cut to cutter diameter ratio (ar/D) of $\frac{3}{4}$ is maintained (i.e., $\frac{6}{8} = \frac{3}{4}$). However, if a different feed or axial depth of cut is required, or if the ar/D ratio is not equal to $\frac{3}{4}$, the cutting speed must be adjusted for the conditions. The adjusted cutting speed V is calculated from $V = V_{opt} \times F_f \times F_d \times F_{ar}$, where V_{opt} is the lower of the two speeds given in the speed table, and F_f, F_d, and F_{ar} are adjustment factors for feed, axial depth of cut, and radial depth of cut, respectively, obtained from Table 15d (face milling); except, when cutting near the end or edge of the workpiece as in Fig. 4, Table 15c (side milling) is used to obtain F_f.

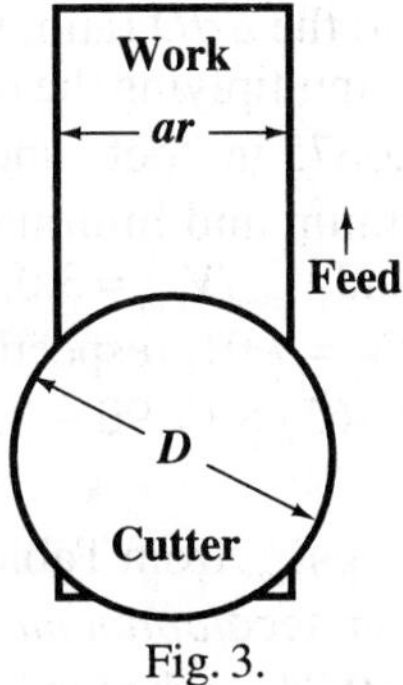

Fig. 3.

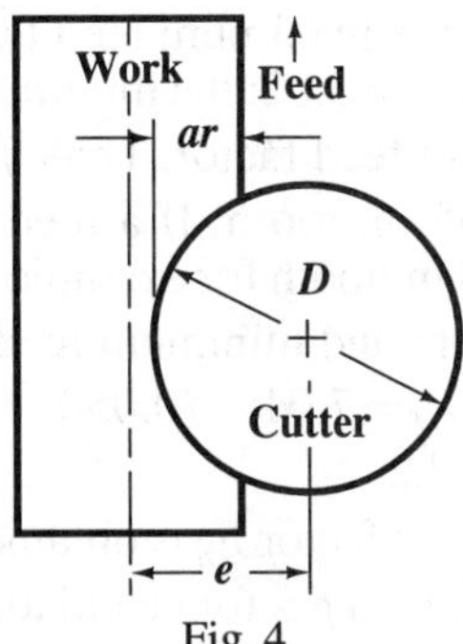

Fig. 4.

In this example, the cutting conditions match the standard conditions specified in the speed table for radial depth of cut to cutter diameter (3 in./4 in.), and depth of cut (0.01 in), but the desired feed of 0.006 in./tooth does not match either of the feeds given in the speed table (0.004 or 0.008). Therefore, the cutting speed must be adjusted for this feed. As with turning, the feed factor F_f is determined by calculating the ratio of the desired feed f to maximum feed f_{opt} from the speed table, and from the ratio V_{avg}/V_{opt} of the two speeds given in the speed table. The feed factor is found at the intersection of the feed ratio row and the speed ratio column in Table 15d. The speed is then obtained using the following equation:

$$\frac{\text{Chosen feed}}{Optimum\text{ feed}} = \frac{f}{f_{opt}} = \frac{0.006}{0.008} = 0.75 \qquad \frac{Average\text{ speed}}{Optimum\text{ speed}} = \frac{V_{avg}}{V_{opt}} = \frac{4755}{2320} \approx 2.0$$

$$F_f = (1.25 + 1.43)/2 = 1.34 \qquad F_d = 1.0 \qquad F_{ar} = 1.0$$

$$V = 2320 \times 1.34 \times 1.0 \times 1.0 = 3109 \text{ fpm, and } 3.82 \times 3109/4 = 2970 \text{ rpm}$$

Example 2, End Milling: What cutting speed should be used for cutting a full slot (i.e., a slot cut from the solid, in one pass, that is the same width as the cutter) in 5140 steel with hardness of 300 Bhn using a 1-inch diameter coated carbide (insert) 0° lead angle end mill, a feed of 0.003 in./tooth, and a 0.2-inch axial depth of cut?

The feed and speed data for end milling 5140 steel, Brinell hardness = 275–325, with a coated carbide tool are given in Table 11 as 15/80 and 8/240 for *optimum* and *average* sets, respectively. The speed adjustment factors for feed and depth of cut for full slot (end milling) are obtained from Table 15b. The calculations are the same as in the previous examples: $f/f_{opt} = 3/15 = 0.2$ and $V_{avg}/V_{opt} = 240/80 = 3.0$, therefore, $F_f = 6.86$ and $F_d = 1.0$. The cutting speed for a 45-minute tool life is $V = 80 \times 6.86 \times 1.0 = 548.8$, approximately 550 ft/min.

Example 3, End Milling: What cutting speed should be used in Example 2 if the radial depth of cut *ar* is 0.02 inch and axial depth of cut is 1 inch?

In end milling, when the radial depth of cut is less than the cutter diameter (as in Fig. 4), first obtain the feed factor F_f from Table 15c, then the axial depth of cut and lead angle factor F_d from Table 15b. The radial depth of cut to cutter diameter ratio ar/D is used in Table 15c to determine the maximum and minimum feeds that guard against tool failure at high feeds and against premature tool wear caused by the tool rubbing against the work at very low feeds. The feed used should be selected so that it falls within the minimum to maximum feed range, and then the feed factor F_f can be determined from the feed factors at minimum and maximum feeds, F_{f1} and F_{f2} as explained below.

The maximum feed f_{max} is found in Table 15c by multiplying the *optimum* feed from the speed table by the maximum feed factor that corresponds to the *ar*/*D* ratio, which in this instance is 0.02/1 = 0.02; the minimum feed f_{min} is found by multiplying the *optimum* feed by the minimum feed factor. Thus, f_{max} = 4.5 × 0.015 = 0.0675 in./tooth and f_{min} = 3.1 × 0.015 = 0.0465 in./tooth. If a feed between these maximum and minimum values is selected, 0.050 in./tooth for example, then for *ar*/*D* = 0.02 and V_{avg}/V_{opt} = 3.0, the feed factors at maximum and minimum feeds are F_{f1} = 7.90 and F_{f2} = 7.01, respectively, and by interpolation, F_f = 7.01 + (0.050 − 0.0465)(0.0675 − 0.0465) × (7.90 − 7.01) = 7.16, approximately 7.2.

The depth of cut factor F_d is obtained from Table 15b, using f_{max} from Table 15c instead of the *optimum* feed f_{opt} for calculating the feed ratio (chosen feed/*optimum* feed). In this example, the feed ratio = chosen feed/f_{max} = 0.050/0.0675 = 0.74, so the feed factor is F_d = 0.93 for a depth of cut = 1.0 inch and 0° lead angle. Therefore, the final cutting speed is 80 × 7.2 × 0.93 = 587 ft/min. Notice that f_{max} obtained from Table 15c was used instead of the *optimum* feed from the speed table, in determining the feed ratio needed to find F_d.

Slit Milling.—The tabular data for slit milling is based on an 8-tooth, 10-degree helix angle cutter with a width of 0.4 inch, a diameter *D* of 4.0 inch, and a depth of cut of 0.6 inch. The given feeds and speeds are valid for any diameters and tool widths, as long as sufficient machine power is available. Adjustments to cutting speeds for other feeds and depths of cut are made using Table 15c or 15d, depending on the orientation of the cutter to the work, as illustrated in Case 1 and Case 2 of Fig. 5. The situation illustrated in Case 1 is approximately equivalent to that illustrated in Fig. 3, and Case 2 is approximately equivalent to that shown in Fig. 4.

Case 1: If the cutter is fed directly into the workpiece, i.e., the feed is perpendicular to the surface of the workpiece, as in cutting off, then Table 15d (face milling) is used to adjust speeds for other feeds. The depth of cut portion of Table 15d is not used in this case (F_d = 1.0), so the adjusted cutting speed $V = V_{opt} \times F_f \times F_{ar}$. In determining the factor F_{ar} from Table 15d, the radial depth of cut *ar* is the length of cut created by the portion of the cutter engaged in the work.

Case 2: If the cutter feed is parallel to the surface of the workpiece, as in slotting or side milling, then Table 15c (side milling) is used to adjust the given speeds for other feeds. In Table 15c, the cutting depth (slot depth, for example) is the radial depth of cut *ar* that is used to determine maximum and minimum allowable feed/tooth and the feed factor F_f. These minimum and maximum feeds are determined in the manner described previously, however, the axial depth of cut factor F_d is not required. The adjusted cutting speed, valid for cutters of any thickness (width), is given by $V = V_{opt} \times F_f$.

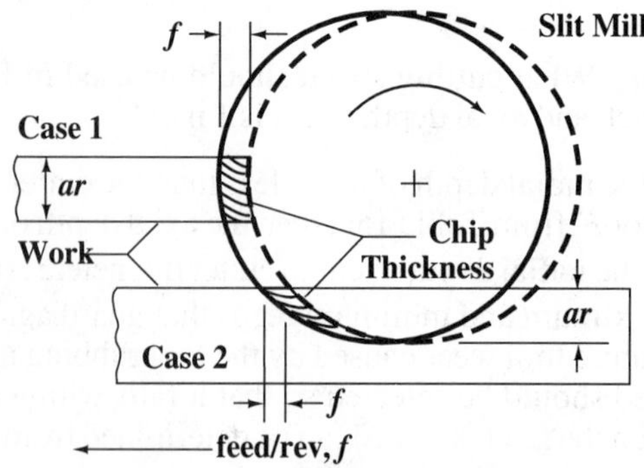

Fig. 5. Determination of Radial Depth of Cut or in Slit Milling

Table 10. Cutting Feeds and Speeds for Milling Aluminum Alloys

Material	Material Condition*		End Milling				Face Milling				Slit Milling			
			HSS		Indexable Insert Uncoated Carbide		Indexable Insert Uncoated Carbide		Polycrystalline Diamond		HSS		Indexable Insert Uncoated Carbide	
			f = feed (0.001 in./tooth), s = speed (ft/min)											
			Opt.	*Avg.*	*Opt.*	*Avg.*	*Opt.*	*Avg.*	*Opt.*	*Avg.*	*Opt.*	*Avg.*	*Opt.*	*Avg.*
All wrought aluminum alloys, 6061-T651, 5000, 6000, 7000 series	CD ST and A	f	15	8	15	8	39	20	8	4	16	8	39	20
All aluminum sand and permanent mold casting alloys	CD ST and A	s	165	850	620	2020	755	1720	3750	8430	1600	4680	840	2390
Aluminum Die-Casting Alloys														
Alloys 308.0 and 319.0	—	f	15	8	15	8	39	20			16	8	39	20
		s	30	100	620	2020	755	1720			160	375	840	2390
Alloys 360.0 and 380.0	—	f	15	8	15	8	39	20	8	4	16	8	39	20
		s	30	90	485	1905	555	1380	3105	7845	145	355	690	2320
Alloys 390.0 and 392.0	—	f					39	20						
		s					220	370						
Alloy 413	—	f			15	8	39	20	8	4			39	20
All other aluminum die-casting alloys	ST and A	s			355	1385	405	665	2320	4755			500	1680
	AC	f	15	8	15	8	39	20	8	4	16	8	39	20
		s	30	90	485	1905	555	1380	3105	7845	145	335	690	2320

Abbreviations designate: A, annealed; AC, as cast; CD, cold drawn; and ST and A, solution treated and aged, respectively.

End Milling: Table data for end milling are based on a 3-tooth, 20-degree helix angle tool with a diameter of 1.0 inch, an axial depth of cut of 0.2 inch, and a radial depth of cut of 1 inch (full slot). Use Table 15b to adjust speeds for other feeds and axial depths of cut, and Table 15c to adjust speeds if the radial depth of cut is less than the tool diameter. Speeds are valid for all tool diameters.

Face Milling: Table data for face milling are based on a 10-tooth, 8-inch diameter face mill, operating with a 15-degree lead angle, $\frac{3}{64}$-inch nose radius, axial depth of cut = 0.1 inch, and radial depth (width) of cut = 6 inches (i.e., width of cut to cutter diameter ratio = $\frac{3}{4}$). These speeds are valid if the cutter axis is above or close to the center line of the workpiece (eccentricity is small). Under these conditions, use Table 15d to adjust speeds for other feeds and axial and radial depths of cut. For larger eccentricity (i.e., when the cutter axis to workpiece center line offset is one half the cutter diameter or more), use the end and side milling adjustment factors (Tables 15b and 15c) instead of the face milling factors.

Slit and Slot Milling: Table data for slit milling are based on an 8-tooth, 10-degree helix angle tool with a cutter width of 0.4 inch, diameter *D* of 4.0 inch, and a depth of cut of 0.6 inch. Speeds are valid for all tool diameters and widths. See the examples in the text for adjustments to the given speeds for other feeds and depths of cut.

Tool life for all tabulated values is approximately 45 minutes; use Table 15e to adjust tool life from 15 to 180 minutes. The combined feed/speed data in this table are based on tool grades (identified in Table 16) as follows: uncoated carbide = 15; diamond = 9.

Table 11. Cutting Feeds and Speeds for Milling Plain Carbon and Alloy Steels

Material	Brinell Hardness	HSS Speed (fpm)		End Milling HSS Opt.	Avg.	End Milling Uncoated Carbide Opt.	Avg.	End Milling Coated Carbide Opt.	Avg.	Face Milling Uncoated Carbide Opt.	Avg.	Face Milling Coated Carbide Opt.	Avg.	Slit Milling Uncoated Carbide Opt.	Avg.	Slit Milling Coated Carbide Opt.	Avg.
				f = feed (0.001 in./tooth), s = speed (ft/min)													
Free-machining plain carbon steels (resulfurized): 1212, 1213, 1215	100–150	140	f s	7 45	4 125	7 465	4 735	7 800	4 1050	39 225	20 335	39 415	20 685	39 265	20 495	39 525	20 830
	150–200	130	f s	7 35	4 100							39 215	20 405				
(Resulfurized): 1108, 1109, 1115, 1117, 1118, 1120, 1126, 1211	100–150	130	f	7	4	7	4	7	4	39	20	39	20	39	20	39	20
	150–200	115	s	30	85	325	565	465	720	140	220	195	365	170	350	245	495
	175–225	115	f s	7 30	4 85									39 185	20 350		
(Resulfurized): 1132, 1137, 1139, 1140, 1144, 1146, 1151	275–325	70															
	325–375	45	f s	7 25	4 70	7 210	4 435	7 300	4 560	39 90	20 170	39 175	20 330	39 90	20 235	39 135	20 325
	375–425	35															
(Leaded): 11L17, 11L18, 12L13, 12L14	100–150	140	f s	7 35	4 100							39 215	20 405				
	150–200	130															
	200–250	110	f s	7 30	4 85							39 185	20 350				
Plain carbon steels: 1006, 1008, 1009, 1010, 1012, 1015, 1016, 1017, 1018, 1019, 1020, 1021, 1022, 1023, 1024, 1025, 1026, 1513, 1514	100–125	110	f s	7 45	4 125	7 465	4 735	7 800	4 1050	39 225	20 335	39 415	20 685	39 265	20 495	39 525	20 830
	125–175	110	f s	7 35	4 100							39 215	20 405				
	175–225	90	f s	7 30	4 85							39 185	20 350				
	225–275	65															

Table 11. *(Continued)* **Cutting Feeds and Speeds for Milling Plain Carbon and Alloy Steels**

Material	Brinell Hardness	HSS Speed (fpm)		End Milling HSS		End Milling Uncoated Carbide		End Milling Coated Carbide		Face Milling Uncoated Carbide		Face Milling Coated Carbide		Slit Milling Uncoated Carbide		Slit Milling Coated Carbide	
				f = feed (0.001 in./tooth), s = speed (ft/min)													
				Opt.	*Avg.*	*Opt.*	*Avg.*	*Opt.*	*Avg.*	*Opt.*	*Avg.*	*Opt.*	*Avg.*	*Opt.*	*Avg.*	*Opt.*	*Avg.*
Plain carbon steels: 1027, 1030, 1033, 1035, 1036, 1037, 1038, 1039, 1040, 1041, 1042, 1043, 1045, 1046, 1048, 1049, 1050, 1052, 1524, 1526, 1527, 1541	125–175	100	**f**	7	4							39	20				
			s	35	100							215	405				
	175–225 225–275	85 70	**f**	7	4							39	20				
			s	30	85							185	350				
	275–325 325–375 375–425	55 35 25	**f**	7	4	7	4	7	4	39	20	39	20	39	20	39	20
			s	25	70	210	435	300	560	90	170	175	330	90	235	135	325
Plain carbon steels: 1055, 1060, 1064, 1065, 1070, 1074, 1078, 1080, 1084, 1086, 1090, 1095, 1548, 1551, 1552, 1561, 1566	125–175 175–225	90 75	**f**	7	4	7	4	7	4	39	20	39	20	39	20	39	20
			s	30	85	325	565	465	720	140	220	195	365	170	350	245	495
	225–275	60	**f**	7	4							39	20				
			s	30	85							185	350				
	275–325 325–375 375–425	45 30 15	**f**	7	4	7	4	7	4	39	20	39	20	39	20	39	20
			s	25	70	210	435	300	560	90	170	175	330	90	235	135	325
Free-machining alloy steels (Resulfurized): 4140, 4150	175–200 200–250	100 90	**f**	15	8	15	8	15	8			39	20	39	20	7	4
			s	7	30	105	270	270	450			295	475	135	305	25	70
	250–300	60	**f**	15	8	15	8	15	8			39	20	39	20	7	4
			s	6	25	50	175	85	255			200	320	70	210	25	70
	300–375 375–425	45 35	**f**	15	8	15	8	15	8			39	20				
			s	5	20	40	155	75	225			175	280				

Table 11. *(Continued)* **Cutting Feeds and Speeds for Milling Plain Carbon and Alloy Steels**

Material	Brinell Hardness	HSS Speed (fpm)		End Milling HSS		End Milling Uncoated Carbide		End Milling Coated Carbide		Face Milling Uncoated Carbide		Face Milling Coated Carbide		Slit Milling Uncoated Carbide		Slit Milling Coated Carbide	
				f = feed (0.001 in./tooth), s = speed (ft/min)													
				Opt.	*Avg.*	*Opt.*	*Avg.*	*Opt.*	*Avg.*	*Opt.*	*Avg.*	*Opt.*	*Avg.*	*Opt.*	*Avg.*	*Opt.*	*Avg.*
Free-machining alloy steels (Leaded): 41L30, 41L40, 41L47, 41L50, 43L47, 51L32, 52L100, 86L20, 86L40	150–200	115	f	7	4	7	4	7	4	39	20	39	20	39	20	39	20
			s	30	85	325	565	465	720	140	220	195	365	170	350	245	495
	200–250	95	f	7	4							39	20				
			s	30	85							185	350				
	250–300 300–375 375–425	70 50 40	f	7	4	7	4	7	4	39	20	39	20	39	20	39	20
			s	25	70	210	435	300	560	90	170	175	330	90	235	135	325
Alloy steels: 4012, 4023, 4024, 4028, 4118, 4320, 4419, 4422, 4427, 4615, 4620, 4621, 4626, 4718, 4720, 4815, 4817, 4820, 5015, 5117, 5120, 6118, 8115, 8615, 8617, 8620, 8622, 8625, 8627, 8720, 8822, 94B17	125–175 175–225	100 90	f	15	8	15	8	15	8			39	20	39	20	39	20
			s	7	30	105	270	220	450			295	475	135	305	265	495
	225–275	60	f	15	8	15	8	15	8			39	20	39	20	39	20
			s	6	25	50	175	85	255			200	320	70	210	115	290
	275–325	50	f	15	8	15	8	15	8			39	20				
			s	5	20	45	170	80	240			190	305				
	325–375 375–425	40 25	f	15	8	15	8	15	8			39	20				
			s	5	20	40	155	75	225			175	280				
Alloy steels: 1330, 1335, 1340, 1345, 4032, 4037, 4042, 4047, 4130, 4135, 4137, 4140, 4142, 4145, 4147, 4150, 4161, 4337, 4340, 50B44, 50B46, 50B50, 50B60, 5130, 5132, 5140, 5145, 5147, 5150, 5160, 51B60, 6150, 81B45, 8630, 8635, 8637, 8640, 8642, 8645, 8650, 8655, 8660, 8740, 9254, 9255, 9260, 9262, 94B30 E51100, E52100: use (HSS speeds)	175–225	75 (65)	f	15	8	15	8	15	8			39	20	39	20	39	20
			s	5	30	105	270	220	450			295	475	135	305	265	495
	225–275	60	f	15	8	15	8	15	8			39	20	39	20	39	20
			s	5	25	50	175	85	255			200	320	70	210	115	290
	275–325	50 (40)	f	15	8	15	8	15	8			39	20				
			s	5	25	45	170	80	240			190	305				
	325–375	35 (30)															
	375–425	20	f	15	8	15	8	15	8			39	20				
			s	5	20	40	155	75	225			175	280				

Table 11. *(Continued)* **Cutting Feeds and Speeds for Milling Plain Carbon and Alloy Steels**

Material	Brinell Hardness	HSS Speed (fpm)		End Milling HSS		End Milling Uncoated Carbide		End Milling Coated Carbide		Face Milling Uncoated Carbide		Face Milling Coated Carbide		Slit Milling Uncoated Carbide		Slit Milling Coated Carbide	
				f = feed (0.001 in./tooth), s = speed (ft/min)													
				Opt.	*Avg.*	*Opt.*	*Avg.*	*Opt.*	*Avg.*	*Opt.*	*Avg.*	*Opt.*	*Avg.*	*Opt.*	*Avg.*	*Opt.*	*Avg.*
Ultra-high-strength steels (not AISI): AMS 6421 (98B37 Mod.), 6422 (98BV40), 6424, 6427, 6428, 6430, 6432, 6433, 6434, 6436, and 6442; 300M, D6ac	220–300	60	**f**			8	4	8	4								
	300–350	45	**s**			165	355	300	480								
	350–400	20	**f**	8	4	8	4					39	20	39	20		
			s	15	45	150	320					130	235	75	175		
	43–52 Rc	—	**f**			5	3							39	20		
			s			20†	55							5	15		
Maraging steels (not AISI): 18% Ni Grades 200, 250, 300, and 350	250–325	50	**f**			8	4	8	4								
			s			165	355	300	480								
	50–52 Rc	—	**f**			5	3							39	20		
			s			20†	55							5	15		
Nitriding steels (not AISI): Nitralloy 125, 135, 135 Mod., 225, and 230, Nitralloy N, Nitralloy EZ, Nitrex 1	200–250	60	**f**	15	8	15	8	15	8			39	20	39	20	39	20
			s	7	30	105	270	220	450			295	475	135	305	265	495
	300–350	25	**f**	15	8	15	8	15	8			39	20				
			s	5	20	40	155	75	225			175	280				

For HSS (high-speed steel) tools in the first speed column only, use Table 15a for recommended feed in inches per tooth and depth of cut.

End Milling: Table data for end milling are based on a 3-tooth, 20-degree helix angle tool with a diameter of 1.0 inch, an axial depth of cut of 0.2 inch, and a radial depth of cut of 1 inch (full slot). Use Table 15b to adjust speeds for other feeds and axial depths of cut, and Table 15c to adjust speeds if the radial depth of cut is less than the tool diameter. Speeds are valid for all tool diameters.

Face Milling: Table data for face milling are based on a 10-tooth, 8-inch diameter face mill, operating with a 15-degree lead angle, $^3/_{64}$-inch nose radius, axial depth of cut = 0.1 inch, and radial depth (width) of cut = 6 inches (i.e., width of cut to cutter diameter ratio = $^3/_4$). These speeds are valid if the cutter axis is above or close to the center line of the workpiece (eccentricity is small). Under these conditions, use Table 15d to adjust speeds for other feeds and axial and radial depths of cut. For larger eccentricity (i.e., when the cutter axis to workpiece center line offset is one half the cutter diameter or more), use the end and side milling adjustment factors (Tables 15b and 15c) instead of the face milling factors.

Slit and Slot Milling: Table data for slit milling are based on an 8-tooth, 10-degree helix angle tool with a cutter width of 0.4 inch, diameter *D* of 4.0 inches, and a depth of cut of 0.6 inch. Speeds are valid for all tool diameters and widths. See the examples in the text for adjustments to the given speeds for other feeds and depths of cut.

Tool life for all tabulated values is approximately 45 minutes; use Table 15e to adjust tool life from 15 to 180 minutes. The combined feed/speed data in this table are based on tool grades (identified in Table 16) as follows: end and slit milling uncoated carbide = 20 except † = 15; face milling uncoated carbide = 19; end, face, and slit milling coated carbide = 10.

Table 12. Cutting Feeds and Speeds for Milling Tool Steels

Material	Brinell Hardness	HSS Speed (fpm)		End Milling: HSS Opt.	End Milling: HSS Avg.	End Milling: Uncoated Carbide Opt.	End Milling: Uncoated Carbide Avg.	End Milling: Coated Carbide Opt.	End Milling: Coated Carbide Avg.	Face Milling: Uncoated Carbide Opt.	Face Milling: Uncoated Carbide Avg.	Face Milling: CBN Opt.	Face Milling: CBN Avg.	Slit Milling: Uncoated Carbide Opt.	Slit Milling: Uncoated Carbide Avg.	Slit Milling: Coated Carbide Opt.	Slit Milling: Coated Carbide Avg.
			f = feed (0.001 in./tooth), s = speed (ft/min)														
Water hardening: W1, W2, W5	150–200	85															
Shock resisting: S1, S2, S5, S6, S7	175–225	55															
Cold work, oil hardening: O1, O2, O6, O7	175–225	50	f	8	4	8	4	8	4	39	20			39	20	39	20
			s	25	70	235	455	405	635	235	385			115	265	245	445
Cold work, high carbon, high chromium: D2, D3, D4, D5, D7	200–250	40															
Cold work, air hardening: A2, A3, A8, A9, A10	200–250	50															
A4, A6	200–250	45	f							39	20						
A7	225–275	40	s							255	385						
Hot work, chromium type: H10, H11, H12, H13, H14, H19	150–200	60															
	200–250	50															
	325–375	30	f	8	4	8	4			39	20			39	20		
			s	15	45	150	320			130	235			75	175		
	48–50 Rc	—	f			5	3					39	20	39	20		
	50–52 Rc	—	s			20†	55					50	135	5†	15		
	52–56 Rc	—															
Hot work, tungsten and molybdenum types: H21, H22, H23, H24, H25, H26, H41, H42, H43	150–200	55	f							39	20						
	200–250	45	s							255	385						
Special-purpose, low alloy: L2, L3, L6	150–200	65	f	8	4	8	4	8	4	39	20			39	20	39	20
Mold: P2, P3, P4, P5, P6 P20, P21	100–150	75	s	25	70	235	455	405	635	235	385			115	265	245	445
	150–200	60															
High-speed steel: M1, M2, M6, M10, T1, T2, T6	200–250	50															
M3-1, M4, M7, M30, M33, M34, M36, M41, M42, M43, M44, M46, M47, T5, T8	225–275	40	f							39	20						
			s							255	385						
T15, M3-2	225–275	30															

For HSS (high-speed steel) tools in the first speed column only, use Table 15a for recommended feed in inches per tooth and depth of cut.

End Milling: Table data for end milling are based on a 3-tooth, 20-degree helix angle tool with a diameter of 1.0 inch, an axial depth of cut of 0.2 inch, and a radial depth of cut of 1 inch (full slot). Use Table 15b to adjust speeds for other feeds and axial depths of cut, and Table 15c to adjust speeds if the radial depth of cut is less than the tool diameter. Speeds are valid for all tool diameters.

Face Milling: Table data for face milling are based on a 10-tooth, 8-inch diameter face mill, operating with a 15-degree lead angle, $\frac{3}{64}$-inch nose radius, axial depth of cut = 0.1 inch, and radial depth (width) of cut = 6 inches (i.e., width of cut to cutter diameter ratio = $\frac{3}{4}$). These speeds are valid if the cutter axis is above or close to the center line of the workpiece (eccentricity is small). Under these conditions, use Table 15d to adjust speeds for other feeds and axial and radial depths of cut. For larger eccentricity (i.e., when the cutter axis to workpiece center line offset is one half the cutter diameter or more), use the end and side milling adjustment factors (Tables 15b and 15c) instead of the face milling factors.

Slit and Slot Milling: Table data for slit milling are based on an 8-tooth, 10-degree helix angle tool with a cutter width of 0.4 inch, diameter *D* of 4.0 inches, and a depth of cut of 0.6 inch. Speeds are valid for all tool diameters and widths. See the examples in the text for adjustments to the given speeds for other feeds and depths of cut.

Tool life for all tabulated values is approximately 45 minutes; use Table 15e to adjust tool life from 15 to 180 minutes. The combined feed/speed data in this table are based on tool grades (identified in Table 16) as follows: uncoated carbide = 20, † = 15; coated carbide = 10; CBN = 1.

Table 13. Cutting Feeds and Speeds for Milling Stainless Steels

Material	Brinell Hardness	HSS		End Milling						Face Milling		Slit Milling			
				HSS		Uncoated Carbide		Coated Carbide		Coated Carbide		Uncoated Carbide		Coated Carbide	
		Speed (fpm)		f = feed (0.001 in./tooth), s = speed (ft/min)											
				Opt.	*Avg.*	*Opt.*	*Avg.*	*Opt.*	*Avg.*	*Opt.*	*Avg.*	*Opt.*	*Avg.*	*Opt.*	*Avg.*
Free-machining stainless steels (Ferritic): 430F, 430FSe	135–185	110	f s	7 30	4 80	7 305	4 780	7 420	4 1240	39 210	20 385	39 120	20 345	39 155	20 475
(Austenitic): 203EZ, 303, 303Se, 303MA, 303Pb, 303Cu, 303 Plus X	135–185 225–275	100 80	f s	7 20	4 55	7 210	4 585					39 75	20 240		
(Martensitic): 416, 416Se, 416 Plus X, 420F, 420FSe, 440F, 440FSe	135–185 185–240 275–325 375–425	110 100 60 30													
Stainless steels (Ferritic): 405, 409, 429, 430, 434, 436, 442, 446, 502	135–185	90	f s	7 30	4 80	7 305	4 780	7 420	4 1240	39 210	20 385	39 120	20 345	39 155	20 475
(Austenitic): 201, 202, 301, 302, 304, 304L, 305, 308, 321, 347, 348	135–185 225–275	75 65	f s	7 20	4 55	7 210	4 585					39 75	20 240		
(Austenitic): 302B, 309, 309S, 310, 310S, 314, 316, 316L, 317, 330	135–185	70													
(Martensitic): 403, 410, 420, 501	135–175 175–225 275–325 375–425	95 85 55 35													

Table 13. *(Continued)* **Cutting Feeds and Speeds for Milling Stainless Steels**

Material	Brinell Hardness	HSS		End Milling						Face Milling		Slit Milling			
				HSS		Uncoated Carbide		Coated Carbide		Coated Carbide		Uncoated Carbide		Coated Carbide	
		Speed (fpm)		f = feed (0.001 in./tooth), s = speed (ft/min)											
				Opt.	Avg.	Opt.	Avg.	Opt.	Avg.	Opt.	Avg.	Opt.	Avg.	Opt.	Avg.
Stainless Steels (Martensitic): 414, 431, Greek Ascoloy, 440A, 440B, 440C	225–275	55–60													
	275–325	45–50													
	375–425	30													
(Precipitation hardening): 15-5PH, 17-4PH, 17-7PH, AF-71, 17-14CuMo, AFC-77, AM-350, AM-355, AM-362, Custom 455, HNM, PH13-8, PH14-8Mo, PH15-7Mo, Stainless W	150–200	60	f	7	4	7	4					39	20		
			s	20	55	210	585					75	240		
	275–325	50													
	325–375	40													
	375–450	25													

For HSS (high-speed steel) tools in the first speed column only, use Table 15a for recommended feed in inches per tooth and depth of cut.

End Milling: Table data for end milling are based on a 3-tooth, 20-degree helix angle tool with a diameter of 1.0 inch, an axial depth of cut of 0.2 inch, and a radial depth of cut of 1 inch (full slot). Use Table 15b to adjust speeds for other feeds and axial depths of cut, and Table 15c to adjust speeds if the radial depth of cut is less than the tool diameter. Speeds are valid for all tool diameters.

Face Milling: Table data for face milling are based on a 10-tooth, 8-inch diameter face mill, operating with a 15-degree lead angle, $\frac{3}{64}$-inch nose radius, axial depth of cut = 0.1 inch, and radial depth (width) of cut = 6 inches (i.e., width of cut to cutter diameter ratio = $\frac{3}{4}$). These speeds are valid if the cutter axis is above or close to the center line of the workpiece (eccentricity is small). Under these conditions, use Table 15d to adjust speeds for other feeds and axial and radial depths of cut. For larger eccentricity (i.e., when the cutter axis to workpiece center line offset is one half the cutter diameter or more), use the end and side milling adjustment factors (Tables 15b and 15c) instead of the face milling factors.

Slit and Slot Milling: Table data for slit milling are based on an 8-tooth, 10-degree helix angle tool with a cutter width of 0.4 inch, diameter *D* of 4.0 inch, and a depth of cut of 0.6 inch. Speeds are valid for all tool diameters and widths. See the examples in the text for adjustments to the given speeds for other feeds and depths of cut.

Tool life for all tabulated values is approximately 45 minutes; use Table 15e to adjust tool life from 15 to 180 minutes. The combined feed/speed data in this table are based on tool grades (identified in Table 16) as follows: uncoated carbide = 20; coated carbide = 10.

Table 14. Cutting Feeds and Speeds for Milling Ferrous Cast Metals

Material	Brinell Hardness	HSS Speed (fpm)		End Milling: HSS Opt.	Avg.	End Milling: Uncoated Carbide Opt.	Avg.	End Milling: Coated Carbide Opt.	Avg.	Face Milling: Uncoated Carbide Opt.	Avg.	Face Milling: Coated Carbide Opt.	Avg.	Face Milling: Ceramic Opt.	Avg.	Face Milling: CBN Opt.	Avg.	Slit Milling: Uncoated Carbide Opt.	Avg.	Slit Milling: Coated Carbide Opt.	Avg.
				f = feed (0.001 in./tooth), s = speed (ft/min)																	
Gray Cast Iron																					
ASTM Class 20	120–150	100																			
ASTM Class 25	160–200	80	f	5	3	5	3			39	20	39	20	39	20	39	20	39	20		
ASTM Class 30, 35, and 40	190–220	70	s	35	90	520	855			140	225	285	535	1130	1630	200	530	205	420		
ASTM Class 45 and 50	220–260	50	f	5	3	5	3			39	20	39	20	39	20	39	20	39	20		
ASTM Class 55 and 60	250–320	30	s	30	70	515	1100			95	160	185	395	845	1220	150	400	145	380		
ASTM Type 1, 1b, 5 (Ni resist)	100–215	50																			
ASTM Type 2, 3, 6 (Ni resist)	120–175	40																			
ASTM Type 2b, 4 (Ni resist)	150–250	30																			
Malleable Iron																					
(Ferritic): 32510, 35018	110–160	110	f	5	3	5	3			39	20	39	20	39	20			39	20		
			s	30	70	180	250			120	195	225	520	490	925			85	150		
(Pearlitic): 40010, 43010, 45006, 45008, 48005, 50005	160–200	80	f	5	3	5	3			39	20	39	20	39	20			39	20		
	200–240	65	s	25	65	150	215			90	150	210	400	295	645			70	125		
(Martensitic): 53004, 60003, 60004	200–255	55																			
(Martensitic): 70002, 70003	220–260	50																			
(Martensitic): 80002	240–280	45																			
(Martensitic): 90001	250–320	25																			
Nodular (Ductile) Iron																					
(Ferritic): 60-40-18, 65-45-12	140–190	75	f	7	4	7	4			39	20	39	20	39	20			39	20		
			s	15	35	125	240			100	155	120	255	580	920			60	135		
(Ferritic-Pearlitic): 80-55-06 {	190–225	60																			
	225–260	50	f	7	4	7	4			39	20	39	20	39	20			39	20		
(Pearlitic-Martensitic): 100-70-03	240–300	40	s	10	30	90	210			95	145	150	275	170	415			40	100		
(Martensitic): 120-90-02	270–330	25																			

Table 14. *(Continued)* **Cutting Feeds and Speeds for Milling Ferrous Cast Metals**

Material	Brinell Hardness	HSS Speed (fpm)		End Milling HSS Opt.	Avg.	End Milling Uncoated Carbide Opt.	Avg.	End Milling Coated Carbide Opt.	Avg.	Face Milling Uncoated Carbide Opt.	Avg.	Face Milling Coated Carbide Opt.	Avg.	Face Milling Ceramic Opt.	Avg.	Face Milling CBN Opt.	Avg.	Slit Milling Uncoated Carbide Opt.	Avg.	Slit Milling Coated Carbide Opt.	Avg.
				f = feed (0.001 in./tooth), s = speed (ft/min)																	
Cast Steels																					
(Low carbon): 1010, 1020	100–150	100	f	7	4	7	4	7	4			39	20					39	20	39	20
	125–175	95	s	25	70	245†	410	420	650			265‡	430					135†	260	245	450
(Medium carbon): 1030, 1040 1050	175–225	80																			
	225–300	60	f	7	4	7	4	7	4			39	20					39	20	39	20
(Low-carbon alloy): 1320, 2315, 2320, 4110, 4120, 4320, 8020, 8620	150–200	85	s	20	55	160†	400	345	560			205‡	340					65†	180	180	370
	200–250	75																			
	250–300	50																			
(Medium-carbon alloy): 1330, 1340, 2325, 2330, 4125, 4130, 4140, 4330, 4340, 8030, 80B30, 8040, 8430, 8440, 8630, 8640, 9525, 9530, 9535	175–225	70	f	7	4	7	4											39	20		
	225–250	65	s	15	45	120†	310											45†	135		
	250–300	50	f							39	20										
	300–350	30	s							25	40										

For HSS (high-speed steel) tools in the first speed column only, use Table 15a for recommended feed in inches per tooth and depth of cut.

End Milling: Table data for end milling are based on a 3-tooth, 20-degree helix angle tool with a diameter of 1.0 inch, an axial depth of cut of 0.2 inch, and a radial depth of cut of 1 inch (full slot). Use Table 15b to adjust speeds for other feeds and axial depths of cut, and Table 15c to adjust speeds if the radial depth of cut is less than the tool diameter. Speeds are valid for all tool diameters.

Face Milling: Table data for face milling are based on a 10-tooth, 8-inch diameter face mill, operating with a 15-degree lead angle, $\frac{3}{64}$-inch nose radius, axial depth of cut = 0.1 inch, and radial depth (width) of cut = 6 inches (i.e., width of cut to cutter diameter ratio = $\frac{3}{4}$). These speeds are valid if the cutter axis is above or close to the center line of the workpiece (eccentricity is small). Under these conditions, use Table 15d to adjust speeds for other feeds and axial and radial depths of cut. For larger eccentricity (i.e., when the cutter axis to workpiece center line offset is one half the cutter diameter or more), use the end and side milling adjustment factors (Tables 15b and 15c) instead of the face milling factors.

Slit and Slot Milling: Table data for slit milling are based on an 8-tooth, 10-degree helix angle tool with a cutter width of 0.4 inch, diameter *D* of 4.0 inches, and a depth of cut of 0.6 inch. Speeds are valid for all tool diameters and widths. See the examples in the text for adjustments to the given speeds for other feeds and depths of cut.

Tool life for all tabulated values is approximately 45 minutes; use Table 15e to adjust tool life from 15 to 180 minutes. The combined feed/speed data in this table are based on tool grades (identified in Table 16) as follows: uncoated carbide = 15 except † = 20; end and slit milling coated carbide = 10; face milling coated carbide = 11 except ‡ = 10. ceramic = 6; CBN = 1.

Table 15a. Recommended Feed in Inches per Tooth (f_t) for Milling with High Speed Steel Cutters

Material	Hardness, HB	End Mills							Plain or Slab Mills	Form Relieved Cutters	Face Mills and Shell End Mills	Slotting and Side Mills
		Depth of Cut, .250 in			Depth of Cut, .050 in							
		Cutter Diam., in			Cutter Diam., in							
		½	¾	1 and up	¼	½	¾	1 and up				
		Feed per Tooth, inch										
Free-machining plain carbon steels	100–185	.001	.003	.004	.001	.002	.003	.004	.003–.008	.005	.004–.012	.002–.008
Plain carbon steels, AISI 1006 to 1030; 1513 to 1522	100–150	.001	.003	.003	.001	.002	.003	.004	.003–.008	.004	.004–.012	.002–.008
	150–200	.001	.002	.003	.001	.002	.002	.003	.003–.008	.004	.003–.012	.002–.008
AISI 1033 to 1095; 1524 to 1566	120–180	.001	.003	.003	.001	.002	.003	.004	.003–.008	.004	.004–.012	.002–.008
	180–220	.001	.002	.003	.001	.002	.002	.003	.003–.008	.004	.003–.012	.002–.008
	220–300	.001	.002	.002	.001	.001	.002	.003	.002–.006	.003	.002–.008	.002–.006
Alloy steels having less than 3% carbon. Typical examples: AISI 4012, 4023, 4027, 4118, 4320 4422, 4427, 4615, 4620, 4626, 4720, 4820, 5015, 5120, 6118, 8115, 8620 8627, 8720, 8820, 8822, 9310, 93B17	125–175	.001	.003	.003	.001	.002	.003	.004	.003–.008	.004	.004–.012	.002–.008
	175–225	.001	.002	.003	.001	.002	.003	.003	.003–.008	.004	.003–.012	.002–.008
	225–275	.001	.002	.003	.001	.001	.002	.003	.002–.006	.003	.003–.008	.002–.006
	275–325	.001	.002	.002	.001	.001	.002	.002	.002–.005	.003	.002–.008	.002–.005
Alloy steels having 3% carbon or more. Typical examples: AISI 1330, 1340, 4032, 4037, 4130, 4140, 4150, 4340, 50B40, 50B60, 5130, 51B60, 6150, 81B45, 8630, 8640, 86B45, 8660, 8740, 94B30	175–225	.001	.002	.003	.001	.002	.003	.004	.003–.008	.004	.003–.012	.002–.008
	225–275	.001	.002	.003	.001	.001	.002	.003	.002–.006	.003	.003–.010	.002–.006
	275–325	.001	.002	.002	.001	.001	.002	.003	.002–.005	.003	.002–.008	.002–.005
	325–375	.001	.002	.002	.001	.001	.002	.002	.002–.004	.002	.002–.008	.002–.005
Tool steel	150–200	.001	.002	.002	.001	.002	.003	.003	.003–.008	.004	.003–.010	.002–.006
	200–250	.001	.002	.002	.001	.002	.002	.003	.002–.006	.003	.003–.008	.002–.005
Gray cast iron	120–180	.001	.003	.004	.002	.003	.004	.004	.004–.012	.005	.005–.016	.002–.010
	180–225	.001	.002	.003	.001	.002	.003	.003	.003–.010	.004	.004–.012	.002–.008
	225–300	.001	.002	.002	.001	.001	.002	.002	.002–.006	.003	.002–.008	.002–.005
Free malleable iron	110–160	.001	.003	.004	.002	.003	.004	.004	.003–.010	.005	.005–.016	.002–.010

Table 15a. *(Continued)* **Recommended Feed in Inches per Tooth (f_t) for Milling with High Speed Steel Cutters**

Material *(Continued)*	Hard-ness, HB	End Mills							Plain or Slab Mills	Form Relieved Cutters	Face Mills and Shell End Mills	Slotting and Side Mills
		Depth of Cut, .250 in			Depth of Cut, .050 in							
		Cutter Diam., in			Cutter Diam., in							
		½	¾	1 and up	¼	½	¾	1 and up				
		Feed per Tooth, inch										
Pearlitic-Martensitic malleable iron	160–200	.001	.003	.004	.001	.002	.003	.004	.003–.010	.004	.004–.012	.002–.018
	200–240	.001	.002	.003	.001	.002	.003	.003	.003–.007	.004	.003–.010	.002–.006
	240–300	.001	.002	.002	.001	.001	.002	.002	.002–.006	.003	.002–.008	.002–.005
Cast steel	100–180	.001	.003	.003	.001	.002	.003	.004	.003–.008	.004	.003–.012	.002–.008
	180–240	.001	.002	.003	.001	.002	.003	.003	.003–.008	.004	.003–.010	.002–.006
	240–300	.001	.002	.002	.005	.002	.002	.002	.002–.006	.003	.003–.008	.002–.005
Zinc alloys (die castings)	…	.002	.003	.004	.001	.003	.004	.006	.003–.010	.005	.004–.015	.002–.012
Copper alloys (brasses & bronzes)	100–150	.002	.004	.005	.002	.003	.005	.006	.003–.015	.004	.004–.020	.002–.010
	150–250	.002	.003	.004	.001	.003	.004	.005	.003–.015	.004	.003–.012	.002–.008
Free cutting brasses & bronzes	80–100	.002	.004	.005	.002	.003	.005	.006	.003–.015	.004	.004–.015	.002–.010
Cast aluminum alloys—as cast	…	.003	.004	.005	.002	.004	.005	.006	.005–.016	.006	.005–.020	.004–.012
Cast aluminum alloys—hardened	…	.003	.004	.005	.002	.003	.004	.005	.004–.012	.005	.005–.020	.004–.012
Wrought aluminum alloys— cold drawn	…	.003	.004	.005	.002	.003	.004	.005	.004–.014	.005	.005–.020	.004–.012
Wrought aluminum alloys—hardened	…	.002	.003	.004	.001	.002	.003	.004	.003–.012	.004	.005–.020	.004–.012
Magnesium alloys	…	.003	.004	.005	.003	.004	.005	.007	.005–.016	.006	.008–.020	.005–.012
Ferritic stainless steel	135–185	.001	.002	.003	.001	.002	.003	.003	.002–.006	.004	.004–.008	.002–.007
Austenitic stainless steel	135–185	.001	.002	.003	.001	.002	.003	.003	.003–.007	.004	.005–.008	.002–.007
	185–275	.001	.002	.003	.001	.002	.002	.002	.003–.006	.003	.004–.006	.002–.007
Martensitic stainless steel	135–185	.001	.002	.002	.001	.002	.003	.003	.003–.006	.004	.004–.010	.002–.007
	185–225	.001	.002	.002	.001	.002	.002	.003	.003–.006	.004	.003–.008	.002–.007
	225–300	.0005	.002	.002	.0005	.001	.002	.002	.002–.005	.003	.002–.006	.002–.005
Monel	100–160	.001	.003	.004	.001	.002	.003	.004	.002–.006	.004	.002–.008	.002–.006

Table 15b. End Milling (Full Slot) Speed Adjustment Factors for Feed, Depth of Cut, and Lead Angle

Cutting Speed, $V = V_{opt} \times F_f \times F_d$																	
Ratio of Chosen Feed to *Optimum* Feed	Ratio of the two cutting speeds (*average*/*optimum*) given in the tables V_{avg}/V_{opt}							Depth of Cut and Lead Angle									
								1 in	(25.4 mm)	0.4 in	(10.2 mm)	0.2 in	(5.1 mm)	0.1 in	(2.4 mm)	0.04 in	(1.0 mm)
	1.00	1.25	1.50	2.00	2.50	3.00	4.00	0°	45°	0°	45°	0°	45°	0°	45°	0°	45°
	Feed Factor, F_f							Depth of Cut and Lead Angle Factor, F_d									
1.00	1.0	1.0	1.0	1.0	1.0	1.0	1.0	0.91	1.36	0.94	1.38	1.00	0.71	1.29	1.48	1.44	1.66
0.90	1.00	1.06	1.09	1.14	1.18	1.21	1.27	0.91	1.33	0.94	1.35	1.00	0.72	1.26	1.43	1.40	1.59
0.80	1.00	1.12	1.19	1.31	1.40	1.49	1.63	0.92	1.30	0.95	1.32	1.00	0.74	1.24	1.39	1.35	1.53
0.70	1.00	1.18	1.30	1.50	1.69	1.85	2.15	0.93	1.26	0.95	1.27	1.00	0.76	1.21	1.35	1.31	1.44
0.60	1.00	1.20	1.40	1.73	2.04	2.34	2.89	0.94	1.22	0.96	1.25	1.00	0.79	1.18	1.28	1.26	1.26
0.50	1.00	1.25	1.50	2.00	2.50	3.00	4.00	0.95	1.17	0.97	1.18	1.00	0.82	1.14	1.21	1.20	1.21
0.40	1.00	1.23	1.57	2.29	3.08	3.92	5.70	0.96	1.11	0.97	1.12	1.00	0.86	1.09	1.14	1.13	1.16
0.30	1.00	1.14	1.56	2.57	3.78	5.19	8.56	0.98	1.04	0.99	1.04	1.00	0.91	1.04	1.07	1.05	1.09
0.20	1.00	0.90	1.37	2.68	4.49	6.86	17.60	1.00	0.85	1.00	0.95	1.00	0.99	0.97	0.93	0.94	0.88
0.10	1.00	0.44	0.80	2.08	4.26	8.00	20.80	1.05	0.82	1.00	0.81	1.00	1.50	0.85	0.76	0.78	0.67

For HSS (high-speed steel) tool speeds in the first speed column of Tables 10 through 14, use Table 15a to determine appropriate feeds and depths of cut.

Cutting feeds and speeds for end milling given in Tables 11 through 14 (except those for high-speed steel in the first speed column) are based on milling a 0.20-inch deep full slot (i.e., radial depth of cut = end mill diameter) with a 1-inch diameter, 20-degree helix angle, 0-degree lead angle end mill. For other depths of cut (axial), lead angles, or feed, use the two feed/speed pairs from the tables and calculate the ratio of desired (new) feed to *optimum* feed (largest of the two feeds are given in the tables), and the ratio of the two cutting speeds (V_{avg}/V_{opt}). Find the feed factor F_f at the intersection of the feed ratio row and the speed ratio column in the left half of the Table. The depth of cut factor F_d is found in the same row as the feed factor, in the right half of the table under the column corresponding to the depth of cut and lead angle. The adjusted cutting speed can be calculated from $V = V_{opt} \times F_f \times F_d$, where V_{opt} is the smaller (*optimum*) of the two speeds from the speed table (from the left side of the column containing the two feed/speed pairs). See the text for examples.

If the radial depth of cut is less than the cutter diameter (i.e., for cutting less than a full slot), the feed factor F_f in the previous equation and the maximum feed f_{max} must be obtained from Table 15c. The axial depth of cut factor F_d can then be obtained from this table using f_{max} in place of the *optimum* feed in the feed ratio. Also see the footnote to Table 15c.

Table 15c. End, Slit, and Side Milling Speed Adjustment Factors for Radial Depth of Cut

Ratio of Radial Depth of Cut to Diameter	Cutting Speed, $V = V_{opt} \times F_f \times F_d$													
	Maximum Feed/Tooth Factor	V_{avg}/V_{opt}						Maximum Feed/Tooth Factor	V_{avg}/V_{opt}					
		1.25	1.50	2.00	2.50	3.00	4.00		1.25	1.50	2.00	2.50	3.00	4.00
		Feed Factor F_f at Maximum Feed per Tooth, F_{f1}							Feed Factor F_f at Minimum Feed per Tooth, F_{f2}					
1.00	1.00	1.00	1.00	1.00	1.00	1.00	1.00	0.70	1.18	1.30	1.50	1.69	1.85	2.15
0.75	1.00	1.15	1.24	1.46	1.54	1.66	1.87	0.70	1.24	1.48	1.93	2.38	2.81	3.68
0.60	1.00	1.23	1.40	1.73	2.04	2.34	2.89	0.70	1.24	1.56	2.23	2.95	3.71	5.32
0.50	1.00	1.25	1.50	2.00	2.50	3.00	4.00	0.70	1.20	1.58	2.44	3.42	4.51	6.96
0.40	1.10	1.25	1.55	2.17	2.83	3.51	4.94	0.77	1.25	1.55	2.55	3.72	5.08	8.30
0.30	1.35	1.20	1.57	2.28	3.05	3.86	5.62	0.88	1.23	1.57	2.64	4.06	5.76	10.00
0.20	1.50	1.14	1.56	2.57	3.78	5.19	8.56	1.05	1.40	1.56	2.68	4.43	6.37	11.80
0.10	2.05	0.92	1.39	2.68	4.46	6.77	13.10	1.44	0.92	1.29	2.50	4.66	7.76	17.40
0.05	2.90	0.68	1.12	2.50	4.66	7.75	17.30	2.00	0.68	1.12	2.08	4.36	8.00	20.80
0.02	4.50	0.38	0.71	1.93	4.19	7.90	21.50	3.10	0.38	0.70	1.38	3.37	7.01	22.20

This table is for side milling, end milling when the radial depth of cut (width of cut) is less than the tool diameter (i.e., less than full slot milling), and slit milling when the feed is parallel to the work surface (slotting). The radial depth of cut to diameter ratio is used to determine the recommended maximum and minimum values of feed/tooth, which are found by multiplying the feed/tooth factor from the appropriate column above (maximum or minimum) by $feed_{opt}$ from the speed tables. For example, given two feed/speed pairs $^7/_{15}$ and $^4/_{45}$ for end milling cast, medium-carbon, alloy steel, and a radial depth of cut to diameter ratio ar/D of 0.10 (a 0.05-inch width of cut for a ½-inch diameter end mill, for example), the maximum feed $f_{max} = 2.05 \times 0.007 = 0.014$ in./tooth and the minimum feed $f_{min} = 1.44 \times 0.007 = 0.010$ in./tooth. The feed selected should fall in the range between f_{min} and f_{max}. The feed factor F_d is determined by interpolating between the feed factors F_{f1} and F_{f2} corresponding to the maximum and minimum feed per tooth, at the appropriate ar/D and speed ratio. In the example given, $ar/D = 0.10$ and $V_{avg}/V_{opt} = 45/15 = 3$, so the feed factor F_{f1} at the maximum feed per tooth is 6.77, and the feed factor F_{f2} at the minimum feed per tooth is 7.76. If a working feed of 0.012 in./tooth is chosen, the feed factor F_f is half way between 6.77 and 7.76 or by formula, $F_f = F_{f1} + (feed - f_{min})/(f_{max} - f_{min}) \times (f_{f2} - f_{f1}) = 6.77 + (0.012 - 0.010)/(0.014 - 0.010) \times (7.76 - 6.77) = 7.27$. The cutting speed is $V = V_{opt} \times F_f \times F_d$, where F_d is the depth of cut and lead angle factor from Table 15b that corresponds to the feed ratio (chosen feed)/f_{max}, not the ratio (chosen feed)/*optimum* feed. For a feed ratio = 0.012/0.014 = 0.86 (chosen feed/f_{max}), depth of cut = 0.2 inch and lead angle = 45°, the depth of cut factor F_d in Table 15b is between 0.72 and 0.74. Therefore, the final cutting speed for this example is $V = V_{opt} \times F_f \times F_d = 15 \times 7.27 \times 0.73 = 80$ ft/min.

Slit and Side Milling: This table only applies when feed is parallel to the work surface, as in slotting. If feed is perpendicular to the work surface, as in cutting off, obtain the required speed-correction factor from Table 15d (face milling). The minimum and maximum feeds/tooth for slit and side milling are determined in the manner described above, however, the axial depth of cut factor F_d is not required. The adjusted cutting speed, valid for cutters of any thickness (width), is given by $V = V_{opt} \times F_f$. Examples are given in the text.

Table 15d. Face Milling Speed Adjustment Factors for Feed, Depth of Cut, and Lead Angle

Cutting Speed $V = V_{opt} \times F_f \times F_d \times F_{ar}$																								
Ratio of Chosen Feed to *Optimum* Feed	Ratio of the two cutting speeds (*average*/*optimum*) given in the tables V_{avg}/V_{opt}							Depth of Cut, inch (mm), and Lead Angle										Ratio of Radial Depth of Cut/Cutter Diameter, *ar*/*D*						
								1 in (25.4 mm)		0.4 in (10.2 mm)		0.2 in (5.1 mm)		0.1 in (2.4 mm)		0.04 in (1.0 mm)								
	1.00	1.10	1.25	1.35	1.50	1.00	2.00	15°	45°	15°	45°	15°	45°	15°	45°	15°	45°	1.00	0.75	0.50	0.40	0.30	0.20	0.10
	Feed Factor, F_f							Depth of Cut Factor, F_d										Radial Depth of Cut Factor, F_{ar}						
1.00	1.0	1.0	1.0	1.0	1.0	1.0	1.0	0.78	1.11	0.94	1.16	0.90	1.10	1.00	1.29	1.47	1.66	0.72	1.00	1.53	1.89	2.43	3.32	5.09
0.90	1.00	1.02	1.05	1.07	1.09	1.10	1.12	0.78	1.10	0.94	1.16	0.90	1.09	1.00	1.27	1.45	1.58	0.73	1.00	1.50	1.84	2.24	3.16	4.69
0.80	1.00	1.03	1.09	1.10	1.15	1.20	1.25	0.80	1.10	0.94	1.14	0.91	1.08	1.00	1.25	1.40	1.52	0.75	1.00	1.45	1.73	2.15	2.79	3.89
0.70	1.00	1.05	1.13	1.22	1.22	1.32	1.43	0.81	1.09	0.95	1.14	0.91	1.08	1.00	1.24	1.39	1.50	0.75	1.00	1.44	1.72	2.12	2.73	3.77
0.60	1.00	1.08	1.20	1.25	1.35	1.50	1.66	0.81	1.09	0.95	1.13	0.92	1.08	1.00	1.23	1.38	1.48	0.76	1.00	1.42	1.68	2.05	2.61	3.52
0.50	1.00	1.10	1.25	1.35	1.50	1.75	2.00	0.81	1.09	0.95	1.13	0.92	1.08	1.00	1.23	1.37	1.47	0.76	1.00	1.41	1.66	2.02	2.54	3.39
0.40	1.00	1.09	1.28	1.44	1.66	2.03	2.43	0.82	1.08	0.95	1.12	0.92	1.07	1.00	1.21	1.34	1.43	0.78	1.00	1.37	1.60	1.90	2.34	2.99
0.30	1.00	1.06	1.32	1.52	1.85	2.42	3.05	0.84	1.07	0.96	1.11	0.93	1.06	1.00	1.18	1.30	1.37	0.80	1.00	1.32	1.51	1.76	2.10	2.52
0.20	1.00	1.00	1.34	1.60	2.07	2.96	4.03	0.86	1.06	0.96	1.09	0.94	1.05	1.00	1.15	1.24	1.29	0.82	1.00	1.26	1.40	1.58	1.79	1.98
0.10	1.00	0.80	1.20	1.55	2.24	3.74	5.84	0.90	1.04	0.97	1.06	0.96	1.04	1.00	1.10	1.15	1.18	0.87	1.00	1.16	1.24	1.31	1.37	1.32

For HSS (high-speed steel) tool speeds in the first speed column, use Table 15a to determine appropriate feeds and depths of cut.

Tabular feeds and speeds data for face milling in Tables 11 through 14 are based on a 10-tooth, 8-inch diameter face mill, operating with a 15-degree lead angle, 3⁄64-inch cutter insert nose radius, axial depth of cut = 0.1 inch, and radial depth (width) of cut = 6 inches (i.e., width of cut to cutter diameter ratio = 3⁄4). For other depths of cut (radial or axial), lead angles, or feed, calculate the ratio of desired (new) feed to *optimum* feed (largest of the two feeds given in the speed table), and the ratio of the two cutting speeds (V_{avg}/V_{opt}). Use these ratios to find the feed factor F_f at the intersection of the feed ratio row and the speed ratio column in the left third of the table. The depth of cut factor F_d is found in the same row as the feed factor, in the center third of the table, in the column corresponding to the depth of cut and lead angle. The radial depth of cut factor F_{ar} is found in the same row as the feed factor, in the right third of the table, in the column corresponding to the radial depth of cut to cutter diameter ratio *ar*/*D*. The adjusted cutting speed can be calculated from $V = V_{opt} \times F_f \times F_d \times F_{ar}$, where V_{opt} is the smaller (*optimum*) of the two speeds from the speed table (from the left side of the column containing the two feed/speed pairs).

The cutting speeds as calculated above are valid if the cutter axis is centered above or close to the center line of the workpiece (eccentricity is small). For larger eccentricity (i.e., the cutter axis is offset from the center line of the workpiece by about one-half the cutter diameter or more), use the adjustment factors from Tables 15b and 15c (end and side milling) instead of the factors from this table. Use Table 15e to adjust end and face milling speeds for increased tool life up to 180 minutes.

Slit and Slot Milling: Tabular speeds are valid for all tool diameters and widths. Adjustments to the given speeds for other feeds and depths of cut depend on the circumstances of the cut. *Case 1:* If the cutter is fed directly into the workpiece, i.e., the feed is perpendicular to the surface of the workpiece, as in cutting off, then this table (face milling) is used to adjust speeds for other feeds. The depth of cut factor is not used for slit milling ($F_d = 1.0$), so the adjusted cutting speed $V = V_{opt} \times F_f \times F_{ar}$. For determining the factor F_{ar}, the radial depth of cut *ar* is the length of cut created by the portion of the cutter engaged in the work. *Case 2:* If the cutter is fed parallel to the surface of the workpiece, as in slotting, then Tables 15b and 15c are used to adjust the given speeds for other feeds. See Fig. 5.

Table 15e. Tool Life Adjustment Factors for Face Milling, End Milling, Drilling, and Reaming

Tool Life, T (minutes)	Face Milling with Carbides and Mixed Ceramics			End Milling with Carbides and HSS			Twist Drilling and Reaming with HSS		
	f_s	f_m	f_l	f_s	f_m	f_l	f_s	f_m	f_l
15	1.69	1.78	1.87	1.10	1.23	1.35	1.11	1.21	1.30
45	1.00	1.00	1.00	1.00	1.00	1.00	1.00	1.00	1.00
90	0.72	0.70	0.67	0.94	0.89	0.83	0.93	0.89	0.85
180	0.51	0.48	0.45	0.69	0.69	0.69	0.87	0.80	0.72

The feeds and speeds given in Tables 11 through 14 and Tables 17 through 23 (except for HSS speeds in the first speed column) are based on a 45-minute tool life. To adjust the given speeds to obtain another tool life, multiply the adjusted cutting speed for the 45-minute tool life V_{45} by the tool life factor from this table according to the following rules: for small feeds, where feed ≤ $\frac{1}{2} f_{opt}$, the cutting speed for the desired tool life T is $V_T = f_s \times V_{15}$; for medium feeds, where $\frac{1}{2} f_{opt}$ < feed < $\frac{3}{4} f_{opt}$, $V_T = f_m \times V_{15}$; and for larger feeds, where $\frac{3}{4} f_{opt}$ ≤ feed ≤ f_{opt}, $V_T = f_l \times V_{15}$. Here, f_{opt} is the largest (*optimum*) feed of the two feed/speed values given in the speed tables or the maximum feed f_{max} obtained from Table 15c, if that table was used in calculating speed adjustment factors.

Table 16. Cutting Tool Grade Descriptions and Common Vendor Equivalents

Grade Description	Tool Identification Code	Approximate Vendor Equivalents			
		Sandvik Coromant	Kennametal	Seco	Valenite
Cubic boron nitride	1	CB50	KD050	CBN2 0	VC721
Ceramics	2	CC620	K060	480	—
	3	CC650	K090	480	Q32
	4 (Whiskers)	CC670	KYON2500	—	—
	5 (Sialon)	CC680	KYON2000	480	—
	6	CC690	KYON3000	—	Q6
Cermets	7	CT515	KT125	CM	VC605
	8	CT525	KT150	CR	VC610
Polycrystalline	9	CD10	KD100	PAX20	VC727
Coated carbides	10	GC-A	—	—	—
	11	GC3015	KC910	TP100	SV310
	12	GC235	KC9045	TP300	SV235
	13	GC4025	KC9025	TP200	SV325
	14	GC415	KC950	TP100	SV315
Uncoated carbides	15	H13A	K8, K4H	883	VC2
	16	S10T	K420, K28	CP20	VC7
	17	S1P	K45	CP20	VC7
	18	S30T	—	CP25	VC5
	19	S6	K21, K25	CP50	VC56
	20	SM30	KC710	CP25	VC35M

See Table 2 on page 757 and the section *Cemented Carbides and Other Hard Materials* for more detailed information on cutting tool grades.

The identification codes in column two correspond to the grade numbers given in the footnotes to Tables 1 to 4b, 6 to 14, and 17 to 23.

Using the Feed and Speed Tables for Drilling, Reaming, and Threading.—The first two speed columns in Tables 17 through 23 give traditional Handbook speeds for drilling and reaming. The following material can be used for selecting feeds for use with the traditional speeds.

The remaining columns in Tables 17 through 23 contain combined feed/speed data for drilling, reaming, and threading, organized in the same manner as in the turning and milling tables. Operating at the given feeds and speeds is expected to result in a tool life of approximately 45 minutes, except for indexable insert drills, which have an expected tool life of approximately 15 minutes per edge. Examples of using this data follow.

Adjustments to HSS drilling speeds for feed and diameter are made using Table 22; Table 5a is used for adjustments to indexable insert drilling speeds, where one-half the drill diameter D is used for the depth of cut. Tool life for HSS drills, reamers, and thread chasers and taps may be adjusted using Table 15e and for indexable insert drills using Table 5b.

The feed for drilling is governed primarily by the size of the drill and by the material to be drilled. Other factors that also affect selection of the feed are the workpiece configuration, the rigidity of the machine tool and the workpiece setup, and the length of the chisel edge. A chisel edge that is too long will result in a very significant increase in the thrust force, which may cause large deflections to occur on the machine tool and drill breakage.

For ordinary twist drills, the feed rate used is 0.001 to 0.003 in /rev for drills smaller than ⅛ in, 0.002 to 0.006 in./rev for ⅛- to ¼-in drills; 0.004 to 0.010 in./rev for ¼- to ½-in drills; 0.007 to 0.015 in./rev for ½- to 1-in drills; and, 0.010 to 0.025 in./rev for drills larger than 1 inch.

The lower values in the feed ranges should be used for hard materials such as tool steels, superalloys, and work-hardening stainless steels; the higher values in the feed ranges should be used to drill soft materials such as aluminum and brass.

Example 1, Drilling: Determine the cutting speed and feed for use with HSS drills in drilling 1120 steel.

Table 15a gives two sets of feed and speed parameters for drilling 1120 steel with HSS drills. These sets are 16⁄50 and 8⁄95, i.e., 0.016 in./rev feed at 50 ft/min and 0.008 in./rev at 95 fpm, respectively. These feed/speed sets are based on a 0.6-inch diameter drill. Tool life for either of the given feed/speed settings is expected to be approximately 45 minutes.

For different feeds or drill diameters, the cutting speeds must be adjusted and can be determined from $V = V_{opt} \times F_f \times F_d$, where V_{opt} is the minimum speed for this material given in the speed table (50 fpm in this example) and F_f and F_d are the adjustment factors for feed and diameter, respectively, found in Table 22.

Table 17. Feeds and Speeds for Drilling, Reaming, and Threading Plain Carbon and Alloy Steels

Material	Brinell Hardness	Drilling	Reaming		Drilling				Reaming		Threading	
		HSS			HSS		Indexable Insert Coated Carbide		HSS		HSS	
		Speed (fpm)			f = feed (0.001 in./rev), s = speed (ft/min)							
					Opt.	*Avg.*	*Opt.*	*Avg.*	*Opt.*	*Avg.*	*Opt.*	*Avg.*
Free-machining plain carbon steels (Resulfurized): 1212, 1213, 1215	100–150	120	80	f s	21 55	11 125	8 310	4 620	36 140	18 185	83 140	20 185
	150–200	125	80									
(Resulfurized): 1108, 1109, 1115, 1117, 1118, 1120, 1126, 1211	100–150	110	75	f s	16 50	8 95	8 370	4 740	27 105	14 115	83 90	20 115
	150–200	120	80									
	175–225	100	65	f s			8 365	4 735				
(Resulfurized): 1132, 1137, 1139, 1140, 1144, 1146, 1151	275–325	70	45									
	325–375	45	30									
	375–425	35	20									
(Leaded): 11L17, 11L18, 12L13, 12L14	100–150	130	85									
	150–200	120	80									
	200–250	90	60	f s			8 365	4 735				
Plain carbon steels: 1006, 1008, 1009, 1010, 1012, 1015, 1016, 1017, 1018, 1019, 1020, 1021, 1022, 1023, 1024, 1025, 1026, 1513, 1514	100–125	100	65	f s	21 55	11 125	8 310	4 620	36 140	18 185	83 140	20 185
	125–175	90	60									
	175–225	70	45	f s			8 365	4 735				
	225–275	60	40									
Plain carbon steels: 1027, 1030, 1033, 1035, 1036, 1037, 1038, 1039, 1040, 1041, 1042, 1043, 1045, 1046, 1048, 1049, 1050, 1052, 1524, 1526, 1527, 1541	125–175	90	60									
	175–225	75	50									
	225–275	60	40	f s			8 365	4 735				
	275–325	50	30									
	325–375	35	20									
	375–425	25	15									

Table 17. *(Continued)* **Feeds and Speeds for Drilling, Reaming, and Threading Plain Carbon and Alloy Steels**

Material	Brinell Hardness	Drilling	Reaming		Drilling				Reaming		Threading	
		HSS			HSS		Indexable Insert Coated Carbide		HSS		HSS	
		Speed (fpm)			f = feed (0.001 in./rev), s = speed (ft/min)							
					Opt.	*Avg.*	*Opt.*	*Avg.*	*Opt.*	*Avg.*	*Opt.*	*Avg.*
Plain carbon steels *(Continued)*: 1055, 1060, 1064, 1065, 1070, 1074, 1078, 1080, 1084, 1086, 1090, 1095, 1548, 1551, 1552, 1561, 1566	125–175	85	55	**f**	16	8	8	4	27	14	83	20
	175–225	70	45	**s**	50	95	370	740	105	115	90	115
	225–275	50	30	**f**			8	4				
				s			365	735				
	275–325	40	25									
	325–375	30	20									
	375–425	15	10									
Free-machining alloy steels (Resulfurized): 4140, 4150	175–200	90	60	**f**	16	8	8	4	26	13	83	20
	200–250	80	50	**s**	75	140	410	685	150	160	125	160
	250–300	55	30	**f**			8	4				
				s			355	600				
	300–375	40	25	**f**			8	4				
	375–425	30	15	**s**			310	525				
(Leaded): 41L30, 41L40, 41L47, 41L50, 43L47, 51L32, 52L100, 86L20, 86L40	150–200	100	65	**f**	16	8	8	4	27	14	83	20
				s	50	95	370	740	105	115	90	115
	200–250	90	60	**f**			8	4				
				s			365	735				
	250–300	65	40									
	300–375	45	30									
	375–425	30	15									
Alloy steels: 4012, 4023, 4024, 4028, 4118, 4320, 4419, 4422, 4427, 4615, 4620, 4621, 4626, 4718, 4720, 4815, 4817, 4820, 5015, 5117, 5120, 6118, 8115, 8615, 8617, 8620, 8622, 8625, 8627, 8720, 8822, 94B17	125–175	85	55	**f**	16	8	8	4	26	13	83	20
	175–225	70	45	**s**	75	140	410	685	150	160	125	160
	225–275	55	35	**f**			8	4				
				s			355	600				
	275–325	50	30	**f**	11	6	8	4	19	10	83	20
				s	50	85	335	570	95	135	60	95
	325–375	35	25	**f**			8	4				
	375–425	25	15	**s**			310	525				

Table 17. *(Continued)* **Feeds and Speeds for Drilling, Reaming, and Threading Plain Carbon and Alloy Steels**

Material	Brinell Hardness	Drilling	Reaming		Drilling				Reaming		Threading	
		HSS			HSS		Indexable Insert Coated Carbide		HSS		HSS	
		Speed (fpm)			f = feed (0.001 in./rev), s = speed (ft/min)							
					Opt.	*Avg.*	*Opt.*	*Avg.*	*Opt.*	*Avg.*	*Opt.*	*Avg.*
Alloy steels: 1330, 1335, 1340, 1345, 4032, 4037, 4042, 4047, 4130, 4135, 4137, 4140, 4142, 4145, 4147, 4150, 4161, 4337, 4340, 50B44, 50B46, 50B50, 50B60, 5130, 5132, 5140, 5145, 5147, 5150, 5160, 51B60, 6150, 81B45, 8630, 8635, 8637, 8640, 8642, 8645, 8650, 8655, 8660, 8740, 9254, 9255, 9260, 9262, 94B30	175–225	75 (60)	50 (40)	f	16	8	8	4	26	13	83	20
				s	75	140	410	685	150	160	125	160
	225–275	60 (50)	40 (30)	f			8	4				
				s			355	600				
	275–325	45 (35)	30 (25)	f	11	6	8	4	19	10	83	20
				s	50	85	335	570	95	135	60	95
E51100, E52100: use (HSS speeds)	325–375	30 (30)	15 (20)	f			8	4				
	375–425	20 (20)	15 (10)	s			310	525				
Ultra-high-strength steels (not AISI): AMS 6421 (98B37 Mod.), 6422 (98BV40), 6424, 6427, 6428, 6430, 6432, 6433, 6434, 6436, and 6442; 300M, D6ac	220–300	50	30	f			8	4				
	300–350	35	20	s			325	545				
	350–400	20	10	f			8	4				
				s			270	450				
Maraging steels (not AISI): 18% Ni Grade 200, 250, 300, and 350	250–325	50	30	f			8	4				
				s			325	545				
Nitriding steels (not AISI): Nitralloy 125, 135, 135 Mod., 225, and 230, Nitralloy N, Nitralloy EZ, Nitrex I	200–250	60	40	f	16	8	8	4	26	13	83	20
				s	75	140	410	685	150	160	125	160
	300–350	35	20	f			8	4				
				s			310	525				

The two leftmost speed columns in this table contain traditional Handbook speeds for drilling and reaming with HSS steel tools. The section Feed Rates for Drilling and Reaming contains useful information concerning feeds to use in conjunction with these speeds.

HSS Drilling and Reaming: The combined feed/speed data for drilling are based on a 0.60-inch diameter HSS drill with standard drill point geometry (2-flute with 118° tip angle). Speed adjustment factors in Table 22 are used to adjust drilling speeds for other feeds and drill diameters. Examples of using this data are given in the text. The given feeds and speeds for reaming are based on an 8-tooth, $\frac{25}{32}$-inch diameter, 30° lead angle reamer, and a 0.008-inch radial depth of cut. For other feeds, the correct speed can be obtained by interpolation using the given speeds if the desired feed lies in the recommended range (between the given values of *optimum* and *average* feed). If a feed lower than the given *average* value is chosen, the speed should be maintained at the corresponding *average* speed (i.e., the highest of the two speed values given). The cutting speeds for reaming do not require adjustment for tool diameters for standard ratios of radical depth of cut to reamer diameter (i.e., f_d = 1.00). Speed adjustment factors to modify tool life are found in Table 15e.

Indexable Insert Drilling: The feed/speed data for indexable insert drilling are based on a tool with two cutting edges, an insert nose radius of $\frac{3}{64}$ inch, a 10-degree lead angle, and diameter $D = 1$ inch. Adjustments to cutting speed for feed and depth of cut are made using Table 5aAdjustment Factors) using a depth of cut of $D/2$, or one-half the drill diameter. Expected tool life at the given feeds and speeds is approximately 15 minutes for short hole drilling (i.e., where maximum hole depth is about $2D$ or less). Speed adjustment factors to increase tool life are found in Table 5b.

Tapping and Threading: The data in this column are intended for use with thread chasers and for tapping. The feed used for tapping and threading must be equal to the lead (feed = lead = pitch) of the thread being cut. The two feed/speed pairs given for each material, therefore, are representative speeds for two thread pitches, 12 and 50 threads per inch (1/0.083 = 12, and 1/0.020 = 50). Tool life is expected to be approximately 45 minutes at the given feeds and speeds. When cutting fewer than 12 threads per inch (pitch ≥ 0.08 inch), use the lower (*optimum*) speed; for cutting more than 50 threads per inch (pitch ≤ 0.02 inch), use the larger (*average*) speed; and, in the intermediate range between 12 and 50 threads per inch, interpolate between the given *average* and *optimum* speeds.

The combined feed/speed data in this table are based on tool grades (identified in Table 16) as follows: coated carbide = 10.

Example 2, Drilling: If the 1120 steel of Example 1 is to be drilled with a 0.60-inch drill at a feed of 0.012 in./rev, what is the cutting speed in ft/min? Also, what spindle rpm of the drilling machine is required to obtain this cutting speed?

To find the feed factor F_d in Table 22, calculate the ratio of the desired feed to the *optimum* feed and the ratio of the two cutting speeds given in the speed tables. The desired feed is 0.012 in./rev and the *optimum* feed, as explained above is 0.016 in./rev, therefore, $feed/f_{opt} = 0.012/0.016 = 0.75$ and $V_{avg}/V_{opt} = 95/50 = 1.9$, approximately 2.

The feed factor F_f is found at the intersection of the feed ratio row and the speed ratio column. $F_f = 1.40$ corresponds to about halfway between 1.31 and 1.50, which are the feed factors that correspond to $V_{avg}/V_{opt} = 2.0$ and $feed/f_{opt}$ ratios of 0.7 and 0.8, respectively. F_d, the diameter factor, is found on the same row as the feed factor (halfway between the 0.7 and 0.8 rows, for this example) under the column for drill diameter = 0.60 inch. Because the speed table values are based on a 0.60-inch drill diameter, $F_d = 1.0$ for this example, and the cutting speed is $V = V_{opt} \times F_f \times F_d = 50 \times 1.4 \times 1.0 = 70$ ft/min. The spindle speed in rpm is $N = 12 \times V/(\pi \times D) = 12 \times 70/(3.14 \times 0.6) = 445$ rpm.

Example 3, Drilling: Using the same material and feed as in the previous example, what cutting speeds are required for 0.079-inch and 4-inch diameter drills? What machine rpm is required for each?

Because the feed is the same as in the previous example, the feed factor is $F_f = 1.40$ and does not need to be recalculated. The diameter factors are found in Table 22 on the same row as the feed factor for the previous example (about halfway between the diameter factors corresponding to $feed/f_{opt}$ values of 0.7 and 0.8) in the column corresponding to drill diameters 0.079 and 4.0 inches, respectively. Results of the calculations are summarized below.

Drill diameter = 0.079 inch	*Drill diameter = 4.0 inches*
$F_f = 1.40$	$F_f = 1.40$
$F_d = (0.34 + 0.38)/2 = 0.36$	$F_d = (1.95 + 1.73)/2 = 1.85$
$V = 50 \times 1.4 \times 0.36 = 25.2$ fpm	$V = 50 \times 1.4 \times 1.85 = 129.5$ fpm
$12 \times 25.2/(3.14 \times 0.079) = 1219$ rpm	$12 \times 129.5/(3.14 \times 4) = 124$ rpm

Drilling Difficulties: A drill split at the web is evidence of too much feed or insufficient lip clearance at the center due to improper grinding. Rapid wearing away of the extreme outer corners of the cutting edges indicates that the speed is too high. A drill chipping or breaking out at the cutting edges indicates that either the feed is too heavy or the drill has been ground with too much lip clearance. Nothing will "check" a high-speed steel drill quicker than to turn a stream of cold water on it after it has been heated while in use. It is equally bad to plunge it in cold water after the point has been heated in grinding. The small checks or cracks resulting from this practice will eventually chip out and cause rapid wear or breakage. Insufficient speed in drilling small holes with hand feed greatly increases the risk of breakage, especially at the moment the drill is breaking through the farther side of the work, due to the operator's inability to gage the feed when the drill is running too slowly.

Small drills have heavier webs and smaller flutes in proportion to their size than do larger drills, so breakage due to clogging of chips in the flutes is more likely to occur. When drilling holes deeper than three times the diameter of the drill, it is advisable to withdraw the drill (peck feed) at intervals to remove the chips and permit coolant to reach the tip of the drill.

Drilling Holes in Glass: The simplest method of drilling holes in glass is to use a standard, tungsten-carbide-tipped masonry drill of the appropriate diameter, in a gun-drill. The edges of the carbide in contact with the glass should be sharp. Kerosene or other liquid may be used as a lubricant, and a light force is maintained on the drill until just before the point breaks through. The hole should then be started from the other side if possible, or a very light force applied for the remainder of the operation, to prevent excessive breaking of material from the sides of the hole. As the hard particles of glass are abraded, they accumulate and act to abrade the hole, so it may be advisable to use a slightly smaller drill than the required diameter of the finished hole.

Alternatively, for holes of medium and large size, use brass or copper tubing, having an outside diameter equal to the size of hole required. Revolve the tube at a peripheral speed of about 100 feet per minute, and use carborundum (80 to 100 grit) and light machine oil between the end of the pipe and the glass. Insert the abrasive under the drill with a thin piece of soft wood, to avoid scratching the glass. The glass should be supported by a felt or rubber cushion, not much larger than the hole to be drilled. If practicable, it is advisable to drill about halfway through, then turn the glass over, and drill down to meet the first cut. Any fin that may be left in the hole can be removed with a round second-cut file wetted with turpentine.

Smaller-diameter holes may also be drilled with triangular-shaped cemented carbide drills that can be purchased in standard sizes. The end of the drill is shaped into a long tapering triangular point. The other end of the cemented carbide bit is brazed onto a steel shank. A glass drill can be made to the same shape from hardened drill rod or an old three-cornered file. The location at which the hole is to be drilled is marked on the workpiece. A dam of putty or glazing compound is built up on the work surface to contain the cutting fluid, which can be either kerosene or turpentine mixed with camphor. Chipping on the back edge of the hole can be prevented by placing a scrap plate of glass behind the area to be drilled and drilling into the backup glass. This procedure also provides additional support to the workpiece and is essential for drilling very thin plates. The hole is usually drilled with an electric hand drill. When the hole is being produced, the drill should be given a small circular motion using the point as a fulcrum, thereby providing a clearance for the drill in the hole.

Very small round or intricately shaped holes and narrow slots can be cut in glass by the ultrasonic machining process or by the abrasive jet cutting process.

Table 18. Feeds and Speeds for Drilling, Reaming, and Threading Tool Steels

Material	Brinell Hardness	Drilling	Reaming		Drilling				Reaming		Threading	
		HSS			HSS		Indexable Insert Uncoated Carbide		HSS		HSS	
		Speed (fpm)			f = feed (0.001 in./rev), s = speed (ft/min)							
					Opt.	*Avg.*	*Opt.*	*Avg.*	*Opt.*	*Avg.*	*Opt.*	*Avg.*
Water hardening: W1, W2, W5	150–200	85	55									
Shock resisting: S1, S2, S5, S6, S7	175–225	50	35									
Cold work (oil hardening): O1, O2, O6, O7	175–225	45	30									
(High carbon, high chromium): D2, D3, D4, D5, D7	200–250	30	20									
(Air hardening): A2, A3, A8, A9, A10	200–250	50	35	**f** **s**	15 45	7 85	8 360	4 605	24 90	12 95	83 75	20 95
A4, A6	200–250	45	30									
A7	225–275	30	20									
Hot work (chromium type): H10, H11, H12, H13, H14, H19	150–200	60	40									
	200–250	50	30									
	325–375	30	20	**f** **s**			8 270	4 450				
(Tungsten type): H21, H22, H23, H24, H25, H26	150–200	55	35									
	200–250	40	25									
(Molybdenum type): H41, H42, H43	150–200	45	30									
	200–250	35	20									
Special-purpose, low alloy: L2, L3, L6	150–200	60	40									
Mold steel: P2, P3, P4, P5, P6	100–150	75	50	**f** **s**	15 45	7 85	8 360	4 605	24 90	12 95	83 75	20 95
P20, P21	150–200	60	40									
High-speed steel: M1, M2, M6, M10, T1, T2, T6	200–250	45	30									
M3-1, M4, M7, M30, M33, M34, M36, M41, M42, M43, M44, M46, M47, T5, T8	225–275	35	20									
T15, M3-2	225–275	25	15									

See the footnote to Table 17 for instructions concerning the use of this table. The combined feed/speed data in this table are based on tool grades (identified in Table 16) as follows: coated carbide = 10.

Table 19. Feeds and Speeds for Drilling, Reaming, and Threading Stainless Steels

Material	Brinell Hardness	Drilling HSS Speed (fpm)	Reaming HSS Speed (fpm)		Drilling HSS Opt.	Drilling HSS Avg.	Drilling Indexable Insert Coated Carbide Opt.	Drilling Indexable Insert Coated Carbide Avg.	Reaming HSS Opt.	Reaming HSS Avg.	Threading HSS Opt.	Threading HSS Avg.
					f = feed (0.001 in./rev), s = speed (ft/min)							
Free-machining stainless steels (Ferritic): 430F, 430FSe	135–185	90	60	f	15	7	8	4	24	12	83	20
				s	25	45	320	540	50	50	40	51
(Austenitic): 203EZ, 303, 303Se, 303MA, 303Pb, 303Cu, 303 Plus X	135–185	85	55	f	15	7	8	4	24	12	83	20
	225–275	70	45	s	20	40	250	425	40	40	35	45
(Martensitic): 416, 416Se, 416 Plus X, 420F, 420FSe, 440F, 440FSe	135–185	90	60									
	185–240	70	45									
	275–325	40	25									
	375–425	20	10									
Stainless steels (Ferritic): 405, 409, 429, 430, 434	135–185	65	45	f	15	7	8	4	24	12	83	20
				s	25	45	320	540	50	50	40	51
(Austenitic): 201, 202, 301, 302, 304, 304L, 305, 308, 321, 347, 348	135–185	55	35									
	225–275	50	30	f	15	7	8	4	24	12	83	20
(Austenitic): 302B, 309, 309S, 310, 310S, 314, 316	135–185	50	30	s	20	40	250	425	40	40	35	45
(Martensitic): 403, 410, 420, 501	135–175	75	50									
	175–225	65	45									
	275–325	40	25									
	375–425	25	15									
(Martensitic): 414, 431, Greek Ascoloy	225–275	50	30									
	275–325	40	25									
	375–425	25	15									
(Martensitic): 440A, 440B, 440C	225–275	45	30									
	275–325	40	25									
	375–425	20	10									
(Precipitation hardening): 15–5PH, 17–4PH, 17–7PH, AF–71, 17–14CuMo, AFC–77, AM–350, AM–355, AM–362, Custom 455, HNM, PH13–8, PH14–8Mo, PH15–7Mo, Stainless W	150–200	50	30	f	15	7	8	4	24	12	83	20
				s	20	40	250	425	40	40	35	45
	275–325	45	25									
	325–375	35	20									
	375–450	20	10									

See the footnote to Table 17 for instructions concerning the use of this table. The combined feed/speed data in this table are based on tool grades (identified in Table 16) as follows: coated carbide = 10.

Table 20. Feeds and Speeds for Drilling, Reaming, and Threading Ferrous Cast Metals

Material	Brinell Hardness	Drilling	Reaming		Drilling						Reaming		Threading	
		HSS			HSS		Indexable Carbide Insert: Uncoated		Indexable Carbide Insert: Coated		HSS		HSS	
		Speed (fpm)			f = feed (0.001 in./rev), s = speed (ft/min)									
					Opt.	*Avg.*	*Opt.*	*Avg.*	*Opt.*	*Avg.*	*Opt.*	*Avg.*	*Opt.*	*Avg.*
ASTM Class 20	120–150	100	65											
ASTM Class 25	160–200	90	60	**f**	16	8	11	6	11	6	26	13	83	20
ASTM Class 30, 35, and 40	190–220	80	55	**s**	80	90	85	180	235	485	85	65	90	80
ASTM Class 45 and 50	220–260	60	40	**f**	13	6	11	6	11	6	21	10	83	20
ASTM Class 55 and 60	250–320	30	20	**s**	50	50	70	150	195	405	50	30	55	45
ASTM Type 1, 1b, 5 (Ni resist)	100–215	50	30											
ASTM Type 2, 3, 6 (Ni resist)	120–175	40	25											
ASTM Type 2b, 4 (Ni resist)	150–250	30	20											
Malleable Iron														
(Ferritic): 32510, 35018	110–160	110	75	**f**	19	10	11	6	11	6	30	16	83	20
				s	80	100	85	180	270	555	95	80	100	85
(Pearlitic): 40010, 43010, 45006, 45008, 48005, 50005	160–200	80	55	**f**	14	7			11	6	22	11	83	20
	200–240	70	45	**s**	65	65			235	485	65	45	70	60
(Martensitic): 53004, 60003, 60004	200–255	55	35											
(Martensitic): 70002, 70003	220–260	50	30											
(Martensitic): 80002	240–280	45	30											
(Martensitic): 90001	250–320	25	15											
Nodular (Ductile) Iron														
(Ferritic): 60-40-18, 65-45-12	140–190	100	65	**f**	17	9	11	6	11	6	28	14	83	20
				s	70	80	85	180	235	485	80	60	80	70

Table 20. *(Continued)* **Feeds and Speeds for Drilling, Reaming, and Threading Ferrous Cast Metals**

Material	Brinell Hardness	Drilling	Reaming		Drilling						Reaming		Threading	
							Indexable Carbide Insert							
		HSS			HSS		Uncoated		Coated		HSS		HSS	
		Speed (fpm)			f = feed (0.001 in./rev), s = speed (ft/min)									
					Opt.	*Avg.*	*Opt.*	*Avg.*	*Opt.*	*Avg.*	*Opt.*	*Avg.*	*Opt.*	*Avg.*
(Martensitic): 120-90-02	270–330	25	15											
	330–400	10	5											
(Ferritic-Pearlitic): 80-55-06	190–225	70	45	f	13	6	11	6	11	6	21	11	83	20
	225–260	50	30	s	60	60	70	150	195	405	55	40	60	55
(Pearlitic-Martensitic): 100-70-03	240–300	40	25											
Cast Steels														
(Low carbon): 1010, 1020	100–150	100	65	f	18	9					29	15	83	20
				s	35	70					75	85	65	85
(Medium carbon): 1030, 1040, 1050	125–175	90	60											
	175–225	70	45											
	225–300	55	35	f	15	7			8	4	24	12	83	20
(Low-carbon alloy): 1320, 2315, 2320, 4110, 4120, 4320, 8020, 8620	150–200	75	50	s	35	60			195†	475	65	70	55	70
	200–250	65	40											
	250–300	50	30											
(Medium-carbon alloy): 1330, 1340, 2325, 2330, 4125, 4130, 4140, 4330, 4340, 8030, 80B30, 8040, 8430, 8440, 8630, 8640, 9525, 9530, 9535	175–225	70	45	f					8	4				
	225–250	60	35	s					130†	315				
	250–300	45	30											
	300–350	30	20											
	350–400	20	10											

See the footnote to Table 17 for instructions concerning the use of this table. The combined feed/speed data in this table are based on tool grades (identified in Table 16) as follows: uncoated = 15; coated carbide = 11, † = 10.

Table 21. Feeds and Speeds for Drilling, Reaming, and Threading Light Metals

Material	Brinell Hardness	Drilling	Reaming		Drilling				Reaming		Threading	
		HSS			HSS		Indexable Insert Uncoated Carbide		HSS		HSS	
		Speed (fpm)			f = feed (0.001 in./rev), s = speed (ft/min)							
					Opt.	*Avg.*	*Opt.*	*Avg.*	*Opt.*	*Avg.*	*Opt.*	*Avg.*
All wrought aluminum alloys, 6061-T651, 5000, 6000, 7000 series	CD	400	400									
	ST and A	350	350	f	31	16	11	6	52	26	83	20
All aluminum sand and permanent mold casting alloys	AC	500	500	s	390	580	3235	11370	610	615	635	565
	ST and A	350	350									
Aluminum Die-Casting Alloys												
Alloys 308.0 and 319.0	—	—	—	f	23	11	11	6	38	19	83	20
				s	110	145	945	3325	145	130	145	130
Alloys 360.0 and 380.0	—	—	—	f	27	14	11	6	45	23	83	20
				s	90	125	855	3000	130	125	130	115
Alloys 390.0 and 392.0	AC	300	300									
	ST and A	70	70									
Alloys 413	—	—	—	f	24	12	11	6	40	20	83	20
	ST and A	45	40	s	65	85	555	1955	85	80	85	80
All other aluminum die-casting alloys	AC	125	100	f	27	14	11	6	45	23	83	20
				s	90	125	855	3000	130	125	130	115
Magnesium Alloys												
All wrought magnesium alloys	A,CD,ST and A	500	500									
All cast magnesium alloys	A,AC, ST and A	450	450									

Abbreviations designate: A, annealed; AC, as cast; CD, cold drawn; and ST and A, solution treated and aged, respectively. See the footnote to Table 17 for instructions concerning the use of this table. The combined feed/speed data in this table are based on tool grades (identified in Table 16) as follows; uncoated carbide = 15.

Table 22. Feed and Diameter Speed Adjustment Factors for HSS Twist Drills and Reamers

	Cutting Speed, $V = V_{opt} \times F_f \times F_d$															
Ratio of Chosen Feed to *Optimum* Feed	Ratio of the two cutting speeds (*average/optimum*) given in the tables V_{avg}/V_{opt}							Tool Diameter								
	1.00	1.25	1.50	2.00	2.50	3.00	4.00	0.08 in (2 mm)	0.15 in (4 mm)	0.25 in (6 mm)	0.40 in (10 mm)	0.60 in (15 mm)	1.00 in (25 mm)	2.00 in (50 mm)	3.00 in (75 mm)	4.00 in (100 mm)
	Feed Factor, F_f							Diameter Factor, F_d								
1.00	1.00	1.00	1.00	1.00	1.00	1.00	1.00	0.30	0.44	0.56	0.78	1.00	1.32	1.81	2.11	2.29
0.90	1.00	1.06	1.09	1.14	1.18	1.21	1.27	0.32	0.46	0.59	0.79	1.00	1.30	1.72	1.97	2.10
0.80	1.00	1.12	1.19	1.31	1.40	1.49	1.63	0.34	0.48	0.61	0.80	1.00	1.27	1.64	1.89	1.95
0.70	1.00	1.15	1.30	1.50	1.69	1.85	2.15	0.38	0.52	0.64	0.82	1.00	1.25	1.52	1.67	1.73
0.60	1.00	1.23	1.40	1.73	2.04	2.34	2.89	0.42	0.55	0.67	0.84	1.00	1.20	1.46	1.51	1.54
0.50	1.00	1.25	1.50	2.00	2.50	3.00	5.00	0.47	0.60	0.71	0.87	1.00	1.15	1.30	1.34	1.94
0.40	1.00	1.23	1.57	2.29	3.08	3.92	5.70	0.53	0.67	0.77	0.90	1.00	1.10	1.17	1.16	1.12
0.30	1.00	1.14	1.56	2.57	3.78	5.19	8.56	0.64	0.76	0.84	0.94	1.00	1.04	1.02	0.96	0.90
0.20	1.00	0.90	1.37	2.68	4.49	6.86	17.60	0.83	0.92	0.96	1.00	1.00	0.96	0.81	0.73	0.66
0.10	1.00	1.44	0.80	2.08	4.36	8.00	20.80	1.29	1.26	1.21	1.11	1.00	0.84	0.60	0.46	0.38

This table is specifically for use with the combined feed/speed data for HSS twist drills in Tables 17 through 23; use Tables 5a and 5b to adjust speed and tool life for indexable insert drilling with carbides. The combined feed/speed data for HSS twist drilling are based on a 0.60-inch diameter HSS drill with standard drill point geometry (2-flute with 118° tip angle). To adjust the given speeds for different feeds and drill diameters, use the two feed/speed pairs from the tables and calculate the ratio of desired (new) feed to *optimum* feed (largest of the two feeds from the speed table), and the ratio of the two cutting speeds V_{avg}/V_{opt}. Use the values of these ratios to find the feed factor F_f at the intersection of the feed ratio row and the speed ratio column in the left half of the table. The diameter factor F_d is found in the same row as the feed factor, in the right half of the table, under the column corresponding to the drill diameter. For diameters not given, interpolate between the nearest available sizes. The adjusted cutting speed can be calculated from $V = V_{opt} \times F_f \times F_d$, where V_{opt} is the smaller (*optimum*) of the two speeds from the speed table (from the left side of the column containing the two feed/speed pairs). Tool life using the selected feed and the adjusted speed should be approximately 45 minutes. Speed adjustment factors to modify tool life are found in Table 15e.

Table 23. Feeds and Speeds for Drilling and Reaming Copper Alloys

Group 1
Architectural bronze(C38500); Extra-high-leaded brass (C35600); Forging brass (C37700); Free-cutting phosphor bronze (B-2) (C54400); Free-cutting brass (C36000); Free-cutting Muntz metal (C37000); High-leaded brass (C33200, C34200); High-leaded brass tube (C35300); Leaded commercial bronze (C31400); Leaded naval brass (C48500); Medium-leaded brass (C34000)
Group 2
Aluminum brass, arsenical (C68700); Cartridge brass, 70% (C26000); High-silicon bronze, B (C65500); Admiralty brass (inhibited) (C44300, C44500); Jewelry bronze, 87.5% (C22600); Leaded Muntz metal (C36500, C36800); Leaded nickel silver (C79600); Low brass, 80% (C24000); Low-leaded brass (C33500); Low-silicon bronze, B (C65100); Manganese bronze, A (C67500); Muntz metal, 60% (C28000); Nickel silver, 55–18 (C77000); Red brass, 85% (C23000); Yellow brass (C26800)
Group 3
Aluminum bronze, D (C61400); Beryllium copper (C17000, C17200, C17500); Commercial bronze, 90% (C22000); Copper nickel, 10% (C70600); Copper nickel, 30% (C71500);Electrolytic tough-pitch copper (C11000); Gilding, 95% (C21000); Nickel silver, 65–10 (C74500); Nickel silver, 65–12 (C75700); Nickel silver, 65–15 (C75400); Nickel silver, 65–18 (C75200); Oxygen-free copper (C10200); Phosphor bronze, 1.25% (C50200); Phosphor bronze, 10% D (C52400); Phosphor bronze, 5% A (C51000); Phosphor bronze, 8% C (C52100); Phosphorus deoxidized copper (C12200)

Alloy Description and UNS Alloy Numbers	Material Condition	Drilling	Reaming		Drilling				Reaming	
		HSS			HSS		Indexable Insert Uncoated Carbide		HSS	
		Speed (fpm)			f = feed (0.001 in./rev), s = speed (ft/min)					
					Opt.	*Avg.*	*Opt.*	*Avg.*	*Opt.*	*Avg.*
Wrought Alloys										
Group 1	A	160	160	f	21	11	11	6	36	18
	CD	175	175	s	210	265	405	915	265	230
Group 2	A	120	110	f	24	12	11	6	40	20
	CD	140	120	s	100	130	205	455	130	120
Group 3	A	60	50	f	23	11	11	6	38	19
	CD	65	60	s	155	195	150	340	100	175

Abbreviations designate: A, annealed; CD, cold drawn. The two leftmost speed columns in this table contain traditional Handbook speeds for HSS steel tools. The text contains information concerning feeds to use in conjunction with these speeds.

HSS Drilling and Reaming: The combined feed/speed data for drilling and Table 22 are used to adjust drilling speeds for other feeds and drill diameters. Examples are given in the text. The given feeds and speeds for reaming are based on an 8-tooth, $^{25}/_{32}$-inch diameter, 30° lead angle reamer, and a 0.008-inch radial depth of cut. For other feeds, the correct speed can be obtained by interpolation using the given speeds if the desired feed lies in the recommended range (between the given values of *optimum* and *average* feed). The cutting speeds for reaming do not require adjustment for tool diameter as long as the radial depth of cut does not become too large. Speed adjustment factors to modify tool life are found in Table 15e.

Indexable Insert Drilling: The feed/speed data for indexable insert drilling are based on a tool with two cutting edges, an insert nose radius of $^{3}/_{64}$ inch, a 10-degree lead angle, and diameter *D* of 1 inch. Adjustments for feed and depth of cut are made using Table 5a (Turning Speed Adjustment Factors) using a depth of cut of D/2, or one-half the drill diameter. Expected tool life at the given feeds and speeds is 15 minutes for short hole drilling (i.e., where hole depth is about *2D* or less). Speed adjustment factors to increase tool life are found in Table 5b. The combined feed/speed data in this table are based on tool grades (identified in Table 16) as follows: uncoated carbide = 15.

Using the Feed and Speed Tables for Tapping and Threading.—The feed used in tapping and threading is always equal to the pitch of the screw thread being formed. The threading data contained in the tables for drilling, reaming, and threading (Tables 17 through 23) are primarily for tapping and thread chasing, and do not apply to thread cutting with single-point tools.

The threading data in Tables 17 through 23 give two sets of feed (pitch) and speed values, for 12 and 50 threads/inch, but these values can be used to obtain the cutting speed for any other thread pitches. If the desired pitch falls between the values given in the tables, i.e., between 0.020 inch (50 tpi) and 0.083 inch (12 tpi), the required cutting speed is obtained by interpolation between the given speeds. If the pitch is less than 0.020 inch (more than 50 tpi), use the *average* speed, i.e., the largest of the two given speeds. For pitches greater than 0.083 inch (fewer than 12 tpi), the *optimum* speed should be used. Tool life using the given feed/speed data is intended to be approximately 45 minutes, and should be about the same for threads between 12 and 50 threads per inch.

Example: Determine the cutting speed required for tapping 303 stainless steel with a ½–20 coated HSS tap.

The two feed/speed pairs for 303 stainless steel, in Table 19, are 83/35 (0.083 in./rev at 35 fpm) and 20/45 (0.020 in./rev at 45 fpm). The pitch of a ½–20 thread is 1/20 = 0.05 inch, so the required feed is 0.05 in./rev. Because 0.05 is between the two given feeds (Table 19), the cutting speed can be obtained by interpolation between the two given speeds as follows:

$$V = 35 + \frac{0.05 - 0.02}{0.083 - 0.02}(45 - 35) = 40 \text{ fpm}$$

The cutting speed for coarse-pitch taps must be lower than for fine-pitch taps with the same diameter. Usually, the difference in pitch becomes more pronounced as the diameter of the tap becomes larger and slight differences in the pitch of smaller-diameter taps have little significant effect on the cutting speed. Unlike all other cutting tools, the feed per revolution of a tap cannot be independently adjusted—it is always equal to the lead of the thread and is always greater for coarse pitches than for fine pitches. Furthermore, the thread form of a coarse-pitch thread is larger than that of a fine-pitch thread; therefore, it is necessary to remove more metal when cutting a coarse-pitch thread.

Taps with a long chamfer, such as starting or tapper taps, can cut faster in a short hole than short chamfer taps, such as plug taps. In deep holes, however, short chamfer or plug taps can run faster than long chamfer taps. Bottoming taps must be run more slowly than either starting or plug taps. The chamfer helps to start the tap in the hole. It also functions to involve more threads, or thread form cutting edges, on the tap in cutting the thread in the hole, thus reducing the cutting load on any one set of thread form cutting edges. In so doing, more chips and thinner chips are produced that are difficult to remove from deeper holes. Shortening the chamfer length causes fewer thread form cutting edges to cut, thereby producing fewer and thicker chips that can easily be disposed of. Only one or two sets of thread form cutting edges are cut on bottoming taps, causing these cutting edges to assume a heavy cutting load and produce very thick chips.

Spiral-pointed taps can operate at a faster cutting speed than taps with normal flutes. These taps are made with supplementary angular flutes on the end that push the chips ahead of the tap and prevent the tapped hole from becoming clogged with chips. They are used primarily to tap open or through holes although some are made with shorter supplementary flutes for tapping blind holes.

The tapping speed must be reduced as the percentage of full thread to be cut is increased. Experiments have shown that the torque required to cut a 100 per cent thread form is more than twice that required to cut a 50 per cent thread form. An increase in the percentage of full thread will also produce a greater volume of chips.

The tapping speed must be lowered as the length of the hole to be tapped is increased. More friction must be overcome in turning the tap and more chips accumulate in the hole.

It will be more difficult to apply the cutting fluid at the cutting edges and to lubricate the tap to reduce friction. This problem becomes greater when the hole is being tapped in a horizontal position.

Cutting fluids have a very great effect on the cutting speed for tapping. Although other operating conditions when tapping frequently cannot be changed, a free selection of the cutting fluid usually can be made. When planning the tapping operation, the selection of a cutting fluid warrants a very careful consideration and perhaps an investigation.

Taper threaded taps, such as pipe taps, must be operated at a slower speed than straight thread taps with a comparable diameter. All the thread form cutting edges of a taper threaded tap that are engaged in the work cut and produce a chip, but only those cutting edges along the chamfer length cut on straight thread taps. Pipe taps often are required to cut the tapered thread from a straight hole, adding to the cutting burden.

The machine tool used for the tapping operation must be considered in selecting the tapping speed. Tapping machines and other machines that are able to feed the tap at a rate of advance equal to the lead of the tap, and that have provisions for quickly reversing the spindle, can be operated at high cutting speeds. On machines where the feed of the tap is controlled manually—such as on drill presses and turret lathes—the tapping speed must be reduced to allow the operator to maintain safe control of the operation.

There are other special considerations in selecting the tapping speed. Very accurate threads are usually tapped more slowly than threads with a commercial grade of accuracy. Thread forms that require deep threads for which a large amount of metal must be removed, producing a large volume of chips, require special techniques and slower cutting speeds. Acme, buttress, and square threads, therefore, are generally cut at lower speeds.

Cutting Speed for Broaching.—Broaching offers many advantages in manufacturing metal parts, including high production rates, excellent surface finishes, and close dimensional tolerances. These advantages are not derived from the use of high cutting speeds; they are derived from the large number of cutting teeth that can be applied consecutively in a given period of time, from their configuration and precise dimensions, and from the width or diameter of the surface that can be machined in a single stroke. Most broaching cutters are expensive in their initial cost and are expensive to sharpen. For these reasons, a long tool life is desirable, and to obtain a long tool life, relatively slow cutting speeds are used. In many instances, slower cutting speeds are used because of the limitations of the machine in accelerating and stopping heavy broaching cutters. At other times, the available power on the machine places a limit on the cutting speed that can be used; i.e., the cubic inches of metal removed per minute must be within the power capacity of the machine.

The cutting speeds for high-speed steel broaches range from 3 to 50 feet per minute, although faster speeds have been used. In general, the harder and more difficult to machine materials are cut at a slower cutting speed and those that are easier to machine are cut at a faster speed. Some typical recommendations for high-speed steel broaches are: AISI 1040, 10 to 30 fpm; AISI 1060, 10 to 25 fpm; AISI 4140, 10 to 25 fpm; AISI 41L40, 20 to 30 fpm; 201 austenitic stainless steel, 10 to 20 fpm; Class 20 gray cast iron, 20 to 30 fpm; Class 40 gray cast iron, 15 to 25 fpm; aluminum and magnesium alloys, 30 to 50 fpm; copper alloys, 20 to 30 fpm; commercially pure titanium, 20 to 25 fpm; alpha and beta titanium alloys, 5 fpm; and the superalloys, 3 to 10 fpm. Surface broaching operations on gray iron castings have been conducted at a cutting speed of 150 fpm, using indexable insert cemented carbide broaching cutters. In selecting the speed for broaching, the cardinal principle of the performance of all metal cutting tools should be kept in mind; i.e., increasing the cutting speed may result in a proportionately larger reduction in tool life, and reducing the cutting speed may result in a proportionately larger increase in the tool life. When broaching most materials, a suitable cutting fluid should be used to obtain a good surface finish and a better tool life. Gray cast iron can be broached without using a cutting fluid although some shops prefer to use a soluble oil.

ESTIMATING SPEEDS AND MACHINING POWER

Estimating Planer Cutting Speeds.—Whereas most planers of modern design have a means of indicating the speed at which the table is traveling, or cutting, many older planers do not. Thus, the following formulas are useful for planers that do not have a means of indicating the table or cutting speed. It is not practicable to provide a formula for calculating the exact cutting speed at which a planer is operating because the time to stop and start the table when reversing varies greatly. The formulas below will, however, provide a reasonable estimate.

$$V_c \cong S_c L$$

$$S_c \cong \frac{V_c}{L}$$

where V_c = cutting speed; fpm or m/min

S_c = number of cutting strokes per minute of planer table

L = length of table cutting stroke; ft or m

Cutting Speed for Planing and Shaping.—The traditional HSS cutting tool speeds in Tables through 4b and Tables 8 through 9 can be used for planing and shaping. The feed and depth of cut factors in Tables 5c should also be used, as explained previously. Very often, other factors relating to the machine or the setup will require a reduction in the cutting speed used on a specific job.

Cutting Time for Turning, Boring, and Facing.—The time required to turn a length of metal can be determined by the following formula in which T = time in minutes, L = length of cut in inches, f = feed in inches per revolution, and N = lathe spindle speed in revolutions per minute.

$$T = \frac{L}{fN}$$

When making job estimates, the time required to load and to unload the workpiece on the machine, and the machine handling time, must be added to the cutting time for each length cut to obtain the floor-to-floor time.

Planing Time.—The approximate time required to plane a surface can be determined from the following formula in which T = time in minutes, L = length of stroke in feet, V_c = cutting speed in feet per minute, V_r = return speed in feet per minute; W = width of surface to be planed in inches, F = feed in inches, and 0.025 = approximate reversal time factor per stroke in minutes for most planers:

$$T = \frac{W}{F}\left[L \times \left(\frac{1}{V_c} + \frac{1}{V_r}\right) + 0.025\right]$$

Speeds for Metal-Cutting Saws.—The following speeds and feeds for solid-tooth, high-speed-steel, circular, metal-cutting saws are recommended by Saws International, Inc. (sfpm = surface feet per minute = 3.142 × blade diameter in inches × rpm of saw shaft ÷ 12).

Speeds for Turning Unusual Materials.—*Slate*, on account of its peculiarly stratified formation, is rather difficult to turn, but if handled carefully, can be machined in an ordinary lathe. The cutting speed should be about the same as for cast iron. A sheet of fiber or pressed paper should be interposed between the chuck or steadyrest jaws and the slate, to protect the latter. Slate rolls must not be centered and run on the tailstock. A satisfactory method of supporting a slate roll having journals at the ends is to bore a piece of lignum vitae to receive the turned end of the roll, and center it for the tailstock spindle.

Rubber can be turned at a peripheral speed of 200 feet per minute, although it is much easier to grind it with an abrasive wheel that is porous and soft. For cutting a rubber roll in

Speeds, Feeds, and Tooth Angles for Sawing Various Materials

Materials	Front Rake Angle α (deg)	Back Rake Angle β (deg)	Stock Diameters (inches)			
			¼–¾	¾–1½	1½–2½	2½–3½
Aluminum	24	12	6500 sfpm 100 in./min	6200 sfpm 85 in./min	6000 sfpm 80 in./min	5000 sfpm 75 in./min
Light Alloys with Cu, Mg, and Zn	22	10	3600 sfpm 70 in./min	3300 sfpm 65 in./min	3000 sfpm 63 in./min	2600 sfpm 60 in./min
Light Alloys with High Si	20	8	650 sfpm 16 in./min	600 sfpm 16 in./min	550 sfpm 14 in./min	550 sfpm 12 in./min
Copper	20	10	1300 sfpm 24 in./min	1150 sfpm 24 in./min	1000 sfpm 22 in./min	800 sfpm 22 in./min
Bronze	15	8	1300 sfpm 24 in./min	1150 sfpm 24 in./min	1000 sfpm 22 in./min	800 sfpm 20 in./min
Hard Bronze	10	8	400 sfpm 6.3 in./min	360 sfpm 6 in./min	325 sfpm 5.5 in./min	300 sfpm 5.1 in./min
Cu-Zn Brass	16	8	2000 sfpm 43 in./min	2000 sfpm 43 in./min	1800 sfpm 39 in./min	1800 sfpm 35 in./min
Gray Cast Iron	12	8	82 sfpm 4 in./min	75 sfpm 4 in./min	72 sfpm 3.5 in./min	66 sfpm 3 in./min
Carbon Steel	20	8	160 sfpm 6.3 in./min	150 sfpm 5.9 in./min	150 sfpm 5.5 in./min	130 sfpm 5.1 in./min
Medium Hard Steel	18	8	100 sfpm 5.1 in./min	100 sfpm 4.7 in./min	80 sfpm 4.3 in./min	80 sfpm 4.3 in./min
Hard Steel	15	8	66 sfpm 4.3 in./min	66 sfpm 4.3 in./min	60 sfpm 4 in./min	57 sfpm 3.5 in./min
Stainless Steel	15	8	66 sfpm 2 in./min	63 sfpm 1.75 in./min	60 sfpm 1.75 in./min	57 sfpm 1.5 in./min

two, the ordinary parting tool should not be used, but a tool shaped like a knife; such a tool severs the rubber without removing any material.

Gutta percha can be turned as easily as wood, but the tools must be sharp and a good soap-and-water lubricant used.

Copper can be turned easily at 200 feet per minute.

Limestone such as is used in the construction of pillars for balconies, etc., can be turned at 150 feet per minute, and the formation of ornamental contours is quite easy. *Marble* is a treacherous material to turn. It should be cut with a tool such as would be used for brass, but at a speed suitable for cast iron. It must be handled very carefully to prevent flaws in the surface.

The foregoing speeds are for high-speed steel tools. Tools tipped with tungsten carbide are adapted for cutting various non-metallic products which cannot be machined readily with steel tools, such as slate, marble, synthetic plastic materials, etc. In drilling slate and marble, use flat drills; and for plastic materials, tungsten-carbide-tipped twist drills. Cutting speeds ranging from 75 to 150 feet per minute have been used for drilling slate (without coolant) and a feed of 0.025 inch per revolution for drills ¾ and 1 inch in diameter.

Estimating Machining Power.—Knowledge of the power required to perform machining operations is useful when planning new machining operations, for optimizing existing machining operations, and to develop specifications for new machine tools that are to be acquired. The available power on any machine tool places a limit on the size of the cut that it can take. When much metal must be removed from the workpiece it is advisable to estimate the cutting conditions that will utilize the maximum power on the machine. Many machining operations require only light cuts to be taken for which the machine obviously has ample power; in this event, estimating the power required is a wasteful effort. Conditions in different shops may vary and machine tools are not all designed alike, so some variations between the estimated results and those obtained on the job are to be expected. However, by using the methods provided in this section a reasonable estimate of the power required can be made, which will suffice in most practical situations.

The measure of power in customary inch units is the horsepower; in SI metric units it is the kilowatt, which is used for both mechanical and electrical power. The power required to cut a material depends upon the rate at which the material is being cut and upon an experimentally determined power constant, K_p, which is also called the unit horsepower, unit power, or specific power consumption. The power constant is equal to the horsepower required to cut a material at a rate of one cubic inch per minute; in SI metric units the power constant is equal to the power in kilowatts required to cut a material at a rate of one cubic centimeter per second, or 1000 cubic millimeters per second (1 cm^3 = 1000 mm^3). Different values of the power constant are required for inch and for metric units, which are related as follows: to obtain the SI metric power constant, multiply the inch power constant by 2.73; to obtain the inch power constant, divide the SI metric power constant by 2.73. Values of the power constant in Tables 24, 30, and 25 can be used for all machining operations except drilling and grinding. Values given are for sharp tools.

Table 24. Power Constants, K_p, for Ferrous Cast Metals, Using Sharp Cutting Tools

Material	Brinell Hardness Number	K_p Inch Units	K_p SI Metric Units	Material	Brinell Hardness Number	K_p Inch Units	K_p SI Metric Units
Gray Cast Iron	100–120	0.28	0.76	Malleable Iron			
	120–140	0.35	0.96	Ferritic	150–175	0.42	1.15
	140–160	0.38	1.04	Pearlitic	175–200	0.57	1.56
	160–180	0.52	1.42		200–250	0.82	2.24
	180–200	0.60	1.64		250–300	1.18	3.22
	200–220	0.71	1.94				
	220–240	0.91	2.48	Cast Steel	150–175	0.62	1.69
					175–200	0.78	2.13
Alloy Cast Iron	150–175	0.30	0.82		200–250	0.86	2.35
	175–200	0.63	1.72	…	…	…	…
	200–250	0.92	2.51	…	…	…	…

The value of the power constant is essentially unaffected by the cutting speed, the depth of cut, and the cutting tool material. Factors that do affect the value of the power constant, and thereby the power required to cut a material, include the hardness and microstructure of the work material, the feed rate, the rake angle of the cutting tool, and whether the cutting edge of the tool is sharp or dull. Values are given in the power constant tables for different material hardness levels, whenever this information is available. Feed factors for the power constant are given in Table 25. All metal cutting tools wear but a worn cutting edge requires more power to cut than a sharp cutting edge.

Factors to provide for tool wear are given in Table 26. In this table, the extra-heavy-duty category for milling and turning occurs only on operations where the tool is allowed to wear more than a normal amount before it is replaced, such as roll turning. The effect of the rake angle usually can be disregarded. The rake angle for which most of the data in the power constant tables are given is positive 14 degrees. Only when the deviation from this angle is large is it necessary to make an adjustment. Using a rake angle that is more positive reduces the power required approximately 1 per cent per degree; using a rake angle that is more negative increases the power required; again approximately 1 per cent per degree.

Many indexable insert cutting tools are formed with an integral chip breaker or other cutting edge modifications, which have the effect of reducing the power required to cut a material. The extent of this effect cannot be predicted without a test of each design. Cutting fluids will also usually reduce the power required, when operating in the lower range of cutting speeds. Again, the extent of this effect cannot be predicted because each cutting fluid exhibits its own characteristics.

Table 25. Feed Factors, *C*, for Power Constants

Inch Units				SI Metric Units			
Feed in.[a]	*C*	Feed in.[a]	*C*	Feed mm[b]	*C*	Feed mm[b]	*C*
0.001	1.60	0.014	0.97	0.02	1.70	0.35	0.97
0.002	1.40	0.015	0.96	0.05	1.40	0.38	0.95
0.003	1.30	0.016	0.94	0.07	1.30	0.40	0.94
0.004	1.25	0.018	0.92	0.10	1.25	0.45	0.92
0.005	1.19	0.020	0.90	0.12	1.20	0.50	0.90
0.006	1.15	0.022	0.88	0.15	1.15	0.55	0.88
0.007	1.11	0.025	0.86	0.18	1.11	0.60	0.87
0.008	1.08	0.028	0.84	0.20	1.08	0.70	0.84
0.009	1.06	0.030	0.83	0.22	1.06	0.75	0.83
0.010	1.04	0.032	0.82	0.25	1.04	0.80	0.82
0.011	1.02	0.035	0.80	0.28	1.01	0.90	0.80
0.012	1.00	0.040	0.78	0.30	1.00	1.00	0.78
0.013	0.98	0.060	0.72	0.33	0.98	1.50	0.72

[a] Turning—in./rev; milling—in./tooth: planing and shaping—in./stroke; broaching—in./tooth.
[b] Turning—mm/rev; milling—mm/tooth: planing and shaping—mm/stroke; broaching—mm/tooth.

Table 26. Tool Wear Factors, *W*

Type of Operation		*W*
For all operations with sharp cutting tools		1.00
Turning:	Finish turning (light cuts)	1.10
	Normal rough and semifinish turning	1.30
	Extra-heavy-duty rough turning	1.60–2.00
Milling:	Slab milling	1.10
	End milling	1.10
	Light and medium face milling	1.10–1.25
	Extra-heavy-duty face milling	1.30–1.60
Drilling:	Normal drilling	1.30
	Drilling hard-to-machine materials and drilling with a very dull drill	1.50
Broaching:	Normal broaching	1.05–1.10
	Heavy-duty surface broaching	1.20–1.30

For planing and shaping, use values given for turning.

The machine tool transmits the power from the driving motor to the workpiece, where it is used to cut the material. The effectiveness of this transmission is measured by the machine tool efficiency factor, E. Average values of this factor are given in Table 28. Formulas for calculating the metal removal rate, Q, for different machining operations are given in Table 29. These formulas are used together with others given below. The following formulas can be used with either customary inch or with SI metric units.

$$P_c = K_p CQW \tag{1}$$

$$P_m = \frac{P_c}{E} = \frac{K_p CQW}{E} \tag{2}$$

where P_c = power at the cutting tool; hp, or kW
P_m = power at the motor; hp, or kW
K_p = power constant (see Tables 24, 30, and 25)
Q = metal removal rate; in. ^{3}/min. or cm^3/s (see Table 29)
C = feed factor for power constant (see Table 25)
W = tool wear factor (see Table 26)
E = machine tool efficiency factor (see Table 28)
V = cutting speed, fpm, or m/min
N = cutting speed, rpm
f = feed rate for turning; in./rev. or mm/rev
f = feed rate for planing and shaping; in./stroke, or mm/stroke
f_t = feed per tooth; in./tooth, or mm/tooth
f_m = feed rate; in./min. or mm/min
d_t = maximum depth of cut per tooth: in., or mm
d = depth of cut; in., or mm
n_t = number of teeth on milling cutter

Table 27. Power Constant, K_p, for High-Temperature Alloys, Tool Steel, Stainless Steel, and Nonferrous Metals, Using Sharp Cutting Tools

Material	Brinell Hardness Number	K_p Inch Units	K_p Metric Units	Material	Brinell Hardness Number	K_p Inch Units	K_p Metric Units
High-Temperature Alloys				Stainless Steel	150-175	0.60	1.64
A286	165	0.82	2.24	Stainless Steel	175-200	0.72	1.97
A286	285	0.93	2.54	Stainless Steel	200-250	0.88	2.40
Chromoloy	200	0.78	3.22	Zinc Die Cast Alloys	…	0.25	0.68
Chromoloy	310	1.18	3.00	Copper (pure)	…	0.91	2.48
Inco 700	330	1.12	3.06	Brass			
Inco 702	230	1.10	3.00	Hard	…	0.83	2.27
Hastelloy-B	230	1.10	3.00	Medium	…	0.50	1.36
M-252	230	1.10	3.00	Soft	…	0.25	0.68
M-252	310	1.20	3.28	Leaded	…	0.30	0.82
Ti-150A	340	0.65	1.77				
U-500	375	1.10	3.00	Bronze			
				Hard	…	0.91	2.48
Monel Metal	…	1.00	2.73	Medium	…	0.50	1.36
Tool Steel	175-200	0.75	2.05	Aluminum			
Tool Steel	200-250	0.88	2.40	Cast	…	0.25	0.68
Tool Steel	250-300	0.98	2.68	Rolled (hard)	…	0.33	0.90
Tool Steel	300-350	1.20	3.28				
Tool Steel	350-400	1.30	3.55	Magnesium Alloys	…	0.10	0.27

n_c = number of teeth engaged in work

w = width of cut; in., or mm

Table 28. Machine Tool Efficiency Factors, E

Type of Drive	E	Type of Drive	E
Direct Belt Drive	0.90	Geared Head Drive	0.70–0.80
Back Gear Drive	0.75	Oil-Hydraulic Drive	0.60–0.90

Example: A 180–200 Bhn AISI shaft is to be turned on a geared head lathe using a cutting speed of 350 fpm (107 m/min), a feed rate of 0.016 in./rev (0.40 mm/rev), and a depth of cut of 0.100 inch (2.54 mm). Estimate the power at the cutting tool and at the motor, using both the inch and metric data.

Inch units:

$K_p = 0.62$ (from Table 30)

$C = 0.94$ (from Table 25)

$W = 1.30$ (from Table 26)

$E = 0.80$ (from Table 28)

$Q = 12\ Vfd = 12 \times 350 \times 0.016 \times 0.100$ (from Table 29)

$Q = 6.72$ in.^{3}/min

Table 29. Formulas for Calculating the Metal Removal Rate, Q

Operation	Metal Removal Rate	
	For Inch Units Only Q = in.³/min	For SI Metric Units Only Q = cm³/s
Single-Point Tools (Turning, Planing, and Shaping)	$12Vfd$	$\frac{V}{60}fd$
Milling	$f_m wd$	$\frac{f_m wd}{60,000}$
Surface Broaching	$12Vwn_c d_t$	$\frac{V}{60}un_c d_t$

$$P_c = K_p CQW = 0.62 \times 0.94 \times 6.72 \times 1.30 = 5 \text{ hp}$$

$$P_m = \frac{P_c}{E} = \frac{5}{0.80} = 6.25 \text{ hp}$$

SI metric units:

$K_p = 1.60$ (from Table 24)
$C = 0.94$ (from Table 25)
$W = 1.30$ (from Table 26)
$E = 0.80$ (from Table 30)

$$Q = \frac{V}{60}fd = \frac{107}{60} \times 0.40 \times 2.54 \text{ (from Table 29)}$$

$$= 1.81 \text{ cm}^3/\text{s}$$

$$P_c = K_p CQW = 1.69 \times 0.94 \times 1.81 \times 1.30 = 3.74 \text{ kW}$$

$$P_m = \frac{P_c}{E} = \frac{3.74}{0.80} = 4.675 \text{ kW}$$

Whenever possible the maximum power available on a machine tool should be used when heavy cuts must be taken.

The cutting conditions for utilizing the maximum power should be selected in the following order: 1) select the maximum depth of cut that can be used; 2) select the maximum feed rate that can be used; and 3) estimate the cutting speed that will utilize the maximum power available on the machine. This sequence is based on obtaining the longest tool life of the cutting tool and at the same time obtaining as much production as possible from the machine.

The life of a cutting tool is most affected by the cutting speed, then by the feed rate, and least of all by the depth of cut. The maximum metal removal rate that a given machine is capable of machining from a given material is used as the basis for estimating the cutting speed that will utilize all the power available on the machine.

Example: A 0.125 inch deep cut is to be taken on a 200–210 Bhn AISI 1050 steel part using a 10 hp geared head lathe. The feed rate selected for this job is 018 in./rev. Estimate the cutting speed that will utilize the maximum power available on the lathe.

$K_p = 0.85$ (From Table 30)
$C = 0.92$ (From Table 25)

$W = 1.30$ (From Table 26)

$E = 0.80$ (From Table 28)

$$Q_{max} = \frac{P_m E}{K_p C W} = \frac{10 \times 0.80}{0.85 \times 0.92 \times 1.30} \qquad \left(P_m = \frac{K_p C Q W}{E}\right)$$

$$= 7.87 \text{ in.}^3\text{/min}$$

$$V = \frac{Q_{max}}{12 f d} = \frac{7.87}{12 \times 0.018 \times 0.125} \qquad (Q = 12 V f d)$$

$$= 290 \text{ fpm}$$

Example: A 160-180 Bhn gray iron casting that is 6 inches wide is to have ⅛ inch stock removed on a 10 hp milling machine, using an 8 inch diameter, 10 tooth, indexable insert cemented carbide face milling cutter. The feed rate selected for this cutter is 0.012 in./tooth, and all the stock (0.125 in.) will be removed in one cut. Estimate the cutting speed that will utilize the maximum power available on the machine.

$K_p = 0.52$ (From Table 30)

$C = 1.00$ (From Table 25)

$W = 1.20$ (From Table 26)

$E = 0.80$ (From Table 27)

$$Q_{max} = \frac{P_m E}{K_p C W} = \frac{10 \times 0.80}{0.52 \times 1.00 \times 1.20} = 12.82 \text{ in.}^3\text{/min} \qquad \left(P_m = \frac{K_p C Q W}{E}\right)$$

$$f_m = \frac{Q_{max}}{wd} = \frac{12.82}{6 \times 0.125} = 17 \text{ in./min} \qquad (Q = f_m w d)$$

$$N = \frac{f_{max}}{f_t n_t} = \frac{17}{0.012 \times 10} = 140 \text{ rpm} \qquad (f_m = f_t n_t N)$$

$$V = \frac{\pi D N}{12} = \frac{\pi \times 8 \times 140}{12} = 293 \text{ fpm} \qquad \left(N = \frac{12V}{\pi D}\right)$$

Estimating Drilling Thrust, Torque, and Power.—Although the lips of a drill cut metal and produce a chip in the same manner as the cutting edges of other metal cutting tools, the chisel edge removes the metal by means of a very complex combination of extrusion and cutting. For this reason a separate method must be used to estimate the power required for drilling. Also, it is often desirable to know the magnitude of the thrust and the torque required to drill a hole. The formulas and tabular data provided in this section are based on information supplied by the National Twist Drill Division of Regal-Beloit Corp. The values in Tables 31 through 34 are for sharp drills and the tool wear factors are given in Table 26. For most ordinary drilling operations 1.30 can be used as the tool wear factor. When drilling most difficult-to-machine materials and when the drill is allowed to become very dull, 1.50 should be used as the value of this factor. It is usually more convenient to measure the web thickness at the drill point than the length of the chisel edge; for this reason, the approximate w/d ratio corresponding to each c/d ratio for a correctly ground drill is provided in Table 32. For most standard twist drills the c/d ratio is 0.18, unless the drill has been ground short or the web has been thinned. The c/d ratio of split point drills is 0.03. The formulas given below can be used for spade drills, as well as for twist drills. Separate formulas are required for use with customary inch units and for SI metric units.

Table 30. Power Constants, K_p, for Wrought Steels, Using Sharp Cutting Tools

Material	Brinell Hardness Number	K_p Inch Units	K_p SI Metric Units
Plain Carbon Steels			
All Plain Carbon Steels	80–100	0.63	1.72
	100–120	0.66	1.80
	120–140	0.69	1.88
	140–160	0.74	2.02
	160–180	0.78	2.13
	180–200	0.82	2.24
	200–220	0.85	2.32
	220–240	0.89	2.43
	240–260	0.92	2.51
	260–280	0.95	2.59
	280–300	1.00	2.73
	300–320	1.03	2.81
	320–340	1.06	2.89
	340–360	1.14	3.11
Free Machining Steels			
AISI 1108, 1109, 1110, 1115, 1116, 1117, 1118, 1119, 1120, 1125, 1126, 1132	100–120	0.41	1.12
	120–140	0.42	1.15
	140–160	0.44	1.20
	160–180	0.48	1.31
	180–200	0.50	1.36
AISI 1137, 1138, 1139, 1140, 1141, 1144, 1145, 1146, 1148, 1151	180–200	0.51	1.39
	200–220	0.55	1.50
	220–240	0.57	1.56
	240–260	0.62	1.69
Alloy Steels			
AISI 4023, 4024, 4027, 4028, 4032, 4037, 4042, 4047, 4137, 4140, 4142, 4145, 4147, 4150, 4340, 4640, 4815, 4817, 4820, 5130, 5132, 5135, 5140, 5145, 5150, 6118, 6150, 8637, 8640, 8642, 8645, 8650, 8740	140–160	0.62	1.69
	160–180	0.65	1.77
	180–200	0.69	1.88
	200–220	0.72	1.97
	220–240	0.76	2.07
	240–260	0.80	2.18
	260–280	0.84	2.29
	280–300	0.87	2.38
	300–320	0.91	2.48
	320–340	0.96	2.62
	340–360	1.00	2.73
AISI 4130, 4320, 4615, 4620, 4626, 5120, 8615, 8617, 8620, 8622, 8625, 8630, 8720	140–160	0.56	1.53
	160–180	0.59	1.61
	180–200	0.62	1.69
	200–220	0.65	1.77
	220–240	0.70	1.91
	240–260	0.74	2.02
	260–280	0.77	2.10
	280–300	0.80	2.18
	300–320	0.83	2.27
	320–340	0.89	2.43
AISI 1330, 1335, 1340, E52100	160–180	0.79	2.16
	180–200	0.83	2.27
	200–220	0.87	2.38
	220–240	0.91	2.48
	240–260	0.95	2.59
	260–280	1.00	2.73

Table 31. Work Material Factor, K_d, for Drilling with a Sharp Drill

Work Material			Work Material Constant, K_d
AISI 1117 (Resulfurized free machining mild steel)			12,000
Steel, 200 Bhn			24,000
Steel, 300 Bhn			31,000
Steel, 400 Bhn			34,000
Cast Iron, 150 Bhn			14,000
Most Aluminum Alloys			7,000
Most Magnesium Alloys			4,000
Most Brasses			14,000
Leaded Brass			7,000
Austenitic Stainless Steel (Type 316)			24,000[a] for Torque
			35,000[a] for Thrust
Titanium Alloy T16A	4V	$40R_c$	18,000[a] for Torque
			29,000[a] for Thrust
René 41	$40R_c$		40,000[ab] min.
Hastelloy-C			30,000[a] for Torque
			37,000[a] for Thrust

[a] Values based upon a limited number of tests.
[b] Will increase with rapid wear.

Table 32. Chisel Edge Factors for Torque and Thrust

c/d	Approx. w/d	Torque Factor A	Thrust Factor B	Thrust Factor J	c/d	Approx. w/d	Torque Factor A	Thrust Factor B	Thrust Factor J
0.03	0.025	1.000	1.100	0.001	0.18	0.155	1.085	1.355	0.030
0.05	0.045	1.005	1.140	0.003	0.20	0.175	1.105	1.380	0.040
0.08	0.070	1.015	1.200	0.006	0.25	0.220	1.155	1.445	0.065
0.10	0.085	1.020	1.235	0.010	0.30	0.260	1.235	1.500	0.090
0.13	0.110	1.040	1.270	0.017	0.35	0.300	1.310	1.575	0.120
0.15	0.130	1.080	1.310	0.022	0.40	0.350	1.395	1.620	0.160

For drills of standard design, use $c/d = .18$.
For split point drills, use $c/d = .03$.
c/d = Length of Chisel Edge ÷ Drill Diameter.
w/d = Web Thickness at Drill Point ÷ Drill Diameter.

For inch units only:

$$T = 2k_d F_f F_T BW + K_d d^2 JW \tag{3}$$

$$M = K_d F_f F_M AW \tag{4}$$

$$P_c = MN/63.025 \tag{5}$$

For SI metric units only:

$$T = 0.05\, K_d F_f F_T BW + 0.007\, K_d d^2 JW \qquad (6)$$

$$M = \frac{K_d F_f F_M AW}{40{,}000} = 0.000025\, K_d F_f F_M AW \qquad (7)$$

$$P_c = MN/9550 \qquad (8)$$

Use with either inch or metric units:

$$P_m = \frac{P_c}{E} \qquad (9)$$

where P_c = Power at the cutter; hp, or kW
P_m = Power at the motor; hp, or kW
M = Torque; in. lb, or N.m
T = Thrust; lb, or N
K_d = Work material factor (See Table 31)
F_f = Feed factor (See Table 33)
F_T = Thrust factor for drill diameter (See Table 34)
F_M = Torque factor for drill diameter (See Table 34)
A = Chisel edge factor for torque (See Table 32)
B = Chisel edge factor for thrust (See Table 32)
J = Chisel edge factor for thrust (See Table 32)
W = Tool wear factor (See Table 26)
N = Spindle speed; rpm
E = Machine tool efficiency factor (See Table 28)
D = Drill diameter; in., or mm
c = Chisel edge length; in., or mm (See Table 32)
w = Web thickness at drill point; in., or mm (See Table 32)

Table 33. Feed Factors, F_f, for Drilling

Inch Units				SI Metric Units			
Feed, in./rev	F_f	Feed, in./rev	F_f	Feed, mm/rev	F_f	Feed, mm/rev	F_f
0.0005	0.0023	0.012	0.029	0.01	0.025	0.30	0.382
0.001	0.004	0.013	0.031	0.03	0.060	0.35	0.432
0.002	0.007	0.015	0.035	0.05	0.091	0.40	0.480
0.003	0.010	0.018	0.040	0.08	0.133	0.45	0.528
0.004	0.012	0.020	0.044	0.10	0.158	0.50	0.574
0.005	0.014	0.022	0.047	0.12	0.183	0.55	0.620
0.006	0.017	0.025	0.052	0.15	0.219	0.65	0.708
0.007	0.019	0.030	0.060	0.18	0.254	0.75	0.794
0.008	0.021	0.035	0.068	0.20	0.276	0.90	0.919
0.009	0.023	0.040	0.076	0.22	0.298	1.00	1.000
0.010	0.025	0.050	0.091	0.25	0.330	1.25	1.195

Table 34. Drill Diameter Factors: F_T for Thrust; F_M for Torque

Inch Units						SI Metric Units					
Drill Dia., in.	F_T	F_M	Drill Dia., in.	F_T	F_M	Drill Dia., mm	F_T	F_M	Drill Dia., mm	F_T	F_M
0.063	0.110	0.007	0.875	0.899	0.786	1.60	1.46	2.33	22.00	11.86	260.8
0.094	0.151	0.014	0.938	0.950	0.891	2.40	2.02	4.84	24.00	12.71	305.1
0.125	0.189	0.024	1.000	1.000	1.000	3.20	2.54	8.12	25.50	13.34	340.2
0.156	0.226	0.035	1.063	1.050	1.116	4.00	3.03	12.12	27.00	13.97	377.1
0.188	0.263	0.049	1.125	1.099	1.236	4.80	3.51	16.84	28.50	14.58	415.6
0.219	0.297	0.065	1.250	1.195	1.494	5.60	3.97	22.22	32.00	16.00	512.0
0.250	0.330	0.082	1.375	1.290	1.774	6.40	4.42	28.26	35.00	17.19	601.6
0.281	0.362	0.102	1.500	1.383	2.075	7.20	4.85	34.93	38.00	18.36	697.6
0.313	0.395	0.124	1.625	1.475	2.396	8.00	5.28	42.22	42.00	19.89	835.3
0.344	0.426	0.146	1.750	1.565	2.738	8.80	5.96	50.13	45.00	21.02	945.8
0.375	0.456	0.171	1.875	1.653	3.100	9.50	6.06	57.53	48.00	22.13	1062
0.438	0.517	0.226	2.000	1.741	3.482	11.00	6.81	74.90	50.00	22.86	1143
0.500	0.574	0.287	2.250	1.913	4.305	12.50	7.54	94.28	58.00	25.75	1493
0.563	0.632	0.355	2.500	2.081	5.203	14.50	8.49	123.1	64.00	27.86	1783
0.625	0.687	0.429	2.750	2.246	6.177	16.00	9.19	147.0	70.00	29.93	2095
0.688	0.741	0.510	3.000	2.408	7.225	17.50	9.87	172.8	76.00	31.96	2429
0.750	0.794	0.596	3.500	2.724	9.535	19.00	10.54	200.3	90.00	36.53	3293
0.813	0.847	0.689	4.000	3.031	12.13	20.00	10.98	219.7	100.00	39.81	3981

Example: A standard ⅞ inch drill is to drill steel parts having a hardness of 200 Bhn on a drilling machine having an efficiency of 0.80. The spindle speed to be used is 350 rpm and the feed rate will be 0.008 in./rev. Calculate the thrust, torque, and power required to drill these holes:

$K_d = 24{,}000$ (From Table 31)

$F_f = 0.021$ (From Table 33)

$F_T = 0.899$ (From Table 34)

$F_M = 0.786$ (From Table 34)

$A = 1.085$ (From Table 32)

$B = 1.355$ (From Table 32)

$J = 0.030$ (From Table 32)

$W = 1.30$ (From Table 26)

$$T = 2K_dF_fF_TBW + K_dd^2JW$$
$$= 2 \times 24{,}000 \times 0.21 \times 0.899 \times 1.355 \times 1.30 + 24{,}000 \times 0.875^2 \times 0.030 \times 1.30$$
$$= 2313 \text{ lb}$$

$$M = K_dF_fF_mAW$$
$$= 24{,}000 \times 0.021 \times 0.786 \times 1.085 \times 1.30 = 559 \text{ in. lb}$$

$$P_c = \frac{MN}{63{,}025} = \frac{559 \times 350}{63{,}025} = 3.1 \text{ hp} \qquad P_m = \frac{P_c}{E} = \frac{3.1}{0.80} = 3.9 \text{ hp}$$

Twist drills are generally the most highly stressed of all metal cutting tools. They must not only resist the cutting forces on the lips, but also the drill torque resulting from these forces and the very large thrust force required to push the drill through the hole. Therefore, often when drilling smaller holes, the twist drill places a limit on the power used and for very large holes, the machine may limit the power.

MACHINING ECONOMETRICS

Tool Wear And Tool Life Relationships

Tool wear.—Tool-life is defined as the cutting time to reach a predetermined wear, called the tool wear criterion. The size of tool wear criterion depends on the grade used, usually a tougher grade can be used at bigger flank wear. For finishing operations, where close tolerances are required, the wear criterion is relatively small. Other alternative wear criteria are a predetermined value of the surface roughness, or a given depth of the crater which develops on the rake face of the tool. The most appropriate wear criteria depends on cutting geometry, grade, and materials.

Tool-life is determined by assessing the time — the tool-life — at which a given predetermined flank wear is reached, 0.25, 0.4, 0.6, 0.8 mm etc. Fig. 1 depicts how flank wear varies with cutting time (approximately straight lines in a semi-logarithmic graph) for three combinations of cutting speeds and feeds. Alternatively, these curves may represent how variations of machinability impact on tool-life, when cutting speed and feed are constant. All tool wear curves will sooner or later bend upwards abruptly and the cutting edge will break, i.e., catastrophic failure as indicated by the white arrows in Fig. 1.

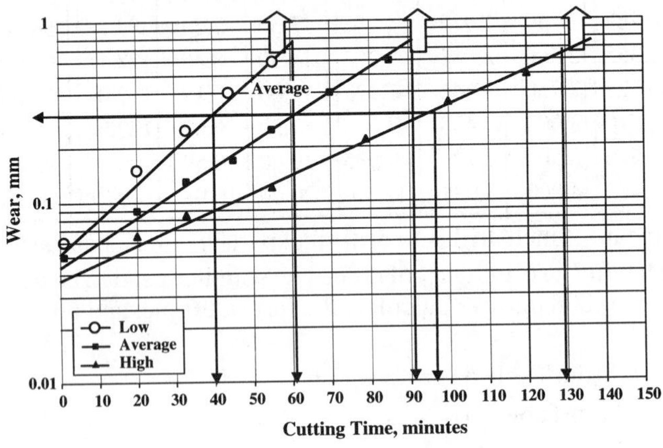

Fig. 1. Flank Wear as a Function of Cutting Time

The maximum deviation from the average tool-life 60 minutes in Fig. 1 is assumed to range between 40 and 95 minutes, i.e. −33% and +58% variation. The positive deviation from the average (longer than expected tool-life) is not important, but the negative one (shorter life) is, as the edge may break before the scheduled tool change after 60 minutes, when the flank wear is 0.6 mm.

It is therefore important to set the wear criterion at a safe level such that tool failures due to "normal" wear become negligible. This is the way machinability variations are mastered.

Equivalent Chip Thickness (*ECT*).—*ECT* combines the four basic turning variables, depth of cut, lead angle, nose radius and feed per revolution into one basic parameter. For all other metal cutting operations such as drilling, milling and grinding, additional variables such as number of teeth, width of cut, and cutter diameter are included in the parameter *ECT*. In turning, milling, and drilling, according to the *ECT* principle, when the product of feed times depth of cut is constant the tool-life is constant no matter how the depth of cut or feed is selected, provided that the cutting speed and cutting edge length are maintained constant. By replacing the geometric parameters with *ECT*, the number of tool-life tests to evaluate cutting parameters can be reduced considerably, by a factor of 4 in turning, and in milling by a factor of 7 because radial depth of cut, cutter diameter and number of teeth are additional parameters.

The introduction of the *ECT* concept constitutes a major simplification when predicting tool-life and calculating cutting forces, torque, and power. *ECT* was first presented in 1931 by Professor R. Woxen, who both theoretically and experimentally proved that *ECT* is a basic metal cutting parameter for high-speed cutting tools. Dr. Colding later proved that the concept also holds for carbide tools, and extended the calculation of *ECT* to be valid for cutting conditions when the depth of cut is smaller than the tool nose radius, or for round inserts. Colding later extended the concept to all other metal cutting operations, including the grinding process.

The definition of *ECT* is:

$$ECT = \frac{Area}{CEL} \text{ (mm or inch)}$$

where A = cross sectional area of cut (approximately = feed × depth of cut), (mm^2 or $inch^2$)

CEL = cutting edge length (tool contact rubbing length), (mm or inch), see Fig.9.

An exact value of A is obtained by the product of *ECT* and *CEL*. In turning, milling, and drilling, *ECT* varies between 0.05 and 1 mm, and is always less than the feed/rev or feed/tooth; its value is usually about 0.7 to 0.9 times the feed.

Example 1: For a feed of 0.8 mm/rev, depth of cut a = 3 mm, and a cutting edge length $CEL = 4\ mm^2$, the value of *ECT* is approximately $ECT = 0.8 \times 3 \div 4 = 0.6$ mm.

The product of *ECT*, *CEL*, and cutting speed *V* (m/min or ft/min) is equal to the metal removal rate, *MRR*, which is measured in terms of the volume of chips removed per minute:

$$MRR = 1000V \times Area = 1000V \times ECT \times CEL \text{ mm}^3\text{/min}$$
$$= V \times Area \text{ cm}^3\text{/min or inch}^3\text{/min}$$

The specific metal removal rate *SMRR* is the metal removal rate per mm cutting edge length *CEL*, thus:

$$SMMR = 1000V \times ECT \text{ mm}^3\text{/min/mm}$$
$$= V \times ECT \text{ cm}^3\text{/min/mm or inch}^3\text{/min/inch}$$

Example 2: Using above data and a cutting speed of $V = 250$ m/min specific metal removal rate becomes $SMRR = 0.6 \times 250 = 150$ (cm^3/min/mm).

ECT in Grinding: In grinding *ECT* is defined as in the other metal cutting processes, and is approximately equal to $ECT = Vw \times ar \div V$, where Vw is the work speed, ar is the depth of cut, and $A = Vw \times ar$. Wheel life is constant no matter how depth ar, or work speed Vw, is selected at V = constant (usually the influence of grinding contact width can be neglected). This translates into the same wheel life as long as the specific metal removal rate is constant, thus:

$$SMMR = 1000Vw \times ar \text{ mm}^3\text{/min/mm}$$

In grinding, *ECT* is much smaller than in the other cutting processes, ranging from about 0.0001 to 0.001 mm (0.000004 to 0.00004 inch). The grinding process is described in a separate chapter *GRINDING FEEDS AND SPEEDS* starting on page 1120.

Tool-life Relationships.—Plotting the cutting times to reach predetermined values of wear typically results in curves similar to those shown in Fig. 2 (cutting time versus cutting speed at constant feed per tooth) and Fig. 3 (cutting time versus feed per tooth at constant cutting speed). These tests were run in 1993 with mixed ceramics turn-milling hard steel, 82 R_C, at the Technische Hochschule Darmstadt.

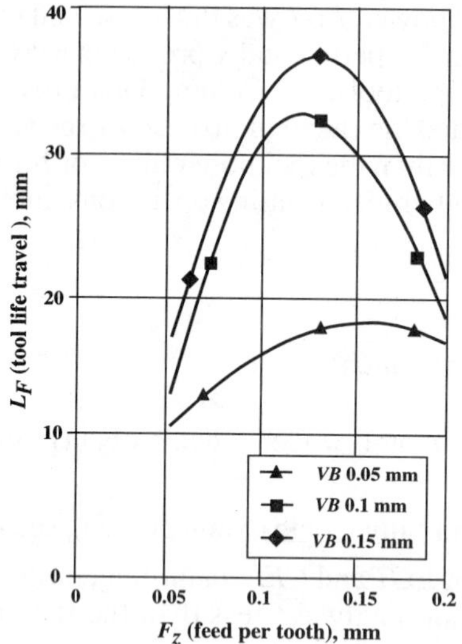

Fig. 2. Influence of feed per tooth on cutting time

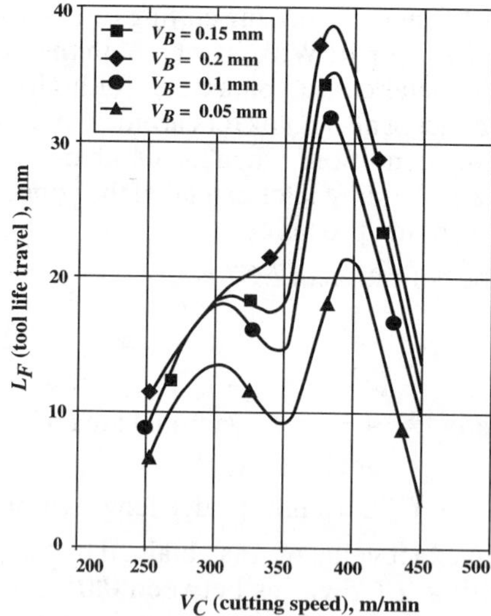

Fig. 3. Influence of cutting speed on tool-life

Tool-life has a maximum value at a particular setting of feed and speed. Economic and productive cutting speeds always occur on the right side of the curves in Figs. 2 and 4, which are called Taylor curves, represented by the so called Taylor's equation.

The variation of tool-life with feed and speed constitute complicated relationships, illustrated in Figs. 6a, 6b, and 6c.

Taylor's Equation.—Taylor's equation is the most commonly used relationship between tool-life *T*, and cutting speed *V*. It constitutes a straight line in a log-log plot, one line for each feed, nose radius, lead angle, or depth of cut, mathematically represented by:

$$V \times T^n = C \tag{1a}$$

where n = is the slope of the line

C = is a constant equal to the cutting speed for $T = 1$ minute

By transforming the equation to logarithmic axes, the Taylor lines become straight lines with slope = n. The constant C is the cutting speed on the horizontal (V) axis at tool-life T = 1 minute, expressed as follows

$$\ln V + n \times \ln T = \ln C \tag{1b}$$

For different values of feed or *ECT*, log-log plots of Equation (1a) form approximately straight lines in which the slope decreases slightly with a larger value of feed or *ECT*. In practice, the Taylor lines are usually drawn parallel to each other, i.e., the slope n is assumed to be constant.

Fig. 4 illustrates the Taylor equation, tool-life T versus cutting speed V, plotted in log-log coordinates, for four values of $ECT = 0.1, 0.25, 0.5$ and 0.7 mm.

In Fig. 4, starting from the right, each $T–V$ line forms a generally straight line that bends off and reaches its maximum tool-life, then drops off with decreasing speed (see also Figs. 2 and 3. When operating at short tool-lives, approximately when T is less than 5 minutes, each line bends a little so that the cutting speed for 1 minute life becomes less than the value calculated by constant C.

The Taylor equation is a very good approximation of the right hand side of the real tool-life curve (slightly bent). The portion of the curve to the left of the maximum tool-life gives shorter and shorter tool-lives when decreasing the cutting speed starting from the point of maximum tool-life. Operating at the maximum point of maximum tool-life, or to the left of it, causes poor surface finish, high cutting forces, and sometimes vibrations.

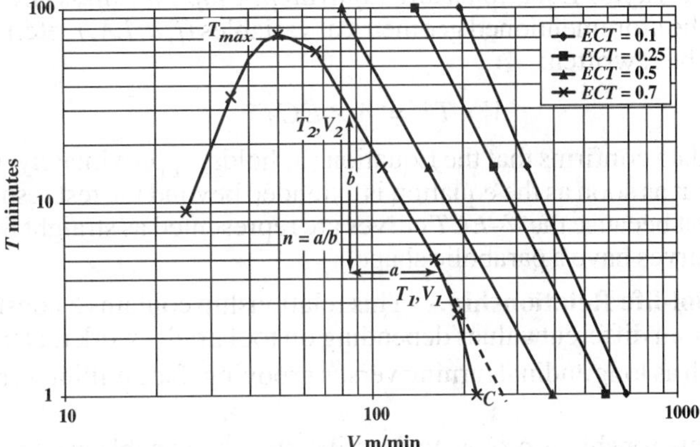

Fig. 4. Definition of slope *n* and constant C in Taylor's equation

Evaluation of Slope *n*, and Constant C.—When evaluating the value of the Taylor slope based on wear tests, care must be taken in selecting the tool-life range over which the slope is measured, as the lines are slightly curved.

The slope *n* can be found in three ways:

- Calculate *n* from the formula $n = (\ln C - \ln V)/\ln T$, reading the values of C and *V* for any value of *T* in the graph.
- Alternatively, using two points on the line, (V_1, T_1) and (V_2, T_2), calculate *n* using the relationship $V_1 \times T_1^n = V_2 \times T_2^n$. Then, solving for *n*,

$$n = \frac{\ln(V_1/V_2)}{\ln(T_2/T_1)}$$

- Graphically, *n* may be determined from the graph by measuring the distances "a" and "b" using a mm scale, and *n* is the ratio of *a* and *b*, thus, *n* = a/b

Example: Using Fig. 4, and a given value of *ECT*= 0.7 mm, calculate the slope and constant of the Taylor line.

On the Taylor line for *ECT*= 0.7, locate points corresponding to tool-lives $T_1 = 15$ minutes and $T_2 = 60$ minutes. Read off the associated cutting speeds as, approximately, $V_1 =$ 110 m/min and $V_2 = 65$ m/min.

The slope *n* is then found to be $n = \ln(110/65)/\ln(60/15) = 0.38$

The constant C can be then determined using the Taylor equation and either point (T_1, V_1) or point (T_2, V_2), with equivalent results, as follows:

$$C = V \times T^n = 110 \times 15^{0.38} = 65 \times 60^{0.38} = 308 \text{ m/min (1027 fpm)}$$

The Generalized Taylor Equation.—The above calculated slope and constant C define tool-life at one particular value of feed *f*, depth of cut *a*, lead angle *LA*, nose radius *r*, and other relevant factors.

The generalized Taylor equation includes these parameters and is written

$$T^n = A \times f^m \times a^p \times LA^q \times r^s \quad (2)$$

where *A* = area; and, *n, m, p, q,* and *s* = constants.

There are two problems with the generalized equation: 1) a great number of tests have to be run in order to establish the constants *n, m, p, q, s,* etc.; and 2) the accuracy is not very good because Equation (2) yields straight lines when plotted versus *f, a, LA*, and *r,* when in reality, they are parabolic curves..

The Generalized Taylor Equation using Equivalent Chip Thickness (ECT): Due to the compression of the aforementioned geometrical variables (f, a, LA, r, etc.) into ECT, Equation (2) can now be rewritten:

$$V \times T^n = A \times ECT^m \tag{3}$$

Experimental data confirms that the Equation (3) holds, approximately, within the range of the test data, but as soon as the equation is extended beyond the test results, the error can become very great because the *V–ECT* curves are represented as straight lines by Equation (3)and the real curves have a parabolic shape.

The Colding Tool-life Relationship.—This relationship contains 5 constants H, K, L, M, and N_0, which attain different values depending on tool grade, work material, and the type of operation, such as longitudinal turning versus grooving, face milling versus end milling, etc.

This tool-life relationship is proven to describe, with reasonable accuracy, how tool-life varies with ECT and cutting speed for any metal cutting and grinding operation. It is expressed mathematically as follows either as a generalized Taylor equation (4a), or, in logarithmic coordinates (4b):

$$V \times T^{(N_0 - L \times \ln ECT)} \times ECT^{\left(-\frac{H}{2M} + \frac{\ln ECT}{4M}\right)} = e^{\left(K - \frac{H}{4M}\right)} \tag{4a}$$

$$y = K - \frac{x - H}{4M} - z(N_0 - L_x) \tag{4b}$$

where $x = \ln ECT$ $y = \ln V$ $z = \ln T$

M = the vertical distance between the maximum point of cutting speed (ECT_H, V_H) for $T = 1$ minute and the speed V_G at point (ECT_G, V_G), as shown in Fig. 5.

$2M$ = the horizontal distance between point (ECT_H, V_G) and point (V_G, ECT_G)

H and K = the logarithms of the coordinates of the maximum speed point (ECT_H, V_H) at tool-life $T = 1$ minute, thus $H = \ln(ECT_H)$ and $K = \ln(V_H)$

N_0 and L = the variation of the Taylor slope n with ECT: $n = N_0 - L \times \ln(ECT)$

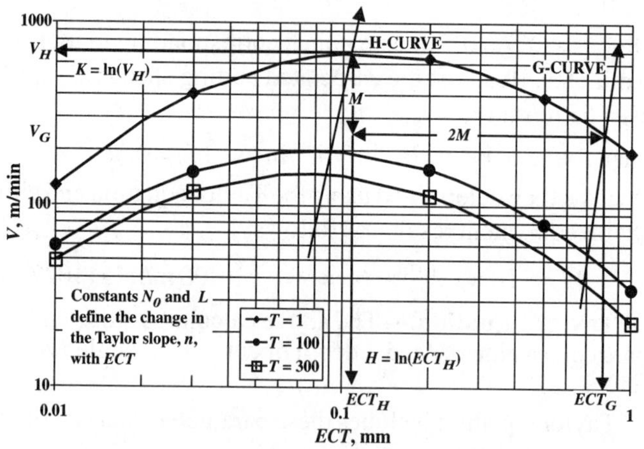

Fig. 5. Definitions of the constants H, K, L, M, and N_0 for tool-life equation in the V-ECT plane with tool-life constant

The constants L and N_0 are determined from the slopes n_1 and n_2 of two Taylor lines at ECT_1 and ECT_2, and the constant M from 3 *V–ECT* values at any constant tool-life. Constants H and K are then solved using the tool-life equation with the above-calculated values of L, N_0 and M.

The *G*- and *H*-curves.—The *G*-curve defines the longest possible tool-life for any given metal removal rate, *MRR*, or specific metal removal rate, *SMRR*. It also defines the point where the total machining cost is minimum, after the economic tool-life T_E, or optimal tool-life T_O, has been calculated, see *Optimization Models, Economic Tool-life when Feed is Constant* starting on page 1073.

The tool-life relationship is depicted in the 3 planes: *T–V*, where *ECT* is the plotted parameter (the Taylor plane); *T–ECT*, where *V* is plotted; and, *V–ECT*, where *T* is a parameter. The latter plane is the most useful because the optimal cutting conditions are more readily understood when viewing in the *V–ECT* plane. Figs. 6a, 6b, and 6c show how the tool-life curves look in these 3 planes in log-log coordinates.

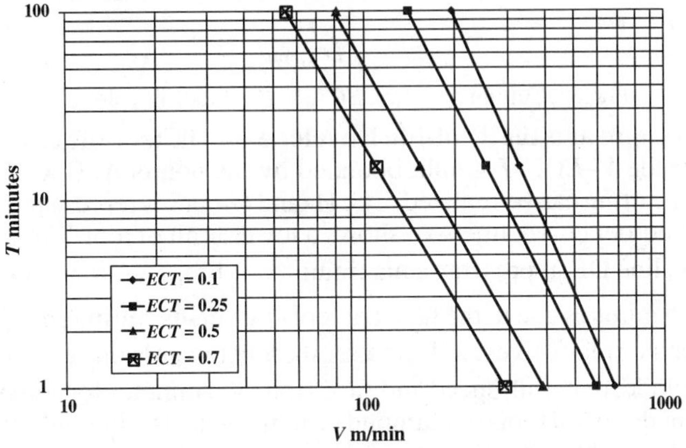

Fig. 6a. Tool-life vs. cutting sped *T–V*, *ECT* plotted

Fig. 6a shows the Taylor lines, and Fig. 6b illustrates how tool-life varies with *ECT* at different values of cutting speed, and shows the *H*-curve. Fig. 6c illustrates how cutting speed varies with *ECT* at different values of tool-life. The *H*- and *G*-curves are also drawn in Fig. 6c.

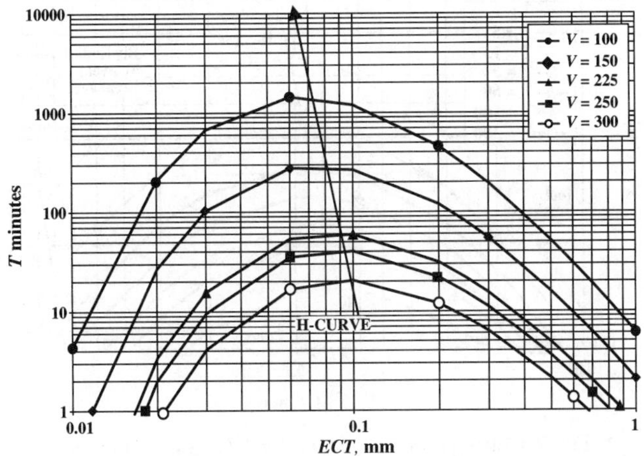

Fig. 6b. Tool-life vs. *ECT*, *T–ECT*, cutting speed plotted

A simple and practical method to ascertain that machining is not done to the left of the *H*-curve is to examine the chips. When *ECT* is too small, about 0.03-0.05 mm, the chips tend to become irregular and show up more or less as dust.

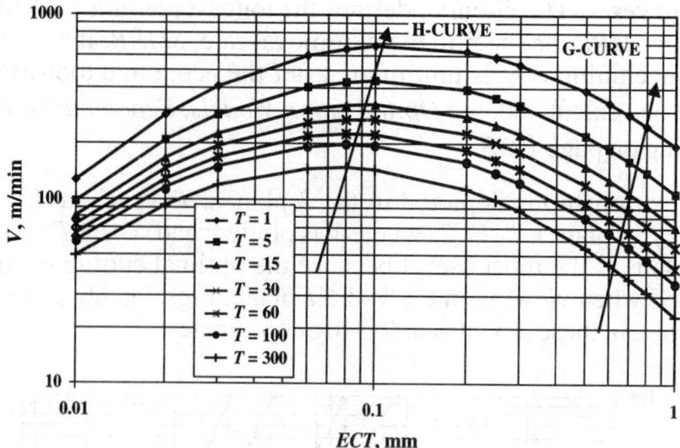

Fig. 6c. Cutting speed vs. *ECT*, *V–ECT*, tool-life plotted

The *V–ECT–T* Graph and the Tool-life Envelope.— The tool-life envelope, in Fig. 7, is an area laid over the *V–ECT–T* graph, bounded by the points A, B, C, D, and E, within which successful cutting can be realized. The H- and *G*-curves represent two borders, lines $\overline{AE}$ and $\overline{BC}$. The border curve, line $\overline{AB}$, shows a lower limit of tool-life, $T_{MIN} = 5$ minutes, and border curve, line $\overline{DE}$, represents a maximum tool-life, $T_{MAX} = 300$ minutes.

T_{MIN} is usually 5 minutes due to the fact that tool-life versus cutting speed does not follow a straight line for short tool-lives; it decreases sharply towards one minute tool-life. T_{MAX} varies with tool grade, material, speed and *ECT* from 300 minutes for some carbide tools to 10000 minutes for diamond tools or diamond grinding wheels, although systematic studies of maximum tool-lives have not been conducted.

Sometimes the metal cutting system cannot utilize the maximum values of the *V–ECT–T* envelope, that is, cutting at optimum *V–ECT* values along the *G*-curve, due to machine power or fixture constraints, or vibrations. Maximum ECT values, ECT_{MAX}, are related to the strength of the tool material and the tool geometry, and depend on the tool grade and material selection, and require a relatively large nose radius.

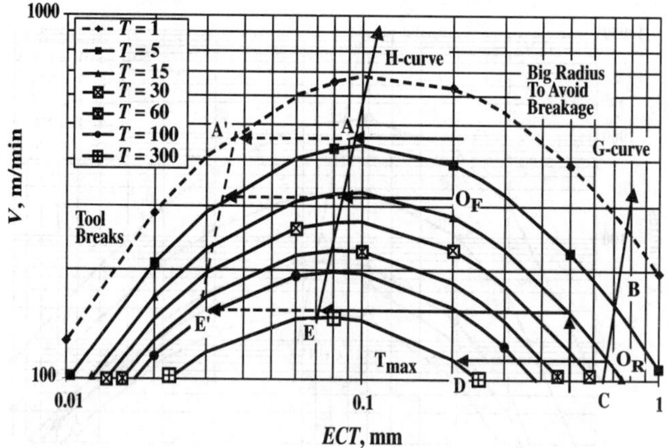

Fig. 7. Cutting speed vs. *ECT*, *V–ECT*, tool-life plotted

Minimum *ECT* values, ECT_{MIN}, are defined by the conditions at which surface finish suddenly deteriorates and the cutting edge begins rubbing rather than cutting. These conditions begin left of the *H*-curve, and are often accompanied by vibrations and built-up edges on the tool. If feed or *ECT* is reduced still further, excessive tool wear with sparks and tool breakage, or melting of the edge occurs. For this reason, values of *ECT* lower than approximately 0.03 mm should not be allowed. In Fig. 7, the ECT_{MIN} boundary is indicated by contour line $\overline{\text{A}'\text{E}'}$.

In milling the minimum feed/tooth depends on the ratio *ar*/*D*, of radial depth of cut *ar*, and cutter diameter *D*. For small *ar*/*D* ratios, the chip thickness becomes so small that it is necessary to compensate by increasing the feed/tooth. See *High-speed Machining Econometrics* starting on page 1085 for more on this topic.

Fig. 7 demonstrates, in principle, minimum cost conditions for roughing at point $\mathbf{O_R}$, and for finishing at point $\mathbf{O_F}$, where surface finish or tolerances have set a limit. Maintaining the speed at $\mathbf{O_R}$, 125 m/min, and decreasing feed reaches a maximum tool-life = 300 minutes at $ECT = 0.2$, and a further decrease of feed will result in shorter lives.

Similarly, starting at point X ($V = 150, ECT = 0.5, T = 15$) and reducing feed, the *H*-curve will be reached at point E ($ECT = 0.075, T = 300$). Continuing to the left, tool-life will decrease and serious troubles occur at point E′ ($ECT = 0.03$).

Starting at point $\mathbf{O_F}$ ($V = 300, ECT = 0.2, T = 15$) and reducing feed the *H*-curve will be reached at point E ($ECT = 0.08, T = 15$). Continuing to the left, life will decrease and serious troubles occur at $ECT = 0.03$.

Starting at point X ($V = 400, ECT = 0.2, T = 5$) and reducing feed the *H*-curve will be reached at point E ($ECT = 0.09, T = 7$). Continuing to the left, life will decrease and serious troubles occur at point A′ ($ECT = 0.03$), where $T = 1$ minute.

Cutting Forces and Chip Flow Angle.—There are three cutting forces, illustrated in Fig. 8, that are associated with the cutting edge with its nose radius *r*, depth of cut *a*, lead angle *LA*, and feed per revolution *f*, or in milling feed per tooth f_z. There is one drawing for roughing and one for finishing operations.

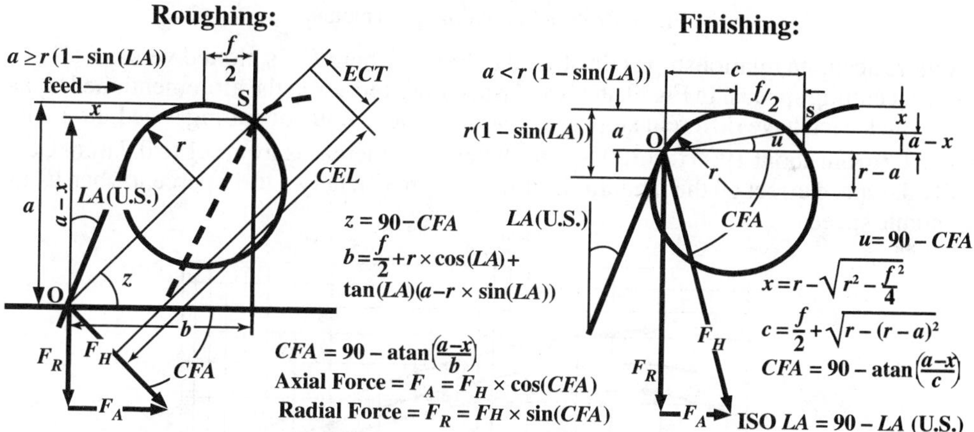

Fig. 8. Definitions of equivalent chip thickness, *ECT*, and chip flow angle, *CFA*.

The cutting force F_C, or tangential force, is perpendicular to the paper plane. The other two forces are the feed or axial force F_A, and the radial force F_R directed towards the work piece. The resultant of F_A and F_R is called F_H. When finishing, F_R is bigger than F_A, while in roughing F_A is usually bigger than F_R. The direction of F_H, measured by the chip flow angle *CFA*, is perpendicular to the rectangle formed by the cutting edge length *CEL* and *ECT* (the product of *ECT* and CEL constitutes the cross sectional area of cut, *A*). The important task of determining the direction of F_H, and calculation of F_A and F_R, are shown in the formulas given in the Fig. 8.

The method for calculating the magnitudes of F_H, F_A, and F_R is described in the following. The first thing is to determine the value of the cutting force F_C. Approximate formulas to calculate the tangential cutting force, torque and required machining power are found in the section *ESTIMATING SPEEDS AND MACHINING POWER* starting on page 1044.

Specific Cutting Force, Kc: The specific cutting force, or the specific energy to cut, *Kc*, is defined as the ratio between the cutting force F_C and the chip cross sectional area, *A*. thus, $Kc = F_C \div A$ N/mm^2.

The value of *Kc* decreases when *ECT* increases, and when the cutting speed *V* increases. Usually, *Kc* is written in terms of its value at *ECT* = 1, called Kc_1, and neglecting the effect of cutting speed, thus $Kc = Kc_1 \times ECT^B$, where *B* = slope in log-log coordinates

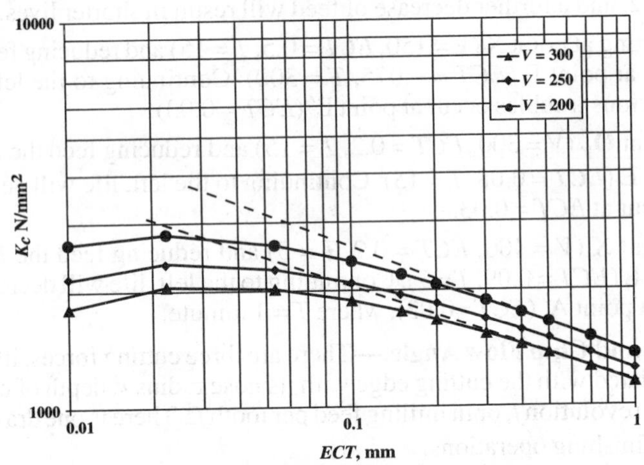

Fig. 9. *Kc* vs. *ECT*, cutting speed plotted

A more accurate relationship is illustrated in Fig. 9, where *Kc* is plotted versus *ECT* at 3 different cutting speeds. In Fig. 9, the two dashed lines represent the aforementioned equation, which each have different slopes, *B*. For the middle value of cutting speed, *Kc* varies with *ECT* from about 1900 to 1300 N/mm^2 when *ECT* increases from 0.1 to 0.7 mm. Generally the speed effect on the magnitude of *Kc* is approximately 5 to 15 percent when using economic speeds.

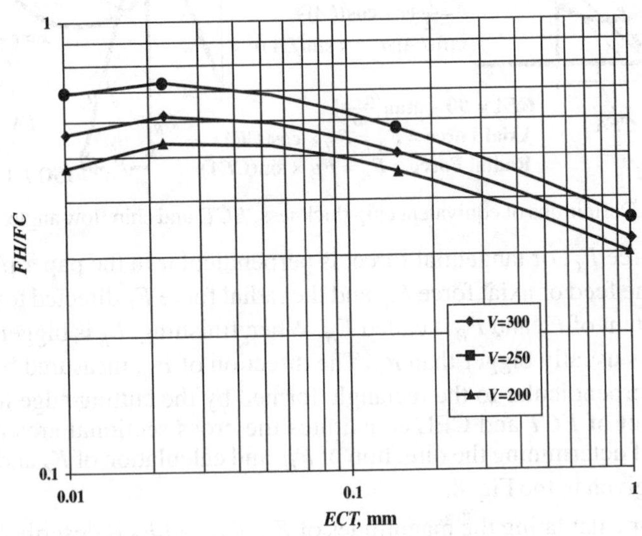

Fig. 10. F_H/F_C vs. *ECT*, cutting speed plotted

Determination of Axial, F_A, and Radial, F_R, Forces: This is done by first determining the resultant force F_H and then calculating F_A and F_R using the Fig. 8 formulas. F_H is derived from the ratio F_H/F_C, which varies with *ECT* and speed in a fashion similar to *Kc*. Fig. 10 shows how this relationship may vary.

As seen in Fig. 10, F_H/F_C is in the range 0.3 to 0.6 when *ECT* varies from 0.1 to 1 mm, and speed varies from 200 to 250 m/min using modern insert designs and grades. Hence, using reasonable large feeds F_H/F_C is around 0.3 – 0.4 and when finishing about 0.5 – 0.6.

Example: Determine F_A and F_R, based on the chip flow angle *CFA* and the cutting force F_C, in turning.

Using a value of $Kc = 1500$ N/mm^2 for roughing, when $ECT = 0.4$, and the cutting edge length $CEL = 5$ mm, first calculate the area $A = 0.4 \times 5 = 2$ mm^2. Then, determine the cutting force $F_C = 2 \times 1500 = 3000$ Newton, and an approximate value of $F_H = 0.5 \times 3000 = 1500$ Newton.

Using a value of $Kc = 1700$ N/mm^2 for finishing, when $ECT = 0.2$, and the cutting edge length $CEL = 2$ mm, calculate the area $A = 0.2 \times 2 = 0.4$ mm^2. The cutting force $F_C = 0.4 \times 1700 = 680$ Newton and an approximate value of $F_H = 0.35 \times 680 = 238$ Newton.

Fig. 8 can be used to estimate *CFA* for rough and finish turning. When the lead angle *LA* is 15 degrees and the nose radius is relatively large, an estimated value of the chip flow angle becomes about 30 degrees when roughing, and about 60 degrees in finishing. Using the formulas for F_A and F_R relative to F_H gives:

Roughing:

$F_A = F_H \times \cos(CFA) = 1500 \times \cos 30 = 1299$ Newton

$F_R = F_H \times \sin(CFA) = 1500 \times \sin 30 = 750$ Newton

Finishing:

$F_A = F_H \times \cos(CFA) = 238 \times \cos 60 = 119$ Newton

$F_R = F_H \times \sin(CFA) = 238 \times \sin 60 = 206$ Newton

The force ratio F_H/F_C also varies with the tool rake angle and increases with negative rakes. In grinding, F_H is much larger than the grinding cutting force F_C; generally F_H/F_C is approximately 2 to 4, because grinding grits have negative rakes of the order –35 to –45 degrees.

Forces and Tool-life.—Forces and tool life are closely linked. The ratio F_H/F_C is of particular interest because of the unique relationship of F_H/F_C with tool-life.

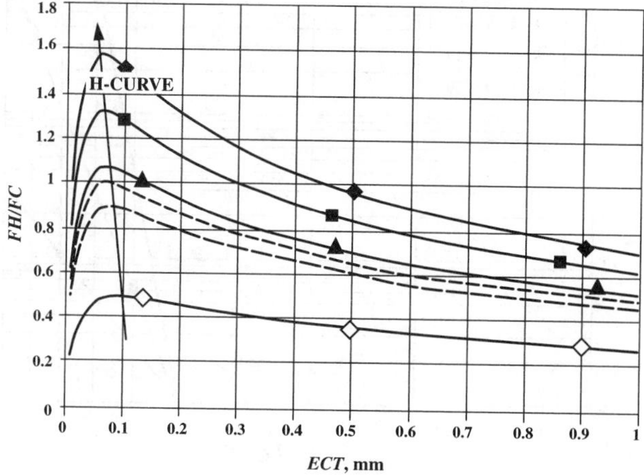

Fig. 11a. F_H/F_C vs. *ECT*

The results of extensive tests at Ford Motor Company are shown in Figs. 11a and 11b, where F_H/F_C and tool-life T are plotted versus ECT at different values of cutting speed V. For any constant speed, tool-life has a maximum at approximately the same values of ECT as has the function F_H/F_C.

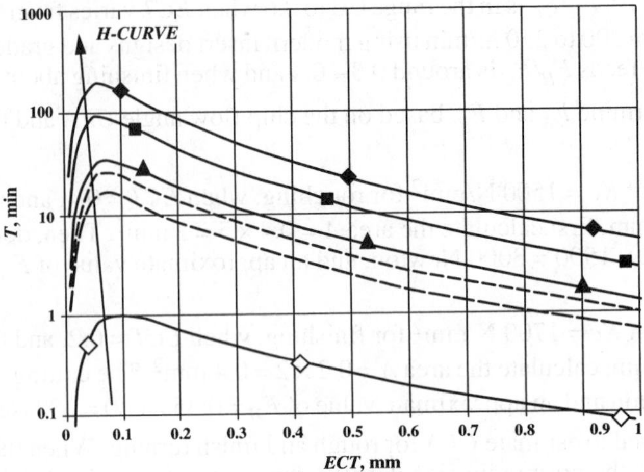

Fig. 11b. Tool-life vs. ECT

The Force Relationship: Similar tests performed elsewhere confirm that the F_H/F_C function can be determined using the 5 tool-life constants (H, K, M, L, N_0) introduced previously, and a new constant (L_F/L).

$$\ln\left(\frac{1}{a}\cdot\frac{F_H}{F_C}\right) = \frac{K - y - \dfrac{(x-H)^2}{4M}}{\dfrac{L_F}{L}(N_0 - Lx)} \tag{5}$$

The constant a depends on the rake angle; in turning a is approximately 0.25 to 0.5 and L_F/L is 10 to 20. F_C attains it maximum values versus ECT along the H-curve, when the tool-life equation has maxima, and the relationships in the three force ratio planes look very similar to the tool-life functions shown in the tool-life planes in Figs. 6a, 6b, and 6c.

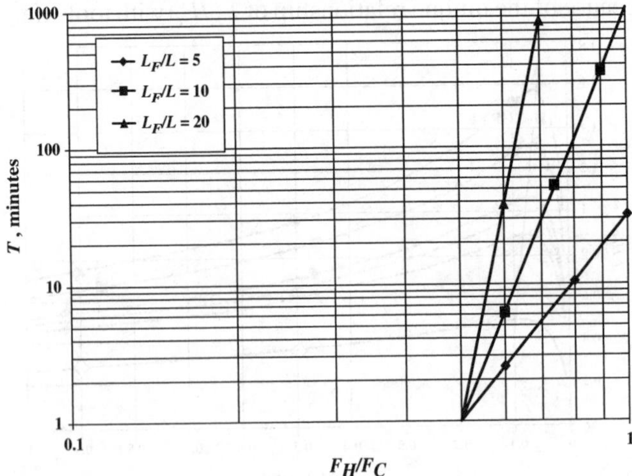

Fig. 12. Tool-life vs. F_H/F_C

Tool-life varies with F_H/F_C with a very simple formula according to Equation (5) as follows:

$$T = \left(\frac{F_H}{aF_C}\right)^{\frac{L_F}{L}}$$

where L is the constant in the tool-life equation, Equation (4a) or (4b), and L_F is the corresponding constant in the force ratio equation, Equation (5). In Fig. 12 this function is plotted for $a = 0.5$ and for $L_F/L = 5$, 10, and 20.

Accurate calculations of aforementioned relationships require elaborate laboratory tests, or better, the design of a special test and follow-up program for parts running in the ordinary production. A software machining program, such as Colding International Corp. *COMP* program can be used to generate the values of all 3 forces, torque and power requirements both for sharp and worn tools

Surface Finish R_a and Tool-life.—It is well known that the surface finish in turning decreases with a bigger tool nose radius and increases with feed; usually it is assumed that R_a increases with the square of the feed per revolution, and decreases inversely with increasing size of the nose radius. This formula, derived from simple geometry, gives rise to great errors. In reality, the relationship is more complicated because the tool geometry must taken into account, and the work material and the cutting conditions also have a significant influence.

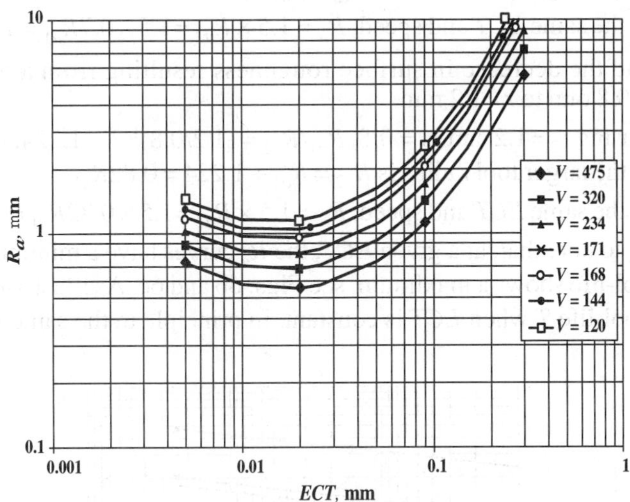

Fig. 13. R_a vs. *ECT*, nose radius *r* constant

Fig. 13 shows surface finish R_a versus *ECT* at various cutting speeds for turning cast iron with carbide tools and a nose radius $r = 1.2$ mm. Increasing the cutting speed leads to a smaller R_a value.

Fig. 14 shows how the finish improves when the tool nose radius, *r*, increases at a constant cutting speed (168 m/min) in cutting nodular cast iron.

In Fig. 15, R_a is plotted versus *ECT* with cutting speed *V* for turning a 4310 steel with carbide tools, for a nose radius $r = 1.2$ mm, illustrating that increasing the speed also leads to a smaller R_a value for steel machining.

A simple rule of thumb for the effect of increasing nose radius *r* on decreasing surface finish R_a, regardless of the ranges of *ECT* or speeds used, albeit within common practical values, is as follows. In finishing,

$$\frac{R_{a1}}{R_{a2}} = \left(\frac{r_2}{r_1}\right)^{0.5} \tag{6}$$

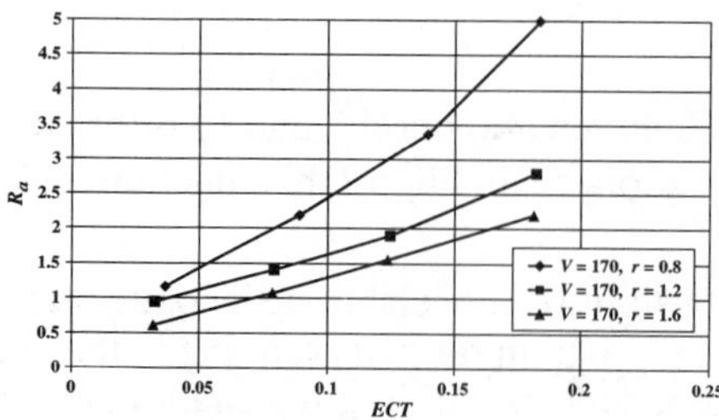

Fig. 14. R_a vs. *ECT*, cutting speed constant, nose radius *r* varies

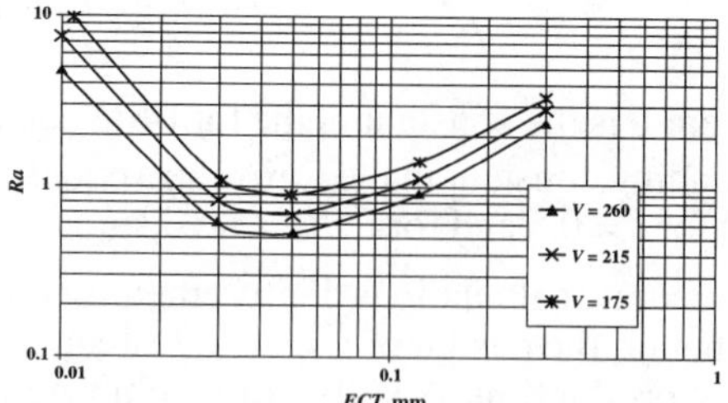

Fig. 15. R_a vs. *ECT*, cutting speed and nose radius *r* constant

In roughing, multiply the finishing values found using Equation (6) by 1.5, thus, $R_{a\,(\mathrm{Rough})} = 1.5 \times R_{a\,(\mathrm{Finish})}$ for each *ECT* and speed.

Example 1: Find the decrease in surface roughness resulting from a tool nose radius change from $r = 0.8$ mm to $r = 1.6$ mm in finishing. Also, find the comparable effect in roughing.

For finishing, using $r_2 = 1.6$ and $r_1 = 0.8$, $R_{a1}/R_{a2} = (1.6/0.8)^{0.5} = 1.414$, thus, the surface roughness using the larger tool radius is $R_{a2} = R_{a1} \div 1.414 = 0.7R_{a1}$

In roughing, at the same *ECT* and speed, $R_a = 1.5 \times R_{a2} = 1.5 \times 0.7R_{a1} = 1.05R_{a1}$

Example 2: Find the decrease in surface roughness resulting from a tool nose radius change from $r = 0.8$ mm to $r = 1.2$ mm

For finishing, using $r_2 = 1.2$ and $r_1 = 0.8$, $R_{a1}/R_{a2} = (1.2/0.8)^{0.5} = 1.224$, thus, the surface roughness using the larger tool radius is $R_{a2} = R_{a1} \div 1.224 = 0.82R_{a1}$

In roughing, at the same *ECT* and speed, $R_a = 1.5 \times R_{a2} = 1.5 \times 0.82R_{a1} = 1.23R_{a1}$

It is interesting to note that, at a given *ECT*, the R_a curves have a minimum, see Figs. 13 and 15, while tool-life shows a maximum, see Figs. 6b and 6c. As illustrated in Fig. 16, R_a increases with tool-life *T* when *ECT* is constant, in principle in the same way as does the force ratio.

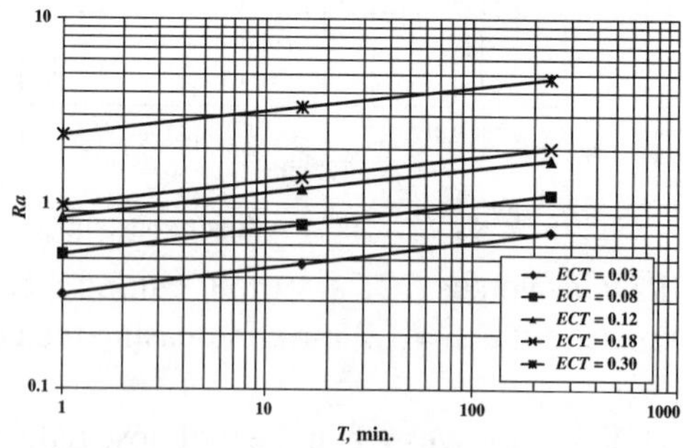

Fig. 16. R_a vs. *T*, holding *ECT* constant

The Surface Finish Relationship: R_a is determined using the same type of mathematical relationship as for tool-life and force calculations:

$$y = K_{Ra} - \frac{x - H_{Ra}{}^2}{4M_{Ra}} - (N_{0Ra} - L_{Ra})\ln(R_a)$$

where K_{RA}, H_{RA}, M_{RA}, N_{ORA}, and L_{RA} are the 5 surface finish constants.

Shape of Tool-life Relationships for Turning, Milling, Drilling and Grinding Operations—Overview.—A summary of the general shapes of tool-life curves (*V–ECT–T* graphs) for the most common machining processes, including grinding, is shown in double logarithmic coordinates in Fig. 17a through Fig. 17h.

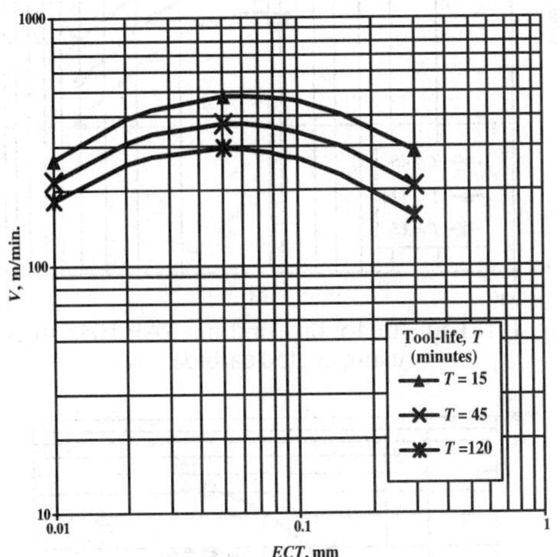

Fig. 17a. Tool-life for turning cast iron using coated carbide

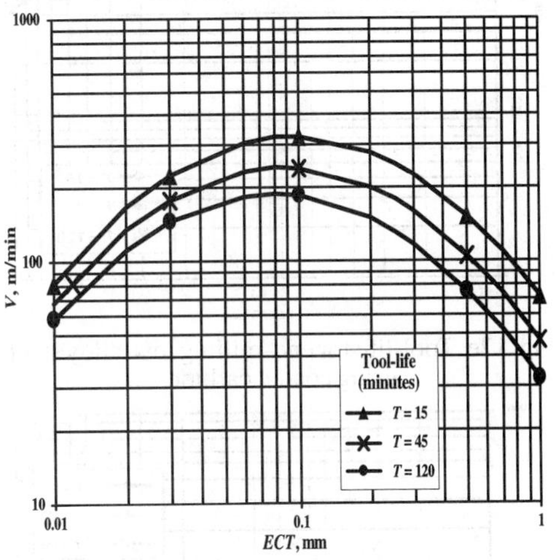

Fig. 17b. Tool-life for turning low-alloy steel using coated carbide

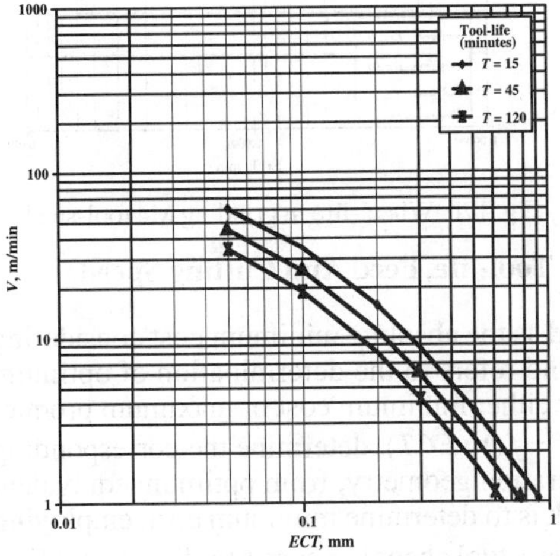

Fig. 17c. Tool-life for end-milling AISI 4140 steel using high-speed steel

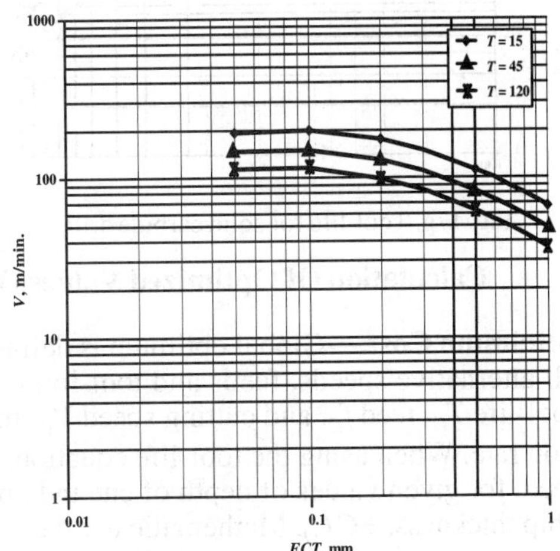

Fig. 17d. Tool-life for end-milling low-allow steel using uncoated carbide

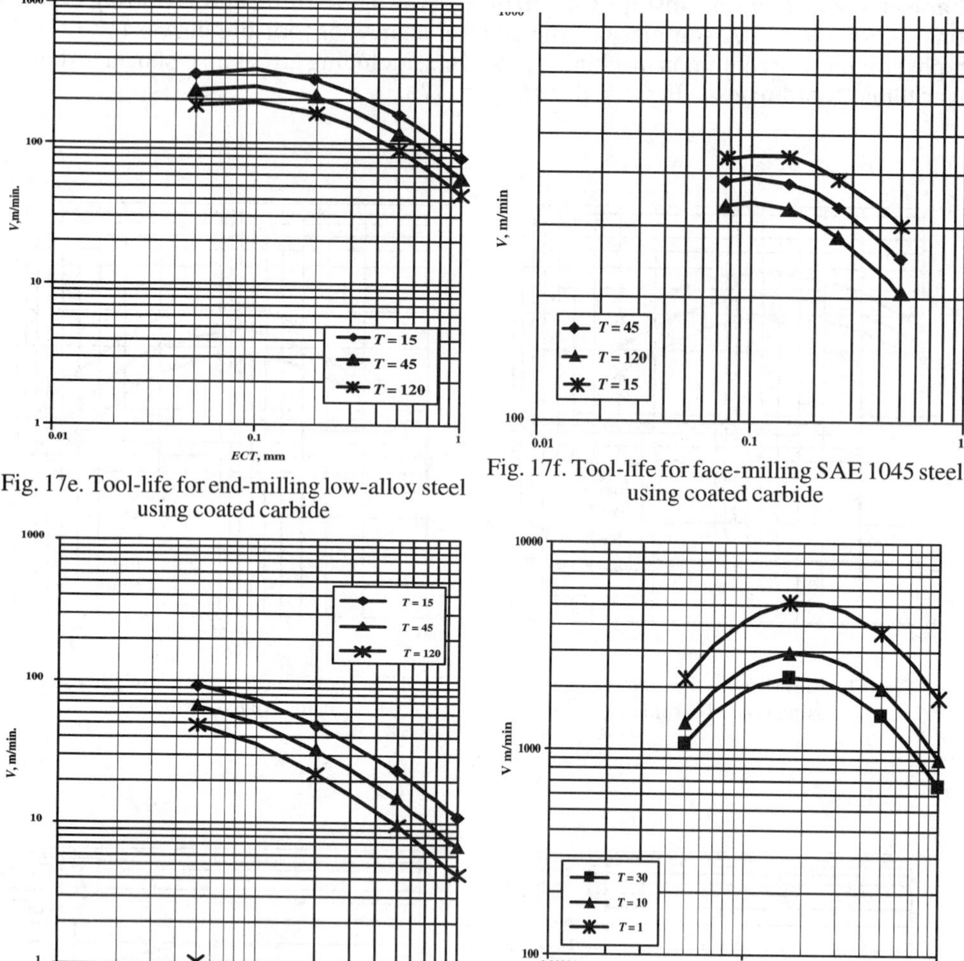

Fig. 17e. Tool-life for end-milling low-alloy steel using coated carbide

Fig. 17f. Tool-life for face-milling SAE 1045 steel using coated carbide

Fig. 17g. Tool-life for solid carbide drill

Fig. 17h. Wheel-life in grinding M4 tool-steel

Calculation Of Optimized Values Of Tool-life, Feed And Cutting Speed

Minimum Cost.—Global optimum is defined as the absolute minimum cost considering all alternative speeds, feeds and tool-lives, and refers to the determination of optimum tool-life T_O, feed f_O, and cutting speed V_O, for either minimum cost or maximum production rate. When using the tool-life equation, $T = f(V, ECT)$, determine the corresponding feed, for given values of depth of cut and operation geometry, from optimum equivalent chip thickness, ECT_O. Mathematically the task is to determine minimum cost, employing the cost function C_{TOT} = cost of machining time + tool changing cost + tooling cost. Minimum cost optima occur along the so-called *G*-curve, identified in Fig. 6c.

Another important factor when optimizing cutting conditions involves choosing the proper cost values for cost per edge C_E, replacement time per edge T_{RPL}, and not least, the hourly rate H_R that should be applied. H_R is defined as the portion of the hourly shop rate that is applied to the operations and machines in question. If optimizing all operations in the portion of the shop for which H_R is calculated, use the full rate; if only one machine is involved, apply a lower rate, as only a portion of the general overhead rate should be used, otherwise the optimum, and anticipated savings, are erroneous.

Production Rate.—The production rate is defined as the cutting time or the metal removal rate, corrected for the time required for tool changes, but neglecting the cost of tools.

The result of optimizing production rate is a shorter tool-life, higher cutting speed, and a higher feed compared to minimum cost optimization, and the tooling cost is considerably higher. Production rates optima also occur along the *G*-curve.

The Cost Function.—There are a number of ways the total machining cost C_{TOT} can be plotted, for example, versus feed, *ECT*, tool-life, cutting speed or other parameter. In Fig. 18a, cost for a face milling operation is plotted versus cutting time, holding feed constant, and using a range of tool-lives, *T*, varying from 1 to 240 minutes.

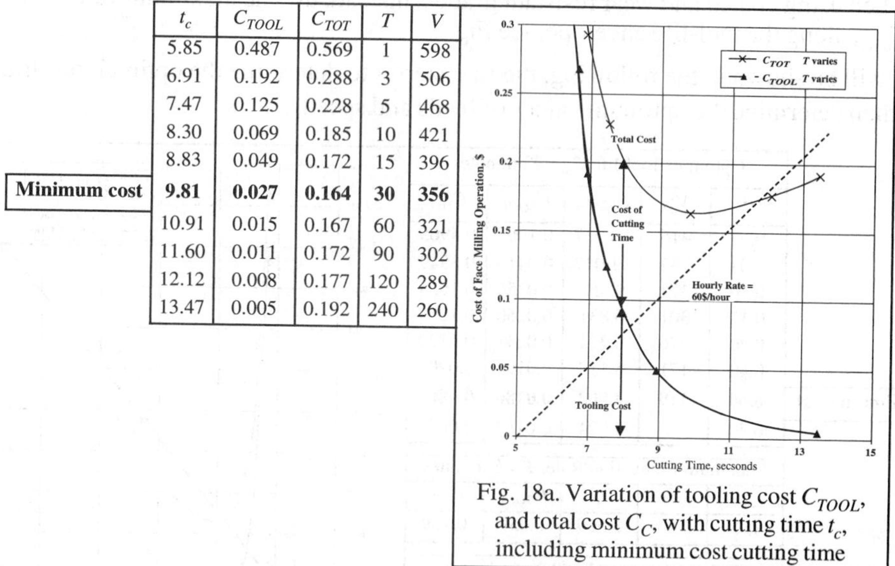

	t_c	C_{TOOL}	C_{TOT}	T	V
	5.85	0.487	0.569	1	598
	6.91	0.192	0.288	3	506
	7.47	0.125	0.228	5	468
	8.30	0.069	0.185	10	421
	8.83	0.049	0.172	15	396
Minimum cost	**9.81**	**0.027**	**0.164**	**30**	**356**
	10.91	0.015	0.167	60	321
	11.60	0.011	0.172	90	302
	12.12	0.008	0.177	120	289
	13.47	0.005	0.192	240	260

Fig. 18a. Variation of tooling cost C_{TOOL}, and total cost C_C, with cutting time t_c, including minimum cost cutting time

The tabulated values show the corresponding cutting speeds determined from the tool-life equation, and the influence of tooling on total cost. Tooling cost, C_{TOOL} = sum of tool cost + cost of replacing worn tools, decreases the longer the cutting time, while the total cost, C_{TOT}, has a minimum at around 10 seconds of cutting time. The dashed line in the graph represents the cost of machining time: the product of hourly rate H_R, and the cutting time t_c divided by 60. The slope of the line defines the value of H_R.

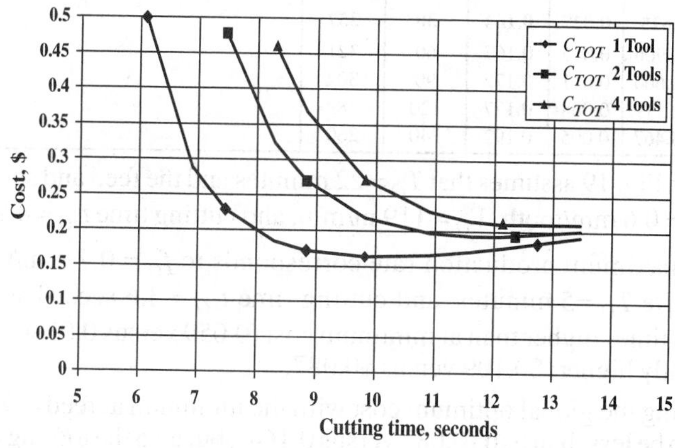

Fig. 18b. Total cost vs. cutting time for simultaneously cutting with 1, 2, and 4 tools

The cutting time for minimum cost varies with the ratio of tooling cost and H_R. Minimum cost moves towards a longer cutting time (longer tool-life) when either the price of the tooling increases, or when several tools cut simultaneously on the same part. In Fig. 18b, this is exemplified by running 2 and 4 cutters simultaneously on the same work piece, at the same feed and depth of cut, and with a similar tool as in Fig. 18a. As the tooling cost goes up 2 and 4 times, respectively, and H_R is the same, the total costs curves move up, but also moves to the right, as do the points of minimum cost and optimal cutting times. This means that going somewhat slower, with more simultaneously cutting tools, is advantageous.

Global Optimum.—Usually, global optimum occurs for large values of feed, heavy roughing, and in many cases the cutting edge will break trying to apply the large feeds required. Therefore, true optima cannot generally be achieved when roughing, in particular when using coated and wear resistant grades; instead, use the maximum values of feed, ECT_{max}, along the tool-life envelope, see Fig. 7.

As will be shown in the following, the first step is to determine the optimal tool-life T_O, and then determine the optimum values of feeds and speeds.

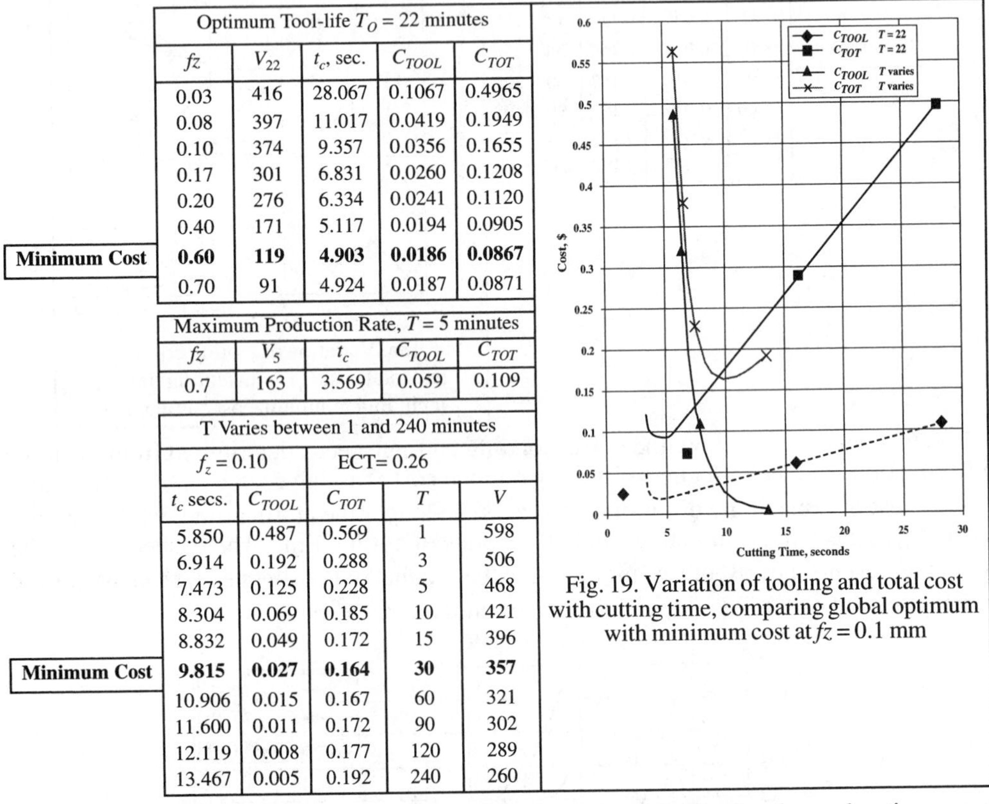

	Optimum Tool-life T_O = 22 minutes				
	fz	V_{22}	t_c, sec.	C_{TOOL}	C_{TOT}
	0.03	416	28.067	0.1067	0.4965
	0.08	397	11.017	0.0419	0.1949
	0.10	374	9.357	0.0356	0.1655
	0.17	301	6.831	0.0260	0.1208
	0.20	276	6.334	0.0241	0.1120
	0.40	171	5.117	0.0194	0.0905
Minimum Cost	**0.60**	**119**	**4.903**	**0.0186**	**0.0867**
	0.70	91	4.924	0.0187	0.0871

Maximum Production Rate, T = 5 minutes				
fz	V_5	t_c	C_{TOOL}	C_{TOT}
0.7	163	3.569	0.059	0.109

	T Varies between 1 and 240 minutes				
	f_z = 0.10		ECT= 0.26		
	t_c secs.	C_{TOOL}	C_{TOT}	T	V
	5.850	0.487	0.569	1	598
	6.914	0.192	0.288	3	506
	7.473	0.125	0.228	5	468
	8.304	0.069	0.185	10	421
	8.832	0.049	0.172	15	396
Minimum Cost	**9.815**	**0.027**	**0.164**	**30**	**357**
	10.906	0.015	0.167	60	321
	11.600	0.011	0.172	90	302
	12.119	0.008	0.177	120	289
	13.467	0.005	0.192	240	260

Fig. 19. Variation of tooling and total cost with cutting time, comparing global optimum with minimum cost at fz = 0.1 mm

The example in Fig. 19 assumes that T_O = 22 minutes and the feed and speed optima were calculated as f_O = 0.6 mm/tooth, V_O = 119 m/min, and cutting time t_{cO} = 4.9 secs.

The point of maximum production rate corresponds to f_O = 0.7 mm/tooth, V_O = 163 m/min, at tool-life T_O =5 minutes, and cutting time t_{cO} = 3.6 secs. The tooling cost is approximately 3 times higher than at minimum cost (0.059 versus 0.0186), while the piece cost is only slightly higher: $0.109 versus $0.087.

When comparing the global optimum cost with the minimum at feed = 0.1 mm/tooth the graph shows it to be less than half (0.087 versus 0.164), but also the tooling cost is about 1/3 lower (0.0186 versus 0.027). The reason why tooling cost is lower depends on the tooling cost term $t_c \times C_E/T$ (see *Calculation of Cost of Cutting and Grinding Operations* on page 1078). In this example, cutting times t_c= 4.9 and 9.81 seconds, at T = 22 and 30 minutes respectively, and the ratios are proportional to 4.9/22 = 0.222 and 9.81/30 = 0.327 respectively.

The portions of the total cost curve for shorter cutting times than at minimum corresponds to using feeds and speeds right of the *G*-curve, and those on the other side are left of this curve.

Optimization Models, Economic Tool-life when Feed is Constant.—Usually, optimization is performed versus the parameters tool-life and cutting speed, keeping feed at a constant value. The cost of cutting as function of cutting time is a straight line with the slope = H_R = hourly rate. This cost is independent of the values of tool change and tooling. Adding the cost of tool change and tooling, gives the variation of total cutting cost which shows a minimum with cutting time that corresponds to an economic tool-life, T_E. Economic tool-life represents a local optima (minimum cost) at a given constant value of feed, feed/tooth, or *ECT*.

Using the Taylor Equation: $V \times T = C$ and differentiating C_{TOT} with respect to T yields:

Economic tool-life:

$T_E = T_V \times (1/n - 1)$, minutes

Economic cutting speed:

$V_E = C/T_E^n$, m/min, or sfm

In these equations, n and C are constants in the Taylor equation for the given value of feed. Values of Taylor slopes, n, are estimated using the speed and feed Tables 1 through 23 starting on page 996 and handbook Table 5b on page 1004 for turning, and Table 15e on page 1028 for milling and drilling; and T_V is the equivalent tooling-cost time. $T_V = T_{RPL} + 60 \times C_E \div H_R$, minutes, where T_{RPL} = time for replacing a worn insert, or a set of inserts in a milling cutter or inserted drill, or a twist drill, reamer, thread chaser, or tap. T_V is described in detail, later; C_E = cost per edge, or set of edges, or cost per regrind including amortized price of tool; and H_R = hourly shop rate, or that rate that is impacted by the changes of cutting conditions .

In two dimensions, Fig. 20a shows how economic tool-life varies with feed per tooth. In this figure, the equivalent tooling-cost time T_V is constant, however the Taylor constant n varies with the feed per tooth.

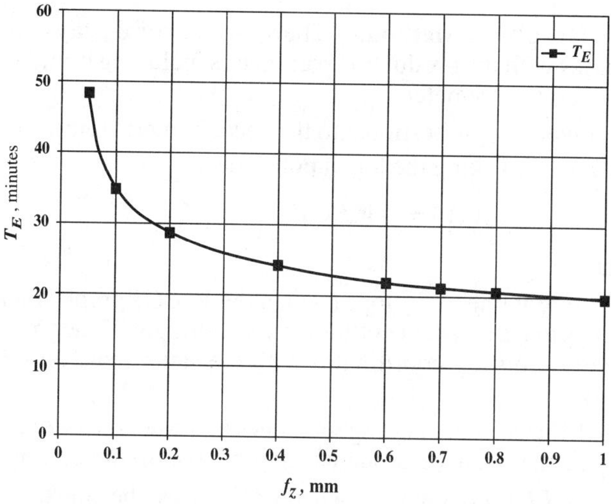

Fig. 20a. Economic tool-life, T_E vs. feed per tooth, fz

Economic tool-life increases with greater values of T_V, either when T_{RPL} is longer, or when cost per edge C_E is larger for constant H_R, or when H_R is smaller and T_{RPL} and C_E are unchanged. For example, when using an expensive machine (which makes H_R bigger) the value of T_V gets smaller, as does the economic tool-life, $T_E = T_V \times (1/n - 1)$. Reducing T_E results in an increase in the economic cutting speed, V_E. This means raising the cutting speed, and illustrates the importance, in an expensive system, of utilizing the equipment better by using more aggressive machining data.

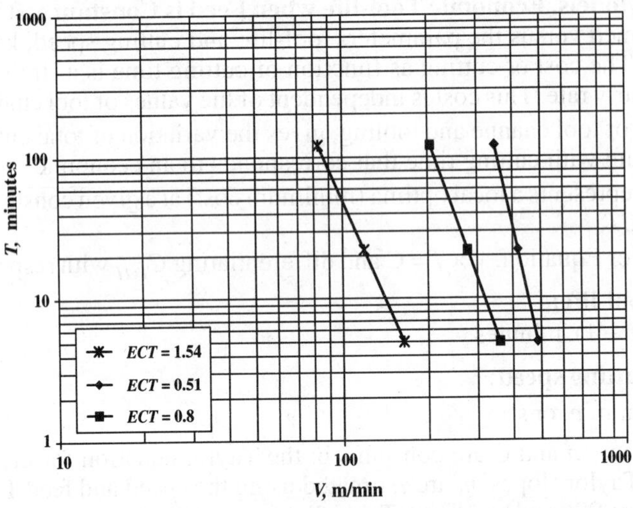

Fig. 20b. Tool-life vs. cutting speed, constant *ECT*

As shown in Fig. 20a for a face milling operation, economic tool-life T_E varies considerably with feed/tooth f_z, in spite of the fact that the Taylor lines have only slightly different slopes (ECT= 0.51, 0.6, 1.54), as shown in Fig. 20b. The calculation is based on the following cost data: T_V= 6, hourly shop rate H_R = \$60/hour, cutter diameter D = 125 mm with number of teeth z = 10, and radial depth of cut ar = 40 mm.

The conclusion relating to the determination of economic tool-life is that both hourly rate H_R and slope n must be evaluated with reasonable accuracy in order to arrive at good values. However, the method shown will aid in setting the trend for general machining economics evaluations.

Global Optimum, Graphical Method.—There are several ways to demonstrate in graphs how cost varies with the production parameters including optimal conditions. In all cases, tool-life is a crucial parameter.

Cutting time t_c is inversely proportional to the specific metal removal rate, $SMRR = V \times ECT$, thus, $1/t_c = V \times ECT$. Taking the log of both sides,

$$\ln V = -\ln ECT - \ln t_c + C \qquad (7)$$

where C is a constant.

Equation (7) is a straight line with slope (– 1) in the *V–ECT* graph when plotted in a log-log graph. This means that a constant cutting time is a straight 45-degree line in the *V–ECT* graph, when plotted in log-log coordinates with the same scale on both axis (a square graph).

The points at which the constant cutting time lines (at 45 degrees slope) are tangent to the tool-life curves define the *G*-curve, along which global optimum cutting occurs.

Note: If the ratio *a*/*CEL* is not constant when *ECT* varies, the constant cutting time lines are not straight, but the cutting time deviation is quite small in most cases.

In the *V–ECT* graph, Fig. 21, 45-degree lines have been drawn tangent to each tool-life curve: T=1, 5, 15, 30, 60, 100 and 300 minutes. The tangential points define the *G*-curve, and the 45-degree lines represent different constant cutting times: 1, 2, 3, 10 minutes, etc. Following one of these lines and noting the intersection points with the tool-life curves T = 1, 5, etc., many different speed and feed combinations can be found that will give the same cutting time. As tool-life gets longer (tooling cost is reduced), *ECT* (feed) increases but the cutting speed has to be reduced.

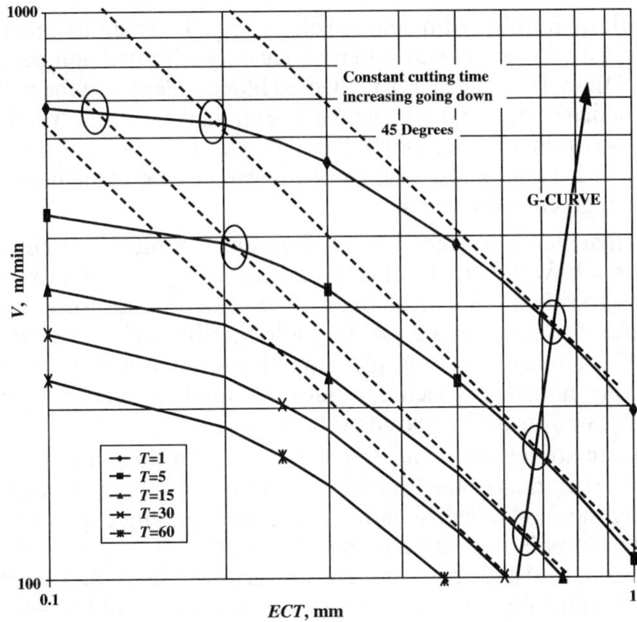

Fig. 21. Constant cutting time in the *V-ECT* plane, tool-life constant

Global Optimum, Mathematical Method.—Global optimization is the search for extremum of C_{TOT} for the three parameters: *T*, *ECT*, and *V*. The results, in terms of the tool-life equation constants, are:

Optimum tool-life:

$$T_O = T_V \times \left(\frac{1}{n_O} - 1\right)$$

$$n_O = 2M \times (L \times \ln T_O)^2 + 1 - N_0 + L \times (2M + H)$$

where n_O = slope at optimum *ECT*.

The same approach is used when searching for maximum production rate, but without the term containing tooling cost.

Optimum cutting speed:

$$V_O = e^{-M + K + (H \times L - N_0) \times \ln T_O + M \times L^2 \times (\ln T_O)^2}$$

Optimum *ECT*:

$$ECT_O = e^{H + 2M \times (L \times \ln(T_O) + 1)}$$

Global optimum is not reached when face milling for very large feeds, and C_{TOT} decreases continually with increasing feed/tooth, but can be reached for a cutter with many teeth, say 20 to 30. In end milling, global optimum can often be achieved for big feeds and for 3 to 8 teeth.

Determination Of Machine Settings And Calculation Of Costs

Based on the rules and knowledge presented in Chapters 1 and 2, this chapter demonstrates, with examples, how machining times and costs are calculated.

Additional formulas are given, and the speed and feed tables given in *SPEED AND FEED TABLES* starting on page 991 should be used. Finally the selection of feeds, speeds and tool-lives for optimized conditions are described with examples related to turning, end milling, and face milling.

There are an infinite number of machine settings available in the machine tool power train producing widely different results. In practice only a limited number of available settings are utilized. Often, feed is generally selected independently of the material being cut, however, the influence of material is critical in the choice of cutting speed. The tool-life is normally not known or directly determined, but the number of pieces produced before the change of worn tools is better known, and tool-life can be calculated using the formula for piece cutting time t_c given in this chapter.

It is well known that increasing feeds or speeds reduces the number of pieces cut between tool changes, but not how big are the changes in the basic parameter tool-life. Therefore, there is a tendency to select "safe" data in order to get a long tool-life. Another common practice is to search for a tool grade yielding a longer life using the current speeds and feeds, or a 10–20% increase in cutting speed while maintaining the current tool-life. The reason for this old-fashioned approach is the lack of knowledge about the opportunities the metal cutting process offers for increased productivity.

For example, when somebody wants to calculate the cutting time, he/she can select a value of the feed rate (product of feed and rpm), and easily find the cutting time by dividing cutting distance by the feed rate. The number of pieces obtained out of a tool is a guesswork, however. This problem is very common and usually the engineers find desired tool-lives after a number of trial and error runs using a variety of feeds and speeds. If the user is not well familiar with the material cut, the tool-life obtained could be any number of seconds or minutes, or the cutting edge might break.

There are an infinite number of feeds and speeds, giving the same feed rate, producing equal cutting time. The same cutting time per piece t_c is obtained independent of the selection of feed/rev f and cutting speed V, (or rpm), as long as the feed rate F_R remains the same: $F_R = f_1 \times \text{rpm}_1 = f_2 \times \text{rpm}_2 = f_3 \times \text{rpm}_3 \ldots$, etc. However, the number of parts before tool change N_{ch} will vary considerably including the tooling cost c_{tool} and the total cutting cost c_{tot}.

The dilemma confronting the machining-tool engineer or the process planner is how to set feeds and speeds for either desired cycle time, or number of parts between tool changes, while balancing the process versus other operations or balancing the total times in one cell with another. These problems are addressed in this section.

Nomenclature

f = feed/rev or tooth, mm f_E = economic feed f_O = optimum feed
T = tool-life, minutes T_E = economic tool-life T_O = optimum tool-life
V = cutting speed, m/min V_E = economic cutting speed V_O = optimum cutting speed, m/min

Similarly, economic and optimum values of:

c_{tool} = piece cost of tooling, \$ C_{TOOL} = cost of tooling per batch, \$
c_{tot} = piece total cost of cutting, \$ C_{TOT} = total cost of cutting per batch, \$
F_R = feed rate measured in the feeding direction, mm/rev
N = batch size
N_{ch} = number of parts before tool change
t_c = piece cutting time, minutes T_C = cutting time per batch, minutes
t_{cyc} = piece cycle time, minutes T_{CYC} = cycle time before tool change, minutes

t_i = idle time (tool "air" motions during cycle), minutes
z = cutter number of teeth

The following variables are used for calculating the per batch cost of cutting:

C_C = cost of cutting time per batch, $
C_{CH} = cost of tool changes per batch, $
C_E = cost per edge, for replacing or regrinding, $
H_R = hourly rate, $
T_V = equivalent tooling-cost time, minutes
T_{RPL} = time for replacing worn edge(s), or tool for regrinding, minutes

Note: In the list above, when two variables use the same name, one in capital letters and one lower case, T_C and t_c for example, the variable name in capital letters refers to batch processing and lowercase letters to per piece processing, such as $T_C = N_{ch} \times t_c$, $C_{TOT} = N_{ch} \times c_{tot}$, etc.

Formulas Valid For All Operation Types Including Grinding

Calculation of Cutting Time and Feed Rate

Feed Rate:

$F_R = f \times \text{rpm}$ (mm/min), where f is the feed in mm/rev along the feeding direction, rpm is defined in terms of work piece or cutter diameter D in mm, and cutting speed V in m/min, as follows:

$$\text{rpm} = \frac{1000V}{\pi D} = \frac{318V}{D}$$

Cutting time per piece:

Note: Constant cutting time is a straight 45-degree line in the *V–ECT* graph, along which tool-life varies considerably, as is shown in Chapter 2.

$$t_c = \frac{Dist}{F_R} = \frac{Dist}{f \times \text{rpm}} = \frac{Dist \times \pi D}{1000V \times f}$$

where the units of distance cut *Dist*, diameter *D*, and feed *f* are mm, and *V* is in m/min.

In terms of *ECT*, cutting time per piece, t_c, is as follows:

$$t_c = \frac{Dist \times \pi D}{1000V} \times \frac{a}{CEL \times ECT}$$

where a = depth of cut, because feed × cross sectional chip area $= f \times a = CEL \times ECT$.

Example 3, Cutting Time: Given $Dist$ = 105 mm, D = 100 mm, f = 0.3 mm, V = 300 m/min, rpm = 700, F_R = 210 mm/min, find the cutting time.

Cutting time = t_c = 105 × 3.1416 × 100 ÷ (1000 × 300 × 0.3) = 0.366 minutes = 22 seconds

Scheduling of Tool Changes

Number of parts before tool change:

$N_{ch} = T \div t_c$

Cycle time before tool change:

$T_{CYC} = N_{ch} \times (t_c + t_i)$, where $t_{cyc} = t_c + t_i$, where t_c = cutting time per piece, t_i = idle time per piece

Tool-life:

$T = N_{ch} \times t_c$

Example 4: Given tool-life T = 90 minutes, cutting time t_c = 3 minutes, and idle time t_i = 3 minutes, find the number of parts produced before a tool change is required and the time until a tool change is required.

Number of parts before tool change = N_{ch} = 90/3 = 30 parts.

Cycle time before tool change = T_{CYC} = 30 × (3 + 3) = 180 minutes

Example 5: Given cutting time, t_c = 1 minute, idle time t_i = 1 minute, N_{ch} = 100 parts, calculate the tool-life T required to complete the job without a tool change, and the cycle time before a tool change is required.

Tool-life = $T = N_{ch} \times t_c$ = 100 × 1 = 100 minutes.

Cycle time before tool change = T_{CYC} = 100 × (1 + 1) = 200 minutes.

Calculation of Cost of Cutting and Grinding Operations.—When machining data varies, the cost of cutting, tool changing, and tooling will change, but the costs of idle and slack time are considered constant.

Cost of Cutting per Batch:

$C_C = H_R \times T_C/60$

T_C = cutting time per batch = (number of parts) × t_c, minutes, or when determining time for tool change $T_{Cch} = N_{ch} \times t_c$ minutes = cutting time before tool change.

t_c = Cutting time/part, minutes

H_R = Hourly Rate

Cost of Tool Changes per Batch:

$$C_{CH} = \frac{H_R}{60} \times T_C \times \frac{T_{RPL}}{T} \qquad \frac{\$}{min} \cdot min = \$$$

where T = tool-life, minutes, and T_{RPL} = time for replacing a worn edge(s), or tool for regrinding, minutes

Cost of Tooling per Batch:

Including cutting tools and holders, but without tool changing costs,

$$C_{TOOL} = \frac{H_R}{60} \times T_C \times \frac{\frac{60 C_E}{H_R}}{T} \qquad \frac{\$}{min} \cdot min \cdot \frac{\frac{min}{hr} \cdot \$ \cdot \frac{hr}{\$}}{min} = \$$$

Cost of Tooling + Tool Changes per Batch:

Including cutting tools, holders, and tool changing costs,

$$(C_{TOOL} + C_{CH}) = \frac{H_R}{60} \times T_C \times \frac{T_{RPL} + \frac{60 C_E}{H_R}}{T}$$

Total Cost of Cutting per Batch:

$$C_{TOT} = \frac{H_R}{60} \times T_C \left(1 + \frac{T_{RPL} + \frac{60 C_E}{H_R}}{T} \right)$$

Equivalent Tooling-cost Time, T_V:

The two previous expressions can be simplified by using $T_V = T_{RPL} + \frac{60 C_E}{H_R}$

thus:

$$(C_{TOOL} + C_{CH}) = \frac{H_R}{60} \times T_C \times \frac{T_V}{T}$$

$$C_{TOT} = \frac{H_R}{60} \times T_C\left(1 + \frac{T_V}{T}\right)$$

C_E = cost per edge(s) is determined using two alternate formulas, depending on whether tools are reground or inserts are replaced:

Cost per Edge, Tools for Regrinding

$$C_E = \frac{\text{cost of tool} + (\text{number of regrinds} \times \text{cost/regrind})}{1 + \text{number of regrinds}}$$

Cost per Edge, Tools with Inserts:

$$C_E = \frac{\text{cost of insert(s)}}{\text{number of edges per insert}} + \frac{\text{cost of cutter body}}{\text{cutter body life in number of edges}}$$

Note: In practice allow for insert failures by multiplying the insert cost by 4/3, that is, assuming only 3 out of 4 edges can be effectively used.

Example 6, Cost per Edge–Tools for Regrinding: Use the data in the table below to calculate the cost per edge(s) C_E, and the equivalent tooling-cost time T_V, for a drill.

Time for cutter replacement T_{RPL}, minute	Cutter Price, $	Cost per regrind, $	Number of regrinds	Hourly shop rate, $	Batch size	Taylor slope, n	Economic cutting time, t_{cE} minute
1	40	6	5	50	1000	0.25	1.5

Using the cost per edge formula for reground tools, $C_E = (40 + 5 \times 6) \div (1 + 5) = \$6.80$

When the hourly rate is \$50/hr, $T_V = T_{RPL} + \frac{60 C_E}{H_R} = 1 + \frac{60(6.8)}{50} = 9.16\text{minutes}$

Calculate economic tool-life using $T_E = T_V \times \left(\frac{1}{n} - 1\right)$ thus, $T_E = 9.17 \times (1/0.25 - 1) =$ $9.16 \times 3 = 27.48$ minutes.

Having determined, elsewhere, the economic cutting time per piece to be $t_{cE} = 1.5$ minutes, for a batch size = 1000 calculate:

Cost of Tooling + Tool Change per Batch:

$$(C_{TOOL} + C_{CH}) = \frac{H_R}{60} \times T_C \times \frac{T_V}{T} = \frac{50}{60} \times 1000 \times 1.5 \times \frac{9.16}{27.48} = \$\ 417$$

Total Cost of Cutting per Batch:

$$C_{TOT} = \frac{H_R}{60} \times T_C\left(1 + \frac{T_V}{T}\right) = \frac{50}{60} \times 1000 \times 1.5 \times \left(1 + \frac{9.16}{27.48}\right) = \$\ 1617$$

Example 7, Cost per Edge–Tools with Inserts: Use data from the table below to calculate the cost of tooling and tool changes, and the total cost of cutting.

For face milling, multiply insert price by safety factor 4/3 then calculate the cost per edge: $C_E = 10 \times (5/3) \times (4/3) + 750/500 = 23.72$ per set of edges

When the hourly rate is \$50, equivalent tooling-cost time is $T_V = 2 + 23.72 \times 60/50 =$ 30.466 minutes (first line in table below). The economic tool-life for Taylor slope $n =$ 0.333 would be $T_E = 30.466 \times (1/0.333 - 1) = 30.466 \times 2 = 61$ minutes.

When the hourly rate is \$25, equivalent tooling-cost time is $T_V = 2 + 23.72 \times 60/25 =$ 58.928 minutes (second line in table below). The economic tool-life for Taylor slope $n =$ 0.333 would be $T_E = 58.928 \times (1/0.333 - 1) = 58.928 \times 2 = 118$ minutes.

Time for replacement of inserts T_{RPL}, minutes	Number of inserts	Price per insert	Edges per insert	Cutter Price	Edges per cutter	Cost per set of edges, C_E	Hourly shop rate	T_V minutes
Face mill								
2	10	5	3	750	500	23.72	50	30.466
2	10	5	3	750	500	23.72	25	58.928
End mill								
1	3	6	2	75	200	4.375	50	6.25
Turning								
1	1	5	3	50	100	2.72	30	6.44

With above data for the face mill, and after having determined the economic cutting time as $t_{cE} = 1.5$ minutes, calculate for a batch size = 1000 and \$50 per hour rate:

Cost of Tooling + Tool Change per Batch:

$$(C_{TOOL} + C_{CH}) = \frac{H_R}{60} \times T_C \times \frac{T_V}{T} = \frac{50}{60} \times 1000 \times 1.5 \times \frac{30.466}{61} = \$\ 624$$

Total Cost of Cutting per Batch:

$$C_{TOT} = \frac{H_R}{60} \times T_C\left(1 + \frac{T_V}{T}\right) = \frac{50}{60} \times 1000 \times 1.5 \times \left(1 + \frac{30.466}{61}\right) = \$\ 1874$$

Similarly, at the \$25/hour shop rate, $(C_{TOOL} + C_{CH})$ and C_{TOT} are \$312 and \$937, respectively.

Example 8, Turning: Production parts were run in the shop at feed/rev = 0.25 mm. One series was run with speed $V_1 = 200$ m/min and tool-life was $T_1 = 45$ minutes. Another was run with speed $V_2 = 263$ m/min and tool-life was $T_2 = 15$ minutes. Given idle time $t_i = 1$ minute, cutting distance $Dist$ =1000 mm, work diameter $D = 50$ mm.

First, calculate Taylor slope, n, using Taylor's equation $V_1 \times T_1^{n} = V_2 \times T_2^{n}$, as follows:

$$n = \ln\frac{V_1}{V_2} \div \ln\frac{T_2}{T_1} = \ln\frac{200}{263} \div \ln\frac{15}{45} = 0.25$$

Economic tool-life T_E is next calculated using the equivalent tooling-cost time T_V, as described previously. Assuming a calculated value of $T_V = 4$ minutes, then T_E can be calculated from

$$T_E = T_V \times \left(\frac{1}{n} - 1\right) = 4 \times \left(\frac{1}{0.25} - 1\right) = 12 \text{ minutes}$$

Economic cutting speed, V_E can be found using Taylor's equation again, this time using the economic tool-life, as follows,

$$V_{E1} \times (T_E)^n = V_2 \times (T_2)^n$$

$$V_{E1} = V_2 \times \left(\frac{T_2}{T_E}\right)^n = 263 \times \left(\frac{15}{12}\right)^{0.25} = 278 \text{ m/min}$$

Using the process data, the remaining economic parameters can be calculated as follows:

Economic spindle rpm, $\text{rpm}_E = (1000V_E)/(\pi D) = (1000 \times 278)/(3.1416 \times 50) = 1770$ rpm

Economic feed rate, $F_{RE} = f \times \text{rpm}_E = 0.25 \times 1770 = 443$ mm/min

Economic cutting time, $t_{cE} = Dist/F_{RE} = 1000/443 = 2.259$ minutes

Economic number of parts before tool change, $N_{chE} = T_E \div t_{cE} = 12 \div 2.259 = 5.31$ parts

Economic cycle time before tool change, $T_{CYCE} = N_{chE} \times (t_c + t_i) = 5.31 \times (2.259 + 1) =$ 17.3 minutes.

Variation Of Tooling And Total Cost With The Selection Of Feeds And Speeds

It is a well-known fact that tool-life is reduced when either feed or cutting speed is increased. When a higher feed/rev is selected, the cutting speed must be decreased in order to maintain tool-life. However, a higher feed rate (feed rate = feed/rev × rpm, mm/min) can result in a longer tool-life if proper cutting data are applied. Optimized cutting data require accurate machinability databases and a computer program to analyze the options. Reasonably accurate optimized results can be obtained by selecting a large feed/rev or tooth, and then calculating the economic tool-life T_E. Because the cost versus feed or *ECT* curve is shallow around the true minimum point, i.e., the global optimum, the error in applying a large feed is small compared with the exact solution.

Once a feed has been determined, the economic cutting speed V_E can be found by calculating the Taylor slope, and the time/cost calculations can be completed using the formulas described in last section.

The remainder of this section contains examples useful for demonstrating the required procedures. Global optimum may or may not be reached, and tooling cost may or may not be reduced, compared to currently used data. However, the following examples prove that significant time and cost reductions are achievable in today's industry.

Note: Starting values of reasonable feeds in mm/rev can be found in the Handbook speed and feed tables, see *Principal Feeds and Speeds Tables* on page 991, by using the f_{avg} values converted to mm as follows: feed (mm/rev) = feed (inch/rev) × 25.4 (mm/inch), thus 0.001 inch/rev = 0.001× 25.4 = 0.0254 mm/rev. When using speed and feed Tables 1 through 23, where feed values are given in thousandths of inch per revolution, simply multiply the given feed by 25.4/1000 = 0.0254, thus feed (mm/rev) = feed (0.001 inch/rev) × 0.0254 (mm/ 0.001inch).

Example 9, Converting Handbook Feed Values From Inches to Millimeters: Handbook tables give feed values f_{opt} and f_{avg} for 4140 steel as 17 and 8 × (0.001 inch/rev) = 0.017 and 0.009 inch/rev, respectively. Convert the given feeds to mm/rev.

feed = 0.017 × 25.4 = 17 × 0.0254 = 0.4318 mm/rev

feed = 0.008 × 25.4 = 9 × 0.0254 = 0.2032 mm/rev

Example 10, Using Handbook Tables to Find the Taylor Slope and Constant: Calculate the Taylor slope and constant, using cutting speed data for 4140 steel in Table 1 starting on page 996, and for ASTM Class 20 grey cast iron using data from Table 4a on page 1002, as follows:

For the 175–250 Brinell hardness range, and the hard tool grade,

$$n = \frac{\ln(V_1/V_2)}{\ln(T_2/T_1)} = \frac{\ln(525/705)}{\ln(15/45)} = 0.27 \qquad C = V_1 \times (T_1)^n = 1467$$

For the 175–250 Brinell hardness range, and the tough tool grade,

$$n = \frac{\ln(V_1/V_2)}{\ln(T_2/T_1)} = \frac{\ln(235/320)}{\ln(15/45)} = 0.28 \qquad C = V_1 \times (T_1)^n = 1980$$

For the 300–425 Brinell hardness range, and the hard tool grade,

$$n = \frac{\ln(V_1/V_2)}{\ln(T_2/T_1)} = \frac{\ln(330/440)}{\ln(15/45)} = 0.26 \qquad C = V_1 \times (T_1)^n = 2388$$

For the 300–425 Brinell hardness range, and the tough tool grade,

$$n = \frac{\ln(V_1/V_2)}{\ln(T_2/T_1)} = \frac{\ln(125/175)}{\ln(15/45)} = 0.31 \qquad C = V_1 \times (T_1)^n = 1324$$

For ASTM Class 20 grey cast iron, using hard ceramic,

$$n = \frac{\ln(V_1/V_2)}{\ln(T_2/T_1)} = \frac{\ln(1490/2220)}{\ln(15/45)} = 0.36 \qquad C = V_1 \times (T_1)^n = 5932$$

Selection of Optimized Data.—Fig. 22 illustrates cutting time, cycle time, number of parts before a tool change, tooling cost, and total cost, each plotted versus feed for a constant tool-life. Approximate minimum cost conditions can be determined using the formulas previously given in this section.

First, select a large feed/rev or tooth, and then calculate economic tool-life T_E, and the economic cutting speed V_E, and do all calculations using the time/cost formulas as described previously.

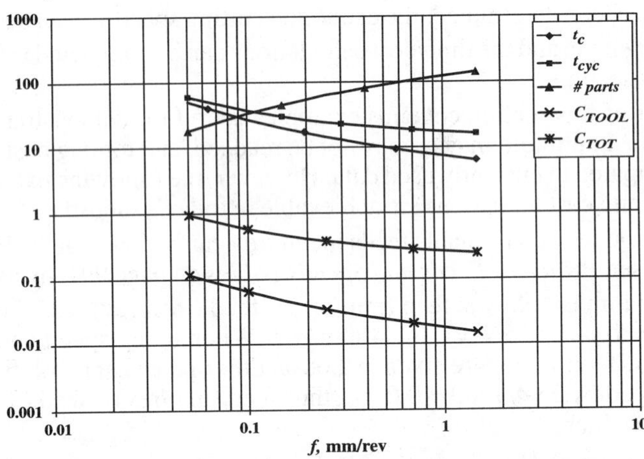

Fig. 22. Cutting time, cycle time, number of parts before tool change, tooling cost, and total cost vs. feed for tool-life = 15 minutes, idle time = 10 s, and batch size = 1000 parts

Example 11, Step by Step Procedure: Turning – Facing out: 1) Select a big feed/rev, in this case $f = 0.9$ mm/rev (0.035 inch/rev). A Taylor slope n is first determined using the Handbook tables and the method described in Example 10. In this example, use $n = 0.35$.

2) Calculate T_V from the tooling cost parameters:

If cost of insert = \$7.50; edges per insert = 2; cost of tool holder = \$100; life of holder = 100 insert sets; and for tools with inserts, allowance for insert failures = cost per insert by 4/3, assuming only 3 out of 4 edges can be effectively used.

Then, cost per edge = C_E is calculated as follows:

$$C_E = \frac{\text{cost of insert(s)}}{\text{number of edges per insert}} + \frac{\text{cost of cutter body}}{\text{cutter body life in number of edges}}$$

$$= \frac{7.50}{4/3 \times 2} + \frac{100}{100} = \$6.00$$

The time for replacing a worn edge of the facing insert $= T_{RPL} = 2.24$ minutes. Assuming an hourly rate $H_R = \$50$/hour, calculate the equivalent tooling-cost time T_V

$$T_V = T_{RPL} + 60 \times C_E/H_R = 2.24 + 60 \times 6/50 = 8.24 \text{ minutes.}$$

3) Determine economic tool-life T_E

$$T_E = T_V \times (1/n - 1) = T_E = T_V \times (1/n - 1) = 8.24 \times (1/0.35 - 1) = 15 \text{ minutes}$$

4) Determine economic cutting speed using the Handbook tables using the method shown in Example 10,

$$V_E = C \times T_E{}^n \text{ m/min} = C \times T_E{}^n = 280 \times 15^{-0.35} = 109 \text{ m/min}$$

5) Determine cost of tooling per batch (cutting tools, holders and tool changing) then total cost of cutting per batch:

$$C_{TOOL} = H_R \times T_C \times (C_E/T)/60$$

$$(C_{TOOL}+C_{CH}) = H_R \times T_C \times ((T_{RPL}+C_E/T)/60$$

$$C_{TOT} = H_R \times T_C(1 + (T_{RPL}+C_E)/T).$$

Example 12, Face Milling – Minimum Cost : This example demonstrates how a modern firm, using the formulas previously described, can determine optimal data. It is here applied to a face mill with 10 teeth, milling a 1045 type steel, and the radial depth versus the cutter diameter is 0.8. The *V–ECT–T* curves for tool-lives 5, 22, and 120 minutes for this operation are shown in Fig. 23a.

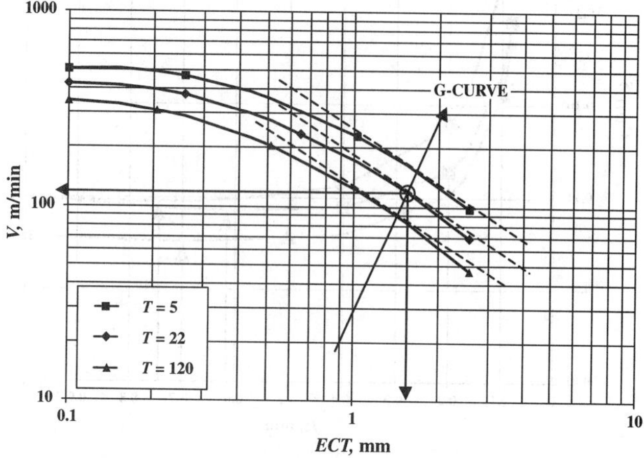

Fig. 23a. Cutting speed vs. ECT, tool-life constant

The global cost minimum occurs along the *G*-curve, see Fig. 6c and Fig. 23a, where the 45-degree lines defines this curve. Optimum *ECT* is in the range 1.5 to 2 mm.

For face and end milling operations, $ECT = z \times f_z \times ar/D \times aa/CEL \div \pi$. The ratio aa/CEL = 0.95 for lead angle $LA = 0$, and for $ar/D = 0.8$ and 10 teeth, using the formula to calculate the feed/tooth range gives for $ECT = 1.5, f_z = 0.62$ mm and for $ECT = 2, f_z = 0.83$ mm.

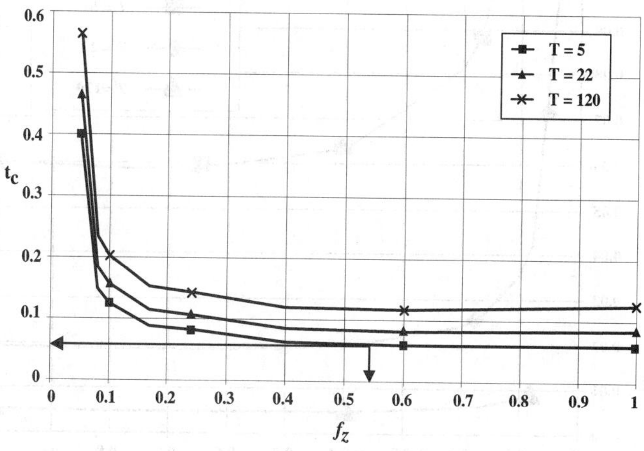

Fig. 23b. Cutting time per part vs. feed per tooth

Using computer simulation, the minimum cost occurs approximately where Fig. 23a indicates it should be. Total cost has a global minimum at f_z around 0.6 to 0.7 mm and a speed of around 110 m/min. *ECT* is about 1.9 mm and the optimal cutter life is $T_O = 22$ minutes. Because it may be impossible to reach the optimum feed value due to tool breakage, the maximum practical feed f_{max} is used as the optimal value. The difference in costs between a global optimum and a practical minimum cost condition is negligible, as shown in Figs. 23c and 23e. A summary of the results are shown in Figs. 23a through 23e, and Table 1.

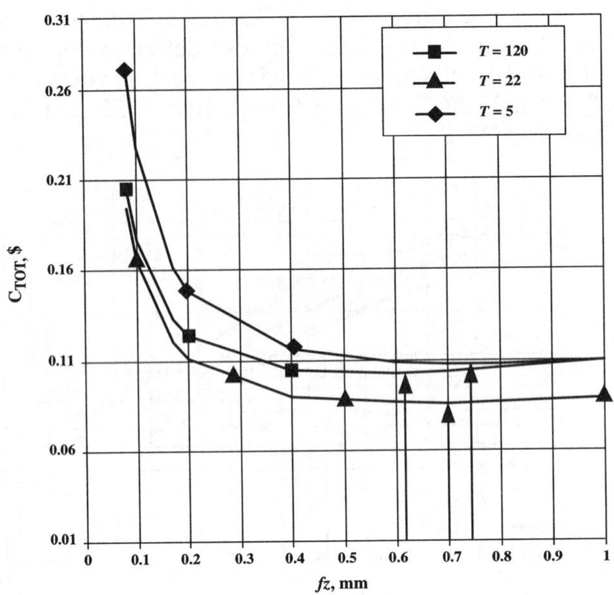

Fig. 23c. Total cost vs. feed/tooth

When plotting cutting time/part, t_c, versus feed/tooth, f_z, at $T = 5$, 22, 120 in Figs. 23b, tool-life $T = 5$ minutes yields the shortest cutting time, but total cost is the highest; the minimum occurs for f_z about 0.75 mm, see Figs. 23c. The minimum for $T = 120$ minutes is about 0.6 mm and for $T_O = 22$ minutes around 0.7 mm.

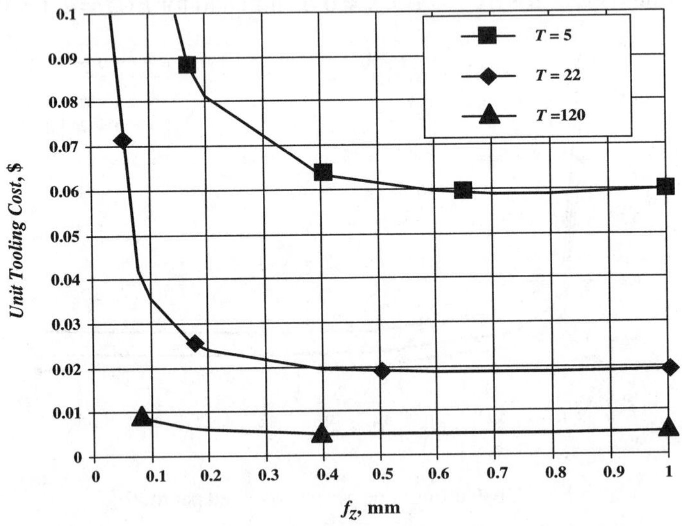

Fig. 23d. Tooling cost versus feed/tooth

Fig. 23d shows that tooling cost drop off quickly when increasing feed from 0.1 to 0.3 to 0.4 mm, and then diminishes slowly and is almost constant up to 0.7 to 0.8 mm/tooth. It is generally very high at the short tool-life 5 minutes, while tooling cost of optimal tool-life 22 minutes is about 3 times higher than when going slow at T=120 minutes.

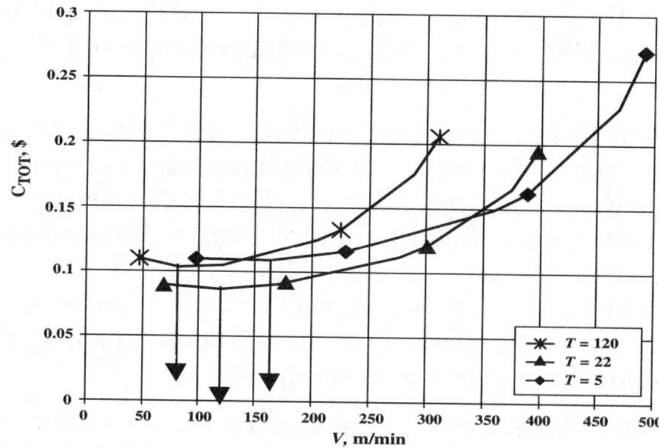

Fig. 23e. Total cost vs. cutting speed at 3 constant tool-lives, feed varies

The total cost curves in Fig. 24e. were obtained by varying feed and cutting speed in order to maintain constant tool-lives at 5, 22 and 120 minutes. Cost is plotted as a function of speed *V* instead of feed/tooth. Approximate optimum speeds are $V = 150$ m/min at $T = 5$ minutes, $V = 180$ m/min at $T = 120$ minutes, and the global optimum speed is $V_O = 110$ m/min for $T_O = 22$ minutes.

Table 1 displays the exact numerical values of cutting speed, tooling cost and total cost for the selected tool-lives of 5, 22, and 120 minutes, obtained from the software program.

Table 1. Face Milling, Total and Tooling Cost versus *ECT*, Feed/tooth *fz*, and Cutting Speed *V*, at Tool-lives 5, 22, and 120 minutes

		$T = 5$ minutes			$T = 22$ minutes			$T = 120$ minutes		
fz	*ECT*	*V*	C_{TOT}	C_{TOOL}	*V*	C_{TOT}	C_{TOOL}	*V*	C_{TOT}	C_{TOOL}
0.03	0.08	489	0.72891	0.39759	416	0.49650	0.10667	344	0.49378	0.02351
0.08	0.21	492	0.27196	0.14834	397	0.19489	0.04187	311	0.20534	0.00978
0.10	0.26	469	0.22834	0.12455	374	0.16553	0.03556	289	0.17674	0.00842
0.17	0.44	388	0.16218	0.08846	301	0.12084	0.02596	225	0.13316	0.00634
0.20	0.51	359	0.14911	0.08133	276	0.11204	0.02407	205	0.12466	0.00594
0.40	1.03	230	0.11622	0.06339	171	0.09051	0.01945	122	0.10495	0.00500
0.60	1.54	164	0.10904	0.05948	119	0.08672	0.01863	83	0.10301	0.00491
0.70	1.80	141	0.10802	0.05892	102	0.08665	0.01862	70	0.10393	0.00495
0.80	2.06	124	0.10800	0.05891	89	0.08723	0.01874	60	0.10547	0.00502
1.00	2.57	98	0.10968	0.05982	69	0.08957	0.01924	47	0.10967	0.00522

High-speed Machining Econometrics

High-speed Machining – No Mystery.—This section describes the theory and gives the basic formulas for any milling operation and high-speed milling in particular, followed by several examples on high-speed milling econometrics. These rules constitute the basis on which selection of milling feed factors is done. Selection of cutting speeds for general milling is done using the Handbook Table 10 through 14, starting on page 1013.

High-speed machining is no mystery to those having a good knowledge of metal cutting. Machining materials with very good machinability, such as low-alloyed aluminum, have for ages been performed at cutting speeds well below the speed values at which these materials should be cut. Operating at these low speeds often results in built-up edges and poor surface finish, because the operating conditions selected are on the wrong side of the Taylor curve, i.e. to the left of the *H*-curve representing maximum tool-life values (see Fig. 4 on page 1059).

In the 1950's it was discovered that cutting speed could be raised by a factor of 5 to 10 when hobbing steel with HSS cutters. This is another example of being on the wrong side of the Taylor curve.

One of the first reports on high-speed end milling using high-speed steel (HSS) and carbide cutters for milling 6061-T651 and A356-T6 aluminum was reported in a study funded by Defense Advanced Research Project Agency (DARPA). Cutting speeds of up to 4400 m/min (14140 fpm) were used. Maximum tool-lives of 20 through 40 minutes were obtained when the feed/tooth was 0.2 through 0.25 mm (0.008 to 0.01 inch), or measured in terms of *ECT* around 0.07 to 0.09 mm. Lower or higher feed/tooth resulted in shorter cutter lives. The same types of previously described curves, namely *T–ECT* curves with maximum tool-life along the *H*-curve, were produced.

When examining the influence of *ECT*, or feed/rev, or feed/tooth, it is found that too small values cause chipping, vibrations, and poor surface finish. This is caused by inadequate (too small) chip thickness, and as a result the material is not cut but plowed away or scratched, due to the fact that operating conditions are on the wrong (left) side of the tool-life versus *ECT* curve (*T-ECT* with constant speed plotted).

There is a great difference in the thickness of chips produced by a tooth traveling through the cutting arc in the milling process, depending on how the center of the cutter is placed in relation to the workpiece centerline, in the feed direction. Although end and face milling cut in the same way, from a geometry and kinematics standpoint they are in practice distinguished by the cutter center placement away from, or close to, the work centerline, respectively, because of the effect of cutter placement on chip thickness. This is the criteria used to distinguishing between the end and face milling processes in the following.

Depth of Cut/Cutter Diameter, ar/D is the ratio of the radial depth of cut *ar* and the cutter diameter *D*. In face milling when the cutter axis points approximately to the middle of the work piece axis, eccentricity is close to zero, as illustrated in Figs. 3 and 4, page 1011, and Fig. 5 on page 1012. In end milling, $ar/D = 1$ for full slot milling.

Mean Chip Thickness, hm is a key parameter that is used to calculate forces and power requirements in high-speed milling. If the mean chip thickness hm is too small, which may occur when feed/tooth is too small (this holds for all milling operations), or when *ar/D* decreases (this holds for ball nose as well as for straight end mills), then cutting occurs on the left (wrong side) of the tool-life versus ECT curve, as illustrated in Figs. 6b and 6c.

In order to maintain a given chip thickness in end milling, the feed/tooth has to be increased, up to 10 times for very small *ar/D* values in an extreme case with no run out and otherwise perfect conditions. A 10 times increase in feed/tooth results in 10 times bigger feed rates (F_R) compared to data for full slot milling (valid for $ar/D = 1$), yet maintain a given chip thickness. The cutter life at any given cutting speed will not be the same, however.

Increasing the number of teeth from say 2 to 6 increases equivalent chip thickness *ECT* by a factor of 3 while the mean chip thickness *hm* remains the same, but does not increase the feed rate to 30 (3×10) times bigger, because the cutting speed must be reduced. However, when the *ar/D* ratio matches the number of teeth, such that one tooth enters when the second tooth leaves the cutting arc, then $ECT = hm$. Hence, *ECT* is proportional to the number of teeth. Under ideal conditions, an increase in number of teeth *z* from 2 to 6 increases the feed rate by, say, 20 times, maintaining tool-life at a reduced speed. In practice about 5 times greater feed rates can be expected for small *ar/D* ratios (0.01 to 0.02), and up to 10 times with 3 times as many teeth. So, high-speed end milling is no mystery.

Chip Geometry in End and Face Milling.—Fig. 24 illustrates how the chip forming process develops differently in face and end milling, and how mean chip thickness *hm* varies with the angle of engagement *AE*, which depends on the *ar/D* ratio. The pertinent chip geometry formulas are given in the text that follows.

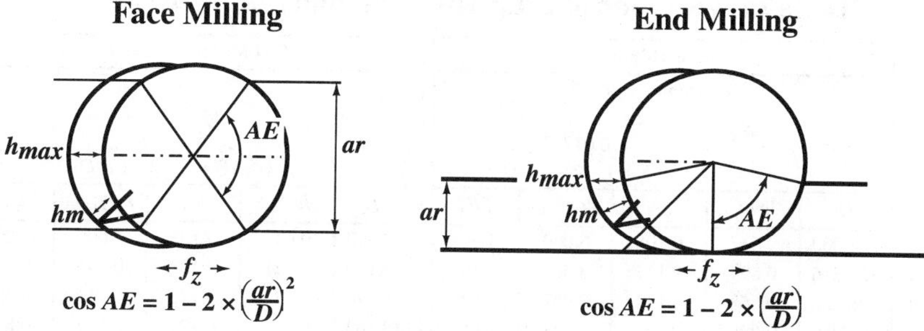

Fig. 24. Comparison of face milling and end milling geometry

High-speed end milling refers to values of *ar*/*D* that are less than 0.5, in particular to *ar*/*D* ratios which are considerably smaller. When *ar*/*D* = 0.5 (*AE* = 90 degrees) and diminishing in end milling, the chip thickness gets so small that poor cutting action develops, including plowing or scratching. This situation is remedied by increasing the feed/tooth, as shown in Table 2a as an increasing f_z/f_{z0} ratio with decreasing *ar*/*D*. For end milling, the f_z/f_{z0} feed ratio is 1.0 for *ar*/*D* = 1 and also for *ar*/*D* = 0.5. In order to maintain the same *hm* as at *ar*/*D* = 1, the feed/tooth should be increased, by a factor of 6.38 when *ar*/*D* is 0.01 and by more than 10 when *ar*/*D* is less than 0.01. Hence high-speed end milling could be said to begin when *ar*/*D* is less than 0.5

In end milling, the ratio f_z/f_{z0} = 1 is set at *ar*/*D* = 1.0 (full slot), a common value in vendor catalogs and handbooks, for *hm* = 0.108 mm.

The face milling chip making process is exactly the same as end milling when face milling the side of a work piece and *ar*/*D* = 0.5 or less. However, when face milling close to and along the work centerline (eccentricity is close to zero) chip making is quite different, as shown in Fig. 24. When *ar*/*D* = 0.74 (*AE* = 95 degrees) in face milling, the f_z/f_{z0} ratio = 1 and increases up to 1.4 when the work width is equal to the cutter diameter (*ar*/*D* = 1). The face milling f_z/f_{z0} ratio continues to diminish when the *ar*/*D* ratio decreases below *ar*/*D* = 0.74, but very insignificantly, only about 11 percent when *ar*/*D* = 0.01.

In face milling f_z/f_{z0} = 1 is set at *ar*/*D* = 0.74, a common value recommended in vendor catalogs and handbooks, for *hm* = 0.151 mm.

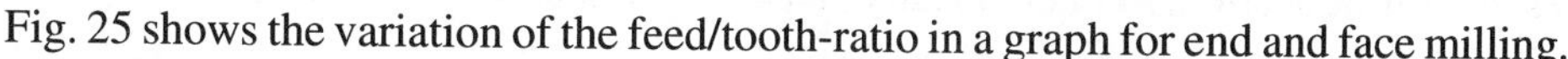
Fig. 25 shows the variation of the feed/tooth-ratio in a graph for end and face milling.

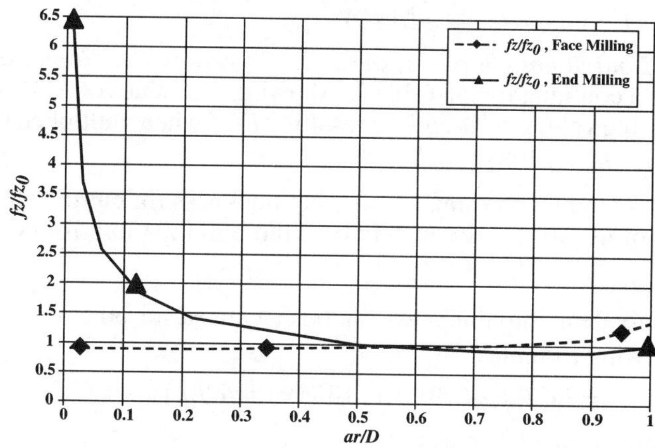

Fig. 25. Feed/tooth versus *ar*/*D* for face and end milling

Table 2a. Variation of Chip Thickness and f_z/f_{z0} with ar/D

	Face Milling					End Milling (straight)				
	ecentricity $e = 0$ $z = 8$ $f_{z0} = 0.017$ $\cos AE = 1 - 2 \times (ar/D)^2$					$z = 2$ $f_{z0} = 0.017$ $\cos AE = 1 - 2 \times (ar/D)$				
ar/D	AE	hm/f_z	hm	ECT/hm	f_z/f_{z0}	AE	hm/f_z	hm	ECT/hm	f_z/f_{z0}
1.0000	180.000	0.637	0.108	5.000	1.398	180.000	0.637	0.108	1.000	1.000
0.9000	128.316	0.804	0.137	3.564	1.107	143.130	0.721	0.122	0.795	0.884
0.8000	106.260	0.863	0.147	2.952	1.032	126.870	0.723	0.123	0.711	0.881
0.7355	94.702	0.890	0.151	2.631	1.000	118.102	0.714	0.122	0.667	0.892
0.6137	75.715	0.929	0.158	1.683	0.958	103.144	0.682	0.116	0.573	0.934
0.5000	60.000	0.162	0.932	0.216	0.202	90.000	0.674	0.115	0.558	1.000
0.3930	46.282	0.973	0.165	1.028	0.915	77.643	0.580	0.099	0.431	1.098
0.2170	25.066	0.992	0.169	0.557	0.897	55.528	0.448	0.076	0.308	1.422
0.1250	14.361	0.997	0.170	0.319	0.892	41.410	0.346	0.059	0.230	1.840
0.0625	7.167	0.999	0.170	0.159	0.891	28.955	0.247	0.042	0.161	2.574
0.0300	3.438	1.000	0.170	0.076	0.890	19.948	0.172	0.029	0.111	3.694
0.0100	1.146	1.000	0.170	0.025	0.890	11.478	0.100	0.017	0.064	6.377
0.0010	0.115	1.000	0.000	0.000	0.890	3.624	0.000	0.000	0.000	20.135

In Table 2a, a standard value f_{z0} = 0.17 mm/tooth (commonly recommended average feed) was used, but the f_z/f_{z0} values are independent of the value of feed/tooth, and the previously mentioned relationships are valid whether f_{z0} = 0.17 or any other value.

In both end and face milling, hm = 0.108 mm for f_{z0} = 0.17mm when ar/D =1. When the f_z/f_{z0} ratio = 1, hm = 0.15 for face milling, and 0.108 in end milling both at ar/D = 1 and 0.5. The tabulated data hold for perfect milling conditions, such as, zero run-out and accurate sharpening of all teeth and edges.

Mean Chip Thickness *hm* and Equivalent Chip Thickness *ECT*.—The basic formula for equivalent chip thickness ECT for any milling process is:

$ECT = f_z \times z/\pi \times (ar/D) \times aa/CEL$, where f_z = feed/tooth, z = number of teeth, D = cutter diameter, ar = radial depth of cut, aa = axial depth of cut, and CEL = cutting edge length. As a function of mean chip thickness hm:

$ECT = hm \times (z/2) \times (AE/180)$, where AE = angle of engagement.

Both terms are exactly equal when one tooth engages as soon as the preceding tooth leaves the cutting section. Mathematically, $hm = ECT$ when $z = 360/AE$; thus:

for face milling, $AE = \arccos(1 - 2 \times (ar/D)^2)$

for end milling, $AE = \arccos(1 - 2 \times (ar/D))$

Calculation of Equivalent Chip Thickness (ECT) versus Feed/tooth and Number of teeth.: Table 2b is a continuation of Table 2a, showing the values of ECT for face and end milling for decreasing values ar/D, and the resulting ECT when multiplied by the f_z/f_{z0} ratio f_{z0} = 0.17 (based on hm = 0.108).

Small ar/D ratios produce too small mean chip thickness for cutting chips. In practice, minimum values of hm are approximately 0.02 through 0.04 mm for both end and face milling.

Formulas.— Equivalent chip thickness can be calculated for other values of f_z and z by means of the following formulas:

Face milling: $ECT_F = ECT_{0F} \times (z/8) \times (f_z/0.17) \times (aa/CEL)$

or, if ECT_F is known calculate f_z using:

$f_z = 0.17 \times (ECT_F/ECT_{0F}) \times (8/z) \times (CEL/aa)$

Table 2b. Variation of *ECT*, Chip Thickness and f_z/f_{z0} with *ar/D*

	Face Milling				End Milling (straight)			
ar/D	*hm*	*fz/f*$_{z0}$	*ECT*	ECT_0 corrected for *fz/f*$_{z0}$	*hm*	*fz/f*$_{z0}$	*ECT*	ECT_0 corrected for *fz/f*$_{z0}$
1.0000	0.108	1.398	0.411	0.575	0.108	1.000	0.103	0.103
0.9000	0.137	1.107	0.370	0.410	0.122	0.884	0.093	0.082
0.8080	0.146	1.036	0.332	0.344	0.123	0.880	0.083	0.073
0.7360	0.151	1.000	0.303	0.303	0.121	0.892	0.076	0.067
0.6137	0.158	0.958	0.252	0.242	0.116	0.934	0.063	0.059
0.5900	0.159	0.952	0.243	0.231	0.115	0.945	0.061	0.057
0.5000	0.162	0.932	0.206	0.192	0.108	1.000	0.051	0.051
0.2170	0.169	0.897	0.089	0.080	0.076	1.422	0.022	0.032
0.1250	0.170	0.892	0.051	0.046	0.059	1.840	0.013	0.024
0.0625	0.170	0.891	0.026	0.023	0.042	2.574	0.006	0.017
0.0300	0.170	0.890	0.012	0.011	0.029	3.694	0.003	0.011
0.0100	0.170	0.890	0.004	0.004	0.017	6.377	0.001	0.007
0.0010	0.170	0.890	0.002	0.002	0.005	20.135	0.001	0.005

In face milling, the approximate values of $aa/CEL = 0.95$ for lead angle $LA = 0°$ (90° in the metric system); for other values of LA, $aa/CEL = 0.95 \times \sin(LA)$, and $0.95 \times \cos(LA)$ in the metric system.

Example, Face Milling: For a cutter with $D = 250$ mm and $ar = 125$ mm, calculate ECT_F for $f_z = 0.1$, $z = 12$, and $LA = 30$ degrees. First calculate $ar/D = 0.5$, and then use Table 2b and find $ECT_{0F} = 0.2$.

Calculate ECT_F with above formula:

$ECT_F = 0.2 \times (12/8) \times (0.1/0.17) \times 0.95 \times \sin 30 = 0.084$ mm.

End milling: $ECT_E = ECT_{0E} \times (z/2) \times (f_z/0.17) \times (aa/CEL)$,

or if ECT_E is known calculate f_z from:

$f_z = 0.17 \times (ECT_E/ECT_{0E}) \times (2/z)) \times (CEL/aa)$

The approximate values of $aa/CEL = 0.95$ for lead angle $LA = 0°$ (90° in the metric system).

Example, High-speed End Milling: For a cutter with $D = 25$ mm and $ar = 3.125$ mm, calculate ECT_E for $f_z = 0.1$ and $z = 6$. First calculate $ar/D = 0.125$, and then use Table 2b and find $ECT_{0E} = 0.0249$.

Calculate ECT_E with above formula:

$ECT_E = 0.0249 \times (6/2) \times (0.1/0.17) \times 0.95 \times 1 = 0.042$ mm.

Example, High-speed End Milling: For a cutter with $D = 25$ mm and $ar = 0.75$ mm, calculate ECT_E for $f_z = 0.17$ and $z = 2$ and 6. First calculate $ar/D = 0.03$, and then use Table 2b and find $f_z/f_{z0} = 3.694$

Then, $f_z = 3.694 \times 0.17 = 0.58$ mm/tooth and $ECT_E = 0.0119 \times 0.95 = 0.0113$ mm and $0.0357 \times 0.95 = 0.0339$ mm for 2 and 6 teeth respectively. These cutters are marked HS2 and HS6 in Figs. 26a, 26d, and 26e.

Example, High-speed End Milling: For a cutter with $D = 25$ mm and $ar = 0.25$ mm, calculate ECT_E for $f_z = 0.17$ and $z = 2$ and 6. First calculate $ar/D = 0.01$, and then use Table 2b and find $ECT_{0E} = 0.0069$ and 0.0207 for 2 and 6 teeth respectively. When obtaining such small values of ECT, there is a great danger to be far on the left side of the H-curve, at least when there are only 2 teeth. Doubling the feed would be the solution if cutter design and material permit.

Example, Full Slot Milling: For a cutter with $D = 25$ mm and $ar = 25$ mm, calculate ECT_E for $f_z = 0.17$ and $z = 2$ and 6. First calculate $ar/D = 1$, and then use Table 2b and find $ECT_E =$

$0.108 \times 0.95 = 0.103$ and $3 \times 0.108 \times 0.95 = 0.308$ for 2 and 6 teeth, respectively. These cutters are marked SL2 and SL6 in Figs. 26a, 26d, and 26e.

Physics behind *hm* and *ECT*, Forces and Tool-life (*T*).—The *ECT* concept for all metal cutting and grinding operations says that the more energy put into the process, by increasing feed/rev, feed/tooth, or cutting speed, the life of the edge decreases. When increasing the number of teeth (keeping everything else constant) the work and the process are subjected to a higher energy input resulting in a higher rate of tool wear.

In high-speed milling when the angle of engagement *AE* is small the contact time is shorter compared to slot milling ($ar/D = 1$) but the chip becomes shorter as well. Maintaining the same chip thickness as in slot milling has the effect that the energy consumption to remove the chip will be different. Hence, maintaining a constant chip thickness is a good measure when calculating cutting forces (keeping speed constant), but not when determining tool wear. Depending on cutting conditions the wear rate can either increase or decrease, this depends on whether cutting occurs on the left or right side of the *H*-curve.

Fig. 26a shows an example of end milling of steel with coated carbide inserts, where cutting speed *V* is plotted versus *ECT* at 5, 15, 45 and 180 minutes tool-lives. Notice that the *ECT* values are independent of *ar/D* or number of teeth or feed/tooth, or whether f_z or f_{z0} is used, as long as the corresponding f_z/f_{z0}-ratio is applied to determine ECT_E. The result is one single curve per tool-life. Had cutting speed been plotted versus f_{z0}, *ar/D*, or *z* values (number of teeth), several curves would be required at each constant tool-life, one for each of these parameters This illustrates the advantage of using the basic parameter *ECT* rather than f_z, or *hm*, or *ar/D* on the horizontal axis.

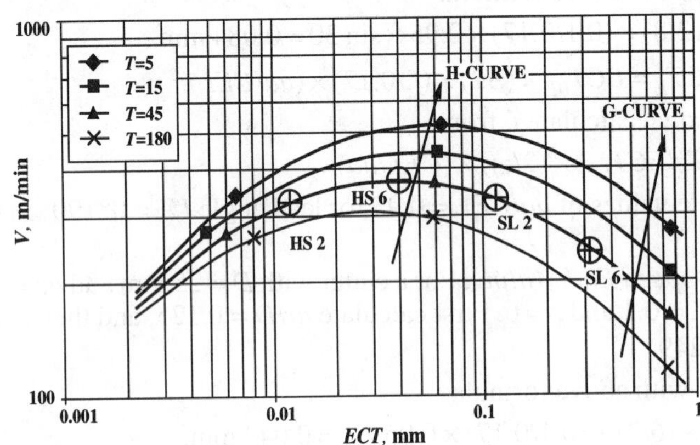

Fig. 26a. Cutting speed vs. ECT, tool-life plotted, for end milling

Example: The points (HS2, HS6) and (SL2, SL6) on the 45-minute curve in Fig. 26a relate to the previous high-speed and full slot milling examples for 2 and 6 teeth, respectively.

Running a slot at $f_{z0} = 0.17$ mm/tooth ($hm = 0.108$, $ECT_E = 0.103$ mm) with 2 teeth and for a tool-life 45 minutes, the cutting speed should be selected at $V = 340$ m/min at point SL2 and for six teeth ($hm = 0.108$ mm, $ECT_E = 0.308$) at $V = 240$ m/min at point SL6.

When high-speed milling for $ar/D = 0.03$ at $f_z = 3.394 \times 0.17 = 0.58$ mm/tooth = 0.58 mm/tooth, *ECT* is reduced to 0.011 mm ($hm = 0.108$) the cutting speed is 290 m/min to maintain $T = 45$ minutes, point HS2. This point is far to the left of the *H*-curve in Fig.26b, but if the number of teeth is increased to 6 ($ECT_E = 3 \times 0.103 = 0.3090$), the cutting speed is 360 m/min at $T = 45$ minutes and is close to the *H*-curve, point HS6. Slotting data using 6 teeth are on the right of this curve at point SL6, approaching the *G*-curve, but at a lower slotting speed of 240 m/min.

Depending on the starting f_z value and on the combination of cutter grade - work material, the location of the *H*-curve plays an important role when selecting high-speed end milling data.

Feed Rate and Tool-life in High-speed Milling, Effect of *ECT* and Number of Teeth.—Calculation of feed rate is done using the formulas in previously given:

Feed Rate:

$F_R = z \times f_z \times \text{rpm}$, where $z \times f_z = f$ (feed/rev of cutter). Feed is measured along the feeding direction.

rpm = 1000 × V/3.1416/D, where D is diameter of cutter.

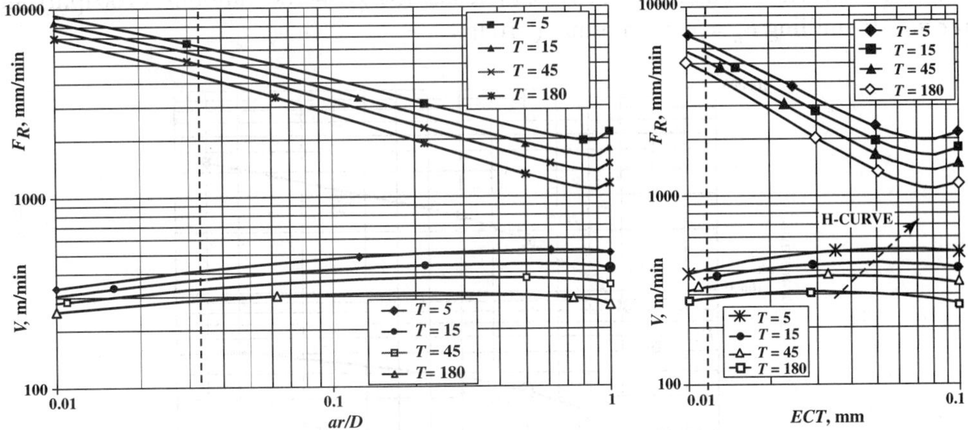

Fig. 26b. High speed feed rate and cutting speed versus *ar*/*D* at *T* = 5, 15, 45, and 180 minutes

Fig. 26c. High speed feed rate and cutting speed versus *ECT*, *ar*/*D* plotted at *T* = 5, 15, 45, and 180 minutes

Fig. 26b shows the variation of feed rate F_R plotted versus *ar*/*D* for tool-lives 5, 15, 45 and 180 minutes with a 25 mm diameter cutter and 2 teeth. Fig. 26c shows the variation of feed rate F_R when plotted versus *ECT*. In both graphs the corresponding cutting speeds are also plotted. The values for *ar*/*D* = 0.03 in Fig. 26b correspond to *ECT* = 0.011 in Fig. 26c.

Feed rates have minimum around values of *ar*/*D* = 0.8 and *ECT*=0.75 and not along the *H*-curve. This is due to the fact that the f_z/f_{z0} ratio to maintain a mean chip thickness = 0.108 mm changes F_R in a different proportion than the cutting speed.

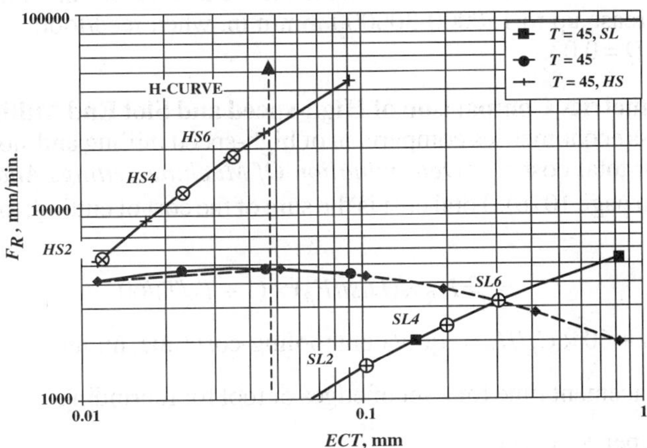

Fig. 26d. Feed rate versus *ECT* comparison of slot milling (*ar*/*D* = 1) and high-speed milling at (*ar*/*D* = 0.03) for 2, 4, and 6 teeth at *T* = 45 minutes

A comparison of feed rates for full slot (ar/D = 1) and high-speed end milling (ar/D = 0.03 and $f_z = 3.69 \times f_{z0} = 0.628$ mm) for tool-life 45 minutes is shown in Fig. 26d. The points SL2, SL4, SL6 and HS2, HS4, HS6, refer to 2, 4, and 6 teeth (2 to 6 teeth are commonly used in practice). Feed rate is also plotted versus number of teeth z in Fig. 26e, for up to 16 teeth, still at $f_z = 0.628$ mm.

Comparing the effect of using 2 versus 6 teeth in high-speed milling shows that feed rates increase from 5250 mm/min (413 ipm) up to 18000 mm/min (1417ipm) at 45 minutes tool-life. The effect of using 2 versus 6 teeth in full slot milling is that feed rate increases from 1480 mm/min (58 ipm) up to 3230 mm/min (127 ipm) at tool-life 45 minutes. If 16 teeth could be used at $ar/D = 0.03$, the feed rate increases to $F_R = 44700$ mm/min (1760 ipm), and for full slot milling $F_R = 5350$ mm/min (210 ipm).

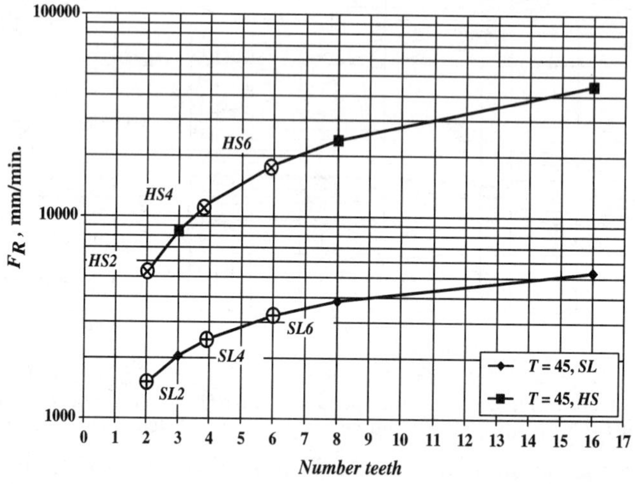

Fig. 26e. Feed rate versus number of teeth comparison of slot milling ($ar/D = 1$) and high-speed milling at ($ar/D = 0.03$) for 2, 4, and 6 teeth at $T = 45$ minutes

Comparing the feed rates that can be obtained in steel cutting with the one achieved in the earlier referred DARPA investigation, using HSS and carbide cutters milling 6061-T651 and A356-T6 aluminum, it is obvious that aluminium end milling can be run at 3 to 6 times higher feed rates. This requires 3 to 6 times higher spindle speeds (cutter diameter 25 mm, radial depth of cut ar = 12.5 mm, 2 teeth). Had these tests been run with 6 teeth, the feed rates would increase up to 150000-300000 mm/min, when feed/tooth = 3.4 × 0.25 = 0.8 mm/tooth at $ar/D = 0.03$.

Process Econometrics Comparison of High-speed and Slot End Milling.—When making a process econometrics comparison of high-speed milling and slot end milling use the formulas for total cost c_{tot} (*Determination Of Machine Settings And Calculation Of Costs* starting on page 1076). Total cost is the sum of the cost of cutting, tool changing, and tooling:

$$c_{tot} = H_R \times (Dist/F_R) \times (1 + T_V/T)/60$$

where $T_V = T_{RPL} + 60 \times C_E/H_R$ = equivalent tooling-cost time, minutes

T_{RPL} = replacement time for a set of edges or tool for regrinding

C_E = cost per edge(s)

H_R = hourly rate, $

Fig. 27. compares total cost c_{tot}, using the end milling cutters of the previous examples, for full slot milling with high-speed milling at *ar*/*D* =0.03, and versus *ECT* at *T* =45 minutes.

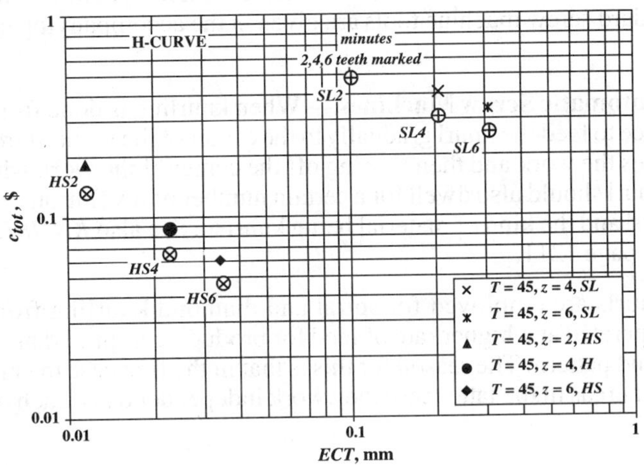

Fig. 27. Cost comparison of slot milling (*ar*/*D* = 1) and high-speed milling at (*ar*/*D* = 0.03) for 2, 4, and 6 teeth at *T* = 45 minutes

The feed/tooth for slot milling is $f_{z0} = 0.17$ and for high-speed milling at *ar*/*D* = 0.03 the feed is $f_z = 3.69 \times f_{z0} = 0.628$ mm.

The calculations for total cost are done according to above formula using tooling cost at T_V = 6, 10, and 14 minutes, for z = 2, 4, and 6 teeth respectively. The distance cut is *Dist* = 1000 mm. Full slot milling costs are,

at feed rate $F_R = 3230$ and $z = 6$

$$c_{tot} = 50 \times (1000/3230) \times (1 + 14/45)/60 = \$0.338 \text{ per part}$$

at feed rate $F_R = 1480$ and $z = 2$

$$c_{tot} = 50 \times (1000/1480) \times (1 + 6/45)/60 = \$0.638 \text{ per part}$$

High-speed milling costs,

at F_R=18000, z=6

$$c_{tot} = 50 \times (1000/18000) \times (1 + 14/45)/60 = \$0.0606 \text{ per part}$$

at F_R= 5250, z=2

$$c_{tot} = 50 \times (1000/5250) \times (1 + 6/45)/60 = \$0.208.$$

The cost reduction using high-speed milling compared to slotting is enormous. For high-speed milling with 2 teeth, the cost for high-speed milling with 2 teeth is 61 percent (0.208/0.338) of full slot milling with 6 teeth (z = 6). The cost for high-speed milling with 6 teeth is 19 percent (0.0638/0.338) of full slot for z = 6.

Aluminium end milling can be run at 3 to 6 times lower costs than when cutting steel. Costs of idle (non-machining) and slack time (waste) are not considered in the example. These data hold for perfect milling conditions such as zero run-out and accurate sharpening of all teeth and edges.

SCREW MACHINE FEEDS AND SPEEDS

Feeds and Speeds for Automatic Screw Machine Tools.—Approximate feeds and speeds for standard screw machine tools are given in the accompanying table.

Knurling in Automatic Screw Machines.—When knurling is done from the cross slide, it is good practice to feed the knurl gradually to the center of the work, starting to feed when the knurl touches the work and then passing off the center of the work with a quick rise of the cam. The knurl should also dwell for a certain number of revolutions, depending on the pitch of the knurl and the kind of material being knurled. See also *KNURLS AND KNURLING* starting on page 1211.

When two knurls are employed for spiral and diamond knurling from the turret, the knurls can be operated at a higher rate of feed for producing a spiral than they can for producing a diamond pattern. The reason for this is that in the first case the knurls work in the same groove, whereas in the latter case they work independently of each other.

Revolutions Required for Top Knurling.—The depth of the teeth and the feed per revolution govern the number of revolutions required for top knurling from the cross slide. If R is the radius of the stock, d is the depth of the teeth, c is the distance the knurl travels from the point of contact to the center of the work at the feed required for knurling, and r is the radius of the knurl; then

$$c = \sqrt{(R+r)^2 - (R+r-d)^2}$$

For example, if the stock radius R is $\frac{5}{32}$ inch, depth of teeth d is 0.0156 inch, and radius of knurl r is 0.3125 inch, then

$$c = \sqrt{(0.1562 + 0.3125)^2 - (0.1562 + 0.3125 - 0.0156)^2}$$
$$= 0.120 \text{ inch} = \text{cam rise required}$$

Assume that it is required to find the number of revolutions to knurl a piece of brass $\frac{5}{16}$ inch in diameter using a 32 pitch knurl. The included angle of the teeth for brass is 90 degrees, the circular pitch is 0.03125 inch, and the calculated tooth depth is 0.0156 inch. The distance c (as determined in the previous example) is 0.120 inch. Referring to the accompanying table of feeds and speeds, the feed for top knurling brass is 0.005 inch per revolution. The number of revolutions required for knurling is, therefore, $0.120 \div 0.005 =$ 24 revolutions. If conditions permit, the higher feed of 0.008 inch per revolution given in the table may be used, and 15 revolutions are then required for knurling.

Cams for Threading.—The table *Spindle Revolutions and Cam Rise for Threading* on page 1097 gives the revolutions required for threading various lengths and pitches and the corresponding rise for the cam lobe. To illustrate the use of this table, suppose a set of cams is required for threading a screw to the length of $\frac{3}{8}$ inch in a Brown & Sharpe machine. Assume that the spindle speed is 2400 revolutions per minute; the number of revolutions to complete one piece, 400; time required to make one piece, 10 seconds; pitch of the thread, $\frac{1}{32}$ inch or 32 threads per inch. By referring to the table, under 32 threads per inch, and opposite $\frac{3}{8}$ inch (length of threaded part), the number of revolutions required is found to be 15 and the rise required for the cam, 0.413 inch.

Approximate Cutting Speeds and Feeds for Standard Automatic Screw Machine Tools—Brown and Sharpe

Tool	Cut: Width or Depth, Inches	Cut: Dia. of Hole, Inches	Brass[a]: Feed, Inches per Rev.	Mild or Soft Steel: Feed, Inches per Rev.	Mild or Soft Steel: Surface Speed, Feet per Min., Carbon Tools	Mild or Soft Steel: Surface Speed, Feet per Min., H.S.S. Tools	Tool Steel, 0.80–1.00% C: Feed, Inches per Rev.	Tool Steel, 0.80–1.00% C: Surface Speed, Feet per Min., Carbon Tools	Tool Steel, 0.80–1.00% C: Surface Speed, Feet per Min., H.S.S. Tools
Boring tools	0.005	…	…	0.008	50	110	0.004	30	60
Box tools, roller rest, Single chip finishing	1/32	…	0.012	0.010	70	150	0.005	40	75
	1/16	…	0.010	0.008	70	150	0.004	40	75
	1/8	…	0.008	0.007	70	150	0.003	40	75
	3/16	…	0.008	0.006	70	150	0.002	40	75
	1/4	…	0.006	0.005	70	150	0.0015	40	75
Finishing	0.005	…	0.010	0.010	70	150	0.006	40	75
Center drills	…	Under 1/8	0.003	0.0015	50	110	0.001	30	75
	…	Over 1/8	0.006	0.0035	50	110	0.002	30	75
Cutoff tools: Angular	…	…	0.0015	0.0006	80	150	0.0004	50	85
Cutoff tools: Circular	3/64–1/8	…	0.0035	0.0015	80	150	0.001	50	85
Cutoff tools: Straight	1/16–1/8	…	0.0035	0.0015	80	150	0.001	50	85
Stock diameter under 1/8 in.	…	…	0.002	0.0008	80	150	0.0005	50	85
Dies: Button	…	…	…	…	30	…	…	14	…
Dies: Chaser	…	…	…	…	30	40	…	16	20
Drills, twist cut	…	0.02	0.0014	0.001	40	60	0.0006	30	45
	…	0.04	0.002	0.0014	40	60	0.0008	30	45
	…	1/16	0.004	0.002	40	60	0.0012	30	45
	…	3/32	0.006	0.0025	40	60	0.0016	30	45
	…	1/8	0.009	0.0035	40	75	0.002	30	60
	…	3/16	0.012	0.004	40	75	0.003	30	60
	…	1/4	0.014	0.005	40	75	0.003	30	60
	…	5/16	0.016	0.005	40	75	0.0035	30	60
	…	3/8–5/8	0.016	0.006	40	85	0.004	30	60
Form tools, circular	1/8	…	0.002	0.0009	80	150	0.0006	50	85
	1/4	…	0.002	0.0008	80	150	0.0005	50	85
	3/8	…	0.0015	0.0007	80	150	0.0004	50	85
	1/2	…	0.0012	0.0006	80	150	0.0004	50	85
	5/8	…	0.001	0.0005	80	150	0.0003	50	85
	3/4	…	0.001	0.0005	80	150	0.0003	50	85
	1	…	0.001	0.0004	80	150	…	…	…

Approximate Cutting Speeds and Feeds for Standard Automatic Screw Machine Tools—Brown and Sharpe *(Continued)*

Tool		Cut: Width or Depth, Inches	Cut: Dia. of Hole, Inches	Brass[a]: Feed, Inches per Rev.	Mild or Soft Steel: Feed, Inches per Rev.	Mild or Soft Steel: Surface Speed, Feet per Min., Carbon Tools	Mild or Soft Steel: Surface Speed, Feet per Min., H.S.S. Tools	Tool Steel, 0.80–1.00% C: Feed, Inches per Rev.	Tool Steel: Surface Speed, Feet per Min., Carbon Tools	Tool Steel: Surface Speed, Feet per Min., H.S.S. Tools
Hollow mills and balance turning tools {	Turned diam. under 5/32 in. {	1/32	…	0.012	0.010	70	150	0.008	40	85
		1/16	…	0.010	0.009	70	150	0.006	40	85
	Turned diam. over 5/32 in. {	1/32	…	0.017	0.014	70	150	0.010	40	85
		1/16	…	0.015	0.012	70	150	0.008	40	85
		1/8	…	0.012	0.010	70	150	0.008	40	85
		3/16	…	0.010	0.008	70	150	0.006	40	85
		1/4	…	0.009	0.007	70	150	0.0045	40	85
Knee tools		1/32	…	…	0.010	70	150	0.008	40	85
Knurling tools {	Turret {	On	…	0.020	0.015	150	…	0.010	105	…
		Off	…	0.040	0.030	150	…	0.025	105	…
	Side or swing {	…	…	0.004	0.002	150	…	0.002	105	…
		…	…	0.006	0.004	150	…	0.003	105	…
	Top {	…	…	0.005	0.003	150	…	0.002	105	…
		…	…	0.008	0.006	150	…	0.004	105	…
Pointing and facing tools		…	…	0.001	0.0008	70	150	0.0005	40	80
		…	…	0.0025	0.002	70	150	0.0008	40	80
Reamers and bits		0.003 – 0.004	1/8 or less	0.010 – 0.007	0.008 – 0.006	70	105	0.006 – 0.004	40	60
		0.004 – 0.008	1/8 or over	0.010	0.010	70	105	0.006 – 0.008	40	60
Recessing tools {	End cut {	…	… {	0.001	0.0006	70	150	0.0004	40	75
		…	…	0.005	0.003	70	150	0.002	40	75
	Inside cut {	1/16–1/8	… {	0.0025	0.002	70	105	0.0015	40	60
			…	0.0008	0.0006	70	105	0.0004	40	60
Swing tools, forming		1/8	…	0.002	0.0007	70	150	0.0005	40	85
		1/4	…	0.0012	0.0005	70	150	0.0003	40	85
		3/8	…	0.001	0.0004	70	150	0.0002	40	85
		1/2	…	0.0008	0.0003	70	150	0.0002	40	85
Turning, straight and taper[b]		1/32	…	0.008	0.006	70	150	0.0035	40	85
		1/16	…	0.006	0.004	70	150	0.003	40	85
		1/8	…	0.005	0.003	70	150	0.002	40	85
		3/16	…	0.004	0.0025	70	150	0.0015	40	85
Taps		…	…	…	…	25	30	…	12	15

[a] Use maximum spindle speed on machine.

[b] For taper turning use feed slow enough for greatest depth depth of cut.

Spindle Revolutions and Cam Rise for Threading

Length of Threaded Portion, Inch	Number of Threads per Inch														
	80	72	64	56	48	40	36	32	30	28	24	20	18	16	14
	First Line: Revolutions of Spindle for Threading. Second Line: Rise on Cam for Threading, Inch														
1⁄16	9.50	9.00	8.50	8.00	6.00	5.50	5.50	5.00	5.00	5.00	3.00	…	…	…	…
	0.107	0.113	0.120	0.129	0.110	0.121	0.134	0.138	0.147	0.157	0.106	…	…	…	…
1⁄8	14.50	13.50	12.50	11.50	9.00	8.00	7.00	7.00	7.00	6.50	4.50	4.00	3.50	3.50	…
	0.163	0.169	0.176	0.185	0.165	0.176	0.171	0.193	0.205	0.204	0.159	0.170	0.165	0.186	…
3⁄16	19.50	18.00	16.50	15.00	12.00	10.50	10.00	9.00	8.50	8.50	6.00	5.50	5.00	4.50	4.00
	0.219	0.225	0.232	0.241	0.220	0.231	0.244	0.248	0.249	0.267	0.213	0.234	0.236	0.239	0.243
1⁄4	24.50	23.508	20.50	18.50	15.00	13.00	12.00	11.00	10.50	10.00	7.50	6.50	6.00	5.50	5.00
	0.276	0.294	0.288	0.297	0.275	0.286	0.293	0.303	0.308	0.314	0.266	0.276	0.283	0.292	0.304
5⁄16	29.50	27.00	24.50	22.00	18.00	15.50	14.50	13.00	12.50	12.00	9.00	8.00	7.00	6.50	6.00
	0.332	0.338	0.345	0.354	0.340	0.341	0.354	0.358	0.367	0.377	0.319	0.340	0.330	0.345	0.364
3⁄8	34.50	31.50	28.50	25.50	21.00	18.00	16.50	15.00	14.50	13.50	10.50	9.00	8.50	7.50	7.00
	0.388	0.394	0.401	0.410	0.385	0.396	0.403	0.413	0.425	0.424	0.372	0.383	0.401	0.398	0.425
7⁄16	39.50	36.00	32.50	29.00	24.00	20.50	19.00	17.00	16.00	15.50	12.00	10.50	9.50	8.50	7.50
	0.444	0.450	0.457	0.466	0.440	0.451	0.464	0.468	0.469	0.487	0.425	0.446	0.448	0.451	0.455
1⁄2	44.50	40.50	36.50	32.50	27.00	23.00	21.00	19.00	18.00	17.00	13.50	11.50	10.50	9.50	8.50
	0.501	0.506	0.513	0.522	0.495	0.506	0.513	0.523	0.528	0.534	0.478	0.489	0.496	0.504	0.516
9⁄16	49.50	45.00	40.50	36.00	30.00	25.50	23.50	21.00	20.00	19.00	15.00	13.00	11.50	10.50	9.50
	0.559	0.563	0.570	0.579	0.550	0.561	0.574	0.578	0.587	0.597	0.531	0.553	0.543	0.558	0.577
5⁄8	54.50	49.50	44.50	39.50	33.00	28.00	25.50	23.00	22.00	20.50	16.50	14.00	13.00	11.50	10.50
	0.613	0.619	0.626	0.635	0.605	0.616	0.623	0.633	0.645	0.644	0.584	0.595	0.614	0.611	0.637
11⁄16	59.50	54.00	48.50	43.00	36.00	30.50	28.00	25.00	23.50	22.50	18.00	15.50	14.00	12.50	11.00
	0.679	0.675	0.682	0.691	0.660	0.671	0.684	0.688	0.689	0.707	0.638	0.659	0.661	0.664	0.668
3⁄4	64.50	58.50	52.50	46.50	39.00	33.00	30.00	27.00	25.50	24.00	19.50	16.50	15.00	13.50	12.00
	0.726	0.731	0.738	0.747	0.715	0.726	0.733	0.743	0.748	0.754	0.691	0.701	0.708	0.717	0.728

Threading cams are often cut on a circular milling attachment. When this method is employed, the number of minutes the attachment should be revolved for each 0.001 inch rise, is first determined. As 15 spindle revolutions are required for threading and 400 for completing one piece, that part of the cam surface required for the actual threading operation equals 15 ÷ 400 = 0.0375, which is equivalent to 810 minutes of the circumference. The total rise, through an arc of 810 minutes is 0.413 inch, so the number of minutes for each 0.001 inch rise equals 810 ÷ 413 = 1.96 or, approximately, two minutes. If the attachment is graduated to read to five minutes, the cam will be fed laterally 0.0025 inch each time it is turned through five minutes of arc.

Practical Points on Cam and Tool Design.—The following general rules are given to aid in designing cams and special tools for automatic screw machines, and apply particularly to Brown and Sharpe machines:

1) Use the highest speeds recommended for the material used that the various tools will stand.

2) Use the arrangement of circular tools best suited for the class of work.

3) Decide on the quickest and best method of arranging the operations before designing the cams.

4) Do not use turret tools for forming when the cross-slide tools can be used to better advantage.

5) Make the shoulder on the circular cutoff tool large enough so that the clamping screw will grip firmly.

6) Do not use too narrow a cutoff blade.

7) Allow 0.005 to 0.010 inch for the circular tools to approach the work and 0.003 to 0.005 inch for the cutoff tool to pass the center.

8) When cutting off work, the feed of the cutoff tool should be decreased near the end of the cut where the piece breaks off.

9) When a thread is cut up to a shoulder, the piece should be grooved or necked to make allowance for the lead on the die. An extra projection on the forming tool and an extra amount of rise on the cam will be needed.

10) Allow sufficient clearance for tools to pass one another.

11) Always make a diagram of the cross-slide tools in position on the work when difficult operations are to be performed; do the same for the tools held in the turret.

12) Do not drill a hole the depth of which is more than 3 times the diameter of the drill, but rather use two or more drills as required. If there are not enough turret positions for the extra drills needed, make provision for withdrawing the drill clear of the hole and then advancing it into the hole again.

13) Do not run drills at low speeds. Feeds and speeds recommended in the table starting on page 1095 should be followed as far as is practicable.

14) When the turret tools operate farther in than the face of the chuck, see that they will clear the chuck when the turret is revolved.

15) See that the bodies of all turret tools will clear the side of the chute when the turret is revolved.

16) Use a balance turning tool or a hollow mill for roughing cuts.

17) The rise on the thread lobe should be reduced so that the spindle will reverse when the tap or die holder is drawn out.

18) When bringing another tool into position after a threading operation, allow clearance before revolving the turret.

19) Make provision to revolve the turret rapidly, especially when pieces are being made in from three to five seconds and when only a few tools are used in the turret. It is sometimes desirable to use two sets of tools.

20) When using a belt-shifting attachment for threading, clearance should be allowed, as it requires extra time to shift the belt.

21) When laying out a set of cams for operating on a piece that requires to be slotted, cross-drilled or burred, allowance should be made on the lead cam so that the transferring arm can descend and ascend to and from the work without coming in contact with any of the turret tools.

22) Always provide a vacant hole in the turret when it is necessary to use the transferring arm.

23) When designing special tools allow as much clearance as possible. Do not make them so that they will just clear each other, as a slight inaccuracy in the dimensions will often cause trouble.

24) When designing special tools having intricate movements, avoid springs as much as possible, and use positive actions.

Stock for Screw Machine Products.—The amount of stock required for the production of 1000 pieces on the automatic screw machine can be obtained directly from the table *Stock Required for Screw Machine Products*. To use this table, add to the length of the work the width of the cut-off tool blade; then the number of feet of material required for 1000 pieces can be found opposite the figure thus obtained, in the column headed "Feet per 1000 Parts." Screw machine stock usually comes in bars 10 feet long, and in compiling this table an allowance was made for chucking on each bar.

The table can be extended by using the following formula, in which

F = number of feet required for 1000 pieces

L = length of piece in inches

W = width of cut-off tool blade in inches

$$F = (L + W) \times 84$$

The amount to add to the length of the work, or the width of the cut-off tool, is given in the following, which is standard in a number of machine shops:

Diameter of Stock, Inches	Width of Cut-off Tool Blade, Inches
0.000–0.250	0.045
0.251–0.375	0.062
0.376–0.625	0.093
0.626–1.000	0.125
1.001–1.500	0.156

It is sometimes convenient to know the weight of a certain number of pieces, when estimating the price. The weight of round bar stock can be found by means of the following formulas, in which

W = weight in pounds

D = diameter of stock in inches

F = length in feet

For brass stock: $W = D^2 \times 2.86 \times F$

For steel stock: $W = D^2 \times 2.675 \times F$

For iron stock: $W = D^2 \times 2.65 \times F$

Stock Required for Screw Machine Products

The table gives the amount of stock, in feet, required for 1000 pieces, when the length of the finished part plus the thickness of the cut-off tool blade is known. Allowance has been made for chucking. To illustrate, if length of cut-off tool and work equals 0.140 inch, 11.8 feet of stock is required for the production of 1000 parts.

Length of Piece and Cut-Off Tool	Feet per 1000 Parts	Length of Piece and Cut-Off Tool	Feet per 1000 Parts	Length of Piece and Cut-Off Tool	Feet per 1000 Parts	Length of Piece and Cut-Off Tool	Feet per 1000 Parts
0.050	4.2	0.430	36.1	0.810	68.1	1.380	116.0
0.060	5.0	0.440	37.0	0.820	68.9	1.400	117.6
0.070	5.9	0.450	37.8	0.830	69.7	1.420	119.3
0.080	6.7	0.460	38.7	0.840	70.6	1.440	121.0
0.090	7.6	0.470	39.5	0.850	71.4	1.460	122.7
0.100	8.4	0.480	40.3	0.860	72.3	1.480	124.4
0.110	9.2	0.490	41.2	0.870	73.1	1.500	126.1
0.120	10.1	0.500	42.0	0.880	73.9	1.520	127.7
0.130	10.9	0.510	42.9	0.890	74.8	1.540	129.4
0.140	11.8	0.520	43.7	0.900	75.6	1.560	131.1
0.150	12.6	0.530	44.5	0.910	76.5	1.580	132.8
0.160	13.4	0.540	45.4	0.920	77.3	1.600	134.5
0.170	14.3	0.550	46.2	0.930	78.2	1.620	136.1
0.180	15.1	0.560	47.1	0.940	79.0	1.640	137.8
0.190	16.0	0.570	47.9	0.950	79.8	1.660	139.5
0.200	16.8	0.580	48.7	0.960	80.7	1.680	141.2
0.210	17.6	0.590	49.6	0.970	81.5	1.700	142.9
0.220	18.5	0.600	50.4	0.980	82.4	1.720	144.5
0.230	19.3	0.610	51.3	0.990	83.2	1.740	146.2
0.240	20.2	0.620	52.1	1.000	84.0	1.760	147.9
0.250	21.0	0.630	52.9	1.020	85.7	1.780	149.6
0.260	21.8	0.640	53.8	1.040	87.4	1.800	151.3
0.270	22.7	0.650	54.6	1.060	89.1	1.820	152.9
0.280	23.5	0.660	55.5	1.080	90.8	1.840	154.6
0.290	24.4	0.670	56.3	1.100	92.4	1.860	156.3
0.300	25.2	0.680	57.1	1.120	94.1	1.880	158.0
0.310	26.1	0.690	58.0	1.140	95.8	1.900	159.7
0.320	26.9	0.700	58.8	1.160	97.5	1.920	161.3
0.330	27.7	0.710	59.7	1.180	99.2	1.940	163.0
0.340	28.6	0.720	60.5	1.200	100.8	1.960	164.7
0.350	29.4	0.730	61.3	1.220	102.5	1.980	166.4
0.360	30.3	0.740	62.2	1.240	104.2	2.000	168.1
0.370	31.1	0.750	63.0	1.260	105.9	2.100	176.5
0.380	31.9	0.760	63.9	1.280	107.6	2.200	184.9
0.390	32.8	0.770	64.7	1.300	109.2	2.300	193.3
0.400	33.6	0.780	65.5	1.320	110.9	2.400	201.7
0.410	34.5	0.790	66.4	1.340	112.6	2.500	210.1
0.420	35.3	0.800	67.2	1.360	114.3	2.600	218.5

Band Saw Blade Selection.—The primary factors to consider in choosing a saw blade are: the pitch, or the number of teeth per inch of blade; the tooth form; and the blade type (material and construction). Tooth pitch selection depends on the size and shape of the work, whereas tooth form and blade type depend on material properties of the workpiece and on economic considerations of the job.

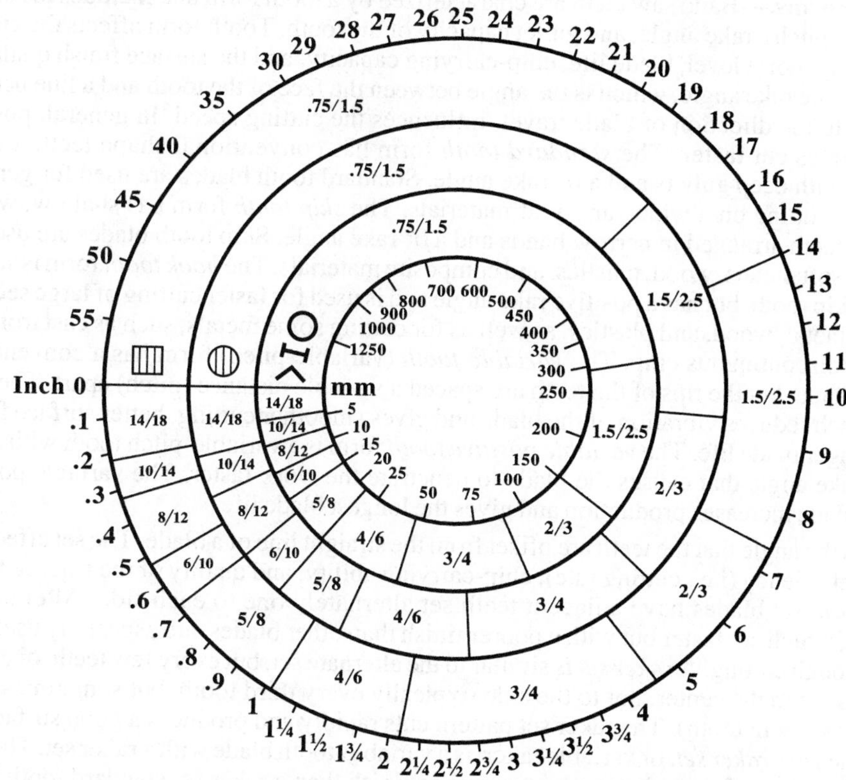

Courtesy of American Saw and Manufacturing Company

The tooth selection chart above is a guide to help determine the best blade pitch for a particular job. The tooth specifications in the chart are standard variable-pitch blade sizes as specified by the Hack and Band Saw Association. The variable-pitch blades listed are designated by two numbers that refer to the approximate maximum and minimum tooth pitch. A 4⁄6 blade, for example, has a maximum tooth spacing of approximately $\frac{1}{4}$ inch and a minimum tooth spacing of about $\frac{1}{6}$ inch. Blades are available, from most manufacturers, in sizes within about ±10 per cent of the sizes listed.

To use the chart, locate the length of cut in inches on the outside circle of the table (for millimeters use the inside circle) and then find the tooth specification that aligns with the length, on the ring corresponding to the material shape. The length of cut is the distance that any tooth of the blade is in contact with the work as it passes once through the cut. For cutting solid round stock, use the diameter as the length of cut and select a blade from the ring with the solid circle. When cutting angles, channels, I-beams, tubular pieces, pipe, and hollow or irregular shapes, the length of cut is found by dividing the cross-sectional area of the cut by the distance the blade needs to travel to finish the cut. Locate the length of cut on the outer ring (inner ring for mm) and select a blade from the ring marked with the angle, I-beam, and pipe sections.

Example: A 4-inch pipe with a 3-inch inside diameter is to be cut. Select a variable pitch blade for cutting this material.

The area of the pipe is $\pi/4 \times (4^2 - 3^2) = 5.5$ in.^{2} The blade has to travel 4 inches to cut through the pipe, so the average length of cut is 5.5/4 = 1.4 inches. On the tooth selection wheel, estimate the location of 1.4 inches on the outer ring, and read the tooth specification from the ring marked with the pipe, angle, and I-beam symbols. The chart indicates that a 4/6 variable-pitch blade is the preferred blade for this cut.

Tooth Forms.—Band saw teeth are characterized by a tooth form that includes the shape, spacing (pitch), rake angle, and gullet capacity of the tooth. Tooth form affects the cutting efficiency, noise level, blade life, chip-carrying capacity, and the surface finish quality of the cut. The rake angle, which is the angle between the face of the tooth and a line perpendicular to the direction of blade travel, influences the cutting speed. In general, positive rake angles cut faster. The *standard tooth* form has conventional shape teeth, evenly spaced with deep gullets and a 0° rake angle. Standard tooth blades are used for general-purpose cutting on a wide variety of materials. The *skip tooth* form has shallow, widely spaced teeth arranged in narrow bands and a 0° rake angle. Skip tooth blades are used for cutting soft metals, wood, plastics, and composite materials. The *hook tooth* form is similar to the skip tooth, but has a positive rake angle and is used for faster cutting of large sections of soft metal, wood, and plastics, as well as for cutting some metals, such as cast iron, that form a discontinuous chip. The *variable-tooth* (variable-pitch) form has a conventional tooth shape, but the tips of the teeth are spaced a variable distance (pitch) apart. The variable pitch reduces vibration of the blade and gives smoother cutting, better surface finish, and longer blade life. The *variable positive tooth* form is a variable-pitch tooth with a positive rake angle that causes the blade to penetrate the work faster. The variable positive tooth blade increases production and gives the longest blade life.

Set is the angle that the teeth are offset from the straight line of a blade. The set affects the blade efficiency (i.e., cutting rate), chip-carrying ability, and quality of the surface finish. *Alternate set* blades have adjacent teeth set alternately one to each side. Alternate set blades, which cut faster but with a poorer finish than other blades, are especially useful for rapid rough cutting. A *raker set is* similar to the alternate set, but every few teeth, one of the teeth is set to the center, not to the side (typically every third tooth, but sometimes every fifth or seventh tooth). The raker set pattern cuts rapidly and produces a good surface finish. The *vari-raker set,* or variable raker, is a variable-tooth blade with a raker set. The vari-raker is quieter and produces a better surface finish than a raker set standard tooth blade. *Wavy set* teeth are set in groups, alternately to one side, then to the other. Both wavy set and vari-raker set blades are used for cutting tubing and other interrupted cuts, but the blade efficiency and surface finish produced are better with a vari-raker set blade.

Types of Blades.—The most important band saw blade types are carbon steel, bimetal, carbide tooth, and grit blades made with embedded carbide or diamond. *Carbon steel blades* have the lowest initial cost, but they may wear out faster. Carbon steel blades are used for cutting a wide variety of materials, including mild steels, aluminum, brass, bronze, cast iron, copper, lead, and zinc, as well as some abrasive materials such as cork, fiberglass, graphite, and plastics. *Bimetal blades* are made with a high-speed steel cutting edge that is welded to a spring steel blade back. Bimetal blades are stronger and last longer, and they tend to produce straighter cuts because the blade can be tensioned higher than carbon steel blades. Because bimetal blades last longer, the cost per cut is frequently lower than when using carbon steel blades. Bimetal blades are used for cutting all ferrous and nonferrous metals, a wide range of shapes of easy to moderately machinable material, and solids and heavy wall tubing with moderate to difficult machinability. *Tungsten carbide blades* are similar to bimetal blades but have tungsten carbide teeth welded to the blade back. The welded teeth of carbide blades have greater wear and high-temperature resistance than either carbon steel or bimetal blades and produce less tooth vibration, while giving smoother, straighter, faster, and quieter cuts requiring less feed force. Carbide blades are used on tough alloys such as cobalt, nickel- and titanium-based alloys, and for nonferrous materials such as aluminum castings, fiberglass, and graphite. The *carbide grit blade* has tungsten carbide grit metallurgically bonded to either a gulleted (serrated) or toothless steel band. The blades are made in several styles and grit sizes. Both carbide grit and diamond grit blades are used to cut materials that conventional (carbon and bimetal) blades are unable to cut such as: fiberglass, reinforced plastics, composite materials, carbon and graphite, aramid fibers, plastics, cast iron, stellites, high-hardness tool steels, and superalloys.

Band Saw Speed and Feed Rate.—The band speed necessary to cut a particular material is measured in feet per minute (fpm) or in meters per minute (m/min), and depends on material characteristics and size of the workpiece. Typical speeds for a bimetal blade cutting 4-inch material with coolant are given in the speed selection table that follows. For other size materials or when cutting without coolant, adjust speeds according to the instructions at the bottom of the table.

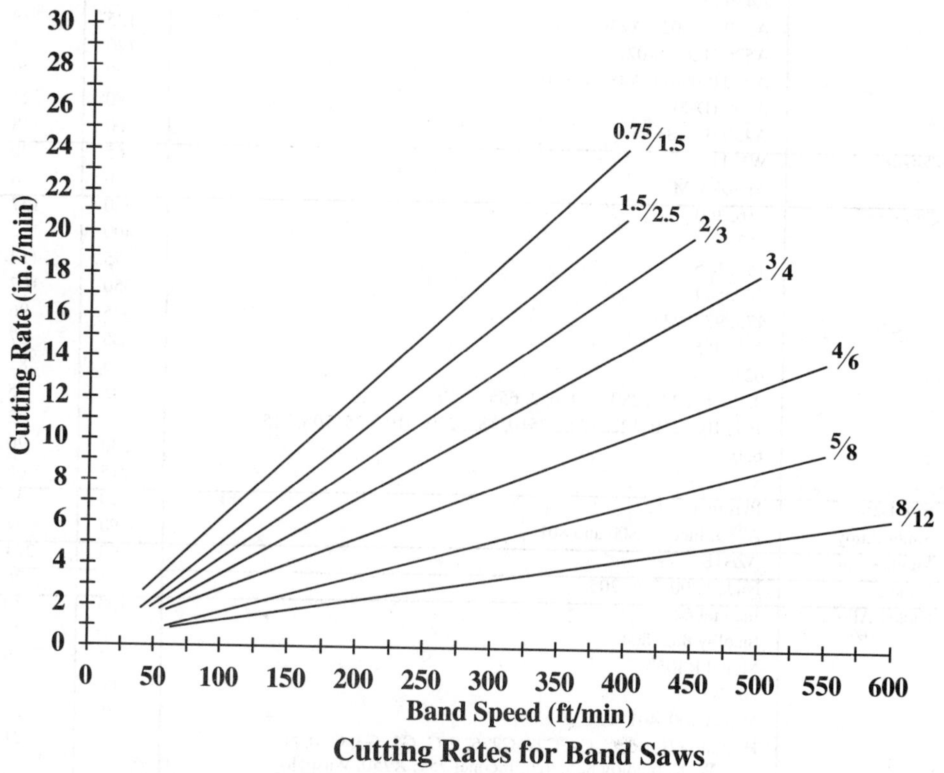

Cutting Rates for Band Saws

The feed or cutting rate, usually measured in square inches or square meters per minute, indicates how fast material is being removed and depends on the speed and pitch of the blade, not on the workpiece material. The graph above, based on material provided by American Saw and Mfg., gives approximate cutting rates (in.^{2}/min) for various variable-pitch blades and cutting speeds. Use the value from the graph as an initial starting value and then adjust the feed based on the performance of the saw. The size and character of the chips being produced are the best indicators of the correct feed force. Chips that are curly, silvery, and warm indicate the best feed rate and band speed. If the chips appear burned and heavy, the feed is too great, so reduce the feed rate, the band speed, or both. If the chips are thin or powdery, the feed rate is too low, so increase the feed rate or reduce the band speed. The actual cutting rate achieved during a cut is equal to the area of the cut divided by the time required to finish the cut. The time required to make a cut is equal to the area of the cut divided by the cutting rate in square inches per minute.

Bimetal Band Saw Speeds for Cutting 4-Inch Material with Coolant

Material	Category (AISI/SAE)	Speed (fpm)	Speed (m/min)
Aluminum Alloys	1100, 2011, 2017, 2024, 3003, 5052, 5086, 6061, 6063, 6101, 6262, 7075	500	152
Cast Iron	A536 (60-40-18)	360	110
	A47	300	91
	A220 (50005), A536 (80-55-06)	240	73
	A48 (20 ksi)	230	70
	A536 (100-70-03)	185	56
	A48 (40 ksi)	180	55
	A220 (60004)	170	52
	A436 (1B)	150	46
	A220 (70003)	145	44
	A436 (2)	140	43
	A220 (80002), A436 (2B)	125	38
	A536 (120-90-02)	120	37
	A220 (90001), A48 (60 ksi)	100	30
	A439 (D-2)	80	24
	A439 (D-2B)	60	18
Cobalt	WF-11	65	20
	Astroloy M	60	18
Copper	356, 360	450	137
	353	400	122
	187, 1452	375	114
	380, 544	350	107
	173, 932, 934	315	96
	330, 365	285	87
	623, 624	265	81
	230, 260, 272, 280, 464, 632, 655	245	75
	101, 102, 110, 122, 172, 17510, 182, 220, 510, 625, 706, 715	235	72
	630	230	70
	811	215	66
Iron Base Super Alloy	Pyromet X-15	120	37
	A286, Incoloy 800 and 801	90	27
Magnesium	AZ31B	900	274
Nickel	Nickel 200, 201, 205	85	26
Nickel Alloy	Inconel 625	100	30
	Incoloy 802, 804	90	27
	Monel R405	85	26
	20CB3	80	24
	Monel 400, 401	75	23
	Hastelloy B, B2, C, C4, C22, C276, F, G, G2, G3, G30, N, S, W, X, Incoloy 825, 926, Inconel 751, X750, Waspaloy	70	21
	Monel K500	65	20
	Incoloy 901, 903, Inconel 600, 718, Ni-Span-C902, Nimonic 263, Rene 41, Udimet 500	60	18
	Nimonic 75	50	15
Stainless Steel	416, 420	190	58
	203EZ, 430, 430F, 4302	150	46
	303, 303PB, 303SE, 410, 440F, 30323	140	43
	304	120	37
	414, 30403	115	35
	347	110	34
	316, 31603	100	30
	Greek Ascoloy	95	29
	18-18-2, 309, Ferralium	90	27
	15-5PH, 17-4PH, 17-7PH, 2205, 310, AM350, AM355, Custom 450, Custom 455, PH13-8Mo, PH14-8Mo, PH15-7Mo	80	24
	22-13-5, Nitronic 50, 60	60	18

Bimetal Band Saw Speeds for Cutting 4-Inch Material with Coolant *(Continued)*

Material	Category (AISI/SAE)	Speed (fpm)	Speed (m/min)
Steel	12L14	425	130
	1213, 1215	400	122
	1117	340	104
	1030	330	101
	1008, 1015, 1020, 1025	320	98
	1035	310	94
	1018, 1021, 1022, 1026, 1513, A242 Cor-Ten A	300	91
	1137	290	88
	1141, 1144, 1144 Hi Stress	280	85
	41L40	275	84
	1040, 4130, A242 Cor-Ten B, (A36 Shapes)	270	82
	1042, 1541, 4140, 4142	250	76
	8615, 8620, 8622	240	73
	W-1	225	69
	1044, 1045, 1330, 4340, E4340, 5160, 8630	220	67
	1345, 4145, 6150	210	64
	1060, 4150, 8640, A-6, O-1, S-1	200	61
	H-11, H-12, H-13, L-6, O-6	190	58
	1095	185	56
	A-2	180	55
	E9310	175	53
	300M, A-10, E52100, HY-80, HY-100	160	49
	S-5	140	43
	S-7	125	38
	M-1	110	34
	HP 9-4-20, HP 9-4-25	105	32
	M-2, M-42, T1	100	30
	D-2	90	27
	T-15	70	21
Titanium	Pure, Ti-3Al-8V-6Cr-4Mo-4Z, Ti-8Mo-8V-2Fe-3Al	80	24
	Ti-2Al-11Sn-5Zr-1Mo, Ti-5Al-2.5Sn, Ti-6Al-2Sn-4Zr-2Mo	75	23
	Ti-6Al-4V	70	21
	Ti-7Al-4Mo, Ti-8Al-1Mo-1V	65	20

The speed figures given are for 4-in. material (length of cut) using a 3⁄4 variable-tooth bimetal blade and cutting fluid. For cutting dry, reduce speed 30–50%; for carbon steel band saw blades, reduce speed 50%. For other cutting lengths: increase speed 15% for ¼-in. material (10/14 blade); increase speed 12% for ¾-in. material (6/10 blade); increase speed 10% for 1¼-in. material (4/6 blade); decrease speed 12% for 8-in. material (2/3 blade).

Table data are based on material provided by LENOX Blades, American Saw & Manufacturing Co.

Example: Find the band speed, the cutting rate, and the cutting time if the 4-inch pipe of the previous example is made of 304 stainless steel.

The preceding blade speed table gives the band speed for 4-inch 304 stainless steel as 120 fpm (feet per minute). The average length of cut for this pipe (see the previous example) is 1.4 inches, so increase the band saw speed by about 10 per cent (see footnote on) to 130 fpm to account for the size of the piece. On the cutting rate graph above, locate the point on the 4/6 blade line that corresponds to the band speed of 130 fpm and then read the cutting rate from the left axis of the graph. The cutting rate for this example is approximately 4 in.^{2}/min. The cutting time is equal to the area of the cut divided by the cutting rate, so cutting time = 5.5/4 = 1.375 minutes.

Band Saw Blade Break-In.—A new band saw blade must be broken in gradually before it is allowed to operate at its full recommended feed rate. Break-in relieves the blade of residual stresses caused by the manufacturing process so that the blade retains its cutting ability longer. Break-in requires starting the cut at the material cutting speed with a low feed rate and then gradually increasing the feed rate over time until enough material has been cut. A blade should be broken in with the material to be cut.

To break in a new blade, first set the band saw speed at the recommended cutting speed for the material and start the first cut at the feed indicated on the starting feed rate graph below. After the saw has penetrated the work to a distance equal to the width of the blade, increase the feed slowly. When the blade is about halfway through the cut, increase the feed again slightly and finish the cut without increasing the feed again. Start the next and each successive cut with the same feed rate that ended the previous cut, and increase the feed rate slightly again before the blade reaches the center of the cut. Repeat this procedure until the area cut by the new blade is equal to the total area required as indicated on the graph below. At the end of the break-in period, the blade should be cutting at the recommended feed rate, otherwise adjusted to that rate.

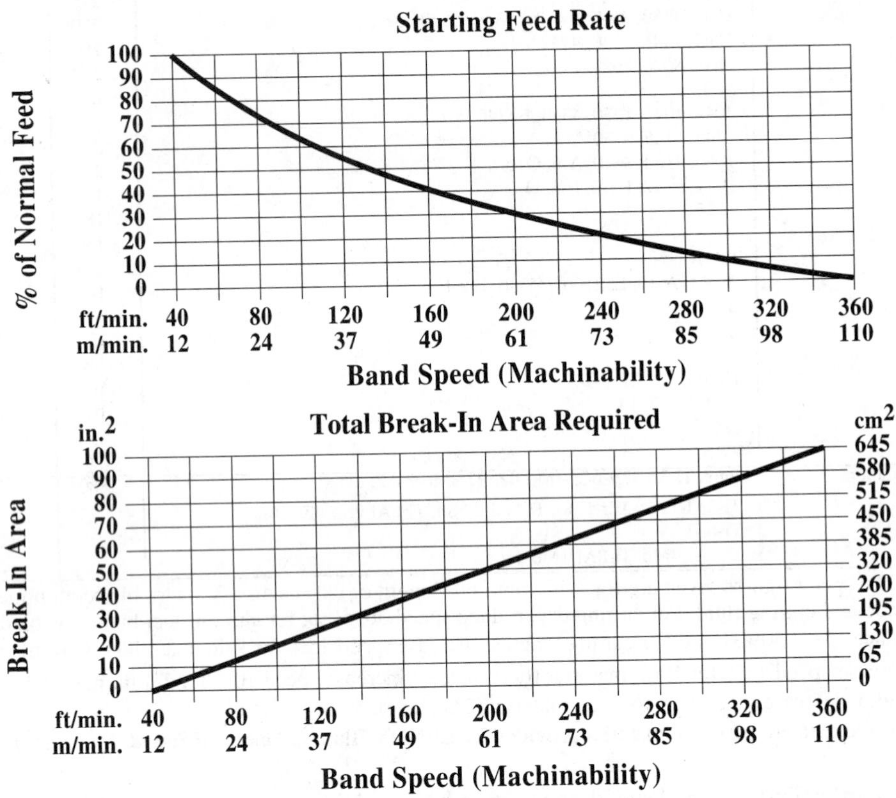

Cutting Fluids for Machining

The goal in all conventional metal-removal operations is to raise productivity and reduce costs by machining at the highest practical speed consistent with long tool life, fewest rejects, and minimum downtime, and with the production of surfaces of satisfactory accuracy and finish. Many machining operations can be performed "dry," but the proper application of a cutting fluid generally makes possible: higher cutting speeds, higher feed rates, greater depths of cut, lengthened tool life, decreased surface roughness, increased dimensional accuracy, and reduced power consumption. Selecting the proper cutting fluid for a specific machining situation requires knowledge of fluid functions, properties, and limitations. Cutting fluid selection deserves as much attention as the choice of machine tool, tooling, speeds, and feeds.

To understand the action of a cutting fluid it is important to realize that almost all the energy expended in cutting metal is transformed into heat, primarily by the deformation of the metal into the chip and, to a lesser degree, by the friction of the chip sliding against the tool face. With these factors in mind it becomes clear that the primary functions of any cutting fluid are: cooling of the tool, workpiece, and chip; reducing friction at the sliding contacts; and reducing or preventing welding or adhesion at the contact surfaces, which forms the "built-up edge" on the tool. Two other functions of cutting fluids are flushing away chips from the cutting zone and protecting the workpiece and tool from corrosion.

The relative importance of the functions is dependent on the material being machined, the cutting tool and conditions, and the finish and accuracy required on the part. For example, cutting fluids with greater lubricity are generally used in low-speed machining and on most difficult-to-cut materials. Cutting fluids with greater cooling ability are generally used in high-speed machining on easier-to-cut materials.

Types of Cutting and Grinding Fluids.—In recent years a wide range of cutting fluids has been developed to satisfy the requirements of new materials of construction and new tool materials and coatings.

There are four basic types of cutting fluids; each has distinctive features, as well as advantages and limitations. Selection of the right fluid is made more complex because the dividing line between types is not always clear. Most machine shops try to use as few different fluids as possible and prefer fluids that have long life, do not require constant changing or modifying, have reasonably pleasant odors, do not smoke or fog in use, and, most important, are neither toxic nor cause irritation to the skin. Other issues in selection are the cost and ease of disposal.

The major divisions and subdivisions used in classifying cutting fluids are:

Cutting Oils, including straight and compounded mineral oils plus additives.

Water-Miscible Fluids, including emulsifiable oils; chemical or synthetic fluids; and semichemical fluids.

Gases.

Paste and Solid Lubricants.

Since the cutting oils and water-miscible types are the most commonly used cutting fluids in machine shops, discussion will be limited primarily to these types. It should be noted, however, that compressed air and inert gases, such as carbon dioxide, nitrogen, and Freon, are sometimes used in machining. Paste, waxes, soaps, graphite, and molybdenum disulfide may also be used, either applied directly to the workpiece or as an impregnant in the tool, such as in a grinding wheel.

Cutting Oils.—Cutting oils are generally compounds of mineral oil with the addition of animal, vegetable, or marine oils to improve the wetting and lubricating properties. Sulfur, chlorine, and phosphorous compounds, sometimes called extreme pressure (EP) additives, provide for even greater lubricity. In general, these cutting oils do not cool as well as water-miscible fluids.

Water-Miscible Fluids.—*Emulsions or soluble oils* are a suspension of oil droplets in water. These suspensions are made by blending the oil with emulsifying agents (soap and soaplike materials) and other materials. These fluids combine the lubricating and rust-prevention properties of oil with water's excellent cooling properties. Their properties are affected by the emulsion concentration, with "lean" concentrations providing better cooling but poorer lubrication, and with "rich" concentrations having the opposite effect. Additions of sulfur, chlorine, and phosphorus, as with cutting oils, yield "extreme pressure" (EP) grades.

Chemical fluids are true solutions composed of organic and inorganic materials dissolved in water. Inactive types are usually clear fluids combining high rust inhibition, high cooling, and low lubricity characteristics with high surface tension. Surface-active types include wetting agents and possess moderate rust inhibition, high cooling, and moderate lubricating properties with low surface tension. They may also contain chlorine and/or sulfur compounds for extreme pressure properties.

Semichemical fluids are combinations of chemical fluids and emulsions. These fluids have a lower oil content but a higher emulsifier and surface-active-agent content than emulsions, producing oil droplets of much smaller diameter. They possess low surface tension, moderate lubricity and cooling properties, and very good rust inhibition. Sulfur, chlorine, and phosphorus also are sometimes added.

Selection of Cutting Fluids for Different Materials and Operations.—The choice of a cutting fluid depends on many complex interactions including the machinability of the metal; the severity of the operation; the cutting tool material; metallurgical, chemical, and human compatibility; fluid properties, reliability, and stability; and finally cost. Other factors affect results. Some shops standardize on a few cutting fluids which have to serve all purposes. In other shops, one cutting fluid must be used for all the operations performed on a machine. Sometimes, a very severe operating condition may be alleviated by applying the "right" cutting fluid manually while the machine supplies the cutting fluid for other operations through its coolant system. Several voluminous textbooks are available with specific recommendations for the use of particular cutting fluids for almost every combination of machining operation and workpiece and tool material. In general, when experience is lacking, it is wise to consult the material supplier and/or any of the many suppliers of different cutting fluids for advice and recommendations. Another excellent source is the Machinability Data Center, one of the many information centers supported by the U.S. Department of Defense. While the following recommendations represent good practice, they are to serve as a guide only, and it is not intended to say that other cutting fluids will not, in certain specific cases, also be effective.

Steels: Caution should be used when using a cutting fluid on steel that is being turned at a high cutting speed with cemented carbide cutting tools. See *Application of Cutting Fluids to Carbides* later. Frequently this operation is performed dry. If a cutting fluid is used, it should be a soluble oil mixed to a consistency of about 1 part oil to 20 to 30 parts water. A sulfurized mineral oil is recommended for reaming with carbide tipped reamers although a heavy-duty soluble oil has also been used successfully.

The cutting fluid recommended for machining steel with high speed cutting tools depends largely on the severity of the operation. For ordinary turning, boring, drilling, and milling on medium and low strength steels, use a soluble oil having a consistency of 1 part oil to 10 to 20 parts water. For tool steels and tough alloy steels, a heavy-duty soluble oil having a consistency of 1 part oil to 10 parts water is recommended for turning and milling. For drilling and reaming these materials, a light sulfurized mineral-fatty oil is used. For tough operations such as tapping, threading, and broaching, a sulfochlorinated mineral-fatty oil is recommended for tool steels and high-strength steels, and a heavy sulfurized mineral-fatty oil or a sulfochlorinated mineral oil can be used for medium- and low-strength steels. Straight sulfurized mineral oils are often recommended for machining tough, stringy low carbon steels to reduce tearing and produce smooth surface finishes.

Stainless Steel: For ordinary turning and milling a heavy-duty soluble oil mixed to a consistency of 1 part oil to 5 parts water is recommended. Broaching, threading, drilling, and reaming produce best results using a sulfochlorinated mineral-fatty oil.

Copper Alloys: Most brasses, bronzes, and copper are stained when exposed to cutting oils containing active sulfur and chlorine; thus, sulfurized and sulfochlorinated oils should not be used. For most operations a straight soluble oil, mixed to 1 part oil and 20 to 25 parts water is satisfactory. For very severe operations and for automatic screw machine work a mineral-fatty oil is used. A typical mineral-fatty oil might contain 5 to 10 per cent lard oil with the remainder mineral oil.

Monel Metal: When turning this material, an emulsion gives a slightly longer tool life than a sulfurized mineral oil, but the latter aids in chip breakage, which is frequently desirable.

Aluminum Alloys: Aluminum and aluminum alloys are frequently machined dry. When a cutting fluid is used it should be selected for its ability to act as a coolant. Soluble oils mixed to a consistency of 1 part oil to 20 to 30 parts water can be used. Mineral oil-base cutting fluids, when used to machine aluminum alloys, are frequently cut back to increase their viscosity so as to obtain good cooling characteristics and to make them flow easily to cover the tool and the work. For example, a mineral-fatty oil or a mineral plus a sulfurized fatty oil can be cut back by the addition of as much as 50 per cent kerosene.

Cast Iron: Ordinarily, cast iron is machined dry. Some increase in tool life can be obtained or a faster cutting speed can be used with a chemical cutting fluid or a soluble oil mixed to consistency of 1 part oil and 20 to 40 parts water. A soluble oil is sometimes used to reduce the amount of dust around the machine.

Magnesium: Magnesium may be machined dry, or with an air blast for cooling. A light mineral oil of low acid content may be used on difficult cuts. Coolants containing water should not be used on magnesium because of the danger of releasing hydrogen caused by reaction of the chips with water. Proprietary water-soluble oil emulsions containing inhibitors that reduce the rate of hydrogen generation are available.

Grinding: Soluble oil emulsions or emulsions made from paste compounds are used extensively in precision grinding operations. For cylindrical grinding, 1 part oil to 40 to 50 parts water is used. Solution type fluids and translucent grinding emulsions are particularly suited for many fine-finish grinding applications. Mineral oil-base grinding fluids are recommended for many applications where a fine surface finish is required on the ground surface. Mineral oils are used with vitrified wheels but are not recommended for wheels with rubber or shellac bonds. Under certain conditions the oil vapor mist caused by the action of the grinding wheel can be ignited by the grinding sparks and explode. To quench the grinding spark a secondary coolant line to direct a flow of grinding oil below the grinding wheel is recommended.

Broaching: For steel, a heavy mineral oil such as sulfurized oil of 300 to 500 Saybolt viscosity at 100 degrees F can be used to provide both adequate lubricating effect and a dampening of the shock loads. Soluble oil emulsions may be used for the lighter broaching operations.

Cutting Fluids for Turning, Milling, Drilling and Tapping.—The following table, *Cutting Fluids Recommended for Machining Operations*, gives specific cutting oil recommendations for common machining operations.

Soluble Oils: Types of oils paste compounds that form emulsions when mixed with water: Soluble oils are used extensively in machining both ferrous and non-ferrous metals when the cooling quality is paramount and the chip-bearing pressure is not excessive. Care should be taken in selecting the proper soluble oil for precision grinding operations. Grinding coolants should be free from fatty materials that tend to load the wheel, thus affecting the finish on the machined part. Soluble coolants should contain rust preventive constituents to prevent corrosion.

Base Oils: Various types of highly sulfurized and chlorinated oils containing inorganic, animal, or fatty materials. This "base stock" usually is "cut back" or blended with a lighter oil, unless the chip-bearing pressures are high, as when cutting alloy steel. Base oils usually have a viscosity range of from 300 to 900 seconds at 100 degrees F.

Mineral Oils: This group includes all types of oils extracted from petroleum such as paraffin oil, mineral seal oil, and kerosene. Mineral oils are often blended with base stocks, but they are generally used in the original form for light machining operations on both free-machining steels and non-ferrous metals. The coolants in this class should be of a type that has a relatively high flash point. Care should be taken to see that they are nontoxic, so that they will not be injurious to the operator. The heavier mineral oils (paraffin oils) usually have a viscosity of about 100 seconds at 100 degrees F. Mineral seal oil and kerosene have a viscosity of 35 to 60 seconds at 100 degrees F.

Cutting Fluids Recommended for Machining Operations

Material to be Cut	Turning	Milling
Aluminum[a]	Mineral Oil with 10 Per cent Fat (or) Soluble Oil	Soluble Oil (96 Per Cent Water) (or) Mineral Seal Oil (or) Mineral Oil
Alloy Steels[b]	25 Per Cent Sulfur base Oil[b] with 75 Per Cent Mineral Oil	10 Per Cent Lard Oil with 90 Per Cent Mineral Oil
Brass	Mineral Oil with 10 Per Cent Fat	Soluble Oil (96 Per Cent Water)
Tool Steels and Low-carbon Steels	25 Per Cent Lard Oil with 75 Per Cent Mineral Oil	Soluble Oil
Copper	Soluble Oil	Soluble Oil
Monel Metal	Soluble Oil	Soluble Oil
Cast Iron[c]	Dry	Dry
Malleable Iron	Soluble Oil	Soluble Oil
Bronze	Soluble Oil	Soluble Oil
Magnesium[d]	10 Per Cent Lard Oil with 90 Per Cent Mineral Oil	Mineral Seal Oil
Material to be Cut	**Drilling**	**Tapping**
Aluminum[e]	Soluble Oil (75 to 90 Per Cent Water) (or) 10 Per Cent Lard Oil with 90 Per Cent Mineral Oil	Lard Oil (or) Sperm Oil (or) Wool Grease (or) 25 Per Cent Sulfur-base Oil[b] Mixed with Mineral Oil
Alloy Steels[b]	Soluble Oil	30 Per Cent Lard Oil with 70 Per Cent Mineral Oil
Brass	Soluble Oil (75 to 90 Per Cent Water) (or) 30 Per Cent Lard Oil with 70 Per Cent Mineral Oil	10 to 20 Per Cent Lard Oil with Mineral Oil
Tool Steels and Low-carbon Steels	Soluble Oil	25 to 40 Per Cent Lard Oil with Mineral Oil (or) 25 Per Cent Sulfur-base Oil[b] with 75 Per Cent Mineral Oil
Copper	Soluble Oil	Soluble Oil
Monel Metal	Soluble Oil	25 to 40 Per Cent Lard Oil Mixed with Mineral Oil (or) Sulfur-base Oil[b] Mixed with Mineral Oil
Cast Iron[c]	Dry	Dry (or) 25 Per Cent Lard Oil with 75 Per Cent Mineral Oil
Malleable Iron	Soluble Oil	Soluble Oil
Bronze	Soluble Oil	20 Per Cent Lard Oil with 80 Per Cent Mineral Oil
Magnesium[d]	60-second Mineral Oil	20 Per Cent Lard Oil with 80 Per Cent Mineral Oil

[a] In machining aluminum, several varieties of coolants may be used. For rough machining, where the stock removal is sufficient to produce heat, water soluble mixtures can be used with good results to dissipate the heat. Other oils that may be recommended are straight mineral seal oil; a 50–50 mixture of mineral seal oil and kerosene; a mixture of 10 per cent lard oil with 90 per cent kerosene; and a 100-second mineral oil cut back with mineral seal oil or kerosene.

[b] The sulfur-base oil referred to contains 4½ per cent sulfur compound. Base oils are usually dark in color. As a rule, they contain sulfur compounds resulting from a thermal or catalytic refinery process. When so processed, they are more suitable for industrial coolants than when they have had such compounds as flowers of sulfur added by hand. The adding of sulfur compounds by hand to the coolant reservoir is of temporary value only, and the non-uniformity of the solution may affect the machining operation.

[c] A soluble oil or low-viscosity mineral oil may be used in machining cast iron to prevent excessive metal dust.

[d] When a cutting fluid is needed for machining magnesium, low or nonacid mineral seal or lard oils are recommended. Coolants containing water should not be used because of the fire danger when magnesium chips react with water, forming hydrogen gas.

[e] Sulfurized oils ordinarily are not recommended for tapping aluminum; however, for some tapping operations they have proved very satisfactory, although the work should be rinsed in a solvent right after machining to prevent discoloration.

Application of Cutting Fluids to Carbides.—Turning, boring, and similar operations on lathes using carbides are performed dry or with the help of soluble oil or chemical cutting fluids. The effectiveness of cutting fluids in improving tool life or by permitting higher cutting speeds to be used, is less with carbides than with high-speed steel tools. Furthermore, the effectiveness of the cutting fluid is reduced as the cutting speed is increased. Cemented carbides are very sensitive to sudden changes in temperature and to temperature gradients within the carbide. Thermal shocks to the carbide will cause thermal cracks to form near the cutting edge, which are a prelude to tool failure. An unsteady or interrupted flow of the coolant reaching the cutting edge will generally cause these thermal cracks. The flow of the chip over the face of the tool can cause an interruption to the flow of the coolant reaching the cutting edge even though a steady stream of coolant is directed at the tool. When a cutting fluid is used and frequent tool breakage is encountered, it is often best to cut dry. When a cutting fluid must be used to keep the workpiece cool for size control or to allow it to be handled by the operator, special precautions must be used. Sometimes applying the coolant from the front and the side of the tool simultaneously is helpful. On lathes equipped with overhead shields, it is very effective to apply the coolant from below the tool into the space between the shoulder of the work and the tool flank, in addition to applying the coolant from the top. Another method is not to direct the coolant stream at the cutting tool at all but to direct it at the workpiece above or behind the cutting tool.

The danger of thermal cracking is great when milling with carbide cutters. The nature of the milling operation itself tends to promote thermal cracking because the cutting edge is constantly heated to a high temperature and rapidly cooled as it enters and leaves the workpiece. For this reason, carbide milling operations should be performed dry.

Lower cutting-edge temperatures diminish the danger of thermal cracking. The cutting-edge temperatures usually encountered when reaming with solid carbide or carbide-tipped reamers are generally such that thermal cracking is not apt to occur except when reaming certain difficult-to-machine metals. Therefore, cutting fluids are very effective when used on carbide reamers. Practically every kind of cutting fluid has been used, depending on the job material encountered. For difficult surface-finish problems in holes, heavy duty soluble oils, sulfurized mineral-fatty oils, and sulfochlorinated mineral-fatty oils have been used successfully. On some work, the grade and the hardness of the carbide also have an effect on the surface finish of the hole.

Cutting fluids should be applied where the cutting action is taking place and at the highest possible velocity without causing splashing. As a general rule, it is preferable to supply from 3 to 5 gallons per minute for each single-point tool on a machine such as a turret lathe or automatic. The temperature of the cutting fluid should be kept below 110 degrees F. If the volume of fluid used is not sufficient to maintain the proper temperature, means of cooling the fluid should be provided.

Cutting Fluids for Machining Magnesium.—In machining magnesium, it is the general but not invariable practice in the United States to use a cutting fluid. In other places, magnesium usually is machined dry except where heat generated by high cutting speeds would not be dissipated rapidly enough without a cutting fluid. This condition may exist when, for example, small tools without much heat-conducting capacity are employed on automatics.

The cutting fluid for magnesium should be an anhydrous oil having, at most, a very low acid content. Various mineral-oil cutting fluids are used for magnesium.

Occupational Exposure To Metalworking Fluids

The term *metalworking fluids* (MWFs) describes coolants and lubricants used during the fabrication of products from metals and metal substitutes. These fluids are used to prolong the life of machine tools, carry away debris, and protect or treat the surfaces of the material being processed. MWFs reduce friction between the cutting tool and work surfaces, reduce wear and galling, protect surface characteristics, reduce surface adhesion or welding, carry away generated heat, and flush away swarf, chips, fines, and residues. Table 1 describes the four different classes of metal working fluids:

Table 1. Classes of Metalworking fluids (MWFs)

MWF	Description	Dilution factor
Straight oil (neat oil or cutting oil)	Highly refined petroleum oils (lubricant-base oils) or other animal, marine, vegetable, or synthetic oils used singly or in combination with or without additives. These are lubricants, or function to improve the finish on the metal cut, and prevent corrosion.	none
Soluble oil (emulsifiable oil)	Combinations of 30% to 85% highly refined, high-viscosity lubricant-base oils and emulsifiers that may include other performance additives. Soluble oils are diluted with water before use at ratios of parts water.	1 part concentrate to 5 to 40 parts water
Semisynthetic	Contain smaller amounts of severely refined lubricant-base oil (5 to 30% in the concentrate), a higher proportion of emulsifiers that may include other performance additives, and 30 to 50% water.	1 part concentrate to 10 to 40 parts water
Synthetic[a]	Contain no petroleum oils and may be water soluble or water dispersible. The simplest synthetics are made with organic and inorganic salts dissolved in water. Offer good rust protection and heat removal but usually have poor lubricating ability. May be formulated with other performance additives. Stable, can be made bioresistant.	1 part concentrate to 10 to 40 parts water

[a] Over the last several decades major changes in the U.S. machine tool industry have increased the consumption of MWFs. Specifically, the use of synthetic MWFs increased as tool and cutting speeds increased.

Occupational Exposures to Metal Working Fluids (MWFs).—Workers can be exposed to MWFs by inhalation of aerosols (mists) or by skin contact resulting in an increased risk of respiratory (lung) and skin disease. Health effects vary based on the type of MWF, route of exposure, concentration, and length of exposure.

Skin contact usually occurs when the worker dips his/her hands into the fluid, floods the machine tool, or handling parts, tools, equipment or workpieces coated with the fluid, without the use of personal protective equipment such as gloves and apron. Skin contact can also result from fluid splashing onto worker from the machine if guarding is absent or inadequate.

Inhalation exposures result from breathing MWF mist or aerosol. The amount of mist generated (and the severity of the exposure) depends on a variety of factors: the type of MWF and its application process; the MWF temperature; the specific machining or grinding operation; the presence of splash guarding; and the effectiveness of the ventilation system. In general, the exposure will be higher if the worker is in close proximity to the machine, the operation involves high tool speeds and deep cuts, the machine is not enclosed, or if ventilation equipment was improperly selected or poorly maintained. In addition, high-pressure and/or excessive fluid application, contamination of the fluid with tramp oils, and improper fluid selection and maintenance will tend to result in higher exposure.

Each MWF class consists of a wide variety of chemicals used in different combinations and the risk these chemicals pose to workers may vary because of different manufacturing processes, various degrees of refining, recycling, improperly reclaimed chemicals, different degrees of chemical purity, and potential chemical reactions between components.

Exposure to hazardous contaminants in MWFs may present health risks to workers. Contamination may occur from: process chemicals and ancillary lubricants inadvertently introduced; contaminants, metals, and alloys from parts being machined; water and cleaning agents used for routine housekeeping; and, contaminants from other environmental sources at the worksite. In addition, bacterial and fungal contaminants may metabolize and degrade the MWFs to hazardous end-products as well as produce endotoxins.

The improper use of biocides to manage microbial growth may result in potential health risks. Attempts to manage microbial growth solely with biocides may result in the emergence of biocide-resistant strains from complex interactions that may occur among different member species or groups within the population. For example, the growth of one species, or the elimination of one group of organisms may permit the overgrowth of another. Studies also suggest that exposure to certain biocides can cause either allergic or contact dermatitis.

Fluid Selection, Use, and Application.— The MWFs selected should be as nonirritating and nonsensitizing as possible while remaining consistent with operational requirements. Petroleum-containing MWFs should be evaluated for potential carcinogenicity using ASTM Standard D1687-95, "Determining Carcinogenic Potential of Virgin Base Oils in Metalworking Fluids". If soluble oil or synthetic MWFs are used, ASTM Standard E1497-94, "Safe Use of Water-Miscible Metalworking Fluids" should be consulted for safe use guidelines, including those for product selection, storage, dispensing, and maintenance. To minimize the potential for nitrosamine formation, nitrate-containing materials should not be added to MWFs containing ethanolamines.

Many factors influence the generation of MWF mists, which can be minimized through the proper design and operation of the MWF delivery system. ANSI Technical Report B11 TR2-1997, "Mist Control Considerations for the Design, Installation and Use of Machine Tools Using Metalworking Fluids" provides directives for minimizing mist and vapor generation. These include minimizing fluid delivery pressure, matching the fluid to the application, using MWF formulations with low oil concentrations, avoiding contamination with tramp oils, minimizing the MWF flow rate, covering fluid reservoirs and return systems where possible, and maintaining control of the MWF chemistry. Also, proper application of MWFs can minimize splashing and mist generation. Proper application includes: applying MWFs at the lowest possible pressure and flow volume consistent with provisions for adequate part cooling, chip removal, and lubrication; applying MWFs at the tool/workpiece interface to minimize contact with other rotating equipment; ceasing fluid delivery when not performing machining; not allowing MWFs to flow over the unprotected hands of workers loading or unloading parts; and using mist collectors engineered for the operation and specific machine enclosures.

Properly maintained filtration and delivery systems provide cleaner MWFs, reduce mist, and minimize splashing and emissions. Proper maintenance of the filtration and delivery systems includes: the selection of appropriate filters; ancillary equipment such as chip handling operations, dissolved air-flotation devices, belt-skimmers, chillers or plate and frame heat exchangers, and decantation tanks; guard coolant return trenches to prevent dumping of floor wash water and other waste fluids; covering sumps or coolant tanks to prevent contamination with waste or garbage (e.g., cigarette butts, food, etc.); and, keeping the machine(s) clean of debris. Parts washing before machining can be an important part of maintaining cleaner MWFs.

Since all additives will be depleted with time, the MWF and additives concentrations should be monitored frequently so that components and additives can be made up as needed. The MWF should be maintained within the pH and concentration ranges recommended by the formulator or supplier. MWF temperature should be maintained at the lowest practical level to slow the growth of microorganisms, reduce water losses and changes in viscosity, and–in the case of straight oils–reduce fire hazards.

Fluid Maintenance.—Drums, tanks, or other containers of MWF concentrates should be stored appropriately to protect them from outdoor weather conditions and exposure to low or high temperatures. Extreme temperature changes may destabilize the fluid concentrates, especially in the case of concentrates mixed with water, and cause water to seep into unopened drums encouraging bacterial growth. MWFs should be maintained at as low a temperature as is practical. Low temperatures slow the growth of microorganisms, reduce water losses and change in viscosity, and in the case of straight oils, reduce the fire hazard risks.

To maintain proper MWF concentrations, neither water nor concentrate should be used to top off the system. The MWF mixture should be prepared by first adding the concentrate to the clean water (in a clean container) and then adding the emulsion to that mixture in the coolant tank. MWFs should be mixed just before use; large amounts should not be stored, as they may deteriorate before use.

Personal Protective Clothing: Personal protective clothing and equipment should always be worn when removing MWF concentrates from the original container, mixing and diluting concentrate, preparing additives (including biocides), and adding MWF emulsions, biocides, or other potentially hazardous ingredients to the coolant reservoir. Personal protective clothing includes eye protection or face shields, gloves, and aprons which do not react with but shed MWF ingredients and additives.

System Service: Coolant systems should be regularly serviced, and the machines should be rigorously maintained to prevent contamination of the fluids by tramp oils (e.g., hydraulic oils, gear box oils, and machine lubricants leaking from the machines or total loss slideway lubrication). Tramp oils can destabilize emulsions, cause pumping problems, and clog filters. Tramp oils can also float to the top of MWFs, effectively sealing the fluids from the air, allowing metabolic products such as volatile fatty acids, mercaptols, scatols, ammonia, and hydrogen sulfide are produced by the anaerobic and facultative anaerobic species growing within the biofilm to accumulate in the reduced state.

When replacing the fluids, thoroughly clean all parts of the system to inhibit the growth of microorganisms growing on surfaces. Some bacteria secrete layers of slime that may grow in stringy configurations that resemble fungal growth. Many bacteria secrete polymers of polysaccharide and/or protein, forming a glycocalyx which cements cells together much as mortar holds bricks. Fungi may grow as masses of hyphae forming mycelial mats. The attached community of microorganisms is called a biofilm and may be very difficult to remove by ordinary cleaning procedures.

Biocide Treatment: Biocides are used to maintain the functionality and efficacy of MWFs by preventing microbial overgrowth. These compounds are often added to the stock fluids as they are formulated, but over time the biocides are consumed by chemical and biological demands Biocides with a wide spectrum of biocidal activity should be used to suppress the growth of the widely diverse contaminant population. Only the concentration of biocide needed to meet fluid specifications should be used since overdosing could lead to skin or respiratory irritation in workers, and under-dosing could lead to an inadequate level of microbial control.

Ventilation Systems: The ventilation system should be designed and operated to prevent the accumulation or recirculation of airborne contaminants in the workplace. The ventilation system should include a positive means of bringing in at least an equal volume of air from the outside, conditioning it, and evenly distributing it throughout the exhausted area.

Exhaust ventilation systems function through suction openings placed near a source of contamination. The suction opening or exhaust hood creates and air motion sufficient to overcome room air currents and any airflow generated by the process. This airflow cap-tures the contaminants and conveys them to a point where they can either be discharged or removed from the airstream. Exhaust hoods are classified by their position relative to the process as canopy, side draft, down draft or enclosure. ANSI Technical Report B11 TR 2-1997 contains guidelines for exhaust ventilation of machining and grinding operations. Enclosures are the only type of exhaust hood recommended by the ANSI committee. They consist of physical barriers between the process and the worker's environment. Enclosures can be further classified by the extent of the enclosure: close capture (enclosure of the point of operation, total enclosure (enclosure of the entire machine), or tunnel enclosure (continuous enclosure over several machines).

If no fresh make up air is introduced into the plant, air will enter the building through open doors and windows, potentially causing cross-contamination of all process areas. Ideally, all air exhausted from the building should be replaced by tempered air from an uncontaminated location. By providing a slight excess of make up air in relatively clean areas and s slight deficit of make up air in dirty areas, cross-contamination can be reduced. In addition, this air can be channeled directly to operator work areas, providing the cleanest possible work environment. Ideally, this fresh air should be supplied in the form of a low-velocity air shower (<100 ft/min to prevent interference with the exhaust hoods) directly above the worker.

Protective Clothing and Equipment: Engineering controls are used to reduce worker exposure to MWFs. But in the event of airborne exposures that exceed the NIOSH REL or dermal contact with the MWFs, the added protection of chemical protective clothing (CPC) and respirators should be provided. Maintenance staff may also need CPC because their work requires contact with MWFs during certain operations. All workers should be trained in the proper use and care of CPC. After any item of CPC has been in routine use, it should be examined to ensure that its effectiveness has not been compromised.

Selection of the appropriate respirator depends on the operation, chemical components, and airborne concentrations in the worker's breathing zone. Table 2. lists the NIOSH- recommended respiratory protection for workers exposed to MWF aerosol.

Respiratory Protection for Workers Exposed to MWF Aerosols[*]

Concentration of MWF aerosol (mg/m^3)	Minimum respiratory protection[a]
#0.5 mg/m^3 (1 × REL)[b]	No respiratory protection required for healthy workers[c]
#5.0 mg/m^3 (10 × REL)	Any air-purifying, half-mask respirator including a disposable respirator[d,e] equipped with any P- or R-series particulate filter (P95, P99, P100, R95, R99, or R100) number
#12.5 mg/m^3 (25 × REL)	Any powered, air-purifying respirator equipped with a hood or helmet and a HEPA filter[f]

[a] Respirators with higher assigned protection factors (APFs) may be substituted for those with lower APFs [NIOSH 1987a].

[b] APF times the NIOSH REL for total particulate mass. The APF [NIOSH 1987b] is the minimum anticipated level of protection provided by each type of respirator.

[c] See text for recommendations regarding workers with asthma and for other workers affected by MWF aerosols.

[d] A respirator that should be discarded after the end of the manufacturer's recommended period of use or after a noticeable increase in breathing resistance or when physical damage, hygiene considerations, or other warning indicators render the respirator unsuitable for further use.

[e] An APF of 10 is assigned to disposable particulate respirators if they have been properly fitted.

[f] High-efficiency particulate air filter. When organic vapors are a potential hazard during metalworking operations, a combination particulate and organic vapor filter is necessary.

[*] Only NIOSH/MSHA-approved or NIOSH-approved (effective date July 10, 1995) respiratory equipment should be used.

MACHINING NONFERROUS METALS

Machining Aluminum.—Some of the alloys of aluminum have been machined successfully without any lubricant or cutting compound, but some form of lubricant is desirable to obtain the best results. For many purposes, a soluble cutting oil is good.

Tools for aluminum and aluminum alloys should have larger relief and rake angles than tools for cutting steel. For high-speed steel turning tools the following angles are recommended: relief angles, 14 to 16 degrees; back rake angle, 5 to 20 degrees; side rake angle, 15 to 35 degrees. For very soft alloys even larger side rake angles are sometimes used. High silicon aluminum alloys and some others have a very abrasive effect on the cutting tool. While these alloys can be cut successfully with high-speed-steel tools, cemented carbides are recommended because of their superior abrasion resistance. The tool angles recommended for cemented carbide turning tools are: relief angles, 12 to 14 degrees; back rake angle, 0 to 15 degrees; side rake angle, 8 to 30 degrees.

Cut-off tools and necking tools for machining aluminum and its alloys should have from 12 to 20 degrees back rake angle and the end relief angle should be from 8 to 12 degrees. Excellent threads can be cut with single-point tools in even the softest aluminum. Experience seems to vary somewhat regarding the rake angle for single-point thread cutting tools. Some prefer to use a rather large back and side rake angle although this requires a modification in the included angle of the tool to produce the correct thread contour. When both rake angles are zero, the included angle of the tool is ground equal to the included angle of the thread. Excellent threads have been cut in aluminum with zero rake angle thread-cutting tools using large relief angles, which are 16 to 18 degrees opposite the front side of the thread and 12 to 14 degrees opposite the back side of the thread. In either case, the cutting edges should be ground and honed to a keen edge. It is sometimes advisable to give the face of the tool a few strokes with a hone between cuts when chasing the thread to remove any built-up edge on the cutting edge.

Fine surface finishes are often difficult to obtain on aluminum and aluminum alloys, particularly the softer metals. When a fine finish is required, the cutting tool should be honed to a keen edge and the surfaces of the face and the flank will also benefit by being honed smooth. Tool wear is inevitable, but it should not be allowed to progress too far before the tool is changed or sharpened. A sulphurized mineral oil or a heavy-duty soluble oil will sometimes be helpful in obtaining a satisfactory surface finish. For best results, however, a diamond cutting tool is recommended. Excellent surface finishes can be obtained on even the softest aluminum and aluminum alloys with these tools.

Although ordinary milling cutters can be used successfully in shops where aluminum parts are only machined occasionally, the best results are obtained with coarse-tooth, large helix-angle cutters having large rake and clearance angles. Clearance angles up to 10 to 12 degrees are recommended. When slab milling and end milling a profile, using the peripheral teeth on the end mill, climb milling (also called down milling) will generally produce a better finish on the machined surface than conventional (or up) milling. Face milling cutters should have a large axial rake angle. Standard twist drills can be used without difficulty in drilling aluminum and aluminum alloys although high helix-angle drills are preferred. The wide flutes and high helix-angle in these drills helps to clear the chips. Sometimes split-point drills are preferred. Carbide tipped twist drills can be used for drilling aluminum and its alloys and may afford advantages in some production applications. Ordinary hand and machine taps can be used to tap aluminum and its alloys although spiral-fluted ground thread taps give superior results. Experience has shown that such taps should have a right-hand ground flute when intended to cut right-hand threads and the helix angle should be similar to that used in an ordinary twist drill.

Machining Magnesium.—Magnesium alloys are readily machined and with relatively low power consumption per cubic inch of metal removed. The usual practice is to employ high cutting speeds with relatively coarse feeds and deep cuts. Exceptionally fine finishes can be obtained so that grinding to improve the finish usually is unnecessary. The horsepower normally required in machining magnesium varies from 0.15 to 0.30 per cubic inch per minute. While this value is low, especially in comparison with power required for cast iron and steel, the total amount of power for machining magnesium usually is high because of the exceptionally rapid rate at which metal is removed.

Carbide tools are recommended for maximum efficiency, although high-speed steel frequently is employed. Tools should be designed so as to dispose of chips readily or without excessive friction, by employing polished chip-bearing surfaces, ample chip spaces, large clearances, and small contact areas. *Keen-edged tools should always be used.*

Feeds and Speeds for Magnesium: Speeds ordinarily range up to 5000 feet per minute for rough- and finish-turning, up to 3000 feet per minute for rough-milling, and up to 9000 feet per minute for finish-milling. For rough-turning, the following combinations of speed in feet per minute, feed per revolution, and depth of cut are recommended: Speed 300 to 600 feet per minute — feed 0.030 to 0.100 inch, depth of cut 0.5 inch; speed 600 to 1000 — feed 0.020 to 0.080, depth of cut 0.4; speed 1000 to 1500 — feed 0.010 to 0.060, depth of cut 0.3; speed 1500 to 2000 — feed 0.010 to 0.040, depth of cut 0.2; speed 2000 to 5000 — feed 0.010 to 0.030, depth of cut 0.15.

Lathe Tool Angles for Magnesium: The true or actual rake angle resulting from back and side rakes usually varies from 10 to 15 degrees. Back rake varies from 10 to 20, and side rake from 0 to 10 degrees. Reduced back rake may be employed to obtain better chip breakage. The back rake may also be reduced to from 2 to 8 degrees on form tools or other broad tools to prevent chatter.

Parting Tools: For parting tools, the back rake varies from 15 to 20 degrees, the front end relief 8 to 10 degrees, the side relief measured perpendicular to the top face 8 degrees, the side relief measured in the plane of the top face from 3 to 5 degrees.

Milling Magnesium: In general, the coarse-tooth type of cutter is recommended. The number of teeth or cutting blades may be one-third to one-half the number normally used; however, the two-blade fly-cutter has proved to be very satisfactory. As a rule, the land relief or primary peripheral clearance is 10 degrees followed by secondary clearance of 20 degrees. The lands should be narrow, the width being about 3⁄64 to 1⁄16 inch. The rake, which is positive, is about 15 degrees.

For rough-milling and speeds in feet per minute up to 900 — feed, inch per tooth, 0.005 to 0.025, depth of cut up to 0.5; for speeds 900 to 1500 — feed 0.005 to 0.020, depth of cut up to 0.375; for speeds 1500 to 3000 — feed 0.005 to 0.010, depth of cut up to 0.2.

Drilling Magnesium: If the depth of a hole is less than five times the drill diameter, an ordinary twist drill with highly polished flutes may be used. The included angle of the point may vary from 70 degrees to the usual angle of 118 degrees. The relief angle is about 12 degrees. The drill should be kept sharp and the outer corners rounded to produce a smooth finish and prevent burr formation. For deep hole drilling, use a drill having a helix angle of 40 to 45 degrees with large polished flutes of uniform cross-section throughout the drill length to facilitate the flow of chips. A pyramid-shaped "spur" or "pilot point" at the tip of the drill will reduce the "spiraling or run-off."

Drilling speeds vary from 300 to 2000 feet per minute with feeds per revolution ranging from 0.015 to 0.050 inch.

Reaming Magnesium: Reamers up to 1 inch in diameter should have four flutes; larger sizes, six flutes. These flutes may be either parallel with the axis or have a negative helix angle of 10 degrees. The positive rake angle varies from 5 to 8 degrees, the relief angle from 4 to 7 degrees, and the clearance angle from 15 to 20 degrees.

Tapping Magnesium: Standard taps may be used unless Class 3B tolerances are required, in which case the tap should be designed for use in magnesium. A high-speed steel concentric type with a ground thread is recommended. The concentric form, which eliminates the radial thread relief, prevents jamming of chips while the tap is being backed out of the hole. The positive rake angle at the front may vary from 10 to 25 degrees and the "heel rake angle" at the back of the tooth from 3 to 5 degrees. The chamfer extends over two to three threads. For holes up to ¼ inch in diameter, two-fluted taps are recommended; for sizes from ½ to ¾ inch, three flutes; and for larger holes, four flutes. Tapping speeds ordinarily range from 75 to 200 feet per minute, and mineral oil cutting fluid should be used.

Threading Dies for Magnesium: Threading dies for use on magnesium should have about the same cutting angles as taps. Narrow lands should be used to provide ample chip space. Either solid or self-opening dies may be used. The latter type is recommended when maximum smoothness is required. Threads may be cut at speeds up to 1000 feet per minute.

Grinding Magnesium: As a general rule, magnesium is ground dry. The highly inflammable dust should be formed into a sludge by means of a spray of water or low-viscosity mineral oil. Accumulations of dust or sludge should be avoided. For surface grinding, when a fine finish is desirable, a low-viscosity mineral oil may be used.

Machining Zinc Alloy Die-Castings.—Machining of zinc alloy die-castings is mostly done without a lubricant. For particular work, especially deep drilling and tapping, a lubricant such as lard oil and kerosene (about half and half) or a 50-50 mixture of kerosene and machine oil may be used to advantage. A mixture of turpentine and kerosene has been been found effective on certain difficult jobs.

Reaming: In reaming, tools with six straight flutes are commonly used, although tools with eight flutes irregularly spaced have been found to yield better results by one manufacturer. Many standard reamers have a land that is too wide for best results. A land about 0.015 inch wide is recommended but this may often be ground down to around 0.007 or even 0.005 inch to obtain freer cutting, less tendency to loading, and reduced heating.

Turning: Tools of high-speed steel are commonly employed although the application of Stellite and carbide tools, even on short runs, is feasible. For steel or Stellite, a positive top rake of from 0 to 20 degrees and an end clearance of about 15 degrees are commonly recommended. Where side cutting is involved, a side clearance of about 4 degrees minimum is recommended. With carbide tools, the end clearance should not exceed 6 to 8 degrees and the top rake should be from 5 to 10 degrees positive. For boring, facing, and other lathe operations, rake and clearance angles are about the same as for tools used in turning.

Machining Monel and Nickel Alloys.—These alloys are machined with high-speed steel and with cemented carbide cutting tools. High-speed steel lathe tools usually have a back rake of 6 to 8 degrees, a side rake of 10 to 15 degrees, and relief angles of 8 to 12 degrees. Broad-nose finishing tools have a back rake of 20 to 25 degrees and an end relief angle of 12 to 15 degrees. In most instances, standard commercial cemented-carbide tool holders and tool shanks can be used which provide an acceptable tool geometry. Honing the cutting edge lightly will help if chipping is encountered.

The most satisfactory tool materials for machining Monel and the softer nickel alloys, such as Nickel 200 and Nickel 230, are M2 and T5 for high-speed steel and crater resistant grades of cemented carbides. For the harder nickel alloys such as K Monel, Permanickel, Duranickel, and Nitinol alloys, the recommended tool materials are T15, M41, M42, M43, and for high-speed steel, M42. For carbides, a grade of crater resistant carbide is recommended when the hardness is less than 300 Bhn, and when the hardness is more than 300 Bhn, a grade of straight tungsten carbide will often work best, although some crater resistant grades will also work well.

A sulfurized oil or a water-soluble oil is recommended for rough and finish turning. A sulfurized oil is also recommended for milling, threading, tapping, reaming, and broaching. Recommended cutting speeds for Monel and the softer nickel alloys are 70 to 100 fpm for high-speed steel tools and 200 to 300 fpm for cemented carbide tools. For the harder nickel alloys, the recommended speed for high-speed steel is 40 to 70 fpm for a hardness up to 300 Bhn and for a higher hardness, 10 to 20 fpm; for cemented carbides, 175 to 225 fpm when the hardness is less than 300 Bhn and for a higher hardness, 30 to 70 fpm.

Nickel alloys have a high tendency to work harden. To minimize work hardening caused by machining, the cutting tools should be provided with adequate relief angles and positive rake angles. Furthermore, the cutting edges should be kept sharp and replaced when dull to prevent burnishing of the work surface. The depth of cut and feed should be sufficiently large to ensure that the tool penetrates the work without rubbing.

Machining Copper Alloys.—Copper alloys can be machined by tooling and methods similar to those used for steel, but at higher surface speeds. Machinability of copper alloys is discussed in the tables starting on page 526 and starting on page 533. Machinability is based on a rating of 100 per cent for the free-cutting alloy C35000, which machines with small, easily broken chips. As with steels, copper alloys containing lead have the best machining properties, with alloys containing tin, and lead, having machinability ratings of 80 and 70 per cent. Tellurium and sulphur are added to copper alloys to increase machinability with minimum effect on conductivity. Lead additions are made to facilitate machining, as their effect is to produce easily broken chips.

Copper alloys containing silicon, aluminum, manganese and nickel become progressively more difficult to machine, and produce long, stringy chips, the latter alloys having only 20 per cent of the machinability of the free-cutting alloys. Although copper is frequently machined dry, a cooling compound is recommended. Other lubricants that have been used include tallow for drilling, gasoline for turning, and beeswax for threading.

Machining Hard Rubber.—Tools suitable for steel may be used for hard rubber, with no top or side rake angles and 10 to 20 deg. clearance angles, of high speed steel or tungsten carbide. Without coolant, surface speeds of about 200 ft./min. are recommended for turning, boring and facing, and may be increased to 300 surface ft./min. with coolant.

Drilling of hard rubber requires high speed steel drills of 35 to 40 deg. helix angle to obtain maximum cutting speeds and drill life. Feed rates for drilling range up to 0.015 in./rev. Deep-fluted taps are best for threading hard rubber, and should be 0.002 to 0.005 in. oversize if close tolerances are to be held. Machine oil is used for a lubricant. Hard rubber may be sawn with band saws having 5 to 10 points per inch, running at about 3000 ft./min. or cut with abrasive wheels. Use of coolant in grinding rubber gives a smoother finish.

Piercing and blanking of sheet rubber is best performed with the rubber or dies heated. Straightening of the often-distorted blanks may be carried out by dropping them into a pan of hot water.

GRINDING FEEDS AND SPEEDS

Grinding data are scarcely available in handbooks, which usually recommend a small range of depths and work speeds at constant wheel speed, including small variations in wheel and work material composition. Wheel life or grinding stiffness are seldom considered.

Grinding parameter recommendations typically range as follows:

- Wheel speeds are usually recommended in the 1200 to 1800 m/min (4000 to 6000 fpm) range, or in rare cases up to 3600 m/min (12000 fpm)
- Work speeds are in the range 20 to 40 m/min (70 to 140 fpm); and, depths of cut of 0.01 to 0.025 mm (0.0004 to 0.001 inch) for roughing, and around 0.005 mm (.0002 in.) for finish grinding.
- Grit sizes for roughing are around 46 to 60 for easy-to-grind materials, and for difficult-to-grind materials higher such as 80 grit. In finishing, a smaller grit size (higher grit number) is recommended. Internal grinding grit sizes for small holes are approximately 100 to 320.
- Specific metal removal rate, *SMRR*, represents the rate of material removal per unit of wheel contact width and are commonly recommended from 200 to 500 mm^3/mm width/min (0.3 to 0.75 in^3/inch width/min).
- Grinding stiffness is a major variable in determining wheel-life and spark-out time. A typical value of system stiffness in outside-diameter grinding, for 10:1 length/diameter ratio, is approximately $K_{ST} = 30–50$ N/μm. System stiffness K_{ST} is calculated from the stiffness of the part, K_w and the machine and fixtures, K_m. Machine values can be obtained from manufacturers, or can be measured using simple equipment along with the part stiffness.
- Generally a lower wheel hardness (soft wheel) is recommended when the system stiffness is poor or when a better finish is desired.

Basic Rules

The wheel speed *V* and *equivalent chip thickness* $ECT = SMRR \div V \div 1000$ are the primary parameters that determine wheel-life, forces and surface finish in grinding. The following general rules and recommendations, using *ECT*, are based on extensive laboratory and industry tests both in Europe and USA. The relationships and shapes of curves pertaining to grinding tool-life, grinding time, and cost are similar to those of any metal cutting operation such as turning, milling and drilling.

In turning and milling, the *ECT* theory says that if the product of feed times depth of cut is constant, the tool-life is constant no matter how the depth of cut or feed is varied, provided that the cutting speed and cutting edge length are maintained constant.

In grinding, wheel-life *T* remains constant for constant cutting speed *V*, regardless of how depth of cut a_r or work speed V_w are selected as long as the specific metal removal rate $SMMR = V_w \times a_r$ is held constant (neglecting the influence of grinding contact width).

ECT is much smaller in grinding than in milling, ranging from about 0.0001 to 0.001 mm (0.000004 to 0.00004 inch). See the section *Machining Econometrics* starting on page 1 for a detailed explanation of the role of *ECT* in conventional machining.

Wheel life *T* and *Grinding Ratio*.—A commonly used measure of relative wheel-life in grinding is the *grinding ratio* that is used to compare grindability when varying grinding wheel composition and work material properties under otherwise constant cutting conditions.

The *grinding ratio* is defined as the slope of the wear curve versus metal removal rate: *grinding ratio* = $MRR \div W^*$, where *MRR* is the metal removal rate, and W^* is the volume wheel wear at which the wheel has to be dressed. The grinding ratio is not a measure of wheel-life, but a relationship between *grinding ratio* and wheel-life *T* can be obtained from the formula *grinding ratio* = *SMRR* × *T* ÷ *W**, where *SMRR* (specific metal removal rate) is determined from *MRR* = *SMRR* × *T* or from *ECT* = *SMRR* ÷ *V* ÷ 1000.

Thus, *grinding ratio* = 1000 × *ECT* × *V* × *T* ÷ *W**, and *T* = *grinding ratio* × *W** ÷ (1000 × *ECT* × *V*), provided that the wheel wear criterion *W** is valid for all data combinations.

Example 1: If *W** in one test is found to be 500 mm^3 for *ECT* = 0.00033 mm and *V* = 3600 m/min, and *grinding ratio* = 10, then wheel-life will vary with measured grinding ratios, wheel speed, and *ECT* as follows: *T* =500 × *grinding ratio* ÷ (*V* × *ECT*) = 4.2 minutes.

In the remainder of this section the *grinding ratio* will not used, and wheel-life is expressed in terms of *ECT* or *SMRR* and wheel speed *V*.

***ECT* in Grinding.**—In turning and milling, *ECT* is defined as the volume of chips removed per unit cutting edge length per revolution of the work or cutter. In milling specifically, *ECT* is defined as the ratio of (number of teeth z × feed per tooth f_z × radial depth of cut a_r × and axial depth of cut a_a) and (cutting edge length *CEL* divided by πD), where *D* is the cutter diameter, thus,

$$ECT = \frac{\pi D z f_z a_r a_a}{CEL}$$

In grinding, the same definition of *ECT* applies if we replace the number of teeth with the average number of grits along the wheel periphery, and replace the feed per tooth by the average feed per grit. This definition is not very practical, however, and *ECT* is better defined by the ratio of the specific metal removal rate *SMMR*, and the wheel speed *V*. Thus, *ECT* = 1000 × *SMRR* ÷ *V*. Keeping *ECT* constant when varying *SMRR* requires that the wheel speed must be changed proportionally.

In milling and turning *ECT* can also be redefined in terms of *SMRR* divided by the work and the cutter speeds, respectively, because *SMRR* is proportional to the feed rate F_R.

Work Speed and Depth of Cut Selection: Work speed V_w is determined by dividing *SMMR* by the depth of cut a_r, or by using the graph in Fig. 1.

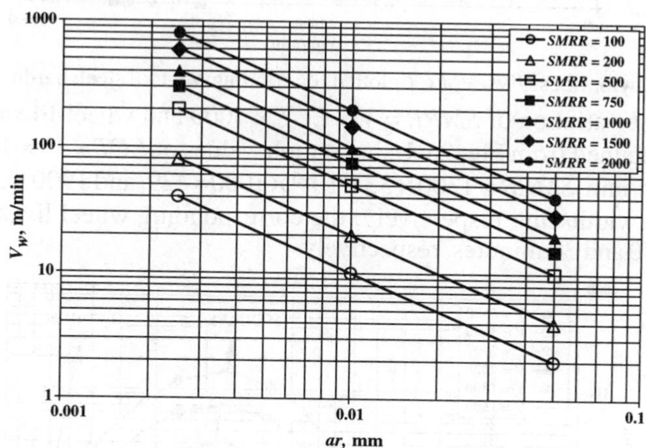

Fig. 1. Work speed V_w vs. depth of cut a_r

Referring to Fig. 1, for depths of cuts of 0.01 and 0.0025 mm, a specific metal removal rate *SMMR* = 1000 mm^3/mm width/min is achieved at work speeds of 100 and 400 m/min, respectively, and for *SMMR* =100 mm^3/mm width/min at work speeds of 10 and 40 m/min, respectively.

Unfortunately, the common use of low values of work speed (20 to 40 m/min) in finishing cause thermal surface damage, disastrous in critical parts such as aircraft components. As the grains slide across the work they generate surface heat and fatigue-type loading may cause residual tensile stresses and severe surface cracks. Proper finish grinding conditions are obtained by increasing the work speed 5 to 10 times higher than the above recommendations indicate. These higher work speeds will create compressive stresses that are not detrimental to the surface. The by-product of higher work speeds is much higher *SMRR* values and thereby much shorter grinding times. Compressive stresses are also obtained by reducing the depth of cut a_r.

Wheel Life Relationships and Optimum Grinding Data.—Figs. 2a, 2b, and 2c show, in three planes, the 3-dimensional variation of wheel-life *T* with wheel speed *V* and *ECT* when grinding a hardened tool steel. Fig. 2a depicts wheel-life versus wheel speed (the *T–V* plane) with constant *ECT* appearing as approximately straight lines when plotted in log-log coordinates.

In grinding, the wheel-life variation follows curves similar to those obtained for conventional metal cutting processes, including a bend-off of the Taylor lines (*T–V* graph) towards shorter life and lower cutting speeds when a certain maximum life is achieved for each value of *ECT*. In the two other planes (*T–ECT*, and *V–ECT*) we usually find smooth curves in which the maximum values of wheel-life are defined by points along a curve called the *H*–curve.

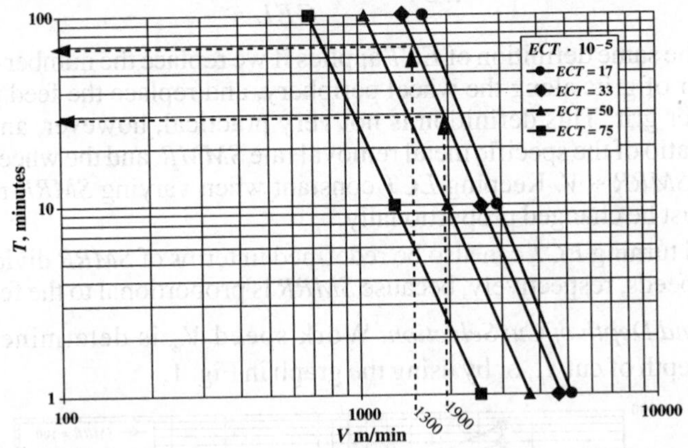

Fig. 2a. Taylor lines: *T* vs. *V*, *ECT* plotted for grinding M4 tool steel, hardness Rc 64

Example 2: The variation of $SMRR = V \times ECT \times 1000$ and wheel-life at various wheel speeds can be obtained from Fig. 2a. Using sample values of $ECT = 33 \times 10^{-5}$ mm and $V =$ 1300 and 1900 m/min, $SMRR = 1300 \times 33 \times 10^{-5} \times 1000 = 429$, and $1900 \times 33 \times 10^{-5} \times 1000 = 627$ mm^3/mm width/min, respectively; the corresponding wheel lives are read off as approximately 70 and 30 minutes, respectively.

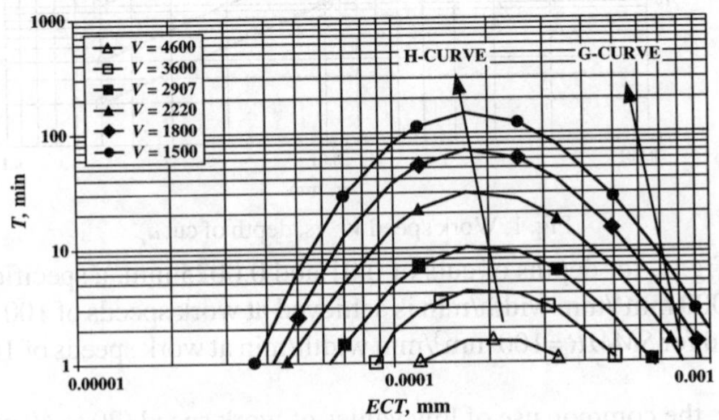

Fig. 2b. *T* vs. *ECT*, *V* plotted

Fig. 2b depicts wheel-life *T* versus *ECT* with constant wheel speed *V* shown as curves plotted in log-log coordinates, similar to those for the other cutting operations.

Example 3: Fig. 2b shows that maximum values of wheel-life occur along the *H*-curve. For the 3 speeds 1800, 2700, and 3600 m/min, maximum wheel lives are approximately 70, 14 and 4 minutes, respectively, at *ECT* around 17×10^{-5} through 20×10^{-5} mm along the *H*-curve. Left and right of the *H*-curve wheel-life is shorter.

Fig. 2c depicts wheel speed *V* versus *ECT* with wheel-life *T* parameter shown as curves in log-log coordinates, similar to those for the other cutting operations, with the characteristic *H*- and *G*-curves.

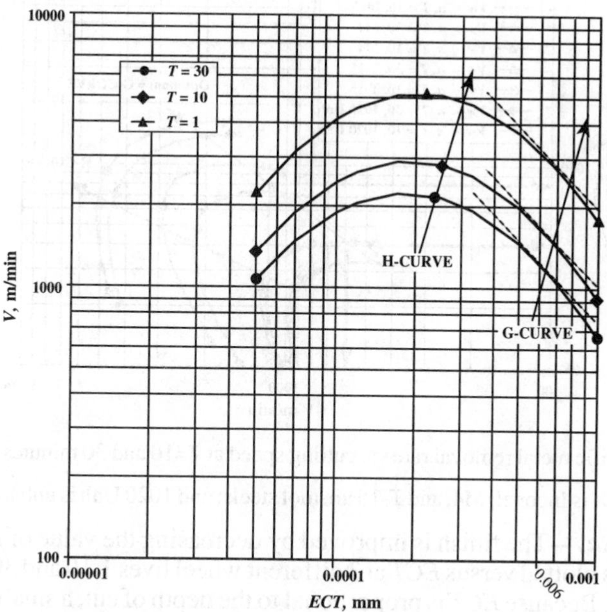

Fig. 2c. *V* vs. *ECT*, *T* plotted

Optimum grinding data for roughing occur along the *G*-curve, which is determined from the *V-ECT* graph by drawing 45-degree lines tangent to the *T*-curves, as shown in Fig. 2c, and drawing a line (the *G*-curve) through the points of tangency on the respective *T*-curves, thus the location and direction of the *G*-curve is determined. Globally optimum data correspond to the *T*-curve for which wheel-life is calculated using the corresponding equivalent tooling-cost time, T_V, calculated from $T_V = T_{RPL} + 60 \times C_E \div H_R$, minutes, where T_{RPL} is the time required to replace wheel, C_E = cost per wheel dressing = wheel cost + cost per dressing, and H_R is the hourly rate.

Minimum cost conditions occur along the *G*-curve; if optimum life T_O was determined at either 10 or 30 minutes then V_O = 1500 and 1100 m/min, respectively, and *ECT* is around $65 - 70 \times 10^{-5}$ mm in both cases. The corresponding optimum values of *SMRR* are $1000 \times 1500 \times 67 \times 10^{-5} = 1000$ and $1000 \times 1100 \times 67 \times 10^{-5} = 740$ mm³/min/mm wheel contact width (1.5 to 1.1in³/in/min).

Using Fig. 1 we find optimum work speeds for depths of cut a_r = 0.01 and 0.005 mm to be V_w = 100 and 75 m/min, and 200 and 150 m/min (330 and 250 fpm, and 660 and 500 fpm) respectively for 10- and 30-minute wheel-life.

These high work speeds are possible using proper dressing conditions, high system stiffness, good grinding fluid quality and wheel composition.

Fig. 3 shows the variation of specific metal removal rate with wheel speed for several materials and a range of *ECT*s at 10- and 30-minutes wheel-life. *ECT* decreases when moving to the left and down along each curve. The two curves for unhardened 1020 steel have the largest values of *SMRR*, and represent the most productive grinding conditions, while the heat resistant alloy Inconel yields the least productive grinding conditions. Each branch attains a maximum *SMRR* along the *G*-curve (compare with the same curve in the V–*ECT* graph, Fig. 2c) and a maximum speed region along the *H*-curve. When the *SMRR*-values are lower than the *H*-curve the *ECT* values for each branch decrease towards the bottom of the graph, then the speed for constant wheel-life must be reduced due to the fact that the *ECT* values are to the left of their respective *H*-curves in *V–ECT* graphs.

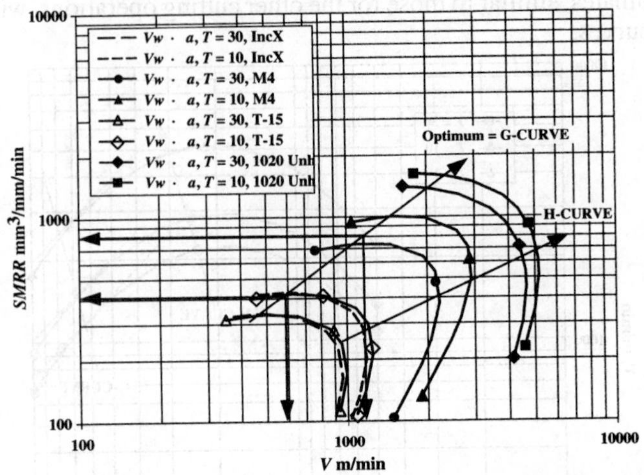

Fig. 3. Specific metal removal rate vs. cutting speed at *T*=10 and 30 minutes wheel life

In the figure, IncX is Inconel; M4, and T-15 are tool steels; and 1020 Unh is unhardened 1020 steel.

Surface Finish, *Ra*.—The finish is improved by decreasing the value of *ECT* as shown in Fig. 4, where *Ra* is plotted versus *ECT* at 3 different wheel lives 1,10 and 30 minutes at constant wheel speed. Because *ECT* is proportional to the depth of cut, a smaller depth of cut is favorable for reducing surface roughness when the work speed is constant.

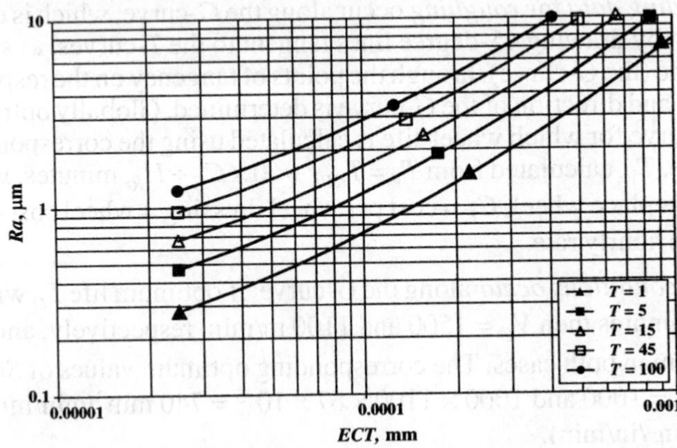

Fig. 4. Surface finish, *Ra* vs. *ECT*, wheel-life *T* plotted

In Fig. 5, *Ra* is plotted versus wheel-life at 5 different *ECT*'s. Both Figs. 4 and 5 illustrate that a shorter life improves the surface finish, which means that either an increased wheel speed (wheel-life decreases) at constant *ECT*, or a smaller *ECT* at constant speed (wheel-life increases), will result in an improved finish. For a required surface finish, *ECT* and wheel-life have to be selected appropriately in order to also achieve an optimum grinding time or cost. In cylindrical grinding a reduction of side feed f_s improves *Ra* as well.

In terms of specific metal removal rate, reducing *SMRR* will improve the surface finish R_A.

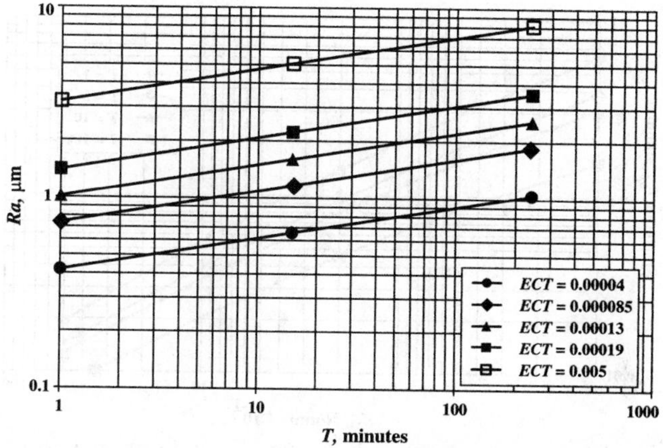

Fig. 5. Surface roughness, *Ra* vs. wheel life *T*, *ECT* plotted

Example 4, Specific Metal Removal Rates and Work Speeds in Rough and Finish Grinding: The tabulated values in the following table indicate that a decreasing *ECT* combined with a higher wheel speed for 10 minutes wheel-life will decrease the metal removal rate and thereby increasing the grinding time. This change is accompanied by a better finish in both roughing and finishing operations. Note the high work speeds when finishing.

	Tool Life $T = 10$ minutes		Roughing Depth $a_r = 0.025$ mm	Finishing Depth $a_r = 0.0025$ mm
ECT mm	Wheel speed V_{10} m/min	Removal Rate $SMRR_{10}$ mm³/mm/ min	Work speed V_w m/min	
0.00050	1970	985	39	390
0.00033	2580	850	34	340
0.00017	2910	500	20	200

The grit size, however, is a major parameter. Fig. 6, shows that a high wheel speed, combined with a small grit size, say 320 Mesh, can achieve R_A values as small as 0.03 microns.

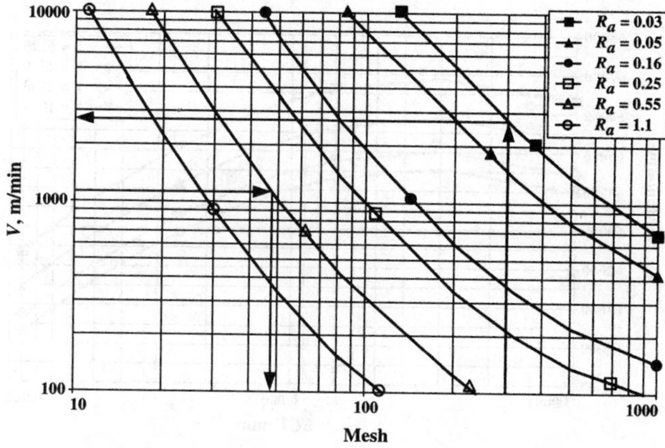

Fig. 6. Wheel speed vs. wheel mesh, *Ra* plotted

Spark-out Time.—Fig. 7 shows how spark-out time varies with system stiffness. As with surface finish, when wheel-life is short (high wear rate) the spark-out time decreases.

Equivalent Diameter (Work Conformity) Factor: The difference in curvature of the work and wheel in the contact region, determined by the equivalent diameter or work conformity formula, is an important factor for calculating spark-out time and forces, but has a

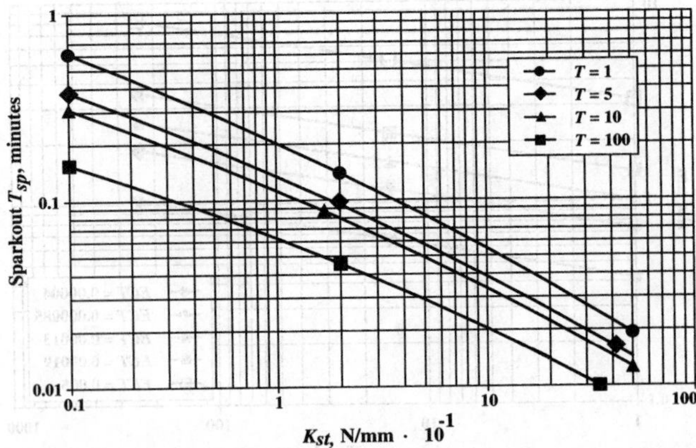

Fig. 7. Sparkout time vs. system stiffness, wheel-life *T* plotted

negligible influence on wheel-life. Therefore, an equivalent diameter, $D_e = D/(1 \pm (D/D_w)$, with the minus sign for internal grinding and the plus sign for external grinding operations, is used to consider the effect of conformity when using internal and external grinding with varying work and wheel diameters. D_e is equal to the wheel diameter in surface grinding (work flat); in internal grinding, the wheel conforms closely to the work and D_e is therefore larger than in external grinding.

Grinding Cutting Forces, Torque and Power.—Formulas to calculate the tangential cutting force, torque and required machining power are found in *Estimating Machining Power* on page 1046, but the values of K_c, specific cutting force or specific energy, are approximately 30 to 40 times higher in grinding than in turning, milling and drilling. This is primarily due to the fact that the *ECT* values in grinding are 1000 to 10000 times smaller, and also due to the negative rake angles of the grit. Average grinding rake angles are around −35 to −45 degrees. K_c for grinding unhardened steel is around 50000 to 70000 N/mm^2 and up to 150000 to 200000 N/mm^2 for hardened steels and heat resistant alloys. The grinding cutting forces are relatively small because the chip area is very small.

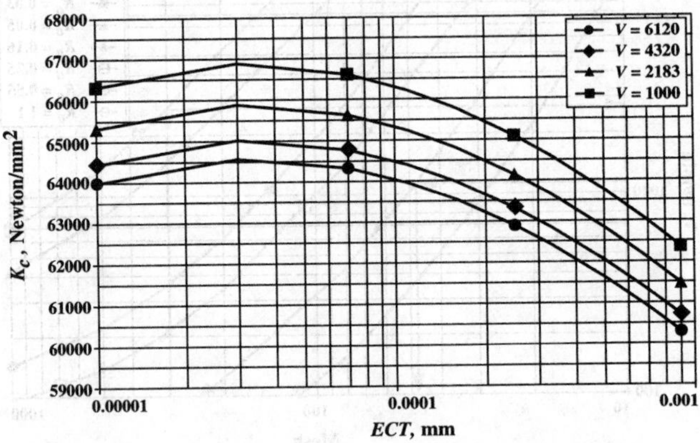

Fig. 8. Specific grinding force *Kc* vs. *ECT*; *V* plotted

As in the other metal cutting operations, the forces vary with *ECT* and to a smaller extent with the wheel speed *V*. An example is shown in Fig. 8, where K_c, specific cutting force, is plotted versus *ECT* at wheel speeds between 1000 and 6000 m/min. The material is medium unhardened carbon steel ground by an aluminum oxide wheel. The impact of wheel speed is relatively small (2 to 5% lower with increasing speed).

Example 5: Find the cutting force when ECT = 0.00017 mm, the cutting edge length (width of cut) CEL is 10 mm, and K_c=150000 N/mm^2.

The chip area is $ECT \times CEL$ = 0.0017 mm^2. For K_c=150000, the cutting force is 0.0017 × 150000 = 255 Newton.

Another difference compared to turning is the influence of the negative rake angles, illustrated by the ratio of F_H/F_C, where F_H is the normal force and F_C the tangential grinding force acting in the wheel speed direction. F_H is much larger than the grinding cutting force, generally F_H/F_C ratio is approximately 2 to 4. An example is shown in Fig. 9, where F_H/F_C, is plotted versus ECT at wheel speeds between 1000 and 6000 m/min, under the same conditions as in Fig. 8.

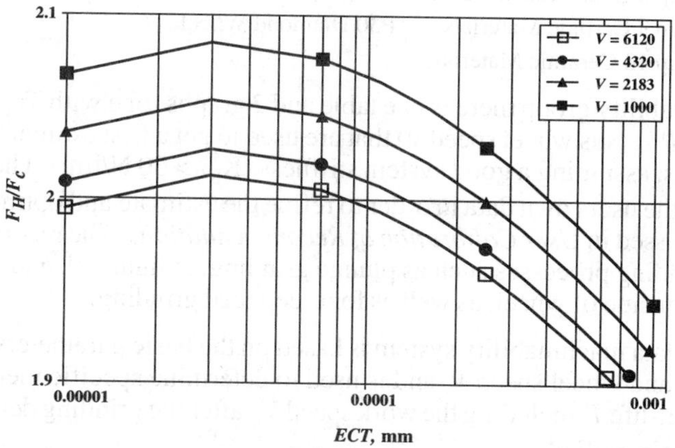

Fig. 9. F_H/F_C vs. ECT; cutting speed plotted

In both Fig. 8 and Fig. 9, it is apparent that both K_c and F_H/F_C attain maximum values for given small values of ECT, in this case approximately ECT = 0.00005 mm. This fact illustrates that forces and wheel-life are closely linked; for example, wheel speed has a maximum for constant wheel-life at approximately the same values of ECT shown in the two graphs (compare with the trends illustrated in Figs. 2a, 2b, 2c, and 3). As a matter of fact, force relationships obey the same type of relationships as those of wheel-life. Colding's force relationship uses the same 5 constants as the tool life equation, but requires values for the specific cutting force at ECT = 0.001 and an additional constant, obtained by a special data base generator. This requires more elaborate laboratory tests, or better, the design of a special test and follow-up program for parts running in the ordinary production.

Grinding Data Selection Including Wheel Life

The first estimate of machine settings is based on dividing work materials into 10 groups, based on grindability, as given in Table 1. Compositions of these work materials are found in the Handbook in the section *STANDARD STEELS* starting on page 403.

Grinding wheel nomenclature is described in *American National Standard Grinding Wheel Markings* starting on page 1141. The wheel compositions are selected according to the grade recommendations in the section *The Selection of Grinding Wheels* starting on page 1142. Grinding fluid recommendations are given in *Cutting Fluids for Machining* starting on page 1106.

Note: Maximum wheel speeds should always be checked using the safety standards in the section *Safe Operating Speeds* starting on page 1171, because the recommendations will sometimes lead to speeds above safety levels.

The material in this section is based on the use of a typical standard wheel composition such as 51-A-46-L-5-V-23, with wheel grade (wheel hardness) = L or above, and mesh (grit size) = 46 or above.

Table 1. Grindability Groups

Group	Examples
Group 1 Unhardened Steels	
Group 2 Stainless Steels	SAE 30201–30347, 51409–51501
Group 3 Cast Iron	
Group 4 Tool Steels	M1, M8, T1, H, O, L, F, 52100
Group 5 Tool Steels	M2, T2, T5, T6, D2, H41, H42, H43, M50
Group 6 Tool Steels	M3, M4, T3, D7
Group 7 Tool Steels	T15, M15
Group 8 Heat Resistant Steels	Inconel, Rene etc.
Group 9 Carbide Materials	P30 Diamond Wheel
Group 10 Ceramic Materials	

For each grindability group there is one table and 2 graphs (one with Taylor lines and the other with *SMRR* versus wheel speed *V*) that are used to get a first estimate of standardized machine settings, assuming a good system stiffness ($K_{ST} > 30$ N/μm). These data are then calibrated with the users own data in order to refine the estimate and optimize the grinding process, as discussed in *User Calibration of Recommendations*. The recommendations are valid for all grinding processes such as plunge grinding, cylindrical, and surface grinding with periphery or side of wheel, as well as for creep feed grinding.

The grinding data machinability system is based on the basic parameters equivalent chip thickness *ECT*, and wheel speed *V*, and is used to determine specific metal removal rates *SMRR* and wheel-life *T*, including the work speed V_w after the grinding depths for roughing and finishing are specified.

For each material group, the grinding data machinability system consists of *T–V* Taylor lines in log-log coordinates for 3 wheel speeds at wheel lives of 1, 10 and 100 minutes wheel-life with 4 different values of equivalent chip thickness *ECT*. The wheel speeds are designated V_1, V_{10}, and V_{100} respectively. In each table the corresponding specific metal removal rates *SMRR* are also tabulated and designated as $SMRR_1$, $SMRR_{10}$ and $SMRR_{100}$ respectively. The user can select any value of *ECT* and interpolate between the Taylor lines. These curves look the same in grinding as in the other metal cutting processes and the slope is set at n = 0.26, so each Taylor line is formulated by $V \times T^{0.26} = C$, where C is a constant tabulated at four *ECT* values, $ECT = 17, 33, 50$ and 75×10^{-5} mm, for each material group. Hence, for each value of *ECT*, $V_1 \times 1^{0.26} = V_{10} \times 10^{0.26} = V_{100} \times 100^{0.26} = C$.

Side Feed, Roughing and Finishing.—In cylindrical grinding, the side feed, $f_s = C \times$ Width, does not impact on the values in the tables, but on the feed rate F_R, where the fraction of the wheel width C is usually selected for roughing and in finishing operations, as shown in the following table.

Work Material	Roughing, C	Finishing, C
Unhardened Steel	2⁄3–3⁄4	1⁄3–3⁄8
Stainless Steel	1⁄2	1⁄4
Cast Iron	3⁄4	3⁄8
Hardened Steel	1⁄2	1⁄4

Finishing: The depth of cut in rough grinding is determined by the allowance and usually set at $a_r = 0.01$ to 0.025 mm. The depth of cut for finishing is usually set at $a_r = 0.0025$ mm and accompanied by higher wheel speeds in order to improve surface finish. However, the most important criterion for critical parts is to increase the work speed in order to avoid thermal damage and surface cracks. In cylindrical grinding, a reduction of side feed f_s

improves R_a as well. Small grit sizes are very important when very small finishes are required. See Figs. 4, 5, and 6 for reference.

Terms and Definitions

a_a = depth of cut

a_r = radial depth of cut, mm

C = fraction of grinding wheel width

CEL = cutting edge length, mm

C_U = Taylor constant

D = wheel diameter, mm

$DIST$ = grinding distance, mm

d_w = work diameter, mm

ECT = equivalent chip thickness = $f(a_r, V, V_w, f_s)$, mm

$$= 1 \div (V \div V_w \div a_r + 1 \div f_s) = \frac{V_w f_s (a_r + 1)}{V}$$

= approximately $V_w \times a_r \div V = SMRR \div V \div 1000$

$= z \times f_z \times a_r \times a_a \div CEL \div (\pi D)$ mm

F_R = feed rate, mm/min

$= f_s \times RPM_w$ for cylindrical grinding

$= f_i \times RPM_w$ for plunge (in-feed) grinding

f_i = in-feed in plunge grinding, mm/rev of work

f_s = side feed or engaged wheel width in cylindrical grinding = $C \times Width$ = a_a approximately equal to the cutting edge length CEL

Grindingratio = $MRR \div W^* = SMRR \times T \div W^* = 1000 \times ECT \times V \times T \div W^*$

MRR = metal removal rate = $SMRR \times T = 1000 \times f_s \times a_r \times V_w$ mm^3/min

$SMRR$ = specific metal removal rate obtained by dividing MRR by the engaged wheel width ($C \times Width$) = $1000 \times a_r \times V_w$ mm^3/mm width/min

Note: 100 mm^3/mm/min = 0.155 in^3/in/min, and 1 in^3/in/min = 645.16 mm^3/mm/min

T, T_U = wheel-life = *Grinding ratio* $\times W \div (1000 \times ECT \times V)$ minutes

t_c = grinding time per pass = $DIST \div F_R$ min

$= DIST \div F_R + t_{sp}$ (min) when spark-out time is included

= # Strokes × ($DIST \div F_R + t_{sp}$) (min) when spark-out time and strokes are included

t_{sp} = spark-out time, minutes

V, V_U = wheel speed, m/min

V_w, V_{wU} = work speed = $SMRR \div 1000 \div a_r$ m/min

W^* = volume wheel wear, mm^3

$Width$ = wheel width (mm)

RPM = wheel speed = $1000 \times V \div D \div \pi$ rpm

RPM_w = work speed = $1000 \times V_w \div D_w \div \pi$ rpm

Relative Grindability.—An overview of grindability of the data base, which must be based on a constant wheel wear rate, or wheel-life, is demonstrated using 10 minutes wheel-life shown in Table 2.

Table 2. Grindability Overview

Material Group	$ECT \times 10^{-5}$	V_{10}	$SMRR_{10}$	V_w Roughing Depth $a_r = 0.025$	V_w *Finishing Depth* $a_r = 0.0025$
1 Unhardened	33	3827	1263	50	500
2 Stainless	33	1080	360	15	150
3 Cast Iron	33	4000	1320	53	530
4 Tool Steel	33	3190	1050	42	420
5 Tool Steel	33	2870	950	38	380
6 Tool Steel	33	2580	850	35	350
7Tool Steel	33	1080	360	15	150
8 Heat resistant	33	1045	345	14	140
9 Carbide with Diamond Wheel	5	$V_{600} = 1200$	$SMRR_{600} = 50$	2	20
10 Ceramics with Diamond Wheel	5	$V_{600} = 411$	$SMRR_{600} = 21$	0.84	84

Procedure to Determine Data.—The following wheel-life recommendations are designed for 4 values of ECT = 0.00017, 0.00033, 0.00050 and 0.00075 mm (shown as 17, 33, 50 and 75 in the tables). Lower values of ECT than 0.00010 mm (0.000004 in.) are not recommended as these may lie to the left of the *H*-curve.

The user selects any one of the ECT values, or interpolates between these, and selects the wheel speed for 10 or 100 minutes life, denoted by V_{10} and V_{100}, respectively. For other desired wheel lives the wheel speed can be calculated from the tabulated Taylor constants C and $n = 0.26$ as follows:

$(V \times T_{(desired)})^{0.26}$ = C, the value of which is tabulated for each ECT value. C is the value of cutting speed V at $T = 1$ minute, hence is the same as for the speed V_1 (V_1×1^0.26 =C)

$$V_{10} = C \div 10^{0.26} = C \div 1.82$$

$$V_{100} = C \div 100^{0.26} = C \div 3.31.$$

Example 6: A tool steel in material group 6 with ECT = 0.00033, has constant C= 4690, V_{10} = 2578 m/min, and V_{100} = 1417 m/min. From this information, find the wheel speed for desired wheel-life of T = 15 minutes and T = 45 minutes

For T = 15 minutes we get $V_{15} = 4690 \div 15^{0.26} = 2319$ m/min (7730 fpm) and for T = 45 minutes $V_{45} = 4690 \div 45^{0.26} = 1743$ m/min (5810 fpm).

The Tables are arranged in 3 sections:

1. Speeds V_{10} and V_1 = Constant CST(standard) for 4 ECT values 0.00017, 0.00033, 0.00050 and 0.00075 mm. Values C_U and V_{10U} refer to user calibration of the standard values in each material group, explained in the following.

2. Speeds V_{100} (first row of 3), V_{10} and V_1 (last in row) corresponding to wheel lives 100, 10 and 1 minutes, for 4 ECT values 0.00017, 0.00033, 0.00050 and 0.00075 mm.

3. Specific metal removal rates $SMRR_{100}$, $SMRR_{10}$ and $SMRR_1$ corresponding to wheel lives 100, 10 and 1 minutes, for the 4 ECT values 0.00017, 0.00033, 0.00050, and 0.00075 mm

The 2 Graphs show: wheel life versus wheel speed in double logarithmic coordinates (Taylor lines); and, *SMRR* versus wheel speed in double logarithmic coordinates for 4 ECT values: 0.00017, 0.00033, 0.00050 and 0.00075 mm.

Table 1. Group 1—Unhardened Steels

Tool Life T (min)	ECT = 0.00017 mm		ECT = 0.00033 mm		ECT = 0.00050 mm		ECT = 0.00075 mm	
	Constant C = 8925		Constant C = 6965		Constant C = 5385		Constant C = 3885	
	V_T	*SMRR*	V_T	*SMRR*	V_T	*SMRR*	V_T	*SMRR*
100	2695	460	2105	695	1625	815	1175	880
10	**4905**	**835**	**3830**	**1265**	**2960**	**1480**	**2135**	**1600**
1	8925	1520	6965	2300	5385	2695	3885	2915

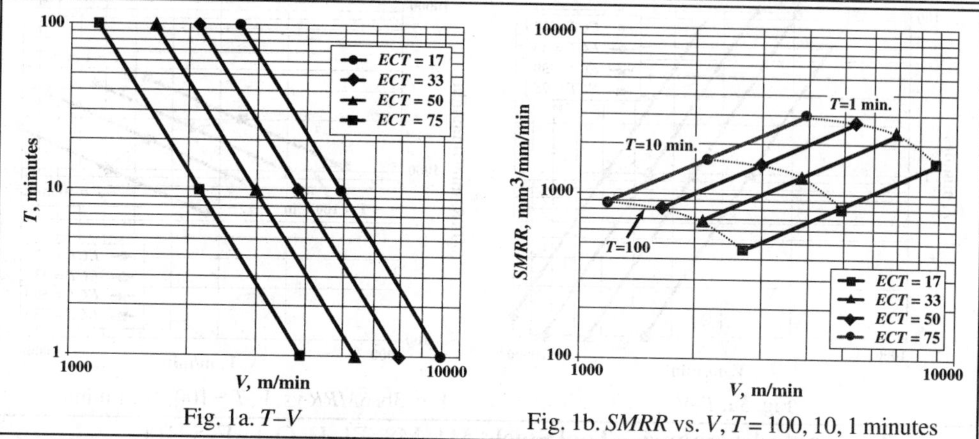

Fig. 1a. *T–V*

Fig. 1b. *SMRR* vs. *V*, *T* = 100, 10, 1 minutes

Table 2. Group 2—Stainless Steels SAE 30201 – 30347, SAE 51409 – 51501

Tool Life T (min)	ECT = 0.00017 mm		ECT = 0.00033 mm		ECT = 0.00050 mm		ECT = 0.00075 mm	
	Constant C = 2270		Constant C = 1970		Constant C = 1505		Constant C = 1010	
	V_T	*SMRR*	V_T	*SMRR*	V_T	*SMRR*	V_T	*SMRR*
100	685	115	595	195	455	225	305	230
10	**1250**	**210**	**1080**	**355**	**825**	**415**	**555**	**415**
1	2270	385	1970	650	1505	750	1010	760

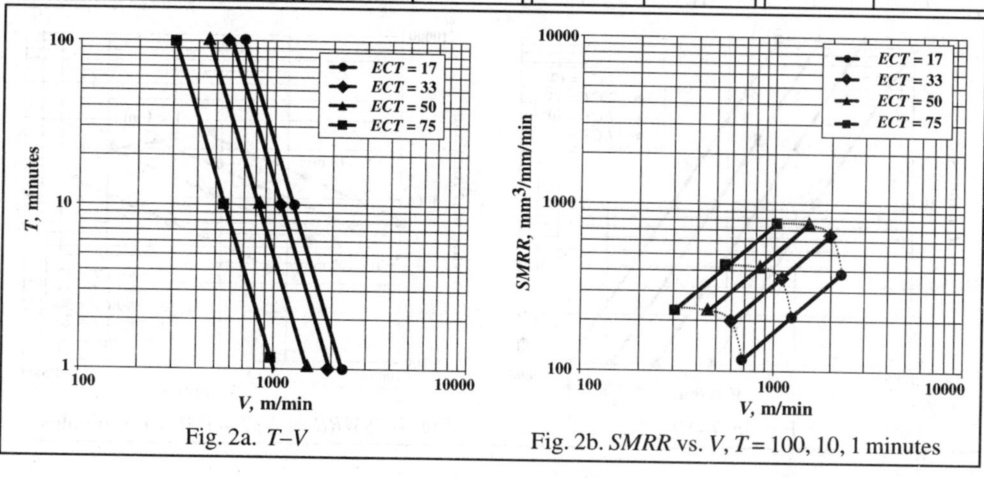

Fig. 2a. *T–V*

Fig. 2b. *SMRR* vs. *V*, *T* = 100, 10, 1 minutes

Table 3. Group 3—Cast Iron

Tool Life T (min)	ECT = 0.00017 mm		ECT = 0.00033 mm		ECT = 0.00050 mm		ECT = 0.00075 mm	
	Constant C = 10710		Constant C = 8360		Constant C = 6465		Constant C = 4665	
	V_T	$SMRR$	V_T	$SMRR$	V_T	$SMRR$	V_T	$SMRR$
100	3235	550	2525	835	1950	975	1410	1055
10	**5885**	**1000**	**4595**	**1515**	**3550**	**1775**	**2565**	**1920**
1	10710	1820	8360	2760	6465	3230	4665	3500

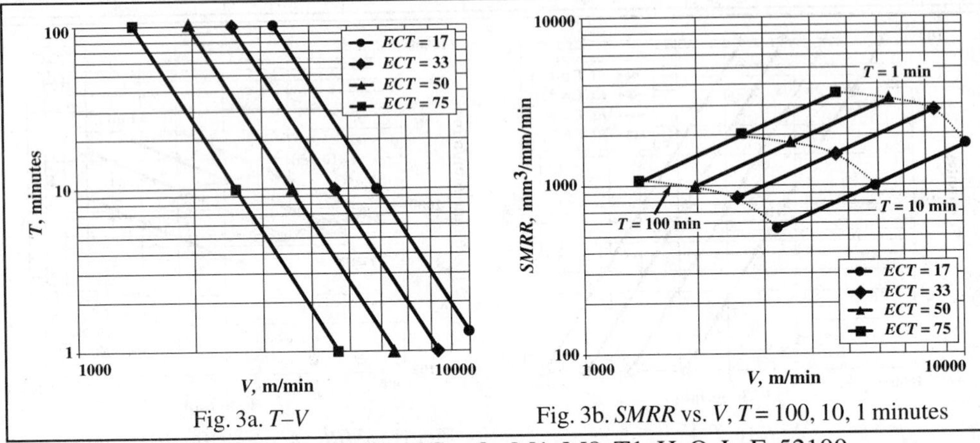

Fig. 3a. T–V

Fig. 3b. $SMRR$ vs. V, T = 100, 10, 1 minutes

Table 4. Group 4—Tool Steels, M1, M8, T1, H, O, L, F, 52100

Tool Life T (min)	ECT = 0.00017 mm		ECT = 0.00033 mm		ECT = 0.00050 mm		ECT = 0.00075 mm	
	Constant C = 7440		Constant C = 5805		Constant C = 4490		Constant C = 3240	
	V_T	$SMRR$	V_T	$SMRR$	V_T	$SMRR$	V_T	$SMRR$
100	2245	380	1755	580	1355	680	980	735
10	**4090**	**695**	**3190**	**1055**	**2465**	**1235**	**1780**	**1335**
1	7440	1265	5805	1915	4490	2245	3240	2430

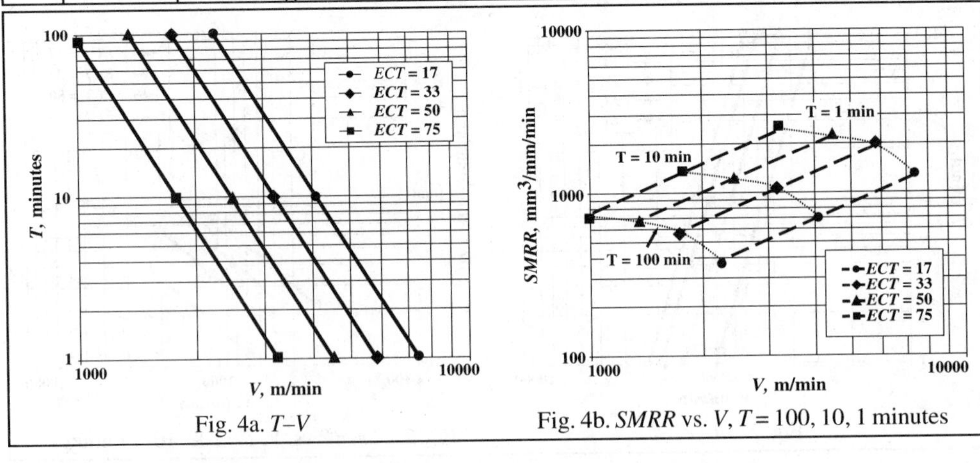

Fig. 4a. T–V

Fig. 4b. $SMRR$ vs. V, T = 100, 10, 1 minutes

Table 5. Group 5—Tool Steels, M2, T2, T5, T6, D2, D5, H41, H42, H43, M50

Tool Life T (min)	ECT = 0.00017 mm		ECT = 0.00033 mm		ECT = 0.00050 mm		ECT = 0.00075 mm	
	Constant C = 6695		Constant C = 5224		Constant C = 4040		Constant C = 2915	
	V_T	*SMRR*	V_T	*SMRR*	V_T	*SMRR*	V_T	*SMRR*
100	2020	345	1580	520	1220	610	880	660
10	**3680**	**625**	**2870**	**945**	**2220**	**1110**	**1600**	**1200**
1	6695	1140	5225	1725	4040	2020	2915	2185

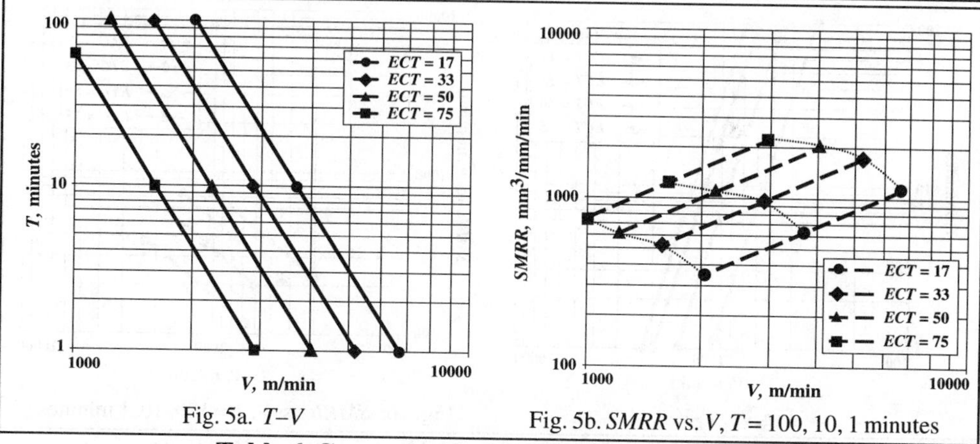

Fig. 5a. *T–V* Fig. 5b. *SMRR* vs. *V*, *T* = 100, 10, 1 minutes

Table 6. Group 6—Tool Steels, M3, M4, T3, D7

Tool Life T (min)	ECT = 0.00017 mm		ECT = 0.00033 mm		ECT = 0.00050 mm		ECT = 0.00075 mm	
	Constant C = 5290		Constant C = 4690		Constant C = 3585		Constant C = 2395	
	V_T	*SMRR*	V_T	*SMRR*	V_T	*SMRR*	V_T	*SMRR*
100	1600	270	1415	465	1085	540	725	540
10	**2910**	**495**	**2580**	**850**	**1970**	**985**	**1315**	**985**
1	5290	900	4690	1550	3585	1795	2395	1795

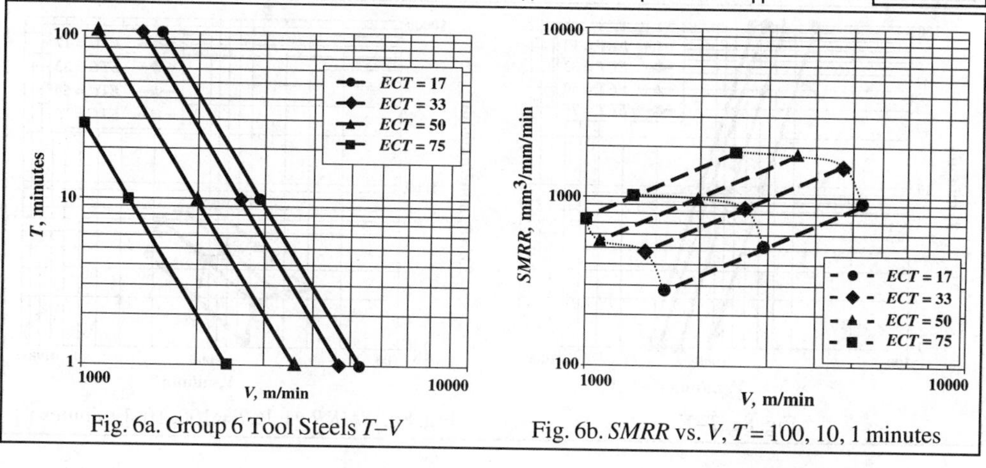

Fig. 6a. Group 6 Tool Steels *T–V* Fig. 6b. *SMRR* vs. *V*, *T* = 100, 10, 1 minutes

Table 7. Group 7—Tool Steels, T15, M15

Tool Life T (min)	ECT = 0.00017 mm		ECT = 0.00033 mm		ECT = 0.00050 mm		ECT = 0.00075 mm	
	Constant C = 2270		Constant C = 1970		Constant C = 1505		Constant C = 1010	
	V_T	$SMRR$	V_T	$SMRR$	V_T	$SMRR$	V_T	$SMRR$
100	685	115	595	195	455	225	305	230
10	**1250**	**210**	**1080**	**355**	**825**	**415**	**555**	**415**
1	2270	385	1970	650	1505	750	1010	760

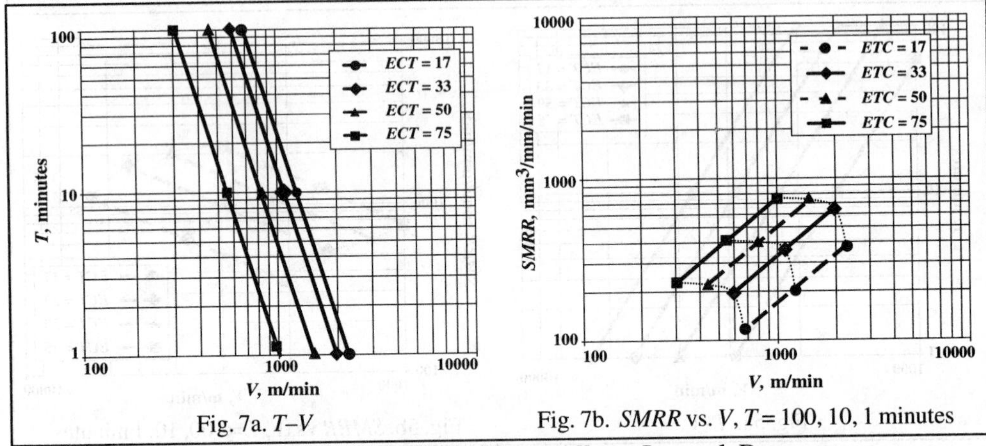

Fig. 7a. T–V Fig. 7b. *SMRR* vs. V, T = 100, 10, 1 minutes

Table 8. Group 8—Heat Resistant Alloys, Inconel, Rene, etc.

Tool Life T (min)	ECT = 0.00017 mm		ECT = 0.00033 mm		ECT = 0.00050 mm		ECT = 0.00075 mm	
	Constant C = 2150		Constant C = 1900		Constant C = 1490		Constant C = 1035	
	V_T	$SMRR$	V_T	$SMRR$	V_T	$SMRR$	V_T	$SMRR$
100	650	110	575	190	450	225	315	235
10	**1185**	**200**	**1045**	**345**	**820**	**410**	**570**	**425**
1	2150	365	1900	625	1490	745	1035	780

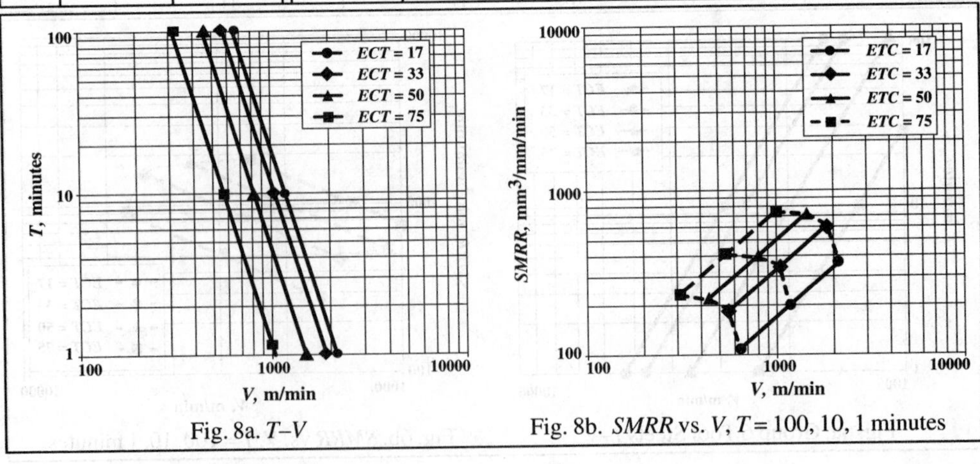

Fig. 8a. T–V Fig. 8b. *SMRR* vs. V, T = 100, 10, 1 minutes

Table 9. Group 9—Carbide Materials, Diamond Wheel

Tool Life *T* (min)	*ECT* = 0.00002 mm		*ECT* = 0.00003 mm		*ECT* = 0.00005 mm		*ECT* = 0.00008 mm	
	Constant C = 9030		Constant C = 8030		Constant C = 5365		Constant C = 2880	
	V_T	*SMRR*	V_T	*SMRR*	V_T	*SMRR*	V_T	*SMRR*
4800	1395	30	1195	35	760	40	390	30
600	2140	45	1855	55	1200	60	625	50
10	4960	100	4415	130	2950	145	1580	125

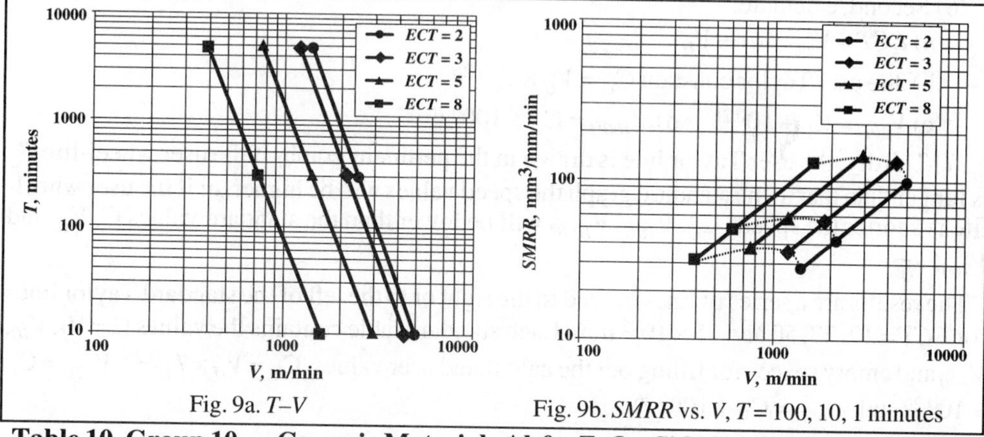

Fig. 9a. *T–V* Fig. 9b. *SMRR* vs. *V*, *T* = 100, 10, 1 minutes

Table 10. Group 10 — Ceramic Materials Al_2O_3, ZrO_2, SiC, Si_3N_4, Diamond Wheel

Tool Life *T* (min)	*ECT* = 0.00002 mm		*ECT* = 0.00003 mm		*ECT* = 0.00005 mm		*ECT* = 0.00008 mm	
	Constant C = 2460		Constant C = 2130		Constant C = 1740		Constant C = 1420	
	V_T	*SMRR*	V_T	*SMRR*	V_T	*SMRR*	V_T	*SMRR*
4800	395	8	335	10	265	13	210	17
600	595	12	510	15	410	20	330	25
10	1355	25	1170	35	955	50	780	60

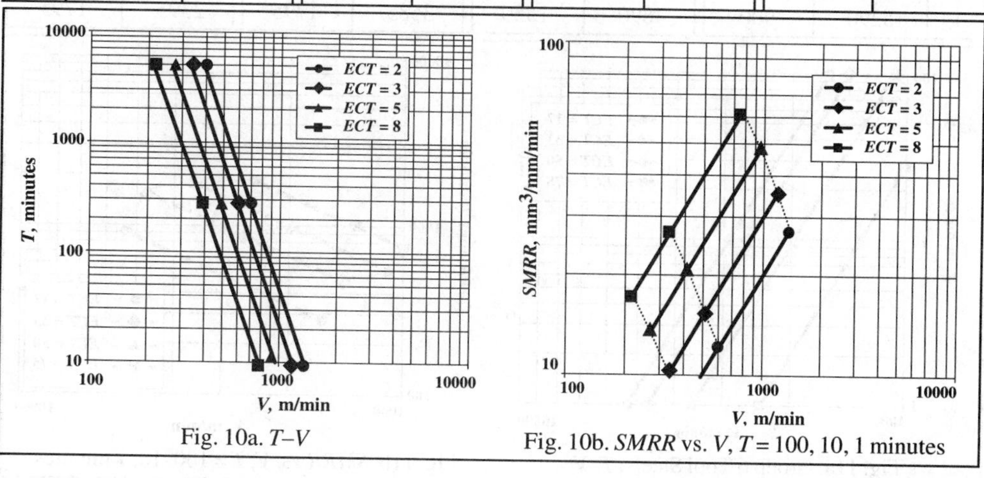

Fig. 10a. *T–V* Fig. 10b. *SMRR* vs. *V*, *T* = 100, 10, 1 minutes

User Calibration of Recommendations

It is recommended to copy or redraw the standard graph for any of the material groups before applying the data calibration method described below. The method is based on the user's own experience and data. The procedure is described in the following and illustrated in Table 11 and Fig. 12.

Only one shop data set is needed to adjust all four Taylor lines as shown below. The required shop data is the user's wheel-life T_U obtained at the user's wheel speed V_U, the user's work speed V_{wU}, and depth of cut a_r.

25) First the user finds out which wheel-life T_U was obtained in the shop, and the corresponding wheel speed V_U, depth of cut a_r and work speed V_{wU}.

26) Second, calculate:

a) $ECT = V_{wU} \times ar \div V_U$

b) the user Taylor constant $C_U = V_U \times T_U^{0.26}$

c) $V_{10U} = C_U \div 10^{0.26}$ d) $V_{100U} = C_U \div 100^{0.26}$

27) Thirdly, the user Taylor line is drawn in the pertinent graph. If the user wheel-life T_U is longer than that in the standard graph the speed values will be higher, or if the user wheel-life is shorter the speeds C_U, V_{10U}, V_{100U} will be lower than the standard values C, V_{10} and V_{100}.

The results are a series of lines moved to the right or to the left of the standard Taylor lines for $ECT = 17, 33, 50$ and 75×10^{-5} mm. Each standard table contains the values $C = V_1$, V_{10}, V_{100} and empty spaces for filling out the calculated user values: $C_U = V_U \times T_U^{0.26}$, $V_{10U} = C_U \div 10^{0.26}$ and $V_{100U} = C_U \div 100^{0.26}$.

Example 7: Assume the following test results on a Group 6 material: user speed is V_U = 1800 m/min, wheel-life T_U = 7 minutes, and ECT = 0.00017 mm. The Group 6 data is repeated below for convenience.

Standard Table Data, Group 6 Material

Tool Life *T* (min)	*ECT* = 0.00017 mm		*ECT* = 0.00033 mm		*ECT* = 0.00050 mm		*ECT* = 0.00075 mm	
	Constant C = 5290		Constant C = 4690		Constant C = 3585		Constant C = 2395	
	V_T	*SMRR*	V_T	*SMRR*	V_T	*SMRR*	V_T	*SMRR*
100	1600	270	1415	465	1085	540	725	540
10	**2910**	**495**	**2580**	**850**	**1970**	**985**	**1315**	**985**
1	5290	900	4690	1550	3585	1795	2395	1795

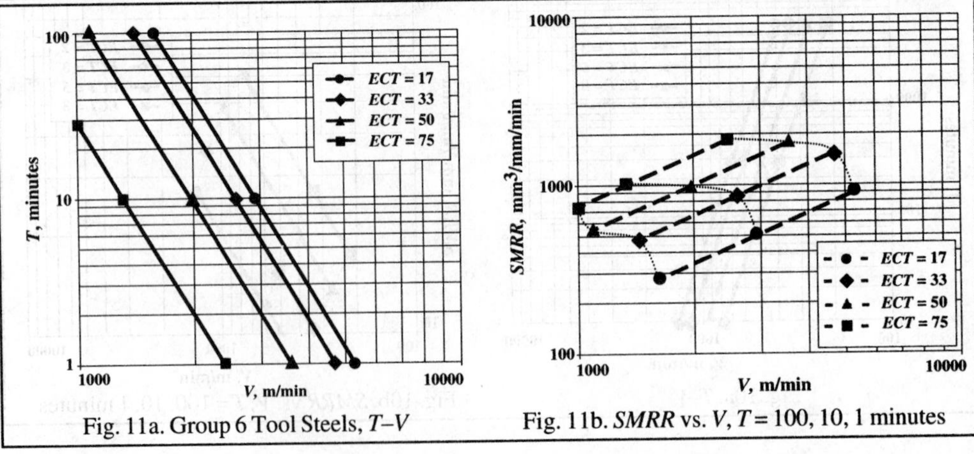

Fig. 11a. Group 6 Tool Steels, *T–V*

Fig. 11b. *SMRR* vs. *V*, *T* = 100, 10, 1 minutes

Calculation Procedure

1) Calculate V_{1U}, V_{10U}, V_{100U} and $SMRR_{1U}$, $SMRR_{10U}$, $SMRR_{100U}$ for $ECT = 0.00017$ mm

a) V_{1U} = the user Taylor constant $C_U = V_U \times T_U^{0.26} = 1800 \times 7^{0.26} = 2985$ m/min, and $SMRR_{1U} = 1000 \times 2985 \times 0.00017 = 507$ mm³/mm width/min

b) $V_{10U} = C_U \div 10^{0.26} = 2985 \div 10^{0.26} = 1640$ m/min, and $SMRR_{10U} = 1000 \times 1640 \times 0.00017 = 279$ mm³/mm width/min

c) $V_{100U} = C_U \div 100^{0.26} = 2985 \div 100^{0.26} = 900$ m/min, and $SMRR_{100U} = 1000 \times 900 \times 0.00017 = 153$ mm³/mm width/min

2) For $ECT = 0.00017$ mm, calculate the ratio of user Taylor constant to standard Taylor constant from the tables = $C_U \div C_{ST} = C_U \div V_1 = 2985 \div 5290 = 0.564$ (see Table 6 for the value of $C_{ST} = V_1$ at $ECT = 0.00017$ mm).

3) For ECT = 0.00033, 0.00050, and 0.00075 mm calculate the user Taylor constants from $C_U = C_{ST} \times$ (the ratio calculated in step 2) = $V_1 \times 0.564 = V_{1U}$. Then, calculate V_{10U} and V_{100U} and $SMRR_{1U}$, $SMRR_{10U}$, $SMRR_{100U}$ using the method in items 1b) and 1c) above.

a) For $ECT = 0.00033$ mm

$V_{1U} = C_U = 4690 \times 0.564 = 2645$ m/min

$V_{10U} = C_U \div 10^{0.26} = 2645 \div 10^{0.26} = 1455$ m/min

$V_{100U} = C_U \div 100^{0.26} = 2645 \div 100^{0.26} = 800$ m/min

$SMRR_{1U}$, $SMRR_{10U}$, and $SMRR_{100U}$ = 876, 480, and 264 mm³/mm width/min

b) For $ECT = 0.00050$ mm

$V_{1U} = C_U = 3590 \times 0.564 = 2025$ m/min

$V_{10U} = C_U \div 10^{0.26} = 2025 \div 10^{0.26} = 1110$ m/min

$V_{100U} = C_U \div 100^{0.26} = 2025 \div 100^{0.26} = 610$ m/min

$SMRR_{1U}$, $SMRR_{10U}$, and $SMRR_{100U}$ = 1013, 555, and 305 mm³/mm width/min

c) For $ECT = 0.00075$ mm

$V_{1U} = C_U = 2395 \times 0.564 = 1350$ m/min

$V_{10U} = C_U \div 10^{0.26} = 1350 \div 10^{0.26} = 740$ m/min

$V_{100U} = C_U \div 100^{0.26} = 1350 \div 100^{0.26} = 405$ m/min

$SMRR_{1U}$, $SMRR_{10U}$, and $SMRR_{100U}$ = 1013, 555, and 305 mm³/mm width/min

Thus, the wheel speed for any desired wheel-life at a given ECT can be calculated from $V = C_U \div T^{0.26}$. For example, at $ECT = 0.00050$ mm and desired tool-life $T = 9$, $V_9 = 2025 \div 9^{0.26} = 1144$ m/min. The corresponding specific metal removal rate is $SMRR = 1000 \times 1144 \times 0.0005 = 572$ mm³/mm width/min (0.886 in³/inch width/min).

Table 11. User Calculated Data, Group 6 Material

Tool Life T (min)	ECT = 0.00017 mm		ECT = 0.00033 mm		ECT = 0.00050 mm		ECT = 0.00075 mm	
	User Constant C_U = 2985		User Constant C_U = 2645		User Constant C_U = 2025		User Constant C_U = 1350	
	V_T	$SMRR$	V_T	$SMRR$	V_T	$SMRR$	V_T	$SMRR$
100	900	153	800	264	610	305	405	305
10	**1640**	**279**	**1455**	**480**	**1110**	**555**	**740**	**555**
1	2985	507	2645	876	2025	1013	1350	1013

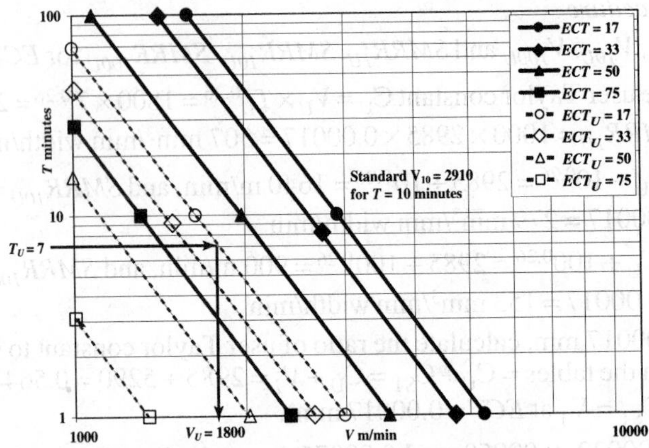

Fig. 12. Calibration of user grinding data to standard Taylor Lines
User Input: $V_U = 1800$ m/min, $T_U = 7$ minutes, $ECT = 0.00017$ mm

Optimization.— As shown, a global optimum occurs along the *G*-curve, in selected cases for values of *ECT* around 0.00075, i.e. at high metal removal rates as in other machining operations. It is recommended to use the simple formula for economic life: $T_E = 3 \times T_V$ minutes. $T_V = T_{RPL} + 60 \times C_E \div H_R$, minutes, where T_{RPL} is the time required to replace wheel, C_E = cost per wheel dressing = wheel cost + cost per dressing, and H_R is the hourly rate.

In grinding, values of T_V range between 2 and 5 minutes in conventional grinders, which means that the economic wheel lives range between 6 and 15 minutes indicating higher metal removal rates than are commonly used. When wheels are sharpened automatically after each stroke as in internal grinding, or when grits are continually replaced as in abrasive grinding (machining), T_V may be less than one minute. This translates into wheel lives around one minute in order to achieve minimum cost grinding.

Grinding Cost, Optimization and Process Planning: More accurate results are obtained when the firm collects and systemizes the information on wheel lives, wheel and work speeds, and depths of cut from production runs. A computer program can be used to plan the grinding process and apply the rules and formulas presented in this chapter. A complete grinding process planning program, such as that developed by Colding International Corporation, can be used to optimize machine settings for various feed-speed preferences corresponding wheel-life requirements, minimum cost or maximum production rate grinding, required surface finish and sparkout time; machine and fixture requirements based on the grinding forces, torque and power for sharp and worn grinding wheels; and, detailed time and cost analysis per part and per batch including wheel dressing and wheel changing schedules.

Table 12 summarizes the time and cost savings per batch as it relates to tool life. The sensitivity of how grinding parameters are selected is obvious. Minimum cost conditions yield a 51% reduction of time and 44% reduction of cost, while maximum production rate reduces total time by 65% but, at the expense of heavy wheel consumption (continuous dressing), cost by only 18%.

Table 12. Wheel Life vs. Cost

Preferences	Time per Batch, minutes	Cost per Batch, $		Reduction from Long Life,%	
		Tooling	Total Cost	Time	Cost
Long Life	2995	39	2412	—	—
Economic Life	2433	252	2211	19	8
Minimum Cost	1465	199	1344	51	44
Max Production Rate	1041	1244	1980	65	18

GRINDING AND OTHER ABRASIVE PROCESSES

Processes and equipment discussed under this heading use abrasive grains for shaping workpieces by means of machining or related methods. Abrasive grains are hard crystals either found in nature or manufactured. The most commonly used materials are aluminum oxide, silicon carbide, cubic boron nitride and diamond. Other materials such as garnet, zirconia, glass and even walnut shells are used for some applications. Abrasive products are used in three basic forms by industry:

A) *Bonded* to form a solid shaped tool such as disks (the basic shape of grinding wheels), cylinders, rings, cups, segments, or sticks to name a few.

B) *Coated* on backings made of paper or cloth, in the form of sheets, strips, or belts.

C) *Loose,* held in some liquid or solid carrier (for lapping, polishing, tumbling), or propelled by centrifugal force, air, or water pressure against the work surface (blast cleaning).

The applications for abrasive processes are multiple and varied. They include:

A) *Cleaning* of surfaces, also the coarse removal of excess material—such as rough off-hand grinding in foundries to remove gates and risers.

B) *Shaping,* such as in form grinding and tool sharpening.

C) *Sizing,* a general objective, but of primary importance in precision grinding.

D) *Surface finish improvement,* either primarily as in lapping, honing, and polishing or as a secondary objective in other types of abrasive processes.

E) *Separating,* as in cut-off or slicing operations.

The main field of application of abrasive processes is in metalworking, because of the capacity of abrasive grains to penetrate into even the hardest metals and alloys. However, the great hardness of the abrasive grains also makes the process preferred for working other hard materials, such as stones, glass, and certain types of plastics. Abrasive processes are also chosen for working relatively soft materials, such as wood, rubber, etc., for such reasons as high stock removal rates, long-lasting cutting ability, good form control, and fine finish of the worked surface.

Grinding Wheels

Abrasive Materials.— In earlier times, only natural abrasives were available. From about the beginning of this century, however, manufactured abrasives, primarily silicon carbide and aluminum oxide, have replaced the natural materials; even natural diamonds have been almost completely supplanted by synthetics. Superior and controllable properties, and dependable uniformity characterize the manufactured abrasives.

Both silicon carbide and aluminum oxide abrasives are very hard and brittle. This brittleness, called friability, is controllable for different applications. Friable abrasives break easily, thus forming sharp edges. This decreases the force needed to penetrate into the work material and the heat generated during cutting. Friable abrasives are most commonly used for precision and finish grinding. Tough abrasives resist fracture and last longer. They are used for rough grinding, snagging, and off-hand grinding.

As a general rule, although subject to variation:

1) Aluminum oxide abrasives are used for grinding plain and alloyed steel in a soft or hardened condition.

2) Silicon carbide abrasives are selected for cast iron, nonferrous metals, and nonmetallic materials.

3) Diamond is the best type of abrasive for grinding cemented carbides. It is also used for grinding glass, ceramics, and hardened tool steel.

4) Cubic Boron Nitride (CBN) is known by several trade names including Borazon (General Electric Co.), ABN (De Beers), Sho-bon (Showa-Denko), and Elbor (USSR). CBN is a synthetic superabrasive used for grinding hardened steels and wear-resistant superalloys. (See *Cubic Boron Nitride (CBN)* starting on page 982.) CBN grinding wheels have long lives and can maintain close tolerances with superior surface finishes.

Bond Properties and Grinding Wheel Grades.—The four main types of bonds used for grinding wheels are the vitrified, resinoid, rubber, and metal.

Vitrified bonds are used for more than half of all grinding wheels made, and are preferred because of their strength and other desirable qualities. Being inert, glass-like materials, vitrified bonds are not affected by water or by the chemical composition of different grinding fluids. Vitrified bonds also withstand the high temperatures generated during normal grinding operations. The structure of vitrified wheels can be controlled over a wide range of strength and porosity. Vitrified wheels, however, are more sensitive to impact than those made with organic bonds.

Resinoid bonds are selected for wheels subjected to impact, or sudden loads, or very high operating speeds. They are preferred for snagging, portable grinder uses, or roughing operations. The higher flexibility of this type of bond—essentially a filled thermosetting plastic—helps it withstand rough treatment.

Rubber bonds are even more flexible than the resinoid type, and for that reason are used for producing a high finish and for resisting sudden rises in load. Rubber bonded wheels are commonly used for wet cut-off wheels because of the nearly burr-free cuts they produce, and for centerless grinder regulating wheels to provide a stronger grip and more reliable workpiece control.

Metal bonds are used in CBN and diamond wheels. In metal bonds produced by electrodeposition, a single layer of superabrasive material (diamond or CBN) is bonded to a metal core by a matrix of metal, usually nickel. The process is so controlled that about 30–40 per cent of each abrasive particle projects above the deposited surface, giving the wheel a very aggressive and free-cutting action. With proper use, such wheels have remarkably long lives. When dulled, or worn down, the abrasive can be stripped off and the wheel renewed by a further deposit process. These wheels are also used in electrical discharge grinding and electrochemical grinding where an electrically conductive wheel is needed.

In addition to the basic properties of the various bond materials, each can also be applied in different proportions, thereby controlling the grade of the grinding wheel.

Grinding wheel grades commonly associated with hardness, express the amount of bond material in a grinding wheel, and hence the strength by which the bond retains the individual grains.

During grinding, the forces generated when cutting the work material tend to dislodge the abrasive grains. As the grains get dull and if they don't fracture to resharpen themselves, the cutting forces will eventually tear the grains from their supporting bond. For a "soft" wheel the cutting forces will dislodge the abrasive grains before they have an opportunity to fracture. When a "hard" wheel is used, the situation is reversed. Because of the extra bond in the wheel the grains are so firmly held that they never break loose and the wheel becomes glazed. During most grinding operations it is desirable to have an intermediate wheel where there is a continual slow wearing process composed of both grain fracture and dislodgement.

The grades of the grinding wheels are designated by capital letters used in alphabetical order to express increasing "hardness" from A to Z.

Grinding Wheel Structure.—The individual grains, which are encased and held together by the bond material, do not fill the entire volume of the grinding wheel; the intermediate open space is needed for several functional purposes such as heat dissipation, coolant application, and particularly, for the temporary storage of chips. It follows that the spacing of the grains must be greater for coarse grains which cut thicker chips and for large contact areas within which the chips have to be retained on the surface of the wheel before being disposed of. On the other hand, a wide spacing reduces the number of grains that contact the work surface within a given advance distance, thereby producing a coarser finish.

In general, denser structures are specified for grinding hard materials, for high-speed grinding operations, when the contact area is narrow, and for producing fine finishes and/or accurate forms. Wheels with open structure are used for tough materials, high stock removal rates, and extended contact areas, such as grinding with the face of the wheel. There are, however, several exceptions to these basic rules, an important one being the grinding of parts made by powder metallurgy, such as cemented carbides; although they represent one of the hardest industrial materials, grinding carbides requires wheels with an open structure.

Most kinds of general grinding operations, when carried out with the periphery of the wheel, call for medium spacing of the grains. The structure of the grinding wheels is expressed by numerals from 1 to 16, ranging from dense to open. Sometimes, "induced porosity" is used with open structure wheels. This term means that the grinding wheel manufacturer has placed filler material (which later burns out when the wheel is fired to vitrify the bond) in the grinding wheel mix. These fillers create large "pores" between grain clusters without changing the total volume of the "pores" in the grinding wheel. Thus, an A46-H12V wheel and an A46H12VP wheel will contain the same amounts of bond, abrasive, and air space. In the former, a large number of relatively small pores will be distributed throughout the wheel. The latter will have a smaller number of larger pores.

American National Standard Grinding Wheel Markings.—ANSI Standard B74.13-1990" Markings for Identifying Grinding Wheels and Other Bonded Abrasives," applies to grinding wheels and other bonded abrasives, segments, bricks, sticks, hones, rubs, and other shapes that are for removing material, or producing a desired surface or dimension. It does not apply to specialities such as sharpening stones and provides only a standard system of markings. Wheels having the same standard markings but made by different wheel manufacturers may not—and probably will not—produce exactly the same grinding action. This desirable result cannot be obtained because of the impossibility of closely correlating any measurable physical properties of bonded abrasive products in terms of their grinding action.

Symbols for designating diamond and cubic boron wheel compositions are given on page 1166.

Sequence of Markings.—The accompanying illustration taken from ANSI B74.13-1990 shows the makeup of a typical wheel or bonded abrasive marking.

	1		2		3		4		5		6
Prefix	**Abrasive Type**		**Grain Size**		**Grade**		**Structure**		**Bond Type**		**Manufacturer's Record**
51	**A**	–	**36**	–	**L**	–	**5**	–	**V**	–	**23**

The meaning of each letter and number in this or other markings is indicated by the following complete list.

1) *Abrasive Letters:* The letter (A) is used for aluminum oxide, (C) for silicon carbide, and (Z) for aluminum zirconium. The manufacturer may designate some particular type in any one of these broad classes, by using his own symbol as a prefix (example, 51).

2) *Grain Size:* The grain sizes commonly used and varying from coarse to very fine are indicated by the following numbers: 8, 10, 12, 14, 16, 20, 24, 30, 36, 46, 54, 60,70, 80, 90, 100, 120, 150, 180, and 220. The following additional sizes are used occasionally: 240, 280, 320, 400, 500, and 600. The wheel manufacturer may add to the regular grain number an additional symbol to indicate a special grain combination.

3) *Grade:* Grades are indicated by letters of the alphabet from A to Z in all bonds or processes. Wheel grades from A to Z range from soft to hard.

4) *Structure:* The use of a structure symbol is optional. The structure is indicated by Nos. 1 to 16 (or higher, if necessary) with progressively higher numbers indicating a progressively wider grain spacing (more open structure).

5) *Bond or Process:* Bonds are indicated by the following letters: V, vitrified; S, silicate; E, shellac or elastic; R, rubber; RF, rubber reinforced; B, resinoid (synthetic resins); BF, resinoid reinforced; O, oxychloride.

6) *Manufacturer's Record:* The sixth position may be used for manufacturer's private factory records; this is optional.

American National Standard Shapes and Sizes of Grinding Wheels.—The ANSI Standard B74.2-1982 which includes shapes and sizes of grinding wheels, gives a wide variety of grinding wheel shape and size combinations. These are suitable for the majority of applications. Although grinding wheels can be manufactured to shapes and dimensions different from those listed, it is advisable, for reasons of cost and inventory control, to avoid using special shapes and sizes, unless technically warranted.

Standard shapes and size ranges as given in this Standard together with typical applications are shown in Table for inch dimensions and in Table for metric dimensions.

The operating surface of the grinding wheel is often referred to as the wheel face. In the majority of cases it is the periphery of the grinding wheel which, when not specified otherwise, has a straight profile. However, other face shapes can also be supplied by the grinding wheel manufacturers, and also reproduced during usage by appropriate truing. ANSI B74.2-1982 standard offers 13 different shapes for grinding wheel faces, which are shown in Table 2.

The Selection of Grinding Wheels.—In selecting a grinding wheel, the determining factors are the composition of the work material, the type of grinding machine, the size range of the wheels used, and the expected grinding results, in this approximate order.

The Norton Company has developed, as the result of extensive test series, a method of grinding wheel recommendation that is more flexible and also better adapted to taking into consideration pertinent factors of the job, than are listings based solely on workpiece categories. This approach is the basis for Tables 3 through 6, inclusive. Tool steels and constructional steels are considered in the detailed recommendations presented in these tables.

Table 3 assigns most of the standardized tool steels to five different grindability groups. The AISI-SAE tool steel designations are used.

After having defined the grindability group of the tool steel to be ground, the operation to be carried out is found in the first column of Table . The second column in this table distinguishes between different grinding wheel size ranges, because wheel size is a factor in determining the contact area between wheel and work, thus affecting the apparent hardness of the grinding wheel. Distinction is also made between wet and dry grinding.

Finally, the last two columns define the essential characteristics of the recommended types of grinding wheels under the headings of first and second choice, respectively. Where letters are used *preceding* A, the standard designation for aluminum oxide, they indicate a degree of friability different from the regular, thus: SF = semi friable (Norton equivalent 16A) and F = friable (Norton equivalent 33A and 38A). The suffix P, where applied, expresses a degree of porosity that is more open than the regular.

Table 1a. Standard Shapes and Inch Size Ranges of Grinding Wheels
ANSI B74.2-1982

Applications	Size Ranges of Principal Dimensions, Inches		
	D = Dia.	T = Thick.	H = Hole
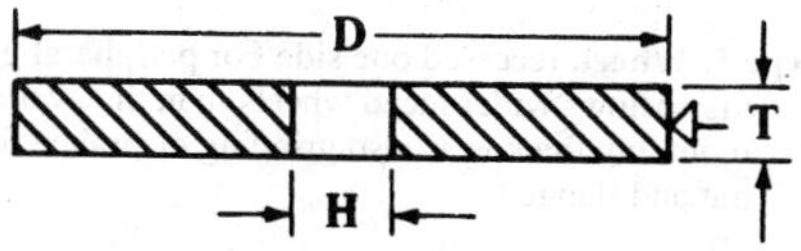 D T H	**Type 1. Straight Wheel** For peripheral grinding.		
CUTTING OFF (Organic bonds only)	1 to 48	$\frac{1}{64}$ to $\frac{3}{8}$	$\frac{1}{16}$ to 6
CYLINDRICAL GRINDING Between centers	12 to 48	$\frac{1}{2}$ to 6	5 to 20
CYLINDRICAL GRINDING Centerless grinding wheels	14 to 30	1 to 20	5 or 12
CYLINDRICAL GRINDING Centerless regulating wheels	8 to 14	1 to 12	3 to 6
INTERNAL GRINDING	$\frac{1}{4}$ to 4	$\frac{1}{4}$ to 2	$\frac{3}{32}$ to $\frac{7}{8}$
OFFHAND GRINDING Grinding on the periphery			
General purpose	6 to 36	$\frac{1}{2}$ to 4	$\frac{1}{2}$ to 3
For wet tool grinding only	30 or 36	3 or 4	20
SAW GUMMING (F-type face)	6 to 12	$\frac{1}{4}$ to $1\frac{1}{2}$	$\frac{1}{2}$ to $1\frac{1}{4}$
SNAGGING Floor stand machines	12 to 24	1 to 3	$1\frac{1}{4}$ to $2\frac{1}{2}$
SNAGGING Floor stand machines (Organic bond, wheel speed over 6500 sfpm)	20 to 36	2 to 4	6 or 12
SNAGGING Mechanical grinders (Organic bond, wheel speed up to 16,500 sfpm)	24	2 to 3	12
SNAGGING Portable machines	3 to 8	$\frac{1}{4}$ to 1	$\frac{3}{8}$ to $\frac{5}{8}$
SNAGGING Portable machines (Reinforced organic bond, 17,000 sfpm)	6 or 8	$\frac{3}{4}$ or 1	1
SNAGGING Swing frame machines	12 to 24	2 to 3	$3\frac{1}{2}$ to 12
SURFACE GRINDING Horizontal spindle machines	6 to 24	$\frac{1}{2}$ to 6	$1\frac{1}{4}$ to 12
TOOL GRINDING Broaches, cutters, mills, reamers, taps, etc.	6 to 10	$\frac{1}{4}$ to $\frac{1}{2}$	$\frac{5}{8}$ to 5
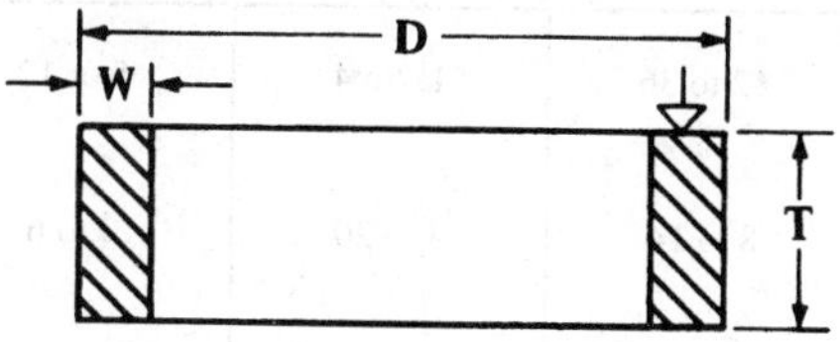 D W T	**Type 2. Cylindrical Wheel** Side grinding wheel — mounted on the diameter; may also be mounted in a chuck or on a plate.		
			W = Wall
SURFACE GRINDING Vertical spindle machines	8 to 20	4 or 5	1 to 4

Table 1a. *(Continued)* **Standard Shapes and Inch Size Ranges of Grinding Wheels** *ANSI B74.2-1982*

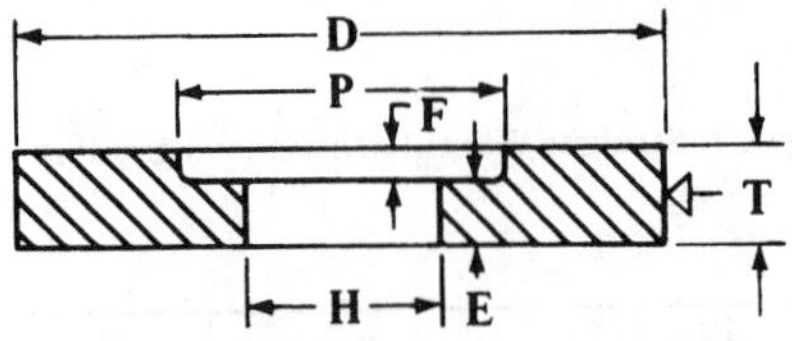

Applications	Size Ranges of Principal Dimensions, Inches		
	D = Dia.	T = Thick.	H = Hole
Type 5. Wheel, recessed one side For peripheral grinding. Allows wider faced wheels than the available mounting thickness, also grinding clearance for the nut and flange.			
CYLINDRICAL GRINDING Between centers	12 to 36	$1\frac{1}{2}$ to 4	5 or 12
CYLINDRICAL GRINDING Centerless regulating wheel	8 to 14	3 to 6	3 or 5
INTERNAL GRINDING	$\frac{3}{8}$ to 4	$\frac{3}{8}$ to 2	$\frac{1}{8}$ to $\frac{7}{8}$
SURFACE GRINDING Horizontal spindle machines	7 to 24	$\frac{3}{4}$ to 6	$1\frac{1}{4}$ to 12
Type 6. Straight-Cup Wheel Side grinding wheel, in whose dimensioning the wall thickness (W) takes precedence over the diameter of the recess. Hole is $\frac{5}{8}$-11UNC-2B threaded for the snagging wheels and $\frac{1}{2}$ or $1\frac{1}{4}''$ for the tool grinding wheels.			W = Wall
SNAGGING Portable machines, organic bond only.	4 to 6	2	$\frac{3}{4}$ to $1\frac{1}{2}$
TOOL GRINDING Broaches, cutters, mills, reamers, taps, etc.	2 to 6	$1\frac{1}{4}$ to 2	$\frac{5}{16}$ or $\frac{3}{8}$
Type 7. Wheel, recessed two sides Peripheral grinding. Recesses allow grinding clearance for both flanges and also narrower mounting thickness than overall thickness.			
CYLINDRICAL GRINDING Between centers	12 to 36	$1\frac{1}{2}$ to 4	5 or 12
CYLINDRICAL GRINDING Centerless regulating wheel	8 to 14	4 to 20	3 to 6
SURFACE GRINDING Horizontal spindle machines	12 to 24	2 to 6	5 to 12

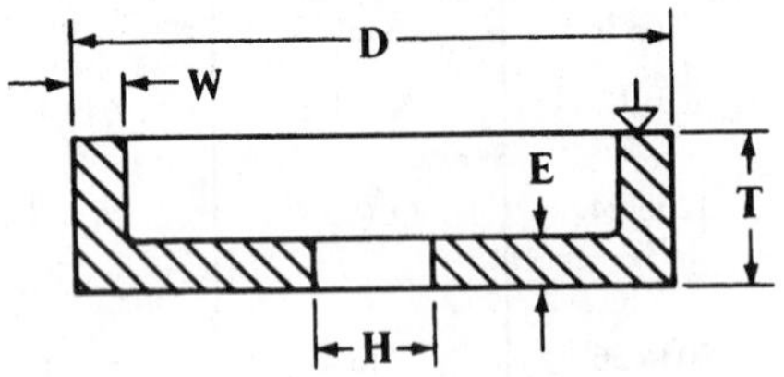

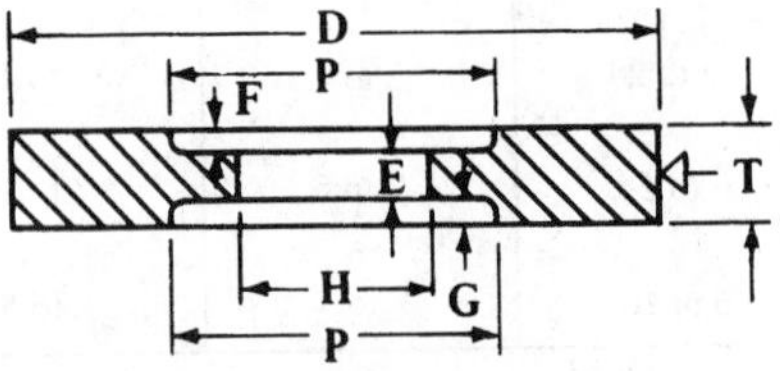

Table 1a. *(Continued)* **Standard Shapes and Inch Size Ranges of Grinding Wheels**
ANSI B74.2-1982

Applications	Size Ranges of Principal Dimensions, Inches		
	D = Dia.	T = Thick.	H = Hole
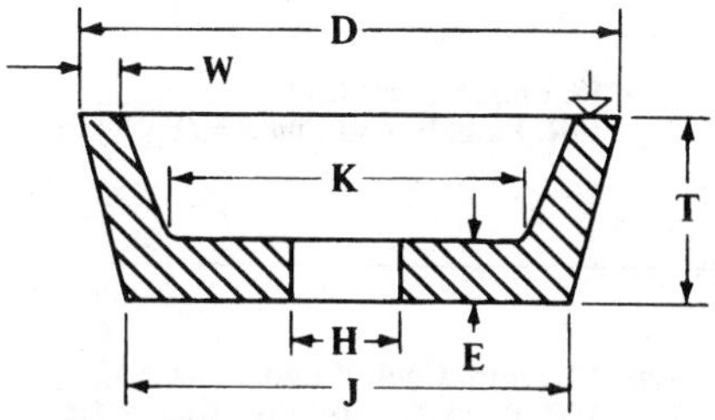 D, W, K, T, H, E, J	**Type 11. Flaring-Cup Wheel** Side grinding wheel with wall tapered outward from the back; wall generally thicker in the back.		
SNAGGING Portable machines, organic bonds only, threaded hole	4 to 6	2	⅝-11 UNC-2B
TOOL GRINDING Broaches, cutters, mills, reamers, taps, etc.	2 to 5	1 ¼ to 2	½ to 1 ¼
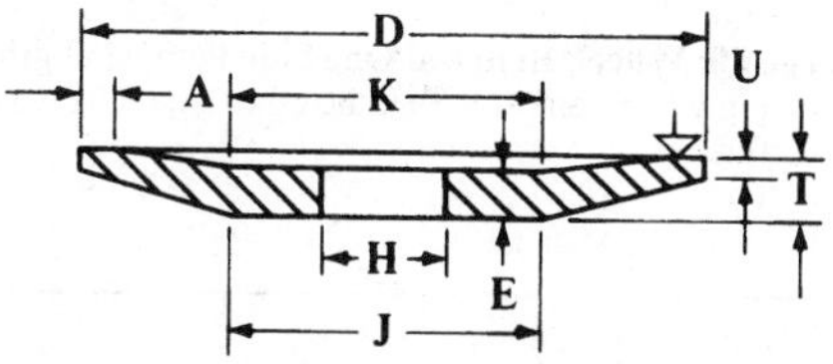 D, U, A, K, T, H, E, J	**Type 12. Dish Wheel** Grinding on the side or on the U-face of the wheel, the U-face being always present in this type.		
TOOL GRINDING Broaches, cutters, mills, reamers, taps, etc.	3 to 8	½ or ¾	½ to 1 ¼
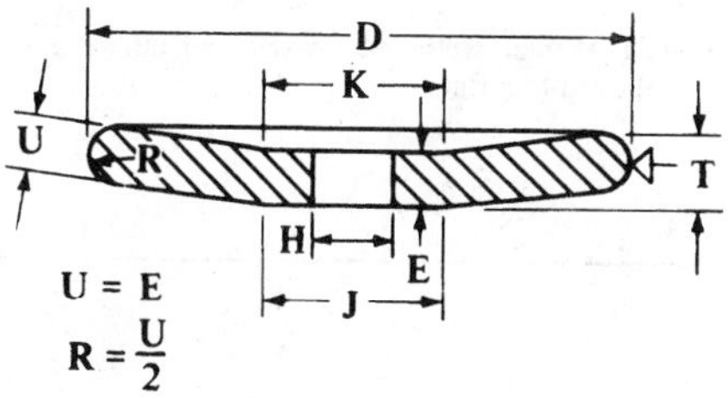 D, K, U, R, T, H, E, J U = E $R = \frac{U}{2}$	**Type 13. Saucer Wheel** Peripheral grinding wheel, resembling the shape of a saucer, with cross section equal throughout.		
SAW GUMMING Saw tooth shaping and sharpening	8 to 12	½ to 1 ¾ *U & E* ¼ to 1½	¾ to 1 ¼
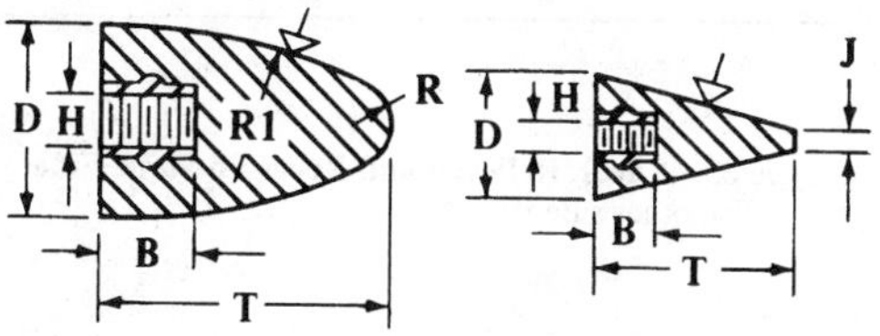 D, H, R1, R, B, T, J	**Type 16. Cone, Curved Side** **Type 17. Cone, Straight Side, Square Tip** **Type 17R. Cone, Straight Side, Round Tip** (Tip Radius $R = J/2$)		
SNAGGING Portable machine, threaded holes	1¼ to 3	2 to 3½	⅜-24UNF-2B to ⅝-11UNC-2B

Table 1a. *(Continued)* **Standard Shapes and Inch Size Ranges of Grinding Wheels**
ANSI B74.2-1982

Applications	Size Ranges of Principal Dimensions, Inches		
	D = Dia.	T = Thick.	H = Hole
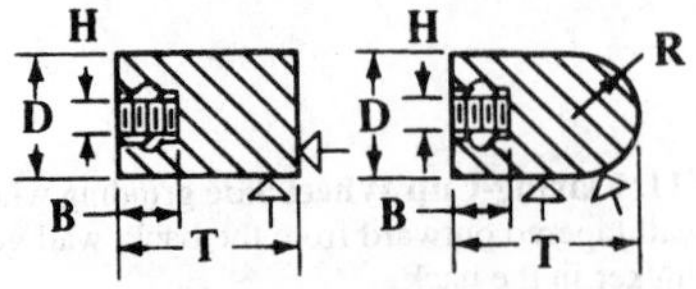 **Type 18. Plug, Square End** **Type 18R. Plug, Round End** $R = D/2$			
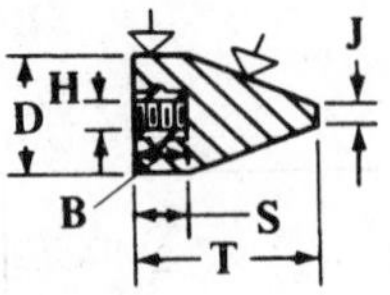 **Type 19. Plugs, Conical End, Square Tip** **Type 19R. Plugs, Conical End, Round Tip** (Tip Radius $R = J/2$)			
SNAGGING Portable machine, threaded holes	1¼ to 3	2 to 3½	⅜-24UNF-2B to ⅝-11UNC-2B
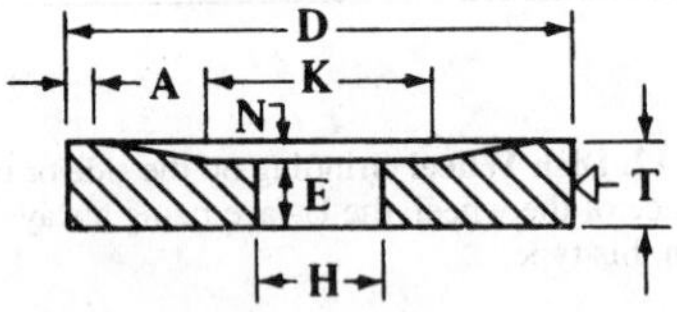 **Type 20. Wheel, Relieved One Side** Peripheral grinding wheel, one side flat, the other side relieved to a flat.			
CYLINDRICAL GRINDING Between centers	12 to 36	¾ to 4	5 to 20
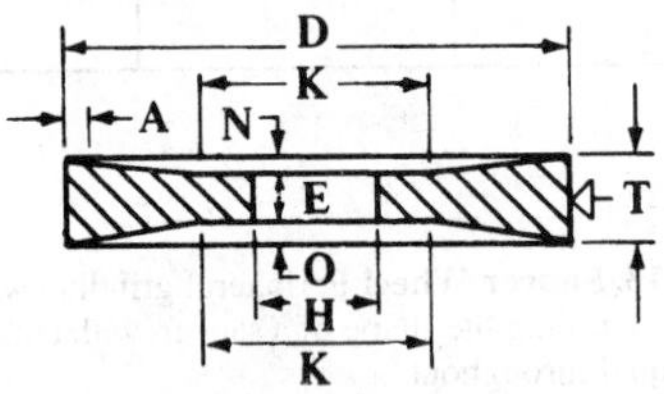 **Type 21. Wheel, Relieved Two Sides** Both sides relieved to a flat.			
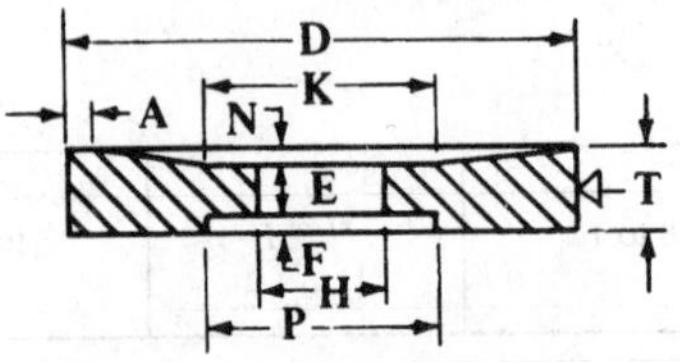 **Type 22. Wheel, Relieved One Side, Recessed Other Side** One side relieved to a flat.			
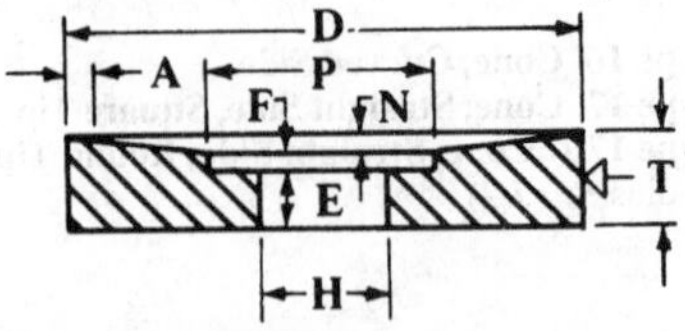 **Type 23. Wheel, Relieved and Recessed Same Side** The other side is straight.			
CYLINDRICAL GRINDING Between centers, with wheel periphery	20 to 36	2 to 4	12 or 20

Table 1a. *(Continued)* **Standard Shapes and Inch Size Ranges of Grinding Wheels** *ANSI B74.2-1982*

Applications	Size Ranges of Principal Dimensions, Inches		
	D = Dia.	T = Thick.	H = Hole
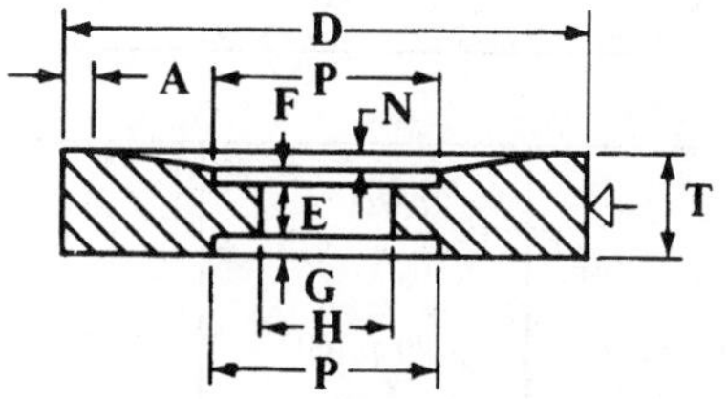 **Type 24. Wheel, Relieved and Recessed One Side, Recessed Other Side** One side recessed, the other side is relieved to a recess.			
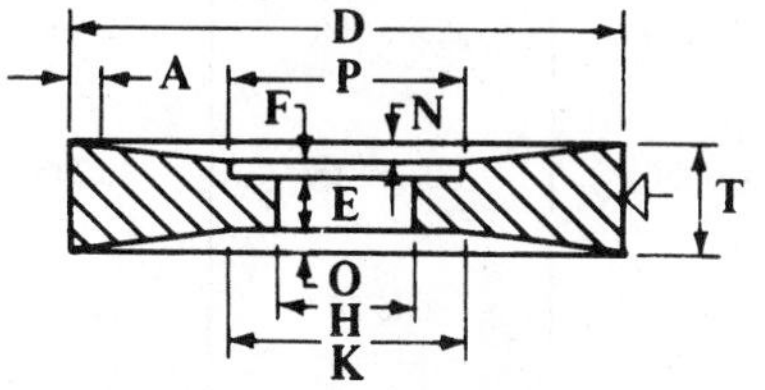 **Type 25. Wheel, Relieved and Recessed One Side, Relieved Other Side** One side relieved to a flat, the other side relieved to a recess.			
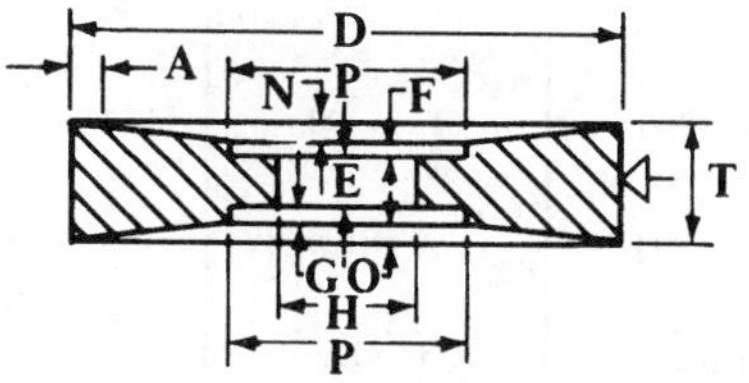 **Type 26. Wheel, Relieved and Recessed Both Sides**			
CYLINDRICAL GRINDING Between centers, with the periphery of the wheel	20 to 36	2 to 4	12 or 20
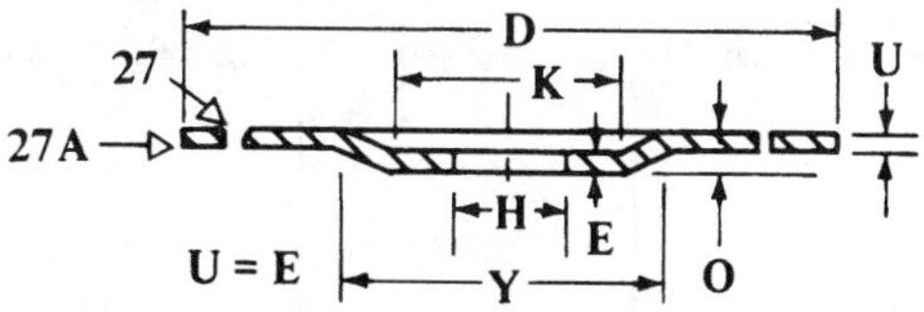 **TYPES 27 & 27A. Wheel, Depressed Center** 27. *Portable Grinding:* Grinding normally done by contact with work at approx. a 15° angle with face of the wheel. 27A. *Cutting-off:* Using the periphery as grinding face.			
CUTTING OFF Reinforced organic bonds only	16 to 30	$U = E$ = 5⁄32 to 1⁄4	1 or 1 1⁄2
SNAGGING Portable machine	3 to 9	U = Uniform thick. 1⁄8 to 3⁄8	3⁄8 or 7⁄8
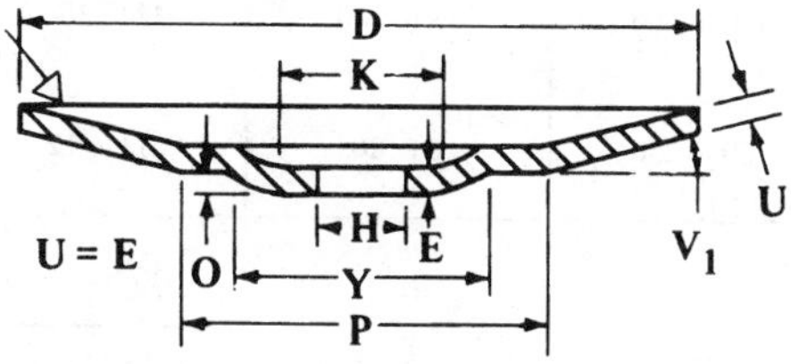 **Type 28. Wheel, Depressed Center (Saucer Shaped Grinding Face)** Grinding at approx. 15° angle with wheel face.			
SNAGGING Portable machine	7 or 9	U = Uniform thickness 1⁄4	7⁄8

Throughout table large open-head arrows indicate grinding surfaces.

Table 1b. Standard Shapes and Metric Size Ranges of Grinding Wheels
ANSI B74.2-1982

Applications	Size Ranges of Principal Dimensions, Millimeters		
	D = Diam.	*T* = Thick.	*H* = Hole
Type 1. Straight Wheel[a]			
CUTTING OFF (nonreinforced and reinforced organic bonds only)	150 to 1250	0.8 to 10	16 to 152.4
CYLINDRICAL GRINDING Between centers	300 to 1250	20 to 160	127 to 508
CYLINDRICAL GRINDING Centerless grinding wheels	350 to 750	25 to 500	127 or 304.8
CYLINDRICAL GRINDING Centerless regulating wheels	200 to 350	25 to 315	76.2 to 152.4
INTERNAL GRINDING	6 to 100	6 to 50	2.5 to 25
OFFHAND GRINDING Grinding on the periphery			
General purpose	150 to 900	13 to 100	20 to 76.2
For wet tool grinding only	750 or 900	80 or 100	508
SAW GUMMING (F-type face)	150 to 300	6 to 40	32
SNAGGING Floor stand machines	300 to 600	25 to 80	32 to 76.2
SNAGGING Floor stand machines(organic bond, wheel speed over 33 meters per second)	500 to 900	50 to 100	152.4 or 304.8
SNAGGING Mechanical grinders (organic bond, wheel speed up to 84 meters per second)	600	50 to 80	304.8
SNAGGING Portable machines	80 to 200	6 to 25	10 to 16
SNAGGING Swing frame machines (organic bond)	300 to 600	50 to 80	88.9 to 304.8
SURFACE GRINDING Horizontal spindle machines	150 to 600	13 to 160	32 to 304.8
TOOL GRINDING Broaches, cutters, mills, reamers, taps, etc.	150 to 250	6 to 20	32 to 127
Type 2. Cylindrical Wheel[a]			
			W = Wall
SURFACE GRINDING Vertical spindle machines	200 to 500	100 or 125	25 to 100

Table 1b. *(Continued)* **Standard Shapes and Metric Size Ranges of Grinding Wheels** *ANSI B74.2-1982*

Applications	Size Ranges of Principal Dimensions, Millimeters		
	D = Diam.	T = Thick.	H = Hole
Type 5. Wheel, recessed one side[a]			
CYLINDRICAL GRINDING Between centers	300 to 900	40 to 100	127 or 304.8
CYLINDRICAL GRINDING Centerless regulating wheels	200 to 350	80 to 160	76.2 or 127
INTERNAL GRINDING	10 to 100	10 to 50	3.18 to 25
Type 6. Straight-Cup Wheel[a]			
			W = Wall
SNAGGING Portable machines, organic bond only (hole is ⅝-11 UNC-2B)	100 to 150	50	20 to 40
TOOL GRINDING Broaches, cutters, mills, reamers, taps, etc. (Hole is 13 to 32 mm)	50 to 150	32 to 50	8 or 10
Type 7. Wheel, recessed two sides[a]			
CYLINDRICAL GRINDING Between centers	300 to 900	40 to 100	127 or 304.8
CYLINDRICAL GRINDING Centerless regulating wheels	200 to 350	100 to 500	76.2 to 152.4
Type 11. Flaring-Cup Wheel[a]			
SNAGGING Portable machines, organic bonds only, threaded hole	100 to 150	50	⅝-11 UNC-2B
TOOL GRINDING Broaches, cutters, mills, reamers, taps, etc.	50 to 125	32 to 50	13 to 32
Type 12. Dish Wheel[a]			
TOOL GRINDING Broaches, cutters, mills, reamers, taps, etc.	80 to 200	13 or 20	13 to 32
Type 27 and 27A. Wheel, depressed center[a]			
CUTTING OFF Reinforced organic bonds only	400 to 750	$U = E = 6$	25.4 or 38.1
SNAGGING Portable machines	80 to 230	$U = E = 3.2$ to 10	9.53 or 22.23

[a] See Table 1a for diagrams and descriptions of each wheel type.
All dimensions in millimeters.

Table 2. Standard Shapes of Grinding Wheel Faces *ANSI B74.2-1982*

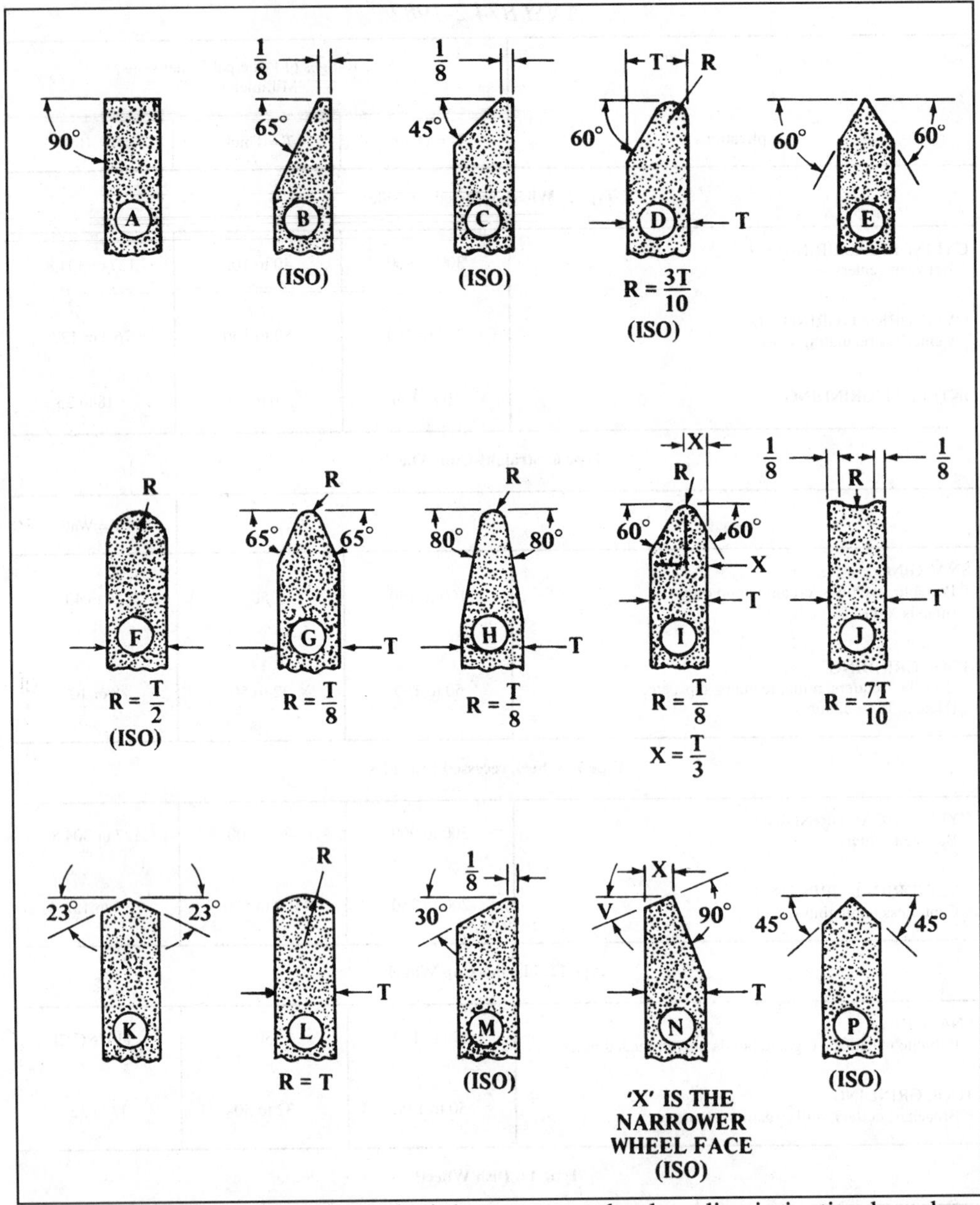

Recommendations, similar in principle, yet somewhat less discriminating have been developed by the Norton Company for *constructional steels*. These materials can be ground either in their original state (soft) or in their after-hardened state (directly or following carburization). Constructional steels must be distinguished from structural steels which are used primarily by the building industry in mill shapes, without or with a minimum of machining.

Constructional steels are either plain carbon or alloy type steels assigned in the AISI-SAE specifications to different groups, according to the predominant types of alloying elements. In the following recommendations no distinction is made because of different compositions since that factor generally, has a minor effect on grinding wheel choice in constructional steels. However, separate recommendations are made for soft (Table 5) and hardened (Table 6) constructional steels. For the relatively rare instance where the use of a single type of wheel for both soft and hardened steel materials is considered more important than the selection of the best suited types for each condition of the work materials, Table 5 lists "All Around" wheels in its last column.

For applications where cool cutting properties of the wheel are particularly important, Table 6 lists, as a second alternative, porous-type wheels. The sequence of choices as presented in these tables does not necessarily represent a second, or third best; it can also apply to conditions where the first choice did not provide optimum results and by varying slightly the composition of the grinding wheel, as indicated in the subsequent choices, the performance experience of the first choice might be improved.

Table 3. Classification of Tool Steels by their Relative Grindability

Relative Grindability Group	AISI-SAE Designation of Tool Steels
GROUP 1—Any area of work surface	W1, W2, W5
	S1, S2, S4, S5, S6, S7
High grindability tool and die steels	O1, O2, O6, O7
(Grindability index greater than 12)	H10, H11, H12, H13, H14
	L2, L6
GROUP 2—Small area of work surface	H19, H20, H21, H22, H23, H24, H26
(*as found in tools*)	P6, P20, P21
	T1, T7, T8
Medium grindability tool and die steels	M1, M2, M8, M10, M33, M50
(Grindability index 3 to 12)	D1, D2, D3, D4, D5, D6
	A2, A4, A6, A8, A9, A10
GROUP 3—Small area of work surface	T4, T5, T6, T8
(*as found in tools*)	M3, M6, M7, M34, M36, M41, M42, M46, M48, M52, M62
Low grindability tool and die steels	D2, D5
(Grindability index between 1.0 and 3)	A11
GROUP 4—Large area of work surface	
(*as found in dies*)	All steels found in Groups 2 and 3
Medium and low grindability tool and die steels	
(Grindability index between 1.0 and 12)	
GROUP 5—Any area of work surface	D3, D4, D7
	M4
Very low grindability tool and die steels	A7
(Grindability index less than 1.0)	T15

Table 4. Grinding Wheel Recommendations for Hardened Tool Steels According to their Grindability

Operation	Wheel or Rim Diameter, Inches	First-Choice Specifications	Second-Choice Specifications
Group 1 Steels			
Surfacing			
Surfacing wheels	14 and smaller	Wet FA46-I8V	SFA46-G12VP
	14 and smaller	Dry FA46-H8V	FA46-F12VP
	Over 14	Wet FA36-I8V	SFA36-I8V
Segments or Cylinders	1½ rim or less	Wet FA30-H8V	FA30-F12VP
Cups	¾ rim or less	Wet FA36-H8V	FA46-F12VP
	(for rims wider than 1½ inches, go one grade softer in available specifications)		
Cutter sharpening			
Straight wheel	…	Wet FA46-K8V	FA60-K8V
	…	Dry FA46-J8V	FA46-H12VP
Dish shape	…	Dry FA60-J8V	FA60-H12VP
Cup shape	…	Dry FA46-L8V	FA60-H12VP
	…	Wet SFA46-L5V	SFA60-L5V
Form tool grinding	8 and smaller	Wet FA60-L8V to FA100-M7V	
	8 and smaller	Dry FA60-K8V to FA100-L8V	
	10 and larger	Wet FA60-L8V to FA80-M6V	
Cylindrical	14 and smaller	Wet SFA60-L5V	…
	16 and larger	Wet SFA60-M5V	…
Centerless	…	Wet SFA60-M5V	…
Internal			
Production grinding	Under ½	Wet SPA80-N6V	SFA80-N7V
	½ to 1	Wet SFA60-M5V	SFA60-M6V
	Over 1 to 3	Wet SFA54-L5V	SFA54-L6V
	Over 3	Wet SFA46-L5V	SFA46-K5V
Tool room grinding	Under ½	Dry FA80-L6V	SFA80-L7V
	½ to 1	Dry FA70-K7V	SFA70-K7V
	Over 1 to 3	Dry FA60-J8V	FA60-H12VP
	Over 3	Dry FA46-J8V	FA54-H12VP
Group 2 Steels			
Surfacing			
Straight wheels	14 and smaller	Wet FA46-I8V	FA46-G12VP
	14 and smaller	Dry FA46-H8V	FA46-F12VP
	Over 14	Wet FA46-H8V	SFA46-I8V
Segments or Cylinders	1½ rim or less	Wet FA30-G8V	FA36-E12VP
Cups	¾ rim or less	Wet FA36-H8V	FA46-F12VP
	(for rims wider than 1½ inches, go one grade softer in available specifications)		

Table 4. *(Continued)* **Grinding Wheel Recommendations for Hardened Tool Steels According to their Grindability**

Operation	Wheel or Rim Diameter, Inches	First-Choice Specifications	Second-Choice Specifications
Cutter sharpening			
Straight wheel	…	Wet FA46-L5V	FA60-K8V
	…	Dry FA46-J8V	FA60-H12VP
Dish shape	…	Dry FA60-J5V	FA60-G12VP
Cup shape	…	Dry FA46-K5V	FA60-G12VP
	…	Wet FA46-L5V	FA60-J8V
Form tool grinding	8 and smaller	Wet FA60-K8V to FA120-L8V	
	8 and smaller	Dry FA80-K8V to FA150-K8V	
	10 and larger	Wet FA60-K8V to FA120-L8V	
Cylindrical	14 and less	Wet FA60-L5V	SFA60-L5V
	16 and larger	Wet FA60-K5V	SFA60-K5V
Centerless	…	Wet FA60-M5V	SFA60-M5V
Internal			
Production grinding	Under ½	Wet FA80-L6V	SFA80-L6V
	½ to 1	Wet FA70-K5V	SFA70-K5V
	Over 1 to 3	Wet FA60-J8V	SFA60-J7V
	Over 3	Wet FA54-J8V	SFA54-J8V
Tool room grinding	Under ½	Dry FA80-I8V	SFA80-K7V
	½ to 1	Dry FA70-J8V	SFA70-J7V
	Over 1 to 3	Dry FA60-I8V	FA60-G12VP
	Over 3	Dry FA54-I8V	FA54-G12VP
Group 3 Steels			
Surfacing			
Straight wheels	14 and smaller	Wet FA60-I8V	FA60-G12VP
	14 and smaller	Dry FA60-H8V	FA60-F12VP
	Over 14	Wet FA60-H8V	SFA60-I8V
Segments or Cylinders	1½ rim or less	Wet FA46-G8V	FA46-E12VP
Cups	¾ rim or less	Wet FA46-G8V	FA46-E12VP
	(for rims wider than 1½ inches, go one grade softer in available specifications)		
Cutter grinding			
Straight wheel	…	Wet FA46-J8V	FA60-J8V
	…	Dry FA46-I8V	FA46-G12VP
Dish shape	…	Dry FA60-H8V	FA60-F12VP
Cup shape	…	Dry FA46-I8V	FA60-F12VP
	…	Wet FA46-J8V	FA60-J8V
Form tool grinding	8 and smaller	Wet FA80-K8V to FA150-L9V	
	8 and smaller	Dry FA100-J8V to FA150-K8V	
	10 and larger	Wet FA80-J8V to FA150-J8V	

Table 4. *(Continued)* **Grinding Wheel Recommendations for Hardened Tool Steels According to their Grindability**

Operation	Wheel or Rim Diameter, Inches	First-Choice Specifications	Second-Choice Specifications
Cylindrical	14 and less	Wet FA80-L5V	SFA80-L6V
	16 and larger	Wet FA60-L6V	SFA60-K5V
Centerless	…	Wet FA60-L5V	SFA60-L5V
Internal			
Production grinding	Under ½	Wet FA90-L6V	SFA90-L6V
	½ to 1	Wet FA80-L6V	SFA80-L6V
	Over 1 to 3	Wet FA70-K5V	SFA70-K5V
	Over 3	Wet FA60-J5V	SFA60-J5V
Tool room grinding	Under ½	Dry FA90-K8V	SFA90-K7V
	½ to 1	Dry FA80-J8V	SFA80-J7V
	Over 1 to 3	Dry FA70-I8V	SFA70-G12VP
	Over 3	Dry FA60-I8V	SFA60-G12VP
Group 4 Steels			
Surfacing			
Straight wheels	14 and smaller	Wet FA60-I8V	C60-JV
	14 and smaller	Wet FA60-H8V	C60-IV
	Over 14	Wet FA46-H8V	C60-HV
Segments	1 ½ rim or less	Wet FA46-G8V	C46-HV
Cylinders	1 ½ rim or less	Wet FA46-G8V	C60-HV
Cups	¾ rim or less	Wet FA46-G6V	C60-IV
	(for rims wider than 1 ½ inches, go one grade softer in available specifications)		
Form tool grinding	8 and smaller	Wet FA60-J8V to FA150-K8V	
	8 and smaller	Dry FA80-I8V to FA180-J8V	
	10 and larger	Wet FA60-J8V to FA150-K8V	
Cylindrical	14 and less	Wet FA80-K8V	C60-KV
	16 and larger	Wet FA60-J8V	C60-KV
Internal			
Production grinding	Under ½	Wet FA90-L8V	C90-LV
	½ to 1	Wet FA80-K5V	C80-KV
	Over 1 to 3	Wet FA70-J8V	C70-JV
	Over 3	Wet FA60-I8V	C60-IV
Tool room grinding	Under ½	Dry FA90-K8V	C90-KV
	½ to 1	Dry FA80-J8V	C80-JV
	Over 1 to 3	Dry FA70-I8V	C70-IV
	Over 3	Dry FA60-H8V	C60-HV

Table 4. *(Continued)* **Grinding Wheel Recommendations for Hardened Tool Steels According to their Grindability**

Operation	Wheel or Rim Diameter, Inches	First-Choice Specifications	Second-Choice Specifications	Third-Choice Specifications
Group 5 Steels				
Surfacing				
Straight wheels	14 and smaller	Wet SFA60-H8V	FA60-E12VP	C60-IV
	14 and smaller	Dry SFA80-H8V	FA80-E12VP	C80-HV
	Over 14	Wet SFA60-H8V	FA60-E12VP	C60-HV
Segments or Cylinders	1 ½ rim or less	Wet SFA46-G8V	FA46-E12VP	C46-GV
Cups	¾ rim or less	Wet SFA60-G8V	FA60-E12VP	C60-GV
	(for rims wider than 1 ½ inches, go one grade softer in available specifications)			
Cutter grinding				
Straight wheels	…	Wet SFA60-I8V	SFA60-G12VP	…
	…	Dry SFA60-H8V	SFA80-F12VP	…
Dish shape	…	Dry SFA80-H8V	SFA80-F12VP	…
Cup shape	…	Dry SFA60-I8V	SFA60-G12VP	…
	…	Wet SFA60-J8V	SFA60-H12VP	…
Form tool grinding	8 and smaller	Wet FA80-J8V to FA180-J9V		…
	8 and smaller	Dry FA100-I8V to FA220-J9V		…
	10 and larger	Wet FA80-J8V to FA180-J9V		…
Cylindrical	14 and less	Wet FA80-J8V	C80-KV	FA80-H12VP
	16 and larger	Wet FA80-I8V	C80-KV	FA80-G12VP
Centerless	…	Wet FA80-J5V	C80-LV	…
Internal				
Production grinding	Under ½	Wet FA100-L8V	C90-MV	…
	½ to 1	Wet FA90-K8V	C80-LV	…
	Over 1 to 3	Wet FA80-J8V	C70-KV	FA80-H12VP
	Over 3	Wet FA70-I8V	C60-JV	FA70-G12VP
Tool room grinding	Under ½	Dry FA100-K8V	C90-KV	…
	½ to 1	Dry FA90-J8V	C80-JV	…
	Over 1 to 3	Dry FA80-I8V	C70-IV	FA80-G12VP
	Over 3	Dry FA70-I8V	C60-IV	FA70-G12VP

Table 5. Grinding Wheel Recommendations for Constructional Steels (Soft)

Grinding Operation	Wheel or Rim Diameter, Inches	First Choice	Alternate Choice (Porous type)	All-Around Wheel
Surfacing				
Straight wheels	14 and smaller	Wet FA46-J8V	FA46-H12VP	FA46-J8V
	14 and smaller	Dry FA46-I8V	FA46-H12VP	FA46-I8V
	Over 14	Wet FA36-J8V	FA36-H12VP	FA36-J8V
Segments	1½ rim or less	Wet FA24-H8V	FA30-F12VP	FA24-H8V
Cylinders	1½ rim or less	Wet FA24-I8V	FA30-G12VP	FA24-H8V
Cups	¾ rim or less	Wet FA24-H8V	FA30-F12VP	FA30-H8V
		(for wider rims, go one grade softer)		
Cylindrical	14 and smaller	Wet SFA60-M5V	…	SFA60-L5V
	16 and larger	Wet SFA54-M5V	…	SFA54-L5V
Centerless	…	Wet SFA54-N5V	…	SFA60-M5V
Internal	Under ½	Wet SFA60-M5V	…	SFA80-L6V
	½ to 1	Wet SFA60-L5V	…	SFA60-K5V
	Over 1 to 3	Wet SFA54-K5V	…	SFA54-J5V
	Over 3	Wet SFA46-K5V	…	SFA46-J5V

Table 6. Grinding Wheel Recommendations for Constructional Steels (Hardened or Carburized)

Grinding Operation	Wheel or Rim Diameter, Inches	First Choice	Alternate Choice (Porous Type)
Surfacing			
Straight wheels	14 and smaller	Wet FA46-I8V	FA46-G12VP
	14 and smaller	Dry FA46-H8V	FA46-F12VP
	Over 14	Wet FA36-I8V	FA36-G12VP
Segments or Cylinders	1½ rim or less	Wet FA30-H8V	FA36-F12VP
Cups	¾ rim or less	Wet FA36-H8V	FA46-F12VP
		(for wider rims, go one grade softer)	
Forms and Radius Grinding	8 and smaller	Wet FA60-L7V to FA100-M8V	
	8 and smaller	Dry FA60-K8V to FA100-L8V	
	10 and larger	Wet FA60-L7V to FA80-M7V	
Cylindrical			
Work diameter			
1 inch and smaller	14 and smaller	Wet SFA80-L6V	…
Over 1 inch	14 and smaller	Wet SFA80-K5V	…
1 inch and smaller	16 and larger	Wet SFA60-L5V	…
Over 1 inch	16 and larger	Wet SFA60-L5V	…
Centerless	…	Wet SFA80-M6V	…
Internal	Under ½	Wet SFA80-N6V	…
	½ to 1	Wet SFA60-M5V	…
	Over 1 to 3	Wet SFA54-L5V	…
	Over 3	Wet SFA46-K5V	…
	Under ½	Dry FA80-L6V	…
	½ to 1	Dry FA70-K8V	…
	Over 1 to 3	Dry FA60-J8V	FA60-H12VP
	Over 3	Dry FA46-J8V	FA54-H12VP

Cubic Boron Nitride (CBN) Grinding Wheels.—Although CBN is not quite as hard, strong, and wear-resistant as a diamond, it is far harder, stronger, and more resistant to wear than aluminum oxide and silicon carbide. As with diamond, CBN materials are available in different types for grinding workpieces of 50 Rc and above, and for superalloys of 35 Rc and harder. Microcrystalline CBN grinding wheels are suitable for grinding mild steels, medium-hard alloy steels, stainless steels, cast irons, and forged steels. Wheels with larger mesh size grains (up to 20/30), now available, provide for higher rates of metal removal.

Special types of CBN are produced for resin, vitrified, and electrodeposited bonds. Wheel standards and nomenclature generally conform to those used for diamond wheels (page 1163), except that the letter **B** instead of **D** is used to denote the type of abrasive. Grinding machines for CBN wheels are generally designed to take full advantage of the ability of CBN to operate at high surface speeds of 9,000–25,000 sfm. CBM is very responsive to changes in grinding conditions, and an increase in wheel speed from 5,000 to 10,000 sfm can increase wheel life by a factor of 6 or more. A change from a water-based coolant to a coolant such as a sulfochlorinated or sulfurized straight grinding oil can increase wheel life by a factor of 10 or more.

Machines designed specifically for use with CBN grinding wheels generally use either electrodeposited wheels or have special trueing systems for other CBN bond wheels, and are totally enclosed so they can use oil as a coolant. Numerical control systems are used, often running fully automatically, including loading and unloading. Machines designed for CBN grinding with electrodeposited wheels are extensively used for form and gear grinding, special systems being used to ensure rapid mounting to exact concentricity and truth in running, no trueing or dressing being required. CBN wheels can produce workpieces having excellent accuracy and finish, with no trueing or dressing for the life of the wheel, even over many hours or days of production grinding of hardened steel components.

Resin-, metal-, and vitrified-bond wheels are used extensively in production grinding, in standard and special machines. Resin-bonded wheels are used widely for dry tool and cutter resharpening on conventional hand-operated tool and cutter grinders. A typical wheel for such work would be designated 11V9 cup type, 100/120 mesh, 75 concentration, with a $\frac{1}{16}$ or $\frac{1}{8}$ in. rim section. Special shapes of resin-bonded wheels are used on dedicated machines for cutting tool manufacture. These types of wheels are usually self-dressing, and allow full machine control of the operation without the need for an operator to see, hear, or feel the action.

Metal-bonded CBN wheels are usually somewhat cheaper than those using other types of bond because only a thin layer of abrasive is present. Metal bonding is also used in manufacture of CBN honing stones. Vitrified-bond CBN wheels are a recent innovation, and high-performance bonds are still being developed. These wheels are used for grinding cams, internal diameters, and bearing components, and can be easily redressed.

An important aspect of grinding with CBN and diamond wheels is reduced heating of the workpiece, thought to result from their superior thermal conductivity compared with aluminum oxide, for instance. CBN and diamond grains also are harder, which means that they stay sharp longer than aluminum oxide grains. The superior ability to absorb heat from the workpiece during the grinding process reduces formation of untempered martensite in the ground surface, caused by overheating followed by rapid quenching. At the same time, a higher compressive residual stress is induced in the surface, giving increased fatigue resistance, compared with the tensile stresses found in surfaces ground with aluminum oxide abrasives. Increased fatigue resistance is of particular importance for gear grinding, especially in the root area.

Variations from General Grinding Wheel Recommendations.—Recommendations for the selection of grinding wheels are usually based on average values with regard to both operational conditions and process objectives. With variations from such average values, the composition of the grinding wheels must be adjusted to obtain optimum results. Although it is impossible to list and to appraise all possible variations and to define their effects on the selection of the best suited grinding wheels, some guidance is obtained from experience. The following tabulation indicates the general directions in which the characteristics of the initially selected grinding wheel may have to be altered in order to approach optimum performance. Variations in a sense opposite to those shown will call for wheel characteristic changes in reverse.

Conditions or Objectives	Direction of Change
To increase cutting rate	Coarser grain, softer bond, higher porosity
To retain wheel size and/or form	Finer grain, harder bond
For small or narrow work surface	Finer grain, harder bond
For larger wheel diameter	Coarser grain
To improve finish on work	Finer grain, harder bond, or resilient bond
For increased work speed or feed rate	Harder bond
For increased wheel speed	Generally, softer bond, except for high-speed grinding, which requires a harder bond for added wheel strength
For interrupted or coarse work surface	Harder bond
For thin walled parts	Softer bond
To reduce load on the machine drive motor	Softer bond

Dressing and Truing Grinding Wheels.—The perfect grinding wheel operating under ideal conditions will be self sharpening, i.e., as the abrasive grains become dull, they will tend to fracture and be dislodged from the wheel by the grinding forces, thereby exposing new, sharp abrasive grains. Although in precision machine grinding this ideal sometimes may be partially attained, it is almost never attained completely. Usually, the grinding wheel must be dressed and trued after mounting on the precision grinding machine spindle and periodically thereafter.

Dressing may be defined as any operation performed on the face of a grinding wheel that improves its cutting action. Truing is a dressing operation but is more precise, i.e., the face of the wheel may be made parallel to the spindle or made into a radius or special shape. Regularly applied truing is also needed for accurate size control of the work, particularly in automatic grinding. The tools and processes generally used in grinding wheel dressing and truing are listed and described in Table .

Table 1. Tools and Methods for Grinding Wheel Dressing and Truing

Designation	Description	Application
Rotating Hand Dressers	Freely rotating discs, either star-shaped with protruding points or discs with corrugated or twisted perimeter, supported in a fork-type handle, the lugs of which can lean on the tool rest of the grinding machine.	Preferred for bench- or floor-type grinding machines; also for use on heavy portable grinders (snagging grinders) where free-cutting proper ties of the grinding wheel are primarily sought and the accuracy of the trued profile is not critical.
Abrasive Sticks	Made of silicon carbide grains with a hard bond. Applied directly or supported in a handle. Less frequently abrasive sticks are also made of boron carbide.	Usually hand held and use limited to smaller-size wheels. Because it also shears the grains of the grinding wheel, or preshaping, prior to final dressing with, e.g., a diamond.

Table 1. *(Continued)* **Tools and Methods for Grinding Wheel Dressing and Truing**

Designation	Description	Application
Abrasive Wheels (Rolls)	Silicon carbide grains in a hard vitrified bond are cemented on ball-bearing mounted spindles. Use either as hand tools with handles or rigidly held in a supporting member of the grinding machine. Generally freely rotating; also available with adjustable brake for diamond wheel dressing.	Preferred for large grinding wheels as a diamond saver, but also for improved control of the dressed surface characteristics. By skewing the abrasive dresser wheel by a few degrees out of parallel with the grinding wheel axis, the basic crushing action is supplemented with wiping and shearing, thus producing the desired degree of wheel surface smoothness.
Single-Point Diamonds	A diamond stone of selected size is mounted in a steel nib of cylindrical shape with or without head, dimensioned to fit the truing spindle of specific grinding machines. Proper orientation and retainment of the diamond point in the setting is an important requirement.	The most widely used tool for dressing and truing grinding wheels in precision grinding. Permits precisely controlled dressing action by regulating infeed and cross feed rate of the truing spindle when the latter is guided by cams or templates for accurate form truing.
Single-Point Form Truing Diamonds	Selected diamonds having symmetrically located natural edges with precisely lapped diamond points, controlled cone angles and vertex radius, and the axis coinciding with that of the nib.	Used for truing operations requiring very accurately controlled, and often steeply inclined wheel profiles, such as are needed for thread and gear grinding, where one or more diamond points participate in generating the resulting wheel periphery form. Dependent on specially designed and made truing diamonds and nibs.
Cluster-Type Diamond Dresser	Several, usually seven, smaller diamond stones are mounted in spaced relationship across the working surface of the nib. In some tools, more than a single layer of such clusters is set at parallel levels in the matrix, the deeper positioned layer becoming active after the preceding layer has worn away.	Intended for straight-face dressing and permits the utilization of smaller, less expensive diamond stones. In use, the holder is canted at a 3° to 10° angle, bringing two to five points into contact with the wheel. The multiple-point contact permits faster cross feed rates during truing than may be used with single-point diamonds for generating a specific degree of wheel-face finish.
Impregnated Matrix-Type Diamond Dressers	The operating surface consists of a layer of small, randomly distributed, yet rather uniformly spaced diamonds that are retained in a bond holding the points in an essentially common plane. Supplied either with straight or canted shaft, the latter being used to cancel the tilt of angular truing posts.	For the truing of wheel surfaces consisting of a single or several flat elements. The nib face should be held tangent to the grinding wheel periphery or parallel with a flat working surface. Offers economic advantages where technically applicable because of using less expensive diamond splinters presented in a manner permitting efficient utilization.
Form- Generating Truing Devices	Swiveling diamond holder post with adjustable pivot location, arm length, and swivel arc, mounted on angularly adjustable cross slides with controlled traverse movement, permits the generation of various straight and circular profile elements, kept in specific mutual locations.	Such devices are made in various degrees of complexity for the positionally controlled interrelation of several different profile elements. Limited to regular straight and circular sections, yet offers great flexibility of setup, very accurate adjustment, and unique versatility for handling a large variety of frequently changing profiles.

Table 1. *(Continued)* **Tools and Methods for Grinding Wheel Dressing and Truing**

Designation	Description	Application
Contour-Duplicating Truing Devices	The form of a master, called cam or template, shaped to match the profile to be produced on the wheel, or its magnified version, is translated into the path of the diamond point by means of mechanical linkage, a fluid actuator, or a pantograph device.	Preferred single-point truing method for profiles to be produced in quantities warranting the making of special profile bars or templates. Used also in small- and medium-volume production when the complexity of the profile to be produced excludes alternate methods of form generation.
Grinding Wheel Contouring by Crush Truing	A hardened steel or carbide roll, which is free to rotate and has the desired form of the workpiece, is fed gradually into the grinding wheel, which runs at slow speed. The roll will, by crushing action, produce its reverse form in the wheel. Crushing produces a free-cutting wheel face with sharp grains.	Requires grinding machines designed for crush truing, having stiff spindle bearings, rigid construction, slow wheel speed for truing, etc. Due to the cost of crush rolls and equipment, the process is used for repetitive work only. It is one of the most efficient methods for precisely duplicating complex wheel profiles that are capable of grinding in the 8-microinch AA range. Applicable for both surface and cylindrical grinding.
Rotating Diamond Roll-Type Grinding Wheel Truing	Special rolls made to agree with specific profile specifications have their periphery coated with a large number of uniformly distributed diamonds, held in a matrix into which the individual stones are set by hand (for larger diamonds) or bonded by a plating process (for smaller elements).	The diamond rolls must be rotated by an air, hydraulic, or electric motor at about one-fourth of the grinding wheel surface speed and in opposite direction to the wheel rotation. Whereas the initial costs are substantially higher than for single-point diamond truing the savings in truing time warrants the method's application in large-volume production of profile-ground components.
Diamond Dressing Blocks	Made as flat blocks for straight wheel surfaces, are also available for radius dressing and profile truing. The working surface consists of a layer of electroplated diamond grains, uniformly distributed and capable of truing even closely toleranced profiles.	For straight wheels, dressing blocks can reduce dressing time and offer easy installation on surface grinders, where the blocks mount on the magnetic plate. Recommended for small- and medium-volume production for truing intricate profiles on regular surface grinders, because the higher pressure developed in crush dressing is avoided.

Guidelines for Truing and Dressing with Single-Point Diamonds.— The diamond nib should be canted at an angle of 10 to 15 degrees in the direction of the wheel rotation and also, if possible, by the same amount in the direction of the cross feed traverse during the truing (see diagram). The dragging effect resulting from this "angling," combined with the occasional rotation of the diamond nib in its holder, will prolong the diamond life by limiting the extent of wear facets and will also tend to produce a pyramid shape of the diamond tip. The diamond may also be set to contact the wheel at about 1⁄8 to 1⁄4 inch below its centerline.

Depth of Cut: This amount should not exceed 0.001 inch per pass for general work, and will have to be reduced to 0.0002 to 0.0004 inch per pass for wheels with fine grains used for precise finishing work.

Diamond crossfeed rate: This value may be varied to some extent depending on the required wheel surface: faster crossfeed for free cutting, and slower crossfeed for producing fine finishes. Such variations, however, must always stay within the limits set by the grain size of the wheel. Thus, the advance rate of the truing diamond per wheel revolution should not exceed the diameter of a grain or be less than half of that rate. Consequently, the diamond crossfeed must be slower for a large wheel than for a smaller wheel having the same grain size number.

Typical crossfeed values for frequently used grain sizes are given in Table 2.

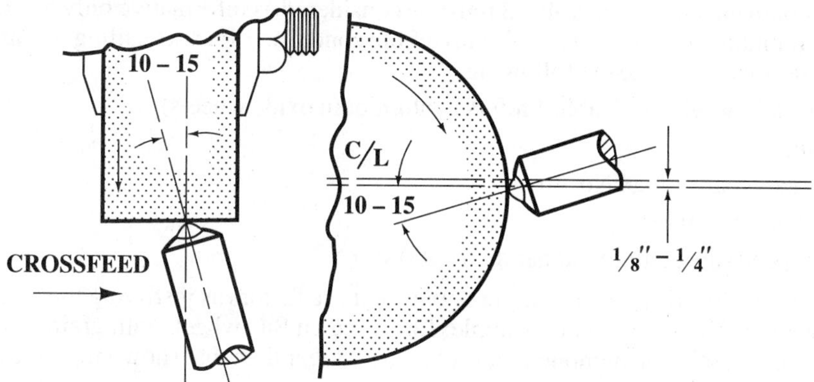

Table 2. Typical Diamond Truing and Crossfeeds

Grain Size	30	36	46	50
Crossfeed per Wheel Rev., in.	0.014–0.024	0.012–0.019	0.008–0.014	0.007–0.012
Grain Size	60	80	120	…
Crossfeed per Wheel Rev., in.	0.006–0.010	0.004–0.007	0.0025–0.004	…

These values can be easily converted into the more conveniently used inch-per-minute units, simply by multiplying them by the rpm of the grinding wheel.

Example: For a 20-inch diameter wheel, Grain No. 46, running at 1200 rpm: Crossfeed rate for roughing-cut truing—approximately 17 ipm, for finishing-cut truing—approximately 10 ipm

Coolant should be applied before the diamond comes into contact with the wheel and must be continued in generous supply while truing.

The speed of the grinding wheel should be at the regular grinding rate, or not much lower. For that reason, the feed wheels of centerless grinding machines usually have an additional speed rate higher than functionally needed, that speed being provided for wheel truing only.

The initial approach of the diamond to the wheel surface must be carried out carefully to prevent sudden contact with the diamond, resulting in penetration in excess of the selected depth of cut. It should be noted that the highest point of a worn wheel is often in its center portion and not at the edge from which the crossfeed of the diamond starts.

The general conditions of the truing device are important for best truing results and for assuring extended diamond life. A rigid truing spindle, well-seated diamond nib, and firmly set diamond point are mandatory. Sensitive infeed and smooth traverse movement at uniform speed also must be maintained.

Resetting of the diamond point.: Never let the diamond point wear to a degree where the grinding wheel is in contact with the steel nib. Such contact can damage the setting of the diamond point and result in its loss. Expert resetting of a worn diamond can repeatedly add to its useful life, even when applied to lighter work because of reduced size.

Size Selection Guide for Single-Point Truing Diamonds.—There are no rigid rules for determining the proper size of the diamond for any particular truing application because of the very large number of factors affecting that choice. Several of these factors are related to the condition, particularly the rigidity, of the grinding machine and truing device, as well as to such characteristics of the diamond itself as purity, crystalline structure, etc. Although these factors are difficult to evaluate in a generally applicable manner, the expected effects of several other conditions can be appraised and should be considered in the selection of the proper diamond size.

The recommended sizes in Table 3 must be considered as informative only and as representing minimum values for generally favorable conditions. Factors calling for larger diamond sizes than listed are the following:

Silicon carbide wheels (Table 3 refers to aluminum oxide wheels)

Dry truing

Grain sizes coarser than No. 46

Bonds harder than M

Wheel speed substantially higher than 6500 sfm.

It is advisable to consider any single or pair of these factors as justifying the selection of one size larger diamond. As an example: for truing an SiC wheel, with grain size No. 36 and hardness P, select a diamond that is two sizes larger than that shown in Table 3 for the wheel size in use.

Table 3. Recommended Minimum Sizes for Single-Point Truing Diamonds

Diamond Size in Carats[a]	Index Number (Wheel Dia. × Width in Inches)	Examples of Max. Grinding Wheel Dimensions	
		Diameter	Width
0.25	3	4	0.75
0.35	6	6	1
0.50	10	8	1.25
0.60	15	10	1.50
0.75	21	12	1.75
1.00	30	12	2.50
1.25	48	14	3.50
1.50	65	16	4.00
1.75	80	20	4.00
2.00	100	20	5.00
2.50	150	24	6.00
3.00	200	24	8.00
3.50	260	30	8.00
4.00	350	36	10.00

[a] One carat equals 0.2 gram.

Single-point diamonds are available as loose stones, but are preferably procured from specialized manufacturers supplying the diamonds set into steel nibs. Expert setting, comprising both the optimum orientation of the stone and its firm retainment, is mandatory for assuring adequate diamond life and satisfactory truing. Because the holding devices for truing diamonds are not yet standardized, the required nib dimensions vary depending on the make and type of different grinding machines. Some nibs are made with angular heads, usually hexagonal, to permit occasional rotation of the nib either manually, with a wrench, or automatically.

Diamond Wheels

Diamond Wheels.—A diamond wheel is a special type of grinding wheel in which the abrasive elements are diamond grains held in a bond and applied to form a layer on the operating face of a non-abrasive core. Diamond wheels are used for grinding very hard or highly abrasive materials. Primary applications are the grinding of cemented carbides, such as the sharpening of carbide cutting tools; the grinding of glass, ceramics, asbestos, and cement products; and the cutting and slicing of germanium and silicon.

Shapes of Diamond Wheels.—The industry-wide accepted Standard (ANSI B74.3-1974) specifies ten basic diamond wheel core shapes which are shown in Table 1 with the applicable designation symbols. The applied diamond abrasive layer may have different cross-sectional shapes. Those standardized are shown in Table 2. The third aspect which is standardized is the location of the diamond section on the wheel as shown by the diagrams in Table . Finally, modifications of the general core shape together with pertinent designation letters are given in Table 4.

The characteristics of the wheel shape listed in these four tables make up the components of the standard designation symbol for diamond wheel shapes. An example of that symbol with arbitrarily selected components is shown in Fig. 1.

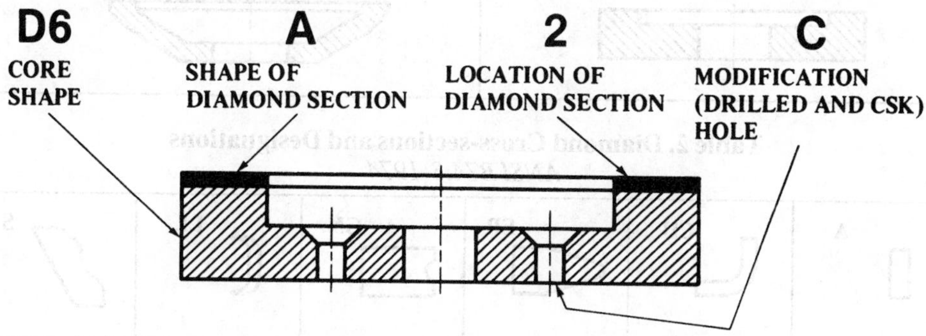

Fig. 1. A Typical Diamond Wheel Shape Designation Symbol

An explanation of these components is as follows:

Basic Core Shape: This portion of the symbol indicates the basic shape of the core on which the diamond abrasive section is mounted. The shape is actually designated by a number. The various core shapes and their designations are given in Table 1.

Diamond Cross-Section Shape: This, the second component, consisting of one or two letters, denotes the cross-sectional shape of the diamond abrasive section. The various shapes and their corresponding letter designations are given in Table 2.

Diamond Section Location: The third component of the symbol consists of a number which gives the location of the diamond section, i.e., periphery, side, corner, etc. An explanation of these numbers is shown in Table 3.

Modification: The fourth component of the symbol is a letter designating some modification, such as drilled and counterbored holes for mounting or special relieving of diamond section or core. This modification position of the symbol is used only when required. The modifications and their designations are given in Table 4.

Table 1. Diamond Wheel Core Shapes and Designations *ANSI B74.3-1974*

1	9
2	11 LESS THAN 90° AND OVER 45°
3	12 45° OR LESS
4	14
6	15

Table 2. Diamond Cross-sections and Designations *ANSI B74.3-1974*

A	CH	ER	GN	LL	S
AH	D	ET	H	M	U
B	DD	F	J	P	V
BT	E	FF	K	Q	Y
C	EE	G	L	QQ	

Table 3. Designations for Location of Diamond Section on Diamond Wheel *ANSI B74.3-1974*

Designation No. and Location	Description	Illustration
1 — Periphery	The diamond section shall be placed on the periphery of the core and shall extend the full thickness of the wheel. The axial length of this section may be greater than, equal to, or less than the depth of diamond, measured radially. A hub or hubs shall not be considered as part of the wheel thickness for this definition.	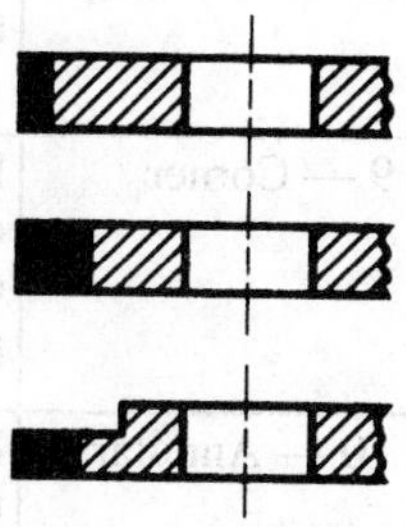
2 — Side	The diamond section shall be placed on the side of the wheel and the length of the diamond section shall extend from the periphery toward the center. It may or may not include the entire side and shall be greater than the diamond depth measured axially. It shall be on that side of the wheel which is commonly used for grinding purposes.	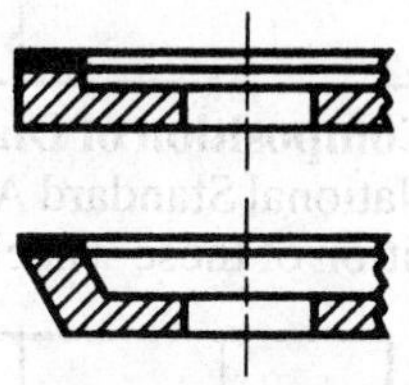
3 — Both Sides	The diamond sections shall be placed on both sides of the wheel and shall extend from the periphery toward the center. They may or may not include the entire sides, and the radial length of the diamond section shall exceed the axial diamond depth.	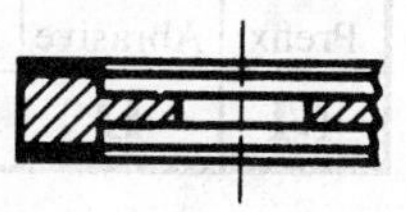
4 — Inside Bevel or Arc	This designation shall apply to the general wheel types 2, 6, 11, 12, and 15 and shall locate the diamond section on the side wall. This wall shall have an angle or arc extending from a higher point at the wheel periphery to a lower point toward the wheel center.	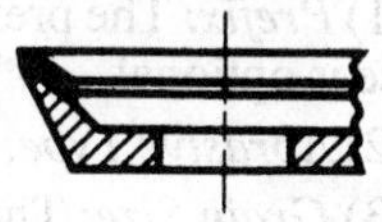
5 — Outside Bevel or Arc	This designation shall apply to the general wheel types, 2, 6, 11, and 15 and shall locate the diamond section on the side wall. This wall shall have an angle or arc extending from a lower point at the wheel periphery to a higher point toward the wheel center.	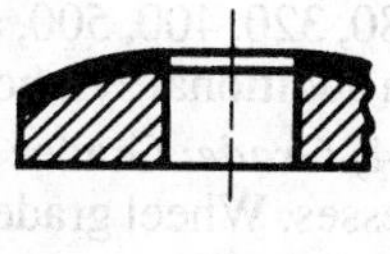
6 — Part of Periphery	The diamond section shall be placed on the periphery of the core but shall not extend the full thickness of the wheel and shall not reach to either side.	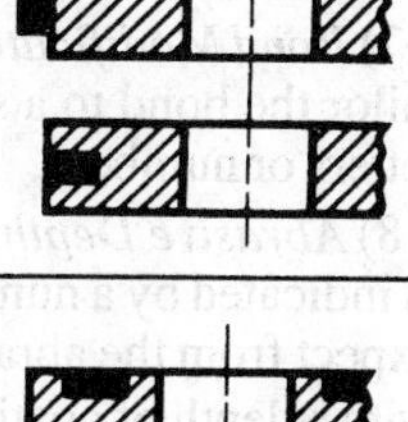
7 — Part of Side	The diamond section shall be placed on the side of the core and shall not extend to the wheel periphery. It may or may not extend to the center.	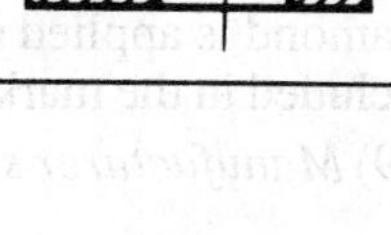

Table 3. *(Continued)* **Designations for Location of Diamond Section on Diamond Wheel** *ANSI B74.3-1974*

Designation No. and Location	Description	Illustration
8 — Throughout	Designates wheels of solid diamond abrasive section without cores.	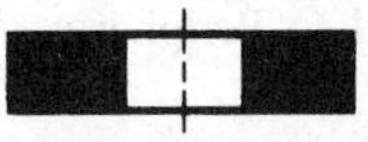
9 — Corner	Designates a location which would commonly be considered to be on the periphery except that the diamond section shall be on the corner but shall not extend to the other corner.	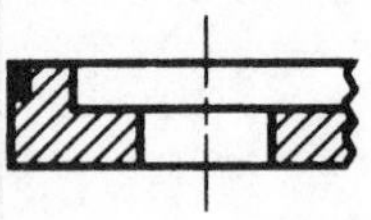
10 — Annular	Designates a location of the diamond abrasive section on the inner annular surface of the wheel.	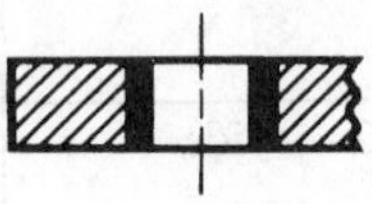

Composition of Diamond and Cubic Boron Nitride Wheels.—According to American National Standard ANSI B74.13-1990, a series of symbols is used to designate the composition of these wheels. An example is shown below.

Prefix	Abrasive	Grain Size	Grade	Concentration	Bond Type	Bond Modification	Depth of Abrasive	Manufacturer's Identification Symbol
M	D	120	R	100	B	56	1/8	*

Fig. 2. Designation Symbols for Composition of Diamond and Cubic Boron Nitride Wheels

The meaning of each symbol is indicated by the following list:

1) *Prefix:* The prefix is a manufacturer's symbol indicating the exact kind of abrasive. Its use is optional.

2) *Abrasive Type:* The letter (B) is used for cubic boron nitride and (D) for diamond.

3) *Grain Size:* The grain sizes commonly used and varying from coarse to very fine are indicated by the following numbers: 8, 10, 12, 14, 16, 20, 24, 30, 36, 46, 54, 60, 70, 80, 90, 100, 120, 150, 180, and 220. The following additional sizes are used occasionally: 240, 280, 320, 400, 500, and 600. The wheel manufacturer may add to the regular grain number an additional symbol to indicate a special grain combination.

4) *Grade:* Grades are indicated by letters of the alphabet from A to Z in all bonds or processes. Wheel grades from A to Z range from soft to hard.

5) *Concentration:* The concentration symbol is a manufacturer's designation. It may be a number or a symbol.

6) *Bond:* Bonds are indicated by the following letters: B, resinoid; V, vitrified; M, metal.

7) *Bond Modification:* Within each bond type a manufacturer may have modifications to tailor the bond to a specific application. These modifications may be identified by either letters or numbers.

8) *Abrasive Depth:* Abrasive section depth, in inches or millimeters (inches illustrated), is indicated by a number or letter which is the amount of total dimensional wear a user may expect from the abrasive portion of the product. Most diamond and CBN wheels are made with a depth of coating on the order of 1/16 in., 1/8 in., or more as specified. In some cases the diamond is applied in thinner layers, as thin as one thickness of diamond grains. The L is included in the marking system to identify a layered type product.

9) *Manufacturer's Identification Symbol:* The use of this symbol is optional.

Table 4. Designation Letters for Modifications of Diamond Wheels
ANSI B74.3-1974

Designation Letter[a]	Description	Illustration
B — Drilled and Counterbored	Holes drilled and counterbored in core.	
C — Drilled and Countersunk	Holes drilled and countersunk in core.	
H — Plain Hole	Straight hole drilled in core.	
M — Holes Plain and Threaded	Mixed holes, some plain, some threaded, are in core.	
P — Relieved One Side	Core relieved on one side of wheel. Thickness of core is less than wheel thickness.	
R — Relieved Two Sides	Core relieved on both sides of wheel. Thickness of core is less than wheel thickness.	
S — Segmented-Diamond Section	Wheel has segmental diamond section mounted on core. (Clearance between segments has no bearing on definition.)	
SS — Segmental and Slotted	Wheel has separated segments mounted on a slotted core.	
T — Threaded Holes	Threaded holes are in core.	
Q — Diamond Inserted	Three surfaces of the diamond section are partially or completely enclosed by the core.	
V — Diamond Inverted	Any diamond cross section, which is mounted on the core so that the interior point of any angle, or the concave side of any arc, is exposed shall be considered inverted. *Exception:* Diamond cross section AH shall be placed on the core with the concave side of the arc exposed.	

[a] Y — Diamond Inserted and Inverted. See definitions for Q and V.

The Selection of Diamond Wheels.—Two general aspects must be defined: (a) The shape of the wheel, also referred to as the basic wheel type and (b) The specification of the abrasive portion.

Table 5. General Diamond Wheel Recommendations for Wheel Type and Abrasive Specification

Typical Applications or Operation	Basic Wheel Type	Abrasive Specification
Single Point Tools (offhand grinding)	D6A2C	*Rough:* MD100-N100-B1/8 *Finish:* MD220-P75-B1/8
Single Point Tools (machine ground)	D6A2H	*Rough:* MD180-J100-B1/8 *Finish:* MD320-L75-B1/8
Chip Breakers	D1A1	MD150-R100-B1/8
Multitooth Tools and Cutters (face mills, end mills, reamers, broaches, etc.)		
Sharpening and Backing off	D11V9	Rough: MD100-R100-B1/8 Combination: MD150-R100-B1/8 *Finish:* MD220-R100-B1/8
Fluting	D12A2	MD180-N100-B1/8
Saw Sharpening	D12A2	MD180-R100-B1/8
Surface Grinding (horizontal spindle)	D1A1	*Rough:* MD120-N100-B1/8 *Finish:* MD240-P100-B1/8
Surface Grinding (vertical spindle)	D2A2T	MD80-R75-B1/8
Cylindrical or Centertype Grinding	D1A1	MD120-P100-B1/8
Internal Grinding	D1A1	MD150-N100-B1/8
Slotting and Cutoff	D1A1R	MD150-R100-B1/4
Lapping	Disc	MD400-L50-B1/16
Hand Honing	DH1, DH2	*Rough:* MD220-B1/16 *Finish:* MD320-B1/6

General recommendations for the dry grinding, with resin bond diamond wheels, of most grades of cemented carbides of average surface to ordinary finishes at normal rates of metal removal with average size wheels, as published by Cincinnati Milacron, are listed in Table 5.

A further set of variables are *the dimensions of the wheel,* which must be adapted to the available grinding machine and, in some cases, to the configuration of the work.

The general abrasive specifications in Table 5 may be modified to suit operating conditions by the following suggestions:

Use softer wheel grades for harder grades of carbides, for grinding larger areas or larger or wider wheel faces.

Use harder wheel grades for softer grades of carbides, for grinding smaller areas, for using smaller and narrower face wheels and for light cuts.

Use fine grit sizes for harder grades of carbides and to obtain better finishes.

Use coarser grit sizes for softer grades of carbides and for roughing cuts.

Use higher diamond concentration for harder grades of carbides, for larger diameter or wider face wheels, for heavier cuts, and for obtaining better finish.

Guidelines for the Handling and Operation of Diamond Wheels.—Grinding machines used for grinding with diamond wheels should be of the precision type, in good service condition, with true running spindles and smooth slide movements.

Mounting of Diamond Wheels: Wheel mounts should be used which permit the precise centering of the wheel, resulting in a runout of less than 0.001 inch axially and 0.0005 inch radially. These conditions should be checked with a 0.0001-inch type dial indicator. Once mounted and centered, the diamond wheel should be retained on its mount and stored in that condition when temporarily removed from the machine.

Truing and Dressing: Resinoid bonded diamond wheels seldom require dressing, but when necessary a soft silicon carbide stick may be hand-held against the wheel. Peripheral and cup type wheels may be sharpened by grinding the cutting face with a 60 to 80 grit silicon carbide wheel. This can be done with the diamond wheel mounted on the spindle of the machine, and with the silicon carbide wheel driven at a relatively slow speed by a specially designed table-mounted grinder or by a small table-mounted tool post grinder. The diamond wheel can be mounted on a special arbor and ground on a lathe with a tool post grinder; peripheral wheels can be ground on a cylindrical grinder or with a special brake-controlled truing device with the wheel mounted on the machine on which it is used. Cup and face type wheels are often lapped on a cast iron or glass plate using a 100 grit silicon carbide abrasive. Care must be used to lap the face parallel to the back, otherwise they must be ground to restore parallelism. Peripheral diamond wheels can be trued and dressed by grinding a silicon carbide block or a special diamond impregnated bronze block in a manner similar to surface grinding. Conventional diamonds must not be used for truing and dressing diamond wheels.

Speeds and Feeds in Diamond Grinding.—General recommendations are as follows:

Wheel Speeds: The generally recommended wheel speeds for diamond grinding are in the range of 5000 to 6000 surface feet per minute, with this upper limit as a maximum to avoid harmful "overspeeding." Exceptions from that general rule are diamond wheels with coarse grains and high concentration (100 per cent) where the wheel wear in dry surface grinding can be reduced by lowering the speed to 2500–3000 sfpm. However, this lower speed range can cause rapid wheel breakdown in finer grit wheels or in those with reduced diamond concentration.

Work Speeds: In diamond grinding, work rotation and table traverse are usually established by experience, adjusting these values to the selected infeed so as to avoid excessive wheel wear.

Infeed per Pass: Often referred to as downfeed and usually a function of the grit size of the wheel. The following are general values which may be increased for raising the productivity, or lowered to improve finish or to reduce wheel wear.

Wheel Grit Size Range	Infeed per Pass
100 to 120	0.001 inch
150 to 220	0.0005 inch
250 and finer	0.00025 inch

GRINDING WHEEL SAFETY

Safety in Operating Grinding Wheels.—Grinding wheels, although capable of exceptional cutting performance due to hardness and wear resistance, are prone to damage caused by improper handling and operation. Vitrified wheels, comprising the major part of grinding wheels used in industry, are held together by an inorganic bond which is actually a type of pottery product and therefore brittle and breakable. Although most of the organic bond types are somewhat more resistant to shocks, it must be realized that all grinding wheels are conglomerates of individual grains joined by a bond material whose strength is limited by the need of releasing the dull, abrasive grains during use.

It must also be understood that during the grinding process very substantial forces act on the grinding wheel, including the centrifugal force due to rotation, the grinding forces resulting from the resistance of the work material, and shocks caused by sudden contact with the work. To be able to resist these forces, the grinding wheel must have a substantial minimum strength throughout that is well beyond that needed to hold the wheel together under static conditions.

Finally, a damaged grinding wheel can disintegrate during grinding, liberating dormant forces which normally are constrained by the resistance of the bond, thus presenting great hazards to both operator and equipment.

To avoid breakage of the operating wheel and, should such a mishap occur, to prevent damage or injury, specific precautions must be applied. These safeguards have been formulated into rules and regulations and are set forth in the American National Standard ANSI B7.1-1988, entitled the American National Standard Safety Requirements for the Use, Care, and Protection of Abrasive Wheels.

Handling, Storage and Inspection.—Grinding wheels should be hand carried, or transported, with proper support, by truck or conveyor. A grinding wheel must not be rolled around on its periphery.

The storage area, positioned not far from the location of the grinding machines, should be free from excessive temperature variations and humidity. Specially built racks are recommended on which the smaller or thin wheels are stacked lying on their sides and the larger wheels in an upright position on two-point cradle supports consisting of appropriately spaced wooden bars. Partitions should separate either the individual wheels, or a small group of identical wheels. Good accessibility to the stored wheels reduces the need of undesirable handling.

Inspection will primarily be directed at detecting visible damage, mostly originating from handling and shipping. Cracks which are not obvious can usually be detected by "ring testing," which consists of suspending the wheel from its hole and tapping it with a nonmetallic implement. Heavy wheels may be allowed to rest vertically on a clean, hard floor while performing this test. A clear metallic tone, a "ring", should be heard; a dead sound being indicative of a possible crack or cracks in the wheel.

Machine Conditions.—The general design of the grinding machines must ensure safe operation under normal conditions. The bearings and grinding wheel spindle must be dimensioned to withstand the expected forces and ample driving power should be provided to ensure maintenance of the rated spindle speed. For the protection of the operator, stationary machines used for dry grinding should have a provision made for connection to an exhaust system and when used for off-hand grinding, a work support must be available.

Wheel guards are particularly important protection elements and their material specifications, wall thicknesses and construction principles should agree with the Standard's specifications. The exposure of the wheel should be just enough to avoid interference with the grinding operation. The need for access of the work to the grinding wheel will define the boundary of guard opening, particularly in the direction of the operator.

Grinding Wheel Mounting.—The mass and speed of the operating grinding wheel makes it particularly sensitive to imbalance. Vibrations that result from such conditions are harmful to the machine, particularly the spindle bearings, and they also affect the ground surface, i.e., wheel imbalance causes chatter marks and interferes with size control. Grinding wheels are shipped from the manufacturer's plant in a balanced condition, but retaining the balanced state after mounting the wheel is quite uncertain. Balancing of the mounted wheel is thus required, and is particularly important for medium and large size wheels, as well as for producing acccurate and smooth surfaces. The most common way of balancing mounted wheels is by using balancing flanges with adjustable weights. The wheel and balancing flanges are mounted on a short balancing arbor, the two concentric and round stub ends of which are supported in a balancing stand.

Such stands are of two types: J) the parallel straight-edged, which must be set up precisely level; and K) the disk type having two pairs of ball bearing mounted overlapping disks, which form a V for containing the arbor ends without hindering the free rotation of the wheel mounted on that arbor.

The wheel will then rotate only when it is out of balance and its heavy spot is not in the lowest position. Rotating the wheel by hand to different positions will move the heavy spot, should such exist, from the bottom to a higher location where it can reveal its presence by causing the wheel to turn. Having detected the presence and location of the heavy spot, its effect can be cancelled by displacing the weights in the circular groove of the flange until a balanced condition is accomplished.

Flanges are commonly used means for holding grinding wheels on the machine spindle. For that purpose, the wheel can either be mounted directly through its hole or by means of a sleeve which slips over a tapered section of the machine spindle. Either way, the flanges must be of equal diameter, usually not less than one-third of the new wheel's diameter. The purpose is to securely hold the wheel between the flanges without interfering with the grinding operation even when the wheel becomes worn down to the point where it is ready to be discarded. Blotters or flange facings of compressible material should cover the entire contact area of the flanges.

One of the flanges is usually fixed while the other is loose and can be removed and adjusted along the machine spindle. The movable flange is held against the mounted grinding wheel by means of a nut engaging a threaded section of the machine spindle. The sense of that thread should be such that the nut will tend to tighten as the spindle revolves. In other words, to remove the nut, it must be turned in the direction that the spindle revolves when the wheel is in operation.

Safe Operating Speeds.—Safe grinding processes are predicated on the proper use of the previously discussed equipment and procedures, and are greatly dependent on the application of adequate operating speeds.

The Standard establishes maximum speeds at which grinding wheels can be operated, assigning the various types of wheels to several classification groups. Different values are listed according to bond type and to wheel strength, distinguishing between low, medium and high strength wheels.

For the purpose of general information, the accompanying table shows an abbreviated version of the Standard's specification. However, for the governing limits, the authoritative source is the manufacturer's tag on the wheel which, particularly for wheels of lower strength, might specify speeds below those of the table.

All grinding wheels of 6 inches or greater diameter must be test run in the wheel manufacturer's plant at a speed that for all wheels having operating speeds in excess of 5000 sfpm is 1.5 times the maximum speed marked on the tag of the wheel.

The table shows the permissible wheel speeds in surface feet per minute (sfpm) units, whereas the tags on the grinding wheels state, for the convenience of the user, the maximum operating speed in revolutions per minute (rpm). The sfpm unit has the advantage of remaining valid for worn wheels whose rotational speed may be increased to the applicable sfpm value. The conversion from either one to the other of these two kinds of units is a matter of simple calculation using the formulas:

$$\text{sfpm} = \text{rpm} \times \frac{D}{12} \times \pi$$

or

$$\text{rpm} = \frac{\text{sfpm} \times 12}{D \times \pi}$$

Where D = maximum diameter of the grinding wheel, in inches. Table 2, showing the conversion values from surface speed into rotational speed, can be used for the direct reading of the rpm values corresponding to several different wheel diameters and surface speeds.

Special Speeds: Continuing progress in grinding methods has led to the recognition of certain advantages that can result from operating grinding wheels above, sometimes even higher than twice, the speeds considered earlier as the safe limits of grinding wheel operations. Advantages from the application of high speed grinding are limited to specific processes, but the Standard admits, and offers code regulations for the use of wheels at special high speeds. These regulations define the structural requirements of the grinding machine and the responsibilities of the grinding wheel manufacturers, as well as of the users. High speed grinding should not be applied unless the machines, particularly guards, spindle assemblies, and drive motors, are suitable for such methods. Also, appropriate grinding wheels expressly made for special high speeds must be used and, of course, the maximum operating speeds indicated on the wheel's tag must never be exceeded.

Portable Grinders.—The above discussed rules and regulations, devised primarily for stationary grinding machines apply also to portable grinders. In addition, the details of various other regulations, specially applicable to different types of portable grinders are discussed in the Standard, which should be consulted, particularly for safe applications of portable grinding machines.

Table 1. Maximum Peripheral Speeds for Grinding Wheels Based on ANSI B7.1–1988

Classification No.	Types of Wheels[a]	Maximum Operating Speeds, sfpm, Depending on Strength of Bond	
		Inorganic Bonds	Organic Bonds
1	Straight wheels — Type 1, except classifications 6, 7, 9, 10, 11, and 12 below Type 4[b] — Taper Side Wheels Types 5, 7, 20, 21, 22, 23, 24, 25, 26 Dish wheels — Type 12 Saucer wheels — Type 13 Cones and plugs — Types 16, 17, 18, 19	5,500 to 6,500	6,500 to 9,500
2	Cylinder wheels — Type 2 Segments	5,000 to 6,000	5,000 to 7,000
3	Cup shape tool grinding wheels — Types 6 and 11 (for fixed base machines)	4,500 to 6,000	6,000 to 8,500
4	Cup shape snagging wheels — Types 6 and 11 (for portable machines)	4,500 to 6,500	6,000 to 9,500
5	Abrasive disks	5,500 to 6,500	5,500 to 8,500
6	Reinforced wheels — except cutting-off wheels (depending on diameter and thickness)	…	9,500 to 16,000
7	Type 1 wheels for bench and pedestal grinders, Types 1 and 5 also in certain sizes for surface grinders	5,500 to 7,550	6,500 to 9,500
8	Diamond and cubic boron nitride wheels Metal bond Steel centered cutting off	to 6,500 to 12,000 to 16,000	to 9,500 … to 16,000
9	Cutting-off wheels — Larger than 16-inch diameter (incl. reinforced organic)	…	9,500 to 14,200
10	Cutting-off wheels — 16-inch diameter and smaller (incl. reinforced organic)	…	9,500 to 16,000
11	Thread and flute grinding wheels	8,000 to 12,000	8,000 to 12,000
12	Crankshaft and camshaft grinding wheels	5,500 to 8,500	6,500 to 9,500

[a] See Tables and Tables starting on page 1148.

[b] Non-standard shape. For snagging wheels, 16 inches and larger — Type 1, internal wheels — Types 1 and 5, and mounted wheels, see ANSI B7.1–1988. Under no conditions should a wheel be operated faster than the maximum operating speed established by the manufacturer.

Values in this table are for general information only.

Table 2. Revolutions per Minute for Various Grinding Speeds and Wheel Diameters (Based on B7.1–1988)

Wheel Diameter, Inch	Peripheral (Surface) Speed, Feet per Minute															Wheel Diameter, Inch
	4,000	4,500	5,000	5,500	6,000	6,500	7,000	7,500	8,000	8,500	9,000	9,500	10,000	12,000	14,000	16,000	
	Revolutions per Minute																
1	15,279	17,189	19,099	21,008	22,918	24,828	26,738	28,648	30,558	32,468	34,377	36,287	38,197	45,837	53,476	61,115	1
2	7,639	8,594	9,549	10,504	11,459	12,414	13,369	14,324	15,279	16,234	17,189	18,144	19,099	22,918	26,738	30,558	2
3	5,093	5,730	6,366	7,003	7,639	8,276	8,913	9,549	10,186	10,823	11,459	12,096	12,732	15,279	17,825	20,372	3
4	3,820	4,297	4,775	5,252	5,730	6,207	6,685	7,162	7,639	8,117	8,594	9,072	9,549	11,459	13,369	15,279	4
5	3,056	3,438	3,820	4,202	4,584	4,966	5,348	5,730	6,112	6,494	6,875	7,257	7,639	9,167	10,695	12,223	5
6	2,546	2,865	3,183	3,501	3,820	4,138	4,456	4,775	5,093	5,411	5,730	6,048	6,366	7,639	8,913	10,186	6
7	2,183	2,456	2,728	3,001	3,274	3,547	3,820	4,093	4,365	4,638	4,911	5,184	5,457	6,548	7,639	8,731	7
8	1,910	2,149	2,387	2,626	2,865	3,104	3,342	3,581	3,820	4,058	4,297	4,536	4,775	5,730	6,685	7,639	8
9	1,698	1,910	2,122	2,334	2,546	2,759	2,971	3,183	3,395	3,608	3,820	4,032	4,244	5,093	5,942	6,791	9
10	1,528	1,719	1,910	2,101	2,292	2,483	2,674	2,865	3,056	3,247	3,438	3,629	3,820	4,584	5,348	6,112	10
12	1,273	1,432	1,592	1,751	1,910	2,069	2,228	2,387	2,546	2,706	2,865	3,024	3,183	3,820	4,456	5,093	12
14	1,091	1,228	1,364	1,501	1,637	1,773	1,910	2,046	2,183	2,319	2,456	2,592	2,728	3,274	3,820	4,365	14
16	955	1,074	1,194	1,313	1,432	1,552	1,671	1,790	1,910	2,029	2,149	2,268	2,387	2,865	3,342	3,820	16
18	849	955	1,061	1,167	1,273	1,379	1,485	1,592	1,698	1,804	1,910	2,016	2,122	2,546	2,971	3,395	18
20	764	859	955	1,050	1,146	1,241	1,337	1,432	1,528	1,623	1,719	1,814	1,910	2,292	2,674	3,056	20
22	694	781	868	955	1,042	1,129	1,215	1,302	1,389	1,476	1,563	1,649	1,736	2,083	2,431	2,778	22
24	637	716	796	875	955	1,035	1,114	1,194	1,273	1,353	1,432	1,512	1,592	1,910	2,228	2,546	24
26	588	661	735	808	881	955	1,028	1,102	1,175	1,249	1,322	1,396	1,469	1,763	2,057	2,351	26
28	546	614	682	750	819	887	955	1,023	1,091	1,160	1,228	1,296	1,364	1,637	1,910	2,183	28
30	509	573	637	700	764	828	891	955	1,019	1,082	1,146	1,210	1,273	1,528	1,783	2,037	30
32	477	537	597	657	716	776	836	895	955	1,015	1,074	1,134	1,194	1,432	1,671	1,910	32
34	449	506	562	618	674	730	786	843	899	955	1,011	1,067	1,123	1,348	1,573	1,798	34
36	424	477	531	584	637	690	743	796	849	902	955	1,008	1,061	1,273	1,485	1,698	36
38	402	452	503	553	603	653	704	754	804	854	905	955	1,005	1,206	1,407	1,608	38
40	382	430	477	525	573	621	668	716	764	812	859	907	955	1,146	1,337	1,528	40
42	364	409	455	500	546	591	637	682	728	773	819	864	909	1,091	1,273	1,455	42
44	347	391	434	477	521	564	608	651	694	738	781	825	868	1,042	1,215	1,389	44
46	332	374	415	457	498	540	581	623	664	706	747	789	830	996	1,163	1,329	46
48	318	358	398	438	477	517	557	597	637	676	716	756	796	955	1,114	1,273	48
53	288	324	360	396	432	468	504	541	577	613	649	685	721	865	1,009	1,153	53
60	255	286	318	350	382	414	446	477	509	541	573	605	637	764	891	1,019	60
72	212	239	265	292	318	345	371	398	424	451	477	504	531	637	743	849	72

Cylindrical Grinding

Cylindrical grinding designates a general category of various grinding methods that have the common characteristic of rotating the workpiece around a fixed axis while grinding outside surface sections in controlled relation to that axis of rotation.

The form of the part or section being ground in this process is frequently cylindrical, hence the designation of the general category. However, the shape of the part may be tapered or of curvilinear profile; the position of the ground surface may also be perpendicular to the axis; and it is possible to grind concurrently several surface sections, adjacent or separated, of equal or different diameters, located in parallel or mutually inclined planes, etc., as long as the condition of a common axis of rotation is satisfied.

Size Range of Workpieces and Machines: Cylindrical grinding is applied in the manufacture of miniature parts, such as instrument components and, at the opposite extreme, for grinding rolling mill rolls weighing several tons. Accordingly, there are cylindrical grinding machines of many different types, each adapted to a specific work-size range. Machine capacities are usually expressed by such factors as maximum work diameter, work length and weight, complemented, of course, by many other significant data.

Plain, Universal, and Limited-Purpose Cylindrical Grinding Machines.—The plain cylindrical grinding machine is considered the basic type of this general category, and is used for grinding parts with cylindrical or slightly tapered form.

The universal cylindrical grinder can be used, in addition to grinding the basic cylindrical forms, for the grinding of parts with steep tapers, of surfaces normal to the part axis, including the entire face of the workpiece, and for internal grinding independently or in conjunction with the grinding of the part's outer surfaces. Such variety of part configurations requiring grinding is typical of work in the tool room, which constitutes the major area of application for universal cylindrical grinding machines.

Limited-purpose cylindrical grinders are needed for special work configurations and for high-volume production, where productivity is more important than flexibility of adaptation. Examples of limited-purpose cylindrical grinding machines are crankshaft and camshaft grinders, polygonal grinding machines, roll grinders, etc.

Traverse or Plunge Grinding.—In traverse grinding, the machine table carrying the work performs a reciprocating movement of specific travel length for transporting the rotating workpiece along the face of the grinding wheel. At each or at alternate stroke ends, the wheel slide advances for the gradual feeding of the wheel into the work. The length of the surface that can be ground by this method is generally limited only by the stroke length of the machine table. In large roll grinders, the relative movement between work and wheel is accomplished by the traverse of the wheel slide along a stationary machine table.

In plunge grinding, the machine table, after having been set, is locked and, while the part is rotating, the wheel slide continually advances at a preset rate, until the finish size of the part is reached. The width of the grinding wheel is a limiting factor of the section length that can be ground in this process. Plunge grinding is required for profiled surfaces and for the simultaneous grinding of multiple surfaces of different diameters or located in different planes.

When the configuration of the part does not make use of either method mandatory, the choice may be made on the basis of the following general considerations: traverse grinding usually produces a better finish, and the productivity of plunge grinding is generally higher.

Work Holding on Cylindrical Grinding Machines.—The manner in which the work is located and held in the machine during the grinding process determines the configuration of the part that can be adapted for cylindrical grinding and affects the resulting accuracy of the ground surface. The method of work holding also affects the attainable production rate, because the mounting and dismounting of the part can represent a substantial portion of the total operating time.

Whatever method is used for holding the part on cylindrical types of grinding machines, two basic conditions must be satisfied: 1) the part should be located with respect to its correct axis of rotation; and 2) the work drive must cause the part to rotate, at a specific speed, around the established axis.

The lengthwise location of the part, although controlled, is not too critical in traverse grinding; however, in plunge grinding, particularly when shoulder sections are also involved, it must be assured with great accuracy.

Table 1 presents a listing, with brief discussions, of work-holding methods and devices that are most frequently used in cylindrical grinding.

Table 1. Work-Holding Methods and Devices for Cylindrical Grinding

Designation	Description	Discussion
Centers, nonrotating ("dead"), with drive plate	Headstock with nonrotating spindle holds the center. Around the spindle, an independently supported sleeve carries the drive plate for rotating the work. Tailstock for opposite center.	The simplest method of holding the work between two opposite centers is also the potentially most accurate, as long as correctly prepared and located center holes are used in the work.
Centers, driving type	Word held between two centers obtains its rotation from the concurrently applied drive by the live headstock spindle and live tailstock spindle.	Eliminates the drawback of the common center-type grinding with driver plate, which requires a dog attached to the workpiece. Driven spindles permit the grinding of the work up to both ends.
Chuck, geared, or cam-actuated	Two, three, or four jaws moved radially through mechanical elements, hand-, or power-operated, exert concentrically acting clamping force on the workpiece.	Adaptable to workpieces of different configurations and within a generally wide capacity of the chuck. Flexible in uses that, however, do not include high-precision work.
Chuck, diaphragm	Force applied by hand or power of a flexible diaphragm causes the attached jaws to deflect temporarily for accepting the work, which is held when force is released.	Rapid action and flexible adaptation to different work configurations by means of special jaws offer varied uses for the grinding of disk-shaped and similar parts.
Collets	Holding devices with externally or internally acting clamping force, easily adaptable to power actuation, assuring high centering accuracy.	Limited to parts with previously machined or ground holding surfaces, because of the small range of clamping movement of the collet jaws.
Face plate	Has four independently actuated jaws, any or several of which may be used, or entirely removed, using the base plate for supporting special clamps.	Used for holding bulky parts, or those of awkward shape, which are ground in small quantities not warranting special fixtures.
Magnetic plate	Flat plates, with pole distribution adapted to the work, are mounted on the spindle like chucks and may be used for work with the locating face normal to the axis.	Applicable for light cuts such as are frequent in tool making, where the rapid clamping action and easy access to both the O.D. and the exposed face are sometimes of advantage.
Steady rests	Two basic types are used: (a) the two-jaw type supporting the work from the back (back rest), leaving access by the wheel; (b) the three-jaw type (center rest).	A complementary work-holding device, used in conjunction with primary work holders, to provide additional support, particularly to long and/or slender parts.
Special fixtures	Single-purpose devices, designed for a particular workpiece, primarily for providing special locating elements.	Typical workpieces requiring special fixturing are, as examples, crankshafts where the holding is combined with balancing functions; or internal gears located on the pitch circle of the teeth for O.D. grinding.

Selection of Grinding Wheels for Cylindrical Grinding.—For cylindrical grinding, as for grinding in general, the primary factor to be considered in wheel selection is the work material. Other factors are the amount of excess stock and its rate of removal (speeds and feeds), the desired accuracy and surface finish, the ratio of wheel and work diameter, wet or dry grinding, etc. In view of these many variables, it is not practical to set up a complete list of grinding wheel recommendations with general validity. Instead, examples of recommendations embracing a wide range of typical applications and assuming common practices are presented in Table 2. This is intended as a guide for the starting selection of grinding-wheel specifications which, in case of a not entirely satisfactory performance, can be refined subsequently. The content of the table is a version of the grinding-wheel recommendations for cylindrical grinding by the Norton Company using, however, non-proprietary designations for the abrasive types and bonds.

Table 2. Wheel Recommendations for Cylindrical Grinding

Material	Wheel Marking	Material	Wheel Marking
Aluminum	SFA46-18V	Forgings	A46-M5V
Armatures (laminated)	SFA100-18V	Gages (plug)	SFA80-K8V
Axles (auto & railway)	A54-M5V	General-purpose grinding	SFA54-L5V
Brass	C36-KV	Glass	BFA220-011V
Bronze		Gun barrels	
Soft	C36-KV	Spotting and O.D.	BFA60-M5V
Hard	A46-M5V	Nitralloy	
Bushings (hardened steel)	BFA60-L5V	Before nitriding	A60-K5V
Bushings (cast iron)	C36-JV	After nitriding	
Cam lobes (cast alloy)		Commercial finish	SFA60-18V
Roughing	BFA54-N5V	High finish	C100-1V
Finishing	A70-P6B	Reflective finish	C500-19E
Cam lobes (hardened steel)		Pistons (aluminum)	SFA46-18V
Roughing	BFA54-L5V	(cast iron)	C36-KV
Finishing	BFA80-T8B	Plastics	C46-JV
Cast iron	C36-JV	Rubber	
Chromium plating		Soft	SFA20-K5B
Commercial finish	SFA60-J8V	Hard	C36-KB
High finish	A150-K5E	Spline shafts	SFA60-N5V
Reflective finish	C500-I9E	Sprayed metal	C60-JV
Commutators (copper)	C60-M4E	Steel	
Crankshafts (airplane)		Soft	
Pins	BFA46-K5V	1 in. dia. and smaller	SFA60-M5V
Bearings	A46-L5V	over 1 in dia.	SFA46-L5V
Crankshafts (automotive pins and bearings)		Hardened	
		1 in. dia. and smaller	SFA80-L8V
Finishing	A54-N5V	over 1 in. dia.	SFA60-K5V
Roughing & finishing	A54-O5V	300 series stainless	SFA46-K8V
Regrinding	A54-M5V	Stellite	BFA46-M5V
Regrinding, sprayed metal	C60-JV	Titanium	C60-JV
		Valve stems (automative)	BFA54-N5V
Drills	BFA54-N5V	Valve tappets	BFA54-M5V

Note: Prefixes to the standard designation "A" of aluminum oxide indicate modified abrasives as follows:

BFA = Blended friable (a blend of regular and friable).

SFA = Semifriable.

Operational Data for Cylindrical Grinding.—In cylindrical grinding, similarly to other metalcutting processes, the applied speed and feed rates must be adjusted to the operational conditions as well as to the objectives of the process. Grinding differs, however, from other types of metalcutting methods in regard to the cutting speed of the tool which, in grinding, is generally not a variable; it should be maintained at, or close to the optimum rate, commonly 6500 feet per minute peripheral speed.

In establishing the proper process values for grinding, of prime consideration are the work material, its condition (hardened or soft), and the type of operation (roughing or finishing). Other influencing factors are the characteristics of the grinding machine (stability, power), the specifications of the grinding wheel, the material allowance, the rigidity and balance of the workpiece, as well as several grinding process conditions, such as wet or dry grinding, the manner of wheel truing, etc.

Variables of the cylindrical grinding process, often referred to as *grinding data,* comprise the speed of work rotation (measured as the surface speed of the work); the infeed (in inches per pass for traverse grinding, or in inches per minute for plunge grinding); and, in the case of traverse grinding, the speed of the reciprocating table movement (expressed either in feet per minute, or as a fraction of the wheel width for each revolution of the work).

For the purpose of starting values in setting up a cylindrical grinding process, a brief listing of basic data for common cylindrical grinding conditions and involving frequently used materials, is presented in Table 3.

Table 3. Basic Process Data for Cylindrical Grinding

Traverse Grinding						
Work Material	Material Condition	Work Surface Speed, fpm	Infeed, Inch/Pass		Traverse for Each Work Revolution, In Fractions of the Wheel Width	
			Roughing	Finishing	Roughing	Finishing
Plain Carbon Steel	Annealed	100	0.002	0.0005	½	⅙
	Hardened	70	0.002	0.0003 to 0.0005	¼	⅛
Alloy Steel	Annealed	100	0.002	0.0005	½	⅙
	Hardened	70	0.002	0.0002 to 0.0005	¼	⅛
Tool Steel	Annealed	60	0.002	0.0005 max.	½	⅙
	Hardened	0.002	0.002	0.0001 to 0.0005	¼	⅛
Copper Alloys	Annealed or Cold Drawn	100	0.002	0.0005 max.	⅓	⅙
Aluminum Alloys	Cold Drawn or Solution Treated	150	0.002	0.0005 max.	⅓	⅙

Plunge Grinding		
Work Material	Infeed per Revolution of the Work, Inch	
	Roughing	Finishing
Steel, soft	0.0005	0.0002
Plain carbon steel, hardened	0.0002	0.000050
Alloy and tool steel, hardened	0.0001	0.000025

These data, which are, in general, considered conservative, are based on average operating conditions and may be modified subsequently,

reducing the values in case of unsatisfactory quality of the grinding or the occurrence of failures;

increasing the rates for raising the productivity of the process, particularly for rigid workpieces, substantial stock allowance, etc.

High-Speed Cylindrical Grinding.—The maximum peripheral speed of the wheels in regular cylindrical grinding is generally 6500 feet per minute; the commonly used grinding wheels and machines are designed to operate efficiently at this speed. Recently, efforts were made to raise the productivity of different grinding methods, including cylindrical grinding, by increasing the peripheral speed of the grinding wheel to a substantially higher than traditional level, such as 12,000 feet per minute or more. Such methods are designated by the distinguishing term of high-speed grinding.

For high-speed grinding, special grinding machines have been built with high dynamic stiffness and static rigidity, equipped with powerful drive motors, extra-strong spindles and bearings, reinforced wheel guards, etc., and using grinding wheels expressly made and tested for operating at high peripheral speeds. The higher stock-removal rate accomplished by high-speed grinding represents an advantage when the work configuration and material permit, and the removable stock allowance warrants its application.

CAUTION: High-speed grinding must *not* be applied on standard types of equipment, such as general types of grinding machines and regular grinding wheels. Operating grinding wheels, even temporarily, at higher than approved speed constitutes a grave safety hazard.

Areas and Degrees of Automation in Cylindrical Grinding.—Power drive for the work rotation and for the reciprocating table traverse are fundamental machine movements that, once set for a certain rate, will function without requiring additional attention. Loading and removing the work, starting and stopping the main movements, and applying infeed by hand wheel are carried out by the operator on cylindrical grinding machines in their basic degree of mechanization. Such equipment is still frequently used in tool room and jobbing-type work.

More advanced levels of automation have been developed for cylindrical grinders and are being applied in different degrees, particularly in the following principal respects:

A) *Infeed,* in which different rates are provided for rapid approach, roughing and finishing, followed by a spark-out period, with presetting of the advance rates, the cutoff points, and the duration of time-related functions.

B) *Automatic cycling* actuated by a single lever to start work rotation, table reciprocation, grinding-fluid supply, and infeed, followed at the end of the operation by wheel slide retraction, the successive stopping of the table movement, the work rotation, and the fluid supply.

C) *Table traverse dwells* (tarry) in the extreme positions of the travel, over preset periods, to assure uniform exposure to the wheel contact of the entire work section.

D) *Mechanized work loading,* clamping, and, after termination of the operation, unloading, combined with appropriate work-feeding devices such as indexing-type drums.

E) *Size control* by in-process or post-process measurements. Signals originated by the gage will control the advance movement or cause automatic compensation of size variations by adjusting the cutoff points of the infeed.

F) *Automatic wheel dressing* at preset frequency, combined with appropriate compensation in the infeed movement.

G) *Numerical control* obviates the time-consuming setups for repetitive work performed on small- or medium-size lots. As an application example: shafts with several sections of different lengths and diameters can be ground automatically in a single operation, grinding the sections in consecutive order to close dimensional limits, controlled by an in-process gage, which is also automatically set by means of the program.

The choice of the grinding machine functions to be automated and the extent of automation will generally be guided by economic considerations, after a thorough review of the available standard and optional equipment. Numerical control of partial or complete cycles is being applied to modern cylindrical and other grinding machines.

Cylindrical Grinding Troubles and Their Correction.—Troubles that may be encountered in cylindrical grinding may be classified as work defects (chatter, checking, burning, scratching, and inaccuracies), improperly operating machines (jumpy infeed or traverse), and wheel defects (too hard or soft action, loading, glazing, and breakage). The Landis Tool Company has listed some of these troubles, their causes, and corrections as follows:

Chatter.—Sources of chatter include: 1) faulty coolant; 2) wheel out of balance; 3) wheel out of round; 4) wheel too hard; 5) improper dressing; 6) faulty work support or rotation; 7) improper operation; 8) faulty traverse; 9) work vibration; 10) outside vibration transmitted to machine; 11) interference; 12) wheel base; and 13) headstock.

Suggested procedures for correction of these troubles are:

1) *Faulty coolant:* Clean tanks and lines. Replace dirty or heavy coolant with correct mixture.

2) *Wheel out of balance:* Rebalance on mounting before and after dressing. Run wheel without coolant to remove excess water. Store a removed wheel on its side to keep retained water from causing a false heavy side. Tighten wheel mounting flanges. Make sure wheel center fits spindle.

3) *Wheel out of round:* True before and after balancing. True sides to face.

4) *Wheel too hard:* Use coarser grit, softer grade, more open bond. See *Wheel Defects* on page 1183.

5) *Improper dressing:* Use sharp diamond and hold rigidly close to wheel. It must not overhang excessively. Check diamond in mounting.

6) *Faulty work support or rotation:* Use sufficient number of work rests and adjust them more carefully. Use proper angles in centers of work. Clean dirt from footstock spindle and be sure spindle is tight. Make certain that work centers fit properly in spindles.

7) *Improper operation:* Reduce rate of wheel feed.

8) *Faulty traverse:* See *Uneven Traverse or Infeed of Wheel Head* on page 1182.

9) *Work vibration:* Reduce work speed. Check workpiece for balance.

10) *Outside vibration transmitted to machine:* Check and make sure that machine is level and sitting solidly on foundation. Isolate machine or foundation.

11) *Interference:* Check all guards for clearance.

12) *Wheel base:* Check spindle bearing clearance. Use belts of equal lengths or uniform cross-section on motor drive. Check drive motor for unbalance. Check balance and fit of pulleys. Check wheel feed mechanism to see that all parts are tight.

13) *Headstock:* Put belts of same length and cross-section on motor drive; check for correct work speeds. Check drive motor for unbalance. Make certain that headstock spindle is not loose. Check work center fit in spindle. Check wear of face plate and jackshaft bearings.

Spirals on Work (traverse lines with same lead on work as rate of traverse).— Sources of spirals include: 1) machine parts out of line; and 2) truing.

Suggested procedures for correction of these troubles are:

1) *Machine parts out of line:* Check wheel base, headstock, and footstock for proper alignment.

2) *Truing:* Point truing tool down 3 degrees at the workwheel contact line. Round off wheel edges.

Check Marks on Work.—Sources of check marks include: 1) improper operation; 2) improper heat treatment; 3) improper size control; 4) improper wheel; and 5) improper dressing.

Suggested procedures for correction of these troubles are:

1) *Improper operation:* Make wheel act softer. See *Wheel Defects*. Do not force wheel into work. Use greater volume of coolant and a more even flow. Check the correct positioning of coolant nozzles to direct a copious flow of clean coolant at the proper location.

2) *Improper heat treatment:* Take corrective measures in heat-treating operations.

3) *Improper size control:* Make sure that engineering establishes reasonable size limits. See that they are maintained.

4) *Improper wheel:* Make wheel act softer. Use softer-grade wheel. Review the grain size and type of abrasive. A finer grit or more friable abrasive or both may be called for.

5) *Improper dressing:* Check that the diamond is sharp, of good quality, and well set. Increase speed of the dressing cycle. Make sure diamond is not cracked.

Burning and Discoloration of Work.—Sources of burning and discoloration are: 1) improper operation; and 2) improper wheel.

Suggested procedures for correction of these troubles are:

1) *Improper operation:* Decrease rate of infeed. Don't stop work while in contact with wheel.

2) *Improper wheel:* Use softer wheel or obtain softer effect. See *Wheel Defects*. Use greater volume of coolant.

Isolated Deep Marks on Work.—Source of trouble is an unsuitable wheel. Use a finer wheel and consider a change in abrasive type.

Fine Spiral or Thread on Work.—Sources of this trouble are: 1) improper operation; and 2) faulty wheel dressing.

Suggested procedures for corrections of these troubles are:

1) *Improper operation:* Reduce wheel pressure. Use more work rests. Reduce traverse with respect to work rotation. Use different traverse rates to break up pattern when making numerous passes. Prevent edge of wheel from penetrating by dressing wheel face parallel to work.

2) *Faulty wheel dressing:* Use slower or more even dressing traverse. Set dressing tool at least 3 degrees down and 30 degrees to the side from time to time. Tighten holder. Don't take too deep a cut. Round off wheel edges. Start dressing cut from wheel edge.

Narrow and Deep Regular Marks on Work.—Source of trouble is that the wheel is too coarse. Use finer grain size.

Wide, Irregular Marks of Varying Depth on Work.—Source of trouble is too soft a wheel. Use a harder grade wheel. See *Wheel Defects*.

Widely Spaced Spots on Work.—Sources of trouble are oil spots or glazed areas on wheel face. Balance and true wheel. Keep oil from wheel face.

Irregular "Fish-tail" Marks of Various Lengths and Widths on Work.—Source of trouble is dirty coolant. Clean tank frequently. Use filter for fine finish grinding. Flush wheel guards after dressing or when changing to finer wheel.

Wavy Traverse Lines on Work.—Source of trouble is wheel edges. Round off. Check for loose thrust on spindle and correct if necessary.

Irregular Marks on Work.—Cause is loose dirt. Keep machine clean.

Deep, Irregular Marks on Work.—Source of trouble is loose wheel flanges. Tighten and make sure blotters are used.

Isolated Deep Marks on Work.—Sources of trouble are: 1) grains pull out; coolant too strong; 2) coarse grains or foreign matter in wheel face; and 3) improper dressing.

Respective suggested procedures for corrections of these troubles are: 1) decrease soda content in coolant mixture; 2) dress wheel; and 3) use sharper dressing tool.

Brush wheel after dressing with stiff bristle brush.

Grain Marks on Work.—Sources of trouble are: 1) improper finishing cut; 2) grain sizes of roughing and finishing wheels differ too much; 3) dressing too coarse; and 4) wheel too coarse or too soft.

Respective suggested procedures for corrections of these troubles are: start with high work and traverse speeds; finish with high work speed and slow traverse, letting wheel "spark-out" completely; finish out better with roughing wheel or use finer roughing wheel; use shallower and slower cut; and use finer grain size or harder-grade wheel.

Inaccuracies in Work.—Work out-of-round, out-of-parallel, or tapered.

Sources of trouble are: 1) misalignment of machine parts; 2) work centers; 3) improper operation; 4) coolant; 5) wheel; 6) improper dressing; 7) spindle bearings; and 8) work.

Suggested procedures for corrections of these troubles are:

1) *Misalignment of machine parts:* Check headstock and tailstock for alignment and proper clamping.

2) *Work centers:* Centers in work must be deep enough to clear center point. Keep work centers clean and lubricated. Check play of footstock spindle and see that footstock spindle is clean and tightly seated. Regrind work centers if worn. Work centers must fit taper of work-center holes. Footstock must be checked for proper tension.

3) *Improper operation:* Don't let wheel traverse beyond end of work. Decrease wheel pressure so work won't spring. Use harder wheel or change feeds and speeds to make wheel act harder. Allow work to "spark-out." Decrease feed rate. Use proper number of work rests. Allow proper amount of tarry. Workpiece must be balanced if it is an odd shape.

4) *Coolant:* Use greater volume of coolant.

5) *Wheel:* Rebalance wheel on mounting before and after truing.

6) *Improper dressing:* Use same positions and machine conditions for dressing as in grinding.

7) *Spindle bearings:* Check clearance.

8) *Work:* Work must come to machine in reasonably accurate form.

Inaccurate Work Sizing (when wheel is fed to same position, it grinds one piece to correct size, another oversize, and still another undersize).—Sources of trouble are: 1) improper work support or rotation; 2) wheel out of balance; 3) loaded wheel; 4) improper infeed; 5) improper traverse; 6) coolant; 7) misalignment; and 8) work.

Suggested procedures for corrections of these troubles are:

1) *Improper work support or rotation:* Keep work centers clean and lubricated. Regrind work-center tips to proper angle. Be sure footstock spindle is tight. Use sufficient work rests, properly spaced.

2) *Wheel out of balance:* Balance wheel on mounting before and after truing.

3) *Loaded wheel:* See *Wheel Defects*.

4) *Improper infeed:* Check forward stops of rapid feed and slow feed. When readjusting position of wheel base by means of the fine feed, move the wheel base back after making the adjustment and then bring it forward again to take up backlash and relieve strain in feed-up parts. Check wheel spindle bearings. Don't let excessive lubrication of wheel base slide cause "floating." Check and tighten wheel feed mechanism. Check parts for wear. Check pressure in hydraulic system. Set infeed cushion properly. Check to see that pistons are not sticking.

5) *Improper traverse:* Check traverse hydraulic system and the operating pressure. Prevent excessive lubrication of carriage ways with resultant "floating" condition. Check to see if carriage traverse piston rods are binding. Carriage rack and driving gear must not bind. Change length of tarry period.

6) *Coolant:* Use greater volume of clean coolant.

7) *Misalignment:* Check level and alignment of machine.

8) *Work:* Workpieces may vary too much in length, permitting uneven center pressure.

Uneven Traverse or Infeed of Wheel Head.—Sources of uneven traverse or infeed of wheel head are: carriage and wheel head, hydraulic system, interference, unbalanced conditions, and wheel out of balance. Suggested procedures for correction of these troubles are:

1) *Carriage and wheel head:* Ways may be scored. Be sure to use recommended oil for both lubrication and hydraulic system. Make sure ways are not so smooth that they press out oil film. Check lubrication of ways. Check wheel feed mechanism, traverse gear, and carriage rack clearance. Prevent binding of carriage traverse cylinder rods.

2) *Hydraulic systems:* Remove air and check pressure of hydraulic oil. Check pistons and valves for oil leakage and for gumminess caused by incorrect oil. Check worn valves or pistons that permit leakage.

3) *Interference:* Make sure guard strips do not interfere.

4) *Unbalanced conditions:* Eliminate loose pulleys, unbalanced wheel drive motor, uneven belts, or high spindle keys.

5) *Wheel out of balance:* Balance wheel on mounting before and after truing.

Wheel Defects.—When *wheel is acting too hard,* such defects as glazing, some loading, lack of cut, chatter, and burning of work result.

Suggested procedures for correction of these faults are: 1) Increase work and traverse speeds as well as rate of in-feed; 2) decrease wheel speed, diameter, or width; 3) dress more sharply; 4) use thinner coolant; 5) don't tarry at end of traverse; 6) select softer wheel grade and coarser grain size; 7) avoid gummy coolant; and 8) on hardened work select finer grit, more fragile abrasive or both to get penetration. Use softer grade.

When *wheel is acting too soft,* such defects as wheel marks, tapered work, short wheel life, and not-holding-cut result.

Suggested procedures for correction of these faults are: 1) Decrease work and traverse speeds as well as rate of in-feed; 2) increase wheel speed, diameter, or width; 3) dress with little in-feed and slow traverse; 4) use heavier coolants; 5) don't let wheel run off work at end of traverse; and 6) select harder wheel or less fragile grain or both.

Wheel Loading and Glazing.—Sources of the trouble of wheel loading or glazing are:

1) Incorrect wheel; 2) improper dress; 3) faulty operation; 4) faulty coolant; and 5) gummy coolant.

Suggested procedures for correction of these faults are:

1) *Incorrect wheel:* Use coarser grain size, more open bond, or softer grade.

2) *Improper dressing:* Keep wheel sharp with sharp dresser, clean wheel after dressing, use faster dressing traverse, and deeper dressing cut.

3) *Faulty operation:* Control speeds and feeds to soften action of wheel. Use less in-feed to prevent loading; more in-feed to stop glazing.

4) *Faulty coolant:* Use more, cleaner and thinner coolant, and less oily coolant.

5) *Gummy coolant:* To stop wheel glazing, increase soda content and avoid the use of soluble oils if water is hard. In using soluble oil coolant with hard water a suitable conditioner or "softener" should be added.

Wheel Breakage.—Suggested procedures for the correction of a radial break with three or more pieces are: 1) Reduce wheel speed to or below rated speed; 2) mount wheel properly, use blotters, tight arbors, even flange pressure and be sure to keep out dirt between flange and wheel; 3) use plenty of coolant to prevent over-heating; 4) use less in-feed; and 5) don't allow wheel to become jammed on work.

A radial break with two pieces may be caused by excessive side strain. To prevent an irregular wheel break, don't let wheel become jammed on work; don't allow striking of wheel; and never use wheels that have been damaged in handling. In general, do not use a wheel that is too tight on the arbor since the wheel is apt to break when started. Prevent excessive hammering action of wheel. Follow rules of the American National Standard Safety Requirements for the Use, Care, and Protection of Abrasive Wheels (ANSI B7.1-1978).

Centerless Grinding

In centerless grinding the work is supported on a work rest blade and is between the grinding wheel and a regulating wheel. The regulating wheel generally is a rubber bonded abrasive wheel. In the normal grinding position the grinding wheel forces the work downward against the work rest blade and also against the regulating wheel. The latter imparts a uniform rotation to the work giving it its same peripheral speed which is adjustable.

The higher the work center is placed above the line joining the centers of the grinding and regulating wheels the quicker the rounding action. Rounding action is also increased by a high work speed and a slow rate of traverse (if a through-feed operation). It is possible to have a higher work center when using softer wheels, as their use gives decreased contact pressures and the tendency of the workpiece to lift off the work rest blade is lessened.

Long rods or bars are sometimes ground with their centers below the line-of-centers of the wheels to eliminate the whipping and chattering due to slight bends or kinks in the rods or bars, as they are held more firmly down on the blade by the wheels.

There are three general methods of centerless grinding which may be described as through-feed, in-feed, and end-feed methods.

Through-feed Method of Grinding.—The through-feed method is applied to straight cylindrical parts. The work is given an axial movement by the regulating wheel and passes between the grinding and regulating wheels from one side to the other. The rate of feed depends upon the diameter and speed of the regulating wheel and its inclination which is adjustable. It may be necessary to pass the work between the wheels more than once, the number of passes depending upon such factors as the amount of stock to be removed, the roundness and straightness of the unground work, and the limits of accuracy required.

The work rest fixture also contains adjustable guides on either side of the wheels that directs the work to and from the wheels in a straight line.

In-feed Method of Centerless Grinding.—When parts have shoulders, heads or some part larger than the ground diameter, the in-feed method usually is employed. This method is similar to "plungecut" form grinding on a center type of grinder. The length or sections to be ground in any one operation are limited by the width of the wheel. As there is no axial feeding movement, the regulating wheel is set with its axis approximately parallel to that of the grinding wheel, there being a slight inclination to keep the work tight against the end stop.

End-feed Method of Grinding.—The end-feed method is applied only to taper work. The grinding wheel, regulating wheel, and the work rest blade are set in a fixed relation to each other and the work is fed in from the front mechanically or manually to a fixed end stop. Either the grinding or regulating wheel, or both, are dressed to the proper taper.

Automatic Centerless Grinding.—The grinding of relatively small parts may be done automatically by equipping the machine with a magazine, gravity chute, or hopper feed, provided the shape of the part will permit using these feed mechanisms.

Internal Centerless Grinding.—Internal grinding machines based upon the centerless principle utilize the outside diameter of the work as a guide for grinding the bore which is concentric with the outer surface. In addition to straight and tapered bores, interrupted and "blind" holes can be ground by the centerless method. When two or more grinding operations such as roughing and finishing must be performed on the same part, the work can be rechucked in the same location as often as required.

Centerless Grinding Troubles.—A number of troubles and some corrective measures compiled by a manufacturer are listed here for the through-feed and in-feed methods of centerless grinding.

Chattermarks: are caused by having the work center too high above the line joining the centers of the grinding and regulating wheels; using too hard or too fine a grinding wheel; using too steep an angle on the work support blade; using too thin a work support blade; "play" in the set-up due to loosely clamped members; having the grinding wheel fit loosely on the spindle; having vibration either transmitted to the machine or caused by a defective drive in the machine; having the grinding wheel out-of-balance; using too heavy a stock removal; and having the grinding wheel or the regulating wheel spindles not properly adjusted.

Feed lines or spiral marks: in through-feed grinding are caused by too sharp a corner on the exit side of the grinding wheel which may be alleviated by dressing the grinding wheel to a slight taper about ½ inch from the edge, dressing the edge to a slight radius, or swiveling the regulating wheel a bit.

Scored work: is caused by burrs, abrasive grains, or removed material being imbedded in or fused to the work support blade. This condition may be alleviated by using a coolant with increased lubricating properties and if this does not help a softer grade wheel should be used.

Work not ground round: may be due to the work center not being high enough above the line joining the centers of the grinding and regulating wheels. Placing the work center higher and using a softer grade wheel should help to alleviate this condition.

Work not ground straight: in through-feed grinding may be due to an incorrect setting of the guides used in introducing and removing the work from the wheels, and the existence of convex or concave faces on the regulating wheel. For example, if the work is tapered on the front end, the work guide on the entering side is deflected toward the regulating wheel. If tapered on the back end, then the work guide on the exit side is deflected toward the regulating wheel. If both ends are tapered, then both work guides are deflected toward the regulating wheel. The same barrel-shaped pieces are also obtained if the face of the regulating wheel is convex at the line of contact with the work. Conversely, the work would be ground with hollow shapes if the work guides were deflected toward the grinding wheel or if the face of the regulating wheel were concave at the line of contact with the work. The use of a warped work rest blade may also result in the work not being ground straight and the blade should be removed and checked with a straight edge.

In in-feed grinding, in order to keep the wheel faces straight which will insure straightness of the cylindrical pieces being ground, the first item to be checked is the straightness and the angle of inclination of the work rest blade. If this is satisfactory then one of three corrective measures may be taken: the first might be to swivel the regulating wheel to compensate for the taper, the second might be to true the grinding wheel to that angle that will give a perfectly straight workpiece, and the third might be to change the inclination of the regulating wheel (this is true only for correcting very slight tapers up to 0.0005 inch).

Difficulties in sizing: the work in in-feed grinding are generally due to a worn in-feed mechanism and may be overcome by adjusting the in-feed nut.

Flat spots: on the workpiece in in-feed grinding usually occur when grinding heavy work and generally when the stock removal is light. This condition is due to insufficient driving power between the work and the regulating wheel which may be alleviated by equipping the work rest with a roller that exerts a force against the workpiece; and by feeding the workpiece to the end stop using the upper slide.

Surface Grinding

The term surface grinding implies, in current technical usage, the grinding of surfaces which are essentially flat. Several methods of surface grinding, however, are adapted and used to produce surfaces characterized by parallel straight line elements in one direction, while normal to that direction the contour of the surface may consist of several straight line sections at different angles to each other (e.g., the guideways of a lathe bed); in other cases the contour may be curved or profiled (e.g., a thread cutting chaser).

Advantages of Surface Grinding.—Alternate methods for machining work surfaces similar to those produced by surface grinding are milling and, to a much more limited degree, planing. Surface grinding, however, has several advantages over alternate methods that are carried out with metal-cutting tools. Examples of such potential advantages are as follows:

1) Grinding is applicable to very hard and/or abrasive work materials, without significant effect on the efficiency of the stock removal.

2) The desired form and dimensional accuracy of the work surface can be obtained to a much higher degree and in a more consistent manner.

3) Surface textures of very high finish and—when the appropriate system is utilized—with the required lay, are generally produced.

4) Tooling for surface grinding as a rule is substantially less expensive, particularly for producing profiled surfaces, the shapes of which may be dressed into the wheel, often with simple devices, in processes that are much more economical than the making and the maintenance of form cutters.

5) Fixturing for work holding is generally very simple in surface grinding, particularly when magnetic chucks are applicable, although the mechanical holding fixture can also be simpler, because of the smaller clamping force required than in milling or planing.

6) Parallel surfaces on opposite sides of the work are produced accurately, either in consecutive operations using the first ground surface as a dependable reference plane or, simultaneously, in double face grinding, which usually operates without the need for holding the parts by clamping.

7) Surface grinding is well adapted to process automation, particularly for size control, but also for mechanized work handling in the large volume production of a wide range of component parts.

Principal Systems of Surface Grinding.—Flat surfaces can be ground with different surface portions of the wheel, by different arrangements of the work and wheel, as well as by different interrelated movements. The various systems of surface grinding, with their respective capabilities, can best be reviewed by considering two major distinguishing characteristics:

1) *The operating surface of the grinding wheel,* which may be the periphery or the face (the side);

2) *The movement of the work during the process,* which may be traverse (generally reciprocating) or rotary (continuous), depending on the design of a particular category of surface grinders.

The accompanying table provides a concise review of the principal surface grinding systems, defined by the preceding characteristics. It should be noted that many surface grinders are built for specific applications, and do not fit exactly into any one of these major categories.

Selection of Grinding Wheels for Surface Grinding.—The most practical way to select a grinding wheel for surface grinding is to base the selection on the work material. Table 1a gives the grinding wheel recommendations for Types 1, 5, and 7 straight wheels used on reciprocating and rotary table surface grinders with horizontal spindles. Table 1b gives the grinding wheel recommendations for Type 2 cylinder wheels, Type 6 cup wheels, and wheel segments used on vertical spindle surface grinders.

The last letters (two or three) that may follow the bond designation V (vitrified) or B (resinoid) refer to: 1) bond modification, "BE" being especially suitable for surface grinding; 2) special structure, "P" type being distinctively porous; and 3) for segments made of 23A type abrasives, the term 12VSM implies porous structure, and the letter "P" is not needed.

Table 1a. Grinding Wheel Recommendations for Surface Grinding—Using Straight Wheel Types 1, 5, and 7

	Horizontal-spindle, reciprocating-table surface grinders	
Material	Wheels less than 16 inches in diameter	Wheels 16 inches in diameter and over
Cast iron	37C36-K8V or 23A46-I8VBE	23A36-I8VBE
Nonferrous metal	37C36-K8V	37C36-K8V
Soft steel	23A46-J8VBE	23A36-J8VBE

Table 1a. *(Continued)* **Grinding Wheel Recommendations for Surface Grinding—Using Straight Wheel Types 1, 5, and 7**

Horizontal-spindle, reciprocating-table surface grinders		
Material	Wheels less than 16 inches in diameter	Wheels 16 inches in diameter and over
Hardened steel—broad contact	32A46-H8VBE or 32A60-F12VBEP	32A36-H8VBE or 32A36-F12VBEP
Hardened steel—narrow contact or interrupted cut	32A46-I8VBE	32A36-J8VBE
General-purpose wheel	23A46-H8VBE	23A36-I8VBE
Cemented carbides	Diamond wheels[a]	Diamond wheels[a]

[a] General diamond wheel recommendations are listed in Table 5 on page 1168.

Horizontal-spindle, rotary-table surface grinders	
Material	Wheels of any diameter
Cast iron	37C36-K8V or 23A46-I8VBE
Nonferrous metals	37C36-K8V
Soft steel	23A46-J8VBE
Hardened steel—broad contact	32A46-I8VBE
Hardened steel—narrow contact or interrupted cut	32A46-J8VBE
General-purpose wheel	23A46-I8VBE
Cemented carbides—roughing	Diamond wheels[a]

Courtesy of Norton Company

Table 1b. Grinding Wheel Recommendations for Surface Grinding—Using Type 2 Cylinder Wheels, Type 6 Cup Wheels, and Wheel Segments

Material	Type 2 Cylinder Wheels	Type 6 Cup Wheels	Wheel Segments
High tensile cast iron and nonferrous metals	37C24-HKV	37C24-HVK	37C24-HVK
Soft steel, malleable cast iron, steel castings, boiler plate	23A24-I8VBE or 23A30-G12VBEP	23A24-I8VBE	23A24-I8VSM or 23A30-H12VSM
Hardened steel—broad contact	32A46-G8VBE or 32A36-E12VBEP	32A46-G8VBE or 32A60-E12VBEP	32A36-G8VBE or 32A46-E12VBEP
Hardened steel—narrow contact or interrupt cut	32A46-H8VBE	32A60-H8VBE	32A46-G8VBE or 32A60-G12VBEP
General-purpose use	23A30-H8VBE or 23A30-E12VBEP	…	23A30-H8VSM or 23A30-G12VSM

The wheel markings in the tables are those used by the Norton Co., complementing the basic standard markings with Norton symbols. The complementary symbols used in these tables, that is, those preceding the letter designating A (aluminum oxide) or C (silicon carbide), indicate the special type of basic abrasive that has the friability best suited for particular work materials. Those preceding A (aluminum oxide) are

57—a versatile abrasive suitable for grinding steel in either a hard or soft state.

38—the most friable abrasive.

32—the abrasive suited for tool steel grinding.

23—an abrasive with intermediate grinding action, and

19—the abrasive produced for less heat-sensitive steels.

Those preceding C (silicon carbide) are

37—a general application abrasive, and

39—an abrasive for grinding hard cemented carbide.

Principal Systems of Surface Grinding — Diagrams

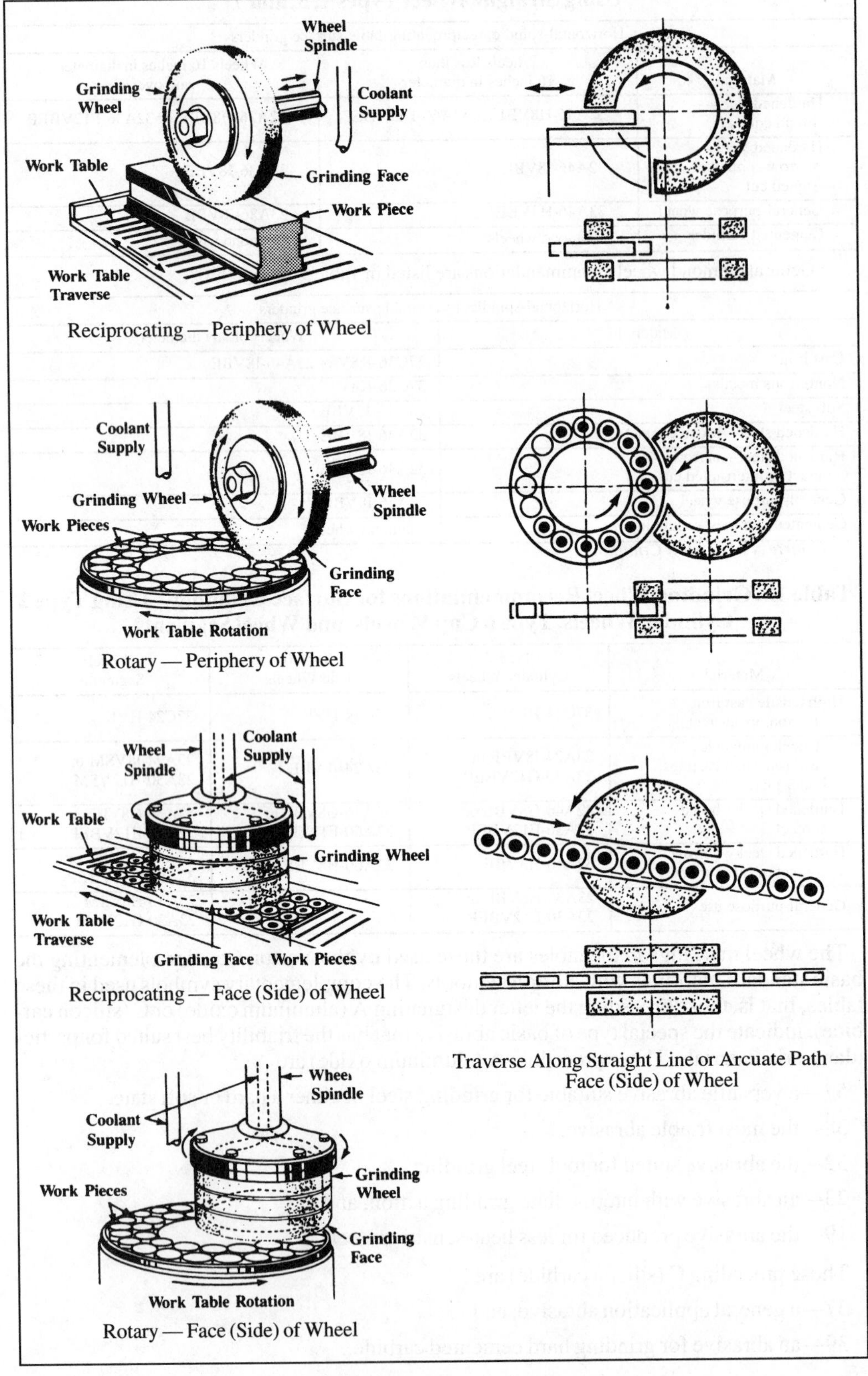

Reciprocating — Periphery of Wheel

Rotary — Periphery of Wheel

Reciprocating — Face (Side) of Wheel

Traverse Along Straight Line or Arcuate Path — Face (Side) of Wheel

Rotary — Face (Side) of Wheel

Principal Systems of Surface Grinding—Principles of Operation

Effective Grinding Surface—Periphery of Wheel Movement of Work—Reciprocating
Work is mounted on the horizontal machine table that is traversed in a reciprocating movement at a speed generally selected from a steplessly variable range. The transverse movement, called cross feed of the table or of the wheel slide, operates at the end of the reciprocating stroke and assures the gradual exposure of the entire work surface, which commonly exceeds the width of the wheel. The depth of the cut is controlled by the downfeed of the wheel, applied in increments at the reversal of the transverse movement.
Effective Grinding Surface—Periphery of Wheel Movement of Work—Rotary
Work is mounted, usually on the full-diameter magnetic chuck of the circular machine table that rotates at a preset constant or automatically varying speed, the latter maintaining an approximately equal peripheral speed of the work surface area being ground. The wheelhead, installed on a cross slide, traverses over the table along a radial path, moving in alternating directions, toward and away from the center of the table. Infeed is by vertical movement of the saddle along the guideways of the vertical column, at the end of the radial wheelhead stroke. The saddle contains the guideways along which the wheelhead slide reciprocates.
Effective Grinding Surface—Face (Side) of Wheel Movement of Work—Reciprocating
Operation is similar to the reciprocating table-type peripheral surface grinder, but grinding is with the face, usually with the rim of a cup-shaped wheel, or a segmental wheel for large machines. Capable of covering a much wider area of the work surface than the peripheral grinder, thus frequently no need for cross feed. Provides efficient stock removal, but is less adaptable than the reciprocating table-type peripheral grinder.
Effective Grinding Surface—Face (Side) of Wheel Movement of Work—Rotary
The grinding wheel, usually of segmental type, is set in a position to cover either an annular area near the periphery of the table or, more commonly, to reach beyond the table center. A large circular magnetic chuck generally covers the entire table surface and facilitates the mounting of workpieces, even of fixtures, when needed. The uninterrupted passage of the work in contact with the large wheel face permits a very high rate of stock removal and the machine, with single or double wheelhead, can be adapted also to automatic operation with continuous part feed by mechanized work handling.
Effective Grinding Surface—Face (Side) of Wheel Movement of Work—Traverse Along Straight or Arcuate Path
Operates with practically the entire face of the wheel, which is designated as an abrasive disc (hence "disc grinding") because of its narrow width in relation to the large diameter. Built either for one or, more frequently, for two discs operating with opposed faces for the simultaneous grinding of both sides of the workpiece. The parts pass between the operating faces of the wheel (a) pushed-in and retracted by the drawerlike movement of a feed slide; (b) in an arcuate movement carried in the nests of a rotating feed wheel; (c) nearly diagonally advancing along a rail. Very well adapted to fully mechanized work handling.

Process Data for Surface Grinding.—In surface grinding, similarly to other metal-cutting processes, the speed and feed rates that are applied must be adjusted to the operational conditions as well as to the objectives of the process. Grinding differs, however, from other types of metal cutting methods in regard to the cutting speed of the tool; the peripheral speed of the grinding wheel is maintained within a narrow range, generally 5500 to 6500 surface feet per minute. Speed ranges different from the common one are used in particular processes which require special wheels and equipment.

Table 2. Basic Process Data for Peripheral Surface Grinding on Reciprocating Table Surface Grinders

Work Material	Hardness	Material Condition	Wheel Speed, fpm	Table Speed, fpm	Downfeed, in. per pass Rough	Downfeed, in. per pass Finish	Crossfeed per pass, fraction of wheel width
Plain carbon steel	52 Rc max.	Annealed, Cold drawn	5500 to 6500	50 to 100	0.003	0.0005 max.	¼
	52 to 65 Rc	Carburized and/or quenched and tempered	5500 to 6500	50 to 100	0.003	0.0005 max.	$^1/_{10}$
Alloy steels	52 Rc max.	Annealed or quenched and tempered	5500 to 6500	50 to 100	0.003	0.001 max.	¼
	52 to 65 Rc	Carburized and/or quenched and tempered	5500 to 6500	50 to 100	0.003	0.0005 max.	$^1/_{10}$
Tool steels	150 to 275 Bhn	Annealed	5500 to 6500	50 to 100	0.002	0.0005 max.	⅕
	56 to 65 Rc	Quenched and tempered	5500 to 6500	50 to 100	0.002	0.0005 max.	$^1/_{10}$
Nitriding steels	200 to 350 Bhn	Normalized, annealed	5500 to 6500	50 to 100	0.003	0.001 max.	¼
	60 to 65 Rc	Nitrided	5500 to 6500	50 to 100	0.003	0.0005 max.	$^1/_{10}$
Cast steels	52 Rc max.	Normalized, annealed	5500 to 6500	50 to 100	0.003	0.001 max.	¼
	Over 52 Rc	Carburized and/or quenched and tempered	5500 to 6500	50 to 100	0.003	0.0005 max.	$^1/_{10}$
Gray irons	52 Rc max.	As cast, annealed, and/or quenched and tempered	5000 to 6500	50 to 100	0.003	0.001 max.	⅓
Ductile irons	52 Rc max.	As cast, annealed or quenched and tempered	5500 to 6500	50 to 100	0.003	0.001 max.	⅕
Stainless steels, martensitic	135 to 235 Bhn	Annealed or cold drawn	5500 to 6500	50 to 100	0.002	0.0005 max.	¼
	Over 275 Bhn	Quenched and tempered	5500 to 6500	50 to 100	0.001	0.0005 max.	⅛
Aluminum alloys	30 to 150 Bhn	As cast, cold drawn or treated	5500 to 6500	50 to 100	0.003	0.001 max.	⅓

In establishing the proper process values for grinding, of prime consideration are the work material, its condition, and the type of operation (roughing or finishing). Table 2 gives basic process data for peripheral surface grinding on reciprocating table surface grinders. For different work materials and hardness ranges data are given regarding table speeds, downfeed (infeed) rates and cross feed, the latter as a function of the wheel width.

Common Faults and Possible Causes in Surface Grinding.—Approaching the ideal performance with regard to both the quality of the ground surface and the efficiency of surface grinding, requires the monitoring of the process and the correction of conditions adverse to the attainment of that goal.

Table 3. Common Faults and Possible Causes in Surface Grinding

CAUSES	FAULTS	WORK DIMENSION			METALLURGICAL DEFECTS		SURFACE QUALITY				WHEEL CONDITION			WORK RETAINMENT	
		Work not flat	Work not parallel	Poor size holding	Burnishing of work	Burning or checking	Feed lines	Chatter marks	Scratches on surface	Poor finish	Wheel loading	Wheel glazing	Rapid wheel wear	Not firmly seated	Work sliding on chuck
WORK CONDITION	Heat treat stresses	X	..	..	..	..	..	..	..	..	..	..	..	..	..
	Work too thin	X	X	..	..	..	..	..	..	..	..	..	..	..	..
	Work warped	X	..	..	..	..	..	..	..	..	..	..	..	X	..
	Abrupt section changes	X	X	..	..	..	..	..	..	..	..	..	..	..	..
GRINDING WHEEL	Grit too fine	..	..	..	X	X	..	..	..	..	X	X	..	..	..
	Grit too coarse	..	..	..	..	..	..	..	..	X	..	..	..	..	..
	Grade too hard	X	..	..	X	X	..	X	..	..	X	X	..	..	..
	Grade too soft	..	..	X	..	..	..	X	X	..	..	..	X	..	..
	Wheel not balanced	..	..	..	..	..	..	X	..	..	..	..	..	..	..
	Dense structure	..	..	..	..	..	..	..	..	..	X	X	..	..	..
TOOLING AND COOLANT	Improper coolant	..	..	..	..	..	..	..	..	..	X	..	..	..	..
	Insufficient coolant	X	X	..	X	X	..	..	..	..	..	X	..	..	..
	Dirty coolant	..	..	..	..	..	..	..	X	..	X	..	..	..	..
	Diamond loose or chipped	X	X	..	..	..	..	..	X	..	..	..	..	..	..
	Diamond dull	..	..	X	..	..	..	..	..	X	X	X	..	..	..
	No or poor magnetic force	..	..	X	..	..	..	..	X	..	..	..	..	X	X
	Chuck surface worn or burred	X	X	..	..	..	..	..	X	..	..	..	..	X	..
MACHINE AND SETUP	Chuck not aligned	X	X	..	..	..	..	..	..	..	..	..	..	..	..
	Vibrations in machine	..	..	..	..	..	..	X	..	..	..	..	..	..	..
	Plane of movement out of parallel	X	X	..	..	..	..	..	..	..	..	..	..	..	..
OPERATIONAL CONDITIONS	Too low work speed	..	..	..	..	..	..	..	..	..	X	..	..	..	..
	Too light feed	..	..	..	..	..	..	..	..	..	..	X	..	..	..
	Too heavy cut	X	..	..	X	..	..	..	..	X	..	..	..	..	..
	Chuck retained swarf	X	X	..	..	..	..	..	..	..	..	..	..	..	..
	Chuck loading improper	X	X	..	..	..	..	..	..	..	..	..	..	..	X
	Insufficient blocking of parts	..	..	..	..	..	..	..	X	..	..	..	..	..	X
	Wheel runs off the work	..	X	X	..	X	..	..	..	..	..	..	..	..	..
	Wheel dressing too fine	X	..	..	..	..	..	..	..	..	..	..	X	..	..
	Wheel edge not chamfered	..	..	..	..	..	X	..	..	..	..	..	..	..	..
	Loose dirt under guard	..	..	..	..	..	..	..	X	..	..	..	..	..	..

Defective, or just not entirely satisfactory surface grinding may have any one or more of several causes. Exploring and determining the cause for eliminating its harmful effects is facilitated by knowing the possible sources of the experienced undesirable performance. Table 3, associating the common faults with their possible causes, is intended to aid in determining the actual cause, the correction of which should restore the desired performance level.

While the table lists the more common faults in surface grinding, and points out their frequent causes, other types of improper performance and/or other causes, in addition to those indicated, are not excluded.

Offhand Grinding

Offhand grinding consists of holding the wheel to the work or the work to the wheel and grinding to broad tolerances and includes such operations as certain types of tool sharpening, weld grinding, snagging castings and other rough grinding. Types of machines that are used for rough grinding in foundries are floor- and bench-stand machines. Wheels for these machines vary from 6 to 30 inches in diameter. Portable grinding machines (electric, flexible shaft, or air-driven) are used for cleaning and smoothing castings.

Many rough grinding operations on castings can be best done with shaped wheels, such as cup wheels (including plate mounted) or cone wheels, and it is advisable to have a good assortment of such wheels on hand to do the odd jobs the best way.

Floor- and Bench-Stand Grinding.—The most common method of rough grinding is on double-end floor and bench stands. In machine shops, welding shops, and automotive repair shops, these grinders are usually provided with a fairly coarse grit wheel on one end for miscellaneous rough grinding and a finer grit wheel on the other end for sharpening tools. The pressure exerted is a very important factor in selecting the proper grinding wheel. If grinding is to be done mostly on hard sharp fins, then durable, coarse and hard wheels are required, but if grinding is mostly on large gate and riser pads, then finer and softer wheels should be used for best cutting action.

Portable Grinding.—Portable grinding machines are usually classified as air grinders, flexible shaft grinders, and electric grinders. The electric grinders are of two types; namely, those driven by standard 60 cycle current and so-called high-cycle grinders. Portable grinders are used for grinding down and smoothing weld seams; cleaning metal before welding; grinding out imperfections, fins and parting lines in castings and smoothing castings; grinding punch press dies and patterns to proper size and shape; and grinding manganese steel castings.

Wheels used on portable grinders are of three bond types; namely, resinoid, rubber, and vitrified. By far the largest percentage is resinoid. Rubber bond is used for relatively thin wheels and where a good finish is required. Some of the smaller wheels such as cone and plug wheels are vitrified bonded.

Grit sizes most generally used in wheels from 4 to 8 inches in diameter are 16, 20, and 24. In the still smaller diameters, finer sizes are used, such as 30, 36, and 46.

The particular grit size to use depends chiefly on the kind of grinding to be done. If the work consists of sharp fins and the machine has ample power, a coarse grain size combined with a fairly hard grade should be used. If the job is more in the nature of smoothing or surfacing and a fairly good finish is required, then finer and softer wheels are called for.

Swing-Frame Grinding.—This type of grinding is employed where a considerable amount of material is to be removed as on snagging large castings. It may be possible to remove 10 times as much material from steel castings using swing-frame grinders as with portable grinders; and 3 times as much material as with high-speed floor-stand grinders.

The largest field of application for swing-frame machines is on castings which are too heavy to handle on a floor stand; but often it is found that comparatively large gates and

risers on smaller castings can be ground more quickly with swing-frame grinders, even if fins and parting lines have to be ground on floor stands as a second operation.

In foundries, the swing-frame machines are usually suspended from a trolley on a jib that can be swung out of the way when placing the work on the floor with the help of an overhead crane. In steel mills when grinding billets, a number of swing-frame machines are usually suspended from trolleys on a line of beams which facilitate their use as required.

The grinding wheels used on swing-frame machines are made with coarser grit sizes and harder grades than wheels used on floor stands for the same work. The reason is that greater grinding pressures can be obtained on the swing-frame machines.

Mounted Wheels and Mounted Points.—These wheels and points are used in hard-to-get-at places and are available with a vitrified bond. The wheels are available with aluminum oxide or silicon carbide abrasive grains. The aluminum oxide wheels are used to grind tough and tempered die steels and the silicon carbide wheels, cast iron, chilled iron, bronze, and other non-ferrous metals.

The illustrations on pages 1205 and 1206 give the standard shapes of mounted wheels and points as published by the Grinding Wheel Institute. A note about the maximum operating speed for these wheels is given at the bottom of the first page of illustrations. Metric sizes are given on page 1204.

Abrasive Belt Grinding

Abrasive belts are used in the metalworking industry for removing stock, light cleaning up of metal surfaces, grinding welds, deburring, breaking and polishing hole edges, and finish grinding of sheet steel. The types of belts that are used may be coated with aluminum oxide (the most common coating) for stock removal and finishing of all alloy steels, high-carbon steel, and tough bronzes; and silicon carbide for use on hard, brittle, and low-tensile strength metals which would include aluminum and cast irons.

Table 1 is a guide to the selection of the proper abrasive belt, lubricant, and contact wheel. This table is entered on the basis of the material used and type of operation to be done and gives the abrasive belt specifications (type of bonding andabrasive grain size and material), the range of speeds at which the belt may best be operated, the type of lubricant to use, and the type and hardness of the contact wheel to use. Table serves as a guide in the selection of contact wheels. This table is entered on the basis of the type of contact wheel surface and the contact wheel material. The table gives the hardness and/or density, the type of abrasive belt grinding for which the contact wheel is intended, the character of the wheel action and such comments as the uses, and hints for best use. Both tables are intended only as guides for general shop practice; selections may be altered to suit individual requirements.

There are three types of abrasive belt grinding machines. One type employs a contact wheel behind the belt at the point of contact of the workpiece to the belt and facilitates a high rate of stock removal. Another type uses an accurate parallel ground platen over which the abrasive belt passes and facilitates the finishing of precision parts. A third type which has no platens or contact wheel is used for finishing parts having uneven surfaces or contours. In this type there is no support behind the belt at the point of contact of the belt with the workpiece. Some machines are so constructed that besides grinding against a platen or a contact wheel the workpiece may be moved and ground against an unsupported portion of the belt, thereby in effect making it a dual machine.

Although abrasive belts at the time of their introduction were used dry, since the advent of the improved waterproof abrasive belts, they have been used with coolants, oil-mists, and greases to aid the cutting action. The application of a coolant to the area of contact retards loading, resulting in a cool, free cutting action, a good finish and a long belt life.

Table 1. Guide to the Selection and Application of Abrasive Belts

Material	Type of Operation	Abrasive Belt[a]	Grit	Belt Speed, fpm	Type of Grease Lubricant	Contact Wheel Type	Contact Wheel Durometer Hardness
Hot-and Cold-Rolled Steel	Roughing	R/R Al_2O_3	24–60	4000–65000	Light-body or none	Cog-tooth, serrated rubber	70–90
	Polishing	R/G or R/R Al_2O_3	80–150	4500–7000	Light-body or none	Plain or serrated rubber, sectional or finger-type cloth wheel, free belt	20–60
	Fine Polishing	R/G or electro-coated Al_2O_3 cloth	180–500	4500–7000	Heavy or with abrasive compound	Smooth-faced rubber or cloth	20–40
Stainless Steel	Roughing	R/R Al_2O_3	50–80	3500–5000	Light-body or none	Cog-tooth, serrated rubber	70–90
	Polishing	R/G or R/R Al_2O_3	80–120	4000–5500	Light-body or none	Plain or serrated rubber, sectional or finger-type cloth wheel, free belt	30–60
	Fine Pol.	Closed-coat SiC	150–280	4500–5500	Heavy or oil mist	Smooth-faced rubber or cloth	20–40
Aluminum, Cast or Fabricated	Roughing	R/R SiC or Al_2O_3	24–80	5000–6500	Light	Cog-tooth, serrated rubber	70–90
	Polishing	R/G SiC or Al_2O_3	100–180	4500–6500	Light	Plain or serrated rubber, sectional or finger-type cloth wheel, free belt	30–50
	Fine Polishing	Closed-coat SiC or electro-coated Al_2O_3	220–320	4500–6500	Heavy or with abrasive compound	Plain faced rubber, finger-type cloth or free belt	20–50
Copper Alloys or Brass	Roughing	R/R SiC or Al_2O_3	36–80	2200–4500	Light-body	Cog-tooth, serrated rubber	70–90
	Polishing	Closed-coat SiC or electro-coated Al_2O_3 or R/G SiC or Al_2O_3	100–150	4000–6500	Light-body	Plain or serrated rubber, sectional or finger-type cloth wheel, free belt	30–50
	Fine Polishing	Closed-coat SiC or electro-coated Al_2O_3	180–320	4000–6500	Light or with abrasive compound	Same as for polishing	20–30
Non-ferrous Die-castings	Roughing	R/R SiC or Al_2O_3	24–80	4500–6500	Light-body	Hard wheel depending on application	50–70
	Polishing	R/G SiC or Al_2O_3	100–180	4500–6500	Light-body	Plain rubber, cloth or free belt	30–50
	Fine Polishing	Electro-coated Al_2O_3 or closed-coat SiC	220–320	4500–6500	Heavy or with abrasive compound	Plain or finger-type cloth wheel, or free belt	20–30
Cast Iron	Roughing	R/R Al_2O_3	24–60	2000–4000	None	Cog-tooth, serrated rubber	70–90
	Polishing	R/R Al_2O_3	80–150	4000–5500	None	Serrated rubber	30–70
	Fine Polishing	R/R Al_2O_3	120–240	4000–5500	Light-body	Smooth-faced rubber	30–40
Titanium	Roughing	R/R SiC or Al_2O_3	36–50	700–1500	Sulfur-chlorinated	Small-diameter, cog-tooth serrated rubber	70–80
	Polishing	R/R SiC	60–120	1200–2000	Light-body	Standard serrated rubber	50
	Fine Pol.	R/R SiC	120–240	1200–2000	Light-body	Smooth-faced rubber or cloth	20–40

[a] R/R indicates that both the making and sizing bond coats are resin. R/G indicates that the making coat is glue and the sizing coat is resin. The abbreviations Al_2O_3 for aluminum oxide and SiC for silicon carbide are used. Almost all R/R and R/G Al_2O_3 and SiC belts have a heavy-drill weight cloth backing. Most electro-coated Al_2O_3 and closed-coat SiC belts have a jeans weight cloth backing.

Table 2. Guide to the Selection and Application of Contact Wheels

Surface	Material	Hardness and Density	Purposes	Wheel Action	Comments
Cog-tooth	Rubber	70 to 90 durometer	Roughing	Fast cutting, allows long belt life.	For cutting down projections on castings and weld beads.
Standard serrated	Rubber	40 to 50 durometer, medium density	Roughing	Leaves rough- to medium-ground surface.	For smoothing projections and face defects.
X-shaped serrations	Rubber	20 to 50 durometer	Roughing and polishing	Flexibility of rubber allows entry into contours. Medium polishing, light removal.	Same as for standard serrated wheels but preferred for soft non-ferrous metals.
Plain face	Rubber	20 to 70 durometer	Roughing and polishing	Plain wheel face allows controlled penetration of abrasive grain. Softer wheels give better finishes.	For large or small flat faces.
Flat flexible	Compressed canvas	About nine densities from very hard to very soft	Roughing and polishing	Hard wheels can remove metal, but not as quickly as cog-tooth rubber wheels. Softer wheels polish well.	Good for medium-range grinding and polishing.
Flat flexible	Solid sectional canvas	Soft, medium, and hard	Polishing	Uniform polishing. Avoids abrasive pattern on work. Adjusts to contours. Can be performed for contours.	A low-cost wheel with uniform density at the face. Handles all types of polishing.
Flat flexible	Buff section canvas	Soft	Contour polishing	For fine polishing and finishing.	Can be widened or narrowed by adding or removing sections. Low cost.
Flat flexible	Sponge rubber inserts	5 to 10 durometer, soft	Polishing	Uniform polishing and finishing. Polishes and blends contours.	Has replaceable segments. Polishes and blends contours. Segments allow density changes.
Flexible	Fingers of canvas attached to hub	Soft	Polishing	Uniform polishing and finishing.	For polishing and finishing.
Flat flexible	Rubber segments	Varies in hardness	Roughing and polishing	Grinds or polishes depending on density and hardness of inserts.	For portable machines. Uses replaceable segments that save on wheel costs and allow density changes.
Flat flexible	Inflated rubber	Air pressure controls hardness	Roughing and polishing	Uniform finishing.	Adjusts to contours.

Abrasive Cutting

Abrasive cut-off wheels are used for cutting steel, brass and aluminum bars and tubes of all shapes and hardnesses, ceramics, plastics, insulating materials, glass and cemented carbides. Originally a tool or stock room procedure, this method has developed into a high-speed production operation. While the abrasive cut-off machine and cut-off wheel can be said to have revolutionized the practice of cutting-off materials, the metal saw continues to be the more economical method for cutting-off large cross-sections of certain materials. However, there are innumerable materials and shapes that can be cut with much greater speed and economy by the abrasive wheel method. On conventional chop-stroke abrasive cutting machines using 16-inch diameter wheels, 2-inch diameter bar stock is the maximum size that can be cut with satisfactory wheel efficiency, but bar stock up to 6 inches in diameter can be cut efficiently on oscillating-stroke machines. Tubing up to 3½ inches in diameter can also be cut efficiently.

Abrasive wheels are commonly available in four types of bonds: Resinoid, rubber, shellac and fiber or fabric reinforced. In general, resinoid bonded cut-off wheels are used for dry cutting where burrs and some burn are not objectionable and rubber bonded wheels are used for wet cutting where cuts are to be smooth, clean and free from burrs. Shellac bonded wheels have a soft, free cutting quality which makes them particularly useful in the tool room where tool steels are to be cut without discoloration. Fiber reinforced bonded wheels are able to withstand severe flexing and side pressures and fabric reinforced bonded wheels which are highly resistant to breakage caused by extreme side pressures, are fast cutting and have a low rate of wear.

The types of abrasives available in cut-off wheels are: Aluminum oxide, for cutting steel and most other metals; silicon carbide, for cutting non-metallic materials such as carbon, tile, slate, ceramics, etc.; and diamond, for cutting cemented carbides. The method of denoting abrasive type, grain size, grade, structure and bond type by using a system of markings is the same as for grinding wheels (see page 1141). Maximum wheel speeds given in the American National Standard Safety Requirements for The Use, Care, and Protection of Abrasive Wheels (ANSI B7.1-1988) range from 9500 to 14,200 surface feet per minute for organic bonded cut-off wheels larger than 16 inches in diameter and from 9500 to 16,000 surface feet per minute for organic bonded cut-off wheels 16 inches in diameter and smaller. Maximum wheel speeds specified by the manufacturer should never be exceeded even though they may be lower than those given in the B7.1.

There are four basic types of abrasive cutting machines: Chop-stroke, oscillating stroke, horizontal stroke and work rotating. Each of these four types may be designed for dry cutting or for wet cutting (includes submerged cutting).

The accompanying table based upon information made available by The Carborundum Co. gives some of the probable causes of cutting off difficulties that might be experienced when using abrasive cut-off wheels.

Probable Causes of Cutting-Off Difficulties

Difficulty	Probable Cause
Angular Cuts and Wheel Breakage	(1) Inadequate clamping which allows movement of work while the wheel is in the cut. The work should be clamped on both sides of the cut. (2) Work vise higher on one side than the other causing wheel to be pinched. (3) Wheel vibration resulting from worn spindle bearings. (4) Too fast feeding into the cut when cutting wet.
Burning of Stock	(1) Insufficient power or drive allowing wheel to stall. (2) Cuts too heavy for grade of wheel being used. (3) Wheel fed through the work too slowly. This causes a heating up of the material being cut. This difficulty encountered chiefly in dry cutting.
Excessive Wheel Wear	(1) Too rapid cutting when cutting wet. (2) Grade of wheel too hard for work, resulting in excessive heating and burning out of bond. (3) Inadequate coolant supply in wet cutting. (4) Grade of wheel too soft for work. (5) Worn spindle bearings allowing wheel vibration.
Excessive Burring	(1) Feeding too slowly when cutting dry. (2) Grit size in wheel too coarse. (3) Grade of wheel too hard. (4) Wheel too thick for job.

Honing Process

The hone-abrading process for obtaining cylindrical forms with precise dimensions and surfaces can be applied to internal cylindrical surfaces with a wide range of diameters such as engine cylinders, bearing bores, pin holes, etc. and also to some external cylindrical surfaces.

The process is used to: 1) eliminate inaccuracies resulting from previous operations by generating a true cylindrical form with respect to roundness and straightness within minimum dimensional limits; 2) generate final dimensional size accuracy within low tolerances, as may be required for interchangeability of parts; 3) provide rapid and economical stock removal consistent with accomplishment of the other results; and 4) generate surface finishes of a specified degree of surface smoothness with high surface quality.

Amount and Rate of Stock Removal.—Honing may be employed to increase bore diameters by as much as 0.100 inch or as little as 0.001 inch. The amount of stock removed by the honing process is entirely a question of processing economy. If other operations are performed before honing then the bulk of the stock should be taken off by the operation that can do it most economically. In large diameter bores that have been distorted in heat treating, it may be necessary to remove as much as 0.030 to 0.040 inch from the diameter to make the bore round and straight. For out-of-round or tapered bores, a good "rule of thumb" is to leave twice as much stock (on the diameter) for honing as there is error in the bore. Another general rule is: For bores over one inch in diameter, leave 0.001 to 0.0015 inch stock per inch of diameter. For example, 0.002 to 0.003 inch of stock is left in two-inch bores and 0.010 to 0.015 inch in ten-inch bores. Where parts are to be honed for finish only, the amount of metal to be left for removing tool marks may be as little as 0.0002 to 0.015 inch on the diameter.

In general, the honing process can be employed to remove stock from bore diameters at the rate of 0.009 to 0.012 inch per minute on cast-iron parts and from 0.005 to 0.008 inch per minute on steel parts having a hardness of 60 to 65 Rockwell C. These rates are based on parts having a length equal to three or four times the diameter. Stock has been removed from long parts such as gun barrels, at the rate of 65 cubic inches per hour. Recommended honing speeds for cast iron range from 110 to 200 surface feet per minute of rotation and from 50 to 110 lineal feet per minute of reciprocation. For steel, rotating surface speeds range from 50 to 110 feet per minute and reciprocation speeds from 20 to 90 lineal feet per minute. The exact rotation and reciprocation speeds to be used depend upon the size of the work, the amount and characteristics of the material to be removed and the quality of the finish desired. In general, the harder the material to be honed, the lower the speed. Interrupted bores are usually honed at faster speeds than plain bores.

Formula for Rotative Speeds.—Empirical formulas for determining rotative speeds for honing have been developed by the Micromatic Hone Corp. These formulas take into consideration the type of material being honed, its hardness and its surface characteristics; the abrasive area; and the type of surface pattern and degree of surface roughness desired. Because of the wide variations in material characteristics, abrasives available, and types of finishes specified, these formulas should be considered as a guide only in determining which of the available speeds (pulley or gear combinations) should be used for any particular application.

The formula for rotative speed, S, in surface feet per minute is:

$$S = \frac{K \times D}{W \times N}$$

The formula for rotative speed in revolutions per minute is:

$$\text{R.P.M} = \frac{R}{W \times N}$$

where, K and R are factors taken from the table on the following page, D is the diameter of the bore in inches, W is the width of the abrasive stone or stock in inches, and N is the number of stones.

Although the actual speed of the abrasive is the resultant of both the rotative speed and the reciprocation speed, this latter quantity is seldom solved for or used. The reciprocation speed is not determined empirically but by testing under operating conditions. Changing the reciprocation speed affects the dressing action of the abrasive stones, therefore, the reciprocation speed is adjusted to provide for a desired surface finish which is usually a well lubricated bearing surface that will not scuff.

Table of Factors for Use in Rotative Speed Formulas

Character of Surface[a]	Material	Hardness[b] Soft		Medium		Hard	
		Factors					
		K	*R*	*K*	*R*	*K*	*R*
Base Metal	Cast Iron	110	420	80	300	60	230
	Steel	80	300	60	230	50	190
Dressing Surface	Cast Iron	150	570	110	420	80	300
	Steel	110	420	80	300	60	230
Severe Dressing	Cast Iron	200	760	150	570	110	420
	Steel	150	570	110	420	80	300

[a] The character of the surface is classified according to its effect on the abrasive; *Base Metal* being a honed, ground or fine bored section that has little dressing action on the grit; *Dressing Surface* being a rough bored, reamed or broached surface or any surface broken by cross holes or ports; *Severe Dressing* being a surface interrupted by keyways, undercuts or burrs that dress the stones severely. If over half of the stock is to be removed after the surface is cleaned up, the speed should be computed using the *Base Metal* factors for *K* and *R*.

[b] Hardness designations of soft, medium and hard cover the following ranges on the Rockwell "C" hardness scale, respectively: 15 to 45, 45 to 60 and 60 to 70.

Possible Adjustments for Eliminating Undesirable Honing Conditions

Undesirable Condition	Adjustment Required to Correct Condition[a] Abrasive[b] Friability	Grain Size	Hardness	Structure	Other Feed Pressure	Reciprocation	R.P.M.	Runout Time	Stroke Length
Abrasive Glazing	+	– –	– –	+	+ +	+ +	– –	–	0
Abrasive Loading	0	– –	–	–	+ +	+	– –	0	0
Too Rough Surface Finish	0	+ +	+ +	–	–	–	+ +	+	0
Too Smooth Surface Finish	0	– –	– –	+	+	+	– –	–	0
Poor Stone Life	–	+	+ +	–	–	–	+	0	0
Slow Stock Removal	+	– –	–	+	+ +	+ +	– –	0	0
Taper — Large at Ends	0	0	0	0	0	0	0	0	–
Taper — Small at Ends	0	0	0	0	0	0	0	0	+

[a] The + and + + symbols generally indicate that there should be an increase or addition while the – and – – symbols indicate that there should be a reduction or elimination. In each case, the double symbol indicates that the contemplated change would have the greatest effect. The 0 symbol means that a change would have no effect.

[b] For the abrasive adjustments the + and + + symbols indicate a more friable grain, a finer grain, a harder grade or a more open structure and the – and – – symbols just the reverse.

Compiled by Micromatic Hone Corp.

Abrasive Stones for Honing.—Honing stones consist of aluminum oxide, silicon carbide, CBN or diamond abrasive grits, held together in stick form by a vitrified clay, resinoid or metal bond. CBN metal-bond stones are particularly suitable and widely used for honing. The grain and grade of abrasive to be used in any particular honing operation depend upon the quality of finish desired, the amount of stock to be removed, the material being honed and other factors.

The following general rules may be followed in the application of abrasive for honing:

1) Silicon-carbide abrasive is commonly used for honing cast iron, while aluminum-oxide abrasive is generally used on steel; 2) The harder the material being honed, the softer the abrasive stick used; 3) A rapid reciprocating speed will tend to make the abrasive cut fast because the dressing action on the grits will be severe; and 4) To improve the finish, use a finer abrasive grit, incorporate more multi-direction action, allow more "runout" time after honing to size, or increase the speed of rotation.

Surface roughnesses ranging from less than 1 micro-inch r.m.s. to a relatively coarse roughness can be obtained by judicious choice of abrasive and honing time but the most common range is from 3 to 50 micro-inches r.m.s.

Adjustments for Eliminating Undesirable Honing Conditions.—The accompanying table indicates adjustments that may be made to correct certain undesirable conditions encountered in honing. Only one change should be made at a time and its effect noted before making other adjustments.

Tolerances.—For bore diameters above 4 inches the tolerance of honed surfaces with respect to roundness and straightness ranges from 0.0005 to 0.001 inch; for bore diameters from 1 to 4 inches, 0.0003 to 0.0005 inch; and for bore diameters below 1 inch, 0.00005 to 0.0003 inch.

Laps and Lapping

Material for Laps.—Laps are usually made of soft cast iron, copper, brass or lead. In general, the best material for laps to be used on very accurate work is soft, close-grained cast iron. If the grinding, prior to lapping, is of inferior quality, or an excessive allowance has been left for lapping, copper laps may be preferable. They can be charged more easily and cut more rapidly than cast iron, but do not produce as good a finish. Whatever material is used, the lap should be softer than the work, as, otherwise, the latter will become charged with the abrasive and cut the lap, the order of the operation being reversed. A common and inexpensive form of lap for holes is made of lead which is cast around a tapering steel arbor. The arbor usually has a groove or keyway extending lengthwise, into which the lead flows, thus forming a key that prevents the lap from turning. When the lap has worn slightly smaller than the hole and ceases to cut, the lead is expanded or stretched a little by the driving in of the arbor. When this expanding operation has been repeated two or three times, the lap usually must be trued or replaced with a new one, owing to distortion.

The tendency of lead laps to lose their form is an objectionable feature. They are, however, easily molded, inexpensive, and quickly charged with the cutting abrasive. A more elaborate form for holes is composed of a steel arbor and a split cast-iron or copper shell which is sometimes prevented from turning by a small dowel pin. The lap is split so that it can be expanded to accurately fit the hole being operated upon. For hardened work, some toolmakers prefer copper to either cast iron or lead. For holes varying from ¼ to ½ inch in diameter, copper or brass is sometimes used; cast iron is used for holes larger than ½ inch in diameter. The arbors for these laps should have a taper of about ¼ or ⅜ inch per foot. The length of the lap should be somewhat greater than the length of the hole, and the thickness of the shell or lap proper should be from ⅛ to ⅙ its diameter.

External laps are commonly made in the form of a ring, there being an outer ring or holder and an inner shell which forms the lap proper. This inner shell is made of cast iron, copper, brass or lead. Ordinarily the lap is split and screws are provided in the holder for adjustment. The length of an external lap should at least equal the diameter of the work, and might well be longer. Large ring laps usually have a handle for moving them across the work.

Laps for Flat Surfaces.—Laps for producing plane surfaces are made of cast iron. In order to secure accurate results, the lapping surface must be a true plane. A flat lap that is used for roughing or "blocking down" will cut better if the surface is scored by narrow grooves. These are usually located about ½ inch apart and extend both lengthwise and crosswise, thus forming a series of squares similar to those on a checker-board. An abrasive of No. 100 or 120 emery and lard oil can be used for charging the roughing lap. For finer work, a lap having an unscored surface is used, and the lap is charged with a finer abrasive. After a lap is charged, all loose abrasive should be washed off with gasoline, for fine work, and when lapping, the surface should be kept moist, preferably with kerosene. Gasoline will cause the lap to cut a little faster, but it evaporates so rapidly that the lap soon becomes dry and the surface caked and glossy in spots. Loose emery should not be applied while lapping, for if the lap is well charged with abrasive in the beginning, is kept well moistened and not crowded too hard, it will cut for a considerable time. The pressure upon the work should be just enough to insure constant contact. The lap can be made to cut only so fast, and if excessive pressure is applied it will become "stripped" in places. The causes of scratches are: Loose abrasive on the lap; too much pressure on the work, and poorly graded abrasive. To produce a perfectly smooth surface free from scratches, the lap should be charged with a very fine abrasive.

Grading Abrasives for Lapping.—For high-grade lapping, abrasives can be evenly graded as follows: A quantity of flour-emery or other abrasive is placed in a heavy cloth bag, which is gently tapped, causing very fine particles to be sifted through. When a sufficient quantity has been obtained in this way, it is placed in a dish of lard or sperm oil. The largest particles will then sink to the bottom and in about one hour the oil should be poured into another dish, care being taken not to disturb the sediment at the bottom. The oil is then allowed to stand for several hours, after which it is poured again, and so on, until the desired grade is obtained.

Charging Laps.—To charge a flat cast-iron lap, spread a very thin coating of the prepared abrasive over the surface and press the small cutting particles into the lap with a hard steel block. There should be as little rubbing as possible. When the entire surface is apparently charged, clean and examine for bright spots; if any are visible, continue charging until the entire surface has a uniform gray appearance. When the lap is once charged, it should be used without applying more abrasive until it ceases to cut. If a lap is over-charged and an excessive amount of abrasive is used, there is a rolling action between the work and lap which results in inaccuracy. The surface of a flat lap is usually finished true, prior to charging, by scraping and testing with a standard surface-plate, or by the well-known method of scraping-in three plates together, in order to secure a plane surface. In any case, the bearing marks or spots should be uniform and close together. These spots can be blended by covering the plates evenly with a fine abrasive and rubbing them together. While the plates are being ground in, they should be carefully tested and any high spots which may form should be reduced by rubbing them down with a smaller block.

To charge cylindrical laps for internal work, spread a thin coating of prepared abrasive over the surface of a hard steel block, preferably by rubbing lightly with a cast-iron or copper block; then insert an arbor through the lap and roll the latter over the steel block, pressing it down firmly to embed the abrasive into the surface of the lap. For external cylindrical laps, the inner surface can be charged by rolling-in the abrasive with a hard steel roller that is somewhat smaller in diameter than the lap. The taper cast-iron blocks which are sometimes used for lapping taper holes can also be charged by rolling-in the abrasive, as previously described; there is usually one roughing and one finishing lap, and when charging the former, it may be necessary to vary the charge in accordance with any error which might exist in the taper.

Rotary Diamond Lap.—This style of lap is used for accurately finishing very small holes, which, because of their size, cannot be ground. While the operation is referred to as lapping, it is, in reality, a grinding process, the lap being used the same as a grinding wheel. Laps employed for this work are made of mild steel, soft material being desirable because it can be charged readily. Charging is usually done by rolling the lap between two hardened steel plates. The diamond dust and a little oil is placed on the lower plate, and as the lap revolves, the diamond is forced into its surface. After charging, the lap should be washed in benzine. The rolling plates should also be cleaned before charging with dust of a finer grade. It is very important not to force the lap when in use, especially if it is a small size. The lap should just make contact with the high spots and gradually grind them off. If a diamond lap is lubricated with kerosene, it will cut freer and faster. These small laps are run at very high speeds, the rate depending upon the lap diameter. Soft work should never be ground with diamond dust because the dust will leave the lap and charge the work.

When using a diamond lap, it should be remembered that such a lap will not produce sparks like a regular grinding wheel; hence, it is easy to crowd the lap and "strip" some of the diamond dust. To prevent this, a sound intensifier or "harker" should be used. This is placed against some stationary part of the grinder spindle, and indicates when the lap touches the work, the sound produced by the slightest contact being intensified.

Grading Diamond Dust.—The grades of diamond dust used for charging laps are designated by numbers, the fineness of the dust increasing as the numbers increase. The diamond, after being crushed to powder in a mortar, is thoroughly mixed with high-grade olive oil. This mixture is allowed to stand five minutes and then the oil is poured into another receptacle. The coarse sediment which is left is removed and labeled No. 0, according to one system. The oil poured from No. 0 is again stirred and allowed to stand ten minutes, after which it is poured into another receptacle and the sediment remaining is labeled No. 1. This operation is repeated until practically all of the dust has been recovered from the oil, the time that the oil is allowed to stand being increased as shown by the following table. This is done in order to obtain the smaller particles that require a longer time for precipitation:

To obtain No. 1 — 10 minutes	To obtain No. 4 — 2 hours
To obtain No. 2 — 30 minutes	To obtain No. 5 — 10 hours
To obtain No. 3 — 1 hour	To obtain No. 6 — until oil is clear

The No. 0 or coarse diamond which is obtained from the first settling is usually washed in benzine, and re-crushed unless very coarse dust is required. This No. 0 grade is sometimes known as "ungraded" dust. In some places the time for settling, in order to obtain the various numbers, is greater than that given in the table.

Cutting Properties of Laps and Abrasives.—In order to determine the cutting properties of abrasives when used with different lapping materials and lubricants, a series of tests was conducted, the results of which were given in a paper by W. A. Knight and A. A. Case, presented before the American Society of Mechanical Engineers. In connection with these tests, a special machine was used, the construction being such that quantitative results could be obtained with various combinations of abrasive, lubricant, and lap material. These tests were confined to surface lapping.

It was not the intention to test a large variety of abrasives, three being selected as representative; namely, Naxos emery, carborundum, and alundum. Abrasive No. 150 was used in each case, and seven different lubricants, five different pressures, and three different lap materials were employed. The lubricants were lard oil, machine oil, kerosene, gasoline, turpentine, alcohol, and soda water.

These tests indicated throughout that there is, for each different combination of lap and lubricant, a definite size of grain that will give the maximum amount of cutting. With all the tests, except when using the two heavier lubricants, some reduction in the size of the grain below that used in the tests (No. 150) seemed necessary before the maximum rate of cutting was reached. This reduction, however, was continuous and soon passed below that which gave the maximum cutting rate.

Cutting Qualities with Different Laps.—The surfaces of the steel and cast-iron laps were finished by grinding. The hardness of the different laps, as determined by the scleroscope was, for cast-iron, 28; steel, 18; copper, 5. The total amount ground from the test-pieces with each of the three laps showed that, taking the whole number of tests as a standard, there is scarcely any difference between the steel and cast iron, but that copper has somewhat better cutting qualities, although, when comparing the laps on the basis of the highest and lowest values obtained with each lap, steel and cast iron are as good for all practical purposes as copper, when the proper abrasive and lubricant are used.

Wear of Laps.—The wear of laps depends upon the material from which they are made and the abrasive used. The wear on all laps was about twice as fast with carborundum as with emery, while with alundum the wear was about one and one-fourth times that with emery. On an average, the wear of the copper lap was about three times that of the cast-iron lap. This is not absolute wear, but wear in proportion to the amount ground from the test-pieces.

Lapping Abrasives.—As to the qualities of the three abrasives tested, it was found that carborundum usually began at a lower rate than the other abrasives, but, when once started, its rate was better maintained. The performance gave a curve that was more nearly a straight line. The charge or residue as the grinding proceeded remained cleaner and sharper and did not tend to become pasty or mucklike, as is so frequently the case with emery. When using a copper lap, carborundum shows but little gain over the cast-iron and steel laps, whereas, with emery and alundum, the gain is considerable.

Effect of Different Lapping Lubricants.—The action of the different lubricants, when tested, was found to depend upon the kind of abrasive and the lap material.

Lard and Machine Oil The test showed that lard oil, without exception, gave the higher rate of cutting, and that, in general, the initial rate of cutting is higher with the lighter lubricants, but falls off more rapidly as the test continues. The lowest results were obtained with machine oil, when using an emery-charged, cast-iron lap. When using lard oil and a carborundum-charged steel lap, the highest results were obtained.

Gasoline and Kerosene On the cast-iron lap, gasoline was superior to any of the lubricants tested. Considering all three abrasives, the relative value of gasoline, when applied to the different laps, is as follows: Cast iron, 127; copper, 115; steel, 106. Kerosene, like gasoline, gives the best results on cast iron and the poorest on steel. The values obtained by carborundum were invariably higher than those obtained with emery, except when using gasoline and kerosene on a copper lap.

Turpentine and Alcohol Turpentine was found to do good work with carborundum on any lap. With emery, turpentine did fair work on the copper lap, but, with the emery on cast-iron and steel laps, it was distinctly inferior. Alcohol gives the lowest results with emery on the cast-iron and steel laps.

Soda Water Soda water gives medium results with almost any combination of lap and abrasives, the best work being on the copper lap and the poorest on the steel lap. On the cast-iron lap, soda water is better than machine or lard oil, but not so good as gasoline or kerosene. Soda water when used with alundum on the copper lap, gave the highest results of any of the lubricants used with that particular combination.

Lapping Pressures.—Within the limits of the pressures used, that is, up to 25 pounds per square inch, the rate of cutting was found to be practically proportional to the pressure. The higher pressures of 20 and 25 pounds per square inch are not so effective on the copper lap as on the other materials.

Wet and Dry Lapping.—With the "wet method" of using a surface lap, there is a surplus of oil and abrasive on the surface of the lap. As the specimen being lapped is moved over it, there is more or less movement or shifting of the abrasive particles. With the "dry method," the lap is first charged by rubbing or rolling the abrasive into its surface. All surplus oil and abrasive are then washed off, leaving a clean surface, but one that has embedded uniformly over it small particles of the abrasive. It is then like the surface of a very fine oilstone and will cut away hardened steel that is rubbed over it. While this has been termed the dry method, in practice, the lap surface is kept moistened with kerosene or gasoline.

Experiments on dry lapping were carried out on the cast-iron, steel, and copper laps used in the previous tests, and also on one of tin made expressly for the purpose. Carborundum alone was used as the abrasive and a uniform pressure of 15 pounds per square inch was applied to the specimen throughout the tests. In dry lapping, much depends upon the manner of charging the lap. The rate of cutting decreased much more rapidly after the first 100 revolutions than with the wet method. Considering the amounts ground off during the first 100 revolutions, and the best result obtained with each lap taken as the basis of comparison, it was found that with a tin lap, charged by rolling No. 150 carborundum into the surface, the rate of cutting, when dry, approached that obtained with the wet method. With the other lap materials, the rate with the dry method was about one-half that of the wet method.

Summary of Lapping Tests.—The initial rate of cutting does not greatly differ for different abrasives. There is no advantage in using an abrasive coarser than No. 150. The rate of cutting is practically proportional to the pressure. The wear of the laps is in the following proportions: cast iron, 1.00; steel, 1.27; copper, 2.62. In general, copper and steel cut faster than cast iron, but, where permanence of form is a consideration, cast iron is the superior metal. Gasoline and kerosene are the best lubricants to use with a cast-iron lap. Machine and lard oil are the best lubricants to use with copper or steel laps. They are, however, least effective on a cast-iron lap. In general, wet lapping is from 1.2 to 6 times as fast as dry lapping, depending upon the material of the lap and the manner of charging.

Portable Grinding Tools

Circular Saw Arbors.—ANSI Standard B107.4-1982 "Driving and Spindle Ends for Portable Hand, Air, and Air Electric Tools" calls for a round arbor of ⅝-inch diameter for nominal saw blade diameters of 6 to 8.5 inches, inclusive, and a ¾-inch diameter round arbor for saw blade diameters of 9 to 12 inches, inclusive.

Spindles for Geared Chucks.—Recommended threaded and tapered spindles for portable tool geared chucks of various sizes are as given in the following table:

Recommended Spindle Sizes

Chuck Sizes, Inch	Recommended Spindles	
	Threaded	Taper[a]
3⁄16 and ¼ Light	⅜–24	1
¼ and 5⁄16 Medium	⅜–24 or ½–20	2 Short
⅜ Light	⅜–24 or ½–20	2
⅜ Medium	½–20 or ⅝–16	2
½ Light	½–20 or ⅝–16	33
½ Medium	⅝–16 or ¾–16	6
⅝ and ¾ Medium	⅝–16 or ¾–16	3

[a] Jacobs number.

Vertical and Angle Portable Tool Grinder Spindles.—The ⅝–11 spindle with a length of 1⅛ inches shown on page 1209 is designed to permit the use of a jam nut with threaded cup wheels. When a revolving guard is used, the length of the spindle is measured from the wheel bearing surface of the guard. For unthreaded wheels with a ⅞-inch hole, a safety sleeve nut is recommended. The unthreaded wheel with ⅝-inch hole is not recommended because a jam nut alone may not resist the inertia effect when motor power is cut off.

Standard Shapes and Metric Sizes of Mounted Wheels and Points

ANSI B74.2-1982

Abrasive Shape No.[a]	Abrasive Shape Size		Abrasive Shape No.[a]	Abrasive Shape Size	
	Diameter	Thickness		Diameter	Thickness
A 1	20	65	A 24	6	20
A 3	22	70	A 25	25	…
A 4	30	30	A 26	16	…
A 5	20	28	A 31	35	26
A 11	21	45	A 32	25	20
A 12	18	30	A 34	38	10
A 13	25	25	A 35	25	10
A 14	18	22	A 36	40	10
A 15	6	25	A 37	30	6
A 21	25	25	A 38	25	25
A 23	20	25	A 39	20	20
B 41	16	16	B 97	3	10
B 42	13	20	B 101	16	18
B 43	6	8	B 103	16	5
B 44	5.6	10	B 104	8	10
B 51	11	20	B 111	11	18
B 52	10	20	B 112	10	13
B 53	8	16	B 121	13	…
B 61	20	8	B 122	10	…
B 62	13	10	B 123	5	…
B 71	16	3	B 124	3	…
B 81	20	5	B 131	13	13
B 91	13	16	B 132	10	13
B 92	6	6	B 133	10	10
B 96	3	6	B 135	6	13
W 144	3	6	W 196	16	26
W 145	3	10	W 197	16	50
W 146	3	13	W 200	20	3
W 152	5	6	W 201	20	6
W 153	5	10	W 202	20	10
W 154	5	13	W 203	20	13
W 158	6	3	W 204	20	20
W 160	6	6	W 205	20	25
W 162	6	10	W 207	20	40
W 163	6	13	W 208	20	50
W 164	6	20	W 215	25	3
W 174	10	6	W 216	25	6
W 175	10	10	W 217	25	10
W 176	10	13	W 218	25	13
W 177	10	20	W 220	25	25
W 178	10	25	W 221	25	40
W 179	10	30	W 222	25	50
W 181	13	1.5	W 225	30	6
W 182	13	3	W 226	30	10
W 183	13	6	W 228	30	20
W 184	13	10	W 230	30	30
W 185	13	13	W 232	30	50
W 186	13	20	W 235	40	6
W 187	13	25	W 236	40	13
W 188	13	40	W 237	40	25
W 189	13	50	W 238	40	40
W 195	16	20	W 242	50	25

[a] See shape diagrams on pages 1205 and 1206.

All dimensions are in millimeters.

Standard Shapes and Inch Sizes of Mounted Wheels and Points

ANSI B74.2-1982 — **1**

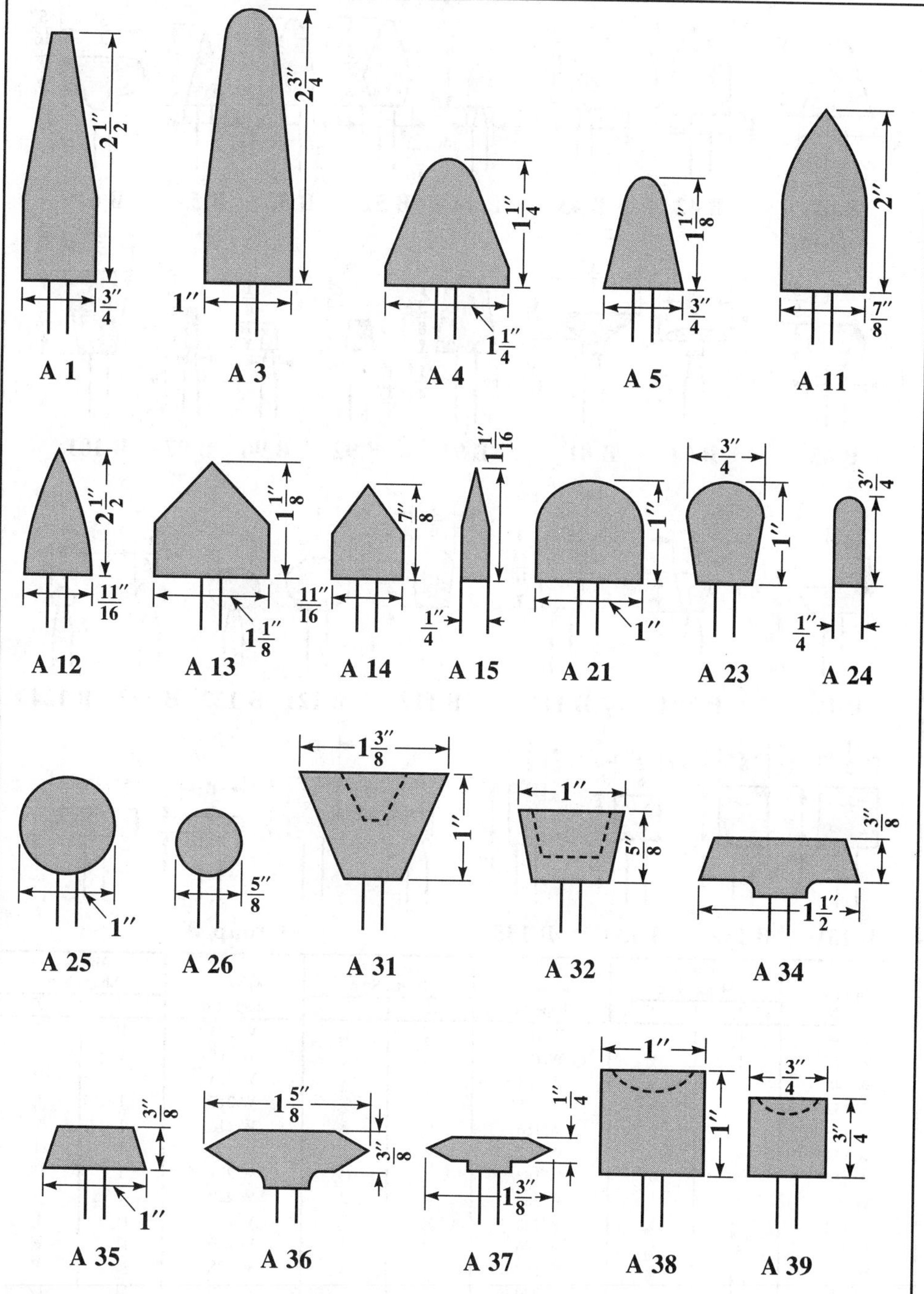

The maximum speeds of mounted vitrified wheels and points of average grade range from about 38,000 to 152,000 rpm for diameters of 1 inch down to ¼ inch. However, the safe operating speed usually is limited by the critical speed (speed at which vibration or whip tends to become excessive) which varies according to wheel or point dimensions, spindle diameter, and overhang.

Standard Shapes and Inch Sizes of Mounted Wheels and Points *ANSI B74.2-1982* — 2

B 41 B 42 B 43 B 44 B 51 B 52 B 53 B 61

B 62 B 71 B 81 B 91 B 92 B 96 B 97 B 101

B 103 B 104 B 111 B 112 B 121 B 122 B 123 B 124

B 131 B 132 B 133 B 135 Group W

Abrasive Shape No.	Abrasive Shape Size		Abrasive Shape No.	Abrasive Shape Size		Abrasive Shape No.	Abrasive Shape Size	
	D	*T*		*D*	*T*		*D*	*T*
W 144	⅛	¼	W 182	½	⅛	W 208	¾	2
W 145	⅛	⅜	W 183	½	¼	W 215	1	⅛
W 146	⅛	½	W 184	½	⅜	W 216	1	¼
W 152	3⁄16	¼	W 185	½	½	W 217	1	⅜
W 153	3⁄16	⅜	W 186	½	¾	W 218	1	½
W 154	3⁄16	½	W 187	½	1	W 220	1	1
W 158	¼	⅛	W 188	½	1½	W 221	1	1½
W 160	¼	¼	W 189	½	2	W 222	1	2
W 162	¼	⅜	W 195	⅝	¾	W 225	1¼	¼
W 163	¼	½	W 196	⅝	1	W 226	1¼	⅜
W 164	¼	¾	W 197	⅝	2	W 228	1¼	¾
W 174	⅜	¼	W 200	¾	⅛	W 230	1¼	1¼
W 175	⅜	⅜	W 201	¾	¼	W 232	1¼	2
W 176	⅜	½	W 202	¾	⅜	W 235	1½	¼
W 177	⅜	¾	W 203	¾	½	W 236	1½	½
W 178	⅜	1	W 204	¾	¾	W 237	1½	1
W 179	⅜	1¼	W 205	¾	1	W 238	1½	1½
W 181	½	1⁄16	W 207	¾	1½	W 242	2	1

Straight Grinding Wheel Spindles for Portable Tools.—Portable grinders with pneumatic or induction electric motors should be designed for the use of organic bond wheels rated 9500 feet per minute. Light-duty electric grinders may be designed for vitrified wheels rated 6500 feet per minute. Recommended maximum sizes of wheels of both types are as given in the following table:

Recommended Maximum Grinding Wheel Sizes for Portable Tools

Spindle Size	Maximum Wheel Dimensions			
	9500 fpm		6500 fpm	
	Diameter D	Thickness T	Diameter D	Thickness T
⅜-24 × 1⅛	2½	½	4	½
½-13 × 1¾	4	¾	5	¾
⅝-11 × 2⅛	8	1	8	1
⅝-11 × 3⅛	6	2	…	…
⅝-11 × 3⅛	8	1½	…	…
¾-10 × 3¼	8	2	…	…

Minimum T with the first three spindles is about ⅛ inch to accommodate cutting off wheels. Flanges are assumed to be according to ANSI B7.1 and threads to ANSI B1.1.

American Standard Threaded and Tapered Spindles for Portable Air and Electric Tools *ASA B5.38-1958*

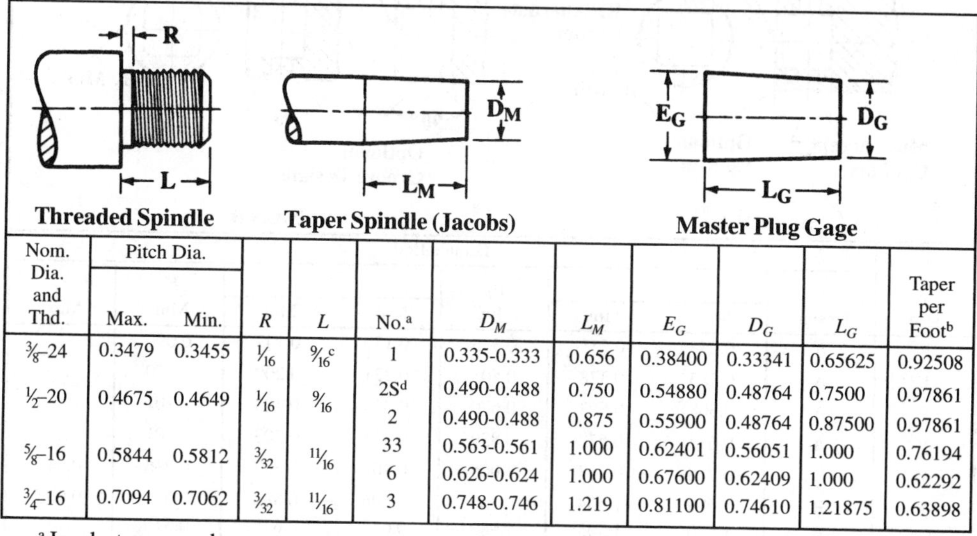

Nom. Dia. and Thd.	Pitch Dia. Max.	Pitch Dia. Min.	R	L	No.[a]	D_M	L_M	E_G	D_G	L_G	Taper per Foot[b]
⅜-24	0.3479	0.3455	1/16	9/16[c]	1	0.335-0.333	0.656	0.38400	0.33341	0.65625	0.92508
½-20	0.4675	0.4649	1/16	9/16	2S[d]	0.490-0.488	0.750	0.54880	0.48764	0.7500	0.97861
					2	0.490-0.488	0.875	0.55900	0.48764	0.87500	0.97861
⅝-16	0.5844	0.5812	3/32	11/16	33	0.563-0.561	1.000	0.62401	0.56051	1.000	0.76194
					6	0.626-0.624	1.000	0.67600	0.62409	1.000	0.62292
¾-16	0.7094	0.7062	3/32	11/16	3	0.748-0.746	1.219	0.81100	0.74610	1.21875	0.63898

[a] Jacobs taper number.

[b] Calculated from E_G, D_G, L_G for the master plug gage.

[c] Also 7/16 inch.

[d] 2S stands for 2 Short.

All dimensions in inches.

Threads are per inch and right-hand.

Tolerances: On R, plus or minus 1/64 inch; on L, plus 0.000, minus 0.030 inch.

American Standard Square Drives for Portable Air and Electric Tools
ASA B5.38-1958

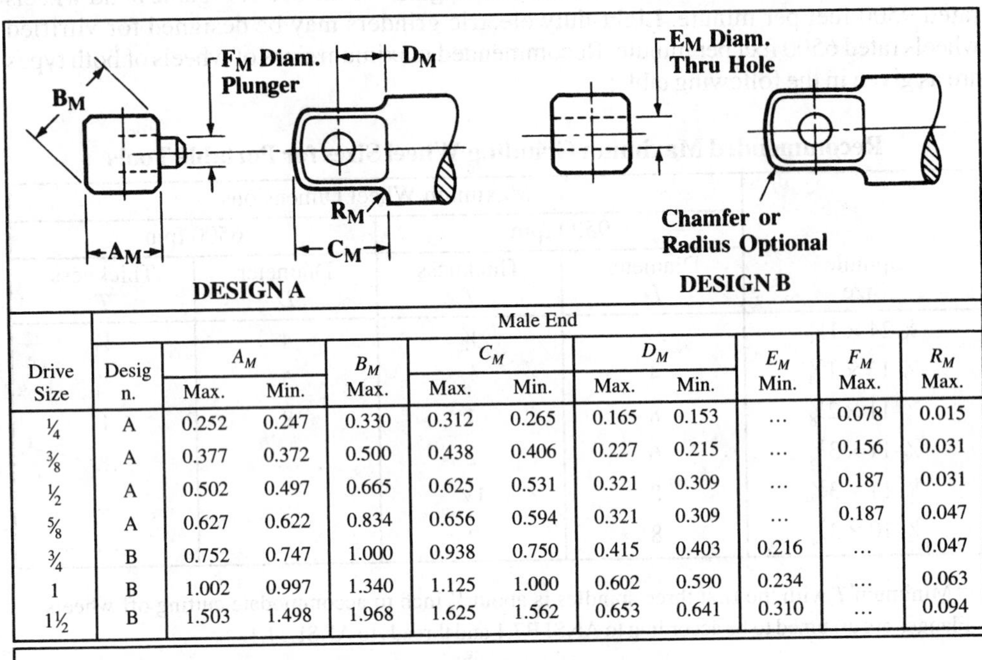

Drive Size	Desig n.	A_M Max.	A_M Min.	B_M Max.	C_M Max.	C_M Min.	D_M Max.	D_M Min.	E_M Min.	F_M Max.	R_M Max.
		Male End									
¼	A	0.252	0.247	0.330	0.312	0.265	0.165	0.153	…	0.078	0.015
⅜	A	0.377	0.372	0.500	0.438	0.406	0.227	0.215	…	0.156	0.031
½	A	0.502	0.497	0.665	0.625	0.531	0.321	0.309	…	0.187	0.031
⅝	A	0.627	0.622	0.834	0.656	0.594	0.321	0.309	…	0.187	0.047
¾	B	0.752	0.747	1.000	0.938	0.750	0.415	0.403	0.216	…	0.047
1	B	1.002	0.997	1.340	1.125	1.000	0.602	0.590	0.234	…	0.063
1½	B	1.503	1.498	1.968	1.625	1.562	0.653	0.641	0.310	…	0.094

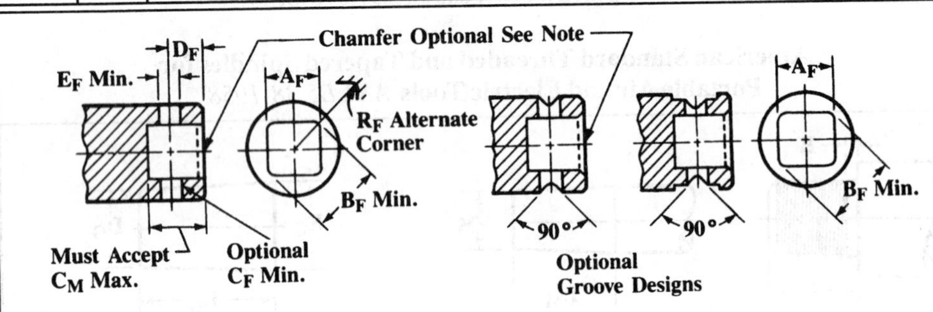

Drive Size	Design	A_F Max.	A_F Min.	B_F Min.	D_F Max.	D_F Min.	E_F Min.	R_F Max.
		Female End						
¼	A	0.258	0.253	0.335	0.159	0.147	0.090	…
⅜	A	0.383	0.378	0.505	0.221	0.209	0.170	…
½	A	0.508	0.503	0.670	0.315	0.303	0.201	…
⅝	A	0.633	0.628	0.839	0.315	0.303	0.201	…
¾	B	0.758	0.753	1.005	0.409	0.397	0.216	0.047
1	B	1.009	1.004	1.350	0.596	0.584	0.234	0.062
1½	B	1.510	1.505	1.983	0.647	0.635	0.310	0.125

All dimensions in inches.

Incorporating fillet radius (R_M) at shoulder of male tang precludes use of minimum diameter cross-hole in socket (E_F), unless female drive end is chamfered (shown as optional).

If female drive end is not chamfered, socket cross-hole diameter (E_F) is increased to compensate for fillet radius R_M, max.

Minimum clearance across flats male to female is 0.001 inch through ¾-inch size; 0.002 inch in 1- and 1½-inch sizes. For impact wrenches A_M should be held as close to maximum as practical.

C_F, min. for both designs A and B should be equal to C_M, max.

American Standard Abrasion Tool Spindles for Portable Air and Electric Tools

ASA B5.38-1958

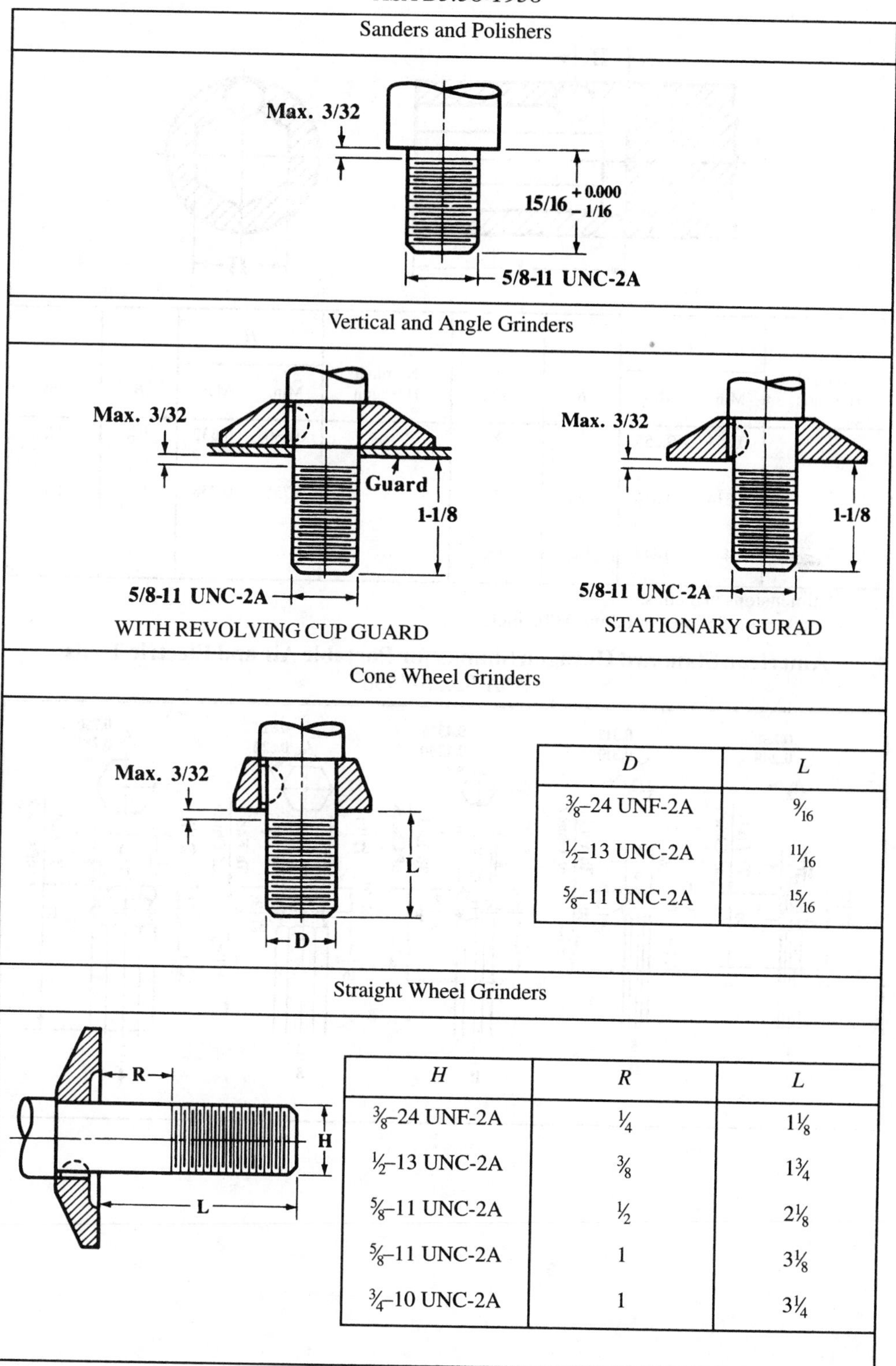

D	*L*
3/8–24 UNF-2A	9/16
1/2–13 UNC-2A	11/16
5/8–11 UNC-2A	15/16

H	*R*	*L*
3/8–24 UNF-2A	1/4	1 1/8
1/2–13 UNC-2A	3/8	1 3/4
5/8–11 UNC-2A	1/2	2 1/8
5/8–11 UNC-2A	1	3 1/8
3/4–10 UNC-2A	1	3 1/4

All dimensions in inches. Threads are right-hand.

American Standard Hexagonal Chucks for Portable Air and Electric Tools

ASA B5.38-1958

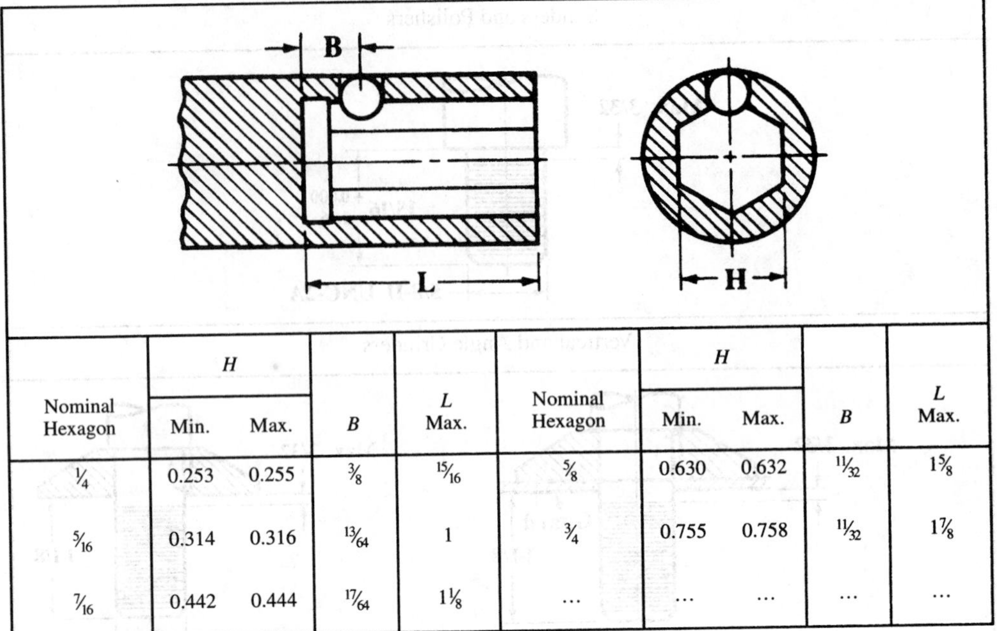

Nominal Hexagon	H Min.	H Max.	B	L Max.	Nominal Hexagon	H Min.	H Max.	B	L Max.
1⁄4	0.253	0.255	3⁄8	15⁄16	5⁄8	0.630	0.632	11⁄32	1 5⁄8
5⁄16	0.314	0.316	13⁄64	1	3⁄4	0.755	0.758	11⁄32	1 7⁄8
7⁄16	0.442	0.444	17⁄64	1 1⁄8	…	…	…	…	…

All dimensions in inches.
Tolerances on *B* is plus or minus 0.005 inch.

American Standard Hexagon Shanks for Portable Air and Electric Tools

ASA B5.38-1958

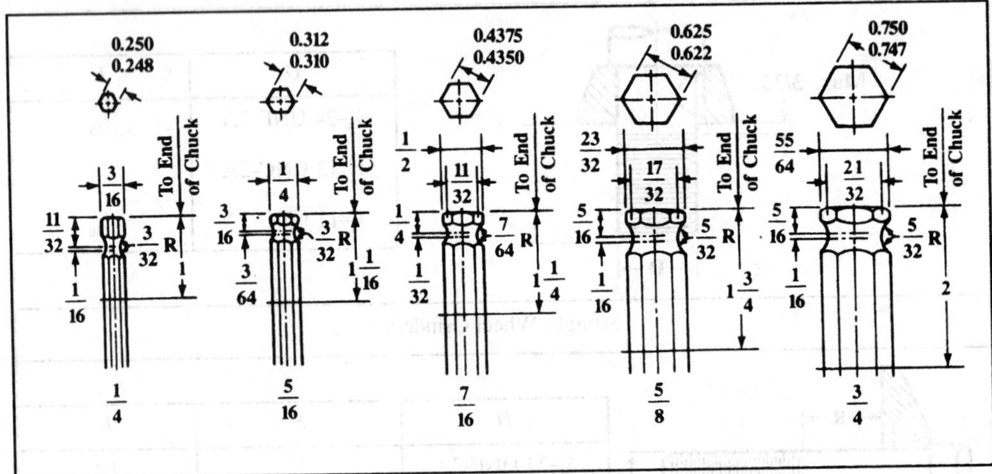

KNURLS AND KNURLING

ANSI Standard Knurls and Knurling.—The ANSI/ASME Standard B94.6-1984 covers knurling tools with standardized diametral pitches and their dimensional relations with respect to the work in the production of straight, diagonal, and diamond knurling on cylindrical surfaces having teeth of uniform pitch parallel to the cylinder axis or at a helix angle not exceeding 45 degrees with the work axis.

These knurling tools and the recommendations for their use are equally applicable to general purpose and precision knurling. The advantage of this ANSI Standard system is the provision by which good tracking (the ability of teeth to mesh as the tool penetrates the work blank in successive revolutions) is obtained by tools designed on the basis of diametral pitch instead of TPI (teeth per inch) when used with work blank diameters that are multiples of 1⁄64 inch for 64 and 128 diametral pitch or 1⁄32 inch for 96 and 160 diametral pitch. The use of knurls and work blank diameters which will permit good tracking should improve the uniformity and appearance of knurling, eliminate the costly trial and error methods, reduce the failure of knurling tools and production of defective work, and decrease the number of tools required. Preferred sizes for cylindrical knurls are given in Table 1 and detailed specifications appear in Table 2.

Table 1. ANSI Standard Preferred Sizes for Cylindrical Type Knurls
ANSI/ASME B94.6-1984

Nominal Outside Diameter D_{nt}	Width of Face F	Diameter of Hole A	Standard Diametral Pitches, P			
			64	96	128	160
			Number of Teeth, N_t, for Standard Pitches			
1⁄2	3⁄16	3⁄16	32	48	64	80
5⁄8	1⁄4	1⁄4	40	60	80	100
3⁄4	3⁄8	1⁄4	48	72	96	120
7⁄8	3⁄8	1⁄4	56	84	112	140
Additional Sizes for Bench and Engine Lathe Tool Holders						
5⁄8	5⁄16	7⁄32	40	60	80	100
3⁄4	5⁄8	1⁄4	48	72	96	120
1	3⁄8	5⁄16	64	96	128	160

The 96 diametral pitch knurl should be given preference in the interest of tool simplification. Dimensions D_{nt}, F, and A are in inches.

Table 2. ANSI Standard Specifications for Cylindrical Knurls with Straight or Diagonal Teeth *ANSI/ASME B94.6-1984*

Diametral Pitch P	Nominal Diameter, D_{nt}: 1⁄2	5⁄8	3⁄4	7⁄8	1	Tracking Correction Factor Q	Tooth Depth, h, + 0.0015, − 0.0000: Straight	Diagonal	Radius at Root R
	Major Diameter of Knurl, D_{ot}, +0.0000, −0.0015								
64	0.4932	0.6165	0.7398	0.8631	0.9864	0.0006676	0.024	0.021	0.0070 0.0050
96	0.4960	0.6200	0.7440	0.8680	0.9920	0.0002618	0.016	0.014	0.0060 0.0040
128	0.4972	0.6215	0.7458	0.8701	0.9944	0.0001374	0.012	0.010	0.0045 0.0030
160	0.4976	0.6220	0.7464	0.8708	0.9952	0.00009425	0.009	0.008	0.0040 0.0025

All dimensions except diametral pitch are in inches.

Approximate angle of space between sides of adjacent teeth for both straight and diagonal teeth is 80 degrees. The permissible eccentricity of teeth for all knurls is 0.002 inch maximum (total indicator reading).

Number of teeth in a knurl equals diametral pitch multiplied by nominal diameter.

Diagonal teeth have 30-degree helix angle, ψ.

The term *Diametral Pitch* applies to the quotient obtained by dividing the total number of teeth in the circumference of the work by the basic blank diameter; in the case of the knurling tool it would be the total number of teeth in the circumference divided by the *nominal* diameter. In the Standard the diametral pitch and number of teeth are always measured in a transverse plane which is perpendicular to the axis of rotation for diagonal as well as straight knurls and knurling.

Cylindrical Knurling Tools.—The cylindrical type of knurling tool comprises a tool holder and one or more knurls. The knurl has a centrally located mounting hole and is provided with straight or diagonal teeth on its periphery. The knurl is used to reproduce this tooth pattern on the work blank as the knurl and work blank rotate together.

**Formulas for Cylindrical Knurls*

$$P = \text{diametral pitch of knurl} = N_t \div D_{nt} \quad (1)$$

$$D_{nt} = \text{nominal diameter of knurl} = N_t \div P \quad (2)$$

$$N_t = \text{no. of teeth on knurl} = P \times D_{nt} \quad (3)$$

$$^*P_{nt} = \text{circular pitch on nominal diameter} = \pi \div P \quad (4)$$

$$^*P_{ot} = \text{circular pitch on major diameter} = \pi D_{ot} \div N_t \quad (5)$$

$$D_{ot} = \text{major diameter of knurl} = D_{nt} - (N_t Q \div \pi) \quad (6)$$

$$Q = P_{nt} - P_{ot} = \text{tracking correction factor in Formula} \quad (7)$$

Tracking Correction Factor Q: Use of the preferred pitches for cylindrical knurls, Table 2, results in good tracking on all fractional work-blank diameters which are multiples of $\frac{1}{64}$ inch for 64 and 128 diametral pitch, and $\frac{1}{32}$ inch for 96 and 160 diametral pitch; an indication of good tracking is evenness of marking on the work surface during the first revolution of the work.

The many variables involved in knurling practice require that an empirical correction method be used to determine what actual circular pitch is needed at the major diameter of the knurl to produce good tracking and the required circular pitch on the workpiece. The empirical tracking correcton factor, Q, in Table 2 is used in the calculation of the major diameter of the knurl, Formula (6).

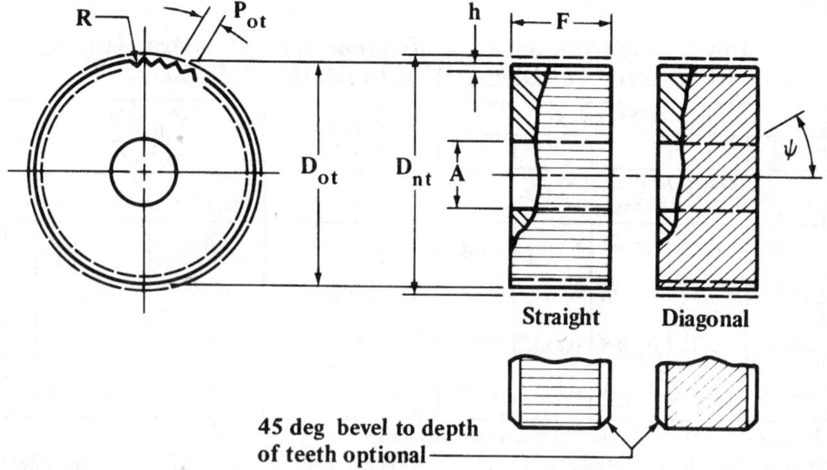

Cylindrical Knurl

* *Note:* For diagonal knurls, P_{nt} and P_{ot} are the transverse circular pitches which are measured in the plane perpendicular to the axis of rotation.

Flat Knurling Tools.—The flat type of tool is a knurling die, commonly used in reciprocating types of rolling machines. Dies may be made with either single or duplex faces having either straight or diagonal teeth. No preferred sizes are established for flat dies.

Flat Knurling Die with Straight Teeth:

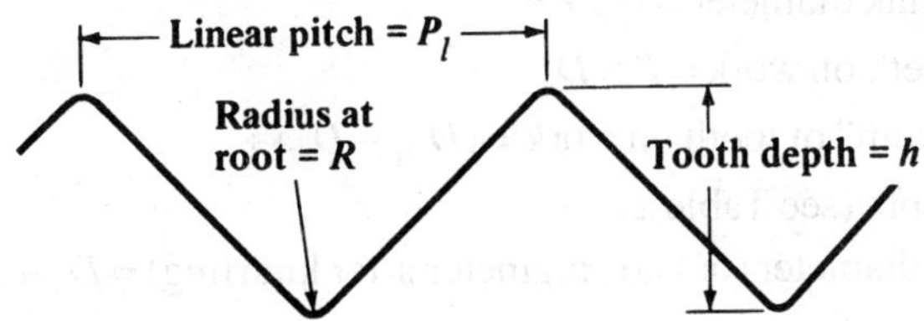

R = radius at root

P = diametral pitch = $N_w \div D_w$ (8)

D_w = work blank (pitch) diameter = $N_w \div P$ (9)

N_w = number of teeth on work = $P \times D_w$ (10)

h = tooth depth

Q = tracking correction factor (see Table 2)

P_l = linear pitch on die

= circular pitch on work pitch diameter = $P - Q$ (11)

Table 3. ANSI Standard Specifications for Flat Knurling Dies
ANSI/ASME B94.6-1984

Diametral Pitch, P	Linear Pitch,[a] P_l	Tooth Depth, h Straight	Tooth Depth, h Diagonal	Radius at Root, R	Diametral Pitch, P	Linear Pitch,[a] P_l	Tooth Depth, h Straight	Tooth Depth, h Diagonal	Radius at Root, R
64	0.0484	0.024	0.021	0.0070 0.0050	128	0.0244	0.012	0.010	0.0045 0.0030
96	0.0325	0.016	0.014	0.0060 0.0040	160	0.0195	0.009	0.008	0.0040 0.0025

[a] The linear pitches are theoretical. The exact linear pitch produced by a flat knurling die may vary slightly from those shown depending upon the rolling condition and the material being rolled.

All dimensions except diametral pitch are in inches.

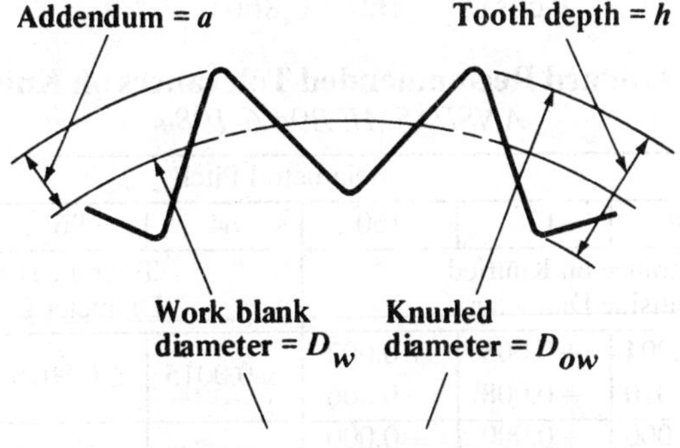

Teeth on Knurled Work

Formulas Applicable to Knurled Work.—The following formulas are applicable to knurled work with straight, diagonal, and diamond knurling.

Formulas for Straight or Diagonal Knurling with Straight or Diagonal Tooth Cylindrical Knurling Tools Set with Knurl Axis Parallel with Work Axis:

$$P = \text{diametral pitch} = N_w \div D_w \quad (12)$$

$$D_w = \text{work blank diameter} = N_w \div P \quad (13)$$

$$N_w = \text{no. of teeth on work} = P \times D_w \quad (14)$$

$$a = \text{"addendum" of tooth on work} = (D_{ow} - D_w) \div 2 \quad (15)$$

$$h = \text{tooth depth (see Table 2)}$$

$$D_{ow} = \text{knurled diameter (outside diameter after knurling)} = D_w + 2a \quad (16)$$

Formulas for Diagonal and Diamond Knurling with Straight Tooth Knurling Tools Set at an Angle to the Work Axis:

If, ψ = angle between tool axis and work axis

P = diametral pitch on tool

P_ψ = diametral pitch produced on work blank (as measured in the transverse plane) by setting tool axis at an angle ψ with respect to work blank axis

D_w = diameter of work blank; and

N_w = number of teeth produced on work blank (as measured in the transverse plane)

then, $P_\psi = P\cos\psi$ (17)

and, $N = D_w P\cos\psi$ (18)

For example, if 30 degree diagonal knurling were to be produced on 1-inch diameter stock with a 160 pitch straight knurl:

$$N_w = D_w P\cos 30^\circ = 1.000 \times 160 \times 0.86603 = 138.56 \text{ teeth}$$

Good tracking is theoretically possible by changing the helix angle as follows to correspond to a whole number of teeth (138):

$$\cos\psi = N_w \div D_w P = 138 \div (1 \times 160) = 0.8625$$

$$\psi = 30\tfrac{1}{2} \text{ degrees, approximately}$$

Whenever it is more practical to machine the stock, good tracking can be obtained by reducing the work blank diameter as follows to correspond to a whole number of teeth (138):

$$D_w = \frac{N_w}{P\cos\psi} = \frac{138}{160 \times 0.866} = 0.996 \text{ inch}$$

Table 4. ANSI Standard Recommended Tolerances on Knurled Diameters
ANSI/ASME B94.6-1984

Tolerance Class	Diametral Pitch							
	64	96	128	160	64	96	128	160
	Tolerance on Knurled Outside Diameter				Tolerance on Work-Blank Diameter Before Knurling			
I	+0.005 −0.012	+0.004 −0.010	+0.003 −0.008	+0.002 −0.006	±0.0015	±0.0010	±0.0007	±0.0005
II	+0.000 −0.010	+0.000 −0.009	+0.000 −0.008	+0.000 −0.006	±0.0015	±0.0010	±0.0007	±0.0005
III	+0.000 −0.006	+0.000 −0.005	+0.000 −0.004	+0.000 −0.003	+0.000 −0.0015	+0.0000 −0.0010	+0.000 −0.0007	+0.0000 −0.0005

Recommended Tolerances on Knurled Outside Diameters.—The recommended applications of the tolerance classes shown in Table 4 are as follows:

Class I: Tolerances in this classification may be applied to straight, diagonal and raised diamond knurling where the knurled outside diameter of the work need not be held to close dimensional tolerances. Such applications include knurling for decorative effect, grip on thumb screws, and inserts for moldings and castings.

Class II: Tolerances in this classification may be applied to straight knurling only and are recommended for applications requiring closer dimensional control of the knurled outside diameter than provided for by Class I tolerances.

Class III: Tolerances in this classification may be applied to straight knurling only and are recommended for applications requiring closest possible dimensional control of the knurled outside diameter. Such applications include knurling for close fits.

Note: The width of the knurling should not exceed the diameter of the blank, and knurling wider than the knurling tool cannot be produced unless the knurl starts at the end of the work.

Marking on Knurls and Dies.—Each knurl and die should be marked as follows: *a.* when straight to indicate its diametral pitch; *b.* when diagonal, to indicate its diametral pitch, helix angle, and hand of angle.

Concave Knurls.—The radius of a concave knurl should not be the same as the radius of the piece to be knurled. If the knurl and the work are of the same radius, the material compressed by the knurl will be forced down on the shoulder *D* and spoil the appearance of the work. A design of concave knurl is shown in the accompanying illustration, and all the important dimensions are designated by letters. To find these dimensions, the pitch of the knurl required must be known, and also, approximately, the throat diameter *B*. This diameter must suit the knurl holder used, and be such that the circumference contains an even number of teeth with the required pitch. When these dimensions have been decided upon, all the other unknown factors can be found by the following formulas: Let R = radius of piece to be knurled; r = radius of concave part of knurl; C = radius of cutter or hob for cutting the teeth in the knurl; B = diameter over concave part of knurl (throat diameter); A = outside diameter of knurl; d = depth of tooth in knurl; P = pitch of knurl (number of teeth per inch circumference); p = circular pitch of knurl; then $r = R + \frac{1}{2}d$; $C = r + d$; $A = B + 2r - (3d + 0.010 \text{ inch})$; and $d = 0.5 \times p \times \cot \alpha/2$, where α is the included angle of the teeth.

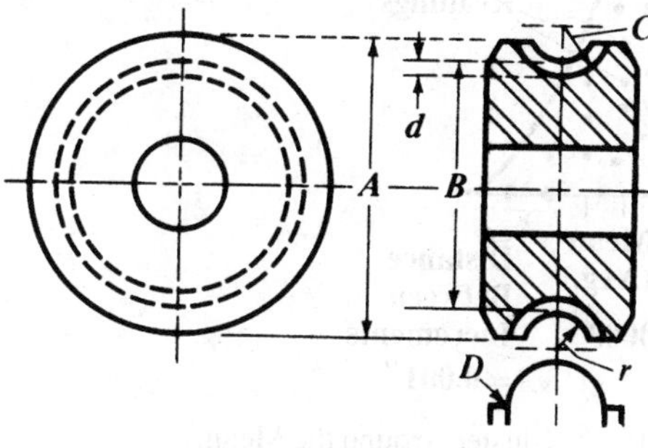

As the depth of the tooth is usually very slight, the throat diameter *B* will be accurate enough for all practical purposes for calculating the pitch, and it is not necessary to take into consideration the pitch circle. For example, assume that the pitch of a knurl is 32, that the throat diameter *B* is 0.5561 inch, that the radius *R* of the piece to be knurled is $\frac{1}{16}$ inch, and that the angle of the teeth is 90 degrees; find the dimensions of the knurl. Using the notation given:

$$p = \frac{1}{P} = \frac{1}{32} = 0.03125 \text{ inch} \qquad d = 0.5 \times 0.03125 \times \cot 45° = 0.0156 \text{ inch}$$

$$r = \frac{1}{16} + \frac{0.0156}{2} = 0.0703 \text{ inch} \qquad C = 0.0703 + 0.0156 = 0.0859 \text{ inch}$$

$$A = 0.5561 + 0.1406 - (0.0468 + 0.010) = 0.6399 \text{ inch}$$

MACHINE TOOL ACCURACY

Accuracy, Repeatability, and Resolution: In machine tools, accuracy is the maximum spread in measurements made of slide movements during successive runs at a number of target points, as discussed below. Repeatability is the spread of the normal curve at the target point that has the largest spread. A rule of thumb says that repeatability is approximately half the accuracy value, or twice as good as the accuracy, but this rule is somewhat nullified due to the introduction of error-compensation features on NC machines. Resolution refers to the smallest units of measurement that the system (controller plus servo) can recognize. Resolution is an electronic/electrical term and the unit is usually smaller than either the accuracy or the repeatability. Low values for resolution are usually, though not necessarily, applied to machines of high accuracy. In addition to high cost, a low-resolution-value design usually has a low maximum feed rate and the use of such designs is usually restricted to applications requiring high accuracy.

*Positioning Accuracy:*The positioning accuracy of a numerically controlled machine tool refers to the ability of an NC machine to place the tip of a tool at a preprogrammed target. Although no metal cutting is involved, this test is very significant for a machine tool and the cost of an NC machine will rise almost geometrically with respect to its positioning accuracy. Care, therefore, should be taken when deciding on the purchase of such a machine, to avoid paying the premium for unneeded accuracy but instead to obtain a machine that will meet the tolerance requirements for the parts to be produced.

Accuracy can be measured in many ways. A tool tip on an NC machine could be moved, for example, to a target point whose *X*-coordinate is 10.0000 inches. If the move is along the *X*-axis, and the tool tip arrives at a point that measures 10.0001 inches, does this mean that the machine has an accuracy of 0.0001 inch? What if a repetition of this move brought the tool tip to a point measuring 10.0003 inches, and another repetition moved the tool to a point that measured 9.9998 inches? In practice, it is expected that there would be a scattering or distribution of measurements and some kind of averaging is normally used.

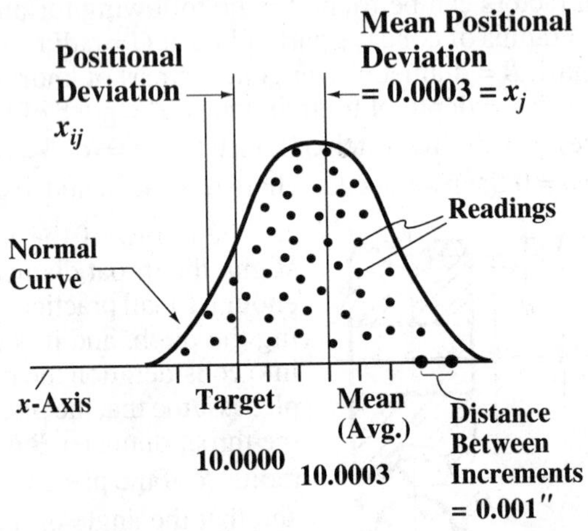

Fig. 1. In a Normal Distribution, Plotted Points Cluster Around the Mean.

Although averaging the results of several runs is an improvement over a single run, the main problem with averaging is that it does not consider the extent or width of the spread of readings. For example, if one measurement to the 10.0000-inch target is 9.9000 inches and another is 10.1000 inches, the difference of the two readings is 0.2000 inch, and the accuracy is poor. However, the readings average a perfect 10 inches. Therefore, the average and the spread of several readings must both be considered in determining the accuracy.

Plotting the results of a large number of runs generates a *normal distribution curve,* as shown in Fig. 1. In this example, the readings are plotted along the *X*-axis in increments of

0.0001 inch (0.0025 mm). Usually, five to ten such readings are sufficient. The distance of any one reading from the target is called the *positional deviation* of the point. The distance of the mean, or average, for the normal distribution from the target is called the *mean positional deviation.*

The spread for the normal curve is determined by a mathematical formula that calculates the distance from the mean that a certain percentage of the readings fall into. The mathematical formula used calculates one *standard deviation,* which represents approximately 32 per cent of the points that will fall within the normal curve, as shown in Fig. 2. One standard deviation is also called one sigma, or 1σ. Plus or minus one sigma (±1σ) represents 64 per cent of all the points under the normal curve. A wider range on the curve, ±2σ, means that 95.44 per cent of the points are within the normal curve, and ±3σ means that 99.74 per cent of the points are within the normal curve. If an infinite number of runs were made, almost all the measurements would fall within the ±3σ range.

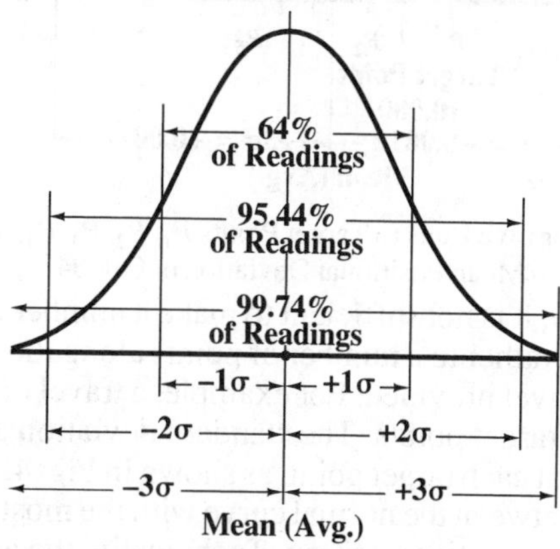

Fig. 2. Percentages of Points Falling in the ±1σ (64%), ±2σ (95.44%), and ±3σ (99.74%) Ranges

The formula for calculating one standard deviation is

$$1\sigma = \sqrt{\frac{1}{n-1}\sum_{i=1}^{n}(X_{ij}-\bar{X}_j)^2}$$

where n = number of runs to the target; i = identification for any one run; X_{ij} = positional deviation for any one run (see Fig. 1); and, $\bar{X}_j$ = mean positional deviation (see Fig. 1).

The bar over the $\bar{X}$ in the formula indicates that the value is the mean or average for the normal distribution.

Example: From Fig. 3, five runs were made at a target point that is 10.0000 inches along the X-axis and the positional deviations for each run were:
$x_{1j} = -0.0002$, $x_{2j} = +0.0002$, $x_{3j} = +0.0005$, $x_{4j} = +0.0007$, and $x_{5j} = +0.0008$ inch. The algebraic total of these five runs is +0.0020, and the mean positional deviation $= \bar{X}_j = 0.0020/5 = 0.0004$.

The calculations for one standard deviation are:

$$1\sigma = \sqrt{\frac{1}{n-1}[(X_{1j}-\bar{X}_j)^2 + (X_{2j}-\bar{X}_j)^2 + (X_{3j}-\bar{X}_j)^2 + (X_{4j}-\bar{X}_j)^2 + (X_{5j}-\bar{X}_j)^2]}$$

$$1\sigma = \sqrt{\frac{1}{5-1}[(-0.0002-0.0004)^2 + (0.0002-0.0004)^2 \\ (0.0005-0.0004)^2 + (0.0007-0.0004)^2 + (0.0008-0.0004)^2]}$$

$$= \sqrt{\frac{1}{4}(0.00000066)} = \sqrt{0.17\times10^{-6}} = 0.0004$$

Three sigma variations or 3σ, is 3 times sigma, equal to 0.0012 for the example.

If an infinite number of trials were made to the target position of 10.0000 inches for the ongoing example, 99.74 per cent of the points would fall between 9.9992 and 10.0016 inches, giving a spread of ± 3σ, or 0.0024 inch. This spread alone is not considered as the *accuracy* but rather the repeatability for the target point 10.0000.

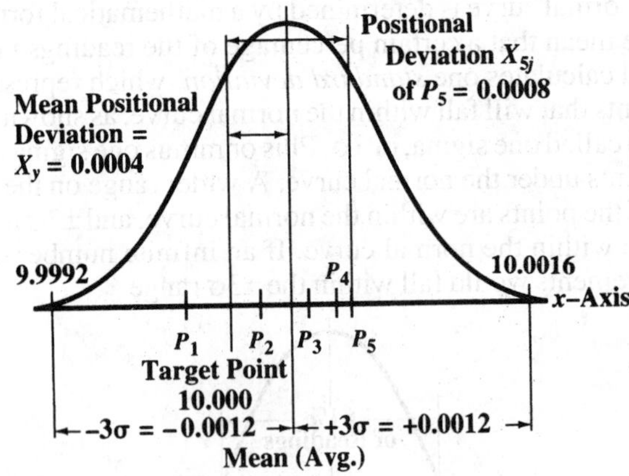

Fig. 3. Readings for Five Runs to Target Points P_1, P_2, P_3, P_4, and P_5 Result in a Mean Positional Deviation of 0.0004

To calculate the accuracy, it is not sufficient to make a number of runs to one target point along a particular axis, but rather to a number of points along the axis, the number depending on the length of axis travel provided. For example, a travel of about 3 ft requires 5, and a travel of 6 ft requires 10 target points. The standard deviation and spread for the normal curve must be determined at each target point, as shown in Fig. 4. The *accuracy* for the axis would then be the spread between the normal curve with the most negative position and the normal curve with the most positive position. Technically, the accuracy is a spread rather than a ± figure, but it is often referred to as a ± figure and it may be assumed that a ±0.003, for expediency, is equal to a spread of 0.006.

The above description for measuring accuracy considers unidirectional approaches to target points. Bidirectional movements (additional movements to the same target point from either direction) will give different results, mostly due to backlash in the lead-screw, though backlash is small with ballnut leadscrews. Measurements made with bidirectional movements will show greater spreads and somewhat less accuracy than will unidirectional movements.

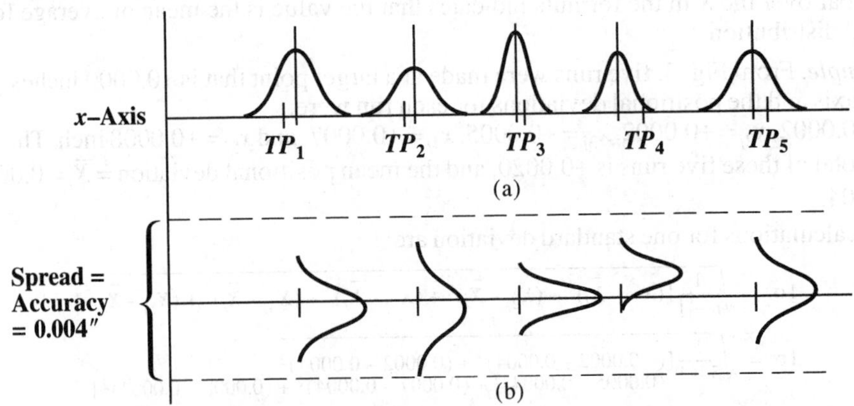

Fig. 4. Two Ways of Plotting Five Target Point Spreads

Rules for determining accuracy were standardized in guidelines last revised by the Association for Manufacturing Technology (AMT) in 1972. Some European machine tool builders use the VDI/DGQ 3441 (German) guidelines, which are similar to those of the

AMT in that normal distributions are used and a number of target points are selected along an axis. Japanese standards JIS-B-6201, JIS-B-6336, and JIS-B-6338 are somewhat simpler and consider only the spread of the readings, so that the final accuracy figure may be almost double that given by the AMT or VDI methods. The International Standards Organization (ISO), in 1988, issued ISO 230-2, which follows the procedures discussed above, but is somewhat less strict than the AMT recommendations. Table 1 lists some types of NC machines and the degree of accuracy that they normally provide.

Table 1. Degrees of Accuracy Expected with NC Machine Tools

Type of NC Machine	Accuracy	
	inches	mm
Large boring machines or boring mills	0.0010–0.0020	0.025–0.050
Small milling machines	0.0006–0.0010	0.015–0.025
Large machining centers	0.0005–0.0008	0.012–0.020
Small and medium-sized machining centers	0.0003–0.0006	0.008–0.015
Lathes, slant bed, small and medium sizes	0.0002–0.0005	0.006–0.012
Lathes, small precision	0.0002–0.0003	0.004–0.008
Horizontal jigmill	0.0002–0.0004	0.004–0.010
Vertical jig boring machines	0.0001–0.0002	0.002–0.005
Vertical jig grinding machines	0.0001–0.0002	0.002–0.005
Cylindrical grinding machines, small to medium sizes	0.00004–0.0003	0.001–0.007
Diamond turning lathes	0.00002–0.0001	0.0005–0.003

Significance of Accuracy: Numerically controlled machines are generally considered to be more accurate and more consistent in their movements than their conventional counterparts. CNC controllers have improved the accuracy by providing the ability to compensate for mechanical inaccuracies. Thus, compensation for errors in the lead-screw, parallelism and squareness of the machine ways, and for the effects of heating can be made automatically on NC machines. Some machine tool types are expected to be more accurate than others; for instance, grinding machines are more accurate than milling machines, and lathes for diamond turning are more accurate than normal slant-bed lathes.

Accuracy of machine tools depends on temperature, air pressure, local vibrations, and humidity. ISO standard 230-2 requires that, where possible, the ambient temperature for conducting such tests be held between 67.1 and 68.9 degrees F (19.5 and 20.5 degrees C).

Autocollimation: Checks on movements of slides and spindles, and alignment and other characteristics of machine tools are performed with great accuracy by means of an autocollimator, which is an optical, noncontact, angle-measuring instrument. Flatness, straightness, perpendicularity, and runout can also be checked by autocollimation. The instrument is designed to project a beam of light from a laser or an incandescent bulb onto an optically flat mirror. When the light beam is reflected back to the instrument, the distance traveled by the beam, also deviations from a straight line, can be detected by the projector and calculated electronically or measured by the scale.

Autocollimators have a small angular measuring range and are usually calibrated in arc-seconds. One arc-second is an angle of 4.85 millionths of an inch (0.00000485 in.) per inch of distance from the vertex, and is often rounded to 5 millionths of an inch per inch. Angles can also be described in terms of radians and 1 arc-second is equal to 4.85 microradians, or 0.0000573 deg.

In practice, the interferometer or autocollimator is fixed to a rigid structure and the optical mirror, which should have a flatness of one-quarter wavelength of the light used (see page 696), is fixed to the workpiece to be measured. The initial reading is taken, and then the workpiece is moved to another position. Readings of movement can be made to within a few millionths of an inch. Angular displacements, corresponding to successive positions, of about 1 arc-second can be taken from most autocollimators, in azimuth or elevation or a combination of the two. Generally, the line width of the reticle limits the accuracy of reading such instruments.

Laser interferometers are designed to allow autocollimation readings to be taken by a photodetector instead of the eye, and some designs can measure angles to 0.001 arc-second, closer than is required for most machine shop applications. Output from an electronic autocollimator is usually transferred to a computer for recording or analysis if required. The computer calculates, lists, and plots the readings for the target points automatically, under control of the inspection program.

A typical plot from such a setup is seen in Fig. 5, where the central line connects the averages for the normal distributions at each target point. The upper line connects the positive outer limits and the lower line the negative outer limits for the normal distributions. The normal spread, indicating the accuracy of positioning, is 0.00065 inch (0.016 mm), for the *Y*-axis along which the measurements were taken.

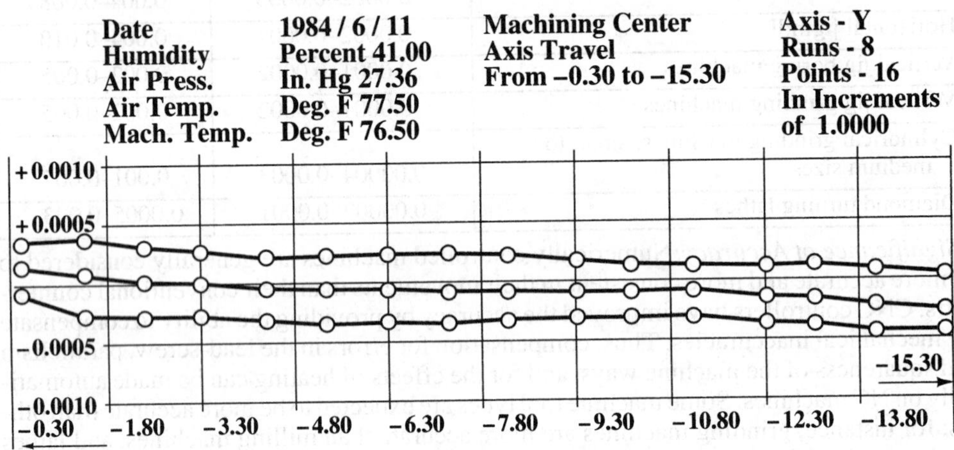

Fig. 5. Laser Interferometer Plots of Movements of Slides on a Large Horizontal Machining Center Showing an Accuracy of 0.00065 inch (0.016 mm) for the *y* Axis

Effect of Machine Accuracy on Part Tolerances

Part tolerances are usually shown on prints, usually in a control block to ANSI Standard 14.5M-1994 (see *Geometric Dimensioning and Tolerancing* starting on page 606.) Table shows some part tolerance symbols that relate to machine tool positioning accuracy. The accuracy of a part is affected by machine and cutting tool dynamics, alignment, fixture accuracy, operator settings, and accuracies of the cutting tools, holders, and collets, but the positioning accuracy of the machine probably has the greatest influence. Spindle rotation accuracy, or runout, also has a large influence on part accuracy.

The ratio of the attainable part accuracy to the no-load positioning accuracy can vary from 1.7:1 to 8.31:1, depending on the type of cutting operation. For instance, making a hole by drilling, followed by a light boring or reaming operation, produces a quite accurate result in about the 1.7:1 range, whereas contour milling on hard material could be at the higher end of the range. A good average for part accuracy versus machine positioning accuracy is 3.3:1, which means that the part accuracy is 3.3 times the positioning accuracy.

Table 2. Symbols and Feature Control Frames *ANSI 14.5M-1994*

Symbol	Characteristic	Meaning of Characteristic	Relationship to the Machine Tool
⌖	Position	The allowable true position tolerance of a feature from a datum (assume feature to be a drilled hole). Feature control block might appear as: | ⌖ | ∅ **0.005** | **A** | A is the datum, which can be another surface, another hole, or other feature	Assume tolerance is 0.005 mm. Machine positioning accuracy would be at least $0.005 \times 0.707 = 0.0035$ mm even if it is assumed that the hole accuracy is the same as the positioning accuracy. Machine could be milling, drilling, or machining center. **y−axis** **True Position Tolerance Zone** **x−axis** **45**
	Position	Assume feature to be a turned circumference, the axis of which has to be within a tolerance to another feature. Feature control block would appear as follows if feature A were the axis of hole 1: | ⌖ | ∅ **0.005** | **A** |	**Center (axis) for Hole 2** **True Position Tolerance Zone 2 (0.005 mm)** **Center (axis) for Hole 1** **Hole 2**

Table 2. Symbols and Feature Control Frames *ANSI 14.5M-1994*

Symbol	Characteristic	Meaning of Characteristic	Relationship to the Machine Tool
○	Roundness	The roundness tolerance establishes a band. Tolerance band	This tolerance would apply to turning and would be the result of radial spindle runout.
⌀	Diameter	Usually expressed as a ± tolerance attached to the dimension.	Diametral accuracy of the part would depend on the positioning accuracy of the cross-slide of lathe or grinder. Positioning accuracy would be from ½ to ¼ of part accuracy, depending chiefly on the rigidity of the tool, depth of cut, and material being cut.
⌓	Profile of a surface	Specifies a uniform boundary, along a true profile. Tolerance 0.005 Datum A Feature control block might appear as: ⌓ | ⌀ 0.005 | A	Affected by positioning accuracy of machine. There would be side and/or end forces on the tool so expect part to machine positioning accuracy to be high, say, 5:1

Table 2. Symbols and Feature Control Frames *ANSI 14.5M-1994*

Symbol	Characteristic	Meaning of Characteristic	Relationship to the Machine Tool
//	Parallelism	A feature (surface) parallel to a datum plane or datum axis. **Tolerance 0.010** **Datum A** Feature control block might appear as: **// | ∅ 0.010 | A**	Affected by positioning accuracy, machine alignment, and fixturing.
◎	Concentricity	Applies to turning. The axis of the feature must lie within the tolerance zone of another axis. **Tolerance 0.010** **A** **Datum A** Feature control block might appear as follows: **◎ | ∅ 0.005 | A**	Affected by positioning accuracy, most likely along Z axis.

Table 2. Symbols and Feature Control Frames *ANSI 14.5M-1994*

Symbol	Characteristic	Meaning of Characteristic	Relationship to the Machine Tool
↗	Runout	Applies to the runout (both radial and axial) of a circular feature at any one position around the circumference or flat, perpendicular to the axis. **Runout at a Point (Radial)** **Runout at a Point (Axial)** Runout at a Point (Radial)	Radial runout on part is not affected by spindle radial runout unless whole machine is untrue. Axial runout on part is affected by axial runout on machine. Feature would normally be perpendicular to datum. Feature control block might appear as: ↗ | ⌀ **0.005** | **A**
⌰	Total runout	Similar to runout but applies to total surface and therefore consider both radial and axial runout.	Would be affected by either radial or axial runout, or both, machine misalignment, or setup.
⊥	Perpendicularity	A feature is perpendicular to a datum plane or axis. **Tolerance Zone**	Affected principally by misalignment of machine or fixturing.

NUMERICAL CONTROL

Introduction.—The Electronic Industries Association (EIA) defines numerical control as "a system in which actions are controlled by the direct insertion of numerical data at some point." More specifically, numerical control, or NC as it will be called here, involves machines controlled by electronic systems designed to accept numerical data and other instructions, usually in a coded form. These instructions may come directly from some source such as a punched tape, a floppy disk, directly from a computer, or from an operator.

The key to the success of numerical control lies in its flexibility. To machine a different part, it is only necessary to "play" a different tape. NC machines are more productive than conventional equipment and consequently produce parts at less cost even when the higher investment is considered. NC machines also are more accurate and produce far less scrap than their conventional counterparts. By 1985, over 110,000 NC machine tools were operating in the United States. Over 80 per cent of the dollars being spent on the most common types of machine tools, namely, drilling, milling, boring, and turning machines, are going into NC equipment.

NC is a generic term for the whole field of numerical control and encompasses a complete field of endeavor. Sometimes CNC, which stands for Computer Numerical Control and applies only to the control system, is used erroneously as a replacement term for NC. Albeit a monumental development, use of the term CNC should be confined to installations where the older hardware control systems have been replaced.

Metal cutting is the most popular application, but NC is being applied successfully to other equipment, including punch presses, EDM wire cutting machines, inspection machines, laser and other cutting and torching machines, tube bending machines, and sheet metal cutting and forming machines.

State of the CNC Technology Today.—Early numerical control machines were ordinary machines retrofitted with controls and motors to drive tools and tables. The operations performed were the same as the operations were on the machines replaced. Over the years, NC machines began to combine additional operations such as automatically changing tools and workpieces. The structure of the machines has been strengthened to provide more rigid platforms. These changes have resulted in a class of machine that can outperform its predecessors in both speed and accuracy. Typical capabilities of a modern machining center are accuracy better than ±0.00035 inch; spindle speeds in the range up to 25,000 rpm or more, and increasing; feed rates up to 400 inches per minute and increasing; tool change times hovering between 2 and 4 seconds and decreasing. Specialized machines have been built that can achieve accuracy better than one millionth (0.000001) of an inch.

Computer numerical control of machines has undergone a great deal of change in the last decade, largely as a result of rapid increases in computer capability. Development of new and improved materials for tooling and bearings, improvements in tool geometry, and the added structural stiffness of the new machines have made it possible to perform cutting operations at speeds and feeds that were formerly impossible to attain.

Numerical Control vs. Manual Operations.—The initial cost of a CNC machine is generally much higher than a manual machine of the same nominal capacity, and the higher initial cost leads to a higher overall cost of the machine per hour of its useful life. However, the additional cost of a CNC machine has to be considered against potential savings that the machine may make possible. Some of the individual factors that make NC and CNC machining attractive are considered below.

Labor is usually one of the highest costs in the production of a part, but the labor rate paid to a CNC machine operator may be lower than the rate paid to the operator of conventional machines. This statement is particularly true when there is a shortage of operators with specialized skills necessary for setting up and operating a manual machine. However, it should not be assumed that skilled CNC machine operators are not needed because most CNCs have manual overrides that allow the operator to adjust feeds and speeds and to manually edit or enter programs as necessary. Also, skilled setup personnel and operators are likely to promote better production rates and higher efficiency in the shop. In addition, the labor rate for setting up and operating a CNC machine can sometimes be divided between two or more machines, further reducing the labor costs and cost per part produced.

The quantity and quality requirements for an order of parts often determines what manufacturing process will be used to produce them. CNC machines are probably most effective when the jobs call for a small to medium number of components that require a wide range of operations to be performed. For example, if a large number of parts are to be machined and the allowable tolerances are large, then manual or automatic fixed-cycle machines may be the most viable process. But, if a large quantity of high quality parts with strict tolerances are required, then a CNC machine will probably be able to produce the parts for the lowest cost per piece because of the speed and accuracy of CNC machines. Moreover, if the production run requires designing and making a lot of specialized form tools, cams, fixtures, or jigs, then the economics of CNC machining improves even more because much of the preproduction work is not required by the nature of the CNC process.

CNC machines can be effective for producing one-of-a-kind jobs if the part is complicated and requires a lot of different operations that, if done manually, would require specialized setups, jigs, fixtures, etc. On the other hand, a single component requiring only one or two setups might be more practical to produce on a manual machine, depending on the tolerances required. When a job calls for a small to medium number of components that require a wide range of operations, CNC is usually preferable. CNC machines are also especially well suited for batch jobs where small numbers of components are produced from an existing part program, as inventory is needed. Once the part program has been tested, a batch of the parts can be run whenever necessary. Design changes can be incorporated by changing the part program as required. The ability to process batches also has an additional benefit of eliminating large inventories of finished components.

CNC machining can help reduce machine idle time. Surveys have indicated that when machining on manual machines, the average time spent on material removal is only about 40 per cent of the time required to complete a part. On particularly complicated pieces, this ratio can drop to as low as 10 per cent or even less. The balance of the time is spent on positioning the tool or work, changing tools, and similar activities. On numerically controlled machines, the metal removal time frequently has been found to be in excess of 70 per cent of the total time spent on the part. CNC nonmachining time is lower because CNC machines perform quicker tool changes and tool or work positioning than manual machines. CNC part programs require a skilled programmer and cost additional preproduction time, but specialized jigs and fixtures that are frequently required with manual machines are not usually required with CNC machines, thereby reducing setup time and cost considerably.

Additional advantages of CNC machining are reduced lead time; improved cutting efficiency and longer tool life, as a result of better control over the feeds and speeds; improved quality and consistently accurate parts, reduced scrap, and less rework; lower inspection costs after the first part is produced and proven correct; reduced handling of parts because more operations can be performed per setup; and faster response to design changes because most part changes can be made by editing the CNC program.

Numerical Control Standards.—Standards for NC hardware and software have been developed by many organizations, and copies of the latest standards may be obtained from the following: Electronic Industries Association (EIA), 2001 Pennsylvania Avenue NW, Washington, DC 20006 (EIA and ANSI/EIA); American Society of Mechanical Engineers (ASME), 345 East 47th Street, New York, NY 10017 (ANSI/ASME); American National Standards Institute (ANSI), II West 42nd Street, New York, NY 10017 (ANSI, ANSI/EIA, ANSI/ASME, and ISO); National Standards Association, Inc. (NSA), 1200 Quince Orchard Boulevard, Gaithersburg, MD 20878; NMTBA The Association for Manufacturing Technology, 7901 Westpark Drive, McLean, VA 22102. Some of the standards and their contents are listed briefly in the accompanying table.

Numerical Control Standards

Standard Title	Description
ANSI/CAM-I 101-1990	Dimensional Measuring Interface Specification
ANSI/ASME B5.50	V-Flange Tool Shanks for Machining Centers with Automatic Tool Changers
ANSI/ASME B5.54-1992	Methods for Performance Evaluation of Computer Numerically Controlled Machining Centers
ANSI/ASME B89.1.12M	Methods for Performance Evaluation of Coordinate Measuring Machines
ANSI/EIA 227-A	1-inch Perforated Tape
ANSI/EIA 232-D	Interface Between Data Terminal Equipment and Data Circuit-Terminating Equipment Employing Serial Binary Data Interchange
ANSI/EIA 267-B	Axis and Motion Nomenclature for Numerically Controlled Machines
ANSI/EIA 274-D	Interchangeable Variable Block Data Format for Positioning, Contouring and Contouring/Positioning Numerically Controlled Machines
ANSI/EIA 358-B	Subset of American National Standarde Code for Information Interchange for Numerical Machine Control Perforated Tape
ANSI/EIA 408	Interface Between NC Equipment and Data Terminal Equipment Employing Parallel Binary Data Interchange
ANSI/EIA 423-A	Electrical Characteristics of Unbalanced Voltage Digital Interface Circuits
ANSI/EIA 431	Electrical Interface Between Numerical Control and Machine Tools
ANSI/EIA 441	Operator Interface Function of Numerical Controls
ANSI/EIA 449	General Purpose 37-position and 9-position Interface for Data Terminal Equipment and Data Circuit-Terminating Equipment Employing Serial Binary Data Interchange
ANSI/EIA 484	Electrical and Mechanical Interface Characteristics and Line Control Protocol Using Communication Control Characters for Serial Data Link between a Direct Numerical Control System and Numerical Control Equipment Employing Asynchronous Full Duplex Transmission
ANSI/EIA 491-A -1990	Interface between a Numerical Control Unit and Peripheral Equipment Employing Asynchronous Binary Data Interchange over Circuits having EIA-423-A Electrical Characteristics
ANSI/EIA 494	32-bit Binary CL Interchange (BCL) Input Format for Numerically Controlled Machines
EIA AB_3-D	Glossary of Terms for Numerically Controlled Machines
EIA Bulletin 12	Application Notes on Interconnection between Interface Circuits Using RS-449 and RS-232-C
ANSI X 3.94	Programming Aid for Numerically Controlled Manufacturing
ANSI X 3.37	Programming Language APT
ANSI X 3.20	1-inch Perforated Tape Take-up Reels for Information Interchange
ANSI X 3.82	One-sided Single Density Unformatted 5.25 inch Flexible Disc Cartridges

Numerical Control Standards *(Continued)*

Standard Title	Description
ISO 841	Numerical Control of Machines—Axis and Motion Nomenclature
ISO 2806	Numerical Control of Machines—Bilingual Vocabulary
ISO 2972	Numerical Control of Machines—Symbols
ISO 3592	Numerical Control of Machines—Numerical Control Processor Output, Logical Structure and Major Words
ISO 4336	Numerical Control of Machines—Specification of Interface Signals between the Numerical Control Unit and the Electrical Equipment of a Numerically Controlled Machine
ISO 4343	Numerical Control of Machines—NC Processor Output— Minor Elements of 2000-type Records (Post Processor Commands)
ISO TR 6132	Numerical Control of Machines—Program Format and Definition of Address Words—Part 1: Data Format for Positioning, Line Motion and Contouring Control Systems
ISO 230-1	Geometric Accuracy of Machines Operating Under No-Load or Finishing Conditions
ISO 230-2	Determination of Accuracy and Repeatability of Positioning of Numerically Controlled Machine Tools
NAS 911	Numerically Controlled Skin/Profile Milling Machines
NAS 912	Numerically Controlled Spar Milling Machines
NAS 913	Numerically Controlled Profiling and Contouring Milling Machines
NAS 914	Numerically Controlled Horizontal Boring, Drilling and Milling Machines
NAS 960	Numerically Controlled Drilling Machines
NAS 963	Computer Numerically Controlled Vertical and Horizontal Jig Boring Machines
NAS 970	Basic Tool Holders for Numerically Controlled Machine Tools
NAS 971	Precision Numerically Controlled Measuring/Inspection Machines
NAS 978	Numerically Controlled Machining Centers
NAS 990	Numerically Controlled Composite Filament Tape Laying Machines
NAS 993	Direct Numerical Control System
NAS 994	Adaptive Control System for Numerically Controlled Milling Machines
NAS 995	Specification for Computerized Numerical Control (CNC)
NMTBA	Common Words as They Relate to Numerical Control Software
NMTBA	Definition and Evaluation of Accuracy and Repeatability of Numerically Controlled Machine Tools
NMTBA	Numerical Control Character Code Cross Reference Chart
NMTBA	Selecting an Appropriate Numerical Control Programming Method
NEMA 1A1	Industrial Cell Controller Classification Concepts and Selection Guide

Programmable Controller.—Frequently referred to as a PC or PLC (the latter term meaning Programmable Logic Controller), a programmable controller is an electronic unit or small computer. PLCs are used to control machinery, equipment, and complete processes, and to assist CNC systems in the control of complex NC machine tools and flexible manufacturing modules and cells. In effect, PLCs are the technological replacements for electrical relay systems.

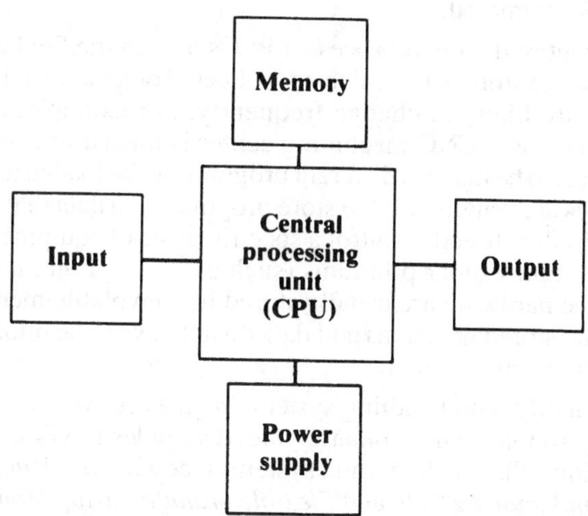

Fig. 1. Programmable Controllers' Four Basic Elements

As shown in Fig. 1, a PLC is composed of four basic elements: the equipment for handling input and output (I/O) signals, the central processing unit (CPU), the power supply, and the memory. Generally, the CPU is a microprocessor and the brain of the PLC. Early PLCs used hardwired special-purpose electronic logic circuits, but most PLCs now being offered are based on microprocessors and have far more logic and control capabilities than was possible with hardwired systems. The CPU scans the status of the input devices continuously, correlates these inputs with the control logic in the memory, and produces the appropriate output responses needed to control the machine or equipment.

Input to a PLC is either discrete or continuous. Discrete inputs may come from push buttons, micro switches, limit switches, photocells, proximity switches or pressure switches, for instance. Continuous inputs may come from sources such as thermocouples, potentiometers, or voltmeters. Outputs from a PLC normally are directed to actuating hardware such as solenoids, solenoid valves, and motor starters. The function of a PLC is to examine the status of an input or set of inputs and, based on this status, actuate or regulate an output or set of outputs.

Digital control logic and sensor input signals are stored in the memory as a series of binary numbers (zeros and ones). Each memory location holds only one "bit" (either 0 or 1) of binary information; however, most of the data in a PLC are used in groups of 8 bits, or bytes. A word is a group of bytes that is operated on at one time by the PLC. The word size in modern PLCs ranges from 8 to 32 bits (1 to 4 bytes), depending on the design of the PLC. In general, the larger the word size that a system is able to operate on (that is, to work on at one time), the faster the system is going to perform. New systems are now beginning to appear that can operate on 64 bits of information at a time.

There are two basic categories of memory: volatile and nonvolatile. Volatile memory loses the stored information when the power is turned off, but nonvolatile memory retains its logic even when power is cut off. A backup battery must be used if the information stored in volatile memory is to be retained. There are six commonly used types of memory. Of these six, random-access memory (RAM) is the most common type because it is the easiest to program and edit. RAM is also the only one of the six common types that is volatile memory. The five nonvolatile memory types are: core memory, read-only memory (ROM), programmable read-only memory (PROM), electronically alterable programmable read-only memory (EAPROM), and electronically erasable programmable read-only memory (EEPROM). EEPROMs are becoming more popular due to their relative ease of programming and their nonvolatile characteristic. ROM is often used as a generic term to refer to the general class of read-only memory types and to indicate that this type of memory is not usually reprogrammed.

More than 90 per cent of the microprocessor PLCs now in the field use RAM memory. RAM is primarily used to store data, which are collected or generated by a process, and to store programs that are likely to change frequently. For example, a part program for machining a workpiece on a CNC machining center is loaded into and stored in RAM. When a different part is to be made, a different program can be loaded in its place. The nonvolatile memory types are usually used to store programs and data that are not expected to be changed. Programs that directly control a specific piece of equipment and contain specific instructions that allow other programs (such as a part program stored in RAM) to access and operate the hardware are usually stored in nonvolatile memory or ROM. The benefit of ROM is that stored programs and data do not have to be reloaded into the memory after the power has been turned off.

PLCs are used primarily with handling systems such as conveyors, automatic retrieval and storage systems, robots, and automatic guided vehicles (AGV), such as are used in flexible manufacturing cells, modules, and systems (see *Flexible Manufacturing Systems (FMS)*, *Flexible Manufacturing Cell*, and *Flexible Manufacturing Module*). PLCs are also to be found in applications as diverse as combustion chamber control, chemical process control, and printed-circuit-board manufacturing.

Types of Programmable Controllers

Type	No. of I/Os	General Applications	Math Capability
Mini	32	Replaces relays, timers, and counters.	Yes
Micro	32–64	Replaces relays, timers, and counters.	Yes
Small	64–128	Replaces relays, timers, and counters. Used for materials handling, and some process control.	Yes
Medium	128–512	Replaces relays, timers, and counters. Used for materials handling, process control, and data collection.	Yes
Large	512+	Replaces relays, timers, and counters. Master control for other PLCs and cells and for generation of reports. High-level network capability	Yes

Types of PLCs may be divided into five groups consisting of micro, mini, small, medium, and large according to the number of I/Os, functional capabilities, and memory capacity. The smaller the number of I/Os and memory capacity, and the fewer the functions, the simpler the PLC. Micro and mini PLCs are usually little more than replacements for relay systems, but larger units may have the functional capabilities of a small computer and be able to handle mathematical functions, generate reports, and maintain high-level communications.

The preceding guidelines have some gray areas because mini, micro, and small PLCs are now available with large memory sizes and functional capacities normally reserved for medium and large PLCs. The accompanying table compares the various types of PLCs and their applications.

Instructions that are input to a PLC are called programs. Four major programming languages are used with PLCs, comprising ladder diagrams, Boolean mnemonics, functional blocks, and English statements. Some PLC systems even support high-level programming languages such as BASIC and PASCAL. Ladder diagrams and Boolean mnemonics are the basic control-level languages. Functional blocks and English statements are considered high-level languages. Ladder diagrams were used with electrical relay systems before these systems were replaced by PLCs and are still the most popular programming method, so they will be discussed further.

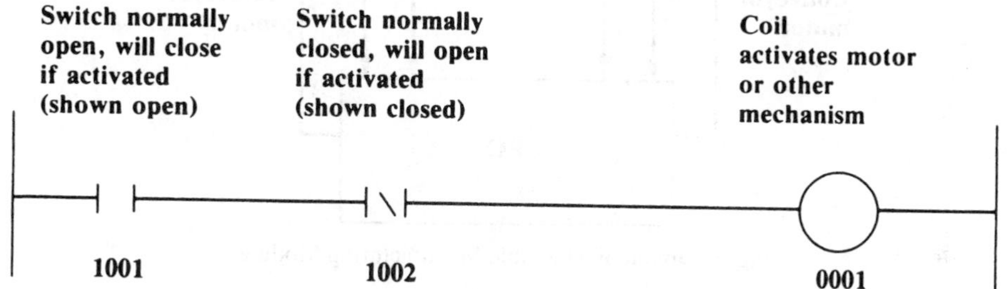

Fig. 2. One Rung on a Ladder Diagram

A ladder diagram consists of symbols, or ladder logic elements, that represent relay contacts or switches and other elements in the control system. One of the more basic symbols represents a normally open switch and is described by the symbol ┤ ├. Another symbol is the normally closed switch, described by the symbol ┤\├. When the normally open switch is activated, it will close, and when the normally closed switch is activated, it will open. Fig. 2 shows one rung (line) on a ladder diagram. Switch 1001 is normally open and switch 1002 is closed. A symbol for a coil (0001) is shown at the right. If switch 1001 is actuated, it will close. If switch 1002 is not activated, it will stay closed. With the two switches closed, current will flow through the line and energize coil 0001. The coil will activate some mechanism such as an electric motor, a robot, or an NC machine tool, for instance.

As an example, Fig. 3 shows a flexible manufacturing module (FMM), consisting of a turning center (NC lathe), an infeed conveyor, an outfeed conveyor, a robot that moves workpieces between the infeed conveyor, the turning center, and the outfeed conveyor, and a PLC. The arrowed lines show the signals going to and coming from the PLC.

Fig. 4 shows a ladder diagram for a PLC that would control the operations of the FMM by:

1) Activating the infeed conveyor to move the workpiece to a position where the robot can pick it up

2) Activating the robot to pick up the workpiece and load it into the chuck on the NC lathe

3) Activating the robot to remove the finished workpiece and place it on the outfeed conveyor

4) Activating the outfeed conveyor to move the workpiece to the next operation

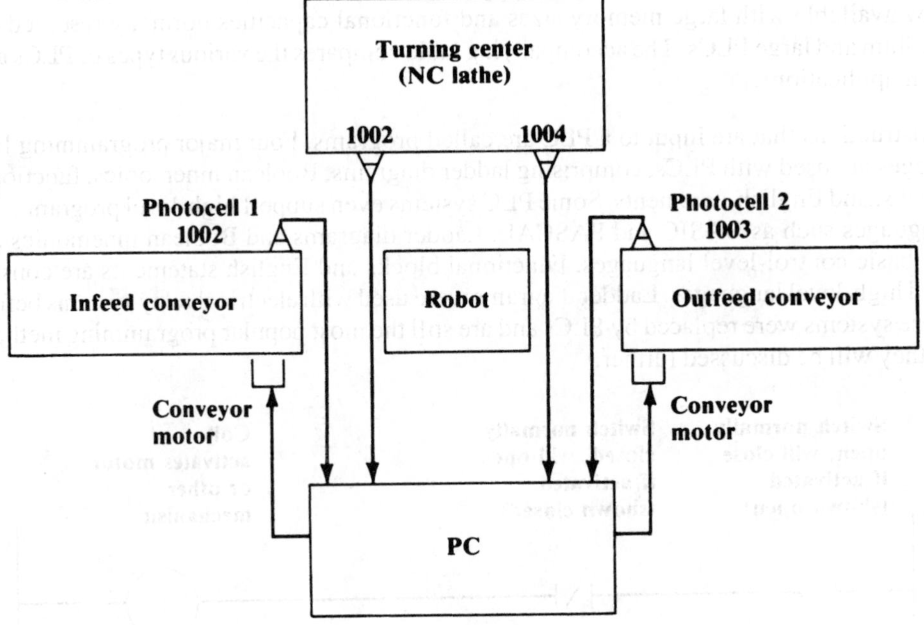

Fig. 3. Layout of a Flexible Manufacturing Module

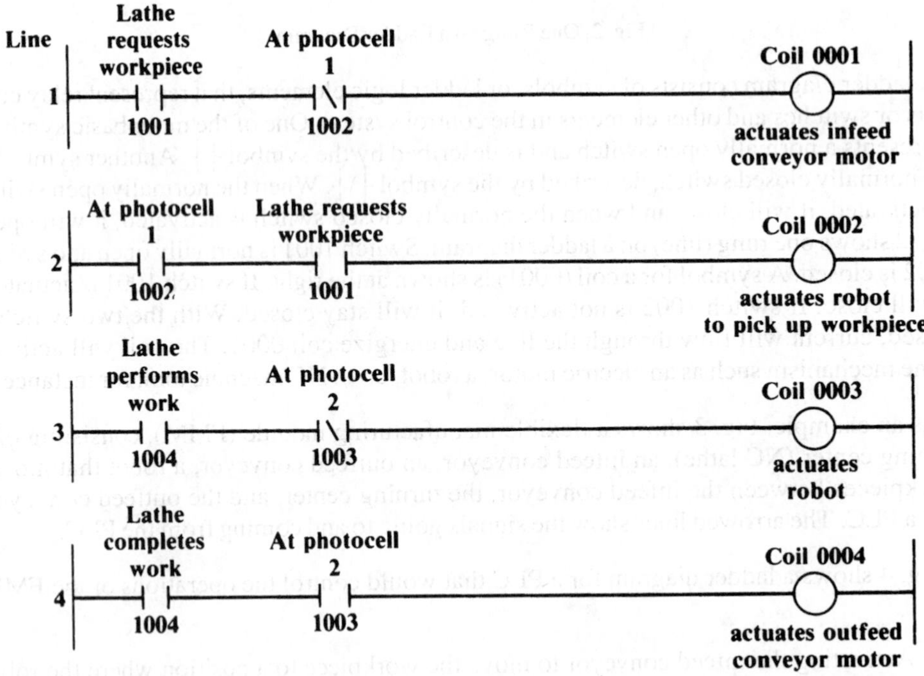

Fig. 4. Portion of a Typical Ladder Diagram for Control of a Flexible Manufacturing Module Including a Turning Center, Conveyors, a Robot, and a Programmable Controller

In Rung 1 of Fig. 4, a request signal for a workpiece from the NC lathe closes the normally open switch 1001. Switch 1002 will remain closed if photocell 1 is not activated, i.e., if it does not detect a workpiece. The signal therefore closes the circuit, energizes the coil, and starts the conveyor motor to bring the next workpiece into position for the robot to grasp.

In Rung 2, switch 1002 (which has been changed in the program of the PLC from a normally closed to a normally open switch) closes when it is activated as photocell 1 detects the workpiece. The signal thus produced, together with the closing of the now normally open switch 1001, energizes the coil, causing the robot to pick up the workpiece from the infeed conveyor.

In Rung 3, switch 1004 on the lathe closes when processing of the part is completed and it is ready to be removed by the robot. Photocell 2 checks to see if there is a space on the conveyor to accept the completed part. If no part is seen by photocell 2, switch 1003 will remain closed, and with switch 1004 closed, the coil will be energized, activating the robot to transfer the completed part to the outfeed conveyor.

Rung 4 shows activation of the output conveyor when a part is to be transferred. Normally open switch 1004 was closed when processing of the part was completed. Switch 1003 (which also was changed from a normally closed to a normally open switch by the program) closes if photocell 2 detects a workpiece. The circuit is then closed and the coil is energized, starting the conveyor motor to move the workpiece clear to make way for the succeeding workpiece.

Closed-Loop System.— Also referred to as a servo or feedback system, a closed-loop system is a control system that issues commands to the drive motors of an NC machine. The system then compares the results of these commands as measured by the movement or location of the machine component, such as the table or spindlehead. The feedback devices normally used for measuring movement or location of the component are called resolvers, encoders, Inductosyns, or optical scales. The resolver, which is a rotary analog mechanism, is the least expensive, and has been the most popular since the first NC machines were developed. Resolvers are normally connected to the lead-screws of NC machines. Linear measurement is derived from monitoring the angle of rotation of the leadscrew and is quite accurate.

Encoders also are normally connected to the leadscrew of the NC machine, and measurements are in digital form. Pulses, or a binary code in digital form, are generated by rotation of the encoder, and represent turns or partial turns of the leadscrew. These pulses are well suited to the digital NC system, and encoders have therefore become very popular with such systems. Encoders generally are somewhat more expensive than resolvers.

The Inductosyn (a trade name of Farrand Controls, Inc.) also produces analog signals, but is attached to the slide or fixed part of a machine to measure the position of the table, spindlehead, or other component. The Inductosyn provides almost twice the measurement accuracy of the resolver, but is considerably more expensive, depending on the length of travel to be measured.

Optical scales generally produce information in digital form and, like the Inductosyn, are attached to the slide or fixed part of the machine. Optical scale measurements are more accurate than either resolvers or encoders and, because of their digital nature, are well suited to the digital computer in a CNC system. Like the Inductosyn, optical scales are more costly than either resolvers or encoders.

Open-Loop System.— A control system that issues commands to the drive motors of an NC machine and has no means of assessing the results of these commands is known as an open-loop system. In such a system, no provision is made for feedback of information concerning movement of the slide(s), or rotation of the leadscrew(s). Stepping motors are popular as drives for open-loop systems.

Adaptive Control.— Measuring performance of a process and then adjusting the process to obtain optimum performance is called adaptive control. In the machine tool field, adaptive control is a means of adjusting the feed and/or speed of the cutting tool, based on sensor feedback information, to maintain optimum cutting conditions. A typical arrangement is seen in Fig. 5. Adaptive control is used primarily for cutting higher-strength materials such as titanium, although the concept is applicable to the cutting of any material. The costs of the sensors and software have restricted wider use of the feature.

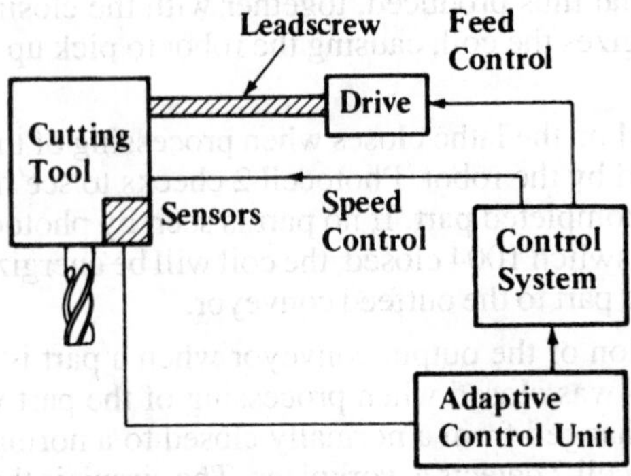

Fig. 5.

The sensors used for adaptive control are generally mounted on the machine drive shafts, tools, or even built into the drive motor. Typically, sensors are used to provide information such as the temperature at the tip of the cutting tool and the cutting force exerted by the tool. The information measured by the sensors is used by the control system computer to analyze the cutting process and adjust the feeds and speeds of the machine to maximize the material removal rate or to optimize another process variable such as surface finish. For the computer to effectively evaluate the process in real time (i.e., while cutting is in progress), details such as maximum allowable tool temperature, maximum allowable cutting force, and information about the drive system need to be integrated into the computer program monitoring the cutting process.

Adaptive control can be used to detect worn, broken, or dull tooling. Ordinarily, the adaptive control system monitors the cutting process to keep the process variables (cutting speed and feed rate, for example) within the proper range. Because the force required to machine a workpiece is lowest when the tool is new or recently resharpened, a steady increase in cutting force during a machining operation, assuming that the feed remains the same, is an indication that the tool is becoming dull (temperature may increase as well). Upon detecting cutting forces that are greater than a predetermined maximum allowable force, the control system causes the feed rate, the cutting speed, or both to be adjusted to maintain the cutting force within allowable limits. If the cutting force cannot be maintained without causing the speed and/or feed rate to be adjusted outside its allowable limits, the machine will be stopped, indicating that the tool is too dull and must be resharpened or replaced.

On some systems, the process monitoring equipment can interface directly with the machine control system, as discussed above. On other systems, the adaptive control is implemented by a separate monitoring system that is independent of the machine control system. These systems include instrumentation to monitor the operations of the machine tool, but do not have the capability to directly change operating parameters, such as feeds and speeds. In addition, this type of control does not require any modification of the existing part programs for control of the machine.

Flexible Manufacturing Systems (FMS).—A flexible manufacturing system (FMS) is a computer-controlled machining arrangement that can perform a variety of continuous metal-cutting operations on a range of components without manual intervention. The objective of such a system is to produce components at the lowest possible cost, especially components of which only small quantities are required. Flexibility, or the ability to switch from manufacture of one type of component to another, or from one type of machining to another, without interrupting production, is the prime requirement of such a system. In general, FMS are used for production of numbers of similar parts between 200 and 2000, although larger quantities are not uncommon. An FMS involves almost all the departments in a company, including engineering, methods, tooling and part programming, planning and scheduling, purchasing, sales and customer service, accounting, maintenance, and quality control. Initial costs of an FMS are estimated as being borne (percentages in parentheses) by machine tools (46.2), materials handling systems (7.7), tooling and fixtures (5.9), pallets (1.9), computer hardware (3.7), computer software (2.2), wash stations (2.8), automatic storage and retrieval systems (6.8), coolant and chip systems (2.4), spares (2), and others (18.4).

FMS are claimed to bring reductions in direct labor (80–90), production planning and control (65), and inspection (70). Materials handling and shop supervision are reduced, and individual productivity is raised. In the materials field, savings are made in tooling (35), scrap and rework (65), and floor space (50). Inventory is reduced and many other costs are avoided. Intangible savings claimed to result from FMS include reduced tooling changeover time, ability to produce complex parts, to incorporate engineering changes more quickly and efficiently than with other approaches, and to make special designs, so that a company can adapt quickly to changing market conditions. Requirements for spare parts with good fit are easily met, and the lower costs combine with higher quality to improve market share. FMS also are claimed to improve morale among workers, leading to higher productivity, with less paper work and more orderly shop operations. Better control of costs and improved cost data help to produce more accurate forecasts of sales and manpower requirements. Response to surges in demand and more economical materials ordering are other advantages claimed with FMS.

Completion of an FMS project is said to average 57 months, including 20 months from the time of starting investigations to the placing of the purchase order. A further 13 months are needed for delivery and a similar period for installation. Debugging and building of production takes about another 11 months before production is running smoothly. FMS are expensive, requiring large capital outlays and investments in management time, software, engineering, and shop support. Efficient operation of FMS also require constant workflow because gaps in the production cycle are very costly.

Flexible Manufacturing Cell.—A flexible manufacturing cell usually consists of two or three NC machines with some form of pallet-changing equipment or an industrial robot. Prismatic-type parts, such as would be processed on a machining center, are usually handled on pallets. Cylindrical parts, such as would be machined on an NC lathe, usually are handled with an overhead type of robot. The cell may be controlled by a computer, but is often run by programmable controllers. The systems can be operated without attendants, but the mixture of parts usually must be less than with a flexible manufacturing system (FMS).

Flexible Manufacturing Module.—A flexible manufacturing module is defined as a single machining center (or turning center) with some type of automatic materials handling equipment such as multiple pallets for machining centers, or robots for manipulating cylindrical parts and chucks for turning centers. The entire module is usually controlled by one or more programmable logic controllers.

Axis Nomenclature.—To distinguish among the different motions, or axes, of a machine tool, a system of letter addresses has been developed. A letter is assigned, for example, to the table of the machine, another to the saddle, and still another to the spindle head. These letter addresses, or axis designations, are necessary for the electronic control system to assign movement instructions to the proper machine element. The assignment of these letter addresses has been standardized on a worldwide basis and is contained in three standards, all of which are in agreement. These standards are EIA RS-267-B, issued by the Electronics Industries Association; AIA NAS-938, issued by the Aerospace Industries Association; and ISO/R 841, issued by the International Organization for Standardization.

The standards are based on a "right-hand rule," which describes the orientation of the motions as well as whether the motions are positive or negative. If a right hand is laid palm up on the table of a vertical milling machine, as shown in Fig. 1, for example, the thumb will point in the positive *X*-direction, the forefinger in the positive *Y*-direction, and the erect middle finger in the positive *Z*-direction, or up. The direction signs are based on the motion of the cutter relative to the workpiece. The movement of the table shown in Fig. 2 is therefore positive, even though the table is moving to the left, because the motion of the cutter relative to the workpiece is to the right, or in the positive direction. The motions are considered from the part programmer's viewpoint, which assumes that the cutter always moves around the part, regardless of whether the cutter or the part moves. The right-hand rule also holds with a horizontal-spindle machine and a vertical table, or angle plate, as shown in Fig. 3. Here, spindle movement back and away from the angle plate, or workpiece, is a positive *Z*-motion, and movement toward the angle plate is a negative *Z*-motion.

Rotary motions also are governed by a right-hand rule, but the fingers are joined and the thumb is pointed in the positive direction of the axis. Fig. 4 shows the designations of the rotary motions about the three linear axes, *X*, *Y*, and *Z*. Rotary motion about the *X*-axis is designated as *A*; rotary motion about the *Y*-axis is *B*; and rotary motion about the *Z*-axis is *C*. The fingers point in the positive rotary directions. Movement of the rotary table around the *Y*-axis shown in Fig. 4 is a *B* motion and is common with horizontal machining centers. Here, the view is from the spindle face looking toward the rotary table. Referring, again, to linear motions, if the spindle is withdrawn axially from the work, the motion is a positive *Z*. A move toward the work is a negative Z.

When a second linear motion is parallel to another linear motion, as with the horizontal boring mill seen in Fig. 5, the horizontal motion of the spindle, or quill, is designated as *Z* and a parallel motion of the angle plate is *W*. A movement parallel to the *X*-axis is *U* and a movement parallel to the *Y*-axis is *V*. Corresponding motions are summarized as follows:

Linear	Rotary	Linear and Parallel
X	*A*	*U*
Y	*B*	*V*
Z	*C*	*W*

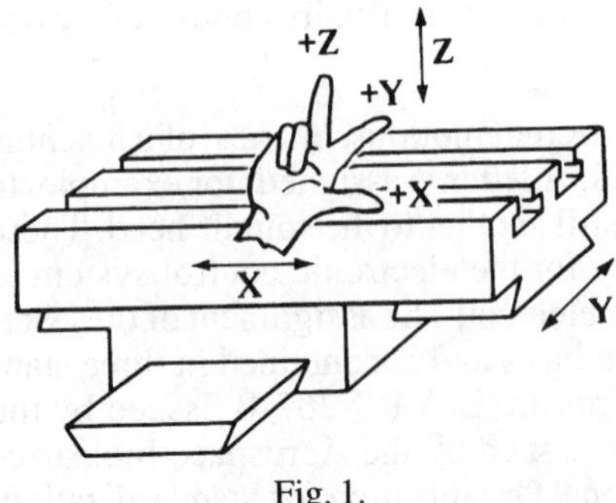
+Z
Z
+Y
+X
X
Y
Fig. 1.

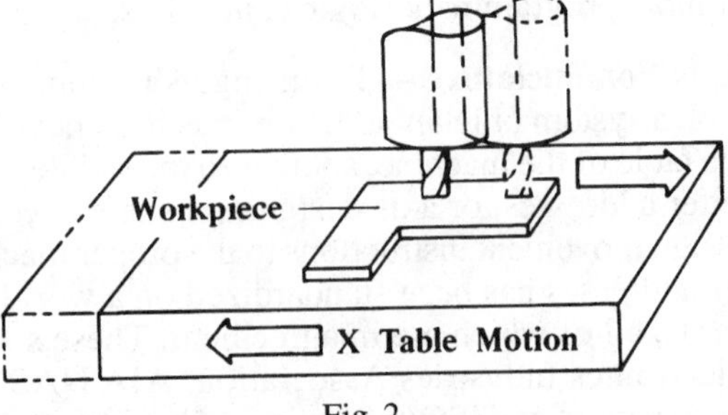
Table movement to the left in effect moves the cutter to the right, relative to the workpiece, or in the + direction
Workpiece
X Table Motion
Fig. 2.

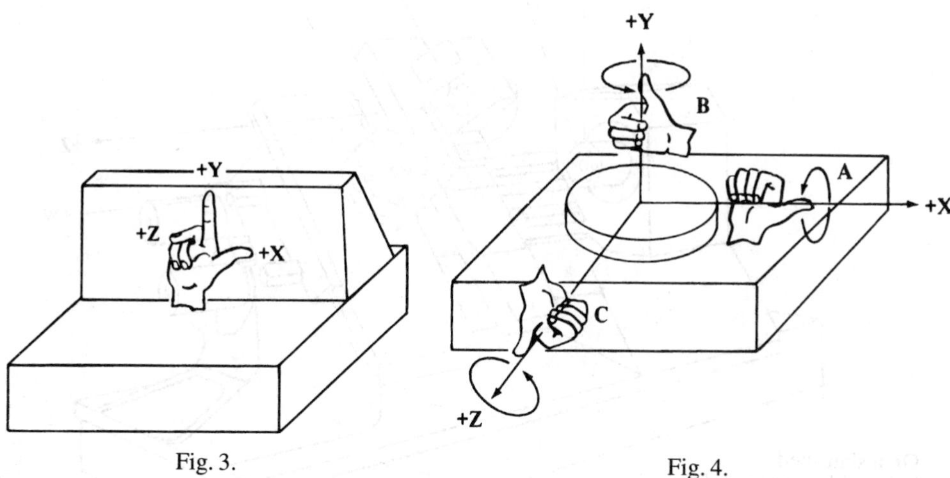

Fig. 3.

Fig. 4.

Axis designations for a lathe are shown in Fig. 6. Movement of the cross-slide away from the workpiece, or the centerline of the spindle, is noted as a plus *X*. Movement toward the workpiece is a minus *X*. The middle finger points in the positive *Z*-direction; therefore, movement away from the headstock is positive and movement toward the headstock is negative. Generally, there is no *Y*-movement.

The machine shown in Fig. 6 is of conventional design, but most NC lathes look more like that shown in Fig. 7. The same right-hand rule applies to this four-axis lathe, on which each turret moves along its own two independent axes. Movement of the outside-diameter or upper turret, up and away from the workpiece, or spindle centerline, is a positive *X*-motion, and movement toward the workpiece is a negative *X*-motion. The same rules apply to the *U*-movement of the inside-diameter, or boring, turret. Movement of the lower turret parallel to the *Z*-motion of the outside-diameter turret is called the *W*-motion. A popular lathe configuration is to have both turrets on one slide, giving a two-axis system rather than the four-axis system shown. *X*-and *Z*-motions may be addressed for either of the two heads. Upward movement of the boring head therefore is a positive *X*-motion.

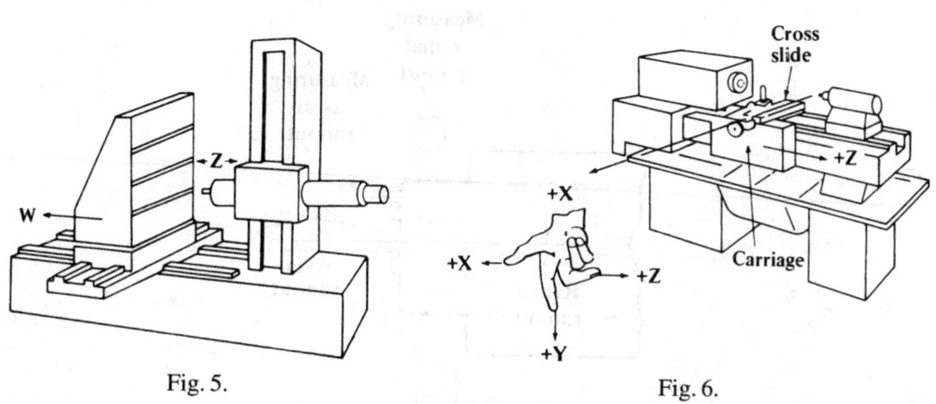

Fig. 5.

Fig. 6.

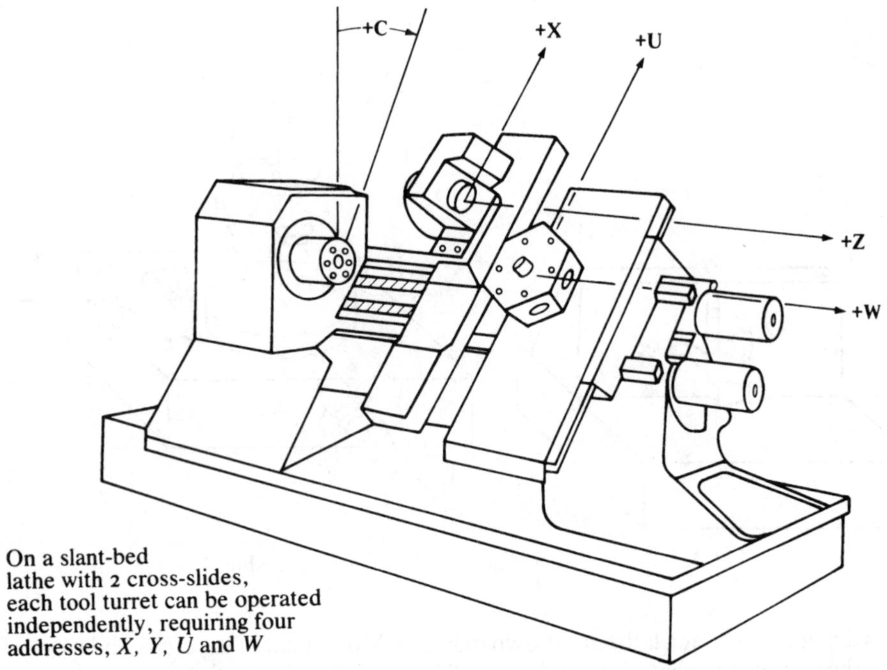

Fig. 7.

Axis nomenclature for other machine configurations is shown in Fig. 9. The letters with the prime notation (e.g., *X′*, *Y′*, *Z′*, *W′*, *A′*, and *B′*) mean that the motion shown is positive, because the movement of the cutter with respect to the work is in a positive direction. In these instances, the workpiece is moving rather than the cutter.

Total Indicator Reading (TIR).—Total indicator reading is used as a measure of the range of machine tool error. TIR is particularly useful for describing the error in a machine tool spindle, referred to as runout. As shown in Fig. 8, there are two types of runout: axial and radial, which can be measured with a dial indicator. Axial runout refers to the wobble of a spindle and is measured at the spindle face. Radial runout is the range of movement of the spindle centerline and is measured on the side of the spindle or quill.

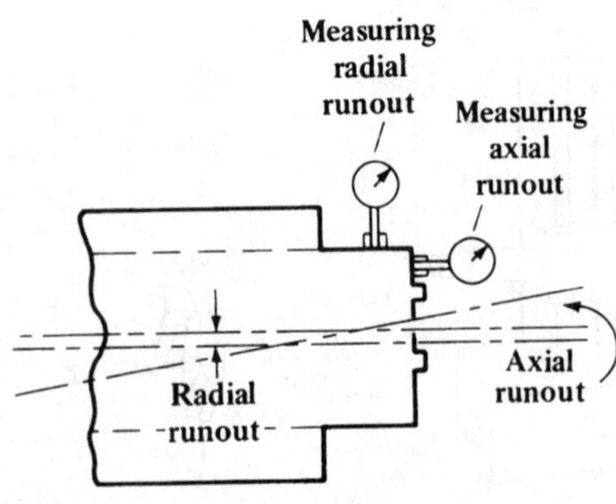

Fig. 8.

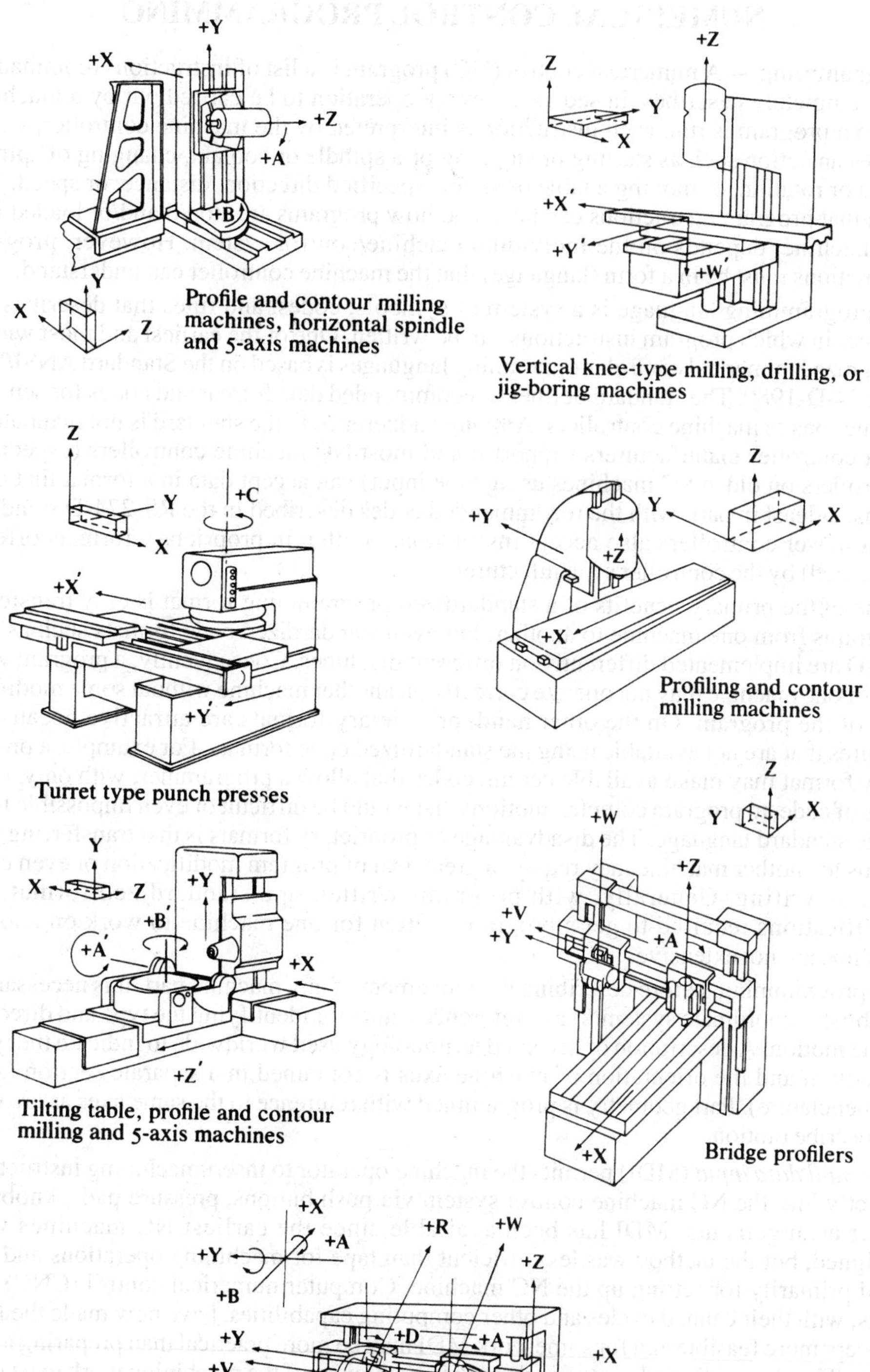
+Y
+X
+Z
+A
+B
Y
X
Z
Profile and contour milling machines, horizontal spindle and 5-axis machines
+Z
Z
Y
X
+X′
+Y′
+W′
Vertical knee-type milling, drilling, or jig-boring machines
Z
Y
X
+C
+X′
+Y
Turret type punch presses
Z
Y
X
+Y
+Z
Profiling and contour milling machines
+X
Y
X
Z
+Y
+B′
+A′
+X
+Z′
Tilting table, profile and contour milling and 5-axis machines
Z
Y
X
+W
+Z
+V
+Y
+A
+X
Bridge profilers
+X
+R
+W
+A
+Y
+Z
+B
+Y
+D
+A
+V
+X
+B
Gantry profilers

Fig. 9.

NUMERICAL CONTROL PROGRAMMING

Programming.—A numerical control (NC) program is a list of instructions (commands) that completely describes, in sequence, every operation to be carried out by a machine. When a program is run, each instruction is interpreted by the machine controller, which causes an action such as starting or stopping of a spindle or coolant, changing of spindle speed or rotation, or moving a table or slide a specified direction, distance, or speed. The form that program instructions can take, and how programs are stored and/or loaded into the machine, depends on the individual machine/control system. However, program instructions must be in a form (language) that the machine controller can understand.

A programming language is a system of symbols, codes, and rules that describes the manner in which program instructions can be written. One of the earliest and most widely recognized numerical control programming languages is based on the Standard ANSI/EIA RS-274-D-1980. The standard defines a recommended data format and codes for sending instructions to machine controllers. Although adherence to the standard is not mandatory, most controller manufacturers support it and most NC machine controllers (especially controllers on older NC machines using tape input) can accept data in a format that conforms, at least in part, with the recommended codes described in the RS-274-D standard. Most newer controllers also accept instructions written in proprietary formats offered (specified) by the controller's manufacturer.

One of the primary benefits of a standardized programming format is easy transfer of programs from one machine to another, but even standardized code formats such as RS-274-D are implemented differently on different machines. Consequently, a program written for one machine may not operate correctly on another machine without some modification of the program. On the other hand, proprietary formats are attractive because of features that are not available using the standardized code formats. For example, a proprietary format may make available certain codes that allow a programmer, with only a few lines of code, to program complex motions that would be difficult or even impossible to do in the standard language. The disadvantage of proprietary formats is that transferring programs to another machine may require a great deal of program modification or even complete rewriting. Generally, with programs written in a standardized format, the modifications required to get a program written for one machine to work on another machine are not extensive.

In programming, before describing the movement of any machine part, it is necessary to establish a coordinate system(s) as a reference frame for identifying the type and direction of the motion. A description of accepted terminology used worldwide to indicate the types of motion and the orientation of machine axes is contained in a separate section (Axis Nomenclature). Part geometry is programmed with reference to the same axes as are used to describe motion.

Manual data input (MDI) permits the machine operator to insert machining instructions directly into the NC machine control system via push buttons, pressure pads, knobs, or other arrangements. MDI has been available since the earliest NC machines were designed, but the method was less efficient than tape for machining operations and was used primarily for setting up the NC machine. Computer numerical control (CNC) systems, with their canned cycles and other computing capabilities, have now made the MDI concept more feasible and for some work MDI may be more practical than preparing a program. The choice depends very much on the complexity of the machining work to be done and, to a lesser degree, on the skill of the person who prepares the program.

Conversational part programming is a form of MDI that requires the operator or programmer to answer a series of questions displayed on the control panel of the CNC. The operator replies to questions that describe the part, material, tool and machine settings, and machining operations by entering numbers that identify the material, blank size and thickness or diameter, tool definitions, and other required data. Depending on capability, some controls can select the required spindle speed and feed rate automatically by using a materials look-up table; other systems request the appropriate feed and speed data. Tool motions needed to machine a part are described by selecting a linear or circular motion programming mode and entering endpoint and intersection coordinates of lines and radius, diameter, tangent points, and directions of arcs and circles (with some controllers, intersection and tangent points are calculated automatically). Machined elements such as holes, slots, and bolt circles are entered by selecting the appropriate tool and describing its action, or with "canned routines" built into the CNC to perform specific machining operations. On some systems, if a feature is once described, it can be copied and/or moved by: translation (copy and/or move), rotation about a point, mirror image (copy and rotate about an axis), and scaling (copy and change size). On many systems, as each command is entered, a graphic image of the part or operation gives a visual check that the program is producing the intended results. When all the necessary data have been entered, the program is constructed and can be run immediately or saved on tape, floppy disk, or other storage media for later use.

Conversational programming gives complete control of machine operations to the shop personnel, taking advantage of the experience and practical skills of the machine operator/programmer. Control systems that provide conversational programming usually include many built-in routines (fixed or canned cycles) for commonly used machining operations and may also have routines for specialized operations. Built-in routines speed programming because one command may replace many lines of program code that would take considerable time to write. Some built-in cycles allow complex machining operations to be programmed simply by specifying the final component profile and the starting stock size, handling such details as developing tool paths, depth of cut, number of roughing passes, and cutter speed automatically. On turning machines, built-in cycles for reducing diameters, chamfer and radius turning, and cutting threads automatically are common. Although many CNC machines have a conversational programming mode, the programming methods used and the features available are not standardized. Some control systems cannot be programmed from the control panel while another program is running (i.e., while a part is being machined), but those systems that can be thus programmed are more productive because programming does not require the machine to be idle. Conversational programming is especially beneficial In reducing programming time in shops that do most of their part programming from the control panel of the machine.

Manual part programming describes the preparation of a part program by manually writing the part program in word addressed format. In the past, this method implied programming without using a computer to determine tool paths, speeds and feeds, or any of the calculations normally required to describe the geometry of a part. Today, however, computers are frequently used for writing and storing the program on disk, as well as for calculations required to program the part. Manual part programming consists of writing codes, in a format appropriate to the machine controller, that instruct the controller to perform a specific action. The most widely accepted form of coding the instructions for numerically controlled machines uses the codes and formats suggested in the ANSI/EIA RS-274-D-1980, standard. This type of programming is sometimes called G-code programming, referring to a commonly used word address used in the RS-274-D standard. Basic details of programming in this format, using the various codes available, are discussed in the next section (G-Code Programming).

Computer-assisted part programming (CAPP) uses a computer to help in the preparation of the detailed instructions for operating an NC machine. In the past, defining a curve or complicated surface profile required a series of complex calculationsto describe the features in intimate detail. However, with the introduction of the microprocessor as an integral part of the CNC machine, the process of defining many complex shapes has been reduced to the simple task of calling up a canned cycle to calculate the path of the cutter. Most new CNC systems have some graphic programming capability, and many use graphic images of the part "drawn" on a computer screen. The part programmer moves a cutter about the part to generate the part program or the detailed block format instructions required by the control system. Machining instructions, such as the speed and feed rate, are entered via the keyboard. Using the computer as an assistant is faster and far more accurate than the manual part programming method.

Computer-assisted part programming methods generally can be characterized as either language-based or graphics-based, the distinction between the two methods being primarily in the manner by which the tool paths are developed. Some modern-language-based programming systems, such as Compact II, use interactive alphanumeric input so that programming errors are detected as soon as they are entered. Many of these programming systems are completely integrated with computer graphics and display an image of the part or operation as soon as an instruction is entered. The language-based programming systems are usually based on, or are a variation of, the APT programming language, which is discussed separately within this section (APT Programming).

The choice between computer-assisted part programming and manual part programming depends on the complexity of the part (particularly its geometry) and how many parts need to be programmed. The more complicated the part, the more benefit to be gained by CAPP, and if many parts are to be programmed, even if they are simple ones, the benefits of a computer-aided system are substantial. If the parts are not difficult to program but involve much repetition, computer-assisted part programming may also be preferred. If parts are to be programmed for several different control systems, a high-level part programming language such as APT will make writing the part programs easier. Because almost all machines have some deviations from standard practices, and few control systems use exactly the same programming format, a higher-level language allows the programmer to concentrate primarily on part geometry and machining considerations. The postprocessors (see *Postprocessors* below) for the individual control systems accommodate most of the variations in the programming required. The programmer only needs to write the program; the postprocessor deals with the machine specifics.

Graphical programming involves building a two- or three-dimensional model of a part on a computer screen by graphically defining the geometric shapes and surfaces of the part using the facilities of a CAD program. In many cases, depending on features of the CAD software package, the same computer drawing used in the design and drafting stage of a project can also be used to generate the program to produce the part. The graphical entities, such as holes, slots, and surfaces, are linked with additional information required for the specific machining operations needed. Most of the cutter movements (path of the cutter), such as those needed for the generation of pockets and lathe roughing cuts, are handled automatically by the computer. The program may then sort the various machining operations into an efficient sequence so that all operations that can be performed with a particular tool are done together, if possible. The output of graphical part programming is generally an alphanumeric part programming language output file, in a format such as an APT or Compact II file.

The part programming language file can be manually checked, and modified, as necessary before being run, and to help detect errors, many graphics programming systems also include some form of part verification software that simulates machining the part on the computer screen. Nongraphic data, such as feed rates, spindle speeds and coolant on/off, must be typed in by the part programmer or entered from acomputer data base at the appropriate points in the program, although some programs prompt for this information when needed. When the part program language file is run or compiled, the result is a center line data (CL data) file describing the part. With most computer-aided part programming output files, the CL data file needs to be processed through a postprocessor (see *Postprocessors* below) to tailor the final code produced to the actual machine being used. Postprocessor output is in a form that can be sent directly to the control system, or can be saved on tape or magnetic media and transferred to the machine tool when necessary. The graphic image of the part and the alphanumeric output files are saved in separate files so that either can be edited in the future if changes in the part become necessary. Revised files must be run and processed again for the part modifications to be included in the part program. Software for producing part programs is discussed further in the CAD/CAM section.

Postprocessors.— A postprocessor is computer software that contains a set of computer instructions designed to tailor the cutter center line location data (CL data), developed by a computerized part programming language, to meet the requirements of a particular machine tool/system combination. Generally, when a machine tool is programmed in a graphical programming environment or any high-level language such as APT, a file is created that describes all movements required of a cutting tool to make the part. The file thus created is run, or compiled, and the result is a list of coordinates (CL data) that describes the successive positions of the cutter relative to the origin of the machine's coordinate system. The output of the program must be customized to fit the input requirements of the machine controller that will receive the instructions. Cutter location data must be converted into a format recognized by the control system, such as G codes and M codes, or into another language or proprietary format recognized by the controller. Generally, some instructions are also added or changed by the programmer at this point.

The lack of standardization among machine tool control systems means that almost all computerized part programming languages require a postprocessor to translate the computer-generated language instructions into a form that the machine controller recognizes. Postprocessors are software and are generally prepared for a fee by the machine tool builder, the control system builder, a third party vendor, or by the user.

G-Code Programming

Programs written to operate numerical control (NC) machines with control systems that comply with the ANSI/EIA RS-274-D-1980, Standard consist of a series of data blocks, each of which is treated as a unit by the controller and contains enough information for a complete command to be carried out by the machine. Each block is made up of one or more words that indicate to the control system how its corresponding action is to be performed. A word is an ordered set of characters, consisting of a letter plus some numerical digits, that triggers a specific action of a machine tool. The first letter of the word is called the letter address of the word, and is used to identify the word to the control system. For example, X is the letter address of a dimension word that requires a move in the direction of the X-axis, Y is the letter address of another dimension word; and F is the letter address of the feed rate. The assigned letter addresses and their meanings, as listed in ANSI/EIA RS-274-D, are shown in Table 1.

Format Classification.— The *format classification sheet* completely describes the format requirements of a control system and gives other important information required to program a particular control including: the type of machine, the format classification shorthand and format detail, a listing of specific letter address codes recognized by the system (for example, G-codes: G01, G02, G17, etc.) and the range of values the available codes may take (S range: 10 to 1800 rpm, for example), an explanation of any codes not specifically assigned by the Standard, and any other unique features of the system.

The format classification shorthand is a nine- or ten-digit code that gives the type of system, the number of motion and other words available, the type and format of dimensional data required by the system, the number of motion control channels, and the number of numerically controlled axes of the system. The *format detail* verysuccinctly summarizes details of the machine and control system. This NC shorthand gives the letter address words and word lengths that can be used to make up a block. The format detail defines the basic features of the control system and the type of machine tool to which it refers. For example, the format detail

Table 1. Letter Addresses Used in Numerical Control

Letter Address	Description	Refers to
A	Angular dimension about the *X*-axis. Measured in decimal parts of a degree	Axis nomenclature
B	Angular dimension about the *Y*-axis. Measured in decimal parts of a degree	Axis nomenclature
C	Angular dimension about the *Z*-axis. Measured in decimal parts of a degree	Axis nomenclature
D	Angular dimension about a special axis, or third feed function, or tool function for selection of tool compensation	Axis nomenclature
E	Angular dimension about a special axis or second feed function	Axis nomenclature
F	Feed word (code)	Feed words
G	Preparatory word (code)	Preparatory words
H	Unassigned	
I	Interpolation parameter or thread lead parallel to the *X*-axis	Circular interpolation and threading
J	Interpolation parameter or thread lead parallel to the *Y*-axis	Circular interpolation and threading
K	Interpolation parameter or thread lead parallel to the *Z*-axis	Circular interpolation and threading
L	Unassigned	
M	Miscellaneous or auxilliary function	Miscellaneous functions
N	Sequence number	Sequence number
O	Sequence number for secondary head only	Sequence number
P	Third rapid-traverse dimension or tertiary-motion dimension parallel to *X*	Axis nomenclature
Q	Second rapid-traverse dimension or tertiary-motion dimension parallel to *Y*	Axis nomenclature
R	First rapid-traverse dimension or tertiary-motion dimension parallel to *Z* or radius for constant surface-speed calculation	Axis nomenclature
S	Spindle-speed function	Spindle speed
T	Tool function	Tool function
U	Secondary-motion dimension parallel to *X*	Axis nomenclature
V	Secondary-motion dimension parallel to *Y*	Axis nomenclature
W	Secondary-motion dimension parallel to *Z*	Axis nomenclature
X	Primary *X*-motion dimension	Axis nomenclature
Y	Primary *Y*-motion dimension	Axis nomenclature
Z	Primary *Z*-motion dimension	Axis nomenclature

N4G2X + 24Y + 24Z + 24B24I24J24F31T4M2

specifies that the NC machine is a machining center (has *X*-, *Y*-, and *Z*-axes) and a tool changer with a four-digit tool selection code (T4); the three linear axes are programmed with two digits before the decimal point and four after the decimal point (X + 24Y + 24Z + 24) and can be positive or negative; probably has a horizontal spindle and rotary table (B24

= rotary motion about the *Y*-axis); has circular interpolation (I24J24); has a feed rate range in which there are three digits before and one after the decimal point (F31); and can handle a four-digit sequence number (N4), two-digit G-words (G2), and two-digit miscellaneous words (M2). The sequence of letter addresses in the format detail is also the sequence in which words with those addresses should appear when used in a block.

The information given in the format shorthand and format detail is especially useful when programs written for one machine are to be used on different machines. Programs that use the variable block data format described in RS-274-D can be used interchangeably on systems that have the same format classification, but for complete program compatibility between machines, other features of the machine and control system must also be compatible, such as the relationships of the axes and the availability of features and control functions.

Control systems differ in the way that the numbers may be written. Most newer CNC machines accept numbers written in a decimal-point format, however, some systems require numbers to be in a fixed-length format that does not use an explicit decimal point. In the latter case, the control system evaluates a number based on the number of digits it has, including zeros. *Zero suppression* in a control system is an arrangement that allows zeros before the first significant figure to be dropped (leading zero suppression) or allows zeros after the last significant figure to be dropped (trailing zero suppression). An *X*-axis movement of 05.3400, for example, could be expressed as 053400 if represented in the full field format, 53400 (leading zero suppression), or 0534 (trailing zero suppression). With decimal-point programming, the above number is expressed simply as 5.34. To ensure program compatibility between machines, all leading and trailing zeros should be included in numbers unless decimal-point programming is used.

Sequence Number (N-Word).—A block normally starts with a sequence number that identifies the block within the part program. Most control systems use a four-digit sequence number allowing step numbers up to N9999. The numbers are usually advanced by fives or tens in order to leave spaces for additional blocks to be inserted later if required. For example, the first block in a program would be N0000, the next block N0005; the next N0010; and so on. The slash character, /, placed in a block, before the sequence number, is called an *optional stop* and causes the block to be skipped over when actuated by the operator. The block that is being worked on by the machine is often displayed on a digital readout so that the operator may know the precise operation being performed.

Preparatory Word (G-Word).—A preparatory word (also referred to as a preparatory function or G-code) consists of the letter address G and usually two digits. The preparatory word is placed at the beginning of a block, normally following the sequence number. Most newer CNC machines allow more than one G-code to be used in a single block, although many of the older systems do not. To ensure compatability with older machines and with the RS-274-D Standard, only one G-code per block should be used.

The G-word indicates to the control system how to interpret the remainder of theblock. For example, G01 refers to linear interpolation and indicates that the words following in the block will move the cutter in a straight line. The G02 code indicates that the words following in the block will move the cutter in a clockwise circular path. A G-word can completely change the normal meaning of other words in a block. For example, X is normally a dimension word that describes a distance or position in the *X*-direction. However, if a block contains the G04 word, which is the code for a dwell, the X word represents the time, in seconds, that the machine is to dwell.

The majority of G-codes are designated as modal, which means that once used, the code remains in effect for succeeding blocks unless it is specifically changed or canceled. Therefore, it is not necessary to include modal G-codes in succeeding blocks except to change or cancel them. Unless a G-code is modal, it is only effective within its designated block for the operation it defines. Table , G-Code Addresses, lists standardized G-code addresses and modality.

Table 2. G-Code Addresses

Code		Description	Code		Description
G00	ab*	Rapid traverse, point to point (M,L)	G34	ab*	Thread cutting, increasing lead (L)
G01	abc	Linear interpolation (M,L)	G35	abc	Thread cutting, decreasing lead (L)
G02	abc	Circular interpolation — clockwise movement (M,L)	G36-G39	ab	Permanently unassigned
G03	abc	Circular interpolation—counter-clockwise movement (M,L)	G36	c	Used for automatic acceleration and deceleration when the blocks are short (M,L)
G04	ab	Dwell—a programmed time delay (M,L)	G37, G37.1, G37.2, G37.3 G37.4		Used for tool gaging (M,L)
G05	ab	Unassigned			
G06	abc	Parabolic interpolation (M,L)	G38		Used for probing to measure the diameter and center of a hole (M)
G07	c	Used for programming with cylindrical diameter values (L)	G38.1		Used with a probe to measure the parallelness of a part with respect to an axis (M)
G08	ab	Programmed acceleration (M,L). [d] Also for lathe programming with cylindrical diameter values	G39, G39.1		Generates a nonprogrammed block to improve cycle time and corner cutting quality when used with cutter compensation (M)
G09	ab	Programmed deceleration (M,L). [d] Used to stop the axis movement at a precise location (M,L)	G39		Tool tip radius compensation used with linear generated block (L)
G10–G12	ab	Unassigned. [d]Sometimes used for machine lock and unlock devices	G39.1		Tool tip radius compensation used used with circular generated block (L)
G13–G16	ac	Axis selection (M,L)	G40	abc	Cancel cutter compensation/ offset (M)
G13–G16	b	Unassigned	G41	abc	Cutter compensation, left (M)
G13		Used for computing lines and circle intersections (M,L)	G42	abc	Cutter compensation, right (M)
G14, G14.1	c	Used for scaling (M,L)	G43	abc	Cutter offset, inside corner (M,L)
G15–G16	c	Polar coordinate programming (M)	G44	abc	Cutter offset, outside corner (M,L)
G15, G16.1	c	Cylindrical interpolation—C axis (L)	G45–G49	ab	Unassigned
G16.2	c	End face milling—C axis (L)	G50–G59	a	Reserved for adaptive control (M,L)
G17–G19	abc	*X-Y, X-Z, Y-Z* plane selection, respectively (M,L)	G50	bb	Unassigned
G20		Unassigned	G50.1	c	Cancel mirror image (M,L)
G22–G32	ab	Unassigned	G51.1	c	Program mirror image (M,L)
G22–G23	c	Defines safety zones in which the machine axis may not enter (M,L)	G52	b	Unassigned
G22.1, G233.1	c	Defines safety zones in which the cutting tool may not exit (M,L)	G52		Used to offset the axes with respect to the coordinate zero point (see G92) (M,L)
G24	c	Single-pass rough-facing cycle (L)	G53	bc	Datum shift cancel
G27–G29		Used for automatically moving to and returning from home position (M,L)	G53	c	Call for motion in the machine coordinate system (M,L)
			G54–G59	bc	Datum shifts (M,L)
G30		Return to an alternate home position (M,L)	G54–G59.3	c	Allows for presetting of work coordinate systems (M,L)
G31, G31.1, G31.2, G31.3, G31.4		External skip function, moves an axis on a linear path until an external signal aborts the move (M,L)	G60–G62	abc	Unassigned
G33	abc	Thread cutting, constant lead (L)			

Table 2. *(Continued)* **G-Code Addresses**

Code		Description	Code		Description
G61	c	Modal equivalent of G09 except that rapid moves are not taken to a complete stop before the next motion block is executed (M,L)	G80	abc	Cancel fixed cycles
G62	c	Automatic corner override, reduces the feed rate on an inside corner cut (M,L)	G81	abc	Drill cycle, no dwell and rapid out (M,L)
G63	a	Unassigned	G82	abc	Drill cycle, dwell and rapid out (M,L)
G63	bc	Tapping mode (M,L)	G83	abc	Deep hole peck drilling cycle (M,L)
G64–G69	abc	Unassigned	G84	abc	Right-hand tapping cycle (M,L)
G64	c	Cutting mode, usually set by the system installer (M,L)	G84.1	c	Left-hand tapping cycle (M,L)
G65	c	Calls for a parametric macro (M,L)	G85	abc	Boring cycle, no dwell, feed out (M,L)
G66	c	Calls for a parametric macro. Applies to motion blocks only (M,L)	G86	abc	Boring cycle, spindle stop, rapid out (M,L)
G66.1	c	Same as G66 but applies to all blocks (M,L)	G87	abc	Boring cycle, manual retraction (M,L)
G67	c	Stop the modal parametric macro (see G65, G66, G66.1) (M,L)	G88	abc	Boring cycle, spindle stop, manual retraction (M,L)
G68	c	Rotates the coordinate system (i.e., the axes) (M)	G88.1		Pocket milling (rectangular and circular), roughing cycle (M)
G69	c	Cancel axes rotation (M)	G88.2		Pocket milling (rectangular and circular), finish cycle (M)
G70	abc	Inch programming (M,L)	G88.3		Post milling, roughs out material around a specified area (M)
G71	abc	Metric programming (M,L)	G88.4		Post milling, finish cuts material around a post (M)
G72	ac	Circular interpolation CW (three-dimensional) (M)	G88.5		Hemisphere milling, roughing cycle (M)
G72	b	Unassigned	G88.6		Hemisphere milling, finishing cycle (M)
G72	c	Used to perform the finish cut on a turned part along the Z-axis after the roughing cuts initiated under G73, G74, or G75 codes (L)	G89	abc	Boring cycle, dwell and feed out (M,L)
G73	b	Unassigned	G89.1		Irregular pocket milling, roughing cycle (M)
G73	c	Deep hole peck drilling cycle (M); OD and ID roughing cycle, running parallel to the Z-axis (L)	G89.2		Irregular pocket milling, finishing cycle (M)
G74	ac	Cancel multiquadrant circular interpolation (M,L)	G90	abc	Absolute dimension input (M,L)
G74	bc	Move to home position (M,L)	G91	abc	Incremental dimension input (M,L)
G74	c	Left-hand tapping cycle (M)	G92	abc	Preload registers, used to shift the coordinate axes relative to the current tool position (M,L)
G74		Rough facing cycle (L)	G93	abc	Inverse time feed rate (velocity/distance) (M,L)
G75	ac	Multiquadrant circular interpolation (M,L)	G94	c	Feed rate in inches or millimeters per minute (ipm or mpm) (M,L)
G75	b	Unassigned	G95	abc	Feed rate given directly in inches or millimeters per revolution (ipr or mpr) (M,L)
G75		Roughing routine for castings or forgings (L)	G96	abc	Maintains a constant surface speed, feet (meters) per minute (L)
G76–G79	ab	Unassigned	G97	abc	Spindle speed programmed in rpm (M,L)

Table 2. *(Continued)* **G-Code Addresses**

Code	Description	Code		Description
		G98–99	ab	Unassigned

[a] Adheres to ANSI/EIA RS-274-D;

[b] Adheres to ISO 6983/1,2,3 Standards; where both symbols appear together, the ANSI/EIA and ISO standard codes are comparable;

[c] This code is modal. All codes that are not identified as modal are nonmodal, when used according to the corresponding definition.

[d] Indicates a use of the code that does not conform with the Standard.

Symbols following a description: (M) indicates that the code applies to a mill or machining center; (L) indicates that the code applies to turning machines; (M,L) indicates that the code applies to both milling and turning machines.

Codes that appear more than once in the table are codes that are in common use, but are not defined by the Standard or are used in a manner that is different than that designated by the Standard (e.g., see G61).

Most systems that support the RS-274-D Standard codes do not use all the codes available in the Standard. Unassigned G-words in the Standard are often used by builders of machine tool control systems for a variety of special purposes, sometimes leading to confusion as to the meanings of unassigned codes. Even more confusing, some builders of systems and machine tools use the less popular standardized codes for other than the meaning listed in the Standard. For these reasons, machine code written specifically for one machine/controller will not necessarily work correctly on another machine controller without modification.

Dimension words contain numerical data that indicate either a distance or a position. The dimension units are selected by using G70 (inch programming) or G71 (metric programming) code. G71 is canceled by a G70 command, by miscellaneous functions M02 (end of program), or by M30 (end of data). The dimension words immediately follow the G-word in a block and on multiaxis machines should be placed in the following order: X, Y, Z, U, V, W, P, Q, R, A, B, C, D, and E.

Absolute programming (G90) is a method of defining the coordinate locations of points to which the cutter (or workpiece) is to move based on the fixed machine zero point. In Fig. 1, the $X-Y$ coordinates of P1 are $X=1.0$, $Y=0.5$ and the coordinates of P2 are $X=2.0$, $Y=1.1$. To indicate the movement of the cutter from one point to another when using the absolute coordinate system, only the coordinates of the destination point P2 are needed.

Incremental programming (G91) is a method of identifying the coordinates of a particular location in terms of the distance of the new location from the current location. In the example shown in Fig. 2, a move from P1 to P2 is written as X + 1.0, Y + 0.6. If there is no movement along the *Z*-axis, *Z* is zero and normally is not noted. An $X-Y$ incremental move from P2 to P3 in Fig. 2 is written as X + 1.0, Y − 0.7.

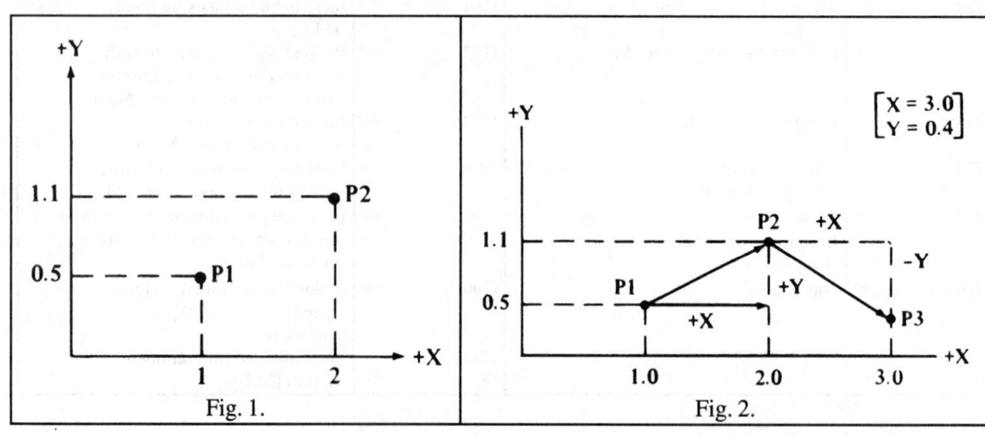

Fig. 1.

Fig. 2.

Most CNC systems offer both absolute and incremental part programming. The choice is handled by G-code G90 for absolute programming and G91 for incremental programming. G90 and G91 are both modal, so they remain in effect until canceled.

The G92 word is used to preload the registers in the control system with desired values. A common example is the loading of the axis-position registers in the control system for a lathe. Fig. 3 shows a typical home position of the tool tip with respect to the zero point on the machine. The tool tip here is registered as being 15.0000 inches in the *Z*-direction and 4.5000 inches in the *X*-direction from machine zero. No movement of the tool is required. Although it will vary with different control system manufacturers, the block to accomplish the registration shown in Fig. 3 will be approximately:

N0050 G92 X4.5 Z15.0

Miscellaneous Functions (M-Words).—Miscellaneous functions, or M-codes, also referred to as auxiliary functions, constitute on-off type commands. M functions are used to control actions such as starting and stopping of motors, turning coolant on and off, changing tools, and clamping and unclamping parts. M functions are made up of the letter M followed by a two-digit code. Table lists the standardized M-codes, however, the functions available will vary from one control system to another. Most systems provide fewer M functions than the complete list and may use some of the unassigned codes to provide additional functions that are not covered by the Standard. If an M-code is used in a block, it follows the T-word and is normally the last word in the block.

Table 3. Miscellaneous Function Words from *ANSI/EIA RS-274-D*

Code	Description
M00	Automatically *stops* the machine. The operator must push a button to continue with the remainder of the program.
M01	An *optional stop* acted upon only when the operator has previously signaled for this command by pushing a button. The machine will automatically stop when the control system senses the M01 code.
M02	This *end-of-program* code stops the machine when all commands in the block are completed. May include rewinding of tape.
M03	Start *spindle rotation* in a *clockwise* direction—looking out from the spindle face.
M04	Start *spindle rotation* in a *counterclockwise* direction—looking out from the spindle face.
M05	*Stop* the spindle in a normal and efficient manner.
M06	Command to *change a tool* (or tools) manually or automatically. Does not cover tool selection, as is possible with the T-words.
M07 to M08	M07 (coolant 2) and M08 (coolant 1) are codes to *turn on coolant.* M07 may control *flood* coolant and M08 *mist* coolant.
M09	Shuts off the coolant.
M10 to M11	M10 applies to automatic *clamping* of the machine slides, workpiece, fixture spindle, etc. M11 is an unclamping code.
M12	An inhibiting code used to synchronize multiple sets of axes, such as a four-axis lathe having two independently operated heads (turrets).
M13	Starts *CW spindle* motion and *coolant on* in the same command.
M14	Starts *CCW spindle* motion and *coolant on* in the same command.
M15 to M16	Rapid traverse of feed motion in either the +(M15) or −(M16) direction.
M17 to M18	Unassigned.
M19	Oriented spindle stop. Causes the spindle to stop at a predetermined angular position.
M20 to M29	Permanently unassigned.

Table 3. *(Continued)* **Miscellaneous Function Words from** *ANSI/EIA RS-274-D*

Code	Description
M30	An *end-of-tape* code similar to M02, but M30 will also rewind the tape; also may switch automatically to a second tape reader.
M31	A command known as *interlock bypass* for temporarily circumventing a normally provided interlock.
M32 to M35	Unassigned.
M36 to M39	Permanently unassigned.
M40 to M46	Used to signal gear changes if required at the machine; otherwise, unassigned.
M47	Continues program execution from the start of the program unless inhibited by an interlock signal.
M48 to M49	M49 deactivates a manual spindle or feed override and returns the parameter to the programmed value; M48 cancels M49.
M50 to M57	Unassigned.
M58 to M59	Holds the rpm constant at the value in use when M59 is initiated; M58 cancels M59.
M60 to M89	Unassigned.
M90 to M99	Reserved for use by the machine user.

Feed Function (F-Word).—F-word stands for feed-rate word or feed rate. The meaning of the feed word depends on the system of units in use and the feed mode. For example, F15 could indicate a feed rate of 0.15 inch (or millimeter) per revolution or 15 inches (or millimeters) per minute, depending on whether G70 or G71 is used to indicate inch or metric programming and whether G94 or G95 is used to specify feed rate expressed as inches (or mm) per minute or revolution. The G94 word is used to indicate inches/minute (ipm) or millimeters/minute (mmpm) and G95 is used for inches/revolution (ipr) or millimeters/revolution (mmpr). The default system of units is selected by G70 (inch programming) or G71 (metric programming) prior to using the feed function. The feed function is modal, so it stays in effect until it is changed by setting a new feed rate. In a block, the feed function is placed immediately following the dimension word of the axis to which it applies or immediately following the last dimension word to which it applies if it is used for more than one axis.

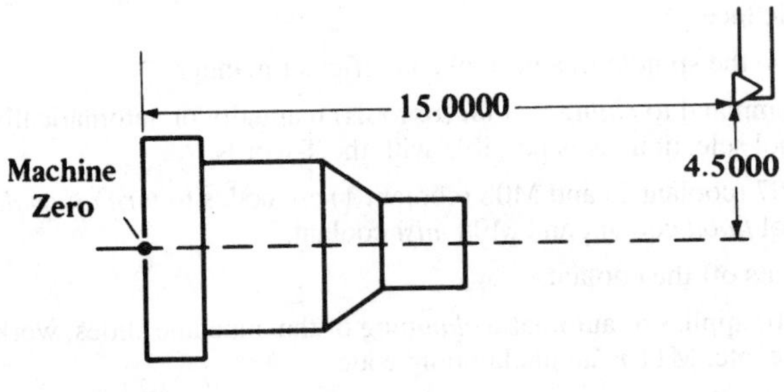

Fig. 3.

In turning operations, when G95 is used to set a constant feed rate per revolution, the spindle speed is varied to compensate for the changing diameter of the work — the spindle speed increases as the working diameter decreases. To prevent the spindle speed from increasing beyond a maximum value, the S-word, see *Spindle Function (S-Word)*, is used to specify the maximum allowable spindle speed before issuing the G95 command. If the spindle speed is changed after the G95 is used, the feed rate is also changed accordingly. If G94 is used to set a constant feed per unit of time (inches or millimeters per minute), changes in the spindle speed do not affect the feed rate.

Feed rates expressed in inches or millimeters per revolution can be converted to feed rates in inches or millimeters per minute by multiplying the feed rate by the spindle speed in revolutions per minute: feed/minute = feed/revolution × spindle speed in rpm. Feed rates for milling cutters are sometimes given in inches or millimeters per tooth. To convert feed per tooth to feed per revolution, multiply the feed rate per tooth by the number of cutter teeth: feed/revolution = feed/tooth × number of teeth.

For certain types of cuts, some systems require an inverse-time *feed command* that is the reciprocal of the time in minutes required to complete the block of instructions. The feed command is indicated by a G93 code followed by an F-word value found by dividing the feed rate, in inches (millimeters) or degrees per minute, by the distance moved in the block: feed command = feed rate/distance = (distance/time)/distance = 1/time.

Feed-rate override refers to a control, usually a rotary dial on the control system panel, that allows the programmer or operator to override the programmed feed rate. Feed-rate override does not change the program; permanent changes can only be made by modifying the program. The range of override typically extends from 0 to 150 per cent of the programmed feed rate on CNC machines; older hardwired systems are more restrictive and most cannot be set to exceed 100 per cent of the preset rate.

Spindle Function (S-Word).—An S-word specifies the speed of rotation of the spindle. The spindle function is programmed by the address S followed by the number of digits specified in the format detail (usually a four-digit number). Two G-codes control the selection of spindle speed input: G96 selects a constant cutting speed in surface feet per minute (sfm) or meters per minute (mpm) and G97 selects a constant spindle speed in revolutions per minute (rpm).

In turning, a constant spindle speed (G97) is applied for threading cycles and for machining parts in which the diameter remains constant. Feed rate can be programmed with either G94 (inches or millimeters per minute) or G95 (inches or millimeters per revolution) because each will result in a constant cutting speed to feed relationship.

G96 is used to select a constant cutting speed (i.e., a constant surface speed) for facing and other cutting operations in which the diameter of the workpiece changes. The spindle speed is set to an initial value specified by the S-word and then automatically adjusted as the diameter changes so that a constant surface speed is maintained. The control system adjusts spindle speed automatically, as the working diameter of the cutting tool changes, decreasing spindle speed as the working diameter increasesor increasing spindle speed as the working diameter decreases. When G96 is used for a constant cutting speed, G95 in a succeeding block maintains a constant feed rate per revolution.

Speeds given in surface feet or meters per minute can be converted to speeds in revolutions per minute (rpm) by the formulas:

$$\text{rpm} = \frac{\text{sfm} \times 12}{\pi \times d} \qquad \text{rpm} = \frac{\text{mpm} \times 1000}{\pi \times d}$$

where d is the diameter, in inches or millimeters, of the part on a lathe or of the cutter on a milling machine; and π is equal to 3.14159.

Tool Function (T-Word).—The T-word calls out the tool that is to be selected on a machining center or lathe having an automatic tool changer or indexing turret. On machines without a tool changer, this word causes the machine to stop and request a tool change. This word also specifies the proper turret face on a lathe. The word usually is accompanied by several numbers, as in T0101, where the first pair of numbers refers to the tool number (and carrier or turret if more than one) and the second pair of numbers refers to the tool offset number. Therefore, T0101 refers to tool 1, offset 1.

Information about the tools and the tool setups is input to the CNC system in the form of a *tool data table.* Details of specific tools are transferred from the table to the part program via the T-word. The tool nose radius of a lathe tool, for example, is recorded in the tool data table so that the necessary tool path calculations can be made by the CNC system. The miscellaneous code M06 can also be used to signal a tool change, either manually or automatically.

Compensation for variations in the tool nose radius, particularly on turning machines, allows the programmer to program the part geometry from the drawing and have the tool follow the correct path in spite of variations in the tool nose shape. Typical of the data required, as shown in Fig. 4, are the nose radius of the cutter, the *X* and *Z* distances from the gage point to some fixed reference point on the turret, and the orientation of the cutter (tool tip orientation code), as shown in Fig. 5. Details of nose radius compensation for numerical control is given in a separate section (Indexable Insert Holders for NC).

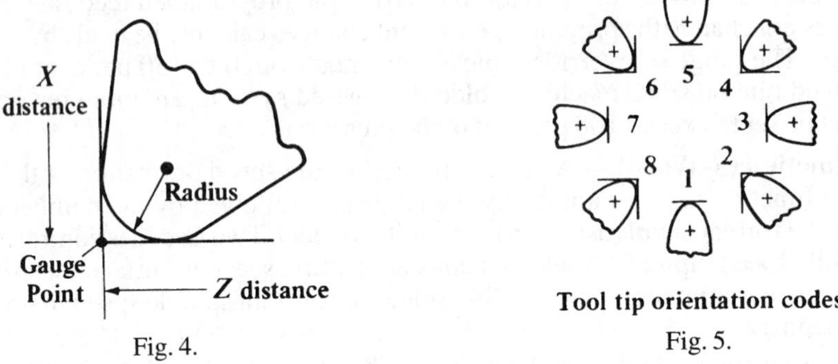

Fig. 4.

Fig. 5.

Tool offset, also called cutter offset, is the amount of cutter adjustment in a direction parallel to the axis of a tool. Tool offset allows the programmer to accommodate the varying dimensions of different tooling by assuming (for the sake of the programming) that all the tools are identical. The actual size of the tool is totally ignored by the programmer who programs the movement of the tools to exactly follow the profile of theworkpiece shape. Once tool geometry is loaded into the tool data table and the cutter compensation controls of the machine activated, the machine automatically compensates for the size of the tools in the programmed movements of the slide. In gage length programming, the tool length and tool radius or diameter are included in the program calculations. Compensation is then used only to account for minor variations in the setup dimensions and tool size.

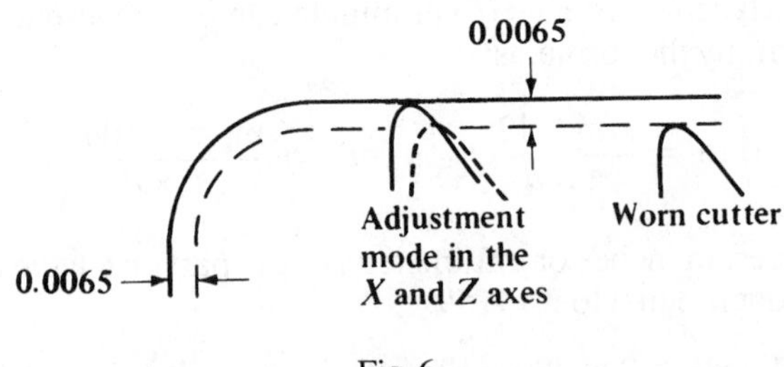

Fig. 6.

Customarily, the tool offset is used in the beginning of a program to initialize each individual tool. Tool offset also allows the machinist to correct for conditions, such as tool wear, that would cause the location of the cutting edge to be different from the programmed location. For example, owing to wear, the tool tip in Fig. 6 is positioned a distance of 0.0065 inch from the location required for the work to be done. To compensate for this wear, the operator (or part programmer), by means of the CNC control panel, adjusts the tool tip with reference to the *X*- and *Z*-axes, moving the tool closer to the work by

0.0065 inch throughout its traverse. The tool offset number causes the position of the cutter to be displaced by the value assigned to that offset number.

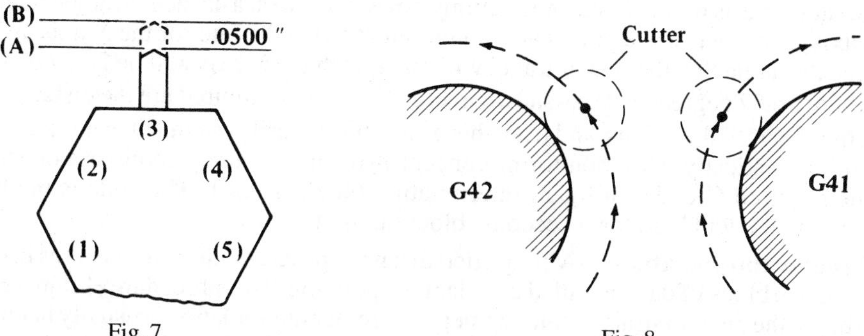

Fig. 7. Fig. 8.

Changes to the programmed positions of cutting tool tip(s) can be made by *tool length offset* programs included in the control system. A dial or other means is generally provided on milling, drilling, and boring machines, and machining centers, allowing the operator or part programmer to override the programmed axial, or *Z*-axis, position. This feature is particularly helpful when setting the lengths of tools in their holders or setting a tool in a turret, as shown in Fig. 7, because an exact setting is not necessary. The tool can be set to an approximate length and the discrepancy eliminated by the control system.

The amount of offset may be determined by noting the amount by which the cutter is moved manually to a fixed point on the fixture or on the part, from the programmed *Z*-axis location. For example, in Fig. 7, the programmed *Z*-axis motion results in the cutter being moved to position A, whereas the required location for the tool is at B. Rather than resetting the tool or changing the part program, the tool length offset amount of 0.0500 inch is keyed into the control system. The 0.0500-inch amount is measured by moving the cutter tip manually to position B and reading the distance moved on the readout panel. Thereafter, every time that cutter is brought into the machining position, the programmed *Z*-axis location will be overridden by 0.0500 inch.

Manual adjustment of the cutter center path to correct for any variance between nominal and actual cutter radius is called *cutter compensation.* The net effect is to move the path of the center of the cutter closer to, or away from, the edge of the workpiece, as shown in Fig. 8. The compensation may also be handled via a tool data table.

When cutter compensation is used, it is necessary to include in the program a G41 code if the cutter is to be to the left of the part and a G42 code if to the right of the part, as shown in Fig. 8. A G40 code cancels cutter compensation. Cutter compensation with earlier hardwire systems was expensive, very limited, and usually held to ±0.0999 inch. The range for cutter compensation with CNC control systems can go as high as ±999.9999 inches, although adjustments of this magnitude are unlikely to be required.

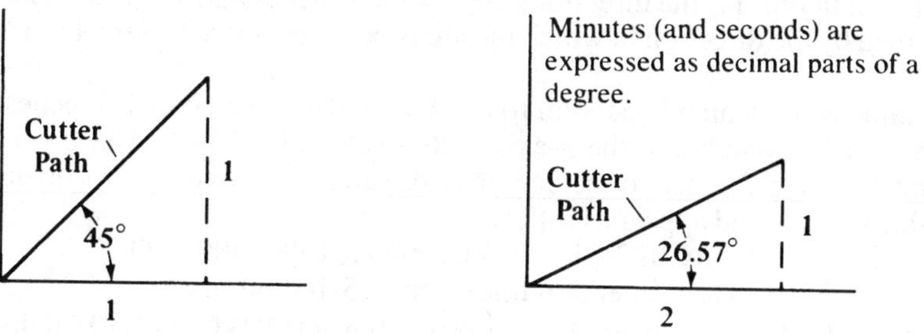

Fig. 9.

Linear Interpolation.—The ability of the control system to guide the workpiece along a straight-line path at an angle to the slide movements is called linear interpolation. Movements of the slides are controlled through simultaneous monitoring of pulses by the control system. For example, if monitoring of the pulses for the *X*-axis of a milling machine is at the same rate as for the *Y*-axis, the cutting tool will move at a 45-degree angle relative to the *X*-axis. However, if the pulses are monitored at twice the rate for the *X*-axis as for the *Y*-axis, the angle that the line of travel will make with the *X*-axis will be 26.57 degrees (tangent of 26.57 degrees = ½), as shown in Fig. 9. The data required are the distances traveled in the *X*- and *Y*-directions, and from these data, the control system will generate the straight line automatically. This monitoring concept also holds for linear motions along three axes. The required G-code for linear interpolation blocks is G01. The code is modal, which means that it will hold for succeeding blocks until it is changed.

Circular Interpolation.—A simplified means of programming circular arcs in one plane, using one block of data, is called circular interpolation. This procedure eliminates the need to break the arc into straight-line segments. Circular interpolation is usually handled in one plane, or two dimensions, although three-dimensional circular interpolation is described in the Standards. The plane to be used is selected by a G or preparatory code. In Fig. 10, G17 is used if the circle is to be formed in the *X–Y* plane, G18 if in the *X–Z* plane, and G19 if in the *Y–Z* plane. Often the control system is preset for

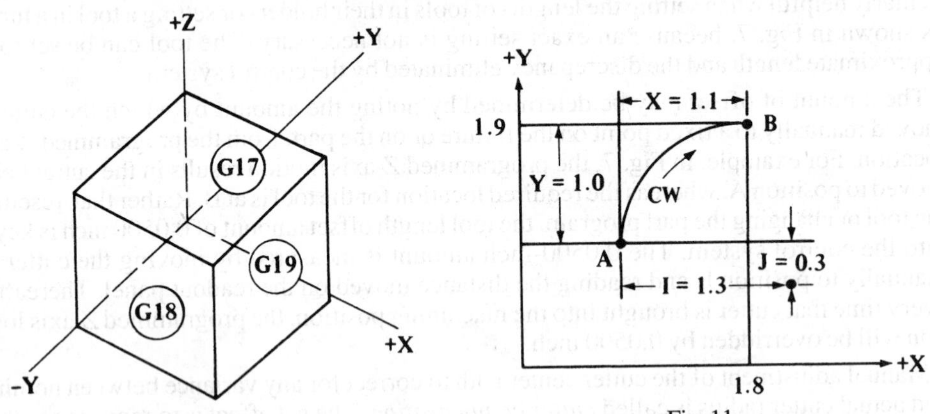

Fig. 10. Fig. 11.

the circular interpolation feature to operate in only one plane (e.g., the *X–Y* plane for milling machines or machining centers or the *X–Z* plane for lathes), and for these machines, the G-codes are not necessary.

A circular arc may be described in several ways. Originally, the RS-274 Standard specified that, with incremental programming, the block should contain:

1) A G-code describing the direction of the arc, G02 for clockwise (CW), and G03 for counterclockwise (CCW).

2) Directions for the component movements around the arc parallel to the axes. In the example shown in Fig. 11, the directions are $X = +1.1$ inches and $Y = +1.0$ inch. The signs are determined by the direction in which the arc is being generated. Here, both *X* and *Y* are positive.

3) The I dimension, which is parallel to the *X*-axis with a value of 1.3 inches, and the J dimension, which is parallel to the *Y*-axis with a value of 0.3 inch. These values, which locate point A with reference to the center of the arc, are called offset dimensions. The block for this work would appear as follows:

N0025 G02 X011000 Y010000 I013000 J003000

(The sequence number, N0025, is arbitrary.)

The block would also contain the plane selection (i.e., G17, G18, or G19), if this selection is not preset in the system. Most of the newer control systems allow duplicate words in the same block, but most of the older systems do not. In these older systems, it is necessary to insert the plane selection code in a separate and prior block, for example, N0020 G17.

Another stipulation in the Standard is that the arc is limited to one quadrant. Therefore, four blocks would be required to complete a circle. Four blocks would also be required to complete the arc shown in Fig. 12, which extends into all four quadrants.

When utilizing absolute programming, the coordinates of the end point are described. Again from Fig. 11, the block, expressed in absolute coordinates, appears as:

N0055 G02 X01800 Y019000 I013000 J003000

where the arc is continued from a previous block; the starting point for the arc in this block would be the end point of the previous block.

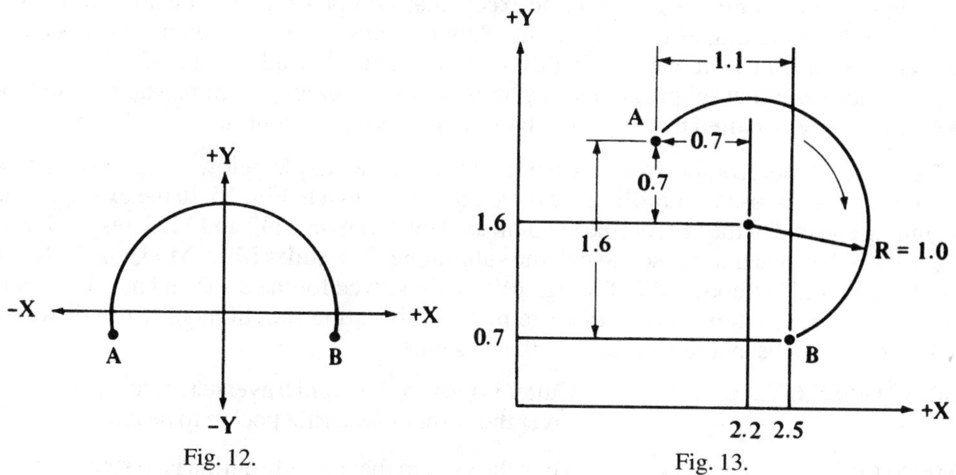

Fig. 12. Fig. 13.

The Standard still contains the format discussed, but simpler alternatives have been developed. The latest version of the Standard (RS-274-D) allows *multiple quadrant programming* in one block, by inclusion of a G75 word. In the absolute-dimension mode (G90), the coordinates of the arc center are specified. In the incremental-dimension mode (G91), the signed (plus or minus) incremental distances from the beginning point of the arc to the arc center are given. Most system builders have introduced some variations on this format. One system builder utilizes the center and the end point of the arc when in an absolute mode, and might describe the block for going from A to B in Fig. 13 as:

N0065 G75 G02 X2.5 Y0.7 I2.2 J1.6

The I and the J words are used to describe the coordinates of the arc center. Decimal-point programming is also used here. A block for the same motion when programmed incrementally might appear as:

N0075 G75 G02 X1.1 Y – 1.6 I0.7 J0.7

This approach is more in conformance with the RS-274-D Standard in that the *X* and *Y* values describe the displacement between the starting and ending points (points A and B), and the I and J indicate the offsets of the starting point from the center. Another and even more convenient way of formulating a circular motion block is to note the coordinates of the ending point and the radius of the arc. Using absolute programming, the block for the motion in Fig. 13 might appear as:

N0085 G75 G02 X2.5 Y0.7 R10.0

The starting point is derived from the previous motion block. Multiquadrant circular interpolation is canceled by a G74 code.

Helical and Parabolic Interpolation.—Helical interpolation is used primarily for milling large threads and lubrication grooves, as shown in Fig. 14. Generally, helical interpolation involves motion in all three axes (*X, Y, Z*) and is accomplished by using circular interpolation (G02 or G03) while changing the third dimension. Parabolic interpolation

(G06) is simultaneous and coordinated control of motion-such that the resulting cutter path describes part of a parabola. The RS-274-D Standard provides further details.

Subroutine.— A subroutine is a set of instructions or blocks that can be inserted into a program and repeated whenever required. Parametric subroutines permit letters or symbols to be inserted into the program in place of numerical values (see *Parametric Expressions and Macros*). Parametric subroutines can be called during part programming and values assigned to the letters or symbols. This facility is particularly helpful when dealing with families of parts.

A subprogram is similar to a subroutine except that a subprogram is not wholly contained within another program, as is a subroutine. Subprograms are used when it is necessary to perform the same task frequently, in different programs. The advantage of subprograms over subroutines is that subprograms may be called by any other program, whereas the subroutine can only be called by the program that contains the subroutine.

There is no standard subroutine format; however, the example below is typical of a program that might be used for milling the three pockets shown in Fig. 15. In the example, the beginning and end of the subroutine are indicated by the codes M92 and M93, respectively, and M94 is the code that is used to call the subroutine. The codes M92, M93, and M94 are not standardized (M-codes M90 through M99 are reserved for the user) and may be different from control system to control system. The subroutine functions may use different codes or may not be available at all on other systems.

N0010 G00 X.6 Y.85	Cutter is moved at a rapid traverse rate to a position over the corner of the first pocket to be cut.
N0020 M92	Tells the system that the subroutine is to start in the next block.
N0030 G01 Z–.25 F2.0	Cutter is moved axially into the workpiece 0.25 inch at 2.0 ipm.
N0040 X.8	Cutter is moved to the right 0.8 inch.
N0050 Y.2	Cutter is moved laterally up 0.2 inch.
N0060 X–.8	Cutter is moved to the left 0.8 inch.
N0070 Y.2	Cutter is moved laterally up 0.2 inch.

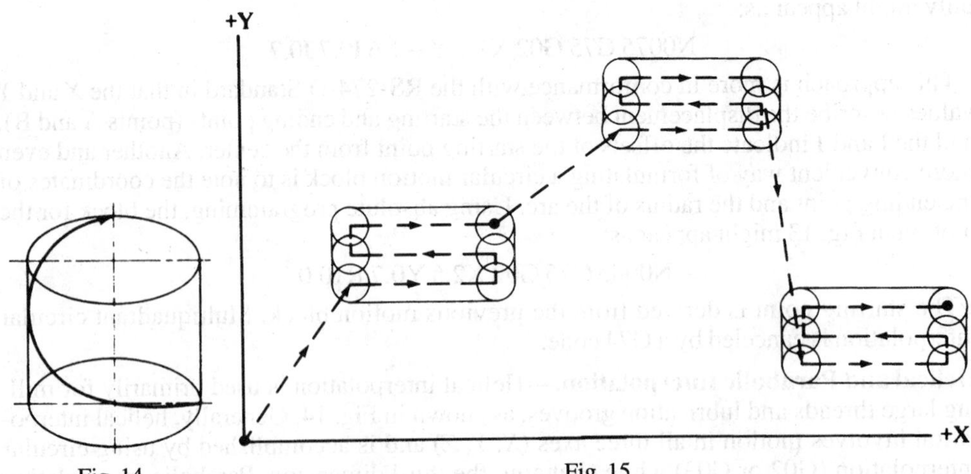

Fig. 14. Fig. 15.

N0080 X.8	Cutter is moved to the right 0.8 inch.
N0090 G00 Z.25 M93	Cutter is moved axially out of pocket at rapid traverse rate. Last block of subroutine is signaled by word M93.
N0100 X.75 Y.5	Cutter is moved to bottom left-hand corner of second pocket at rapid traverse rate.
N0110 M94 N0030	Word M94 calls for repetition of the subroutine that starts at sequence number N0030 and ends at sequence number N0090.
N0120 G00 X.2 Y–I.3	After the second pocket is cut by repetition of sequence numbers N0030 through N0090, the cutter is moved to start the third pocket.
N0130 M94 N0030	Repetition of subroutine is called for by word M94 and the third pocket is cut.

Parametric Expressions and Macros.—Parametric programming is a method whereby a variable or replaceable parameter representing a value is placed in the machining code instead of using the actual value. In this manner, a section of code can be used several or many times with different numerical values, thereby simplifying the programming and reducing the size of the program. For example, if the values of X and Y in lines N0040 to N0080 of the previous example are replaced as follows:

N0040 X#1

N0050 Y#2

N0060 X#3

N0070 Y#4

then the subroutine starting at line N0030 is a parametric subroutine. That is, the numbers following the # signs are the variables or parameters that will be replaced with actual values when the program is run. In this example, the effect of the program changes is to allow the same group of code to be used for milling pockets of different sizes. If on the other hand, lines N0010, N0100, and N0120 of the original example were changed in a similar manner, the effect would be to move the starting location of each of the slots to the location specified by the replaceable parameters.

Before the program is run, the values that are to be assigned to each of the parameters or variables are entered as a list at the start of the part program in this manner:

#1 = .8

#2 = .2

#3 = .8

#4 = .2

All that is required to repeat the same milling process again, but this time creating a different size pocket, is to change the values assigned to each of the parameters #1, #2, #3, and #4 as necessary. Techniques for using parametric programming are not standardized and are not recognized by all control systems. For this reason, consult the programming manual of the particular system for specific details.

As with a parametric subroutine, macro describes a type of program that can be recalled to allow insertion of finite values for letter variables. The difference between a macro and a parametric subroutine is minor. The term macro normally applies toa source program that is used with computer-assisted part programming; the parametric subroutine is a feature of the CNC system and can be input directly into that system.

Conditional Expressions.—It is often useful for a program to make a choice between two or more options, depending on whether or not a certain condition exists. A program can contain one or more blocks of code that are not needed every time the program is run, but are needed some of the time. For example, refer to the previous program for milling three slots. An occasion arises that requires that the first and third slots be milled, but not the second one. If the program contained the following block of code, the machine could be easily instructed to skip the milling of the second slot:

N0095 IF [#5 EQ 0] GO TO N0120

In this block, #5 is the name of a variable; EQ is a conditional expression meaning *equals;* and GO TO is a branch statement meaning resume execution of the program at the following line number. The block causes steps N0100 and N0110 of the program to be skipped if the value of #5 (a dummy variable) is set equal to zero. If the value assigned to #5 is any number other than zero, the expression (#5 EQ 0) is not true and the remaining instructions in block N0095 are not executed. Program execution continues with the next step, N0100, and the second pocket is milled. For the second pocket to be milled, parameter #5 is initialized at the beginning of the program with a statement such as #5 = 1 or #5 = 2. Initializing #5 = 0 guarantees that the pocket is not machined. On control systems that automatically initialize all variables to zero whenever the system is reset or a program is loaded, the second slot will not be machined unless the #5 is assigned a nonzero value each time the program is run.

Other conditional expressions are: NE = not equal to; GT = greater than; LT = less than; GE = greater than or equal to; and LE = less than or equal to. As with parametric expressions, conditional expressions may not be featured on all machines and techniques and implementation will vary. Therefore, consult the control system programming manual for the specific command syntax.

Fixed (Canned) Cycles.—Fixed (canned) cycles comprise sets of instructions providing for a preset sequence of events initiated by a single command or a block of data. Fixed cycles generally are offered by the builder of the control system or machine tool as part of the software package that accompanies the CNC system. Limited numbers of canned cycles began to appear on hardwire control systems shortly before their demise. The canned cycles offered generally consist of the standard G-codes covering driling, boring, and tapping operations, plus options that have been developed by the system builder such as thread cutting and turning cycles. (See *Thread Cutting* and *Turning Cycles*.) Some standard canned cycles included in RS-274-D are shown herewith. A block of data that might be used to generate the cycle functions is also shown above each illustration. Although the G-codes for the functions are standardized, the other words in the block and the block format are not, and different control system builders have different arrangements. The blocks shown are reasonable examples of fixed cycles and do not represent those of any particular system builder.

The G81 block for a simple drilling cycle is:

N_____G81 X_____Y_____C_____D_____F_____EOB

N_____X_____Y_____EOB

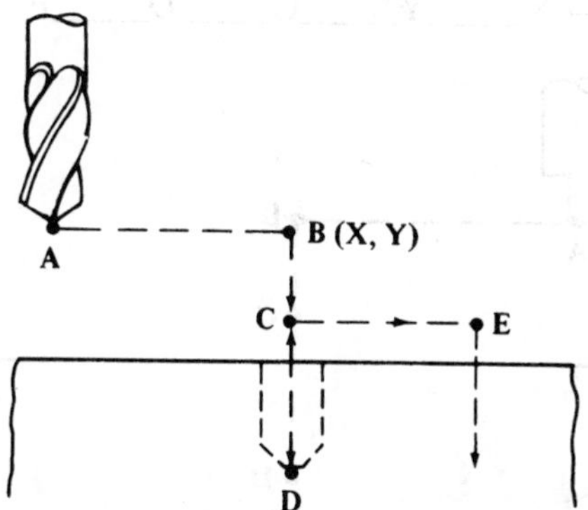

This G81 drilling cycle will move the drill point from position A to position B and then down to C at a rapid traverse rate; the drill point will next be fed from C to D at the programmed feed rate, then returned to C at the rapid traverse rate. If the cycle is to be repeated at a subsequent point, such as point E in the illustration, it is necessary Only to give the required X and Y coordinates. This repetition capability is typical of canned cycles.

The G82 block for a spotfacing or drilling cycle with a dwell is:

N_____G82 X_____Y_____C_____D_____T_____F_____EOB

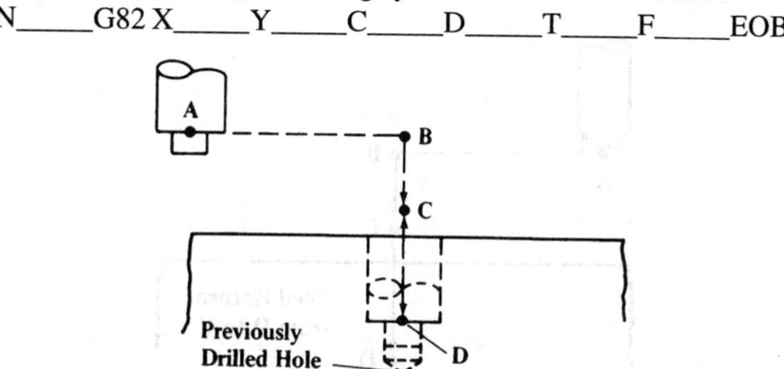

This G82 code produces a cycle that is very similar to the cycle of the G81 code except for the dwell period at point D. The dwell period allows the tool to smooth out the bottom of the counterbore or spotface. The time for the dwell, in seconds, is noted as a T-word.

The G83 block for a peck-drilling cyle is:

N_____G83 X_____Y_____C_____D_____K_____F_____EOB

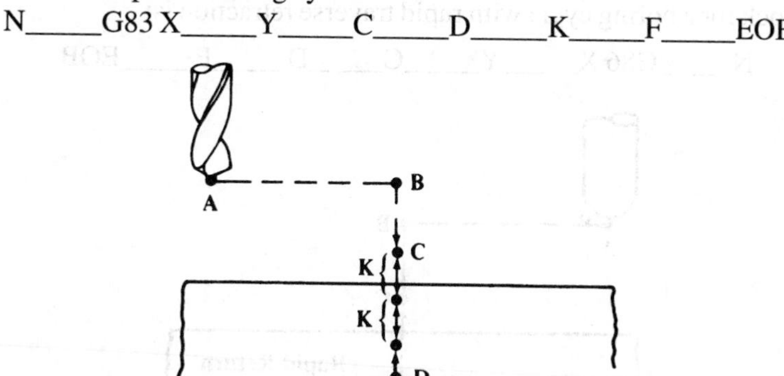

In the G83 peck-drilling cycle, the drill is moved from point A to point B and then to point C at the rapid traverse rate; the drill is then fed the incremental distance K, followed by rapid return to C. Down feed again at the rapid traverse rate through the distance K is next, after which the drill is fed another distance K. The drill is thenrapid traversed back to C, followed by rapid traverse for a distance of K + K; down feed to D follows before the drill is rapid traversed back to C, to end the cycle.

The G84 block for a tapping cycle is:

N_____G84 X_____Y_____C_____D_____F_____EOB

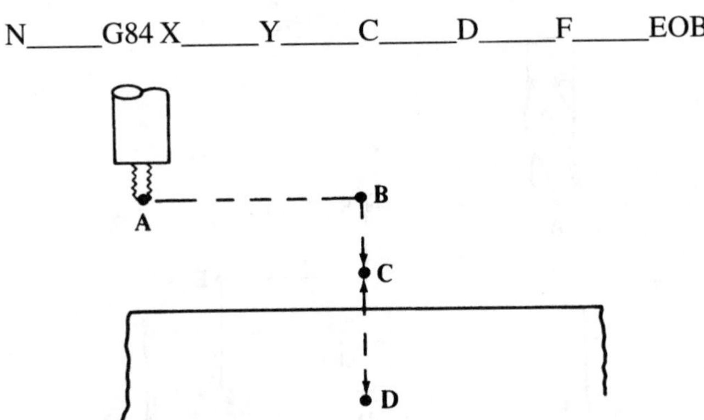

The G84 canned tapping cycle starts with the end of the tap being moved from point A to point B and then to point C at the rapid traverse rate. The tap is then fed to point D, reversed, and moved back to point C.

The G85 block for a boring cycle with tool retraction at the feed rate is:

N_____G85 X_____Y_____C_____D_____F_____EOB

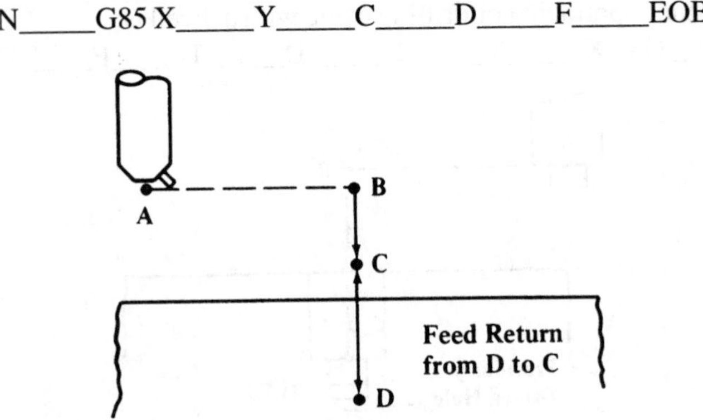

In the G85 boring cycle, the tool is moved from point A to point B and then to point C at the rapid traverse rate. The tool is next fed to point D and then, while still rotating, is moved back to point C at the same feed rate.

The G86 block for a boring cycle with rapid traverse retraction is:

N_____G86 X_____Y_____C_____D_____F_____EOB

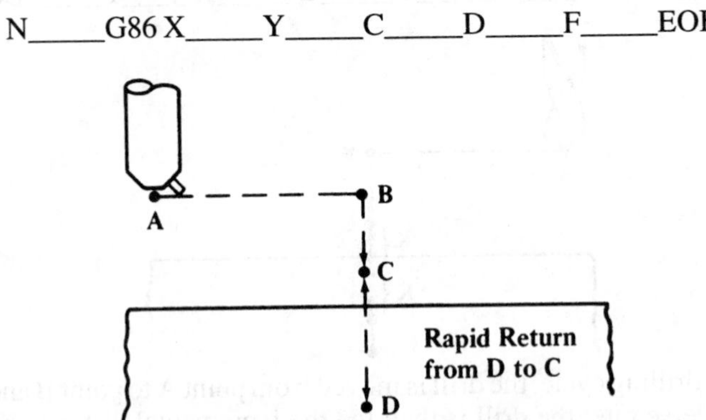

The G86 boring cycle is similar to the G85 cycle except that the tool is withdrawn at the rapid traverse rate.

The G87 block for a boring cycle with manual withdrawal of the tool is:

N_____G87 X_____Y_____C_____D_____F_____EOB

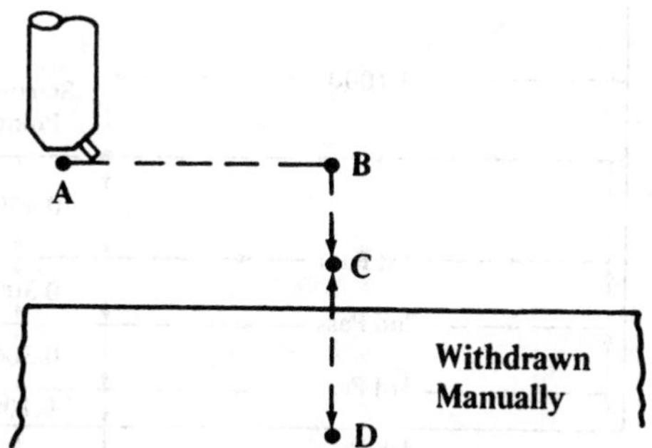

In the G87 canned boring cycle, the cutting tool is moved from A to B and then to C at the rapid traverse rate. The tool is then fed to D. The cycle is identical to the other boring cycles except that the tool is withdrawn manually.

The G88 block for a boring cycle with dwell and manual withdrawal is:

N_____G88 X_____Y_____C_____D_____T_____F_____EOB

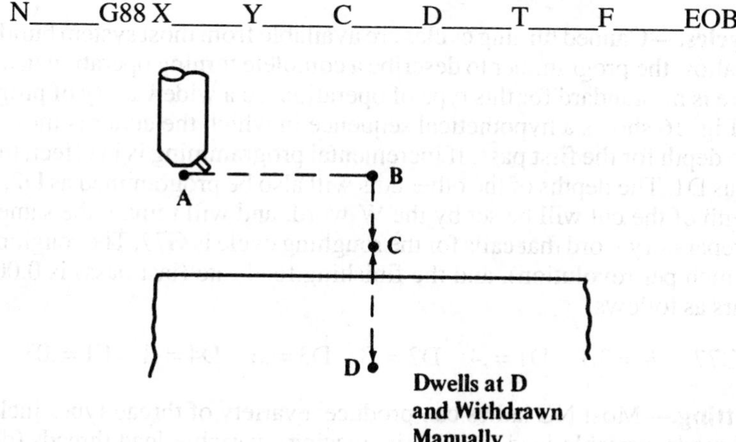

In the G88 dwell cycle, the tool is moved from A to B to C at the rapid traverse rate and then fed at the prescribed feed rate to D. The tool dwells at D, then stops rotating and is withdrawn manually.

The G89 block for a boring cycle with dwell and withdrawal at the feed rate is:

N_____G89 X_____Y_____C_____D_____T_____F_____EOB

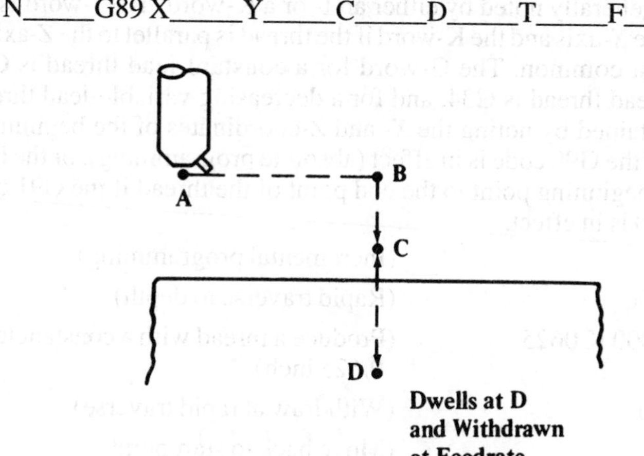

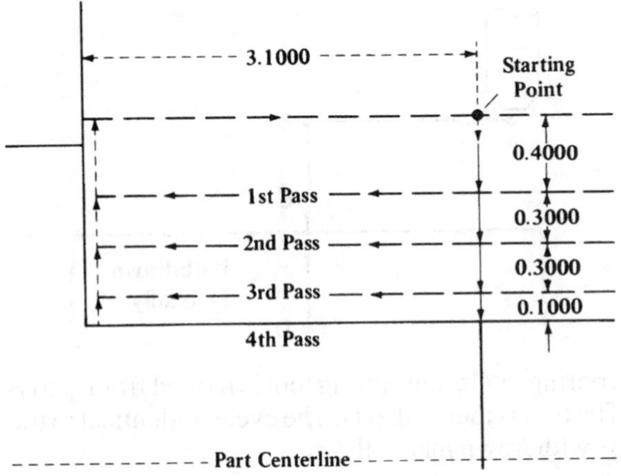

Fig. 16.

Turning Cycles.—Canned turning cycles are available from most system builders and are designed to allow the programmer to describe a complete turning operation in one or a few blocks. There is no standard for this type of operation, so a wide variety of programs have developed. Fig. 16 shows a hypothetical sequence in which the cutter is moved from the start point to depth for the first pass. If incremental programming is in effect, this distance is specified as D1. The depths of the other cuts will also be programmed as D2, D3, and so on. The length of the cut will be set by the W-word, and will remain the same with each pass. The preparatory word that calls for the roughing cycle is G77. The roughing feed rate is 0.03 ipr (inch per revolution), and the finishing feed rate (last pass) is 0.005 ipr. The block appears as follows:

N0054 G77 W = 3.1 D1 = .4 D2 = .3 D3 = .3 D4 = .1 F1 = .03 F2 = .005

Thread Cutting.—Most NC lathes can produce a variety of thread types including constant-lead threads, variable-lead threads (increasing), variable-lead threads (decreasing), multiple threads, taper threads, threads running parallel to the spindle axis, threads (spiral groove) perpendicular to the spindle axis, and threads containing a combination of the preceding. Instead of the feed rate, the lead is specified in the threading instruction block, so that the feed rate is made consistent with, and dependent upon, the selected speed (rpm) of the spindle.

The thread lead is generally noted by either an I- or a K-word. The I-word is used if the thread is parallel to the *X*-axis and the K-word if the thread is parallel to the *Z*-axis, the latter being by far the most common. The G-word for a constant-lead thread is G33, for an increasing variable-lead thread is G34, and for a decreasing variable-lead thread is G35. Taper threads are obtained by noting the *X*- and *Z*-coordinates of the beginning and end points of the thread if the G90 code is in effect (absolute programming), or the incremental movement from the beginning point to the end point of the thread if the G91 code (incremental programming) is in effect.

N0001 G91	(Incremental programming)
N0002 G00 X–.1000	(Rapid traverse to depth)
N0003 G33 Z–1.0000 K.0625	(Produce a thread with a constant lead of 0.625 inch)
N0004 G00 X.1000	(Withdraw at rapid traverse)
N0005 Z1.0000	(Move back to start point)

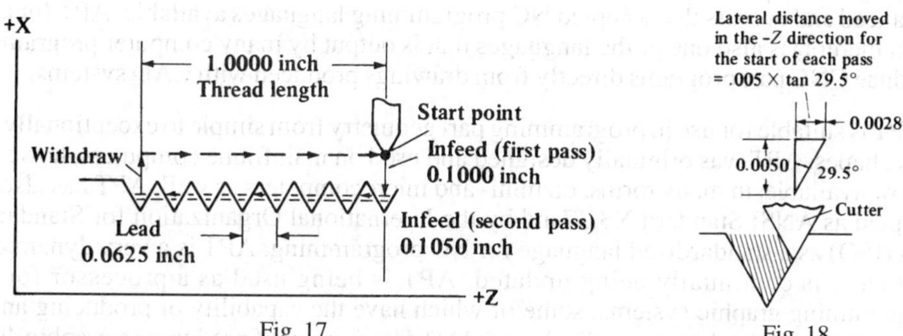

Fig. 17. Fig. 18.

Multiple threads are specified by a code in the block that spaces the start of the threads equally around the cylinder being threaded. For example, if a triple thread is to be cut, the threads will start 120 degrees apart. Typical single-block thread cutting utilizing a plunge cut is illustrated in Fig. 17 and shows two passes. The passes areidentical except for the distance of the plunge cut. Builders of control systems and machine tools use different codewords for threading, but those shown below can be considered typical. For clarity, both zeros and decimal points are shown.

The only changes in the second pass are the depth of the plunge cut and the withdrawal. The blocks will appear as follows:

N0006 X − .1050
N0007 G33 Z − 1.0000 K.0625
N0008 G00 X.1050
N0009 Z1.000

Compound thread cutting, rather than straight plunge thread cutting, is possible also, and is usually used on harder materials. As illustrated in Fig. 18, the starting point for the thread is moved laterally in the -Z direction by an amount equal to the depth of the cut times the tangent of an angle that is slightly less than 30 degrees. The program for the second pass of the example shown in Fig. 18 is as follows:

N0006 X − .1050 Z − .0028
N0007 G33 Z − 1.0000 K.0625
N0008 G00 X.1050
N0009 Z1.0000

Fixed (canned), one-block cycles also have been developed for CNC systems to produce the passes needed to complete a thread. These cycles may be offered by the builder of the control system or machine tool as standard or optional features. Subroutines also can generally be prepared by the user to accomplish the same purpose (see Subroutine). A one-block fixed threading cycle might look something like:

N0048 G98 X − .2000 Z − 1.0000 D.0050 F.0010

where G98 = preparatory code for the threading cycle
X−.2000 = total distance from the starting point to the bottom of the thread
Z−1.0000 = length of the thread
D.0050 = depths of successive cuts
F.0010 = depth(s) of the finish cut(s)

APT Programming

APT.—APT stands for Automatically Programmed Tool and is one of many computer languages designed for use with NC machine tools. The selection of a computer-assisted part-programming language depends on the type and complexity of the parts being machined more than on any other factor. Although some of the other languages may be easier to use, APT has been chosen to be covered in this book because it is a nonproprietary language in the public domain, has the broadest range of capability, and is one of the most advanced and universally accepted NC programming languages available. APT (or a variation thereof) is also one of the languages that is output by many computer programs that produce CNC part programs directly from drawings produced with CAD systems.

APT is suitable for use in programming part geometry from simple to exceptionally complex shapes. APT was originally designed and used on mainframe computers, however, it is now available, in many forms, on mini- and microcomputers as well. APT has also been adopted as ANSI Standard X3.37and by the International Organization for Standardization (ISO) as a standardized language for NC programming. APT is a very dynamic program and is continually being updated. APT is being used as a processor for part-programming graphic systems, some of which have the capability of producing an APT program from a graphic screen display or CAD drawing and of producing a graphic display on the CAD system from an APT program.

APT is a high-level programming language. One difference between APT and the ANSI/EIA RS-274-D (G-codes) programming format discussed in the last section is that APT uses English like words and expressions to describe the motion of the tool or workpiece. APT has the capability of programming the machining of parts in up to five axes, and also allows computations and variables to be included in the programming statements so that a whole family of similar parts can be programmed easily. This section describes the general capabilities of the APT language and includes a ready reference guide to the basic geometry and motion statements of APT, which is suitable for use in programming the machining of the majority of cubic type parts involving two-dimensional movements. Some of the three-dimensional geometry capability of APT and a description of its five-dimensional capability are also included.

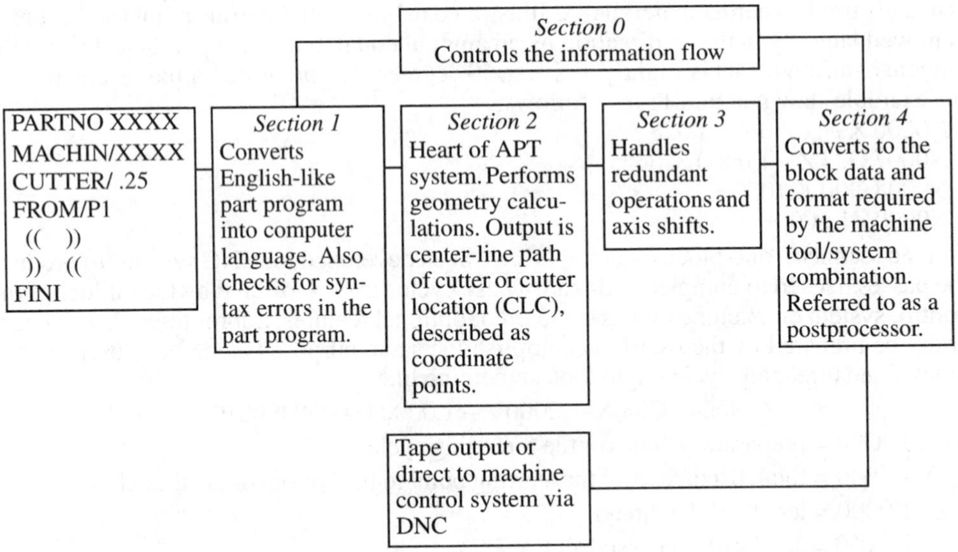

As shown in Fig. 1, the APT system can be thought of comprising the input program, the five sections 0 through IV, and the output program. The input program shown on the left progresses through the first four sections and all four are controlled by the fifth, section 0. Section IV, the postprocessor, is the software package that is added to sections II and III to customize the output and produce the necessary program format (including the G-words, M-words, etc.) so that the coded instructions will be recognizable by the control system. The postprocessor is software that is separate from the main body of the APT program, but for purposes of discussion, it may be easier to consider it as a unit within the APT program.

APT Computational Statements.—Algebraic and trigonometric functions and computations can be performed with the APT system as follows:

Arithmetic Form	APT Form	Arithmetic Form	APT Form	Arithmetic Form	APT Form
25×25	25*25	25^2	25**2	cos θ	COSF(θ)
$25 \div 25$	25/25	25^n	25**n	tan θ	TANF(θ)
$25 + 25$	25 + 25	$\sqrt{25}$	SQRTF (25)	arctan .5000	ATANF(.5)
$25 - 25$	25 − 25	sin θ	SINF(θ)		

Computations may be used in the APT system in two ways. One way is to let a factor equal the computation and then substitute the factor in a statement; the other is to put the computation directly into the statement. The following is a series of APT statements illustrating the first approach.

P1 = POINT/0,0,1

T = (25*2/3 + (3**2 − 1))

P2 = POINT/T,0,0

The second way would be as follows;

P1 = POINT/0,0,1

P2 = POINT/(25*2/3 + (3**2 − 1)),0,0

Note: The parentheses have been used as they would be in an algebraic formula so that the calculations will be carried out in proper sequence. The operations within the inner parentheses would be carried out first. It is important for the total number of left-hand parentheses to equal the total number of right-hand parentheses; otherwise, the program will fail.

APT Geometry Statements.—Before movements around the geometry of a part can be described, the geometry must be defined. For example, in the statement GOTO/P1, the computer must know where P1 is located before the statement can be effective. P1 therefore must be described in a geometry statement, prior to its use in the motion statement GOTO/P1. The simplest and most direct geometry statement for a point is

P1 = POINT/X ordinate, Y ordinate, Z ordinate

If the *Z* ordinate is zero and the point lies on the *X*–*Y* plane, the *Z* location need not be noted. There are other ways of defining the position of a point, such as at the intersection of two lines or where a line is tangent to a circular arc. These alternatives are described below, together with ways to define lines and circles. Referring to the preceding statement, P1 is known as a symbol. Any combination of letters and numbers may be used as a symbol providing the total does not exceed six characters and at least one of them is a letter. MOUSE2 would be an acceptable symbol, as would CAT3 or FRISBE. However, it is sensible to use symbols that help define the geometry. For example, C1 or CIR3 would be good symbols for a circle. A good symbol for a vertical line would be VL5.

Next, and after the equal sign, the particular geometry is noted. Here, it is a POINT. This word is a vocabulary word and must be spelled exactly as prescribed. Throughout, the designers of APT have tried to use words that are as close to English as possible. A slash follows the vocabulary word and is followed by a specific description of the particular geometry, such as the coordinates of the point P1. A usable statement for P1 might appear as P1 = POINT/1,5,4. The 1 would be the *X* ordinate; the 5, the *Y* ordinate; and the 4, the *Z* ordinate.

Lines as calculated by the computer are infinitely long, and circles consist of 360 degrees. As the cutter is moved about the geometry under control of the motion statements, the lengths of the lines and the amounts of the arcs are "cut" to their proper size. (Some of the geometry statements shown in the accompanying illustrations for defining POINTS, LINES, CIRCLES, TABULATED CYLINDERS, CYLINDERS, CONES, and SPHERES, in the APT language, may not be included in some two-dimensional [ADAPT] systems.)

Points

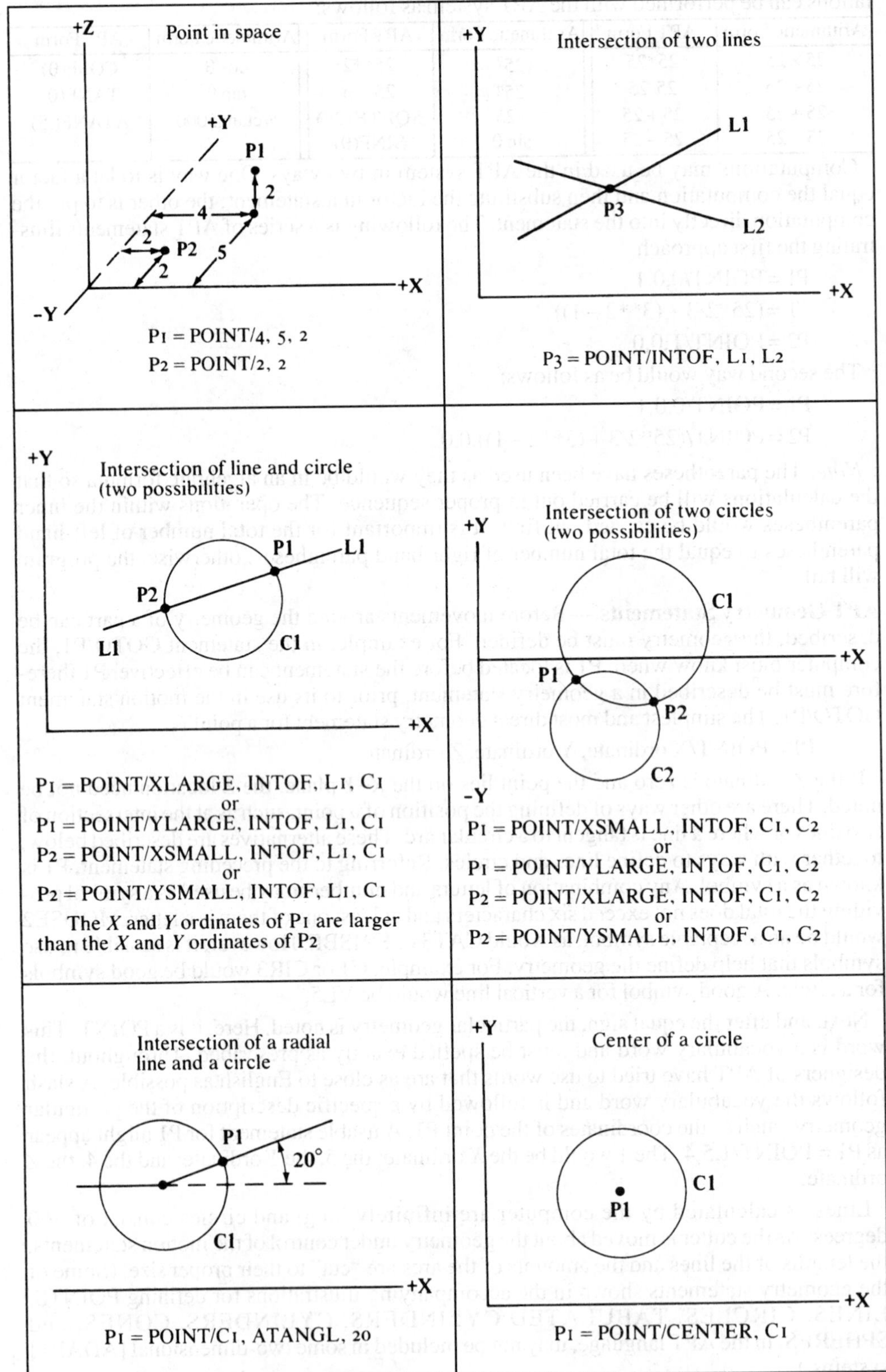
Point in space
+Z
+Y
-Y
+X
P1
P2
2
4
2
5
2
P1 = POINT/4, 5, 2
P2 = POINT/2, 2
Intersection of two lines
+Y
+X
L1
L2
P3
P3 = POINT/INTOF, L1, L2
Intersection of line and circle
(two possibilities)
+Y
+X
P1
P2
L1
C1
P1 = POINT/XLARGE, INTOF, L1, C1
or
P1 = POINT/YLARGE, INTOF, L1, C1
P2 = POINT/XSMALL, INTOF, L1, C1
or
P2 = POINT/YSMALL, INTOF, L1, C1
The X and Y ordinates of P1 are larger
than the X and Y ordinates of P2
Intersection of two circles
(two possibilities)
+Y
-Y
+X
C1
C2
P1
P2
P1 = POINT/XSMALL, INTOF, C1, C2
or
P1 = POINT/YLARGE, INTOF, C1, C2
P2 = POINT/XLARGE, INTOF, C1, C2
or
P2 = POINT/YSMALL, INTOF, C1, C2
Intersection of a radial
line and a circle
P1
20°
C1
+X
P1 = POINT/C1, ATANGL, 20
Center of a circle
+Y
+X
C1
P1
P1 = POINT/CENTER, C1

Lines

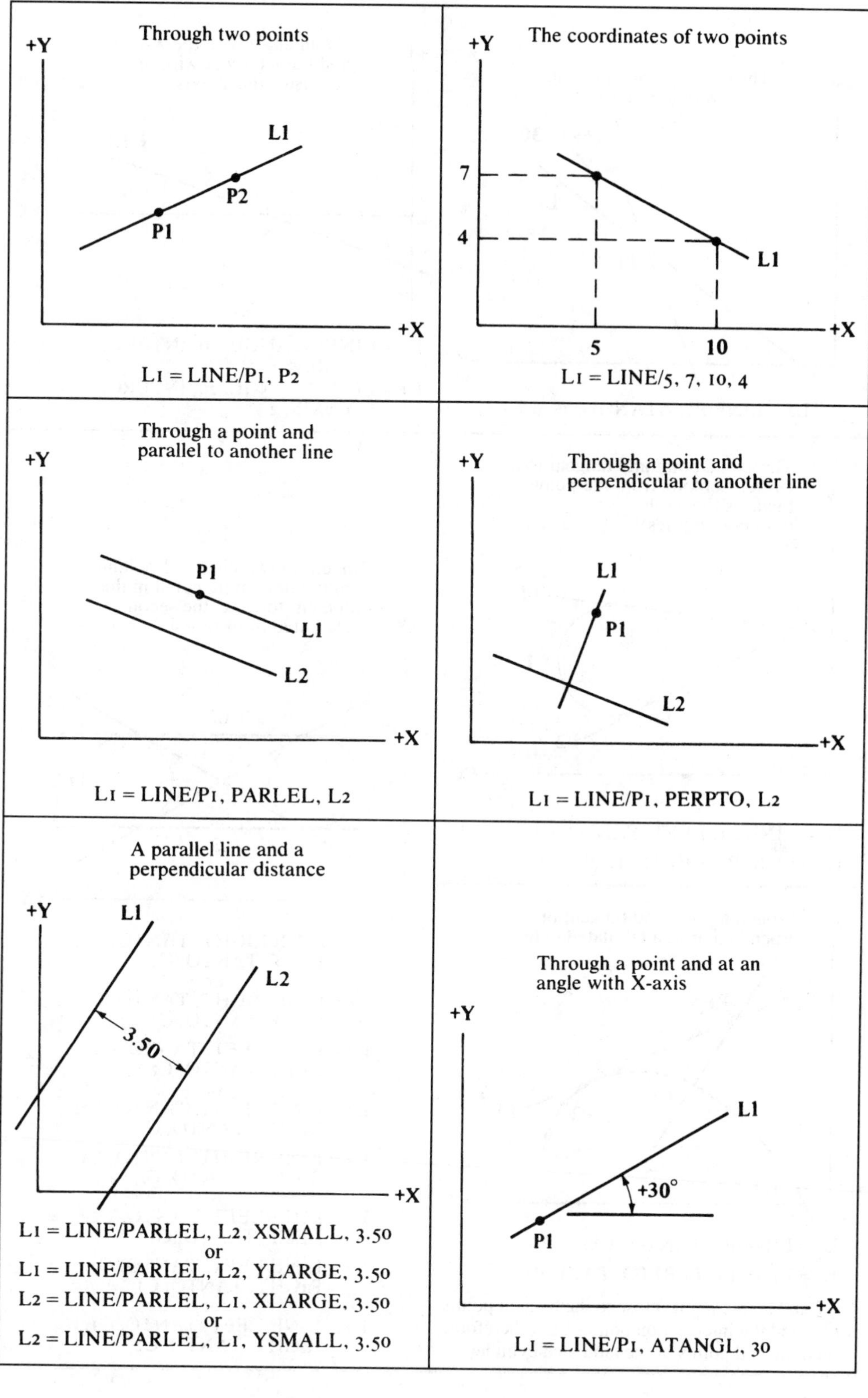
Through two points
+Y
L1
P2
P1
+X
L1 = LINE/P1, P2
The coordinates of two points
+Y
7
4
L1
5
10
+X
L1 = LINE/5, 7, 10, 4
Through a point and
parallel to another line
+Y
P1
L1
L2
+X
L1 = LINE/P1, PARLEL, L2
Through a point and
perpendicular to another line
+Y
L1
P1
L2
+X
L1 = LINE/P1, PERPTO, L2
A parallel line and a
perpendicular distance
+Y
L1
L2
3.50
+X
L1 = LINE/PARLEL, L2, XSMALL, 3.50
or
L1 = LINE/PARLEL, L2, YLARGE, 3.50
L2 = LINE/PARLEL, L1, XLARGE, 3.50
or
L2 = LINE/PARLEL, L1, YSMALL, 3.50
Through a point and at an
angle with X-axis
+Y
L1
+30°
P1
+X
L1 = LINE/P1, ATANGL, 30

Lines *(Continued)*

Through a point and at an angle with another line

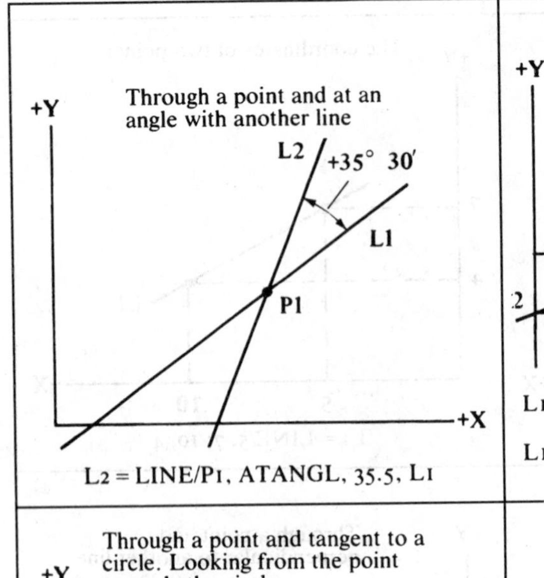

L2 = LINE/P1, ATANGL, 35.5, L1

By an angle with the X-axis and the intercept with the X-axis or the Y-axis

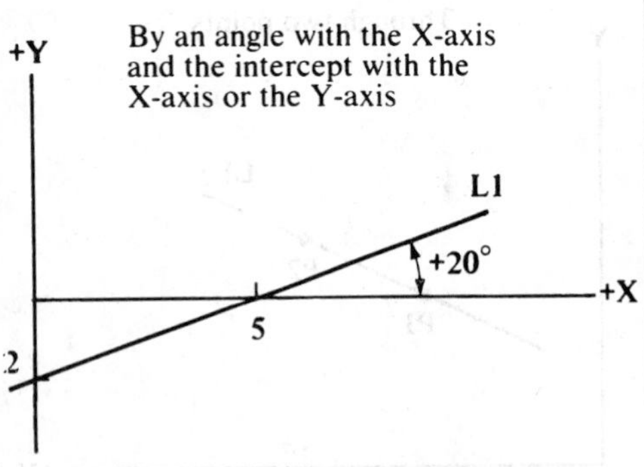

L1 = LINE/ATANGL, 20, INTERC, XAXIS, 5

L1 = LINE/ATANGL, 20, INTERC, YAXIS, 2

Through a point and tangent to a circle. Looking from the point towards the circle (two possibilities)

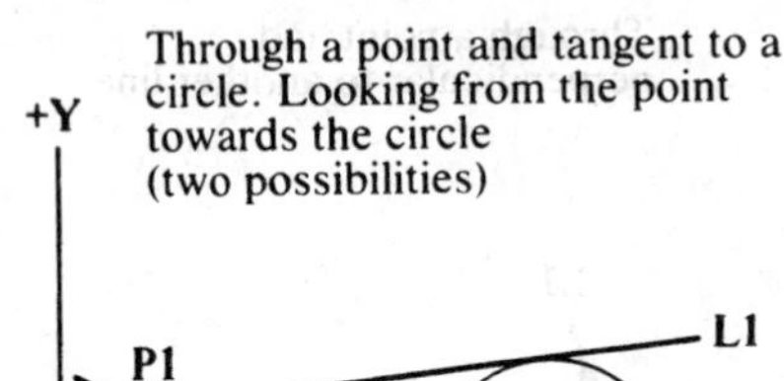

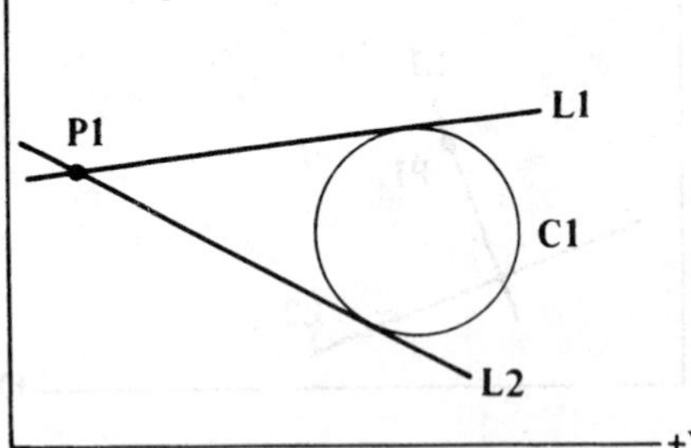

L1 = LINE/P1, LEFT, TANTO, C1

L2 = LINE/P1, RIGHT, TANTO, C1

Through a point and tangent or perpendicular to a tabulated cylinder

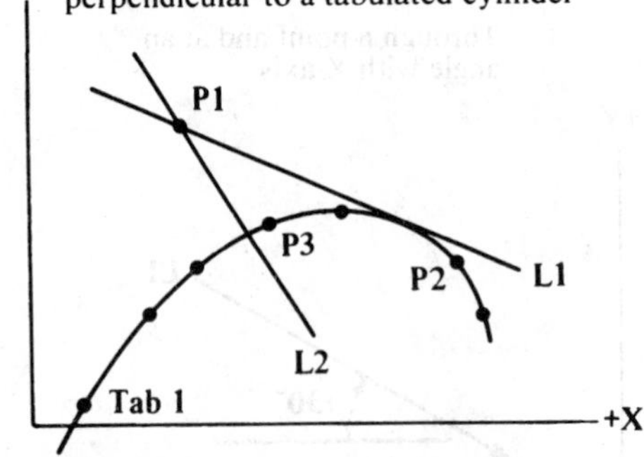

L1 = LINE/P1, TANTO, TAB1, P3

L1 = LINE/P1, PERPTO, TAB1, P3

P2 and P3 are points close to the tangent points of L1 and the intersection point of L2, therefore cannot be end points of the tabulated cylinder

Tangent to two circles. Looking from the first circle noted in the statement towards the second circle noted (four possibilities)

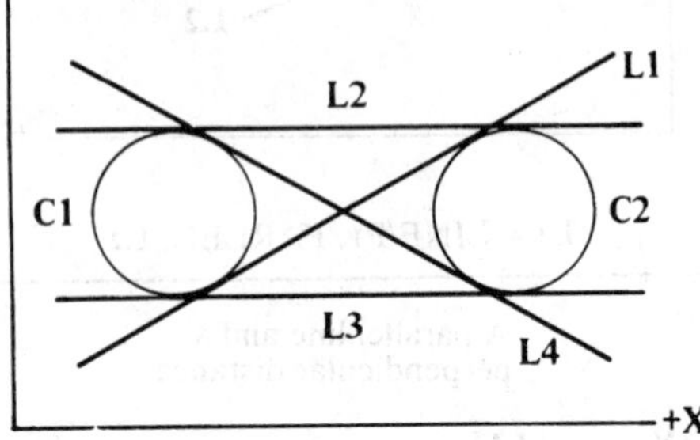

L1 = LINE/RIGHT, TANTO, C2, LEFT, TANTO, C1

or

L1 = LINE/RIGHT, TANTO, C1, LEFT, TANTO, C2

L2 = LINE/LEFT, TANTO, C1, LEFT, TANTO, C2

or

L2 = LINE/RIGHT, TANTO, C2, RIGHT, TANTO, C1

L3 = LINE/RIGHT, TANTO, C1, RIGHT, TANTO, C2

or

L3 = LINE/LEFT, TANTO, C2, LEFT, TANTO, C1

L4 = LINE/LEFT, TANTO, C2, RIGHT, TANTO, C1

or

L4 = LINE/LEFT, TANTO, C1, RIGHT, TANTO, C2

Circles

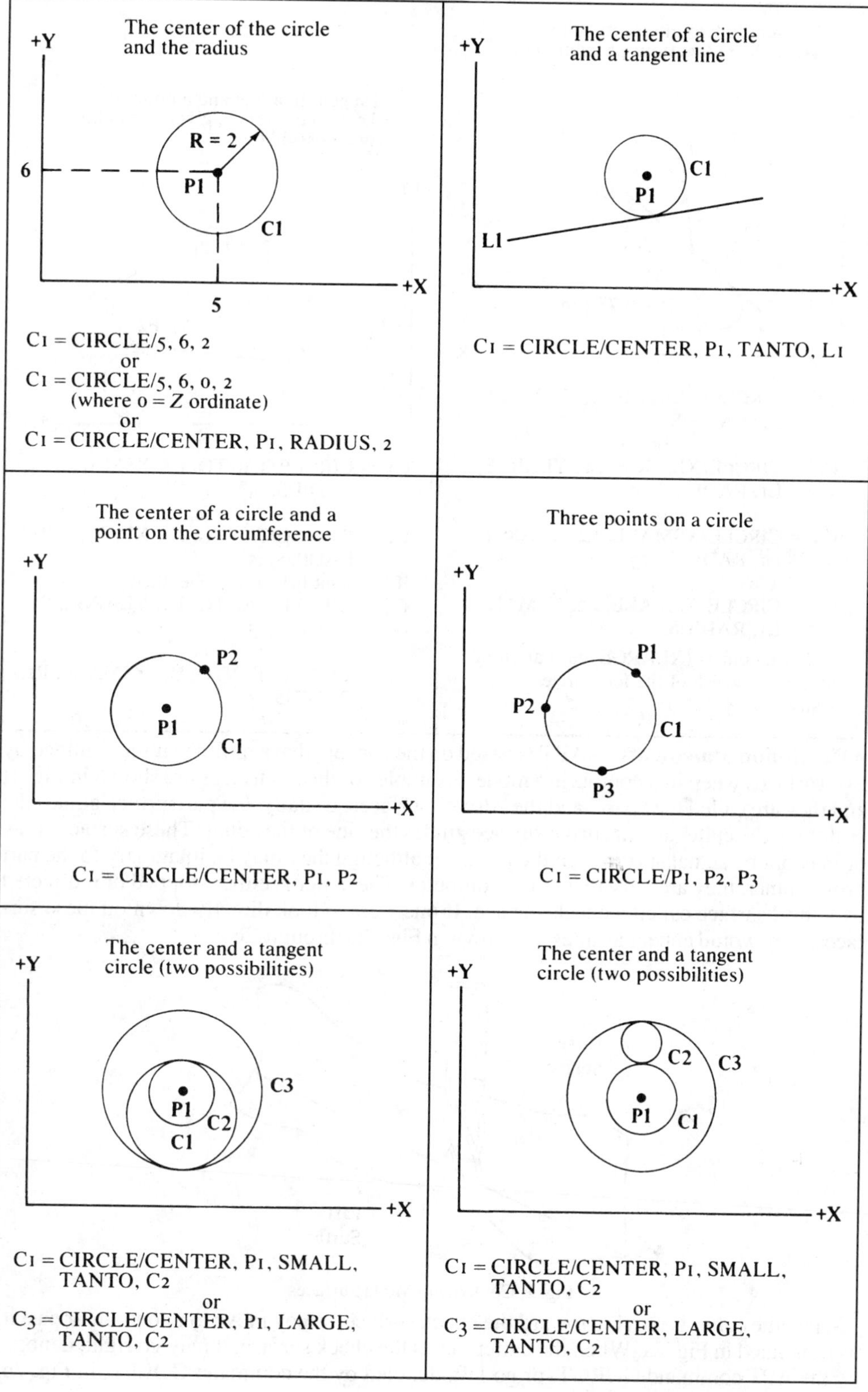
The center of the circle and the radius
+Y
6
R = 2
P1
C1
+X
5
C1 = CIRCLE/5, 6, 2
or
C1 = CIRCLE/5, 6, 0, 2
(where 0 = Z ordinate)
or
C1 = CIRCLE/CENTER, P1, RADIUS, 2
The center of a circle and a tangent line
+Y
C1
P1
L1
+X
C1 = CIRCLE/CENTER, P1, TANTO, L1
The center of a circle and a point on the circumference
+Y
P2
P1
C1
+X
C1 = CIRCLE/CENTER, P1, P2
Three points on a circle
+Y
P1
P2
C1
P3
+X
C1 = CIRCLE/P1, P2, P3
The center and a tangent circle (two possibilities)
+Y
C3
P1
C2
C1
+X
C1 = CIRCLE/CENTER, P1, SMALL, TANTO, C2
or
C3 = CIRCLE/CENTER, P1, LARGE, TANTO, C2
The center and a tangent circle (two possibilities)
+Y
C2
C3
P1
C1
+X
C1 = CIRCLE/CENTER, P1, SMALL, TANTO, C2
or
C3 = CIRCLE/CENTER, LARGE, TANTO, C2

Circles

Tangent to two intersecting lines, given the radius (four possibilities)

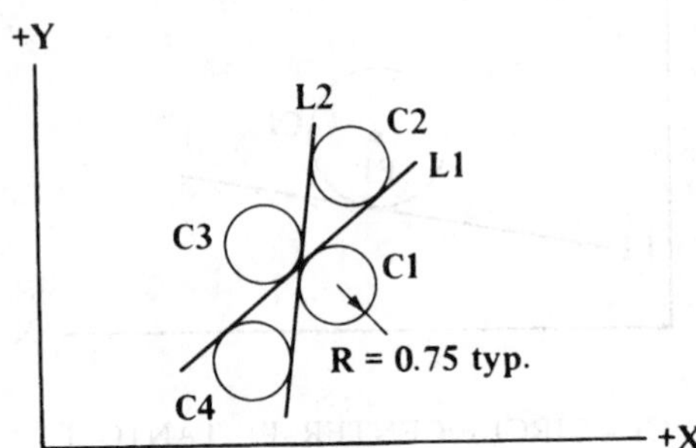

C1 = CIRCLE/XLARGE, L2, YSMALL, L1, RADIUS, .75

or

C2 = CIRCLE/XLARGE, L2, YLARGE, L1, RADIUS, .75

or

C3 = CIRCLE/XSMALL, L2, YLARGE, L1, RADIUS, .75

or

C4 = CIRCLE/XSMALL, L2, YSMALL, L1, RADIUS, .75

The modifiers (XLARGE, etc.) are used to indicate which of the four circles is wanted.

Tangent to a line and a point on the circumference, given the radius (two possibilities)

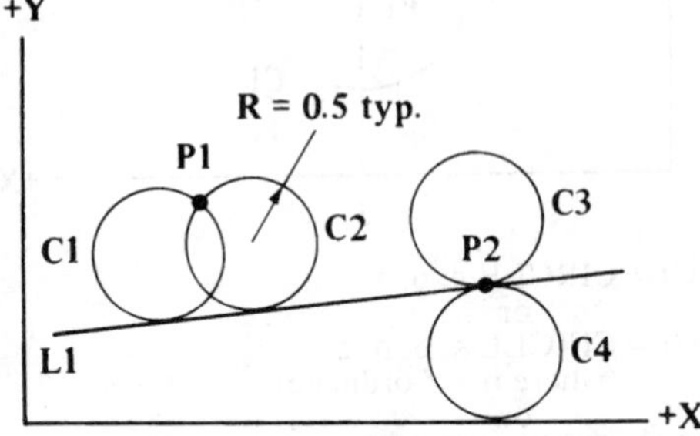

C1 = CIRCLE/TANTO, L1, XSMALL, P1, RADIUS, .5

or

C2 = CIRCLE/TANTO, L1, XLARGE, P1, RADIUS, .5

if the point lies on the line, then:

C3 = CIRCLE/TANTO, L1, YLARGE, P2, RADIUS, .5

or

C4 = CIRCLE/TANTO, L1, YSMALL, P2, RADIUS, .5

APT Motion Statements.—APT is based on the concept that a milling cutter is guided by two surfaces when in a contouring mode. Examples of these surfaces are shown in Fig. 1, and they are called the "part" and the "drive" surfaces. Usually, the partsurface guides the bottom of the cutter and the drive surface guides the side of the cutter. These surfaces may or may not be actual surfaces on the part, and although they may be imaginary to the part programmer, they are very real to the computer. The cutter is either stopped or redirected by a third surface called a check surface. If one were to look directly down on these surfaces, they would appear as lines, as shown in Figs. 2a through 2c.

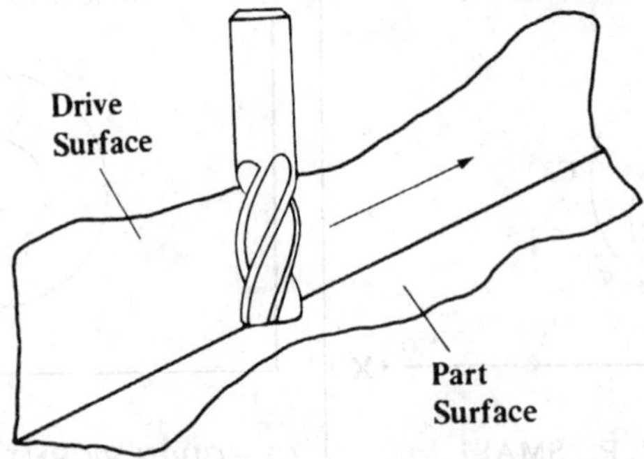

Fig. 1. Contouring Mode Surfaces

When the cutter is moving toward the check surface, it may move to it, onto it, or past it, as illustrated in Fig. 2a. When the cutter meets the check surface, it may go right, denoted by the APT command GORGT, or go left, denoted by the command GOLFT, in Fig. 2b.

Alternatively, the cutter may go forward, instructed by the command GOFWD, as in Fig. 2c. The command GOFWD is used when the cutter is moving either onto or off a tangent circular arc. These code instructions are part of what are called motion commands. Fig. 3 shows a cutter moving along a drive surface, L1, toward a check surface, L2. When it arrives at L2, the cutter will make a right turn and move along L2 and past the new check surface L3. Note that L2 changes from a check surface to a drive surface the moment the cutter begins to move along it. The APT motion statement for this move is:

GORGT/L2,PAST,L3

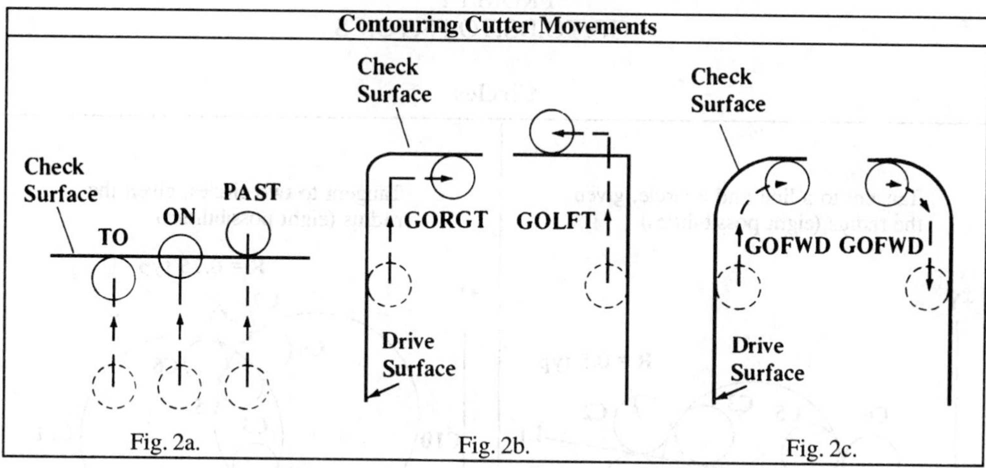

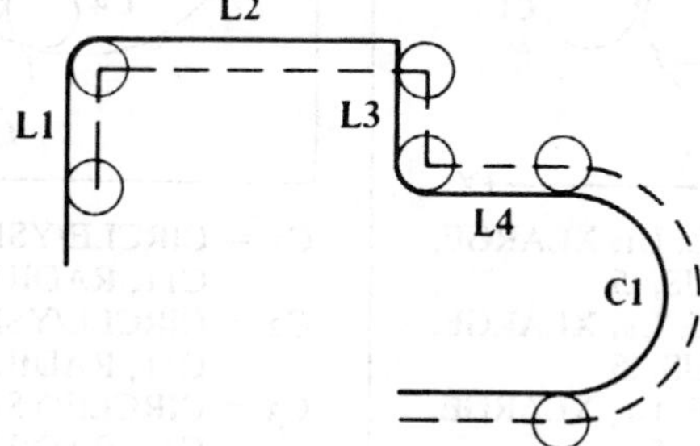

Fig. 3. Motion Statements for Movements Around a Workpiece

Still referring to Fig. 3, the cutter moves along L3 until it comes to L4. L3 now becomes the drive surface and L4 the check surface. The APT statement is:

GORGT/L3,TO,L4

The next statement is:

GOLFT/L4,TANTO,C1

Even though the cutter is moving to the right, it makes a left turn if one is looking in the direction of travel of the cutter. In writing the motion statements, the part programmers must imagine they are steering the cutter. The drive surface now becomes L4 and the check surface, C1. The next statement will therefore be:

GOFWD/C1,TANTO,L5

This movement could continue indefinitely, with the cutter being guided by the drive, part, and check surfaces.

Start-Up Statements: For the cutter to move along them, it must first be brought into contact with the three guiding surfaces by means of a start-up statement. There are three different start-up statements, depending on how many surfaces are involved.

A three-surface start-up statement is one in which the cutter is moved to the drive, part, and check surfaces, as seen in Fig. 4a. A two-surface start-up is one in which the cutter is moved to the drive and part surfaces, as in Fig. 4b. A one-surface start-up is one in which the cutter is moved to the drive surface and the X–Y plane, where $Z = 0$, as in Fig. 4c. With the two- and one-surface start-up statements, the cutter moves in the most direct path, or perpendicular to the surfaces. Referring to Fig. 4a(three-surface start-up), the move is initiated from a point P1. The two statements that will move the cutter from P1 to the three surfaces are:

```
FROM/P1
GO/TO,DS,TO,PS,TO,CS
```

Circles

Tangent to a line and a circle, given the radius (eight possibilities)

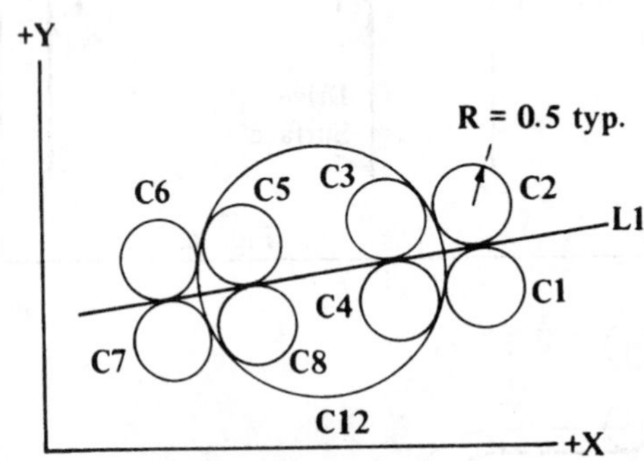

C1 = CIRCLE/YSMALL, L1, XLARGE, OUT, C12, RADIUS, .5
C2 = CIRCLE/YLARGE, L1, XLARGE, OUT, C12, RADIUS, .5
C3 = CIRCLE/YLARGE, L1, XLARGE, IN, C12, RADIUS, .5
C4 = CIRCLE/YSMALL, L1, XLARGE, IN, C12, RADIUS, .5
C5 = CIRCLE/YLARGE, L1, XSMALL, OUT, C12, RADIUS, .5
C6 = CIRCLE/XLARGE, L1, XSMALL, OUT, C12, RADIUS, .5
C7 = CIRCLE/YSMALL, L1, XSMALL, OUT, C12, RADIUS, .5
C8 = CIRCLE/YSMALL, L1, XSMALL, IN, C12, RADIUS, .5

Recommendations:

1. Note which side of line circle is on (e.g., YSMALL, L1).
2. Note whether the circle being defined is inside (IN), or outside (OUT), the known circle.
3. Of the two remaining circles, note whether the circle to be defined is XLARGE, XSMALL, or YLARGE or YSMALL, to arrive at the second modifier in the statement.

Tangent to two circles, given the radius (eight possibilities)

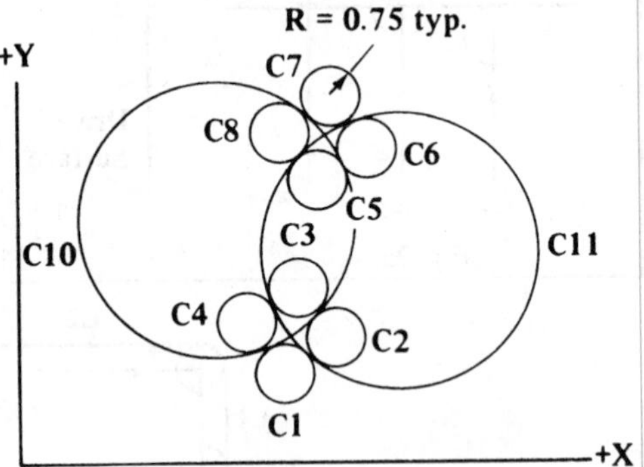

C1 = CIRCLE/YSMALL, OUT, C10, OUT, C11, RADIUS, .75
C2 = CIRCLE/YSMALL, OUT, C10, IN, C11, RADIUS, .75
C3 = CIRCLE/YSMALL, IN, C10, IN, C11, RADIUS, .75
C4 = CIRCLE/YSMALL, IN, C10, OUT, C11, RADIUS, .75
C5 = CIRCLE/YLARGE, IN, C10, IN, C11, RADIUS, .75
C6 = CIRCLE/YLARGE, OUT, C10, IN, C11, RADIUS, .75
C7 = CIRCLE/YLARGE, OUT, C10, OUT, C11, RADIUS, .75
C8 = CIRCLE/YLARGE, IN, C10, OUT, C11, RADIUS, .75

Recommendations

1. Apply IN, OUT modifiers.
2. Apply XLARGE, etc., modifiers.

DS is used as the symbol for the Drive Surface; PS as the symbol for the Part Surface; and CS as the symbol for the Check Surface. The surfaces must be denoted in this sequence. The drive surface is the surface that the cutter will move along after coming in contact with the three surfaces. The two statements applicable to the two-surface start-up (Fig. 4b) are:

FROM/P1
GO/TO,DS,TO,PS

The one-surface start-up (Fig. 4c) is:

FROM/P1
GO/TO,DS

Planes

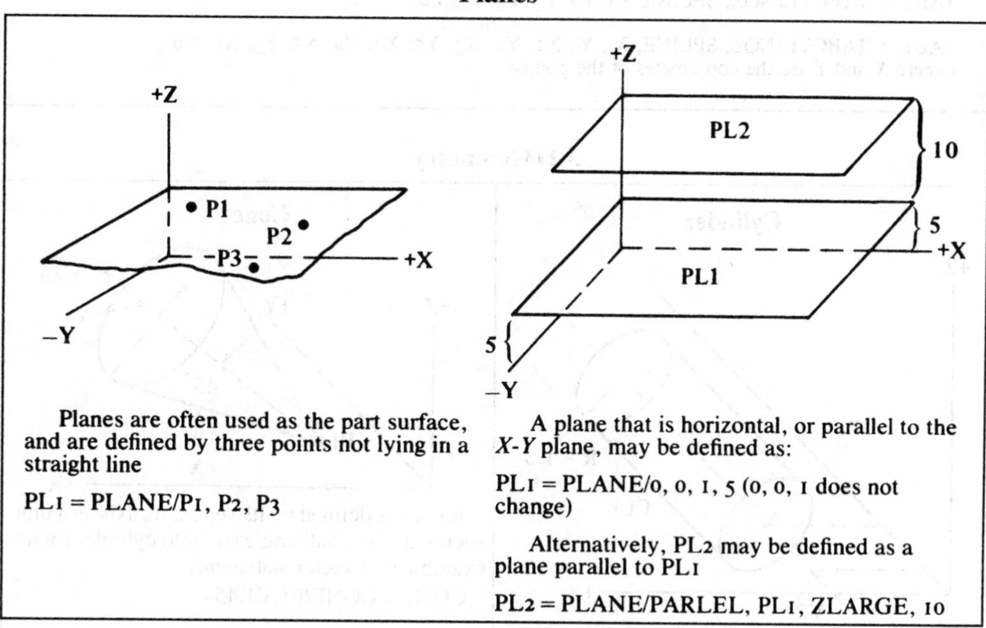

Planes are often used as the part surface, and are defined by three points not lying in a straight line

PL1 = PLANE/P1, P2, P3

A plane that is horizontal, or parallel to the *X-Y* plane, may be defined as:

PL1 = PLANE/0, 0, 1, 5 (0, 0, 1 does not change)

Alternatively, PL2 may be defined as a plane parallel to PL1

PL2 = PLANE/PARLEL, PL1, ZLARGE, 10

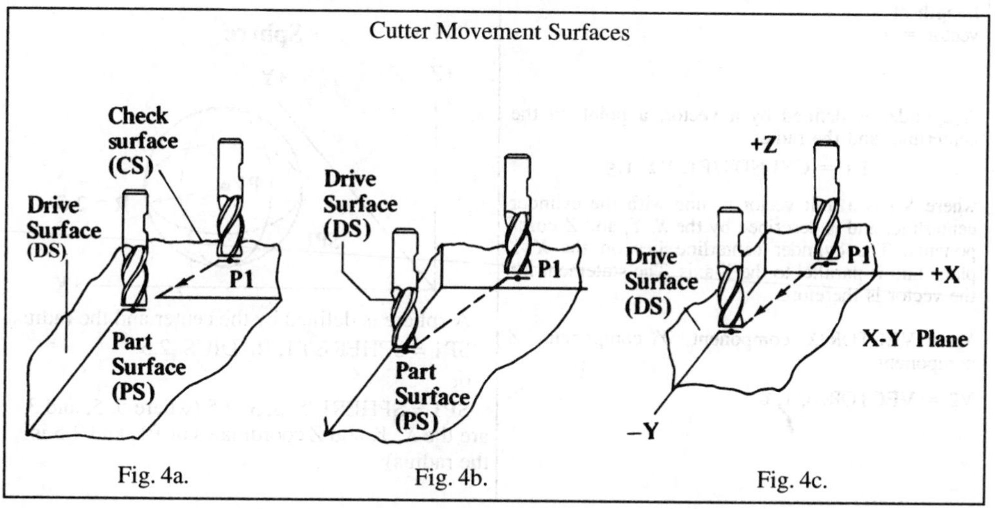

Fig. 4a. Fig. 4b. Fig. 4c.

Tabulated Cylinder

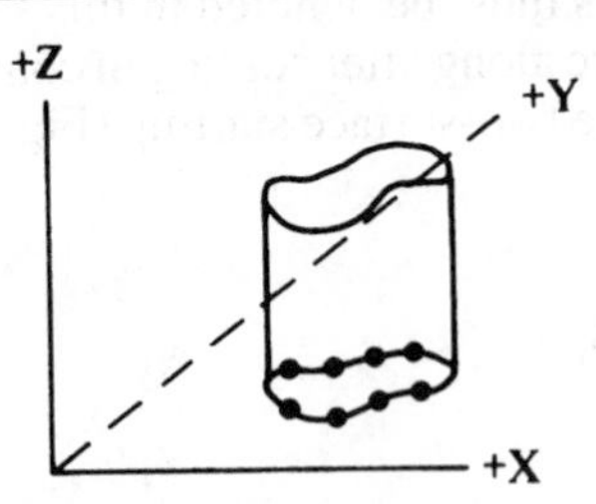

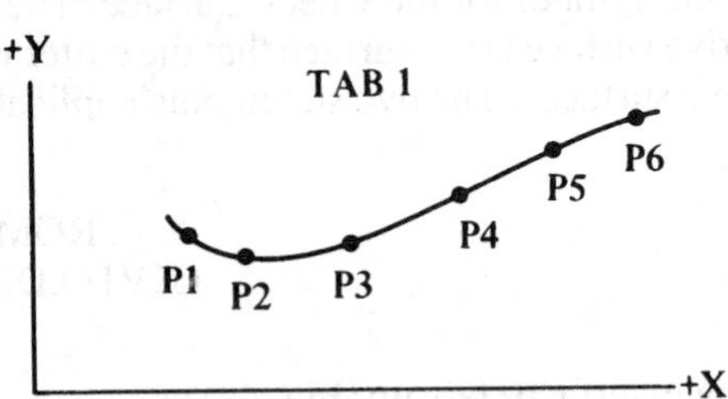

A tabulated cylinder is the line that is formed when an irregular cylinder intersects a plane. The plane intersected in the figure at the left is the *X–Y* plane.

A section of the line can be defined by a series of points on the line, as seen at the right. This line is called a TABCYL. The line must pass through all the points, therefore, it is best not to use too many. The statement to the computer would read:

TAB1 = TABCYL/NOZ, SPLINE, P1, P2, P3, P4, P5, P6

or

TAB1 = TABCYL/NOZ, SPLINE, X., Y., X2, Y2, X3, Y3, X4, Y4, X5, Y5, X6, Y6
(where *X* and *Y* are the coordinates of the points)

3-D Geometry

Cylinder

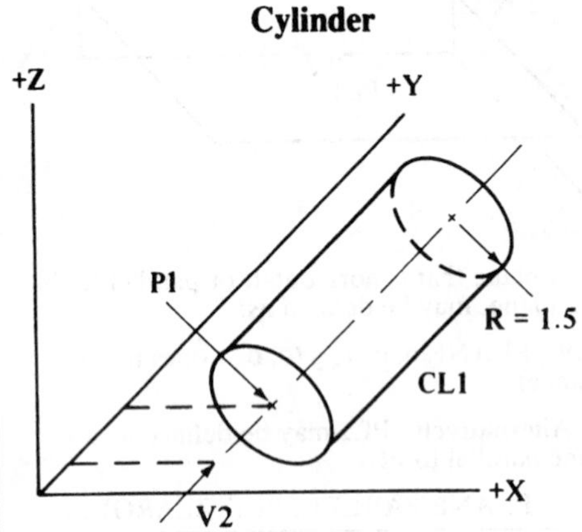

Length of vector = 1

A cylinder is defined by a vector, a point on the centerline, and the radius

CL1 = CYLNDR/P1, V2, 1.5

where V2 is a unit vector in line with the cylinder centerline, and is described by the *X*, *Y*, and *Z* components. The cylinder centerline lies on the *X–Y* plane and is parallel to the *Y*-axis. The statement for the vector is therefore:

V2 = VECTOR/*X* component, *Y* component, *Z* component

V2 = VECTOR/0, 1, 0

Cone

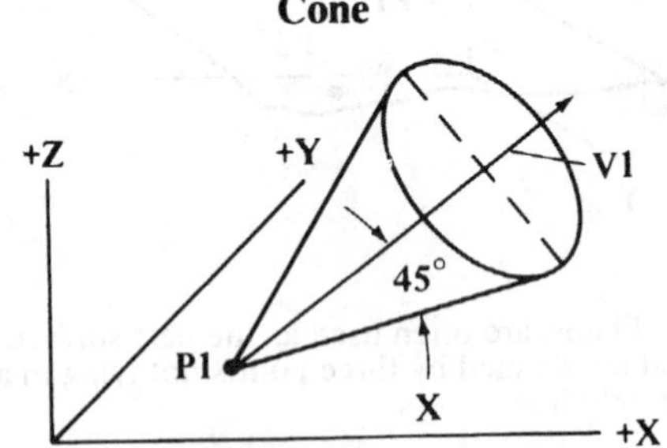

A cone is defined by its vertex, its axis as a unit vector, and the half angle (refer to cylinder for an example of a vector statement)

CON1 = CONE/P1,V1,45

Sphere

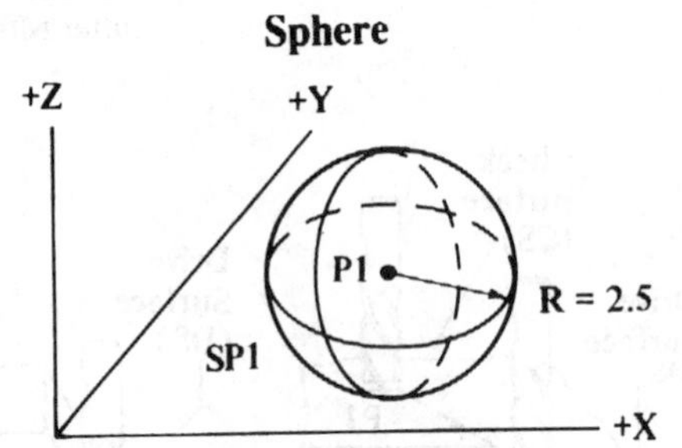

A sphere is defined by the center and the radius

SP1 = SPHERE/P1, RADIUS, 2.5

or

SP1 = SPHERE/5, 5, 3, 2.5 (where 5, 5, and 3 are the *X*, *Y*, and *Z* coordinates or P1, and 2.5 is the radius)

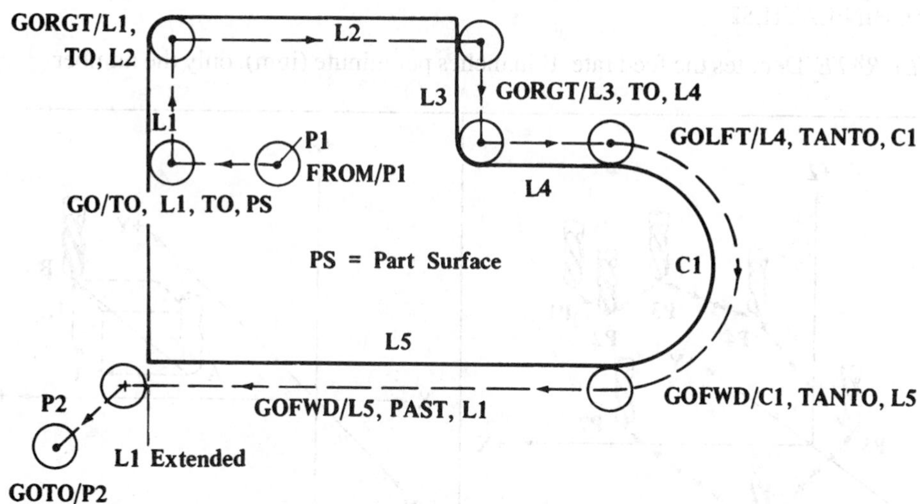

Fig. 5. A Completed Two-Surface Start-Up

Note that, in all three motion statements, the slash mark (/) lies between the GO and the TO. When the cutter is moving to a point rather than to surfaces, such as in a start-up, the statement is GOTO/ rather than GO/TO. A two-surface start-up, Fig. 3, when completed, might appear as shown in Fig. 5, which includes the motion statements needed. The motion statements, as they would appear in a part program, are shown at the left, below:

```
FROM/P1                    FROM/P1
GO/TO,L1,TO,PS             GOTO/P2
GORGT/L1,TO,L2             GOTO/P3
GORGT/L2,PAST,L3           GOTO/P4
GORGT/L3,TO,L4             GOTO/P5
GOLFT/L4,TANTO,C1          GOTO/P6
GOFWD/C1,TANTO,L5          GOTO/P7
GOFWD/L5,PAST,L1
GOTO/P2
```

GOTO statements can move the cutter throughout the range of the machine, as shown in Fig. 6. APT statements for such movements are shown at the right in the preceding example. The cutter may also be moved incrementally, as shown in Fig. 7. Here, the cutter is to move 2 inches in the + X direction, 1 inch in the + Y direction, and 1.5 inches in the + Z direction. The incremental move statement (indicated by DLTA) is:

GODLTA/2,1,1.5

The first position after the slash is the X movement; the second the Y movement, and the third, the Z movement.

Five-Axis Machining: Machining on five axes is achieved by causing the APT program to generate automatically a unit vector that is normal to the surface being machined, as shown in Fig. 8. The vector would be described by its X, Y, and Z components. These components, along with the X, Y, and Z coordinate positions of the tool tip, are fed into the postprocessor, which determines the locations and angles for the machine tool head and/or table.

APT Postprocessor Statements.—Statements that refer to the operation of the machine rather than to the geometry of the part or the motion of the cutter about the part are called postprocessor statements. APT postprocessor statements have been standardized internationally. Some common statements and an explanation of their meaning follow:

MACHIN/ Specifies the postprocessor that is to be used. Every postprocessor has an identity code, and this code must follow the slash mark (/). For example: MACHIN/LATH,82

FEDRATE/ Denotes the feed rate. If in inches per minute (ipm), only the number

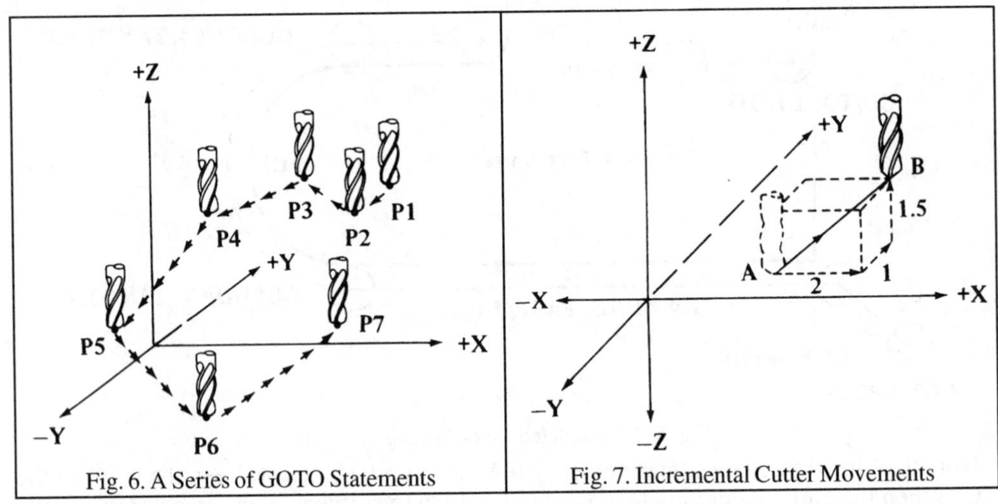

Fig. 6. A Series of GOTO Statements

Fig. 7. Incremental Cutter Movements

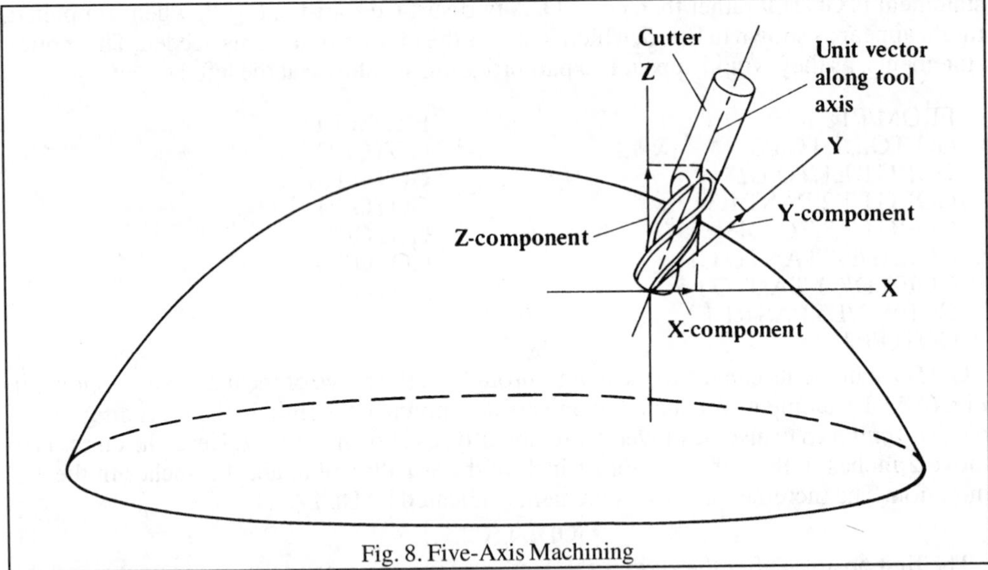

Fig. 8. Five-Axis Machining

need be shown. If in inches per revolution (ipr), IPR must be shown, for example: FEDRAT/.005,IPR

RAPID Means rapid traverse and applies only to the statement that immediately follows it

SPINDL/ Refers to spindle speed. If in revolutions per minute (rpm), only the number need be shown. If in surface feet per minute (sfm), the letters SFM need to be shown, for example: SPINDL/ 100SFM

COOLNT/ Means cutting fluid and can be subdivided into: COOLNT/ON, COOLNT/MIST, COOLNT/FLOOD, COOLNT/OFF

TURRET/ Used to call for a selected tool or turret position

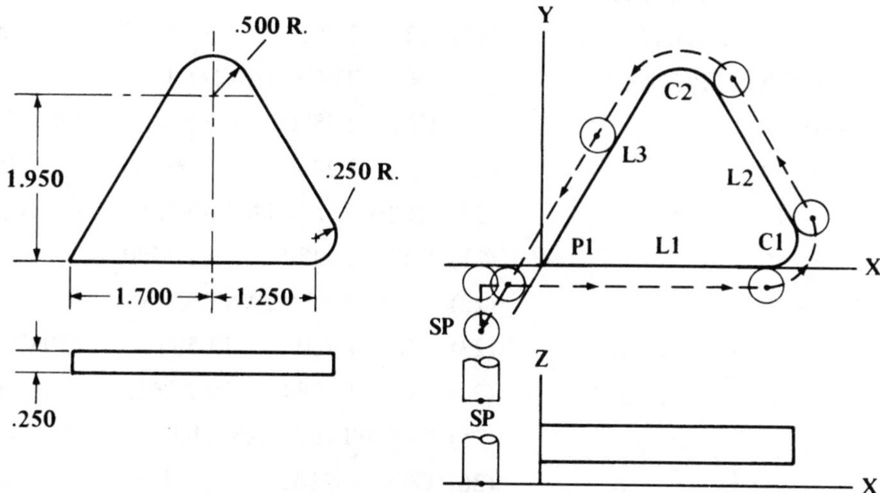

Fig. 9. Symbols for Geometrical Elements

CYCLE/ Specifies a cycle operation such as a drilling or boring cycle. An example of a drilling cycle is: CYCLE/DRILL,RAPTO,.45,FEDTO,0,IPR,.004. The next statement might be GOTO/PI and the drill will then move to P1 and perform the cycle operation. The cycle will repeat until the CYCLE/OFF statement is read

END Stops the machine but does not turn off the control system

APT Example Program.— A dimensioned drawing of a part and a drawing with the symbols for the geometry elements are shown in Fig. 9. A complete APT program for this part, starting with the statement PARTNO 47F36542 and ending with FINI, is shown at the left below.

(1) PARTNO	(1) PARTNO
(2) CUTTER/.25	(2) CUTTER/.25
(3) FEDRAT/5	(3) FEDRAT/5
(4) SP = POINT/–.5, –.5, .75	(4) SP = POINT/–.5, –.5, .75
(5) P1 = POINT/0, 0, 1	(5) P1 = POINT/0, 0, 1
(6) L1 = LINE/P1, ATANGL, 0	(6) L1 = LINE/P1, ATANGL, 0
(7) C1 = CIRCLE/(1.700 + 1.250), .250, .250	(7) C1 = CIRCLE/(1.700 + 1.250), .250, .250
(8) C2 = CIRCLE/1.700, 1.950, .5	(8) C2 = CIRCLE/1.700, 1.950, .5
(9) L2 = LINE/RIGHT, TANTO, C1, RIGHT, TANTO, C2	(9) L2 = LINE/RIGHT, TANTO, C1, RIGHT, TANTO, C2
(10) L3 = LINE/P1, LEFT, TANTO, C2	(10) L3 = LINE/P1, LEFT, TANTO, C2
(11) FROM/SP	(11) FROM/SP
(12) GO/TO, L1	(12) FRO M –.500 –.5000 .7500
(13) GORGT/L1, TANTO, C1	(13) GO/TO/, L1
(14) GOFWD/C1, TANTO, L2	(14) GT –.5000 –.1250 .0000

```
(15) GOFWD/L2, TANTO, C2
(16) GOFWD/C2, TANTO, L3
(17) GOFWD/L3, PAST, L1
(18) GOTO/SP
(19) FINI
```

```
(15) GORGT/L1, TANTO, C1
(16) GT     2.9500     -.1250      .0000
(17) GOFWD/C1, TANTO, L2
(18) CIR    2.9500     .2500       .3750 CCLW
(19)        3.2763     .4348       .0000
(20) GOFWD/L2, TANTO, C2
(21) GT     2.2439     2.2580      .0000
(22) GOFWD/C2, TANTO, L3
(23) CIR    1.700      1.9500      .6250 CCLW
(24)        1.1584     2.2619      .0000
(25) GOFWD/L3, PAST, L1
(26) GT     -.2162     -.1250      .0000
(27) GOTO/SP
(28) GT     -.5000     -.5000      .7500
(29) FINI
```

The numbers at the left of the statements are for reference purposes only, and are not part of the program. The cutter is set initially at a point represented by the symbol SP, having coordinates $X = -0.5$, $Y = -0.5$, $Z = 0.75$, and moves to L1 (extended) with a one-surface start-up so that the bottom of the cutter rests on the X–Y plane. The cutter then moves counterclockwise around the part, past L1 (extended), and returns to SP. The coordinates of P1 are $X = 0$, $Y = 0$, and $Z = 1$.

Referring to the numbers at the left of the program:

(1) PARTNO must begin every program. Any identification can follow.

(2) The diameter of the cutter is specified. Here it is 0.25 inch.

(3) The feed rate is given as 5 inches per minute, which is contained in a postprocessor statement.

(4)–(10) Geometry statements.

(11)–(18) Motion statements.

(19) All APT programs end with FINI.

A computer printout from section II of the APT program is shown at the right, above. This program was run on a desktop personal computer. Lines (1) through (10) repeat the geometry statements from the original program. The motion statements are also repeated, and below each motion statement are shown the X, Y, and Z coordinates of the end points of the center-line (CL) movements for the cutter. Two lines of data follow those for the circular movements. For example, Line (18), which follows Line (17), GOFWD/C1,TANTO,L2, describes the X coordinate of the center of the arc, 2.9500, the Y coordinate of the center of the arc, 0.2500, and the radius of the arc required to be traversed by the cutter.

This radius is that of the arc shown on the part print, plus the radius of the cutter ($0.2500 + 0.1250 = 0.3750$). Line (18) also shows that the cutter is traveling in a counterclockwise (CCLW) motion. A circular motion is described in Lines (22), (23), and (24). Finally, the cutter is directed to return to the starting point, SP, and this command is noted in Line (27). The X, Y, and Z coordinates of SP are shown in Line (28).

APT for Turning.—In its basic form, APT is not a good program for turning. Although APT is probably the most suitable program for three-, four-, and five-axis machining, it is awkward for the simple two-axis geometry required for lathe operations. To overcome this problem, preprocessors have been developed especially for lathe part programming. The statements in the lathe program are automatically converted to basic APT statements in the computer and processed by the regular APT processor. An example of a lathe program, based on the APT processor and made available by the McDonnell Douglas Automation Co., is shown below. The numbers in parentheses are not part of the program, but are used only for reference. Fig. 10 shows the general set-up for the part, and Fig. 11 shows an enlarged view of the part profile with dimensions expressed along what would be the *X*- and *Y*-axes on the part print.

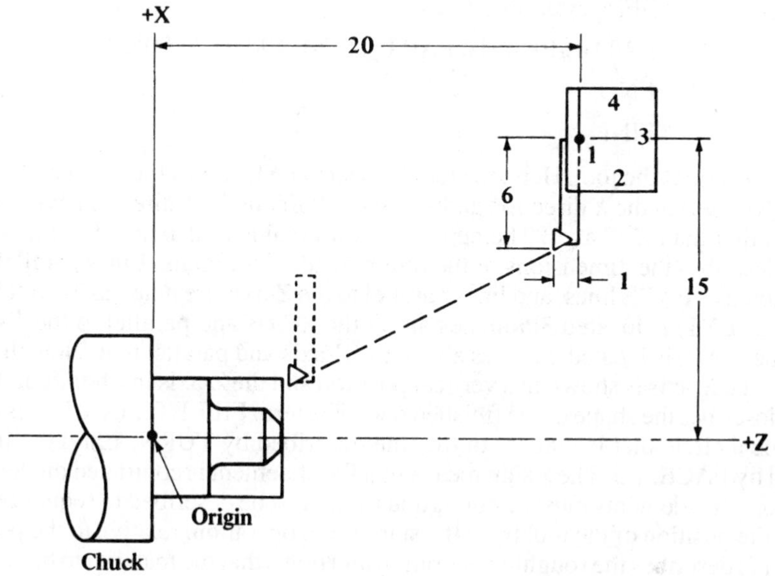

Fig. 10. Setup for APT Turning

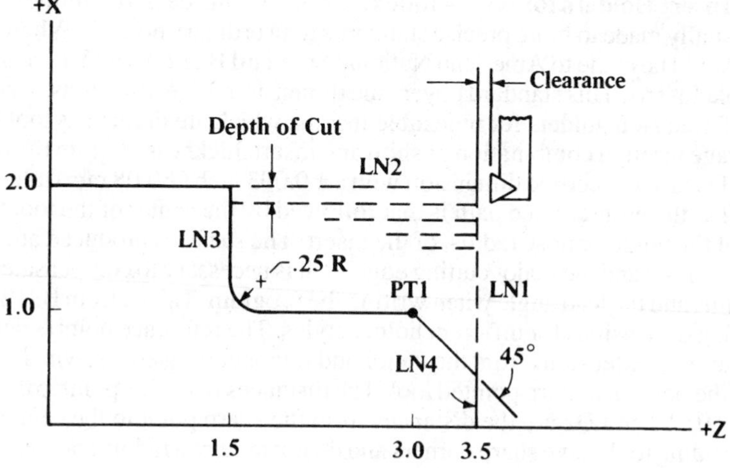

Fig. 11.

```
(1)  PARTNO LATHE EXAMPLE
(2)  MACHIN/MODEL LATHE
(3)          T1 = TOOL/FACE, 1, XOFF, –1, YOFF, –6, RADIUS, .031
(4)  BLANK1 = SHAPE/FACE, 3.5, TURN, 2
(5)     PART1 = SHAPE/FACE, 3.5, TAPER, 3.5, .5, ATANGL, – 45, TURN, 1,$
                FILLET, .25 FACE, 1.5 TURN, 2
(6)             FROM/(20–1), (15–6)
(7)             LATHE/ROUGH, BLANK1, PART1, STEP, .1, STOCK, .05,$
                SFM, 300, IPR, .01, T1
(8)             LATHE/FINISH, PART1, SFM, 400, IPR, .005, T1
(9)             END
(10)            FINI
```

Line (3) describes the tool. Here, the tool is located on face 1 of the turret and its tip is –1 inch "off" (offset) in the *X* direction and –6 inches "off" in the *Y* direction, when considering *X–Y* rather than *X–Z* axes. The cutting tool tip radius is also noted in this statement. Line (4) describes the dimensions of the rough material, or blank. Lines parallel to the *X*-axis are noted as FACE lines, and lines parallel to the *Z*-axis are noted as TURN lines. The FACE line (LN1) is located 3.5 inches along the *Z*-axis and parallel to the *X*-axis. The TURN line (LN2) is located 2 inches above the *Z*-axis and parallel to it. Note that in Figs. 10 and 11, the *X*-axis is shown in a vertical position and the *Z*-axis in a horizontal position. Line (5) describes the shape of the finished part. The term FILLET is used in this statement to describe a circle that is tangent to the line described by TURN, 1 and the line that is described by FACE, 1.5. The $ sign means that the statement is continued on the next line. These geometry elements must be contiguous and must be described in sequence. Line (6) specifies the position of the tool tip at the start of the operation, relative to the point of origin. Line (7) describes the roughing operation and notes that the material to be roughed out lies between BLANK1 and PART1; that the STEP, or depth of roughing cuts, is to be 0.1 inch; that 0.05 inch is to be left for the finish cut; that the speed is to be 300 sfm and the feed rate is to be 0.01 ipr; and that the tool to be used is identified by the symbol T1. Line (8) describes the finish cut, which is to be along the contour described by PART1.

Indexable Insert Holders for NC.—Indexable insert holders for numerical control lathes are usually made to more precise standards than ordinary holders. Where applicable, reference should be made to American National Standard B212.3-1986, Precision Holders for Indexable Inserts. This standard covers the dimensional specifications, styles, and designations of precision holders for indexable inserts, which are defined as tool holders that locate the gage insert (a combination of shim and insert thicknesses) from the back or front and end surfaces to a specified dimension with a ±0.003 inch (±0.08 mm) tolerance. In NC programming, the programmed path is that followed by the center of the tool tip, which is the center of the point, or nose radius, of the insert. The surfaces produced are the result of the path of the nose and the major cutting edge, so it is necessary to compensate for the nose or point radius and the lead angle when writing the program. Table , from B212.3, gives the compensating dimensions for different holder styles. The reference point is determined by the intersection of extensions from the major and minor cutting edges, which would be the location of the point of a sharp pointed tool. The distances from this point to the nose radius are *L*1 and *D*1; *L*2 and *D*2 are the distances from the sharp point to the center of the nose radius. Threading tools have sharp corners and do not require a radius compensation. Other dimensions of importance in programming threading tools are also given in Table 2; the data were developed by Kennametal, Inc.

Table 1. Insert Radius Compensation *ANSI B212.3-1986*

Style		Rad.	*L*-1	*L*-2	*D*-1	*D*-2
Square Profile						
		Turning 15° Lead Angle				
B Style[a] Also Applies to R Style	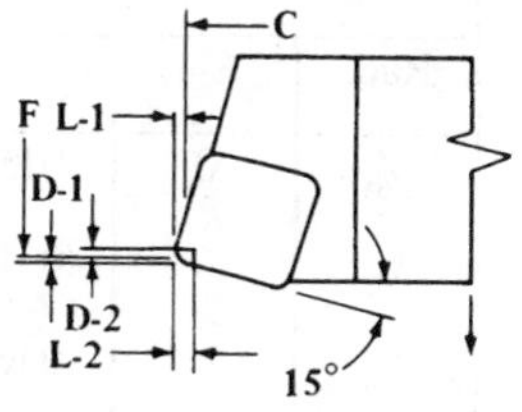	1/64	.0035	.0191	.0009	.0110
		1/32	.0070	.0383	.0019	.0221
		3/64	.0105	.0574	.0028	.0331
		1/16	.0140	.0765	.0038	.0442
		Turning 45° Lead Angle				
D Style[a]; Also Applies to S Style	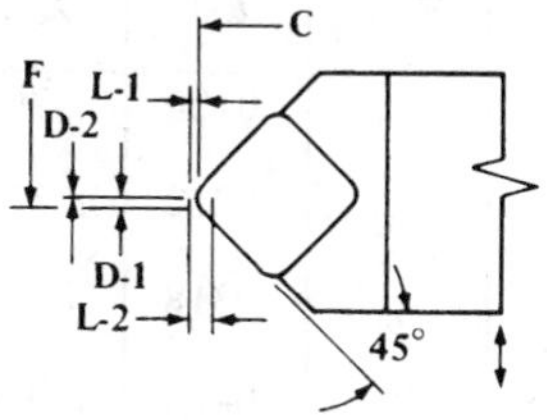	1/64	.0065	.0221	.0065	0
		1/32	.0129	.0442	.0129	0
		3/64	.0194	.0663	.0194	0
		1/16	.0259	.0884	.0259	0
		Facing 15° Lead Angle				
K Style[a];	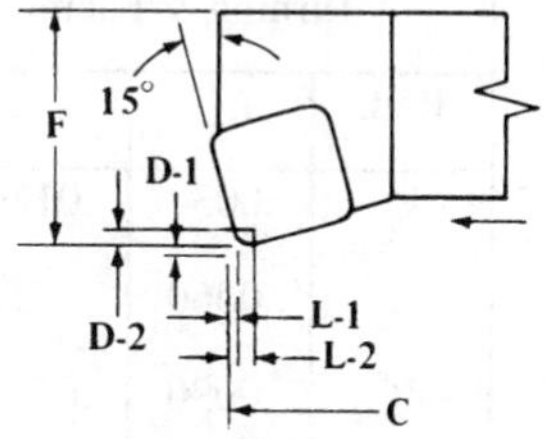	1/64	.0009	.0110	.0035	.0191
		1/32	.0019	.0221	.0070	.0383
		3/64	.0028	.0331	.0105	.0574
		1/16	.0038	.0442	.0140	.0765
Triangle Profile						
		Turning 0° Lead Angle				
G Style[a];	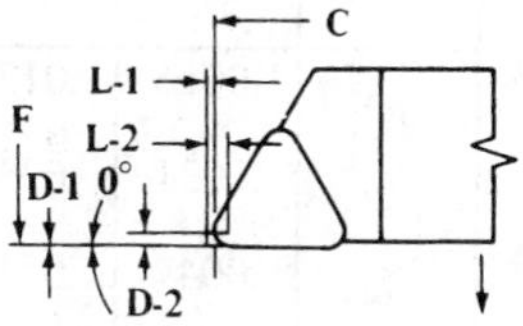	1/64	.0114	.0271	0	.0156
		1/32	.0229	.0541	0	.0312
		3/64	.0343	.0812	0	.0469
		1/16	.0458	.1082	0	.0625
		Turning and Facing 15° Lead Angle				
B Style[a]; Also Applies to R Style	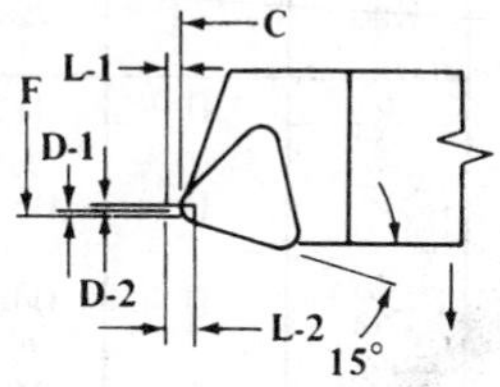	1/64	.0146	.0302	.0039	.0081
		1/32	.0291	.0604	.0078	.0162
		3/64	.0437	.0906	.0117	.0243
		1/16	.0582	.1207	.0156	.0324

Table 1. *(Continued)* **Insert Radius Compensation** *ANSI B212.3-1986*

Triangle Profile *(continued)*

F Style[a];

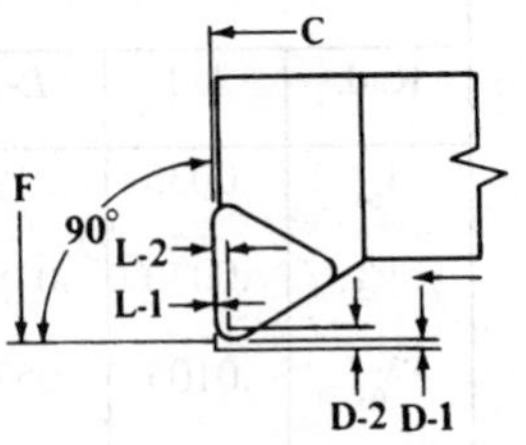

Facing 90° Lead Angle				
Rad.	*L*-1	*L*-2	*D*-1	*D*-2
1/64	0	.0156	.0114	.0271
1/32	0	.0312	.0229	.0541
3/64	0	.0469	.0343	.0812
1/16	0	.0625	.0458	.1082

J Style[a];

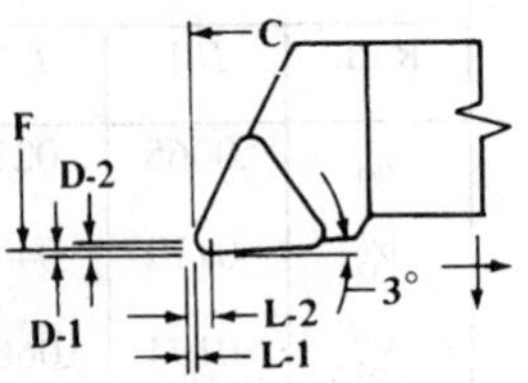

Turning & Facing 3° Lead Angle				
Rad.	*L*-1	*L*-2	*D*-1	*D*-2
1/64	.0106	.0262	.0014	.0170
1/32	.0212	.0524	.0028	.0340
3/64	.0318	.0786	.0042	.0511
1/16	.0423	.1048	.0056	.0681

80° Diamond Profile

G Style[a];

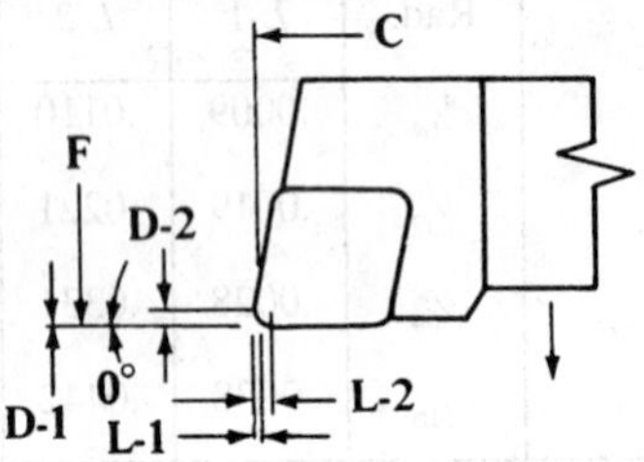

Turning & Facing 0° Lead Angle				
Rad.	*L*-1	*L*-2	*D*-1	*D*-2
1/64	.0030	.0186	0	.0156
1/32	.0060	.0312	0	.0312
3/64	.0090	.0559	0	.0469
1/16	.0120	.0745	0	.0625

L Style[a];

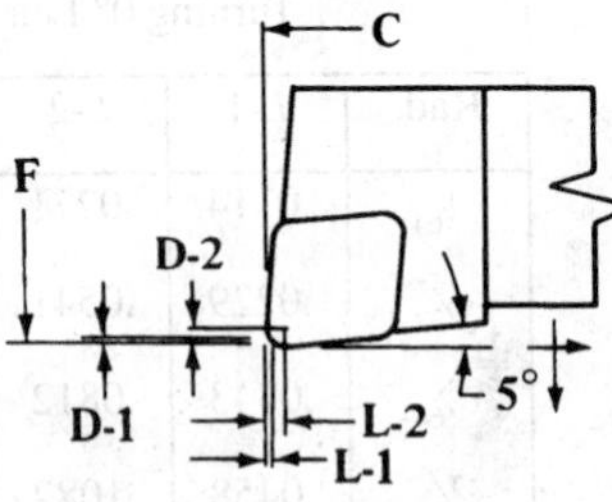

Turning & Facing 5° Reverse Lead Angle				
Rad.	*L*-1	*L*-2	*D*-1	*D*-2
1/64	.0016	.0172	.0016	.0172
1/32	.0031	.0344	.0031	.0344
3/64	.0047	.0516	.0047	.0516
1/16	.0062	.0688	.0062	.0688

F Style[a];

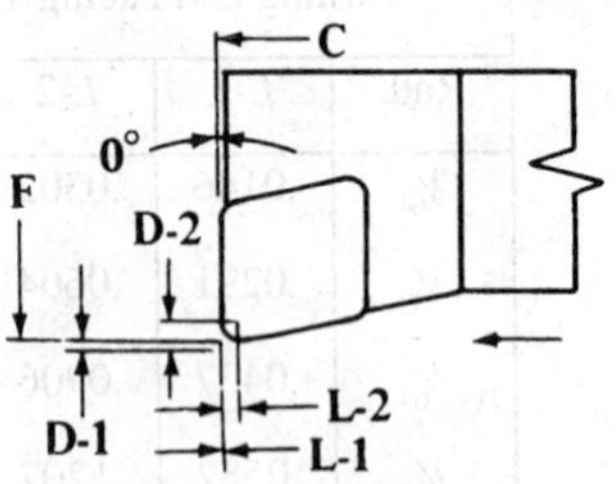

Facing 0° Lead Angle				
Rad.	*L*-1	*L*-2	*D*-1	*D*-2
1/64	0	.0156	.0030	.0186
1/32	0	.0312	.0060	.0372
3/64	0	.0469	.0090	.0559
1/16	0	.0625	.0120	.0745

Table 1. *(Continued)* **Insert Radius Compensation** *ANSI B212.3-1986*

Style	Rad.	*L*-1	*L*-2	*D*-1	*D*-2
80° Diamond Profile *(continued)*					
	Turning 15° Lead Angle				
R Style[a];	1/64	.0011	.0167	.0003	.0117
	1/32	.0022	.0384	.0006	.0234
	3/64	.0032	.0501	.0009	.0351
	1/16	.0043	.0668	.0012	.0468
	Facing 15° Lead Angle				
K Style[a];	1/64	.0003	.0117	.0011	.0167
	1/32	.0006	.0234	.0022	.0334
	3/64	.0009	.0351	.0032	.0501
	1/16	.0012	.0468	.0043	.0668
55° Profile					
	Profiling 3° Reverse Lead Angle				
J Style[a];	1/64	.0135	.0292	.0015	.0172
	1/32	.0271	.0583	.0031	.0343
	3/64	.0406	.0875	.0046	.0519
	1/16	.0541	.1166	.0062	.0687
35° Profile					
	Profiling 3° Reverse Lead Angle				
J Style[a]; Negative rake holders have 6° back rake and 6° side rake	1/64	.0330	.0487	.0026	.0182
	1/32	.0661	.0973	.0051	.0364
	3/64	.0991	.1460	.0077	.0546
	1/16	.1322	.1947	.0103	.0728
	Profiling 5° Lead Angle				
L Style[a];	1/64	.0324	.0480	.0042	.0198
	1/32	.0648	.0360	.0086	.0398
	3/64	.0971	.1440	.0128	.0597
	1/16	.1205	.1920	.0170	.0795

[a] *L*-1 and *D*-1 over sharp point to nose radius; and *L*-2 and *D*-2 over sharp point to center of nose radius. The *D*-1 dimension for the *B, E, D, M, P, S, T*, and *V* style tools are over the sharp point of insert to a sharp point at the intersection of a line on the lead angle on the cutting edge of the insert and the *C* dimension. The *L*-1 dimensions on *K* style tools are over the sharp point of insert to sharp point intersection of lead angle and *F* dimensions.

All dimensions are in inches.

Table 2. Threading Tool Insert Radius Compensation for NC Programming

Threading						
Insert Size	T	R	U	Y	X	Z
2	5/32 Wide	.040	.075	.040	.024	.140
3	3/16 Wide	.046	.098	.054	.031	.183
4	1/4 Wide	.053	.128	.054	.049	.239
5	3/8 Wide	.099	.190	…	…	…

Buttress Threading | 29° Acme | 60° V-Threading

Z | X | R | Y | 3° U T

NTB-B | NTB-A | NA | NTF | NT

All dimensions are given in inches.

Courtesy of Kennametal, Inc.

The *C* and *F* characters are tool holder dimensions other than the shank size. In all instances, the *C* dimension is parallel to the length of the shank and the *F* dimension is parallel to the side dimension; actual dimensions must be obtained from the manufacturer. For all *K* style holders, the *C* dimension is the distance from the end of the shank to the tangent point of the nose radius and the end cutting edge of the insert. For all other holders, the *C* dimension is from the end of the shank to a tangent to the nose radius of the insert. The *F* dimension on all B, D, E, M, P, and V style holders is measured from the back side of the shank to the tangent point of the nose radius and the side cutting edge of the insert. For all A, F, G, J, K, and L style holders, the *F* dimension is the distance from the back side of the shank to the tangent of the nose radius of the insert. In all these designs, the nose radius is the standard radius corresponding to those given in the paragraph *Cutting Point Configuration* on page 732.

V-Flange Tool Shanks and Retention Knobs.—Dimensions of ANSI B5.18-1972 standard tool shanks and corresponding spindle noses are detailed on pages 913 through pages 910, and are suitable for spindles used in milling and associated machines. Corresponding equipment for higher-precision numerically controlled machines, using retention knobs instead of drawbars, is usually made to the ANSI/ASME B5.50-1985 standard.

Essential Dimensions of V-Flange Tool Shanks ***ANSI/ASME B5.50-1985***

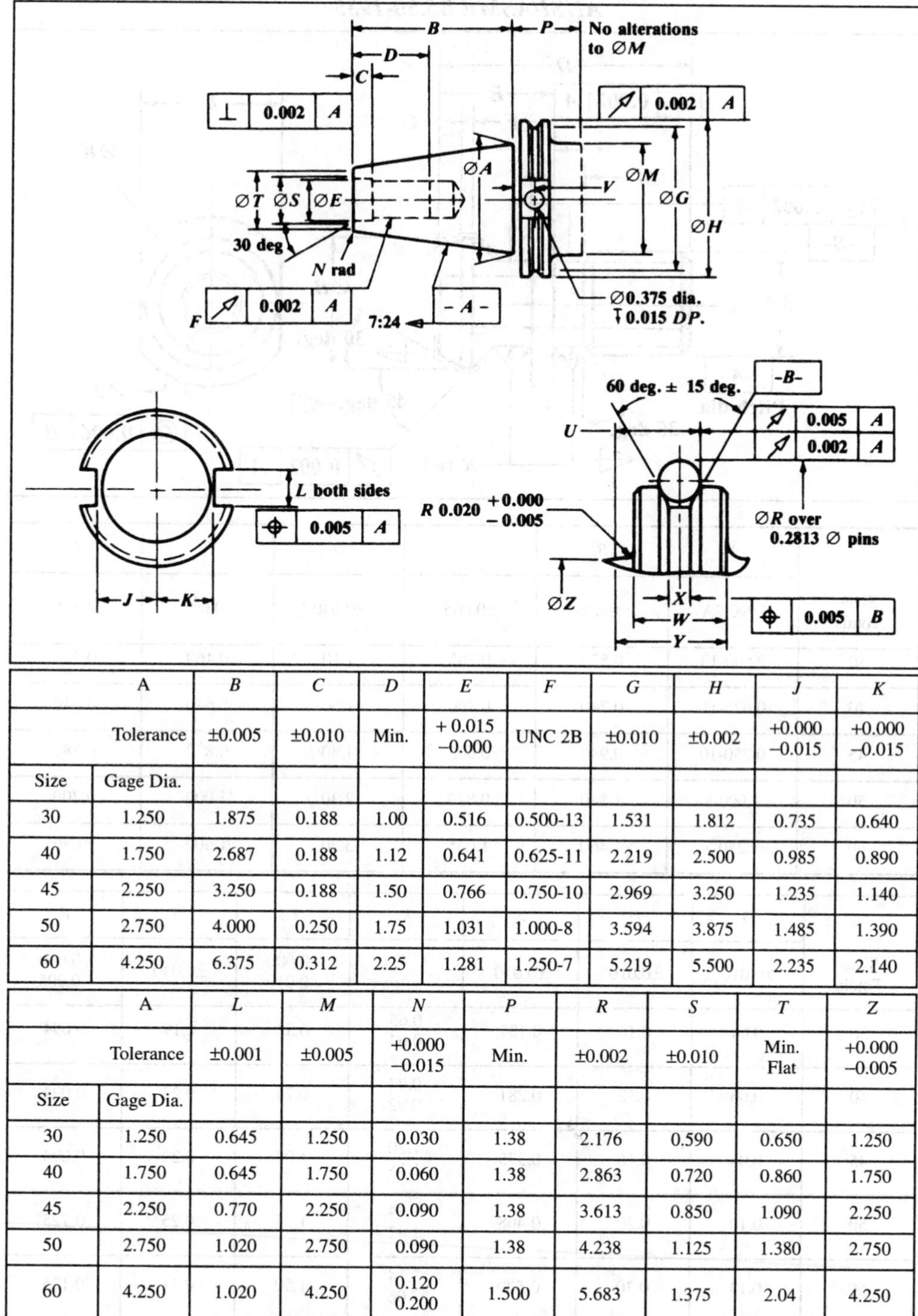

A		*B*	*C*	*D*	*E*	*F*	*G*	*H*	*J*	*K*
Tolerance		±0.005	±0.010	Min.	+0.015 −0.000	UNC 2B	±0.010	±0.002	+0.000 −0.015	+0.000 −0.015
Size	Gage Dia.									
30	1.250	1.875	0.188	1.00	0.516	0.500-13	1.531	1.812	0.735	0.640
40	1.750	2.687	0.188	1.12	0.641	0.625-11	2.219	2.500	0.985	0.890
45	2.250	3.250	0.188	1.50	0.766	0.750-10	2.969	3.250	1.235	1.140
50	2.750	4.000	0.250	1.75	1.031	1.000-8	3.594	3.875	1.485	1.390
60	4.250	6.375	0.312	2.25	1.281	1.250-7	5.219	5.500	2.235	2.140

A		*L*	*M*	*N*	*P*	*R*	*S*	*T*	*Z*
Tolerance		±0.001	±0.005	+0.000 −0.015	Min.	±0.002	±0.010	Min. Flat	+0.000 −0.005
Size	Gage Dia.								
30	1.250	0.645	1.250	0.030	1.38	2.176	0.590	0.650	1.250
40	1.750	0.645	1.750	0.060	1.38	2.863	0.720	0.860	1.750
45	2.250	0.770	2.250	0.090	1.38	3.613	0.850	1.090	2.250
50	2.750	1.020	2.750	0.090	1.38	4.238	1.125	1.380	2.750
60	4.250	1.020	4.250	0.120 0.200	1.500	5.683	1.375	2.04	4.250

Notes: Taper tolerance to be 0.001 in. in 12 in. applied in direction that increases rate of taper. Geometric dimensions symbols are to ANSI Y14.5M-1982. Dimensions are in inches. Deburr all sharp edges. Unspecified fillets and radii to be 0.03 ± 0.010*R*, or 0.03 ± 0.010 × 45 degrees. Data for size 60 are not part of Standard. For all sizes, the values for dimensions *U* (tol. ± 0.005) are 0.579: for *V* (tol. ± 0.010), 0.440; for *W* (tol. ± 0.002), 0.625; for *X* (tol. ± 0.005), 0.151; and for *Y* (tol. ± 0.002), 0.750.

Essential Dimensions of V-Flange Tool Shank Retention Knobs
ANSI/ASME B5.50-1985

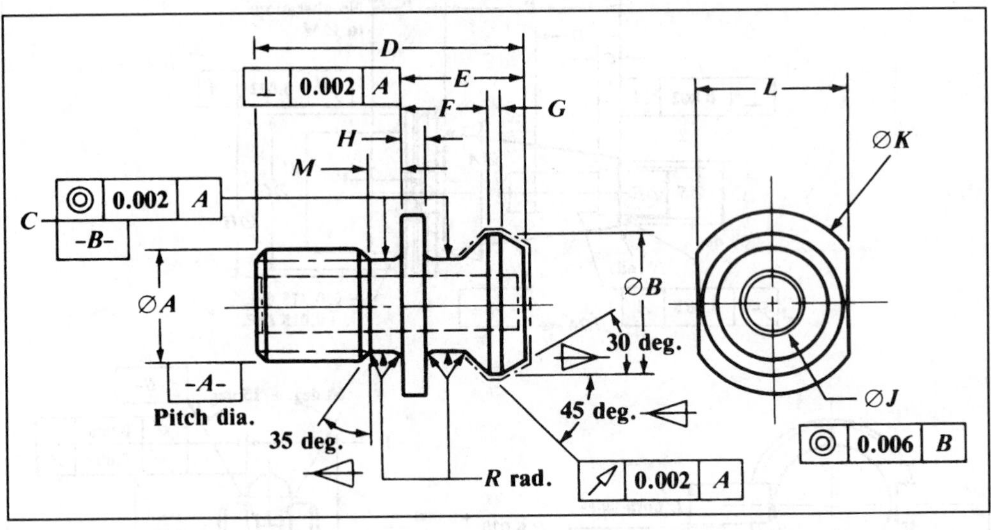

Size/ Totals	A	B	C	D	E	F
	UNC 2A	±0.005	±0.005	±0.040	±0.005	±0.005
30	0.500-13	0.520	0.385	1.10	0.460	0.320
40	0.625-11	0.740	0.490	1.50	0.640	0.440
45	0.750-10	0.940	0.605	1.80	0.820	0.580
50	1.000-8	1.140	0.820	2.30	1.000	0.700
60	1.250-7	1.460	1.045	3.20	1.500	1.080

Size/ Totals	G	H	J	K	L	M	R
	±0.010	±0.010	±0.010		+0.000 −0.010	±0.040	+0.010 −0.005
30	0.04	0.10	0.187	0.65 0.64	0.53	0.19	0.094
40	0.06	0.12	0.281	0.94 0.92	0.75	0.22	0.094
45	0.08	0.16	0.375	1.20 1.18	1.00	0.22	0.094
50	0.10	0.20	0.468	1.44 1.42	1.25	0.25	0.125
60	0.14	0.30	0.500	2.14 2.06	1.50	0.31	0.125

Notes: Dimensions are in inches. Material: low-carbon steel. Heat treatment: carburize and harden to 0.016 to 0.028 in. effective case depth. Hardness of noted surfaces to be Rockwell 56-60; core hardness Rockwell C35-45. Hole *J* shall not be carburized. Surfaces *C* and *R* to be free from tool marks. Deburr all sharp edges. Geometric dimension symbols are to

ANSI Y14.5M-1982. Data for size 60 are not part of Standard.

CAD/CAM

CAD/CAM.—CAD in engineering means computer-aided design using a computer graphics system to develop mechanical, electrical/electronic, and architectural designs. A second D (CADD) is sometimes added (computer-aided drafting and design) and simply indicates a mechanical drafting or drawing program. CAD technology is the foundation for a wide variety of engineering, design, drafting, analysis, and manufacturing activities. Often a set of drawings initially developed in the design phase of a project is also used for analyzing and optimizing the design, creating mechanical drawings of parts and assemblies and for generating NC/CNC part programs that control machining operations. Formerly, after a component had been designed with CAD, the design was passed to a part programmer who developed a program for machining the components, either manually or directly on the computer (graphic) screen, but the process often required redefining and reentering part geometry. This procedure is often regarded as the CAM part of CAD/CAM, although CAM (for computer-aided manufacturing) has a much broader meaning and involves the computer in many other manufacturing activities such as factory simulation and planning analyses. Improvements in the speed and capability of computers, operating systems, and programs (including, but not limited to CAD) have simplified the process of integrating the manufacturing process and passing drawings (revised, modified, and translated, as necessary) through the design, analysis, simulation, and manufacturing stages.

A CAD drawing is a graphic representation of part geometry data stored in a drawing database file. The drawing database generally contains the complete list of entity (line, arc, etc.) and coordinate information required to build the CAD drawing, and additional information that may be required to define solid surfaces and other model characteristics. The format of data in a drawing file depends on the CAD program used to create the file. Generally, drawings are not directly interchangeable between drawing programs, however, drawings created in one system can usually be translated into an intermediate format or file type, such as DXF, that allows some of the drawing information to be exchanged between different programs. Translation frequently results in some loss of detail or loss of other drawing information because the various drawing programs do not all have the same features. The section Drawing Exchange Standards covers some of the available methods of transferring drawing data between different CAD programs.

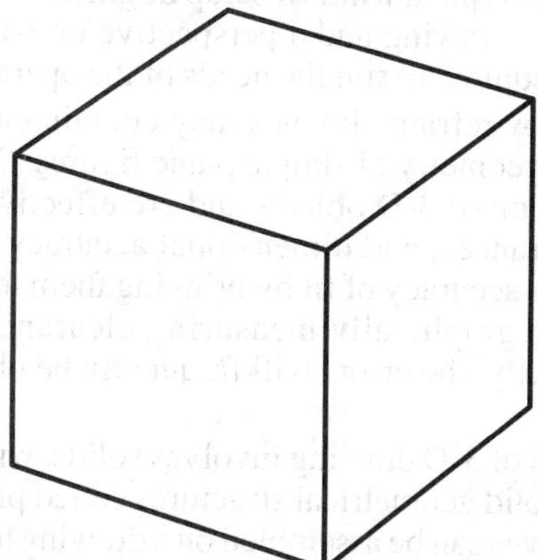

Fig. 1. Simple Wireframe Cube with Hidden Lines Automatically Removed

The simplest CAD drawings are two-dimensional and conform to normal engineering drafting practice showing orthographic (front, top, and side views, for example), exploded, isometric, or other views of a component. Depending on the complexity of the part and machining requirements, two-dimensional drawings are often sufficient for use in developing NC/CNC part programs. If a part can be programmed within a two-dimensional

CAD framework, a significant cost saving may be realized because 3-D drawings require considerably more time, drawing skill, and experience to produce than 2-D drawings.

Wireframes are the simplest two- and three-dimensional forms of drawing images and are created by defining all edges of a part and, where required, lines defining surfaces. Wireframe drawing elements consist primarily of lines and arcs that can be used in practically any combination. A wireframe drawing of a cube, as in Fig. 1, consists of 12 lines of equal length (some are hidden and thus not shown), each perpendicular to the others. Information about the interior of the cube and the character of the surfaces is not included in the drawing. With such a system, if a 1-inch cube is drawn and a 0.5-inch cylinder is required to intersect the cube's surface at the center of one of its faces, the intersection points cannot be determined because nothing is known about the area between the edges. A wireframe model of this type is ambiguous if the edges overlap or do not meet where they should. Hidden-line removal can be used to indicate the relative elevations of the drawing elements, but normally a drawing cannot be edited when hidden lines have been removed. Hidden lines can be shown dashed or can be omitted from the view.

Two-dimensional drawing elements, such as lines, arcs, and circles, are constructed by directly or indirectly specifying point coordinates, usually x and y, that identify the location, size, and orientation of the entities. Three-dimensional drawings are also made up of a collection of lines, arcs, circles, and other drawing elements and are stored in a similar manner. A third point coordinate, z, indicates the elevation of a point in 3-D drawings. On the drawing screen, working in the x-y plane, the elevation is commonly thought of as the distance of a point or object into the screen (away from the observer) or out of the viewing screen (toward the observer). Coordinate axes are oriented according to the right-hand rule: If the fingers of the right hand point in the direction from the positive x-axis to the positive y-axis, the thumb of the right hand points in the direction of the positive z-axis.

Assigning a thickness (or extruding) to objects drawn in two dimensions quickly gives some 3-D characteristics to an object and can be used to create simple prismatic 3-D shapes, such as cubes and cylinders. Usually, the greatest difficulty in creating 3-D drawings is in picking and visualizing the three-dimensional points in a two-dimensional workspace (the computer display screen). To assist in the selection of 3-D points, many CAD programs use a split or windowed screen drawing area that can simultaneously show different views of a drawing. Changes made in the current or active window are reflected in each of the other windows. A typical window setup might show three orthogonal (mutually perpendicular) views of the drawing and a perspective or 3-D view. Usually, the views shown can be changed as required to suit the needs of the operator.

If carefully constructed, wireframe images may contain enough information to completely define the external geometry of simple plane figures. Wireframe images are especially useful for visualization of 3-D objects and are effectively used during the design process to check fits, clearances, and dimensional accuracy. Parts designed to be used together can be checked for accuracy of fit by bringing them together in a drawing, superimposing the images, and graphically measuring clearances. If the parts have been designed or drawn incorrectly, the errors will frequently be obvious and appropriate corrections can be made.

A more complicated level of 3-D drawing involves solids, with sections of the part being depicted on the screen as solid geometrical structures called primitives, such as cylinders, spheres, and cubes. Primitives can be assembled on a drawing to show more complex parts. Three distinct forms of image may be generated by 3-D systems, although not all systems make use of all three.

Surface Images: A surface image defines not only the edges of the part, but also the "skin" of each face or surface. For the example mentioned previously, the intersection for the 0.5-inch cylinder would be calculated and drawn in position. Surface models are necessary for designing free-form objects such as automotive body panels and plastics injection moldings used in consumer goods. For a surface model, the computer must be provided with much more information about the part in addition to the *x*, *y*, *z* coordinates defining each point, as in a wireframe. This information may include tangent vectors, surface normals, and weighting that determines how much influence one point has on another, twists, and other mathematical data that define abstract curves, for instance. Fig. 2 shows a typical 3-D surface patch.

Shaded images may be constructed using simulated light sources, reflections, colors, and textures to make renderings more lifelike. Surface images are sometimes ambiguous, with surfaces that overlap or miss each other entirely. Information about the interior of the part, such as the center of gravity or the volume, also may not be available, depending on the CAD package.

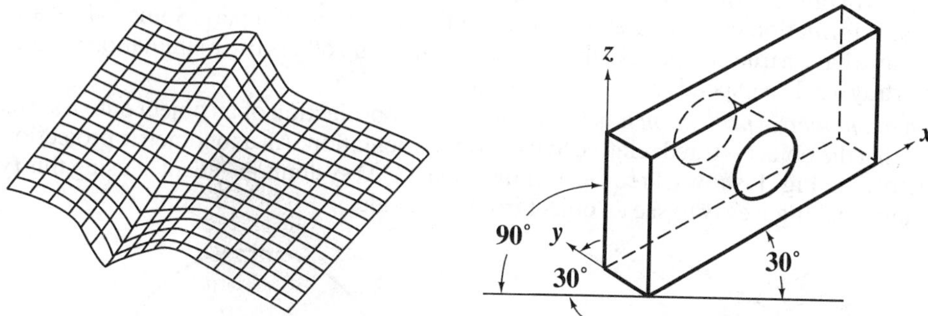

Fig. 2. A 3-D Surface Patch

Fig. 3. Isometric Drawing Showing Orientation of Principle Drawing Axes

Solid Images: A solid image is the ultimate electronic representation of a part, containing all the necessary information about edges, surfaces, and the interior. Most solid-imaging programs can calculate volume, center of mass, centroid, and moment of inertia. Several methods are available for building a solid model. One method is to perform Boolean operations on simple shapes such as cylinders, cones, cubes, and blocks. Boolean operations are used to union (join), difference (subtract one from another), and intersect (find the common volume between two objects). Thus, making a hole in a part requires subtracting a cylinder from a rectangular block. This type of program is called constructive solid geometry (CSG).

The boundary representation type of imaging program uses profiles of 2-D shapes that it extrudes, rotates, and otherwise translates in 3-D space to create the required solid. Sometimes combinations of the above two programs are used to attain a blend of flexibility, accuracy, and performance. For more precision, greatly increased time is needed for calculations, so compromises sometimes are needed to maintain reasonable productivity. Solid images may be sliced or sectioned on the screen to provide a view of the interior. This type of image is also useful for checking fit and assembly of one part with another.

Solid images provide complete, unambiguous representation of a part, but the programs require large amounts of computer memory. Each time a Boolean operation is performed, the list of calculations that must be done to define the model becomes longer, so that computation time increases.

Drawing Projections.—Several different techniques are used to display objects on paper or a computer screen to give an accurate three-dimensional appearance. Several of these methods are commonly used in CAD drawings.

Isometric drawings, as in Fig. 3, can be used to good effect for visualizing a part because they give the impression of a 3-D view and are often much faster to create. Isometric drawings are created in 2-D space, with the *x*- and *y*-axes being inclined at 30 degrees to the horizontal, as shown in Fig. 3, and the *z*-axis as vertical. Holes and cylinders in isometric drawings become elliptical. Because of the orientation of the *x*-, *y*-, and *z*-axes, the true length of lines may not be accurately represented in isometric drawings and dimensions should not be taken directly from a print. Some CAD programs have a special set of predefined drawing axes to facilitate creating isometric drawings.

In *parallel projections,* lines that are parallel in an object, assembly, or part being portrayed remain parallel in the drawing. Parallel projections show 3-D objects in a dimensionally correct manner, so that relative and scaled dimensions may be taken directly from a drawing. However, drawings may not appear as realistic as isometric or perspective drawings.

A characteristic of *perspective drawings* is that parallel lines converge (see Fig. 4) so that objects that are farther away from the observer appear smaller. Perspective drawing techniques are used in some three-dimensional drawings to convey the true look of an object, or group of objects. Because objects in perspective drawings are not drawn to scale, dimensional information cannot be extracted from the drawings of a part. Some 3-D drawing packages have a true perspective drawing capability, and others use a simulation technique to portray a 3-D perspective.

An *axonometric projection* is a 3-D perpendicular projection of an object onto a surface, such that the object is tilted relative to its normal orientation. An axonometric projection of a cube, as in Fig. 1, shows three faces of the cube. CAD systems are adept at using this type of view, making it easy to see an object from any angle.

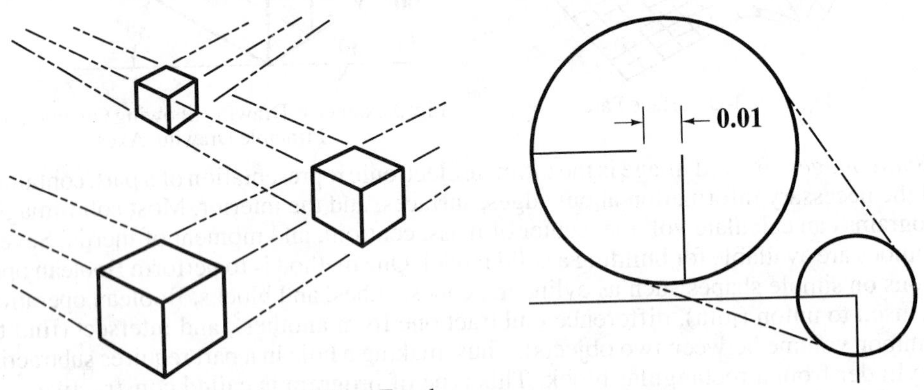

Fig. 4. Perspective Drawing of Three Equal-Size Cubes and Construction Lines

Fig. 5. A Common Positioning Error

Drawing Tips and Traps.—Images sometimes appear correct on the screen but contain errors that show up when the drawing is printed or used to produce NC/CNC part programs. In Fig. 5, the two lines within the smaller circle appear to intersect at a corner, but when the view of the intersection is magnified, as in the larger circle, it is clear that the lines actually do not touch. Although an error of this type may not be easily visible, other parts placed in the drawing relative to this part will be out of position.

A common problem that shows up in plotting, but is difficult to detect on the screen, comes from placing lines in the same spot. When two or more lines occupy exactly the same location on the screen, there is usually no noticeable effect on the display. However, when the drawing is plotted, each line is plotted separately, causing the single line visible to become thicker and darker. Likewise, if a line that appears continuous on the screen is actually made up of several segments, plotting the line will frequently result in a broken, marred, or blotted appearance to the line because the individual segments are plotted separately, and at different times. To avoid these problems and to get cleaner looking plots, replace segmented lines with single lines and avoid constructions that place one line directly on top of another.

Exact decimal values should be used when entering point coordinates from the keyboard, if possible; fractional sizes should be entered as fractions, not truncated decimals. For example, $\frac{5}{16}$ should be entered as 0.3125 or $\frac{5}{16}$, not 0.313. Accumulated rounding errors and surprises later on when parts do not fit are thus reduced. Drawing dimensions, on the other hand, should not have more significant digits or be more precise than necessary. Unnecessary precision in dimensioning leads to increased difficulty in the production stage because the part has to be made according to the accuracy indicated on the drawing.

Snap and *object snap* commands make selecting lines, arcs, circles, or other drawing entities faster, easier, and more accurate when picking and placing objects on the screen. Snap permits only points that are even multiples of the snap increment to be selected by the pointer. A ⅛-inch snap setting, for example, will allow points to be picked at exactly ⅛-inch intervals. Set the *snap increment* to the smallest distance increment (1 in., ¼ in., 1 ft., etc.) being used in the area of the drawing under construction and reset the snap increment frequently, if necessary. The snap feature can be turned off during a command to override the setting or to select points at a smaller interval than the snap increment allows. Some systems permit setting a different snap value for each coordinate axis.

The *object snap* selection mode is designed to select points on a drawing entity according to predefined characteristics of the entity. For example, if end-point snap is in effect, picking a point anywhere along a line will select the end point of the line nearest the point picked. Object snap modes include point, intersection, midpoint, center and quadrants of circles, tangency point (allows picking a point on an arc or circle that creates a tangent to a line), and perpendicular point (picks a point that makes a perpendicular from the base point to the object selected). When two or more object snap modes are used together, the nearest point that meets the selection criteria will be chosen. Using object snap will greatly reduce the frequency of the type of problem shown in Fig. 5.

Copy: Once drawn, avoid redrawing the same object. It is almost always faster to copy and modify a drawing than to draw it again. The basic copy commands are: copy, array, offset, and mirror. Use these, along with move and rotate and the basic editing commands, to modify existing objects. Copy and move should be the most frequently used commands. If possible, create just one instance of a drawing object and then copy and move it to create others.

To create multiple copies of an object, use the *copy, multiple* feature to copy selected objects as many times as required simply by indicating the destination points. The *array* command makes multiple copies of an object according to a regular pattern. The rectangular array produces rows and columns, and the polar array puts the objects into a circular pattern, such as in a bolt circle. *Offset* copies an entity and places the new entity a specified distance from the original and is particularly effective at placing parallel lines and curves, and for creating concentric copies of closed shapes. *Mirror* creates a mirror image copy of an object, and is useful for making right- and left-hand variations of an object as well as for copying objects from one side of an assembly to the other. In some CAD programs, a system variable controls whether text is mirrored along with other objects.

Many manufacturers distribute drawings of their product lines in libraries of CAD drawings, usually as DXF files, that can be incorporated into existing drawings. The suitability of such drawings depends on the CAD program and drawing format being used, the skill of the technician who created the drawings, and the accuracy of the drawings. A typical example, Fig. 6, shows a magnetically coupled actuator drawing distributed by Tol-O-Matic, Inc. Libraries of frequently used drawing symbols and blocks are also available from commercial sources.

Create Blocks of Frequently Used Objects: Once created, complete drawings or parts of drawings can be saved and later recalled, as needed, into another drawing. Such objects can be scaled, copied, stretched, mirrored, rotated, or otherwise modified without changing the original. When shapes are initially drawn in unit size (i.e., fitting within a 1×1 square) and saved, they can be inserted into any drawing and scaled very easily. One or more individual drawing elements can be saved as a group element, or *block,* that can be manipulated in a drawing as a single element. Block properties vary, depending on the drawing program, but are among the most powerful features of CAD. Typically, blocks are uniquely named and, as with simple objects, may be saved in a file on the disk. Blocks are ideal for creating libraries of frequently used drawing symbols. Blocks can be copied, moved, scaled very easily, rotated, arrayed, and inserted as many times as is required in a drawing and manipulated as one object. When scaled, each object within the block is also scaled to the same degree.

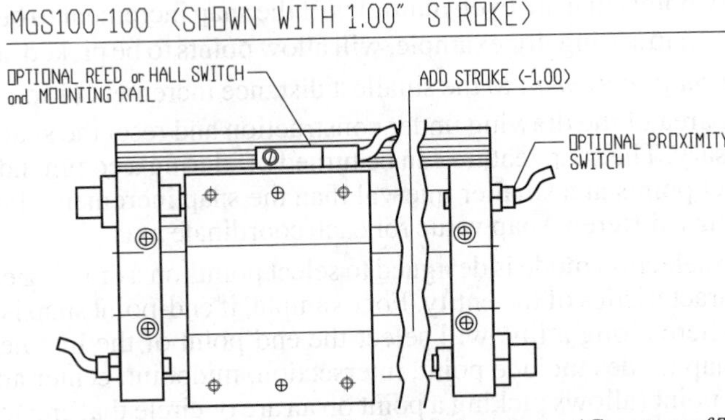

Fig. 6. Manufacturer's Drawing of a Magnetically Coupled Actuator (*Courtesy of Tol-O-Matic, Inc.*)

When a family of parts is to be drawn, create and block a single drawing of the part that fits within a unit cube of convenient size, such as 1 × 1 × 1. When the block is inserted in a drawing, it is scaled appropriately in the *x*-, *y*-, and *z*-directions. For example, ⅜-inch bolts can be drawn 1 inch long in the *x*-direction and ⅜-inch in diameter in the *y*-*z* plane. If a 5-inch bolt is needed, insert the "bolt" block with a scale of 5 in the *x*-direction and a scale of 1 in the *y*- and *z*-directions.

Once blocked, the individual components of a block (lines, arcs, circles, surfaces, and text, for example) cannot be individually changed or edited. To edit a block, a copy (instance) of the block must be *exploded* (unblocked) to divide it into its original components. Once exploded, all the individual elements of the block (except other blocks) can be edited. When the required changes have been made, the block must be redefined (redeclared as a block by giving it a name and identifying its components). If the block is redefined using the same name, any previous references to the block in the drawing will be updated to match the redefined block. For example, an assembly drawing is needed that shows a mechanical frame with 24 similar control panels attached to it. Once one of the panels is drawn and defined as a block (using the name *PANEL*, for instance), the block can be inserted (or copied) into the drawing 24 times. Later, if changes need to be made to the panel design, one instance of the block *PANEL* can be exploded, modified, and redefined with the name *PANEL*. When *PANEL* is redefined, every other copy of the *PANEL* block in the drawing is also redefined, so every copy of *PANEL* in the drawing is updated. On the other hand, if the block was redefined with a different name, say, *PANEL1*, existing copies of *PANEL* remain unchanged. When redefining a block that already exists in the drawing, be sure to use the same insertion point that was used for the original definition of the block; otherwise, the positions of existing blocks with the same name will be changed.

Use of Text Attributes to Request Drawing Information Automatically: Text attributes are a useful method for attaching textual information to a particular part or feature of a drawing. An attribute is basically a text variable that has a name and can be assigned a value. Attributes are created by defining attribute characteristics such as a name, location in the drawing, text size and style, and default value. The attribute value is assigned when the attribute is inserted into a drawing as part of a block.

Fig. 7 shows two views of a title block for size A to C drawing sheets. The upper figure includes the title block dimensions (included only for reference) and the names and locations of the attributes (COMPANY, TITLE1, TITLE2, etc.). When a block containing text attributes is inserted in a drawing, the operator is asked to enter the value of each attribute.

To create this title block, first draw the frame of the title block and define the attributes (name, location and default value for: company name and address, drawing titles [2 lines], drawing size, drawing number, revision number, scale, and sheet number). Finally, create and name a block containing the title frame and all the attribute definitions (do not include the dimensions).

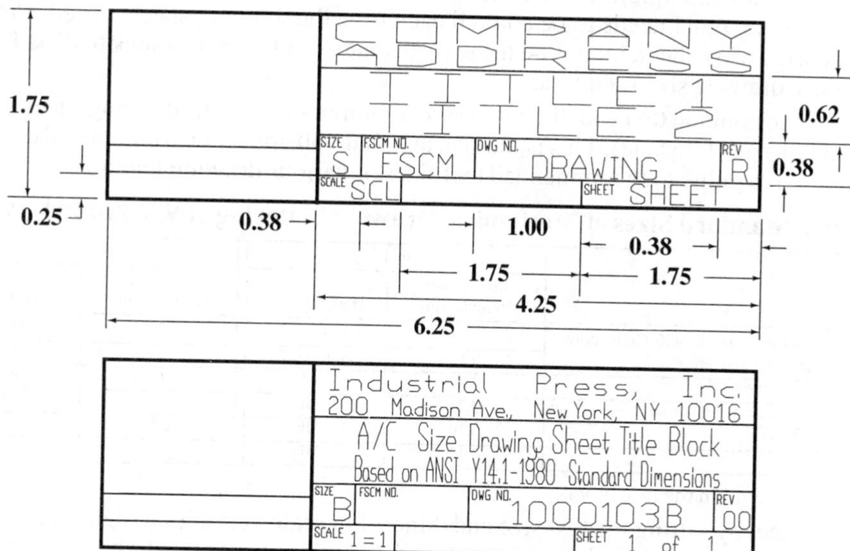

Fig. 7. Title Block for A to C Size Drawing Sheets Showing the Placement of Text Attributes. The Lower Figure Shows the Completed Block

When the block is inserted into a drawing, the operator is asked to enter the attribute values (such as company name, drawing title, etc.), which are placed into the title block at the predetermined location. The lower part of Fig. 7 shows a completed title block as it might appear inserted in a drawing. A complete drawing sheet could include several additional blocks, such as a sheet frame, a revision block, a parts list block, and any other supplementary blocks needed. Some of these blocks, such as the sheet frame, title, and parts list blocks, might be combined into a single block that could be inserted into a drawing at one time.

Define a Default Drawing Configuration: Drawing features that are commonly used in a particular type of drawing can be set up in a template file so that frequently used settings, such as text and dimension styles, text size, drawing limits, initial view, and other default settings, are automatically set up when a new drawing is started. Different configurations can be defined for each frequently used drawing type, such as assembly, parts, or printed circuit drawings. When creating a new drawing, use one of the template files as a pattern or open a template file and use it to create the new drawing, saving it with a new name.

Scaling Drawings: Normally, for fast and accurate drawing, it is easiest to draw most objects full scale, or with a 1:1 scale. This procedure greatly simplifies creation of the initial drawing, and ensures accuracy, because scale factors do not need to be calculated. If it becomes necessary to fit a large drawing onto a small drawing sheet (for example, to fit a 15 × 30 inch assembly onto a 11 × 17 inch, B-sized, drawing sheet), the drawing sheet can be scaled larger to fit the assembly size. Likewise, large drawing sheets can be scaled down to fit small drawings. The technique takes some practice, but it permits the drawing assembly to be treated full scale. If editing is required at a later date (to move something or add a hole in a particular location, for example), changes can be made without rescaling and dimensions can be taken directly from the unscaled drawing on the computer.

Scaling Text on Drawing Sheets: It is usually desirable that text, dimensions, and a few other features on drawings stay a consistent size on each sheet, even when the drawing size is very different. The following procedure ensures that text and dimensions (other features as well, if desired) will be the same size, from drawing to drawing without resorting to scaling the drawing to fit onto the drawing sheet.

Create a drawing sheet having the exact dimensions of the actual sheet to be output (A, B, C, D, or E size, for example). Use text attributes, such as the title block illustrated in Fig. 7, to include any text that needs to be entered each time the drawing sheet is used. Create a block of the drawing sheet, including the text attributes, and save the block to disk. Repeat for each size drawing sheet required.

Establish the nominal text and dimension size requirements for the drawing sheet when it is plotted full size (1:1 scale). This is the size text that will appear on a completed drawing. Use Table 1 as a guide to recommended text sizes of various drawing features.

Table 1. Standard Sizes of Mechanical Drawing Lettering ***ANSI Y14.2M–1992***

Use For	Inch		Metric	
	Min. Letter Heights, (in)	Drawing Size	Min. Letter Heights, (mm)	Drawing Size
Drawing title, drawing size, CAGE Code, drawing number, and revision letter[a]	0.24	D, E, F, H, J, K	6	A0, A1
	0.12	A, B, C, G	3	A2, A3, A4
Section and view letters	0.24	All	6	All
Zone letters and numerals in borders	0.24	All	6	All
Drawing block headings	0.10	All	2.5	All
All other characters	0.12	All	3	All

[a] When used within the title block.

Test the sheet by setting the text size and dimension scale variables to their nominal values (established above) and place some text and dimensions onto the drawing sheet. Plot a copy of the drawing sheet and check that text and dimensions are the expected size.

To use the drawing sheet, open a drawing to be placed on the sheet and insert the sheet block into the drawing. Scale and move the sheet block to locate the sheet relative to the drawing contents. When scaling the sheet, try to use whole-number scale factors (3:1, 4:1, etc.), if possible; this will make setting text size and dimension scale easier later on. Set the text-size variable equal to the nominal text size multiplied by the drawing sheet insertion scale (for example, for 0.24 text height on a drawing sheet scaled 3:1, the text-size variable will be set to $3 \times 0.24 = 0.72$). Likewise, set the dimension-scale variable equal to the nominal dimension size multiplied by the drawing sheet insertion scale.

Once the text size and dimensions scale variables have been set, enter all the text and dimensions into the drawing. If text of another size is needed, multiply the *new* nominal text size by the sheet scale to get the actual size of the text to use in the drawing.

Use Appropriate Detail: Excessive detail may reduce the effectiveness of the drawing, increase the drawing time on individual commands and the overall time spent on a drawing, and reduce performance and speed of the CAD program. Whenever possible, symbolic drawing elements should be used to represent more complicated parts of a drawing unless the appearance of that particular component is essential to the drawing.

Drawing everything to scale often serves no purpose but to complicate a drawing and increase drawing time. The importance of detail depends on the purpose of a drawing, but detail in one drawing is unnecessary in another. For example, the slot size of a screw head (length and width) varies with almost every size of screw. If the purpose of a drawing is to show the type and location of the hardware, a symbolic representation of a screw is usually all that is required. The same is generally true of other screw heads, bolt threads, bolt head diameters and width across the flats, wire diameters, and many other hardware features.

Drawing Exchange Standards.—The ability to transfer working data between different CAD, CAD/CAM, design analysis, and NC/CNC programs is one of the most important requirements of engineering drawing programs. Once an engineer, designer, draftsman, or machinist enters relevant product data into his or her machine (computer or machine tool), the information defining the characteristics of the product should be available to the others involved in the project without recreating or reentering it. In view of manufacturing goals of reducing lead time and increasing productivity, concurrent engineering, and improved product performance, interchangeable data are a critical component in a CAD/CAM program. Depending on the requirements of a project, it may be entirely possible to transfer most if not all of the necessary product drawings from one drawing system to another.

IGES stands for Initial Graphics Exchange Specification and is a means of exchanging or converting drawings and CAD files for use in a different computer graphics system. The concept is shown diagrammatically in Fig. 8. Normally, a drawing prepared on the computer graphics system supplied by company A would have to be redrawn before it would operate on the computer graphics system supplied by company B. However, with IGES, the drawing can be passed through a software package called a preprocessor that converts it into a standardized IGES format that can be stored on a magnetic disk. A postprocessor at company B is then used to convert the standard IGES format to that required for their graphics system. Both firms would be responsible for purchasing or developing their own preprocessors and postprocessors, to suit their own machines and control systems. Almost all the major graphics systems manufacturing companies today either have or are developing IGES preprocessor and postprocessor programs to convert software from one system to another.

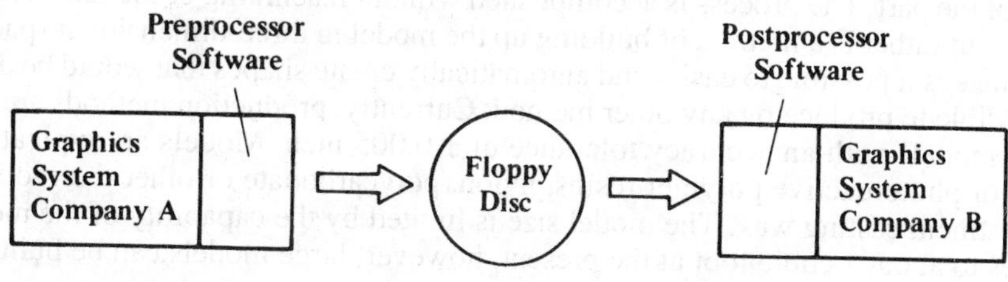

Fig. 8.

DXF stands for Drawing Exchange Format and is a pseudo-standard file format used for exchanging drawings and associated information between different CAD and design analysis programs. Nearly all two- and three-dimensional CAD programs support some sort of drawing exchange through the use of DXF files, and most can read and export DXF files. There are, however, differences in the drawing features supported and the manner in which the DXF files are handled by each program. For example, if a 3-D drawing is exported in the DXF format and imported into a 2-D CAD program, some loss of information results because all the 3-D features are not supported by the 2-D program, so that most attempts to make a transfer between such programs fail completely. Most common drawing entities (lines, arcs, etc.) will transfer successfully, although other problems may occur. For example, drawing entities that are treated as a single object in an original drawing (such as blocks, hatch patterns, and symbols) may be divided into hundreds of individual components when converted into a DXF file. Consequently, such a drawing may become much more difficult to edit after it is transferred to another drawing program.

ASCII stands for American Standard Code for Information Interchange. ASCII is a code system that describes the manner in which character-based information is stored in a computer system. Files stored in the ASCII format can be transferred easily between computers, even those using different operating systems. Although ASCII is not a drawing file format, many CAD drawing formats (DXF and IGES, for example) are ASCII files. In these files, the drawing information is stored according to a specific format using ASCII characters. ASCII files are often referred to as pure text files because they can be read and edited by simple text editors.

HPGL, for Hewlett-Packard Graphics Language, is a format that was first developed for sending vector- (line-) based drawing information to pen plotters. The format is commonly used for sending drawing files to printers and plotters for printing. Because HPGL is a character-based format (ASCII), it can be transferred between computers easily. Normally, devices that recognize the HPGL format can print the files without using the program on which the file (a drawing, for example) was created.

STL is a CAD drawing format that is primarily used to send CAD drawings to *rapid automated prototyping* machines. STL is a mnemonic abbreviation for stereo-lithography, the technique that is used to create three-dimensional solid models directly from computer-generated drawings and for which the drawing format was originally developed. Most prototyping machines use 3-D CAD drawing files in STL format to create a solid model of the part represented by a drawing.

STEP stands for Standard for Exchange of Product Model Data and is a series of existing and proposed ISO standards written to allow access to all the data that surround a product. It extends the IGES idea of providing a geometric data transfer to include all the other data that would need to be communicated about a product over its lifetime, and facilitates the use and accessibility of the product data. Although STEP is a new standard, software tools have been developed for converting data from the IGES to STEP format and from STEP to IGES.

Rapid Automated Prototyping.—Rapid automated prototyping is a method of quickly creating an accurate three-dimensional physical model directly from a computerized conception of the part. The process is accomplished without machining or the removal of any material, but rather is a method of building up the model in three-dimensional space. The process makes it possible to easily and automatically create shapes that would be difficult or impossible to produce by any other method. Currently, production methods are able to produce models with an accuracy tolerance of ± 0.005 inch. Models are typically constructed of photoreactive polymer resins, nylon, polycarbonate or other thermoplastics, and investment casting wax. The model size is limited by the capability of the modeling machines to about 1 cubic foot at the present, however, large models can be built in sections and glued or otherwise fastened together.

Much of the work and a large part of the cost associated with creating a physical model by rapid prototyping are in the initial creation of the CAD model. The model needs to be a 3-D design model, built using wireframe, surface, or solid CAD modeling techniques. Many full-featured CAD systems support translation of drawing files into the STL format, which is the preferred file format for downloading CAD models to rapid prototyping machines. CAD programs without STL file format capability can use the IGES or DXF file format. This process can be time-consuming and expensive because additional steps may have to be taken by the service bureau to recreate features lost in converting the IGES or DXF file into STL format. If the design file has to be edited by a service bureau to recreate surfaces lost in the translation, unwanted changes to the model may occur, unnoticed. The safest route is to create a CAD model and export it directly into the STL format, leaving little chance for unexpected errors. Reverse STL generators are also available that will display a file saved in STL format or convert it into a form that can be imported into a CAD program.

DNC.—DNC stands for Direct Numerical Control and refers to a method of controlling numerical control machines from a remote location by means of a link to a computer or computer network. In its simplest form, DNC consists of one NC or CNC machine linked by its serial port to a computer. The computer may be used to develop and store CNC part programs and to transfer part programs to the machine as required. DNC links are normally two-directional, meaning that the NC/CNC can be operated from a computer terminal and the computer can be operated or ordered to supply data to the NC/CNC from the machine's control panel.

The number of machines that can be connected to a DNC network depends on the network's capability; in theory, any number of machines can be attached, and controlled. The type of network depends on the individual DNC system, but most industry standard network protocols are supported, so DNC nodes can be connected to existing networks very easily. Individual NC/CNC machines on a network can be controlled locally, from a network terminal in another building, or even from a remote location miles away through phone or leased lines.

Machinery Noise.—Noise from machinery or other mechanical systems can be controlled to some degree in the design or development stage if quantified noise criteria are provided the designer. Manufacturers and consumers may also use the same information in deciding whether the noise generated by a particular machine will be acceptable for a specific purpose.

Such criteria for noise may be classified into three types: 1) those relating to the degree of interference with speech communications; 2) those relating to physiological damage to humans, especially their hearing; and 3) those relating to psychological disturbances in people exposed to noise.

Sound Level Specifications: Noise criteria generally are specified in some system of units representing sound levels. One commonly used system specifies sound levels in units called decibels on the "A" scale, written dBA. The dBA scale designates a sound level system weighted to match human hearing responses to various frequencies and loudness. For example, to permit effective speech communication, typical criteria for indoor maximum noise levels are: meeting and conference rooms, 42 dBA; private offices and small meeting rooms, 38 to 47 dBA; supervisors' offices and reception rooms, 38 to 52 dBA; large offices and cafeterias, 42 to 52 dBA; laboratories, drafting rooms, and general office areas, 47 to 56 dBA; maintenance shops, computer rooms, and washrooms, 52 to 61 dBA; control and electrical equipment rooms, 56 to 66 dBA; and manufacturing areas and foremen's offices, 66 dBA. Similarly, there are standards and recommendations for daily permissible times of exposure at various steady sound levels to avoid hearing damage. For a working shift of 8 hours, a steady sound level of 90 dBA is the maximum generally permitted, with marked reduction in allowable exposure times for higher sound levels.*

Measuring Machinery Noise.—The noise level produced by a single machine can be measured by using a standard sound level meter of the handheld type set to the dBA scale. However, when other machines are running at the same time, or when there are other background noises, the noise of the machine cannot be measured directly. In such cases, two measurements, taken as follows, can be used to calculate the noise level of the individual machine. The meter should be held at arm's length and well away from any bystanders to avoid possible significant error up to 5 dBA.

Step 1. At the point of interest, measure the total noise, T, in decibels; that is, measure the noise of the shop and the machine in question when all machines are running; Step 2. Turn off the machine in question and measure B, the remaining background noise level; Step 3. Calculate M, the noise of the machine alone, $M = 10\log_{10}[10^{(T/10)} - 10^{(B/10)}]$.

Example 1: With a machine running, the sound level meter reads 51 decibels as the total shop noise T; and with the machine shut off the meter reads 49 decibels as the remaining background noise B. What is the noise level M of the machine alone? $M = 10\log_{10}[10^{(51/10)} - 10^{(49/10)}] = 46.7$ decibels dBA.

Example 2: If in Example 1 the remaining background noise level B was 41 decibels instead of 49, what is the noise level of the machine alone? $M = 10\log_{10}[10^{(51/10)} - 10^{(41/10)}] = 50.5$ decibels dBA.

Note: From this example it is evident that when the background noise level B is approximately 10 or more decibels lower than the total noise level T measured at the machine in question, then the background noise does not contribute significantly to the sound level at the machine and, for practical purposes, $M = T$ and no calculation is required.

* After April 1983, if employee noise exposures equal or exceed an 8-hour, time-weighted average sound level of 85 dB, OSHA requires employers to administer an effective hearing conservation program.

MANUFACTURING PROCESSES

PUNCHES, DIES, AND PRESS WORK

ELECTRICAL DISCHARGE MACHINING

IRON AND STEEL CASTINGS

IRON AND STEEL CASTINGS (cont.)

METAL JOINING, CUTTING, AND SURFACING

METAL JOINING, CUTTING, AND SURFACING (cont.)

WELDING

WELDING (cont.)

MANUFACTURING PROCESSES

PUNCHES, DIES, AND PRESS WORK

Clearance between Punches and Dies.—The amount of clearance between a punch and die for blanking and perforating is governed by the thickness and kind of stock to be operated upon. For thin material, the punch should be a close sliding fit to prevent ragged edges, but for heavier stock, there should be some clearance. The clearance between the punch and die in cutting heavy material reduces the pressure required for the punching operation and the danger of breaking the punch.

Meaning of the Term "Clearance".—There is a difference of opinion among diemakers as to the method of designating clearance. The prevailing practice of fifteen firms specializing in die work is as follows: Ten of these firms define clearance as the space between the punch and die on *one side,* or one-half the difference between the punch and die sizes. The remaining five firms consider clearance as the total difference between the punch and die sizes; for example, if the die is round, clearance equals die diameter minus punch diameter. The advantage of designating clearance as the space on each side is particularly evident with dies of irregular form or of angular shape. Although the practice of designating clearance as the difference between the punch and die diameters may be satisfactory for round dies, it leads to confusion when the dies are of unsymmetrical forms. The term "clearance" should not be used in specifications without indicating clearly just what it means. According to one die manufacturer, the term "cutting clearance" is used to indicate the space between the punch and die on each side, and the term "die clearance" refers to the angular clearance provided below the cutting edge so that the parts will fall easily through the die. The term "clearance" as here used means the space on one side only; hence, for round dies, clearance equals die radius minus punch radius.

Clearances Generally Allowed.—For brass and soft steel, most dies are given a clearance on one side equal to the stock thickness multiplied by 0.05 or 0.06; but one-half of this clearance is preferred for some classes of work, and a clearance equal to the stock thickness multiplied by 0.10 may give the cleanest fracture for certain other operations such as punching holes in ductile steel boiler plate.

Where Clearance Is Applied.—Whether clearance is deducted from the diameter of the punch or added to the diameter of the die depends upon the nature of the work. If a blank of given size is required, the die is made to that size and the punch is made smaller. Inversely, when holes of a given size are required, the punch is made to the diameter wanted and the die is made larger. Therefore, for blanking to a given size, the clearance is deducted from the size of the punch, and for perforating, the clearance is added to the size of the die.

Effect of Clearance on Working Pressure.—Clearance affects not only the smoothness of the fracture, but also the pressure required for punching or blanking. This pressure is greatest when the punch diameter is small compared to the thickness of the stock. In one test, for example, a punching pressure of about 32,000 pounds was required to punch ¾-inch holes into 5⁄16-inch mild steel plate when the clearance was about 10 per cent. With a clearance of about 4.5 per cent, the pressure increased to 33,000 pounds and a clearance of 2.75 per cent resulted in a pressure of 34,500 pounds.

Soft ductile metal requires more clearance than hard metal, although it has been common practice to increase the clearance for harder metals. In punching holes in fairly hard steel, a clean fracture was obtained with a clearance of only 0.03 times stock thickness.

Angular Clearance for Dies.—The angular clearance ordinarily used in a blanking die varies from 1 to 2 degrees, although dies intended for producing a comparatively small number of blanks are sometimes given a clearance angle of 4 or 5 degrees to facilitate making the die quickly. When large numbers of blanks are required, a clearance of about 1 degree is used.

There are two methods of giving clearance to dies: In one method, the clearance extends to the top face of the die; and in the other, there is a space about ⅛ inch below the cutting edge that is left practically straight, or having a very small amount of clearance.

For very soft metal, such as soft, thin brass, the first method is employed, but for harder material, such as hard brass, steel, etc., it is better to have a very small clearance for a short distance below the cutting edge. When a die is made in this way, thousands of blanks can be cut with little variation in their size, as grinding the die face will not enlarge the hole to any appreciable extent.

Lubricants for Press Work.—Blanking dies used for carbon and low-alloy steels are often run with only residual mill lubricant, but will last longer if lightly oiled. Higher alloy and stainless steels require thicker lubricants. Kerosene is usually used with aluminum. Lubricant thickness needs to be about 0.0001 in. and can be obtained with about 1 pint of fluid to cover 500 sq. ft of material. During successive strokes, metal debris adheres to the punch and may accelerate wear, but damage may be reduced by application of the lubricant to the sheet or strip by means of rollers or spray. High-speed blanking may require heavier applications or a continuous airless spraying of oil. For sheet thicker than ⅛ in. and for stainless steel, high-pressure lubricants containing sulfurs and chlorines are often used.

Shallow drawing and forming of steel can be done with low-viscosity oils and soap solutions, but deeper draws require light- to medium-viscosity oils containing fats and such active elements as sulfur or phosphorus, and mineral fillers such as chalk or mica. Deep drawing often involves ironing or thinning of the walls by up to 35 per cent, and thick oils containing high proportions of chemically active compounds are used. Additives used in drawing compounds are selected for their ability to maintain a physical barrier between the tool surfaces and the metal being shaped. Dry soaps and polymer films are frequently used for these purposes. Aluminum can be shallow drawn with oils of low to medium viscosity, and for deep drawing, tallow may be added, also wax or soap suspensions for very large reductions.

Annealing Drawn Shells.—When drawing steel, iron, brass, or copper, annealing is necessary after two or three draws have been made, because the metal is hardened by the drawing process. For steel and brass, anneal between alternate reductions, at least. Tin plate or stock that cannot be annealed without spoiling the finish must ordinarily be drawn to size in one or two operations. Aluminum can be drawn deeper and with less annealing than the other commercial metals, provided the proper grade is used. If it is necessary to anneal aluminum, it should be heated in a muffle furnace, care being taken to see that the temperature does not exceed 700 degrees F.

Drawing Brass.—When drawing brass shells or cup-shaped articles, it is usually possible to make the depth of the first draw equal to the diameter of the shell. By heating brass to a temperature just below what would show a dull red in a dark room, it is possible to draw difficult shapes, otherwise almost impossible, and to produce shapes with square corners.

Drawing Rectangular Shapes.—When square or rectangular shapes are to be drawn, the radius of the corners should be as large as possible, because defects usually occur in the corners when drawing. Moreover, the smaller the radius, the less the depth that can be obtained in the first draw.

The maximum depths that can be drawn with corners of a given radii are approximately as follows: With a radius of 3⁄32 to 3⁄16 inch, depth of draw, 1 inch; radius 3⁄16 to ⅜ inch, depth 1½ inches; radius ⅜ to ½ inch, depth 2 inches; and radius ½ to ¾ inch, depth 3 inches.

These figures are taken from actual practice and can doubtless be exceeded slightly when using metal prepared for the process. If the box needs to be quite deep and the radius is quite small, two or more drawing operations will be necessary.

Speeds and Pressures for Presses.—The speeds for presses equipped with cutting dies depend largely upon the kind of material being worked, and its thickness. For punching and shearing ordinary metals not over ¼ inch thick, the speeds usually range between 50 and 200 strokes per minute, 100 strokes per minute being a fair average. For punching metal over ¼ inch thick, geared presses with speeds ranging from 25 to 75 strokes per minute are commonly employed.

The cutting pressures required depend upon the shearing strength of the material, and the actual area of the surface being severed. For round holes, the pressure required equals the circumference of the hole × the thickness of the stock × the shearing strength.

To allow for some excess pressure, the tensile strength may be substituted for the shearing strength; the tensile strength for these calculations may be roughly assumed as follows: Mild steel, 60,000; wrought iron, 50,000; bronze, 40,000; copper, 30,000; aluminum, 20,000; zinc, 10,000; and tin and lead, 5,000 pounds per square inch.

Pressure Required for Punching.—The formula for the force in tons required to punch a circular hole in sheet steel is $\pi DST/2000$, where S = the shearing strength of the material in lb/in.^{2}, T = thickness of the steel in inches, and 2000 is the number of lb in 1 ton. An approximate formula is DT × 80, where D and T are the diameter of the hole and the thickness of the steel, respectively, both in inches, and 80 is a factor for steel. The result is the force in tons.

Example: Find the pressure required to punch a hole, 2 inches in diameter, through ¼-in. thick steel. By applying the approximate formula, $2 \times \frac{1}{4} \times 80 = 40$ tons.

If the hole is not circular, replace the hole diameter with the value of one-third of the perimeter of the hole to be punched.

Example: Find the pressure required to punch a 1-inch square hole in ¼-in. thick steel. The total length of the hole perimeter is 4 in. and one-third of 4 in. is 1 ⅓ in., so the force is $1\frac{1}{3} \times \frac{1}{4} \times 80 = 26\frac{2}{3}$ tons.

The corresponding factor for punching holes in brass is 65 instead of 80. So, to punch a hole measuring 1 by 2 inches in ¼-in. thick brass sheet, the factor for hole size is the perimeter length $6 \div 3 = 2$, and the formula is $2 \times \frac{1}{4} \times 65 = 32\frac{1}{2}$ tons.

Shut Height of Press.—The term "shut height," as applied to power presses, indicates the die space when the slide is at the bottom of its stroke and the slide connection has been adjusted upward as far as possible. The "shut height" is the distance from the lower face of the slide, either to the top of the bed or to the top of the bolster plate, there being two methods of determining it; hence, this term should always be accompanied by a definition explaining its meaning. According to one press manufacturer, the safest plan is to define "shut height" as the distance from the top of the bolster to the bottom of the slide, with the stroke down and the adjustment up, because most dies are mounted on bolster plates of standard thickness, and a misunderstanding that results in providing too much die space is less serious than having insufficient die space. It is believed that the expression "shut height" was applied first to dies rather than to presses, the shut height of a die being the distance from the bottom of the lower section to the top of the upper section or punch, excluding the shank, and measured when the punch is in the lowest working position.

Diameters of Shell Blanks.—The diameters of blanks for drawing plain cylindrical shells can be obtained from the table on the following pages, which gives a very close approximation for thin stock. The blank diameters given in this table are for sharp-cornered shells and are found by the following formula:

$$D = \sqrt{d^2 + 4dh} \qquad (1)$$

where D = diameter of flat blank; d = diameter of finished shell; and h = height of finished shell.

Example: If the diameter of the finished shell is to be 1.5 inches, and the height, 2 inches, the trial diameter of the blank would be found as follows:

$$D = \sqrt{1.5^2 + 4 \times 1.5 \times 2} = \sqrt{14.25} = 3.78 \text{ inches}$$

For a round-cornered cup, the following formula, in which r equals the radius of the corner, will give fairly accurate diameters, provided the radius does not exceed, say, ¼ the height of the shell:

$$D = \sqrt{d^2 + 4dh} - r \quad (2)$$

These formulas are based on the assumption that the thickness of the drawn shell is the same as the original thickness of the stock, and that the blank is so proportioned that its area will equal the area of the drawn shell. This method of calculating the blank diameter is quite accurate for thin material, when there is only a slight reduction in the thickness of the metal incident to drawing; but when heavy stock is drawn and the thickness of the finished shell is much less than the original thickness of the stock, the blank diameter obtained from Formula (1) or (2) will be too large, because when the stock is drawn thinner, there is an increase in area. When an appreciable reduction in thickness is to be made, the blank diameter can be obtained by first determining the "mean height" of the drawn shell by the following formula. This formula is only approximately correct, but will give results sufficiently accurate for most work:

$$M = \frac{ht}{T} \quad (3)$$

where M = approximate mean height of drawn shell; h = height of drawn shell; t = thickness of shell; and T = thickness of metal before drawing.

After determining the mean height, the blank diameter for the required shell diameter is obtained from the table previously referred to, the mean height being used instead of the actual height.

Example: Suppose a shell 2 inches in diameter and 3 ¾ inches high is to be drawn, and that the original thickness of the stock is 0.050 inch, and the thickness of drawn shell, 0.040 inch. To what diameter should the blank be cut? By using Formula (3) to obtain the mean height:

$$M = \frac{ht}{T} = \frac{3.75 \times 0.040}{0.050} = 3 \text{ inches}$$

According to the table, the blank diameter for a shell 2 inches in diameter and 3 inches high is 5.29 inches. Formula (3) is accurate enough for all practical purposes, unless the reduction in the thickness of the metal is greater than about one-fifth the original thickness. When there is considerable reduction, a blank calculated by this formula produces a shell that is too long. However, the error is in the right direction, as the edges of drawn shells are ordinarily trimmed.

If the shell has a rounded corner, the radius of the corner should be deducted from the figures given in the table. For example, if the shell referred to in the foregoing example had a corner of ¼-inch radius, the blank diameter would equal 5.29 − 0.25 = 5.04 inches.

Another formula that is sometimes used for obtaining blank diameters for shells, when there is a reduction in the thickness of the stock, is as follows:

$$D = \sqrt{a^2 + (a^2 - b^2)\frac{h}{t}} \quad (4)$$

Diameters of Blanks for Drawn Cylindrical Shells

Diam. of Shell	Height of Shell																			
	¼	½	¾	1	1 ¼	1 ½	1 ¾	2	2 ¼	2 ½	2 ¾	3	3 ¼	3 ½	3 ¾	4	4 ½	5	5 ½	6
¼	0.56	0.75	0.90	1.03	1.14	1.25	1.35	1.44	1.52	1.60	1.68	1.75	1.82	1.89	1.95	2.01	2.14	2.25	2.36	2.46
½	0.87	1.12	1.32	1.50	1.66	1.80	1.94	2.06	2.18	2.29	2.40	2.50	2.60	2.69	2.78	2.87	3.04	3.21	3.36	3.50
¾	1.14	1.44	1.68	1.89	2.08	2.25	2.41	2.56	2.70	2.84	2.97	3.09	3.21	3.33	3.44	3.54	3.75	3.95	4.13	4.31
1	1.41	1.73	2.00	2.24	2.45	2.65	2.83	3.00	3.16	3.32	3.46	3.61	3.74	3.87	4.00	4.12	4.36	4.58	4.80	5.00
1 ¼	1.68	2.01	2.30	2.56	2.79	3.01	3.21	3.40	3.58	3.75	3.91	4.07	4.22	4.37	4.51	4.64	4.91	5.15	5.39	5.62
1 ½	1.94	2.29	2.60	2.87	3.12	3.36	3.57	3.78	3.97	4.15	4.33	4.50	4.66	4.82	4.98	5.12	5.41	5.68	5.94	6.18
1 ¾	2.19	2.56	2.88	3.17	3.44	3.68	3.91	4.13	4.34	4.53	4.72	4.91	5.08	5.26	5.41	5.58	5.88	6.17	6.45	6.71
2	2.45	2.83	3.16	3.46	3.74	4.00	4.24	4.47	4.69	4.90	5.10	5.29	5.48	5.66	5.83	6.00	6.32	6.63	6.93	7.21
2 ¼	2.70	3.09	3.44	3.75	4.04	4.31	4.56	4.80	5.03	5.25	5.46	5.66	5.86	6.05	6.23	6.41	6.75	7.07	7.39	7.69
2 ½	2.96	3.36	3.71	4.03	4.33	4.61	4.87	5.12	5.36	5.59	5.81	6.02	6.22	6.42	6.61	6.80	7.16	7.50	7.82	8.14
2 ¾	3.21	3.61	3.98	4.31	4.62	4.91	5.18	5.44	5.68	5.92	6.15	6.37	6.58	6.79	6.99	7.18	7.55	7.91	8.25	8.58
3	3.46	3.87	4.24	4.58	4.90	5.20	5.48	5.74	6.00	6.25	6.48	6.71	6.93	7.14	7.35	7.55	7.94	8.31	8.66	9.00
3 ¼	3.71	4.13	4.51	4.85	5.18	5.48	5.77	6.04	6.31	6.56	6.80	7.04	7.27	7.49	7.70	7.91	8.31	8.69	9.06	9.41
3 ½	3.97	4.39	4.77	5.12	5.45	5.77	6.06	6.34	6.61	6.87	7.12	7.36	7.60	7.83	8.05	8.26	8.67	9.07	9.45	9.81
3 ¾	4.22	4.64	5.03	5.39	5.73	6.05	6.35	6.64	6.91	7.18	7.44	7.69	7.92	8.16	8.38	8.61	9.03	9.44	9.83	10.20
4	4.47	4.90	5.29	5.66	6.00	6.32	6.63	6.93	7.21	7.48	7.75	8.00	8.25	8.49	8.72	8.94	9.38	9.80	10.20	10.58
4 ¼	4.72	5.15	5.55	5.92	6.27	6.60	6.91	7.22	7.50	7.78	8.05	8.31	8.56	8.81	9.04	9.28	9.72	10.15	10.56	10.96
4 ½	4.98	5.41	5.81	6.19	6.54	6.87	7.19	7.50	7.79	8.08	8.35	8.62	8.87	9.12	9.37	9.60	10.06	10.50	10.92	11.32
4 ¾	5.22	5.66	6.07	6.45	6.80	7.15	7.47	7.78	8.08	8.37	8.65	8.92	9.18	9.44	9.69	9.93	10.40	10.84	11.27	11.69
5	5.48	5.92	6.32	6.71	7.07	7.42	7.75	8.06	8.37	8.66	8.94	9.22	9.49	9.75	10.00	10.25	10.72	11.18	11.62	12.04
5 ¼	5.73	6.17	6.58	6.97	7.33	7.68	8.02	8.34	8.65	8.95	9.24	9.52	9.79	10.05	10.31	10.56	11.05	11.51	11.96	12.39
5 ½	5.98	6.42	6.84	7.23	7.60	7.95	8.29	8.62	8.93	9.23	9.53	9.81	10.08	10.36	10.62	10.87	11.37	11.84	12.30	12.74
5 ¾	6.23	6.68	7.09	7.49	7.86	8.22	8.56	8.89	9.21	9.52	9.81	10.10	10.38	10.66	10.92	11.18	11.69	12.17	12.63	13.08
6	6.48	6.93	7.35	7.75	8.12	8.49	8.83	9.17	9.49	9.80	10.10	10.39	10.68	10.95	11.23	11.49	12.00	12.49	12.96	13.42

In this formula, D = blank diameter; a = outside diameter; b = inside diameter; t = thickness of shell at bottom; and h = depth of shell. This formula is based on the volume of the metal in the drawn shell. It is assumed that the shells are cylindrical, and no allowance is made for a rounded corner at the bottom, or for trimming the shell after drawing. To allow for trimming, add the required amount to depth h. When a shell is of irregular cross-section, if its weight is known, the blank diameter can be determined by the following formula:

$$D = 1.1284\sqrt{\frac{W}{wt}} \tag{5}$$

where D = blank diameter in inches; W = weight of shell; w = weight of metal per cubic inch; and t = thickness of the shell.

In the construction of dies for producing shells, especially of irregular form, a common method to be used is to make the drawing tool first. The actual blank diameter then can be determined by trial. One method is to cut a trial blank as near to size and shape as can be estimated. The outline of this blank is then scribed on a flat sheet, after which the blank is drawn. If the finished shell shows that the blank is not of the right diameter or shape, a new trial blank is cut either larger or smaller than the size indicated by the line previously scribed, this line acting as a guide. If a model shell is available, the blank diameter can also be determined as follows:

First, cut a blank somewhat large, and from the same material used for making the model; then, reduce the size of the blank until its weight equals the weight of the model.

Depth and Diameter Reductions of Drawn Cylindrical Shells.—The depth to which metal can be drawn in one operation depends upon the quality and kind of material, its thickness, the slant or angle of the dies, and the amount that the stock is thinned or "ironed" in drawing. A general rule for determining the depth to which cylindrical shells can be drawn in one operation is as follows: The depth or length of the first draw should never be greater than the diameter of the shell. If the shell is to have a flange at the top, it may not be practicable to draw as deeply as is indicated by this rule, unless the metal is extra good, because the stock is subjected to a higher tensile stress, owing to the larger blank needed to form the flange. According to another rule, the depth given the shell on the first draw should equal one-third the diameter of the blank. Ordinarily, it is possible to draw sheet steel of any thickness up to ¼ inch, so that the diameter of the first shell equals about six-tenths of the blank diameter. When drawing plain shells, the amount that the diameter is reduced for each draw must be governed by the quality of the metal and its susceptibility to drawing. The reduction for various thicknesses of metal is about as follows:

Approximate thickness of sheet steel	1⁄16	1⁄8	3⁄16	1⁄4	5⁄16
Possible reduction in diameter for each succeeding step, per cent	20	15	12	10	8

For example, if a shell made of 1⁄16-inch stock is 3 inches in diameter after the first draw, it can be reduced 20 per cent on the next draw, and so on until the required diameter is obtained. These figures are based upon the assumption that the shell is annealed after the first drawing operation, and at least between every two of the following operations. Necking operations—that is, the drawing out of a short portion of the lower part of the cup into a long neck—may be done without such frequent annealings. In double-action presses, where the inside of the cup is supported by a bushing during drawing, the reductions possible may be increased to 30, 24, 18, 15, and 12 per cent, respectively. (The latter figures may also be used for brass in single-action presses.)

When a hole is to be pierced at the bottom of a cup and the remaining metal is to be drawn after the hole has been pierced or punched, always pierce from the opposite direction to that in which the stock is to be drawn after piercing. It may be necessary to machine the metal around the pierced hole to prevent the starting of cracks or flaws in the subsequent drawing operations.

The foregoing figures represent conservative practice and it is often possible to make greater reductions than are indicated by these figures, especially when using a good drawing metal. Taper shells require smaller reductions than cylindrical shells, because the metal tends to wrinkle if the shell to be drawn is much larger than the punch. The amount that the stock is "ironed" or thinned out while being drawn must also be considered, because a reduction in gage or thickness means greater force will be exerted by the punch against the bottom of the shell; hence the amount that the shell diameter is reduced for each drawing operation must be smaller when much ironing is necessary. The extent to which a shell can be ironed in one drawing operation ranges between 0.002 and 0.004 inch per side, and should not exceed 0.001 inch on the final draw, if a good finish is required.

Allowances for Bending Sheet Metal.—In bending steel, brass, bronze, or other metals, the problem is to find the length of straight stock required for each bend; these lengths are added to the lengths of the straight sections to obtain the total length of the material before bending.

If L = length in inches, of straight stock required before bending; T = thickness in inches; and R = inside radius of bend in inches:

For 90-degree bends in soft brass and soft cooper see Table 1 or:

$$L = (0.55 \times T) + (1.57 \times R) \tag{1}$$

For 90-degree bends in half-hard copper and brass, soft steel, and aluminum see Table 2 or:

$$L = (0.64 \times T) + (1.57 \times R) \tag{2}$$

For 90-degree bends in bronze, hard cooper, cold-rolled steel, and spring steel see Table 3 or:

$$L = (0.71 \times T) + (1.57 \times R) \tag{3}$$

Angle of Bend Other Than 90 Degrees: For angles other than 90 degrees, find length L, using tables or formulas, and multiply L by angle of bend, in degrees, divided by 90 to find length of stock before bending. In using this rule, note that *angle of bend* is the angle through which the material has actually been bent; hence, it is not always the angle as given on a drawing. To illustrate, in Fig. 1, the angle on the drawing is 60 degrees, but the angle of bend A is 120 degrees (180 − 60 = 120); in Fig. 2, the angle of bend A is 60 degrees; in Fig. 3, angle A is 90 − 30 = 60 degrees. Formulas (1), (2), and (3) are based on extensive experiments of the Westinghouse Electric Co. They apply to parts bent with simple tools or on the bench, where limits of ± 1⁄64 inch are specified. If a part has two or more bends of the same radius, it is, of course, only necessary to obtain the length required for one of the bends and then multiply by the number of bends, to obtain the total allowance for the bent sections.

Table 1. Lengths of Straight Stock Required for 90-Degree Bends in Soft Copper and Soft Brass

Radius R of Bend, Inches	Thickness T of Material, Inch												
	1/64	1/32	3/64	1/16	5/64	3/32	1/8	5/32	3/16	7/32	1/4	9/32	5/16
1/32	0.058	0.066	0.075	0.083	0.092	0.101	0.118	0.135	0.152	0.169	0.187	0.204	0.221
3/64	0.083	0.091	0.100	0.108	0.117	0.126	0.143	0.160	0.177	0.194	0.212	0.229	0.246
1/16	0.107	0.115	0.124	0.132	0.141	0.150	0.167	0.184	0.201	0.218	0.236	0.253	0.270
3/32	0.156	0.164	0.173	0.181	0.190	0.199	0.216	0.233	0.250	0.267	0.285	0.302	0.319
1/8	0.205	0.213	0.222	0.230	0.239	0.248	0.265	0.282	0.299	0.316	0.334	0.351	0.368
5/32	0.254	0.262	0.271	0.279	0.288	0.297	0.314	0.331	0.348	0.365	0.383	0.400	0.417
3/16	0.303	0.311	0.320	0.328	0.337	0.346	0.363	0.380	0.397	0.414	0.432	0.449	0.466
7/32	0.353	0.361	0.370	0.378	0.387	0.396	0.413	0.430	0.447	0.464	0.482	0.499	0.516
1/4	0.401	0.409	0.418	0.426	0.435	0.444	0.461	0.478	0.495	0.512	0.530	0.547	0.564
9/32	0.450	0.458	0.467	0.475	0.484	0.493	0.510	0.527	0.544	0.561	0.579	0.596	0.613
5/16	0.499	0.507	0.516	0.524	0.533	0.542	0.559	0.576	0.593	0.610	0.628	0.645	0.662
11/32	0.549	0.557	0.566	0.574	0.583	0.592	0.609	0.626	0.643	0.660	0.678	0.695	0.712
3/8	0.598	0.606	0.615	0.623	0.632	0.641	0.658	0.675	0.692	0.709	0.727	0.744	0.761
13/32	0.646	0.654	0.663	0.671	0.680	0.689	0.706	0.723	0.740	0.757	0.775	0.792	0.809
7/16	0.695	0.703	0.712	0.720	0.729	0.738	0.755	0.772	0.789	0.806	0.824	0.841	0.858
15/32	0.734	0.742	0.751	0.759	0.768	0.777	0.794	0.811	0.828	0.845	0.863	0.880	0.897
1/2	0.794	0.802	0.811	0.819	0.828	0.837	0.854	0.871	0.888	0.905	0.923	0.940	0.957
9/16	0.892	0.900	0.909	0.917	0.926	0.935	0.952	0.969	0.986	1.003	1.021	1.038	1.055
5/8	0.990	0.998	1.007	1.015	1.024	1.033	1.050	1.067	1.084	1.101	1.119	1.136	1.153
11/16	1.089	1.097	1.106	1.114	1.123	1.132	1.149	1.166	1.183	1.200	1.218	1.235	1.252
3/4	1.187	1.195	1.204	1.212	1.221	1.230	1.247	1.264	1.281	1.298	1.316	1.333	1.350
13/16	1.286	1.294	1.303	1.311	1.320	1.329	1.346	1.363	1.380	1.397	1.415	1.432	1.449
7/8	1.384	1.392	1.401	1.409	1.418	1.427	1.444	1.461	1.478	1.495	1.513	1.530	1.547
15/16	1.481	1.489	1.498	1.506	1.515	1.524	1.541	1.558	1.575	1.592	1.610	1.627	1.644
1	1.580	1.588	1.597	1.605	1.614	1.623	1.640	1.657	1.674	1.691	1.709	1.726	1.743
1 1/16	1.678	1.686	1.695	1.703	1.712	1.721	1.738	1.755	1.772	1.789	1.807	1.824	1.841
1 1/8	1.777	1.785	1.794	1.802	1.811	1.820	1.837	1.854	1.871	1.888	1.906	1.923	1.940
1 3/16	1.875	1.883	1.892	1.900	1.909	1.918	1.935	1.952	1.969	1.986	2.004	2.021	2.038
1 1/4	1.972	1.980	1.989	1.997	2.006	2.015	2.032	2.049	2.066	2.083	2.101	2.118	2.135

Table 2. Lengths of Straight Stock Required for 90-Degree Bends in Half-Hard Brass and Sheet Copper, Soft Steel, and Aluminum

Radius *R* of Bend, Inches	Thickness *T* of Material, Inch												
	1⁄64	1⁄32	3⁄64	1⁄16	5⁄64	3⁄32	1⁄8	5⁄32	3⁄16	7⁄32	1⁄4	9⁄32	5⁄16
1⁄32	0.059	0.069	0.079	0.089	0.099	0.109	0.129	0.149	0.169	0.189	0.209	0.229	0.249
3⁄64	0.084	0.094	0.104	0.114	0.124	0.134	0.154	0.174	0.194	0.214	0.234	0.254	0.274
1⁄16	0.108	0.118	0.128	0.138	0.148	0.158	0.178	0.198	0.218	0.238	0.258	0.278	0.298
3⁄32	0.157	0.167	0.177	0.187	0.197	0.207	0.227	0.247	0.267	0.287	0.307	0.327	0.347
1⁄8	0.206	0.216	0.226	0.236	0.246	0.256	0.276	0.296	0.316	0.336	0.356	0.376	0.396
5⁄32	0.255	0.265	0.275	0.285	0.295	0.305	0.325	0.345	0.365	0.385	0.405	0.425	0.445
3⁄16	0.305	0.315	0.325	0.335	0.345	0.355	0.375	0.395	0.415	0.435	0.455	0.475	0.495
7⁄32	0.354	0.364	0.374	0.384	0.394	0.404	0.424	0.444	0.464	0.484	0.504	0.524	0.544
1⁄4	0.403	0.413	0.423	0.433	0.443	0.453	0.473	0.493	0.513	0.533	0.553	0.573	0.593
9⁄32	0.452	0.462	0.472	0.482	0.492	0.502	0.522	0.542	0.562	0.582	0.602	0.622	0.642
5⁄16	0.501	0.511	0.521	0.531	0.541	0.551	0.571	0.591	0.611	0.631	0.651	0.671	0.691
11⁄32	0.550	0.560	0.570	0.580	0.590	0.600	0.620	0.640	0.660	0.680	0.700	0.720	0.740
3⁄8	0.599	0.609	0.619	0.629	0.639	0.649	0.669	0.689	0.709	0.729	0.749	0.769	0.789
13⁄32	0.648	0.658	0.668	0.678	0.688	0.698	0.718	0.738	0.758	0.778	0.798	0.818	0.838
7⁄16	0.697	0.707	0.717	0.727	0.737	0.747	0.767	0.787	0.807	0.827	0.847	0.867	0.887
15⁄32	0.746	0.756	0.766	0.776	0.786	0.796	0.816	0.836	0.856	0.876	0.896	0.916	0.936
1⁄2	0.795	0.805	0.815	0.825	0.835	0.845	0.865	0.885	0.905	0.925	0.945	0.965	0.985
17⁄32	0.844	0.854	0.864	0.874	0.884	0.894	0.914	0.934	0.954	0.974	0.994	1.014	1.034
9⁄16	0.894	0.904	0.914	0.924	0.934	0.944	0.964	0.984	1.004	1.024	1.044	1.064	1.084
5⁄8	0.992	1.002	1.012	1.022	1.032	1.042	1.062	1.082	1.102	1.122	1.42	1.162	1.182
11⁄16	1.090	1.100	1.110	1.120	1.130	1.140	1.160	1.180	1.200	1.220	1.240	1.260	1.280
3⁄4	1.188	1.198	1.208	1.218	1.228	1.238	1.258	1.278	1.298	1.318	1.338	1.358	1.378
13⁄16	1.286	1.296	1.306	1.316	1.326	1.336	1.356	1.376	1.396	1.416	1.436	1.456	1.476
7⁄8	1.384	1.394	1.404	1.414	1.424	1.434	1.454	1.474	1.494	1.514	1.534	1.554	1.574
15⁄16	1.483	1.493	1.503	1.513	1.523	1.553	1.553	1.573	1.693	1.613	1.633	1.653	1.673
1	1.581	1.591	1.601	1.611	1.621	1.631	1.651	1.671	1.691	1.711	1.731	1.751	1.771
1 1⁄16	1.697	1.689	1.699	1.709	1.719	1.729	1.749	1.769	1.789	1.809	1.829	1.849	1.869
1 1⁄8	1.777	1.787	1.797	1.807	1.817	1.827	1.847	1.867	1.887	1.907	1.927	1.947	1.967
1 3⁄16	1.875	1.885	1.895	1.905	1.915	1.925	1.945	1.965	1.985	1.005	2.025	2.045	2.065
1 1⁄4	1.973	1.983	1.993	1.003	2.013	2.023	2.043	2.063	2.083	2.103	2.123	2.143	2.163

Table 3. Lengths of Straight Stock Required for 90-Degree Bends in Hard Copper, Bronze, Cold-Rolled Steel, and Spring Steel

Radius *R* of Bend, Inches	Thickness *T* of Material, Inch												
	1⁄64	1⁄32	3⁄64	1⁄16	5⁄64	3⁄32	1⁄8	5⁄32	3⁄16	7⁄32	1⁄4	9⁄32	5⁄16
1⁄32	0.060	0.071	0.082	0.093	0.104	0.116	0.138	0.160	0.182	0.204	0.227	0.249	0.271
3⁄64	0.085	0.096	0.107	0.118	0.129	0.141	0.163	0.185	0.207	0.229	0.252	0.274	0.296
1⁄16	0.109	0.120	0.131	0.142	0.153	0.165	0.187	0.209	0.231	0.253	0.276	0.298	0.320
3⁄32	0.158	0.169	0.180	0.191	0.202	0.214	0.236	0.258	0.280	0.302	0.325	0.347	0.369
1⁄8	0.207	0.218	0.229	0.240	0.251	0.263	0.285	0.307	0.329	0.351	0.374	0.396	0.418
5⁄32	0.256	0.267	0.278	0.289	0.300	0.312	0.334	0.356	0.378	0.400	0.423	0.445	0.467
3⁄16	0.305	0.316	0.327	0.338	0.349	0.361	0.383	0.405	0.427	0.449	0.472	0.494	0.516
7⁄32	0.355	0.366	0.377	0.388	0.399	0.411	0.433	0.455	0.477	0.499	0.522	0.544	0.566
1⁄4	0.403	0.414	0.425	0.436	0.447	0.459	0.481	0.503	0.525	0.547	0.570	0.592	0.614
9⁄32	0.452	0.463	0.474	0.485	0.496	0.508	0.530	0.552	0.574	0.596	0.619	0.641	0.663
5⁄16	0.501	0.512	0.523	0.534	0.545	0.557	0.579	0.601	0.623	0.645	0.668	0.690	0.712
11⁄32	0.551	0.562	0.573	0.584	0.595	0.607	0.629	0.651	0.673	0.695	0.718	0.740	0.762
3⁄8	0.600	0.611	0.622	0.633	0.644	0.656	0.678	0.700	0.722	0.744	0.767	0.789	0.811
13⁄32	0.648	0.659	0.670	0.681	0.692	0.704	0.726	0.748	0.770	0.792	0.815	0.837	0.859
7⁄16	0.697	0.708	0.719	0.730	0.741	0.753	0.775	0.797	0.819	0.841	0.864	0.886	0.908
15⁄32	0.736	0.747	0.758	0.769	0.780	0.792	0.814	0.836	0.858	0.880	0.903	0.925	0.947
1⁄2	0.796	0.807	0.818	0.829	0.840	0.852	0.874	0.896	0.918	0.940	0.963	0.985	1.007
9⁄16	0.894	0.905	0.916	0.927	0.938	0.950	0.972	0.994	1.016	1.038	1.061	1.083	1.105
5⁄8	0.992	1.003	1.014	1.025	1.036	1.048	1.070	1.092	1.114	1.136	1.159	1.181	1.203
11⁄16	1.091	1.102	1.113	1.124	1.135	1.147	1.169	1.191	1.213	1.235	1.258	1.280	1.302
3⁄4	1.189	1.200	1.211	1.222	1.233	1.245	1.267	1.289	1.311	1.333	1.356	1.378	1.400
13⁄16	1.288	1.299	1.310	1.321	1.332	1.344	1.366	1.388	1.410	1.432	1.455	1.477	1.499
7⁄8	1.386	1.397	1.408	1.419	1.430	1.442	1.464	1.486	1.508	1.530	1.553	1.575	1.597
15⁄16	1.483	1.494	1.505	1.516	1.527	1.539	1.561	1.583	1.605	1.627	1.650	1.672	1.694
1	1.582	1.593	1.604	1.615	1.626	1.638	1.660	1.682	1.704	1.726	1.749	1.771	1.793
1 1⁄16	1.680	1.691	1.702	1.713	1.724	1.736	1.758	1.780	1.802	1.824	1.847	1.869	1.891
1 1⁄8	1.779	1.790	1.801	1.812	1.823	1.835	1.857	1.879	1.901	1.923	1.946	1.968	1.990
1 3⁄16	1.877	1.888	1.899	1.910	1.921	1.933	1.955	1.977	1.999	2.021	2.044	2.066	2.088
1 1⁄4	1.974	1.985	1.996	2.007	2.018	2.030	2.052	2.074	2.096	2.118	2.141	2.163	2.185

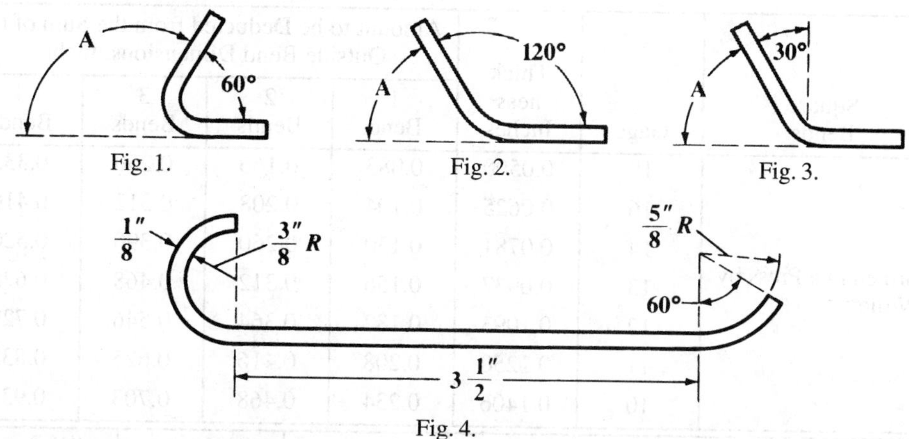

Fig. 4.

Example, Showing Application of Formulas: Find the length before bending of the part illustrated by Fig. 4. Soft steel is to be used.

For bend at left-hand end (180-degree bend)

$$L = [(0.64 \times 0.125) + (1.57 \times 0.375)] \times \frac{180}{90} = 1.338$$

For bend at right-hand end (60-degree bend)

$$L = [(0.64 \times 0.125) + (1.57 \times 0.625)] \times \frac{60}{90} = 0.707$$

Total length before bending = 3.5 + 1.338 + 0.707 = 5.545 inches

Other Bending Allowance Formulas.—When bending sheet steel or brass, add from ⅓ to ½ of the thickness of the stock, for *each bend,* to the sum of the inside dimensions of the finished piece, to get the length of the straight blank. The harder

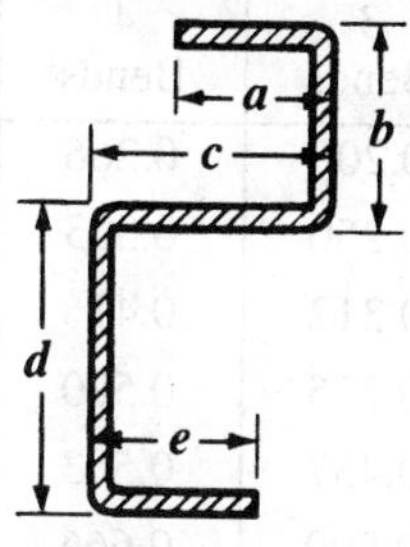

the material the greater the allowance (⅓ of the thickness is added for soft stock and ½ of the thickness for hard material). The data given in the table, *Allowances for Bends in Sheet Metal,* refer more particularly to the bending of sheet metal for counters, bank fittings and general office fixtures, for which purpose it is not absolutely essential to have the sections of the bends within very close limits. Absolutely accurate data for this work cannot be deduced, as the stock varies considerably as to hardness, etc. The figures given apply to sheet steel, aluminum, brass and bronze. Experience has demonstrated that for the semisquare corners, such as are formed in a V-die, the amount to be deducted from the sum of the outside bend dimensions, as shown in the accompanying illustration by the sum of the letters from a to e, is as follows: $X = 1.67\ BG$, where X = the amount to be deducted; B = the number of bends; and G = the decimal equivalent of the gage. The values of X for different gages and numbers of bends are given in the table. Its application may be illustrated by an example: A strip having two bends is to have outside dimensions of 2, 1½ and 2 inches, and is made of stock 0.125 inch thick. The sum of the outside dimensions is thus 5½ inches, and from the table the amount to be deducted is found to be 0.416; hence the blank will be 5.5 − 0.416 = 5.084 inches long.

The lower part of the table applies to square bends which are either drawn through a block of steel made to the required shape, or else drawn through rollers in a drawbench. The pressure applied not only gives a much sharper corner, but it also elongates the material more than in the V-die process. In this case, the deduction is $X = 1.33\ BG$.

Allowances for Bends in Sheet Metal

Square Bends	Gage	Thickness Inches	Amount to be Deducted from the Sum of the Outside Bend Dimensions, Inches						
			1 Bend	2 Bends	3 Bends	4 Bends	5 Bends	6 Bends	7 Bends
Formed in a Press by a V-die	18	0.0500	0.083	0.166	0.250	0.333	0.416	0.500	0.583
	16	0.0625	0.104	0.208	0.312	0.416	0.520	0.625	0.729
	14	0.0781	0.130	0.260	0.390	0.520	0.651	0.781	0.911
	13	0.0937	0.156	0.312	0.468	0.625	0.781	0.937	1.093
	12	0.1093	0.182	0.364	0.546	0.729	0.911	1.093	1.276
	11	0.1250	0.208	0.416	0.625	0.833	1.041	1.250	1.458
	10	0.1406	0.234	0.468	0.703	0.937	1.171	1.406	1.643

Square Bends	Gage	Thickness Inches	Amount to be Deducted from the Sum of the Outside Bend Dimensions, Inches						
			1 Bend	2 Bends	3 Bends	4 Bends	5 Bends	6 Bends	7 Bends
Rolled or Drawn in a Draw-bench	18	0.0500	0.066	0.133	0.200	0.266	0.333	0.400	0.466
	16	0.0625	0.083	0.166	0.250	0.333	0.416	0.500	0.583
	14	0.0781	0.104	0.208	0.312	0.416	0.521	0.625	0.729
	13	0.0937	0.125	0.250	0.375	0.500	0.625	0.750	0.875
	12	0.1093	0.145	0.291	0.437	0.583	0.729	0.875	1.020
	11	0.1250	0.166	0.333	0.500	0.666	0.833	1.000	1.166
	10	0.1406	0.187	0.375	0.562	0.750	0.937	1.125	1.312

Fine Blanking

The process called fine blanking uses special presses and tooling to produce flat components from sheet metal or plate, with high dimensional accuracy. According to Hydrel A. G., Romanshorn, Switzerland, fine-blanking presses can be powered hydraulically or mechanically, or by a combination of these methods, but they must have three separate and distinct movements. These movements serve to clamp the work material, to perform the blanking operation, and to eject the finished part from the tool. Forces of 1.5–2.5 times those used in conventional stamping are needed for fine blanking, so machines and tools must be designed and constructed accordingly. In mechanical fine-blanking presses the clamping and ejection forces are exerted hydraulically. Such presses generally are of toggle-type design and are limited to total forces of up to about 280 tons. Higher forces generally require all-hydraulic designs. These presses are also suited to embossing, coining, and impact extrusion work.

Cutting elements of tooling for fine blanking generally are made from 12 per cent chromium steel, although high speed steel and tungsten carbide also are used for long runs or improved quality. Cutting clearances between the intermediate punch and die are usually held between 0.0001 and 0.0003 in. The clamping elements are sharp projections of 90-degree V-section that follow the outline of the workpiece and that are incorporated into each tool as part of the stripper plate with thin material and also as part of the die plate when material thicker than 0.15 in. is to be blanked. Pressure applied to the elements containing the V-projections prior to the blanking operation causes the sharp edges to enter the material surface, preventing sideways movement of the blank. The pressure applied as the projections bite into the work surface near the contour edges also squeezes the material, causing it to flow toward the cutting edges, reducing the usual rounding effect at the cut edge. When small details such as gear teeth are to be produced, V-projections are often used on both sides of the work, even with thin materials, to enhance the flow effect. With suitable tooling, workpieces can be produced with edges that are perpendicular to top and bottom surfaces within 0.004 in. on thicknesses of 0.2 in., for instance. V-projection dimensions for various material thicknesses are shown in the table.

Fine-blanked edges are free from the fractures that result from conventional tooling, and can have surface finishes down to 80 μin. Ra with suitable tooling. Close tolerances can be held on inner and outer forms, and on hole center distances. Flatness of fine-blanked components is better than that of parts made by conventional methods, but distortion may occur with thin materials due to release of internal stresses. Widths must be slightly greater than are required for conventional press working. Generally, the strip width must be 2–3 times the thickness, plus the width of the part measured transverse to the feed direction. Other factors to be considered are shape, material quality, size and shape of the V-projection in relation to the die outline, and spacing between adjacent blanked parts. Holes and slots can be produced with ratios of width to material thickness down to 0.7, compared with the 1:1 ratio normally specified for conventional tooling. Operations such as countersinking, coining, and bending up to 60 degrees can be incorporated in fine-blanking tooling.

The cutting force in lb exerted in fine blanking is 0.9 times the length of the cut in inches times the material thickness in inches, times the tensile strength in $lb_f/in.^2$. Pressure in lb exerted by the clamping element(s) carrying the V-projections is calculated by multiplying the length of the V-projection, which depends on its shape, in inches by its height (h), times the material tensile strength in $lb_f/in.^2$, times an empirical factor f. Factor f has been determined to be 2.4–4.4 for a tensile strength of 28,000–113,000 $lb_f/in.^2$. The clamping pressure is approximately 30 per cent of the cutting force, calculated as above. Dimensions and positioning of the V-projection(s) are related to the material thickness, quality, and tensile strength. A small V-projection close to the line of cut has about the same effect as a large V-projection spaced away from the cut. However, if the V-projection is too close to the cut, it may move out of the material at the start of the cutting process, reducing its effectiveness.

Positioning the V-projection at a distance from the line of cut increases both material and blanking force requirements. Location of the V-projection relative to the line of cut also affects tool life.

Dimensions for V-projections Used in Fine-Blanking Tools

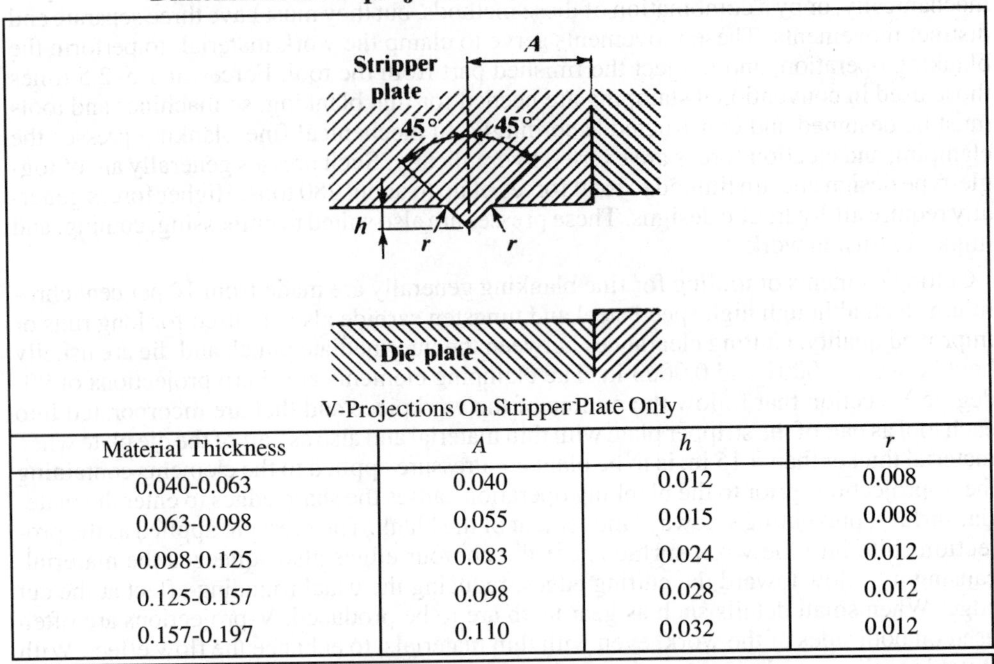

V-Projections On Stripper Plate Only

Material Thickness	*A*	*h*	*r*
0.040-0.063	0.040	0.012	0.008
0.063-0.098	0.055	0.015	0.008
0.098-0.125	0.083	0.024	0.012
0.125-0.157	0.098	0.028	0.012
0.157-0.197	0.110	0.032	0.012

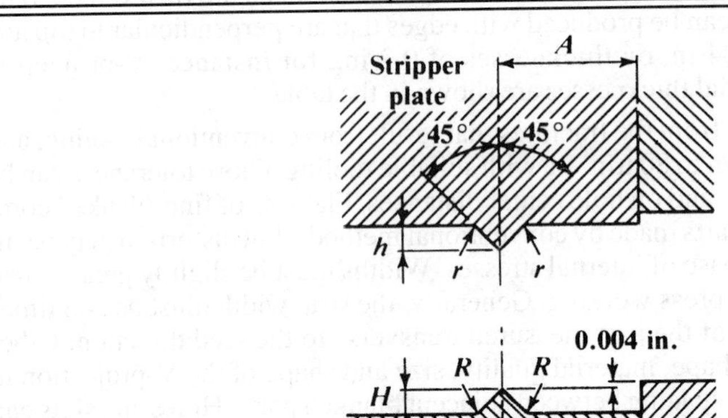

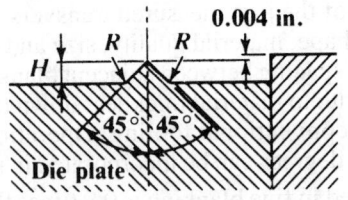

V-Projections On Both Stripper and Die Plate

Material Thickness	*A*	*H*	*R*	*h*	*r*
0.157–0.197	0.098	0.032	0.032	0.020	0.008
0.197–0.248	0.118	0.040	0.040	0.028	0.008
0.248–0.315	0.138	0.047	0.047	0.032	0.008
0.315–0.394	0.177	0.060	0.060	0.040	0.020
0.394–0.492	0.217	0.070	0.080	0.047	0.020
0.492–0.630	0.276	0.087	0.118	0.063	0.020

All units are in inches.

Steel Rule Dies

Steel rule dies (or knife dies) were patented by Robert Gair in 1879, and, as the name implies, have cutting edges made from steel strips of about the same proportions as the steel strips used in making graduated rules for measuring purposes. According to J. A. Richards, Sr., of the J. A. Richards Co., Kalamazoo, MI, a pioneer in the field, these dies were first used in the printing and shoemaking industries for cutting out shapes in paper, cardboard, leather, rubber, cork, felt, and similar soft materials. Steel rule dies were later adopted for cutting upholstery material for the automotive and other industries, and for cutting out simple to intricate shapes in sheet metal, including copper, brass, and aluminum. A typical steel rule die, partially cut away to show the construction, is shown in Fig. 1, and is designed for cutting a simple circular shape. Such dies generally cost 25 to 35 per cent of the cost of conventional blanking dies, and can be produced in much less time. The die shown also cuts a rectangular opening in the workpiece, and pierces four holes, all in one press stroke.

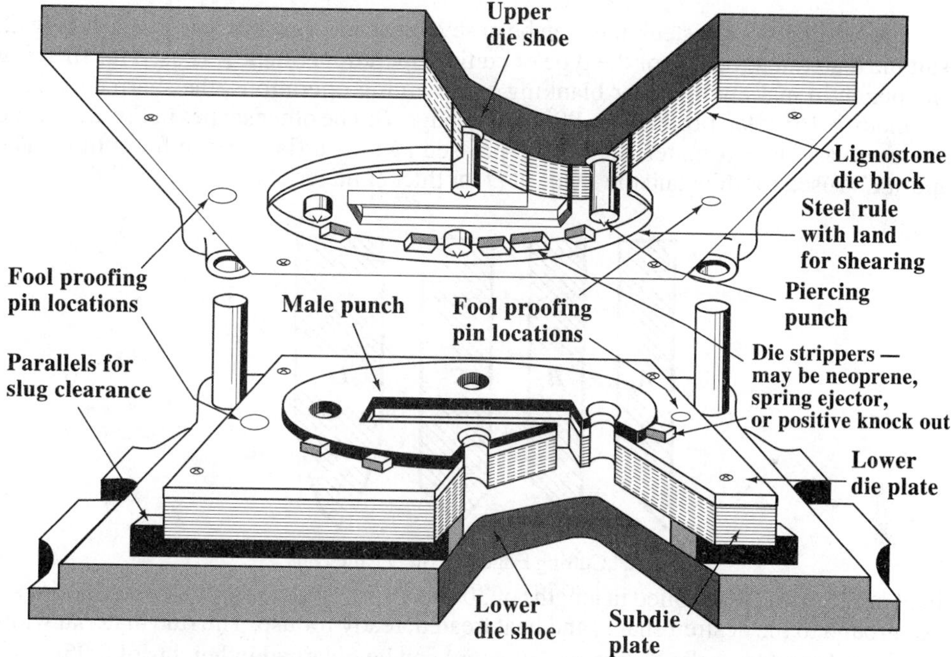

Fig. 1. Steel Rule Die for Cutting a Circular Shape, Sectioned to Show the Construction

The die blocks that hold the steel strips on edge on the press platen or in the die set may be made from plaster, hot lead or type metal, or epoxy resin, all of which can be poured to shape. However, the material most widely used for light work is ¾-in. thick, five- or seven-ply maple or birch wood. Narrow slots are cut in this wood with a jig saw to hold the strips vertically. Where greater forces are involved, as with operations on metal sheets, the blocks usually are made from Lignostone densified wood or from metal. In the ¾-in. thickness mostly used, medium- and high-density grades of Lignostone are available. The ¾-in. thickness is made from about 35 plies of highly compressed lignite wood, bonded with phenolformaldehyde resin, which imparts great density and strength. The material is made in thicknesses up to 6 in., and in various widths and lengths.

Steel rule die blocks can carry punches of various shapes to pierce holes in the stock, also projections designed to form strengthening ribs and other shapes in material such as aluminum, at the same time as the die cuts the component to shape. Several dies can be combined or nested, and operated together in a large press, to produce various shapes simultaneously from one sheet of material.

As shown in Fig. 1, the die steel is held in the die block slot on its edge, usually against the flat platen of a die set attached to the moving slide of the press. The sharp, free end of the rule faces toward the workpiece, which is supported by the face of the other die half. This other die half may be flat or may have a punch attached to it, as shown, and it withstands the pressure exerted in the cutting or forming action when the press is operated. The closed height of the die is adjusted to permit the cutting edge to penetrate into the material to the extent needed, or, if there is a punch, to carry the cutting edges just past the punch edges for the cutting operation. After the sharp edge has penetrated it, the material often clings to the sides of the knife. Ejector inserts made from rubber, combinations of cork and rubber, and specially compounded plastics material, or purpose-made ejectors, either spring- or positively actuated, are installed in various positions alongside the steel rules and the punch. These ejectors are compressed as the dies close, and when the dies open, they expand, pushing the material clear of the knives or the punch.

The cutting edges of the steel rules can be of several shapes, as shown in profile in Fig. 2, to suit the material to be cut, or the type of cutting operation. Shape *A* is used for shearing in the punch in making tools for blanking and piercing operations, the sharp edge later being modified to a flat, producing a 90° cutting edge, *B*. The other shapes in Fig. 2 are used for cutting various soft materials that are pressed against a flat surface for cutting. The shape at *C* is used for thin, and the shape at *D* for thicker materials.

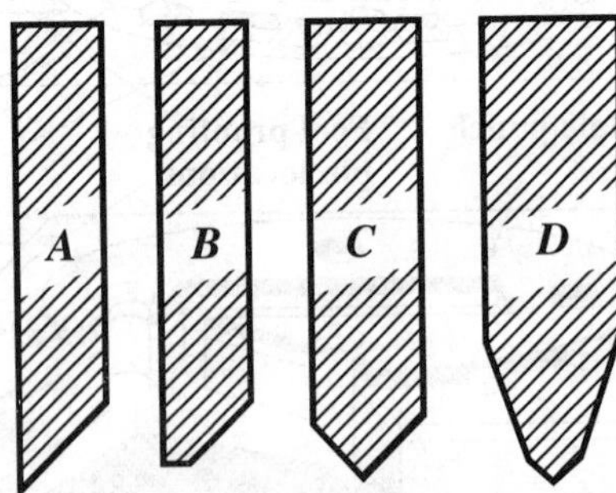

Fig. 2. Cutting Edges for Steel Rule Dies

Steel rule die steel is supplied in lengths of 30 and 50 in., or in coils of any length, with the edges ground to the desired shape, and heat treated, ready for use. The rule material width is usually referred to as the height, and material can be obtained in heights of 0.95, 1, $1\frac{1}{8}$, $1\frac{1}{4}$, and $1\frac{1}{2}$ in. Rules are available in thicknesses of 0.055, 0.083, 0.11, 0.138, 0.166, and 0.25 in. (4 to 18 points in printers' measure of 72 points = 1 in.). Generally, stock thicknesses of 0.138 or 0.166 in. (10 and 12 points) are preferred, the thinner rules being used mainly for dies requiring intricate outlines. The stock can be obtained in soft or hard temper. The standard edge bevel is 46°, but bevels of 40 to 50° can be used. Thinner rule stock is easiest to form to shape and is often used for short runs of 50 pieces or thereabouts. The thickness and hardness of the material to be blanked also must be considered when choosing rule thickness.

Making of Steel Rule Dies.—Die making begins with a drawing of the shape required. Saw cutting lines may be marked directly on the face of the die block in a conventional layout procedure using a height gage, or a paper drawing may be pasted to or drawn on the die board. Because paper stretches and shrinks, Mylar or other nonshrink plastics sheets may be preferred for the drawing. A hole is drilled off the line to allow a jig saw to be inserted, and jig saw or circular saw cuts are then made under manual control along the drawing lines to produce the slots for the rules. Jig saw blades are available in a range of sizes to suit various thicknesses of rule and for sawing medium-density Lignostone, a speed of 300 strokes/min is recommended, the saw having a stroke of about 2 in. To make sure the rule thickness to be used will be a tight fit in the slot, trials are usually carried out on scrap pieces of die block before cuts are made on a new block.

During slot cutting, the saw blade must always be maintained vertical to the board being cut, and magnifying lenses are often used to keep the blade close to the line. Carbide or carbide-tipped saw blades are recommended for clean cuts as well as for long life. To keep any "islands" (such as the center of a circle) in position, various places in the sawn line are cut to less than full depth for lengths of ¼ to ½ in., and to heights of ⅝ to ¾ in. to bridge the gaps. Slots of suitable proportions must be provided in the steel rules, on the sides away from the cutting edges, to accommodate these die block bridges.

Rules for steel rule dies are bent to shape to fit the contours called for on the drawing by means of small, purpose-built bending machines, fitted with suitable tooling. For bends of small radius, the tooling on these machines is arranged to perform a peening or hammering action to force the steel rule into close contact with the radius-forming component of the machine so that quite small radii, as required for jig saw puzzles, for instance, can be produced with good accuracy. Some forms are best made in two or more pieces, then joined by welding or brazing. The edges to be joined are mitered for a perfect fit, and are clamped securely in place for joining. Electrical resistance or a gas heating torch is used to heat the joint. Wet rags are applied to the steel at each side of the joint to keep the material cool and the hardness at the preset level, as long as possible.

When shapes are to be blanked from sheet metal, the steel rule die is arranged with flat, 90° edges (*B*, in Fig. 2), which cut by pushing the work past a close-fitting counter-punch. This counterpunch, shown in Fig. 1, may be simply a pad of steel or other material, and has an outline corresponding to the shape of the part to be cut. Sometimes the pad may be given a gradual, slight reduction in height to provide a shearing action as the moving tool pushes the work material past the pad edges. As shown in Fig. 1, punches can be incorporated in the die to pierce holes, cut slots, or form ribs and other details during the blanking operation. These punches are preferably made from high-carbon, high-vanadium, alloy steel, heat treated to Rc 61 to 63, with the head end tempered to Rc 45 to 50.

Heat treatment of the high-carbon-steel rules is designed to produce a hardness suited to the application. Rules in dies for cutting cartons and similar purposes, with mostly straight cuts, are hardened to Rc 51 to 58. For dies requiring many intricate bends, lower-carbon material is used, and is hardened to Rc 38 to 45. And for dies to cut very intricate shapes, a steel in dead-soft condition with hardness of about Rb 95 is recommended. After the intricate bends are made, this steel must be carburized before it is hardened and tempered. For this material, heat treatment uses an automatic cycle furnace, and consists of carburizing in a liquid compound heated to 1500°F and quenching in oil, followed by "tough" tempering at 550°F and cooling in the furnace.

After the hardened rule has been reinstalled in the die block, the tool is loaded into the press and the sharp die is used with care to shear the sides of the pad to match the die contours exactly. A close fit, with clearances of about half those used in conventional blanking dies, is thus ensured between the steel rule and the punch. Adjustments to the clearances can be made at this point by grinding the die steel or the punch. After the adjustment work is done, the sharp edges of the rule steel are ground flat to produce a land of about $1/64$ in. wide (*A* in Fig. 2), for the working edges of the die. Clearances for piercing punches should be similar to those used on conventional piercing dies.

ELECTRICAL DISCHARGE MACHINING

Generally called EDM, electrical discharge machining uses an electrode to remove metal from a workpiece by generating electric sparks between conducting surfaces. The two main types of EDM are termed sinker or plunge, used for making mold or die cavities, and wire, used to cut shapes such as are needed for stamping dies. For die sinking, the electrode usually is made from copper or graphite and is shaped as a positive replica of the shape to be formed on or in the workpiece. A typical EDM sinker machine, shown diagrammatically in Fig. 1, resembles a vertical milling machine, with the electrode attached to the vertical slide. The slide is moved down and up by an electronic, servo-controlled drive unit that controls the spacing between the electrode and the workpiece on the table. The table can be adjusted in three directions, often under numerical control, to positions that bring a workpiece surface to within 0.0005 to 0.030 in. from the electrode surface, where a spark is generated.

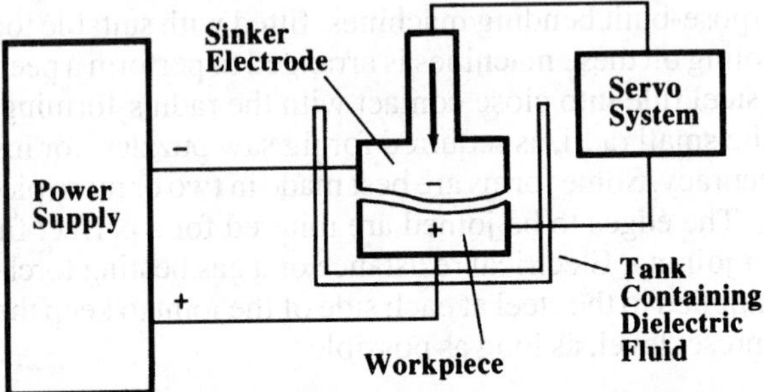

Fig. 1. Sinker or Plunge Type EDM Machines Are Used to Sink Cavities in Molds and Dies

Wire EDM, shown diagrammatically in Fig. 2, are numerically controlled and somewhat resemble a bandsaw with the saw blade replaced by a fine brass or copper wire, which forms the electrode. This wire is wound off one reel, passed through tensioning and guide rollers, then through the workpiece and through lower guide rollers before being wound onto another reel for storage and eventual recycling. One set of guide rollers, usually the lower, can be moved on two axes at 90 degrees apart under numerical control to adjust the angle of the wire when profiles of varying angles are to be produced. The table also is movable in two directions under numerical control to adjust the position of the workpiece relative to the wire. Provision must be made for the cut-out part to be supported when it is freed from the workpiece so that it does not pinch and break the wire.

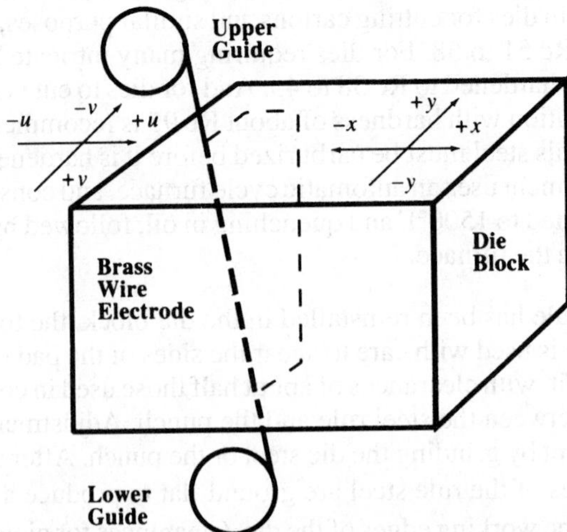

Fig. 2. Wire Type EDM Machines Are Used to Cut Stamping Die Profiles.

EDM applied to grinding machines is termed EDG. The process uses a graphite wheel as an electrode, and wheels can be up to 12 in. in diameter by 6 in. wide. The wheel periphery is dressed to the profile required on the workpiece and the wheel profile can then be transferred to the workpiece as it is traversed past the wheel, which rotates but does not touch the work. EDG machines are highly specialized and are mainly used for producing complex profiles on polycrystaline diamond cutting tools and for shaping carbide tooling such as form tools, thread chasers, dies, and crushing rolls.

EDM Terms*

Anode: The positive terminal of an electrolytic cell or battery. In EDM, incorrectly applied to the tool or electrode.

Barrel effect: In wire EDM, a condition where the center of the cut is wider than the entry and exit points of the wire, due to secondary discharges caused by particles being pushed to the center by flushing pressure from above and beneath the workpiece.

Capacitor: An electrical component that stores an electric charge. In some EDM power supplies, several capacitors are connected across the machining gap and the current for the spark comes directly from the capacitors when they are discharged.

Cathode: The negative terminal in an electrolytic cell or battery. In EDM incorrectly applied to the workpiece.

Colloidal suspension: Particles suspended in a liquid that are too fine to settle out. In EDM, the tiny particles produced in the sparking action form a colloidal suspension in the dielectric fluid.

Craters: Small cavities left on an EDM surface by the sparking action. Also known as pits.

Dielectric filter : A filter that removes particles from 5 μm (0.00020 in.) down to as fine as 1 μm (0.00004 in) in size, from dielectric fluid.

Dielectric fluid : The non-conductive fluid that circulates between the electrode and the workpiece to provide the dielectric strength across which an arc can occur, to act as a coolant to solidify particles melted by the arc, and to flush away the solidified particles.

Dielectric strength: In EDM, the electrical potential (voltage) needed to break down (ionize) the dielectric fluid in the gap between the electrode and the workpiece.

Discharge channel: The conductive pathway formed by ionized dielectric and vapor between the electrode and the workpiece.

Dither: A slight up and down movement of the machine ram and attached electrode, used to improve cutting stability.

Duty cycle: The percentage of a pulse cycle during which the current is turned on (on time), relative to the total duration of the cycle.

EDG: Electrical discharge grinding using a machine that resembles a surface grinder but has a wheel made from electrode material. Metal is removed by an EDM process rather than by grinding.

Electrode growth: A plating action that occurs at certain low-power settings, whereby workpiece material builds up on the electrode, causing an increase in size.

Electrode wear: Amount of material removed from the electrode during the EDM process. This removal can be end wear or corner wear, and is measured linearly or volumetrically but is most often expressed as end wear per cent, measured linearly.

Electro-forming: An electro-plating process used to make metal EDM electrodes.

Energy: Measured in joules, is the equivalent of volt-coulombs or volt-ampere- seconds.

Farad: Unit of electrical capacitance, or the energy-storing capacity of a capacitor.

* *Source:* Hansvedt Industries

Gap: The closest point between the electrode and the workpiece where an electrical discharge will occur. (See *Overcut*)

Gap current: The average amperage flowing across the machining gap.

Gap voltage: The voltage across the gap while current is flowing. The voltage across the electrode/workpiece before current flows is called the open gap voltage. Heat-affected zone. The layer below the recast layer, which has been subjected to elevated temperatures that have altered the properties of the workpiece metal.

Ion: An atom or group of atoms that has lost or gained one or more electrons and is therefore carrying a positive or negative electrical charge, and is described as being ionized.

Ionization: The change in the dielectric fluid that is subjected to a voltage potential whereby it becomes electrically conductive, allowing it to conduct the arc.

Low-wear: An EDM process in which the volume of electrode wear is between 2 and 15 per cent of the volume of workpiece wear. Normal negative polarity wear ratios are 15 to 40 per cent.

Negative electrode: When the electrode voltage potential is negative with respect to the workpiece.

No-wear: An EDM process in which electrode wear is virtually eliminated and the wear ratio is usually less than 2 per cent by volume.

Orbit: A programmable motion between the electrode and the workpiece, produced by a feature built in to the machine, or an accessory, that produces a cavity or hole larger than the electrode. The path can be planetary (circular), vectorial, or polygonal (trace). These motions can often be performed in sequence, and combined with x-axis movement of the electrode.

Overcut: The distance between one side of an electrode and the adjacent wall of the workpiece cavity.

Overcut taper: The difference between the overcut dimensions at the top (entrance) and at the bottom of the cavity.

Plasma: A superheated, highly ionized gas that forms in the discharge channel due to the applied voltage.

Positive electrode: The electrode voltage potential is positive with respect to the workpiece. is the opposite of this condition.

Power parameters: A set of power supply, servo, electrode material, workpiece material, and flushing settings that are selected to produce a desired metal removal rate and surface finish.

Quench: The rapid cooling of the EDM surface by the dielectric fluid, which is partially responsible for metallurgical changes in the recast layer and in the heat- affected zone.

Recast layer: A layer created by the solidification of molten metal on the workpiece surface after it has been melted by the EDM process.

Secondary discharge: A discharge that occurs as conductive particles are carried out along the side of the electrode by the dielectric fluid.

Spark in: A method of locating an electrode with respect to the workpiece, using high frequency, low amperage settings so that there is no cutting action. The electrode is advanced toward the workpiece until contact is indicated and this point is used as the basis for setting up the job.

Spark out: A technique used in orbiting, which moves the electrode in the same path until sparking ceases.

Square wave: An electrical wave shape generated by a solid state power supply.

Stroke: The distance the ram travels under servo control.

UV axis: A mechanism that provides for movement of the upper head of a wire EDM machine to allow inclined surfaces to be generated.

White layer: The surface layer of an EDM cut that is affected by the heat generated during the process. The characteristics of the layer depend on the material, and may be extremely hard martensite or an annealed layer.

Wire EDM: An EDM machine or process in which the electrode is a continuously unspooling, conducting wire that moves in preset patterns in relation to the workpiece.

Wire guide: A replaceable precision round diamond insert, sized to match the wire, that guides the wire at the entrance and exit points of a wire cut.

Wire speed: The rate at which the wire is fed axially through the workpiece (not the rate at which cutting takes place), adjusted so that clean wire is maintained in the cut but slow enough to minimize waste.

The EDM Process.—During the EDM process, energy from the sparks created between the electrode and the workpiece is dissipated by the melting and vaporizing of the workpiece material preferentially, only small amounts of material being lost from the electrode. When current starts to flow between the electrode and the work, the dielectric fluid in the small area in which the gap is smallest, and in which the spark will occur, is transformed into a plasma of hydrogen, carbon, and various oxides. This plasma forms a conducting passageway, consisting of ionized or electrically charged particles, through which the spark can form between the electrode and the workpiece. After current starts to flow, to heat and vaporize a tiny area, the striking voltage is reached, the voltage drops, and the field of ionized particles loses its energy, so that the spark can no longer be sustained. As the voltage then begins to rise again with the increase in resistance, the electrical supply is cut off by the control, causing the plasma to implode and creating a low-pressure pulse that draws in dielectric fluid to flush away metallic debris and cool the impinged area. Such a cycle typically lasts a few microseconds (millionths of a second, or μs), and is repeated continuously in various places on the workpiece as the electrode is moved into the work by the control system.

Flushing: An insulating dielectric fluid is made to flow in the space between the workpiece and the electrode to prevent premature spark discharge, cool the workpiece and the electrode, and flush away the debris. For sinker machines, this fluid is paraffin, kerosene, or a silicon-based dielectric fluid, and for wire machines, the dielectric fluid is usually deionized water. The dielectric fluid can be cooled in a heat exchanger to prevent it from rising above about 100°F, at which cooling efficiency may be reduced. The fluid must also be filtered to remove workpiece particles that would prevent efficient flushing of the spark gaps. Care must be taken to avoid the possibility of entrapment of gases generated by sparking. These gases may explode, causing danger to life, breaking a valuable electrode or workpiece, or causing a fire.

Flushing away of particles generated during the process is vital to successful EDM operations. A secondary consideration is the heat transferred to the side walls of a cavity, which may cause the workpiece material to expand and close in around the electrode, leading to formation of dc arcs where conductive particles are trapped. Flushing can be done by forcing the fluid to pass through the spark gap under pressure, by sucking it through the gap, or by directing a side nozzle to move the fluid in the tank surrounding the workpiece. In pressure flushing, fluid is usually pumped through strategically placed holes in the electrode or in the workpiece. Vacuum flushing is used when side walls must be accurately formed and straight, and is seldom needed on numerically controlled machines because the table can be programmed to move the workpiece sideways.

Flushing needs careful consideration because of the forces involved, especially where fluid is pumped or sucked through narrow passageways, and large hydraulic forces can easily be generated. Excessively high pressures can lead to displacement of the electrode, the workpiece, or both, causing inaccuracy in the finished product. Many low-pressure flushing holes are preferable to a few high-pressure holes. Pressure-relief valves in the system are recommended.

Electronic Controls: The electrical circuit that produces the sparks between the electrode and the workpiece is controlled electronically, the length of the extremely short on and off periods being matched by the operator or the programmer to the materials of the electrode and the workpiece, the dielectric, the rate of flushing, the speed of metal removal, and the quality of surface finish required. The average current flowing between the electrode and the workpiece is shown on an ammeter on the power source, and is the determining factor in machining time for a specific operation. The average spark gap voltage is shown on a voltmeter.

EDM machines can incorporate provision for orbiting the electrode so that flushing is easier, and cutting is faster and increased on one side. Numerical control can also be used to move the workpiece in relation to the electrode with the same results. Numerical control can also be used for checking dimensions and changing electrodes when necessary. The clearance on all sides between the electrode and the workpiece, after the machining operation, is called the overcut or overburn. The overcut becomes greater with increases in the on time, the spark energy, or the amperage applied, but its size is little affected by voltage changes. Allowances must be made for overcut in the dimensioning of electrodes. Sidewall encroachment and secondary discharge can take up parts of these allowances, and electrodes must always be made smaller to avoid making a cavity or hole too large.

Polarity: Polarity can affect processing speed, finish, wear, and stability of the EDM operation. On sinker machines, the electrode is generally, made positive to protect the electrode from excessive wear and preserve its dimensional accuracy. This arrangement removes metal at a slower rate than electrode negative, which is mostly used for high-speed metal removal with graphite electrodes. Negative polarity is also used for machining carbides, titanium, and refractory alloys using metallic electrodes. Metal removal with graphite electrodes can be as much as 50 per cent faster with electrode negative polarity than with electrode positive, but negative polarity results in much faster electrode wear, so it is generally restricted to electrode shapes that can be redressed easily.

Newer generators can provide less than 1 per cent wear with either copper or graphite electrodes during roughing operations. Roughing is typically done with a positive-polarity electrode using elevated on times. Some electrodes, particularly micrograin graphites, have a high resistance to wear. Fine-grain, high-density graphites provide better wear characteristics than coarser, less dense grades, and copper-tungsten resists wear better than pure copper electrodes.

Machine Settings: For vertical machines, a rule of thumb for power selection on graphite and copper electrodes is 50 to 65 amps per square inch of electrode engagement. For example, an electrode that is ½ in. square might use $0.5 \times 0.5 \times 50 = 12.5$ amps. Although each square inch of electrode surface may be able to withstand higher currents, lower settings should be used with very large jobs or the workpiece may become overheated and it may be difficult to clean up the recast layer. Lower amperage settings are required for electrodes that are thin or have sharp details. The voltage applied across the arc gap between the electrode and the workpiece is ideally about 35 volts, but should be as small as possible to maintain stability of the process.

Spark Frequency: Spark frequency is the number of times per second that the current is switched on and off. Higher frequencies are used for finishing operations and for work on cemented carbide, titanium, and copper alloys. The frequency of sparking affects the surface finish produced, low frequencies being used with large spark gaps for rapid metal removal with a rough finish, and higher frequencies with small gaps for finer finishes. High frequency usually increases, and low frequency reduces electrode wear.

The Duty Cycle: Electronic units on modern EDM machines provide extremely close control of each stage in the sparking cycle, down to millionths of a second (μs). A typical EDM cycle might last 100 μs. Of this time, the current might be on for 40 μs and off for 60 μs. The relationship between the lengths of the on and off times is called the duty cycle and it indicates the degree of efficiency of the operation. The duty cycle states the on time as a percentage of the total cycle time and in the previous example it is 40 per cent. Although reducing the off time will increase the duty cycle, factors such as flushing efficiency, electrode and workpiece material, and dielectric condition control the minimum off time. Some EDM units incorporate sensors and fuzzy logic circuits that provide for adaptive control of cutting conditions for unattended operation. Efficiency is also reported as the amount of metal removed, expressed as in.^{3}/hr.

In the EDM process, work is done only during the on time, and the longer the on time, the more material is removed in each sparking cycle. Roughing operations use extended on time for high metal-removal rates, resulting in fewer cycles per second, or lower frequency. The resulting craters are broader and deeper so that the surface is rougher and the heat-affected zone (HAZ) on the workpiece is deeper. With positively charged electrodes, the spark moves from the electrode toward the workpiece and the maximum material is removed from the workpiece. However, every spark takes a minute particle from the electrode so that the electrode also is worn away. Finishing electrodes tend to wear much faster than roughing electrodes because more sparks are generated in unit time.

The part of the cycle needed for reionizing the dielectric (the off time) greatly affects the operating speed. Although increasing the off time slows the process, longer off times can increase stability by providing more time for the ejected material to be swept away by the flow of the dielectric fluid, and for deionization of the fluid, so that erratic cycling of the servo-mechanisms that advance and retract the electrode is avoided. In any vertical EDM operation, if the overcut, wear, and finish are satisfactory, machining speed can best be adjusted by slowly decreasing the off time setting in small increments of 1 to 5 μs until machining becomes erratic, then returning to the previous stable setting. As the off time is decreased, the machining gap or gap voltage will slowly fall and the working current will rise. The gap voltage should not be allowed to drop below 35 to 40 volts.

Metal Removal Rates (MRR): Amounts of metal removed in any EDM process depend largely on the length of the on time, the energy/spark, and the number of sparks/second. The following data were provided by Poco Graphite, Inc., in their *EDM Technical Manual*. For a typical roughing operation using electrode positive polarity on high-carbon steel, a 67 per cent duty cycle removed 0.28 in.^{3}/hr. For the same material, a 50 per cent duty cycle removed 0.15 in.^{3}/hr, and a 33 per cent duty cycle for finishing removed 0.075 in.^{3}/hr.

In another example, shown in the top data row in Table 1, a 40 per cent duty cycle with a frequency of 10 kHz and peak current of 50 amps was run for 5 minutes of cutting time. Metal was removed at the rate of 0.8 in.^{3}/hr with electrode wear of 2.5 per cent and a surface finish of 400 μin. R_a. When the on and off times in this cycle were halved, as shown in the second data row in Table 1, the duty cycle remained at 40 per cent, but the frequency doubled to 20 kHz. The result was that the peak current remained unaltered, but with only half the on time the MRR was reduced to 0.7 in.^{3}/hr, the electrode wear increased to 6.3 per cent, and the surface finish improved to 300 μin. R_a. The third and fourth rows in Table 1 show other variations in the basic cycle and the results.

Table 1. Effect of Electrical Control Adjustments on EDM Operations

On Time (μs)	Off Time (μs)	Frequency (kHz)	Peak Current (Amps)	Metal Removal Rate (in.^{3}/hr)	Electrode Wear (%)	Surface Finish (μ in. R_a)
40	60	10	50	0.08	2.5	400
20	30	20	50	0.7	6.3	300
40	10	20	50	1.2	1.4	430
40	60	10	25	0.28	2.5	350

The Recast Layer: One drawback of the EDM process when used for steel is the recast layer, which is created wherever sparking occurs. The oil used as a dielectric fluid causes the EDM operation to become a random heat-treatment process in which the metal surface is heated to a very high temperature, then quenched in oil. The heat breaks down the oil into hydrocarbons, tars, and resins, and the molten metal draws out the carbon atoms and traps them in the resolidified metal to form the very thin, hard, brittle surface called the recast layer that covers the heat-affected zone (HAZ). This recast layer has a white appearance and consists of particles of material that have been melted by the sparks, enriched with carbon, and drawn back to the surface or retained by surface tension. The recast layer is harder than the parent metal and can be as hard as glass, and must be reduced or removed by vapor blasting with glass beads, polishing, electrochemical or abrasive flow machining, after the shaping process is completed, to avoid cracking or flaking of surface layers that may cause failure of the part in service.

Beneath the thin recast layer, the HAZ, in steel, consists of martensite that usually has been hardened by the heating and cooling sequences coupled with the heat-sink cooling effect of a thick steel workpiece. This martensite is hard and its rates of expansion and contraction are different from those of the parent metal. If the workpiece is subjected to heating and cooling cycles in use, the two layers are constantly stressed and these stresses may cause formation of surface cracks. The HAZ is usually much deeper in a workpiece cut on a sinker than on a wire machine, especially after roughing, because of the increased heating effect caused by the higher amounts of energy applied.

The depth of the HAZ depends on the amperage and the length of the on time, increasing as these values increase, to about 0.012 to 0.015 in. deep. Residual stress in the HAZ can range up to 650 N/mm^2. The HAZ cannot be removed easily, so it is best avoided by programming the series of cuts taken on the machine so that most of the HAZ produced by one cut is removed by the following cut. If time is available, cut depth can be reduced gradually until the finishing cuts produce an HAZ having a thickness of less than 0.0001 in.

Workpiece Materials.—Most homogeneous materials used in metalworking can be shaped by the EDM process. Some data on typical workpiece materials are given in Table 2. Sintered materials present some difficulties caused by the use of a cobalt or other binder used to hold the carbide or other particles in the matrix. The binder usually melts at a lower temperature than the tungsten, molybdenum, titanium, or other carbides, so it is preferentially removed by the sparking sequence and the carbide particles are thus loosened and freed from the matrix. The structures of sintered materials based on tungsten, cobalt, and molybdenum require higher EDM frequencies with very short on times, so that there is less danger of excessive heat buildup, leading to melting. Copper-tungsten electrodes are recommended for EDM of tungsten carbides. When used with high frequencies for powdered metals, graphite electrodes often suffer from excessive wear.

Workpieces of aluminum, brass, and copper should be processed with metallic electrodes of low melting points such as copper or copper-tungsten. Workpieces of carbon and stainless steel that have high melting points should be processed with graphite electrodes.

Table 2. Characteristics of Common Workpiece Materials for EDM

Material	Specific Gravity	Melting Point		Vaporization Temperature		Conductivity (Silver = 100)
		°F	°C	°F	°C	
Aluminum	2.70	1220	660	4442	2450	63.00
Brass	8.40	1710	930	…		…
Cobalt	8.71	2696	1480	5520	2900	16.93
Copper	8.89	1980	1082	4710	2595	97.61
Graphite	2.07	N/A		6330	3500	70.00
Inconel	…	2350	1285	…		…
Magnesium	1.83	1202	650	2025	1110	39.40
Manganese	7.30	2300	1260	3870	2150	15.75
Molybdenum	10.20	4748	2620	10,040	5560	17.60
Nickel	8.80	2651	1455	4900	2730	12.89
Carbon Steel	7.80	2500	1371	…		12.00
Tool Steel	…	2730	1500	…		…
Stainless Steel	…	2750	1510	…		…
Titanium	4.50	3200	1700	5900	3260	13.73
Tungsten	18.85	6098	3370	10,670	5930	14.00
Zinc	6.40	790	420	1663	906	26.00

The melting points and specific gravities of the electrode material and of the workpiece should preferably be similar.

Electrode Materials.—Most EDM electrodes are made from graphite, which provides a much superior rate of metal removal than copper because of the ability of graphite to resist thermal damage. Graphite has a density of 1.55 to 1.85 g/cm^3, lower than most metals. Instead of melting when heated, graphite sublimates, that is, it changes directly from a solid to a gas without passing through the liquid stage. Sublimation of graphite occurs at a temperature of 3350°C (6062°F). EDM graphite is made by sintering a compressed mixture of fine graphite powder (1 to 100 micron particle size) and coal tar pitch in a furnace. The open structure of graphite means that it is eroded more rapidly than metal in the EDM process. The electrode surface is also reproduced on the surface of the workpiece. The sizes of individual surface recesses may be reduced during sparking when the work is moved under numerical control of workpiece table movements.

The fine grain sizes and high densities of graphite materials that are specially made for high-quality EDM finishing provide high wear resistance, better finish, and good reproduction of fine details, but these fine grades cost more than graphite of larger grain sizes and lower densities. Premium grades of graphite cost up to five times as much as the least expensive and about three times as much as copper, but the extra cost often can be justified by savings during machining or shaping of the electrode.

Graphite has a high resistance to heat and wear at lower frequencies, but will wear more rapidly when used with high frequencies or with negative polarity. Infiltrated graphites for EDM electrodes are also available as a mixture of copper particles in a graphite matrix, for applications where good machinability of the electrode is required. This material presents a trade-off between lower arcing and greater wear with a slower metal-removal rate, but costs more than plain graphite.

EDM electrodes are also made from copper, tungsten, silver-tungsten, brass, and zinc, which all have good electrical and thermal conductivity. However, all these metals have melting points below those encountered in the spark gap, so they wear rapidly. Copper with 5 per cent tellurium, added for better machining properties, is the most commonly used metal alloy. Tungsten resists wear better than brass or copper and is more rigid when used for thin electrodes but is expensive and difficult to machine. Metal electrodes, with their more even surfaces and slower wear rates, are often preferred for finishing operations on work that requires a smooth finish. In fine-finishing operations, the arc gap between the surfaces of the electrode and the workpiece is very small and there is a danger of dc arcs being struck, causing pitting of the surface. This pitting is caused when particles dislodged from a graphite electrode during fine-finishing cuts are not flushed from the gap. If struck by a spark, such a particle may provide a path for a continuous discharge of current that will mar the almost completed work surface.

Some combinations of electrode and workpiece material, electrode polarity, and likely amounts of corner wear are listed in Table 3. Corner wear rates indicate the ability of the electrode to maintain its shape and reproduce fine detail. The column headed Capacitance refers to the use of capacitors in the control circuits to increase the impact of the spark without increasing the amperage. Such circuits can accomplish more work in a given time, at the expense of surface-finish quality and increased electrode wear.

Table 3. Types of Electrodes Used for Various Workpiece Materials

Electrode	Electrode Polarity	Workpiece Material	Corner Wear (%)	Capacitance
Copper	+	Steel	2–10	No
Copper	+	Inconel	2–10	No
Copper	+	Aluminum	<3	No
Copper	−	Titanium	20–40	Yes
Copper	−	Carbide	35–60	Yes
Copper	−	Copper	34–45	Yes
Copper	−	Copper-tungsten	40–60	Yes
Copper-tungsten	+	Steel	1–10	No
Copper-tungsten	−	Copper	20–40	Yes
Copper-tungsten	−	Copper-tungsten	30–50	Yes
Copper-tungsten	−	Titanium	15–25	Yes
Copper-tungsten	−	Carbide	35–50	Yes
Graphite	+	Steel	<1	No
Graphite	−	Steel	30–40	No
Graphite	+	Inconel	<1	No
Graphite	−	Inconel	30–40	No
Graphite	+	Aluminum	<1	No
Graphite	−	Aluminum	10–20	No
Graphite	−	Titanium	40–70	No
Graphite	−	Copper	N/A	Yes

Electrode Wear: Wear of electrodes can be reduced by leaving the smallest amounts of finishing stock possible on the workpiece and using no-wear or low-wear settings to remove most of the remaining material so that only a thin layer remains for finishing with the redressed electrode. The material left for removal in the finishing step should be only slightly more than the maximum depth of the craters left by the previous cut. Finishing operations should be regarded as only changing the quality of the finish, not removing metal or sizing. Low power with very high frequencies and minimal amounts of offset for each finishing cut are recommended.

On manually adjusted machines, fine finishing is usually carried out by several passes of a full-size finishing electrode. Removal of a few thousandths of an inch from a cavity with such an arrangement requires the leading edge of the electrode to recut the cavity over the entire vertical depth. By the time the electrode has been sunk to full depth, it is so worn that precision is lost. This problem sometimes can be avoided on a manual machine by use of an orbiting attachment that will cause the electrode to traverse the cavity walls, providing improved speed, finish, and flushing, and reducing corner wear on the electrode.

Selection of Electrode Material: Factors that affect selection of electrode material include metal-removal rate, wear resistance (including volumetric, corner, end, and side, with corner wear being the greatest concern), desired surface finish, costs of electrode manufacture and material, and characteristics of the material to be machined. A major factor is the ability of the electrode material to resist thermal damage, but the electrode's density, the polarity, and the frequencies used are all important factors in wear rates. Copper melts at about 1085°C (1985°F) and spark-gap temperatures must generally exceed 3800°C (6872°F), so use of copper may be made unacceptable because of its rapid wear rates. Graphites have good resistance to heat and wear at low frequencies, but will wear more with high frequency, negative polarity, or a combination of these.

Making Electrodes.—Electrodes made from copper and its alloys can be machined conventionally by lathes, and milling and grinding machines, but copper acquires a burr on run-off edges during turning and milling operations. For grinding copper, the wheel must often be charged with beeswax or similar material to prevent loading of the surface. Flat grinding of copper is done with wheels having open grain structures (46-J, for instance) to contain the wax and to allow room for the soft, gummy, copper chips. For finish grinding, wheels of at least 60 and up to 80 grit should be used for electrodes requiring sharp corners and fine detail. These wheels will cut hot and load up much faster, but are necessary to avoid rapid breakdown of sharp corners.

Factors to be considered in selection of electrode materials are: the electrode material cost cost/in^3; the time to manufacture electrodes; difficulty of flushing; the number of electrodes needed to complete the job; speed of the EDM; amount of electrode wear during EDM; and workpiece surface-finish requirements.

Copper electrodes have the advantage over graphite in their ability to be discharge-dressed in the EDM, usually under computer numerical control (CNC). The worn electrode is engaged with a premachined dressing block made from copper-tungsten or carbide. The process renews the original electrode shape, and can provide sharp, burr-free edges. Because of its higher vaporization temperature and wear resistance, discharge dressing of graphite is slow, but graphite has the advantage that it can be machined conventionally with ease.

Machining Graphite: Graphites used for EDM are very abrasive, so carbide tools are required for machining them. The graphite does not shear away and flow across the face of the tool as metal does, but fractures or is crushed by the tool pressure and floats away as a fine powder or dust. Graphite particles have sharp edges and, if allowed to mix with the machine lubricant, will form an abrasive slurry that will cause rapid wear of machine guiding surfaces. The dust may also cause respiratory problems and allergic reactions, especially if the graphite is infiltrated with copper, so an efficient exhaust system is needed for machining.

Compressed air can be used to flush out the graphite dust from blind holes, for instance, but provision must be made for vacuum removal of the dust to avoid hazards to health and problems with wear caused by the hard, sharp-edged particles. Air velocities of at least 500 ft/min are recommended for flushing, and of 2000 ft/min in collector ducts to prevent settling out. Fluids can also be used, but small-pore filters are needed to keep the fluid clean. High-strength graphite can be clamped or chucked tightly but care must be taken to avoid crushing. Collets are preferred for turning because of the uniform pressure they apply to the workpiece. Sharp corners on electrodes made from less dense graphite are liable to chip or break away during machining.

For conventional machining of graphite, tools of high-quality tungsten carbide or polycrystaline diamond are preferred and must be kept sharp. Recommended cutting speeds for high-speed steel tools are 100 to 300, tungsten carbide 500 to 750, and polycrystaline diamond, 500 to 2000 surface ft/min. Tools for turning should have positive rake angles and nose radii of $\frac{1}{64}$ to $\frac{1}{32}$ in. Depths of cut of 0.015 to 0.020 in. produce a better finish than light cuts such as 0.005 in. because of the tendency of graphite to chip away rather than flow across the tool face. Low feed rates of 0.005 in./rev for rough- and 0.001 to 0.003 in./rev for finish-turning are preferred. Cutting off is best done with a tool having an angle of 20°.

For bandsawing graphite, standard carbon steel blades can be run at 2100 to 3100 surface ft/min. Low power feed rates should be used to avoid overloading the teeth and the feed rate should be adjusted until the saw has a very slight speed up at the point of breakthrough. Milling operations require rigid machines, short tool extensions, and firm clamping of parts. Milling cutters will chip the exit side of the cut, but chipping can be reduced by use of sharp tools, positive rake angles, and low feed rates to reduce tool pressure. Feed/tooth for two-flute end mills is 0.003 to 0.005 in. for roughing and 0.001 to 0.003 in. for finishing.

Standard high-speed steel drills can be used for drilling holes but will wear rapidly, causing holes that are tapered or undersized, or both. High-spiral, tungsten carbide drills should be used for large numbers of holes over $\frac{1}{16}$ in. diameter, but diamond-tipped drills will last longer. Pecking cycles should be used to clear dust from the holes. Compressed air can be passed through drills with through coolant holes to clear dust. Feed rates for drilling are 0.0015 to 0.002 in./rev for drills up to $\frac{1}{32}$, 0.001 to 0.003 in./rev for $\frac{1}{32}$- to $\frac{1}{8}$-in. drills, and 0.002 to 0.005 in./rev for larger drills. Standard taps without fluid are best used for through holes, and for blind holes, tapping should be completed as far as possible with a taper tap before the bottoming tap is used.

For surface grinding of graphite, a medium (60) grade, medium-open structure, vitreous-bond, green-grit, silicon-carbide wheel is most commonly used. The wheel speed should be 5300 to 6000 surface ft/min, with traversing feed rates at about 56 ft/min. Roughing cuts are taken at 0.005 to 0.010 in./pass, and finishing cuts at 0.001 to 0.003 in./pass. Surface finishes in the range of 18 to 32 μin. R_a are normal, and can be improved by longer spark-out times and finer grit wheels, or by lapping. Graphite can be centerless ground using a silicon-carbide, resinoid-bond work wheel and a regulating wheel speed of 195 ft/min.

Wire EDM, orbital abrading, and ultrasonic machining are also used to shape graphite electrodes. Orbital abrading uses a die containing hard particles to remove graphite, and can produce a fine surface finish. In ultrasonic machining, a water-based abrasive slurry is pumped between the die attached to the ultrasonic transducer and the graphite workpiece on the machine table. Ultrasonic machining is rapid and can reproduce small details down to 0.002 in. in size, with surface finishes down to 8 μin. R_a. If coolants are used, the graphite should be dried for 1 hour at over 400°F (but not in a microwave oven) to remove liquids before used.

Wire EDM.—In the wire EDM process, with deionized water as the dielectric fluid, carbon is extracted from the recast layer, rather than added to it. When copper-base wire is used, copper atoms migrate into the recast layer, softening the surface slightly so that wire-cut surfaces are sometimes softer than the parent metal. On wire EDM machines, very high amperages are used with very short on times, so that the heat-affected zone (HAZ) is quite shallow. With proper adjustment of the on and off times, the depth of the HAZ can be held below 1 micron (0.00004 in.).

The cutting wire is used only once, so that the portion in the cut is always cylindrical and has no spark-eroded sections that might affect the cut accuracy. The power source controls the electrical supply to the wire and to the drive motors on the table to maintain the preset arc gap within 0.1 micron (0.000004 in.) of the programmed position. On wire EDM machines, the water used as a dielectric fluid is deionized by a deionizer included in the cooling system, to improve its properties as an insulator. Chemical balance of the water is also important for good dielectric properties.

Drilling Holes for Wire EDM: Before an aperture can be cut in a die plate, a hole must be provided in the workpiece. Such holes are often "drilled" by EDM, and the wire threaded through the workpiece before starting the cut. The "EDM drill" does not need to be rotated, but rotation will help in flushing and reduce electrode wear. The EDM process can drill a hole 0.04 in. in diameter through 4-in. thick steel in about 3 minutes, using an electrode made from brass or copper tubing. Holes of smaller diameter can be drilled, but the practical limit is 0.012 in. because of the overcut, the lack of rigidity of tubing in small sizes, and the excessive wear on such small electrodes. The practical upper size limit on holes is about 0.12 in. because of the comparatively large amounts of material that must be eroded away for larger sizes. However, EDM is commonly used for making large or deep holes in such hard materials as tungsten carbide. For instance, a 0.2-in. hole has been made in carbide 2.9 in. thick in 49 minutes by EDM. Blind holes are difficult to produce with accuracy, and must often be made with cut-and-try methods.

Deionized water is usually used for drilling and is directed through the axial hole in the tubular electrode to flush away the debris created by the sparking sequence. Because of the need to keep the extremely small cutting area clear of metal particles, the dielectric fluid is often not filtered but is replaced continuously by clean fluid that is pumped from a supply tank to a disposal tank on the machine.

Wire Electrodes: Wire for EDM generally is made from yellow brass containing copper 63 and zinc 37 per cent, with a tensile strength of 50,000 to 145,000 $lb_f/in.^2$, and may be from 0.002 to 0.012 in. diameter.

In addition to yellow brass, electrode wires are also made from brass alloyed with aluminum or titanium for tensile strengths of 140,000 to 160,000 $lb_f/in.^2$. Wires with homogeneous, uniform electrolytic coatings of alloys such as brass or zinc are also used. Zinc is favored as a coating on brass wires because it gives faster cutting and reduced wire breakage due to its low melting temperature of 419°C, and vaporization temperature of 906°C. The layer of zinc can boil off while the brass core, which melts at 930°C, continues to deliver current.

Some wires for EDM are made from steel for strength, with a coating of brass, copper, or other metal. Most wire machines use wire negative polarity (the wire is negative) because the wire is constantly renewed and is used only once, so wear is not important. Important qualities of wire for EDM include smooth surfaces, free from nicks, scratches and cracks, precise diameters to ±0.00004 in. for drawn and ±0.00006 in. for plated, high tensile strength, consistently good ductility, uniform spooling, and good protective packaging.

IRON AND STEEL CASTINGS

Cast irons and cast steels encompass a large family of ferrous alloys, which, as the name implies, are cast to shape rather than being formed by working in the solid state. In general, cast irons contain more than 2 per cent carbon and from 1 to 3 per cent silicon. Varying the balance between carbon and silicon, alloying with different elements, and changing melting, casting, and heat-treating practices can produce a broad range of properties. In most cases, the carbon exists in two forms: free carbon in the form of graphite and combined carbon in the form of iron carbide (cementite). Mechanical and physical properties depend strongly on the shape and distribution of the free graphite and the type of matrix surrounding the graphite particles.

The four basic types of cast iron are white iron, gray iron, malleable iron, and ductile iron. In addition to these basic types, there are other specific forms of cast iron to which special names have been applied, such as chilled iron, alloy iron, and compacted graphite cast iron.

Gray Cast Iron.—Gray cast iron may easily be cast into any desirable form and it may also be machined readily. It usually contains from 1.7 to 4.5 per cent carbon, and from 1 to 3 per cent silicon. The excess carbon is in the form of graphite flakes and these flakes impart to the material the dark-colored fracture which gives it its name. Gray iron castings are widely used for such applications as machine tools, automotive cylinder blocks, cast-iron pipe and fittings and agricultural implements.

The American National Standard Specifications for Gray Iron Castings—ANSI/ASTM A48-76 groups the castings into two categories. Gray iron castings in Classes 20A, 20B, 20C, 25A, 25B, 25C, 30A, 30B, 30C, 35A, 35B, and 35C are characterized by excellent machinability, high damping capacity, low modulus of elasticity, and comparative ease of manufacture. Castings in Classes 40B, 40C, 45B, 45C, 50B, 50C, 60B, and 60C are usually more difficult to machine, have lower damping capacity, a higher modulus of elasticity, and are more difficult to manufacture. The prefix number is indicative of the minimum tensile strength in pounds per square inch, i.e., 20 is 20,000 psi, 25 is 25,000 psi, 30 is 30,000 psi, etc.

High-strength iron castings produced by the Meehanite-controlled process may have various combinations of physical properties to meet different requirements. In addition to a number of general engineering types, there are heat-resisting, wear-resisting and corrosion-resisting Meehanite castings.

White Cast Iron.—When nearly all of the carbon in a casting is in the combined or cementite form, it is known as white cast iron. It is so named because it has a silvery-white fracture. White cast iron is very hard and also brittle; its ductility is practically zero. Castings of this material need particular attention with respect to design since sharp corners and thin sections result in material failures at the foundry. These castings are less resistant to impact loading than gray iron castings, but they have a compressive strength that is usually higher than 200,000 pounds per square inch as compared to 65,000 to 160,000 pounds per square inch for gray iron castings. Some white iron castings are used for applications that require maximum wear resistance but most of them are used in the production of malleable iron castings.

Chilled Cast Iron.—Many gray iron castings have wear-resisting surfaces of white cast iron. These surfaces are designated by the term "chilled cast iron" since they are produced in molds having metal chills for cooling the molten metal rapidly. This rapid cooling results in the formation of cementite and white cast iron.

Alloy Cast Iron.—This term designates castings containing alloying elements such as nickel, chromium, molybdenum, copper, and manganese in sufficient amounts to appreciably change the physical properties. These elements may be added either to increase the strength or to obtain special properties such as higher wear resistance, corrosion resistance, or heat resistance. Alloy cast irons are used extensively for such parts as automotive cylinders, pistons, piston rings, crankcases, brake drums; for certain machine tool castings, for certain types of dies, for parts of crushing and grinding machinery, and for application where the casting must resist scaling at high temperatures. Machinable alloy cast irons having tensile strengths up to 70,000 pounds per square inch or even higher may be produced.

Malleable-iron Castings.—Malleable iron is produced by the annealing or graphitization of white iron castings. The graphitization in this case produces temper carbon which is graphite in the form of compact rounded aggregates. Malleable castings are used for many industrial applications where strength, ductility, machinability, and resistance to shock are important factors. In manufacturing these castings, the usual procedure is to first produce a hard, brittle white iron from a charge of pig iron and scrap. These hard white-iron castings are then placed in stationary batch-type furnaces or car-bottom furnaces and the graphitization (malleablizing) of the castings is accomplished by means of a suitable annealing heat treatment. During this annealing period the temperature is slowly (50 hours) increased to as much as 1650 or 1700 degrees F, after which time it is slowly (60 hours) cooled. The American National Standard Specifications for Malleable Iron Castings—ANSI/ASTM A47-77 specifies the following grades and their properties: No. 32520, having a minimum tensile strength of 50,000 pounds per square inch, a minimum yield strength of 32,500 psi., and a minimum elongation in 2 inches of 10 per cent; and No. 35018, having a minimum tensile strength of 53,000 psi., a minimum yield strength of 35,000 psi., and a minimum elongation in 2 inches of 18 per cent.

Cupola Malleable Iron: Another method of producing malleable iron involves initially the use of a cupola or a cupola in conjunction with an air furnace. This type of malleable iron, called cupola malleable iron, exhibits good fluidity and will produce sound castings. It is used in the making of pipe fittings, valves, and similar parts and possesses the useful property of being well suited to galvanizing. The American National Standard Specifications for Cupola Malleable Iron — ANSI/ASTM 197-79 calls for a minimum tensile strength of 40,000 pounds per square inch; a minimum yield strength of 30.000 psi.; and a minimum elongation in 2 inches of 5 per cent.

Pearlitic Malleable Iron: This type of malleable iron contains some combined carbon in various forms. It may be produced either by stopping the heat treatment of regular malleable iron during production before the combined carbon contained therein has all been transformed to graphite or by reheating regular malleable iron above the transformation range. Pearlitic malleable irons exhibit a wide range of properties and are used in place of steel castings or forgings or to replace malleable iron when a greater strength or wear resistance is required. Some forms are made rigid to resist deformation while others will undergo considerable deformation before breaking. This material has been used in axle housings, differential housings, camshafts, and crankshafts for automobiles; machine parts; ordnance equipment; and tools. Tension test requirements of pearlitic malleable iron castings called for in ASTM Specification A 220–79 are given in the accompanying table.

Tension Test Requirements of Pearlitic Malleable Iron Castings *ASTM A 220-79*

Casting Grade Numbers		40010	45008	45006	50005	60004	70003	80002	90001
Min. Tensile Strength	1000s Lbs. per Sq. In.	60	65	65	70	80	85	95	105
Min. Yield Strength		40	45	45	50	60	70	80	90
Min. Elong. in 2 In., Per Cent		10	8	6	5	4	3	2	1

Ductile Cast Iron.—A distinguishing feature of this widely used type of cast iron, also known as spheroidal graphite iron or nodular iron, is that the graphite is present in ball-like form instead of in flakes as in ordinary gray cast iron. The addition of small amounts of magnesium- or cerium-bearing alloys together with special processing produces this spheroidal graphite structure and results in a casting of high strength and appreciable ductility. Its toughness is intermediate between that of cast iron and steel, and its shock resistance is comparable to ordinary grades of mild carbon steel. Melting point and fluidity are similar to those of the high-carbon cast irons. It exhibits good pressure tightness under high stress and can be welded and brazed. It can be softened by annealing or hardened by normalizing and air cooling or oil quenching and drawing.

Five grades of this iron are specified in ASTM A 536-80—Standard Specification for Ductile Iron Castings. The grades and their corresponding matrix microstructures and heat treatments are as follows: Grade 60-40-18, ferritic, may be annealed; Grade 65-45-12, mostly ferritic, as-cast or annealed; Grade 80-55-06, ferritic/pearlitic, as-cast; Grade 100-70-03, mostly pearlitic, may be normalized; Grade 120-90-02, martensitic, oil quenched and tempered. The grade nomenclature identifies the minimum tensile strength, on per cent yield strength, and per cent elongation in 2 inches. Thus, Grade 60–40–18 has a minimum tensile strength of 60,000 psi, a minimum 0.2 per cent yield strength of 40,000 psi, and minimum elongation in 2 inches of 18 per cent. Several other types are commercially available to meet specific needs. The common grades of ductile iron can also be specified by only Brinell hardness, although the appropriate microstructure for the indicated hardness is also a requirement. This method is used in SAE Specification J434C for automotive castings and similar applications. Other specifications not only specify tensile properties, but also have limitations in composition. Austenitic types with high nickel content, high corrosion resistance, and good strength at elevated temperatures, are specified in ASTM A439-80.

Ductile cast iron can be cast in molds containing metal chills if wear-resisting surfaces are desired. Hard carbide areas will form in a manner similar to the forming of areas of chilled cast iron in gray iron castings. Surface hardening by flame or induction methods is also feasible. Ductile cast iron can be machined with the same ease as gray cast iron. It finds use as crankshafts, pistons, and cylinder heads in the automotive industry; forging hammer anvils, cylinders, guides, and control levers in the heavy machinery field; and wrenches, clamp frames, face-plates, chuck bodies, and dies for forming metals in the tool and die field. The production of ductile iron castings involves complex metallurgy, the use of special melting stock, and close process control. The majority of applications of ductile iron have been made to utilize its excellent mechanical properties in combination with the castability, machinability, and corrosion resistance of gray iron.

Steel Castings.—Steel castings are especially adapted for machine parts that must withstand shocks or heavy loads. They are stronger than either wrought iron, cast iron, or malleable iron and are very tough. The steel used for making steel castings may be produced either by the open-hearth, electric arc, side-blow converter, or electric induction methods. The raw materials used are steel scrap, pig iron, and iron ore, the materials and their proportions varying according to the process and the type of furnace used. The open-hearth method is used when large tonnages are continually required while a small electric furnace might be used for steels of widely differing analyses, which are required in small lot production. The high frequency induction furnace is used for small quantity production of expensive steels of special composition such as high-alloy steels. Steel castings are used for such parts as hydroelectric turbine wheels, forging presses, gears, railroad car frames, valve bodies, pump casings, mining machinery, marine equipment, engine casings, etc.

Steel castings can generally be made from any of the many types of carbon and alloy steels produced in wrought form and respond similarly to heat treatment; they also do not exhibit directionality effects that are typical of wrought steel. Steel castings are classified into two general groups: carbon steel and alloy steel.

Table 1. Mechanical Properties of Steel Casings
For general information only. Not for use as design or specification limit values.

Tensile Strength, Lbs. per Sq. In.	Yield Point, Lbs. per Sq. In.	Elongation in 2 In., Per Cent	Brinell Hardness Number	Type of Heat Treatment	Application Indicating Properties
Structural Grades of Carbon Steel Castings					
60,000	30,000	32	120	Annealed	Low electric resistivity. Desirable magnetic properties. Carburizing and case hardening grades. Weldability.
65,000	35,000	30	130	Normalized	Good weldability. Medium strength with good machinability and high ductility.
70,000	38,000	28	140	Normalized	
80,000	45,000	26	160	Normalized and tempered	High strength carbon steels with good machinability, toughness and good fatigue resistance.
85,000	50,000	24	175	Normalized and tempered	
100,000	70,000	20	200	Quenched and tempered	Wear resistance. Hardness.
Engineering Grades of Low Alloy Steel Castings					
70,000	45,000	26	150	Normalized and tempered	Good weldability. Medium strength with high toughness and good machinability. For high temperature service.
80,000	50,000	24	170	Normalized and tempered	
90,000	60,000	22	190	Normalized and tempered[a]	Certain steels of these classes have good high temperature properties and deep hardening properties. Toughness.
100,000	68,000	20	209	Normalized and tempered[a]	
110,000	85,000	20	235	Quenched and tempered	Impact resistance. Good low temperature properties for certain steels. Deep hardening. Good combination of strength and toughness.
120,000	95,000	16	245	Quenched and tempered	
150,000	125,000	12	300	Quenched and tempered	Deep hardening. High strength. Wear and fatigue resistance.
175,000	148,000	8	340	Quenched and tempered	High strength and hardness. Wear resistance. High fatigue resistance.
200,000	170,000	5	400	Quenched and tempered	

[a] Quench and temper heat treatments may also be employed for these classes.

The values listed above have been compiled by the Steel Founders' Society of America as those normally expected in the production of steel castings. The castings are classified according to tensile strength values which are given in the first column. Specifications covering steel castings are prepared by the American Society for Testing and Materials, the Association of American Railroads, the Society of Automotive Engineers, the United States Government (Federal and Military Specifications), etc. These specifications appear in publications issued by these organizations.

Carbon Steel Castings.—Carbon steel castings may be designated as low-carbon medium-carbon, and high-carbon. Low-carbon steel castings have a carbon content of less than 0.20 per cent (most are produced in the 0.16 to 0.19 per cent range). Other elements present are: manganese, 0.50 to 0.85 per cent; silicon, 0.25 to 0.70 per cent; phosphorus, 0.05 per cent max.; and sulfur, 0.06 per cent max. Their tensile strengths (annealed condition) range from 40,000 to 70,000 pounds per square inch. Medium-carbon steel castings have a carbon content of from 0.20 to 0.50 per cent. Other elements present are: manganese, 0.50 to 1.00 per cent; silicon, 0.20 to 0.80 per cent; phosphorus, 0.05 per cent max.; and sulfur, 0.06 per cent max. Their tensile strengths range from 65,000 to 105,000 pounds per square inch depending, in part, upon heat treatment. High-carbon steel castings have a carbon content of more than 0.50 per cent and also contain: manganese, 0.50 to 1.00 per cent; silicon, 0.20 to 0.70 per cent; and phosphorus and sulfur, 0.05 per cent max. each. Fully annealed high-carbon steel castings exhibit tensile strengths of from 95,000 to 125,000 pounds per square inch. See Table 1 for grades and properties of carbon steel castings.

Table 2. Nominal Chemical Composition and Mechanical Properties of Heat-Resistant Steel Castings *ASTM A297-81*

Grade	Nominal Chemical Composition, Per Cent[a]	Tensile Strength, min		0.2 Per Cent Yield Strength, min		Per Cent Elongation in 2 in., or 50 mm, min.
		ksi	MPa	ksi	MPa	
HF	19 Chromium, 9 Nickel	70	485	35	240	25
HH	25 Chromium, 12 Nickel	75	515	35	240	10
HI	28 Chromium, 15 Nickel	70	485	35	240	10
HK	25 Chromium, 20 Nickel	65	450	35	240	10
HE	29 Chromium, 9 Nickel	85	585	40	275	9
HT	15 Chromium, 35 Nickel	65	450	…	…	4
HU	19 Chromium, 39 Nickel	65	450	…	…	4
HW	12 Chromium, 60 Nickel	60	415	…	…	…
HX	17 Chromium, 66 Nickel	60	415	…	…	…
HC	28 Chromium	55	380	…	…	…
HD	28 Chromium, 5 Nickel	75	515	35	240	8
HL	29 Chromium, 20 Nickel	65	450	35	240	10
HN	20 Chromium, 25 Nickel	63	435	…	…	8
HP	26 Chromium, 35 Nickel	62.5	430	34	235	4.5

[a] Remainder is iron.

ksi = kips per square inch = 1000s of pounds per square inch; MPa = megapascals.

Alloy Steel Castings.—Alloy cast steels are those in which special alloying elements such as manganese, chromium, nickel, molybdenum, vanadium have been added in sufficient quantities to obtain or increase certain desirable properties. Alloy cast steels are comprised of two groups—the low-alloy steels with their alloy content totaling less than 8 per cent and the high-alloy steels with their alloy content totaling 8 per cent or more. The addition of these various alloying elements in conjunction with suitable heat-treatments, makes it possible to secure steel castings having a wide range of properties. The three accompanying tables give information on these steels. The lower portion of Table 1 gives the engineering grades of low-alloy cast steels grouped according to tensile strengths and gives properties normally expected in the production of steel castings. Tables 2 and 3 give the standard designations and nominal chemical composition ranges of high-alloy castings which may be classified according to heat or corrosion resistance. The grades given in these tables are recognized in whole or in part by the Alloy Casting Institute (ACI), the American Society for Testing and Materials (ASTM), and the Society of Automotive Engineers (SAE).

Table 3. Nominal Chemical Composition and Mechanical Properties of Corrosion-Resistant Steel Castings *ASTM A743-81a*

Grade	Nominal Chemical Composition, Per Cent[a]	Tensile Strength, min		0.2% Yield Strength, min		Per Cent Elongation in 2 in., or 50 mm, min	Per Cent Reduction of Area, min
		ksi	MPa	ksi	MPa		
CF-8	9 Chromium, 9 Nickel	70[b]	485[b]	30[b]	205[b]	35	…
CG-12	22 Chromium, 12 Nickel	70	485	28	195	35	…
CF-20	19 Chromium, 9 Nickel	70	485	30	205	30	…
CF-8M	19 Chromium, 10 Nickel, with Molybdenum	70	485	30	205	30	…
CF-8C	19 Chromium, 10 Nickel with Niobium	70	485	30	205	30	…
CF-16 and CF-16Fa	19 Chromium, 9 Nickel, Free Machining	70	485	30	205	25	…
CH-20 and CH-10	25 Chromium, 12 Nickel	70	485	30	205	30	…
CK-20	25 Chromium, 20 Nickel	65	450	28	195	30	…
CE-30	29 Chromium, 9 Nickel	80	550	40	275	10	…
CA-15 and CA-15M	12 Chromium	90	620	65	450	18	30
CB-30	20 Chromium	65	450	30	205	…	…
CC-50	28 Chromium	55	380	…	…	…	…
CA-40	12 Chromium	100	690	70	485	15	25
CF-3	19 Chromium, 9 Nickel	70	485	30	205	35	…
CF-3M	19 Chromium, 10 Nickel, with Molybdenum	70	485	30	205	30	…
CG6MMN	Chromium-Nickel-Manganese -Molybdenum	75	515	35	240	30	…
CG-8M	19 Chromium, 11 Nickel, with Molybdenum	75	520	35	240	25	…
CN-7M	20 Chromium, 29 Nickel, with Copper and Molybdenum	62	425	25	170	35	…
CN-7MS	19 Chromium, 24 Nickel, with Copper and Molybdenum	70	485	30	205	35	…
CW-12M	Nickel, Molybdenum, Chromium	72	495	46	315	4.0	…
CY-40	Nickel, Chromium, Iron	70	485	28	195	30	…
CZ-100	Nickel Alloy	50	345	18	125	10	…
M-35-1	Nickel-Copper Alloy	65	450	25	170	25	…
M-35-2	Nickel-Copper Alloy	65	450	30	205	25	…
CA-6NM	12 Chromium, 4 Nickel	110	755	80	550	15	35
CD-4MCu	25 Chromium, 5 Nickel, 2 Molybdenum, 3 Copper	100	690	70	485	16	…
CA-6N	11 Chromium, 7 Nickel	140	965	135	930	15	50

[a] Remainder is iron.

[b] For low ferrite or non-magnetic castings of this grade, the following values shall apply: tensile strength, min, 65 ksi (450 MPa); yield point, min, 28 ksi (195 MPa).

The specifications committee of the Steel Founders Society issues a *Steel Castings Handbook* with supplements. Supplement 1 provides design rules and data based on the fluidity and solidification of steel, mechanical principles involved in production of molds and cores, cleaning of castings, machining, and functionality and weight aspects. Data and examples are included to show how these rules are applied. Supplement 2 summarizes the standard steel castings specification issued by the ASTM SAE, Assoc. of Am. Railroads (AAR), Am. Bur of Shipping (ABS), and Federal authorities, and provides guidance as to their applications. Information is included for carbon and alloy cast steels, high alloy cast steels, and centrifugally cast steel pipe. Details are also given of standard test methods for steel castings, including mechanical, non-destructive (visual, liquid penetrant, magnetic particle, radiographic, and ultrasonic), and testing of qualifications of welding procedures and personnel. Other supplements cover such subjects as tolerances, drafting practices, properties, repair and fabrication welding, of carbon, low alloy and high alloy castings, foundry terms, and hardenability and heat treatment.

Austenitic Manganese Cast Steel: Austenitic manganese cast steel is an important high-alloy cast steel which provides a high degree of shock and wear resistance. Its composition normally falls within the following ranges: carbon, 1.00 to 1.40 per cent; manganese, 10.00 to 14.00 per cent; silicon, 0.30 to 1.00 per cent; sulfur, 0.06 per cent max.; phosphorus, 0.10 per cent, max. In the as-cast condition, austenitic manganese steel is quite brittle. In order to strengthen and toughen the steel, it is heated to between 1830 and 1940 degrees F and quenched in cold water. Physical properties of quenched austenitic manganese steel that has been cast to size are as follows: tensile strength, 80,000 to 100,000 pounds per square inch; shear strength (single shear), 84,000 pounds per square inch; elongation in 2 inches, 15 to 35 per cent; reduction in area, 15 to 35 per cent; and Brinell hardness number, 180 to 220. When cold worked, the surface of such a casting increases to a Brinell hardness of from 450 to 550. In many cases the surfaces are cold worked to maximum hardness to assure immediate hardness in use. Heat-treated austenitic manganese steel is machined only with great difficulty since it hardens at and slightly ahead of the point of contact of the cutting tool. Grinding wheels mounted on specially adapted machines are used for boring, planing, keyway cutting, and similar operations on this steel. Where grinding cannot be employed and machining must be resorted to, high-speed tool steel or cemented carbide tools are used with heavy, rigid equipment and slow, steady operation. In any event, this procedure tends to be both tedious and expensive. Austenitic manganese cast steel can be arc-welded with manganese-nickel steel welding rods containing from 3 to 5 per cent nickel, 10 to 15 per cent manganese, and, usually, 0.60 to 0.80 per cent carbon.

Casting of Metals

Molten metals are shaped by pouring (casting) into a mold of the required form, which they enter under gravity, centrifugal force, or various degrees of pressure. Molds are made of refractory materials like sand, plaster, graphite, or metal. Sand molds are formed around a pattern or replica of the part to be made, usually of wood though plastics or metal may be used when large numbers of molds for similar parts are to be made.

Green-sand molding is used for most sand castings, sand mixed with a binder being packed around the pattern by hand, with power tools, or in a vibrating machine which may also exert a compressive force to pack the grains more closely. The term "green-sand" implies that the binder is not cured by heating or chemical reactions. The pattern is made in two "halves," which usually are attached to opposite sides of a flat plate. Shaped bars and other projections are fastened to the plate to form connecting channels and funnels in the sand for entry of the molten metal into the casting cavities. The sand is supported at the plate edges by a box-shaped frame or flask, with locating tabs that align the two mold halves when they are later assembled for the pouring operation.

Hollows and undercut surfaces in the casting are produced by cores, also made from sand, that are placed in position before the mold is closed, and held in place by tenons in grooves (called prints) formed in the sand by pattern projections. An *undercut surface* is one from which the pattern cannot be withdrawn in a straight line, so must be formed by a core in the mold. When the poured metal has solidified, the frame is removed and the sand falls or is cleaned off, leaving the finished casting(s) ready to be cut from the runners.

Gray iron is easily cast in complex shapes in green-sand and other molds and can be machined readily. The iron usually contains carbon, 1.7–4.5, and silicon, 1–3 per cent by weight. Excess carbon in the form of graphite flakes produces the gray surface from which the name is derived, when a casting is fractured.

Shell molding: invented by a German engineer, Croning, uses a resin binder to lock the grains of sand in a ¼- to ⅜-in.-thick layer of sand/resin mixture, which adheres to a heated pattern plate after the mass of the mixture has been dumped back into the container. The hot resin quickly hardens enough to make the shell thus formed sufficiently rigid to be removed from the pattern, producing a half mold. The other half mold is produced on another plate by the same method. Pattern projections form runner channels, basins, core prints, and locating tenons in each mold half. Cores are inserted to form internal passages and undercuts. The shell assembly is placed in a molding box and supported with some other material such as steel shot or a coarse sand, when the molten metal is to be poured in. Some shell molds are strong enough to be filled without backup, and the two mold halves are merely clamped together for metal to be poured in to make the casting(s).

V-Process is a method whereby dry, unbonded sand is held to the shape of a pattern by a vacuum. The pattern is provided with multiple vent passages that terminate in various positions all over its surface, and are connected to a common plenum chamber. A heat-softened, 0.002–0.005-in.-thick plastics film is draped over the pattern and a vacuum of 200–400 mm of mercury is applied to the chamber, sucking out the air beneath the film so that the plastics is drawn into close contact with the pattern. A sand box or flask with walls that also contain hollow chambers and a flat grid that spans the central area is placed on the pattern plate to confine the dry unbonded sand that is allowed to fall through the grid on to the pattern.

After vibration to compact the sand around the pattern, a former is used to shape a sprue cup into the upper surface of the sand, connecting with a riser on the pattern, and the top surface of the sand is covered with a plastics film that extends over the flask sides. The hollow chambers in the flask walls are then connected to the vacuum source. The vacuum is sufficient to hold the sand grains in their packed condition between the plastics films above and beneath, firmly in the shape defined by the pattern, so that the flask and the sand half-mold can be lifted from the pattern plate. Matching half molds made by these procedures are assembled into a complete mold, with cores inserted if needed. With both mold halves still held by vacuum, molten metal is poured through the sprue cup into the mold, the plastics film between the mold surfaces being melted and evaporated by the hot metal. After solidification, the vacuum is released and the sand, together with the casting(s), falls from the mold flasks. The castings emerge cleanly, and the sand needs only to be cooled before reuse.

Permanent mold, or gravity die, casting is mainly used for nonferrous metals and alloys. The mold (or die) is usually iron or steel, or graphite, and is cooled by water channels or by air jets on the outer surfaces. Cavity surfaces in metal dies are coated with a thin layer of heat-resistant material. The mold or die design is usually in two halves, although many multiple-part molds are in use, with loose sand or metal cores to form undercut surfaces. The cast metal is simply poured into a funnel formed in the top of the mold, although elaborate tilting mechanisms are often used to control the passage of metal into (and emergence of air from) the remote portions of die cavities.

Because the die temperature varies during the casting cycle, its dimensions vary correspondingly. The die is opened and ejectors push the casting(s) out as soon as its temperature is low enough for sufficient strength to build up. During the period after solidification and before ejection, cooling continues but shrinkage of the casting(s) is restricted by the die. The alloy being cast must be sufficiently ductile to accommodate these restrictions without fracturing. An alloy that tears or splits during cooling in the die is said to be *hot short* and cannot be cast in rigid molds. Dimensions of the casting(s) at shop temperatures will be related to the die temperature and the dimensions at ejection. Rules for casting shrinkage that apply to friable (sand) molds do not hold for rigid molds. Designers of metal molds and dies rely on temperature-based calculations and experience in evolving shrinkage allowances.

Low-pressure casting uses mold or die designs similar to those for gravity casting. The container (crucible) for the molten metal has provision for an airtight seal with the mold, and when gas or air pressure (6–10 lb/in.^{2}) is applied to the bath surface inside the crucible, the metal is forced up a hollow refractory tube (stalk) projecting from the die underside. This stalk extends below the bath level so that metal entering the die is free from oxides and impurities floating on the surface. The rate of filling is controlled so that air can be expelled from the die by the entering metal. With good design and control, high-quality, nonporous castings are made by both gravity and low-pressure methods, though the extra pressure in low-pressure die casting may increase the density and improve the reproduction of fine detail in the die.

Squeeze casting uses a metal die, of which one half is clamped to the bed of a large (usually) hydraulic press and the other to the vertically moving ram of the press. Molten metal is poured into the lower die and the upper die is brought down until the die is closed. The amount of metal in the die is controlled to produce a slight overflow as the die closes to ensure complete filling of the cavity. The heated dies are lubricated with graphite and pressures up to 25 tons per square inch may be applied by the press to squeeze the molten metal into the tiniest recesses in the die. When the press is opened, the solidified casting is pushed out by ejectors.

Finishing Operations for Castings

Removal of Gates and Risers from Castings.— After the molten iron or steel has solidified and cooled, the castings are removed from their molds, either manually or by placing them on vibratory machines and shaking the sand loose from the castings. The gates and risers that are not broken off in the shake-out are removed by impact, sawing, shearing, or burning-off methods. In the impact method, a hammer is used to knock off the gates and risers. Where the possibility exists that the fracture would extend into the casting itself, the gates or risers are first notched to assure fracture in the proper place. Some risers have a necked-down section at which the riser breaks off when struck. Sprue-cutter machines are also used to shear off gates. These machines facilitate the removal of a number of small castings from a central runner. Band saws, power saws. and abrasive cut-off wheel machines are also used to remove gates and risers. The use of band saws permits following the contour of the casting when removing unwanted appendages. Abrasive cut-off wheels are used when the castings are too hard or difficult to saw. Oxyacetylene cutting torches are used to cut off gates and risers and to gouge out or remove surface defects on castings. These torches are used on steel castings where the gates and risers are of a relatively large size. Surface defects are subsequently repaired by conventional welding methods.

Any unwanted material in the form of fins, gates, and riser pads that come above the casting surface, chaplets, parting-line flash, etc., is removed by chipping with pneumatic hammers, or by grinding with such equipment as floor or bench-stand grinders, portable grinders, and swing-frame grinders.

Blast Cleaning of Castings.— Blast cleaning of castings is performed to remove adhering sand, to remove cores, to improve the casting appearance, and to prepare the castings for their final finishing operation, which includes painting, machining, or assembling. Scale produced as a result of heat treating can also be removed. A variety of machines are used to handle all sizes of casting. The methods employed include blasting with sand, metal shot, or grit; and hydraulic cleaning or tumbling. In blasting, sharp sand, shot, or grit is carried by a stream of compressed air or water or by centrifugal force (gained as a result of whirling in a rapidly rotating machine) and directed against the casting surface by means of nozzles. The operation is usually performed in cabinets or enclosed booths. In some setups the castings are placed on a revolving table and the abrasive from the nozzles that are either mechanically or hand-held is directed against all the casting surfaces. Tumbling machines are also employed for cleaning, the castings being placed in large revolving drums together with slugs, balls, pins, metal punchings, or some abrasive, such as sandstone or granite chips, slag, silica, sand, or pumice. Quite frequently, the tumbling and blasting methods are used together, the parts then being tumbled and blasted simultaneously. Castings may also be cleaned by hydroblasting. This method uses a water-tight room in which a mixture of water and sand under high pressure is directed at the castings by means of nozzles. The action of the water and sand mixture cleans the castings very effectively.

Heat Treatment of Steel Castings.—Steel castings can be heat treated to bring about diffusion of carbon or alloying elements, softening, hardening, stress-relieving, toughening, improved machinability, increased wear resistance, and removal of hydrogen entrapped at the surface of the casting. Heat treatment of steel castings of a given composition follows closely that of wrought steel of similar composition. For discussion of types of heat treatment refer to the "Heat Treatment of Steel" section of this Handbook.

Estimating Casting Weight.—Where no pattern or die has yet been made, as when preparing a quotation for making a casting, the weight of a cast component can be estimated with fair accuracy by calculating the volume of each of the casting features, such as box- or rectangular-section features, cylindrical bosses, housings, ribs, and other parts, and adding them together. Several computer programs, also measuring mechanisms that can be applied to a drawing, are available to assist with these calculations. When the volume of metal has been determined it is necessary only to multiply by the unit weight of the alloy to be used, to arrive at the weight of the finished casting. The cost of the metal in the finished casting can then be estimated by multiplying the weight in lb by the cost/lb of the alloy. Allowances for melting losses, and for the extra metal used in risers and runners, and the cost of melting and machining may also be added to the cost/lb. Estimates of the costs of pattern- or die-making, molding, pouring and finishing of the casting(s), may also be added, to complete the quotation estimate.

Pattern Materials—Shrinkage, Draft, and Finish Allowances

Woods for Patterns.—Woods commonly used for patterns are white pine, mahogany, cherry, maple, birch, white wood, and fir. For most patterns, white pine is considered superior because it is easily worked, readily takes glue and varnish, and is fairly durable. For medium- and small-sized patterns, especially if they are to be used extensively, a harder wood is preferable. Mahogany is often used for patterns of this class, although many prefer cherry. As mahogany has a close grain, it is not as susceptible to atmospheric changes as a wood of coarser grain. Mahogany is superior in this respect to cherry, but is more expensive. In selecting cherry, never use young timber. Maple and birch are employed quite extensively, especially for turned parts, as they take a good finish. White wood is sometimes substituted for pine, but it is inferior to the latter in being more susceptible to atmospheric changes.

Selection of Wood.—It is very important to select well-seasoned wood for patterns; that is, it should either be kiln-dried or kept 1 or 2 years before using, the time depending upon the size of the lumber. During the seasoning or drying process, the moisture leaves the wood cells and the wood shrinks, the shrinkage being almost entirely across the grain rather than in a lengthwise direction. Naturally, after this change takes place, the wood is less liable to warp, although it will absorb moisture in damp weather. Patterns also tend to absorb moisture from the damp sand of molds, and to minimize troubles from this source they are covered with varnish. Green or water-soaked lumber should not be put in a drying room, because the ends will dry out faster than the rest of the log, thus causing cracks. In a log, there is what is called "sap wood" and "heart wood." The outer layers form the sap wood, which is not as firm as the heart wood and is more likely to warp; hence, it should be avoided, if possible.

Pattern Varnish.—Patterns intended for repeated use are varnished to protect them against moisture, especially when in the damp molding sand. The varnish used should dry quickly to give a smooth surface that readily draws from the sand. Yellow shellac varnish is generally used. It is made by dissolving gum shellac in grain alcohol. Wood alcohol is sometimes substituted, but is inferior. The color of the varnish is commonly changed for covering core prints, in order that the prints may be readily distinguished from the body of the pattern. Black shellac varnish is generally used. At least three coats of varnish should be applied to patterns, the surfaces being rubbed down with sandpaper after applying the preliminary coats, in order to obtain a smooth surface.

Shrinkage Allowances.—The shrinkage allowances ordinarily specified for patterns to compensate for the contraction of castings in cooling are as follows: cast iron, $\frac{3}{32}$ to $\frac{1}{8}$ inch per foot; common brass, $\frac{3}{16}$ inch per foot; yellow brass, $\frac{7}{32}$ inch per foot; bronze, $\frac{5}{32}$ inch per foot; aluminum, $\frac{1}{8}$ to $\frac{5}{32}$ inch per foot; magnesium, $\frac{1}{8}$ to $\frac{11}{64}$ inch per foot; steel, $\frac{3}{16}$ inch per foot. These shrinkage allowances are approximate values only because the exact allowance depends upon the size and shape of the casting and the resistance of the mold to the normal contraction of the casting during cooling. It is, therefore, possible that more than one shrinkage allowance will be required for different parts of the same pattern. Another factor that affects shrinkage allowance is the molding method, which may vary to such an extent from one foundry to another, that different shrinkage allowances for each would have to be used for the same pattern. For these reasons it is recommended that patterns be made at the foundry where the castings are to be produced to eliminate difficulties due to lack of accurate knowledge of shrinkage requirements.

An example of how casting shape can affect shrinkage allowance is given in the Steel Castings Handbook. In this example a straight round steel bar required a shrinkage allowance of approximately $\frac{9}{32}$ inch per foot. The same bar but with a large knob on each end required a shrinkage allowance of only $\frac{3}{16}$ inch per foot. A third steel bar with large flanges at each end required a shrinkage allowance of only $\frac{7}{64}$ inch per foot. This example would seem to indicate that the best practice in designing castings and making patterns is to obtain shrinkage values from the foundry that is to make the casting because there can be no fixed allowances.

Metal Patterns.—Metal patterns are especially adapted to molding machine practice, owing to their durability and superiority in retaining the required shape. The original master pattern is generally made of wood, the casting obtained from the wood pattern being finished to make the metal pattern. The materials commonly used are brass, cast iron, aluminum, and steel. Brass patterns should have a rather large percentage of tin, to improve the casting surface. Cast iron is generally used for large patterns because it is cheaper than brass and more durable. Cast-iron patterns are largely used on molding machines. Aluminum patterns are light but they require large shrinkage allowances. White metal is sometimes used when it is necessary to avoid shrinkage. The gates for the mold may be cast or made of sheet brass. Some patterns are made of vulcanized rubber, especially for light match-board work.

Obtaining Weight of Casting from Pattern Weight.—To obtain the approximate weight of a casting, multiply the weight of the pattern by the factor given in the accompanying table. For example, if the weight of a white-pine pattern is 4 pounds what is the weight of a solid cast-iron casting obtained from that pattern? Casting weight = 4 × 16 = 64 pounds. If the casting is cored, fill the core-boxes with dry sand, and multiply the weight of the sand by one of the following factors: For cast iron, 4; for brass, 4.65; for aluminum, 1.4. Then subtract the product of the sand weight and the factor just given from the weight of the solid casting, to obtain the weight of the cored casting. The weight of wood varies considerably, so the results obtained by the use of the table are only approximate, the factors being based on the average weight of the woods listed. For metal patterns, the results may be more accurate.

Factors for Obtaining Weight of Casting from Pattern Weight

Pattern Material	Factors				
	Cast Iron	Aluminum	Copper	Zinc	Brass, 70% Copper, 30% Zinc
White pine	16.00	5.70	19.60	15.00	19.00
Mahogany, Honduras	12.00	4.50	14.70	11.50	14.00
Cherry	10.50	3.80	13.00	10.00	12.50
Cast Iron	1.00	0.35	1.22	0.95	1.17
Aluminum	2.85	1.00	3.44	2.70	3.30

Die Casting

Die casting is a method of producing finished castings by forcing molten metal into a hard metal die, which is arranged to open after the metal has solidified so that the casting can be removed. The die-casting process makes it possible to secure accuracy and uniformity in castings, and machining costs are either eliminated altogether or are greatly reduced. The greatest advantage of the die-casting process is that parts are accurately and often completely finished when taken from the die. When the dies are properly made, castings may be accurate within 0.001 inch or even less and a limit of 0.002 or 0.003 inch per inch of casting dimension can be maintained on many classes of work.

Die castings are used extensively in the manufacture of such products as cash registers, meters, time-controlling devices, small housings, washing machines, and parts for a great variety of mechanisms. Lugs and gear teeth are cast in place and both external and internal screw threads can be cast. Holes can be formed within about 0.001 inch of size and the most accurate bearings require only a finish-reaming operation. Figures and letters may be cast sunken or in relief on wheels for counting or printing devices, and with ingenious die designs, many shapes that formerly were believed too intricate for die casting are now produced successfully by this process.

Die casting uses hardened steel molds (dies) into which the molten metal is injected at high speed, reaching pressures up to 10 tons/in.^{2}, force being applied by a hydraulically actuated plunger moving in a cylindrical pressure chamber connected to the die cavity(s). If the plan area of the casting and its runner system cover 50 in.^{2}, the total power applied is 10 tons/in.^{2} of pressure on the metal × 50 in.^{2} of projected area, producing a force of 500 tons, and the die-casting machine must hold the die shut against this force. Massive toggle mechanisms stretch the heavy (6-in. diameter) steel tie bars through about 0.045 in. on a typical (500-ton) machine to generate this force. Although the die is hot, metal entering the die cavity is cooled quickly, producing layers of rapidly chilled, dense material about 0.015 in. thick in the metal having direct contact with the die cavity surfaces. Because the high injection forces allow castings to be made with thin walls, these dense layers form a large proportion of the total wall thickness, producing high casting strength. This phenomenon is known as the skin effect, and should be taken into account when considering the tensile strengths and other properties measured in (usually thicker) test bars.

As to the limitations of the die-casting process it may be mentioned that the cost of dies is high, and, therefore, die casting is economical only when large numbers of duplicate parts are required. The stronger and harder metals cannot be die cast, so that the process is not applicable for casting parts that must necessarily be made of iron or steel, although special alloys have been developed for die casting that have considerable tensile and compressive strength.

Many die castings are produced by the hot-chamber method in which the pressure chamber connected to the die cavity is immersed permanently in the molten metal and is automatically refilled through a hole that is uncovered as the (vertical) pressure plunger moves back after filling the die. This method can be used for alloys of low melting point and high fluidity such as zinc, lead, tin, and magnesium. Other alloys requiring higher pressure, such as brass, or that can attack and dissolve the ferrous pressure chamber material, such as aluminum, must use the slower cold-chamber method with a water-cooled (horizontal) pressure chamber outside the molten metal.

Porosity.—Molten metal injected into a die cavity displaces most of the air, but some of the air is trapped and is mixed with the metal. The high pressure applied to the metal squeezes the pores containing the air to very small size, but subsequent heating will soften the casting so that air in the surface pores can expand and cause blisters. Die castings are seldom solution heat treated or welded because of this blistering problem. The chilling effect of the comparatively cold die causes the outer layers of a die casting to be dense and relatively free of porosity. Vacuum die casting, in which the cavity atmosphere is evacuated before metal is injected, is sometimes used to reduce porosity. Another method is to displace the air by filling the cavity with oxygen just prior to injection. The oxygen is burned by the hot metal so that porosity does not occur.

When these special methods are not used, machining depths must be limited to 0.020–0.035 inch if pores are not to be exposed, but as-cast accuracy is usually good enough for only light finishing cuts to be needed. Special pore-sealing techniques must be used if pressure tightness is required.

Designing Die Castings.—Die castings are best designed with uniform wall thicknesses (to reduce cooling stresses) and cores of simple shapes (to facilitate extraction from the die). Heavy sections should be avoided or cored out to reduce metal concentrations that may attract trapped gases and cause porosity concentrations. Designs should aim at arranging for metal to travel through thick sections to reach thin ones if possible. Because of the high metal injection pressures, conventional sand cores cannot be used, so cored holes and apertures are made by metal cores that form part of the die. Small and slender cores are easily bent or broken, so should be avoided in favor of piercing or drilling operations on the finished castings. Ribbing adds strength to thin sections, and fillets should be used on all inside corners to avoid high stress concentrations in the castings. Sharp outside corners should be avoided. Draft allowances on a die casting are usually from 0.5 to 1.5 degrees per side to permit the castings to be pushed off cores or out of the cavity.

Alloys Used for Die Casting.—The alloys used in modern die-casting practice are based on aluminum, zinc, and copper, with small numbers of castings also being made from magnesium-, tin-, and lead-based alloys.

Aluminum-Base Alloys.—Aluminum-base die-casting alloys are used more extensively than any other base metal alloy because of their superior strength combined with ease of castability. Linear shrinkage of aluminum alloys on cooling is about 12.9 to 15.5×10^{-6} in./in.-°F. Casting temperatures are of the order of 1200 deg. F. Most aluminum die castings are produced in aluminum-silicon-copper alloys such as the Aluminum Association (AA) No. 380 (ASTM SC84A; UNS A038000), containing silicon 7.5 to 9.5 and copper 3 to 4 per cent. Silicon increases fluidity for complete die filling, but reduces machinability, and copper adds hardness but reduces ductility in aluminum alloys. A less-used alloy having slightly greater fluidity is AA No. 384 (ASTM SC114A; UNS A03840) containing silicon 10.5 to 12.0 and copper 3.0 to 4.5 per cent. For marine applications, AA 360 (ASTM 100A; UNS A03600) containing silicon 9 to 10 and copper 0.6 per cent is recommended, the copper content being kept low to reduce susceptibility to corrosion in salt atmospheres.

The tensile strengths of AA 380, 384, and 360 alloys are 47,000, 48,000, and 46,000 lb/in.^{2}, respectively. Although 380, 384, and 360 are the most widely used die-castable alloys, several other aluminum alloys are used for special applications. For instance, the AA 390 alloy, with its high silicon content (16 to 18 per cent), is used for internal combustion engine cylinder castings, to take advantage of the good wear resistance provided by the hard silicon grains. No. 390 alloy also contains 4 to 5 per cent copper, and has a hardness of 120 Brinell with low ductility, and a tensile strength of 41,000 lb/in.^{2}.

Zinc-Base Alloys.—In the molten state, zinc is extremely fluid and can therefore be cast into very intricate shapes. The metal also is plentiful and has good mechanical properties. Zinc die castings can be made to closer dimensional limits and with thinner walls than aluminum. Linear shrinkage of these alloys on cooling is about 9 to 13×10^{-6} in./in.-°F. The low casting temperatures (750–800 deg. F) and the hot-chamber process allow high production rates with simple automation. Zinc die castings can be produced with extremely smooth surfaces, lending themselves well to plating and other finishing methods. The established zinc alloys numbered 3, 5 and 7 [ASTM B86 (AG40A; UNS Z33520), AG41A (UNS Z35531), and AG40B (UNS Z33522)] each contains 3.5 to 4.3 per cent of aluminum, which adds strength and hardness, plus carefully controlled amounts of other elements. Recent research has brought forward three new alloys of zinc containing 8, 12, and 27 per cent of aluminum, which confer tensile strength of 50,000–62,000 lb/in.^{2} and hardness approaching that of cast iron (105–125 Brinell). These alloys can be used for gears and racks, for instance, and as housings for shafts that run directly in reamed or bored holes, with no need for bearing bushes.

Copper-Base Alloys.—Brass alloys are used for plumbing, electrical, and marine components where resistance to corrosion must be combined with strength and wear resistance. With the development of the cold-chamber casting process, it became possible to make die castings from several standard alloys of copper and zinc such as yellow brass (ASTM B176-Z30A; UNS C85800) containing copper 58, zinc 40, tin 1, and lead 1 per cent. Tin and lead are included to improve corrosion resistance and machinability, respectively, and this alloy has a tensile strength of 45,000 lb/in^2. Silicon brass (ASTM B176-ZS331A; UNS C87800) with copper 65 and zinc 34 per cent also contains 1 per cent silicon, giving it more fluidity for castability and with higher tensile strength (58,000 psi) and better resistance to corrosion. High silicon brass or tombasil (ASTM B176-ZS144A), containing copper 82, zinc 14, and silicon 4 per cent, has a tensile strength of 70,000 lb/in.2 and good wear resistance, but at the expense of machinability.

Magnesium-Base Alloys.—Light weight combined with good mechanical properties and excellent damping characteristics are principal reasons for using magnesium die castings. Magnesium has a low specific heat and does not dissolve iron so it may be die cast by the cold- or hot-chamber methods. For the same reasons, die life is usually much longer than for aluminum. The lower specific heat and more rapid solidification make production about 50 per cent faster than with aluminum. To prevent oxidation, an atmosphere of CO_2 and air, containing about 0.5 per cent of SF_6 gas, is used to exclude oxygen from the surface of the molten metal. The most widely used alloy is AZ91D (ASTM B94; UNS 11916), a high-purity alloy containing aluminum 9 and zinc 0.7 per cent, and having a yield strength of 23,000 lb/in.^{2} (Table 7a on page 559). AZ91D has a corrosion rate similar to that of 380 aluminum (see *Aluminum-Base Alloys*).

Tin-Base Alloys.—In this group tin is alloyed with copper, antimony, and lead. SAE Alloy No. 10 contains, as the principal ingredients, in percentages, tin, 90; copper, 4 to 5; antimony, 4 to 5; lead, maximum, 0.35. This high-quality babbitt mixture is used for main-shaft and connecting-rod bearings or bronze-backed bearings in the automotive and aircraft industries. SAE No. 110 contains tin, 87.75; antimony, 7.0 to 8.5; copper, maximum, 2.25 to 3.75 per cent and other constituents the same as No. 10. SAE No. 11, which contains a little more copper and antimony and about 4 per cent less tin than No. 10, is also used for bearings or other applications requiring a high-class tin-base alloy. These tin-base compositions are used chiefly for automotive bearings but they are also used for milking machines, soda fountains, syrup pumps, and similar apparatus requiring resistance against the action of acids, alkalies, and moisture.

Lead-Base Alloys.—These alloys are employed usually where a cheap noncorrosive metal is needed and strength is relatively unimportant. Such alloys are used for parts of lead-acid batteries, for automobile wheel balancing weights, for parts that must withstand the action of strong mineral acids and for parts of X-ray apparatus. SAE Composition No. 13 contains (in percentages) lead, 86; antimony, 9.25 to 10.75; tin, 4.5 to 5.5 per cent. SAE Specification No. 14 contains less lead and more antimony and copper. The lead content is 76; antimony, 14 to 16; and tin, 9.25 to 10.75 per cent. Alloys Nos. 13 and 14 are inexpensive owing to the high lead content and may be used for bearings that are large and subjected to light service.

Dies for Die-Casting Machines.—Dies for die-casting machines are generally made of steel although cast iron and nonmetallic materials of a refractory nature have been used, the latter being intended especially for bronze or brass castings, which, owing to their comparatively high melting temperatures, would damage ordinary steel dies. The steel most generally used is a low-carbon steel. Chromium-vanadium and tungsten steels are used for aluminum, magnesium, and brass alloys, when dies must withstand relatively high temperatures.

Making die-casting dies requires considerable skill and experience. Dies must be so designed that the metal will rapidly flow to all parts of the impression and at the same time allow the air to escape through shallow vent channels, 0.003 to 0.005 inch deep, cut into the parting of the die. To secure solid castings, the gates and vents must be located with reference to the particular shape to be cast. Shrinkage is another important feature, especially on accurate work. The amount usually varies from 0.002 to 0.007 inch per inch, but to determine the exact shrinkage allowance for an alloy containing three or four elements is difficult except by experiment.

Die-Casting Bearing Metals in Place.—Practically all the metals that are suitable for bearings can be die cast in place. Automobile connecting rods are an example of work to which this process has been applied sucessfully. After the bearings are cast in place, they are finished by boring or reaming. The best metals for the bearings, and those that also can be die cast most readily, are the babbitts containing about 85 per cent tin with the remainder copper and antimony. These metals should not contain over 9 per cent copper. The copper constitutes the hardening element in the bearing. A recommended composition for a high-class bearing metal is 85 per cent tin, 10 per cent antimony, and 5 per cent copper. The antimony may vary from 7 to 10 per cent and the copper from 5 to 8 per cent. To reduce costs, some bearing metals use lead instead of tin. One bearing alloy contains from 95 to 98 per cent lead. The die-cast metal becomes harder upon seasoning a few days. In die-casting bearings, the work is located from the bolt holes that are drilled previous to die casting. It is important that the bolt holes be drilled accurately with relation to the remainder of the machined surfaces.

Injection Molding of Metal.—The die casting and injection molding processes have been combined to make possible the injection molding of many metal alloys by mixing powdered metal, of 5 to 10 μm (0.0002 to 0.0004 in.) particle size with thermoplastic binders. These binders are chosen for maximum flow characteristics to ensure that the mixture can penetrate to the most remote parts of the die/mold cavities. Moderate pressures and temperatures are used for the injection molding of these mixtures, and the molded parts harden as they cool so that they can be removed as solids from the mold. Shrinkage allowances for the cavities are greater than are required for the die casting process, because the injection molded parts are subject to a larger shrinkage (10 to 35 per cent) after removal from the die, due to evaporation of the binder and consolidation of the powder.

Binder removal may take several days because of the need to avoid distortion, and when it is almost complete the molded parts are sintered in a controlled atmosphere furnace at high temperatures to remove the remaining binder and consolidate the powdered metal component that remains. Density can thus be increased to about 95 per cent of the density of similar material produced by other processes. Tolerances are similar to those in die casting, and some parts are sized by a coining process for greater accuracy. The main limitation of the process is size, parts being restricted to about a 1.5-in. cube.

Precision Investment Casting

Investment casting is a highly developed process that is capable of great casting accuracy and can form extremely intricate contours. The process may be utilized when metals are too hard to machine or otherwise fabricate; when it is the only practical method of producing a part; or when it is more economical than any other method of obtaining work of the quality required. Precision investment casting is especially applicable in producing either exterior or interior contours of intricate form with surfaces so located that they could not be machined readily if at all. The process provides efficient, accurate means of producing such parts as turbine blades, airplane, or other parts made from alloys that have high melting points and must withstand exceptionally high temperatures, and many other products. The accuracy and finish of precision investment castings may either eliminate machining entirely or reduce it to a minimum. The quantity that may be produced economically may range from a few to thousands of duplicate parts.

Investment casting uses an expendable pattern, usually of wax or injection-molded plastics. Several wax replicas or patterns are usually joined together or to bars of wax that are shaped to form runner channels in the mold. Wax shapes that will produce pouring funnels also are fastened to the runner bars. The mold is formed by dipping the wax assembly (tree) into a thick slurry containing refractory particles. This process is known as investing. After the coating has dried, the process is repeated until a sufficient thickness of material has been built up to form a one-piece mold shell. Because the mold is in one piece, undercuts, apertures, and hollows can be produced easily. As in shell molding, this invested shell is baked to increase its strength, and the wax or plastics pattern melts and runs out or evaporates (lost-wax casting). Some molds are backed up with solid refractory material that is also dried and baked to increase the strength. Molds for lighter castings are often treated similarly to shell molds described before. Filling of the molds may take place in the atmosphere, in a chamber filled with inert gas or under vacuum, to suit the metal being cast.

Materials That May Be Cast.—The precision investment process may be applied to a wide range of both ferrous and nonferrous alloys. In industrial applications, these include alloys of aluminum and bronze, Stellite, Hastelloys, stainless and other alloy steels, and iron castings, especially where thick and thin sections are encountered. In producing investment castings, it is possible to control the process in various ways so as to change the porosity or density of castings, obtain hardness variations in different sections, and vary the corrosion resistance and strength by special alloying.

General Procedure in Making Investment Castings.—Precision investment casting is similar in principle to the "lost-wax" process that has long been used in manufacturing jewelry, ornamental pieces, and individual dentures, inlays, and other items required in dentistry, which is not discussed here. When this process is employed, both the pattern and mold used in producing the casting are destroyed after each casting operation, but they may both be replaced readily. The "dispensable patterns" (or cluster of duplicate patterns) is first formed in a permanent mold or die and is then used to form the cavity in the mold or "investment" in which the casting (or castings) is made. The investment or casting mold consists of a refractory material contained within a reinforcing steel flask. The pattern is made of wax, plastics, or a mixture of the two. The material used is evacuated from the investment to form a cavity (without parting lines) for receiving the metal to be cast. Evacuation of the pattern (by the application of sufficient heat to melt and vaporize it) and the use of a master mold or die for reproducing it quickly and accurately in making duplicate castings are distinguishing features of this casting process. Modern applications of the process include many developments such as variations in the preparation of molds, patterns, investments, etc., as well as in the casting procedure. Application of the process requires specialized knowledge and experience.

Master Mold for Making Dispensable Patterns.—Duplicate patterns for each casting operation are made by injecting the wax, plastics, or other pattern material into a master mold or die that usually is made either of carbon steel or of a soft metal alloy. Rubber, alloy steels, and other materials may also be used. The mold cavity commonly is designed to form a cluster of patterns for multiple castings. The mold cavity is not, as a rule, an exact duplicate of the part to be cast because it is necessary to allow for shrinkage and perhaps to compensate for distortion that might affect the accuracy of the cast product. In producing master pattern molds there is considerable variation in practice. One general method is to form the cavity by machining; another is by pouring a molten alloy around a master pattern that usually is made of monel metal or of a high-alloy stainless steel. If the cavity is not machined, a master pattern is required. Sometimes, a sample of the product itself may be used as a master pattern, when, for example, a slight reduction in size due to shrinkage is not objectionable. The dispensable pattern material, which may consist of waxes, plastics, or a combination of these materials, is injected into the mold by pressure, by gravity, or by the centrifugal method. The mold is made in sections to permit removal of the dispensable pattern. The mold while in use may be kept at the correct temperature by electrical means, by steam heating, or by a water jacket.

Shrinkage Allowances for Patterns.—The shrinkage allowance varies considerably for different materials. In casting accurate parts, experimental preliminary casting operations may be necessary to determine the required shrinkage allowance and possible effects of distortion. Shrinkage allowances, in inches per inch, usually average about 0.022 for steel, 0.012 for gray iron, 0.016 for brass, 0.012 to 0.022 for bronze, 0.014 for aluminum and magnesium alloys. (See also *Shrinkage Allowances* on page 1340.)

Casting Dimensions and Tolerances.—Generally, dimensions on investment castings can be held to ±0.005 in. and on specified dimensions to as low as ±0.002 in. Many factors, such as the grade of refractory used for the initial coating on the pattern, the alloy composition, and the pouring temperature, affect the cast surface finish. Surface discontinuities on the as-cast products therefore can range from 30 to 300 microinches in height.

Investment Materials.—For investment casting of materials having low melting points, a mixture of plaster of Paris and powdered silica in water may be used to make the molds, the silica forming the refractory and the plaster acting as the binder. To cast materials having high melting points, the refractory may be changed to sillimanite, an alumina-silicate material having a low coefficient of expansion that is mixed with powdered silica as the binder. Powdered silica is then used as the binder. The interior surfaces of the mold are reproduced on the casting so, when fine finishes are needed, a first coating of fine sillimanite sand and a silicon ester such as ethyl silicate with a small amount of piperidine, is applied and built up to a thickness of about 0.06 in. This investment is covered with a coarser grade of refractory that acts to improve bonding with the main refractory coatings, before the back up coatings are applied.

With light castings, the invested material may be used as a shell, without further reinforcement. With heavy castings the shell is placed in a larger container which may be of thick waxed paper or card, and further slurry is poured around it to form a thicker mold of whatever proportions are needed to withstand the forces generated during pouring and solidification. After drying in air for several hours, the invested mold is passed through an oven where it is heated to a temperature high enough to cause the wax to run out. When pouring is to take place, the mold is pre-heated to between 700 and 1000°C, to get rid of any remaining wax, to harden the binder and prepare for pouring the molten alloy. Pouring metal into a hot mold helps to ensure complete filling of intricate details in the castings. Pouring may be done under gravity, under a vacuum under pressure, or with a centrifuge. When pressure is used, attention must be paid to mold permeability to ensure gases can escape as the metal enters the cavities.

Casting Operations.—The temperature of the flask for casting may range all the way from a chilled condition up to 2000 degrees F or higher, depending upon the metal to be cast, the size and shape of the casting or cluster, and the desired metallurgical conditions. During casting, metals are nearly always subjected to centrifugal force vacuum, or other pressure. The procedure is governed by the kind of alloy, the size of the investment cavity, and its contours or shape.

Investment Removal.—When the casting has solidified, the investment material is removed by destroying it. Some investments are soluble in water, but those used for ferrous castings are broken by using pneumatic tools, hammers, or by shot or abrasive blasting and tumbling to remove all particles. Gates, sprues, and runners may be removed from the castings by an abrasive cutting wheel or a band saw according to the shape of the cluster and machinability of the material.

Accuracy of Investment Castings.—The accuracy of precision investment castings may, in general, compare favorably with that of many machined parts. The overall tolerance varies with the size of the work, the kind of metal and the skill and experience of the operators. Under normal conditions, tolerances may vary from ±0.005 or ±0.006 inch per inch, down to ±0.0015 to ±0.002 inch per inch, and even smaller tolerances are possible on very small dimensions. Where tolerances applying to a lengthwise dimension must be smaller than would be normal for the casting process, the casting gate may be placed at one end to permit controlling the length by a grinding operation when the gate is removed.

Casting Weights and Sizes.—Investment castings may vary in weight from a fractional part of an ounce up to 75 pounds or more. Although the range of weights representing the practice of different firms specializing in investment casting may vary from about ½ pound up to 10 or 20 pounds, a practical limit of 10 or 15 pounds is common. The length of investment castings ordinarily does not exceed 12 or 15 inches, but much longer parts may be cast. It is possible to cast sections having a thickness of only a few thousandths of an inch, but the preferred minimum thickness, as a general rule, is about 0.020 inch for alloys of high castability and 0.040 inch for alloys of low castability.

Design for Investment Casting.—As with most casting processes, best results from investment casting are achieved when uniform wall thicknesses between 0.040 and 0.375 in. are used for both cast components and channels forming runners in the mold. Gradual transition from thick to thin sections is also desirable. It is important that molten metal should not have to pass through a thin section to fill a thick part of the casting. Thin edges should be avoided because of the difficulty of producing them in the wax pattern. Fillets should be used in all internal corners to avoid stress concentrations that usually accompany sharp angles. Thermal contraction usually causes distortion of the casting, and should be allowed for if machining is to be minimized. Machining allowances vary from 0.010 in. on small, to 0.040 in. on large parts. With proper arrangement of castings in the mold, grain size and orientation can be controlled and directional solidification can often be used to advantage to ensure desired physical properties in the finished components.

Casting Milling Cutters by Investment Method.—Possible applications of precision investment casting in tool manufacture and in other industrial applications are indicated by its use in producing high-speed steel milling cutters of various forms and sizes. Removal of the risers, sand blasting to improve the appearance, and grinding the cutting edges are the only machining operations required. The bore is used as cast. Numerous tests have shown that the life of these cutters compares favorably with high-speed steel cutters made in the usual way.

Extrusion of Metals

The Basic Process.—Extrusion is a metalworking process used to produce long, straight semifinished products such as bars, tubes, solid and hollow sections, wire and strips by squeezing a solid slug of metal, either cast or wrought, from a closed container through a die. An analogy to the process is the dispensing of toothpaste from a collapsible tube.

During extrusion, compressive and shear, but no tensile, forces are developed in the stock, thus allowing the material to be heavily deformed without fracturing. The extrusion process can be performed at either room or high temperature, depending on the alloy and method. Cross sections of varying complexity can also be produced, depending on the materials and dies used.

In the specially constructed presses used for extrusion, the load is transmitted by a ram through an intermediate dummy block to the stock. The press container is usually fitted with a wear-resistant liner and is constructed to withstand high radial loads. The die stack consists of the die, die holder, and die backer, all of which are supported in the press end housing or platen, which resists the axial loads.

The following are characteristics of different extrusion methods and presses: 1) The movement of the extrusion relative to the ram. In "direct extrusion," the ram is advanced toward the die stack; in "indirect extrusion," the die moves down the container bore;

2) The position of the press axis, which is either horizontal or vertical; 3) The type of drive, which is either hydraulic or mechanical; and 4) The method of load application, which is either conventional or hydrostatic.

In forming a hollow extrusion, such as a tube, a mandrel integral with the ram is pushed through the previously pierced raw billet.

Cold Extrusion: Cold extrusion has often been considered a separate process from hot extrusion; however, the only real difference is that cold or only slightly warm billets are used as starting stock. Cold extrusion is not limited to certain materials; the only limiting factor is the stresses in the tooling. In addition to the soft metals such as lead and tin, aluminum alloys, copper, zirconium, titanium, molybdenum, beryllium, vanadium, niobium, and steel can be extruded cold or at low deformation temperatures. Cold extrusion has many advantages, such as no oxidation or gas/metal reactions; high mechanical properties due to cold working if the heat of deformation does not initiate recrystallization; narrow tolerances; good surface finish if optimum lubrication is used; fast extrusion speeds can be used with alloys subject to hot shortness.

Examples of cold extruded parts are collapsible tubes, aluminum cans, fire extinguisher cases, shock absorber cylinders, automotive pistons, and gear blanks.

Hot Extrusion: Most hot extrusion is performed in horizontal hydraulic presses rated in size from 250 to 12,000 tons. The extrusions are long pieces of uniform cross sections, but complex cross sections are also produced. Most types of alloys can be hot extruded.

Owing to the temperatures and pressures encountered in hot extrusion, the major problems are the construction and the preservation of the equipment. The following are approximate temperature ranges used to extrude various types of alloys: magnesium, 650–850 degrees F; aluminum, 650–900 degrees F; copper, 1200–2000 degrees F; steel, 2200–2400 degrees F; titanium, 1300–2100 degrees F; nickel 1900–2200 degrees F; refractory alloys, up to 4000 degrees F. In addition, pressures range from as low as 5000 to over 100,000 psi. Therefore, lubrication and protection of the chamber, ram, and die are generally required. The use of oil and graphite mixtures is often sufficient at the lower temperatures; while at higher temperatures, glass powder, which becomes a molten lubricant, is used.

Extrusion Applications: The stress conditions in extrusion make it possible to work materials that are brittle and tend to crack when deformed by other primary metalworking processes. The most outstanding feature of the extrusion process, however, is its ability to produce a wide variety of cross-sectional configurations; shapes can be extruded that have complex, nonuniform, and nonsymmetrical sections that would be difficult or impossible to roll or forge. Extrusions in many instances can take the place of bulkier, more costly assemblies made by welding, bolting, or riveting. Many machining operations may also be reduced through the use of extruded sections. However, as extrusion temperatures increase, processing costs also increase, and the range of shapes and section sizes that can be obtained becomes narrower.

While many asymmetrical shapes are produced, symmetry is the most important factor in determining extrudability. Adjacent sections should be as nearly equal as possible to permit uniform metal flow through the die. The length of their protruding legs should not exceed 10 times their thickness.

The size and weight of extruded shapes are limited by the section configuration and properties of the material extruded. The maximum size that can be extruded on a press of a given capacity is determined by the "circumscribing circle," which is defined as the smallest diameter circle that will enclose the shape. This diameter controls the die size, which in turn is limited by the press size. For instance, the larger presses are generally capable of extruding aluminum shapes with a 25-in.-diam circumscribing circle and steel and titanium shapes with about 22-in.-diam circle.

The minimum cross-sectional area and minimum thickness that can be extruded on a given size press are dependent on the properties of the material, the extrusion ratio (ratio of the cross-sectional area of the billet to the extruded section), and the complexity of shape. As a rule thicker sections are required with increased section size.

The following table gives the approximate minimum cross section and minimum thickness of some commonly extruded metals.

Material	Minimum Cross Section (sq in.)	Minimum Thickness (in.)
Carbon and alloy steels	0.40	0.120
Stainless steels	0.45-0.70	0.120-0.187
Titanium	0.50	0.150
Aluminum	<0.40	0.040
Magnesium	<0.40	0.040

Extruded shapes minimize and sometimes eliminate the need for machining; however, they do not have the dimensional accuracy of machined parts. Smooth surfaces with finishes better than 30 μin. rms are attainable in magnesium and aluminum; an extruded finish of 125 μin. rms is generally obtained with most steels and titanium alloys. Minimum corner and fillet radii of $\frac{1}{64}$ in. are preferred for aluminum and magnesium alloys; while for steel, minimum corner radii of 0.030 in. and fillet radii of 0.125 in. are typical.

Extrusion of Tubes: In tube extrusion, the metal passes through a die, which determines its outer diameter, and around a central mandrel, which determines its inner diameter. Either solid or hollow billets may be used, with the solid billet being used most often. When a solid billet is extruded, the mandrel must pierce the billet by pushing axially through it before the metal can pass through the annular gap between the die and the mandrel. Special presses are used in tube extrusion to increase the output and improve the quality compared to what is obtained using ordinary extrusion presses. These special hydraulic presses independently control ram and mandrel positioning and movement.

Powder Metallurgy

Powder metallurgy is a process whereby metal parts in large quantities can be made by the compressing and sintering of various powdered metals such as brass, bronze, aluminum, and iron. Compressing of the metal powder into the shape of the part to be made is done by accurately formed dies and punches in special types of hydraulic or mechanical presses. The "green" compressed pieces are then sintered in an atmosphere controlled furnace at high temperatures, causing the metal powder to be bonded together into a solid mass. A subsequent sizing or pressing operation and supplementary heat treatments may also be employed. The physical properties of the final product are usually comparable to those of cast or wrought products of the same composition. Using closely controlled conditions, steel of high hardness and tensile strength has also been made by this process.

Any desired porosity from 5 to 50 per cent can be obtained in the final product. Large quantities of porous bronze and iron bearings, which are impregnated with oil for self-lubrication, have been made by this process. Other porous powder metal products are used for filtering liquids and gases. Where continuous porosity is desired in the final product, the voids between particles are kept connected or open by mixing one per cent of zinc stearate or other finely powdered metallic soap throughout the metal powder before briquetting and then boiling this out in a low temperature baking before the piece is sintered.

The dense type of powdered metal products include refractory metal wire and sheet, cemented carbide tools, and electrical contact materials (products which could not be made as satisfactorily by other processes) and gears or other complex shapes which might also have been made by die casting or the precise machining of wrought or cast metal.

Advantages of Powder Metallurgy.—Parts requiring irregular curves, eccentrics, radial projections, or recesses often can be produced only by powder metallurgy. Parts that require irregular holes, keyways, flat sides, splines or square holes that are not easily machined, can usually be made by this process. Tapered holes and counter-bores are easily produced. Axial projections can be formed but the permissible size depends on the extent to which the powder will flow into the die recesses. Projections not more than one-quarter the length of the part are practicable. Slots, grooves, blind holes, and recesses of varied depths are also obtainable.

Limiting Factors in Powdered Metal Process.—The number and variety of shapes that may be obtained are limited by lack of plastic flow of powders, i.e., the difficulty with which they can be made to flow around corners. Tolerances in diameter usually cannot be held closer than 0.001 inch and tolerances in length are limited to 0.005 inch. This difference in diameter and length tolerances may be due to the elasticity of the powder and spring of the press.

Factors Affecting Design of Briquetting Tools.—High-speed steel is recommended for dies and punches and oil-hardening steel for strippers and knock-outs. One manufacturer specifies dimensional tolerances of 0.0002 inch and super-finished surfaces for these tools. Because of the high pressures employed and the abrasive character of certain refractory materials used in some powdered metal composition, there is frequently a tendency toward severe wear of dies and punches. In such instances, carbide inserts, chrome plating, or highly resistant die steels are employed. With regard to the shape of the die, corner radii, fillets, and bevels should be used to avoid sharp corners. Feather edges, threads, and reentrant angles are usually impracticable. The making of punches and dies is particularly exacting because allowances must be made for changes in dimensions due to growth after pressing and shrinkage or growth during sintering.

Flame Spraying Process

In this process, the forerunner of which was called the metal spraying process, metals, alloys, ceramics, and cermets are deposited on metallic or other surfaces. The object may be to build up worn or undersize parts, provide wear-resisting or corrosion-resisting surfaces, correct defective castings, etc.

Different types of equipment are available that provide the means of depositing the coatings on the surfaces. In one, wire is fed automatically through the nozzle of the spray gun; then a combustible gas, oxygen and compressed air serve to melt and blow the atomized metal against the surface to be coated. The gas usually used is acetylene but other gases may be used. Any desired thickness of metal may be deposited and the metals include steels, ranging from low to high carbon content, various brass and bronze compositions, babbitt metal, tin, zinc, lead, nickel, copper, and aluminum. The movement of the spray gun, in covering a given surface, is controlled either mechanically or by hand. In enlarging worn or undersize shafts, spindles, etc., it is common practice to clamp the gun in a lathe toolholder and use the feed mechanism to traverse the gun at a uniform rate while the metal is being deposited upon the rotating workpiece. The spraying operation may be followed by machining or grinding to obtain a more precise dimension.

Some typical production applications using the wire process are the coating of automotive exhaust valves, refinishing of transfer ink rollers for the printing industry and the rebuilding of worn truck clutch plates. Other production applications include the metallizing of glass meter box windows, the spraying of aluminum onto cloth gauze to produce electrolytic condenser plates, and the spraying of zinc or copper for coating ceramic insulators.

With another type of equipment, metal, refractory, and ceramic powder are used instead of wire. Ordinarily this equipment employs the use of two gases, oxygen and a fuel gas. The fuel gas is usually acetylene but in some instances hydrogen may be used. When hand-held, a small reservoir supplies the powder to the equipment but a larger reservoir is used for lathe-mounted equipment or for large-scale production work. The four basic types of coating powders used with this equipment are ceramics, oxidation-resistant metals and alloys, self-bonding alloys, and alloys for fused coatings. These powders are used to produce wear-resistant, corrosion-resistant, heat-resistant, and electrically conductive coatings.

Still other equipment employs the use of plasma flame with which vapors of materials are raised to a higher energy level than the ordinary gaseous state. Its use raises the temperature ceiling and provides a controlled atmosphere by permitting employment of an inert or chemically inactive gas so that chemical action, such as oxidation, during the heating and application of the spray material can be controlled. The temperatures that can be obtained with commercially available plasma equipment often exceed 30,000 degrees F but for most plasma flame spray processes the temperature range of from 12,000 to 20,000 degrees F is optimum. Plasma flame spray materials include alumina, zirconia, tungsten, molybdenum, tantalum, copper, aluminum, carbides, and nickel-base alloys.

Regardless of the equipment used, what is important is the proper preparation of the surface that will receive the sprayed coating. Preparation activities include the degreasing or solvent cleaning of the surface, undercutting of the surface to provide room for the proper coating thickness, abrasive or grit blasting the substrate to provide a roughened surface, grooving (in the case of flat surfaces) or rough threading (in the case of cylindrical work) the surface to be coated, preheating the base metal. Methods of obtaining a bond between the sprayed material and the substrate are: heating the base, roughening the base, or spraying a "self-bonding" material onto a smooth surface; however, heating alone is seldom used in machine element work as the elevated temperatures required to obtain the proper bond causes problems of warpage and surface corrosion.

METAL JOINING, CUTTING, AND SURFACING

Metals may be joined without using fasteners by employing soldering, brazing, and welding. Soldering involves the use of a non-ferrous metal whose melting point is below that of the base metal and in all cases below 800 degrees F. Brazing entails the use of a non-ferrous filler metal with a melting point below that of the base metal but above 800 degrees F. In fusion welding, abutting metal surfaces are made molten, are joined in the molten state, and then allowed to cool. The use of a filler metal and the application of pressure are considered to be optional in the practice of fusion welding.

Soldering

Soldering employs lead- or tin-base alloys with melting points below 800 degrees F and is commonly referred to as soft soldering. Use of hard solders, silver solders and spelter solders which have silver, copper, or nickel bases and have melting points above 800 degrees F is known as brazing. Soldering is used to provide a convenient joint that does not require any great mechanical strength. It is used in a great many instances in combination with mechanical staking, crimping or folding, the solder being used only to seal against leakage or to assure electrical contact. The accompanying table gives some of the properties and uses of various solders that are generally available.

Forms Available.—Soft solders can be obtained in bar, cake, wire, pig, slab ingot, ribbon, segment, powder, and foil-form for various uses to which they are put. In bar form they are commonly used for hand soldering. The pigs, ingots, and slabs are used in operations that employ melting kettles. The ribbon, segment, powder and foil forms are used for special applications and the cake form is used for wiping. Wire forms are either solid or they contain acid or rosin cores for fluxing. These wire forms, both solid and core containing, are used in hand and automatic machine applications. Prealloyed powders, suspended in a fluxing medium, are frequently applied by brush and, upon heating, consistently wet the solderable surfaces to produce a satisfactory joint.

Fluxes for Soldering.—The surfaces of the metals being joined in the soldering operation must be clean in order to obtain an efficient joint. Fluxes clean the surfaces of the metal in the joint area by removing the oxide coating present, keep the area clean by preventing formation of oxide films, and lower the surface tension of the solder thereby increasing its wetting properties. Rosin, tallow, and stearin are mild fluxes which prevent oxidation but are not too effective in removing oxides present. Rosin is used for electrical applications since the residue is non-corrosive and non-conductive. Zinc chloride and ammonium chloride (sal ammoniac), used separately or in combination, are common fluxes that remove oxide films readily. The residue from these fluxes may in time cause trouble, due to their corrosive effects, if they are not removed or neutralized. Washing with water containing about 5 ounces of sodium citrate (for non-ferrous soldering) or 1 ounce of trisodium phosphate (for ferrous and non-ferrous soldering) per gallon followed by a clear water rinse or washing with commercial water-soluble detergents are methods of inactivating and removing this residue.

Methods of Application.—Solder is applied using a soldering iron, a torch, a solder bath, electric induction or resistance heating, a stream of hot neutral gas or by wiping. Clean surfaces which are hot enough to melt the solder being applied or accept molten solder are necessary to obtain a good clean bond. Pelts being soldered should be free of oxides, dirt, oil, and scale. Scraping and the use of abrasives as well as fluxes are resorted to for preparing surfaces for soldering. The procedures followed in soldering aluminum, magnesium and stainless steel differ somewhat from conventional soldering techniques and are indicated in the material which follows

Properties of Soft Solder Alloys *Appendix, ASTM:B 32-70*

Nominal Composition[a] Per Cent				Specific Gravity[b]	Melting Ranges,[c] Degrees Fahrenheit		Uses
Sn	Pb	Sb	Ag		Solidus	Liquidus	
70	30	…	…	8.32	361	378	For coating metals.
63	37	…	…	8.40	361	361	As lowest melting solder for dip and hand soldering methods.
60	40	…	…	8.65	361	374	"Fine Solder." For general purposes, but particularly where the temperature requirements are critical.
50	50	…	…	8.85	361	421	For general purposes. Most popular of all.
45	55	…	…	8.97	361	441	For automobile radiator cores and roofing seams.
40	60	…	…	9.30	361	460	Wiping solder for joining lead pipes and cable sheaths. For automobile radiator cores and heating units.
35	65	…	…	9.50	361	477	General purpose and wiping solder.
30	70	…	…	9.70	361	491	For machine and torch soldering.
25	75	…	…	10.00	361	511	For machine and torch soldering.
20	80	…	…	10.20	361	531	For coating and joining metals. For filling dents or seams in automobile bodies.
15	85	…	…	10.50	440[d]	550	For coating and joining metals.
10	90	…	…	10.80	514[d]	570	For coating and joining metals.
5	95	…	…	11.30	518	594	For coating and joining metals.
40	58	2	…	9.23	365	448	Same uses as (50-50) tin-lead but not recommended for use on galvanized iron.
35	63.2	1.8	…	9.44	365	470	For wiping and all uses except on galvanized iron.
30	68.4	1.6	…	9.65	364	482	For torch soldering or machine soldering, except on galvanized iron.
25	73.7	1.3	…	9.96	364	504	For torch and maching soldering, except on galvanized iron.
20	79	1	…	10.17	363	517	For machine soldering and coating of metals, tipping, and like uses, but not recommended for use on galvanized iron.
95	…	5	…	7.25	452	464	For joints on copper in electrical, plumbing and heating work.
…	97.5	…	2.5	11.35	579	579	For use on copper, brass, and similar metals with torch heating. Not recommended in humid environments due to its known susceptibility to corrosion.
1	97.5	…	1.5	11.28	588	588	For use on copper, brass, and similar metals with torch heating.

[a] Abbreviations of alloying elements are as follows: Sn, tin; Pb, lead; Sb, antimony; and Ag, silver.

[b] The specific gravity multiplied by 0.0361 equals the density in pounds per cubic inch.

[c] The alloys are completely solid below the lower point given, designated "solidus," and completely liquid above the higher point given, designated "liquidus." In the range of temperatures between these two points the alloys are partly solid and partly liquid.

[d] For some engineering design purposes, it is well to consider these alloys as having practically no mechanical strength above 360 degrees F.

Soldering Aluminum: Two properties of aluminum which tend to make it more difficult to solder are its high thermal conductivity and the tenacity of its ever-present oxide film. Aluminum soldering is performed in a temperature range of from 550 to 770 degrees F, compared to 375 to 400 degrees F temperature range for ordinary metals, because of the metal's high thermal conductivity. Two methods can be used, one using flux and one using abrasion. The method employing flux is most widely used and is known as flow soldering. In this method flux dissolves the aluminum oxide and keeps it from re-forming. The flux should be fluid at soldering temperatures so that the solder can displace it in the joint. In the friction method the oxide film is mechanically abraded with a soldering iron, wire brush, or multi-toothed tool while being covered with molten solder. The molten solder keeps the oxygen in the atmosphere from reacting with the newly-exposed aluminum surface; thus wetting of the surface can take place.

The alloys that are used in soldering aluminum generally contain from 50 to 75 per cent tin with the remainder zinc.

The following aluminum alloys are listed in order of ease of soldering: commercial and high-purity aluminum; wrought alloys containing not more than 1 per cent manganese or magnesium; and finally the heat-treatable alloys which are the most difficult.

Cast and forged aluminum parts are not generally soldered.

Soldering Magnesium: Magnesium is not ordinarily soldered to itself or other metals. Soldering is generally used for filling small surface defects, voids or dents in castings or sheets where the soldered area is not to be subjected to any load. Two solders can be used: one with a composition of 60 per cent cadmium, 30 per cent zinc, and 10 per cent tin has a melting point of 315 degrees F; the other has a melting point of 500 degrees F and has a nominal composition of 90 per cent cadmium and 10 per cent zinc.

The surfaces to be soldered are cleaned to a bright metallic luster by abrasive methods before soldering. The parts are preheated with a torch to the approximate melting temperature of the solder being used. The solder is applied and the surface under the molten solder is rubbed vigorously with a sharp pointed tool or wire brush. This action results in the wetting of the magnesium surface. To completely wet the surface, the solder is kept molten and the rubbing action continued. The use of flux is not recommended.

Soldering Stainless Steel: Stainless steel is somewhat more difficult to solder than other common metals. This is true because of a tightly adhering oxide film on the surface of the metal and because of its low thermal conductivity. The surface of the stainless steel must be thoroughly cleaned. This can be done by abrasion or by clean white pickling with acid. Muriatic (hydrochloric) acid saturated with zinc or combinations of this mixture and 25 per cent additional muriatic acid, or 10 per cent additional acetic acid, or 10 to 20 per cent additional water solution of orthophosphoric acid may all be used as fluxes for soldering stainless steel. Tin-lead solder can be used successfully. Because of the low thermal conductivity of stainless steel, a large soldering iron is needed to bring the surfaces to the proper temperature. The proper temperature is reached when the solder flows freely into the area of the joint. Removal of the corrosive flux is important in order to prevent joint failure. Soap and water or a suitable commercial detergent may be used to remove the flux residue.

Ultrasonic Fluxless Soldering.—This more recently introduced method of soldering makes use of ultrasonic vibrations which facilitates the penetration of surface films by the molten solder thus eliminating the need for flux. The equipment offered by one manufacturer consists of an ultrasonic generator, ultrasonic soldering head which includes a transducer coupling, soldering tip, tip heater, and heating platen. Metals that can be soldered by this method include aluminum, copper, brass, silver, magnesium, germanium, and silicon.

Brazing

Brazing is a metal joining process which uses a non-ferrous filler metal with a melting point below that of the base metals but above 800 degrees F. The filler metal wets the base metal when molten in a manner similar to that of a solder and its base metal. There is a slight diffusion of the filler metal into the hot, solid base metal or a surface alloying of the base and filler metal. The molten metal flows between the close-fitting metals because of capillary forces.

Filler Metals for Brazing Applications.— Brazing filler metals have melting points that are lower than those of the base metals being joined and have the ability when molten to flow readily into closely fitted surfaces by capillary action. The commonly used brazing metals may be considered as grouped into the seven standard classifications shown in Table . These are aluminum-silicon; copper-phosphorus; silver; nickel; copper and copper-zinc; magnesium; and precious metals.

The solidus and liquidus are given in the table instead of the melting and flow points in order to avoid confusion. The solidus is the highest temperature at which the metal is completely solid or, in other words, the temperature above which the melting starts. The liquidus is the lowest temperature at which the metal is completely liquid, that is, the temperature below which the solidification starts.

Fluxes for Brazing.— In order to obtain a sound joint the surfaces in and adjacent to the joint must be free from dirt, oil, and oxides or other foreign matter at the time of brazing. Cleaning may be achieved by chemical or mechanical means. Some of the mechanical means employed are filing, grinding, scratch brushing and machining. The chemical means include the use of trisodium phosphate, carbon tetrachloride, and trichloroethylene for removing oils and greases.

Fluxes are used mainly to prevent the formation of oxides and to remove any oxides on the base and filler metals. They also promote free flow of the filler metal during the course of the brazing operation.

They are made available in the following forms: powders; pastes or solutions; gases or vapors; and as coatings on the brazing rods.

In the powder form a flux can be sprinkled along the joint, provided that the joint has been preheated sufficiently to permit the sprinkled flux to adhere and not be blown away by the torch flame during brazing. A thin paste or solution is easily applied and when spread on evenly, with no bare spots, gives a very satisfactory flux coating. Gases or vapors are used in controlled atmosphere furnace brazing where large amounts of assemblies are mass-brazed. Coatings on the brazing rods protect the filler metal from becoming oxidized and eliminate the need for dipping rods into the flux, but it is recommended that flux be applied to the base metal since it may become oxidized in the heating operation. No matter which flux is used, however, it performs its task only if it is chemically active at the brazing temperature.

Chemical compounds incorporated into brazing fluxes include borates (sodium, potassium, lithium, etc.), fused borax, fluoborates (potassium, sodium, etc.), fluorides (sodium, potassium, lithium, etc.), chlorides (sodium, potassium, lithium), acids (boric, calcined boric acid), alkalies (potassium hydroxide, sodium hydroxide), wetting agents, and water (either as water of crystallization or as an addition for paste fluxes). The accompanying Table 3 provides a guide which will aid in the selection of brazing fluxes that are available commercially.

Table 1. Brazing Filler Metals [Based on Specification and Appendix of AWS (American Welding Society) A5.8–81]

AWS Classification[a]	Nominal Composition,[b] Per Cent						Temperature, Degrees F			Standard Form[c]	Uses
	Ag	Cu	Zn	Al	Ni	Other	Solidus	Liquidus	Brazing Range		
BAlSi-2	…	…	…	92.5	…	Si, 7.5	1070	1135	1110–1150	7	For joining the following aluminum alloys: 1060, EC, 1100, 3003, 3004, 5005, 5050, 6053, 6061, 6062, 6063, 6951 and cast alloys A612 and C612. All of these filler metals are suitable for furnace and dip brazing. BAlSi-3, -4 and -5 are suitable for torch brazing. Used with lap and tee joints rather than butt joints. Joint clearances run from .006 to .025 inch.
BAlSi-3	…	4	…	86	…	Si, 10	970	1085	1160–1120	2, 3, 5	
BAlSi-4	…	…	…	88	…	Si, 12	1070	1080	1080–1120	2, 3, 4, 5	
BAlSi-5	…	…	…	90	…	Si, 10	1070	1095	1090–1120	7	
BAlSi-6	…	…	…	90	…	Si, 7.5 Mg. 2.5	1038	1125	1110–1150	7	BAlSi-6 through -11 are vacuum brazing filler metals. Magnesium is present as an O_2 getter. When used in vacuum, solidus & liquidus temperatures are different from those shown.
BAlSi-7	…	…	…	88.5	…	Si, 10 Mg, 1.5	1038	1105	1090–1120	7	
BAlSi-8	…	…	…	86.5	…	Si, 12 Mg, 1.5	1038	1075	1080–1120	2, 7	
BAlSi-9	…	…	…	87	…	Si, 12 Mg, 0.3	1044	1080	1080–1120	7	
BAlSi-10	…	…	…	86.5	…	Si, 11 Mg, 2.5	1038	1086	1080–1120	2	
BAlSi-11	…	…	…	88.4	…	Si, 10 Mg, 1.5 Bi, 0.1	1038	1105	1090–1120	7	
BCuP-1	…	95	…	…	…	P, 5	1310	1695	1450–1700	1	For joining copper and its alloys with some limited use on silver, tungsten and molybdenum. Not for use on ferrous or nickel-base alloys. Are used for cupro-nickels but caution should be exercised when nickel content is greater than 30 per cent. Suitable for all brazing processes. Lap joints recommended but butt joints may be used. Clearances used range from .001 to .005 inch.
BCuP-2	…	93	…	…	…	P, 7	1310	1460	1350–1550	2, 3, 4	
BCuP-3	5	89	…	…	…	P, 6	1190	1485	1300–1500	2, 3, 4	
BCuP-4	6	87	…	…	…	P, 7	1190	1335	1300–1450	2, 3, 4	
BCuP-5	15	80	…	…	…	P, 5	1190	1475	1300–1500	1, 2, 3, 4	
BCuP-6	2	91	…	…	…	P, 7	1190	1450	1350–1500	2, 3, 4	
BCuP-7	5	88	…	…	…	P, 6.8	1190	1420	1300–1500	2, 3, 4	

[a] These classifications contain chemical symbols preceded by "B" which stands for brazing filler metal.

[b] These are nominal compositions. Trace elements may be present in small amounts and are not shown. Abbreviations used are: Ag, silver; Cu, copper; Zn, zinc; Al, aluminum; Ni, nickel; Ot, other; Si, silicon; P, phosphorus; Cd, cadmium; Sn, tin; Li, lithium; Cr, chromium; B, boron; Fe, iron; O, oxygen; Mg, magnesium; W, tungsten; Pd, palladium; and Au, gold.

[c] Numbers specify standard forms as follows: 1, strip; 2, wire; 3, rod; 4, powder; 5, sheet; 6, paste; 7, clad sheet or strip; and 8, transfer tape.

Table 2. Brazing Filler Metals [Based on Specification and Appendix of AWS (American Welding Society) A5.8–81]

AWS Classification[a]	Nominal Composition,[b] Per Cent						Temperature, Degrees F			Standard Form[c]	Uses
	Ag	Cu	Zn	Al	Ni	Other	Solidus	Liquidus	Brazing Range		
BAg-1	45	15	16	…	…	Cd, 24	1125	1145	1145–1400	1, 2, 4	For joining most ferrous and nonferrous metals except aluminum and magnesium. These filler metals have good brazing properties and are suitable for preplacement in the joint or for manual feeding into the joint. All methods of heating may be used. Lap joints are generally used; however, butt joints may be used. Joint clearances of .002 to .005 inch are recommended. Flux is generally required.
BAg-1a	50	15.5	16.5	…	…	Cd, 18	1160	1175	1175–1400	1, 2, 4	
BAg-2	35	26	21	…	…	Cd, 18	1125	1295	1295–1550	1, 2, 4, 7	
BAg-2a	30	27	23	…	…	Cd, 20	1125	1310	1310–1550	1, 2, 4	
BAg-3	50	15.5	15.5	…	3	Cd, 16	1170	1270	1270–1500	1, 2, 4, 7	
BAg-4	40	30	28	…	2	…	1240	1435	1435–1650	1, 2	
BAg-5	45	30	25	…	…	…	1250	1370	1370–1550	1, 2	
BAg-6	50	34	16	…	…	…	1270	1425	1425–1600	1, 2	
BAg-7	56	22	17	…	…	Sn, 5	1145	1205	1205–1400	1, 2	
BAg-8	72	28	…	…	…	…	1435	1435	1435–1650	1, 2, 4	
BAg-8a	72	27.8	…	…	…	Li, 2.	1410	1410	1410–1600	1, 2	
BAg-13	54	40	5	…	1	…	1325	1575	1575–1775	1, 2	
BAg-13a	56	42	…	…	2	…	1420	1640	1600–1800	1, 2	
BAg-18	60	30	…	…	…	Sn, 10	1115	1325	1325–1550	1, 2	
BAg-19	92.5	7.3	…	…	…	Li, 2	1435	1635	1610–1800	1, 2	
BAg-20	30	38	32	…	…	…	1250	1410	1410–1600	1, 2, 4	
BAg-21	63	28.5	…	…	2.5	Sn, 6	1275	1475	1475–1650	1, 2, 4	
BAg-22	49	16	23	…	4.5	Mn, 7.5	1260	1290	1290–1525	1, 2, 4, 7	
BAg-23	85	…	…	…	…	Mn, 15	1760	1780	1780–1900	1, 2, 4	
BAg-24	50	20	28	…	2	…	1220	1305	1305–1550	1, 2	
BAg-25	20	40	35	…	…	Mn, 5	1360	1455	1455–1555	2, 4	
BAg-26	25	38	33	…	2	Mn, 2	1305	1475	1475–1600	1, 2, 4, 7	
BAg-27	25	35	26.5	…	…	Cd, 13.5	1125	1375	1375–1575	1, 2, 4	
BAg-28	40	30	28	…	…	Sn, 2	1200	1310	1310–1550	1, 2, 4	

AWS Classification[a]	Nominal Composition,[b] Per Cent						Temperature, Degrees F			Standard Form[c]	Uses
	Ni	Cu	Cr	B	Si	Other	Solidus	Liquidus	Brazing Range		
BNi-1	74	…	14	3.5	4	Fe, 4.5	1790	1900	1950–2200	1, 2, 3, 4, 8	For brazing AISI 300 and 400 series stainless steels, and nickel- and cobalt-base alloys. Particularly suited to vacuum systems and vacuum tube applications because of their very low vapor pressure. The limiting element is chromium in those alloys in which it is employed. Special brazing procedures required with filler metal containing manganese.
BNi-2	82.5	…	7	3	4.5	Fe, 3	1780	1830	1850–2150	1, 2, 3, 4, 8	
BNi-3	91	…	…	3	4.5	Fe, 1.5	1800	1900	1850–2150	1, 2, 3, 4, 8	
BNi-4	93.5	…	…	1.5	3.5	Fe, 1.5	1800	1950	1850–2150	1, 2, 3, 4, 8	
BNi-5	71	…	19	…	10	…	1975	2075	2100–2200	1, 2, 3, 4, 8	
BNi-6	89	…	…	…	…	P, 11	1610	1610	1700–1875	1, 2, 3, 4, 8	
BNi-7	77	…	13	…	…	P, 10	1630	1630	1700–1900	1, 2, 3, 4, 8	
BNi-8	65.5	4.5	…	…	7	Mn, 23	1800	1850	1850–2000	1, 2, 3, 4, 8	

Table 2. *(Continued)* **Brazing Filler Metals [Based on Specification and Appendix of AWS (American Welding Society) A5.8–81]**

AWS Classification[a]	Nominal Composition,[b] Per Cent						Temperature, Degrees F			Standard Form[c]	Uses
	Ag	Cu	Zn	Al	Ni	Other	Solidus	Liquidus	Brazing Range		
BCu-1	…	100	…	…	…	…	1980	1980	2000–2100	1, 2	For joining various ferrous and nonferrous metals. They can also be used with various brazing processes. Avoid overheating the Cu-Zn alloys. Lap and butt joints are commonly used.
BCu-1a	…	99	…	…	…	Ot, 1	1980	1980	2000–2100	4	
BCu-2	…	86.5	…	…	…	O, 13.5	1980	1980	2000–2100	6	
RBCuZn-A	…	59	…	…	…	Zn, 41	1630	1650	1670–1750	1, 2, 3	
RBCuZn-C	…	58	…	…	0.1	Zn, 40 Fe, 0.7 Mn, 0.3 Sn, 1	1590	1630	1670–1750	2	
RBCuZn-D	10	48	…	…	0.2	Zn, 42	1690	1715	1720–1800	1, 2, 3	
BCuZn-E	…	50	…	…	…	Zn, 50	1595	1610	1610–1725	1, 2, 3, 4, 5	
BCuZn-F	…	50	…	…	…	Zn, 46.5 Sn, 3.5	1570	1580	1580–1700	1, 2, 3, 4, 5	
BCuZn-G	…	70	…	…	…	Zn, 30	1680	1750	1750–1850	1, 2, 3, 4, 5	
BCuZn-H	…	80	…	…	…	Zn, 20	1770	1830	1830–1950	1, 2, 3, 4, 5	
BMg-1	…	…	…	…	…	d	830	1100	1120–1160	2, 3	BMg-1 is used for joining AZ10A, K1A, and M1A magnesium-base metals.
BAu-1	…	63	…	…	…	Au, 37	1815	1860	1860–2000	1, 2, 4	For brazing of iron, nickel, and cobalt-base metals where resistance to oxidation or corrosion is required. Low rate of interaction with base metal facilitates use on thin base metals. Used with induction, furnace, or resistance heating in a reducing atmosphere or in a vacuum and with no flux. For other applications, a borax-boric acid flux is used.
BAu-2	…	20.5	…	…	…	Au, 79.5	1635	1635	1635–1850	1, 2, 4	
BAu-3	3	62.5	…	…	…	Au, 34.5	1785	1885	1885–1995	1, 2, 4	
BAu-4	18.5	…	…	…	…	Au, 81.5	1740	1740	1740–1840	1, 2, 4	
BAu-5	36	…	…	…	…	Au, 30 Pd, 34	2075	2130	2130–2250	1, 2, 4	
BAu-6	22	…	…	…	…	Au, 70 Pd, 8	1845	1915	1915–2050	1, 2, 4	
BCo-1	17	…	…	…	8	Cr, 19 W, 4 B, 0.8 C, 0.4 Co, 59	2050	2100	2100–2250	1, 3, 4, 8	Generally used for high temperature properties and compatability with cobalt-base metals.

[a] These classifications contain chemical symbols preceded by "B" which stands for brazing filler metal.

[b] These are nominal compositions. Trace elements may be present in small amounts and are not shown. Abbreviations used are: Ag, silver; Cu, copper; Zn, zinc; Al, aluminum; Ni, nickel; Ot, other; Si, silicon; P, phosphorus; Cd, cadmium; Sn, tin; Li, lithium; Cr, chromium; B, boron; Fe, iron; O, oxygen; Mg, magnesium; W, tungsten; Pd, palladium; and Au, gold.

[c] Numbers specify standard forms as follows: 1, strip; 2, wire; 3, rod; 4, powder; 5, sheet; 6, paste; 7, clad sheet or strip; and 8, transfer tape.

[d] Al, 9; Zn, 2; Mg, 89.

Table 3. Guide to Selection of Brazing Filler Metals and Fluxes

Base Metals Being Brazed	Filler Metals Recommended[a]	Flux				
		American Welding Society Brazing Flux Type No.	Effective Temperature Range, Degrees F	Ingredients Contained Therein	Form Supplied In	Method of Use[b]
All brazeable aluminum alloys	BA1Si	1	700 to 1190	Chlorides Fluorides	Powder	1, 2 3, 4
All brazeable magnesium alloys	BMg	2	900 to 1200	Chlorides Fluorides	Powder	3, 4
Alloys such as alumi-num-bronze; aluminum-brass con-taining addi-tions of aluminum of 0.5 per cent or more	BCuZn BCuP	4[c]	1050 to 1800	Chlorides Fluorides Borates Wetting agent	Paste or Powder	1, 2, 3
Titanium and zirconium in base alloys	BAg	6	700 to 1600	Chlorides Fluorides Wetting agent	Paste or Powder	1, 2, 3
Any other brazeable alloys not listed above	All brazing filler metals except BA1Si and BMg	3	700 to 2000	Boric acid Borates Fluorides Fluoborates (must contain fluorine com-pound) Wetting agent	Paste, Powder, or Liquid	1, 2, 3
	All brazing filler metals except BA1Si, BMg, and BAg 1 through BAg 7	5	1000 to 2200	Borax Boric acid Borates Wetting agent *No fluorine in any form*	Paste, Powder, or Liquid	1, 2, 3

[a] Abbreviations used in this column are as follows: B, brazing filler metal; Al, aluminum; Si, silicon; Mg, magnesium; Cu, copper; Zn, zinc; P, phosphorus; and Ag, silver.

[b] Explanation of numbering system used is as follows: 1—dry powder is sprinkled in joint region; 2—heated metal filler rod is dipped into powder or paste; 3—flux is mixed with alcohol, water, monochlorobenzene, etc., to form a paste or slurry; 4—flux is used molten in a bath.

[c] Types 1 and 3 fluxes, alone or in combination, may be used with some of these base metals also.

Methods of Steadying Work for Brazing.—Pieces to be joined by brazing after being properly jointed may be held in a stable position by means of clamping devices, spot welds, or mechanical means such as crimping, staking, or spinning. When using clamping devices care must be taken to avoid the use of devices containing springs for applying pressure because springs tend to lose their properties under the influence of heat. Care must also be taken to be sure that the clamping devices are no larger than is necessary for strength considerations, because a large metal mass in contact with the base metal near the brazing area would tend to conduct heat away from the area too quickly and result in an inefficient braze. Thin sections that are to be brazed are frequently held together by spot welds. It must be remembered that these spot welds may interfere with the flow of the molten brazing alloy and appropriate steps must be taken to be sure that the alloy is placed where it can flow into all portions of the joint.

Methods of Supplying Heat for Brazing.—The methods of supplying heat for brazing form the basis of the classification of the different brazing methods and are given as follows.

Torch or Blowpipe Brazing: Air-gas, oxy-acetylene, air-acetylene, and oxy-other fuel gas blowpipes are used to bring the areas of the joint and the filler material to the proper heat for brazing. The flames should generally be neutral or slightly reducing but in some instances some types of bronze welding require a slightly oxidizing flame.

Dip Brazing: Baths of molten alloy, covered with flux, or baths of molten salts are used for dip brazing. The parts to be brazed are first assembled, usually with the aid of jigs, and are dipped into the molten metal, then raised and allowed to drain. The molten alloy enters the joint by capillary action. When the salt bath is used, the filler metal is first inserted between the parts being joined, or, in the form of wire, is wrapped around the area of the joint. The brazing metal melts and flows into the joint, again by capillary action.

Furnace Brazing: Furnaces that are heated electrically or by gas or oil with auxiliary equipment that maintains a reducing or protective atmosphere and controlled temperatures therein are used for brazing large numbers of units, usually without flux.

Resistance Brazing: Heat is supplied by means of hot or incandescent electrodes. The heat is produced by the resistance of the electrodes to the flow of electricity and the filler metal is frequently used as an insert between the parts being joined.

Induction Brazing: Parts to be joined are heated by being placed near a coil carrying an electric current. Eddy current losses of the induced electric current are dissipated in the form of heat raising the temperature of the work to a point higher than the melting point of the brazing alloy. This method is both quick and clean.

Vacuum Furnace Brazing: Cold-wall vacuum furnaces, with electrical-resistance radiant heaters, and pumping systems capable of evacuating a conditioned chamber to moderate vacuum (about 0.01 micron) in 5 minutes are recommended for vacuum brazing. Metals commonly brazed in vacuum are the stainless steels, heat-resistant alloys, titanium, refractory metals, and aluminum. Fluxes and filler metals containing alloying elements with low boiling points or high vapor pressure are no t used.

Brazing Symbol Application.—ANSI/AWS A2.4.-79 symbols for brazing are also used for welding with the exception of the symbol for a scarf joint (see the diagram at the top of the following page, and the symbol for a scarf joint in the table *Basic Weld Symbols* on page 1394, for applications of brazing symbols). The second, third and fourth figures from the top of the next page show how joint clearances are indicated. If no special joint preparation is required, only the arrow is used with the brazing process indicated In the tail.

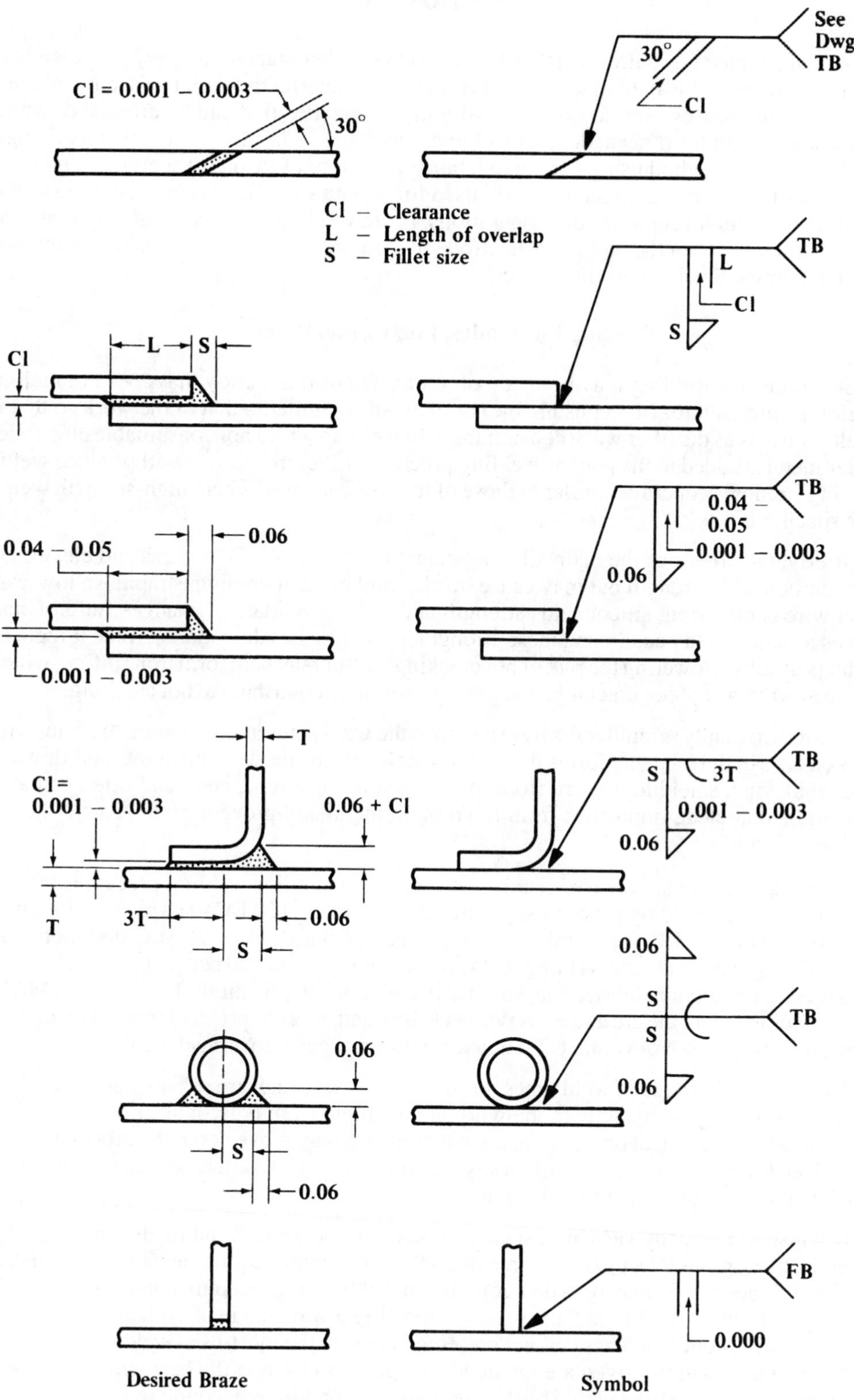

Typical Applications of Standard Brazing Symbols

WELDING

Welding of metals requires that they be heated to a molten state so that they fuse together. A filler wire or rod is held in the heated zone to add material that will replace metal consumed by the process and to produce a slightly raised area that can be dressed down to make a level surface if needed. Most welding operations today use an electric arc, though the autogenous method using a torch that burns a mixture of (usually) acetylene and oxygen gases to heat the components is still used for certain work. Lasers are also used as the heating medium for some welding operations. In arc welding, a low-voltage, high-current arc is struck between the end of an electrode in a holder and the work, generating intense heat that immediately melts tile surface.

Welding Electrodes, Fluxes, and Processes

Electrodes for welding may be made of a tungsten or other alloy that does not melt at welding temperatures (nonconsumable) or of an alloy similar to that of the work so that it melts and acts as the filler wire (consumable). In welding with a nonconsumable electrode, filler metal is added to the pool as welding proceeds. Filler metals that will produce welds having strength properties similar to those of the work are used where high-strength welds are specified.

Briefly, the effects of the main alloying elements in welding filler wires and electrodes are: carbon adds strength but may cause brittle weld metal if cooling is rapid, so low-carbon wire is preferred; silicon adds strength and reduces oxidation, changes fluidity, and gives a flatter weld bead; manganese strengthens and assists deoxidation, plus it reduces effects of sulfur, lowering the risk of hot cracking; sulfur may help form iron sulfide, which increases the risk of hot cracking; and phosphorus, may contribute to hot cracking.

Fluxes in (usually) granular form are added to the weld zone, as coatings on the filler wire or as a core in the tube that forms the (consumable) electrode. The flux melts and flows in the weld zone, shielding the arc from the oxygen in the atmosphere, and often contains materials that clean impurities from the molten metal and prevent grain growth during recrystallization.

Processes.—There are approximately 100 welding and allied welding processes but the four manual arc welding processes: gas metal arc welding (GMAW) (which is also commonly known as MIG for metal inert gas), flux-cored arc (FCAW), shielded metal arc (SMAW), gas tungsten arc welding (GTAW), account for over 90 per cent of the arc welding used in production, fabrication, structural, and repair applications. FCAW and SMAW use fluxes to shield the arc and FCAW uses fluxes and gases to protect the weld from oxygen and nitrogen. GMAW and GTAW use mixtures of gases to protect the weld.

There are two groups of weld types, groove and fillet, which are self-explanatory. Each type of weld may be made with the work at any angle from horizontal (flat) to inverted (overhead). In a vertical orientation, the electrode tip may move down the groove or fillet (vertical down), or up (vertical up). In any weld other than flat, skill is needed to prevent the molten metal falling from the weld area.

Because of the many variables, such as material to be welded and its thickness, equipment, fluxes, gases, electrodes, degree of skill, and strength requirements for the finished welds, it is not practicable to set up a complete list of welding recommendations that would have general validity. Instead, examples embracing a wide range of typical applications, and assuming common practices, are presented here for the most-used welding processes. The recommendations given are intended as a guide to finding the best approach to any welding job, and are to be varied by the user to fit the conditions encountered in the specific welding situation.

Gas Metal Arc Welding (GMAW)

The two most cost-effective manual arc welding processes are GMAW and FCAW. These two welding processes are used with more than 50 per cent of the arc welding consumable electrodes purchased. Gas metal arc welding modes extend from short-circuit welding, where the consumable electrode wire is melted into the molten pool in a rapid succession of short circuits during which the arc is extinguished, to pulsed and regular spray transfer, where a stream of fine drops and vaporized weld metal is propelled across the continuous arc gap by electromagnetic forces in the arc.

GMAW is the most-used welding process and the two most common GMAW low-carbon steel electrodes used for production welding in North America are the E70S-3 and E70S-6 from the ANSI/AWS Standard A5 series of specifications for arc welding. The E70S-3 contains manganese and silicon as deoxidants and is mainly used for welding low-carbon steels, using argon mixtures as shielding gases. The wire used in the E70S-6 electrodes has more silicon than wire used for the E70S-3 electrodes, and is preferred where straight CO_2 or argon mixes are used as the shielding gas or if the metal to be welded is contaminated. The deoxidizing properties of the E70S-6 electrode also may be beneficial for high-current, deep-penetration welds, and welds in which higher than normal impact-strength properties are required.

E80S-D2 wire contains more manganese and silicon, plus 0.5 per cent molybdenum for welding such steels as AISI 4130, and steels for high-temperature service. The argon + CO_2 mixture is preferred to exert the influence of argon's inertness over the oxidizing action of CO_2. E70S-2 electrodes contain aluminum, titanium, and zirconium to provide greater deoxidation action and are valuable for welding contaminated steel plate.

When the GMAW welding process is used for galvanized steels, minute welding cracks may be caused by the reaction of the zinc coating on the work with silicon in the electrode. Galvanized steel should be welded with an electrode having the lowest possible silicon content such as the E70S-3. For welding low-carbon and low-alloy steels with conventional argon mixture shielding gases, there is little difference between the E70S-3 and E70S-6.

Electrode Diameters.—One of the most important welding decisions is selecting the optimum GMAW electrode diameter. Selection of electrode diameters should be based on the material thickness, as shown for carbon and stainless steels in Table 1, the compatibility of the electrode current requirements with the material thickness, the mode of weld metal transfer, and the deposition rate potential shown in Table 2. The two most popular GMAW electrode sizes are 0.035 in. (1.0 mm) and 0.045 in. (1.2 mm). Diameters of electrodes used for GMAW exert a strong influence on cost of welding. Table 2 also shows how the weld deposition rate varies in short-circuit and spray transfer modes in welding carbon and stainless steels.

Table 1. GMAW Electrode Sizes for Welding Carbon and Stainless Steels

	Electrode Diameter			
Material Thickness	0.030 in. (0.8 mm)	0.035 in. (1.0 mm)	0.045 in. (1.2 mm)	0.062 in. (1.6 mm)
25 to 21 gage (0.020 to 0.032 in.)	yes	…	…	…
20 gage to 1⁄4 in. (0.036 to 0.25 in.)	…	yes	…	…
3⁄16 to 7⁄16 in. flat and horizontal	…	…	yes	…
1⁄2 in. and up	…	…	…	yes

The table is based on suitability of the electrode size to mode of weld transfer, material thickness, and cost effectiveness. If a smaller electrode size is selected, the lower deposition rates could increase welding costs by 20 to 60 per cent.

Table 2. Typical Maximum GMAW Deposition Rates for Carbon and Stainless Steels. Constant-Voltage 450-amp Power Source and Standard Wire Feeder

Weld transfer mode	Electrode Diameter			
	0.030 in. (0.8 mm)	0.035 in. (1.0 mm)	0.045 in. (1.2 mm)	0.062 in. (1.6 mm)
Short circuit	5 lb/h (2.3 kg/h)	7 lb/h (3.2 kg/h)	9 lb/h (4 kg/h)	…
Spray transfer	9 lb/h (4 kg/h)	11 lb/h (5 kg/h)	19 lb/h (8.6 kg/h)	21 lb/h (9.5 kg/h)

For the lowest-cost welds with GMAW electrodes larger than 0.030 in. in diameter, the power source should provide a minimum of 350 amps. The compatibility of the optimum current range of the 0.035-in. (1.0-mm) electrode and its deposition potential make it the first choice for welding of 20 gage to ¼ in. (0.88 to 6.4 mm) thicknesses. For welding thinner sheet metals of 25 to 21 gage, the optimum electrode diameter is 0.030 in. (0.8 mm). The 0.045-in. (1.2-mm) electrode is the most practical choice for spray transfer applications on materials over ¼ in. (6.4 mm) thick and thicker.

As an example, when welding ¼-in. (6.4-mm) thick steel, with 100 per cent arc-on time and a labor cost of $15/h, the deposition rate with a 0.035-in. (0.9-mm) electrode is approximately 11 lb/h (5 kg/h). The labor cost per lb at $15/h ÷ 11 lb/h = $1.36/lb ($3.00/kg). If an electrode of 0.045-in. (1.2-mm) diameter is used for the same application, the deposition rate is 16 lb/h (7.2 kg/h) and at a $15/h labor rate, the cost of weld metal deposited = $15/h ÷ 16 lb/h = $0.93/lb ($2.00/kg). The 0.045-in. diameter electrode would also cost less per pound than a smaller wire, and the weld time with the 0.045-in. electrode would be reduced, so less shielding gas also would be consumed.

GMAW Welding of Sheet Steel.—In GMAW, the short-circuit transfer mode is used to weld carbon steel, low-alloy steel, and stainless steel sheet of 24 gage (0.023 in., or 0.6 mm) to 11 gage (0.12 in., or 3 mm). The most common gage sizes welded with short-circuit transfer are 20 gage to 11 gage (0.88 to 3 mm) and the best GMAW electrode for these thin, sheet metal gages is the 0.035-in. (1-mm) diameter electrode. The short-circuit current requirements for these operations are typically 50 to 200 amps with voltages in the range of 14 to 22 volts. The optimum short-circuit voltage for the majority of applications is 16 to 18 volts.

Shielding Gases for Welding Carbon and Low-Alloy Steels.—With more than 40 GMAW gas mixtures available for welding carbon steels, low-alloy steels, and stainless steels, selection is often confusing. Reactive oxygen and carbon dioxide (CO_2) are added to argon to stabilize the arc and add energy to the weld. CO_2 can provide more energy to the weld than oxygen. As the CO_2 content in a shielding gas mixture is increased to certain levels, the voltage requirements are increased. Argon + oxygen mixtures will require lower voltages than mixtures containing argon with 10 to 25 per cent CO_2. Helium may also be added to argon if increased weld energy is required.

Shielding Gases for Short-Circuit Welding of Carbon Steels.—GMAW short-circuit transfer (SCT) is used mainly for welding thin metals of less than 10 gage, and gaps. With the SCT mode of weld metal transfer, the arc short circuits many times each second. The numerous short circuits switch the arc energy on and off. The short circuits and low current cause the transferred weld to freeze rapidly. Short-circuit transfer on carbon steel gage metals thicker than ¹⁄₁₆ in. (1.6 mm) requires a shielding gas that will provide substantial weld energy. For these applications, argon with 15–25 per cent CO_2 is recommended.

If short-circuit transfer is used on metals thinner than 18 gage (0.047 in., 1.2 mm), melt-through and distortion often occur. Melt-through and distortion can be reduced on very thin-gage carbon and low-alloy steels by using a shielding gas that provides less weld energy than argon + 15 to 25 per cent CO_2 mixes. Argon + oxygen mixtures can utilize lower voltages to sustain the arc. Argon mixed with 2 to 5 per cent oxygen is a practical mixture for thin carbon steel of less than 16 gage, where there is sensitivity to heat.

Shielding Gases for Spray Transfer Welding of Carbon Steels.—With GMAW spray transfer, all traditional argon gas mixtures will provide spatter-free spray weld transfer, depending on the electrode diameter and welding parameters used. The electrode diameter and the electrode current density influence the formation of the weld metal to be transferred. For example, with a 0.035-in. diameter electrode using a mixture containing argon 75 + CO_2 25 per cent, a small globular weld droplet is formed on the end of the electrode tip in the conventional spray transfer parameter range. With the same gas mixture, a 0.045-in. (1.2-mm) diameter electrode, and current above 330 amps, the globular formation disappears and the metal transfers in the spray mode.

Spatter potential stemming from shielding gas, with 0.035-in. (1.0-mm) and smaller diameter electrodes can be controlled by reducing the CO_2 content in the argon mixture to less than 21 per cent. Each different shielding gas will primarily influence the open arc spray transfer mode by variations in the weld energy provided through the welding voltage requirements.

Gas selection in spray transfer must be given careful consideration. In welding of clean cold-rolled carbon steel or low-alloy steel less than ⅜ in. (9.5 mm) thick, the energy potential of the arc is less important than it is for welding of steels thicker than ¼ in. (13 mm) or steels with mill scale. The energy level of the arc is also a key factor in welding steels for which higher than normal impact properties are specified.

A simple, practical multipurpose gas mixture for carbon and low-alloy steels is argon + 15 to 20 per cent CO_2, and a mixture of argon + 17 per cent CO_2 would be ideal. This two-part argon/CO_2 mixture provides higher weld energy than two-component argon + CO_2 mixtures having less than 10 per cent CO_2, argon + oxygen mixtures, or argon + CO_2 + oxygen tri-component mixtures. The argon + 17 per cent CO_2 mixture will provide an arc slightly less sensitive to mill scale than the other mixtures mentioned.

The argon + 17 per cent CO_2 mixture also has practical benefits in that it provides sufficient weld energy for all GMAW short-circuit and spray transfer applications with cylinder or bulk gases. The argon + 17 per cent CO_2 mixture may also be used for all-position FCAW electrodes in welding carbon steels, low-alloy steels, and stainless steels.

Shielding Gases for GMAW Welding of Stainless Steels.—The major problems encountered when using GMAW on stainless steels of thinner than 14 gage include controlling potential melt-through, controlling distortion, and black oxidation on the weld surface. These three welding problems have a common denominator, which is heat. The key to welding thin stainless steel is to minimize the potential heat when welding, by appropriate choice of gas mixture.

A popular gas mixture that is often recommended for GMAW welding of thin-gage stainless steel is the three-part helium gas mixture containing helium 90 + argon 7.5 + CO_2 2.5 per cent. In contrast to gas mixtures without helium, the helium tri-mixture requires the use of higher voltages to sustain the arc, which adds unnecessary heat to the heat-sensitive thin-gage welds.

A practical and lower-cost alternative for GMAW short-circuit transfer on stainless steels is an argon mixture with 2 to 4 per cent CO_2. The argon + CO_2 mixture allows use of lower voltages than is practical with argon/helium mixtures, and the lower voltages resulting from the argon + CO_2 mixture will help to reduce distortion and oxidation, and decrease the melt-through potential. The mixture that works with short-circuit transfer is also a logical practical choice for spray transfer welding of stainless steel because it is less oxidizing than argon/oxygen mixtures. Table 3 provides practical gas mixture recommendations for specific applications.

Table 3. Shielding Gases for Welding Carbon Steels and Stainless Steels

Application	Gas mixtures					
	Argon + Oxygen	Argon + CO_2 + Oxygen	Argon + 2–4% CO_2	Argon + 6–10% CO_2	Argon + 13–20% CO_2	Argon + 25% CO_2
Short-circuit melt-through problems; less than 20 gage	1	1	1	1	2	3
Short-circuit 18 to 11 gage	…	…	…	…	1	1
Spray if mill scale or surface problems; carbon steels	…	…	…	…	1	2
Spray if low energy required; carbon steel	1	1	1	1	…	…
Spray, best impact strengths, lowest porosity; carbon steels	…	…	…	…	1	…
Best single gas mixture for carbon steels	…	…	…	…	1	…
Short-circuit; stainless steels	…	…	1	…	…	…
Spray; stainless steels	2	…	1	…	…	…
Best single gas mixture for stainless and duplex steels	…	…	1	…	…	…

Preferred choice of shielding gas is 1, followed by 2 and 3.

For GMAW spray transfer welding of stainless steels thicker than 11 gage, the traditional GMAW shielding gas has been argon 98 + oxygen 2 per cent. The argon + oxygen mixture provides excellent, stable, spray transfer, but the oxygen promotes oxidation, leaving the weld with a black surface. To reduce the oxidation, the 2 per cent oxygen can be replaced with the less oxidizing 2–4 per cent CO_2.

Shielding Gases for GMAW Welding of Aluminum.—For GMAW welding of aluminum, helium is added to argon to provide additional weld energy, increasing penetration width, and reducing porosity potential. A gas mixture that has worked well in practice and can be used on the majority of aluminum applications is argon + 25 to 35 per cent helium. Mixtures with higher helium content, of 50 to 90 per cent, require voltages and flow rates that may be excessive for many established aluminum applications.

Welding Controls.—The two primary controls for welding with GMAW are the electrode wire feed control on the wire feeder and the voltage control on the power source. As shown in Fig. 1, these controls typically consist of switches and knobs but do not have the scales, seen enlarged at the upper left, that indicate combinations of wire feed rate, wire gage, volts, and amps. These scales have been added here to allow clearer explanation of the functioning of the wire feed control.

The typical wire feed unit provides maximum feed rates of 600 to 800 in./min. The scale surrounding the setting knob on a wire feed control unit usually has only 10 unnumbered graduations, somewhat like the hour markers on a clock face. On most machines, each of these graduations represents an adjustment of the feed rate of approximately 70 in./min. For each increase in the wire feed rate of 70 in./min, depending on the voltage, the welding current increases by approximately 20 to 40 amps, depending on the wire diameter and wire feed positions.

In Fig. 1, a black sector has been drawn in on the wire feed rate adjustment knob to indicate the range of wire feed rates usable with the gas mixture and the electrode diameter (gage) specified. The wire feed and voltage settings shown are for welding thin-gage carbon, low-alloy, or stainless steels with a 0.030- or 0.035-in. (0.8- or 1-mm) diameter electrode. The left edge of the sector on the wire feed knob is set to the eight o'clock position, corresponding to 70 in./min. The optimum voltage for this wire feed rate is 15. If a setting is too low, the knob is turned to the second (nine o'clock) or third (ten o'clock) position to increase the current. The voltage typically increases or decreases by 1 volt for each graduation of the wire feed quadrant.

The short-circuit transfer current range of 50 to 200 amps corresponds to a wire feed rate of 70 to 420 in./min, and is typically found between the eight and one o'clock positions on the scale, as indicated by the black sector on the knob in Fig. 1.

Diagrammatic quadrants have been added at the left in Fig. 1, to show the material thickness, voltage, and current that correspond to the setting of the wire feed rate adjustment knob. Optimum settings are easily made for short-circuit welding of sheet metals. When using a 0.030-in. (0.8-mm) or 0.035-in. (1-mm) diameter GMAW electrode, for instance, to weld 16-gage carbon or stainless steel with a conventional 200-to 450-amp constant-voltage power source and wire feeder, the wire feed control is set to the ten o'clock position for a feed rate of 210 in./min. With digital wire feed units, the short-circuit current range is typically between 100 and 400 in./min, so a good starting point is to set the wire feeder at 210 in./min. The welding voltage is set to 17.

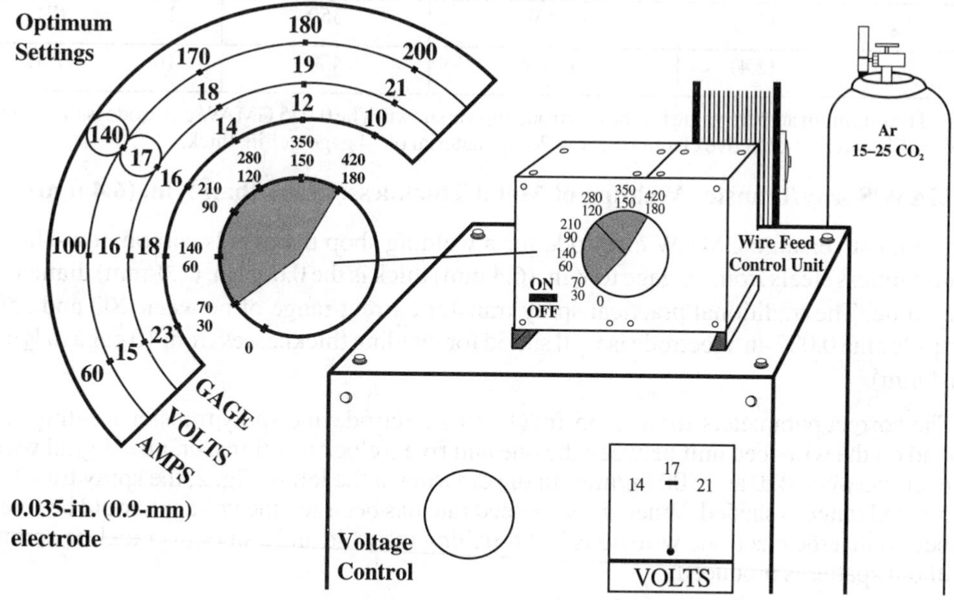

Fig. 1. Wire Feed Settings for Short-Circuit Welding of Carbon, Low-Alloy, and Stainless Steel Sheet.

Many welders set their parameters by an established mark on the equipment or by the sound of the arc as the weld is being made. The sound of the arc, influenced by the optimum current and voltage set, should be a consistent, smooth, crackling noise. If the SCT sound is harsh, the voltage should be increased slightly. If the sound is soft, the voltage should be decreased in volt increments until the sound becomes a smooth crackle. For welding metals thicker than 16 gage but less than 10 gage, the wire feed control should be moved to the eleven o'clock position (280 in./min), and the voltage reset to 18.

Welding of thicknesses less than 16 gage should be started with the wire feed control at the nine o'clock position (140 in./min) and the voltage control set to 16. The parameters discussed above apply when using argon mixtures containing 15 to 25 per cent CO_2.

GMAW Spray Transfer.—In the spray transfer mode, spatter is often caused by the voltage being set so low that the electrode runs into the weld, resulting in expulsion of molten metal from the weld pool. GMAW spray transfer is normally used for welding carbon, low-alloy, and stainless steels of a minimum thickness of ⅛ in. (3.2 mm).

In Table 4, typical deposition rates with a 0.045-in. (1.2-mm) carbon steel electrode are compared with rates for larger carbon steel GMAW and flux-cored electrodes. These welds are typically carried out in the flat and horizontal positions. The practical GMAW electrode diameters commonly used for spray transfer are 0.035-in. (1-mm), 0.045-in. (1.2-mm), and 0.062-in. (1.6-mm) diameter. The most cost-effective GMAW electrode that also has the greatest range of applications on metals over 3⁄16 in. thick is the 0.045-in. (1.2-mm) diameter size.

Table 4. Typical Deposition Rates for Carbon Steel Welding Electrodes

Electrode Diameter				Deposit Rates	
in.	(mm)	Electrode Type	Amperage[a]	lb/h	(kg/h)
0.035	(1.0)	GMAW	350	11	(5)
0.045	(1.2)	GMAW	380	13	(6)
0.062	(1.6)	GMAW	400	14	(6.4)
1⁄16	(1.6)	FCAW	350	15	(7)
3⁄32	(2.4)	FCAW	450	16	(7.3)

[a] The optimum ampere value for the electrode type is shown. The 0.045 GMAW electrode is the most versatile and cost-effective electrode for welding material of 14 gage to 1 in. thick.

GMAW Spray Transfer Welding of Metal Thicknesses Less than ¼ in. (6.4 mm).—

The most versatile GMAW electrode for a welding shop that welds carbon, low-alloy, and stainless steels from 20 gage to ¼ in. (6.4 mm) thick is the 0.035-in. (1.0-mm) diameter electrode. The traditional practical spray transfer current range of between 200 and 350 amps for the 0.035-in. electrode is well suited for welding thicknesses from 10 gage to ¼ in. (6.4 mm).

The correct parameters for a 0.035-in. (1-mm) electrode and spray transfer welding are found on the wire feed unit between the one and five o'clock positions, or, on a digital wire feeder, between 420 and 700 in./min. In the drawing at the left in Fig. 2, the spray transfer wire feed range is shaded. When the wire feed rate has been set, the voltage should be fine-tuned so that the electrode wire tip is just touching the weld and a smooth crackling sound without spatter is produced.

An optimum single spray transfer mode current setting for a 0.035-in. (1-mm) diameter electrode for most welding applications is approximately 280 amps with the wire feed set at the three o'clock position for 560 in./min. Manual or high-speed mechanized welds on material of 10 gage to ¼ in. thick can be made at the three o'clock wire feed position with only an adjustment for voltage, which should be set initially at 31 volts, when using an argon + CO_2 mixture.

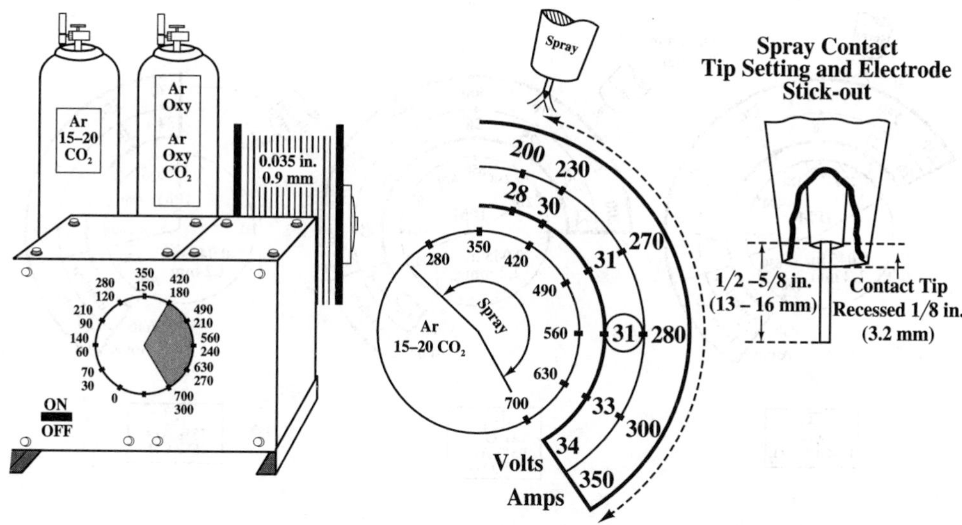

Fig. 2. GMAW Spray Transfer Parameters with 0.035-in. (0.9-mm) Diameter Electrodes

GMAW Spray Transfer for Metal Thicknesses ¼ in. (6.4 mm) and Up.—The 0.45-in. (1.2-mm) diameter is the most cost-effective GMAW electrode for spray transfer welding of carbon, low-alloy, and stainless steels ¼ in. and thicker. A 7⁄16-in. (11.2-mm) single-pass, no-weave, fillet weld can be produced with this electrode. If larger single-pass welds are required, use of flux-cored electrodes should be considered.

A 400-amp power source is a practical cost-effective unit to use with the 0.045-in. diameter electrode. Globular spray transfer, obtained at the ten o'clock position on the wire feed adjustment knob, starts at current levels of approximately 230 amps and requires a wire feed rate of approximately 210 in./min (90 mm/s). Most spray applications are carried out in the higher-energy, deeper-penetrating 270- to 380-amp range, or between twelve and two o'clock wire feed positions giving 350 to 490 in./rain (150 to 210 mm/s). In this range, in which there is minimum weld spatter, the weld deposits are in the form of minute droplets and vaporized weld metal.

The quadrants at the top in Fig. 3 show some typical settings for feed rate, voltage, and current, with different shielding gases. An ideal starting point with a 0.045-in. (1.2-mm) diameter electrode is to set the wire feed rate knob at the one o'clock position, or 420 in./min, at which rate the current drawn, depending on the power source used, should be about 320 to 350 amps. The best starting voltage for the 0.045-in. (1.2-mm) electrode is 30 volts. The arc length should then be set as indicated in Fig. 4. With current over 400 amps at 560 in./min, the 0.045-in. diameter electrode may produce a turbulent weld puddle and a digging arc, which can lead to lack of fusion, porosity, and cracks.

GMAW Spray Transfer with 0.062-in. (1.6-mm) Diameter Electrodes.—Electrode wire of 0.062-in. (1.6-mm) diameter is the largest size in normal use and is often chosen for its high deposition rates. Due to the high-current requirements for the spray transfer mode, use of these thicker electrodes is generally restricted to metal thicknesses of ½ in. (13 mm) and thicker. The high-current requirement reduces ease of welding. This electrode size is suitable for mechanized welding in which fillet welds greater than ⅜ in. (9.6 mm) are required.

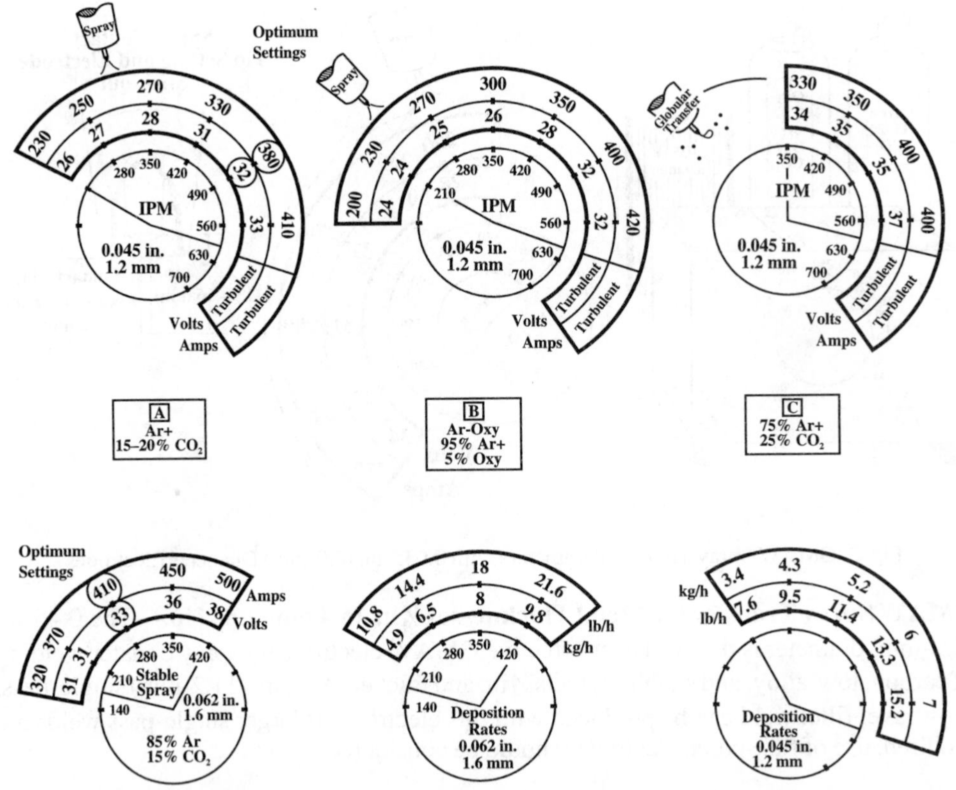

Fig. 3. GMAW Spray Transfer Parameters for Various Electrodes and Gases.

As indicated at the lower left in Fig. 3, the current range for 0.062-in. (1.6-mm) electrodes is narrow and most welds are made in the range of 360 to 420 amps, or between the ten and eleven o'clock positions on the wire feed control unit for 210 to 280 in./min (90 to 120 mm/s). The quadrants at the lower center and lower right in Fig. 3 show deposition rates in lb/h and kg/h for 0.045-in. (1.2-mm) and 0.062-in. (1.6-mm) diameter electrodes.

Some optimum settings for GMAW welding with a mixture of argon + 15 to 20 per cent CO_2 gases are given in Table 5.

Table 5. Optimum Settings for GMAW with Argon + 15–20 per cent CO_2

Diameters in.	mm	Wire Feed Rates in./min	m/min	Amps	Volts
0.035	1.0				
short circuit		210	5.3	140	17
spray transfer		560	14.2	280	29–30
0.045	1.2				
short circuit		210	5.3	190	18
spray transfer		420	10.7	380	30–31
0.052	1.4				
spray transfer		280	7.1	370	31–32
0.062	1.6				
spray transfer		280	7.1	410	31–32

Note: If argon + oxygen gas mixtures are used, voltage should be lowered by 1 to 4 volts for the spray transfer mode. The faster the weld travel speed, the lower the voltage required.

Spray Transfer Voltage.—The usual setting for spray transfer welding with commonly used electrode diameters is between 25 and 35 volts (see Fig. 4A). To set the optimum voltage for GMAW spray transfer, set the voltage initially so that it is too high, usually between 30 and 35 volts. With excess voltage, there should be a visible gap between the tip of the electrode and the weld, and the arc sound should be free from crackle. With the sequence shown in Fig. 4, the voltage should now be reduced until a consistent smooth crackle sound is produced. If the voltage is lowered too much, the electrode will run into the weld, making a harsh crackling sound, and the resulting weld expulsion will cause spatter.

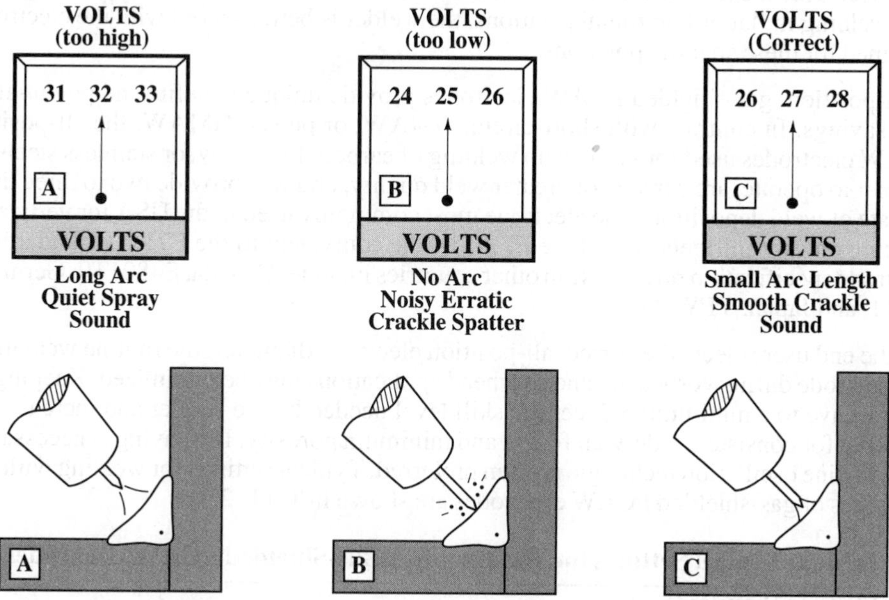

Fig. 4. Setting Optimum Voltage for GMAW Spray Transfer Welding.

Flux-Cored Arc Welding

FCAW welding offers unique benefits for specific applications, but flux-cored consumable electrodes cost more than the solid electrodes used in gas metal arc welding, so users need to be aware of FCAW benefits and disadvantages compared with those of GMAW welding. Generally, flux-cored electrodes designed for use without a shielding gas are intended for welding outdoors. Most indoor FCAW welding is done with gas-shielded FCAW welding electrode wire. Some Standards for gas-shielded FCAW electrodes for various countries are listed in Table 1.

Table 1. Standards for Gas-Shielded, Flux-Cored Welding Electrodes

Steel Type	Country	Standard
Low-Carbon Steels	USA	AWS A5.20
	Canada	CSA W48.5
	Japan	JIS Z3313
	Germany	DIN 8559
Low-Alloy Steels	USA	AWS A5.29
	Canada	CSA W48.3-M
	United Kingdom	BS 639-2492
Stainless Steels	USA	AWS A5.22

All-Position, Gas-Shielded Electrodes.—The term "all-position" does not necessarily mean that these electrodes are the best choice for all positions. Also, flux-cored electrodes may meet all standard specifications, but there will inevitably be subtle differences in weld transfer characteristics and recommended current, voltage, and other settings between electrodes made by different manufacturers. The chemistry and slag of the electrodes developed for welding in the flat and horizontal positions (E70T-X) typically provide superior results when they are used for flat and horizontal applications where the surface conditions of the plate are suspect or large, deep-penetrating welds are required. All-position electrodes are intended for, and best used in, vertical and overhead welds. For extensive welding in flat or horizontal positions, the welder is better served with the electrodes designed for these specific positions.

All-position, gas-shielded FCAW electrodes provide unique benefits and potential for cost savings. In contrast with short-circuit GMAW, or pulsed GMAW, the all-position FCAW electrodes used for vertical up welding of carbon, low-alloy, or stainless steels are simpler to operate, are capable of greater weld quality, and will provide two to three times the rate of weld deposition. The electrode most commonly used in the USA for vertical up welding on carbon steels is the type E71T-1. The equivalent to the E71T-1 standardized electrode specification now in use in other countries include: Canada, E4801T9; Germany, SGR1; and Japan, YFW 24.

If the end user selects the correct all-position electrode diameter, the routine weaving of the electrode during vertical up and overhead applications may be minimized. Keeping the weld weave to a minimum reduces the skill level needed by the welder and increases the potential for consistent side-wall fusion and minimum porosity. If weaving is necessary, a straight-line oscillation technique is often preferred. Typical settings for welding with various sizes of gas-shielded FCAW electrodes are shown in Table 2.

Table 2. Typical Settings for Welding with Gas-Shielded FCAW Electrodes

Electrode Diameter in.	(mm)	Vertical Up Welds		Flat and Horizontal Welds	
0.035	(1)	Feed rate Current Voltage	450 ipm 165 amps 28 volts	Feed rate Current Voltage	630 ipm 250 amps 30 volts
0.045	(1.2)	Feed rate Current Voltage	350 ipm 200 amps 25 volts	Feed rate Current Voltage	560 ipm 280 amps 26 Volts
0.052	(1.4)	Feed rate Current Voltage	240 ipm 200 amps 25 volts	Feed rate Current Voltage	520 ipm 300 amps 30 volts
0.062	(1.6)	Feed rate Current Voltage	210 ipm 240 amps 25 volts	Feed rate Current Voltage	350 ipm 340 amps 29 volts
3⁄32	(2.4)	…		Feed rate Current Voltage	210 ipm 460 amps 32 volts

Material Condition and Weld Requirements.—Practical considerations for selecting a gas-shielded, flux-cored electrode depend on the material condition and weld requirements. FCAW electrodes are beneficial if the surface of the material to be welded is contaminated with mill scale, rust, oil, or paint; the fillet weld size is to be over 3⁄8 in. (9.6 mm) wide (a GMAW single-pass fillet weld with an electrode size of 0.045 in. is typically 3⁄8 in. wide); the weld is vertical up, or overhead; the required impact strengths and other mechanical properties are above normal levels; crack resistance needs to be high; and increased penetration is required.

Selecting an FCAW Electrode.—Selection of FCAW electrodes is simplified by matching the characteristics of flux-cored types with the material and weld requirements listed above. Once the correct electrode type is selected, the next step is to choose the optimum size. In selecting an all-position, flux-cored electrode for vertical up or overhead welding, the steel thickness is the prime consideration. Selecting the optimum electrode diameter allows the high current capability of the electrode to be fully used to attain maximum deposition rates and allows use of the highest penetrating current without concern for excessive heat-related problems during welding. When used in the optimum current range, the deposited filler metal matches the required amount of filler metal for the specific size of the weld, determined by the plate thickness.

The following suggestions and recommendations are made for FCAW welding of carbon, low-alloy, and stainless steels having flat surfaces. For vertical up welds on steels of thicknesses from ⅛ to $^3/_{16}$ in. (3.2 to 4.8 mm) and for vertical up welds on pipe in the thickness range of ¼ to ½ in. (6.4 to 13 mm), consider the 0.035-in. (1.0-mm) diameter E71T-1 electrode. For vertical up welds on steels in the range of ¼ to ⅜ in. (6.4 to 9.6 mm) thickness, consider the 0.045-in. (1.2-mm) diameter E71T-1 electrode or, in nonheat-sensitive applications, the 0.062-in. (1.6-mm) diameter E71T-1 electrode. For vertical up welds on steels of over ⅜ in. (9.6 mm) thickness, consider the 0.062-in. (1.6-mm) diameter E71T-1 electrodes for optimum deposition rates. For flat and horizontal welds on steels of ⅜ to ¾ in. (9.6 to 19 mm) thickness, consider the $^1/_{16}$-in. (1.6-mm) diameter E70T-X electrodes. For flat and horizontal welds on steels over ¾ in. (19 mm) in thickness, consider the $^3/_{32}$-in. (2.4-mm) E70T-X electrodes.

FCAW Welding of Low-Carbon Steels.—Low-carbon steel is usually called carbon steel or mild steel. The most-used FCAW electrode for welding carbon steels in the flat or horizontal welding positions is the type E70T-1. which is suited to welding of reasonably clean steel using single-pass or multi-pass welds. Type E70T-2 has added deoxidizers and is suited to surfaces with mill scale or other contamination. This type is used when no more than two layers of weld are to be applied. Type E70T-5 is used for single-pass or multi-pass welds where superior impact properties or improved crack resistance are required. The E70T-X electrodes typically range in size from 0.045 to $^3/_{32}$ in. (1.2 to 2.4 mm) in diameter. Type E71T-1 all-position electrodes are available in diameters of 0.035 in. (1 mm) to 0.062 in. (1.6 mm). With the FCAW process, multi-pass welds are defined as a condition where three or more weld passes are placed on top of each other.

Settings for Gas-Shielded, All-Position, FCAW Electrodes.—The optimum setting range (volts and amps) for vertical up welding with all-position FCAW electrodes is rather narrow. The welder usually obtains the greatest degree of weld puddle control at the recommended low to medium current settings. The electrode manufacturers' recommended current range for an E71T-1 electrode of 0.045-in. (1.2-mm) diameter for vertical up welding may be approximately 130 to 250 amps. Using the 0.045-in. (1.2-mm) diameter electrode at 250 amps for a vertical up weld in ¼-in. (6.4-mm) thick steel, the welder may find that after 3 to 4 inches (75–100 mm) of weld, the weld heat built up in the steel being welded is sufficient to make the weld puddle fluidity increasingly difficult to control. Reducing the current to 160–220 amps will make it possible to maintain control over the weld puddle.

A typical optimum setting for a vertical up weld with an E71T-1, 0.035-in. (1.2-mm) diameter, all-position electrode is as follows. First, set the wire feed rate. If the wire feeder maximum rate is 650 to 750 in./min, the setting mark on the adjustment knob should be set between the one and two o'clock positions on the dial to obtain a feed rate of 450 in./min. If the wire feeder has a digital readout, the rate setting should be the same. At the 450-in./min setting, the welding current with the 0.035-in. all-position electrode should be optimized at between 160 and 170 amps. The welding voltage should be set at 27 to 28 volts with the electrode tip just touching the weld. If there is a gap causing the weld puddle to become too fluid, the voltage should be lowered. If the electrode runs into the weld, causing spatter, the voltage needs to be increased.

With the above conditions, welding steel of ⅛ to ¼ in. thickness will deposit 5 to 7 lb/h (2.2 to 3 kg/h). The 0.035-in. electrode is also ideal for welding steel pipe with wall thicknesses of less than ½ in. (13 mm). The thickness of the pipe after bevelling controls the size of electrode to be used. The 0.035-in. (6.4-mm) electrode can produce a ¼ -in. (6.4-mm) vertical up fillet weld on such a pipe without weaving.

Contact Tip Recess.—The dimension labeled contact tip recess in Fig. 2, and indicated as ⅛ in., should be about ½ in. (13 mm) for a minimum electrode extension of ¾ in. (19 mm), for FCAW welding. This dimension is critical for obtaining high-quality welds with all-position electrodes because they have a fast-freezing slag and operate with low to medium current and voltage. If the recess dimension is less than the optimum, the voltage may be lower than the minimum recommended, and if the settings are less than the minimum, the fast-freezing slag may solidify too rapidly, causing excess porosity or "worm tracks" on the weld surface.

The recommended length of electrode extension for all-position FCAW, E71T-1 electrodes is ¾ to 1 in. (19 to 25 mm). The size of this extension not only affects the minimum required parameters, but a long electrode extension also ensures preheating of the electrode and allows lower current to be used. Preheating the electrode is further beneficial as it reduces moisture on the electrode surface, and in the electrode flux. When a change is made from the GMAW to the FCAW process, welders must be aware of the influence on weld quality of the electrode extension in the FCAW process.

Porosity and Worm Tracks.—As mentioned above, porosity and worm tracks typically result from a combination of incorrect electrode extension, incorrect welding settings, humidity, electrode moisture, refill scale, rust, paint, oils, or poor welding technique. Where humidity levels are high, potential for porosity and worm tracks increases. The FCAW process is less sensitive to mill scale than the GMAW spray transfer mode but mill scale will often cause excess weld porosity. The best way to avoid the effects of mill scale, rust, oil, and surface contaminants is to grind the area to be welded.

Another way to reduce porosity is to keep weaving to a minimum. If the correct size flux-cored electrode is used, weaving can be kept to a minimum for most flux-cored applications. The forehand technique produces the best weld bead surface on fillet weld beads up to ¾ in. (19 mm) steel thickness in the flat and horizontal weld positions. On larger single-pass fillet welds, the backhand technique is beneficial because the voltage directed at the weld provides additional weld puddle control to the fluid welds. The backhand technique used for flat and horizontal welds produces a more convex weld bead, reduces potential for porosity, and increases penetration.

If porosity or worm tracks occur, the prime solution is in weld practices that increase heat at the weld, but the following remedies can also be tried. Grind clean the surface to be welded; use recommended electrode extensions; increase current (wire feed rate) decrease voltage; use the backhand welding technique; slow down travel speed, consider use of a different electrode formulation containing increased deoxidizers, avoid weaving; change from argon + CO_2 mixture to straight CO_2; and provide a protective cover to keep the electrode spool clean and dry.

Welding with 0.045-in. (1.2-mm) Diameter All-Position Electrodes.—Fig. 5 shows wire feed settings for welding of steel with 0.045-in. (1.2-mm) diameter, E71T-1 all-position electrodes using a mixture of argon + 15 to 25 per cent CO_2 as the shielding gas, and an electrode extension of ¾ in. (18 mm). Parameters for vertical up welding, shown at the left in Fig. 5, include setting the wire feed rate at the twelve o'clock position, or about 350 in./min, using 200 to 190 amps, and setting the voltage between 24 and 25 volts. Optimum parameters for flat welding, shown at the right in Fig. 5 include setting the wire feed rate at three o'clock position, or 560 in./min (240 mm/s), and 270 amps at 25 to 27 volts.

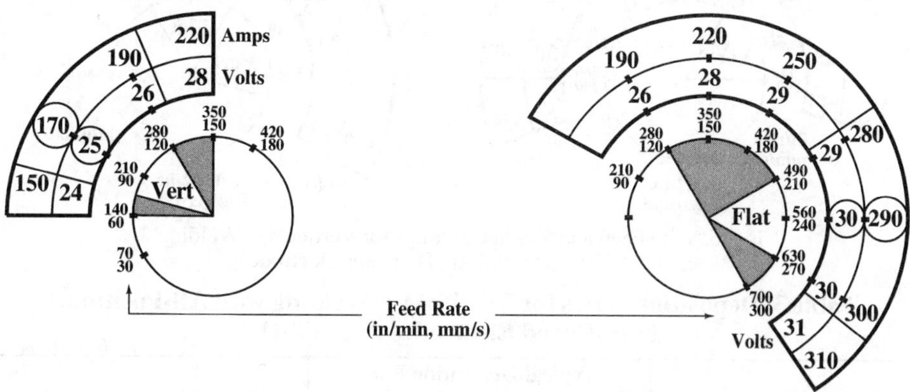

Fig. 5. Wire Feed and Voltage Settings for FCAW Welding with 0.045-in. (1.2-mm) Diameter E71T-1 Electrodes. Optimum settings are circled.

Welding with 0.052-in. (1.3-mm) Diameter All-Position Electrodes.—Settings for vertical up and flat welding with all-position E71T-1 electrodes of 0.052-in. (1.3-mm) diameter are seen at the left in Fig. 6. These electrodes are suited to welding steel having thicknesses of ¼ in. (6 mm) and thicker. For vertical up welding, the wire feed rate is set between the ten and eleven o'clock positions, or 250 in./min (106 mm/s), with about 200 amps at 25 volts. Flat welding with these electrodes is best done with the wire feed rate set between the two and three o'clock positions, or 490 to 560 in./min (207 to 237 mm/s), giving approximately 300 amps at 28 volts.

Settings for all-position E71T-1 electrodes of 0.062-in. (1.6-mm) diameter, shown at the right in Fig. 6, for vertical up welding are just before the ten o'clock position, or 190 in./min, giving 230 to 240 amps with voltage adjusted to 24–25 volts. For flat welding, the wire feed is set to the twelve o'clock position, giving 340–350 amps with a voltage of 29–30 volts.

High-Deposition, All-Position Electrodes.—Vertical up weld deposition rates of 10 to 14 lb/h can be achieved with the E71T-1, 0.062-in. (1.6-mm) and 0.045-in. (1.2-mm) flux-cored electrodes. Settings are shown in Fig. 6 for E71T-1, FCAW electrodes of 0.052- and 0.062-in. (1.4- and 1.6-mm) diameter. These electrodes are suited to applications in which the steel thickness is ¼ in. and thicker, and are the most cost-effective diameter for all-position welds on carbon and stainless steels of ¼ in. (6.35 mm) thickness and thicker. In contrast, vertical up welds, using GMAW or SMAW, may deposit an average of 2 to 4 lb/h (1 to 2 kg/h). Deposition rates are based on welding 60 minutes of each hour. Pulsed GMAW provides deposition rates of 3 to 6 lb/h (1.3-2.7 kg/h). Average rates for vertical up welding with all-position, flux-cored electrodes are shown in Table 3.

The average deposition rates in Table 4 are to be expected with FCAW electrodes available today. Special electrodes are also available that are specifically designed to provide higher deposition rates. A typical manual welder, welding on steel of ¼ to ⅜ in. thickness for 30 minutes of each hour with an all-position flux-cored 0.062-in. (1.6-mm) or 0.045-in. (1.2-mm) diameter electrode would deposit about 4–5 lb/h.

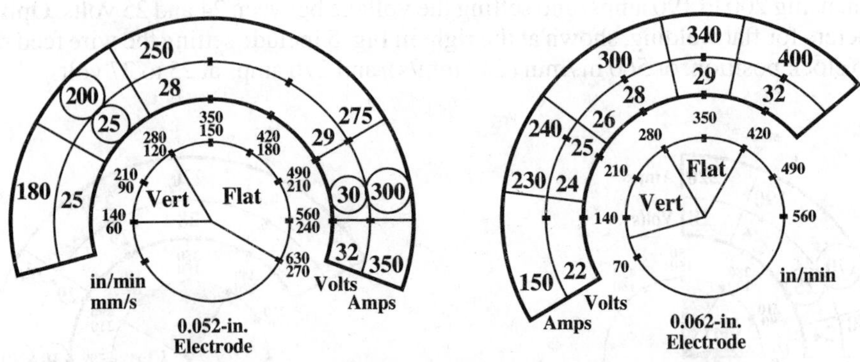

Fig. 6. Wire Feed and Voltage Settings for Vertical Up Welding with 0.052- and 0.062-in. Diameter Electrodes

Table 3. Deposition Rates for Vertical Up Welding with All-Position, Flux-Cored Electrodes (ET71T-1)

Electrode Diameter		Typical Deposition Rate Range		Average Deposition Rate	
in.	(mm)	lb/h	(kg/h)	lb/h	(kg/h)
0.035	(1)	2.7–6.5	(1.2–3)	5	(2.3)
0.045	(1.2)	5–11	(2.3–5)	8	(3.6)
0.052	(1.4)	4–8	(1.8–3.6)	6.5	(3)
0.062	(1.6)	4–11	(1.8–5)	8.5	(4)

Table 4. Average Deposition Rates for Flat and Horizontal Welds

Process	Electrode Size in.	(mm)	Cost-Effective Current Range (amps)	Optimum Current (amps)	Deposition Rate lb/h	(kg/h)
GMAW spray transfer	0.035	(1)	250–350	285	9	(4)
	0.045	(1.2)	300–400	385	13	(5.9)
	0.052	(1.4)	350–470	410	11	(5)
	0.062	(1.6)	375–500	450	17	(7.7)
FCAW	0.045	(1.2)	225–310	300	14	(6.4)
	0.052	(1.4)	260–350	310	15	(6.8)
	0.062	(1.6)	300–400	340	15	(6.8)
	3⁄32	(2.4)	380–560	460	17	(7.7)

The average deposition rates of pulsed GMAW and FCAW for vertical up welds are similar for applications where the steel thickness is 1⁄8 in. (3.2 mm) or less. On steels thicker than 1⁄8 in., where the current may be increased, and larger-diameter all-position FCAW electrodes may be used, deposition rates will be much greater than with pulsed GMAW. Compared with GMAW electrodes for pulsed welding, FCAW all-position electrodes require less costly equipment, less welding skill, and have potential for increased weld fusion with less porosity than with GMAW pulsed techniques.

Electrode Diameters and Deposition Rates.—A cost-effective welding shop can achieve deposition rates on flat and horizontal welds of 12 to 15 lb/h (5 to 7 kg/h) with both the GMAW 0.045-in. wire and the 0.062-in. flux-cored wire electrodes, without welder discomfort, and with welds of consistent quality.

The first consideration in selecting the optimum size of gas-shielded FCAW E70T-X electrode for manual flat and horizontal welds on steels thicker than 1⁄4 in. (6.4 mm) is the current requirements needed to achieve deposition rates of 12 to 15 lb/h (5 to 7 kg/h).

Large-size electrodes of $\frac{3}{32}$-in. (2.4-mm) diameter require 500 amps or more to attain optimum deposition rates. These $\frac{3}{32}$-in. diameter electrodes are often used with power sources in the 300–400 amp range, but even when the power source provides 500 to 600 amps, welding is often performed at the low end of the electrode's current requirements. With the large, $\frac{3}{32}$-in. diameter electrodes, welder appeal is low, smoke is often excessive, and deposition rates are often only comparable with smaller, easier-to-operate FCAW electrodes.

Typical deposition rates for flat and horizontal welds with various electrode sizes and weld settings are shown in Table 4. In connection with this table, it may be noted that high deposition rates in welding steel plate thicker than $\frac{1}{4}$ in. require use of currents above the minimum shown for the various sizes of electrodes. The optimum current requirements for the most popular electrode sizes indicate that a 450-amp power source is the most suitable for welding steel of more than $\frac{1}{4}$ in. thickness. The two most cost-effective and versatile consumables for thin and thick steel sections are the 0.045 in. for GMAW and the 0.062 in. for FCAW electrodes.

The approach to a welding application is critical to achievement of optimum weld quality at minimum cost. In many applications, minimal consideration is given to weld costs. Half of every man-hour of welding in many shops could be saved with selection of the correct electrode diameter used with optimum parameter settings. A practical point that is often overlooked in selection of FCAW electrodes is that the larger the electrode diameter, the more restricted is the application thickness range. Large FCAW electrodes such as the $\frac{3}{32}$ in. (2.4 mm) are neither suitable nor cost-effective for the common steel thickness range of $\frac{1}{4}$ to $\frac{1}{2}$ in. (6.4 to 13 mm). Smaller FCAW electrodes such as the $\frac{1}{16}$-in. (1.6-mm) diameter, are suitable for both thin and thick applications. A $\frac{1}{6}$-in. diameter FCAW electrode used in the 300- to 350-amp range provides excellent deposition rates with superior welder appeal and negligible smoke.

Large-diameter $\frac{3}{32}$-in. (2.4-mm) electrodes are popular for manual applications. However, from a practical point of view, this electrode size is often better suited to mechanized high-current welding in which the high currents required for optimum deposition rates may be safely used without health risks. Use of an electrode at 60 to 80 per cent of its welding current capability indicates that the correct diameter electrode has been selected for the application. When an electrode is used at its maximum-current capability, it shows that the next size larger electrode should be preferred, and when the low end of the current capability is in use, the electrode selected is typically too large.

The 0.062-in. (1.6-mm) Diameter, E70T-X Electrode.—The 0.062-in. (1.6-mm) diameter FCAW electrode is the most practical size of its type and will provide excellent deposition rate potential with a practical current range and the broadest application range. Settings for the common $\frac{1}{16}$-in. diameter E70T-1 electrode are shown in Fig. 7. With the GMAW process, a $\frac{3}{8}$- to $\frac{7}{16}$-in. (9.6- to 11-mm) minimum-weave fillet weld is typically the maximum size that can be made in a single pass. The 0.062-in. ($\frac{1}{16}$-in, 1.6-mm) FCAW electrode can easily produce a $\frac{3}{4}$-in. (19-mm), nonweave, single-pass fillet weld. This size of electrode is also a practical choice for welding steel of $\frac{1}{4}$ in. (6.4 mm) or greater thickness. From a cost perspective, FCAW consumable electrodes should be used whenever the GMAW process is not suitable.

The average deposition efficiency of a flux-cored electrode is 85 per cent, which means that for every 100 lb or kg of electrode material used, 85 lb or kg ends up as weld material. In contrast, the average deposition efficiency of a GMAW electrode used with argon mixtures and correct equipment settings should be a minimum of 99 per cent.

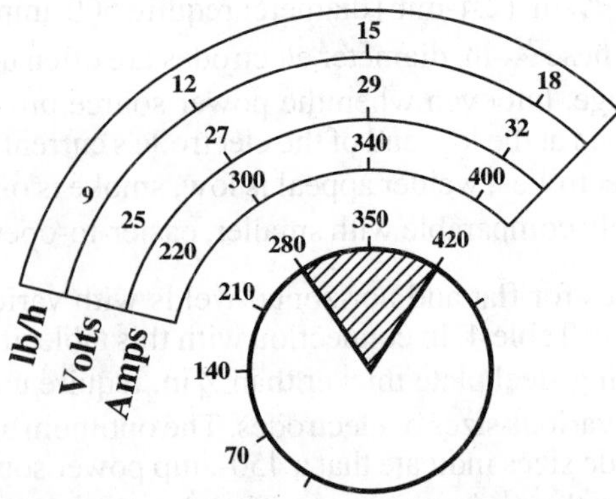

Fig. 7. Settings for $\frac{1}{16}$-in. (1.6-mm) FCAW, E70T-X Electrodes.

Shielding Gases and FCAW Electrodes.—The E70T-X flux-cored electrodes that are recommended for flat and horizontal welds use CO_2 gas shielding. Because of new OSHA welding smoke restrictions, manufacturers of FCAW electrodes now provide E70T-X consumable electrodes that can be used with less reactive argon + CO_2 mixtures to reduce smoke levels. The fast-freezing slag, all-position, E71T-1 flux-cored electrodes can use either CO_2 or argon + 15 to 25 per cent CO_2 mixtures for welding carbon, low-alloy, or stainless steels. The argon + CO_2 mixture is often selected because it provides the highest energy from a reactive gas mixture with a compatible voltage range.

Instead of CO_2, welders often prefer the arc characteristics, lower smoke levels, and lower voltage requirements of the argon + CO_2 mixtures for all-position welding. However, if lower reactive argon mixtures such as argon + oxygen, or argon with less than 13 per cent CO_2, are used, the weld voltage requirements and the arc plasma energy are reduced, adding to the possibility of changing the mechanical properties significantly, increasing the porosity, and raising the potential for forming worm tracks.

Shielded Metal Arc Welding

With the shielded metal arc welding (SMAW) process, commonly known as stick welding, it is most important to select an electrode that is suited to the application. For welding austenitic stainless or high-alloy steels, the electrode is first selected to match the mechanical and chemical requirements of the metal to be welded. Secondary requirements such as the welding position, penetration potential, deposition capabilities, and ease of slag removal are then considered. Many electrodes for SMAW welding of low- to medium-carbon steels have unique characteristics making them the most suitable and cost-effective for a specific welding application.

In interpreting the ANSI/AWS Standard specification code for SMAW electrodes shown in Table 1, for example, E60XX, the E stands for a low-carbon steel, metal arc welding electrode. The next two digits, such as 60 or 70, indicate the approximate tensile strength of the weld deposit in thousands of psi.

Of the last two digits, the first indicates the usability as follows: 1 = usable in all welding positions; 2 = usable for flat or horizontal positions; and 3 = usable in flat position only.

The final digit, combined with the above, indicates the type of flux coating, as shown in Table 1.

British Standard BS 639:1986 defines requirements for covered carbon- and carbon-manganese-steel electrodes for manual metal arc welding, depositing weld metal having a tensile strength of not more than 650 N/mm^2. Appendix A of this standard lists minimum mandatory and optional characteristics of these electrodes. The extensive classifications provide for electrodes to be rated for strength, toughness, and covering (STC), with codes such as E 51 5 4 BB [160 3 0 H]. In this series, E indicates that the electrode is covered and is for manual metal arc welding. The next two digits (51) indicate the strength (tensile, yield, and elongation) properties. The next digits (5 and 4) give the temperatures at which minimum average impact strengths of 28J (at −40°C) and 47J (at −30°C), using Charpy V-notch test specimens, are required. The next group is for the covering and the BB stands for basic, high efficiency. Other letters are B for basic; C for cellulosic; R for rutile, RR for rutile, heavy coated; and S for other.

Table 1. Significance of Digits in ANSI/AWS A5.18-1979 Standard

Third and Fourth Digits	Flux Type and Characteristics, SMAW Electrodes
10	High-cellulose coating bonded with sodium silicate. Deep penetration, energetic spray-type arc. All-positional, DCEP[a] only
11	Similar to 10 but bonded with potassium silicate to permit use with AC or DCEP
12	High-rutile coating, bonded with sodium silicate. Quiet arc, medium penetration, all-positional, AC or DCEN
13	Similar to 12 but bonded with sodium silicate and with easily ionized materials added. Gives steady arc on low voltage. All-positional, AC or DCEN
14	Similar to 12 with addition of medium amount of iron powder. All-positional, AC or DC
15	Lime-fluoride coating (basic low-hydrogen) bonded with sodium silicate. All-positional. For welding high-tensile steels. DCEP only
16	Similar to 15 but bonded with potassium silicate. AC or DCEP
18	Similar to 15 but with addition of iron powder. All-positional, AC or DC
20	High iron-oxide coating bonded with sodium silicate. Flat or HV positions. Good X-ray quality. AC or DC
24	Heavy coating containing high percentage of iron powder for fast deposition rates. Flat and horizontal positions only. AC or DC
27	Very heavy coating with ingredients similar to 20 and high percentage of iron powder. Flat or horizontal positions. High X-ray quality. AC or DC
28	Similar to 18 but heavier coating and suited for use in flat and HV positions only. AC or DC
30	High-iron-oxide-type coating but produces less fluid slag than 20. For use in flat position only (primarily narrow-groove butt welds). Good X-ray quality. AC or DC

[a] DC = direct current, AC = alternating current, EP = electrode positive, EN = electrode negative.

The letters in brackets are optional, and the first group indicates the efficiency, which is the ratio of the mass of weld metal to the mass of nominal diameter core wire consumed with the largest diameter electrode, rounded up to the nearest multiple of 10. The next digit (3) is the maker's advice for the position(s) to be used. Codes for this category include 1, all positions; 2, all positions except vertical down; 3, flat, and for fillet welds, horizontal/vertical; 4, flat; 5, flat, vertical/down; and for fillet welds, horizontal/vertical; and 9, other. The digit at (0), which may have numbers from 0 to 9, shows the polarity, and the minimum open-circuit voltage to be used for that electrode. A 0 here indicates that the electrode is not suited for use with AC. The (H) is included only for hydrogen-controlled electrodes that will deposit not more than 15 ml of diffusible hydrogen for each 100 g of deposited weld metal. The corresponding ISO Standard for BS 639 is ISO 2560. Low-alloy steel electrodes and chromium and chromium nickel steel electrodes are covered in BS 2493 and BS 2926.

The most common electrodes used for the SMAW process are the AWS types E60XX and E70XX. SMAW welding electrode Standards are issued by the American Welding Society (AWS), the British Standards Institute (BS), Canada (CSA), Germany (DIN), and Japan (JIS) and are shown in Table 2.

AWS E60XX Electrodes.—Characteristics of the E60XX electrodes influence the weld position capability, ease of slag removal, penetration potential, weld travel speed capability, and weld deposition rates. These electrodes are designed for welding low-carbon steels and they provide welds with typical tensile strength in the range of 58,000 to 65,000 lb_f/in^2, depending on the specific electrode utilized, the base metal condition and chemistry, and the amount of weld dilution. In selecting an electrode for SMAW welding, knowing that the mechanical and chemical requirements have been matched, it is necessary to choose electrodes with characteristics that influence the features required, as shown in Table 2.

Recommended current ranges, shown in Table 4 for the various sizes of AWS E60XX electrodes most commonly used for welding carbon steel, will give optimum results with SMAW electrodes. An ideal starting point for the current setting for any SMAW electrode diameter is in the middle of the range. The current ranges shown are average values taken from literature of electrode manufacturers in three different countries.

Table 2. Characteristics of SMAW Welding Electrodes Made to Standards of Various Countries

Standard	Description
AWS E6010 CSA E41010 BS E4343C10 DIN E4343C4 JIS	Designed for welding pipe and general structures. Excellent for all-position and vertical down welding. Slag is light and easy to remove. Deep, penetrating arc. Low deposition rates. Polarity DC + (electrode positive).
AWS E6011 CSA E41011 BS E4343C13 DIN E4343C4 JIS D4311	Similar to E6010 but modified to allow use of AC. Excellent for welding sheet metal corner joints vertical down. Polarity AC or DC + (electrode positive).
AWS 6012 CSA E41012 BS E4332R12 DIN E4332R(C) JIS D4313	Designed for welding sheet metal and light structural steels. Medium penetration suitable for gaps or where minimum weld dilution is needed. Ideal for flat, horizontal, or vertical down welding. Will weld faster than the E6010-11 electrode. Polarity AC or DC—(electrode negative).
AWS E6013 CSA E41013 BS E4332R21 DIN E4332R3 JIS D4313	Excellent AC or DC—performance. All-position. Shallow penetration. Good choice for low open-circuit welding machines. AC or DC both excellent on thin structural applications. Polarity AC or DC (DC both polarities).
AWS 6027 CSA 41027 BSE4343A13035 DIN 4343AR11 JIS D4327	Iron powder is added to the flux to provide higher deposition rates. Ideal for multipass groove and fillet welding in flat and horizontal positions. Polarity AC or DC (both polarities).

Table 3. Diameters of AWS E6010/E6011 SMAW Electrodes for Welding Low-Carbon Steel Sheet Metal

SWG of Sheet Metal to be Welded	Electrode Diameter		Current Starting Level
	in.	(mm)	(amps)
18	$\frac{3}{32}$	(2.5)	45–60
16–14	$\frac{1}{8}$	(3.2)	80–110
12	$\frac{5}{32}$	(4)	125–135
10	$\frac{3}{16}$	(5)	135–150

Table 4. Current Ranges for AWS E60XX SMAW Electrodes

Electrode Diameter in.	(mm)	E6010/E6011 (amps)	E6012 (amps)	E6013 (amps)	E6027 (amps)
1⁄16	(1.6)	…	25–50	20–40	…
3⁄32	(2.5)	40–75	40–100	50–100	…
1⁄8	(3.2)	75–130	85–140	75–135	120–180
5⁄32	(4)	90–170	115–185	110–185	155–245
3⁄16	(5)	135–220	145–240	150–235	200–300
1⁄4	(6.4)	205–325	250–390	240–340	300–410
5⁄16	(8)	260–420	290–480	310–425	370–480

Table 3 shows approximate current requirements for AWS E60XX electrodes for welding sheet metal carbon steels. The current ranges specified vary slightly with different electrode manufacturers. For welding sheet metal start at the low end of the given current requirements with electrodes of 3⁄16-in (5-mm) diameter or smaller. For metals thicker than 10 gage (0.134 in.), start in the center of the current range, then adjust to suit. A high DC current may result in arc blow, and improved results may then be obtained with AC.

For welding thicker materials, a good starting setting is in the middle of the current range shown in Table 3. In welding material less than 1⁄4-in. (6.4-mm) thick, vertically, with an E6010 electrode, try a 1⁄8-in. (3.2-mm) electrode at 90 to 100 amps. For welding thicknesses between 3⁄16 and 5⁄16 in. (5 and 8 mm) with the E6010 electrode, vertically, try the 5⁄16-in. (8-mm) diameter electrode at 100 to 125 amps. For thicknesses of 3⁄8 to 1 in. (9.5 to 25 mm), try a 3⁄16-in. (5-mm) diameter electrode at 155 to 165 amps.

AWS E70XX Electrodes.—Information on the most commonly used AWS E70XX electrodes is given in Table 6. For critical welding applications, low-hydrogen electrodes are typically used. It is most important that manufacturers' instructions regarding storage requirements for keeping low-hydrogen electrodes free from moisture are followed. Current ranges for welding low-carbon steel sheet metal with E70XX electrodes of diameters from 3⁄32 to 3⁄16 in. (2.5 to 5 mm) are shown in Table 5. The optimum starting point is in the middle of the current range indicated.

Table 5. Current Ranges for SMAW E70XX Welding Electrodes

Electrode Diameter in.	(mm)	E7014 (amps)	E7018 (amps)	E7024 (amps)
3⁄32	(2.5)	75–120	70–105	85–135
1⁄8	(3.2)	110–155	110–160	130–180
5⁄32	(4)	145–210	150–215	175–240
3⁄16	(5)	190–280	180–275	230–315
7⁄32	(5.5)	255–335	255–350	280–370
1⁄4	(6.4)	330–415	295–360	325–450
5⁄16	(8)	380–490	370–480	390–530

In using AWS E7018 electrodes for vertical up welding of plate thicknesses of 3⁄16 to 5⁄16 in. (5 to 8 mm), try a 1⁄8-in. (3.2-mm) diameter electrode. For vertical up welding of thicknesses greater than $\frac{5}{16}$ in. (8 mm), try a $\frac{5}{32}$-in. (4-mm) electrode. With AWS E7018 electrodes, to make horizontal fillet welds in plate thicknesses of 10 swg (0.135 in., 3.4 mm), try a $\frac{3}{16}$-in. (5-mm) electrode, for $\frac{1}{4}$-in. (6.4-mm) plate, try the $\frac{7}{32}$-in. (5.5-mm) electrode, and for steel plate thicker than $\frac{1}{4}$ in., try the $\frac{1}{4}$-in. (6.4-mm) diameter electrode.

Table 6. Characteristics of AWS Electrodes for SMAW Welding

Standard	Description
AWS E7014 CSA E48014 BS E5121RR11011 DIN E5121RR8 JIS D4313	An iron-powder, all-position electrode for shallow penetration. Excellent for vertical down and applications with poor fit. Similar to AWS E6012-E6013 with added iron powder. For welding mild and low-alloy steels. Polarity AC or DC, + or –.
AWS E7018 CSA E48018 BS E5154B11026(H) DIN E5154B(R)10 JIS D5016	An iron-powder, low-hydrogen, all-position electrode. Excellent for rigid, highly stressed structures of low- to medium-carbon steel. Can also be used for welding mild and high-strength steels, high-carbon steels, and alloy steels. Polarity AC or DC + reverse polarity.
AWS 7024 CSA E48024 BS E5122RR13034 DIN E5122RR11 JIS D4324	An iron-powder electrode with low hydrogen, usable in all positions. Excellent for high-amperage, large, fillet welds in flat and horizontal positions. Polarity AC or DC, + or –.
AWS E7028 CSA E48028 BS E514B12036(H) DIN E5143B(R)12 JIS D5026	An iron-powder, low-hydrogen electrode suitable for horizontal fillets and grooved flat position welding. Higher deposition rates. More cost-effective than the AWS E7018 electrode. Polarity AC or DC + reverse polarity.

The E7024 electrode is suggested for horizontal fillet welds. For 10-gage (0.135-in, 3.4-mm) material, try the $\frac{1}{8}$-in. (3.2-mm) diameter electrode; for above 10-gage to $\frac{3}{16}$-in. (5-mm) material, try the $\frac{5}{32}$-in. (4-mm) diameter electrode. For plate of $\frac{3}{16}$- to $\frac{1}{4}$-in. thickness, try the $\frac{3}{16}$-in. size, and for plate thicker than $\frac{1}{4}$ in., try the $\frac{1}{4}$-in. (6.4-mm) electrode.

Gas Tungsten Arc Welding

Often called TIG (for tungsten inert gas) welding, gas tungsten arc welding (GTAW) uses a nonconsumable tungsten electrode with a gas shield, and was, until the development of plasma arc welding (PAW), the most versatile of all common manual welding processes. Plasma arc welding is a modified GTAW process. In contrast to GTAW, plasma arc welding has less sensitivity to arc length variations, superior low-current arc stability, greater potential tungsten life, and the capability for single-pass, full-penetration welds on thick sections.

In examining a potential welding application, the three primary considerations are: achieving a quality weld, ease of welding, and cost. Selecting the optimum weld process becomes more complex as sophisticated electronic technology is applied to conventional welding equipment and consumable electrodes. Rapid advances in gas metal arc and PAW welding power source technology, and the development of many new flux-cored electrodes, have made selection of the optimum welding process or weld consumable more difficult. When several manual welding processes are available, the logical approach in considering GTAW for production welding is to first examine whether the job can be welded by gas metal arc or flux-cored methods.

GTAW Welding Current.—A major benefit offered by GTAW, compared with GMAW, FCAW, or SMAW, is the highly concentrated, spatter-free, inert heat from the tungsten arc, which is beneficial for many applications. The GTAW process can use any of three types of welding current, including: direct-current straight polarity, electrode negative (DC–), direct-current reverse polarity, electrode positive (DC+), alternating current with high frequency for arc stabilization (ACHF). Each of the different current types provides benefits that can be used for a specific application.

GTAW Direct-Current Straight Polarity (DC–): The most common GTAW current is straight polarity, where the electrode is connected to the negative terminal on the power source and the ground is connected to the positive terminal. Gas tungsten arc welding is used with inert gases such as argon, and argon + helium to weld most metals. During a DC– straight-polarity weld, electrons flow from the negative tungsten electrode tip and pass through the electric field in the arc plasma to the positive workpiece, as shown in Fig. 1.

Plasma is a high-temperature, ionized, gaseous column that is formed when electrons in the arc collide with the shielding gas molecules. The gas atoms lose one or more electrons, leaving them positively charged. The electrons and the resulting plasma are concentrated at the electrode tip, where they cause the plasma pressure to be at its greatest. The electron density thins out as the electrons travel from the straight-polarity, negatively charged, tungsten electrode across the open arc. As the electrons traverse the arc to the work, the resulting arc column width increases slightly, controlled in part by the electromagnetic forces generated by the current. With the increase in the arc column width, the density and pressure of the plasma decrease. The electrons collide with the work, liberating much heat. The downward pressure of the plasma is exerted against the surface of the weld pool. The gas ions in the plasma are positively charged and greater in mass than the electrons.

In DC–, straight-polarity welding, the positive gas ions are drawn to the negative electrode. The electron flow to the weld ensures that most of the arc heat is generated at the positive work side of the arc. This current setup provides maximum penetration potential, as indicated in Fig. 1. With DC–, straight polarity, the tungsten electrode can carry a higher current and operate at lower temperatures than with the other current arrangements.

Direct-Current Reverse Polarity (DC+): With direct-current positive polarity (DC+), the tungsten electrode is connected to the power-source positive terminal so that the electrons flow from the negative work to the positive electrode. As illustrated in Fig. 2, the electrons impinging on the electrode tip reverse the direction of the heat concentration that occurs with straight polarity, as described above. Approximately two-thirds of the heat generated with DC+ reverse polarity is at the electrode tip, and the electrode becomes very hot, even with low current levels. DC+ reverse polarity requires large-diameter electrodes.

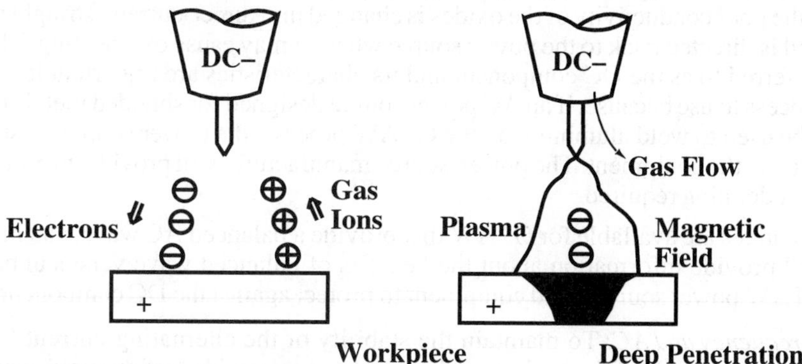

Fig. 1. Straight Polarity (DC–) Provides Highest Electrode Current Capacity and Deepest Penetration Potential.

In the current range of 100 to 150 amps, DC+ reverse polarity requires a ¼-in. (6.4-mm) diameter electrode. This larger electrode produces a weld puddle almost twice as wide as that produced by a 120-amp, $\frac{1}{16}$-in. (1.6-mm) diameter, DC– straight polarity electrode. Most of the heat is generated at the electrode tip with DC+ reverse polarity, so penetration is much less than with DC– straight polarity. With DC+ reverse polarity, the positive gas ions in the arc plasma are drawn to the negative workpiece where they bombard and break up the surface oxides that form on metals such as aluminum and magnesium. However, the best welding method for aluminum and magnesium is to use alternating current (AC), which combines the benefits of DC– straight and DC+ reverse polarity.

Alternating Current (AC): The surface oxides formed on metals such as aluminum and magnesium disturb the arc and reduce the weld quality. Welding of these metals requires DC+ reverse or AC polarity to break up the surface oxides. An alternating current (AC) cycle consists of one-half cycle of straight polarity and one-half cycle of reverse polarity. With alternating current, the cleaning action benefits of the reverse-polarity arc can be combined with the electrode current-carrying capacity of the straight-polarity arc. In welding aluminum and magnesium, the half cycles of AC polarity may become unbalanced. During the AC cycle, the reverse electrode-positive portion of the cycle is restricted by the oxides on the surfaces of these materials. The surface oxides are poor conductors and make it difficult for the electrons generated by the reverse-polarity part of the cycle to flow from the work to the electrode tip, but they do not upset the straight polarity in which the electrons flow from the electrode to the work.

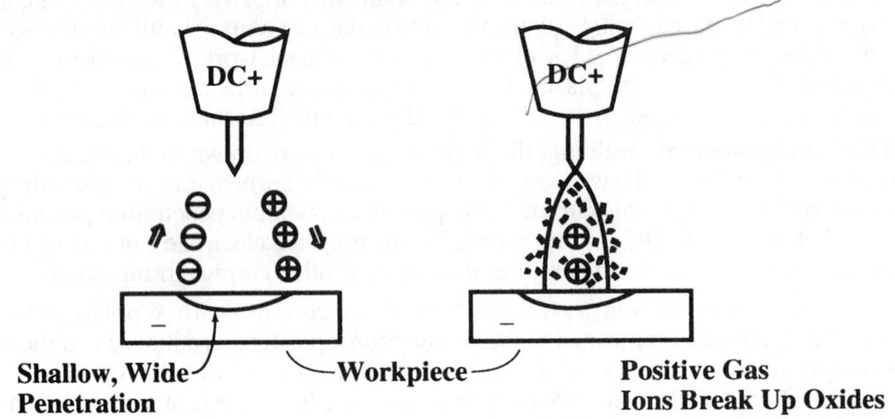

Fig. 2. Direct-Current (DC+) Reverse Polarity Provides a Shallow, Wide, Weld Pool.

DC Component: The part of the reverse-polarity cycle of alternating current (AC) that is upset by the poor conductivity of the oxides is changed into direct-current, straight polarity (DC–) and is directed back to the power source where it may cause overheating. The feedback is referred to as the DC component and its characteristics are important in deciding which process to use because, if an AC power source designed for shielded metal arc welding is to be used to weld aluminum by the GTAW process, the power source must be derated to protect the equipment. The power-source manufacturer will provide information on the level of derating required.

Power sources are available for GTAW that provide a balanced AC wave, and manufacturers will provide information about the benefits of balanced wave versus unbalanced wave, GTAW power sources, and equipment to protect against the DC component.

High Frequency and AC: To maintain the stability of the alternating-current (AC) arc when the positive cycle of the arc is upset by the aluminum oxide, and to avoid contamination of the tungsten electrode, high-frequency current is used to assist in arc ignition during each AC cycle. In direct-current, straight-polarity (DC–) welding of carbon and stainless steels, the high-frequency current is typically selected by the HF arc start-only switch. During AC welding of steels without oxide problems, the HF switch may be left on the arc start-only setting. When AC is used on aluminum, magnesium, or other metals with poor electron-conductive oxides, the HF switch should be moved to the continuous setting.

High-frequency current is also beneficial in that it promotes gas ionization. The more positively charged molecules produced, the more cleaning action takes place in the direct-current, reverse-polarity (DCRP) cycle.

Selecting the Tungsten Electrode Type.—Use of the correct tungsten electrode composition is vital to producing good-quality GTAW welds. Tungsten has the highest melting temperature of all metals. Pure tungsten provides a low-current capacity and requires addition of such alloying elements as thorium or zirconium to increase the current-carrying capability. The electrode diameter and the electrode tip configuration also require consideration as both have a great influence on the performance and application potential of GTAW welding.

Table 8 shows typical compositions of commonly used GTAW tungsten electrode materials from the American Welding Society AWS A5.12 Standard. New electrode compositions have been designed that utilize other alloys and rare-earth metals. These electrodes are designed for longer lives in both GTAW and plasma welding.

Pure Tungsten: Pure tungsten electrode material provides good arc stability with alternating current (AC). Tungsten has low current capacity and low resistance to electrode contamination. Pure tungsten is good for low-amperage welding of aluminum and magnesium alloys. On medium- to high-current ferrous applications, there is a potential for tungsten inclusions in the weld.

With DC, the current capacity of pure tungsten is lower than with the alloyed tungsten electrodes. During AC welding, a molten ball shape forms at the pure tungsten electrode tip, and this formation is desirable for welding aluminum.

Table 7. Selection of Gas Tungsten Arc Welding (GTAW) Electrodes

Base Metal	Electrode	Current	Recommendations
Carbon, low-alloy, stainless, and nickel steels	Thoriated	DCEN	Use EWZr electrodes with AC on thin materials
Aluminum	Zirconium or pure tungsten	AC	Use EWZr on critical applications
Aluminum	Thoriated zirconium	DCEP	Use EWZr or EWP electrodes with DCEP on thin sections
Copper and copper alloys	Thoriated	DCEN	Use EWZr or EWP with AC on thin sections
Magnesium	Zirconium	AC	Use DCEP with same electrode on thin sections
Titanium	Thoriated	DCEN	…

Table 8. Common Tungsten Electrode Compositions

Classification	Color	Tungsten (%)	Thorium Oxide (%)	Zirconium Oxide (%)
EWP	Green	99.50	…	…
EWTh-1	Yellow	98.50	0.8–1.2	…
EWTh-2	Red	97.50	1.7–2.2	…
EWTh-3	Blue	98.95	0.35–0.55	…
EWZr	Brown	99.20	…	0.15–0.4

In the classification column, E = electrode; W = tungsten; P = pure; Th = thoriated (thorium oxide); Zr = zirconiated (zirconium oxide). The colors are codes used by manufacturers to identify the material. Tungsten percentages are minimum requirements. The EWTh-3 is also called striped tungsten because it is made with a strip of thoriated material along the length. This electrode needs to be preheated by striking an arc to melt the tip, providing for the thorium and the tungsten to combine before welding is started.

The electrode recommendations in Table 7 are a guide to attaining good-quality GTAW welds from the venous types of polarities available.

Electrode and Current Selection.— Tables 9 and 10 show approximate current recommendations for common electrode types and diameters. The GTAW electrode size should be selected so that its midrange current provides the energy required for the intended application. If the electrode is too thin, excess current may be required, causing the electrode to wear too quickly or melt and contaminate the weld. If the electrodes used are found to be constantly at the top end of the current range, a change should be made to the next larger size. Tables 11 and 12 show recommended sizes of electrodes and filler metal rods or wires for welding various thicknesses of carbon, low-alloy, and stainless steels and aluminum.

Table 9. Recommended Current Ranges for Thoriated GTAW Electrodes

Electrode		Current Range (amps)
1⁄16 in.	(1.6 mm)	60–150
3⁄32 in.	(2.4 mm)	150–250
1⁄8 in.	(3.2 mm)	250–400
5⁄32 in.	(4 mm)	400–500

The electrode selected must suit the application and the current capacity of the power source.

Table 10. Current Ranges for EWP and EWZr GTAW Electrodes

Electrode		Ampere Range AC Balanced		Ampere Range AC Unbalanced	
in.	(mm)	EWP	EWZr	EWP	EWZr
1⁄16	(1.6)	30–80	60–120	50–100	70–150
3⁄32	(2.4)	60–130	100–180	100–160	140–235
1⁄8	(3.2)	100–180	160–250	150–210	225–325
5⁄32	(4)	160–240	200–320	200–275	300–400

Table 11. Electrode and Current Recommendations for Carbon, Low-Alloy, and Stainless Steels

Material Thickness		Electrode Diameter		Filler Rod Diameter		Current Range (amps)
in.	(mm)	in.	(mm)	in.	(mm)	DCEN EWTh
1⁄16	(1.6)	1⁄16	(1.6)	1⁄16	(1.6)	60–100
1⁄8	(3.2)	3⁄32	(2.4)	3⁄32	(2.4)	150–170
3⁄16	(4.8)	3⁄32	(2.4)	1⁄8	(3.2)	180–220
1⁄4	(6.4)	1⁄8	(3.2)	5⁄32	(7.2)	260–300

Note: The shielding gas is argon at 15 to 20 cu ft/h (CFH). For stainless steel, reduce the current by approximately 10 per cent.

Table 12. Recommendations for GTAW Welding of Aluminum with EWP Electrodes Using AC and High-Frequency Current

Material Thickness		Electrode Diameter		Filler Rod Diameter		AC Current Range
in.	(mm)	in.	(mm)	in.	(mm)	(amps)
1⁄16	(1.6)	1⁄16	(1.6)	1⁄16	(1.6)	40–70
1⁄8	(3.2)	3⁄32	(2.4)	3⁄32	(2.4)	70–125
3⁄16	(4.8)	1⁄8	(3.2)	1⁄8	(3.2)	110–170
1⁄4	(6.4)	5⁄32	(4)	3⁄16	(4.8)	170–220

Thoriated Electrodes: In contrast with the pure EWP electrodes, thoriated electrodes have a higher melting temperature and up to about 50 per cent more current-carrying capacity, with superior arc starting and arc stability. These electrodes are typically the first choice for critical DC welding applications, but do not have the potential to maintain a rounded ball shape at the tip. The best welding mode for these electrodes is with the tip ground to a tapered or fine point.

Zirconiated Electrodes: Tungsten electrodes with zirconium are practical for critical applications and have less sensitivity to contamination and superior current capacity than pure tungsten electrodes.

Protecting and Prolonging Electrode Life: To improve tungsten electrode life, the tip should be tapered in accordance with the manufacturer's recommendations. There must also be preflow, postflow shielding gas coverage to protect the electrode before and after the weld. When possible, high frequency should be used to avoid scratch starts, which contaminate the electrode. The shortest possible electrode extension should be employed, to avoid the possibility of the electrode touching the filler or weld metal. The grinding wheel used to sharpen the tungsten must not be contaminated from grinding other metals or with dirt.

Filler Metals.—Specifications covering composition and mechanical properties for GTAW filler metal are published by the American Welding Society under the following classifications: A5.7, copper and copper alloys; A5.9, chromium and chromium nickel; A5.10, aluminum; A5.14, nickel; A5.16, titanium; A5.18, carbon steels; A5.19, magnesium; and A5.28, low-alloy steels.

Filler metals must be kept dry and clean if they are to be used satisfactorily.

Shielding Gases.—Inert gases such as argon, and argon + helium mixtures are most commonly used for GTAW. Helium provides greater thermal conductivity and additional arc voltage potential than argon, and is normally added to argon when more weld energy is required for improved penetration and increased mechanized welding travel speeds. Argon gas mixtures containing 30 to 75 per cent helium provide benefits for manual welding of aluminum over ⅜ in. (9.6 mm) thick; mechanized welding of aluminum where high speeds are required; mechanized welding of carbon and stainless steels where good penetration is needed; mechanized welding of stainless steel where good penetration and faster speeds are required; and for copper of ¼ in. (6.4 mm) thickness and thicker.

Shielding gas purity for GTAW welding is important. Welding-grade argon is supplied at a purity of at least 99.996 per cent and helium is produced to a minimum purity of 99.995 per cent. However, shielding gases may be contaminated due to poor cylinder filling practices. If impure gas is suspected, the following test is suggested. With the HF and power on, create an arc without welding and hold the arc for about 30 seconds. Examine the electrode tip for signs of unusual coloration, oxidation, or contamination, which result from impurities in the shielding gas.

Plasma Arc Welding (PAW)

When an electric current passes between two electrodes through certain gases, the energy of the gas molecules is increased so that they accelerate and collide with each other more often. With increases in energy, the binding forces between the nuclei and the electrons are exceeded, and electrons are released from the nuclei. The gas now consists of neutral molecules, positively charged atoms, and negatively charged electrons. The plasma gas is said to be ionized, so that it is capable of conducting electric current. Plasma forms in all welding arcs but in plasma arc welding it is generated by a series of events that begins with inert gas passing through the welding torch nozzle. High-frequency current is then generated between the tungsten electrode (cathode) and the torch nozzle (anode), forming a low-current pilot arc. The ionized path of this nontransferred arc is then transferred from the tungsten electrode to the work, and a preset plasma current is generated.

The above sequence of events provides the ionized path for the plasma current between the electrode and the work so that arcing between the electrode and the nozzle ceases. (Nontransferred arcs may be used for metal spraying or nonmetallic welds.) Forcing the ionized gas through the small orifice in the nozzle increases both the level of ionization and the arc velocity, and arc temperatures between 30,000 and 50,000°F (16,650 and 27,770°C) are generated.

Gases for Plasma Arc Welding

Welding Gases.—Argon is the preferred gas for plasma arc welding (PAW) as it is easily ionized and the plasma column formed by argon can be sustained by a low voltage. The low thermal conductivity of argon produces a plasma column with a narrow, concentrated hot core surrounded by a cooler outer zone. Argon plasmas are suited to welding steel up to ⅛ in. (3.2 mm). For thicker materials, requiring a hotter arc and using higher current melt-in technique, a mixture of argon 25 + helium 75 per cent may be used. Additions of helium and hydrogen to the gas mixture improve heat transfer, reduce porosity, and increase weld travel speed. For welding materials thinner than ⅛ in. thick by the plasma gas keyhole method (full penetration welds), gases may contain up to 15 per cent hydrogen with the remainder argon. Good results are obtained with argon + 5 per cent hydrogen in welding stainless and nickel steels over ⅛ in. thick.

Shielding Gases.—A shielding gas is needed to protect the narrow plasma arc column and the weld pool, and generally is provided by mixtures of argon, argon + hydrogen, argon + helium, or argon + O_2 + CO_2, depending on compatibility with the material being welded. Shielding gas flow rates vary from 5 to 35 cu ft/h (2.4 to 17 l/min). However, if argon is used for both plasma and shielding, the plasma gas will become less concentrated. The normally tight plasma arc column will expand in contact with the colder shielding gas, reducing ionization and thus concentration and intensity of the plasma column. With no shielding gas, the tight column is unaffected by the surrounding oxygen and nitrogen of the atmosphere, which are not easily ionized.

Hydrogen is added to the shielding gas when welding low-alloy steels of less than $1/16$ in. (1.6 mm) thickness, or stainless and nickel steels, with many benefits. The hydrogen molecules dissociate in contact with the arc at temperatures of about 7,000°F (3,870°C) and the energy thus created is released when the hydrogen molecules recombine on contact with the work surface. The diatomic molecular action creates a barrier around the plasma, maintaining column stiffness. Hydrogen in the shielding gas combines with oxygen in the weld zone, releasing it into the atmosphere and keeping the weld clean. Hydrogen reduces the surface tension of the weld pool, increasing fluidity, and the added energy increases penetration.

Helium mixed with the argon shielding gas is beneficial for all metals as it increases the ionization potential, allowing use of higher voltages that give increased welding temperatures. Flow rates are in the range of 15 to 50 cu ft/h (7 to 24 l/min). Arc-starting efficiency is reduced with pure helium, but adding 25 per cent of argon helps both arc starting and stability. Helium additions of 25 to 75 per cent are made to obtain increased thermal benefits.

Argon + CO_2 shielding gas mixtures are beneficial in fusion welding of carbon steels. A mixture of argon with 20 to 30 per cent CO_2 improves weld fluidity. Shielding gas mixtures of argon + CO_2 with an argon + 5 per cent hydrogen plasma should be considered for welding carbon steel of $1/16$ to ¼ in. thickness. Steels with higher amounts of carbon have higher heat conductivity and need application of more heat than is needed with stainless steels. Manufacturers usually make recommendations on types of gas mixtures to use with their equipment.

PAW Welding Equipment.— The PAW process uses electrode negative (DCEN) polarity in a current range from 25 to 400 amps, and equipment is offered by many manufacturers. Solid-state inverter units are available with nonmechanical contactors. Most PAW units contain a high-frequency generator, a small DC power supply, controls for welding and shielding gas mixtures, and a torch coolant control. A weld sequencer is recommended, especially for keyhole mode welding, but it is also useful in automated fusion welding. The sequencer provides control of up-slope and down-slope conditions for gas mixtures and current, so that it is possible to make welds without run-on and run-off tabs, as is necessary with circumferential welds.

Generally, plasma arc torches are liquid-cooled using deionized water in the coolant lines to the torch to avoid effects of electrolysis. Electrodes are usually tungsten with 2 per cent thorium. If the welding shop already has a constant-current power supply and a coolant recirculator, plasma arc welding may be used by addition of a pilot arc welding console and a torch.

Applications.— Fusion welding is the main use for plasma arc welding. The process is used for high-volume, repetitive, high-duty cycle, manual and automated operations on lap, flange, butt, and corner fusion welds, in all positions. Joint design for materials less than 0.01 in. (0.254 mm) thick may require a flange type joint for rigidity and to allow use of extra, weld metal reinforcement. Filler metal may be added during fusion welding, and automated hot or cold wire feeders can be used. Fusion welding uses a soft, less-restricted arc with low gas flows, and the current level may vary from approximately 25 to 200 amps. The soft arc is obtained by setting the end of the tungsten electrode level with the face of the torch nozzle, in which position lower currents and gas flows are required. With these conditions, the weld bead is slightly wider than a bead produced with a recessed electrode.

Low-Current Plasma Fusion Welding: With the reduced consumption of gas and electric current, the low-current plasma fusion welding method is ideal for welding metals down to 0.001 in. (0.025 mm) in thickness, as the low-current plasma pilot arc allows arcs to be started consistently with currents of less than 1 amp. With currents below 1 amp, the pilot arc is usually left in the continuous mode to maintain the arc. In the conditions described, arc stability is improved and the process is much less sensitive to variations in the distance of the torch from the workpiece. Given this height tolerance, setting up is simplified, and with the smaller torches required, it is often easier to see the weld pool than with the GTAW process. Some plasma welding units incorporate gas flow meters that are designed for low flow rates, and currents in the range of 0.1 to 15 amps can be selected.

Low-current plasma arc welding is more economical than other gas tungsten arc welding methods, especially with solid-state inverter systems and smaller torches. The process is useful for sealing type welds where joint access is good, and for welding components of office furniture, household items, electronic and aerospace parts, metallic screening, and thin-wall tubing.

Keyhole mode welding describes a method whereby abutting edges of two plates are melted simultaneously, forming a vapor capillary (or keyhole) and the resulting molten-walled hole moves along the joint line. This method requires the end of the tungsten electrode to be positioned well back inside the torch nozzle to produce a high-velocity, restricted arc column with sufficient energy to pierce the workpiece. This mode is also used for the plasma cutting process, but the major difference is that welding uses very low plasma flow rates of the order of 1 to 3 cu ft/h (0.5 to 1.4 l/min) for work thicknesses of $\frac{1}{16}$ to $\frac{5}{32}$ in. (1.6 to 4 mm). These low rates avoid unwanted displacement of the weld metal. After the arc pierces the workpiece, the torch moves along the weld line and the thin layer of molten metal is supported by surface tension as it flows to the rear of the line of movement, where it solidifies and forms the weld.

As it passes through the keyhole, the high-velocity plasma gas column flushes the molten weld pool and carries away trapped gases and contaminants that otherwise would be trapped in the weld. Plasma arc keyhole welding is affected less by surface and internal defects in the work material than is the GTAW process. Most metals that can be welded by the gas tungsten arc method can be plasma arc welded with the conventional DC electrode, negative keyhole method, except aluminum, which requires a variable polarity keyhole method.

Plasma keyhole welding is usually automated because it requires consistent travel speed and torch height above the work. A typical operation is welding steel with square abutting edges (no bevels) in thicknesses of 0.09 to 0.375 in. (2.3 to 9 mm), where 100 per cent penetration in a single pass is required. Producing square-groove butt welds in materials thicker than ½ in. by the plasma arc keyhole process requires some edge preparation and several filler passes. The finished weld is uniformly narrow and the even distribution of heat means that distortion is minimized.

Welding Aluminum.—The variable polarity plasma arc (VPPA) process was developed for welding metals that form an oxide skin, such as aluminum. Electrode negative (straight) polarity is necessary for the plasma arc to provide sufficient heat to the workpiece and minimize heat buildup in the tungsten electrode. With electrode negative polarity, electrons move rapidly from the negative cathode tungsten electrode to the positive anode workpiece, generating most of the heat in the workpiece. Because of the oxide skin on aluminum, however, straight polarity produces an erratic arc, poor weld fluidity, and an irregularly shaped weld bead. The oxide skin must be broken up if the metal flow is to be controlled, and this breakup is effected by a power supply that constantly switches from negative to positive polarity.

A typical cycle uses a 20-ms pulse of electrode negative polarity and a 3-ms pulse of electrode positive polarity. The pulses are generated as square waves and the positive (cleaning) pulse is set at 30 to 80 amps higher than the negative pulse for greater oxide-breaking action. The tenacious oxide skin is thus broken constantly and the rapid cycle changes result in optimum cathode cleaning with minimum deterioration of the tungsten electrode and consistent arc stability. Varying polarity has advantages in both gas metal arc and plasma arc welding, but with the keyhole process it allows single-pass, square-groove, full-penetration welds in materials up to ½ in. (12.7 mm) thick.

The VPPA process ensures extremely low levels of porosity in weld areas in aluminum. VPPA welding is often used in the vertical up position for aluminum because it provides superior control of root reinforcement, which tends to be excessive when welding is done in the flat position. Pulsing in the VPPA process when welding aluminum of ⅛ to ¼ in. thickness in the flat position gives satisfactory root profiles. Pulsing gives improved arc control in keyhole welds in both ferrous and nonferrous metals and is beneficial with melt-in fusion welding of thin materials as it provides better control of heat input to the workpiece.

Plasma Arc Surface Coating

Plasma Arc Surfacing uses an arc struck between the electrode and the workpiece, or transferred arc, to apply coatings of other metals or alloys to the workpiece surface. This high-temperature process produces homogeneous welds in which the ionized plasma gas stream melts both the work surface and a stream of powdered alloy or filler wire fed into the arc. Dilution of the base metal can be held below 5 per cent if required. With arc temperatures between 25,000 and 50,000°F (14,000 and 28,000°C), deposition occurs rapidly, and a rate of 15 lb/h (6.8 kg/h) of powdered alloy is not unusual. Deposition from wire can be performed at rates up to 28 lb/h (12.7 kg/h), much higher than with oxygen/fuel or gas metal/arc methods.

In the nontransferred arc process used for coating of surfaces, the arc is struck between the electrode and the torch nozzle, so that it does not attach to the work surface. This process is sometimes called metal spraying, and is used for building up surfaces for hard facing, and for application of anticorrosion and barrier layers. Argon is frequently used as the plasma gas. As the coating material in the form of powder or wire enters the plasma, it is melted thoroughly by the plasma column and is propelled toward the work at high velocity to form a mechanical bond with the work surface. Some 500 different powder combinations are available for this process, so that a variety of requirements can be fulfilled, and deposition rates up to 100 lb/h (45 kg/h) can be achieved.

The plasma arc process allows parts to be modified or recovered if worn, and surfaces with unique properties can be provided on new or existing components. Low levels of porosity in the deposited metal can be achieved. Metal spraying can be performed manually or automatically, and its use depends primarily on whether a mechanical bond is acceptable. Other factors include the volume of parts to be treated, the time needed for the process and for subsequent finishing, the quality requirements for the finished parts, rejection rates, and costs of consumable materials and energy.

Some systems are available that can use either metal powder or wire as the spray material, and can be operated at higher voltage settings that result in longer plasma arc lengths at temperatures over 10,000°F (5,537°C). With these systems, the plasma velocity is increased to about 12,000 ft/s (3,658 m/s), giving an extremely dense coating with less than 1 per cent porosity. Current ranges of 30 to 500 amps are available, and nitrogen is frequently used as the plasma gas, coupled with CO_2, nitrogen, or compressed air as the shielding gas. Gas flow rates are between 50 and 350 cu ft/h (24 and 165 l/min). Large or small surface areas can be coated at low cost, with minimum heat input, if other aspects of the process are compatible with the product being made.

Plasma Arc Cutting.—Higher current and gas flow rates than for plasma arc welding are used for the plasma arc cutting (PAC) process, which operates on DC straight polarity, and uses a transferred arc to melt through the material to be cut. The nozzle is positioned close to the work surface and the velocity of the plasma jet is greatly increased by a restricting nozzle orifice so that it blows away the metal as it is melted to make the cut. The higher energy level makes the process much faster than cutting with an oxygen/fuel torch on cutting steel of less than ½ in. thick, but the process produces kerfs with some variation in the width and in the bevel angle, affecting the precision of the part. Some of the molten metal may recast itself on the edges of the cut and may be difficult to remove.

Factors that affect plasma cutting include the type and pressure of the gas, its flow pattern, the current, the size and shape of the nozzle orifice, and its closeness to the work surface. To reduce noise and fumes, mechanized plasma arc cutting is often performed with the workpiece submerged in water. Oxidation of cut surfaces is almost nonexistent with the underwater method.

Precision Plasma Arc Cutting.—A later development of the above process uses a magnetic field in the cutter head to stabilize the plasma arc by means of Lorentz forces that cause it to spin faster and tighter on the electrode tip. The magnetic field also confines the spinning plasma so that a narrower kerf is produced without adverse effect on cutting speed. Results from this process are somewhat comparable with those from laser cutting and, with numerical control of machine movements, it is used for production of small batches of blanks for stamping and similar applications. With galvanized and aluminized steel, edges are clean and free from burrs, but some slag may cling to edges of mild-steel parts.

Electron-Beam (EB) Welding

Heat for melting of metals in electron-beam welding is obtained by generating electrons, concentrating them into a beam, and accelerating them to between 30 and 70 per cent of the speed of light, using voltages between 25 and 200 kV. The apparatus used is called an electron-beam gun, and it is provided with electrical coils to focus and deflect the beam as needed for the welding operation. Energy input depends on the number of electrons impinging on the work in unit time, their velocity, the degree of concentration of the beam, and the traveling speed of the workpiece being welded. Some 6.3×10^{15} electrons/s are generated in a 1-mA current stream. With beam diameters of 0.01 to 0.03 in. (0.25 to 0.76 mm), beam power can reach 100 kW and power density can be as high as 10^7 W/in^2 (1.55×10^4 W/mm^2), higher than most arc welding levels.

At these power densities, an electron beam can penetrate steel up to 4 in thick and form a vapor capillary or keyhole, as described earlier. Although patterns can be traced by deflecting the beam, the method used in welding is to move the electron gun or the workpiece. A numerical control, or computer numerical control, program is used because of the accuracy required to position the narrow beam in relation to the weld line.

Equipment is available for electron-beam welding under atmospheric pressure or at various degrees of vacuum. The process is most efficient (produces the narrowest width and deepest penetration welds) at high levels of vacuum, of the order of 10^{-6} to 10^{-3} torr or lower (standard atmospheric pressure is about 760 torr, or 760 mm of mercury), so that a vacuum chamber large enough to enclose the work is needed. Operation in a vacuum minimizes contamination of the molten weld material by oxygen and nitrogen. Gases produced during welding are also extracted rapidly by the vacuum pump so that welding of reactive metals is eased. However, the pumping time and the size of many workpieces restrict the use of high-vacuum enclosures.

At atmospheric pressures, scattering of the beam electrons by gas molecules is increased in relation to the number of stray molecules and the distance traveled, so that penetration depth is less and the beam spread is greater. In the atmosphere, the gun-to-work distance must be less than about 1.5 in. (38 mm). Electron-beam welding at atmospheric pressure requires beam-accelerating voltages above 150 kV, but lower values can be used with a protective gas. Helium is preferred because it is lighter than air and permits greater penetration. Argon, which is heavier than air and allows less penetration, can also be used to prevent contamination.

Required safety precautions, such as radiation shields to guard workers against the effects of X-rays when the electron beam strikes the work, are essential when electron-beam welding is done at atmospheric pressure. Such barriers are usually built into enclosures that are designed specifically for electron-beam welding in a partial vacuum. Adequate ventilation is also required to remove ozone and other gases generated when the process is used in the atmosphere.

Carbon, low-alloy, and stainless steels; high-temperature and refractory alloys; copper and aluminum alloys can be electron-beam welded, and single-pass, reasonably square, butt welds can be made in materials up to 1 in. (25.4 mm) thick at good speeds with nonvacuum equipment rated at 60 kW. Edges of thick material to be electron-beam welded require precision machining to provide good joint alignment and minimize the joint gap. Dissimilar metals usually may be welded without problems.

Because of the heat-sink effect, electron-beam welds solidify and cool very rapidly, causing cracking in certain materials such as low-ferrite stainless steel. Although capital costs for electron-beam welding are generally higher than for other methods, welding of large numbers of parts and the high welding travel rates make the process competitive.

Weld and Welding Symbols

American National Standard Weld and Welding Symbols.—Graphical symbols for welding provide a means of conveying complete welding information from the designer to the welder by means of drawings. The symbols and their method of use (examples of which are given in the table following this section) are part of the American National Standard ANSI/AWS A2.4-79 sponsored by the American Welding Society.

In the Standard a distinction is made between the terms *weld symbol* and *welding symbol*. Weld symbols, shown in the table *Basic Weld Symbols*, are ideographs used to indicate the type of weld desired, whereas welding symbol denotes a symbol made up of as many as eight elements conveying explicit welding instructions.

The eight elements which may appear in a welding symbol are: reference line; arrow; basic weld symbols; dimensions and other data; supplementary symbols; finish symbols; tail and specification; and process or other reference.

The standard location of elements of a welding symbol are shown in Fig. 1.

Reference Line: This is the basis of the welding symbol. All other elements are oriented with respect to this line. The arrow is affixed to one end and a tail, when necessary, is affixed to the other.

Arrow: This connects the reference line to one side of the joint in the case of groove, fillet, flange, and flash or upset welding symbols. This side of the joint is known as the *arrow side* of the joint. The opposite side is known as the *other side* of the joint. In the case of plug, slot, projection, and seam welding symbols, the arrow connects the reference line to the outer surface of one of the members of the joint at the center line of the weld. In this case the member to which the arrow points is the *arrow side* member: the other member is the *other side* member. In the case of bevel and J-groove weld symbols, a two-directional arrow pointing toward a member indicates that the member is to be chamfered.

Basic Weld Symbols: These designate the type of welding to be performed. The basic symbols which are shown in the table *Basic Weld Symbols* are placed approximately in the center of the reference line, either above or below it or on both sides of it as shown in Fig. 1. Welds on the arrow side of the joint are shown by placing the weld symbols on the side of the reference line towards the reader (lower side). Welds on the other side of the joint are shown by placing the weld symbols on the side of the reference line away from the reader (upper side).

Supplementary Symbols: These convey additional information relative to the extent of the welding, where the welding is to be performed, and the contour of the weld bead. The "weld-all-around" and "field" symbols are placed at the end of the reference line at the base of the arrow as shown in Fig. 2 and the table *Supplementary Weld Symbols*.

Dimensions: These include the size, length, spacing, etc., of the weld or welds. The size of the weld is given to the left of the basic weld symbol and the length to the right. If the length is followed by a dash and another number, this number indicates the center-to-center spacing of intermittent welds. Other pertinent information such as groove angles, included angle of countersink for plug welds and the designation of the number of spot or projection welds are also located above or below the weld symbol. The number designating the number of spot or projection welds is always enclosed in parentheses.

Contour and Finish Symbols: The contour symbol is placed above or below the weld symbol. The finish symbol always appears above or below the contour symbol (see Fig. 1).

The following finish symbols indicate the method, not the degrees of finish: C—chipping; G—grinding; M—machining; R—rolling; and H—hammering.

For indication of surface finish refer to the Surface Texture section.

Tail: The tail which appears on the end of the reference line opposite to the arrow end is used when a specification, process, or other reference is made in the welding symbol. When no specification, process, or other reference is used with a welding symbol, the tail may be omitted.

Table 1. Basic Weld Symbols

Groove Weld Symbols							
Square	Scarf[a]	V	Bevel	U	J	Flare V	Flare bevel

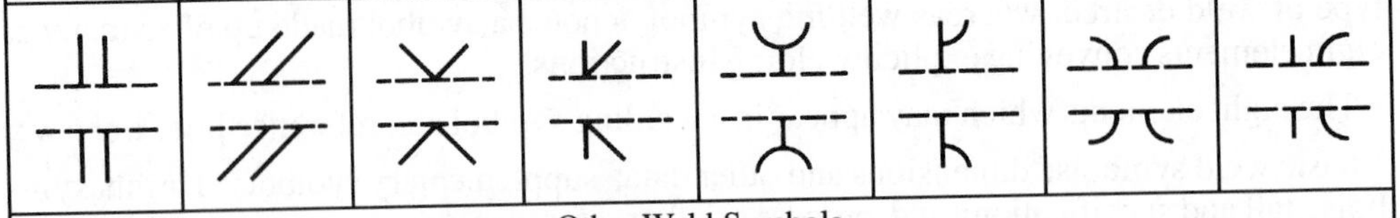

Other Weld Symbols							
Fillet	Plug or slot	Spot or projection	Seam	Back or backing	Surfacing	Flange	
						Edge	Corner

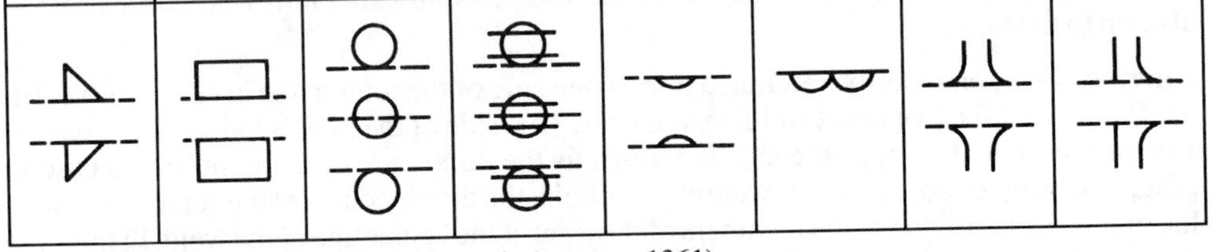

[a] This scarf symbol used for brazing only (see page 1361).

For examples of basic weld symbol applications see starting on page 1397.

Table 2. Supplementary Weld Symbols

Weld all around	Field weld	Melt-thru	Backing or spacer material	Countour		
				Flush	Convex	Concave
			*			

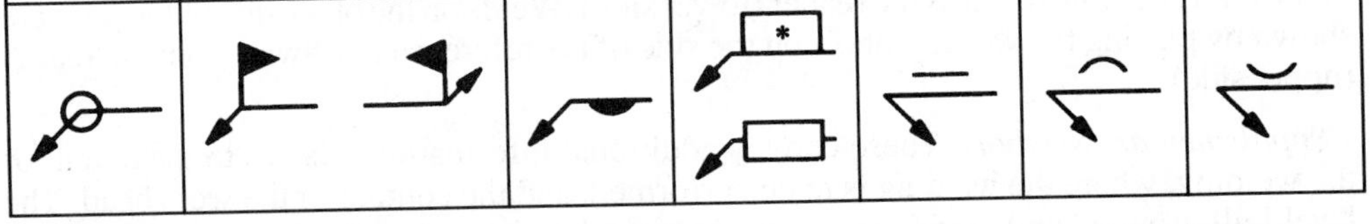

Melt-Thru Symbol: The melt-thru symbol is used only where 100 per cent joint or member penetration plus reinforcement are required.

Specification, Process, or Other Designation: These are placed in the tail of the welding symbol and are in accordance with the American National Standard. They do not have to be used if a note is placed on the drawing indicating that the welding is to be done to some specification or that instructions are given elsewhere as to the welding procedure to be used.

Letter Designations: American National Standard letter designations for welding and allied processes are shown in the table on page 1396.

Further Information: For complete information concerning welding specification by the use of standard symbols, reference should be made to American National Standard ANSI/AWS A2.4-79, which may be obtained from either the American National Standards Institute or the American Welding Society listed below.

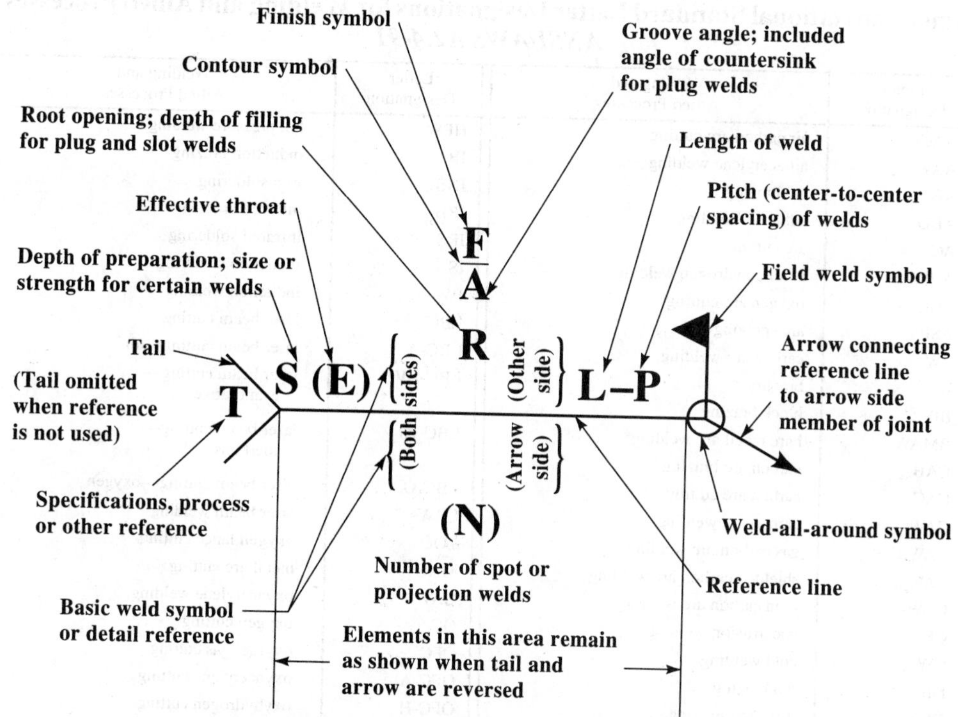

Fig. 1. Standard Location of Elements of a Welding Symbol

Welding Codes, Rules, Regulations, and Specifications.—Codes recommending procedures for obtaining specified results in the welding of various structures have been established by societies, institutes, bureaus, and associations, as well as state and federal departments.

The latest codes, rules, etc., may be obtained from these agencies, whose names and addresses are listed as follows: PV = Pressure Vessels; P = Piping; T = Tanks; SB = Structural and Bridges; S = Ships; AC = Aircraft Construction; and EWM = Electrical Welding Machinery.

Air Force/LGM, Department of the Air Force, Washington, DC 20330. (AC)

American Bureau of Shipping, 2 World Trade Center, NY, NY 10048. (S)

American Institute of Steel Cons., 1 E. Wacker Drive, Chicago, IL 60601. (SB)

American National Stds Inst., 11 West 42nd St. NY, NY 10036. (PV, P, EWM)

American Petroleum Institute, 1220 L St., NW, Washington, DC 20005. (PV)

American Soc. of Mechanical Engineers, 345 E. 47th St., NY, NY 10017. (PV)

American Welding Society, 550 LeJeune Road, Miami, FL 33135. (T, S, SB, AC)

Fed. Aviation Admin., Dept. of Transportation, Washington, DC 20590. (AC)

Insurance Services Office, 7 World Trade Center, NY, NY 10048. (PV)

Lloyd's Register of Shipping, 17 Battery Place, NY, NY 10004. (S)

Mechanical Contractors Assn., 1385 Piccard Drive, Rockville, MD. (P)

National Electrical Mfrs. Assn., 2101 L St. NW, Washington, DC 20037. (EWM)

Naval Facilities Eng'g Command, 200 Stovall St., Alexandria, VA 22332. (SB)

Naval Ship Engineering Center, Dept. of the Navy, Hyattsville, MD 20782. (PV, S)

U.S. Government Printing Office, Washington, DC 20402. (PV)

American National Standard Letter Designations for Welding and Allied Processes ANSI/AWS A2.4-91

Letter Designation	Welding and Allied Processes	Letter Designation	Welding and Allied Processes
AAC	air carbon arc cutting	HPW	hot pressure welding
AAW	air acetylene welding	IB	induction brazing
AB	arc brazing	INS	iron soldering
ABD	adhesive bonding	IRB	infrared brazing
AC	arc cutting	IRS	infrared soldering
AHW	atomic hydrogen welding	IS	induction soldering
AOC	oxygen arc cutting	IW	induction welding
ASP	arc spraying	LBC	laser beam cutting
AW	carbon arc welding	LBC-A	laser beam cutting—air
B	brazing	LBC-EV	laser beam cutting—evaporative
BB	block brazing	LBC-IG	laser beam cutting—inert gas
BMAW	bare metal arc welding	LBC-O	laser beam cutting—oxygen
CAB	carbon arc brazing	LBW	laser beam welding
CAC	carbon arc cutting	LOC	oxygen lance cutting
CAW	carbon arc welding	MAC	metal arc cutting
CAW-G	gas carbon arc welding	OAW	oxyacetylene welding
CAW-S	shielded carbon arc welding	OC	oxygen cutting
CAW-T	twin carbon arc welding	OFC	oxyfuel gas cutting
CEW	coextrusion welding	OFC-A	oxyacetylene cutting
CW	cold welding	OFC-H	oxyhydrogen cutting
DB	dip brazing	OFC-N	oxynatural gas cutting
DFB	diffusion brazing	OFC-P	oxypropane cutting
DFW	diffusion welding	OFW	oxyfuel gas cutting
DS	dip soldering	OHW	oxyhydrogen welding
EBC	electron beam cutting	PAC	plasma arc cutting
EBW	electron beam welding	PAW	plasma arc welding
EBW-HV	electron beam welding—high vacuum	PEW	percussion welding
EBW-MV	electron beam welding—medium vacuum	PGW	pressure gas welding
EBW-NV	electron beam welding—nonvacuum	POC	metal powder cutting
EGW	electrogas welding	PSP	plasma spraying
ESW	electroslag welding	PW	projection welding
EXW	explosion welding	RB	resistance brazing
FB	furnace brazing	RS	resistance soldering
FCAW	flux-cored arc welding	RSEW	resistance seam welding
FLB	flow brazing	RSEW-HF	resistance seam welding—high frequency
FLOW	flow welding	RSEW-I	resistance seam welding—induction
FLSP	flame spraying	RSW	resistance spot welding
FOC	chemical flux cutting	ROW	roll welding
FOW	forge welding	RW	resistance welding
FRW	friction welding	S	soldering
FS	furnace soldering	SAW	submerged arc welding
FW	flash welding	SAW-S	series submerged arc welding
GMAC	gas metal arc cutting	SMAC	shielded metal arccutting
GMAW	gas metal arc welding	SMAW	shielded metal arc welding
GMAW-P	gas metal arc welding—pulsed arc	SSW	solid state welding
GMAW-S	gas metal arc welding—short-circuiting arc	SW	stud arc welding
GTAC	gas tungsten arc cutting		
GTAW	gas tungsten arc welding		
GTAW-P	gas tungsten arc welding—pulsed arc		

Application of American National Standard Welding Symbols

Desired Weld	Symbol	Symbol Meaning
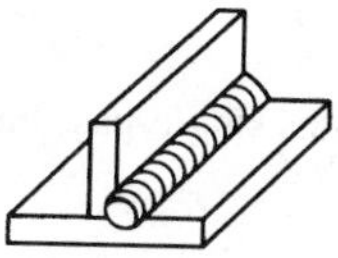	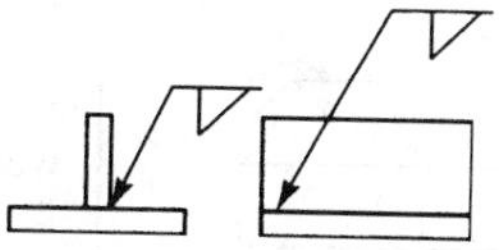	Symbol indicates fillet weld on *arrow side* of the joint.
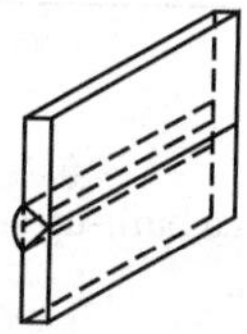	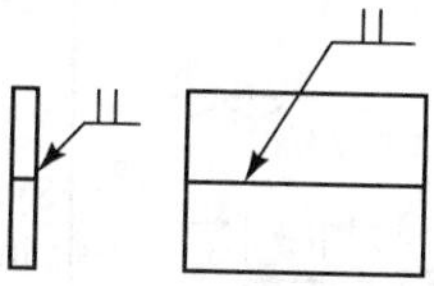	Symbol indicates square-groove weld on *other side* of the joint.
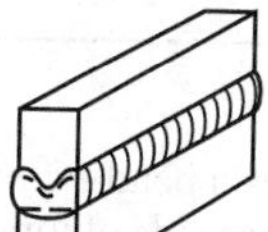	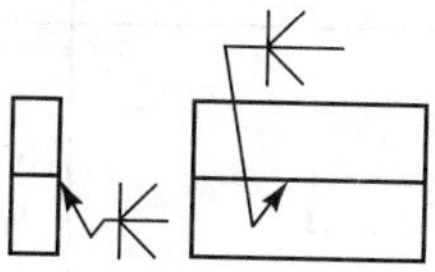	Symbol indicates bevel-groove weld on both sides of joint. Breaks in arrow indicate bevels on upper member of joint. Breaks in arrows are used on symbols designating bevel and J-groove welds.
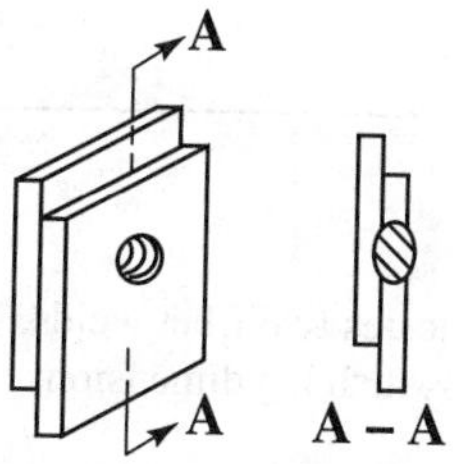 A A A – A	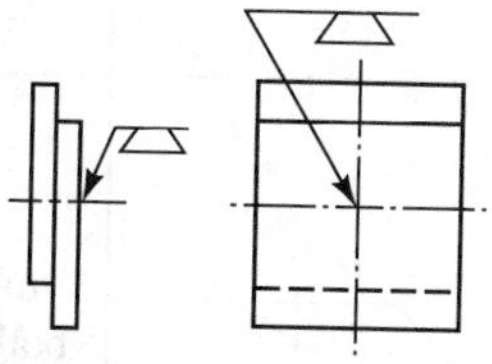	Symbol indicates plug weld on *arrow side* of joint.
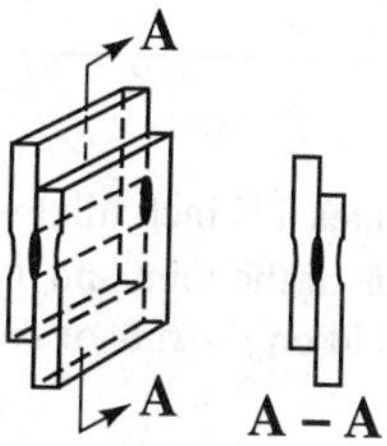 A A A – A	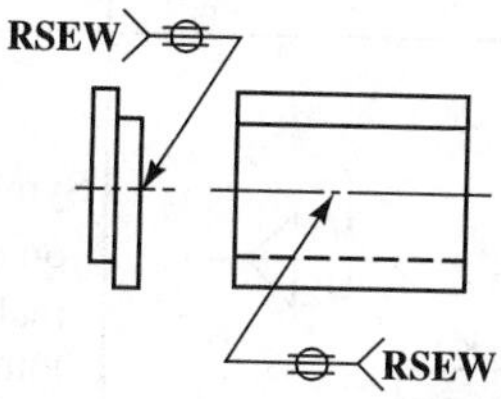 RSEW RSEW	Symbol indicates resistance-seam weld. Weld symbol appears on both sides of reference line pointing up the fact that *arrow* and *other side* of joint references have no significance.
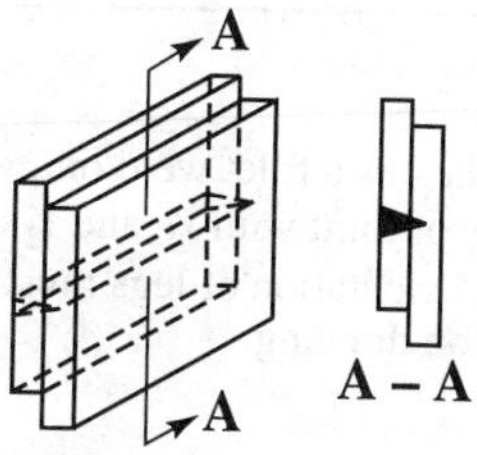 A A A – A	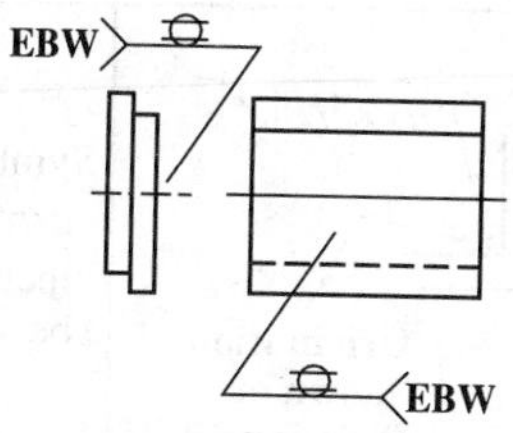 EBW EBW	Symbol indicates electron beam seam weld on *other side* of joint.

Application of American National Standard Welding Symbols *(Continued)*

Desired Weld	Symbol	Symbol Meaning
Groove Weld Made Before Welding Other Side **Back Weld**		Symbol indicates single-pass back weld.
2 1/8 3 2	1/8 2 3 2	Symbol indicates a built-up surface ⅛ inch thick.
		Symbol indicates a bead-type back weld on the *other side* of joint, and a J-groove grooved horizontal member (shown by break in arrow) and fillet weld on *arrow side* of the joint.
1/2 1/2 1/2	1/2	Symbol indicates two fillet welds, both with ½-inch leg dimensions.
1/4 1/2	1/4 1/2	Symbol indicates a ½-inch fillet weld on *arrow side* of the joint and a ¼-inch fillet weld on *far side* of the joint.
1/4 1/2	(1/4 × 1/2) **Orientation Shown on Drawing**	Symbol indicates a fillet weld on *arrow side* of joint with ¼- and ½-inch legs. Orientation of legs must be shown on drawing.

Application of American National Standard Welding Symbols *(Continued)*

Desired Weld	Symbol	Symbol Meaning
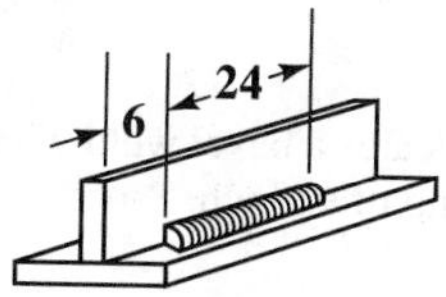	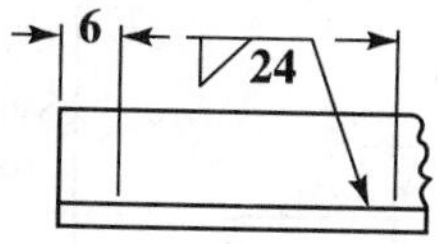	Symbol indicates a 24-inch long fillet weld on the *arrow side* of the joint.
Locate Welds at Ends of Joint 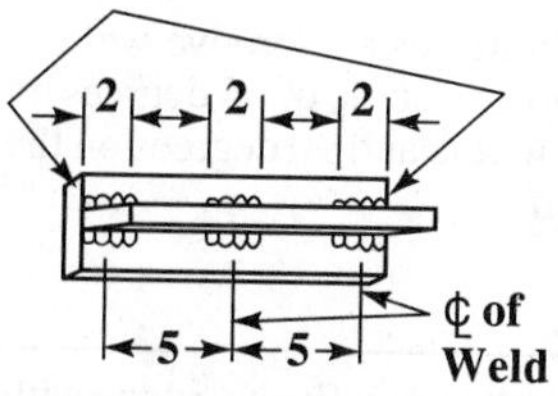	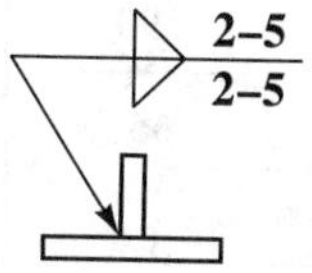	Symbol indicates a series of intermittent fillet welds each 2 inches long and spaced 5 inches apart on centers directly opposite each other on both sides of the joint.
Locate Welds at Ends of Joint 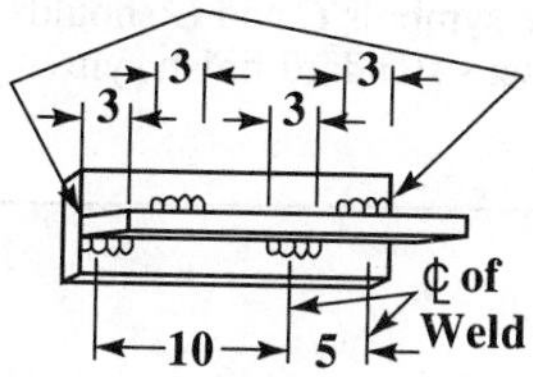	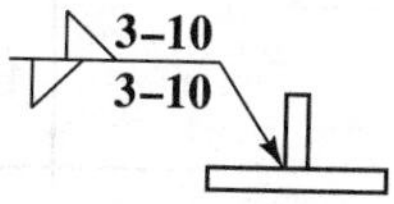	Symbol indicates a series of intermittent fillet welds each 3 inches long and spaced 10 inches apart on centers. The centers of the welds on one side of the joint are displaced from those on the other.
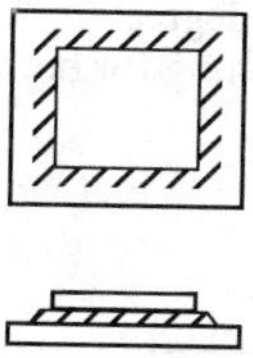	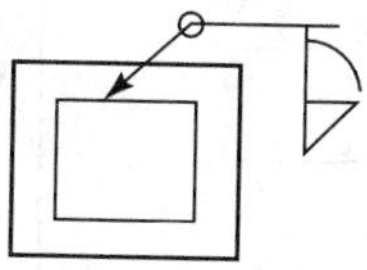	Symbol indicates a fillet weld around the perimeter of the member.
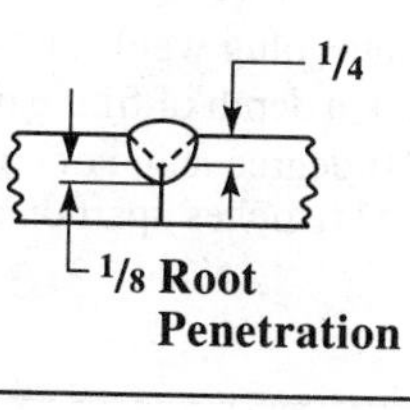	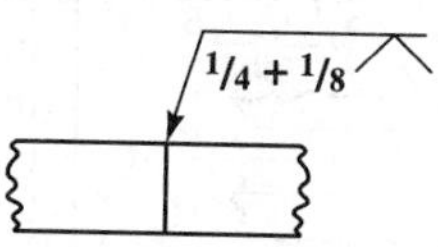	Symbol indicates a ¼-inch V-groove weld with a ⅛-inch root penetration.
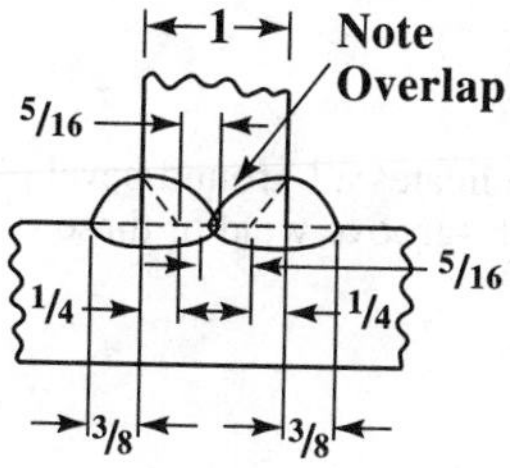	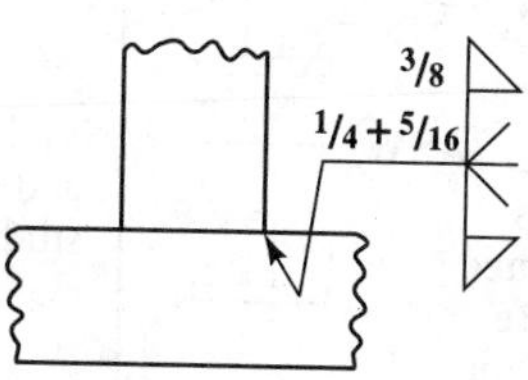	Symbol indicates a ¼-inch bevel weld with a 5⁄16-inch root penetration plus a subsequent ⅜-inch fillet weld.

Application of American National Standard Welding Symbols *(Continued)*

Desired Weld	Symbol	Symbol Meaning
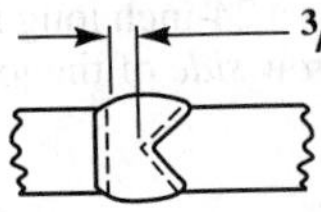 3/16	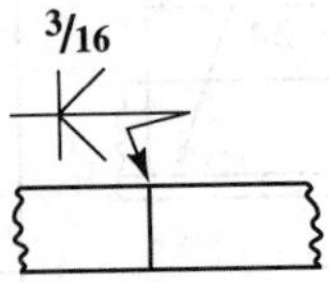 3/16	Symbol indicates a bevel weld with a root opening of $\frac{3}{16}$ inch.
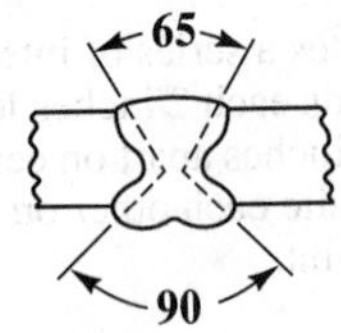 65 90	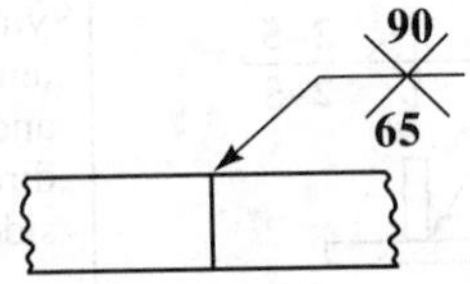 90 65	Symbol indicates a V-groove weld with a groove angle of 65 degrees on the *arrow side* and 90 degrees on the *other side*.
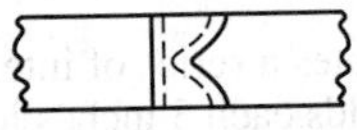	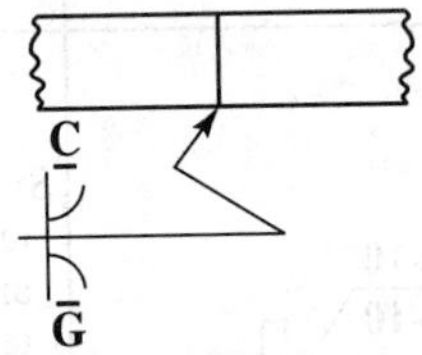 C G	Symbol indicates a flush surface with the reinforcement removed by chipping on the *other side* of the joint and a smooth grind on the *arrow side*. The symbols *C* and *G* should be the user's standard finish symbols.
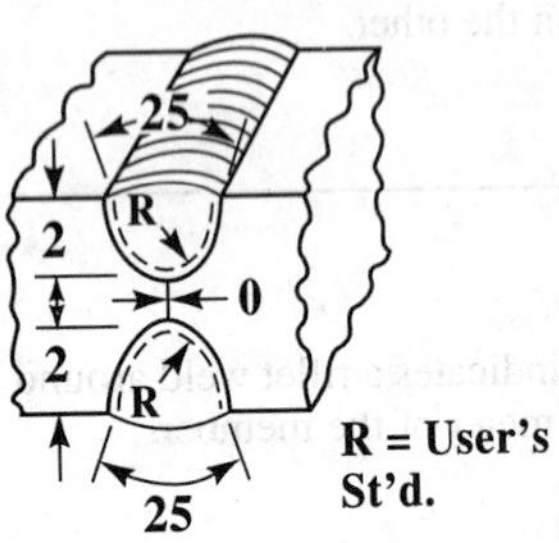 25 R 2 0 2 R 25 R = User's St'd.	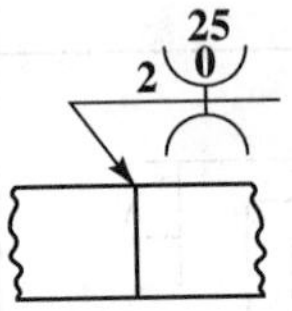 25 2 0	Symbol indicates a 2-inch U-groove weld with a 25-degree groove angle and no root opening for both sides of the joint.
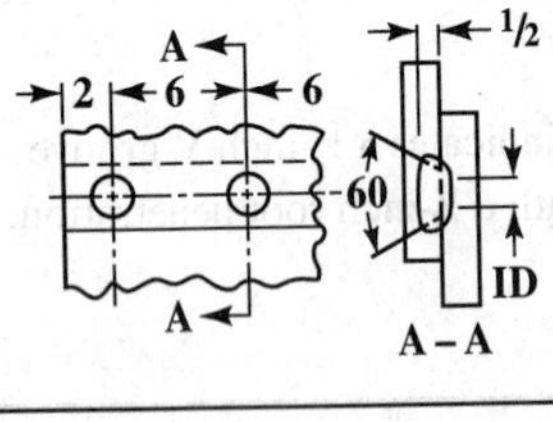 A 2 6 6 A 1/2 60 ID A – A	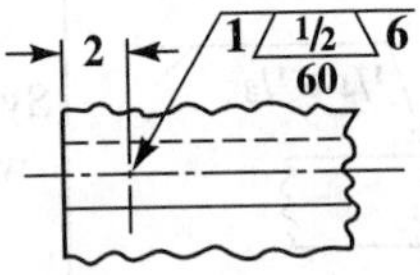 2 1 1/2 6 60	Symbol indicates plug welds of 1-inch diameter, a depth of filling of $\frac{1}{2}$ inch and a 60-degree angle of countersink spaced 6 inches apart on centers.
Preparation 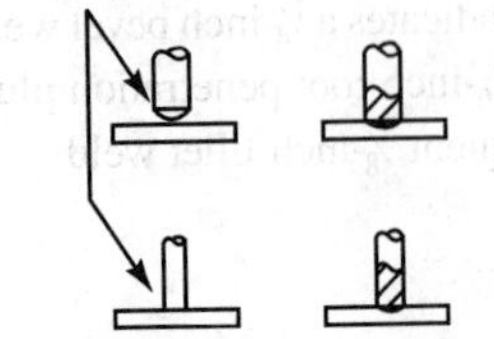	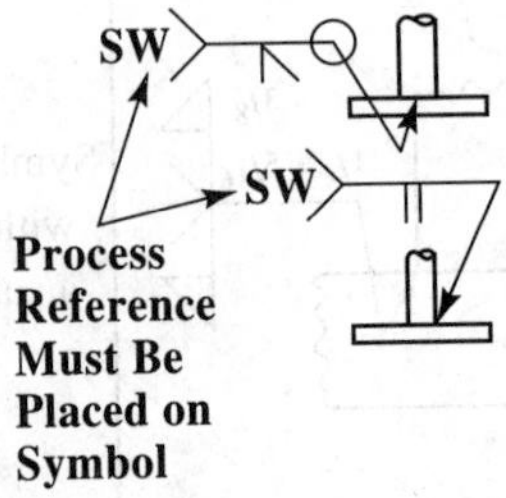 SW SW **Process Reference Must Be Placed on Symbol**	Symbol indicates all-around bevel and square-groove weld of these studs.

Application of American National Standard Welding Symbols *(Continued)*

Desired Weld	Symbol	Symbol Meaning
Min. Acceptable Shear Strength 200 lb/in. in. A A A–A	EBW 200	Symbol indicates an electron beam seam weld with a minimum acceptable joint strength of 200 pounds per lineal inch.
A 0.10″ 0.10″ A A–A	0.10″ (4) EBW	Symbol indicates four 0.10-inch diameter electron beam spot welds located at random.
		Symbol indicates a fillet weld on the *other side* of joint and a flare-bevel-groove weld and a fillet weld on the *arrow side* of the joint.
A A A–A	GTAW GTAW	Symbol indicates gas tungsten-arc seam weld on *arrow side* of joint.
		Symbol indicates edge-flange weld on *arrow side* of joint and flare-V-groove weld on *other side* of joint.
		Symbol indicates melt-thru weld. By convention, this symbol is placed on the opposite side of the reference line from the corner-flange symbol.

Pipe Welding

Pipe Welding.—Welding of (usually steel) pipe is commonly performed manually, with the pipe joint stationary, or held in a fixture whereby rotation can be used to keep the weld location in a fixed, downhand, position. Alternatively, pipe may need to be welded on site, without rotation, and the welder then has to exert considerable skill to produce a satisfactory, pressure-tight joint. Before welding stationary pipe, a welder must be proficient in welding in the four basic positions: 1G flat, 2G horizontal, 3G vertical and 4G overhead, depicted at the top in Figs. 1a, 1b, and 1c.

Positioning of Joint Components in Pipe Welding

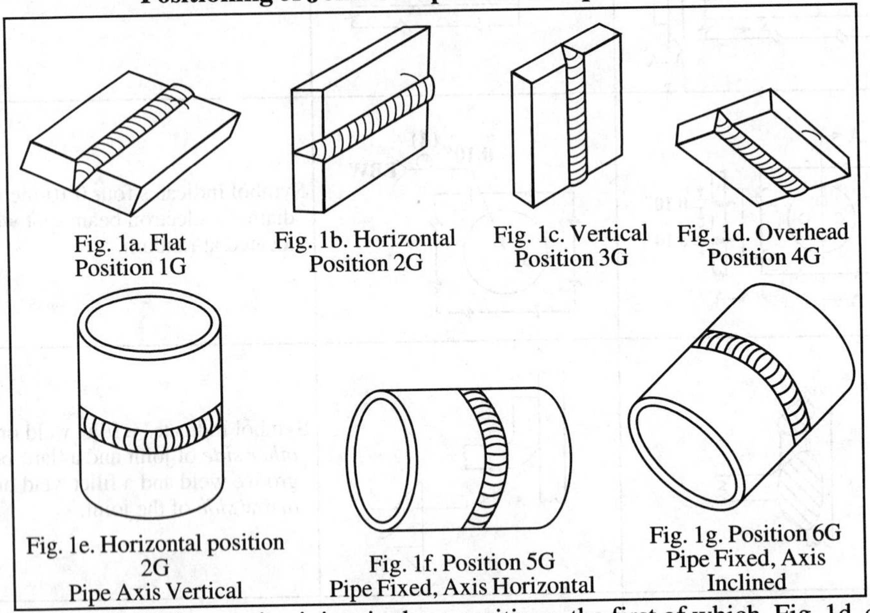

Fig. 1a. Flat Position 1G

Fig. 1b. Horizontal Position 2G

Fig. 1c. Vertical Position 3G

Fig. 1d. Overhead Position 4G

Fig. 1e. Horizontal position 2G Pipe Axis Vertical

Fig. 1f. Position 5G Pipe Fixed, Axis Horizontal

Fig. 1g. Position 6G Pipe Fixed, Axis Inclined

At the bottom are shown pipe joints in three positions, the first of which, Fig. 1d, corresponds to the 2G horizontal (non-rotational) position in the upper row. The remaining two are respectively 5G, Fig. 1e, that represents pipe with the weld in a fixed vertical (non-rotational) position; and 6G, Fig. 1f, that typifies pipe to be welded at an angle and not rotated during welding.

For satisfactory pipe welding, consideration must be given to the chemical composition and thickness of the metal to be welded; selection of a suitable electrode material composition and size; determination of the current, voltage and wire feed rate to be used; preparation of the joint or edges of the pipes; and ways of holding the pipes in the positions needed while welding is carried out. High-quality tack welds, each about 1.5 inches (38 mm) long, and projecting about $\frac{1}{16}$ inch (1.6 mm) beyond the inner wall of the pipe, are usually made to hold the parts of the assembly in position during welding.

SMAW (stick) welding was used almost exclusively for pipe welding until the advent of MIG welding with its potential for much greater rates of deposition. It cannot be emphasized too strongly that practices suitable for SMAW cannot be transferred to MIG welding, for which greater expertise is required if satisfactory welds are to be produced. MIG short-circuit, globular, and spray transfer, and pulsed MIG, with flux or metal-cored consumables (electrodes) can now all be used for pipe welding. Use of all-position, flux-cored, MIG consumables in particular, can reduce skill requirements, improve weld quality, and hold down costs in pipe welding.

Among the important items involved in the change to the MIG process is the automatic wire feeder. With today's wire feeding equipment, an increase of one increment on the dial, say from the 9 to the 10 o'clock position, can increase the wire feed rate by 70 in/min.

As an example, such an increase could raise the weld current from 110 to 145 amps and the weld voltage from 16 to 17, resulting in an increase of 40 per cent in the energy supplied to the weld. Another vital parameter is the amount that the wire sticks out from the contact tip. In low-parameter, short-circuit welding, a small change in the wire stick-out can alter the energy supplied to the weld by 20 to 30 per cent.

Root passes: Whatever welding process is selected, the most important step in pipe welding, as in other types of welding, is the root pass, which helps to determine the degree of penetration of the weld metal, and affects the amount of lack of fusion in the finished weld. During the root pass, the action of the arc in the weld area should reshape the gap between the adjacent sides of the joint into a pear-shaped opening, often called a "keyhole." As the work proceeds, this keyhole opening is continuously being filled, on the trailing side of the weld, by the metal being deposited from the electrode. The keyhole travels along with the weld so that the root pass produces a weld that penetrates slightly through the inner wall of the pipe.

MIG short-circuit root welding of carbon steel pipe requires a gap of $\frac{5}{32} \pm \frac{1}{32}$ (4 ± 0.8 mm), between the ends of the pipe, and the width of the root faces (at the base of the bevels) should be $\frac{1}{16}$ to $\frac{3}{32}$ inch (1.6 to 2.4 mm). The recommended bevel angle for MIG pipe welding is 40E (80E included angle) and the maximum root gap is $\frac{3}{16}$ inch (4.8 mm). The root pass in 1G welds should be made in the vertical-down direction with the electrode held between the 2 and 3 o'clock positions. When an 0.035-inch (0.9-mm) diameter E70S-3 MIG wire is used with the above root dimensions, weaving is not needed for the root pass except when welding over tack welds.

Fill passes: In welding carbon steel pipe in the 1G position with an 0.035-inch diameter electrode wire, MIG short circuit fill passes should use a minimum of 135 amps and be done in the vertical-up position. Fill passes should deposit a maximum thickness of no more than $\frac{1}{8}$ inch (3.2 mm). Inclusion of CO_2 gas in the mixture will improve weld fusion. With flux-cored electrodes, the minimum amount of wire stick out is $\frac{3}{4}$ inch (19 mm). Weld fusion can be improved in welding pipe of 0.4 inch (10 mm) wall thickness and thicker by preheating the work to a temperature between 400 and 500°F.

Horizontal Pipe Welding: In 1G welds (see Fig. 1a), the pipe should be rotated in the direction that moves the solidifying area away from the wire tip, to minimize penetration and resulting breakthrough. Welding of pipe in the 2G, horizontal position is made more difficult by the tendency for the molten metal to drip from the weld pool. Such dripping may cause an excessively large keyhole to form during the root-welding pass, and in subsequent passes electrode metal may be lost. Metal may also be lost from the edge of the upper pipe, causing an undercut at that side of the weld.

Vertical-down welding: With the pipe axis horizontal (as in the 5G position in Fig. 1f), vertical-down welding is usually started at the top or 12 o'clock location, and proceeds until the 6 o'clock location is reached. Welding then starts again at the 12 o'clock location and continues in the opposite direction until the 6 o'clock location is reached. Vertical-down welding is mainly used for thin-walled, low-carbon steel pipe of $\frac{1}{8}$ to $\frac{5}{16}$ inch (3.2 to 7.9 mm) wall thickness, which has low heat-retaining capacity so that the weld metal cools slowly, producing a soft and ductile structure. The slow rate of cooling also permits faster weld deposition, and, when several beads are deposited, causes an annealing effect that may refine the entire weld structure.

Vertical-up welding: In the 5G position, vertical-up welding normally begins at the 6 o'clock location and continues up to the 12 o'clock location, the weld then being completed by starting at the 6 o'clock location on the other side of the pipe and traversing up to the 12 o'clock location again. Vertical-up welding is more suited to pipe with thick walls and to alloy steels. However, the greater heat sink effect of the heavy-walled pipe may result in a faster cooling rate and embrittlement of the material, especially in alloy steels.

The cooling rate can be reduced by slowing the rate of traverse and depositing a heavier bead of metal, both facilitated by welding in the vertical-up direction.

Using a thicker electrode and higher current for thicker-walled pipe to reduce the number of beads required may result in dripping from the molten puddle of metal. Defects such as pin holes, lack of fusion, and cold lap, may then appear in the weld. Vertical-up welding of pipe in the 5G, fixed, horizontal position, Fig. 1f, used for thick-walled pipe, is probably the most difficult for a welder, but once mastered will form the basis for other methods of pipe welding. Starting at the 6 o'clock location, the arc for the root pass is struck overhead, with the electrode at an angle of 5 to 10° from the vertical, on the joint, not on the tack weld. A long arc should be maintained for a short-period while weaving the electrode to pre-heat the area ahead of the weld. Only small amounts of filler metal will be transferred while this long arc is maintained in the overhead position. The electrode tip is then advanced to establish the correct arc length and held in position long enough for the keyhole to form before starting to lay down the root bead, moving up toward the 12 o'clock location.

Thin-wall pipe: The optimum globular/spray parameters for welding rotated, (1G position) thin-wall pipe of less than 12 inch diameter are 0.035-inch electrode wire fed at 380 to 420 in/min with a protective gas mixture of argon 80 to 85, CO_2 15 to 20 per cent, and current of 190 to 210 amps. These conditions will provide deposition rates of about 6 lb/hr (3 kg/hr).

Use of Flux-cored Electrodes.—Small diameter, flux cored electrodes developed in the eighties are still a rarity in many pipe welding shops, but flux cored welding can produce consistent, high-quality, low-cost welds on carbon steel or stainless pipe. Flux cored E71T-1, 035-inch (1 mm) diameter wire provides a continuous, medium energy, open arc, with a practical current range of 135 to 165 amps for welding pipe. This current range is similar to the optimum MIG short-circuit current range, and is 25 to 30 percent less current than the minimum open arc spray transfer current for an 0.035-inch diameter MIG wire.

In contrast to MIG short circuit welding, FCAW works with an open arc and no short circuits. The FCAW arc energy is continuous, and, in contrast to short-circuit transfer, provides increased weld fusion potential. The weld metal from the flux-cored tubular wire is transferred from the periphery and the center of the wire, resulting in broad coverage of the weld. The plasma in the flux cored arc is wider than MIG plasma, and the flux-cored arc is less focused and easier to control than the MIG spray arc.

Open arc, gas shielded, flux-cored welding can produce spray type transfer at lower currents than open arc MIG spray transfer. With FCAW, the current density is high because the electrode wire cross-sectional area is less than that of the same size MIG solid wire due to the central core of flux. This higher density provides for improved weld penetration potential. The FCAW process produces slag, which serves as a mold to hold the fluid molten metal in place, an ideal arrangement for vertical-up and overhead welds.

All position, flux-cored wires require less operator skill for vertical-up and overhead welds than MIG, SMAW, and TIG processes. Fill passes can also be completed in 30 to 50 percent less time with all-position, flux-cored wires than with MIG short circuit and SMAW wires.

For good quality FCAW, welders need to know the best root and bevel dimensions, and the importance of maintaining those dimensions for continuous weld fusion; the preferred direction of pipe rotation; the diameter of flux cored electrode best suited for welding thin wall pipe; the optimum parameter range for that electrode on 1G and 5G welds; the preferred amount of wire stick out (typically 0.7 inch or 18 mm); and how to fine tune the voltage. When flux-cored welding is to be used for the fill passes, MIG short circuit welding is recommended for the root welds to reduce the possibility of slag from the flux being trapped in the weld. Higher weld deposit rates are provided with flux-cored, vertical-up welding, and there is the temptation to weld faster with a process that's easy to use. Conservative wire feed setting are recommended unless the high deposition rates are shown to provide consistent weld fusion. Wire feed settings should allow the welder time to control and direct the weave into the critical groove locations.

Complete Weld Fusion.—It is essential that new weld metal deposits be completely fused with the pipe components, and with metal laid down in successive passes. Factors that can prevent complete fusing are too numerous to list here. Some basic rules that, if followed, will improve weld fusion and quality in MIG welding in the 1G and 5G positions are:

1) The maximum gap at the root should be $\frac{3}{16}$ inch (5 mm)

2) The root land should be $\frac{1}{16}$ to $\frac{3}{32}$ inch (1.6 to 2.4 mm) wide

3) A bevel angle of 80° inclusive should be used for MIG and flux-cored welding of pipe to provide width for weaving and improve fusion.

4) An 0.035-in MIG electrode should have a minimum short circuit current of 135 amps for fill passes

5) Tack and root welds should be made in the vertical-down position.

6) Tack welds should be about 1.5 in (38 mm) long by $\frac{1}{16}$ to $\frac{3}{32}$ inch (1.6 to 2.4 mm) thick.

7) Short circuit fill passes should be made in the vertical-up position.

8) With flux-cored electrodes, a minimum of 0.7 inch (18 mm) wire stick out from the contact tip must be maintained.

9) Current and voltage must be related to the pipe wall thickness

10) Argon+25 per cent CO_2 is recommended for short circuit welding of pipe roots.

11) Use of undiluted CO_2 gas will improve MIG weld fusion in fill passes because of the "digging" action of the arc, and the increased weld energy

12) With pipe wall thicknesses of 0.4 in (10 mm) or greater, preheating to between 400 and 500° F (205 and 260°C), will help make fusion complete

Other Methods.—Pulsed MIG is a viable alternative to flux-cored for all-position welds on 5G pipe, but requires more costly equipment. The pulsed MIG process however, has few advantages over conventional MIG and flux-cored when the latter are used correctly. Pulsed MIG may have some advantage on mechanized 5G welds, and on welding of stainless steel, pipe in the 5G position.

Metal-cored electrode wire also has few advantages for pipe welds because they work best with low-energy gas welds, which cancels out the increased current density claimed for them.

On most manual pipe welds, the welder needs time to control and direct the weave to ensure even heating and avoid lack of fusion. Satisfactory welds are often performed at travel speeds of 4 to 12 in/min giving deposit rates of 3 to 5 lb/hr.

Pipe Welding Procedure

Because of the variety of parameter combinations that can be used in pipe welding, it is suggested that charts be prepared and displayed in welding booths to remind welders of the basic settings to be used. Examples of such charts for tack, root, fill and cover passes, are included in what follows:

FCAW 5G (Non-rotated) MIG Welding of Thick-walled, Carbon-steel Pipes, Procedure for Root Welding.—This procedure can be applied to most pipe sizes, and should be given special consideration for 5G (non-rotated) welds on carbon steel pipe with $\frac{3}{8}$ inch (10 mm) wall thickness and thicker.

Pipe and Weld Data

Pipe bevel included angle = 80° *Root face land* = 3/32 ± $\frac{1}{32}$ inch (2.4 ± 0.8 mm)

Root gap between faces = $\frac{5}{32}$ ± $\frac{1}{32}$ inch (4 ± 0.8 mm)

Electrode for root weld = 0.035-inch (0.9 mm) diameter, E70S-3 flux-cored.

Gas = argon with 15–25% CO_2

Gasflowrates = 30 to 40 cubic ft/hr

Set wire feeder to 210–280 in/min(10 to 11 o'clock position on many feeders) for current of 140–170 amps, 17–18 volts.

Wire extension: For MIG root weld, set contact tip to stick outside the nozzle, $\frac{1}{16}$ to $\frac{1}{8}$ inch (1.6 to 3 mm). Maintain $\frac{3}{8}$ to $\frac{5}{8}$ inch (10 to 16 mm) maximum wire stick out from contact tip.

Tack Welding Procedures for FCAW 5G Pipe Welds: Make tack welds 1.5 to 2 inches (38 to 50 mm) long. After welding, grind full length of tack to thickness of approximately $\frac{1}{16}$ inch (1.6 mm). Feather tack ends back $\frac{3}{8}$ to $\frac{1}{2}$ inches (9.5 to 13 mm).

On pipes of less than 6 inches (15 cm) outside diameter, use three tack welds, equally spaced, starting at 12 o'clock.

On pipes over 6 inches outside diameter use 4 tack welds. Locate tack welds at 12, 3, 6, and 9 o'clock.

Root Welding Procedures for FCAW 5G Pipe Welds: Root weld MIG vertical-down. Weld sequence: 12 to 3, 9 to 6, 3 to 6, and 12 to 9 o'clock positions.

Start and finish MIG root welds at tack centers. Use slight weave oscillation over tacks. No weave necessary if $\frac{1}{8}$- to $\frac{5}{32}$-inch root gap is maintained. Weaving may be required if root gap is less than $\frac{1}{8}$ inch (3 mm). Weaving is also beneficial for root welds between 7 and 6 o'clock, and between 5 and 6 o'clock. After each root pass, blend the starts and stops back to the original tack thickness.

To complete the root, ensure that the weld stops and starts on the last tack, and that the root weld center is ground flat or slightly concave. Remove any slag islands.

FCAW 5G (Non-rotated) MIG Welding of Thick-walled, Carbon-steel Pipes, Procedure for Fill and Cover Welds.—This procedure can be applied to most common pipe sizes, and should be given special consideration for 5G (non-rotated) welds on carbon steel pipe with $\frac{3}{8}$ inch (10 mm) wall thickness and thicker.

Pipe and Weld Data

Electrodeforfill and coverpasses = 0.035 inch (0.9 mm) diameter, 71T-1 flux-cored

Gas = argon with 15–25% CO_2

Gasflowrates = 30 to 40 cubic ft/hr

Set an initial wire feed rate of 350 to 450 in/min (12 to 1 o'clock position on typical wire feed unit), 135–165 amps, 25–28 volts. Alternatively, use a wire feed setting of 350 in/min (12 o'clock on wire feed unit), which should result in about 135–145 amps, 25–26 volts. If the weld pool and weld heat build up permit, increase the wire feed rate to 380 in/min (between the 12 and 1 o'clock positions), 150 amps, 27 volts. Try also a wire feed setting of 420 in/min (1 o'clock on the wire feeder), 165 amps, 28 volts. Determine the low and maximum wire feed rates to be used by examination of the weld fusion obtained in sectioned test samples.

Wire extension: Adjust contact tip so it is recessed $\frac{1}{2}$ inch within the nozzle to provide a total wire stick out from the contact tip of 0.7 to 1 inch (18 to 25 mm).

Fill and Cover Pass Procedures for FCAW 5G Pipe Welds: Weld vertical-up. If the pipe diameter allows the fill pass to be made in two passes, start at the 7 o'clock position and weld to the 1 o'clock position. This approach is preferable to starting and finishing on the root tacks. Starting at the 7 o'clock position will ensure that optimum weld energy is achieved as the first pass welds over the initial 6 o'clock root tack location. Use the grinder to feather the first 1 inch (25 mm) of the weld start and stop of the first pass, before applying the second vertical-up weld pass. Use a slight weave action for the fill pass.

Remove all flux-cored slag between weld passes. Make sure no fill pass is greater in depth than $\frac{1}{8}$ inch (3 mm). Use a straight weave across the root face. At the bevel edge use a slight upward motion with the gun. The motion should be no greater than the wire diameter. Then use a slight back step for added bevel fusion and to avoid undercuts.

Leave 1⁄32 to 1⁄16 inch (0.8 to 1.6 mm) of the groove depth to provide for the optimum cover pass profile. The bevel edge will act as a guide for the cover pass weld. If more weld fusion is required for pipe thicker than 3⁄8 inch (10 mm), after the root weld is complete, preheat the pipe to between 400 and 600°F (200–300°C) before welding. Preheating is typically not necessary for a cover pass.

For pipe diameters on which the welder needs more than two passes for the vertical-up welds, the recommended sequence for vertical-up welding is:

1) First pass, weld from the 7 to the 4 o'clock position. Start with a slight forehand nozzle angle. At the 4 o'clock position, the gun should be at the same angle as the pipe.
2) Second pass, weld from the 10 to the 1 o'clock position, then grind all stops and start again at the 1-inch (25 mm) position.
3) Third pass, weld from the 4 to the 1 o'clock position.
4) Fourth pass, weld from the 7 to the 10 o'clock position.

FCAW 5G (Non-rotated) Welding of Thin-Walled Carbon Steel Pipes, Procedure for Root, Fill and Cover Pass Welding.—This procedure can be applied to most common pipe sizes, and should be given special consideration for 5G (non-rotated) welding of carbon steel pipe with wall thicknesses up to 3⁄8 inch (10 mm).

Pipe and Weld Data

Electrode for root weld = 0.035 inch diameter, E70S-6 flux cored.

Gas = argon with 15–25% CO_2

Gas flow rates = 30 to 40 cubic ft/hr

Root Welding Procedure for 5G Welds: Use root welding data from *Root Welding Procedures for FCAW 5G Pipe Welds*, above.

Fill and Cover Pass Procedures for 5G Welds: Use MIG short-circuit, vertical-up for fill and cover passes. Electrode wire and gas, same as for root weld.

Weld vertical-up. If the vertical-up fill pass can be made in two passes, weld from the 7 to the 1 o'clock position, to avoid starting and finishing on the root tacks. Starting just past 6 o'clock ensures that optimum weld energy is achieved as the first pass welds over the initial 6 o'clock root tack location. Feather 1 inch (25 mm) of the weld start and stop on the first pass with the grinder before applying the second vertical-up weld pass. Use a slight weave action.

Use MIG short-circuit wire feed, 200–230 in/min 125–135 amps, 19–22 volts. Start at optimum 210 in/min (10 o'clock on the wire feeder) for 130 amps, 21–22 volts. Fine tune voltage by listening to arc sound to obtain a consistent rapid crackle sound.

Electrode sticks out 1⁄2 to 5⁄8 inch, contact tip flush with nozzle end.

Remove MIG surface slag islands between weld passes. No fill pass should be thicker than 1⁄8 inch (3 mm). Use straight weave across the root face. At the bevel, use a slight upward motion with the gun. The motion should be no greater than the wire diameter. Then use a slight back step for added bevel fusion and to avoid possibility of undercut.

For the cover pass, leave 1⁄32 to 1⁄16 inch of the groove depth for the optimum cover pass profile. The bevel edge will act as a guide for the cover pass weld.

If more weld fusion is required after the root is complete and between fill passes, pre-heat pipe to 200 to 400°F.

For pipe diameters on which more than two passes are required for the circumference the weld sequence is:

1) First pass, weld from the 7 to the 4 o'clock position. Start with a slight forehand nozzle angle. At the 4 o'clock position the gun should point straight at the joint.

2) Second pass, weld from the 10 to the 1 o'clock position, then grind all stops and starts for at least 1 inch (25 mm).
3) Third pass, weld from the 4 to the 1 o'clock position.
4) Fourth pass, weld from the 7 to the 10 o'clock position.

Nondestructive Testing

Nondestructive Testing.—Nondestructive testing (NDT) is aimed at examination of a component or assembly, usually for surface or internal cracks or other nonhomogeneities, to determine the structure, or to measure thickness, by some means that will not impair its use for the intended purpose. Traditional methods include use of radiography, ultrasonic vibration, dye penetrants, magnetic particles, acoustic emission, leakage, and eddy currents. These methods are simple to use but some thought needs to be given to their application and to interpretation of the results. Space limitations preclude a full discussion of NDT here, but the nature of the welding process makes these methods particularly useful, so some information on use of NDT for testing welds is given below.

Nondestructive Testing Symbol Application.—The application of nondestructive testing symbols is also covered in American National Standard ANSI/AWS 2.4–79.

Basic Testing Symbols: These are shown in the following table.

American National Standard Basic Symbols for Nondestructive Testing
ANSI/AWS 2.4–79

Symbol	Type of Test	Symbol	Type of Test
AET	Acoustic Emission	PT	Penetrant
ET	Eddy Current	PRT	Proof
LT	Leak	RT	Radiographic
MT	Magnetic Particle	UT	Ultrasonic
NRT	Neutron Radiographic	VT	Visual

Testing Symbol Elements: The testing symbol consists of the following elements:

Reference Line
- Arrow
- Basic Testing Symbol
- Test-all-around Symbol
- (N) Number of Tests
- Test in Field

Tail
- Specification or other reference

The standard location of the testing symbol elements are shown in the following figure.

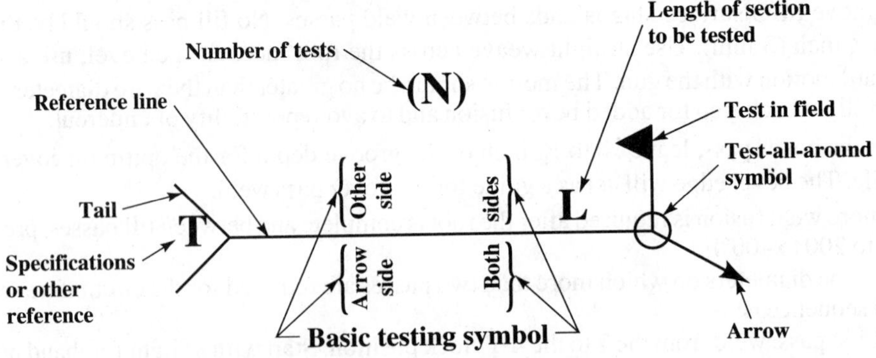

Locations of Testing Symbol Elements

The arrow connects the reference line to the part to be tested. The side of the part to which the arrow points is considered to be the *arrow side*. The side opposite the arrow side is considered to be the *other side*.

Location of Testing Symbol: Tests to be made on the arrow side of the part are indicated by the basic testing symbol on the side of the reference line toward the reader.

Tests to be made on the other side of the part are indicated by the basic testing symbol on the side of the reference line away from the reader.

To specify where only a certain length of a section is to be considered, the actual length or percentage of length to be tested is shown to the right of the basic test symbol. To specify the number of tests to be taken on a joint or part, the number of tests is shown in parentheses.

Tests to be made on both sides of the part are indicated by test symbols on both sides of the reference line. Where nondestructive symbols have no arrow or other significance, the testing symbols are centered in the reference line.

Combination of Symbols: Nondestructive basic testing symbols may be combined and nondestructive and welding symbols may be combined.

Direction of Radiation: When specified, the direction of radiation may be shown in conjunction with the radiographic or neutron radiographic basic testing symbols by means of a radiation symbol located on the drawing at the desired angle.

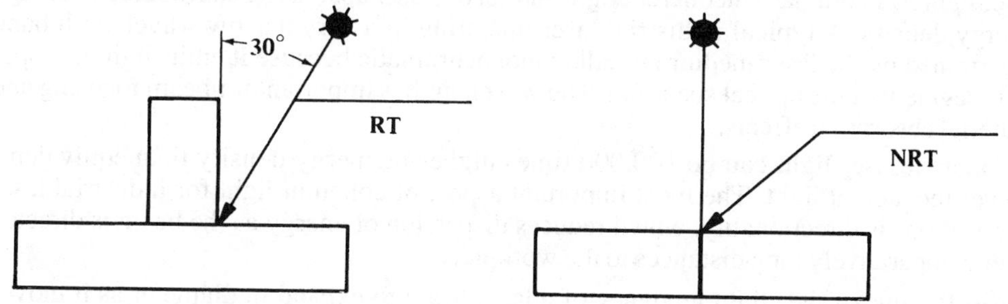

Tests Made All Around the Joint: To specify tests to be made all around a joint a circular test-all-around symbol is used.

Areas of Revolution: For nondestructive testing of areas of revolution, the area is indicated by the test-all-around symbol and appropriate dimensions.

Plane Areas: The area to be examined is enclosed by straight broken lines having a small circle around the angle apex at each change in direction.

LASERS

Lasers are used for cutting, welding, drilling, surface treatment, and marking. The word laser stands for Light Amplification by Stimulated Emission of Radiation, and a laser is a unit that produces optical-frequency radiation in intense, controllable quantities of energy. When directed against the surface of a material, this quantity of energy is high enough to cause a localized effect. Heating by a laser is controlled to produce only the desired result in a specific area, ensuring low part distortion.

The four basic components of a laser, shown in Fig. 1, are an amplifying medium, a means to excite this medium, mirrors arranged to form an optical resonator, and an output transmission device to cause beam energy to exit from the laser. The laser output wavelength is controlled by the type of amplifying medium used. The most efficient industrial lasers use optical excitation or electrical discharge to stimulate the medium and start the lasing action.

Solid-state lasers, in which the medium is a solid crystal of an optically pure material such as glass or yttrium aluminum garnet (YAG) doped with neodymium (Nd), are excited by a burst of light from a flashlamp(s) arranged in a reflective cavity that acts to concentrate the excitation energy into the crystal. Neodymium lasers emit radiation at 1.06 μm (1 μm = 0.00004 in.), in the near infrared portion of the spectrum.

The carbon dioxide (CO_2) laser uses a gaseous mixture of helium, nitrogen, and carbon dioxide. The gas molecules are energized by an electric discharge between strategically placed cathodes and anodes. The light produced by CO_2 lasers has a wavelength of approximately 10.6 μm.

Laser Light.—The characteristics of light emitted from a laser are determined by the medium and the design of the optical resonator. Photons traveling parallel to the optical axis are amplified and the design provides for a certain portion of this light energy to be transmitted from the resonator. This amplifier/resonator action determines the wavelength and spatial distribution of the laser light.

The transmitted laser light beam is monochromatic (one color) and coherent (parallel rays), with low divergence and high brightness, characteristics that distinguish coherent laser light from ordinary incoherent light and set the laser apart as a beam source with high energy density. A typical industrial laser operating in a very narrow wavelength band determined by the laser medium is called monochromatic because it emits light in a specific segment of the optical spectrum. The wavelength is important for beam focusing and material absorption effects.

Coherent laser light can be 100,000 times higher in energy density than equivalent-power incoherent light. The most important aspect of coherent light for industrial laser applications is directionality. which reduces dispersion of energy as the beam is directed over comparatively long distances to the workpiece.

Laser Beams.—The slight tendency of a laser beam to expand in diameter as it moves away from its source is called beam divergence, and is important in determining the size of the spot where it is focused on the work surface. The beam-divergence angle for high-power lasers used in processing industrial materials is larger than the diffraction-limited value because the divergence angle tends to increasewith increasing laser output power. The amount of divergence thus is a major factor in concentration of energy in the work.

The power emitted per unit area per unit solid angle is called brightness. Because the laser can produce very high levels of power in very narrowly collimated beams, it is a source of high brightness energy. This brightness factor is a major characteristic of solid-state lasers. Other important beam characteristics in industrial lasers include spatial mode and depth of focus. Ideally, the output beam of the laser selected should have a mode structure, divergence, and wavelength sufficient to process the application in optimum time and with a minimum of heat input. A beam-quality factor, M^2, is commonly used to define the productive performance of a laser. This factor is a measure of the ratio between the spot diameter of a given laser to that of a theoretically perfect beam. Beam quality is expressed as "times diffraction" and is always greater than 1. For CO_2 lasers at the 1-kW level, $M^2 = 1.5$, and for YAG lasers at 500 W, $M^2 = 12.0$ is typical.

The mode of a laser beam is described by the power distribution profile over its cross-section. Called transverse modes, these profiles are represented by the term TEM_{mn}, where TEM stands for transverse electromagnetic, and m and n are small integers indicating that power distribution is bell-shaped (Gaussian) TEM_{00}, or donut-shaped TEM_{01}*.

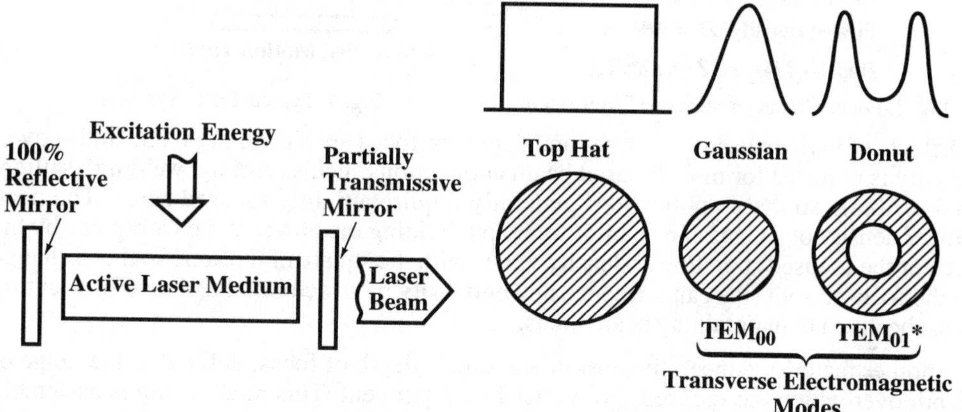

Fig. 1. Basic Components of a Laser

Fig. 2. Spatial Intensity Distribution in Laser Beams

Fig. 2 shows various transverse electromagnetic modes commonly used in materials processing applications. For such applications, it is helpful to determine the peak and total power generated by the laser. Diffraction in Gaussian beams is inherently limited and other modes of operation may have larger beam-divergence angles, causing less power to be delivered to the workpiece. The selection process for industrial applications should consider only those lasers that produce the lowest-order mode beam, in a Gaussian-shaped energy profile (see Fig. 2), with a narrow beam divergence. Solid-state lasers do not meet all these criteria, and with high-power CO_2 lasers, it is sometimes necessary to compromise because of reduced output power, large physical size, and complexity of the laser design.

Although a laser with a TEM_{00} output beam is preferred for optimum performance, the application may not always require such a beam. For example, many CO_2 laser cutting operations are performed with a TEM_{01}* beam and welding is often done with a mixture of each of these modes. Lasers can be operated in three temporal modes, continuous wave (CW), pulsed, and superpulsed (called Q-switched for YAG lasers), depending on the materials being processed.

The smallest focused spot diameter that will provide the highest energy intensity can be produced by a TEM_{00} laser. The fundamental mode output of CO_2 lasers is limited to 2500 watts. Complex spatial patterns are often caused by inhomogeneities in solid-state laser crystals and are controlled by insertion of apertures that greatly reduce output power. However, standard lasers suit the needs of most industrial users as beam divergence is only one factor in laser design.

Beam Focusing.—The diameter of a focused laser beam spot can be estimated by multiplying the published beam divergence value by the focal length of the lensor by the relationship of the wavelength to the unfocused beam diameter. Thus, the beam from a CO_2 laser operating at a 10.6-μm wavelength, using the same focal length lens, will produce a focused spot ten times larger than the beam from a Nd:YAG laser operating at a 1.06-μm wavelength.

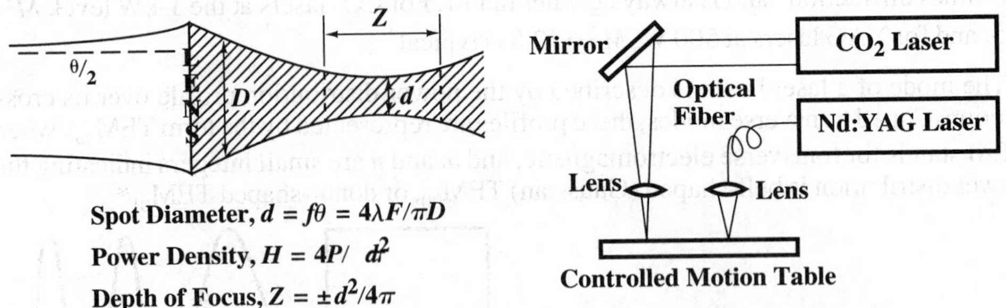

Fig. 3. Focus Characteristics of a Laser Beam.

Fig. 4. Typical Laser Systems.

Effects of various beam spot sizes and depths of focus are shown in Fig. 3. High-power density is required for most focused beam applications such as cutting, welding, drilling, and scribing, so these applications generally require a tightly focused beam. The peak power density of a Gaussian beam is found by dividing the power at the workpiece by the area of the focused spot. Power density varies with the square of the area, so that a change in the focused spot size can influence power density by a factor of 4 and careful attention must be given to maintaining beam focus.

Another factor of concern in laser processing is depth of focus, defined as the range of depth over which the focused spot varies by ±5 per cent. This relationship is extremely important in cutting sheet metal, where it is affected by variations in surface flatness. Cutting heads that adapt automatically to maintain constant surface-to-nozzle spacing are used to reduce this effect.

Types of Industrial Lasers.—Specific types of lasers are suited to specific applications, and Fig. 1 lists the most common lasers used in processing typical industrial materials. Solid-state lasers are typically used for drilling, cutting, spot and seam welding, and marking on thin sheet metal. CO_2 lasers are used to weld, cut, surface treat, and mark both metals and nonmetals. For example, CO_2 lasers are suited to ceramic scribing and Nd:YAG lasers for drilling turbine blades. Factors that affect suitability include wavelength, power density, and spot size. Some applications can use more than one laser type. Cutting sheet metal, an established kilowatt-level CO_2 laser application, can also be done with kilowatt-level Nd:YAG lasers. For some on-line applications that require multiaxis beam motion, the Nd:YAG laser may have advantages in close coupling the laser beam to the workpiece through fiber optics.

Table 1. Common Industrial Laser Applications

Type	Wavelength (μm)	Operating Mode	Power Range (watts)	Applications
Nd:YAG	1.06	Pulsed	10–2,000	A, B, D, E, F
Nd: YAG	1.06	Continuous	500–3,000	A, B, C
Nd: YAG	1.06	Q-switched	5–150	D, E, F
CO_2	10.6	Pulsed	5–3,000	A, B, D, E
CO_2	10.6	Superpulsed	1,000–5,000	A
CO_2	10.6	Continuous	100–25,000	A, B, C

Applications: A = cutting, B = welding, C = surface treatment, D =drilling, E = marking, F = micro-machining.

Industrial Laser Systems.—The laser should be located as close as possible to the workpiece to minimize beam-handling problems. Ability to locate the beam source away from its power supply and ancillary equipment, and to arrange the beam source at an angle to the workpiece allows the laser to be used in many automatic and numerically controlled set ups. Fig. 4 shows typical laser system arrangements.

Lasers require power supplies and controllers for lasers are usually housed in industrial grade enclosures suited to factory floor conditions. Because the laser is a relatively inefficient converter of electrical energy to electromagnetic energy (light), the waste heat from the beam source must be removed by heat exchangers located away from the processing area. Flowing gas CO_2 lasers require a source of laser gas, used to make up any volume lost in the normal recycling process. Gas can be supplied from closely linked tanks or piped from remote bulk storage.

Delivery of a high-quality beam from the laser to the workpiece often requires subsystems that change the beam path by optical means or cause the beam to be directed along two or more axes. Five-axis beam motion systems, for example, using multiple optical elements to move the beam in X, Y, Z, and rotation/tilt, are available.

Solid-state laser beams can be transmitted through flexible optical fibers. If there is no beam motion, the workpiece must be moved. The motion systems used can be as simple as an XY or rotary table, or as complex as a multistation, dual-feed table. Hybrid systems offer a combination of beam and workpiece motion and are frequently used in multiaxis cutting applications. All motions are controlled by an auxiliary unit such as a CNC, NC, paper tape, or programmable controller. Newer types of controllers interface with the beam source to control the entire process. Gas jet nozzles, wire feed, or seam tracking equipment are often used, and processing may be monitored and controlled by signals from height sensors, ionized by-product (plasma) detectors, and other systems.

Safety.—Safety for lasers is covered in ANSI Z136.1-1993: Safe Use Of Lasers. Most industrial lasers require substantial electrical input at high-voltage and -amperage conditions. Design of the beam source and the associated power supply should be to accepted industry electrical standards. Protective shielding is advised where an operator could interact, physically, with the laser beam, and would be similar to safety shields provided on other industrial equipment.

Radiation from a laser is intense light concentrated in tight bundles of energy. The high energy density and selective absorption characteristics of the laser beam have the potential to cause serious damage to the eye. For this reason, direct viewing of the beam from the laser should be restricted. Safety eyewear is commercially available to provide protection for each type of laser used. Certain lasers, such as the 1.06-μm solid-state units, should be arranged in a system such that workers are shielded from direct and indirect radiation. Other types of lasers, such as the 10.6-μm CO_2 laser, when operated without shielding, should meet industry standards for maximum permissible exposure levels. Much information is published on laser radiation safety, so that the subject is highly documented. Laser suppliers are very familiar with local regulations and are a good source for prepurchase information. Certain materials, notably many plastics compositions, when vaporized, will produce potentially harmful fumes.Precautionary measures such as workstation exhaust systems typically handle this problem.

Laser Beam/Material Interaction.—Industrial lasers fall into categories of effectiveness because the absorption of laser light by industrial materials depends on the specific wavelength. However, at room temperature, CO_2 laser light at 10.6 μm wavelength is fully absorbed by most organic and inorganic nonmetals.

Both CO_2 and YAG can be used in metalworking applications, although YAG laser light at 1.06 μm is absorbed to a higher degree in metals. Compensation for the lower absorption of CO: light by metals is afforded by high-energy-density beams, which create small amounts of surface temperature change that tend to increase the beam-coupling coefficient.

At CO_2 power densities in excess of 10^6 W/cm^2, effective absorptivity in metals approaches that of nonmetals. Above certain temperatures, metals will absorb more infrared energy. In steel at 400°C, for instance, the absorption rate is increased by 50 per cent. In broad-area beam processing, where the energy density (10^4 W/cm^2) is low, some form of surface coating may be required to couple the beam energy into a metal surface.

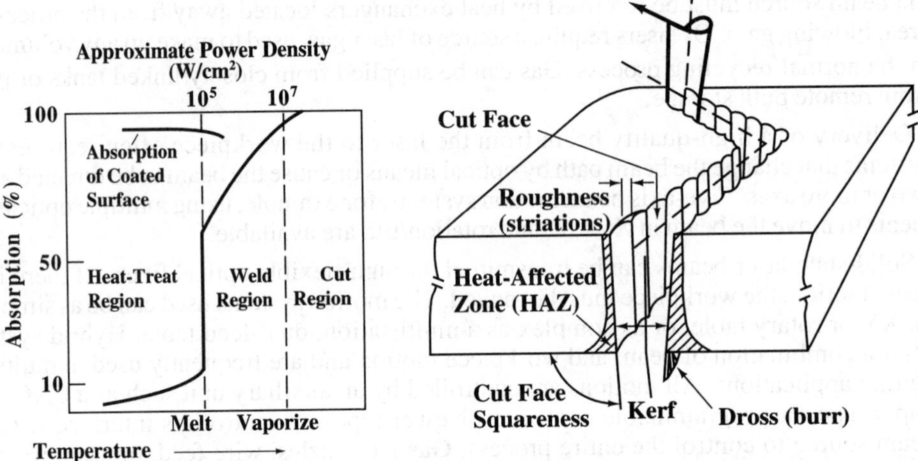

Fig. 5. Laser Energy Absorption Intensity vs. Temperature

Fig. 6. Factors in Laser Cutting.

Thermal Properties of Workpieces.—When a laser beam is coupled to a workpiece, initial conversion of energy to work, in the form of heat, is confined to a very thin layer (100-200 Ångstroms) of surface material. The absorbed energy converted to heat will change the physical state of the workpiece, and depending on the energy intensity of the beam, a material will heat, melt, or vaporize. Fig. 5 shows percentage of energy absorption versus temperature for various phase changes in materials.

Heating, melting, and vaporization of a material by laser radiation depends on the thermal conductivity and specific heat of the material. The heating rate is inversely proportional to the specific heat per unit volume, so that the important factor for heat flow is the thermal diffusivity of the work material. This value determines how rapidly a material will accept and conduct thermal energy, and a high thermal diffusivity will allow a greater depth of fusion penetration with less risk of thermal cracking.

Heat produced by a laser in surface layers is rapidly quenched into the material and the complementary cooling rate is also rapid. In some metals, the rate is 10^6 C°/s. This rapid cooling results in minimum residual heat effects, due to the slower thermal diffusivity of heat spreading from the processed area. However, rapid cooling may produce undesired effects in some metals. Cooling that is too rapid prevents chemical mixing and may result in brittle welds.

Thermosetting plastics are specifically sensitive to reheating, which may produce a gummy appearance or a charred, ashlike residue. Generally, the sensitivity of a material to heat from a laser is as apparent as with any other localized heating process. Any literature describing the behavior of materials when exposed to heat will apply to laser processing.

Cutting Metal with Lasers

The energy in a laser beam is absorbed by the surface of the impinged material, and the energy is converted into work in the form of heat, which raises the temperature to the melting or vaporization point. A jet of gas is arranged to expel excess molten metal and vapor from the molten area. Moving the resulting molten-walled hole along a path with continuous or rapidly pulsed beam power produces a cut. The width of this cut (kerf), the quality of the cut edges, and the appearance of the underside of the cut (where the dross collects) are determined by choice of laser, beam quality, delivered power, and type of motion employed (beam, workpiece, or combination). Fig. 6 identifies the factors involved in producing a high-quality cut.

Power versus penetration and cutting rate are essentially straight-line functions for most ferrous metals cut with lasers. A simple relationship states that process depth is proportional to power and inversely proportional to speed. Thus, for example, doubling power will double penetration depth. The maximum possible thickness that can be cut is, therefore, a function of power, cutting rate, and compromise on cut quality. Currently, 25 mm (1 in.) is considered the maximum thickness of steel alloys that can be cut. The most economically efficient range of thicknesses is up to 12.5 mm (0.49 in.).

Metals reflect laser light at increasing percentages with increasing wavelength. The high-energy densities generated by high-power CO_2 lasers overcome these reflectivity effects. Shorter-wavelength lasers such as Nd:YAG do not suffer these problems because more of their beam energy is absorbed.

Beam Assistance Techniques.—In cutting ferrous alloys, a jet of oxygen concentric with the laser beam is directed against the heated surface of the metal. The heat of the molten puddle of steel produced by the laser power causes the oxygen to combine with the metal, so that the jet burns through the entire thickness of the steel. This melt ablation process also uses the gas pressure to eject the molten metal from the cut kerf. Control of the gas pressure, shape of the gas stream, and positioning of the gas nozzle orifice above the metal surface are critical factors. A typical gas jet nozzle is shown in Fig. 7. Cutting highly alloyed steels, such as stainless steel, is done with pulsed CO_2 laser beams. High-pressure gas jets with the nozzle on the surface of the metal and nonoxidizing gas assistance can be used to minimize or eliminate clinging dross.

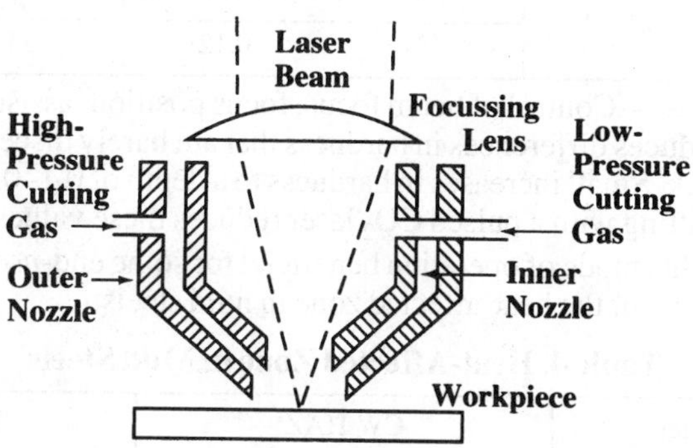

Fig. 7. Laser Gas Cutting Nozzle for Steel.

The narrow kerf produced by the laser allows cut patterns to be nested as close as one beam diameter apart, and sharply contoured and profiled cuts can be made, even in narrow angle locations. For this type of work and for other reasons, confining the kerf width to a dimension equal to, or slightly greater than, the diameter of the laser beam is important. Kerf width is a function of beam quality, focus, focus position, gas pressure, gas nozzle to surface spacing, and processing rate. Table 2 shows typical kerf widths.

Table 2. Typical Kerf Widths in CO_2 Laser Cutting

Material	Thickness		Kerf	
	mm	in.	mm	in.
Carbon Steel	1.5	0.06	0.05	0.002
	2.25	0.09	0.12	0.005
	3.12	0.12	0.2	0.008
	6.25	0.25	0.3	0.012
Aluminum	2.25	0.09	0.25	0.01
Plastics	<4.0	<0.16	2 × beam diameter	

Cut Edge Roughness.—Cutting with a continuous-wave (CW) output CO_2 laser can produce surface roughness values of 8–15 μm (315–590 μin) in 1.6-mm (0.063-in) cold-rolled steel and 30-35 μm (1180–1380 μin) in mild steel. Surface roughness of 30–50 μm (1180–1970 μin) in thin-gage stainless steel sheets is routine when using oxygen to assist cutting. Table 3 lists some surface roughness values.

Table 3. Surface Roughness Values for Laser Cutting with Oxygen

Material	Thickness		Surface Finish	
	mm	in.	μm	μin
Stainless Steel	1	0.04	30	1200
	2	0.08	35	1400
	3	0.12	50	2000
Cold-Rolled Steel	1	0.04	8	320
	2	0.08	10	400
	3	0.12	15	600
Mild Steel	1	0.04	30	1200
	2	0.08	30	1200
	3	0.12	35	1400

Heat-Affected Zones.—Control of beam focus, focus position, assist gas conditions, and processing rates produces differences in hardness that are barely discernible in steels up to 2 mm (0.078 in.) thick. Small increases in hardness to a depth of 0.1–0.2 mm (0.004–0.008 in.) are common. Cutting with a pulsed CO_2 laser reduces these values to less than 0.1 mm (0.004 in.), making this mode of operation beneficial for some end-use applications. Table 4 shows typical values for the heat-affected zone in mild steels.

Table 4. Heat-Affected Zone in Mild Steels

Material Thickness		CW HAZ		Pulsed HAZ	
mm	in.	mm	in.	mm	in.
4	0.157	0.50	0.020	0.15	0.006
3	0.118	0.37	0.015	0.15	0.006
2	0.078	0.10	0.004	0.12	0.005
1	0.039	0.75	0.030	0.07	0.003

Rates for laser cutting of metals are typically reported as data developed under ideal conditions, that is, in a controlled development laboratory environment using technician-operated equipment. Rates achieved on the shop floor using semiskilled system operators to produce complicated shapes may vary dramatically from published data. Speed versus thickness for cutting steel is shown in Fig. 8 for 1000- and 1500-W power levels, and cutting performance for several other metals of various thicknesses is shown in Table 5. Cutting rate data for pulsed and CW Nd:YAG lasers for steel also are shown in Fig. 8, and for Nd:YAG cutting of other metals in Table 5.

Table 5. CO_2 and Nd: YAG Cutting Speeds for Nonferrous Metals

Material	CO_2 (1500 watts)				Nd: YAG				
	Thickness		Speed		Thickness		Speed		Power
	mm	in.	m/min	ft/min	mm	in.	m/min	ft/min	watts
Copper	1	0.04	2.25	7.4	…		…		…
	2	0.08	0.75	2.5	…		…		…
	3	0.12	0.35	1.15	…		…		…
Aluminum	1	0.04	8	26.2	1.5	0.06	2.5	8.2	1000
	2	0.08	4	13.1	2.5	0.1	1.0	3.3	1000
	3	0.12	1.5	4.9	3.5	0.14	0.5	1.6	1000
Titanium	1	0.04	6	19.7	0.4	0.016	1.0	3.3	150
	2	0.08	3	9.8	…		…		…
Tungsten	…		…		0.08	0.003	0.03	0.1	250
Brass	1	0.04	3	9.8	…		…		…
	2	0.08	1.5	4.9	…		…		…
Hastalloy	2.5	0.1	2.8	9.2	…		…		…
Hastalloy X	…		…		0.08	0.003	0.5	1.6	150
Inconel 718	4	0.16	1.1	3.6	…		…		…

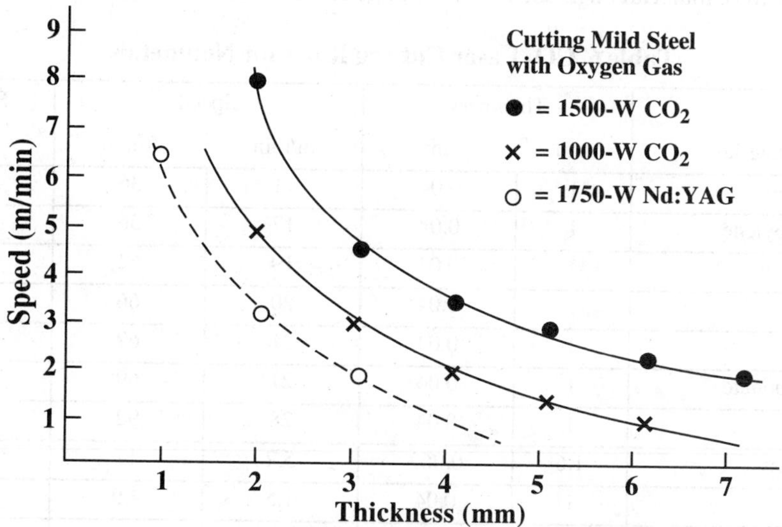

Fig. 8. Typical Cutting Rates for CO_2 and YAG Lasers.

Cutting of Nonmetals.—Laser cutting of nonmetals has three requirements: a focused beam of energy at a wavelength that will be absorbed easily by the material so that melting or vaporization can occur; a concentric jet of gas, usually compressed air, to remove the by-products from the cut area; and a means to generate cuts in straight or curved outlines. Residual thermal effects resulting from the process present a greater problem than in cutting of metals and limit applications of lasers in nonmetal processing.

When subjected to a laser beam, paper, wood, and other cellular materials undergo vaporization caused by combustion. The cutting speed depends on laser power, material thickness, and water and air content of the material. Thermoplastic polymer materials are cut by melting and gas jet expulsion of the melted material from the cut area. The cutting speed is governed by laser power, material thickness, and pressure of gas used to eject the displaced material.

Polymers that may be cut by combustion or chemical degradation include the thermosetting plastics, for example, epoxies and phenolics. Cutting speed is determined by the laser power and is higher for thermosets than for other polymers due to the phase change to vapor.

Composite materials are generally easy to cut, but the resulting cut may not be of the highest quality, depending on the heat sensitivity of the composite materials. High-pressure cutting processes such as fluid jets have proven to be more effective than lasers for cutting many composite materials.

Nonmetal cutting processes require moderate amounts of power, so the only limitation on cut thickness is the quality of the cut. In practice, the majority of cutting applications are to materials less than 12 mm thick. Cutting rates for some commonly used nonmetals are shown in Table 6. Nonmetal cutting applications require a gas jet to remove molten, vaporized, or chemically degraded matter from the cut area.

Compressed air is used for many plastics cutting applications because it is widely available and cheap to produce, so it is a small cost factor in nonmetal cutting. A narrow kerf is a feature of nonmetal cutting, and it is especially important in the cutting of compactly nested parts such as those produced in cutting of fabrics. Nonmetals react in a variety of ways to laser-generated heat, so that it is difficult to generalize on edge roughness, but thermally sensitive materials will usually show edge effects.

Table 6. CO_2 Laser Cutting Rates for Nonmetals

	Thickness		Speed		Power
Material	mm	in	m/min	ft/min	watts
Polythene	1	0.04	11	36	500
Polypropylene	1	0.04	17	56	500
Polystyrene	1	0.04	19	62	500
Nylon	1	0.04	20	66	500
ABS	1	0.04	21	69	500
Polycarbonate	1	0.04	21	69	500
PVC	1	0.04	28	92	500
Fiberglass	1.6	0.063	5.2	17	450
Glass	1	0.04	1.5	4.9	500
Alumina	1	0.04	1.4	4.6	500
Hardwood	10	0.39	2.6	8.5	500
Plywood	12	0.47	4.8	15.7	1000
Cardboard	4.6	0.18	9.0	29.5	350

Welding with Lasers

Laser Welding Theory.—Conversion of absorbed laser energy into heat causes metals to undergo a phase change from solid to liquid and, as energy is removed, back to solid. This fusion welding process is used to produce selective area spot welds or linear continuous seam welds. The two types of laser welding processes, conduction and deep penetration, or keyhole, are shown in Fig. 9.

Conduction welding: relies on the thermal diffusivity characteristics of the metal to conduct heat into the joint area. By concentrating heat into the focused beam diameter and programming this heat input for short time periods, more heat is conducted into the joint than is radiated outward from the joint. Conduction welds are generally used for spot welding and partial penetration seam welding.

Deep penetration keyhole welding: is produced by beam energy converted to heat that causes a hole to be produced through the thickness of the metal. Vapor pressure of evaporated metal holds a layer of molten metal in place against the hole wall.

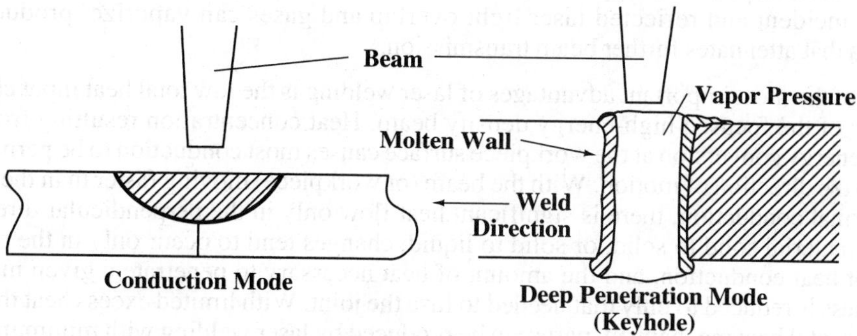

Fig. 9. Types of Laser Welds.

Movement of the hole, by beam or workpiece motion, causes the molten metal to flow around the hole and solidify behind the beam interaction point. The resolidified metal has a different structure than the base metal. Maximum practical penetration limits are approximately 25 mm (2 in.) with today's available laser power technology.

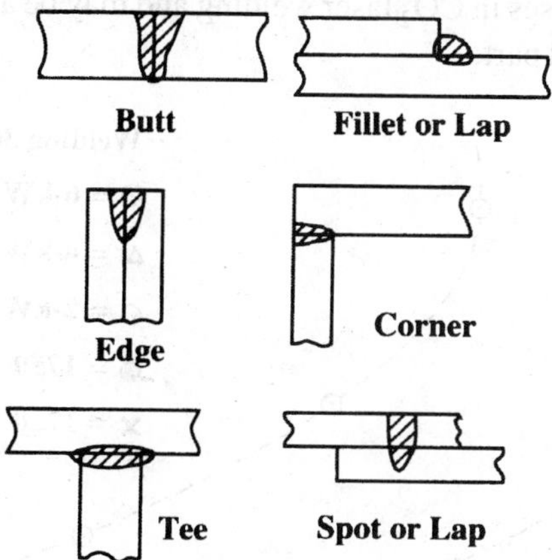

Fig. 10. Examples of Laser Weld Joint Designs.

If the physical change from solid to liquid to solid does not produce a ductile fusion zone, and if the brittleness of the resolidified metal cannot be reduced easily by postweld annealing, then the laser welding process, as with other fusion welding processes, may not be viable. If the metal-to-metal combination does not produce an effective weld, other considerations such as filler metal additions to modify fusion zone chemistry should be considered.

Welded Joint Design.—For optimum results, the edges of parts to be laser beam welded should be in close contact. When a part is being designed and there is a choice of welding process, designers should design joints and joint tolerances to the optimum for laser welding. Fig. 10 shows suitable joint designs for the laser fusion welding process. Joint tolerances are one of the more important parameters influencing part weldability, and for corner, tee, and lap joints, gaps should be not more than 25 per cent of the thickness of the thinnest section. For butt and edge joints, the percentage is reduced to 10. Addition of filler metal to compensate for large joint gaps is becoming popular.

Welding Rates.—The information presented in Fig. 11 for welding with CO_2 and Nd:YAG lasers should be considered as typical for the specific lasers shown and is for use in optimum conditions. These data are provided only as guidelines.

Processing Gas.—The proper choice of processing gas is important for both conduction and keyhole welding. Gases that ionize easily should not be used to shield the beam/material interaction point. Energy intensities of 10^6 W/cm^2 or higher can occur in the zone where incident and reflected laser light overlap and gases can vaporize, producing a plasma that attenuates further beam transmission.

One of the most important advantages of laser welding is the low total heat input characteristic of the focused high-energy density beam. Heat concentration resulting from the beam energy conversion at the workpiece surface causes most conduction to be perpendicular to the direction of motion. With the beam (or workpiece) moving faster than the speed of thermal conduction, there is significant heat flow only in the perpendicular direction. Thus, material solid to solid, or solid to liquid, changes tend to occur only in the narrow path of heat conduction, and the amount of heat necessary to penetrate a given material thickness is reduced to only that needed to fuse the joint. With limited excess heat through the low total heat mechanism, parts can be produced by laser welding with minimum thermal distortion.

Helium is the ideal gas for laser welding, but other gases such as CO_2 and argon have been used. Neither CO_2 nor argon produces a clean, perfectly smooth weld, but weld integrity seems sufficient to suggest them as alternatives. The cost of welding assistance gas can be greater than for laser gases in CO_2 laser welding and may be a significant factor in manufacturing cost per welded part.

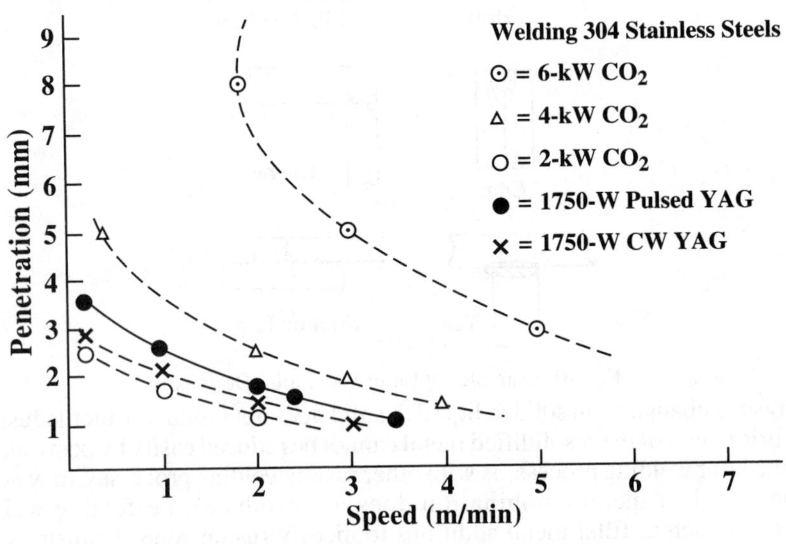

Fig. 11. Rates for CO_2 and Nd:YAG Laser Welding.

Drilling with Lasers

Laser Drilling Theory.—Laser drilling is performed by direct, percussive, and trepanning methods that produce holes of increasing quality respectively, using increasingly more sophisticated equipment. The drilling process occurs when the localized heating of the material by a focused laser beam raises the surface temperature above the melting temperature for metal or, for nonmetals, above the vaporization temperature.

Direct Drilling.—The single-pulse, single-hole process is called direct drilling. The process hole size is determined by the thermal characteristics of the material, the beam spot size, the power density, the beam quality, and the focus location. Of these parameters, beam quality, in terms of beam divergence, is an important criterion because of its effect on the hole size. Single-pulse drilled holes are usually limited to a depth of 1.5 mm (0.06 in.) in metals and up to 8 mm (0.315 in.) in nonmetals. Maximum hole diameter for pulsed solid-state laser metal drilling is in the 0.5-to 0.75-mm (0.02- to 0.03-in) range, and CO_2 direct drilling can produce holes up to 1.0 mm (0.04 in.) in diameter. The aspect ratio (depth to midhole diameter) is typically under 10:1 in metals and for many nonmetals it can be 15:1. Hole taper is usually present in direct drilling of metals. The amount of taper (entrance hole to exit hole diameter change) can be as much as 25 per cent in many metals. Direct drilling produces a recast layer with a depth of about 0.1 mm (0.004 in.). Diameter tolerances are ±10 per cent for the entrance hole, depending on beam quality and assist gas pressure.

Percussive Drilling.—Firing a rapid sequence of pulses produces a hole of higher quality than direct drilling in metal thicknesses up to 25 mm (1 in.). This process is known as percussive drilling. Multiple pulses may be necessary, depending on the metal thickness. Typical results using percussion drilled holes are: maximum depth achievable, 25 mm (1 in.); maximum hole diameter, 1.5 mm (0.06 in.); aspect ratio, 50:1; recast layer, 0.5 mm (0.02 in.); taper under 10 per cent; and hole diameter tolerance ±5 per cent.

Trepanning.—To improve hole quality, some companies use the trepanning method to cut a hole. In this process, a focused beam is moved around the circumference of the hole to be drilled by a rotating mirror assembly. The closeness of spacing of the beam pulses that need to be overlapped to produce the hole depends on the quality requirements. Typical results are: maximum hole depth, 10 mm (0.39 in.); maximum hole diameter, 2.5 mm (0.1 in.); and recast layer thickness, 25 μm (985 μin).

Drilling Rates.—Laser drilling is a fast process but is very dependent on the above-mentioned process factors. It is difficult to generalize on laser drilling rates because of the large number of combinations of material, hole diameter, depth, number of holes per part, and part throughput. With Nd:YAG lasers, direct drilling rates of 1 ms are typical.

Heat Treatment with Lasers

The defocused beam from a CO_2 laser impinging on a metal surface at room temperature will have 90 per cent or more of its power reflected. In steels, the value is about 93 per cent. Compared with focused beam processing, which uses power densities greater than 10^5 W/cm^2, the power density of laser beams designed for heat treatment, at less than 10^4 W/cm^2, is insufficient to overcome reflectivity effects. Therefore, the metal surface needs to be prepared by one of several processes that will enhance absorption characteristics. Surface roughening can be used to produce tiny craters that can trap portions of the beam long enough to raise the surface temperature to a point where more beam energy is absorbed. Coating the metal surface is a common expedient. Black enamel paint is easy to apply and the laser beam causes the enamel to vaporize, leaving a clean surface.

The absorbed laser beam energy, converted to heat, raises the temperature of the metal in the beam pattern for as long as the beam remains in one place. The length of the dwell time is used to control the depth of the heat treatment and is an extremely effective means for control of case depth in hardening.

Materials Applicability.—Hardenable ferrous metals, such as medium- and high-carbon steels, tool steels, low-alloy steels and cast irons, and steels with fine-carbide dispersion, are good candidates for laser heat treating. Marginally hardenable metals include annealed carbon steels, spheroidized carbon steels, mild-carbon steels (0.2 per cent C), and ferritic nodular cast irons. Low-carbon steels (<0.1 per cent C), austenitic stainless steels, and nonferrous alloys and metals are not hardenable.

The effect of the metal microstructure on depth of hardening is an important factor. Cast iron, with a graphite and tempered martensite structure, presents a low carbon-diffusion distance that favors deep-hardened cases. The same is true for steel with a tempered martensite or bainite structure. On the other hand, cast iron with a graphite/ferrite structure and spheroidized iron (Fe_3C plus ferrite) structures have large carbon-diffusion patterns and therefore produce very shallow or no case depths.

Hardening Rates.—Laser hardening is typically slower than conventional techniques such as induction heating. However, by limiting the area to be hardened, the laser can prove to be cost-effective through the elimination of residual heat effects that cause part distortion. A typical hardening rate is 130 cm^2/min. (20 in^2/min.) for a 1-mm (0.039-in) case depth in 4140 steel.

Cladding with Lasers

In laser cladding, for applying a coating of a hard metal to a softer alloy, for instance, a shaped or defocused laser beam is used to heat either preplaced or gravity-fed powdered alloys. The cladding alloy melts and flows across the surface of the substrate, rapidly solidifying when laser power is removed. Control of laser power, beam or part travel speed, clad thickness, substrate thickness, powder feed rate, and shielding gas are process variables that are determined for each part.

Many of the alloys currently used in plasma arc or metal inert gas cladding techniques can be used with the laser cladding process. Among these materials, Stellites, Colmonoys, and other alloys containing carbides are included, plus Inconel, Triballoy, Fe-Cr-C-X alloys, and tungsten and titanium carbides.

Controlled minimal dilution may be the key technical advantage of the laser cladding process. Dilution is defined as the total volume of the surface layer contributed by melting of the substrate, and it increases with increasing power, but decreases with either increasing travel speed or increasing beam width transverse to the direction of travel. Tests comparing laser dilution to other cladding techniques show the laser at <2 per cent compared to 5–15 per cent for plasma arc and 20–25 per cent for stick electrode processes.

The laser cladding process results in a dense, homogenous, nonporous clad layer that is metallurgically bonded to the substrate. These qualities are in contrast to the mechanically bonded, more porous layer produced by other methods.

Marking with Lasers

Laser marking technology can be divided into two groups; those that produce a repetitive mark are listed as mask marking, and those that involve rapid changes of mark characteristics are classified as scanned beam marking. The amount of data that can be marked in a unit of time (writing speed) depends on laser energy density, galvanometer speed, computer control, and the dimensions of the mark. Heat-type marks have been made at rates up to 2500 mm/s (100 in/s) and engraved marks at rates of 500–800 mm/s (20–30 in/s). Writing fields are of various sizes, but a typical field measures 100 × 100 mm (4 × 4 in.).

Mask Marking.—In mask marking, the beam from a CO_2 laser is projected through a reflective mask that passes beam energy only through uncoated areas. The beam energy is reimaged by a wide field lens onto the material's surface where the absorbed heat changes the molecular structure of the material to produce a visible mark. Examples are clouding PVC or acrylics, effecting a change in a colored surface (usually by adjusting proportions of pigment dyes), or by ablating a surface layer to expose a sublayer of a different color.

CO_2 lasers can be pulsed at high rates and have produced legible marks at line speeds of 20,000 marks/h. These lasers produce energy densities in the 1–20 J/cm^2 range, which corresponds to millions of watts/cm^2 of power density and allows marking to be performed in areas covering 0.06 to 6 cm^2. The minimum width of an individual line is 0.1 mm (0.004 in.). Mask marking is done by allowing the beam from a laser to be projected through a mask containing the mark to be made. Reimaging the beam by optics onto the workpiece causes a visible change in the material, resulting in a permanent mark. Mask marking is used for materials that are compatible with the wavelength of the laser used.

Scanned-Beam Marking.—Focusing a pulsed laser beam to a small diameter concentrates the power and produces high-energy density that will cause a material to change its visual character. Identified by several names (spot, stroke, pattern generation, or engraving), this application is best known as scanned beam.

In the scanned-beam method, the beam from a pulsed YAG or CO_2 laser is directed onto the surface of a part by a controlled mirror oscillation that changes the beam path in a preprogrammed manner. The programming provides virtually unlimited choice of patterns to be traced on the part. The pulsed laser output can be sequenced with beam manipulation to produce a continuous line or a series of discrete spots that visually suggest a pattern (dot matrix).

The energy density in the focused beam is sufficient to produce a physical or chemical change in most materials. For certain highly reflective metals, such as aluminum, better results are obtained by pretreating the surface (anodizing). Not all scanned beam applications result in removal of base metal. Some remove only a coating or produce a discoloration, caused by heating, that serves as a mark.

Hard Facing

Hard facing is a method of adding a coating, edge, or point, of a metal or alloy capable of resisting abrasion, corrosion, heat, or impact, to a metal component. The process can be applied equally well to new parts or old worn parts. The most common welding methods used to apply hard-facing materials include the oxyacetylene gas, shielded-metal arc, submerged arc, plasma arc, and inert-gas-shielded arc (consuming and nonconsuming electrode). Such coatings can also be applied by a spraying process, using equipment designed to handle the coating material in the form of a wire or a powder.

Hard-Facing Materials.—The first thing to be considered in the selection of a hard-facing material is the type of service the part in question is to undergo. Other considerations include machinability, cost of hard-facing material, porosity of the deposit, appearance in use, and ease of application. Only generalized information can be given here to guide the selection of a material as the choice is dependent upon experience with a particular type of service. Generally, the greater the hardness of the facing material, the greater is its resistance to abrasion and shock or impact wear. Many hardenable materials may be used for hard facing such as carbon steels, low-alloy steels, medium-alloy steels, and medium-high alloys but none of these is outstanding. Some of the materials that might be considered to be preferable are high-speed steel, austenitic manganese steel, austenitic high-chromium iron, cobalt-chromium alloy, copper-base alloy, and nickel-chromium-boron alloy.

High-Speed Steels.—These steels are available in the form of welding rods (RFe5) and electrodes (EFe5) for hard facing where hardness is required at service temperatures up to 1100 degrees F and where wear resistance and toughness are also required. Typical surfacing operations are done on cutting tools, shear blades, reamers, forming dies, shearing dies, guides, ingot tongs, and broaches using these metals.

Hardness: These steels have a hardness of 55 to 60 on the Rockwell C scale in the as-welded condition and a hardness of 30 Rockwell C in the annealed condition. At a temperature of 1100 degrees F, the as-deposited hardness of 60 Rockwell C falls off very slowly to 47 Rockwell C. At about 1200 degrees F, the maximum Rockwell C hardness is 30.

Resistance Properties: As deposited, the alloys can withstand only medium impact, but when tempered, the impact resistance is increased appreciably. Deposits of these alloys will oxidize readily because of their high molybdenum content but can withstand atmospheric corrosion. They do not withstand liquid corrosives.

Other Properties or Characteristics: The metals are well suited for metal-to-metal wear especially at elevated temperatures. They retain their hardness at elevated temperatures and can take a high polish. For machining, these alloys must first be annealed. Full hardness may be regained by a subsequent heat treatment of the metal.

Austenitic Manganese Steels.—These metals are available in the form of electrodes (EFeMn) for hard facing when dealing with metal-to-metal wear and impact. Uses include facing rock-crushing equipment and railway frogs and crossings.

Hardness: Hardness of the as-deposited metals are 170 to 230 Bhn, but they can be work-hardened to 450 to 550 Bhn very readily. For all practical purposes, these metals have no hot hardness as they become brittle when reheated above 500 to 600 degrees F.

Resistance Properties: These metals have high impact resistance. Their corrosion and oxidation resistance are similar to those of ordinary carbon steels. Their resistance to abrasion is only mediocre compared with hard abrasives like quartz.

Other Properties or Characteristics: The yield strength of the deposited metal in compression is low, but any compressive deformation rapidly raises it until plastic flow ceases. This property is an asset in impact wear situations. Machining is difficult with ordinary tools and equipment; finished surfaces are usually ground.

Austenitic High-Chromium Irons.—These metals are available in rod (RFeCr-A) and electrode (EFeCr-A) form and are used for facing agricultural machinery parts, coke chutes, steel mill guides, sand-blasting equipment, and brick-making machinery.

Hardness: The as-welded deposit ranges in hardness from 51 to 62 Rockwell C. Under impact, the deposit work hardens somewhat, but the resulting deformation also leads to cracking and impact service is therefore avoided. Hot hardness decreases slowly at temperatures up to 800 and 900 degrees F. At 900 degrees F, the instantaneous hardness is 43 Rockwell C. In 3 minutes under load, the hardness drops to 37 Rockwell C. At 1200 degrees F, the instantaneous hardness is 5 Rockwell C. The decrease in hardness during hot testing is practically recovered on cooling to ambient temperatures.

Resistance Properties: Deposits will withstand only light impact without cracking. Dynamic compression stresses above 60,000 pounds per square inch should be avoided. These metals exhibit good oxidation resistance up to 1800 degrees F and can be considered for hot wear applications where hot plasticity is not objectionable. They are not very resistant to corrosion from liquids and will rust in moist air, but are more stable than ordinary iron and steel. Resistance to low-stress scratching is outstanding and is related to the amount of hard carbides present. However, under high-stress grinding abrasion, performance is only mediocre and they are not deemed suitable for such service.

Other Properties or Characteristics: The deposited metals have a yield strength (0.1 per cent offset) of between 80,000 and 140,000 pounds per square inch in compression and an ultimate strength of from 150,000 to 280,000 pounds per square inch. Their tensile strength is low and therefore tension uses are avoided in design. These deposits are considered to be commercially unmachinable and are also very difficult to grind. When ground, a grinding wheel of aluminum oxide abrasive with a 24-grit size and a hard (Q) and medium-spaced resinoid bond is recommended for off-hand high-speed work and a slightly softer (P) vitrified bond for off-hand low-speed work.

Cobalt-Base Alloys.—These metals are available in both rod (RCoCr) and electrode (ECoCr) form and are frequently used to surface the contact surfaces of exhaust valves in aircraft, truck, and bus engines. Other uses include parts such as valve trim in steam engines, and on pump shafts, where conditions of corrosion and erosion are encountered. Several metals with a greater carbon content are available (CoCr-B, CoCr-C) and are used in applications requiring greater hardness and abrasion resistance but where impact resistance is not mandatory or expected to be a factor.

Hardness: Hardness ranges on the Rockwell C scale for gas-welded deposits are as follows: CoCr-A, 38 to 47; CoCr-B, 45 to 49; and CoCr-C, 48 to 58. For arc-welded deposits, hardness ranges (Rockwell C) as follows: CoCr-A, 23 to 47; CoCr-B, 34 to 47; and CoCr-C, 43 to 58. The values for arc-weld deposits depend for the most part on the base metal dilution. The greater the dilution, the lower the hardness. Many surfacing alloys are softened permanently by heating to elevated temperatures, however, these metals are exceptional. They do exhibit lower hardness values when hot but return to their approximate original hardness values upon cooling. Elevated-temperature strength and hardness are outstanding properties of this group. Their use at 1200 degrees F and above is considered advantageous but between 1000 and 1200 degrees F, their advantages are not definitely established, and at temperatures below 1000 degrees F, other surfacing metals may prove better.

Resistance Properties: In the temperature range from 1000 to 1200 degrees F, weld deposits of these metals have a great resistance to creep. Tough martensitic steel deposits are considered superior to cobalt-base deposits in both flow resistance and toughness. The chromium in the deposited metal promotes the formation of a thin, tightly adherent scale that provides a scaling resistance to combustion products of internal combustion engines, including deposits from leaded fuels. These metals are corrosion-resistant in such media as air, food, and certain acids. It is advisable to conduct field tests to determine specific corrosion resistance for the application being considered.

Other Properties or Characteristics: Deposits are able to take a high polish and have a low coefficient of friction and therefore are well suited for metal-to-metal wear resistance. Machining of these deposits is difficult; the difficulty increases in proportion to the increase in carbon content. CoCr-A alloys are preferably machined with sintered carbide tools. CoCr-C deposits are finished by grinding.

Copper-Base Alloys.—These metals are available in rod (RCuA1-A2, RCuA1-B, RCuA1-C, RCuA1-D, RCuA1-E, RCuSi-A, RCuSn, RCuSn-D, RCuSn-E, and RCuZn-E) and electrode (ECuA1-A2, ECuA1-B, ECuA1-C, ECuA1-D, ECuA1-E, ECuSi, ECuSn-A, ECuSn-C, ECuSn-E, and ECuZn-E) forms and are used in depositing overlays and inlays for bearing, corrosion-resistant, and wear-resistant surfaces. The CuA1-A2 rods and electrodes are used for surfacing bearing surfaces between the hardness ranges of 130 to 190 Bhn as well as for corrosion-resistant surfaces. The CuA1-B and CuA1-C rods and electrodes are used for surfacing bearing surfaces of hardness ranges 140 to 290 Bhn. The CuA1-D and CuA1-E rods and electrodes are used on bearing and wear-resistant surfaces requiring the higher hardnesses of 230 to 390 Bhn such as are found on gears, cams, wear plates, and dies. The copper-tin (CuSn) metals are used where a lower hardness is required for surfacing, for corrosion-resistant surfaces, and sometimes for wear-resistant applications.

Hardness: Hardness of a deposit depends upon the welding process employed and the manner of depositing the metal. Deposits made by the inert-gas metal-arc process (both consumable and nonconsumable electrode) will be higher in hardness than deposits made with the gas, metal-arc, and carbon-arc processes because lower losses of aluminum, tin, silicon, and zinc are achieved due to the better shielding from oxidation. Copper-base alloys are not recommended for use at elevated temperatures because their hardness and mechanical properties decrease consistently as the temperature goes above 400 degrees F.

Resistance Properties: The highest impact resistance of the copper-base alloy metals is exhibited by CuA1-A2 deposits. As the aluminum content increases, the impact resistance decreases markedly. CuSi weld deposits have good impact properties. CuSn metals as deposited have low impact resistance and CuZn-E deposits have a very low impact resistance. Deposits of the CuA1 filler metals form a protective oxide coating upon exposure to the atmosphere. Oxidation resistance of CuSi deposits is fair and that of CuSn deposits are comparable to pure copper. With the exception of the CuSn-E and CuZn-E alloys, these metals are widely used to resist many acids, mild alkalies, and salt water. Copper-base alloy deposits are not recommended for use where severe abrasion is encountered in service. CuA1 filler metals are used to overlay surfaces subjected to excessive wear from metal-to-metal contact such as gears, cams, sheaves, wear plates, and dies.

Other Properties or Characteristics: All copper-base alloy metals are used for overlays and inlays for bearing surfaces with the exception of the CuSi metals. Metals selected for bearing surfaces should have a Brinell hardness of 50 to 75 units below that of the mating metal surface. Slight porosity is generally acceptable in bearing service as a porous deposit is able to retain oil for lubricating purposes. CuA1 deposits in compression have elastic limits ranging from 25,000 TO 65,000 lb/in.^{2} and ultimate strengths of 120,000 to 171,000 lb/in.^{2} The elastic limit and ultimate strength of CuSi deposits in compression are 22,000 lb/in.^{2} and 60,000 lb/in.^{2}, respectively. CuZn-E deposits in compression have an elastic limit of only about 5000 lb/in.^{2} and an ultimate strength of 20,000 lb/in.^{2} All copper-base alloy deposits can be machined.

Nickel-Chromium-Boron Alloys.—These metals are available in both rod (RNiCr) and electrode (ENiCr) form and their deposits have good metal-to-metal wear resistance, good low-stress, scratch-abrasion resistance, corrosion resistance, and retention of hardness at elevated temperatures. These properties make the alloys suitable for use on seal rings, cement pump screws, valves, screw conveyors, and cams. Three different formulations of these metals are recognized (NiCr-A, NiCr-B, and NiCr-C).

Hardness: Hardness of the deposited NiCr-A from rods range from 35 to 40 Rockwell C; of NiCr-B rods, 45 to 50 Rockwell C; of NiCr-C rods, 56 to 62 Rockwell C. Hardness of the deposited NiCr-A from electrodes ranges from 24 to 35 Rockwell C; of NiCr-B from electrodes, 30 to 45 Rockwell C; and of NiCr-C electrodes, 35 to 56. The lower hardness values and greater ranges of hardness values of the electrode deposits are attributed to the dilution of deposit and base metals. Hot Rockwell C hardness values of NiCr-A electrode deposits range from 30 to 19 in the temperature range from 600 to 1000 degrees F from instantaneous loading to a 3-minute loading interval. NiCr-A rod deposits range from 34 to 24 in the same temperature range and under the same load conditions. Hot Rockwell C hardness values of NiCr-B electrode deposits range from 41 to 26 in the temperature range from 600 to 1000 degrees F from instantaneous loading to a 3-minute loading interval. NiCr-B rod deposits range from 46 to 37 in the same temperature range and under the same load conditions. Hot Rockwell C hardness values of NiCr-C electrode deposits range from 49 to 31 in the temperature range from 600 to 1000 degrees F from instantaneous loading to a 3-minute loading interval. NiCr-C rod deposits range from 55 to 40 in the same temperature range and under the same load conditions.

Resistance Properties: Deposits of these metal alloys will withstand light impact fairly well. When plastic deformation occurs, cracks are more likely to appear in the NiCr-C deposit than in the NiCr-A and NiCr-B deposits. NiCr deposits are oxidation-resistant up to 1800 degrees F. Their use above 1750 degrees F is not recommended because fusion may begin near this temperature. NiCr deposits are completely resistant to atmospheric, steam, salt water, and salt spray corrosion and to the milder acids and many common corrosive chemicals. It is advisable to conduct field tests when a corrosion application is contemplated. These metals are not recommended for high-stress grinding abrasion. NiCr deposits have good metal-to-metal wear resistance, take a high polish under wearing conditions, and are particularly resistant to galling. These properties are especially evident in the NiCr-C alloy.

Other Properties or Characteristics: In compression, these alloys have an elastic limit of 42,000 lb/in.^{2} Their yield strength in compression is 92,000 lb/in.^{2} (0.01 per cent offset), 150,000 lb/in.^{2} (0.10 per cent offset), and 210,000 lb/in.^{2} (0.20 per cent offset). Deposits of NiCr filler metals may be machined with tungsten carbide tools using slow speeds, light feeds, and heavy tool shanks. They are also finished by grinding using a soft-to-medium vitrified silicon carbide wheel.

Chromium Plating.—Chromium plating is an electrolytic process of depositing chromium on metals either as a protection against corrosion or to increase the surface-wearing qualities. The value of chromium-plating plug and ring gages has probably been more thoroughly demonstrated than any other single application of this treatment. Chromium-plated gages not only wear longer, but when worn, the chromium may be removed and the gage replated and reground to size.

In general, chromium-plated tools have operated well, giving greatly improved performance on nearly all classes of materials such as brass, bronze, copper, nickel, aluminum, cast iron, steel, plastics, asbestos compositions, and similar materials. Increased cutting life has been obtained with chromium-plated drills, taps, reamers, files, broaches, tool tips, saws, thread chasers, and the like. Dies for stamping, drawing, hot forging, die casting, and for molding plastics materials have shown greatly increased life after being plated with hard chromium.

Special care is essential in grinding and lapping tools preparatory to plating the cutting edges, because the chromium deposit is influenced materially by the grain structure and hardness of the base metal. The thickness of the plating may vary from 0.0001 to 0.001 or 0.002 inch, the thicker platings being used to build up undersize tools such as taps and reamers. A common procedure in the hard chromium plating of tools, as well as for parts to be salvaged by depositing chromium to increase diameters, is as follows:

1) Degrease with solvent;

2) Mount the tools on racks;

3) Clean in an anodic alkali bath held at a temperature of 82 degrees C for from 3 to 5 minutes;

4) Rinse in boiling water;

5) Immerse in a 20 per cent hydrochloric acid solution for 2 to 3 seconds;

6) Rinse in cold water;

7) Rinse in hot water;

8) Etch in a reverse-current chromic acid bath for 2 to 5 minutes;

9) Place work immediately in the chromium plating bath; and

10) Remove hydrogen embrittlement, if necessary, by immersing the plated tools for 2 hours in an oil bath maintained at 177 degrees C.

Chromium has a very low coefficient of friction. The static coefficient of friction for steel on chromium-plated steel is 0.17, and the sliding coefficient of friction is 0.16. This value may be compared with the static coefficient of friction for steel on steel of 0.30 and a sliding coefficient of friction of 0.20. The static coefficient of friction for steel on babbitt is 0.25, and the sliding coefficient of friction 0.20, whereas for chromium-plated steel on babbitt, the static coefficient of friction is 0.15, and the sliding coefficient of friction is 0.13. These figures apply to highly polished bearing surfaces. Articles that are to be chromium plated in order to resist frictional wear should be highly polished before plating so that full advantage can be taken of the low coefficient of friction that is characteristic of chromium. Chromium resists attack by almost all organic and inorganic compounds, except muriatic and sulfuric acids. The melting point of chromium is 2930 degrees F, and it remains bright up to 1200 degrees F. Above 1200 degrees F, a light adherent oxide forms and does not readily become detached. For this reason, chromium has been used successfully for protecting articles that must resist high temperatures, even above 2000 degrees F.

Cutting Metals with an Oxidizing Flame

The oxyhydrogen and oxyacetylene flames are especially adapted to cutting metals. When iron or steel is heated to a high temperature, it has a great affinity for oxygen and readily combines with it to form various oxides, and causing the metal to be disintegrated and burned with great rapidity. The metal-cutting or burning torch operates on this principle. A torch tip is designed to preheat the metal, which is then burned or oxidized by a jet of pure oxygen. The kerf or path left by the flame is suggestive of a saw cut when the cutting torch has been properly adjusted and used. The traversing motion of the torch along the work may be controlled either by hand or mechanically.

The Cutting Torch.—The ordinary cutting torch consists of a heating jet using oxygen and acetylene, oxygen and hydrogen, or, in fact, any other gas that, when combined with oxygen, will produce sufficient heat. By the use of this heating jet, the metal is first brought to a sufficiently high temperature, and an auxiliary jet of pure oxygen is then turned onto the red-hot metal, and the action just referred to takes place. Some cutting torches have a number of preheating flame ports surrounding the central oxygen port, so that a preheating flame will precede the oxygen regardless of the direction in which the torch is moved. This arrangement has been used to advantage in mechanically guided torches. The rate of cutting varies with the thickness of the steel, the size of the tip, and the oxygen pressure.

Adjustment and Use of Cutting Torch.—When using the cutting torch for the cutting of steel plate, the preheating flame first comes into contact with the edge of the plate and quickly raises it to a white-hot temperature. The oxygen valve is then opened, and as the pure oxygen comes into contact with the heated metal, the latter is burned or oxidized.

Metals That Can Be Cut.—Metals such as wrought iron and steels of comparatively low-carbon content can be cut readily with the cutting torch. High-carbon steels may be cut successfully if preheated to a temperature that depends somewhat on the carbon content. The higher the carbon content, the greater the degree of preheating required. A black heat is sufficient for ordinary tool steel, but a low red heat may be required for some alloy tool steels. Brass and bronze plates have been cut by interposing them between steel plates.

Cutting Stainless Steel.—Stainless steel can be cut readily by the flux-injection method. The elements that give stainless steels their desirable properties produce oxides that reduce the flame cutting operation to a slow melting-away process when the conventional oxy-acetylene cutting equipment is used. By injecting a suitable flux directly into the stream of cutting oxygen before it enters the torch, the obstructing oxides can be removed. Portable flux feeding units are designed to inject a predetermined amount of the flux powder. The rate of flux flow is accurately regulated by a vibrator type of dispenser with rheostat control. The flux-injection method is applicable either to machine cutting or to a hand-controlled torch. The operating procedure and speed of cutting are practically the same as in cutting mild steel.

Cutting Cast Iron.—The cutting of cast iron with the oxyacetylene torch is practicable, although it cannot be cut as readily as steel. The ease of cutting seems to depend largely on the physical character of the cast iron, very soft cast iron being more difficult to cut than harder varieties. The cost is much higher than that for cutting the same thickness of steel, because of the larger preheating flame necessary and the larger oxygen consumption. In spite of this extra cost, however, this method is often economical. The slag from a cast-iron cut contains considerable melted cast iron, whereas in steel, the slag is practically free from particles of the metal, indicating that cast-iron cutting is partly a melting operation. Increased speed and decreased cost often can be obtained by feeding a steel rod, about ¼ inch in diameter, into the top of the cut, beneath the torch tip. This rod furnishes a large amount of slag that flows over the cut and increases the temperature of the cast iron. Special tips are used because of the larger amounts of heat and oxygen required.

Mechanically Guided Torches.—Cutting torches used for cutting openings in plates or blocks or for cutting parts to some definite outline are often guided mechanically or by numerical control. Torches guided by pantograph mechanisms are especially adapted for tracing the outline to be cut from a pattern or drawing. Other designs are preferable for straight-line cutting and one type is designed for circular cutting.

Cutting Steel Castings.—When cutting steel castings, care should be taken to prevent burning pockets in the metal when the flame strikes a blowhole. If a blowhole is penetrated, the molten oxide will splash into the cavity and the flame will be diverted. The presence of the blowhole is generally indicated by excessive sparks. The operator should immediately move the torch back along the cut and direct it at an angle so as to strike the metal beneath the blowhole and burn it away if possible beyond the cavity. Cutting in the normal position then may be resumed.

Thickness of Metal That Can Be Cut.—The maximum thickness of metal that can be cut by these high-temperature flames depends largely upon the gases used and the pressure of the oxygen, which may be as high as 150 lb/in.^{2} The thicker the metal, the higher the pressure required. When using an oxyacetylene flame, it might be practicable to cut iron or steel up to 12 or 14 inches in thickness, whereas an oxyhydrogen flame has been used to cut steel plates 24 inches thick. The oxyhydrogen flame will cut thicker material principally because it is longer than the oxyacetylene flame and can penetrate to the full depth of the cut, thus keeping all the oxide in a molten condition so that it can be easily blown out by the oxygen cutting jet. A mechanically guided torch will cut thick material more satisfactorily than a hand-guided torch, because the flame is directed straight into the cut and does not wobble, as it tends to do when the torch is held by hand. With any flame, the cut is less accurate and the kerf wider, as the thickness of the metal increases. When cutting light material, the kerf might be $\frac{1}{16}$ inch wide, whereas for heavy stock, it might be $\frac{1}{4}$ or $\frac{3}{8}$ inch wide.

Arc Cutting of Metals

Arc Cutting.—According to the *Procedure Handbook of Arc-Welding Design & Practice,* published by The Lincoln Electric Co., a steel may be cut easily, and with great accuracy by means of the oxyacetylene torch. All metals, however, do not cut as easily as steel. Cast iron, stainless steels, manganese steels, and nonferrous materials are not as readily cut and shaped with the oxyacetylene cutting process because of their reluctance to oxidize. For these materials, arc cutting is often used to good advantage.

The cutting of steel is a chemical action. The oxygen combines readily with the iron to form iron oxide. In cast iron, this action is hindered by the presence of carbon in graphite form. Thus, cast iron cannot be cut as readily as steel; higher temperatures are necessary and cutting is slower. In steel, the action starts at bright red heat, whereas in cast iron, the temperature must be nearer to the melting point to obtain a sufficient reaction.

Plasma Cutting of Metals.—Because of its convenience and ability for close control, much precision cutting of outlines in sheet steel today is carried out with the plasma cutting torch, as discussed on page 1391 in the section on welding.

FILES AND BURS

Definitions of File Terms.—The following file terms apply to hand files but not to rotary files and burs.

Axis: Imaginary line extending the entire length of a file equidistant from faces and edges.

Back: The convex side of a file having the same or similar cross-section as a half-round file.

Bastard Cut: A grade of file coarseness between coarse and second cut of American pattern files and rasps.

Blank: A file in any process of manufacture before being cut.

Blunt: A file whose cross-sectional dimensions from point to tang remain unchanged.

Coarse Cut: The coarsest of all American pattern file and rasp cuts.

Coarseness: Term describing the relative number of teeth per unit length, the coarsest having the least number of file teeth per unit length; the smoothest, the most. American pattern files and rasps have four degrees of coarseness: coarse, bastard, second and smooth. Swiss pattern files usually have seven degrees of coarseness: 00, 0, 1, 2, 3, 4, 6 (from coarsest to smoothest). Curved tooth files have three degrees of coarseness: standard, fine and smooth.

Curved Cut: File teeth which are made in curved contour across the file blank.

Cut: Term used to describe file teeth with respect to their coarseness or their character (single, double, rasp, curved, special).

Double Cut: A file tooth arrangement formed by two series of cuts, namely the overcut followed, at an angle, by the upcut.

Edge: Surface joining faces of a file. May have teeth or be smooth.

Face: Widest cutting surface or surfaces that are used for filing.

Heel or Shoulder: That portion of a file that abuts the tang.

Hopped: A term used among file makers to represent a very wide skip or spacing between file teeth.

Length: The distance from the heel to the point.

Overcut: The first series of teeth put on a double-cut file.

Point: The front end of a file; the end opposite the tang.

Rasp Cut: A file tooth arrangement of round-topped teeth, usually not connected, that are formed individually by means of a narrow, punch-like tool.

Re-cut: A worn-out file which has been re-cut and re-hardened after annealing and grinding off the old teeth.

Safe Edge: An edge of a file that is made smooth or uncut, so that it will not injure that portion or surface of the workplace with which it may come in contact during filing.

Second Cut: A grade of file coarseness between bastard and smooth of American pattern files and rasps.

Set: To blunt the sharp edges or corners of file blanks before and after the overcut is made, in order to prevent weakness and breakage of the teeth along such edges or corners when the file is put to use.

Shoulder or Heel: See *Heel or Shoulder.*

Single Cut: A file tooth arrangement where the file teeth are composed of single unbroken rows of parallel teeth formed by a single series of cuts.

Smooth Cut: An American pattern file and rasp cut that is smoother than second cut.

Tang: The narrowed portion of a file which engages the handle.

Upcut: The series of teeth superimposed on the overcut, and at an angle to it, on a double-cut file.

File Characteristics.—Files are classified according to their shape or cross-section and according to the pitch or spacing of their teeth and the nature of the cut.

Cross-section and Outline: The cross-section may be quadrangular, circular, triangular, or some special shape. The outline or contour may be tapered or blunt. In the former, the point is more or less reduced in width and thickness by a gradually narrowing section that extends for one-half to two-thirds of the length. In the latter the cross-section remains uniform from tang to point.

Cut: The character of the teeth is designated as single, double, rasp or curved. The *single cut file* (or *float* as the coarser cuts are sometimes called) has a single series of parallel teeth extending across the face of the file at an angle of from 45 to 85 degrees with the axis of the file. This angle depends upon the form of the file and the nature of the work for which it is intended. The single cut file is customarily used with a light pressure to produce a smooth finish. The *double cut file* has a multiplicity of small pointed teeth inclining toward the point of the file arranged in two series of diagonal rows that cross each other. For general work, the angle of the first series of rows is from 40 to 45 degrees and of the second from 70 to 80 degrees. For *double cut finishing files* the first series has an angle of about 30 degrees and the second, from 80 to 87 degrees. The second, or upcut, is almost always deeper than the first or overcut. Double cut files are usually employed, under heavier pressure, for fast metal removal and where a rougher finish is permissible. The *rasp* is formed by raising a series of individual rounded teeth from the surface of the file blank with a sharp narrow, punch-like cutting tool and is used with a relatively heavy pressure on soft substances for fast removal of material. The curved tooth file has teeth that are in the form of parallel arcs extending across the face of the file, the middle portion of each arc being closest to the point of the file. The teeth are usually single cut and are relatively coarse. They may be formed by steel displacement but are more commonly formed by milling.

With reference to coarseness of cut the terms *coarse, bastard, second* and *smooth cuts* are used, the coarse or bastard files being used on the heavier classes of work and the second or smooth cut files for the finishing or more exacting work. These degrees of coarseness are only comparable when files of the same length are compared, as the number or teeth per inch of length decreases as the length of the file increases. The number of teeth per inch varies considerably for different sizes and shapes and for files of different makes. The coarseness range for the curved tooth files is given as standard, fine and smooth. In the case of Swiss pattern files, a series of numbers is used to designate coarseness instead of names; Nos. 00, 0, 1, 2, 3, 4 and 6 being the most common with No. 00 the coarsest and No. 6 the finest.

Classes of Files.—There are five main classes of files: mill or saw files; machinists' files; curved tooth files; Swiss pattern files; and rasps. The first two classes are commonly referred to as American pattern files.

Mill or Saw Files: These are used for sharpening mill or circular saws, large crosscut saws; for lathe work; for draw filing; for filing brass and bronze; and for smooth filing generally.

Cantsaw files 1) have an obtuse isosceles triangular section, a blunt outline, are single cut and are used for sharpening saws having "M"-shaped teeth and teeth of less than 60-degree angle. *Crosscut files*; 2) have a narrow triangular section with short side rounded, a blunt outline, are single cut and are used to sharpen crosscut saws. The rounded portion is used to deepen the gullets of saw teeth and the sides are used to sharpen the teeth themselves. *Double ender files*; 3) have a triangular section, are tapered from the middle to both ends, are tangless are single cut and are used reversibly for sharpening saws. The *mill file*; 4) itself, is usually single cut, tapered in width, and often has two square cutting edges in addition to the cutting sides. Either or both edges may be rounded, however, forfiling the gullets saw teeth. The *blunt mill file* has a uniform rectangular cross-section from tip to tang. The *triangular saw files* or *taper saw files*; 5) have an equilateral triangular section, are tapered, are single cut and are used for filing saws with 60-degree angle teeth. They come in taper, slim taper, extra slim taper and double extra slim taper thicknesses *Blunt triangular* and *blunt hand saw files* are without taper *Web saw files*; 6) have a diamond-shaped section, a blunt outline, are single cut and are used for sharpening pulpwood or web saws. *Machinists' Files:* These files are used throughout industry where metal must be removed rapidly and finish is of secondary importance. Except for certain exceptions in the round and half-round shapes, all are double cut. *Flat files*; 7) have a rectangular section, are tapered in width and thickness, are cut on both sides and edges and are used for general utility work. *Half round files*; 8) have a circular segmental section, are tapered in width and thickness, have their flat side double cut, their rounded side mostly double but sometimes single cut, and are used to file rounded holes, concave corners, etc. in general filing work. *Hand files*;

9) are similar to flat files but taper in thickness only. One edge is uncut or "safe." *Knife files*; 10) have a "knife-blade" section, are tapered in width only, are double cut, and are used by tool and die makers on work having acute angles. *Machinist's General Purpose files* have a rectangular section, are tapered and have single cut teeth divided by angular serrations which produce short cutting edges. These edges help stock removal but still leave a smooth finish and are suitable for use on various materials including aluminum, bronze, cast iron, malleable iron, mild steels and annealed tool steels. *Pillar files*; 11) are similar to hand files but are thicker and not as wide. *Round files*; 12) have a circular section, are tapered, single cut, and are generally used to file circular openings or curved surfaces. *Square files*; 13) have a square section, are tapered, and are used for filing slots, keyways and for general surface filing where a heavier section is preferred. *Three square files*; 14) have an equilateral triangular section and are tapered on all sides. They are double cut and have sharp corners as contrasted with taper triangular files which are single cut and have somewhat rounded corners. They are used for filing accurate internal angles, for clearing out square corners, and for filing taps and cutters. *Warding files*; and 15) have a rectangular section, and taper in width to a narrow point. They are used for general narrow space filing. *Wood files* are made in the same sections as flat and half round files but with coarser teeth especially suited for working on wood..

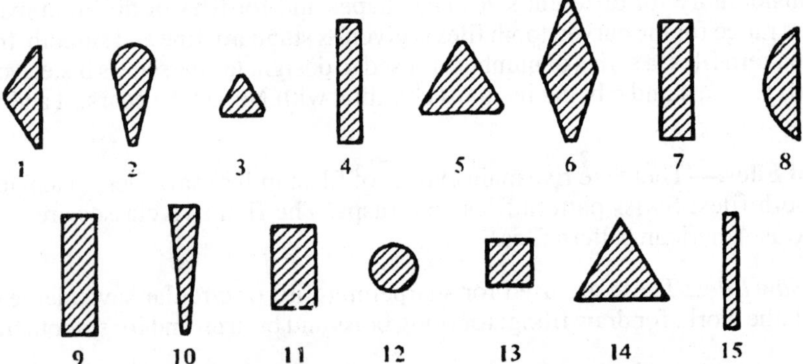

Curved Tooth Files: Regular curved tooth files are made in both rigid and flexible forms. The rigid type has either a tang for a conventional handle or is made plain with a hole at each end for mounting in a special holder. The flexible type is furnished for use in special holders only. The curved tooth files come in standard fine and smooth cuts and in parallel flat, square, pillar, pillar narrow, half round and shell types. A special curved tooth file is available with teeth divided by long angular serrations. The teeth are cut in an "off center" arc. When moved across the work toward one edge of the file a fast cutting action is provided; when moved toward the other edge, a smoothing action; thus the file is made to serve a dual purpose.

Swiss Pattern Files: These are used by tool and die makers, model makers and delicate instrument parts finishers. They are made to closer tolerances than the conventional American pattern files although with similar cross-sections. The points of the Swiss pattern files are smaller, the tapers are longer and they are available in much finer cuts. They are primarily finishing tools for removing burrs left from previous finishing operations truing up narrow grooves, notches and keyways, cleaning out corners and smoothing small parts. For very fine work, *round* and *square handled needle files,* available in numerous cross-sectional shapes in overall lengths from 4 to 7 ¾ inches, are used. Die sinkers use *die sinkers files* and *die sinkers rifflers.* The files, also made in many different cross-sectional shapes, are 3½ inches in length and are available in the cut Nos. 0, 1, 2, and 4. The rifflers are from 5½ to 6¾ inches long, have cutting surfaces on either end, and come in numerous cross-sectional shapes in cut Nos. 0, 2, 3, 4 and 6. These rifflers are used by die makers for getting into corners, crevices, holes and contours of intricate dies and molds. Used in the same fashion as die sinkers rifflers, *silversmiths rifflers,* that have a much heavier cross-section, are available in lengths from 6 ⅞ to 8 inches and in cuts Nos. 0, 1, 2, and 3. *Blunt machine files* in Cut Nos. 00, 0, and 2 for use in ordinary and bench filing machines are available in many different cross-sectional shapes, in lengths from 3 to 8 inches.

Rasps: Rasps are employed for work on relatively soft substances such as wood, leather, and lead where fast removal or material is required. They come in rectangular and half round cross-sections, the latter with and without a sharp edge.

Special Purpose Files: Falling under one of the preceding five classes of files, but modified to meet the requirements of some particular function, are a number of special purpose files. The *long angle lathe file* is used for filing work that is rotating in a lathe. The long tooth angle provides a clean shear, eliminates drag or tear and is self-clearing. This file has safe or uncut edges to protect shoulders of the work which are not to be filed. The *foundry file* has especially sturdy teeth with heavy set edges for the snagging of castings—the removing of fins, sprues, and other projections. The *die casting file* has extra strong teeth on corners and edges as well as sides for working on die castings of magnesium, zinc, or aluminum alloys. A special file for stainless steel is designed to stand up under the abrasive action of stainless steel alloys. *Aluminum rasps* and *files* are designed to eliminate clogging. A special tooth construction is used in one type of aluminum tile which breaks up the filings, allows the file to clear itself and overcomes chatter. A *brass file* is designed so that with a little pressure the sharp, high-cut teeth bite deep while with less pressure, their short uncut angle produces a smoothing effect. The *lead float* has coarse, single cut teeth at almost right angles to the file axis. These shear away the metal under ordinary pressure and produce a smoothing effect under light pressure. The *shear tooth file* has a coarse single cut with a long angle for soft metals or alloys, plastics, hard rubber and wood. *Chain saw files* are designed to sharpen all types of chain saw teeth. These files come in round, rectangular, square and diamond-shaped sections. The round and square sectioned files have either double or single cut teeth, the rectangular files have single cut teeth and the diamond-shaped files have double cut teeth.

Effectiveness of Rotary Files and Burs.—There it very little difference in the efficiency of rotary files or burs when used in electric tools and when used in air tools, provided the speeds have been reasonably well selected. Flexible-shaft and other machines used as a source of power for these tools have a limited number of speeds which govern the revolutions per minute at which the tools can be operated.

The carbide bur may be used on hard or soft materials with equally good results. The principle difference in construction of the carbide bur is that its teeth or flutes are provided with a negative rather than a radial rake. Carbide burs are relatively brittle, and must be treated more carefully than ordinary burs. They should be kept cutting freely, in order to prevent too much pressure, which might result in crumbling of the cutting epics.

At the same speeds, both high-speed steel and carbide burs remove approximately the same amount of metal. However, when carbide burs are used at their most efficient speeds, the rate of stock removal may be as much as four times that of ordinary burs. In certain cases, speeds much higher than those shown in the table can be used. It has been demonstrated that a carbide bur will last up to 100 times as long as a high-speed steel bur of corresponding size and shape.

Approximate Speeds of Rotary Files and Burs

Tool Diam., Inches	Medium Cut, High-Speed Steel Bur or File		
	Mild Steel	Cast Iron	Bronze
	Speed, Revolutions per Minute		
1§8	4600	7000	15,000
1§4	3450	5250	11,250
3§8	2750	4200	9000
1§2	2300	3500	7500
5§8	2000	3100	6650
3§4	1900	2900	6200
7§8	1700	2600	5600
1	1600	2400	5150
1⅛	1500	2300	4850
1¼	1400	2100	4500

Tool Diam., Inches	Medium Cut, High-Speed Steel Bur or File		Carbide Bur	
	Speed, Revolutions Per Minute		Medium Cut	Fine Cut
	Aluminum	Magnesium	Any Material	
1§8	20,000	30,000	45,000	30,000
1§4	15,000	22,500	30,000	20,000
3§8	12,000	18,000	24,000	16,000
1§2	10,000	15,000	20,000	13,350
5§8	8900	13,350	18,000	12,000
3§4	8300	12,400	16,000	10,650
7§8	7500	11,250	14,500	9650
1	6850	10,300	13,000	8650
1⅛	6500	9750	…	…
1¼	6000	9000	…	…

As recommended by the Nicholson File Company.

Power Brush Finishing

Power brush finishing is a production method of metal finishing that employs wire, elastomer bonded wire, or non-metallic (cord, natural fiber or synthetic) brushing wheels in automatic machines, semi-automatic machines and portable air tools to smooth or roughen surfaces, remove surface oxidation and weld scale or remove burrs.

Description of Brushes.—Brushes work in the following ways: the wire points of a brush can be considered to act as individual culling tools so that the brush, in effect, is a multiple-tipped cutting tool. The fill material, as it is rotated, contacts the surface of the work and imparts an impact action which produces a coldworking effect. The type of finish produced depends upon the wheel material, wheel speed, and how the wheel is applied.

Brushes differ in the following ways 1) fill material (wire—carbon steel, stainless steel; synthetic; Tampico; and cord); 2) length of fill material (or trim); and 3) the density of the fill material.

To aid in wheel selection and use, the accompanying table made up from information supplied by *The Osborn Manufacturing Company* lists the characteristics and mayor uses of brushing wheels.

Use of Brushes.—The brushes should be located so as to bring the full face of the brush in contact with the work. Full face contact is necessary to avoid grooving the brush. Operations that are set up with the brush face not in full contact with the work require some provision for dressing the brush face. When the tips of a brush, used with full face contact, become dull during use with subsequent loss of working clearance, reconditioning and resharpening is necessary. This is accomplished simply and efficiently by alternately reversing the direction of rotation during use.

Deburring and Producing a Radius on the Tooth Profile of Gears.—The brush employed for deburring and producing a radius on the tooth profile of gears is a short trim, dense, wire-fill radial brush. The brush should be set up so as to brush across the edge as shown in Fig. 1A. Line contact brushing, as shown in Fig. 1B should be avoided because the Crisis face will wear non-uniformly; and the wire points, being flexible, tend to flare to the side, thus minimizing the effectiveness of the brushing operation. When brushing gears, the brushes are spaced and contact the tooth profile on the center line of the gear as shown in Fig. 2. This facilitates using brush reversal to maintain the wire brushing points at their maximum cutting efficiency.

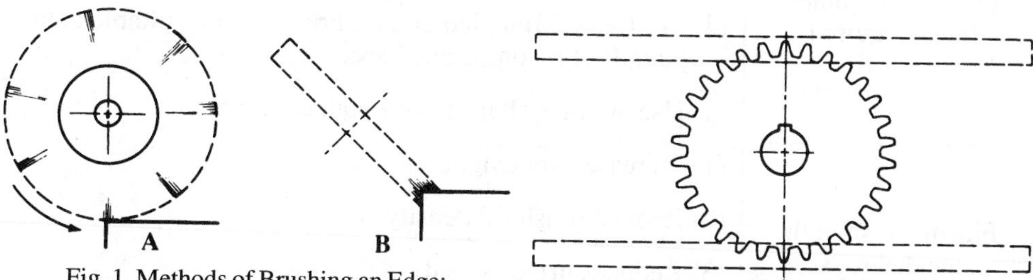

Fig. 1. Methods of Brushing an Edge; (A) Correct, (B) Incorrect

Fig. 2. Setup for Deburring Gears

The setup for brushing spline bores differs from brushing gears in that the brushes are located off-center, as illustrated in Fig. 3. When helical gears are brushed, it is sometimes necessary to favor the acute side of the gear tooth to develop a generous radius prior to shaving. This can be accomplished by locating the brushes as shown In Fig. 4. Elastomer bonded wire-filled brushes are used for deburring fine pitch gears. These brushes remove the burrs without leaving any secondary roll. The use of bonded brushes is necessary when the gears are not shaved after hobbing or gear shaping.

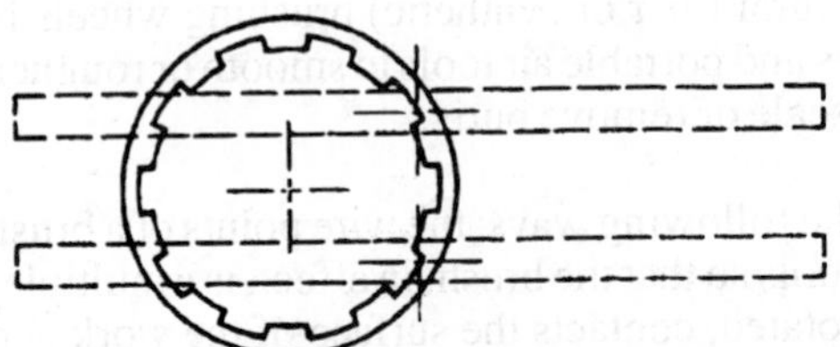

Fig. 3. Setup for Brushing Broached Splines

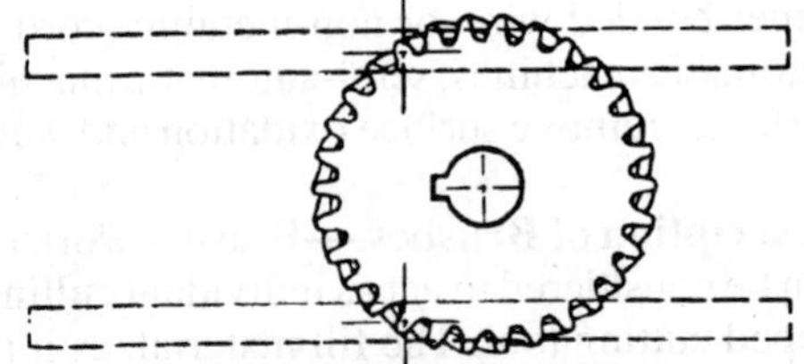

Fig. 4. Setup for Finishing Helical Gears

Adjustments for Eliminating Undesirable Conditions in Power Brush Finishing

Undesirable Condition	Possible Adjustments for Eliminating Condition
Brush works too slowly	(1) Decrease trim length and increase fill density. (2) Increase filament diameter. (3) Increase surface speed by increasing R.P.M. or outside diameter.
Brush works too fast	(1) Reduce filament diameter. (2) Reduce surface speed by reducing R.P.M. or outside diameter. (3) Reduce fill density. (4) Increase trim length.
Action of brush peens burr to adjacent surface	(1) Decrease trim length and increase fill density. (2) If wire brush tests indicate metal too ductile (burr is peened rather than removed), change to nonmetallic brush such as a treated Tampico brush used with a burring compound.
Finer or smoother finish required	(1) Decrease trim length and increase fill density. (2) Decrease filament diameter. (3) Try treated Tampico or cord brushes with suitable compounds at recommended speeds. (4) Use auxiliary buffing compound with brush.
Finish too smooth and lustrous	(1) Increase trim length. (2) Reduce brush fill density. (3) Reduce surface speed. (4) Increase filament diameter.
Brushing action not sufficiently uniform	(1) Devise hand-held or mechanical fixture or machine which will avoid irregular off-hand manipulation. (2) Increase trim length and decrease fill density.

Characteristics and Applications of Brushes Used in Power Finishing

Brush Type	Description	Operating Speed Range, sfpm	Uses	Remarks
Radial, short trim dense wire fill	Develops very little impact action but maximum cutting action.	6500	Removal of burrs from gear teeth and sprockets. Produces blends and radii at juncture of intersecting surfaces.	Brush should be set up so as to brush across any edge. Reversal of rotation needed to maintain maximum cutting efficiency of brush points.
Radial, medium to long trim twisted knot wire fill	Normally used singly and on portable tools. Brush is versatile and provides high impact action.	7500–9500 for high speeds. 1200 for slow speeds.	For cleaning welds in the automotive and pipeline industries. Also for cleaning surfaces prior to painting, stripping rubber flash from molded products and cleaning mesh-wire conveyor belts.	Surface speed plays an important role since at low speeds the brush is very flexible and at high speeds it is extremely hard and fast cutting.
Radial, medium to long trim crimped wire fill	With the 4- to 8-inch diameter brush, part is hand held. With the 10- to 15-inch diameter brush, part is held by machine.	4500–6000	Serves as utility tool on bench grinder for removing feather grinding burrs, machining burrs, and for cleaning and producing a satin or matte finish.	Good for hand held parts as brush is soft enough to conform to irregular surfaces and hard-to-reach areas. Smaller diameter brushes are not recommended for high-production operations.
Radial, sectional, nonmetallic fill (treated and untreated Tampico or cord)	Provides means for improving finish or improving surface for plating. Works best with grease base deburring or buffing compound.	5500–6500 7500 for polishing	For producing radii and improving surface finish. Removes the sharp peaks that fixed abrasives leave on a surface so that surface will accept a uniform plating. Polishing marks and draw marks can be successfully blended.	Brush is selective to an edge which means that it removes metal from an edge but not from adjoining surfaces. It will produce a very uniform radius without peening or rolling any secondary metal.
Radial, wide-face, nonmetallic fill (natural fibers or synthetics)	Can be used with flow-through mounting which facilitates feeding of cold water and hot alkaline solutions through brush face to prevent buildup.	750–1200 for cleaning steel. 600 when used with slurries	For cleaning steel. Used in electrolytic tinplate lines, continuous galvanizing and annealing lines, and cold reduction lines. Used to produce dull or matte-type finishes on stainless steel and synthetics.	Speeds above 3600 sfpm will not appreciably improve operation as brush wear will be excessive. Avoid excessive pressures. Ammeters should be installed in drive-motor circuit to indicate brushing pressure.

Characteristics and Applications of Brushes Used in Power Finishing *(Continued)*

Brush Type	Description	Operating Speed Range, sfpm	Uses	Remarks
Radial, wide face, metallic fill	This brush is made to customer's specifications. It is dynamically balanced at the speed at which it will operate.	2000–4000	Removes buildup of aluminum oxide from work rolls in aluminum mill. Removes lime or magnesium coatings from certain types of steel. Burnishes hot-dipped galvanized steel to produce a minimum spangled surface.	Each brush should have its own drive. An ammeter should be present in drive-motor circuit to measure brushing pressure. If strip is being brushed, a steel backup roll should be opposite the brush roll.
Radial, wide face, strip (interrupted brush face)	Performs cleaning operations that would cause a solid face brush to become loaded and unusable.	When cleaning conveyor belts, brush speed is 2 to 3 times that of conveyer belt.	Need for cleaning rubber and fabric conveyor belts of carry-back material which would normally foul snubber pulley and return idlers.	Designed for medium- to light-duty work. Brush face does not load.
Radial, Cup, Flared End, and Straight End, wire fill elastomer bonded	Extremely fast cutting with maximum operator safety. No loss of wire through fatigue. Always has uniform face.	3600–9000	For removing oxide weld scale, burrs, and insulation from wire.	Periodic reversing of brush direction will result in a brush life ten times greater than non-bonded wheels. Fast cutting action necessitates precise holding of part with respect to brush.
Cup, twisted knot wire fill	Fast cutting wheel used on portable tools to clean welds, scale, rust, and other oxides.	8000–10,000 4500–6500 for deburring and producing a radius around periphery of holes.	Used in shipyards and in structural steel industry. For cleaning outside diameter of pipe and removing burrs and producing radii on heat exchanger tube sheets and laminations for stator cores.	Fast acting brush cleans large areas economically. Setup time is short.
Radial, wire or treated Tampico or cord	For use with standard centerless grinders. Brush will not remove metal from a cylindrical surface. Parts must be ground to size before brushing.	…	For removing feather grinding burrs and improving surface finish. Parts of 24 microinches can be finished down to 15 to 10 microinches. Parts of 10 to 12 microinches can be finished down to 7 to 4 microinches.	Follows centerless grinding principles, except that accuracy in pressure and adjustment is not critical. A machine no longer acceptable for grinding can be used for brushing.

Polishing and Buffing

The terms "polishing" and "buffing" are sometimes applied to similar classes of work in different plants, but according to approved usage of the terms, there is the following distinction: Polishing is any operation performed with wheels having abrasive glued to the working surfaces, whereas buffing is done with wheels having the abrasive applied loosely instead of imbedding it into glue; moreover, buffing is not so harsh an operation as ordinary polishing, and it is commonly utilized to obtain very fine surfaces having a "grainless finish."

Polishing Wheels.—The principal materials from which polishing wheels are made are wood, leather, canvas, cotton cloth, plastics, felt, paper, sheepskin, impregnated rubber, canvas composition, and wool. Leather and canvas are the materials most commonly used in polishing wheel construction. Wooden wheels covered with material to which emery or some other abrasive is glued are employed extensively for polishing flat surfaces, especially when good edges must be maintained. Cloth wheels are made in various ways; wheels having disks that are cemented together are very hard and used for rough, coarse work, whereas those having sewn disks are made of varying densities by sewing together a larger or smaller number of disks into sections and gluing then.

Wheels in which the disks are held together by thread or metal stitches and which are not stiffened by the use of glue usually require metal side plates to support the canvas disks. Muslin wheels are made from sewed or stapled buffs glued together, but the outer edges of a wheel frequently are left open or free from glue to provide an open face of any desired depth. Wool felt wheels are flexible and resilient, and the density may be varied by sewing two or more disks together and then cementing to form a wheel. Solid felt wheels are quite popular for fine finishing but have little value as general utility wheels. Paper wheels are made from strawboard paper disks and are cemented together under pressure to form a very hard wheel for rough work. Softer wheels are similarly made from felt paper. The "compress" canvas wheel has a cushion of polishing material formed by pieces of leather, canvas, or felt, that are held in a crosswise radial position by two side plates attached to the wheel hub. This cushion of polishing material may be varied in density to suit the requirements; it may be readily shaped to conform to the curvature of the work and this shape can be maintained. Sheepskin polishing wheels and paper wheels are little used.

Polishing Operations and Abrasives.—Polishing operations on such parts as chisels, hammers, screwdrivers, wrenches, and similar parts that are given a fine finish but are not plated, usually require four operations, which are "roughing," "dry fining," "greasing," and "coloring." The roughing is frequently regarded as a solid grinding wheel job. Sometimes there are two steps to the greasing operation—rough and fine greasing. For some hardware, such as the cheaper screwdrivers, wrenches, etc., the operations of roughing and dry fining are considered sufficient. For knife blades and cutlery, the roughing operation is performed with solid grinding wheels and the polishing is known as fine or blue glazing, but these terms are never used when referring to the polishing of hardware parts, plumbers' supplies, etc. A term used in finishing German silver, white metal, and similar materials is "sand-buffing," which, in distinction from the ordinary buffing operation that is used only to produce a very high finish, actually removes considerable metal, as in rough polishing or flexible grinding. For sand-buffing, pumice and other abrasive powders are loosely applied.

Aluminum oxide abrasives are widely used for polishing high-tensile-strength metals such as carbon and alloy steels, tough iron, and nonferrous alloys. Silicon carbide abrasives are recommended for hard, brittle substances such as grey iron, cemented carbide tools, and materials of low tensile strength such as brass, aluminum, and copper.

Buffing Wheels.—Buffing wheels are manufactured from disks (either whole or pieced) of bleached or unbleached cotton or woolen cloth, and they are used as the agent for carrying abrasive powders, such as tripoli, crocus, rouge, lime, etc., which are mixed with waxes or greases as a bond. There are two main classes of buffs, one of which is known as the "pieced-sewed" buffs, and is made from various weaves and weights of cloth. The other is the "full-disk" buffs, which are made from specially woven material. Bleached cloth is harder and stiffer than unbleached cloth, and is used for the faster cutting buffs. Coarsely woven unbleached cloth is recommended for highly colored work on soft metals, and the finer woven unbleached cloths are better adapted for harder metals. When working at the usual speed, a stiff buff is not suitable for "cutting down" soft metal or for light plated ware, but is used on harder metals and for heavy nickel-plated articles.

Speed of Polishing Wheels.—The proper speed for polishing is governed to some extent by the nature of the work, but for ordinary operations, the polishing wheel should have a peripheral speed of about 7500 ft/min. If run at a lower speed, the work tends to tear the polishing material from the wheel too readily, and the work is not as good in quality. Muslin, felt, or leather polishing wheels having wood or iron centers should be run at peripheral speeds varying from 300 to 7000 ft/min. It is rarely necessary to exceed 6000 ft/min, and for most purposes, 4000 ft/min is sufficient. If the wheels are kept in good condition, in perfect balance, and are suitably mounted on substantial buffing lathes, they can be used safely at speeds within the limits given. However, manufacturers' recommendations concerning wheel speeds should be followed, where they apply.

Grain Numbers of Emery.—The numbers commonly used in designating the different grains of emery, corundum, and other abrasives are 10, 12, 14, 16, 18, 20, 24, 30, 36, 40, 46, 54, 60, 70, 80, 90, 100, 120, 150, 180, and 200, ranging from coarse to fine, respectively. These numbers represent the number of meshes per linear inch in the grading sieve. An abrasive finer than No. 200 is known as "flour" and the degree of fineness is designated by the letters CF, F, FF, FFF, FFFF, and PCF or SF, ranging from coarse to fine. The methods of grading flour-emery adopted by different manufacturers do not exactly agree, the letters differing somewhat for the finer grades. Again, manufacturers' recommendations should be followed.

Grades of Emery Cloth.—The coarseness of emery cloth is indicated by letters and numbers corresponding to the grain number of the loose emery used in the manufacture of the cloth. The letters and numbers for grits ranging from fine to coarse are as follows: FF, F, 120, 100, 90, 80, 70, 60, 54, 46, and 40. For large work roughly filed, use coarse cloth such as numbers 46 or 54, and then finer grades to obtain the required polish. If the work has been carefully filed, a good polish can be obtained with numbers 60 and 90 cloth, and a brilliant polish can be achieved by finishing with number 120 and flour-emery.

Mixture for Cementing Emery Cloth to a Lapping Wheel.—Many proprietary adhesives are available for application of emery cloth to the periphery of a buffing or lapping wheel, and generally are supplied with application instructions. In the absence of such instructions, clean the wheel thoroughly before applying the adhesive, and then rub the emery cloth down so as to exclude all air from between the surface of the wheel and the cloth.

Etching and Etching Fluids

Etching Fluids for Different Metals.—A common method of etching names or simple designs upon steel is to apply a thin, even coating of beeswax or some similar substance which will resist acid; then mark the required lines or letters in the wax with a sharp-pointed scriber, thus exposing the steel (where the wax has been removed by the scriber point) to the action of an acid, which is finally applied. To apply a very thin coating of beeswax, place the latter in a silk cloth, warm the piece to be etched, and tub the pad over it. Regular coach varnish is also used instead of wax, as a "resist."

An etching fluid ordinarily used for carbon steel consists of nitric acid, 1 part; water, 4 parts. It may be necessary to vary the amount of water, as the exact proportion depends upon the carbon content and whether the steel is hard or soft. For hard steel, use nitric acid, 2 parts; acetic acid, 1 part. For high-speed steel, nickel or brass, use nitro-hydrochloric acid (nitric, 1 part; hydrochloric, 4 parts). For high-speed steel it is sometimes better to add a little more nitric acid. For etching bronze, use nitric acid, 100 parts; muriatic acid, 5 parts. For brass, nitric acid, 16 parts; water, 160 parts, dissolve 6 parts potassium chlorate in 100 parts of water; then mix the two solutions and apply.

A fluid which may be used either for producing a frosted effect or for deep etching (depending upon the time it is allowed to act) is composed of 1 ounce sulphate of copper (blue vitriol); ¼ ounce alum; ½ teaspoonful of salt; 1 gill of vinegar, and 20 drops of nitric acid. For aluminum, use a solution composed of alcohol, 4 ounces; acetic acid, 6 ounces; antimony chloride, 4 ounces; water, 40 ounces.

Various acid-resisting materials are used for covering the surfaces of steel rules etc., prior to marking off the lines on a graduating machine. When the graduation lines are fine and very closely spaced, as on machinists' scales which are divided into hundredths or sixty-fourths, it is very important to use a thin resist that will cling to the metal and prevent any under-cutting of the acid: the resist should also enable fine lines to be drawn without tearing or crumbling as the tool passes through it. One resist that has been extensively used is composed of about 50 per cent of asphaltum, 25 per cent of beeswax, and, in addition, a small percentage of Burgundy pitch, black pitch, and turpentine. A thin covering of this resisting material is applied to the clean polished surface to be graduated and, after it is dry, the work is ready for the graduating machine. For some classes of work, paraffin is used for protecting the surface surrounding the graduation lines which are to be etched. The method of application consists in melting the paraffin and raising its temperature high enough so that it will flow freely; then the work is held at a slight angle and the paraffin is poured on its upper surface. The melted paraffin forms a thin protective coating.

Conversion Coatings and the Coloring of Metals

Conversion Coatings.—Conversion coatings are thin, adherent chemical compounds that are produced on metallic surfaces by chemical or electrochemical treatment. These coatings are insoluble, passive, and protective, and are divided into two basic systems: oxides or mixtures of oxides with other compounds, usually chromates or phosphates. Conversion coatings are used for corrosion protection, as an adherent paint base; and for decorative purposes because of their inherent color and because they can absorb dyes and colored sealants.

Conversion coatings are produced in three or four steps. First there is a pretreatment, which often involves mechanical surface preparation followed by decreasing and/or chemical or electrochemical cleaning or etching. Then thermal, chemical, or electrochemical surface conversion processes take place in acid or alkaline solutions applied by immersion spraying, or brushing. A post treatment follows, which includes rinsing and drying, and may also include sealing or dyeing. If coloring is the main purpose of the coating, then oiling, waxing, or lacquering may be required.

Passivation of Copper.—The blue-green patina that forms on copper alloys during atmospheric exposure is a passivated film; i.e., it prevents corrosion. This patina may be produced artificially or its growth may be accelerated by a solution of ammonium sulfate, 6 pounds; copper sulfate, 3 ounces; ammonia (technical grade, 0.90 specific gravity), 1.34 fluid ounces; and water, 6.5 gallons. This solution is applied as a fine spray to a chemically cleaned surface and is allowed to dry between each of five or six applications. In about 6 hours a patina somewhat bluer than natural begins to develop and continues after exposure to weathering.

Small copper parts can be coated with a passivated film by immersion in or brushing with a solution consisting of the following weight proportions: copper, 30; nitric acid, concentrated, 60; acetic acid (6%), 600; ammonium chloride, 11; and ammonium hydroxide (technical grade, 0.90 specific gravity), 20. To prepare the solution, the copper is dissolved in the nitric acid before the remaining chemicals are added, and the solution is allowed to stand for several days before use. A coating of linseed oil is applied to the treated parts.

Coloring of Copper Alloys.—Metals are colored to enhance their appearance, to produce an undercoat for an organic finish, or to reduce light reflection. Copper alloys can be treated to produce a variety of colors, with the final color depending on the base metal composition, the coloring solution's composition, the immersion time, and the operator's skill. Cleaning is an important part of the pretreatment; nitric and sulfuric acid solutions are used to remove oxides and to activate the surface.

The following solutions are used to color alloys that contain 85 per cent or more of copper. A dark red color is produced by immersing the parts in molten potassium nitrate, at 1200–1300°F, for up to 20 seconds, followed by a hot water quench. The parts must then be lacquered. A steel black color can be obtained by immersing the parts in a 180°F solution of arsenious oxide (white arsenic), 4 ounces; hydrochloric acid (1.16 specific gravity), 8 fluid ounces; and water, 1 gallon. The parts are immersed until a uniform color is obtained; they are scratch brushed while wet, and then dried and lacquered. A light brown color is obtained using a room-temperature solution of barium sufate, 0.5 ounce; ammonium carbonate, 0.25 ounce; and water, 1 gallon.

The following solutions are used to color alloys that contain less than 85 per cent copper. To color brass black, parts are placed in an oblique tumbling barrel made of stainless steel and covered with 3 to 5 gallons of water. Three ounces of copper sulfate and 6 ounces of sodium thiosulfate are dissolved in warm water and added to the barrel's contents. After tumbling for 15 to 30 minutes to obtain the finish, the solution is drained from the barrel, and the parts are washed thoroughly in clean water, dried in sawdust or air-blasted and, if necessary, lacquered. To produce a blue-black color, the parts are immersed in a 130–175°F solution of copper carbonate 1 pound; ammonium hydroxide (0.89 specific gravity), 1 quart; and water, 3 quarts. Excess copper carbonate should be present. The proper color is obtained in 1 minute. To color brass a hardware green, immerse the parts in a 160°F solution of ferric nitrate, 1 ounce; sodium thiosulfate, 6 ounces; and water, 1 gallon. To color brass a light brown, immerse the parts in a 195–212°F solution of potassium chlorate, 5.5 ounces; nickel sulfate, 2.75 ounces; copper sulfate, 24 ounces; and water, 1 gallon.

Post treatment: The treated parts should be scratch brushed to remove any excess or loose deposits. A contrast of colors may be obtained by brushing with a slurry of fine pumice, hand nabbing with an abrasive paste, mass finishing, or buffing to remove the color from the highlights. In order to prolong the life of parts used for outdoor decorative purposes, a clear lacquer should be applied. Parts intended for indoor purposes are often used without additional protection.

Coloring of Iron and Steel.—Thin black oxide coatings are applied to steel by immersing the parts to be coated in a boiling solution of sodium hydroxide and mixtures of nitrates nitrites. These coatings serve as paint bases and, in some cases, as final finishes. When the coatings are impregnated with oil or wax, they furnish fairly good corrosion resistance. These finishes are relatively inexpensive compared to other coatings.

Phosphate Coatings: Phosphate coatings are applied to iron and steel parts by reacting them with a dilute solution of phosphoric acid and other chemicals. The surface of the metal is converted into an integral, mildly protective layer of insoluble crystalline phosphate. Small items are coated in tumbling barrels; large items are spray coated on conveyors.

The three types of phosphate coatings in general use are zinc, iron, and manganese. Zinc phosphate coatings vary from light to dark gray. The color depends on the carbon content and pretreatment of the steel's surface, as well as the composition of the solution. Zinc phosphate coatings are generally used as a base for paint or oil, as an aid in cold working, for increased wear resistance, or for rustproofing. Iron phosphate coatings were the first type to be used; they produce dark gray coatings and their chief application is as a paint base. Manganese phosphate coatings are usually dark gray; however, since they are used almost exclusively as an oil base, for break in and to prevent galling, they become black in appearance.

In general, stainless steels and certain alloy steels cannot be phosphated. Most cast irons and alloy steels accept coating with various degrees of difficulty depending on alloy content.

Anodizing Aluminum Alloys.—In the anodizing process, the aluminum object to be treated is immersed as the anode in an acid electrolyte, and a direct current is applied. Oxidation of the surface occurs, producing a greatly thickened, hard, porous film of aluminum oxide. The object is then immersed in boiling water to seal the porosity and render the film impermeable. Before sealing, the film can be colored by impregnation with dyes or pigments. Special electrolytes may also be used to produce colored anodic films directly in the anodizing bath. The anodic coatings are used primarily for corrosion protection and abrasion resistance, and as a paint base.

The three principal types of anodizing processes are: chromic, in which the active agent is chromic acid; sulfuric, in which the active agent is sulfuric acid, and hard anodizing, in which sulfuric acid is used by itself or with additives in a low-temperature electrolyte bath. Most of the anodic coatings range in thickness from 0.2 to 0.7 mil. The hard anodizing process can produce coatings up to 2 mils. The chromic acid coating is less brittle than the sulfuric, and, since the chromic electrolyte does not attack aluminum, it does not present a corrosion problem when it is trapped in crevices. The chromic coating is less resistant to abrasion than the sulfuric, but it cannot be used with alloys containing more than 5 per cent copper due to corrosion of the base metal.

Chemical Conversion Coatings for Aluminum: Chemical conversion coatings for aluminum alloys are adherent surface layers of low volubility oxide, phosphate, or chromate compounds produced by the reaction of the metal surface with suitable reagents. The conversion coatings are much thinner and softer than anodic coatings but they are less expensive and serve as an excellent paint base.

Magnesium Alloys.—Chemical treatment of magnesium alloys is used to provide a paint base and to improve corrosion resistance. The popular conversion "dip" coatings are chrome pickle and dichromate treatments, and they are very thin. Anodic coatings are thicker and harder, and, after sealing, give the same protection against corrosion, although painting is still desirable.

Titanium Alloys.—Chemical conversion coatings are used on titanium alloys to improve lubricity by acting as a base for the retention of lubricants. The coatings are applied by immersion, spraying, or brushing. A popular coating bath is an aqueous solution of phosphates, fluorides, and hydrofluoric acid. The coating is composed primarily of titanium and potassium fluorides and phosphates.

Plating

Surface Coatings.—The following is a list of military plating and coating specifications.

Anodize (Chromic and Sulfuric), MIL-A-8625F: Conventional Types I, IB, and II anodic coatings are intended to improve surface corrosion protection under severe conditions or as a base for paint systems. Coatings can be colored with a large variety of dyes and pigments. Class 1 is non-dyed; Class 2 dyed. Color is to be specified on the contract. Prior to dying or sealing, coatings shall meet the weight requirements.

Type I and IB coatings should be used on fatigue critical components (due to thinness of coating). Type I unless otherwise specified shall not be applied to aluminum alloys with over 5% copper or 7% silicon or total alloying constituents over 7.5%. Type IC is a mineral or mixed mineral/organic acid that anodizes. It provides a non-chromate alternative for Type I and IB coatings where corrosion resistance, paint adhesion, and fatigue resistance are required. Type IIB is a thin sulfuric anodizing coating for use as non-chromate alternatives for Type I and IB coatings where corrosion resistance, paint adhesion, and fatigue resistance are required. Be sure to specify the class of anodic coating and any special sealing requirements.

Types I, IB, IC, and IIB shall have a thickness between 0.00002 and 0.0007 in. Type II shall have a thickness between 0.0007 and 0.0010 in.

Black Chrome, MIL-C-14538C: A hard, non-reflective, abrasion, heat and corrosion resistant coating approximately 0.0002 in. thick. Provides limited corrosion protection, but added protection can be obtained by specifying underplate such as nickel. Color is a dull dark gray, approaching black and may be waxed or oiled to darken.

Black chromium has poor throwing power, and conforming anodes are necessary for intricate shapes. Apply coating after heat treating and all mechanical operations are performed. Steel parts with hardness in excess of 40 Rc shall be stress relieved prior to plating by baking one hour or more (300 to 500°F) and baked after plating (375°F ± 25°F) for 3 hours.

Black Oxide Coating, MIL-C-13924C: A uniform, mostly decorative black coating for ferrous metals used to decrease light reflection. Only very limited corrosion protection under mild corrosion conditions. Black oxide coatings should normally be given a supplementary treatment.

Used for moving parts that cannot tolerate the dimensional change of a more corrosion resistant finish. Use alkaline oxidizing for wrought iron, cast and malleable irons, plain carbon, low alloy steel and corrosion resistant steel alloys. Alkaline-chromite oxidizing may be used on certain corrosion resistant steel alloys tempered at less than 900°F Salt oxidizing is suitable for corrosion resistant steel alloys that are tempered at 900°F or higher.

Cadmium, QQ-P-416F: Cadmium plating is required to be smooth, adherent, uniform in appearance, free from blisters, pits, nodules, burning, and other defects when examined visually without magnification. Unless otherwise specified in the engineering drawing or procurement documentation, the use of brightening agents in the plating solution to modify luster is prohibited on components with a specified heat treatment of 180 ksi minimum tensile strength (or 40 Rc) and higher. Either a bright (not caused by brightening agents) or dull luster shall be acceptable. Baking on Types II and III shall be done prior to application of supplementary coatings. For Classes 1, 2, and 3 the minimum thicknesses shall be 0.0005, 0.0003, and 0.0002 in. respectively.

Type I is to be used as plated. Types II and III require supplementary chromate and phosphate treatment respectively. Chromate treatment required for type II may be colored iridescent bronze to brown including olive drab, yellow and forest green. Type II is recommended for corrosion resistance. Type III is used as a paint base and is excellent for plating stainless steels that are to be used in conjunction with aluminum to prevent galvanic corrosion. For Types II and III the minimum cadmium thickness requirement shall be met after the supplementary treatment.

Chemical Films, MIL-C-5541E: The materials that qualify produce coatings that range in color from clear to iridescent yellow or brown. Inspection difficulties may arise with clear coatings because of their invisibility.

Class 1A chemical conversion coatings are intended to provide corrosion prevention when left unpainted as well as to improve adhesion of paint finish systems on aluminum and aluminum alloys. May be used on tanks, tubings, and component structures where paint finishes are not required for the exterior surfaces but are required for the interior surfaces.

Class 3 chemical conversion coatings are intended for use as a corrosive film for electrical and electronic applications where lower resistant contacts are required. The primary difference between Class 1A and Class 3 coating is thickness.

Chemical Finish: Black, MIL-F-495E: A uniform black corrosion retardant for copper. Coating has no abrasion resistance. Used to blacken color and reduce gloss on copper-alloy surfaces other than food service and water supply items. Also used as a base for subsequent coatings such as lacquer, varnish, oil, and wax.

Chrome, QQ-C-320B: Has excellent hardness, wear resistance, and erosion resistance. In addition chrome has a low coefficient of friction, is resistant to heat, and can be rendered porous for lubrication purposes.

Types I and II have bright and satin appearances respectively.

Class 1 is used as plating for corrosion protection and Class 2, for engineering plating. Class 1 and 2 both shall have a minimum thickness of 0.00001 in. on all visible surfaces. If thickness is not specified use 0.002 in.

Class 2a will be plated to specified dimensions or processed to specified dimensions after plating. Class 2b will be used on parts below 40 Rc and subject to static loads or designed for limited life under dynamic loads. Class 2c will be used on parts below 40 Rc and designed for unlimited life under dynamic loads. Class 2d parts have hardness of 40 Rc or above, which are subject to static loads or designed for unlimited life under dynamic loads. Class 2e parts have hardness of 40 Rc or above, which are designed for unlimited life under dynamic loads.

All coated steel parts having a hardness of Rc 36 and higher shall be baked at a minimum of 375°F ± 25°F per the following conditions.With a tensile strength of 160-180 (ksi), the time at temperature will be 3 hr.; at 181-220 ksi, the time will be 8 hr.; and at 221 ksi and above, the time will be 12 hr.

Copper, MIL-C-14550B: Has good corrosion resistance when used as an undercoat. A number of copper processes are available, each designed for a specific purpose such as, to improve brightness (to eliminate the need for buffing), high speed (for electro-forming), and fine grain (to prevent case-hardening).

All steel parts having a hardness of Rc 35 and higher shall be baked at 375°F ± 25°F for 24 hours, within four hours after plating to provide hydrogen embrittlement relief. Plated springs and other parts subject to flexure shall not be flexed prior to baking operations.

Class 0 will have a thickness 0.001 - 0.005 in. and is used for heat treatment stop-off; Class 1 is 0.001 in. and is used to provide carburizing shield, also for plated through printed circuit boards. Class 2 is 0.0005 in. thick and is used as an undercoat for nickel and other platings. Class 3 is 0.0002 in. thick and is used to prevent basis metal migration into tin (prevents poisoning solderability). Class 4 is 0.0001 in. thick.

Tin Lead, MIL-P-81728A: It has excellent solderability. Either a matte or bright luster is acceptable. For electronics components, use only parts with a matte or flow brightened finish.

For brightened electronic components, the maximum thickness will be 0.0003 in. Tin 50 to 70% by weight and with a lead remainder, 0.0003-0.0005 in.

Magnesium Process, MIL-M-3171C: Process #1-A chrome pickle treatment for magnesium. Color varies from matte gray to yellow-red. Has only fair corrosion resistance (< 24 hours, 20% salt spray resistance).

#7-A dichromate treatment for magnesium. Color varies from light brown to gray depending on alloy. Only fair corrosion resistance (< 24 hours, 20% salt spray resistance).

#9-A galvanic anodize treatment for magnesium. Produces a dark brown to black coating. Designed to give a protective film on alloys which do not react to Dow No. 7 treatment. Only fair corrosion resistance (< 24 hours, 20% salt spray resistance).

Type/Class	Thickness (in.)	Comments
Type 1	Removes metal. (approx. 0.0006 for wrought, less for die castings.) No dimensional change	Used for protecting magnesium during shipment, storage and machining. Can be used as a paint base. NOTE: Must remove Type I coating before applying Type III and Type IV treatments.
Type III	…	*Note:* precleaning and pickling may result in dimensional changes due to metal loss.
Type IV	No dimensional change	Can be used as a paint base, and is applicable to all magnesium alloys. Used where optical properties (black) are required on close tolerance parts. NOTE: Precleaning and pickling may result in dimensional changes due to metal loss.

Magnesium Anodic Treatment, MIL-M-45202C: The HAE anodic finish is probably the hardest coating currently available for magnesium. It exhibits stability at high temperatures and has good dielectric strength. It serves as an excellent paint base. It requires resin seal or paint for maximum corrosion protection.

Type/Class	Typical Thickness	Comments
Type I, Light coating.		
Class A	0.2 mil	Tan coating (HAE)
Grade 1	…	Without post treatment (dyed)
Grade 2	…	With biflouride-dichromate post treatment
Class C	0.3 mil	Light green coating (Dow #17)
Type II, Heavy coating		
Class A	1.5 mil	Hard brown coating (HAE)
Grade 1	…	Without post treatment
Grade 3	…	With biflouride-dichromate post treatment
Grade 4	…	With biflouride-dichromate post treatment including moist heat aging
Grade 5		With double application of biflouride-dichromate post treatment including moist heat aging.
Class D	1.2 mil	Dark green coating (Dow #17)

Coatings range from thin clear to light gray-green, to thick dark-green coatings. The clear coatings are used as a base for subsequent clear lacquers or paints to produce a final appearance similar to clear anodizing on aluminum. The light gray-green coatings are used in most applications which are to be painted. The thick, dark-green coating offers the best combination of abrasion resistance, protective value and paint base characteristics.

Electroless Nickel, AMS 2404C, AMS 2405B, AMS 2433B: Is typically used as a coating to provide a hard-ductile, wear-resistant, and corrosion-resistant surface for operation in service up to 1000°F, to provide uniform build-up on complex shapes.

AMS 2404C, is deposited directly on the basis metal without a flash coating of other metal, unless otherwise specified. AMS 2405B, is deposited directly on the basis metal except where parts fabricated from corrosion resistant steels or alloys where a "strike" coating of nickel or other suitable metal is required, unless otherwise specified. AMS 2433B, is a type of electroless nickel typically used to enhance the solderability of surfaces, but usage is not limited to such applications. Generally, the plate shall be placed directly on the basis metal. However, aluminum alloys shall be zinc immersion coated per ASTM B253 followed by copper flash; corrosion resistant steels and nickel and cobalt alloys or other basis metals may use a nickel or copper flash undercoat when the purchaser permits.

Electroless Nickel Preparation: Parts having a hardness higher than Rc 40 and have been machined or ground after heat treatment shall be suitably stress-relieved before cleaning and plating.

After treatment, parts having a hardness of Rc 33 and over shall be heated to 375°F ± 15°F for three hours. If such treatment is injurious to the parts, bake at 275°F ± 15°F for four hours.

Electroless Nickel, Low-Phosphorous

Note: If permitted by drawing, the maximum hardness and wear resistance are obtained by heating parts for 30-60 minutes, preferably in an inert atmosphere, at 750°F ± 15°F except aluminum parts shall be baked at 450°F ± 15°F for four hours. If such heating is not specified, bake at 375°F ± 15°F for three hours. If this treatment is injurious to parts or assemblies, bake at 275°F for five hours.

Plating: nickel-thallium-boron (Electroless Deposition) and nickel-boron (Electroless Deposition)

Preparation: All fabrication-type operations shall be completed.

Post-treatment: Cold worked or heat treated parts and aluminum alloys and other parts requiring special thermal treatment shall be post treated as agreed upon by purchaser and vendor. Other plated parts within four hours after plating shall be heat treated for 90 ± 10 minutes at 675°F ± 15°F.

Electropolishing, (No MIL-SPEC No.): This process electrolytically removes or diminishes scratches, burrs, and unwanted sharp edges from most metals. Finishes from satin to mirror-bright are produced by controlling time, temperature, or both.

Typically the thickness loss is 0.0002 in. This process is not recommended for close tolerance surfaces.

Gold, MIL-G-45204C: Has a yellow to orange color depending on the proprietary process used. Will range from matte to bright finish depending on basis metal. It has good corrosive resistance and a high tarnish resistance. It provides a low contact resistance, is a good conductor of electricity, and has excellent solderability. If the hardness grade for the gold coating is not specified, Type I shall be furnished at a hardness of Grade A, and Type II furnished at a hardness of Grade C.

For soldering, a thin pure soft gold coating is preferred. A minimum and maximum thickness 0.00005 and 0.00010 in., respectively, shall be plated.

Unless otherwise specified, gold over silver underplate combinations shall be excluded from electronics hardware. Silver or copper plus silver may not be used as an underplate unless required by the item specification. When gold is applied to brass bronze or beryllium copper, or a copper plate or strike, an antidiffusion underplate such as nickel shall be applied.

Type I is 99.7% gold minimum (Grades A, B, or C); Type II is 99.0% (Grades B, C, or D); and Type III is 99.9% (Grade A only).

Grade A is 90 Knoop maximum; Grade B is 91-129 Knoop; Grade C is 130-200 Knoop; and Grade D is 201 Knoop and over.

Class 00 has a thickness of 0.00002 in. minimum; Class 0, 0.00003 in.; Class 1, 0.00005 in.; Class 2, 0.0001 in.; Class 3, 0.0002 in.; Class 4, 0.0003 in.; Class 5, 0.0005 in.; and Class 6, 0.0015 in.

Hard Anodize, MIL-A-8625F: The color will vary from light tan to black depending on alloy and thickness. Can be dyed in darker colors depending on the thickness. Coating penetrates base metal as much as builds up on the surface. The term thickness includes both the buildup and penetration. It provides a very hard ceramic type coating. Abrasion resistance will vary with alloy and thickness of coating. Has good dielectric properties.

Do not seal coatings where the main function is to obtain maximum abrasion or wear resistance. When used for exterior applications requiring corrosion resistance but permitting reduced abrasion, the coating shall be sealed (boiling deionized water or hot 5% sodium dichromate solution, or other suitable chemical solutions).

Type III will have a thickness specified on the contract or applicable drawing. If not specified use a nominal thickness of 0.002 in. Hard coatings may vary in thickness from 0.0005 - 0.0045 in.

Class 1 shall be not dyed or pigmented. Class 2 shall be dyed and the color specified on the contract. The process can be controlled to very close thickness tolerances. Where maximum serviceability or special properties are required, consult metal finisher for best alloy choice. Thick coatings (those over 0.004 in.) will tend to break down sharp edges. Can be used as an electrical insulation coating. "Flash" hard anodize may be used instead of conventional anodize for corrosion resistance and may be more economical in conjunction with other hard anodized areas.

Lubrication, Solid Film MIL-L-46010D: The Military Plating Specification establishes the requirements for three types of heat cured solid film lubricants that are intended to reduce wear and prevent galling, corrosion, and seizure of metals. For use on aluminum, copper, steel, stainless steel, titanium, and chromium, and nickel bearing surfaces.

Types I, II, and III have a thicknesses of 0.008 - 0.013 mm. No single reading less than 0.005 mm or greater than 0.018 mm.

Type I has a curing temperature of 150 ± 15°C and an endurance life of 250 minutes; Type II, 204 ± 15°C and 450 minutes; and Type III is a low volatile organic compound (VOC) content lubricant with cure cycles of 150 ± 15°C for two hours, or 204 ± 15°C for one hour with an endurance life of 450 minutes. Color 1 has a natural product color and Color 2 has a black color.

Nickel, QQ-N-290A: There is a nickel finish for almost any need. Nickel can be deposited soft, hard-dull, or bright, depending on process used and conditions employed in plating. Thus, hardness can range from 150-500 Vickers. Nickel can be similar to stainless steel in color, or can be a dull gray (almost white) color. Corrosion resistance is a function of thickness. Nickel has a low coefficient of thermal expansion.

All steel parts having a tensile strength of 220,000 or greater shall not be a nickel plate without specific approval of procuring agency.

Class 1 is used for corrosion protection. Plating shall be applied over an underplating of copper or yellow brass on zinc and zinc based alloys. In no case, shall the copper underplate be substituted for any part of the specified nickel thickness. Class 2 is used in engineering applications.

Grade A has a thickness of 0.0016 in.; Grade B, 0.0012 in.; Grade C, 0.001 in.; Grade D, 0.0008 in.; Grade E, 0.0006 in.; Grade F, 0.0004 in.; and Grade G, 0.002 in.

Palladium, MIL-P-45209B: A gray, dense deposit good for undercoats. Has good wear characteristics, corrosion resistance, catalytic properties, and good conductivity.

The thickness shall be 0.00005 in. unless otherwise specified.

Steel springs and other steel parts subject to flexure or repeated impact and of hardness greater than Rc 40 shall be heated to 375°F ± 25°F for three hours after plating.

Passivate, QQ-P-35C: Intended to improve the corrosion resistance of parts made from austentic, ferritic, and martensitic corrosion-resistant steels of the 200, 300, and 400 series and precipitation hardened corrosion resistant steels. 440C grades may be exempt from passivation treatments of the procuring activity.

Type II is a medium temperature nitric acid solution with sodium dichromate additive. Type VI, a low temperature nitric acid solution; Type VII, a medium temperature nitric acid solution; and Type VIII, a medium temperature high concentration nitric acid solution.

Phosphate Coating: Light, TT-C-490D: This specification covers cleaning methods and pretreatment processes.

Methods /Types	Typical Thickness (in.)	Comments
Cleaning Methods		
	–	Light coating for use as a paint base.
Method I	…	Mechanical or abrasive cleaning (for ferrous surfaces only).
Method II	…	Used for solvent cleaning.
Method III	…	Used for hot alkalines (for ferrous surfaces only).
Method IV	…	Emulsion.
Method V	…	Used for alkaline derusting (for ferrous surfaces only).
Method VI	…	Phosphoric acid.
Pretreatment Coatings		
Type I	…	Zinc phosphate. Class 1-spray application: Class 2A and 2B-Immersion or Dip application
Type II	…	Aqueous Iron Phosphate
Type III	0.0003–0.0005	Is an organic pretreatment coating
Type IV	…	Non-aqueous iron phosphate
Type V	…	Zinc phosphate

Type I is intended as a general all-purpose pretreatment prior to painting. Type II and IV are intended primarily for use where metal parts are to be formed after painting. Type III is intended for use where size and shape preclude using Type I, II, or IV and where items containing mixed metal components are assembled prior to treatment.

Phosphate Coating: Heavy, DOD-P-16232-F: The primary differences are that Type M is used as a heavy manganese phosphate coating for corrosion and wear resistance and Type Z is used as a Zinc phosphate coating.

Type M has a thickness from 0.0002-0.0004 in. and Type Z, 0.0002-0.0006 in. Class 1, for both types has a supplementary preservative treatment or coating as specified; Class 2, has a supplementary treatment with lubricating oil; and Class 3, no supplementary treatment is required. For Type M, Class 4 is chemically converted (may be dyed to color as specified) with no supplementary coating or supplementary coating as specified. For Type Z, Class 4 is the same as Class 3.

This coating is for medium and low alloy steels. The coatings range from gray to black in color. The "heavy" phosphate coatings covered by this specification are intended as a base for holding/retaining supplemental coatings which provide the major portion of the corrosion resistance. "Light" phosphate coatings used for a paint base are covered by other specifications. Heavy zinc phosphate coatings may be used when paint and supplemental oil coatings are required on various parts or assemblies.

Rhodium, MIL-R-46085B: Rhodium is metallic and similar to stainless steel in color, has excellent corrosion and abrasion resistance, is almost as hard as chromium, and has a high reflectivity. Thicker coatings of Rhodium are very brittle.

Class/Types	Thickness (in.)	Comments
Type I	–	Over nickel, silver, gold, or platinum.
Type II	–	Over other metals, requires nickel undercoat.
Class 1	0.000002	Used on silver for tarnish resistance.
Class 2	0.00001	
Class 3	0.00002	Applications range from electronic to nose cones -wherever wear, corrosion resist solderability and reflectivity are important.
Class 4	0.00010	
Class 5	0.00025	

Parts having a hardness of Rc 33 or above shall be baked at 375°F for three hours prior to cleaning. Parts having hardness of 40 Rc and above shall be baked within four hours after plating at 375°F for three hours.

Silver, QQ-S-365D: Silver has an increasing use in both decorative and engineering fields, including electrical and electronic fields.

Silver is white matte to very bright in appearance. Has good corrosion resistance, depending on base metal and will tarnish easily. Its hardness varies from about 90-135 Brinell depending on process and plating conditions. Solderability is excellent, but decreases with age. Silver is the best conductor of electricity. Has excellent lubricity and smear characteristics for antigalling uses on static seals, bushing, etc. Stress relief steel parts at a minimum 375°F ± 25°F or more prior to cleaning and plating if they contain or are suspected of having damaging residual tensile stresses.

All types and grades will have a minimum thickness of 0.0005 in. unless otherwise specified. Type I is matte, Type II is semi-bright, and Type III is bright. Grade A has a chromate post-treatment to improve tarnish resistance. In contrast Grade B has no supplementary treatment.

Tin, MIL-T-10727C: There are two different types of coating methods used, electrodeposited (based on Use ASTM B545 standard specification for electrodeposited coatings of tin) and hot dipped.

Thickness as specified on drawing (thickness is not part of the specification) is 0.0001-0.0025 in., flash for soldering; 0.0002-0.0004 in., to prevent galling and seizing; 0.0003 in. minimum, where corrosion resistance is important; and 0.0002-0.0006 to prevent formation of case during nitriding.

Color is a gray-white color in plated condition. Tin is soft, but very ductile. It has good corrosion resistance, and has excellent solderability. Tin is not good for low temperature applications.

If a bright finish is desired to be used in lieu of fused tin, specify Bright Tin plate. Thickness can exceed that of fused tin and deposit shows excellent corrosion resistance and solderability.

Vacuum Cadmium, MIL-C-8837B: Is used primarily to provide corrosion resistance to ferrous parts free from hydrogen contamination and possible embrittlement. Recommended on steels with a strength of 2.2x105 psi or above.

Coating is applied after all machining, brazing, welding, and forming has been completed. Prior to coating, all steel parts shall be stress relieved by baking at 375°F ± 25 for three hours if suspected of having residual tensile stresses. Immediately prior to coating, lightly dry abrasive blast areas are to be coated.

Type I shall be as plated; and Types II and III require supplementary chromate and phosphate treatments respectively.

Classes 1, 2, and 3 have thicknesses of 0.0005, 0.0003, and 0.0002 in. respectively.

Cadmium coating shall not be used, if in service, temperature reaches 450°F.

A salt spray test is required for type II and is 96 hours.

Zinc, ASTM-B633: This specification covers requirements for electrodeposited zinc coatings applied to iron or steel articles to protect them from corrosion. It does not cover zinc-coated wire or sheets.

Type I will be as plated; Type II will have colored chromate conversion coatings; Type III will have colorless chromate conversion coatings; and Type IV will have phosphate conversion coatings.

High strength steels (tensile strength over 1700 Mpa) shall not be electroplated.

Stress relief: All parts with ultimate tensile strength 1000 Mpa and above at minimum 190°C for three hours or more before cleaning and plating.

Hydrogen embrittlement relief: All electroplated parts 1200 Mpa or higher shall be baked at 190°C for three hours or more within four hours after electroplating.

Corrosion Resistance Requirements	
Types	Test Period Hr.
II	96
III	12

FASTENERS

NAILS, SPIKES AND WOOD SCREWS

RIVETS AND RIVETED JOINTS

TORQUE AND TENSION IN FASTENERS

BOLTS, SCREWS, NUTS, AND WASHERS

METRIC FASTENERS

METRIC FASTENERS (cont.)

BRITISH FASTENERS

BRITISH FASTENERS (cont.)

MACHINE SCREWS

CAP AND SET SCREWS

SELF-THREADING SCREWS

T- Slots, Bolts, and Nuts

PINS AND STUDS

RETAINING RINGS

WING NUTS, WING SCREWS AND THUMB SCREWS

NAILS, SPIKES AND WOOD SCREWS

Standard Wire Nails and Spikes (Size, Length and Approximate Number to Pound)

Size of Nail	Length, Inches	Common Wire Nails and Brads: Gage	Num/lb	Flooring Brads: Gage	Num/lb	Fence Nails: Gage	Num/lb	Casing, Smooth and Barbed Box: Gage	Num/lb	Finishing Nails: Gage	Num/lb
2 *d*	1	15	876	…	…	…	…	15 ½	1010	16 ½	1351
3 *d*	1 ¼	14	568	…	…	…	…	14 ½	635	15 ½	807
4 *d*	1 ½	12 ½	316	…	…	…	…	14	473	15	584
5 *d*	1 ¾	12 ½	271	…	…	10	142	14	406	15	500
6 *d*	2	11 ½	181	11	157	10	124	12 ½	236	13 ½	309
7 *d*	2 ¼	11 ½	161	11	139	9	92	12 ½	210	13	238
8 *d*	2 ½	10 ¼	106	10	99	9	82	11 ½	145	12 ½	189
9 *d*	2 ¾	10 ¼	96	10	90	8	62	11 ½	132	12 ½	172
10 *d*	3	9	69	9	69	7	50	10 ½	94	11 ½	121
12 *d*	3 ¼	9	64	8	54	6	40	10 ½	87	11 ½	113
16 *d*	3 ½	8	49	7	43	5	30	10	71	11	90
20 *d*	4	6	31	6	31	4	23	9	52	10	62
30 *d*	4 ½	5	24	…	…	…	…	9	46	…	…
40 *d*	5	4	18	…	…	…	…	8	35	…	…
50 *d*	5 ½	3	16	…	…	…	…	…	…	…	…
60 *d*	6	2	11	…	…	…	…	…	…	…	…

Size and Length		Hinge Nails, Heavy		Hinge Nails, Light		Clinch Nails		Barbed Car Nails, Heavy		Barbed Car Nails, Light	
2 *d*	1	…	…	…	…	14	710	…	…	…	…
3 *d*	1 ¼	…	…	…	…	13	429	…	…	…	…
4 *d*	1 ½	3	50	6	82	12	274	10	165	12	274
5 *d*	1 ¾	3	38	6	62	12	235	9	118	10	142
6 *d*	2	3	30	6	50	11	157	9	103	10	124
7 *d*	2 ¼	00	12	3	25	11	139	8	76	9	92
8 *d*	2 ½	00	11	3	23	10	99	8	69	9	82
9 *d*	2 ¾	00	10	3	22	10	90	7	54	8	62
10 *d*	3	00	9	3	19	9	69	7	50	8	57
12 *d*	3 ¼	…	…	…	…	9	62	6	42	7	50
16 *d*	3 ½	…	…	…	…	8	49	6	35	7	43
20 *d*	4	…	…	…	…	7	37	5	26	6	31
30 *d*	4 ½	…	…	…	…	…	…	5	24	6	28
40 *d*	5	…	…	…	…	…	…	4	18	5	21
50 *d*	5 ½	…	…	…	…	…	…	3	15	4	17
60 *d*	6	…	…	…	…	…	…	3	13	4	15

Size and Length		Boat Nails, Heavy		Boat Nails, Light		Slating Nails		Spikes: Size and Length		Gage	No. to Lb
2 *d*	1	…	…	…	…	12	411				
3 *d*	1 ¼	…	…	…	…	10 ½	225				
4 *d*	1 ½	¼	44	$^3/_{16}$	82	10 ½	187				
5 *d*	1 ¾	…	…	…	…	10	142	10 *d*	3	6	41
6*d*	2	¼	32	$^3/_{16}$	62	9	103	12 *d*	3 ¼	6	38
7 *d*	2 ¼	…	…	…	…	…	…	16 *d*	3 ½	5	30
8 *d*	2 ½	¼	26	$^3/_{16}$	50	…	…	20 *d*	4	4	23
9 *d*	2 ¾	…	…	…	…	…	…	30 *d*	4 ½	3	17
10 *d*	3	⅜	14	¼	22	…	…	40 *d*	5	2	13
12 *d*	3 ¼	⅜	13	¼	20	…	…	50 *d*	5 ½	1	10
16 *d*	3 ½	⅜	12	¼	18	…	…	60 *d*	6	1	8
20 *d*	4	⅜	10	¼	16	…	…	…	7	0	7
30 *d*	4 ½	…	…	…	…	…	…	…	8	00	6
40 *d*	5	…	…	…	…	…	…	…	9	00	5
50 *d*	5 ½	…	…	…	…	…	…	…	10	⅜	4
60 *d*	6	…	…	…	…	…	…	…	12	⅜	3

ANSI Flat, Pan, and Oval Head Wood Screws *ANSI B18.6.1-1981 (R1997)*

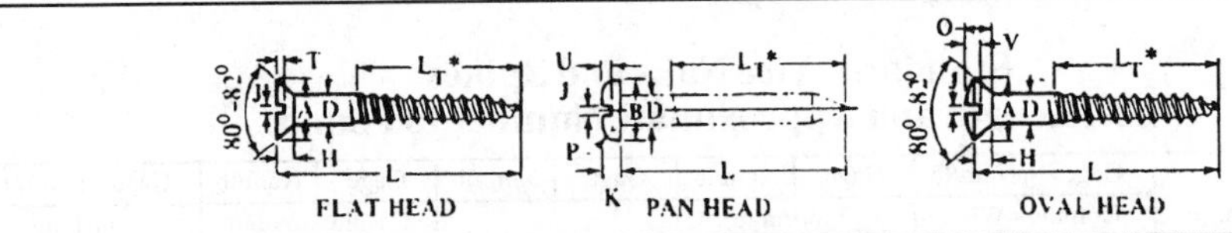

Nominal Size	Threads per inch	D[a] Basic Diam. of Screw	J Width of Slot Max.	J Min.	A Head Diameter Max., Sharp Edge	A Min., Edge Rounded or Flat	B Head Diameter Max.	B Min.	P Head Radius Max.	H Height of Head Ref.
0	32	.060	.023	.016	.119	.099	.116	.104	.020	.035
1	28	.073	.026	.019	.146	.123	.142	.130	.025	.043
2	26	.086	.031	.023	.172	.147	.167	.155	.035	.051
3	24	.099	.035	.027	.199	.171	.193	.180	.037	.059
4	22	.112	.039	.031	.225	.195	.219	.205	.042	.067
5	20	.125	.043	.035	.252	.220	.245	.231	.044	.075
6	18	.138	.048	.039	.279	.244	.270	.256	.046	.083
7	16	.151	.048	.039	.305	.268	.296	.281	.049	.091
8	15	.164	.054	.045	.332	.292	.322	.306	.052	.100
9	14	.177	.054	.045	.358	.316	.348	.331	.056	.108
10	13	.190	.060	.050	.385	.340	.373	.357	.061	.116
12	11	.216	.067	.056	.438	.389	.425	.407	.078	.132
14	10	.242	.075	.064	.507	.452	.492	.473	.087	.153
16	9	.268	.075	.064	.544	.485	.528	.508	.094	.164
18	8	.294	.084	.072	.635	.568	.615	.594	.099	.191
20	8	.320	.084	.072	.650	.582	.631	.608	.121	.196
24	7	.372	.094	.081	.762	.685	.740	.716	.143	.230

[a] *Diameter Tolerance:* Equals + 0.004 in. and – 0.007 in. for cut threads. For rolled thread body diameter tolerances, see ANSI 18.6.1-1981 (R1991).

Nominal Size	Threads per Inch	O Tot. Hgt. of Head Max.	O Min.	K Height of Head Max.	K Min.	T Depth of Slot Max.	T Min.	U Depth of Slot Max.	U Min.	V Depth of Slot Max.	V Min.
0	32	.056	.041	.039	.031	.015	.010	.022	.014	.030	.025
1	28	.068	.052	.046	.038	.019	.012	.027	.018	.038	.031
2	26	.080	.063	.053	.045	.023	.015	.031	.022	.045	.037
3	24	.092	.073	.060	.051	.027	.017	.035	.027	.052	.043
4	22	.104	.084	.068	.058	.030	.020	.040	.030	.059	.049
5	20	.116	.095	.075	.065	.034	.022	.045	.034	.067	.055
6	18	.128	.105	.082	.072	.038	.024	.050	.037	.074	.060
7	16	.140	.116	.089	.079	.041	.027	.054	.041	.081	.066
8	15	.152	.126	.096	.085	.045	.029	.058	.045	.088	.072
9	14	.164	.137	.103	.092	.049	.032	.063	.049	.095	.078
10	13	.176	.148	.110	.099	.053	.034	.068	.053	.103	.084
12	11	.200	.169	.125	.112	.060	.039	.077	.061	.117	.096
14	10	.232	.197	.144	.130	.070	.046	.087	.070	.136	.112
16	9	.248	.212	.153	.139	.075	.049	.093	.074	.146	.120
18	8	.290	.249	.178	.162	.083	.054	.106	.085	.171	.141
20	8	.296	.254	.182	.166	.090	.059	.108	.087	.175	.144
24	7	.347	.300	.212	.195	.106	.070	.124	.100	.204	.168

All dimensions in inches. The edge of flat and oval head screws may be flat or rounded. Wood screws are also available with Types I, IA, and II recessed heads. Consult the standard for recessed head dimensions. *The length of the thread, L_T, on wood screws having cut threads shall be equivalent to approximately two-thirds of the nominal length of the screw. For rolled threads, L_T shall be at least four times the basic screw diameter or two-thirds of the nominal screw length, whichever is greater. Screws of nominal lengths that are too short to accommodate the minimum thread length shall have threads extending as close to the underside of the head as practicable.

Pilot Hole Drill Sizes for Wood Screws

Work Material	Wood Screw Size 2	4	6	8	10	12	14
Hardwood	3⁄64	1⁄16	5⁄64	3⁄32	7⁄64	1⁄8	9⁄64
Softwood	1⁄32	3⁄64	1⁄16	5⁄64	3⁄32	7⁄64	1⁄8

RIVETS AND RIVETED JOINTS

Classes and Types of Riveted Joints.—Riveted joints may be classified by application as: 1) pressure vessel; 2) structural; and 3) machine member.

For information and data concerning joints for pressure vessels such as boilers, reference should be made to standard sources such as the ASME Boiler Code. The following sections will cover only structural and machine-member riveted joints.

Basically there are two kinds of riveted joints, the *lap-joint* and the *butt-joint.* In the ordinary *lap-joint*, the plates overlap each other and are held together by one or more rows of rivets. In the *butt-joint*, the plates being joined are in the same plane and are joined by means of a cover plate or butt strap, which is riveted to both plates by one or more rows of rivets. The term *single riveting* means one row of rivets in a lap-joint or one row on each side of a butt-joint; *double riveting* means two rows of rivets in a lap-joint or two rows on each side of the joint in butt riveting. Joints are also triple and quadruple riveted. Lap-joints may also be made with inside or outside cover plates. Types of lap and butt joints are illustrated in the table on starting on page 1461.

General Considerations of a Riveted Joint.—Factors to be considered in the design or specification of a riveted joint are: type of joint; spacing of rivets; type and size of rivet; type and size of hole; and rivet material.

Spacing of Rivets: The spacing between rivet centers is called *pitch* and between row center lines, *back pitch* or *transverse pitch.* The distance between centers of rivets nearest each other in adjacent rows is called *diagonal pitch.* The distance from the edge of the plate to the center line of the nearest row of rivets is called *margin.*

Examination of a riveted joint made up of several rows of rivets will reveal that after progressing along the joint a given distance, the rivet pattern or arrangement is repeated. (For a butt joint, the length of a *repeating section* is usually equal to the *long pitch* or pitch of the rivets in the outer row, that is the row farthest from the edge of the joint.) For structural and machine-member joints, the proper pitch may be determined by making the tensile strength of the plate over the length of the repeating section, that is the distance between rivets in the outer row, equal to the total shear strength of the rivets in the repeating section. Minimum pitch and diagonal pitch are also governed by the clearance required for the hold-on (Dolly bar) and rivet set. Dimensions for different sizes of hold-ons and rivet sets are given in the table on page 1467.

When fastening thin plate, it is particularly important to maintain accurate spacing to avoid buckling.

Size and Type of Rivets: The rivet diameter d commonly falls between $d = 1.2\sqrt{t}$ and $d = 1.4\sqrt{t}$, where t is the thickness of the plate. Dimensions for various types of American Standard large (½-inch diameter and up) rivets and small solid rivets are shown in tables that follow. It may be noted that countersunk heads are not as strong as other types.

Size and Type of Hole: Rivet holes may be punched, punched and reamed, or drilled. Rivet holes are usually made $\frac{1}{16}$ inch larger in diameter than the nominal diameter of the rivet although in some classes of work in which the rivet is driven cold, as in automatic machine riveting, the holes are reamed to provide minimum clearance so that the rivet fills the hole completely.

When holes are punched in heavy steel plate, there may be considerable loss of strength unless the holes are reamed to remove the inferior metal immediately surrounding them. This results in the diameter of the punched hole being increased by from $\frac{1}{16}$ to $\frac{1}{8}$ inch. Annealing after punching tends to restore the strength of the plate in the vicinity of the holes.

Rivet Material: Rivets for structural and machine-member purposes are usually made of wrought iron or soft steel, but for aircraft and other applications where light weight or resistance to corrosion is important, copper, aluminum alloy, Monel, Inconel, etc., may be used as rivet material.

Failure of Riveted Joints.—Rivets may fail by:

1) Shearing through one cross-section (single shear)

2) Shearing through two cross-sections (double shear)

3) Crushing

Plates may fail by:

4) Shearing along two parallel lines extending from opposite sides of the rivet hole to the edge of the plate

5) Tearing along a single line from middle of rivet hole to edge of plate

6) Crushing

7) Tearing between adjacent rivets (tensile failure) in the same row or in adjacent rows

Types 4 and 5 failures are caused by rivets being placed too close to the edge of the plate. These types of failure are avoided by placing the center of the rivet at a minimum of one and one-half times the rivet diameter away from the edge.

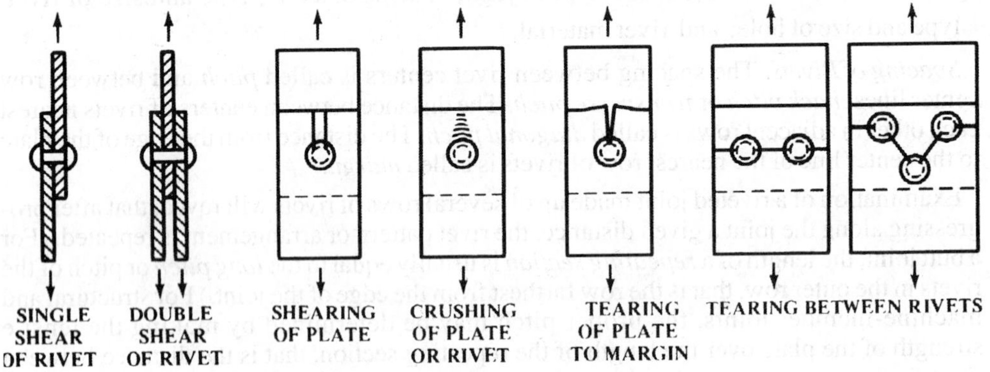

Types of Rivet and Plate Failure

Failure due to tearing on a diagonal between rivets in adjacent rows when the pitch is four times the rivet diameter or less is avoided by making the transverse pitch one and three-quarters times the rivet diameter.

Theoretical versus Actual Riveted Joint Failure.—If it is assumed that the rivets are placed the suggested distance from the edge of the plate and each row the suggested distance from another row, then the failure of a joint is most likely to occur as a result of shear failure of the rivets, bearing failure (crushing) of the plate or rivets, or tensile failure of the plate, alone or in combination depending on the makeup of the joints.

Joint failure in actuality is more complex than this. Rivets do not undergo pure shear especially in lap-joints where rivets are subjected to single shear. The rivet, in this instance, would be subject to a combination of tensile and shearing stresses and it would fail because of combined stresses, not a single stress. Furthermore, the shearing stress is usually considered to be distributed evenly over the cross-section, which is also not the case.

Rivets that are usually driven hot contract on cooling. This contraction in the length of the rivet draws the plates together and sets up a stress in the rivet estimated to be equal in magnitude to the yield point of the rivet steel. The contraction in the diameter of the rivet results in a little clearance between the rivetand the hole in the plate. The tightness in the plates caused by the contraction in length of the rivet gives rise to a condition in which quite a sizeable frictional force would have to be overcome before the plates would slip over one another and subject the rivets to a shearing force. It is European practice to design joints for resistance to this slipping. It has been found, however, that the strength-basis designs obtained in American and English practice are not very different from European designs.

Design of Riveted Joints.—In the design of riveted joints, a simplified treatment is frequently used in which the following assumptions are made:

1) The load is carried equally by the rivets.

2) No combined stresses act on a rivet to cause failure.

3) The shearing stress in a rivet is uniform across the cross-section under question.

4) The load that would cause failure in single shear would have to be doubled to cause failure in double shear.

5) The bearing stress of rivet and plate is distributed equally over the projected area of the rivet.

6) The tensile stress is uniform in the section of metal between the rivets.

Allowable Stresses.—The design stresses for riveted joints are usually set by codes, practices, or specifications. The American Institute of Steel Construction issues specifications for the design, fabrication, and erection of structural steel for buildings in which the allowable stress permitted in tension for structural steel and rivets is specified at 20,000 pounds per square inch, the allowable bearing stress for rivets is 40,000 psi in double shear and 32,000 psi in single shear, and the allowable shearing stress for rivets is 15,000 psi. The American Society of Mechanical Engineers in its Boiler Code lists the following ultimate stresses: tensile, 55,000 psi; shearing, 44,000 psi; compressive or bearing, 95,000 psi. The design stresses usually are one-fifth of these, that is tensile, 11,000 psi; shearing, 8800 psi; compressive or bearing, 19,000 psi. In machine design work, values close to these or somewhat lower are commonly used.

Analysis of Joint Strength.—The following examples and strength analyses of riveted joints are based on the six assumptions previously outlined for a simplified treatment.

Example 1: Consider a 12-inch section of single-riveted lap-joint made up with plates of ¼-inch thickness and six rivets, ⅝ inch in diameter. Assume that rivet holes are 1⁄16 inch larger in diameter than the rivets. In this joint, the entire load is transmitted from one plate to the other by means of the rivets. Each plate and the six rivets carry the entire load. The safe tensile load L and the efficiency η may be determined in the following way: Design stresses of 8500 psi for shear, 20,000 psi for bearing, and 10,000 psi for tension are arbitrarily assigned and it is assumed that the rivets will not tear or shear through the plate to the edge of the joint.

A) The safe tensile load L based on single shear of the rivets is equal to the number of rivets n times the cross-sectional area of one rivet A_r times the allowable shearing stress S_s or

$$L = n \times A_r \times S_s$$

$$L = 6 \times \frac{\pi}{4}(0.625)^2 \times 8500$$

$$L = 15,647 \text{ pounds}$$

B) The safe tensile load L based on bearing stress is equal to the number of rivets n times the projected bearing area of the rivet A_b (diameter times thickness of plate) times the allowable bearing stress S_c or $L = n \times A_b \times S_c = 6 \times (0.625 \times 0.25) \times 20{,}000 = 18{,}750$ pounds.

C) The safe load L based on the tensile stress is equal to the net cross-sectional area of the plate between rivet holes A_p times the allowable tensile stress S_t or

$L = A_p \times S_t = 0.25[12 - 6(0.625 + 0.0625)] \times 10{,}000 = 19{,}688$ pounds.

The safe tensile load for the joint would be the least of the three loads just computed or 15,647 pounds and the efficiency η would be equal to this load divided by the tensile strength of the section of plate under consideration, if it were unperforated or

$$\eta = \frac{15,647}{12 \times 0.25 \times 10,000} \times 100 = 52.2 \text{ per cent}$$

Example 2: Under consideration is a 12-inch section of double-riveted butt-joint with main plates ½ inch thick and two cover plates each 5⁄16 inch thick. There are 3 rivets in the inner row and 2 on the outer and their diameters are ⅞ inch. Assume that the diameter of the rivet holes is 1⁄16 inch larger than that of the rivets. The rivets are so placed that the main plates will not tear diagonally from one rivet row to the others nor will they tear or fail in shear out to their edges. The safe tensile load L and the efficiency η may be determined in the following way: Design stresses for 8500 psi for shear, 20,000 psi for bearing, and 10,000 psi for tension are arbitrarily assigned.

A) The safe tensile load L based on double shearing of the rivets is equal to the number of rivets n times the number of shearing planes per rivet times the cross-sectional area of one rivet A_r times the allowable shearing stress S_s or $L = n \times 2 \times A_r \times S_s = 5 \times 2 \times \frac{\pi}{4}(0.875)^2 \times 8500 = 51,112$ pounds

B) The safe tensile load L based on bearing stress is equal to the number of rivets n times the projected bearing area of the rivet A_b (diameter times thickness of plate) times the allowable bearing stress S_c or $L = n \times A_b \times S_c = 5 \times (0.875 \times 0.5) \times 20,000 = 43,750$ pounds.

(Cover plates are not considered since their combined thickness is ¼ inch greater than the main plate thickness.)

C) The safe tensile load L based on the tensile stress is equal to the net cross-sectional area of the plate between the two rivets in the outer row A_p times the allowable tensile stress S_t or $L = A_p \times S_t = 0.5[12 - 2(0.875 + 0.0625)] \times 10,000 = 50,625$ pounds.

In completing the analysis, the sum of the load that would cause tearing between rivets in the three-hole section plus the load carried by the two rivets in the two-hole section is also investigated. The sum is necessary because if the joint is to fail, it must fail at both sections simultaneously. The least safe load that can be carried by the two rivets of the two-hole section is based on the bearing stress (see the foregoing calculations).

1) The safe tensile load L based on the bearing strength of two rivets of the two-hole section is $L = n \times A_b \times S_c = 2 \times (0.875 \times 0.5) \times 20,000 = 17,500$ pounds.

2) The safe tensile load L based on the tensile strength of the main plate between holes in the three-hole section is $L \times A_p \times S_t = 0.5[12 - 3(0.875 + 0.0625)] \times 10,000 = 45,938$ pounds.

The total safe tensile load based on this combination is 17,500 + 45,938 = 63,438 pounds, which is greater than any of the other results obtained.

The safe tensile load for the joint would be the least of the loads just computed or 43,750 pounds and the efficiency η would be equal to this load divided by the tensile strength of the section of plate under consideration, if it were unperforated or

$$\eta = \frac{43,750}{0.5 \times 12 \times 10,000} \times 100 = 72.9 \text{ per cent}$$

A riveted joint may fail by shearing through the rivets (single or double shear), crushing the rivets, tearing the plate between the rivets, crushing the plate or by a combination of two or more of the foregoing causes. Rivets placed too close to the edge of the plate may tear or shear the plate out to the edge but this type of failure is avoided by placing the center of the rivet 1.5 times the rivet diameter away from the edge.

The efficiency of a riveted joint is equal to the strength of the joint divided by the strength of the unriveted plate, expressed as a percentage.

In the following formulas, let

d = diameter of holes
t = thickness of plate
t_c= thickness of cove plates
p = Pitch of inner row of rivets

P = Pitch of outer row of ribets
S_s = Shear stress for rivets
S_t=Tensile stress for plates
S_c= Compressive or bearing stress for rivets or plates

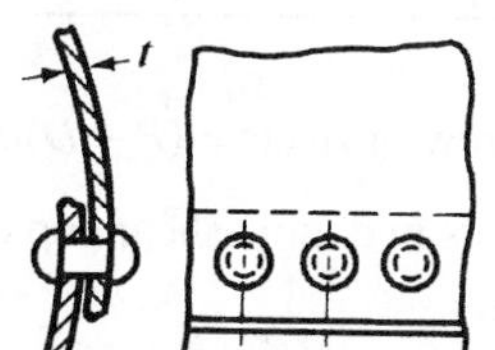

For Single-Riveted Lap-Joint

(1) Resistance to shearing one rivet = $\frac{\pi d^2}{4} S_s$

(2) Resistance to tearing plate between rivets = $(p - D)tS_t$

(3) Resistance to crushing rivet or plate = dtS_c

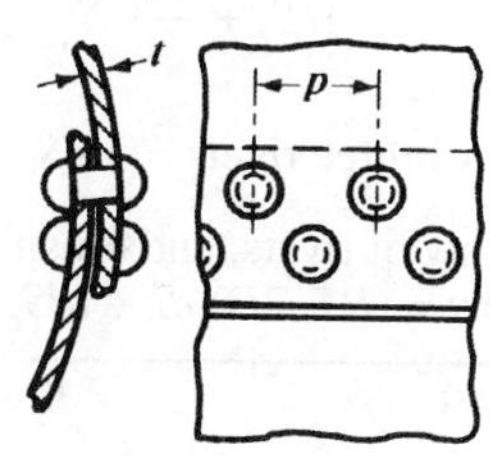

Double-Riveted Lap-Joint

(1) Resistance to shearing two rivets = $\frac{2\pi d^2}{4} S_s$

(2) Resistance to tearing between two rivets = $(p - D)tS_t$

(3) Resistance to crushing in front of two rivets = $2dtS_c$

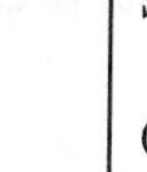

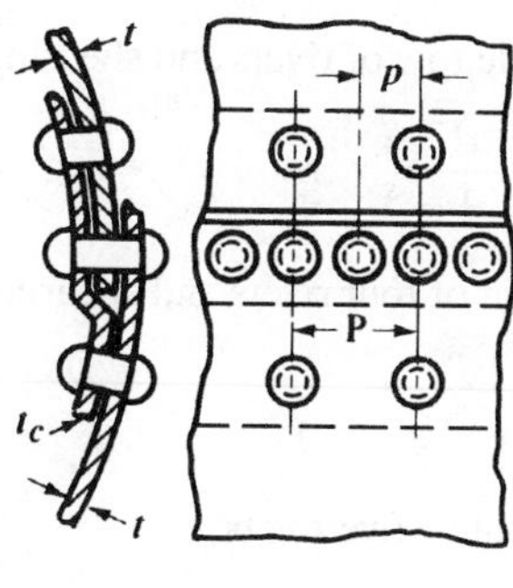

Single-Riveted Lap-Joint with Inside Cover Plate

(1) Resistance to tearing between outer row of rivets $= (P - D)tS_t$

(2) Resistance to tearing between inner row of rivets, and shearing outer row of rivets = $(P - 2D)tS_t + \frac{\pi d^2}{4} S_s$

(3) Resistance to shearing three rivets = $\frac{3\pi d^2}{4} S_s$

(4) Resistance to crushing in front of three rivets = $3tdS_c$

(5) Resistance to tearing at inner row of rivets, and crushing in front of one rivet in outer row $= (P - 2D)tS_t + tdS_c$

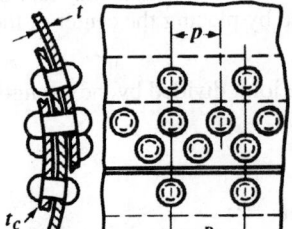

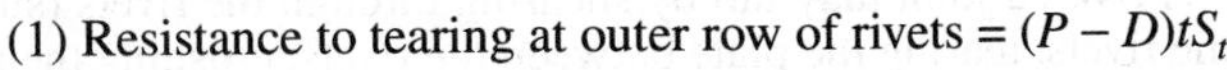

Double-Riveted Lap-Joint with Inside Cover Plate

(1) Resistance to tearing at outer row of rivets = $(P - D)tS_t$

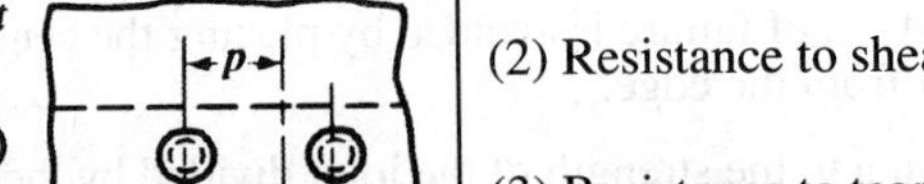

(2) Resistance to shearing four rivets = $\dfrac{4\pi d^2}{4} S_s$

(3) Resistance to tearing at inner row and shearing outer row of rivets = $(P - 1\tfrac{1}{2}D)tS_t + \dfrac{\pi d^2}{4} S_s$

(4) Resistance to crushing in front of four rivets = $4tdS_c$

(5) Resistance to tearing at inner row of rivets, and crushing in front of one rivet = $\left(P - 1\dfrac{1}{2}D\right)tS_t + tdS_c$

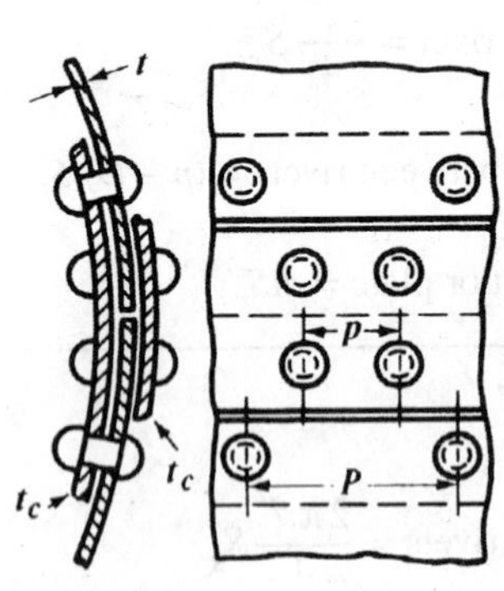

Double-Riveted Butt-Joint

(1) Resistance to tearing at outer row of rivets = $(P - D)tS_t$

(2) Resistance to shearing two rivets in double shear and one in single shear = $\dfrac{5\pi d^2}{4} S_s$

(3) Resistance to tearing at inner row of rivets and shearing one rivet of the outer row = $(P - 2D)tS_t + \dfrac{\pi d^2}{4} S_s$

(4) Resistance to crushing in front of three rivets = $3tdS_c$

(5) Resistance to tearing at inner row of rivets, and crushing in front of one rivet in outer row = $(P - 2D)\ tS_t + tdS_c$

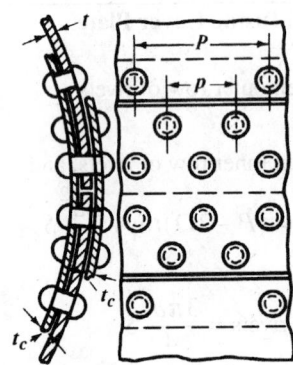

Triple-Riveted Butt-Joint

(1) Resistance to tearing at outer row of rivets = $(P - D)tS_t$

(2) Resistance to shearing four rivets in double shear and one in single shear = $\dfrac{9\pi d^2}{4} S_s$

(3) Resistance to tearing at middle row of rivets and shearing one rivet = $(P - 2D)tS_t + \dfrac{\pi d^2}{4} S_s$

(4) Resistance to crushing in front of four rivets and shearing one rivet = $4dtS_c + \dfrac{\pi d^2}{4} S_s$

(5) Resistance to crushing in front of five rivets = $4dtS_c + dt_cS_c$

Dimensions are usually specified in inches and stresses in pounds per square inch. See page 1459 for a discussion of allowable stresses that may be used in calculating the strengths given by the formulas. The design stresses are usually set by codes, practices, or specifications.

Rivet Lengths for Forming Round and Countersunk Heads

Grip in Inches	To Form Round Head — Diameter of Rivet in Inches: ½	⅝	¾	⅞	1	1⅛	1¼	Grip in Inches	To Form Countersunk Head — Diameter of Rivet in Inches: ½	⅝	¾	⅞	1	1⅛	1¼
	Length of Rivet in Inches								Length of Rivet in Inches						
½	1⅝	1⅞	1⅞	2	2⅛	…	…	½	1	1	1⅛	1¼	1¼	…	…
⅝	1¾	2	2	2⅛	2¼	…	…	⅝	1⅛	1¼	1¼	1⅜	1⅜	…	…
¾	1⅞	2⅛	2⅛	2¼	2⅜	…	…	¾	1⅜	1⅜	1⅜	1½	1½	…	…
⅞	2	2¼	2¼	2⅜	2½	…	…	⅞	1½	1½	1½	1⅝	1⅝	…	…
1	2¼	2⅜	2⅜	2½	2⅝	2¾	2⅞	1	1⅝	1⅝	1⅝	1¾	1¾	1⅞	1⅞
1⅛	2⅜	2½	2½	2⅝	2¾	2⅞	3	1⅛	1¾	1¾	1⅞	1⅞	1⅞	2	2
1¼	2½	2⅝	2⅝	2¾	2⅞	3	3⅛	1¼	2	2	2	2	2	2⅛	2⅛
1⅜	2⅝	2¾	2¾	2⅞	3	3⅛	3¼	1⅜	2⅛	2⅛	2⅛	2¼	2¼	2⅜	2⅜
1½	2⅞	3	3	3⅛	3¼	3⅜	3½	1½	2¼	2¼	2¼	2⅜	2⅜	2½	2½
1⅝	3	3⅛	3⅛	3¼	3⅜	3½	3½	1⅝	2⅜	2⅜	2⅜	2½	2½	2⅝	2⅝
1¾	3⅛	3¼	3¼	3½	3⅝	3¾	3¾	1¾	2⅝	2⅝	2⅝	2⅝	2⅝	2¾	2¾
1⅞	3¼	3⅜	3⅜	3⅝	3¾	3⅞	3⅞	1⅞	2¾	2¾	2¾	2¾	2¾	2⅞	2⅞
2	3½	3½	3⅝	3¾	3⅞	4	4	2	2⅞	2⅞	2⅞	2⅞	2⅞	3	3
2⅛	3⅝	3⅝	3¾	3⅞	4	4⅛	4⅛	2⅛	3⅛	3	3	3	3	3⅛	3⅛
2¼	3¾	3⅞	3⅞	4	4⅛	4¼	4¼	2¼	3¼	3⅛	3⅛	3⅛	3¼	3¼	3¼
2⅜	4	4	4	4⅛	4¼	4⅜	4⅜	2⅜	3⅜	3⅜	3⅜	3⅜	3⅜	3⅜	3⅜
2½	4⅛	4⅛	4⅛	4¼	4⅜	4½	4½	2½	3½	3½	3½	3½	3½	3⅝	3⅝
2⅝	4¼	4¼	4¼	4⅜	4½	4⅝	4⅝	2⅝	3¾	3⅝	3⅝	3⅝	3⅝	3¾	3¾
2¾	4⅜	4⅜	4⅜	4½	4⅝	4¾	4¾	2¾	3⅞	3¾	3¾	3¾	3¾	3⅞	3⅞
2⅞	4⅝	4⅝	4⅝	4⅝	4¾	4⅞	5	2⅞	4	3⅞	3⅞	3⅞	3⅞	4	4
3	…	4¾	4¾	4⅞	5	5	5⅛	3	…	4⅛	4⅛	4⅛	4⅛	4⅛	4⅛
3⅛	…	4⅞	4⅞	5	5⅛	5¼	5¼	3⅛	…	4¼	4¼	4¼	4¼	4¼	4¼
3¼	…	5	5	5⅛	5¼	5⅜	5⅜	3¼	…	4⅜	4⅜	4⅜	4⅜	4⅜	4⅜
3⅜	…	5⅛	5⅛	5¼	5⅜	5½	5½	3⅜	…	4½	4½	4½	4½	4½	4½
3½	…	5⅜	5⅜	5⅜	5½	5⅝	5⅝	3½	…	4⅝	4⅝	4⅝	4⅝	4⅝	4⅝
3⅝	…	5½	5½	5½	5⅝	5¾	5¾	3⅝	…	4¾	4¾	4¾	4¾	4⅞	4⅞
3¾	…	5⅝	5⅝	5⅝	5¾	5⅞	5⅞	3¾	…	5	5	5	5	5	5
3⅞	…	5¾	5¾	5¾	5⅞	6	6	3⅞	…	5⅛	5⅛	5⅛	5⅛	5⅛	5⅛
4	…	…	5⅞	6	6	6⅛	6¼	4	…	…	5¼	5¼	5¼	5¼	5¼
4⅛	…	…	6	6⅛	6¼	6⅜	6⅜	4⅛	…	…	5⅜	5⅜	5⅜	5⅜	5⅜
4¼	…	…	6⅛	6¼	6½	6½	6½	4¼	…	…	5½	5½	5½	5½	5½
4⅜	…	…	6⅜	6½	6½	6⅝	6⅝	4⅜	…	…	5⅝	5⅝	5⅝	5⅝	5⅝
4½	…	…	6½	6⅝	6⅝	6¾	6¾	4½	…	…	5¾	5¾	5¾	5¾	5¾
4⅝	…	…	6⅝	6¾	6¾	6¾	6⅞	4⅝	…	…	6	6	6	6	6
4¾	…	…	6¾	6⅞	6⅞	7	7	4¾	…	…	6⅛	6⅛	6⅛	6⅛	6⅛
4⅞	…	…	6⅞	7	7	7⅛	7⅛	4⅞	…	…	6¼	6¼	6¼	6¼	6¼
5	…	…	…	7⅛	7⅛	7¼	7¼	5	…	…	…	6⅜	6⅜	6⅜	6⅜
5⅛	…	…	…	7¼	7¼	7⅜	7⅜	5⅛	…	…	…	6½	6½	6½	6½
5¼	…	…	…	7⅜	7⅜	7½	7½	5¼	…	…	…	6⅝	6⅝	6⅝	6⅝
5⅜	…	…	…	7⅝	7⅝	7¾	7⅝	5⅜	…	…	…	6¾	6¾	6¾	6¾
5½	…	…	…	7¾	7¾	7⅞	7⅞	5½	…	…	…	6⅞	6⅞	6⅞	6⅞
5⅝	…	…	…	7⅞	7⅞	8	8	5⅝	…	…	…	7	7	7	7
5¾	…	…	…	8	8	8⅛	8⅛	5¾	…	…	…	7¼	7¼	7¼	7¼
5⅞	…	…	…	8⅛	8⅛	8¼	8¼	5⅞	…	…	…	7⅜	7⅜	7⅜	7⅜

As given by the American Institute of Steel Construction. Values may vary from standard practice of individual fabricators and should be checked against the fabricator's standard.

Standard Rivets

Standards for rivets published by the American National Standards Institute and the British Standards Institution are as follows:

American National Standard Large Rivets.—The types of rivets covered by this standard (ANSI B18.1.2-1972 (R1995)) are shown on pages 1465 and 1466. It may be noted, however, that when specified, the swell neck included in this standard is applicable to all standard large rivets except the flat countersunk head and oval countersunk head types. Also shown are the hold-on (dolly bar) and rivet set impression dimensions (see page 1467). All standard large rivets have fillets under the head not exceeding an 0.062-inch radius. The length tolerances for these rivets are given as follows: through 6 inches in length, ½- and ⅝-inch diameters, ±0.03 inch; ¾- and ⅞-inch diameters, ±0.06-inch; and 1- through 1¾-inch diameters, ±0.09 inch. For rivets over 6 inches in length, ½- and ⅝-inch diameters, ±0.06 inch; ¾- and ⅞-inch diameters, ±0.12 inch; and 1- through 1¾-inch diameters, ±0.19 inch. Steel and wrought iron rivet materials appear in ASTM Specifications A31, A131, A152, and A502.

American National Standard Small Solid Rivets.—The types of rivets covered by this standard (ANSI/ASME B18.1.1-1972 (R1995)) are shown on pages 1468 through 1470. In addition, the standard gives the dimensions of 60-degree flat countersunk head rivets used to assemble ledger plates and guards for mower cutter bars, but these are not shown. As the heads of standard rivets are not machined or trimmed, the circumference may be somewhat irregular and edges may be rounded or flat. Rivets other than countersunk types are furnished with a definite fillet under the head, whose radius should not exceed 10 per cent of the maximum shank diameter or 0.03 inch, whichever is the smaller. With regard to head dimensions, tolerances shown in the dimensional tables are applicable to rivets produced by the normal cold heading process. Unless otherwise specified, rivets should have plain sheared ends that should be at right angles within 2 degrees to the axis of the rivet and be reasonably flat. When so specified by the user, rivets may have the standard header points shown on page 1468. Rivets may be made of ASTM Specification A31, Grade A steel; or may adhere to SAE Recommended Practice, Mechanical and Chemical Requirements for Nonthreaded Fasteners—SAE J430, Grade 0. When specified, rivets may be made of other materials.

ANSI/ASME B18.1.3M-1983 (R1995), Metric Small Solid Rivets, provides data for small, solid rivets with flat, round, and flat countersunk heads in metric dimensions. The main series of rivets has body diameters, in millimeters, of 1.6, 2, 2.5, 3, 4, 5, 6, 8, 10, and 12. A secondary series (nonpreferred) consists of sizes, 1, 1.2, 1.4, 3.5, 7, 9, and 11 millimeters.

British Standard Small Rivets for General Purposes.—Dimensions of small rivets for general purposes are given in British Standard 641:1951 and are shown in the table on page 1472. In addition, the standard lists the standard lengths of these rivets, gives the dimensions of washers to be used with countersunk head rivets (140°), indicates that the rivets may be made from mild steel, copper, brass, and a range of aluminum alloys and pure aluminum specified in B.S. 1473, and gives the dimensions of Coopers' flat head rivets ½ inch in diameter and below, in an appendix. In all types of rivets, except those with countersunk heads, there is a small radius or chamfer at the junction of the head and the shank.

British Standard Dimensions of Rivets (½ to 1¾ inch diameter).—The dimensions of rivets covered in BS 275:1927 (obsolescent) are given on page 1471 and do not apply to boiler rivets. With regard to this standard the terms "nominal diameter" and "standard diameter" are synonymous. The term "tolerance" refers to the variation from the nominal diameter of the rivet and not to the difference between the diameter under the head and the diameter near the point.

American National Standard Large Rivets — 1 *ANSI B18.1.2-1972 (R1995)*

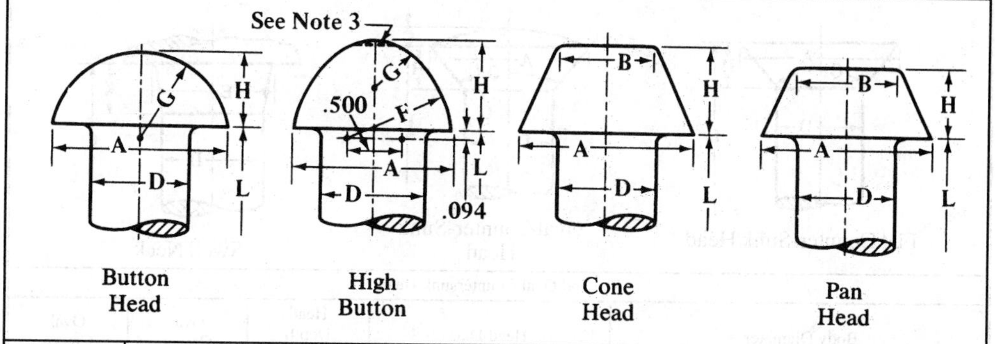

Nom. Body Dia. D^a	Head Dia. *A*		Height *H*		Head Dia. *A*		Height *H*	
	Mfd. Note 1	Driven Note 2	Mfd. Note 1	Driven Note 2	Mfd. Note 1	Driven Note 2	Mfd. Note 1	Driven Note 2
	Button Head				High Button Head (Acorn)			
½	0.875	0.922	0.375	0.344	0.781	0.875	0.500	0.375
⅝	1.094	1.141	0.469	0.438	0.969	1.062	0.594	0.453
¾	1.312	1.375	0.562	0.516	1.156	1.250	0.688	0.531
⅞	1.531	1.594	0.656	0.609	1.344	1.438	0.781	0.609
1	1.750	1.828	0.750	0.688	1.531	1.625	0.875	0.688
1⅛	1.969	2.062	0.844	0.781	1.719	1.812	0.969	0.766
1¼	2.188	2.281	0.938	0.859	1.906	2.000	1.062	0.844
1⅜	2.406	2.516	1.031	0.953	2.094	2.188	1.156	0.938
1½	2.625	2.734	1.125	1.031	2.281	2.375	1.250	1.000
1⅝	2.844	2.969	1.219	1.125	2.469	2.562	1.344	1.094
1¾	3.062	3.203	1.312	1.203	2.656	2.750	1.438	1.172
	Cone Head				Pan Head			
½	0.875	0.922	0.438	0.406	0.800	0.844	0.350	0.328
⅝	1.094	1.141	0.547	0.516	1.000	1.047	0.438	0.406
¾	1.312	1.375	0.656	0.625	1.200	1.266	0.525	0.484
⅞	1.531	1.594	0.766	0.719	1.400	1.469	0.612	0.578
1	1.750	1.828	0.875	0.828	1.600	1.687	0.700	0.656
1⅛	1.969	2.063	0.984	0.938	1.800	1.891	0.788	0.734
1¼	2.188	2.281	1.094	1.031	2.000	2.094	0.875	0.812
1⅜	2.406	2.516	1.203	1.141	2.200	2.312	0.962	0.906
1½	2.625	2.734	1.312	1.250	2.400	2.516	1.050	0.984
1⅝	2.844	2.969	1.422	1.344	2.600	2.734	1.138	1.062
1¾	3.062	3.203	1.531	1.453	2.800	2.938	1.225	1.141

[a] Tolerance for diameter of body is plus and minus from nominal and for ½-in. size equals +0.020, –0.022; for sizes ⅝ to 1-in., incl., equals +0.030, –0.025; for sizes 1⅛ and 1¼-in. equals +0.035; –0.027; for sizes 1⅜ and 1½-in. equals +0.040, –0.030; for sizes 1⅝ and 1¾-in. equals +0.040, –0.037.

All dimensions are given in inches.

Note 1. Basic dimensions of head as manufactured.

Note 2. Dimensions of manufactured head after driving and also of driven head.

Note 3. Slight flat permissible within the specified head-height tolerance.

The following formulas give the basic dimensions for manufactured shapes: *Button Head,* $A = 1.750D$; $H = 0.750D$; $G = 0.885D$. *High Button Head,* $A = 1.500D + 0.031$; $H = 0.750D + 0.125$; $F = 0.750D + 0.281$; $G = 0.750D - 0.281$. *Cone Head,* $A = 1.750D$; $B = 0.938D$; $H = 0.875D$. *Pan Head,* $A = 1.600D$; $B = 1.000D$; $H = 0.700D$. Length *L* is measured parallel to the rivet axis, from the extreme end to the bearing surface plane for flat bearing surface head type rivets, or to the intersection of the head top surface with the head diameter for countersunk head-type rivets.

American National Standard Large Rivets — 2 *ANSI B18.1.2-1972 (R1995)*

Flat Counter-Sunk Head

Oval Counter-Sunk Head

Swell Neck

Flat and Oval Countersunk Head								
Body Diameter *D*				Head Dia. *A*		Head Depth *H*	Oval Crown height[a] *C*	Oval Crown Radius[a] *G*
Nominal[a]		Max.	Min.	Max.[b]	Min.[c]	Ref.		
½	0.500	0.520	0.478	0.936	0.872	0.260	0.095	1.125
⅝	0.625	0.655	0.600	1.194	1.112	0.339	0.119	1.406
¾	0.750	0.780	0.725	1.421	1.322	0.400	0.142	1.688
⅞	0.875	0.905	0.850	1.647	1.532	0.460	0.166	1.969
1	1.000	1.030	0.975	1.873	1.745	0.520	0.190	2.250
1⅛	1.125	1.160	1.098	2.114	1.973	0.589	0.214	2.531
1¼	1.250	1.285	1.223	2.340	2.199	0.650	0.238	2.812
1⅜	1.375	1.415	1.345	2.567	2.426	0.710	0.261	3.094
1½	1.500	1.540	1.470	2.793	2.652	0.771	0.285	3.375
1⅝	1.625	1.665	1.588	3.019	2.878	0.831	0.309	3.656
1¾	1.750	1.790	1.713	3.262	3.121	0.901	0.332	3.938

[a] Basic dimension as manufactured. For tolerances see table footnote on page 1465.
[b] Sharp edged head.
[c] Rounded or flat edged irregularly shaped head (heads are not machined or trimmed).

Swell Neck[a]						
Body Diameter *D*				Diameter Under Head *E*		Neck Length *K*[a]
Nominal[a]		Max.	Min.	Max. (Basic)	Min.	
½	0.500	0.520	0.478	0.563	0.543	0.250
⅝	0.625	0.655	0.600	0.688	0.658	0.312
¾	0.750	0.780	0.725	0.813	0.783	0.375
⅞	0.875	0.905	0.850	0.938	0.908	0.438
1	1.000	1.030	0.975	1.063	1.033	0.500
1⅛	1.125	1.160	1.098	1.188	1.153	0.562
1¼	1.250	1.285	1.223	1.313	1.278	0.625
1⅜	1.375	1.415	1.345	1.438	1.398	0.688
1½	1.500	1.540	1.470	1.563	1.523	0.750
1⅝	1.625	1.665	1.588	1.688	1.648	0.812
1¾	1.750	1.790	1.713	1.813	1.773	0.875

[a] The swell neck is applicable to all standard forms of large rivets except the flat countersunk and oval countersunk head types.

All dimensions are given in inches.

The following formulas give basic dimensions for manufactured shapes: *Flat Countersunk Head,* $A = 1.810D$; $H = 1.192(\text{Max } A - D)/2$; included angle Q of head = 78 degrees. *Oval Countersunk Head,* $A = 1.810D$; $H = 1.192(\text{Max } A - D)/2$; included angle of head = 78 degrees. *Swell Neck,* $E = D + 0.063$; $K = 0.500D$. Length L is measured parallel to the rivet axis, from the extreme end to the bearing surface plane for flat bearing surface head-type rivets, or to the intersection of the head top surface with the head diameter for countersunk head-type rivets.

American National Standard Dimensions for Hold-On (Dolly Bar) and Rivet Set Impression *ANSI B18.1.2-1972 (R1995)*

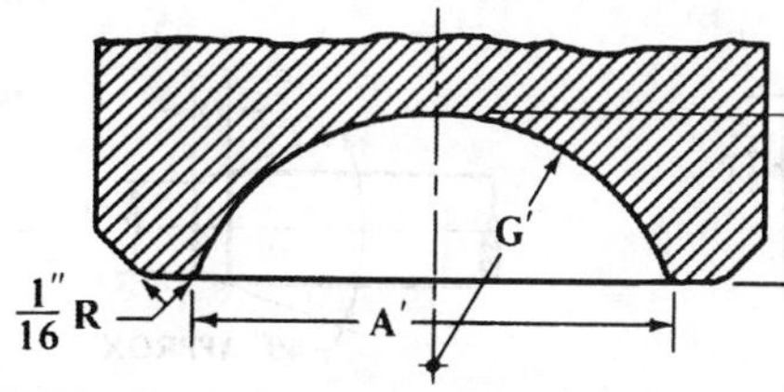

Button Head

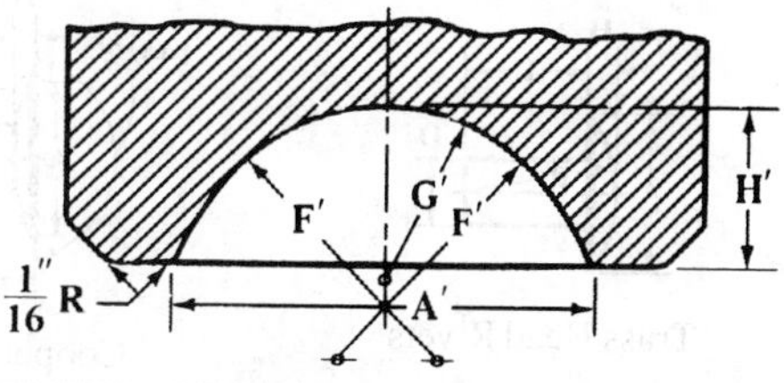

High Button Head

Rivet Body Dia.	Button Head			High Button Head			
	A′	*H′*	*G′*	*A′*	*H′*	*F′*	*G′*
½	0.906	0.312	0.484	0.859	0.344	0.562	0.375
⅝	1.125	0.406	0.594	1.047	0.422	0.672	0.453
¾	1.344	0.484	0.719	1.234	0.500	0.797	0.531
⅞	1.578	0.562	0.844	1.422	0.578	0.922	0.609
1	1.812	0.641	0.953	1.609	0.656	1.031	0.688
1⅛	2.031	0.719	1.078	1.797	0.719	1.156	0.766
1¼	2.250	0.797	1.188	1.984	0.797	1.266	0.844
1⅜	2.469	0.875	1.312	2.172	0.875	1.406	0.938
1½	2.703	0.953	1.438	2.344	0.953	1.500	1.000
1⅝	2.922	1.047	1.547	2.531	1.031	1.641	1.094
1¾	3.156	1.125	1.672	2.719	1.109	1.750	1.172

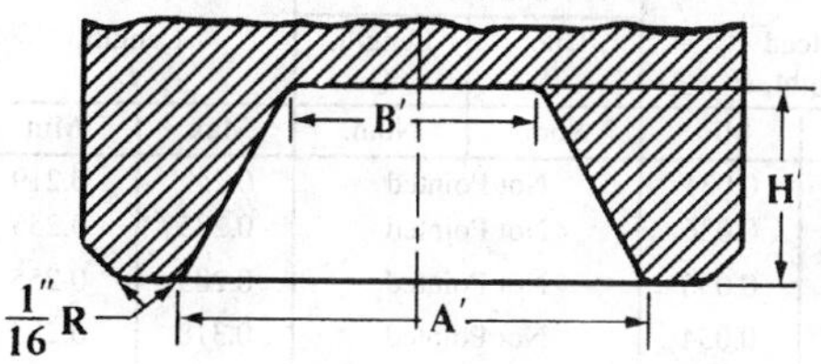

Cone Head

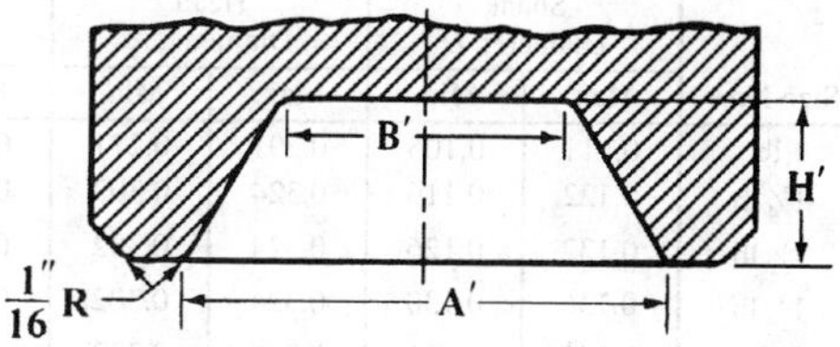

Pan Head

Rivet Body Dia.	Cone Head		
	A′	*B′*	*H′*
½	0.891	0.469	0.391
⅝	1.109	0.594	0.484
¾	1.328	0.703	0.578
⅞	1.562	0.828	0.688
1	1.781	0.938	0.781
1⅛	2.000	1.063	0.875
1¼	2.219	1.172	0.969
1⅜	2.453	1.297	1.078
1½	2.672	1.406	1.172
1⅝	2.891	1.531	1.266
1¾	3.109	1.641	1.375

Rivet Body Dia.	Pan Head		
	A′	*B′*	*H′*
½	0.812	0.500	0.297
⅝	1.031	0.625	0.375
¾	1.234	0.750	0.453
⅞	1.438	0.875	0.531
1	1.641	1.000	0.609
1⅛	1.844	1.125	0.688
1¼	2.047	1.250	0.766
1⅜	2.250	1.375	0.844
1½	2.453	1.500	0.906
1⅝	2.656	1.625	0.984
1¾	2.875	1.750	1.063

All dimensions are given in inches.

Table 1. American National Standard Small Solid Rivets
ANSI/ASME B18.1.1-1972 (R1995) and Appendix

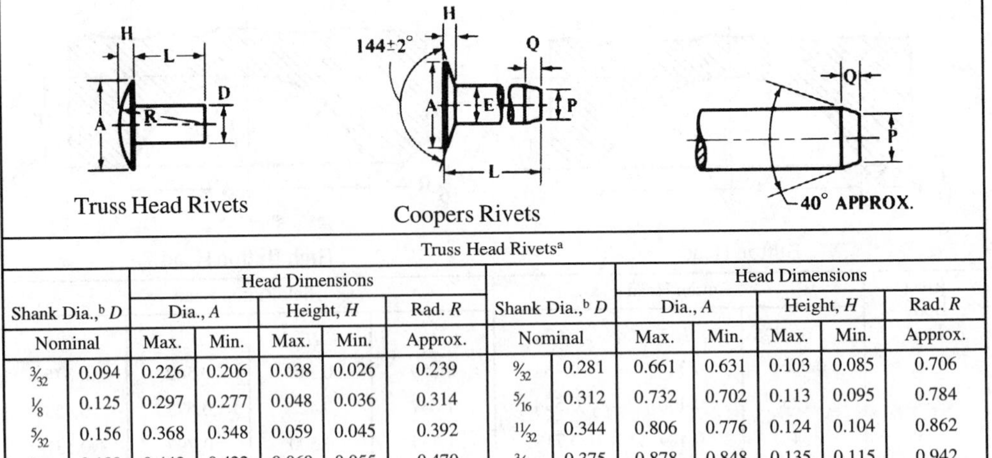

Truss Head Rivets

Coopers Rivets

Truss Head Rivets[a]

Shank Dia.,[b] D		Head Dimensions					Shank Dia.,[b] D		Head Dimensions				
		Dia., A		Height, H		Rad. R			Dia., A		Height, H		Rad. R
Nominal		Max.	Min.	Max.	Min.	Approx.	Nominal		Max.	Min.	Max.	Min.	Approx.
3/32	0.094	0.226	0.206	0.038	0.026	0.239	9/32	0.281	0.661	0.631	0.103	0.085	0.706
1/8	0.125	0.297	0.277	0.048	0.036	0.314	5/16	0.312	0.732	0.702	0.113	0.095	0.784
5/32	0.156	0.368	0.348	0.059	0.045	0.392	11/32	0.344	0.806	0.776	0.124	0.104	0.862
3/16	0.188	0.442	0.422	0.069	0.055	0.470	3/8	0.375	0.878	0.848	0.135	0.115	0.942
7/32	0.219	0.515	0.495	0.080	0.066	0.555	13/32	0.406	0.949	0.919	0.145	0.123	1.028
1/4	0.250	0.590	0.560	0.091	0.075	0.628	7/16	0.438	1.020	0.990	0.157	0.135	1.098

[a] Length tolerance of rivets is + or − .016 inch. Approximate proportions of rivets: $A = 2.300 \times D$, $H = 0.330 \times D$, $R = 2.512 \times D$.

[b] Tolerances on the nominal shank diameter in inches are given for the following body diameter ranges: 3/32 to 5/32, plus 0.002, minus 0.004; 3/16 to 1/4, plus 0.003, minus 0.006; 9/32 to 11/32, plus 0.004, minus 0.008; and 3/8 to 7/16, plus 0.005, minus 0.010.

Coopers Rivets

Size No.[a]	Shank Diameter, D		Head Diameter, A		Head Height, H		Point Dimensions Dia., P	Point Dimensions Length, Q	Length, L	
	Max.	Min.	Max.	Min.	Max.	Min.	Nom.	Nom.	Max.	Min.
1 lb	0.111	0.105	0.291	0.271	0.045	0.031	Not Pointed		0.249	0.219
1 1/4 lb	0.122	0.116	0.324	0.302	0.050	0.036	Not Pointed		0.285	0.255
1 1/2 lb	0.132	0.126	0.324	0.302	0.050	0.036	Not Pointed		0.285	0.255
1 3/4 lb	0.136	0.130	0.324	0.302	0.052	0.034	Not Pointed		0.318	0.284
2 lb	0.142	0.136	0.355	0.333	0.056	0.038	Not Pointed		0.322	0.288
3 lb	0.158	0.152	0.386	0.364	0.058	0.040	0.123	0.062	0.387	0.353
4 lb	0.168	0.159	0.388	0.362	0.058	0.040	0.130	0.062	0.418	0.388
5 lb	0.183	0.174	0.419	0.393	0.063	0.045	0.144	0.062	0.454	0.420
6 lb	0.206	0.197	0.482	0.456	0.073	0.051	0.160	0.094	0.498	0.457
7 lb	0.223	0.214	0.513	0.487	0.076	0.054	0.175	0.094	0.561	0.523
8 lb	0.241	0.232	0.546	0.516	0.081	0.059	0.182	0.094	0.597	0.559
9 lb	0.248	0.239	0.578	0.548	0.085	0.063	0.197	0.094	0.601	0.563
10 lb	0.253	0.244	0.578	0.548	0.085	0.063	0.197	0.094	0.632	0.594
12 lb	0.263	0.251	0.580	0.546	0.086	0.060	0.214	0.094	0.633	0.575
14 lb	0.275	0.263	0.611	0.577	0.091	0.065	0.223	0.094	0.670	0.612
16 lb	0.285	0.273	0.611	0.577	0.089	0.063	0.223	0.094	0.699	0.641
18 lb	0.285	0.273	0.642	0.608	0.108	0.082	0.230	0.125	0.749	0.691
20 lb	0.316	0.304	0.705	0.671	0.128	0.102	0.250	0.125	0.769	0.711
3/8 in.	0.380	0.365	0.800	0.762	0.136	0.106	0.312	0.125	0.840	0.778

[a] Size numbers in pounds refer to the approximate weight of 1000 rivets.

When specified American National Standard Small Solid Rivets may be obtained with points. Point dimensions for belt and coopers rivets are given in the accompanying tables. Formulas for calculating point dimensions of other rivets are given with the diagram alongside

All dimensions in inches except where otherwise noted.

Table 2. American National Standard Small Solid Rivets

ANSI/ASME B18.1.1-1972 (R1995)

Tinners Rivets

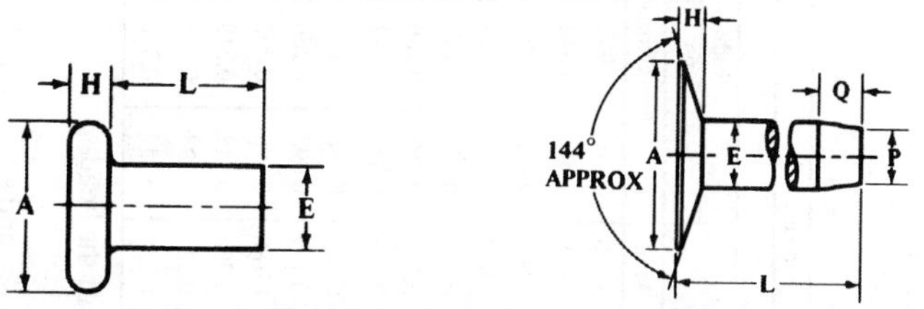

Belt Rivets

Tinners Rivets									
Size No.[a]	Shank Diameter, *E*		Head Dia., *A*		Head Height, *H*		Length, *L*		
	Max.	Min.	Max.	Min.	Max.	Min.	Nom.	Max.	Min.
6 oz.	0.081	0.075	0.213	0.193	0.028	0.016	1/8	0.135	0.115
8 oz.	0.091	0.085	0.225	0.205	0.036	0.024	5/32	0.166	0.146
10 oz.	0.097	0.091	0.250	0.230	0.037	0.025	11/64	0.182	0.162
12 oz.	0.107	0.101	0.265	0.245	0.037	0.025	3/16	0.198	0.178
14 oz.	0.111	0.105	0.275	0.255	0.038	0.026	3/16	0.198	0.178
1 lb	0.113	0.107	0.285	0.265	0.040	0.028	13/64	0.213	0.193
1 1/4 lb	0.122	0.116	0.295	0.275	0.045	0.033	7/32	0.229	0.209
1 1/2 lb	0.132	0.126	0.316	0.294	0.046	0.034	15/64	0.244	0.224
1 3/4 lb	0.136	0.130	0.331	0.309	0.049	0.035	1/4	0.260	0.240
2 lb	0.146	0.140	0.341	0.319	0.050	0.036	17/64	0.276	0.256
2 1/2 lb	0.150	0.144	0.311	0.289	0.069	0.055	9/32	0.291	0.271
3 lb	0.163	0.154	0.329	0.303	0.073	0.059	5/16	0.323	0.303
3 1/2 lb	0.168	0.159	0.348	0.322	0.074	0.060	21/64	0.338	0.318
4 lb	0.179	0.170	0.368	0.342	0.076	0.062	11/32	0.354	0.334
5 lb	0.190	0.181	0.388	0.362	0.084	0.070	3/8	0.385	0.365
6 lb	0.206	0.197	0.419	0.393	0.090	0.076	25/64	0.401	0.381
7 lb	0.223	0.214	0.431	0.405	0.094	0.080	13/32	0.416	0.396
8 lb	0.227	0.218	0.475	0.445	0.101	0.085	7/16	0.448	0.428
9 lb	0.241	0.232	0.490	0.460	0.103	0.087	29/64	0.463	0.443
10 lb	0.241	0.232	0.505	0.475	0.104	0.088	15/32	0.479	0.459
12 lb	0.263	0.251	0.532	0.498	0.108	0.090	1/2	0.510	0.490
14 lb	0.288	0.276	0.577	0.543	0.113	0.095	33/64	0.525	0.505
16 lb	0.304	0.292	0.597	0.563	0.128	0.110	17/32	0.541	0.521
18 lb	0.347	0.335	0.706	0.668	0.156	0.136	19/32	0.603	0.583

[a] Size numbers refer to the approximate weight of 1000 rivets.

Belt Rivets[a]								
Size No.[b]	Shank Diameter, *E*		Head Dia., *A*		Head Height, *H*		Point Dimensions	
							Dia., *P*	Length, *Q*
	Max.	Min.	Max.	Min.	Max.	Min.	Nominal	Nominal
14	0.085	0.079	0.260	0.240	0.042	0.030	0.065	0.078
13	0.097	0.091	0.322	0.302	0.051	0.039	0.073	0.078
12	0.111	0.105	0.353	0.333	0.054	0.040	0.083	0.078
11	0.122	0.116	0.383	0.363	0.059	0.045	0.097	0.078
10	0.136	0.130	0.417	0.395	0.065	0.047	0.109	0.094
9	0.150	0.144	0.448	0.426	0.069	0.051	0.122	0.094
8	0.167	0.161	0.481	0.455	0.072	0.054	0.135	0.094
7	0.183	0.174	0.513	0.487	0.075	0.056	0.151	0.125
6	0.206	0.197	0.606	0.580	0.090	0.068	0.165	0.125
5	0.223	0.214	0.700	0.674	0.105	0.083	0.185	0.125
4	0.241	0.232	0.921	0.893	0.138	0.116	0.204	0.141

[a] Length tolerance on belt rivets is plus 0.031 inch, minus 0 inch.

[b] Size number refers to the Stub's iron wire gage number of the stock used in the shank of the rivet.

All dimensions in inches.

Note: American National Standard Small Solid Rivets may be obtained with or without points. Point proportions are given in the diagram in Table 1.

Table 3. American National Standard Small Solid Rivets *ANSI/ASME B18.1.1-1972 (R1995) and Appendix*

Shank Diameter				Flat Head				Flat Countersunk Head			Button Head					Pan Head						
				Head Dimensions																		
D		*E*		Dia., *A*		Height, *H*		Dia., *A* Sharp		Height[a] *H*	Dia., *A*		Height, *H*		Radius, *R*	Dia., *A*		Height, *H*		Radii *R1*	*R2*	*R3*
Nominal		Max.	Min.	Max.	Min.	Max.	Min.	Max.[b]	Min.[c]	Ref.	Max.	Min.	Max.	Min.	Approx.	Max.	Min.	Max.	Min.	Approximate		
1/16	0.062	0.064	0.059	0.140	0.120	0.027	0.017	0.118	0.110	0.027	0.122	0.102	0.052	0.042	0.055	0.118	0.098	0.040	0.030	0.019	0.052	0.217
3/32	0.094	0.096	0.090	0.200	0.180	0.038	0.026	0.176	0.163	0.040	0.182	0.162	0.077	0.065	0.084	0.173	0.153	0.060	0.048	0.030	0.080	0.326
1/8	0.125	0.127	0.121	0.260	0.240	0.048	0.036	0.235	0.217	0.053	0.235	0.215	0.100	0.088	0.111	0.225	0.205	0.078	0.066	0.039	0.106	0.429
5/32	0.156	0.158	0.152	0.323	0.301	0.059	0.045	0.293	0.272	0.066	0.290	0.268	0.124	0.110	0.138	0.279	0.257	0.096	0.082	0.049	0.133	0.535
3/16	0.188	0.191	0.182	0.387	0.361	0.069	0.055	0.351	0.326	0.079	0.348	0.322	0.147	0.133	0.166	0.334	0.308	0.114	0.100	0.059	0.159	0.641
7/32	0.219	0.222	0.213	0.453	0.427	0.080	0.065	0.413	0.384	0.094	0.405	0.379	0.172	0.158	0.195	0.391	0.365	0.133	0.119	0.069	0.186	0.754
1/4	0.250	0.253	0.244	0.515	0.485	0.091	0.075	0.469	0.437	0.106	0.460	0.430	0.196	0.180	0.221	0.444	0.414	0.151	0.135	0.079	0.213	0.858
9/32	0.281	0.285	0.273	0.579	0.545	0.103	0.085	0.528	0.491	0.119	0.518	0.484	0.220	0.202	0.249	0.499	0.465	0.170	0.152	0.088	0.239	0.963
5/16	0.312	0.316	0.304	0.641	0.607	0.113	0.095	0.588	0.547	0.133	0.572	0.538	0.243	0.225	0.276	0.552	0.518	0.187	0.169	0.098	0.266	1.070
11/32	0.344	0.348	0.336	0.705	0.667	0.124	0.104	0.646	0.602	0.146	0.630	0.592	0.267	0.247	0.304	0.608	0.570	0.206	0.186	0.108	0.292	1.176
3/8	0.375	0.380	0.365	0.769	0.731	0.135	0.115	0.704	0.656	0.159	0.684	0.646	0.291	0.271	0.332	0.663	0.625	0.225	0.205	0.118	0.319	1.286
13/32	0.406	0.411	0.396	0.834	0.790	0.146	0.124	0.763	0.710	0.172	0.743	0.699	0.316	0.294	0.358	0.719	0.675	0.243	0.221	0.127	0.345	1.392
7/16	0.438	0.443	0.428	0.896	0.852	0.157	0.135	0.823	0.765	0.186	0.798	0.754	0.339	0.317	0.387	0.772	0.728	0.261	0.239	0.137	0.372	1.500

[a] Given for reference purposes only. Variations in this dimension are controlled by the head and shank diameters and the included angle of the head.

[b] Tabulated maximum values calculated on basic diameter of rivet and 92° included angle extended to a sharp edge.

[c] Minimum of rounded or flat-edged irregular-shaped head. Rivet heads are not machined or trimmed and the circumference may be irregular and edges rounded or flat.

All dimensions in inches. Length tolerance of all rivets is plus or minus 0.016 inch. Approximate proportions of rivets: flat head, $A = 2.00 \times D$, $H = 0.33\ D$; flat countersunk head, $A = 1.850 \times D$, $H = 0.425 \times D$; button head, $A = 1.750 \times D$, $H = 0.750 \times D$, $R = 0.885 \times D$; pan head, $A = 1.720 \times D$, $H = 0.570 \times D$, $R1 = 0.314 \times D$, $R2 = 0.850 \times D$, $R3 = 3.430 \times D$.

Note: ANSI Small Solid Rivets may be obtained with or without points. Point proportions are given in the diagram in Table 1.

Head Dimensions and Diameters of British Standard Rivets

BS 275:1927 (obsolescent) This standard does not apply to Boiler Rivets

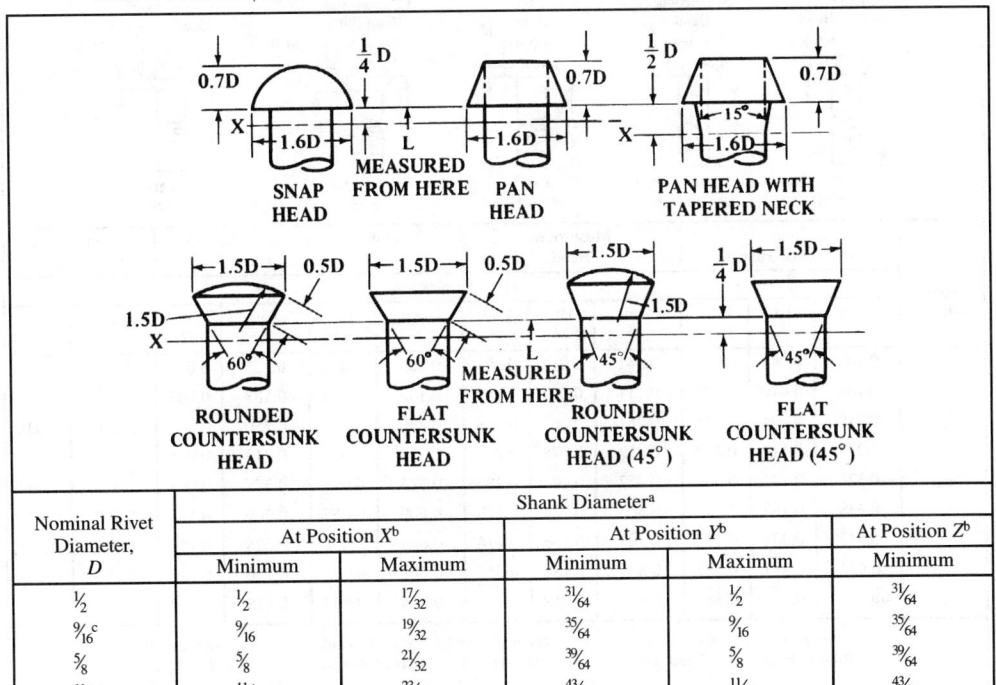

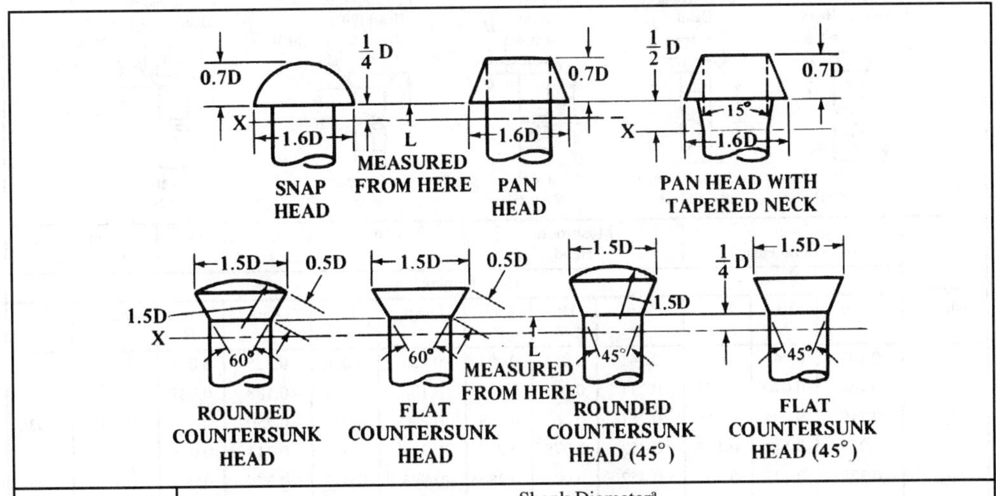

Nominal Rivet Diameter, *D*	Shank Diameter[a]				
	At Position *X*[b]		At Position *Y*[b]		At Position *Z*[b]
	Minimum	Maximum	Minimum	Maximum	Minimum
½	½	17⁄32	31⁄64	½	31⁄64
9⁄16[c]	9⁄16	19⁄32	35⁄64	9⁄16	35⁄64
⅝	⅝	21⁄32	39⁄64	⅝	39⁄64
11⁄16[c]	11⁄16	23⁄32	43⁄64	11⁄16	43⁄64
¾	¾	25⁄32	47⁄64	¾	47⁄64
13⁄16[c]	13⁄16	27⁄32	51⁄64	13⁄16	51⁄64
⅞	⅞	29⁄32	55⁄64	⅞	55⁄64
15⁄16[c]	15⁄16	31⁄32	59⁄64	15⁄16	59⁄64
1	1	1 1⁄32	63⁄64	1	63⁄64
1 1⁄16[c]	1 1⁄16	1 3⁄32	1 3⁄64	1 1⁄16	1 3⁄64
1⅛	1⅛	1 5⁄32	1 7⁄64	1⅛	1 7⁄64
1 3⁄16[c]	1 3⁄16	1 7⁄32	1 11⁄64	1 3⁄16	1 11⁄64
1¼	1¼	1 9⁄32	1 15⁄64	1¼	1 15⁄64
1 5⁄16[c]	1 5⁄16	1 11⁄32	1 19⁄64	1 5⁄16	1 19⁄64
1⅜	1⅜	1 13⁄32	1 23⁄64	1⅜	1 23⁄64
1 7⁄16[c]	1 7⁄16	1 15⁄32	1 27⁄64	1 7⁄16	1 27⁄64
1½	1½	1 17⁄32	1 31⁄64	1½	1 31⁄64
1 9⁄16[c]	1 9⁄16	1 19⁄32	1 35⁄64	1 9⁄16	1 35⁄64
1⅝	1⅝	1 21⁄32	1 39⁄64	1⅝	1 39⁄64
1 11⁄16[c]	1 11⁄16	1 23⁄32	1 43⁄64	1 11⁄16	1 43⁄64
1¾	1¾	1 25⁄32	1 47⁄64	1¾	1 47⁄64

[a] Tolerances of the rivet diameter are as follows: at position *X*, plus 1⁄32 inch, minus zero; at position *Y*, plus zero, minus 1⁄64 inch; at position *Z*, minus 1⁄64 inch but in no case shall the difference between the diameters at positions *X* and *Y* exceed 1⁄32 inch, nor shall the diameter of the shank between positions *X* and *Y* be less than the minimum diameter specified at position *Y*.

[b] The location of positions *Y* and *Z* are as follows: Position *Y* is located ½*D* from the end of the rivet for rivet lengths 5 diameters long and under. For longer rivets, position *Y* is located 4½*D* from the head of the rivet. Position *Z* (found only on rivets longer than 5*D*) is located ½*D* from the end of the rivet.

[c] At the recommendation of the British Standards Institution, these sizes are to be dispensed with wherever possible.

All dimensions that are tabulated are given in inches.

British Standard Small Rivets for General Purposes *BS 641:1951 (obsolescent)*

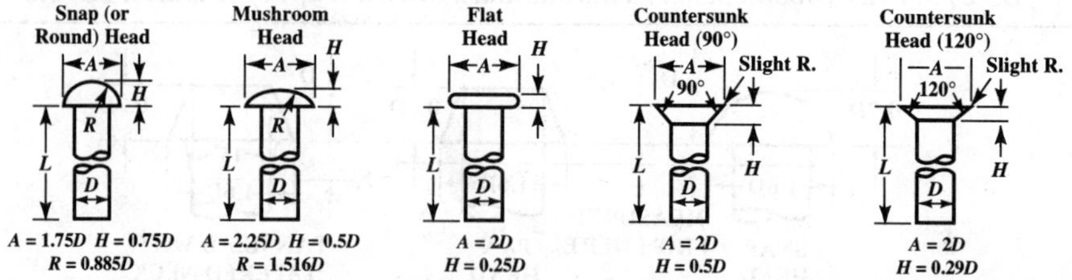

Snap (or Round) Head: $A = 1.75D$ $H = 0.75D$ $R = 0.885D$

Mushroom Head: $A = 2.25D$ $H = 0.5D$ $R = 1.516D$

Flat Head: $A = 2D$ $H = 0.25D$

Countersunk Head (90°): $A = 2D$ $H = 0.5D$

Countersunk Head (120°): $A = 2D$ $H = 0.29D$

Nom. Diam. D	Snap (or Round Head)			Mushroom Head			Flat Head		Countersunk Head (90°)		Countersunk Head (120°)	
	Head Dimensions											
	Diam. A	Ht. H	Rad. R	Diam. A	Ht. H	Rad. R	Diam. A	Ht. H	Diam. A	Ht. H	Diam. A	Ht. H
1⁄16	0.109	0.047	0.055	0.141	0.031	0.095	0.125	0.016	0.125	0.031	…	…
3⁄32	0.164	0.070	0.083	0.211	0.047	0.142	0.188	0.023	0.188	0.047	…	…
1⁄8	0.219	0.094	0.111	0.281	0.063	0.189	0.250	0.031	0.250	0.063	0.250	0.036
5⁄32	0.273	0.117	0.138	0.352	0.078	0.237	0.313	0.039	0.313	0.078	…	…
3⁄16	0.328	0.141	0.166	0.422	0.094	0.284	0.375	0.047	0.375	0.094	0.375	0.054
1⁄4	0.438	0.188	0.221	0.563	0.125	0.379	0.500	0.063	0.500	0.125	0.500	0.073
5⁄16	0.547	0.234	0.277	0.703	0.156	0.474	0.625	0.078	0.625	0.156	0.625	0.091
3⁄8	0.656	0.281	0.332	0.844	0.188	0.568	0.750	0.094	0.750	0.188	0.750	0.109
7⁄16	0.766	0.328	0.387	0.984	0.219	0.663	0.875	0.109	0.875	0.219	…	…

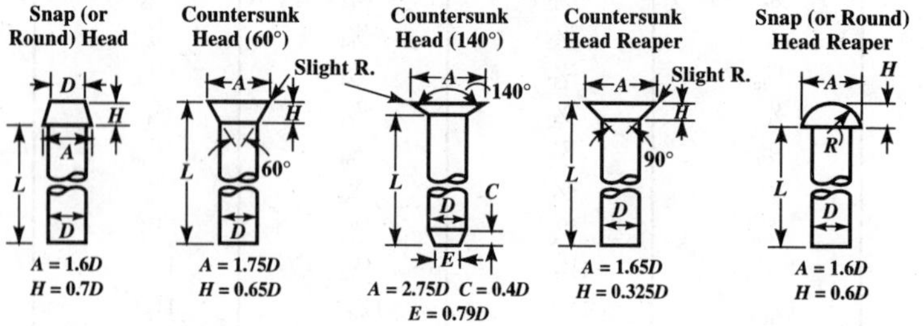

Snap (or Round) Head: $A = 1.6D$ $H = 0.7D$

Countersunk Head (60°): $A = 1.75D$ $H = 0.65D$

Countersunk Head (140°): $A = 2.75D$ $C = 0.4D$ $E = 0.79D$

Countersunk Head Reaper: $A = 1.65D$ $H = 0.325D$

Snap (or Round) Head Reaper: $A = 1.6D$ $H = 0.6D$

Nominal Diameter D		Pan Head		Countersunk Head (60°)		Countersunk Head (140°)			Countersunk Head Reaper		Snap (or Round) Head Reaper	
		Head Dimensions										
Inch	Gage No.[a]	Diam. A	Ht. H	Diam. A	Ht. H	Diam. A	Ht. C	Diam. E	Diam. A	Ht. H	Diam. A	Ht. H
0.104	12	…	…	…	…	0.286	0.042	0.082	…	…	…	…
0.116	11	…	…	…	…	0.319	0.046	0.092	…	…	…	…
0.128	10	…	…	…	…	0.352	0.051	0.101	…	…	…	…
0.144	9	…	…	…	…	0.396	0.058	0.114	…	…	…	…
0.160	8	…	…	…	…	0.440	0.064	0.126	…	…	…	…
0.176	7	…	…	…	…	0.484	0.070	0.139	…	…	…	…
3⁄16	…	0.300	0.131	0.328	0.122	…	…	…	…	…	…	…
0.192	6	…	…	…	…	0.528	0.077	0.152	0.317	0.062	0.307	0.115
0.202	…	…	…	…	…	…	…	…	0.333	0.066	0.323	0.121
0.212	5	…	…	…	…	0.583	0.085	0.167	0.350	0.069	0.339	0.127
0.232	4	…	…	…	…	0.638	0.093	0.183	0.383	0.075	0.371	0.139
1⁄4	…	0.400	0.175	0.438	0.162	0.688	0.100	0.198	…	…	…	…
0.252	3	…	…	…	…	…	…	…	0.416	0.082	0.403	0.151
5⁄16	…	0.500	0.219	0.547	0.203	0.859	0.125	0.247	…	…	…	…
3⁄8	…	0.600	0.263	0.656	0.244	10.031	0.150	0.296	…	…	…	…
7⁄16	…	0.700	0.306	0.766	0.284	…	…	…	…	…	…	…

[a] Gage numbers are British Standard Wire Gage (S.W.G.) numbers.

All dimensions in inches unless specified otherwise0.

British Standard Rivets for General Engineering.—Dimensions in metric units of rivets for general engineering purposes are given in this British Standard, BS 4620;1970, which is based on ISO Recommendation ISO/R 1051. The snap head rivet dimensions of 14 millimeters and above are taken from the German Standard DIN 124, Round Head Rivets for Steel Structures. The shapes of heads have been restricted to those in common use in the United Kingdom. Table 2 shows the rivet dimensions. Table 1 shows a tentative range of preferred nominal lengths as given in an appendix to the Standard. It is stated that these lengths will be reviewed in the light of usage. The rivets are made by cold or hot forging methods from mild steel, copper, brass, pure aluminum, aluminum alloys, or other suitable metal. It is stated that the radius under the head of a rivet shall run smoothly into the face of the head and shank without step or discontinuity.

In this Standard, the following definitions apply: 1) *Nominal diameter:* The diameter of the shank; 2) *Nominal length of rivets other than countersunk or raised countersunk rivets:* The length from the underside of the head to the end of the shank; 3) *Nominal length of countersunk and raised countersunk rivets:* The distance from the periphery of the head to the end of the rivet measured parallel to the axis of the rivet; and 4) *Manufactured head:* The head on the rivet as received from the manufacturer.

Table 1. Tentative Range of Lengths for Rivets *Appendix to BS 4620:1970 (1998)*

Nom. Shank Dia.	Nominal Length																				
	3	4	5	6	8	10	12	14	16	(18)	20	(22)	25	(28)	30	(32)	35	(38)	40	45	..
1	X	X	X	X	X	X	X	X	X	..	X	..	..	..	..	..	..	..	..	..	..
1.2	X	X	X	X	X	X	X	X	X	..	X	..	..	..	..	..	..	..	..	..	..
1.6	X	X	X	X	X	X	X	X	X	X	..	X	X	..	..	..	..	..	..	..	..
2	X	X	X	X	X	X	X	X	X	..	X	X	X	..	..	..	..	..	..	..	..
2.5	X	X	X	X	X	X	X	X	X	X	..	X	X	..	..	..	..	..	..	..	..
3	..	X	X	X	X	X	X	X	X	X	X	X	X	..	..	..	..	..	..	..	..
(3.5)	..	..	..	..	..	..	..	..	..	..	..	..	..	..	..	..	..	..	..	..	..
4	..	..	..	X	X	X	X	X	X	X	..	X	X	..	..	..	..	..	..	..	..
5	..	..	..	X	X	X	X	X	X	X	X	X	X	X	X	..	X	X	..	..	..
6	..	..	..	X	X	X	X	X	X	X	X	X	X	X	X	..	X	X	..	X	..
Nom. Shank Dia.	Nominal Length																				
	10	12	14	16	(18)	20	(22)	25	(28)	30	(32)	35	(38)	40	45	50	55	60	65	70	75
(7)	..	..	..	..	..	..	..	..	..	..	..	..	..	..	..	..	..	..	..	..	..
8	X	X	X	X	X	X	X	X	X	X	..	X	X	..	X	X	..	..	..	..	..
10	..	..	X	X	X	X	X	X	X	X	..	X	X	X	X	X	..	..	..	..	..
12	..	..	..	..	..	..	..	X	..	X	..	X	..	X	..	X	X	..	..	..	X
(14)	..	..	..	..	..	..	..	..	..	..	..	..	..	..	..	..	..	..	..	..	..
16	..	..	..	..	..	..	..	..	..	..	..	..	..	X	X	X	X	X	X	..	X
Nom. Shank Dia.	Nominal Length																				
	45	50	55	60	65	70	75	80	85	90	(95)	100	(105)	110	(115)	120	(125)	130	140	150	160
(18)	..	..	..	..	..	..	..	..	..	..	..	..	..	..	..	..	..	..	..	..	..
20	X	..	..	X	..	X	..	X	..	..	..	..	..	..	..	..	..	..	..	..	..
(22)	..	..	..	..	..	..	..	..	..	..	..	..	..	..	..	..	..	..	..	..	..
24	..	..	..	..	X	..	X	..	X	X	..	X	..	..	..	..	..	..	..	..	..
(27)	..	..	..	..	..	..	..	..	..	..	..	..	..	..	..	..	..	..	..	..	..
30	..	..	..	..	..	..	..	..	X	X	..	X	..	X	..	X	..	..	..	..	..
(33)	..	..	..	..	..	..	..	..	..	..	..	..	..	..	..	..	..	..	..	..	..
36	..	..	..	..	..	..	..	..	..	..	..	X	..	X	..	X	..	X	..	..	..
(39)	..	..	..	..	..	..	..	..	..	..	..	..	..	..	..	X	..	X	X	X	X

All dimensions are in millimeters.

Note: Sizes and lengths shown in parenthesis are nonpreferred and should be avoided if possible.

Table 2. British Standard Rivets for General Engineering Purposes
BS 4620:1970 (1998)

60° Csk. and Raised Csk. Head — 60° ± 2½° — **60° Csk. Head** — 90° ± 2½° — **Snap Head** — **Universal Head** — **Flat Head**

***K = 0.43 d (for ref. only)**   **†K = 0.5 d (for ref. only)**

Hot Forged Rivets									
		60° Csk. and Raised Csk. Head		Snap Head		Universal Head			
		Head Dimensions							
Nom. Shank Diam. *d*	Tol. on Dia. *d*	Nom. Dia. *D*	Height of Raise *W*	Nom. Dia. *D*	Nom. Depth *K*	Nom. Dia. *D*	Nom. Depth *K*	Rad. *R*	Rad. *r*
(14)		21	2.8	22	9	28	5.6	42	8.4
16	±0.43	24	3.2	25	10	32	6.4	48	9.6
(18)		27	3.6	28	11.5	36	7.2	54	11
20		30	4.0	32	13	40	8.0	60	12
(22)	±0.52	33	4.4	36	14	44	8.8	66	13
24		36	4.8	40	16	48	9.6	72	14
(27)		40	5.4	43	17	54	10.8	81	16
30		45	6.0	48	19	60	12.0	90	18
(33)	±0.62	50	6.6	53	21	66	13.2	99	20
36		55	7.2	58	23	72	14.4	108	22
(39)		59	7.8	62	25	78	15.6	117	23

Cold Forged Rivets										
		90° Csk. Head	Snap Head		Universal Head				Flat Head	
		Head Dimensions								
Nom. Shank Dia. *d*	Tol. on Dia. *d*	Nom. Dia. *D*	Nom. Dia. *D*	Nom. Dia. *K*	Nom. Dia. *D*	Nom. Depth *K*	Rad. *R*	Rad. *r*	Nom. Dia. *D*	Nom. Depth *K*
1		2	1.8	0.6	2	0.4	3.0	0.6	2	0.25
1.2		2.4	2.1	0.7	2.4	0.5	3.6	0.7	2.4	0.3
1.6	±0.07	3.2	2.8	1.0	3.2	0.6	4.8	1.0	3.2	0.4
2		4	3.5	1.2	4	0.8	6.0	1.2	4	0.5
2.5		5	4.4	1.5	5	1.0	7.5	1.5	5	0.6
3		6	5.3	1.8	6	1.2	9.0	1.8	6	0.8
(3.5)		7	6.1	2.1	7	1.4	10.5	2.1	7	0.9
4	±0.09	8	7	2.4	8	1.6	12	2.4	8	1.0
5		10	8.8	3.0	10	2.0	15	3.0	10	1.3
6		12	10.5	3.6	12	2.4	18	3.6	12	1.5
(7)		14	12.3	4.2	14	2.8	21	4.2	14	1.8
8	±0.11	16	14	4.8	16	3.2	24	4.8	16	2
10		20	18	6.0	20	4.0	30	6	20	2.5
12		24	21	7.2	24	4.8	36	7.2	…	…
(14)	±0.14	…	25	8.4	28	5.6	42	8.4	…	…
16		…	28	9.6	32	6.4	48	9.6	…	…

All dimensions are in millimeters. Sizes shown in parentheses are nonpreferred.

TORQUE AND TENSION IN FASTENERS

Tightening Bolts: Bolts are often tightened by applying torque to the head or nut, which causes the bolt to stretch. The stretching results in bolt tension or preload, which is the force that holds a joint together. Torque is relatively easy to measure with a torque wrench, so it is the most frequently used indicator of bolt tension. Unfortunately, a torque wrench does not measure bolt tension accurately, mainly because it does not take friction into account. The friction depends on bolt, nut, and washer material, surface smoothness, machining accuracy, degree of lubrication, and the number of times a bolt has been installed. Fastener manufacturers often provide information for determining torque requirements for tightening various bolts, accounting for friction and other effects. If this information is not available, the methods described in what follows give general guidelines for determining how much tension should be present in a bolt, and how much torque may need to be applied to arrive at that tension.

High preload tension helps keep bolts tight, increases joint strength, creates friction between parts to resist shear, and improves the fatigue resistance of bolted connections. The recommended preload F_i, which can be used for either static (stationary) or fatigue (alternating) applications, can be determined from: $F_i = 0.75 \times A_t \times S_p$ for reusable connections, and $F_i = 0.9 \times A_t \times S_p$ for permanent connections. In these formulas, F_i is the bolt preload, A_t is the tensile stress area of the bolt, and S_p is the proof strength of the bolt. Determine A_t from screw-thread tables or by means of formulas in this section. Proof strength S_p of commonly used ASTM and SAE steel fasteners is given in this section and in the section on metric screws and bolts for those fasteners. For other materials, an approximate value of proof strength can be obtained from: $S_p = 0.85 \times S_y$, where S_y is the yield strength of the material. Soft materials should not be used for threaded fasteners.

Once the required preload has been determined, one of the best ways to be sure that a bolt is properly tensioned is to measure its tension directly with a strain gage. Next best is to measure the change in length (elongation) of the bolt during tightening, using a micrometer or dial indicator. Each of the following two formulas calculates the required change in length of a bolt needed to make the bolt tension equal to the recommended preload. The change in length δ of the bolt is given by:

$$\delta = F_i \times \frac{A_d \times l_t + A_t \times l_d}{A_d \times A_t \times E} \quad (1) \qquad \text{or} \qquad \delta = \frac{F_i \times l}{A \times E} \quad (2)$$

In Equation (1), F_i is the bolt preload; A_d is the major-diameter area of the bolt; A_t is the tensile-stress area of the bolt; E is the bolt modulus of elasticity; l_t is the length of the threaded portion of the fastener within the grip; and l_d is the length of the unthreaded portion of the grip. Here, the grip is defined as the total thickness of the clamped material. Equation (2) is a simplified formula for use when the area of the fastener is constant, and gives approximately the same results as Equation (1). In Equation (2), l is the bolt length; A is the bolt area; and δ, F_i, and E are as described before.

If measuring bolt elongation is not possible, the torque necessary to tighten the bolt must be estimated. If the recommended preload is known, use the following general relation for the torque: $T = K \times F_i \times d$, where T is the wrench torque, K is a constant that depends on the bolt material and size, F_i is the preload, and d is the nominal bolt diameter. A value of K = 0.2 may be used in this equation for mild-steel bolts in the size range of ¼ to 1 inch. For other steel bolts, use the following values of K: nonplated black finish, 0.3; zinc-plated, 0.2; lubricated, 0.18; cadmium-plated, 0.16. Check with bolt manufacturers and suppliers for values of K to use with bolts of other sizes and materials.

The proper torque to use for tightening bolts in sizes up to about ½ inch may also be determined by trial. Test a bolt by measuring the amount of torque required to fracture it (use bolt, nut, and washers equivalent to those chosen for the real application). Then, use a tightening torque of about 50 to 60 per cent of the fracture torque determined by the test. The tension in a bolt tightened using this procedure will be about 60 to 70 per cent of the elastic limit (yield strength) of the bolt material.

The table that follows can be used to get a rough idea of the torque necessary to properly tension a bolt by using the bolt diameter d and the coefficients b and m from the table; the approximate tightening torque T in ft-lb for the listed fasteners is obtained by solving the equation $T = 10^{b+m \log d}$. This equation is approximate, for use with unlubricated fasteners as supplied by the mill. See the notes at the end of the table for more details on using the equation.

Wrench Torque $T = 10^{b+m \log d}$ for Steel Bolts, Studs, and Cap Screws (*see notes*)

Fastener Grade(s)	Bolt Diameter d (in.)	m	b
SAE 2, ASTM A307	¼ to 3	2.940	2.533
SAE 3	¼ to 3	3.060	2.775
ASTM A-449, A-354-BB, SAE 5	¼ to 3	2.965	2.759
ASTM A-325[a]	½ to 1½	2.922	2.893
ASTM A-354-BC	¼ to ⅝	3.046	2.837
SAE 6, SAE 7	¼ to 3	3.095	2.948
SAE 8	¼ to 3	3.095	2.983
ASTM A-354-BD, ASTM A490[a]	⅜ to 1¾	3.092	3.057
Socket Head Cap Screws	¼ to 3	3.096	3.014

[a] Values for permanent fastenings on steel structures.

Usage: Values calculated using the preceding equation are for standard, unplated industrial fasteners as received from the manufacturer; for cadmium-plated cap screws, multiply the torque by 0.9; for cadmium-plated nuts and bolts, multiply the torque by 0.8; for fasteners used with special lubricants, multiply the torque by 0.9; for studs, use cap screw values for equivalent grade.

Preload for Bolts in Loaded Joints.—The following recommendations are based on MIL-HDBK-60, a subsection of FED-STD-H28, Screw Thread Standards for Federal Service. Generally, bolt preload in joints should be high enough to maintain joint members in contact and in compression. Loss of compression in a joint may result in leakage of pressurized fluids past compression gaskets, loosening of fasteners under conditions of cyclic loading, and reduction of fastener fatigue life.

The relationship between fastener fatigue life and fastener preload is illustrated by Fig. 1. An axially loaded bolted joint in which there is no bolt preload is represented by line OAB, that is, the bolt load is equal to the joint load. When joint load varies between P_a and P_b, the bolt load varies accordingly between P_{Ba} and P_{Bb}. However, if preload P_{B1} is applied to the bolt, the joint is compressed and bolt load changes more slowly than the joint load (indicated by line P_{B1}A, whose slope is less than line OAB) because some of the load is absorbed as a reduction of compression in the joint. Thus, the axial load applied to the joint varies between $P_{Ba'}$ and $P_{Bb'}$ as joint load varies between P_a and P_b. This condition results in a considerable reduction in cyclic bolt-load variation and thereby increases the fatigue life of the fastener.

Preload for Bolts In Shear.—In shear-loaded joints, with members that slide, the joint members transmit shear loads to the fasteners in the joint and the preload must be sufficient to hold the joint members in contact. In joints that do not slide (i.e., there is no relative motion between joint members), shear loads are transmitted within the joint by frictional forces that mainly result from the preload. Therefore, preload must be great enough for the resulting friction forces to be greater than the applied shear force. With high applied shear loads, the shear stress induced in the fastener during application of the preload must also be considered in the bolted-joint design. Joints with combined axial and shear loads must be analyzed to ensure that the bolts will not fail in either tension or shear.

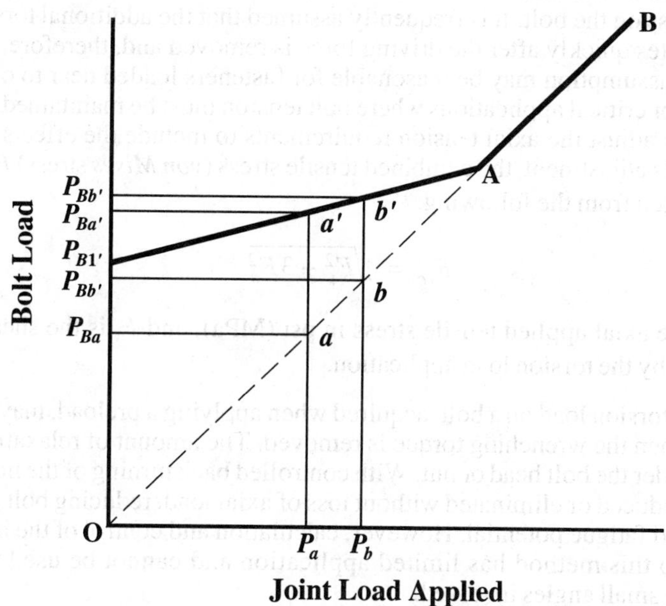

Fig. 1. Bolt Load in a Joint with Applied Axial Load

General Application of Preload.—Preload values should be based on joint requirements, as outlined before. Fastener applications are generally designed for maximum utilization of the fastener material; that is to say, the fastener size is the minimum required to perform its function and a maximum safe preload is generally applied to it. However, if a low-strength fastener is replaced by one of higher strength, for the sake of convenience or standardization, the preload in the replacement should not be increased beyond that required in the original fastener.

To utilize the maximum amount of bolt strength, bolts are sometimes tightened to or beyond the yield point of the material. This practice is generally limited to ductile materials, where there is considerable difference between the yield strength and the ultimate (breaking) strength, because low-ductility materials are more likely to fail due to unexpected overloads when preloaded to yield. Joints designed for primarily static load conditions that use ductile bolts, with a yield strain that is relatively far from the strain at fracture, are often preloaded above the yield point of the bolt material. Methods for tightening up to and beyond the yield point include tightening by feel without special tools, and the use of electronic equipment designed to compare the applied torque with the angular rotation of the fastener and detect changes that occur in the elastic properties of fasteners at yield.

Bolt loads are maintained below the yield point in joints subjected to cyclic loading and in joints using bolts of high-strength material where the yield strain is close to the strain at fracture. For these conditions, the maximum preloads generally fall within the following ranges: 50 to 80 per cent of the minimum tensile ultimate strength; 75 to 90 per cent of the minimum tensile yield strength or proof load; or 100 per cent of the observed proportional limit or onset of yield.

Bolt heads, driving recesses (in socket screws, for example), and the juncture of head and shank must be sufficiently strong to withstand the preload and any additional stress encountered during tightening. There must also be sufficient thread to prevent stripping (generally, at least three fully engaged threads). Materials susceptible to stress-corrosion cracking may require further preload limitations.

Preload Adjustments.—Preloads may be applied directly by axial loading or indirectly by turning of the nut or bolt. When preload is applied by turning of nuts or bolts, a torsion load component is added to the desired axial bolt load. This combined loading increases the tensile stress on the bolt. It is frequently assumed that the additional torsion load component dissipates quickly after the driving force is removed and, therefore, can be largely ignored. This assumption may be reasonable for fasteners loaded near to or beyond yield strength, but for critical applications where bolt tension must be maintained below yield, it is important to adjust the axial tension requirements to include the effects of the preload torsion. For this adjustment, the combined tensile stress (*von Mises* stress) F_{tc} in psi (MPa) can be calculated from the following:

$$F_{tc} = \sqrt{F_t^2 + 3F_s^2} \qquad (3)$$

where F_t is the axial applied tensile stress in psi (MPa), and F_s is the shear stress in psi (MPa) caused by the torsion load application.

Some of the torsion load on a bolt, acquired when applying a preload, may be released by springback when the wrenching torque is removed. The amount of relaxation depends on the friction under the bolt head or nut. With controlled back turning of the nut, the torsional load may be reduced or eliminated without loss of axial load, reducing bolt stress and lowering creep and fatigue potential. However, calculation and control of the back-turn angle is difficult, so this method has limited application and cannot be used for short bolts because of the small angles involved.

For relatively soft work-hardenable materials, tightening bolts in a joint slightly beyond yield will work-harden the bolt to some degree. Back turning of the bolt to the desired tension will reduce embedment and metal flow and improve resistance to preload loss.

The following formula for use with single-start Unified inch screw threads calculates the combined tensile stress, F_{tc}:

$$F_{tc} = F_t\sqrt{1 + 3\left(\frac{1.96 + 2.31\mu}{1 - 0.325P/d_2} - 1.96\right)^2} \qquad (4)$$

Single-start UNJ screw threads in accordance with MIL-S-8879 have a thread stress diameter equal to the bolt pitch diameter. For these threads, F_{tc} can be calculated from:

$$F_{tc} = F_t\sqrt{1 + 3\left(\frac{0.637P}{d_2} + 2.31\mu\right)^2} \qquad (5)$$

where μ is the coefficient of friction between threads, P is the thread pitch ($P = 1/n$, and n is the number of threads per inch), and d_2 is the bolt-thread pitch diameter in inches. Both Equations (2) and (3) are derived from Equation (1); thus, the quantity within the radical ($\sqrt{\ }$) represents the proportion of increase in axial bolt tension resulting from preload torsion. In these equations, tensile stress due to torsion load application becomes most significant when the thread friction, μ, is high.

Coefficients of Friction for Bolts and Nuts.—Table 1 gives examples of coefficients of friction that are frequently used in determining torque requirements. Dry threads, indicated by the words "None added" in the Lubricant column, are assumed to have some residual machine oil lubrication. Table 1 values are not valid for threads that have been cleaned to remove all traces of lubrication because the coefficient of friction of these threads may be very much higher unless a plating or other film is acting as a lubricant.

Table 1. Coefficients of Friction of Bolts and Nuts

Bolt/Nut Materials	Lubricant	Coefficient of Friction, μ ± 20%
Steel[a]	Graphite in petrolatum or oil	0.07
	Molybdenum disulfide grease	0.11
	Machine oil	0.15
Steel,[a] cadmium-plated	None added	0.12
Steel,[a] zinc-plated	None added	0.17
Steel[a]/bronze	None added	0.15
Corrosion-resistant steel or nickel-base alloys/silver-plated materials	None added	0.14
Titanium/steel[a]	Graphite in petrolatum	0.08
Titanium	Molybdenum disulfide grease	0.10

[a] "Steel" includes carbon and low-alloy steels but not corrosion-resistant steels.

Where two materials are separated by a slash (/), either may be the bolt material; the other is the nut material.

Preload Relaxation.—Local yielding, due to excess bearing stress under nuts and bolt heads (caused by high local spots, rough surface finish, and lack of perfect squareness of bolt and nut bearing surfaces), may result in preload relaxation after preloads are first applied to a bolt. Bolt tension also may be unevenly distributed over the threads in a joint, so thread deformation may occur, causing the load to be redistributed more evenly over the threaded length. Preload relaxation occurs over a period of minutes to hours after the application of the preload, so retightening after several minutes to several days may be required. As a general rule, an allowance for loss of preload of about 10 per cent may be made when designing a joint.

Increasing the resilience of a joint will make it more resistant to local yielding, that is, there will be less loss of preload due to yielding. When practical, a joint-length to bolt-diameter ratio of 4 or more is recommended (e.g., a ¼-inch bolt and a 1-inch or greater joint length). Through bolts, far-side tapped holes, spacers, and washers can be used in the joint design to improve the joint-length to bolt-diameter ratio.

Over an extended period of time, preload may be reduced or completely lost due to vibration; temperature cycling, including changes in ambient temperature; creep; joint load; and other factors. An increase in the initial bolt preload or the use of thread-locking methods that prevent relative motion of the joint may reduce the problem of preload relaxation due to vibration and temperature cycling. Creep is generally a high-temperature effect, although some loss of bolt tension can be expected even at normal temperatures. Harder materials and creep-resistant materials should be considered if creep is a problem or high-temperature service of the joint is expected.

The mechanical properties of fastener materials vary significantly with temperature, and allowance must be made for these changes when ambient temperatures range beyond 30 to 200°F. Mechanical properties that may change include tensile strength, yield strength, and modulus of elasticity. Where bolts and flange materials are generically dissimilar, such as carbon steel and corrosion-resistant steel or steel and brass, differences in thermal expansion that might cause preload to increase or decrease must be taken into consideration.

Methods of Applying and Measuring Preload.—Depending on the tightening method, the accuracy of preload application may vary up to 25 per cent or more. Care must be taken to maintain the calibration of torque and load indicators. Allowance should be made for uncertainties in bolt load to prevent overstressing the bolts or failing to obtain sufficient preload. The method of tensioning should be based on the required accuracy and relative costs.

The most common methods of bolt tension control are indirect because it is usually difficult or impractical to measure the tension produced in each fastener during assembly. Table 2 lists the most frequently used methods of applying bolt preload and the approximate accuracy of each method. For many applications, fastener tension can be satisfactorily controlled within certain limits by applying a known torque to the fastener. Laboratory tests have shown that whereas a satisfactory torque tension relationship can be established for a given set of conditions, a change of any of the variables, such as fastener material, surface finish, and the presence or absence of lubrication, may severely alter the relationship. Because most of the applied torque is absorbed in intermediate friction, a change in the surface roughness of the bearing surfaces or a change in the lubrication will drastically affect the friction and thus the torque tension relationship. Regardless of the method or accuracy of applying the preload, tension will decrease in time if the bolt, nut, or washer seating faces deform under load, if the bolt stretches or creeps under tensile load, or if cyclic loading causes relative motion between joint members.

Table 2. Accuracy of Bolt Preload Application Methods

Method	Accuracy	Method	Accuracy
By feel	±35%	Computer-controlled wrench	
Torque wrench	±25%	below yield (turn-of-nut)	±15%
Turn-of-nut	±15%	yield-point sensing	±8%
Preload indicating washer	±10%	Bolt elongation	±3–5%
Strain gages	±1%	Ultrasonic sensing	±1%

Tightening methods using power drivers are similar in accuracy to equivalent manual methods.

Elongation Measurement.—Bolt elongation is directly proportional to axial stress when the applied stress is within the elastic range of the material. If both ends of a bolt are accessible, a micrometer measurement of bolt length made before and after the application of tension will ensure the required axial stress is applied. The elongation δ in inches (mm) can be determined from the formula $\delta = F_t \times L_B \div E$, given the required axial stress F_t in psi (MPa), the bolt modulus of elasticity E in psi (MPa), and the effective bolt length L_B in inches (mm). L_B, as indicated in Fig. 2, includes the contribution of bolt area and ends (head and nut) and is calculated from:

$$L_B = \left(\frac{d_{ts}}{d}\right)^2 \times \left(L_s + \frac{H_B}{2}\right) + L_J - L_S + \frac{H_N}{2} \qquad (6)$$

where d_{ts} is the thread stress diameter, d is the bolt diameter, L_s is the unthreaded length of the bolt shank, L_j is the overall joint length, H_B is the height of the bolt head, and H_N is the height of the nut.

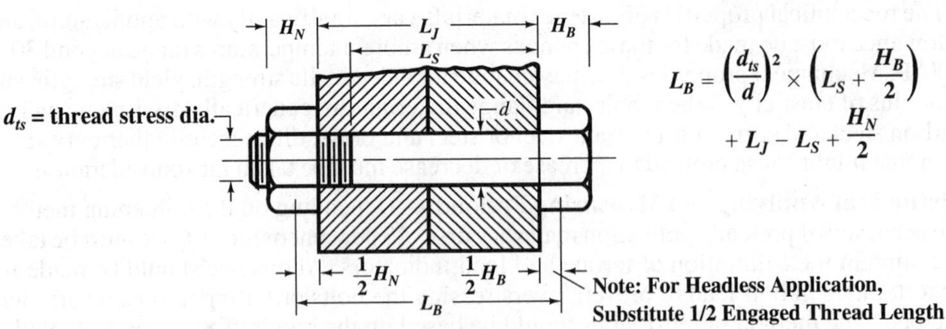

Fig. 2. Effective Length Applicable in Elongation Formulas

The micrometer method is most easily and accurately applied to bolts that are essentially uniform throughout the bolt length, that is, threaded along the entire length or that have only a few threads in the bolt grip area. If the bolt geometry is complex, such as tapered or stepped, the elongation is equal to the sum of the elongations of each section with allowances made for transitional stresses in bolt head height and nut engagement length.

The direct method of measuring elongation is practical only if both ends of a bolt are accessible. Otherwise, if the diameter of the bolt or stud is sufficiently large, an axial hole can be drilled, as shown in Fig. 3, and a micrometer depth gage or other means used to determine the change in length of the hole as the fastener is tightened. A similar method uses a special indicating bolt that has a blind axial hole containing a pin fixed at the bottom. The pin is usually made flush with the bolt head surface before load application. As the bolt is loaded, the elongation causes the end of the pin to move below the reference surface. The displacement of the pin can be converted directly into unit stress by means of a calibrated gage. In some bolts of this type, the pin is set a distance above the bolt so that the pin is flush with the bolt head when the required axial load is reached.

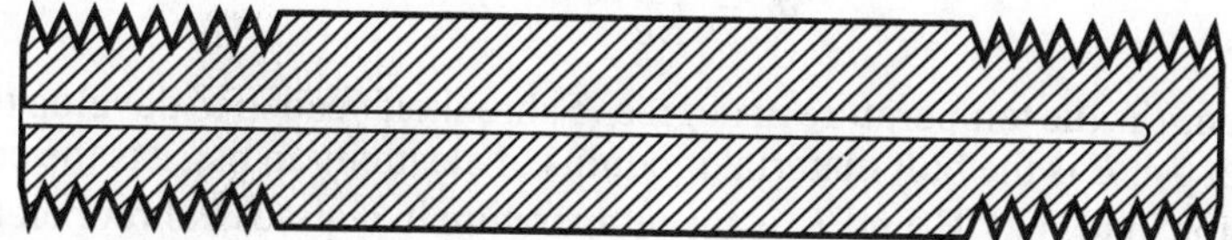

Fig. 3. Hole Drilled to Measure Elongation When One End of Stud or Bolt Is Not Accessible

The *ultrasonic method* of measuring elongation uses a sound pulse, generated at one end of a bolt, that travels the length of a bolt, bounces off the far end, and returns to the sound generator in a measured period of time. The time required for the sound pulse to return depends on the length of the bolt and the speed of sound in the bolt material. The speed of sound in the bolt depends on the material, the temperature, and the stress level. The ultrasonic measurement system can compute the stress, load, or elongation of the bolt at any time by comparing the pulse travel time in the loaded and unstressed conditions. In a similar method, measuring round-trip transit times of longitudinal and shear wave sonic pulses allows calculation of tensile stress in a bolt without consideration of bolt length. This method permits checking bolt tension at any time and does not require a record of the ultrasonic characteristics of each bolt at zero load.

To ensure consistent results, the ultrasonic method requires that both ends of the bolt be finished square to the bolt axis. The accuracy of ultrasonic measurement compares favorably with strain gage methods, but is limited by sonic velocity variations between bolts of the same material and by corrections that must be made for unstressed portions of the bolt heads and threads.

The *turn-of-nut method* applies preload by turning a nut through an angle that corresponds to a given elongation. The elongation of the bolt is related to the angle turned by the formula: $\delta_B = \theta \times l \div 360$, where δ_B is the elongation in inches (mm), θ is the turn angle of the nut in degrees, and l is the lead of the thread helix in inches (mm). Substituting $F_t \times L_B \div E$ for elongation δ_B in this equation gives the turn-of-nut angle required to attain preload F_t:

$$\theta = 360\frac{F_t L_B}{El} \qquad (7)$$

where L_B is given by Equation (6), and E is the modulus of elasticity.

Accuracy of the turn-of-nut method is affected by elastic deformation of the threads, by roughness of the bearing surfaces, and by the difficulty of determining the starting point for measuring the angle. The starting point is usually found by tightening the nut enough to seat the contact surfaces firmly, and then loosening it just enough to release any tension and twisting in the bolt. The nut-turn angle will be different for each bolt size, length, material, and thread lead. The preceding method of calculating the nut-turn angle also requires elongation of the bolt without a corresponding compression of the joint material. The turn-of-nut method, as just outlined, is not valid for joints with compressible gaskets or other soft material, or if there is a significant deformation of the nut and joint material relative to that of the bolt. The nut-turn angle would then have to be determined empirically using a simulated joint and a tension-measuring device.

The Japanese Industrial Standards (JIS) Handbook, *Fasteners and Screw Threads,* indicates that the turn-of-nut tightening method is applicable in both elastic and plastic region tightening. Refer to JIS B 1083 for more detail on this subject.

Heating: causes a bolt to expand at a rate proportional to its coefficient of expansion. When a hot bolt and nut are fastened in a joint and cooled, the bolt shrinks and tension is developed. The temperature necessary to develop an axial stress, F_t, (when the stress is below the elastic limit) can be found as follows:

$$T = \frac{F_t}{Ee} + T_o \qquad (8)$$

In this equation, T is the temperature in degrees Fahrenheit needed to develop the axial tensile stress F_t in psi, E is the bolt material modulus of elasticity in psi, e is the coefficient of linear expansion in in./in.-°F, and T_o is the temperature in degrees Fahrenheit to which the bolt will be cooled. $T - T_o$ is, therefore, the temperature change of the bolt. In finite-element simulations, heating and cooling are frequently used to preload mesh elements in tension or compression. Equation (8) can be used to determine required temperature changes in such problems.

Example: A tensile stress of 40,000 psi is required for a steel bolt in a joint operating at 70°F. If E is 30×10^6 psi and e is 6.2×10^{-6} in./in.-°F, determine the temperature of the bolt needed to develop the required stress on cooling.

$$T = \frac{40{,}000}{(30 \times 10^6)(6.2 \times 10^{-6})} + 70 = 285°\text{F}$$

In practice, the bolt is heated slightly above the required temperature (to allow for some cooling while the nut is screwed down) and the nut is tightened snugly. Tension develops as the bolt cools. In another method, the nut is tightened snugly on the bolt, and the bolt is heated in place. When the bolt has elongated sufficiently, as indicated by inserting a thickness gage between the nut and the bearing surface of the joint, the nut is tightened. The bolt develops the required tension as it cools; however, preload may be lost if the joint temperature increases appreciably while the bolt is being heated.

Calculating Thread Tensile-Stress Area.—The tensile-stress area for Unified threads is based on a diameter equivalent to the mean of the pitch and minor diameters. The pitch and the minor diameters for Unified screw threads can be found from the major (nominal) diameter, d, and the screw pitch, $P = 1/n$, where n is the number of threads per inch, by use of the following formulas: the pitch diameter $d_p = d - 0.649519 \times P$; the minor diameter $d_m = d - 1.299038 \times P$. The tensile stress area, A_s, for Unified threads can then be found as follows:

$$A_s = \frac{\pi}{4}\left(\frac{d_m + d_p}{2}\right)^2 \qquad (9)$$

UNJ threads in accordance with MIL-S-8879 have a tensile thread area that is usually considered to be at the basic bolt pitch diameter, so for these threads, $A_s = \pi d_p^2/4$. The tensile stress area for Unified screw threads is smaller than this area, so the required tightening torque for UNJ threaded bolts is greater than for an equally stressed Unified threaded bolt in an equivalent joint. To convert tightening torque for a Unified fastener to the equivalent torque required with a UNJ fastener, use the following relationship:

$$\mathrm{UNJ}_{\mathrm{torque}} = \left(\frac{d \times n - 0.6495}{d \times n - 0.9743}\right)^2 \times \mathrm{Unified}_{\mathrm{torque}} \qquad (10)$$

where d is the basic thread major diameter, and n is the number of threads per inch.

The tensile stress area for metric threads is based on a diameter equivalent to the mean of the pitch diameter and a diameter obtained by subtracting ⅙ the height of the fundamental thread triangle from the external-thread minor diameter. The Japanese Industrial Standard JIS B 1082 (see also ISO 898/1) defines the stress area of metric screw threads as follows:

$$A_s = \frac{\pi}{4}\left(\frac{d_2 + d_3}{2}\right)^2 \qquad (11)$$

In Equation (11), A_s is the stress area of the metric screw thread in mm²; d_2 is the pitch diameter of the external thread in mm, given by $d_2 = d - 0.649515 \times P$; and d_3 is defined by $d_3 = d_1 - H/6$. Here, d is the nominal bolt diameter; P is the thread pitch; $d_1 = d - 1.082532 \times P$ is the minor diameter of the external thread in mm; and $H = 0.866025 \times P$ is the height of the fundamental thread triangle. Substituting the formulas for d_2 and d_3 into Equation (11) results in $A_s = 0.7854(d - 0.9382P^2)$.

The stress area, A_s, of Unified threads in mm² is given in JIS B 1082 as:

$$A_s = 0.7854\left(d - \frac{0.9743}{n} \times 25.4\right)^2 \qquad (12)$$

Relation between Torque and Clamping Force.—The Japanese Industrial Standard JIS B 1803 defines fastener tightening torque T_f as the sum of the bearing surface torque T_w and the shank (threaded) portion torque T_s. The relationship between the applied tightening torque and bolt preload F_{ft} is as follows: $T_f = T_s + T_w = K \times F_f \times d$. In the preceding, d is the nominal diameter of the screw thread, and K is the torque coefficient defined as follows:

$$K = \frac{1}{2d}\left(\frac{P}{\pi} + \mu_s d_2 \sec\alpha' + \mu_w D_w\right) \qquad (13)$$

where P is the screw thread pitch; μ_s is the coefficient of friction between threads; d_2 is the pitch diameter of the thread; μ_w is the coefficient of friction between bearing surfaces; D_w is the equivalent diameter of the friction torque bearing surfaces; and α' is the flank angle at the ridge perpendicular section of the thread ridge, defined by $\tan\alpha' = \tan\alpha\cos\beta$, where α is the thread half angle (30°, for example), and β is the thread helix, or lead, angle. β can be found from $\tan\beta = l \div 2\pi r$, where l is the thread lead, and r is the thread radius (i.e., one-half the nominal diameter d). When the bearing surface contact area is circular, D_w can be obtained as follows:

$$D_w = \frac{2}{3} \times \frac{D_o^3 - D_i^3}{D_o^2 - D_i^2} \qquad (14)$$

where D_o and D_i are the outside and inside diameters, respectively, of the bearing surface contact area.

The torques attributable to the threaded portion of a fastener, T_s, and bearing surfaces of a joint, T_w, are as follows:

$$T_s = \frac{F_f}{2}\left(\frac{P}{\pi} + \mu_s d_2 \sec\alpha'\right) \qquad (15) \qquad\qquad T_w = \frac{F_f}{2}\mu_w D_w \qquad (16)$$

where F_f, P, μ, d_2, α', μ_w, and D_w are as previously defined.

Tables 3 and 4 give values of torque coefficient K for coarse- and fine-pitch metric screw threads corresponding to various values of μ_s and μ_w. When a fastener material yields according to the shearing-strain energy theory, the torque corresponding to the yield clamping force (see Fig. 4) is $T_{fy} = K \times F_{fy} \times d$, where the yield clamping force F_{fy} is given by:

$$F_{fy} = \frac{\sigma_y A_s}{\sqrt{1 + 3\left[\frac{2}{d_A}\left(\frac{P}{\pi} + \mu_s d_2 \sec\alpha'\right)\right]^2}} \quad (17)$$

Table 3. Torque Coefficients K for Metric Hexagon Head Bolt and Nut Coarse Screw Threads

Between Threads, μ_s	Coefficient of Friction Between Bearing Surfaces, μ_w									
	0.08	0.10	0.12	0.15	0.20	0.25	0.30	0.35	0.40	0.45
0.08	0.117	0.130	0.143	0.163	0.195	0.228	0.261	0.293	0.326	0.359
0.10	0.127	0.140	0.153	0.173	0.206	0.239	0.271	0.304	0.337	0.369
0.12	0.138	0.151	0.164	0.184	0.216	0.249	0.282	0.314	0.347	0.380
0.15	0.153	0.167	0.180	0.199	0.232	0.265	0.297	0.330	0.363	0.396
0.20	0.180	0.193	0.206	0.226	0.258	0.291	0.324	0.356	0.389	0.422
0.25	0.206	0.219	0.232	0.252	0.284	0.317	0.350	0.383	0.415	0.448
0.30	0.232	0.245	0.258	0.278	0.311	0.343	0.376	0.409	0.442	0.474
0.35	0.258	0.271	0.284	0.304	0.337	0.370	0.402	0.435	0.468	0.500
0.40	0.285	0.298	0.311	0.330	0.363	0.396	0.428	0.461	0.494	0.527
0.45	0.311	0.324	0.337	0.357	0.389	0.422	0.455	0.487	0.520	0.553

Values in the table are average values of torque coefficient calculated using: Equations (13) and (14) for K and D_w; diameters d of 4, 5, 6, 8, 10, 12, 16, 20, 24, 30, and 36 mm; and selected corresponding pitches P and pitch diameters d_2 according to JIS B 0205 (ISO 724) thread standard. Dimension D_i was obtained for a Class 2 fit without chamfer from JIS B 1001, Diameters of Clearance Holes and Counterbores for Bolts and Screws (equivalent to ISO 273-1979). The value of D_o was obtained by multiplying the reference dimension from JIS B 1002, width across the flats of the hexagon head, by 0.95.

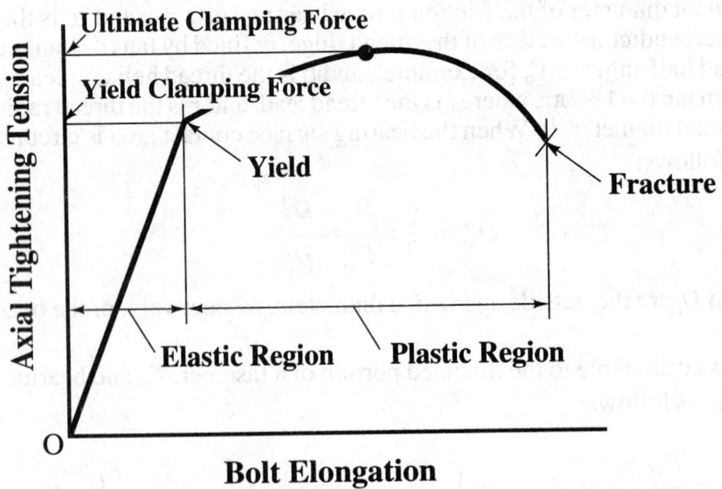

Fig. 4. The Relationship between Bolt Elongation and Axial Tightening Tension

Table 4. Torque Coefficients *K* for Metric Hexagon Head Bolt and Nut Fine-Screw Threads

Between Threads, μ_s	Coefficient of Friction Between Bearing Surfaces, μ_w									
	0.08	0.10	0.12	0.15	0.20	0.25	0.30	0.35	0.40	0.45
0.08	0.106	0.118	0.130	0.148	0.177	0.207	0.237	0.267	0.296	0.326
0.10	0.117	0.129	0.141	0.158	0.188	0.218	0.248	0.278	0.307	0.337
0.12	0.128	0.140	0.151	0.169	0.199	0.229	0.259	0.288	0.318	0.348
0.15	0.144	0.156	0.168	0.186	0.215	0.245	0.275	0.305	0.334	0.364
0.20	0.171	0.183	0.195	0.213	0.242	0.272	0.302	0.332	0.361	0.391
0.25	0.198	0.210	0.222	0.240	0.270	0.299	0.329	0.359	0.389	0.418
0.30	0.225	0.237	0.249	0.267	0.297	0.326	0.356	0.386	0.416	0.445
0.35	0.252	0.264	0.276	0.294	0.324	0.353	0.383	0.413	0.443	0.472
0.40	0.279	0.291	0.303	0.321	0.351	0.381	0.410	0.440	0.470	0.500
0.45	0.306	0.318	0.330	0.348	0.378	0.408	0.437	0.467	0.497	0.527

Values in the table are average values of torque coefficient calculated using Equations (13) and (14) for K and D_w; diameters d of 8, 10, 12, 16, 20, 24, 30, and 36 mm; and selected respective pitches P and pitch diameters d_2 according to JIS B 0207 thread standard (ISO 724). Dimension D_i was obtained for a Class 1 fit without chamfer from JIS B 1001, Diameters of Clearance Holes and Counterbores for Bolts and Screws (equivalent to ISO 273-1979). The value of D_o was obtained by multiplying the reference dimension from JIS B 1002 (small type series), width across the flats of the hexagon head, by 0.95.

In Equation (17), σ_y is the yield point or proof stress of the bolt, A_s is the stress area of the thread, and $d_A = (4A_s/\pi)^{1/2}$ is the diameter of a circle having an area equal to the stress area of the thread. The other variables have been identified previously.

Example: Find the torque required to tighten a 10-mm coarse-threaded ($P = 1.5$) grade 8.8 bolt to yield assuming that both the thread- and bearing-friction coefficients are 0.12.

Solution: From Equation (17), calculate F_{fy} and then solve $T_{fy} = KF_{fy}d$ to obtain the torque required to stress the bolt to the yield point.

$$\sigma_y = 800 \text{ N/mm}^2 \text{ (MPa) (minimum, based on 8.8 grade rating)}$$
$$A_s = 0.7854(10 - 0.9382 \times 1.5)^2 = 57.99 \text{ mm}^2$$
$$d_A = (4A_s/\pi)^{1/2} = 8.6 \text{ mm}$$
$$d_2 = 9.026 \text{ mm (see JIS B 0205 or ISO 724)}$$

Find α' from $\tan\alpha' = \tan\alpha\cos\beta$ using:

$$\alpha = 30°;\ \tan\beta = l \div 2\pi r;\ l = P = 1.5;\ \text{and } r = d \div 2 = 5 \text{ mm}$$
$$\tan\beta = 1.5 \div 10\pi = 0.048, \text{ therefore } \beta = 2.73°$$
$$\tan\alpha' = \tan\alpha\cos\beta = \tan 30° \times \cos 2.73° = 0.577, \text{ and } \alpha' = 29.97°$$

Solving Equation (17) gives the yield clamping force as follows:

$$F_{fy} = \frac{800 \times 58.0}{\sqrt{1 + 3\left[\frac{2}{8.6}\left(\frac{1.5}{\pi} + 0.12 \times 9.026 \times \sec 29.97°\right)\right]^2}} = 38{,}075 \text{ N}$$

K can be determined from Tables 3 (coarse thread) and Tables 4 (fine thread) or from Equations (13) and (14). From Table 3, for μ_s and μ_w equal to 0.12, $K = 0.164$. The yield-point tightening torque can then be found from $T_{fy} = K \times F_{fy} \times d = 0.164 \times 38{,}075 \times 10 = 62.4 \times 10^3$ N-mm = 62.4 N-m.

Obtaining Torque and Friction Coefficients.—Given suitable test equipment, the torque coefficient K and friction coefficients between threads μ_s or between bearing surfaces μ_w can be determined experimentally as follows: Measure the value of the axial tightening tension and the corresponding tightening torque at an arbitrary point in the 50 to 80 per cent range of the bolt yield point or proof stress (for steel bolts, use the minimum value of the yield point or proof stress multiplied by the stress area of the bolt). Repeat this test several times and average the results. The tightening torque may be considered as the sum of the torque on the threads plus the torque on the bolt head- or nut-to-joint bearing surface. The torque coefficient can be found from $K = T_f \div F_f \times d$, where F_f is the measured axial tension, and T_f is the measured tightening torque.

To measure the coefficient of friction between threads or bearing surfaces, obtain the total tightening torque and that portion of the torque due to the thread or bearing surface friction. If only tightening torque and the torque on the bearing surfaces can be measured, then the difference between these two measurements can be taken as the thread-tightening torque. Likewise, if only the tightening torque and threaded-portion torque are known, the torque due to bearing can be taken as the difference between the known torques. The coefficients of friction between threads and bearing surfaces, respectively, can be obtained from the following:

$$\mu_s = \frac{2T_s \cos\alpha'}{d_2 F_f} - \cos\alpha' \tan\beta \quad (18) \qquad \mu_w = \frac{2T_w}{D_w F_f} \quad (19)$$

As before, T_s is the torque attributable to the threaded portion of the screw, T_w is the torque due to bearing, D_w is the equivalent diameter of friction torque on bearing surfaces according to Equation (14), and F_f is the measured axial tension.

Torque-Tension Relationships.—Torque is usually applied to develop an axial load in a bolt. To achieve the desired axial load in a bolt, the torque must overcome friction in the threads and friction under the nut or bolt head. In Fig. 5, the axial load P_B is a component of the normal force developed between threads. The normal-force component perpendicular to the thread helix is $P_{N\beta}$ and the other component of this force is the torque load $P_B \tan\beta$ that is applied in tightening the fastener. Assuming the turning force is applied at the pitch diameter of the thread, the torque T_1 needed to develop the axial load is $T_1 = P_B \times \tan\beta \times d_2/2$. Substituting $\tan\beta = l \div \pi d_2$ into the previous expression gives $T_1 = P_B \times l \div 2\pi$.

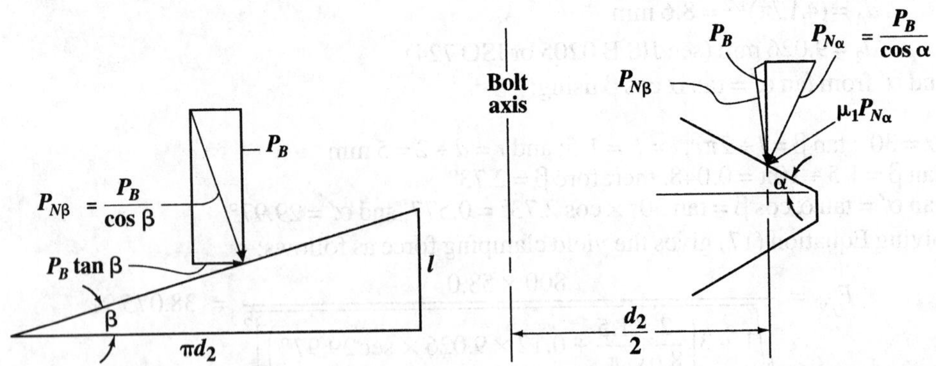

Fig. 5. Free Body Diagram of Thread Helix Forces

Fig. 6. Thread Friction Force

In Fig. 6, the normal-force component perpendicular to the thread flanks is $P_{N\alpha}$. With a coefficient of friction μ_1 between the threads, the friction load is equal to $\mu_1 P_{N\alpha}$, or $\mu_1 P_B \div \cos\alpha$. Assuming the force is applied at the pitch diameter of the thread, the torque T_2 to overcome thread friction is given by:

$$T_2 = \frac{d_2 \mu_1 P_B}{2\cos\alpha} \quad (20)$$

With the coefficient of friction μ_2 between a nut or bolt-head pressure face and a component face, as in Fig. 7, the friction load is equal to $\mu_2 P_B$. Assuming the force is applied midway between the nominal (bolt) diameter d and the pressure-face diameter b, the torque T_3 to overcome the nut or bolt underhead friction is:

$$T_3 = \frac{d+b}{4}\mu_2 P_B \tag{21}$$

The total torque, T, required to develop axial bolt load, P_B, is equal to the sum of the torques T_1, T_2, and T_3 as follows:

$$T = P_B\left(\frac{l}{2\pi} + \frac{d_2\mu_1}{2\cos\alpha} + \frac{(d+b)\mu_2}{4}\right) \tag{22}$$

For a fastener system with 60° threads, $\alpha = 30°$ and d_2 is approximately $0.92d$. If no loose washer is used under the rotated nut or bolt head, b is approximately $1.5d$ and Equation (22) reduces to:

$$T = P_B[0.159 \times l + d(0.531\mu_1 + 0.625\mu_2)] \tag{23}$$

In addition to the conditions of Equation (23), if the thread and bearing friction coefficients, μ_1 and μ_2, are equal (which is not necessarily so), then $\mu_1 = \mu_2 = \mu$, and the previous equation reduces to:

$$T = P_B(0.159l + 1.156\mu d) \tag{24}$$

Example: Estimate the torque required to tighten a UNC ½-13 grade 8 steel bolt to a preload equivalent to 55 per cent of the minimum tensile bolt strength. Assume that the bolt is unplated and both the thread and bearing friction coefficients equal 0.15.

Solution: The minimum tensile strength for SAE grade 8 bolt material is 150,000 psi (from page 1488). To use Equation (24), find the stress area of the bolt using Equation (9) with $P = 1/13$, $d_m = d - 1.2990P$, and $d_p = d - 0.6495P$, and then calculate the necessary preload, P_B, and the applied torque, T.

$$A_s = \frac{\pi}{4}\left(\frac{0.4500 + 0.4001}{2}\right)^2 = 0.1419 \text{ in.}^2$$

$$P_B = \sigma_{allow} \times A_s = 0.55 \times 150{,}000 \times 0.1419 = 11{,}707 \text{ lb}_f$$

$$T = 11{,}707\left(\frac{0.159}{13} + 1.156 \times 0.15 \times 0.500\right) = 1158 \text{ lb-in.} = 96.5 \text{ lb-ft}$$

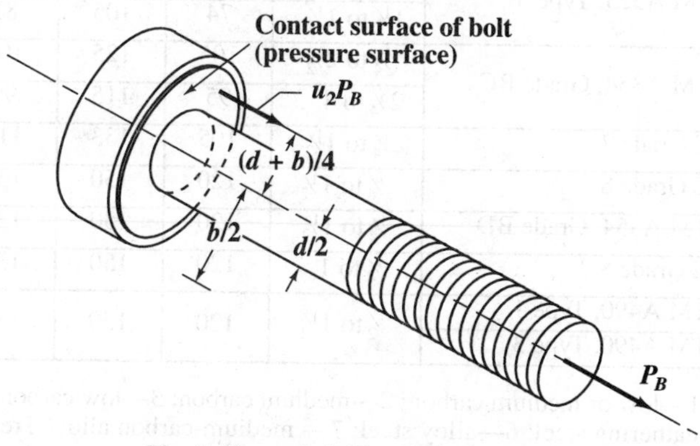

Fig. 7. Nut or Bolt Head Frictin Force

Grade Marks and Material Properties for Bolts and Screws.—Bolts, screws, and other fasteners are marked on the head with a symbol that identifies the grade of the fastener. The grade specification establishes the minimum mechanical properties that the fastener must meet. Additionally, industrial fasteners must be stamped with a registered head mark that identifies the manufacturer. The grade identification table identifies the grade markings and gives mechanical properties for some commonly used ASTM and SAE steel fasteners. Metric fasteners are identified by property grade marks, which are specified in ISO and SAE standards. These marks are discussed with metric fasteners.

Grade Identification Marks and Mechanical Properties of Bolts and Screws

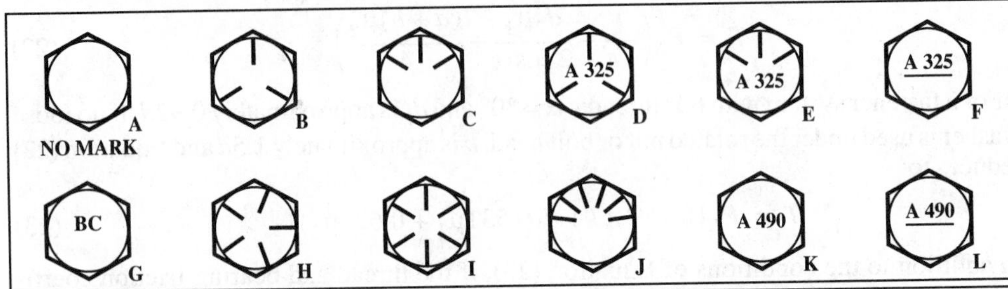

Identifier	Grade	Size (in.)	Min. Strength (10^3 psi) Proof	Tensile	Yield	Material & Treatment
A	SAE Grade 1	¼ to 1½	33	60	36	1
	ASTM A307	¼ to 1½	33	60	36	3
	SAE Grade 2	¼ to ¾	55	74	57	1
		⅞ to 1½	33	60	36	
	SAE Grade 4	¼ to 1½	65	115	100	2, a
B	SAE Grade 5	¼ to 1	85	120	92	2,b
	ASTM A449	1⅛ to 1½	74	105	81	
	ASTM A449	1¾ to 3	55	90	58	
C	SAE Grade 5.2	¼ to 1	85	120	92	4, b
D	ASTM A325, Type 1	½ to 1	85	120	92	2,b
		1⅛ to 1½	74	105	81	
E	ASTM A325, Type 2	½ to 1	85	120	92	4, b
		1⅛ to 1½	74	105	81	
F	ASTM A325, Type 3	½ to 1	85	120	92	5, b
		1⅛ to 1½	74	105	81	
G	ASTM A354, Grade BC	¼ to 2½	105	125	109	5,b
		2¾ to 4	95	115	99	
H	SAE Grade 7	¼ to 1½	105	133	115	7, b
I	SAE Grade 8	¼ to 1½	120	150	130	7, b
	ASTM A354, Grade BD	¼ to 1½	120	150	130	6, b
J	SAE Grade 8.2	¼ to 1	120	150	130	4, b
K	ASTM A490, Type 1	½ to 1½	120	150	130	6, b
L	ASTM A490, Type 3					5, b

Material Steel: 1—low or medium carbon; 2—medium carbon; 3—low carbon; 4—low-carbon martensite; 5—weathering steel; 6—alloy steel; 7— medium-carbon alloy. Treatment: a—cold drawn; b—quench and temper.

Detecting Counterfeit Fasteners.—Fasteners that have markings identifying them as belonging to a specific grade or property class are counterfeit if they do not meet the standards established for that class. Counterfeit fasteners may break unexpectedly at smaller loads than expected. Generally, these fasteners are made from the wrong material or they are not properly strengthened during manufacture. Either way, counterfeit fasteners can lead to dangerous failures in assemblies. The law now requires testing of fasteners used in some critical applications. Detection of counterfeit fasteners is difficult because the counterfeits look genuine. The only sure way to determine if a fastener meets its specification is to test it. However, reputable distributors will assist in verifying the authenticity of the fasteners they sell. For important applications, fasteners can be checked to determine whether they perform according to the standard. Typical laboratory checks used to detect fakes include testing hardness, elongation, and ultimate loading, and a variety of chemical tests.

Mechanical Properties and Grade Markings of Nuts.—Three grades of hex and square nuts designated Grades 2, 5, and 8 are specified by the SAE J995 standard covering nuts in the ¼- to 1½-inch diameter range. Grades 2, 5, and 8 nuts roughly correspond to the SAE specified bolts of the same grade. Additional specifications are given for miscellaneous nuts such as hex jam nuts, hex slotted nuts, heavy hex nuts, etc. Generally speaking, use nuts of a grade equal to or greater than the grade of the bolt being used. Grade 2 nuts are not required to be marked, however, all Grades 5 and 8 nuts in the ¼- to 1½-inch range must be marked in one of three ways: Grade 5 nuts may be marked with a dot on the face of the nut and a radial or circumferential mark at 120° counterclockwise from the dot; or a dot at one corner of the nut and a radial line at 120° clockwise from the nut, or one notch at each of the six corners of the nut. Grade 8 nuts may be identified by a dot on the face of the nut with a radial or circumferential mark at 60° counterclockwise from the dot; or a dot at one corner of the nut and a radial line at 60° clockwise from the nut, or two notches at each of the six corners of the nut.

Working Strength of Bolts.—When the nut on a bolt is tightened, an initial tensile load is placed on the bolt that must be taken into account in determining its safe working strength or external load-carrying capacity. The total load on the bolt theoretically varies from a maximum equal to the sum of the initial and external loads (when the bolt is absolutely rigid and the parts held together are elastic) to a minimum equal to either the initial or external loads, whichever is the greater (where the bolt is elastic and the parts held together are absolutely rigid). No material is absolutely rigid, so in practice the total load values fall somewhere between these maximum and minimum limits, depending upon the relative elasticity of the bolt and joint members.

Some experiments made at Cornell University to determine the initial stress due to tightening nuts on bolts sufficiently to make a packed joint steam-tight showed that experienced mechanics tighten nuts with a pull roughly proportional to the bolt diameter. It was also found that the stress due to nut tightening was often sufficient to break a ½-inch (12.7-mm) bolt, but not larger sizes, assuming that the nut is tightened by an experienced mechanic. It may be concluded, therefore, that bolts smaller than ⅝ inch (15.9 mm) should not be used for holding cylinder heads or other parts requiring a tight joint. As a result of these tests, the following empirical formula was established for the working strength of bolts used for packed joints or joints where the elasticity of a gasket is greater than the elasticity of the studs or bolts.

$$W = S_t(0.55d^2 - 0.25d)$$

In this formula, W = working strength of bolt or permissible load, in pounds, after allowance is made for initial load due to tightening; S_t = allowable working stress in tension, pounds per square inch; and d = nominal outside diameter of stud or bolt, inches. A somewhat more convenient formula, and one that gives approximately the same results, is

$$W = S_t(A - 0.25d)$$

In this formula, W, S_t, and d are as previously given, and A = area at the root of the thread, square inches.

Example: What is the working strength of a 1-inch bolt that is screwed tightly in a packed joint when the allowable working stress is 10,000 psi?

$$W = 10{,}000(0.55 \times 1 - 0.25 \times 1) \;= 3000 \text{ pounds approx.}$$

Formulas for Stress Areas and Lengths of Engagement of Screw Threads.— The critical areas of stress of mating screw threads are: 1) The effective cross-sectional area, or tensile-stress area, of the external thread; 2) the shear area of the external thread, which depends principally on the minor diameter of the tapped hole; and 3) the shear area of the internal thread, which depends principally on the major diameter of the external thread. The relation of these three stress areas to each other is an important factor in determining how a threaded connection will fail, whether by breakage in the threaded section of the screw (or bolt) or by stripping of either the external or internal thread.

If failure of a threaded assembly should occur, it is preferable for the screw to break rather than have either the external or internal thread strip. In other words, the length of engagement of mating threads should be sufficient to carry the full load necessary to break the screw without the threads stripping.

If mating internal and external threads are manufactured of materials having equal tensile strengths, then to prevent stripping of the external thread, the length of engagement should be not less than that given by Formula (1):

$$L_e = \frac{2 \times A_t}{3.1416 K_n \max[\tfrac{1}{2} + 0.57735 n(E_s \min - K_n \max)]} \qquad (1)$$

In this formula, the factor of 2 means that it is assumed that the area of the screw in shear must be twice the tensile-stress area to attain the full strength of the screw (this value is slightly larger than required and thus provides a small factor of safety against stripping); L_e = length of engagement, in inches; n = number of threads per inch; K_n max = maximum minor diameter of internal thread; E_s min = minimum pitch diameter of external thread for the class of thread specified; and A_t = tensile-stress area of screw thread given by Formula (2a) or (2b) or the thread tables for Unified threads, Tables 4a through 4k starting on page page 1740, which are based on Formula (2a).

For steels of up to 100,000 psi ultimate tensile strength,

$$A_t = 0.7854\left(D - \frac{0.9743}{n}\right)^2 \qquad (2a)$$

For steels of over 100,000 psi ultimate tensile strength,

$$A_t = 3.1416\left(\frac{E_s \min}{2} - \frac{0.16238}{n}\right)^2 \qquad (2b)$$

In these formulas, D = basic major diameter of the thread and the other symbols have the same meanings as before.

Stripping of Internal Thread: If the internal thread is made of material of lower strength than the external thread, stripping of the internal thread may take place before the screw breaks. To determine whether this condition exists, it is necessary to calculate the factor J for the relative strength of the external and internal threads given by Formula (3):

$$J = \frac{A_s \times \text{tensile strength of external thread material}}{A_n \times \text{tensile strength of internal thread material}} \qquad (3)$$

If J is less than or equal to 1, the length of engagement determined by Formula (1) is adequate to prevent stripping of the internal thread; if J is greater than 1, the required length of engagement Q to prevent stripping of the internal thread is obtained by multiplying the length of engagement L_e, Formula (1), by J:

$$Q = JL_e \tag{4}$$

In Formula (3), A_s and A_n are the shear areas of the external and internal threads, respectively, given by Formulas (5) and (6):

$$A_s = 3.1416nL_eK_n\max\left[\frac{1}{2n} + 0.57735(E_s\min - K_n\max)\right] \tag{5}$$

$$A_n = 3.1416nL_eD_s\min\left[\frac{1}{2n} + 0.57735(D_s\min - E_n\max)\right] \tag{6}$$

In these formulas, n = threads per inch; L_e = length of engagement from Formula (1); K_n max = maximum minor diameter of internal thread; E_s min = minimum pitch diameter of the external thread for the class of thread specified; D_s min = minimum major diameter of the external thread; and E_n max = maximum pitch diameter of internal thread.

Load to Break Threaded Portion of Screws and Bolts.—The direct tensile load P to break the threaded portion of a screw or bolt (assuming that no shearing or torsional stresses are acting) can be determined from the following formula:

$$P = SA_t$$

where P = load in pounds to break screw; S = ultimate tensile strength of material of screw or bolt in pounds per square inch; and A_t = tensile-stress area in square inches from Formula (2a), (2b), or from the screw thread tables.

Lock Wire Procedure Detail.—Wire ties are frequently used as a locking device for bolted connections to prevent loosening due to vibration and loading conditions, or tampering. The use of safety wire ties is illustrated in Figs. 1 and 2 below. The illustrations assume the use of right-hand threaded fasteners and the following additional rules apply:

1) No more that three (3) bolts may be tied together; 2) Bolt heads may be tied as shown only when the female thread receiver is captive; 3) Pre-drilled nuts may be tied in a fashion similar to that illustrated with the following conditions. a) Nuts must be heat-treated; and b) Nuts are factory drilled for use with lock wire.

4) Lock wire must fill a minimum of 75% of the drilled hole provided for the use of lock wire; and 5) Lock wire must be aircraft quality stainless steel of 0.508 mm (0.020 inch) diameter, 0.8128 mm (0.032 inch) diameter, or 1.067 mm 0.042 inch) diameter. Diameter of lock wire is determined by the thread size of the fastener to be safe-tied. a) Thread sizes of 6 mm (0.25 inch) and smaller use 0.508mm (0.020 inch) wire; b) Thread sizes of 6 mm (0.25 inch) to 12 mm (0.5 inch) use 0.8128 mm (0.032 inch) wire; c) Thread sizes > 12 mm (0.5 inch) use 1.067 mm (0.042 inch) wire; and d) The larger wire may be used in smaller bolts in cases of convenience, but smaller wire must not be used in larger fastener sizes.

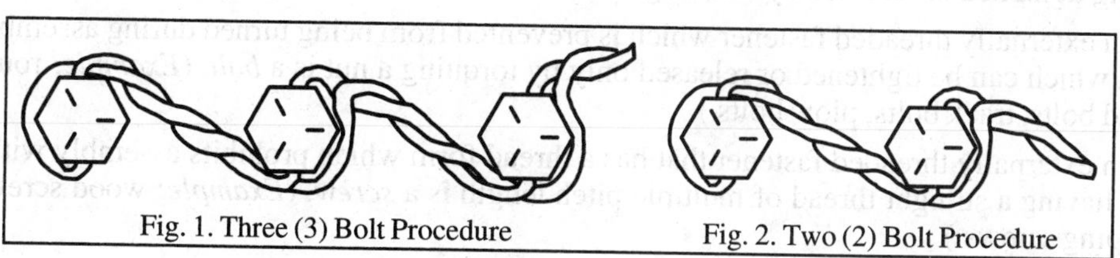

Fig. 1. Three (3) Bolt Procedure

Fig. 2. Two (2) Bolt Procedure

BOLTS, SCREWS, NUTS, AND WASHERS

Dimensions of bolts, screws, nuts, and washers used in machine construction are given here. For data on thread forms, see *Square and Hex Bolts, Screws, and Nuts* in "Screw Thread Systems" section.

American Square and Hexagon Bolts, Screws, and Nuts.—The 1941 American Standard ASA B18.2 covered head dimensions only. In 1952 and 1955 the Standard was revised to cover the entire product. Some bolt and nut classifications were simplified by elimination or consolidation in agreements reached with the British and Canadians. In 1965 ASA B18.2 was redesignated into two standards: B18.2.1 covering square and hexagon bolts and screws including hexagon cap screws and lag screws and B18.2.2 covering square and hexagon nuts. In B18.2.1-1965, hexagon head cap screws and finished hexagon bolts were consolidated into a single product heavy semifinished hexagon bolts and heavy finished hexagon bolts were consolidated into a single product; regular semifinished hexagon bolts were eliminated; a new tolerance pattern for all bolts and screws and a positive identification procedure for determining whether an externally threaded product should be designated as a bolt or screw were established. Also included in this standard are heavy hexagon bolts and heavy hexagon structural bolts. In B18.2.2-1965, regular semifinished nuts were discontinued; regular hexagon and heavy hexagon nuts in sizes ¼ through 1 inch, finished hexagon nuts in sizes larger than 1½ inches, washer-faced semifinished style of finished nuts in sizes ⅝-inch and smaller and heavy series nuts in sizes $\frac{7}{16}$-inch and smaller were eliminated.

Further revisions and refinements include the addition of askew head bolts and hex head lag screws and the specifying of countersunk diameters for the various hex nuts. Heavy hex structural bolts and heavy hex nuts were moved to a new structural applications standard. Additionally, B18.2.1 has been revised to allow easier conformance to Public Law 101-592. All these changes are reflected in ANSI/ASME B18.2.1-1996, and ANSI/ASME B18.2.2-1987 (R1999).

Unified Square and Hexagon Bolts, Screws, and Nuts.—Items that are recognized in the Standard as "unified" dimensionally with British and Canadian standards are shown in bold-face in certain tables.

The other items in the same tables are based on formulas accepted and published by the British for sizes outside the ranges listed in their standards which, as a matter of information, are BS 1768:1963 (obsolescent) for Precision (Normal Series) Unified Hexagon Bolts, Screws, Nuts (UNC and UNF Threads) and B.S. 1769 and amendments for Black (Heavy Series) Unified Hexagon Bolts, etc. Tolerances applied to comparable dimensions of American and British Unified bolts and nuts may differ because of rounding off practices and other factors.

Differentiation between Bolt and Screw.—A bolt is an externally threaded fastener designed for insertion through holes in assembled parts, and is normally intended to be tightened or released by torquing a nut.

A screw is an externally threaded fastener capable of being inserted into holes in assembled parts, of mating with a preformed internal thread or forming its own thread and of being tightened or released by torquing the head.

An externally threaded fastener which is prevented from being turned during assembly, and which can be tightened or released only by torquing a nut is a *bolt.* (*Example:* round head bolts, track bolts, plow bolts.)

An externally threaded fastener that has a thread form which prohibits assembly with a nut having a straight thread of multiple pitch length is a *screw.* (*Example:* wood screws, tapping screws.)

An externally threaded fastener that must be assembled with a nut to perform its intended service is a *bolt*. (*Example:* heavy hex structural bolt.)

An externally threaded fastener that must be torqued by its head into a tapped or other preformed hole to perform its intended service is a *screw*. (*Example:* square head set screw.)

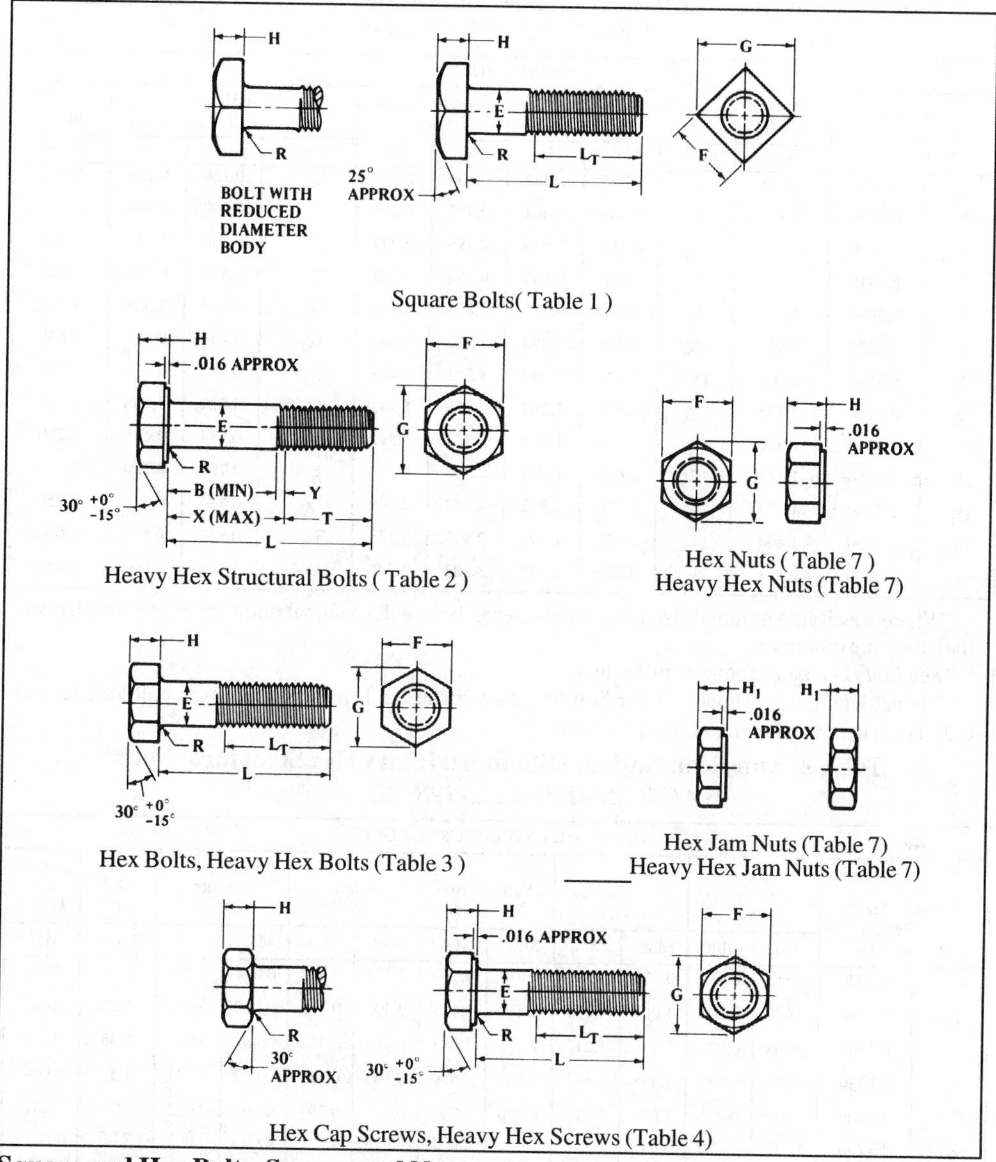

Square and Hex Bolts, Screws, and Nuts.—The dimensions for square and hex bolts and screws given in the following tables have been taken from American National Standard ANSI/ASME B18.2.1-1996 and for nuts from American National Standard ANSI/ASME B18.2.2-1987 (R1999) Reference should be made to these Standards for information or data not found in the following text and tables:

Designation: Bolts and screws should be designated by the following data in the sequence shown: nominal size (fractional and decimal equivalent); threads per inch (omit for lag screws); product length for bolts and screws (fractional or two-place decimal equivalent); product name; material, including specification, where necessary; and protective finish, if required. Examples: (1) 3/8-16 × 1 1/2 Square Bolt, Steel, Zinc Plated; (2) 1/2-13 × 3 Hex Cap Screw, SAE Grade 8 Steel; and (3) .75 × 5.00 Hex Lag Screw, Steel. (4) 1/2-13 Square Nut, Steel, Zinc Plated; (5) 3/4-16 Heavy Hex Nut, SAE J995 Grade 5 Steel; and (6) 1000-8 Hex Thick Slotted Nut, ASTM F594 (Alloy Group 1) Corrosion-Resistant Steel.

Table 1. American National Standard and Unified Standard Square Bolts *ANSI/ASME B18.2.1-1996*

SQUARE BOLTS											
Nominal Size[a] or Basic Product Dia.		Body Dia.[b] E	Width Across Flats F			Width Across Corners G		Head Height H			Thread Length[c] L_T
		Max.	Basic	Max.	Min.	Max.	Min.	Basic	Max.	Min.	Nom.
1/4	**0.2500**	**0.260**	**3/8**	**0.375**	**0.362**	**0.530**	**0.498**	**11/64**	**0.188**	**0.156**	0.750
5/16	**0.3125**	**0.324**	**1/2**	**0.500**	**0.484**	**0.707**	**0.665**	**13/64**	**0.220**	**0.186**	0.875
3/8	**0.3750**	**0.388**	**9/16**	**0.562**	**0.544**	**0.795**	**0.747**	**1/4**	**0.268**	**0.232**	1.000
7/16	**0.4375**	**0.452**	**5/8**	**0.625**	**0.603**	**0.884**	**0.828**	**19/64**	**0.316**	**0.278**	1.125
1/2	**0.5000**	**0.515**	**3/4**	**0.750**	**0.725**	**1.061**	**0.995**	**21/64**	**0.348**	**0.308**	1.250
5/8	**0.6250**	**0.642**	**15/16**	**0.938**	**0.906**	**1.326**	**1.244**	**27/64**	**0.444**	**0.400**	1.500
3/4	**0.7500**	**0.768**	**1 1/8**	**1.125**	**1.088**	**1.591**	**1.494**	**1/2**	**0.524**	**0.476**	1.750
7/8	**0.8750**	**0.895**	**1 5/16**	**1.312**	**1.269**	**1.856**	**1.742**	**19/32**	**0.620**	**0.568**	2.000
1	**1.0000**	**1.022**	**1 1/2**	**1.500**	**1.450**	**2.121**	**1.991**	**21/32**	**0.684**	**0.628**	2.250
1 1/8	**1.1250**	**1.149**	**1 11/16**	**1.688**	**1.631**	**2.386**	**2.239**	**3/4**	**0.780**	**0.720**	2.500
1 1/4	**1.2500**	**1.277**	**1 7/8**	**1.875**	**1.812**	**2.652**	**2.489**	**27/32**	**0.876**	**0.812**	2.750
1 3/8	**1.3750**	**1.404**	**2 1/16**	**2.602**	**1.994**	**2.917**	**2.738**	**29/32**	**0.940**	**0.872**	3.000
1 1/2	**1.5000**	**1.531**	**2 1/4**	**2.250**	**2.175**	**3.182**	**2.986**	**1**	**1.036**	**0.964**	3.250

[a] Where specifying nominal size in decimals, zeros before the decimal point and in the fourth decimal place are omitted.

[b] See *Body Diameter* footnote in Table 3.

[c] Thread lengths, L_T, shown are for bolt lengths 6 inches and shorter. For longer bolt lengths add 0.250 inch to thread lengths shown.

Table 2. American National Standard Heavy Hex Structural Bolts *ANSI/ASME B18.2.1-1981 (R1992)*[a]

HEAVY HEX STRUCTURAL BOLTS													
Nominal Size[a] or Basic Product Dia.		Body Dia. E		Width Across Flats F		Width Across Corners G		Height H		Radius of Fillet R		Thrd. Lgth. L_T	Transition Thrd. Y
		Max.	Min.	Max.	Min.	Max.	Min.	Max.	Min.	Max.	Min.	Basic	Max.
1/2	0.5000	0.515	0.482	0.875	0.850	1.010	0.969	0.323	0.302	0.031	0.009	1.00	0.19
5/8	0.6250	0.642	0.605	1.062	1.031	1.227	1.175	0.403	0.378	0.062	0.021	1.25	0.22
3/4	0.7500	0.768	0.729	1.250	1.212	1.443	1.383	0.483	0.455	0.062	0.021	1.38	0.25
7/8	0.8750	0.895	0.852	1.438	1.394	1.660	1.589	0.563	0.531	0.062	0.031	1.50	0.28
1	1.0000	1.022	0.976	1.625	1.575	1.876	1.796	0.627	0.591	0.093	0.062	1.75	0.31
1 1/8	1.1250	1.149	1.098	1.812	1.756	2.093	2.002	0.718	0.658	0.093	0.062	2.00	0.34
1 1/4	1.2500	1.277	1.223	2.000	1.938	2.309	2.209	0.813	0.749	0.093	0.062	2.00	0.38
1 3/8	1.3750	1.404	1.345	2.188	2.119	2.526	2.416	0.878	0.810	0.093	0.062	2.25	0.44
1 1/2	1.5000	1.531	1.470	2.375	2.300	2.742	2.622	0.974	0.902	0.093	0.062	2.25	0.44

[a] Heavy hex structural bolts have been removed from the latest version, ANSI/ASME B18.2.1-1996. The table has been included for reference.

All dimensions are in inches. **Bold type shows bolts unified dimensionally with British and Canadian Standards.** Threads, when rolled, shall be Unified Coarse, Fine, or 8-thread series (UNRC, UNRF, or 8 UNR Series), Class 2A. Threads produced by other methods may be Unified Coarse, Fine, or 8-thread series (UNC, UNF, or 8 UN Series), Class 2A.

Table 3. American National Standard and Unified Standard Hex and Heavy Hex Bolts *ANSI/ASME B18.2.1-1996*

Nominal Size[a] or Basic Dia.	Full Size Body Dia. *E*	Width Across Flats *F*			Width Across Corners *G*		Head Height *H*			Thread Length[b] L_T
	Max.	Basic	Max.	Min.	Max.	Min.	Basic	Max.	Min.	Nom.
HEX BOLTS										
1/4 0.2500	**0.260**	**7/16**	**0.438**	**0.425**	**0.505**	**0.484**	**11/64**	**0.188**	**0.150**	0.750
5/16 0.3125	**0.324**	**1/2**	**0.500**	**0.484**	**0.577**	**0.552**	**7/32**	**0.235**	**0.195**	0.875
3/8 0.3750	**0.388**	**9/16**	**0.562**	**0.544**	**0.650**	**0.620**	**1/4**	**0.268**	**0.226**	1.000
7/16 0.4375	**0.452**	**5/8**	**0.625**	**0.603**	**0.722**	**0.687**	**19/64**	**0.316**	**0.272**	1.125
1/2 0.5000	**0.515**	**3/4**	**0.750**	**0.725**	**0.866**	**0.826**	**11/32**	**0.364**	**0.302**	1.250
5/8 0.6250	**0.642**	**15/16**	**0.938**	**0.906**	**1.083**	**1.033**	**27/64**	**0.444**	**0.378**	1.500
3/4 0.7500	**0.768**	**1 1/8**	**1.125**	**1.088**	**1.299**	**1.240**	**1/2**	**0.524**	**0.455**	1.750
7/8 0.8750	**0.895**	**1 5/16**	**1.312**	**1.269**	**1.516**	**1.447**	**37/64**	**0.604**	**0.531**	2.000
1 1.0000	**1.022**	**1 1/2**	**1.500**	**1.450**	**1.732**	**1.653**	**43/64**	**0.700**	**0.591**	2.250
1 1/8 1.1250	**1.149**	**1 11/16**	**1.688**	**1.631**	**1.949**	**1.859**	**3/4**	**0.780**	**0.658**	2.500
1 1/4 1.2500	**1.277**	**1 7/8**	**1.875**	**1.812**	**2.165**	**2.066**	**27/32**	**0.876**	**0.749**	2.750
1 3/8 1.3750	**1.404**	**2 1/16**	**2.062**	**1.994**	**2.382**	**2.273**	**29/32**	**0.940**	**0.810**	3.000
1 1/2 1.5000	**1.531**	**2 1/4**	**2.250**	**2.175**	**2.598**	**2.480**	**1**	**1.036**	**0.902**	3.250
1 3/4 1.7500	**1.785**	**2 5/8**	**2.625**	**2.538**	**3.031**	**2.893**	**1 5/32**	**1.196**	**1.054**	3.750
2 2.000	**2.039**	**3**	**3.000**	**2.900**	**3.464**	**3.306**	**1 11/32**	**1.388**	**1.175**	4.250
2 1/4 2.2500	2.305	3 3/8	3.375	3.262	3.897	3.719	1 1/2	1.548	1.327	4.750
2 1/2 2.5000	2.559	3 3/4	3.750	3.625	4.330	4.133	1 21/32	1.708	1.479	5.250
2 3/4 2.7500	2.827	4 1/8	4.125	3.988	4.763	4.546	1 13/16	1.869	1.632	5.750
3 3.0000	3.081	4 1/2	4.500	4.350	5.196	4.959	2	2.060	1.815	6.250
3 1/4 3.2500	3.335	4 7/8	4.875	4.712	5.629	5.372	2 3/16	2.251	1.936	6.750
3 1/2 3.5000	3.589	5 1/4	5.250	5.075	6.062	5.786	2 5/16	2.380	2.057	7.250
3 3/4 3.7500	3.858	5 5/8	5.625	5.437	6.495	6.198	2 1/2	2.572	2.241	7.750
4 4.0000	4.111	6	6.000	5.800	6.928	6.612	2 11/16	2.764	2.424	8.250
HEAVY HEX BOLTS										
1/2 0.5000	**0.515**	**7/8**	**0.875**	**0.850**	**1.010**	**0.969**	11/32	0.364	0.302	1.250
5/8 0.6250	**0.642**	**1 1/16**	**1.062**	**1.031**	**1.227**	**1.175**	27/64	0.444	0.378	1.500
3/4 0.7500	**0.768**	**1 1/4**	**1.250**	**1.212**	**1.443**	**1.383**	1/2	0.524	0.455	1.750
7/8 0.8750	**0.895**	**1 7/16**	**1.438**	**1.394**	**1.660**	**1.589**	37/64	0.604	0.531	2.000
1 1.0000	**1.022**	**1 5/8**	**1.625**	**1.575**	**1.876**	**1.796**	43/64	0.700	0.591	2.250
1 1/8 1.1250	**1.149**	**1 13/16**	**1.812**	**1.756**	**2.093**	**2.002**	3/4	0.780	0.658	2.500
1 1/4 1.2500	**1.277**	**2**	**2.000**	**1.938**	**2.309**	**2.209**	27/32	0.876	0.749	2.750
1 3/8 1.3750	**1.404**	**2 3/16**	**2.188**	**2.119**	**2.526**	**2.416**	29/32	0.940	0.810	3.000
1 1/2 1.5000	**1.531**	**2 3/8**	**23.75**	**2.300**	**2.742**	**2.622**	1	1.036	0.902	3.250
1 3/4 1.7500	**1.785**	**2 3/4**	**2.750**	**2.662**	**3.175**	**3.035**	1 5/32	1.196	1.054	3.750
2 2.0000	**2.039**	3 1/8	**3.125**	**3.025**	**3.608**	**3.449**	1 11/32	1.388	1.175	4.250
2 1/4 2.2500	2.305	3 1/2	3.500	3.388	4.041	3.862	1 1/2	1.548	1.327	4.750
2 1/2 2.5000	2.559	3 7/8	3.875	3.750	4.474	4.275	1 21/32	1.708	1.479	5.250
2 3/4 2.7500	2.827	4 1/4	4.250	4.112	4.907	4.688	1 13/16	1.869	1.632	5.750
3 3.0000	3.081	4 5/8	4.625	4.475	5.340	5.102	2	2.060	1.815	6.250

[a] *Nominal Size:* Where specifying nominal size in decimals, zeros preceding the decimal point and in the fourth decimal place are omitted.

[b] Thread lengths, L_T, shown are for bolt lengths 6 inches and shorter. For longer bolt lengths add 0.250 inch to thread lengths shown.

All dimensions are in inches.

Bold type shows bolts unified dimensionally with British and Canadian Standards.

Threads: Threads, when rolled, are Unified Coarse, Fine, or 8-thread series (UNRC, UNRF, or 8 UNR Series), Class 2A. Threads produced by other methods may be Unified Coarse, Fine or 8-thread series (UNC, UNF, or 8 UN Series), Class 2A.

Body Diameter: Bolts may be obtained in "reduced diameter body." Where "reduced diameter body" is specified, the body diameter may be reduced to approximately the pitch diameter of the thread. A shoulder of full body diameter under the head may be supplied at the option of the manufacturer.

Material: Unless otherwise specified, chemical and mechanical properties of steel bolts conform to ASTM A307, Grade A. Other materials are as agreed upon by manufacturer and purchaser.

Table 4. American National Standard and Unified Standard Heavy Hex Screws and Hex Cap Screws *ANSI/ASME B18.2.1-1996*

Nominal Size[a] or Basic Product Dia.		Body Dia. E		Width Across Flats F			Width Across Corners G		Height H			Thread Length[b] L_T
		Max.	Min.	Basic	Max.	Min.	Max.	Min.	Basic	Max.	Min.	Basic
HEAVY HEX SCREWS												
1/2	0.5000	0.5000	0.482	7/8	0.875	.0850	1.010	0.969	5/16	0.323	0.302	1.250
5/8	0.6250	0.6250	0.605	1 1/16	1.062	1.031	1.227	1.175	25/64	0.403	0.378	1.500
3/4	0.7500	0.7500	0.729	1 1/4	1.250	1.212	1.443	1.383	15/32	0.483	0.455	1.750
7/8	0.8750	0.8750	0.852	1 7/16	1.438	1.394	1.660	1.589	35/64	0.563	0.531	2.000
1	1.0000	1.0000	0.976	1 5/8	1.625	1.575	1.876	1.796	39/64	0.627	0.591	2.250
1 1/8	1.1250	1.1250	1.098	1 13/16	1.812	1.756	2.093	2.002	11/16	0.718	0.658	2.500
1 1/4	1.2500	1.2500	1.223	2	2.000	1.938	2.309	2.209	25/32	0.813	0.749	2.750
1 3/8	1.3750	1.3750	1.345	2 3/16	2.188	2.119	2.526	2.416	27/32	0.878	0.810	3.000
1 1/2	1.5000	1.5000	1.470	2 3/8	2.375	2.300	2.742	2.622	15/16	0.974	0.902	3.250
1 3/4	1.7500	1.7500	1.716	2 3/4	2.750	2.662	3.175	3.035	1 3/32	1.134	1.054	3.750
2	2.0000	2.0000	1.964	3 1/8	3.125	3.025	3.608	3.449	1 7/32	1.263	1.175	4.250
2 1/4	2.2500	2.2500	2.214	3 1/2	3.500	3.388	4.041	3.862	1 3/8	1.423	1.327	5.000[c]
2 1/2	2.5000	2.5000	2.461	3 7/8	3.875	3.750	4.474	4.275	1 17/32	1.583	1.479	5.500[c]
2 3/4	2.7500	2.7500	2.711	4 1/4	4.250	41.112	4.907	4.688	1 11/16	1.744	1.632	6.000[c]
3	3.0000	3.0000	2.961	4 5/8	4.625	4.475	5.340	5.102	1 7/8	1.935	1.815	6.500[c]
HEX CAP SCREWS (Finished Hex Bolts)												
1/4	0.2500	0.2500	0.2450	7/16	0.438	0.428	0.505	0.488	5/32	0.163	0.150	0.750
5/16	0.3125	0.3125	0.3065	1/2	0.500	0.489	0.577	0.557	13/64	0.211	0.195	0.875
3/8	0.3750	0.3750	0.3690	9/16	0.562	0.551	0.650	0.628	15/64	0.243	0.226	1.000
7/16	0.4375	0.4375	0.4305	5/8	0.625	0.612	0.722	0.698	9/32	0.291	0.272	1.125
1/2	0.5000	0.5000	0.4930	3/4	0.750	0.736	0.866	0.840	5/16	0.323	0.302	1.250
9/16	0.5625	0.5625	0.5545	13/16	0.812	0.798	0.938	0.910	23/64	0.371	0.348	1.375
5/8	0.6250	0.6250	0.6170	15/16	0.938	0.922	1.083	1.051	25/64	0.403	0.378	1.500
3/4	0.7500	0.7500	0.7410	1 1/8	1.125	1.100	1.299	1.254	15/32	0.483	0.455	1.750
7/8	0.8750	0.8750	0.8660	1 5/16	1.312	1.285	1.516	1.465	35/64	0.563	0.531	2.000
1	1.0000	1.0000	0.9900	1 1/2	1.500	1.469	1.732	1.675	39/64	0.627	0.591	2.250
1 1/8	1.1250	1.1250	1.1140	1 11/16	1.688	1.631	1.949	1.859	11/16	0.718	0.658	2.500
1 1/4	1.2500	1.2500	1.2390	1 7/8	1.875	1.812	2.165	2.066	25/35	0.813	0.749	2.750
1 3/8	1.3750	1.3750	1.3630	2 1/16	2.062	1.994	2.382	2.273	27/32	0.878	0.810	3.000
1 1/2	1.5000	1.5000	1.4880	2 1/4	2.250	2.175	2.598	2.480	15/16	0.974	0.902	3.250
1 3/4	1.7500	1.7500	1.7380	2 5/8	2.625	2.538	3.031	2.893	1 3/32	1.134	1.054	3.750
2	2.0000	2.0000	1.9880	3	3.000	2.900	3.464	3.306	1 7/32	1.263	1.175	4.250
2 1/4	2.2500	2.2500	2.2380	3 3/8	3.375	3.262	3.897	3.719	1 3/8	1.423	1.327	5.000[c]
2 1/2	2.5000	2.5000	2.4880	3 3/4	3.750	3.625	4.330	4.133	1 17/32	1.583	1.479	5.500[c]
2 3/4	2.7500	2.7500	2.7380	4 1/8	4.125	3.988	4.763	4.546	1 11/16	1.744	1.632	6.000[c]
3	3.0000	3.0000	2.9880	4 1/2	4.500	4.350	5.196	4.959	1 7/8	1.935	1.815	6.500[c]

[a] *Nominal Size:* Where specifying nominal size in decimals, zeros preceding the decimal and in the fourth decimal place are omitted.

[b] Thread lengths, L_T, shown are for bolt lengths 6 inches and shorter. For longer bolt lengths add 0.250 inch to thread lengths shown.

[c] Thread lengths, L_T, shown are for bolt lengths over 6 inches.

All dimensions are in inches.

Unification: **Bold type indicates product features unified dimensionally with British and Canadian Standards.** Unification of fine thread products is limited to sizes 1 inch and smaller.

Bearing Surface: Bearing surface is flat and washer faced. Diameter of bearing surface is equal to the maximum width across flats within a tolerance of minus 10 per cent.

Threads Series: Threads, when rolled, are Unified Coarse, Fine, or 8-thread series (UNRC, UNRF, or 8 UNR Series), Class 2A. Threads produced by other methods shall preferably be UNRC, UNRF or 8 UNR but, at manufacturer's option, may be Unified Coarse, Fine or 8-thread series (UNC, UNF, or 8 UN Series), Class 2A.

Material: Chemical and mechanical properties of steel screws normally conform to Grades 2, 5, or 8 of SAE J429, ASTM A449 or ASTM A354 Grade BD. Where specified, screws may also be made from brass, bronze, corrosion-resisting steel, aluminum alloy or other materials.

Table 5. American National Standard Square Lag Screws *ANSI/ASME B18.2.1-1996*

Nominal Size[a] or Basic Product Dia.		Body or Shoulder Dia. E		Width Across Flats F			Width Across Corners G		Height H			Shoulder Length S	Radius of Fillet R	Thds. per Inch	Thread Dimensions			
		Max.	Min.	Basic	Max.	Min.	Max.	Min.	Basic	Max.	Min.	Min.	Max.		Pitch P	Flat at Root B	Depth of Thd. T	Root Dia. D_1
No. 10	0.1900	0.199	0.178	9/32	0.281	0.271	0.398	0.372	1/8	0.140	0.110	0.094	0.03	11	0.091	0.039	0.035	0.120
1/4	0.2500	0.260	0.237	3/8	0.375	0.362	0.530	0.498	11/64	0.188	0.156	0.094	0.03	10	0.100	0.043	0.039	0.173
5/16	0.3125	0.324	0.298	1/2	0.500	0.484	0.707	0.665	13/64	0.220	0.186	0.125	0.03	9	0.111	0.048	0.043	0.227
3/8	0.3750	0.388	0.360	9/16	0.562	0.544	0.795	0.747	1/4	0.268	0.232	0.125	0.03	7	0.143	0.062	0.055	0.265
7/16	0.4375	0.452	0.421	5/8	0.625	0.603	0.884	0.828	19/64	0.316	0.278	0.156	0.03	7	0.143	0.062	0.055	0.328
1/2	0.5000	0.515	0.482	3/4	0.750	0.725	1.061	0.995	21/64	0.348	0.308	0.156	0.03	6	0.167	0.072	0.064	0.371
5/8	0.6250	0.642	0.605	15/16	0.938	0.906	1.326	1.244	27/64	0.444	0.400	0.312	0.06	5	0.200	0.086	0.077	0.471
3/4	0.7500	0.768	0.729	1 1/8	1.125	1.088	1.591	1.494	1/2	0.524	0.476	0.375	0.06	4 1/2	0.222	0.096	0.085	0.579
7/8	0.8750	0.895	0.852	1 5/16	1.312	1.269	1.856	1.742	19/42	0.620	0.568	0.375	0.06	4	0.250	0.108	0.096	0.683
1	1.0000	1.022	0.976	1 1/2	1.500	1.450	2.121	1.991	21/32	0.684	0.628	0.625	0.09	3 1/2	0.286	0.123	0.110	0.780
1 1/8	1.1250	1.149	1.098	1 11/16	1.688	1.631	2.386	2.239	3/4	0.780	0.720	0.625	0.09	3 1/4	0.308	0.133	0.119	0.887
1 1/4	1.2500	1.277	1.223	1 7/8	1.875	1.812	2.652	2.489	27/32	0.876	0.812	0.625	0.09	3 1/4	0.308	0.133	0.119	1.012

[a] When specifying decimal nominal size, zeros before decimal point and in fourth decimal place are omitted.

All dimensions in inches.

Minimum thread length is 1/2 length of screw plus 0.50 inch, or 6.00 inches, whichever is shorter. Screws too short for the formula thread length shall be threaded as close to the head as practicable.

Thread formulas: Pitch = 1 ÷ thds. per inch. Flat at root = 0.4305 × pitch. Depth of single thread = 0.385 × pitch.

Table 6. American National Standard Hex Lag Screws *ANSI/ASME B18.2.1-1996*

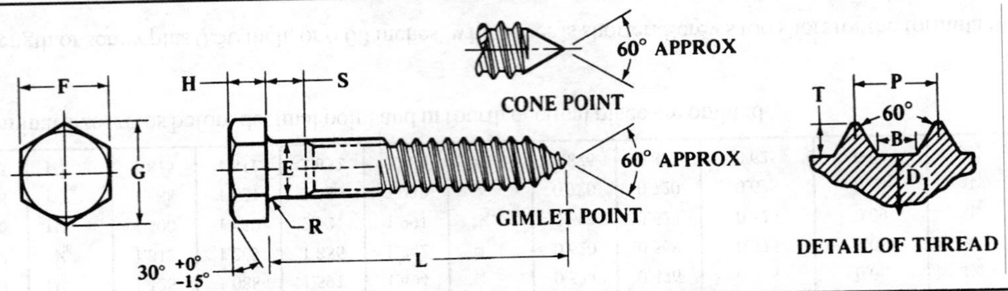

Nominal Size[a] or Basic Product Dia.		Body or Shoulder Dia. *E*		Width Across Flats *F*			Width Across Corners *G*		Height *H*			Shoulder Length *S*	Radius of Fillet *R*	Thds. per Inch	Thread Dimensions			
		Max.	Min.	Basic	Max.	Min.	Max.	Min.	Basic	Max.	Min.	Min.	Max.		Pitch *P*	Flat at Root *B*	Depth of Thd. *T*	Root Dia. D_1
No. 10	0.1900	0.199	0.178	9⁄32	0.281	0.271	0.323	0.309	1⁄8	0.140	0.110	0.094	0.03	11	0.091	0.039	0.035	0.120
1⁄4	0.2500	0.260	0.237	3⁄8	0.438	0.425	0.505	0.484	11⁄64	0.188	0.150	0.094	0.03	10	0.100	0.043	0.039	0.173
5⁄16	0.3125	0.324	0.298	1⁄2	0.500	0.484	0.577	0.552	7⁄32	0.235	0.195	0.125	0.03	9	0.111	0.048	0.043	0.227
3⁄8	0.3750	0.388	0.360	9⁄16	0.562	0.544	0.650	0.620	1⁄4	0.268	0.226	0.125	0.03	7	0.143	0.062	0.055	0.265
7⁄16	0.4375	0.452	0.421	5⁄8	0.625	0.603	0.722	0.687	19⁄64	0.316	0.272	0.156	0.03	7	0.143	0.062	0.055	0.328
1⁄2	0.5000	0.515	0.482	3⁄4	0.750	0.725	0.866	0.826	11⁄32	0.364	0.302	0.156	0.03	6	0.167	0.072	0.064	0.371
5⁄8	0.6250	0.642	0.605	15⁄16	0.938	0.906	1.083	1.033	27⁄64	0.444	0.378	0.312	0.06	5	0.200	0.086	0.077	0.471
3⁄4	0.7500	0.768	0.729	1 1⁄8	1.125	1.088	1.299	1.240	1⁄2	0.524	0.455	0.375	0.06	4 1⁄2	0.222	0.096	0.085	0.579
7⁄8	0.8750	0.895	0.852	1 5⁄16	1.312	1.269	1.516	1.447	37⁄64	0.604	0.531	0.375	0.06	4	0.250	0.108	0.096	0.683
1	1.0000	1.022	0.976	1 1⁄2	1.500	1.450	1.732	1.653	43⁄64	0.700	0.591	0.625	0.09	3 1⁄2	0.286	0.123	0.110	0.780
1 1⁄8	1.1250	1.149	1.098	1 11⁄16	1.688	1.631	1.949	1.859	3⁄4	0.780	0.658	0.625	0.09	3 1⁄4	0.308	0.133	0.119	0.887
1 1⁄4	1.2500	1.277	1.223	1 7⁄8	1.875	1.812	2.165	2.066	27⁄32	0.876	0.749	0.625	0.09	3 1⁄4	0.308	0.133	0.119	1.012

[a] When specifying decimal nominal size, zeros before decimal point and in fourth decimal place are omitted.

All dimensions in inches.

Minimum thread length is 1⁄2 length of screw plus 0.50 inch, or 6.00 inches, whichever is shorter. Screws too short for the formula thread length shall be threaded as close to the head as practicable.

Thread formulas: Pitch = 1 ÷ thds. per inch. Flat at root = 0.4305 × pitch. Depth of single thread = 0.385 × pitch.

Table 7. American National Standard and Unified Standard Hex Nuts and Jam Nuts and Heavy Hex Nuts and Jam *ANSI/ASME B18.2.2-1987 (R1999)*

Nominal Size or Basic Major Dia. of Thread		Width Across Flats F			Width Across Corners G		Thickness, Nuts H			Thickness, Jam Nuts H_1		
		Basic	Max.	Min.	Max.	Min.	Basic	Max.	Min.	Basic	Max.	Min.
Hex Nuts and Hex Jam Nuts												
1/4	**0.2500**	**7/16**	**0.438**	**0.428**	**0.505**	**0.488**	**7/32**	**0.226**	**0.212**	**5/32**	**0.163**	**0.150**
5/16	**0.3125**	**1/2**	**0.500**	**0.489**	**0.577**	**0.557**	**17/64**	**0.273**	**0.258**	**3/16**	**0.195**	**0.180**
3/8	**0.3750**	**9/16**	**0.562**	**0.551**	**0.650**	**0.628**	**21/64**	**0.337**	**0.320**	**7/32**	**0.227**	**0.210**
7/16	**0.4375**	**11/16**	**0.688**	**0.675**	**0.794**	**0.768**	**3/8**	**0.385**	**0.365**	**1/4**	**0.260**	**0.240**
1/2	**0.5000**	**3/4**	**0.750**	**0.736**	**0.866**	**0.840**	**7/16**	**0.448**	**0.427**	**5/16**	**0.323**	**0.302**
9/16	**0.5625**	**7/8**	**0.875**	**0.861**	**1.010**	**0.982**	**31/64**	**0.496**	**0.473**	**5/16**	**0.324**	**0.301**
5/8	**0.6250**	**15/16**	**0.938**	**0.922**	**1.083**	**1.051**	**35/64**	**0.559**	**0.535**	**3/8**	**0.387**	**0.363**
3/4	**0.7500**	**1 1/8**	**1.125**	**1.088**	**1.299**	**1.240**	**41/64**	**0.665**	**0.617**	**27/64**	**0.446**	**0.398**
7/8	**0.8750**	**1 5/16**	**1.312**	**1.269**	**1.516**	**1.447**	**3/4**	**0.776**	**0.724**	**31/64**	**0.510**	**0.458**
1	**1.0000**	**1 1/2**	**1.500**	**1.450**	**1.732**	**1.653**	**55/64**	**0.887**	**0.831**	**35/64**	**0.575**	**0.519**
1 1/8	**1.1250**	**1 11/16**	**1.688**	**1.631**	**1.949**	**1.859**	**31/32**	**0.999**	**0.939**	**39/64**	**0.639**	**0.579**
1 1/4	**1.2500**	**1 7/8**	**1.875**	**1.812**	**2.165**	**2.066**	**1 1/16**	**1.094**	**1.030**	**23/32**	**0.751**	**0.687**
1 3/8	**1.3750**	**2 1/16**	**2.062**	**1.994**	**2.382**	**2.273**	**1 11/64**	**1.206**	**1.138**	**25/32**	**0.815**	**0.747**
1 1/2	**1.5000**	**2 1/4**	**2.250**	**2.175**	**2.598**	**2.480**	**1 9/32**	**1.317**	**1.245**	**27/32**	**0.880**	**0.808**
Heavy Hex Nuts and Heavy Hex Jam Nuts												
1/4	0.2500	1/2	0.500	0.488	0.577	0.556	15/64	0.250	0.218	11/64	0.188	0.156
5/16	0.3125	9/16	0.562	0.546	0.650	0.622	19/64	0.314	0.280	13/64	0.220	0.186
3/8	0.3750	11/16	0.688	0.669	0.794	0.763	23/64	0.377	0.341	15/64	0.252	0.216
7/16	0.4375	3/4	0.750	0.728	0.866	0.830	27/64	0.441	0.403	17/64	0.285	0.247
1/2	**0.5000**	**7/8**	**0.875**	**0.850**	**1.010**	**0.969**	**31/64**	**0.504**	**0.464**	**19/64**	**0.317**	**0.277**
9/16	0.5625	15/16	0.938	0.909	1.083	1.037	35/64	0.568	0.526	21/64	0.349	0.307
5/8	**0.6250**	**1 1/16**	**1.062**	**1.031**	**1.227**	**1.1175**	**39/64**	**0.631**	**0.587**	**23/64**	**0.381**	**0.337**
3/4	**0.7500**	**1 1/4**	**1.250**	**1.212**	**1.443**	**1.382**	**47/64**	**0.758**	**0.710**	**27/64**	**0.446**	**0.398**
7/8	**0.8750**	**1 7/16**	**1.438**	**1.394**	**1660**	**1.589**	**55/64**	**0.885**	**0.833**	**31/64**	**0.510**	**0.458**
1	**1.0000**	**1 5/8**	**1.625**	**1.575**	**1.876**	**1.796**	**63/64**	**1.012**	**0.956**	**35/64**	**0.575**	**0.519**
1 1/8	**1.1250**	**1 13/16**	**1.812**	**1.756**	**2.093**	**2.002**	**1 7/64**	**1.139**	**1.079**	**39/64**	**0.639**	**0.579**
1 1/4	**1.2500**	**2**	**2.000**	**1.938**	**2.309**	**2.209**	**1 7/32**	**1.251**	**1.187**	**23/32**	**0.751**	**0.687**
1 3/8	**1.3750**	**2 3/16**	**2.188**	**2.119**	**2.526**	**2.416**	**1 11/32**	**1.378**	**1.310**	**25/32**	**0.815**	**0.747**
1 1/2	**1.5000**	**2 3/8**	**2.375**	**2.300**	**2.742**	**2.622**	**1 15/32**	**1.505**	**1.433**	**27/32**	**0.880**	**0.808**
1 5/8	1.6250	2 9/16	2.562	2.481	2.959	2.828	1 19/32	1.632	1.556	29/32	0.944	0.868
1 3/4	**1.7500**	**2 3/4**	**2.750**	**2.662**	**3.175**	**3.035**	**1 23/32**	**1.759**	**1.679**	**31/32**	**1.009**	**0.929**
1 7/8	1.8750	2 15/16	2.938	2.844	3.392	3.242	1 27/32	1.886	1.802	1 1/32	1.073	0.989
2	**2.0000**	**3 1/8**	**3.125**	**3.025**	**3.608**	**3.449**	**1 31/32**	**2.013**	**1.925**	**1 3/32**	**1.138**	**1.050**
2 1/4	2.2500	3 1/2	3.500	3.388	4.041	3.862	2 13/64	2.251	2.155	1 13/64	1.251	1.155
2 1/2	2.5000	3 7/8	3.875	3.750	4.474	4.275	2 29/64	2.505	2.401	1 29/64	1.505	1.401
2 3/4	2.7500	4 1/4	4.250	4.112	4.907	4.688	2 45/64	2.759	2.647	1 37/64	1.634	1.522
3	3.0000	4 5/8	4.625	4.475	5.340	5.102	2 61/64	3.013	2.893	1 45/64	1.763	1.643
3 1/4	3.2500	5	5.000	4.838	5.774	5.515	3 3/16	3.252	3.124	1 13/16	1.876	1.748
3 1/2	3.5000	5 3/8	5.375	5.200	6.207	5.928	3 7/16	3.506	3.370	1 15/16	2.006	1.870
3 3/4	3.7500	5 3/4	5.750	5.562	6.640	6.341	3 11/16	3.760	3.616	2 1/16	2.134	1.990
4	4.0000	6 1/8	6.125	5.925	7.073	6.755	3 15/16	4.014	3.862	2 3/16	2.264	2.112

All dimensions are in inches.

Bold type shows nuts unified dimensionally with British and Canadian Standards.

Threads are Unified Coarse-, Fine-, or 8-thread series (UNC, UNF or 8UN), Class 2B. Unification of fine-thread nuts is limited to sizes 1 inch and under.

Table 8. American National Standard and Unified Standard Hex Flat Nuts and Flat Jam Nuts and Heavy Hex Flat Nuts and Flat Jam Nuts

ANSI/ASME B18.2.2-1987 (R1999)

Nominal Size or Basic Major Dia. of Thread		Width Across Flats F			Width Across Corners G		Thickness, Flat Nuts H			Thickness, Flat Jam Nuts H_1		
		Basic	Max.	Min.	Max.	Min.	Basic	Max.	Min.	Basic	Max.	Min.
Hex Flat Nuts and Hex Flat Jam Nuts												
1 1/8	**1.1250**	**1 11/16**	**1.688**	**1.631**	**1.949**	**1.859**	**1**	**1.030**	**0.970**	**5/8**	**0.655**	**0.595**
1 1/4	**1.2500**	**1 7/8**	**1.875**	**1.812**	**2.165**	**2.066**	**1 3/32**	**1.126**	**1.062**	**3/4**	**0.782**	**0.718**
1 3/8	**1.3750**	**2 1/16**	**2.062**	**1.994**	**2.382**	**2.273**	**1 13/64**	**1.237**	**1.169**	**13/16**	**0.846**	**0.778**
1 1/2	**1.5000**	**2 1/4**	**2.250**	**2.175**	**2.598**	**2.480**	**1 5/16**	**1.348**	**1.276**	**7/8**	**0.911**	**0.839**
Heavy Hex Flat Nuts and Heavy Hex Flat Jam Nuts												
1 1/8	**1.1250**	**1 13/16**	**1.812**	**1.756**	**2.093**	**2.002**	**1 1/8**	**1.155**	**1.079**	**5/8**	**0.655**	**0.579**
1 1/4	**1.2500**	**2**	**2.000**	**1.938**	**2.309**	**2.209**	**1 1/4**	**1.282**	**1.187**	**3/4**	**0.782**	**0.687**
1 3/8	**1.3750**	**2 3/16**	**2.188**	**2.119**	**2.526**	**2.416**	**1 3/8**	**1.409**	**1.310**	**13/16**	**0.846**	**0.747**
1 1/2	**1.5000**	**2 3/8**	**2.375**	**2.300**	**2.742**	**2.622**	**1 1/2**	**1.536**	**1.433**	**7/8**	**0.911**	**0.808**
1 3/4	**1.7500**	**2 3/4**	**2.750**	**2.662**	**3.175**	**3.035**	**1 3/4**	**1.790**	**1.679**	**1**	**1.040**	**0.929**
2	**2.0000**	**3 1/8**	**3.125**	**3.025**	**3.608**	**3.449**	**2**	**2.044**	**1.925**	**1 1/8**	**1.169**	**1.050**
2 1/4	2.2500	3 1/2	3.500	3.388	4.041	3.862	2 1/4	2.298	2.155	1 1/4	1.298	1.155
2 1/2	2.5000	3 7/8	3.875	3.750	4.474	4.275	2 1/2	2.552	2.401	1 1/2	1.552	1.401
2 3/4	2.7500	4 1/4	4.250	4.112	4.907	4.688	2 3/4	2.806	2.647	1 5/8	1.681	1.522
3	3.0000	4 5/8	4.625	4.475	5.340	5.102	3	3.060	2.893	1 3/4	1.810	1.643
3 1/4	3.2500	5	5.000	4.838	5.774	5.515	3 1/4	3.314	3.124	1 7/8	1.939	1.748
3 1/2	3.5000	5 3/8	5.375	5.200	6.207	5.928	3 1/2	3.568	3.370	2	2.068	1.870
3 3/4	3.7500	5 3/4	5.750	5.562	6.640	6.341	3 3/4	3.822	3.616	2 1/8	2.197	1.990
4	4.0000	6 1/8	6.125	5.925	7.073	6.755	4	4.076	3.862	2 1/4	2.326	2.112

All dimensions are in inches.

Bold type indicates nuts unified dimensionally with British and Canadian Standards.

Threads are Unified Coarse-thread series (UNC), Class 2B.

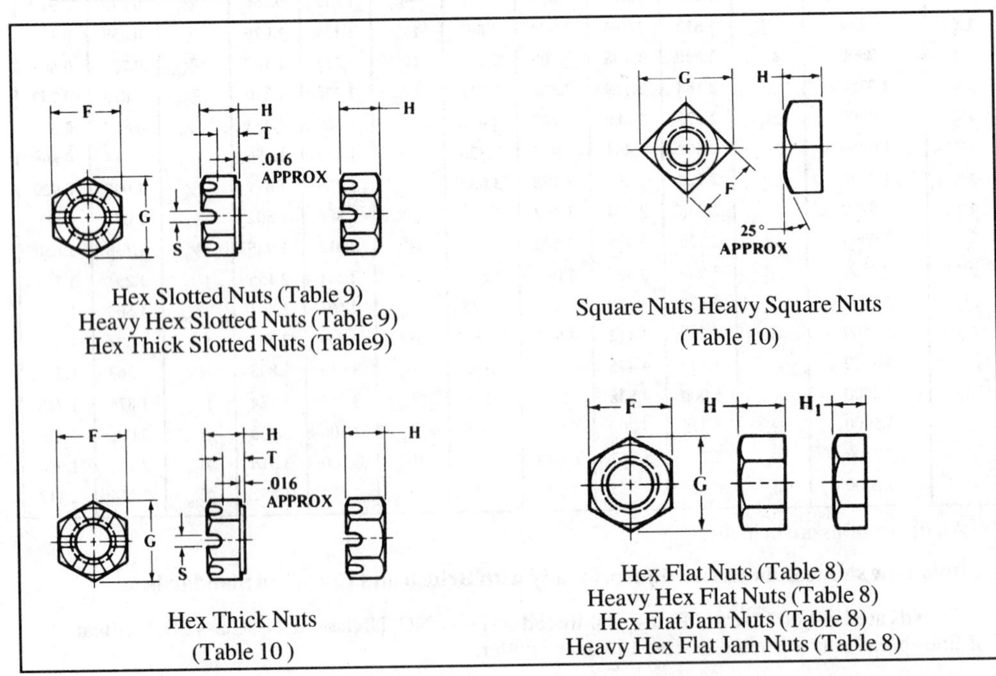

Hex Slotted Nuts (Table 9)
Heavy Hex Slotted Nuts (Table 9)
Hex Thick Slotted Nuts (Table9)

Square Nuts Heavy Square Nuts (Table 10)

Hex Thick Nuts (Table 10)

Hex Flat Nuts (Table 8)
Heavy Hex Flat Nuts (Table 8)
Hex Flat Jam Nuts (Table 8)
Heavy Hex Flat Jam Nuts (Table 8)

Table 9. American National and Unified Standard Hex Slotted Nuts, Heavy Hex Slotted Nuts, and Hex Thick Slotted Nuts *ANSI/ASME B18.2.2-1987 (R1999)*

Nominal Size or Basic Major Dia. of Thread		Width Across Flats F			Width Across Corners G		Thickness H			Unslotted Thickness T		Width of Slot S	
		Basic	Max.	Min.	Max.	Min.	Basic	Max.	Min.	Max.	Min.	Max.	Min.
Hex Slotted Nuts													
1/4	**0.2500**	**7/16**	**0.438**	**0.428**	**0.505**	**0.488**	**7/32**	**0.226**	**0.212**	0.14	0.12	0.10	0.07
5/16	**0.3125**	**1/2**	**0.500**	**0.489**	**0.577**	**0.577**	**17/64**	**0.273**	**0.258**	0.18	0.16	0.12	0.09
3/8	**0.3750**	**9/16**	**0.562**	**0.551**	**0.650**	**0.628**	**21/64**	**0.337**	**0.320**	0.21	0.19	0.15	0.12
7/16	**0.4375**	**11/16**	**0.688**	**0.675**	**0.794**	**0.768**	**3/8**	**0.385**	**0.365**	0.23	0.21	0.15	0.12
1/2	**0.5000**	**3/4**	**0.750**	**0.736**	**0.866**	**0.840**	**7/16**	**0.448**	**0.427**	0.29	0.27	0.18	0.15
9/16	**0.5625**	**7/8**	**0.875**	**0.861**	**1.010**	**0.982**	**31/64**	**0.496**	**0.473**	0.31	0.29	0.18	0.15
5/8	**0.6250**	**15/16**	**0.938**	**0.922**	**1.083**	**1.051**	**35/64**	**0.559**	**0.535**	0.34	0.32	0.24	0.18
3/4	**0.7500**	**1 1/8**	**1.125**	**1.088**	**1.299**	**1.240**	**41/64**	**0.665**	**0.617**	0.40	0.38	0.24	0.18
7/8	**0.8750**	**1 5/16**	**1.312**	**1.269**	**1.516**	**1.447**	**3/4**	**0.776**	**0.724**	0.52	0.49	0.24	0.18
1	**1.0000**	**1 1/2**	**1.500**	**1.450**	**1.732**	**1.653**	**55/64**	**0.887**	**0.831**	0.59	0.56	0.30	0.24
1 1/8	**1.1250**	**1 11/16**	**1.688**	**1.631**	**1.949**	**1.859**	**31/32**	**0.999**	**0.939**	0.64	0.61	0.33	0.24
1 1/4	**1.2500**	**1 7/8**	**1.875**	**1.812**	**2.165**	**2.066**	**1 1/16**	**1.094**	**1.030**	0.70	0.67	0.40	0.31
1 3/8	**1.3750**	**2 1/16**	**2.062**	**1.994**	**2.382**	**2.273**	**1 11/64**	**1.206**	**1.138**	0.82	0.78	0.40	0.31
1 1/2	**1.5000**	**2 1/4**	**2.250**	**2.175**	**2.598**	**2.480**	**1 9/32**	**1.317**	**1.245**	0.86	0.82	0.46	0.37
Heavy Hex Slotted Nuts													
1/4	0.2500	1/2	0.500	0.488	0.577	0.556	15/64	0.250	0.218	0.15	0.13	0.10	0.07
5/16	0.3125	9/16	0.562	0.546	0.650	0.622	19/64	0.314	0.280	0.21	0.19	0.12	0.09
3/8	0.3750	11/16	0.688	0.669	0.794	0.763	23/64	0.377	0.341	0.24	0.22	0.15	0.12
7/16	0.4375	3/4	0.750	0.728	0.866	0.830	27/64	0.441	0.403	0.28	0.26	0.15	0.12
1/2	**0.5000**	**7/8**	**0.875**	**0.850**	**1.010**	**0.969**	**31/64**	**0.504**	**0.464**	0.34	0.32	0.18	0.15
9/16	0.5625	15/16	0.938	0.909	1.083	1.037	35/64	0.568	0.526	0.37	0.35	0.18	0.15
5/8	**0.6250**	**1 1/16**	**1.062**	**1.031**	**1.227**	**1.175**	**39/64**	**0.631**	**0.587**	0.40	0.38	0.24	0.18
3/4	**0.7500**	**1 1/4**	**1.250**	**1.212**	**1.443**	**1.382**	**47/64**	**0.758**	**0.710**	0.49	0.47	0.24	0.18
7/8	**0.8750**	**1 7/16**	**1.438**	**1.394**	**1.660**	**1.589**	**55/64**	**0.885**	**0.833**	0.62	0.59	0.24	0.18
1	**1.0000**	**1 5/8**	**1.625**	**1.575**	**1.876**	**1.796**	**63/64**	**1.012**	**0.956**	0.72	0.69	0.30	0.24
1 1/8	**1.1250**	**1 13/16**	**1.812**	**1.756**	**2.093**	**2.002**	**1 7/64**	**1.139**	**1.079**	0.78	0.75	0.33	0.24
1 1/4	**1.2500**	**2**	**2.000**	**1.938**	**2.309**	**2.209**	**1 7/32**	**1.251**	**1.187**	0.86	0.83	0.40	0.31
1 3/8	**1.3750**	**2 3/16**	**2.188**	**2.119**	**2.526**	**2.416**	**1 11/32**	**1.378**	**1.310**	0.99	0.95	0.40	0.31
1 1/2	**1.5000**	**2 3/8**	**2.375**	**2.300**	**2.742**	**2.622**	**1 15/32**	**1.505**	**1.433**	1.05	1.01	0.46	0.37
1 3/4	**1.7500**	**2 3/4**	**2.750**	**2.662**	**3.175**	**3.035**	**1 23/32**	**1.759**	**1.679**	1.24	1.20	0.52	0.43
2	**2.0000**	**3 1/8**	**3.125**	**3.025**	**3.608**	**3.449**	**1 31/32**	**2.013**	**1.925**	1.43	1.38	0.52	0.43
2 1/4	2.2500	3 1/2	3.500	3.388	4.041	3.862	2 13/64	2.251	2.155	1.67	1.62	0.52	0.43
2 1/2	2.5000	3 7/8	3.875	3.750	4.474	4.275	2 29/64	2.505	2.401	1.79	1.74	0.64	0.55
2 3/4	2.7500	4 1/4	4.250	4.112	4.907	4.688	2 45/64	2.759	2.647	2.05	1.99	0.64	0.55
3	3.0000	4 5/8	4.625	4.475	5.340	5.102	2 61/64	3.013	2.893	2.23	2.17	0.71	0.62
3 1/4	3.2500	5	5.000	4.838	5.774	5.515	3 3/16	3.252	3.124	2.47	2.41	0.71	0.62
3 1/2	3.5000	5 3/8	5.375	5.200	6.207	5.928	3 7/16	3.506	3.370	2.72	2.65	0.71	0.62
3 3/4	3.7500	5 3/4	5.750	5.562	6.640	6.341	3 11/16	3.760	3.616	2.97	2.90	0.71	0.62
4	4.0000	6 1/8	6.125	5.925	7.073	6.755	3 15/16	4.014	3.862	3.22	3.15	0.71	0.62
Hex Thick Slotted Nuts													
1/4	**0.2500**	**7/16**	**0.438**	**0.428**	**0.505**	**0.488**	**9/32**	**0.288**	**0.274**	0.20	0.18	0.10	0.07
5/16	**0.3125**	**1/2**	**0.500**	**0.489**	**0.577**	**0.557**	**21/64**	**0.336**	**0.320**	0.24	0.22	0.12	0.09
3/8	**0.3750**	**9/16**	**0.562**	**0.551**	**0.650**	**0.628**	**13/32**	**0.415**	**0.398**	0.29	0.27	0.15	0.12
7/16	**0.4375**	**11/16**	**0.688**	**0.675**	**0.794**	**0.768**	**29/24**	**0.463**	**0.444**	0.31	0.29	0.15	0.12
1/2	**0.5000**	**3/4**	**0.750**	**0.736**	**0.866**	**0.840**	**9/16**	**0.573**	**0.552**	0.42	0.40	0.18	0.15
9/16	**0.5625**	**7/8**	**0.875**	**0.861**	**1.010**	**0.982**	**39/64**	**0.621**	**0.598**	0.43	0.41	0.18	0.15
5/8	**0.6250**	**15/16**	**0.938**	**0.922**	**1.083**	**1.051**	**23/32**	**0.731**	**0.706**	0.51	0.49	0.24	0.18
3/4	**0.7500**	**1 1/8**	**1.125**	**1.088**	**1.299**	**1.240**	**13/16**	**0.827**	**0.798**	0.57	0.55	0.24	0.18
7/8	**0.8750**	**1 5/16**	**1.312**	**1.269**	**1.516**	**1.447**	**29/32**	**0.922**	**0.890**	0.67	0.64	0.24	0.18
1	**1.0000**	**1 1/2**	**1.500**	**1.450**	**1.732**	**1.653**	**1**	**1.018**	**0.982**	0.73	0.70	0.30	0.24
1 1/8	**1.1250**	**1 11/16**	**1.688**	**1.631**	**1.949**	**1.859**	**1 5/32**	**1.176**	**1.136**	0.83	0.80	0.33	0.24
1 1/4	**1.2500**	**1 7/8**	**1.875**	**1.812**	**2.165**	**2.066**	**1 1/4**	**1.272**	**1.228**	0.89	0.86	0.40	0.31
1 3/8	**1.3750**	**2 1/16**	**2.062**	**1.994**	**2.382**	**2.273**	**1 3/8**	**1.399**	**1.351**	1.02	0.98	0.40	0.31
1 1/2	**1.5000**	**2 1/4**	**2.250**	**2.175**	**2.598**	**2.480**	**1 1/2**	**1.526**	**1.474**	1.08	1.04	0.46	0.37

All dimensions are in inches.

Bold type indicates nuts unified dimensionally with British and Canadian Standards.

Threads are Unified Coarse-, Fine-, or 8-thread series (UNC, UNF, or 8UN), Class 2B.

Unification of fine-thread nuts is limited to sizes 1 inch and under.

Table 10. American National and Unified Standard Square Nuts and Heavy Square Nuts and American National Standard Hex Thick Nuts

ANSI/ASME B18.2.2-1987 (R1999)

Nominal Size or Basic Major Dia. of Thread		Width Across Flats *F*			Width Across Corners *G*		Thickness *H*		
		Basic	Max.	Min.	Max.	Min.	Basic	Max.	Min.
Square Nuts[a]									
1/4	**0.2500**	**7/16**	**0.438**	**0.425**	**0.619**	**0.554**	**7/32**	**0.235**	**0.203**
5/16	**0.3125**	**9/16**	**0.562**	**0.547**	**0.795**	**0.721**	**17/64**	**0.283**	**0.249**
3/8	**0.3750**	**5/8**	**0.625**	**0.606**	**0.884**	**0.802**	**21/64**	**0.346**	**0.310**
7/16	**0.4375**	**3/4**	**0.750**	**0.728**	**1.061**	**0.970**	**3/8**	**0.394**	**0.356**
1/2	**0.5000**	**13/16**	**0.812**	**0.788**	**1.149**	**1.052**	**7/16**	**0.458**	**0.418**
5/8	**0.6250**	**1**	**1.000**	**0.969**	**1.414**		**35/64**	**0.569**	**0.525**
3/4	**0.7500**	**1 1/8**	**1.125**	**1.088**	**1.591**	**1.464**	**21/32**	**0.680**	**0.632**
7/8	**0.8750**	**1 5/16**	**1.312**	**1.269**	**1.856**	**1.712**	**49/64**	**0.792**	**0.740**
1	**1.0000**	**1 1/2**	**1.500**	**1.450**	**2.121**	**1.961**	**7/8**	**0.903**	**0.847**
1 1/8	**1.1250**	**1 11/16**	**1.688**	**1.631**	**2.386**	**2.209**	**1**	**1.030**	**0.970**
1 1/4	**1.2500**	**1 7/8**	**1.875**	**1.812**	**2.652**	**2.458**	**1 3/32**	**1.126**	**1.062**
1 3/8	**1.3750**	**2 1/16**	**2.062**	**1.994**	**2.917**	**2.708**	**1 13/64**	**1.237**	**1.169**
1 1/2	**1.5000**	**2 1/4**	**2.250**	**2.175**	**3.182**	**2.956**	**1 5/16**	**1.348**	**1.276**
Heavy Square Nuts[a]									
1/4	0.2500	1/2	0.500	0.488	0.707	0.640	1/4	0.266	0.218
5/16	0.3125	9/16	0.562	0.546	0.795	0.720	5/16	0.330	0.280
3/8	0.3750	11/16	0.688	0.669	0.973	0.889	3/8	0.393	0.341
7/16	0.4375	3/4	0.750	0.728	1.060	0.970	7/16	0.456	0.403
1/2	0.5000	7/8	0.875	0.850	1.237	1.137	1/2	0.520	0.464
5/8	0.6250	1 1/16	1.062	1.031	1.503	1.386	5/8	0.647	0.587
3/4	0.7500	1 1/4	1.250	1.212	1.768	1.635	3/4	0.774	0.710
7/8	0.8750	1 7/16	1.438	1.394	2.033	1.884	7/8	0.901	0.833
1	1.0000	1 5/8	1.625	1.575	2.298	2.132	1	1.028	0.956
1 1/8	1.1250	1 13/16	1.812	1.756	2.563	2.381	1 1/8	1.155	1.079
1 1/4	1.2500	2	2.000	1.938	2.828	2.631	1 1/4	1.282	1.187
1 3/8	1.3750	2 3/16	2.188	2.119	3.094	2.879	1 3/8	1.409	1.310
1 1/2	1.5000	2 3/8	2.375	2.300	3.359	3.128	1 1/2	1.536	1.433
Hex Thick Nuts[b]									
1/4	0.2500	7/16	0.438	0.428	0.505	0.488	9/32	0.288	0.274
5/16	0.3125	1/2	0.500	0.489	0.577	0.557	21/64	0.336	0.320
3/8	0.3750	9/16	0.562	0.551	0.650	0.628	13/32	0.415	0.398
7/16	0.4375	11/16	0.688	0.675	0.794	0.768	29/64	0.463	0.444
1/2	0.5000	3/4	0.750	0.736	0.866	0.840	9/16	0.573	0.552
9/16	0.5625	7/8	0.875	0.861	1.010	0.982	39/64	0.621	0.598
5/8	0.6250	15/16	0.938	0.922	1.083	1.051	23/32	0.731	0.706
3/4	0.7500	1 1/8	1.125	1.088	1.299	1.240	13/16	0.827	0.798
7/8	0.8750	1 5/16	1.312	1.269	1.516	1.447	29/32	0.922	0.890
1	1.0000	1 1/2	1.500	1.450	1.732	1.653	1	1.018	0.982
1 1/8	1.1250	1 11/16	1.688	1.631	1.949	1.859	1 5/32	1.176	1.136
1 1/4	1.2500	1 7/8	1.875	1.812	2.165	2.066	1 1/4	1.272	1.228
1 3/8	1.3750	2 1/16	2.062	1.994	2.382	2.273	1 3/8	1.399	1.351
1 1/2	1.5000	2 1/4	2.250	2.175	2.598	2.480	1 1/2	1.526	1.474

[a] Coarse-thread series, Class 2B.

[b] Unified Coarse-, Fine-, or 8-thread series (8 UN), Class 2B.

All dimensions are in inches.

Bold type indicates nuts unified dimensionally with British and Canadian Standards.

Low and High Crown (Blind, Acorn) Nuts *SAE Recommended Practice J483a*

Low Crown

Nom. Size[a] or Basic Major Dia. of Thread		Width Across Flats, *F*			Width Across Corners, *G*		Body Dia., *A*	Over-all Hgt., *H*	Hexa-gon Hgt., *Q*	Nose Rad., *R*	Body Rad., *S*	Drill Dep., *T*	Full Thd., *U*
		Max.	(Basic)	Min.	Max.	Min.						Max.	Min.
6	0.1380	5/16	0.3125	0.302	0.361	0.344	0.30	0.34	0.16	0.08	0.17	0.25	0.16
8	0.1640	5/16	0.3125	0.302	0.361	0.344	0.30	0.34	0.16	0.08	0.17	0.25	0.16
10	0.1900	3/8	0.3750	0.362	0.433	0.413	0.36	0.41	0.19	0.09	0.22	0.28	0.19
12	0.2160	3/8	0.3750	0.362	0.433	0.413	0.36	0.41	0.19	0.09	0.22	0.31	0.22
1/4	0.2500	7/16	0.4375	0.428	0.505	0.488	0.41	0.47	0.22	0.11	0.25	0.34	0.25
5/16	0.3125	1/2	0.5000	0.489	0.577	0.557	0.47	0.53	0.25	0.12	0.28	0.41	0.31
3/8	0.3750	9/16	0.5625	0.551	0.650	0.628	0.53	0.62	0.28	0.14	0.33	0.45	0.38
7/16	0.4375	5/8	0.6250	0.612	0.722	0.698	0.59	0.69	0.31	0.16	0.36	0.52	0.44
1/2	0.5000	3/4	0.7500	0.736	0.866	0.840	0.72	0.81	0.38	0.19	0.42	0.59	0.50
9/16	0.5625	7/8	0.8750	0.861	1.010	0.982	0.84	0.94	0.44	0.22	0.50	0.69	0.56
5/8	0.6250	15/16	0.9375	0.922	1.083	1.051	0.91	1.00	0.47	0.23	0.53	0.75	0.62
3/4	0.7500	1 1/16	1.0625	1.045	1.227	1.191	1.03	1.16	0.53	0.27	0.59	0.88	0.75
7/8	0.8750	1 1/4	1.2500	1.231	1.443	1.403	1.22	1.36	0.62	0.31	0.70	1.00	0.88
1	1.0000	1 7/16	1.4375	1.417	1.660	1.615	1.41	1.55	0.72	0.36	0.81	1.12	1.00
1 1/8	1.1250	1 5/8	1.6250	1.602	1.876	1.826	1.59	1.75	0.81	0.41	0.92	1.31	1.12
1 1/4	1.2500	1 13/16	1.8125	1.788	2.093	2.038	1.78	1.95	0.91	0.45	1.03	1.44	1.25

High Crown

Nom. Size[a] or Basic Major Dia. of Thread		Width Across Flats, *F*			Width Across-Corners, *G*		Body Dia., *A*	Over-all Hgt., *H*	Hexa-gon Hgt. *Q*	Nose Rad., *R*	Body Rad., *S*	Drill Dep., *T*	Full Thd., *U*
		Max.	(Basic)	Min.	Max.	Min.						Max.	Min.
6	0.1380	5/16	0.3125	0.302	0.361	0.344	0.30	0.42	0.17	0.05	0.25	0.28	0.19
8	0.1640	5/16	0.3125	0.302	0.361	0.344	0.30	0.42	0.17	0.05	0.25	0.28	0.19
10	0.1900	3/8	0.3750	0.362	0.433	0.413	0.36	0.52	0.20	0.06	0.30	0.34	0.25
12	0.2160	3/8	0.3750	0.362	0.433	0.413	0.36	0.52	0.20	0.06	0.30	0.38	0.28
1/4	0.2500	7/16	0.4375	0.428	0.505	0.488	0.41	0.59	0.23	0.06	0.34	0.41	0.31
5/16	0.3125	1/2	0.5000	0.489	0.577	0.557	0.47	0.69	0.28	0.08	0.41	0.47	0.38
3/8	0.3750	9/16	0.5625	0.551	0.650	0.628	0.53	0.78	0.31	0.09	0.44	0.56	0.47
7/16	0.4375	5/8	0.6250	0.612	0.722	0.698	0.59	0.88	0.34	0.09	0.50	0.62	0.53
1/2	0.5000	3/4	0.7500	0.736	0.866	0.840	0.72	1.03	0.42	0.12	0.59	0.75	0.62
9/16	0.5625	7/8	0.8750	0.861	1.010	0.982	0.84	1.19	0.48	0.16	0.69	0.81	0.69
5/8	0.6250	15/16	0.9375	0.922	1.083	1.051	0.91	1.28	0.53	0.16	0.75	0.91	0.78
3/4	0.7500	1 1/16	1.0625	1.045	1.227	1.191	1.03	1.45	0.59	0.17	0.84	1.06	0.94
7/8	0.8750	1 1/4	1.12500	1.231	1.443	1.403	1.22	1.72	0.70	0.20	0.98	1.22	1.09
1	1.0000	1 7/16	1.4375	1.417	1.660	1.615	1.41	1.97	0.81	0.23	1.14	1.38	1.25
1 1/8	1.1250	1 5/8	1.6250	1.602	1.876	1.826	1.59	2.22	0.92	0.27	1.28	1.59	1.41
1 1/4	1.2500	1 13/16	1.8125	1.788	2.093	2.038	1.78	2.47	1.03	0.28	1.44	1.75	1.56

[a] When specifying a nominal size in decimals, any zero in the fourth decimal place is omitted.

All dimensions are in inches. Threads are Unified Standard Class 2B, UNC or UNF Series.

Hex High and Hex Slotted High Nuts *SAE Standard J482a*

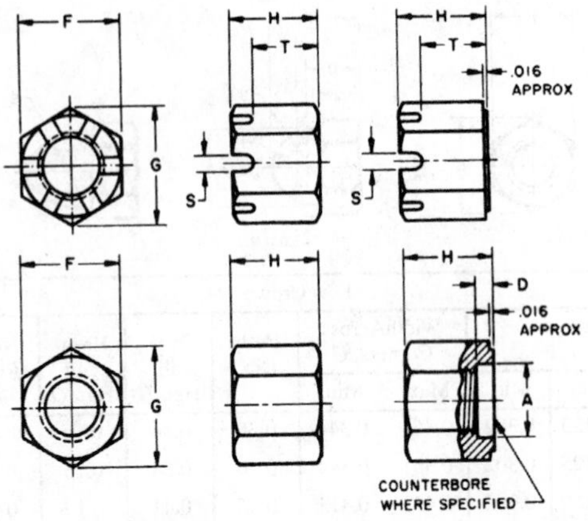

Nominal Size[a] or Basic Major Diameter of Thread		Width Across Flats, F			Width Across Corners, G		Slot Width, S	
		Basic	Max.	Min.	Max.	Min.	Min.	Max.
1/4	0.2500	7/16	0.4375	0.428	0.505	0.488	0.07	0.10
5/16	0.3125	1/2	0.5000	0.489	0.577	0.557	0.09	0.12
3/8	0.3750	9/16	0.5625	0.551	0.650	0.628	0.12	0.15
7/16	0.4375	11/16	0.6875	0.675	0.794	0.768	0.12	0.15
1/2	0.5000	3/4	0.7500	0.736	0.866	0.840	0.15	0.18
9/16	0.5625	7/8	0.8750	0.861	1.010	0.982	0.15	0.18
5/8	0.6250	15/16	0.9375	0.922	1.083	1.051	0.18	0.24
3/4	0.7500	1 1/8	1.1250	1.088	1.299	1.240	0.18	0.24
7/8	0.8750	1 5/16	1.3125	1.269	1.516	1.447	0.18	0.24
1	1.0000	1 1/2	1.5000	1.450	1.732	1.653	0.24	0.30
1 1/8	1.1250	1 11/16	1.6875	1.631	1.949	1.859	0.24	0.33
1 1/4	1.2500	1 7/8	1.8750	1.812	2.165	2.066	0.31	0.40

Nominal Size[a] or Basic Major Diameter of Thread		Thickness, H			Unslotted Thickness, T		Counterbore (Optional)	
		Basic	Max.	Min.	Max.	Min.	Dia., A	Depth, D
1/4	0.2500	3/8	0.382	0.368	0.29	0.27	0.266	0.062
5/16	0.3125	29/64	0.461	0.445	0.37	0.35	0.328	0.078
3/8	0.3750	1/2	0.509	0.491	0.38	0.36	0.391	0.094
7/16	0.4375	39/64	0.619	0.599	0.46	0.44	0.453	0.109
1/2	0.5000	21/32	0.667	0.645	0.51	0.49	0.516	0.125
9/16	0.5625	49/64	0.778	0.754	0.59	0.57	0.594	0.141
5/8	0.6250	27/32	0.857	0.831	0.63	0.61	0.656	0.156
3/4	0.7500	1	1.015	0.985	0.76	0.73	0.781	0.188
7/8	0.8750	1 5/32	1.172	1.140	0.92	0.89	0.906	0.219
1	1.0000	1 5/16	1.330	1.292	1.05	1.01	1.031	0.250
1 1/8	1.1250	1 1/2	1.520	1.480	1.18	1.14	1.156	0.281
1 1/4	1.2500	1 11/16	1.710	1.666	1.34	1.29	1.281	0.312

[a] When specifying a nominal size in decimals, any zero in the fourth decimal place is omitted.

All dimensions are in inches. Threads are Unified Standard Class 2B, UNC or UNF Series.

American National Standard Round Head and Round Head Square Neck Bolts *ANSI/ASME B18.5 –1990*

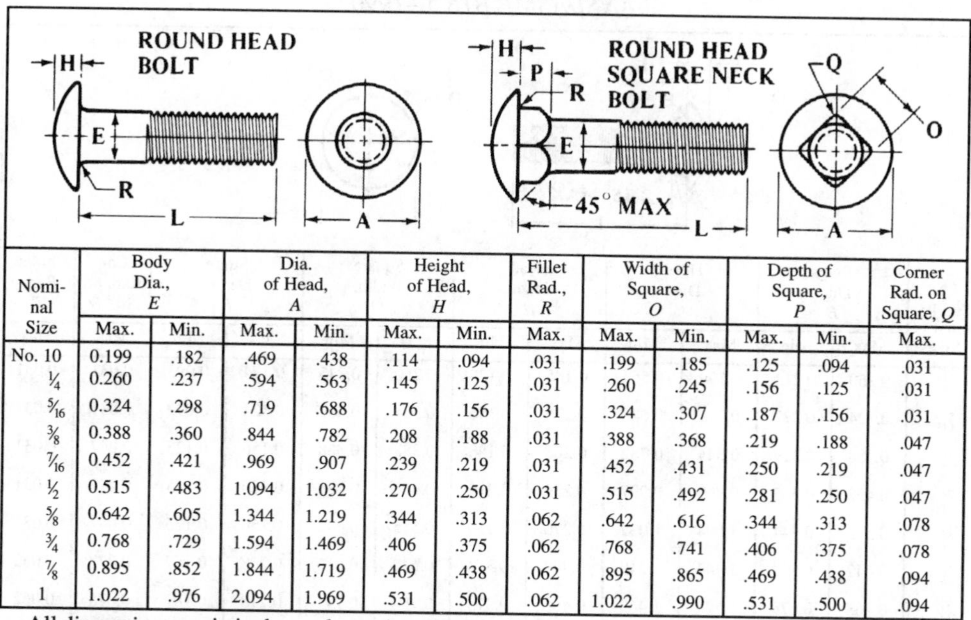

Nominal Size	Body Dia., *E*		Dia. of Head, *A*		Height of Head, *H*		Fillet Rad., *R*	Width of Square, *O*		Depth of Square, *P*		Corner Rad. on Square, *Q*
	Max.	Min.	Max.	Min.	Max.	Min.	Max.	Max.	Min.	Max.	Min.	Max.
No. 10	0.199	.182	.469	.438	.114	.094	.031	.199	.185	.125	.094	.031
¼	0.260	.237	.594	.563	.145	.125	.031	.260	.245	.156	.125	.031
5/16	0.324	.298	.719	.688	.176	.156	.031	.324	.307	.187	.156	.031
⅜	0.388	.360	.844	.782	.208	.188	.031	.388	.368	.219	.188	.047
7/16	0.452	.421	.969	.907	.239	.219	.031	.452	.431	.250	.219	.047
½	0.515	.483	1.094	1.032	.270	.250	.031	.515	.492	.281	.250	.047
⅝	0.642	.605	1.344	1.219	.344	.313	.062	.642	.616	.344	.313	.078
¾	0.768	.729	1.594	1.469	.406	.375	.062	.768	.741	.406	.375	.078
⅞	0.895	.852	1.844	1.719	.469	.438	.062	.895	.865	.469	.438	.094
1	1.022	.976	2.094	1.969	.531	.500	.062	1.022	.990	.531	.500	.094

All dimensions are in inches unless otherwise specified.

Threads are Unified Standard, Class 2A, UNC Series, in accordance with ANSI B1.1. For threads with additive finish, the maximum diameters of Class 2A shall apply before plating or coating, whereas the basic diameters (Class 2A maximum diameters plus the allowance) shall apply to a bolt after plating or coating.

Bolts are designated in the sequence shown: nominal size (number, fraction or decimal equivalent); threads per inch; nominal length (fraction or decimal equivalent); product name; material; and protective finish, if required.

i.e.:½-13 × 3 Round Head Square Neck Bolt, Steel .375-16 × 2.50 Step Bolt, Steel, Zinc Plated

American National Standard T-Head Bolts *ANSI/ASME B18.5 –1990*

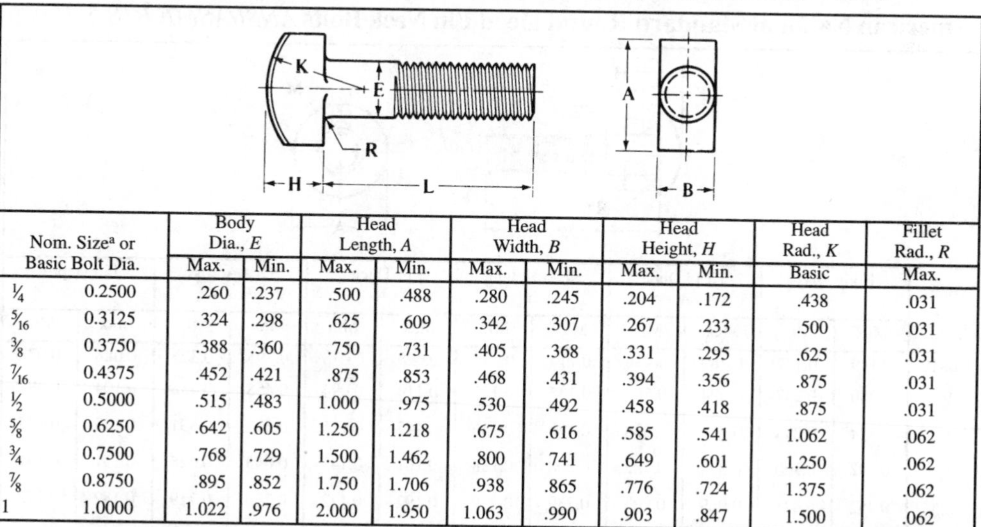

Nom. Size[a] or Basic Bolt Dia.		Body Dia., *E*		Head Length, *A*		Head Width, *B*		Head Height, *H*		Head Rad., *K*	Fillet Rad., *R*
		Max.	Min.	Max.	Min.	Max.	Min.	Max.	Min.	Basic	Max.
¼	0.2500	.260	.237	.500	.488	.280	.245	.204	.172	.438	.031
5/16	0.3125	.324	.298	.625	.609	.342	.307	.267	.233	.500	.031
⅜	0.3750	.388	.360	.750	.731	.405	.368	.331	.295	.625	.031
7/16	0.4375	.452	.421	.875	.853	.468	.431	.394	.356	.875	.031
½	0.5000	.515	.483	1.000	.975	.530	.492	.458	.418	.875	.031
⅝	0.6250	.642	.605	1.250	1.218	.675	.616	.585	.541	1.062	.062
¾	0.7500	.768	.729	1.500	1.462	.800	.741	.649	.601	1.250	.062
⅞	0.8750	.895	.852	1.750	1.706	.938	.865	.776	.724	1.375	.062
1	1.0000	1.022	.976	2.000	1.950	1.063	.990	.903	.847	1.500	.062

[a] Where specifying nominal size in decimals, zeros preceding the decimal point and in the fourth decimal place are omitted. For information as to threads and method of bolt designation, see footnotes to preceding table.

All dimensions are given in inches.

American National Standard Round Head Short Square Neck Bolts
ANSI/ASME B18.5 –1990

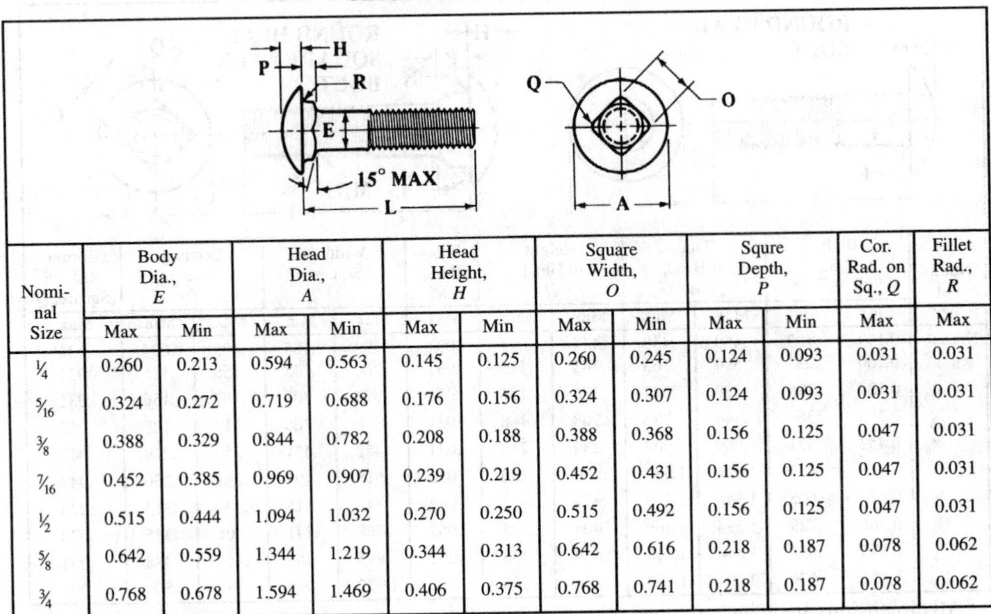

Nominal Size	Body Dia., *E*		Head Dia., *A*		Head Height, *H*		Square Width, *O*		Squre Depth, *P*		Cor. Rad. on Sq., *Q*	Fillet Rad., *R*
	Max	Min	Max	Min	Max	Min	Max	Min	Max	Min	Max	Max
¼	0.260	0.213	0.594	0.563	0.145	0.125	0.260	0.245	0.124	0.093	0.031	0.031
5⁄16	0.324	0.272	0.719	0.688	0.176	0.156	0.324	0.307	0.124	0.093	0.031	0.031
⅜	0.388	0.329	0.844	0.782	0.208	0.188	0.388	0.368	0.156	0.125	0.047	0.031
7⁄16	0.452	0.385	0.969	0.907	0.239	0.219	0.452	0.431	0.156	0.125	0.047	0.031
½	0.515	0.444	1.094	1.032	0.270	0.250	0.515	0.492	0.156	0.125	0.047	0.031
⅝	0.642	0.559	1.344	1.219	0.344	0.313	0.642	0.616	0.218	0.187	0.078	0.062
¾	0.768	0.678	1.594	1.469	0.406	0.375	0.768	0.741	0.218	0.187	0.078	0.062

All dimensions are given in inches.

Threads are Unified Standard, Class 2A, UNC Series, in accordance with ANSI B1.1. For threads with additive finish, the maximum diameters of Class 2A apply before plating or coating, whereas the basic diameters (Class 2A maximum diameters plus the allowance) apply to a bolt after plating or coating.

Bolts are designated in the sequence shown: nominal size (number, fraction or decimal equivalent); threads per inch; nominal length (fraction or decimal equivalent); product name; material; and protective finish, if required.

i.e.,½-13 × 3 Round Head Short Square Neck Bolt, Steel .375-16 × 2.50 Round Head Short Square Neck Bolt, Steel, Zinc Plated

American National Standard Round Head Fin Neck Bolts *ANSI/ASME B18.5-1990*

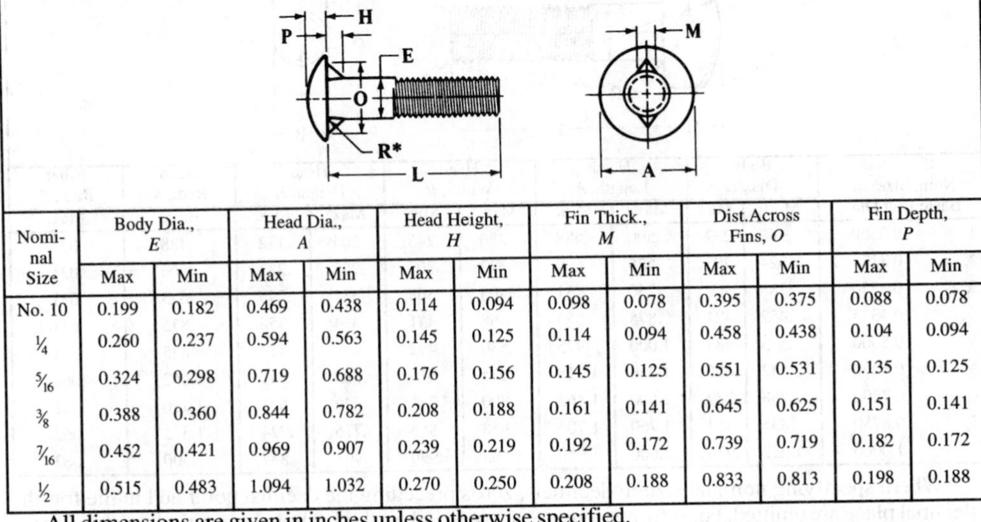

Nominal Size	Body Dia., *E*		Head Dia., *A*		Head Height, *H*		Fin Thick., *M*		Dist.Across Fins, *O*		Fin Depth, *P*	
	Max	Min	Max	Min	Max	Min	Max	Min	Max	Min	Max	Min
No. 10	0.199	0.182	0.469	0.438	0.114	0.094	0.098	0.078	0.395	0.375	0.088	0.078
¼	0.260	0.237	0.594	0.563	0.145	0.125	0.114	0.094	0.458	0.438	0.104	0.094
5⁄16	0.324	0.298	0.719	0.688	0.176	0.156	0.145	0.125	0.551	0.531	0.135	0.125
⅜	0.388	0.360	0.844	0.782	0.208	0.188	0.161	0.141	0.645	0.625	0.151	0.141
7⁄16	0.452	0.421	0.969	0.907	0.239	0.219	0.192	0.172	0.739	0.719	0.182	0.172
½	0.515	0.483	1.094	1.032	0.270	0.250	0.208	0.188	0.833	0.813	0.198	0.188

All dimensions are given in inches unless otherwise specified.

*Maximum fillet radius *R* is 0.031 inch for all sizes.

For information as to threads and method of bolt designation, see foonotes to the preceding table.

American National Standard Round Head Ribbed Neck Bolts *ANSI/ASME B18.5–1990*

Nominal Size[a] or Basic Bolt Diameter		Body Diameter, *E*		Head Diameter, *A*		Head Height, *H*		Head to Ribs, *M*: For Lengths of ⅞ in. and Shorter	Head to Ribs, *M*: For Lengths of 1 in. and Longer	Number of Ribs, *N*	Dia. Over Ribs, *O*	Depth Over Ribs, *P*: For Lengths of ⅞ in. and Shorter	Depth Over Ribs, *P*: For Lengths of 1 in. and 1⅛ in.	Depth Over Ribs, *P*: For Lengths of ¼ in. and Longer	Fillet Radius, *R*
		Max	Min	Max	Min	Max	Min	±0.031[b]		Approx	Min	±0.031			Max[c]
No. 10	0.1900	0.199	0.182	0.469	0.438	0.114	0.094	0.031†	0.063	9	0.210	0.250	0.407	0.594	0.031
¼	0.2500	0.260	0.237	0.594	0.563	0.145	0.125	0.031†	0.063	10	0.274	0.250	0.407	0.594	0.031
5⁄16	0.3125	0.324	0.298	0.719	0.688	0.176	0.156	0.031†	0.063	12	0.340	0.250	0.407	0.594	0.031
⅜	0.3750	0.388	0.360	0.844	0.782	0.208	0.188	0.031†	0.063	12	0.405	0.250	0.407	0.594	0.031
7⁄16	0.4375	0.452	0.421	0.969	0.907	0.239	0.219	0.031†	0.063	14	0.470	0.250	0.407	0.594	0.031
½	0.5000	0.515	0.483	1.094	1.032	0.270	0.250	0.031†	0.063	16	0.534	0.250	0.407	0.594	0.031
⅝	0.6250	0.642	0.605	1.344	1.219	0.344	0.313	0.094	0.094	19	0.660	0.313	0.438	0.625	0.062
¾	0.7500	0.768	0.729	1.594	1.469	0.406	0.375	0.094	0.094	22	0.785	0.313	0.438	0.625	0.062

[a] Where specifying nominal size in decimals, zeros preceding decimal and in the fourth decimal place shall be omitted.
[b] Tolerance on the No. 10 through ½ in. sizes for nominal lengths ⅞ in. and shorter shall be +0.031 and −0.000.
[c] The minimum radius is one half of the value shown.

All dimensions are given in inches unless otherwise specified.

For information as to threads and method of designating bolts, see following table.

American National Standard Step and 114 Degree Countersunk Square Neck Bolts

ANSI/ASME B18.5–1990

Nominal Size	Step & 114° Countersunk Bolts					Step Bolts							114° Countersunk Square Neck Bolts						
	Body Dia., *E*		Corner Rad. on Square, *Q*	Width of Square, *O*		Depth of Square, *P*		Dia. of Head, *A*		Height of Head, *H*		Fillet Radius, *R*	Depth of Square, *P*		Dia. of Head, *A*		Flat on Head, *F*	Height of Head, *H*	
	Max.	Min.	Max.	Min.	Min.	Max.	Min.	Max.	Min.	Max.	Min.	Max.	Max.	Min.	Max.	Min.	Min.	Max.	Min.
No. 10	0.199	0.182	0.031	0.199	0.185	0.125	0.094	0.656	0.625	0.114	0.094	0.031	0.125	0.094	0.548	0.500	0.015	0.131	0.112
¼	0.260	0.237	0.031	0.260	0.245	0.156	0.125	0.844	0.813	0.145	0.125	0.031	0.156	0.125	0.682	0.625	0.018	0.154	0.135
5⁄16	0.324	0.298	0.031	0.324	0.307	0.187	0.156	1.031	1.000	0.176	0.156	0.031	0.219	0.188	0.821	0.750	0.023	0.184	0.159
⅜	0.388	0.360	0.047	0.388	0.368	0.219	0.188	1.219	1.188	0.208	0.188	0.031	0.250	0.219	0.960	0.875	0.027	0.212	0.183
7⁄16	0.452	0.421	0.047	0.452	0.431	0.250	0.219	1.406	1.375	0.239	0.219	0.031	0.281	0.250	1.093	1.000	0.030	0.235	0.205
½	0.515	0.483	0.047	0.515	0.492	0.281	0.250	1.594	1.563	0.270	0.250	0.031	0.312	0.281	1.233	1.125	0.035	0.265	0.229
⅝[a]	.642	0.605	0.078	0.642	0.616	…	…	…	…	…	…	…	0.406	0.375	1.495	1.375	0.038	0.316	0.272
¾[a]	0.768	0.729	0.078	0.768	0.741	…	…	…	…	…	…	…	0.500	0.469	10.754	1.625	0.041	0.368	0.314

[a] These sizes pertain to 114 degree countersunk square neck bolts only. Dimensions given in last seven columns to the right are for these bolts only.

All dimensions are in inches unless otherwise specified.

Threads are Unified Standard, Class 2A, UNC Series, in accordance with ANSI B1.1. For threads with additive finish, the maximum diameters of Class 2A shall apply before plating or coating, whereas the basic diameters (Class 2A maximum diameters plus the allowance) shall apply to a bolt after plating or coating.

Bolts are designated in the sequence shown: nominal size (number, fraction or decimal equivalent); threads per inch; nominal length (fraction or decimal equivalent); product name; material; and protective finish, if required. For example

½-13 × 3 Round Head Square Neck Bolt, Steel .375-16 × 2.50 Step Bolt, Steel, Zinc Plated

American National Standard Countersunk Bolts and Slotted Countersunk Bolts

ANSI/ASME B18.5–1990

Nominal Size[a] or Basic Bolt Diameter		Body Diameter, *E* Max	Body Diameter, *E* Min	Head Diameter, *A* Max Edge Sharp	Head Diameter, *A* Min Edge Sharp	Head Diameter, *A* Absolute Min Edge Rounded or Flat	Flat on Min Dia., Head, *F*[b] Max
¼	0.2500	0.260	0.237	0.493	0.477	0.445	0.018
5/16	0.3125	0.324	0.298	0.618	0.598	0.558	0.023
⅜	0.3750	0.388	0.360	0.740	0.715	0.668	0.027
7/16	0.4375	0.452	0.421	0.803	0.778	0.726	0.030
½	0.5000	0.515	0.483	0.935	0.905	0.845	0.035
⅝	0.6250	0.642	0.605	1.169	1.132	1.066	0.038
¾	0.7500	0.768	0.729	1.402	1.357	1.285	0.041
⅞	0.8750	0.895	0.852	1.637	1.584	1.511	0.042
1	1.0000	1.022	0.976	1.869	1.810	1.735	0.043
1⅛	1.1250	1.149	1.098	2.104	2.037	1.962	0.043
1¼	1.2500	1.277	1.223	2.337	2.262	2.187	0.043
1⅜	1.3750	1.404	1.345	2.571	2.489	2.414	0.043
1½	1.5000	1.531	1.470	2.804	2.715	2.640	0.043

Nom. Size or Basic Bolt Dia.		Head Height, *H* Max[c]	Head Height, *H* Min[d]	Slot Width, *J* Max	Slot Width, *J* Min	Slot Depth, *T* Max	Slot Depth, *T* Min
¼	0.2500	0.150	0.131	0.075	0.064	0.068	0.045
5/16	0.3125	0.189	0.164	0.084	0.072	0.086	0.057
⅜	0.3750	0.225	0.196	0.094	0.081	0.103	0.068
7/16	0.4375	0.226	0.196	0.094	0.081	0.103	0.068
½	0.5000	0.269	0.233	0.106	0.091	0.103	0.068
⅝	0.6250	0.336	0.292	0.133	0.116	0.137	0.091
¾	0.7500	0.403	0.349	0.149	0.131	0.171	0.115
⅞	0.8750	0.470	0.408	0.167	0.147	0.206	0.138
1	1.0000	0.537	0.466	0.188	0.166	0.240	0.162
1⅛	1.1250	0.604	0.525	0.196	0.178	0.257	0.173
1¼	1.2500	0.671	0.582	0.211	0.193	0.291	0.197
1⅜	1.3750	0.738	0.641	0.226	0.208	0.326	0.220
1½	1.5000	0.805	0.698	0.258	0.240	0.360	0.244

[a] Where specifying size in decimals, zeros preceding decimal and in fourth decimal place are omitted.

[b] Flat on minimum diameter head calculated on minimum sharp and absolute minimum head diameters and 82° head angle.

[c] Maximum head height calculated on maximum sharp head diameter, basic bolt diameter, and 78° head angle.

[d] Minimum head height calculated on minimum sharp head diameter, basic bolt diameter, and 82° head angle.

All dimensions are given in inches.

For thread information and method of bolt designation see foonotes to previous table.

Heads are unslotted unless otherwise specified. For slot dimensions see Table 1 in Slotted Head Cap Screw section.

Wrench Openings for Nuts *ANSI/ASME B18.2.2-1987 (R1999), Appendix*

Max. [a] Width Across Flats of Nut	Wrench Opening[b] Min.	Wrench Opening[b] Max.	Max. [a] Width Across Flats of Nut	Wrench Opening[b] Min.	Wrench Opening[b] Max.	Max.[a] Width Across Flats of Nut	Wrench Opening[b] Min.	Wrench Opening[b] Max.
$\frac{5}{32}$	0.158	0.163	$1\frac{1}{4}$	1.257	1.267	$2\frac{15}{16}$	2.954	2.973
$\frac{3}{16}$	0.190	0.195	$1\frac{5}{16}$	1.320	1.331	*3*	*3.016*	*3.035*
$\frac{7}{32}$	0.220	0.225	*$1\frac{3}{8}$*	*1.383*	*1.394*	$3\frac{1}{8}$	3.142	3.162
$\frac{1}{4}$	0.252	0.257	$1\frac{7}{16}$	1.446	1.457	*$3\frac{3}{8}$*	*3.393*	*3.414*
$\frac{9}{32}$	0.283	0.288	$1\frac{1}{2}$	1.508	1.520	$3\frac{1}{2}$	3.518	3.540
$\frac{5}{16}$	0.316	0.322	$1\frac{5}{8}$	1.634	1.646	*$3\frac{3}{4}$*	*3.770*	*3.793*
$\frac{11}{32}$	0.347	0.353	$1\frac{11}{16}$	1.696	1.708	$3\frac{7}{8}$	3.895	3.918
$\frac{3}{8}$	0.378	0.384	$1\frac{13}{16}$	1.822	1.835	*$4\frac{1}{8}$*	*4.147*	*4.172*
$\frac{7}{16}$	0.440	0.446	$1\frac{7}{8}$	1.885	1.898	$4\frac{1}{4}$	4.272	4.297
$\frac{1}{2}$	0.504	0.510	2	2.011	2.025	$4\frac{1}{2}$	4.524	4.550
$\frac{9}{16}$	0.556	0.573	$2\frac{1}{16}$	2.074	2.088	$4\frac{5}{8}$	4.649	4.676
$\frac{5}{8}$	0.629	0.636	$2\frac{3}{16}$	2.200	2.215	*$4\frac{7}{8}$*	*4.900*	*4.928*
$\frac{11}{16}$	0.692	0.699	$2\frac{1}{4}$	2.262	2.277	5	5.026	5.055
$\frac{3}{4}$	0.755	0.763	$2\frac{3}{8}$	2.388	2.404	*$5\frac{1}{4}$*	*5.277*	*5.307*
$\frac{13}{16}$	0.818	0.826	*$2\frac{7}{16}$*	*2.450*	*2.466*	$5\frac{3}{8}$	5.403	5.434
$\frac{7}{8}$	0.880	0.888	$2\frac{9}{16}$	2.576	2.593	*$5\frac{5}{8}$*	*5.654*	*5.686*
$\frac{15}{16}$	0.944	0.953	*$2\frac{5}{8}$*	*2.639*	*2.656*	$5\frac{3}{4}$	5.780	5.813
1	1.006	1.015	$2\frac{3}{4}$	2.766	2.783	*6*	*6.031*	6.157
$1\frac{1}{16}$	1.068	1.077	*$2\frac{13}{16}$*	*2.827*	*2.845*	$6\frac{1}{8}$	6.065	6.192
$1\frac{1}{8}$	1.132	1.142						

[a] Wrenches are marked with the "Nominal Size of Wrench," which is equal to the basic or maximum width across flats of the corresponding nut. Minimum wrench opening is (1.005W + 0.001). Tolerance on wrench opening is (0.005W + 0.004) from minimum, where W equals nominal size of wrench.

[b] Openings for $\frac{5}{32}$ to $\frac{3}{8}$ widths from old ASA B18.2-1960 and italic values are from former ANSI B18.2.2-1972.

All dimensions given in inches.

Wrench Clearance Dimensions.—Wrench clearances are given in Tables 1 and Tables 2. They are based on a wrench opening corresponding to the dimensions across the flats of the fastener. The listed values were obtained from a composite study of the alloy steel wrenches that are commercially available and military specifications. They are suitable for general use as minimum requirements.

Table 1. Wrench Clearances for Box Wrench—12 Point
From SAE Aeronautical Drafting Manual

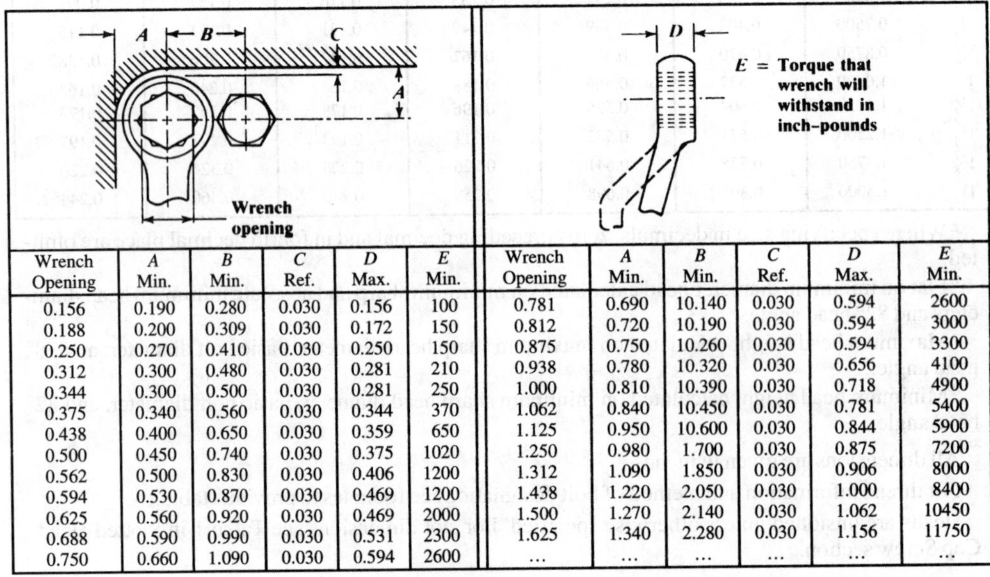

Wrench Opening	A Min.	B Min.	C Ref.	D Max.	E Min.	Wrench Opening	A Min.	B Min.	C Ref.	D Max.	E Min.
0.156	0.190	0.280	0.030	0.156	100	0.781	0.690	10.140	0.030	0.594	2600
0.188	0.200	0.309	0.030	0.172	150	0.812	0.720	10.190	0.030	0.594	3000
0.250	0.270	0.410	0.030	0.250	150	0.875	0.750	10.260	0.030	0.594	3300
0.312	0.300	0.480	0.030	0.281	210	0.938	0.780	10.320	0.030	0.656	4100
0.344	0.300	0.500	0.030	0.281	250	1.000	0.810	10.390	0.030	0.718	4900
0.375	0.340	0.560	0.030	0.344	370	1.062	0.840	10.450	0.030	0.781	5400
0.438	0.400	0.650	0.030	0.359	650	1.125	0.950	10.600	0.030	0.844	5900
0.500	0.450	0.740	0.030	0.375	1020	1.250	0.980	1.700	0.030	0.875	7200
0.562	0.500	0.830	0.030	0.406	1200	1.312	1.090	1.850	0.030	0.906	8000
0.594	0.530	0.870	0.030	0.469	1200	1.438	1.220	2.050	0.030	1.000	8400
0.625	0.560	0.920	0.030	0.469	2000	1.500	1.270	2.140	0.030	1.062	10450
0.688	0.590	0.990	0.030	0.531	2300	1.625	1.340	2.280	0.030	1.156	11750
0.750	0.660	1.090	0.030	0.594	2600	…	…	…	…	…	…

Table 2. Wrench Clearances for Open End Engineers Wrench 15° and Socket Wrench (Regular Length)
From SAE Aeronautical Drafting Manual; © Society of Automotive Engineers, Inc.

H = Thickness of wrench head
J = Torque that wrench will withstand in inch–pounds
P = Torque that wrench will withstand in inch–pounds
* = Does not include allowance for torque device
All dimensions in inches except where otherwise noted

Wrench Open-ing	Open End Engineers Wrench 15° A Min.	B Max.	C Min.	D Min.	E Min.	F Max.	G Ref.	H Max.	J Min.	Socket (Regular Length) K Min.	L Ref.	Q = .250 M Max.	Q = .250 N Max.	Q = .250 P Min.	Q = .375 M Max.	Q = .375 N Max.	Q = .375 P Min.	Q = .500 M Max.	Q = .500 N Max.	Q = .500 P Min.	Q = .750 M Max.	Q = .750 N Max.	Q = .750 P Min.	Wrench Opening
.156	.220	.250	.390	.160	.250	.200	.030	.094	25	…	…	…	…	…	…	…	…	…	…	…	…	…	…	…
.188	.250	.280	.430	.190	.270	.230	.030	.172	40	.370	.030	1.000	.510	125	…	…	…	…	…	…	…	…	…	.188
.250	.280	.340	.530	.270	.310	.310	.030	.172	60	.470	.030	1.000	.510	200	1.250	.690	250	…	…	…	…	…	…	.250
.312	.380	.470	.660	.280	.390	.390	.050	.203	125	.550	.030	1.000	.510	300	1.250	.690	400	…	…	…	…	…	…	.312
.344	.420	.500	.750	.340	.450	.450	.050	.203	175	.580	.030	1.000	.519	450	1.250	.690	675	…	…	…	…	…	…	.344
.375	.420	.500	.780	.360	.450	.520	.050	.219	250	.620	.030	1.000	.580	550	1.250	.690	900	1.500	.880	1600	…	…	…	.375
.438	.470	.590	.890	.420	.520	.640	.050	.250	375	.750	.030	1.000	.683	550	1.250	.880	1250	1.500	.940	1700	…	…	…	.438
.500	.520	.640	1.000	.470	.580	.660	.050	.266	490	.810	.030	1.000	.692	600	1.250	.880	1450	1.500	.940	2000	…	…	…	.500
.562	.590	.770	1.130	.520	.660	.700	.050	.297	700	.870	.030	…	…	…	1.250	.932	1600	1.500	.940	2700	…	…	…	.562
.594	.640	.830	1.210	.530	.700	.700	.050	.344	800	.920	.030	…	…	…	1.250	.963	1750	1.562	.970	3000	…	…	…	.594
.625	.640	.830	1.230	.550	.700	.700	.050	.344	935	.950	.030	…	…	…	1.250	.995	2000	1.562	1.000	3600	…	…	…	.625
.688	.770	.920	1.470	.660	.880	.800	.060	.375	1250	1.030	.030	…	…	…	1.250	1.058	2000	1.562	1.065	4300	…	…	…	.688
.750	.770	.920	1.510	.670	.880	.800	.060	.375	1500	1.120	.030	…	…	…	1.250	1.120	2000	1.562	1.130	5000	…	…	…	.750
.781	.830	.950	1.550	.690	.890	.840	.060	.375	1615	1.150	.030	…	…	…	1.250	1.126	2000	1.625	1.130	5000	…	…	…	.781
.812	.910	1.120	1.660	.720	.970	.860	.060	.406	1710	1.200	.030	…	…	…	1.250	1.213	2000	1.625	1.222	5000	…	…	…	.812
.875	.970	1.150	1.810	.800	1.060	.910	.060	.438	2250	1.280	.030	…	…	…	…	…	…	1.750	1.285	5000	…	…	…	.875
.938	.970	1.150	1.850	.810	1.060	.950	.060	.438	2750	1.370	.030	…	…	…	…	…	…	1.750	1.410	5000	…	…	…	.938
1.000	1.050	1.230	2.000	.880	1.160	1.060	.060	.500	3250	1.470	.030	…	…	…	…	…	…	1.750	1.410	5000	…	…	…	1.000
1.062	1.090	1.250	2.100	.970	1.200	1.200	.080	.500	3500	1.550	.030	…	…	…	…	…	…	1.844	1.505	5000	…	…	…	1.062
1.125	1.140	1.370	2.210	1.000	1.270	1.230	.080	.500	4000	1.610	.030	…	…	…	…	…	…	1.938	1.567	5000	…	…	…	1.125
1.250	1.270	1.420	2.440	1.080	1.390	1.310	.080	.562	5250	1.890	.030	…	…	…	…	…	…	2.000	1.723	5000	2.375	1.855	7250	1.250
1.312	1.390	1.690	2.630	1.170	1.520	1.340	.080	.562	6000	1.980	.030	…	…	…	…	…	…	…	…	…	2.500	1.920	8000	1.312
1.438	1.470	1.720	2.800	1.250	1.590	1.340	.090	.641	7500	2.140	.030	…	…	…	…	…	…	…	…	…	2.625	2.075	9550	1.438
1.500	1.470	1.720	2.840	1.270	1.590	1.450	.090	.641	8250	2.200	.030	…	…	…	…	…	…	…	…	…	2.625	2.170	10450	1.500
1.625	1.560	1.880	3.100	1.380	1.750	1.560	.090	.641	9000	2.390	.030	…	…	…	…	…	…	…	…	…	2.750	2.325	11750	1.625

Table 1a. American National Standard Type A Plain Washers— Preferred Sizes *ANSI/ASME B18.22.1-1965 (R1998)*

Nominal Washer Size[a]		Series	Inside Diameter			Outside Diameter			Thickness		
			Basic	Tolerance Plus	Tolerance Minus	Basic	Tolerance Plus	Tolerance Minus	Basic	Max.	Min.
—	—		0.078	0.000	0.005	0.188	0.000	0.005	0.020	0.025	0.016
—	—		0.094	0.000	0.005	0.250	0.000	0.005	0.020	0.025	0.016
—	—		0.125	0.008	0.005	0.312	0.008	0.005	0.032	0.040	0.025
No. 6	0.138		0.156	0.008	0.005	0.375	0.015	0.005	0.049	0.065	0.036
No. 8	0.164		0.188	0.008	0.005	0.438	0.015	0.005	0.049	0.065	0.036
No. 10	0.190		0.219	0.008	0.005	0.500	0.015	0.005	0.049	0.065	0.036
3/16	0.188		0.250	0.015	0.005	0.562	0.015	0.005	0.049	0.065	0.036
No. 12	0.216		0.250	0.015	0.005	0.562	0.015	0.005	0.065	0.080	0.051
1/4	0.250	N	0.281	0.015	0.005	0.625	0.015	0.005	0.065	0.080	0.051
1/4	0.250	W	0.312	0.015	0.005	0.734[b]	0.015	0.007	0.065	0.080	0.051
5/16	0.312	N	0.344	0.015	0.005	0.688	0.015	0.007	0.065	0.080	0.051
5/16	0.312	W	0.375	0.015	0.005	0.875	0.030	0.007	0.083	0.104	0.064
3/8	0.375	N	0.406	0.015	0.005	0.812	0.015	0.007	0.065	0.080	0.051
3/8	0.375	W	0.438	0.015	0.005	1.000	0.030	0.007	0.083	0.104	0.064
7/16	0.438	N	0.469	0.015	0.005	0.922	0.015	0.007	0.065	0.080	0.051
7/16	0.438	W	0.500	0.015	0.005	1.250	0.030	0.007	0.083	0.104	0.064
1/2	0.500	N	0.531	0.015	0.005	1.062	0.030	0.007	0.095	0.121	0.074
1/2	0.500	W	0.562	0.015	0.005	1.375	0.030	0.007	0.109	0.132	0.086
9/16	0.562	N	0.594	0.015	0.005	1.156[b]	0.030	0.007	0.095	0.121	0.074
9/16	0.562	W	0.625	0.015	0.005	1.469[b]	0.030	0.007	0.109	0.132	0.086
5/8	0.625	N	0.656	0.030	0.007	1.312	0.030	0.007	0.095	0.121	0.074
5/8	0.625	W	0.688	0.030	0.007	1.750	0.030	0.007	0.134	0.160	0.108
3/4	0.750	N	0.812	0.030	0.007	1.469	0.030	0.007	0.134	0.160	0.108
3/4	0.750	W	0.812	0.030	0.007	2.000	0.030	0.007	0.148	0.177	0.122
7/8	0.875	N	0.938	0.030	0.007	1.750	0.030	0.007	0.134	0.160	0.108
7/8	0.875	W	0.938	0.030	0.007	2.250	0.030	0.007	0.165	0.192	0.136
1	1.000	N	1.062	0.030	0.007	2.000	0.030	0.007	0.134	0.160	0.108
1	1.000	W	1.062	0.030	0.007	2.500	0.030	0.007	0.165	0.192	0.136
1 1/8	1.125	N	1.250	0.030	0.007	2.250	0.030	0.007	0.134	0.160	0.108
1 1/8	1.125	W	1.250	0.030	0.007	2.750	0.030	0.007	0.165	0.192	0.136
1 1/4	1.250	N	1.375	0.030	0.007	2.500	0.030	0.007	0.165	0.192	0.136
1 1/4	1.250	W	1.375	0.030	0.007	3.000	0.030	0.007	0.165	0.192	0.136
1 3/8	1.375	N	1.500	0.030	0.007	2.750	0.030	0.007	0.165	0.192	0.136
1 3/8	1.375	W	1.500	0.045	0.010	3.250	0.045	0.010	0.180	0.213	0.153
1 1/2	1.500	N	1.625	0.030	0.007	3.000	0.030	0.007	0.165	0.192	0.136
1 1/2	1.500	W	1.625	0.045	0.010	3.500	0.045	0.010	0.180	0.213	0.153
1 5/8	1.625		1.750	0.045	0.010	3.750	0.045	0.010	0.180	0.213	0.153
1 3/4	1.750		1.875	0.045	0.010	4.000	0.045	0.010	0.180	0.213	0.153
1 7/8	1.875		2.000	0.045	0.010	4.250	0.045	0.010	0.180	0.213	0.153
2	2.000		2.125	0.045	0.010	4.500	0.045	0.010	0.180	0.213	0.153
2 1/4	2.250		2.375	0.045	0.010	4.750	0.045	0.010	0.220	0.248	0.193
2 1/2	2.500		2.625	0.045	0.010	5.000	0.045	0.010	0.238	0.280	0.210
2 3/4	2.750		2.875	0.065	0.010	5.250	0.065	0.010	0.259	0.310	0.228
3	3.000		3.125	0.065	0.010	5.500	0.065	0.010	0.284	0.327	0.249

[a] Nominal washer sizes are intended for use with comparable nominal screw or bolt sizes.

[b] The 0.734-inch, 1.156-inch, and 1.469-inch outside diameters avoid washers which could be used in coin operated devices.

All dimensions are in inches.

Preferred sizes are for the most part from series previously designated "Standard Plate" and "SAE." Where common sizes existed in the two series, the SAE size is designated "N" (narrow) and the Standard Plate "W" (wide). These sizes as well as all other sizes of Type A Plain Washers are to be ordered by ID, OD, and thickness dimensions.

Additional selected sizes of Type A Plain Washers are shown in Table 1b.

Table 1b. American National Standard Type A Plain Washers — Additional Selected Sizes *ANSI/ASME B18.22.1-1965 (R1998)*

Inside Diameter			Outside Diameter			Thickness		
	Tolerance			Tolerance				
Basic	Plus	Minus	Basic	Plus	Minus	Basic	Max.	Min.
0.094	0.000	0.005	0.219	0.000	0.005	0.020	0.025	0.016
0.125	0.000	0.005	0.250	0.000	0.005	0.022	0.028	0.017
0.156	0.008	0.005	0.312	0.008	0.005	0.035	0.048	0.027
0.172	0.008	0.005	0.406	0.015	0.005	0.049	0.065	0.036
0.188	0.008	0.005	0.375	0.015	0.005	0.049	0.065	0.036
0.203	0.008	0.005	0.469	0.015	0.005	0.049	0.065	0.036
0.219	0.008	0.005	0.438	0.015	0.005	0.049	0.065	0.036
0.234	0.008	0.005	0.531	0.015	0.005	0.049	0.065	0.036
0.250	0.015	0.005	0.500	0.015	0.005	0.049	0.065	0.036
0.266	0.015	0.005	0.625	0.015	0.005	0.049	0.065	0.036
0.312	0.015	0.005	0.875	0.015	0.007	0.065	0.080	0.051
0.375	0.015	0.005	0.734[a]	0.015	0.007	0.065	0.080	0.051
0.375	0.015	0.005	1.125	0.015	0.007	0.065	0.080	0.051
0.438	0.015	0.005	0.875	0.030	0.007	0.083	0.104	0.064
0.438	0.015	0.005	1.375	0.030	0.007	0.083	0.104	0.064
0.500	0.015	0.005	1.125	0.030	0.007	0.083	0.104	0.064
0.500	0.015	0.005	1.625	0.030	0.007	0.083	0.104	0.064
0.562	0.015	0.005	1.250	0.030	0.007	0.109	0.132	0.086
0.562	0.015	0.005	1.875	0.030	0.007	0.109	0.132	0.086
0.625	0.015	0.005	1.375	0.030	0.007	0.109	0.132	0.086
0.625	0.015	0.005	2.125	0.030	0.007	0.134	0.160	0.108
0.688	0.030	0.007	1.469[a]	0.030	0.007	0.134	0.160	0.108
0.688	0.030	0.007	2.375	0.030	0.007	0.165	0.192	0.136
0.812	0.030	0.007	1.750	0.030	0.007	0.148	0.177	0.122
0.812	0.030	0.007	2.875	0.030	0.007	0.165	0.192	0.136
0.938	0.030	0.007	2.000	0.030	0.007	0.165	0.192	0.136
0.938	0.030	0.007	3.375	0.045	0.010	0.180	0.213	0.153
1.062	0.030	0.007	2.250	0.030	0.007	0.165	0.192	0.136
1.062	0.045	0.010	3.875	0.045	0.010	0.238	0.280	0.210
1.250	0.030	0.007	2.500	0.030	0.007	0.165	0.192	0.136
1.375	0.030	0.007	2.750	0.030	0.007	0.165	0.192	0.136
1.500	0.045	0.010	3.000	0.045	0.010	0.180	0.213	0.153
1.625	0.045	0.010	3.250	0.045	0.010	0.180	0.213	0.153
1.688	0.045	0.010	3.500	0.045	0.010	0.180	0.213	0.153
1.812	0.045	0.010	3.750	0.045	0.010	0.180	0.213	0.153
1.938	0.045	0.010	4.000	0.045	0.010	0.180	0.213	0.153
2.062	0.045	0.010	4.250	0.045	0.010	0.180	0.213	0.153

[a] The 0.734-inch and 1.469-inch outside diameters avoid washers which could be used in coin operated devices.

All dimensions are in inches.

The above sizes are to be ordered by ID, OD, and thickness dimensions.

Preferred Sizes of Type A Plain Washers are shown in Table 1a.

ANSI Standard Plain Washers.—The Type A plain washers were originally developed in a light, medium, heavy and extra heavy series. These series have been discontinued and the washers are now designated by their nominal dimensions.

The Type B plain washers are available in a narrow, regular and wide series with proportions designed to distribute the load over larger areas of lower strength materials.

Plain washers are made of ferrous or non-ferrous metal, plastic or other material as specified. The tolerances indicated in the tables are intended for metal washers only.

Table 2. American National Standard Type B Plain Washers —

Nominal Washer Size[a]		Series[b]	Inside Diameter			Outside Diameter			Thickness		
			Basic	Tolerance		Basic	Tolerance		Basic	Max.	Min.
				Plus	Minus		Plus	Minus			
		N	0.068	0.000	0.005	0.125	0.000	0.005	0.025	0.028	0.022
No. 0	0.060	R	0.068	0.000	0.005	0.188	0.000	0.005	0.025	0.028	0.022
		W	0.068	0.000	0.005	0.250	0.000	0.005	0.025	0.028	0.022
		N	0.084	0.000	0.005	0.156	0.000	0.005	0.025	0.028	0.022
No. 1	0.073	R	0.084	0.000	0.005	0.219	0.000	0.005	0.025	0.028	0.022
		W	0.084	0.000	0.005	0.281	0.000	0.005	0.032	0.036	0.028
		N	0.094	0.000	0.005	0.188	0.000	0.005	0.025	0.028	0.022
No. 2	0.086	R	0.094	0.000	0.005	0.250	0.000	0.005	0.032	0.036	0.028
		W	0.094	0.000	0.005	0.344	0.000	0.005	0.032	0.036	0.028
		N	0.109	0.000	0.005	0.219	0.000	0.005	0.025	0.028	0.022
No. 3	0.099	R	0.109	0.000	0.005	0.312	0.000	0.005	0.032	0.036	0.028
		W	0.109	0.008	0.005	0.406	0.008	0.005	0.040	0.045	0.036
		N	0.125	0.000	0.005	0.250	0.000	0.005	0.032	0.036	0.028
No. 4	0.112	R	0.125	0.008	0.005	0.375	0.008	0.005	0.040	0.045	0.036
		W	0.125	0.008	0.005	0.438	0.008	0.005	0.040	0.045	0.036
		N	0.141	0.000	0.005	0.281	0.000	0.005	0.032	0.036	0.028
No. 5	0.125	R	0.141	0.008	0.005	0.406	0.008	0.005	0.040	0.045	0.036
		W	0.141	0.008	0.005	0.500	0.008	0.005	0.040	0.045	0.036
		N	0.156	0.000	0.005	0.312	0.000	0.005	0.032	0.036	0.028
No. 6	0.138	R	0.156	0.008	0.005	0.438	0.008	0.005	0.040	0.045	0.036
		W	0.156	0.008	0.005	0.562	0.008	0.005	0.040	0.045	0.036
		N	0.188	0.008	0.005	0.375	0.008	0.005	0.040	0.045	0.036
No. 8	0.164	R	0.188	0.008	0.005	0.500	0.008	0.005	0.040	0.045	0.036
		W	0.188	0.008	0.005	0.625	0.015	0.005	0.063	0.071	0.056
		N	0.203	0.008	0.005	0.406	0.008	0.005	0.040	0.045	0.036
No. 10	0.190	R	0.203	0.008	0.005	0.562	0.008	0.005	0.040	0.045	0.036
		W	0.203	0.008	0.005	0.734[c]	0.015	0.007	0.063	0.071	0.056
		N	0.234	0.008	0.005	0.438	0.008	0.005	0.040	0.045	0.036
No. 12	0.216	R	0.234	0.008	0.005	0.625	0.015	0.005	0.063	0.071	0.056
		W	0.234	0.008	0.005	0.875	0.015	0.007	0.063	0.071	0.056
		N	0.281	0.015	0.005	0.500	0.015	0.005	0.063	0.071	0.056
1⁄4	0.250	R	0.281	0.015	0.005	0.734[c]	0.015	0.007	0.063	0.071	0.056
		W	0.281	0.015	0.005	1.000	0.015	0.007	0.063	0.071	0.056
		N	0.344	0.015	0.005	0.625	0.015	0.005	0.063	0.071	0.056
5⁄16	0.312	R	0.344	0.015	0.005	0.875	0.015	0.007	0.063	0.071	0.056
		W	0.344	0.015	0.005	1.125	0.015	0.007	0.063	0.071	0.056
		N	0.406	0.015	0.005	0.734[c]	0.015	0.007	0.063	0.071	0.056
3⁄8	0.375	R	0.406	0.015	0.005	1.000	0.015	0.007	0.063	0.071	0.056
		W	0.406	0.015	0.005	1.250	0.030	0.007	0.100	0.112	0.090
		N	0.469	0.015	0.005	0.875	0.015	0.007	0.063	0.071	0.056
7⁄16	0.438	R	0.469	0.015	0.005	1.125	0.015	0.007	0.063	0.071	0.056
		W	0.469	0.015	0.005	1.469[c]	0.030	0.007	0.100	0.112	0.090
		N	0.531	0.015	0.005	1.000	0.015	0.007	0.063	0.071	0.056
1⁄2	0.500	R	0.531	0.015	0.005	1.250	0.030	0.007	0.100	0.112	0.090
		W	0.531	0.015	0.005	1.750	0.030	0.007	0.100	0.112	0.090
		N	0.594	0.015	0.005	1.125	0.015	0.007	0.063	0.071	0.056
9⁄16	0.562	R	0.594	0.015	0.005	1.469[c]	0.030	0.007	0.100	0.112	0.090
		W	0.594	0.015	0.005	2.000	0.030	0.007	0.100	0.112	0.090
		N	0.656	0.030	0.007	1.250	0.030	0.007	0.100	0.112	0.090
5⁄8	0.625	R	0.656	0.030	0.007	1.750	0.030	0.007	0.100	0.112	0.090
		W	0.656	0.030	0.007	2.250	0.030	0.007	0.160	0.174	0.146
		N	0.812	0.030	0.007	1.375	0.030	0.007	0.100	0.112	0.090
3⁄4	0.750	R	0.812	0.030	0.007	2.000	0.030	0.007	0.100	0.112	0.090
		W	0.812	0.030	0.007	2.500	0.030	0.007	0.160	0.174	0.146
		N	0.938	0.030	0.007	1.469[c]	0.030	0.007	0.100	0.112	0.090
7⁄8	0.875	R	0.938	0.030	0.007	2.250	0.030	0.007	0.160	0.174	0.146
		W	0.938	0.030	0.007	2.750	0.030	0.007	0.160	0.174	0.146
		N	1.062	0.030	0.007	1.750	0.030	0.007	0.100	0.112	0.090
1	1.000	R	1.062	0.030	0.007	2.500	0.030	0.007	0.160	0.174	0.146
		W	1.062	0.030	0.007	3.000	0.030	0.007	0.160	0.174	0.146
		N	1.188	0.030	0.007	2.000	0.030	0.007	0.100	0.112	0.090
1 1⁄8	1.125	R	1.188	0.030	0.007	2.750	0.030	0.007	0.160	0.174	0.146
		W	1.188	0.030	0.007	3.250	0.030	0.007	0.160	0.174	0.146

Table 2. *(Continued)* **American National Standard Type B Plain Washers —**

Nominal Washer Size[a]		Series[b]	Inside Diameter			Outside Diameter			Thickness		
			Basic	Tolerance		Basic	Tolerance		Basic	Max.	Min.
				Plus	Minus		Plus	Minus			
		N	1.312	0.030	0.007	2.250	0.030	0.007	0.160	0.174	0.146
1¼	1.250	R	1.312	0.030	0.007	3.000	0.030	0.007	0.160	0.174	0.146
		W	1.312	0.045	0.010	3.500	0.045	0.010	0.250	0.266	0.234
		N	1.438	0.030	0.007	2.500	0.030	0.007	0.160	0.174	0.146
1⅜	1.375	R	1.438	0.030	0.007	3.250	0.030	0.007	0.160	0.174	0.146
		W	1.438	0.045	0.010	3.750	0.045	0.010	0.250	0.266	0.234
		N	1.562	0.030	0.007	2.750	0.030	0.007	0.160	0.174	0.146
1½	1.500	R	1.562	0.045	0.010	3.500	0.045	0.010	0.250	0.266	0.234
		W	1.562	0.045	0.010	4.000	0.045	0.010	0.250	0.266	0.234
		N	1.750	0.030	0.007	3.000	0.030	0.007	0.160	0.174	0.146
1⅝	1.625	R	1.750	0.045	0.010	3.750	0.045	0.010	0.250	0.266	0.234
		W	1.750	0.045	0.010	4.250	0.045	0.010	0.250	0.266	0.234
		N	1.875	0.030	0.007	3.250	0.030	0.007	0.160	0.174	0.146
1¾	1.750	R	1.875	0.045	0.010	4.000	0.045	0.010	0.250	0.266	0.234
		W	1.875	0.045	0.010	4.500	0.045	0.010	0.250	0.266	0.234
		N	2.000	0.045	0.010	3.500	0.045	0.010	0.250	0.266	0.234
1⅞	1.875	R	2.000	0.045	0.010	4.250	0.045	0.010	0.250	0.266	0.234
		W	2.000	0.045	0.010	4.750	0.045	0.010	0.250	0.266	0.234
		N	2.125	0.045	0.010	3.750	0.045	0.010	0.250	0.266	0.234
2	2.000	R	2.125	0.045	0.010	4.500	0.045	0.010	0.250	0.266	0.234
		W	2.125	0.045	0.010	5.000	0.045	0.010	0.250	0.266	0.234

[a] Nominal washer sizes are intended for use with comparable nominal screw or bolt sizes.

[b] N indicates Narrow; R, Regular; and W, Wide Series.

[c] The 0.734-inch and 1.469-inch outside diameter avoids washers which could be used in coin operated devices.

All dimensions are in inches.

Inside and outside diameters shall be concentric within at least the inside diameter tolerance.

Washers shall be flat within 0.005-inch for basic outside diameters up through 0.875-inch and within 0.010 inch for larger outside diameters.

For 2¼-, 2½- , 2¾-, and 3-inch sizes see ANSI/ASME B18.22.1-1965 (R1998).

American National Standard Helical Spring and Tooth Lock Washers ANSI/ASME B18.21.1-1994.—This standard covers helical spring lock washers of carbon steel; boron steel; corrosion resistant steel, Types 302 and 305; aluminum-zinc alloy; phosphor-bronze; silicon-bronze; and K-Monel; in various series. Tooth lock washers of carbon steel having internal teeth, external teeth, and both internal and external teeth, of two constructions, designated as Type A and Type B. Washers intended for general industrial application are also covered. American National Standard Lock Washers (Metric Series) ANSI/ASME B18.21.2M-1994 covers metric sizes for helical spring and tooth lock washers.

Helical spring lock washers: These washers are used to provide: 1) good bolt tension per unit of applied torque for tight assemblies; 2) hardened bearing surfaces to create uniform torque control; 3) uniform load distribution through controlled radii—section—cut-off; and 4) protection against looseness resulting from vibration and corrosion.

Nominal washer sizes are intended for use with comparable nominal screw or bolt sizes. These washers are designated by the following data in the sequence shown: Product name; nominal size (number, fraction or decimal equivalent); series; material;and protective finish, if required. For example: Helical Spring Lock Washer, 0.375 Extra Duty, Steel, Phosphate Coated.

Helical spring lock washers are available in four series: Regular, heavy, extra duty and hi-collar as given in Tables 2 and 1. Helical spring lock washers made of materials other than carbon steel are available in the regular series as given in Table 2.

Table 1. American National Standard Hi-Collar Helical Spring Lock Washers *ANSI/ASME B18.21.1-1994*

Nominal Washer Size		Inside Diameter		Outside Diameter	Washer Section	
					Width	Thickness[a]
		Min.	Max.	Max.	Min.	Min.
No. 4	0.112	0.114	0.120	0.173	0.022	0.022
No. 5	0.125	0.127	0.133	0.202	0.030	0.030
No. 6	0.138	0.141	0.148	0.216	0.030	0.030
No. 8	0.164	0.167	0.174	0.267	0.042	0.047
No. 10	0.190	0.193	0.200	0.294	0.042	0.047
1⁄4	0.250	0.252	0.260	0.363	0.047	0.078
5⁄16	0.3125	0.314	0.322	0.457	0.062	0.093
3⁄8	0.375	0.377	0.385	0.550	0.076	0.125
7⁄16	0.4375	0.440	0.450	0.644	0.090	0.140
1⁄2	0.500	0.502	0.512	0.733	0.103	0.172
5⁄8	0.625	0.628	0.640	0.917	0.125	0.203
3⁄4	0.750	0.753	0.765	1.105	0.154	0.218
7⁄8	0.875	0.878	0.890	1.291	0.182	0.234
1	1.000	1.003	1.015	1.478	0.208	0.250
1 1⁄8	1.125	1.129	1.144	1.663	0.236	0.313
1 1⁄4	1.250	1.254	1.272	1.790	0.236	0.313
1 3⁄8	1.375	1.379	1.399	2.031	0.292	0.375
1 1⁄2	1.500	1.504	1.524	2.159	0.292	0.375
1 3⁄4	1.750	1.758	1.778	2.596	0.383	0.469
2	2.000	2.008	2.028	2.846	0.383	0.469
2 1⁄4	2.250	2.262	2.287	3.345	0.508	0.508
2 1⁄2	2.500	2.512	2.537	3.559	0.508	0.508
2 3⁄4	2.750	2.762	2.787	4.095	0.633	0.633
3	3.000	3.012	3.037	4.345	0.633	0.633

[a] Mean section thickness = (inside thickness + outside thickness) ÷ 2.

Table 2. American National Standard Helical Spring Lock Washers *ANSI/ASME B18.21.1-1994*

Nominal Washer Size		Inside Diameter, A		Regular			Heavy			Extra Duty		
		Max.	Min.	O.D., B Max.	Section Width, W	Section Thickness, T[a]	O.D., B Max.	Section Width, W	Section Thickness, T[a]	O.D., B Max.	Section Width, W	Section Thickness, T[a]
No. 2	0.086	0.094	0.088	0.172	0.035	0.020	0.182	0.040	0.025	0.208	0.053	0.027
No. 3	0.099	0.107	0.101	0.195	0.040	0.025	0.209	0.047	0.031	0.239	0.062	0.034
No. 4	0.112	0.120	0.114	0.209	0.040	0.025	0.223	0.047	0.031	0.253	0.062	0.034
No. 5	0.125	0.133	0.127	0.236	0.047	0.031	0.252	0.055	0.040	0.300	0.079	0.045
No. 6	0.138	0.148	0.141	0.250	0.047	0.031	0.266	0.055	0.040	0.314	0.079	0.045
No. 8	0.164	0.174	0.167	0.293	0.055	0.040	0.307	0.062	0.047	0.375	0.096	0.057
No. 10	0.190	0.200	0.193	0.334	0.062	0.047	0.350	0.070	0.056	0.434	0.112	0.068
No. 12	0.216	0.227	0.220	0.377	0.070	0.056	0.391	0.077	0.063	0.497	0.130	0.080
¼	0.250	0.060	0.252	0.487	0.109	0.062	0.489	0.110	0.077	0.533	0.132	0.084
5/16	0.3125	0.322	0.314	0.583	0.125	0.078	0.293	0.130	0.097	0.619	0.143	0.108
⅜	0.375	0.385	0.377	0.680	0.141	0.094	0.688	0.145	0.115	0.738	0.170	0.123
7/16	0.4375	0.450	0.440	0.776	0.156	0.109	0.784	0.160	0.133	0.836	0.186	0.143
½	0.500	0.512	0.502	0.869	0.171	0.125	0.879	0.176	0.151	0.935	0.204	0.162
9/16	0.5625	0.574	0.564	0.965	0.188	0.141	0.975	0.193	0.170	1.035	0.223	0.182
⅝	0.625	0.641	0.628	1.073	0.203	0.156	1.087	0.210	0.189	1.151	0.242	0.202
11/16	0.6875	0.704	0.691	1.170	0.219	0.172	1.186	0.227	0.207	1.252	0.260	0.221
¾	0.750	0.766	0.753	1.265	0.234	0.188	1.285	0.244	0.226	1.355	0.279	0.241
13/16	0.8125	0.832	0.816	1.363	0.250	0.203	1.387	0.262	0.246	1.458	0.298	0.261
⅞	0.875	0.894	0.878	1.459	0.266	0.219	1.489	0.281	0.266	1.571	0.322	0.285
15/16	0.9375	0.958	0.941	1.556	0.281	0.234	1.590	0.298	0.284	1.684	0.345	0.308
1	1.000	1.024	1.003	1.656	0.297	0.250	1.700	0.319	0.306	1.794	0.366	0.330
1 1/16	1.0625	1.087	1.066	1.751	0.312	0.266	1.803	0.338	0.326	1.905	0.389	0.352
1⅛	1.125	1.153	1.129	1.847	0.328	0.281	1.903	0.356	0.345	2.013	0.411	0.375
1 3/16	1.1875	1.217	1.192	1.943	0.344	0.297	2.001	0.373	0.364	2.107	0.431	0.396
1¼	1.250	1.280	1.254	2.036	0.359	0.312	2.104	0.393	0.384	2.222	0.452	0.417
1 5/16	1.3125	1.344	1.317	2.133	0.375	0.328	2.203	0.410	0.403	2.327	0.472	0.438
1⅜	1.375	1.408	1.379	2.219	0.391	0.344	2.301	0.427	0.422	2.429	0.491	0.458
1 7/16	1.4375	1.472	1.442	2.324	0.406	0.359	2.396	0.442	0.440	2.530	0.509	0.478
1½	1.500	1.534	1.504	2.419	0.422	0.375	2.491	0.458	0.458	2.627	0.526	0.496

[a] T = mean section thickness $= (t_i + t_o) \div 2$.

All dimensions are given in inches.*See ANSI/ASME B18.21.1-1994 standard for sizes over 1½ to 3, inclusive, for regular and heavy helical spring lock washers and over 1½ to 2, inclusive, for extra-duty helical spring lock washers.

When carbon steel helical spring lock washers are to be hot-dipped galvanized for use with hot-dipped galvanized bolts or screws, they are to be coiled to limits onto inch in excess of those specified in Tables 2 and 1 for minimum inside diameter and maximum outside diameter. Galvanizing washers under ¼ inch nominal size are not recommended.

Tooth lock washers: These washers serve to lock fasteners, such as bolts and nuts, to the component parts of an assembly, or increase the friction between the fasteners and the assembly. They are designated in a manner similar to helical spring lock washers, and are available in carbon steel. Dimensions are given in Tables 3 and 4.

Table 3. American National Standard Internal-External Tooth Lock Washers
ANSI/ASME B18.21.1-1994

All dimensions are given in inches except whole numbers under "Size"

TYPE A TYPE B

Size	A Inside Diameter Max.	A Inside Diameter Min.	B Outside Diameter Max.	B Outside Diameter Min.	C Thickness Max.	C Thickness Min.
No. 4	.123	.115	.475	.460	.021	.016
			.510	.495	.021	.017
			.610	.580	.021	.017
No. 6	.150	.141	.510	.495	.028	.023
			.610	.580	.028	.023
			.690	.670	.028	.023
No. 8	.176	.168	.610	.580	.034	.028
			.690	.670	.034	.028
			.760	.740	.034	.028
No. 10	.204	.195	.610	.580	.034	.028
			.690	.670	.040	.032
			.760	.740	.040	.032
			.900	.880	.040	.032
No. 12	.231	.221	.690	.670	.040	.032
			.760	.725	.040	.032
			.900	.880	.040	.032
			.985	.965	.045	.037
¼	.267	.256	.760	.725	.040	.032
			.900	.880	.045	.037
			.985	.965	.045	.037
			1.070	1.045	.045	.037

Size	A Inside Diameter Max.	A Inside Diameter Min.	B Outside Diameter Max.	B Outside Diameter Min.	C Thickness Max.	C Thickness Min.
5⁄16	.332	.320	.900	.865	.040	.032
			.985	.965	.045	.037
			1.070	1.045	.050	.042
			1.155	1.130	.050	.042
⅜	.398	.384	.985	.965	.045	.037
			1.070	1.045	.050	.042
			1.070	1.130	.050	.042
			1.260	1.220	.050	.042
7⁄16	.464	.448	1.070	1.045	.050	.042
			1.155	1.130	.050	.042
			1.260	1.220	.055	.047
			1.315	1.290	.055	.047
½	.530	.512	1.260	1.220	.055	.047
			1.315	1.290	.055	.047
			1.410	1.380	.060	.052
			1.620	1.590	.067	.059
9⁄16	.596	.576	1.315	1.290	.055	.047
			1.430	1.380	.060	.052
			1.620	1.590	.067	.059
			1.830	1.797	.067	.059
⅝	.663	.640	1.410	1.380	.060	.052
			1.620	1.590	.067	.059
			1.830	1.797	.067	.059
			1.975	1.935	.067	.059

Table 4. American National Standard Internal and External Tooth Lock Washers *ANSI/ASME B18.21.1-1994*

TYPE A TYPE B
Internal Tooth

TYPE A TYPE B
External Tooth

TYPE A 80°-82° TYPE B 80°-82°
Countersunk External Tooth

Internal Tooth Lock Washers

	Size	#2	#3	#4	#5	#6	#8	#10	#12	1/4	5/16	3/8	7/16	1/2	9/16	5/8	11/16	3/4	13/16	7/8	1	1 1/8	1 1/4
A	Max	0.095	0.109	0.123	0.136	0.150	0.176	0.204	0.231	0.267	0.332	0.398	0.464	0.530	0.596	0.663	0.728	0.795	0.861	0.927	1.060	1.192	1.325
	Min	0.089	0.102	0.115	0.129	0.141	0.168	0.195	0.221	0.256	0.320	0.384	0.448	0.512	0.576	0.640	0.704	0.769	0.832	0.894	1.019	1.144	1.275
B	Max	0.200	0.232	0.270	0.280	0.295	0.340	0.381	0.410	0.478	0.610	0.692	0.789	0.900	0.985	1.071	1.166	1.245	1.315	1.410	1.637	1.830	1.975
	Min	0.175	0.215	0.245	0.255	0.275	0.325	0.365	0.394	0.460	0.594	0.670	0.740	0.867	0.957	1.045	1.130	1.220	1.290	1.364	1.590	1.799	1.921
C	Max	0.015	0.019	0.019	0.021	0.021	0.023	0.025	0.025	0.028	0.034	0.040	0.040	0.045	0.045	0.050	0.050	0.055	0.055	0.060	0.067	0.067	0.067
	Min	0.010	0.012	0.015	0.017	0.017	0.018	0.020	0.020	0.023	0.028	0.032	0.032	0.037	0.037	0.042	0.042	0.047	0.047	0.052	0.059	0.059	0.059

External Tooth Lock Washers

	Size	#2	#3	#4	#5	#6	#8	#10	#12	1/4	5/16	3/8	7/16	1/2	9/16	5/8	11/16	3/4	13/16	7/8	1	1 1/8	1 1/4
A	Max	…	0.109	0.123	0.136	0.150	0.176	0.204	0.231	0.267	0.332	0.398	0.464	0.530	0.596	0.663	0.728	0.795	0.861	0.927	1.060	…	…
	Min	…	0.102	0.115	0.129	0.141	0.168	0.195	0.221	0.256	0.320	0.384	0.448	0.513	0.576	0.641	0.704	0.768	0.833	0.897	1.025	…	…
B	Max	…	0.235	0.260	0.285	0.320	0.381	0.410	0.475	0.510	0.610	0.694	0.760	0.900	0.985	1.070	1.155	1.260	1.315	1.410	1.620	…	…
	Min	…	0.220	0.245	0.270	0.305	0.365	0.395	0.460	0.494	0.588	0.670	0.740	0.880	0.960	1.045	1.130	1.220	1.290	1.380	1.590	…	…
C	Max	…	0.015	0.019	0.019	0.022	0.023	0.025	0.028	0.028	0.034	0.040	0.040	0.045	0.045	0.050	0.050	0.055	0.055	0.060	0.067	…	…
	Min	…	0.012	0.014	0.015	0.016	0.018	0.020	0.023	0.023	0.028	0.032	0.032	0.037	0.037	0.042	0.042	0.047	0.047	0.052	0.059	…	…

Heavy Internal Tooth Lock Washers

	Size	1/4	5/6	3/8	7/16	1/2	9/16	5/8	3/4	7/8
A	Max	0.267	0.332	0.398	0.464	0.530	0.596	0.663	0.795	0.927
	Min	0.256	0.320	0.384	0.448	0.512	0.576	0.640	0.768	0.894
B	Max	0.536	0.607	0.748	0.858	0.924	1.034	1.135	1.265	1.447
	Min	0.500	0.590	0.700	0.800	0.880	0.990	1.100	1.240	1.400
C	Max	0.045	0.050	0.050	0.067	0.067	0.067	0.067	0.084	0.084
	Min	0.035	0.040	0.042	0.050	0.055	0.055	0.059	0.070	0.075

Countersunk External Tooth Lock Washers[a]

	Size	#4	#6	#8	#10	#12	1/4	#16	5/16	3/8	7/16	1/2
A	Max	0.123	0.150	0.177	0.205	0.231	0.267	0.287	0.333	0.398	0.463	0.529
	Min	0.113	0.140	0.167	0.195	0.220	0.255	0.273	0.318	0.383	0.448	0.512
C	Max	0.019	0.021	0.021	0.025	0.025	0.025	0.028	0.028	0.034	0.045	0.045
	Min	0.015	0.017	0.017	0.020	0.020	0.020	0.023	0.023	0.028	0.037	0.037
D	Max	0.065	0.092	0.105	0.099	0.128	0.128	0.147	0.192	0.255	0.270	0.304
	Min	0.050	0.082	0.088	0.083	0.118	0.113	0.137	0.165	0.242	0.260	0.294

[a] Starting with #4, approx. O.D.'s are: 0.213, 0.289, 0.322, 0.354, 0.421 0.454, 0.505, 0.599, 0.765, 0.867, and 0.976.

All dimensions are given in inches.

METRIC FASTENERS

A number of American National Standards covering metric bolts, screws, nuts, and washers have been established in cooperation with the Department of Defense in such a way that they could be used by the Government for procurement purposes. Extensive information concerning these metric fasteners is given in the following text and tables, but for additional manufacturing and acceptance specifications reference should be made to the respective Standards which may be obtained by nongovernmental agencies from the American National Standards Institute, 11 West 42nd Street, New York, N.Y. 10036. These Standards are:

ANSI B18.2.3.1M-1979 (R1989) Metric Hex Cap Screws	Table 1
ANSI B18.2.3.2M-1979 (R1989) Metric Formed Hex Screws	Table 2
ANSI B18.2.3.3M-1979 (R1989) Metric Heavy Hex Screws	Table 3
ANSI B18.2.3.4M-1984 Metric Hex Flange Screws	Table 7
ANSI B18.2.3.5M-1979 (R1989) Metric Hex Bolts	Table 12
ANSI B18.2.3.6M-1979 (R1989) Metric Heavy Hex Bolts	Table 10
ANSI B18.2.3.7M-1979 (R1989) Metric Heavy Hex Structural Bolts	Table 11
ANSI B18.2.3.8M-1981 (R1991) Metric Hex Lag Screws	Table 5
ANSI B18.2.3.9M-1984 Metric Heavy Hex Flange Screws	Table 6
ANSI B18.5.2.2M-1982 Metric Round Head Square Neck Bolts	Table 9
ANSI B18.3.1M-1986 Socket Head Cap Screws (Metric Series)	Table 23
ANSI B18.2.4.1M-1979 (R1989) Metric Hex Nuts, Style 1	Table 26
ANSI B18.2.4.2M-1979 (R1989) Metric Hex Nuts, Style 2	Table 26
ANSI B18.2.4.3M-1979 (R1989) Metric Slotted Hex Nuts	Table 27
ANSI B18.2.4.4M-1982 Metric Hex Flange Nuts	Table 28
ANSI B18.2.4.5M-1979 Metric Hex Jam Nuts	Table 31
ANSI B18.2.4.6M-1979 (R1990) Metric Heavy Hex Nuts	Table 31
ANSI B18.16.3M-1998 Prevailing-Torque Metric Hex Nuts	Table 29
ANSI B18.16.3M-1998 Prevailing-Torque Metric Hex Flange Nuts	Table 30
ANSI B18.22M-1981 (R1990) Metric Plain Washers	Table 32

Manufacturers should be consulted concerning which items and sizes are in stock production.

Comparison with ISO Standards.—American National Standards for metric bolts, screws and nuts have been coordinated to the extent possible with the comparable ISO Standards or proposed Standards. The dimensional differences between the ANSI and the comparable ISO Standards or proposed Standards are few, relatively minor, and none will affect the functional interchangeability of bolts, screws, and nuts manufactured to the requirements of either.

Where no comparable ISO Standard had been developed, as was the case when the ANSI Standards for Metric Heavy Hex Screws, Metric Heavy Hex Bolts, and Metric Hex Lag Screws were adopted, nominal diameters, thread pitches, body diameters, widths across flats, head heights, thread lengths, thread dimensions, and nominal lengths are in accord with ISO Standards for related hex head screws and bolts. At the time of ANSI adoption (1982) there was no ISO Standard for round head square neck bolts.

The following functional characteristics of hex head screws and bolts are in agreement between the respective ANSI Standard and the comparable ISO Standard or proposed Standard: diameters and thread pitches, body diameters, widths across flats (see exception below), bearing surface diameters (except for metric hex bolts), flange diameters (for metric hex flange screws), head heights, thread lengths, thread dimensions, and nominal lengths.

Table 1. American National Standard Metric Hex Cap Screws
ANSI/ASME B18.2.3.1M-1979 (R1995)

K L
K_1 L_g
S L_s B (REF)
C X (REF)
E
D_w–D_s
INCOMPLETE THREAD
OPTIONAL POINT CONSTRUCTION
15°-30°
FILLET
PROPERTY CLASS AND MANUFACTURER'S IDENTIFICATION TO APPEAR ON TOP OF HEAD

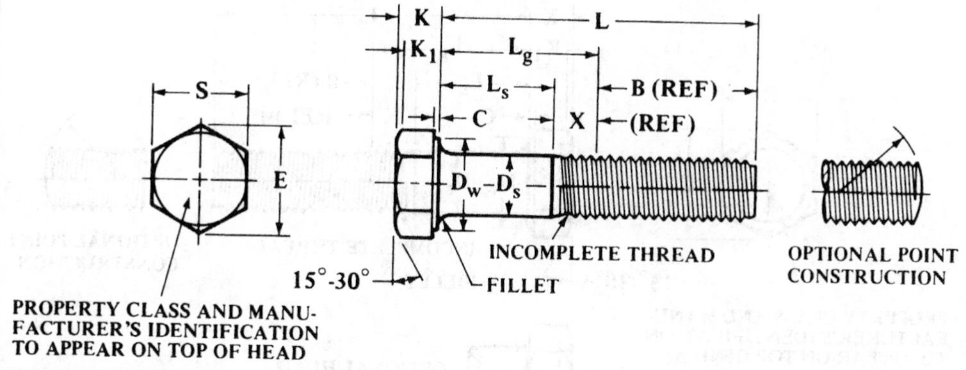

Nominal Screw Dia., *D*, and Thread Pitch	Body Dia., D_s		Width Across Flats, *S*		Width Across Corners, *E*		Head Height, *K*		Wrenching Height, K_1	Washer Face Thick., *C*		Washer Face Dia., D_w
	Max	Min	Max	Min	Max	Min	Max	Min	Min	Max	Min	Min
M5 × 0.8	5.00	4.82	8.00	7.78	9.24	8.79	3.65	3.35	2.4	0.5	0.2	7.0
M6 × 1	6.00	5.82	10.00	9.78	11.55	11.05	4.15	3.85	2.8	0.5	0.2	8.9
M8 × 1.25	8.00	7.78	13.00	12.73	15.01	14.38	5.50	5.10	3.7	0.6	0.3	11.6
[a]M10 × 1.5	10.00	9.78	15.00	14.73	17.32	16.64	6.63	6.17	4.5	0.6	0.3	13.6
M10 × 1.5	10.00	9.78	16.00	15.73	18.48	17.77	6.63	6.17	4.5	0.6	0.3	14.6
M12 × 1.75	12.00	11.73	18.00	17.73	20.78	20.03	7.76	7.24	5.2	0.6	0.3	16.6
M14 × 2	14.00	13.73	21.00	20.67	24.25	23.35	9.09	8.51	6.2	0.6	0.3	19.6
M16 × 2	16.00	15.73	24.00	23.67	27.71	26.75	10.32	9.68	7.0	0.8	0.4	22.5
M20 × 2.5	20.00	19.67	30.00	29.16	34.64	32.95	12.88	12.12	8.8	0.8	0.4	27.7
M24 × 3	24.00	23.67	36.00	35.00	41.57	39.55	15.44	14.56	10.5	0.8	0.4	33.2
M30 × 3.5	30.00	29.67	46.00	45.00	53.12	50.85	19.48	17.92	13.1	0.8	0.4	42.7
M36 × 4	36.00	35.61	55.00	53.80	63.51	60.79	23.38	21.62	15.8	0.8	0.4	51.1
M42 × 4.5	42.00	41.38	65.00	62.90	75.06	71.71	26.97	25.03	18.2	1.0	0.5	59.8
M48 × 5	48.00	47.38	75.00	72.60	86.60	82.76	31.07	28.93	21.0	1.0	0.5	69.0
M56 × 5.5	56.00	55.26	85.00	82.20	98.15	93.71	36.20	33.80	24.5	1.0	0.5	78.1
M64 × 6	64.00	63.26	95.00	91.80	109.70	104.65	41.32	38.68	28.0	1.0	0.5	87.2
M72 × 6	72.00	71.26	105.00	101.40	121.24	115.60	46.45	43.55	31.5	1.2	0.6	96.3
M80 × 6	80.00	79.26	115.00	111.00	132.72	126.54	51.58	48.42	35.0	1.2	0.6	105.4
M90 × 6	90.00	89.13	130.00	125.50	150.11	143.07	57.74	54.26	39.2	1.2	0.6	119.2
M100 × 6	100.00	99.13	145.00	140.00	167.43	159.60	63.90	60.10	43.4	1.2	0.6	133.0

[a] This size with width across flats of 15 mm is not standard. Unless specifically ordered, M10 hex cap screws with 16 mm width across flats will be furnished.

All dimensions are in millimeters.

Basic thread lengths, *B*, are the same as given in Table 12.

Transition thread length, *X*, includes the length of incomplete threads and tolerances on grip gaging length and body length. It is intended for calculation purposes.

For additional manufacturing and acceptance specifications, reference should be made to the ANSI/ASME B18.2.3.1M-1979 (R1995).

Table 2. American National Standard Metric Formed Hex Screws
ANSI/ASME B18.2.3.2M-1979 (R1995)

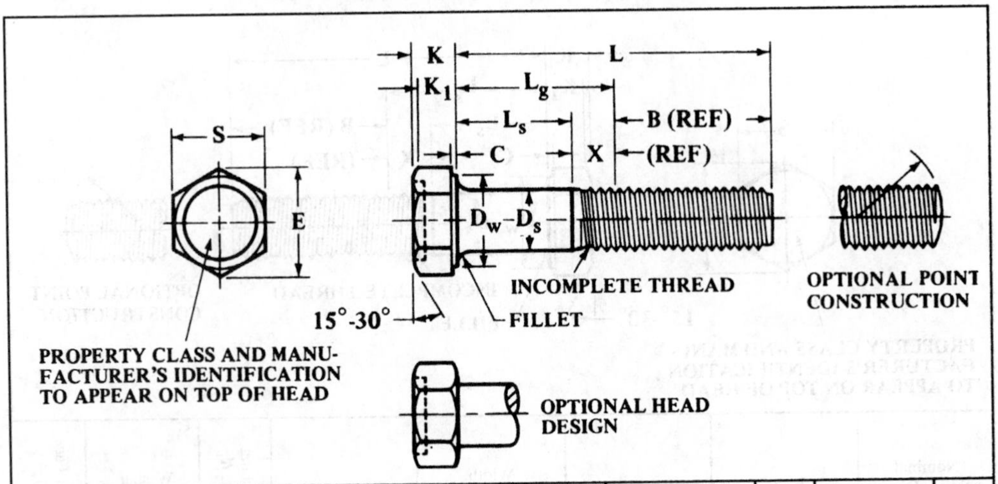

K
L
K_1
L_g
S
L_s
B (REF)
C
X
(REF)
E
D_w–D_s
INCOMPLETE THREAD
OPTIONAL POINT CONSTRUCTION
15°-30°
FILLET
PROPERTY CLASS AND MANUFACTURER'S IDENTIFICATION TO APPEAR ON TOP OF HEAD
OPTIONAL HEAD DESIGN

Nominal Screw Dia., *D*, and Thread Pitch	Body Dia., D_s		Width Across Flats, *S*		Width Across Corners, *E*		Head Height, *K*		Wrenching Height, K_1	Washer Face Thick., *C*		Washer Face Dia., D_w
	Max	Min	Max	Min	Max	Min	Max	Min	Min	Max	Min	Max
M5 × 0.8	5.00	4.82	8.00	7.64	9.24	8.56	3.65	3.35	2.4	0.5	0.2	6.9
M6 × 1	6.00	5.82	10.00	9.64	11.55	10.80	4.15	3.85	2.0	0.5	0.2	8.9
M8 × 1.25	8.00	7.78	13.00	12.57	15.01	14.08	5.50	5.10	3.7	0.6	0.3	11.6
[a]M10 × 1.5	10.00	9.78	15.00	14.57	17.32	16.32	6.63	6.17	4.5	0.6	0.3	13.6
M10 × 1.5	10.00	9.78	16.00	15.57	18.48	17.43	6.63	6.17	4.5	0.6	0.3	14.6
M12 × 1.75	12.00	11.73	18.00	17.57	20.78	19.68	7.76	7.24	5.2	0.6	0.3	16.6
M14 × 2	14.00	13.73	21.00	20.16	24.25	22.58	9.09	8.51	6.2	0.6	0.3	19.6
M16 × 2	16.00	15.73	24.00	23.16	27.71	25.94	10.32	9.68	7.0	0.8	0.4	22.5
M20 × 2.5	20.00	19.67	30.00	29.16	34.64	32.66	12.88	12.12	8.8	0.8	0.4	27.7
M24 × 3	24.00	23.67	36.00	35.00	41.57	39.20	15.44	14.56	10.5	0.8	0.4	33.2

[a] This size with width across flats of 15 mm is not standard. Unless specifically ordered, M10 formed hex screws with 16 mm width across flats will be furnished.

All dimensions are in millimeters.

†Basic thread lengths, *B*, are the same as given in Table 12.

‡Transition thread length, *X*, includes the length of incomplete threads and tolerances on the grip gaging length and body length. It is intended for calculation purposes.

For additional manufacturing and acceptance specifications, reference should be made to the Standard.

Socket head cap screws ANSI B18.3.1M-1986 are functionally interchangeable with screws which conform to ISO R861-1968 or ISO 4762-1977. However, the thread lengths specified in the ANSI Standard are equal to or longer than required by either ISO Standard. Consequently the grip lengths also vary on screws where the North American thread length practice differs. Minor variations in head diameter, head height, key engagement and wall thickness are due to diverse tolerancing practice and will be found documented in the ANSI Standard.

One exception with respect to width across flats for metric hex cap screws, formed hex screws, and hex bolts is the M10 size. These are currently being produced in the United States with a width across flats of 15 mm. This size, however, is not an ISO Standard. Unless these M10 screws and bolts with 15 mm width across flats are specifically ordered, the M10 size with 16 mm across flats will be furnished.

Table 3. American National Standard Metric Heavy Hex Screws
ANSI B18.2.3.3M-1979 (R1995)

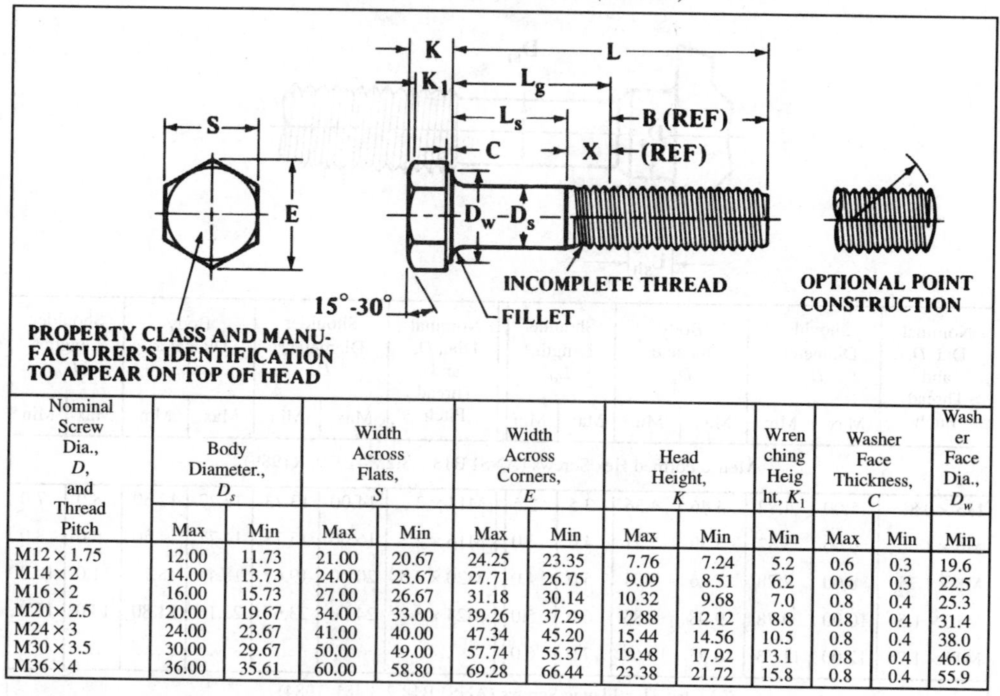

Nominal Screw Dia., *D*, and Thread Pitch	Body Diameter., D_s		Width Across Flats, *S*		Width Across Corners, *E*		Head Height, *K*		Wrenching Height, K_1	Washer Face Thickness, *C*		Washer Face Dia., D_w
	Max	Min	Max	Min	Max	Min	Max	Min	Min	Max	Min	Min
M12 × 1.75	12.00	11.73	21.00	20.67	24.25	23.35	7.76	7.24	5.2	0.6	0.3	19.6
M14 × 2	14.00	13.73	24.00	23.67	27.71	26.75	9.09	8.51	6.2	0.6	0.3	22.5
M16 × 2	16.00	15.73	27.00	26.67	31.18	30.14	10.32	9.68	7.0	0.8	0.4	25.3
M20 × 2.5	20.00	19.67	34.00	33.00	39.26	37.29	12.88	12.12	8.8	0.8	0.4	31.4
M24 × 3	24.00	23.67	41.00	40.00	47.34	45.20	15.44	14.56	10.5	0.8	0.4	38.0
M30 × 3.5	30.00	29.67	50.00	49.00	57.74	55.37	19.48	17.92	13.1	0.8	0.4	46.6
M36 × 4	36.00	35.61	60.00	58.80	69.28	66.44	23.38	21.72	15.8	0.8	0.4	55.9

All dimensions are in millimeters.

Basic thread lengths, *B*, are the same as given in Table 12.

Transition thread length, *X*, includes the length of incomplete threads and tolerances on grip gaging length and body length. It is intended for calculation purposes.

For additional manufacturing and acceptance specifications, reference should be made to the Standard.

ANSI letter symbols designating dimensional characteristics are in accord with those used in ISO Standards except capitals have been used for data processing convenience instead of the lower case letters used in the ISO Standards.

Metric Screw and Bolt Diameters.—Metric screws and bolts are furnished with full diameter body within the limits shown in the respective dimensional tables, or are threaded to the head (see *Metric Screw and Bolt Thread Lengths*) unless the purchaser specifies "reduced body diameter." Metric formed hex screws (Table 4), hex flange screws (Table 4), hex bolts (Table 4), heavy hex bolts (Table 4), hex lag screws (Table 5), heavy hex flange screws (Table 6), and round head square neck bolts (Table 8) may be obtained with reduced diameter body, if so specified; however, formed hex screws, hex flange screws, heavy hex flange screws, hex bolts, or heavy hex bolts with nominal lengths shorter than 4*D*, where *D* is the nominal diameter, are not recommended. Metric formed hex screws, hex flange screws, heavy hex flange screws, and hex lag screws with reduced body diameter will be furnished with a shoulder under the head. For metric hex bolts and heavy hex bolts this is optional with the manufacturer.

For bolts and lag screws there may be a reasonable swell, fin, or die seam on the body adjacent to the head not exceeding the nominal bolt diameter by: 0.50 mm for M5, 0.65 mm for M6, 0.75 mm for M8 through M14, 1.25 mm for M16, 1.50 mm for M20 through M30, 2.30 mm for M36 through M48, 3.00 mm for M56 through M72, and 4.80 mm for M80 through M100.

Table 4. American National Standard Metric Hex Screws and Bolts — Reduced Body Diameters

Nominal Dia., D, and Thread Pitch	Shoulder Diameter,[a] D_s		Body Diameter, D_{si}		Shoulder Length,[a] L_{sh}		Nominal Dia., D, and Thread Pitch	Shoulder Diameter,[a] D_s		Body Diameter, D_{si}		Shoulder Length,[a] L_{sh}	
	Max	Min	Max	Min	Max	Min		Max	Min	Max	Min	Max	Min
Metric Formed Hex Screws (ANSI B18.2.3.2M–1979, R1989)													
M5 × 0.8	5.00	4.82	4.46	4.36	3.5	2.5	M14 × 2	14.00	13.73	12.77	12.50	8.0	7.0
M6 × 1	6.00	5.82	5.39	5.21	4.0	3.0	M16 × 2	16.00	15.73	14.77	14.50	9.0	8.0
M8 × 1.25	8.00	7.78	7.26	7.04	5.0	4.0	M20 × 2.5	20.00	19.67	18.49	18.16	11.0	10.0
M10 × 1.5	10.00	9.78	9.08	8.86	6.0	5.0	M24 × 3	24.00	23.67	22.13	21.80	13.0	12.0
M12 × 1.75	12.00	11.73	10.95	10.68	7.0	6.0	…	…	…	…	…	…	…
Metric Hex Flange Screws (ANSI B18.2.3.4M–1984)													
M5 × 0.8	5.00	4.82	4.46	4.36	3.5	2.5	M12 × 1.75	12.00	11.73	10.95	10.68	7.0	6.0
M6 × 1	6.00	5.82	5.39	5.21	4.0	3.0	M14 × 2	14.00	13.73	12.77	12.50	8.0	7.0
M8 × 1.25	8.00	7.78	7.26	7.04	5.0	4.0	M16 × 2	16.00	15.73	14.77	14.50	9.0	8.0
M10 × 1.5	10.00	9.78	9.08	8.86	6.0	5.0	…	…	…	…	…	…	…
Metric Hex Bolts (ANSI B18.2.3.5M–1979, R1989)													
M5 × 0.8	5.48	4.52	4.46	4.36	3.5	2.5	M14 × 2	14.70	13.30	12.77	12.50	8.0	7.0
M6 × 1	6.48	5.52	5.39	5.21	4.0	3.0	M16 × 2	16.70	15.30	14.77	14.50	9.0	8.0
M8 × 1.25	8.58	7.42	7.26	7.04	5.0	4.0	M20 × 2.5	20.84	19.16	18.49	18.16	11.0	10.0
M10 × 1.5	10.58	9.42	9.08	8.86	6.0	5.0	M24 × 3	24.84	23.16	22.13	21.80	13.0	12.0
M12 × 1.75	12.70	11.30	10.95	10.68	7.0	6.0	…	…	…	…	…	…	…
Metric Heavy Hex Bolts (ANSI B18.2.3.6M–1979, R1989)													
M12 × 1.75	12.70	11.30	10.95	10.68	7.0	6.0	M20 × 2.5	20.84	19.16	18.49	18.16	11.0	10.0
M14 × 2	14.70	13.30	12.77	12.50	8.0	7.0	M24 × 3	24.84	23.16	22.13	21.80	13.0	12.0
M16 × 2	16.70	15.30	14.77	14.50	9.0	8.0	…	…	…	…	…	…	…
Metric Heavy Hex Flange Screws (ANSI B18.2.3.9M–1984)													
M10 × 1.5	10.00	9.78	9.08	8.86	6.0	5.0	M16 × 2	16.00	15.73	14.77	14.50	9.0	8.0
M12 × 1.75	12.00	11.73	10.95	10.68	7.0	6.0	M20 × 2.5	20.00	19.67	18.49	18.16	11.0	10.0
M14 × 2	14.00	13.73	12.77	12.50	8.0	7.0	…	…	…	…	…	…	…

[a] Shoulder is mandatory for formed hex screws, hex flange screws, and heavy hex flange screws. Shoulder is optional for hex bolts and heavy hex bolts.

All dimensions are in millimeters.

Table 5. American National Standard Metric Hex Lag Screws

ANSI B18.2.3.8M–1981, R1991

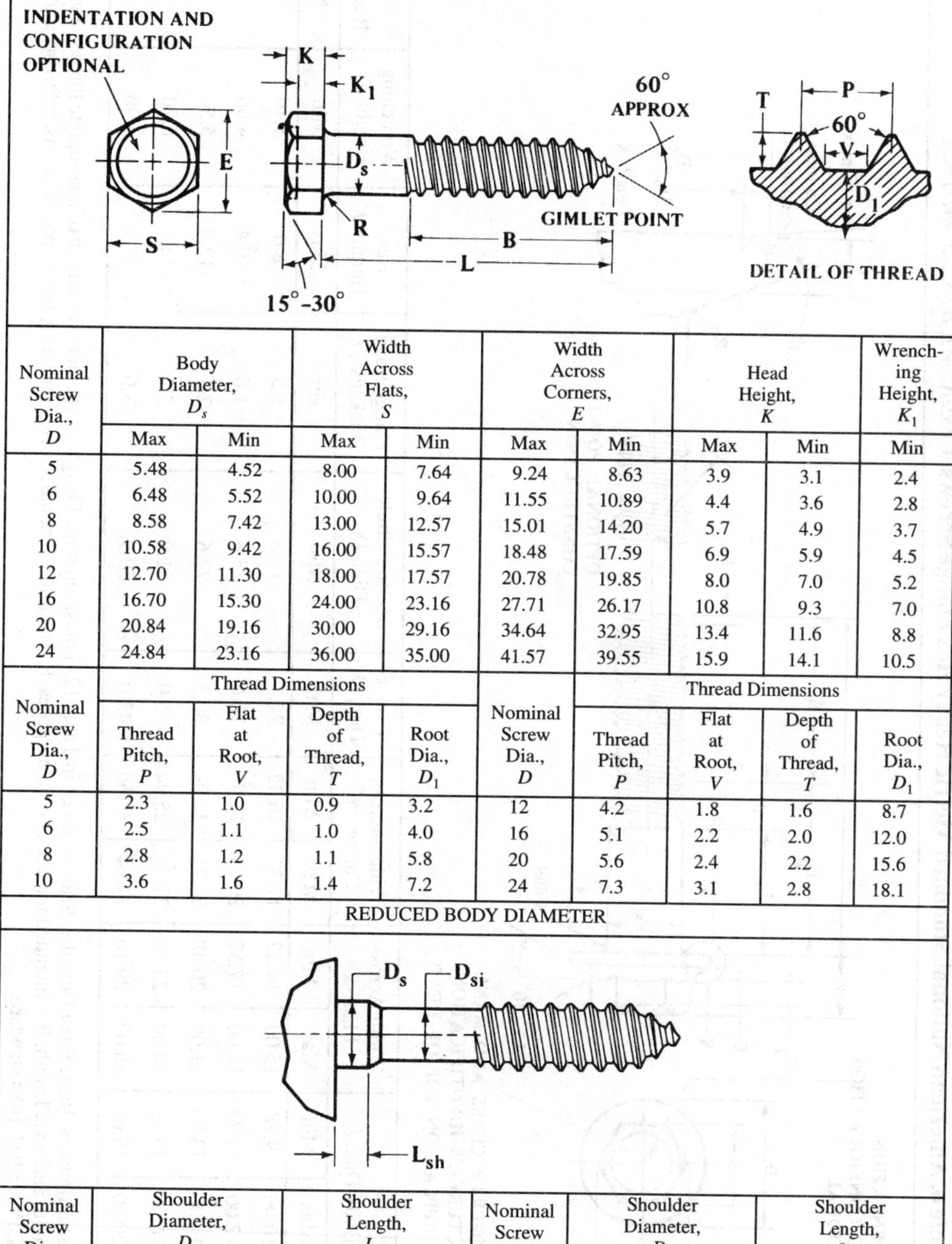

Nominal Screw Dia., D	Body Diameter, D_s		Width Across Flats, S		Width Across Corners, E		Head Height, K		Wrenching Height, K_1
	Max	Min	Max	Min	Max	Min	Max	Min	Min
5	5.48	4.52	8.00	7.64	9.24	8.63	3.9	3.1	2.4
6	6.48	5.52	10.00	9.64	11.55	10.89	4.4	3.6	2.8
8	8.58	7.42	13.00	12.57	15.01	14.20	5.7	4.9	3.7
10	10.58	9.42	16.00	15.57	18.48	17.59	6.9	5.9	4.5
12	12.70	11.30	18.00	17.57	20.78	19.85	8.0	7.0	5.2
16	16.70	15.30	24.00	23.16	27.71	26.17	10.8	9.3	7.0
20	20.84	19.16	30.00	29.16	34.64	32.95	13.4	11.6	8.8
24	24.84	23.16	36.00	35.00	41.57	39.55	15.9	14.1	10.5

Nominal Screw Dia., D	Thread Dimensions				Nominal Screw Dia., D	Thread Dimensions			
	Thread Pitch, P	Flat at Root, V	Depth of Thread, T	Root Dia., D_1		Thread Pitch, P	Flat at Root, V	Depth of Thread, T	Root Dia., D_1
5	2.3	1.0	0.9	3.2	12	4.2	1.8	1.6	8.7
6	2.5	1.1	1.0	4.0	16	5.1	2.2	2.0	12.0
8	2.8	1.2	1.1	5.8	20	5.6	2.4	2.2	15.6
10	3.6	1.6	1.4	7.2	24	7.3	3.1	2.8	18.1

REDUCED BODY DIAMETER

Nominal Screw Dia., D	Shoulder Diameter, D_s		Shoulder Length, L_{sh}		Nominal Screw Dia., D	Shoulder Diameter, D_s		Shoulder Length, L_{sh}	
	Max	Min	Max	Min		Max	Min	Max	Min
5	5.48	4.52	3.5	2.5	12	12.70	11.30	7.0	6.0
6	6.48	5.52	4.0	3.0	16	16.70	15.30	9.0	8.0
8	8.58	7.42	5.0	4.0	20	20.84	19.16	11.0	10.0
10	10.58	9.42	6.0	5.0	24	24.84	23.16	13.0	12.0

All dimensions are in millimeters. Reduced body diameter, D_{si}, is the blank diameter before rolling. Shoulder is mandatory when body diameter is reduced.

Table 6. American National Standard Metric Heavy Hex Flange Screws *ANSI/ASME B18.2.3.9M-1984*

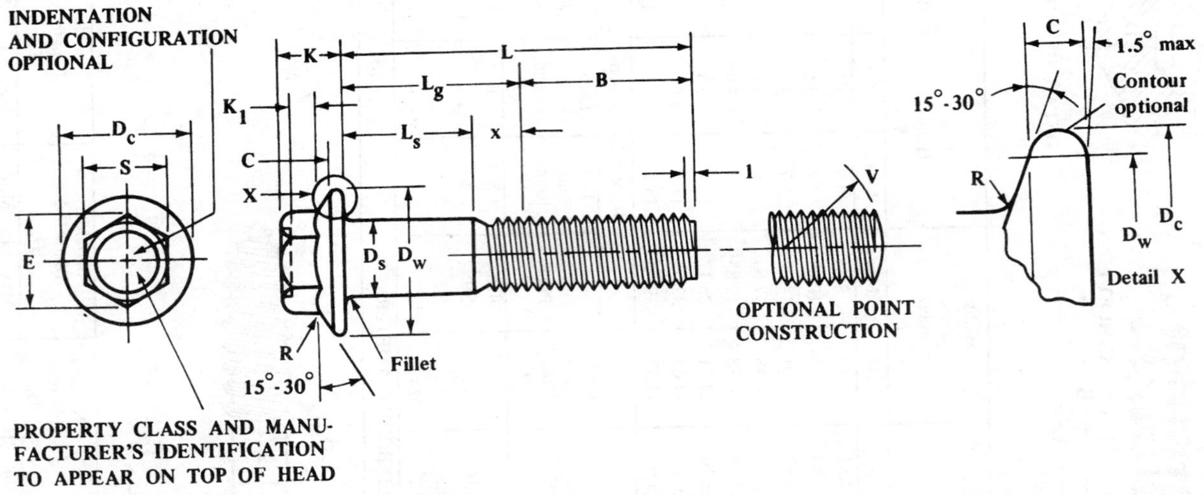

Nominal Screw Dia., D, and Thread Pitch	Body Dia., D_s		Width Across Flats, S		Width Across Corners, E		Flange Dia., D_c	Bearing Circle Dia., D_w	Flange Edge Thickness, C	Head Height, K	Wrenching Height K_1	Fillet Radius, R
	Max	Min	Max	Min	Max	Min	Max	Min	Min	Max	Min	Max
M10 × 1.5	10.00	9.78	15.00	14.57	17.32	16.32	22.3	19.6	1.5	8.6	3.70	0.6
M12 × 1.75	12.00	11.73	18.00	17.57	20.78	19.68	26.6	23.8	1.8	10.4	4.60	0.7
M14 × 2	14.00	13.73	21.00	20.48	24.25	22.94	30.5	27.6	2.1	12.4	5.50	0.9
M16 × 2	16.00	15.73	24.00	23.16	27.71	25.94	35.0	31.9	2.4	14.1	6.20	1.0
M20 × 2.5	20.00	19.67	30.00	29.16	34.64	32.66	43.0	39.9	3.0	17.7	7.90	1.2

All dimensions are in millimeters. Basic thread lengths, B, are as given in Table 12. Transition thread length, x, includes the length of incomplete threads and tolerances on grip gaging length and body length. It is intended for calculation purposes. For additional manufacturing and acceptance specifications, reference should be made to ANSI/ASME B18.2.3.9M-1984 standard.

Table 7. American National Standard Metric Hex Flange Screws
ANSI/ASME B18.2.3.4M-1984

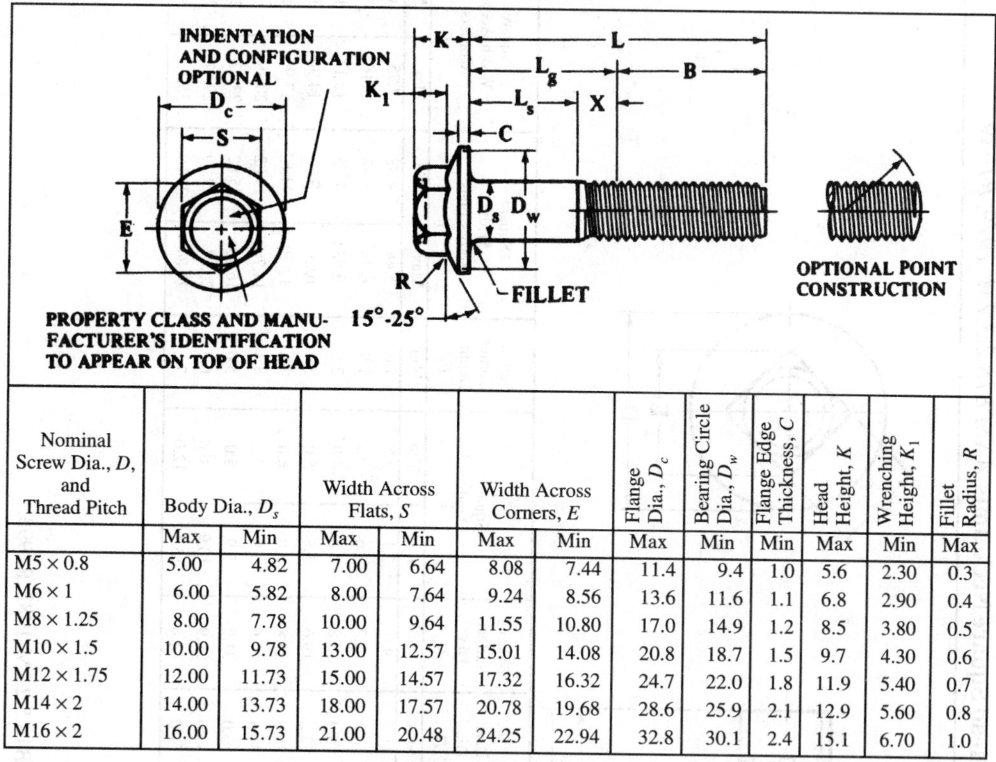

Nominal Screw Dia., *D*, and Thread Pitch	Body Dia., D_s		Width Across Flats, *S*		Width Across Corners, *E*		Flange Dia., D_c	Bearing Circle Dia., D_w	Flange Edge Thickness, *C*	Head Height, *K*	Wrenching Height, K_1	Fillet Radius, *R*
	Max	Min	Max	Min	Max	Min	Max	Min	Min	Max	Min	Max
M5 × 0.8	5.00	4.82	7.00	6.64	8.08	7.44	11.4	9.4	1.0	5.6	2.30	0.3
M6 × 1	6.00	5.82	8.00	7.64	9.24	8.56	13.6	11.6	1.1	6.8	2.90	0.4
M8 × 1.25	8.00	7.78	10.00	9.64	11.55	10.80	17.0	14.9	1.2	8.5	3.80	0.5
M10 × 1.5	10.00	9.78	13.00	12.57	15.01	14.08	20.8	18.7	1.5	9.7	4.30	0.6
M12 × 1.75	12.00	11.73	15.00	14.57	17.32	16.32	24.7	22.0	1.8	11.9	5.40	0.7
M14 × 2	14.00	13.73	18.00	17.57	20.78	19.68	28.6	25.9	2.1	12.9	5.60	0.8
M16 × 2	16.00	15.73	21.00	20.48	24.25	22.94	32.8	30.1	2.4	15.1	6.70	1.0

All dimensions are in millimeters. Basic thread lengths, *B*, are the same as given in Table 12. Transition thread length, *x*, includes the length of incomplete threads and tolerances on grip gaging length and body length. This dimension is intended for calculation purposes only. For additional manufacturing and acceptance specifications, reference should be made to ANSI/ASME B18.2.3.4M-1984 standard.

Table 8. American National Standard Metric Round Head Square Neck Bolts Reduced Body Diameters *ANSI/ASME B18.5.2.2M-1982, R1993*

Nominal Bolt Dia., *D* and Thread Pitch	Diameter of Reduced Body D_r		Nominal Bolt Dia., *D* and Thread Pitch	Diameter of Reduced Body D_r	
	Max	Min		Max	Min
M5 × 0.8	5.00	4.36	M14 × 2	14.00	12.50
M6 × 1	6.00	5.21	M16 × 2	16.00	14.50
M8 × 1.25	8.00	7.04	M20 × 2.5	20.00	18.16
M10 × 1.5	10.00	8.86	M24 × 3	24.00	21.80
M12 × 1.75	12.00	10.68	…	…	…

All dimensions are in millimeters.

Table 9. American National Standard Metric Round Head Square Neck Bolts *ANSI B18.2.3.7M–1979, R1989*

Nominal Bolt Dia., *D* and Thread Pitch	Diameter of Full Body, D_s		Head Radius, (R_k)	Head height, *K*		Head Edge Thickness, *C*		Head Dia., D_c	Bearing Surface Dia., D_w	Square Depth, *F*		Square Corner Depth, F_1	Square Width Across Flats, *V*		Square Width Across Corners, *E*	
	Max	Min	Ref.	Max	Min	Max	Min	Max	Min	Max	Min	Min	Max	Min	Max	Min
M5 × 0.8	5.48	4.52	8.8	3.1	2.5	1.8	1.0	11.8	9.8	3.1	2.5	1.6	5.48	4.88	7.75	6.34
M6 × 1	6.48	5.52	10.7	3.6	3.0	1.9	1.1	14.2	12.2	3.6	3.0	1.9	6.48	5.88	9.16	7.64
M8 × 1.25	8.58	7.42	12.5	4.8	4.0	2.2	1.2	18.0	15.8	4.8	4.0	2.5	8.58	7.85	12.13	10.20
M10 × 1.5	10.58	9.42	15.5	5.8	5.0	2.5	1.5	22.3	19.6	5.8	5.0	3.2	10.58	9.85	14.96	12.80
M12 × 1.75	12.70	11.30	19.0	6.8	6.0	2.8	1.8	26.6	23.8	6.8	6.0	3.8	12.70	11.82	17.96	15.37
M14 × 2	14.70	13.30	21.9	7.9	7.0	3.3	2.1	30.5	27.6	7.9	7.0	4.4	14.70	13.82	20.79	17.97
M16 × 2	16.70	15.30	25.5	8.9	8.0	3.6	2.4	35.0	31.9	8.9	8.0	5.0	16.70	15.82	23.62	20.57
M20 × 2.5	20.84	19.16	31.9	10.9	10.0	4.2	3.0	43.0	39.9	10.9	10.0	6.3	20.84	19.79	29.47	25.73
M24 × 3	24.84	23.16	37.9	13.1	12.0	5.1	3.6	51.0	47.6	13.1	12.0	7.6	24.84	23.79	35.13	30.93

All dimensions are in millimeters.

† L_g is the grip gaging length which controls the length of thread *B*.

‡ *B* is the basic thread length and is a reference dimension (see Table 13).

For additional manufacturing and acceptance specifications, see ANSI/ASME B18.5.2.2M-1982, R1993.

Table 10. ANSI Heavy Hex Bolts *ANSI B18.2.3.6M–1979, R1989*

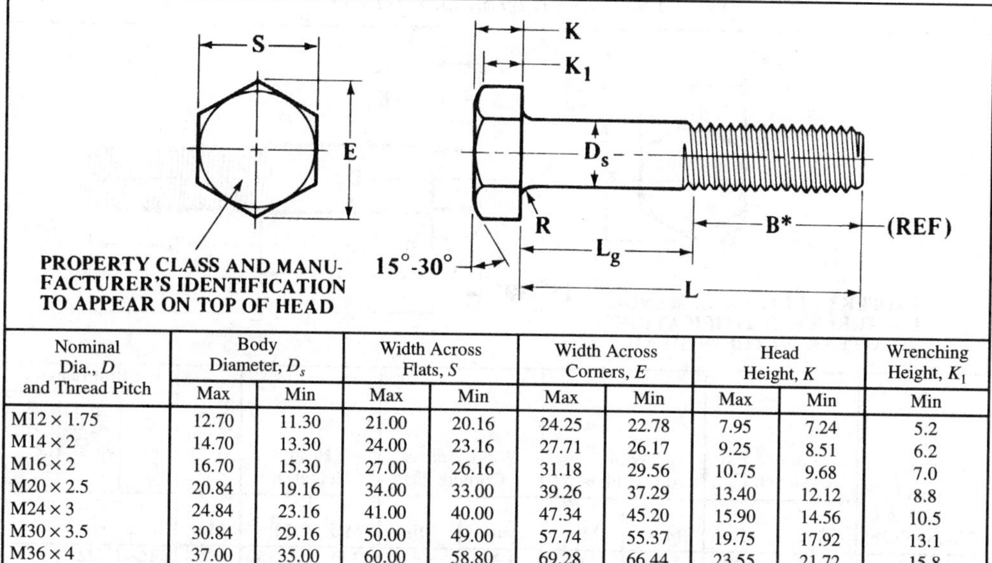

Nominal Dia., D and Thread Pitch	Body Diameter, D_s		Width Across Flats, S		Width Across Corners, E		Head Height, K		Wrenching Height, K_1
	Max	Min	Max	Min	Max	Min	Max	Min	Min
M12 × 1.75	12.70	11.30	21.00	20.16	24.25	22.78	7.95	7.24	5.2
M14 × 2	14.70	13.30	24.00	23.16	27.71	26.17	9.25	8.51	6.2
M16 × 2	16.70	15.30	27.00	26.16	31.18	29.56	10.75	9.68	7.0
M20 × 2.5	20.84	19.16	34.00	33.00	39.26	37.29	13.40	12.12	8.8
M24 × 3	24.84	23.16	41.00	40.00	47.34	45.20	15.90	14.56	10.5
M30 × 3.5	30.84	29.16	50.00	49.00	57.74	55.37	19.75	17.92	13.1
M36 × 4	37.00	35.00	60.00	58.80	69.28	66.44	23.55	21.72	15.8

All dimensions are in millimeters.*Basic thread lengths, B, are the same as given in Table 12.For additional manufacturing and acceptance specifications, reference should be made to the ANSI B18.2.3.6M–1979, R1989 standard.

Table 11. ANSI Metric Heavy Hex Structural Bolts *ANSI B18.2.3.7M–1979, R1989*

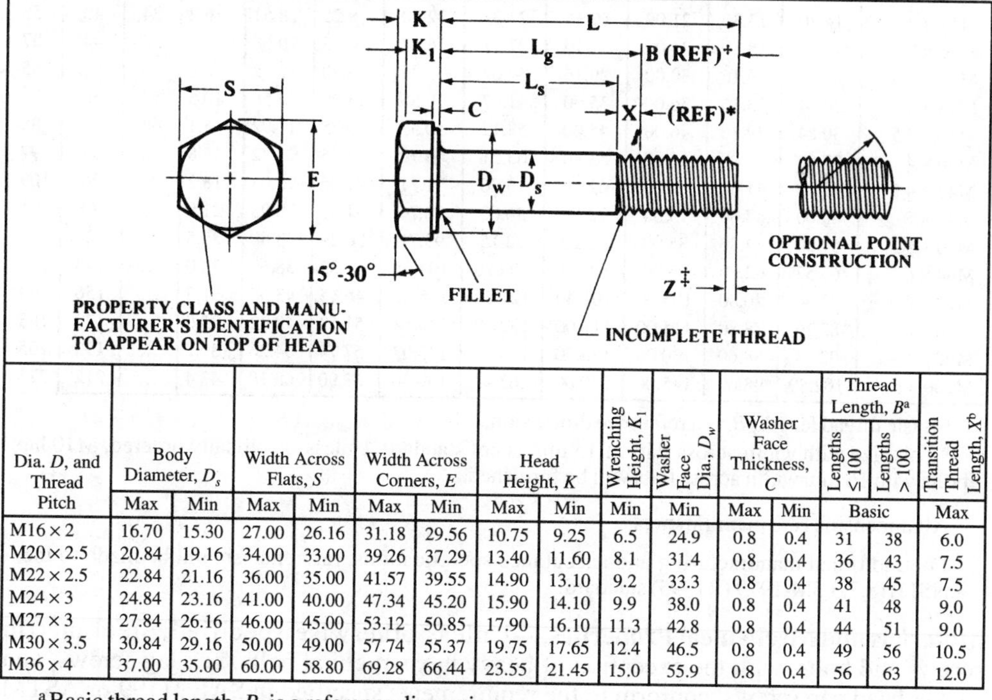

Dia. D, and Thread Pitch	Body Diameter, D_s		Width Across Flats, S		Width Across Corners, E		Head Height, K		Wrenching Height, K_1	Washer Face Dia., D_w	Washer Face Thickness, C		Thread Length, B[a] Lengths ≤ 100	Thread Length, B[a] Lengths > 100	Transition Thread Length, X[b]
	Max	Min	Max	Min	Max	Min	Max	Min	Min	Min	Max	Min	Basic		Max
M16 × 2	16.70	15.30	27.00	26.16	31.18	29.56	10.75	9.25	6.5	24.9	0.8	0.4	31	38	6.0
M20 × 2.5	20.84	19.16	34.00	33.00	39.26	37.29	13.40	11.60	8.1	31.4	0.8	0.4	36	43	7.5
M22 × 2.5	22.84	21.16	36.00	35.00	41.57	39.55	14.90	13.10	9.2	33.3	0.8	0.4	38	45	7.5
M24 × 3	24.84	23.16	41.00	40.00	47.34	45.20	15.90	14.10	9.9	38.0	0.8	0.4	41	48	9.0
M27 × 3	27.84	26.16	46.00	45.00	53.12	50.85	17.90	16.10	11.3	42.8	0.8	0.4	44	51	9.0
M30 × 3.5	30.84	29.16	50.00	49.00	57.74	55.37	19.75	17.65	12.4	46.5	0.8	0.4	49	56	10.5
M36 × 4	37.00	35.00	60.00	58.80	69.28	66.44	23.55	21.45	15.0	55.9	0.8	0.4	56	63	12.0

[a] Basic thread length, B, is a reference dimension.

[b] Transition thread length, X, includes the length of incomplete threads and tolerances on grip gaging length and body length. It is intended for calculation purposes.

All dimensions are in millimeters.

For additional manufacturing and acceptance specifications, reference should be made to the ANSI B18.2.3.7M-1979 (R1995) standard.

Table 12. American National Standard Metric Hex Bolts
ANSI/ASME B18.2.3.5M-1989

PROPERTY CLASS AND MANUFACTURER'S IDENTIFICATION TO APPEAR ON TOP OF HEAD

Nominal Bolt Dia., D and Thread Pitch	Body Diameter, D_s		Width Across Flats, S		Width Across Corners, E		Head Height, K		Wrenching Height, K_1	For Bolt Lengths ≤ 125 mm	> 125 and ≤ 200 mm	> 200 mm
	Max	Min	Max	Min	Max	Min	Max	Min	Min	Basic Thread Length,[a] B		
M5 × 0.8	5.48	4.52	8.00	7.64	9.24	8.63	3.58	3.35	2.4	16	22	35
M6 × 1	6.19	5.52	10.00	9.64	11.55	10.89	4.38	3.55	2.8	18	24	37
M8 × 1.25	8.58	7.42	13.00	12.57	15.01	14.20	5.68	5.10	3.7	22	28	41
[b]M10 × 1.5	10.58	9.42	15.00	14.57	17.32	16.46	6.85	6.17	4.5	26	32	45
M10 × 1.5	10.58	9.42	16.00	15.57	18.48	17.59	6.85	6.17	4.5	26	32	45
M12 × 1.75	12.70	11.30	18.00	17.57	20.78	19.85	7.95	7.24	5.2	30	36	49
M14 × 2	14.70	13.30	21.00	20.16	24.25	22.78	9.25	8.51	6.2	34	40	53
M16 × 2	16.70	15.30	24.00	23.16	27.71	26.17	10.75	9.68	7.0	38	44	57
M20 × 2.5	20.84	19.16	30.00	29.16	34.64	32.95	13.40	12.12	8.8	46	52	65
M24 × 3	24.84	23.16	36.00	35.00	41.57	39.55	15.90	14.56	10.5	54	60	73
M30 × 3.5	30.84	29.16	46.00	45.00	53.12	50.55	19.75	17.92	13.1	66	72	85
M36 × 4	37.00	35.00	55.00	53.80	63.51	60.79	23.55	21.72	15.8	78	84	97
M42 × 4.5	43.00	41.00	65.00	62.90	75.06	71.71	27.05	25.03	18.2	90	96	109
M48 × 5	49.00	47.00	75.00	72.60	86.60	82.76	31.07	28.93	21.0	102	108	121
M56 × 5.5	57.20	54.80	85.00	82.20	98.15	93.71	36.20	33.80	24.5	…	124	137
M64 × 6	65.52	62.80	95.00	91.80	109.70	104.65	41.32	38.68	28.0	…	140	153
M72 × 6	73.84	70.80	105.00	101.40	121.24	115.60	46.45	43.55	31.5	…	156	169
M80 × 6	82.16	78.80	115.00	111.00	132.79	126.54	51.58	48.42	35.0	…	172	185
M90 × 6	92.48	88.60	130.00	125.50	150.11	143.07	57.74	54.26	39.2	…	192	205
M100 × 6	102.80	98.60	145.00	140.00	167.43	159.60	63.90	60.10	43.4	…	212	225

[a] Basic thread length, B, is a reference dimension.

[b] This size with width across flats of 15 mm is not standard. Unless specifically ordered, M10 hex bolts with 16 mm width across flats will be furnished.

All dimensions are in millimeters.

For additional manufacturing and acceptance specifications, reference should be made to the ANSI B18.2.3.5M-1979 (R1995) standard.

Materials and Mechanical Properties.—Unless otherwise specified, steel metric screws and bolts, with the exception of heavy hex structural bolts, hex lag screws, and socket head cap screws, conform to the requirements specified in SAE J1199 or ASTM F568. Steel heavy hex structural bolts conform to ASTM A325M or ASTM A490M. Alloy steel socket head cap screws conform to ASTM 574M, property class 12.9, where the numeral 12 represents approximately one-hundredth of the minimum tensile strength in megapascals and the decimal .9 approximates the ratio of the minimum yield stress to the minimum tensile stress. This is in accord with ISO designation practice. Screws and bolts of other materials, and all materials for hex lag bolts, have properties as agreed upon by the purchaser and the manufacturer.

Except for socket head cap screws, metric screws and bolts are furnished with a natural (as processed) finish, unplated or uncoated unless otherwise specified.

Alloy steel socket head cap screws are furnished with an oiled black oxide coating (thermal or chemical) unless a protective plating or coating is specified by the purchaser.

Metric Screw and Bolt Designation.—Metric screws and bolts with the exception of socket head cap screws are designated by the following data, preferably in the sequence shown: product name, nominal diameter and thread pitch (except for hex lag screws), nominal length, steel property class or material identification, and protective coating, if required.

Example: Hex cap screw, M10 × 1.5 × 50, class 9.8, zinc plated

Heavy hex structural bolt, M24 × 3 × 80, ASTM A490M

Hex lag screw, 6 × 35, silicon bronze.

Socket head cap screws (metric series) are designated by the following data in the order shown: ANSI Standard number, nominal size, thread pitch, nominal screw length, name of product (may be abbreviated SHCS), material and property class (alloy steel screws are supplied to property class 12.9 as specified in ASTM A574M: corrosion-resistant steel screws are specified to the property class and material requirements in ASTM F837M), and protective finish, if required.

Example: B18.3.1M—6 × 1 × 20 Hexagon Socket Head Cap Screw, Alloy Steel

B18.3.1M—10 × 1.5 × 40 SHCS, Alloy Steel Zinc Plated.

Metric Screw and Bolt Thread Lengths.—The length of thread on metric screws and bolts (except for metric lag screws) is controlled by the grip gaging length, L_g max. This is the distance measured parallel to the axis of the screw or bolt, from under the head bearing surface to the face of a noncounterbored or noncountersunk standard GO thread ring gage assembled by hand as far as the thread will permit. The maximum grip gaging length, as calculated and rounded to one decimal place, is equal to the nominal screw length, L, minus the basic thread length, B, or in the case of socket head cap screws, minus the minimum thread length L_T. B and L_T are reference dimensions intended for calculation purposes only and will be found in Tables 12 and 14, respectively.

Table 13. Basic Thread Lengths for Metric Round Head Square Neck Bolts ***ANSI/ASME B18.5.2.2M-1982, R1993***

Nom. Bolt Dia., D and Thread Pitch	Bolt Length, L			Nom. Bolt Dia., D and Thread Pitch	Bolt Length, L		
	≤ 125	> 125 and ≤ 200	> 200		≤ 125	> 125 and ≤ 200	> 200
	Basic Thread Length, B				Basic Thread length, B		
M5 × 0.8	16	22	35	M14 × 2	34	40	53
M6 × 1	18	24	37	M16 × 2	38	44	57
M8 × 1.25	22	28	41	M20 × 2.5	46	52	65
M10 × 1.5	26	32	45	M24 × 3	54	60	73
M12 × 1.75	30	36	49	…	…	…	…

All dimensions are in millimeters

Basic thread length B is a reference dimension intended for calculation purposes only.

Table 14. Socket Head Cap Screws (Metric Series)—Length of Complete Thread
ANSI/ASME B18.3.1M-1986

Nominal Size	Length of Complete Thread, L_T	Nominal Size	Length of Complete Thread, L_T	Nominal Size	Length of Complete Thread, L_T
M1.6	15.2	M6	24.0	M20	52.0
M2	16.0	M8	28.0	M24	60.0
M2.5	17.0	M10	32.0	M30	72.0
M3	18.0	M12	36.0	M36	84.0
M4	20.0	M14	40.0	M42	96.0
M5	22.0	M16	44.0	M48	108.0

Grip length, L_G equals screw length, L, minus L_T. Total length of thread L_{TT} equals L_T plus 5 times the pitch of the coarse thread for the respective screw size. Body length L_B equals L minus L_{TT}.

The minimum thread length for hex lag screws is equal to one-half the nominal screw length plus 12 mm, or 150 mm, whichever is shorter. Screws too short for this formula to apply are threaded as close to the head as practicable.

Metric Screw and Bolt Diameter-Length Combinations.—For a given diameter, the recommended range of lengths of metric cap screws, formed hex screws, heavy hex screws, hex flange screws, and heavy hex flange screws can be found in Table 16, for heavy hex structural bolts in Table 17, for hex lag screws in Table 15, for round head square neck bolts in Table 18, and for socket head cap screws in Table 19. No recommendations for diameter-length combinations are given in the Standards for hex bolts and heavy hex bolts.

Hex bolts in sizes M5 through M24 and heavy hex bolts in sizes M12 through M24 are standard only in lengths longer than 150 mm or 10*D*, whichever is shorter. When shorter lengths of these sizes are ordered, hex cap screws are normally supplied in place of hex bolts and heavy hex screws in place of heavy hex bolts. Hex bolts in sizes M30 and larger and heavy hex bolts in sizes M30 and M36 are standard in all lengths; however, at manufacturer's option, hex cap screws may be substituted for hex bolts and heavy hex screws for heavy hex bolts for any diameter-length combination.

Table 15. Recommended Diameter-Length Combinations for Metric Hex Lag Screws *ANSI B18.2.3.8M-1981 (R1999)*

Nominal Length, L	Nominal Screw Diameter								Nominal Length, L	Nominal Screw Diameter				
	5	6	8	10	12	16	20	24		10	12	16	20	24
8	X	…	…	…	…	…	…	…	90	X	X	X	X	X
10	X	X	…	…	…	…	…	…	100	X	X	X	X	X
12	X	X	X	…	…	…	…	…	110	…	X	X	X	X
14	X	X	X	…	…	…	…	…	120	…	X	X	X	X
16	X	X	X	X	…	…	…	…	130	…	…	X	X	X
20	X	X	X	X	X	…	…	…	140	…	…	X	X	X
25	X	X	X	X	X	X	…	…	150	…	…	X	X	X
30	X	X	X	X	X	X	X	…	160	…	…	X	X	X
35	X	X	X	X	X	X	X	X	180	…	…	…	X	X
40	X	X	X	X	X	X	X	X	200	…	…	…	X	X
45	X	X	X	X	X	X	X	X	220	…	…	…	…	X
50	X	X	X	X	X	X	X	X	240	…	…	…	…	X
60	…	X	X	X	X	X	X	X	260	…	…	…	…	X
70	…	…	X	X	X	X	X	X	280	…	…	…	…	X
80	…	…	X	X	X	X	X	X	300	…	…	…	…	X

All dimensions are in millimeters.

Recommended diameter-length combinations are indicated by the symbol X.

Table 16. Rec'd Diameter-Length Combinations for Metric Hex Cap Screws, Formed Hex and Heavy Hex Screws, Hex Flange and Heavy Hex Flange Screws

Nominal Length[a]	Diameter—Pitch										
	M5 ×0.8	M6 ×1	M8 ×1.25	M10 ×1.5	M12 ×1.75	M14 ×2	M16 ×2	M20 ×2.5	M24 ×3	M30 ×3.5	M36 ×4
8	X	…	…	…	…	…	…	…	…	…	…
10	X	X	…	…	…	…	…	…	…	…	…
12	X	X	X	…	…	…	…	…	…	…	…
14	X	X	X	X[b]	…	…	…	…	…	…	…
16	X	X	X	X	X[b]	X[b]	…	…	…	…	…
20	X	X	X	X	X	X	…	…	…	…	…
25	X	X	X	X	X	X	X	…	…	…	…
30	X	X	X	X	X	X	X	X	…	…	…
35	X	X	X	X	X	X	X	X	X	…	…
40	X	X	X	X	X	X	X	X	X	X	…
45	X	X	X	X	X	X	X	X	X	X	…
50	X	X	X	X	X	X	X	X	X	X	X
(55)	…	X	X	X	X	X	X	X	X	X	X
60	…	X	X	X	X	X	X	X	X	X	X
(65)	…	…	X	X	X	X	X	X	X	X	X
70	…	…	X	X	X	X	X	X	X	X	X
(75)	…	…	X	X	X	X	X	X	X	X	X
80	…	…	X	X	X	X	X	X	X	X	X
(85)	…	…	…	X	X	X	X	X	X	X	X
90	…	…	…	X	X	X	X	X	X	X	X
100	…	…	…	X	X	X	X	X	X	X	X
110	…	…	…	…	X	X	X	X	X	X	X
120	…	…	…	…	X	X	X	X	X	X	X
130	…	…	…	…	…	X	X	X	X	X	X
140	…	…	…	…	…	X	X	X	X	X	X
150	…	…	…	…	…	…	X	X	X	X	X
160	…	…	…	…	…	…	X	X	X	X	X
(170)	…	…	…	…	…	…	…	X	X	X	X
180	…	…	…	…	…	…	…	X	X	X	X
(190)	…	…	…	…	…	…	…	X	X	X	X
200	…	…	…	…	…	…	…	X	X	X	X
220	…	…	…	…	…	…	…	…	X	X	X
240	…	…	…	…	…	…	…	…	X	X	X
260	…	…	…	…	…	…	…	…	…	X	X
280	…	…	…	…	…	…	…	…	…	X	X
300	…	…	…	…	…	…	…	…	…	X	X

[a] Lengths in parentheses are not recommended. Recommended lengths of formed hex screws, hex flange screws, and heavy hex flange screws do not extend above 150 mm. Recommended lengths of heavy hex screws do not extend below 20 mm. Standard sizes for government use. Recommended diameter-length combinations are indicated by the symbol X. Screws with lengths above heavy cross lines are threaded full length.

[b] Does not apply to hex flange screws and heavy hex flange screws.

All dimensions are in millimeters.

For available diameters of each type of screw, see respective dimensional table.

Table 17. Recommended Diameter-Length Combinations for Metric Heavy Hex Structural Bolts

Nominal Length, *L*	Nominal Diameter and Thread Pitch						
	M16 × 2	M20 × 2.5	M22 × 2.5	M24 × 3	M27 × 3	M30 × 3.5	M36 × 4
45	X	…	…	…	…	…	…
50	X	X	…	…	…	…	…
55	X	X	X	…	…	…	…
60	X	X	X	X	…	…	…
65	X	X	X	X	X	…	…
70	X	X	X	X	X	X	…
75	X	X	X	X	X	X	…
80	X	X	X	X	X	X	X
85	X	X	X	X	X	X	X
90	X	X	X	X	X	X	X
95	X	X	X	X	X	X	X
100	X	X	X	X	X	X	X
110	X	X	X	X	X	X	X
120	X	X	X	X	X	X	X
130	X	X	X	X	X	X	X
140	X	X	X	X	X	X	X
150	X	X	X	X	X	X	X
160	X	X	X	X	X	X	X
170	X	X	X	X	X	X	X
180	X	X	X	X	X	X	X
190	X	X	X	X	X	X	X
200	X	X	X	X	X	X	X
210	X	X	X	X	X	X	X
220	X	X	X	X	X	X	X
230	X	X	X	X	X	X	X
240	X	X	X	X	X	X	X
250	X	X	X	X	X	X	X
260	X	X	X	X	X	X	X
270	X	X	X	X	X	X	X
280	X	X	X	X	X	X	X
290	X	X	X	X	X	X	X
300	X	X	X	X	X	X	X

All dimensions are in millimeters.

Recommended diameter-length combinations are indicated by the symbol X.

Bolts with lengths above the heavy cross lines are threaded full length.

Table 18. Recommended Diameter-Length Combinations for Metric Round Head Square Neck Bolts

Nominal Length,[a] *L*	Nominal Diameter and Thread Pitch								
	M5 × 0.8	M6 × 1	M8 × 1.25	M10 × 1.5	M12 × 1.75	M14 × 2	M16 × 2	M20 × 2.5	M24 × 3
10	X	…	…	…	…	…	…	…	…
12	X	X	…	…	…	…	…	…	…
(14)	X	X	…	…	…	…	…	…	…
16	X	X	X	…	…	…	…	…	…
20	X	X	X	X	…	…	…	…	…
25	X	X	X	X	X	…	…	…	…
30	X	X	X	X	X	X	X	…	…
35	X	X	X	X	X	X	X	…	…
40	X	X	X	X	X	X	X	X	…
45	X	X	X	X	X	X	X	X	X
50	X	X	X	X	X	X	X	X	X
(55)	…	X	X	X	X	X	X	X	X
60	…	X	X	X	X	X	X	X	X
(65)	…	…	X	X	X	X	X	X	X
70	…	…	X	X	X	X	X	X	X
(75)	…	…	X	X	X	X	X	X	X
80	…	…	X	X	X	X	X	X	X
(85)	…	…	…	X	X	X	X	X	X

Table 18. *(Continued)* **Recommended Diameter-Length Combinations for Metric Round Head Square Neck Bolts**

Nominal Length,[a] *L*	Nominal Diameter and Thread Pitch								
	M5 × 0.8	M6 × 1	M8 × 1.25	M10 × 1.5	M12 × 1.75	M14 × 2	M16 × 2	M20 × 2.5	M24 × 3
90	…	…	…	X	X	X	X	X	X
100	…	…	…	X	X	X	X	X	X
110	…	…	…	…	X	X	X	X	X
120	…	…	…	…	X	X	X	X	X
130	…	…	…	…	…	X	X	X	X
140	…	…	…	…	…	X	X	X	X
150	…	…	…	…	…	…	X	X	X
160	…	…	…	…	…	…	X	X	X
(170)	…	…	…	…	…	…	…	X	X
180	…	…	…	…	…	…	…	X	X
(190)	…	…	…	…	…	…	…	X	X
200	…	…	…	…	…	…	…	X	X
220	…	…	…	…	…	…	…	…	X
240	…	…	…	…	…	…	…	…	X

[a] Bolts with lengths above the heavy cross lines are threaded full length. Lengths in ( ) are not recommended.

All dimensions are in millimeters. Recommended diameter-length combinations are indicated by the symbol X. Standard sizes for government use.

Table 19. Diameter-Length Combinations for Socket Head Cap Screws (Metric Series)

Nominal Length, *L*	Nominal Size													
	M1.6	M2	M2.5	M3	M4	M5	M6	M8	M10	M12	M14	M16	M20	M24
20	X	X												
25	X	X	X	X										
30	X	X	X	X	X									
35	…	X	X	X	X	X	X							
40	…	X	X	X	X	X	X							
45	…	…	X	X	X	X	X	X						
50	…	…	X	X	X	X	X	X						
55	…	…	…	X	X	X	X	X	X					
60	…	…	…	X	X	X	X	X	X	X				
65	…	…	…	X	X	X	X	X	X	X	X			
70	…	…	…	…	X	X	X	X	X	X	X	X		
80	…	…	…	…	X	X	X	X	X	X	X	X		
90	…	…	…	…	…	X	X	X	X	X	X	X	X	
100	…	…	…	…	…	X	X	X	X	X	X	X	X	X
110	…	…	…	…	…	…	X	X	X	X	X	X	X	X
120	…	…	…	…	…	…	X	X	X	X	X	X	X	X
130	…	…	…	…	…	…	…	X	X	X	X	X	X	X
140	…	…	…	…	…	…	…	X	X	X	X	X	X	X
150	…	…	…	…	…	…	…	X	X	X	X	X	X	X
160	…	…	…	…	…	…	…	X	X	X	X	X	X	X
180	…	…	…	…	…	…	…	…	X	X	X	X	X	X
200	…	…	…	…	…	…	…	…	X	X	X	X	X	X
220	…	…	…	…	…	…	…	…	…	X	X	X	X	X
240	…	…	…	…	…	…	…	…	…	X	X	X	X	X
260	…	…	…	…	…	…	…	…	…	…	X	X	X	X
300	…	…	…	…	…	…	…	…	…	…	…	X	X	X

All dimensions are in millimeters. Screws with lengths above heavy cross lines are threaded full length. Diameter-length combinations are indicated by the symbol X. Standard sizes for government use. In addition to the lengths shown, the following lengths are standard: 3, 4, 5, 6, 8, 10, 12, and 16 mm. No diameter-length combinations are given in the Standard for these lengths. Screws larger than M24 with lengths equal to or shorter than L_{TT} (see Table 14 footnote) are threaded full length.

Metric Screw and Bolt Clearance Holes.—Clearance holes for screws and bolts with the exception of hex lag screws, socket head cap screws, and round head square neck bolts are given in Table 20. Clearance holes for round head square neck bolts are given in Table 8 and drill and counterbore sizes for socket head cap screws are given in Table 21.

Table 20. Recommended Clearance Holes for Metric Hex Screws and Bolts

Nominal Dia., D and Thread Pitch	Clearance Hole Dia., Basic, D_h			Nominal Dia., D and Thread Pitch	Clearance Hole Dia., Basic, D_h		
	Close	Normal, Preferred	Loose		Close	Normal, Preferred	Loose
M5 × 0.8	5.3	5.5	5.8	M30 × 3.5	31.0	33.0	35.0
M6 × 1	6.4	6.6	7.0	M36 × 4	37.0	39.0	42.0
M8 × 1.25	8.4	9.0	10.0	M42 × 4.5	43.0	45.0	48.0
M10 × 1.5	10.5	11.0	12.0	M48 × 5	50.0	52.0	56.0
M12 × 1.75	13.0	13.5	14.5	M56 × 5.5	58.0	62.0	66.0
M14 × 2	15.0	15.5	16.5	M64 × 6	66.0	70.0	74.0
M16 × 2	17.0	17.5	18.5	M72 × 6	74.0	78.0	82.0
M20 × 2.5	21.0	22.0	24.0	M80 × 6	82.0	86.0	91.0
M22 × 2.5[a]	23.0	24.0	26.0	M90 × 6	93.0	96.0	101.0
M24 × 3	25.0	26.0	28.0	M100 × 6	104.0	107.0	112.0
M27 × 3[a]	28.0	30.0	32.0	…	…	…	…

[a] Applies only to heavy hex structural bolts.

All dimensions are in millimeters.

Does not apply to hex lag screws, hex socket head cap screws, or round head square neck bolts.

Normal Clearance: This is preferred for general purpose applications and should be specified unless special design considerations dictate the need for either a close or loose clearance hole.

Close Clearance: This should be specified only where conditions such as critical alignment of assembled parts, wall thickness or other limitations necessitate use of a minimum hole. When close clearance holes are specified, special provision (e.g. countersinking) must be made at the screw or bolt entry side to permit proper seating of the screw or bolt head.

Loose Clearance: This should be specified only for applications where maximum adjustment capability between components being assembled is necessary.

Recommended Tolerances: The clearance hole diameters given in this table are minimum size. Recommended tolerances are: for screw or bolt diameter M5, +0.2 mm; for M6 through M16, +0.3 mm; for M20 through M42, +0.4 mm; for M48 through M72, +0.5 mm; and for M80 through M100, +0.6 mm.

Table 21. Drill and Counterbore Sizes for Metric Socket Head Cap Screws

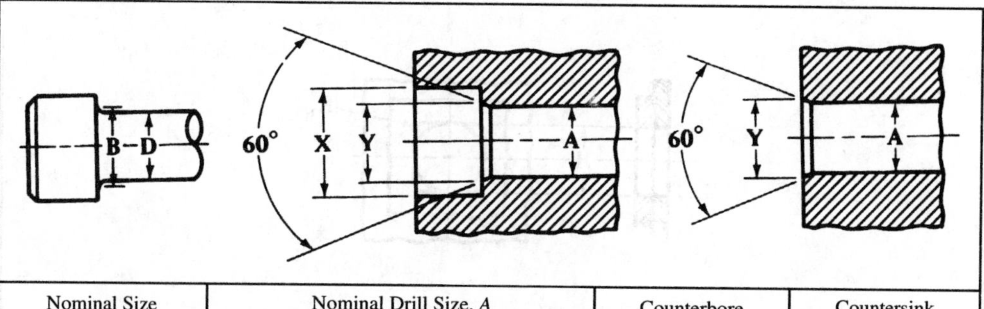

Nominal Size or Basic Screw Diameter	Nominal Drill Size, *A*		Counterbore Diameter, *X*	Countersink Diameter,[a] *Y*
	Close Fit[b]	Normal Fit[c]		
M1.6	1.80	1.95	3.50	2.0
M2	2.20	2.40	4.40	2.6
M2.5	2.70	3.00	5.40	3.1
M3	3.40	3.70	6.50	3.6
M4	4.40	4.80	8.25	4.7
M5	5.40	5.80	9.75	5.7
M6	6.40	6.80	11.25	6.8
M8	8.40	8.80	14.25	9.2
M10	10.50	10.80	17.25	11.2
M12	12.50	12.80	19.25	14.2
M14	14.50	14.75	22.25	16.2
M16	16.50	16.75	25.50	18.2
M20	20.50	20.75	31.50	22.4
M24	24.50	24.75	37.50	26.4
M30	30.75	31.75	47.50	33.4
M36	37.00	37.50	56.50	39.4
M42	43.00	44.00	66.00	45.6
M48	49.00	50.00	75.00	52.6

[a] *Countersink:* It is considered good practice to countersink or break the edges of holes which are smaller than *B* Max. (see Table 23) in parts having a hardness which approaches, equals, or exceeds the screw hardness. If such holes are not countersunk, the heads of screws may not seat properly or the sharp edges on holes may deform the fillets on screws, thereby making them susceptible to fatigue in applications involving dynamic loading. The countersink or corner relief, however, should not be larger than is necessary to ensure that the fillet on the screw is cleared. Normally, the diameter of countersink does not have to exceed *B* Max. Countersinks or corner reliefs in excess of this diameter reduce the effective bearing area and introduce the possibility of embedment where the parts to be fastened are softer than the screws or of brinnelling or flaring the heads of the screws where the parts to be fastened are harder than the screws.

[b] *Close Fit:* The close fit is normally limited to holes for those lengths of screws which are threaded to the head in assemblies where only one screw is to be used or where two or more screws are to be used and the mating holes are to be produced either at assembly or by matched and coordinated tooling.

[c] *Normal Fit:* The normal fit is intended for screws of relatively long length or for assemblies involving two or more screws where the mating holes are to be produced by conventional tolerancing methods. It provides for the maximum allowable eccentricity of the longest standard screws and for certain variations in the parts to be fastened, such as: deviations in hole straightness, angularity between the axis of the tapped hole and that of the hole for shank, differences in center distances of the mating holes, etc.

All dimensions are in millimeters.

Table 22. Recommended Clearance Holes for Metric Round Head Square Neck Bolts

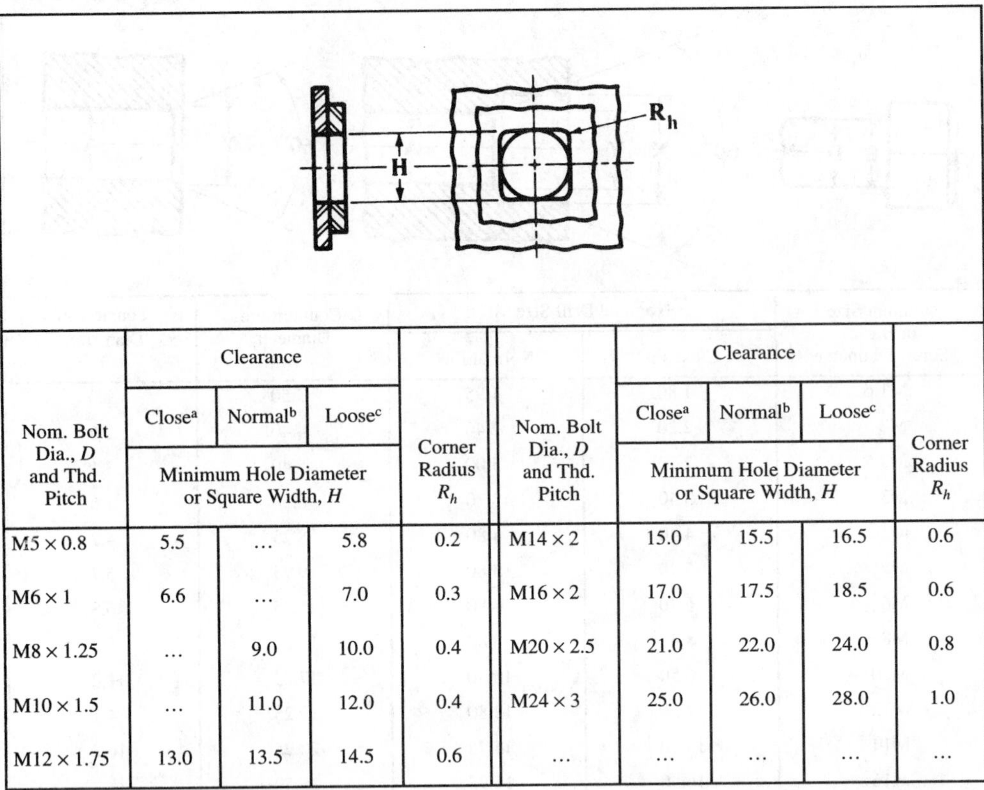

Nom. Bolt Dia., D and Thd. Pitch	Clearance			Corner Radius R_h	Nom. Bolt Dia., D and Thd. Pitch	Clearance			Corner Radius R_h
	Close[a]	Normal[b]	Loose[c]			Close[a]	Normal[b]	Loose[c]	
	Minimum Hole Diameter or Square Width, H					Minimum Hole Diameter or Square Width, H			
M5 × 0.8	5.5	…	5.8	0.2	M14 × 2	15.0	15.5	16.5	0.6
M6 × 1	6.6	…	7.0	0.3	M16 × 2	17.0	17.5	18.5	0.6
M8 × 1.25	…	9.0	10.0	0.4	M20 × 2.5	21.0	22.0	24.0	0.8
M10 × 1.5	…	11.0	12.0	0.4	M24 × 3	25.0	26.0	28.0	1.0
M12 × 1.75	13.0	13.5	14.5	0.6	…	…	…	…	…

[a] *Close Clearance:* Close clearance should be specified only for square holes in very thin and/or soft material, or for slots, or where conditions such as critical alignment of assembled parts, wall thickness, or other limitations necessitate use of a minimal hole. Allowable swell or fins on the bolt body and/or fins on the corners of the square neck may interfere with close clearance round or square holes.

[b] *Normal Clearance:* Normal clearance hole sizes are preferred for general purpose applications and should be specified unless special design considerations dictate the need for either a close or loose clearance hole.

[c] *Loose Clearance:* Loose clearance hole sizes should be specified only for applications where maximum adjustment capability between components being assembled is necessary. Loose clearance square hole or slots may not prevent bolt turning during wrenching.

All dimensions are in millimeters.

Metric Screw and Bolt Thread Series.—Unless otherwise specified, metric screws and bolts, except for hex lag screws, are furnished with metric coarse threads conforming to the dimensions for general purpose threads given in ANSI B1.13M (see *Metric Screw and Bolt Diameter-Length Combinations*Metric Screw Threads in Index). Except for socket head cap screws, the tolerance class is 6g, which applies to plain finish (unplated or uncoated) screws or bolts and to plated or coated screws or bolts before plating or coating. For screws with additive finish, the 6g diameters may be exceeded by the amount of the allowance, i.e. the basic diameters apply to the screws or bolts after plating or coating. For socket head cap screws, the tolerance class is 4g6g, but for plated screws, the allowance g may be consumed by the thickness of plating so that the maximum limit of size after plating is tolerance class 4h6h. Thread limits are in accordance with ANSI B1.13M. Metric hex lag screws have a special thread which is covered in Table 5.

Metric Screw and Bolt Identification Symbols.—Screws and bolts are identified on the top of the head by property class symbols and manufacturer's identification symbol.

Table 23. American National Standard Socket Head Cap Screws—Metric Series
ANSI/ASME B18.3.1M-1986

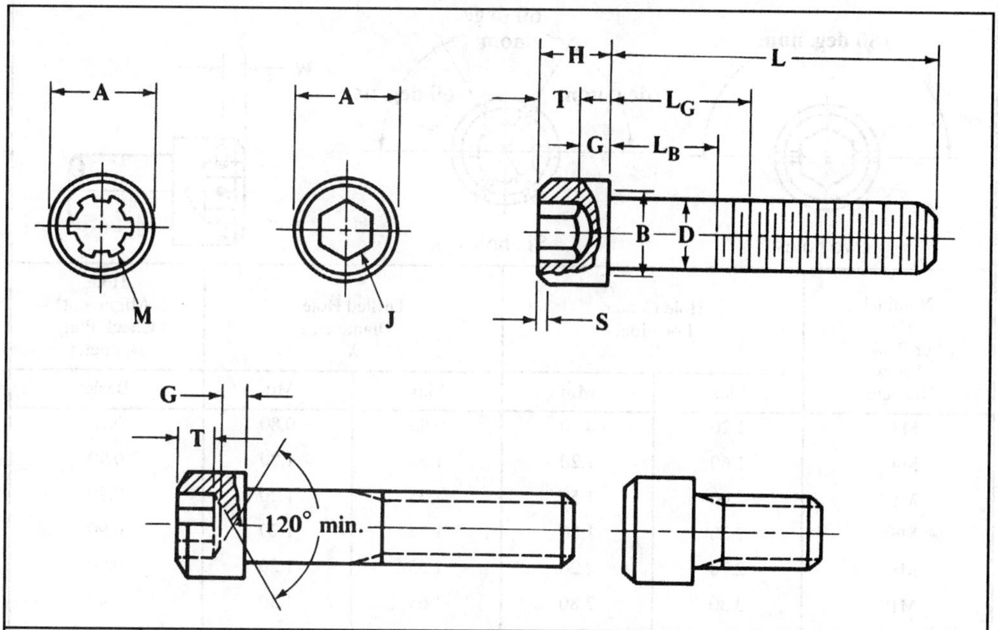

Nom. Size and Thread Pitch	Body Diameter, *D*		Head Diameter *A*		Head Height *H*		Chamfer or Radius *S*	Hexagon Socket Size[a] *J*	Spline Socket Size[a] *M*	Key Engagement *T*	Transition Dia. *B*[a]
	Max	Min	Max	Min	Max	Min	Max	Nom.	Nom.	Min	Max
M1.6 × 0.35	1.60	1.46	3.00	2.87	1.60	1.52	0.16	1.5	1.829	0.80	2.0
M2 × 0.4	2.00	1.86	3.80	3.65	2.00	1.91	0.20	1.5	1.829	1.00	2.6
M2.5 × 0.45	2.50	2.36	4.50	4.33	2.50	2.40	0.25	2.0	2.438	1.25	3.1
M3 × 0.5	3.00	2.86	5.50	5.32	3.00	2.89	0.30	2.5	2.819	1.50	3.6
M4 × 0.7	4.00	3.82	7.00	6.80	4.00	3.88	0.40	3.0	3.378	2.00	4.7
M5 × 0.8	5.00	4.82	8.50	8.27	5.00	4.86	0.50	4.0	4.648	2.50	5.7
M6 × 1	6.00	5.82	10.00	9.74	6.00	5.85	0.60	5.0	5.486	3.00	6.8
M8 × 1.25	8.00	7.78	13.00	12.70	8.00	7.83	0.80	6.0	7.391	4.00	9.2
M10 × 1.5	10.00	9.78	16.00	15.67	10.00	9.81	1.00	8.0	…	5.00	11.2
M12 × 1.75	12.00	11.73	18.00	17.63	12.00	11.79	1.20	10.0	…	6.00	14.2
M14 × 2[b]	14.00	13.73	21.00	20.60	14.00	13.77	1.40	12.0	…	7.00	16.2
M16 × 2	16.00	15.73	24.00	23.58	16.00	15.76	1.60	14.0	…	8.00	18.2
M20 × 2.5	20.00	19.67	30.00	29.53	20.00	19.73	2.00	17.0	…	10.00	22.4
M24 × 3	24.00	23.67	36.00	35.48	24.00	23.70	2.40	19.0	…	12.00	26.4
M30 × 3.5	30.00	29.67	45.00	44.42	30.00	29.67	3.00	22.0	…	15.00	33.4
M36 × 4	36.00	35.61	54.00	53.37	36.00	35.64	3.60	27.0	…	18.00	39.4
M42 × 4.5	42.00	41.61	63.00	62.31	42.00	41.61	4.20	32.0	…	21.00	45.6
M48 × 5	48.00	47.61	72.00	71.27	48.00	47.58	4.80	36.0	…	24.00	52.6

[a] See also Table 25.

[b] The M14 × 2 size is not recommended for use in new designs.

All dimensions are in millimeters

L_G is grip length and L_B is body length (see Table 14).

For length of complete thread, see Table 14.

For additional manufacturing and acceptance specifications, see ANSI/ASME B18.3.1M-1986.

Table 24. Drilled Head Dimensions for Metric Hex Socket Head Cap Screws

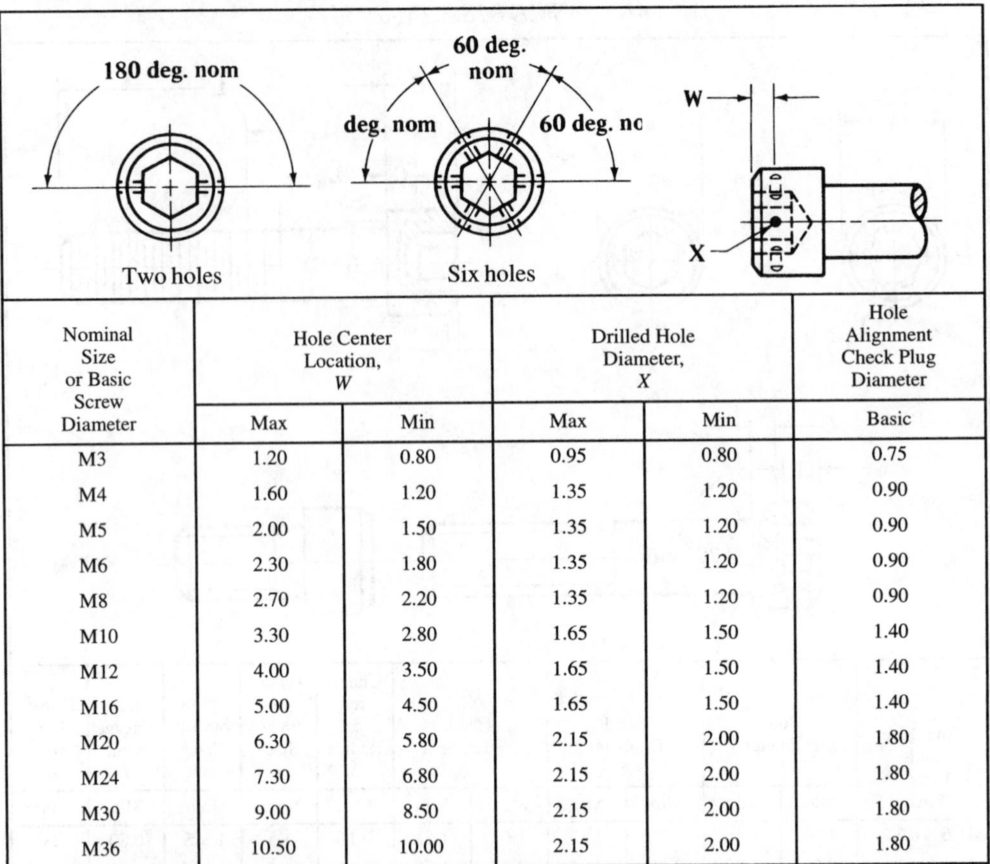

Nominal Size or Basic Screw Diameter	Hole Center Location, *W*		Drilled Hole Diameter, *X*		Hole Alignment Check Plug Diameter
	Max	Min	Max	Min	Basic
M3	1.20	0.80	0.95	0.80	0.75
M4	1.60	1.20	1.35	1.20	0.90
M5	2.00	1.50	1.35	1.20	0.90
M6	2.30	1.80	1.35	1.20	0.90
M8	2.70	2.20	1.35	1.20	0.90
M10	3.30	2.80	1.65	1.50	1.40
M12	4.00	3.50	1.65	1.50	1.40
M16	5.00	4.50	1.65	1.50	1.40
M20	6.30	5.80	2.15	2.00	1.80
M24	7.30	6.80	2.15	2.00	1.80
M30	9.00	8.50	2.15	2.00	1.80
M36	10.50	10.00	2.15	2.00	1.80

All dimensions are in millimeters.

Drilled head metric hexagon socket head cap screws normally are not available in screw sizes smaller than M3 nor larger than M36. The M3 and M4 nominal screw sizes have two drilled holes spaced 180 degrees apart. Nominal screw sizes M5 and larger have six drilled holes spaced 60 degrees apart unless the purchaser specifies two drilled holes. The positioning of holes on opposite sides of the socket should be such that the hole alignment check plug will pass completely through the head without any deflection. When so specified by the purchaser, the edges of holes on the outside surface of the head will be chamfered 45 degrees to a depth of 0.30 to 0.50 mm.

Table 25. American National Standard Hexagon and Spline Sockets for Socket Head Cap Screws—Metric Series *ANSI/ASME B18.3.1M-1986*

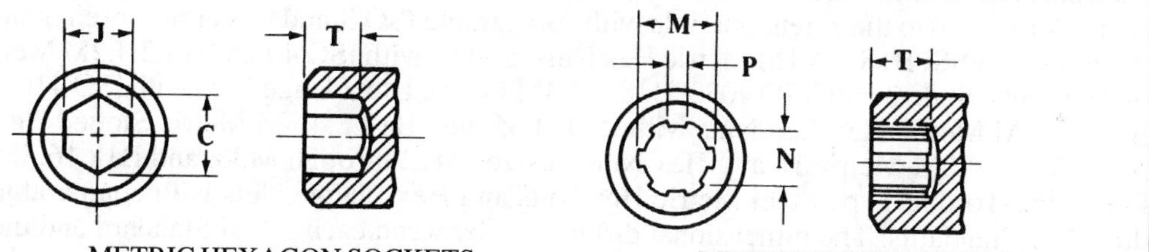

METRIC HEXAGON SOCKETS

See Table 23

METRIC SPLINE SOCKET

See Table 23

Nominal Hexagon Socket Size	Socket Width Across Flats, *J*		Socket Width Across Corners, *C*	Nominal Hexagon Socket Size	Socket Width Across Flats, *J*		Socket Width Across Corners, *C*
Metric Hexagon Sockets							
	Max	Min	Min		Max	Min	Min
1.5	1.545	1.520	1.73	12	12.146	12.032	13.80
2	2.045	2.020	2.30	14	14.159	14.032	16.09
2.5	2.560	2.520	2.87	17	17.216	17.050	19.56
3	3.071	3.020	3.44	19	19.243	19.065	21.87
4	4.084	4.020	4.58	22	22.319	22.065	25.31
5	5.084	5.020	5.72	24	24.319	24.065	27.60
6	6.095	6.020	6.86	27	27.319	27.065	31.04
8	8.115	8.025	9.15	32	32.461	32.080	36.80
10	10.127	10.025	11.50	36	36.461	36.080	41.38

Nominal Spline Socket Size	Socket Major Diameter, *M*		Socket Minor Diameter, *N*		Width of Tooth, *P*	
Metric Spline Sockets[a]						
	Max	Min	Max	Min	Max	Min
1.829	1.8796	1.8542	1.6256	1.6002	0.4064	0.3810
2.438	2.4892	2.4638	2.0828	2.0320	0.5588	0.5334
2.819	2.9210	2.8702	2.4892	2.4384	0.6350	0.5842
3.378	3.4798	3.4290	2.9972	2.9464	0.7620	0.7112
4.648	4.7752	4.7244	4.1402	4.0894	0.9906	0.9398
5.486	5.6134	5.5626	4.8260	4.7752	1.2700	1.2192
7.391	7.5692	7.5184	6.4516	6.4008	1.7272	2.6764

[a] The tabulated dimensions represent direct metric conversions of the equivalent inch size spline sockets shown in American National Standard Socket Cap, Shoulder and Set Screws — Inch Series ANSI B18.3. Therefore, the spline keys and bits shown therein are applicable for wrenching the corresponding size metric spline sockets.

Metric Nuts

The American National Standards covering metric nuts have been established in cooperation with the Department of Defense in such a way that they could be used by the Government for procurement purposes. Extensive information concerning these nuts is given in the following text and tables, but for more complete manufacturing and acceptance specifications, reference should be made to the respective Standards, which may be obtained by non-governmental agencies from the American National Standards Institute, 11 West

42nd Street, New York, N.Y. 10036. Manufacturers should be consulted concerning items and sizes which are in stock production.

Comparison with ISO Standards.—American National Standards for metric nuts have been coordinated to the extent possible with comparable ISO Standards or proposed Standards, thus: ANSI B18.2.4.1M Metric Hex Nuts, Style 1 with ISO 4032; B18.2.4.2M Metric Hex Nuts, Style 2 with ISO 4033; B18.2.4.4M Metric Hex Flange Nuts with ISO 4161; B18.2.4.5M Metric Hex Jam Nuts with ISO 4035; and B18.2.4.3M Metric Slotted Hex Nuts, B18.2.4.6M Metric Heavy Hex Nuts in sizes M12 through M36, and B18.16.3M Prevailing-Torque Type Steel Metric Hex Nuts and Hex Flange Nuts with comparable draft ISO Standards. The dimensional differences between each ANSI Standard and the comparable ISO Standard or draft Standard are very few, relatively minor, and none will affect the interchangeability of nuts manufactured to the requirements of either.

At its meeting in Varna, May 1977, ISO/TC2 studied several technical reports analyzing design considerations influencing determination of the best series of widths across flats for hex bolts, screws, and nuts. A primary technical objective was to achieve a logical ratio between under head (nut) bearing surface area (which determines the magnitude of compressive stress on the bolted members) and the tensile stress area of the screw thread (which governs the clamping force that can be developed by tightening the fastener). The series of widths across flats in the ANSI Standards agree with those which were selected by ISO/TC2 to be ISO Standards.

One exception for width across flats of metric hex nuts, styles 1 and 2, metric slotted hex nuts, metric hex jam nuts, and prevailing-torque metric hex nuts is the M10 size. These nuts in M10 size are currently being produced in the United States with a width across flats of 15 mm. This width, however, is not an ISO Standard. Unless these M10 nuts with width across flats of 15 mm are specifically ordered, the M10 size with 16 mm width across flats will be furnished.

In ANSI Standards for metric nuts, letter symbols designating dimensional characteristics are in accord with those used in ISO Standards, except capitals have been used for data processing convenience instead of lower case letters used in ISO Standards.

Metric Nut Tops and Bearing Surfaces.—Metric hex nuts, styles 1 and 2, slotted hex nuts, and hex jam nuts are double chamfered in sizes M16 and smaller and in sizes M20 and larger may either be double chamfered or have a washer-faced bearing surface and a chamfered top at the option of the manufacturer. Metric heavy hex nuts are optional either way in all sizes. Metric hex flange nuts have a flange bearing surface and a chamfered top and prevailing-torque type metric hex nuts have a chamfered bearing surface. Prevailing-torque type metrix hex flange nuts have a flange bearing surface. All types of metric nuts have the tapped hole countersunk on the bearing face and metric slotted hex nuts, hex flange nuts, and prevailing-torque type hex nuts and hex flange nuts may be countersunk on the top face.

Materials and Mechanical Properties.—Nonheat-treated carbon steel metric hex nuts, style 1 and slotted hex nuts conform to material and property class requirements specified for property class 5 nuts; hex nuts, style 2 and hex flange nuts to property class 9 nuts; hex jam nuts to property class 04 nuts, and nonheat-treated carbon and alloy steel heavy hex nuts to property classes 5, 9, 8S, or 8S3 nuts; all as covered in ASTM A563M. Carbon steel metric hex nuts, style 1 and slotted hex nuts that have specified heat treatment conform to material and property class requirements specified for property class 10 nuts; hex nuts, style 2 to property class 12 nuts; hex jam nuts to property class 05 nuts; hex flange nuts to property classes 10 and 12 nuts; and carbon or alloy steel heavy hex nuts to property classes 10S, 10S3, or 12 nuts, all as covered in ASTM A563M. Carbon steel prevailing-torque type hex nuts and hex flange nuts conform to mechanical and property class requirements as given in ANSI B18.16.1M.

Table 26. American National Standard Metric Hex Nuts, Styles 1 and 2
ANSI/ASME B18.2.4.1M and B18.2.4.2M-1979 (R1995)

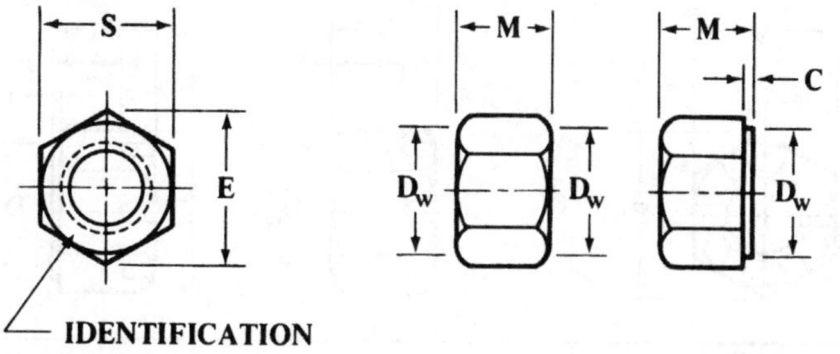

Nominal Nut Dia. and Thread Pitch	Width Across Flats, S		Width Across Corners, E		Thickness, M		Bearing Face Dia., D_w	Washer Face Thickness, C	
	Max	Min	Max	Min	Max	Min	Min	Max	Min
Metric Hex Nuts — Style 1									
M1.6 × 0.35	3.20	3.02	3.70	3.41	1.30	1.05	2.3	…	…
M2 × 0.4	4.00	3.82	4.62	4.32	1.60	1.35	3.1	…	…
M2.5 × 0.45	5.00	4.82	5.77	5.45	2.00	1.75	4.1	…	…
M3 × 0.5	5.50	5.32	6.35	6.01	2.40	2.15	4.6	…	…
M3.5 × 0.6	6.00	5.82	6.93	6.58	2.80	2.55	5.1	…	…
M4 × 0.7	7.00	6.78	8.08	7.66	3.20	2.90	6.0	…	…
M5 × 0.8	8.00	7.78	9.24	8.79	4.70	4.40	7.0	…	…
M6 × 1	10.00	9.78	11.55	11.05	5.20	4.90	8.9	…	…
M8 × 1.25	13.00	12.73	15.01	14.38	6.80	6.44	11.6	…	…
[a]M10 × 1.5	15.00	14.73	17.32	16.64	9.1	8.7	13.6	0.6	0.3
M10 × 1.5	16.00	15.73	18.48	17.77	8.40	8.04	14.6	…	…
M12 × 1.75	18.00	17.73	20.78	20.03	10.80	10.37	16.6	…	…
M14 × 2	21.00	20.67	24.25	23.36	12.80	12.10	19.4	…	…
M16 × 2	24.00	23.67	27.71	26.75	14.80	14.10	22.4	…	…
M20 × 2.5	30.00	29.16	34.64	32.95	18.00	16.90	27.9	0.8	0.4
M24 × 3	36.00	35.00	41.57	39.55	21.50	20.20	32.5	0.8	0.4
M30 × 3.5	46.00	45.00	53.12	50.85	25.60	24.30	42.5	0.8	0.4
M36 × 4	55.00	53.80	63.51	60.79	31.00	29.40	50.8	0.8	0.4
Metric Hex Nuts — Style 2									
M3 × 0.5	5.50	5.32	6.35	6.01	2.90	2.65	4.6	…	…
M3.5 × 0.6	6.00	5.82	6.93	6.58	3.30	3.00	5.1	…	…
M4 × 0.7	7.00	6.78	8.08	7.66	3.80	3.50	5.9	…	…
M5 × 0.8	8.00	7.78	9.24	8.79	5.10	4.80	6.9	…	…
M6 × 1	10.00	9.78	11.55	11.05	5.70	5.40	8.9	…	…
M8 × 1.25	13.00	12.73	15.01	14.38	7.50	7.14	11.6	…	…
[a]M10 × 1.5	15.00	14.73	17.32	16.64	10.0	9.6	13.6	0.6	0.3
M10 × 1.5	16.00	15.73	18.48	17.77	9.30	8.94	14.6	…	…
M12 × 1.75	18.00	17.73	20.78	20.03	12.00	11.57	16.6	…	…
M14 × 2	21.00	20.67	24.25	23.35	14.10	13.40	19.6	…	…
M16 × 2	24.00	23.67	27.71	26.75	16.40	15.70	22.5	…	…
M20 × 2.5	30.00	29.16	34.64	32.95	20.30	19.00	27.7	0.8	0.4
M24 × 3	36.00	35.00	41.57	39.55	23.90	22.60	33.2	0.8	0.4
M30 × 3.5	46.00	45.00	53.12	50.85	28.60	27.30	42.7	0.8	0.4
M36 × 4	55.00	53.80	63.51	60.79	34.70	33.10	51.1	0.8	0.4

[a] This size with width across flats of 15 mm is not standard. Unless specifically ordered, M10 hex nuts with 16 mm width across flats will be furnished.

All dimensions are in millimeters.

Table 27. American National Standard Metric Slotted Hex Nuts
ANSI B18.2.4.4M-1982 (R1999)

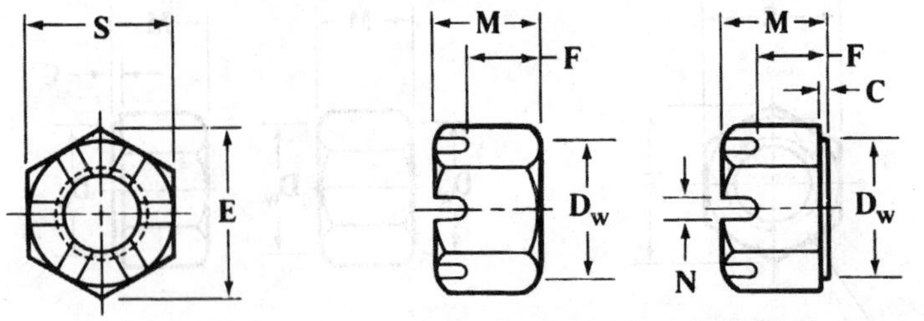

Nominal Nut Dia. and Thread Pitch	Width Across Flats, *S*		Width Across Corners, *E*		Thickness, *M*		Bearing Face Dia., D_w	Unslotted Thickness, *F*		Width of Slot, *N*		Washer Face Thickness *C*	
	Max	Min	Max	Min	Max	Min	Min	Max	Min	Max	Min	Max	Min
M5 × 0.8	8.00	7.78	9.24	8.79	5.10	4.80	6.9	3.2	2.9	2.0	1.4	…	…
M6 × 1	10.00	9.78	11.55	11.05	5.70	5.40	8.9	3.5	3.2	2.4	1.8	…	…
M8 × 1.25	13.00	12.73	15.01	14.38	7.50	7.14	11.6	4.4	4.1	2.9	2.3	…	…
[a]M10 × 1.5	15.00	14.73	17.32	16.64	10.0	9.6	13.6	5.7	5.4	3.4	2.8	0.6	0.3
M10 × 1.5	16.00	15.73	18.48	17.77	9.30	8.94	14.6	5.2	4.9	3.4	2.8	…	…
M12 × 1.75	18.00	17.73	20.78	20.03	12.00	11.57	16.6	7.3	6.9	4.0	3.2	…	…
M14 × 2	21.00	20.67	24.25	23.35	14.10	13.40	19.6	8.6	8.0	4.3	3.5	…	…
M16 × 2	24.00	23.67	27.71	26.75	16.40	15.70	22.5	9.9	9.3	5.3	4.5	…	…
M20 × 2.5	30.00	29.16	34.64	32.95	20.30	19.00	27.7	13.3	12.2	5.7	4.5	0.8	0.4
M24 × 3	36.00	35.00	41.57	39.55	23.90	22.60	33.2	15.4	14.3	6.7	5.5	0.8	0.4
M30 × 3.5	46.00	45.00	53.12	50.85	28.60	27.30	42.7	18.1	16.8	8.5	7.0	0.8	0.4
M36 × 4	55.00	53.80	63.51	60.79	34.70	33.10	51.1	23.7	22.4	8.5	7.0	0.8	0.4

[a] This size with width across flats of 15 mm is not standard. Unless specifically ordered, M10 slotted hex nuts with 16 mm width across flats will be furnished.

All dimensions are in millimeters.

Metric nuts of other materials, such as stainless steel, brass, bronze, and aluminum alloys, have properties as agreed upon by the manufacturer and purchaser. Properties of nuts of several grades of non-ferrous materials are covered in ASTM F467M.

Unless otherwise specified, metric nuts are furnished with a natural (unprocessed) finish, unplated or uncoated.

Metric Nut Thread Series.—Metric nuts have metric coarse threads with class 6H tolerances in accordance with ANSI B1.13M (see *Metric Screw and Bolt Diameter-Length Combinations*Metric Screw Threads in index). For prevailing-torque type metric nuts this condition applies before introduction of the prevailing torque feature. Nuts intended for use with externally threaded fasteners which are plated or coated with a plating or coating thickness (e.g., hot dip galvanized) requiring overtapping of the nut thread to permit assembly, have over-tapped threads in conformance with requirements specified in ASTM A563M.

Table 28. American National Standard Metric Hex Flange Nuts
ANSI B18.2.4.4M-1982 (R1999)

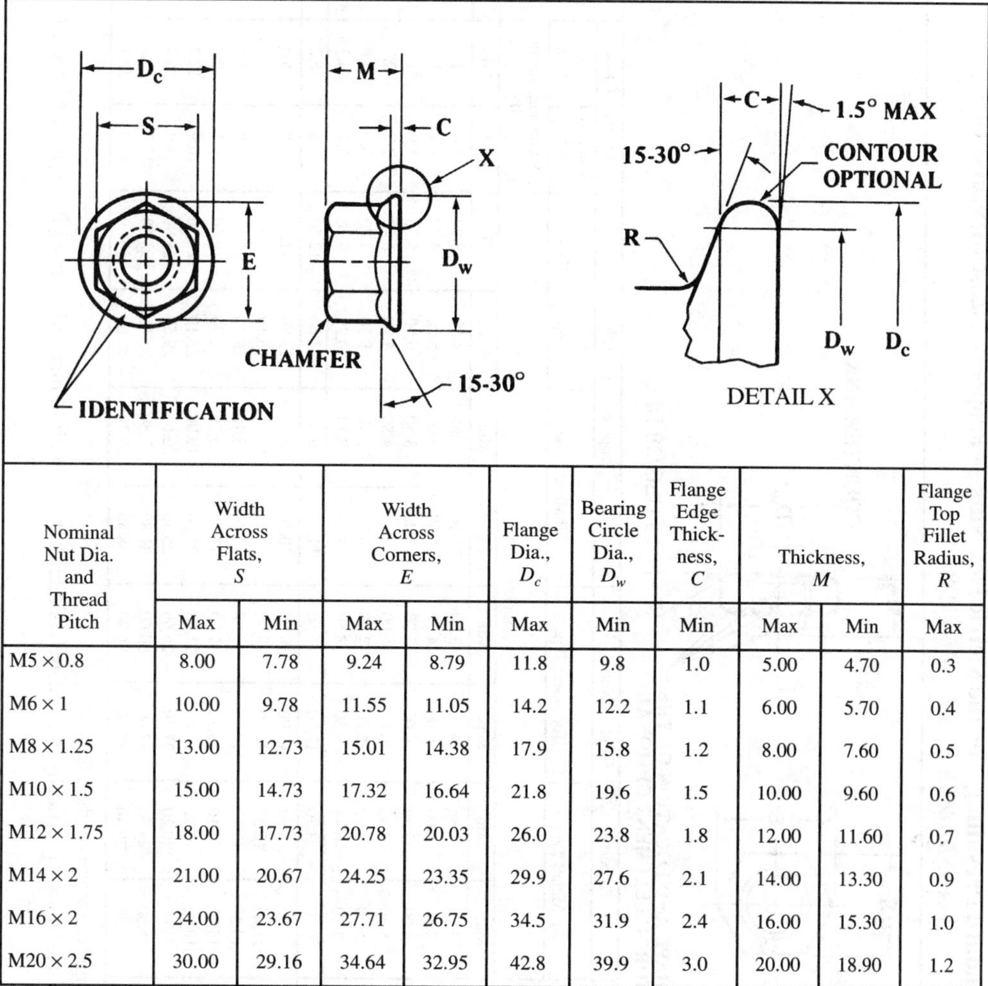

Nominal Nut Dia. and Thread Pitch	Width Across Flats, *S*		Width Across Corners, *E*		Flange Dia., D_c	Bearing Circle Dia., D_w	Flange Edge Thick-ness, *C*	Thickness, *M*		Flange Top Fillet Radius, *R*
	Max	Min	Max	Min	Max	Min	Min	Max	Min	Max
M5 × 0.8	8.00	7.78	9.24	8.79	11.8	9.8	1.0	5.00	4.70	0.3
M6 × 1	10.00	9.78	11.55	11.05	14.2	12.2	1.1	6.00	5.70	0.4
M8 × 1.25	13.00	12.73	15.01	14.38	17.9	15.8	1.2	8.00	7.60	0.5
M10 × 1.5	15.00	14.73	17.32	16.64	21.8	19.6	1.5	10.00	9.60	0.6
M12 × 1.75	18.00	17.73	20.78	20.03	26.0	23.8	1.8	12.00	11.60	0.7
M14 × 2	21.00	20.67	24.25	23.35	29.9	27.6	2.1	14.00	13.30	0.9
M16 × 2	24.00	23.67	27.71	26.75	34.5	31.9	2.4	16.00	15.30	1.0
M20 × 2.5	30.00	29.16	34.64	32.95	42.8	39.9	3.0	20.00	18.90	1.2

All dimensions are in millimeters.

Types of Metric Prevailing-Torque Type Nuts.—There are three basic designs for prevailing-torque type nuts:

1) All-metal, one-piece construction nuts which derive their prevailing-torque characteristics from controlled distortion of the nut thread and/or body.

2) Metal nuts which derive their prevailing-torque characteristics from addition or fusion of a nonmetallic insert, plug. or patch in their threads.

3) Top insert, two-piece construction nuts which derive their prevailing-torque characteristics from an insert, usually a full ring of non-metallic material, located and retained in the nut at its top surface.

The first two designs are designated in Tables 29 and 30 as "all-metal" type and the third design as "top-insert" type.

Table 29. American National Standard Prevailing-Torque Metric Hex Nuts — Property Classes 5, 9, and 10
ANSI/AMSE B18.16.3M-1998

NOTE: SIZE, SHAPE AND LOCATION OF THE PREVAILING-TORQUE ELEMENT OPTIONAL

Nominal Nut Dia. and Thread Pitch	Width Across Flats, *S*		Width Across Corners, *E*		Thickness, *M*: Property Classes 5 and 10 Nuts, All Metal[a] Type		Property Classes 5 and 10 Nuts, Top Insert Type		Property Class 9 Nuts, All Metal Type		Property Class 9 Nuts, Top Insert Type		Wrenching Height, M_1: Property Class 5 and 10 Nuts	Property Class 9 Nuts	Bearing Face Dia., D_w
	Max	Min	Max	Min	Max	Min	Max	Min	Max	Min	Max	Min	Min	Min	Min
M3 × 0.5	5.50	5.32	6.35	6.01	3.10	2.65	4.50	3.90	3.10	2.65	4.50	3.90	1.4	1.4	4.6
M3.5 × 0.6	6.00	5.82	6.93	6.58	3.50	3.00	5.00	4.30	3.50	3.00	5.00	4.30	1.7	1.7	5.1
M4 × 0.7	7.00	6.78	8.08	7.66	4.00	3.50	6.00	5.30	4.00	3.50	6.00	5.30	1.9	1.9	5.9
M5 × 0.8	8.00	7.78	9.24	8.79	5.30	4.80	6.80	6.00	5.30	4.80	7.20	6.40	2.7	2.7	6.9
M6 × 1	10.00	9.78	11.55	11.05	5.90	5.40	8.00	7.20	6.70	5.40	8.50	7.70	3.0	3.0	8.9
M8 × 1.25	13.00	12.73	15.01	14.38	7.10	6.44	9.50	8.50	8.00	7.14	10.20	9.20	3.7	4.3	11.6
[b]M10 × 1.5	15.00	14.73	17.32	16.64	9.70	8.70	12.50	11.50	11.20	9.60	13.50	12.50	5.6	6.2	13.6
M10 × 1.5	16.00	15.73	18.48	17.77	9.00	8.04	11.90	10.90	10.50	8.94	12.80	11.80	4.8	5.6	14.6
M12 × 1.75	18.00	17.73	20.78	20.03	11.60	10.37	14.90	13.90	13.30	11.57	16.10	15.10	6.7	7.7	16.6
M14 × 2	21.00	20.67	24.25	23.35	13.20	12.10	17.00	15.80	15.40	13.40	18.30	17.10	7.8	8.9	19.6
M16 × 2	24.00	23.67	27.71	26.75	15.20	14.10	19.10	17.90	17.90	15.70	20.70	19.50	9.1	10.5	22.5
M20 × 2.5	30.00	29.16	34.64	32.95	19.00	16.90	22.80	21.50	21.80	19.00	25.10	23.80	10.9	12.7	27.7
M24 × 3	36.00	35.00	41.57	39.55	23.00	20.20	27.10	25.60	26.40	22.60	29.50	28.00	13.0	15.1	33.2
M30 × 3.5	46.00	45.00	53.12	50.85	26.90	24.30	32.60	30.60	31.80	27.30	35.60	33.60	15.7	18.2	42.7
M36 × 4	55.00	53.80	63.51	60.79	32.50	29.40	38.90	36.90	38.50	33.10	42.60	40.60	19.0	22.1	51.1

[a] Also includes metal nuts with non-metallic inserts, plugs, or patches in their threads.

[b] This size with width across flats of 15 mm is not standard. Unless specifically ordered, M10 slotted hex nuts with 16 mm width across flats will be furnished.

All dimensions are in millimeters.

Table 30. American National Standard Prevailing-Torque Metric Hex Flange Nuts *ANSI B18.16.3M-1998*

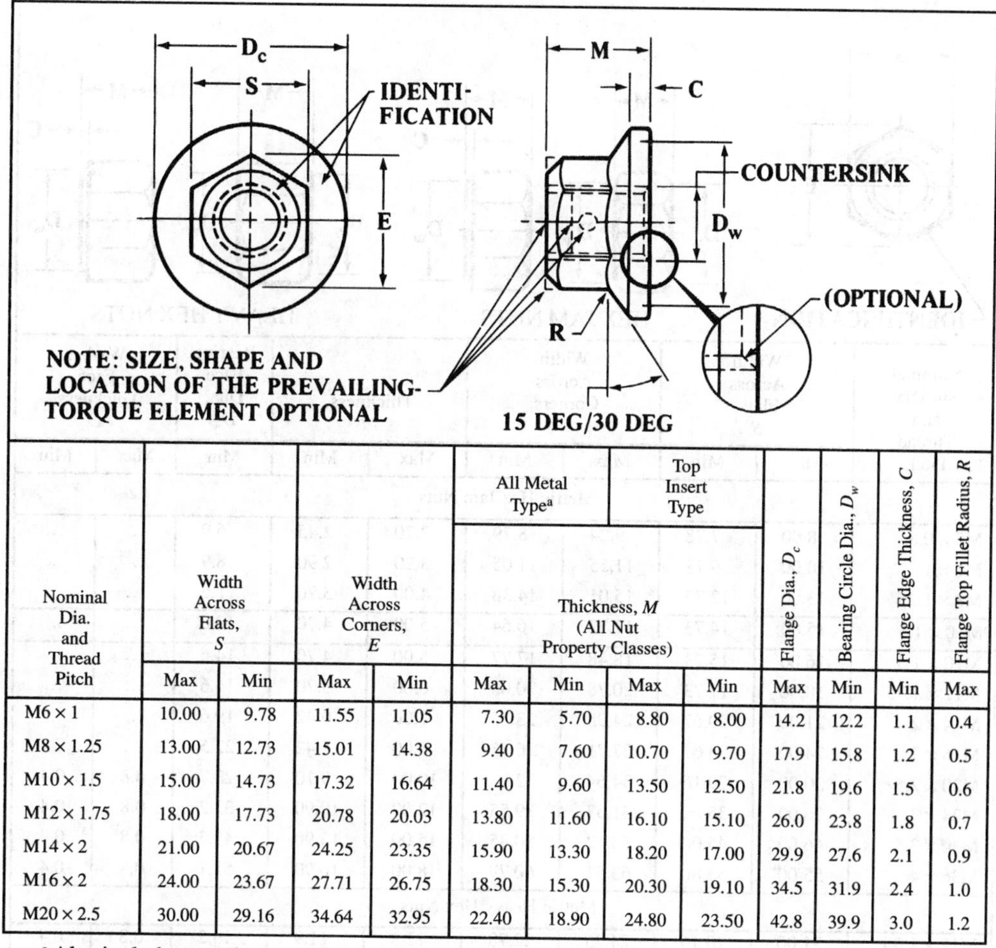

Nominal Dia. and Thread Pitch	Width Across Flats, *S*		Width Across Corners, *E*		Thickness, *M* (All Nut Property Classes): All Metal Type[a]		Top Insert Type		Flange Dia., D_c	Bearing Circle Dia., D_w	Flange Edge Thickness, *C*	Flange Top Fillet Radius, *R*
	Max	Min	Max	Min	Max	Min	Max	Min	Max	Min	Min	Max
M6 × 1	10.00	9.78	11.55	11.05	7.30	5.70	8.80	8.00	14.2	12.2	1.1	0.4
M8 × 1.25	13.00	12.73	15.01	14.38	9.40	7.60	10.70	9.70	17.9	15.8	1.2	0.5
M10 × 1.5	15.00	14.73	17.32	16.64	11.40	9.60	13.50	12.50	21.8	19.6	1.5	0.6
M12 × 1.75	18.00	17.73	20.78	20.03	13.80	11.60	16.10	15.10	26.0	23.8	1.8	0.7
M14 × 2	21.00	20.67	24.25	23.35	15.90	13.30	18.20	17.00	29.9	27.6	2.1	0.9
M16 × 2	24.00	23.67	27.71	26.75	18.30	15.30	20.30	19.10	34.5	31.9	2.4	1.0
M20 × 2.5	30.00	29.16	34.64	32.95	22.40	18.90	24.80	23.50	42.8	39.9	3.0	1.2

[a] Also includes metal nuts with nonmetallic inserts, plugs, or patches in their threads.

All dimensions are in millimeters.

Metric Nut Identification Symbols.—Carbon steel hex nuts, styles 1 and 2, hex flange nuts, and carbon and alloy steel heavy hex nuts are marked to identify the property class and manufacturer in accordance with requirements specified in ASTM A563M. The aforementioned nuts when made of other materials, as well as slotted hex nuts and hex jam nuts, are marked to identify the property class and manufacturer as agreed upon by manufacturer and purchaser. Carbon steel prevailing-torque type hex nuts and hex flange nuts are marked to identify property class and manufacturer as specified in ANSI B18.16.1M. Prevailing-torque type nuts of other materials are identified as agreed upon by the manufacturer and purchaser.

Metric Nut Designation.—Metric nuts are designated by the following data, preferably in the sequence shown: product name, nominal diameter and thread pitch, steel property class or material identification, and protective coating, if required. (Note: It is common practice in ISO Standards to omit thread pitch from the product designation when the nut threads are the metric coarse thread series, e.g., M10 stands for M10 × 1.5).

Example: Hex nut, style 1, M10 × 1.5, ASTM A563M class 10, zinc plated

Heavy hex nut, M20 × 2.5, silicon bronze, ASTM F467, grade 651

Slotted hex nut, M20, ASTM A563M class 10.

Table 31. American National Standard Metric Hex Jam Nuts and Heavy Hex Nuts *ANSI B18.2.4.5M* **and** *B18.2.4.6M-1979 (R1998)*

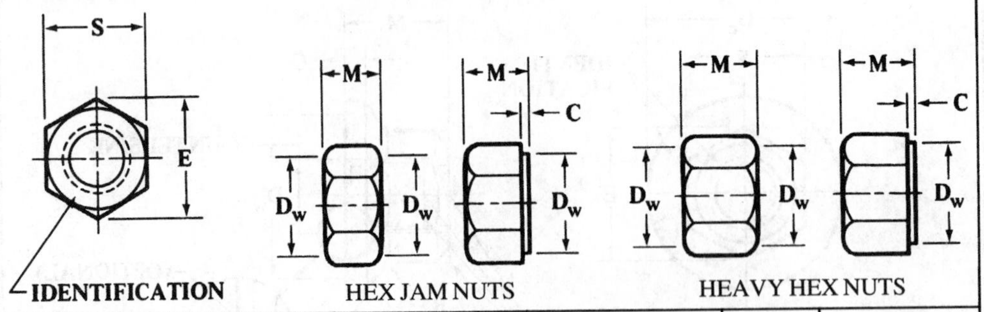

Nominal Nut Dia. and Thread Pitch	Width Across Flats, *S*		Width Across Corners, *E*		Thickness, *M*		Bearing Face Dia., D_w	Washer Face Thickness, *C*	
	Max	Min	Max	Min	Max	Min	Min	Max	Min
Metric Hex Jam Nuts									
M5 × 0.8	8.00	7.78	9.24	8.79	2.70	2.45	6.9	…	…
M6 × 1	10.00	9.78	11.55	11.05	3.20	2.90	8.9	…	…
M8 × 1.25	13.00	12.73	15.01	14.38	4.00	3.70	11.6	…	…
[a]M10 × 1.5	15.00	14.73	17.32	16.64	5.00	4.70	13.6	…	…
M10 × 1.5	16.00	15.73	18.48	17.77	5.00	4.70	14.6	…	…
M12 × 1.75	18.00	17.73	20.78	20.03	6.00	5.70	16.6	…	…
M14 × 2	21.00	20.67	24.25	23.35	7.00	6.42	19.6	…	…
M16 × 2	24.00	23.67	27.71	26.75	8.00	7.42	22.5	…	…
M20 × 2.5	30.00	29.16	34.64	32.95	10.00	9.10	27.7	0.8	0.4
M24 × 3	36.00	35.00	41.57	39.55	12.00	10.90	33.2	0.8	0.4
M30 × 3.5	46.00	45.00	53.12	50.85	15.00	13.90	42.7	0.8	0.4
M36 × 4	55.00	53.80	63.51	60.79	18.00	16.90	51.1	0.8	0.4
Metric Heavy Hex Nuts									
M12 × 1.75	21.00	20.16	24.25	22.78	12.3	11.9	19.2	0.8	0.4
M14 × 2	24.00	23.16	27.71	26.17	14.3	13.6	22.0	0.8	0.4
M16 × 2	27.00	26.16	31.18	29.56	17.1	16.4	24.9	0.8	0.4
M20 × 2.5	34.00	33.00	39.26	37.29	20.7	19.4	31.4	0.8	0.4
M22 × 2.5	36.00	35.00	41.57	39.55	23.6	22.3	33.3	0.8	0.4
M24 × 3	41.00	40.00	47.34	45.20	24.2	22.9	38.0	0.8	0.4
M27 × 3	46.00	45.00	53.12	50.85	27.6	26.3	42.8	0.8	0.4
M30 × 3.5	50.00	49.00	57.74	55.37	30.7	29.1	46.6	0.8	0.4
M36 × 4	60.00	58.80	69.28	66.44	36.6	35.0	55.9	0.8	0.4
M42 × 4.5	70.00	67.90	80.83	77.41	42.0	40.4	64.5	1.0	0.5
M48 × 5	80.00	77.60	92.38	88.46	48.0	46.4	73.7	1.0	0.5
M56 × 5.5	90.00	87.20	103.92	99.41	56.0	54.1	82.8	1.0	0.5
M64 × 6	100.00	96.80	115.47	110.35	64.0	62.1	92.0	1.0	0.5
M72 × 6	110.00	106.40	127.02	121.30	72.0	70.1	101.1	1.2	0.6
M80 × 6	120.00	116.00	138.56	132.24	80.0	78.1	110.2	1.2	0.6
M90 × 6	135.00	130.50	155.88	148.77	90.0	87.8	124.0	1.2	0.6
M100 × 6	150.00	145.00	173.21	165.30	100.0	97.8	137.8	1.2	0.6

[a] This size with width across flats of 15 mm is not standard. Unless specifically ordered, M10 hex jam nuts with 16 mm width across flats will be furnished.

All dimensions are in millimeters.

Metric Washers

Metric Plain Washers.—American National Standard ANSI B18.22M-1981 (R1990) covers general specifications and dimensions for flat, round-hole washers, both soft (as fabricated) and hardened, intended for use in general-purpose applications. Dimensions are given in the following table. Manufacturers should be consulted for current information on stock sizes.

Comparison with ISO Standards.—The washers covered by this ANSI Standard are nominally similar to those covered in various ISO documents. Outside diameters were selected, where possible, from ISO/TC2/WG6/N47 "General Plan for Plain Washers for Metric Bolts, Screws, and Nuts." The thicknesses given in the ANSI Standard are similar to the nominal ISO thicknesses, however the tolerances differ. Inside diameters also differ.

ISO metric washers are currently covered in ISO 887, "Plain Washers for Metric Bolts, Screws, and Nuts – General Plan."

Types of Metric Plain Washers.—Soft (as fabricated) washers are generally available in nominal sizes 1.6 mm through 36 mm in a variety of materials. They are normally used in low-strength applications to distribute bearing load, to provide a uniform bearing surface, and to prevent marring of the work surface.

Hardened steel washers are normally available in sizes 6 mm through 36 mm in the narrow and regular series. They are intended primarily for use in high-strength joints to minimize embedment, to provide a uniform bearing surface, and to bridge large clearance holes and slots.

Metric Plain Washer Materials and Finish.—Soft (as fabricated) washers are made of nonhardened steel unless otherwise specified by the purchaser. Hardened washers are made of through-hardened steel tempered to a hardness of 38 to 45 Rockwell C.

Unless otherwise specified, washers are furnished with a natural (as fabricated) finish, unplated or uncoated with a light film of oil or rust inhibitor.

Metric Plain Washer Designation.—When specifying metric plain washers, the designation should include the following data in the sequence shown: description, nominal size, series, material type, and finish, if required.

Example: Plain washer, 6 mm, narrow, soft, steel, zinc plated
Plain washer, 10 mm, regular, hardened steel.

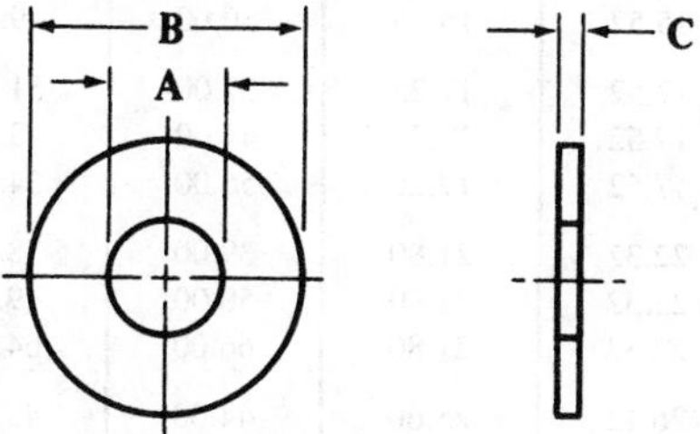

Table 32. American National Standard Metric Plain Washers
ANSI B18.22M-1981, R1990

Nominal Washer Size[a]	Washer Series	Inside Diameter, A		Outside Diameter, B		Thickness, C	
		Max	Min	Max	Min	Max	Min
1.6	Narrow	2.09	1.95	4.00	3.70	0.70	0.50
	Regular	2.09	1.95	5.00	4.70	0.70	0.50
	Wide	2.09	1.95	6.00	5.70	0.90	0.60
2	Narrow	2.64	2.50	5.00	4.70	0.90	0.60
	Regular	2.64	2.50	6.00	5.70	0.90	0.60
	Wide	2.64	2.50	8.00	7.64	0.90	0.60

Table 32. *(Continued)* **American National Standard Metric Plain Washers** *ANSI B18.22M-1981, R1990*

Nominal Washer Size[a]	Washer Series	Inside Diameter, *A*		Outside Diameter, *B*		Thickness, *C*	
		Max	Min	Max	Min	Max	Min
2.5	Narrow	3.14	3.00	6.00	5.70	0.90	0.60
	Regular	3.14	3.00	8.00	7.64	0.90	0.60
	Wide	3.14	3.00	10.00	9.64	1.20	0.80
3	Narrow	3.68	3.50	7.00	6.64	0.90	0.60
	Regular	3.68	3.50	10.00	9.64	1.20	0.80
	Wide	3.68	3.50	12.00	11.57	1.40	1.00
3.5	Narrow	4.18	4.00	9.00	8.64	1.20	0.80
	Regular	4.18	4.00	10.00	9.64	1.40	1.00
	Wide	4.18	4.00	15.00	14.57	1.75	1.20
4	Narrow	4.88	4.70	10.00	9.64	1.20	0.80
	Regular	4.88	4.70	12.00	11.57	1.40	1.00
	Wide	4.88	4.70	16.00	15.57	2.30	1.60
5	Narrow	5.78	5.50	11.00	10.57	1.40	1.00
	Regular	5.78	5.50	15.00	14.57	1.75	1.20
	Wide	5.78	5.50	20.00	19.48	2.30	1.60
6	Narrow	6.87	6.65	13.00	12.57	1.75	1.20
	Regular	6.87	6.65	18.80	18.37	1.75	1.20
	Wide	6.87	6.65	25.40	24.88	2.30	1.60
8	Narrow	9.12	8.90	18.80[b]	18.37[b]	2.30	1.60
	Regular	9.12	8.90	25.40[b]	24.48[b]	2.30	1.60
	Wide	9.12	8.90	32.00	31.38	2.80	2.00
10	Narrow	11.12	10.85	20.00	19.48	2.30	1.60
	Regular	11.12	10.85	28.00	27.48	2.80	2.00
	Wide	11.12	10.85	39.00	38.38	3.50	2.50
12	Narrow	13.57	13.30	25.40	24.88	2.80	2.00
	Regular	13.57	13.30	34.00	33.38	3.50	2.50
	Wide	13.57	13.30	44.00	43.38	3.50	2.50
14	Narrow	15.52	15.25	28.00	27.48	2.80	2.00
	Regular	15.52	15.25	39.00	38.38	3.50	2.50
	Wide	15.52	15.25	50.00	49.38	4.00	3.00
16	Narrow	17.52	17.25	32.00	31.38	3.50	2.50
	Regular	17.52	17.25	44.00	43.38	4.00	3.00
	Wide	17.52	17.25	56.00	54.80	4.60	3.50
20	Narrow	22.32	21.80	39.00	38.38	4.00	3.00
	Regular	22.32	21.80	50.00	49.38	4.60	3.50
	Wide	22.32	21.80	66.00	64.80	5.10	4.00
24	Narrow	26.12	25.60	44.00	43.38	4.60	3.50
	Regular	26.12	25.60	56.00	54.80	5.10	4.00
	Wide	26.12	25.60	72.00	70.80	5.60	4.50
30	Narrow	33.02	32.40	56.00	54.80	5.10	4.00
	Regular	33.02	32.40	72.00	70.80	5.60	4.50
	Wide	33.02	32.40	90.00	88.60	6.40	5.00
36	Narrow	38.92	38.30	66.00	64.80	5.60	4.50
	Regular	38.92	38.30	90.00	88.60	6.40	5.00
	Wide	38.92	38.30	110.00	108.60	8.50	7.00

[a] Nominal washer sizes are intended for use with comparable screw and bolt sizes.

[b] The 18.80/18.37 and 25.40/24.48 mm outside diameters avoid washers which could be used in coin-operated devices.

All dimensions are in millimeters.

BRITISH FASTENERS

British Standard Square and Hexagon Bolts, Screws and Nuts.—Important dimensions of precision hexagon bolts, screws and nuts (B.S.W. and B.S.F. threads) as covered by British Standard 1083:1965 are given in Tables 1 and 2. The use of fasteners in this standard will decrease as fasteners having Unified inch and ISO metric threads come into increasing use. Dimensions of Unified precision hexagon bolts, screws and nuts (UNC and UNF threads) are given in BS 1768:1963 (obsolescent); of Unified black hexagon bolts, screws and nuts (UNC and UNF threads) in BS 1769:1951 (obsolescent); and of Unified black square and hexagon bolts, screws and nuts (UNC and UNF threads) in BS 2708:1956 (withdrawn). Unified nominal and basic dimensions in these British Standards are the same as the comparable dimensions in the American Standards, but the tolerances applied to these basic dimensions may differ because of rounding-off practices and other factors. For Unified dimensions of square and hexagon bolts and nuts as given in ANSI/ASME B18.2.1-1996 and ANSI/ASME B18.2.2-1987 (R1999) see Tables through 4, , 8 and starting on page 1494. ISO metric precision hexagon bolts, screws and nuts are specified in the British Standard BS 3692:1967 (obsolescent) (see *British Standard ISO Metric Precision Hexagon Bolts, Screws and Nuts* starting on page 1559), and ISO metric black hexagon bolts, screws and nuts are covered by British Standard BS 4190:1967 (obsolescent).

British Standard Screwed Studs.—General purpose screwed studs are covered in British Standard 2693: Part 1:1956. The aim in this standard is to provide for a stud having tolerances which would not render it expensive to manufacture and which could be used in association with standard tapped holes for most purposes. Provision has been made for the use of both Unified Fine threads, Unified Coarse threads, British Standard Fine threads, and British Standard Whitworth threads as shown in the table on page 1554.

Designations: The *metal end* of the stud is the end which is screwed into the component. The *nut end* is the end of the screw of the stud which is not screwed into the component. The *plain portion* of the stud is the unthreaded length.

Recommended Fitting Practices for Metal End of Stud: It is recommended that holes tapped to Class 3B limits (see starting on page 1716) in accordance with B.S. 1580 "Unified Screw Threads" or to Close Class limits in accordance with B.S. 84 "Screw Threads of Whitworth Form" as appropriate, be used in association with the metal end of the stud specified in this standard. Where fits are not critical, however, holes may be tapped to Class 2B limits (see table on page 1716) in accordance with B.S. 1580 or Normal Class limits in accordance with B.S. 84.

It is recommended that the B.A. stud specified in this standard be associated with holes tapped to the limits specified for nuts in B.S. 93, 1919 edition. Where fits for these studs are not critical, holes may be tapped to limits specified for nuts in the current edition of B.S. 93.

In general, it will be found that the amount of oversize specified for the studs will produce a satisfactory fit in conjunction with the standard tapping as above. Even when interference is not present, locking will take place on the thread runout which has been carefully controlled for this purpose. Where it is considered essential to assure a true interference fit, higher grade studs should be used. It is recommended that standard studs be used even under special conditions where selective assembly may be necessary.

British Standard Whitworth (B.S.W.) and Fine (B.S.F.) Precision Hexagon Bolts, Screws, and Nuts

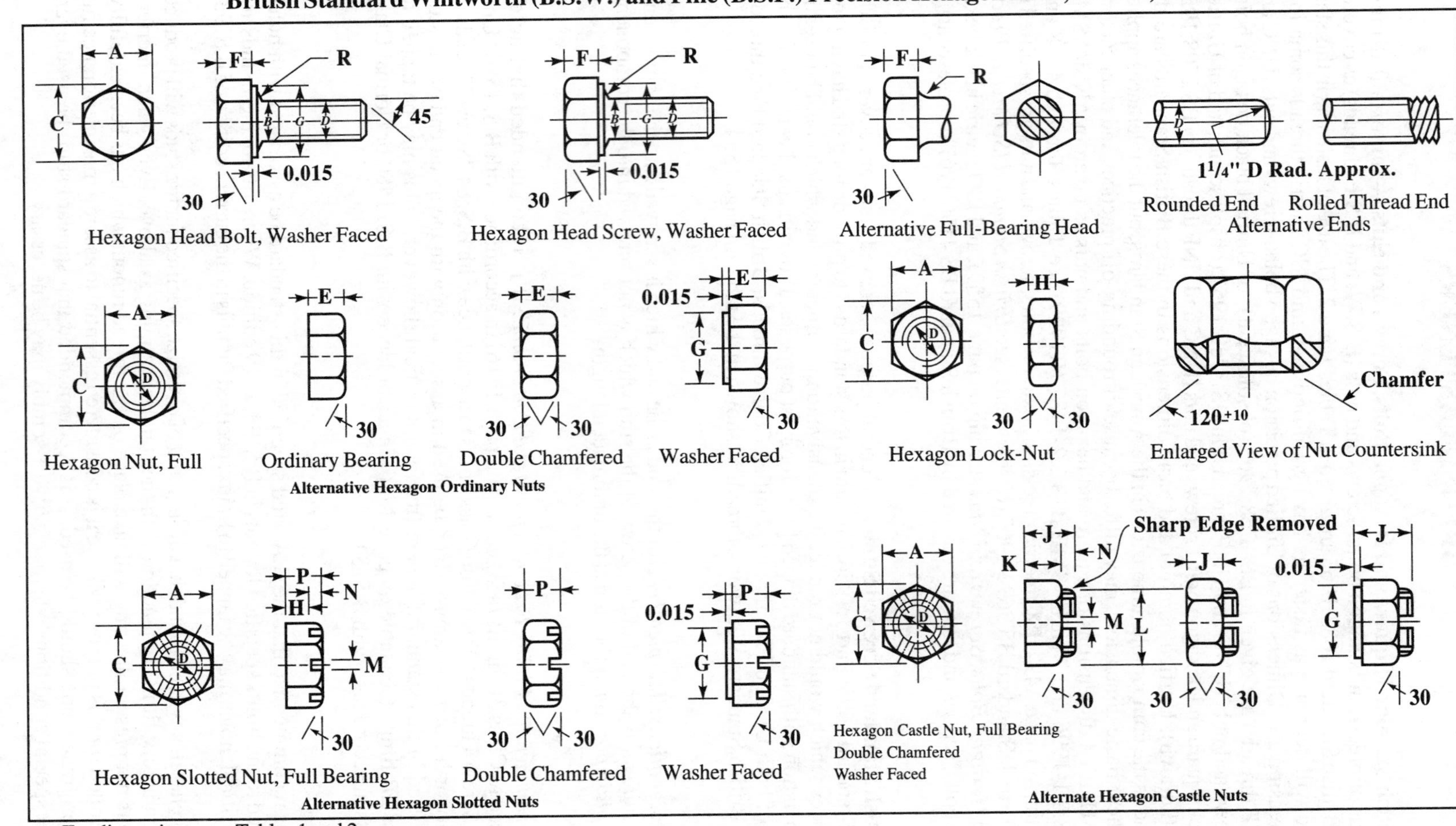

For dimensions, see Tables 1 and 2.

Table 1. British Standard Whitworth (B.S.W.) and Fine (B.S.F.) Precision Hexagon Slotted and Castle Nuts *BS 1083:1965 (obsolescent)*

Nominal Size *D*	Number of Threads per Inch		Bolts, Screws, and Nuts					Bolts and Screws						Nuts			
			Width			Diameter of Washer Face *G*		Radius Under Head *R*		Diameter of Unthreaded Portion of Shank *B*		Thickness		Thickness			
			Across Flats *A*		Across Corners *C*							Head *F*		Ordinary *E*		Lock *H*	
	B.S.W.	B.S.F.	Max.	Min.[a]	Max.	Max.	Min.	Max.	Min.	Max.	Min.	Max.	Min.	Max.	Min.	Max.	Min.
1/4	20	26	0.445	0.438	0.51	0.428	0.418	0.025	0.015	0.2500	0.2465	0.176	0.166	0.200	0.190	0.185	0.180
5/16	18	22	0.525	0.518	0.61	0.508	0.498	0.025	0.015	0.3125	0.3090	0.218	0.208	0.250	0.240	0.210	0.200
3/8	16	20	0.600	0.592	0.69	0.582	0.572	0.025	0.015	0.3750	0.3715	0.260	0.250	0.312	0.302	0.260	0.250
7/16	14	18	0.710	0.702	0.82	0.690	0.680	0.025	0.015	0.4375	0.4335	0.302	0.292	0.375	0.365	0.275	0.265
1/2	12	16	0.820	0.812	0.95	0.800	0.790	0.025	0.015	0.5000	0.4960	0.343	0.333	0.437	0.427	0.300	0.290
9/16	12	16	0.920	0.912	1.06	0.900	0.890	0.045	0.020	0.5625	0.5585	0.375	0.365	0.500	0.490	0.333	0.323
5/8	11	14	1.010	1.000	1.17	0.985	0.975	0.045	0.020	0.6250	0.6190	0.417	0.407	0.562	0.552	0.375	0.365
3/4	10	12	1.200	1.190	1.39	1.175	1.165	0.045	0.020	0.7500	0.7440	0.500	0.480	0.687	0.677	0.458	0.448
7/8	9	11	1.300	1.288	1.50	1.273	1.263	0.065	0.040	0.8750	0.8670	0.583	0.563	0.750	0.740	0.500	0.490
1	8	10	1.480	1.468	1.71	1.453	1.443	0.095	0.060	1.0000	0.9920	0.666	0.636	0.875	0.865	0.583	0.573
1 1/8	7	9	1.670	1.640	1.93	1.620	1.610	0.095	0.060	1.1250	1.1170	0.750	0.710	1.000	0.990	0.666	0.656
1 1/4	7	9	1.860	1.815	2.15	1.795	1.785	0.095	0.060	1.2500	1.2420	0.830	0.790	1.125	1.105	0.750	0.730
1 3/8[b]	…	8	2.050	2.005	2.37	1.985	1.975	0.095	0.060	1.3750	1.3650	0.920	0.880	1.250	1.230	0.833	0.813
1 1/2	6	8	2.220	2.175	2.56	2.155	2.145	0.095	0.060	1.5000	1.4900	1.000	0.960	1.375	1.355	0.916	0.896
1 3/4	5	7	2.580	2.520	2.98	2.495	2.485	0.095	0.060	1.7500	1.7400	1.170	1.110	1.625	1.605	1.083	1.063
2	4.5	7	2.760	2.700	3.19	2.675	2.665	0.095	0.060	2.0000	1.9900	1.330	1.270	1.750	1.730	1.166	1.146

[a] When bolts from 1/4 to 1 inch are hot forged, the tolerance on the width across flats shall be two and a half times the tolerance shown in the table and shall be unilaterally minus from maximum size. For dimensional notation, see diagram on page 1552.

[b] Noted standard with B.S.W. thread.

All dimensions in inches except where otherwise noted.

Table 2. British Standard Whitworth (B.S.W.) and Fine (B.S.F.) Precision Hexagon Slotted and Castle Nuts *BS 1083:1965 (obsolescent)*

Nominal Size *D*	Number of Threads per Inch		Slotted Nuts: Thickness *P*		Slotted Nuts: Lower Face to Bottom of Slot *H*		Castle Nuts: Total Thickness *J*		Castle Nuts: Lower Face to Bottom of Slot *K*		Castle Nuts: Castellated Portion Diameter *L*		Slotted and Castle Nuts: Slots Width *M*		Slotted and Castle Nuts: Slots Depth *N*
	B.S.W.	B.S.F.	Max.	Min.	Max.	Min.	Max.	Min.	Max.	Min.	Max.	Min.	Max.	Min.	Approx.
1/4	20	26	0.200	0.190	0.170	0.160	0.290	0.280	0.200	0.190	0.430	0.425	0.100	0.090	0.090
5/16	18	22	0.250	0.240	0.190	0.180	0.340	0.330	0.250	0.240	0.510	0.500	0.100	0.090	0.090
3/8	16	20	0.312	0.302	0.222	0.212	0.402	0.392	0.312	0.302	0.585	0.575	0.100	0.090	0.090
7/16	14	18	0.375	0.365	0.235	0.225	0.515	0.505	0.375	0.365	0.695	0.685	0.135	0.125	0.140
1/2	12	16	0.437	0.427	0.297	0.287	0.577	0.567	0.437	0.427	0.805	0.795	0.135	0.125	0.140
9/16	12	16	0.500	0.490	0.313	0.303	0.687	0.677	0.500	0.490	0.905	0.895	0.175	0.165	0.187
5/8	11	14	0.562	0.552	0.375	0.365	0.749	0.739	0.562	0.552	0.995	0.985	0.175	0.165	0.187
3/4	10	12	0.687	0.677	0.453	0.443	0.921	0.911	0.687	0.677	1.185	1.165	0.218	0.208	0.234
7/8	9	11	0.750	0.740	0.516	0.506	0.984	0.974	0.750	0.740	1.285	1.265	0.218	0.208	0.234
1	8	10	0.875	0.865	0.595	0.585	1.155	1.145	0.875	0.865	1.465	1.445	0.260	0.250	0.280
1 1/8	7	9	1.000	0.990	0.720	0.710	1.280	1.270	1.000	0.990	1.655	1.635	0.260	0.250	0.280
1 1/4	7	9	1.125	1.105	0.797	0.777	1.453	1.433	1.125	1.105	1.845	1.825	0.300	0.290	0.328
1 3/8[a]	…	8	1.250	1.230	0.922	0.902	1.578	1.558	1.250	1.230	2.035	2.015	0.300	0.290	0.328
1 1/2	6	8	1.375	1.355	1.047	1.027	1.703	1.683	1.375	1.355	2.200	2.180	0.300	0.290	0.328
1 3/4	5	7	1.625	1.605	1.250	1.230	2.000	1.980	1.625	1.605	2.555	2.535	0.343	0.333	0.375
2	4.5	7	1.750	1.730	1.282	1.262	2.218	2.198	1.750	1.730	2.735	2.715	0.426	0.416	0.468

[a] Not standard with B.S.W. thread. For widths across flats, widths across corners, and diameter of washer face see Table 1. For dimensional notation, see diagram on page 1552.

All dimensions in inches except where otherwise noted.

Table 1. British Standard ISO Metric Precision Hexagon Bolts, Screws and Nuts

BS 3692:1967 (obsolescent)

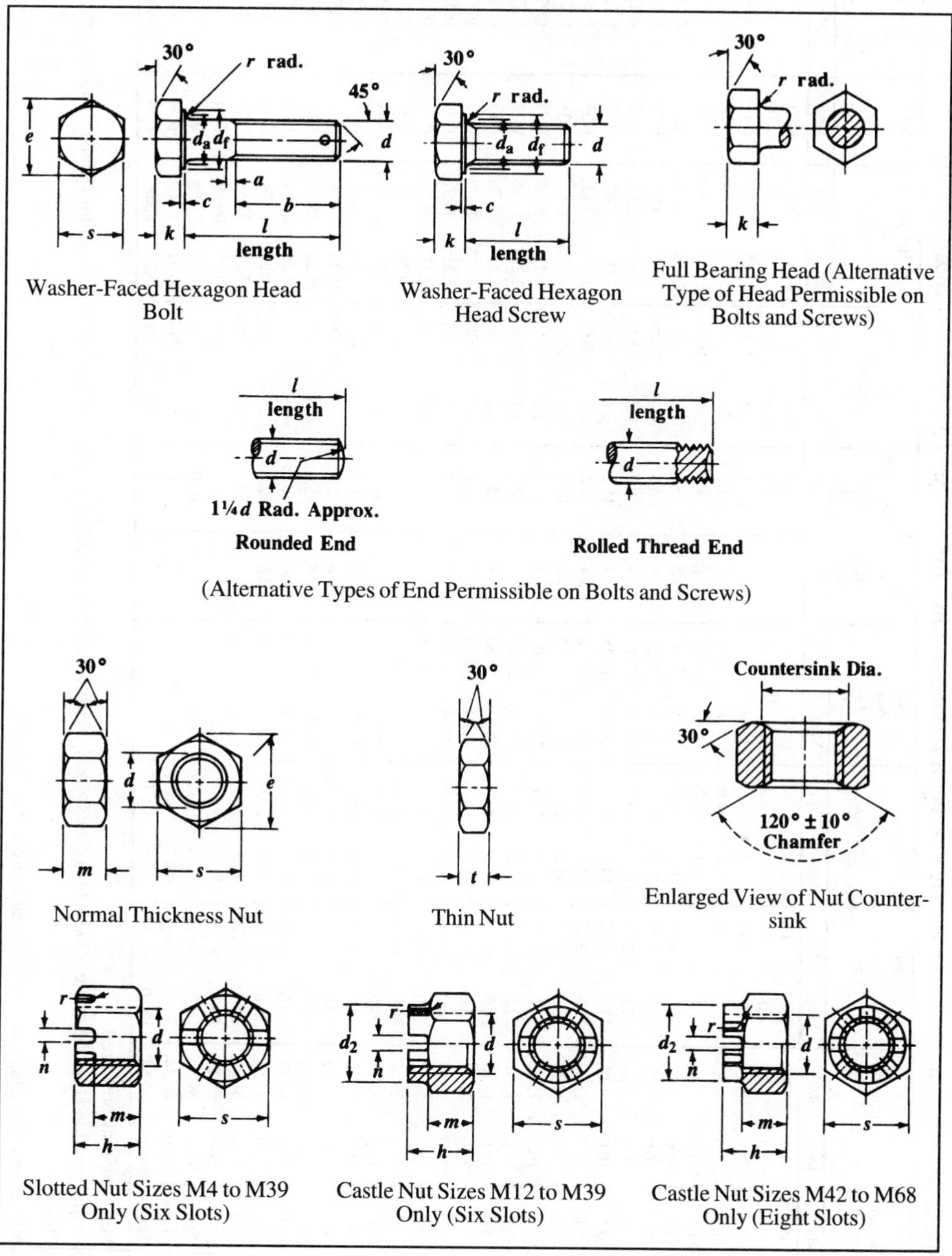

Washer-Faced Hexagon Head Bolt

Washer-Faced Hexagon Head Screw

Full Bearing Head (Alternative Type of Head Permissible on Bolts and Screws)

(Alternative Types of End Permissible on Bolts and Screws)

Normal Thickness Nut

Thin Nut

Enlarged View of Nut Countersink

Slotted Nut Sizes M4 to M39 Only (Six Slots)

Castle Nut Sizes M12 to M39 Only (Six Slots)

Castle Nut Sizes M42 to M68 Only (Eight Slots)

Table 2. British Standard ISO Metric Precision Hexagon Bolts and Screws *BS 3692:1967 (obsolescent)*

Nom. Size and Thread Dia.[a] d	Pitch of Thread (Coarse Pitch-Series)	Thread Runout a	Dia. of Unthreaded Shank d		Width Across Flats s		Width Across Corners e		Dia. of Washer Face d_t		Depth of Washer Face c	Transition Dia.[b] d_a	Radius Under Head[b] r		Height of Head k		Eccentricity of Head	Eccentricity of Shank and Split Pin Hole to the Thread
		Max.	Max.	Min.	Max.	Min.	Max.	Min.	Max.	Min.		Max.	Max.	Min.	Max.	Min.	Max.	Max.
M1.6	0.35	0.8	1.6	1.46	3.2	3.08	3.7	3.48	…	…	…	2.0	0.2	0.1	1.225	0.975	0.18	0.14
M2	0.4	1.0	2.0	1.86	4.0	3.88	4.6	4.38	…	…	…	2.6	0.3	0.1	1.525	1.275	0.18	0.14
M2.5	0.45	1.0	2.5	2.36	5.0	4.88	5.8	5.51	…	…	…	3.1	0.3	0.1	2.125	1.875	0.18	0.14
M3	0.5	1.2	3.0	2.86	5.5	5.38	6.4	6.08	5.08	4.83	0.1	3.6	0.3	0.1	2.125	1.875	0.18	0.14
M4	0.7	1.6	4.0	3.82	7.0	6.85	8.1	7.74	6.55	6.30	0.1	4.7	0.35	0.2	2.925	2.675	0.22	0.18
M5	0.8	2.0	5.0	4.82	8.0	7.85	9.2	8.87	7.55	7.30	0.2	5.7	0.35	0.2	3.650	3.35	0.22	0.18
M6	1	2.5	6.0	5.82	10.0	9.78	11.5	11.05	9.48	9.23	0.3	6.8	0.4	0.25	4.15	3.85	0.22	0.18
M8	1.25	3.0	8.0	7.78	13.0	12.73	15.0	14.38	12.43	12.18	0.4	9.2	0.6	0.4	5.65	5.35	0.27	0.22
M10	1.5	3.5	10.0	9.78	17.0	16.73	19.6	18.90	16.43	16.18	0.4	11.2	0.6	0.4	7.18	6.82	0.27	0.22
M12	1.75	4.0	12.0	11.73	19.0	18.67	21.9	21.10	18.37	18.12	0.4	14.2	1.1	0.6	8.18	7.82	0.33	0.27
(M14)	2	5.0	14.0	13.73	22.0	21.67	25.4	24.49	21.37	21.12	0.4	16.2	1.1	0.6	9.18	8.82	0.33	0.27
M16	2	5.0	16.0	15.73	24.0	23.67	27.7	26.75	23.27	23.02	0.4	18.2	1.1	0.6	10.18	9.82	0.33	0.27
(M18)	2.5	6.0	18.0	17.73	27.0	26.67	31.2	30.14	26.27	26.02	0.4	20.2	1.1	0.6	12.215	11.785	0.33	0.27
M20	2.5	6.0	20.0	19.67	30.0	29.67	34.6	33.53	29.27	28.80	0.4	22.4	1.2	0.8	13.215	12.785	0.33	0.33
(M22)	2.5	6.0	22.0	21.67	32.0	31.61	36.9	35.72	31.21	30.74	0.4	24.4	1.2	0.8	14.215	13.785	0.39	0.33
M24	3	7.0	24.0	23.67	36.0	35.38	41.6	39.98	34.98	34.51	0.5	26.4	1.2	0.8	15.215	14.785	0.39	0.33
(M27)	3	7.0	27.0	26.67	41.0	40.38	47.3	45.63	39.98	39.36	0.5	30.4	1.7	1.0	17.215	16.785	0.39	0.33
M30	3.5	8.0	30.0	29.67	46.0	45.38	53.1	51.28	44.98	44.36	0.5	33.4	1.7	1.0	19.26	18.74	0.39	0.33
(M33)	3.5	8.0	33.0	32.61	50.0	49.38	57.7	55.80	48.98	48.36	0.5	36.4	1.7	1.0	21.26	20.74	0.39	0.39
M36	4	10.0	36.0	35.61	55.0	54.26	63.5	61.31	53.86	53.24	0.5	39.4	1.7	1.0	23.26	22.74	0.46	0.39
(M39)	4	10.0	39.0	38.61	60.0	59.26	69.3	66.96	58.86	58.24	0.6	42.4	1.7	1.0	25.26	24.74	0.46	0.39
M42	4.5	11.0	42.0	41.61	65.0	64.26	75.1	72.61	63.76	63.04	0.6	45.6	1.8	1.2	26.26	25.74	0.46	0.39
(M45)	4.5	11.0	45.0	44.61	70.0	69.26	80.8	78.26	68.76	68.04	0.6	48.6	1.8	1.2	28.26	27.74	0.46	0.39
M48	5	12.0	48.0	47.61	75.0	74.26	86.6	83.91	73.76	73.04	0.6	52.6	2.3	1.6	30.26	29.74	0.46	0.39
(M52)	5	12.0	52.0	51.54	80.0	79.26	92.4	89.56	…	…	…	56.6	2.3	1.6	33.31	32.69	0.46	0.46
M56	5.5	19.0	56.0	55.54	85.0	84.13	98.1	95.07	…	…	…	63.0	3.5	2.0	35.31	34.69	0.54	0.46
(M60)	5.5	19.0	60.0	59.54	90.0	89.13	103.9	100.72	…	…	…	67.0	3.5	2.0	38.31	37.69	0.54	0.46
M64	6	21.0	64.0	63.54	95.0	94.13	109.7	106.37	…	…	…	71.0	3.5	2.0	40.31	39.69	0.54	0.46
(M68)	6	21.0	68.0	67.54	100.0	99.13	115.5	112.02	…	…	…	75.0	3.5	2.0	43.31	42.69	0.54	0.46

[a] Sizes shown in parentheses are non-preferred.

[b] A true radius is not essential provided that the curve is smooth and lies wholly within the maximum radius, determined from the maximum transitional diameter, and the minimum radius specified.

All dimensions are in millimeters. For illustration of bolts and screws see Table 1.

Table 3. British Standard ISO Metric Precision Hexagon Nuts and Thin Nuts *BS 3692:1967 (obsolescent)*

Nominal Size and Thread Diameter[a] *d*	Pitch of Thread (Coarse Pitch Series)	Width Across Flats *s*		Width Across Corners *e*		Thickness of Normal Nut *m*		Tolerance on Squareness of Thread to Face of Nut[b]	Eccentricity of Hexagon	Thickness of Thin Nut *t*	
		Max.	Min.	Max.	Min.	Max.	Min.	Max.	Max.	Max.	Min.
M1.6	0.35	3.20	3.08	3.70	3.48	1.30	1.05	0.05	0.14	…	…
M2	0.4	4.00	3.88	4.60	4.38	1.60	1.35	0.06	0.14	…	…
M2.5	0.45	5.00	4.88	5.80	5.51	2.00	1.75	0.08	0.14	…	…
M3	0.5	5.50	5.38	6.40	6.08	2.40	2.15	0.09	0.14	…	…
M4	0.7	7.00	6.85	8.10	7.74	3.20	2.90	0.11	0.18	…	…
M5	0.8	8.00	7.85	9.20	8.87	4.00	3.70	0.13	0.18	…	…
M6	1	10.00	9.78	11.50	11.05	5.00	4.70	0.17	0.18	…	…
M8	1.25	13.00	12.73	15.00	14.38	6.50	6.14	0.22	0.22	5.0	4.70
M10	1.5	17.00	16.73	19.60	18.90	8.00	7.64	0.29	0.22	6.0	5.70
M12	1.75	19.00	18.67	21.90	21.10	10.00	9.64	0.32	0.27	7.0	6.64
(M14)	2	22.00	21.67	25.4	24.49	11.00	10.57	0.37	0.27	8.0	7.64
M16	2	24.00	23.67	27.7	6.75	13.00	12.57	0.41	0.27	8.0	7.64
(M18)	2.5	27.00	26.67	31.20	30.14	15.00	14.57	0.46	0.27	9.0	8.64
M20	2.5	30.00	29.67	34.60	33.53	16.00	15.57	0.51	0.33	9.0	8.64
(M22)	2.5	32.00	31.61	36.90	35.72	18.00	17.57	0.54	0.33	10.0	9.64
M24	3	36.00	35.38	41.60	39.98	19.00	18.48	0.61	0.33	10.0	9.64
(M27)	3	41.00	40.38	47.3	45.63	22.00	21.48	0.70	0.33	12.0	11.57
M30	3.5	46.00	45.38	53.1	51.28	24.00	23.48	0.78	0.33	12.0	11.57
(M33)	3.5	50.00	49.38	57.70	55.80	26.00	25.48	0.85	0.39	14.0	13.57
M36	4	55.00	54.26	63.50	61.31	29.00	28.48	0.94	0.39	14.0	13.57
(M39)	4	60.00	59.26	69.30	66.96	31.00	30.38	1.03	0.39	16.0	15.57
M42	4.5	65.00	64.26	75.10	72.61	34.00	33.38	1.11	0.39	16.0	15.57
(M45)	4.5	70.00	69.26	80.80	78.26	36.00	35.38	1.20	0.39	18.0	17.57
M48	5	75.00	74.26	86.60	83.91	38.00	37.38	1.29	0.39	18.0	17.57
(M52)	5	80.00	79.26	92.40	89.56	42.00	41.38	1.37	0.46	20.0	19.48
M56	5.5	85.00	84.13	98.10	95.07	45.00	44.38	1.46	0.46	…	…
(M60)	5.5	90.00	89.13	103.90	100.72	48.00	47.38	1.55	0.46	…	…
M64	6	95.00	94.13	109.70	106.37	51.00	50.26	1.63	0.46	…	…
(M68)	6	100.00	99.13	115.50	112.02	54.00	53.26	1.72	0.46	…	…

[a] Sizes shown in parentheses are non-preferred.

[b] As measured with the nut squareness gage described in the text and illustrated in Appendix A of the Standard and a feeler gage.

All dimensions are in millimeters. For illustration of hexagon nuts and thin nuts see Table 1.

Table 4. British Standard ISO Metric Precision Hexagon Slotted Nuts and Castle Nuts *BS 3692:1967 (obsolescent)*

Nominal Size and Thread Diameter[a] d	Width Across Flats s		Width Across Corners e		Diameter d_2		Thickness h		Lower Face of Nut to Bottom of Slot m		Width of Slot n		Radius (0.25 n) r	Eccentricity of the Slots
	Max.	Min.	Max.	Min.	Max.	Min.	Max.	Min.	Max.	Min.	Max.	Min.	Min.	Max.
M4	7.00	6.85	8.10	7.74	…	…	5	4.70	3.2	2.90	1.45	1.2	0.3	0.18
M5	8.00	7.85	9.20	8.87	…	…	6	5.70	4.0	3.70	1.65	1.4	0.35	0.18
M6	10.00	9.78	11.50	11.05	…	…	7.5	7.14	5	4.70	2.25	2	0.5	0.18
M8	13.00	12.73	15.00	14.38	…	…	9.5	9.14	6.5	6.14	2.75	2.5	0.625	0.22
M10	17.00	16.73	19.60	18.90	…	…	12	11.57	8	7.64	3.05	2.8	0.70	0.22
M12	19.00	18.67	21.90	21.10	17	16.57	15	14.57	10	9.64	3.80	3.5	0.875	0.27
(M14)	22.00	21.67	25.4	24.49	19	18.48	16	15.57	11	10.57	3.80	3.5	0.875	0.27
M16	24.00	23.67	27.7	26.75	22	21.48	19	18.48	13	12.57	4.80	4.5	1.125	0.27
(M18)	27.00	26.67	31.20	30.14	25	24.48	21	20.48	15	14.57	4.80	4.5	1.125	0.27
M20	30.00	29.67	34.60	33.53	28	27.48	22	21.48	16	15.57	4.80	4.5	1.125	0.33
(M22)	32.00	31.61	36.90	35.72	30	29.48	26	25.48	18	17.57	5.80	5.5	1.375	0.33
M24	36.00	35.38	41.60	39.98	34	33.38	27	26.48	19	18.48	5.80	5.5	1.375	0.33
(M27)	41.00	40.38	47.3	45.63	38	37.38	30	29.48	22	21.48	5.80	5.5	1.375	0.33
M30	46.00	45.38	53.1	51.28	42	41.38	33	32.38	24	23.48	7.36	7	1.75	0.33
(M33)	50.00	49.38	57.70	55.80	46	45.38	35	34.38	26	25.48	7.36	7	1.75	0.39
M36	55.00	54.26	63.50	61.31	50	49.38	38	37.38	29	28.48	7.36	7	1.75	0.39
(M39)	60.00	59.26	69.30	66.96	55	54.26	40	39.38	31	30.38	7.36	7	1.75	0.39
M42	65.00	64.26	75.10	72.61	58	57.26	46	45.38	34	33.38	9.36	9	2.25	0.39
(M45)	70.00	69.26	80.80	78.26	62	61.26	48	47.38	36	35.38	9.36	9	2.25	0.39
M48	75.00	74.26	86.60	83.91	65	64.26	50	49.38	38	37.38	9.36	9	2.25	0.39
(M52)	80.00	79.26	92.40	89.56	70	69.26	54	53.26	42	41.38	9.36	9	2.25	0.46
M56	85.00	84.13	98.10	95.07	75	74.26	57	56.26	45	44.38	9.36	9	2.25	0.46
(M60)	90.00	89.13	103.90	100.72	80	79.26	63	62.26	48	47.38	11.43	11	2.75	0.46
M64	95.00	94.13	109.70	106.37	85	84.13	66	65.26	51	50.26	11.43	11	2.75	0.46
(M68)	100.00	99.13	115.50	112.02	90	89.13	69	68.26	54	53.26	11.43	11	2.75	0.46

[a] Sizes shown in parentheses are non-preferred.

All dimensions are in millimeters. For illustration of hexagon slotted nuts and castle nuts see Table 1.

After several years of use of BS 2693:Part 1:1956 (obsolescent), it was recognized that it would not meet the requirements of all stud users. The thread tolerances specified could result in clearance of interference fits because locking depended on the run-out threads. Thus, some users felt that true interference fits were essential for their needs. As a result, the British Standards Committee has incorporated the Class 5 interference fit threads specified in American Standard ASA B1.12 into the BS 2693:Part 2:1964, "Recommendations for High Grade Studs."

British Standard ISO Metric Precision Hexagon Bolts, Screws and Nuts.— This British Standard BS 3692:1967 (obsolescent) gives the general dimensions and tolerances of precision hexagon bolts, screws and nuts with ISO metric threads in diameters from 1.6 to 68 mm. It is based on the following ISO recommendations and draft recommendations: R 272, R 288, DR 911, DR 947, DR 950, DR 952 and DR 987. Mechanical properties are given only with respect to carbon or alloy steel bolts, screws and nuts, which are not to be used for special applications such as those requiring weldability, corrosion resistance or ability to withstand temperatures above 300°C or below − 50°C. The dimensional requirements of this standard also apply to non-ferrous and stainless steel bolts, screws and nuts.

Finish: Finishes may be dull black which results from the heat-treating operation or may be bright finish, the result of bright drawing. Other finishes are possible by mutual agreement between purchaser and producer. It is recommended that reference be made to BS 3382 "Electroplated Coatings on Threaded Components" in this respect.

General Dimensions: The bolts, screws and nuts conform to the general dimensions given in Tables 1, 2, 3 and 4.

Nominal Lengths of Bolts and Screws: The nominal length of a bolt or screw is the distance from the underside of the head to the extreme end of the shank including any chamfer or radius. Standard nominal lengths and tolerances thereon are given in Table 5.

Table 5. British Standard ISO Metric Bolt and Screw Nominal Lengths
BS 3692:1967 (obsolescent)

Nominal Length[a] *l*	Tolerance	Nominal Length[a] *l*	Tolerance	Nominal Length[a] *l*	Tolerance	Nominal Length[a] *l*	Tolerance
5	± 0.24	30	± 0.42	90	± 0.70	200	± 0.925
6	± 0.24	(32)	± 0.50	(95)	± 0.70	220	± 0.925
(7)	± 0.29	35	± 0.50	100	± 0.70	240	± 0.925
8	± 0.29	(38)	± 0.50	(105)	± 0.70	260	± 1.05
(9)	± 0.29	40	± 0.50	110	± 0.70	280	± 1.05
10	± 0.29	45	± 0.50	(115)	± 0.70	300	± 1.05
(11)	± 0.35	50	± 0.50	120	± 0.70	325	± 1.15
12	± 0.35	55	± 0.60	(125)	± 0.80	350	± 1.15
14	± 0.35	60	± 0.60	130	± 0.80	375	± 1.15
16	± 0.35	65	± 0.60	140	± 0.80	400	± 1.15
(18)	± 0.35	70	± 0.60	150	± 0.80	425	± 1.25
20	± 0.42	75	± 0.60	160	± 0.80	450	± 1.25
(22)	± 0.42	80	± 0.60	170	± 0.80	475	± 1.25
25	± 0.42	85	± 0.70	180	± 0.80	500	± 1.25
(28)	± 0.42	…	…	190	± 0.925	…	…

[a] Nominal lengths shown in parentheses are non-preferred.
All dimensions are in millimeters.

Bolt and Screw Ends: The ends of bolts and screws may be finished with either a 45-degree chamfer to a depth slightly exceeding the depth of thread or a radius approximately equal to 1¼ times the nominal diameter of the shank. With rolled threads, the lead formed at the end of the bolt by the thread rolling operation may be regarded as providing the necesssary chamfer to the end; the end being reasonably square with the center line of the shank.

Screw Thread Form: The form of thread and diameters and associated pitches of standard ISO metric bolts, screws, and nuts are in accordance with BS 3643:Part 1:1981 (1998), "Principles and Basic Data" The screw threads are made to the tolerances for the medium class of fit (6*H*/6*g*) as specified in BS 3643:Part 2:1981 (1998), "Specification for Selected Limits of Size."

Length of Thread on Bolts: The length of thread on bolts is the distance from the end of the bolt (including any chamfer or radius) to the leading face of a screw ring gage which has been screwed as far as possible onto the bolt by hand. Standard thread lengths of bolts are $2d + 6$ mm for a nominal length of bolt up to and including 125 mm, $2d + 12$ mm for a nominal bolt length over 125 mm up to and including 200 mm, and $2d + 25$ mm for a nominal bolt length over 200 mm. Bolts that are too short for minimum thread lengths are threaded as screws and designated as screws. The tolerance on bolt thread lengths are plus two pitches for all diameters.

Length of Thread on Screws: Screws are threaded to permit a screw ring gage being screwed by hand to within a distance from the underside of the head not exceeding two and a half times the pitch for diameters up to and including 52 mm and three and a half times the pitch for diameters over 52 mm.

Angularity and Eccentricity of Bolts, Screws and Nuts: The axis of the thread of the nut is square to the face of the nut subject to the "squareness tolerance" given in Table 3.

In gaging, the nut is screwed by hand onto a gage, having a truncated taper thread, until the thread of the nut is tight on the thread of the gage. A sleeve sliding on a parallel extension of the gage, which has a face of diameter equal to the minimum distance across the flats of the nut and exactly at 90 degrees to the axis of the gage, is brought into contact with the leading face of the nut. With the sleeve in this position, it should not be possible for a feeler gage of thickness equal to the "squareness tolerance" to enter anywhere between the leading nut face and sleeve face.

The hexagon flats of bolts, screws and nuts are square to the bearing face, and the angularity of the head is within the limits of 90 degrees, plus or minus 1 degree. The eccentricity of the hexagon flats of nuts relative to the thread diameter should not exceed the values given in Table 3 and the eccentricity of the head relative to the width across flats and eccentricity between the shank and thread of bolts and screws should not exceed the values given in Table 2.

Chamfering, Washer Facing and Countersinking: Bolt and screw heads have a chamfer of approximately 30 degrees on their upper faces and, at the option of the manufacturer, a washer face or full bearing face on the underside. Nuts are countersunk at an included angle of 120 degrees plus or minus 10 degrees at both ends of the thread. The diameter of the countersink should not exceed the nominal major diameter of the thread plus 0.13 mm up to and including 12 mm diameter, and plus 0.25 mm above 12 mm diameter. This stipulation does not apply to slotted, castle or thin nuts.

Strength Grade Designation System for Steel Bolts and Screws: This Standard includes a strength grade designation system consisting of two figures. The first figure is one tenth of the minimum tensile strength in kgf/mm², and the second figure is one tenth of the ratio between the minimum yield stress (or stress at permanent set limit, $R_{0.2}$) and the minimum tensile strength, expressed as a percentage. For example with the strength designation grade 8.8, the first figure 8 represents ⅒ the minimum tensile strength of 80 kgf/mm² and the second figure 8 represents ⅒ the ratio

$$\frac{\text{stress at permanent set limit } R_{0.2}\%}{\text{minimum tensile strength}} = \frac{1}{10} \times \frac{64}{80} \times \frac{100}{1}$$

the numerical values of stress and strength being obtained from the accompanying table.

Strength Grade Designations of Steel Bolts and Screws

Strength Grade Designation	4.6	4.8	5.6	5.8	6.6	6.8	8.8	10.9	12.9	14.9
Tensile Strength (R_m), Min.	40	40	50	50	60	60	80	100	120	140
Yield Stress (R_e), Min.	24	32	30	40	36	48	…	…	…	…
Stress at Permanent Set Limit ($R_{0.2}$), Min.	…	…	…	…	…	…	64	90	108	126
All stress and strength values are in kgf/mm² units.										

Strength Grade Designation System for Steel Nuts: The strength grade designation system for steel nuts is a number which is one-tenth of the specified proof load stress in kgf/mm². The proof load stress corresponds to the minimum tensile strength of the highest grade of bolt or screw with which the nut can be used.

Strength Grade Designations of Steel Nuts

Strength Grade Designation	4	5	6	8	12	14
Proof Load Stress (kgf/mm²)	40	50	60	80	120	140

Recommended Bolt and Nut Combinations

Grade of Bolt	4.6	4.8	5.6	5.8	6.6	6.8	8.8	10.9	12.9	14.9
Recommended Grade of Nut	4	4	5	5	6	6	8	12	12	14
Note: Nuts of a higher strength grade may be substituted for nuts of a lower strength grade.										

Marking: The marking and identification requirements of this Standard are only mandatory for steel bolts, screws and nuts of 6 mm diameter and larger; manufactured to strength grade designations 8.8 (for bolts or screws) and 8 (for nuts) or higher. Bolts and screws are identified as ISO metric by either of the symbols "ISO M" or "M", embossed or indented on top of the head. Nuts may be indented or embossed by alternative methods depending on their method of manufacture.

Designation: Bolts 10 mm diameter, 50 mm long manufactured from steel of strength grade 8.8, would be designated:

"Bolts M10 × 50 to BS 3692 — 8.8."

Brass screws 8 mm diameter, 20 mm long would be designated:

"Brass screws M8 × 20 to BS 3692."

Nuts 12 mm diameter, manufactured from steel of strength grade 6, cadmium plated could be designated:

"Nuts M12 to BS 3692 — 6, plated to BS 3382: Part 1."

Miscellaneous Information: The Standard also gives mechanical properties of steel bolts, screws and nuts [i.e., tensile strengths; hardnesses (Brinell, Rockwell, Vickers); stresses (yield, proof load); etc.], material and manufacture of steel bolts, screws and nuts; and information on inspection and testing. Appendices to the Standard give information on gaging; chemical composition; testing of mechanical properties; examples of marking of bolts, screws and nuts; and a table of preferred standard sizes of bolts and screws, to name some.

British Standard General Purpose Studs *BS 2693:Part 1:1956 (obsolescent)*

Limits for End Screwed into Component (All threads except B.A.)													
Nom. Dia. *D*	Major Dia.	Thds. per In.	Major Dia.	Effective Diameter		Minor Diameter		Thds. per In.	Major Dia.	Effective Diameter		Minor Dia.	
	Max.		Min.	Max.	Min.	Max.	Min.		Min.	Max.	Min.	Max.	Min.
UN THREADS		UNF THREADS						UNC THREADS					
1/4	0.2500	28	0.2435	0.2294	0.2265	0.2088	0.2037	20	0.2419	0.2201	0.2172	0.1913	0.1849
5/16	0.3125	24	0.3053	0.2883	0.2852	0.2643	0.2586	18	0.3038	0.2793	0.2762	0.2472	0.2402
3/8	0.3750	24	0.3678	0.3510	0.3478	0.3270	0.3211	16	0.3656	0.3375	0.3343	0.3014	0.2936
7/16	0.4375	20	0.4294	0.4084	0.4050	0.3796	0.3729	14	0.4272	0.3945	0.3911	0.3533	0.3447
1/2	0.5000	20	0.4919	0.4712	0.4675	0.4424	0.4356	13	0.4891	0.4537	0.4500	0.4093	0.4000
9/16	0.5625	18	0.5538	0.5302	0.5264	0.4981	0.4907	12	0.5511	0.5122	0.5084	0.4641	0.4542
5/8	0.6250	18	0.6163	0.5929	0.5889	0.5608	0.5533	11	0.6129	0.5700	0.5660	0.5175	0.5069
3/4	0.7500	16	0.7406	0.7137	0.7094	0.6776	0.6693	10	0.7371	0.6893	0.6850	0.6316	0.6200
7/8	0.8750	14	0.8647	0.8332	0.8286	0.7920	0.7828	9	0.8611	0.8074	0.8028	0.7433	0.7306
1	1.0000	12	0.9886	0.9510	0.9459	0.9029	0.8925	8	0.9850	0.9239	0.9188	0.8517	0.8376
1 1/8	1.1250	12	1.1136	1.0762	1.0709	1.0281	1.0176	7	1.1086	1.0375	1.0322	0.9550	0.9393
1 1/4	1.2500	12	1.2386	1.2014	1.1959	1.1533	1.1427	7	1.2336	1.1627	1.1572	1.0802	1.0644
1 3/8	1.3750	12	1.3636	1.3265	1.3209	1.2784	1.2677	6	1.3568	1.2723	1.2667	1.1761	1.1581
1 1/2	1.5000	12	1.4886	1.4517	1.4459	1.4036	1.3928	6	1.4818	1.3975	1.3917	1.3013	1.2832
B.S.THREADS		B.S.F. THREADS						B.S.W. THREADS					
1/4	0.2500	26	0.2455	0.2280	0.2251	0.2034	0.1984	20	0.2452	0.2206	0.2177	0.1886	0.1831
5/16	0.3125	22	0.3077	0.2863	0.2832	0.2572	0.2517	18	0.3073	0.2798	0.2767	0.2442	0.2383
3/8	0.3750	20	0.3699	0.3461	0.3429	0.3141	0.3083	16	0.3695	0.3381	0.3349	0.0981	0.2919
7/16	0.4375	18	0.4320	0.4053	0.4019	0.3697	0.3635	14	0.4316	0.3952	0.3918	0.3495	0.3428
1/2	0.5000	16	0.4942	0.4637	0.4600	0.4237	0.4172	12	0.4937	0.4503	0.4466	0.3969	0.3897
9/16	0.5625	16	0.5566	0.5263	0.5225	0.4863	0.4797	12	0.5560	0.5129	0.5091	0.4595	0.4521
5/8	0.6250	14	0.6187	0.5833	0.5793	0.5376	0.5305	11	0.6183	0.5708	0.5668	0.5126	0.5050
3/4	0.7500	12	0.7432	0.7009	0.6966	0.6475	0.6398	10	0.7428	0.6903	0.6860	0.6263	0.6182
7/8	0.8750	11	0.8678	0.8214	0.8168	0.7632	0.7551	9	0.8674	0.8085	0.8039	0.7374	0.7288
1	1.0000	10	0.9924	0.9411	0.9360	0.8771	0.8686	8	0.9920	0.9251	0.9200	0.8451	0.8360
1 1/8	1.1250	9	1.1171	1.0592	1.0539	0.9881	0.9792	7	1.1164	1.0388	1.0335	0.9473	0.9376
1 1/4	1.2500	9	1.2419	1.1844	1.1789	1.1133	1.1042	7	1.2413	1.1640	1.1585	1.0725	1.0627
1 3/8	1.3750	8	1.3665	1.3006	1.2950	1.2206	1.2110	0..	…	…	…	…	…
1 1/2	1.5000	8	1.4913	1.4258	1.4200	1.3458	1.3360	6	1.4906	1.3991	1.3933	1.2924	1.2818

Limits for End Screwed into Component (B.A. Threads)[a]							
Designation No.	Pitch	Major Diameter		Effective Diameter		Minor Diameter	
		Max.	Min.	Max.	Min.	Max.	Min.
2	0.8100 mm 0.03189 in.	4.700 mm 0.1850 in.	4.580 mm 0.1803 in.	4.275 mm 0.1683 in.	4.200 mm 0.1654 in.	3.790 mm 0.1492 in.	3.620 mm 0.1425 in.
4	0.6600 mm 0.2598 in.	3.600 mm 0.1417 in.	3.500 mm 0.1378 in.	3.260 mm 0.1283 in.	3.190 mm 0.1256 in.	2.865 mm 0.1128 in.	2.720 mm 0.1071 in.

[a] Approximate inch equivalents are shown below the dimensions given in mm.

Minimum Nominal Lengths of Studs[a]								
Nom. Stud. Dia.	For Thread Length (Component End) of		Nom. Stud. Dia.	For Thread Length (Component End) of		Nom. Stud Dia.	For Thread Length (Component End) of	
	1*D*	1.5*D*		1*D*	1.5*D*		1*D*	1.5*D*
1/4	7/8	1	9/16	2	2 3/8	1 1/8	4	4 5/8
5/16	1 1/8	1 3/8	5/8	2 1/4	2 5/8	1 1/4	4 3/4	5 1/2
3/8	1 3/8	1 5/8	3/4	2 5/8	3	1 3/8	5	5 3/4
7/16	1 5/8	1 7/8	7/8	3 1/8	3 5/8	1 1/2	5 1/4	6
1/2	1 3/4	2	1	3 1/2	4	…	…	…

[a] The standard also gives preferred and standard lengths of studs: *Preferred* lengths of studs: 7/8, 1, 1 1/8, 1 1/4, 1 3/8, 1 1/2, 1 3/4, 2, 2 1/4, 2 1/2, 2 3/4, 3, 3 1/4, 3 1/2 and for lengths above 3 1/2 the preferred increment is 1/2. *Standard* lengths of studs: 7/8, 1, 1 1/8, 1 1/4, 1 3/8, 1 1/2, 1 5/8, 1 3/4, 1 7/8, 2, 2 1/8, 2 1/4, 2 3/8, 2 1/2, 2 5/8, 2 3/4, 2 7/8, 3, 3 1/8, 3 1/4, 3 3/8, 3 1/2 and for lengths above 3 1/2 the standard increment is 1/4.

All dimensions are in inches except where otherwise noted.

See page 1786 for interference-fit threads.

British Standard Single Coil Rectangular Section Spring Washers; Metric Series — Types B and BP *BS 4464:1969 (1998)*

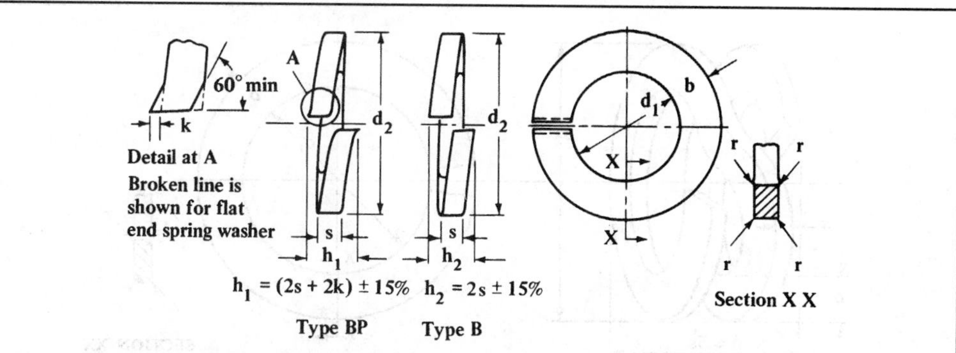

Nom. Size & Thread Dia., d	Inside Dia., d_1		Width, b	Thickness, s	Outside Dia., d_2 Max	Radius, r Max	k (Type BP Only)
	Max	Min					
M1.6	1.9	1.7	0.7 ± 0.1	0.4 ± 0.1	3.5	0.15	…
M2	2.3	2.1	0.9 ± 0.1	0.5 ± 0.1	4.3	0.15	…
(M2.2)	2.5	2.3	1.0 ± 0.1	0.6 ± 0.1	4.7	0.2	…
M2.5	2.8	2.6	1.0 ± 0.1	0.6 ± 0.1	5.0	0.2	…
M3	3.3	3.1	1.3 ± 0.1	0.8 ± 0.1	6.1	0.25	…
(M3.5)	3.8	3.6	1.3 ± 0.1	0.8 ± 0.1	6.6	0.25	0.15
M4	4.35	4.1	1.5 ± 0.1	0.9 ± 0.1	7.55	0.3	0.15
M5	5.35	5.1	1.8 ± 0.1	1.2 ± 0.1	9.15	0.4	0.15
M6	6.4	6.1	2.5 ± 0.15	1.6 ± 0.1	11.7	0.5	0.2
M8	8.55	8.2	3 ± 0.15	2 ± 0.1	14.85	0.65	0.3
M10	10.6	10.2	3.5 ± 0.2	2.2 ± 0.15	18.0	0.7	0.3
M12	12.6	12.2	4 ± 0.2	2.5 ± 0.15	21.0	0.8	0.4
(M14)	14.7	14.2	4.5 ± 0.2	3 ± 0.15	24.1	1.0	0.4
M16	16.9	16.3	5 ± 0.2	3.5 ± 0.2	27.3	1.15	0.4
(M18)	19.0	18.3	5 ± 0.2	3.5 ± 0.2	29.4	1.15	0.4
M20	21.1	20.3	6 ± 0.2	4 ± 0.2	33.5	1.3	0.4
(M22)	23.3	22.4	6 ± 0.2	4 ± 0.2	35.7	1.3	0.4
M24	25.3	24.4	7 ± 0.25	5 ± 0.2	39.8	1.65	0.5
(M27)	28.5	27.5	7 ± 0.25	5 ± 0.2	43.0	1.65	0.5
M30	31.5	30.5	8 ± 0.25	6 ± 0.25	48.0	2.0	0.8
(M33)	34.6	33.5	10 ± 0.25	6 ± 0.25	55.1	2.0	0.8
M36	37.6	36.5	10 ± 0.25	6 ± 0.25	58.1	2.0	0.8
(M39)	40.8	39.6	10 ± 0.25	6 ± 0.25	61.3	2.0	0.8
M42	43.8	42.6	12 ± 0.25	7 ± 0.25	68.3	2.3	0.8
(M45)	46.8	45.6	12 ± 0.25	7 ± 0.25	71.3	2.3	0.8
M48	50.0	48.8	12 ± 0.25	7 ± 0.25	74.5	2.3	0.8
(M52)	54.1	52.8	14 ± 0.25	8 ± 0.25	82.6	2.65	1.0
M56	58.1	56.8	14 ± 0.25	8 ± 0.25	86.6	2.65	1.0
(M60)	62.3	60.9	14 ± 0.25	8 ± 0.25	90.8	2.65	1.0
M64	66.3	64.9	14 ± 0.25	8 ± 0.25	93.8	2.65	1.0
(M68)	70.5	69.0	14 ± 0.25	8 ± 0.25	99.0	2.65	1.0

All dimensions are given in millimeters. Sizes shown in parentheses are non-preferred, and are not usually stock sizes.

British Standard Double Coil Rectangular Section Spring Washers; Metric Series — Type D *BS 4464:1969 (1998)*

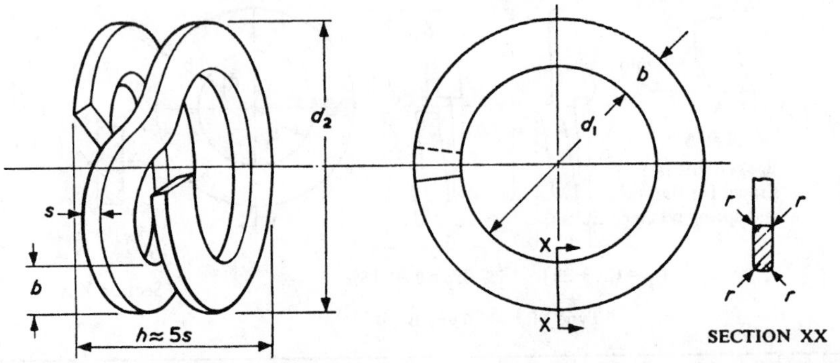

Nom. Size, d	Inside Dia., d_1		Width, b	Thickness, s	O.D., d_2 Max	Radius, r Max
	Max	Min				
M2	2.4	2.1	0.9 ± 0.1	0.5 ± 0.05	4.4	0.15
(M2.2)	2.6	2.3	1.0 ± 0.1	0.6 ± 0.05	4.8	0.2
M2.5	2.9	2.6	1.2 ± 0.1	0.7 ± 0.1	5.5	0.23
M3.0	3.6	3.3	1.2 ± 0.1	0.8 ± 0.1	6.2	0.25
(M3.5)	4.1	3.8	1.6 ± 0.1	0.8 ± 0.1	7.5	0.25
M4	4.6	4.3	1.6 ± 0.1	0.8 ± 0.1	8.0	0.25
M5	5.6	5.3	2 ± 0.1	0.9 ± 0.1	9.8	0.3
M6	6.6	6.3	3 ± 0.15	1 ± 0.1	12.9	0.33
M8	8.8	8.4	3 ± 0.15	1.2 ± 0.1	15.1	0.4
M10	10.8	10.4	3.5 ± 0.20	1.2 ± 0.1	18.2	0.4
M12	12.8	12.4	3.5 ± 0.2	1.6 ± 0.1	20.2	0.5
(M14)	15.0	14.5	5 ± 0.2	1.6 ± 0.1	25.4	0.5
M16	17.0	16.5	5 ± 0.2	2 ± 0.1	27.4	0.65
(M18)	19.0	18.5	5 ± 0.2	2 ± 0.1	29.4	0.65
M20	21.5	20.8	5 ± 0.2	2 ± 0.1	31.9	0.65
(M22)	23.5	22.8	6 ± 0.2	2.5 ± 0.15	35.9	0.8
M24	26.0	25.0	6.5 ± 0.2	3.25 ± 0.15	39.4	1.1
(M27)	29.5	28.0	7 ± 0.25	3.25 ± 0.15	44.0	1.1
M30	33.0	31.5	8 ± 0.25	3.25 ± 0.15	49.5	1.1
(M33)	36.0	34.5	8 ± 0.25	3.25 ± 0.15	52.5	1.1
M36	40.0	38.0	10 ± 0.25	3.25 ± 0.15	60.5	1.1
(M39)	43.0	41.0	10 ± 0.25	3.25 ± 0.15	63.5	1.1
M42	46.0	44.0	10 ± 0.25	4.5 ± 0.2	66.5	1.5
M48	52.0	50.0	10 ± 0.25	4.5 ± 0.2	72.5	1.5
M56	60.0	58.0	12 ± 0.25	4.5 ± 0.2	84.5	1.5
M64	70.0	67.0	12 ± 0.25	4.5 ± 0.2	94.5	1.5

All dimensions are given in millimeters. Sizes shown in parentheses are non-preferred, and are not usually stock sizes. The free height of double coil washers before compression is normally approximately five times the thickness but, if required, washers with other free heights may be obtained by arrangement with manufacturer.

British Standard Single Coil Square Section Spring Washers; Metric Series — Type A-1 *BS 4464:1969 (1998)*

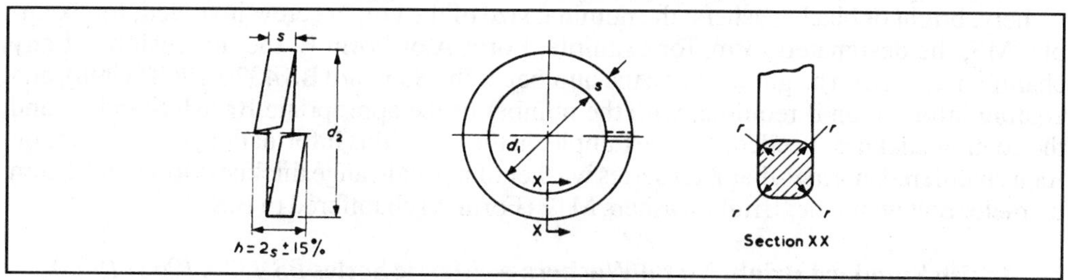

British Standard Single Coil Square Section Spring Washers; Metric Series — Type A-2 *BS 4464:1969 (1998)*

Nom. Size, *d*	Inside Dia., d_1		Thickness & Width, *s*	O.D., d_2	Radius, *r*
	Max	Min		Max	Max
M3	3.3	3.1	1 ± 0.1	5.5	0.3
(M3.5)	3.8	3.6	1 ± 0.1	6.0	0.3
M4	4.35	4.1	1.2 ± 0.1	6.95	0.4
M5	5.35	5.1	1.5 ± 0.1	8.55	0.5
M6	6.4	6.1	1.5 ± 0.1	9.6	0.5
M8	8.55	8.2	2 ± 0.1	12.75	0.65
M10	10.6	10.2	2.5 ± 0.15	15.9	0.8
M12	12.6	12.2	2.5 ± 0.15	17.9	0.8
(M14)	14.7	14.2	3 ± 0.2	21.1	1.0
M16	16.9	16.3	3.5 ± 0.2	24.3	1.15
(M18)	19.0	18.3	3.5 ± 0.2	26.4	1.15
M20	21.1	20.3	4.5 ± 0.2	30.5	1.5
(M22)	23.3	22.4	4.5 ± 0.2	32.7	1.5
M24	25.3	24.4	5 ± 0.2	35.7	1.65
(M27)	28.5	27.5	5 ± 0.2	38.9	1.65
M30	31.5	30.5	6 ± 0.2	43.9	2.0
(M33)	34.6	33.5	6 ± 0.2	47.0	2.0
M36	37.6	36.5	7 ± 0.25	52.1	2.3
(M39)	40.8	39.6	7 ± 0.25	55.3	2.3
M42	43.8	42.6	8 ± 0.25	60.3	2.65
(M45)	46.8	45.6	8 ± 0.25	63.3	2.65
M48	50.0	48.8	8 ± 0.25	66.5	2.65

All dimensions are in millimeters. Sizes shown in parentheses are nonpreferred and are not usually stock sizes.

British Standard for Metric Series Metal Washers.—BS 4320:1968 (1998) specifies bright and black metal washers for general engineering purposes.

Bright Metal Washers: These washers are made from either CS4 cold-rolled strip steel BS 1449:Part 3B or from CZ 108 brass strip B.S. 2870: 1980, both in the hard condition. However, by mutual agreement between purchaser and supplier, washers may be made available with the material in any other condition, or they may be made from another material, or may be coated with a protective or decorative finish to some appropriate British Standard. Washers are reasonably flat and free from burrs and are normally supplied unchamfered. They may, however, have a 30-degree chamfer on one edge of the external diameter. These washers are made available in two size categories, normal and large diameter, and in two thicknesses, normal (Form A or C) and light (Form B or D). The thickness of a light-range washer is from ½ to ⅔ the thickness of a normal range washer.

Black Metal Washers: These washers are made from mild steel, and can be supplied in three size categories designated normal, large, and extra large diameters. The normal-diameter series is intended for bolts ranging from M5 to M68 (Form E washers), the large-diameter series for bolts ranging from M8 to M39 (Form F washers), and the extra large series for bolts from M5 to M39 (Form G washers). A protective finish can be specified by the purchaser in accordance with any appropriate British Standard.

Washer Designations: The Standard specifies the details that should be given when ordering or placing an inquiry for washers. These details are the general description, namely, bright or black washers; the nominal size of the bolt or screw involved, for example, M5; the designated form, for example, Form A or Form E; the dimensions of any chamfer required on bright washers; the number of the Standard BS 4320:1968 (1998), and coating information if required, with the number of the appropriate British Standard and the coating thickness needed. As an example, in the use of this information, the designation for a chamfered, normal-diameter series washer of normal-range thickness to suit a 12-mm diameter bolt would be: Bright washers M12 (Form A) chamfered to B.S. 4320.

British Standard Bright Metal Washers — Metric Series *BS 4320:1968 (1998)*

NORMAL DIAMETER SIZES												
Nominal Size of Bolt or Screw	Inside Diameter			Outside Diameter			Thickness					
							Form A (Normal Range)			Form B (Light Range)		
	Nom	Max	Min	Nom	Max	Min	Nom	Max	Min	Nom	Max	Min
M 1.0	1.1	1.25	1.1	2.5	2.5	2.3	0.3	0.4	0.2	…	…	…
M 1.2	1.3	1.45	1.3	3.0	3.0	2.8	0.3	0.4	0.2	…	…	…
(M 1.4)	1.5	1.65	1.5	3.0	3.0	2.8	0.3	0.4	0.2	…	…	…
M 1.6	1.7	1.85	1.7	4.0	4.0	3.7	0.3	0.4	0.2	…	…	…
M 2.0	2.2	2.35	2.2	5.0	5.0	4.7	0.3	0.4	0.2	…	…	…
(M 2.2)	2.4	2.55	2.4	5.0	5.0	4.7	0.5	0.6	0.4	…	…	…
M 2.5	2.7	2.85	2.7	6.5	6.5	6.2	0.5	0.6	0.4	…	…	…
M3	3.2	3.4	3.2	7	7	6.7	0.5	0.6	0.4	…	…	…
(M 3.5)	3.7	3.9	3.7	7	7	6.7	0.5	0.6	0.4	…	…	…
M4	4.3	4.5	4.3	9	9	8.7	0.8	0.9	0.7	…	…	…
(M 4.5)	4.8	5.0	4.8	9	9	8.7	0.8	0.9	0.7	…	…	…
M 5	5.3	5.5	5.3	10	10	9.7	1.0	1.1	0.9	…	…	…
M 6	6.4	6.7	6.4	12.5	12.5	12.1	1.6	1.8	1.4	0.8	0.9	0.7
(M 7)	7.4	7.7	7.4	14	14	13.6	1.6	1.8	1.4	0.8	0.9	0.7
M 8	8.4	8.7	8.4	17	17	16.6	1.6	1.8	1.4	1.0	1.1	0.9
M 10	10.5	10.9	10.5	21	21	20.5	2.0	2.2	1.8	1.25	1.45	1.05
M 12	13.0	13.4	13.0	24	24	23.5	2.5	2.7	2.3	1.6	1.80	1.40
(M 14)	15.0	15.4	15.0	28	28	27.5	2.5	2.7	2.3	1.6	1.8	1.4
M 16	17.0	17.4	17.0	30	30	29.5	3.0	3.3	2.7	2.0	2.2	1.8
(M 18)	19.0	19.5	19.0	34	34	33.2	3.0	3.3	2.7	2.0	2.2	1.8
M 20	21	21.5	21	37	37	36.2	3.0	3.3	2.7	2.0	2.2	1.8
(M 22)	23	23.5	23	39	39	38.2	3.0	3.3	2.7	2.0	2.2	1.8
M24	25	25.5	25	44	44	43.2	4.0	4.3	3.7	2.5	2.7	2.3
(M 27)	28	28.5	28	50	50	49.2	4.0	4.3	3.7	2.5	2.7	2.3
M30	31	31.6	31	56	56	55.0	4.0	4.3	3.7	2.5	2.7	2.3
(M 33)	34	34.6	34	60	60	59.0	5.0	5.6	4.4	3.0	3.3	2.7
M 36	37	37.6	37	66	66	65.0	5.0	5.6	4.4	3.0	3.3	2.7
(M 39)	40	40.6	40	72	72	71.0	6.0	6.6	5.4	3.0	3.3	2.7
LARGE DIAMETER SIZES												
Nominal Size of Bolt or Screw	Inside Diameter			Outside Diameter			Thickness					
							Form C (Normal Range)			Form D (Light Range)		
	Nom	Max	Min	Nom	Max	Min	Nom	Max	Min	Nom	Max	Min
M 4	4.3	4.5	4.3	10.0	10.0	9.7	0.8	0.9	0.7	…	…	…
M 5	5.3	5.5	5.3	12.5	12.5	12.1	1.0	1.1	0.9	…	…	…
M 6	6.4	6.7	6.4	14	14	13.6	1.6	1.8	1.4	0.8	0.9	0.7
M 8	8.4	8.7	8.4	21	21	20.5	1.6	1.8	1.4	1.0	1.1	0.9
M 10	10.5	10.9	10.5	24	24	23.5	2.0	2.2	1.8	1.25	1.45	1.05
M 12	13.0	13.4	13.0	28	28	27.5	2.5	2.7	2.3	1.6	1.8	1.4
(M 14)	15.0	15.4	15	30	30	29.5	2.5	2.7	2.3	1.6	1.8	1.4
M 16	17.0	17.4	17	34	34	33.2	3.0	3.3	2.7	2.0	2.2	1.8
(M 18)	19.0	19.5	19	37	37	36.2	3.0	3.3	2.7	2.0	2.2	1.8
M 20	21	21.5	21	39	39	38.2	3.0	3.3	2.7	2.0	2.2	1.8
(M 22)	23	23.5	23	44	44	43.2	3.0	3.3	2.7	2.0	2.2	1.8
M 24	25	25.5	25	50	50	49.2	4.0	4.3	3.7	2.5	2.7	2.3
(M 27)	28	28.5	28	56	56	55	4.0	4.3	3.7	2.5	2.7	2.3
M 30	31	31.6	31	60	60	59	4.0	4.3	3.7	2.5	2.7	2.3
(M 33)	34	34.6	34	66	66	65	5.0	5.6	4.4	3.0	3.3	2.7
M 36	37	37.6	37	72	72	71	5.0	5.6	4.4	3.0	3.3	2.7
(M 39)	40	40.6	40	77	77	76	6.0	6.6	5.4	3.0	3.3	2.7

All dimensions are in millimeters.

Nominal bolt or screw sizes shown in parentheses are nonpreferred.

British Standard Black Metal Washers — Metric Series *BS 4320:1968 (1998)*

Nom Bolt or Screw Size	Inside Diameter			Outside Diameter			Thickness		
	Nom	Max	Min	Nom	Max	Min	Nom	Max	Min
NORMAL DIAMETER SIZES (Form E)									
M 5	5.5	5.8	5.5	10.0	10.0	9.2	1.0	1.2	0.8
M 6	6.6	7.0	6.6	12.5	12.5	11.7	1.6	1.9	1.3
(M 7)	7.6	8.0	7.6	14.0	14.0	13.2	1.6	1.9	1.3
M 8	9.0	9.4	9.0	17	17	16.2	1.6	1.9	1.3
M 10	11.0	11.5	11.0	21	21	20.2	2.0	2.3	1.7
M 12	14	14.5	14	24	24	23.2	2.5	2.8	2.2
(M 14)	16	16.5	16	28	28	27.2	2.5	2.8	2.2
M 16	18	18.5	18	30	30	29.2	3.0	3.6	2.4
(M 18)	20	20.6	20	34	34	32.8	3.0	3.6	2.4
M 20	22	22.6	22	37	37	35.8	3.0	3.6	2.4
(M 22)	24	24.6	24	39	39	37.8	3.0	3.6	2.4
M 24	26	26.6	26	44	44	42.8	4	4.6	3.4
(M 27)	30	30.6	30	50	50	48.8	4	4.6	3.4
M 30	33	33.8	33	56	56	54.5	4	4.6	3.4
(M 33)	36	36.8	36	60	60	58.5	5	6.0	4.0
M 36	39	39.8	39	66	66	64.5	5	6.0	4.0
(M 39)	42	42.8	42	72	72	70.5	6	7.0	5.0
M 42	45	45.8	45	78	78	76.5	7	8.2	5.8
(M 45)	48	48.8	48	85	85	83	7	8.2	5.8
M 48	52	53	52	92	92	90	8	9.2	6.8
(M 52)	56	57	56	98	98	96	8	9.2	6.8
M 56	62	63	62	105	105	103	9	10.2	7.8
(M 60)	66	67	66	110	110	108	9	10.2	7.8
M 64	70	71	70	115	115	113	9	10.2	7.8
(M 68)	74	75	74	120	120	118	10	11.2	8.8
LARGE DIAMETER SIZES (Form F)									
M 8	9	9.4	9.0	21	21	20.2	1.6	1.9	1.3
M 10	11	11.5	11	24	24	23.2	2	2.3	1.7
M 12	14	14.5	14	28	28	27.2	2.5	2.8	2.2
(M 14)	16	16.5	16	30	30	29.2	2.5	2.8	2.2
M 16	18	18.5	18	34	34	32.8	3	3.6	2.4
(M 18)	20	20.6	20	37	37	35.8	3	3.6	2.4
M 20	22	22.6	22	39	39	37.8	3	3.6	2.4
(M 22)	24	24.6	24	44	44	42.8	3	3.6	2.4
M 24	26	26.6	26	50	50	48.8	4	4.6	3.4
(M 27)	30	30.6	30	56	56	54.5	4	4.6	3.4
M 30	33	33.8	33	60	60	58.5	4	4.6	3.4
(M 33)	36	36.8	36	66	66	64.5	5	6.0	4
M 36	39	39.8	39	72	72	70.5	5	6.0	4
(M 39)	42	42.8	42	77	77	75.5	6	7	5
EXTRA LARGE DIAMETER SIZES (Form G)									
M 5	5.5	5.8	5.5	15	15	14.2	1.6	1.9	1.3
M 6	6.6	7.0	6.6	18	18	17.2	2	2.3	1.7
(M 7)	7.6	8.0	7.6	21	21	20.2	2	2.3	1.7
M 8	9	9.4	9.0	24	24	23.2	2	2.3	1.7
M 10	11	11.5	11.0	30	30	29.2	2.5	2.8	2.2
M 12	14	14.5	14.0	36	36	34.8	3	3.6	2.4
(M 14)	16	16.5	16.0	42	42	40.8	3	3.6	2.4
M 16	18	18.5	18	48	48	46.8	4	4.6	3.4
(M 18)	20	20.6	20	54	54	52.5	4	4.6	3.4
M 20	22	22.6	22	60	60	58.5	5	6.0	4
(M 22)	24	24.6	24	66	66	64.5	5	6.0	4
M 24	26	26.6	26	72	72	70.5	6	7	5
(M 27)	30	30.6	30	81	81	79	6	7	5
M 30	33	33.8	33	90	90	88	8	9.2	6.8
(M 33)	36	36.8	36	99	99	97	8	9.2	6.8
M 36	39	39.8	39	108	108	106	10	11.2	8.8
(M39)	42	42.8	42	117	117	115	10	11.2	8.8

All dimensions are in millimeters.

Nominal bolt or screw sizes shown in parentheses are nonpreferred.

MACHINE SCREWS

American National Standard Machine Screws and Machine Screw Nuts.—This Standard ANSI B18.6.3 covers both slotted and recessed head machine screws. Dimensions of various types of slotted machine screws, machine screw nuts, and header points are given in Tables 1 through 12. The Standard also covers flat trim head, oval trim head and drilled fillister head machine screws and gives cross recess dimensions and gaging dimensions for all types of machine screw heads. Information on metric machine screws B18.6.7M is given beginning on page 1577.

Threads: Except for sizes 0000, 000, and 00, machine screw threads may be either Unified Coarse (UNC) and Fine thread (UNF) Class 2A (see *American Standard for Unified Screw Threads* starting on page 1712) or UNRC and UNRF Series, at option of manufacturer. Thread dimensions for sizes 0000, 000, and 00 are given in Table 7 on page 1573.

Threads for hexagon machine screw nuts may be either UNC or UNF, Class 2B, and for square machine screw nuts are UNC Class 2B.

Length of thread: Machine screws of sizes No. 5 and smaller with nominal lengths equal to 3 diameters and shorter have full form threads extending to within 1 pitch (thread) of the bearing surface of the head, or closer, if practicable. Nominal lengths greater than 3 diameters, up to and including 1⅛ inch, have full form threads extending to within two pitches (threads) of the bearing surface of the head, or closer, if practicable. Unless otherwise specified, screws of longer nominal length have a minimum length of full form thread of 1.00 inch.Machine screws of sizes No. 6 and larger with nominal length equal to 3 diameters and shorter have full form threads extending to within 1 pitch (thread) of the bearing surface of the head, or closer, if practicable. Nominal lengths greater than 3 diameters, up to and including 2 inches, have full form threads extending to within 2 pitches (threads) of the bearing surface of the head, or closer, if practicable. Screws of longer nominal length, unless otherwise specified, have a minimum length of full form thread of 1.50 inches.

Table 1. Square and Hexagon Machine Screw Nuts *ANSI B18.6.3-1972 (R1991)*

F H F H G G_1 30 30 Optional; See Note

Nom. Size	Basic Dia.	Basic F	Max. F	Min. F	Max. G	Min. G	Max. G_1	Min. G_1	Max. H	Min. H
0	0.0600	5⁄32	0.156	0.150	0.221	0.206	0.180	0.171	0.050	0.043
1	0.0730	5⁄32	0.156	0.150	0.221	0.206	0.180	0.171	0.050	0.043
2	0.0860	3⁄16	0.188	0.180	0.265	0.247	0.217	0.205	0.066	0.057
3	0.0990	3⁄16	0.188	0.180	0.265	0.247	0.217	0.205	0.066	0.057
4	0.1120	1⁄4	0.250	0.241	0.354	0.331	0.289	0.275	0.098	0.087
5	0.1250	5⁄16	0.312	0.302	0.442	0.415	0.361	0.344	0.114	0.102
6	0.1380	5⁄16	0.312	0.302	0.442	0.415	0.361	0.344	0.114	0.102
8	0.1640	11⁄32	0.344	0.332	0.486	0.456	0.397	0.378	0.130	0.117
10	0.1900	3⁄8	0.375	0.362	0.530	0.497	0.433	0.413	0.130	0.117
12	0.2160	7⁄16	0.438	0.423	0.619	0.581	0.505	0.482	0.161	0.148
1⁄4	0.2500	7⁄16	0.438	0.423	0.619	0.581	0.505	0.482	0.193	0.178
5⁄16	0.3125	9⁄16	0.562	0.545	0.795	0.748	0.650	0.621	0.225	0.208
3⁄8	0.3750	5⁄8	0.625	0.607	0.884	0.833	0.722	0.692	0.257	0.239

All dimensions in inches. Hexagon machine screw nuts have tops flat and chamfered. Diameter of top circle should be the maximum width across flats within a tolerance of minus 15 per cent. Bottoms are flat but may be chamfered if so specified. Square machine screw nuts have tops and bottoms flat without chamfer.

Diameter of body: The diameter of machine screw bodies is not less than Class 2A thread minimum pitch diameter nor greater than the basic major diameter of the thread. Cross-recessed trim head machine screws not threaded to the head have an 0.062 in. minimum length shoulder under the head with diameter limits as specified in the dimensional tables in the standard.

Designation: Machine screws are designated by the following data in the sequence shown: Nominal size (number, fraction, or decimal equivalent); threads per inch; nominal length (fraction or decimal equivalent); product name, including head type and driving provision; header point, if desired; material; and protective finish, if required. For example:

¼ – 20 × 1¼ Slotted Pan Head Machine Screw, Steel, Zinc Plated

6 – 32 × ¾ Type IA Cross Recessed Fillister Head Machine Screw, Brass

Machine screw nuts are designated by the following data in the sequence shown: Nominal size (number, fraction, or decimal equivalent); threads per inch; product name; material; and protective finish, if required. For example:

10 – 24 Hexagon Machine Screw Nut, Steel, Zinc Plated

0.138 – 32 Square Machine Screw Nut, Brass

Table 2. American National Standard Slotted 100-Degree Flat Countersunk Head Machine Screws *ANSI B18.6.3-1972 (R1977)*

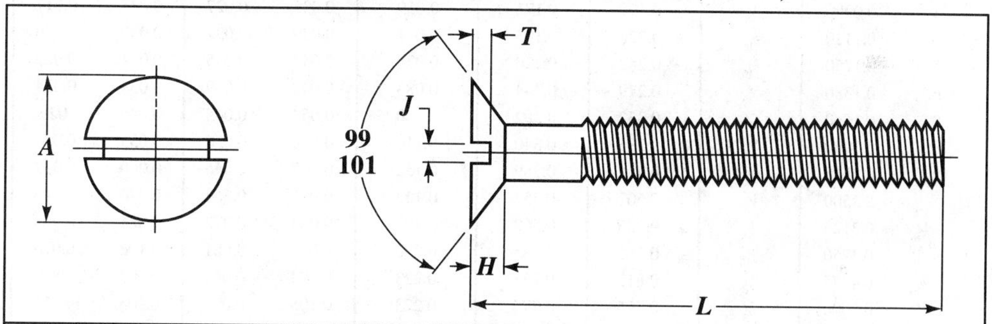

Nominal Size[a] or Basic Screw Dia.		Head Dia., *A*		Head Height, *H*	Slot Width, *J*		Slot Depth, *T*	
		Max., Edge Sharp	Min., Edge Rounded or Flat	Ref.	Max.	Min.	Max.	Min.
0000	0.0210	0.043	0.037	0.009	0.008	0.005	0.008	0.004
000	0.0340	0.064	0.058	0.014	0.012	0.008	0.011	0.007
00	0.0470	0.093	0.085	0.020	0.017	0.010	0.013	0.008
0	0.0600	0.119	0.096	0.026	0.023	0.016	0.013	0.008
1	0.0730	0.146	0.120	0.031	0.026	0.019	0.016	0.010
2	0.0860	0.172	0.143	0.037	0.031	0.023	0.019	0.012
3	0.0990	0.199	0.167	0.043	0.035	0.027	0.022	0.014
4	0.1120	0.225	0.191	0.049	0.039	0.031	0.024	0.017
6	0.1380	0.279	0.238	0.060	0.048	0.039	0.030	0.022
8	0.1640	0.332	0.285	0.072	0.054	0.045	0.036	0.027
10	0.1900	0.385	0.333	0.083	0.060	0.050	0.042	0.031
¼	0.2500	0.507	0.442	0.110	0.075	0.064	0.055	0.042
5/16	0.3125	0.635	0.556	0.138	0.084	0.072	0.069	0.053
⅜	0.3750	0.762	0.670	0.165	0.094	0.081	0.083	0.065

[a] When specifying nominal size in decimals, zeros preceding the decimal point and in the fourth decimal place are omitted.

All dimensions are in inches.

Table 3. American National Standard Slotted Flat Countersunk Head and Close Tolerance 100-Degree Flat Countersunk Head Machine Screws *ANSI B18.6.3-1972 (R1991)*

SLOTTED FLAT COUNTERSUNK HEAD TYPE

Nominal Size[a] or Basic Screw Dia.		Max., L[b]	Head Dia., A Max., Edge Sharp	Head Dia., A Min., Edge[c]	Head Height, H Ref.	Slot Width, J Max.	Slot Width, J Min.	Slot Depth, T Max.	Slot Depth, T Min.
0000	0.0210	....	0.043	0.037	0.011	0.008	0.004	0.007	0.003
000	0.0340	....	0.064	0.058	0.016	0.011	0.007	0.009	0.005
00	0.0470	....	0.093	0.085	0.028	0.017	0.010	0.014	0.009
0	0.0600	1/8	0.119	0.099	0.035	0.023	0.016	0.015	0.010
1	0.0730	1/8	0.146	0.123	0.043	0.026	0.019	0.019	0.012
2	0.0860	1/8	0.172	0.147	0.051	0.031	0.023	0.023	0.015
3	0.0990	1/8	0.199	0.171	0.059	0.035	0.027	0.027	0.017
4	0.1120	3/16	0.225	0.195	0.067	0.039	0.031	0.030	0.020
5	0.1250	3/16	0.252	0.220	0.075	0.043	0.035	0.034	0.022
6	0.1380	3/16	0.279	0.244	0.083	0.048	0.039	0.038	0.024
8	0.1640	1/4	0.332	0.292	0.100	0.054	0.045	0.045	.029
10	0.1900	5/16	0.385	0.340	0.116	0.060	0.050	0.053	0.034
12	0.2160	3/8	0.438	0.389	0.132	0.067	0.056	0.060	0.039
1/4	0.2500	7/16	0.507	0.452	0.153	0.075	0.064	0.070	0.046
5/16	0.3125	1/2	0.635	0.568	0.191	0.084	0.072	0.088	0.058
3/8	0.3750	9/16	0.762	0.685	0.230	0.094	0.081	0.106	0.070
7/16	0.4375	5/8	0.812	0.723	0.223	0.094	0.081	0.103	.066
1/2	0.5000	3/4	0.875	0.775	0.223	0.106	0.091	0.103	0.065
9/16	0.5625	...	1.000	0.889	0.260	0.118	0.102	0.120	0.077
5/8	0.6250	...	1.125	1.002	0.298	0.133	0.116	0.137	0.088
3/4	0.7500	...	1.375	1.230	0.372	0.149	0.131	0.171	0.111

[a] When specifying nominal size in decimals, zeros preceding the decimal point and in the fourth decimal place are omitted.

[b] These lengths or shorter are undercut.

[c] May be rounded or flat.

CLOSE TOLERANCE 100-DEGREE FLAT COUNTERSUNK HEAD TYPE

Nominal Size[a] or Basic Screw Dia.		Head Diameter, A Max., Edge Sharp	Head Diameter, A Min., Edge[c]	Head Height, H Ref.	Slot Width, J Max.	Slot Width, J Min.	Slot Depth, T Max.	Slot Depth, T Min.
4	0.1120	0.225	0.191	0.049	0.039	0.031	0.024	0.017
6	0.1380	0.279	0.238	0.060	0.048	0.039	0.030	0.022
8	0.1640	0.332	0.285	0.072	0.054	0.045	0.036	0.027
10	0.1900	0.385	0.333	0.083	0.060	0.050	0.042	0.031
1/4	0.2500	0.507	0.442	0.110	0.075	0.064	0.055	0.042
5/16	0.3125	0.635	0.556	0.138	0.084	0.072	0.069	0.053
3/8	0.3750	0.762	0.670	0.165	0.094	0.081	0.083	0.065
7/16	0.4375	0.890	0.783	0.193	0.094	0.081	0.097	0.076
1/2	0.5000	1.017	0.897	0.221	0.106	0.091	0.111	0.088
9/16	0.5625	1.145	1.011	0.249	0.118	0.102	0.125	0.099
5/8	0.6250	1.272	1.124	0.276	0.133	0.116	0.139	0.111

All dimensions are in inches.

Table 4. American National Standard Slotted Undercut Flat Countersunk Head and Plain and Slotted Hex Washer Head Machine Screws *ANSI B18.6.3-1972 (R1991)*

SLOTTED UNDERCUT FLAT COUNTERSUNK HEAD TYPE

Nominal Size[a] or Basic Screw Dia.		Max., L[b]	Head Dia., A		Head Height, H		Slot Width, J		Slot Depth, T	
			Max., Edge Sharp	Min., Edge Rnded. or Flat	Max.	Min.	Max.	Min.	Max.	Min.
0	0.0600	1/8	0.119	0.099	0.025	0.018	0.023	0.016	0.011	0.007
1	0.0730	1/8	0.146	0.123	0.031	0.023	0.026	0.019	0.014	0.009
2	0.0860	1/8	0.172	0.147	0.036	0.028	0.031	0.023	0.016	0.011
3	0.0990	1/8	0.199	0.171	0.042	0.033	0.035	0.027	0.019	0.012
4	0.1120	3/16	0.225	0.195	0.047	0.038	0.039	0.031	0.022	0.014
5	0.1250	3/16	0.252	0.220	0.053	0.043	0.043	0.035	0.024	0.016
6	0.1380	3/16	0.279	0.244	0.059	0.048	0.048	0.039	0.027	0.017
8	0.1640	1/4	0.332	0.292	0.070	0.058	0.054	0.045	0.032	0.021
10	0.1900	5/16	0.385	0.340	0.081	0.068	0.060	0.050	0.037	0.024
12	0.2160	3/8	0.438	0.389	0.092	0.078	0.067	0.056	0.043	0.028
1/4	0.2500	7/16	0.507	0.452	0.107	0.092	0.075	0.064	0.050	0.032
5/16	0.3125	1/2	0.635	0.568	0.134	0.116	0.084	0.072	0.062	0.041
3/8	0.3750	9/16	0.762	0.685	0.161	0.140	0.094	0.081	0.075	0.049
7/16	0.4375	5/8	0.812	0.723	0.156	0.133	0.094	0.081	0.072	0.045
1/2	0.5000	3/4	0.875	0.775	0.156	0.130	0.106	0.091	0.072	0.046

[a] When specifying nominal size in decimals, zeros preceding the decimal point and in the fourth decimal place are omitted.

[b] These lengths or shorter are undercut.

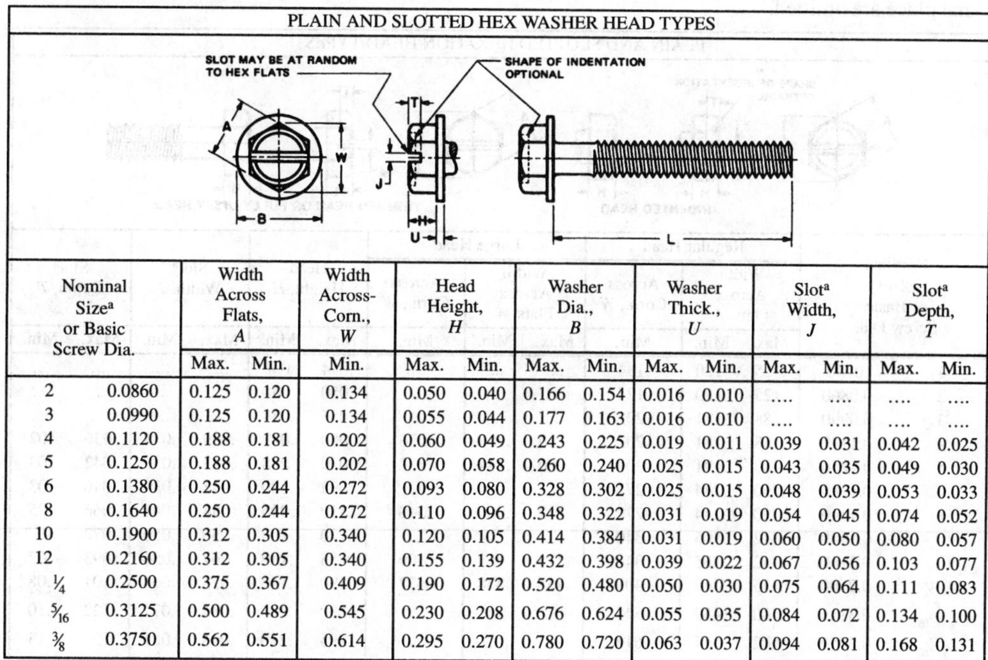

PLAIN AND SLOTTED HEX WASHER HEAD TYPES

Nominal Size[a] or Basic Screw Dia.		Width Across Flats, A		Width Across-Corn., W	Head Height, H		Washer Dia., B		Washer Thick., U		Slot[a] Width, J		Slot[a] Depth, T	
		Max.	Min.	Min.	Max.	Min.	Max.	Min.	Max.	Min.	Max.	Min.	Max.	Min.
2	0.0860	0.125	0.120	0.134	0.050	0.040	0.166	0.154	0.016	0.010	....	....	....	....
3	0.0990	0.125	0.120	0.134	0.055	0.044	0.177	0.163	0.016	0.010	....	....	....	....
4	0.1120	0.188	0.181	0.202	0.060	0.049	0.243	0.225	0.019	0.011	0.039	0.031	0.042	0.025
5	0.1250	0.188	0.181	0.202	0.070	0.058	0.260	0.240	0.025	0.015	0.043	0.035	0.049	0.030
6	0.1380	0.250	0.244	0.272	0.093	0.080	0.328	0.302	0.025	0.015	0.048	0.039	0.053	0.033
8	0.1640	0.250	0.244	0.272	0.110	0.096	0.348	0.322	0.031	0.019	0.054	0.045	0.074	0.052
10	0.1900	0.312	0.305	0.340	0.120	0.105	0.414	0.384	0.031	0.019	0.060	0.050	0.080	0.057
12	0.2160	0.312	0.305	0.340	0.155	0.139	0.432	0.398	0.039	0.022	0.067	0.056	0.103	0.077
1/4	0.2500	0.375	0.367	0.409	0.190	0.172	0.520	0.480	0.050	0.030	0.075	0.064	0.111	0.083
5/16	0.3125	0.500	0.489	0.545	0.230	0.208	0.676	0.624	0.055	0.035	0.084	0.072	0.134	0.100
3/8	0.3750	0.562	0.551	0.614	0.295	0.270	0.780	0.720	0.063	0.037	0.094	0.081	0.168	0.131

[a] Unless otherwise specified, hexagon washer head machine screws are not slotted.

All dimensions are in inches.

Table 5. American National Standard Slotted Truss Head and Plain and Slotted Hexagon Head Machine Screws *ANSI B18.6.3-1972 (R1991)*

SLOTTED TRUSS HEAD TYPE

Nominal Size[a] or Basic Screw Dia.		Head Dia., *A*		Head Height, *H*		Head Radius, *R*	Slot Width, *J*		Slot Depth, *T*	
		Max.	Min.	Max.	Min.	Max.	Max.	Min.	Max.	Min.
0000	0.0210	0.049	0.043	0.014	0.010	0.032	0.009	0.005	0.009	0.005
000	0.0340	0.077	0.071	0.022	0.018	0.051	0.013	0.009	0.013	0.009
00	0.0470	0.106	0.098	0.030	0.024	0.070	0.017	0.010	0.018	0.012
0	0.0600	0.131	0.119	0.037	0.029	0.087	0.023	0.016	0.022	0.014
1	0.0730	0.164	0.149	0.045	0.037	0.107	0.026	0.019	0.027	0.018
2	0.0860	0.194	0.180	0.053	0.044	0.129	0.031	0.023	0.031	0.022
3	0.0990	0.226	0.211	0.061	0.051	0.151	0.035	0.027	0.036	0.026
4	0.1120	0.257	0.241	0.069	0.059	0.169	0.039	0.031	0.040	0.030
5	0.1250	0.289	0.272	0.078	0.066	0.191	0.043	0.035	0.045	0.034
6	0.1380	0.321	0.303	0.086	0.074	0.211	0.048	0.039	0.050	0.037
8	0.1640	0.384	0.364	0.102	0.088	0.254	0.054	0.045	0.058	0.045
10	0.1900	0.448	0.425	0.118	0.103	0.283	0.060	0.050	0.068	0.053
12	0.2160	0.511	0.487	0.134	0.118	0.336	0.067	0.056	0.077	0.061
1/4	0.2500	0.573	0.546	0.150	0.133	0.375	0.075	0.064	0.087	0.070
5/16	0.3125	0.698	0.666	0.183	0.162	0.457	0.084	0.072	0.106	0.085
3/8	0.3750	0.823	0.787	0.215	0.191	0.538	0.094	0.081	0.124	0.100
7/16	0.4375	0.948	0.907	0.248	0.221	0.619	0.094	0.081	0.142	0.116
1/2	0.5000	1.073	1.028	0.280	0.250	0.701	0.106	0.091	0.161	0.131
9/16	0.5625	1.198	1.149	0.312	0.279	0.783	0.118	0.102	0.179	0.146
5/8	0.6250	1.323	1.269	0.345	0.309	0.863	0.133	0.116	0.196	0.162
3/4	0.7500	1.573	1.511	0.410	0.368	1.024	0.149	0.131	0.234	0.182

[a] Where specifying nominal size in decimals, zeros preceding decimal points and in the fourth decimal place are omitted.

PLAIN AND SLOTTED HEXAGON HEAD TYPES

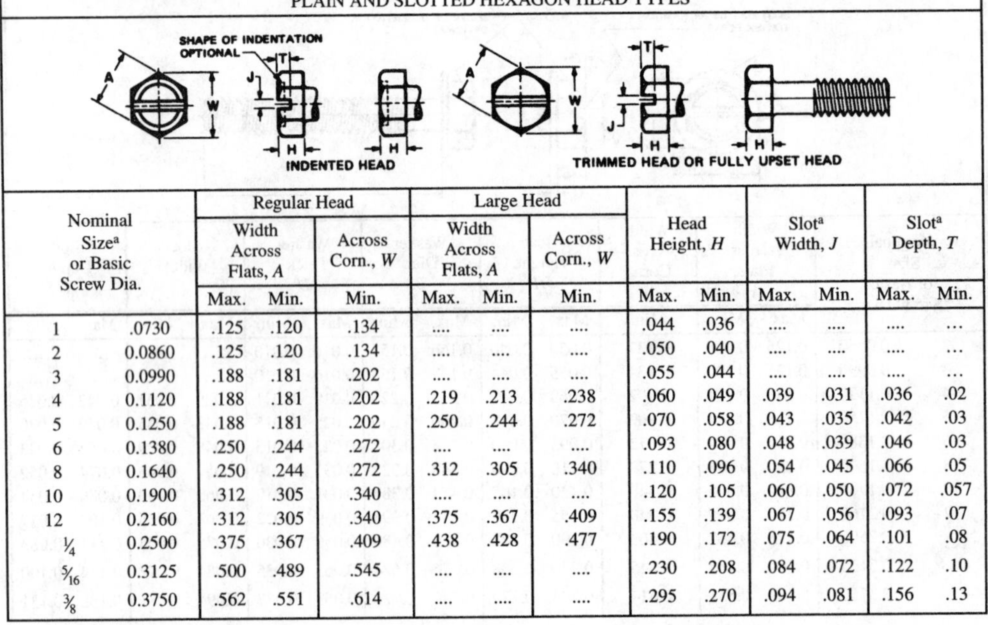

Nominal Size[a] or Basic Screw Dia.		Regular Head Width Across Flats, *A*		Regular Head Across Corn., *W*	Large Head Width Across Flats, *A*		Large Head Across Corn., *W*	Head Height, *H*		Slot[a] Width, *J*		Slot[a] Depth, *T*	
		Max.	Min.	Min.	Max.	Min.	Min.	Max.	Min.	Max.	Min.	Max.	Min.
1	.0730	.125	.120	.134	....	....	....	.044	.036	....	....	....	...
2	0.0860	.125	.120	.134	....	....	....	.050	.040	....	....	....	...
3	0.0990	.188	.181	.202	....	....	....	.055	.044	....	....	....	...
4	0.1120	.188	.181	.202	.219	.213	.238	.060	.049	.039	.031	.036	.02
5	0.1250	.188	.181	.202	.250	.244	.272	.070	.058	.043	.035	.042	.03
6	0.1380	.250	.244	.272	....	....	....	.093	.080	.048	.039	.046	.03
8	0.1640	.250	.244	.272	.312	.305	.340	.110	.096	.054	.045	.066	.05
10	0.1900	.312	.305	.340	....	....	....	.120	.105	.060	.050	.072	.057
12	0.2160	.312	.305	.340	.375	.367	.409	.155	.139	.067	.056	.093	.07
1/4	0.2500	.375	.367	.409	.438	.428	.477	.190	.172	.075	.064	.101	.08
5/16	0.3125	.500	.489	.545	....	....	....	.230	.208	.084	.072	.122	.10
3/8	0.3750	.562	.551	.614	....	....	....	.295	.270	.094	.081	.156	.13

[a] Unless otherwise specified, hexagon head machine screws are not slotted.

All dimensions are in inches.

Table 6. American National Standard Slotted Pan Head Machine Screws *ANSI B18.6.3-1972 (R1991)*

Nominal Size[a] or Basic Screw Dia.		Head Dia., *A*		Head Height, *H*		Head Radius, *R*	Slot Width, *J*		Slot Depth, *T*	
		Max.	Min.	Max.	Min.	Max.	Max.	Min.	Max.	Min.
0000	0.0210	.042	.036	.016	.010	.007	.008	.004	.008	.004
000	0.0340	.066	.060	.023	.017	.010	.012	.008	.012	.008
00	0.0470	.090	.082	.032	.025	.015	.017	.010	.016	.010
0	0.0600	.116	.104	.039	.031	.020	.023	.016	.022	.014
1	0.0730	.142	.130	.046	.038	.025	.026	.019	.027	.018
2	0.0860	.167	.155	.053	.045	.035	.031	.023	.031	.022
3	0.0990	.193	.180	.060	.051	.037	.035	.027	.036	.026
4	0.1120	.219	.205	.068	.058	.042	.039	.031	.040	.030
5	0.1250	.245	.231	.075	.065	.044	.043	.035	.045	.034
6	0.1380	.270	.256	.082	.072	.046	.048	.039	.050	.037
8	0.1640	.322	.306	.096	.085	.052	.054	.045	.058	.045
10	0.1900	.373	.357	.110	.099	.061	.060	.050	.068	.053
12	0.2160	.425	.407	.125	.112	.078	.067	.056	.077	.061
¼	0.2500	.492	.473	.144	.130	.087	.075	.064	.087	.070
5/16	0.3125	.615	.594	.178	.162	.099	.084	.072	.106	.085
⅜	0.3750	.740	.716	.212	.195	.143	.094	.081	.124	.100
7/16	0.4375	.863	.837	.247	.228	.153	.094	.081	.142	.116
½	0.5000	.987	.958	.281	.260	.175	.106	.091	.161	.131
9/16	0.5625	1.041	1.000	.315	.293	.197	.118	.102	.179	.146
⅝	0.6250	1.172	1.125	.350	.325	.219	.133	.116	.197	.162
¾	0.7500	1.435	1.375	.419	.390	.263	.149	.131	.234	.192

[a] Where specifying nominal size in decimals, zeros preceding decimal and in the fourth decimal place are omitted.

All dimensions are in inches.

Table 7. Nos. 0000, 000 and 00 Threads *ANSI B18.6.3-1972 (R1991) Appendix*

Nominal Size[a] and Threads Per Inch	Series Designat.	External[b]							Internal[c]				
		Class	Major Diameter		Pitch Diameter			Minor Dia.	Class	Pitch Diameter			Major Dia.
			Max.	Min.	Max.	Min.	Tol.			Min.	Max.	Tol.	Min.
0000-160 or 0.0210-160	NS	2	.0210	.0195	.0169	.0158	.0011	.0128	2	.0169	.0181	.0012	.0210
000-120 or 0.0340-120	NS	2	.0340	.0325	.0286	0.272	.0014	.0232	2	.0286	.0300	.0014	.034
00-90 or 0.0470-90	NS	2	.0470	.0450	.0398	.0382	.0016	.0326	2	.0398	.0414	.0016	.047
00-96 or 0.0470-96	NS	2	.0470	.0450	.0402	.0386	.0016	.0334	2	.0402	.0418	.0016	.047

[a] Where specifying nominal size in decimals, zeros preceding decimal and in the fourth decimal place are omitted.

[b] There is no allowance provided on the external threads.

[c] The minor diameter limits for internal threads are not specified, they being determined by the amount of thread engagement necessary to satisfy the strength requirements and tapping performance in the intended application.

All dimensions are in inches.

Table 8. American National Standard Slotted Fillister and Slotted Drilled Fillister Head Machine Screws *ANSI B18.6.3-1972 (R1991)*

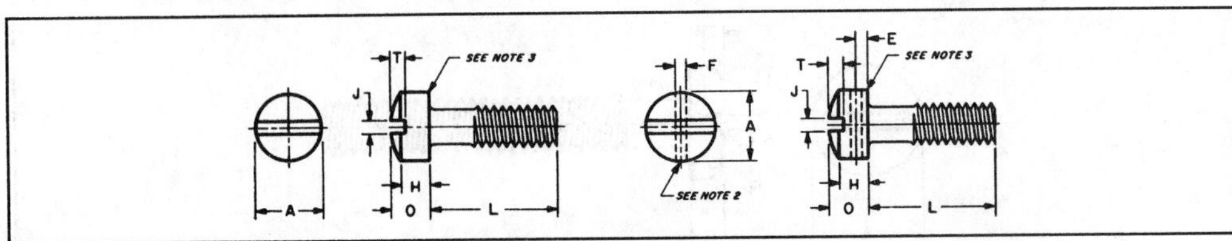

SLOTTED FILLISTER HEAD TYPE

Nominal Size[1] or Basic Screw Dia.		Head Dia., *A*		Head Side Height, *H*		Total Head Height, *O*		Slot Width, *J*		Slot Depth, *T*	
		Max.	Min.	Max.	Min.	Max.	Min.	Max	Min.	Max.	Min.
0000	0.0210	.038	.032	.019	.011	.025	.15	.008	.004	.012	.006
000	0.0340	.059	.053	.029	.021	.035	.027	.012	.006	.017	.011
00	0.0470	.082	.072	.037	.028	.047	.039	.017	.010	.022	.015
0	0.0600	.096	.083	.043	.038	.055	.047	.023	.016	.025	.015
1	0.0730	.118	.104	.053	.045	.066	.058	.026	.019	.031	.020
2	0.0860	.140	.124	.062	.053	.083	.066	.031	.023	.037	.025
3	0.0990	.161	.145	.070	.061	.095	.077	.035	.027	.043	.030
4	0.1120	.183	.166	.079	.069	.107	.088	.039	.031	.048	.035
5	0.1250	.205	.187	.088	.078	.120	.100	.043	.035	.054	.040
6	0.1380	.226	.208	.096	.086	.132	.111	.048	.039	.060	.045
8	0.1640	.270	.250	.113	.102	.156	.133	.054	.045	.071	.054
10	0.1900	.313	.292	.130	.118	.180	.156	.060	.050	.083	.064
12	0.2160	.357	.334	.148	.134	.205	.178	.067	.056	.094	.074
¼	0.2500	.414	.389	.170	.155	.237	.207	.075	.064	.109	.087
5⁄16	0.3125	.518	.490	.211	.194	.295	.262	.084	.072	.137	.110
⅜	0.3750	.622	.590	.253	.233	.355	.315	.094	.081	.164	.133
7⁄16	0.4375	.625	.589	.265	.242	.368	.321	.094	.081	.170	.135
½	0.5000	.750	.710	.297	.273	.412	.362	.106	.091	.190	.151
9⁄16	0.5625	.812	.768	.336	.308	.466	.410	.118	.102	.214	.172
⅝	0.6250	.875	.827	.375	.345	.521	.461	.133	.116	.240	.193
¾	0.7500	1.000	.945	.441	.406	.612	.542	.149	.131	.281	.226

SLOTTED DRILLED FILLISTER HEAD TYPE

Nominal Size[1] or Basic Screw Dia.		Head Dia., *A*		Head Side Height, *H*		Total Head Height, *O*		Slot Width, *J*		Slot Depth, *T*		Drilled Hole Locat., *E*	Drilled Hole. Dia., *F*
		Max.	Min.	Max.	Min.	Max.	Min.	Max.	Min.	Max.	Min.	Basic	Basic
2	0.0860	.140	.124	.062	.055	.083	.070	.031	.023	.030	.022	.026	.031
3	0.0990	.161	.145	.070	.064	.095	.082	.035	.027	.034	.026	.030	.037
4	0.1120	.183	.166	.079	.072	.107	.094	.039	.031	.038	.030	.035	.037
5	0.1250	.205	.187	.088	.081	.120	.106	.043	.035	.042	.033	.038	.046
6	0.1380	.226	.208	.096	.089	.132	.118	.048	.039	.045	.035	.043	.046
8	0.1640	.270	.250	.113	.106	.156	.141	.054	.045	.065	.054	.043	.046
10	0.1900	.313	.292	.130	.123	.180	.165	.060	.050	.075	.064	.043	.046
12	0.2160	.357	.334	.148	.139	.205	.188	.067	.056	.087	.074	.053	.046
¼	0.2500	.414	.389	.170	.161	.237	.219	.075	.064	.102	.087	.062	.062
5⁄16	0.3125	.518	.490	.211	.201	.295	.276	.084	.072	.130	.110	.078	.070
⅜	0.3750	.622	.590	.253	.242	.355	.333	.094	.081	.154	.134	.094	.070

All dimensions are in inches.

[1]Where specifying nominal size in decimals, zeros preceding decimal points and in the fourth decimal place are omitted.

[2]Drilled hole shall be approximately perpendicular to the axis of slot and may be permitted to break through bottom of the slot. Edges of the hole shall be free from burrs.

[3]A slight rounding of the edges at periphery of head is permissible provided the diameter of the bearing circle is equal to no less than 90 per cent of the specified minimum head diameter.

Table 9. American National Standard Slotted Oval Countersunk Head Machine Screws *ANSI B18.6.3-1972 (R1991)*

Nominal Size[a]	or Basic Screw Dia.	Max L^b	Head Dia., A: Max., Edge Sharp	Head Dia., A: Min., Edge Rnded. or Flat	Head Side Height, H, Ref.	Total Head Height, O: Max.	Total Head Height, O: Min.	Slot Width, J: Max.	Slot Width, J: Min.	Slot Depth, T: Max.	Slot Depth, T: Min.
00	0.0470	...	.093	.085	.028	.042	.034	.017	.010	.023	.016
0	0.0600	1/8	.119	.099	.035	.056	.041	.023	.016	.030	.025
1	0.0730	1/8	.146	.123	.043	.068	.052	.026	.019	.038	.031
2	0.0860	1/8	.172	.147	.051	.080	.063	.031	.023	.045	.037
3	0.0990	1/8	.199	.171	.059	.092	.073	.035	.027	.052	.043
4	0.1120	3/16	.225	.195	.067	.104	.084	.039	.031	.059	.049
5	0.1250	3/16	.252	.220	.075	.116	.095	.043	.035	.067	.055
6	0.1380	3/16	.279	.244	.083	.128	.105	.048	.039	.074	.060
8	0.1640	1/4	.332	.292	.100	.152	.126	.054	.045	.088	.072
10	0.1900	5/16	.385	.340	.116	.176	.148	.060	.050	.103	.084
12	0.2160	3/8	.438	.389	.132	.200	.169	.067	.056	.117	.096
1/4	0.2500	7/16	.507	.452	.153	.232	.197	.075	.064	.136	.112
5/16	0.3125	1/2	.635	.568	.191	.290	.249	.084	.072	.171	.141
3/8	0.3750	9/16	.762	.685	.230	.347	.300	.094	.081	.206	.170
7/16	0.4375	5/8	.812	.723	.223	.345	.295	.094	.081	.210	.174
1/2	0.5000	3/4	.875	.775	.223	.354	.299	.106	.091	.216	.176
9/16	0.5625	...	1.000	.889	.260	.410	.350	.118	.102	.250	.207
5/8	0.6250	...	1.125	1.002	.298	.467	.399	.133	.116	.285	.235
3/4	0.7500	...	1.375	1.230	.372	.578	.497	.149	.131	.353	.293

[a] When specifying nominal size in decimals, zeros preceding decimal points and in the fourth decimal place are omitted.

[b] These lengths or shorter are undercut.

All dimensions are in inches.

Table 10. American National Standard Header Points for Machine Screws before Threading *ANSI B18.6.3-1972 (R1991)*

Nom. Size	Threads per Inch	Max. P	Min. P	Max. L
2	56	0.057	0.050	1/2
	64	0.060	0.053	1/2
4	40	0.074	0.065	1/2
	48	0.079	0.070	1/2
5	40	0.086	0.076	1/2
	44	0.088	0.079	1/2
6	32	0.090	0.080	3/4
	40	0.098	0.087	3/4
8	32	0.114	0.102	1
	36	0.118	0.106	1
10	24	0.125	0.112	1 1/4
	32	0.138	0.124	1 1/4
12	24	0.149	0.134	1 3/8
	28	0.156	0.141	1 3/8
1/4	20	0.170	0.153	1 1/2
	28	0.187	0.169	1 1/2
5/16	18	0.221	0.200	1 1/2
	24	0.237	0.215	1 1/2
3/8	16	0.270	0.244	1 1/2
	24	0.295	0.267	1 1/2
7/16	14	0.316	0.287	1 1/2
	20	0.342	0.310	1 1/2
1/2	13	0.367	0.333	1 1/2
	20	0.399	0.362	1 1/2

All dimensions in inches. Edges of point may be rounded and end of point need not be flat nor perpendicular to shank. Machine screws normally have plain sheared ends but when specified may have header points, as shown above.

Table 11. American National Standard Slotted Binding Head and Slotted Undercut Oval Countersunk Head Machine Screws *ANSI B18.6.3-1972 (R1991)*

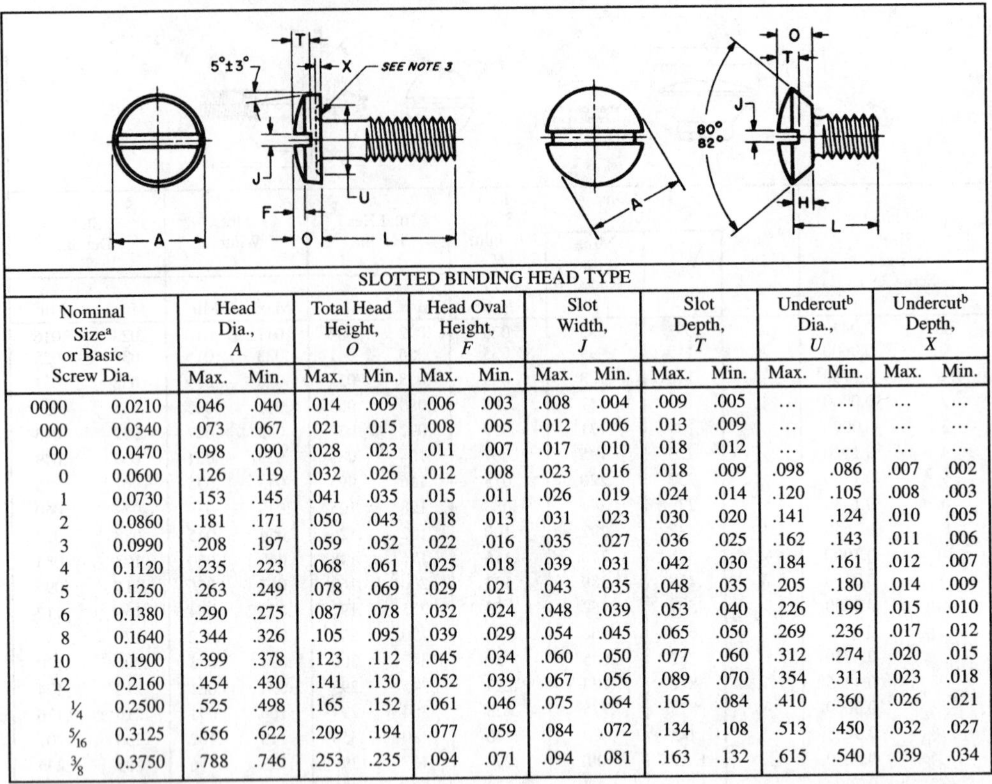

SLOTTED BINDING HEAD TYPE																
Nominal Size[a] or Basic Screw Dia.		Head Dia., *A*		Total Head Height, *O*		Head Oval Height, *F*		Slot Width, *J*		Slot Depth, *T*		Undercut[b] Dia., *U*		Undercut[b] Depth, *X*	
		Max.	Min.	Max.	Min.	Max.	Min.	Max.	Min.	Max.	Min.	Max.	Min.	Max.	Min.
0000	0.0210	.046	.040	.014	.009	.006	.003	.008	.004	.009	.005	…	…	…	…
000	0.0340	.073	.067	.021	.015	.008	.005	.012	.006	.013	.009	…	…	…	…
00	0.0470	.098	.090	.028	.023	.011	.007	.017	.010	.018	.012	…	…	…	…
0	0.0600	.126	.119	.032	.026	.012	.008	.023	.016	.018	.009	.098	.086	.007	.002
1	0.0730	.153	.145	.041	.035	.015	.011	.026	.019	.024	.014	.120	.105	.008	.003
2	0.0860	.181	.171	.050	.043	.018	.013	.031	.023	.030	.020	.141	.124	.010	.005
3	0.0990	.208	.197	.059	.052	.022	.016	.035	.027	.036	.025	.162	.143	.011	.006
4	0.1120	.235	.223	.068	.061	.025	.018	.039	.031	.042	.030	.184	.161	.012	.007
5	0.1250	.263	.249	.078	.069	.029	.021	.043	.035	.048	.035	.205	.180	.014	.009
6	0.1380	.290	.275	.087	.078	.032	.024	.048	.039	.053	.040	.226	.199	.015	.010
8	0.1640	.344	.326	.105	.095	.039	.029	.054	.045	.065	.050	.269	.236	.017	.012
10	0.1900	.399	.378	.123	.112	.045	.034	.060	.050	.077	.060	.312	.274	.020	.015
12	0.2160	.454	.430	.141	.130	.052	.039	.067	.056	.089	.070	.354	.311	.023	.018
¼	0.2500	.525	.498	.165	.152	.061	.046	.075	.064	.105	.084	.410	.360	.026	.021
5/16	0.3125	.656	.622	.209	.194	.077	.059	.084	.072	.134	.108	.513	.450	.032	.027
3/8	0.3750	.788	.746	.253	.235	.094	.071	.094	.081	.163	.132	.615	.540	.039	.034

[a] Where specifying nominal size in decimals, zeros preceding decimal points and in the fourth decimal place are omitted.

[b] Unless otherwise specified, slotted binding head machine screws are not undercut.

SLOTTED UNDERCUT OVAL COUNTERSUNK HEAD TYPES											
Nominal Size[a] or Basic Screw Dia.		Max. *L*[a]	Head Dia., *A*		Head Side Height, *H*	Total Head Height, *O*		Slot Width, *J*		Slot Depth, *T*	
			Max., Edge Sharp	Min., Edge Rnded. or Flat	Ref.	Max.	Min.	Max.	Min.	Max.	Min.
0	0.0600	1/8	.119	.099	.025	.046	.033	.023	.016	.028	.022
1	0.0730	1/8	.146	.123	.031	.056	.042	.026	.019	.034	.027
2	0.0860	1/8	.172	.147	.036	.065	.050	.031	.023	.040	.033
3	0.0990	1/8	.199	.171	.042	.075	.059	.035	.027	.047	.038
4	0.1120	3/16	.225	.195	.047	.084	.067	.039	.031	.053	.043
5	0.1250	3/16	.252	.220	.053	.094	.076	.043	.035	.059	.048
6	0.1380	3/16	.279	.244	.059	.104	.084	.048	.039	.065	.053
8	0.1640	¼	.332	.292	.070	.123	.101	.054	.045	.078	.064
10	0.1900	5/16	.385	.340	.081	.142	.118	.060	.050	.090	.074
12	0.2160	3/8	.438	.389	.092	.161	.135	.067	.056	.103	.085
¼	0.2500	7/16	.507	.452	.107	.186	.158	.075	.064	.119	.098
5/16	0.3125	½	.635	.568	.134	.232	.198	.084	.072	.149	.124
3/8	0.3750	9/16	.762	.685	.161	.278	.239	.094	.081	.179	.149
7/16	0.4375	5/8	.812	.723	.156	.279	.239	.094	.081	.184	.154
½	0.5000	¾	.875	.775	.156	.288	.244	.106	.091	.204	.169

[a] These lengths or shorter are undercut.

All dimensions are in inches.

Table 12. Slotted Round Head Machine Screws

ANSI B18.6.3-1972 (R1991) Appendix

Nominal Size[a] or Basic Screw Dia.		Head Diameter, *A*		Head Height, *H*		Slot Width, *J*		Slot Depth, *T*	
		Max.	Min.	Max.	Min.	Max.	Min.	Max.	Min.
0000	0.0210	.041	.035	.022	.016	.008	.004	.017	.013
000	0.0340	.062	.056	.031	.025	.012	.008	.018	.012
00	0.0470	.089	.080	.045	.036	.017	.010	.026	.018
0	0.0600	.113	.099	.053	.043	.023	.016	.039	.029
1	0.0730	.138	.122	.061	.051	.026	.019	.044	.033
2	0.0860	.162	.146	.069	.059	.031	.023	.048	.037
3	0.0990	.187	.169	.078	.067	.035	.027	.053	.040
4	0.1120	.211	.193	.086	.075	.039	.031	.058	.044
5	0.1250	.236	.217	.095	.083	.043	.035	.063	.047
6	0.1380	.260	.240	.103	.091	.048	.039	.068	.051
8	0.1640	.309	.287	.120	.107	.054	.045	.077	.058
10	0.1900	.359	.334	.137	.123	.060	.050	.087	.065
12	0.2160	.408	.382	.153	.139	.067	.056	.096	.073
1/4	0.2500	.472	.443	.175	.160	.075	.064	.109	.082
5/16	0.3125	.590	.557	.216	.198	.084	.072	.132	.099
3/8	0.3750	.708	.670	.256	.237	.094	.081	.155	.117
7/16	0.4375	.750	.707	.328	.307	.094	.081	.196	.148
1/2	0.5000	.813	.766	.355	.332	.106	.091	.211	.159
9/16	0.5625	.938	.887	.410	.385	.118	.102	.242	.183
5/8	0.6250	1.000	.944	.438	.411	.133	.116	.258	.195
3/4	0.7500	1.250	1.185	.547	.516	.149	.131	.320	.242

[a] When specifying nominal size in decimals, zeros preceding decimal point and in the fourth decimal place are omitted.

All dimensions are in inches.

Not recommended, use Pan Head machine screws.

ANSI Cross References for Machine Screws and Metric Machine Screw

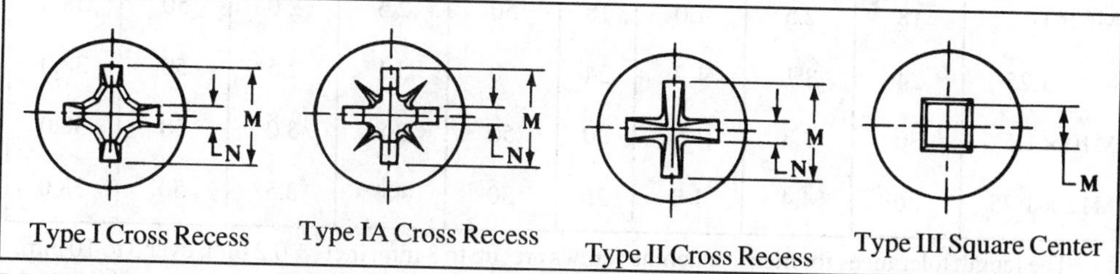

Type I Cross Recess Type IA Cross Recess Type II Cross Recess Type III Square Center

Machine Screw Cross Recesses.—Four cross recesses, Types I, IA, II, and III, may be used in lieu of slots in machine screw heads. Dimensions for recess diameter *M*, width *N*, and depth *T* (not shown above) together with recess penetration gaging depths are given in American National Standard ANSI B18.6.3-1972 (R1991) for machine screws, and in ANSI/ASME B18.6.7M-1985 for metric machine screws.

American National Standard Metric Machine Screws.—This Standard B18.6.7M covers metric flat and oval countersunk and slotted and recessed pan head machine screws and metric hex head and hex flange head machine screws. Dimensions are given in Tables 1 through 4 and 6.

Table 1. American National Standard Thread Lengths for Metric Machine Screws

ANSI/ASME B18.6.7M-1985

Pan, Hex, and Hex Flange Head Screws

Flat and Oval Countersunk Head Screws

Heat-Treated Recessed Flat Countersunk Head Screws

Nominal Screw Size and Thread Pitch	L	L_{US}	L_U	L		L_{US}	L_U	L	B
	Nominal Screw Length Equal to or Shorter than[a]	Unthreaded Length[b]		Nominal Screw Length[a]		Unthreaded Length[b]		Nominal Screw Length Longer than[a]	Full Form Thread Length[c]
		Max[d]	Max[e]	Over	To and Including	Max[d]	Max[e]		Min
M2 × 0.4	6	1.0	0.4	6	30	1.0	0.8	30	25.0
M2.5 × 0.45	8	1.1	0.5	8	30	1.1	0.9	30	25.0
M3 × 0.5	9	1.2	0.5	9	30	1.2	1.0	30	25.0
M3.5 × 0.6	10	1.5	0.6	10	50	1.5	1.2	50	38.0
M4 × 0.7	12	1.8	0.7	12	50	1.8	1.4	50	38.0
M5 × 0.8	15	2.0	0.8	15	50	2.0	1.6	50	38.0
M6 × 1	18	2.5	1.0	18	50	2.5	2.0	50	38.0
M8 × 1.25	24	3.1	1.2	24	50	3.1	2.5	50	38.0
M10 × 1.5	30	3.8	1.5	30	50	3.8	3.0	50	38.0
M12 × 1.75	36	4.4	1.8	36	50	4.4	3.5	50	38.0

[a] The length tolerances for metric machine screws are: up to 3 mm, incl., ± 0.2 mm; over 3 to 10 mm, incl., ± 0.3 mm; over 10 to 16 mm, incl., ± 0.4 mm; over 16 to 50 mm, incl., ± 0.5 mm; over 50 mm, ± 1.0 mm.

[b] Unthreaded lengths L_U and L_{US} represent the distance, measured parallel to the axis of screw, from the underside of the head to the face of a nonchamfered or noncounterbored standard GO thread ring gage assembled by hand as far as the thread will permit.

[c] Refer to the illustrations for respective screw head styles.

[d] The L_{US} values apply only to heat treated recessed flat countersunk head screws.

[e] The L_U values apply to all screws except heat treated recessed flat countersunk head screws.

All dimensions in millimeters.

Table 2. American National Standard Slotted, Cross and Square Recessed Flat Countersunk Head Metric Machine Screws *ANSI/ASME B18.6.7M-1985*

Nominal Screw Size and Thread Pitch	Slotted and Style A		Style B															
	D_S		D_{SH}[a]		D_S	L_{SH}[a]		D_K			K	R		N		T	
	Body Diameter		Body and Shoulder Diameter	Shoulder Diameter	Body Diameter	Shoulder Length		Head Diameter: Theoretical Sharp		Head Diameter: Actual	Head Height	Underhead Fillet Radius		Slot Width		Slot Depth	
	Max	Min	Max	Min	Min	Max	Min	Max	Min	Min	Max Ref	Max	Min	Max	Min	Max	Min
M2 × 0.4[b]	2.00	1.65	2.00	1.86	1.65	0.50	0.30	4.4	4.1	3.5	1.2	0.8	0.4	0.7	0.5	0.6	0.4
M2.5 × 0.45	2.50	2.12	2.50	2.36	2.12	0.55	0.35	5.5	5.1	4.4	1.5	1.0	0.5	0.8	0.6	0.7	0.5
M3 × 0.5	3.00	2.58	3.00	2.86	2.58	0.60	0.40	6.3	5.9	5.2	1.7	1.2	0.6	1.0	0.8	0.9	0.6
M3.5 × 0.6	3.50	3.00	3.50	3.32	3.00	0.70	0.50	8.2	7.7	6.9	2.3	1.4	0.7	1.2	1.0	1.2	0.9
M4 × 0.7	4.00	3.43	4.00	3.82	3.43	0.80	0.60	9.4	8.9	8.0	2.7	1.6	0.8	1.5	1.2	1.3	1.0
M5 × 0.8	5.00	4.36	5.00	4.82	4.36	0.90	0.70	10.4	9.8	8.9	2.7	2.0	1.0	1.5	1.2	1.4	1.1
M6 × 1	6.00	5.21	6.00	5.82	5.21	1.10	0.90	12.6	11.9	10.9	3.3	2.4	1.2	1.9	1.6	1.6	1.2
M8 × 1.25	8.00	7.04	8.00	7.78	7.04	1.40	1.10	17.3	16.5	15.4	4.6	3.2	1.6	2.3	2.0	2.3	1.8
M10 × 1.5	10.00	8.86	10.00	9.78	8.86	1.70	1.30	20.0	19.2	17.8	5.0	4.0	2.0	2.8	2.5	2.6	2.0

[a] All recessed head heat-treated steel screws of property class 9.8 or higher strength have the Style B head form. Recessed head screws other than those specifically designated to be Style B have the Style A head form. The underhead shoulder on the Style B head form is mandatory and all other head dimensions are common to both the Style A and Style B head forms.

[b] This size is not specified for Type III square recessed flat countersunk heads; Type II cross recess is not specified for any size.

All dimensions in millimeters.

For dimension *B*, see Table 1.

For dimension *L*, see Table 8.

Table 3. American National Standard Slotted, Cross and Square Recessed Oval Countersunk Head Metric Machine Screws *ANSI/ASME B18.6.7M-1985*

Nominal Screw Size and Thread Pitch	D_S Body Diameter		D_K Head Diameter			K	F	R_F	R		N		T	
			Theoretical Sharp		Actual	Head Side Height	Raised Head Height	Head Top Radius	Underhead Fillet Radius		Slot Width		Slot Depth	
	Max	Min	Max	Min	Min	Max Ref	Max	Approx	Max	Min	Max	Min	Max	Min
M2 × 0.4[a]	2.00	1.65	4.4	4.1	3.5	1.2	0.5	5.0	0.8	0.4	0.7	0.5	1.0	0.8
M2.5 × 0.45	2.50	2.12	5.5	5.1	4.4	1.5	0.6	6.6	1.0	0.5	0.8	0.6	1.2	1.0
M3 × 0.5	3.00	2.58	6.3	5.9	5.2	1.7	0.7	7.4	1.2	0.6	1.0	0.8	1.5	1.2
M3.5 × 0.6	3.50	3.00	8.2	7.7	6.9	2.3	0.8	10.9	1.4	0.7	1.2	1.0	1.7	1.4
M4 × 0.7	4.00	3.43	9.4	8.9	8.0	2.7	1.0	11.6	1.6	0.8	1.5	1.2	1.9	1.6
M5 × 0.8	5.00	4.36	10.4	9.8	8.9	2.7	1.2	11.9	2.0	1.0	1.5	1.2	2.4	2.0
M6 × 1	6.00	5.21	12.6	11.9	10.9	3.3	1.4	14.9	2.4	1.2	1.9	1.6	2.8	2.4
M8 × 1.25	8.00	7.04	17.3	16.5	15.4	4.6	2.0	19.7	3.2	1.6	2.3	2.0	3.7	3.2
M10 × 1.5	10.00	8.86	20.0	19.2	17.8	5.0	2.3	22.9	4.0	2.0	2.8	2.5	4.4	3.8

[a] This size is not specified for Type III square recessed oval countersunk heads; Type II cross recess is not specified for any size.

All dimensions in millimeters.

For dimension *B*, see Table 1.

For dimension *L*, see Table 8.

Table 4. American National Standard Slotted and Cross and Square Recessed Pan Head Metric Machine Screws
ANSI/ASME B18.6.7M-1985

Nominal Screw Size and Thread Pitch	D_s		D_K		Slotted			Cross and Square Recess			D_A	R	N		T	W
	Body Diameter		Head Diameter		K Head Height		R_1 Head Radius	K Head Height		R_1 Head Radius	Underhead Fillet Transition Dia	Underhead Fillet Radius	Slot Width		Slot Depth	Unslotted Head Thickness
	Max	Min	Max	Min	Max	Min	Max	Max	Min	Ref	Max	Min	Max	Min	Min	Min
M2 × 0.4[a]	2.00	1.65	4.0	3.7	1.3	1.1	0.8	1.6	1.4	3.2	2.6	0.1	0.7	0.5	0.5	0.4
M2.5 × 0.45	2.50	2.12	5.0	4.7	1.5	1.3	1.0	2.1	1.9	4.0	3.1	0.1	0.8	0.6	0.6	0.5
M3 × 0.5	3.00	2.58	5.6	5.3	1.8	1.6	1.2	2.4	2.2	5.0	3.6	0.1	1.0	0.8	0.7	0.7
M3.5 × 0.6	3.50	3.00	7.0	6.6	2.1	1.9	1.4	2.6	2.3	6.0	4.1	0.1	1.2	1.0	0.8	0.8
M4 × 0.7	4.00	3.43	8.0	7.6	2.4	2.2	1.6	3.1	2.8	6.5	4.7	0.2	1.5	1.2	1.0	0.9
M5 × 0.8	5.00	4.36	9.5	9.1	3.0	2.7	2.0	3.7	3.4	8.0	5.7	0.2	1.5	1.2	1.2	1.2
M6 × 1	6.00	5.21	12.0	11.5	3.6	3.3	2.5	4.6	4.3	10.0	6.8	0.3	1.9	1.6	1.4	1.4
M8 × 1.25	8.00	7.04	16.0	15.5	4.8	4.5	3.2	6.0	5.6	13.0	9.2	0.4	2.3	2.0	1.9	1.9
M10 × 1.5	10.00	8.86	20.0	19.4	6.0	5.7	4.0	7.5	7.1	16.0	11.2	0.4	2.8	2.5	2.4	2.4

[a] This size not specified for Type III square recessed pan heads; Type II cross recess is not specified for any size.

All dimensions in millimeters.

For dimension *B*, see Table 1.

For dimension *L*, see Table 8.

Table 5. American National Standard Header Points for Metric Machine Screws Before Threading *ANSI/ASME B18.6.7M-1985*

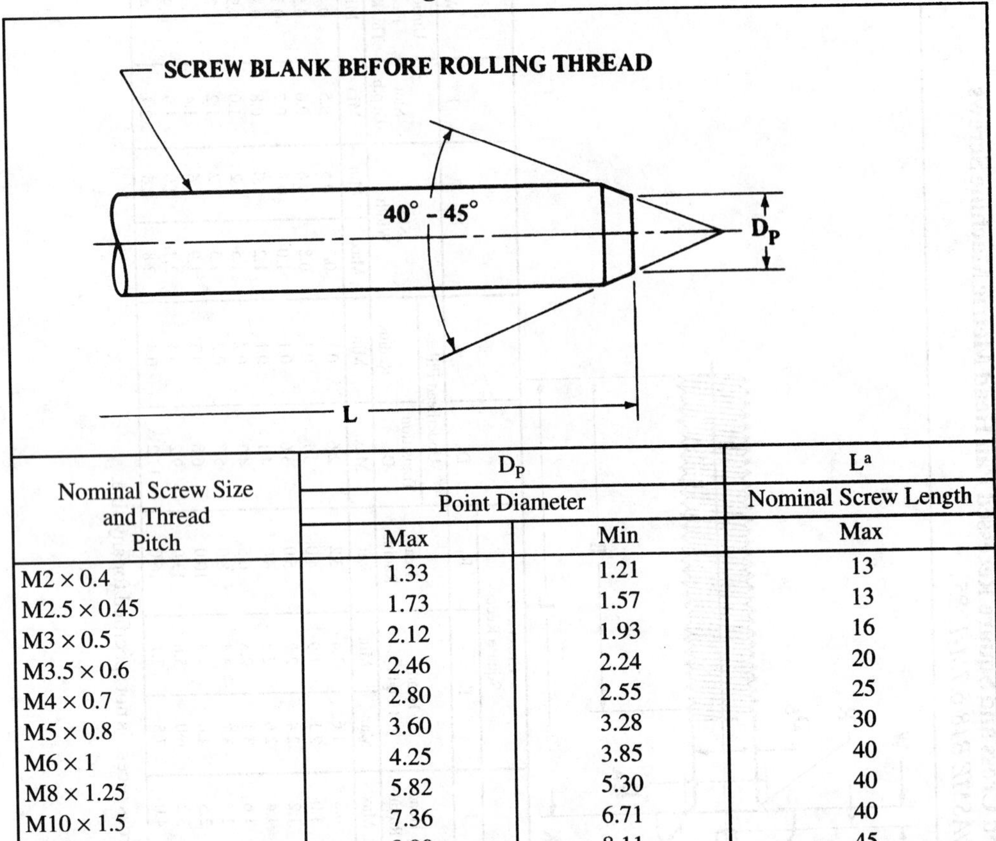

Nominal Screw Size and Thread Pitch	D_P Point Diameter		L^a Nominal Screw Length
	Max	Min	Max
M2 × 0.4	1.33	1.21	13
M2.5 × 0.45	1.73	1.57	13
M3 × 0.5	2.12	1.93	16
M3.5 × 0.6	2.46	2.24	20
M4 × 0.7	2.80	2.55	25
M5 × 0.8	3.60	3.28	30
M6 × 1	4.25	3.85	40
M8 × 1.25	5.82	5.30	40
M10 × 1.5	7.36	6.71	40
M12 × 1.75	8.90	8.11	45

[a] Header points apply to these nominal lengths or shorter. The pointing of longer lengths may require machining to the dimensions specified.

All dimensions in millimeters.

The edge of the point may be rounded and the end of point need not be flat nor perpendicular to the axis of screw shank.

Threads: Threads for metric machine screws are coarse M profile threads, as given in ANSI B1.13M (see page 1755), unless otherwise specified.

Length of Thread: The lengths of threads on metric machine screws are given in Table 1 for the applicable screw type, size, and length.

Diameter of Body: The body diameters of metric machine screws are within the limits specified in the dimensional tables (Tables 3 through 4 and 6).

Designation: Metric machine screws are designated by the following data in the sequence shown: Nominal size and thread pitch; nominal length; product name, including head type and driving provision; header point if desired; material (including property class, if steel); and protective finish, if required. For example:

M8 × 1.25 × 30 Slotted Pan Head Machine Screw, Class 4.8 Steel, Zinc Plated

M3.5 × 0.6 × 20 Type IA Cross Recessed Oval Countersunk Head Machine Screw, Header Point, Brass

It is common ISO practice to omit the thread pitch from the product size designation when screw threads are the metric coarse thread series, e.g., M10 stands for M10 × 1.5.

Table 6. American National Standard Hex and Hex Flange Head Metric Machine Screws *ANSI/ASME B18.6.7M-1985*

Hex Head

Shape of Indentation Optional

Nominal Screw Size and Thread Pitch	D_S Body Diameter		S[a] Hex Width Across Flats		E[a] Hex Width Across Corners	K Head Height		D_A Underhead Fillet Transition Dia	R Underhead Fillet Radius
	Max	Min	Max	Min	Min	Max	Min	Max	Min
M2 × 0.4	2.00	1.65	3.20	3.02	3.38	1.6	1.3	2.6	0.1
M2.5 × 0.45	2.50	2.12	4.00	3.82	4.28	2.1	1.8	3.1	0.1
M3 × 0.5	3.00	2.58	5.00	4.82	5.40	2.3	2.0	3.6	0.1
M3.5 × 0.6	3.50	3.00	5.50	5.32	5.96	2.6	2.3	4.1	0.1
M4 × 0.7	4.00	3.43	7.00	6.78	7.59	3.0	2.6	4.7	0.2
M5 × 0.8	5.00	4.36	8.00	7.78	8.71	3.8	3.3	5.7	0.2
M6 × 1	6.00	5.21	10.00	9.78	10.95	4.7	4.1	6.8	0.3
M8 × 1.25	8.00	7.04	13.00	12.73	14.26	6.0	5.2	9.2	0.4
M10 × 1.5	10.00	8.86	16.00	15.73	17.62	7.5	6.5	11.2	0.4
M12 × 1.75	12.00	10.68	18.00	17.73	19.86	9.0	7.8	13.2	0.4
M10 × 1.5[b]	10.00	8.86	15.00	14.73	16.50	7.5	6.5	11.2	0.4

[a] Dimensions across flats and across corners of the head are measured at the point of maximum metal. Taper of sides of head (angle between one side and the axis) shall not exceed 2° or 0.10 mm, whichever is greater, the specified width across flats being the large dimension.

[b] The contour of the edge at periphery of flange is optional provided the minimum flange thickness is maintained at the minimum flange diameter. The top surface of flange may be straight or slightly rounded (convex) upward.

Table 7. American National Standard Hex and Hex Flange Head Metric Machine Screws *ANSI/ASME B18.6.7M-1985*

Hex Flange Head

Nominal Screw Size and Thread Pitch	D_S Body Diameter Max	D_S Body Diameter Min	S^a Hex Width Across Flats Max	S^a Hex Width Across Flats Min	E^a Hex Width Across Corners Min	D_C Flange Diameter Max	D_C Flange Diameter Min	K Overall Head Height	K_1 Hex Height Min	C^b Flange Edge Thickness Min	R_1 Flange Top Fillet Radius Max	D_A Underhead Fillet Max Transition Dia	R Underhead Fillet Min Radius
M2 × 0.4	2.00	1.65	3.00	2.84	3.16	4.5	4.1	2.2	1.3	0.3	0.1	2.6	0.1
M2.5 × 0.45	2.50	2.12	3.20	3.04	3.39	5.4	5.0	2.7	1.6	0.3	0.2	3.1	0.1
M3 × 0.5	3.00	2.58	4.00	3.84	4.27	6.4	5.9	3.2	1.9	0.4	0.2	3.6	0.1
M3.5 × 0.6	3.50	3.00	5.00	4.82	5.36	7.5	6.9	3.8	2.4	0.5	0.2	4.1	0.1
M4 × 0.7	4.00	3.43	5.50	5.32	5.92	8.5	7.8	4.3	2.8	0.6	0.2	4.7	0.2
M5 × 0.8	5.00	4.36	7.00	6.78	7.55	10.6	9.8	5.4	3.5	0.7	0.3	5.7	0.2
M6 × 1	6.00	5.21	8.00	7.78	8.66	12.8	11.8	6.7	4.2	1.0	0.4	6.8	0.3
M8 × 1.25	8.00	7.04	10.00	9.78	10.89	16.8	15.5	8.6	5.6	1.2	0.5	9.2	0.4
M10 × 1.5	10.00	8.86	13.00	12.72	14.16	21.0	19.3	10.7	7.0	1.4	0.6	11.2	0.4
M12 × 1.75	12.00	10.68	15.00	14.72	16.38	24.8	23.3	13.7	8.4	1.8	0.7	13.2	0.4

[a] Dimensions across flats and across corners of the head are measured at the point of maximum metal. Taper of sides of head (angle between one side and the axis) shall not exceed 2° or 0.10 mm, whichever is greater, the specified width across flats being the large dimension.

[b] The contour of the edge at periphery of flange is optional provided the minimum flange thickness is maintained at the minimum flange diameter. The top surface of flange may be straight or slightly rounded (convex) upward.

All dimensions in millimeters.

A slight rounding of all edges of the hexagon surfaces of indented hex heads is permissible provided the diameter of the bearing circle is not less than the equivalent of 90 per cent of the specified minimum width across flats dimension.

Heads may be indented, trimmed, or fully upset at the option of the manufacturer.

The M10 size screws having heads with 15 mm width across flats are not ISO Standard. Unless M10 size screws with 15 mm width across flats are specifically ordered, M10 size screws with 16 mm width across flats shall be furnished.

For dimension *B*, see Table 1.

For dimension *L*, see Table 8.

Table 8. Recommended Nominal Screw Lengths for Metric Machine Screws

Nominal Screw Length	Nominal Screw Size									
	M2	M2.5	M3	M3.5	M4	M5	M6	M8	M10	M12
2.5	PH									
3	A	PH								
4	A	A	PH							
5	A	A	A	PH	PH					
6	A	A	A	A	A	PH				
8	A	A	A	A	A	A	A			
10	A	A	A	A	A	A	A	A		
13	A	A	A	A	A	A	A	A	A	
16	A	A	A	A	A	A	A	A	A	H
20	A	A	A	A	A	A	A	A	A	H
25		A	A	A	A	A	A	A	A	H
30			A	A	A	A	A	A	A	H
35				A	A	A	A	A	A	H
40					A	A	A	A	A	H
45						A	A	A	A	H
50						A	A	A	A	H
55							A	A	A	H
60							A	A	A	H
65								A	A	H
70								A	A	H
80								A	A	H
90									A	H

All dimensions in millimeters.

[1]The nominal screw lengths included between the heavy lines are recommended for the respective screw sizes and screw head styles as designated by the symbols.

A — Signifies screws of all head styles covered in this standard.

P — Signifies pan head screws.

H — Signifies hex and hex flange head screws.

Table 9. Clearance Holes for Metric Machine Screws
***ANSI/ASME B18.6.7M-1985* Appendix**

Nominal Screw Size	Basic Clearance Hole Diameter[a]		
	Close Clearance[b]	Normal Clearance (Preferred)[b]	Loose Clearance[b]
M2	2.20	2.40	2.60
M2.5	2.70	2.90	3.10
M3	3.20	3.40	3.60
M3.5	3.70	3.90	4.20
M4	4.30	4.50	4.80
M5	5.30	5.50	5.80
M6	6.40	6.60	7.00
M8	8.40	9.00	10.00
M10	10.50	11.00	12.00
M12	13.00	13.50	14.50

[a] The values given in this table are minimum limits. The recommended plus tolerances are as follows: for clearance hole diameters over 1.70 to and including 5.80 mm, plus 0.12, 0.20, and 0.30 mm for close, normal, and loose clearances, respectively; for clearance hole diameters over 5.80 to 14.50 mm, plus 0.18, 0.30, and 0.45 mm for close, normal, and loose clearances, respectively.

[b] Normal clearance hole sizes are preferred. Close clearance hole sizes are for situations such as critical alignment of assembled components, wall thickness, or other limitations which necessitate the use of a minimal hole. Countersinking or counterboring at the fastener entry side may be necessary for the proper seating of the head. Loose clearance hole sizes are for applications where maximum adjustment capability between the components being assembled is necessary.

All dimensions in millimeters.

British Machine Screws.—Many of these classifications of fasteners are covered in British Standards B.S. 57:1951, "B.A. Screws, Bolts and Nuts"; BS 450:1958 (obsolescent), "Machine Screws and Machine Screw Nuts (BSW and BSF Threads)"; B.S. 1981:1953, "Unified Machine Screws and Machine Screw Nuts"; BS 2827:1957 (obsolescent):1957, "Machine Screw Nuts, Pressed Type (B.A. and Whitworth Form Threads)"; B.S. 3155:1960, "American Machine Screws and Nuts in Sizes Below ¼ inch Diameter"; and BS 4183:1967 (obsolescent), "Machine Screws and Machine Screw Nuts, Metric Series." At a conference organized by the British Standards Institution in 1965 at which the major sectors of British industry were represented, a policy statement was approved that urged British firms to regard the traditional screw thread systems—Whitworth, B.A. and BSF—as obsolescent, and to make the internationally-agreed ISO metric thread their first choice (with ISO Unified thread as second choice) for all future designs. It is recognized that some sections of British industry already using ISO inch (Unified) screw threads may find it necessary, for various reasons, to continue with their use for some time: Whitworth and B.A. threads should, however, be superseded by ISO metric threads in preference to an intermediate change to ISO inch threads. Fasteners covered by B.S. 57, B.S. 450 and BS 2827:1957 (obsolescent) eventually would be superseded and replaced by fasteners specified by B.S. 4183.

British Standard Whitworth (BSW) and Fine (BSF) Machine Screws.—British Standard BS 450:1958 (obsolescent) covers machine screws and nuts with British Standard Whit-worth and British Standard Fine threads. All the various heads in common use in both slotted and recessed forms are covered. Head shapes are shown on page 1595 and dimensions on page 1598. It is intended that this standard will eventually be superseded by B.S. 4183, "Machine Screws and Machine Screw Nuts, Metric Series."

British Standard Machine Screws and Machine Screw Nuts, Metric Series.—British Standard BS 4183:1967 (obsolescent) gives dimensions and tolerances for: countersunk head, raised countersunk head, and cheese head slotted head screws in a diameter range from M1 (1 mm) to M20 (20 mm); pan head slotted head screws in a diameter range from M2.5 (2.5 mm) to M10 (10 mm); countersunk head and raised countersunk head recessed head screws in a diameter range from M2.5 (2.5 mm) to M12 (12 mm); pan head recessed head screws in a diameter range from M2.5 (2.5 mm) to M10 (10 mm); and square and hexagon machine screw nuts in a diameter range from M1.6 (1.6 mm) to M10 (10 mm). Mechanical properties are also specified for steel, brass and aluminum alloy machine screws and machine screw nuts in this standard.

Material: The materials from which the screws and nuts are manufactured have a tensile strength not less than the following: steel, 40 kgf/mm^2 (392 N/mm^2); brass, 32 kgf/mm^2 (314 N/mm^2); and aluminum alloy, 32 kgf/mm^2 (314 N/mm^2). The unit, kgf/mm^2 is in accordance with ISO DR 911 and the unit in parentheses has the relationship, 1 kgf = 9.80665 Newtons. These minimum strengths are applicable to the finished products. Steel machine screws conform to the requirements for strength grade designation 4.8. The strength grade designation system for machine screws consists of two figures, the first is $\frac{1}{10}$ of the minimum tensile strength in kgf/mm^2, the second is $\frac{1}{10}$ of the ratio between the yield stress and the minimum tensile strength expressed as a percentage: $\frac{1}{10}$ minimum tensile strength of 40 kgf/mm^2 gives the symbol "4"; $\frac{1}{10}$ ratio $\frac{\text{yield stress}}{\text{minimum tensile strength}}\% = \frac{1}{10} \times \frac{32}{40} \times 100/1 =$ "8"; giving the strength grade designation "4.8." Multiplication of these two figures gives the minimum yield stress in kgf/mm^2.

Coating of Screws and Nuts: It is recommended that the coating comply with the appropriate part of BS 3382. "Electroplated Coatings on Threaded Components."

Screw Threads: Screw threads are ISO metric coarse pitch series threads in accordance with B.S. 3643. "ISO Metric Screw Threads," Part 1, "Thread Data and Standard Thread Series." The external threads used for screws conform to tolerance Class 6g limits (medium fit) as given in B.S. 3643, "ISO Metric Screw Threads," Part 2, "Limits and Tolerances for Coarse Pitch Series Threads." The internal threads used for nuts conform to tolerance Class 6H limits (medium fit) as given in B.S. 3643: Part 2.

Nominal Lengths of Screws: For countersunk head screws the nominal length is the distance from the upper surface of the head to the extreme end of the shank, including any chamfer, radius, or cone point. For raised countersunk head screws the nominal length is the distance from the upper surface of the head (excluding the raised portion) to the extreme end of the shank, including any chamfer, radius, or cone point. For pan and cheese head screws the nominal length is the distance from the underside of the head to the extreme end of the shank, including any chamfer, radius, or cone point. Standard nominal lengths and tolerances are given in Table 5.

Lengths of Thread on Screws: The length of thread is the distance from the end of the screw (including any chamfer, radius, or cone point) to the leading face of a nut without countersink which has been screwed as far as possible onto the screw by hand. The minimum thread length is shown in the following table:

Nominal Thread Dia., d[a]	M1	M1.2	(M1.4)	M1.6	M2	(M2.2)	M2.5	M3	(M3.5)	M4
Thread Length b (Min.)	b	b	b	15	16	17	18	19	20	22
Nominal Thread Dia., d[a]	**(M4.5)**	**M5**	**M6**	**M8**	**M10**	**M12**	**(M14)**	**M16**	**(M18)**	**M20**
Thread Length b (Min.)	24	25	28	34	40	46	52	58	64	70

[a] Items shown in parentheses are non-preferred.

[b] Threaded up to the head.

All dimensions are in millimeters.

Screws of nominal thread diameter M1, M1.2 and M1.4 and screws of larger diameters that are too short for the above thread lengths are threaded as far as possible up to the head.

In these screws the length of unthreaded shank under the head does not exceed 1½ pitches for lengths up to twice the diameter and 2 pitches for longer lengths, and is defined as the distance from the leading face of a nut that has been screwed as far as possible onto the screw by hand to: 1) the junction of the basic major diameter and the countersunk portion of the head on countersunk and raised countersunk heads; and 2) the underside of the head on other types of heads.

Diameter of Unthreaded Shank on Screws: The diameter of the unthreaded portion of the shank on screws is not greater than the basic major diameter of the screw thread and not less than the minimum effective diameter of the screw thread. The diameter of the unthreaded portion of shank is closely associated with the method of manufacture; it will generally be nearer the major diameter of the thread for turned screws and nearer the effective diameter for those produced by cold heading.

Radius Under the Head of Screws: The radius under the head of pan and cheese head screws runs smoothly into the face of the head and shank without any step or discontinuity. A true radius is not essential providing that the curve is smooth and lies wholly within the maximum radius. Any radius under the head of countersunk head screws runs smoothly into the conical bearing surface of the head and the shank without any step or discontinuity. The radius values given in Tables 1 and 2 are regarded as the maximum where the shank diameter is equal to the major diameter of the thread and minimum where the shank diameter is approximately equal to the effective diameter of the thread.

Table 1. British Standard Slotted Countersunk Head Machine Screws—Metric Series *BS 4183:1967 (obsolescent)*

Nominal Size d[a]	Head Diameter D		Head Height k		Radius r[b]	Thread Length b	Thread Run-out a	Flushness Tolerance[c]	Slot Width n		Slot Depth t	
	Max. (Theor. Sharp) $2d$	Min. $1.75d$	Max. $0.5d$	Min. $0.45d$		Min.	Max. $2p$[d]	Max.	Max.	Min.	Max. $0.3d$	Min. $0.2d$
M1	2.00	1.75	0.50	0.45	0.1	[e]	0.50	....	0.45	0.31	0.30	0.20
M1.2	2.40	2.10	0.60	0.54	0.1	[e]	0.50	....	0.50	0.36	0.36	0.24
(M1.4)	2.80	2.45	0.70	0.63	0.1	[e]	0.60	....	0.50	0.36	0.42	0.28
M1.6	3.20	2.80	0.80	0.72	0.1	15.0	0.70	....	0.60	0.46	0.48	0.32
M2.0	4.00	3.50	1.00	0.90	0.1	16.0	0.80	....	0.70	0.56	0.60	0.40
(M2.2)	4.40	3.85	1.10	0.99	0.1	17.0	0.90	....	0.80	0.66	0.66	0.44
M2.5	5.00	4.38	1.25	1.12	0.1	18.0	0.90	0.10	0.80	0.66	0.75	0.50
M3	6.00	5.25	1.50	1.35	0.1	19.0	1.00	0.12	1.00	0.86	0.90	0.60
(M3.5)	7.00	6.10	1.75	1.57	0.2	20.0	1.20	0.13	1.00	0.86	1.05	0.70
M4	8.00	7.00	2.00	1.80	0.2	22.0	1.40	0.15	1.20	1.06	1.20	0.80
(M4.5)	9.00	7.85	2.25	2.03	0.2	24.0	1.50	0.17	1.20	1.06	1.35	0.90
M5	10.00	8.75	2.50	2.25	0.2	25.0	1.60	0.19	1.51	1.26	1.50	1.00
M6	12.00	10.50	3.00	2.70	0.25	28.0	2.00	0.23	1.91	1.66	1.80	1.20
M8	16.00	14.00	4.00	3.60	0.4	34.0	2.50	0.29	2.31	2.06	2.40	1.60
M10	20.00	17.50	5.00	4.50	0.4	40.0	3.00	0.37	2.81	2.56	3.00	2.00
M12	24.00	21.00	6.00	5.40	0.6	46.0	3.50	0.44	3.31	3.06	3.60	2.40
(M14)	28.00	24.50	7.00	6.30	0.6	52.0	4.00	0.52	3.31	3.06	4.20	2.80
M16	32.00	28.00	8.00	7.20	0.6	58.0	4.00	0.60	4.37	4.07	4.80	3.20
(M18)	36.00	31.50	9.00	8.10	0.6	64.0	5.00	0.67	4.37	4.07	5.40	3.60
M20	40.00	35.00	10.00	9.00	0.8	70.0	5.00	0.75	5.37	5.07	6.00	4.00

[a] Nominal sizes shown in parentheses are non-preferred.

[b] See *Radius Under the Head of Screws* description in text.

[c] See *Dimensions of 90-Degree Countersunk Head Screws* description in text.

[d] See text following table in *Lengths of Thread on Screws* description in text.

[e] Threaded up to head.

All dimensions are given in millimeters. For dimensional notation, see diagram on page 1590. Recessed head screws are also standard and are available. For dimensions see British Standard.

Table 2. British Standard Slotted Raised Countersunk Head Machine Screws—Metric Series *BS 4183:1967 (obsolescent)*

Nominal Size d^a	Head Diameter D		Head Height k		Radius Under Head r^b	Thread Length b	Thread Run-out a	Height of Raised Portion f	Head Radius R	Slot Width n		Slot Depth t	
	Max. (Theor. Sharp) $2d$	Min. $1.75d$	Max. $0.5d$	Min. $0.45d$		Min.	Max. $2p^c$	Nom. $0.25d$	Nom.	Max.	Min.	Max. $0.5d$	Min. $0.4d$
M1	2.00	1.75	0.50	0.45	0.1	[d]	0.50	0.25	2.0	0.45	0.31	0.50	0.40
M1.2	2.40	2.10	0.60	0.54	0.1	[d]	0.50	0.30	2.5	0.50	0.36	0.60	0.48
(M1.4)	2.80	2.45	0.70	0.63	0.1	[d]	0.60	0.35	2.5	0.50	0.36	0.70	0.56
M1.6	3.20	2.80	0.80	0.72	0.1	15.0	0.70	0.40	3.0	0.60	0.46	0.80	0.64
M2.0	4.00	3.50	1.00	0.90	0.1	16.0	0.80	0.50	4.0	0.70	0.56	1.00	0.80
(M2.2)	4.40	3.85	1.10	0.99	0.1	17.0	0.90	0.55	4.0	0.80	0.66	1.10	0.88
M2.5	5.00	4.38	1.25	1.12	0.1	18.0	0.90	0.60	5.0	0.80	0.66	1.25	1.00
M3	6.00	5.25	1.50	1.35	0.1	19.0	1.00	0.75	6.0	1.00	0.86	1.50	1.20
(M3.5)	7.00	6.10	1.75	1.57	0.2	20.0	1.20	0.90	6.0	1.00	0.86	1.75	1.40
M4	8.00	7.00	2.00	1.80	0.2	22.0	1.40	1.00	8.0	1.20	1.06	2.00	1.60
(M4.5)	9.00	7.85	2.25	2.03	0.2	24.0	1.50	1.10	8.0	1.20	1.06	2.25	1.80
M5	10.00	8.75	2.50	2.25	0.2	25.0	1.60	1.25	10.0	1.51	1.26	2.50	2.00
M6	12.00	10.50	3.00	2.70	0.25	28.0	2.00	1.50	12.0	1.91	1.66	3.00	2.40
M8	16.00	14.00	4.00	3.60	0.4	34.0	2.50	2.00	16.0	2.31	2.06	4.00	3.20
M10	20.00	17.50	5.00	4.50	0.4	40.0	3.00	2.50	20.0	2.81	2.56	5.00	4.00
M12	24.00	21.00	6.00	5.40	0.6	46.0	3.50	3.00	25.0	3.31	3.06	6.00	4.80
(M14)	28.00	24.50	7.00	6.30	0.6	52.0	4.00	3.50	25.0	3.31	3.06	7.00	5.60
M16	32.00	28.00	8.00	7.20	0.6	58.0	4.00	4.00	32.0	4.37	4.07	8.00	6.40
(M18)	36.00	31.50	9.00	8.10	0.6	64.0	5.00	4.50	32.0	4.37	4.07	9.00	7.20
M20	40.00	35.00	10.00	9.00	0.8	70.0	5.00	5.00	40.0	5.37	5.07	10.00	8.00

[a] Nominal sizes shown in parentheses are non-preferred.
[b] See *Radius Under the Head of Screws* description in text.
[c] See text following table in *Lengths of Thread on Screws* description in text.
[d] Threaded up to head.

All dimensions are given in millimeters. For dimensional notation see diagram on page 1590. Recessed head screws are also standard and available. For dimensions see British Standard.

Ends of Screws: When screws are made with rolled threads, the "lead" formed by the thread rolling operation is normally regarded as providing the necessary chamfer and no other machining is necessary. The ends of screws with cut threads are normally finished with a chamfer conforming to the dimension in Fig. 1a through Fig. 1d. At the option of the manufacturer, the ends of screws smaller than M6 (6-mm diameter) may be finished with a radius approximately equal to 1¼ times the nominal diameter of the shank. When cone point ends are required, they should have the dimensions given in Fig. 1a through Fig. 1d.

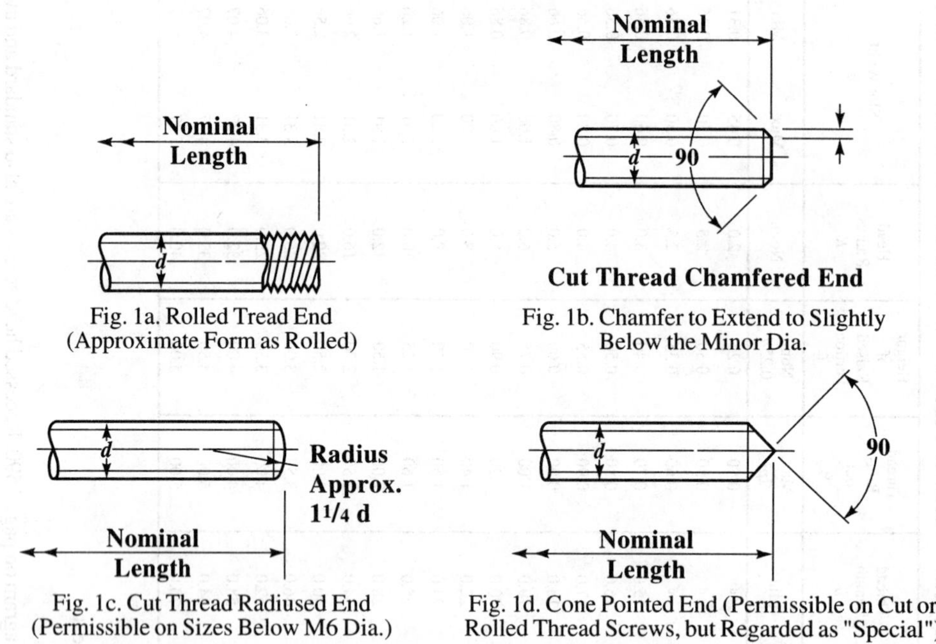

Fig. 1a. Rolled Tread End (Approximate Form as Rolled)

Fig. 1b. Chamfer to Extend to Slightly Below the Minor Dia.

Fig. 1c. Cut Thread Radiused End (Permissible on Sizes Below M6 Dia.)

Fig. 1d. Cone Pointed End (Permissible on Cut or Rolled Thread Screws, but Regarded as "Special")

Dimensions of 90-Degree Countersunk Head Screws: One of the appendices to this British Standard states that countersunk head screws should fit into the countersunk hole with as great a degree of flushness as possible. To achieve this condition, it is necessary for the dimensions of both the head of the screw and the countersunk hole to be controlled within prescribed limits. The maximum or design size of the head is controlled by a theoretical diameter to a sharp corner and the minimum head angle of 90 degrees. The minimum head size is controlled by a minimum head diameter, the maximum head angle of 92 degrees and a flushness tolerance (see Fig. 2, page 1591). The edge of the head may be flat or rounded, as shown in Fig. 3 on page 1591.

British Standard Machine Screws and Machine Screw Nuts—Metric Series

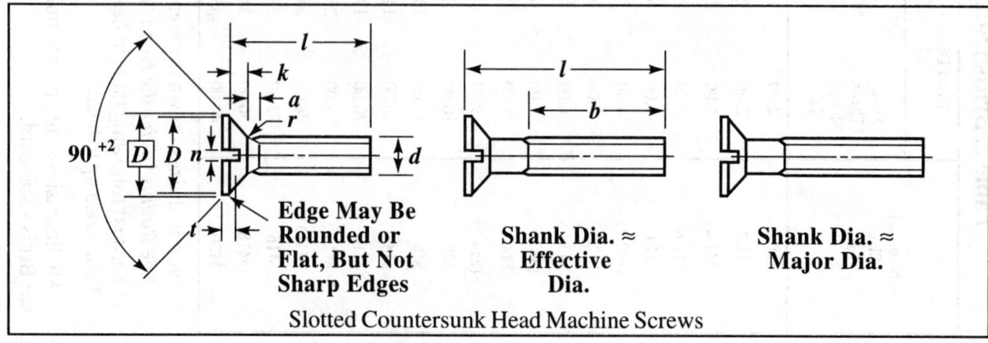

Slotted Countersunk Head Machine Screws

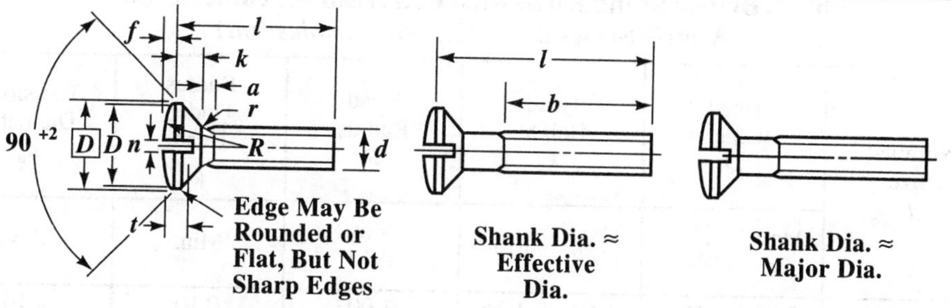

Edge May Be Rounded or Flat, But Not Sharp Edges
Shank Dia. ≈ Effective Dia.
Shank Dia. ≈ Major Dia.

Slotted Raised Countersunk Head Machine Screws

0 TO 5
Shank Dia. ≈ Effective Dia.
Shank Dia. ≈ Major Dia.

Slotted Pan Head Machine Screws

0 TO 5
Shank Dia. ≈ Effective Dia.
Shank Dia. ≈ Major Dia.

Slotted Cheese Head Machine Screws

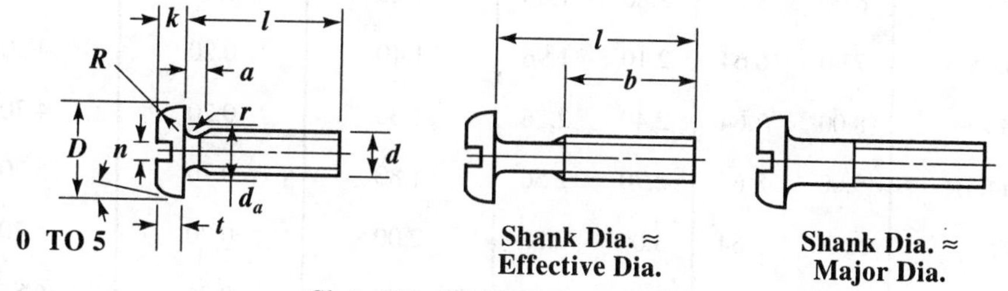

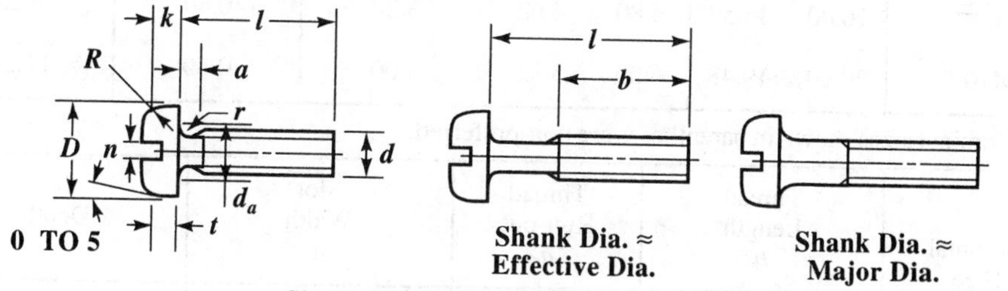

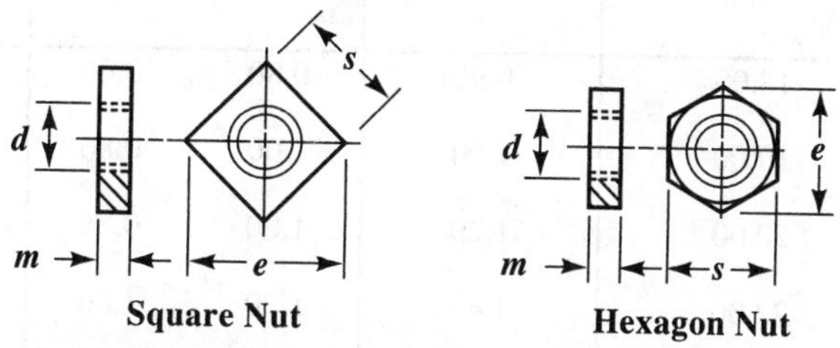

Square Nut
Hexagon Nut

Machine Screw Nuts, Pressed Type, Square and Hexagon

For dimensions, see Tables 1 through 5.

90°
Theoretical Sharp
Minimum Head Dia.
92°
Flushness Tolerance (Max.)
Minimum Envelope
Maximum Envelope

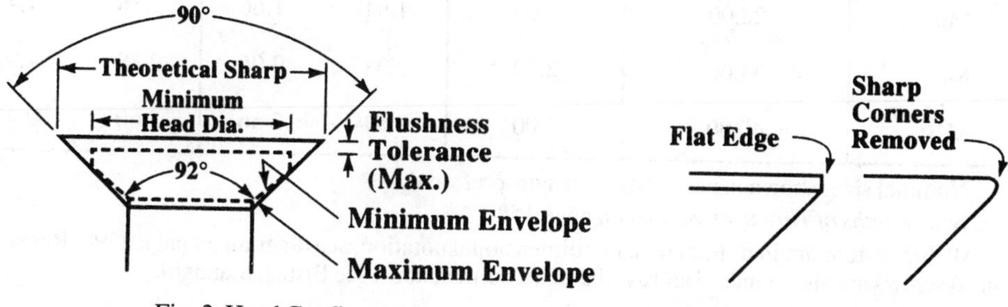

Fig. 2. Head Configuration

Flat Edge
Sharp Corners Removed

Fig. 3. Edge Configuration

Table 3. British Standard Slotted Pan Head Machine Screws— Metric Series *BS 4183:1967 (obsolescent)*

Nominal Size d[a]	Head Diameter D		Head Height k		Head Radius R	Radius Under Head r	Transition Diameter d_a
	Max. $2d$	Min.	Max. $0.6d$	Min.	Max $0.4d$	Min.	Max.
M2.5	5.00	4.70	1.50	1.36	1.00	0.10	3.10
M3	6.00	5.70	1.80	1.66	1.20	0.10	3.60
(M3.5)	7.00	6.64	2.10	1.96	1.40	0.20	4.30
M4	8.00	7.64	2.40	2.26	1.60	0.20	4.70
(M4.5)	9.00	8.64	2.70	2.56	1.80	0.20	5.20
M5	10.00	9.64	3.00	2.86	2.00	0.20	5.70
M6	12.00	11.57	3.60	3.42	2.50	0.25	6.80
M8	16.00	15.57	4.80	4.62	3.20	0.40	9.20
M10	20.00	19.48	6.00	5.82	4.00	0.40	11.20

[a] Nominal sizes shown in parentheses are non-preferred.

Nominal Size d[a]	Thread Length b	Thread Run-out a	Slot Width n		Slot Depth t	
	Min.	Max. $2p$[b]	Max.	Min.	Max. $0.6k$	Min. $0.4k$
M2.5	18.00	0.90	0.80	0.66	0.90	0.60
M3	19.00	1.00	1.00	0.86	1.08	0.72
(M3.5)	20.00	1.20	1.00	0.86	1.26	0.84
M4	22.00	1.40	1.20	1.06	1.44	0.96
(M4.5)	24.00	1.50	1.20	1.06	1.62	1.08
M5	25.00	1.60	1.51	1.26	1.80	1.20
M6	28.00	2.00	1.91	1.66	2.16	1.44
M8	34.00	2.50	2.31	2.06	2.88	1.92
M10	40.00	3.00	2.81	2.56	3.60	2.40

[a] Nominal sizes shown in parentheses are non-preferred.

[b] See *Lengths of Thread on Screws* on page 1587.

All dimensions are in millimeters. For dimensional notation, see diagram on page 1590. Recessed head screws are also standard and available. For dimensions, see British Standard.

Table 4. British Standard Slotted Cheese Head Machine Screws—Metric Series *BS 4183:1967 (obsolescent)*

Nominal Size d^a	Head Diameter D		Head Height k		Radius r^b	Transition Diameter d_a	Thread Length b	Thread Run-out a	Slot Width n		Slot Depth t	
	Max.	Min.	Max.	Min.	Min.	Max.	Min.	Max. [c]	Max.	Min.	Max.	Min.
M1	2.00	1.75	0.70	0.56	0.10	1.30	[b]	0.50	0.45	0.31	0.44	0.30
M1.2	2.30	2.05	0.80	0.66	0.10	1.50	[b]	0.50	0.50	0.36	0.49	0.35
(M1.4)	2.60	2.35	0.90	0.76	0.10	1.70	[b]	0.60	0.50	0.36	0.60	0.40
M1.6	3.00	2.75	1.00	0.86	0.10	2.00	15.00	0.70	0.60	0.46	0.65	0.45
M2	3.80	3.50	1.30	1.16	0.10	2.60	16.00	0.80	0.70	0.56	0.85	0.60
(M2.2)	4.00	3.70	1.50	1.36	0.10	2.80	17.00	0.90	0.80	0.66	1.00	0.70
M2.5	4.50	4.20	1.60	1.46	0.10	3.10	18.00	0.90	0.80	0.66	1.00	0.70
M3	5.50	5.20	2.00	1.86	0.10	3.60	19.00	1.00	1.00	0.86	1.30	0.90
(M3.5)	6.00	5.70	2.40	2.26	0.10	4.10	20.00	1.20	1.00	0.86	1.40	1.00
M4	7.00	6.64	2.60	2.46	0.20	4.70	22.00	1.40	1.20	1.06	1.60	1.20
(M4.5)	8.00	7.64	3.10	2.92	0.20	5.20	24.00	1.50	1.20	1.06	1.80	1.40
M5	8.50	8.14	3.30	3.12	0.20	5.70	25.00	1.60	1.51	1.26	2.00	1.50
M6	10.00	9.64	3.90	3.72	0.25	6.80	28.00	2.00	1.91	1.66	2.30	1.80
M8	13.00	12.57	5.00	4.82	0.40	9.20	34.00	2.50	2.31	2.06	2.80	2.30
M10	16.00	15.57	6.00	5.82	0.40	11.20	40.00	3.00	2.81	2.56	3.20	2.70
M12	18.00	17.57	7.00	6.78	0.60	14.20	46.00	3.50	3.31	3.06	3.80	3.20
(M14)	21.00	20.48	8.00	7.78	0.60	16.20	52.00	4.00	3.31	3.06	4.20	3.60
M16	24.00	23.48	9.00	8.78	0.60	18.20	58.00	4.00	4.37	4.07	4.60	4.00
(M18)	27.00	26.48	10.00	9.78	0.60	20.20	64.00	5.00	4.37	4.07	5.10	4.50
M20	30.00	29.48	11.00	10.73	0.80	22.40	70.00	5.00	5.27	5.07	5.60	5.00

[a] Nominal sizes shown in parentheses are non-preferred.

[b] Threaded up to head.

[c] See text following table in *Lengths of Thread on Screws* description in text.

All dimensions are given in millimeters. For dimensional notation, see diagram on page 1590.

General Dimensions: The general dimensions and tolerances for screws and nuts are given in the accompanying tables. Although slotted screw dimensions are given, recessed head screws are also standard and available. Dimensions of recessed head screws are given in BS 4183:1967 (obsolescent).

Table 5. British Standard Machine Screws and Nuts — Metric Series *BS 4183:1967 (obsolescent)*

Concentricity Tolerances

IT 13 IT 13

Countersunk & Raised Countersunk Heads

Pan & Cheese Heads

Slot to Head **Head to Shank**

Nominal Size d^a	Head to Shank and Slot to Head (IT 13) Countersunk, Raised Csk.and Pan Heads	Head to Shank and Slot to Head (IT 13) Cheese Heads
M1	0.14	0.14
M1.2	0.14	0.14
(M1.4)	0.14	0.14
M1.6	0.18	0.14
M2	0.18	0.18
(M2.2)	0.18	0.18
M2.5	0.18	0.18
M3	0.18	0.18
(M3.5)	0.22	0.18
M4	0.22	0.22
(M4.5)	0.22	0.22
M5	0.22	0.22
M6	0.27	0.22
M8	0.27	0.27
M10	0.33	0.27
M12	0.33	0.27
(M14)	0.33	0.33
M16)	0.39	0.33
(M18)	0.39	0.33
M20	0.39	0.33

Nominal Lengths and Tolerances on Length for Machine Screws

Nominal Length[a]	Tolerance	Nominal-Length[a]	Tolerance
1.5	±0.12	45	±0.50
2	±0.12	50	±0.060
2.5	±0.20	55	±0.60
3	±0.20	60	±0.60
4	±0.24	65	±0.60
5	±0.24	70	±0.60
6	±0.24	75	±0.60
(7)	±0.29	80	±0.60
8	±0.29	85	±0.70
(9)	±0.29	90	
10	±0.29	(95)	±0.70
(11)	±0.35	100	±0.70
12	±0.35	(105)	±0.70
14	±0.35	110	±0.70
16	±0.35	(115)	±0.70
(18)	±0.35	120	±0.70
20	±0.42	(125)	±0.70
(22)	±0.42	130	±0.80
25	±0.42	140	±0.80
(28)	±0.42	150	
30	±0.42	160	±0.80
(38)	±0.50	190	±0.925
40	±0.50	200	±0.925

Dimensions of Machine Screw Nuts, Pressed Type, Square and Hexagon

Nominal Size d^a	Width Across Flats s Max.	Width Across Flats s Min.	Width Across Corners e Square
M1.6	3.2	3.02	4.5
M2	4.0	3.82	5.7
(M2.2)	4.5	4.32	6.4
(M2.5)	5.0	4.82	7.1
M3	5.5	5.32	7.8
(M3.5)	6.0	5.82	8.5
M4	7.0	6.78	9.9
M5	8.0	7.78	11.3
M6	10.0	9.78	14.1
M8	13.0	12.73	18.4
M10	17.0	16.73	24.0

Nominal Size d^a	Width Across Corners e	Thickness m Max.	Thickness m Min
M1.6	3.7	1.0	0.75
M2	4.6	1.2	0.95
(M2.2)	5.2	1.2	0.95
M2.5	5.8	1.6	1.35
M3	6.4	1.6	1.35
(M3.5)	6.9	2.0	1.75
M4	8.1	2.0	1.75
M5	9.2	2.5	2.25
M6	11.5	3.0	2.75
M8	15.0	4.0	3.70
M10	19.6	5.0	4.70

[a] Nominal sizes and lengths shown in parentheses are non-preferred.

All dimensions are given in millimeters. For dimensional notation, see diagram on page 1590.

British Unified Machine Screws and Nuts.—British Standard B.S. 1981:1953 covers certain types of machine screws and machine screw nuts for which agreement has been reached with the United States and Canada as to general dimensions for interchangeability. These types are: countersunk, raised-countersunk, pan, and raised-cheese head screws with slotted or recessed heads; small hexagon head screws; and precision and pressed nuts. All have Unified threads. Head shapes are shown on page 1595 and dimensions are given on page 1597.

Identification: As revised by Amendment No. 1 in February 1955, this standard now requires that the above-mentioned screws and nuts that conform to this standard should have a distinguishing feature applied to identify them as Unified. All *recessed head screws* are to be identified as Unified by a groove in the form of four arcs of a circle in the upper surface of the head. All *hexagon head screws* are to be identified as Unified by: 1) a circular recess in the upper surface of the head; 2) a continuous line of circles indented on one or more of the flats of the hexagon and parallel to the screw axis; and 3) at least two contiguous circles indented on the upper surface of the head. All *machine screw* nuts of the pressed type shall be identified as Unified by means of the application of a groove indented in one face of the nut approximately midway between the major diameter of the thread and flats of the square or hexagon. *Slotted head screws* shall be identified as Unified either by a circular recess or by a circular platform or raised portion on the upper surface of the head. *Machine screw nuts* of the *precision type* shall be identified as Unified by either a groove indented on one face of the front approximately midway between the major diameter of the thread and the flats of the hexagon or a continuous line of circles indented on one or more of the flats of the hexagon and parallel to the nut axis.

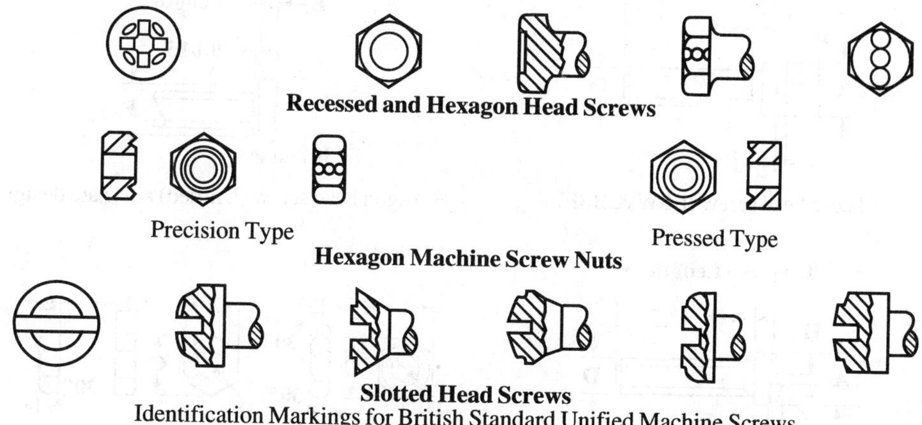

Identification Markings for British Standard Unified Machine Screws

British Standard Machine Screws and Nuts

BS 450:1958 (obsolescent) and B.S. 1981:1953

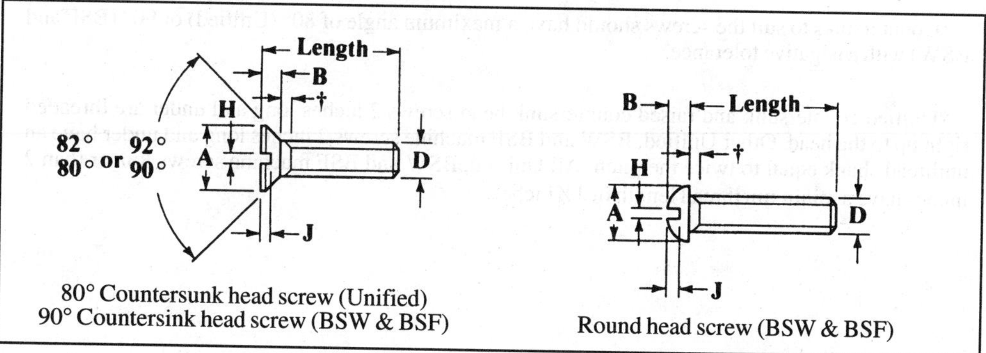

80° Countersunk head screw (Unified)
90° Countersink head screw (BSW & BSF)

Round head screw (BSW & BSF)

British Standard Machine Screws and Nuts

BS 450:1958 (obsolescent) and B.S. 1981:1953

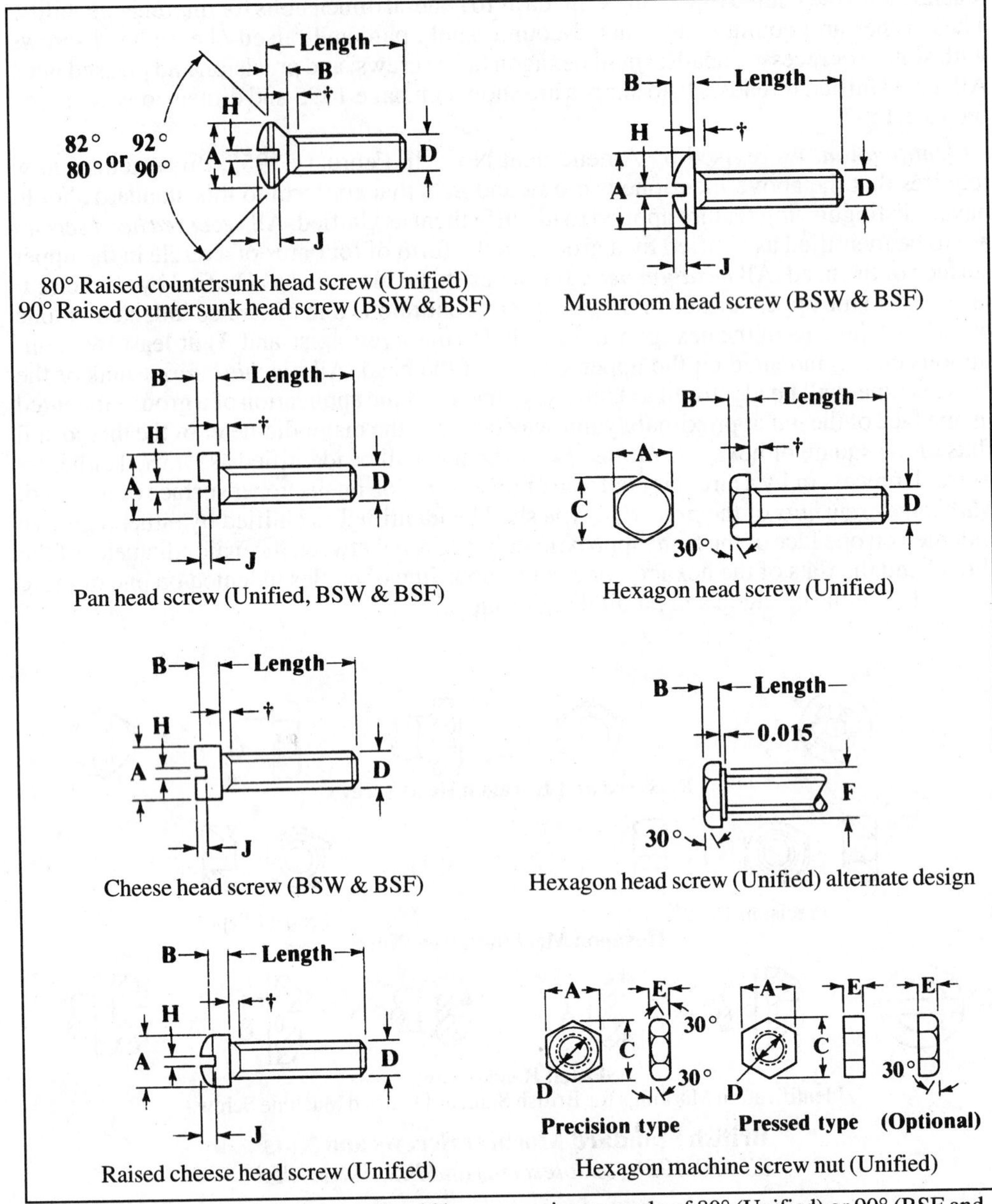

80° Raised countersunk head screw (Unified)
90° Raised countersunk head screw (BSW & BSF)

Mushroom head screw (BSW & BSF)

Pan head screw (Unified, BSW & BSF)

Hexagon head screw (Unified)

Cheese head screw (BSW & BSF)

Hexagon head screw (Unified) alternate design

Raised cheese head screw (Unified)

Hexagon machine screw nut (Unified)

*Countersinks to suit the screws should have a maximum angle of 80° (Unified) or 90° (BSF and BSW) with a negative tolerance.

†Unified countersunk and raised countersunk head screws 2 inches long and under are threaded right up to the head. Other Unified, BSW and BSF machine screws 2 inches long and under have an unthread shank equal to twice the pitch. All Unified, BSW and BSF machine screws longer than 2 inches have a minimum thread length of 1¾ inches.

British Standard Unified Machine Screws and Nuts *B.S. 1981:1953*

Nom.Size of Screw	Basic Dia. *D*	Threads per Inch UNC	UNF	Dia. of Head *A* Max.	Min.	Depth of Head *B* Max.	Min.	Width of Slot *H* Max.	Min.	Depth of Slot *J*
80° Countersunk Head Screws[a,b]										
4	0.112	40	…	0.211	0.194	0.067	…	0.039	0.031	0.025
6	0.138	32	…	0.260	0.242	0.083	…	0.048	0.039	0.031
8	0.164	32	…	0.310	0.291	0.100	…	0.054	0.045	0.037
10	0.190	24[c]	32	0.359	0.339	0.116	…	0.060	0.050	0.044
1/4	0.250	20	28	0.473	0.450	0.153	…	0.075	0.064	0.058
5/16	0.3125	18	24	0.593	0.565	0.191	…	0.084	0.072	0.073
3/8	0.375	16	24	0.712	0.681	0.230	…	0.094	0.081	0.086
7/16	0.4375	14	20	0.753	0.719	0.223	…	0.094	0.081	0.086
1/2	0.500	13	20	0.808	0.770	0.223	…	0.106	0.091	0.086
5/8	0.625	11	18	1.041	0.996	0.298	…	0.133	0.116	0.113
3/4	0.750	10	16	1.275	1.223	0.372	…	0.149	0.131	0.141
Pan Head Screws[b]										
4	0.112	40	…	0.219	0.205	0.068	0.058	0.039	0.031	0.036
6	0.138	32	…	0.270	0.256	0.082	0.072	0.048	0.039	0.044
8	0.164	32	…	0.322	0.306	0.096	0.085	0.054	0.045	0.051
10	0.190	24[c]	32	0.373	0.357	0.110	0.099	0.060	0.050	0.059
1/4	0.250	20	28	0.492	0.473[d]	0.144	0.130	0.075	0.064	0.079
5/16	0.3125	18	24	0.615	0.594	0.178	0.162	0.084	0.072	0.101
3/8	0.375	16	24	0.740	0.716	0.212	0.195	0.094	0.081	0.122
7/16	0.4375	14	20	0.863	0.838	0.247	0.227	0.094	0.081	0.133
1/2	0.500	13	20	0.987	0.958	0.281	0.260	0.106	0.091	0.152
5/8	0.625	11	18	1.125	1.090	0.350	0.325	0.133	0.116	0.189
3/4	0.750	10	16	1.250	1.209	0.419	0.390	0.149	0.131	0.226
Raised Cheese-Head Screws[b]										
4	0.112	40	…	0.183	0.166	0.107	0.088	0.039	0.031	0.042
6	0.138	32	…	0.226	0.208	0.132	0.111	0.048	0.039	0.053
8	0.164	32	…	0.270	0.250	0.156	0.133	0.054	0.045	0.063
10	0.190	24[c]	32	0.313	0.292	0.180	0.156	0.060	0.050	0.074
1/4	0.250	20	28	0.414	0.389	0.237	0.207	0.075	0.064	0.098
5/16	0.3125	18	24	0.518	0.490	0.295	0.262	0.084	0.072	0.124
3/8	0.375	16	24	0.622	0.590	0.355	0.315	0.094	0.081	0.149
7/16	0.4375	14	20	0.625	0.589	0.368	0.321	0.094	0.081	0.153
1/2	0.500	13	20	0.750	0.710	0.412	0.362	0.106	0.091	0.171
5/8	0.625	11	18	0.875	0.827	0.521	0.461	0.133	0.116	0.217
3/4	0.750	10	16	1.000	0.945	0.612	0.542	0.149	0.131	0.254

[a] All dimensions, except *J*, given for the No. 4 to 3/8-inch sizes, incl., also apply to all the 80° Raised Countersunk Head Screws given in the Standard.

[b] Also available with recessed heads.

[c] Non-preferred.

[d] By arrangement may also be 0.468.

Nom. Size	Basic Dia. *D*	Threads per Inch UNC	UNF	Width Across Flats *A* Max.	Min.	Corners *C* Max.	H'd Depth *B* Nut Thick. *E* Max.	Min.	Wash. Face Dia. *F* Max.	Min.
Hexagon Head Screws										
4	0.112	40	…	0.1875	0.1835	0.216	0.060	0.055	0.183	0.173
6	0.138	32	…	0.2500	0.2450	0.289	0.080	0.074	0.245	0.235
8	0.164	32	…	0.2500	0.2450	0.289	0.110	0.104	0.245	0.235
10	0.190	24[c]	32	0.3125	0.3075	0.361	0.120	0.113	0.307	0.297
Hexagon Machine Screw Nuts—Precision Type										
4	0.112	40	…	0.1875	0.1835	0.216	0.098	0.087	…	…
6	0.138	32	…	0.2500	0.2450	0.269	0.114	0.102	…	…
8	0.164	32	…	0.3125	0.3075	0.361	0.130	0.117	…	…
10	0.190	24[c]	…	0.3125	0.3075	0.361	0.130	0.117	…	…
Hexagon Machine Screw Nuts—Pressed Type										
4	0.112	40	…	0.2500	0.2410	0.289	0.087	0.077	…	…
6	0.138	32	…	0.3125	0.3020	0.361	0.114	0.102	…	…
8	0.164	32	…	0.3438	0.3320	0.397	0.130	0.117	…	…
10	0.190	24[c]	32	0.3750	0.3620	0.433	0.130	0.117	…	…
1/4	0.250	20	28	0.4375	0.4230	0.505	0.193	0.178	…	…
5/16	0.3125	18	24	0.5625	0.5450	0.649	0.225	0.208	…	…
3/8	0.375	16	24	0.6250	0.6070	0.722	0.257	0.239	…	…

All dimensions in inches. See page 1595 for a pictorial representation and letter dimensions.

British Standard Whitworth (BSW) and Fine (BSF) Machine Screws
BS 450:1958 (obsolescent)

	Nom. Size of Screw	Basic Dia. D	Threads per Inch		Dia. of Head A		Depth of Head B		Width of Slot H		Depth of Slot J
			BSW	BSF	Max.	Min.	Max.	Min.	Max.	Min.	
90° Countersunk Head Screws[a,b]	1/8	0.1250	40	…	0.219	0.201	0.056	…	0.039	0.032	0.027
	3/16	0.1875	24	32[c]	0.328	0.307	0.084	…	0.050	0.042	0.041
	7/32	0.2188	…	28[c]	0.383	0.360	0.098	…	0.055	0.046	0.048
	1/4	0.2500	20	26	0.438	0.412	0.113	…	0.061	0.051	0.055
	5/16	0.3125	18	22	0.547	0.518	0.141	…	0.071	0.061	0.069
	3/8	0.3750	16	20	0.656	0.624	0.169	…	0.082	0.072	0.083
	7/16	0.4375	14	18	0.766	0.729	0.197	…	0.093	0.082	0.097
	1/2	0.5000	12	16	0.875	0.835	0.225	…	0.104	0.092	0.111
	9/16	0.5625	12[c]	16[c]	0.984	0.941	0.253	…	0.115	0.103	0.125
	5/8	0.6250	11	14	1.094	1.046	0.281	…	0.126	0.113	0.138
	3/4	0.7500	10	12	1.312	1.257	0.338	…	0.148	0.134	0.166
Round Head Screws[b]	1/8	0.1250	40	…	0.219	0.206	0.087	0.082	0.039	0.032	0.048
	3/16	0.1875	24	32[c]	0.328	0.312[d]	0.131	0.124	0.050	0.042	0.072
	7/32	0.2188	…	28[c]	0.383	0.365	0.153	0.145	0.055	0.046	0.084
	1/4	0.2500	20	26	0.438	0.417	0.175	0.165	0.061	0.051	0.096
	5/16	0.3125	18	22	0.547	0.524	0.219	0.207	0.071	0.061	0.120
	3/8	0.3750	16	20	0.656	0.629	0.262	0.249	0.082	0.072	0.144
	7/16	0.4375	14	18	0.766	0.735	0.306	0.291	0.093	0.082	0.168
	1/2	0.5000	12	16	0.875	0.840	0.350	0.333	0.104	0.092	0.192
	9/16	0.5625	12[c]	16[c]	0.984	0.946	0.394	0.375	0.115	0.103	0.217
	5/8	0.6250	11	14	1.094	1.051	0.437	0.417	0.126	0.113	0.240
	3/4	0.7500	10	12	1.312	1.262	0.525	0.500	0.148	0.134	0.288
Pan Head Screws[b]	1/8	0.1250	40	…	0.245	0.231	0.075	0.065	0.039	0.032	0.040
	3/16	0.1875	24	32[c]	0.373	0.375	0.110	0.099	0.050	0.042	0.061
	7/32	0.2188	…	28[c]	0.425	0.407	0.125	0.112	0.055	0.046	0.069
	1/4	0.2500	20	26	0.492	0.473[e]	0.144	0.130	0.061	0.051	0.078
	5/16	0.3125	18	22	0.615	0.594	0.178	0.162	0.071	0.061	0.095
	3/8	0.3750	16	20	0.740	0.716	0.212	0.195	0.082	0.072	0.112
	7/16	0.4375	14	18	0.863	0.838	0.247	0.227	0.093	0.082	0.129
	1/2	0.5000	12	16	0.987	0.958	0.281	0.260	0.104	0.092	0.145
	9/16	0.5625	12[c]	16[c]	1.031	0.999	0.315	0.293	0.115	0.103	0.162
	5/8	0.6250	11	14	1.125	1.090	0.350	0.325	0.126	0.113	0.179
	3/4	0.7500	10	12	1.250	1.209	0.419	0.390	0.148	0.134	0.213
Cheese Head Screws[b]	1/8	0.1250	40	…	0.188	0.180	0.087	0.082	0.039	0.032	0.039
	3/16	0.1875	24	32[c]	0.281	0.270	0.131	0.124	0.050	0.042	0.059
	7/32	0.2188	…	28[c]	0.328	0.315	0.153	0.145	0.055	0.046	0.069
	1/4	0.2500	20	26	0.375	0.360	0.175	0.165	0.061	0.051	0.079
	5/16	0.3125	18	22	0.469	0.450	0.219	0.207	0.071	0.061	0.098
	3/8	0.3750	16	20	0.562	0.540	0.262	0.249	0.082	0.072	0.118
	7/16	0.4375	14	18	0.656	0.630	0.306	0.291	0.093	0.082	0.138
	1/2	0.5000	12	16	0.750	0.720	0.350	0.333	0.104	0.092	0.157
	9/16	0.5625	12[c]	16[c]	0.844	0.810	0.394	0.375	0.115	0.103	0.177
	5/8	0.6250	11	14	0.938	0.900	0.437	0.417	0.126	0.113	0.197
	3/4	0.7500	10	12	1.125	1.080	0.525	0.500	0.148	0.134	0.236
Mushroom Head Screws[b]	1/8	0.1250	40	…	0.289	0.272	0.078	0.066	0.043	0.035	0.040
	3/16	0.1875	24	32[c]	0.448	0.425	0.118	0.103	0.060	0.050	0.061
	1/4	0.2500	20	26	0.573	0.546	0.150	0.133	0.075	0.064	0.079
	5/16	0.3125	18	22	0.698	0.666	0.183	0.162	0.084	0.072	0.096
	3/8	0.3750	16	20	0.823	0.787	0.215	0.191	0.094	0.081	0.112

[a] All dimensions, except *J*, given for the 1/8-through 3/8-inch sizes also apply to all the 90° Raised Countersunk Head Screw dimensions given in the Standard.

[b] These screws are also available with recessed heads; dimensions of recess are not given here but may be found in the Standard.

[c] Non-preferred size; avoid use whenever possible.

[d] By arrangement may also be 0.309.

[e] By arrangement may also be 0.468.

All dimensions in inches.

See diagram on page 1595 for a pictorial representation of screws and letter dimensions.

CAP AND SET SCREWS

Slotted Head Cap Screws.—American National Standard ANSI/ASME B18.6.2-1998 is intended to cover the complete general and dimensional data for the various styles of slotted head cap screws as well as square head and slotted headless set screws (see page 1606). Reference should be made to this Standard for information or data not found in the following text or tables.

Length of Thread: The length of complete (full form) thread on cap screws is equal to twice the basic screw diameter plus 0.250 in. with a plus tolerance of 0.188 in. or an amount equal to 2½ times the pitch of the thread, whichever is greater. Cap screws of lengths too short to accommodate the minimum thread length have full form threads extending to within a distance equal to 2½ pitches (threads) of the head.

Designation: Slotted head cap screws are designated by the following data in the sequence shown: Nominal size (fraction or decimal equivalent); threads per inch; screw length (fraction or decimal equivalent); product name; material; and protective finish, if required. Examples: ½-13 × 3 Slotted Round Head Cap Screw, SAE Grade 2 Steel, Zinc Plated. .750-16 × 2.25 Slotted Flat Countersunk Head Cap Screw, Corrosion Resistant Steel.

Table 1. American National Standard Slotted Flat Countersunk Head Cap Screws ***ANSI/ASME B18.6.2-1998***

Nominal Size[a] or Basic Screw Diam.		Body Diam., *E*		Head Dia., *A*		Head Hgt., *H*	Slot Width, *J*		Slot Depth, *T*		Filet Rad., *U*
				Edge Sharp	Edge Rnd'd. or Flat						
		Max.	Min.	Max.	Min.	Ref.	Max.	Min.	Max.	Min.	Max.
¼	0.2500	.2500	.2450	.500	.452	.140	.075	.064	.068	.045	.100
5/16	0.3125	.3125	.3070	.625	.567	.177	.084	.072	.086	.057	.125
⅜	0.3750	.3750	.3690	.750	.682	.210	.094	.081	.103	.068	.150
7/16	0.4375	.4375	.4310	.812	.736	.210	.094	.081	.103	.068	.175
½	0.5000	.5000	.4930	.875	.791	.210	.106	.091	.103	.068	.200
9/16	0.5625	.5625	.5550	1.000	.906	.244	.118	.102	.120	.080	.225
⅝	0.6250	.6250	.6170	1.125	1.020	.281	.133	.116	.137	.091	.250
¾	0.7500	.7500	.7420	1.375	1.251	.352	.149	.131	.171	.115	.300
⅞	0.8750	.8750	.8660	1.625	1.480	.423	.167	.147	.206	.138	.350
1	1.0000	1.0000	.9900	1.875	1.711	.494	.188	.166	.240	.162	.400
1⅛	1.1250	1.1250	1.1140	2.062	1.880	.529	.196	.178	.257	.173	.450
1¼	1.2500	1.2500	1.2390	2.312	2.110	.600	.211	.193	.291	.197	.500
1⅜	1.3750	1.3750	1.3630	2.562	2.340	.665	.226	.208	.326	.220	.550
1½	1.5000	1.5000	1.4880	2.812	2.570	.742	.258	.240	.360	.244	.600

[a] When specifying a nominal size in decimals, the zero preceding the decimal point is omitted as is any zero in the fourth decimal place.

All dimensions are in inches.

Threads: Threads are Unified Standard Class 2A; UNC, UNF and 8 UN Series or UNRC, UNRF, and 8 UNR Series.

Table 2. American National Standard Slotted Round Head Cap Screws *ANSI/ASME B18.6.2-1998*

Nom. Size[a] or Basic Screw Diameter		Body Diameter, E		Head Diameter, A		Head Height, H		Slot Width, J		Slot Depth, T	
		Max.	Min.	Max.	Min.	Max.	Min.	Max.	Min.	Max.	Min.
1/4	0.2500	.2500	.2450	.437	.418	.191	.175	.075	.064	.117	.097
5/16	0.3125	.3125	.3070	.562	.540	.245	.226	.084	.072	.151	.126
3/8	0.3750	.3750	.3690	.625	.603	.273	.252	.094	.081	.168	.138
7/16	0.4375	.4375	.4310	.750	.725	.328	.302	.094	.081	.202	.167
1/2	0.5000	.5000	.4930	.812	.786	.354	.327	.106	.091	.218	.178
9/16	0.5625	.5625	.5550	.937	.909	.409	.378	.118	.102	.252	.207
5/8	0.6250	.6250	.6170	1.000	.970	.437	.405	.133	.116	.270	.220
3/4	0.7500	.7500	.7420	1.250	1.215	.546	.507	.149	.131	.338	.278

[a] When specifying a nominal size in decimals, the zero preceding the decimal point is omitted as is any zero in the fourth decimal place.

All dimensions are in inches.

Fillet Radius, U: For fillet radius see foonote to table below.

Threads: Threads are Unified Standard Class 2A; UNC, UNF and 8 UN Series or UNRC, UNRF and 8 UNR Series.

Table 3. American National Standard Slotted Fillister Head Cap Screws *ANSI/ASME B18.6.2-1998*

Nom. Size[a] or Basic Screw Dia.		Body Dia.., E		Head Dia.., A		Head Side Height, H		Total Head Height, O		Slot Width, J		Slot Depth, T	
		Max.	Min.	Max.	Min.	Max.	Min.	Max.	Min.	Max.	Min.	Max.	Min.
1/4	0.2500	.2500	.2450	.375	.363	.172	.157	.216	.194	.075	.064	.097	.077
5/16	0.3125	.3125	.3070	.437	.424	.203	.186	.253	.230	.084	.072	.115	.090
3/8	0.3750	.3750	.3690	.562	.547	.250	.229	.314	.284	.094	.081	.142	.112
7/16	0.4375	.4375	.4310	.625	.608	.297	.274	.368	.336	.094	.081	.168	.133
1/2	0.5000	.5000	.4930	.750	.731	.328	.301	.413	.376	.106	.091	.193	.153
9/16	0.5625	.5625	.5550	.812	.792	.375	.346	.467	.427	.118	.102	.213	.168
5/8	0.6250	.6250	.6170	.875	.853	.422	.391	.521	.478	.133	.116	.239	.189
3/4	0.7500	.7500	.7420	1.000	.976	.500	.466	.612	.566	.149	.131	.283	.223
7/8	0.8750	.8750	.8660	1.125	1.098	.594	.556	.720	.668	.167	.147	.334	.264
1	1.0000	1.0000	.9900	1.312	1.282	.656	.612	.803	.743	.188	.166	.371	.291

[a] When specifying nominal size in decimals, the zero preceding the decimal point is omitted as is any zero in the fourth decimal place.

All dimensions are in inches.

Fillet Radius, U: The fillet radius is as follows: For screw sizes 1/4 to 3/8 incl., .031 max. and .016 min.; 7/16 to 9/16, incl., .047 max., .016 min.; and for 5/8 to 1, incl., .062 max., .031 min.

Threads: Threads are Unified Standard Class 2A; UNC, UNF and 8 UN Series or UNRC, UNRF and 8 UNR Series.

Table 4. American National Standard Hexagon and Spline Socket Head Cap Screws *ANSI/ASME B18.3-1998*

Nominal Size	Body Diameter, D Max	Body Diameter, D Min	Head Diameter, A Max	Head Diameter, A Min	Head Height, H Max	Head Height, H Min	Spline Socket Size M Nom	Hex. Socket Size J Nom	Fillet Ext. F Max	Key Engagement[a] T
0	0.0600	0.0568	0.096	0.091	0.060	0.057	0.060	0.050	0.007	0.025
1	0.0730	0.0695	0.118	0.112	0.073	0.070	0.072	1/16 0.062	0.007	0.031
2	0.0860	0.0822	0.140	0.134	0.086	0.083	0.096	5/64 0.078	0.008	0.038
3	0.0990	0.0949	0.161	0.154	0.099	0.095	0.096	5/64 0.078	0.008	0.044
4	0.1120	0.1075	0.183	0.176	0.112	0.108	0.111	3/32 0.094	0.009	0.051
5	0.1250	0.1202	0.205	0.198	0.125	0.121	0.111	3/32 0.094	0.010	0.057
6	0.1380	0.1329	0.226	0.218	0.138	0.134	0.133	7/64 0.109	0.010	0.064
8	0.1640	0.1585	0.270	0.262	0.164	0.159	0.168	9/64 0.141	0.012	0.077
10	0.1900	0.1840	0.312	0.303	0.190	0.185	0.183	5/32 0.156	0.014	0.090
1/4	0.2500	0.2435	0.375	0.365	0.250	0.244	0.216	3/16 0.188	0.014	0.120
5/16	0.3125	0.3053	0.469	0.457	0.312	0.306	0.291	1/4 0.250	0.017	0.151
3/8	0.3750	0.3678	0.562	0.550	0.375	0.368	0.372	5/16 0.312	0.020	0.182
7/16	0.4375	0.4294	0.656	0.642	0.438	0.430	0.454	3/8 0.375	0.023	0.213
1/2	0.5000	0.4919	0.750	0.735	0.500	0.492	0.454	3/8 0.375	0.026	0.245
5/8	0.6250	0.6163	0.938	0.921	0.625	0.616	0.595	1/2 0.500	0.032	0.307
3/4	0.7500	0.7406	1.125	1.107	0.750	0.740	0.620	5/8 0.625	0.039	0.370
7/8	0.8750	0.8647	1.312	1.293	0.875	0.864	0.698	3/4 0.750	0.044	0.432
1	1.0000	0.9886	1.500	1.479	1.000	0.988	0.790	3/4 0.750	0.050	0.495
1 1/8	1.1250	1.1086	1.688	1.665	1.125	1.111	…	7/8 0.875	0.055	0.557
1 1/4	1.2500	1.2336	1.875	1.852	1.250	1.236	…	7/8 0.875	0.060	0.620
1 3/8	1.3750	1.3568	2.062	2.038	1.375	1.360	…	1 1.000	0.065	0.682
1 1/2	1.5000	1.4818	2.250	2.224	1.500	1.485	…	1 1.000	0.070	0.745
1 3/4	1.7500	1.7295	2.625	2.597	1.750	1.734	…	1 1/4 1.250	0.080	0.870
2	2.0000	1.9780	3.000	2.970	2.000	1.983	…	1 1/2 1.500	0.090	0.995
2 1/4	2.2500	2.2280	3.375	3.344	2.250	2.232	…	1 3/4 1.750	0.100	1.120
2 1/2	2.5000	2.4762	3.750	3.717	2.500	2.481	…	1 3/4 1.750	0.110	1.245
2 3/4	2.7500	2.7262	4.125	4.090	2.750	2.730	…	2 2.000	0.120	1.370
3	3.0000	2.9762	4.500	4.464	3.000	2.979	…	2 1/4 2.250	0.130	1.495
3 1/4	3.2500	3.2262	4.875	4.837	3.250	3.228	…	2 1/4 2.250	0.140	1.620
3 1/2	3.5000	3.4762	5.250	5.211	3.500	3.478	…	2 3/4 2.750	0.150	1.745
3 3/4	3.7500	3.7262	5.625	5.584	3.750	3.727	…	2 3/4 2.750	0.160	1.870
4	4.0000	3.9762	6.000	5.958	4.000	3.976	…	3 3.000	0.170	1.995

[a] (Key engagement depths are minimum.)

All dimensions in inches. The body length L_B of the screw is the length of the unthreaded cylindrical portion of the shank. The length of thread, L_T, is the distance from the extreme point to the last complete (full form) thread. Standard length increments for screw diameters up to 1 inch are 1/16 inch for lengths 1/8 through 1/4 inch, 1/8 inch for lengths 1/4 through 1 inch, 1/4 inch for lengths 1 through 3 1/2 inches, 1/2 inch for lengths 3 1/2 through 7 inches, 1 inch for lengths 7 through 10 inches and for diameters over 1 inch are 1/2 inch for lengths 1 through 7 inches, 1 inch for lengths 7 through 10 inches, and 2 inches for lengths over 10 inches. Heads may be plain or knurled, and chamfered to an angle *E* of 30 to 45 degrees with the surface of the flat. The thread conforms to the Unified Standard with radius root, Class 3A UNRC and UNRF for screw sizes No. 0 through 1 inch inclusive, Class 2A UNRC and UNRF for over 1 inch through 1 1/2 inches inclusive, and Class 2A UNRC for larger sizes. Socket dimensions are given in Table 11. For details not shown, including materials, see ANSI/ASME B18.3-1998.

Table 5. Drill and Counterbore Sizes For Socket Head Cap Screws (1960 Series)

Nominal Size or Basic Screw Diameter		Nominal Drill Size				Counterbore Diameter	Countersink Diameter[a]
		Close Fit[b]		Normal Fit[c]			
		Number or Fractional Size	Decimal Size	Number or Fractional Size	Decimal Size		
		A				B	C
0	0.0600	51	0.067	49	0.073	1/8	0.074
1	0.0730	46	0.081	43	0.089	5/32	0.087
2	0.0860	3/32	0.094	36	0.106	3/16	0.102
3	0.0990	36	0.106	31	0.120	7/32	0.115
4	0.1120	1/8	0.125	29	0.136	7/32	0.130
5	0.1250	9/64	0.141	23	0.154	1/4	0.145
6	0.1380	23	0.154	18	0.170	9/32	0.158
8	0.1640	15	0.180	10	0.194	5/16	0.188
10	0.1900	5	0.206	2	0.221	3/8	0.218
1/4	0.2500	17/64	0.266	9/32	0.281	7/16	0.278
5/16	0.3125	21/64	0.328	11/32	0.344	17/32	0.346
3/8	0.3750	25/64	0.391	13/32	0.406	5/8	0.415
7/16	0.4375	29/64	0.453	15/32	0.469	23/32	0.483
1/2	0.5000	33/64	0.516	17/32	0.531	13/16	0.552
5/8	0.6250	41/64	0.641	21/32	0.656	1	0.689
3/4	0.7500	49/64	0.766	25/32	0.781	1 3/16	0.828
7/8	0.8750	57/64	0.891	29/32	0.906	1 3/8	0.963
1	1.0000	1 1/64	1.016	1 1/32	1.031	1 5/8	1.100
1 1/4	1.2500	1 9/32	1.281	1 5/16	1.312	2	1.370
1 1/2	1.5000	1 17/32	1.531	1 9/16	1.562	2 3/8	1.640
1 3/4	1.7500	1 25/32	1.781	1 13/16	1.812	2 3/4	1.910
2	2.0000	2 1/32	2.031	2 1/16	2.062	3 1/8	2.180

[a] *Countersink:* It is considered good practice to countersink or break the edges of holes which are smaller than (D Max + $2F$ Max) in parts having a hardness which approaches, equals or exceeds the screw hardness. If such holes are not countersunk, the heads of screws may not seat properly or the sharp edges on holes may deform the fillets on screws thereby making them susceptible to fatigue in applications involving dynamic loading. The countersink or corner relief, however, should not be larger than is necessary to insure that the fillet on the screw is cleared.

[b] *Close Fit:* The close fit is normally limited to holes for those lengths of screws which are threaded to the head in assemblies where only one screw is to be used or where two or more screws are to be used and the mating holes are to be produced either at assembly or by matched and coordinated tooling.

[c] *Normal Fit:* The normal fit is intended for screws of relatively long length or for assemblies involving two or more screws where the mating holes are to be produced by conventional tolerancing methods. It provides for the maximum allowable eccentricity of the longest standard screws and for certain variations in the parts to be fastened, such as: deviations in hole straightness, angularity between the axis of the tapped hole and that of the hole for the shank, differences in center distances of the mating holes, etc.

All dimensions in inches.

Source: Appendix to American National Standard ANSI/ASME B18.3-1998.

Table 6. American National Standard Hexagon and Spline Socket Flat Countersunk Head Cap Screws *ANSI/ASME B18.3-1998*

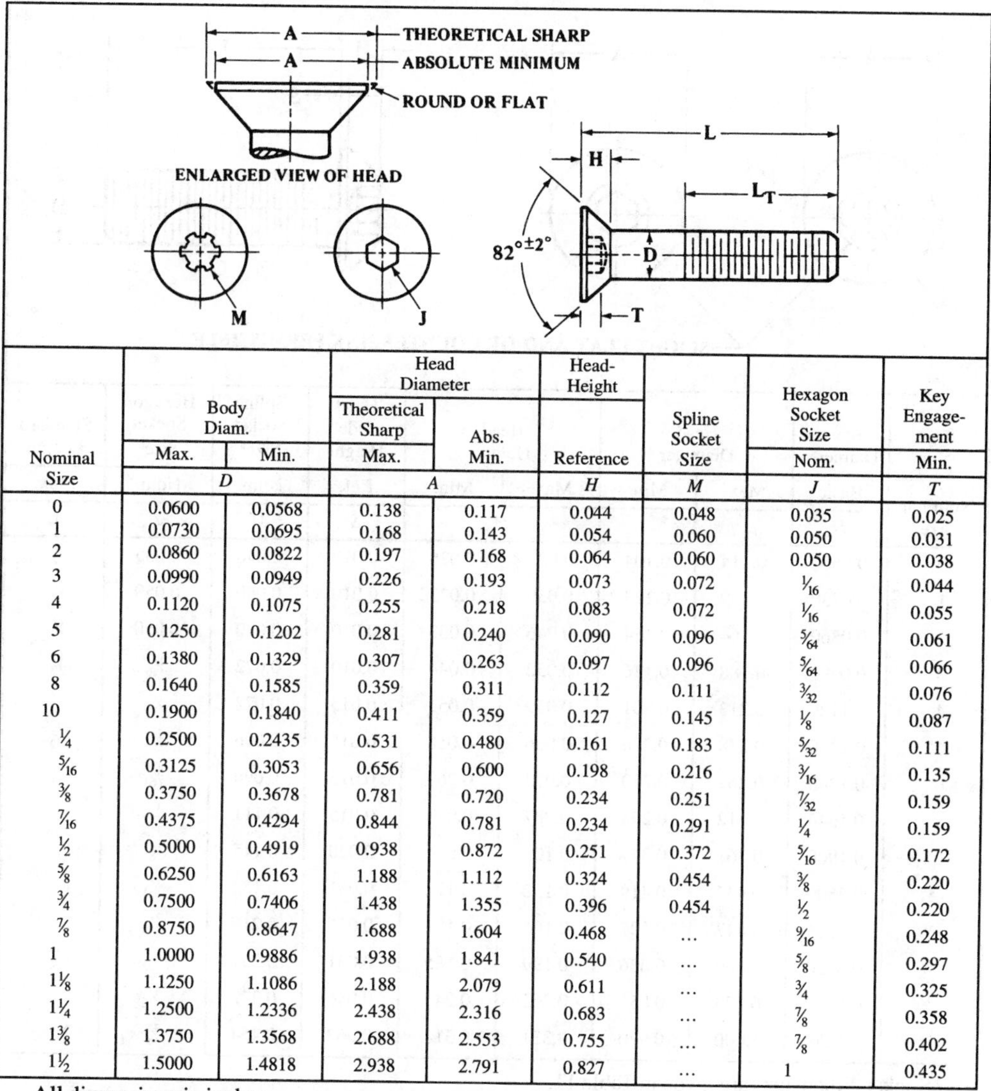

Nominal Size	Body Diam. Max.	Body Diam. Min.	Head Diameter Theoretical Sharp Max.	Head Diameter Abs. Min.	Head-Height Reference	Spline Socket Size	Hexagon Socket Size Nom.	Key Engagement Min.
	D		A		H	M	J	T
0	0.0600	0.0568	0.138	0.117	0.044	0.048	0.035	0.025
1	0.0730	0.0695	0.168	0.143	0.054	0.060	0.050	0.031
2	0.0860	0.0822	0.197	0.168	0.064	0.060	0.050	0.038
3	0.0990	0.0949	0.226	0.193	0.073	0.072	1⁄16	0.044
4	0.1120	0.1075	0.255	0.218	0.083	0.072	1⁄16	0.055
5	0.1250	0.1202	0.281	0.240	0.090	0.096	5⁄64	0.061
6	0.1380	0.1329	0.307	0.263	0.097	0.096	5⁄64	0.066
8	0.1640	0.1585	0.359	0.311	0.112	0.111	3⁄32	0.076
10	0.1900	0.1840	0.411	0.359	0.127	0.145	1⁄8	0.087
1⁄4	0.2500	0.2435	0.531	0.480	0.161	0.183	5⁄32	0.111
5⁄16	0.3125	0.3053	0.656	0.600	0.198	0.216	3⁄16	0.135
3⁄8	0.3750	0.3678	0.781	0.720	0.234	0.251	7⁄32	0.159
7⁄16	0.4375	0.4294	0.844	0.781	0.234	0.291	1⁄4	0.159
1⁄2	0.5000	0.4919	0.938	0.872	0.251	0.372	5⁄16	0.172
5⁄8	0.6250	0.6163	1.188	1.112	0.324	0.454	3⁄8	0.220
3⁄4	0.7500	0.7406	1.438	1.355	0.396	0.454	1⁄2	0.220
7⁄8	0.8750	0.8647	1.688	1.604	0.468	…	9⁄16	0.248
1	1.0000	0.9886	1.938	1.841	0.540	…	5⁄8	0.297
1 1⁄8	1.1250	1.1086	2.188	2.079	0.611	…	3⁄4	0.325
1 1⁄4	1.2500	1.2336	2.438	2.316	0.683	…	7⁄8	0.358
1 3⁄8	1.3750	1.3568	2.688	2.553	0.755	…	7⁄8	0.402
1 1⁄2	1.5000	1.4818	2.938	2.791	0.827	…	1	0.435

All dimensions in inches.

The body of the screw is the unthreaded cylindrical portion of the shank where not threaded to the head; the shank being the portion of the screw from the point of juncture of the conical bearing surface and the body to the flat of the point. The length of thread L_T is the distance measured from the extreme point to the last complete (full form) thread.

Standard length increments of No. 0 through 1-inch sizes are as follows: 1⁄16 inch for nominal screw lengths of 1⁄8 through 1⁄4 inch; 1⁄8 inch for lengths of 1⁄4 through 1 inch; 1⁄4 inch for lengths of 1 inch through 3 1⁄2 inches; 1⁄2 inch for lengths of 3 1⁄2 through 7 inches; and 1 inch for lengths of 7 through 10 inches, incl. For screw sizes over 1 inch, length increments are: 1⁄2 inch for nominal screw lengths of 1 inch through 7 inches; 1 inch for lengths of 7 through 10 inches; and 2 inches for lengths over 10 inches.

Threads shall be Unified external threads with radius root; Class 3A UNRC and UNRF series for sizes No. 0 through 1 inch and Class 2A UNRC and UNRF series for sizes over 1 inch to 1 1⁄2 inches, incl.

For manufacturing details not shown, including materials, see American National Standard ANSI/ASME B18.3-1998 Socket dimensions are given in Table 11.

Table 7. American National Standard Hexagon Socket and Spline Socket Button Head Cap Screws ***ANSI/ASME B18.3-1998***

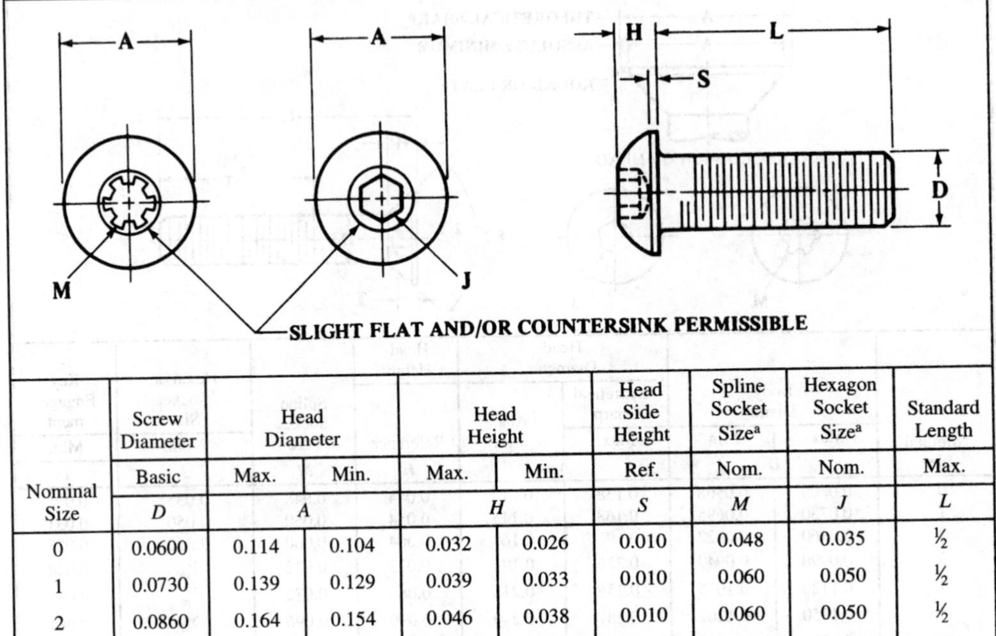

Nominal Size	Screw Diameter	Head Diameter		Head Height		Head Side Height	Spline Socket Size[a]	Hexagon Socket Size[a]	Standard Length
	Basic	Max.	Min.	Max.	Min.	Ref.	Nom.	Nom.	Max.
	D	*A*		*H*		*S*	*M*	*J*	*L*
0	0.0600	0.114	0.104	0.032	0.026	0.010	0.048	0.035	½
1	0.0730	0.139	0.129	0.039	0.033	0.010	0.060	0.050	½
2	0.0860	0.164	0.154	0.046	0.038	0.010	0.060	0.050	½
3	0.0990	0.188	0.176	0.052	0.044	0.010	0.072	1/16	½
4	0.1120	0.213	0.201	0.059	0.051	0.015	0.072	1/16	½
5	0.1250	0.238	0.226	0.066	0.058	0.015	0.096	5/64	½
6	0.1380	0.262	0.250	0.073	0.063	0.015	0.096	5/64	⅝
8	0.1640	0.312	0.298	0.087	0.077	0.015	0.111	3/32	¾
10	0.1900	0.361	0.347	0.101	0.091	0.020	0.145	⅛	1
¼	0.2500	0.437	0.419	0.132	0.122	0.031	0.183	5/32	1
5/16	0.3125	0.547	0.527	0.166	0.152	0.031	0.216	3/16	1
⅜	0.3750	0.656	0.636	0.199	0.185	0.031	0.251	7/32	1¼
½	0.5000	0.875	0.851	0.265	0.245	0.046	0.372	5/16	2
⅝	0.6250	1.000	0.970	0.331	0.311	0.062	0.454	⅜	2

[a] Socket dimensions are given in Table 11.

All dimensions in inches.

These cap screws have been designed and recommended for light fastening applications. They are not suggested for use in critical high-strength applications where socket head cap screws should normally be used.

Standard length increments for socket button head cap screws are as follows: 1/16 inch for nominal screw lengths of ⅛ through ¼ inch, ⅛ inch for nominal screw lengths of ¼ through 1 inch, and ¼ inch for nominal screw lengths of 1 inch through 2 inches. Tolerances on lengths are –0.03 inch for lengths up to 1 inch inclusive. For lengths from 1 through 2 inches, inclusive, length tolerances are –0.04 inch.

The thread conforms to the Unified standard, Class 3A, with radius root, UNRC and UNRF.

To prevent interference, American National Standard ANSI/ASME B18.3.4M-1986 gives metric dimensional and general requirements for a lower head profile hexagon socket button head cap screw. Because of its design, wrenchability and other design factors are reduced; therefore, B18.3.4M should be reviewed carefully. Available only in metric sizes and with metric threads.

For manufacturing details, including materials, not shown, see American National Standard ANSI/ASME B18.3-1998

Table 8. American National Standard Hexagon Socket Head Shoulder Screws *ANSI/ASME B18.3-1998*

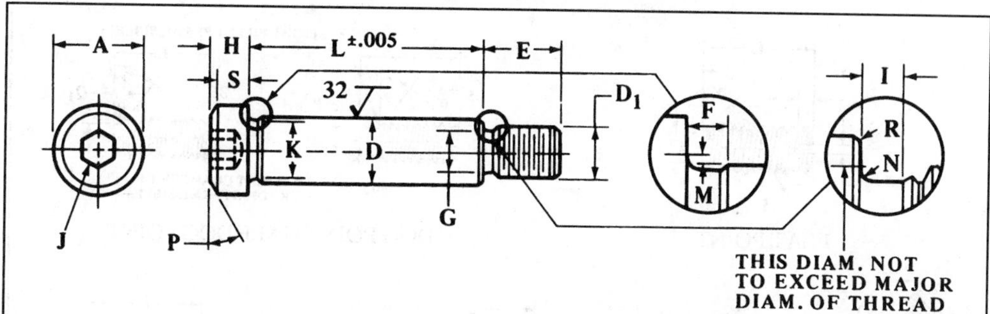

Nominal Size	Shoulder Diameter		Head Diameter		Head Height		Head Side Height	Nominal Thread Size	Thread Length
	Max.	Min.	Max.	Min.	Max.	Min.	Min.		
	D		*A*		*H*		*S*	D_1	*E*
¼	0.2480	0.2460	0.375	0.357	0.188	0.177	0.157	10–24	0.375
5⁄16	0.3105	0.3085	0.438	0.419	0.219	0.209	0.183	¼-20	0.438
⅜	0.3730	0.3710	0.562	0.543	0.250	0.240	0.209	5⁄16-18	0.500
½	0.4980	0.4960	0.750	0.729	0.312	0.302	0.262	⅜-16	0.625
⅝	0.6230	0.6210	0.875	0.853	0.375	0.365	0.315	½-13	0.750
¾	0.7480	0.7460	1.000	0.977	0.500	0.490	0.421	⅝-11	0.875
1	0.9980	0.9960	1.312	1.287	0.625	0.610	0.527	¾-10	1.000
1¼	1.2480	1.2460	1.750	1.723	0.750	0.735	0.633	⅞-9	1.125
1½	1.4980	1.4960	2.125	2.095	1.000	0.980	0.842	1⅛-7	1.500
1¾	1.7480	1.7460	2.375	2.345	1.125	1.105	0.948	1¼-7	1.750
2	1.9980	1.9960	2.750	2.720	1.250	1.230	1.054	1½-6	2.000

Nominal Size	Thread Neck Diameter		Thread Neck Width	Shoulder Neck Diam.	Shoulder Neck Width	Thread Neck Fillet		Head Fillet Extension Above *D*	Hexagon Socket Size
	Max.	Min.	Max.	Min.	Max.	Max.	Min.	Max.	Nom.
	G		*I*	*K*	*F*	*N*		*M*	*J*
¼	0.142	0.133	0.083	0.227	0.093	0.023	0.017	0.014	⅛
5⁄16	0.193	0.182	0.100	0.289	0.093	0.028	0.022	0.017	5⁄32
⅜	0.249	0.237	0.111	0.352	0.093	0.031	0.025	0.020	3⁄16
½	0.304	0.291	0.125	0.477	0.093	0.035	0.029	0.026	¼
⅝	0.414	0.397	0.154	0.602	0.093	0.042	0.036	0.032	5⁄16
¾	0.521	0.502	0.182	0.727	0.093	0.051	0.045	0.039	⅜
1	0.638	0.616	0.200	0.977	0.125	0.055	0.049	0.050	½
1¼	0.750	0.726	0.222	1.227	0.125	0.062	0.056	0.060	⅝
1½	0.964	0.934	0.286	1.478	0.125	0.072	0.066	0.070	⅞
1¾	1.089	1.059	0.286	1.728	0.125	0.072	0.066	0.080	1
2	1.307	1.277	0.333	1.978	0.125	0.102	0.096	0.090	1¼

All dimensions are in inches. The shoulder is the enlarged, unthreaded portion of the screw. Standard length increments for shoulder screws are: ⅛ inch for nominal screw lengths of ¼ through ¾ inch; ¼ inch for lengths above ¾ through 5 inches; and ½ inch for lengths over 5 inches. The thread conforms to the Unified Standard Class 3A, UNC. Hexagon socket sizes for the respective shoulder screw sizes are the same as for set screws of the same nominal size (see Table 7) except for shoulder screw size 1 inch, socket size is ½ inch, for screw size 1 ½ inches, socket size is ⅞ inch, and for screw size 2 inches, socket size is 1 ¼ inches. For details not shown, including materials, see ANSI/ASME B18.3-1998.

Table 9. American National Standard Slotted Headless Set Screws *ANSI/ASME B18.6.2-1998*

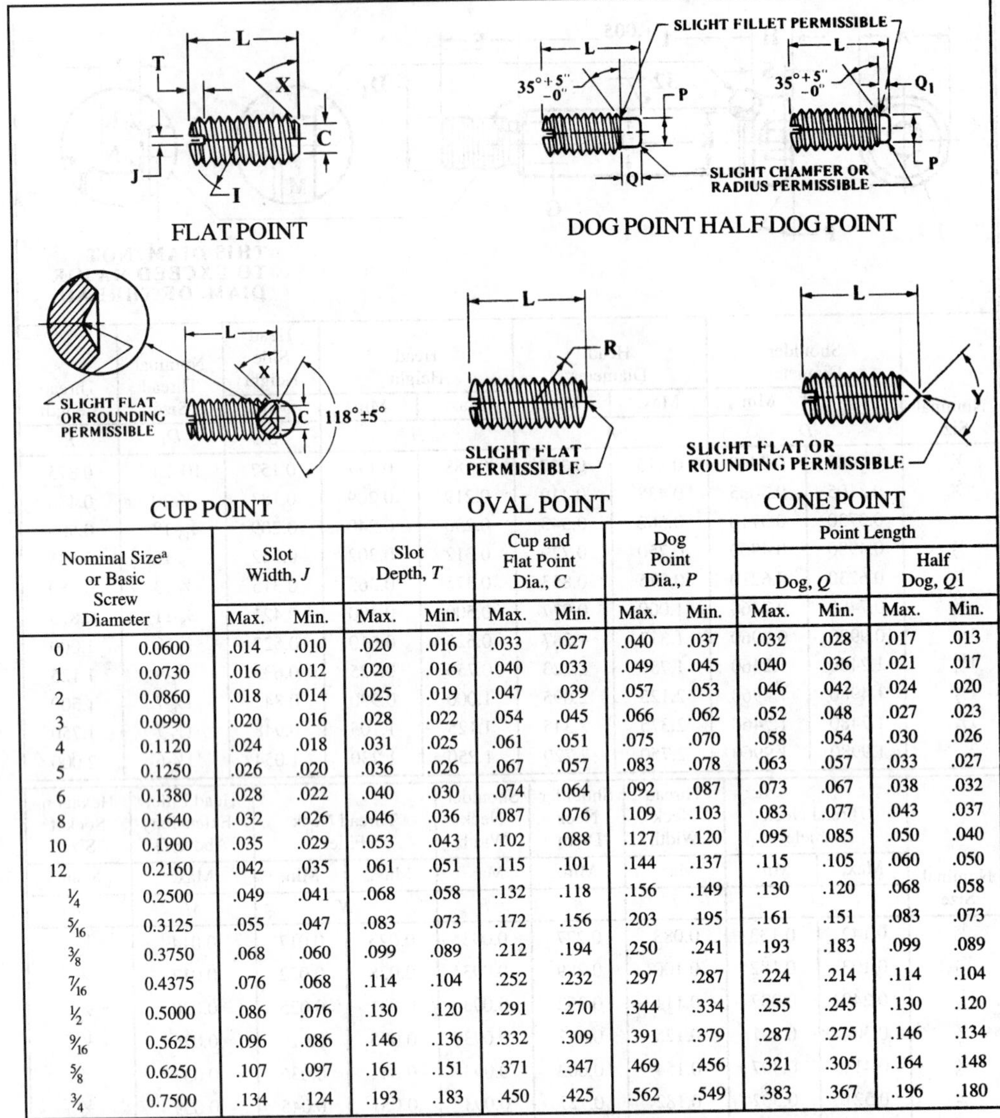

Nominal Size[a] or Basic Screw Diameter		Slot Width, *J*		Slot Depth, *T*		Cup and Flat Point Dia., *C*		Dog Point Dia., *P*		Point Length Dog, *Q*		Point Length Half Dog, *Q1*	
		Max.	Min.	Max.	Min.	Max.	Min.	Max.	Min.	Max.	Min.	Max.	Min.
0	0.0600	.014	.010	.020	.016	.033	.027	.040	.037	.032	.028	.017	.013
1	0.0730	.016	.012	.020	.016	.040	.033	.049	.045	.040	.036	.021	.017
2	0.0860	.018	.014	.025	.019	.047	.039	.057	.053	.046	.042	.024	.020
3	0.0990	.020	.016	.028	.022	.054	.045	.066	.062	.052	.048	.027	.023
4	0.1120	.024	.018	.031	.025	.061	.051	.075	.070	.058	.054	.030	.026
5	0.1250	.026	.020	.036	.026	.067	.057	.083	.078	.063	.057	.033	.027
6	0.1380	.028	.022	.040	.030	.074	.064	.092	.087	.073	.067	.038	.032
8	0.1640	.032	.026	.046	.036	.087	.076	.109	.103	.083	.077	.043	.037
10	0.1900	.035	.029	.053	.043	.102	.088	.127	.120	.095	.085	.050	.040
12	0.2160	.042	.035	.061	.051	.115	.101	.144	.137	.115	.105	.060	.050
¼	0.2500	.049	.041	.068	.058	.132	.118	.156	.149	.130	.120	.068	.058
5⁄16	0.3125	.055	.047	.083	.073	.172	.156	.203	.195	.161	.151	.083	.073
⅜	0.3750	.068	.060	.099	.089	.212	.194	.250	.241	.193	.183	.099	.089
7⁄16	0.4375	.076	.068	.114	.104	.252	.232	.297	.287	.224	.214	.114	.104
½	0.5000	.086	.076	.130	.120	.291	.270	.344	.334	.255	.245	.130	.120
9⁄16	0.5625	.096	.086	.146	.136	.332	.309	.391	.379	.287	.275	.146	.134
⅝	0.6250	.107	.097	.161	.151	.371	.347	.469	.456	.321	.305	.164	.148
¾	0.7500	.134	.124	.193	.183	.450	.425	.562	.549	.383	.367	.196	.180

[a] When specifying a nominal size in decimals a zero preceding the decimal point or any zero in the fourth decimal place is omitted.

All dimensions are in inches.

Crown Radius, I: The crown radius has the same value as the basic screw diameter to three decimal places.

Oval Point Radius, R: Values of the oval point radius according to nominal screw size are: For a screw size of 0, a radius of .045; 1, .055; 2, .064; 3, .074; 4, .084; 5, .094; 6, .104; 8, .123; 10, .142; 12, .162; ¼, .188; 5⁄16, .234; ⅜, .281; 7⁄16, .328; ½, .375; 9⁄16, .422; ⅝, .469; and for ¾, .562.

Cone Point Angle, Y: The cone point angle is 90° ± 2° for the following nominal lengths, or longer, shown according to screw size: For nominal size 0, a length of 5⁄64; 1, 3⁄32; 2, 7⁄64; 3, ⅛; 4, 5⁄32; 5, 3⁄16; 6, 3⁄16; 8, ¼; 10, ¼; 12, 5⁄16; ¼, 5⁄16; 5⁄16, ⅜; ⅜, 7⁄16; 7⁄16, ½; ½, 9⁄16; 9⁄16, ⅝; ⅝, ¾; and for ¾, ⅞. For shorter screws, the cone point angle is 118° ± 2°.

Point Angle X: The point angle is 45°, + 5°, − 0°, for screws of nominal lengths, or longer, as given just above for cone point angle, and 30°, min. for shorter screws.

Threads: are Unified Standard Class 2A; UNC and UNF Series or UNRC and UNRF Series.

Table 10. American National Standard Hexagon and Spline Socket Set Screw Optional Cup Points ***ANSI/ASME B18.3-1998***

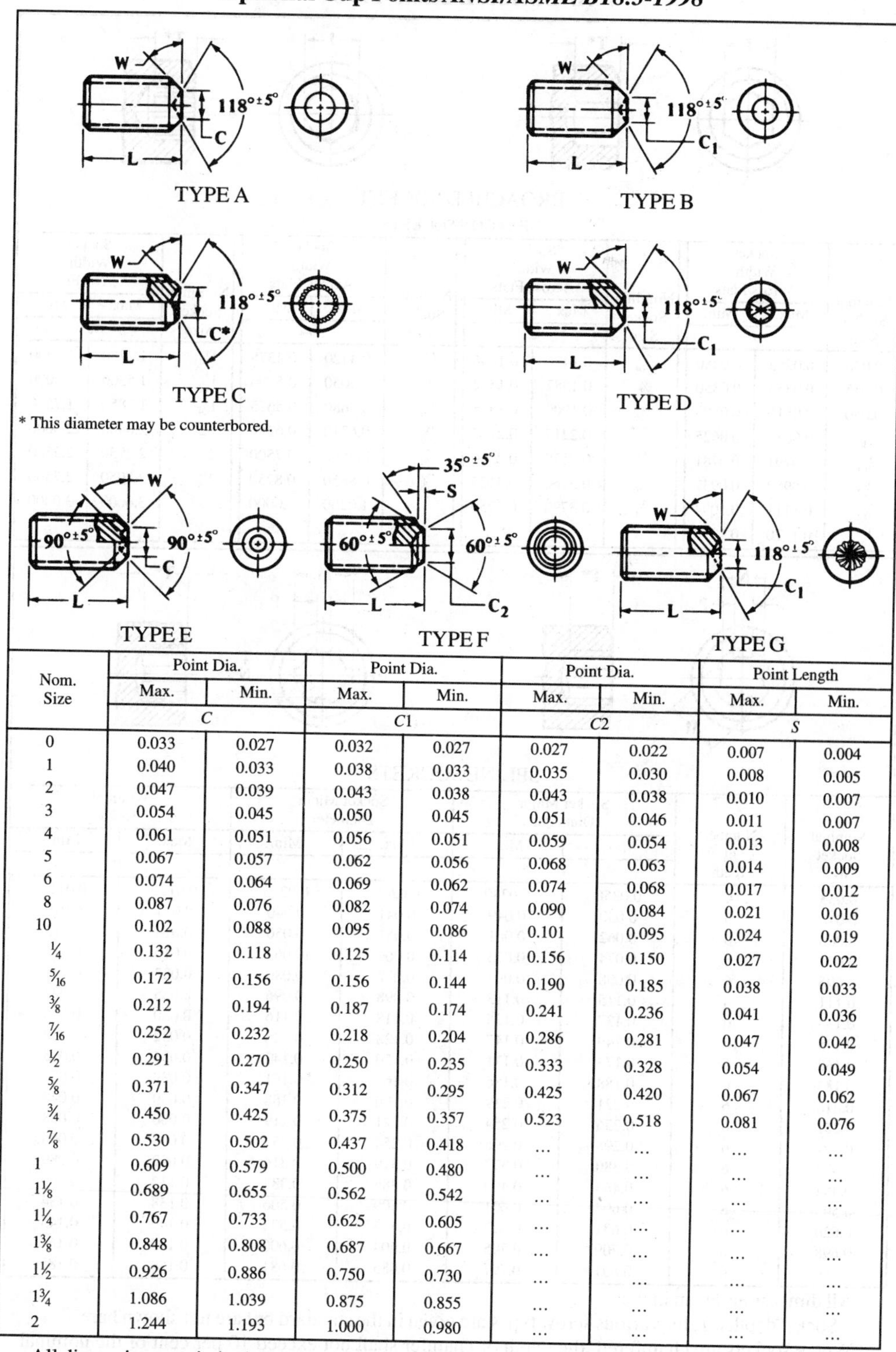

* This diameter may be counterbored.

Nom. Size	Point Dia. Max.	Point Dia. Min.	Point Dia. Max.	Point Dia. Min.	Point Dia. Max.	Point Dia. Min.	Point Length Max.	Point Length Min.
	C		*C1*		*C2*		*S*	
0	0.033	0.027	0.032	0.027	0.027	0.022	0.007	0.004
1	0.040	0.033	0.038	0.033	0.035	0.030	0.008	0.005
2	0.047	0.039	0.043	0.038	0.043	0.038	0.010	0.007
3	0.054	0.045	0.050	0.045	0.051	0.046	0.011	0.007
4	0.061	0.051	0.056	0.051	0.059	0.054	0.013	0.008
5	0.067	0.057	0.062	0.056	0.068	0.063	0.014	0.009
6	0.074	0.064	0.069	0.062	0.074	0.068	0.017	0.012
8	0.087	0.076	0.082	0.074	0.090	0.084	0.021	0.016
10	0.102	0.088	0.095	0.086	0.101	0.095	0.024	0.019
1/4	0.132	0.118	0.125	0.114	0.156	0.150	0.027	0.022
5/16	0.172	0.156	0.156	0.144	0.190	0.185	0.038	0.033
3/8	0.212	0.194	0.187	0.174	0.241	0.236	0.041	0.036
7/16	0.252	0.232	0.218	0.204	0.286	0.281	0.047	0.042
1/2	0.291	0.270	0.250	0.235	0.333	0.328	0.054	0.049
5/8	0.371	0.347	0.312	0.295	0.425	0.420	0.067	0.062
3/4	0.450	0.425	0.375	0.357	0.523	0.518	0.081	0.076
7/8	0.530	0.502	0.437	0.418	…	…	…	…
1	0.609	0.579	0.500	0.480	…	…	…	…
1 1/8	0.689	0.655	0.562	0.542	…	…	…	…
1 1/4	0.767	0.733	0.625	0.605	…	…	…	…
1 3/8	0.848	0.808	0.687	0.667	…	…	…	…
1 1/2	0.926	0.886	0.750	0.730	…	…	…	…
1 3/4	1.086	1.039	0.875	0.855	…	…	…	…
2	1.244	1.193	1.000	0.980	…	…	…	…

All dimensions are in inches.

The cup point types shown are those available from various manufacturers.

Table 11. American National Standard Hexagon and Spline Sockets *ANSI/ASME B18.3-1998*

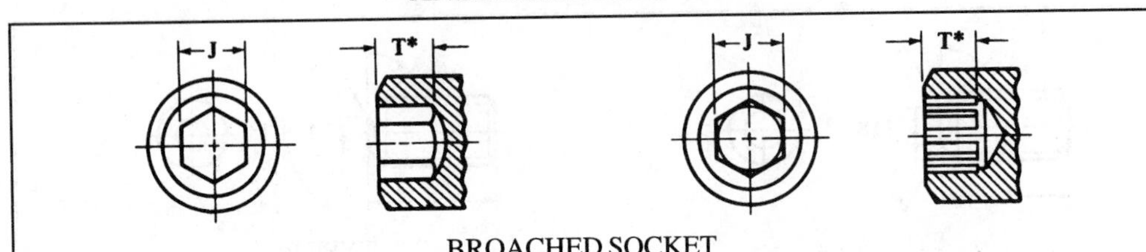

BROACHED SOCKET

HEXAGON SOCKETS

Nominal Socket Size	Socket Width Across Flats J Max.	Socket Width Across Flats J Min.	Nominal Socket Size	Socket Width Across Flats J Max.	Socket Width Across Flats J Min.	Nominal Socket Size	Socket Width Across Flats J Max.	Socket Width Across Flats J Min.	Nominal Socket Size	Socket Width Across Flats J Max.	Socket Width Across Flats J Min.
0.028	0.0285	0.0280	9⁄64	0.1426	0.1406	7⁄16	0.4420	0.4375	1¼	1.2750	1.2500
0.035	0.0355	0.0350	5⁄32	0.1587	0.1562	½	0.5050	0.5000	1½	1.5300	1.5000
0.050	0.0510	0.0500	3⁄16	0.1900	0.1875	9⁄16	0.5680	0.5625	1¾	1.7850	1.7500
1⁄16	0.0635	0.0625	7⁄32	0.2217	0.2187	5⁄8	0.6310	0.6250	2	2.0400	2.0000
5⁄64	0.0791	0.0781	¼	0.2530	0.2500	¾	0.7570	0.7500	2¼	2.2950	2.2500
3⁄32	0.0952	0.0937	5⁄16	0.3160	0.3125	7⁄8	0.8850	0.8750	2¾	2.8050	2.7500
7⁄64	0.1111	0.1094	3⁄8	0.3790	0.3750	1	1.0200	1.0000	3	3.0600	3.0000
1⁄8	0.1270	0.1250	…	…	…	…	…	…	…	…	…

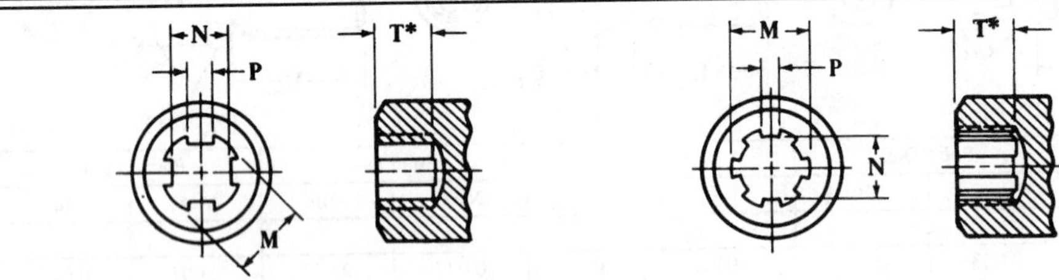

SPLINE SOCKETS

Nominal Socket Size	Number of Teeth	Socket Major Diameter M Max.	Socket Major Diameter M Min.	Socket Minor Diameter N Max.	Socket Minor Diameter N Min.	Width of Tooth P Max.	Width of Tooth P Min.
0.033	4	0.0350	0.0340	0.0260	0.0255	0.0120	0.0115
0.048	6	0.050	0.049	0.041	0.040	0.011	0.010
0.060	6	0.062	0.061	0.051	0.050	0.014	0.013
0.072	6	0.074	0.073	0.064	0.063	0.016	0.015
0.096	6	0.098	0.097	0.082	0.080	0.022	0.021
0.111	6	0.115	0.113	0.098	0.096	0.025	0.023
0.133	6	0.137	0.135	0.118	0.116	0.030	0.028
0.145	6	0.149	0.147	0.128	0.126	0.032	0.030
0.168	6	0.173	0.171	0.150	0.147	0.036	0.033
0.183	6	0.188	0.186	0.163	0.161	0.039	0.037
0.216	6	0.221	0.219	0.190	0.188	0.050	0.048
0.251	6	0.256	0.254	0.221	0.219	0.060	0.058
0.291	6	0.298	0.296	0.254	0.252	0.068	0.066
0.372	6	0.380	0.377	0.319	0.316	0.092	0.089
0.454	6	0.463	0.460	0.386	0.383	0.112	0.109
0.595	6	0.604	0.601	0.509	0.506	0.138	0.134
0.620	6	0.631	0.627	0.535	0.531	0.149	0.145
0.698	6	0.709	0.705	0.604	0.600	0.168	0.164
0.790	6	0.801	0.797	0.685	0.681	0.189	0.185

All dimensions are in inches.

* Socket depths, *T*, for various screw types are given in the standard but are not shown here.

Where sockets are chamfered, the depth of chamfer shall not exceed 10 per cent of the nominal socket size for sizes up to and including 1⁄16 inch for hexagon sockets and 0.060 for spline sockets, and 7.5 per cent for larger sizes.

Table 12. American National Standard Square Head Set Screws *ANSI/ASME B18.6.2-1998*

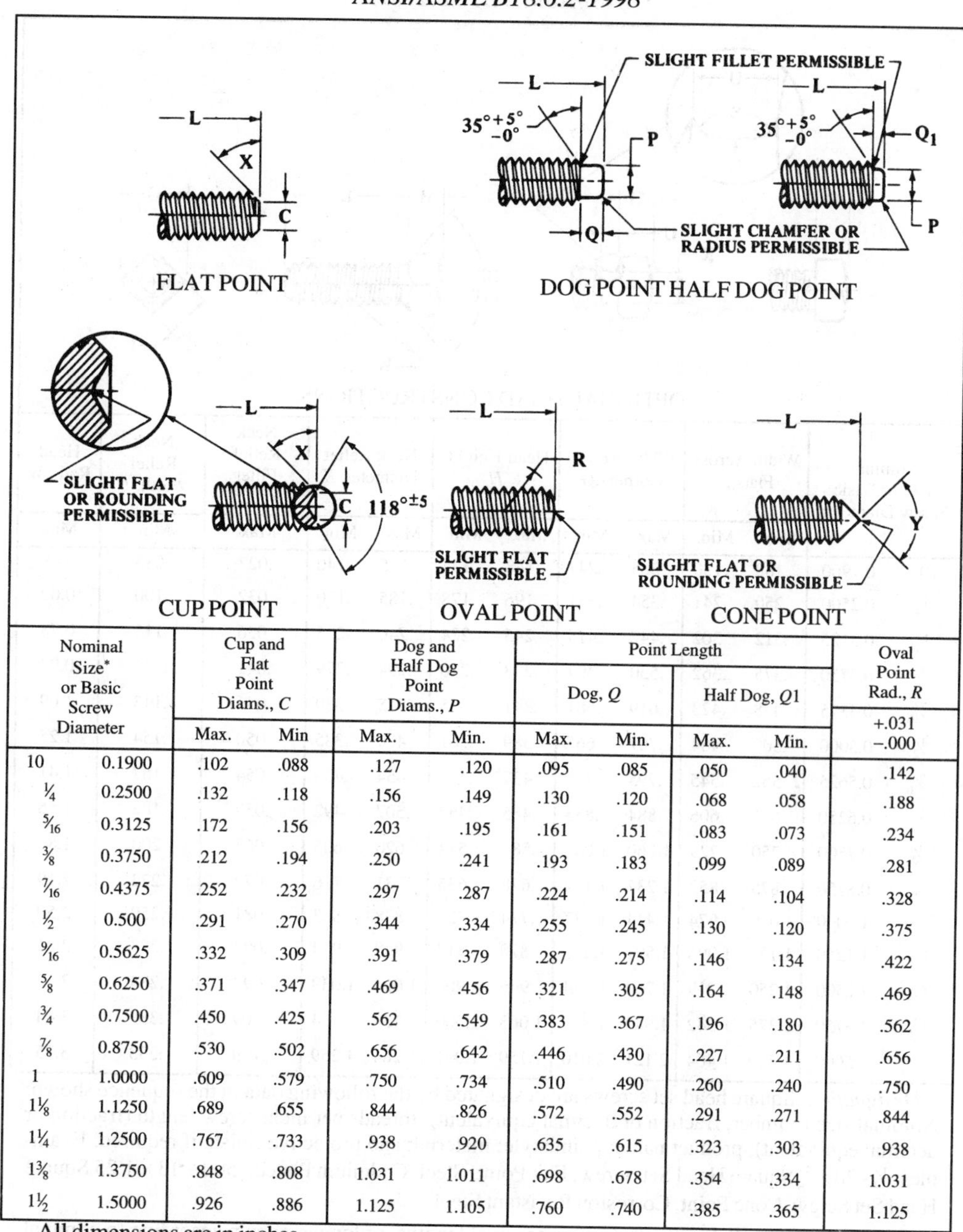

Nominal Size* or Basic Screw Diameter		Cup and Flat Point Diams., *C*		Dog and Half Dog Point Diams., *P*		Point Length				Oval Point Rad., *R*
						Dog, *Q*		Half Dog, *Q1*		
		Max.	Min	Max.	Min.	Max.	Min.	Max.	Min.	+.031 −.000
10	0.1900	.102	.088	.127	.120	.095	.085	.050	.040	.142
¼	0.2500	.132	.118	.156	.149	.130	.120	.068	.058	.188
5⁄16	0.3125	.172	.156	.203	.195	.161	.151	.083	.073	.234
⅜	0.3750	.212	.194	.250	.241	.193	.183	.099	.089	.281
7⁄16	0.4375	.252	.232	.297	.287	.224	.214	.114	.104	.328
½	0.500	.291	.270	.344	.334	.255	.245	.130	.120	.375
9⁄16	0.5625	.332	.309	.391	.379	.287	.275	.146	.134	.422
⅝	0.6250	.371	.347	.469	.456	.321	.305	.164	.148	.469
¾	0.7500	.450	.425	.562	.549	.383	.367	.196	.180	.562
⅞	0.8750	.530	.502	.656	.642	.446	.430	.227	.211	.656
1	1.0000	.609	.579	.750	.734	.510	.490	.260	.240	.750
1⅛	1.1250	.689	.655	.844	.826	.572	.552	.291	.271	.844
1¼	1.2500	.767	.733	.938	.920	.635	.615	.323	.303	.938
1⅜	1.3750	.848	.808	1.031	1.011	.698	.678	.354	.334	1.031
1½	1.5000	.926	.886	1.125	1.105	.760	.740	.385	.365	1.125

All dimensions are in inches.

**Threads:* Threads are Unified Standard Class 2A; UNC, UNF and 8 UN Series or UNRC, UNRF and 8 UNR Series.

Length of Thread: Square head set screws have complete (full form) threads extending over that portion of the screw length which is not affected by the point. For the respective constructions, threads extend into the neck relief, to the conical underside of head, or to within one thread (as measured with a thread ring gage) from the flat underside of the head. Threads through angular or crowned portions of points have fully formed roots with partial crests.

* When specifying a nominal size in decimals, the zero preceding the decimal point is omitted as is any zero in the fourth decimal place.

Table 13. American National Standard Square Head Set Screws
ANSI/ASME B18.6.2-1998

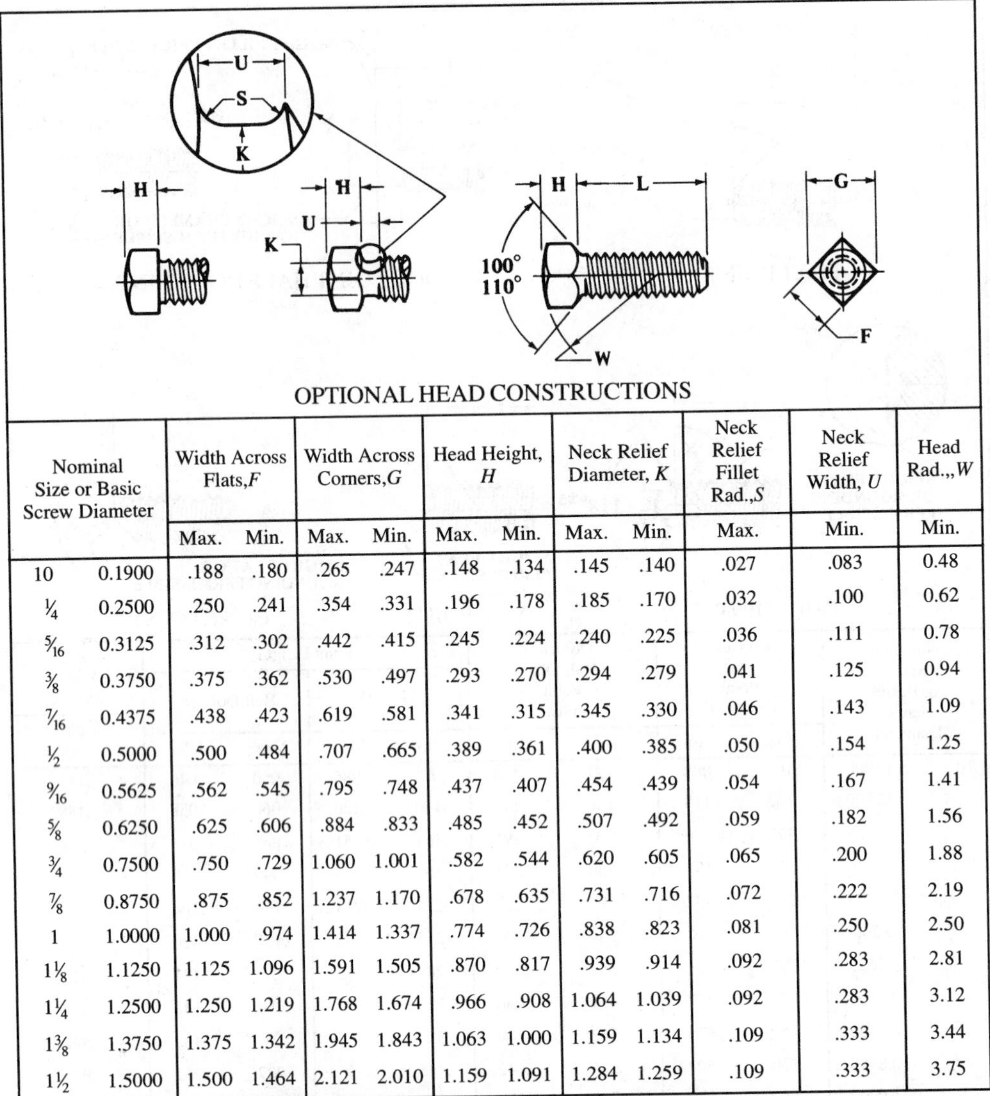

OPTIONAL HEAD CONSTRUCTIONS

Nominal Size or Basic Screw Diameter		Width Across Flats, *F*		Width Across Corners, *G*		Head Height, *H*		Neck Relief Diameter, *K*		Neck Relief Fillet Rad., *S*	Neck Relief Width, *U*	Head Rad., *W*
		Max.	Min.	Max.	Min.	Max.	Min.	Max.	Min.	Max.	Min.	Min.
10	0.1900	.188	.180	.265	.247	.148	.134	.145	.140	.027	.083	0.48
¼	0.2500	.250	.241	.354	.331	.196	.178	.185	.170	.032	.100	0.62
5⁄16	0.3125	.312	.302	.442	.415	.245	.224	.240	.225	.036	.111	0.78
⅜	0.3750	.375	.362	.530	.497	.293	.270	.294	.279	.041	.125	0.94
7⁄16	0.4375	.438	.423	.619	.581	.341	.315	.345	.330	.046	.143	1.09
½	0.5000	.500	.484	.707	.665	.389	.361	.400	.385	.050	.154	1.25
9⁄16	0.5625	.562	.545	.795	.748	.437	.407	.454	.439	.054	.167	1.41
⅝	0.6250	.625	.606	.884	.833	.485	.452	.507	.492	.059	.182	1.56
¾	0.7500	.750	.729	1.060	1.001	.582	.544	.620	.605	.065	.200	1.88
⅞	0.8750	.875	.852	1.237	1.170	.678	.635	.731	.716	.072	.222	2.19
1	1.0000	1.000	.974	1.414	1.337	.774	.726	.838	.823	.081	.250	2.50
1⅛	1.1250	1.125	1.096	1.591	1.505	.870	.817	.939	.914	.092	.283	2.81
1¼	1.2500	1.250	1.219	1.768	1.674	.966	.908	1.064	1.039	.092	.283	3.12
1⅜	1.3750	1.375	1.342	1.945	1.843	1.063	1.000	1.159	1.134	.109	.333	3.44
1½	1.5000	1.500	1.464	2.121	2.010	1.159	1.091	1.284	1.259	.109	.333	3.75

Designation: Square head set screws are designated by the following data in the sequence shown: Nominal size (number, fraction or decimal equivalent); threads per inch; screw length (fraction or decimal equivalent); product name; point style; material; and protective finish, if required. Examples: ¼ - 20 × ¾ Square Head Set Screw, Flat Point, Steel, Cadmium Plated. .500 – 13 × 1.25 Square Head Set Screw, Cone Point, Corrosion Resistant Steel.

Cone Point Angle, Y: For the following nominal lengths, or longer, shown according to nominal size, the cone point angle is 90° ± 2°: For size No. 10, ¼; ¼, 5⁄16; 5⁄16, ⅜; ⅜, 7⁄16; 7⁄16, ½; ½, 9⁄16; 9⁄16, ⅝; ⅝, ¾; ¾, ⅞; ⅞, 1; 1, 1⅛; 1⅛, 1¼; 1¼, 1½; 1⅜, 1⅝; and for 1½, 1¾. For shorter screws the cone point angle is 118° ± 2°.

Point Types: Unless otherwise specified, square head set screws are supplied with cup points. Cup points as furnished by some manufacturers may be externally or internally knurled. Where so specified by the purchaser, screws have cone, dog, half-dog, flat or oval points as given on the following page.

Point Angle, X: The point angle is 45°, + 5°, – 0° for screws of the nominal lengths, or longer, given just above for cone point angle, and 30° min. for shorter lengths.

Table 14. Applicability of Hexagon and Spline Keys and Bits

Nominal Key or Bit Size		Cap Screws 1960 Series	Flat Countersunk Head Cap Screws	Button Head Cap Screws	Shoulder Screws	Set Screws
		Nominal Screw Sizes				
HEXAGON KEYS AND BITS						
	0.028	…	…	…	…	0
	0.035	…	0	0	…	1 & 2
	0.050	0	1 & 2	1 & 2	…	3 & 4
1/16	0.062	1	3 & 4	3 & 4	…	5 & 6
5/64	0.078	2 & 3	5 & 6	5 & 6	…	8
3/32	0.094	4 & 5	8	8	…	10
7/64	0.109	6	…	…	…	…
1/8	0.125	…	10	10	1/4	1/4
9/64	0.141	8	…	…	…	…
5/32	0.156	10	1/4	1/4	5/16	5/16
3/16	0.188	1/4	5/16	5/16	3/8	3/8
7/32	0.219	…	3/8	3/8	…	7/16
1/4	0.250	5/16	7/16	…	1/2	1/2
5/16	0.312	3/8	1/2	1/2	5/8	5/8
3/8	0.375	7/16 & 1/2	5/8	5/8	3/4	3/4
7/16	0.438	…	…	…	…	…
1/2	0.500	5/8	3/4	…	1	7/8
9/16	0.562	…	7/8	…	…	1 & 1 1/8
5/8	0.625	3/4	1	…	1 1/4	1 1/4 & 1 3/8
3/4	0.750	7/8 & 1	1 1/8	…	…	1 1/2
7/8	0.875	1 1/8 & 1 1/4	1 1/4 & 1 3/8	…	1 1/2	…
1	1.000	1 3/8 & 1 1/2	1 1/2	…	1 3/4	1 3/4 & 2
1 1/4	1.250	1 3/4	…	…	2	…
1 1/2	1.500	2	…	…	…	…
1 3/4	1.750	2 1/4 & 2 1/2	…	…	…	…
2	2.000	2 3/4	…	…	…	…
2 1/4	2.250	3 & 3 1/4	…	…	…	…
2 3/4	2.750	3 1/2 & 3 3/4	…	…	…	…
3	3.000	4	…	…	…	…
SPLINE KEYS AND BITS						
	0.033	…	…	…	…	0 & 1
	0.048	…	0	0	…	2 & 3
	0.060	0	1 & 2	1 & 2	…	4
	0.072	1	3 & 4	3 & 4	…	5 & 6
	0.096	2 & 3	5 & 6	5 & 6	…	8
	0.111	4 & 5	8	8	…	10
	0.133	6	…	…	…	…
	0.145	…	10	10	…	1/4
	0.168	8	…	…	…	…
	0.183	10	1/4	1/4	…	5/16
	0.216	1/4	5/16	5/16	…	3/8
	0.251	…	3/8	3/8	…	7/16
	0.291	5/16	7/16	…	…	1/2
	0.372	3/8	1/2	1/2	…	5/8
	0.454	7/16 & 1/2	5/8 & 3/4	5/8	…	3/4
	0.595	5/8	…	…	…	7/8
	0.620	3/4	…	…	…	…
	0.698	7/8	…	…	…	…
	0.790	1	…	…	…	…

Source: Appendix to American National StandardANSI/ASME B18.3-1998.

Table 15. American National Standard Hexagon and Spline Socket Set Screws
ANSI/ASME B18.3-1998

Flat Point

Oval Point

Cone Point

Half Dog

Cup Point

For optional cup points and their dimensions see Table 10.

Nominal Size or Basic Screw Diameter		Socket Size Hex. Nom.	Socket Size Spl. Nom.	Cup and Flat Point Diameters Max.	Cup and Flat Point Diameters Min.	Half Dog Point Dia. Max.	Half Dog Point Lgth. Max.	Oval Point Radius Basic	Min. Key Engagement Depth Hex.	Min. Key Engagement Depth Spl.	Lgth. Limit for Angle
		J	M	C		P	Q	R	T_H[a]	T_S[a]	Y[b]
0	0.0600	0.028	0.033	0.033	0.027	0.040	0.017	0.045	0.050	0.026	0.09
1	0.0730	0.035	0.033	0.040	0.033	0.049	0.021	0.055	0.060	0.035	0.09
2	0.0860	0.035	0.048	0.047	0.039	0.057	0.024	0.064	0.060	0.040	0.13
3	0.0990	0.050	0.048	0.054	0.045	0.066	0.027	0.074	0.070	0.040	0.13
4	0.1120	0.050	0.060	0.061	0.051	0.075	0.030	0.084	0.070	0.045	0.19
5	0.1250	1/16	0.072	0.067	0.057	0.083	0.033	0.094	0.080	0.055	0.19
6	0.1380	1/16	0.072	0.074	0.064	0.092	0.038	0.104	0.080	0.055	0.19
8	0.1640	5/64	0.096	0.087	0.076	0.109	0.043	0.123	0.090	0.080	0.25
10	0.1900	3/32	0.111	0.102	0.088	0.127	0.049	0.142	0.100	0.080	0.25
1/4	0.2500	1/8	0.145	0.132	0.118	0.156	0.067	0.188	0.125	0.125	0.31
5/16	0.3125	5/32	0.183	0.172	0.156	0.203	0.082	0.234	0.156	0.156	0.38
3/8	0.3750	3/16	0.216	0.212	0.194	0.250	0.099	0.281	0.188	0.188	0.44
7/16	0.4375	7/32	0.251	0.252	0.232	0.297	0.114	0.328	0.219	0.219	0.50
1/2	0.5000	1/4	0.291	0.291	0.270	0.344	0.130	0.375	0.250	0.250	0.57
5/8	0.6250	5/16	0.372	0.371	0.347	0.469	0.164	0.469	0.312	0.312	0.75
3/4	0.7500	3/8	0.454	0.450	0.425	0.562	0.196	0.562	0.375	0.375	0.88
7/8	0.8750	1/2	0.595	0.530	0.502	0.656	0.227	0.656	0.500	0.500	1.00
1	1.0000	9/16	…	0.609	0.579	0.750	0.260	0.750	0.562	…	1.13
1 1/8	1.1250	9/16	…	0.689	0.655	0.844	0.291	0.844	0.562	…	1.25
1 1/4	1.2500	5/8	…	0.767	0.733	0.938	0.323	0.938	0.625	…	1.50
1 3/8	1.3750	5/8	…	0.848	0.808	1.031	0.354	1.031	0.625	…	1.63
1 1/2	1.5000	3/4	…	0.926	0.886	1.125	0.385	1.125	0.750	…	1.75
1 3/4	1.7500	1	…	1.086	1.039	1.312	0.448	1.321	1.000	…	2.00
2	2.0000	1	…	1.244	1.193	1.500	0.510	1.500	1.000	…	2.25

[a] Reference should be made to the Standard for shortest optimum nominal lengths to which the minimum key engagement depths T_H and T_S apply.

[b] Cone point angle Y is 90 degrees plus or minus 2 degrees for these nominal lengths or longer and 118 degrees plus or minus 2 degrees for shorter nominal lengths.

All dimensions are in inches. The thread conforms to the Unified Standard, Class 3A, UNC and UNF series. The socket depth T is included in the Standard and some are shown here. The nominal length L of all socket type set screws is the total or overall length. For nominal screw lengths of 1/16 through 3/16 inch (0 through 3 sizes incl.) the standard length increment is 0.06 inch; for lengths 1/8 through 1 inch the increment is 1/8 inch; for lengths 1 through 2 inches the increment is 1/4 inch; for lengths 2 through 6 inches the increment is 1/2 inch; for lengths 6 inches and longer the increment is 1 inch. Socket dimensions are given in Table 11.

Length Tolerance: The allowable tolerance on length L for all set screws of the socket type is ±0.01 inch for set screws up to 5/8 inch long; ±0.02 inch for screws over 5/8 to 2 inches long; ±0.03 inch for screws over 2 to 6 inches long and ± 0.06 inch for screws over 6 inches long. Socket dimensions are given in Table 11.

For manufacturing details, including materials, not shown, see American National Standard ANSI/ASME B18.3-1998.

British Standard Hexagon Socket Screws — Metric Series.—The first five parts of British Standard BS 4168: 1981 provide specifications for hexagon socket head cap screws and hexagon socket set screws.

Hexagon Socket Head Cap Screws: The dimensional data in Table 1 are based upon BS 4168: Part 1: 1981. These screws are available in stainless steel and alloy steel, the latter having class 12.9 properties as specified in BS 6104:Part 1. When ordering these screws, the designation "Hexagon socket head cap screw BS 4168 M5 × 20-12.9" would mean, as an example, a cap screw having a thread size of d = M5, nominal length l = 20 mm, and property class 12.9. Alloy steel cap screws are furnished with a black oxide finish (thermal or chemical); stainless steel cap screws with a plain finish. Combinations of thread size, nominal length, and length of thread are shown in Table 2; the screw threads in these combinations are in the ISO metric coarse pitch series specified in BS 3643 with tolerances in the 5g6g class. (See Metric Screw Threads in Index.)

Hexagon Socket Set Screws: Part 2 of B.S. 4168:1981 specifies requirements for hexagon socket set screws with fiat point having ISO metric threads, and diameters from 1.6 mm up to and including 24 mm. The dimensions of these set screws along with those of cone-point, dog-point, and cup-point set screws in accord, respectively, with Parts 3, 4, and 5 of the Standard are given in Table 3 and the accompanying illustration. All of these set screws are available in either steel processed to mechanical properties class 45H B.S. 6104:Part 3; or stainless steel processed to mechanical properties described in B.S. 6105. Steel set screws are furnished with black oxide (thermal or chemical) finish; stainless steel set screws are furnished plain. The tolerances applied to the threads of these set screws are for ISO product grade A, based on ISO 4759/1-1978 "Tolerances for fasteners — Part 1: Bolts, screws, and nuts with thread diameters greater than or equal to 1.6 mm and less than or equal to 150 mm and product grades A, B, and C."

Hexagon socket set screws are designated by the type, the thread size, nominal length, and property class. As an example, for a flat-point set screw of thread size d = M6, nominal length l = 12 mm, and property class 45H:

Hexagon socket set screw flat point BS 4168 M6 × 12-45H

British Standard Hexagon Socket Countersunk and Button Head Screws — Metric Series: British Standard BS 4168:1967 provides a metric series of hexagon socket countersunk and button head screws. The dimensions of these screws are given in Table 4. The revision of this Standard will constitute Parts 6 and 8 of BS 4168.

British Standards for Mechanical Properties of Fasteners: B.S. 6104: Part 1:1981 specifies mechanical properties for bolts, screws, and studs with nominal diameters up to and including 39 mm of any triangular ISO thread and made of carbon or alloy steel. It does not apply to set screws and similar threaded fasteners. Part 2 of this Standard specifies the mechanical properties of set screws and similar fasteners, not under tensile stress, in the range from M1.6 up to and including M39 and made of carbon or alloy steel.

B.S. 6105:1981 provides specifications for bolts, screws, studs, and nuts made from austenitic, ferritic, and martensitic grades of corrosion-resistant steels. This Standard applies only to fastener components after completion of manufacture with nominal diameters from M1.6 up to and including M39. These Standards are not described further here. Copies may be obtained from the British Standards Institution, 2 Park Street, London W1A 2BS and also from the American National Standards Institute, 11 West 42nd Street, New York, N.Y. 10036.

Table 1. British Standard Hexagon Socket Head Cap Screws—Metric Series

BS 4168:Part 1:1981 (obsolescent)

Nominal Size,[a] d	Body Diameter, D		Head Diameter, A			Head Height, H		Hexagon Socket Size, J[b]	Key Engagement, K	Wall Thickness, W	Fillet Rad., F	Fillet Diam., d_a
	Max	Min	Max[c]	Max[d]	Min	Max	Min	Nom	Min	Min	Min	Max
M1.6	1.6	1.46	3	3.14	2.86	1.6	1.46	1.5	0.7	0.55	0.1	2
M2	2	1.86	3.8	3.98	3.62	2	1.86	1.5	1	0.55	0.1	2.6
M2.5	2.5	2.36	4.5	4.68	4.32	2.5	2.36	2	1.1	0.85	0.1	3.1
M3	3	2.86	5.5	5.68	5.32	3	2.86	2.5	1.3	1.15	0.1	3.6
M4	4	3.82	7	7.22	6.78	4	3.82	3	2	1.4	0.2	4.7
M5	5	4.82	8.5	8.72	8.28	5	4.82	4	2.5	1.9	0.2	5.7
M6	6	5.82	10	10.22	9.78	6	5.70	5	3	2.3	0.25	6.8
M8	8	7.78	13	13.27	12.73	8	7.64	6	4	3.3	0.4	9.2
M10	10	9.78	16	16.27	15.73	10	9.64	8	5	4	0.4	11.2
M12	12	11.73	18	18.27	17.73	12	11.57	10	6	4.8	0.6	14.2
(M14)	14	13.73	21	21.33	20.67	14	13.57	12	7	5.8	0.6	16.2
M16	16	15.73	24	24.33	23.67	16	15.57	14	8	6.8	0.6	18.2
M20	20	19.67	30	30.33	29.67	20	19.48	17.	10	8.6	0.8	22.4
M24	24	23.67	36	36.39	35.61	24	23.48	19	12	10.4	0.8	26.4
M30	30	29.67	45	45.39	44.61	30	29.48	22	15.5	13.1	1	33.4
M36	36	35.61	54	54.46	53.54	36	35.38	27	19	15.3	1	39.4

[a] The size shown in ( ) is non-preferred.

[b] See Table 2 for min/max.

[c] For plain heads.

[d] For knurled heads.

All dimensions are given in millimeters.

Table 2. British Standard Hexagon Socket Screws — Metric Series *BS 4168:Part 1:1981 (obsolescent)*

Dimensions of Hexagon Sockets

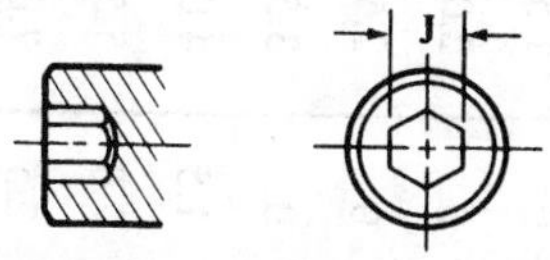

Nominal Socket Size	Socket Width Across Flats, J		Nominal Socket Size	Socket Width Across Flats, J	
	Max.	Min.		Max.	Min.
1.5	1.545	1.52	6	6.095	6.02
2.0	2.045	2.02	8	8.115	8.025
2.5	2.56	2.52	10	10.115	10.025
3	3.08	3.02	12	12.142	12.032
4	4.095	4.02	14	14.142	14.032
5	5.095	5.02	17	17.23	17.05
…	…	…	19	19.275	19.065

Association of Nominal and Thread Lengths for Each Thread Size

Nominal Length, L	Nominal Thread Size, D							
	M1.6	M2	M2.5	M3	M4	M5	M6	M8
2.5								
3								
4								
5								
6								
8								
10								
12								
16								
20								
25								
30								
35								
40								
45								
50								
55								
60								
65								
70								
80	…	…	…	…	…	…	…	…
…	…	…	…	…	…	…	…	…
…								
b (ref)	15	16	17	18	20	22	24	28

Nominal Length, L	Nominal Thread Size, D							
	M10	M12	(M14)	M16	M20	M24	M30	M36
16								
20								
25								
30								
35								
40								
45								
50								
55								
60								
65								
70								
80								
90								
100								
110								
120								
130								
140								
150								
160								
180								
200								
b (ref)	32	36	40	44	52	60	72	84

All dimensions are in millimeters.

The popular lengths are those between the stepped solidlines. Lengths above the dashed lines are threaded to the head within 3 pitch lengths (3P). Lengths below the dashed lines have values of L_g and L_s (see Table 1) given by the formulas: $L_g$max. = L nom. – b ref., and $L_s$min. = $L_g$max. – 5P.

Table 3. British Standard Hexagon Socket Set Screws — Metric Series
BS 4168:Parts 2, 3, 4, and 5:1994

Nom. Size, d	Pitch, P	Socket Size, s	Depth of Key Engagement, t[a]		Range of Popular Lengths				Length of Dog on Dog Point Screws[b]					End Diameters			
					Flat Point	Cone Point	Dog Point	Cup Point	Short Dog, z		Long Dog, z			Flat Point, d_z	Cone Point d_t	Dog Point, d_p	Cup Point, d_z
		nom	min	min	l	l	l	l	min	max	min	max	b	max	max	max	max
M1.6	0.35	0.7	0.7	1.5	2–8	2–8	2–8	2–8	0.4	0.65	0.8	1.05	2.5	0.8	0	0.8	0.8
M2	0.4	0.9	0.8	1.7	2–10	2–10	2.5–10	2–10	0.5	0.75	1.0	1.25	3.0	1.0	0	1.0	1.0
M2.5	0.45	1.3	1.2	2.0	2–12	2.5–12	3–12	2–12	0.63	0.88	1.25	1.5	4	1.5	0	1.5	1.2
M3	0.5	1.5	1.2	2.0	2–16	2.5–16	4–16	2.5–16	0.75	1.0	1.5	1.75	5	2.0	0	2.0	1.4
M4	0.7	2.0	1.5	2.5	2.5–20	3–20	5–20	3–20	1.0	1.25	2.0	2.25	6	2.5	0	2.5	2.0
M5	0.8	2.5	2.0	3.0	3–25	4–25	6–25	4–25	1.25	1.5	2.5	2.75	6	3.5	0	3.5	2.5
M6	1.0	3.0	2.0	3.5	4–30	5–30	8–30	5–30	1.5	1.75	3.0	3.25	8	4.0	1.5	4.0	3.0
M8	1.25	4.0	3.0	5.0	5–40	6–40	8–40	6–40	2.0	2.25	4.0	4.3	10	5.5	2.0	5.5	5.0
M10	1.5	5.0	4.0	6.0	6–50	8–50	10–50	8–50	2.5	2.75	5.0	5.3	12	7.0	2.5	7.0	6.0
M12	1.75	6.0	4.8	8.0	8–60	10–60	12–60	10–60	3.0	3.25	6.0	6.3	16	8.5	3.0	8.5	8.0
M16	2.0	8.0	6.4	10.0	10–60	12–60	16–60	12–60	4.0	4.3	8.0	8.36	20	12.0	4.0	12.0	10.0
M20	2.5	10.0	8.0	12.0	12–60	16–60	20–60	16–60	5.0	5.3	10.0	10.36	25	15.0	5.0	15.0	14.0
M24	3.0	12.0	10.0	15.0	16–60	20–60	25–60	20–60	6.0	6.3	12.0	12.43	30	18.0	6.0	18.0	16.0

[a] The smaller of the two t min. values applies to certain short-length set screws. These short-length screws are those whose length is approximately equal to the diameter of the screw. The larger t min. values apply to longer-length screws.

[b] A dog point set screw having a nominal length equal to or less than the length shown in the (*) column of the table is supplied with length z shown in the short dog column. For set screws of lengths greater than shown in the (*) column, z for long dogs applies.

All dimensions are in millimeters. For dimensional notation, see diagram, page 1618.

Table 4. British Standard Hexagon Socket Countersunk and Button Head Screws — Metric Series *B.S. 4168:1967*

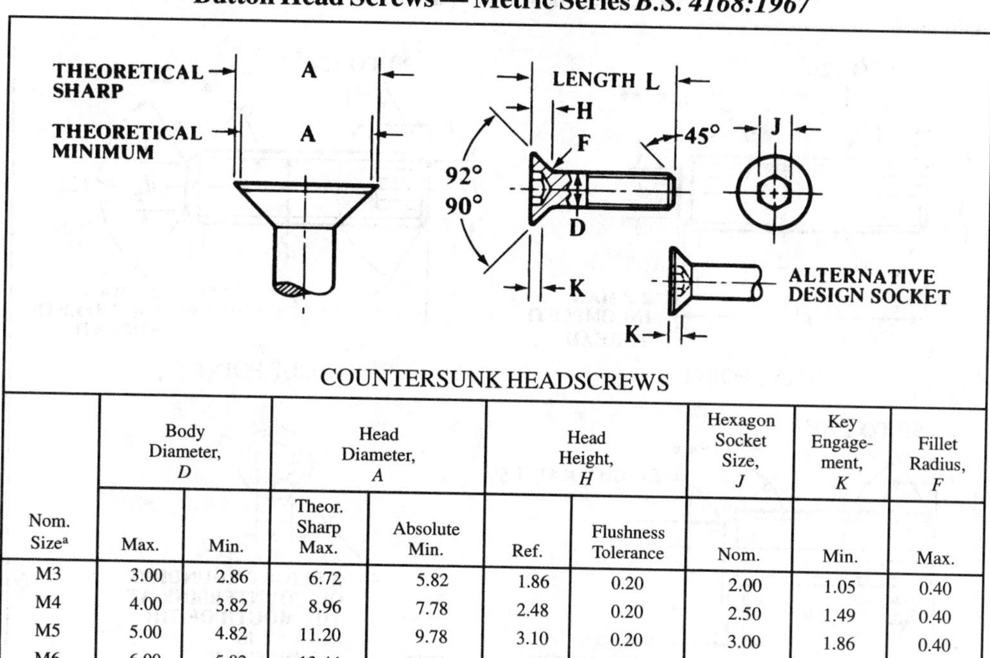

COUNTERSUNK HEADSCREWS

Nom. Size[a]	Body Diameter, *D*		Head Diameter, *A*		Head Height, *H*		Hexagon Socket Size, *J*	Key Engage-ment, *K*	Fillet Radius, *F*
	Max.	Min.	Theor. Sharp Max.	Absolute Min.	Ref.	Flushness Tolerance	Nom.	Min.	Max.
M3	3.00	2.86	6.72	5.82	1.86	0.20	2.00	1.05	0.40
M4	4.00	3.82	8.96	7.78	2.48	0.20	2.50	1.49	0.40
M5	5.00	4.82	11.20	9.78	3.10	0.20	3.00	1.86	0.40
M6	6.00	5.82	13.44	11.73	3.72	0.20	4.00	2.16	0.60
M8	8.00	7.78	17.92	15.73	4.96	0.24	5.00	2.85	0.70
M10	10.00	9.78	22.40	19.67	6.20	0.30	6.00	3.60	0.80
M12	12.00	11.73	26.88	23.67	7.44	0.36	8.00	4.35	1.10
(M14)	14.00	13.73	30.24	26.67	8.12	0.40	10.00	4.65	1.10
M16	16.00	15.73	33.60	29.67	8.80	0.45	10.00	4.89	1.10
(M18)	18.00	17.73	36.96	32.61	9.48	0.50	12.00	5.25	1.10
M20	20.00	19.67	40.32	35.61	10.16	0.54	12.00	5.45	1.10

[a] Sizes shown in parentheses are non-preferred.

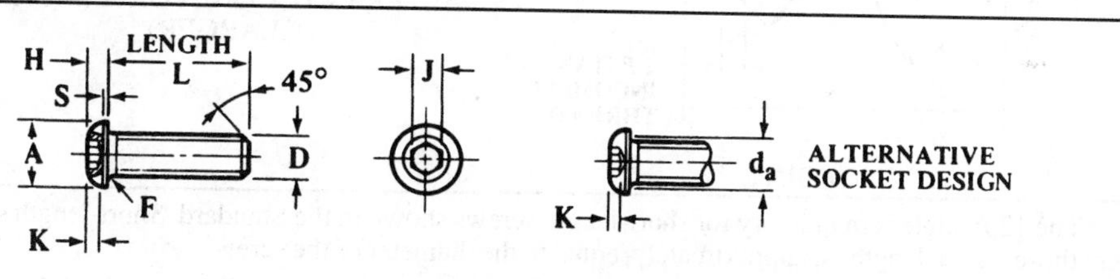

BUTTON HEADSCREWS

Nom. Size, *D*	Head Diameter, *A*		Head Height, *H*		Head Side Height, *S*	Hexagon Socket Size, *J*	Key Engage-ment, *K*	Fillet Radius *F*	Fillet Radius d_a
	Max.	Min.	Max.	Min.	Ref.	Nom.	Min.	Min.	Max.
M3	5.50	5.32	1.60	1.40	0.38	2.00	1.04	0.10	3.60
M4	7.50	7.28	2.10	1.85	0.38	2.50	1.30	0.20	4.70
M5	9.50	9.28	2.70	2.45	0.50	3.00	1.56	0.20	5.70
M6	10.50	10.23	3.20	2.95	0.80	4.00	2.08	0.25	6.80
M8	14.00	13.73	4.30	3.95	0.80	5.00	2.60	0.40	9.20
M10	18.00	17.73	5.30	4.95	0.80	6.00	3.12	0.40	11.20
M12	21.00	20.67	6.40	5.90	0.80	8.00	4.16	0.60	14.20

All dimensions are given in millimeters.

British Standard Hexagon Socket Set Screws — Metric Series
BS 4168:Parts 2, 3, 4, and 5:1994

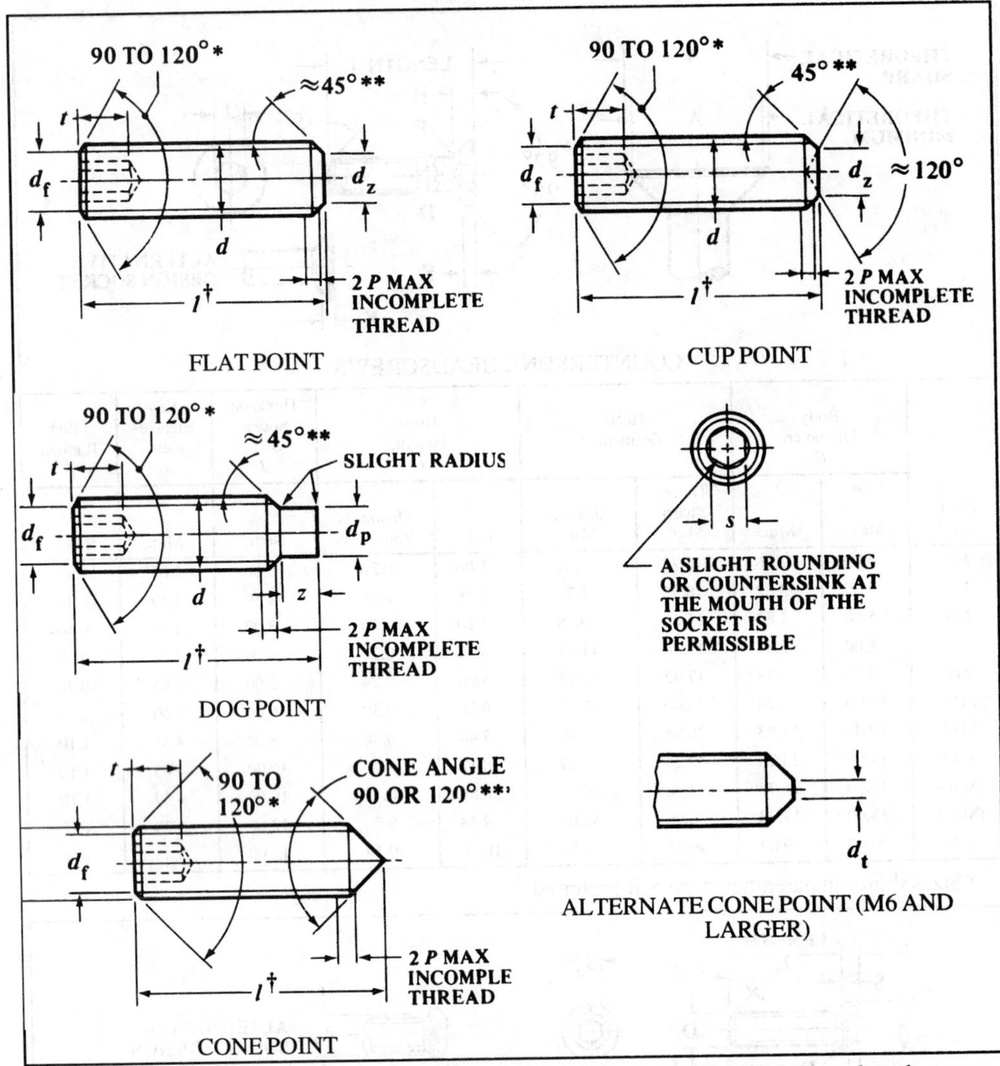

*The 120° angle is mandatory for short-length screws shown in the Standard. Short-length screws are those whose length is, approximately, equal to the diameter of the screw.

**The 45° angle applies only to that portion of the point below the root diameter, *df*, of the thread.

***The cone angle applies only to the portion of the point below the root diameter, *df*, of the thread and shall be 120° for certain short lengths listed in the Standard. All other lengths have a 90° cone angle.

†The popular length ranges of these set screws are listed in Table 3. These lengths have been selected from the following nominal lengths: 2, 2.5, 3, 4, 6, 8, 10, 12, 16, 20, 25,30, 35, 40, 45, 50, 55, and 60 millimeters.

Holding Power of Set-screws.—While the amount of power a set-screw of given size will transmit without slipping (when used for holding a pulley, gear, or other part from turning relative to a shaft) varies somewhat according to the physical properties of both set-screw and shaft and other variable factors, experiments have shown that the safe holding force in pounds for different diameters of set-screws should be approximately as follows: For ¼-inch diameter set-screws the safe holding force is 100 pounds, for ⅜-inch diameter set-screws the safe holding force is 250 pounds, for ½-inch diameter set-screws the safe holding force is 500 pounds, for ¾-inch diameter set-screws the safe holding force is 1300 pounds, and for 1-inch diameter set-screws the safe holding force is 2500 pounds.

The power or torque that can be safely transmitted by a set-screw may be determined from the formulas, $P = (DNd^{2.3}) \div 50$; or $T = 1250Dd^{2.3}$ in which P is the horsepower transmitted; T is the torque in inch-pounds transmitted; D is the shaft diameter in inches; N is the speed of the shaft in revolutions per minute; and d is the diameter of the set-screw in inches.

Example:— How many ½-inch diameter set-screws would be required to transmit 3 horsepower at a shaft speed of 1000 rpm if the shaft diameter is 1 inch?

Using the first formula given above, the power transmitted by a single ½-inch diameter set-screw is determined: $P = [1 \times 1000 \times (\frac{1}{2})^{2.3}] \div 50 = 4.1$ hp. Therefore a single ½-inch diameter set-screw is sufficient.

Example:— In the previous example, how many ⅜-inch diameter set-screws would be required? $P = [1 \times 1000 \times (\frac{3}{8})^{2.3}] \div 50 = 2.1$ hp. Therefore two ⅜-inch diameter set-screws are required.

Table 5. British Standard Whitworth (BSW) and British Standard Fine (BSF) Bright Square Head Set-Screws (With Flat Chamfered Ends)

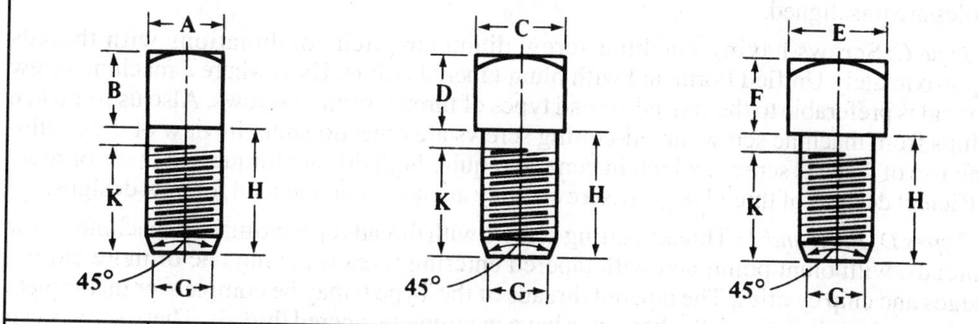

Nominal Size and Max. Diam., Inches	Number of Threads per Inch		No. 1 Standard		No. 2 Standard		No. 3 Standard	
	BSW	BSF	Width Across Flats *A*	Depth of Head *B*	Width Across Flats *C*	Depth of Head *D*	Width Across Flats *E*	Depth of Head *F*
¼	20	26	0.250	0.250	0.313	0.250	0.375	0.250
5⁄16	18	22	0.313	0.313	0.375	0.313	0.438	0.313
⅜	16	20	0.375	0.375	0.438	0.375	0.500	0.375
7⁄16	14	18	0.438	0.438	0.500	0.438	0.625	0.438
½	12	16	0.500	0.500	0.563	0.500	0.750	0.500
⅝	11	14	0.625	0.625	0.750	0.625	0.875	0.625
¾	10	12	0.750	0.750	0.875	0.750	1.000	0.750
⅞	9	11	0.875	0.875	1.000	0.875	1.125	0.875
1	8	10	1.000	1.000	1.125	1.000	1.250	1.000

* Depth of Head *B*, *D* and *F* same as for Width Across Flats, No. 1 Standard.

Dimensions *A*, *B*, *C*, *D*, *E*, and *F* are in inches.

SELF-THREADING SCREWS

ANSI Standard Sheet Metal, Self-Tapping, and Metallic Drive Screws.—Table 1 shows the various types of "self-tapping" screw threads covered by the ANSI B18.6.4-1981 (R1991) standard. (Metric thread forming and thread cutting tapping screws are discussed beginning on page 1635). ANSI designations are also shown. Types A, AB, B, BP and C when turned into a hole of proper size form a thread by a displacing action. Types D, F, G, T, BF and BT when turned into a hole of proper size form a thread by a cutting action. Type U when driven into a hole of proper size forms a series of multiple threads by a displacing action. These screws have the following descriptions and applications:

Type A: Spaced-thread screw with gimlet point primarily for use in light sheet metal, resin-impregnated plywood, and asbestos compositions. This type is no longer recommended. Use Type AB in new designs and whenever possible substitute for Type A in existing designs.

Type AB: Spaced-thread screw with same pitches as Type B but with gimlet point, primarily for similar uses as for Type A.

Type B: Spaced-thread screw with a blunt point with pitches generally somewhat finer than Type A. Used for thin metal, non-ferrous castings, plastics, resin-impregnated plywood, and asbestos compositions.

Type BP: Spaced-thread screw, the same as Type B but having a conical point extending beyond incomplete entering threads. Used for piercing fabrics or in assemblies where holes are misaligned.

Type C: Screws having machine screw diameter-pitch combinations with threads approximately Unified Form and with blunt tapered points. Used where a machine screw thread is preferable to the spaced-thread types of thread forming screws. Also useful when chips from machine screw thread-cutting screws are objectionable. In view of the declining use of Type C screws, which in general require high driving torques, in favor of more efficient designs of thread tapping screws, they are not recommended for new designs.

Types D, F, G, and T: Thread-cutting screws with threads approximating machine screw threads, with blunt point, and with tapered entering threads having one or more cutting edges and chip cavities. The tapered threads of the Type F may be complete or incomplete at the producer's option; all other types have incomplete tapered threads. These screws can be used in materials such as aluminum, zinc, and lead die-castings; steel sheets and shapes; cast iron; brass; and plastics.

Types BF and BT: Thread-cutting screws with spaced threads as in Type B, with blunt points, and one or more cutting grooves. Used in plastics, asbestos, and other similar compositions.

Type U: Multiple-threaded drive screw with large helix angle, having a pilot point, for use in metal and plastics. This screw is forced into the work by pressure and is intended for making permanent fastenings.

ANSI Standard Head Types for Tapping and Metallic Drive Screws: Many of the head types used with "self-tapping" screw threads are similar to the head types of American National Standard machine screws shown in the section with that heading.

Round Head: The round head has a semi-elliptical top surface and a flat bearing surface. Because of the superior slot driving characteristics of pan head screws over round head screws, and the overlap in dimensions of cross recessed pan heads and round heads, it is recommended that pan head screws be used in new designs and wherever possible substituted in existing designs.

Undercut Flat and Oval Countersunk Heads: For short lengths, 82-degree and oval countersunk head tapping screws have heads undercut to 70 per cent of normal side height to afford greater length of thread on the screws.

Flat Countersunk Head: The flat countersunk head has a flat top surface and a conical bearing surface with a head angle for one design of approximately 82 degrees and for another design of approximately 100 degrees. Because of its limited usage and in the interest of curtailing product varieties, the 100-degree flat countersunk head is considered nonpreferred.

Oval Countersunk Head: The oval countersunk head has a rounded top surface and a conical bearing surface with a head angle of approximately 82 degrees.

Flat and Oval Countersunk Trim Heads: Flat and oval countersunk trim heads are similar to the 82-degree flat and oval countersunk heads except that the size of head for a given size screw is one (large trim head) or two (small trim head) sizes smaller than the regular flat and oval countersunk head size. Oval countersunk trim heads have a definite radius where the curved top surface meets the conical bearing surface. Trim heads are furnished only in cross recessed types.

Pan Head: The slotted pan head has a flat top surface rounded into cylindrical sides and a flat bearing surface. The recessed pan head has a rounded top and a flat bearing surface. This head type is now preferred to the round head.

Fillister Head: The fillister head has a rounded top surface, cylindrical sides, and a flat bearing surface.

Hex Head: The hex head has a flat or indented top surface, six flat sides, and a flat bearing surface. Because the slotted hex head requires a secondary operation in manufacture which often results in burrs at the extremity of the slot that interfere with socket wrench engagement and the wrenching capability of the hex far exceeds that of the slot, it is not recommended for new designs.

Hex Washer Head: The hex washer head has an indented top surface and six flat sides formed integrally with a flat washer that projects beyond the sides and provides a flat bearing surface. Because the slotted hex washer head requires a secondary operation in manufacture which often results in burrs at the extremity of the slot that often interferes with socket wrench engagement and because the wrenching capability of the hex far exceeds that of the slot in the indented head, it is not recommended for new designs.

Truss Head: The truss head has a low rounded top surface with a flat bearing surface, the diameter of which for a given screw size is larger than the diameter of the corresponding round head. In the interest of product simplification and recognizing that the truss head is an inherently weak design, it is not recommended for new designs.

Method of Designation.—Tapping screws are designated by the following data in the sequence shown: Nominal size (number, fraction or decimal equivalent); threads per inch; nominal length (fraction or decimal equivalent); point type; product name, including head type and driving provision; material; and protective finish, if required.
Examples:

¼–14 × 1½ Type AB Slotted Pan Head Tapping Screw, Steel, Nickel Plated

6–32 × ¾ Type T, Type 1A Cross Recessed Pan Head Tapping Screw, Corrosion Resistant Steel

0.375–16 × 1.50 Type D, Washer Head Tapping Screw, Steel

Metallic Drive Screws: Type U metallic drive screws are designated by the following data in the sequence shown: Nominal size (number, fraction, or decimal equivalent); nominal length (fraction or decimal equivalent); product name, including head type; material; and protective finish, if required. Examples:

10 × 5⁄16 Round Head Metallic Drive Screw, Steel

0.312 × 0.50 Round Head Metallic Drive Screw, Steel, Zinc Plated

Table 1. ANSI Standard Threads and Points for Thread Forming Self-Tapping Screws *ANSI B18.6.4-1981 (R1991)*

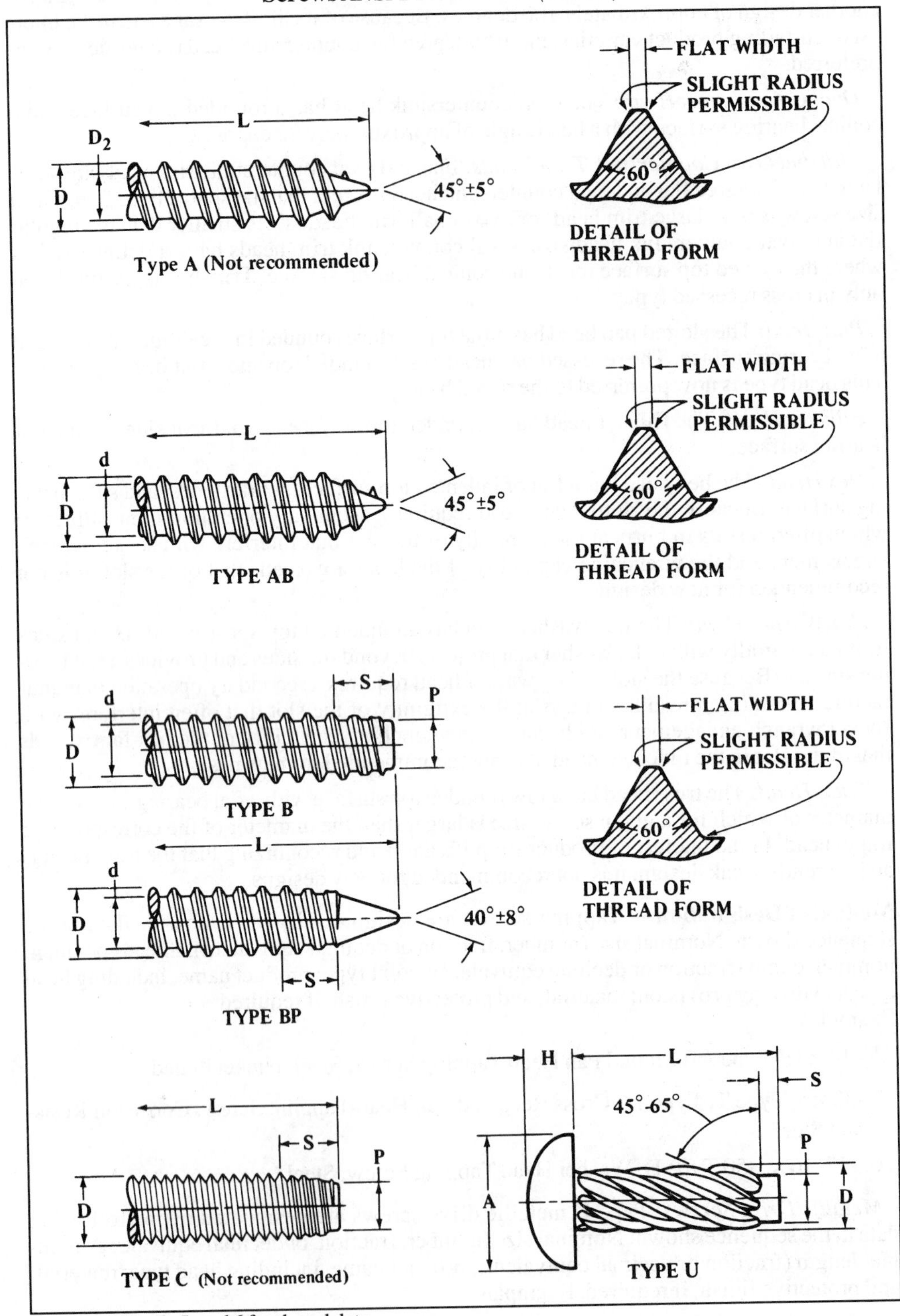

See Tables 3, 5, and 6 for thread data.

Table 2. ANSI Standard Threads and Points for Thread Cutting Self-Tapping Screws *ANSI B18.6.4-1981 (R1991)*

L
S
d
D
P
TYPE BF
FLAT WIDTH
SLIGHT RADIUS PERMISSIBLE
60°
DETAIL OF THREAD FORM
TYPE BT
TYPE D
TYPE F
TYPE G
TYPE T

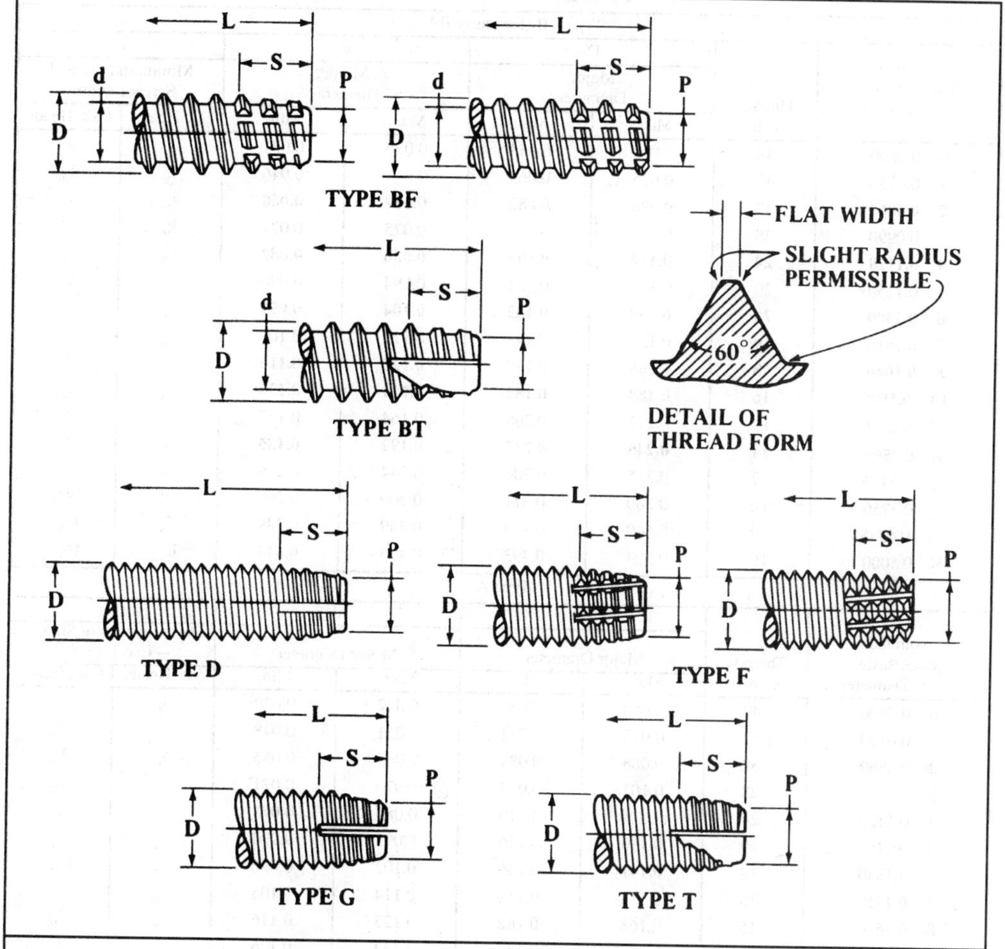

See Tables 5 and for thread data.

Cross Recesses.—Type I cross recess has a large center opening, tapered wings, and blunt bottom, with all edges relieved or rounded. Type IA cross recess has a large center opening, wide straight wings, and blunt bottom, with all edges relieved or rounded. Type II consists of two intersecting slots with parallel sides converging to a slightly truncated apex at the bottom of the recess. Type III has a square center opening, slightly tapered side walls, and a conical bottom, with top edges relieved or rounded.

Table 3. ANSI Standard Cross Recesses for Self-Tapping Screws *ANSI B18.6.4-1981 (R1991)* **and Metric Thread Forming and Thread Cutting Tapping Screws** *ANSI/ASME B18.6.5M-1986*

N
M
TYPE I
TYPE IA
TYPE II
TYPE III

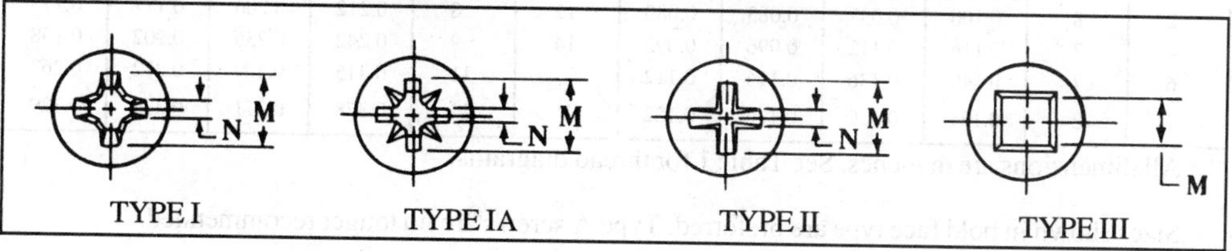

Table 4. ANSI Standard Thread and Point Dimensions for Types AB, A and U Thread Forming Tapping Screws *ANSI B18.6.4-1981 (R1991)*

Type AB (Formerly BA)							
		D		d		L	
Nominal Size or Basic Screw Diameter	Threads per inch	Major Diameter		Minor Diameter		Minimum Practical Screw Lengths	
		Max.	Min.	Max.	Min.	90° Heads	Csk. Heads
0 0.0600	48	0.060	0.054	0.036	0.033	1⁄8	5⁄32
1 0.0730	42	0.075	0.069	0.049	0.046	5⁄32	3⁄16
2 0.0860	**32**	**0.088**	**0.082**	**0.064**	**0.060**	3⁄16	7⁄32
3 0.0990	28	0.101	0.095	0.075	0.071	3⁄16	1⁄4
4 0.1120	**24**	**0.114**	**0.108**	**0.086**	**0.082**	7⁄32	9⁄32
5 0.1250	20	0.130	0.123	0.094	0.090	1⁄4	5⁄16
6 0.1380	**20**	**0.139**	**0.132**	**0.104**	**0.099**	9⁄32	11⁄32
7 0.1510	19	0.154	0.147	0.115	0.109	5⁄16	3⁄8
8 0.1640	**18**	**0.166**	**0.159**	**0.122**	**0.116**	5⁄16	3⁄8
10 0.1900	**16**	**0.189**	**0.182**	**0.141**	**0.135**	3⁄8	7⁄16
12 0.2160	14	0.215	0.208	0.164	0.157	7⁄16	21⁄32
1⁄4 0.2500	**14**	**0.246**	**0.237**	**0.192**	**0.185**	1⁄2	19⁄32
5⁄16 0.3125	12	0.315	0.306	0.244	0.236	5⁄8	3⁄4
3⁄8 0.3750	12	0.380	0.371	0.309	0.299	3⁄4	29⁄32
7⁄16 0.4375	10	0.440	0.429	0.359	0.349	7⁄8	1 1⁄32
1⁄2 0.5000	10	0.504	0.493	0.423	0.413	1	1 5⁄32
Type A							
		D		d		L	
Nominal Size[a] Basic Screw Diameter	Threads per inch	Major Diameter		Minor Diameter		These Lengths or Shorter —Use Type AB	
		Max.	Min.	Max.	Min.	90° Heads	Csk. Heads
0 0.0600	40	0.060	0.057	0.042	0.039	1⁄8	3⁄16
1 0.0730	32	0.075	0.072	0.051	0.048	1⁄8	3⁄16
2 0.0860	32	0.088	0.084	0.061	0.056	5⁄32	3⁄16
3 0.0990	28	0.101	0.097	0.076	0.071	3⁄16	7⁄32
4 0.1120	24	0.114	0.110	0.083	0.078	3⁄16	1⁄4
5 0.1250	20	0.130	0.126	0.095	0.090	3⁄16	1⁄4
6 0.1380	18	0.141	0.136	0.102	0.096	1⁄4	5⁄16
7 0.1510	16	0.158	0.152	0.114	0.108	5⁄16	3⁄8
8 0.1640	15	0.168	0.162	0.123	0.116	3⁄8	7⁄16
10 0.1900	12	0.194	0.188	0.133	0.126	3⁄8	1⁄2
12 0.2160	11	0.221	0.215	0.162	0.155	7⁄16	9⁄16
14 0.2420	10	0.254	0.248	0.185	0.178	1⁄2	5⁄8
16 0.2680	10	0.280	0.274	0.197	0.189	9⁄16	3⁄4
18 0.2940	9	0.306	0.300	0.217	0.209	5⁄8	13⁄16
20 0.3200	9	0.333	0.327	0.234	0.226	11⁄16	13⁄16
24 0.3720	9	0.390	0.383	0.291	0.282	3⁄4	1

[a] Where specifying nominal size in decimals, zeros preceding decimal and in fourth place are omitted.

Type U Metallic Drive Screws											
Nom. Size	No. of Starts	Out. Dia.		Pilot Dia.		Nom. Size	No. of Starts	Out. Dia.		Pilot Dia.	
		Max.	Min.	Max.	Min.			Max.	Min.	Max.	Min.
00	6	0.060	0.057	0.049	0.046	8	8	0.167	0.162	0.136	0.132
0	6	0.075	0.072	0.063	0.060	10	8	0.182	0.177	0.150	0.146
2	8	0.100	0.097	0.083	0.080	12	8	0.212	0.206	0.177	0.173
4	7	0.116	0.112	0.096	0.092	14	9	0.242	0.236	0.202	0.198
6	7	0.140	0.136	0.116	0.112	5⁄16	11	0.315	0.309	0.272	0.267
7	8	0.154	0.150	0.126	0.122	3⁄8	12	0.378	0.371	0.334	0.329

All dimensions are in inches. See Table 1 for thread diagrams.

Sizes shown in bold face type are preferred. Type A screws are no longer recommended.

Table 5. ANSI Standard Thread and Point Dimensions for B and BP Thread Forming and BF and BT Thread Cutting Tapping Screws *ANSI B18.6.4-1981 (R1991)*

THREAD FORMING TYPES B AND BP													
		D		d		P		S		L			
Nominal Size[a] or Basic Screw Diameter	Thds per Inch[b]	Major Diameter		Minor Diameter		Point Diameter[c]		Point Taper Length[d]		Minimum Practical Nominal Screw Lengths			
										Type B		Type BP	
		Max	Min	Max	Min	Max	Min	Max	Min	90° Heads	Csk Heads	90° Heads	Csk Heads
0 0.0600	48	0.060	0.054	0.036	0.033	0.031	0.027	0.042	0.031	1/8	1/8	5/32	3/16
1 0.0730	42	0.075	0.069	0.049	0.046	0.044	0.040	0.048	0.036	1/8	5/32	3/16	7/32
2 0.0860	32	0.088	0.082	0.064	0.060	0.058	0.054	0.062	0.047	5/32	3/16	1/4	9/32
3 0.0990	28	0.101	0.095	0.075	0.071	0.068	0.063	0.071	0.054	3/16	7/32	9/32	5/16
4 0.1120	24	0.114	0.108	0.086	0.082	0.079	0.074	0.083	0.063	3/16	1/4	5/16	11/32
5 0.1250	20	0.130	0.123	0.094	0.090	0.087	0.082	0.100	0.075	7/32	9/32	11/32	13/32
6 0.1380	20	0.139	0.132	0.104	0.099	0.095	0.089	0.100	0.075	1/4	9/32	3/8	7/16
7 0.1510	19	0.154	0.147	0.115	0.109	0.105	0.099	0.105	0.079	1/4	5/16	13/32	15/32
8 0.1640	18	0.166	0.159	0.122	0.116	0.112	0.106	0.111	0.083	9/32	11/32	7/16	1/2
10 0.1900	16	0.189	0.182	0.141	0.135	0.130	0.123	0.125	0.094	5/16	3/8	1/2	19/32
12 0.2160	14	0.215	0.208	0.164	0.157	0.152	0.145	0.143	0.107	11/32	7/16	9/16	21/32
1/4 0.2500	14	0.246	0.237	0.192	0.185	0.179	0.171	0.143	0.107	3/8	1/2	21/32	3/4
5/16 0.3125	12	0.315	0.306	0.244	0.236	0.230	0.222	0.167	0.125	15/32	19/32	27/32	31/32
3/8 0.3750	12	0.380	0.371	0.309	0.299	0.293	0.285	0.167	0.125	17/32	11/16	15/16	1 1/8
7/16 0.4375	10	0.440	0.429	0.359	0.349	0.343	0.335	0.200	0.150	5/8	25/32	1 1/8	1 1/4
1/2 0.5000	10	0.504	0.493	0.423	0.413	0.407	0.399	0.200	0.150	11/16	27/32	1 1/4	1 13/32

[a] Where specifying nominal size in decimals, zeros preceding decimal and in the fourth decimal place shall be omitted.

[b] The width of flat at crest of thread shall not exceed 0.004 inch for sizes up to No. 8, inclusive, and 0.006 inch for larger sizes.

[c] Point diameters specified apply to screw threads before roll threading.

[d] Points of screws are tapered and fluted or slotted. The flute on Type BT screws has an included angle of 90 to 95 degrees and the thread cutting edge is located above the axis of the screw. Flutes and slots extend through first full form thread beyond taper except for Type BF screw on which tapered threads may be complete at manufacturer's option and flutes may be one pitch short of first full form thread.

THREAD CUTTING TYPES BF AND BT[d]											
		D		d		P		S		L	
Nominal Size[a] or Basic Screw Diameter	Thds per Inch[b]	Major Diameter		Minor Diameter		Point Diameter[c]		Point Taper Length[d]		Minimum Practical Nominal Screw Lengths	
		Max	Min	Max	Min	Max	Min	Max	Min	90° Heads	Csk Heads
0 0.0600	48	0.060	0.054	0.036	0.033	0.031	0.027	0.042	0.031	1/8	1/8
1 0.0730	42	0.075	0.069	0.049	0.046	0.044	0.040	0.048	0.036	1/8	5/32
2 0.0860	32	0.088	0.082	0.064	0.060	0.058	0.054	0.062	0.047	5/32	3/16
3 0.0990	28	0.101	0.095	0.075	0.071	0.068	0.063	0.071	0.054	3/16	7/32
4 0.1120	24	0.114	0.108	0.086	0.082	0.079	0.074	0.083	0.063	3/16	1/4
5 0.1250	20	0.130	0.123	0.094	0.090	0.087	0.082	0.100	0.075	7/32	9/32
6 0.1380	20	0.139	0.132	0.104	0.099	0.095	0.089	0.100	0.075	1/4	9/32
7 0.1510	19	0.154	0.147	0.115	0.109	0.105	0.099	0.105	0.079	1/4	5/16
8 0.1640	18	0.166	0.159	0.122	0.116	0.112	0.106	0.111	0.083	9/32	11/32
10 0.1900	16	0.189	0.182	0.141	0.135	0.130	0.123	0.125	0.094	5/16	3/8
12 0.2160	14	0.215	0.208	0.164	0.157	0.152	0.145	0.143	0.107	11/32	7/16
1/4 0.2500	14	0.246	0.237	0.192	0.185	0.179	0.171	0.143	0.107	3/8	1/2
5/16 0.3125	12	0.315	0.306	0.244	0.236	0.230	0.222	0.167	0.125	15/32	19/32
3/8 0.3750	12	0.380	0.371	0.309	0.299	0.293	0.285	0.167	0.125	17/32	11/16
7/16 0.4375	10	0.440	0.429	0.359	0.349	0.343	0.335	0.200	0.150	5/8	25/32
1/2 0.5000	10	0.504	0.493	0.423	0.413	0.407	0.399	0.200	0.150	11/16	27/32

All dimensions are in inches. See Tables 1 and 2 for thread diagrams.

Table 6. Thread and Point Dimensions for Type C Thread Forming Tapping Screws (ANSI B18.6.4–1981, R1991 Appendix)

Nominal Size[a] or Basic Screw Diameter	Threads per inch	D		P		S				L			
		Major Diameter		Point Diameter[b]		Point Taper Length[c]				Determinant Lengths for Point Taper[c]		Minimum Practical Nominal Screw Lengths	
						For Short Screws		For Long Screws					
		Max	Min	Max	Min	Max	Min	Max	Min	90° Heads	Csk Heads	90° Heads	Csk Heads
2 0.0860	56	0.0860	0.0813	0.068	0.061	0.062	0.045	0.080	0.062	5/32	3/16	5/32	3/16
2 0.0860	64	0.0860	0.0816	0.070	0.064	0.055	0.039	0.070	0.055	1/8	3/16	1/8	5/32
3 0.0990	48	0.0990	0.0938	0.078	0.070	0.073	0.052	0.094	0.073	3/16	7/32	5/32	7/32
3 0.0990	56	0.0990	0.0942	0.081	0.074	0.062	0.045	0.080	0.062	5/32	3/16	5/32	3/16
4 0.1120	40	0.1120	0.1061	0.087	0.078	0.088	0.062	0.112	0.088	7/32	1/4	3/16	1/4
4 0.1120	48	0.1120	0.1068	0.091	0.083	0.073	0.052	0.094	0.073	3/16	7/32	5/32	7/32
5 0.1250	40	0.1250	0.1191	0.100	0.091	0.088	0.062	0.112	0.088	7/32	9/32	3/16	1/4
5 0.1250	44	0.1250	0.1195	0.102	0.094	0.080	0.057	0.102	0.080	3/16	1/4	3/16	1/4
6 0.1380	32	0.1380	0.1312	0.107	0.096	0.109	0.078	0.141	0.109	1/4	5/16	1/4	5/16
6 0.1380	40	0.1380	0.1321	0.113	0.104	0.088	0.062	0.112	0.088	7/32	9/32	3/16	1/4
8 0.1640	32	0.1640	0.1571	0.132	0.122	0.109	0.078	0.141	0.109	1/4	11/32	1/4	5/16
8 0.1640	36	0.1640	0.1577	0.136	0.126	0.097	0.069	0.125	0.097	7/32	5/16	7/32	9/32
10 0.1900	24	0.1900	0.1818	0.148	0.135	0.146	0.104	0.188	0.146	11/32	7/16	5/16	13/32
10 0.1900	32	0.1900	0.1831	0.158	0.148	0.109	0.078	0.141	0.109	1/4	11/32	1/4	5/16
12 0.2160	24	0.2160	0.2078	0.174	0.161	0.146	0.104	0.188	0.146	11/32	7/16	5/16	13/32
12 0.2160	28	0.2160	0.2085	0.180	0.168	0.125	0.089	0.161	0.125	5/16	13/32	9/32	3/8
1/4 0.2500	20	0.2500	0.2408	0.200	0.184	0.175	0.125	0.225	0.175	13/32	17/32	3/8	1/2
1/4 0.2500	28	0.2500	0.2425	0.214	0.202	0.125	0.089	0.161	0.125	5/16	13/32	9/32	3/8
5/16 0.3125	18	0.3125	0.3026	0.257	0.239	0.194	0.139	0.250	0.194	15/32	19/32	7/16	9/16
5/16 0.3125	24	0.3125	0.3042	0.271	0.257	0.146	0.104	0.188	0.146	11/32	15/32	5/16	15/32
3/8 0.3750	16	0.3750	0.3643	0.312	0.293	0.219	0.156	0.281	0.219	1/2	11/16	15/32	5/8
3/8 0.3750	24	0.3750	0.3667	0.333	0.319	0.146	0.104	0.188	0.146	11/32	1/2	5/16	1/2
7/16 0.4375	14	0.4375	0.4258	0.366	0.344	0.250	0.179	0.321	0.250	19/32	3/4	9/16	23/32
7/16 0.4375	20	0.4375	0.4281	0.387	0.371	0.175	0.125	0.225	0.175	13/32	9/16	3/8	17/32
1/2 0.5000	13	0.5000	0.4876	0.423	0.399	0.269	0.192	0.346	0.269	5/8	25/32	19/32	3/4
1/2 0.5000	20	0.5000	0.4906	0.450	0.433	0.175	0.125	0.225	0.175	13/32	9/16	3/8	17/32

[a] Where specifying nominal size in decimals, zeros preceding decimal and in the fourth decimal place shall be omitted.

[b] The tabulated values apply to screw blanks before roll threading.

[c] Screws of these nominal lengths and shorter shall have point taper length specified above for short screws. Longer lengths shall have point taper length specified for long screws.

All dimensions are in inches. See Table 1 for thread diagrams. Type C is not recommended for new designs.

Tapered threads shall have unfinished crests.

Table 7. ANSI Standard Thread and Point Dimensions for Types D, F, G, and T Thread Cutting Tapping Screws *ANSI B18.6.4-1981 (R1991)*

Nominal Size[a] or Basic Screw Diameter	Threads per inch	D Major Diameter Max	D Major Diameter Min	P Point Diameter[b] Max	P Point Diameter[b] Min	S Point Taper Length[c] For Short Screws Max	S For Short Screws Min	S For Long Screws Max	S For Long Screws Min	L Determinant Lengths for Point Taper[c] 90° Heads	L Determinant Lengths Csk Heads	L Minimum Practical Nominal Screw Lengths 90° Heads	L Minimum Practical Csk Heads
2 0.0860	56	0.0860	0.0813	0.068	0.061	0.062	0.045	0.080	0.062	5/32	3/16	5/32	3/16
2 0.0860	64	0.0860	0.0816	0.070	0.064	0.055	0.039	0.070	0.055	1/8	3/16	1/8	5/32
3 0.0990	48	0.0990	0.0938	0.078	0.070	0.073	0.052	0.094	0.073	3/16	7/32	5/32	7/32
3 0.0990	56	0.0990	0.0942	0.081	0.074	0.062	0.045	0.080	0.062	5/32	3/16	5/32	3/16
4 0.1120	40	0.1120	0.1061	0.087	0.078	0.088	0.062	0.112	0.088	7/32	1/4	3/16	1/4
4 0.1120	48	0.1120	0.1068	0.091	0.083	0.073	0.052	0.094	0.073	3/16	7/32	5/32	7/32
5 0.1250	40	0.1250	0.1191	0.100	0.091	0.088	0.062	0.112	0.088	7/32	9/32	3/16	1/4
5 0.1250	44	0.1250	0.1195	0.102	0.094	0.080	0.057	0.102	0.080	3/16	1/4	3/16	1/4
6 0.1380	32	0.1380	0.1312	0.107	0.096	0.109	0.078	0.141	0.109	1/4	5/16	1/4	5/16
6 0.1380	40	0.1380	0.1321	0.113	0.104	0.088	0.062	0.112	0.088	7/32	9/32	3/16	1/4
8 0.1640	32	0.1640	0.1571	0.132	0.122	0.109	0.078	0.141	0.109	1/4	11/32	1/4	5/16
8 0.1640	36	0.1640	0.1577	0.136	0.126	0.097	0.069	0.125	0.097	7/32	5/16	7/32	9/32
10 0.1900	24	0.1900	0.1818	0.148	0.135	0.146	0.104	0.188	0.146	11/32	7/16	5/16	13/32
10 0.1900	32	0.1900	0.1831	0.158	0.148	0.109	0.078	0.141	0.109	1/4	11/32	1/4	5/16
12 0.2160	24	0.2160	0.2078	0.174	0.161	0.146	0.104	0.188	0.146	11/32	7/16	5/16	13/32
12 0.2160	28	0.2160	0.2085	0.180	0.168	0.125	0.089	0.161	0.125	5/16	13/32	9/32	3/8
1/4 0.2500	20	0.2500	0.2408	0.200	0.184	0.175	0.125	0.225	0.175	13/32	17/32	3/8	1/2
1/4 0.2500	28	0.2500	0.2425	0.214	0.202	0.125	0.089	0.161	0.125	5/16	13/32	9/32	3/8
5/16 0.3125	18	0.3125	0.3026	0.257	0.239	0.194	0.139	0.250	0.194	15/32	19/32	7/16	9/16
5/16 0.3125	24	0.3125	0.3042	0.271	0.257	0.146	0.104	0.188	0.146	11/32	15/32	5/16	15/32
3/8 0.3750	16	0.3750	0.3643	0.312	0.293	0.219	0.156	0.281	0.219	1/2	11/16	15/32	5/8
3/8 0.3750	24	0.3750	0.3667	0.333	0.319	0.146	0.104	0.188	0.146	11/32	1/2	5/16	1/2
7/16 0.4375	14	0.4375	0.4258	0.366	0.344	0.250	0.179	0.321	0.250	19/32	3/4	9/16	23/32
7/16 0.4375	20	0.4375	0.4281	0.387	0.371	0.175	0.125	0.225	0.175	13/32	9/16	3/8	17/32
1/2 0.5000	13	0.5000	0.4876	0.423	0.399	0.269	0.192	0.346	0.269	5/8	25/32	19/32	3/4
1/2 0.5000	20	0.5000	0.4906	0.450	0.433	0.175	0.125	0.225	0.175	13/32	9/16	3/8	17/32

[a] Where specifying nominal size in decimals, zeros preceding decimal and in the fourth decimal place shall be omitted.

[b] The tabulated values apply to screw blanks before roll threading.

[c] Screws of these nominal lengths and shorter shall have point taper length specified above for short screws. Longer lengths shall have point taper length specified for long screws.

All dimensions are in inches. See Table 2 for thread diagrams.

Type "Type D" otherwise designated "Type 1."

Type "Type T" otherwise designated "Type 23."

Table 8. Approximate Hole Sizes for Type A Steel Thread Forming Screws

In Steel, Stainless Steel, Monel Metal, Brass, and Aluminum Sheet Metal									
		Hole Size					Hole Size		
Screw Size	Metal Thick-ness	Pierced or Extruded	Drilled or Clean Punched	Drill Size	Screw Size	Metal Thick-ness	Pierced or Extruded	Drilled or Clean Punched	Drill Size
4	0.015	…	0.086	44	8	0.024	0.136	0.125	1/8
	0.018	…	0.086	44		0.030	0.136	0.125	1/8
	0.024	0.098	0.094	42		0.036	0.136	0.125	1/8
	0.030	0.098	0.094	42		0.048	0.136	0.128	30
	0.036	0.098	0.098	40	10	0.018	…	0.136	29
6	0.015	…	0.104	37		0.024	0.157	0.136	29
	0.018	…	0.104	37		0.030	0.157	0.136	29
	0.024	0.111	0.104	37		0.036	0.157	0.136	29
	0.030	0.111	0.104	37		0.048	0.157	0.149	25
	0.036	0.111	0.106	36	12	0.024	…	0.161	20
7	0.015	…	0.116	32		0.030	0.185	0.161	20
	0.018	…	0.116	32		0.036	0.185	0.161	20
	0.024	0.120	0.116	32		0.048	0.185	0.161	20
	0.030	0.120	0.116	32	14	0.024	…	0.185	13
	0.036	0.120	0.116	32		0.030	0.209	0.189	12
	0.048	0.120	0.120	31		0.036	0.209	0.191	11
8	0.018	…	0.125	1/8		0.048	0.209	0.196	9

In Plywood (Resin Impregnated)						In Asbestos Compositions					
				Penetration in Blind Holes						Penetration in Blind Holes	
Screw Size	Hole Size	Drill Size	Min. Mat'l Thickness	Min.	Max.	Screw Size	Hole Size	Drill Size	Min. Mat'l Thickness	Min.	Max.
4	0.098	40	0.188	0.250	0.750	4	0.094	42	0.188	0.250	0.750
6	0.110	35	0.188	0.250	0.750	6	0.106	36	0.188	0.250	0.750
7	0.128	30	0.250	0.312	0.750	7	0.125	1/8	0.250	0.312	0.750
8	0.140	28	0.250	0.312	0.750	8	0.136	29	0.250	0.312	0.750
10	0.170	18	0.312	0.375	1.000	10	0.161	20	0.312	0.375	1.000
12	0.189	12	0.312	0.375	1.000	12	0.185	13	0.312	0.375	1.000
14	0.228	1	0.438	0.500	1.000	14	0.213	3	0.438	0.500	1.000

Type A is not recommended, use Type AB.

See footnote at bottom of Table 9.

Table 9. Approximate Hole Sizes for Type C Steel Thread Forming Screws

In Sheet Steel											
Screw Size	Metal Thick-ness	Hole Size	Drill Size	Screw Size	Metal Thick-ness	Hole Size	Drill Size	Screw Size	Metal Thick-ness	Hole Size	Drill Size
4–40	0.037	0.094	42	10-24	0.037	0.154	23	1/4-20	0.037	0.221	2
	0.048	0.094	42		0.048	0.161	20		0.048	0.221	2
	0.062	0.096	41		0.062	0.166	19		0.062	0.228	1
	0.075	0.100	39		0.075	0.170	18		0.075	0.234	A
	0.105	0.102	38		0.105	0.173	17		0.105	0.234	A
	0.134	0.102	38		0.134	0.177	16		0.134	0.236	6mm
6–32	0.037	0.113	33	10–32	0.037	0.170	18	1/4-28	0.037	0.224	5.7mm
	0.048	0.116	32		0.048	0.170	18		0.048	0.228	1
	0.062	0.116	32		0.062	0.170	18		0.062	0.232	5.9mm
	0.075	0.122	3.1mm		0.075	0.173	17		0.075	0.234	A
	0.105	0.125	1/8		0.105	0.177	16		0.105	0.238	B
	0.134	0.125	1/8		0.134	0.177	16		0.134	0.238	B
8–32	0.037	0.136	29	12–24	0.037	0.189	12	5/16-18	0.037	0.290	L
	0.048	0.144	27		0.048	0.194	10		0.048	0.290	L
	0.062	0.144	27		0.062	0.194	10		0.062	0.290	L
	0.075	0.147	26		0.075	0.199	8		0.075	0.295	M
	0.105	0.150	25		0.105	0.199	8		0.105	0.295	M
	0.134	0.150	25		0.134	0.199	8		0.134	0.295	M

All dimensions are in inches except drill sizes. It may be necessary to vary the hole size to suit a particular application.

Type C is not recommended for new designs.

Table 10. Approximate Pierced or Extruded Hole Sizes for Types AB, B, and BP Steel Thread Forming Screws

Screw Size	Metal Thickness	Pierced or Extruded Hole Size	Screw Size	Metal Thickness	Pierced or Extruded Hole Size	Screw Size	Metal Thickness	Pierced or Extruded Hole Size
In Steel, Stainless Steel, Monel Metal, and Brass Sheet Metal								
4	0.015	0.086	7	0.024	0.120	10	0.030	0.157
	0.018	0.086		0.030	0.120		0.036	0.157
	0.024	0.098		0.036	0.120		0.048	0.157
	0.030	0.098		0.048	0.120	12	0.024	0.185
	0.036	0.098	8	0.018	0.136		0.030	0.185
6	0.015	0.111		0.024	0.136		0.036	0.185
	0.018	0.111		0.030	0.136		0.048	0.185
	0.024	0.111		0.036	0.136	¼	0.030	0.209
	0.030	0.111		0.048	0.136		0.036	0.209
	0.036	0.111	10	0.018	0.157		0.048	0.209
7	0.018	0.120		0.024	0.157	…	…	…
In Aluminum Alloy Sheet Metal								
4	0.024	0.086	6	0.048	0.111	8	0.036	0.136
	0.030	0.086	7	0.024	0.120		0.048	0.136
	0.036	0.086		0.030	0.120	10	0.024	0.157
	0.048	0.086		0.036	0.120		0.030	0.157
6	0.024	0.111		0.048	0.120		0.036	0.157
	0.030	0.111	8	0.024	0.136		0.048	0.157
	0.036	0.111		0.030	0.136	…	…	…

All dimensions are in inches except whole number screw and drill sizes.

Since conditions differ widely, it may be necessary to vary the hole size to suit a particular application.

Table 11. Drilled Hole Sizes for Types AB, B, and BP Steel Thread Forming Screws

Screw Size	Hole Size	Drill Size	Min. Mat'l Thickness	Penetration in Blind Holes		Screw Size	Hole Size	Drill Size	Min. Mat'l Thickness	Penetration in Blind Holes	
				Min.	Max.					Min.	Max.
In Plywood (Resin Impregnated)						In Asbestos Compositions					
2	0.073	49	0.125	0.188	0.500	2	0.076	48	0.125	0.188	0.500
4	0.100	39	0.188	0.250	0.625	4	0.101	38	0.188	0.250	0.625
6	0.125	⅛	0.188	0.250	0.625	6	0.120	31	0.188	0.250	0.625
7	0.136	29	0.188	0.250	0.750	7	0.136	29	0.250	0.312	0.750
8	0.144	27	0.188	0.250	0.750	8	0.147	26	0.312	0.375	0.750
10	0.173	17	0.250	0.312	1.000	10	0.166	19	0.312	0.375	1.000
12	0.194	10	0.312	0.375	1.000	12	0.196	9	0.312	0.375	1.000
¼	0.228	1	0.312	0.375	1.000	¼	0.228	1	0.438	0.500	1.000
In Aluminum, Magnesium, Zinc, Brass, and Bronze Castings[a]						In Phenol Formaldehyde Plastics[a]					
2	0.078	47	…	0.125	…	2	0.078	47	…	0.188	…
4	0.104	37	…	0.188	…	4	0.100	39	…	0.250	…
6	0.128	30	…	0.250	…	6	0.128	30	…	0.250	…
7	0.144	27	…	0.250	…	7	0.136	29	…	0.250	…
8	0.152	24	…	0.250	…	8	0.150	25	…	0.312	…
10	0.177	16	…	0.250	…	10	0.177	16	…	0.312	…
12	0.199	8	…	0.281	…	12	0.199	8	…	0.375	…
¼	0.234	15/64	…	0.312	…	¼	0.234	15/64	…	0.375	…
In Cellulose Acetate and Nitrate, and Acrylic and Styrene Resins[a]											
2	0.078	47	…	0.188	…	8	0.144	27	…	0.312	…
4	0.094	42	…	0.250	…	10	0.170	18	…	0.312	…
6	0.120	31	…	0.250	…	12	0.191	11	…	0.375	…
7	0.128	30	…	0.250	…	¼	0.221	2	…	0.375	…

[a] Data below apply to Types B and BP only.

All dimensions are in inches except whole number screw and drill sizes.

Since conditions differ widely, it may be necessary to vary the hole size to suit a particular application.

Table 12. Approximate Drilled or Clean-Punched Hole Sizes for Types AB, B, and BP Steel Thread Forming Screws

Screw Size	Metal Thick-ness	Hole Size	Drill Size	Screw Size	Metal Thick-ness	Hole Size	Drill Size	Screw Size	Metal Thick-ness	Hole Size	Drill Size
In Steel, Stainless Steel, Monel Metal, and Brass Sheet Metal											
2	0.015	0.064	52	7	0.018	0.116	32	10	0.125	0.170	18
	0.018	0.064	52		0.024	0.116	32		0.135	0.170	18
	0.024	0.067	51		0.030	0.116	32		0.164	0.173	17
	0.030	0.070	50		0.036	0.116	32	12	0.024	0.166	19
	0.036	0.073	49		0.048	0.120	31		0.030	0.166	19
	0.048	0.073	49		0.060	0.128	30		0.036	0.166	19
	0.060	0.076	48		0.075	0.136	29		0.048	0.170	18
4	0.015	0.086	44		0.105	0.140	28		0.060	0.177	16
	0.018	0.086	44	8	0.024	0.125	⅛		0.075	0.182	14
	0.024	0.089	43		0.030	0.125	⅛		0.105	0.185	13
	0.030	0.094	42		0.036	0.125	⅛		0.125	0.196	9
	0.036	0.094	42		0.048	0.128	30		0.135	0.196	9
	0.048	0.096	41		0.060	0.136	29		0.164	0.201	7
	0.060	0.100	39		0.075	0.140	28	¼	0.030	0.194[a]	10[a]
	0.075	0.102	38		0.105	0.150	25		0.036	0.194[a]	10[a]
6	0.015	0.104	37		0.125	0.150	25		0.048	0.194[a]	10[a]
	0.018	0.104	37		0.135	0.152	24		0.060	0.199[a]	8[a]
	0.024	0.106	36	10	0.024	0.144	27		0.075	0.204[a]	6[a]
	0.030	0.106	36		0.030	0.144	27		0.105	0.209	4
	0.036	0.110	35		0.036	0.147	26		0.125	0.228	1
	0.048	0.111	34		0.048	0.152[a]	24[a]		0.135	0.228	1
	0.060	0.116	32		0.060	0.152[a]	24[a]		0.164	0.234	$^{15}/_{64}$
	0.075	0.120	31		0.075	0.157	22		0.187	0.234	$^{15}/_{64}$
	0.105	0.128	30		0.105	0.161	20		0.194	0.234	$^{15}/_{64}$
In Aluminum Alloy Sheet Metal											
2	0.024	0.064	52	7	0.060	0.120	31	10	0.164	0.159	21
	0.030	0.064	52		0.075	0.128	30		0.200 to 0.375	0.166	19
	0.036	0.064	52		0.105	0.136	29	12	0.048	0.161	20
	0.048	0.067	51		0.128 to 0.250	0.136	29		0.060	0.166	19
	0.060	0.070	50	8	0.030	0.116	32		0.075	0.173	17
4	0.030	0.086	44		0.036	0.120	31		0.105	0.180	15
	0.036	0.086	44		0.048	0.128	30		0.125	0.182	14
	0.048	0.086	44		0.060	0.136	29		0.135	0.182	14
	0.060	0.089	43		0.075	0.140	28		0.164	0.189	12
	0.075	0.089	43		0.105	0.147	26		0.200 to 0.375	0.196	9
	0.105	0.094	42		0.125	0.147	26	¼	0.060	0.199	8
6	0.030	0.104	37		0.135	0.149	25		0.075	0.201	7
	0.036	0.104	37		0.162 to 0.375	0.152	24		0.105	0.204	6
	0.048	0.104	37	10	0.036	0.144	27		0.125	0.209	4
	0.060	0.106	36		0.048	0.144	27		0.135	0.209	4
	0.75	0.110	35		0.060	0.144	27		0.164	0.213	3
	0.105	0.111	34		0.075	0.147	26		0.187	0.213	3
	0.128 to 250	120	31		0.105	0.147	26		0.194	0.221	2
7	0.030	0.113	33		0.125	0.154	23		0.200 to 0.375	228	1
	0.036	0.113	33		0.135	0.154	23				
	0.048	0.116	32								

Table 12. *(Continued)* **Approximate Drilled or Clean-Punched Hole Sizes for Types AB, B, and BP Steel Thread Forming Screws**

Screw Size	Metal Thick-ness	Hole Size	Drill Size	Screw Size	Metal Thick-ness	Hole Size	Drill Size	Screw Size	Metal Thick-ness	Hole Size	Drill Size
In Steel, Stainless Steel, Monel Metal, and Brass Sheet Metal											
10	0.018	0.144	27	¼	0.018	0.196	9	¼	0.048	0.205	5
10	0.048	0.149	25	¼	0.024	0.196	9	¼	0.060	0.228	1
10	0.060	0.154	23	¼	0.030	0.196	9	¼	0.075	0.232	5.9 mm
…	…	…	…	¼	0.036	0.196	9	…	…	…	…

[a] For Types B and BP only; for Type AB see concluded Table following.
All dimensions are in inches except numbered screw and drill sizes.

Since conditions differ widely, it may be necessary to vary the hole size to suit a particular application. Hole sizes for metal thicknesses above 0.075 inch are for Types B and BP only.

Table 13. Approximate Hole Sizes for Types D, F, G, and T Steel Thread Cutting Screws in Sheet Metals

Screw Size	Thick-ness	Steel		Aluminum Alloy	
		Hole Size	Drill Size	Hole Size	Drill Size
2–56	0.050	0.073	49	0.070	50
	0.060	0.073	49	0.073	49
	0.083	0.073	49	0.073	49
	0.109	0.073	49	0.073	49
	0.125	0.076	48	0.073	49
	0.140	0.076	48	0.073	49
3–48	0.050	0.081	46	0.078	⁵⁄₆₄
	0.060	0.081	46	0.081	46
	0.083	0.082	45	0.082	45
	0.109	0.086	44	0.082	45
	0.125	0.086	44	0.082	45
	0.140	0.086	44	0.086	44
	0.187	0.089	43	0.086	44
4–40	0.050	0.089	43	0.089	43
	0.060	0.089	43	0.089	43
	0.083	0.094	42	0.089	43
	0.109	0.096	41	0.094	42
	0.125	0.098	40	0.094	42
	0.140	0.098	40	0.094	³⁄₃₂
	0.187	0.102	38	0.098	40
5–40	0.050	0.106	36	0.102	38
	0.060	0.106	36	0.102	38
	0.083	0.106	36	0.104	37
	0.109	0.106	36	0.104	37
	0.125	0.109	⁷⁄₆₄	0.106	36
	0.140	0.110	35	0.106	36
	0.187	0.116	32	0.110	35
	0.250	0.116	32	0.113	33
6–32	0.050	0.110	35	0.109	⁷⁄₆₄
	0.060	0.113	33	0.109	⁷⁄₆₄
	0.083	0.116	32	0.111	34
	0.109	0.116	32	0.113	33
	0.125	0.116	32	0.116	32
	0.140	0.120	31	0.116	32
	0.187	0.125	⅛	0.120	31
	0.250	0.125	⅛	0.125	⅛
8–32	0.050	0.136	29	0.136	29
	0.060	0.140	28	0.136	29
	0.083	0.140	28	0.136	29
	0.109	0.144	27	0.140	28
	0.125	0.144	27	0.140	28
	0.140	0.147	26	0.144	27

Screw Size	Thick-ness	Steel		Aluminum Alloy	
		Hole Size	Drill Size	Hole Size	Drill Size
8–32	0.187	0.150	25	0.147	26
	0.250	0.150	25	0.150	25
	0.312	0.150	25	0.150	25
10–24	0.050	0.152	24	0.150	25
	0.060	0.154	23	0.152	24
	0.083	0.161	20	0.154	23
	0.109	0.161	20	0.157	22
	0.125	0.166	19	0.159	21
	0.140	0.170	18	0.161	20
	0.187	0.173	17	0.166	19
	0.250	0.173	17	0.172	¹¹⁄₆₄
	0.312	0.173	17	0.173	17
	0.375	0.173	17	0.173	17
10–32	0.050	0.159	21	0.161	20
	0.060	0.166	19	0.161	20
	0.083	0.166	19	0.161	20
	0.109	0.170	18	0.166	19
	0.125	0.170	18	0.166	19
	0.140	0.170	18	0.166	19
	0.187	0.177	16	0.172	¹¹⁄₆₄
	0.250	0.177	16	0.177	16
	0.312	0.177	16	0.177	16
	0.375	0.177	16	0.177	16
12–24	0.060	0.180	15	0.177	16
	0.083	0.182	14	0.180	15
	0.109	0.188	³⁄₁₆	0.182	14
	0.125	0.191	11	0.185	13
	0.140	0.191	11	0.188	³⁄₁₆
	0.187	0.199	8	0.191	11
	0.250	0.199	8	0.199	8
	0.312	0.199	8	0.199	8
	0.375	0.199	8	0.199	8
	0.500	0.199	8	0.199	8
¼–20	0.083	0.213	3	0.206	5
	0.109	0.219	⁷⁄₃₂	0.209	4
	0.125	0.221	2	0.213	3
	0.140	0.221	2	0.213	3
	0.187	0.228	1	0.221	2
	0.250	0.228	1	0.228	1
	0.312	0.228	1	0.228	1
	0.375	0.228	1	0.228	1
	0.500	0.228	1	0.228	1

Table 13. *(Continued)* Approximate Hole Sizes for Types D, F, G, and T Steel Thread Cutting Screws in Sheet Metals

Screw Size	Thick-ness	Steel		Aluminum Alloy		Screw Size	Thick-ness	Steel		Aluminum Alloy	
		Hole Size	Drill Size	Hole Size	Drill Size			Hole Size	Drill Size	Hole Size	Drill Size
1/4–28	0.083	0.221	2	0.219	7/32	5/16–24	0.187	0.295	M	0.290	L
	0.109	0.228	1	0.221	2		0.250	0.295	M	0.295	M
	0.125	0.228	1	0.221	2		0.312	0.295	M	0.295	M
	0.140	0.234	A	0.221	2		0.375	0.295	M	0.295	M
	0.187	0.234	15/64	0.228	1		0.500	0.295	M	0.295	M
	0.250	0.234	15/64	0.234	15/64	3/8–16	0.125	0.339	R	0.328	21/64
	0.312	0.234	15/64	0.234	15/64		0.140	0.339	R	0.332	Q
	0.375	0.234	15/64	0.234	15/64		0.187	0.348	S	0.339	R
	0.500	0.234	15/64	0.234	15/64		0.250	0.358	T	0.348	S
5/16–18	0.109	0.277	J	0.266	H		0.312	0.358	T	0.348	S
	0.125	0.277	J	0.272	I		0.375	0.358	T	0.348	S
	0.140	0.281	9/32	0.272	I		0.500	0.358	T	0.348	S
	0.187	0.290	L	0.281	K	3/8–24	0.125	0.348	S	0.344	11/32
	0.250	0.290	L	0.290	L		0.140	0.348	S	0.344	11/32
	0.312	0.290	L	0.290	L		0.187	0.358	T	0.348	S
	0.375	0.290	L	0.290	L		0.250	0.358	T	0.358	T
	0.500	0.290	L	0.290	L		0.312	0.358	T	0.358	T
5/16–24	0.109	0.290	L	0.281	K		0.375	0.358	T	0.358	T
	0.125	0.290	L	0.281	9/32		0.500	0.358	T	0.358	T
	0.140	0.290	L	0.281	9/32		…	…	…	…	…

All dimensions are in inches except numbered drill and screw sizes. It may be necessary to vary the hole size to suit a particular application.

Table 14. Approximate Hole Sizes for Types D, F, G, and T Steel Thread Cutting Screws in Sheet Metals

Screw Size	Thick-ness	Cast Iron		Zinc and Aluminum[a]		Screw Size	Thick-ness	Cast Iron		Zinc and Aluminum[a]	
		Hole Size	Drill Size	Hole Size	Drill Size			Hole Size	Drill Size	Hole Size	Drill Size
2–56	0.050	0.076	48	0.073	49	5–40	0.083	0.113	33	0.106	36
	0.060	0.076	48	0.073	49		0.109	0.113	33	0.110	35
	0.083	0.076	48	0.076	48		0.125	0.116	32	0.110	35
	0.109	0.078	5/64	0.076	48		0.140	0.116	32	0.110	35
	0.125	0.078	5/64	0.076	48		0.187	0.116	32	0.111	34
	0.140	0.078	5/64	0.076	48		0.250	0.116	32	0.113	33
3–48	0.050	0.089	43	0.082	45	6–32	0.050	0.120	31	0.116	32
	0.060	0.089	43	0.082	45		0.060	0.120	31	0.120	31
	0.083	0.089	43	0.082	45		0.083	0.125	1/8	0.120	31
	0.109	0.089	43	0.086	44		0.109	0.125	1/8	0.120	31
	0.125	0.089	43	0.089	43		0.125	0.125	1/8	0.120	31
	0.140	0.094	42	0.089	43		0.140	0.125	1/8	0.120	31
	0.187	0.094	42	0.089	43		0.187	0.128	30	0.120	31
4–40	0.050	0.100	39	0.090	41		0.250	0.128	30	0.120	31
	0.060	0.100	39	0.096	41	8–32	0.050	0.147	26	0.144	27
	0.083	0.102	38	0.096	41		0.060	0.150	25	0.144	27
	0.109	0.102	38	0.096	41		0.083	0.150	25	0.144	27
	0.125	0.102	38	0.100	39		0.109	0.150	25	0.144	27
	0.140	0.102	38	0.100	39		0.125	0.150	25	0.147	26
	0.187	0.104	37	0.100	39		0.140	0.150	25	0.147	26
5–40	0.050	0.111	34	0.106	36		0.187	0.154	23	0.147	26
	0.060	0.111	34	0.106	36		0.250	0.154	23	0.150	25
							0.312	0.154	23	0.150	25

Table 14. *(Continued)* **Approximate Hole Sizes for Types D, F, G, and T Steel Thread Cutting Screws in Sheet Metals**

Screw Size	Thick-ness	Cast Iron Hole Size	Cast Iron Drill Size	Zinc and Aluminum[a] Hole Size	Zinc and Aluminum[a] Drill Size
10–24	0.050	0.170	18	0.161	20
	0.060	0.170	18	0.166	19
	0.083	0.172	11/64	0.166	19
	0.109	0.173	17	0.166	19
	0.125	0.173	17	0.166	19
	0.140	0.173	17	0.166	19
	0.187	0.177	16	0.170	18
	0.250	0.177	16	0.170	18
	0.312	0.177	16	0.172	11/64
	0.375	0.177	16	0.172	11/64
10–32	0.050	0.173	17	0.170	18
	0.060	0.173	17	0.170	18
	0.083	0.177	16	0.172	11/64
	0.109	0.177	16	0.172	11/64
	0.125	0.177	16	0.172	11/64
	0.140	0.177	16	0.172	11/64
	0.187	0.180	15	0.172	11/64
	0.250	0.180	15	0.173	17
	0.312	0.180	15	0.173	17
	0.375	0.180	15	0.177	16
12–24	0.060	0.196	9	0.189	12
	0.083	0.199	8	0.191	11
	0.109	0.199	8	0.191	11
	0.125	0.199	8	0.191	11
	0.140	0.199	8	0.194	10
	0.187	0.203	13/64	0.194	10
	0.250	0.204	6	0.196	9
	0.312	0.204	6	0.196	9
	0.375	0.204	6	0.199	8
	0.500	0.204	6	0.199	8
1/4–20	0.083	0.228	1	0.219	7/32
	0.109	0.228	1	0.219	7/32
	0.125	0.228	1	0.221	2
	0.140	0.228	1	0.221	2
	0.187	0.234	15/64	0.221	2
	0.250	0.234	15/64	0.228	1
	0.312	0.234	15/64	0.228	1
	0.375	0.234	15/64	0.228	1
	0.500	0.234	15/64	0.228	1

Screw Size	Thick-ness	Cast Iron Hole Size	Cast Iron Drill Size	Zinc and Aluminum[a] Hole Size	Zinc and Aluminum[a] Drill Size
1/4–28	0.083	0.234	A	0.228	1
	0.109	0.234	15/64	0.228	1
	0.125	0.234	15/64	0.228	1
	0.140	0.234	15/64	0.228	1
	0.187	0.238	B	0.228	1
	0.250	0.238	B	0.234	A
	0.312	0.238	B	0.234	A
	0.375	0.238	B	0.234	15/64
	0.500	0.238	B	0.234	15/64
5/16–18	0.109	0.290	L	0.277	J
	0.125	0.290	L	0.281	K
	0.140	0.290	L	0.281	K
	0.187	0.295	M	0.281	9/32
	0.250	0.295	M	0.281	9/32
	0.312	0.295	M	0.290	L
	0.375	0.295	M	0.290	L
	0.500	0.295	M	0.290	L
5/16–24	0.109	0.295	M	0.290	L
	0.125	0.295	M	0.290	L
	0.140	0.295	M	0.290	L
	0.187	0.302	N	0.290	L
	0.250	0.302	N	0.290	L
	0.312	0.302	N	0.295	M
	0.375	0.302	N	0.295	M
	0.500	0.302	N	0.295	M
3/8–16	0.125	0.348	S	0.339	R
	0.140	0.348	S	0.339	R
	0.187	0.348	S	0.339	R
	0.250	0.348	S	0.344	11/32
	0.312	0.348	S	0.344	11/32
	0.375	0.348	S	0.348	S
	0.500	0.348	S	0.348	S
3/8–24	0.125	0.358	T	0.348	S
	0.140	0.358	T	0.348	S
	0.187	0.358	T	0.348	S
	0.250	0.358	T	0.358	T
	0.312	0.358	T	0.358	T
	0.375	0.358	T	0.358	T
	0.500	0.358	T	0.358	T

[a] Die Castings

Screw Size	Phenol Formaldehyde[a]				Cellulose Acetate, Cellulose Nitrate, Acrylic Resin, and Styrene Resin[a]			
	Hole Size	Drill Size	Depth of Penetration		Hole Size	Drill Size	Depth of Penetration	
			Min	Max			Min	Max
2–56	0.078	5/64	0.219	0.375	0.076	48	0.219	0.375
3–48	0.089	43	0.219	0.375	0.086	44	0.219	0.375
4–40	0.098	40	0.250	0.312	0.093	42	0.250	0.312
5–40	0.113	33	0.250	0.438	0.110	35	0.250	0.438
6–32	0.116	32	0.250	0.312	0.116	32	0.250	0.312
8–32	0.144	27	0.312	0.500	0.144	27	0.312	0.500
10–24	0.161	20	0.375	0.500	0.161	20	0.375	0.500
10–32	0.166	19	0.375	0.500	0.166	19	0.375	0.500
1/4–20	0.228	1	0.375	0.625	0.228	1	0.375	1.000

[a] Plastics

For footnotes see Table .

Table 15. Approximate Hole Sizes for Types BF and BT Steel Thread Cutting Screws in Cast Metals

In Die Cast Zinc and Aluminum							
Screw Size	Thickness	Hole Size	Drill Size	Screw Size	Thickness	Hole Size	Drill Size
2	0.060	0.073	49	10	0.125	0.166	19
	0.083	0.073	49		0.140	0.166	19
	0.109	0.076	48		0.188	0.166	19
	0.125	0.076	48		0.250	0.170	18
	0.140	0.076	48		0.312	0.172	11⁄64
3	0.060	0.086	44		0.375	0.172	11⁄64
	0.083	0.086	44	12	0.125	0.191	11
	0.109	0.086	44		0.140	0.191	11
	0.125	0.086	44		0.188	0.191	11
	0.140	0.089	43		0.250	0.196	9
	0.188	0.089	43		0.312	0.196	9
4	0.109	0.098	40		0.375	0.196	9
	0.125	0.100	39	1⁄4	0.125	0.221	2
	0.140	0.100	39		0.140	0.221	2
	0.188	0.100	39		0.188	0.221	2
	0.250	0.102	38		0.250	0.228	1
5	0.109	0.111	34		0.312	0.228	1
	0.125	0.111	34		0.375	0.228	1
	0.140	0.113	33	5⁄16	0.125	0.281	K
	0.188	0.113	33		0.140	0.281	K
	0.250	0.116	32		0.188	0.281	K
6	0.125	0.120	31		0.250	0.281	K
	0.140	0.120	31		0.312	0.290	L
	0.188	0.120	31		0.375	0.290	L
	0.250	0.125	1⁄8	3⁄8	0.125	0.344	11⁄32
	0.312	0.125	1⁄8		0.140	0.344	11⁄32
8	0.125	0.149	25		0.188	0.344	11⁄32
	0.140	0.149	25		0.250	0.344	11⁄32
	0.188	0.149	25		0.312	0.348	S
	0.250	0.152	24		0.375	0.348	S
	0.312	0.152	24		…	…	…

All dimensions are in inches except numbered drill and screw sizes. It may be necessary to vary the hole size to suit a particular application.

Table 16. Approximate Hole Size for Types BF and BT Steel Thread Cutting Screws in Plastics

	Phenol Formaldehyde				Cellulose Acetate, Cellulose Nitrate, Acrylic Resin and Styrene Resin			
			Depth of Penetration				Depth of Penetration	
Screw Size	Hole Size	Drill Size	Min	Max	Hole Size	Drill Size	Min	Max
2	0.078	5⁄64	0.094	0.250	0.076	48	0.094	0.250
3	0.089	43	0.125	0.312	0.089	43	0.125	0.312
4	0.104	37	0.125	0.312	0.100	39	0.125	0.312
5	0.116	32	0.188	0.375	0.113	33	0.188	0.375
6	0.125	1⁄8	0.188	0.375	0.120	31	0.188	0.375
8	0.147	26	0.250	0.500	0.144	27	0.250	0.500
10	0.170	18	0.312	0.625	0.166	19	0.312	0.625
12	0.194	10	0.375	0.625	0.189	12	0.375	0.625
1⁄4	0.228	1	0.375	0.750	0.221	2	0.375	0.750

For footnotes see above table.

Table 17. Approximate Hole Sizes for Type U Hardened Steel Metallic Drive Screws

In Ferrous and Non-Ferrous Castings, Sheet Metals, Plastics, Plywood (Resin-Impregnated) and Fiber								
Screw Size	Hole Size	Drill Size	Screw Size	Hole Size	Drill Size	Screw Size	Hole Size	Drill Size
00	.052	55	6	.120	31	12	.191	11
0	.067	51	7	.136	29	14	.221	2
2	.086	44	8	.144	27	5⁄16	.295	M
4	.104	37	10	.161	20	3⁄8	.358	T

All dimensions are in inches except whole number screw and drill sizes and letter drill sizes.

Table 18. ANSI Standard Torsional Strength Requirements for Tapping Screws
ANSI B18.6.4-1981 (R1991)

Nom. Screw Size	Type A	Types AB,B,BF, BP,andBT	Types C, D, F, G, and T		Nom-Screw Size	Type A	Types AB, B, BF, BP, and BT	Types C, D, F, G, and T	
			Coarse Thread	Fine Thread				Coarse Thread	Fine Thread
2	4	4	5	6	1⁄4	…	142	140	179
3	9	9	9	10	16	152	…	…	…
4	12	13	13	15	18	196	…	…	…
5	18	18	18	20	5⁄16	…	290	306	370
6	24	24	23	27	20	250	…	…	…
7	30	30	…	…	24	492	…	…	…
8	39	39	42	47	3⁄8	…	590	560	710
10	48	56	56	74	7⁄16	…	620	700	820
12	83	88	93	108	1⁄2	…	1020	1075	1285
14	125	…	…	…	…	…	…	…	…

Torsional strength data are in pound-inches.

Self-tapping Thread Inserts.—Self-tapping screw thread inserts are essentially hard bushings with internal and external threads. The internal threads conform to Unified and American standard classes 2B and 3B, depending on the type of insert used. The external thread has cutting edges on the end that provide the self-tapping feature. These inserts may be used in magnesium, aluminum, cast iron, zinc, plastics, and other materials. Self-tapping inserts are made of case-hardened carbon steel, stainless steel, and brass, the brass type being designed specifically for installation in wood.

Screw Thread Inserts.—Screw thread inserts are helically formed coils of diamond-shaped stainless steel or phosphor bronze wire that screw into a threaded hole to form a mating internal thread for a screw or stud. These inserts provide a convenient means of repairing stripped-out threads and are also used to provide stronger threads in soft materials such as aluminum, zinc die castings, wood, magnesium, etc. than can be obtained by direct tapping of the base metal involved.

According to the Heli-Coil Corp., conventional design practice in specifying boss diameters or edge distances can usually be applied since the major diameter of a hole tapped to receive a thread insert is not much larger than the major diameter of thread the insert provides.

Screw thread inserts are available in thread sizes from 4–40 to 1½–6 inch National and Unified Coarse Thread Series and in 6–40 to 1½–12 sizes in the fine-thread series. When used in conjunction with appropriate taps and gages, screw thread inserts will meet requirements of 2, 2B, 3, and 3B thread classes.

ANSI Standard Metric Thread Forming and Thread Cutting Tapping Screws.—

Table 1 shows the various types of metric thread forming and thread cutting screw threads covered by the standard ANSI/ASME B18.6.5M-1986. The designations of the American National Standards Institute are shown.

Table 1. ANSI Standard Threads and Points for Metric Thread Forming and Thread Cutting Tapping Screws *ANSI/ASME B18.6.5M-1986*

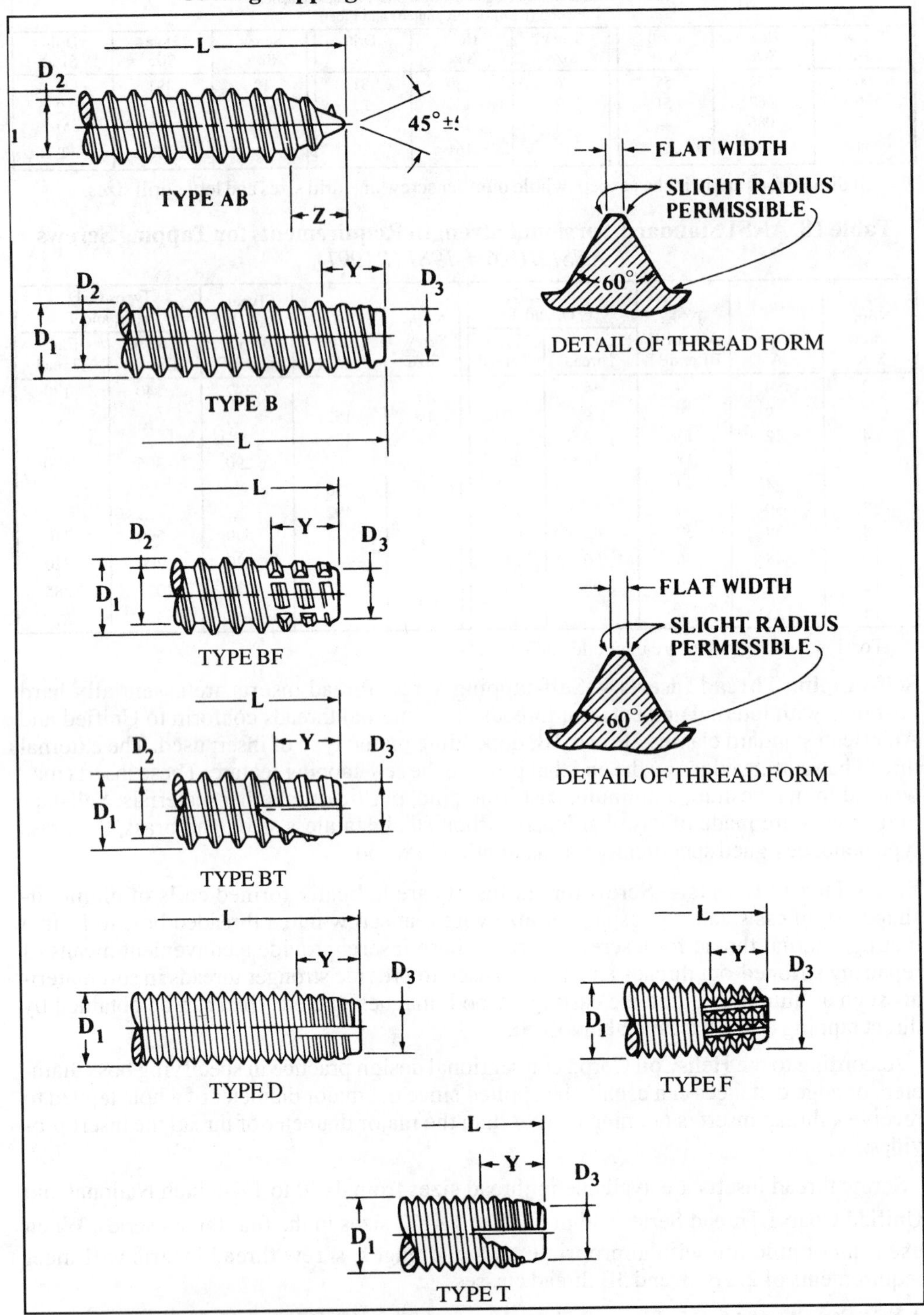

See Tables 3 and 4 for thread data.

Thread Forming Tapping Screws: These types are generally for application in materials where large internal stresses are permissible or desirable, to increase resistance to loosening. These screws have the following descriptions and applications:

Type AB: Spaced thread screw with gimlet point primarily intended for use in thin metal, resin impregnated plywood, and asbestos compositions.

Type B: Spaced thread screw with a blunt point that has tapered entering threads with unfinished crests and same pitches as Type AB. Used for thin metal, nonferrous castings, resin impregnated plywood, certain resilient plastics, and asbestos compositions.

Thread Cutting Tapping Screws: These screws are generally for application in materials where disruptive internal stresses are undesirable or where excessive driving torques are encountered with thread forming tapping screws. These screws have the following descriptions and applications:

Types BF and BT: Spaced threads with blunt point and tapered entering threads having unfinished crests, as on Type B, with one or more cutting edges or chip cavities, intended for use in plastics, asbestos compositions, and other similar materials.

Types D, F, and T: Tapping screws with threads of machine screw diameter-pitch combinations (metric coarse thread series) approximating a 60 degree basic thread form (not necessarily conforming to any standard thread profile) with a blunt point and tapered entering threads with unfinished crests and having one or more cutting edges and chip cavities, intended for use in materials such as aluminum, zinc, and lead die castings; steel sheets and shapes; cast iron; brass; and plastics.

ANSI Standard Head Types for Metric Thread Forming and Cutting Tapping Screws.—The head types covered by ANSI/ASME B18.6.5M-1986 include those commonly applicable to metric tapping screws and are described as follows:

Flat Countersunk Head: The flat countersunk head has a flat top surface and a conical bearing surface with a head angle of 90 to 92 degrees.

Oval Countersunk Head: The oval countersunk head has a rounded top surface and a conical bearing surface with a head angle of 90 to 92 degrees.

Pan Head: The slotted pan head has a flat top surface rounding into cylindrical sides and a flat bearing surface. The recessed pan head has a rounded top surface blending into cylindrical sides and a flat bearing surface.

Hex Head: The hex head has a flat or indented top surface, six flat sides, and a flat bearing surface.

Hex Flange Head: The hex flange head has a flat or indented top surface and six flat sides formed integrally with a frustroconical or slightly rounded (convex) flange that projects beyond the sides and provides a flat bearing surface.

Method of Designation.—Metric tapping screws are designated with the following data, preferably in the sequence shown: Nominal size; thread pitch; nominal length; thread and point type; product name, including head style and driving provision; material; and protective finish, if required.

Examples:

6.3 × 1.8 × 30 Type AB, Slotted Pan Head Tapping Screw, Steel, Zinc Plated

6 × 1 × 20 Type T, Type 1A Cross Recessed Pan Head Tapping Screw, Corrosion Resistant Steel

4.2 × 1.4 × 13 Type BF, Type 1 Cross Recessed Oval Countersunk Head Tapping Screw, Steel, Chromium Plated

10 × 1.5 × 40 Type D, Hex Flange Head Tapping Screw, Steel

Table 2. Recommended Nominal Screw Lengths for Metric Tapping Screws *ANSI/ASME B18.6.5M-1986*

Nominal Screw Length	Nominal Screw Size for Types AB, B, BF, and BT									
	2.2	-	2.9	3.5	4.2	4.8	5.5	6.3	8	9.5
	Nominal Screw Size for Types D, F, and T									
	2	2.5	3	3.5	4	5	-	6	8	10
4	PH	PH								
5	PH	PH								
6	A	A	PH							
8	A	A	A	PH	PH					
10	A	A	A	A	A	PH				
13	A	A	A	A	A	A	A	PH		
16		A	A	A	A	A	A	A	PH	
20				A	A	A	A	A	A	PH
25				A	A	A	A	A	A	A
30						A	A	A	A	A
35						A	A	A	A	A
40							A	A	A	A
45									A	A
50									A	A
55										A
60										A

Table 3. ANSI Standard Thread and Point Dimensions for Types AB and B Metric Thread Forming Tapping Screws *ANSI/ASME B18.6.5M-1986*

Nominal Screw Size and Thread Pitch[a]	Basic Screw Diameter	Basic Thread Pitch	D_1 Thread Major Diameter		D_2 Thread Minor Diameter		D_3 Point Diameter[b]		Y Point Taper Length Type B[c]		Z Point Length Factor TypeAB	L Min. Practical Nominal Screw Length[d]			
												Type AB		Type B	
	Ref[e]	Ref[e]	Max	Min	Max	Min	Max	Min	Max	Min	Ref[f]	Note 7	Note 8	Note 7	Note 8
2.2 × 0.8	2.184	0.79	2.24	2.10	1.63	1.52	1.47	1.37	1.6	1.2	2.0	4	6	4	5
2.9 × 1	2.845	1.06	2.90	2.76	2.18	2.08	2.01	1.88	2.1	1.6	2.6	6	7	5	7
3.5 × 1.3	3.505	1.27	3.53	3.35	2.64	2.51	2.41	2.26	2.5	1.9	3.2	7	9	6	8
4.2 × 1.4	4.166	1.41	4.22	4.04	3.10	2.95	2.84	2.69	2.8	2.1	3.7	8	10	7	10
4.8 × 1.6	4.826	1.59	4.80	4.62	3.58	3.43	3.30	3.12	3.2	2.4	4.3	9	12	8	11
5.5 × 1.8	5.486	1.81	5.46	5.28	4.17	3.99	3.86	3.68	3.6	2.7	5.0	11	14	9	12
6.3 × 1.8	6.350	1.81	6.25	6.03	4.88	4.70	4.55	4.34	3.6	2.7	6.0	12	16	10	13
8 × 2.1	7.938	2.12	8.00	7.78	6.20	5.99	5.84	5.64	4.2	3.2	7.5	16	20	12	17
9.5 × 2.1	9.525	2.12	9.65	9.43	7.85	7.59	7.44	7.24	4.2	3.2	8.0	19	24	14	19

[a] The body diameter (unthreaded portion) is not less than the minimum minor diameter nor greater than the maximum major diameter of the thread.

[b] The tabulated values shall apply to screw blanks prior to roll threading.

[c] The tabulated maximum limits are equal to approximately two times the thread pitch.

[d] Lengths shown are theoretical minimums and are intended to assist the user in the selection of appropriate short screw lengths. Refer to Table 2 for recommended diameter-length combinations.

[e] Basic screw diameter and basic thread pitch shall be used for calculation purposes wherever these factors appear in formulations for dimensions.

[f] The minimum effective grip length on Type AB tapping screws shall be determined by subtracting the point length factor from the minimum screw length.

All dimensions are in millimeters. See Table 1 for thread diagrams.

[7] Pan, hex, and hex flange heads.

[8] Flat and oval countersunk heads.

Table 4. ANSI Standard Thread and Point Dimensions for Types BF, BT, D, F, and T Metric Thread Cutting Tapping Screws *ANSI/ASME B18.6.5M-1986*

Types BF and BT												
			D1		D2		D3		Y		L	
Nominal Screw Size and Thread Pitch	Basic Screw Diameter	Basic Thread Pitch	Thread Major Diameter		Thread Minor Diameter		Point Diameter[a]		Point Taper Length Type B[b]		Minimal Practical Nominal Screw Length[c]	
	Ref[d]	Ref[d]	Max	Min	Max	Min	Max	Min	Max	Min	Pan, Hex and Hex Flange Heads	Flat and Oval Csunk Heads
2.2 × 0.8	2.184	0.79	2.24	2.10	1.63	1.52	1.47	1.37	1.6	1.2	4	5
2.9 × 1	2.845	1.06	2.90	2.76	2.18	2.08	2.01	1.88	2.1	1.6	5	7
3.5 × 1.3	3.505	1.27	3.53	3.35	2.64	2.51	2.41	2.26	2.5	1.9	6	8
4.2 × 1.4	4.166	1.41	4.22	4.04	3.10	2.95	2.84	2.69	2.8	2.1	7	10
4.8 × 1.6	4.826	1.59	4.80	4.62	3.58	3.43	3.30	3.12	3.2	2.4	8	11
5.5 × 1.8	5.486	1.81	5.46	5.28	4.17	3.99	3.86	3.68	3.6	2.7	9	12
6.3 × 1.8	6.350	1.81	6.25	6.03	4.88	4.70	4.55	4.34	3.6	2.7	10	13
8 × 2.1	7.938	2.12	8.00	7.78	6.20	5.99	5.84	5.64	4.2	3.2	12	17
9.5 × 2.1	9.525	2.12	9.65	9.43	7.85	7.59	7.44	7.24	4.2	3.2	14	19

[a] The tabulated values apply to screw blanks prior to roll threading.

[b] The tabulated maximum limits are equal to approximately two times the thread pitch.

[c] Lengths shown are theoretical minimums and are intended to assist in the selection of appropriate short screw lengths. See Table 2 for recommended length-diameter combinations. For Types D, F, and T, shorter screws are available with the point length reduced to the limits tabulated for short screws.

[d] Basic screw diameter and basic thread pitch are used for calculation purposes whenever these factors appear in formulations for dimensions.

Types D, F, T												
	D1		D3		Ds	Y					L	
Nominal Screw Size and Thread Pitch	Thread Major Diameter		Point Diameter[a]		Body Diameter[a]	Point Taper Length					Minimum Practical Nominal Screw Length[c]	
						For Short Screws		For Long Screws[b]			Pan, Hex and Hex Flange Heads	Flat and Oval Csunk Heads
	Max	Min	Max	Min	Min	Max	Min	Max	Min			
2 × 0.4	2.00	1.88	1.45	1.39	1.65	1.4	1.0	1.8	1.4		4	5
2.5 × 0.45	2.50	2.37	1.88	1.82	2.12	1.6	1.1	2.0	1.6		4	6
3 × 0.5	3.00	2.87	2.32	2.26	2.58	1.8	1.3	2.3	1.8		5	6
3.5 × 0.6	3.50	3.35	2.68	2.60	3.00	2.1	1.5	2.7	2.1		5	8
4 × 0.7	4.00	3.83	3.07	2.97	3.43	2.5	1.8	3.2	2.5		6	9
5 × 0.8	5.00	4.82	3.94	3.84	4.36	2.8	2.0	3.6	2.8		7	10
6 × 1	6.00	5.79	4.69	4.55	5.21	3.5	2.5	4.5	3.5		9	12
8 × 1.25	8.00	7.76	6.40	6.24	7.04	4.4	3.1	5.6	4.4		11	16
10 × 1.5	10.00	9.73	8.08	7.88	8.86	5.3	3.8	6.8	5.3		13	18

[a] Minimum limits for body diameter (unthreaded portion) are tabulated for convenient reference. For Types BF and BT, the body diameter is not less than the minimum minor diameter nor greater than the maximum major diameter of the thread.

[b] Long screws are screws of nominal lengths equal to or longer than those listed under L.

All dimensions are in millimeters. See Table 1 for thread diagrams.

Material and Heat Treatment.—Tapping screws are normally fabricated from carbon steel and are suitably processed to meet the performance and test requirements outlined in the standard, B18.6.5M. Tapping screws may also be made from corrosion resistant steel, Monel, brass, and aluminum alloys. The materials, properties, and performance characteristics applicable to such screws should be mutually agreed upon between the manufacturer and the purchaser.

Table 5. Clearance Holes for Metric Tapping Screws
ANSI/ASME B18.6.5M-1986 **Appendix**

Nominal Screw Size and Thread Pitch	Basic Clearance Hole Diameter[a]			Nominal Screw Size and Thread Pitch	Basic Clearance Hole Diameter[a]		
	Close Clearance[b]	Normal Clearance (Preferred)[b]	Loose Clearance[b]		Close Clearance[b]	Normal Clearance (Preferred)[b]	Loose Clearance[b]
	Types AB, B, BF, and BT				Types D, F, and T		
2.2 ×0.8	2.40	2.60	2.80	2 ×0.4	2.20	2.40	2.60
2.9 ×1	3.10	3.30	3.50	2.5 ×0.45	2.70	2.90	3.10
3.5 ×1.3	3.70	3.90	4.20	3 ×0.5	3.20	3.40	3.60
4.2 ×1.4	4.50	4.70	5.00	3.5 ×0.6	3.70	3.90	4.20
4.8 ×1.6	5.10	5.30	5.60	4 ×0.7	4.30	4.50	4.80
5.5 ×1.8	5.90	6.10	6.50	5 ×0.8	5.30	5.50	5.80
6.3×1.8	6.70	6.90	7.30	6 ×1	6.40	6.60	7.00
8 ×2.1	8.40	9.00	10.00	8 ×1.25	8.40	9.00	10.00
9.5 ×2.1	10.00	10.50	11.50	10 ×1.5	10.50	11.00	12.00

[a] The values given in this table are minimum limits. The recommended plus tolerances are as follows: for clearance hole diameters over 1.70 to and including 5.80 mm, plus 0.12, 0.20, and 0.30 mm for close, normal, and loose clearances, respectively; over 5.80 to and including 14.50 mm, plus 0.18, 0.30, and 0.45 mm for close, normal, and loose clearances, respectively.

[b] Normal clearance hole sizes are preferred. Close clearance hole sizes are for situations such as critical alignment of assembled components, wall thickness, or other limitations that necessitate the use of a minimal hole. Countersinking or counterboring at the fastener entry side may be necessary for the proper seating of the head. Loose clearance hole sizes are for applications where maximum adjustment capability between the components being assembled is necessary.

All dimensions are in millimeters.

Approximate Installation Hole Sizes for Metric Tapping Screws.—The approximate hole sizes given in Tables 7 through 9 provide general guidance in selecting holes for installing the respective types of metric thread forming and thread cutting tapping screws in various commonly used materials. Types AB, B, BF, and BT metric tapping screws are covered in these tables; hole sizes for Types D, F, and T metric thread cutting tapping screws are still under development.

Table 6. Approximate Pierced or Extruded Hole Sizes for Steel Types AB and B Metric Thread Forming Tapping Screws

Nominal Screw Size and Thread Pitch	Metal Thickness	Hole Size	Nominal Screw Size and Thread Pitch	Metal Thickness	Hole Size	Nominal Screw Size and Thread Pitch	Metal Thickness	Hole Size
In Steel, Stainless Steel, Monel, and Brass Sheet Metal								
2.9 × 1	0.38	2.18	4.2 × 1.4	0.46	3.45	5.5 × 1.8	0.61	4.70
	0.46	2.18		0.61	3.45		0.76	4.70
	0.61	2.49		0.76	3.45		0.91	4.70
	0.76	2.49		0.91	3.45		1.22	4.70
	0.91	2.49		1.22	3.45		…	…
3.5 × 1.3	0.38	2.82	4.8 × 1.6	0.46	3.99	6.3 × 1.8	0.76	5.31
	0.46	2.82		0.61	3.99		0.91	5.31
	0.61	2.82		0.76	3.99		1.22	5.31
	0.76	2.82		0.91	3.99		…	…
	0.91	2.82		1.22	3.99		…	…
In Aluminum Alloy								
2.9 × 1	0.61	2.18	3.5 × 1.3	0.91	2.82	4.8 × 1.6	0.61	3.99
	0.76	2.18		1.22	2.82		0.76	3.99
	0.91	2.18	4.2 × 1.4	0.61	3.45		0.91	3.99
	1.22	2.18		0.76	3.45		1.22	3.99
3.5 × 1.3	0.61	2.82		0.91	3.45		…	…
	0.76	2.82		1.22	3.45		…	…

All dimensions are in millimeters.

Table 7. Approximate Drilled or Clean-Punched Hole Sizes for Steel Type AB Metric Thread Forming Tapping Screws in Sheet Metal

Nominal Screw Size and Thread Pitch	Metal Thick-ness	Hole Size	Drill Size[a]	Nominal Screw Size and Thread Pitch	Metal Thick-ness	Hole Size	Drill Size[a]	Nominal Screw Size and Thread Pitch	Metal Thick-ness	Hole Size	Drill Size[a]
In Steel, Stainless Steel, Monel, and Brass Sheet Metal											
2.2 × 0.8	0.38	1.63	52	3.5 × 1.3	0.61	2.69	36	4.8 × 1.6	1.22	3.78	25
	0.46	1.63	52		0.76	2.69	36		1.52	3.91	23
	0.61	1.70	51		0.91	2.79	35		1.90	3.99	22
	0.76	1.78	50		1.22	2.82	34	5.5 × 1.8	0.46	…	…
	0.91	1.85	49		1.52	2.95	32		0.61	4.22	19
	1.22	1.85	49		1.90	3.05	31		0.76	4.22	19
	1.52	1.93	48	4.2 × 1.4	0.46	…	…		0.91	4.22	19
2.9 × 1	0.38	2.18	44		0.61	3.18	…		1.22	4.32	18
	0.46	2.18	44		0.76	3.18	…		1.52	4.50	16
	0.61	2.26	43		0.91	3.18	…		1.90	4.62	14
	0.76	2.39	42		1.22	3.25	30	6.3 × 1.8	0.46	4.98	9
	0.91	2.39	42		1.52	3.45	29		0.61	4.98	9
	1.22	2.44	41		1.90	3.56	28		0.76	4.98	9
	1.52	2.54	39	4.8 × 1.6	0.46	3.66	27		0.91	4.98	9
	1.90	2.59	38		0.61	3.66	27		1.22	5.21	W
3.5 × 1.3	0.38	2.64	37		0.76	3.66	27		1.52	5.79	1
	0.46	2.64	37		0.91	3.73	26		1.90	5.89	…
In Aluminum Alloy Sheet Metal											
2.2 × 0.8	0.38	…	…	3.5 × 1.3	0.61	…	…	4.8 × 1.6	1.22	3.66	27
	0.46	…	…		0.76	2.64	37		1.52	3.66	27
	0.61	1.63	52		0.91	2.64	37		1.90	3.73	26
	0.76	1.63	52		1.22	2.64	37	5.5 × 1.8	0.46	…	…
	0.91	1.63	52		1.52	2.69	36		0.61	…	…
	1.22	1.70	51		1.90	2.79	35		0.76	…	…
	1.52	1.78	50	4.2 × 1.4	0.46	…	…		0.91	…	…
2.9 × 1	0.38	…	…		0.61	…	…		1.22	4.09	20
	0.46	…	…		0.76	2.95	32		1.52	4.22	19
	0.61	…	…		0.91	3.05	31		1.90	4.39	17
	0.76	2.18	44		1.22	3.25	30	6.3 × 1.8	0.46	…	…
	0.91	2.18	44		1.52	3.45	29		0.61	…	…
	1.22	2.18	44		1.90	3.56	28		0.76	…	…
	1.52	2.26	43	4.8 × 1.6	0.46	…	…		0.91	…	…
	1.90	2.26	43		0.61	…	…		1.22	…	…
3.5 × 1.3	0.38	…	…		0.76	…	…		1.52	5.05	8
	0.46	…	…		0.91	3.66	27		1.90	5.11	7

[a] Customary drill size references have been retained where the metric hole diameters are direct conversions of their decimal inch equivalents.

All dimensions are in millimeters except drill sizes.

Table 8. Approximate Hole Sizes for Steel Type AB Metric Thread Forming Tapping Screws in Plywoods and Asbestos

Nominal Screw Size and Thread Pitch	Hole Size	Drill Size[a]	Min Mat'l Thickness	Penetration in Blind Holes Min	Penetration in Blind Holes Max	Hole Size	Drill Size[a]	Min Mat'l Thickness	Penetration in Blind Holes Min	Penetration in Blind Holes Max
	In Plywood (Resin Impregnated)					In Asbestors Compositions				
2.2 ×0.8	1.85	49	3.18	4.78	12.70	1.93	48	3.18	4.78	12.70
2.9 ×1	2.54	39	4.78	6.35	15.88	2.57	38	4.78	6.35	15.88
3.5 ×1.3	3.18	…	4.78	6.35	15.88	3.05	31	4.78	6.35	15.88
4.2 ×1.4	3.66	27	4.78	6.35	19.05	3.73	26	7.92	9.52	19.05
4.8 ×1.6	4.39	17	6.35	7.92	25.40	4.22	19	7.92	9.52	25.40
5.5 ×1.8	4.93	10	7.92	9.52	25.40	4.98	9	7.92	9.52	25.40
6.3 ×1.8	5.79	1	7.92	9.52	25.40	5.79	1	11.13	12.70	25.40

[a] Customary drill size references have been retained where the metric hole diameters are direct conversions of their decimal inch equivalents.

All dimensions are in millimeters except drill sizes.

Table 9. Approximate Hole Sizes for Steel Type B Metric Thread Forming Tapping Screws in Plywoods, Asbestos, and Plastics

Nominal Screw Sizeand Thread Pitch	Hole Size	Drill Size[a]	Min Mat'l Thick-ness	Penetration in Blind Holes Min	Penetration in Blind Holes Max	Nominal Screw Size and Thread Pitch	Hole Size	Drill Size[a]	Min Mat'l Thick-ness	Penetration in Blind Holes Min	Penetration in Blind Holes Max
In Plywood (Resin Impregnated)											
2.2 × 0.8	1.85	49	3.18	4.78	12.70	4.8 × 1.6	4.39	17	6.35	7.92	25.40
2.9 × 1	2.54	39	4.78	6.35	15.88	5.5 × 1.8	4.93	10	7.92	9.52	25.40
3.5 × 1.3	3.18	…	4.78	6.35	15.88	6.3 × 1.8	5.79	1	7.92	9.52	25.40
4.2 × 1.4	3.66	27	4.78	6.35	19.05	…	…	…	…	…	…

[a] Customary drill size references have been retained where the metric hole diameters are direct conversions of their decimal inch equivalents.

Nominal Screw Size and Thread Pitch	Hole Size	Drill Size[a]	Min Mat'l Thickness	Penetration in Blind Holes Min	Penetration in Blind Holes Max
In Asbestos Compositions					
2.2 × 0.8	1.93	48	3.18	4.78	12.70
2.9 × 1	2.57	38	4.78	6.35	15.88
3.5 × 1.3	3.05	31	4.78	6.35	15.88
4.2 × 1.4	3.73	26	7.92	9.52	19.05
4.8 × 1.6	4.22	19	7.92	9.52	25.40
5.5 × 1.8	4.98	9	7.92	9.52	25.40
6.3 × 1.8	5.79	1	11.13	12.70	25.40

Nominal Screw Size and Thread Pitch	Hole Size	Drill Size[a]	Min Penetration in Blind Holes	Hole Size	Drill Size[a]	Min Penetration in Blind Holes
	In Phenol Formaldehyde			In Cellulose Acetate & Nitrate, Acrylic and Styrene Resins		
2.2 × 0.8	1.98	47	4.78	1.98	47	4.78
2.9 × 1	2.54	39	6.35	2.39	42	6.35
3.5 × 1.3	3.25	30	6.35	3.05	32	6.35
4.2 × 1.4	3.81	25	7.92	3.66	27	7.92
4.8 × 1.6	4.50	16	7.92	4.32	18	7.92
5.5 × 1.8	5.05	8	9.52	4.85	11	9.52
6.3 × 1.8	5.94	…	9.52	5.61	2	9.52

All dimensions are in millimeters except drill sizes.

Table 10. Approximate Drilled or Clean-Punched Hole Sizes for Steel Type B Metric Thread Forming Tapping Screws in Sheet Metal and Cast Metals

In Steel, Stainless Steel, Monel, and Brass Sheet Metal

Nominal Screw Size and Thread Pitch	Metal Thick-ness	Hole Size	Drill Size[a]
2.2 × 0.8	0.38	1.63	52
	0.46	1.63	52
	0.61	1.70	51
	0.76	1.78	50
	0.91	1.85	49
	1.22	1.85	49
	1.52	1.93	48
2.9 × 1	0.38	2.18	44
	0.46	2.18	44
	0.61	2.26	43
	0.76	2.39	42
	0.91	2.39	42
	1.22	2.44	41
	1.52	2.54	39
	1.90	2.59	38
3.5 × 1.3	0.38	2.64	37
	0.46	2.64	37
	0.61	2.69	36
	0.76	2.69	36
	0.91	2.79	35
	1.22	2.82	34
	1.52	2.95	32

Nominal Screw Size and Thread Pitch	Metal Thick-ness	Hole Size	Drill Size[a]
3.5 × 1.3	1.90	3.05	31
	2.67	3.25	30
4.2 × 1.4	0.61	3.18	…
	0.76	3.18	…
	0.91	3.18	…
	1.22	3.25	30
	1.52	3.45	29
	1.90	3.56	28
	2.67	3.81	25
	3.18	3.81	25
	3.43	3.86	24
4.8 × 1.6	0.61	3.66	27
	0.76	3.66	27
	0.91	3.73	26
	1.22	3.86	24
	1.52	3.86	24
	1.90	3.99	22
	2.67	4.09	20
	3.18	4.32	18
	3.43	4.32	18
	4.17	4.39	17

Nominal Screw Size and Thread Pitch	Metal Thick-ness	Hole Size	Drill Size[a]
5.5 × 1.8	0.61	4.22	19
	0.76	4.22	19
	0.91	4.22	19
	1.22	4.32	18
	1.52	4.50	16
	1.90	4.62	14
	2.67	4.70	13
	3.18	4.98	9
	3.43	4.98	9
	4.17	5.11	7
6.3 × 1.8	0.76	4.93	10
	0.91	4.93	10
	1.22	4.93	10
	1.52	5.05	8
	1.90	5.18	6
	2.67	5.31	4
	3.18	5.79	1
	3.43	5.79	1
	4.17	5.94	…
	4.75	5.94	…
	4.93	5.94	…

In Aluminum Alloy Sheet Metal

Nominal Screw Size and Thread Pitch	Metal Thick-ness	Hole Size	Drill Size[a]
2.2 × 0.8	0.61	1.63	52
	0.76	1.63	52
	0.91	1.63	52
	1.22	1.70	51
	1.52	1.78	50
2.9 × 1	0.76	2.18	44
	0.91	2.18	44
	1.22	2.18	44
	1.52	2.26	43
	1.90	2.26	43
	2.67	2.39	42
3.5 × 1.3	0.76	2.64	37
	0.91	2.64	37
	1.22	2.64	37
	1.52	2.69	36
	1.90	2.79	35
	2.67	2.82	34
	3.25 to 6.25	3.05	31

Nominal Screw Size and Thread Pitch	Metal Thick-ness	Hole Size	Drill Size[a]
4.2 × 1.4	0.76	2.95	32
	0.91	3.05	31
	1.22	3.25	30
	1.52	3.45	29
	1.90	3.56	28
	2.67	3.73	26
	3.18	3.73	26
	3.43	3.78	25
	4.11 to 9.52	3.86	24
4.8 × 1.6	0.91	3.66	27
	1.22	3.66	27
	1.52	3.66	27
	1.90	3.73	26
	2.67	3.73	26
	3.18	3.91	23
	3.43	3.91	23
	4.17	4.04	21
	5.08 to 9.52	4.22	19

Nominal Screw Size and Thread Pitch	Metal Thick-ness	Hole Size	Drill Size[a]
5.5 × 1.8	1.22	4.09	20
	1.52	4.22	19
	1.90	4.39	17
	2.67	4.57	15
	3.18	4.62	14
	3.43	4.62	14
	4.17	4.80	12
	5.08 to 9.52	4.98	9
6.3 × 1.8	1.52	5.05	8
	1.90	5.11	7
	2.67	5.18	6
	3.18	5.31	4
	3.43	5.31	4
	4.17	5.41	3
	4.75	5.41	3
	4.93	5.61	2
	5.08 to 9.52	5.79	1

[a] Customary drill size references have been retained where the metric hole diameters are direct conversions of their decimal inch equivalents.

In Aluminum, Magnesium, Zinc, Brass, and Bronze Cast Metals							
Nominal Screw Size and Thread Pitch	Hole Size	Drill Size[a]	Min Penetration in Blind Holes	Nominal Screw Size and Thread Pitch	Hole Size	Drill Size[a]	Min Penetration in Blind Holes
2.2 × 0.8	1.98	47	3.18	4.8 × 1.6	4.50	16	6.35
2.9 × 1	2.64	37	4.78	5.5 × 1.8	5.05	8	7.14
3.5 × 1.3	3.25	30	6.35	6.3 × 1.8	5.94	4	7.92
4.2 × 1.4	3.86	24	6.35	…	…	…	…

All dimensions are in millimeters, except drill sizes.

Table 11. Approximate Hole Sizes for Steel Types BF and BT Metric Thread Cutting Tapping Screws for Cast Metals and Plastics

Nominal Screw Size and Thread Pitch	Material Thickness	Hole Size	Drill Size[a]	Nominal Screw Size and Thread Pitch	Material Thickness	Hole Size	Drill Size[a]
In Die Cast Zinc and Aluminum							
2.2 × 0.8	1.52	1.85	49	3.5 × 1.3	3.18	3.05	31
	2.11	1.85	49		3.56	3.05	31
	2.77	1.93	48		4.78	3.05	31
	3.18	1.93	48		6.35	3.18	…
	3.56	1.93	48		7.92	3.18	…
2.9 × 1	2.77	2.49	40	4.2 × 1.4	3.18	3.78	25
	3.18	2.54	39		3.56	3.78	25
	3.56	2.54	39		4.78	3.78	25
	4.78	2.54	39		6.35	3.86	24
	6.35	2.59	38		7.92	3.86	24
4.8 × 1.6	3.18	4.22	19	6.3 × 1.8	6.35	5.79	1
	3.56	4.22	19		7.92	5.79	1
	4.78	4.22	19		9.52	5.79	1
	6.35	4.32	18	8 × 2.1	3.18	7.14	K
	7.92	4.37	…		3.56	7.14	K
	9.52	4.37	…		4.78	7.14	K
5.5 × 1.8	3.18	4.85	11		6.35	7.14	K
	3.56	4.85	11		7.92	7.37	L
	4.78	4.85	11		9.52	7.37	L
	6.35	4.98	9	9.5 × 2.1	3.18	8.74	…
	7.92	4.98	9		3.56	8.74	…
	9.52	4.98	9		4.78	8.74	…
6.3 × 1.8	3.18	5.61	2		6.35	8.74	…
	3.56	5.61	2		7.92	8.84	S
	4.78	5.61	2		9.52	8.84	S

[a] Customary drill size references have been retained where the metric hole sizes are direct conversions of their decimal inch equivalents.

Nominal Screw Size and Thread Pitch	Hole Size	Drill Size[a]	Depth of Penetration	
			Min	Max
In Phenol Formaldehyde				
2.2 × 0.8	1.98	…	2.39	6.35
2.9 × 1	2.64	37	3.18	7.92
3.5 × 1.3	3.18	…	4.78	9.52
4.2 × 1.4	3.73	26	6.35	12.70
4.8 × 1.6	4.32	18	7.92	15.88
5.5 × 1.8	4.93	10	9.52	15.88
6.3 × 1.8	5.79	1	9.52	19.05
In Cellulose Acetate and Nitrate, Acrylic and Styrene Resins				
2.2 × 0.8	1.93	48	2.39	6.35
2.9 × 1	2.54	39	3.18	7.92
3.5 × 1.3	3.05	31	4.78	9.52
4.2 × 1.4	3.66	27	6.35	12.70
4.8 × 1.6	4.22	19	7.92	15.88
5.5 × 1.8	4.80	12	9.52	15.88
6.3 × 1.8	5.61	2	9.52	19.05

All dimensions are in millimeters except drill sizes.

The finish (plating or coating) on metric tapping screws and the material composition and hardness of the mating component are factors that affect assembly torques in individual applications. Although the recommended installation hole sizes given in Tables 7 through 9 were based on the use of plain unfinished carbon steel metric tapping screws, experience has shown that the specified holes are also suitable for screws having most types of commercial finishes. However, owing to various finishes providing different degrees of lubricity, some adjustment of installation torques may be necessary to suit individual applications. Also, where exceptionally heavy finishes are involved or screws are to be assembled into materials of higher hardness, some deviation from the specified hole sizes may be required to provide optimum assembly. The necessity and extent of such deviations can best be determined by experiment in the particular assembly environment.

Table 1. American National Standard T-Slots *ANSI/ASME B5.1M-1985 (R1998)*

T-SLOTS

Basic DimensionS

Suggested Approximate Dimensions For Rounding Or Breaking Of Corners

Nominal T-Bolt Size[a]		Width of Throat A_1[b]		Headspace Dimensions: Width B_1				Headspace Dimensions: Depth C_1				Depth of Throat D_1				Rounding or Breaking of Corners[c]					
				inch		mm		inch		mm		inch		mm		inch			mm		
inch	mm	inch	mm	min	max	min	max	min	max	min	max	min	max	min	max	R_1 max	W_1 max	U_1 max	R_1 max	W_1 max	U_1 max
	4		5			10	11			3	3.5			4.5	7				0.5	0.8	0.8
	5		6			11	12.5			5	6			5	8				0.5	0.8	0.8
0.250	6	0.282	8	0.500	0.562	14.5	16	0.203	0.234	7	8	0.125	0.375	7	11	0.02	0.02	0.03	0.5	0.8	0.8
0.312	8	0.344	10	0.594	0.656	16	18	0.234	0.266	7	8	0.156	0.438	9	14	0.02	0.03	0.03	0.5	0.8	0.8
0.375	10	0.438	12	0.719	0.781	19	21	0.297	0.328	8	9	0.219	0.562	11	17	0.02	0.03	0.03	0.5	0.8	0.8
0.500	12	0.562	14	0.906	0.969	23	25	0.359	0.391	9	11	0.312	0.688	12	19	0.02	0.03	0.03	0.5	0.8	0.8
0.625	16	0.688	18	1.188	1.250	30	32	0.453	0.484	12	14	0.438	0.875	16	24	0.03	0.03	0.05	0.8	0.8	1.3
0.750	20	0.812	22	1.375	1.469	37	40	0.594	0.625	16	18	0.562	1.062	20	29	0.03	0.03	0.05	0.8	0.8	1.3
1.000	24	1.062	28	1.750	1.844	46	50	0.781	0.828	20	22	0.750	1.250	26	36	0.03	0.06	0.05	0.8	1.5	1.3
1.250	30	1.312	36	2.125	2.219	56	60	1.031	1.094	25	28	1.000	1.562	33	46	0.03	0.06	0.05	0.8	1.5	1.3
1.500	36	1.562	42	2.562	2.656	68	72	1.281	1.344	32	35	1.250	1.938	39	53	0.03	0.06	0.05	0.8	1.5	1.3
	42		48			80	85			36	40			44	59				1.5	2.5	2
	48		54			90	95			40	44			50	66				1.5	2.5	2

[a] Width of tongue (tenon) to be used with the above T-Slots will be found in the complete standard, B5.1M.

[b] Throat dimensions are basic. When slots are intended to be used for holding only, tolerances can be 0.0 + 0.010 inch or H12 Metric (ISO/R286); when intended for location, tolerance can be 0.0 + 0.001 inch or H8 Metric (see page 648).

[c] Corners of T-Slots may be square or may be rounded or broken to the indicated maximum dimensions at the manufacturer's option.

For the dimensions of tongue seats, inserted tongues, and solid tongues refer to the complete standard, B5.1M.

Table 2. American National Standard T-Bolts *ANSI/ASME B5.1M-1985 (R1998)*

T-BOLTS

Nominal T-Bolt Size and Thread A_2[ab]		Bolt Head Dimensions												Rounding of Corners[c]			
		Width Across Flats B_2				Width Across Corners		Height C_2				R_2		W_2	
inch	metric	inch		mm		inch	mm	inch		mm		inch	mm	inch	mm
UNC-2A	ISO[d]	max	min	max	min	max	max	max	min	max	min	max	max	max	max
	M4			9	8.5		12.7			2.5	2.1		0.3		0.5
	M5			10	9.5		14.1			4	3.6		0.3		0.5
0.250–20	M6	0.469	0.438	13	12	0.663	18.4	0.156	0.141	6	5.6	0.02	0.5	0.03	0.8
0.312–18	M8	0.562	0.531	15	14	0.796	21.2	0.188	0.172	6	5.6	0.02	0.5	0.03	0.8
0.375–16	M10	0.688	0.656	18	17	0.972	25.5	0.250	0.234	7	6.6	0.02	0.5	0.03	0.8
0.500–13	M12	0.875	0.844	22	21	1.238	31.1	0.312	0.297	8	7.6	0.02	0.5	0.06	1.5
0.625–11	M16	1.125	1.094	28	27	1.591	39.6	0.406	0.391	10	9.6	0.03	0.8	0.06	1.5
0.750–10	M20	1.312	1.281	34	33	1.856	48.1	0.531	0.500	14	13.2	0.03	0.8	0.06	1.5
1.000–8	M24	1.688	1.656	43	42	2.387	60.8	0.688	0.656	18	17.2	0.03	0.8	0.06	1.5
1.250–7	M30	2.062	2.031	53	52	2.917	75	0.938	0.906	23	22.2	0.03	0.8	0.06	1.5
1.500–6	M36	2.500	2.469	64	63	3.536	90.5	1.188	1.156	28	27.2	0.03	0.8	0.06	1.5
	M42			75	74		106.1			32	30.5		1		2
	M48			85	84		120.2			36	34.5		1		2

[a] For inch tolerances for thread diameters of bolts or studs and for threads see page 1716.

[b] T-slots to be used with these bolts will be found in Table 1.

[c] Corners of T-bolts may be square or may be rounded or broken to the indicated maximum dimensions at the manufacturer's option.

[d] Metric thread grade and tolerance position is 5g 6g (see page 1764).

Table 3. American National Standard T-Nuts *ANSI/ASME B5.1M-1985 (R1998)*

T-NUTS

Nominal T-Bolt Size[a]		Width of Tongue A_3				Tap for Stud[b] E_3		Width of Nut B_3				Height of Nut C_3				Total Thickness Including Tongue[c] K_3		Length of Nut[c] L_3		Rounding of Corners			
		inch		mm		inch	mm	inch		mm		inch		mm						R_3		W_3	
																				inch	mm	inch	mm
inch	mm	max	min	max	min	UNC-3B	ISO[d]	max	min	max	min	max	min	max	min	inch	mm	inch	mm	max	max	max	max
	4		…		…		…		…	…	…			…	…		…		…		…		…
	5		…		…		…		…	…	…			…	…		…		…		…		…
0.250	6	…	…	…	…	…	…	…	…	…	…	…	…	…	…	…	…	…	…	…	…	…	…
0.312	8	0.330	0.320	8.7	8.5	0.250–20	M6	0.562	0.531	15	14	0.188	0.172	6	5.6	0.281	9	0.562	18	0.02	0.5	0.03	0.8
0.375	10	0.418	0.408	11	10.75	0.312–18	M8	0.688	0.656	18	17	0.250	0.234	7	6.6	0.375	10.5	0.688	20	0.02	0.5	0.03	0.8
0.500	12	0.543	0.533	13.5	13.25	0.375–1	6M10	0.875	0.844	22	21	0.312	0.297	8	7.6	0.531	12	0.875	23	0.02	0.5	0.06	1.5
0.625	16	0.668	0.658	17.25	17	0.500–13	M12	1.125	1.094	28	27	0.406	0.391	10	9.6	0.625	15	1.125	27	0.03	0.8	0.06	1.5
0.750	20	0.783	0.773	20.5	20.25	0.625–11	M16	1.312	1.281	34	33	0.531	0.500	14	13.2	0.781	21	1.312	35	0.03	0.8	0.06	1.5
1.000	24	1.033	1.018	26.5	26	0.750–10	M20	1.688	1.656	43	42	0.688	0.656	18	17.2	1.000	27	1.688	46	0.03	0.8	0.06	1.5
1.250	30	1.273	1.258	33	32.5	1.000–8	M24	2.062	2.031	53	52	0.938	0.906	23	22.2	1.312	34	2.062	53	0.03	0.8	0.06	1.5
1.500	36	1.523	1.508	39.25	38.75	1.250–7	M30	2.500	2.469	64	63	1.188	1.156	28	27.2	1.625	42	2.500	65	0.03	0.8	0.06	1.5
	42			46.75	46.25		M36			75	74			32	30.5		48		75		1		2
	48			52.5	51.75		M42			85	84			36	34.5		54		85		1		2

[a] T-slot dimensions to fit the above nuts will be found in Table 1.
[b] For tolerances of inch threads see page 1716.
[c] No tolerances are given for "Total Thickness" or "Nut Length" as they need not be held to close limits.
[d] Metric tapped thread grade and tolerance position is 5H (see page 1764).

PINS AND STUDS

Dowel-Pins.—Dowel-pins are used either to retain parts in a fixed position or to preserve alignment. Under normal conditions a properly fitted dowel-pin is subjected solely to shearing strain, and this strain occurs only at the junction of the surfaces of the two parts which are being held by the dowel-pin. It is seldom necessary to use more than two dowel-pins for holding two pieces together and frequently one is sufficient. For parts that have to be taken apart frequently, and where driving out of the dowel-pins would tend to wear the holes, and also for very accurately constructed tools and gages that have to be taken apart, or that require to be kept in absolute alignment, the taper dowel-pin is preferable. The taper dowel-pin is most commonly used for average machine work, but the straight type is given the preference on tool and gage work, except where extreme accuracy is required, or where the tool or gage is to be subjected to rough handling.

The size of the dowel-pin is governed by its application. For locating nests, gage plates, etc., pins from ⅛ to 3⁄16 inch in diameter are satisfactory. For locating dies, the diameter of the dowel-pin should never be less than ¼ inch; the general rule is to use dowel-pins of the same size as the screws used in fastening the work. The length of the dowel-pin should be about one and one-half to two times its diameter in each plate or part to be doweled.

When hardened cylindrical dowel-pins are inserted in soft parts, ream the hole about 0.001 inch smaller than the dowel-pin. If the doweled parts are hardened, grind (or lap) the hole 0.0002 to 0.0003 inch under size. The hole should be ground or lapped straight, that is, without taper or "bell-mouth."

American National Standard Cotter Pins *ANSI B18.8.1-1972 (R1994)*

Nom. Size	Dia. A^a & Width B Max.	Wire Width B Min.	Head Dia. C Min.	Prong Length D Min.	Hole Size	Nom. Size	Dia. A^a & Width B Max.	Wire Width B Min.	Head Dia. C Min.	Prong Length D Min.	Hole Size
1⁄32	0.032	0.022	0.06	0.01	0.047	3⁄16	0.176	0.137	0.38	0.09	0.203
3⁄64	0.048	0.035	0.09	0.02	0.062	7⁄32	0.207	0.161	0.44	0.10	0.234
1⁄16	0.060	0.044	0.12	0.03	0.078	¼	0.225	0.176	0.50	0.11	0.266
5⁄64	0.076	0.057	0.16	0.04	0.094	5⁄16	0.280	0.220	0.62	0.14	0.312
3⁄32	0.090	0.069	0.19	0.04	0.109	⅜	0.335	0.263	0.75	0.16	0.375
7⁄64	0.104	0.080	0.22	0.05	0.125	7⁄16	0.406	0.320	0.88	0.20	0.438
⅛	0.120	0.093	0.25	0.06	0.141	½	0.473	0.373	1.00	0.23	0.500
9⁄64	0.134	0.104	0.28	0.06	0.156	⅝	0.598	0.472	1.25	0.30	0.625
5⁄32	0.150	0.116	0.31	0.07	0.172	¾	0.723	0.572	1.50	0.36	0.750

[a] Tolerances are: –0.004 inch for the 1⁄32- to 3⁄16-inch sizes, incl.; –0.005 inch for the 7⁄32- to 5⁄16-inch sizes, incl.; –0.006 inch for the ⅜- to ½-inch sizes, incl.; and –0.008 inch for the ⅝- and ¾-inch sizes. Note: Tolerances for length are: up to 1 inch ± 0.030 inch, over 1 inch ±0.060 inch.

All dimensions are in inches.

American National Standard Clevis Pins *ANSI B18.8.1-1972 (R1994)*

Nom.Size (Basic Pin Dia.)	Shank Dia.*A* Max	Head Dia.*B* Max.[a]	Head Hgt.*C* Max.[b]	Head Chamfer *D* Nom.[c]	Hole Dia.*E* Max.[d]	Point Dia.*F* Max.[e]	Pin Lgth.*G* Basic[f]	Head to HoleCenter *H* Max.[g]	Point Length *L* Max.	Point Length *L* Min.	Cotter Pin Size for Hole
3/16	0.186	0.32	0.07	0.02	0.088	0.15	0.58	0.504	0.055	0.035	1/16
1/4	0.248	0.38	0.10	0.03	0.088	0.21	0.77	0.692	0.055	0.035	1/16
5/16	0.311	0.44	0.10	0.03	0.119	0.26	0.94	0.832	0.071	0.049	3/32
3/8	0.373	0.51	0.13	0.03	0.119	0.33	1.06	0.958	0.071	0.049	3/32
7/16	0.436	0.57	0.16	0.04	0.119	0.39	1.19	1.082	0.071	0.049	3/32
1/2	0.496	0.63	0.16	0.04	0.151	0.44	1.36	1.223	0.089	0.063	1/8
5/8	0.621	0.82	0.21	0.06	0.151	0.56	1.61	1.473	0.089	0.063	1/8
1/4	0.746	0.94	0.26	0.07	0.182	0.68	1.91	1.739	0.110	0.076	5/32
7/8	0.871	1.04	0.32	0.09	0.182	0.80	2.16	1.989	0.110	0.076	5/32
1	0.996	1.19	0.35	0.10	0.182	0.93	2.41	2.239	0.110	0.076	5/32

[a] Tolerance is –0.05 inch.

[b] Tolerance is –0.02 inch.

[c] Tolerance is ±0.01 inch.

[d] Tolerance is –0.015 inch.

[e] Tolerance is –0.01 inch.

[f] Lengths tabulated are intended for use with standard clevises, without spacers. When other lengths are required, it is recommended that they be limited wherever possible to nominal lengths in 0.06-inch increments.

[g] Tolerance is –0.020 inch.

All dimensions are in inches.

British Standard for Metric Series Dowel Pins.—Steel parallel dowel pins specified in British Standard 1804:Part 2:1968 are divided into three grades which provide different degrees of pin accuracy.

Grade 1 is a precision ground pin made from En 32A or En 32B low carbon steel (BS 970) or from high carbon steel to BS 1407 or BS 1423. Pins below 4 mm diameter are unhardened. Those of 4 mm diameter and above are hardened to a minimum of 750 HV 30 in accordance with BS 427, but if they are made from steels to BS 1407 or BS 1423 then the hardness shall be within the range 600 to 700 HV 30, in accordance with BS 427. The values of other hardness scales may be used in accordance with BS 860.

Grade 2 is a ground pin made from any of the steels used for Grade 1. The pins are normally supplied unhardened, unless a different condition is agreed on between the purchaser and supplier.

Grade 3 pins are made from En 1A free cutting steel (BS 970) and are supplied with a machined, bright rolled or drawn finish. They are normally supplied unhardened unless a different condition is agreed on between the purchaser and supplier.

Pins of any grade may be made from different steels in accordance with BS 970, by mutual agreement between the purchaser and manufacturer. If steels other than those in the standard range are used, the hardness of the pins shall also be decided on by mutual agreement between purchaser and supplier. As shown in the illustration at the head of the accompanying table, one end of each pin is chamfered to provide a lead. The other end may be similarly chamfered, or domed.

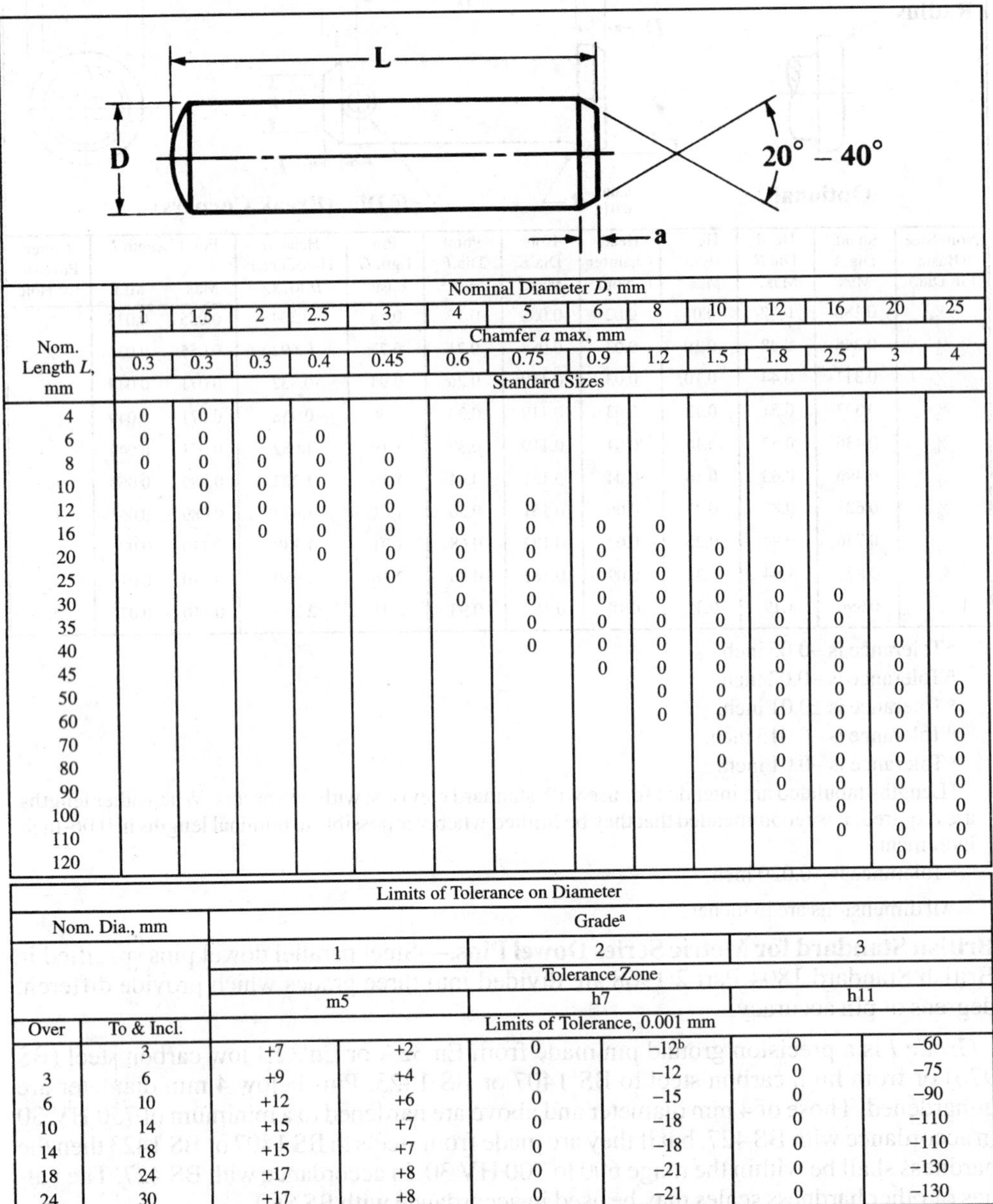

Nom. Length *L*, mm	Nominal Diameter *D*, mm													
	1	1.5	2	2.5	3	4	5	6	8	10	12	16	20	25
	Chamfer *a* max, mm													
	0.3	0.3	0.3	0.4	0.45	0.6	0.75	0.9	1.2	1.5	1.8	2.5	3	4
	Standard Sizes													
4	0	0												
6	0	0	0	0										
8	0	0	0	0	0									
10	0	0	0	0	0	0								
12		0	0	0	0	0	0							
16			0	0	0	0	0	0	0					
20				0	0	0	0	0	0	0				
25					0	0	0	0	0	0	0			
30						0	0	0	0	0	0	0		
35							0	0	0	0	0	0		
40							0	0	0	0	0	0	0	
45								0	0	0	0	0	0	
50									0	0	0	0	0	0
60									0	0	0	0	0	0
70										0	0	0	0	0
80										0	0	0	0	0
90											0	0	0	0
100												0	0	0
110												0	0	0
120													0	0

Limits of Tolerance on Diameter							
Nom. Dia., mm		Grade[a]					
		1		2		3	
		Tolerance Zone					
		m5		h7		h11	
Over	To & Incl.	Limits of Tolerance, 0.001 mm					
	3	+7	+2	0	−12[b]	0	−60
3	6	+9	+4	0	−12	0	−75
6	10	+12	+6	0	−15	0	−90
10	14	+15	+7	0	−18	0	−110
14	18	+15	+7	0	−18	0	−110
18	24	+17	+8	0	−21	0	−130
24	30	+17	+8	0	−21	0	−130

[a] The limits of tolerance for grades 1 and 2 dowel pins have been chosen to provide satisfactory assembly when used in standard reamed holes (H7 and H8 tolerance zones). If the assembly is not satisfactory, refer to B.S. 1916: Part 1, Limits and Fits for Engineering, and select a different class of fit.

[b] This tolerance is larger than that given in BS 1916, and has been included because the use of a closer tolerance would involve precision grinding by the manufacturer, which is uneconomic for a grade 2 dowel pin.

The tolerance limits on the overall length of all grades of dowel pin up to and including 50 mm long are +0.5, −0.0 mm, and for pins over 50 mm long are +0.8, −0.0 mm. The Standard specifies that the roughness of the cylindrical surface of grades 1 and 2 dowel pins, when assessed in accordance with BS 1134, shall not be greater than 0.4 μm CLA (16 CLA).

Table 1. American National Standard Hardened Ground Machine Dowel Pins *ANSI/ASME B18.8.2-1995*

Nominal Size[a] or Nominal Pin Diameter		Pin Diameter, A						Point Diameter, B		Crown Height, C	Crown Radius, R	Range of Preferred Lengths,[b] L	Single Shear Load, for Carbon or Alloy Steel, Calculated lb	Suggested Hole Diameter[c]	
		Standard Series Pins			Oversize Series Pins										
		Basic	Max	Min	Basic	Max	Min	Max	Min	Max	Min			Max	Min
1/16	0.0625	0.0627	0.0628	0.0626	0.0635	0.0636	0.0634	0.058	0.048	0.020	0.008	3/16–3/4	400	0.0625	0.0620
5/64[d]	0.0781	0.0783	0.0784	0.0782	0.0791	0.0792	0.0790	0.074	0.064	0.026	0.010	…	620	0.0781	0.0776
3/32	0.0938	0.0940	0.0941	0.0939	0.0948	0.0949	0.0947	0.089	0.079	0.031	0.012	5/16–1	900	0.0937	0.0932
1/8	0.1250	0.1252	0.1253	0.1251	0.1260	0.1261	0.1259	0.120	0.110	0.041	0.016	3/8–2	1,600	0.1250	0.1245
5/32[d]	0.1562	0.1564	0.1565	0.1563	0.1572	0.1573	0.1571	0.150	0.140	0.052	0.020	…	2,500	0.1562	0.1557
3/16	0.1875	0.1877	0.1878	0.1876	0.1885	0.1886	0.1884	0.180	0.170	0.062	0.023	1/2–2	3,600	0.1875	0.1870
1/4	0.2500	0.2502	0.2503	0.2501	0.2510	0.2511	0.2509	0.240	0.230	0.083	0.031	1/2–2 1/2	6,400	0.2500	0.2495
5/16	0.3125	0.3127	0.3128	0.3126	0.3135	0.3136	0.3134	0.302	0.290	0.104	0.039	1/2–2 1/2	10,000	0.3125	0.3120
3/8	0.3750	0.3752	0.3753	0.3751	0.3760	0.3761	0.3759	0.365	0.350	0.125	0.047	1/2–3	14,350	0.3750	0.3745
7/16	0.4375	0.4377	0.4378	0.4376	0.4385	0.4386	0.4384	0.424	0.409	0.146	0.055	7/8–3	19,550	0.4375	0.4370
1/2	0.5000	0.5002	0.5003	0.5001	0.5010	0.5011	0.5009	0.486	0.471	0.167	0.063	3/4, 1–4	25,500	0.5000	0.4995
5/8	0.6250	0.6252	0.6253	0.6251	0.6260	0.6261	0.6259	0.611	0.595	0.208	0.078	1 1/4–5	39,900	0.6250	0.6245
3/4	0.7500	0.7502	0.7503	0.7501	0.7510	0.7511	0.7509	0.735	0.715	0.250	0.094	1 1/2–6	57,000	0.7500	0.7495
7/8	0.8750	0.8752	0.8753	0.8751	0.8760	0.8761	0.8759	0.860	0.840	0.293	0.109	2, 2 1/2–6	78,000	0.8750	0.8745
1	1.0000	1.0002	1.0003	1.0001	1.0010	1.0011	1.0009	0.980	0.960	0.333	0.125	2, 2 1/2–5, 6	102,000	1.0000	0.9995

[a] Where specifying nominal size as basic diameter, zeros preceding decimal and in the fourth decimal place are omitted.

[b] Lengths increase in 1/16-inch steps up to 3/8 inch, in 1/8-inch steps from 3/8 inch to 1 inch, in 1/4-inch steps from 1 inch to 2 1/2 inches, and in 1/2-inch steps above 2 1/2 inches. Tolerance on length is ±0.010 inch.

[c] These hole sizes have been commonly used for press fitting Standard Series machine dowel pins into materials such as mild steels and cast iron. In soft materials such as aluminum or zinc die castings, hole size limits are usually decreased by 0.0005 inch to increase the press fit.

[d] Nonpreferred sizes, not recommended for use in new designs.

All dimensions are in inches.

If a dowel pin is driven into a blind hole where no provision is made for releasing air, the worker assembling the pin may be endangered, and damage may be caused to the associated component, or stresses may be set up. The appendix of the Standard describes one method of overcoming this problem by providing a small flat surface along the length of a pin to permit the release of air.

For purposes of marking, the Standard states that each package or lot of dowel pins shall bear the manufacturer's name or trademark, the BS number, and the grade of pin.

American National Standard Hardened Ground Machine Dowel Pins.—Hardened ground machine dowel pins are furnished in two diameter series: Standard Series having basic diameters 0.0002 inch over the nominal diameter, intended for initial installations; and Oversize Series having basic diameters 0.001 inch over the nominal diameter, intended for replacement use.

Preferred Lengths and Sizes: The preferred lengths and sizes in which these pins are normally available are given in Table 1. Other sizes and lengths are produced as required by the purchaser.

Effective Length: The effective length, L_e, must not be less than 75 per cent of the overall length of the pin.

Shear Strength: Single shear strength values are listed in Table 1. Prior versions of ANSI/ASME B18.8.2-1995 had listed double shear load minimum values and had specified a minimum single shear strength of 130,000 psi. See ANSI/ASME B18.8.2-1995, Appendix B for a description of the double shear test.

Designation: These pins are designated by the following data in the sequence shown: Product name (noun first), including pin series, nominal pin diameter (fraction or decimal equivalent), length (fraction or decimal equivalent), material, and protective finish, if required.

Examples: Pins, Hardened Ground Machine Dowel — Standard Series, 3⁄8 × 1 1⁄2, Steel, Phosphate Coated.

Pins, Hardened Ground Machine Dowel — Oversize Series, 0.625 × 2.500, Steel

Installation Precaution: Pins should not be installed by striking or hammering and when installing with a press, a shield should be used and safety glasses worn.

American National Standard Hardened Ground Production Dowel Pins.—Hardened ground production dowel pins have basic diameters that are 0.0002 inch over the nominal pin diameter.

Preferred Lengths and Sizes: The preferred lengths and sizes in which these pins are available are given in Table 2. Other sizes and lengths are produced as required by the purchaser.

Shear Strength: Single shear strength values are listed in Table 2. Prior versions of ANSI/ASME B18.8.2-1995 had listed double shear load minimum values and had specified a minimum single shear strength of 102,000 psi. See ANSI/ASME B18.8.2-1995, Appendix B for a description of the double shear test.

Ductility: These standard pins are sufficiently ductile to withstand being pressed into holes 0.0005 inch smaller than the nominal pin diameter in hardened steel without cracking or shattering.

Designation: These pins are designated by the following data in the sequence shown: Product name (noun first), nominal pin diameter (fraction or decimal equivalent), length (fraction or decimal equivalent), material, and protective finish, if required.

Examples: Pins, Hardened Ground Production Dowel, 1⁄8 × 3⁄4, Steel, Phosphate Coated

Pins, Hardened Ground Production Dowel, 0.375 × 1.500, Steel

Table 2. American National Standard Hardened Ground Production Dowel Pins
ANSI/ASME B18.8.2-1995

Nominal Size[a] or Nominal Pin Diameter		Pin Diameter, *A*			Corner Radius, *R*		Range of Preferred Lengths,[b] *L*	Single Shear Load, Calculated,lb	Suggested Hole Diameter[c]	
		Basic	Max	Min	Max	Min			Max	Min
1/16	0.0625	0.0627	0.0628	0.0626	0.020	0.010	3/16–1	395	0.0625	0.0620
3/32	0.0938	0.0939	0.0940	0.0938	0.020	0.010	3/16–2	700	0.0937	0.0932
7/64	0.1094	0.1095	0.1096	0.1094	0.020	0.010	3/16–2	950	0.1094	0.1089
1/8	0.1250	0.1252	0.1253	0.1251	0.020	0.010	3/16–2	1,300	0.1250	0.1245
5/32	0.1562	0.1564	0.1565	0.1563	0.020	0.010	3/16–2	2,050	0.1562	0.1557
3/16	0.1875	0.1877	0.1878	0.1876	0.020	0.010	3/16–2	2,950	0.1875	0.1870
7/32	0.2188	0.2189	0.2190	0.2188	0.020	0.010	1/4–2	3,800	0.2188	0.2183
1/4	0.2500	0.2502	0.2503	0.2501	0.020	0.010	1/4–1 1/2, 1 3/4, 2–2 1/2	5,000	0.2500	0.2495
5/16	0.3125	0.3127	0.3128	0.3126	0.020	0.010	5/16–1 1/2, 1 3/4, 2–2 1/2	8,000	0.3125	0.3120
3/8	0.3750	0.3752	0.3753	0.3751	0.020	0.010	3/8–1 1/2, 1 3/4, 2–3	11,500	0.3750	0.3745

[a] Where specifying nominal pin size in decimals, zeros preceding decimal and in the fourth decimal place are omitted.

[b] Lengths increase in 1/16-inch steps up to 1 inch, in 1/8-inch steps from 1 inch to 2 inches and then are 2 1/4, 2 1/2, and 3 inches.

[c] These hole sizes have been commonly used for press fitting production dowel pins into materials such as mild steels and cast iron. In soft materials such as aluminum or zinc die castings, hole size limits are usually decreased by 0.0005 inch to increase the press fit.

All dimensions are in inches.

American National Standard Unhardened Ground Dowel Pins.—Unhardened ground dowel pins are normally produced by grinding the outside diameter of commercial wire or rod material to size. Consequently, the maximum diameters of the pins, as specified in Table 3, are below the minimum commercial stock sizes by graduated amounts from 0.0005 inch on the 1/16-inch nominal pin size to 0.0028 inch on the 1-inch nominal pin size.

Preferred Lengths and Sizes: The preferred lengths and sizes in which unhardened ground pins are normally available are given in Table 3. Other sizes and lengths are produced as required by the purchaser.

Shear Strength: These pins must have a single shear strength of 64,000 psi minimum for pins made from steel and 40,000 psi minimum for pins made from brass and must be capable of withstanding the minimum double shear loads given in Table 3 when tested in accordance with the procedure outlined in ANSI/ASME B18.8.2-1995, Appendix B.

Designation: These pins are designated by the following data in the order shown: Product name (noun first), nominal pin diameter (fraction or decimal equivalent), length (fraction or decimal equivalent), material, and protective finish, if required.

Examples: Pins, Unhardened Ground Dowel, 1/8 × 3/4, Steel

Pins, Unhardened Ground Dowel, 0.250 × 2.500, Steel, Zinc Plated

Table 3. American National Standard Unhardened Ground Dowel Pins
ANSI/ASME B18.8.2-1995

Nominal Size[a] or Basic Pin Diameter		Pin Diameter, A		Chamfer Length, C		Range of Preferred Lengths,[b] L	Suggested Hole Diameter[c]		Double Shear Load Min, lb.	
		Max	Min	Max	Min		Max	Min	Carbon-Steel	Brass
1/16	0.0625	0.0600	0.0595	0.025	0.005	1/4–1	0.0595	0.0580	350	220
3/32	0.0938	0.0912	0.0907	0.025	0.005	1/4–1 1/2	0.0907	0.0892	820	510
[d]7/64	0.1094	0.1068	0.1063	0.025	0.005	…	0.1062	0.1047	1,130	710
1/8	0.1250	0.1223	0.1218	0.025	0.005	1/4–2	0.1217	0.1202	1,490	930
5/32	0.1562	0.1535	0.1530	0.025	0.005	1/4–2	0.1528	0.1513	2,350	1,470
3/16	0.1875	0.1847	0.1842	0.025	0.005	1/4–2	0.1840	0.1825	3,410	2,130
7/32	0.2188	0.2159	0.2154	0.025	0.005	1/4–2	0.2151	0.2136	4,660	2,910
1/4	0.2500	0.2470	0.2465	0.025	0.005	1/4–1 1/2, 1 3/4, 2–2 1/2	0.2462	0.2447	6,120	3,810
5/16	0.3125	0.3094	0.3089	0.040	0.020	5/16–1 1/2, 1 3/4, 2–2 1/2	0.3085	0.3070	9,590	5,990
3/8	0.3750	0.3717	0.3712	0.040	0.020	3/8–1 1/2, 1 3/4, 2–2 1/2	0.3708	0.3693	13,850	8,650
7/16	0.4375	0.4341	0.4336	0.040	0.020	7/16–5/8, 3/4, 7/8–1 1/2, 1 3/4, 2–2 1/2	0.4331	0.4316	18,900	11,810
1/2	0.5000	0.4964	0.4959	0.040	0.020	1/2, 5/8, 3/4, 7/8, 1–1 1/2, 1 3/4, 2–3	0.4954	0.4939	24,720	15,450
5/8	0.6250	0.6211	0.6206	0.055	0.035	5/8, 3/4, 7/8, 1–1 1/2, 1 3/4, 2, 2 1/2–4	0.6200	0.6185	38,710	24,190
3/4	0.7500	0.7458	0.7453	0.055	0.035	3/4, 7/8, 1, 1 1/4, 1 1/2, 1 3/4, 2, 2 1/2–4	0.7446	0.7431	55,840	34,900
7/8	0.8750	0.8705	0.8700	0.070	0.050	7/8, 1, 1 1/4, 1 1/2, 1 3/4, 2, 2 1/2–4	0.8692	0.8677	76,090	47,550
1	1.0000	0.9952	0.9947	0.070	0.050	1, 1 1/4, 1 1/2, 1 3/4, 2, 2 1/2–4	0.9938	0.9923	99,460	62,160

[a] Where specifying pin size in decimals, zeros preceding decimal and in the fourth decimal place are omitted.

[b] Lengths increase in 1/16-inch increments from 1/4 to 1 inch, in 1/8-inch increments from 1 inch to 2 inches, and in 1/4-inch increments from 2 to 2 1/2 inches, and in 1/2-inch increments from 2 1/2 to 4 inches.

[c] These hole sizes have been found to be satisfactory for press fitting pins into mild steel and cast and malleable irons. In soft materials such as aluminum alloys or zinc die castings, hole size limits are usually decreased by 0.0005 inch to increase the press fit.

[d] Nonpreferred size, not recommended for use in new designs.

All dimensions are in inches.

American National Standard Straight Pins.—The diameter of both chamfered and square end straight pins is that of the commercial wire or rod from which the pins are made. The tolerances shown in Table 4 are applicable to carbon steel and some deviations in the diameter limits may be necessary for pins made from other materials.

Table 4. American National Standard Chamfered and Square End Straight Pins *ANSI/ASME B18.8.2-1995*

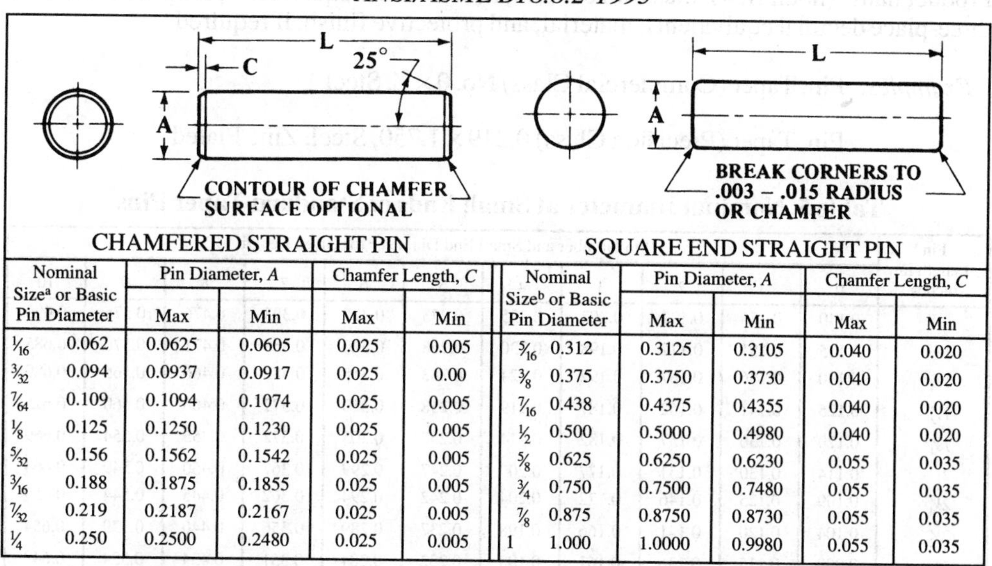

CHAMFERED STRAIGHT PIN — SQUARE END STRAIGHT PIN

Nominal Size[a] or Basic Pin Diameter	Pin Diameter, *A*		Chamfer Length, *C*		Nominal Size[b] or Basic Pin Diameter	Pin Diameter, *A*		Chamfer Length, *C*	
	Max	Min	Max	Min		Max	Min	Max	Min
1/16 0.062	0.0625	0.0605	0.025	0.005	5/16 0.312	0.3125	0.3105	0.040	0.020
3/32 0.094	0.0937	0.0917	0.025	0.00	3/8 0.375	0.3750	0.3730	0.040	0.020
7/64 0.109	0.1094	0.1074	0.025	0.005	7/16 0.438	0.4375	0.4355	0.040	0.020
1/8 0.125	0.1250	0.1230	0.025	0.005	1/2 0.500	0.5000	0.4980	0.040	0.020
5/32 0.156	0.1562	0.1542	0.025	0.005	5/8 0.625	0.6250	0.6230	0.055	0.035
3/16 0.188	0.1875	0.1855	0.025	0.005	3/4 0.750	0.7500	0.7480	0.055	0.035
7/32 0.219	0.2187	0.2167	0.025	0.005	7/8 0.875	0.8750	0.8730	0.055	0.035
1/4 0.250	0.2500	0.2480	0.025	0.005	1 1.000	1.0000	0.9980	0.055	0.035

[a] Where specifying nominal size in decimals, zeros preceding decimal point are omitted.
[b] Where specifying nominal size in decimals, zeros preceding decimal point are omitted.

All dimensions are in inches.

Length Increments: Lengths are as specified by the purchaser; however, it is recommended that nominal pin lengths be limited to increments of not less than 0.062 inch.

Material: Straight pins are normally made from cold drawn steel wire or rod having a maximum carbon content of 0.28 per cent. Where required, pins may also be made from corrosion resistant steel, brass, or other metals.

Designation: Straight pins are designated by the following data, in the sequence shown: Product name (noun first), nominal size (fraction or decimal equivalent), material, and protective finish, if required.

Examples: Pin, Chamfered Straight, 1/8 × 1.500, Steel

Pin, Square End Straight, 0.250 × 2.250, Steel, Zinc Plated

American National Standard Taper Pins.—Taper pins have a uniform taper over the pin length with both ends crowned. Most sizes are supplied in commercial and precision classes, the latter having generally tighter tolerances and being more closely controlled in manufacture.

Diameters: The major diameter of both commercial and precision classes of pins is the diameter of the large end and is the basis for pin size. The diameter at the small end is computed by multiplying the nominal length of the pin by the factor 0.02083 and subtracting the result from the basic pin diameter. See also Table 5.

Taper: The taper on commercial class pins is 0.250 ± 0.006 inch per foot and on the precision class pins is 0.250 ± 0.004 inch per foot of length.

Materials: Unless otherwise specified, taper pins are made from SAE 1211 steel or cold drawn SAE 1212 or 1213 steel or equivalents, and no mechanical property requirements apply.

Hole Sizes: Under most circumstances, holes for taper pins require taper reaming. Sizes and lengths of taper pins for which standard reamers are available are given in Table 6. Drilling specifications for taper pins are given below.

Designation: Taper pins are designated by the following data in the sequence shown: Product name (noun first), class, size number (or decimal equivalent), length (fraction or three-place decimal equivalent), material, and protective finish, if required.

Examples: Pin, Taper (Commercial Class) No. 0 × ¾, Steel

Pin, Taper (Precision Class) 0.219 × 1.750, Steel, Zinc Plated

Table 5. Nominal Diameter at Small Ends of Standard Taper Pins

Pin Length in inches	Pin Number and Small End Diameter for Given Length										
	0	1	2	3	4	5	6	7	8	9	10
¾	0.140	0.156	0.177	0.203	0.235	0.273	0.325	0.393	0.476	0.575	0.690
1	0.135	0.151	0.172	0.198	0.230	0.268	0.320	0.388	0.471	0.570	0.685
1¼	0.130	0.146	0.167	0.192	0.224	0.263	0.315	0.382	0.466	0.565	0.680
1½	0.125	0.141	0.162	0.187	0.219	0.258	0.310	0.377	0.460	0.560	0.675
1¾	0.120	0.136	0.157	0.182	0.214	0.252	0.305	0.372	0.455	0.554	0.669
2	0.114	0.130	0.151	0.177	0.209	0.247	0.299	0.367	0.450	0.549	0.664
2¼	0.109	0.125	0.146	0.172	0.204	0.242	0.294	0.362	0.445	0.544	0.659
2½	0.104	0.120	0.141	0.166	0.198	0.237	0.289	0.356	0.440	0.539	0.654
2¾	0.099	0.115	0.136	0.161	0.193	0.232	0.284	0.351	0.434	0.534	0.649
3	0.094	0.110	0.131	0.156	0.188	0.227	0.279	0.346	0.429	0.528	0.643
3¼	…	…	…	0.151	0.182	0.221	0.273	0.340	0.424	0.523	0.638
3½	…	…	…	0.146	0.177	0.216	0.268	0.335	0.419	0.518	0.633
3¾	…	…	…	0.141	0.172	0.211	0.263	0.330	0.414	0.513	0.628
4	…	…	…	0.136	0.167	0.206	0.258	0.326	0.409	0.508	0.623
4¼	…	…	…	0.131	0.162	0.201	0.253	0.321	0.403	0.502	0.617
4½	…	…	…	0.125	0.156	0.195	0.247	0.315	0.398	0.497	0.612
5	…	…	…	…	0.146	0.185	0.237	0.305	0.389	0.487	0.602
5½	…	…	…	…	…	…	…	0.294	0.377	0.476	0.591
6	…	…	…	…	…	…	…	0.284	0.367	0.466	0.581

Drilling Specifications for Taper Pins.—When helically fluted taper pin reamers are used, the diameter of the through hole drilled prior to reaming is equal to the diameter at the small end of the taper pin. (See Table 5.) However, when straight fluted taper reamers are to be used, it may be necessary, for long pins, to step drill the hole before reaming, the number and sizes of the drills to be used depending on the depth of the hole (pin length).

To determine the number and sizes of step drills required: Find the length of pin to be used at the top of the chart on page 1657 and follow this length down to the intersection with that heavy line which represents the size of taper pin (see taper pin numbers at the right-hand end of each heavy line). If the length of pin falls between the first and second dots, counting from the left, only one drill is required. Its size is indicated by following the nearest horizontal line from the point of intersection (of the pin length) on the heavy line over to the drill diameter values at the left. If the intersection of pin length comes between the second and third dots, then two drills are required. The size of the smaller drill then corresponds to the intersection of the pin length and the heavy line and the larger is the corresponding drill diameter for the intersection of one-half this length with the heavy line. Should the pin length fall between the third and fourth dots, three drills are required. The smallest drill will have a diameter corresponding to the intersection of the total pin length with the heavy line, the next in size will have a diameter corresponding to the intersection of two-thirds of this length with the heavy line and the largest will have a diameter corresponding to the intersection of one-third of this length with the heavy line. Where the intersection falls between two drill sizes, use the smaller.

Chart to Facilitate Selection of Number and Sizes of Drills for Step-Drilling Prior to Taper Reaming

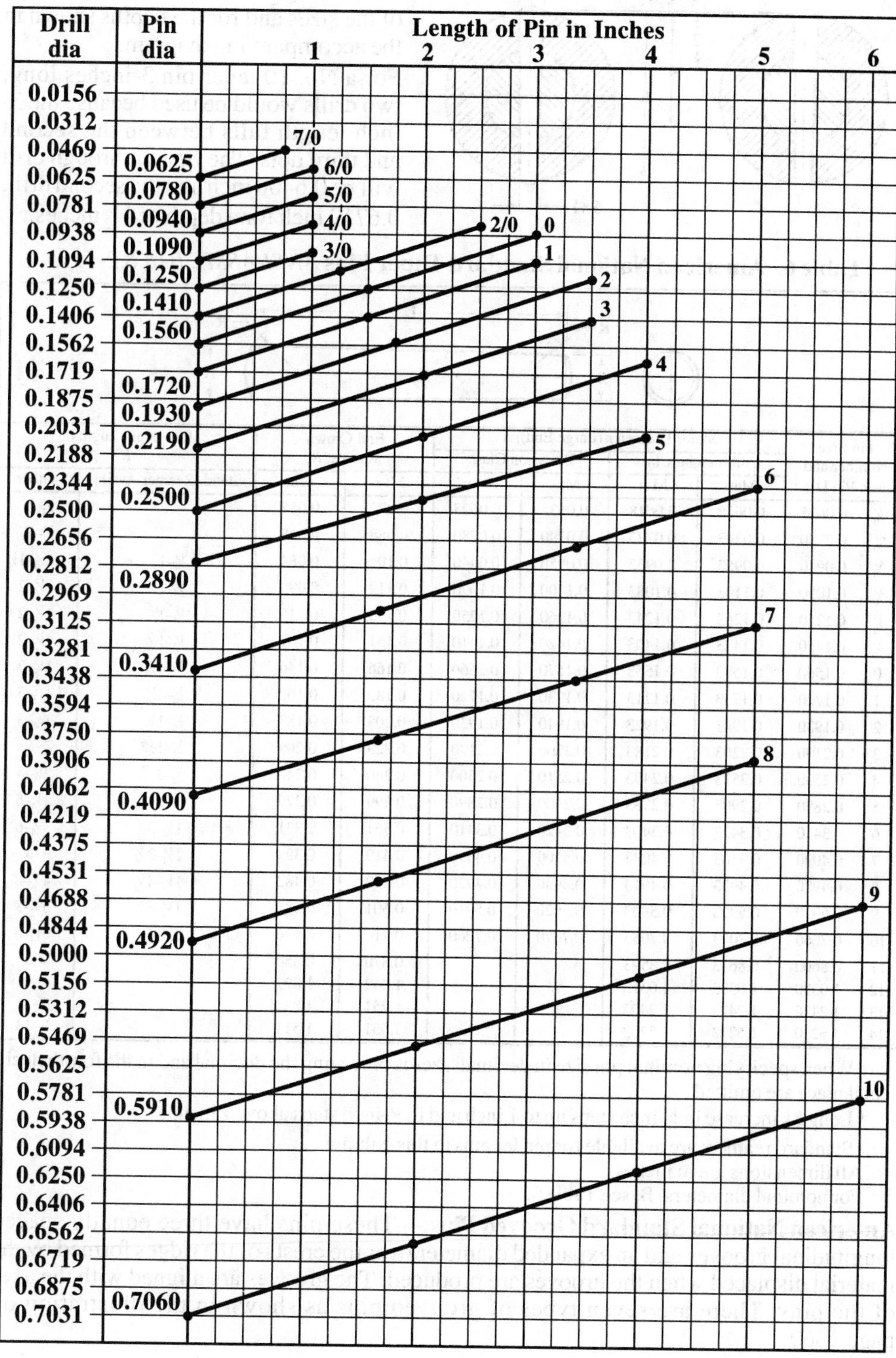

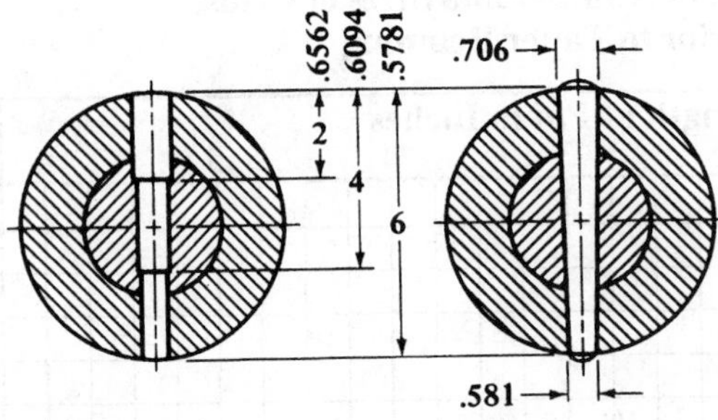

Examples: For a No. 10 taper pin 6-inches long, three drills would be used, of the sizes and for the depths shown in the accompanying diagram.

For a No. 10 taper pin 3-inches long, two drills would be used because the 3-inch length falls between the second and third dots. The first or through drill will be 0.6406 inch and the second drill, 0.6719 inch for a depth of 1½ inches.

Table 6. American National Standard Taper Pins *ANSI/ASME B18.8.2-1995*

Pin Size Number and Basic Pin Dia.[a]		Major Diameter (Large End), *A*				End Crown Radius, *R*		Range of Lengths,[b] *L*	
		Commercial Class		Precision Class					
		Max	Min	Max	Min	Max	Min	Stand. Reamer Avail.[c]	Other
7/0	0.0625	0.0638	0.0618	0.0635	0.0625	0.072	0.052	…	¼–1
6/0	0.0780	0.0793	0.0773	0.0790	0.0780	0.088	0.068	…	¼–½
5/0	0.0940	0.0953	0.0933	0.0950	0.0940	0.104	0.084	¼–1	1¼, 1½
4/0	0.1090	0.1103	0.1083	0.1100	0.1090	0.119	0.099	¼–1	1¼–2
3/0	0.1250	0.1263	0.1243	0.1260	0.1250	0.135	0.115	¼–1	1¼–2
2/0	0.1410	0.1423	0.1403	0.1420	0.1410	0.151	0.131	½–1¼	1½–2½
0	0.1560	0.1573	0.1553	0.1570	0.1560	0.166	0.146	½–1¼	1½–3
1	0.1720	0.1733	0.1713	0.1730	0.1720	0.182	0.162	¾–1¼	1½–3
2	0.1930	0.1943	0.1923	0.1940	0.1930	0.203	0.183	¾–1½	1¾–3
3	0.2190	0.2203	0.2183	0.2200	0.2190	0.229	0.209	¾–1¾	2–4
4	0.2500	0.2513	0.2493	0.2510	0.2500	0.260	0.240	¾–2	2¼–4
5	0.2890	0.2903	0.2883	0.2900	0.2890	0.299	0.279	1–2½	2¾–6
6	0.3410	0.3423	0.3403	0.3420	0.3410	0.351	0.331	1¼–3	3¼–6
7	0.4090	0.4103	0.4083	0.4100	0.4090	0.419	0.399	1¼–3¾	4–8
8	0.4920	0.4933	0.4913	0.4930	0.4920	0.502	0.482	1¼–4½	4¾–8
9	0.5910	0.5923	0.5903	0.5920	0.5910	0.601	0.581	1¼–5¼	5½–8
10	0.7060	0.7073	0.7053	0.7070	0.7060	0.716	0.696	1½–6	6¼–8
11	0.8600	0.8613	0.8593	…	…	0.870	0.850	…	2–8
12	1.0320	1.0333	1.0313	…	…	1.042	1.022	…	2–9
13	1.2410	1.2423	1.2403	…	…	1.251	1.231	…	3–11
14	1.5210	1.5223	1.5203	…	…	1.531	1.511	…	3–13

[a] When specifying nominal pin size in decimals, zeros preceding the decimal and in the fourth decimal place are omitted.

[b] Lengths increase in ⅛-inch steps up to 1 inch and in ¼-inch steps above 1 inch.

[c] Standard reamers are available for pin lengths in this column.

All dimensions are in inches.

For nominal diameters, B, see Table 5.

American National Standard Grooved Pins.—These pins have three equally spaced longitudinal grooves and an expanded diameter over the crests of the ridges formed by the material displaced when the grooves are produced. The grooves are aligned with the axes of the pins. There are seven types of grooved pins as shown in the illustration on page 1660.

Standard Sizes and Lengths: The standard sizes and lengths in which grooved pins are normally available are given in Table 7.

Materials: Grooved pins are normally made from cold drawn low carbon steel wire or rod. Where additional performance is required, carbon steel pins may be supplied surface hardened and heat treated to a hardness consistent with the performance requirements. Pins may also be made from alloy steel, corrosion resistant steel, brass, Monel and other non-ferrous metals having chemical properties as agreed upon between manufacturer and purchaser.

Performance Requirements: Grooved pins are required to withstand the minimum double shear loads given in Table 7 for the respective materials shown, when tested in accordance with the Double Shear Testing of Pins as set forth in ANSI/ASME B18.8.2-1995, Appendix B.

Hole Sizes: To obtain maximum product retention under average conditions, it is recommended that holes for the installation of grooved pins be held as close as possible to the limits shown in Table 7. The minimum limits correspond to the drill size, which is the same as the basic pin diameter. The maximum limits are generally suitable for length-diameter ratios of not less than 4 to 1 nor greater than 10 to 1. For smaller length-to-diameter ratios, the hole should be held closer to the minimum limits where retention is critical. Conversely for larger ratios where retention requirements are less important, it may be desirable to increase the hole diameters beyond the maximum limits shown.

Designation: Grooved pins are designated by the following data in the sequence shown: Product name (noun first) including type designation, nominal size (number, fraction or decimal equivalent), length (fraction or decimal equivalent), material, including specification or heat treatment where necessary, protective finish, if required.

Examples: Pin, Type A Grooved, $\frac{3}{32} \times \frac{3}{4}$, Steel, Zinc Plated

Pin, Type F Grooved, 0.250×1.500, Corrosion Resistant Steel

American National Standard Grooved T-Head Cotter Pins and Round Head Grooved Drive Studs.—The cotter pins have a T-head and the studs a round head. Both pins and studs have three equally spaced longitudinal grooves and an expanded diameter over the crests of the raised ridges formed by the material displaced when the grooves are formed.

Standard Sizes and Lengths: The standard sizes and range of standard lengths are given in Tables 8 and 9.

Material: Unless otherwise specified these pins are made from low carbon steel. Where so indicated by the purchaser they may be made from corrosion resistant steel, brass or other non-ferrous alloys.

Hole Sizes: To obtain optimum product retention under average conditions, it is recommended that holes for the installation of grooved T-head cotter pins and grooved drive studs be held as close as possible to the limits tabulated. The minimum limits given correspond to the drill size, which is equivalent to the basic shank diameter. The maximum limits shown are generally suitable for length-diameter ratios of not less than 4 to 1 and not greater than 10 to 1. For smaller length-to-diameter ratios, the holes should be held closer to minimum limits where retention is critical. Conversely, for larger length-to-diameter ratios or where retention requirements are not essential, it may be desirable to increase the hole diameter beyond the maximum limits shown.

Designation: Grooved T-head cotter pins and round head grooved drive studs are designated by the following data, in the order shown: Product name (noun first), nominal size (number, fraction or decimal equivalent), length (fraction or decimal equivalent), material including specification or heat treatment where necessary, and protective finish, if required.

Examples: Pin, Grooved T-Head Cotter, $\frac{1}{4} \times 1\frac{1}{4}$, Steel, Zinc Plated

Drive Stud, Round Head Grooved, No. $10 \times \frac{1}{2}$, Corrosion Resistant Steel

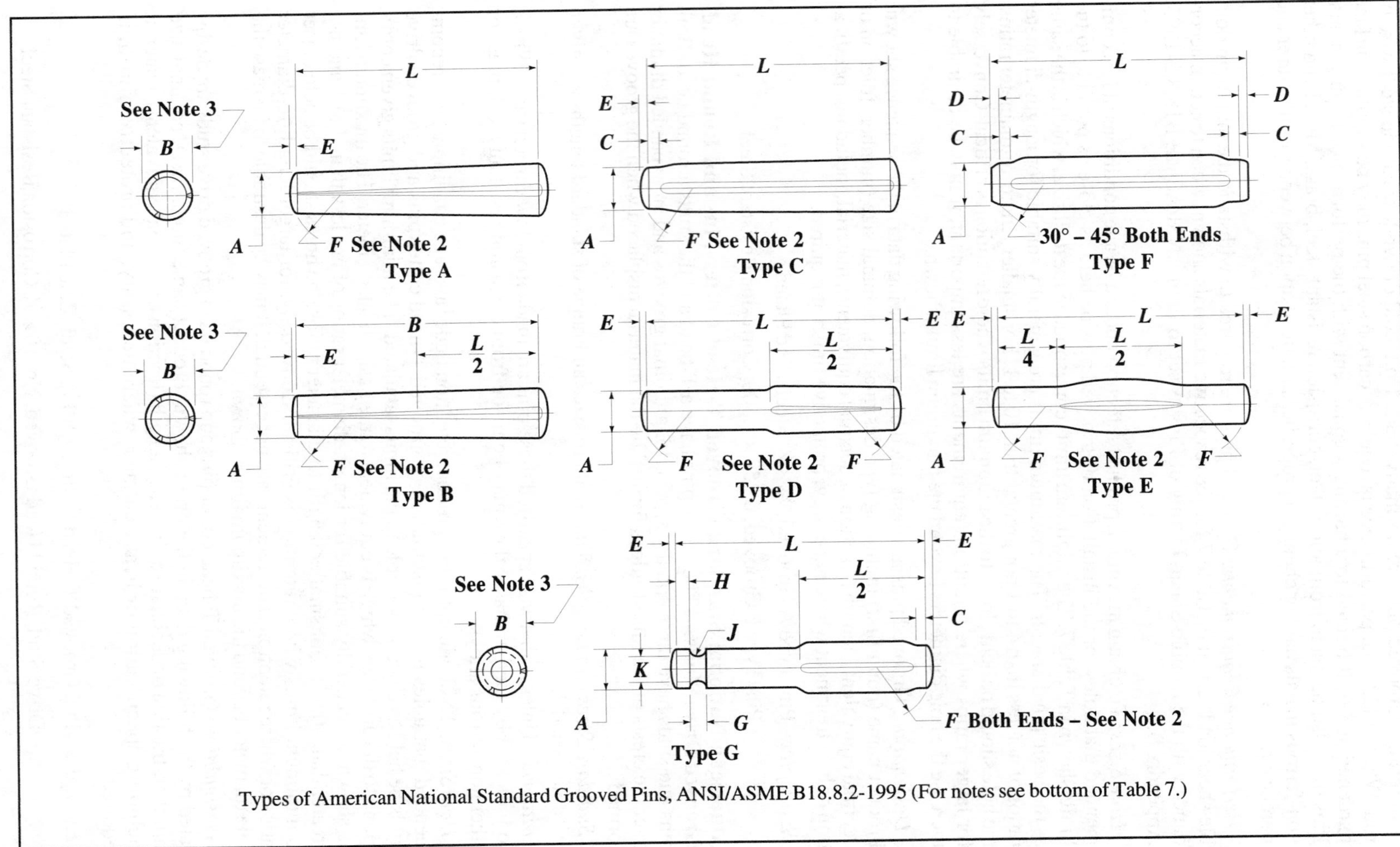

Types of American National Standard Grooved Pins, ANSI/ASME B18.8.2-1995 (For notes see bottom of Table 7.)

Table 7. American National Standard Grooved Pins *ANSI/ASME B18.8.2-1995*

Nominal Size or Basic Pin Diameter		Pin Diameter,[a] A		Pilot Length, C	Chamfer Length,[b] D	Crown Height,[b] E		Crown Radius,[b] F		Neck Width, G		Shoulder Length, H		Neck Radius, J	Neck Diameter, K		Range of Standard Lengths[c]
		Max	Min	Ref	Min	Max	Min	Max	Min	Max	Min	Max	Min	Ref	Max	Min	
1/32[d]	0.0312	0.0312	0.0302	0.015	…	…	…	…	…	…	…	…	…	…	…	…	1/8–1/2
3/64[d]	0.0469	0.0469	0.0459	0.031	…	…	…	…	…	…	…	…	…	…	…	…	1/8–5/8
1/16	0.0625	0.0625	0.0615	0.031	0.016	0.0115	0.0015	0.088	0.068	…	…	…	…	…	…	…	1/8–1
5/64[d]	0.0781	0.0781	0.0771	0.031	0.016	0.0137	0.0037	0.104	0.084	…	…	…	…	…	…	…	1/4–1
3/32	0.0938	0.0938	0.0928	0.031	0.016	0.0141	0.0041	0.135	0.115	0.038	0.028	0.041	0.031	0.016	0.067	0.057	1/4–1 1/4
7/64[d]	0.1094	0.1094	0.1074	0.031	0.016	0.0160	0.0060	0.150	0.130	0.038	0.028	0.041	0.031	0.016	0.082	0.072	1/4–1 1/4
1/8	0.1250	0.1250	0.1230	0.031	0.016	0.0180	0.0080	0.166	0.146	0.069	0.059	0.041	0.031	0.031	0.088	0.078	1/4–1 1/2
5/32	0.1563	0.1563	0.1543	0.062	0.031	0.0220	0.0120	0.198	0.178	0.069	0.059	0.057	0.047	0.031	0.109	0.099	3/8–2
3/16	0.1875	0.1875	0.1855	0.062	0.031	0.0230	0.0130	0.260	0.240	0.069	0.059	0.057	0.047	0.031	0.130	0.120	3/8–2 1/4
7/32	0.2188	0.2188	0.2168	0.062	0.031	0.0270	0.0170	0.291	0.271	0.101	0.091	0.072	0.062	0.031	0.151	0.141	1/2–3
1/4	0.2500	0.2500	0.2480	0.062	0.031	0.0310	0.0210	0.322	0.302	0.101	0.091	0.072	0.062	0.047	0.172	0.162	1/2–3 1/4
5/16	0.3125	0.3125	0.3105	0.094	0.047	0.0390	0.0290	0.385	0.365	0.132	0.122	0.104	0.094	0.062	0.214	0.204	5/8–3 1/2
3/8	0.3750	0.3750	0.3730	0.094	0.047	0.0440	0.0340	0.479	0.459	0.132	0.122	0.135	0.125	0.062	0.255	0.245	3/4–4 1/4
7/16	0.4375	0.4375	0.4355	0.094	0.047	0.0520	0.0420	0.541	0.521	0.195	0.185	0.135	0.125	0.094	0.298	0.288	7/8–4 1/2
1/2	0.5000	0.5000	0.4980	0.094	0.047	0.0570	0.0470	0.635	0.615	0.195	0.185	0.135	0.125	0.094	0.317	0.307	1–4 1/2

[a] For expanded diameters, *B*, see ANSI/ASME B18.8.2-1995.

[b] Pins in 1/32- and 3/64-inch sizes of any length and all sizes of 1/4-inch nominal length or shorter are not crowned or chamfered.

[c] Standard lengths increase in 1/8-inch steps from 1/8 to 1 inch, and in 1/4-inch steps above 1 inch. Standard lengths for the 1/32-, 3/64-, 1/16-, and 5/64-inch sizes and the 1/4-inch length for the 3/32-, 7/64-, and 1/8-inch sizes do not apply to Type G grooved pins.

[d] Non-stock items, not recommended for new designs.

Pin Material	Nominal Pin Size														
	1/32	3/64	1/16	5/64	3/32	7/64	1/8	5/32	3/16	7/32	1/4	5/16	3/8	7/16	1/2
Steels	Double Shear Load, Min, lb														
Low Carbon	100	220	410	620	890	1,220	1,600	2,300	3,310	4,510	5,880	7,660	11,000	15,000	19,600
Alloy (R_c 40 – 48 hardness)	180	400	720	1,120	1,600	2,180	2,820	4,520	6,440	8,770	11,500	17,900	26,000	35,200	46,000
Corrosion Resistant	140	300	540	860	1,240	1,680	2,200	3,310	4,760	6,480	8,460	12,700	18,200	24,800	32,400
Brass	60	140	250	390	560	760	990	1,540	2,220	3,020	3,950	6,170	9,050	12,100	15,800
	Recommended Hole Sizes for Unplated Pins (The minimum drill size is the same as the pin size. See also text on page 1659.)														
Maximum Diameter	0.0324	0.0482	0.0640	0.0798	0.0956	0.1113	0.1271	0.1587	0.1903	0.2219	0.2534	0.3166	0.3797	0.4428	0.5060
Minimum Diameter	0.0312	0.0469	0.0625	0.0781	0.0938	0.1094	0.1250	0.1563	0.1875	0.2188	0.2500	0.3125	0.3750	0.4375	0.5000

All dimensions are in inches.

Table 8. American National Standard Grooved T-Head Cotter Pins
ANSI/ASME B18.8.2-1995

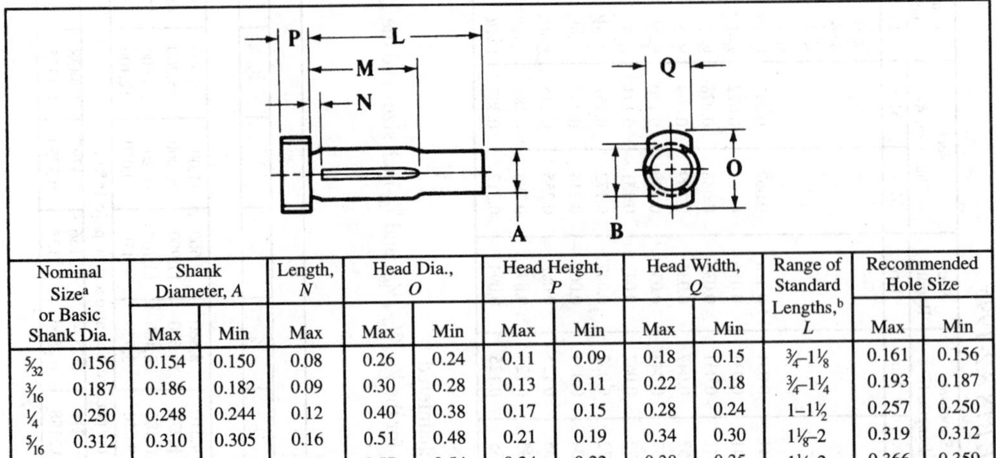

Nominal Size[a] or Basic Shank Dia.		Shank Diameter, *A*		Length, *N*	Head Dia., *O*		Head Height, *P*		Head Width, *Q*		Range of Standard Lengths,[b] *L*	Recommended Hole Size	
		Max	Min	Max	Max	Min	Max	Min	Max	Min		Max	Min
5/32	0.156	0.154	0.150	0.08	0.26	0.24	0.11	0.09	0.18	0.15	3/4–1 1/8	0.161	0.156
3/16	0.187	0.186	0.182	0.09	0.30	0.28	0.13	0.11	0.22	0.18	3/4–1 1/4	0.193	0.187
1/4	0.250	0.248	0.244	0.12	0.40	0.38	0.17	0.15	0.28	0.24	1–1 1/2	0.257	0.250
5/16	0.312	0.310	0.305	0.16	0.51	0.48	0.21	0.19	0.34	0.30	1 1/8–2	0.319	0.312
23/64	0.359	0.358	0.353	0.18	0.57	0.54	0.24	0.22	0.38	0.35	1 1/4–2	0.366	0.359
1/2	0.500	0.498	0.493	0.25	0.79	0.76	0.32	0.30	0.54	0.49	2–3	0.508	0.500

[a] When specifying nominal size in decimals, zeros preceding decimal point and in the fourth decimal place are omitted.

[b] Lengths increase in 1/8-inch steps from 3/4 to 1 1/4 inch and in 1/4-inch steps above 1 1/4 inches. For groove length, *M*, dimensions see ANSI/ASME B18.8.2-1995.

All dimensions are in inches.

For expanded diameter, *B*, dimensions, see ANSI/ASME B18.8.2-1995.

Table 9. American National Standard Round Head Grooved Drive Studs
ANSI/ASME B18.8.2-1995

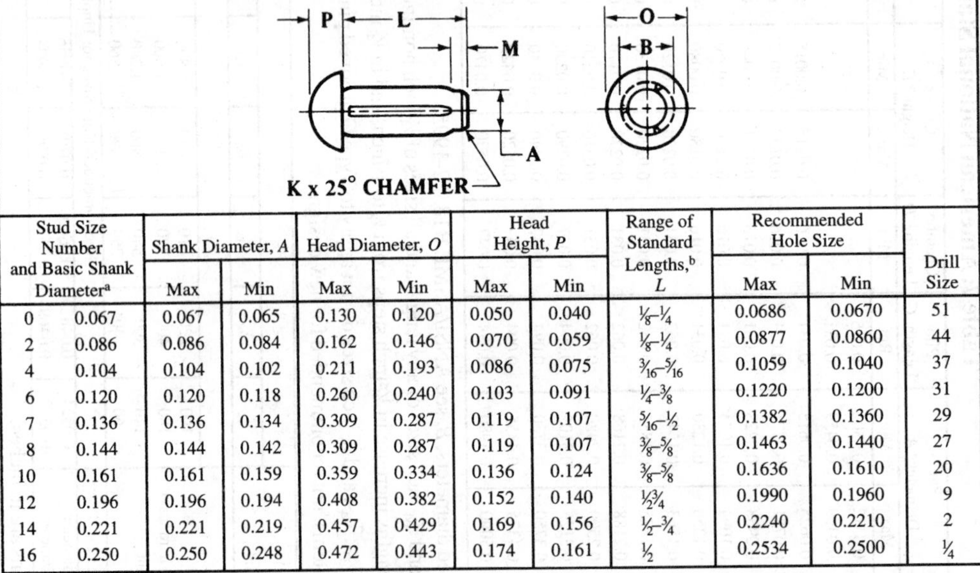

Stud Size Number and Basic Shank Diameter[a]		Shank Diameter, *A*		Head Diameter, *O*		Head Height, *P*		Range of Standard Lengths,[b] *L*	Recommended Hole Size		Drill Size
		Max	Min	Max	Min	Max	Min		Max	Min	
0	0.067	0.067	0.065	0.130	0.120	0.050	0.040	1/8–1/4	0.0686	0.0670	51
2	0.086	0.086	0.084	0.162	0.146	0.070	0.059	1/8–1/4	0.0877	0.0860	44
4	0.104	0.104	0.102	0.211	0.193	0.086	0.075	3/16–5/16	0.1059	0.1040	37
6	0.120	0.120	0.118	0.260	0.240	0.103	0.091	1/4–3/8	0.1220	0.1200	31
7	0.136	0.136	0.134	0.309	0.287	0.119	0.107	5/16–1/2	0.1382	0.1360	29
8	0.144	0.144	0.142	0.309	0.287	0.119	0.107	3/8–5/8	0.1463	0.1440	27
10	0.161	0.161	0.159	0.359	0.334	0.136	0.124	3/8–5/8	0.1636	0.1610	20
12	0.196	0.196	0.194	0.408	0.382	0.152	0.140	1/2 3/4	0.1990	0.1960	9
14	0.221	0.221	0.219	0.457	0.429	0.169	0.156	1/2–3/4	0.2240	0.2210	2
16	0.250	0.250	0.248	0.472	0.443	0.174	0.161	1/2	0.2534	0.2500	1/4

[a] Where specifying nominal size in decimals, zeros preceding decimal point and in the fourth decimal place are omitted.
[b] Lengths increase in 1/16-inch steps from 1/8 to 3/8 inch and in 1/8-inch steps above 3/8 inch.

All dimensions are in inches.
For pilot length, *M*, and expanded diameter, *B*, dimensions see ANSI/ASME B18.8.2-1995.

Table 10. American National Standard Slotted Type Spring Pins
ANSI/ASME B18.8.2-1995

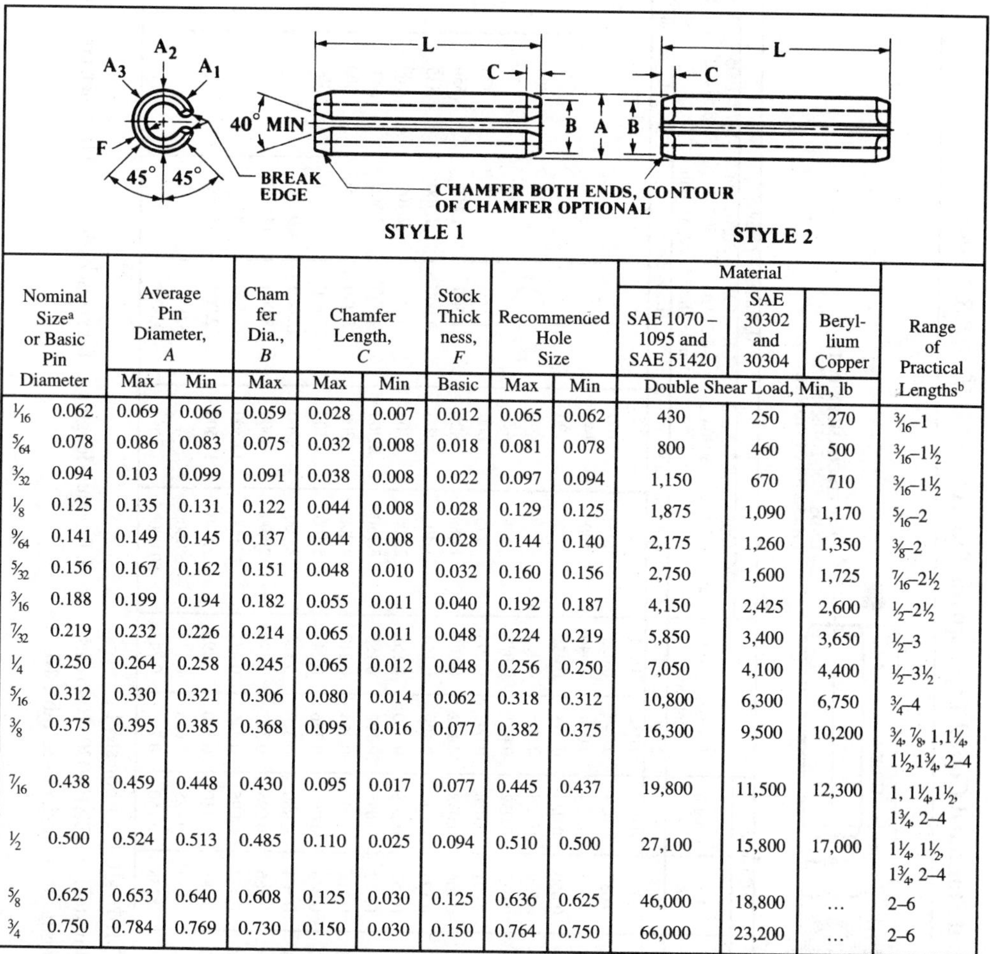

Nominal Size[a] or Basic Pin Diameter		Average Pin Diameter, *A*		Chamfer Dia., *B*	Chamfer Length, *C*		Stock Thickness, *F*	Recommended Hole Size		Material: SAE 1070–1095 and SAE 51420	Material: SAE 30302 and 30304	Material: Beryllium Copper	Range of Practical Lengths[b]
		Max	Min	Max	Max	Min	Basic	Max	Min	Double Shear Load, Min, lb			
1/16	0.062	0.069	0.066	0.059	0.028	0.007	0.012	0.065	0.062	430	250	270	3/16–1
5/64	0.078	0.086	0.083	0.075	0.032	0.008	0.018	0.081	0.078	800	460	500	3/16–1 1/2
3/32	0.094	0.103	0.099	0.091	0.038	0.008	0.022	0.097	0.094	1,150	670	710	3/16–1 1/2
1/8	0.125	0.135	0.131	0.122	0.044	0.008	0.028	0.129	0.125	1,875	1,090	1,170	5/16–2
9/64	0.141	0.149	0.145	0.137	0.044	0.008	0.028	0.144	0.140	2,175	1,260	1,350	3/8–2
5/32	0.156	0.167	0.162	0.151	0.048	0.010	0.032	0.160	0.156	2,750	1,600	1,725	7/16–2 1/2
3/16	0.188	0.199	0.194	0.182	0.055	0.011	0.040	0.192	0.187	4,150	2,425	2,600	1/2–2 1/2
7/32	0.219	0.232	0.226	0.214	0.065	0.011	0.048	0.224	0.219	5,850	3,400	3,650	1/2–3
1/4	0.250	0.264	0.258	0.245	0.065	0.012	0.048	0.256	0.250	7,050	4,100	4,400	1/2–3 1/2
5/16	0.312	0.330	0.321	0.306	0.080	0.014	0.062	0.318	0.312	10,800	6,300	6,750	3/4–4
3/8	0.375	0.395	0.385	0.368	0.095	0.016	0.077	0.382	0.375	16,300	9,500	10,200	3/4, 7/8, 1, 1 1/4, 1 1/2, 1 3/4, 2–4
7/16	0.438	0.459	0.448	0.430	0.095	0.017	0.077	0.445	0.437	19,800	11,500	12,300	1, 1 1/4, 1 1/2, 1 3/4, 2–4
1/2	0.500	0.524	0.513	0.485	0.110	0.025	0.094	0.510	0.500	27,100	15,800	17,000	1 1/4, 1 1/2, 1 3/4, 2–4
5/8	0.625	0.653	0.640	0.608	0.125	0.030	0.125	0.636	0.625	46,000	18,800	…	2–6
3/4	0.750	0.784	0.769	0.730	0.150	0.030	0.150	0.764	0.750	66,000	23,200	…	2–6

[a] Where specifying nominal size in decimals, zeros preceding decimal point are omitted.

[b] Length increments are 1/16 inch from 1/8 to 1 inch; 1/8 from 1 inch to 2 inches; and 1/4 inch from 2 inches to 6 inches.

All dimensions are in inches.

American National Standard Spring Pins.—These pins are made in two types: one type has a slot throughout its length; the other is shaped into a coil.

Preferred Lengths and Sizes: The preferred lengths and sizes in which these pins are normally available are given in Tables 10 and 11.

Materials: Spring pins are normally made from SAE 1070–1095 carbon steel, SAE 6150H alloy steel, SAE types 51410 through 51420, 30302 and 30304 corrosion resistant steels, and beryllium copper alloy, heat treated or cold worked to attain the hardness and performance characteristics set forth in ANSI/ASME B18.8.2-1995.

Designation: Spring pins are designated by the following data in the sequence shown:

Examples: Pin, Coiled Spring, 1/4 × 1 1/4, Standard Duty, Steel, Zinc Plated

Pin, Slotted Spring, 1/2 × 3, Steel, Phosphate Coated

Table 11. American National Standard Coiled Type Spring Pins *ANSI/ASME B18.8.2-1995*

BREAK EDGE
SWAGED CHAMFER BOTH ENDS. CONTOUR OF CHAMFER OPTIONAL

Nominal Size or Basic Pin Diameter	Pin Diameter, A						Chamfer		Recommended Hole Size		SAE Material Number					
	Standard Duty		Heavy Duty		Light Duty		Dia., B	Length, C			1070–1095 and 51420	30302 and 30304	1070–1095 and 51420	30302 and 30304	1070–1095 and 51420	30302 and 30304
											Double Shear Load, Min, lb					
	Max	Min	Max	Min	Max	Min	Max	Ref	Max	Min	Standard Duty		Heavy Duty		Light Duty	
1/32 0.031	0.035	0.033	…	…	…	…	0.029	0.024	0.032	0.031	90[a]	65	…	…	…	…
0.039	0.044	0.041	…	…	…	…	0.037	0.024	0.040	0.039	135[a]	100	…	…	…	…
3/64 0.047	0.052	0.049	…	…	…	…	0.045	0.024	0.048	0.046	190[a]	145	…	…	…	…
0.052	0.057	0.054	…	…	…	…	0.050	0.024	0.053	0.051	250[a]	190	…	…	…	…
1/16 0.062	0.072	0.067	0.070	0.066	0.073	0.067	0.059	0.028	0.065	0.061	330	265	475	360	205	160
5/64 0.078	0.088	0.083	0.086	0.082	0.089	0.083	0.075	0.032	0.081	0.077	550	425	800	575	325	250
3/32 0.094	0.105	0.099	0.103	0.098	0.106	0.099	0.091	0.038	0.097	0.093	775	600	1,150	825	475	360
7/64 0.109	0.120	0.114	0.118	0.113	0.121	0.114	0.106	0.038	0.112	0.108	1,050	825	1,500	1,150	650	500
1/8 0.125	0.138	0.131	0.136	0.130	0.139	0.131	0.121	0.044	0.129	0.124	1,400	1,100	2,000	1,700	825	650
5/32 0.156	0.171	0.163	0.168	0.161	0.172	0.163	0.152	0.048	0.160	0.155	2,200	1,700	3,100	2,400	1,300	1,000
3/16 0.188	0.205	0.196	0.202	0.194	0.207	0.196	0.182	0.055	0.192	0.185	3,150	2,400	4,500	3,500	1,900	1,450
7/32 0.219	0.238	0.228	0.235	0.226	0.240	0.228	0.214	0.065	0.224	0.217	4,200	3,300	5,900	4,600	2,600	2,000
1/4 0.250	0.271	0.260	0.268	0.258	0.273	0.260	0.243	0.065	0.256	0.247	5,500	4,300	7,800	6,200	3,300	2,600
5/16 0.312	0.337	0.324	0.334	0.322	0.339	0.324	0.304	0.080	0.319	0.308	8,700	6,700	12,000	9,300	5,200	4,000
3/8 0.375	0.403	0.388	0.400	0.386	0.405	0.388	0.366	0.095	0.383	0.370	12,600	9,600	18,000	14,000	…	…
7/16 0.438	0.469	0.452	0.466	0.450	0.471	0.452	0.427	0.095	0.446	0.431	17,000	13,300	23,500	18,000	…	…
1/2 0.500	0.535	0.516	0.532	0.514	0.537	0.516	0.488	0.110	0.510	0.493	22,500	17,500	32,000	25,000	…	…
5/8 0.625	0.661	0.642	0.658	0.640	…	…	0.613	0.125	0.635	0.618	35,000[b]	…	48,000[b]	…	…	…
3/4 0.750	0.787	0.768	0.784	0.766	…	…	0.738	0.150	0.760	0.743	50,000[b]	…	70,000[b]	…	…	…

[a] Sizes 1/32 inch through 0.052 inch are not available in SAE 1070–1095 carbon steel.

[b] Sizes 5/8 inch and larger are produced from SAE 6150H alloy steel, not SAE 1070–1095 carbon steel. Practical lengths, *L*, for sizes 1/32 through 0.052 inch are 1/8 through 5/8 inch and for the 7/64-inch size, 1/4 through 1 3/4 inches. For lengths of other sizes see Table 10.

All dimensions are in inches.

RETAINING RINGS

Retaining Rings.—The purpose of a retaining ring is to act as an artificial shoulder that will retain an object in a housing (internal ring), as shown in Fig. 1, or on a shaft (external ring). Two types of retaining ring are common, the stamped ring and the spiral-wound ring. The stamped type of retaining ring, or snap ring, is stamped from tempered sheet metal and has a nonuniform cross-section. The typical spiral-wound retaining ring has a uniform cross-section and is made up of two or more turns of coiled, spring-tempered steel, although one-turn spiral-wound rings are common. Spiral-wound retaining rings provide a continuous gapless shoulder to a housing or shaft. Most stamped rings can only be installed at or near the end of a shaft or housing. The spiral-wound design generally requires installation from the end of a shaft or housing. Both types, stamped and spiral, are usually installed into grooves on the shaft or housing.

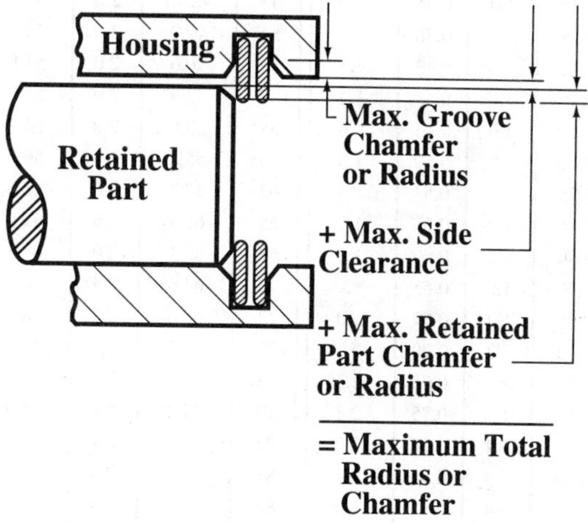

Fig. 1. Typical Retaining Ring Installation Showing Maximum Total Radius or Chamfer (*Courtesy Spirolox Retaining Rings*)

In the section that follows, Tables 1 through 6 give dimensions and data on general-purpose tapered and reduced cross-section metric retaining rings (stamped type) covered by ANSI B27.7M-1977, R1983. Tables 1 and 4 cover Type 3AM1 tapered external retaining rings, Tables 2 and 5 cover Type 3BM1 tapered internal rings, and Tables 3 and 6 cover Type 3CM1 reduced cross-section external rings. Tables 7 through 10 cover inch sizes of internal and external spiral retaining rings corresponding to MIL-R-27426 Types A (external) and B (internal), Class 1 (medium duty) and Class 2 (heavy duty). Tables 11 through 17 cover stamped retaining rings in inch sizes.

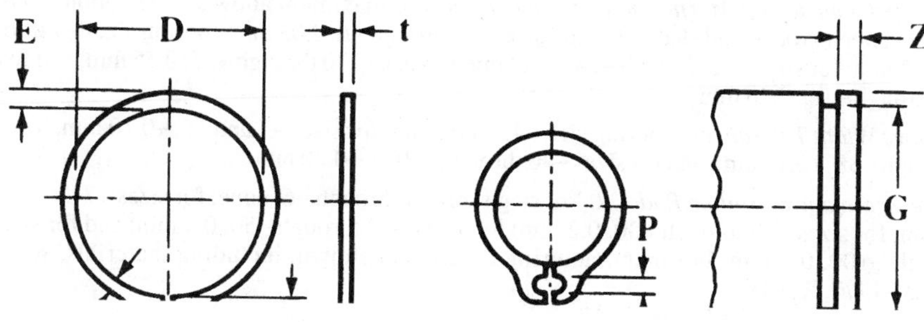

Table 1. American National Standard Metric Tapered Retaining Rings — Basic External Series — 3AM1 *ANSI B27.7M-1977, R1983*

Shaft Diam.	Ring		Groove				Shaft Diam	Ring		Groove			
	Free Diam.	Thick ness	Diam.	Width	Depth	Edge Margin		Free Diam.	Thick ness	Diam.	Width	Depth	Edge Margin
S	*D*	*t*	*G*	*W*	*d* ref	*Z* min	*S*	*D*	*t*	*G*	*W*	*d* ref	*Z* min
4	3.60	0.25	3.80	0.32	0.1	0.3	36	33.25	1.3	33.85	1.4	1.06	3.2
5	4.55	0.4	4.75	0.5	0.13	0.4	38	35.20	1.3	35.8	1.4	1.10	3.3
6	5.45	0.4	5.70	0.5	0.15	0.5	40	36.75	1.6	37.7	1.75	1.15	3.4
7	6.35	0.6	6.60	0.7	0.20	0.6	42	38.80	1.6	39.6	1.75	1.20	3.6
8	7.15	0.6	7.50	0.7	0.25	0.8	43	39.65	1.6	40.5	1.75	1.25	3.8
9	8.15	0.6	8.45	0.7	0.28	0.8	45	41.60	1.6	42.4	1.75	1.30	3.9
10	9.00	0.6	9.40	0.7	0.30	0.9	46	42.55	1.6	43.3	1.75	1.35	4.0
11	10.00	0.6	10.35	0.7	0.33	1.0	48	44.40	1.6	45.2	1.75	1.40	4.2
12	10.85	0.6	11.35	0.7	0.33	1.0	50	46.20	1.6	47.2	1.75	1.40	4.2
13	11.90	0.9	12.30	1.0	0.35	1.0	52	48.40	2.0	49.1	2.15	1.45	4.3
14	12.90	0.9	13.25	1.0	0.38	1.2	54	49.9	2.0	51.0	2.15	1.50	4.5
15	13.80	0.9	14.15	1.0	0.43	1.3	55	50.6	2.0	51.8	2.15	1.60	4.8
16	14.70	0.9	15.10	1.0	0.45	1.4	57	52.9	2.0	53.8	2.15	1.60	4.8
17	15.75	0.9	16.10	1.0	0.45	1.4	58	53.6	2.0	54.7	2.15	1.65	4.9
18	16.65	1.1	17.00	1.2	0.50	1.5	60	55.8	2.0	56.7	2.15	1.65	4.9
19	17.60	1.1	17.95	1.2	0.53	1.6	62	57.3	2.0	58.6	2.15	1.70	5.1
20	18.35	1.1	18.85	1.2	0.58	1.7	65	60.4	2.0	61.6	2.15	1.70	5.1
21	19.40	1.1	19.80	1.2	0.60	1.8	68	63.1	2.0	64.5	2.15	1.75	5.3
22	20.30	1.1	20.70	1.2	0.65	1.9	70	64.6	2.4	66.4	2.55	1.80	5.4
23	21.25	1.1	21.65	1.2	0.67	2.0	72	66.6	2.4	68.3	2.55	1.85	5.5
24	22.20	1.1	22.60	1.2	0.70	2.1	75	69.0	2.4	71.2	2.55	1.90	5.7
25	23.10	1.1	23.50	1.2	0.75	2.3	78	72.0	2.4	74.0	2.55	2.00	6.0
26	24.05	1.1	24.50	1.2	0.75	2.3	80	74.2	2.4	75.9	2.55	2.05	6.1
27	24.95	1.3	25.45	1.4	0.78	2.3	82	76.4	2.4	77.8	2.55	2.10	6.3
28	25.80	1.3	26.40	1.4	0.80	2.4	85	78.6	2.4	80.6	2.55	2.20	6.6
30	27.90	1.3	28.35	1.4	0.83	2.5	88	81.4	2.8	83.5	2.95	2.25	6.7
32	29.60	1.3	30.20	1.4	0.90	2.7	90	83.2	2.8	85.4	2.95	2.30	6.9
34	31.40	1.3	32.00	1.4	1.00	3.0	95	88.1	2.8	90.2	2.95	2.40	7.2
35	32.30	1.3	32.90	1.4	1.05	3.1	100	92.5	2.8	95.0	2.95	2.50	7.5

All dimensions are in millimeters. Sizes –4, –5, and –6 are available in beryllium copper only.

These rings are designated by series symbol and shaft diameter, thus: for a 4 mm diameter shaft, 3AM1-4; for a 20 mm diameter shaft, 3AM1-20; etc.

Ring Free Diameter Tolerances: For ring sizes –4 through –6, +0.05, –0.10 mm; for sizes –7 through –12, +0.05, –0.15 mm; for sizes –13 through – 26, +0.15, –0.25 mm; for sizes –27 through –38, +0.25, –0.40 mm; for sizes –40 through –50, +0.35, –0.50 mm; for sizes –52 through –62, +0.35, –0.65 mm; and for sizes –65 through –100, +0.50, –0.75 mm.

Groove Diameter Tolerances: For ring sizes –4 through –6, –0.08 mm; for sizes –7 through –10, –0.10 mm; for sizes –11 through –15, –0.12 mm; for sizes –16 through –26, –0.15 mm; for sizes –27 through – 36, –0.20 mm; for sizes –38 through –55, –0.30 mm; and for sizes –57 through –100, –0.40 mm.

Groove Diameter F.I.M. (full indicator movement) or maximum allowable deviation of concentricity between groove and shaft: For ring sizes –4 through –6, 0.03 mm; for ring sizes –7 through –12, 0.05 mm; for sizes –13 through –28, 0.10 mm; for sizes –30 through –55, 0.15 mm; and for sizes –57 through – 00, 0.20 mm.

Groove Width Tolerances: For ring size –4, +0.05 mm; for sizes –5 and –6, +0.10 mm, for sizes –7 through –38, +0.15 mm; and for sizes –40 through – 100, +0.20 mm.

Groove Maximum Bottom Radii, R: For ring sizes –4 through –6, none; for sizes –7 through –18, 0.1 mm; for sizes –19 through –30, 0.2 mm; for sizes –32 through –50, 0.3 mm; and for sizes –52 through –100, 0.4 mm. For manufacturing details not shown, including materials, see ANSI B27.7M-1977, R1983.

Table 2. American National Standard Metric Tapered Retaining Rings — Basic Internal Series — 3BM1 *ANSI B27.7M-1977, R1983*

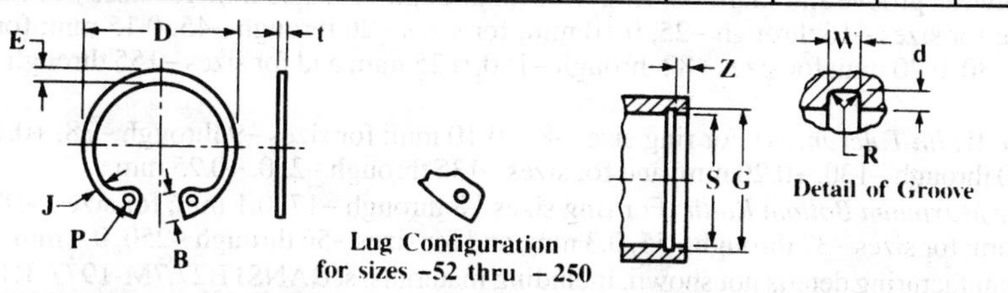

Lug Configuration for sizes –52 thru – 250

Detail of Groove

Shaft Diam.	Ring		Groove				Shaft Diam.	Ring		Groove			
	Free Diam.	Thickness	Diam.	Width	Depth	Edge Margin		Free Diam.	Thickness	Diam.	Width	Depth	Edge Margin
S	*D*	*t*	*G*	*W*	*d* ref	*Z* min	*S*	*D*	*t*	*G*	*W*	*d* ref	*Z* min
8	8.80	0.4	8.40	0.5	0.2	0.6	65	72.2	2.4	69.0	2.55	2.00	6.0
9	10.00	0.6	9.45	0.7	0.23	0.7	68	75.7	2.4	72.2	2.55	2.10	6.3
10	11.10	0.6	10.50	0.7	0.25	0.8	70	77.5	2.4	74.4	2.55	2.20	6.6
11	12.20	0.6	11.60	0.7	0.3	0.9	72	79.6	2.4	76.5	2.55	2.25	6.7
12	13.30	0.6	12.65	0.7	0.33	1.0	75	83.3	2.4	79.7	2.55	2.35	7.1
13	14.25	0.9	13.70	1.0	0.35	1.1	78	86.8	2.8	82.8	2.95	2.40	7.2
14	15.45	0.9	14.80	1.0	0.40	1.2	80	89.1	2.8	85.0	2.95	2.50	7.5
15	16.60	0.9	15.85	1.0	0.43	1.3	82	91.1	2.8	87.2	2.95	2.60	7.8
16	17.70	0.9	16.90	1.0	0.45	1.4	85	94.4	2.8	90.4	2.95	2.70	8.1
17	18.90	0.9	18.00	1.0	0.50	1.5	88	97.9	2.8	93.6	2.95	2.80	8.4
18	20.05	0.9	19.05	1.0	0.53	1.6	90	100.0	2.80	95.7	2.95	2.85	8.6
19	21.10	0.9	20.10	1.0	0.55	1.7	92	102.2	2.8	97.8	2.95	2.90	8.7
20	22.25	0.9	21.15	1.0	0.57	1.7	95	105.6	2.8	101.0	2.95	3.00	9.0
21	23.30	0.9	22.20	1.0	0.60	1.8	98	109.0	2.8	104.2	2.95	3.10	9.3
22	24.40	1.1	23.30	1.2	0.65	1.9	100	110.7	2.8	106.3	2.95	3.15	9.5
23	25.45	1.1	24.35	1.2	0.67	2.0	102	112.4	2.8	108.4	2.95	3.20	9.6
24	26.55	1.1	25.4	1.2	0.70	2.1	105	115.8	2.8	111.5	2.95	3.25	9.8
25	27.75	1.1	26.6	1.2	0.80	2.4	108	119.2	2.8	114.6	2.95	3.30	9.9
26	28.85	1.1	27.7	1.2	0.85	2.6	110	120.8	2.8	116.7	2.95	3.35	10.1
27	29.95	1.3	28.8	1.4	0.90	2.7	115	126.0	2.8	121.9	2.95	3.45	10.4
28	31.10	1.3	29.8	1.4	0.90	2.7	120	132.4	2.8	127.0	2.95	3.50	10.5
30	33.40	1.3	31.9	1.4	0.95	2.9	125	137.1	2.8	132.1	2.95	3.55	10.7
32	35.35	1.3	33.9	1.4	0.95	2.9	130	142.5	2.8	137.2	2.95	3.60	10.8
34	37.75	1.3	36.1	1.4	1.05	3.2	135	148.5	3.2	142.3	3.40	3.65	11.0
35	38.75	1.3	37.2	1.4	1.10	3.3	140	154.1	3.2	147.4	3.40	3.70	11.1
36	40.00	1.3	38.3	1.4	1.15	3.5	145	159.5	3.2	152.5	3.40	3.75	11.3
37	41.05	1.3	39.3	1.4	1.15	3.5	150	164.5	3.2	157.6	3.40	3.80	11.4
38	42.15	1.3	40.4	1.4	1.20	3.6	155	168.8	3.2	162.7	3.40	3.85	11.6
40	44.25	1.6	42.4	1.75	1.20	3.6	160	175.1	4.0	167.8	4.25	3.90	11.7
42	46.60	1.6	44.5	1.75	1.25	3.7	165	180.3	4.0	172.9	4.25	3.95	11.9
45	49.95	1.6	47.6	1.75	1.30	3.9	170	185.6	4.0	178.0	4.25	4.00	12.0
46	51.05	1.6	48.7	1.75	1.35	4.0	175	191.3	4.0	183.2	4.25	4.10	12.3
47	52.15	1.6	49.8	1.75	1.40	4.2	180	196.6	4.0	188.4	4.25	4.20	12.6
48	53.30	1.6	50.9	1.75	1.45	4.3	185	202.7	4.8	193.6	5.10	4.30	12.9
50	55.35	1.6	53.1	1.75	1.55	4.6	190	207.7	4.8	198.8	5.10	4.40	13.2
52	57.90	2.0	55.3	2.15	1.65	5.0	200	217.8	4.8	209.0	5.10	4.50	13.5
55	61.10	2.0	58.4	2.15	1.70	5.1	210	230.3	4.8	219.4	5.10	4.70	14.1
57	63.25	2.0	60.5	2.15	1.75	5.3	220	240.5	4.8	230.0	5.10	5.00	15.0
58	64.4	2.0	61.6	2.15	1.80	5.4	230	251.4	4.8	240.6	5.10	5.30	15.9
60	66.8	2.0	63.8	2.15	1.90	5.7	240	262.3	4.8	251.0	5.10	5.50	16.5
62	68.6	2.0	65.8	2.15	1.90	5.7	250	273.3	4.8	261.4	5.10	5.70	17.1
63	69.9	2.0	66.9	2.15	1.95	5.9	…	…	…	…	…	…	…

All dimensions are in millimeters.

These rings are designated by series symbol and shaft diameter, thus: for a 9 mm diameter shaft, 3BM1-9; for a 22 mm diameter shaft, 3BM1-22; etc.

Ring Free Diameter Tolerances: For ring sizes –8 through –20, +0.25, –0.13 mm; for sizes –21 through –26, +0.40, –0.25 mm; for sizes –27 through –38, +0.65, –0.50 mm; for sizes –40 through –50, +0.90, –0.65 mm; for sizes –52 through –75, +1.00, –0.75 mm; for sizes –78 through –92, +1.40,

−1.40 mm; for sizes −95 through −155, +1.65, −1.65 mm; for sizes −160 through −180, +2.05, −2.05 mm; and for sizes −185 through −250, +2.30, −2.30 mm.

Groove Diameter Tolerances: For ring sizes −8 and −9, +0.06 mm; for sizes −10 through −18, +0.10 mm; for sizes −19 through − 28, +0.15 mm; for sizes −30 through −50, +0.20 mm; for sizes − 52 through −98, +0.30; for sizes −100 through −160, +0.40 mm; and for sizes −165 through −250, +0.50 mm.

Groove Diameter F.I.M. (full indicator movement) or maximum allowable deviation of concentricity between groove and shaft: For ring sizes −8 through −10, 0.03 mm; for sizes −11 through −15, 0.05 mm; for sizes −16 through −25, 0.10 mm; for sizes −26 through −45, 0.15 mm; for sizes −46 through −80, 0.20 mm; for sizes −82 through −150, 0.25 mm; and for sizes −155 through −250, 0.30 mm.

Groove Width Tolerances: For ring size −8, +0.10 mm; for sizes −9 through −38, +0.15 mm; for sizes −40 through −130, +0.20 mm; and for sizes −135 through −250, +0.25 mm.

Groove Maximum Bottom Radii: For ring sizes −8 through −17, 0.1 mm; for sizes −18 through − 30, 0.2 mm; for sizes −32 through −55, 0.3 mm; and for sizes −56 through −250, 0.4 mm.

For manufacturing details not shown, including materials, see ANSI B27.7M-1977, R1983.

Table 3. American National Standard Metric Reduced Cross Section Retaining Rings — E Ring External Series —3CM1 *ANSI B27.7M-1977, R1983*

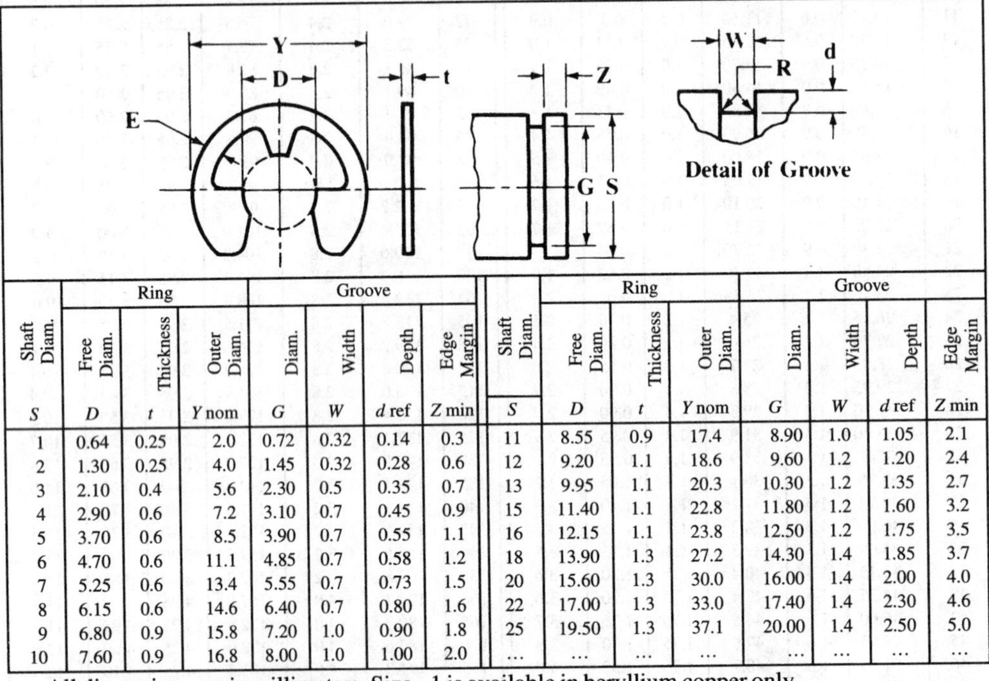

Shaft Diam.	Ring			Groove				Shaft Diam.	Ring			Groove			
	Free Diam.	Thickness	Outer Diam.	Diam.	Width	Depth	Edge Margin		Free Diam.	Thickness	Outer Diam.	Diam.	Width	Depth	Edge Margin
S	D	t	Y nom	G	W	d ref	Z min	S	D	t	Y nom	G	W	d ref	Z min
1	0.64	0.25	2.0	0.72	0.32	0.14	0.3	11	8.55	0.9	17.4	8.90	1.0	1.05	2.1
2	1.30	0.25	4.0	1.45	0.32	0.28	0.6	12	9.20	1.1	18.6	9.60	1.2	1.20	2.4
3	2.10	0.4	5.6	2.30	0.5	0.35	0.7	13	9.95	1.1	20.3	10.30	1.2	1.35	2.7
4	2.90	0.6	7.2	3.10	0.7	0.45	0.9	15	11.40	1.1	22.8	11.80	1.2	1.60	3.2
5	3.70	0.6	8.5	3.90	0.7	0.55	1.1	16	12.15	1.1	23.8	12.50	1.2	1.75	3.5
6	4.70	0.6	11.1	4.85	0.7	0.58	1.2	18	13.90	1.3	27.2	14.30	1.4	1.85	3.7
7	5.25	0.6	13.4	5.55	0.7	0.73	1.5	20	15.60	1.3	30.0	16.00	1.4	2.00	4.0
8	6.15	0.6	14.6	6.40	0.7	0.80	1.6	22	17.00	1.3	33.0	17.40	1.4	2.30	4.6
9	6.80	0.9	15.8	7.20	1.0	0.90	1.8	25	19.50	1.3	37.1	20.00	1.4	2.50	5.0
10	7.60	0.9	16.8	8.00	1.0	1.00	2.0	…	…	…	…	…	…	…	…

All dimensions are in millimeters. Size −1 is available in beryllium copper only.

These rings are designated by series symbol and shaft diameter, thus: for a 2 mm diameter shaft, 3CM1-2; for a 13 mm shaft, 3CMI -13; etc.

Ring Free Diameter Tolerances: For ring sizes − 1 through −7, +0.03, −0.08 mm; for sizes −8 through −13, +0.05, −0.10 mm; and for sizes −15 through −25, +0.10, −0.15 mm.

Groove Diameter Tolerances: For ring sizes −1 and −2, −0.05 mm; for sizes −3 through −6, −0.08; for sizes −7 through −11, −0.10 mm; for sizes −12 through −18, − 0.15 mm; and for sizes −20 through −25, − 0.20 mm.

Groove Diameter F.I.M. (Full Indicator Movement) or maximum allowable deviation of concentricity between groove and shaft: For ring sizes −1 through −3, 0.04 mm; for −4 through −6, 0.05 mm; for −7 through −10, 0.08 mm; for −11 through −25, 0.10 mm.

Groove Width Tolerances: For ring sizes − 1 and −2, +0.05 mm; for size −3, +0.10 mm; and for sizes −4 through − 25, +0.15 mm.

Groove Maximum Bottom Radii: For ring sizes −1 and −2, 0.05 mm; for −3 through −7, 0.15 mm; for −8 through −13, 0.25 mm; and for −15 through −25, 0.4 mm.

For manufacturing details not shown, including materials, see ANSI B27.7M-1977, R1983.

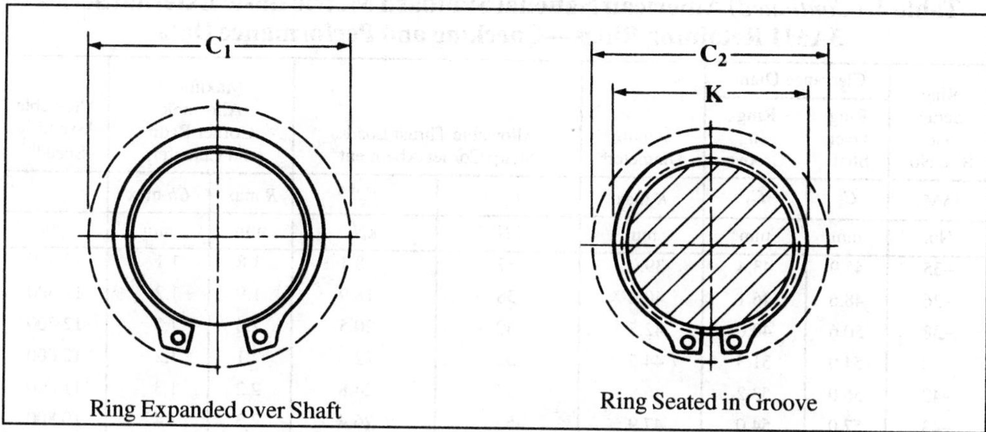

Ring Expanded over Shaft

Ring Seated in Groove

Table 4. American National Standard Metric Basic External Series 3AM1 Retaining Rings—Checking and Performance Data

Ring Series and Size No.	Clearance Diam. Ring Over Shaft	Ring in Groove	Gaging Diameter[a]	Allowable Thrust Loads Sharp Corner Abutment		Maximum Allowable Corner Radii and Chamfers		Allowable Assembly Speed[b]
3AM1	C_1	C_2	*K* max	P_r[c]	P_g[d]	*R* max	*Ch* max	…
No.	mm	mm	mm	kN	kN	mm	mm	rpm
−4[a]	7.0	6.8	4.90	0.6	0.2	0.35	0.25	70 000
−5[a]	8.2	7.9	5.85	1.1	0.3	0.35	0.25	70 000
−6[a]	9.1	8.8	6.95	1.4	0.4	0.35	0.25	70 000
−7	12.3	11.8	8.05	2.6	0.7	0.45	0.3	60 000
−8	13.6	13.0	9.15	3.1	1.0	0.5	0.35	55 000
−9	14.5	13.8	10.35	3.5	1.2	0.6	0.35	48 000
−10	15.5	14.7	11.50	3.9	1.5	0.7	0.4	42 000
−11	16.4	15.6	12.60	4.3	1.8	0.75	0.45	38 000
−12	17.4	16.6	13.80	4.7	2.0	0.8	0.45	34 000
−13	19.7	18.8	15.05	7.5	2.2	0.8	0.5	31 000
−14	20.7	19.7	15.60	8.1	2.6	0.9	0.5	28 000
−15	21.7	20.6	17.20	8.7	3.2	1.0	0.6	27 000
−16	22.7	21.6	18.35	9.3	3.5	1.1	0.6	25 000
−17	23.7	22.6	19.35	9.9	4.0	1.1	0.6	24 000
−18	26.2	25.0	20.60	16.0	4.4	1.2	0.7	23 000
−19	27.2	25.9	21.70	16.9	4.9	1.2	0.7	21 500
−20	28.2	26.8	22.65	17.8	5.7	1.2	0.7	20 000
−21	29.2	27.7	23.80	18.6	6.2	1.3	0.7	19 000
−22	30.3	28.7	24.90	19.6	7.0	1.3	0.8	18 500
−23	31.3	29.6	26.00	20.5	7.6	1.3	0.8	18 000
−24	34.1	32.4	27.15	21.4	8.2	1.4	0.8	17 500
−25	35.1	33.3	28.10	22.3	9.2	1.4	0.8	17 000
−26	36.0	34.2	29.25	23.2	9.6	1.5	0.9	16 500
−27	37.8	35.9	30.35	28.4	10.3	1.5	0.9	16 300
−28	38.8	36.9	31.45	28.4	11.0	1.6	1.0	15 800
−30	40.8	38.8	33.6	31.6	12.3	1.6	1.0	15 000
−32	42.8	40.7	35.9	33.6	14.1	1.7	1.0	14 800
−34	44.9	42.5	37.9	36	16.7	1.7	1.1	14 000

Table 4. *(Continued)* **American National Standard Metric Basic External Series 3AM1 Retaining Rings—Checking and Performance Data**

Ring Series and Size No.	Clearance Diam. Ring Over Shaft	Clearance Diam. Ring in Groove	Gaging Diameter[a]	Allowable Thrust Loads Sharp Corner Abutment		Maximum Allowable Corner Radii and Chamfers		Allowable Assembly Speed[b]
3AM1	C_1	C_2	*K* max	P_r[c]	P_g[d]	*R* max	*Ch* max	…
No.	mm	mm	mm	kN	kN	mm	mm	rpm
−35	45.9	43.4	39.0	37	18.1	1.8	1.1	13 500
−36	48.6	46.1	40.2	38	18.9	1.9	1.2	13 300
−38	50.6	48.0	42.5	40	20.5	2.0	1.2	12 700
−40	54.0	51.3	44.5	52	22.6	2.1	1.2	12 000
−42	56.0	53.2	46.9	54	24.8	2.2	1.3	11 000
−43	57.0	54.0	47.9	55	26.4	2.3	1.4	10 800
−45	59.0	55.9	50.0	58	28.8	2.3	1.4	10 000
−46	60.0	56.8	50.9	59	30.4	2.4	1.4	9 500
−48	62.4	59.1	53.0	62	33	2.4	1.4	8 800
−50	64.4	61.1	55.2	64	35	2.4	1.4	8 000
−52	67.6	64.1	57.4	84	37	2.5	1.5	7 700
−54	69.6	66.1	59.5	87	40	2.5	1.5	7 500
−55	70.6	66.9	60.4	89	44	2.5	1.5	7 400
−57	72.6	68.9	62.7	91	45	2.6	1.5	7 200
−58	73.6	69.8	63.6	93	46	2.6	1.6	7 100
−60	75.6	71.8	65.8	97	49	2.6	1.6	7 000
−62	77.6	73.6	67.9	100	52	2.7	1.6	6 900
−65	80.6	76.6	71.2	105	54	2.8	1.7	6 700
−68	83.6	79.5	74.5	110	58	2.9	1.7	6 500
−70	88.1	83.9	76.4	136	62	2.9	1.7	6 400
−72	90.1	85.8	78.5	140	65	2.9	1.7	6 200
−75	93.1	88.7	81.7	147	69	3.0	1.8	5 900
−78	95.4	92.1	84.6	151	76	3.0	1.8	5 600
−80	97.9	93.1	87.0	155	80	3.1	1.9	5 400
−82	100.0	95.1	89.0	159	84	3.2	1.9	5 200
−85	103.0	97.9	92.1	165	91	3.2	1.9	5 000
−88	107.0	100.8	95.1	199	97	3.2	1.9	4 800
−90	109.0	103.6	97.1	204	101	3.2	1.9	4 500
−95	114.0	108.6	102.7	215	112	3.4	2.1	4 350
−100	119.5	113.7	108.0	227	123	3.5	2.1	4 150

[a] For checking when ring is seated in groove.

[b] These values have been calculated for steel rings.

[c] These values apply to rings made from SAE 1060–1090 steels and PH 15-7 Mo stainless steel used on shafts hardened to R_c 50 minimum, with the exception of sizes −4, −5, and −6 which are supplied in beryllium copper only. Values for other sizes made from beryllium copper can be calculated by multiplying the listed values by 0.75. The values listed include a safety factor of 4.

[d] These values are for all standard rings used on low carbon steel shafts. They include a safety factor of 2.

Maximum allowable assembly loads with *R* max or *Ch* max are: For rings sizes −4, 0.2 kN; for sizes −5 and −6, 0.5 kN; for sizes −7 through −12, 2.1 kN; for sizes −13 through −17, 4.0 kN; for sizes −18 through −26, 6.0 kN; for sizes −27 through −38, 8.6 kN; for sizes −40 through − 50, 13.2 kN; for sizes −52 through −68, 22.0 kN; for sizes −70 through −85, 32 kN; and for sizes −88 through −100, 47 kN.

Source: Appendix to American National Standard ANSI B27.7M-1977, R1983.

Table 5. American National Standard Metric Basic Internal Series 3BMI Retaining Rings — Checking and Performance Data

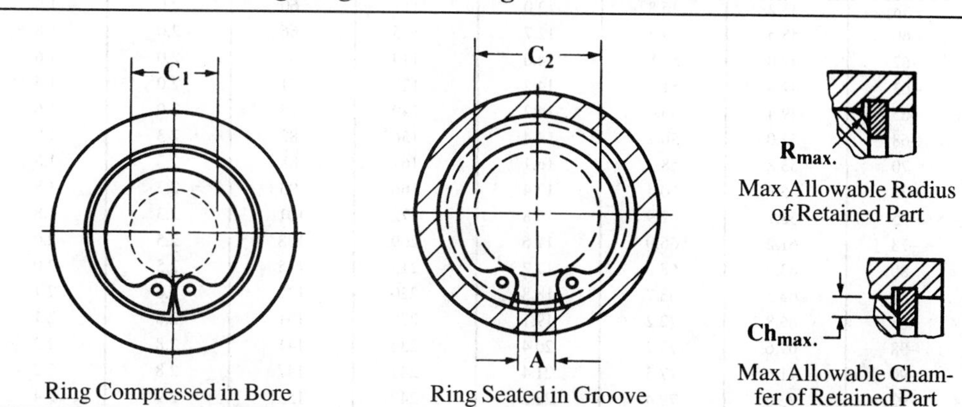

Ring Compressed in Bore | Ring Seated in Groove | Max Allowable Radius of Retained Part | Max Allowable Chamfer of Retained Part

Ring Series and Size No.	Clearance Diam. Ring in Bore	Clearance Diam. Ring in Groove	Gaging Diameter[a]	Allowable Thrust Loads Sharp Corner Abutment		Maximum Allowable Corner Radii and Chamfers	
3BMI	C_1	C_2	*A* min	P_r[b]	P_g[c]	*R* max	*Ch* max
No.	mm	mm	mm	kN	kN	mm	mm
−8	4.4	4.8	1.40	2.4	1.0	0.4	0.3
−9	4.6	5.0	1.50	4.4	1.2	0.5	0.35
−10	5.5	6.0	1.85	4.9	1.5	0.5	0.35
−11	5.7	6.3	1.95	5.4	2.0	0.6	0.4
−12	6.7	7.3	2.25	5.8	2.4	0.6	0.4
−13	6.8	7.5	2.35	8.9	2.6	0.7	0.5
−14	6.9	7.7	2.65	9.7	3.2	0.7	0.5
−15	7.9	8.7	2.80	10.4	3.7	0.7	0.5
−16	8.8	9.7	2.80	11.0	4.2	0.7	0.5
−17	9.8	10.8	3.35	11.7	4.9	0.75	0.6
−18	10.3	11.3	3.40	12.3	5.5	0.75	0.6
−19	11.4	12.5	3.40	13.1	6.0	0.8	0.65
−20	11.6	12.7	3.8	13.7	6.6	0.9	0.7
−21	12.6	13.8	4.2	14.5	7.3	0.9	0.7
−22	13.5	14.8	4.3	22.5	8.3	0.9	0.7
−23	14.5	15.9	4.9	23.5	8.9	1.0	0.8
−24	15.5	16.9	5.2	24.8	9.7	1.0	0.8
−25	16.5	18.1	6.0	25.7	11.6	1.0	0.8
−26	17.5	19.2	5.7	26.8	12.7	1.2	1.0
−27	17.4	19.2	5.9	33	14.0	1.2	1.0
−28	18.2	20.0	6.0	34	14.6	1.2	1.0
−30	20.0	21.9	6.0	37	16.5	1.2	1.0
−32	22.0	23.9	7.3	39	17.6	1.2	1.0
−34	24.0	26.1	7.6	42	20.6	1.2	1.0
−35	25.0	27.2	8.0	43	22.3	1.2	1.0
−36	26.0	28.3	8.3	44	23.9	1.2	1.0
−37	27.0	29.3	8.4	45	24.6	1.2	1.0
−38	28.0	30.4	8.6	46	26.4	1.2	1.0
−40	29.2	31.6	9.7	62	27.7	1.7	1.3
−42	29.7	32.2	9.0	65	30.2	1.7	1.3
−45	32.3	34.9	9.6	69	33.8	1.7	1.3
−46	33.3	36.0	9.7	71	36	1.7	1.3
−47	34.3	37.1	10.0	72	38	1.7	1.3
−48	35.0	37.9	10.5	74	40	1.7	1.3
−50	36.9	40.0	12.1	77	45	1.7	1.3
−52	38.6	41.9	11.7	99	50	2.0	1.6
−55	40.8	44.2	11.9	105	54	2.0	1.6
−57	42.2	45.7	12.5	109	58	2.0	1.6

Table 5. *(Continued)* **American National Standard Metric Basic Internal Series 3BMI Retaining Rings — Checking and Performance Data**

−58	43.2	46.8	13.0	111	60	2.0	1.6
−60	45.5	49.3	12.7	115	66	2.0	1.6
−62	47.0	50.8	14.0	119	68	2.0	1.6
−63	47.8	51.7	14.2	120	71	2.0	1.6
−65	49.4	53.4	14.2	149	75	2.0	1.6
−68	52.0	56.2	14.4	156	82	2.3	1.8
−70	53.8	58.2	16.1	161	88	2.3	1.8
−72	55.9	60.4	17.4	166	93	2.3	1.8
−75	58.2	62.9	16.8	172	101	2.3	1.8
−78	61.2	66.0	17.6	209	108	2.5	2.0
−80	63.0	68.0	17.2	215	115	2.5	2.0
−82	63.5	68.7	18.8	220	122	2.6	2.1
−85	66.8	72.2	19.1	228	131	2.6	2.1
−88	69.6	75.2	20.4	236	141	2.8	2.2
−90	71.6	77.3	21.4	241	147	2.8	2.2
−92	73.6	79.4	22.2	247	153	2.9	2.4
−95	76.7	82.7	22.6	255	164	3.0	2.5
−98	78.3	84.5	22.6	263	174	3.0	2.5
−100	80.3	86.6	24.1	269	181	3.1	2.5
−102	82.2	88.6	25.5	273	187	3.2	2.6
−105	85.1	91.6	26.0	281	196	3.3	2.6
−108	88.1	94.7	26.4	290	205	3.5	2.7
−110	88.4	95.1	27.5	295	212	3.6	2.8
−115	93.2	100.1	29.4	309	227	3.7	2.9
−120	98.2	105.2	27.2	321	241	3.9	3.1
−125	103.1	110.2	30.3	335	255	4.0	3.2
−130	108.0	115.2	31.0	349	269	4.0	3.2
−135	110.4	117.7	30.4	415	283	4.3	3.4
−140	115.3	122.7	30.4	429	298	4.3	3.4
−145	120.4	127.9	31.6	444	313	4.3	3.4
−150	125.3	132.9	33.5	460	327	4.3	3.4
−155	130.4	138.1	37.0	475	343	4.3	3.4
−160	133.8	141.6	35.0	613	359	4.5	3.6
−165	138.7	146.6	33.1	632	374	4.6	3.7
−170	143.6	151.6	38.2	651	390	4.6	3.7
−175	146.0	154.2	37.7	670	403	4.8	3.8
−180	151.4	159.8	39.0	690	434	5.0	4.0
−185	154.7	163.3	37.3	851	457	5.1	4.1
−190	159.5	168.3	35.0	873	480	5.3	4.3
−200	169.2	178.2	43.9	919	517	5.4	4.3
−210	177.5	186.9	40.6	965	566	5.8	4.6
−220	184.1	194.1	38.3	1000	608	6.1	4.9
−230	194.0	204.6	49.0	1060	686	6.3	5.1
−240	200.4	211.4	45.4	1090	725	6.6	5.3
−250	210.0	221.4	53.0	1150	808	6.7	5.4

[a] For checking when ring is seated in groove.

[b] These values apply to rings made from SAE 1060-1090 steels and PH 15-7 Mo stainless steel used in bores hardened to R_c 50 minimum. Values for rings made from beryllium copper can be calculated by multiplying the listed values by 0.75. The values listed include a safety factor of 4.

[c] These values are for standard rings used in low carbon steel bores. They include a safety factor of 2. Maximum allowable assembly loads for *R* max or *Ch* max are: For ring size −8, 0.8 kN; for sizes −9 through −12, 2.0 kN; for sizes −13 through −21, 4.0 kN; for sizes −22 through −26, 7.4 kN; for sizes −27 through −38, 10.8 kN; for sizes −40 through −50, 17.4 kN; for sizes −52 through −63, 27.4 kN; for size −65, 42.0 kN; for sizes −68 through −72, 39 kN; for sizes −75 through −130, 54 kN; for sizes −135 through −155, 67 kN; for sizes −160 through −180, 102 kN; and for sizes −185 through −250, 151 kN.

Source: Appendix to American National Standard ANSI B27.7M-1977, R1983.

Table 6. American National Standard Metric E-Type External Series 3CM1 Retaining Rings — Checking and Performance Data

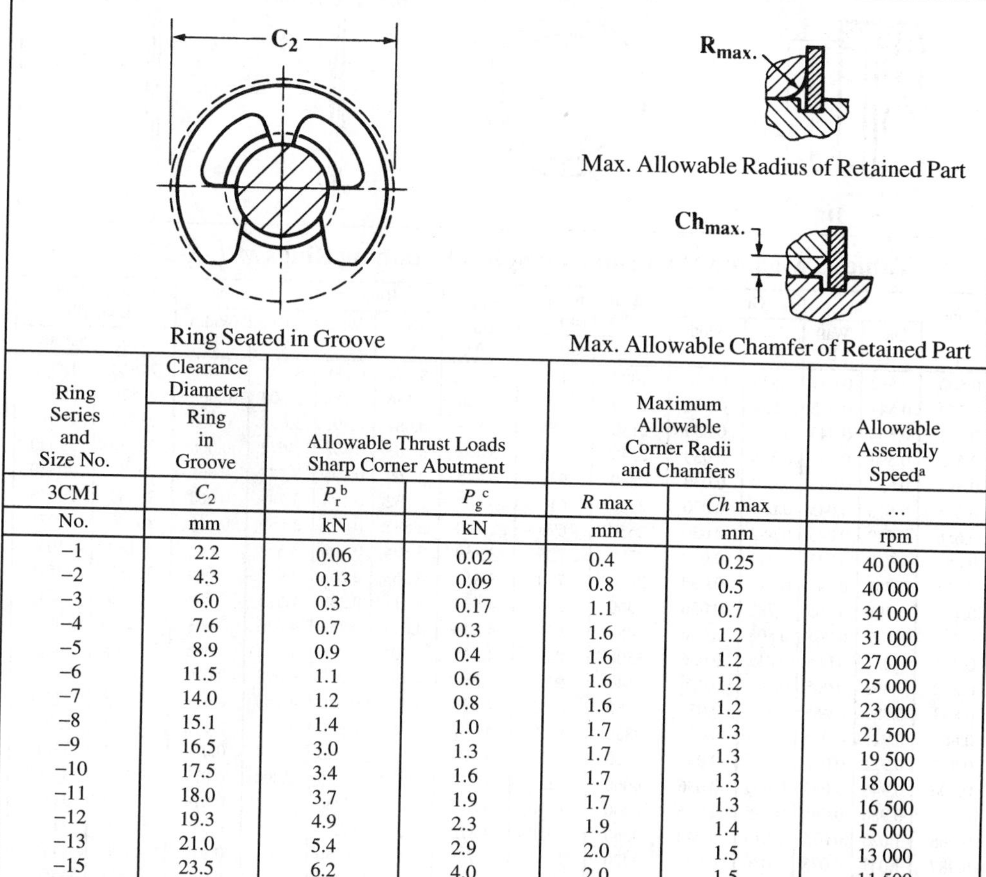

Ring Series and Size No.	Clearance Diameter: Ring in Groove	Allowable Thrust Loads Sharp Corner Abutment		Maximum Allowable Corner Radii and Chamfers		Allowable Assembly Speed[a]
3CM1	C_2	P_r[b]	P_g[c]	*R* max	*Ch* max	…
No.	mm	kN	kN	mm	mm	rpm
−1	2.2	0.06	0.02	0.4	0.25	40 000
−2	4.3	0.13	0.09	0.8	0.5	40 000
−3	6.0	0.3	0.17	1.1	0.7	34 000
−4	7.6	0.7	0.3	1.6	1.2	31 000
−5	8.9	0.9	0.4	1.6	1.2	27 000
−6	11.5	1.1	0.6	1.6	1.2	25 000
−7	14.0	1.2	0.8	1.6	1.2	23 000
−8	15.1	1.4	1.0	1.7	1.3	21 500
−9	16.5	3.0	1.3	1.7	1.3	19 500
−10	17.5	3.4	1.6	1.7	1.3	18 000
−11	18.0	3.7	1.9	1.7	1.3	16 500
−12	19.3	4.9	2.3	1.9	1.4	15 000
−13	21.0	5.4	2.9	2.0	1.5	13 000
−15	23.5	6.2	4.0	2.0	1.5	11 500
−16	24.5	6.6	4.5	2.0	1.5	10 000
−18	27.9	8.7	5.4	2.1	1.6	9 000
−20	30.7	9.8	6.5	2.2	1.7	8 000
−22	33.7	10.8	8.1	2.2	1.7	7 000
−25	37.9	12.2	10.1	2.4	1.9	5 000

[a] These values have been calculated for steel rings.

[b] These values apply to rings made from SAE 1060-1090 steels and PH 15-7 Mo stainless steel used on shafts hardened to R_c 50 minimum, with the exception of size −1 which is supplied in beryllium copper only. Values for other sizes made from beryllium copper can be calculated by multiplying the listed values by 0.75. The values listed include a safety factor of 4.

[c] These values apply to all standard rings used on low carbon steel shafts. They include a safety factor of 2.

Maximum allowable assembly loads with *R* max or *Ch* max are as follows:

Ring Size No.	Maximum Allowable Load,kN	Ring Size No.	Maximum AllowableLoad, kN	Ring Size No.	Maximum AllowableLoad, kN
−1	0.06	−8	1.4	−16	6.6
−2	0.13	−9	3.0	−18	8.7
−3	0.3	−10	3.4	−20	9.8
−4	0.7	−11	3.7	−22	10.8
−5	0.9	−12	4.9	−25	12.2
−6	1.1	−13	5.4	…	…
−7	1.2	−15	6.2	…	…

Source: Appendix to American National Standard ANSI B27.7M-1977, R1983.

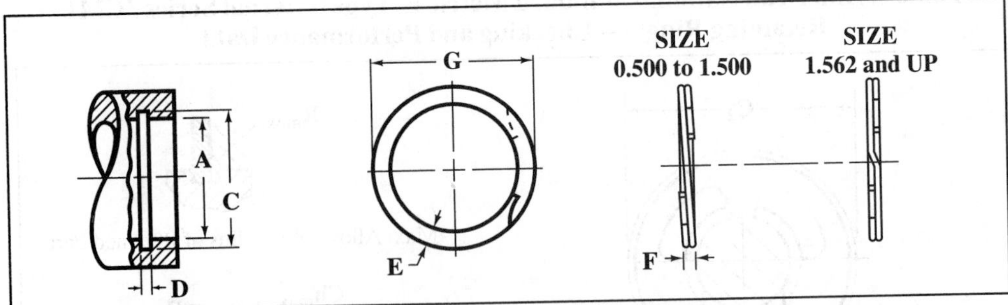

Table 7. Medium Duty Internal Spiral Retaining Rings *MIL-R-27426*

Bore Dia. A	Ring		Groove		Static Thrust Load (lb)		Bore Dia. A	Ring		Groove		Static Thrust Load (lb)	
	Dia. G	Wall E	Dia. C	Width D	Ring	Groove		Dia. G	Wall E	Dia. C	Width D	Ring	Groove
0.500	0.532	0.045	0.526	0.030	2000	405	3.437	3.574	0.188	3.543	0.068	27660	18240
0.512	0.544	0.045	0.538	0.030	2050	420	3.500	3.636	0.188	3.606	0.068	28170	18575
0.531	0.564	0.045	0.557	0.030	2130	455	3.543	3.684	0.198	3.653	0.068	28520	19515
0.562	0.594	0.045	0.588	0.030	2250	495	3.562	3.703	0.198	3.672	0.068	28670	19620
0.594	0.626	0.045	0.619	0.030	2380	535	3.625	3.769	0.198	3.737	0.068	29180	20330
0.625	0.658	0.045	0.651	0.030	2500	610	3.687	3.832	0.198	3.799	0.068	29680	20675
0.656	0.689	0.045	0.682	0.030	2630	670	3.740	3.885	0.198	3.852	0.068	30100	20975
0.687	0.720	0.045	0.713	0.030	2750	725	3.750	3.894	0.198	3.862	0.068	30180	21030
0.718	0.751	0.045	0.744	0.030	2870	790	3.812	3.963	0.208	3.930	0.068	30680	22525
0.750	0.790	0.065	0.782	0.036	3360	800	4.437	4.611	0.238	4.573	0.068	35710	30215
0.777	0.817	0.065	0.808	0.036	3480	835	4.500	4.674	0.238	4.636	0.068	36220	30645
0.781	0.821	0.065	0.812	0.036	3500	840	4.527	4.701	0.238	4.663	0.068	36440	30830
0.812	0.853	0.065	0.843	0.036	3640	915	4.562	4.737	0.238	4.698	0.079	36720	31065
0.843	0.889	0.065	0.880	0.036	3780	1155	4.625	4.803	0.250	4.765	0.079	43940	32420
0.866	0.913	0.065	0.903	0.036	3880	1250	4.687	4.867	0.250	4.827	0.079	44530	32855
0.875	0.922	0.065	0.912	0.036	3920	1250	4.724	4.903	0.250	4.864	0.079	44880	33115
0.906	0.949	0.065	0.939	0.036	4060	1335	4.750	4.930	0.250	4.890	0.079	45130	33300
0.938	0.986	0.065	0.975	0.036	4200	1430	4.812	4.993	0.250	4.952	0.079	45710	33735
0.968	1.025	0.075	1.015	0.042	4340	1950	4.875	5.055	0.250	5.015	0.079	46310	34175
0.987	1.041	0.075	1.030	0.042	4420	1865	4.921	5.102	0.250	5.061	0.079	46750	34495
1.000	1.054	0.075	1.043	0.042	4480	1910	4.937	5.122	0.250	5.081	0.079	46900	35595
1.023	1.078	0.075	1.066	0.042	5470	1660	5.000	5.185	0.250	5.144	0.079	47500	36050
1.031	1.084	0.075	1.074	0.042	5510	1650	5.118	5.304	0.250	5.262	0.079	48620	36905
1.062	1.117	0.075	1.104	0.042	5680	1745	5.125	5.311	0.250	5.269	0.079	48690	36955
1.093	1.147	0.075	1.135	0.042	5840	1820	5.250	5.436	0.250	5.393	0.079	49880	37590
1.125	1.180	0.075	1.167	0.042	6010	1935	5.375	5.566	0.250	5.522	0.079	51050	39565
1.156	1.210	0.075	1.198	0.042	6180	2020	5.500	5.693	0.250	5.647	0.079	52250	40485
1.188	1.249	0.085	1.236	0.048	7380	2115	5.511	5.703	0.250	5.658	0.079	52350	40565
1.218	1.278	0.085	1.266	0.048	7570	2195	5.625	5.818	0.250	5.772	0.079	53440	41405
1.250	1.312	0.085	1.298	0.048	7770	2510	5.708	5.909	0.250	5.861	0.079	54230	43730
1.281	1.342	0.085	1.329	0.048	7960	2425	5.750	5.950	0.250	5.903	0.079	54630	44050
1.312	1.374	0.085	1.360	0.048	8150	2532	5.875	6.077	0.250	6.028	0.079	55810	45010
1.343	1.408	0.085	1.395	0.048	8340	2875	5.905	6.106	0.250	6.058	0.079	56100	45240
1.375	1.442	0.095	1.427	0.048	8540	3070	6.000	6.202	0.312	6.153	0.079	57000	45965
1.406	1.472	0.095	1.458	0.048	8740	3180	6.125	6.349	0.312	6.297	0.094	69500	52750
1.437	1.504	0.095	1.489	0.048	8930	3330	6.250	6.474	0.312	6.422	0.094	70920	53825
1.456	1.523	0.095	1.508	0.048	9050	3410	6.299	6.524	0.312	6.471	0.094	71480	54250
1.468	1.535	0.095	1.520	0.048	9120	3460	6.375	6.601	0.312	6.547	0.094	72340	54905
1.500	1.567	0.095	1.552	0.048	9320	3605	6.500	6.726	0.312	6.672	0.094	73760	55980
1.562	1.634	0.108	1.617	0.056	10100	3590	6.625	6.863	0.312	6.807	0.094	75180	60375
1.574	1.649	0.108	1.633	0.056	10180	3640	6.692	6.931	0.312	6.874	0.094	75940	60985
1.625	1.701	0.108	1.684	0.056	10510	3875	6.750	6.987	0.312	6.932	0.094	76590	61515
1.653	1.730	0.108	1.712	0.056	10690	4020	6.875	7.114	0.312	7.057	0.094	78010	62655
1.687	1.768	0.118	1.750	0.056	10910	4510	7.000	7.239	0.312	7.182	0.094	79430	63790
1.750	1.834	0.118	1.813	0.056	11310	4895	7.086	7.337	0.312	7.278	0.094	80410	68125
1.813	1.894	0.118	1.875	0.056	11720	5080	7.125	7.376	0.312	7.317	0.094	80850	68500

Table 7. *(Continued)* **Medium Duty Internal Spiral Retaining Rings** *MIL-R-27426*

Bore Dia. A	Ring Dia. G	Ring Wall E	Groove Dia. C	Groove Width D	Static Thrust Load (lb) Ring	Static Thrust Load (lb) Groove	Bore Dia. A	Ring Dia. G	Ring Wall E	Groove Dia. C	Groove Width D	Static Thrust Load (lb) Ring	Static Thrust Load (lb) Groove
1.850	1.937	0.118	1.917	0.056	11960	5735	7.250	7.501	0.312	7.442	0.094	82270	69700
1.875	1.960	0.118	1.942	0.056	12120	5825	7.375	7.628	0.312	7.567	0.094	83690	70900
1.938	2.025	0.118	2.005	0.056	12530	6250	7.480	7.734	0.312	7.672	0.094	84880	71910
2.000	2.091	0.128	2.071	0.056	12930	7090	7.500	7.754	0.312	7.692	0.094	85110	72105
2.047	2.138	0.128	2.118	0.056	13230	7275	7.625	7.890	0.312	7.827	0.094	86520	77125
2.062	2.154	0.128	2.132	0.056	13330	7225	7.750	8.014	0.312	7.952	0.094	87940	78390
2.125	2.217	0.128	2.195	0.056	13740	7450	7.875	8.131	0.312	8.077	0.094	89360	79655
2.165	2.260	0.138	2.239	0.056	14000	8020	8.000	8.266	0.312	8.202	0.094	90780	80920
2.188	2.284	0.138	2.262	0.056	14150	8105	8.250	8.528	0.375	8.462	0.094	93620	87575
2.250	2.347	0.138	2.324	0.056	14550	8335	8.267	8.546	0.375	8.479	0.094	93810	87755
2.312	2.413	0.138	2.390	0.056	14950	9030	8.454	8.744	0.375	8.676	0.094	96040	89850
2.375	2.476	0.138	2.453	0.056	15350	9275	8.500	8.780	0.375	8.712	0.094	96450	90230
2.437	2.543	0.148	2.519	0.056	15760	10005	8.750	9.041	0.375	8.972	0.094	99290	97265
2.440	2.546	0.148	2.522	0.056	15780	10015	8.858	9.151	0.375	9.080	0.094	100520	98465
2.500	2.606	0.148	2.582	0.056	16160	10625	9.000	9.293	0.375	9.222	0.094	102130	100045
2.531	2.641	0.148	2.617	0.056	16360	10900	9.055	9.359	0.375	9.287	0.094	102750	105190
2.562	2.673	0.148	2.648	0.056	16560	11030	9.250	9.555	0.375	9.482	0.094	104960	107455
2.625	2.736	0.148	2.711	0.056	16970	11305	9.448	9.755	0.375	9.680	0.094	107210	109755
2.677	2.789	0.158	2.767	0.056	17310	12065	9.500	9.806	0.375	9.732	0.094	107800	110360
2.688	2.803	0.158	2.778	0.056	17380	12115	9.750	10.068	0.375	9.992	0.094	110640	118145
2.750	2.865	0.158	2.841	0.056	17780	12530	10.000	10.320	0.375	10.242	0.094	113470	121175
2.813	2.929	0.158	2.903	0.056	18190	12675	10.250	10.582	0.375	10.502	0.094	116310	129340
2.834	2.954	0.168	2.928	0.056	18320	13340	10.500	10.834	0.375	10.752	0.094	119150	132490
2.875	2.995	0.168	2.969	0.056	18590	13530	10.750	11.095	0.375	11.012	0.094	121980	141030
2.937	3.058	0.168	3.031	0.056	18990	13825	11.000	11.347	0.375	11.262	0.094	124820	144310
2.952	3.073	0.168	3.046	0.056	19090	13890	3.875	4.025	0.208	3.993	0.068	30680	22525
3.000	3.122	0.168	3.096	0.068	24150	14420	3.938	4.089	0.208	4.056	0.068	31700	23265
3.062	3.186	0.168	3.158	0.068	24640	14720	4.000	4.157	0.218	5.124	0.068	32190	24835
3.125	3.251	0.178	3.223	0.068	25150	15335	4.063	4.222	0.218	4.187	0.068	32700	25225
3.149	3.276	0.178	3.247	0.068	25340	15450	4.125	4.284	0.218	4.249	0.068	33200	25610
3.187	3.311	0.178	3.283	0.068	25650	15640	4.188	4.347	0.218	4.311	0.068	33710	25795
3.250	3.379	0.178	3.350	0.068	26160	16270	4.250	4.416	0.228	4.380	0.068	34210	27665
3.312	3.446	0.188	3.416	0.068	26660	17245	4.312	4.479	0.228	4.442	0.068	34710	28065
3.346	3.479	0.188	3.450	0.068	26930	17425	4.330	4.497	0.228	4.460	0.068	34850	28185
3.375	3.509	0.188	3.479	0.068	27160	17575	4.375	4.543	0.228	4.505	0.068	32210	28475

Source: Spirolox Retaining Rings, RR Series. All dimensions are in inches. Depth of groove $d = (C - A)/2$. Standard material: carbon spring steel (SAE 1070-1090).

Ring Thickness, F: For shaft sizes 0.500 through 0.718, 0.025; for sizes 0.750 through 0.938, 0.031; for sizes 0.968 through 1.156, 0.037; for sizes 1.188 through 1.500, 0.043; for sizes 1.562 through 2.952, 0.049; for sizes 3.000 through 4.562, 0.061; for sizes 4.625 through 6.000, 0.072; for sizes 6.125 through 11.000, 0.086.

Ring Free Diameter Tolerances: For housing sizes 0.500 through 1.031, +0.013, −0.000; for sizes 1.062 through 1.500, +0.015, −0.000; for sizes 1.562 through 2.047, +0.020, −0.000; for sizes 2.062 through 3.000, +0.025, −0.000; for sizes 3.062 through 4.063, +0.030, −0.000; for sizes 4.125 through 5.125, +0.035, −0.000; for sizes 5.250 through 6.125, +0.045, −0.000; for sizes 6.250 through 7.125, +0.055, −0.000; for sizes 7.250 through 11.000, +0.065, −0.000.

Ring Thickness Tolerances: Thickness indicated is for unplated rings; add 0.002 to upper thickness tolerance for plated rings. For housing sizes 0.500 through 1.500, ±0.002; for sizes 1.562 through 4.562, ±0.003; for sizes 4.625 through 11.000, ±0.004.

Groove Diameter Tolerances: For housing sizes 0.500 through 0.750, ±0.002; for sizes 0.777 through 1.031, ±0.003; for sizes 1.062 through 1.500, ±0.004; for sizes 1.562 through 2.047, ±0.005; for sizes 2.062 through 5.125, ±0.006; for sizes 5.250 through 6.000, ±0.007; for sizes 6.125 through 11.000, ±0.008.

Groove Width Tolerances: For housing sizes 0.500 through 1.156, +0.003, −0.000; for sizes 1.188 through 2.952, +0.004, −0.000; for sizes 3.000 through 6.000, +0.005, −0.000; for sizes 6.125 through 11.000, +0.006, −0.000.

Table 8. Medium Duty External Spiral Retaining Rings *MIL-R-27426*

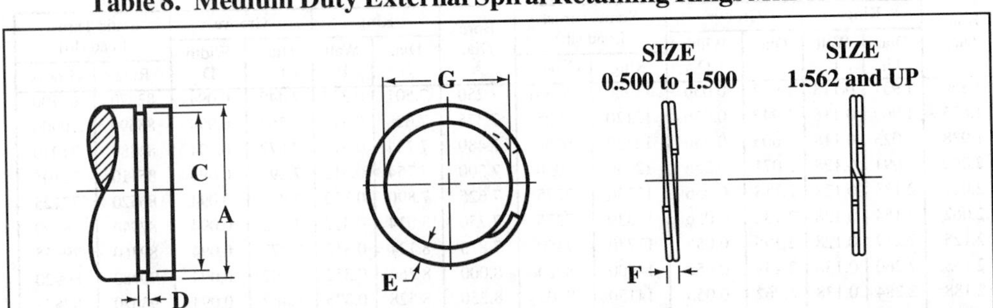

Shaft Dia. A	Ring Dia. G	Ring Wall E	Groove Dia. C	Groove Width D	Static Thrust Load (lb) Ring	Static Thrust Load (lb) Groove	Shaft Dia. A	Ring Dia. G	Ring Wall E	Groove Dia. C	Groove Width D	Static Thrust Load (lb) Ring	Static Thrust Load (lb) Groove
0.500	0.467	0.045	0.474	0.030	2000	550	3.343	3.210	0.188	3.239	0.068	26910	17410
0.531	0.498	0.045	0.505	0.030	2130	640	3.375	3.242	0.188	3.271	0.068	27160	17570
0.551	0.518	0.045	0.525	0.030	2210	700	3.437	3.301	0.188	3.331	0.068	27660	18240
0.562	0.529	0.045	0.536	0.030	2250	730	3.500	3.363	0.188	3.394	0.068	28170	18580
0.594	0.561	0.045	0.569	0.030	2380	740	3.543	3.402	0.198	3.433	0.068	28520	19510
0.625	0.585	0.055	0.594	0.030	2500	970	3.562	3.422	0.198	3.452	0.068	28670	19620
0.656	0.617	0.055	0.625	0.030	2630	1020	3.625	3.483	0.198	3.515	0.068	29180	19970
0.669	0.629	0.055	0.638	0.030	2680	1040	3.687	3.543	0.198	3.575	0.068	29680	20680
0.687	0.647	0.055	0.656	0.030	2750	1060	3.740	3.597	0.198	3.628	0.068	30100	20970
0.718	0.679	0.055	0.687	0.030	2870	1110	3.750	3.606	0.198	3.638	0.068	30180	21030
0.750	0.710	0.065	0.719	0.036	3360	1100	3.812	3.668	0.198	3.700	0.068	30680	21380
0.781	0.741	0.065	0.750	0.036	3500	1210	3.875	3.724	0.208	3.757	0.068	31190	22890
0.812	0.771	0.065	0.781	0.036	3640	1260	3.938	3.784	0.208	3.820	0.068	31700	23270
0.843	0.803	0.065	0.812	0.036	3780	1310	4.000	3.842	0.218	3.876	0.068	32190	24840
0.875	0.828	0.065	0.838	0.036	3920	1620	4.063	3.906	0.218	3.939	0.068	32700	25230
0.906	0.860	0.065	0.869	0.036	4060	1680	4.125	3.967	0.218	4.000	0.068	33200	25820
0.937	0.889	0.065	0.900	0.036	4200	1740	4.134	3.975	0.218	4.010	0.068	33270	25670
0.968	0.916	0.075	0.925	0.042	5180	2080	4.188	4.030	0.218	4.058	0.068	33710	27260
0.984	0.930	0.075	0.941	0.042	5260	2120	4.250	4.084	0.228	4.120	0.068	34210	27660
1.000	0.946	0.075	0.957	0.042	5350	2150	4.312	4.147	0.218	4.182	0.068	34710	28070
1.023	0.968	0.075	0.980	0.042	5470	2200	4.331	4.164	0.218	4.200	0.068	34860	28410
1.031	0.978	0.075	0.988	0.042	5510	2220	4.375	4.208	0.218	4.245	0.068	35210	28480
1.062	1.007	0.075	1.020	0.042	5680	2230	4.437	4.271	0.218	4.307	0.068	35710	28880
1.093	1.040	0.075	1.051	0.042	5840	2300	4.500	4.326	0.238	4.364	0.068	36220	30640
1.125	1.070	0.075	1.083	0.042	6010	2370	4.562	4.384	0.250	4.422	0.079	43340	31980
1.156	1.102	0.075	1.114	0.042	6180	2430	4.625	4.447	0.250	4.485	0.079	43940	32420
1.188	1.127	0.085	1.140	0.048	7380	2850	4.687	4.508	0.250	4.457	0.079	44530	32860
1.218	1.159	0.085	1.170	0.048	7570	2930	4.724	4.546	0.250	4.584	0.079	44880	33120
1.250	1.188	0.085	1.202	0.048	7770	3000	4.750	4.571	0.250	4.610	0.079	45130	33300
1.281	1.221	0.085	1.233	0.048	7960	3080	4.812	4.633	0.250	4.672	0.079	45710	33730
1.312	1.251	0.095	1.264	0.048	8150	3150	4.875	4.695	0.250	4.735	0.079	46310	34170
1.343	1.282	0.095	1.295	0.048	8340	3230	4.937	4.757	0.250	4.797	0.079	46900	34610
1.375	1.308	0.095	1.323	0.048	8540	3580	5.000	4.820	0.250	4.856	0.079	47500	36050
1.406	1.340	0.095	1.354	0.048	8740	3660	5.118	4.934	0.250	4.974	0.079	48620	36900
1.437	1.370	0.095	1.385	0.048	8930	3740	5.125	4.939	0.250	4.981	0.079	48690	36950
1.468	1.402	0.095	1.416	0.048	9120	3820	5.250	5.064	0.250	5.107	0.079	49880	37590
1.500	1.433	0.095	1.448	0.048	9320	3910	5.375	5.187	0.250	5.228	0.079	51060	39560
1.562	1.490	0.108	1.507	0.056	10100	4300	5.500	5.308	0.250	5.353	0.079	52250	40480
1.575	1.503	0.108	1.520	0.056	10190	4340	5.511	5.320	0.250	5.364	0.079	52350	40560
1.625	1.549	0.108	1.566	0.056	10510	4800	5.625	5.433	0.250	5.478	0.079	53440	41400

Table 8. *(Continued)* **Medium Duty External Spiral Retaining Rings** *MIL-R-27426*

1.687	1.610	0.118	1.628	0.056	10910	4980	5.750	5.550	0.250	5.597	0.079	54630	44050
1.750	1.673	0.118	1.691	0.056	11310	5170	5.875	5.674	0.250	5.722	0.079	55810	45010
1.771	1.690	0.118	1.708	0.056	11450	5590	5.905	5.705	0.250	5.752	0.079	56100	45240
1.813	1.730	0.118	1.749	0.056	11720	5810	6.000	5.798	0.250	5.847	0.079	57000	45970
1.875	1.789	0.128	1.808	0.056	12120	6290	6.125	5.903	0.312	5.953	0.094	69500	52750
1.938	1.844	0.128	1.861	0.056	12530	7470	6.250	6.026	0.312	6.078	0.094	70920	53830
1.969	1.882	0.128	1.902	0.056	12730	6610	6.299	6.076	0.312	6.127	0.094	71480	54250
2.000	1.909	0.128	1.992	0.056	12930	7110	6.375	6.152	0.312	6.203	0.094	72340	54900
2.062	1.971	0.128	2.051	0.056	13330	7870	6.500	6.274	0.312	6.328	0.094	73760	55980
2.125	2.029	0.128	2.082	0.056	13740	7990	6.625	6.390	0.312	6.443	0.094	75180	60380
2.156	2.060	0.138	2.091	0.056	13940	8020	6.750	6.513	0.312	6.568	0.094	76590	61515
2.188	2.070	0.138	2.113	0.056	14150	8220	6.875	6.638	0.312	6.693	0.094	78010	62650
2.250	2.092	0.138	2.176	0.056	14550	8340	7.000	6.761	0.312	6.818	0.094	79430	63790
2.312	2.153	0.138	2.234	0.056	14950	9030	7.125	6.877	0.312	6.933	0.094	80850	68500
2.362	2.211	0.138	2.284	0.056	15270	9230	7.250	6.999	0.312	7.058	0.094	82270	69700
2.375	2.273	0.138	2.297	0.056	15350	9280	7.375	7.125	0.312	7.183	0.094	83690	70900
2.437	2.331	0.148	2.355	0.056	15760	10000	7.500	7.250	0.312	7.308	0.094	85110	72100
2.500	2.394	0.148	2.418	0.056	16160	10260	7.625	7.363	0.312	7.423	0.094	86520	77120
2.559	2.449	0.148	2.473	0.056	16540	11020	7.750	7.486	0.312	7.548	0.094	87940	78390
2.562	2.452	0.148	2.476	0.056	16560	11030	7.875	7.611	0.312	7.673	0.094	89360	79650
2.625	2.514	0.148	2.539	0.056	16970	11300	8.000	7.734	0.312	7.798	0.094	90780	80920
2.688	2.572	0.158	2.597	0.056	17380	12250	8.250	7.972	0.375	8.038	0.094	93620	87580
2.750	2.635	0.158	2.660	0.056	17780	12390	8.500	8.220	0.375	8.288	0.094	96450	90230
2.813	2.696	0.168	2.722	0.056	18190	12820	8.750	8.459	0.375	8.528	0.094	99290	97270
2.875	2.755	0.168	2.781	0.056	18590	13530	9.000	8.707	0.375	8.778	0.094	102130	100050
2.937	2.817	0.168	2.843	0.056	18990	13820	9.250	8.945	0.375	9.018	0.094	104960	107560
2.952	2.831	0.168	2.858	0.056	19090	13890	9.500	9.194	0.375	9.268	0.094	107800	110360
3.000	2.877	0.168	2.904	0.068	24150	14420	9.750	9.432	0.375	9.508	0.094	110640	118150
3.062	2.938	0.168	2.966	0.068	24640	14720	10.000	9.680	0.375	9.758	0.094	113470	121180
3.125	3.000	0.178	3.027	0.068	25150	15335	10.250	9.918	0.375	9.998	0.094	116310	129340
3.149	3.023	0.178	3.051	0.068	25340	15450	10.500	10.166	0.375	10.248	0.094	119150	132490
3.187	3.061	0.178	3.089	0.068	25650	15640	10.750	10.405	0.375	10.488	0.094	121980	141030
3.250	3.121	0.178	3.150	0.068	26160	16270	11.000	10.653	0.375	10.738	0.094	124820	144310
3.312	3.180	0.188	3.208	0.068	26660	17250							

Source: Spirolox Retaining Rings, RS Series. All dimensions are in inches. Depth of groove $d = (A - C)/2$. Standard material: carbon spring steel (SAE 1070–1090).

Ring Thickness, F: For shaft sizes 0.500 through 0.718, 0.025; for sizes 0.750 through 0.937, 0.031; for sizes 0.968 through 1.156, 0.037; for sizes 1.188 through 1.500, 0.043; for sizes 1.562 through 2.952, 0.049; for sizes 3.000 through 4.500, 0.061; for sizes 4.562 through 6.000, 0.072; for sizes 6.125 through 11.000, 0.086.

Ring Free Diameter Tolerances: For shaft sizes 0.500 through 1.031, +0.000, + 0.000, −0.013; for sizes 1.062 through 1.500, +0.000, −0.015; for sizes 1.562 through 2.125, +0.000, −0.020; for sizes 2.156 through 2.688, +0.000, −0.025; for sizes 2.750 through 3.437, +0.000, −0.030; for sizes 3.500 through 5.125, +0.000, −0.040; for sizes 5.250 through 6.125, +0.000, −0.050; for sizes 6.250 through 7.375, +0.000, −0.060; for sizes 7.500 through 11.000, +0.000, −0.070.

Ring Thickness Tolerances: Thickness indicated is for unplated rings; add 0.002 to upper tolerance for plated rings. For shaft sizes 0.500 through 1.500, ± 0.002; for sizes 1.562 through 4.500, ± 0.003; for sizes 4.562 through 11.000, ± 0.004.

Groove Diameter Tolerances: For shaft sizes 0.500 through 0.562, ±0.002; for sizes 0.594 through 1.031, ± 0.003; for sizes 1.062 through 1.500, ± 0.004; for sizes 1.562 through 2.000, ± 0.005; for sizes 2.062 through 5.125, ±0.006; for sizes 5.250 through 6.000, ± 0.007; for sizes 6.125 through 11.000, ± 0.008.

Groove Width Tolerances: For shaft sizes 0.500 through 1.156, +0.003, −0.000; for sizes 1.188 through 2.952, +0.004, −0.000; for sizes 3.000 through 6.000, +0.005, −0.000; for sizes 6.125 through 11.000, +0.006, −0.000.

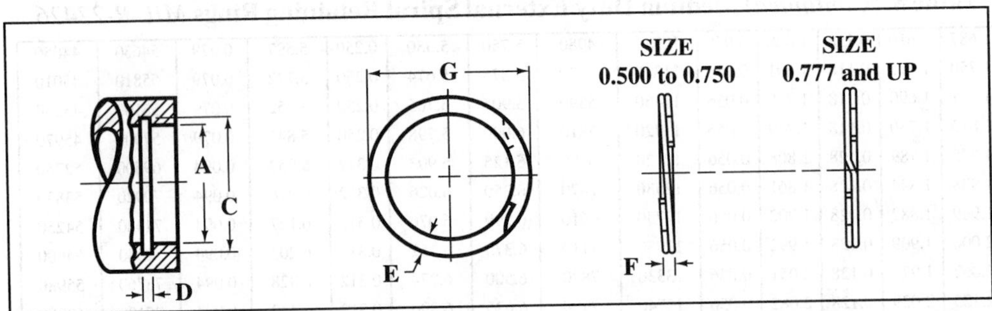

Table 9. Heavy Duty Internal Spiral Retaining Rings *MIL-R-27426*

Bore Dia. A	Ring Dia. G	Ring Wall E	Groove Dia. C	Groove Width D	Static Thrust Load (lb) Ring	Static Thrust Load (lb) Groove	Bore Dia. A	Ring Dia. G	Ring Wall E	Groove Dia. C	Groove Width D	Static Thrust Load (lb) Ring	Static Thrust Load (lb) Groove
0.500	0.538	0.045	0.530	0.039	2530	310	3.543	3.781	0.281	3.755	0.120	49420	28250
0.512	0.550	0.045	0.542	0.039	2590	325	3.562	3.802	0.281	3.776	0.120	49680	28815
0.562	0.605	0.055	0.596	0.039	2840	455	3.625	3.868	0.281	3.841	0.120	50560	30160
0.625	0.675	0.055	0.655	0.039	3160	655	3.750	4.002	0.312	3.974	0.120	52310	33720
0.688	0.743	0.065	0.732	0.039	3480	965	3.875	4.136	0.312	4.107	0.120	54050	37250
0.750	0.807	0.065	0.796	0.039	3790	1065	3.938	4.203	0.312	4.174	0.120	54930	39045
0.777	0.836	0.075	0.825	0.046	4720	1026	4.000	4.270	0.312	4.240	0.120	55790	41025
0.812	0.873	0.075	0.862	0.046	4930	1150	4.125	4.369	0.312	4.339	0.120	57540	38495
0.866	0.931	0.075	0.920	0.046	5260	1395	4.250	4.501	0.312	4.470	0.120	59280	41955
0.875	0.943	0.085	0.931	0.046	5310	1520	4.330	4.588	0.312	4.556	0.120	60400	44815
0.901	0.972	0.085	0.959	0.046	5470	1675	4.500	4.768	0.312	4.735	0.120	62770	50290
0.938	1.013	0.085	1.000	0.046	5690	1925	4.625	4.899	0.312	4.865	0.120	64510	54155
1.000	1.080	0.085	1.066	0.046	6070	2310	4.750	5.030	0.312	4.995	0.120	66260	58270
1.023	1.105	0.085	1.091	0.046	6210	2480	5.000	5.297	0.312	5.260	0.120	69740	65095
1.062	1.138	0.103	1.130	0.056	7010	1940	5.250	5.559	0.350	5.520	0.139	83790	68315
1.125	1.205	0.103	1.197	0.056	7420	2280	5.375	5.690	0.350	5.650	0.139	85780	72840
1.188	1.271	0.103	1.262	0.056	7840	2615	5.500	5.810	0.350	5.770	0.139	87780	74355
1.250	1.339	0.103	1.330	0.056	8250	3110	5.750	6.062	0.350	6.020	0.139	91770	77735
1.312	1.406	0.118	1.396	0.056	8650	3650	6.000	6.314	0.350	6.270	0.139	95760	81120
1.375	1.471	0.118	1.461	0.056	9070	4075	6.250	6.576	0.380	6.530	0.174	122520	80655
1.439	1.539	0.118	1.528	0.056	9490	4670	6.500	6.838	0.380	6.790	0.174	127420	90295
1.456	1.559	0.118	1.548	0.056	9600	4890	6.625	6.974	0.380	6.925	0.174	129870	92060
1.500	1.605	0.118	1.594	0.056	9900	5275	6.750	7.105	0.380	7.055	0.174	132320	102475
1.562	1.675	0.128	1.658	0.068	12780	4840	7.000	7.366	0.380	7.315	0.174	137220	110410
1.625	1.742	0.128	1.725	0.068	13290	5415	7.250	7.628	0.418	7.575	0.209	170370	103440
1.653	1.772	0.128	1.755	0.068	13520	5695	7.500	7.895	0.418	7.840	0.209	176240	115780
1.688	1.810	0.128	1.792	0.068	13810	6070	7.750	8.157	0.418	8.100	0.209	182120	127270
1.750	1.876	0.128	1.858	0.068	14320	7635	8.000	8.419	0.418	8.360	0.209	187990	139370
1.812	1.940	0.128	1.922	0.068	14820	7305	8.250	8.680	0.437	8.620	0.209	193870	152695
1.850	1.981	0.158	1.962	0.068	15130	7960	8.500	8.942	0.437	8.880	0.209	199740	161735
1.875	2.008	0.158	1.989	0.068	15340	8305	8.750	9.209	0.437	9.145	0.209	205620	173065
1.938	2.075	0.158	2.056	0.068	15850	9125	9.000	9.471	0.437	9.405	0.209	211490	182515
2.000	2.142	0.158	2.122	0.068	16360	10040	9.250	9.737	0.437	9.669	0.209	217370	194070
2.062	2.201	0.168	2.186	0.086	21220	8280	9.500	10.000	0.500	9.930	0.209	223240	204550

Table 9. *(Continued)* **Heavy Duty Internal Spiral Retaining Rings** *MIL-R-27426*

Bore Dia. A	Ring Dia. G	Ring Wall E	Groove Dia. C	Groove Width D	Static Thrust Load (lb) Ring	Static Thrust Load (lb) Groove	Bore Dia. A	Ring Dia. G	Ring Wall E	Groove Dia. C	Groove Width D	Static Thrust Load (lb) Ring	Static Thrust Load (lb) Groove
2.125	2.267	0.168	2.251	0.086	21870	8935	9.750	10.260	0.500	10.189	0.209	229120	214325
2.188	2.334	0.168	2.318	0.086	22520	9745	10.000	10.523	0.500	10.450	0.209	234990	225330
2.250	2.399	0.168	2.382	0.086	23160	10455	10.250	10.786	0.500	10.711	0.209	240870	236605
2.312	2.467	0.200	2.450	0.086	23790	11700	10.500	11.047	0.500	10.970	0.209	246740	247110
2.357	2.535	0.200	2.517	0.086	24440	12715	10.750	11.313	0.500	11.234	0.209	252620	260530
2.440	2.602	0.200	2.584	0.086	25110	13550	11.000	11.575	0.500	11.495	0.209	258490	272645
2.500	2.667	0.200	2.648	0.086	25730	14640	11.250	11.838	0.500	11.756	0.209	264360	285040
2.531	2.700	0.200	2.681	0.086	26050	15185	11.500	12.102	0.562	12.018	0.209	270240	298285
2.562	2.733	0.225	2.714	0.103	29940	12775	11.750	12.365	0.562	12.279	0.209	276120	311240
2.625	2.801	0.225	2.781	0.103	30680	13780	12.000	12.628	0.562	12.540	0.209	281990	324475
2.688	2.868	0.225	2.848	0.103	31410	14775	12.250	12.891	0.562	12.801	0.209	287860	337980
2.750	2.934	0.225	2.914	0.103	32140	15790	12.500	13.154	0.562	13.063	0.209	293740	352390
2.813	3.001	0.225	2.980	0.103	32870	16845	12.750	13.417	0.562	13.324	0.209	299610	366460
2.834	3.027	0.225	3.006	0.103	33120	17595	13.000	13.680	0.662	13.585	0.209	305490	380805
2.875	3.072	0.225	3.051	0.103	33600	18505	13.250	13.943	0.662	13.846	0.209	311360	395430
3.000	3.204	0.225	3.182	0.103	35060	20795	13.500	14.207	0.662	14.108	0.209	317240	411000
3.062	3.271	0.281	3.248	0.120	42710	18735	13.750	14.470	0.662	14.369	0.209	323110	426185
3.125	3.338	0.281	3.315	0.120	43590	19865	14.000	14.732	0.662	14.630	0.209	328990	441645
3.157	3.371	0.281	3.348	0.120	44020	20345	14.250	14.995	0.662	14.891	0.209	334860	457380
3.250	3.470	0.281	3.446	0.120	45330	22120	14.500	15.259	0.750	15.153	0.209	340740	474120
3.346	3.571	0.281	3.546	0.120	46670	23905	14.750	15.522	0.750	15.414	0.209	346610	490415
3.469	3.701	0.281	3.675	0.120	48390	26405	15.000	15.785	0.750	15.675	0.209	352490	506990
3.500	3.736	0.281	3.710	0.120	48820	27370							

Source: Spirolox Retaining Rings, RRN Series. All dimensions are in inches. Depth of groove $d = (C - A)/2$. Thickness indicated is for unplated rings; add 0.002 to upper thickness tolerance for plated rings. Standard material: carbon spring steel (SAE 1070–1090).

Ring Thickness, F: For housing sizes 0.500 through 0.750, 0.035; for sizes 0.777 through 1.023, 0.042; for sizes 1.062 through 1.500, 0.050; for sizes 1.562 through 2.000, 0.062; for sizes 2.062 through 2.531, 0.078; for sizes 2.562 through 3.000, 0.093; for sizes 3.062 through 5.000, 0.111; for sizes 5.250 through 7.000, 0.156; for sizes 7.250 through 15.000, 0.187.

Ring Free Diameter Tolerances: For housing sizes 0.500 through 1.500, +0.013, −0.000; for sizes 1.562 through 2.000, +0.020, −0.000; for sizes 2.062 through 2.531, + 0.025, −0.000; for sizes 2.562 through 3.000, +0.030, −0.000; for sizes 3.062 through 5.000, +0.035, −0.000; for sizes 5.250 through 6.000, +0.050, −0.000; for sizes 6.250 through 7.000, +0.055. −0.000; for sizes 7.250 through 10.500, +0.070, −0.000; for sizes 10.750 through 12.750, +0.120, −0.000; for sizes 13.000 through 15.000, +0.140, −0.000.

Ring Thickness Tolerances: For housing sizes 0.500 through 1.500, ± 0.002; for sizes 1.562 through 5.000, ± 0.003; for sizes 5.250 through 6.000, ± 0.004; for sizes 6.250 through 15.000, ± 0.005.

Groove Diameter Tolerances: For housing sizes 0.500 through 0.750, ± 0.002; for sizes 0.777 through 1.023, ± 0.003; for sizes 1.062 through 1.500, ± 0.004; for sizes 1.562 through 2.000, ± 0.005; for sizes 2.062 through 5.000, ± 0.006; for sizes 5.250 through 6.000, ± 0.007; for sizes 6.250 through 10.500, ± 0.008; for sizes 10.750 through 12.500, ± 0.010; for sizes 12.750 through 15.000, ± 0.012.

Groove Width Tolerances: For housing sizes 0.500 through 1.023, +0.003, −0.000; for sizes 1.062 through 2.000, +0.004, −0.000; for sizes 2.062 through 5.000, +0.005, −0.000; for sizes 5.250 through 6.000, +0.006, −0.000; for sizes 6.250 through 7.000, +0.008, −0.000; for sizes 7.250 through 15.000, +0.008, −0.000.

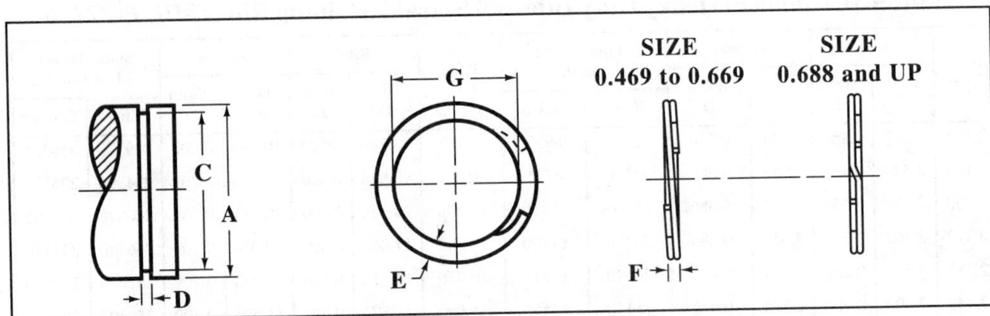

Table 10. Heavy Duty External Spiral Retaining Rings *MIL-R-27426*

Shaft Dia.	Ring		Groove		Static Thrust Load (lb)		Shaft Dia.	Ring		Groove		Static Thrust Load (lb)	
A	Dia. G	Wall E	Dia. C	Width D	Ring	Groove	A	Dia. G	Wall E	Dia. C	Width D	Ring	Groove
0.469	0.439	0.045	0.443	0.029	1880	510	3.500	3.293	0.270	3.316	0.120	48820	32250
0.500	0.464	0.050	0.468	0.039	2530	440	3.543	3.333	0.270	3.357	0.120	49420	33000
0.551	0.514	0.050	0.519	0.039	2790	540	3.625	3.411	0.270	3.435	0.120	50560	34490
0.562	0.525	0.050	0.530	0.039	2840	560	3.687	3.469	0.270	3.493	0.120	51430	35820
0.594	0.554	0.050	0.559	0.039	3000	700	3.750	3.527	0.270	3.552	0.120	52310	37180
0.625	0.583	0.055	0.588	0.039	3160	820	3.875	3.647	0.270	3.673	0.120	54050	39190
0.669	0.623	0.055	0.629	0.039	3380	1070	3.938	3.708	0.270	3.734	0.120	54930	40230
0.688	0.641	0.065	0.646	0.046	4170	960	4.000	3.765	0.270	3.792	0.120	55790	41660
0.750	0.698	0.065	0.704	0.046	4550	1250	4.250	4.037	0.270	4.065	0.120	59280	39370
0.781	0.727	0.065	0.733	0.046	4740	1430	4.375	4.161	0.270	4.190	0.120	61020	40530
0.812	0.756	0.065	0.762	0.046	4930	1620	4.500	4.280	0.270	4.310	0.120	62770	42810
0.875	0.814	0.075	0.821	0.046	5310	2000	4.750	4.518	0.270	4.550	0.120	66260	47570
0.938	0.875	0.075	0.882	0.046	5690	2440	5.000	4.756	0.270	4.790	0.120	69740	52580
0.984	0.919	0.085	0.926	0.046	5970	2790	5.250	4.995	0.350	5.030	0.139	83790	57830
1.000	0.932	0.085	0.940	0.046	6070	2950	5.500	5.228	0.350	5.265	0.139	87780	64720
1.023	0.953	0.085	0.961	0.046	6210	3170	5.750	5.466	0.350	5.505	0.139	91770	70540
1.062	0.986	0.103	0.998	0.056	7010	2810	6.000	5.705	0.350	5.745	0.139	95760	76610
1.125	1.047	0.103	1.059	0.056	7420	2890	6.250	5.938	0.418	5.985	0.174	122520	82930
1.188	1.105	0.103	1.118	0.056	7840	3450	6.500	6.181	0.418	6.225	0.174	127420	89510
1.250	1.163	0.103	1.176	0.056	8250	4110	6.750	6.410	0.418	6.465	0.174	132320	96330
1.312	1.218	0.118	1.232	0.056	8650	4810	7.000	6.648	0.418	6.705	0.174	137220	103400
1.375	1.277	0.118	1.291	0.056	9070	5650	7.250	6.891	0.418	6.942	0.174	142130	111810
1.438	1.336	0.118	1.350	0.056	9490	6340	7.500	7.130	0.437	7.180	0.209	176240	120170
1.500	1.385	0.118	1.406	0.056	9900	7060	7.750	7.368	0.437	7.420	0.209	182120	128060
1.562	1.453	0.128	1.468	0.068	12780	6600	8.000	7.606	0.437	7.660	0.209	187990	136200
1.625	1.513	0.128	1.529	0.068	13290	7330	8.250	7.845	0.437	7.900	0.209	193870	144590
1.687	1.573	0.128	1.589	0.068	13800	8190	8.500	8.083	0.437	8.140	0.209	199740	153220
1.750	1.633	0.128	1.650	0.068	14320	8760	8.750	8.324	0.437	8.383	0.209	205620	160800
1.771	1.651	0.128	1.669	0.068	14490	9040	9.000	8.560	0.500	8.620	0.209	211490	171250
1.812	1.690	0.128	1.708	0.068	14820	9440	9.250	8.798	0.500	8.860	0.209	217370	180640
1.875	1.751	0.158	1.769	0.068	15340	9950	9.500	9.036	0.500	9.100	0.209	223240	190280
1.969	1.838	0.158	1.857	0.068	16110	11040	9.750	9.275	0.500	9.338	0.209	229120	201140
2.000	1.867	0.158	1.886	0.068	16360	11420	10.000	9.508	0.500	9.575	0.209	234990	212810
2.062	1.932	0.168	1.946	0.086	21220	11820	10.250	9.745	0.500	9.814	0.209	240870	223780
2.125	1.989	0.168	2.003	0.086	21870	12980	10.500	9.984	0.500	10.054	0.209	246740	234490
2.156	2.018	0.168	2.032	0.086	22190	13390	10.750	10.221	0.500	10.293	0.209	252620	246000
2.250	2.105	0.168	2.120	0.086	23160	14650	11.000	10.459	0.500	10.533	0.209	258490	257230
2.312	2.163	0.168	2.178	0.086	23790	15510	11.250	10.692	0.500	10.772	0.209	264360	269270
2.375	2.223	0.200	2.239	0.086	24440	16170	11.500	10.934	0.562	11.011	0.209	270240	281590
2.437	2.283	0.200	2.299	0.086	25080	16840	11.750	11.171	0.562	11.250	0.209	276120	294180
2.500	2.343	0.200	2.360	0.086	25730	17530	12.000	11.410	0.562	11.490	0.209	281990	306450
2.559	2.402	0.200	2.419	0.086	26340	17940	12.250	11.647	0.562	11.729	0.209	287860	319580
2.625	2.464	0.200	2.481	0.086	27020	18930	12.500	11.885	0.562	11.969	0.209	293740	332360
2.687	2.523	0.200	2.541	0.086	27650	19640	12.750	12.124	0.562	12.208	0.209	299610	346030
2.750	2.584	0.225	2.602	0.103	32140	20380	13.000	12.361	0.662	12.448	0.209	305490	359330
2.875	2.702	0.225	2.721	0.103	33600	22170	13.250	12.598	0.662	12.687	0.209	311360	373530
2.937	2.760	0.225	2.779	0.103	34320	23240	13.500	12.837	0.662	12.927	0.209	317240	387340

Table 10. *(Continued)* **Heavy Duty External Spiral Retaining Rings** *MIL-R-27426*

Shaft Dia.	Ring		Groove		Static Thrust Load (lb)		Shaft Dia.	Ring		Groove		Static Thrust Load (lb)	
	Dia.	Wall	Dia.	Width				Dia.	Wall	Dia.	Width		
A	G	E	C	D	Ring	Groove	A	G	E	C	D	Ring	Groove
3.000	2.818	0.225	2.838	0.103	35060	24340	13.750	13.074	0.662	13.166	0.209	323110	402090
3.062	2.878	0.225	2.898	0.103	35780	25140	14.000	13.311	0.662	13.405	0.209	328990	417110
3.125	2.936	0.225	2.957	0.103	36520	26290	14.250	13.548	0.662	13.644	0.209	334860	432410
3.156	2.965	0.225	2.986	0.103	36880	26860	14.500	13.787	0.750	13.884	0.209	340740	447250
3.250	3.054	0.225	3.076	0.103	37980	28320	14.750	14.024	0.750	14.123	0.209	346610	463090
3.344	3.144	0.225	3.166	0.103	39080	29800	15.000	14.262	0.750	14.363	0.209	352490	478450
3.437	3.234	0.225	3.257	0.103	40170	30980							

Source: Spirolox Retaining Rings, RSN Series. All dimensions are in inches. Depth of groove $d = (A - C)/2$. Thickness indicated is for unplated rings; add 0.002 to upper tolerance for plated rings. Standard material: carbon spring steel (SAE 1070-1090).

Ring Thickness, F: For shaft size 0.469, 0.025; for sizes 0.500 through 0.669, 0.035; for sizes 0.688 through 1.023, 0.042; for sizes 1.062 through 1.500, 0.050; for sizes 1.562 through 2.000, 0.062; for sizes 2.062 through 2.687, 0.078; for sizes 2.750 through 3.437, 0.093; for sizes 3.500 through 5.000, 0.111; for sizes 5.250 through 6.000, 0.127; for sizes 6.250 through 7.250, 0.156; for sizes 7.500 through 15.000, 0.187.

Ring Free Diameter Tolerances: For shaft sizes 0.469 through 1.500, +0.000, −0.013; for sizes 1.562 through 2.000, +0.000, −0.020; for sizes 2.062 through 2.687, + 0.000, −0.025; for sizes 2.750 through 3.437, +0.000, −0.030; for sizes 3.500 through 5.000, +0.000, −0.035; for sizes 5.250 through 6.000, +0.000, −0.050; for sizes 6.250 through 7.000, +0.000, −0.060; for sizes 7.250 through 10.000, +0.000, −0.070; for sizes 10.250 through 12.500, +0.000, −0.090; for sizes 12.750 through 15.000, +0.000, −0.110.

Ring Thickness Tolerances: For shaft sizes 0.469 through 1.500, ±0.002; for sizes 1.562 through 5.000, ±0.003; for sizes 5.250 through 6.000, ±0.004; for sizes 6.250 through 15.000, ±0.005.

Groove Diameter Tolerances: For shaft sizes 0.469 through 0.562, ±0.002; for sizes 0.594 through 1.023, ±0.003; for sizes 1.062 through 1.500, ±0.004; for sizes 1.562 through 2.000, ±0.005; for sizes 2.062 through 5.000, ±0.006; for sizes 5.250 through 6.000, ±0.007; for sizes 6.250 through 10.000, ±0.008; for sizes 10.250 through 12.500, ±0.010; for sizes 12.750 through 15.000, ±0.012.

Groove Width Tolerances: For shaft sizes 0.469 through 1.023, +0.003, −0.000; for sizes 1.062 through 2.000, +0.004, −0.000; for sizes 2.062 through 5.000, +0.005, −0.000; for sizes 5.250 through 6.000, +0.006; −0.000; for sizes 6.250 through 7.250, + 0.008, −0.000; for sizes 7.500 through 15.000, +0.008, −0.000.

Thrust Load Capacity: The most important criterion in determining which ring is best suited for a specific application is thrust load capacity. The strength of the retaining ring and groove must both be considered when analyzing the thrust load capacity of an application to determine whether the groove or the retaining ring is likely to fail first. When a retaining ring application fails, the fault will usually be with the groove, unless the groove material is of very high strength.

Ring Material: The standard materials for spiral-wound retaining rings are SAE 1070 to 1090 carbon spring steels and 18-8 type 302 stainless steels. The 1070 to 1090 carbon spring steels provide high-strength retaining rings at low cost. Type 302 stainless steel withstands ordinary rusting. Other materials are used for specialized applications, such as the type 316 stainless frequently used in the food industry. For high-temperature use, superalloy A286 rings can be used at up to 900°F and Inconel X-750 at up to 1200°F. Other materials, such as 316 stainless steel, 17-7PH and Inconel stainless steels are sometimes used for special-purpose and custom-made rings. Standard ring are typically supplied uncoated, however, special finishes such as cadmium, phosphate, zinc, or black oxide coatings for carbon spring steel rings and passivation of stainless steel rings are available.

Table 11. Important Dimensions of Inch Series External Retaining Rings
MS 16624

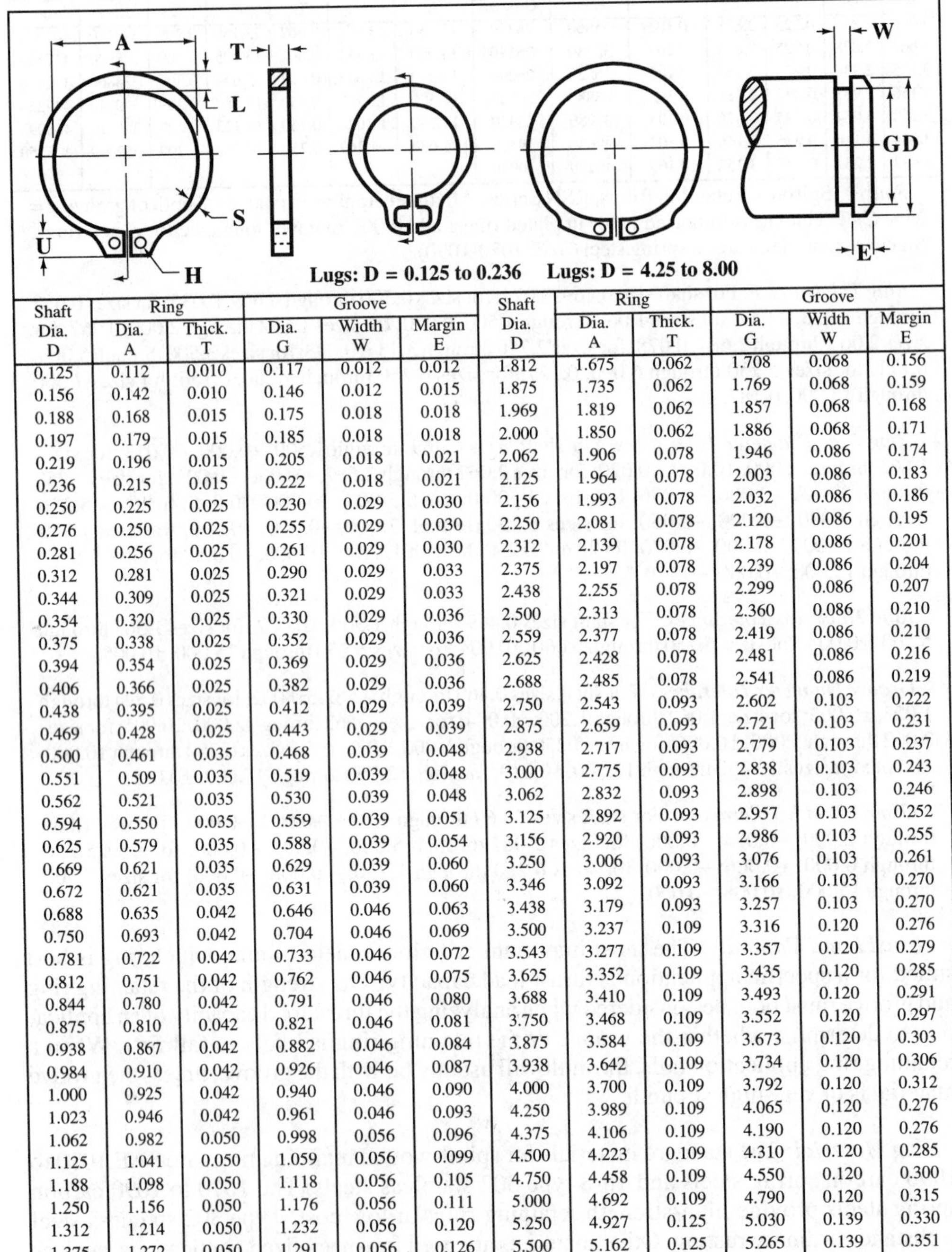

Lugs: D = 0.125 to 0.236 Lugs: D = 4.25 to 8.00

Shaft Dia.	Ring		Groove			Shaft Dia.	Ring		Groove		
	Dia.	Thick.	Dia.	Width	Margin		Dia.	Thick.	Dia.	Width	Margin
D	A	T	G	W	E	D	A	T	G	W	E
0.125	0.112	0.010	0.117	0.012	0.012	1.812	1.675	0.062	1.708	0.068	0.156
0.156	0.142	0.010	0.146	0.012	0.015	1.875	1.735	0.062	1.769	0.068	0.159
0.188	0.168	0.015	0.175	0.018	0.018	1.969	1.819	0.062	1.857	0.068	0.168
0.197	0.179	0.015	0.185	0.018	0.018	2.000	1.850	0.062	1.886	0.068	0.171
0.219	0.196	0.015	0.205	0.018	0.021	2.062	1.906	0.078	1.946	0.086	0.174
0.236	0.215	0.015	0.222	0.018	0.021	2.125	1.964	0.078	2.003	0.086	0.183
0.250	0.225	0.025	0.230	0.029	0.030	2.156	1.993	0.078	2.032	0.086	0.186
0.276	0.250	0.025	0.255	0.029	0.030	2.250	2.081	0.078	2.120	0.086	0.195
0.281	0.256	0.025	0.261	0.029	0.030	2.312	2.139	0.078	2.178	0.086	0.201
0.312	0.281	0.025	0.290	0.029	0.033	2.375	2.197	0.078	2.239	0.086	0.204
0.344	0.309	0.025	0.321	0.029	0.033	2.438	2.255	0.078	2.299	0.086	0.207
0.354	0.320	0.025	0.330	0.029	0.036	2.500	2.313	0.078	2.360	0.086	0.210
0.375	0.338	0.025	0.352	0.029	0.036	2.559	2.377	0.078	2.419	0.086	0.210
0.394	0.354	0.025	0.369	0.029	0.036	2.625	2.428	0.078	2.481	0.086	0.216
0.406	0.366	0.025	0.382	0.029	0.036	2.688	2.485	0.078	2.541	0.086	0.219
0.438	0.395	0.025	0.412	0.029	0.039	2.750	2.543	0.093	2.602	0.103	0.222
0.469	0.428	0.025	0.443	0.029	0.039	2.875	2.659	0.093	2.721	0.103	0.231
0.500	0.461	0.035	0.468	0.039	0.048	2.938	2.717	0.093	2.779	0.103	0.237
0.551	0.509	0.035	0.519	0.039	0.048	3.000	2.775	0.093	2.838	0.103	0.243
0.562	0.521	0.035	0.530	0.039	0.048	3.062	2.832	0.093	2.898	0.103	0.246
0.594	0.550	0.035	0.559	0.039	0.051	3.125	2.892	0.093	2.957	0.103	0.252
0.625	0.579	0.035	0.588	0.039	0.054	3.156	2.920	0.093	2.986	0.103	0.255
0.669	0.621	0.035	0.629	0.039	0.060	3.250	3.006	0.093	3.076	0.103	0.261
0.672	0.621	0.035	0.631	0.039	0.060	3.346	3.092	0.093	3.166	0.103	0.270
0.688	0.635	0.042	0.646	0.046	0.063	3.438	3.179	0.093	3.257	0.103	0.270
0.750	0.693	0.042	0.704	0.046	0.069	3.500	3.237	0.109	3.316	0.120	0.276
0.781	0.722	0.042	0.733	0.046	0.072	3.543	3.277	0.109	3.357	0.120	0.279
0.812	0.751	0.042	0.762	0.046	0.075	3.625	3.352	0.109	3.435	0.120	0.285
0.844	0.780	0.042	0.791	0.046	0.080	3.688	3.410	0.109	3.493	0.120	0.291
0.875	0.810	0.042	0.821	0.046	0.081	3.750	3.468	0.109	3.552	0.120	0.297
0.938	0.867	0.042	0.882	0.046	0.084	3.875	3.584	0.109	3.673	0.120	0.303
0.984	0.910	0.042	0.926	0.046	0.087	3.938	3.642	0.109	3.734	0.120	0.306
1.000	0.925	0.042	0.940	0.046	0.090	4.000	3.700	0.109	3.792	0.120	0.312
1.023	0.946	0.042	0.961	0.046	0.093	4.250	3.989	0.109	4.065	0.120	0.276
1.062	0.982	0.050	0.998	0.056	0.096	4.375	4.106	0.109	4.190	0.120	0.276
1.125	1.041	0.050	1.059	0.056	0.099	4.500	4.223	0.109	4.310	0.120	0.285
1.188	1.098	0.050	1.118	0.056	0.105	4.750	4.458	0.109	4.550	0.120	0.300
1.250	1.156	0.050	1.176	0.056	0.111	5.000	4.692	0.109	4.790	0.120	0.315
1.312	1.214	0.050	1.232	0.056	0.120	5.250	4.927	0.125	5.030	0.139	0.330
1.375	1.272	0.050	1.291	0.056	0.126	5.500	5.162	0.125	5.265	0.139	0.351
1.438	1.333	0.050	1.350	0.056	0.132	5.750	5.396	0.125	5.505	0.139	0.366
1.500	1.387	0.050	1.406	0.056	0.141	6.000	5.631	0.125	5.745	0.139	0.381
1.562	1.446	0.062	1.468	0.068	0.141	6.250	5.866	0.156	5.985	0.174	0.396
1.625	1.503	0.062	1.529	0.068	0.144	6.500	6.100	0.156	6.225	0.174	0.411
1.687	1.560	0.062	1.589	0.068	0.147	6.750	6.335	0.156	6.465	0.174	0.426
1.750	1.618	0.062	1.650	0.068	0.150	7.000	6.570	0.156	6.705	0.174	0.441
1.772	1.637	0.062	1.669	0.068	0.153	7.500	7.009	0.187	7.180	0.209	0.480

Source: Industrial Retaining Rings, 3100 Series. All dimensions are in inches. Depth of groove $d = (D - G)/2$. Thickness indicated is for unplated rings; for most plated rings, the maximum ring thickness will not exceed the minimum groove width (W) minus 0.0002 inch. Standard material: carbon spring steel (SAE 1060-1090).

Ring Free Diameter Tolerances: For shaft sizes 0.125 through 0.250, +0.002, −0.004; for sizes 0.276 through 0.500, +0.002, −0.005; for sizes 0.551 through 1.023, +0.005, −0.010; for sizes 1.062 through 1.500, +0.010, −0.015; for sizes 1.562 through 2.000, +0.013, −0.020; for sizes 2.062 through 2.500, +0.015, −0.025; for sizes 2.559 through 5.000, +0.020, −0.030; for sizes 5.250 through 6.000, +0.020, −0.040; for sizes 6.250 through 6.750, +0.020, −0.050; for sizes 7.000 and 7.500, +0.050, −0.130.

Ring Thickness Tolerances: For shaft sizes 0.125 and 0.156, ±0.001; for sizes 0.188 through 1.500, ±0.002; for sizes 1.562 through 5.000, ±0.003; for sizes 5.250 through 6.000, ±0.004; for sizes 6.250 through 7.500, ±0.005.

Groove Diameter Tolerances: For shaft sizes 0.125 through 0.250, ±0.0015; for sizes 0.276 through 0.562, ±0.002; for sizes 0.594 through 1.023, ±0.003; for sizes 1.062 though 1.500, ±0.004; for sizes 1.562 through 2.000, ±0.005; for sizes 2.062 through 5.000, ±0.006; for sizes 5.250 through 6.000, ±0.007; for sizes 6.250 through 7.500, ±0.008.

Groove Width Tolerances: For shaft sizes 0.125 through 0.236, +0.002, −0.000; for sizes 0.250 through 1.023, +0.003, −0.000; for sizes 1.062 through 2.000, +0.004, −0.000; for sizes 2.062 through 5.000, +0.005, −0.000; for sizes 5.250 through 6.000, +0.006, −0.000; for sizes 6.250 through 7.500, +0.008, −0.000.

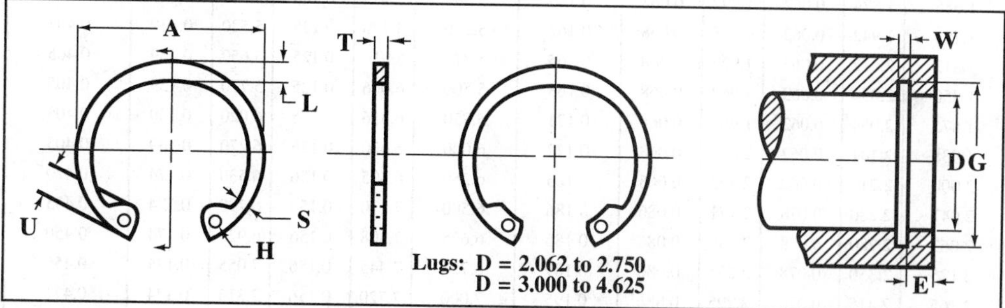

Table 12. Important Dimensions of Inch Series Internal Retaining Rings

Housing Dia. D	Ring Dia. A	Ring Thick. T	Groove Dia. G	Groove Width W	Groove Margin E	Housing Dia. D	Ring Dia. A	Ring Thick. T	Groove Dia. G	Groove Width W	Groove Margin E
0.250	0.280	0.015	0.268	0.018	0.027	2.500	2.775	0.078	2.648	0.086	0.222
0.312	0.346	0.015	0.330	0.018	0.027	2.531	2.775	0.078	2.681	0.086	0.225
0.375	0.415	0.025	0.397	0.029	0.033	2.562	2.844	0.093	2.714	0.103	0.228
0.438	0.482	0.025	0.461	0.029	0.036	2.625	2.910	0.093	2.781	0.103	0.234
0.453	0.498	0.025	0.477	0.029	0.036	2.677	2.980	0.093	2.837	0.103	0.240
0.500	0.548	0.035	0.530	0.039	0.045	2.688	2.980	0.093	2.848	0.103	0.240
0.512	0.560	0.035	0.542	0.039	0.045	2.750	3.050	0.093	2.914	0.103	0.246
0.562	0.620	0.035	0.596	0.039	0.051	2.812	3.121	0.093	2.980	0.103	0.252
0.625	0.694	0.035	0.665	0.039	0.060	2.835	3.121	0.093	3.006	0.103	0.255
0.688	0.763	0.035	0.732	0.039	0.066	2.875	3.191	0.093	3.051	0.103	0.264
0.750	0.831	0.035	0.796	0.039	0.069	2.953	3.325	0.093	3.135	0.103	0.273
0.777	0.859	0.042	0.825	0.046	0.072	3.000	3.325	0.093	3.182	0.103	0.273
0.812	0.901	0.042	0.862	0.046	0.075	3.062	3.418	0.109	3.248	0.120	0.279
0.866	0.961	0.042	0.920	0.046	0.081	3.125	3.488	0.109	3.315	0.120	0.285
0.875	0.971	0.042	0.931	0.046	0.084	3.149	3.523	0.109	3.341	0.120	0.288
0.901	1.000	0.042	0.959	0.046	0.087	3.156	3.523	0.109	3.348	0.120	0.288
0.938	1.041	0.042	1.000	0.046	0.093	3.250	3.623	0.109	3.446	0.120	0.294
1.000	1.111	0.042	1.066	0.046	0.099	3.346	3.734	0.109	3.546	0.120	0.300
1.023	1.136	0.042	1.091	0.046	0.102	3.469	3.857	0.109	3.675	0.120	0.309
1.062	1.180	0.050	1.130	0.056	0.102	3.500	3.890	0.109	3.710	0.120	0.315

Table 12. *(Continued)* **Important Dimensions of Inch Series Internal Retaining Rings**

Housing Dia. D	Ring Dia. A	Ring Thick. T	Groove Dia. G	Groove Width W	Groove Margin E	Housing Dia. D	Ring Dia. A	Ring Thick. T	Groove Dia. G	Groove Width W	Groove Margin E
1.125	1.249	0.050	1.197	0.056	0.108	3.543	3.936	0.109	3.755	0.120	0.318
1.181	1.319	0.050	1.255	0.056	0.111	3.562	3.936	0.109	3.776	0.120	0.321
1.188	1.319	0.050	1.262	0.056	0.111	3.625	4.024	0.109	3.841	0.120	0.324
1.250	1.388	0.050	1.330	0.056	0.120	3.740	4.157	0.109	3.964	0.120	0.336
1.259	1.388	0.050	1.339	0.056	0.120	3.750	4.157	0.109	3.974	0.120	0.336
1.312	1.456	0.050	1.396	0.056	0.126	3.875	4.291	0.109	4.107	0.120	0.348
1.375	1.526	0.050	1.461	0.056	0.129	3.938	4.358	0.109	4.174	0.120	0.354
1.378	1.526	0.050	1.464	0.056	0.129	4.000	4.424	0.109	4.240	0.120	0.360
1.438	1.596	0.050	1.528	0.056	0.135	4.125	4.558	0.109	4.365	0.120	0.360
1.456	1.616	0.050	1.548	0.056	0.138	4.250	4.691	0.109	4.490	0.120	0.360
1.500	1.660	0.050	1.594	0.056	0.141	4.331	4.756	0.109	4.571	0.120	0.360
1.562	1.734	0.062	1.658	0.068	0.144	4.500	4.940	0.109	4.740	0.120	0.360
1.575	1.734	0.062	1.671	0.068	0.144	4.625	5.076	0.109	4.865	0.120	0.360
1.625	1.804	0.062	1.725	0.068	0.150	4.724	5.213	0.109	4.969	0.120	0.366
1.653	1.835	0.062	1.755	0.068	0.153	4.750	5.213	0.109	4.995	0.120	0.366
1.688	1.874	0.062	1.792	0.068	0.156	5.000	5.485	0.109	5.260	0.120	0.390
1.750	1.942	0.062	1.858	0.068	0.162	5.250	5.770	0.125	5.520	0.139	0.405
1.812	2.012	0.062	1.922	0.068	0.165	5.375	5.910	0.125	5.650	0.139	0.405
1.850	2.054	0.062	1.962	0.068	0.168	5.500	6.066	0.125	5.770	0.139	0.405
1.875	2.054	0.062	1.989	0.068	0.171	5.750	6.336	0.125	6.020	0.139	0.405
1.938	2.141	0.062	2.056	0.068	0.177	6.000	6.620	0.125	6.270	0.139	0.405
2.000	2.210	0.062	2.122	0.068	0.183	6.250	6.895	0.156	6.530	0.174	0.420
2.047	2.280	0.078	2.171	0.086	0.186	6.500	7.170	0.156	6.790	0.174	0.435
2.062	2.280	0.078	2.186	0.086	0.186	6.625	7.308	0.156	6.925	0.174	0.450
2.125	2.350	0.078	2.251	0.086	0.189	6.750	7.445	0.156	7.055	0.174	0.456
2.165	2.415	0.078	2.295	0.086	0.195	7.000	7.720	0.156	7.315	0.174	0.471
2.188	2.415	0.078	2.318	0.086	0.195	7.250	7.995	0.187	7.575	0.209	0.486
2.250	2.490	0.078	2.382	0.086	0.198	7.500	8.270	0.187	7.840	0.209	0.510
2.312	2.560	0.078	2.450	0.086	0.207	7.750	8.545	0.187	8.100	0.209	0.525
2.375	2.630	0.078	2.517	0.086	0.213	8.000	8.820	0.187	8.360	0.209	0.540
2.440	2.702	0.078	2.584	0.086	0.216	8.250	9.095	0.187	8.620	0.209	0.555

Source: Industrial Retaining Rings, 3000 Series. All dimensions are in inches. Depth of groove $d = (G - D)/2$. Thickness indicated is for unplated rings. Standard material: carbon spring steel (SAE 1060-1090).

Ring Free Diameter Tolerances: For housing sizes 0.250 through 0.777, +0.010, −0.005; for sizes 0.812 through 1.023, +0.015, − 0.010; for sizes 1.062 through 1.500, + 0.025, − 0.020; for sizes 1.562 through 2.000, +0.035, −0.025; for sizes 2.047 through 3.000, +0.040, −0.030; for sizes 3.062 through 3.625, ±0.055; for sizes 3.740 through 6.000, ±0.065; for sizes 6.250 through 7.000, ±0.080; for sizes 7.250 through 8.250, ±0.090.

Ring Thickness Tolerances: For housing sizes 0.250 through 1.500, ±0.002; for sizes 1.562 through 5.000, ±0.003; for sizes 5.250 through 6.000, ±0.004; for sizes 6.250 through 8.250, ±0.005.

Groove Diameter Tolerances: For housing sizes 0.250 and 0.312, ±0.001; for sizes 0.375 through 0.750, ±0.002; for sizes 0.777 through 1.023 ±0.003; for sizes 1.062 through 1.500, ±0.004; for sizes 1.562 through 2.000, ±0.005; for sizes 2.047 through 5.000 ±0.006; for sizes 5.250 through 6.000, ±0.007; for sizes 6.250 through 8.250, ±0.008.

Groove Width Tolerances: For housing sizes 0.250 and 0.312, +0.002, − 0.000; for sizes 0.375 through 1.023, +0.003, −0.000; for sizes 1.062 through 2.000, +0.004, −0.000; for sizes 2.047 through 5.000, +0.005; −0.000; for sizes 5.250 through 6.000, +0.006, −0.000; for sizes 6.250 through 8.250, +0.008, −0.000.

Table 13. Important Dimensions of Inch Series External Retaining Rings *MS16632*

Shaft Diameter D	Ring			Groove			[a]Static Thrust Load (lb)	
	Free Dia. A	Thickness T	Diameter B	Diameter G	Width W	Margin E	Ring	Groove
0.125	0.102	0.015	0.164	0.106	0.018	0.020	85	40
0.156	0.131	0.015	0.205	0.135	0.018	0.020	110	55
0.188	0.161	0.015	0.245	0.165	0.018	0.022	130	70
0.219	0.187	0.025	0.275	0.193	0.029	0.026	260	100
0.236	0.203	0.025	0.295	0.208	0.029	0.028	280	115
0.250	0.211	0.025	0.311	0.220	0.029	0.030	295	130
0.281	0.242	0.025	0.344	0.247	0.029	0.034	330	170
0.312	0.270	0.025	0.376	0.276	0.029	0.036	370	200
0.375	0.328	0.025	0.448	0.335	0.029	0.040	440	265
0.406	0.359	0.025	0.485	0.364	0.029	0.042	480	300
0.437	0.386	0.025	0.516	0.393	0.029	0.044	515	340
0.500	0.441	0.035	0.581	0.450	0.039	0.050	825	440
0.562	0.497	0.035	0.653	0.507	0.039	0.056	930	550
0.625	0.553	0.035	0.715	0.563	0.039	0.062	1030	690
0.687	0.608	0.042	0.780	0.619	0.046	0.068	1700	820
0.750	0.665	0.042	0.845	0.676	0.046	0.074	1850	985
0.812	0.721	0.042	0.915	0.732	0.046	0.080	2010	1150
0.875	0.777	0.042	0.987	0.789	0.046	0.086	2165	1320
0.937	0.830	0.042	1.054	0.843	0.046	0.094	2320	1550
1.000	0.887	0.042	1.127	0.900	0.046	0.100	2480	1770
1.125	0.997	0.050	1.267	1.013	0.056	0.112	3300	2200
1.188	1.031	0.050	1.321	1.047	0.056	0.140	3500	2900
1.250	1.110	0.050	1.410	1.126	0.056	0.124	3600	2700
1.375	1.220	0.050	1.550	1.237	0.056	0.138	4000	3300
1.500	1.331	0.050	1.691	1.350	0.056	0.150	4400	4000
1.750	1.555	0.062	1.975	1.576	0.068	0.174	6400	5300
2.000	1.777	0.062	2.257	1.800	0.068	0.200	7300	7000

[a] *Thrust Load Safety Factors:* Ring, 4; groove, 2. Groove wall thrust loads are for grooves machined in cold-rolled steel with a tensile yield strength of 45,000 psi; for other shaft materials, the thrust load varies proportionally with the yield strength.

Source: Industrial Retaining Rings, 2000 Series. All dimensions are in inches. Depth of groove $d = (D - G)/2$. Standard material: carbon spring steel (SAE 1060-1090). Thickness indicated is for unplated rings; for most plated rings with shaft sizes less than 1.000 inch, the maximum thickness will not exceed the minimum groove width (W) minus 0.0002 inch; for larger rings, the ring thickness may increase by 0.002 inch.

Groove Maximum Bottom Radii: For shaft diameters less than 0.500 inch, 0.005 inch; for shaft sizes 0.500 through 1.000 inch, 0.010 inch; all larger sizes, 0.015 inch.

Ring Free Diameter Tolerances: For shaft sizes 0.125 through 0.188, +0.002, −0.004; for sizes 0.219 through 0.437, +0.003, −0.005; for sizes 0.500 through 0.625, ±0.006; for sizes 0.687 through 1.000, ±0.007; for sizes 1.125 through 1.500, ±0.008; for sizes 1.750 and 2.000, ±0.010.

Ring Thickness Tolerances: For shaft sizes 0.125 through 1.500, ±0.002; for sizes 1.750 and 2.000, ±0.003.

Groove Diameter Tolerances: For shaft sizes 0.125 through 0.188, ±0.0015; for sizes 0.219 through 0.437, ±0.002; for sizes 0.500 through 1.000, ±0.003; for sizes 1.125 through 1.500, ±0.004; for sizes 1.750 and 2.000, ±0.005.

Groove Width Tolerances: For shaft sizes 0.125 through 0.188, +0.002, −0.000; for sizes 0.219 through 1.000, +0.003, −0.000; for sizes 1.125 through 2.000, +0.004, −0.000.

Table 14. Important Dimensions of Inch Series External Retaining Rings *MS16633*

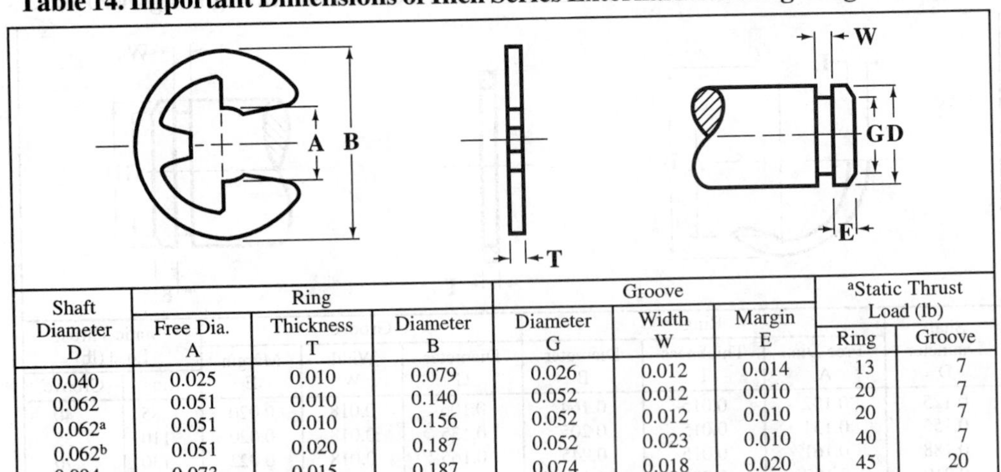

Shaft Diameter	Ring			Groove			[a]Static Thrust Load (lb)	
	Free Dia.	Thickness	Diameter	Diameter	Width	Margin		
D	A	T	B	G	W	E	Ring	Groove
0.040	0.025	0.010	0.079	0.026	0.012	0.014	13	7
0.062	0.051	0.010	0.140	0.052	0.012	0.010	20	7
0.062[a]	0.051	0.010	0.156	0.052	0.012	0.010	20	7
0.062[b]	0.051	0.020	0.187	0.052	0.023	0.010	40	7
0.094	0.073	0.015	0.187	0.074	0.018	0.020	45	20
0.094	0.069	0.015	0.230	0.074	0.018	0.020	45	20
0.110	0.076	0.015	0.375	0.079	0.018	0.030	55	40
0.125	0.094	0.015	0.230	0.095	0.018	0.030	65	45
0.140	0.100	0.015	0.203	0.102	0.018	0.038	70	60
0.140[c]	0.108	0.015	0.250	0.110	0.018	0.030	70	45
0.140[d]	0.102	0.025	0.270	0.105	0.029	0.034	150	55
0.156	0.114	0.025	0.282	0.116	0.029	0.040	165	70
0.172	0.125	0.025	0.312	0.127	0.029	0.044	180	90
0.188	0.145	0.025	0.335	0.147	0.029	0.040	195	90
0.188	0.122	0.025	0.375	0.125	0.029	0.062	195	135
0.218	0.185	0.025	0.437	0.188	0.029	0.030	225	75
0.250	0.207	0.025	0.527	0.210	0.029	0.040	260	115
0.312	0.243	0.025	0.500	0.250	0.029	0.062	325	225
0.375	0.300	0.035	0.660	0.303	0.039	0.072	685	315
0.437	0.337	0.035	0.687	0.343	0.039	0.094	800	485
0.437	0.375	0.035	0.600	0.380	0.039	0.058	800	290
0.500	0.392	0.042	0.800	0.396	0.046	0.104	1100	600
0.625	0.480	0.042	0.940	0.485	0.046	0.140	1370	1040
0.744	0.616	0.050	1.000	0.625	0.056	0.118	1940	1050
0.750	0.616	0.050	1.000	0.625	0.056	0.124	1960	1100
0.750	0.574	0.050	1.120	0.580	0.056	0.170	1960	1500
0.875	0.668	0.050	1.300	0.675	0.056	0.200	2200	2050
0.985	0.822	0.050	1.500	0.835	0.056	0.148	2570	1710
1.000	0.822	0.050	1.500	0.835	0.056	0.164	2620	1900
1.188	1.066	0.062	1.626	1.079	0.068	0.108	3400	1500
1.375	1.213	0.062	1.875	1.230	0.068	0.144	4100	2300

[a] *Thrust Load Safety Factors:* Ring 3; groove, 2.

Source: Industrial Retaining Rings, 1000 Series. All dimensions are in inches. Depth of groove $d = (D - G)/2$. Standard material: carbon spring steel (SAE 1060 –1090). Thickness indicated is for unplated rings; for most plated rings with shaft sizes less than 0.625, the maximum ring thickness will not exceed the minimum groove width (W) minus 0.0002 inch; for larger rings, the thickness may increase by 0.002 inch.

Groove Maximum Bottom Radii: For shaft sizes 0.040 and 0.062, 0.003 inch; for sizes 0.094 through 0.250, 0.005 inch; for sizes 0.312 through 0.437, 0.010 inch; for sizes 0.500 through 1.375, 0.015 inch.

Ring Free Diameter Tolerances: For shaft sizes 0.040 through 0.250, +0.001, –0.003; for sizes 0.312 through 0.500, +0.002, –0.004; for sizes 0.625 through 1.000, +0.003, –0.005; for sizes 1.188 and 1.375, +0.006, –0.010.

Ring Thickness Tolerances: For shaft sizes 0.040 and 0.062[a], ±0.001; for sizes 0.062[b] through 1.000, ±0.002; for sizes 1.188 and 1.375, ±0.003.

Groove Diameter Tolerances: For shaft sizes 0.040 through 0.218, +0.002, –0.000; for sizes 0.250 through 1.000, +0.003, –0.000; for sizes 1.188 and 1.375, +0.005, –0.000.

Grove Width Tolerances: For shaft sizes 0.040 through 0.140[c], +0.002, –0.000; for sizes 0.140[d] through 1.000, +0.003, –0.000; for sizes 1.188 and 1.375, +0.004, –0.000.

Table 15. Dimensions of Inch Series External Retaining Rings *MS3215*

Shaft Diameter D	Ring Free Dia. A	Ring Thickness T	Ring Diameter B	Groove Diameter G	Groove Width W	Groove Margin E	[a]Static Thrust Load (lb) Ring	[a]Static Thrust Load (lb) Groove
0.094	0.072	0.015	0.206	0.074	0.018	0.020	55	13
0.125	0.093	0.015	0.270	0.095	0.018	0.030	75	25
0.156	0.113	0.025	0.335	0.116	0.029	0.040	150	40
0.188	0.143	0.025	0.375	0.147	0.029	0.040	180	50
0.219	0.182	0.025	0.446	0.188	0.029	0.031	215	50
0.250	0.204	0.025	0.516	0.210	0.029	0.040	250	75
0.312	0.242	0.025	0.588	0.250	0.029	0.062	300	135
0.312	0.242	0.035	0.588	0.250	0.039	0.062	420	135
0.375	0.292	0.035	0.660	0.303	0.039	0.072	520	190
0.438	0.332	0.035	0.746	0.343	0.039	0.096	600	285
0.500	0.385	0.042	0.810	0.396	0.046	0.104	820	360
0.562	0.430	0.042	0.870	0.437	0.046	0.124	930	480

[a] *Thrust Load Safety Factors:* Ring, 3; groove, 2.

Source: Industrial Retaining Rings, 1200 Series. All dimensions are in inches. Depth of groove $d = (D - G)/2$. Standard material: carbon spring steel (SAE 1060-1090). Thickness indicated is for unplated rings; for most plated rings the maximum thickness will not exceed the minimum groove width (W) minus 0.0002 inch.

Groove Maximum Bottom Radii: For shaft sizes 0.250 and smaller, 0.005 inch; for sizes 0.312 through 0.438, 0.010 inch; for sizes 0.500 and 0.562, 0.015 inch. *Ring Free Diameter Tolerances:* For shaft sizes 0.094 through 0.156, +0.001, −0.003; for sizes 0.188 through 0.312, ±0.003; for sizes 0.375 through 0.562, ±0.004. *Ring Thickness Tolerances:* For all shaft sizes, ±0.002. *Groove Diameter Tolerances:* For shaft sizes 0.094 through 0.188, +0.002, −0.000; for sizes 0.219 and 0.250, ±0.002; for sizes 0.312 through 0.562, ±0.003. *Groove Width Tolerances:* For shaft sizes 0.094 and 0.125, +0.002, − 0.000; for sizes 0.156 through 0.562, +0.003, −0.000.

The thrust load capacities shown in the tables of this section include safety factors. Usually, a safety factor of 2 is used for groove thrust load calculations when the load is applied through a retained part and groove with both having sharp corners and where the minimum side clearance exists between the retained part and the shaft or bore. Groove thrust load values in the tables of this section are based on these conditions. A safety factor of 3 is usual for calculations of thrust load capacity based on ring shear.

Ideally, the corner of a retained part in contact with a retaining ring should have square corners and contact the ring as closely as possible to the shaft or housing. The tabulated thrust capacities assume that minimum clearances exist between the retained part and shaft or housing, that the groove and retained part have square corners, and that contact between the retained part and the ring occurs close to the shaft or housing. If these conditions apply, the tabulated thrust loads apply. If the application does not meet the previous conditions but the side clearances, radii, and chamfers are less than the maximum total radius or chamfer of Fig. 1, then the thrust load capacity must be reduced by dividing the tabulated value by 2. The maximum total radius is given by $0.5(b - d)$ and the maximum total chamfer by $0.375(b - d)$, where b is the radial wall thickness, and d is the groove depth. The recommended maximum total radius or chamfer specifications are intended to be used as guidelines by the designer, and to ensure the ring application will withstand published and calculated values of static thrust loads.

In analyzing the retaining ring loading conditions, a static, uniformly applied load is usually assumed. Dynamic and eccentric loads, however, are frequently encountered. Eccentric loading occurs when the load is concentrated on a small portion of the ring, such as may be caused by incorrectly machined surfaces, cocking of the retained part, and axial misalignment of parts. Conditions leading to eccentric loading on the ring should be avoided. In addition to the factors that affect the static thrust capacity, applications in which shock or impact loading occurs must be evaluated very carefully and tested in service to assess the effect of the mass and velocity of the retained part striking the ring. Vibration caused by impact loading can also cause the ring to fail if the resonant frequency of the system (retaining ring application) coincides with the resonant frequency of the retaining ring.

Table 16. Dimensions of Inch Series Self-Locking External Retaining Rings

Shaft Diameter		Ring		Optical Groove			[a]Static Thrust Load (lb)	
Min. D	Max. D	Free Dia. A	Thickness T	Diameter G	Width W	Margin E	Ring	Groove
0.078	0.080	0.074	0.025	The use of grooves with these shaft sizes is not suggested.			10	0
0.092	0.096	0.089	0.025				10	0
0.123	0.127	0.120	0.025				20	0
0.134	0.138	0.130	0.025				20	0
0.154	0.158	0.150	0.025				22	0
0.185	0.189	0.181	0.035				25	0
0.248	0.252	0.238	0.035	0.240	0.041	0.030	35	90
0.310	0.316	0.298	0.042	0.303	0.048	0.030	50	110
0.373	0.379	0.354	0.042	0.361	0.048	0.030	55	185
0.434	0.440	0.412	0.050	0.419	0.056	0.030	60	280
0.497	0.503	0.470	0.050	0.478	0.056	0.040	65	390
0.622	0.628	0.593	0.062	0.599	0.069	0.045	85	570
0.745	0.755	0.706	0.062	0.718	0.069	0.050	90	845

[a] *Thrust Load Safety Factors:* Ring, 1; groove, 2.

Source: Industrial Retaining Rings, 7100 Series. All dimensions are in inches. Depth of groove $d = (D - G)/2$. Standard material: carbon spring steel (SAE 1060-1090). Thickness indicated is for unplated rings; for plated, phosphate coated, and stainless steel rings, the maximum ring thickness may be exceeded by 0.002 inch.

Ring Free Diameter Tolerances: For shaft sizes 0.078 through 0.138, +0.002, −0.003; for sizes 0.154 through 0.252, +0.002, −0.004; for sizes 0.310 through 0.440, +0.003, −0.005; for sizes 0.497 through 0.755, +0.004, −0.006. *Ring Thickness Tolerances:* For shaft sizes 0.078 through 0.158, ±0.002; for sizes 0.185 through 0.503, ±0.003; for sizes 0.622 through 0.755, ±0.004. *Groove Diameter Tolerances:* For shaft sizes less than 0.248, grooves are not recommended; for other sizes, grooves are optional. For shaft sizes 0.248 through 0.316, +0.005, −0.0015; for sizes 0.373 through 0.628, +0.001, −0.002; for sizes 0.745 and 0.755, +0.002, −0.003. *Groove Width Tolerances:* For shaft sizes 0.248 through 0.379, +0.003, −0.000; for sizes 0.434 through 0.755, +0.004, −0.000.

Table 17. Inch Series Internal and External Self-Locking Retaining Rings

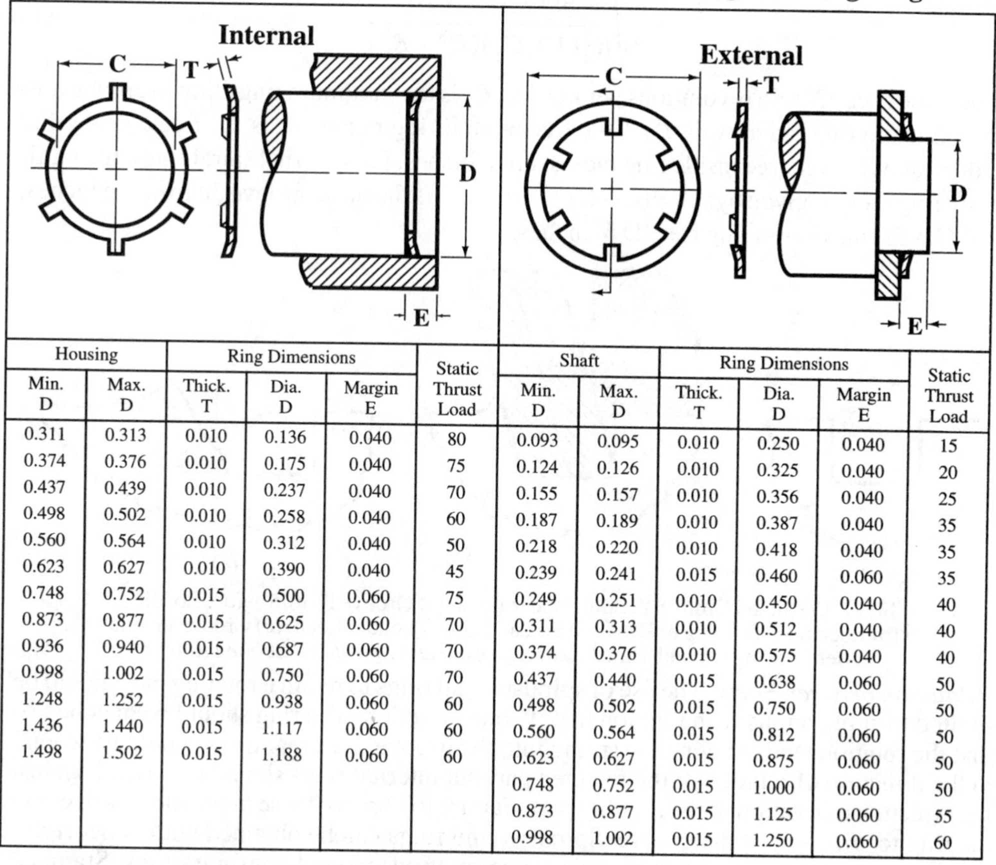

Housing		Ring Dimensions			Static Thrust Load	Shaft		Ring Dimensions			Static Thrust Load
Min. D	Max. D	Thick. T	Dia. D	Margin E		Min. D	Max. D	Thick. T	Dia. D	Margin E	
0.311	0.313	0.010	0.136	0.040	80	0.093	0.095	0.010	0.250	0.040	15
0.374	0.376	0.010	0.175	0.040	75	0.124	0.126	0.010	0.325	0.040	20
0.437	0.439	0.010	0.237	0.040	70	0.155	0.157	0.010	0.356	0.040	25
0.498	0.502	0.010	0.258	0.040	60	0.187	0.189	0.010	0.387	0.040	35
0.560	0.564	0.010	0.312	0.040	50	0.218	0.220	0.010	0.418	0.040	35
0.623	0.627	0.010	0.390	0.040	45	0.239	0.241	0.015	0.460	0.060	35
0.748	0.752	0.015	0.500	0.060	75	0.249	0.251	0.010	0.450	0.040	40
0.873	0.877	0.015	0.625	0.060	70	0.311	0.313	0.010	0.512	0.040	40
0.936	0.940	0.015	0.687	0.060	70	0.374	0.376	0.010	0.575	0.040	40
0.998	1.002	0.015	0.750	0.060	70	0.437	0.440	0.015	0.638	0.060	50
1.248	1.252	0.015	0.938	0.060	60	0.498	0.502	0.015	0.750	0.060	50
1.436	1.440	0.015	1.117	0.060	60	0.560	0.564	0.015	0.812	0.060	50
1.498	1.502	0.015	1.188	0.060	60	0.623	0.627	0.015	0.875	0.060	50
						0.748	0.752	0.015	1.000	0.060	50
						0.873	0.877	0.015	1.125	0.060	55
						0.998	1.002	0.015	1.250	0.060	60

Source: Industrial Retaining Rings, 6000 Series (internal) and 6100 Series (external). All dimensions are in inches, thrust loads are in pounds. Thickness indicated is for unplated rings. Standard material: carbon spring steel (SAE 1060-1090).

Internal Rings: Thrust loads are for rings made of standard material inserted into cold-rolled, low-carbon housing. *Ring Thickness Tolerances:* For housing sizes 0.311 through 0.627, ±0.001; for sizes 0.748 through 1.502, ±0.002. *Ring Diameter Tolerances:* For housing sizes 0.311 through 0.439, ±0.005; for sizes 0.498 through 1.502, ±0.010.

External Rings: Thrust loads are for rings made of standard material installed onto cold-rolled, low-carbon shafts. *Ring Thickness Tolerances:* For shaft sizes 0.093 through 0.220, ±0.001; for size 0.239, ±0.002; for sizes 0.249 through 0.376, ±0.001; for sizes 0.437 through 1.002, ±0.002. *Ring Diameter Tolerances:* For shaft sizes 0.093 through 0.502, ±0.005; for sizes 0.560 through 1.002, ±0.010.

Centrifugal Capacity: Proper functioning of a retaining ring depends on the ring remaining seated on the groove bottom. External rings "cling" to the groove bottom because the ring ID is slightly smaller than the diameter at the bottom of the groove. Ring speed should be kept below the allowable steady-state speed of the ring, or self-locking rings specially designed for high-speed applications should be used, otherwise an external ring can lose its grip on the groove. Applications of large retaining rings that tend to spin in their grooves when subjected to sudden acceleration or deceleration of the retained part can benefit from a ring with more "cling" (i.e., a smaller interior diameter) as long as the stress of installation is within permissible limits. Special rings are also available that lock into a hole in the bottom of the groove, thereby preventing rotation. The following equation can be used to determine the allowable steady-state speed N of an external spiral retaining ring:

$$N = \sqrt{\frac{0.466C_1E^3 \times 10^{12}}{R_n^3(1 + C_1)(R_o^3 - R_i^3)}} \qquad (1)$$

where the speed N is in revolutions per minute, C_1 is the minimum ring cling to groove bottom, E is the ring radial wall, R_n is the free neutral ring radius, R_o is the free outside ring radius, and R_i is the free inside ring radius, all in inches. For external spiral rings, the minimum ring cling is given by: $C_1 = (C - G)/G$, where C is the mean groove diameter in inches, and G is the maximum ring free ID in inches.

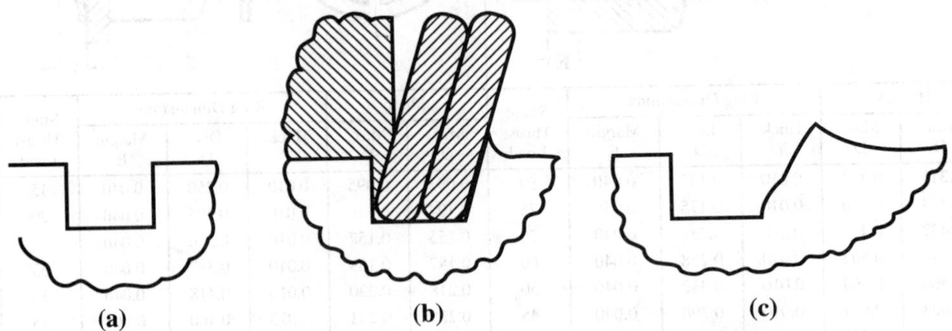

Fig. 2. Localized Groove Yielding under Load. (a) Groove Profile before Loading; (b) Localized Yielding of Retained Part and Groove under Load; (c) Groove Profile after Loading beyond Thrust Capacity (*Courtesy Spirolox Retaining Rings*)

Rotation between Parts: The use of spiral-wound rings to retain a rotating part should be limited to applications with rotation in only one direction. The ring should be matched so that the rotation tends to wind the spring into the groove. External rings should be wound in the direction of rotation of the retained part but internal rings should be wound against the direction of rotation of the rotating part. Failure to observe these precautions will cause the ring to wind out of the groove. Spiral-wound rings can be obtained with either right-hand (normal rotation) or left-hand (reverse rotation) wound configurations. Stamped retaining rings do not have these limitations, and may be used for applications that require rotation of the retained part, regardless of the direction of rotation.

Retaining Ring Failure.—Failure of a retaining ring application can result from failure of the ring itself, failure of the groove, or both. If a ring fails, the cause is likely to be from shearing of the ring. Shear failure occurs when a ring is installed in a groove and loaded by a retained part with both the groove and the retained part having a compressive yield strength greater than 45,000 psi; or when the load is applied through a retained part and groove, both having sharp corners and line-to-line contact; or when the ring is too thin in section compared with its diameter. To examine the possibility of ring shear, the allowable thrust P_s, based on the shear strength of the ring material, is given by

$$P_s = \frac{\pi D t S_s}{K} \qquad (2)$$

where P_s is in lb_f, D is the shaft or housing diameter in inches, t is the ring thickness in inches, S_s is the shear strength of the ring material in lb/in.^{2}, and K is the factor of safety.

Groove Failure: The most common type of groove failure is yielding of the groove material that occurs when the thrust load, applied through the retaining ring against the corner of the groove, exceeds the compressive yield strength of the groove. This yielding of the groove results from a low compressive yield strength of the groove material, and allows the ring to tilt and come out of the groove, as illustrated in Fig. 2(b).

When dishing of a ring occurs as a result of yielding in the groove material, a bending moment across the cross-section of the ring generates a tensile stress that is highest at the interior diameter of the ring. If the maximum stress is greater than the yield strength of the ring material, the ring ID will grow and the ring will become permanently dished in shape. To determine the thrust load capacity of a ring based on groove deformation, the allowable angle of ring deflection must be calculated, then the thrust load based on groove yield can be determined. However, for spiral-wound rings, the thrust load P_G that initiates the onset of groove deformation can be estimated from the following:

$$P_G = \frac{\pi D d S_y}{K} \tag{3}$$

where P_G is given in lb_f, D is the shaft or housing diameter in inches, d is the groove depth in inches, S_y is the yield strength of the groove material, and K is the safety factor. For stamped rings, estimate P_G by multiplying Equation (3) by the fraction of the groove circumference that contacts the ring.

The thrust load capacity of a particular retaining ring application can be increased by changing the workpiece material that houses the groove. Increasing the yield strength of the groove material increases the thrust load capacity of the retaining ring application. However, increasing the strength of the groove material may cause the failure mechanism to shift from groove deformation to ring shear. Therefore, use the lower of the values obtained from Equations (2) and (3) for the allowable thrust load.

Groove Design and Machining: In most applications, grooves are located near the end of a shaft or housing bore to facilitate installation and removal of the rings. The groove is normally located a distance at least two to three times the groove depth from the end of the shaft or bore. If the groove is too close to the end of the shaft or bore, the groove may shear or yield. The following equation can be used to determine the minimum safe distance Y of a groove from the end of a shaft or housing:

$$Y = \frac{K P_t}{\pi D S_c} \tag{4}$$

where K is the factor of safety, P_t is the thrust load on the groove in pounds, S_c is the shear strength of the groove material in psi, and D is the shaft or housing diameter in inches.

A properly designed and machined groove is just as important in a retaining ring application as the ring itself. The walls of grooves should be perpendicular to the shaft or bore diameter; the grooves should have square corners on the top edges, and radii at the bottom, within the tolerances specified by the manufacturers, as shown in Fig. 1 (page 1665). Test data indicate that the ultimate thrust capacity for both static and dynamic loading conditions is greatly affected if these groove requirements are not met. For spiral-wound rings, the maximum bottom groove radius is 0.005 inch for rings up to 1.000 inch free diameter and 0.010 inch for larger rings, internal or external. For stamped rings, the maximum bottom groove radius varies with ring size and style.

Table 18. Retaining Ring Standards

Military	
MIL-R-21248B	MS-16633 Open-type external uniform cross-section
	MS-16634 Open-type external uniform cross-section cylindrically bowed
	MS-3215 Open-type external tapered cross-section
	MS-16632 Crescent-type external
	MS-16625 Internal
	MS-16629 Internal cylindrically bowed
	MS-16624 Closed-type external tapered cross-section

Table 18. Retaining Ring Standards

Military	
MIL-R-21248B	MS-16628 Closed-type external tapered cross-section cylindrically bowed
	MS-16627 Internal inverted
	MS-16626 Closed-type external tapered cross-section
	MS-90707 Self-locking external tapered cross-section
	MS-3217 External heavy-duty tapered cross-section
MIL-R-27426	Uniform cross-section spiral retaining rings, Type 1-External, Type 2-Internal
Acrospace Standard	
AS 3215	Ring, Retaining—Spiral, Internal, Heavy Duty, Stainless Steel
AS 3216	Ring, Retaining—Spiral, External, Heavy Duty, Stainless Steel
AS 3217	Ring, Retaining—Spiral, Internal, Light Duty, Stainless Steel
AS 3218	Ring, Retaining—Spiral, External, Light Duty, Stainless Steel
AS 3219	Ring, Wound, Dimensional and Acceptance Standard for Spiral Wound Retaining Rings
ANSI	
B27.6-1972, R1983	General Purpose Uniform Cross-Section Spiral Retaining Rings
B27.7M-1977, R1983	General Purpose Tapered and Reduced Cross-Section Retaining Rings (Metric)
B27.2M-1977, R1983	General Purpose Metric Tapered and Reduced Cross-Section Retaining Rings
	Type 3DM1—Heavy Duty External Rings
	Type 3EM1—Reinforced "E" Rings
	Type 3FM1—"C" Type Rings
ANSI/SAE	
MA4016	Ring, Retaining—External Spiral Wound, Heavy and Medium Duty, Crescent, Metric
MA4017	Ring, Retaining—External Spiral Wound, Heavy and Medium Duty, Crescent, Metric
MA4020	Ring, Retaining—External Tapered, Type 1, Class 2, AMS 5520, Metric
MA4021	Ring, Retaining—Internal Tapered, Type 1, Class 1, AMS 5520, Metric
MA4029	Ring, Retaining—Internal, Beveled, Tapered, Type 2, Class 1, AMS 5520, Metric
MA4030	Ring, Retaining—External, Reinforced E-Ring, Type 1, Class 3, AMS 5520, Metric
MA4035	Rings, Retaining—Spiral Wound, Uniform Section, Corrosion Resistant, Procurement Specification for, Metric
MA4036	Ring, Retaining—Tapered Width, Uniform Thickness, Corrosion Resistant, Procurement Specification for, Metric
DIN	
DIN 471, 472, 6799, 984, 5417, 7993	Standards for normal and heavy type, internal and external retaining rings and retaining washers
LN 471, 472, 6799	Aerospace standards for internal and external retaining rings

WING NUTS, WING SCREWS AND THUMB SCREWS

Wing Nuts.—A wing nut is a nut having wings designed for manual turning without driver or wrench. As covered by ANSI B18.17-1968 (R1983) wing nuts are classified first, by type on the basis of the method of manufacture; and second, by style on the basis of design characteristics. They consist of:

Type A: Type A wing nuts are cold forged or cold formed solid nuts having wings of moderate height. In some sizes they are produced in regular, light, and heavy series to best suit the requirements of specific applications. Dimensions are given in Table 1.

Table 1. American National Standard Type A Wing Nuts
ANSI B18.17-1968, R1983

Nominal Size or Basic Major Diameter of Thread[a]	Thds. per Inch	Series[b]	Nut Blank Size (Ref)	*A* Wing Spread		*B* Wing Height		*C* Wing Thick.		*D* Between Wings		*E* Boss Diam.		*G* Boss Height	
				Max	Min	Max	Min	Max	Min	Max	Min	Max	Min	Max	Min
3 (0.0990)	48, 56	Hvy.	AA	0.72	0.59	0.41	0.28	0.11	0.07	0.21	0.17	0.33	0.29	0.14	0.10
4 (0.1120)	40, 38	Hvy.	AA	0.72	0.59	0.41	0.28	0.11	0.07	0.21	0.17	0.33	0.29	0.14	0.10
5 (0.1250)	40, 44	**Lgt.**	**AA**	**0.72**	**0.59**	**0.41**	**0.28**	**0.11**	**0.07**	**0.21**	**0.17**	**0.33**	**0.29**	**0.14**	**0.10**
		Hvy.	A	0.91	0.78	0.47	0.34	0.14	0.10	0.27	0.22	0.43	0.39	0.18	0.14
6 (0.1380)	32, 40	Lgt.	AA	0.72	0.59	0.41	0.28	0.11	0.07	0.21	0.17	0.33	0.29	0.14	0.10
		Hvy.	**A**	**0.91**	**0.78**	**0.47**	**0.34**	**0.14**	**0.10**	**0.27**	**0.22**	**0.43**	**0.39**	**0.18**	**0.14**
8 (0.1640)	32, 36	**Lgt.**	**A**	**0.91**	**0.78**	**0.47**	**0.34**	**0.14**	**0.10**	**0.27**	**0.22**	**0.43**	**0.39**	**0.18**	**0.14**
		Hvy.	B	1.10	0.97	0.57	0.43	0.18	0.14	0.33	0.26	0.50	0.45	0.22	0.17
10 (0.1900)	24, 32	**Lgt.**	**A**	**0.91**	**0.78**	**0.47**	**0.34**	**0.14**	**0.10**	**0.27**	**0.22**	**0.43**	**0.39**	**0.18**	**0.14**
		Hvy.	B	1.10	0.97	0.57	0.43	0.18	0.14	0.33	0.26	0.50	0.45	0.22	0.17
12 (0.2160)	24, 28	**Lgt.**	**B**	**1.10**	**0.97**	**0.57**	**0.43**	**0.18**	**0.14**	**0.33**	**0.26**	**0.50**	**0.45**	**0.22**	**0.17**
		Hvy.	C	1.25	1.12	0.66	0.53	0.21	0.17	0.39	0.32	0.58	0.51	0.25	0.20
¼ (0.2500)	20, 28	**Lgt.**	**B**	**1.10**	**0.97**	**0.57**	**0.43**	**0.18**	**0.14**	**0.39**	**0.26**	**0.50**	**0.45**	**0.22**	**0.17**
		Reg.	C	1.25	1.12	0.66	0.53	0.21	0.17	0.39	0.32	0.58	0.51	0.25	0.20
		Hvy.	D	1.44	1.31	0.79	0.65	0.24	0.20	0.48	0.42	0.70	0.64	0.30	0.26
5/16 (0.3125)	18, 24	**Lgt.**	**C**	**1.25**	**1.12**	**0.66**	**0.53**	**0.21**	**0.17**	**0.39**	**0.32**	**0.58**	**0.51**	**0.25**	**0.20**
		Reg.	D	1.44	1.31	0.79	0.65	0.24	0.20	0.48	0.42	0.70	0.64	0.30	0.26
		Hvy.	E	1.94	1.81	1.00	0.87	0.33	0.26	0.65	0.54	0.93	0.86	0.39	0.35
⅜ (0.3750)	16, 24	**Lgt.**	**D**	**1.44**	**1.31**	**0.79**	**0.65**	**0.24**	**0.20**	**0.48**	**0.42**	**0.70**	**0.64**	**0.30**	**0.26**
		Reg.	E	1.94	1.81	1.00	0.87	0.33	0.26	0.65	0.54	0.93	0.86	0.39	0.35
7/16 (0.4375)	14, 20	**Lgt.**	**E**	**1.94**	**1.81**	**1.00**	**0.87**	**0.33**	**0.26**	**0.65**	**0.54**	**0.93**	**0.86**	**0.39**	**0.35**
		Hvy.	F	2.76	2.62	1.44	1.31	0.40	0.34	0.90	0.80	1.19	1.13	0.55	0.51
½ (0.5000)	13, 20	**Lgt.**	**E**	**1.94**	**1.81**	**1.00**	**0.87**	**0.33**	**0.26**	**0.65**	**0.54**	**0.93**	**0.86**	**0.39**	**0.35**
		Hvy.	F	2.76	2.62	1.44	1.31	0.40	0.34	0.90	0.80	1.19	1.13	0.55	0.51
9/16 (0.5625)	12, 18	Hvy.	F	2.76	2.62	1.44	1.31	0.40	0.34	0.90	0.80	1.19	1.13	0.55	0.51
⅝ (0.6250)	11, 18	Hvy.	F	2.76	2.62	1.44	1.31	0.40	0.34	0.90	0.80	1.19	1.13	0.55	0.51
¾ (0.7500)	10, 16	Hvy.	F	2.76	2.62	1.44	1.31	0.40	0.34	0.90	0.80	1.19	1.13	0.55	0.51

[a] Where specifying nominal size in decimals, zeros in the fourth decimal place are omitted.
[b] Lgt. = Light; Hvy. = Heavy; Reg. = Regular. Sizes shown in **bold face** are preferred.

All dimensions in inches.

Type B: Type B wing nuts are hot forged solid nuts available in two wing styles: Style 1, having wings of moderate height; and Style 2, having high wings. Dimensions are given in Table 2.

Table 2. American National Standard Type B Wing Nuts
ANSI B18.17-1968, R1983

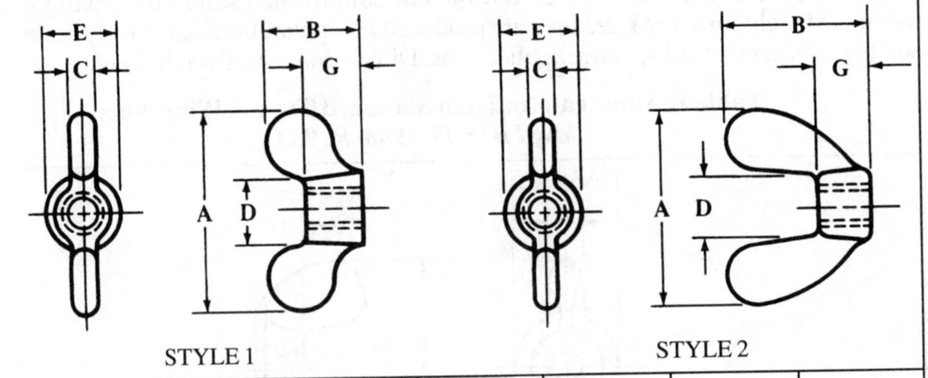

Nominal Size or Basic Major Diameter of Thread[a]	Thds. per Inch	*A* Wing Spread		*B* Wing Height		*C* Wing Thick.		*D* Between Wings		*E* Boss Diam.		*G* Boss Height	
		Max	Min	Max	Min	Max	Min	Max	Min	Max	Min	Max	Min
Type B, Style 1													
5 (0.1250)	40	0.78	0.72	0.36	0.30	0.13	0.10	0.28	0.22	0.31	0.28	0.22	0.16
10 (0.1900)	24	0.97	0.91	0.45	0.39	0.15	0.12	0.34	0.28	0.39	0.36	0.28	0.22
1/4 (0.2500)	20	1.16	1.09	0.56	0.50	0.17	0.14	0.41	0.34	0.47	0.44	0.34	0.28
5/16 (0.3125)	18	1.44	1.38	0.67	0.61	0.18	0.15	0.50	0.44	0.55	0.52	0.41	0.34
3/8 (0.3750)	16	1.72	1.66	0.80	0.73	0.20	0.17	0.59	0.53	0.63	0.60	0.47	0.41
7/16 (0.4375)	14	2.00	1.94	0.91	0.84	0.21	0.18	0.69	0.62	0.71	0.68	0.53	0.47
1/2 (0.5000)	13	2.31	2.22	1.06	0.94	0.23	0.20	0.78	0.69	0.79	0.76	0.62	0.50
9/16 (0.5625)	12	2.59	2.47	1.17	1.05	0.25	0.21	0.88	0.78	0.88	0.84	0.69	0.56
5/8 (0.6250)	11	2.84	2.72	1.31	1.19	0.27	0.23	0.94	0.84	0.96	0.92	0.75	0.62
3/4 (0.7500)	10	3.31	3.19	1.52	1.39	0.29	0.25	1.10	1.00	1.12	1.08	0.88	0.75
Type B, Style 2													
5 (0.1250)	40	0.81	0.75	0.62	0.56	0.12	0.09	0.28	0.22	0.31	0.28	0.22	0.16
10 (0.1900)	24	1.01	0.95	0.78	0.72	0.14	0.11	0.35	0.29	0.39	0.36	0.28	0.22
1/4 (0.2500)	20	1.22	1.16	0.94	0.88	0.16	0.13	0.41	0.35	0.47	0.44	0.34	0.28
5/16 (0.3125)	18	1.43	1.37	1.09	1.03	0.17	0.14	0.48	0.42	0.55	0.52	0.41	0.34
3/8 (0.3750)	16	1.63	1.57	1.25	1.19	0.18	0.15	0.55	0.49	0.63	0.60	0.47	0.41
7/16 (0.4375)	14	1.90	1.84	1.42.	1.36	0.19	0.16	0.62	0.56	0.71	0.68	0.53	0.47
1/2 (0.5000)	13	2.13	2.04	1.58	1.45	0.20	0.17	0.69	0.60	0.79	0.76	0.62	0.50
9/16 (0.5625)	12	2.40	2.28	1.75	1.62	0.22	0.18	0.76	0.67	0.88	0.84	0.69	0.56
5/8 (0.6250)	11	2.60	2.48	1.91	1.78	0.23	0.19	0.83	0.74	0.96	0.92	0.75	0.62
3/4 (0.7500)	10	3.02	2.90	2.22	2.09	0.24	0.20	0.97	0.88	1.12	1.08	0.88	0.75

[a] Where specifying nominal size in decimals, zeros in the fourth decimal place are omitted.

All dimensions in inches.

Table 3. American National Standard Type C Wing Nuts *ANSI B18.17-1968, R1983*

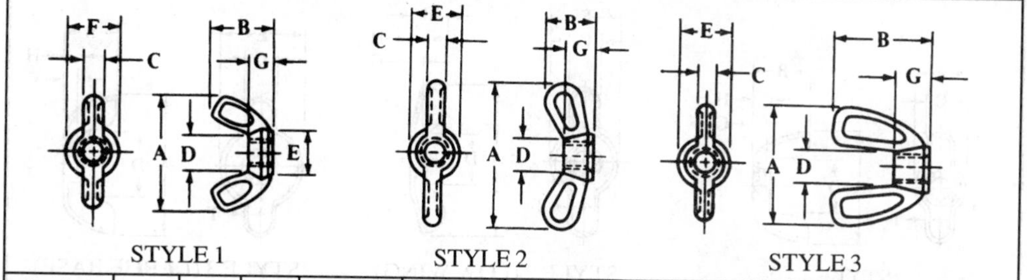

STYLE 1 STYLE 2 STYLE 3

Nominal Size or Basic Major Diameter of Thread[a]	Thds. per Inch	Series	Nut Blank Size (Ref)	*A* Wing Spread		*B* Wing Height		*C* Wing Thick.		*D* Between Wings		*E* Boss Diam.		*F* Boss Diam.		*G* Boss Height	
				Max	Min	Max	Min	Max	Min	Max	Min	Max	Min	Max	Min	Max	Min
Type C, Style 1																	
4 (0.1120)	40	Reg.	AA	0.66	0.64	0.36	0.35	0.11	0.09	0.18	0.16	0.27	0.25	0.32	0.30	0.16	0.14
5 (0.1250)	40	Reg.	AA	0.66	0.64	0.36	0.35	0.11	0.09	0.18	0.16	0.27	0.25	0.32	0.30	0.16	0.14
6 (0.1380)	32	**Reg.**	**AA**	**0.66**	**0.64**	**0.36**	**0.35**	**0.11**	**0.09**	**0.18**	**0.16**	**0.27**	**0.25**	**0.32**	**0.30**	**0.16**	**0.14**
		Hvy.	A	0.85	0.83	0.43	0.42	0.14	0.12	0.29	0.27	0.38	0.36	0.41	0.40	0.20	0.18
8 (0.1640)	32	Reg.	A	0.85	0.83	0.43	0.42	0.14	0.12	0.29	0.27	0.38	0.36	0.41	0.40	0.20	0.18
10 (0.1900)	24, 32	Reg.	A	0.85	0.83	0.43	0.42	0.14	0.12	0.29	0.27	0.38	0.36	0.41	0.40	0.20	0.18
12 (0.2160)	24	**Reg.**	A	**0.85**	**0.83**	**0.43**	**0.42**	**0.14**	**0.12**	**0.29**	**0.27**	**0.38**	**0.36**	**0.41**	**0.40**	**0.20**	**0.18**
		Hvy.	B	1.08	1.05	0.57	0.53	0.16	0.14	0.32	0.30	0.44	0.42	0.48	0.46	0.23	0.21
1/4 (0.2500)	20, 28	Reg.	B	1.08	1.05	0.57	0.53	0.16	0.14	0.32	0.30	0.44	0.42	0.48	0.46	0.23	0.21
5/16 (0.3125)	18, 24	Reg.	C	1.23	1.20	0.64	0.62	0.20	0.18	0.39	0.35	0.50	0.49	0.57	0.55	0.26	0.24
3/8 (0.3750)	16, 24	Reg.	D	1.45	1.42	0.74	0.72	0.23	0.21	0.46	0.42	0.62	0.60	0.69	0.67	0.29	0.27
7/16 (0.4375)	14, 20	**Reg.**	**E**	**1.89**	**1.86**	**0.91**	**0.90**	**0.29**	**0.28**	**0.67**	**0.65**	**0.75**	**0.73**	**0.83**	**0.82**	**0.38**	**0.37**
		Hvy.	EH	1.89	1.86	0.93	0.91	0.34	0.33	0.63	0.62	0.81	0.79	0.89	0.87	0.42	0.40
1/2 (0.5000)	13, 20	**Reg.**	**E**	**1.89**	**1.86**	**0.91**	**0.90**	**0.29**	**0.28**	**0.67**	**0.65**	**0.75**	**0.73**	**0.83**	**0.82**	**0.38**	**0.37**
		Hvy.	EH	1.89	1.86	0.93	0.91	0.34	0.33	0.63	0.62	0.81	0.79	0.89	0.87	0.42	0.40
Type C, Style 2																	
5 (0.1250)	40	…	…	0.82	0.80	0.25	0.23	0.09	0.08	0.21	0.19	0.26	0.24	…	…	0.17	0.15
6 (0.1380)	32	…	…	0.82	0.80	0.25	0.23	0.09	0.08	0.21	0.19	0.26	0.24	…	…	0.17	0.15
8 (0.1640)	32	…	…	1.01	0.99	0.28	0.27	0.11	0.09	0.29	0.28	0.36	0.34	…	…	0.19	0.18
10 (0.1900)	24, 32	…	…	1.01	0.99	0.28	0.27	0.11	0.09	0.29	0.28	0.36	0.34	…	…	0.19	0.18
12 (0.2160)	24	…	…	1.20	1.18	0.32	0.31	0.12	0.11	0.38	0.37	0.44	0.43	…	…	0.22	0.20
1/4 (0.2500)	20	…	…	1.20	1.18	0.32	0.31	0.12	0.11	0.38	0.37	0.44	0.43	…	…	0.22	0.20
5/16 (0.3125)	18	…	…	1.51	1.49	0.36	0.35	0.14	0.12	0.44	0.43	0.51	0.49	…	…	0.24	0.23
3/8 (0.3750)	16	…	…	1.89	1.86	0.58	0.55	0.20	0.17	0.44	0.43	0.63	0.62	…	…	0.37	0.35
Type C, Style 3																	
5 (0.1250)	40	…	…	0.92	0.89	0.70	0.67	0.16	0.15	0.26	0.24	0.38	0.36	…	…	0.25	0.24
6 (0.1380)	32	…	…	0.92	0.89	0.70	0.67	0.16	0.15	0.26	0.24	0.38	0.36	…	…	0.25	0.24
8 (0.1640)	32	…	…	0.92	0.89	0.70	0.67	0.16	0.15	0.26	0.24	0.38	0.36	…	…	0.25	0.24
10 (0.1900)	24, 32	…	…	1.14	1.12	0.85	0.83	0.19	0.17	0.32	0.30	0.44	0.42	…	…	0.29	0.27
12 (0.2160)	24	…	…	1.14	1.12	0.85	0.83	0.19	0.17	0.32	0.30	0.44	0.42	…	…	0.29	0.27
1/4 (0.2500)	20	…	…	1.14	1.12	0.85	0.83	0.19	0.17	0.32	0.30	0.44	0.42	…	…	0.29	0.27
5/16 (0.3125)	18	…	…	1.29	1.27	1.04	1.02	0.23	0.22	0.39	0.36	0.50	0.49	…	…	0.35	0.34
3/8 (0.3750)	16	…	…	1.51	1.49	1.20	1.18	0.27	0.25	0.45	0.42	0.62	0.60	…	…	0.43	0.42

[a] Where specifying nominal size in decimals, zeros in the fourth decimal place are omitted.

All dimensions in inches. Sizes shown in **bold face** are preferred.

Type C: Type C wing nuts are die cast solid nuts and are available in three wing styles: Style 1, having wings of moderate height; Style 2, having low wings; and Style 3, having high wings. In some sizes, the Style 1 nuts are produced in regular, light, and heavy series to best suit the requirements of specific applications. Dimensions are given in Table 3.

Table 4. American National Standard Type D Wing Nuts *ANSI B18.17-1968, R1983*

STYLE 1 STYLE 2 (LOW WING) STYLE 3 (LARGE BASE)

Nominal Size or Basic Major Diameter of Thread[a]	Thds. per Inch	Series[b]	*A* Wing Spread		*B* Wing Height		*C* Wing Thick.		*D* Between Wings	*E* Boss Diam.		*G* Boss Hgt.	*H* Wall Hgt.	*T* Stock Thick.	
			Max	Min	Max	Min	Max	Min	Min	Max	Min	Min	Min	Max	Min
Type D, Style 1															
8 (0.1640)	32, 36	…	0.78	0.72	0.40	0.34	0.18	0.14	0.25	0.41	0.35	0.08	0.12	0.04	0.03
10 (0.1900)	24, 32	…	0.91	0.85	0.47	0.41	0.21	0.17	0.34	0.53	0.47	0.10	0.12	0.04	0.03
12 (0.2160)	24, 28	…	1.09	1.03	0.47	0.41	0.21	0.17	0.34	0.53	0.47	0.10	0.12	0.05	0.04
1/4 (0.2500)	20, 28	…	1.11	1.05	0.50	0.44	0.25	0.21	0.34	0.62	0.56	0.11	0.12	0.05	0.04
5/16 (0.3125)	18, 24	…	1.30	1.24	0.59	0.53	0.30	0.26	0.46	0.73	0.67	0.14	0.18	0.06	0.05
3/8 (0.3750)	16, 24	…	1.41	1.34	0.67	0.61	0.34	0.30	0.69	0.83	0.77	0.16	0.18	0.06	0.05
Type D, Style 2															
5 (0.1250)	40	Reg.	1.03	0.97	0.25	0.19	0.19	0.13	0.30	0.40	0.34	0.07	0.09	0.04	0.03
6 (0.1380)	32	Reg.	1.03	0.97	0.25	0.19	0.19	0.13	0.30	0.40	0.34	0.08	0.09	0.04	0.03
8 (0.1640)	32	Reg.	1.03	0.97	0.25	0.19	0.19	0.13	0.30	0.40	0.34	0.08	0.09	0.04	0.03
10 (0.1900)	24, 32	Reg.	1.40	1.34	0.34	0.28	0.25	0.18	0.32	0.53	0.47	0.09	0.16	0.05	0.04
		Hvy.	1.21	1.16	0.28	0.26	0.31	0.25	0.60	0.61	0.55	0.09	0.13	0.05	0.04
12 (0.2160)	24	Reg.	1.21	1.16	0.28	0.26	0.31	0.25	0.60	0.61	0.55	0.11	0.13	0.05	0.04
1/4 (0.2500)	20	Reg.	1.21	1.16	0.28	0.26	0.31	0.25	0.60	0.61	0.55	0.11	0.13	0.05	0.04
Type D, Style 3															
10 (0.1900)	24, 32	Lgt.	1.31	1.25	0.48	0.42	0.29	0.23	0.47	0.65	0.59	0.08	0.12	0.04	0.03
		Reg.	1.40	1.34	0.53	0.47	0.25	0.19	0.50	0.75	0.69	0.08	0.14	0.04	0.03
12 (0.2160)	24	Reg.	1.28	1.22	0.40	0.34	0.23	0.17	0.59	0.73	0.67	0.11	0.12	0.04	0.03
1/4 (0.2500)	20	Lgt.	1.28	1.22	0.40	0.34	0.23	0.17	0.59	0.73	0.67	0.11	0.12	0.04	0.03
		Reg.	1.78	1.72	0.66	0.60	0.31	0.25	0.70	1.03	0.97	0.14	0.17	0.06	0.04
		Hvy.	1.47	1.40	0.50	0.44	0.37	0.31	0.66	1.03	0.97	0.14	0.14	0.08	0.06
5/16 (0.3125)	18	Reg.	1.78	1.72	0.66	0.60	0.31	0.25	0.70	1.03	0.97	0.14	0.17	0.06	0.04
		Hvy.	1.47	1.40	0.50	0.44	0.37	0.31	0.66	1.03	0.97	0.14	0.14	0.08	0.06

[a] Where specifying nominal size in decimals, zeros in the fourth decimal place are omitted.

[b] Lgt. = Light; Hvy. = Heavy; Reg. = Regular.

All dimensions in inches.

Type D: Type D wing nuts are stamped sheet metal nuts and are available in three styles: Style 1, having wings of moderate height; Style 2, having low wings; and Style 3, having wings of moderate height and a larger bearing surface. In some sizes, Styles 2 and 3 are produced in regular, light, and heavy series to best suit the requirements of specific applications. Dimensions are given in Table 4.

Specification of Wing Nuts.—When specifying wing nuts, the following data should be included in the designation and should appear in the following sequence: nominal size (number, fraction or decimal equivalent), threads per inch, type, style and/or series, material, and finish.

Examples: 10—32 Type A Wing Nut, Regular Series, Steel, Zinc Plated.
0.250—20 Type C Wing Nut, Style 1, Zinc Alloy, Plain.

Threads for Wing Nuts.—Threads are in conformance with the ANSI Standard Unified Thread, Class 2B for all types of wing nuts except type D which have a modified Class 2B thread. Because of the method of manufacture, the minor diameter of the thread in type D nuts may be somewhat larger than the Unified Thread Class 2B maximum but shall in no case exceed the minimum pitch diameter.

Materials and Finish for Wing Nuts.—Types A, B, and D wing nuts are normally supplied as specified by the user in carbon steel, brass or corrosion resistant steel of good quality and adaptable to the manufacturing process. Type C wing nuts are made from die cast zinc alloy. Unless otherwise specified, wing nuts are supplied with a plain (unplated or uncoated) finish.

Wing Screws.—A wing screw is a screw having a wing-shaped head designed for manual turning without a driver or wrench. As covered by ANSI B18.17-1968 (R1983) wing screws are classified first, by type on the basis of the method of manufacture, and second, by style on the basis of design characteristics. They consist of the following:

Type A: Type A wing screws are of two-piece construction having cold formed or cold forged wing portions of moderate height. In some sizes they are produced in regular, light, and heavy series to best suit the requirements of specific applications. Dimensions are given in Table .

Type B: Type B wing screws are of hot forged one-piece construction available in two wing styles: Style 1, having wings of moderate height; and Style 2, having high wings. Dimensions are given in Table .

Type C: Type C wing screws are available in two styles: Style 1, of a one-piece die cast construction having wings of moderate height; and Style 2, of a two-piece construction having a die cast wing portion of moderate height. Dimensions are given in Table 6.

Type D: Type D wing screws are of two-piece welded construction having stamped sheet metal wing portions of moderate height. Dimensions are given in Table 6.

Materials for Wing Screws and Thumb Screws: Type A wing screws are normally supplied in carbon steel with the shank portion case hardened. When so specified, they also may be made from corrosion resistant steel, brass or other materials as agreed upon by the manufacturer and user.

Type B wing screws are normally made from carbon steel but also may be made from corrosion resistant steel, brass or other materials.

Type C, Style 1, wing screws are supplied only in die cast zinc alloy. Type C, Style 2, wing screws have the wing portion made from die cast zinc alloy with the shank portion normally made from carbon steel. Where so specified, the shank portion may be made from corrosion resistant steel, brass or other materials as agreed upon by the manufacturer and user.

Type D wing screws are normally supplied in carbon steel but also may be made from corrosion resistant steel, brass or other materials.

Thumb screws of all types are normally made from a good commercial quality of carbon steel having a maximum ultimate tensile strength of 48,000 psi. Where so specified, carbon steel thumb screws are case hardened. They are also made from corrosion resistant steel, brass, and other materials as agreed upon by the manufacturer and user.

Unless otherwise specified, wing screws and thumb screws are supplied with a plain (unplated or uncoated) finish.

Thumb Screws: A thumb screw is a screw having a flattened head designed for manual turning without a driver or wrench. As covered by ANSI B18.17-1968 (R1983) thumb screws are classified by type on the basis of design characteristics. They consist of the following:

Type A: Type A thumb screws are forged one-piece screws having a shoulder under the head and are available in two series: regular and heavy. Dimensions are given in Table .

Type B: Type B thumb screws are forged one-piece screws without a shoulder and are available in two series: regular and heavy. Dimensions are given in Table .

Table 5. American National Standard Types A and B Wing Screws

ANSI B18.17-1968, R1983

E B L C G A See Footnote 1

SHANK PERMANENTLY INSERTED INTO WING PORTION

TYPE A

E B L C G A See Footnote 1 Style 1

E B L C G A See Footnote 1 Style 2

TYPE B

Nominal Size or Basic Major Diameter[a]	Thds. per Inch	Series[b]	Head Blank size (Ref)	A Wing Spread		B Wing Height		C Wing Thick.		E Boss Diam.		G Boss Height.		L Practical Screw Lengths	
				Max	Min	Max	Min	Max	Min	Max	Min	Max	Min	Max	Min
Type A															
4 (0.1120)	40	Hvy.	AA	0.72	0.59	0.41	0.28	0.11	0.07	0.33	0.29	0.14	0.10	0.75	0.25
6 (0.1380)	32	Lgt.	AA	0.72	0.59	0.41	0.28	0.11	0.07	0.33	0.29	0.14	0.10	} 0.75	0.25
		Hvy.	**A**	**0.91**	**0.78**	**0.47**	**0.34**	**0.14**	**0.10**	**0.43**	**0.39**	**0.18**	**0.14**		
8 (0.1640)	32	**Lgt.**	**A**	**0.91**	**0.78**	**0.47**	**0.34**	**0.14**	**0.10**	**0.43**	**0.39**	**0.18**	**0.14**	} 0.75	0.38
		Hvy.	B	1.10	0.97	0.57	0.43	0.18	0.14	0.50	0.45	0.22	0.17		
10 (0.1900)	24, 32	**Lgt.**	**A**	**0.91**	**0.78**	**0.47**	**0.34**	**0.14**	**0.10**	**0.43**	**0.39**	**0.18**	**0.14**	} 1.00	0.38
		Hvy.	B	1.10	0.97	0.57	0.43	0.18	0.14	0.50	0.45	0.22	0.17		
12 (0.2160)	24	**Lgt.**	**B**	**1.10**	**0.97**	**0.57**	**0.43**	**0.18**	**0.14**	**0.50**	**0.45**	**0.22**	**0.17**	} 1.00	0.38
		Hvy.	C	1.25	1.12	0.66	0.53	0.21	0.17	0.58	0.51	0.25	0.20		
¼ (0.2500)	20	**Lgt.**	**B**	**1.10**	**0.97**	**0.57**	**0.43**	**0.18**	**0.14**	**0.50**	**0.45**	**0.22**	**0.17**		
		Reg.	C	1.25	1.12	0.66	0.53	0.21	0.17	0.58	0.51	0.25	0.20	} 1.50	0.50
		Hvy.	D	1.44	1.31	0.79	0.65	0.24	0.20	0.70	0.64	0.30	0.26		
⁵⁄₁₆ (0.3125)	18	**Lgt.**	**C**	**1.25**	**1.12**	**0.66**	**0.53**	**0.21**	**0.17**	**0.58**	**0.51**	**0.25**	**0.20**		
		Reg.	D	1.44	1.31	0.79	0.65	0.24	0.20	0.70	0.64	0.30	0.26	} 1.50	0.50
		Hvy.	E	1.94	1.81	1.00	0.87	0.33	0.26	0.93	0.86	0.39	0.35		
⅜ (0.3750)	16	**Lgt.**	**D**	**1.44**	**1.31**	**0.79**	**0.65**	**0.24**	**0.20**	**0.70**	**0.64**	**0.30**	**0.26**		
		Reg.	E	1.94	1.81	1.00	0.87	0.33	0.26	0.93	0.86	0.39	0.35	} 2.00	0.75
		Hvy.	F	2.76	2.62	1.44	1.31	0.40	0.34	1.19	1.13	0.55	0.51		
⁷⁄₁₆ (0.4375)	14	**Lgt.**	**E**	**1.94**	**1.81**	**1.00**	0.87	**0.33**	**0.26**	**0.93**	**0.86**	**0.39**	**0.35**	} 4.00	1.00
		Hvy.	F	2.76	2.62	1.44	1.31	0.40	0.34	1.19	1.13	0.55	0.51		
½ (0.5000)	13	**Lgt.**	**E**	**1.94**	**1.81**	**1.00**	**0.87**	**0.33**	**0.26**	**0.93**	**0.86**	**0.39**	**0.35**	} 4.00	1.00
		Hvy.	F	2.76	2.62	1.44	1.31	0.40	0.34	1.19	1.13	0.55	0.51		
⅝ (0.6250)	11	Hvy.	F	2.76	2.62	1.44	1.31	0.40	0.34	1.19	1.13	0.55	0.51	4.00	1.25
Type B, Style 1															
10 (0.1900)	24	…	…	0.97	0.91	0.45	0.39	0.15	0.12	0.39	0.36	0.28	0.22	2.00	0.50
¼ (0.2500)	20	…	…	1.16	1.09	0.56	0.50	0.17	0.14	0.47	0.44	0.34	0.28	3.00	0.50
⁵⁄₁₆ (0.3125)	18	…	…	1.44	1.38	0.67	0.61	0.18	0.15	0.55	0.52	0.41	0.34	3.00	0.50
⅜ (0.3750)	16	…	…	1.72	1.66	0.80	0.73	0.20	0.17	0.63	0.60	0.47	0.41	4.00	0.50
⁷⁄₁₆ (0.4375)	14	…	…	2.00	1.94	0.91	0.84	0.21	0.18	0.71	0.68	0.53	0.47	3.00	1.00
½ (0.5000)	13	…	…	2.31	2.22	1.06	0.94	0.23	0.20	0.79	0.76	0.62	0.50	3.00	1.00
⅝ (0.6250)	11	…	…	2.84	2.72	1.31	1.19	0.27	0.23	0.96	0.92	0.75	0.62	2.50	1.00
Type B, Style 2															
10 (0.1900)	24	…	…	1.01	0.95	0.78	0.72	0.14	0.11	0.39	0.36	0.28	0.22	1.25	0.50
¼ (0.2500)	20	…	…	1.22	1.16	0.94	0.88	0.16	0.13	0.47	0.44	0.34	0.28	2.00	0.50
⁵⁄₁₆ (0.3125)	18	…	…	1.43	1.37	1.09	1.03	0.17	0.14	0.55	0.52	0.41	0.34	2.00	0.50
⅜ (0.3750)	16	…	…	1.63	1.57	1.25	1.19	0.18	0.15	0.63	0.60	0.47	0.41	2.00	0.50

[a] Where specifyin nominal size in decimals, zeros in the fourth decimal place are omitted.

[b] Hvy. = Heavy; Lgt. = Light; Reg. = Regular.

All dimensions in inches. Sizes shown in **bold face** are preferred.

[1] Plain point, unless alternate point from styles shown in Table 8 is specified by user.

Table 6. American National Standard Types C and D Wing Screws
ANSI B18.17-1968, R1983

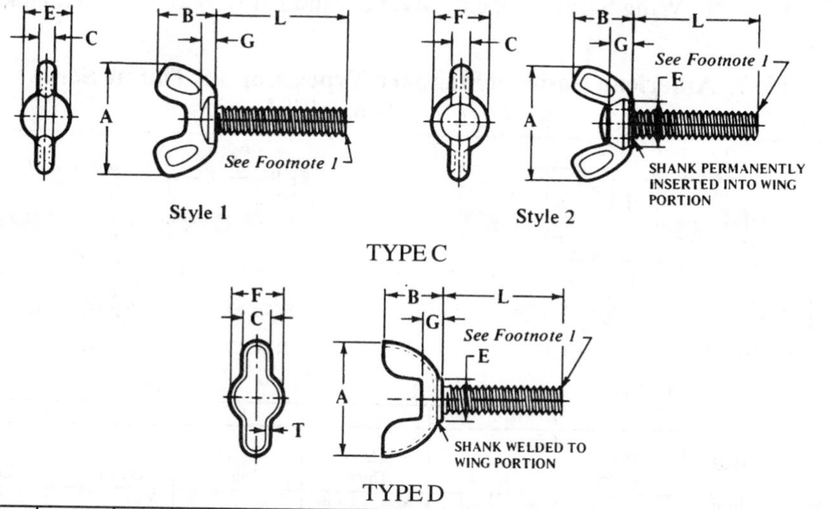

See Footnote 1

SHANK PERMANENTLY INSERTED INTO WING PORTION

Style 1

Style 2

TYPE C

See Footnote 1

SHANK WELDED TO WING PORTION

TYPE D

Nominal Size or Basic Screw Diameter[a]	Thds. per Inch	A Wing Spread		B Wing Height		C Wing Thick.		E Boss Diam.		F Boss Diam.		G Height		L Practical Screw Lengths	
		Max	Min	Max	Min	Max	Min	Max	Min	Max	Min	Max	Min	Max	Min
Type C, Style 1															
6 (0.1380)	32	0.85	0.83	0.45	0.43	0.15	0.12	…	…	0.41	0.39	0.12	0.07	0.75	0.25
8 (0.1640)	32	0.85	0.83	0.45	0.43	0.15	0.12	…	…	0.41	0.39	0.12	0.07	1.00	0.38
10 (0.1900)	24, 32	0.85	0.83	0.45	0.43	0.15	0.12	…	…	0.41	0.39	0.12	0.07	1.25	0.38
¼ (0.2500)	20	1.08	1.05	0.56	0.53	0.17	0.14	…	…	0.46	0.44	0.12	0.07	1.50	0.50
5⁄16 (0.3125)	18	1.23	1.20	0.64	0.62	0.22	0.19	…	…	0.51	0.49	0.14	0.10	1.50	0.50
⅜ (0.3750)	16	1.45	1.42	0.74	0.72	0.24	0.21	…	…	0.63	0.62	0.15	0.12	1.50	0.50
Type C, Style 2															
6 (0.1380)	32	0.85	0.83	0.43	0.42	0.14	0.12	0.38	0.36	0.41	0.40	0.20	0.18	1.00	0.25
8 (0.1640)	32	0.85	0.83	0.43	0.42	0.14	0.12	0.38	0.36	0.41	0.40	0.20	0.18	1.00	0.38
10 (0.1900)	24, 32	0.85	0.83	0.43	0.42	0.14	0.12	0.38	0.36	0.41	0.40	0.20	0.18	2.00	0.38
¼ (0.2500)	20	1.08	1.05	0.57	0.53	0.16	0.14	0.44	0.42	0.48	0.46	0.23	0.21	2.50	0.50
5⁄16 (0.3125)	18	1.23	1.20	0.64	0.62	0.20	0.18	0.50	0.49	0.57	0.55	0.26	0.24	3.00	0.50
⅜ (0.3750)	16	1.45	1.42	0.74	0.72	0.23	0.21	0.62	0.60	0.69	0.67	0.29	0.27	3.00	0.75
7⁄16 (0.4375)	14	1.89	1.86	0.91	0.90	0.29	0.28	0.75	0.73	0.83	0.82	0.38	0.37	4.00	1.00
½ (0.5000)	13	1.89	1.86	0.91	0.90	0.29	0.28	0.75	0.73	0.83	0.82	0.38	0.37	4.00	1.00
Type D															
6 (0.1380)	32	0.78	0.72	0.40	0.34	0.18	0.12	0.35	0.31	0.40	0.34	0.21	0.14	0.75	0.25
8 (0.1640)	32	0.78	0.72	0.40	0.34	0.18	0.12	0.35	0.31	0.40	0.34	0.21	0.14	0.75	0.38
10 (0.1900)	24	0.90	0.84	0.46	0.40	0.21	0.15	0.35	0.31	0.53	0.47	0.22	0.16	1.00	0.38
12 (0.2160)	24	1.09	1.03	0.46	0.40	0.26	0.20	0.44	0.39	0.61	0.55	0.24	0.18	1.00	0.38
¼ (0.2500)	20	1.09	1.03	0.46	0.40	0.26	0.20	0.47	0.43	0.61	0.55	0.24	0.18	1.50	0.50
5⁄16 (0.3125)	18	1.31	1.25	0.62	0.56	0.29	0.23	0.57	0.53	0.68	0.62	0.29	0.23	1.50	0.50
⅜ (0.3750)	16	1.31	1.25	0.62	0.56	0.29	0.23	0.63	0.59	0.68	0.62	0.29	0.23	2.00	0.75

[a] Where specifying nominal size in decimals, zeros in the fourth decimal place are omitted.

All dimensions in inches.

[1] Plain point, unless alternate point from styles shown in Table 8 is specified by user.

Wing Screw and Thumb Screw Designation.—When specifying wing and thumb screws, the following data should be included in the designation and should appear in the following sequence: nominal size (number, fraction or decimal equivalent), threads per inch, length (fractions or decimal equivalents), type, style and/or series, point (if other than plain point), materials, and finish.

Examples: 10—32 × 1¼, Thumb Screw, Type A, Regular, Steel, Zinc Plated.

0.375—16 × 2.00, Wing Screw, Type B, Style 2, Steel, Cadmium Plated.

0.250—20 × 1.50, Wing Screw, Type C, Style 2, Zinc Alloy Wings, Steel Shank, Brass Plated.

Table 7. American National Standard Types A and B Thumb Screws
ANSI B18.17-1968, R1983

See Footnote 1 — REGULAR — HEAVY — TYPE A

See Footnote 1 — REGULAR — HEAVY — TYPE B

Nominal Size or Basic Screw Diameter[a]	Thds. per Inch	A Head Width		B Head Height		C Head Thick.		C′ Head Thick.		E Shoulder Diameter		L Practical Screw Lengths	
		Max	Min	Max	Min	Max	Min	Max	Min	Max	Min	Max	Min
Type A, Regular													
6 (0.1380)	32	0.31	0.29	0.33	0.31	0.05	0.04	…	…	0.25	0.23	0.75	0.25
8 (0.1640)	32	0.36	0.34	0.38	0.36	0.06	0.05	…	…	0.31	0.29	0.75	0.38
10 (0.1900)	24, 32	0.42	0.40	0.48	0.46	0.06	0.05	…	…	0.35	0.32	1.00	0.38
12 (0.2160)	24	0.48	0.46	0.54	0.52	0.06	0.05	…	…	0.40	0.38	1.00	0.38
1/4 (0.2500)	20	0.55	0.52	0.64	0.61	0.07	0.05	…	…	0.47	0.44	1.50	0.50
5/16 (0.3125)	18	0.70	0.67	0.78	0.75	0.09	0.07	…	…	0.59	0.56	1.50	0.50
3/8 (0.3750)	16	0.83	0.80	0.95	0.92	0.11	0.09	…	…	0.76	0.71	2.00	0.75
Type A, Heavy													
10 (0.1900)	24	0.89	0.83	0.84	0.72	0.18	0.16	0.10	0.08	0.33	0.31	2.00	0.50
1/4 (0.2500)	20	1.05	0.99	0.94	0.81	0.24	0.22	0.10	0.08	0.40	0.38	3.00	0.50
5/16 (0.3125)	18	1.21	1.15	1.00	0.88	0.27	0.25	0.11	0.09	0.46	0.44	4.00	0.50
3/8 (0.3750)	16	1.41	1.34	1.16	1.03	0.30	0.28	0.11	0.09	0.55	0.53	4.00	0.50
7/16 (0.4375)	14	1.59	1.53	1.22	1.09	0.36	0.34	0.13	0.11	0.71	0.69	2.50	1.00
1/2 (0.5000)	13	1.81	1.72	1.28	1.16	0.40	0.38	0.14	0.12	0.83	0.81	3.00	1.00
Type B, Regular													
6 (0.1380)	32	0.45	0.43	0.28	0.26	0.08	0.06	0.03	0.02	…	…	1.00	0.25
8 (0.1640)	32	0.51	0.49	0.32	0.30	0.09	0.07	0.04	0.02	…	…	1.00	0.38
10 (0.1900)	24, 32	0.58	0.54	0.39	0.36	0.10	0.08	0.05	0.03	…	…	2.00	0.38
12 (0.2160)	24	0.71	0.67	0.45	0.43	0.11	0.09	0.05	0.03	…	…	2.00	0.38
1/4 (0.2500)	20	0.83	0.80	0.52	0.48	0.16	0.14	0.06	0.03	…	…	2.50	0.50
5/16 (0.3125)	18	0.96	0.91	0.64	0.60	0.17	0.14	0.09	0.06	…	…	3.00	0.50
3/8 (0.3750)	16	1.09	1.03	0.71	0.67	0.22	0.18	0.11	0.08	…	…	3.00	0.75
7/16 (0.4375)	14	1.40	1.35	0.96	0.91	0.27	0.24	0.14	0.11	…	…	4.00	1.00
1/2 (0.5000)	13	1.54	1.46	1.09	1.03	0.33	0.29	0.15	0.11	…	…	4.00	1.00
Type B, Heavy													
10 (0.1900)	24	0.89	0.83	0.78	0.66	0.18	0.16	0.08	0.06	…	…	2.00	0.50
1/4 (0.2500)	20	1.05	0.99	0.81	0.72	0.24	0.22	0.11	0.09	…	…	3.00	0.50
5/16 (0.3125)	18	1.21	1.15	0.88	0.78	0.27	0.25	0.11	0.09	…	…	4.00	0.50
3/8 (0.3750)	16	1.41	1.34	0.94	0.84	0.30	0.28	0.14	0.12	…	…	4.00	0.50
7/16 (0.4375)	14	1.59	1.53	1.00	0.91	0.36	0.34	0.14	0.12	…	…	3.00	1.00
1/2 (0.5000)	13	1.81	1.72	1.09	0.97	0.40	0.38	0.18	0.16	…	…	3.00	1.00

[a] Where specifying nominal size in decimals, zeroes in fourth decimal place are omitted.

All dimensions in inches.

[1] Plain point, unless alternate point from styles shown in Table 8 is specified by user.

Lengths of Wing and Thumb Screws.—The length of wing or thumb screws is measured parallel to the axis of the screw from the intersection of the head or shoulder with the shank to the extreme point of the screw. Standard length increments are as follows: For sizes No. 4 through ¼ inch and for nominal lengths of 0.25 to 0.75 inch, 0.12-inch increments; from 0.75- to 1.50-inch lengths, 0.25-inch increments; and for 1.50- to 3.00-inch lengths, 0.50-inch increments. For sizes 5⁄16 through ½ inch and for 0.50- to 1.50-inch lengths, 0.25-inch increments; for 1.50- to 3.00-inch lengths, 0.50-inch increments; and for 3.00- to 4.00-inch lengths, 1.00-inch increments.

Threads for Wing Screws and Thumb Screws.—Threads for all types of wing screws and thumb screws are in conformance with ANSI Standard Unified Thread, Class 2A. For threads with an additive finish the Class 2A maximum diameters apply to an unplated screw or to a screw before plating, whereas the basic diameters (Class 2A maximum diameters plus the allowance) apply to a screw after plating. All types of wing and thumb screws should have complete (full form) threads extending as close to the head or shoulder as practicable.

Points for Wing and Thumb Screws.—Wing and thumb screws are normally supplied with plain points (sheared ends). Where so specified, these screws may be obtained with cone, cup, dog, flat or oval points as shown in Table 8.

Table 8. American National Standard Alternate Points for Wing and Thumb Screws ***ANSI B18.17-1968, R1983***

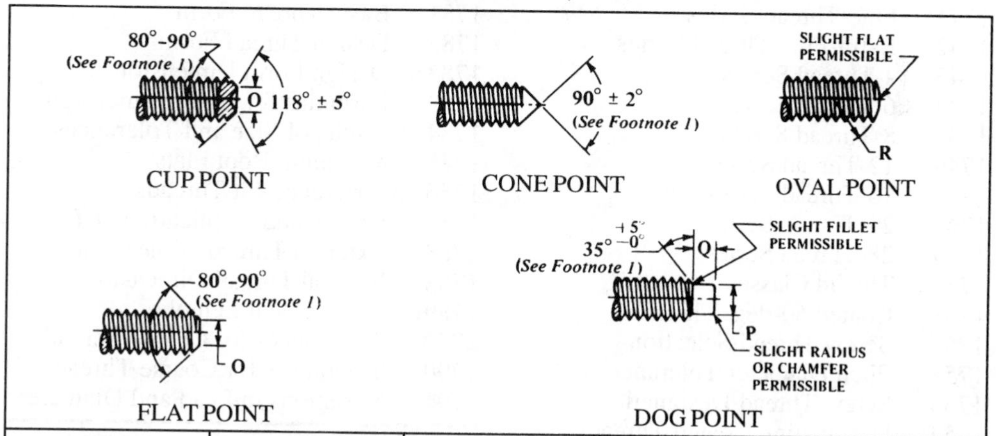

80°–90° (See Footnote 1)
O 118° ± 5°
CUP POINT

90° ± 2° (See Footnote 1)
CONE POINT

SLIGHT FLAT PERMISSIBLE
R
OVAL POINT

80°–90° (See Footnote 1)
O
FLAT POINT

35° +5° −0° (See Footnote 1)
Q
SLIGHT FILLET PERMISSIBLE
P
SLIGHT RADIUS OR CHAMFER PERMISSIBLE
DOG POINT

Nominal Size or Basic Screw Diamter[a]	*O* Cup and Flat Point Diameter		*P* Dog Point[b] Diameter		*Q* Dog Point[b] Length		*R* Oval Point Radius	
	Max	Min	Max	Min	Max	Min	Max	Min
4 (0.1120)	0.061	0.051	0.075	0.070	0.061	0.051	0.099	0.084
6 (0.1380)	0.074	0.064	0.092	0.087	0.075	0.065	0.140	0.109
8 (0.1640)	0.087	0.076	0.109	0.103	0.085	0.075	0.156	0.125
10 (0.1900)	0.102	0.088	0.127	0.120	0.095	0.085	0.172	0.141
12 (0.2160)	0.115	0.101	0.144	0.137	0.115	0.105	0.188	0.156
¼ (0.2500)	0.132	0.118	0.156	0.149	0.130	0.120	0.219	0.188
5⁄16 (0.3125)	0.172	0.156	0.203	0.195	0.161	0.151	0.256	0.234
⅜ (0.3750)	0.212	0.194	0.250	0.241	0.193	0.183	0.312	0.281
7⁄16 (0.4375)	0.252	0.232	0.297	0.287	0.224	0.214	0.359	0.328
½ (0.5000)	0.291	0.270	0.344	0.334	0.255	0.245	0.406	0.375
⅝ (0.6250)	0.371	0.347	0.469	0.456	0.321	0.305	0.500	0.469

[a] Where specifying nominal size in decimals, zeros in the fourth decimal place are omitted.

[b] The axis of dog points shall not be eccentric with the axis of the screw by more than 3 per cent of the basic screw diameter or 0.005 in., whichever is the smaller.

All dimensions in inches.

[1] The external point angles specified shall apply to those portions of the angles which lie below the thread root diameter, it being recognized the angle within the thread profile may be varied due to the manufacturing processes.

THREADS AND THREADING

SCREW THREAD SYSTEMS

UNIFIED SCREW THREADS

METRIC SCREW THREADS

METRIC SCREW THREADS (cont.)

MINIATURE AND INTERFERENCE FIT THREADS

ACME SCREW THREADS

ACME SCREW THREADS (cont.)

BUTTRESS THREADS

BRITISH THREADS

BRITISH THREADS (cont.)

OTHER THREADS

MEASURING SCREW THREADS

THREADS AND THREADING

THREAD ROLLING

THREAD GRINDING

THREAD MILLING

THREAD MILLING (cont.)

SIMPLE, COMPOUND, DIFFERENTIAL, AND BLOCK INDEXING

SCREW THREAD SYSTEMS

Screw Thread Forms.—Of the various screw thread forms which have been developed, the most used are those having symmetrical sides inclined at equal angles with a vertical center line through the thread apex. Present-day examples of such threads would include the Unified, the Whitworth and the Acme forms. One of the early forms was the Sharp V which is now used only occasionally. Symmetrical threads are relatively easy to manufacture and inspect and hence are widely used on mass-produced general-purpose threaded fasteners of all types. In addition to general-purpose fastener applications, certain threads are used to repeatedly move or translate machine parts against heavy loads. For these so-called translation threads a stronger form is required. The most widely used translation thread forms are the square, the Acme, and the buttress. Of these, the square thread is the most efficient, but it is also the most difficult to cut owing to its parallel sides and it cannot be adjusted to compensate for wear. Although less efficient, the Acme form of thread has none of the disadvantages of the square form and has the advantage of being somewhat stronger. The buttress form is used for translation of loads in one direction only because of its non-symmetrical form and combines the high efficiency and strength of the square thread with the ease of cutting and adjustment of the Acme thread.

Sharp V-thread.—The sides of the thread form an angle of 60 degrees with each other. The top and bottom of the thread are, theoretically, sharp, but in practice it is necessary to make the thread with a slight flat. There is no standard adopted for this flat, but it is usually made about one-twenty-fifth of the pitch. If p = pitch of thread, and d = depth of thread, then:

$$d = p \times \cos 30 \text{ deg.} = p \times 0.866 = \frac{0.866}{\text{no. of threads per inch}}$$

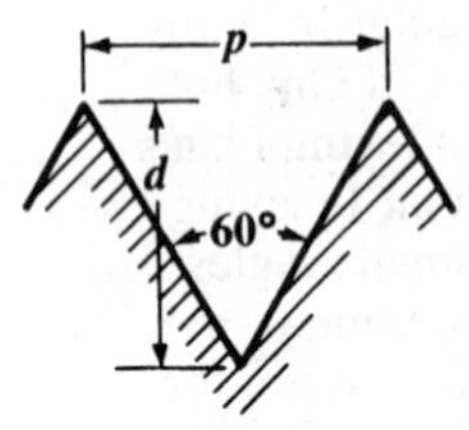

Some modified V-threads, for locomotive boiler taps particularly, have a depth of 0.8 × pitch.

American National and Unified Screw Thread Forms.—The American National form (formerly known as the United States Standard) was used for many years for most screws, bolts, and miscellaneous threaded products produced in the United States. The American National Standard for Unified Screw Threads now in use includes certain modifications of the former standard as is explained on page 1706. The Basic Profile is shown below and is identical for both UN and UNR screw threads. In this figure H is the height of a sharp V-thread.

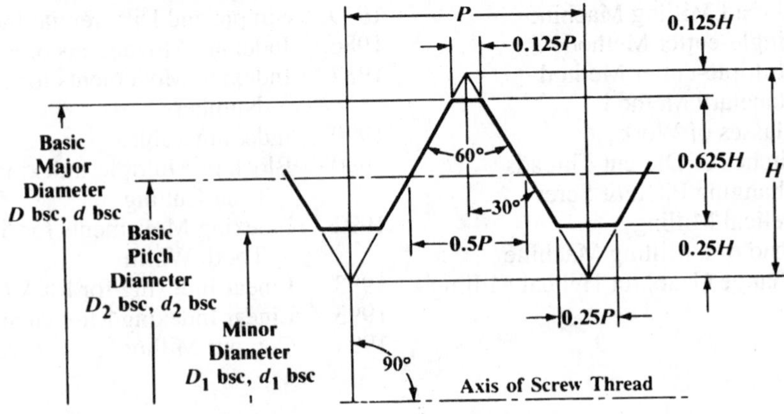

Basic Profile of UN and UNF Screw Threads

Definitions of Screw Threads.— The following definitions are based on American National Standard ANSI/ASME B1.7M-1984 (R1992) "Nomenclature, Definitions, and Letter Symbols for Screw Threads," and refer to both straight and taper threads.

Actual Size: An actual size is a measured size.

Allowance: An allowance is the prescribed difference between the design (maximum material) size and the basic size. It is numerically equal to the absolute value of the ISO term *fundamental deviation.*

Axis of Thread: Thread axis is coincident with the axis of its pitch cylinder or cone.

Basic Profile of Thread: The basic profile of a thread is the cyclical outline, in an axial plane, of the permanently established boundary between the provinces of the external and internal threads. All deviations are with respect to this boundary.

Basic Size: The basic size is that size from which the limits of size are derived by the application of allowances and tolerances.

Bilateral Tolerance: This is a tolerance in which variation is permitted in both directions from the specified dimension.

Black Crest Thread: This is a thread whose crest displays an unfinished cast, rolled, or forged surface.

Blunt Start Thread: "Blunt start" designates the removal of the incomplete thread at the starting end of the thread. This is a feature of threaded parts that are repeatedly assembled by hand, such as hose couplings and thread plug gages, to prevent cutting of hands and crossing of threads. It was formerly known as a Higbee cut.

Chamfer: This is a conical surface at the starting end of a thread.

Class of Thread: The class of a thread is an alphanumerical designation to indicate the standard grade of tolerance and allowance specified for a thread.

Clearance Fit: This is a fit having limits of size so prescribed that a clearance always results when mating parts are assembled at their maximum material condition.

Complete Thread: The complete thread is that thread whose profile lies within the size limits. (See also *Effective Thread* and *Length of Complete Thread.*) *Note:* Formerly in pipe thread terminology this was referred to as "the perfect thread" but that term is no longer considered desirable.

Crest: This is that surface of a thread which joins the flanks of the thread and is farthest from the cylinder or cone from which the thread projects.

Crest Truncation: This is the radial distance between the sharp crest (crest apex) and the cylinder or cone that would bound the crest.

Depth of Thread Engagement: The depth (or height) of thread engagement between two coaxially assembled mating threads is the radial distance by which their thread forms overlap each other.

Design Size: This is the basic size with allowance applied, from which the limits of size are derived by the application of a tolerance. If there is no allowance, the design size is the same as the basic size.

Deviation: Deviation is a variation from an established dimension, position, standard, or value. In ISO usage, it is the algebraic difference between a size (actual, maximum, or minimum) and the corresponding basic size. The term deviation does not necessarily indicate an error. (See also *Error.*)

Deviation, Fundamental (ISO term): For standard threads, the fundamental deviation is the upper or lower deviation closer to the basic size. It is the upper deviation *es* for an external thread and the lower deviation *EI* for an internal thread. (See also *Allowance* and *Tolerance Position.*)

Deviation, Lower (ISO term): The algebraic difference between the minimum limit of size and the basic size. It is designated *EI* for internal and *ei* for external thread diameters.

Deviation, Upper (ISO term): The algebraic difference between the maximum limit of size and the basic size. It is designated *ES* for internal and *es* for external thread diameters.

Dimension: A numerical value expressed in appropriate units of measure and indicated on drawings along with lines, symbols, and notes to define the geometrical characteristic of an object.

Effective Size: See *Pitch Diameter, Functional Diameter.*

Effective Thread: The effective (or useful) thread includes the complete thread, and those portions of the incomplete thread which are fully formed at the root but not at the crest (in taper pipe threads it includes the so-called black crest threads); thus excluding the vanish thread.

Error: The algebraic difference between an observed or measured value beyond tolerance limits, and the specified value.

External Thread: A thread on a cylindrical or conical external surface.

Fit: Fit is the relationship resulting from the designed difference, before assembly, between the sizes of two mating parts which are to be assembled.

Flank: The flank of a thread is either surface connecting the crest with the root. The flank surface intersection with an axial plane is theoretically a straight line.

Flank Angle: The flank angles are the angles between the individual flanks and the perpendicular to the axis of the thread, measured in an axial plane. A flank angle of a symmetrical thread is commonly termed the half-angle of thread.

Flank Diametral Displacement: In a boundary profile defined system, flank diametral displacement is twice the radial distance between the straight thread flank segments of the maximum and minimum boundary profiles. The value of flank diametral displacement is equal to pitch diameter tolerance in a pitch line reference thread system.

Height of Thread: The height (or depth) of thread is the distance, measured radially, between the major and minor cylinders or cones, respectively.

Helix Angle: On a straight thread, the helix angle is the angle made by the helix of the thread and its relation to the thread axis. On a taper thread, the helix angle at a given axial position is the angle made by the conical spiral of the thread with the axis of the thread. The helix angle is the complement of the lead angle. (See also page 1932 for diagram.)

Higbee Cut: See *Blunt Start Thread.*

Imperfect Thread: See *Incomplete Thread.*

Included Angle: See *Thread Angle*.

Incomplete Thread: A threaded profile having either crests or roots or both, not fully formed, resulting from their intersection with the cylindrical or end surface of the work or the vanish cone. It may occur at either end of the thread.

Interference Fit: A fit having limits of size so prescribed that an interference always results when mating parts are assembled.

Internal Thread: A thread on a cylindrical or conical internal surface.

Lead: Lead is the axial distance between two consecutive points of intersection of a helix by a line parallel to the axis of the cylinder on which it lies, i.e., the axial movement of a threaded part rotated one turn in its mating thread.

Lead Angle: On a straight thread, the lead angle is the angle made by the helix of the thread at the pitch line with a plane perpendicular to the axis. On a taper thread, the lead angle at a given axial position is the angle made by the conical spiral of the thread with the perpendicular to the axis at the pitch line.

Lead Thread: That portion of the incomplete thread that is fully formed at the root but not fully formed at the crest that occurs at the entering end of either an external or internal thread.

Left-hand Thread: A thread is a left-hand thread if, when viewed axially, it winds in a counterclockwise and receding direction. Left-hand threads are designated LH.

Length of Complete Thread: The axial length of a thread section having full form at both crest and root but also including a maximum of two pitches at the start of the thread which may have a chamfer or incomplete crests.

Length of Thread Engagement: The length of thread engagement of two mating threads is the axial distance over which the two threads, each having full form at both crest and root, are designed to contact. (See also *Length of Complete Thread.*)

Limits of Size: The applicable maximum and minimum sizes.

Major Clearance: The radial distance between the root of the internal thread and the crest of the external thread of the coaxially assembled designed forms of mating threads.

Major Cone: The imaginary cone that would bound the crests of an external taper thread or the roots of an internal taper thread.

Major Cylinder: The imaginary cylinder that would bound the crests of an external straight thread or the roots of an internal straight thread.

Major Diameter: On a straight thread the major diameter is that of the major cylinder. On a taper thread the major diameter at a given position on the thread axis is that of the major cone at that position. (See also *Major Cylinder* and *Major Cone.*)

Maximum Material Condition: (*MMC*): The condition where a feature of size contains the maximum amount of material within the stated limits of size. For example, minimum internal thread size or maximum external thread size.

Minimum Material Condition: (*Least Material Condition* (*LMC*)): The condition where a feature of size contains the least amount of material within the stated limits of size. For example, maximum internal thread size or minimum external thread size.

Minor Clearance: The radial distance between the crest of the internal thread and the root of the external thread of the coaxially assembled design forms of mating threads.

Minor Cone: The imaginary cone that would bound the roots of an external taper thread or the crests of an internal taper thread.

Minor Cylinder: The imaginary cylinder that would bound the roots of an external straight thread or the crests of an internal straight thread.

Minor Diameter: On a straight thread the minor diameter is that of the minor cylinder. On a taper thread the minor diameter at a given position on the thread axis is that of the minor cone at that position. (See also *Minor Cylinder* and *Minor Cone.*)

Multiple-Start Thread: A thread in which the lead is an integral multiple, other than one, of the pitch.

Nominal Size: Designation used for general identification.

Parallel Thread: See *Screw Thread.*

Partial Thread: See *Vanish Thread.*

Pitch: The pitch of a thread having uniform spacing is the distance measured parallel with its axis between corresponding points on adjacent thread forms in the same axial plane and on the same side of the axis. Pitch is equal to the lead divided by the number of thread starts.

Pitch Cone: The pitch cone is an imaginary cone of such apex angle and location of its vertex and axis that its surface would pass through a taper thread in such a manner as to make the widths of the thread ridge and the thread groove equal. It is, therefore, located equidistantly between the sharp major and minor cones of a given thread form. On a theoretically perfect taper thread, these widths are equal to one-half the basic pitch. (See also *Axis of Thread* and *Pitch Diameter.*)

Pitch Cylinder: The pitch cylinder is an imaginary cylinder of such diameter and location of its axis that its surface would pass through a straight thread in such a manner as to make the widths of the thread ridge and groove equal. It is, therefore, located equidistantly between the sharp major and minor cylinders of a given thread form. On a theoretically perfect thread these widths are equal to one-half the basic pitch. (See also *Axis of Thread* and *Pitch Diameter.*)

Pitch Diameter: On a straight thread the pitch diameter is the diameter of the pitch cylinder. On a taper thread the pitch diameter at a given position on the thread axis is the diameter of the pitch cone at that position. *Note:* When the crest of a thread is truncated beyond the pitch line, the pitch diameter and pitch cylinder or pitch cone would be based on a theoretical extension of the thread flanks.

Pitch Diameter, Functional Diameter: The functional diameter is the pitch diameter of an enveloping thread with perfect pitch, lead, and flank angles and having a specified length of engagement. It includes the cumulative effect of variations in lead (pitch), flank angle, taper, straightness, and roundness. Variations at the thread crest and root are excluded. Other, nonpreferred terms are *virtual diameter, effective size, virtual effective diameter,* and *thread assembly diameter.*

Pitch Line: The generator of the cylinder or cone specified in *Pitch Cylinder* and *Pitch Cone.*

Right-hand Thread: A thread is a fight-hand thread if, when viewed axially, it winds in a clockwise and receding direction. A thread is considered to be right-hand unless specifically indicated otherwise.

Root: That surface of the thread which joins the flanks of adjacent thread forms and is immediately adjacent to the cylinder or cone from which the thread projects.

Root Truncation: The radial distance between the sharp root (root apex) and the cylinder or cone that would bound the root.

Runout: As applied to screw threads, unless otherwise specified, runout refers to circular runout of major and minor cylinders with respect to the pitch cylinder. Circular runout, in accordance with ANSI Y14.5M, controls cumulative variations of circularity and coaxiality. Runout includes variations due to eccentricity and out-of-roundness. The amount of runout is usually expressed in terms of full indicator movement (FIM).

Screw Thread: A screw thread is a continuous and projecting helical ridge usually of uniform section on a cylindrical or conical surface.

Sharp Crest: (*Crest Apex*): The apex formed by the intersection of the flanks of a thread when extended, if necessary, beyond the crest.

Sharp Root: (*Root Apex*): The apex formed by the intersection of the adjacent flanks of adjacent threads when extended, if necessary, beyond the root.

Standoff: The axial distance between specified reference points on external and internal taper thread members or gages, when assembled with a specified torque or under other specified conditions.

Straight Thread: A straight thread is a screw thread projecting from a cylindrical surface.

Taper Thread: A taper thread is a screw thread projecting from a conical surface.

Tensile Stress Area: The tensile stress area is an arbitrarily selected area for computing the tensile strength of an externally threaded fastener so that the fastener strength is consistent with the basic material strength of the fastener. It is typically defined as a function of pitch diameter and/or minor diameter to calculate a circular cross section of the fastener correcting for the notch and helix effects of the threads.

Thread: A thread is a portion of a screw thread encompassed by one pitch. On a single-start thread it is equal to one turn. (See also *Threads per Inch* and *Turns per Inch.*)

Thread Runout: See *Vanish Thread.*

Thread Series: Thread Series are groups of diameter/pitch combinations distinguished from each other by the number of threads per inch applied to specific diameters.

Thread Shear Area: The thread shear area is the total ridge cross-sectional area intersected by a specified cylinder with diameter and length equal to the mating thread engagement. Usually the cylinder diameter for external thread shearing is the minor diameter of the internal thread and for internal thread shearing it is the major diameter of the external thread.

Threads per Inch: The number of threads per inch is the reciprocal of the axial pitch in inches.

Tolerance: The total amount by which a specific dimension is permitted to vary. The tolerance is the difference between the maximum and minimum limits.

Tolerance Class: (*metric*): The tolerance class (metric) is the combination of a tolerance position with a tolerance grade. It specifies the allowance (fundamental deviation), pitch diameter tolerance (flank diametral displacement), and the crest diameter tolerance.

Tolerance Grade: (metric): The tolerance grade (metric) is a numerical symbol that designates the tolerances of crest diameters and pitch diameters applied to the design profiles.

Tolerance Limit: The variation, positive or negative, by which a size is permitted to depart from the design size.

Tolerance Position: (metric): The tolerance position (metric) is a letter symbol that designates the position of the tolerance zone in relation to the basic size. This position provides the allowance (fundamental deviation).

Total Thread: Includes the complete and all the incomplete thread, thus including the vanish thread and the lead thread.

Transition Fit: A fit having limits of size so prescribed that either a clearance or an interference may result when mating parts are assembled.

Turns per Inch: The number of turns per inch is the reciprocal of the lead in inches.

Unilateral Tolerance: A tolerance in which variation is permitted in one direction from the specified dimension.

Vanish Thread: (*Partial Thread, Washout Thread,* or *Thread Runout*): That portion of the incomplete thread which is not fully formed at the root or at crest and root. It is produced by the chamfer at the starting end of the thread forming tool.

Virtual Diameter: See *Pitch Diameter, Functional Diameter.*

Washout Thread: See *Vanish Thread.*

ISO Miniature Screw Threads, Basic Form *ISO/R 1501:1970*

Pitch P	$H = 0.866025P$	$0.554256H = 0.48P$	$0.375H = 0.324760P$	$0.320744H = 0.320744P$	$0.125H = 0.108253P$
0.08	0.069282	0.038400	0.025981	0.022222	0.008660
0.09	0.077942	0.043200	0.029228	0.024999	0.009743
0.1	0.086603	0.048000	0.032476	0.027777	0.010825
0.125	0.108253	0.060000	0.040595	0.034722	0.013532
0.15	0.129904	0.072000	0.048714	0.041666	0.016238
0.175	0.151554	0.084000	0.056833	0.048610	0.018944
0.2	0.173205	0.096000	0.064952	0.055554	0.021651
0.225	0.194856	0.108000	0.073071	0.062499	0.024357
0.25	0.216506	0.120000	0.081190	0.069443	0.027063
0.3	0.259808	0.144000	0.097428	0.083332	0.032476

ISO Miniature Screw Threads, Basic Dimensions *ISO/R 1501:1970*

Nominal Diameter	Pitch P	Major Diameter D, d	Pitch Diameter D_2, d_2	Minor Diameter D_1, d_1
0.30	0.080	0.300000	0.248039	0.223200
0.35	0.090	0.350000	0.291543	0.263600
0.40	0.100	0.400000	0.335048	0.304000
0.45	0.100	0.450000	0.385048	0.354000
0.50	0.125	0.500000	0.418810	0.380000
0.55	0.125	0.550000	0.468810	0.430000
0.60	0.150	0.600000	0.502572	0.456000
0.70	0.175	0.700000	0.586334	0.532000
0.80	0.200	0.800000	0.670096	0.608000
0.90	0.225	0.900000	0.753858	0.684000
1.00	0.250	1.000000	0.837620	0.760000
1.10	0.250	1.100000	0.937620	0.860000
1.20	0.250	1.200000	1.037620	0.960000
1.40	0.300	1.400000	1.205144	1.112000

D and *d* dimensions refer to the nut (internal) and screw (external) threads, respectively.

UNIFIED SCREW THREADS

American Standard for Unified Screw Threads.— American Standard B1.1-1949 was the first American standard to cover those Unified Thread Series agreed upon by the United Kingdom, Canada, and the United States to obtain screw thread interchangeability among these three nations. These Unified threads are now the basic American standard for fastening types of screw threads. In relation to previous American practice, Unified threads have substantially the same thread form and are mechanically interchangeable with the former American National threads of the same diameter and pitch.

The principal differences between the two systems lie in: 1) application of allowances; 2) variation of tolerances with size; 3) difference in amount of pitch diameter tolerance on external and internal threads; and 4) differences in thread designation.

In the Unified system an allowance is provided on both the Classes 1A and 2A external threads whereas in the American National system only the Class I external thread has an allowance. Also, in the Unified system, the pitch diameter tolerance of an internal thread is 30 per cent greater than that of the external thread, whereas they are equal in the American National system.

Revised Standard: The revised screw thread standard ANSI/ASME B1.1-1989 is much the same as that of ANSI B1.1-1982. The latest symbols in accordance with ANSI/ASME B1.7M-1984 (R1992) Nomenclature, are used. Acceptability criteria are described in ANSI/ASME B 1.3M-1986, Screw Thread Gaging Systems for Dimensional Acceptability, Inch or Metric Screw Threads (UN, UNR, UNJ, M, and MJ).

Where the letters U, A or B do not appear in the thread designations, the threads conform to the outdated American National screw threads.

Advantages of Unified Threads: The Unified standard is designed to correct certain production difficulties resulting from the former standard. Often, under the old system, the tolerances of the product were practically absorbed by the combined tool and gage tolerances, leaving little for a working tolerance in manufacture. Somewhat greater tolerances are now provided for nut threads. As contrasted with the old "classes of fit" 1, 2, and 3, for each of which the pitch diameter tolerance on the external and internal threads were equal, the Classes 1B, 2B, and 3B (internal) threads in the new standard have, respectively, a 30 per cent larger pitch diameter tolerance than the 1A, 2A, and 3A (external) threads. Relatively more tolerance is provided for fine threads than for coarse threads of the same pitch. Where previous tolerances were more liberal than required, they were reduced.

Thread Form.— The Design Profiles for Unified screw threads, shown on page 1713, define the maximum material condition for external and internal threads with no allowance and are derived from the Basic Profile, shown on page 1706.

UN External Screw Threads: A flat root contour is specified, but it is necessary to provide for some threading tool crest wear, hence a rounded root contour cleared beyond the $0.25P$ flat width of the Basic Profile is optional.

UNR External Screw Threads: To reduce the rate of threading tool crest wear and to improve fatigue strength of a flat root thread, the Design Profile of the UNR thread has a smooth, continuous, non-reversing contour with a radius of curvature not less than $0.108P$ at any point and blends tangentially into the flanks and any straight segment. At the maximum material condition, the point of tangency is specified to be at a distance not less than $0.625H$ (where H is the height of a sharp V-thread) below the basic major diameter.

UN and UNR External Screw Threads: The Design Profiles of both UN and UNR external screw threads have flat crests. However, in practice, product threads are produced with partially or completely rounded crests. A rounded crest tangent at $0.125P$ flat is shown as an option on page 1713.

UN Internal Screw Thread: In practice it is necessary to provide for some threading tool crest wear, therefore the root of the Design Profile is rounded and cleared beyond the 0.125*P* flat width of the Basic Profile.There is no internal UNR screw thread.

American National Standard Unified Internal and External Screw Thread Design Profiles (Maximum Material Condition) .—

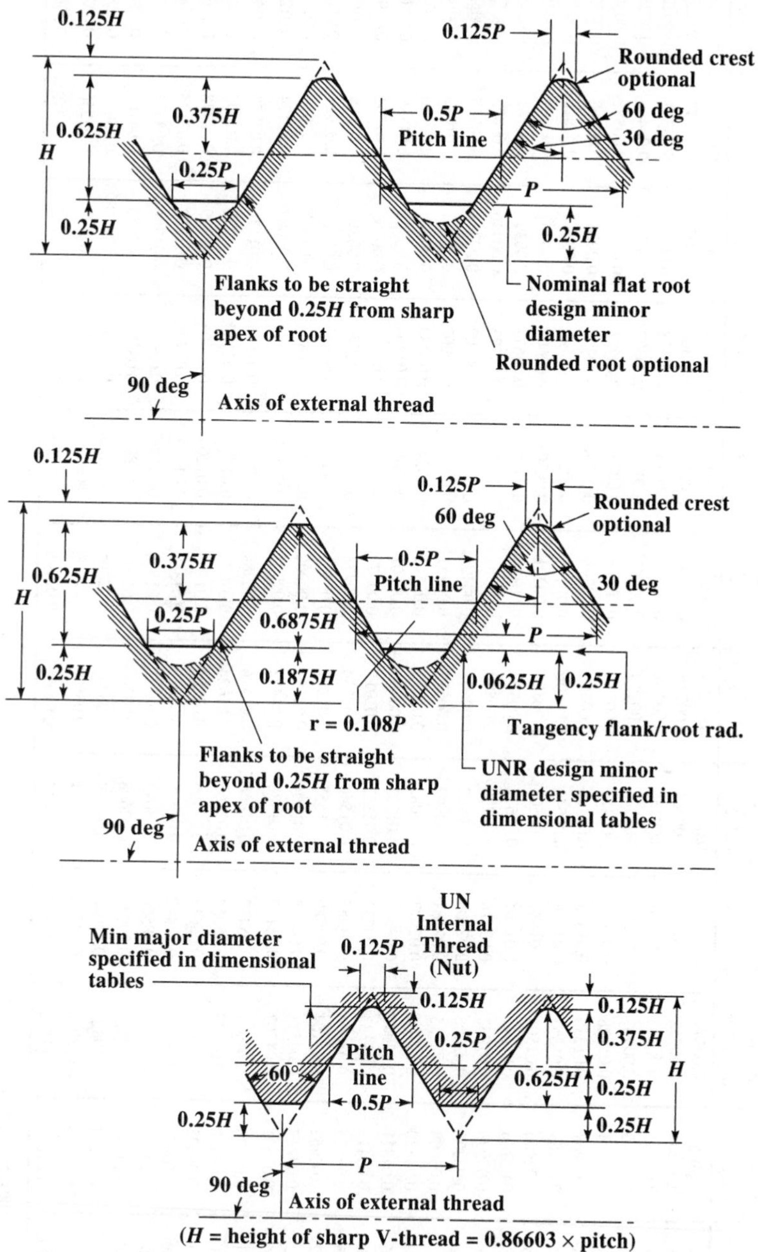

(H = height of sharp V-thread = 0.86603 × pitch)

Thread Series: Thread series are groups of diameter-pitch combinations distinguished from each other by the numbers of threads per inch applied to a specific diameter. The various diameter-pitch combinations of eleven standard series are shown in Table 2. The limits of size of threads in the eleven standard series together with certain selected combinations of diameter and pitch, as well as the symbols for designating the various threads, are given in Table 3.

Table 1. American Standard Unified Inch Screw Thread Form Data

Threads per Inch n	Pitch P	Depth of Sharp V-Thread $0.86603P$	Depth of Int. Thd. and UN Ext. Thd.[a] $0.54127P$	Depth of UNR Ext. Thd. $0.59539P$	Truncation of Ext. Thd. Root $0.21651P$	Truncation of UNR Ext. Thd. Root[b] $0.16238P$	Truncation of Ext. Thd. Crest $0.10825P$	Truncation of Int. Thd. Root $0.10825P$	Truncation of Int. Thd. Crest $0.2165P$	Flat at Ext. Thd. Crest and Int. Thd. Root $0.125P$	Basic Flat at Int. Thd. Crest[c] $0.25P$	Maximum Ext. Thd. Root Radius $0.14434P$	Addendum of Ext. Thd. $0.32476P$
80	0.01250	0.01083	0.00677	0.00744	0.00271	0.00203	0.00135	0.00135	0.00271	0.00156	0.00312	0.00180	0.00406
72	0.01389	0.01203	0.00752	0.00827	0.00301	0.00226	0.00150	0.00150	0.00301	0.00174	0.00347	0.00200	0.00451
64	0.01563	0.01353	0.00846	0.00930	0.00338	0.00254	0.00169	0.00169	0.00338	0.00195	0.00391	0.00226	0.00507
56	0.01786	0.01546	0.00967	0.01063	0.00387	0.00290	0.00193	0.00193	0.00387	0.00223	0.00446	0.00258	0.00580
48	0.02083	0.01804	0.01128	0.01240	0.00451	0.00338	0.00226	0.00226	0.00451	0.00260	0.00521	0.00301	0.00677
44	0.02273	0.01968	0.01230	0.01353	0.00492	0.00369	0.00246	0.00246	0.00492	0.00284	0.00568	0.00328	0.00738
40	0.02500	0.02165	0.01353	0.01488	0.00541	0.00406	0.00271	0.00271	0.00541	0.00312	0.00625	0.00361	0.00812
36	0.02778	0.02406	0.01504	0.01654	0.00601	0.00451	0.00301	0.00301	0.00601	0.00347	0.00694	0.00401	0.00902
32	0.03125	0.02706	0.01691	0.01861	0.00677	0.00507	0.00338	0.00338	0.00677	0.00391	0.00781	0.00451	0.01015
28	0.03571	0.03093	0.01933	0.02126	0.00773	0.00580	0.00387	0.00387	0.00773	0.00446	0.00893	0.00515	0.01160
27	0.03704	0.03208	0.02005	0.02205	0.00802	0.00601	0.00401	0.00401	0.00802	0.00463	0.00926	0.00535	0.01203
24	0.04167	0.03608	0.02255	0.02481	0.00902	0.00677	0.00451	0.00451	0.00902	0.00521	0.01042	0.00601	0.01353
20	0.05000	0.04330	0.02706	0.02977	0.01083	0.00812	0.00541	0.00541	0.01083	0.00625	0.01250	0.00722	0.01624
18	0.05556	0.04811	0.03007	0.03308	0.01203	0.00902	0.00601	0.00601	0.01203	0.00694	0.01389	0.00802	0.01804
16	0.06250	0.05413	0.03383	0.03721	0.01353	0.01015	0.00677	0.00677	0.01353	0.00781	0.01562	0.00902	0.02030
14	0.07143	0.06186	0.03866	0.04253	0.01546	0.01160	0.00773	0.00773	0.01546	0.00893	0.01786	0.01031	0.02320
13	0.07692	0.06662	0.04164	0.04580	0.01655	0.01249	0.00833	0.00833	0.01665	0.00962	0.01923	0.01110	0.02498
12	0.08333	0.07217	0.04511	0.04962	0.01804	0.01353	0.00902	0.00902	0.01804	0.01042	0.02083	0.01203	0.02706
11½	0.08696	0.07531	0.04707	0.05177	0.01883	0.01412	0.00941	0.00941	0.01883	0.01087	0.02174	0.01255	0.02824
11	0.09091	0.07873	0.04921	0.05413	0.01968	0.01476	0.00984	0.00984	0.01968	0.01136	0.02273	0.01312	0.02952
10	0.10000	0.08660	0.05413	0.05954	0.02165	0.01624	0.01083	0.01083	0.02165	0.01250	0.02500	0.01443	0.03248
9	0.11111	0.09623	0.06014	0.06615	0.02406	0.01804	0.01203	0.01203	0.02406	0.01389	0.02778	0.01604	0.03608
8	0.12500	0.10825	0.06766	0.07442	0.02706	0.02030	0.01353	0.01353	0.02706	0.01562	0.03125	0.01804	0.04059
7	0.14286	0.12372	0.07732	0.08506	0.03093	0.02320	0.01546	0.01546	0.03093	0.01786	0.03571	0.02062	0.04639
6	0.16667	0.14434	0.09021	0.09923	0.03608	0.02706	0.01804	0.01804	0.03608	0.02083	0.04167	0.02406	0.05413
5	0.20000	0.17321	0.10825	0.11908	0.04330	0.03248	0.02165	0.02165	0.04330	0.02500	0.05000	0.02887	0.06495
4½	0.22222	0.19245	0.12028	0.13231	0.04811	0.03608	0.02406	0.02406	0.04811	0.02778	0.05556	0.03208	0.07217
4	0.25000	0.21651	0.13532	0.14885	0.05413	0.04059	0.02706	0.02706	0.05413	0.03125	0.06250	0.03608	0.08119

[a] Also depth of thread engagement.

[b] Design profile.

[c] Also basic flat at external UN thread root.

All dimensions are in inches.

Table 2. Diameter-Pitch Combinations for Standard Series of Threads (UN/UNR)

Sizes[a] No. or Inches	Basic Major Dia. Inches	Threads per Inch										
		Series with Graded Pitches			Series with Uniform (Constant) Pitches							
		Coarse UNC	Fine[b] UNF	Extra fine[c] UNEF	4-UN	6-UN	8-UN	12-UN	16-UN	20-UN	28-UN	32-UN
0	0.0600	…	80	Series designation shown indicates the UN thread form; however, the UNR thread form may be specified by substituting UNR in place of UN in all designations for external threads.								
(1)	0.0730	64	72									
2	0.0860	56	64									
(3)	0.0990	48	56									
4	0.1120	40	48									
5	0.1250	40	44	…	…	…	…	…	…	…	…	…
6	0.1380	32	40	…	…	…	…	…	…	…	…	UNC
8	0.1640	32	36	…	…	…	…	…	…	…	…	UNC
10	0.1900	24	32	…	…	…	…	…	…	…	…	UNF
(12)	0.2160	24	28	32	…	…	…	…	…	…	UNF	UNEF
1/4	0.2500	20	28	32	…	…	…	…	…	UNC	UNF	UNEF
5/16	0.3125	18	24	32	…	…	…	…	…	20	28	UNEF
3/8	0.3750	16	24	32	…	…	…	…	UNC	20	28	UNEF
7/16	0.4375	14	20	28	…	…	…	…	16	UNF	UNEF	32
1/2	0.5000	13	20	28	…	…	…	…	16	UNF	UNEF	32
9/16	0.5625	12	18	24	…	…	…	UNC	16	20	28	32
5/8	0.6250	11	18	24	…	…	…	12	16	20	28	32
(11/16)	0.6875	…	…	24	…	…	…	12	16	20	28	32
3/4	0.7500	10	16	20	…	…	…	12	UNF	UNEF	28	32
(13/16)	0.8125	…	…	20	…	…	…	12	16	UNEF	28	32
7/8	0.8750	9	14	20	…	…	…	12	16	UNEF	28	32
(15/16)	0.9375	…	…	20	…	…	…	12	16	UNEF	28	32
1	1.0000	8	12	20	…	…	UNC	UNF	16	UNEF	28	32
(1 1/16)	1.0625	…	…	18	…	…	8	12	16	20	28	…
1 1/8	1.1250	7	12	18	…	…	8	UNF	16	20	28	…
(1 3/16)	1.1875	…	…	18	…	…	8	12	16	20	28	…
1 1/4	1.2500	7	12	18	…	…	8	UNF	16	20	28	…
1 5/16	1.3125	…	…	18	…	…	8	12	16	20	28	…
1 3/8	1.3750	6	12	18	…	UNC	8	UNF	16	20	28	…
(1 7/16)	1.4375	…	…	18	…	6	8	12	16	20	28	…
1 1/2	1.5000	6	12	18	…	UNC	8	UNF	16	20	28	…
(1 9/16)	1.5625	…	…	18	…	6	8	12	16	20	…	…
1 5/8	1.6250	…	…	18	…	6	8	12	16	20	…	…
(1 11/16)	1.6875	…	…	18	…	6	8	12	16	20	…	…
1 3/4	1.7500	5	…	…	…	6	8	12	16	20	…	…
(1 13/16)	1.8125	…	…	…	…	6	8	12	16	20	…	…
1 7/8	1.8750	…	…	…	…	6	8	12	16	20	…	…
(1 15/16)	1.9375	…	…	…	…	6	8	12	16	20	…	…
2	2.0000	4 1/2	…	…	…	6	8	12	16	20	…	…
(2 1/8)	2.1250	…	…	…	…	6	8	12	16	20	…	…
2 1/4	2.2500	4 1/2	…	…	…	6	8	12	16	20	…	…
(2 3/8)	2.3750	…	…	…	…	6	8	12	16	20	…	…
2 1/2	2.5000	4	…	…	UNC	6	8	12	16	20	…	…
(2 5/8)	2.6250	…	…	…	4	6	8	12	16	20	…	…
2 3/4	2.7500	4	…	…	UNC	6	8	12	16	20	…	…
(2 7/8)	2.8750	…	…	…	4	6	8	12	16	20	…	…
3	3.0000	4	…	…	UNC	6	8	12	16	20	…	…
(3 1/8)	3.1250	…	…	…	4	6	8	12	16	…	…	…
3 1/4	3.2500	4	…	…	UNC	6	8	12	16	…	…	…
(3 3/8)	3.3750	…	…	…	4	6	8	12	16	…	…	…
3 1/2	3.5000	4	…	…	UNC	6	8	12	16	…	…	…
(3 5/8)	3.6250	…	…	…	4	6	8	12	16	…	…	…
3 3/4	3.7500	4	…	…	UNC	6	8	12	16	…	…	…
(3 7/8)	3.8750	…	…	…	4	6	8	12	16	…	…	…
4	4.0000	4	…	…	UNC	6	8	12	16	…	…	…

[a] Sizes shown in parentheses are secondary sizes. Primary sizes of 4 1/4, 4 1/2, 4 3/4, 5, 5 1/4, 5 1/2, 5 3/4 and 6 inches also are in the 4, 6, 8, 12, and 16 thread series; secondary sizes of 4 1/8, 4 3/8, 4 5/8, 4 7/8, 5 1/8, 5 3/8, 5 5/8, and 5 7/8 also are in the 4, 6, 8, 12, and 16 thread series.

[b] For diameters over 1 1/2 inches, use 12-thread series.

[c] For diameters over 1 11/16 inches, use 16-thread series.

For UNR thread form substitute UNR for UN for external threads only.

Table 3. Standard Series and Selected Combinations — Unified Screw Threads

Nominal Size, Threads per Inch, and Series Designation[a]	External[b]								Internal[b]					
	Class	Allow-ance	Major Diameter			Pitch Diameter		UNR Minor Dia.,[c] Max (Ref.)	Class	Minor Diameter		Pitch Diameter		Major Diameter
			Max[d]	Min	Min[e]	Max[d]	Min			Min	Max	Min	Max	Min
0–80 UNF	2A	0.0005	0.0595	0.0563	—	0.0514	0.0496	0.0446	2B	0.0465	0.0514	0.0519	0.0542	0.0600
	3A	0.0000	0.0600	0.0568	—	0.0519	0.0506	0.0451	3B	0.0465	0.0514	0.0519	0.0536	0.0600
1–64 UNC	2A	0.0006	0.0724	0.0686	—	0.0623	0.0603	0.0538	2B	0.0561	0.0623	0.0629	0.0655	0.0730
	3A	0.0000	0.0730	0.0692	—	0.0629	0.0614	0.0544	3B	0.0561	0.0623	0.0629	0.0648	0.0730
1–72 UNF	2A	0.0006	0.0724	0.0689	—	0.0634	0.0615	0.0559	2B	0.0580	0.0635	0.0640	0.0665	0.0730
	3A	0.0000	0.0730	0.0695	—	0.0640	0.0626	0.0565	3B	0.0580	0.0635	0.0640	0.0659	0.0730
2–56 UNC	2A	0.0006	0.0854	0.0813	—	0.0738	0.0717	0.0642	2B	0.0667	0.0737	0.0744	0.0772	0.0860
	3A	0.0000	0.0860	0.0819	—	0.0744	0.0728	0.0648	3B	0.0667	0.0737	0.0744	0.0765	0.0860
2–64 UNF	2A	0.0006	0.0854	0.0816	—	0.0753	0.0733	0.0668	2B	0.0691	0.0753	0.0759	0.0786	0.0860
	3A	0.0000	0.0860	0.0822	—	0.0759	0.0744	0.0674	3B	0.0691	0.0753	0.0759	0.0779	0.0860
3–48 UNC	2A	0.0007	0.0983	0.0938	—	0.0848	0.0825	0.0734	2B	0.0764	0.0845	0.0855	0.0885	0.0990
	3A	0.0000	0.0990	0.0945	—	0.0855	0.0838	0.0741	3B	0.0764	0.0845	0.0855	0.0877	0.0990
3–56 UNF	2A	0.0007	0.0983	0.0942	—	0.0867	0.0845	0.0771	2B	0.0797	0.0865	0.0874	0.0902	0.0990
	3A	0.0000	0.0990	0.0949	—	0.0874	0.0858	0.0778	3B	0.0797	0.0865	0.0874	0.0895	0.0990
4–40 UNC	2A	0.0008	0.1112	0.1061	—	0.0950	0.0925	0.0814	2B	0.0849	0.0939	0.0958	0.0991	0.1120
	3A	0.0000	0.1120	0.1069	—	0.0958	0.0939	0.0822	3B	0.0849	0.0939	0.0958	0.0982	0.1120
4–48 UNF	2A	0.0007	0.1113	0.1068	—	0.0978	0.0954	0.0864	2B	0.0894	0.0968	0.0985	0.1016	0.1120
	3A	0.0000	0.1120	0.1075	—	0.0985	0.0967	0.0871	3B	0.0894	0.0968	0.0985	0.1008	0.1120
5–40 UNC	2A	0.0008	0.1242	0.1191	—	0.1080	0.1054	0.0944	2B	0.0979	0.1062	0.1088	0.1121	0.1250
	3A	0.0000	0.1250	0.1199	—	0.1088	0.1069	0.0952	3B	0.0979	0.1062	0.1088	0.1113	0.1250
5–44 UNF	2A	0.0007	0.1243	0.1195	—	0.1095	0.1070	0.0972	2B	0.1004	0.1079	0.1102	0.1134	0.1250
	3A	0.0000	0.1250	0.1202	—	0.1102	0.1083	0.0979	3B	0.1004	0.1079	0.1102	0.1126	0.1250
6–32 UNC	2A	0.0008	0.1372	0.1312	—	0.1169	0.1141	0.1000	2B	0.104	0.114	0.1177	0.1214	0.1380
	3A	0.0000	0.1380	0.1320	—	0.1177	0.1156	0.1008	3B	0.1040	0.1140	0.1177	0.1204	0.1380
6–40 UNF	2A	0.0008	0.1372	0.1321	—	0.1210	0.1184	0.1074	2B	0.111	0.119	0.1218	0.1252	0.1380
	3A	0.0000	0.1380	0.1329	—	0.1218	0.1198	0.1082	3B	0.1110	0.1186	0.1218	0.1243	0.1380
8–32 UNC	2A	0.0009	0.1631	0.1571	—	0.1428	0.1399	0.1259	2B	0.130	0.139	0.1437	0.1475	0.1640
	3A	0.0000	0.1640	0.1580	—	0.1437	0.1415	0.1268	3B	0.1300	0.1389	0.1437	0.1465	0.1640
8–36 UNF	2A	0.0008	0.1632	0.1577	—	0.1452	0.1424	0.1301	2B	0.134	0.142	0.1460	0.1496	0.1640
	3A	0.0000	0.1640	0.1585	—	0.1460	0.1439	0.1309	3B	0.1340	0.1416	0.1460	0.1487	0.1640
10–24 UNC	2A	0.0010	0.1890	0.1818	—	0.1619	0.1586	0.1394	2B	0.145	0.156	0.1629	0.1672	0.1900

Table 3. *(Continued)* **Standard Series and Selected Combinations — Unified Screw Threads**

Nominal Size, Threads per Inch, and Series Designation[a]	External[b]								Internal[b]					
	Class	Allow-ance	Major Diameter			Pitch Diameter		UNR Minor Dia.,[c] Max (Ref.)	Class	Minor Diameter		Pitch Diameter		Major Diameter
			Max[d]	Min	Min[e]	Max[d]	Min			Min	Max	Min	Max	Min
	3A	0.0000	0.1900	0.1828	—	0.1629	0.1604	0.1404	3B	0.1450	0.1555	0.1629	0.1661	0.1900
10–28 UNS	2A	0.0010	0.1890	0.1825	—	0.1658	0.1625	0.1464	2B	0.151	0.160	0.1668	0.1711	0.1900
10–32 UNF	2A	0.0009	0.1891	0.1831	—	1.1688	0.1658	0.1519	2B	0.156	0.164	0.1697	0.1736	0.1900
	3A	0.0000	0.1900	0.1840	—	0.1697	0.1674	0.1528	3B	0.1560	0.1641	0.1697	0.1726	0.1900
10–36 UNS	2A	0.0009	0.1891	0.1836	—	0.1711	0.1681	0.1560	2B	0.160	0.166	0.1720	0.1759	0.1900
10–40 UNS	2A	0.0009	0.1891	0.1840	—	0.1729	0.1700	0.1592	2B	0.163	0.169	0.1738	0.1775	0.1900
10–48 UNS	2A	0.0008	0.1892	0.1847	—	0.1757	0.1731	0.1644	2B	0.167	0.172	0.1765	0.1799	0.1900
10–56 UNS	2A	0.0007	0.1893	0.1852	—	0.1777	0.1752	0.1681	2B	0.171	0.175	0.1784	0.1816	0.1900
12–24 UNC	2A	0.0010	0.2150	0.2078	—	0.1879	0.1845	0.1654	2B	0.171	0.181	0.1889	0.1933	0.2160
	3A	0.0000	0.2160	0.2088	—	0.1889	0.1863	0.1664	3B	0.1710	0.1807	0.1889	0.1922	0.2160
12–28 UNF	2A	0.0010	0.2150	0.2085	—	0.1918	0.1886	0.1724	2B	0.177	0.186	0.1928	0.1970	0.2160
	3A	0.0000	0.2160	0.2095	—	0.1928	0.1904	0.1734	3B	0.1770	0.1857	0.1928	0.1959	0.2160
12–32 UNEF	2A	0.0009	0.2151	0.2091	—	0.1948	0.1917	0.1779	2B	0.182	0.190	0.1957	0.1998	0.2160
	3A	0.0000	0.2160	0.2100	—	0.1957	0.1933	0.1788	3B	0.1820	0.1895	0.1957	0.1988	0.2160
12–36 UNS	2A	0.0009	0.2151	0.2096	—	0.1971	0.1941	0.1821	2B	0.186	0.192	0.1980	0.2019	0.2160
12–40 UNS	2A	0.0009	0.2151	0.2100	—	0.1989	0.1960	0.1835	2B	0.189	0.195	0.1998	0.2035	0.2160
12–48 UNS	2A	0.0008	0.2152	0.2107	—	0.2017	0.1991	0.1904	2B	0.193	0.198	0.2025	0.2059	0.2160
12–56 UNS	2A	0.0007	0.2153	0.2112	—	0.2037	0.2012	0.1941	2B	0.197	0.201	0.2044	0.2076	0.2160
1/4–20 UNC	1A	0.0011	0.2489	0.2367	—	0.2164	0.2108	0.1894	1B	0.196	0.207	0.2175	0.2248	0.2500
	2A	0.0011	0.2489	0.2408	0.2367	0.2164	0.2127	0.1894	2B	0.196	0.207	0.2175	0.2224	0.2500
	3A	0.0000	0.2500	0.2419	—	0.2175	0.2147	0.1905	3B	0.1960	0.2067	0.2175	0.2211	0.2500
1/4–24 UNS	2A	0.0011	0.2489	0.2417	—	0.2218	0.2181	0.1993	2B	0.205	0.215	0.2229	0.2277	0.2500
1/4–27 UNS	2A	0.0010	0.2490	0.2423	—	0.2249	0.2214	0.2049	2B	0.210	0.219	0.2259	0.2304	0.2500
1/4–28 UNF	1A	0.0010	0.2490	0.2392	—	0.2258	0.2208	0.2064	1B	0.211	0.220	0.2268	0.2333	0.2500
	2A	0.0010	0.2490	0.2425	—	0.2258	0.2225	0.2064	2B	0.211	0.220	0.2268	0.2311	0.2500
	3A	0.0000	0.2500	0.2435	—	0.2268	0.2243	0.2074	3B	0.2110	0.2190	0.2268	0.2300	0.2500
1/4–32 UNEF	2A	0.0010	0.2490	0.2430	—	0.2287	0.2255	0.2118	2B	0.216	0.224	0.2297	0.2339	0.2500
	3A	0.0000	0.2500	0.2440	—	0.2297	0.2273	0.2128	3B	0.2160	0.2229	0.2297	0.2328	0.2500
1/4–36 UNS	2A	0.0009	0.2491	0.2436	—	0.2311	0.2280	0.2161	2B	0.220	0.226	0.2320	0.2360	0.2500
1/4–40 UNS	2A	0.0009	0.2491	0.2440	—	0.2329	0.2300	0.2193	2B	0.223	0.229	0.2338	0.2376	0.2500

Table 3. *(Continued)* **Standard Series and Selected Combinations — Unified Screw Threads**

Nominal Size, Threads per Inch, and Series Designation[a]	External[b]								Internal[b]					
	Class	Allowance	Major Diameter			Pitch Diameter		UNR Minor Dia.,[c] Max (Ref.)	Class	Minor Diameter		Pitch Diameter		Major Diameter
			Max[d]	Min	Min[e]	Max[d]	Min			Min	Max	Min	Max	Min
1/4–48 UNS	2A	0.0008	0.2492	0.2447	—	0.2357	0.2330	0.2243	2B	0.227	0.232	0.2365	0.2401	0.2500
1/4–56 UNS	2A	0.0008	0.2492	0.2451	—	0.2376	0.2350	0.2280	2B	0.231	0.235	0.2384	0.2417	0.2500
5/16–18 UNC	1A	0.0012	0.3113	0.2982	—	0.2752	0.2691	0.2452	1B	0.252	0.265	0.2764	0.2843	0.3125
	2A	0.0012	0.3113	0.3026	0.2982	0.2752	0.2712	0.2452	2B	0.252	0.265	0.2764	0.2817	0.3125
	3A	0.0000	0.3125	0.3038	—	0.2764	0.2734	0.2464	3B	0.2520	0.2630	0.2764	0.2803	0.3125
5/16–20 UN	2A	0.0012	0.3113	0.3032	—	0.2788	0.2748	0.2518	2B	0.258	0.270	0.2800	0.2852	0.3125
	3A	0.0000	0.3125	0.3044	—	0.2800	0.2770	0.2530	3B	0.2580	0.2680	0.2800	0.2839	0.3125
5/16–24 UNF	1A	0.0011	0.3114	0.3006	—	0.2843	0.2788	0.2618	1B	0.267	0.277	0.2854	0.2925	0.3125
	2A	0.0011	0.3114	0.3042	—	0.2843	0.2806	0.2618	2B	0.267	0.277	0.2854	0.2902	0.3125
	3A	0.0000	0.3125	0.3053	—	0.2854	0.2827	0.2629	3B	0.2670	0.2754	0.2854	0.2890	0.3125
5/16–27 UNS	2A	0.0010	0.3115	0.3048	—	0.2874	0.2839	0.2674	2B	0.272	0.281	0.2884	0.2929	0.3125
5/16–28 UN	2A	0.0010	0.3115	0.3050	—	0.2883	0.2849	0.2689	2B	0.274	0.282	0.2893	0.2937	0.3125
	3A	0.0000	0.3125	0.3060	—	0.2893	0.2867	0.2699	3B	0.2740	0.2807	0.2893	0.2926	0.3125
5/16–32 UNEF	2A	0.0010	0.3115	0.3055	—	0.2912	0.2880	0.2743	2B	0.279	0.286	0.2922	0.2964	0.3125
	3A	0.0000	0.3125	0.3065	—	0.2922	0.2898	0.2753	3B	0.2790	0.2847	0.2922	0.2953	0.3125
5/16–36 UNS	2A	0.0009	0.3116	0.3061	—	0.2936	0.2905	0.2785	2B	0.282	0.289	0.2945	0.2985	0.3125
5/16–40 UNS	2A	0.0009	0.3116	0.3065	—	0.2954	0.2925	0.2818	2B	0.285	0.291	0.2963	0.3001	0.3125
5/16–48 UNS	2A	0.0008	0.3117	0.3072	—	0.2982	0.2955	0.2869	2B	0.290	0.295	0.2990	0.3026	0.3125
3/8–16 UNC	1A	0.0013	0.3737	0.3595	—	0.3331	0.3266	0.2992	1B	0.307	0.321	0.3344	0.3429	0.3750
	2A	0.0013	0.3737	0.3643	0.3595	0.3331	0.3287	0.2992	2B	0.307	0.321	0.3344	0.3401	0.3750
	3A	0.0000	0.3750	0.3656	—	0.3344	0.3311	0.3005	3B	0.3070	0.3182	0.3344	0.3387	0.3750
3/8–18 UNS	2A	0.0013	0.3737	0.3650	—	0.3376	0.3333	0.3076	2B	0.315	0.328	0.3389	0.3445	0.3750
3/8–20 UN	2A	0.0012	0.3738	0.3657	—	0.3413	0.3372	0.3143	2B	0.321	0.332	0.3425	0.3479	0.3750
	3A	0.0000	0.3750	0.3669	—	0.3425	0.3394	0.3155	3B	0.3210	0.3297	0.3425	0.3465	0.3750
3/8–24 UNF	1A	0.0011	0.3739	0.3631	—	0.3468	0.3411	0.3243	1B	0.330	0.340	0.3479	0.3553	0.3750
	2A	0.0011	0.3739	0.3667	—	0.3468	0.3430	0.3243	2B	0.330	0.340	0.3479	0.3528	0.3750
3/8–24 UNF	3A	0.0000	0.3750	0.3678	—	0.3479	0.3450	0.3254	3B	0.3300	0.3372	0.3479	0.3516	0.3750
3/8–27 UNS	2A	0.0011	0.3739	0.3672	—	0.3498	0.3462	0.3298	2B	0.335	0.344	0.3509	0.3556	0.3750

Table 3. *(Continued)* **Standard Series and Selected Combinations — Unified Screw Threads**

Nominal Size, Threads per Inch, and Series Designation[a]	External[b]								Internal[b]					
			Major Diameter			Pitch Diameter		UNR Minor Dia.,[c] Max (Ref.)		Minor Diameter		Pitch Diameter		Major Diameter
	Class	Allow-ance	Max[d]	Min	Min[e]	Max[d]	Min		Class	Min	Max	Min	Max	Min
3/8–28 UN	2A	0.0011	0.3739	0.3674	—	0.3507	0.3471	0.3313	2B	0.336	0.345	0.3518	0.3564	0.3750
	3A	0.0000	0.3750	0.3685	—	0.3518	0.3491	0.3324	3B	0.3360	0.3426	0.3518	0.3553	0.3750
3/8–32 UNEF	2A	0.0010	0.3740	0.3680	—	0.3537	0.3503	0.3368	2B	0.341	0.349	0.3547	0.3591	0.3750
	3A	0.0000	0.3750	0.3690	—	0.3547	0.3522	0.3378	3B	0.3410	0.3469	0.3547	0.3580	0.3750
3/8–36 UNS	2A	0.0010	0.3740	0.3685	—	0.3560	0.3528	0.3409	2B	0.345	0.352	0.3570	0.3612	0.3750
3/8–40 UNS	2A	0.0009	0.3741	0.3690	—	0.3579	0.3548	0.3443	2B	0.348	0.354	0.3588	0.3628	0.3750
0.390–27 UNS	2A	0.0011	0.3889	0.3822	—	0.3648	0.3612	0.3448	2B	0.350	0.359	0.3659	0.3706	0.3900
7/16–14 UNC	1A	0.0014	0.4361	0.4206	—	0.3897	0.3826	0.3511	1B	0.360	0.376	0.3911	0.4003	0.4375
	2A	0.0014	0.4361	0.4258	0.4206	0.3897	0.3850	0.3511	2B	0.360	0.376	0.3911	0.3972	0.4375
	3A	0.0000	0.4375	0.4272	—	0.3911	0.3876	0.3525	3B	0.3600	0.3717	0.3911	0.3957	0.4375
7/16–16 UN	2A	0.0014	0.4361	0.4267	—	0.3955	0.3909	0.3616	2B	0.370	0.384	0.3969	0.4028	0.4375
	3A	0.0000	0.4375	0.4281	—	0.3969	0.3935	0.3630	3B	0.3700	0.3800	0.3969	0.4014	0.4375
7/16–18 UNS	2A	0.0013	0.4362	0.4275	—	0.4001	0.3958	0.3701	2B	0.377	0.390	0.4014	0.4070	0.4375
7/16–20 UNF	1A	0.0013	0.4362	0.4240	—	0.4037	0.3975	0.3767	1B	0.383	0.395	0.4050	0.4131	0.4375
	2A	0.0013	0.4362	0.4281	—	0.4037	0.3995	0.3767	2B	0.383	0.395	0.4050	0.4104	0.4375
	3A	0.0000	0.4375	0.4294	—	0.4050	0.4019	0.3780	3B	0.3830	0.3916	0.4050	0.4091	0.4375
7/16–24 UNS	2A	0.0011	0.4364	0.4292	—	0.4093	0.4055	0.3868	2B	0.392	0.402	0.4104	0.4153	0.4375
7/16–27 UNS	2A	0.0011	0.4364	0.4297	—	0.4123	0.4087	0.3923	2B	0.397	0.406	0.4134	0.4181	0.4375
7/16–28 UNEF	2A	0.0011	0.4364	0.4299	—	0.4132	0.4096	0.3938	2B	0.399	0.407	0.4143	0.4189	0.4375
	3A	0.0000	0.4375	0.4310	—	0.4143	0.4116	0.3949	3B	0.3990	0.4051	0.4143	0.4178	0.4375
7/16–32 UN	2A	0.0010	0.4365	0.4305	—	0.4162	0.4128	0.3993	2B	0.404	0.411	0.4172	0.4216	0.4375
	3A	0.0000	0.4375	0.4315	—	0.4172	0.4147	0.4003	3B	0.4040	0.4094	0.4172	0.4205	0.4375
1/2–12 UNS	2A	0.0016	0.4984	0.4870	—	0.4443	0.4389	0.3992	2B	0.410	0.428	0.4459	0.4529	0.5000
	3A	0.0000	0.5000	0.4886	—	0.4459	0.4419	0.4008	3B	0.4100	0.4223	0.4459	0.4511	0.5000
1/2–13 UNC	1A	0.0015	0.4985	0.4822	—	0.4485	0.4411	0.4069	1B	0.417	0.434	0.4500	0.4597	0.5000
	2A	0.0015	0.4985	0.4876	0.4822	0.4485	0.4435	0.4069	2B	0.417	0.434	0.4500	0.4565	0.5000
	3A	0.0000	0.5000	0.4891	—	0.4500	0.4463	0.4084	3B	0.4170	0.4284	0.4500	0.4548	0.5000
1/2–14 UNS	2A	0.0015	0.4985	0.4882	—	0.4521	0.4471	0.4135	2B	0.423	0.438	0.4536	0.4601	0.5000
1/2–16 UN	2A	0.0014	0.4986	0.4892	—	0.4580	0.4533	0.4241	2B	0.432	0.446	0.4594	0.4655	0.5000

Table 3. *(Continued)* **Standard Series and Selected Combinations — Unified Screw Threads**

Nominal Size, Threads per Inch, and Series Designation[a]	External[b]								Internal[b]					
	Class	Allow-ance	Major Diameter			Pitch Diameter		UNR Minor Dia.,[c] Max (Ref.)	Class	Minor Diameter		Pitch Diameter		Major Diameter
			Max[d]	Min	Min[e]	Max[d]	Min			Min	Max	Min	Max	Min
	3A	0.0000	0.5000	0.4906	—	0.4594	0.4559	0.4255	3B	0.4320	0.4419	0.4594	0.4640	0.5000
½–18 UNS	2A	0.0013	0.4987	0.4900	—	0.4626	0.4582	0.4326	2B	0.440	0.453	0.4639	0.4697	0.5000
½–20 UNF	1A	0.0013	0.4987	0.4865	—	0.4662	0.4598	0.4392	1B	0.446	0.457	0.4675	0.4759	0.5000
	2A	0.0013	0.4987	0.4906	—	0.4662	0.4619	0.4392	2B	0.446	0.457	0.4675	0.4731	0.5000
	3A	0.0000	0.5000	0.4919	—	0.4675	0.4643	0.4405	3B	0.4460	0.4537	0.4675	0.4717	0.5000
½–24 UNS	2A	0.0012	0.4988	0.4916	—	0.4717	0.4678	0.4492	2B	0.455	0.465	0.4729	0.4780	0.5000
½–27 UNS	2A	0.0011	0.4989	0.4922	—	0.4748	0.4711	0.4548	2B	0.460	0.469	0.4759	0.4807	0.5000
½–28 UNEF	2A	0.0011	0.4989	0.4924	—	0.4757	0.4720	0.4563	2B	0.461	0.470	0.4768	0.4816	0.5000
	3A	0.0000	0.5000	0.4935	—	0.4768	0.4740	0.4574	3B	0.4610	0.4676	0.4768	0.4804	0.5000
½–32 UN	2A	0.0010	0.4990	0.4930	—	0.4787	0.4752	0.4618	2B	0.466	0.474	0.4797	0.4842	0.5000
	3A	0.0000	0.5000	0.4940	—	0.4797	0.4771	0.4628	3B	0.4660	0.4719	0.4797	0.4831	0.5000
9⁄16–12 UNC	1A	0.0016	0.5609	0.5437	—	0.5068	0.4990	0.4617	1B	0.472	0.490	0.5084	0.5186	0.5625
	2A	0.0016	0.5609	0.5495	0.5437	0.5068	0.5016	0.4617	2B	0.472	0.490	0.5084	0.5152	0.5625
	3A	0.0000	0.5625	0.5511	—	0.5084	0.5045	0.4633	3B	0.4720	0.4843	0.5084	0.5135	0.5625
9⁄16–14 UNS	2A	0.0015	0.5610	0.5507	—	0.5146	0.5096	0.4760	2B	0.485	0.501	0.5161	0.5226	0.5625
9⁄16–16 UN	2A	0.0014	0.5611	0.5517	—	0.5205	0.5158	0.4866	2B	0.495	0.509	0.5219	0.5280	0.5625
	3A	0.0000	0.5625	0.5531	—	0.5219	0.5184	0.4880	3B	0.4950	0.5040	0.5219	0.5265	0.5625
9⁄16–18 UNF	1A	0.0014	0.5611	0.5480	—	0.5250	0.5182	0.4950	1B	0.502	0.515	0.5264	0.5353	0.5625
	2A	0.0014	0.5611	0.5524	—	0.5250	0.5205	0.4950	2B	0.502	0.515	0.5264	0.5323	0.5625
	3A	0.0000	0.5625	0.5538	—	0.5264	0.5230	0.4964	3B	0.5020	0.5106	0.5264	0.5308	0.5625
9⁄16–20 UN	2A	0.0013	0.5612	0.5531	—	0.5287	0.5245	0.5017	2B	0.508	0.520	0.5300	0.5355	0.5625
	3A	0.0000	0.5625	0.5544	—	0.5300	0.5268	0.5030	3B	0.5080	0.5162	0.5300	0.5341	0.5625
9⁄16–24 UNEF	2A	0.0012	0.5613	0.5541	—	0.5342	0.5303	0.5117	2B	0.517	0.527	0.5354	0.5405	0.5625
	3A	0.0000	0.5625	0.5553	—	0.5354	0.5325	0.5129	3B	0.5170	0.5244	0.5354	0.5392	0.5625
9⁄16–27 UNS	2A	0.0011	0.5614	0.5547	—	0.5373	0.5336	0.5173	2B	0.522	0.531	0.5384	0.5432	0.5625
9⁄16–28 UN	2A	0.0011	0.5614	0.5549	—	0.5382	0.5345	0.5188	2B	0.524	0.532	0.5393	0.5441	0.5625
	3A	0.0000	0.5625	0.5560	—	0.5393	0.5365	0.5199	3B	0.5240	0.5301	0.5393	0.5429	0.5625
9⁄16–32 UN	2A	0.0010	0.5615	0.5555	—	0.5412	0.5377	0.5243	2B	0.529	0.536	0.5422	0.5467	0.5625
	3A	0.0000	0.5625	0.5565	—	0.5422	0.5396	0.5253	3B	0.5290	0.5344	0.5422	0.5456	0.5625

Table 3. *(Continued)* **Standard Series and Selected Combinations — Unified Screw Threads**

Nominal Size, Threads per Inch, and Series Designation[a]	External[b]								Internal[b]					
	Class	Allow-ance	Major Diameter			Pitch Diameter		UNR Minor Dia.,[c] Max (Ref.)	Class	Minor Diameter		Pitch Diameter		Major Diameter
			Max[d]	Min	Min[e]	Max[d]	Min			Min	Max	Min	Max	Min
5⁄8–11 UNC	1A	0.0016	0.6234	0.6052	—	0.5644	0.5561	0.5152	1B	0.527	0.546	0.5660	0.5767	0.6250
	2A	0.0016	0.6234	0.6113	0.6052	0.5644	0.5589	0.5152	2B	0.527	0.546	0.5660	0.5732	0.6250
	3A	0.0000	0.6250	0.6129	—	0.5660	0.5619	0.5168	3B	0.5270	0.5391	0.5660	0.5714	0.6250
5⁄8–12 UN	2A	0.0016	0.6234	0.6120	—	0.5693	0.5639	0.5242	2B	0.535	0.553	0.5709	0.5780	0.6250
	3A	0.0000	0.6250	0.6136	—	0.5709	0.5668	0.5258	3B	0.5350	0.5463	0.5709	0.5762	0.6250
5⁄8–14 UNS	2A	0.0015	0.6235	0.6132	—	0.5771	0.5720	0.5385	2B	0.548	0.564	0.5786	0.5852	0.6250
5⁄8–16 UN	2A	0.0014	0.6236	0.6142	—	0.5830	0.5782	0.5491	2B	0.557	0.571	0.5844	0.5906	0.6250
	3A	0.0000	0.6250	0.6156	—	0.5844	0.5808	0.5505	3B	0.5570	0.5662	0.5844	0.5890	0.6250
5⁄8–18 UNF	1A	0.0014	0.6236	0.6105	—	0.5875	0.5805	0.5575	1B	0.565	0.578	0.5889	0.5980	0.6250
	2A	0.0014	0.6236	0.6149	—	0.5875	0.5828	0.5575	2B	0.565	0.578	0.5889	0.5949	0.6250
	3A	0.0000	0.6250	0.6163	—	0.5889	0.5854	0.5589	3B	0.5650	0.5730	0.5889	0.5934	0.6250
5⁄8–20 UN	2A	0.0013	0.6237	0.6156	—	0.5912	0.5869	0.5642	2B	0.571	0.582	0.5925	0.5981	0.6250
	3A	0.0000	0.6250	0.6169	—	0.5925	0.5893	0.5655	3B	0.5710	0.5787	0.5925	0.5967	0.6250
5⁄8–24 UNEF	2A	0.0012	0.6238	0.6166	—	0.5967	0.5927	0.5742	2B	0.580	0.590	0.5979	0.6031	0.6250
	3A	0.0000	0.6250	0.6178	—	0.5979	0.5949	0.5754	3B	0.5800	0.5869	0.5979	0.6018	0.6250
5⁄8–27 UNS	2A	0.0011	0.6239	0.6172	—	0.5998	0.5960	0.5798	2B	0.585	0.594	0.6009	0.6059	0.6250
5⁄8–28 UN	2A	0.0011	0.6239	0.6174	—	0.6007	0.5969	0.5813	2B	0.586	0.595	0.6018	0.6067	0.6250
	3A	0.0000	0.6250	0.6185	—	0.6018	0.5990	0.5824	3B	0.5860	0.5926	0.6018	0.6055	0.6250
5⁄8–32 UN	2A	0.0011	0.6239	0.6179	—	0.6036	0.6000	0.5867	2B	0.591	0.599	0.6047	0.6093	0.6250
	3A	0.0000	0.6250	0.6190	—	0.6047	0.6020	0.5878	3B	0.5910	0.5969	0.6047	0.6082	0.6250
11⁄16–12 UN	2A	0.0016	0.6859	0.6745	—	0.6318	0.6264	0.5867	2B	0.597	0.615	0.6334	0.6405	0.6875
	3A	0.0000	0.6875	0.6761	—	0.6334	0.6293	0.5883	3B	0.5970	0.6085	0.6334	0.6387	0.6875
11⁄16–16 UN	2A	0.0014	0.6861	0.6767	—	0.6455	0.6407	0.6116	2B	0.620	0.634	0.6469	0.6531	0.6875
	3A	0.0000	0.6875	0.6781	—	0.6469	0.6433	0.6130	3B	0.6200	0.6284	0.6469	0.6515	0.6875
11⁄16–20 UN	2A	0.0013	0.6862	0.6781	—	0.6537	0.6494	0.6267	2B	0.633	0.645	0.6550	0.6606	0.6875
	3A	0.0000	0.6875	0.6794	—	0.6550	0.6518	0.6280	3B	0.6330	0.6412	0.6550	0.6592	0.6875
11⁄16–24 UNEF	2A	0.0012	0.6863	0.6791	—	0.6592	0.6552	0.6367	2B	0.642	0.652	0.6604	0.6656	0.6875
	3A	0.0000	0.6875	0.6803	—	0.6604	0.6574	0.6379	3B	0.6420	0.6494	0.6604	0.6643	0.6875
11⁄16–28 UN	2A	0.0011	0.6864	0.6799	—	0.6632	0.6594	0.6438	2B	0.649	0.657	0.6643	0.6692	0.6875

Table 3. *(Continued)* **Standard Series and Selected Combinations — Unified Screw Threads**

Nominal Size, Threads per Inch, and Series Designation[a]	External[b]								Internal[b]					
	Class	Allow-ance	Major Diameter			Pitch Diameter		UNR Minor Dia.,[c] Max (Ref.)	Class	Minor Diameter		Pitch Diameter		Major Diameter
			Max[d]	Min	Min[e]	Max[d]	Min			Min	Max	Min	Max	Min
	3A	0.0000	0.6875	0.6810	—	0.6643	0.6615	0.6449	3B	0.6490	0.6551	0.6643	0.6680	0.6875
11⁄16–32 UN	2A	0.0011	0.6864	0.6804	—	0.6661	0.6625	0.6492	2B	0.654	0.661	0.6672	0.6718	0.6875
	3A	0.0000	0.6875	0.6815	—	0.6672	0.6645	0.6503	3B	0.6540	0.6594	0.6672	0.6707	0.6875
3⁄4–10 UNC	1A	0.0018	0.7482	0.7288	—	0.6832	0.6744	0.6291	1B	0.642	0.663	0.6850	0.6965	0.7500
	2A	0.0018	0.7482	0.7353	0.7288	0.6832	0.6773	0.6291	2B	0.642	0.663	0.6850	0.6927	0.7500
	3A	0.0000	0.7500	0.7371	—	0.6850	0.6806	0.6309	3B	0.6420	0.6545	0.6850	0.6907	0.7500
3⁄4–12 UN	2A	0.0017	0.7483	0.7369	—	0.6942	0.6887	0.6491	2B	0.660	0.678	0.6959	0.7031	0.7500
	3A	0.0000	0.7500	0.7386	—	0.6959	0.6918	0.6508	3B	0.6600	0.6707	0.6959	0.7013	0.7500
3⁄4–14 UNS	2A	0.0015	0.7485	0.7382	—	0.7021	0.6970	0.6635	2B	0.673	0.688	0.7036	0.7103	0.7500
3⁄4–16 UNF	1A	0.0015	0.7485	0.7343	—	0.7079	0.7004	0.6740	1B	0.682	0.696	0.7094	0.7192	0.7500
	2A	0.0015	0.7485	0.7391	—	0.7079	0.7029	0.6740	2B	0.682	0.696	0.7094	0.7159	0.7500
	3A	0.0000	0.7500	0.7406	—	0.7094	0.7056	0.6755	3B	0.6820	0.6908	0.7094	0.7143	0.7500
3⁄4–18 UNS	2A	0.0014	0.7486	0.7399	—	0.7125	0.7079	0.6825	2B	0.690	0.703	0.7139	0.7199	0.7500
3⁄4–20 UNEF	2A	0.0013	0.7487	0.7406	—	0.7162	0.7118	0.6892	2B	0.696	0.707	0.7175	0.7232	0.7500
	3A	0.0000	0.7500	0.7419	—	0.7175	0.7142	0.6905	3B	0.6960	0.7037	0.7175	0.7218	0.7500
3⁄4–24 UNS	2A	0.0012	0.7488	0.7416	—	0.7217	0.7176	0.6992	2B	0.705	0.715	0.7229	0.7282	0.7500
3⁄4–24 UNS	2A	0.0012	0.7488	0.7421	—	0.7247	0.7208	0.7047	2B	0.710	0.719	0.7259	0.7310	0.7500
3⁄4–28 UN	2A	0.0012	0.7488	0.7423	—	0.7256	0.7218	0.7062	2B	0.711	0.720	0.7268	0.7318	0.7500
	3A	0.0000	0.7500	0.7435	—	0.7268	0.7239	0.7074	3B	0.7110	0.7176	0.7268	0.7305	0.7500
3⁄4–32 UN	2A	0.0011	0.7489	0.7429	—	0.7286	0.7250	0.7117	2B	0.716	0.724	0.7297	0.7344	0.7500
	3A	0.0000	0.7500	0.7440	—	0.7297	0.7270	0.7128	3B	0.7160	0.7219	0.7297	0.7333	0.7500
13⁄16–12 UN	2A	0.0017	0.8108	0.7994	—	0.7567	0.7512	0.7116	2B	0.722	0.740	0.7584	0.7656	0.8125
	3A	0.0000	0.8125	0.8011	—	0.7584	0.7543	0.7133	3B	0.7220	0.7329	0.7584	0.7638	0.8125
13⁄16–16 UN	2A	0.0015	0.8110	0.8016	—	0.7704	0.7655	0.7365	2B	0.745	0.759	0.7719	0.7782	0.8125
	3A	0.0000	0.8125	0.8031	—	0.7719	0.7683	0.7380	3B	0.7450	0.7533	0.7719	0.7766	0.8125
13⁄16–20 UNEF	2A	0.0013	0.8112	0.8031	—	0.7787	0.7743	0.7517	2B	0.758	0.770	0.7800	0.7857	0.8125
	3A	0.0000	0.8125	0.8044	—	0.7800	0.7767	0.7530	3B	0.7580	0.7662	0.7800	0.7843	0.8125
13⁄16–28 UN	2A	0.0012	0.8113	0.8048	—	0.7881	0.7843	0.7687	2B	0.774	0.782	0.7893	0.7943	0.8125
	3A	0.0000	0.8125	0.8060	—	0.7893	0.7864	0.7699	3B	0.7740	0.7801	0.7893	0.7930	0.8125

Table 3. *(Continued)* **Standard Series and Selected Combinations — Unified Screw Threads**

Nominal Size, Threads per Inch, and Series Designation[a]	External[b]								Internal[b]					
	Class	Allowance	Major Diameter			Pitch Diameter		UNR Minor Dia.,[c] Max (Ref.)	Class	Minor Diameter		Pitch Diameter		Major Diameter
			Max[d]	Min	Min[e]	Max[d]	Min			Min	Max	Min	Max	Min
13/16–32 UN	2A	0.0011	0.8114	0.8054	—	0.7911	0.7875	0.7742	2B	0.779	0.786	0.7922	0.7969	0.8125
	3A	0.0000	0.8125	0.8065	—	0.7922	0.7895	0.7753	3B	0.7790	0.7844	0.7922	0.7958	0.8125
7/8–9 UNC	1A	0.0019	0.8731	0.8523	—	0.8009	0.7914	0.7408	1B	0.755	0.778	0.8028	0.8151	0.8750
	2A	0.0019	0.8731	0.8592	0.8523	0.8009	0.7946	0.7408	2B	0.755	0.778	0.8028	0.8110	0.8750
	3A	0.0000	0.8750	0.8611	—	0.8028	0.7981	0.7427	3B	0.7550	0.7681	0.8028	0.8089	0.8750
7/8–10 UNS	2A	0.0018	0.8732	0.8603	—	0.8082	0.8022	0.7542	2B	0.767	0.788	0.8100	0.8178	0.8750
7/8–12 UN	2A	0.0017	0.8733	0.8619	—	0.8192	0.8137	0.7741	2B	0.785	0.803	0.8209	0.8281	0.8750
	3A	0.0000	0.8750	0.8636	—	0.8209	0.8168	0.7758	3B	0.7850	0.7948	0.8209	0.8263	0.8750
7/8–14 UNF	1A	0.0016	0.8734	0.8579	—	0.8270	0.8189	0.7884	1B	0.798	0.814	0.8286	0.8392	0.8750
	2A	0.0016	0.8734	0.8631	—	0.8270	0.8216	0.7884	2B	0.798	0.814	0.8286	0.8356	0.8750
	3A	0.0000	0.8750	0.8647	—	0.8286	0.8245	0.7900	3B	0.7980	0.8068	0.8286	0.8339	0.8750
7/8–16 UN	2A	0.0015	0.8735	0.8641	—	0.8329	0.8280	0.7900	2B	0.807	0.821	0.8344	0.8407	0.8750
	3A	0.0000	0.8750	0.8656	—	0.8344	0.8308	0.8005	3B	0.8070	0.8158	0.8344	0.8391	0.8750
7/8–18 UNS	2A	0.0014	0.8736	0.8649	—	0.8375	0.8329	0.8075	2B	0.815	0.828	0.8389	0.8449	0.8750
7/8–20 UNEF	2A	0.0013	0.8737	0.8656	—	0.8412	0.8368	0.8142	2B	0.821	0.832	0.8425	0.8482	0.8750
	3A	0.0000	0.8750	0.8669	—	0.8425	0.8392	0.8155	3B	0.8210	0.8287	0.8425	0.8468	0.8750
7/8–24 UNS	2A	0.0012	0.8738	0.8666	—	0.8467	0.8426	0.8242	2B	0.830	0.840	0.8479	0.8532	0.8750
7/8–27 UNS	2A	0.0012	0.8738	0.8671	—	0.8497	0.8458	0.8297	2B	0.835	0.844	0.8509	0.8560	0.8750
7/8–28 UN	2A	0.0012	0.8738	0.8673	—	0.8506	0.8468	0.8312	2B	0.836	0.845	0.8518	0.8568	0.8750
	3A	0.0000	0.8750	0.8685	—	0.8518	0.8489	0.8324	3B	0.8360	0.8426	0.8518	0.8555	0.8750
7/8–32 UN	2A	0.0011	0.8739	0.8679	—	0.8536	0.8500	0.8367	2B	0.841	0.849	0.8547	0.8594	0.8750
	3A	0.0000	0.8750	0.8690	—	0.8547	0.8520	0.8378	3B	0.8410	0.8469	0.8547	0.8583	0.8750
15/16–12 UN	2A	0.0017	0.9358	0.9244	—	0.8817	0.8760	0.8366	2B	0.847	0.865	0.8834	0.8908	0.9375
	3A	0.0000	0.9375	0.9261	—	0.8834	0.8793	0.8383	3B	0.8470	0.8575	0.8834	0.8889	0.9375
15/16–16 UN	2A	0.0015	0.9360	0.9266	—	0.8954	0.8904	0.8615	2B	0.870	0.884	0.8969	0.9034	0.9375
	3A	0.0000	0.9375	0.9281	—	0.8969	0.8932	0.8630	3B	0.8700	0.8783	0.8969	0.9018	0.9375
15/16–20 UNEF	2A	0.0014	0.9361	0.9280	—	0.9036	0.8991	0.8766	2B	0.883	0.895	0.9050	0.9109	0.9375
	3A	0.0000	0.9375	0.9294	—	0.9050	0.9016	0.8780	3B	0.8830	0.8912	0.9050	0.9094	0.9375
15/16–28 UN	2A	0.0012	0.9363	0.9298	—	0.9131	0.9091	0.8937	2B	0.899	0.907	0.9143	0.9195	0.9375

Table 3. *(Continued)* **Standard Series and Selected Combinations — Unified Screw Threads**

Nominal Size, Threads per Inch, and Series Designation[a]	External[b]								Internal[b]					
	Class	Allow-ance	Major Diameter			Pitch Diameter		UNR Minor Dia.,[c] Max (Ref.)	Class	Minor Diameter		Pitch Diameter		Major Diameter
			Max[d]	Min	Min[e]	Max[d]	Min			Min	Max	Min	Max	Min
	3A	0.0000	0.9375	0.9310	—	0.9143	0.9113	0.8949	3B	0.8990	0.9051	0.9143	0.9182	0.9375
15⁄16–32 UN	2A	0.0011	0.9364	0.9304	—	0.9161	0.9123	0.8992	2B	0.904	0.911	0.9172	0.9221	0.9375
	3A	0.0000	0.9375	0.9315	—	0.9172	0.9144	0.9003	3B	0.9040	0.9094	0.9172	0.9209	0.9375
1–8 UNC	1A	0.0020	0.9980	0.9755	—	0.9168	0.9067	0.8492	1B	0.865	0.890	0.9188	0.9320	1.0000
	2A	0.0020	0.9980	0.9830	0.9755	0.9168	0.9100	0.8492	2B	0.865	0.890	0.9188	0.9276	1.0000
	3A	0.0000	1.0000	0.9850	—	0.9188	0.9137	0.8512	3B	0.8650	0.8797	0.9188	0.9254	1.0000
1–10 UNS	2A	0.0018	0.9982	0.9853	—	0.9332	0.9270	0.8792	2B	0.892	0.913	0.9350	0.9430	1.0000
1–12 UNF	1A	0.0018	0.9982	0.9810	—	0.9441	0.9353	0.8990	1B	0.910	0.928	0.9459	0.9573	1.0000
	2A	0.0018	0.9982	0.9868	—	0.9441	0.9382	0.8990	2B	0.910	0.928	0.9459	0.9535	1.0000
	3A	0.0000	1.0000	0.9886	—	0.9459	0.9415	0.9008	3B	0.9100	0.9198	0.9459	0.9516	1.0000
1–14 UNS[f]	1A	0.0017	0.9983	0.9828	—	0.9519	0.9435	0.9132	1B	0.923	0.938	0.9536	0.9645	1.0000
	2A	0.0017	0.9983	0.9880	—	0.9519	0.9463	0.9132	2B	0.923	0.938	0.9536	0.9609	1.0000
	3A	0.0000	1.0000	0.9897	—	0.9536	0.9494	0.9149	3B	0.9230	0.9315	0.9536	0.9590	1.0000
1–16 UN	2A	0.0015	0.9985	0.9891	—	0.9579	0.9529	0.9240	2B	0.932	0.946	0.9594	0.9659	1.0000
	3A	0.0000	1.0000	0.9906	—	0.9594	0.9557	0.9255	3B	0.9320	0.9408	0.9594	0.9643	1.0000
1–18 UNS	2A	0.0014	0.9986	0.9899	—	0.9625	0.9578	0.9325	2B	0.940	0.953	0.9639	0.9701	1.0000
1–20 UNEF	2A	0.0014	0.9986	0.9905	—	0.9661	0.9616	0.9391	2B	0.946	0.957	0.9675	0.9734	1.0000
	3A	0.0000	1.0000	0.9919	—	0.9675	0.9641	0.9405	3B	0.9460	0.9537	0.9675	0.9719	1.0000
1–24 UNS	2A	0.0013	0.9987	0.9915	—	0.9716	0.9674	0.9491	2B	0.955	0.965	0.9729	0.9784	1.0000
1–27 UNS	2A	0.0012	0.9988	0.9921	—	0.9747	0.9707	0.9547	2B	0.960	0.969	0.9759	0.9811	1.0000
1–28 UN	2A	0.0012	0.9988	0.9923	—	0.9756	0.9716	0.9562	2B	0.961	0.970	0.9768	0.9820	1.0000
	3A	0.0000	1.0000	0.9935	—	0.9768	0.9738	0.9574	3B	0.9610	0.9676	0.9768	0.9807	1.0000
1–32 UN	2A	0.0011	0.9989	0.9929	—	0.9786	0.9748	0.9617	2B	0.966	0.974	0.9797	0.9846	1.0000
	3A	0.0000	1.0000	0.9940	—	0.9797	0.9769	0.9628	3B	0.9660	0.9719	0.9797	0.9834	1.0000
1 1⁄16–8 UN	2A	0.0020	1.0605	1.0455	—	0.9793	0.9725	0.9117	2B	0.927	0.952	0.9813	0.9902	1.0625
	3A	0.0000	1.0625	1.0475	—	0.9813	0.9762	0.9137	3B	0.9270	0.9422	0.9813	0.9880	1.0625
1 1⁄16–12 UN	2A	0.0017	1.0608	1.0494	—	1.0067	1.0010	0.9616	2B	0.972	0.990	1.0084	1.0158	1.0625
	3A	0.0000	1.0625	1.0511	—	1.0084	1.0042	0.9633	3B	0.9720	0.9823	1.0084	1.0139	1.0625
1 1⁄16–16 UN	2A	0.0015	1.0610	1.0516	—	1.0204	1.0154	0.9865	2B	0.995	1.009	1.0219	1.0284	1.0625
	3A	0.0000	1.0625	1.0531	—	1.0219	1.0182	0.9880	3B	0.9950	1.0033	1.0219	1.0268	1.0625
1 1⁄16–18 UNEF	2A	0.0014	1.0611	1.0524	—	1.0250	1.0203	0.9950	2B	1.002	1.015	1.0264	1.0326	1.0625

Table 3. *(Continued)* **Standard Series and Selected Combinations — Unified Screw Threads**

Nominal Size, Threads per Inch, and Series Designation[a]	External[b]								Internal[b]					
	Class	Allowance	Major Diameter			Pitch Diameter		UNR Minor Dia.,[c] Max (Ref.)	Class	Minor Diameter		Pitch Diameter		Major Diameter
			Max[d]	Min	Min[e]	Max[d]	Min			Min	Max	Min	Max	Min
	3A	0.0000	1.0625	1.0538	—	1.0264	1.0228	0.9964	3B	1.0020	1.0105	1.0264	1.0310	1.0625
1 1/16–20 UN	2A	0.0014	1.0611	1.0530	—	1.0286	1.0241	1.0016	2B	1.008	1.020	1.0300	1.0359	1.0625
	3A	0.0000	1.0625	1.0544	—	1.0300	1.0266	1.0030	3B	1.0080	1.0162	1.0300	1.0344	1.0625
1 1/16–28 UN	2A	0.0012	1.0613	1.0548	—	1.0381	1.0341	1.0187	2B	1.024	1.032	1.0393	1.0445	1.0625
	3A	0.0000	1.0625	1.0560	—	1.0393	1.0363	1.0199	3B	1.0240	1.0301	1.0393	1.0432	1.0625
1 1/8–7 UNC	1A	0.0022	1.1228	1.0982	—	1.0300	1.0191	0.9527	1B	0.970	0.998	1.0322	1.0463	1.1250
	2A	0.0022	1.1228	1.1064	1.0982	1.0300	1.0228	0.9527	2B	0.970	0.998	1.0322	1.0416	1.1250
	3A	0.0000	1.1250	1.1086	—	1.0322	1.0268	0.9549	3B	0.9700	0.9875	1.0322	1.0393	1.1250
1 1/8–8 UN	2A	0.0021	1.1229	1.1079	1.1004	1.0417	1.0348	0.9741	2B	0.990	1.015	1.0438	1.0528	1.1250
	3A	0.0000	1.1250	1.1100	—	1.0438	1.0386	0.9762	3B	0.9900	1.0047	1.0438	1.0505	1.1250
1 1/8–10 UNS	2A	0.0018	1.1232	1.1103	—	1.0582	1.0520	1.0042	2B	1.017	1.038	1.0600	1.0680	1.1250
1 1/8–12 UNF	1A	0.0018	1.1232	1.1060	—	1.0691	1.0601	1.0240	1B	1.035	1.053	1.0709	1.0826	1.1250
	2A	0.0018	1.1232	1.1118	—	1.0691	1.0631	1.0240	2B	1.035	1.053	1.0709	1.0787	1.1250
	3A	0.0000	1.1250	1.1136	—	1.0709	1.0664	1.0258	3B	1.0350	1.0448	1.0709	1.0768	1.1250
1 1/8–14 UNS	2A	0.0016	1.1234	1.1131	—	1.0770	1.0717	1.0384	2B	1.048	1.064	1.0786	1.0855	1.1250
1 1/8–16 UN	2A	0.0015	1.1235	1.1141	—	1.0829	1.0779	1.0490	2B	1.057	1.071	1.0844	1.0909	1.1250
	3A	0.0000	1.1250	1.1156	—	1.0844	1.0807	1.0505	3B	1.0570	1.0658	1.0844	1.0893	1.1250
1 1/8–18 UNEF	2A	0.0014	1.1236	1.1149	—	1.0875	1.0828	1.0575	2B	1.065	1.078	1.0889	1.0951	1.1250
	3A	0.0000	1.1250	1.1163	—	1.0889	1.0853	1.0589	3B	1.0650	1.0730	1.0889	1.0935	1.1250
1 1/8–20 UN	2A	0.0014	1.1236	1.1155	—	1.0911	1.0866	1.0641	2B	1.071	1.082	1.0925	1.0984	1.1250
	3A	0.0000	1.1250	1.1169	—	1.0925	1.0891	1.0655	3B	1.0710	1.0787	1.0925	1.0969	1.1250
1 1/8–24 UNS	2A	0.0013	1.1237	1.1165	—	1.0966	1.0924	1.0742	2B	1.080	1.090	1.0979	1.1034	1.1250
1 1/8–28 UN	2A	0.0012	1.1238	1.1173	—	1.1006	1.0966	1.0812	2B	1.086	1.095	1.1018	1.1070	1.1250
	3A	0.0000	1.1250	1.1185	—	1.1018	1.0988	1.0824	3B	1.0860	1.0926	1.1018	1.1057	1.1250
1 3/16–8 UN	2A	0.0021	1.1854	1.1704	—	1.1042	1.0972	1.0366	2B	1.052	1.077	1.1063	1.1154	1.1875
	3A	0.0000	1.1875	1.1725	—	1.1063	1.1011	1.0387	3B	1.0520	1.0672	1.1063	1.1131	1.1875
1 3/16–12 UN	2A	0.0017	1.1858	1.1744	—	1.1317	1.1259	1.0866	2B	1.097	1.115	1.1334	1.1409	1.1875
	3A	0.0000	1.1875	1.1761	—	1.1334	1.1291	1.0883	3B	1.0970	1.1073	1.1334	1.1390	1.1875
1 3/16–16 UN	2A	0.0015	1.1860	1.1766	—	1.1454	1.1403	1.1115	2B	1.120	1.134	1.1469	1.1535	1.1875

Table 3. *(Continued)* **Standard Series and Selected Combinations — Unified Screw Threads**

Nominal Size, Threads per Inch, and Series Designation[a]	External[b]								Internal[b]						
	Class	Allowance	Major Diameter			Pitch Diameter		UNR Minor Dia.,[c] Max (Ref.)	Class	Minor Diameter		Pitch Diameter		Major Diameter
			Max[d]	Min	Min[e]	Max[d]	Min			Min	Max	Min	Max	Min
	3A	0.0000	1.1875	1.1781	—	1.1469	1.1431	1.1130	3B	1.1200	1.1283	1.1469	1.1519	1.1875
1 3/16–18 UNEF	2A	0.0015	1.1860	1.1773	—	1.1499	1.1450	1.1199	2B	1.127	1.140	1.1514	1.1577	1.1875
	3A	0.0000	1.1875	1.1788	—	1.1514	1.1478	1.1214	3B	1.1270	1.1355	1.1514	1.1561	1.1875
1 3/16–20 UN	2A	0.0014	1.1861	1.1780	—	1.1536	1.1489	1.1266	2B	1.133	1.145	1.1550	1.1611	1.1875
	3A	0.0000	1.1875	1.1794	—	1.1550	1.1515	1.1280	3B	1.1330	1.1412	1.1550	1.1595	1.1875
1 3/16–28 UN	2A	0.0012	1.1863	1.1798	—	1.1631	1.1590	1.1437	2B	1.149	1.157	1.1643	1.1696	1.1875
	3A	0.0000	1.1875	1.1810	—	1.1643	1.1612	1.1449	3B	1.1490	1.1551	1.1643	1.1683	1.1875
1 1/4–7 UNC	1A	0.0022	1.2478	1.2232	—	1.1550	1.1439	1.0777	1B	1.095	1.123	1.1572	1.1716	1.2500
	2A	0.0022	1.2478	1.2314	1.2232	1.1550	1.1476	1.0777	2B	1.095	1.123	1.1572	1.1668	1.2500
	3A	0.0000	1.2500	1.2336	—	1.1572	1.1517	1.0799	3B	1.0950	1.1125	1.1572	1.1644	1.2500
1 1/4–8 UN	2A	0.0021	1.2479	1.2329	1.2254	1.1667	1.1597	1.0991	2B	1.115	1.140	1.1688	1.1780	1.2500
	3A	0.0000	1.2500	1.2350	—	1.1688	1.1635	1.1012	3B	1.1150	1.1297	1.1688	1.1757	1.2500
1 1/4–10 UNS	2A	0.0019	1.2481	1.2352	—	1.1831	1.1768	1.1291	2B	1.142	1.163	1.1850	1.1932	1.2500
1 1/4–12 UNF	1A	0.0018	1.2482	1.2310	—	1.1941	1.1849	1.1490	1B	1.160	1.178	1.1959	1.2079	1.2500
	2A	0.0018	1.2482	1.2368	—	1.1941	1.1879	1.1490	2B	1.160	1.178	1.1959	1.2039	1.2500
	3A	0.0000	1.2500	1.2386	—	1.1959	1.1913	1.1508	3B	1.1600	1.1698	1.1959	1.2019	1.2500
1 1/4–14 UNS	2A	0.0016	1.2484	1.2381	—	1.2020	1.1966	1.1634	2B	1.173	1.188	1.2036	1.2106	1.2500
1 1/4–16 UN	2A	0.0015	1.2485	1.2391	—	1.2079	1.2028	1.1740	2B	1.182	1.196	1.2094	1.2160	1.2500
	3A	0.0000	1.2500	1.2406	—	1.2094	1.2056	1.1755	3B	1.1820	1.1908	1.2094	1.2144	1.2500
1 1/4–18 UNEF	2A	0.0015	1.2485	1.2398	—	1.2124	1.2075	1.1824	2B	1.190	1.203	1.2139	1.2202	1.2500
	3A	0.0000	1.2500	1.2413	—	1.2139	1.2103	1.1839	3B	1.1900	1.1980	1.2139	1.2186	1.2500
1 1/4–20 UN	2A	0.0014	1.2486	1.2405	—	1.2161	1.2114	1.1891	2B	1.196	1.207	1.2175	1.2236	1.2500
	3A	0.0000	1.2500	1.2419	—	1.2175	1.2140	1.1905	3B	1.1960	1.2037	1.2175	1.2220	1.2500
1 1/4–24 UNS	2A	0.0013	1.2487	1.2415	—	1.2216	1.2173	1.1991	2B	1.205	1.215	1.2229	1.2285	1.2500
1 1/4–28 UN	2A	0.0012	1.2488	1.2423	—	1.2256	1.2215	1.2062	2B	1.211	1.220	1.2268	1.2321	1.2500
	3A	0.0000	1.2500	1.2435	—	1.2268	1.2237	1.2074	3B	1.2110	1.2176	1.2268	1.2308	1.2500
1 5/16–8 UN	2A	0.0021	1.3104	1.2954	—	1.2292	1.2221	1.1616	2B	1.177	1.202	1.2313	1.2405	1.3125
	3A	0.0000	1.3125	1.2975	—	1.2313	1.2260	1.1637	3B	1.1770	1.1922	1.2313	1.2382	1.3125
1 5/16–12 UN	2A	0.0017	1.3108	1.2994	—	1.2567	1.2509	1.2116	2B	1.222	1.240	1.2584	1.2659	1.3125

Table 3. *(Continued)* **Standard Series and Selected Combinations — Unified Screw Threads**

Nominal Size, Threads per Inch, and Series Designation[a]	External[b]								Internal[b]					
	Class	Allowance	Major Diameter			Pitch Diameter		UNR Minor Dia.,[c] Max (Ref.)	Class	Minor Diameter		Pitch Diameter		Major Diameter
			Max[d]	Min	Min[e]	Max[d]	Min			Min	Max	Min	Max	Min
	3A	0.0000	1.3125	1.3011	—	1.2584	1.2541	1.2133	3B	1.2220	1.2323	1.2584	1.2640	1.3125
$1\frac{5}{16}$–16 UN	2A	0.0015	1.3110	1.3016	—	1.2704	1.2653	1.2365	2B	1.245	1.259	1.2719	1.2785	1.3125
	3A	0.0000	1.3125	1.3031	—	1.2719	1.2681	1.2380	3B	1.2450	1.2533	1.2719	1.2769	1.3125
$1\frac{5}{16}$–18 UNEF	2A	0.0015	1.3110	1.3023	—	1.2749	1.2700	1.2449	2B	1.252	1.265	1.2764	1.2827	1.3125
	3A	0.0000	1.3125	1.3038	—	1.2764	1.2728	1.2464	3B	1.2520	1.2605	1.2764	1.2811	1.3125
$1\frac{5}{16}$–20 UN	2A	0.0014	1.3111	1.3030	—	1.2786	1.2739	1.2516	2B	1.258	1.270	1.2800	1.2861	1.3125
	3A	0.0000	1.3125	1.3044	—	1.2800	1.2765	1.2530	3B	1.2580	1.2662	1.2800	1.2845	1.3125
$1\frac{5}{16}$–28 UN	2A	0.0012	1.3113	1.3048	—	1.2881	1.2840	1.2687	2B	1.274	1.282	1.2893	1.2946	1.3125
	3A	0.0000	1.3125	1.3060	—	1.2893	1.2862	1.2699	3B	1.2740	1.2801	1.2893	1.2933	1.3125
$1\frac{3}{8}$–6 UNC	1A	0.0024	1.3726	1.3453	—	1.2643	1.2523	1.1742	1B	1.195	1.225	1.2667	1.2822	1.3750
	2A	0.0024	1.3726	1.3544	1.3453	1.2643	1.2563	1.1742	2B	1.195	1.225	1.2667	1.2771	1.3750
	3A	0.0000	1.3750	1.3568	—	1.2667	1.2607	1.1766	3B	1.1950	1.2146	1.2667	1.2745	1.3750
$1\frac{3}{8}$–8 UN	2A	0.0022	1.3728	1.3578	1.3503	1.2916	1.2844	1.2240	2B	1.240	1.265	1.2938	1.3031	1.3750
	3A	0.0000	1.3750	1.3600	—	1.2938	1.2884	1.2262	3B	1.2400	1.2547	1.2938	1.3008	1.3750
$1\frac{3}{8}$–10 UNS	2A	0.0019	1.3731	1.3602	—	1.3081	1.3018	1.2541	2B	1.267	1.288	1.3100	1.3182	1.3750
$1\frac{3}{8}$–12 UNF	1A	0.0019	1.3731	1.3559	—	1.3190	1.3096	1.2739	1B	1.285	1.303	1.3209	1.3332	1.3750
	2A	0.0019	1.3731	1.3617	—	1.3190	1.3127	1.2739	2B	1.285	1.303	1.3209	1.3291	1.3750
	3A	0.0000	1.3750	1.3636	—	1.3209	1.3162	1.2758	3B	1.2850	1.2948	1.3209	1.3270	1.3750
$1\frac{3}{8}$–14 UNS	2A	0.0016	1.3734	1.3631	—	1.3270	1.3216	1.2884	2B	1.298	1.314	1.3286	1.3356	1.3750
$1\frac{3}{8}$–16 UN	2A	0.0015	1.3735	1.3641	—	1.3329	1.3278	1.2990	2B	1.307	1.321	1.3344	1.3410	1.3750
	3A	0.0000	1.3750	1.3656	—	1.3344	1.3306	1.3005	3B	1.3070	1.3158	1.3344	1.3394	1.3750
$1\frac{3}{8}$–18 UNEF	2A	0.0015	1.3735	1.3648	—	1.3374	1.3325	1.3074	2B	1.315	1.328	1.3389	1.3452	1.3750
	3A	0.0000	1.3750	1.3663	—	1.3389	1.3353	1.3089	3B	1.3150	1.3230	1.3389	1.3436	1.3750
$1\frac{3}{8}$–20 UN	2A	0.0014	1.3736	1.3655	—	1.3411	1.3364	1.3141	2B	1.321	1.332	1.3425	1.3486	1.3750
	3A	0.0000	1.3750	1.3669	—	1.3425	1.3390	1.3155	3B	1.3210	1.3287	1.3425	1.3470	1.3750
$1\frac{3}{8}$–24 UNS	2A	0.0013	1.3737	1.3665	—	1.3466	1.3423	1.3241	2B	1.330	1.340	1.3479	1.3535	1.3750
$1\frac{3}{8}$–28 UN	2A	0.0012	1.3738	1.3673	—	1.3506	1.3465	1.3312	2B	1.336	1.345	1.3518	1.3571	1.3750
	3A	0.0000	1.3750	1.3685	—	1.3518	1.3487	1.3324	3B	1.3360	1.3426	1.3518	1.3558	1.3750
$1\frac{7}{16}$–6 UN	2A	0.0024	1.4351	1.4169	—	1.3268	1.3188	1.2367	2B	1.257	1.288	1.3292	1.3396	1.4375

Table 3. *(Continued)* **Standard Series and Selected Combinations — Unified Screw Threads**

Nominal Size, Threads per Inch, and Series Designation[a]	External[b]								Internal[b]					
	Class	Allow-ance	Major Diameter			Pitch Diameter		UNR Minor Dia.,[c] Max (Ref.)	Class	Minor Diameter		Pitch Diameter		Major Diameter
			Max[d]	Min	Min[e]	Max[d]	Min			Min	Max	Min	Max	Min
	3A	0.0000	1.4375	1.4193	—	1.3292	1.3232	1.2391	3B	1.2570	1.2771	1.3292	1.3370	1.4375
1 7/16–8 UN	2A	0.0022	1.4353	1.4203	—	1.3541	1.3469	1.2865	2B	1.302	1.327	1.3563	1.3657	1.4375
	3A	0.0000	1.4375	1.4225	—	1.3563	1.3509	1.2887	3B	1.3020	1.3172	1.3563	1.3634	1.4375
1 7/16–12 UN	2A	0.0018	1.4357	1.4243	—	1.3816	1.3757	1.3365	2B	1.347	1.365	1.3834	1.3910	1.4375
	3A	0.0000	1.4375	1.4261	—	1.3834	1.3790	1.3383	3B	1.3470	1.3573	1.3834	1.3891	1.4375
1 7/16–16 UN	2A	0.0016	1.4359	1.4265	—	1.3953	1.3901	1.3614	2B	1.370	1.384	1.3969	1.4037	1.4375
	3A	0.0000	1.4375	1.4281	—	1.3969	1.3930	1.3630	3B	1.3700	1.3783	1.3969	1.4020	1.4375
1 7/16–18 UNEF	2A	0.0015	1.4360	1.4273	—	1.3999	1.3949	1.3699	2B	1.377	1.390	1.4014	1.4079	1.4375
	3A	0.0000	1.4375	1.4288	—	1.4014	1.3977	1.3714	3B	1.3770	1.3855	1.4014	1.4062	1.4375
1 7/16–20 UN	2A	0.0014	1.4361	1.4280	—	1.4036	1.3988	1.3766	2B	1.383	1.395	1.4050	1.4112	1.4375
	3A	0.0000	1.4375	1.4294	—	1.4050	1.4014	1.3780	3B	1.3830	1.3912	1.4050	1.4096	1.4375
1 7/16–28 UN	2A	0.0013	1.4362	1.4297	—	1.4130	1.4088	1.3936	2B	1.399	1.407	1.4143	1.4198	1.4375
	3A	0.0000	1.4375	1.4310	—	1.4143	1.4112	1.3949	3B	1.3990	1.4051	1.4143	1.4184	1.4375
1 1/2–6	1A	0.0024	1.4976	1.4703	—	1.3893	1.3772	1.2992	1B	1.320	1.350	1.3917	1.4075	1.5000
	2A	0.0024	1.4976	1.4794	1.4703	1.3893	1.3812	1.2992	2B	1.320	1.350	1.3917	1.4022	1.5000
	3A	0.0000	1.5000	1.4818	—	1.3917	1.3856	1.3016	3B	1.3200	1.3396	1.3917	1.3996	1.5000
1 1/2–8 UN	2A	0.0022	1.4978	1.4828	1.4753	1.4166	1.4093	1.3490	2B	1.365	1.390	1.4188	1.4283	1.5000
	3A	0.0000	1.5000	1.4850	—	1.4188	1.4133	1.3512	3B	1.3650	1.3797	1.4188	1.4259	1.5000
1 1/2–10 UNS	2A	0.0019	1.4981	1.4852	—	1.4331	1.4267	1.3791	2B	1.392	1.413	1.4350	1.4433	1.5000
1 1/2–12 UNF	1A	0.0019	1.4981	1.4809	—	1.4440	1.4344	1.3989	1B	1.410	1.428	1.4459	1.4584	1.5000
	2A	0.0019	1.4981	1.4867	—	1.4440	1.4376	1.3989	2B	1.410	1.428	1.4459	1.4542	1.5000
	3A	0.0000	1.5000	1.4886	—	1.4459	1.4411	1.4008	3B	1.4100	1.4198	1.4459	1.4522	1.5000
1 1/2–14 UNS	2A	0.0017	1.4983	1.4880	—	1.4519	1.4464	1.4133	2B	1.423	1.438	1.4536	1.4608	1.5000
1 1/2–16 UN	2A	0.0016	1.4984	1.4890	—	1.4578	1.4526	1.4239	2B	1.432	1.446	1.4594	1.4662	1.5000
	3A	0.0000	1.5000	1.4906	—	1.4594	1.4555	1.4255	3B	1.4320	1.4408	1.4594	1.4645	1.5000
1 1/2–18 UNEF	2A	0.0015	1.4985	1.4898	—	1.4624	1.4574	1.4324	2B	1.440	1.452	1.4639	1.4704	1.5000
	3A	0.0000	1.5000	1.4913	—	1.4639	1.4602	1.4339	3B	1.4400	1.4480	1.4639	1.4687	1.5000
1 1/2–20 UN	2A	0.0014	1.4986	1.4905	—	1.4661	1.4613	1.4391	2B	1.446	1.457	1.4675	1.4737	1.5000
	3A	0.0000	1.5000	1.4919	—	1.4675	1.4639	1.4405	3B	1.4460	1.4537	1.4675	1.4721	1.5000

Table 3. *(Continued)* **Standard Series and Selected Combinations — Unified Screw Threads**

Nominal Size, Threads per Inch, and Series Designation[a]	External[b]								Internal[b]					
	Class	Allowance	Major Diameter			Pitch Diameter		UNR Minor Dia.,[c] Max (Ref.)	Class	Minor Diameter		Pitch Diameter		Major Diameter
			Max[d]	Min	Min[e]	Max[d]	Min			Min	Max	Min	Max	Min
1½–24 UNS	2A	0.0013	1.4987	1.4915	—	1.4716	1.4672	1.4491	2B	1.455	1.465	1.4729	1.4787	1.5000
1½–28 UN	2A	0.0013	1.4987	1.4922	—	1.4755	1.4713	1.4561	2B	1.461	1.470	1.4768	1.4823	1.5000
	3A	0.0000	1.5000	1.4935	—	1.4768	1.4737	1.4574	3B	1.4610	1.4676	1.4768	1.4809	1.5000
1⁹⁄₁₆–6 UN	2A	0.0024	1.5601	1.5419	—	1.4518	1.4436	1.3617	2B	1.382	1.413	1.4542	1.4648	1.5625
	3A	0.0000	1.5625	1.5443	—	1.4542	1.4481	1.3641	3B	1.3820	1.4021	1.4542	1.4622	1.5625
1⁹⁄₁₆–8 UN	2A	0.0022	1.5603	1.5453	—	1.4791	1.4717	1.4115	2B	1.427	1.452	1.4813	1.4909	1.5625
	3A	0.0000	1.5625	1.5475	—	1.4813	1.4758	1.4137	3B	1.4270	1.4422	1.4813	1.4885	1.5625
1⁹⁄₁₆–12 UN	2A	0.0018	1.5607	1.5493	—	1.5066	1.5007	1.4615	2B	1.472	1.490	1.5084	1.5160	1.5625
	3A	0.0000	1.5625	1.5511	—	1.5084	1.5040	1.4633	3B	1.4720	1.4823	1.5084	1.5141	1.5625
1⁹⁄₁₆–16 UN	2A	0.0016	1.5609	1.5515	—	1.5203	1.5151	1.4864	2B	1.495	1.509	1.5219	1.5287	1.5625
	3A	0.0000	1.5625	1.5531	—	1.5219	1.5180	1.4880	3B	1.4950	1.5033	1.5219	1.5270	1.5625
1⁹⁄₁₆–18 UNEF	2A	0.0015	1.5610	1.5523	—	1.5249	1.5199	1.4949	2B	1.502	1.515	1.5264	1.5329	1.5625
	3A	0.0000	1.5625	1.5538	—	1.5264	1.5227	1.4964	3B	1.5020	1.5105	1.5264	1.5312	1.5625
1⁹⁄₁₆–20 UN	2A	0.0014	1.5611	1.5530	—	1.5286	1.5238	1.5016	2B	1.508	1.520	1.5300	1.5362	1.5625
	3A	0.0000	1.5625	1.5544	—	1.5300	1.5264	1.5030	3B	1.5080	1.5162	1.5300	1.5346	1.5625
1⅝–6 UN	2A	0.0025	1.6225	1.6043	—	1.5142	1.5060	1.4246	2B	1.445	1.475	1.5167	1.5274	1.6250
	3A	0.0000	1.6250	1.6068	—	1.5167	1.5105	1.4271	3B	1.4450	1.4646	1.5167	1.5247	1.6250
1⅝–8 UN	2A	0.0022	1.6228	1.6078	1.6003	1.5416	1.5342	1.4784	2B	1.490	1.515	1.5438	1.5535	1.6250
	3A	0.0000	1.6250	1.6100	—	1.5438	1.5382	1.4806	3B	1.4900	1.5047	1.5438	1.5510	1.6250
1⅝–10 UNS	2A	0.0019	1.6231	1.6102	—	1.5581	1.5517	1.5041	2B	1.517	1.538	1.5600	1.5683	1.6250
1⅝–12 UN	2A	0.0018	1.6232	1.6118	—	1.5691	1.5632	1.5240	2B	1.535	1.553	1.5709	1.5785	1.6250
	3A	0.0000	1.6250	1.6136	—	1.5709	1.5665	1.5258	3B	1.5350	1.5448	1.5709	1.5766	1.6250
1⅝–14 UNS	2A	0.0017	1.6233	1.6130	—	1.5769	1.5714	1.5383	2B	1.548	1.564	1.5786	1.5858	1.6250
1⅝–16 UN	2A	0.0016	1.6234	1.6140	—	1.5828	1.5776	1.5489	2B	1.557	1.571	1.5844	1.5912	1.6250
	3A	0.0000	1.6250	1.6156	—	1.5844	1.5805	1.5505	3B	1.5570	1.5658	1.5844	1.5895	1.6250
1⅝–18 UNEF	2A	0.0015	1.6235	1.6148	—	1.5874	1.5824	1.5574	2B	1.565	1.578	1.5889	1.5954	1.6250
	3A	0.0000	1.6250	1.6163	—	1.5889	1.5852	1.5589	3B	1.5650	1.5730	1.5889	1.5937	1.6250
1⅝–20 UN	2A	0.0014	1.6236	1.6155	—	1.5911	1.5863	1.5641	2B	1.571	1.582	1.5925	1.5987	1.6250
	3A	0.0000	1.6250	1.6169	—	1.5925	1.5889	1.5655	3B	1.5710	1.5787	1.5925	1.5971	1.6250

Table 3. *(Continued)* **Standard Series and Selected Combinations — Unified Screw Threads**

Nominal Size, Threads per Inch, and Series Designation[a]	External[b]								Internal[b]					
	Class	Allow-ance	Major Diameter			Pitch Diameter		UNR Minor Dia.,[c] Max (Ref.)	Class	Minor Diameter		Pitch Diameter		Major Diameter
			Max[d]	Min	Min[e]	Max[d]	Min			Min	Max	Min	Max	Min
1⅝–24 UNS	2A	0.0013	1.6237	1.6165	—	1.5966	1.5922	1.5741	2B	1.580	1.590	1.5979	1.6037	1.6250
1 11/16–6 UN	2A	0.0025	1.6850	1.6668	—	1.5767	1.5684	1.4866	2B	1.507	1.538	1.5792	1.5900	1.6875
	3A	0.0000	1.6875	1.6693	—	1.5792	1.5730	1.4891	3B	1.5070	1.5271	1.5792	1.5873	1.6875
1 11/16–8 UN	2A	0.0022	1.6853	1.6703	—	1.6041	1.5966	1.5365	2B	1.552	1.577	1.6063	1.6160	1.6875
	3A	0.0000	1.6875	1.6725	—	1.6063	1.6007	1.5387	3B	1.5520	1.5672	1.6063	1.6136	1.6875
1 11/16–12 UN	2A	0.0018	1.6857	1.6743	—	1.6316	1.6256	1.5865	2B	1.597	1.615	1.6334	1.6412	1.6875
	3A	0.0000	1.6875	1.6761	—	1.6334	1.6289	1.5883	3B	1.5970	1.6073	1.6334	1.6392	1.6875
1 11/16–16 UN	2A	0.0016	1.6859	1.6765	—	1.6453	1.6400	1.6114	2B	1.620	1.634	1.6469	1.6538	1.6875
	3A	0.0000	1.6875	1.6781	—	1.6469	1.6429	1.6130	3B	1.6200	1.6283	1.6469	1.6521	1.6875
1 11/16–18 UNEF	2A	0.0015	1.6860	1.6773	—	1.6499	1.6448	1.6199	2B	1.627	1.640	1.6514	1.6580	1.6875
	3A	0.0000	1.6875	1.6788	—	1.6514	1.6476	1.6214	3B	1.6270	1.6355	1.6514	1.6563	1.6875
1 11/16–20 UN	2A	0.0015	1.6860	1.6779	—	1.6535	1.6487	1.6265	2B	1.633	1.645	1.6550	1.6613	1.6875
	3A	0.0000	1.6875	1.6794	—	1.6550	1.6514	1.6280	3B	1.6330	1.6412	1.6550	1.6597	1.6875
1¾–5 UNC	1A	0.0027	1.7473	1.7165	—	1.6174	1.6040	1.5092	1B	1.534	1.568	1.6201	1.6375	1.7500
	2A	0.0027	1.7473	1.7268	1.7165	1.6174	1.6085	1.5092	2B	1.534	1.568	1.6201	1.6317	1.7500
	3A	0.0000	1.7500	1.7295	—	1.6201	1.6134	1.5119	3B	1.5340	1.5575	1.6201	1.6288	1.7500
1¾–6 UN	2A	0.0025	1.7475	1.7293	—	1.6392	1.6309	1.5491	2B	1.570	1.600	1.6417	1.6525	1.7500
	3A	0.0000	1.7500	1.7318	—	1.6417	1.6354	1.5516	3B	1.5700	1.5896	1.6417	1.6498	1.7500
1¾–8 UN	2A	0.0023	1.7477	1.7327	1.7252	1.6665	1.6590	1.5989	2B	1.615	1.640	1.6688	1.6786	1.7500
	3A	0.0000	1.7500	1.7350	—	1.6688	1.6632	1.6012	3B	1.6150	1.6297	1.6688	1.6762	1.7500
1¾–10 UNS	2A	0.0019	1.7481	1.7352	—	1.6831	1.6766	1.6291	2B	1.642	1.663	1.6850	1.6934	1.7500
1¾–12 UN	2A	0.0018	1.7482	1.7368	—	1.6941	1.6881	1.6490	2B	1.660	1.678	1.6959	1.7037	1.7500
	3A	0.0000	1.7500	1.7386	—	1.6959	1.6914	1.6508	3B	1.6600	1.6698	1.6959	1.7017	1.7500
1¾–14 UNS	2A	0.0017	1.7483	1.7380	—	1.7019	1.6963	1.6632	2B	1.673	1.688	1.7036	1.7109	1.7500
1¾–16 UN	2A	0.0016	1.7484	1.7390	—	1.7078	1.7025	1.6739	2B	1.682	1.696	1.7094	1.7163	1.7500
	3A	0.0000	1.7500	1.7406	—	1.7094	1.7054	1.6755	3B	1.6820	1.6908	1.7094	1.7146	1.7500
1¾–18 UNS	2A	0.0015	1.7485	1.7398	—	1.7124	1.7073	1.6824	2B	1.690	1.703	1.7139	1.7205	1.7500
1¾–20 UN	2A	0.0015	1.7485	1.7404	—	1.7160	1.7112	1.6890	2B	1.696	1.707	1.7175	1.7238	1.7500
	3A	0.0000	1.7500	1.7419	—	1.7175	1.7139	1.6905	3B	1.6960	1.7037	1.7175	1.7222	1.7500

Table 3. *(Continued)* **Standard Series and Selected Combinations — Unified Screw Threads**

Nominal Size, Threads per Inch, and Series Designation[a]	External[b]								Internal[b]					
	Class	Allow-ance	Major Diameter			Pitch Diameter		UNR Minor Dia.,[c] Max (Ref.)	Class	Minor Diameter		Pitch Diameter		Major Diameter
			Max[d]	Min	Min[e]	Max[d]	Min			Min	Max	Min	Max	Min
1 13/16–6 UN	2A	0.0025	1.8100	1.7918	—	1.7017	1.6933	1.6116	2B	1.632	1.663	1.7042	1.7151	1.8125
	3A	0.0000	1.8125	1.7943	—	1.7042	1.6979	1.6141	3B	1.6320	1.6521	1.7042	1.7124	1.8125
1 13/16–8 UN	2A	0.0023	1.8102	1.7952	—	1.7290	1.7214	1.6614	2B	1.677	1.702	1.7313	1.7412	1.8125
	3A	0.0000	1.8125	1.7975	—	1.7313	1.7256	1.6637	3B	1.6770	1.6922	1.7313	1.7387	1.8125
1 13/16–12 UN	2A	0.0018	1.8107	1.7993	—	1.7566	1.7506	1.7115	2B	1.722	1.740	1.7584	1.7662	1.8125
	3A	0.0000	1.8125	1.8011	—	1.7584	1.7539	1.7133	3B	1.7220	1.7323	1.7584	1.7642	1.8125
1 13/16–16 UN	2A	0.0016	1.8109	1.8015	—	1.7703	1.7650	1.7364	2B	1.745	1.759	1.7719	1.7788	1.8125
	3A	0.0000	1.8125	1.8031	—	1.7719	1.7679	1.7380	3B	1.7450	1.7533	1.7719	1.7771	1.8125
1 13/16–20 UN	2A	0.0015	1.8110	1.8029	—	1.7785	1.7737	1.7515	2B	1.758	1.770	1.7800	1.7863	1.8125
	3A	0.0000	1.8125	1.8044	—	1.7800	1.7764	1.7530	3B	1.7580	1.7662	1.7800	1.7847	1.8125
1 7/8–6 UN	2A	0.0025	1.8725	1.8543	—	1.7642	1.7558	1.6741	2B	1.695	1.725	1.7667	1.7777	1.8750
	3A	0.0000	1.8750	1.8568	—	1.7667	1.7604	1.6766	3B	1.6950	1.7146	1.7667	1.7749	1.8750
1 7/8–8 UN	2A	0.0023	1.8727	1.8577	1.8502	1.7915	1.7838	1.7239	2B	1.740	1.765	1.7938	1.8038	1.8750
	3A	0.0000	1.8750	1.8600	—	1.7938	1.7881	1.7262	3B	1.7400	1.7547	1.7938	1.8013	1.8750
1 7/8–10 UNS	2A	0.0019	1.8731	1.8602	—	1.8081	1.8016	1.7541	2B	1.767	1.788	1.8100	1.8184	1.8750
1 7/8–12 UN	2A	0.0018	1.8732	1.8618	—	1.8191	1.8131	1.7740	2B	1.785	1.803	1.8209	1.8287	1.8750
	3A	0.0000	1.8750	1.8636	—	1.8209	1.8164	1.7758	3B	1.7850	1.7948	1.8209	1.8267	1.8750
1 7/8–14 UNS	2A	0.0017	1.8733	1.8630	—	1.8269	1.8213	1.7883	2B	1.798	1.814	1.8286	1.8359	1.8750
1 7/8–16 UN	2A	0.0016	1.8734	1.8640	—	1.8328	1.8275	1.7989	2B	1.807	1.821	1.8344	1.8413	1.8750
	3A	0.0000	1.8750	1.8656	—	1.8344	1.8304	1.8005	3B	1.8070	1.8158	1.8344	1.8396	1.8750
1 7/8–18 UNS	2A	0.0015	1.8735	1.8648	—	1.8374	1.8323	1.8074	2B	1.815	1.828	1.8389	1.8455	1.8750
1 7/8–20 UN	2A	0.0015	1.8735	1.8654	—	1.8410	1.8362	1.8140	2B	1.821	1.832	1.8425	1.8488	1.8750
	3A	0.0000	1.8750	1.8669	—	1.8425	1.8389	1.8155	3B	1.8210	1.8287	1.8425	1.8472	1.8750
1 15/16–6 UN	2A	0.0026	1.9349	1.9167	—	1.8266	1.8181	1.7365	2B	1.757	1.788	1.8292	1.8403	1.9375
	3A	0.0000	1.9375	1.9193	—	1.8292	1.8228	1.7391	3B	1.7570	1.7771	1.8292	1.8375	1.9375
1 15/16–8 UN	2A	0.0023	1.9352	1.9202	—	1.8540	1.8463	1.7864	2B	1.802	1.827	1.8563	1.8663	1.9375
	3A	0.0000	1.9375	1.9225	—	1.8563	1.8505	1.7887	3B	1.8020	1.8172	1.8563	1.8638	1.9375
1 15/16–12 UN	2A	0.0018	1.9357	1.9243	—	1.8816	1.8755	1.8365	2B	1.847	1.865	1.8834	1.8913	1.9375
	3A	0.0000	1.9375	1.9261	—	1.8834	1.8789	1.8383	3B	1.8470	1.8573	1.8834	1.8893	1.9375

Table 3. *(Continued)* **Standard Series and Selected Combinations — Unified Screw Threads**

Nominal Size, Threads per Inch, and Series Designation[a]	External[b]								Internal[b]					
	Class	Allowance	Major Diameter			Pitch Diameter		UNR Minor Dia.,[c] Max (Ref.)	Class	Minor Diameter		Pitch Diameter		Major Diameter
			Max[d]	Min	Min[e]	Max[d]	Min			Min	Max	Min	Max	Min
1 15/16–16 UN	2A	0.0016	1.9359	1.9265	—	1.8953	1.8899	1.8614	2B	1.870	1.884	1.8969	1.9039	1.9375
	3A	0.0000	1.9375	1.9281	—	1.8969	1.8929	1.8630	3B	1.8700	1.8783	1.8969	1.9021	1.9375
1 15/16–20 UN	2A	0.0015	1.9360	1.9279	—	1.9035	1.8986	1.8765	2B	1.883	1.895	1.9050	1.9114	1.9375
	3A	0.0000	1.9375	1.9294	—	1.9050	1.9013	1.8780	3B	1.8830	1.8912	1.9050	1.9098	1.9375
2–4½ UNC	1A	0.0029	1.9971	1.9641	—	1.8528	1.8385	1.7324	1B	1.759	1.795	1.8557	1.8743	2.0000
	2A	0.0029	1.9971	1.9751	1.9641	1.8528	1.8433	1.7324	2B	1.759	1.795	1.8557	1.8681	2.0000
	3A	0.0000	2.0000	1.9780	—	1.8557	1.8486	1.7353	3B	1.7590	1.7861	1.8557	1.8650	2.0000
2–6 UN	2A	0.0026	1.9974	1.9792	—	1.8891	1.8805	1.7990	2B	1.820	1.850	1.8917	1.9028	2.0000
	3A	0.0000	2.0000	1.9818	—	1.8917	1.8853	1.8016	3B	1.8200	1.8396	1.8917	1.9000	2.0000
2–8 UN	2A	0.0023	1.9977	1.9827	1.9752	1.9165	1.9087	1.8489	2B	1.865	1.890	1.9188	1.9289	2.0000
	3A	0.0000	2.0000	1.9850	—	1.9188	1.9130	1.8512	3B	1.8650	1.8797	1.9188	1.9264	2.0000
2–10 UNS	2A	0.0020	1.9980	1.9851	—	1.9330	1.9265	1.8790	2B	1.892	1.913	1.9350	1.9435	2.0000
2–12 UN	2A	0.0018	1.9982	1.9868	—	1.9441	1.9380	1.8990	2B	1.910	1.928	1.9459	1.9538	2.0000
	3A	0.0000	2.0000	1.9886	—	1.9459	1.9414	1.9008	3B	1.9100	1.9198	1.9459	1.9518	2.0000
2–14 UNS	2A	0.0017	1.9983	1.9880	—	1.9519	1.9462	1.9133	2B	1.923	1.938	1.9536	1.9610	2.0000
2–16 UN	2A	0.0016	1.9984	1.9890	—	1.9578	1.9524	1.9239	2B	1.932	1.946	1.9594	1.9664	2.0000
	3A	0.0000	2.0000	1.9906	—	1.9594	1.9554	1.9255	3B	1.9320	1.9408	1.9594	1.9646	2.0000
2–18 UNS	2A	0.0015	1.9985	1.9898	—	1.9624	1.9573	1.9324	2B	1.940	1.953	1.9639	1.9706	2.0000
2–20 UN	2A	0.0015	1.9985	1.9904	—	1.9660	1.9611	1.9390	2B	1.946	1.957	1.9675	1.9739	2.0000
	3A	0.0000	2.0000	1.9919	—	1.9675	1.9638	1.9405	3B	1.9460	1.9537	1.9675	1.9723	2.0000
2 1/16–16 UNS	2A	0.0016	2.0609	2.0515	—	2.0203	2.0149	1.9864	2B	1.995	2.009	2.0219	2.0289	2.0625
	3A	0.0000	2.0625	2.0531	—	2.0219	2.0179	1.9880	3B	1.9950	2.0033	2.0219	2.0271	2.0625
2 1/8–6 UN	2A	0.0026	2.1224	2.1042	—	2.0141	2.0054	1.9240	2B	1.945	1.975	2.0167	2.0280	2.1250
	3A	0.0000	2.1250	2.1068	—	2.0167	2.0102	1.9266	3B	1.9450	1.9646	2.0167	2.0251	2.1250
2 1/8–8 UN	2A	0.0024	2.1226	2.1076	2.1001	2.0414	2.0335	1.9738	2B	1.990	2.015	2.0438	2.0540	2.1250
	3A	0.0000	2.1250	2.1100	—	2.0438	2.0379	1.9762	3B	1.9900	2.0047	2.0438	2.0515	2.1250
2 1/8–12 UN	2A	0.0018	2.1232	2.1118	—	2.0691	2.0630	2.0240	2B	2.035	2.053	2.0709	2.0788	2.1250
	3A	0.0000	2.1250	2.1136	—	2.0709	2.0664	2.0258	3B	2.0350	2.0448	2.0709	2.0768	2.1250
2 1/8–16 UN	2A	0.0016	2.1234	2.1140	—	2.0828	2.0774	2.0489	2B	2.057	2.071	2.0844	2.0914	2.1250
	3A	0.0000	2.1250	2.1156	—	2.0844	2.0803	2.0505	3B	2.0570	2.0658	2.0844	2.0896	2.1250

Table 3. *(Continued)* **Standard Series and Selected Combinations — Unified Screw Threads**

Nominal Size, Threads per Inch, and Series Designation[a]	External[b]								Internal[b]					
	Class	Allowance	Major Diameter			Pitch Diameter		UNR Minor Dia.,[c] Max (Ref.)	Class	Minor Diameter		Pitch Diameter		Major Diameter
			Max[d]	Min	Min[e]	Max[d]	Min			Min	Max	Min	Max	Min
2 1/8–20 UN	2A	0.0015	2.1235	2.1154	—	2.0910	2.0861	2.0640	2B	2.071	2.082	2.0925	2.0989	2.1250
	3A	0.0000	2.1250	2.1169	—	2.0925	2.0888	2.0655	3B	2.0710	2.0787	2.0925	2.0973	2.1250
2 3/16–16 UNS	2A	0.0016	2.1859	2.1765	—	2.1453	2.1399	2.1114	2B	2.120	2.134	2.1469	2.1539	2.1875
	3A	0.0000	2.1875	2.1781	—	2.1469	2.1428	2.1130	3B	2.1200	2.1283	2.1469	2.1521	2.1875
2 1/4–4 1/2 UNC	1A	0.0029	2.2471	2.2141	—	2.1028	2.0882	1.9824	1B	2.009	2.045	2.1057	2.1247	2.2500
	2A	0.0029	2.2471	2.2251	2.2141	2.1028	2.0931	1.9824	2B	2.009	2.045	2.1057	2.1183	2.2500
	3A	0.0000	2.2500	2.2280	—	2.1057	2.0984	1.9853	3B	2.0090	2.0361	2.1057	2.1152	2.2500
2 1/4–6 UN	2A	0.0026	2.2474	2.2292	—	2.1391	2.1303	2.0490	2B	2.070	2.100	2.1417	2.1531	2.2500
	3A	0.0000	2.2500	2.2318	—	2.1417	2.1351	2.0516	3B	2.0700	2.0896	2.1417	2.1502	2.2500
2 1/4–8 UN	2A	0.0024	2.2476	2.2326	2.2251	2.1664	2.1584	2.0988	2B	2.115	2.140	2.1688	2.1792	2.2500
	3A	0.0000	2.2500	2.2350	—	2.1688	2.1628	2.1012	3B	2.1150	2.1297	2.1688	2.1766	2.2500
2 1/4–10 UNS	2A	0.0020	2.2480	2.2351	—	2.1830	2.1765	2.1290	2B	2.142	2.163	2.1850	2.1935	2.2500
2 1/4–12 UN	2A	0.0018	2.2482	2.2368	—	2.1941	2.1880	2.1490	2B	2.160	2.178	2.1959	2.2038	2.2500
	3A	0.0000	2.2500	2.2386	—	2.1959	2.1914	2.1508	3B	2.1600	2.1698	2.1959	2.2018	2.2500
2 1/4–14 UNS	2A	0.0017	2.2483	2.2380	—	2.2019	2.1962	2.1633	2B	2.173	2.188	2.2036	2.2110	2.2500
2 1/4–16 UN	2A	0.0016	2.2484	2.2390	—	2.2078	2.2024	2.1739	2B	2.182	2.196	2.2094	2.2164	2.2500
	3A	0.0000	2.2500	2.2406	—	2.2094	2.2053	2.1755	3B	2.1820	2.1908	2.2094	2.2146	2.2500
2 1/4–18 UNS	2A	0.0015	2.2485	2.2398	—	2.2124	2.2073	2.1824	2B	2.190	2.203	2.2139	2.2206	2.2500
2 1/4–20 UN	2A	0.0015	2.2485	2.2404	—	2.2160	2.2111	2.1890	2B	2.196	2.207	2.2175	2.2239	2.2500
	3A	0.0000	2.2500	2.2419	—	2.2175	2.2137	2.1905	3B	2.1960	2.2037	2.2175	2.2223	2.2500
2 5/16–16 UNS	2A	0.0017	2.3108	2.3014	—	2.2702	2.2647	2.2363	2B	2.245	2.259	2.2719	2.2791	2.3125
	3A	0.0000	2.3125	2.3031	—	2.2719	2.2678	2.2380	3B	2.2450	2.2533	2.2719	2.2773	2.3125
2 3/8–6 UN	2A	0.0027	2.3723	2.3541	—	2.2640	2.2551	2.1739	2B	2.195	2.226	2.2667	2.2782	2.3750
	3A	0.0000	2.3750	2.3568	—	2.2667	2.2601	2.1766	3B	2.1950	2.2146	2.2667	2.2753	2.3750
2 3/8–8 UN	2A	0.0024	2.3726	2.3576	—	2.2914	2.2833	2.2238	2B	2.240	2.265	2.2938	2.3043	2.3750
	3A	0.0000	2.3750	2.3600	—	2.2938	2.2878	2.2262	3B	2.2400	2.2547	2.2938	2.3017	2.3750
2 3/8–12 UN	2A	0.0019	2.3731	2.3617	—	2.3190	2.3128	2.2739	2B	2.285	2.303	2.3209	2.3290	2.3750
	3A	0.0000	2.3750	2.3636	—	2.3209	2.3163	2.2758	3B	2.2850	2.2948	2.3209	2.3269	2.3750
2 3/8–16 UN	2A	0.0017	2.3733	2.3639	—	2.3327	2.3272	2.2988	2B	2.307	2.321	2.3344	2.3416	2.3750

Table 3. *(Continued)* **Standard Series and Selected Combinations — Unified Screw Threads**

Nominal Size, Threads per Inch, and Series Designation[a]	External[b]								Internal[b]					
	Class	Allowance	Major Diameter			Pitch Diameter		UNR Minor Dia.,[c] Max (Ref.)	Class	Minor Diameter		Pitch Diameter		Major Diameter
			Max[d]	Min	Min[e]	Max[d]	Min			Min	Max	Min	Max	Min
	3A	0.0000	2.3750	2.3656	—	2.3344	2.3303	2.3005	3B	2.3070	2.3158	2.3344	2.3398	2.3750
2⅜–20 UN	2A	0.0015	2.3735	2.3654	—	2.3410	2.3359	2.3140	2B	2.321	2.332	2.3425	2.3491	2.3750
	3A	0.0000	2.3750	2.3669	—	2.3425	2.3387	2.3155	3B	2.3210	2.3287	2.3425	2.3475	2.3750
2 7/16–16 UNS	2A	0.0017	2.4358	2.4264	—	2.3952	2.3897	2.3613	2B	2.370	2.384	2.3969	2.4041	2.4375
	3A	0.0000	2.4375	2.4281	—	2.3969	2.3928	2.3630	3B	2.3700	2.3783	2.3969	2.4023	2.4375
2½–4 UNC	1A	0.0031	2.4969	2.4612	—	2.3345	2.3190	2.1992	1B	2.229	2.267	2.3376	2.3578	2.5000
	2A	0.0031	2.4969	2.4731	2.4612	2.3345	2.3241	2.1992	2B	2.229	2.267	2.3376	2.3511	2.5000
	3A	0.0000	2.5000	2.4762	—	2.3376	2.3298	2.2023	3B	2.2290	2.2594	2.3376	2.3477	2.5000
2½–6 UN	2A	0.0027	2.4973	2.4791	—	2.3890	2.3800	2.2989	2B	2.320	2.350	2.3917	2.4033	2.5000
	3A	0.0000	2.5000	2.4818	—	2.3917	2.3850	2.3016	3B	2.3200	2.3396	2.3917	2.4004	2.5000
2½–8 UN	2A	0.0024	2.4976	2.4826	2.4751	2.4164	2.4082	2.3488	2B	2.365	2.390	2.4188	2.4294	2.5000
	3A	0.0000	2.5000	2.4850	—	2.4188	2.4127	2.3512	3B	2.3650	2.3797	2.4188	2.4268	2.5000
2½–10 UNS	2A	0.0020	2.4980	2.4851	—	2.4330	2.4263	2.3790	2B	2.392	2.413	2.4350	2.4437	2.5000
2½–12 UN	2A	0.0019	2.4981	2.4867	—	2.4440	2.4378	2.3989	2B	2.410	2.428	2.4459	2.4540	2.5000
	3A	0.0000	2.5000	2.4886	—	2.4459	2.4413	2.4008	3B	2.4100	2.4198	2.4459	2.4519	2.5000
2½–14 UNS	2A	0.0017	2.4983	2.4880	—	2.4519	2.4461	2.4133	2B	2.423	2.438	2.4536	2.4612	2.5000
2½–16 UN	2A	0.0017	2.4983	2.4889	—	2.4577	2.4522	2.4238	2B	2.432	2.446	2.4594	2.4666	2.5000
	3A	0.0000	2.5000	2.4906	—	2.4594	2.4553	2.4255	3B	2.4320	2.4408	2.4594	2.4648	2.5000
2½–18 UNS	2A	0.0016	2.4984	2.4897	—	2.4623	2.4570	2.4323	2B	2.440	2.453	2.4639	2.4708	2.5000
2½–20 UN	2A	0.0015	2.4985	2.4904	—	2.4660	2.4609	2.4390	2B	2.446	2.457	2.4675	2.4741	2.5000
	3A	0.0000	2.5000	2.4919	—	2.4675	2.4637	2.4405	3B	2.4460	2.4537	2.4675	2.4725	2.5000
2⅝–6 UN	2A	0.0027	2.6223	2.6041	—	2.5140	2.5050	2.4239	2B	2.445	2.475	2.5167	2.5285	2.6250
	3A	0.0000	2.6250	2.6068	—	2.5167	2.5099	2.4266	3B	2.4450	2.4646	2.5167	2.5255	2.6250
2⅝–8 UN	2A	0.0025	2.6225	2.6075	—	2.5413	2.5331	2.4737	2B	2.490	2.515	2.5438	2.5545	2.6250
	3A	0.0000	2.6250	2.6100	—	2.5438	2.5376	2.4762	3B	2.4900	2.5047	2.5438	2.5518	2.6250
2⅝–12 UN	2A	0.0019	2.6231	2.6117	—	2.5690	2.5628	2.5239	2B	2.535	2.553	2.5709	2.5790	2.6250
	3A	0.0000	2.6250	2.6136	—	2.5709	2.5663	2.5258	3B	2.5350	2.5448	2.5709	2.5769	2.6250
2⅝–16 UN	2A	0.0017	2.6233	2.6139	—	2.5827	2.5772	2.5488	2B	2.557	2.571	2.5844	2.5916	2.6250
	3A	0.0000	2.6250	2.6156	—	2.5844	2.5803	2.5505	3B	2.5570	2.5658	2.5844	2.5898	2.6250

Table 3. *(Continued)* **Standard Series and Selected Combinations — Unified Screw Threads**

Nominal Size, Threads per Inch, and Series Designation[a]	External[b]								Internal[b]					
	Class	Allowance	Major Diameter			Pitch Diameter		UNR Minor Dia.,[c] Max (Ref.)	Class	Minor Diameter		Pitch Diameter		Major Diameter
			Max[d]	Min	Min[e]	Max[d]	Min			Min	Max	Min	Max	Min
2⅝–20 UN	2A	0.0015	2.6235	2.6154	—	2.5910	2.5859	2.5640	2B	2.571	2.582	2.5925	2.5991	2.6250
	3A	0.0000	2.6250	2.6169	—	2.5925	2.5887	2.5655	3B	2.5710	2.5787	2.5925	2.5975	2.6250
2¾–4 UNC	1A	0.0032	2.7468	2.7111	—	2.5844	2.5686	2.4491	1B	2.479	2.517	2.5876	2.6082	2.7500
	2A	0.0032	2.7468	2.7230	2.7111	2.5844	2.5739	2.4491	2B	2.479	2.517	2.5876	2.6013	2.7500
	3A	0.0000	2.7500	2.7262	—	2.5876	2.5797	2.4523	3B	2.4790	2.5094	2.5876	2.5979	2.7500
2¾–6 UN	2A	0.0027	2.7473	2.7291	—	2.6390	2.6299	2.5489	2B	2.570	2.600	2.6417	2.6536	2.7500
	3A	0.0000	2.7500	2.7318	—	2.6417	2.6349	2.5516	3B	2.5700	2.5896	2.6417	2.6506	2.7500
2¾–8 UN	2A	0.0025	2.7475	2.7325	2.7250	2.6663	2.6580	2.5987	2B	2.615	2.640	2.6688	2.6796	2.7500
	3A	0.0000	2.7500	2.7350	—	2.6688	2.6625	2.6012	3B	2.6150	2.6297	2.6688	2.6769	2.7500
2¾–10 UNS	2A	0.0020	2.7480	2.7351	—	2.6830	2.6763	2.6290	2B	2.642	2.663	2.6850	2.6937	2.7500
2¾–12 UN	2A	0.0019	2.7481	2.7367	—	2.6940	2.6878	2.6489	2B	2.660	2.678	2.6959	2.7040	2.7500
	3A	0.0000	2.7500	2.7386	—	2.6959	2.6913	2.6508	3B	2.6600	2.6698	2.6959	2.7019	2.7500
2¾–14 UNS	2A	0.0017	2.7483	2.7380	—	2.7019	2.6961	2.6633	2B	2.673	2.688	2.7036	2.7112	2.7500
2¾–16 UN	2A	0.0017	2.7483	2.7389	—	2.7077	2.7022	2.6738	2B	2.682	2.696	2.7094	2.7166	2.7500
	3A	0.0000	2.7500	2.7406	—	2.7094	2.7053	2.6755	3B	2.6820	2.6908	2.7094	2.7148	2.7500
2¾–18 UNS	2A	0.0016	2.7484	2.7397	—	2.7123	2.7070	2.6823	2B	2.690	2.703	2.7139	2.7208	2.7500
2¾–20 UN	2A	0.0015	2.7485	2.7404	—	2.7160	2.7109	2.6890	2B	2.696	2.707	2.7175	2.7241	2.7500
	3A	0.0000	2.7500	2.7419	—	2.7175	2.7137	2.6905	3B	2.6960	2.7037	2.7175	2.7225	2.7500
2⅞–6 UN	2A	0.0028	2.8722	2.8540	—	2.7639	2.7547	2.6738	2B	2.695	2.725	2.7667	2.7787	2.8750
	3A	0.0000	2.8750	2.8568	—	2.7667	2.7598	2.6766	3B	2.6950	2.7146	2.7667	2.7757	2.8750
2⅞–8 UN	2A	0.0025	2.8725	2.8575	—	2.7913	2.7829	2.7237	2B	2.740	2.765	2.7938	2.8048	2.8750
	3A	0.0000	2.8750	2.8600	—	2.7938	2.7875	2.7262	3B	2.7400	2.7547	2.7938	2.8020	2.8750
2⅞–12 UN	2A	0.0019	2.8731	2.8617	—	2.8190	2.8127	2.7739	2B	2.785	2.803	2.8209	2.8291	2.8750
	3A	0.0000	2.8750	2.8636	—	2.8209	2.8162	2.7758	3B	2.7850	2.7948	2.8209	2.8271	2.8750
2⅞–16 UN	2A	0.0017	2.8733	2.8639	—	2.8327	2.8271	2.7988	2B	2.807	2.821	2.8344	2.8417	2.8750
	3A	0.0000	2.8750	2.8656	—	2.8344	2.8302	2.8005	3B	2.8070	2.8158	2.8344	2.8399	2.8750
2⅞–20 UN	2A	0.0016	2.8734	2.8653	—	2.8409	2.8357	2.8139	2B	2.821	2.832	2.8425	2.8493	2.8750
	3A	0.0000	2.8750	2.8669	—	2.8425	2.8386	2.8155	3B	2.8210	2.8287	2.8425	2.8476	2.8750
3–4 UNC	1A	0.0032	2.9968	2.9611	—	2.8344	2.8183	2.6991	1B	2.729	2.767	2.8376	2.8585	3.0000

Table 3. *(Continued)* **Standard Series and Selected Combinations — Unified Screw Threads**

Nominal Size, Threads per Inch, and Series Designation[a]	External[b]								Internal[b]					
	Class	Allowance	Major Diameter			Pitch Diameter		UNR Minor Dia.,[c] Max (Ref.)	Class	Minor Diameter		Pitch Diameter		Major Diameter
			Max[d]	Min	Min[e]	Max[d]	Min			Min	Max	Min	Max	Min
	2A	0.0032	2.9968	2.9730	2.9611	2.8344	2.8237	2.6991	2B	2.729	2.767	2.8376	2.8515	3.0000
	3A	0.0000	3.0000	2.9762	—	2.8376	2.8296	2.7023	3B	2.7290	2.7594	2.8376	2.8480	3.0000
3–6 UN	2A	0.0028	2.9972	2.9790	—	2.8889	2.8796	2.7988	2B	2.820	2.850	2.8917	2.9038	3.0000
	3A	0.0000	3.0000	2.9818	—	2.8917	2.8847	2.8016	3B	2.8200	2.8396	2.8917	2.9008	3.0000
3–8 UN	2A	0.0026	2.9974	2.9824	2.9749	2.9162	2.9077	2.8486	2B	2.865	2.890	2.9188	2.9299	3.0000
	3A	0.0000	3.0000	2.9850	—	2.9188	2.9124	2.8512	3B	2.8650	2.8797	2.9188	2.9271	3.0000
3–10 UNS	2A	0.0020	2.9980	2.9851	—	2.9330	2.9262	2.8790	2B	2.892	2.913	2.9350	2.9439	3.0000
3–12 UN	2A	0.0019	2.9981	2.9867	—	2.9440	2.9377	2.8989	2B	2.910	2.928	2.9459	2.9541	3.0000
	3A	0.0000	3.0000	2.9886	—	2.9459	2.9412	2.9008	3B	2.9100	2.9198	2.9459	2.9521	3.0000
3–14 UNS	2A	0.0018	2.9982	2.9879	—	2.9518	2.9459	2.9132	2B	2.923	2.938	2.9536	2.9613	3.0000
3–16 UN	2A	0.0017	2.9983	2.9889	—	2.9577	2.9521	2.9238	2B	2.932	2.946	2.9594	2.9667	3.0000
	3A	0.0000	3.0000	2.9906	—	2.9594	2.9552	2.9255	3B	2.9320	2.9408	2.9594	2.9649	3.0000
3–18 UNS	2A	0.0016	2.9984	2.9897	—	2.9623	2.9569	2.9323	2B	2.940	2.953	2.9639	2.9709	3.0000
3–20 UN	2A	0.0016	2.9984	2.9903	—	2.9659	2.9607	2.9389	2B	2.946	2.957	2.9675	2.9743	3.0000
	3A	0.0000	3.0000	2.9919	—	2.9675	2.9636	2.9405	3B	2.9460	2.9537	2.9675	2.9726	3.0000
3⅛–6 UN	2A	0.0028	3.1222	3.1040	—	3.0139	3.0045	2.9238	2B	2.945	2.975	3.0167	3.0289	3.1250
	3A	0.0000	3.1250	3.1068	—	3.0167	3.0097	2.9266	3B	2.9450	2.9646	3.0167	3.0259	3.1250
3⅛–8 UN	2A	0.0026	3.1224	3.1074	—	3.0412	3.0326	2.9736	2B	2.990	3.015	3.0438	3.0550	3.1250
	3A	0.0000	3.1250	3.1100	—	3.0438	3.0374	2.9762	3B	2.9900	3.0047	3.0438	3.0522	3.1250
3⅛–12 UN	2A	0.0019	3.1231	3.1117	—	3.0690	3.0627	3.0239	2B	3.035	3.053	3.0709	3.0791	3.1250
	3A	0.0000	3.1250	3.1136	—	3.0709	3.0662	3.0258	3B	3.0350	3.0448	3.0709	3.0771	3.1250
3⅛–16 UN	2A	0.0017	3.1233	3.1139	—	3.0827	3.0771	3.0488	2B	3.057	3.071	3.0844	3.0917	3.1250
	3A	0.0000	3.1250	3.1156	—	3.0844	3.0802	3.0505	3B	3.0570	3.0658	3.0844	3.0899	3.1250
3¼–4 UNC	1A	0.0033	3.2467	3.2110	—	3.0843	3.0680	2.9490	1B	2.979	3.017	3.0876	3.1088	3.2500
	2A	0.0033	3.2467	3.2229	3.2110	3.0843	3.0734	2.9490	2B	2.979	3.017	3.0876	3.1017	3.2500
	3A	0.0000	3.2500	3.2262	—	3.0876	3.0794	2.9523	3B	2.9790	3.0094	3.0876	3.0982	3.2500
3¼–6 UN	2A	0.0028	3.2472	3.2290	—	3.1389	3.1294	3.0488	2B	3.070	3.100	3.1417	3.1540	3.2500
	3A	0.0000	3.2500	3.2318	—	3.1417	3.1346	3.0516	3B	3.0700	3.0896	3.1417	3.1509	3.2500
3¼–8 UN	2A	0.0026	3.2474	3.2324	3.2249	3.1662	3.1575	3.0986	2B	3.115	3.140	3.1688	3.1801	3.2500
	3A	0.0000	3.2500	3.2350	—	3.1688	3.1623	3.1012	3B	3.1150	3.1297	3.1688	3.1773	3.2500

Table 3. *(Continued)* **Standard Series and Selected Combinations — Unified Screw Threads**

Nominal Size, Threads per Inch, and Series Designation[a]	External[b]								Internal[b]					
	Class	Allowance	Major Diameter			Pitch Diameter		UNR Minor Dia.,[c] Max (Ref.)	Class	Minor Diameter		Pitch Diameter		Major Diameter
			Max[d]	Min	Min[e]	Max[d]	Min			Min	Max	Min	Max	Min
3¼–10 UNS	2A	0.0020	3.2480	3.2351	—	3.1830	3.1762	3.1290	2B	3.142	3.163	3.1850	3.1939	3.2500
3¼–12 UN	2A	0.0019	3.2481	3.2367	—	3.1940	3.1877	3.1489	2B	3.160	3.178	3.1959	3.2041	3.2500
	3A	0.0000	3.2500	3.2386	—	3.1959	3.1912	3.1508	3B	3.1600	3.1698	3.1959	3.2041	3.2500
3¼–14 UNS	2A	0.0018	3.2482	3.2379	—	3.2018	3.1959	3.1632	2B	3.173	3.188	3.2036	3.2113	3.2500
3¼–16 UN	2A	0.0017	3.2483	3.2389	—	3.2077	3.2021	3.1738	2B	3.182	3.196	3.2094	3.2167	3.2500
	3A	0.0000	3.2500	3.2406	—	3.2094	3.2052	3.1755	3B	3.1820	3.1908	3.2094	3.2149	3.2500
3¼–18 UNS	2A	0.0016	3.2484	3.2397	—	3.2123	3.2069	3.1823	2B	3.190	3.203	3.2139	3.2209	3.2500
3⅜–6 UN	2A	0.0029	3.3721	3.3539	—	3.2638	3.2543	3.1737	2B	3.195	3.225	3.2667	3.2791	3.3750
	3A	0.0000	3.3750	3.3568	—	3.2667	3.2595	3.1766	3B	3.1950	3.2146	3.2667	3.2760	3.3750
3⅜–8 UN	2A	0.0026	3.3724	3.3574	—	3.2912	3.2824	3.2236	2B	3.240	3.265	3.2938	3.3052	3.3750
	3A	0.0000	3.3750	3.3600	—	3.2938	3.2872	3.2262	3B	3.2400	3.2547	3.2938	3.3023	3.3750
3⅜–12 UN	2A	0.0019	3.3731	3.3617	—	3.3190	3.3126	3.2739	2B	3.285	3.303	3.3209	3.3293	3.3750
	3A	0.0000	3.3750	3.3636	—	3.3209	3.3161	3.2758	3B	3.2850	3.2948	3.3209	3.3272	3.3750
3⅜–16 UN	2A	0.0017	3.3733	3.3639	—	3.3327	3.3269	3.2988	2B	3.307	3.321	3.3344	3.3419	3.3750
	3A	0.0000	3.3750	3.3656	—	3.3344	3.3301	3.3005	3B	3.3070	3.3158	3.3344	3.3400	3.3750
3½–4 UNC	1A	0.0033	3.4967	3.4610	—	3.3343	3.3177	3.1990	1B	3.229	3.267	3.3376	3.3591	3.5000
	2A	0.0033	3.4967	3.4729	3.4610	3.3343	3.3233	3.1990	2B	3.229	3.267	3.3376	3.3519	3.5000
	3A	0.0000	3.5000	3.4762	—	3.3376	3.3293	3.2023	3B	3.2290	3.2594	3.3376	3.3484	3.5000
3½–6 UN	2A	0.0029	3.4971	3.4789	—	3.3888	3.3792	3.2987	2B	3.320	3.350	3.3917	3.4042	3.5000
	3A	0.0000	3.5000	3.4818	—	3.3917	3.3845	3.3016	3B	3.3200	3.3396	3.3917	3.4011	3.5000
3½–8 UN	2A	0.0026	3.4974	3.4824	3.4749	3.4162	3.4074	3.3486	2B	3.365	3.390	3.4188	3.4303	3.5000
	3A	0.0000	3.5000	3.4850	—	3.4188	3.4122	3.3512	3B	3.3650	3.3797	3.4188	3.4274	3.5000
3½–10 UNS	2A	0.0021	3.4979	3.4850	—	3.4329	3.4260	3.3789	2B	3.392	3.413	3.4350	3.4440	3.5000
3½–12 UN	2A	0.0019	3.4981	3.4867	—	3.4440	3.4376	3.3989	2B	3.410	3.428	3.4459	3.4543	3.5000
	3A	0.0000	3.5000	3.4886	—	3.4459	3.4411	3.4008	3B	3.4100	3.4198	3.4459	3.4522	3.5000
3½–14 UNS	2A	0.0018	3.4982	3.4879	—	3.4518	3.4457	3.4132	2B	3.423	3.438	3.4536	3.4615	3.5000
3½–16 UN	2A	0.0017	3.4983	3.4889	—	3.4577	3.4519	3.4238	2B	3.432	3.446	3.4594	3.4669	3.5000
	3A	0.0000	3.5000	3.4906	—	3.4594	3.4551	3.4255	3B	3.4320	3.4408	3.4594	3.4650	3.5000
3½–18 UNS	2A	0.0017	3.4983	3.4896	—	3.4622	3.4567	3.4322	2B	3.440	3.453	3.4639	3.4711	3.5000

Table 3. *(Continued)* **Standard Series and Selected Combinations — Unified Screw Threads**

Nominal Size, Threads per Inch, and Series Designation[a]	External[b]								Internal[b]					
	Class	Allow-ance	Major Diameter			Pitch Diameter		UNR Minor Dia.,[c] Max (Ref.)	Class	Minor Diameter		Pitch Diameter		Major Diameter
			Max[d]	Min	Min[e]	Max[d]	Min			Min	Max	Min	Max	Min
3⅝–6 UN	2A	0.0029	3.6221	3.6039	—	3.5138	3.5041	3.4237	2B	3.445	3.475	3.5167	3.5293	3.6250
	3A	0.0000	3.6250	3.6068	—	3.5167	3.5094	3.4266	3B	3.4450	3.4646	3.5167	3.5262	3.6250
3⅝–8 UN	2A	0.0027	3.6223	3.6073	—	3.5411	3.5322	3.4735	2B	3.490	3.515	3.5438	3.5554	3.6250
	3A	0.0000	3.6250	3.6100	—	3.5438	3.5371	3.4762	3B	3.4900	3.5047	3.5438	3.5525	3.6250
3⅝–12 UN	2A	0.0019	3.6231	3.6117	—	3.5690	3.5626	3.5239	2B	3.535	3.553	3.5709	3.5793	3.6250
	3A	0.0000	3.6250	3.6136	—	3.5709	3.5661	3.5258	3B	3.5350	3.5448	3.5709	3.5772	3.6250
3⅝–16 UN	2A	0.0017	3.6233	3.6139	—	3.5827	3.5769	3.5488	2B	3.557	3.571	3.5844	3.5919	3.6250
	3A	0.0000	3.6250	3.6156	—	3.5844	3.5801	3.5505	3B	3.5570	3.5658	3.5844	3.5900	3.6250
3¾–4 UNC	1A	0.0034	3.7466	3.7109	—	3.5842	3.5674	3.4489	1B	3.479	3.517	3.5876	3.6094	3.7500
	2A	0.0034	3.7466	3.7228	3.7109	3.5842	3.5730	3.4489	2B	3.479	3.517	3.5876	3.6021	3.7500
	3A	0.0000	3.7500	3.7262	—	3.5876	3.5792	3.4523	3B	3.4790	3.5094	3.5876	3.5985	3.7500
3¾–6 UN	2A	0.0029	3.7471	3.7289	—	3.6388	3.6290	3.5487	2B	3.570	3.600	3.6417	3.6544	3.7500
	3A	0.0000	3.7500	3.7318	—	3.6417	3.6344	3.5516	3B	3.5700	3.5896	3.6417	3.6512	3.7500
3¾–8 UN	2A	0.0027	3.7473	3.7323	3.7248	3.6661	3.6571	3.5985	2B	3.615	3.640	3.6688	3.6805	3.7500
	3A	0.0000	3.7500	3.7350	—	3.6688	3.6621	3.6012	3B	3.6150	3.6297	3.6688	3.6776	3.7500
3¾–10 UNS	2A	0.0021	3.7479	3.7350	—	3.6829	3.6760	3.6289	2B	3.642	3.663	3.6850	3.6940	3.7500
3¾–12 UN	2A	0.0019	3.7481	3.7367	—	3.6940	3.6876	3.6489	2B	3.660	3.678	3.6959	3.7043	3.7500
	3A	0.0000	3.7500	3.7386	—	3.6959	3.6911	3.6508	3B	3.6600	3.6698	3.6959	3.7022	3.7500
3¾–14 UNS	2A	0.0018	3.7482	3.7379	—	3.7018	3.6957	3.6632	2B	3.673	3.688	3.7036	3.7115	3.7500
3¾–16 UN	2A	0.0017	3.7483	3.7389	—	3.7077	3.7019	3.6738	2B	3.682	3.696	3.7094	3.7169	3.7500
	3A	0.0000	3.7500	3.7406	—	3.7094	3.7051	3.6755	3B	3.6820	3.6908	3.7094	3.7150	3.7500
3¾–18 UNS	2A	0.0017	3.7483	3.7396	—	3.7122	3.7067	3.6822	2B	3.690	3.703	3.7139	3.7211	3.7500
3⅞–6 UN	2A	0.0030	3.8720	3.8538	—	3.7637	3.7538	3.6736	2B	3.695	3.725	3.7667	3.7795	3.8750
	3A	0.0000	3.8750	3.8568	—	3.7667	3.7593	3.6766	3B	3.6950	3.7146	3.7667	3.7763	3.8750
3⅞–8 UN	2A	0.0027	3.8723	3.8573	—	3.7911	3.7820	3.7235	2B	3.740	3.765	3.7938	3.8056	3.8750
	3A	0.0000	3.8750	3.8600	—	3.7938	3.7870	3.7262	3B	3.7400	3.7547	3.7938	3.8026	3.8750
3⅞–12 UN	2A	0.0020	3.8730	3.8616	—	3.8189	3.8124	3.7738	2B	3.785	3.803	3.8209	3.8294	3.8750
	3A	0.0000	3.8750	3.8636	—	3.8209	3.8160	3.7758	3B	3.7850	3.7948	3.8209	3.8273	3.8750
3⅞–16 UN	2A	0.0018	3.8732	3.8638	—	3.8326	3.8267	3.7987	2B	3.807	3.821	3.8344	3.8420	3.8750

Table 3. *(Continued)* **Standard Series and Selected Combinations — Unified Screw Threads**

Nominal Size, Threads per Inch, and Series Designation[a]	External[b]								Internal[b]					
	Class	Allowance	Major Diameter			Pitch Diameter		UNR Minor Dia.,[c] Max (Ref.)	Class	Minor Diameter		Pitch Diameter		Major Diameter
			Max[d]	Min	Min[e]	Max[d]	Min			Min	Max	Min	Max	Min
	3A	0.0000	3.8750	3.8656	—	3.8344	3.8300	3.8005	3B	3.8070	3.8158	3.8344	3.8401	3.8750
4–4 UNC	1A	0.0034	3.9966	3.9609	—	3.8342	3.8172	3.6989	1B	3.729	3.767	3.8376	3.8597	4.0000
	2A	0.0034	3.9966	3.9728	3.9609	3.8342	3.8229	3.6989	2B	3.729	3.767	3.8376	3.8523	4.0000
	3A	0.0000	4.0000	3.9762	—	3.8376	3.8291	3.7023	3B	3.7290	3.7594	3.8376	3.8487	4.0000
4–6 UN	2A	0.0030	3.9970	3.9788	—	3.8887	3.8788	3.7986	2B	3.820	3.850	3.8917	3.9046	4.0000
	3A	0.0000	4.0000	3.9818	—	3.8917	3.8843	3.8016	3B	3.8200	3.8396	3.8917	3.9014	4.0000
4–8 UN	2A	0.0027	3.9973	3.9823	3.9748	3.9161	3.9070	3.8485	2B	3.865	3.890	3.9188	3.9307	4.0000
	3A	0.0000	4.0000	3.9850	—	3.9188	3.9120	3.8512	3B	3.8650	3.8797	3.9188	3.9277	4.0000
4–10 UNS	2A	0.0021	3.9979	3.9850	—	3.9329	3.9259	3.8768	2B	3.892	3.913	3.9350	3.9441	4.0000
4–12 UN	2A	0.0020	3.9980	3.9866	—	3.9439	3.9374	3.8988	2B	3.910	3.928	3.9459	3.9544	4.0000
	3A	0.0000	4.0000	3.9886	—	3.9459	3.9410	3.9008	3B	3.9100	3.9198	3.9459	3.9523	4.0000
4–14 UNS	2A	0.0018	3.9982	3.9879	—	3.9518	3.9456	3.9132	2B	3.923	3.938	3.9536	3.9616	4.0000
4–16 UN	2A	0.0018	3.9982	3.9888	—	3.9576	3.9517	3.9237	2B	3.932	3.946	3.9594	3.9670	4.0000
	3A	0.0000	4.0000	3.9906	—	3.9594	3.9550	3.9255	3B	3.9320	3.9408	3.9594	3.9651	4.000

[a] Use UNR designation instead of UN wherever UNR thread form is desired for external use.

[b] Regarding combinations of thread classes, see text on p. 1635.

[c] UN series external thread maximum minor diameter is basic for Class 3A and basic minus allowance for Classes 1A and 2A.

[d] For Class 2A threads having an additive finish the maximum is increased, by the allowance, to the basic size, the value being the same as for Class 3A.

[e] For unfinished hot-rolled material not including standard fasteners with rolled threads.

All dimensions in inches.

Use UNS threads only if Standard Series do not meet requirements (see p. 1635). For sizes above 4 inches see ASME/ANSI B1.1-1989.

Coarse-Thread Series: This series, UNC/UNRC, is the one most commonly used in the bulk production of bolts, screws, nuts and other general engineering applications. It is also used for threading into lower tensile strength materials such as cast iron, mild steel and softer materials (bronze, brass, aluminum, magnesium and plastics) to obtain the optimum resistance to stripping of the internal thread. It is applicable for rapid assembly or disassembly, or if corrosion or slight damage is possible.

Table 4a. Coarse-Thread Series, UNC and UNRC — Basic Dimensions

Sizes No. or Inches	Basic Major Dia., D	Thds. per Inch,	Basic Pitch Dia.,[a] D_2	Minor Diameter		Lead Angle λ at Basic P.D.		Area of Minor Dia. at $D\text{-}2h_b$	Tensile Stress Area[b]
				Ext. Thds.,[c] d_3 (Ref.)	Int. Thds.,[d] D_1				
	Inches	n	Inches	Inches	Inches	Deg.	Min	Sq. In.	Sq. In.
1 (0.073)[e]	**0.0730**	**64**	**0.0629**	**0.0544**	**0.0561**	**4**	**31**	0.00218	0.00263
2 (0.086)	**0.0860**	**56**	**0.0744**	**0.0648**	**0.0667**	**4**	**22**	0.00310	0.00370
3 (0.099)[e]	**0.0990**	**48**	**0.0855**	**0.0741**	**0.0764**	**4**	**26**	0.00406	0.00487
4 (0.112)	**0.1120**	**40**	**0.0958**	**0.0822**	**0.0849**	**4**	**45**	0.00496	0.00604
5 (0.125)	**0.1250**	**40**	**0.1088**	**0.0952**	**0.0979**	**4**	**11**	0.00672	0.00796
6 (0.138)	**0.1380**	**32**	**0.1177**	**0.1008**	**0.1042**	**4**	**50**	0.00745	0.00909
8 (0.164)	**0.1640**	**32**	**0.1437**	**0.1268**	**0.1302**	**3**	**58**	0.01196	0.0140
10 (0.190)	**0.1900**	**24**	**0.1629**	**0.1404**	**0.1449**	**4**	**39**	0.01450	0.0175
12 (0.216)[e]	**0.2160**	**24**	**0.1889**	**0.1664**	**0.1709**	**4**	**1**	0.0206	0.0242
1/4	**0.2500**	**20**	**0.2175**	**0.1905**	**0.1959**	**4**	**11**	0.0269	0.0318
5/16	**0.3125**	**18**	**0.2764**	**0.2464**	**0.2524**	**3**	**40**	0.0454	0.0524
3/8	**0.3750**	**16**	**0.3344**	**0.3005**	**0.3073**	**3**	**24**	0.0678	0.0775
7/16	**0.4375**	**14**	**0.3911**	**0.3525**	**0.3602**	**3**	**20**	0.0933	0.1063
1/2	**0.5000**	**13**	**0.4500**	**0.4084**	**0.4167**	**3**	**7**	0.1257	0.1419
9/16	**0.5625**	**12**	**0.5084**	**0.4633**	**0.4723**	**2**	**59**	0.162	0.182
5/8	**0.6250**	**11**	**0.5660**	**0.5168**	**0.5266**	**2**	**56**	0.202	0.226
3/4	**0.7500**	**10**	**0.6850**	**0.6309**	**0.6417**	**2**	**40**	0.302	0.334
7/8	**0.8750**	**9**	**0.8028**	**0.7427**	**0.7547**	**2**	**31**	0.419	0.462
1	**1.0000**	**8**	**0.9188**	**0.8512**	**0.8647**	**2**	**29**	0.551	0.606
1 1/8	**1.1250**	**7**	**1.0322**	**0.9549**	**0.9704**	**2**	**31**	0.693	0.763
1 1/4	**1.2500**	**7**	**1.1572**	**1.0799**	**1.0954**	**2**	**15**	0.890	0.969
1 3/8	**1.3750**	**6**	**1.2667**	**1.1766**	**1.1946**	**2**	**24**	1.054	1.155
1 1/2	**1.5000**	**6**	**1.3917**	**1.3016**	**1.3196**	**2**	**11**	1.294	1.405
1 3/4	**1.7500**	**5**	**1.6201**	**1.5119**	**1.5335**	**2**	**15**	1.74	1.90
2	**2.0000**	**4 1/2**	**1.8557**	**1.7353**	**1.7594**	**2**	**11**	2.30	2.50
2 1/4	**2.2500**	**4 1/2**	**2.1057**	**1.9853**	**2.0094**	**1**	**55**	3.02	3.25
2 1/2	**2.5000**	**4**	**2.3376**	**2.2023**	**2.2294**	**1**	**57**	3.72	4.00
2 3/4	**2.7500**	**4**	**2.5876**	**2.4523**	**2.4794**	**1**	**46**	4.62	4.93
3	**3.0000**	**4**	**2.8376**	**2.7023**	**2.7294**	**1**	**36**	5.62	5.97
3 1/4	**3.2500**	**4**	**3.0876**	**2.9523**	**2.9794**	**1**	**29**	6.72	7.10
3 1/4	**3.500**	**4**	**3.3376**	**3.2023**	**3.2294**	**1**	**22**	7.92	8.33
3 3/4	**3.7500**	**4**	**3.5876**	**3.4523**	**3.4794**	**1**	**16**	9.21	9.66
4	**4.0000**	**4**	**3.8376**	**3.7023**	**3.7294**	**1**	**11**	10.61	11.08

[a] British: Effective Diameter.

[b] See formula, pages 1482 and 1490.

[c] Design form for UNR threads. (See figure on page 1713.)

[d] Basic minor diameter.

[e] Secondary sizes.

Fine-Thread Series: This series, UNF/UNRF, is suitable for the production of bolts, screws, and nuts and for other applications where the Coarse series is not applicable. External threads of this series have greater tensile stress area than comparable sizes of the

Coarse series. The Fine series is suitable when the resistance to stripping of both external and mating internal threads equals or exceeds the tensile load carrying capacity of the externally threaded member (see p. 1415). It is also used where the length of engagement is short, where a smaller lead angle is desired, where the wall thickness demands a fine pitch, or where finer adjustment is needed.

Table 4b. Fine-Thread Series, UNF and UNRF — Basic Dimensions

Sizes No. or Inches	Basic Major Dia., D	Thds. per Inch,	Basic Pitch Dia.,[a] D_2	Minor Diameter		Lead Angle λ at Basic P.D.		Area of Minor Dia. at $D\text{-}2h_b$	Tensile Stress Area[b]
				Ext. Thds.,[c] d_3 (Ref.)	Int. Thds.,[d] D_1				
	Inches	n	Inches	Inches	Inches	Deg.	Min	Sq. In.	Sq. In.
0 (0.060)	**0.0600**	**80**	**0.0519**	**0.0451**	**0.0465**	**4**	**23**	0.00151	0.00180
1 (0.073)[e]	**0.0730**	**72**	**0.0640**	**0.0565**	**0.0580**	**3**	**57**	0.00237	0.00278
2 (0.086)	**0.0860**	**64**	**0.0759**	**0.0674**	**0.0691**	**3**	**45**	0.00339	0.00394
3 (0.099)[e]	**0.0990**	**56**	**0.0874**	**0.0778**	**0.0797**	**3**	**43**	0.00451	0.00523
4 (0.112)	**0.1120**	**48**	**0.0985**	**0.0871**	**0.0894**	**3**	**51**	0.00566	0.00661
5 (0.125)	**0.1250**	**44**	**0.1102**	**0.0979**	**0.1004**	**3**	**45**	0.00716	0.00830
6 (0.138)	**0.1380**	**40**	**0.1218**	**0.1082**	**0.1109**	**3**	**44**	0.00874	0.01015
8 (0.164)	**0.1640**	**36**	**0.1460**	**0.1309**	**0.1339**	**3**	**28**	0.01285	0.01474
10 (0.190)	**0.1900**	**32**	**0.1697**	**0.1528**	**0.1562**	**3**	**21**	0.0175	0.0200
12 (0.216)[e]	**0.2160**	**28**	**0.1928**	**0.1734**	**0.1773**	**3**	**22**	0.0226	0.258
1/4	**0.2500**	**28**	**0.2268**	**0.2074**	**0.2113**	**2**	**52**	0.0326	0.0364
5/16	**0.3125**	**24**	**0.2854**	**0.2629**	**0.2674**	**2**	**40**	0.0524	0.0580
3/8	**0.3750**	**24**	**0.3479**	**0.3254**	**0.3299**	**2**	**11**	0.0809	0.0878
7/16	**0.4375**	**20**	**0.4050**	**0.3780**	**0.3834**	**2**	**15**	0.1090	0.1187
1/2	**0.5000**	**20**	**0.4675**	**0.4405**	**0.4459**	**1**	**57**	0.1486	0.1599
9/16	**0.5625**	**18**	**0.5264**	**0.4964**	**0.5024**	**1**	**55**	0.189	0.203
5/8	**0.6250**	**18**	**0.5889**	**0.5589**	**0.5649**	**1**	**43**	0.240	0.256
3/4	**0.7500**	**16**	**0.7094**	**0.6763**	**0.6823**	**1**	**36**	0.351	0.373
7/8	**0.8750**	**14**	**0.8286**	**0.7900**	**0.7977**	**1**	**34**	0.480	0.509
1	**1.0000**	**12**	**0.9459**	**0.9001**	**0.9098**	**1**	**36**	0.625	0.663
1 1/8	**1.1250**	**12**	**1.0709**	**1.0258**	**1.0348**	**1**	**25**	0.812	0.856
1 1/4	**1.2500**	**12**	**1.1959**	**1.1508**	**1.1598**	**1**	**16**	1.024	1.073
1 3/8	**1.3750**	**12**	**1.3209**	**1.2758**	**1.2848**	**1**	**9**	1.260	1.315
1 1/2	**1.5000**	**12**	**1.4459**	**1.4008**	**1.4098**	**1**	**3**	1.521	1.581

[a] British: Effective Diameter.
[b] See formula, pages 1482 and 1490.
[c] Design form for UNR threads. (See figure on page 1713.)
[d] Basic minor diameter.
[e] Secondary sizes.

Extra-Fine-Thread Series: This series, UNEF/UNREF, is applicable where even finer pitches of threads are desirable, as for short lengths of engagement and for thin-walled tubes, nuts, ferrules, or couplings. It is also generally applicable under the conditions stated above for the fine threads.

Constant Pitch Series: The various constant-pitch series, UN, with 4, 6, 8, 12, 16, 20, 28 and 32 threads per inch, given in Table 3, offer a comprehensive range of diameter-pitch combinations for those purposes where the threads in the Coarse, Fine, and Extra-Fine series do not meet the particular requirements of the design.

When selecting threads from these constant-pitch series, preference should be given wherever possible to those tabulated in the 8-, 12-, or 16-thread series.

Table 4c. Extra-Fine-Thread Series, UNEF and UNREF — Basic Dimensions

Sizes No. or Inches	Basic Major Dia., D	Thds. per Inch, n	Basic Pitch Dia.,[a] D_2	Minor Diameter Ext. Thds.,[c] d_3 (Ref.)	Minor Diameter Int. Thds.,[d] D_1	Lead Angle λ at Basic P.D.		Area of Minor Dia. at $D - 2h_b$	Tensile Stress Area[b]
	Inches		Inches	Inches	Inches	Deg.	Min	Sq. In.	Sq. In.
12 (0.216)[e]	**0.2160**	**32**	**0.1957**	**0.1788**	**0.1822**	**2**	**55**	0.0242	0.0270
1/4	**0.2500**	**32**	**0.2297**	**0.2128**	**0.2162**	**2**	**29**	0.0344	0.0379
5/16	**0.3125**	**32**	**0.2922**	**0.2753**	**0.2787**	**1**	**57**	0.0581	0.0625
3/8	**0.3750**	**32**	**0.3547**	**0.3378**	**0.3412**	**1**	**36**	0.0878	0.0932
7/16	**0.4375**	**28**	**0.4143**	**0.3949**	**0.3988**	**1**	**34**	0.1201	0.1274
1/2	**0.5000**	**28**	**0.4768**	**0.4574**	**0.4613**	**1**	**22**	0.162	0.170
9/16	**0.5625**	**24**	**0.5354**	**0.5129**	**0.5174**	**1**	**25**	0.203	0.214
5/8	**0.6250**	**24**	**0.5979**	**0.5754**	**0.5799**	**1**	**16**	0.256	0.268
11/16[e]	**0.6875**	**24**	**0.6604**	**0.6379**	**0.6424**	**1**	**9**	0.315	0.329
3/4	**0.7500**	**20**	**0.7175**	**0.6905**	**0.6959**	**1**	**16**	0.369	0.386
13/16[e]	**0.8125**	**20**	**0.7800**	**0.7530**	**0.7584**	**1**	**10**	0.439	0.458
7/8	**0.8750**	**20**	**0.8425**	**0.8155**	**0.8209**	**1**	**5**	0.515	0.536
15/16[e]	**0.9375**	**20**	**0.9050**	**0.8780**	**0.8834**	**1**	**0**	0.598	0.620
1	**1.0000**	**20**	**0.9675**	**0.9405**	**0.9459**	**0**	**57**	0.687	0.711
1 1/16[e]	**1.0625**	**18**	**1.0264**	**0.9964**	**1.0024**	**0**	**59**	0.770	0.799
1 1/8	**1.1250**	**18**	**1.0889**	**1.0589**	**1.0649**	**0**	**56**	0.871	0.901
1 3/16[e]	**1.1875**	**18**	**1.1514**	**1.1214**	**1.1274**	**0**	**53**	0.977	1.009
1 1/4	**1.2500**	**18**	**1.2139**	**1.1839**	**1.1899**	**0**	**50**	1.090	1.123
1 5/16[e]	**1.3125**	**18**	**1.2764**	**1.2464**	**1.2524**	**0**	**48**	1.208	1.244
1 3/8	**1.3750**	**18**	**1.3389**	**1.3089**	**1.3149**	**0**	**45**	1.333	1.370
1 7/16[e]	**1.4375**	**18**	**1.4014**	**1.3714**	**1.3774**	**0**	**43**	1.464	1.503
1 1/2	**1.5000**	**18**	**1.4639**	**1.4339**	**1.4399**	**0**	**42**	1.60	1.64
1 9/16[e]	**1.5625**	**18**	**1.5264**	**1.4964**	**1.5024**	**0**	**40**	1.74	1.79
1 5/8	**1.6250**	**18**	**1.5889**	**1.5589**	**1.5649**	**0**	**38**	1.89	1.94
1 11/16[e]	**1.6875**	**18**	**1.6514**	**1.6214**	**1.6274**	**0**	**37**	2.05	2.10

[a] British: Effective Diameter.

[b] See formula, pages 1482 and 1490.

[c] Design form for UNR threads. (See figure on page 1713.)

[d] Basic minor diameter.

[e] Secondary sizes.

8-Thread Series: The 8-thread series (8-UN) is a uniform-pitch series for large diameters. Although originally intended for high-pressure-joint bolts and nuts, it is now widely used as a substitute for the Coarse-Thread Series for diameters larger than 1 inch.

12-Thread Series: The 12-thread series (12-UN) is a uniform pitch series for large diameters requiring threads of medium-fine pitch. Although originally intended for boiler practice, it is now used as a continuation of the Fine-Thread Series for diameters larger than 1 1/2 inches.

16-Thread Series: The 16-thread series (16-UN) is a uniform pitch series for large diameters requiring fine-pitch threads. It is suitable for adjusting collars and retaining nuts, and also serves as a continuation of the Extra-fine Thread Series for diameters larger than 1 11/16 inches.

4-, 6-, 20-, 28-, and 32-Thread Series: These thread series have been used more or less widely in industry for various applications where the Standard Coarse, Fine or Extra-fine Series were not as applicable. They are now recognized as Standard Unified Thread Series in a specified selection of diameters for each pitch (see Table 2).

Whenever a thread in a constant-pitch series also appears in the UNC, UNF, or UNEF series, the symbols and tolerances for limits of size of UNC, UNF, or UNEF series are applicable, as will be seen in Tables 2 and 3.

Table 4d. 4 – Thread Series, 4 –UN and 4 – UNR — Basic Dimensions

Sizes		Basic Major Dia., D	Basic Pitch Dia.,[a] D_2	Minor Diameter		Lead Angle λ at Basic P.D.		Area of Minor Dia. at $D - 2h_b$	Tensile Stress Area[b]
Primary	Secondary			Ext. Thds.,[c] d_3s (Ref.)	Int. Thds.,[d] D_1				
Inches	Inches	Inches	Inches	Inches	Inches	Deg.	Min.	Sq. In.	Sq. In.
2½[e]		**2.5000**	**2.3376**	**2.2023**	**2.2294**	**1**	**57**	3.72	4.00
	2⅝	**2.6250**	**2.4626**	**2.3273**	**2.3544**	**1**	**51**	4.16	4.45
2¾[e]		**2.7500**	**2.5876**	**2.4523**	**2.4794**	**1**	**46**	4.62	4.93
	2⅞	**2.8750**	**2.7126**	**2.5773**	**2.6044**	**1**	**41**	5.11	5.44
3[e]		**3.0000**	**2.8376**	**2.7023**	**2.7294**	**1**	**36**	5.62	5.97
	3⅛	**3.1250**	**2.9626**	**2.8273**	**2.8544**	**1**	**32**	6.16	6.52
3¼[e]		**3.2500**	**3.0876**	**2.9523**	**2.9794**	**1**	**29**	6.72	7.10
	3⅜	**3.3750**	**3.2126**	**3.0773**	**3.1044**	**1**	**25**	7.31	7.70
3½[e]		**3.5000**	**3.3376**	**3.2023**	**3.2294**	**1**	**22**	7.92	8.33
	3⅝	**3.6250**	**3.4626**	**3.3273**	**3.3544**	**1**	**19**	8.55	9.00
3¾[e]		**3.7500**	**3.5876**	**3.4523**	**3.4794**	**1**	**16**	9.21	9.66
	3⅞	**3.8750**	**3.7126**	**3.5773**	**3.6044**	**1**	**14**	9.90	10.36
4[e]		**4.0000**	**3.8376**	**3.7023**	**3.7294**	**1**	**11**	10.61	11.08
	4⅛	**4.1250**	**3.9626**	**3.8273**	**3.8544**	**1**	**9**	11.34	11.83
4¼		**4.2500**	**4.0876**	**3.9523**	**3.9794**	**1**	**7**	12.10	12.61
	4⅜	**4.3750**	**4.2126**	**4.0773**	**4.1044**	**1**	**5**	12.88	13.41
4½		**4.5000**	**4.3376**	**4.2023**	**4.2294**	**1**	**3**	13.69	14.23
	4⅝	**4.6250**	**4.4626**	**4.3273**	**4.3544**	**1**	**1**	14.52	15.1
4¾		**4.7500**	**4.5876**	**4.4523**	**4.4794**	**1**	**0**	15.4	15.9
	4⅞	**4.8750**	**4.7126**	**4.5773**	**4.6044**	**0**	**58**	16.3	16.8
5		**5.0000**	**4.8376**	**4.7023**	**4.7294**	**0**	**57**	17.2	17.8
	5⅛	**5.1250**	**4.9626**	**4.8273**	**4.8544**	**0**	**55**	18.1	18.7
5¼		**5.2500**	**5.0876**	**4.9523**	**4.9794**	**0**	**54**	19.1	19.7
	5⅜	**5.3750**	**5.2126**	**5.0773**	**5.1044**	**0**	**52**	20.0	20.7
5½		**5.5000**	**5.3376**	**5.2023**	**5.2294**	**0**	**51**	21.0	21.7
	5⅝	**5.6250**	**5.4626**	**5.3273**	**5.3544**	**0**	**50**	22.1	22.7
5¾		**5.7500**	**5.5876**	**5.4523**	**5.4794**	**0**	**49**	23.1	23.8
	5⅞	**5.8750**	**5.7126**	**5.5773**	**5.6044**	**0**	**48**	24.2	24.9
6		**6.0000**	**5.8376**	**5.7023**	**5.7294**	**0**	**47**	25.3	26.0

[a] British: Effective Diameter.
[b] See formula, pages 1482 and 1490.
[c] Design form for UNR threads. (See figure on page 1713).
[d] Basic minor diameter.
[e] These are standard sizes of the UNC series.

Table 4e. 6-Thread Series, 6-UN and 6-UNR—Basic Dimensions

Sizes		Basic Major Dia., D	Basic Pitch Dia.,[a] D_2	Minor Diameter		Lead Angle λ at Basic P.D.		Area of Minor Dia. at $D - 2h_b$	Tensile Stress Area[b]
Primary	Secondary			Ext. Thds.,[c] d_3 (Ref.)	Int. Thds.,[d] D_1				
Inches	Inches	Inches	Inches	Inches	Inches	Deg.	Min.	Sq. In.	Sq. In.
1⅜[e]		1.3750	1.2667	1.1766	1.1946	2	24	1.054	1.155
	1 7/16	1.4375	1.3292	1.2391	1.2571	2	17	1.171	1.277
1½[e]		1.5000	1.3917	1.3016	1.3196	2	11	1.294	1.405
	1 9/16	1.5625	1.4542	1.3641	1.3821	2	5	1.423	1.54
1⅝		1.6250	1.5167	1.4271	1.4446	2	0	1.56	1.68
	1 11/16	1.6875	1.5792	1.4891	1.5071	1	55	1.70	1.83
1¾		1.7500	1.6417	1.5516	1.5696	1	51	1.85	1.98
	1 13/16	1.8125	1.7042	1.6141	1.6321	1	47	2.00	2.14
1⅞		1.8750	1.7667	1.6766	1.6946	1	43	2.16	2.30
	1 15/16	1.9375	1.8292	1.7391	1.7571	1	40	2.33	2.47
2		2.0000	1.8917	1.8016	1.8196	1	36	2.50	2.65
	2⅛	2.1250	2.0167	1.9266	1.9446	1	30	2.86	3.03
2¼		2.2500	2.1417	2.0516	2.0696	1	25	3.25	3.42
	2⅜	2.3750	2.2667	2.1766	2.1946	1	20	3.66	3.85
2½		2.5000	2.3917	2.3016	2.3196	1	16	4.10	4.29
	2⅝	2.6250	2.5167	2.4266	2.4446	1	12	4.56	4.76
2¾		2.7500	2.6417	2.5516	2.5696	1	9	5.04	5.26
	2⅞	2.8750	2.7667	2.6766	2.6946	1	6	5.55	5.78
3		3.0000	2.8917	2.8016	2.8196	1	3	6.09	6.33
	3⅛	3.1250	3.0167	2.9266	2.9446	1	0	6.64	6.89
3¼		3.2500	3.1417	3.0516	3.0696	0	58	7.23	7.49
	3⅜	3.3750	3.2667	3.1766	3.1946	0	56	7.84	8.11
3½		3.5000	3.3917	3.3016	3.3196	0	54	8.47	8.75
	3⅝	3.6250	3.5167	3.4266	3.4446	0	52	9.12	9.42
3¾		3.7500	3.6417	3.5516	3.5696	0	50	9.81	10.11
	3⅞	3.8750	3.7667	3.6766	3.6946	0	48	10.51	10.83
4		4.0000	3.8917	3.8016	3.8196	0	47	11.24	11.57
	4⅛	4.1250	4.0167	3.9266	3.9446	0	45	12.00	12.33
4¼		4.2500	4.1417	4.0516	4.0696	0	44	12.78	13.12
	4⅜	4.3750	4.2667	4.1766	4.1946	0	43	13.58	13.94
4½		4.5000	4.3917	4.3016	4.3196	0	42	14.41	14.78
	4⅝	4.6250	4.5167	4.4266	4.4446	0	40	15.3	15.6
4¾		4.7500	4.6417	4.5516	4.5696	0	39	16.1	16.5
	4⅞	4.8750	4.7667	4.6766	4.6946	0	38	17.0	17.5
5		5.0000	4.8917	4.8016	4.8196	0	37	18.0	18.4
	5⅛	5.1250	5.0167	4.9266	4.9446	0	36	18.9	19.3
5¼		5.2500	5.1417	5.0516	5.0696	0	35	19.9	20.3
	5⅜	5.3750	5.2667	5.1766	5.1946	0	35	20.9	21.3
5½		5.5000	5.3917	5.3016	5.3196	0	34	21.9	22.4
	5⅝	5.6250	5.5167	5.4266	5.4446	0	33	23.0	23.4
5¾		5.7500	5.6417	5.5516	5.5696	0	32	24.0	24.5
	5⅞	5.8750	5.7667	5.6766	5.6946	0	32	25.1	25.6
6		6.0000	5.8917	5.8016	5.8196	0	31	26.3	26.8

[a] British: Effective Diameter.

[b] See formula, pages 1482 and 1490.

[c] Design form for UNR threads. (See figure on page 1713).

[d] Basic minor diameter.

[e] These are standard sizes of the UNC series.

Table 4f. 8-Thread Series, 8-UN and 8-UNR—Basic Dimensions

Sizes		Basic Major Dia., D	Basic Pitch Dia.,[a] D_2	Minor Diameter		Lead Angle λ at Basic P.D.		Area of Minor Dia. at $D - 2h_b$	Tensile Stress Area[b]
Primary	Secondary			Ext.Thds.,[c] d_3 (Ref.)	Int.Thds.,[d] D_1				
Inches	Inches	Inches	Inches	Inches	Inches	Deg.	Min.	Sq. In.	Sq. In.
1[e]		**1.0000**	**0.9188**	**0.8512**	**0.8647**	**2**	**29**	0.551	0.606
	1 1/16	**1.0625**	**0.9813**	**0.9137**	**0.9272**	**2**	**19**	0.636	0.695
1 1/8		**1.1250**	**1.0438**	**0.9792**	**0.9897**	**2**	**11**	0.728	0.790
	1 3/16	**1.1875**	**1.1063**	**1.0387**	**1.0522**	**2**	**4**	0.825	0.892
1 1/4		**1.2500**	**1.1688**	**1.1012**	**1.1147**	**1**	**57**	0.929	1.000
	1 5/16	**1.3125**	**1.2313**	**1.1637**	**1.1772**	**1**	**51**	1.039	1.114
1 3/8		**1.3750**	**1.2938**	**1.2262**	**1.2397**	**1**	**46**	1.155	1.233
	1 7/16	**1.4375**	**1.3563**	**1.2887**	**1.3022**	**1**	**41**	1.277	1.360
1 1/2		**1.5000**	**1.4188**	**1.3512**	**1.3647**	**1**	**36**	1.405	1.492
	1 9/16	**1.5625**	**1.4813**	**1.4137**	**1.4272**	**1**	**32**	1.54	1.63
1 5/8		**1.6250**	**1.5438**	**1.4806**	**1.4897**	**1**	**29**	1.68	1.78
	1 11/16	**1.6875**	**1.6063**	**1.5387**	**1.5522**	**1**	**25**	1.83	1.93
1 3/4		**1.7500**	**1.6688**	**1.6012**	**1.6147**	**1**	**22**	1.98	2.08
	1 13/16	**1.8125**	**1.7313**	**1.6637**	**1.6772**	**1**	**19**	2.14	2.25
1 7/8		**1.8750**	**1.7938**	**1.7262**	**1.7397**	**1**	**16**	2.30	2.41
	1 15/16	**1.9375**	**1.8563**	**1.7887**	**1.8022**	**1**	**14**	2.47	2.59
2		**2.0000**	**1.9188**	**1.8512**	**1.8647**	**1**	**11**	2.65	2.77
	2 1/8	**2.1250**	**2.0438**	**1.9762**	**1.9897**	**1**	**7**	3.03	3.15
2 1/4		**2.2500**	**2.1688**	**2.1012**	**2.1147**	**1**	**3**	3.42	3.56
	2 3/8	**2.3750**	**2.2938**	**2.2262**	**2.2397**	**1**	**0**	3.85	3.99
2 1/2		**2.5000**	**2.4188**	**2.3512**	**2.3647**	**0**	**57**	4.29	4.44
	2 5/8	**2.6250**	**2.5438**	**2.4762**	**2.4897**	**0**	**54**	4.76	4.92
2 3/4		**2.7500**	**2.6688**	**2.6012**	**2.6147**	**0**	**51**	5.26	5.43
	2 7/8	**2.8750**	**2.7938**	**2.7262**	**2.7397**	**0**	**49**	5.78	5.95
3		**3.0000**	**2.9188**	**2.8512**	**2.8647**	**0**	**47**	6.32	6.51
	3 1/8	**3.1250**	**3.0438**	**2.9762**	**2.9897**	**0**	**45**	6.89	7.08
3 1/4		**3.2500**	**3.1688**	**3.1012**	**3.1147**	**0**	**43**	7.49	7.69
	3 3/8	**3.3750**	**3.2938**	**3.2262**	**3.2397**	**0**	**42**	8.11	8.31
3 1/2		**3.5000**	**3.4188**	**3.3512**	**3.3647**	**0**	**40**	8.75	8.96
	3 5/8	**3.6250**	**3.5438**	**3.4762**	**3.4897**	**0**	**39**	9.42	9.64
3 3/4		**3.7500**	**3.6688**	**3.6012**	**3.6147**	**0**	**37**	10.11	10.34
	3 7/8	**3.8750**	**3.7938**	**3.7262**	**3.7397**	**0**	**36**	10.83	11.06
4		**4.0000**	**3.9188**	**3.8512**	**3.8647**	**0**	**35**	11.57	11.81
	4 1/8	**4.1250**	**4.0438**	**3.9762**	**3.9897**	**0**	**34**	12.34	12.59
4 1/4		**4.2500**	**4.1688**	**4.1012**	**4.1147**	**0**	**33**	13.12	13.38
	4 3/8	**4.3750**	**4.2938**	**4.2262**	**4.2397**	**0**	**32**	13.94	14.21
4 1/2		**4.5000**	**4.4188**	**4.3512**	**4.3647**	**0**	**31**	14.78	15.1
	4 5/8	**4.6250**	**4.5438**	**4.4762**	**4.4897**	**0**	**30**	15.6	15.9
4 3/4		**4.7500**	**4.6688**	**4.6012**	**4.6147**	**0**	**29**	16.5	16.8
	4 7/8	**4.8750**	**4.7938**	**4.7262**	**4.7397**	**0**	**29**	17.4	17.7
5		**5.0000**	**4.9188**	**4.8512**	**4.8647**	**0**	**28**	18.4	18.7
	5 1/8	**5.1250**	**5.0438**	**4.9762**	**4.9897**	**0**	**27**	19.3	19.7
5 1/4		**5.2500**	**5.1688**	**5.1012**	**5.1147**	**0**	**26**	20.3	20.7
	5 3/8	**5.3750**	**5.2938**	**5.2262**	**5.2397**	**0**	**26**	21.3	21.7
5 1/2		**5.5000**	**5.4188**	**5.3512**	**5.3647**	**0**	**25**	22.4	22.7
	5 5/8	**5.6250**	**5.5438**	**5.4762**	**5.4897**	**0**	**25**	23.4	23.8
5 3/4		**5.7500**	**5.6688**	**5.6012**	**5.6147**	**0**	**24**	24.5	24.9
	5 7/8	**5.8750**	**5.7938**	**5.7262**	**5.7397**	**0**	**24**	25.6	26.0
6		**6.0000**	**5.9188**	**5.8512**	**5.8647**	**0**	**23**	26.8	27.1

[a] British: Effective Diameter.

[b] See formula, pages 1482 and 1490.

[c] Design form for UNR threads. (See figure on page 1713).

[d] Basic minor diameter.

[e] This is a standard size of the UNC series.

Table 4g. 12-Thread series, 12-UN and 12-UNR—Basic Dimensions

Sizes		Basic Major Dia., D	Basic Pitch Dia.,[a] D_2	Minor Diameter		Lead Angle λ at Basic P.D.		Area of Minor Dia. at $D - 2h_b$	Tensile Stress Area[b]
Primary	Secondary			Ext. Thds.,[c] d_3 (Ref.)	Int. Thds.,[d] D_1				
Inches	Inches	Inches	Inches	Inches	Inches	Deg.	Min.	Sq. In.	Sq. In.
9/16[e]		0.5625	0.5084	0.4633	0.4723	2	59	0.162	0.182
5/8		0.6250	0.5709	0.5258	0.5348	2	40	0.210	0.232
	11/16	0.6875	0.6334	0.5883	0.5973	2	24	0.264	0.289
3/4		0.7500	0.6959	0.6508	0.6598	2	11	0.323	0.351
	13/16	0.8125	0.7584	0.7133	0.7223	2	0	0.390	0.420
7/8		0.8750	0.8209	0.7758	0.7848	1	51	0.462	0.495
	15/16	0.9375	0.8834	0.8383	0.8473	1	43	0.540	0.576
1[e]		1.0000	0.9459	0.9008	0.9098	1	36	0.625	0.663
	1 1/16	1.0625	1.0084	0.9633	0.9723	1	30	0.715	0.756
1 1/8[e]		1.1250	1.0709	1.0258	1.0348	1	25	0.812	0.856
	1 3/16	1.1875	1.1334	1.0883	1.0973	1	20	0.915	0.961
1 1/4[e]		1.2500	1.1959	1.1508	1.1598	1	16	1.024	1.073
	1 5/16	1.3125	1.2584	1.2133	1.2223	1	12	1.139	1.191
1 3/8		1.3750	1.3209	1.2758	1.2848	1	9	1.260	1.315
	1 7/16	1.4375	1.3834	1.3383	1.3473	1	6	1.388	1.445
1 1/2[e]		1.5000	1.4459	1.4008	1.4098	1	3	1.52	1.58
	1 9/16	1.5625	1.5084	1.4633	1.4723	1	0	1.66	1.72
1 5/8		1.6250	1.5709	1.5258	1.5348	0	58	1.81	1.87
	1 11/16	1.6875	1.6334	1.5883	1.5973	0	56	1.96	2.03
1 3/4		1.7500	1.6959	1.6508	1.6598	0	54	2.12	2.19
	1 13/16	1.8125	1.7584	1.7133	1.7223	0	52	2.28	2.35
1 7/8		1.8750	1.8209	1.7758	1.7848	0	50	2.45	2.53
	1 15/16	1.9375	1.8834	1.8383	1.8473	0	48	2.63	2.71
2		2.0000	1.9459	1.9008	1.9098	0	47	2.81	2.89
	2 1/8	2.1250	2.0709	2.0258	2.0348	0	44	3.19	3.28
2 1/4		2.2500	2.1959	2.1508	2.1598	0	42	3.60	3.69
	2 3/8	2.3750	2.3209	2.2758	2.2848	0	39	4.04	4.13
2 1/2		2.5000	2.4459	2.4008	2.4098	0	37	4.49	4.60
	2 5/8	2.6250	2.5709	2.5258	2.5348	0	35	4.97	5.08
2 3/4		2.7500	2.6959	2.6508	2.6598	0	34	5.48	5.59
	2 7/8	2.8750	2.8209	2.7758	2.7848	0	32	6.01	6.13
3		3.0000	2.9459	2.9008	2.9098	0	31	6.57	6.69
	3 1/8	3.1250	3.0709	3.0258	3.0348	0	30	7.15	7.28
3 1/4		3.2500	3.1959	3.1508	3.1598	0	29	7.75	7.89
	3 3/8	3.3750	3.3209	3.2758	3.2848	0	27	8.38	8.52
3 1/2		3.5000	3.4459	3.4008	3.4098	0	26	9.03	9.18
	3 5/8	3.6250	3.5709	3.5258	3.5348	0	26	9.71	9.86
3 3/4		3.7500	3.6959	3.6508	3.6598	0	25	10.42	10.57
	3 7/8	3.8750	3.8209	3.7758	3.7848	0	24	11.14	11.30
4		4.0000	3.9459	3.9008	3.9098	0	23	11.90	12.06
	4 1/8	4.1250	4.0709	4.0258	4.0348	0	22	12.67	12.84
4 1/4		4.2500	4.1959	4.1508	4.1598	0	22	13.47	13.65
	4 3/8	4.3750	4.3209	4.2758	4.2848	0	21	14.30	14.48
4 1/2		4.5000	4.4459	4.4008	4.4098	0	21	15.1	15.3
	4 5/8	4.6250	4.5709	4.5258	4.5348	0	20	16.0	16.2
4 3/4		4.7500	4.6959	4.6508	4.6598	0	19	16.9	17.1
	4 7/8	4.8750	4.8209	4.7758	4.7848	0	19	17.8	18.0
5		5.0000	4.9459	4.9008	4.9098	0	18	18.8	19.0
	5 1/8	5.1250	5.0709	5.0258	5.0348	0	18	19.8	20.0
5 1/4		5.2500	5.1959	5.1508	5.1598	0	18	20.8	21.0
	5 3/8	5.3750	5.3209	5.2758	5.2848	0	17	21.8	22.0
5 1/2		5.5000	5.4459	5.4008	5.4098	0	17	22.8	23.1
	5 5/8	5.6250	5.5709	5.5258	5.5348	0	16	23.9	24.1
5 3/4		5.7500	5.6959	5.6508	5.6598	0	16	25.0	25.2
	5 7/8	5.8750	5.8209	5.7758	5.7848	0	16	26.1	26.4
6		6.0000	5.9459	5.9008	5.9098	0	15	27.3	27.5

[a] British: Effective Diameter.

[b] See formula, pages 1482 and 1490.

[c] Design form for UNR threads. (See figure on page 1713.)

[d] Basic minor diameter.

[e] These are standard sizes of the UNC or UNF Series.

Table 4h. 16-Thread Series, 16-UN and 16-UNR—Basic Dimensions

Sizes		Basic Major Dia., D	Basic Pitch Dia.,[a] D_2	Minor Diameter		Lead Angle λ at Basic P.D.		Area of Minor Dia. at $D - 2h_b$	Tensile Stress Area[b]
Primary	Secondary			Ext. Thds.,[c] d_3 (Ref.)	Int. Thds.,[d] D_1				
Inches	Inches	Inches	Inches	Inches	Inches	Deg.	Min.	Sq. In.	Sq. In.
3⁄8[e]		**0.3750**	**0.3344**	**0.3005**	**0.3073**	**3**	**24**	0.0678	0.0775
7⁄16		**0.4375**	**0.3969**	**0.3630**	**0.3698**	**2**	**52**	0.0997	0.1114
1⁄2		**0.5000**	**0.4594**	**0.4255**	**0.4323**	**2**	**29**	0.1378	0.151
9⁄16		**0.5625**	**0.5219**	**0.4880**	**0.4948**	**2**	**11**	0.182	0.198
5⁄8		**0.6250**	**0.5844**	**0.5505**	**0.5573**	**1**	**57**	0.232	0.250
	11⁄16	**0.6875**	**0.6469**	**0.6130**	**0.6198**	**1**	**46**	0.289	0.308
3⁄4[e]		**0.7500**	**0.7094**	**0.6755**	**0.6823**	**1**	**36**	0.351	0.373
	13⁄16	**0.8125**	**0.7719**	**0.7380**	**0.7448**	**1**	**29**	0.420	0.444
7⁄8		**0.8750**	**0.8344**	**0.8005**	**0.8073**	**1**	**22**	0.495	0.521
	15⁄16	**0.9375**	**0.8969**	**0.8630**	**0.8698**	**1**	**16**	0.576	0.604
1		**1.0000**	**0.9594**	**0.9255**	**0.9323**	**1**	**11**	0.663	0.693
	1 1⁄16	**1.0625**	**1.0219**	**0.9880**	**0.9948**	**1**	**7**	0.756	0.788
1 1⁄8		**1.1250**	**1.0844**	**1.0505**	**1.0573**	**1**	**3**	0.856	0.889
	1 3⁄16	**1.1875**	**1.1469**	**1.1130**	**1.1198**	**1**	**0**	0.961	0.997
1 1⁄4		**1.2500**	**1.2094**	**1.1755**	**1.1823**	**0**	**57**	1.073	1.111
	1 5⁄16	**1.3125**	**1.2719**	**1.2380**	**1.2448**	**0**	**54**	1.191	1.230
1 3⁄8		**1.3750**	**1.3344**	**1.3005**	**1.3073**	**0**	**51**	1.315	1.356
	1 7⁄16	**1.4375**	**1.3969**	**1.3630**	**1.3698**	**0**	**49**	1.445	1.488
1 1⁄2		**1.5000**	**1.4594**	**1.4255**	**1.4323**	**0**	**47**	1.58	1.63
	1 9⁄16	**1.5625**	**1.5219**	**1.4880**	**1.4948**	**0**	**45**	1.72	1.77
1 5⁄8		**1.6250**	**1.5844**	**1.5505**	**1.5573**	**0**	**43**	1.87	1.92
	1 11⁄16	**1.6875**	**1.6469**	**1.6130**	**1.6198**	**0**	**42**	2.03	2.08
1 3⁄4		**1.7500**	**1.7094**	**1.6755**	**1.6823**	**0**	**40**	2.19	2.24
	1 13⁄16	**1.8125**	**1.7719**	**1.7380**	**1.7448**	**0**	**39**	2.35	2.41
1 7⁄8		**1.8750**	**1.8344**	**1.8005**	**1.8073**	**0**	**37**	2.53	2.58
	1 15⁄16	**1.9375**	**1.8969**	**1.8630**	**1.8698**	**0**	**36**	2.71	2.77
2		**2.0000**	**1.9594**	**1.9255**	**1.9323**	**0**	**35**	2.89	2.95
	2 1⁄8	**2.1250**	**2.0844**	**2.0505**	**2.0573**	**0**	**33**	3.28	3.35
2 1⁄4		**2.2500**	**2.2094**	**2.1755**	**2.1823**	**0**	**31**	3.69	3.76
	2 3⁄8	**2.3750**	**2.3344**	**2.3005**	**2.3073**	**0**	**29**	4.13	4.21
2 1⁄2		**2.5000**	**2.4594**	**2.4255**	**2.4323**	**0**	**28**	4.60	4.67
	2 5⁄8	**2.6250**	**2.5844**	**2.5505**	**2.5573**	**0**	**26**	5.08	5.16
2 3⁄4		**2.7500**	**2.7094**	**2.6755**	**2.6823**	**0**	**25**	5.59	5.68
	2 7⁄8	**2.8750**	**2.8344**	**2.8005**	**2.8073**	**0**	**24**	6.13	6.22
3		**3.0000**	**2.9594**	**2.9255**	**2.9323**	**0**	**23**	6.69	6.78
	3 1⁄8	**3.1250**	**3.0844**	**3.0505**	**3.0573**	**0**	**22**	7.28	7.37
3 1⁄4		**3.2500**	**3.2094**	**3.1755**	**3.1823**	**0**	**21**	7.89	7.99
	3 3⁄8	**3.3750**	**3.3344**	**3.3005**	**3.3073**	**0**	**21**	8.52	8.63
3 1⁄2		**3.5000**	**3.4594**	**3.4255**	**3.4323**	**0**	**20**	9.18	9.29
	3 5⁄8	**3.6250**	**3.5844**	**3.5505**	**3.5573**	**0**	**19**	9.86	9.98
3 3⁄4		**3.7500**	**3.7094**	**3.6755**	**3.6823**	**0**	**18**	10.57	10.69
	3 7⁄8	**3.8750**	**3.8344**	**3.8005**	**3.8073**	**0**	**18**	11.30	11.43
4		**4.0000**	**3.9594**	**3.9255**	**3.9323**	**0**	**17**	12.06	12.19
	4 1⁄8	**4.1250**	**4.0844**	**4.0505**	**4.0573**	**0**	**17**	12.84	12.97
4 1⁄4		**4.2500**	**4.2094**	**4.1755**	**4.1823**	**0**	**16**	13.65	13.78
	4 3⁄8	**4.3750**	**4.3344**	**4.3005**	**4.3073**	**0**	**16**	14.48	14.62
4 1⁄2		**4.5000**	**4.4594**	**4.4255**	**4.4323**	**0**	**15**	15.34	15.5
	4 5⁄8	**4.6250**	**4.5844**	**4.5505**	**4.5573**	**0**	**15**	16.2	16.4
4 3⁄4		**4.7500**	**4.7094**	**4.6755**	**4.6823**	**0**	**15**	17.1	17.3
	4 7⁄8	**4.8750**	**4.8344**	**4.8005**	**4.8073**	**0**	**14**	18.0	18.2
5		**5.0000**	**4.9594**	**4.9255**	**4.9323**	**0**	**14**	19.0	19.2
	5 1⁄8	**5.1250**	**5.0844**	**5.0505**	**5.0573**	**0**	**13**	20.0	20.1
5 1⁄4		**5.2500**	**5.2094**	**5.1755**	**5.1823**	**0**	**13**	21.0	21.1
	5 3⁄8	**5.3750**	**5.3344**	**5.3005**	**5.3073**	**0**	**13**	22.0	22.2
5 1⁄2		**5.5000**	**5.4594**	**5.4255**	**5.4323**	**0**	**13**	23.1	23.2
	5 5⁄8	**5.6250**	**5.5844**	**5.5505**	**5.5573**	**0**	**12**	24.1	24.3

Table 4h. *(Continued)* **16-Thread Series, 16-UN and 16-UNR—Basic Dimensions**

Sizes		Basic Major Dia., D	Basic Pitch Dia.,[a] D_2	Minor Diameter		Lead Angle λ at Basic P.D.		Area of Minor Dia. at $D - 2h_b$	Tensile Stress Area[b]
Primary	Secondary			Ext. Thds.,[c] d_3 (Ref.)	Int. Thds.,[d] D_1				
Inches	Inches	Inches	Inches	Inches	Inches	Deg.	Min.	Sq. In.	Sq. In.
5¾		5.7500	5.7094	5.6755	5.6823	0	12	25.2	25.4
	5⅞	5.8750	5.8344	5.8005	5.8073	0	12	26.4	26.5
6		6.0000	5.9594	5.9255	5.9323	0	11	27.5	27.7

[a] British: Effective Diameter.

[b] See formula, pages 1482 and 1490.

[c] Design form for UNR threads. (See figure on page 1713).

[d] Basic minor diamter.

[e] These are standard sizes of the UNC or UNF Series.

Table 4i. 20-Thread Series, 20-UN and 20-UNR—Basic Dimensions

Sizes		Basic Major Dia., D	Basic Pitch Dia.,[a] D_2	Minor Diameter		Lead Angle λ at Basic P.D.		Area of Minor Dia. at $D - 2h_b$	Tensile Stress Area[b]
Primary	Secondary			Ext. Thds.,[c] d_3 (Ref.)	Int. Thds.,[d] D_1				
Inches	Inches	Inches	Inches	Inches	Inches	Deg.	Min.	Sq. In.	Sq. In.
¼[e]		0.2500	0.2175	0.1905	0.1959	4	11	0.0269	0.0318
5/16		0.3125	0.2800	0.2530	0.2584	3	15	0.0481	0.0547
⅜		0.3750	0.3425	0.3155	0.3209	2	40	0.0755	0.0836
7/16[e]		0.4375	0.4050	0.3780	0.3834	2	15	0.1090	0.1187
½[e]		0.5000	0.4675	0.4405	0.4459	1	57	0.1486	0.160
9/16		0.5625	0.5300	0.5030	0.5084	1	43	0.194	0.207
⅝		0.6250	0.5925	0.5655	0.5709	1	32	0.246	0.261
	11/16	0.6875	0.6550	0.6280	0.6334	1	24	0.304	0.320
¾[e]		0.7500	0.7175	0.6905	0.6959	1	16	0.369	0.386
	13/16[e]	0.8125	0.7800	0.7530	0.7584	1	10	0.439	0.458
⅞[e]		0.8750	0.8425	0.8155	0.8209	1	5	0.515	0.536
	15/16[e]	0.9375	0.9050	0.8780	0.8834	1	0	0.0.598	0.620
1[e]		1.0000	0.9675	0.9405	0.9459	0	57	0.687	0.711
	1 1/16	1.0625	1.0300	1.0030	1.0084	0	53	0.782	0.807
1⅛		1.1250	1.0925	1.0655	1.0709	0	50	0.882	0.910
	1 3/16	1.1875	1.1550	1.1280	1.1334	0	47	0.990	1.018
1¼		1.2500	1.2175	1.1905	1.1959	0	45	1.103	1.133
	1 5/16	1.3125	1.2800	1.2530	1.2584	0	43	1.222	1.254
1⅜		1.3750	1.3425	1.3155	1.3209	0	41	1.348	1.382
	1 7/16	1.4375	1.4050	1.3780	1.3834	0	39	1.479	1.51
1½		1.5000	1.4675	1.4405	1.4459	0	37	1.62	1.65
	1 9/16	1.5625	1.5300	1.5030	1.5084	0	36	1.76	1.80
1⅝		1.6250	1.5925	1.5655	1.5709	0	34	1.91	1.95
	1 11/16	1.6875	1.6550	1.6280	1.6334	0	33	2.07	2.11
1¾		1.7500	1.7175	1.6905	1.6959	0	32	2.23	2.27
	1 13/16	1.8125	1.7800	1.7530	1.7584	0	31	2.40	2.44
1⅞		1.8750	1.8425	1.8155	1.8209	0	30	2.57	2.62
	1 15/16	1.9375	1.9050	1.8780	1.8834	0	29	2.75	2.80
2		2.0000	1.9675	1.9405	1.9459	0	28	2.94	2.99
	2⅛	2.1250	2.0925	2.0655	2.0709	0	26	3.33	3.39
2¼		2.2500	2.2175	2.1905	2.1959	0	25	3.75	3.81
	2⅜	2.3750	2.3425	2.3155	2.3209	0	23	4.19	4.25
2½		2.5000	2.4675	2.4405	2.4459	0	22	4.66	4.72
	2⅝	2.6250	2.5925	2.5655	2.5709	0	21	5.15	5.21
2¾		2.7500	2.7175	2.6905	2.6959	0	20	5.66	5.73
	2⅞	2.8750	2.8425	2.8155	2.8209	0	19	6.20	6.27
3		3.0000	2.9675	2.9405	2.9459	0	18	6.77	6.84

[a] British: Effective Diameter.

[b] See formula, pages 1482 and 1490.

[c] Design form for UNR threads. (See figure on page 1713.)

[d] Basic minor diameter.

[e] These are standard sizes of the UNC, UNF, or UNEF Series.

Table 4j. 28-Thread Series, 28-UN and 28-UNR — Basic Dimensions

Sizes		Basic Major Dia., D	Basic Pitch Dia.,[a] D_2	Minor Diameter		Lead Angel λ at Basic P.D.		Area of Minor Dia. at D-$2h_b$	Tensile Stress Area[b]
Primary	Secondary			Ext. Thds.,[c] d_3 (Ref.)	Int. Thds.,[d] D_1				
Inches	Inches	Inches	Inches	Inches	Inches	Deg.	Min.	Sq. In.	Sq. In.
	12 (0.216)[e]	**0.2160**	**0.1928**	**0.1734**	**0.1773**	**3**	**22**	0.0226	0.0258
1/4[e]		**0.2500**	**0.2268**	**0.2074**	**0.2113**	**2**	**52**	0.0326	0,0364
5/16		**0.3125**	**0.2893**	**0.2699**	**0.2738**	**2**	**15**	0.0556	0.0606
3/8		**0.3750**	**0.3518**	**0.3324**	**0.3363**	**1**	**51**	0.0848	0.0909
7/16[e]		**0.4375**	**0.4143**	**0.3949**	**0.3988**	**1**	**34**	0.1201	0.1274
1/2[e]		**0.5000**	**0.4768**	**0.4574**	**0.4613**	**1**	**22**	0.162	0.170
9/16		**0.5625**	**0.5393**	**0.5199**	**0.5238**	**1**	**12**	0.209	0.219
5/8		**0.6250**	**0.6018**	**0.5824**	**0.5863**	**1**	**5**	0.263	0.274
	11/16	**0.6875**	**0.6643**	**0.6449**	**0.6488**	**0**	**59**	0.323	0.335
3/4		**0.7500**	**0.7268**	**0.7074**	**0.7113**	**0**	**54**	0.389	0.402
	13/16	**0.8125**	**0.7893**	**0.7699**	**0.7738**	**0**	**50**	0.461	0.475
7/8		**0.8750**	**0.8518**	**0.8324**	**0.8363**	**0**	**46**	0.539	0.554
	15/16	**0.9375**	**0.9143**	**0.8949**	**0.8988**	**0**	**43**	0.624	0.640
1		**1.0000**	**0.9768**	**0.9574**	**0.9613**	**0**	**40**	0.714	0.732
	1 1/16	**1.0625**	**1.0393**	**1.0199**	**1.0238**	**0**	**38**	0.811	0.830
1 1/8		**1.1250**	**1.1018**	**1.0824**	**1.0863**	**0**	**35**	0.914	0.933
	1 3/16	**1.1875**	**1.1643**	**1.1449**	**1.1488**	**0**	**34**	1.023	1.044
1 1/4		**1.2500**	**1.2268**	**1.2074**	**1.2113**	**0**	**32**	1.138	1.160
	1 5/16	**1.3125**	**1.2893**	**1.2699**	**1.2738**	**0**	**30**	1.259	1.282
1 3/8		**1.3750**	**1.3518**	**1.3324**	**1.3363**	**0**	**29**	1.386	1.411
	1 7/16	**1.4375**	**1.4143**	**1.3949**	**1.3988**	**0**	**28**	1.52	1.55
1 1/2		**1.5000**	**1.4768**	**1.4574**	**1.4613**	**0**	**26**	1.66	1.69

[a] British: Effective Diameter.
[b] See formula, pages 1482 and 1490.
[c] Design form for UNR threads. (See figure on page 1713.)
[d] Basic minor diameter.
[e] These are standard sizes of the UNF or UNEF Series.

Table 4k. 32-Thread Series, 32-UN and 32-UNR — Basic Dimensions

Sizes		Basic Major Dia., D	Basic Pitch Dia.,[a] D_2	Minor Diameter		Lead Angel λ at Basic P.D.		Area of Minor Dia. at D - $2h_b$	Tensile Stress Area[b]
Primary	Secondary			Ext.Thds.,[c] d_3 (Ref.)	Int.Thds.,[d] D_1				
Inches	Inches	Inches	Inches	Inches	Inches	Deg.	Min.	Sq. In.	Sq. In.
6 (0.138)[e]		**0.1380**	**0.1177**	**0.1008**	**0.1042**	**4**	**50**	0.00745	0.00909
8 (0.164)[e]		**0.1640**	**0.1437**	**0.1268**	**0.1302**	**3**	**58**	0.01196	0.0140
10 (0.190)[e]		**0.1900**	**0.1697**	**0.1528**	**0.1562**	**3**	**21**	0.01750	0.0200
	12 (0.216)[e]	**0.2160**	**0.1957**	**0.1788**	**0.1822**	**2**	**55**	0.0242	0.0270
1/4[e]		**0.2500**	**0.2297**	**0.2128**	**0.2162**	**2**	**29**	0.0344	0.0379
5/16[e]		**0.3125**	**0.2922**	**0.2753**	**0.2787**	**1**	**57**	0.0581	0.0625
3/8[e]		**0.3750**	**0.3547**	**0.3378**	**0.3412**	**1**	**36**	0.0878	0.0932
7/16		**0.4375**	**0.4172**	**0.4003**	**0.4037**	**1**	**22**	0.1237	0.1301
1/2		**0.5000**	**0.4797**	**0.4628**	**0.4662**	**1**	**11**	0.166	0.173
9/16		**0.5625**	**0.5422**	**0.5253**	**0.5287**	**1**	**3**	0.214	0.222
5/8		**0.6250**	**0.6047**	**0.5878**	**0.5912**	**0**	**57**	0.268	0.278
	11/16	**0.6875**	**0.6672**	**0.6503**	**0.6537**	**0**	**51**	0.329	0.339
3/4		**0.7500**	**0.7297**	**0.7128**	**0.7162**	**0**	**47**	0.395	0.407
	13/16	**0.8125**	**0.7922**	**0.7753**	**0.7787**	**0**	**43**	0.468	0.480
7/8		**0.8750**	**0.8547**	**0.8378**	**0.8412**	**0**	**40**	0.547	0.560
	15/16	**0.9375**	**0.9172**	**0.9003**	**0.9037**	**0**	**37**	0.632	0.646
1		**1.0000**	**0.9797**	**0.9628**	**0.9662**	**0**	**35**	0.723	0.738

[a] British: Effective Diameter.
[b] See formula, pages 1482 and 1490.
[c] Design form for UNR threads. (See figure on page 1713.)
[d] Basic minor diameter.
[e] These are standard sizes of the UNC, UNF, or UNEF Series.

Fine Threads for Thin-Wall Tubing: Dimensions for a 27-thread series, ranging from ¼- to 1-inch nominal size, also are included in Table 3. These threads are recommended for general use on thin-wall tubing. The minimum length of complete thread is one-third of the basic major diameter plus 5 threads (+ 0.185 in.).

Selected Combinations: Thread data are tabulated in Table 3 for certain additional selected special combinations of diameter and pitch, with pitch diameter tolerances based on a length of thread engagement of 9 times the pitch. The pitch diameter limits are applicable to a length of engagement of from 5 to 15 times the pitch. (This provision should not be confused with the lengths of thread on mating parts, as they may exceed the length of engagement by a considerable amount.) Thread symbols are UNS and UNRS.

Other Threads of Special Diameters, Pitches, and Lengths of Engagement: Thread data for special combinations of diameter, pitch, and length of engagement not included in selected combinations are also given in the Standard but are not given here. Also, when design considerations require non-standard pitches or extreme conditions of engagement not covered by the tables, the allowance and tolerances should be derived from the formulas in the Standard. The thread symbol for such special threads is UNS.

Thread Classes.—Thread classes are distinguished from each other by the amounts of tolerance and allowance. Classes identified by a numeral followed by the letters A and B are derived from certain Unified formulas (not shown here) in which the pitch diameter tolerances are based on increments of the basic major (nominal) diameter, the pitch, and the length of engagement. These formulas and the class identification or symbols apply to all of the Unified threads.

Classes 1A, 2A, and 3A apply to external threads only, and Classes 1B, 2B, and 3B apply to internal threads only. The disposition of the tolerances, allowances, and crest clearances for the various classes is illustrated on pages 1751 and 1752.

Classes 2A and 2B: Classes 2A and 2B are the most commonly used for general applications, including production of bolts, screws, nuts, and similar fasteners.

The maximum diameters of Class 2A (external) uncoated threads are less than basic by the amount of the allowance. The allowance minimizes galling and seizing in high-cycle wrench assembly, or it can be used to accommodate plated finishes or other coating. However, for threads with additive finish, the maximum diameters of Class 2A may be exceeded by the amount of the allowance, for example, the 2A maximum diameters apply to an unplated part or to a part before plating whereas the basic diameters (the 2A maximum diameter plus allowance) apply to a part after plating. The minimum diameters of Class 2B (internal) threads, whether or not plated or coated, are basic, affording no allowance or clearance in assembly at maximum metal limits.

Class 2AG: Certain applications require an allowance for rapid assembly to permit application of the proper lubricant or for residual growth due to high-temperature expansion. In these applications, when the thread is coated and the 2A allowance is not permitted to be consumed by such coating, the thread class symbol is qualified by G following the class symbol.

Classes 3A and 3B: Classes 3A and 3B may be used if closer tolerances are desired than those provided by Classes 2A and 2B. The maximum diameters of Class 3A (external) threads and the minimum diameters of Class 3B (internal) threads, whether or not plated or coated, are basic, affording no allowance or clearance for assembly of maximum metal components.

Classes 1A and 1B: Classes 1A and 1B threads replaced American National Class 1. These classes are intended for ordnance and other special uses. They are used on threaded components where quick and easy assembly is necessary and where a liberal allowance is required to permit ready assembly, even with slightly bruised or dirty threads.

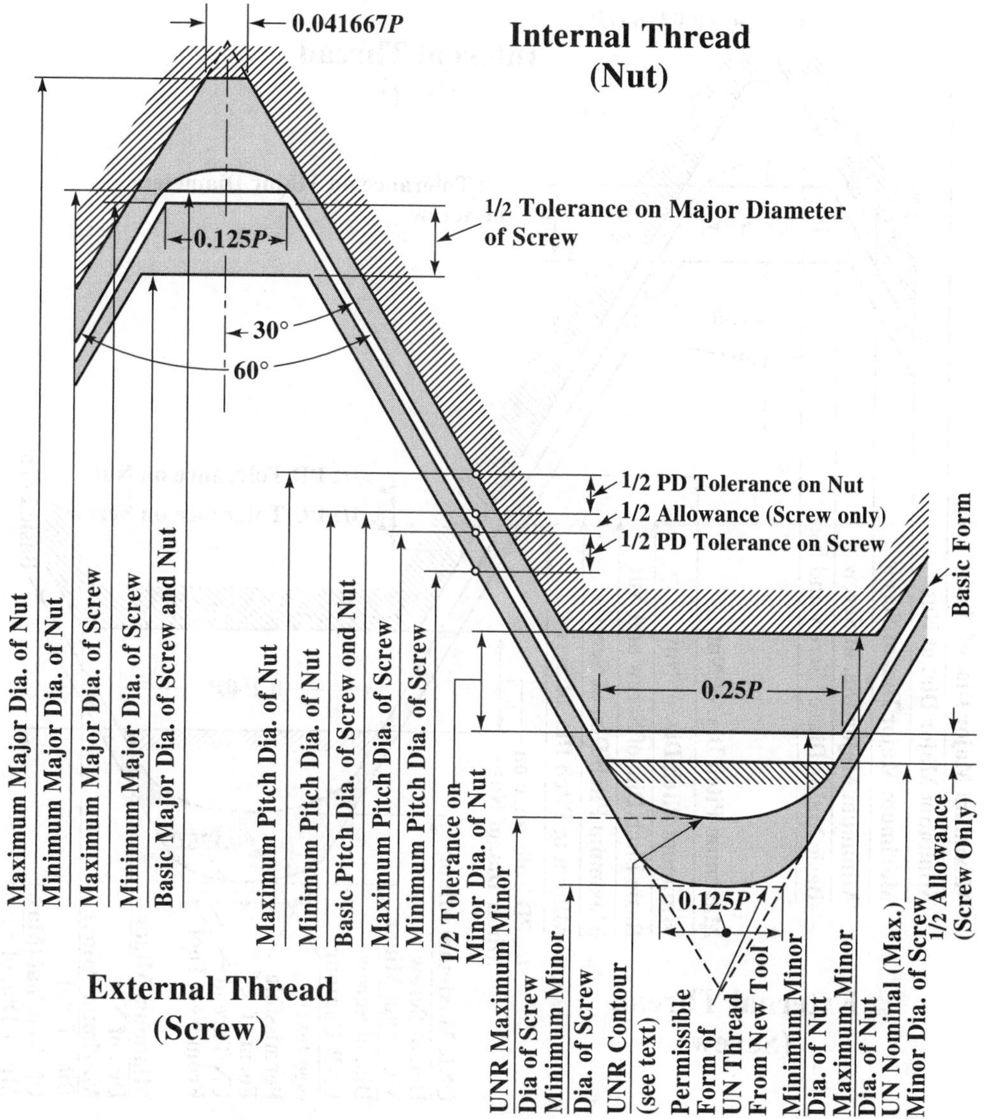

Limits of Size Showing Tolerances, Allowances (Neutral Space), and Crest Clearances for Unified Classes 1A, 2A, 1B, and 2B

Maximum diameters of Class 1A (external) threads are less than basic by the amount of the same allowance as applied to Class 2A. For the intended applications in American practice the allowance is not available for plating or coating. Where the thread is plated or coated, special provisions are necessary. The minimum diameters of Class 1B (internal) threads, whether or not plated or coated, are basic, affording no allowance or clearance for assembly with maximum metal external thread components having maximum diameters which are basic.

Coated 60-deg. Threads.— Although the Standard does not make recommendations for thicknesses of, or specify limits for coatings, it does outline certain principles that will aid mechanical interchangeability if followed whenever conditions permit.

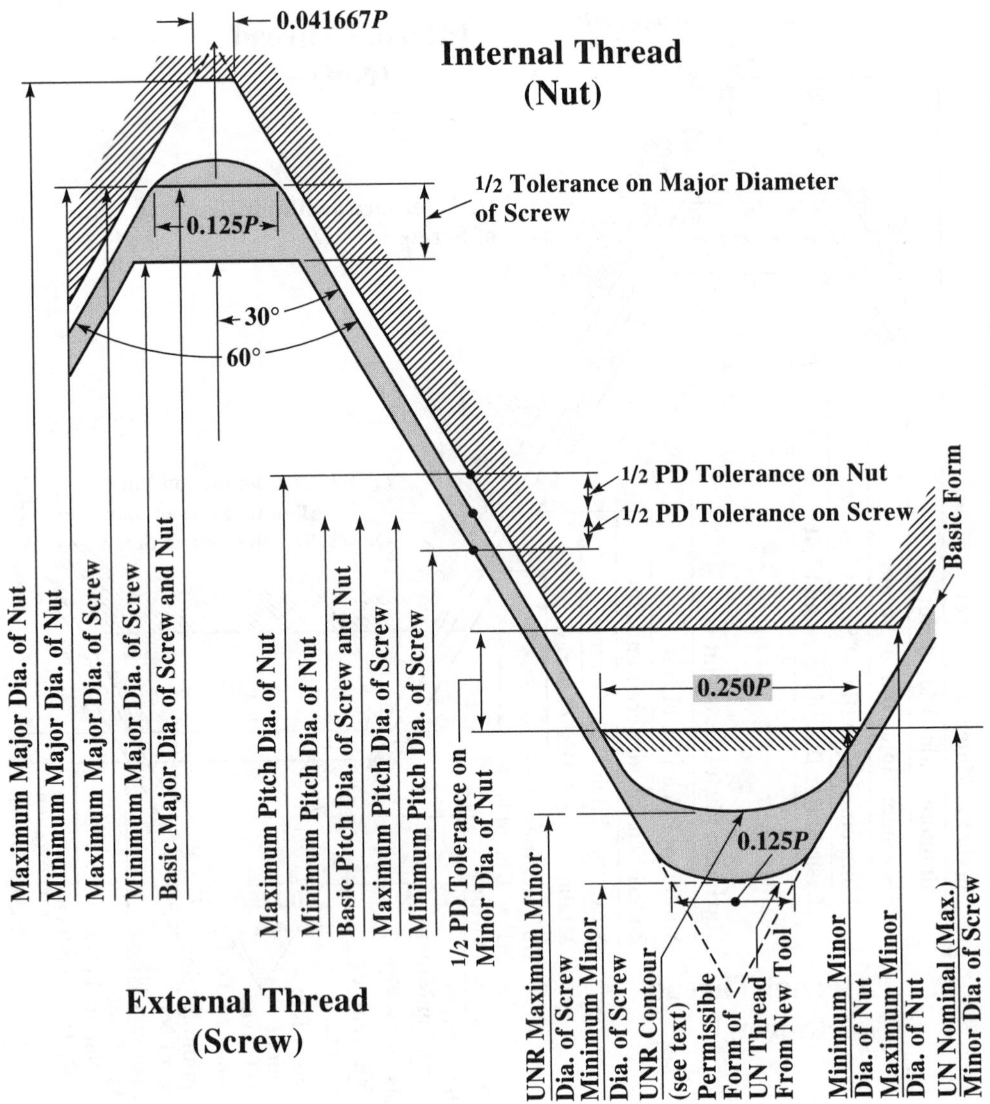

Limits of Size Showing Tolerances and Crest Clearances for Unified Classes 3A and 3B and American National Classes 2 and 3

To keep finished threads within the limits of size established in the Standard, external threads should not exceed basic size after plating and internal threads should not be below basic size after plating. This recommendation does not apply to threads coated by certain commonly used processes such as hot-dip galvanizing where it may not be required to maintain these limits.

Class 2A provides both a tolerance and an allowance. Many thread requirements call for coatings such as those deposited by electro-plating processes and, in general, the 2A allowance provides adequate undercut for such coatings. There may be variations in thickness and symmetry of coating resulting from commercial processes but after plating the threads should be accepted by a basic Class 3A size GO gage and a Class 2A gage as a NOT-GO gage. Class 1A provides an allowance which is maintained for both coated and uncoated product, i.e., it is not available for coating.

Class 3A does not include an allowance so it is suggested that the limits of size before plating be reduced by the amount of the 2A allowance whenever that allowance is adequate.

No provision is made for overcutting internal threads as coatings on such threads are not generally required. Further, it is very difficult to deposit a significant thickness of coating on the flanks of internal threads. Where a specific thickness of coating is required on an internal thread, it is suggested that the thread be overcut so that the thread as coated will be accepted by a GO thread plug gage of basic size.

This Standard ASME/ANSI B1.1-1989 specifies limits of size that pertain whether threads are coated or uncoated. Only in Class 2A threads is an allowance available to accommodate coatings. Thus, in all classes of internal threads and in all Class 1A, 2AG, and 3A external threads, limits of size must be adjusted to provide suitable provision for the desired coating.

For further information concerning dimensional accommodation of coating or plating for 60-degree threads, see Section 7, ASME/ANSI B1.1-1989.

Screw Thread Selection — Combination of Classes.—Whenever possible, selection should be made from Table 2, Standard Series Unified Screw Threads, preference being given to the Coarse- and Fine- thread Series. If threads in the standard series do not meet the requirements of design, reference should be made to the selected combinations in Table 3. The third expedient is to compute the limits of size from the tolerance tables or tolerance increment tables given in the Standard. The fourth and last resort is calculation by the formulas given in the Standard.

The requirements for screw thread fits for specific applications depend on end use and can be met by specifying the proper combinations of thread classes for the components. For example, a Class 2A external thread may be used with a Class 1B, 2B, or 3B internal thread.

Pitch Diameter Tolerances, All Classes.—The pitch diameter tolerances in Table 3 for all classes of the UNC, UNF, 4-UN, 6-UN, and 8-UN series are based on a length of engagement equal to the basic major (nominal) diameter and are applicable for lengths of engagement up to 1½ diameters.

The pitch diameter tolerances used in Table 3 for all classes of the UNEF, 12-UN, 16-UN, 20-UN, 28-UN, and 32-UN series and the UNS series, are based on a length of engagement of 9 pitches and are applicable for lengths of engagement of from 5 to 15 pitches.

Screw Thread Designation.—The basic method of designating a screw thread is used where the standard tolerances or limits of size based on the standard length of engagement are applicable. The designation specifies in sequence the nominal size, number of threads per inch, thread series symbol, thread class symbol, and, finally, gaging system number per ASME/ANSI B1.3M. The nominal size is the basic major diameter and is specified as the fractional diameter, screw number, or their decimal equivalent. Where decimal equivalents are used for size callout, they shall be interpreted as being nominal size designations only and shall have no dimensional significance beyond the fractional size or number designation. The symbol LH is placed after the thread class symbol to indicate a left-hand thread:

Examples:

¼–20 UNC-2A (21) or 0.250–20 UNC-2A (21)

10–32 UNF-2A (22) or 0.190–32 UNF-2A (22)

7⁄16–20 UNRF-2A (23) or 0.4375–20 UNRF-2A (23)

2–12 UN-2A (21) or 2.000–12 UN-2A (21)

¼–20 UNC-3A-LH (21) or 0.250–20 UNC-3A-LH (21)

For uncoated standard series threads these designations may optionally be supplemented by the addition of the pitch diameter limits of size.

Example:

¼–20 UNC-2A (21)

PD 0.2164–0.2127 (Optional for uncoated threads)

Designating Coated Threads.—For coated (or plated) Class 2A external threads, the basic (max) major and basic (max) pitch diameters are given followed by the words AFTER COATING. The major and pitch diameter limits of size before coating are also given followed by the words BEFORE COATING.

Example: ¾–10 UNC-2A (21)

[a]Major dia 0.7500 max }
PD 0.6850 max } AFTER COATING

[b]Major dia 0.7482–0.7353 }
PD 0.6832–0.6773 } BEFORE COATING

[a] Major and PD values are equal to basic and correspond to those in Table 3 for Class 3A.

[b] Major and PD limits are those in Table 3 for Class 2A.

Certain applications require an allowance for rapid assembly, to permit application of a proper lubricant, or for residual growth due to high-temperature expansion. In such applications where the thread is to be coated and the 2A allowance is not permitted to be consumed by such coating, the thread class symbol is qualified by the addition of the letter G (symbol for allowance) following the class symbol, and the maximum major and maximum pitch diameters are reduced below basic size by the amount of the 2A allowance and followed by the words AFTER COATING. This arrangement ensures that the allowance is maintained. The major and pitch diameter limits of size before coating are also given followed by SPL and BEFORE COATING. For information concerning the designating of this and other special coating conditions reference should be made to American National Standard ASME/ANSI B1.1-1989.

Designating UNS Threads.—UNS screw threads which have special combinations of diameter and pitch with tolerance to Unified formulation have the basic form designation set out first followed always by the limits of size.

Designating Multiple Start Threads.—If a screw thread is of multiple start, it is designated by specifying in sequence the nominal size, pitch (in decimals or threads per inch) and lead (in decimals or fractions).

Other Special Designations.—For other special designations including threads with modified limits of size or with special lengths of engagement, reference should be made to American National Standard ASME/ANSI B1.1-1989.

Hole Sizes for Tapping.—Hole size limits for tapping Classes 1B, 2B, and 3B threads of various lengths of engagement are given in the Tapping Section.

Internal Thread Minor Diameter Tolerances.—Internal thread minor diameter tolerances in Table 3 are based on a length of engagement equal to the nominal diameter. For general applications these tolerances are suitable for lengths of engagement up to 1½ diameters. However, some thread applications have lengths of engagement which are greater than 1½ diameters or less than the nominal diameter. For such applications it may be advantageous to increase or decrease the tolerance, respectively, as explained in the Tapping Section.

METRIC SCREW THREADS

American National Standard Metric Screw Threads M Profile.— American National Standard ANSI/ASME B1.13M-1983 (R1995) describes a system of metric threads for general fastening purposes in mechanisms and structures. The standard is in basic agreement with ISO screw standards and resolutions, as of the date of publication, and features detailed information for diameter-pitch combinations selected as to preferred standard sizes. This Standard contains general metric standards for a 60-degree symmetrical screw thread with a basic ISO 68 designated profile.

Application Comparison with Inch Threads.— The metric M profile threads of tolerance class 6H/6g (see page 1762) are intended for metric applications where the inch class 2A/2B have been used. At the minimum material limits, the 6H/6g results in a looser fit than the 2A/2B. Tabular data are also provided for a tighter tolerance fit external thread of class 4g6g which is approximately equivalent to the inch class 3A but with an allowance applied. It may be noted that a 4H5H/4h6h fit is approximately equivalent to class 3A/3B fit in the inch system.

Interchangeability with Other System Threads.— Threads produced to this Standard ANSI/ASME B1.13M are fully interchangeable with threads conforming to other National Standards that are based on ISO 68 basic profile and ISO 965/1 tolerance practices.

Threads produced to this Standard should be mechanically interchangeable with those produced to ANSI B1.18M-1982 (R1987) "Metric Screw Threads for Commercial Mechanical Fasteners—Boundary Profile Defined," of the same size and tolerance class. However, there is a possibility that some parts may be accepted by conventional gages used for threads made to ANSI/ASME B1.13M and rejected by the Double-NOT-GO gages required for threads made to ANSI B1.18M.

Threads produced in accordance with M profile and MJ profile ANSI/ASME B1.21M design data will assemble with each other. However, external MJ threads will encounter interference on the root radii with internal M thread crests when both threads are at maximum material condition.

Definitions.— The following definitions apply to metric screw threads — M profile.

Allowance: The minimum nominal clearance between a prescribed dimension and its basic dimension. Allowance is not an ISO metric screw thread term but it is numerically equal to the absolute value of the ISO term *fundamental deviation.*

Basic Thread Profile: The cyclical outline in an axial plane of the permanently established boundary between the provinces of the external and internal threads. All deviations are with respect to this boundary. (See Fig. 1 and 6.)

Bolt Thread (External Thread): The term used in ISO metric thread standards to describe all external threads. All symbols associated with external threads are designated with lower case letters. This Standard uses the term external threads in accordance with United States practice.

Clearance: The difference between the size of the internal thread and the size of the external thread when the latter is smaller.

Crest Diameter: The major diameter of an external thread and the minor diameter of an internal thread.

Design Profiles: The maximum material profiles permitted for external and internal threads for a specified tolerance class. (See Fig. 2 and 3.)

Deviation: An ISO term for the algebraic difference between a given size (actual, measured, maximum, minimum, etc.) and the corresponding basic size. The term deviation does not necessarily indicate an error.

Fit: The relationship existing between two corresponding external and internal threads with respect to the amount of clearance or interference which is present when they are assembled.

Fundamental Deviation: For Standard threads, the deviation (upper or lower) closer to the basic size. It is the upper deviation, *es*, for an external thread and the lower deviation, *EI*, for an internal thread. (See Fig. 6.)

Limiting Profiles: The limiting M profile for internal threads is shown in Fig. 7. The limiting M profile for external threads is shown in Fig. 8.

Lower Deviation: The algebraic difference between the minimum limit of size and the corresponding basic size.

Nut Thread (Internal Thread): A term used in ISO metric thread standards to describe all internal threads. All symbols associated with internal threads are designated with upper case letters. This Standard uses the term *internal thread* in accordance with United States practice.

Tolerance: The total amount of variation permitted for the size of a dimension. It is the difference between the maximum limit of size and the minimum limit of size (i.e., the algebraic difference between the upper deviation and the lower deviation). The tolerance is an absolute value without sign. Tolerance for threads is applied to the design size in the direction of the minimum material. On external threads the tolerance is applied negatively. On internal threads the tolerance is applied positively.

Tolerance Class: The combination of a tolerance position with a tolerance grade. It specifies the allowance (fundamental deviation) and tolerance for the pitch and major diameters of external threads and pitch and minor diameters of internal threads.

Tolerance Grade: A numerical symbol that designates the tolerances of crest diameters and pitch diameters applied to the design profiles.

Tolerance Position: A letter symbol that designates the position of the tolerance zone in relation to the basic size. This position provides the allowance (fundamental deviation).

Upper Deviation: The algebraic difference between the maximum limit of size and the corresponding basic size.

Basic M Profile.—The basic M thread profile also known as ISO 68 basic profile for metric screw threads is shown in Fig. 1 with associated dimensions listed in Table 3.

Design M Profile for Internal Thread.—The design M profile for the internal thread at maximum material condition is the basic ISO 68 profile. It is shown in Fig. 2 with associated thread data listed in Table 3.

Design M Profile for External Thread.—The design M profile for the external thread at the no allowance maximum material condition is the basic ISO 68 profile except where a rounded root is required. For the standard $0.125P$ minimum radius, the ISO 68 profile is modified at the root with a $0.17783H$ truncation blending into two arcs with radii of $0.125P$ tangent to the thread flanks as shown in Fig. 3 with associated thread data in Table 3.

M Crest and Root Form.—The form of crest at the major diameter of the external thread is flat, permitting corner rounding. The external thread is truncated $0.125H$ from a sharp crest. The form of the crest at the minor diameter of the internal thread is flat. It is truncated $0.25H$ from a sharp crest.

The crest and root tolerance zones at the major and minor diameters will permit rounded crest and root forms in both external and internal threads.

The root profile of the external thread must lie within the "section lined" tolerance zone shown in Fig. 4. For the rounded root thread, the root profile must lie within the "section lined" rounded root tolerance zone shown in Fig. 4. The profile must be a continuous, smoothly blended non-reversing curve, no part of which has a radius of less than $0.125P$, and which is tangential to the thread flank. The profile may comprise tangent flank arcs that are joined by a tangential flat at the root.

The root profile of the internal thread must not be smaller than the basic profile. The maximum major diameter must not be sharp.

General Symbols.—The general symbols used to describe the metric screw thread forms are shown in Table 1.

Table 1. American National Standard Symbols for Metric Threads *ANSI/ASME B1.13M-1983 (R1995)*

Symbol	Explanation
D	Major Diameter Internal Thread
D_1	Minor Diameter Internal Thread
D_2	Pitch Diameter Internal Thread
d	Major Diameter External Thread
d_1	Minor Diameter External Thread
d_2	Pitch Diameter External Thread
d_3	Rounded Form Minor Diameter External Thread
P	Pitch
r	External Thread Root Radius
T	Tolerance
T_{D1}, T_{D2}	Tolerances for D_1, D_2
T_d, T_{d2}	Tolerances for d, d_2
ES	Upper Deviation, Internal Thread [Equals the Allowance (Fundamental Deviation) Plus the Tolerance]. See Fig. 6.
EI	Lower Deviation, Internal Thread Allowance (Fundamental Deviation). See Fig. 6.
G, H	Letter Designations for Tolerance Positions for Lower Deviation, Internal Thread
g, h	Letter Designations for Tolerance Positions for Upper Deviation, External Thread
es	Upper Deviation, External Thread Allowance (Fundamental Deviation). See Fig. 6. In the ISO system es is always negative for an allowance fit or zero for no allowance.
ei	Lower Deviation, External Thread [Equals the Allowance (Fundamental Deviation) Plus the Tolerance]. See Fig. 6. In the ISO system ei is always negative for an allowance fit.
H	Height of Fundamental Triangle
LE	Length of Engagement
LH	Left Hand Thread

Standard M Profile Screw Thread Series.—The standard metric screw thread series for general purpose equipment's threaded components design and mechanical fasteners is a *coarse thread* series. Their diameter/pitch combinations are shown in Table 4. These diameter/pitch combinations are the preferred sizes and should be the first choice as applicable. Additional *fine pitch* diameter/pitch combinations are shown in Table 5.

Table 2. American National Standard General Purpose and Mechanical Fastener Coarse Pitch Metric Thread—M Profile Series *ANSI/ASME B1.13M-1983 (R1995)*

Nom.Size	Pitch	Nom.Size	Pitch	Nom.Size	Pitch	Nom.Size	Pitch
1.6	0.35	6	1	22	2.5[a]	56	5.5
2	0.4	8	1.25	24	3	64	6
2.5	0.45	10	1.5	27	3[a]	72	6
3	0.5	12	1.75	30	3.5	80	6
3.5	0.6	14	2	36	4	90	6
4	0.7	16	2	42	4.5	100	6
5	0.8	20	2.5	48	5	…	…

[a] For high strength structural steel fasteners only.

All dimensions are in millimeters.

Table 3. American National Standard Metric Thread — M Profile Data *ANSI/ASME B1.13M-1983 (R1995)*

Pitch P	Truncation of Internal Thread Root and External Thread Crest $\frac{H}{8}$ 0.108253P	Addendum of Internal Thread and Truncation of Internal Thread $\frac{H}{4}$ 0.216506P	Dedendum of Internal Thread and Addendum External Thread $\frac{3}{8}H$ 0.324760P	Difference[a] $\frac{H}{2}$ 0.433013P	Height of Internal Thread and Depth of Thread Engagement $\frac{5}{8}H$ 0.541266P	Difference[b] 0.711325H 0.616025P	Twice the External Thread Addendum $\frac{3}{4}H$ 0.649519P	Difference[c] $\frac{11}{12}H$ 0.793857P	Height of Sharp V-Thread H 0.8660254P	Double Height of Internal Thread $\frac{5}{4}H$ 1.082532P
0.2	0.02165	0.04330	0.06495	0.08660	0.10825	0.12321	0.12990	0.15877	0.17321	0.21651
0.25	0.02706	0.05413	0.08119	0.10825	0.13532	0.15401	0.16238	0.19846	0.21651	0.27063
0.3	0.03248	0.06495	0.09743	0.12990	0.16238	0.18481	0.19486	0.23816	0.25981	0.32476
0.35	0.03789	0.07578	0.11367	0.15155	0.18944	0.21561	0.22733	0.27785	0.30311	0.37889
0.4	0.04330	0.08660	0.12990	0.17321	0.21651	0.24541	0.25981	0.31754	0.34641	0.43301
0.45	0.04871	0.09743	0.14614	0.19486	0.24357	0.27721	0.29228	0.35724	0.38971	0.48714
0.5	0.05413	0.10825	0.16238	0.21651	0.27063	0.30801	0.32476	0.39693	0.43301	0.64952
0.6	0.06495	0.12990	0.19486	0.25981	0.32476	0.36962	0.38971	0.47631	0.51962	0.64952
0.7	0.07578	0.15155	0.22733	0.30311	0.37889	0.43122	0.45466	0.55570	0.60622	0.75777
0.75	0.08119	0.16238	0.24357	0.32476	0.40595	0.46202	0.48714	0.59539	0.64952	0.81190
0.8	0.08660	0.17321	0.25981	0.34641	0.43301	0.49282	0.51962	0.63509	0.69282	0.86603
1	0.10825	0.21651	0.32476	0.43301	0.54127	0.61603	0.64952	0.79386	0.86603	1.08253
1.25	0.13532	0.27063	0.40595	0.54127	0.67658	0.77003	0.81190	0.99232	1.08253	1.35316
1.5	0.16238	0.32476	0.48714	0.64952	0.81190	0.92404	0.97428	1.19078	1.29904	1.62380
1.75	0.18944	0.37889	0.56833	0.75777	0.94722	1.07804	1.13666	1.38925	1.51554	1.89443
2	0.21651	0.43301	0.64952	0.86603	1.08253	1.23205	1.29904	1.58771	1.73205	2.16506
2.5	0.27063	0.54127	0.81190	1.08253	1.35316	1.54006	1.62380	1.98464	2.16506	2.70633
3	0.32476	0.64652	0.97428	1.29904	1.62380	1.84808	1.94856	2.38157	2.59808	3.24760
3.5	0.37889	0.75777	1.13666	1.51554	1.89443	2.15609	2.27332	2.77850	3.03109	3.78886
4	0.43301	0.86603	1.29904	1.73205	2.16506	2.46410	2.59808	3.17543	3.46410	4.33013
4.5	0.48714	0.97428	1.46142	1.94856	2.43570	2.77211	2.92284	3.57235	3.89711	4.87139
5	0.54127	1.08253	1.62380	2.16506	2.70633	3.08013	3.24760	3.96928	4.33013	5.41266
5.5	0.59539	1.19078	1.78618	2.38157	2.97696	3.38814	3.57235	4.36621	4.76314	5.95392
6	0.64952	1.29904	1.94856	2.59808	3.24760	3.69615	3.89711	4.76314	5.19615	6.49519
8	0.86603	1.73205	2.59808	3.46410	4.33013	4.92820	5.19615	6.35085	6.92820	8.66025

[a] Difference between max theoretical pitch diameter and max minor diameter of external thread and between min theoretical pitch diameter and min minor diameter of internal thread.

[b] Difference between min theoretical pitch diameter and min design minor diameter of external thread for 0.125P root radius.

[c] Difference between max major diameter and max theoretical pitch diameter of internal thread.

All dimensions are in millimeters.

Table 4. American National Standard General Purpose and Mechanical Fastener Coarse Pitch Metric Thread—M Profile Series ***ANSI/ASME B1.13M-1983 (R1995)***

Nom. Size	Pitch	Nom. Size	Pitch	Nom. Size	Pitch	Nom. Size	Pitch
1.6	0.35	6	1	22	2.5[a]	56	5.5
2	0.4	8	1.25	24	3	64	6
2.5	0.45	10	1.5	27	3[a]	72	6
3	0.5	12	1.75	30	3.5	80	6
3.5	0.6	14	2	36	4	90	6
4	0.7	16	2	42	4.5	100	6
5	0.8	20	2.5	48	5	...	...

[a] For high strength structural steel fasteners only.

All dimensions are in millimeters.

Table 5.
American National Standard Fine Pitch Metric Thread—M Profile Series
ANSI/ASME B1.13M-1983 (R1995)

Nom. Size	Pitch			Nom. Size	Pitch		Nom. Size	Pitch		Nom. Size	Pitch
8	1		...	27	...	2	56	...	2	105	2
10	0.75		1.25	30	1.5	2	60	1.5	...	110	2
12	1	1.5[a]	1.25	33	...	2	64	...	2	120	2
14	...		1.5	35	1.5	...	65	1.5	...	130	2
15	1		...	36	...	2	70	1.5	...	140	2
16	...		1.5	39	...	2	72	...	2	150	2
17	1		...	40	1.5	...	75	1.5	...	160	3
18	...		1.5	42	...	2	80	1.5	2	170	3
20	1		1.5	45	1.5	...	85	...	2	180	3
22	...		1.5	48	...	2	90	...	2	190	3
24	...		2	50	1.5	...	95	...	2	200	3
25	1.5		...	55	1.5	...	100	...	2		

[a] Only for wheel studs and nuts.

All dimensions are in millimeters.

Limits and Fits for Metric Screw Threads — M Profile.—The International (ISO) metric tolerance system is based on a system of limits and fits. The limits of the tolerances on the mating parts together with their allowances (fundamental deviations) determine the fit of the assembly. For simplicity the system is described for cylindrical parts (see *British Standard for Metric ISO Limits and Fits* starting on page 657) but in this Standard it is applied to screw threads. Holes are equivalent to internal threads and shafts to external threads.

Basic Size: This is the zero line or surface at assembly where the interface of the two mating parts have a common reference.*

Upper Deviation: This is the algebraic difference between the maximum limit of size and the basic size. It is designated by the French term "écart supérieur" (ES for internal and *es* for external threads).

Lower Deviation: This is the algebraic difference between the minimum limit of size and the basic size. It is designated by the French term "écart inférieur" (*EI* for internal and *ei* for external threads).

* Basic," when used to identify a particular dimension in this Standard, such as basic major diameter, refers to the h/H tolerance position (zero fundamental deviation) value.

Fundamental Deviations (Allowances): These are the deviations which are closest to the basic size. In the accompanying figure they would be *EI* and *es*.

Fits: Fits are determined by the fundamental deviations assigned to the mating parts and may be positive or negative. The selected fits can be clearance, transition, or interference. To illustrate the fits schematically, a zero line is drawn to represent the basic size as shown in Fig. 6. By convention, the external thread lies below the zero line and the internal thread lies above it (except for interference fits). This makes the fundamental deviation negative for the external thread and equal to its upper deviation (*es*). The fundamental deviation is positive for the internal thread and equal to its lower deviation (*EI*).

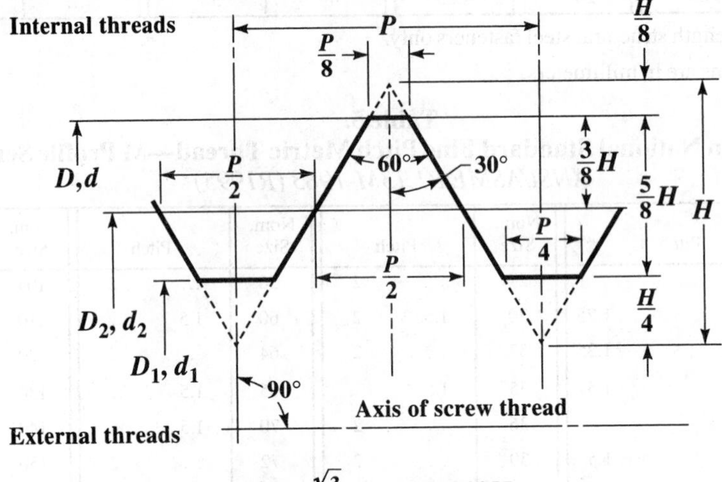

$0.125H - 0.108\,253P$ $0.250H = 0.216\,506P$ $0.375H + 0.324\,760P$ $0.625H - 0.541\,266P$

Fig. 1. Basic M Thread Profile ISO 68 Basic Profile

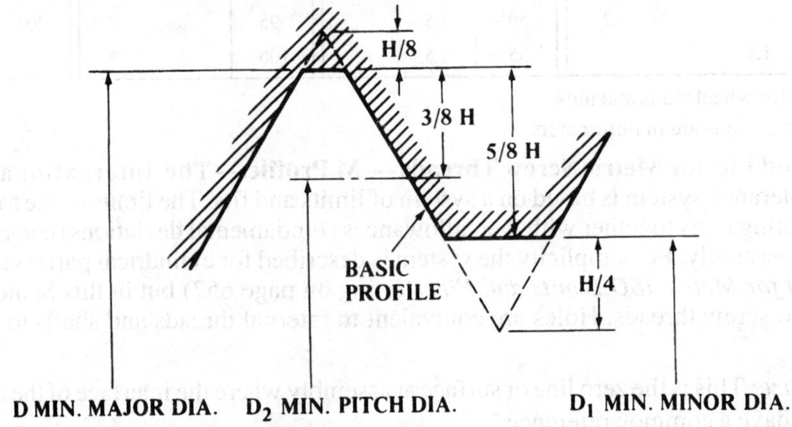

Fig. 2. Internal Thread Design M Profile with No Allowance (Fundamental Deviation) (Maximum Material Condition). For Dimensions see Table 3

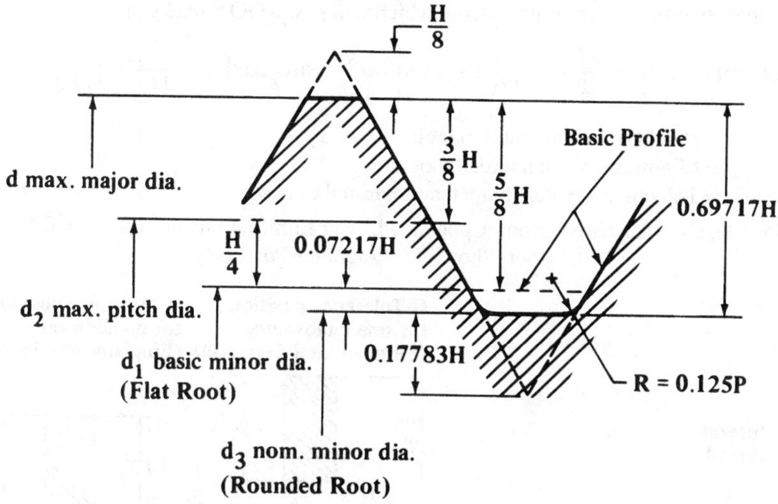

Fig. 3. External Thread Design M Profile with No Allowance (Fundamental Deviation) (Flanks at Maximum Material Condition). For Dimensions see Table 3

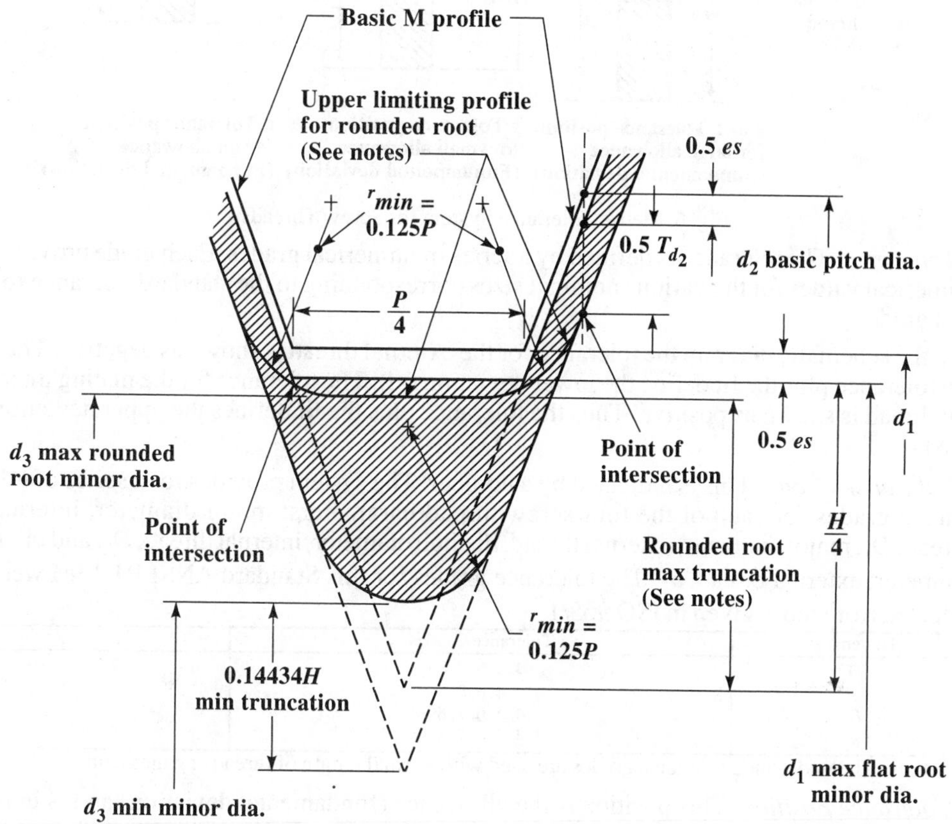

Fig. 4. M Profile, External Thread Root, Upper and Lower Limiting Profiles for $r_{min} = 0.125\ P$ and for Flat Root (Shown for Tolerance Position g)

Notes:

1) "Section lined" portions identify tolerance zone and unshaded portions identify allowance (fundamental deviation).

2) The upper limiting profile for rounded root is not a design profile; rather it indicates the limiting acceptable condition for the rounded root which will pass a GO thread gage.

3) $$\text{Max truncation} = \frac{H}{4} - r_{min}\left(1 - \cos\left[60° - \text{arc}\cos\left(1 - \frac{T_{d2}}{4r_{min}}\right)\right]\right)$$

where H = Height of fundamental triangle
r_{min} = Minimum external thread root radius
T_{d2} = Tolerance on pitch diameter of external threasd

Fig. 5. M Profile, External Thread Root, Upper and Lower Limiting Profiles for $r_{min} = 0.125\,P$ and for Flat Root (Shown for Tolerance Position g)

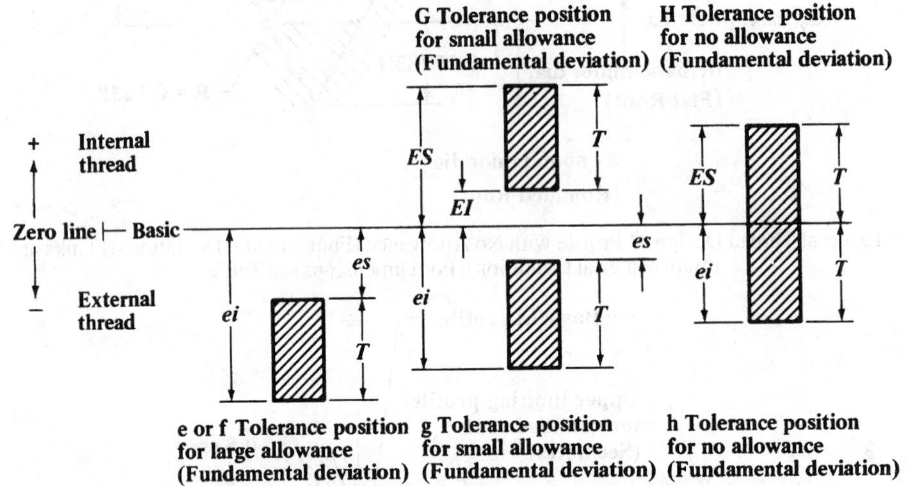

Fig. 6. Metric Tolerance System for Screw Threads

Tolerance: The tolerance is defined by a series of numerical grades. Each grade provides numerical values for the various nominal sizes corresponding to the standard tolerance for that grade.

In the schematic diagram the tolerance for the external thread is shown as negative. Thus the tolerance plus the fit define the lower deviation (*ei*). The tolerance for the mating internal thread is shown as positive. Thus the tolerance plus the fit defines the upper deviation (*ES*).

Tolerance Grade: This is indicated by a number. The system provides for a series of tolerance grades for each of the four screw thread parameters: minor diameter, internal thread, D_1; major diameter, external thread, d; pitch diameter, internal thread, D_2; and pitch diameter, external thread, d_2. The tolerance grades for this Standard ANSI B1.13M were selected from those given in ISO 965/1.

Dimension	Tolerance Grades	Table
D_1	4, 5, <u>6</u>, 7, 8	7
d	4, <u>6</u>, 8	10
D_2	4, 5, <u>6</u>, 7, 8	8
d_2	3, <u>4</u>, 5, <u>6</u>, 7, 8, 9	9

Note: The underlined tolerance grades are used with normal length of thread engagement.

Tolerance Position: This position is the allowance (fundamental deviation) and is indicated by a letter. A capital letter is used for internal threads and a lower case letter for external threads. The system provides a series of tolerance positions for internal and external threads. The underlined letters are used in this Standard:

Internal threads	G, <u>H</u>	Table 7
External threads	e, f, <u>g</u>, h	Table 7

Designations of Tolerance Grade, Tolerance Position, and Tolerance Class: The tolerance grade is given first followed by the tolerance position, thus: 4g or 5H. To designate the tolerance class the grade and position of the pitch diameter is shown first followed by that for the major diameter in the case of the external thread or that for the minor diameter in the case of the internal thread, thus 4g6g for an external thread and 5H6H for an internal thread. If the two grades and positions are identical, it is not necessary to repeat the symbols, thus 4g, alone, stands for 4g4g and 5H, alone, stands for 5H5H.

Lead and Flank Angle Tolerances: For acceptance of lead and flank angles of product screw threads, see Section 10 of ANSI/ASME B1.13M-1983 (R1995).

Short and Long Lengths of Thread Engagement when Gaged with Normal Length Contacts: For short lengths of thread engagement, LE, reduce the pitch diameter tolerance of the external thread by one tolerance grade number. For long lengths of thread engagement, LE, increase the allowance (fundamental deviation) at the pitch diameter of the external thread. Examples of tolerance classes required for normal, short, and long gage length contacts are given in the following table.

For lengths of thread engagement classified as normal, short, and long, see Table 6.

Table 6. American National Standard Length of Metric Thread Engagement *ISO 965/1* **and** *ANSI/ASME B1.13M-1983 (R1995)*

Basic Major Diameter d_{bsc}		Pitch	Length of Thread Engagement			
			Short LE	Normal LE		Long LE
Over	Up to and incl.	*P*	Up to and incl.	Over	Up to and incl.	Over
1.5	2.8	0.2	0.5	0.5	1.5	1.5
		0.25	0.6	0.6	1.9	1.9
		0.35	0.8	0.8	2.6	2.6
		0.4	1	1	3	3
		0.45	1.3	1.3	3.8	3.8
2.8	5.6	0.35	1	1	3	3
		0.5	1.5	1.5	4.5	4.5
		0.6	1.7	1.7	5	5
		0.7	2	2	6	6
		0.75	2.2	2.2	6.7	6.7
		0.8	2.5	2.5	7.5	7.5
5.6	11.2	0.75	2.4	2.4	7.1	7.1
		1	3	3	9	9
		1.25	4	4	12	12
		1.5	5	5	15	15
11.2	22.4	1	3.8	3.8	11	11
		1.25	4.5	4.5	13	13
		1.5	5.6	5.6	16	16
		1.75	6	6	18	18
		2	8	8	24	24
		2.5	10	10	30	30
22.4	45	1	4	4	12	12
		1.5	6.3	6.3	19	19
		2	8.5	8.5	25	25
		3	12	12	36	36
		3.5	15	15	45	45
		4	18	18	53	53
		4.5	21	21	63	63
45	90	1.5	7.5	7.5	22	22
		2	9.5	9.5	28	28
		3	15	15	45	45
		4	19	19	56	56
		5	24	24	71	71
		5.5	28	28	85	85
		6	32	32	95	95
90	180	2	12	12	36	36
		3	18	18	53	53
		4	24	24	71	71
		6	36	36	106	106
180	355	3	20	20	60	60
		4	26	26	80	80
		6	40	40	118	118

All dimensions are in millimeters.

Normal LE	Short LE	Long LE
6g	5g6g	6e6g
4g6g	3g6g	4e6g
6h[a]	5h6h	6g6h
4h6h[a]	3h6h	4g6h

[a] Applies to maximum material functional size (GO thread gage) for plated 6g and 4g6g class threads, respectively.

Coated or Plated Threads: Coating is one or more applications of additive material to the threads, including dry-film lubricants, but excluding soft or liquid lubricants that are readily displaced in assembly or gaging. Plating is included as coating in the Standard. Unless otherwise specified, size limits for standard external tolerance classes 6g and 4g6g apply prior to coating. The external thread allowance may thus be used to accommodate the coating thickness on coated parts, provided that the maximum coating thickness is no more than one-quarter of the allowance. Thus, the thread after coating is subject to acceptance using a basic (tolerance position h) size GO thread gage and tolerance position g thread gage for either minimum material, LO, or NOT-GO. Where the external thread has no allowance or the allowance must be maintained after coating, and for standard internal threads, sufficient allowance must be provided prior to coating to ensure that finished product threads do not exceed the maximum material limits specified. For thread classes with tolerance position H or h, coating allowances in accordance with Table 7 for position G or g, respectively, should be applied wherever possible.

Table 7. American National Standard Allowance (Fundamental Deviation) for Internal and External Metric Threads *ISO 965/1 ANSI/ASME B1.13M-1983 (R1995)*

Pitch	Allowance (Fundamental Deviation)[a]					
	Internal Thread D_2, D_1		External Thread d, d_2			
	G	H	e	f	g	h
P	*EI*	*EI*	*es*	*es*	*es*	*es*
0.2	+0.017	0	…	…	−0.017	0
0.25	+0.018	0	…	…	−0.018	0
0.3	+0.018	0	…	…	−0.018	0
0.35	+0.019	0	…	−0.034	−0.019	0
0.4	+0.019	0	…	−0.034	−0.019	0
0.45	+0.020	0	…	−0.035	−0.020	0
0.5	+0.020	0	−0.050	−0.036	−0.020	0
0.6	+0.021	0	−0.053	−0.036	−0.021	0
0.7	+0.022	0	−0.056	−0.038	−0.022	0
0.75	+0.022	0	−0.056	−0.038	−0.022	0
0.8	+0.024	0	−0.060	−0.038	−0.024	0
1	+0.026	0	−0.060	−0.040	−0.026	0
1.25	+0.028	0	−0.063	−0.042	−0.028	0
1.5	+0.032	0	−0.067	−0.045	−0.032	0
1.75	+0.034	0	−0.071	−0.048	−0.034	0
2	+0.038	0	−0.071	−0.052	−0.038	0
2.5	+0.042	0	−0.080	−0.058	−0.042	0
3	+0.048	0	−0.085	−0.063	−0.048	0
3.5	+0.053	0	−0.090	−0.070	−0.053	0
4	+0.060	0	−0.095	−0.075	−0.060	0
4.5	+0.063	0	−0.100	−0.080	−0.063	0
5	+0.071	0	−0.106	−0.085	−0.071	0
5.5	+0.075	0	−0.112	−0.090	−0.075	0
6	+0.080	0	−0.118	−0.095	−0.080	0

[a] Allowance is the absolute value of fundamental deviation.

All dimensions are in millimeters.

Dimensional Effect of Coating.—On a cylindrical surface, the effect of coating is to change the diameter by twice the coating thickness. On a 60-degree thread, however, since the coating thickness is measured perpendicular to the thread surface while the pitch diameter is measured perpendicular to the thread axis, the effect of a uniformly coated flank on the pitch diameter is to change it by four times the thickness of the coating on the flank.

External Thread with No Allowance for Coating: To determine gaging limits before coating for a uniformly coated thread, decrease: D) maximum pitch diameter by four times maximum coating thickness; E) minimum pitch diameter by four times minimum coating thickness; F) maximum major diameter by two times maximum coating thickness; and G) minimum major diameter by two times minimum coating thickness.

External Thread with Only Nominal or Minimum Thickness Coating: If no coating thickness tolerance is given, it is recommended that a tolerance of plus 50 per cent of the nominal or minimum thickness be assumed.

Then, to determine before coating gaging limits for a uniformly coated thread, decrease:

H) maximum pitch diameter by six times coating thickness; I) minimum pitch diameter by four times coating thickness; J) maximum major diameter by three times coating thickness; and K) minimum major diameter by two times coating thickness.

Adjusted Size Limits: It should be noted that the before coating material limit tolerances are less than the tolerance after coating. This is because the coating tolerance consumes some of the product tolerance. In cases there may be insufficient pitch diameter tolerance available in the before coating condition so that additional adjustments and controls will be necessary.

Strength: On small threads (5 mm and smaller) there is a possibility that coating thickness adjustments will cause base material minimum material conditions which may significantly affect strength of externally threaded parts. Limitations on coating thickness or part redesign may then be necessary.

Internal Threads: Standard internal threads provide no allowance for coating thickness.

To determine before coating, gaging limits for a uniformly coated thread, increase:

L) minimum pitch diameter by four times maximum coating thickness, if specified, or by six times minimum or nominal coating thickness when a tolerance is not specified;

M) maximum pitch diameter by four times minimum or nominal coating thickness;

N) minimum minor diameter by two times maximum coating thickness, if specified, or by three times minimum or nominal coating thickness; and O) maximum minor diameter by two times minimum or nominal coating thickness.

Other Considerations.—It is essential to review all possibilities adequately and consider limitations in the threading and coating production processes before finally deciding on the coating process and the allowance required to accommodate the coating. A no-allowance thread after coating must not transgress the basic profile and is, therefore, subject to acceptance using a basic (tolerance position H/h) size GO thread gage.

Formulas for M Profile Screw Thread Limiting Dimensions.—The limiting dimensions for M profile screw threads are calculated from the following formulas.

Internal Threads: $Min major dia.$ = basic major dia. + EI (Table 7)

$Min pitch dia.$ = basic major dia. − 0.649519P (Table 3) + EI for D_2 (Table 7)

$Max pitch dia.$ = min pitch dia. + T_{D2} (Table 8)

$Max major dia.$ = max pitch dia. + 0.793857P (Table 3)

$Min minor dia.$ = min major dia. − 1.082532P (Table 3)

$Max minor dia.$ = min minor dia. + T_{D1} (Table 9)

Table 8. American National Standard Pitch-Diameter Tolerances of Internal Metric Threads, T_{D2} *ISO 965/1 ANSI/ASME B1.13M-1983 (R1995)*

Basic Major Diameter, *D*			Tolerance Grade				
Over	Up to and incl.	Pitch *P*	4	5	6	7	8
1.5	2.8	0.2	0.042	…	…	…	…
		0.25	0.048	0.060	…	…	…
		0.35	0.053	0.067	0.085	…	…
		0.4	0.056	0.071	0.090	…	…
		0.45	0.060	0.075	0.095	…	…
2.8	5.6	0.35	0.056	0.071	0.090	…	…
		0.5	0.063	0.080	0.100	0.125	…
		0.6	0.071	0.090	0.112	0.140	…
		0.7	0.075	0.095	0.118	0.150	…
		0.75	0.075	0.095	0.118	0.150	…
		0.8	0.080	0.100	0.125	0.160	0.200
5.6	11.2	0.75	0.085	0.106	0.132	0.170	…
		1	0.095	0.118	0.150	0.190	0.236
		1.25	0.100	0.125	0.160	0.200	0.250
		1.5	0.112	0.140	0.180	0.224	0.280
11.2	22.4	1	0.100	0.125	0.160	0.200	0.250
		1.25	0.112	0.140	0.180	0.224	0.280
		1.5	0.118	0.150	0.190	0.236	0.300
		1.75	0.125	0.160	0.200	0.250	0.315
		2	0.132	0.170	0.212	0.265	0.335
		2.5	0.140	0.180	0.224	0.280	0.355
22.4	45	1	0.106	0.132	0.170	0.212	…
		1.5	0.125	0.160	0.200	0.250	0.315
		2	0.140	0.180	0.224	0.280	0.355
		3	0.170	0.212	0.265	0.335	0.425
		3.5	0.180	0.224	0.280	0.355	0.450
		4	0.190	0.236	0.300	0.375	0.475
		4.5	0.200	0.250	0.315	0.400	0.500
45	90	1.5	0.132	0.170	0.212	0.265	0.335
		2	0.150	0.190	0.236	0.300	0.375
		3	0.180	0.224	0.280	0.355	0.450
		4	0.200	0.250	0.315	0.400	0.500
		5	0.212	0.265	0.335	0.425	0.530
		5.5	0.224	0.280	0.355	0.450	0.560
		6	0.236	0.300	0.375	0.475	0.600
90	180	2	0.160	0.200	0.250	0.315	0.400
		3	0.190	0.236	0.300	0.375	0.475
		4	0.212	0.265	0.335	0.425	0.530
		6	0.250	0.315	0.400	0.500	0.630
180	355	3	0.212	0.265	0.335	0.425	0.530
		4	0.236	0.300	0.375	0.475	0.600
		6	0.265	0.335	0.425	0.530	0.670

All dimensions are in millimeters.

Table 9. American National Standard Minor Diameter Tolerances of Internal Metric Threads T_{D1} *ISO 965/1 ANSI/ASME B1.13M-1983 (R1995)*

Pitch *P*	Tolerance Grade				
	4	5	6	7	8
0.2	0.038	…	…	…	…
0.25	0.045	0.056	…	…	…
0.3	0.053	0.067	0.085	…	…
0.35	0.063	0.080	0.100	…	…
0.4	0.071	0.090	0.112	…	…
0.45	0.080	0.100	0.125	…	…
0.5	0.090	0.112	0.140	0.180	…
0.6	0.100	0.125	0.160	0.200	…
0.7	0.112	0.140	0.180	0.224	…
0.75	0.118	0.150	0.190	0.236	…
0.8	0.125	0.160	0.200	0.250	0.315
1	0.150	0.190	0.236	0.300	0.375
1.25	0.170	0.212	0.265	0.335	0.425
1.5	0.190	0.236	0.300	0.375	0.475
1.75	0.212	0.265	0.335	0.425	0.530
2	0.236	0.300	0.375	0.475	0.600
2.5	0.280	0.355	0.450	0.560	0.710
3	0.315	0.400	0.500	0.630	0.800
3.5	0.355	0.450	0.560	0.710	0.900
4	0.375	0.475	0.600	0.750	0.950
4.5	0.425	0.530	0.670	0.850	1.060
5	0.450	0.560	0.710	0.900	1.120
5.5	0.475	0.600	0.750	0.950	1.180
6	0.500	0.630	0.800	1.000	1.250

All dimensions are in millimeters.

External Threads: Maxmajordia. = basic major dia. – *es* (Table 7) (Note that *es* is an absolute value.)

Minmajordia. = max major dia. – T_d (Table 10)

Maxpitchdia. = basic major dia. – 0.649519P (Table 3) – *es* for d_2 (Table 7)

Minpitchdia. = max pitch dia. – T_{d2} (Table 11)

Maxflatform minordia. = max pitch dia. – 0.433013P (Table 3)

Max rounded root minor dia. = max pitch dia. – 2 × max trunc. (See Fig. 4)

Min rounded root minor dia. = min pitch dia. – 0.616025P (Table 3)

Minrootradius = 0.125P

Table 10. American National Standard Major Diameter Tolerances of External Metric Threads, T_d *ISO 965/1 ANSI/ASME B1.13M-1983 (R1995)*

Pitch	Tolerance Grade			Pitch	Tolerance Grade		
P	4	6	8	*P*	4	6	8
0.2	0.036	0.056	…	1.25	0.132	0.212	0.335
0.25	0.042	0.067	…	1.5	0.150	0.236	0.375
0.3	0.048	0.075	…	1.75	0.170	0.265	0.425
0.35	0.053	0.085	…	2	0.180	0.280	0.450
0.4	0.060	0.095	…	2.5	0.212	0.335	0.530
0.45	0.063	0.100	…	3	0.236	0.375	0.600
0.5	0.067	0.106	…	3.5	0.265	0.425	0.670
0.6	0.080	0.125	…	4	0.300	0.475	0.750
0.7	0.090	0.140	…	4.5	0.315	0.500	0.800
0.75	0.090	0.140	…	5	0.335	0.530	0.850
0.8	0.095	0.150	0.236	5.5	0.355	0.560	0.900
1	0.112	0.180	0.280	6	0.375	0.600	0.950

All dimensions are in millimeters.

Table 11. American National Standard Pitch-Diameter Tolerances of External Metric Threads, T_{d2} *ISO 965/1 ANSI/ASME B1.13M-1983 (R1995)*

Basic Major Diameter, *d*		Pitch	Tolerance Grade						
Over	Up to and incl.	*P*	3	4	5	6	7	8	9
1.5	2.8	0.2	0.025	0.032	0.040	0.050	…	…	…
		0.25	0.028	0.036	0.045	0.056	…	…	…
		0.35	0.032	0.040	0.050	0.063	0.080	…	…
		0.4	0.034	0.042	0.053	0.067	0.085	…	…
		0.45	0.036	0.045	0.056	0.071	0.090	…	…
2.8	5.6	0.35	0.034	0.042	0.053	0.067	0.085	…	…
		0.5	0.038	0.048	0.060	0.075	0.095	…	…
		0.6	0.042	0.053	0.067	0.085	0.106	…	…
		0.7	0.045	0.056	0.071	0.090	0.112	…	…
		0.75	0.045	0.056	0.071	0.090	0.112	…	…
		0.8	0.048	0.060	0.075	0.095	0.118	0.150	0.190
5.6	11.2	0.75	0.050	0.063	0.080	0.100	0.125	…	…
		1	0.056	0.071	0.090	0.112	0.140	0.180	0.224
		1.25	0.060	0.075	0.095	0.118	0.150	0.190	0.236
		1.5	0.067	0.085	0.106	0.132	0.170	0.212	0.265
11.2	22.4	1	0.060	0.075	0.095	0.118	0.150	0.190	0.236
		1.25	0.067	0.085	0.106	0.132	0.170	0.212	0.265
		1.5	0.071	0.090	0.112	0.140	0.180	0.224	0.280
		1.75	0.075	0.095	0.118	0.150	0.190	0.236	0.300
		2	0.080	0.100	0.125	0.160	0.200	0.250	0.315
		2.5	0.085	0.106	0.132	0.170	0.212	0.265	0.335
22.4	45	1	0.063	0.080	0.100	0.125	0.160	0.200	0.250
		1.5	0.075	0.095	0.118	0.150	0.190	0.236	0.300
		2	0.085	0.106	0.132	0.170	0.212	0.265	0.335
		3	0.100	0.125	0.160	0.200	0.250	0.315	0.400
		3.5	0.106	0.132	0.170	0.212	0.265	0.335	0.425
		4	0.112	0.140	0.180	0.224	0.280	0.355	0.450
		4.5	0.118	0.150	0.190	0.236	0.300	0.375	0.475
45	90	1.5	0.080	0.100	0.125	0.160	0.200	0.250	0.315
		2	0.090	0.112	0.140	0.180	0.224	0.280	0.355
		3	0.106	0.132	0.170	0.212	0.265	0.335	0.425
		4	0.118	0.150	0.190	0.236	0.300	0.375	0.475
		5	0.125	0.160	0.200	0.250	0.315	0.400	0.500
		5.5	0.132	0.170	0.212	0.265	0.335	0.425	0.530
		6	0.140	0.180	0.224	0.280	0.355	0.450	0.560
90	180	2	0.095	0.118	0.150	0.190	0.236	0.300	0.375
		3	0.112	0.140	0.180	0.224	0.280	0.355	0.450
		4	0.125	0.160	0.200	0.250	0.315	0.400	0.500
		6	0.150	0.190	0.236	0.300	0.375	0.475	0.600
180	355	3	0.125	0.160	0.200	0.250	0.315	0.400	0.500
		4	0.140	0.180	0.224	0.280	0.355	0.450	0.560
		6	0.160	0.200	0.250	0.315	0.400	0.500	0.630

All dimensions are in millimeters.

Tolerance Grade Comparisons.—The approximate ratios of the tolerance grades shown in Tables 9, 8, 11, and 10 in terms of Grade 6 are as follows:

Minor Diameter Tolerance of Internal Thread: T_{D1} *(Table 9):* Grade 4 is 0.63 T_{D1} (6); Grade 5 is 0.8 T_{D1} (6); Grade 7 is 1.25 T_{D1} (6); and Grade 8 is 1.6 T_{D1} (6).

Pitch Diameter Tolerance of Internal Thread: T_{D2} *(Table 8):* Grade 4 is 0.85 T_{d2} (6); Grade 5 is 1.06 T_{d2} (6); Grade 6 is 1.32 T_{d2} (6); Grade 7 is 1.7 T_{d2} (6); and Grade 8 is 2.12 T_{d2} (6). It should be noted that these ratios are in terms of the Grade 6 pitch diameter tolerance for the external thread.

Major Diameter Tolerance of External Thread: T_d *(Table 10):* Grade 4 is 0.63 T_d (6); and Grade 8 is 1.6 T_d (6).

Pitch Diameter Tolerance of External Thread: T_{d2} *(Table 11):* Grade 3 is 0.5 T_{d2} (6); Grade 4 is 0.63 T_{d2} (6); Grade 5 is 0.8 T_{d2} (6); Grade 7 is 1.25 T_{d2} (6); Grade 8 is 1.6 T_{d2} (6); and Grade 9 is 2 T_{d2} (6).

Standard M Profile Screw Threads, Limits of Size.—The limiting M profile for internal threads is shown in Fig. 7 with associated dimensions for standard sizes in Table 12. The limiting M profiles for external threads are shown in Fig. 8 with associated dimensions for standard sizes in Table 13.

Table 12. Internal Metric Thread - M Profile Limiting Dimensions, ANSI/ASME B1.13M-1983 (R1995)

Basic Thread Designation	Toler. Class	Minor Diameter D_1		Pitch Diameter D_2			Major Diameter D	
		Min	Max	Min	Max	Tol	Min	Max[a]
M1.6 × 0.35	6H	1.221	1.321	1.373	1.458	0.085	1.600	1.736
M2 × 0.4	6H	1.567	1.679	1.740	1.830	0.090	2.000	2.148
M2.5 × 0.45	6H	2.013	2.138	2.208	2.303	0.095	2.500	2.660
M3 × 0.5	6H	2.459	2.599	2.675	2.775	0.100	3.000	3.172
M3.5 × 0.6	6H	2.850	3.010	3.110	3.222	0.112	3.500	3.699
M4 × 0.7	6H	3.242	3.422	3.545	3.663	0.118	4.000	4.219
M5 × 0.8	6H	4.134	4.334	4.480	4.605	0.125	5.000	5.240
M6 × 1	6H	4.917	5.153	5.350	5.500	0.150	6.000	6.294
M8 × 1.25	6H	6.647	6.912	7.188	7.348	0.160	8.000	8.340
M8 × 1	6H	6.917	7.153	7.350	7.500	0.150	8.000	8.294
M10 × 1.5	6H	8.376	8.676	9.026	9.206	0.180	10.000	10.396
M10 × 1.25	6H	8.647	8.912	9.188	9.348	0.160	10.000	10.340
M10 × 0.75	6H	9.188	9.378	9.513	9.645	0.132	10.000	10.240
M12 × 1.75	6H	10.106	10.441	10.863	11.063	0.200	12.000	12.453
M12 × 1.5	6H	10.376	10.676	11.026	11.216	0.190	12.000	12.406
M12 × 1.25	6H	10.647	10.912	11.188	11.368	0.180	12.000	12.360
M12 × 1	6H	10.917	11.153	11.350	11.510	0.160	12.000	12.304
M14 × 2	6H	11.835	12.210	12.701	12.913	0.212	14.000	14.501
M14 × 1.5	6H	12.376	12.676	13.026	13.216	0.190	14.000	14.406
M15 × 1	6H	13.917	14.153	14.350	14.510	0.160	15.000	15.304
M16 × 2	6H	13.835	14.210	14.701	14.913	0.212	16.000	16.501
M16 × 1.5	6H	14.376	14.676	15.026	15.216	0.190	16.000	16.406
M17 × 1	6H	15.917	16.153	16.350	16.510	0.160	17.000	17.304
M18 × 1.5	6H	16.376	16.676	17.026	17.216	0.190	18.000	18.406
M20 × 2.5	6H	17.294	17.744	18.376	18.600	0.224	20.000	20.585
M20 × 1.5	6H	18.376	18.676	19.026	19.216	0.190	20.000	20.406
M20 × 1	6H	18.917	19.153	19.350	19.510	0.160	20.000	20.304
M22 × 2.5	6H	19.294	19.744	20.376	20.600	0.224	22.000	22.585
M22 × 1.5	6H	20.376	20.676	21.026	21.216	0.190	22.000	22.406
M24 × 3	6H	20.752	21.252	22.051	22.316	0.265	24.000	24.698
M24 × 2	6H	21.835	22.210	22.701	22.925	0.224	24.000	24.513

Table 12. *(Continued)* **Internal Metric Thread - M Profile Limiting Dimensions, ANSI/ASME B1.13M-1983 (R1995)**

Basic Thread Designation	Toler. Class	Minor Diameter D_1		Pitch Diameter D_2			Major Diameter D	
		Min	Max	Min	Max	Tol	Min	Max[a]
M25 × 1.5	6H	23.376	23.676	24.026	24.226	0.200	25.000	25.416
M27 × 3	6H	23.752	24.252	25.051	25.316	0.265	27.000	27.698
M27 × 2	6H	24.835	25.210	25.701	25.925	0.224	27.000	27.513
M30 × 3.5	6H	26.211	26.771	27.727	28.007	0.280	30.000	30.785
M30 × 2	6H	27.835	28.210	28.701	28.925	0.224	30.000	30.513
M30 × 1.5	6H	28.376	28.676	29.026	29.226	0.200	30.000	30.416
M33 × 2	6H	30.835	31.210	31.701	31.925	0.224	33.000	33.513
M35 × 1.5	6H	33.376	33.676	34.026	34.226	0.200	35.000	35.416
M36 × 4	6H	31.670	32.270	33.402	33.702	0.300	36.000	36.877
M36 × 2	6H	33.835	34.210	34.701	34.925	0.224	36.000	36.513
M39 × 2	6H	36.835	37.210	37.701	37.925	0.224	39.000	39.513
M40 × 1.5	6H	38.376	38.676	39.026	39.226	0.200	40.000	40.416
M42 × 4.5	6H	37.129	37.799	39.077	39.392	0.315	42.000	42.965
M42 × 2	6H	39.835	40.210	40.701	40.925	0.224	42.000	42.513
M45 × 1.5	6H	43.376	43.676	44.026	44.226	0.200	45.000	45.416
M48 × 5	6H	42.587	43.297	44.752	45.087	0.335	48.000	49.057
M48 × 2	6H	45.835	46.210	46.701	46.937	0.236	48.000	48.525
M50 × 1.5	6H	48.376	48.676	49.026	49.238	0.212	50.000	50.428
M55 × 1.5	6H	53.376	53.676	54.026	54.238	0.212	55.000	55.428
M56 × 5.5	6H	50.046	50.796	52.428	52.783	0.355	56.000	57.149
M56 × 2	6H	53.835	54.210	54.701	54.937	0.236	56.000	56.525
M60 × 1.5	6H	58.376	58.676	59.026	59.238	0.212	60.000	60.428
M64 × 6	6H	57.505	58.305	60.103	60.478	0.375	64.000	65.241
M64 × 2	6H	61.835	62.210	62.701	62.937	0.236	64.000	64.525
M65 × 1.5	6H	63.376	63.676	64.026	64.238	0.212	65.000	65.428
M70 × 1.5	6H	68.376	68.676	69.026	69.238	0.212	70.000	70.428
M72 × 6	6H	65.505	66.305	68.103	68.478	0.375	72.000	73.241
M72 × 2	6H	69.835	70.210	70.701	70.937	0.236	72.000	72.525
M75 × 1.5	6H	73.376	73.676	74.026	74.238	0.212	75.000	75.428
M80 × 6	6H	73.505	74.305	76.103	76.478	0.375	80.000	81.241
M80 × 2	6H	77.835	78.210	78.701	78.937	0.236	80.000	80.525
M80 × 1.5	6H	78.376	78.676	79.026	79.238	0.212	80.000	80.428
M85 × 2	6H	82.835	83.210	83.701	83.937	0.236	85.000	85.525
M90 × 6	6H	83.505	84.305	86.103	86.478	0.375	90.000	91.241
M90 × 2	6H	87.835	88.210	88.701	88.937	0.236	90.000	90.525
M95 × 2	6H	92.835	93.210	93.701	93.951	0.250	95.000	95.539
M100 × 6	6H	93.505	94.305	96.103	96.503	0.400	100.000	101.266
M100 × 2	6H	97.835	98.210	98.701	98.951	0.250	100.000	100.539
M105 × 2	6H	102.835	103.210	103.701	103.951	0.250	105.000	105.539
M110 × 2	6H	107.835	108.210	108.701	108.951	0.250	110.000	110.539
M120 × 2	6H	117.835	118.210	118.701	118.951	0.250	120.000	120.539
M130 × 2	6H	127.835	128.210	128.701	128.951	0.250	130.000	130.539
M140 × 2	6H	137.835	138.210	138.701	138.951	0.250	140.000	140.539
M150 × 2	6H	147.835	148.210	148.701	148.951	0.250	150.000	150.539
M160 × 3	6H	156.752	157.252	158.051	158.351	0.300	160.000	160.733
M170 × 3	6H	166.752	167.252	168.051	168.351	0.300	170.000	170.733
M180 × 3	6H	176.752	177.252	178.051	178.351	0.300	180.000	180.733
M190 × 3	6H	186.752	187.252	188.051	188.386	0.335	190.000	190.768
M200 × 3	6H	196.752	197.252	198.051	198.386	0.335	200.000	200.768

[a] This reference dimension is used in design of tools, etc., and is not normally specified. Generally, major diameter acceptance is based upon maximum material condition gaging.

All dimensions are in millimeters.

Table 13. External Metric Thread—M Profile Limiting Dimensions ANSI/ASME B1.13M-1983 (R1995)

Basic Thread Desig.	Toler. Class	Allow. es[a]	Major Diam.[b] d		Pitch Diam.[b] d_2			Minor-Diam., d_1[b]	Minor Diam., d_3[c]
			Max	Min	Max	Min	Tol.	Max	Min
M1.6 × 0.35	6g	0.019	1.581	1.496	1.354	1.291	0.063	1.202	1.075
M1.6 × 0.35	4g6g	0.019	1.581	1.496	1.354	1.314	0.040	1.202	1.098
M2 × 0.4	6g	0.019	1.981	1.886	1.721	1.654	0.067	1.548	1.408
M2 × 0.4	4g6g	0.019	1.981	1.886	1.721	1.679	0.042	1.548	1.433
M2.5 × 0.45	6g	0.020	2.480	2.380	2.188	2.117	0.071	1.993	1.840
M2.5 × 0.45	4g6g	0.020	2.480	2.380	2.188	2.143	0.045	1.993	1.866
M3 × 0.5	6g	0.020	2.980	2.874	2.655	2.580	0.075	2.439	2.272
M3 × 0.5	4g6g	0.020	2.980	2.874	2.655	2.607	0.048	2.439	2.299
M3.5 × 0.6	6g	0.021	3.479	3.354	3.089	3.004	0.085	2.829	2.635
M3.5 × 0.6	4g6g	0.021	3.479	3.354	3.089	3.036	0.053	2.829	2.667
M4 × 0.7	6g	0.022	3.978	3.838	3.523	3.433	0.090	3.220	3.002
M4 × 0.7	4g6g	0.022	3.978	3.838	3.523	3.467	0.056	3.220	3.036
M5 × 0.8	6g	0.024	4.976	4.826	4.456	4.361	0.095	4.110	3.869
M5 × 0.8	4g6g	0.024	4.976	4.826	4.456	4.396	0.060	4.110	3.904
M6 × 1	6g	0.026	5.974	5.794	5.324	5.212	0.112	4.891	4.596
M6 × 1	4g6g	0.026	5.974	5.794	5.324	5.253	0.071	4.891	4.637
M8 × 1.25	6g	0.028	7.972	7.760	7.160	7.042	0.118	6.619	6.272
M8 × 1.25	4g6g	0.028	7.972	7.760	7.160	7.085	0.075	6.619	6.315
M8 × 1	6g	0.026	7.974	7.794	7.324	7.212	0.112	6.891	6.596
M8 × 1	4g6g	0.026	7.974	7.794	7.324	7.253	0.071	6.891	6.637
M10 × 1.5	6g	0.032	9.968	9.732	8.994	8.862	0.132	8.344	7.938
M10 × 1.5	4g6g	0.032	9.968	9.732	8.994	8.909	0.085	8.344	7.985
M10 × 1.25	6g	0.028	9.972	9.760	9.160	9.042	0.118	8.619	8.272
M10 × 1.25	4g6g	0.028	9.972	9.760	9.160	9.085	0.075	8.619	8.315
M10 × 0.75	6g	0.022	9.978	9.838	9.491	9.391	0.100	9.166	8.929
M10 × 0.75	4g6g	0.022	9.978	9.838	9.491	9.428	0.063	9.166	8.966
M12 × 1.75	6g	0.034	11.966	11.701	10.829	10.679	0.150	10.072	9.601
M12 × 1.75	4g6g	0.034	11.966	11.701	10.829	10.734	0.095	10.072	9.656
M12 × 1.5	6g	0.032	11.968	11.732	10.994	10.854	0.140	10.344	9.930
M12 × 1.25	6g	0.028	11.972	11.760	11.160	11.028	0.132	10.619	10.258
M12 × 1.25	4g6g	0.028	11.972	11.760	11.160	11.075	0.085	10.619	10.305
M12 × 1	6g	0.026	11.974	11.794	11.324	11.206	0.118	10.891	10.590
M12 × 1	4g6g	0.026	11.974	11.794	11.324	11.249	0.075	10.891	10.633
M14 × 2	6g	0.038	13.962	13.682	12.663	12.503	0.160	11.797	11.271
M14 × 2	4g6g	0.038	13.962	13.682	12.663	12.563	0.100	11.797	11.331
M14 × 1.5	6g	0.032	13.968	13.732	12.994	12.854	0.140	12.344	11.930
M14 × 1.5	4g6g	0.032	13.968	13.732	12.994	12.904	0.090	12.344	11.980
M15 × 1	6g	0.026	14.974	14.794	14.324	14.206	0.118	13.891	13.590
M15 × 1	4g6g	0.026	14.974	14.794	14.324	14.249	0.075	13.891	13.633
M16 × 2	6g	0.038	15.962	15.682	14.663	14.503	0.160	13.797	13.271
M16 × 2	4g6g	0.038	15.962	15.682	14.663	14.563	0.100	13.797	13.331
M16 × 1.5	6g	0.032	15.968	15.732	14.994	14.854	0.140	14.344	13.930
M16 × 1.5	4g6g	0.032	15.968	15.732	14.994	14.904	0.090	14.344	13.980
M17 × 1	6g	0.026	16.974	16.794	16.324	16.206	0.118	15.891	15.590
M17 × 1	4g6g	0.026	16.974	16.794	16.324	16.249	0.075	15.891	15.633
M18 × 1.5	6g	0.032	17.968	17.732	16.994	16.854	0.140	16.344	15.930
M18 × 1.5	4g6g	0.032	17.968	17.732	16.994	16.904	0.090	16.344	15.980

Table 13. *(Continued)* **External Metric Thread—M Profile Limiting Dimensions ANSI/ASME B1.13M-1983 (R1995)**

Basic Thread Desig.	Toler. Class	Allow. es[a]	Major Diam.[b] d		Pitch Diam.[b] d_2			Minor-Diam., d_1[b]	Minor Diam., d_3[c]
			Max	Min	Max	Min	Tol.	Max	Min
M20 × 2.5	6g	0.042	19.958	19.623	18.334	18.164	0.170	17.252	16.624
M20 × 2.5	4g6g	0.042	19.958	19.623	18.334	18.228	0.106	17.252	16.688
M20 × 1.5	6g	0.032	19.968	19.732	18.994	18.854	0.140	18.344	17.930
M20 × 1.5	4g6g	0.032	19.968	19.732	18.994	18.904	0.090	18.344	17.980
M20 × 1	6g	0.026	19.974	19.794	19.324	19.206	0.118	18.891	18.590
M20 × 1	4g6g	0.026	19.974	19.794	19.324	19.249	0.075	18.891	18.633
M22 × 2.5	6g	0.042	21.958	21.623	20.334	20.164	0.170	19.252	18.624
M22 × 1.5	6g	0.032	21.968	21.732	20.994	20.854	0.140	20.344	19.930
M22 × 1.5	4g6g	0.032	21.968	21.732	20.994	20.904	0.090	20.344	19.980
M24 × 3	6g	0.048	23.952	23.577	22.003	21.803	0.200	20.704	19.955
M24 × 3	4g6g	0.048	23.952	23.557	22.003	21.878	0.125	20.704	20.030
M24 × 2	6g	0.038	23.962	23.682	22.663	22.493	0.170	21.797	21.261
M24 × 2	4g6g	0.038	23.962	23.682	22.663	22.557	0.106	21.797	21.325
M25 × 1.5	6g	0.032	24.968	24.732	23.994	23.844	0.150	23.344	22.920
M25 × 1.5	4g6g	0.032	24.968	24.732	23.994	23.899	0.095	23.344	22.975
M27 × 3	6g	0.048	26.952	26.577	25.003	24.803	0.200	23.704	22.955
M27 × 2	6g	0.038	26.962	26.682	25.663	25.493	0.170	24.797	24.261
M27 × 2	4g6g	0.038	29.962	26.682	25.663	25.557	0.106	24.797	24.325
M30 × 3.5	6g	0.053	29.947	29.522	27.674	27.462	0.212	26.158	25.306
M30 × 3.5	4g6g	0.053	29.947	29.522	27.674	27.542	0.132	26.158	25.386
M30 × 2	6g	0.038	29.962	29.682	28.663	28.493	0.170	27.797	27.261
M30 × 2	4g6g	0.038	29.962	29.682	28.663	28.557	0.106	27.797	27.325
M30 × 1.5	6g	0.032	29.968	29.732	28.994	28.844	0.150	28.344	27.920
M30 × 1.5	4g6g	0.032	29.968	29.732	28.994	28.899	0.095	28.344	27.975
M33 × 2	6g	0.038	32.962	32.682	31.663	31.493	0.170	30.797	30.261
M33 × 2	4g6g	0.038	32.962	32.682	31.663	31.557	0.106	30.797	30.325
M35 × 1.5	6g	0.032	34.968	34.732	33.994	33.844	0.150	33.344	33.920
M36 × 4	6g	0.060	35.940	35.465	33.342	33.118	0.224	31.610	30.654
M36 × 4	4g6g	0.060	35.940	35.465	33.342	33.202	0.140	31.610	30.738
M36 × 2	6g	0.038	35.962	35.682	34.663	34.493	0.170	33.797	33.261
M36 × 2	4g6g	0.038	35.962	35.682	34.663	34.557	0.106	33.797	33.325
M39 × 2	6g	0.038	38.962	38.682	37.663	37.493	0.170	36.797	36.261
M39 × 2	4g6g	0.038	38.962	38.682	37.663	37.557	0.106	36.797	36.325
M40 × 1.5	6g	0.032	39.968	39.732	38.994	38.844	0.150	38.344	37.920
M40 × 1.5	4g6g	0.032	39.968	39.732	38.994	38.899	0.095	38.344	37.975
M42 × 4.5	6g	0.063	41.937	41.437	39.014	38.778	0.236	37.066	36.006
M42 × 4.5	4g6g	0.063	41.937	41.437	39.014	38.864	0.150	37.066	36.092
M42 × 2	6g	0.038	41.962	41.682	40.663	40.493	0.170	39.797	39.261
M42 × 2	4g6g	0.038	41.962	41.682	40.663	40.557	0.106	39.797	39.325
M45 × 1.5	6g	0.032	44.968	44.732	43.994	43.844	0.150	43.344	42.920
M45 × 1.5	4g6g	0.032	44.968	44.732	43.994	43.899	0.095	43.344	42.975
M48 × 5	6g	0.071	47.929	47.399	44.681	44.431	0.250	42.516	41.351
M48 × 5	4g6g	0.071	47.929	47.399	44.681	44.521	0.160	42.516	41.441
M48 × 2	6g	0.038	47.962	47.682	46.663	46.483	0.180	45.797	45.251
M48 × 2	4g6g	0.038	47.962	47.682	46.663	46.551	0.112	45.797	45.319
M50 × 1.5	6g	0.032	49.968	49.732	48.994	48.834	0.160	48.344	47.910
M50 × 1.5	4g6g	0.032	49.968	49.732	48.994	48.894	0.100	48.344	47.970
M55 × 1.5	6g	0.032	54.968	54.732	53.994	53.834	0.160	53.344	52.910

Table 13. *(Continued)* **External Metric Thread—M Profile Limiting Dimensions ANSI/ASME B1.13M-1983 (R1995)**

Basic Thread Desig.	Toler. Class	Allow. es^a	Major Diam.[b] d		Pitch Diam.[b] d_2			Minor-Diam., d_1[b]	Minor Diam., d_3[c]
			Max	Min	Max	Min	Tol.	Max	Min
M55 × 1.5	4g6g	0.032	54.968	54.732	53.994	53.894	0.100	53.344	52.970
M56 × 5.5	6g	0.075	55.925	55.365	52.353	52.088	0.265	49.971	48.700
M56 × 5.5	4g6g	0.075	55.925	55.365	52.353	52.183	0.170	49.971	48.795
M56 × 2	6g	0.038	55.962	55.682	54.663	54.483	0.180	53.797	53.251
M56 × 2	4g6g	0.038	55.962	55.682	54.663	54.551	0.112	53.797	53.319
M60 × 1.5	6	0.032	59.968	59.732	58.994	58.834	0.160	58.344	57.910
M60 × 1.5	4g6g	0.032	59.968	59.732	58.994	58.894	0.100	58.344	57.970
M64 × 6	6g	0.080	63.920	63.320	60.023	59.743	0.280	57.425	56.047
M64 × 6	4g6g	0.080	63.920	63.320	60.023	59.843	0.180	57.425	56.147
M64 × 2	6g	0.038	63.962	63.682	62.663	62.483	0.180	61.797	61.251
M64 × 2	4g6g	0.038	63.962	63.682	62.663	62.551	0.112	61.797	61.319
M65 × 1.5	6g	0.032	64.968	64.732	63.994	63.834	0.160	63.344	62.910
M65 × 1.5	4g6g	0.032	64.968	64.732	63.994	63.894	0.100	63.344	62.970
M70 × 1.5	6g	0.032	69.968	69.732	68.994	68.834	0.160	68.344	67.910
M70 × 1.5	4g6g	0.032	69.968	69.732	68.994	68.894	0.100	68.344	67.970
M72 × 6	6g	0.080	71.920	71.320	68.023	67.743	0.280	65.425	64.047
M72 × 6	4g6g	0.080	71.920	71.320	68.023	67.843	0.180	65.425	64.147
M72 × 2	6g	0.038	71.962	71.682	70.663	70.483	0.180	69.797	69.251
M72 × 2	4g6g	0.038	71.962	71.682	70.663	70.551	0.112	69.797	69.319
M75 × 1.5	6g	0.032	74.968	74.732	73.994	73.834	0.160	73.344	72.910
M75 × 1.5	4g6g	0.032	74.968	74.732	73.994	73.894	0.100	73.344	72.970
M80 × 6	6g	0.080	79.920	79.320	76.023	75.743	0.280	73.425	72.047
M80 × 6	4g6g	0.080	79.920	79.320	76.023	75.843	0.180	73.425	72.147
M80 × 2	6g	0.038	79.962	79.682	78.663	78.483	0.180	77.797	77.251
M80 × 2	4g6g	0.038	79.962	79.682	78.663	78.551	0.112	77.797	77.319
M80 × 1.5	6g	0.032	79.968	79.732	78.994	78.834	0.160	78.344	77.910
M80 × 1.5	4g6g	0.032	79.968	79.732	78.994	78.894	0.100	78.334	77.970
M85 × 2	6g	0.038	84.962	84.682	83.663	83.483	0.180	82.797	82.251
M85 × 2	4g6g	0.038	84.962	84.682	83.663	83.551	0.112	82.797	82.319
M90 × 6	6g	0.080	89.920	89.320	86.023	85.743	0.280	83.425	82.047
M90 × 6	4g6g	0.080	89.920	89.320	86.023	85.843	0.180	83.425	82.147
M90 × 2	6g	0.038	89.962	89.682	88.663	88.483	0.180	87.797	87.251
M90 × 2	4g6g	0.038	89.962	89.682	88.663	88.551	0.112	87.797	87.319
M95 × 2	6g	0.038	94.962	94.682	93.663	93.473	0.190	92.797	92.241
M95 × 2	4g6g	0.038	94.962	94.682	93.663	93.545	0.118	92.797	92.313
M100 × 6	6g	0.080	99.920	99.320	96.023	95.723	0.300	93.425	92.027
M100 × 6	4g6g	0.080	99.920	99.320	96.023	95.833	0.190	93.425	92.137
M100 × 2	6g	0.038	99.962	99.682	98.663	98.473	0.190	97.797	97.241
M100 × 2	4g6g	0.038	99.962	99.682	98.663	98.545	0.118	97.797	97.313
M105 × 2	6g	0.038	104.962	104.682	103.663	103.473	0.190	102.797	102.241
M105 × 2	4g6g	0.038	104.962	104.682	103.663	103.545	0.118	102.797	102.313
M110 × 2	6g	0.038	109.962	109.682	108.663	108.473	0.190	107.797	107.241
M110 × 2	4g6g	0.038	109.962	109.682	108.663	108.545	0.118	107.797	107.313
M120 × 2	6g	0.038	119.962	119.682	118.663	118.473	0.190	117.797	117.241
M120 × 2	4g6g	0.038	119.962	119.682	118.663	118.545	0.118	117.797	117.313
M130 × 2	6g	0.038	129.962	129.682	128.663	128.473	0.190	127.797	127.241
M130 × 2	4g6g	0.038	139.962	139.682	138.663	138.545	0.118	137.797	137.313
M140 × 2	6g	0.038	139.962	139.682	138.663	138.473	0.190	137.797	137.241

Table 13. *(Continued)* **External Metric Thread—M Profile Limiting Dimensions ANSI/ASME B1.13M-1983 (R1995)**

Basic Thread Desig.	Toler. Class	Allow. es[a]	Major Diam.[b] d		Pitch Diam.[b] d_2			Minor-Diam., d_1[b]	Minor Diam., d_3[c]
			Max	Min	Max	Min	Tol.	Max	Min
M140 × 2	4g6g	0.038	139.962	139.682	138.663	138.545	0.118	137.797	137.313
M150 × 2	6g	0.038	149.962	149.682	148.663	148.473	0.190	147.797	147.241
M150 × 2	4g6g	0.038	149.962	149.682	148.663	148.545	0.118	147.797	147.313
M160 × 3	6g	0.048	159.952	159.577	158.003	157.779	0.224	156.704	155.931
M160 × 3	4g6g	0.048	159.952	159.577	158.003	157.863	0.140	156.704	156.015
M170 × 3	6g	0.048	169.952	169.577	168.003	167.779	0.224	166.704	165.931
M170 × 3	4g6g	0.048	169.952	169.577	168.003	167.863	0.140	166.704	166.015
M180 × 3	6g	0.048	179.952	179.577	178.003	177.779	0.224	176.704	175.931
M180 × 3	4g6g	0.048	179.952	179.577	178.003	177.863	0.140	176.704	176.015
M190 × 3	6g	0.048	189.952	189.577	188.003	187.753	0.250	186.704	185.905
M190 × 3	4g6g	0.048	189.952	189.577	188.003	187.843	0.160	186.704	185.995
M200 × 3	6g	0.048	199.952	199.577	198.003	197.753	0.250	196.704	195.905
M200 × 3	4g6g	0.048	199.952	199.577	198.003	197.843	0.160	196.704	195.995

[a] *es* is an absolute value.

[b] (Flat form) For screw threads at maximum limits of tolerance position *h*, add the absolute value *es* to the maximum diameters required. For maximum major diameter this value is the basic thread size listed in Table 12 as Minimum Major Diameter (D_{min}; for maximum pitch diameter this value is the same as listed in Table 12 as Minimum Pitch Diameter ($D_{2\,min}$); and for maximum minor diameter this value is the same as listed in Table 12 as Minimum Minor Diameter ($D_{1\,min}$).

[c] (Rounded form) This reference dimension is used in the design of tools, etc. In dimensioning external threads it is not normally specified. Generally minor diameter acceptance is based upon maximum material condition gaging.

All dimensions are in millimeters.

If the required values are not listed in these tables, they may be calculated using the data in Tables 3, 6, 7, 8, 9, 10, and 11 together with the preceding formulas. If the required data are not included in any of the tables listed above, reference should be made to Sections 6 and 9.3 of ANSI/ASME B1.13M, which gives design formulas.

Metric Screw Thread Designations.—Metric screw threads are identified by the letter (M) for the thread form profile, followed by the nominal diameter size and the pitch expressed in millimeters, separated by the sign (×) and followed by the tolerance class separated by a dash (–) from the pitch.

The simplified international practice for designating coarse pitch M profile metric screw threads is to leave off the pitch. Thus a M14 × 2 thread is designated just M14. However, to prevent misunderstanding, it is mandatory to use the value for pitch in all designations.

Thread acceptability gaging system requirements of ANSI B1.3M may be added to the thread size designation as noted in the examples (numbers in parentheses) or as specified in pertinent documentation, such as the drawing or procurement document.

Unless otherwise specified in the designation, the screw thread is right hand.

Examples: External thread of M profile, right hand: M6 × 1 – 4g6g (22)

Internal thread of M profile, right hand: M6 × 1 – 5H6H (21)

Designation of Left Hand Thread: When a left hand thread is specified, the tolerance class designation is followed by a dash and LH.

Example: M6 × 1 – 5H6H – LH (23)

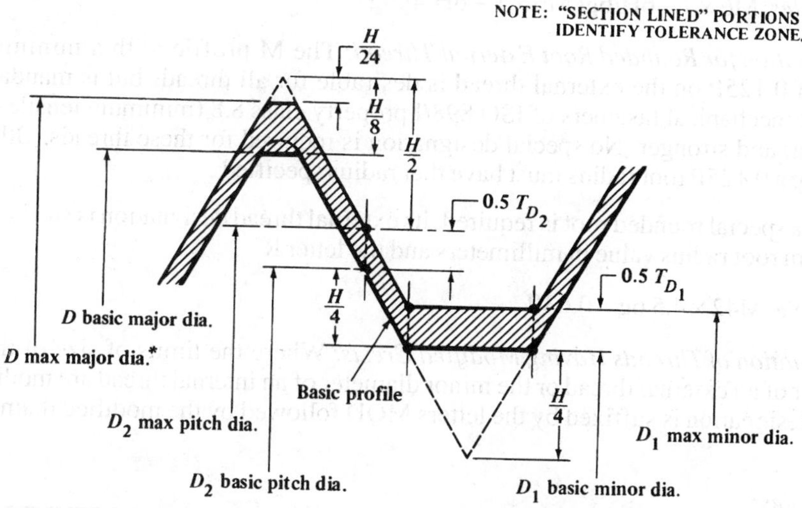

Fig. 7. Internal Thread — Limiting M Profile. Tolerance Position H

*This demension is used in the design of tools, etc. In demsioning internal threads it is not normally specified. Generally, major diameter acceptance is based on maximum material condition gaging.

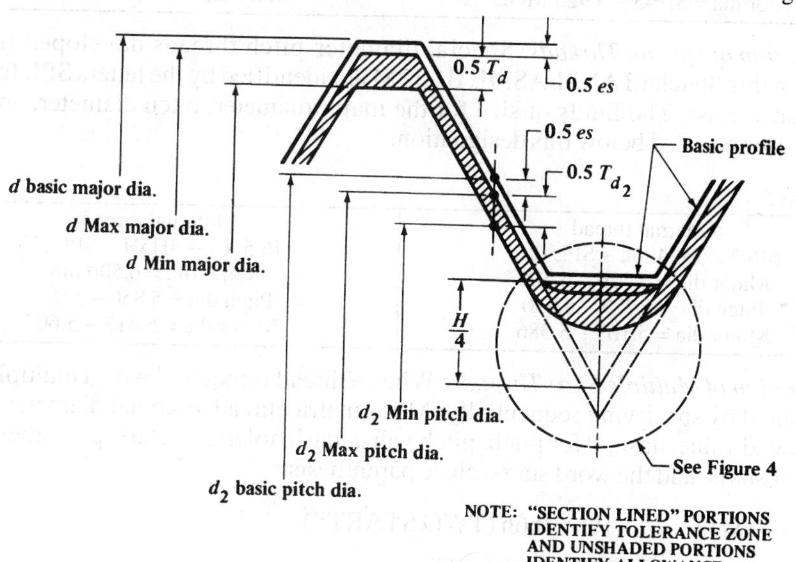

Fig. 8. External Thread — Limiting M Profile. Tolerance Position g

Designation for Identical Tolerance Classes: If the two tolerance class designations for a thread are identical, it is not necessary to repeat the symbols.

Example: M6 × 1 – 6H (21)

Designation Using All Capital Letters: When computer and teletype thread designations use all capital letters, the external or internal thread may need further identification. Thus the tolerance class is followed by the abbreviations EXT or INT in capital letters.

Examples: M6 × 1 – 4G6G EXT; M6 × 1 – 6H INT

Designation for Thread Fit: A fit between mating threads is indicated by the internal thread tolerance class followed by the external thread tolerance class and separated by a slash.

Examples: M6 × 1 – 6H/6g; M6 × 1 – 6H/4g6g

Designation for Rounded Root External Thread: The M profile with a minimum root radius of 0.125P on the external thread is desirable for all threads but is mandatory for threaded mechanical fasteners of ISO 898/I property class 8.8 (minimum tensile strength 800 MPa) and stronger. No special designation is required for these threads. Other parts requiring a 0.125P root radius must have that radius specified.

When a special rounded root is required, its external thread designation is suffixed by the minimum root radius value in millimeters and the letter R.

Example: M42 × 4.5 6g – 0.63R

Designation of Threads Having Modified Crests: Where the limits of size of the major diameter of an external thread or the minor diameter of an internal thread are modified, the thread designation is suffixed by the letters MOD followed by the modified diameter limits.

Examples:

External thread M profile, major diameter reduced 0.075 mm. M6 × 1 – 4h6h MOD Major dia = 5.745 – 5.925 MOD	Internal thread M profile, minor diameter increased 0.075 mm. M6 × 1 – 4H5H MOD Minor dia = 5.101 – 5.291 MOD

Designation of Special Threads: Special diameter-pitch threads developed in accordance with this Standard ANSI/ASME B1.13M are identified by the letters SPL following the tolerance class. The limits of size for the major diameter, pitch diameter, and minor diameter are specified below this designation.

Example:

External thread	Internal thread
M6.5 × 1 – 4h6h – SPL (22)	M6.5 × 1 – 4H5H – SPL (23)
Major dia = 6.320 – 6.500	Major dia = 6.500 min
Pitch dia = 5.779 – 5.850	Pitch dia = 5.850 – 5.945
Minor dia = 5.163 – 5.386	Minor dia = 5.417 – 5.607

Designation of Multiple Start Threads: When a thread is required with a multiple start, it is designated by specifying sequentially: M for metric thread, nominal diameter size, × L for lead, lead value, dash, P for pitch, pitch value, dash, tolerance class, parenthesis, script number of starts, and the word starts, close parenthesis.

Examples: M16 × L4 – P2 – 4h6h (TWO STARTS)

M14 × L6 – P2 – 6H (THREE STARTS)

Designation of Coated or Plated Threads: In designating coated or plated M threads the tolerance class should be specified as after coating or after plating. If no designation of after coating or after plating is specified, the tolerance class applies before coating or plating in accordance with ISO practice. After plating, the thread must not transgress the maximum material limits for the tolerance position H/h.

Examples: M6 × 1 – 6h AFTER COATING or AFTER PLATING

M6 × 1 – 6g AFTER COATING or AFTER PLATING

Where the tolerance position G/g is insufficient relief for the application to hold the threads within product limits, the coating or plating allowance may be specified as the maximum and minimum limits of size for minor and pitch diameters of internal threads or major and pitch diameters for external threads before coating or plating.

Example: Allowance on external thread M profile based on 0.010 mm minimum coating thickness.

M6 × 1 – 4h6h – AFTER COATING

BEFORE COATING

Majordia = 5.780 – 5.940 Pitchdia = 5.239 – 5.290

Metric Screw Threads—MJ Profile.—The MJ screw thread is intended for aerospace metric threaded parts and for other highly stressed applications requiring high fatigue strength, or for "no allowance" applications. The MJ profile thread is a hard metric version similar to the UNJ inch, MIL-S-8879, which has a 0.15*P* to 0.18*P* controlled root radius in the external thread and the internal thread minor diameter truncated to accommodate the external thread maximum root radius.

The American National Standard ANSI/ASME B1.21M-1978 establishes the basic triangular profile for the MJ form of thread; gives a system of designations; lists the standard series of diameter-pitch combinations for diameters from 1.6 to 200 mm; and specifies limiting dimensions and tolerances.

Diameter-Pitch Combinations.—This Standard includes a selected series of diameter-pitch combinations of threads taken from International Standard ISO 261 plus some additional sizes in the constant pitch series. It also includes the standard series of diameter-pitch combinations for aerospace screws, bolts, and nuts as shown below.

American National Standard Thread Series for Aerospace Screws, Bolts, and Nuts
ANSI/ASME B1.21M-1978

Nom.Size	Pitch	Nom.Size	Pitch	Nom.Size	Pitch	Nom.Size	Pitch
1.6	0.35	5	0.8	14	1.5	27	2
2	0.4	6	1	16	1.5	30	2
2.5	0.45	7	1	18	1.5	33	2
3	0.5	8	1	20	1.5	36	2
3.5	0.6	10	1.25	22	1.5	39	2
4	0.7	12	1.25	24	2	…	…

All dimensions are in millimeters.

Tolerances: One tolerance class, 4h6h, is specified in this Standard for all sizes of external threads after processing, including coating or electroplating. The tolerance position h provides no allowance. The pitch diameter tolerance is grade 4 and the major diameter tolerance is grade 6. For coated or plated external threads having pitches of 2 mm or smaller, the tolerance class before processing is applied is 4g6g. The tolerance position g provides an allowance for coating or plating only. For pitches larger than 2 mm, special allowances are provided.

For internal threads, after all processing including coating or plating has been completed, tolerance class 4H6H is specified for sizes 1 through 5 mm and 4H5H for sizes 6 mm and larger. The tolerance position H provides no allowance. The pitch diameter tolerance is grade 4 for all sizes and the minor diameter tolerance is grade 6 for the 5 mm size and smaller and grade 5 for the 6 mm size and larger. For coated or plated internal threads having pitches 2 mm or smaller, the tolerance class is 4G6G for sizes 1 through 5 mm and 4G5G for sizes 6 mm and larger. The tolerance position G provides an allowance for coating or plating.

The above class tolerances are positive for internal threads and negative for external threads, that is, in the direction of minimum material.

Symbols: Standard symbols appearing in the following diagrams are:

D = Basic major diameter of internal thread

D_2 = Basic pitch diameter of internal thread

D_l = Basic minor diameter of internal thread

d = Basic major diameter of external thread

d_2 = Basic pitch diameter of external thread

d_1 = Basic minor diameter of internal thread

d_3 = Diameter to bottom of external thread root radius

H = Height of fundamental triangle P = Pitch

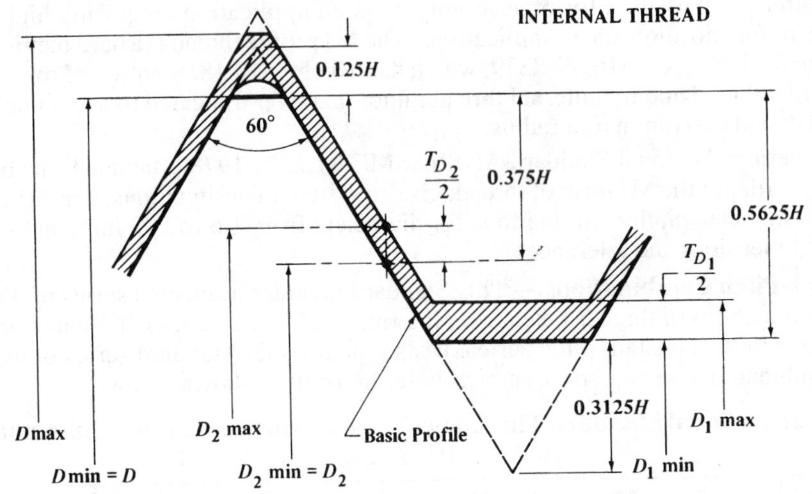

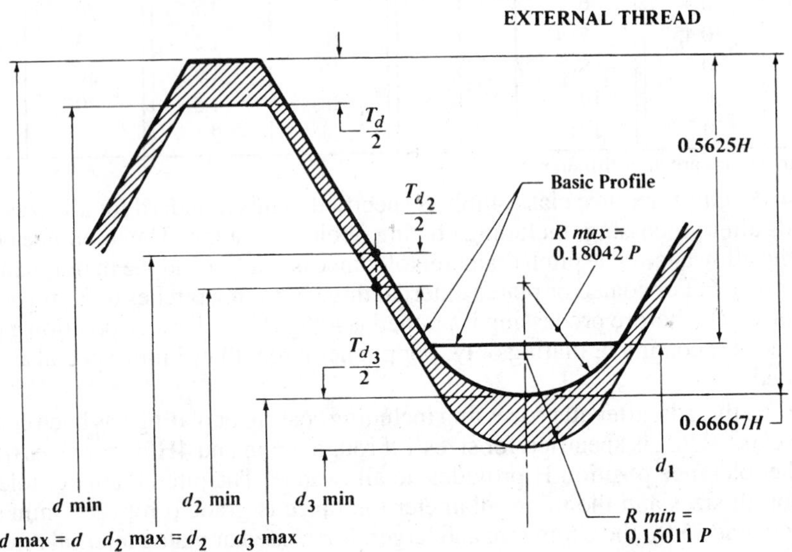

Internal MJ Thread Basic and Design Profiles (Above) and External MJ Thread Basic and Design Profiles (Below) Showing Tolerance Zones

Basic Designations: The aerospace metric screw thread is designated by the letters "MJ" to identify the metric J thread form, followed by the nominal size and pitch in millimeters (separated by the sign "×") and followed by the tolerance class (separated by a dash from the pitch). Unless otherwise specified in the designation, the thread helix is right hand. Example: MJ6 × 1 – 4h6h

For further details concerning limiting dimensions, allowances for coating and plating, modified and special threads, etc., reference should be made to the Standard.

Trapezoidal Metric Thread — Preferred Basic Sizes (DIN 103)

$H = 1.866P$

$h_s = 0.5P + a$

$h_e = 0.5P + a - b$

$h_n = 0.5P + 2a - b$

$h_{as} = 0.25P$

Nom. & Major Diam. of Bolt, D_s	Pitch, P	Pitch Diam., E	Depth of Engagement, h_e	Clearance a	Clearance b	Bolt Minor Diam., K_s	Bolt Depth of Thread, h_s	Nut Major Diam., D_n	Nut Minor Diam., K_n	Nut Depth of Thread, h_n
10	3	8.5	1.25	0.25	0.5	6.5	1.75	10.5	7.5	1.50
12	3	10.5	1.25	0.25	0.5	8.5	1.75	12.5	9.5	1.50
14	4	12	1.75	0.25	0.5	9.5	2.25	14.5	10.5	2.00
16	4	14	1.75	0.25	0.5	11.5	2.25	16.5	12.5	2.00
18	4	16	1.75	0.25	0.5	13.5	2.25	18.5	14.5	2.00
20	4	18	1.75	0.25	0.5	15.5	2.25	20.5	16.5	2.00
22	5	19.5	2	0.25	0.75	16.5	2.75	22.5	18	2.00
24	5	21.5	2	0.25	0.75	18.5	2.75	24.5	20	2.25
26	5	23.5	2	0.25	0.75	20.5	2.75	26.5	22	2.25
28	5	25.5	2	0.25	0.75	22.5	2.75	28.5	24	2.25
30	6	27	2.5	0.25	0.75	23.5	3.25	30.5	25	2.75
32	6	29	2.5	0.25	0.75	25.5	3.25	32.5	27	2.75
36	6	33	2.5	0.25	0.75	29.5	3.25	36.5	31	2.75
40	7	36.5	3	0.25	0.75	32.5	3.75	40.5	34	3.25
44	7	40.5	3	0.25	0.75	36.5	3.75	44.5	38	3.25
48	8	44	3.5	0.25	0.75	39.5	4.25	48.5	41	3.75
50	8	46	3.5	0.25	0.75	41.5	4.25	50.5	43	3.75
52	8	48	3.5	0.25	0.75	43.5	4.25	52.5	45	3.75
55	9	50.5	4	0.25	0.75	45.5	4.75	55.5	47	4.25
60	9	55.5	4	0.25	0.75	50.5	4.75	60.5	52	4.25
65	10	60	4.5	0.25	0.75	54.5	5.25	65.5	56	4.75
70	10	65	4.5	0.25	0.75	59.5	5.25	70.5	61	4.75
75	10	70	4.5	0.25	0.75	64.5	5.25	75.5	66	4.75
80	10	75	4.5	0.25	0.75	69.5	5.25	80.5	71	4.75
85	12	79	5.5	0.25	0.75	72.5	6.25	85.5	74	5.75
90	12	84	5.5	0.25	0.75	77.5	6.25	90.5	79	5.75
95	12	89	5.5	0.25	0.75	82.5	6.25	95.5	84	5.75
100	12	94	5.5	0.25	0.75	87.5	6.25	100.5	89	5.75
110	12	104	5.5	0.25	0.75	97.5	6.25	110.5	99	5.75
120	14	113	6	0.5	1.5	105	7.5	121	108	6.5
130	14	123	6	0.5	1.5	115	7.5	131	118	6.5
140	14	133	6	0.5	1.5	125	7.5	141	128	6.5
150	16	142	7	0.5	1.5	133	8.5	151	136	7.5
160	16	152	7	0.5	1.5	143	8.5	161	146	7.5
170	16	162	7	0.5	1.5	153	8.5	171	156	7.5
180	18	171	8	0.5	1.5	161	9.5	181	164	8.5
190	18	181	8	0.5	1.5	171	9.5	191	174	8.5
200	18	191	8	0.5	1.5	181	9.5	201	184	8.5
210	20	200	9	0.5	1.5	189	10.5	211	192	9.5
220	20	210	9	0.5	1.5	199	10.5	221	202	9.5
230	20	220	9	0.5	1.5	209	10.5	231	212	9.5
240	22	229	10	0.5	1.5	217	11.5	241	220	10.5
250	22	239	10	0.5	1.5	227	11.5	251	230	10.5
260	22	249	10	0.5	1.5	237	11.5	261	240	10.5
270	24	258	11	0.5	1.5	245	12.5	271	248	11.5
280	24	268	11	0.5	1.5	255	12.5	281	258	11.5
290	24	278	11	0.5	1.5	265	12.5	291	268	11.5
300	26	287	12	0.5	1.5	273	13.5	301	276	12.5

All dimensions are in millimeters.

*Roots are rounded to a radius, r, equal to 0.25 mm for pitches of from 3 to 12 mm inclusive and 0.5 mm for pitches of from 14 to 26 mm inclusive for power transmission.

Metric Series Threads — A comparison of Maximum Metal Dimensions of British (B.S. 1095), French (NF E03-104), German (DIN 13), and Swiss (VSM 12003) Systems

Nominal Size and Major Bolt Diam.	Pitch	Pitch Diam.	Bolt				Nut				
			Minor Diameter				Major Diameter			Minor Diameter	
			British	French	German	Swiss	British & German	French	Swiss	French, German& Swiss	British
6	1	5.350	4.863	4.59	4.700	4.60	6.000	6.108	6.100	4.700	4.863
7	1	6.350	5.863	5.59	5.700	5.60	7.000	7.108	7.100	5.700	5.863
8	1.25	7.188	6.579	6.24	6.376	6.25	8.000	8.135	8.124	6.376	6.579
9	1.25	8.188	7.579	7.24	7.376	7.25	9.000	9.135	9.124	7.376	7.579
10	1.5	9.026	8.295	7.89	8.052	7.90	10.000	10.162	10.150	8.052	8.295
11	1.5	10.026	9.295	8.89	9.052	8.90	11.000	11.162	11.150	9.052	9.295
12	1.75	10.863	10.011	9.54	9.726	9.55	12.000	12.189	12.174	9.726	10.011
14	2	12.701	11.727	11.19	11.402	11.20	14.000	14.216	14.200	11.402	11.727
16	2	14.701	13.727	13.19	13.402	13.20	16.000	16.216	16.200	13.402	13.727
18	2.5	16.376	15.158	14.48	14.752	14.50	18.000	18.270	18.250	14.752	15.158
20	2.5	18.376	17.158	16.48	16.752	16.50	20.000	20.270	20.250	16.752	17.158
22	2.5	20.376	19.158	18.48	18.752	18.50	22.000	22.270	22.250	18.752	19.158
24	3	22.051	20.590	19.78	20.102	19.80	24.000	24.324	24.300	20.102[a]	20.590
27	3	25.051	23.590	22.78	23.102	22.80	27.000	27.324	27.300	23.102[b]	23.590
30	3.5	27.727	26.022	25.08	25.454	25.10	30.000	30.378	30.350	25.454	26.022
33	3.5	30.727	29.022	28.08	28.454	28.10	33.000	33.378	33.350	28.454	29.022
36	4	33.402	31.453	30.37	30.804	30.40	36.000	36.432	36.400	30.804	31.453
39	4	36.402	34.453	33.37	33.804	33.40	39.000	39.432	39.400	33.804	34.453
42	4.5	39.077	36.885	35.67	36.154	35.70	42.000	42.486	42.450	36.154	36.885
45	4.5	42.077	39.885	38.67	39.154	38.70	45.000	45.486	45.450	39.154	39.885
48	5	41.752	42.316	40.96	41.504	41.00	48.000	48.540	48.500	41.504	42.316
52	5	48.752	46.316	44.96	45.504	45.00	52.000	52.540	52.500	45.504	46.316
56	5.5	52.428	49.748	48.26	48.856	48.30	56.000	56.594	56.550	48.856	49.748
60	5.5	56.428	53.748	52.26	52.856	52.30	60.000	60.594	60.550	52.856	53.748

[a] The value shown is given in the German Standard; the value in the French Standard is 20.002; and in the Swiss Standard, 20.104.

[b] The value shown is given in the German Standard; the value in the French Standard is 23.002; and in the Swiss Standard, 23.104.

All dimensions are in mm.

MINIATURE AND INTERFERENCE FIT THREADS

American Standard for Unified Miniature Screw Threads.— This American Standard (B1.10-1958, R1988) introduces a new series to be known as Unified Miniature Screw Threads and intended for general purpose fastening screws and similar uses in watches, instruments, and miniature mechanisms. Use of this series is recommended on all new products in place of the many improvised and unsystematized sizes now in existence which have never achieved broad acceptance nor recognition by standardization bodies. The series covers a diameter range from 0.30 to 1.40 millimeters (0.0118 to 0.0551 inch) and thus supplements the Unified and American thread series which begins at 0.060 inch (number 0 of the machine screw series). It comprises a total of fourteen sizes which, together with their respective pitches, are those endorsed by the American-British-Canadian Conference of April 1955 as the basis for a Unified standard among the inch-using countries, and coincide with the corresponding range of sizes in ISO (International Organization for Standardization) Recommendation No. 68. Additionally, it utilizes thread forms which are compatible in all significant respects with both the Unified and ISO basic thread profiles. Thus, threads in this series are interchangeable with the corresponding sizes in both the American-British-Canadian and ISO standardization programs.

Basic Form of Thread: The basic profile by which the design forms of the threads covered by this standard are governed is shown in Table 1. The thread angle is 60 degrees and except for basic height and depth of engagement which are $0.52p$, instead of $0.54127p$, the basic profile for this thread standard is identical with the Unified and American basic thread form. The selection of 0.52 as the exact value of the coefficient for the height of this basic form is based on practical manufacturing considerations and a plan evolved to simplify calculations and achieve more precise agreement between the metric and inch dimensional tables.

Products made to this standard will be interchangeable with products made to other standards which allow a maximum depth of engagement (or combined addendum height) of $0.54127p$. The resulting difference is negligible (only 0.00025 inch for the coarsest pitch) and is completely offset by practical considerations in tapping, since internal thread heights exceeding $0.52p$ are avoided in these (Unified Miniature) small thread sizes in order to reduce excessive tap breakage.

Design Forms of Threads: The design (maximum material) forms of the external and internal threads are shown in Table 2. These forms are derived from the basic profile shown in Table 1 by the application of clearances for the crests of the addenda at the roots of the mating dedendum forms. Basic and design form dimensions are given in Table 3.

Nominal Sizes: The thread sizes comprising this series and their respective pitches are shown in the first two columns of Table 5. The fourteen sizes shown in Table 5 have been systematically distributed to provide a uniformly proportioned selection over the entire range. They are separated alternately into two categories: The sizes shown in bold type are selections made in the interest of simplification and are those to which it is recommended that usage be confined wherever the circumstances of design permit. Where these sizes do not meet requirements the intermediate sizes shown in light type are available.

Table 1. Unified Miniature Screw Threads — Basic Thread Form

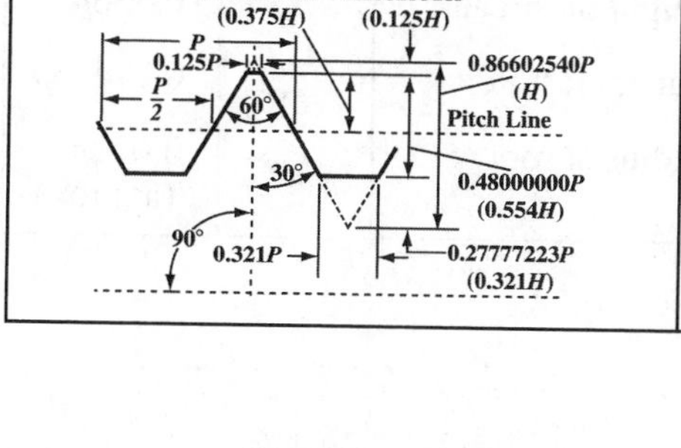

0.32475953P (0.375H)
0.10825318P (0.125H)
P
0.125P
P/2
60°
0.86602540P (H)
Pitch Line
30°
0.48000000P (0.554H)
90°
0.321P
0.27777223P (0.321H)

Thread Element	Symbol	Formula
Formulas for Basic Thread Form Metric units (millimeters) are used in all formulas		
Angle of thread	2α	60°
Half angle of thread	α	30°
Pitch of thread	p	
No. of threads per inch	n	$25.4/p$
Height of sharp V thread	H	0.86603p
Addendum of basic thread	h_{ab}	$0.32476p$
Height of basic thread	h_b	$0.52p$

Table 2. Unified Miniature Screw Threads—Design Thread Form

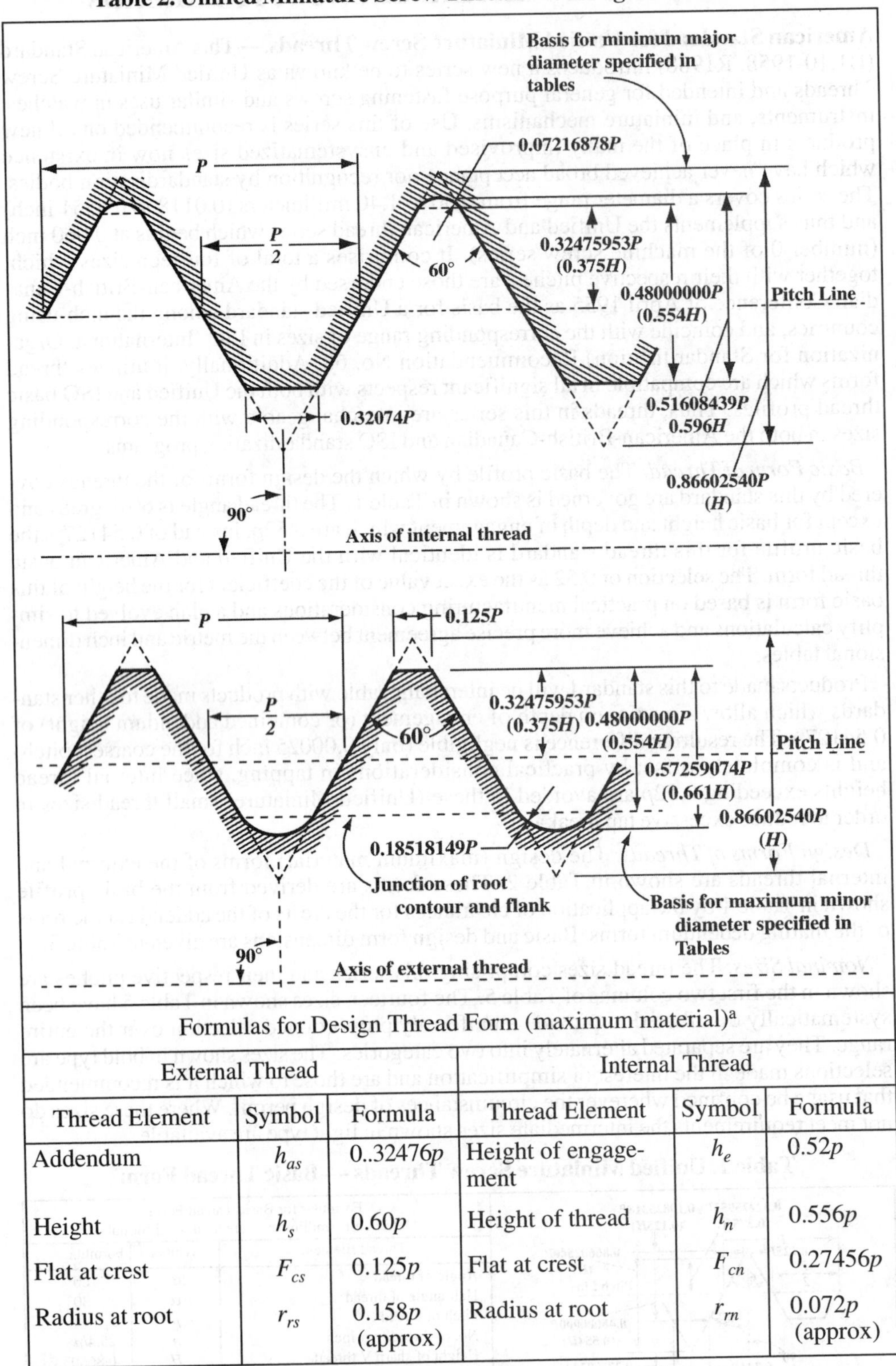

Formulas for Design Thread Form (maximum material)[a]

External Thread			Internal Thread		
Thread Element	Symbol	Formula	Thread Element	Symbol	Formula
Addendum	h_{as}	0..32476p	Height of engagement	h_e	0.52p
Height	h_s	0.60p	Height of thread	h_n	0.556p
Flat at crest	F_{cs}	0.125p	Flat at crest	F_{cn}	0.27456p
Radius at root	r_{rs}	0.158p (approx)	Radius at root	r_{rn}	0.072p (approx)

[a] Metric units (millimeters) are used in all formulas.

Table 3. Unified Miniature Screw Threads—Basic and Design Form Dimensions

Basic Thread Form					External Thread Design Form			Internal Thread Design Form		
Threads per inch n^a	Pitch p	Height of Sharp V $H =$ $0.86603p$	Height $h_b =$ $0.52p$	Addendum $h_{ab} =$ $h_{as} =$ $0.32476p$	Height $h_s =$ $0.60p$	Flat at Crest $F_{cs} =$ $0.125p$	Radius at Root $r_{rs} =$ $0.158p$	Height $h_n =$ $0.556p$	Flat at Crest $F_{cn} =$ $0.27456p$	Radius at Root $r_{rn} =$ $0.072p$
Millimeter Dimensions										
…	.080	.0693	.0416	.0260	.048	.0100	.0126	.0445	.0220	.0058
…	.090	.0779	.0468	.0292	.054	.0112	.0142	.0500	.0247	.0065
…	.100	.0866	.0520	.0325	.060	.0125	.0158	.0556	.0275	.0072
…	.125	.1083	.0650	.0406	.075	.0156	.0198	.0695	.0343	.0090
…	.150	.1299	.0780	.0487	.090	.0188	.0237	.0834	.0412	.0108
…	.175	.1516	.0910	.0568	.105	.0219	.0277	.0973	.0480	.0126
…	.200	.1732	.1040	.0650	.120	.0250	.0316	.1112	.0549	.0144
…	.225	.1949	.1170	.0731	.135	.0281	.0356	.1251	.0618	.0162
…	.250	.2165	.1300	.0812	.150	.0312	.0395	.1390	.0686	.0180
…	.300	.2598	.1560	.0974	.180	.0375	.0474	.1668	.0824	.0216
Inch Dimensions										
$317\frac{1}{2}$	.003150	.00273	.00164	.00102	.00189	.00039	.00050	.00175	.00086	.00023
$282\frac{2}{9}$	.003543	.00307	.00184	.00115	.00213	.00044	.00056	.00197	.00097	.00026
254	.003937	.00341	.00205	.00128	.00236	.00049	.00062	.00219	.00108	.00028
$203\frac{1}{5}$	.004921	.00426	.00256	.00160	.00295	.00062	.00078	.00274	.00135	.00035
$169\frac{1}{3}$	.005906	.00511	.00307	.00192	.00354	.00074	.00093	.00328	.00162	.00043
$145\frac{1}{7}$	.006890	.00597	.00358	.00224	.00413	.00086	.00109	.00383	.00189	.00050
127	.007874	.00682	.00409	.00256	.00472	.00098	.00124	.00438	.00216	.00057
$112\frac{8}{9}$	.008858	.00767	.00461	.00288	.00531	.00111	.00140	.00493	.00243	.00064
$101\frac{3}{5}$	.009843	.00852	.00512	.00320	.00591	.00123	.00156	.00547	.00270	.00071
$84\frac{2}{3}$	.011811	.01023	.00614	.00384	.00709	.00148	.00187	.00657	.00324	.00085

[a] In Tables 5 and 6 these values are shown rounded to the nearest whole number.

Table 4. Unified Miniature Screw Threads — Formulas for Basic and Design Dimensions and Tolerances

Formulas for Basic Dimensions	
D = Basic Major Diameter and Nominal Size in millimeters; p = Pitch in millimeters; E = Basic Pitch Diameter in millimeters = $D - 0.64952p$; and K = Basic Minor Diameter in millimeters = $D - 1.04p$	
Formulas for Design Dimensions (Maximum Material)	
External Thread	Internal Thread
D_s = Major Diameter = D	D_n = Major Diameter = $D + 0.072p$
E_s = Pitch Diameter = E	E_n = Pitch Diameter = E
K_s = Minor Diameter = $D - 1.20p$	K_n = Minor Diameter = K
Formulas for Tolerances on Design Dimensions[a]	
External Thread (−)	Internal Thread (+)
Major Diameter Tol., $0.12p + 0.006$	[b]Major Diameter Tol., $0.168p + 0.008$
Pitch Diameter Tol., $0.08p + 0.008$	Pitch Diameter Tol., $0.08p + 0.008$
[c]Minor Diameter Tol., $0.16p + 0.008$	Minor Diameter Tol., $0.32p + 0.012$

[a] These tolerances are based on lengths of engagement of $\frac{2}{3}D$ to $1\frac{1}{2}D$.

[b] This tolerance establishes the maximum limit of the major diameter of the internal thread. In practice, this limit is applied to the threading tool (tap) and not gaged on the product. Values for this tolerance are, therefore, not given in Table 5.

[c] This tolerance establishes the minimum limit of the minor diameter of the external thread. In practice, this limit is applied to the threading tool and only gaged on the product in confirming new tools. Values for this tolerance are, therefore, not given in Table 5.

Metric units (millimeters) apply in all formulas. Inch tolerances are not derived by direct conversion of the metric values. They are the differences between the rounded off limits of size in inch units.

Table 5. Unified Miniature Screw Threads — Limits of Size and Tolerances

Size Designation[a]	Pitch	External Threads: Major Diam. Max[b]	Major Diam. Min	Pitch Diam. Max[b]	Pitch Diam. Min	Minor Diam. Max[c]	Minor Diam. Min[d]	Internal Threads: Minor Diam. Min[b]	Minor Diam. Max	Pitch Diam. Min[b]	Pitch Diam. Max	Major Diam. Min[e]	Major Diam. Max[d]	Lead Angle at Basic Pitch Diam. deg	Lead Angle min	Sectional Area at Minor Diam. at D — 1.28p
	mm	mm	mm	mm	mm	mm	mm	mm	mm	mm	mm	mm	mm	deg	min	sq mm
0.30 UNM	**0.080**	**0.300**	**0.284**	**0.248**	**0.234**	**0.204**	**0.183**	**0.217**	**0.254**	**0.248**	**0.262**	**0.306**	**0.327**	**5**	**52**	**0.0307**
0.35 UNM	0.090	0.350	0.333	0.292	0.277	0.242	0.220	0.256	0.297	0.292	0.307	0.356	0.380	5	37	0.0433
0.40 UNM	**0.100**	**0.400**	**0.382**	**0.335**	**0.319**	**0.280**	**0.256**	**0.296**	**0.340**	**0.335**	**0.351**	**0.407**	**0.432**	**5**	**26**	**0.0581**
0.45 UNM	0.100	0.450	0.432	0.385	0.369	0.330	0.306	0.346	0.390	0.385	0.401	0.457	0.482	4	44	0.0814
0.50 UNM	**0.125**	**0.500**	**0.479**	**0.419**	**0.401**	**0.350**	**0.322**	**0.370**	**0.422**	**0.419**	**0.437**	**0.509**	**0.538**	**5**	**26**	**0.0908**
0.55 UNM	0.125	0.550	0.529	0.469	0.451	0.400	0.372	0.420	0.472	0.469	0.487	0.559	0.588	4	51	0.1195
0.60 UNM	**0.150**	**0.600**	**0.576**	**0.503**	**0.483**	**0.420**	**0.388**	**0.444**	**0.504**	**0.503**	**0.523**	**0.611**	**0.644**	**5**	**26**	**0.1307**
0.70 UNM	0.175	0.700	0.673	0.586	0.564	0.490	0.454	0.518	0.586	0.586	0.608	0.713	0.750	5	26	0.1780
0.80 UNM	**0.200**	**0.800**	**0.770**	**0.670**	**0.646**	**0.560**	**0.520**	**0.592**	**0.668**	**0.670**	**0.694**	**0.814**	**0.856**	**5**	**26**	**0.232**
0.90 UNM	0.225	0.900	0.867	0.754	0.728	0.630	0.586	0.666	0.750	0.754	0.780	0.916	0.962	5	26	0.294
1.00 UNM	**0.250**	**1.000**	**0.964**	**0.838**	**0.810**	**0.700**	**0.652**	**0.740**	**0.832**	**0.838**	**0.866**	**1.018**	**1.068**	**5**	**26**	**0.363**
1.10 UNM	0.250	1.100	1.064	0.938	0.910	0.800	0.752	0.840	0.932	0.938	0.966	1.118	1.168	4	51	0.478
1.20 UNM	**0.250**	**1.200**	**1.164**	**1.038**	**1.010**	**0.900**	**0.852**	**0.940**	**1.032**	**1.038**	**1.066**	**1.218**	**1.268**	**4**	**23**	**0.608**
1.40 UNM	0.300	1.400	1.358	1.205	1.173	1.040	0.984	1.088	1.196	1.205	1.237	1.422	1.480	4	32	0.811
	Thds. per in.	inch	inch	inch	inch	inch	inch	inch	inch	inch	inch	inch	inch	deg	min	sq in
0.30 UNM	**318**	**0.0118**	**0.0112**	**0.0098**	**0.0092**	**0.0080**	**0.0072**	**0.0085**	**0.0100**	**0.0098**	**0.0104**	**0.0120**	**0.0129**	**5**	**52**	**0.0000475**
0.35 UNM	282	0.0138	0.0131	0.0115	0.0109	0.0095	0.0086	0.0101	0.0117	0.0115	0.0121	0.0140	0.0149	5	37	0.0000671
0.40 UNM	**254**	**0.0157**	**0.0150**	**0.0132**	**0.0126**	**0.0110**	**0.0101**	**0.0117**	**0.0134**	**0.0132**	**0.0138**	**0.0160**	**0.0170**	**5**	**26**	**0.0000901**
0.45 UNM	254	0.0177	0.0170	0.0152	0.0145	0.0130	0.0120	0.0136	0.0154	0.0152	0.0158	0.0180	0.0190	4	44	0.0001262
0.50 UNM	**203**	**0.0197**	**0.0189**	**0.0165**	**0.0158**	**0.0138**	**0.0127**	**0.0146**	**0.0166**	**0.0165**	**0.0172**	**0.0200**	**0.0212**	**5**	**26**	**0.0001407**
0.55 UNM	203	0.0217	0.0208	0.0185	0.0177	0.0157	0.0146	0.0165	0.0186	0.0185	0.0192	0.0220	0.0231	4	51	0.0001852
0.60 UNM	**169**	**0.0236**	**0.0227**	**0.0198**	**0.0190**	**0.0165**	**0.0153**	**0.0175**	**0.0198**	**0.0198**	**0.0206**	**0.0240**	**0.0254**	**5**	**26**	**0.000203**
0.70 UNM	145	0.0276	0.0265	0.0231	0.0222	0.0193	0.0179	0.0204	0.0231	0.0231	0.0240	0.0281	0.0295	5	26	0.000276
0.80 UNM	**127**	**0.0315**	**0.0303**	**0.0264**	**0.0254**	**0.0220**	**0.0205**	**0.0233**	**0.0263**	**0.0264**	**0.0273**	**0.0321**	**0.0337**	**5**	**26**	**0.000360**
0.90 UNM	113	0.0354	0.0341	0.0297	0.0287	0.0248	0.0231	0.0262	0.0295	0.0297	0.0307	0.0361	0.0379	5	26	0.000456
1.00 UNM	**102**	**0.0394**	**0.0380**	**0.0330**	**0.0319**	**0.0276**	**0.0257**	**0.0291**	**0.0327**	**0.0330**	**0.0341**	**0.0401**	**0.0420**	**5**	**26**	**0.000563**
1.10 UNM	102	0.0433	0.0419	0.0369	0.0358	0.0315	0.0296	0.0331	0.0367	0.0369	0.0380	0.0440	0.0460	4	51	0.000741
1.20 UNM	**102**	**0.0472**	**0.0458**	**0.0409**	**0.0397**	**0.0354**	**0.0335**	**0.0370**	**0.0406**	**0.0409**	**0.0420**	**0.0480**	**0.0499**	**4**	**23**	**0.000943**
1.40 UNM	85	0.0551	0.0535	0.0474	0.0462	0.0409	0.0387	0.0428	0.0471	0.0474	0.0487	0.0560	0.0583	4	32	0.001257

[a] Sizes shown in bold type are preferred.

[b] This is also the basic dimension.

[c] This limit, in conjunction with root form shown in Table 2, is advocated for use when optical projection methods of gaging are employed. For mechanical gaging the minimum minor diameter of the internal thread is applied.

[d] This limit is provided for reference only. In practice, the form of the threading tool is relied upon for this limit.

[e] This limit is provided for reference only, and is not gaged. For gaging, the maximum major diameter of the external thread is applied.

Table 6. Unified Miniature Screw Threads— Minimum Root Flats for External Threads

Pitch	No. of Threads Per Inch	Thread Height for Min. Flat at Root 0.64p		Minimum Flat at Root F_{rs} = 0.136p	
mm		mm	Inch	mm	Inch
0.080	318	0.0512	0.00202	0.0109	0.00043
0.090	282	0.0576	0.00227	0.0122	0.00048
0.100	254	0.0640	0.00252	0.0136	0.00054
0.125	203	0.0800	0.00315	0.0170	0.00067
0.150	169	0.0960	0.00378	0.0204	0.00080
0.175	145	0.1120	0.00441	0.0238	0.00094
0.200	127	0.1280	0.00504	0.0272	0.00107
0.225	113	0.1440	0.00567	0.0306	0.00120
0.250	102	0.1600	0.00630	0.0340	0.00134
0.300	85	0.1920	0.00756	0.0408	0.00161

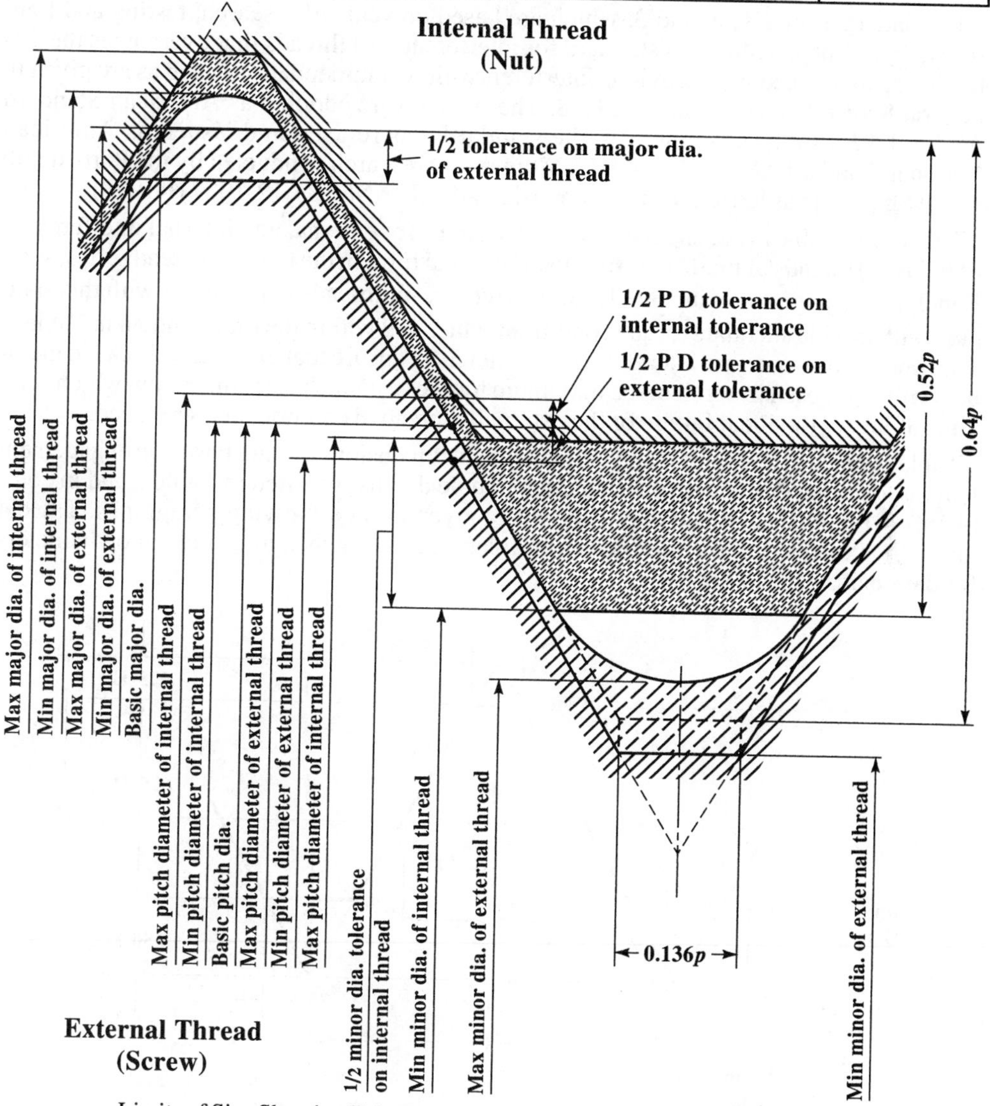

Limits of Size Showing Tolerances and Crest Clearances for UNM Threads

Limits of Size: Formulas used to determine limits of size are given in Table 4; the limits of size are given in Table 5. The diagram on page 1785 illustrates the limits of size and Table 6 gives values for the minimum flat at the root of the external thread shown on the diagram.

Classes of Threads: The standard establishes one class of thread with zero allowance on all diameters. When coatings of a measurable thickness are required, they should be included within the maximum material limits of the threads since these limits apply to both coated and uncoated threads.

Hole Sizes for Tapping: Suggested hole sizes are given in the Tapping Section.

Interference-Fit Threads.—Interference-fit threads are threads in which the externally threaded member is larger than the internally threaded member when both members are in the free state and that, when assembled, become the same size and develop a holding torque through elastic compression, plastic movement of material, or both. By custom, these threads are designated Class 5.

The data in Tables 1, 2, and 3, which are based on years of research, testing and field study, represent an American standard for interference-fit threads that overcomes the difficulties experienced with previous interference-fit recommendations such as are given in Federal Screw Thread Handbook H28. These data were adopted as American Standard ASA B1.12-1963. Subsequently, the standard was revised and issued as American National Standard ANSI B1.12-1972. More recent research conducted by the Portsmouth Naval Shipyard has led to the current revision ASME/ANSI B1.12-1987 (R1998).

The data in Tables 1, 2, and 3 provide dimensions for external and internal interference-fit (Class 5) threads of modified American National form in the Coarse Thread series, sizes ¼ inch to 1½ inches. It is intended that interference-fit threads conforming with this standard will provide adequate torque conditions which fall within the limits shown in Table 3. The minimum torques are intended to be sufficient to ensure that externally threaded members will not loosen in service; the maximum torques establish a ceiling below which seizing, galling, or torsional failure of the externally threaded components is reduced.

Tables 1 and 2 give external and internal thread dimensions and are based on engagement lengths, external thread lengths, and tapping hole depths specified in Table 3 and in compliance with the design and application data given in the following paragraphs. Table 4 gives the allowances and Table 5 gives the tolerances for pitch, major, and minor diameters for the Coarse Thread Series.

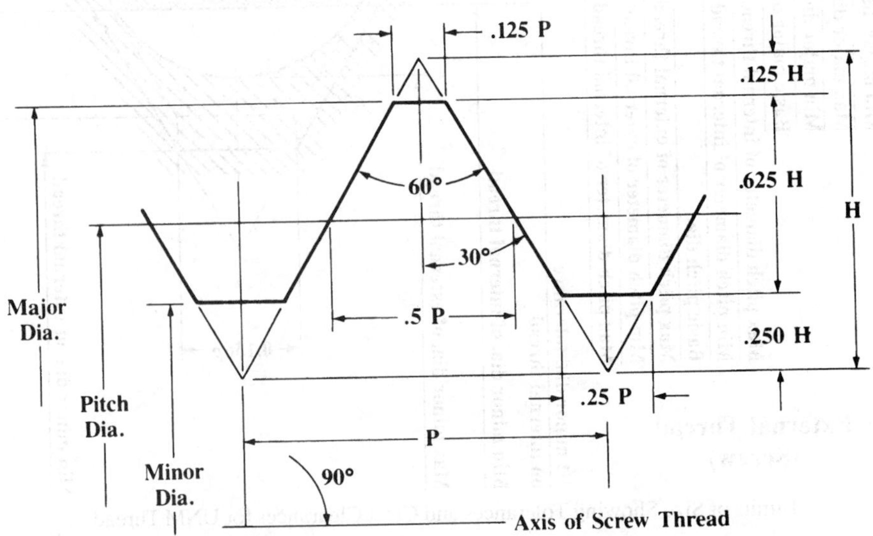

Basic Profile of American National Standard Class 5 Interference Fit Thread

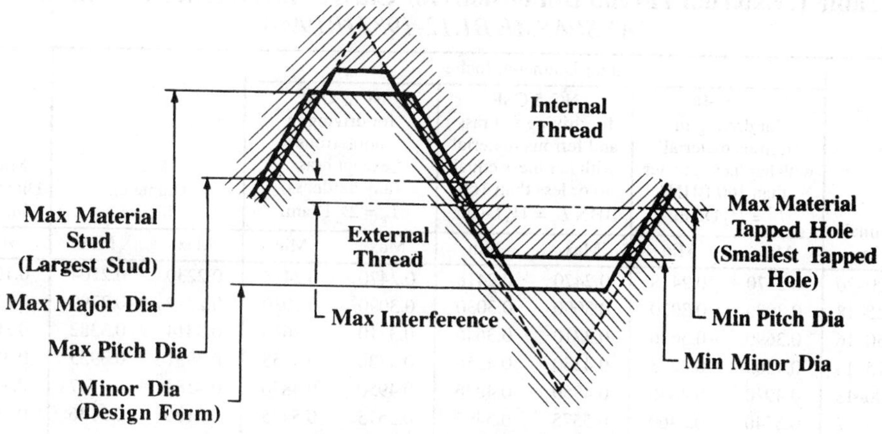

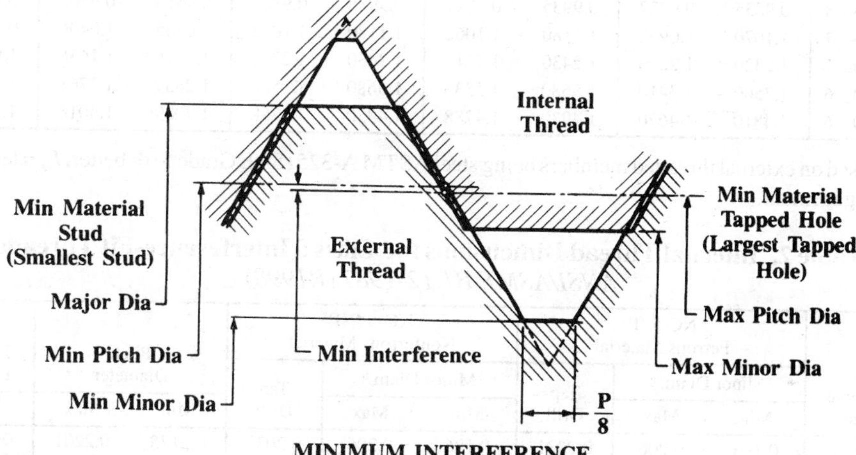

Maximum and Minimum Material Limits for Class 5 Interference-Fit Thread

Design and Application Data for Class 5 Interference-Fit Threads.—Following are conditions of usage and inspection on which satisfactory application of products made to dimensions in Tables 1, 2, and 3 are based.

Thread Designations: The following thread designations provide a means of distinguishing the American Standard Class 5 Threads from the tentative Class 5 and alternate Class 5 threads, specified in Handbook H28. They also distinguish between external and internal American Standard Class 5 Threads.

Class 5 External Threads are designated as follows:

NC-5 HF—For driving in hard ferrous material of hardness over 160 BHN.

NC-5 CSF—For driving in copper alloy and soft ferrous material of 160 BHN or less.

NC-5 ONF—For driving in other nonferrous material (nonferrous materials other than copper alloys), any hardness.

Class 5 Internal Threads are designated as follows:

NC-5 IF—Entire ferrous material range.

NC-5 INF—Entire nonferrous material range.

Table 1. External Thread Dimensions for Class 5 Interference-Fit Threads
ANSI/ASME B1.12-1987 (R1998)

Nominal Size	Major Diameter, Inches						Pitch Diameter, Inches		Minor Diameter, Inches
	NC-5 HF for driving in ferrous material with hardness greater than 160 BHN $L_e = 1\frac{1}{4}$ Diam.		NC-5 CSF for driving in brass and ferrous material with hardness equal to or less than 160 BHN $L_e = 1\frac{1}{4}$ Diam.		NC-5 ONF for driving in nonferrous except brass (any hardness) $L_e = 2\frac{1}{2}$ Diam.				
	Max	Min	Max	Min	Max	Min	Max	Min	Max
0.2500–20	0.2470	0.2418	0.2470	0.2418	0.2470	0.2418	0.2230	0.2204	0.1932
0.3125–18	0.3080	0.3020	0.3090	0.3030	0.3090	0.3030	0.2829	0.2799	0.2508
0.3750–16	0.3690	0.3626	0.3710	0.3646	0.3710	0.3646	0.3414	0.3382	0.3053
0.4375–14	0.4305	0.4233	0.4330	0.4258	0.4330	0.4258	0.3991	0.3955	0.3579
0.5000–13	0.4920	0.4846	0.4950	0.4876	0.4950	0.4876	0.4584	0.4547	0.4140
0.5625–12	0.5540	0.5460	0.5575	0.5495	0.5575	0.5495	0.5176	0.5136	0.4695
0.6250–11	0.6140	0.6056	0.6195	0.6111	0.6195	0.6111	0.5758	0.5716	0.5233
0.7500–10	0.7360	0.7270	0.7440	0.7350	0.7440	0.7350	0.6955	0.6910	0.6378
0.8750– 9	0.8600	0.8502	0.8685	0.8587	0.8685	0.8587	0.8144	0.8095	0.7503
1.0000– 8	0.9835	0.9727	0.9935	0.9827	0.9935	0.9827	0.9316	0.9262	0.8594
1.1250– 7	1.1070	1.0952	1.1180	1.1062	1.1180	1.1062	1.0465	1.0406	0.9640
1.2500– 7	1.2320	1.2200	1.2430	1.2312	1.2430	1.2312	1.1715	1.1656	1.0890
1.3750– 6	1.3560	1.3410	1.3680	1.3538	1.3680	1.3538	1.2839	1.2768	1.1877
1.5000– 6	1.4810	1.4670	1.4930	1.4788	1.4930	1.4788	1.4089	1.4018	1.3127

Based on external threaded members being steel ASTM A-325 (SAE Grade 5) or better. L_e = length of engagement.

Table 2. Internal Thread Dimensions for Class 5 Interference-Fit Threads
ANSI/ASME B1.12-1987 (R1998)

Nominal Size	NC-5 IF Ferrous Material			NC-5 INF Nonferrous Material			Pitch Diameter		Major Diam.
	Minor Diam.[a]		Tap Drill	Minor Diam.[a]		Tap Drill			
	Min	Max		Min	Max		Min	Max	Min
0.2500–20	0.196	0.206	0.2031	0.196	0.206	0.2031	0.2175	0.2201	0.2532
0.3125–18	0.252	0.263	0.2610	0.252	0.263	0.2610	0.2764	0.2794	0.3161
0.3750–16	0.307	0.318	0.3160	0.307	0.318	0.3160	0.3344	0.3376	0.3790
0.4375–14	0.374	0.381	0.3750	0.360	0.372	0.3680	0.3911	0.3947	0.4421
0.5000–13	0.431	0.440	0.4331	0.417	0.429	0.4219	0.4500	0.4537	0.5050
0.5625–12	0.488	0.497	0.4921	0.472	0.485	0.4844	0.5084	0.5124	0.5679
0.6250–11	0.544	0.554	0.5469	0.527	0.540	0.5313	0.5660	0.5702	0.6309
0.7500–10	0.667	0.678	0.6719	0.642	0.655	0.6496	0.6850	0.6895	0.7565
0.8750– 9	0.777	0.789	0.7812	0.755	0.769	0.7656	0.8028	0.8077	0.8822
1.0000– 8	0.890	0.904	0.8906	0.865	0.880	0.8750	0.9188	0.9242	1.0081
1.1250– 7	1.000	1.015	1.0000	0.970	0.986	0.9844	1.0322	1.0381	1.1343
1.2500– 7	1.125	1.140	1.1250	1.095	1.111	1.1094	1.1572	1.1631	1.2593
1.3750– 6	1.229	1.247	1.2344	1.195	1.213	1.2031	1.2667	1.2738	1.3858
1.5000– 6	1.354	1.372	1.3594	1.320	1.338	1.3281	1.3917	1.3988	1.5108

[a] Fourth decimal place is 0 for all sizes.

All dimensions are in inches, unless otherwise specified.

Externally Threaded Products: Points of externally threaded components should be chamfered or otherwise reduced to a diameter below the minimum minor diameter of the thread. The limits apply to bare or metallic coated parts. The threads should be free from excessive nicks, burrs, chips, grit or other extraneous material before driving.

Table 3. Torques, Interferences, and Engagement Lengths for Class 5 Interference-Fit Threads *ANSI/ASME B1.12-1987 (R1998)*

Nominal Size	Interference on Pitch Diameter		Engagement Lengths, External Thread Lengths and Tapped Hole Depths[a]						Torque at 1-¼D Engagement in Ferrous Material	
			In Brass and Ferrous			In Nonferrous Except Brass				
	Max	Min	L_e	T_s	T_h min	L_e	T_s	T_h min	Max, lb-ft	Min, lb-ft
0.2500–20	.0055	.0003	0.312	0.375 + .125 – 0	0.375	0.625	0.688 + .125 – 0	0.688	12	3
0.3125–18	.0065	.0005	0.391	0.469 + .139 – 0	0.469	0.781	0.859 + .139 – 0	0.859	19	6
0.3750–16	.0070	.0006	0.469	0.562 + .156 – 0	0.562	0.938	1.031 + .156 – 0	1.031	35	10
0.4375–14	.0080	.0008	0.547	0.656 + .179 – 0	0.656	1.094	1.203 + .179 – 0	1.203	45	15
0.5000–13	.0084	.0010	0.625	0.750 + .192 – 0	0.750	1.250	1.375 + .192 – 0	1.375	75	20
0.5625–12	.0092	.0012	0.703	0.844 + .208 – 0	0.844	1.406	1.547 + .208 – 0	1.547	90	30
0.6250–11	.0098	.0014	0.781	0.938 + .227 – 0	0.938	1.562	1.719 + .227 – 0	1.719	120	37
0.7500–10	.0105	.0015	0.938	1.125 + .250 – 0	1.125	1.875	2.062 + .250 – 0	2.062	190	60
0.8750– 9	.0016	.0018	1.094	1.312 + .278 – 0	1.312	2.188	2.406 + .278 – 0	2.406	250	90
1.0000– 8	.0128	.0020	1.250	1.500 + .312 – 0	1.500	2.500	2.750 + .312 – 0	2.750	400	125
1.1250– 7	.0143	.0025	1.406	1.688 + .357 – 0	1.688	2.812	3.094 + .357 – 0	3.095	470	155
1.2500– 7	.0143	.0025	1.562	1.875 + .357 – 0	1.875	3.125	3.438 + .357 – 0	3.438	580	210
1.3750– 6	.0172	.0030	1.719	2.062 + .419 – 0	2.062	3.438	3.781 + .419 – 0	3.781	705	250
1.5000– 6	.0172	.0030	1.875	2.250 + .419 – 0	2.250	3.750	4.125 + .419 – 0	4.125	840	325

[a] L_e = Length of engagement. T_s = External thread length of full form thread. T_h = Minimum depth of full form thread in hole.

All dimensions are inches.

Materials for Externally Threaded Products: The length of engagement, depth of thread engagement and pitch diameter in Tables 1, 2, and 3 are designed to produce adequate torque conditions when heat-treated medium-carbon steel products, ASTM A-325 (SAE Grade 5) or better, are used. In many applications, case-carburized and nonheat-treated medium-carbon steel products of SAE Grade 4 are satisfactory. SAE Grades 1 and 2, may be usable under certain conditions. This standard is not intended to cover the use of products made of stainless steel, silicon bronze, brass or similar materials. When such materials are used, the tabulated dimensions will probably require adjustment based on pilot experimental work with the materials involved.

Lubrication: For driving in ferrous material, a good lubricant sealer should be used, particularly in the hole. A non-carbonizing type of lubricant (such as a rubber-in-water dispersion) is suggested. The lubricant must be applied to the hole and it may be applied to the male member. In applying it to the hole, care must be taken so that an excess amount of lubricant will not cause the male member to be impeded by hydraulic pressure in a blind hole. Where sealing is involved, the lubricant selected should be insoluble in the medium being sealed.

For driving, in nonferrous material, lubrication may not be needed. The use of medium gear oil for driving in aluminum is recommended. American research has observed that the minor diameter of lubricated tapped holes in non-ferrous materials may tend to close in, that is, be reduced in driving; whereas with an unlubricated hole the minor diameter may tend to open up.

Driving Speed: This standard makes no recommendation for driving speed. Some opinion has been advanced that careful selection and control of driving speed is desirable to obtain optimum results with various combinations of surface hardness and roughness. Experience with threads made to this standard may indicate what limitations should be placed on driving speeds.

Table 4. Allowances for Coarse Thread Series *ANSI/ASME B1.12-1987 (R1998)*

TPI	Difference between Nom. Size and Max Major Diam of NC-5 HF[a]	Difference between Nom. Size and Max Major Diam. of NC-5 CSF or NC-5 ONF[a]	Difference between Basic Minor Diam. and Min Minor Diam. of NC-5 IF[a]	Difference between Basic Minor Diam. and Min Minor Diam.of NC-5 INF	Max PD Inteference or Neg Allowance, Ext Thread[b]	Difference between Max Minor Diam. and Basic Minor Diam., Ext Thread
20	0.0030	0.0030	0.000	0.000	0.0055	0.0072
18	0.0045	0.0035	0.000	0.000	0.0065	0.0080
16	0.0060	0.0040	0.000	0.000	0.0070	0.0090
14	0.0070	0.0045	0.014	0.000	0.0080	0.0103
13	0.0080	0.0050	0.014	0.000	0.0084	0.0111
12	0.0085	0.0050	0.016	0.000	0.0092	0.0120
11	0.0110	0.0055	0.017	0.000	0.0098	0.0131
10	0.0140	0.0060	0.019	0.000	0.0105	0.0144
9	0.0150	0.0065	0.022	0.000	0.0116	0.0160
8	0.0165	0.0065	0.025	0.000	0.0128	0.0180
7	0.0180	0.0070	0.030	0.000	0.0143	0.0206
6	0.0190	0.0070	0.034	0.000	0.0172	0.0241

[a] The allowances in these columns were obtained from industrial research data.

[b] Negative allowance is the difference between the basic pitch diameter and pitch diameter value at maximum material condition.

All dimensions are in inches.

The difference between basic major diameter and internal thread minimum major diameter is 0.075H and is tabulated in Table 5.

Table 5. Tolerances for Pitch Diameter, Major Diameter, and Minor Diameter for Coarse Thread Series *ANSI/ASME B1.12-1987 (R1998)*

TPI	PD Tolerance for Ext and Int Threads[a]	Major Diam. Tolerance for Ext Thread[b]	Minor Diam. Tolerance for Int Thread NC-5 IF	Minor Diam. Tolerance for Int Thread NC-5 INF[c]	Tolerance 0.075H or 0.065P for Tap Major Diam.
20	0.0026	0.0052	0.010	0.010	0.0032
18	0.0030	0.0060	0.011	0.011	0.0036
16	0.0032	0.0064	0.011	0.011	0.0041
14	0.0036	0.0072	0.008	0.012	0.0046
13	0.0037	0.0074	0.008	0.012	0.0050
12	0.0040	0.0080	0.009	0.013	0.0054
11	0.0042	0.0084	0.010	0.013	0.0059
10	0.0045	0.0090	0.011	0.014	0.0065
9	0.0049	0.0098	0.012	0.014	0.0072
8	0.0054	0.0108	0.014	0.015	0.0093
7	0.0059	0.0118	0.015	0.015	0.0093
6	0.0071	0.0142	0.018	0.018	0.0108

[a] National Class 3 pitch diameter tolerance from ASA B1.1-1960.

[b] Twice the NC-3 pitch diameter tolerance.

[c] National Class 3 minor diameter tolerance from ASA B1.1-1960.

All dimensions are in inches.

Relation of Driving Torque to Length of Engagement: Torques increase directly as the length of engagement and this increase is proportionately more rapid as size increases. The standard does not establish recommended breakloose torques.

Surface Roughness: Surface roughnesss is not a required measurement. Roughness between 63 and 125 μin. Ra is recommended. Surface roughness greater than 125 μin. Ra may encourage galling and tearing of threads. Surfaces with roughness less than 63 μin. Ra may hold insufficient lubricant and wring or weld together.

Lead and Angle Variations: The lead variation values tabulated in Table 6 are the maximum variations from specified lead between any two points not farther apart than the length of the standard GO thread gage. Flank angle variation values tabulated in Table 7 are maximum variations from the basic 30° angle between thread flanks and perpendiculars to the thread axis. The application of these data in accordance with ANSI/ASME B1.3M, the screw thread gaging system for dimensional acceptability, is given in the Standard. Lead variation does not change the volume of displaced metal, but it exerts a cumulative unilateral stress on the pressure side of the thread flank. Control of the difference between pitch diameter size and functional diameter size to within one-half the pitch diameter tolerance will hold lead and angle variables to within satisfactory limits. Both the variations may produce unacceptable torque and faulty assemblies.

Table 6. Maximum Allowable Variations in Lead and Maximum Equivalent Change in Functional Diameter *ANSI/ASME B1.12-1987 (R1998)*

Nominal Size	External and Internal Threads	
	Allowable Variation in Axial Lead (Plus or Minus)	Max Equivalent Change in Functional Diam. (Plus for Ext, Minus for Int)
0.2500–20	0.0008	0.0013
0.3125–18	0.0009	0.0015
0.3750–16	0.0009	0.0016
0.4375–14	0.0010	0.0018
0.5000–13	0.0011	0.0018
0.5625–12	0.0012	0.0020
0.6250–11	0.0012	0.0021
0.7500–10	0.0013	0.0022
0.8750– 9	0.0014	0.0024
1.0000– 8	0.0016	0.0027
1.1250– 7	0.0017	0.0030
1.2500– 7	0.0017	0.0030
1.3750– 6	0.0020	0.0036
1.5000– 6	0.0020	0.0036

All dimensions are in inches.

Note: The equivalent change in functional diameter applies to total effect of form errors.

Maximum allowable variation in lead is permitted only when all other form variations are zero.

For sizes not tabulated, maximum allowable variation in lead is equal to 0.57735 times one-half the pitch diameter tolerance.

Table 7. Maximum Allowable Variation in 30° Basic Half-Angle of External and Internal Screw Threads *ANSI/ASME B1.12-1987 (R1998)*

TPI	Allowable Variation in Half-Angle of Thread (Plus or Minus)	TPI	Allowable Variation in Half-Angle of Thread (Plus or Minus)	TPI	Allowable Variation in Half-Angle of Thread (Plus or Minus)
32	1° 30′	14	0° 55′	8	0° 45′
28	1° 20′	13	0° 55′	7	0° 45′
27	1° 20′	12	0° 50′	6	0° 40′
24	1° 15′	11½	0° 50′	5	0° 40′
20	1° 10′	11	0° 50′	4½	0° 40′
18	1° 05′	10	0° 50′	4	0° 40′
16	1° 00′	9	0° 50′	…	…

ACME SCREW THREADS

American National Standard Acme Screw Threads.—This American National Standard ASME/ANSI B1.5-1988 is a revision of American Standard ANSI B1.5-1977 and provides for two general applications of Acme threads, namely, General Purpose and Centralizing.

The limits and tolerances in this standard relate to single-start Acme threads, and may be used, if considered suitable, for multi-start Acme threads, which provide fast relative traversing motion when this is necessary. For information on additional allowances for multi-start Acme threads, see later section on page 1800.

General Purpose Acme Threads.—Three classes of General Purpose threads, 2G, 3G, and 4G, are provided in the standard, each having clearance on all diameters for free movement, and may be used in assemblies with the internal thread rigidly fixed and movement of the external thread in a direction perpendicular to its axis limited by its bearing or bearings. It is suggested that external and internal threads of the same class be used together for general purpose assemblies, Class 2G being the preferred choice. If less backlash or end play is desired, Classes 3G and 4G are provided. Class 5G is not recommended for new designs.

Thread Form: The accompanying figure shows the thread form of these General Purpose threads, and the formulas accompanying the figure determine their basic dimensions. Table 1 gives the basic dimensions for the most generally used pitches.

Angle of Thread: The angle between the sides of the thread, measured in an axial plane, is 29 degrees. The line bisecting this 29-degree angle shall be perpendicular to the axis of the screw thread.

Thread Series: A series of diameters and associated pitches is recommended in the Standard as preferred. These diameters and pitches have been chosen to meet present needs with the fewest number of items in order to reduce to a minimum the inventory of both tools and gages. This series of diameters and associated pitches is given in Table 3.

Chamfers and Fillets: General Purpose external threads may have the crest corner chamfered to an angle of 45 degrees with the axis to a maximum width of $P/15$, where P is the pitch. This corresponds to a maximum depth of chamfer fiat of $0.0945P$.

Basic Diameters: The max major diameter of the external thread is basic and is the nominal major diameter for all classes. The min pitch diameter of the internal thread is basic and is equal to the basic major diameter minus the basic height of the thread, h. The basic minor diameter is the min minor diameter of the internal thread. It is equal to the basic major diameter minus twice the basic thread height, $2h$.

Length of Engagement: The tolerances specified in this standard are applicable to lengths of engagement not exceeding twice the nominal major diameter.

Major and Minor Diameter Allowances: A minimum diametral clearance is provided at the minor diameter of all external threads by establishing the maximum minor diameter 0.020 inch below the basic minor diameter of the nut for pitches of 10 threads per inch and coarser, and 0.10 inch for finer pitches. A minimum diametral clearance at the major diameter is obtained by establishing the minimum major diameter of the internal thread 0.020 inch above the basic major diameter of the screw for pitches of 10 threads per inch and coarser, and 0.010 inch for finer pitches.

American National Standard General Purpose Acme Thread Form *ASME/ANSI B1.5-1988*, **and Stub Acme Screw Thread Form** *ASME/ANSI B1.8-1988 (R1994)*

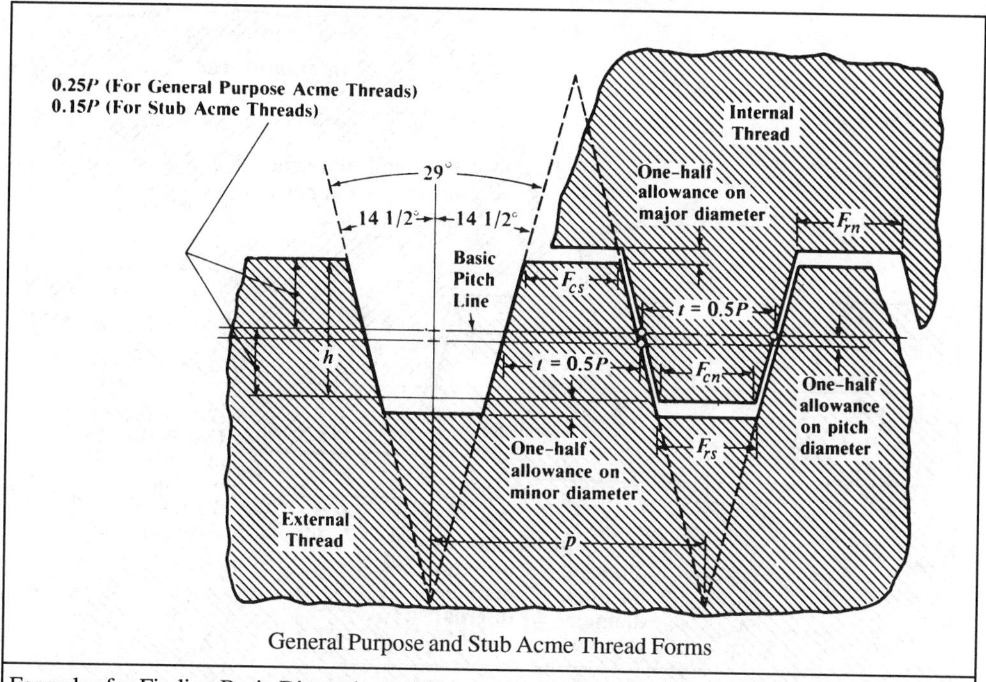

General Purpose and Stub Acme Thread Forms

Formulas for Finding Basic Dimensions of General Purpose Acme and Stub Acme Screw Threads	
General Purpose	Stub Acme Threads
Pitch = P = 1 ÷ No. threads per inch, n	Pitch = P = 1 ÷ No. threads per inch, n
Basic thread height $h = 0.5P$	Basic thread height $h = 0.3P$
Basic thread thickness $t = 0.5P$	Basic thread thickness $t = 0.5P$
Basic flat at crest $F_{cn} = 0.3707P$ (internal thread)	Basic flat at crest $F_{cn} = 0.4224P$ (internal thread)
Basic flat at crest $F_{cs} = 0.3707P - 0.259 \times$ (P.D. allowance on ext. thd.)	Basic flat at crest $F_{cs} = 0.4224P - 0.259 \times$ (pitch dia. allowance on ext. thread)
$F_{rn} = 0.3707P - 0.259 \times$ (major dia. allowance on internal thread)	$F_{rn} = 0.4224P - 0.259 \times$ (major dia. allowance on internal thread)
$F_{rs} = 0.3707P - 0.259 \times$ (minor dia. allowance on ext. thread — pitch dia. allowance on ext. thread)	$F_{rs} = 0.4224P - 0.259 \times$ (minor dia. allowance on ext. thread — pitch dia. allowance on ext. thread)

American National Standard Centralizing Acme Screw Thread Form
ASME/ANSI B1.5-1988

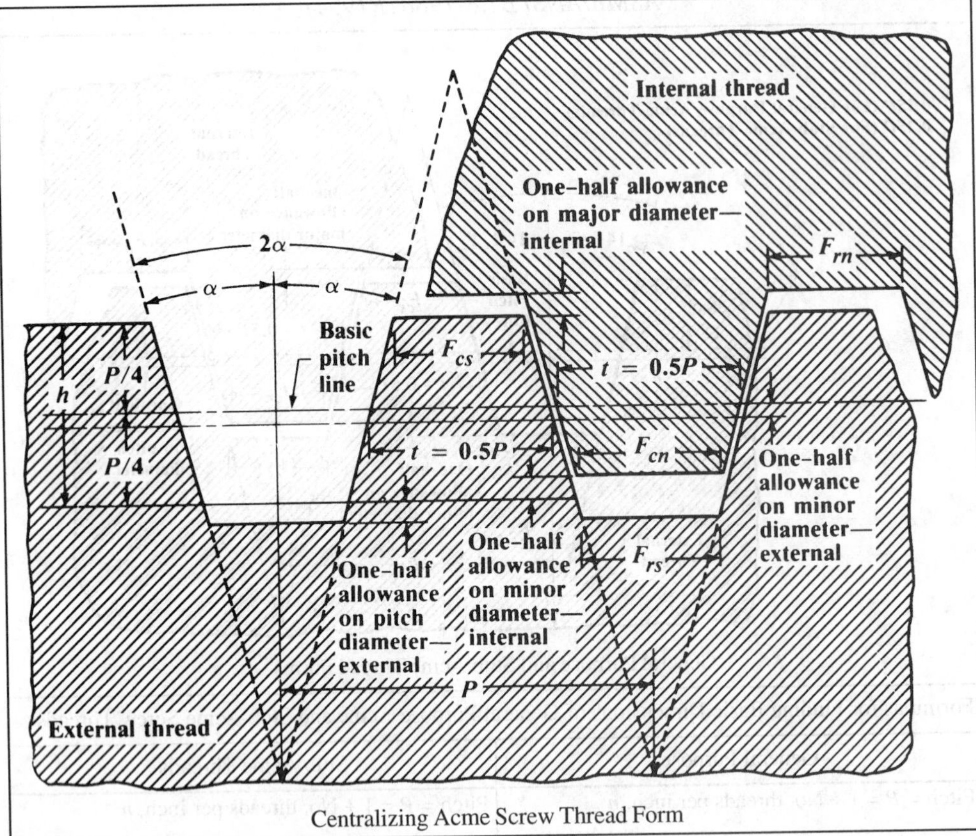

Centralizing Acme Screw Thread Form

Formulas for Finding Basic Dimensions of Centralizing Acme Screw Threads

Pitch = P = 1 ÷ No. threads per inch, n: Basic thread height $h = 0.5P$

Basic thread thickness $t = 0.5P$

Basic flat at crest $F_{cn} = 0.3707P + 0.259 \times$ (minor. dia. allowance on internal threads) (internal thread)

Basic flat at crest $F_{cs} = 0.3707P - 0.259 \times$ (pitch diameter allowance on external thread) (external thread)

$F_{rn} = 0.3707P - 0.259 \times$ (major dia. allowance on internal thread)

$F_{rs} = 0.3707P - 0.259 \times$ (minor dia. allowance on external thread — pitch dia. allowance on external thread)

Stress Area of General Purpose Acme Threads.—For computing the tensile strength of the thread section, the minimum stress area based on the mean of the minimum pitch diameter d_2 min. and the minimum minor diameter d_1 max. of the external thread is used:

$$\text{Stress Area} = 3.1416\left(\frac{d_2 \text{ min.} + d_1 \text{ max.}}{4}\right)^2$$

where d_2 min. and d_1 max. may be computed by Formulas 4 and 6, Table 2a or taken from Table 2b.

Shear Area of General Purpose Acme Threads. For computing the shear area per inch length of engagement of the external thread, the maximum minor diameter of the internal thread D_1 max., and the minimum pitch diameter of the external thread D_2 min., Table 2b or Formulas 12 and 4, Table 2a, are used:

$$\text{Shear Area} = 3.1416 D_1 \text{ max. } [0.5 + n \tan 14\tfrac{1}{2}^\circ (D_2 \text{ min.} - D_1 \text{ max.})]$$

Major and Minor Diameter Tolerances: The tolerance on the external thread major diameter is 0.05P, where P is the pitch, with a minimum of 0.005 inch. The tolerance on the internal thread major diameter is 0.020 inch for 10 threads per inch and coarser and 0.010 for finer pitches. The tolerance on the external thread minor diameter is 1.5 × pitch diameter tolerance. The tolerance on the internal thread minor diameter is 0.05P with a minimum of 0.005 inch.

Pitch Diameter Allowances and Tolerances: Allowances on the pitch diameter of General Purpose Acme threads are given in Table 4. Pitch diameter tolerances are given in Table 5. The ratios of the pitch diameter tolerances of Classes 2G, 3G, and 4G, General Purpose threads are 3.0, 1.4, and 1, respectively.

An increase of 10 per cent in the allowance is recommended for each inch, or fraction thereof, that the length of engagement exceeds two diameters.

Application of Tolerances: The tolerances specified are designed to ensure interchangeability and maintain a high grade of product. The tolerances on diameters of the internal thread are plus, being applied from minimum sizes to above the minimum sizes. The tolerances on diameters of the external thread are minus, being applied from the maximum sizes to below the maximum sizes. The pitch diameter (or thread thickness) tolerances for an external or internal thread of a given class are the same. The thread thickness tolerance is 0.259 times the pitch diameter tolerance.

Limiting Dimensions: Limiting dimensions of General Purpose Acme screw threads in the recommended series are given in Table 2b. These limits are based on the formulas in Table 2a.

For combinations of pitch and diameter other than those in the recommended series, the formulas in Table 2a and the data in Tables 4 and 5 make it possible to readily determine the limiting dimensions required.

A diagram showing the disposition of allowances, tolerances, and crest clearances for General Purpose Acme threads appears on page 1793.

Acme Thread Abbreviations.—The following abbreviations are recommended for use on drawings and in specifications, and on tools and gages:

ACME = Acme threads

G = General Purpose

C = Centralizing

P = pitch

L = lead

LH = left hand

Designation of General Purpose Acme Threads.—The examples listed below are given here to show how General Purpose Acme threads are designated on drawings and tools:

1.750-4 ACME-2G indicates a General Purpose Class 2G Acme thread of 1.750-inch major diameter, 4 threads per inch, single thread, right hand. The same thread, but left hand, is designated 1.750-4 ACME-2G-LH.

2.875-0.4*P*-0.8*L*-ACME-3G indicates a General Purpose Class 3G Acme thread of 2.875-inch major diameter, pitch 0.4 inch, lead 0.8 inch, double thread, right hand.

Table 1. American National Standard General Purpose Acme Screw Thread Form — Basic Dimensions *ASME/ANSI B1.5-1988*

Thds. per Inch n	Pitch, $P = 1/n$	Height of Thread (Basic), $h = P/2$	Total Height of Thread, $h_s = P/2 + \frac{1}{2}$ allowance[a]	Thread Thickness (Basic), $t = P/2$	Width of Flat: Crest of Internal Thread (Basic), $F_{cn} = 0.3707P$	Width of Flat: Root of Internal Thread, F_{rn} $0.3707P - 0.259 \times$ allowance[a]
16	0.06250	0.03125	0.0362	0.03125	0.0232	0.0206
14	0.07143	0.03571	0.0407	0.03571	0.0265	0.0239
12	0.08333	0.04167	0.0467	0.04167	0.0309	0.0283
10	0.010000	0.05000	0.0600	0.05000	0.0371	0.0319
8	0.12500	0.06250	0.0725	0.06250	0.0463	0.0411
6	0.16667	0.08333	0.0933	0.08333	0.0618	0.0566
5	0.20000	0.10000	0.1100	0.10000	0.0741	0.0689
4	0.25000	0.12500	0.1350	0.12500	0.0927	0.0875
3	0.33333	0.16667	0.1767	0.16667	0.1236	0.1184
2½	0.40000	0.20000	0.2100	0.20000	0.1483	0.1431
2	0.50000	0.25000	0.2600	0.25000	0.1853	0.1802
1½	0.66667	0.33333	0.3433	0.33333	0.2471	0.2419
1⅓	0.75000	0.37500	0.3850	0.37500	0.2780	0.2728
1	1.00000	0.50000	0.5100	0.50000	0.3707	0.3655

[a] Allowance is 0.020 inch for 10 threads per inch and coarser, and 0.010 inch for finer threads.

All dimensions are in inches.

Table 2a. American National Standard General Purpose Acme Single-Start Screw Threads — Formulas for Determining Diameters *ASME/ANSI B1.5-1988*

D = Basic Major Diameter and Nominal Size, in Inches.

P = Pitch = 1 ÷ Number of Threads per Inch.

E = Basic Pitch Diameter = $D - 0.5P$

K = Basic Minor Diameter = $D - P$

No.	External Threads (Screws)
1	Major Dia., Max. = D
2	Major Dia., Min. = D *minus* $0.05P$[a] but not less than 0.005.
3	Pitch Dia., Max. = E *minus* allowance from Table 4.
4	Pitch Dia., Min. = Pitch Dia., Max. (Formula 3) *minus* tolerance from Table 5.
5	Minor Dia., Max. = K *minus* 0.020 for 10 threads per inch and coarser and 0.010 for finer-pitches.
6	Minor Dia., Min. = Minor Dia., Max. (Formula 5) *minus* 1.5 × pitch diameter tolerance from Table 5.
	Internal Threads (Nuts)
7	Major Dia., Min. = D *plus* 0.020 for 10 threads per inch and coarser and 0.010 for finer pitches.
8	Major Dia., Max. = Major Dia., Min. (Formula 7) *plus* 0.020 for 10 threads per inch and coarser and 0.010 for finer pitches.
9	Pitch Dia., Min. = E
10	Pitch Dia., Max. = Pitch Dia., Min. (Formula 9) *plus* tolerance from Table 5.
11	Minor Dia., Min. = K
12	Minor Dia., Max. = Minor Dia., Min. (Formula 11) *plus* $0.05P$[a] but not less than 0.005.

[a] If P is between two recommended pitches listed in Table 3, use the coarser of the two pitches in this formula instead of the actual value of P.

Table 2b. Limiting Dimensions of American National Standard General Purpose Acme Single-Start Screw Threads

Limiting Diameters		Nominal Diameter, D											
		1/4	5/16	3/8	7/16	1/2	5/8	3/4	7/8	1	1 1/8	1 1/4	1 3/8
		Threads per Inch[a]											
		16	14	12	12	10	8	6	6	5	5	5	4
		External Threads											
Classes 2G, 3G, and 4G Major Diameter	Max (D)	0.2500	0.3125	0.3750	0.4375	0.5000	0.6250	0.7500	0.8750	1.0000	1.1250	1.2500	1.3750
	Min	0.2450	0.3075	0.3700	0.4325	0.4950	0.6188	0.7417	0.8667	0.9900	1.1150	1.2400	1.3625
Classes 2G, 3G, and 4G Minor Diameter	Max	0.1775	0.2311	0.2817	0.3442	0.3800	0.4800	0.5633	0.6883	0.7800	0.9050	1.0300	1.1050
Class 2G, Minor Diameter	Min	0.1618	0.2140	0.2632	0.3253	0.3594	0.4570	0.5372	0.6615	0.7509	0.8753	0.9998	1.0720
Class 3G, Minor Diameter	Min	0.1702	0.2231	0.2730	0.3354	0.3704	0.4693	0.5511	0.6758	0.7664	0.8912	1.0159	1.0896
Class 4G, Minor Diameter	Min	0.1722	0.2254	0.2755	0.3379	0.3731	0.4723	0.5546	0.6794	0.7703	0.8951	1.0199	1.0940
Class 2G, Pitch Diameter	Max	0.2148	0.2728	0.3284	0.3909	0.4443	0.5562	0.6598	0.7842	0.8920	1.0165	1.1411	1.2406
	Min	0.2043	0.2614	0.3161	0.3783	0.4306	0.5408	0.6424	0.7663	0.8726	0.9967	1.1210	1.2188
Class 3G, Pitch Diameter	Max	0.2158	0.2738	0.3296	0.3921	0.4458	0.5578	0.6615	0.7861	0.8940	1.0186	1.1433	1.2430
	Min	0.2109	0.2685	0.3238	0.3862	0.4394	0.5506	0.6534	0.7778	0.8849	1.0094	1.1339	1.2327
Class 4G, Pitch Diameter	Max	0.2168	0.2748	0.3309	0.3934	0.4472	0.5593	0.6632	0.7880	0.8960	1.0208	1.1455	1.2453
	Min	0.2133	0.2710	0.3268	0.3892	0.4426	0.5542	0.6574	0.7820	0.8895	1.0142	1.1388	1.2380
		Internal Threads											
Classes 2G, 3G, and 4G Major Diameter	Min	0.2600	0.3225	0.3850	0.4475	0.5200	0.6450	0.7700	0.8950	1.0200	1.1450	1.2700	1.3950
	Max	0.2700	0.3325	0.3950	0.4575	0.5400	0.6650	0.7900	0.9150	1.0400	1.1650	1.2900	1.4150
Classes 2G, 3G, and 4G Minor Diameter	Min	0.1875	0.2411	0.2917	0.3542	0.4000	0.5000	0.5833	0.7083	0.8000	0.9250	1.0500	1.1250
	Max	0.1925	0.2461	0.2967	0.3592	0.4050	0.5062	0.5916	0.7166	0.8100	0.9350	1.0600	1.1375
Class 2G, Pitch Diameter	Min	0.2188	0.2768	0.3333	0.3958	0.4500	0.5625	0.6667	0.7917	0.9000	1.0250	1.1500	1.1.2500
	Max	0.2293	0.2882	0.3456	0.4084	0.4637	0.5779	0.6841	0.8096	0.9194	1.0448	1.1701	1.2720
Class 3G, Pitch Diameter	Min	0.2188	0.2768	0.3333	0.3958	0.4500	0.5625	0.6667	0.7917	0.9000	1.0250	1.1500	1.2500
	Max	0.2237	0.2821	0.3391	0.4017	0.4564	0.5697	0.6748	0.8000	0.9091	1.0342	1.1594	1.2603
Class 4G, Pitch Diameter	Min	0.2188	0.2768	0.3333	0.3958	0.4500	0.5625	0.6667	0.7917	0.9000	1.0250	1.1500	1.2500
	Max	0.2223	0.2806	0.3374	0.4000	0.4546	0.5676	0.6725	0.7977	0.9065	1.0316	1.1567	1.2573

[a] All other dimensions are given in inches. The selection of threads per inch is arbitrary and for the purpose of establishing a standard.

Table 2c. Limiting Dimensions of American National Standard General Purpose Acme Single-Start Screw Threads

Limiting Diameters		Nominal Diameter, D										
		1½	1¾	2	2¼	2½	2¾	3	3½	4	4½	5
		Threads per Inch[a]										
		4	4	4	3	3	3	2	2	2	2	2
		External Threads										
Classes 2G, 3G, and 4G Major Diameter	Max (D)	1.5000	1.7500	2.0000	2.2500	2.5000	2.7500	3.0000	3.5000	4.0000	4.5000	5.0000
	Min	1.4875	1.7375	1.9875	2.2333	2.4833	2.7333	2.9750	3.4750	3.9750	4.4750	4.9750
Classes 2G, 3G, and 4G Minor Diameter	Max	1.2300	1.4800	1.7300	1.8967	2.1467	2.3967	2.4800	2.9800	3.4800	3.9800	4.4800
Class 2G, Minor Diameter	Min	1.1965	1.4456	1.6948	1.8572	2.1065	2.3558	2.4326	2.9314	3.4302	3.9291	4.4281
Class 3G, Minor Diameter	Min	1.2144	1.4640	1.7136	1.8783	2.1279	2.3776	2.4579	2.9574	3.4568	3.9563	4.4558
Class 4G, Minor Diameter	Min	1.2189	1.4686	1.7183	1.8835	2.1333	2.3831	2.4642	2.9638	3.4634	3.9631	4.4627
Class 2G, Pitch Diameter	Max	1.3652	1.6145	1.8637	2.0713	2.3207	2.5700	2.7360	3.2350	3.7340	4.2330	4.7319
	Min	1.3429	1.5916	1.8402	2.0450	2.2939	2.5427	2.7044	3.2026	3.7008	4.1991	4.6973
Class 3G, Pitch Diameter	Max	1.3677	1.6171	1.8665	2.0743	2.3238	2.5734	2.7395	3.2388	3.7380	4.2373	4.7364
	Min	1.3573	1.6064	1.8555	2.0620	2.3113	2.5607	2.7248	3.2237	3.7225	4.2215	4.7202
Class 4G, Pitch Diameter	Max	1.3701	1.6198	1.8693	2.0773	2.3270	2.5767	2.7430	3.2425	3.7420	4.2415	4.7409
	Min	1.3627	1.6122	1.8615	2.0685	2.3181	2.5676	2.7325	3.2317	3.7309	4.2302	4.7294
		Internal Threads										
Classes 2G, 3G, and 4G Major Diameter	Min	1.5200	1.7700	2.0200	2.2700	2.5200	2.7700	3.0200	3.5200	4.0200	4.5200	5.0200
	Max	1.5400	1.7900	2.0400	2.2900	2.5400	2.7900	3.0400	3.5400	4.0400	4.5400	5.0400
Classes 2G, 3G, and 4G Minor Diameter	Min	1.2500	1.5000	1.7500	1.9167	2.1667	2.4167	2.5000	3.0000	3.5000	4.0000	4.5000
	Max	1.2625	1.5125	1.7625	1.9334	2.1834	2.4334	2.5250	3.0250	3.5250	4.0250	4.5250
Class 2G, Pitch Diameter	Min	1.3750	1.6250	1.8750	2.0833	2.3333	2.5833	2.7500	3.2500	3.7500	4.2500	4.7500
	Max	1.3973	1.6479	1.8985	2.1096	2.3601	2.6106	2.7816	3.2824	3.7832	4.2839	4.7846
Class 3G, Pitch Diameter	Min	1.3750	1.6250	1.8750	2.0833	2.3333	2.5833	2.7500	3.2500	3.7500	4.2500	4.7500
	Max	1.3854	1.6357	1.8860	2.0956	2.3458	2.5960	2.7647	3.2651	3.7655	4.2658	4.7662
Class 4G, Pitch Diameter	Min	1.3750	1.6250	1.8750	2.0833	2.3333	2.5833	2.7500	3.2500	3.7500	4.2500	4.7500
	Max	1.3824	1.6326	1.8828	2.0921	2.3422	2.5924	2.7605	3.2608	3.7611	4.2613	4.7615

Table 3. General Purpose Acme Single-Start Screw Thread Data *ASME/ANSI B1.5-1988*

Identification		Basic Diameters			Thread Data							
		Classes 2G, 3G, and 4G							Lead Angle at Basic Classes 2G, 3G, and 4G λ			
Nominal Sizes (All Classes)	Threads per Inch,[a] n	Major Diameter, D	Pitch Diameter, $D_2 = D - h$	Minor Diameter, $D_1 = D - 2h$	Pitch, P	Thickness at Pitch Line, $t = P/2$	Basic Height of Thread, $h = P/2$	Basic Width of Flat, $F = 0.3707P$	Deg	Min	Shear Area[b] Class 3G	Stress Area[c] Class 3G
¼	16	0.2500	0.2188	0.1875	0.06250	0.03125	0.03125	0.0232	5	12	0.350	0.0285
5⁄16	14	0.3125	0.2768	0.2411	0.07143	0.03571	0.03571	0.0265	4	42	0.451	0.0474
⅜	12	0.3750	0.3333	0.2917	0.08333	0.04167	0.04167	0.0309	4	33	0.545	0.0699
7⁄16	12	0.4375	0.3958	0.3542	0.08333	0.04167	0.04167	0.0309	3	50	0.660	0.1022
½	10	0.5000	0.4500	0.4000	0.10000	0.05000	0.05000	0.0371	4	3	0.749	0.1287
⅝	8	0.6250	0.5625	0.5000	0.12500	0.06250	0.06250	0.0463	4	3	0.941	0.2043
¾	6	0.7500	0.6667	0.5833	0.16667	0.08333	0.08333	0.0618	4	33	1.108	0.2848
⅞	6	0.8750	0.7917	0.7083	0.16667	0.08333	0.08333	0.0618	3	50	1.339	0.4150
1	5	1.0000	0.9000	0.8000	0.20000	0.10000	0.10000	0.0741	4	3	1.519	0.5354
1⅛	5	1.1250	1.0250	0.9250	0.20000	0.10000	0.10000	0.0741	3	33	1.751	0.709
1¼	5	1.2500	1.1500	1.0500	0.20000	0.10000	0.10000	0.0741	3	10	1.983	0.907
1⅜	4	1.3750	1.2500	1.1250	0.25000	0.12500	0.12500	0.0927	3	39	2.139	1.059
1½	4	1.5000	1.3750	1.2500	0.25000	0.12500	0.12500	0.0927	3	19	2.372	1.298
1¾	4	1.7500	1.6250	1.5000	0.25000	0.12500	0.12500	0.0927	2	48	2.837	1.851
2	4	2.0000	1.8750	1.7500	0.25000	0.12500	0.12500	0.0927	2	26	3.301	2.501
2¼	3	2.2500	2.0833	1.9167	0.33333	0.16667	0.16667	0.1236	2	55	3.643	3.049
2½	3	2.5000	2.3333	2.1667	0.33333	0.16667	0.16667	0.1236	2	36	4.110	3.870
2¾	3	2.7500	2.5833	2.4167	0.33333	0.16667	0.16667	0.1236	2	21	4.577	4.788
3	2	3.0000	2.7500	2.5000	0.50000	0.25000	0.25000	0.1853	3	19	4.786	5.27
3½	2	3.5000	3.2500	3.0000	0.50000	0.25000	0.25000	0.1853	2	48	5.73	7.50
4	2	4.0000	3.7500	3.5000	0.50000	0.25000	0.25000	0.1853	2	26	6.67	10.12
4½	2	4.5000	4.2500	4.0000	0.50000	0.25000	0.25000	0.1853	2	9	7.60	13.13
5	2	5.0000	4.7500	4.5000	0.50000	0.25000	0.25000	0.1853	1	55	8.54	16.53

[a] All other dimensions are given in inches.

[b] Per inch length of engagement of the external thread in line with the minor diameter crests of the internal thread. Figures given are the minimum shear area based on max D_1 and min d_2.

[c] Figures given are the minimum stress area based on the mean of the minimum minor and pitch diameters of the external thread.

See formulas for shear area and stress area on page 1794.

Table 4. American National Standard General Purpose Acme Single-Start Screw Threads — Pitch Diameter Allowances *ASME/ANSI B1.5-1988*

Nominal Size Range[a]		Allowances on External Threads [b]		
Above	To and Including	Class 2G, $0.008\sqrt{D}$	Class 3G, $0.006\sqrt{D}$	Class 4G, $0.004\sqrt{D}$
0	3/16	0.0024	0.0018	0.0012
3/16	5/16	0.0040	0.0030	0.0020
5/16	7/16	0.0049	0.0037	0.0024
7/16	9/16	0.0057	0.0042	0.0028
9/16	11/16	0.0063	0.0047	0.0032
11/16	13/16	0.0069	0.0052	0.0035
13/16	15/16	0.0075	0.0056	0.0037
15/16	1 1/16	0.0080	0.0060	0.0040
1 1/16	1 3/16	0.0085	0.0064	0.0042
1 3/16	1 5/16	0.0089	0.0067	0.0045
1 5/16	1 7/16	0.0094	0.0070	0.0047
1 7/16	1 9/16	0.0098	0.0073	0.0049
1 9/16	1 7/8	0.0105	0.0079	0.0052
1 7/8	2 1/8	0.0113	0.0085	0.0057
2 1/8	2 3/8	0.0120	0.0090	0.0060
2 3/8	2 5/8	0.0126	0.0095	0.0063
2 5/8	2 7/8	0.0133	0.0099	0.0066
2 7/8	3 1/4	0.0140	0.0105	0.0070
3 1/4	3 3/4	0.0150	0.0112	0.0075
3 3/4	4 1/4	0.0160	0.0120	0.0080
4 1/4	4 3/4	0.0170	0.0127	0.0085
4 3/4	5 1/2	0.0181	0.0136	0.0091

[a] The values in columns 3 to 5 are to be used for any size within the range shown in columns 1 and 2. These values are calculated from the mean of the range.

[b] An increase of 10 per cent in the allowance is recommended for each inch, or fraction thereof, that the length of engagement exceeds two diameters.

All dimensions in inches.

It is recommended that the sizes given in Table 3 be used whenever possible.

Allowances for Class $2G_2G$ threads in column 3 also apply to American National Standard Stub Acme threads ASME/ANSI B 1.8-1988.

Multiple Start Acme Threads.—The tabulated diameter-pitch data with allowances and tolerances relate to single-start threads. These data, as tabulated, may be and often are used for two-start Class 2G threads but this usage generally requires reduction of the full working tolerances to provide a greater allowance or clearance zone between the mating threads to assure satisfactory assembly.

When the class of thread requires smaller working tolerances than the 2G class or when threads with 3, 4, or more starts are required, some additional allowances or increased tolerances or both may be needed to ensure adequate working tolerances and satisfactory assembly of mating parts.

It is suggested that the allowances shown in Table 4 be used for all external threads and that allowances be applied to internal threads in the following ratios: for two-start threads, 50 per cent of the allowances shown in the third, fourth, and fifth columns of Table 4; for three-start threads, 75 per cent of these allowances; and for four-start threads, 100 per cent of these same values.

These values will provide for a 0.25-16 ACME-2G thread size, 0.002, 0.003, and 0.004 inch additional clearance for 2-, 3-, and 4-start threads, respectively. For a 5-2 ACME-3G thread size the additional clearances would be 0.0091, 0.0136, and 0.0181 inch, respectively. GO thread plug gages and taps would be increased by these same values. To maintain the same working tolerances on multi-start threads, the pitch diameter of the NOT GO thread plug gage would also be increased by these same values.

For multi-start threads with more than four starts, it is believed that the 100 per cent allowance provided by the above procedures would be adequate as index spacing variables would generally be no greater than on a four-start thread.

Table 5. American National Standard General Purpose Acme Single-Start Screw Threads — Pitch Diameter Tolerances *ASME/ANSI B1.5-1988*

Nom. Dia.,[a] D	Class of Thread 2G	Class of Thread 3G	Class of Thread 4G	Nom. Dia.,[a] D	Class of Thread 2G	Class of Thread 3G	Class of Thread 4G
	Diameter Increment				Diameter Increment		
	$0.006\sqrt{D}$	$0.0028\sqrt{D}$	$0.002\sqrt{D}$		$0.006\sqrt{D}$	$0.0028\sqrt{D}$	$0.002\sqrt{D}$
1/4	.00300	.00140	.00100	1 1/2	.00735	.00343	.00245
5/16	.00335	.00157	.00112	1 3/4	.00794	.00370	.00265
3/8	.00367	.00171	.00122	2	.00849	.00396	.00283
7/16	.00397	.00185	.00132	2 1/4	.00900	.00420	.00300
1/2	.00424	.00198	.00141	2 1/2	.00949	.00443	.00316
5/8	.00474	.00221	.00158	2 3/4	.00995	.00464	.00332
3/4	.00520	.00242	.00173	3	.01039	.00485	.00346
7/8	.00561	.00262	.00187	3 1/2	.01122	.00524	.00374
1	.00600	.00280	.00200	4	.01200	.00560	.00400
1 1/8	.00636	.00297	.00212	4 1/2	.01273	.00594	.00424
1 1/4	.00671	.00313	.00224	5	.01342	.00626	.00447
1 3/8	.00704	.00328	.00235	…	…	…	…
Thds. per Inch, n	Class of Thread 2G	Class of Thread 3G	Class of Thread 4G	Thds. per Inch, n	Class of Thread 2G	Class of Thread 3G	Class of Thread 4G
	Pitch Increment				Pitch Increment		
	$0.030\sqrt{1/n}$	$0.014\sqrt{1/n}$	$0.010\sqrt{1/n}$		$0.030\sqrt{1/n}$	$0.014\sqrt{1/n}$	$0.010\sqrt{1/n}$
16	.00750	.00350	.00250	4	.01500	.00700	.00500
14	.00802	.00374	.00267	3	.01732	.00808	.00577
12	.00866	.00404	.00289	2 1/2	.01897	.00885	.00632
10	.00949	.00443	.00316	2	.02121	.00990	.00707
8	.01061	.00495	.00354	1 1/2	.02449	.01143	.00816
6	.01225	.00572	.00408	1 1/3	.02598	.01212	.00866
5	.01342	.00626	.00447	1	.03000	.01400	.01000

[a] For a nominal diameter between any two tabulated nominal diameters, use the diameter increment for the larger of the two tabulated nominal diameters.

All other dimensions are given in inches.

For any particular size of thread, the pitch diameter tolerance is obtained by adding the *diameter increment* from the upper half of the table to the *pitch increment* from the lower half of the table. *Example:* A 1/4-16 Acme-2G thread has a pitch diameter tolerance of 0.00300 + 0.00750 = 0.0105 inch.

The equivalent tolerance on thread thickness is 0.259 times the pitch diameter tolerance.

Columns for the 2G Class of thread in this table also apply to American National Standard Stub Acme threads, ASME/ANSI B1.8-1988 (R1994).

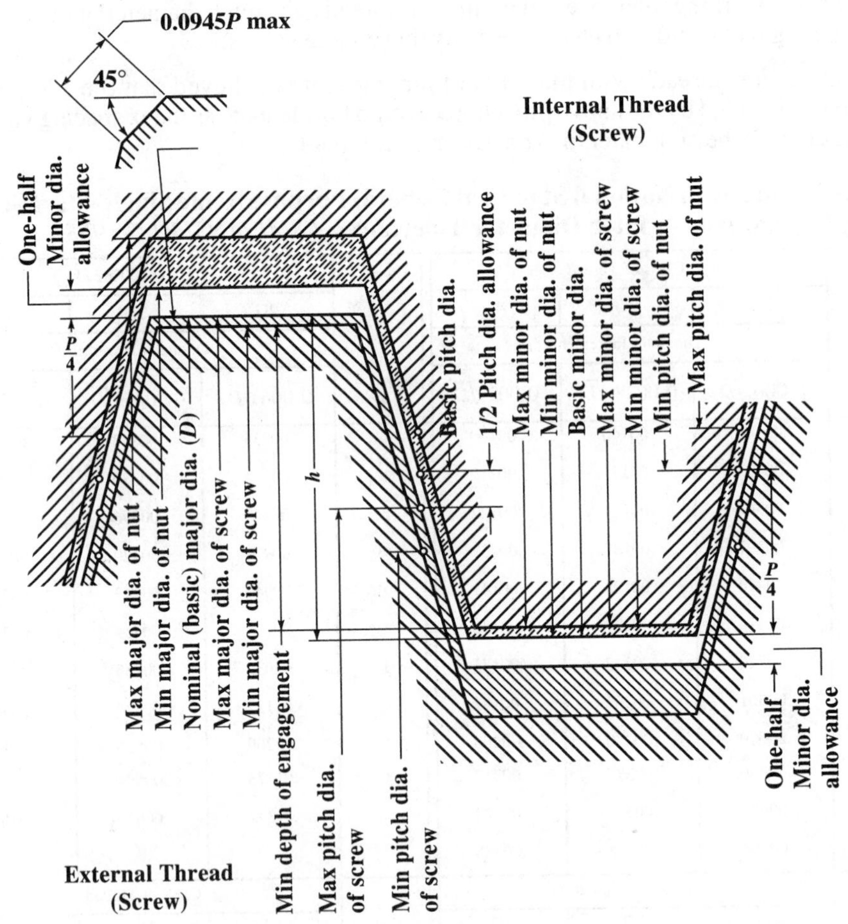

Disposition of Allowances, Tolerances, and Crest Clearances for General Purpose Single-start Acme Threads (All Classes)

In general, for multi-start threads of Classes 2G, 3G, and 4G the percentages would be applied, usually, to allowances for the same class, respectively. However, where exceptionally good control over lead, angle, and spacing variables would produce close to theoretical values in the product, it is conceivable that these percentages could be applied to Class 3G or Class 4G allowances used on Class 2G internally threaded product. Also, these percentages could be applied to Class 4G allowances used on Class 3G internally threaded product. It is not advocated that any change be made in externally threaded products.

Designations for gages or tools for internal threads could cover allowance requirements as follows:

GO and NOT GO thread plug gages for: 2.875-0.4*P*-0.8L-ACME-2G with 50 per cent of the 4G internal thread allowance.

Centralizing Acme Threads.—The three classes of Centralizing Acme threads in American National Standard ASME/ANSI B1.5-1988, designated as 2C, 3C, and 4C, have limited clearance at the major diameters of internal and external threads so that a bearing at the major diameters maintains approximate alignment of the thread axis and prevents wedging on the flanks of the thread. An alternative series having centralizing control on the *minor* diameter is described on page 1816. For any combination of the three classes of threads covered in this standard some end play or backlash will result. Classes 5C and 6C are not recommended for new designs.

Table 6. American National Standard Centralizing Acme Screw Thread Form — Basic Dimensions *ASME/ANSI B1.5-1988*

Thds per Inch, n	Pitch, P	Height of Thread (Basic), $h = P/2$	Total Height of Thread (All External Threads) $h_s = h + \frac{1}{2}$ allowance[a]	Thread Thickness (Basic), $t = P/2$	45-Deg Chamfer Crest of External Threads: Min Depth, $0.05P$	45-Deg Chamfer Crest of External Threads: Min Width of Chamfer-Flat, $0.0707P$	Max Fillet Radius, Root of Tapped Hole, $0.06P$	Fillet Radius at Min or Diameter of Screws Max (All) $0.10P$
16	0.06250	0.03125	0.0362	0.03125	0.0031	0.0044	0.0038	0.0062
14	0.07143	0.03571	0.0407	0.03571	0.0036	0.0050	0.0038	0.0071
12	0.08333	0.04167	0.0467	0.04167	0.0042	0.0059	0.0050	0.0083
10	0.10000	0.05000	0.0600	0.05000	0.0050	0.0071	0.0060	0.0100
8	0.12500	0.06250	0.0725	0.06250	0.0062	0.0088	0.0075	0.0125
6	0.16667	0.08333	0.0933	0.08333	0.0083	0.0119	0.0100	0.0167
5	0.20000	0.10000	0.1100	0.10000	0.0100	0.0141	0.0120	0.0200
4	0.25000	0.12500	0.1350	0.12500	0.0125	0.0177	0.0150	0.0250
3	0.33333	0.16667	0.1767	0.16667	0.0167	0.0236	0.0200	0.0333
2½	0.40000	0.20000	0.2100	0.20000	0.0200	0.0283	0.0240	0.0400
2	0.50000	0.25000	0.2600	0.25000	0.0250	0.0354	0.0300	0.0500
1½	0.66667	0.33333	0.3433	0.33333	0.0330	0.0471	0.0400	0.0667
1⅓	0.75000	0.37500	0.3850	0.37500	0.0380	0.0530	0.0450	0.0750
1	1.00000	0.50000	0.5100	0.50000	0.0500	0.0707	0.0600	0.1000

[a] Allowance is 0.020 inch for 10 or less threads per inch and 0.010 inch for more than 10 threads per inch.

All dimensions in inches. See diagram on page 1794.

Application: These three classes together with the accompanying specifications are for the purpose of ensuring the interchangeable manufacture of Centralizing Acme threaded parts. Each user is free to select the classes best adapted to his particular needs. It is suggested that external and internal threads of the same class be used together for centralizing assemblies, Class 2C providing the maximum end play or backlash. If less backlash or end play is desired, Classes 3C and 4C are provided. The requirement for a centralizing fit is that the sum of the major diameter tolerance plus the major diameter allowance on the internal thread, and the major diameter tolerance on the external thread shall equal or be less than the pitch diameter allowance on the external thread. A Class 2C external thread, which has a larger pitch diameter allowance than either a Class 3C or 4C, can be used interchangeably with a Class 2C, 3C, or 4C internal thread and fulfill this requirement. Similarly, a Class 3C external thread can be used interchangeably with a Class 3C or 4C internal thread, but only a Class 4C internal thread can be used with a Class 4C external thread.

Thread Form: The thread form is the same as the General Purpose Acme Thread and is shown in the figure on page 1793. The formulas accompanying the figure determine the basic dimensions, which are given in Table 6 for the most generally used pitches.

Angle of Thread: The angle between the sides of the thread measured in an axial plane is 29 degrees. The line bisecting this 29-degree angle shall be perpendicular to the axis of the thread.

Chamfers and Fillets: External threads have the crest corners chamfered at an angle of 45 degrees with the axis to a minimum depth of $P/20$ and a maximum depth of $P/15$. These modifications correspond to a minimum width of chamfer flat of $0.0707P$ and a maximum width of $0.0945P$ (see Table 6, columns 6 and 7).

External threads for Classes 2C, 3C, and 4C may have a fillet at the minor diameter not greater than $0.1P$

Thread Series: A series of diameters and pitches is recommended in the Standard as preferred. These diameters and pitches have been chosen to meet present needs with the fewest number of items in order to reduce to a minimum the inventory of both tools and gages. This series of diameters and associated pitches is given in Table 8.

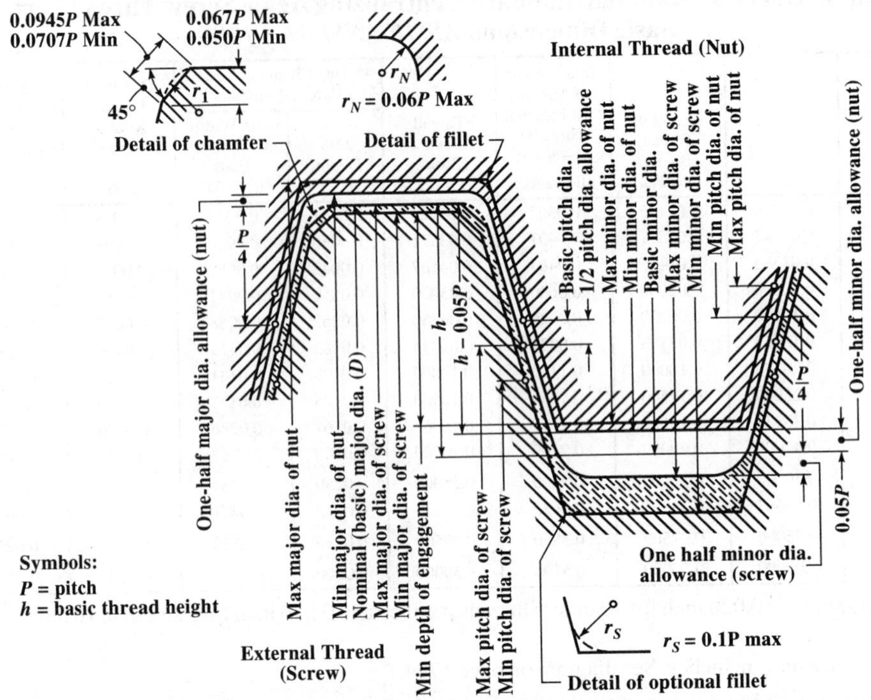

Disposition of Allowances, Tolerances, and Crest Clearances for Centralizing Single-Start Acme Threads—Classes 2C, 3C, and 4C

Table 7a. American National Standard Centralizing Acme Single-Start Screw Threads — Formulas for Determining Diameters *ASME/ANSI B1.5-1988*

D = Nominal Size or Diameter in Inches P = Pitch = 1 ÷ Number of Threads per Inch	
No.	Classes 2C, 3C, and 4C External Threads (Screws)
1	Major Dia., Max = D (Basic).
2	Major Dia., Min = D *minus* tolerance from Table 11, cols. 7, 8, or 10.
3	Pitch Dia., Max = Int. Pitch Dia., Min (Formula 9) *minus* allowance from Table 9, cols. 3, 4, or 5.
5	Minor Dia., Max = D *minus* P *minus* allowance from Table 11, col. 3.
6	Minor Dia., Min = Ext. Minor Dia., Max (Formula 5) *minus* 1.5 × Pitch Dia. tolerance from Table 10.
	Classes 2C, 3C, and 4C Internal Threads (Nuts)
7	Major Dia., Min = D *plus* allowance from Table 11, col. 4.
8	Major Dia., Max = Int. Major Dia., Min (Formula 7) *plus* tolerance from Table 11, cols. 7,9, or 11.
9	Pitch Dia., Min = D *minus* P/2 (Basic).
10	Pitch Dia., Max = Int. Pitch Dia., Min (Formula 9) *plus* tolerance from Table 10.
11	Minor Dia., Min = D *minus* 0.9P.
12	Minor Dia., Max = Int. Minor Dia., Min (Formula 11) *plus* tolerance from Table 11, col. 6.

Table 7b. Limiting Dimensions of American National Standard Centralizing Acme Single-Start Screw Threads, Classes 2C, 3C, and 4C *ASME/ANSI B1.5-1988*

Limiting Diameters		½	⅝	¾	⅞	1	1⅛	1¼	1⅜	1½
		Nominal Diameter, *D*								
		Threads per Inch[a]								
		10	8	6	6	5	5	5	4	4
External Threads										
Classes 2C, 3C, and 4C, Major Diameter	Max	0.5000	0.6250	0.7500	0.8750	1.0000	1.1250	1.2500	1.3750	1.5000
Class 2C, Major Diameter	Min	0.4975	0.6222	0.7470	0.8717	0.9965	1.1213	1.2461	1.3709	1.4957
Class 3C, Major Diameter	Min	0.4989	0.6238	0.7487	0.8736	0.9985	1.1234	1.2483	1.3732	1.4982
Class 4C, Major Diameter	Min	0.4993	0.6242	0.7491	0.8741	0.9990	1.1239	1.2489	1.3738	1.4988
Classes 2C, 3C, and 4C, Minor Diameter	Max	0.3800	0.4800	0.5633	0.6883	0.7800	0.9050	1.0300	1.1050	1.2300
Class 2C, Minor Diameter	Min	0.3594	0.4570	0.5371	0.6615	0.7509	0.8753	0.9998	1.0719	1.1965
Class 3C, Minor Diameter	Min	0.3704	0.4693	0.5511	0.6758	0.7664	0.8912	1.0159	1.0896	1.2144
Class 4C, Minor Diameter	Min	0.3731	0.4723	0.5546	0.6794	0.7703	0.8951	1.0199	1.0940	1.2188
Class 2C, Pitch Diameter	Max	0.4443	0.5562	0.6598	0.7842	0.8920	1.0165	1.1411	1.2406	1.3652
	Min	0.4306	0.5408	0.6424	0.7663	0.8726	0.9967	1.1210	1.2186	1.3429
Class 3C, Pitch Diameter	Max	0.4458	0.5578	0.6615	0.7861	0.8940	1.0186	1.1433	1.2430	1.3677
	Min	0.4394	0.5506	0.6534	0.7778	0.8849	1.0094	1.1339	1.2327	1.3573
Class 4C, Pitch Diameter	Max	0.4472	0.5593	0.6632	0.7880	0.8960	1.0208	1.1455	1.2453	1.3701
	Min	0.4426	0.5542	0.6574	0.7820	0.8895	1.0142	1.1388	1.2380	1.3627
Internal Threads										
Classes 2C, 3C, and 4C, Major Diameter	Min	0.5007	0.6258	0.7509	0.8759	1.0010	1.1261	1.2511	1.3762	1.5012
Classes 2C and 3C, Major Diameter	Max	0.5032	0.6286	0.7539	0.8792	1.0045	1.1298	1.2550	1.3803	1.5055
Class 4C, Major Diameter	Max	0.5021	0.6274	0.7526	0.8778	1.0030	0.1282	1.2533	1.3785	1.5036
Classes 2C, 3C, and 4C, Minor Diameter	Min	0.4100	0.5125	0.6000	0.7250	0.8200	0.9450	0.0700	1.1500	1.2750
	Max	0.04150	0.5187	0.6083	0.7333	0.8300	0.9550	1.0800	1.1625	1.2875
Class 2C, Pitch Diameter	Min	0.4500	0.5625	0.6667	0.7917	0.9000	1.0250	1.1500	1.2500	1.3750
	Max	0.4637	0.5779	0.6841	0.8096	0.9194	1.0448	1.1701	1.2720	1.3973
Class 3C, Pitch Diameter	Min	0.4500	0.5625	0.6667	0.7917	0.9000	1.0250	1.1500	1.2500	1.3750
	Max	0.4564	0.5697	0.6748	0.8000	0.9091	1.0342	1.1594	1.2603	1.3854
Class 4C, Pitch Diameter	Min	0.4500	0.5625	0.6667	0.7917	0.9000	1.0250	1.1500	1.2500	1.3750
	Max	0.4546	0.5676	0.6725	0.7977	0.9065	1.0316	1.1567	1.2573	1.3824

[a] All other dimensions are in inches. The selection of threads per inch is arbitrary and for the purpose of establishing a standard.

Table 7b. *(Continued)* **Limiting Dimensions of American National Standard Centralizing Acme Single-Start Screw Threads, Classes 2C, 3C, and 4C ASME/ANSI B1.5-1988**

Limiting Diameters		Nominal Diameter, *D*									
		1¾	2	2¼	2½	2¾	3	3½12	4	4½	5
		Threads per Inch[a]									
		4	4	3	3	3	2	2	2	2	2
External Threads											
Classes 2C, 3C, and 4C, Major Diameter	Max	1.7500	2.0000	2.2500	2.5000	2.7500	3.0000	3.5000	4.0000	4.5000	5.0000
Class 2C, Major Diameter	Min	1.7454	1.9951	2.2448	2.4945	2.7442	2.9939	3.4935	3.9930	4.4926	4.9922
Class 3C, Major Diameter	Min	1.7480	1.9979	2.2478	2.4976	2.7475	2.9974	3.4972	3.9970	4.4968	4.9966
Class 4C, Major Diameter	Min	1.7487	1.9986	2.2485	2.4984	2.7483	2.9983	3.4981	3.9980	4.4979	4.9978
Classes 2C, 3C, and 4C, Minor Diameter	Max	1.4800	1.7300	1.8967	2.1467	2.3967	2.4800	2.9800	3.4800	3.9800	4.4800
Class 2C, Minor Diameter	Min	1.4456	1.6948	1.8572	2.1065	2.3558	2.4326	2.9314	3.4302	3.9291	4.4281
Class 3C, Minor Diameter	Min	1.4640	1.7136	1.8783	2.1279	2.3776	2.4579	2.9574	3.4568	3.9563	4.4558
Class 4C, Minor Diameter	Min	1.4685	1.7183	1.8835	2.1333	2.3831	2.4642	2.9638	3.4634	3.9631	4.4627
Class 2C, Pitch Diameter	Max	1.6145	1.8637	2.0713	2.3207	2.5700	2.7360	3.2350	3.7340	4.2330	4.7319
	Min	1.5916	1.8402	2.0450	2.2939	2.5427	2.7044	3.2026	3.7008	4.1991	4.6973
Class 3C, Pitch Diameter	Max	1.6171	1.8665	2.0743	2.3238	2.5734	2.7395	3.2388	3.7380	4.2373	4.7364
	Min	1.6064	1.8555	2.0620	2.3113	2.5607	2.7248	3.2237	3.7225	4.2215	4.7202
Class 4C, Pitch Diameter	Max	1.6198	1.8693	2.0773	2.3270	2.5767	2.7430	3.2425	3.7420	4.2415	4.7409
	Min	1.6122	1.8615	2.0685	2.3181	2.5676	2.7325	3.2317	3.7309	4.2302	4.7294
Internal Threads											
Classes 2C, 3C, and 4C, Major Diameter	Min	1.7513	2.0014	2.2515	2.5016	2.7517	3.0017	3.5019	4.0020	4.5021	5.0022
Classes 2C and 3C, Major Diameter	Max	1.7559	2.0063	2.2567	2.5071	2.7575	3.0078	3.5084	4.0090	4.5095	5.0100
Class 4C, Major Diameter	Max	1.7539	2.0042	2.2545	2.5048	2.7550	3.0052	3.5056	4.0060	4.5063	5.0067
Classes 2C, 3C, and 4C, Minor Diameter	Min	1.5250	1.7750	1.9500	2.2000	2.4500	2.5500	3.0500	3.5500	4.0500	4.5500
	Max	1.5375	1.7875	1.9667	2.2167	2.4667	2.5750	3.0750	3.5750	4.0750	4.5750
Class 2C, Pitch Diameter	Min	1.6250	1.8750	2.0833	2.3333	2.5833	2.7500	3.2500	3.7500	4.2500	4.7500
	Max	1.6479	1.8985	2.1096	2.3601	2.6106	2.7816	3.2824	3.7832	4.2839	4.7846
Class 3C, Pitch Diameter	Min	1.6250	1.8750	2.0833	2.3333	2.5833	2.7500	3.2500	3.7500	4.2500	4.7500
	Max	1.6357	1.8860	2.0956	2.3458	2.5960	2.7647	3.2651	3.7655	4.2658	4.7662
Class 4C Pitch Diameter	Min	1.6250	1.8750	2.0833	2.3333	2.5833	2.7500	3.2500	3.7500	4.2500	4.7500
	Max	1.6326	1.8828	2.0921	2.3422	2.5924	2.7605	3.2608	3.7611	4.2613	4.7615

Table 8. American National Standard Centralizing Acme Single-Start Screw Thread Data *ASME/ANSI B1.5-1988*

Identification		Diameters			Thread Data					
Nominal Sizes (All Classes)	Threads per Inch,[a] n	Centralizing, Classes 2C, 3C, and 4C: Basic Major Diameter, D	Pitch Diameter, $D_2 = (D - h)$	Minor Diameter, $D_1 = (D - 2h)$	Pitch, P	Thickness at Pitch Line, $t = P/2$	Basic Height of Thread, $h = P/2$	Basic Width of Flat, $F = 0.3707P$	Lead Angle at Basic Pitch Diameter[a]: Centralizing Classes 2C, 3C, and 4C, λ Deg	Min
¼	16	0.2500	0.2188	0.1875	0.06250	0.03125	0.03125	0.0232	5	12
5/16	14	0.3125	0.2768	0.2411	0.07143	0.03571	0.03571	0.0265	4	42
⅜	12	0.3750	0.3333	0.2917	0.08333	0.04167	0.04167	0.0309	4	33
7/16	12	0.4375	0.3958	0.3542	0.08333	0.04167	0.04167	0.0309	3	50
½	10	0.5000	0.4500	0.4000	0.10000	0.05000	0.05000	0.0371	4	3
⅝	8	0.6250	0.5625	0.5000	0.12500	0.06250	0.06250	0.0463	4	3
¾	6	0.7500	0.6667	0.5833	0.16667	0.08333	0.08333	0.0618	4	33
⅞	6	0.8750	0.7917	0.7083	0.16667	0.08333	0.08333	0.0618	3	50
1	5	1.0000	0.9000	0.8000	0.20000	0.10000	0.10000	0.0741	4	3
1⅛	5	1.1250	1.0250	0.9250	0.20000	0.10000	0.10000	0.0741	3	33
1¼	5	1.2500	1.1500	1.0500	0.20000	0.10000	0.10000	0.0741	3	10
1⅜	4	1.3750	1.2500	1.1250	0.25000	0.12500	0.12500	0.0927	3	39
1½	4	1.5000	1.3750	1.2500	0.25000	0.12500	0.12500	0.0927	3	19
1¾	4	1.7500	1.6250	1.5000	0.25000	0.12500	0.12500	0.0927	2	48
2	4	2.0000	1.8750	1.7500	0.25000	0.12500	0.12500	0.0927	2	26
2¼	3	2.2500	2.0833	1.9167	0.33333	0.16667	0.16667	0.1236	2	55
2½	3	2.5000	2.3333	2.1667	0.33333	0.16667	0.16667	0.1236	2	36
2¾	3	2.7500	2.5833	2.4167	0.33333	0.16667	0.16667	0.1236	2	21
3	2	3.0000	2.7500	2.5000	0.50000	0.25000	0.25000	0.1853	3	19
3½	2	3.5000	3.2500	3.0000	0.50000	0.25000	0.25000	0.1853	2	48
4	2	4.0000	3.7500	3.5000	0.50000	0.25000	0.25000	0.1853	2	26
4½	2	4.5000	4.2500	4.0000	0.50000	0.25000	0.25000	0.1853	2	9
5	2	5.0000	4.7500	4.5000	0.50000	0.25000	0.25000	0.1853	1	55

[a] All other dimensions are given in inches.

Table 9. American National Standard Centralizing Acme Single-Start Screw Threads — Pitch Diameter Allowances *ASME/ANSI B1.5-1988*

Nominal Size Range[a]		Allowances on External Threads[b]		
		Centralizing		
Above	To and Including	Class 2C, $0.008\sqrt{D}$	Class 3C, $0.006\sqrt{D}$	Class 4C, $0.004\sqrt{D}$
0	3/16	0.0024	0.0018	0.0012
3/16	5/16	0.0040	0.0030	0.0020
5/16	7/16	0.0049	0.0037	0.0024
7/16	9/16	0.0057	0.0042	0.0028
9/16	11/16	0.0063	0.0047	0.0032
11/16	13/16	0.0069	0.0052	0.0035
13/16	15/16	0.0075	0.0056	0.0037
15/16	1 1/16	0.0080	0.0060	0.0040
1 1/16	1 3/16	0.0085	0.0064	0.0042
1 3/16	1 5/16	0.0089	0.0067	0.0045
1 5/16	1 7/16	0.0094	0.0070	0.0047
1 7/16	1 9/16	0.0098	0.0073	0.0049
1 9/16	1 7/8	0.0105	0.0079	0.0052
1 7/8	2 1/8	0.0113	0.0085	0.0057
2 1/8	2 3/8	0.0120	0.0090	0.0060
2 3/8	2 5/8	0.0126	0.0095	0.0063
2 5/8	2 7/8	0.0133	0.0099	0.0066
2 7/8	3 1/4	0.0140	0.0105	0.0070
3 1/4	3 3/4	0.0150	0.0112	0.0075
3 3/4	4 1/4	0.0160	0.0120	0.0080
4 1/4	4 3/4	0.0170	0.0127	0.0085
4 3/4	5 1/2	0.0181	0.0136	0.0091

[a] The values in columns 3 to 5 are to be used for any size within the range shown in columns 1 and 2. These values are calculated from the mean of the range.

[b] An increase of 10 per cent in the allowance is recommended for each inch, or fraction thereof, that the length of engagement exceeds two diameters.

All dimensions are given in inches.

It is recommended that the sizes given in Table 8 be used whenever possible.

Basic Diameters: The maximum major diameter of the external thread is basic and is the nominal major diameter for all classes.

The minimum pitch diameter of the internal thread is basic for all classes and is equal to the basic major diameter D minus the basic height of thread, h. The minimum minor diameter of the internal thread for all classes is $0.1P$ above basic.

Length of Engagement: The tolerances specified in this Standard are applicable to lengths of engagement not exceeding twice the nominal major diameter.

Pitch Diameter Allowances: Allowances applied to the pitch diameter of the external thread for all classes are given in Table 9.

Major and Minor Diameter Allowances: A minimum diametral clearance is provided at the minor diameter of all external threads by establishing the maximum minor diameter 0.020 inch below the basic minor diameter for 10 threads per inch and coarser, and 0.010 inch for finer pitches and by establishing the minimum minor diameter of the internal thread $0.1P$ greater than the basic minor diameter.

A minimum diametral clearance at the major diameter is obtained by establishing the minimum major diameter of the internal thread $0.001\sqrt{D}$ above the basic major diameter. These allowances are shown in Table 11.

Table 10. American National Standard Centralizing Acme Single-Start Screw Threads — Pitch Diameter Tolerances *ASME/ANSI B1.5-1988*

Nom. Dia.,[a] D	Class of Thread			Nom. Dia.,[a] D	Class of Thread		
	2C	3C	4C		2C	3C	4C
	Diameter Increment				Diameter Increment		
	$0.006\sqrt{D}$	$0.0028\sqrt{D}$	$0.002\sqrt{D}$		$0.006\sqrt{D}$	$0.0028\sqrt{D}$	$0.002\sqrt{D}$
1/4	.00300	.00140	.00100	1 1/2	.00735	.00343	.00245
5/16	.00335	.00157	.00112	1 3/4	.00794	.00370	.00265
3/8	.00367	.00171	.00122	2	.00849	.00396	.00283
7/16	.00397	.00185	.00132	2 1/4	.00900	.00420	.00300
1/2	.00424	.00198	.00141	2 1/2	.00949	.00443	.00316
5/8	.00474	.00221	.00158	2 3/4	.00995	.00464	.00332
3/4	.00520	.00242	.00173	3	.01039	.00485	.00346
7/8	.00561	.00262	.00187	3 1/2	.01122	.00524	.00374
1	.00600	.00280	.00200	4	.01200	.00560	.00400
1 1/8	.00636	.00297	.00212	4 1/2	.01273	.00594	.00424
1 1/4	.00671	.00313	.00224	5	.01342	.00626	.00447
1 3/8	.00704	.00328	.00235	…	…	…	…
Thds. per Inch, n	Class of Thread			Thds. per Inch, n	Class of Thread		
	2C	3C	4C		2C	3C	4C
	Pitch Increment				Pitch Increment		
	$0.030\sqrt{1/n}$	$0.014\sqrt{1/n}$	$0.010\sqrt{1/n}$		$0.030\sqrt{1/n}$	$0.014\sqrt{1/n}$	$0.010\sqrt{1/n}$
16	.00750	.00350	.00250	4	.01500	.00700	.00500
14	.00802	.00374	.00267	3	.01732	.00808	.00577
12	.00866	.00404	.00289	2 1/2	.01897	.00885	.00632
10	.00949	.00443	.00316	2	.02121	.00990	.00707
8	.01061	.00495	.00354	1 1/2	.02449	.01143	.00816
6	.01225	.00572	.00408	1 1/3	.02598	.01212	.00866
5	.01342	.00626	.00447	1	.03000	.01400	.01000

[a] For a nominal diameter between any two tabulated nominal diameters, use the diameter increment for the larger of the two tabulated nominal diameters.

All dimensions are given in inches.

For any particular size of thread, the pitch diameter tolerance is obtained by adding the *diameter increment* from the upper half of the table to the *pitch increment* from the lower half of the table. *Example:* A 0.250-16-ACME-2C thread has a pitch diameter tolerance of 0.00300 + 0.00750 = 0.0105 inch.

The equivalent tolerance on thread thickness is 0.259 times the pitch diameter tolerance.

Table 11. American National Standard Centralizing Acme Single-Start Screw Threads — Tolerances and Allowances for Major and Minor Diameters *ASME/ANSI B1.5-1988*

Size (Nom.)	Thds[a] per Inch	Allowance From Basic Major and Minor Diameters (All Classes): Minor Diam,[c] All External Threads (Minus)	Allowance — Internal Thread: Major Diam,[d] (Plus $0.0010\sqrt{D}$)	Allowance — Internal Thread: Minor Diam,[c] (Plus $0.1P$)	Toler. on Minor Diam,[b] All Internal Threads, (Plus $0.05P$)	Tolerance on Major Diameter Plus on Internal, Minus on External Threads — Class 2C: External and Internal Threads, $0.0035\sqrt{D}$	Class 3C: External Thread, $0.0015\sqrt{D}$	Class 3C: Internal Thread, $0.0035\sqrt{D}$	Class 4C: External Thread, $0.0010\sqrt{D}$	Class 4C: Internal Thread, $0.0020\sqrt{D}$
¼	16	0.010	0.0005	0.0062	0.0050	0.0017	0.0007	0.0017	0.0005	0.0010
5⁄16	14	0.010	0.0006	0.0071	0.0050	0.0020	0.0008	0.0020	0.0006	0.0011
⅜	12	0.010	0.0006	0.0083	0.0050	0.0021	0.0009	0.0021	0.0006	0.0012
7⁄16	12	0.010	0.0007	0.0083	0.0050	0.0023	0.0010	0.0023	0.0007	0.0013
½	10	0.020	0.0007	0.0100	0.0050	0.0025	0.0011	0.0025	0.0007	0.0014
⅝	8	0.020	0.0008	0.0125	0.0062	0.0028	0.0012	0.0028	0.0008	0.0016
¾	6	0.020	0.0009	0.0167	0.0083	0.0030	0.0013	0.0030	0.0009	0.0017
⅞	6	0.020	0.0009	0.0167	0.0083	0.0033	0.0014	0.0033	0.0009	0.0019
1	5	0.020	0.0010	0.0200	0.0100	0.0035	0.0015	0.0035	0.0010	0.0020
1⅛	5	0.020	0.0011	0.0200	0.0100	0.0037	0.0016	0.0037	0.0011	0.0021
1¼	5	0.020	0.0011	0.0200	0.0100	0.0039	0.0017	0.0039	0.0011	0.0022
1⅜	4	0.020	0.0012	0.0250	0.0125	0.0041	0.0018	0.0041	0.0012	0.0023
1½	4	0.020	0.0012	0.0250	0.0125	0.0043	0.0018	0.0043	0.0012	0.0024
1¾	4	0.020	0.0013	0.0250	0.0125	0.0046	0.0020	0.0046	0.0013	0.0026
2	4	0.020	0.0014	0.0250	0.0125	0.0049	0.0021	0.0049	0.0014	0.0028
2¼	3	0.020	0.0015	0.0333	0.0167	0.0052	0.0022	0.0052	0.0015	0.0030
2½	3	0.020	0.0016	0.0333	0.0167	0.0055	0.0024	0.0055	0.0016	0.0032
2¾	3	0.020	0.0017	0.0333	0.0167	0.0058	0.0025	0.0058	0.0017	0.0033
3	2	0.020	0.0017	0.0500	0.0250	0.0061	0.0026	0.0061	0.0017	0.0035
3½	2	0.020	0.0019	0.0500	0.0250	0.0065	0.0028	0.0065	0.0019	0.0037
4	2	0.020	0.0020	0.0500	0.0250	0.0070	0.0030	0.0070	0.0020	0.0040
4½	2	0.020	0.0021	0.0500	0.0250	0.0074	0.0032	0.0074	0.0021	0.0042
5	2	0.020	0.0022	0.0500	0.0250	0.0078	0.0034	0.0078	0.0022	0.0045

[a] All other dimensions are given in inches. Intermediate pitches take the values of the next coarser pitch listed.
[b] To avoid a complicated formula and still provide an adequate tolerance, the pitch factor is used as a basis, with the minimum tolerance set at 0.005 in.
[c] The minimum clearance at the minor diameter between the internal and external thread is the sum of the values in columns 3 and 5.
[d] The minimum clearance at the major diameter between the internal and external thread is equal to col. 4.

Tolerance on minor diameter of all external threads is 1.5 × pitch diameter tolerance.

Values for intermediate diameters should be calculated from the formulas in column headings, but ordinarily may be interpolated.

Major and Minor Diameter Tolerances: The tolerances on the major and minor diameters of the external and internal threads are listed in Table 11 and are based upon the formulas seven in the column headings.

An increase of 10 per cent in the allowance is recommended for each inch or fraction thereof that the length of engagement exceeds two diameters.

For information on gages for Centralizing Acme threads the Standard ASME/ANSI B1.5 should be consulted.

Pitch Diameter Tolerances: Pitch diameter tolerances for Classes 2C, 3C and 4C for various practicable combinations of diameter and pitch are given in Table 10. The ratios of the pitch diameter tolerances of Classes 2C, 3C, and 4C are 3.0, 1.4, and 1, respectively.

Application of Tolerances: The tolerances specified are such as to insure interchangeability and maintain a high grade of product. The tolerances on the diameters of internal threads are plus, being applied from the minimum sizes to above the minimum sizes. The tolerances on the diameters of external threads are minus, being applied from the maximum sizes to below the maximum sizes. The pitch diameter tolerances for an external or internal thread of a given class are the same

Limiting Dimensions: Limiting dimensions for Centralizing Acme threads in the preferred series of diameters and pitches are given in Table 7b. These limits are based on the formulas in Table 7a.

For combinations of pitch and diameter other than those in the preferred series the formulas in Table 7a and the data in the tables referred to therein make it possible to readily determine the limiting dimension required.

Designation of Centralizing Acme Threads.—The following examples are given to show how these Acme threads are designated on drawings, in specifications, and on tools and gages:

Example, 1.750-6-ACME-4C: Indicates a Centralizing Class 4C Acme thread of 1.750-inch major diameter, 0.1667-inch pitch, single thread, right-hand.

Example, 1.750-6-ACME-4C-LH: Indicates the same thread left-hand.

Example, 2.875-0.4P-0.8L-ACME-3C (Two Start): Indicates a Centralizing Class 3C Acme thread with 2.875-inch major diameter, 0.4-inch pitch, 0.8-inch lead, double thread, right-hand.

Example, 2.500-0.3333P-0.6667L-ACME-4C (Two Start): Indicates a Centralizing Class 4C Acme thread with 2.500-inch nominal major diameter (basic major diameter 2.500 inches), 0.3333-inch pitch, 0.6667-inch lead, double thread, right-hand. The same thread left-hand would have LH at the end of the designation.

American National Standard Stub Acme Threads.—This American National Standard ASME/ANSI B1.8-1988 (R1994) provides a Stub Acme screw thread for those unusual applications where, due to mechanical or metallurgical considerations, a coarse-pitch thread of shallow depth is required. The fit of Stub Acme threads corresponds to the Class 2G General Purpose Acme thread in American National Standard ANSI B1.5-1988. For a fit having less backlash, the tolerances and allowances for Classes 3G or 4G General Purpose Acme threads may be used.

Thread Form: The thread form and basic formulas for Stub Acme threads are given on page 1793 and the basic dimensions in Table 12.

Table 12. American National Standard Stub Acme Screw Thread Form — Basic Dimensions *ASME/ANSI B1.8-1988 (R1994)*

Thds. per Inch[a] n	Pitch, $P = 1/n$	Height of Thread (Basic), $0.3P$	Total Height of Thread, $0.3P + ½$ allowance[b]	Thread Thickness (Basic), $P/2$	Width of Flat: Crest of InternalThread (Basic), $0.4224P$	Width of Flat: Root of Internal Thread, $0.4224P - 0.259 \times$ allowance[b]
16	0.06250	0.01875	0.0238	0.03125	0.0264	0.0238
14	0.07143	0.02143	0.0264	0.03571	0.0302	0.0276
12	0.08333	0.02500	0.0300	0.04167	0.0352	0.0326
10	0.10000	0.03000	0.0400	0.05000	0.0422	0.0370
9	0.11111	0.03333	0.0433	0.05556	0.0469	0.0417
8	0.12500	0.03750	0.0475	0.06250	0.0528	0.0476
7	0.14286	0.04285	0.0529	0.07143	0.0603	0.0551
6	0.16667	0.05000	0.0600	0.08333	0.0704	0.0652
5	0.20000	0.06000	0.0700	0.10000	0.0845	0.0793
4	0.25000	0.07500	0.0850	0.12500	0.1056	0.1004
3½	0.28571	0.08571	0.0957	0.14286	0.1207	0.1155
3	0.33333	0.10000	0.1100	0.16667	0.1408	0.1356
2½	0.40000	0.12000	0.1300	0.20000	0.1690	0.1638
2	0.50000	0.15000	0.1600	0.25000	0.2112	0.2060
1½	0.66667	0.20000	0.2100	0.33333	0.2816	0.2764
1⅓	0.75000	0.22500	0.2350	0.37500	0.3168	0.3116
1	1.00000	0.30000	0.3100	0.50000	0.4224	0.4172

[a] All other dimensions in inches. See diagram, page 1793.

[b] Allowance is 0.020 inch for 10 or less threads per inch and 0.010 inch for more than 10 threads per inch.

Table 13a. American National Standard Stub Acme Single-Start Screw Threads — Formulas for Determining Diameters *ASME/ANSI B1.8-1988 (R1994)*

No.	D = Basic Major Diameter and Nominal Size in Inches D_2 = Basic Pitch Diameter = $D - 0.3P$ D_1 = Basic Minor Diameter = $D - 0.6P$
	External Threads (Screws)
1	Major Dia., Max = D.
2	Major Dia., Min. = D *minus* $0.05P$.
3	Pitch Dia., Max. = D_2 *minus* allowance from column 3, Table 4.
4	Pitch Dia., Min. = Pitch Dia., Max. (Formula 3) *minus* Class 2G tolerance from Table 5.
5	Minor Dia., Max. = D_1 *minus* 0.020 for 10 threads per inch and coarser and 0.010 for finer pitches.
6	Minor Dia., Min. = Minor Dia., Max. (Formula 5) *minus* Class 2G pitch diameter tolerance from Table 5.
	Internal Threads (Nuts)
7	Major Dia., Min. = D *plus* 0.020 for 10 threads per inch and coarser and 0.010 for finer pitches.
8	Major Dia., Max.= Major Dia., Min. (Formula 7) *plus* Class 2G pitch diameter tolerance from Table 5.
9	Pitch Dia., Min. = $D_2 = D - 0.3P$
10	Pitch Dia., Max. = Pitch Dia., Min. (Formula 9) *plus* Class 2G tolerance from Table 5.
11	Minor Dia., Min. = $D_1 = D - 0.6P$
12	Minor Dia., Max = Minor Dia., Min. (Formula 11) *plus* $0.05P$.

Table 13b. Limiting Dimensions for American National Standard Stub Acme Single-Start Screw Threads *ASME/ANSI B1.8-1988 (R1994)*

Limiting Diameters		Nominal Diameter, *D*											
		1/4	5/16	3/8	7/16	1/2	5/8	3/4	7/8	1	1 1/8	1 1/4	1 3/8
		Threads per Inch[a]											
		16	14	12	12	10	8	6	6	5	5	5	4
		External Threads											
Major Dia.	Max (*D*)	0.2500	0.3125	0.3750	0.4375	0.5000	0.6250	0.7500	0.8750	1.0000	1.1250	1.2500	1.3750
	Min	0.2469	0.3089	0.3708	0.4333	0.4950	0.6188	0.7417	0.8667	0.9900	1.1150	1.2400	1.3625
Pitch Dia.	Max	0.2272	0.2871	0.3451	0.4076	0.4643	0.5812	0.6931	0.8175	0.9320	1.0565	1.1811	1.2906
	Min	0.2167	0.2757	0.3328	0.3950	0.4506	0.5658	0.6757	0.7996	0.9126	1.0367	1.1610	1.2686
Minor Dia.	Max	0.2024	0.2597	0.3150	0.3775	0.4200	0.5300	0.6300	0.7550	0.8600	0.9850	1.1100	1.2050
	Min	0.1919	0.2483	0.3027	0.3649	0.4063	0.5146	0.6126	0.7371	0.8406	0.9652	1.0899	1.1830
		Internal Threads											
Major Dia.	Min	0.2600	0.3225	0.3850	0.4475	0.5200	0.6450	0.7700	0.8950	1.0200	1.1450	1.2700	1.3950
	Max	0.2705	0.3339	0.3973	0.4601	0.5337	0.6604	0.7874	0.9129	1.0394	1.1648	1.2901	1.4170
Pitch Dia.	Min	0.2312	0.2911	0.3500	0.4125	0.4700	0.5875	0.7000	0.8250	0.9400	1.0650	1.1900	1.3000
	Max	0.2417	0.3025	0.3623	0.4251	0.4837	0.6029	0.7174	0.8429	0.9594	1.0848	1.2101	1.3220
Minor Dia.	Min	0.2125	0.2696	0.3250	0.3875	0.4400	0.5500	0.6500	0.7750	0.8800	1.0050	1.1300	1.2250
	Max	0.2156	0.2732	0.3292	0.3917	0.4450	0.5562	0.6583	0.7833	0.8900	1.0150	1.1400	1.2375

[a] All other dimensions are given in inches.

Limiting Diameters		Nominal Diameter, *D*										
		1 1/2	1 3/4	2	2 1/4	2 1/2	2 3/4	3	3 1/2	4	4 1/2	5
		Threads per Inch[a]										
		4	4	4	3	3	3	2	2	2	2	2
		External Threads										
Major Dia.	Max (*D*)	1.5000	1.7500	2.0000	2.2500	2.5000	2.7500	3.0000	3.5000	4.0000	4.5000	5.0000
	Min	1.4875	1.7375	1.9875	2.2333	2.4833	2.7333	2.9750	3.4750	3.9750	4.4750	4.9750
Pitch Dia.	Max	1.4152	1.6645	1.9137	2.1380	2.3874	2.6367	2.8360	3.3350	3.8340	4.3330	4.8319
	Min	1.3929	1.6416	1.8902	2.1117	2.3606	2.6094	2.8044	3.3026	3.8008	4.2991	4.7973
Minor Dia.	Max	1.3300	1.5800	1.8300	2.0300	2.2800	2.5300	2.6800	3.1800	3.6800	4.1800	4.6800
	Min	1.3077	1.5571	1.8065	2.0037	2.2532	2.5027	2.6484	3.1476	3.6468	4.1461	4.6454
		Internal Threads										
Major Dia.	Min	1.5200	1.7700	2.0200	2.2700	2.5200	2.7700	3.0200	3.5200	4.0200	4.5200	5.0200
	Max	1.5423	1.7929	2.0435	2.2963	2.5468	2.7973	3.0516	3.5524	4.0532	4.5539	5.0546
Pitch Dia.	Min	1.4250	1.6750	1.9250	2.1500	2.4000	2.6500	2.8500	3.3500	3.8500	4.3500	4.8500
	Max	1.4473	1.6979	1.9485	2.1763	2.4268	2.6773	2.8816	3.3824	3.8832	4.3839	4.8846
Minor Dia.	Min	1.3500	1.6000	1.8500	2.0500	2.3000	2.5500	2.7000	3.2000	3.7000	4.2000	4.7000
	Max	1.3625	1.6125	1.8625	2.0667	2.3167	2.5667	2.7250	3.2250	3.7250	4.2250	4.7250

Allowances and Tolerances: The major and minor diameter allowances for Stub Acme threads are the same as those given for General Purpose Acme threads on page 1792.

Pitch diameter allowances for Stub Acme threads are the same as for Class 2G General Purpose Acme threads and are given in column 3 of Table 4. Pitch diameter tolerances for Stub Acme threads are the same as for Class 2G General Purpose Acme threads and are given in columns 2 and 7 of Table 5.

Limiting Dimensions: Limiting dimensions of American Standard Stub Acme threads may be determined by using the formulas given in Table 13a, or directly from Table 13b. The diagram below shows the limits of size for Stub Acme threads.

Thread Series: A preferred series of diameters and pitches for General Purpose Acme threads (Table 14) is recommended for Stub Acme threads.

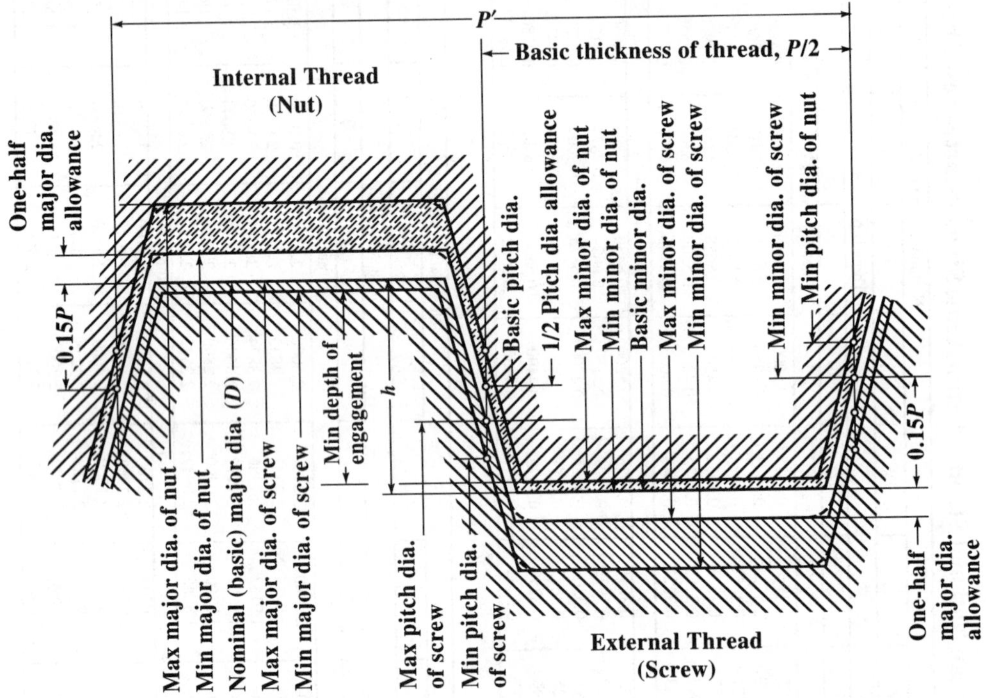

Limits of Size, Allowances, Tolerances, and Crest Clearances for American National Standard Stub Acme Threads

Stub Acme Thread Designations.—The method of designation for Standard Stub Acme threads is illustrated in the following examples: 0.500-20 Stub Acme indicates a ½-inch major diameter, 20 threads per inch, right hand, single thread, Standard Stub Acme thread. The designation 0.500-20 Stub Acme-LH indicates the same thread except that it is left hand.

Alternative Stub Acme Threads.—Since one Stub Acme thread form may not meet the requirements of all applications, basic data for two of the other commonly used forms are included in the appendix of the American Standard for Stub Acme Threads. These so-called Modified Form 1 and Modified Form 2 threads utilize the same tolerances and allowances as Standard Stub Acme threads and have the same major diameter and basic thread thickness at the pitchline (0.5*P*). The basic height of Form 1 threads, *h*, is 0.375*P*; for Form 2 it is 0.250*P*. The basic width of flat at the crest of the internal thread is 0.4030*P* for Form 1 and 0.4353*P* for Form 2.

Table 14. Stub Acme Screw Thread Data *ASME/ANSI B1.8-1988* **(R1994)**

Identification		Basic Diameters			Thread Data					
Nominal Sizes	Threads per Inch,[a] n	Major Diameter, D	Pitch Diameter, $D_2 = D - h$	Minor Diameter, $D_1 = D - 2h$	Pitch, P	Thread Thickness at Pitch Line, $t = P/2$	Basic Thread Height, $h = 0.3P$	Basic Width of Flat, $0.4224P$	Lead Angle at Basic Pitch Diameter	
									Deg	Min
¼	16	0.2500	0.2312	0.2125	0.06250	0.03125	0.01875	0.0264	4	54
5⁄16	14	0.3125	0.2911	0.2696	0.07143	0.03572	0.02143	0.0302	4	28
⅜	12	0.3750	0.3500	0.3250	0.08333	0.04167	0.02500	0.0352	4	20
7⁄16	12	0.4375	0.4125	0.3875	0.08333	0.04167	0.02500	0.0352	3	41
½	10	0.5000	0.4700	0.4400	0.10000	0.05000	0.03000	0.0422	3	52
⅝	8	0.6250	0.5875	0.5500	0.12500	0.06250	0.03750	0.0528	3	52
¾	6	0.7500	0.7000	0.6500	0.16667	0.08333	0.05000	0.0704	4	20
⅞	6	0.8750	0.8250	0.7750	0.16667	0.08333	0.05000	0.0704	3	41
1	5	1.0000	0.9400	0.8800	0.20000	0.10000	0.06000	0.0845	3	52
1⅛	5	1.1250	1.0650	1.0050	0.20000	0.10000	0.06000	0.0845	3	25
1¼	5	1.2500	1.1900	1.1300	0.20000	0.10000	0.06000	0.0845	3	4
1⅜	4	1.3750	1.3000	1.2250	0.25000	0.12500	0.07500	0.1056	3	30
1½	4	1.5000	1.4250	1.3500	0.25000	0.12500	0.07500	0.1056	3	12
1¾	4	1.7500	1.6750	1.6000	0.25000	0.12500	0.07500	0.1056	2	43
2	4	2.0000	1.9250	1.8500	0.25000	0.12500	0.07500	0.1056	2	22
2¼	3	2.2500	2.1500	2.0500	0.33333	0.16667	0.10000	0.1408	2	50
2½	3	2.5000	2.4000	2.3000	0.33333	0.16667	0.10000	0.1408	2	32
2¾	3	2.7500	2.6500	2.5500	0.33333	0.16667	0.10000	0.1408	2	18
3	2	3.0000	2.8500	2.7000	0.50000	0.25000	0.15000	0.2112	3	12
3½	2	3.5000	3.3500	3.2000	0.50000	0.25000	0.15000	0.2112	2	43
4	2	4.0000	3.8500	3.7000	0.50000	0.25000	0.15000	0.2112	2	22
4½	2	4.5000	4.3500	4.2000	0.50000	0.25000	0.15000	0.2112	2	6
5	2	5.0000	4.8500	4.7000	0.50000	0.25000	0.15000	0.2112	1	53

[a] All other dimensions are given in inches.

The pitch diameter and minor diameter for Form 1 threads will be smaller than similar values for the Standard Stub Acme Form and for Form 2 they will be larger owing to the differences in basic thread height h. Therefore, in calculating the dimensions of Form 1 and Form 2 threads using Formulas 1 through 12 in Table 13a, it is only necessary to substitute the following values in applying the formulas: For Form 1, $D_2 = D - 0.375P$, $D_1 = D - 0.75P$; for Form 2, $D_2 = D - 0.25P$, $D_1 = D - 0.5P$.

Thread Designation: These threads are designated in the same manner as Standard Stub Acme threads except for the insertion of either M1 or M2 after "Acme." Thus, 0.500-20 Stub Acme M1 for a Form 1 thread; and 0.500-20 Stub Acme M2 for a Form 2 thread.

Acme Centralizing Threads—Alternative Series with Minor Diameter Centralizing Control.—When Acme centralizing threads are produced in single units or in very small quantities (and principally in sizes larger than the range of commercial taps and dies) where the manufacturing process employs cutting tools (such as lathe cutting), it may be economically advantageous and therefore desirable to have the centralizing control of the mating threads located at the *minor diameters.*

Particularly under the above-mentioned type of manufacturing, the two advantages cited for minor diameter centralizing control over centralizing control at the major diameters of the mating threads are: 1) Greater ease and faster checking of machined thread dimensions. It is much easier to measure the minor diameter (root) of the external thread and the mating minor diameter (crest or bore) of the internal thread than it is to determine the major diameter (root) of the internal thread and the major diameter (crest or turn) of the external thread; and 2) better manufacturing control of the machined size due to greater ease of checking.

In the event that minor diameter centralizing is necessary, recalculate all thread dimensions, reversing major and minor diameter allowances, tolerances, radii, and chamfer.

60-Degree Stub Thread.—Former American Standard B1.3-1941 included a 60-degree stub thread for use where design or operating conditions could be better satisfied by the use of this thread, or other modified threads, than by Acme threads. Data for 60-Degree Stub thread form are given in the accompanying diagram.

60-Degree Stub Thread

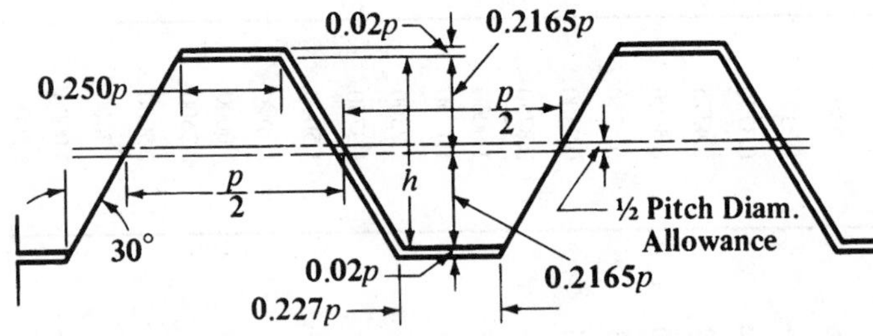

A clearance of at least 0.02 × pitch is added to depth h to produce extra depth, thus avoiding interference with threads of mating part at minor or major diameters.

Basic thread thickness at pitch line = 0.5 × pitch p; basic depth h = 0.433 × pitch; basic width of flat at crest = 0.25 × pitch; width of flat at root of screw thread = 0.227 × pitch; basic pitch diameter = basic major diameter − 0.433 × pitch; basic minor diameter = basic major diameter − 0.866 × pitch.

BUTTRESS THREADS

10-Degree Modified Square Thread.—The included angle between the sides of the thread is 10 degrees (see accompanying diagram). The angle of 10 degrees results in a thread which is the practical equivalent of a "square thread," and yet is capable of economical production. Multiple thread milling cutters and ground thread taps should not be specified for modified square threads of the larger lead angles without consulting the cutting tool manufacturer.

Formulas:

In the following formulas, D = basic major diameter; E = basic pitch diameter; K = basic minor diameter; p = pitch; h = basic depth of thread on screw depth when there is no clearance between root of screw and crest of thread on nut; t = basic thickness of thread at pitch line; F = basic width of flat at crest of screw thread; G = basic width of flat at root of screw thread; C = clearance between root of screw and crest of thread on nut: $E = D - 0.5p$; $K = D - p$; $h = 0.5p$ (see Note); $t = 0.5p$; $F = 0.4563p$; $G = 0.4563p - (0.17 \times C)$. *Note:* A clearance should be added to depth h to avoid interference with threads of mating parts at minor or major diameters.

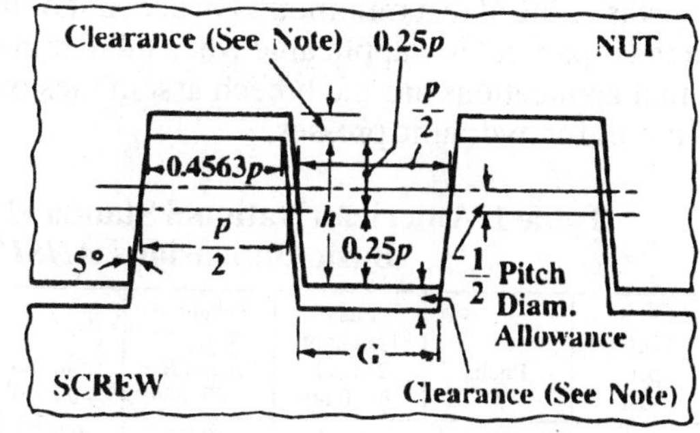

Threads of Buttress Form.—The buttress form of thread has certain advantages in applications involving exceptionally high stresses along the thread axis in one direction only. The contacting flank of the thread, which takes the thrust, is referred to as the *pressure flank* and is so nearly perpendicular to the thread axis that the radial component of the thrust is reduced to a minimum. Because of the small radial thrust, this form of thread is particularly applicable where tubular members are screwed together, as in the case of breech mechanisms of large guns and airplane propeller hubs.

Diagram 1A shows a common form. The front or load-resisting face is perpendicular to the axis of the screw and the thread angle is 45 degrees. According to one rule, the pitch P = 2 × screw diameter ÷ 15. The thread depth d may equal ¾ × pitch, making the flat f = ⅛ × pitch. Sometimes depth d is reduced to ⅔ × pitch, making f = ⅙ × pitch.

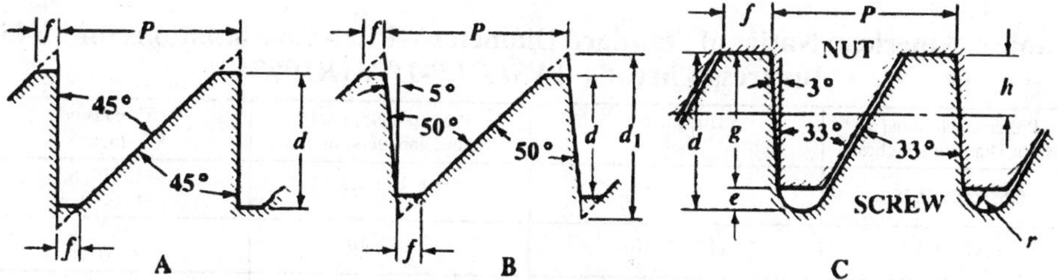

Fig. 1. Buttress Form Threads

The load-resisting side or flank may be inclined an amount (diagram 1B) ranging usually from 1 to 5 degrees to avoid cutter interference in milling the thread. With an angle of 5 degrees and an included thread angle of 50 degrees, if the width of the flat f at both crest and root equals ⅛ × pitch, then the thread depth equals 0.69 × pitch or ¾ d_1.

The saw-tooth form of thread illustrated by diagram 1C is known in Germany as the "Sägengewinde" and in Italy as the "Fillettatura a dente di Sega." Pitches are standardized from 2 millimeters up to 48 millimeters in the German and Italian specifications. The front face inclines 3 degrees from the perpendicular and the included angle is 33 degrees.

The thread depth d for the screw = 0.86777 × pitch P. The thread depth g for the nut = 0.75 × pitch. Dimension $h = 0.341 \times P$. The width f of flat at the crest of the thread on the screw = 0.26384 × pitch. Radius r at the root = 0.12427 × pitch. The clearance space e = 0.11777 × pitch.

American National Standard Buttress Inch Screw Threads.—The buttress form of thread has certain advantages in applications involving exceptionally high stresses along the thread axis in one direction only. As the thrust side (load flank) of the standard buttress thread is made very nearly perpendicular to the thread axis, the radial component of the thrust is reduced to a minimum. On account of the small radial thrust, the buttress form of thread is particularly applicable when tubular members are screwed together. Examples of actual applications are the breech assemblies of large guns, airplane propeller hubs, and columns for hydraulic presses.

Table 1. American National Standard Inch Buttress Screw Threads— Basic Dimensions *ANSI B1.9-1973 (R1992)*

Thds.[a] per Inch	Pitch, p	Basic Height of Thread, $h = 0.6p$	Height of Sharp-V Thread, $H = 0.89064p$	Crest Truncation, $f = 0.14532p$	Height of Thread, h_s or h_n = $0.66271p$	Max. Root Truncation,[b] $s = 0.0826p$	Max. Root Radius,[c] $r = 0.0714p$	Width of Flat at Crest, $F = 0.16316p$
20	0.0500	0.0300	0.0445	0.0073	0.0331	0.0041	0.0036	0.0082
16	0.0625	0.0375	0.0557	0.0091	0.0414	0.0052	0.0045	0.0102
12	0.0833	0.0500	0.0742	0.0121	0.0552	0.0069	0.0059	0.0136
10	0.1000	0.0600	0.0891	0.0145	0.0663	0.0083	0.0071	0.0163
8	0.1250	0.0750	0.1113	0.0182	0.0828	0.0103	0.0089	0.0204
6	0.1667	0.1000	0.1484	0.0242	0.1105	0.0138	0.0119	0.0271
5	0.2000	0.1200	0.1781	0.0291	0.1325	0.0165	0.0143	0.0326
4	0.2500	0.1500	0.2227	0.0363	0.1657	0.0207	0.0179	0.0408
3	0.3333	0.2000	0.2969	0.0484	0.2209	0.0275	0.0238	0.0543
2½	0.4000	0.2400	0.3563	0.0581	0.2651	0.0330	0.0286	0.0653
2	0.5000	0.3000	0.4453	0.0727	0.3314	0.0413	0.0357	0.0816
1½	0.6667	0.4000	0.5938	0.0969	0.4418	0.0551	0.0476	0.1088
1¼	0.8000	0.4800	0.7125	0.1163	0.5302	0.0661	0.0572	0.1305
1	1.0000	0.6000	0.8906	0.1453	0.6627	0.0826	0.0714	0.1632

[a] All other dimensions are in inches.
[b] Minimum root truncation is one-half of maximum.
[c] Minimum root radius is one-half of maximum.

Table 2. American National Standard Diameter—Pitch Combinations for 7°/45° Buttress Threads *ANSI B1.9-1973 (R1992)*

Preferred Nominal Major Diameters,Inches	Threads per Inch[a]	Preferred Nominal Major Diameters, Inches	Threads per Inch[a]
0.5, 0.625, 0.75	(20, 16, 12)	4.5, 5, 5.5, 6	12, 10, 8, (6, 5, 4), 3
0.875, 1.0	(16, 12, 10)	7, 8, 9, 10	10, 8, 6, (5, 4, 3), 2.5, 2
1.25, 1.375, 1.5	16, (12, 10, 8), 6	11, 12, 14, 16	10, 8, 6, 5, (4, 3, 2.5), 2, 1.5, 1.25
1.75, 2, 2.25, 2.5	16, 12, (10, 8, 6), 5, 4	18, 20, 22, 24	8, 6, 5, 4, (3, 2.5, 2), 1.5, 1.25, 1
2.75, 3, 3.5, 4	16, 12, 10, (8, 6, 5), 4		

[a] Preferred pitches are in pitches are in parentheses.

In selecting the form of thread recommended as standard ANSI B1.9-1973 (R1992), manufacture by milling, grinding, rolling, or other suitable means, has been taken into consideration. All dimensions are in inches.

Form of American National Standard 7°/45° Buttress Thread with 0.6*p* Basic Height of Thread Engagement

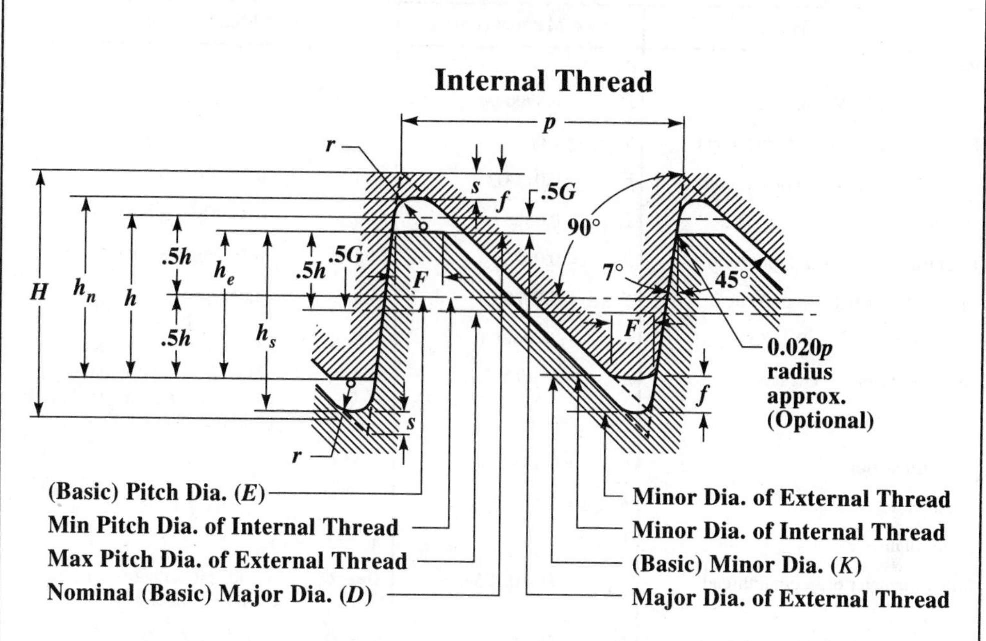

ROUND ROOT EXTERNAL THREAD

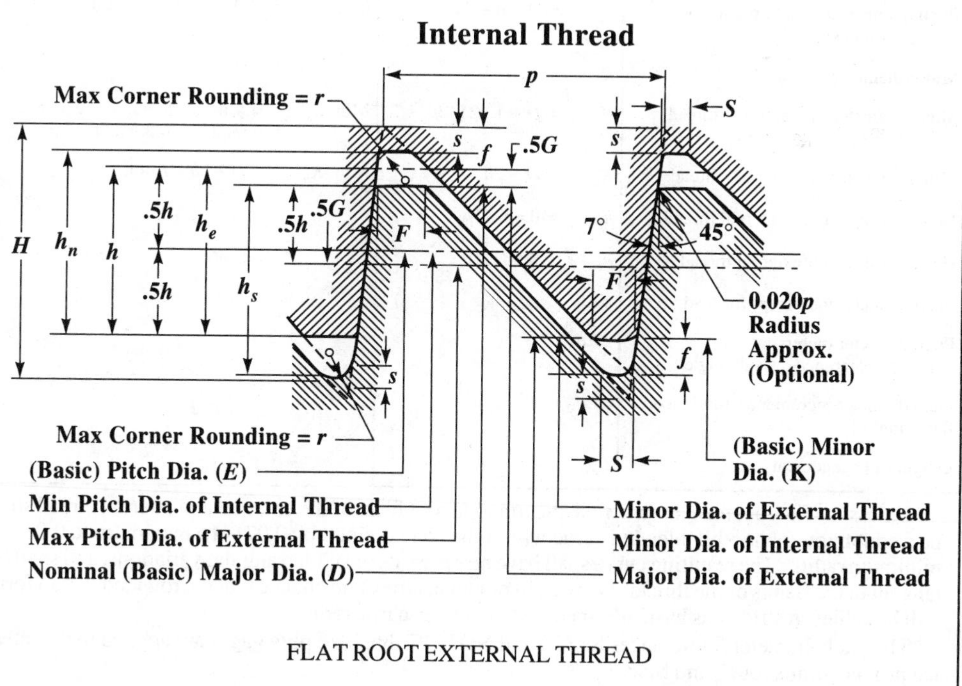

FLAT ROOT EXTERNAL THREAD

Heavy Line Indicates Basic Form

Table 3. American National Standard Buttress Inch Screw Thread Symbols and Form

Thread Element	Max. Material (Basic)		Min. Material	
Pitch	p			
Height of sharp-V thread	H	$= 0.89064p$		
Basic height of thread engagement	h	$= 0.6p$		
Root radius (theoretical)(see footnote [a])	r	$= 0.07141p$	Min. r	$= 0.0357p$
Root truncation	s	$= 0.0826p$	Min. s	$= 0.5$; Max. $s = 0.0413p$
Root truncation for flat root form	s	$= 0.0826p$	Min. s	$= 0.5$; Max. $s = 0.0413p$
Flat width for flat root form	S	$= 0.0928p$	Min. S	$= 0.0464p$
Allowance	G	(see text)		
Height of thread engagement	h_e	$= h - 0.5G$	Min. h_e	$=$ Max. h_e − [0.5 tol. on major dia. external thread + 0.5 tol. on minor dia. internal thread].
Crest truncation	f	$= 0.14532p$		
Crest width	F	$= 0.16316p$		
Major diameter	D			
Major diameter of internal thread	D_n	$= D + 0.12542p$	Max. D_n	= Max. pitch dia.of internal thread + $0.80803p$
Major diameter of external thread	D_s	$= D - G$	Min. D_s	$= D - G - D$ tol.
Pitch diameter	E			
Pitch diameter of internal thread (see footnote [b])	E_n	$= D - h$	Max. E_n	$= D - h + PD$ tol.
Pitch diameter of external thread (see footnote [c])	E_s	$= D - h - G$	Min. E_s	$= D - h - G - PD$ tol.
Minor diameter	K			
Minor diameter of external thread	K_s	$= D - 1.32542p - G$	Min. K_s	= Min. pitch dia. of external thread − $0.80803p$
Minor diameter of internal thread	K_n	$= D - 2h$	Min. K_n	$= D - 2h + K$ tol.
Height of thread of internal thread	h_n	$= 0.66271p$		
Height of thread ofexternal thread	h_s	$= 0.66271p$		
Pitch diameter increment for lead	ΔEl			
Pitch diameter increment for 45° clearance flank angle	$\Delta E\alpha_1$			
Pitch diameter increment for 7° load flank angle	$\Delta E\alpha_2$			
Length of engagement	L_e			

[a] Unless the flat root form is specified, the rounded root form of the external and internal thread shall be a continuous, smoothly blended curve within the zone defined by $0.07141p$ maximum to $0.0357p$ minimum radius. The resulting curve shall have no reversals or sudden angular variations, and shall be tangent to the flanks of the thread. There is, in practice, almost no chance that the rounded thread form will be achieved strictly as basically specified, that is, as a true radius.

[b] The pitch diameter X tolerances for GO and NOT GO threaded plug gages are applied to the internal product limits for E_n and Max. E_n.

[c] The pitch diameter W tolerances for GO and NOT GO threaded setting plug gages are applied to the external product limits for E_s and Min. E_s.

Form of Thread: The form of the buttress thread is shown in the accompanying figure and has the following characteristics:

A) A load flank angle, measured in an axial plane, of 7 degrees from the normal to the axis.

B) A clearance flank angle, measured in an axial plane, of 45 degrees from the normal to the axis.

C) Equal truncations at the crests of the external and internal threads such that the basic height of thread engagement (assuming no allowance) is equal to 0.6 of the pitch.

D) Equal radii, at the roots of the external and internal basic thread forms tangential to the load flank and the clearance flank. (There is, in practice, almost no chance that the thread forms will be achieved strictly as basically specified, that is, as true radii.) When specified, equal flat roots of the external and internal thread may be supplied.

Buttress Thread Tolerances.—Tolerances from basic size on external threads are applied in a minus direction and on internal threads in a plus direction.

Pitch Diameter Tolerances: The following formula is used for determining the pitch diameter product tolerance for Class 2 (standard grade) external or internal threads:

$$\text{PD tolerance} = 0.002\sqrt[3]{D} + 0.00278\sqrt{L_e} + 0.00854\sqrt{p}$$

where D = basic major diameter of external thread (assuming no allowance)

L_e = length of engagement

p = pitch of thread

When the length of engagement is taken as $10p$, the formula reduces to

$$0.002\sqrt[3]{D} + 0.0173\sqrt{p}$$

It is to be noted that this formula relates specifically to Class 2 (standard grade) PD tolerances. Class 3 (precision grade) PD tolerances are two-thirds of Class 2 PD tolerances. Pitch diameter tolerances based on this latter formula, for various diameter pitch combinations, are given in Tables 4 and 5.

Functional Size: Deviations in lead and flank angle of product threads increase the functional size of an external thread and decrease the functional size of an internal thread by the cumulative effect of the diameter equivalents of these deviations. The functional size of all buttress product threads shall not exceed the maximum-material limit.

Tolerances on Major Diameter of External Thread and Minor Diameter of Internal Thread: Unless otherwise specified, these tolerances should be the same as the pitch diameter tolerance for the class used.

Tolerances on Minor Diameter of External Thread and Major Diameter of Internal Thread: It will be sufficient in most instances to state only the maximum minor diameter of the external thread and the minimum major diameter of the internal thread without any tolerance. However, the root truncation from a sharp V should not be greater than $0.0826p$ nor less than $0.0413p$.

Lead and Flank Angle Deviations for Class 2: The deviations in lead and flank angles may consume the entire tolerance zone between maximum and minimum material product limits given in Table 4.

Diameter Equivalents for Variations in Lead and Flank Angles for Class 3: The combined diameter equivalents of variations in lead (including helix deviations), and flank angle for Class 3, shall not exceed 50 percent of the pitch diameter tolerances given in Table 5.

Table 4. American National Standard Class 2 (Standard Grade) Tolerances for Buttress Inch Screw Threads *ANSI B1.9-1973 (R1992)*

Thds. per Inch	Pitch,[a] p Inch	Basic Major Diameter, Inch									Pitch[b] Increment, $0.0173\sqrt{p}$ Inch
		From 0.5 thru 0.7	Over 0.7 thru 1.0	Over 1.0 thru 1.5	Over 1.5 thru 2.5	Over 2.5 thru 4	Over 4 thru 6	Over 6 thru 10	Over 10 thru 16	Over 16 thru 24	
		Tolerance on Major Diameter of External Thread, Pitch Diameter of External and Internal Threads, and Minor Diameter of Internal Thread, Inch									
20	0.0500	.0056	....	....	....	....	....	....	....	....	.00387
16	0.0625	.0060	.0062	.0065	.0068	.0073	....	....	....	....	.00432
12	0.0833	.0067	.0069	.0071	.0075	.0080	.0084	....	....	....	.00499
10	0.1000	....	.0074	.0076	.0080	.0084	.0089	.0095	.0102	....	.00547
8	0.1250	....	....	.0083	.0086	.0091	.0095	.0101	.0108	.0115	.00612
6	0.1667	....	....	.0092	.0096	.0100	.0105	.0111	.0118	.0125	.00706
5	0.2000	....	....	....	.0103	.0107	.0112	.0117	.0124	.0132	.00774
4	0.2500	....	....	....	.0112	.0116	.0121	.0127	.0134	.0141	.00865
3	0.3333	....	....	....	....	....	.0134	.0140	.0147	.0154	.00999
2.5	0.4000	....	....	....	....	....	....	.0149	.0156	.0164	.01094
2.0	0.5000	....	....	....	....	....	....	.0162	.0169	.0177	.01223
1.5	0.6667	....	....	....	....	....	....	....	.0188	.0196	.01413
1.25	0.8000	....	....	....	....	....	....	....	.0202	.0209	.01547
1.0	1.0000	....	....	....	....	....	....	....	....	.0227	.01730
		Diameter Increment,[c] $0.002\sqrt[3]{D}$									
		.00169	.00189	.00215	.00252	.00296	.00342	.00400	.00470	.00543	

[a] For threads with pitches not shown in this table, pitch increment to be used in tolerance formula is to be determined by use of formula P.D. Tolerance $= 0.002\sqrt[3]{D} + 0.00278\sqrt{L_e} + 0.00854\sqrt{p}$, where: D = basic major diameter of external thread (assuming no allowance), L_e = length of engagement, and p = pitch of thread.

[b] When the length of engagement is taken as $10p$, the formula reduces to: $0.002\sqrt[3]{D} + 0.0173\sqrt{p}$

[c] Diameter D, used in diameter increment formula, is based on the average of the range.

Table 5. American National Standard Class 3 (Precision Grade) Tolerances for Buttress Inch Screw Threads *ANSI B1.9-1973 (R1992)*

Threads per Inch	Pitch, p Inch	Basic Major Diameter, Inch								
		From 0.5 thru 0.7	Over 0.7 thru 1.0	Over 1.0 thru 1.5	Over 1.5 thru 2.5	Over 2.5 thru 4	Over 4 thru 6	Over 6 thru 10	Over 10 thru 16	Over 16 thru 24
		Tolerance on Major Diameter of External Thread, Pitch Diameter of External and Internal Threads, and Minor Diameter of Internal Thread, Inch								
20	0.0500	.0037	....	....	....	....	....	....	....	....
16	0.0625	.0040	.0042	.0043	.0046	.0049	....	....	....	....
12	0.0833	.0044	.0046	.0048	.0050	.0053	.0056	....	....	....
10	0.1000	....	.0049	.0051	.0053	.0056	.0059	.0063	.0068	....
8	0.1250	....	....	.0055	.0058	.0061	.0064	.0067	.0072	.0077
6	0.1667	....	....	.0061	.0064	.0067	.0070	.0074	.0078	.0083
5	.02000	....	....	....	.0068	.0071	.0074	.0078	.0083	.0088
4	0.2500	....	....	....	.0074	.0077	.0080	.0084	.0089	.0094
3	.03333	....	....	....	....	....	.0089	.0093	.0098	.0103
2.5	0.4000	....	....	....	....	....	....	.0100	.0104	.0109
2.0	0.5000	....	....	....	....	....	....	.0108	.0113	.0118
1.5	0.6667	....	....	....	....	....	....	....	.0126	.0130
1.25	0.8000	....	....	....	....	....	....	....	.0135	.0139
1.0	1.0000	....	....	....	....	....	....	....	....	.0152

Tolerances on Taper and Roundness: There are no requirements for taper and roundness for Class 2 buttress screw threads.

The major and minor diameters of Class 3 buttress threads shall not taper nor be out of round to the extent that specified limits for major and minor diameter are exceeded. The taper and out-of-roundness of the pitch diameter for Class 3 buttress threads shall not exceed 50 per cent of the pitch-diameter tolerances.

Allowances for Easy Assembly.—An allowance (clearance) should be provided on all external threads to secure easy assembly of parts. The amount of the allowance is deducted from the nominal major, pitch, and minor diameters of the external thread when the maximum material condition of the external thread is to be determined.

The minimum internal thread is basic.

The amount of the allowance is the same for both classes and is equal to the Class 3 pitch-diameter tolerance as calculated by the formulas previously given. The allowances for various diameter-pitch combinations are given in Table 6.

Table 6. American National Standard External Thread Allowances for Classes 2 and 3 Buttress Inch Screw Threads *ANSI B1.9-1973 (R1992)*

Threads per Inch	Pitch, p Inch	Basic Major Diameter, Inch								
		From 0.5 thru 0.7	Over 0.7 thru 1.0	Over 1.0 thru 1.5	Over 1.5 thru 2.5	Over 2.5 thru 4	Over 4 thru 6	Over 6 thru 10	Over 10 thru 16	Over 16 thru 24
		Allowance on Major, Minor and Pitch Diameters of External Thread, Inch								
20	0.0500	.0037	....	....	....	....	....	....	....	....
16	0.0625	.0040	.0042	.0043	.0046	.0049	....	....	....	....
12	0.0833	.0044	.0046	.0048	.0050	.0053	.0056	....	....	....
10	0.1000	....	.0049	.0051	.0053	.0056	.0059	.0063	.0068	....
8	0.1250	....	....	.0055	.0058	.0061	.0064	.0067	.0072	.0077
6	0.1667	....	....	.0061	.0064	.0067	.0070	.0074	.0078	.0083
5	0.2000	....	....	....	.0068	.0071	.0074	.0078	.0083	.0088
4	0.2500	....	....	....	.0074	.0077	.0080	.0084	.0089	.0094
3	0.3333	....	....	....	....	....	.0089	.0093	.0098	.0103
2.5	0.4000	....	....	....	....	....	....	.0100	.0104	.0109
2.0	0.5000	....	....	....	....	....	....	.0108	.0113	.0118
1.5	0.6667	....	....	....	....	....	....	....	.0126	.0130
1.25	0.8000	....	....	....	....	....	....	....	.0135	.0139
1.0	1.0000	....	....	....	....	....	....	....	....	.0152

Example Showing Dimensions for a Typical Buttress Thread.—The dimensions for a 2-inch diameter, 4 threads per inch, Class 2 buttress thread with flank angles of 7 degrees and 45 degrees are

h = basic thread height = 0.1500 (Table 1)

$h_s = h_n$ = height of thread in external and internal threads = 0.1657 (Table 1)

G = pitch-diameter allowance on external thread = 0.0074 (Table 6)

Tolerance on PD of external and internal threads = 0.0112 (Table 4)

Tolerance on major diameter of external thread and minor diameter of internal thread = 0.0112 (Table 4)

Internal Thread

$BasicMajorDiameter = D = 2.0000$

$Min.MajorDiameter = D - 2h + 2h_n = 2.0314$ (see Table 1)

$Min.PitchDiameter = D - h = 1.8500$ (see Table 1)

$Max.PitchDiameter = D - h + PD$ Tolerance = 1.8612 (see Table 4)

$Min.MinorDiameter = D - 2h = 1.7000$ (see Table 1)

$Max.MinorDiameter = D - 2h +$ Minor Diameter Tolerance

= 1.7112 (see Table 4)

External Thread

$MaxMajorDiameter = D - G = 1.9926$ (see Table 6)

$MinMajorDiameter = D - G -$ Major Diameter Tolerance
$= 1.9814$ (see Tables 4 and 6)

$MaxPitchDiameter = D - h - G = 1.8426$ (see Tables 1 and 6)

$MinPitchDiameter = D - h - G - PD$ Tolerance $= 1.8314$ (see Table 4)

$MaxMinorDiameter = D - G - 2h_s = 1.6612$ (see Tables 1 and 6)

Buttress Thread Designations.—When only the designation, BUTT is used, the thread is "pull" type buttress (external thread pulls) with the clearance flank leading and the 7-degree pressure flank following. When the designation, PUSH-BUTT is used, the thread is a push type buttress (external thread pushes) with the 7-degree load flank leading and the 45-degree clearance flank following. Whenever possible this description should be confirmed by a simplified view showing thread angles on the drawing of the product that has the buttress thread.

Standard Buttress Threads: A buttress thread is considered to be standard when: E) opposite flank angles are 7-degrees and 45-degrees; F) basic thread height is 0.6*p*; G) tolerances and allowances are as shown in Tables 4 through 6; and H) length of engagement is 10*p* or less.

Thread Designation Abbreviations: In thread designations on drawings, tools, gages, and in specifications, the following abbreviations and letters are to be used:

BUTT	for buttress thread, pull type
PUSH-BUTT	for buttress thread, push type
LH	for left-hand thread (Absence of LH indicates that the thread is a right-hand thread.)
P	for pitch
L	for lead
A	for external thread
B	for internal thread

Note: Absence of A or B after thread class indicates that designation covers both the external and internal threads.

Le	for length of thread engagement
SPL	for special
FL	for flat root thread
E	for pitch diameter
TPI	for threads per inch
THD	for thread

Designation Sequence for Buttress Inch Screw Threads.—When designating single-start standard buttress threads the nominal size is given first, the threads per inch next, then PUSH if the internal member is to push, but nothing if it is to pull, then the class of thread (2 or 3), then whether external (A) or internal (B), then LH if left-hand, but nothing if right-hand, and finally FL if a flat root thread, but nothing if a radiused root thread; thus, 2.5-8 BUTT-2A indicates a 2.5 inch, 8 threads per inch buttress thread, Class 2 external, right-hand, internal member to pull, with radiused root of thread. The designation 2.5-8 PUSH-BUTT-2A-LH-FL signifies a 2.5 inch size, 8 threads per inch buttress thread with internal member to push, Class 2 external, left-hand, and flat root.

A multiple-start standard buttress thread is similarly designated but the pitch is given instead of the threads per inch, followed by the lead and the number of starts is indicated in parentheses after the class of thread. Thus, 10-0.25P–0.5L – BUTT-3B (2 start) indicates a 10-inch thread with 4 threads per inch, 0.5 inch lead, buttress form with internal member to pull, Class 3 internal, 2 starts, with radiused root of thread.

BRITISH THREADS

British Standard Buttress Threads B.S. 1657: 1950.—Specifications for buttress threads in this standard are similar to those in the American Standard except: 1) A basic depth of thread of 0.4*p* is used instead of 0.6*p*; 2) Sizes below 1 inch are not included; 3) Tolerances on major and minor diameters are the same as the pitch diameter tolerances, whereas in the American Standard separate tolerances are provided; however, provision is made for smaller major and minor diameter tolerances when crest surfaces of screws or nuts are used as datum surfaces, or when the resulting reduction in depth of engagement must be limited; and 4) Certain combinations of large diameters with fine pitches are provided that are not encouraged in the American Standard.

Löwenherz Thread.—The Löwenherz thread has flats at the top and bottom the same as the U.S. standard form, but the angle is 53 degrees 8 minutes. The depth equals 0.75 × the pitch, and the width of the flats at the top and bottom is equal to 0.125 × the pitch. This screw thread is based on the metric system and is used for measuring instruments, especially in Germany.

Löwenherz Thread

Diameter		Pitch, Millimeters	Approximate No. of Threads per Inch	Diameter		Pitch Millimeters	Approximate No. of Threads per Inch
Millimeters	Inches			Millimeters	Inches		
1.0	0.0394	0.25	101.6	9.0	0.3543	1.30	19.5
1.2	0.0472	0.25	101.6	10.0	0.3937	1.40	18.1
1.4	0.0551	0.30	84.7	12.0	0.4724	1.60	15.9
1.7	0.0669	0.35	72.6	14.0	0.5512	1.80	14.1
2.0	0.0787	0.40	63.5	16.0	0.6299	2.00	12.7
2.3	0.0905	0.40	63.5	18.0	0.7087	2.20	11.5
2.6	0.1024	0.45	56.4	20.0	0.7874	2.40	10.6
3.0	0.1181	0.50	50.8	22.0	0.8661	2.80	9.1
3.5	0.1378	0.60	42.3	24.0	0.9450	2.80	9.1
4.0	0.1575	0.70	36.3	26.0	1.0236	3.20	7.9
4.5	0.1772	0.75	33.9	28.0	1.1024	3.20	7.9
5.0	0.1968	0.80	31.7	30.0	1.1811	3.60	7.1
5.5	0.2165	0.90	28.2	32.0	1.2599	3.60	7.1
6.0	0.2362	1.00	25.4	36.0	1.4173	4.00	6.4
7.0	0.2756	1.10	23.1	40.0	1.5748	4.40	5.7
8.0	0.3150	1.20	21.1	…	…	…	…

International Metric Thread System.—The Système Internationale (S.I.) Thread was adopted at the International Congress for the standardization of screw threads held in Zurich in 1898. The thread form is similar to the American standard (formerly U.S. Standard), excepting the depth which is greater. There is a clearance between the root and mating crest fixed at a maximum of $\frac{1}{16}$ the height of the fundamental triangle or 0.054 × pitch. A rounded root profile is recommended. This system formed the basis of the normal metric series of many European countries.

$Depth\, d = 0.7035\, P$ max; $0.6855\, P$ min.

$Flat\, f = 0.125\, P$

$Radius\, r = 0.0633\, P$ max.; $0.054\, P$ min.

$Tap\, drill\, dia$ = major dia.-pitch

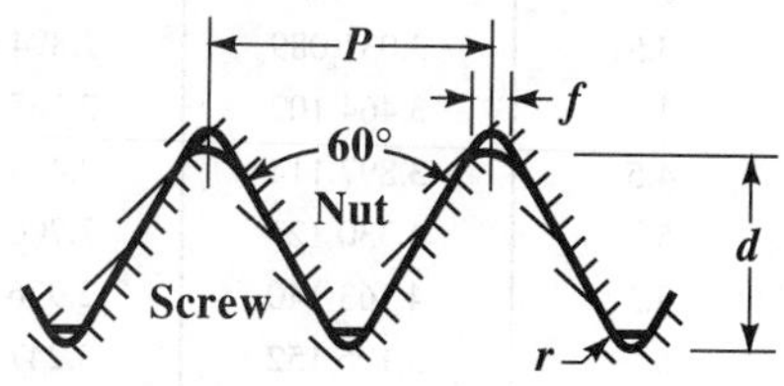

British Standard Unified Screw Threads of UNJ Basic Profile.—This British Standard B.S. 4084: 1978 arises from a request originating from within the British aircraft industry and is based upon specifications for Unified screw threads and American military standard MIL-S-8879.—These UNJ threads, having an enlarged root radius, were introduced for applications requiring high fatigue strength where working stress levels are high, in order to minimize size and weight, as in aircraft engines, airframes, missiles, space vehicles and similar designs where size and weight are critical. To meet these requirements the root radius of external Unified threads is controlled between appreciably enlarged limits, the minor diameter of the mating internal threads being appropriately increased to insure the necessary clearance. The requirement for high strength is further met by restricting the tolerances for UNJ threads to the highest classes, Classes 3A and 3B, of Unified screw threads.

The standard, not described further here, contains both a coarse and a fine pitch series of threads.

British Standard ISO Metric Screw Threads.—BS 3643:Part 1:1981 (1998) provides principles and basic data for ISO metric screw threads. It covers single-start, parallel screw threads of from 1 to 300 millimeters in diameter. Part 2 of the Standard gives the specifications for selected limits of size.

Basic Profile: The ISO basic profile for triangular screw threads is shown in Fig. 1. and basic dimensions of this profile are given in Table 1.

Table 1. British Standard ISO Metric Screw Threads—Basic Profile Dimensions
BS 3643:1981 (1998)

Pitch P	H	$\frac{5}{8}H$	$\frac{3}{8}H$	$H/4$	$H/8$
	0.086603P	0.54127P	0.32476P	0.21651P	0.10825P
0.2	0.173 205	0.108 253	0.064 952	0.043 301	0.021 651
0.25	0.216 506	0.135 316	0.081 190	0.054 127	0.027 063
0.3	0.259 808	0.162 380	0.097 428	0.064 952	0.032 476
0.35	0.303 109	0.189 443	0.113 666	0.075 777	0.037 889
0.4	0.346 410	0.216 506	0.129 904	0.086 603	0.043 301
0.45	0.389 711	0.243 570	0.146 142	0.097 428	0.048 714
0.5	0.433 013	0.270 633	0.162 380	0.108 253	0.054 127
0.6	0.519 615	0.324 760	0.194 856	0.129 904	0.064 952
0.7	0.606 218	0.378 886	0.227 322	0.151 554	0.075 777
0.75	0.649 519	0.405 949	0.243 570	0.162 380	0.081 190
0.8	0.692 820	0.433 013	0.259 808	0.173 205	0.086 603
1	0.866 025	0.541 266	0.324 760	0.216 506	0.108 253
1.25	1.082 532	0.676 582	0.405 949	0.270 633	0.135 316
1.5	1.299 038	0.811 899	0.487 139	0.324 760	0.162 380
1.75	1.515 544	0.947 215	0.568 329	0.378 886	0.189 443
2	1.732 051	1.082 532	0.649 519	0.433 013	0.216 506
2.5	2.165 063	1.353 165	0.811 899	0.541 266	0.270 633
3	2.598 076	1.623 798	0.974 279	0.649 519	0.324 760
3.5	3.031 089	1.894 431	1.136 658	0.757 772	0.378 886
4	3.464 102	2.165 063	1.299 038	0.866 025	0.433 013
4.5	3.897 114	2.435 696	1.461 418	0.974 279	0.487 139
5	4.330 127	2.706 329	1.623 798	1.082 532	0.541 266
5.5	4.763 140	2.976 962	1.786 177	1.190 785	0.595 392
6	5.196 152	3.247 595	1.948 557	1.299 038	0.649 519
8[a]	6.928 203	4.330 127	2.598 076	1.732 051	0.866 025

[a] This pitch is not used in any of the ISO metric standard series.

All dimensions are given in millimeters.

Tolerance System: The tolerance system defines *tolerance classes* in terms of a combination of a *tolerance grade* (figure) and a *tolerance position* (letter). The tolerance position is defined by the distance between the basic size and the nearest end of the tolerance zone, this distance being known as the *fundamental deviation,* EI, in the case of internal threads, and es in the case of external threads. These tolerance positions with respect to the basic size (zero line) are shown in Fig. 2 and fundamental deviations for nut and bolt threads are given in Table 2.

Table 2. Fundamental Deviations for Nut Threads and Bolt Threads

	Nut Thread D_2, D_1		Bolt Thread d, d_2					Nut Thread D_2, D_1		Bolt Thread d, d_2			
	Tolerance Position							Tolerance Position					
	G	H	e	f	g	h		G	H	e	f	g	h
	Fundamental Deviation							Fundamental Deviation					
Pitch *P*	EI	EI	es	es	es	es	Pitch *P*	EI	EI	es	es	es	es
mm	μm	μm	μm	μm	μm	μm	mm	μm	μm	μm	μm	μm	μm
0.2	+17	0	…	…	−17	0	1.25	+28	0	−63	−42	−28	0
0.25	+18	0	…	…	−18	0	1.5	+32	0	−67	−45	−32	0
0.3	+18	0	…	…	−18	0	1.75	+34	0	−71	−48	−34	0
0.35	+19	0	…	−34	−19	0	2	+38	0	−71	−52	−38	0
0.4	+19	0	…	−34	−19	0	2.5	+42	0	−80	−58	−42	0
0.45	+20	0	…	−35	−20	0	3	+48	0	−85	−63	−48	0
0.5	+20	0	−50	−36	−20	0	3.5	+53	0	−90	−70	−53	0
0.6	+21	0	−53	−36	−21	0	4	+60	0	−95	−75	−60	0
0.7	+22	0	−56	−38	−22	0	4.5	+63	0	−100	−80	−63	0
0.75	+22	0	−56	−38	−22	0	5	+71	0	−106	−85	−71	0
0.8	+24	0	−60	−38	−24	0	5.5	+75	0	−112	−90	−75	0
1	+26	0	−60	−40	−26	0	6	+80	0	−118	−95	−80	0

See Figs. 1 and 2 for meaning of symbols.

Tolerance Grades: Tolerance grades specified in the Standard for each of the four main screw thread diameters are as follows:

Minor diameter of nut threads (D_1): tolerance grades 4, 5, 6, 7, and 8.

Major diameter of bolt threads (d): tolerance grades 4, 6, and 8.

Pitch diameter of nut threads (D_2): tolerance grades 4, 5, 6, 7, and 8.

Pitch diameter of bolt threads (d_2): tolerance grades 3, 4, 5, 6, 7, 8, and 9.

Tolerance Positions: Tolerance positions are G and H for nut threads and e, f, g, and h for bolt threads. The relationship of these tolerance position identifying letters to the amount of fundamental deviation is shown in Table 2.

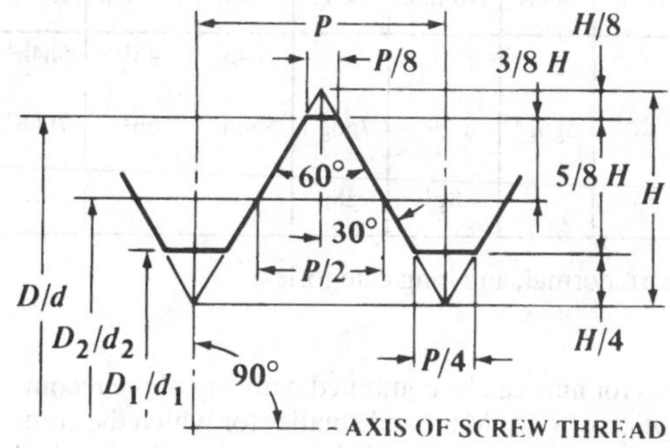

D = maj. diam. of internal thread;
d = maj. diam. of external th
D_2 = pitch diam. of internal thread;
d_2 = pitch diam. of internal thread;
D_1 = minor diam. of internal thread;
d_1 = minor diam. of external thread;
P = Pitch;
H = height of fundamental angle;

Fig. 1. Basic Profile of ISO Metric Thread

Tolerance Classes: To reduce the number of gages and tools, the Standard specifies that the tolerance positions and classes shall be chosen from those listed in Tables 3 and 4 for short, normal, and long lengths of thread engagement. The following rules apply for the choice of tolerance quality: *Fine:* for precision threads when little variation of fit character is needed; *Medium:* for general use; and *Coarse:* for cases where manufacturing difficulties can arise as, for example, when threading hot-rolled bars and long blind holes. If the actual length of thread engagement is unknown, as in the manufacturing of standard bolts, normal is recommended.

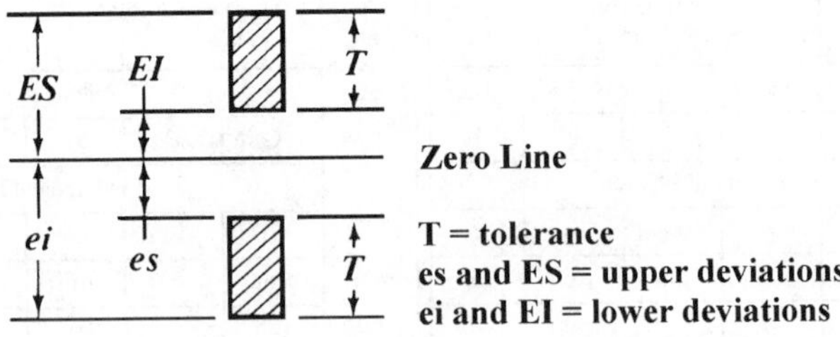

Fig. 2. Tolerance Positions with Respect to Zero Line (Basic Size)

Table 3. Tolerance Classes for Nuts

Tolerance Quality	Tolerance Position G			Tolerance Position H		
	Short	Normal	Long	Short	Normal	Long
Fine	…	…	…	4H[a]	5H[a]	6H[a]
Medium	5G[b]	6G[b]	7G[b]	5H[c]	6H[c,d]	7H[c]
Coarse	…	7G[b]	8G[b]	…	7H[a]	8H[a]

[a] Second choice.
[b] Third choice; these are to be avoided.
[c] First choice.

[d] For commercial nut and bolt threads. See Table 5 for short, normal, and long categories.

Table 4. Tolerance Classes for Bolts

Tolerance Quality	Tolerance Position e			Tolerance Position f			Tolerance Position g			Tolerance Position h		
	Short	Normal	Long	Short	Normal	Long	Short	Normal	Long	Short	Normal	Long
Fine	…	…	…	…	…	…	…	…	…	3h4h[3]	4h[1]	5h4h[3]
Medium	…	6e[1]	7e6e[3]	…	6f[1]	…	5g6g[3]	6g[1,4]	7g6g[3]	5h6h[3]	6h[2]	7h6h[3]
Coarse	…	…	…	…	…	…	…	8g[2]	9g8g[3]	…	…	…

See footnotes to Table 3, and see Table 5 for short, normal, and long categories.

Note: Any of the recommended tolerance classes for nuts can be combined with any of the recommended tolerance classes for bolts with the exception of sizes M1.4 and smaller for which the combination 5H/6h or finer shall be chosen. However, to guarantee a sufficient overlap, the finished components should preferably be made to form the fits H/g, H/h, or G/h.

Table 5. Lengths of Thread Engagements for Short, Normal, and Long Categories

Basic Major Diameter *d*			Short	Normal		Long
			Length of Thread Engagement			
Over	Up to and Incl.	Pitch *P*	Up to and Incl.	Over	Up to and Incl.	Over
0.99	1.4	0.2	0.5	0.5	1.4	1.4
		0.25	0.6	0.6	1.7	1.7
		0.3	0.7	0.7	2	2
1.4	2.8	0.2	0.5	0.5	1.5	1.5
		0.25	0.6	0.6	1.9	1.9
		0.35	0.8	0.8	2.6	2.6
		0.4	1	1	3	3
		0.45	1.3	1.3	3.8	3.8
2.8	5.6	0.35	1	1	3	3
		0.5	1.5	1.5	4.5	4.5
		0.6	1.7	1.7	5	5
		0.7	2	2	6	6
		0.75	2.2	2.2	6.7	6.7
		0.8	2.5	2.5	7.5	7.5
5.6	11.2	0.75	2.4	2.4	7.1	7.1
		1	3	3	9	9
		1.25	4	4	12	12
		1.5	5	5	15	15
11.2	22.4	1	3.8	3.8	11	11
		1.25	4.5	4.5	13	13
		1.5	5.6	5.6	16	16
		1.75	6	6	18	18
		2	8	8	24	24
		2.5	10	10	30	30

Basic Major Diameter *d*			Short	Normal		Long
			Length of Thread Engagement			
Over	Up to and Incl.	Pitch *P*	Up to and Incl.	Over	Up to and Incl.	Over
22.4	45	1	4	4	12	12
		1.5	6.3	6.3	19	19
		2	8.5	8.5	25	25
		3	12	12	36	36
		3.5	15	15	45	45
		4	18	18	53	53
		4.5	21	21	63	63
45	90	1.5	7.5	7.5	22	22
		2	9.5	9.5	28	28
		3	15	15	45	45
		4	19	19	56	56
		5	24	24	71	71
		5.5	28	28	85	85
		6	32	32	95	95
90	180	2	12	12	36	36
		3	18	18	53	53
		4	24	24	71	71
		6	36	36	106	106
180	300	3	20	20	60	60
		4	26	26	80	80
		6	40	40	118	118

All dimensions are given in millimeters.

Design Profiles: The design profiles for ISO metric internal and external screw threads are shown in Fig. 3. These represent the profiles of the threads at their maximum metal condition. It may be noted that the root of each thread is deepened so as to clear the basic flat crest of the other thread. The contact between the thread is thus confined to their sloping flanks. However, for nut threads as well as bolt threads, the actual root contours shall not at any point violate the basic profile.

Designation: Screw threads complying with the requirements of the Standard shall be designated by the letter M followed by values of the nominal diameter and of the pitch, expressed in millimeters, and separated by the sign ×. Example: M6 × 0.75. The absence of the indication of pitch means that a coarse pitch is specified.

The complete designation of a screw thread consists of a designation for the thread system and size, and a designation for the crest diameter tolerance. Each class designation consists of: a figure indicating the tolerance grade; and a letter indicating the tolerance position, capital for nuts, lower case for bolts. If the two class designations for a thread are the same (one for the pitch diameter and one for the crest diameter), it is not necessary to repeat the symbols. As examples, a bolt thread designated M10-6g signifies a thread of 10 mm nominal diameter in the Coarse Thread Series having a tolerance class 6g for both pitch and major diameters. A designation M10 × 1-5g6g signifies a bolt thread of 10 mm nominal diameter having a pitch of 1 mm, a tolerance class 5g for pitch diameter, and a tolerance class 6g for major diameter. A designation M10-6H signifies a nut thread of 10 mm diameter in the Coarse Thread Series having a tolerance class 6H for both pitch and minor diameters.

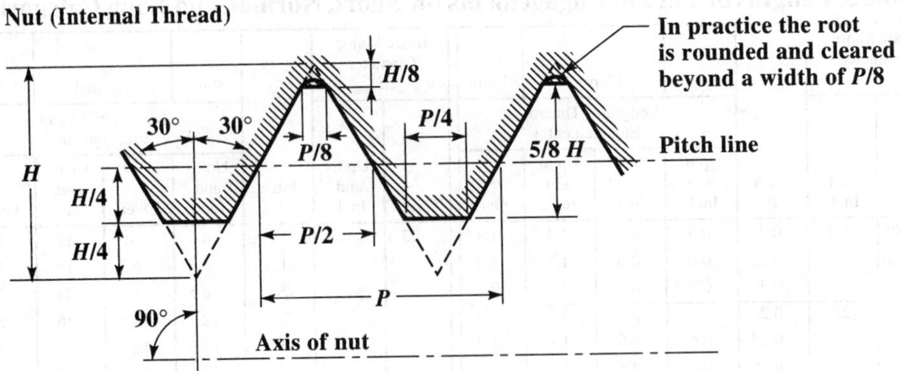

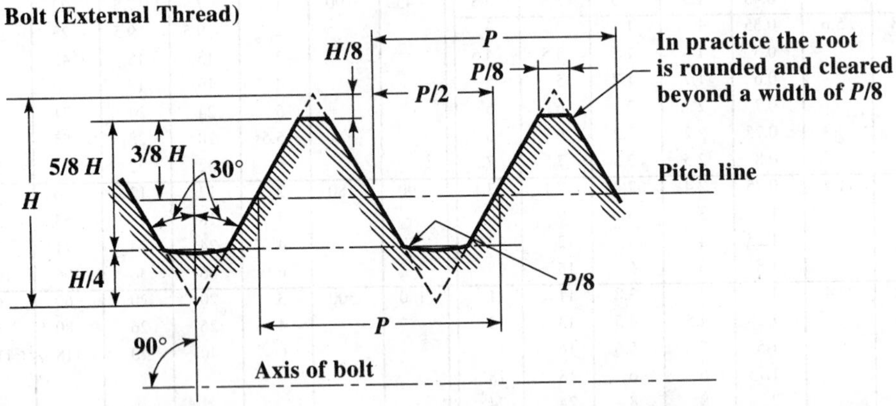

Fig. 3. Maximum Material Profiles for Internal and External Threads

A fit between mating parts is indicated by the nut thread tolerance class followed by the bolt thread tolerance class separated by an oblique stroke. *Examples:* M6-6H/6g and M20 × 2-6H/5g6g. For coated threads, the tolerances apply to the parts before coating, unless otherwise specified. After coating, the actual thread profile shall not at any point exceed the maximum material limits for either tolerance position H or h.

Fundamental Deviation Formulas: The formulas used to calculate the fundamental deviations in Table 2 are:

$EI_G = +(15 + 11P)$

$EI_H = 0$

$es_e = -(50 + 11P)$ except for threads with $P \leq 0.45$ mm

$es_f = -(30 + 11P)$

$es_g = -(15 + 11P)$

and $es_h = 0$

In these formulas, EI and es are expressed in micrometers and P is in millimeters.

Crest Diameter Tolerance Formulas: The tolerances for the major diameter of bolt threads (T_d), grade 6, in Table 7, were calculated from the formula:

$$T_d(6) = 180\sqrt[3]{P^2} - \frac{3.15}{\sqrt{P}}$$

In this formula, $T_d(6)$ is in micrometers and P is in millimeters. For tolerance grades 4 and 8: $T_d(4) = 0.63\,T_d(6)$ and $T_d(8) = 1.6\,T_d(6)$, respectively.

Table 6. British Standard ISO Metric Screw Threads: Limits and Tolerances for Finished Uncoated Threads for Normal Lengths of Engagement *BS 3643: Part 2: 1981*

Nominal Diameter[a]	Pitch		External Threads (Bolts)							Internal Threads (Nuts)[b]					
	Coarse	Fine	Tol. Class	Fund dev.	Major Dia.		Pitch Dia.		Minor Dia	Tol. Class	Major Dia.	Pitch Dia.		Minor Dia	
					Max	Tol(−)	Max	Tol(−)	Min		Min	Max	Tol(−)	Max	Tol(−)
1		0.2	4h	0	1.000	0.036	0.870	0.030	0.717	4H	1.000	0.910	0.040	0.821	0.038
			6g	0.017	0.983	0.056	0.853	0.048	0.682						
	0.25		4h	0	1.000	0.042	0.838	0.034	0.649	4H	1.000	0.883	0.045	0.774	0.045
			6g	0.018	0.982	0.067	0.820	0.053	0.613	5H	1.000	0.894	0.056	0.785	0.056
1.1		0.2	4h	0	1.100	0.036	0.970	0.030	0.817	4H	1.100	1.010	0.040	0.921	0.038
			6g	0.017	1.083	0.056	0.953	0.048	0.782						
	0.25		4h	0	1.100	0.042	0.938	0.034	0.750	4H	1.100	0.983	0.045	0.874	0.045
			6g	0.018	1.082	0.067	0.920	0.053	0.713	5H	1.100	0.994	0.056	0.885	0.056
1.2		0.2	4h	0	1.200	0.036	1.070	0.030	0.917	4H	1.200	1.110	0.040	1.021	0.038
			6g	0.017	1.183	0.056	1.053	0.048	0.882						
	0.25		4h	0	1.200	0.042	1.038	0.034	0.850	4H	1.200	1.083	0.045	0.974	0.045
			6g	0.018	1.182	0.067	1.020	0.053	0.813	5H	1.200	1.094	0.056	0.985	0.056
1.4		0.2	4h	0	1.400	0.036	1.270	0.030	1.117	4H	1.400	1.310	0.040	1.221	0.038
			6g	0.017	1.383	0.056	1.253	0.048	1.082						
	0.3		4h	0	1.400	0.048	1.205	0.036	0.984	4H	1.400	1.253	0.048	1.128	0.053
			6g	0.018	1.382	0.075	1.187	0.056	0.946	5H	1.400	1.265	0.060	1.142	0.067
										6H	1.400	1.280	0.075	1.160	0.085
1.6		0.2	4h	0	1.600	0.036	1.470	0.032	1.315	4H	1.600	1.512	0.042	1.421	0.038
			6g	0.017	1.583	0.056	1.453	0.050	1.280						
	0.35		4h	0	1.600	0.053	1.373	0.040	1.117	4H	1.600	1.426	0.053	1.284	0.063
			6g	0.019	1.581	0.085	1.354	0.063	1.075	5H	1.600	1.440	0.067	1.301	0.080
										6H	1.600	1.458	0.085	1.321	0.100
1.8		0.2	4h	0	1.800	0.036	1.670	0.032	1.515	4H	1.800	1.712	0.042	1.621	0.038
			6g	0.017	1.783	0.056	1.653	0.050	1.480						
	0.35		4h	0	1.800	0.053	1.573	0.040	1.317	4H	1.800	1.626	0.053	1.484	0.063
			6g	0.019	1.781	0.085	1.554	0.063	1.275	5H	1.800	1.640	0.067	1.501	0.080
										6H	1.800	1.658	0.085	1.521	0.100
2		0.25	4h	0	2.000	0.042	1.838	0.036	1.648	4H	2.000	1.886	0.048	1.774	0.045
			6g	0.018	1.982	0.067	1.820	0.056	1.610	5H	2.000	1.898	0.060	1.785	0.056
	0.4		4h	0	2.000	0.060	1.740	0.042	1.452	4H	2.000	1.796	0.056	1.638	0.071
			6g	0.019	1.981	0.095	1.721	0.067	1.408	5H	2.000	1.811	0.071	1.657	0.090
										6H	2.000	1.830	0.090	1.679	0.112
2.2		0.25	4h	0	2.200	0.042	2.038	0.036	1.848	4H	2.200	2.086	0.048	1.974	0.045
			6g	0.018	2.182	0.067	2.020	0.056	1.810	5H	2.200	2.098	0.060	1.985	0.056
	0.45		4h	0	2.200	0.063	1.908	0.045	1.585	4H	2.200	1.968	0.060	1.793	0.080
			6g	0.020	2.180	0.100	1.888	0.071	1.539	5H	2.200	1.983	0.075	1.813	0.100
										6H	2.000	2.003	0.095	1.838	0.125
2.5		0.35	4h	0	2.500	0.053	2.273	0.040	2.017	4H	2.500	2.326	0.053	2.184	0.063
			6g	0.019	2.481	0.085	2.254	0.063	1.975	5H	2.500	2.340	0.067	2.201	0.080
										6H	2.500	2.358	0.085	2.221	0.100
	0.45		4h	0	2.500	0.063	2.208	0.045	1.885	4H	2.500	2.268	0.060	2.093	0.080
			6g	0.020	2.480	0.100	2.188	0.071	1.839	5H	2.500	2.283	0.075	2.113	0.100
										6H	2.500	2.303	0.095	2.138	0.125
3		0.35	4h	0	3.000	0.053	2.773	0.042	2.515	4H	3.000	2.829	0.056	2.684	0.063
			6g	0.019	2.981	0.085	2.754	0.067	2.471	5H	3.000	2.844	0.071	2.701	0.080
										6H	3.000	2.863	0.090	2.721	0.100
	0.5		4h	0	3.000	0.067	2.675	0.048	2.319	5H	3.000	2.755	0.080	2.571	0.112
			6g	0.020	2.980	0.106	2.655	0.075	2.272	6H	3.000	2.775	0.100	2.599	0.140
										7H	3.000	2.800	0.125	2.639	0.180
3.5		0.35	4h	0	3.500	0.053	3.273	0.042	3.015	4H	3.500	3.329	0.056	3.184	0.063
			6g	0.019	3.481	0.085	3.254	0.067	2.971	5H	3.500	3.344	0.071	3.201	0.080
										6H	3.500	3.363	0.090	3.221	0.100
	0.6		4h	0	3.500	0.080	3.110	0.053	2.688	5H	3.500	3.200	0.090	2.975	0.125
			6g	0.021	3.479	0.125	3.089	0.085	2.635	6H	3.500	3.222	0.112	3.010	0.160
										7H	3.500	3.250	0.140	3.050	0.200
4		0.5	4h	0	4.000	0.067	3.675	0.048	3.319	5H	4.000	3.755	0.080	3.571	0.112
			6g	0.020	3.980	0.106	3.655	0.075	3.272	6H	4.000	3.775	0.100	3.599	0.140
										7H	4.000	3.800	0.125	3.639	0.180
	0.7		4h	0	4.000	0.090	3.545	0.056	3.058	5H	4.000	3.640	0.095	3.382	0.140
			6g	0.022	3.978	0.140	3.523	0.090	3.002	6H	4.000	3.663	0.118	3.422	0.180
										7H	4.000	3.695	0.150	3.466	0.224

Table 6. British Standard ISO Metric Screw Threads: Limits and Tolerances for Finished Uncoated Threads for Normal Lengths of Engagement *BS 3643: Part 2: 1981 (Continued)*

Nominal Diameter[a]	Pitch		External Threads (Bolts)							Internal Threads (Nuts)[b]					
	Coarse	Fine	Tol. Class	Fund dev.	Major Dia. Max	Major Dia. Tol(−)	Pitch Dia. Max	Pitch Dia. Tol(−)	Minor Dia Min	Tol. Class	Major Dia. Min	Pitch Dia. Max	Pitch Dia. Tol(−)	Minor Dia Max	Minor Dia Tol(−)
4.5		0.5	4h	0	4.500	0.067	4.175	0.048	3.819	5H	4.500	4.255	0.080	4.071	0.112
			6g	0.020	4.480	0.106	4.155	0.075	3.772	6H	4.500	4.275	0.100	4.099	0.140
										7H	4.500	4.300	0.125	4.139	0.180
	0.75		4h	0	4.500	0.090	4.013	0.056	3.495	5H	4.500	4.108	0.095	3.838	0.150
			6g	0.022	4.478	0.140	3.991	0.090	3.439	6H	4.500	4.131	0.118	3.878	0.190
										7H	4.500	4.163	0.150	3.924	0.236
5		0.5	4h	0	5.000	0.067	4.675	0.048	4.319	5H	5.000	4.755	0.080	4.571	0.112
			6g	0.020	4.980	0.106	4.655	0.075	4.272	6H	5.000	4.775	0.100	4.599	0.140
										7H	5.000	4.800	0.125	4.639	0.180
	0.8		4h	0	5.000	0.095	4.480	0.060	3.927	5H	5.000	4.580	0.100	4.294	0.160
			6g	0.024	4.976	0.150	4.456	0.095	3.868	6H	5.000	4.605	0.125	4.334	0.200
										7H	5.000	4.640	0.160	4.384	0.250
5.5		0.5	4h	0	5.500	0.067	5.175	0.048	4.819	5H	5.500	5.255	0.080	5.071	0.112
			6g	0.020	5.480	0.106	5.155	0.075	4.772	6H	5.500	5.275	0.100	5.099	0.140
										7H	5.500	5.300	0.125	5.139	0.180
6		0.75	4h	0	6.000	0.090	5.513	0.063	4.988	5H	6.000	5.619	0.106	5.338	0.150
			6g	0.022	5.978	0.140	5.491	0.100	4.929	6H	6.000	5.645	0.132	5.378	0.190
										7H	6.000	5.683	0.170	5.424	0.236
	1		4h	0	6.000	0.112	5.350	0.071	4.663	5H	6.000	5.468	0.118	5.107	0.190
			6g	0.026	5.974	0.180	5.324	0.112	4.597	6H	6.000	5.500	0.150	5.153	0.236
			8g	0.026	5.974	0.280	5.324	0.180	4.528	7H	6.000	5.540	0.190	5.217	0.300
7		0.75	4h	0	7.000	0.090	6.513	0.063	5.988	5H	7.000	6.619	0.106	6.338	0.150
			6g	0.022	6.978	0.140	6.491	0.100	5.929	6H	7.000	6.645	0.132	6.378	0.190
										7H	7.000	6.683	0.170	6.424	0.236
	1		4h	0	7.000	0.112	6.350	0.071	5.663	5H	7.000	6.468	0.118	6.107	0.190
			6g	0.026	6.974	0.180	6.324	0.112	5.596	6H	7.000	6.500	0.150	6.153	0.236
			8g	0.026	6.974	0.280	6.324	0.180	5.528	7H	7.000	6.540	0.190	6.217	0.300
8		1	4h	0	8.000	0.112	7.350	0.071	6.663	5H	8.000	7.468	0.118	7.107	0.190
			6g	0.026	7.974	0.180	7.324	0.112	6.596	6H	8.000	7.500	0.150	7.153	0.236
			8g	0.026	7.974	0.280	7.324	0.180	6.528	7H	8.000	7.540	0.190	7.217	0.300
	1.25		4h	0	8.000	0.132	7.188	0.075	6.343	5H	8.000	7.313	0.125	6.859	0.212
			6g	0.028	7.972	0.212	7.160	0.118	6.272	6H	8.000	7.348	0.160	6.912	0.265
			8g	0.028	7.972	0.335	7.160	0.190	6.200	7H	8,000	7.388	0.200	6.982	0.335
9	1.25		4h	0	9.000	0.132	8.188	0.075	7.343	5H	9.000	8.313	0.125	7.859	0.212
			6g	0.028	8.972	0.212	8.160	0.008	7.272	6H	9.000	8.348	0.160	7.912	0.265
			8g	0.028	8.972	0.335	8.160	0.190	7.200	7H	9.000	8.388	0.200	7.982	0.335
10		1.25	4h	0	10.000	0.132	9.188	0.075	8.343	5H	10.000	9.313	0.125	8.859	0.212
			6g	0.028	9.972	0.212	9.160	0.118	8.272	6H	10.000	9.348	0.160	8.912	0.265
			8g	0.028	9.972	0.335	9.160	0.190	8.200	7H	10.000	9.388	0.200	8.982	0.335
	1.5		4h	0	10.000	0.150	9.026	0.085	8.018	5H	10.000	9.166	0.140	8.612	0.236
			6g	0.032	9.968	0.236	8.994	0.132	7.938	6H	10.000	9.206	0.180	8.676	0.300
			8g	0.032	9.968	0.375	8.994	0.212	7.858	7H	10.000	9.250	0.224	8.751	0.375
11	1.5		4h	0	11.000	0.150	10.026	0.085	9.018	5H	11.000	10.166	0.140	9.612	0.236
			6g	0.032	10.968	0.236	9.994	0.132	8.938	6H	11.000	10.206	0.180	9.676	0.300
			8g	0.032	10.968	0.375	9.994	0.212	8.858	7H	11.000	10.250	0.224	9.751	0.375
12		1.25	4h	0	12.000	0.132	11.188	0.085	10.333	5H	12.000	11.328	0.140	10.859	0.212
			6g	0.028	11.972	0.212	11.160	0.132	10.257	6H	12.000	11.398	0.180	10.912	0.265
			8g	0.028	11.972	0.335	11.160	0.212	10.177	7H	12.000	11.412	0.224	10.985	0.335
	1.75		4h	0	12.000	0.170	10.863	0.095	9.692	5H	12.000	11.023	0.160	10.371	0.265
			6g	0.034	11.966	0.265	10.829	0.150	9.602	6H	12.000	11.063	0.200	10.441	0.335
			8g	0.034	11.966	0.425	10.829	0.236	9.516	7H	12.000	11.113	0.250	10.531	0.425
14		1.5	4h	0	14.000	0.150	13.026	0.090	12.012	5H	14.000	13.176	0.150	12.612	0.236
			6g	0.032	13.968	0.236	12.994	0.140	11.930	6H	14.000	13.216	0.190	12.676	0.300
			8g	0.032	13.968	0.375	12.994	0.224	11.846	7H	14.000	13.262	0.236	12.751	0.375
	2		4h	0	14.000	0.180	12.701	0.100	11.369	5H	14.000	12.871	0.170	12.135	0.300
			6g	0.038	13.962	0.280	12.663	0.160	11.271	6H	14.000	12.913	0.212	12.210	0.375
			8g	0.038	13.962	0.450	12.663	0.250	11.181	7H	14.000	12.966	0.265	12.310	0.475

Table 6. British Standard ISO Metric Screw Threads: Limits and Tolerances for Finished Uncoated Threads for Normal Lengths of Engagement *BS 3643: Part 2: 1981 (Continued)*

Nominal Diameter[a]	Pitch		External Threads (Bolts)							Internal Threads (Nuts)[b]					
	Coarse	Fine	Tol. Class	Fund dev.	Major Dia.		Pitch Dia.		Minor Dia	Tol. Class	Major Dia.	Pitch Dia.		Minor Dia	
					Max	Tol(−)	Max	Tol(−)	Min		Min	Max	Tol(−)	Max	Tol(−)
16		1.5	4h	0	16.000	0.150	15.026	0.090	14.012	5H	16.000	15.176	0.150	14.612	0.236
			6g	0.032	15.968	0.236	14.994	0.140	13.930	6H	16.000	15.216	0.190	14.676	0.300
			8g	0.032	15.968	0.375	14.994	0.224	13.846	7H	16.000	15.262	0.236	14.751	0.375
	2		4h	0	16.000	0.180	14.701	0.100	13.369	5H	16.000	14.871	0.170	14.135	0.300
			6g	0.038	15.962	0.280	14.663	0.160	13.271	6H	16.000	14.913	0.212	14.210	0.375
			8g	0.038	15.962	0.450	14.663	0.250	13.181	7H	16.000	14.966	0.265	14.310	0.475
18		1.5	4h	0	18.000	0.150	17.026	0.090	16.012	5H	18.000	17.176	0.150	16.612	0.236
			6g	0.032	17.968	0.236	16.994	0.140	15.930	6H	18.000	17.216	0.190	16.676	0.300
			8g	0.032	17.968	0.375	16.994	0.224	15.846	7H	18.000	17.262	0.236	16.751	0.375
	2.5		4h	0	18.000	0.212	16.376	0.106	14.730	5H	18.000	16.556	0.180	15.649	0.355
			6g	0.042	17.958	0.335	16.334	0.170	14.624	6H	18.000	16.600	0.224	15.774	0.450
			8g	0.042	17.958	0.530	16.334	0.265	14.529	7H	18.000	16.656	0.280	15.854	0.560
20		1.5	4h	0	20.000	0.150	19.026	0.090	18.012	5H	20.000	19.176	0.150	18.612	0.236
			6g	0.032	19.968	0.236	18.994	0.140	17.930	6H	20.000	0.190	0.190	18.676	0.300
			8g	0.032	19.968	0.375	18.994	0.224	17.846	7H	20.000	19.262	0.236	18.751	0.375
	2.5		4h	0	20.000	0.212	18.376	0.106	16.730	5H	20.000	18.556	0.180	17.649	0.355
			6g	0.042	19.958	0.335	18.334	0.170	16.624	6H	20.000	18.600	0.224	17.744	0.450
			8g	0.042	19.958	0.530	18.334	0.265	16.529	7H	20.000	18.650	0.280	17.854	0.560
22		1.5	4h	0	22.000	0.150	21.026	0.090	20.012	5H	22.000	21.176	0.150	20.612	0.236
			6g	0.032	21.968	0.236	20.994	0.140	19.930	6H	22.000	21.216	0.190	20.676	0.300
			8g	0.032	21.968	0.375	20.994	0.224	19.846	7H	22.000	21.262	0.236	20.751	0.375
	2.5		4h	0	22.000	0.212	20.376	0.106	18.730	5H	22.000	20.556	0.180	19.649	0.335
			6g	0.042	21.958	0.335	20.334	0.170	18.624	6H	22.000	20.600	0.224	19.744	0.450
			8g	0.042	21.958	0.530	20.334	0.265	18.529	7H	22.000	20.656	0.280	19.854	0.560
24		2	4h	0	24.000	0.180	22.701	0.106	21.363	5H	24.000	22.881	0.180	22.135	0.300
			6g	0.038	23.962	0.280	22.663	0.170	21.261	6H	24.000	22.925	0.224	22.210	0.375
			8g	0.038	23.962	0.450	22.663	0.265	21.166	7H	24.000	22.981	0.280	22.310	0.475
	3		4h	0	24.000	0.236	22.051	0.125	20.078	5H	24.000	22.263	0.212	21.152	0.400
			6g	0.048	23.952	0.375	22.003	0.200	19.955	6H	24.000	22.316	0.265	21.252	0.500
			8g	0.048	23.952	0.600	22.003	0.315	19.840	7H	24.000	22.386	0.335	21.382	0.630
27		2	4h	0	27.000	0.180	25.701	0.106	24.363	5H	27.000	25.881	0.180	25.135	0.300
			6g	0.038	26.962	0.280	25.663	0.170	24.261	6H	27.000	25.925	0.224	25.210	0.375
			8g	0.038	26.962	0.450	25.663	0.265	24.166	7H	27.000	25.981	0.280	25.310	0.475
	3		4h	0	27.000	0.236	25.051	0.125	23.078	5H	27.000	25.263	0.212	24.152	0.400
			6g	0.048	26.952	0.375	25.003	0.200	22.955	6H	27.000	25.316	0.265	24.252	0.500
			8g	0.048	26.952	0.600	25.003	0.315	22.840	7H	27.000	25.386	0.335	24.382	0.630
30		2	4h	0	30.000	0.180	28.701	0.106	27.363	5H	30.000	28.881	0.180	28.135	0.300
			6g	0.038	29.962	0.280	28.663	0.170	27.261	6H	30.000	27.925	0.224	28.210	0.375
			8g	0.038	29.962	0.450	28.663	0.265	27.166	7H	30.000	28.981	0.280	28.310	0.475
	3.5		4h	0	30.000	0.265	27.727	0.132	25.439	5H	30.000	27.951	0.224	26.661	0.450
			6g	0.053	29.947	0.425	27.674	0.212	25.305	6H	30.000	28.007	0.280	26.771	0.560
			8g	0.053	29.947	0.670	27.674	0.335	25.183	7H	30.000	28.082	0.355	26.921	0.710
33		2	4h	0	33.000	0.180	31.701	0.106	30.363	5H	33.000	31.881	0.180	31.135	0.300
			6g	0.038	32.962	0.280	31.663	0.170	30.261	6H	33.000	31.925	0.224	31.210	0.375
			8g	0.038	32.962	0.450	30.663	0.265	30.166	7H	33.000	31.981	0.280	31.310	0.475
	3.5		4h	0	33.000	0.265	30.727	0.132	28.438	5H	33.000	30.951	0.224	29.661	0.450
			6g	0.053	32.947	0.425	30.674	0.212	28.305	6H	33.000	31.007	0.280	29.771	0.560
			8g	0.053	32.947	0.670	30.674	0.335	28.182	7H	33.000	31.082	0.355	29.921	0.710
36	4		4h	0	36.000	0.300	33.402	0.140	30.798	5H	36.000	33.638	0.236	32.145	0.475
			6g	0.060	35.940	0.475	33.342	0.224	30.654	6H	36.000	33.702	0.300	32.270	0.600
			8g	0.060	35.940	0.750	33.342	0.355	30.523	7H	36.000	33.777	0.375	32.420	0.750
39	4		4h	0	39.000	0.300	36.402	0.140	33.798	5H	39.000	36.638	0.236	35.145	0.475
			6g	0.060	38.940	0.475	36.342	0.224	33.654	6H	39.000	36.702	0.300	35.270	0.600
			8g	0.060	38.940	0.750	36.342	0.355	33.523	7H	39.000	36.777	0.375	35.420	0.750

[a] This table provides coarse- and fine-pitch series data for threads listed in Table 6 for first, second, and third choices. For constant-pitch series and for larger sizes than are shown, refer to the Standard.

[b] The fundamental deviation for internal threads (nuts) is zero for threads in this table.

All dimensions are in millimeters.

The tolerances for the minor diameter of nut threads (T_{D1}), grade 6, in Table 7, were calculated as follows:

For pitches 0.2 to 0.8 mm, T_{D1} (6) = $433P - 190P^{1.22}$.

For pitches 1 mm and coarser, T_{D1} (6) = $230P^{0.7}$.

In these formulas, T_{D1} (6) is in micrometers and P is in millimeters. For tolerance grades 4, 5, 7, and 8: T_{D1} (4) = 0.63 T_{D1} (6); T_{D1} (5) = 0.8 T_{D1} (6); T_{D1} (7) = 1.25 T_{D1} (6); and T_{D1} (8) = 1.6 T_{D1} (6), respectively.

Diameter/Pitch Combinations: Part 1 of BS 3643 provides a choice of diameter/pitch combinations shown here in Table 6. The use of first-choice items is preferred but if necessary, second, then third choice combinations may be selected. If pitches finer than those given in Table 6 are necessary, only the following pitches should be used: 3, 2, 1.5, 1, 0.75, 0.5, 0.35, 0.25, and 0.2 mm. When selecting such pitches it should be noted that there is increasing difficulty in meeting tolerance requirements as the diameter is increased for a given pitch. It is suggested that diameters greater than the following should not be used with the pitches indicated:

Pitch, mm	0.5	0.75	1	1.5	2	3
Maximum Diameter, mm	22	33	80	150	200	300

In cases where it is necessary to use a thread with a pitch larger than 6 mm, in the diameter range of 150 to 300 mm, the 8 mm pitch should be used.

Limits and Tolerances for Finished Uncoated Threads: Part 2 of BS 3643 specifies the fundamental deviations, tolerances, and limits of size for the tolerance classes 4H, 5H, 6H, and 7H for internal threads (nuts) and 4h, 6g, and 8g for external threads (bolts) for coarse-pitch series within the range of 1 to 68 mm; fine-pitch series within the range of 1 to 33 mm; and constant pitch series within the range of 8 to 300 mm diameter.

The data in Table 7 provide the first, second, and third choice combinations shown in Table 6 except that constant-pitch series threads are omitted. For diameters larger than shown in Table 7, and for constant-pitch series data, refer to the Standard.

Table 7. British Standard ISO Metric Screw Threads — Diameter/Pitch Combinations *BS 3643:Part 1:1981 (1998)*

Nominal Diameter						Nominal Diameter			
Choices						Choices			
1st	2nd	3rd	Coarse Pitch	Fine Pitch	Constant Pitch	1st	2nd	3rd	Constant Pitch
1	…	…	0.25	0.2	…	…	…	70	6,4,3,2,1.5
…	1.1	…	0.25	0.2	…	72	…	…	6,4,3,2,1.5
1.2	…	…	0.25	0.2	…	…	…	75	4,3,2,1.5
…	1.4	…	0.3	0.2	…	…	76	…	6,4,3,2,1.5
1.6	…	…	0.35	0.2	…	…	…	78	2
…	1.8	…	0.35	0.2	…	80	…	…	6,4,3,2,1.5
2.0	…	…	0.4	0.25	…	…	…	82	2
…	2.2	…	0.45	0.25	…	…	85	…	6,4,3,2
2.5	…	…	0.45	0.35	…	90	…	…	6,4,3,2
3	…	…	0.5	0.35	…	…	95	…	6,4,3,2
…	3.5	…	0.6	0.35	…	100	…	…	6,4,3,2
4	…	…	0.7	0.5	…	…	105	…	6,4,3,2
…	4.5	…	0.75	0.5	…	110	…	…	6,4,3,2
5	…	…	0.8	0.5	…	…	115	…	6,4,3,2
…	…	5.5	…	(0.5)	…	…	120	…	6,4,3,2

Table 7. *(Continued)* **British Standard ISO Metric Screw Threads — Diameter/Pitch Combinations** *BS 3643:Part 1:1981 (1998)*

Nominal Diameter			Coarse Pitch	Fine Pitch	Constant Pitch	Nominal Diameter			Constant Pitch
Choices						Choices			
1st	2nd	3rd				1st	2nd	3rd	
6	…	…	1	0.75	…	125	…	…	6,4,3,2
…	7	…	1	0.75	…	…	130	…	6,4,3,2
8	…	…	1.25	1	0.75	…	…	135	6,4,3,2
…	…	9	1.25	…	1,0.75	140	…	…	6,4,3,2
10	…	…	1.5	1.25	1,0.75	…	…	145	6,4,3,2
…	…	11	1.5	…	1,0.75	…	150	…	6,4,3,2
12	…	…	1.75	1.25	1.5,1	…	…	155	6,4,3
…	14	…	2	1.5	1.25[a]1	160	…	…	6,4,3
…	…	15	…	…	1.5,1	…	…	165	6,4,3
16	…	…	2	1.5	1	…	170	…	6,4,3
…	…	17	…	…	1.5,1	…	…	175	6,4,3
…	18	…	2.5	1.5	2,1	180	…	…	6,4,3
20	…	…	2.5	1.5	2,1	…	…	185	6,4,3
…	22	…	2.5	1.5	2,1	…	190	…	6,4,3
24	…	…	3	2	1.5,1	…	…	195	6,4,3
…	…	25	…	…	2,1.5,1	200	…	…	6,4,3
…	…	26	…	…	1.5	…	…	205	6,4,3
…	27	…	3	2	1.5,1	…	210	…	6,4,3
…	…	28	…	…	2,1.5,1	…	…	215	6,4,3
30	…	…	3.5	2	(3),1.5,1	220	…	…	6,4,3
…	…	32	…	…	2,1.5	…	…	225	6,4,3
…	33	…	3.5	2	(3),1.5,	…	…	230	6,4,3
…	…	35[b]	…	…	1.5	…	…	235	6,4,3
36	…	…	4	…	3,2,1.5	…	240	…	6,4,3
…	…	38	…	…	1.5	…	…	245	6,4,3
…	39	…	4	…	3,2,1.5	250	…	…	6,4,3
…	…	40	…	…	3,2,1.5	…	…	255	6,4
42	45	…	4.5	…	4,3,2,1.5	…	260	…	6,4
48	…	…	5	…	4,3,2,1.5	…	…	265	6,4
…	…	50	…	…	3,2,1.5	…	…	270	6,4
…	52	…	5	…	4,3,2,1.5	…	…	275	6,4
…	…	55	…	…	4,3,2,1.5	280	…	…	6,4
56	…	…	5.5	…	4,3,2,1.5	…	…	285	6,4
…	…	58	…	…	4,3,2,1.5	…	…	290	6,4
…	60	…	5.5	…	4,3,2,1.5	…	…	295	6,4
…	…	62	…	…	4,3,2,1.5	…	300	…	6,4
64	…	…	6	…	4,3,2,1.5	…	…	…	…
…	…	65	…	…	4,3,2,1.5	…	…	…	…
…	68	…	6	…	4,3,2,1.5	…	…	…	…

[a] Only for spark plugs for engines.

[b] Only for locking nuts for bearings.

All dimensions are given in millimeters.

Pitches in parentheses () are to be avoided as far as possible.

British Standard Whitworth (BSW) and British Standard Fine (BSF) Threads.— The BSW is the Coarse Thread series and the BSF is the Fine Thread series of British Standard 84:1956—Parallel Screw Threads of Whitworth Form. The dimensions given in the tables on the following pages for the major, effective, and minor diameters are, respectively, the maximum limits of these diameters for bolts and the minimum limits for nuts. Formulas for the tolerances on these diameters are given in the table below.

Whitworth Standard Thread Form.—This thread form is used for the British Standard Whitworth (BSW) and British Standard Fine (BSF) screw threads. More recently, both threads have been known as parallel screw threads of Whitworth form.

With standardization of the Unified thread, the Whitworth thread form is expected to be used only for replacements or spare parts. Tables of British Standard Parallel Screw Threads of Whitworth Form will be found on the following pages; tolerance formulas are given in the table below. The form of the thread is shown by the diagram. If

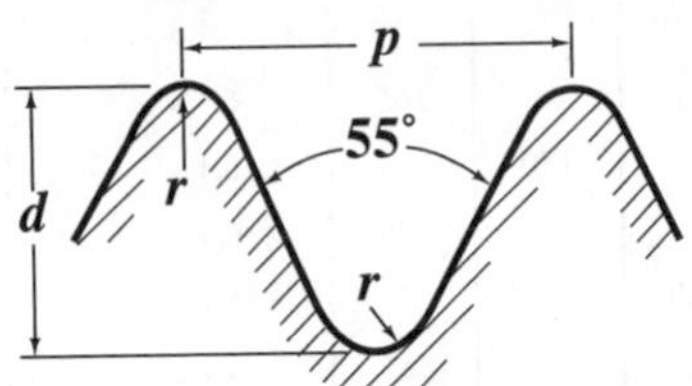

p = pitch, d = depth of thread, r = radius at crest and root, and n = number of threads per inch, then

$$d = \tfrac{1}{3}p \times \cot 27°30' = 0.640327p = 0.640327 \div n$$

$$r = 0.137329p = 0.137329 \div n$$

It is recommended that stainless steel bolts of nominal size ¾ inch and below should not be made to Close Class limits but rather to Medium or Free Class limits. Nominal sizes above ¾ inch should have maximum and minimum limits 0.001 inch smaller than the values obtained from the table.

Tolerance Classes : Close Class bolts. Applies to screw threads requiring a fine snug fit, and should be used only for special work where refined accuracy of pitch and thread form are particularly required. *Medium Class bolts and nuts.* Applies to the better class of ordinary interchangeable screw threads. *Free Class bolts.* Applies to the majority of bolts of ordinary commercial quality. *Normal Class nuts.* Applies to ordinary commercial quality nuts; this class is intended for use with Medium or Free Class bolts.

Table 1. Tolerance Formulas for BSW and BSF Threads

	Class or Fit	Tolerance in inches[a] (+ for nuts, – for bolts)		
		Major Dia.	Effective Dia.	Minor Dia.
Bolts	Close	$\frac{2}{3}T + 0.01\sqrt{p}$	$\frac{2}{3}T$	$\frac{2}{3}T + 0.013\sqrt{p}$
	Medium	$T + 0.01\sqrt{p}$	T	$T + 0.02\sqrt{p}$
	Free	$\frac{3}{2}T + 0.01\sqrt{p}$	$\frac{3}{2}T$	$\frac{3}{2}T + 0.02\sqrt{p}$
Nuts	Close	…	$\frac{2}{3}T$	$0.2p + 0.004$[b]
	Medium	…	T	$0.2p + 0.005$[c]
	Normal	…	$\frac{3}{2}T$	$0.2p + 0.007$[d]

[a] The symbol $T = 0.002\sqrt[3]{D} + 0.003\sqrt{L} + 0.005\sqrt{p}$, where D = major diameter of thread in inches; L = length of engagement in inches; p = pitch in inches. The symbol p signifies pitch.

[b] For 26 threads per inch and finer.

[c] For 24 and 22 threads per inch.

[d] For 20 threads per inch and coarser.

Table 2. Thraeds of Whitworth Form—Basic Dimensions

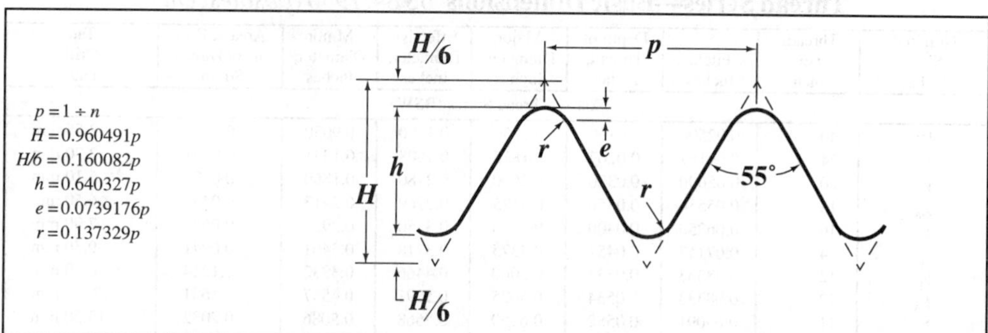

Threads per Inch	Pitch	Triangular Height	Shortening	Depth of Thread	Depth of Rounding	Radius
n	*p*	*H*	*H*/6	*h*	*e*	*r*
72	0.013889	0.013340	0.002223	0.008894	0.001027	0.001907
60	0.016667	0.016009	0.002668	0.010672	0.001232	0.002289
56	0.017857	0.017151	0.002859	0.011434	0.001320	0.002452
48	0.020833	0.020010	0.003335	0.013340	0.001540	0.002861
40	0.025000	0.024012	0.004002	0.016008	0.0011848	0.003433
36	0.027778	0.026680	0.004447	0.017787	0.002053	0.003815
32	0.031250	0.030015	0.005003	0.020010	0.002310	0.004292
28	0.035714	0.034303	0.005717	0.022869	0.002640	0.004905
26	0.038462	0.036942	0.006157	0.024628	0.002843	0.005282
24	0.041667	0.040020	0.006670	0.026680	0.003080	0.005722
22	0.045455	0.043659	0.007276	0.029106	0.003366	0.006242
20	0.050000	0.048025	0.008004	0.032016	0.003696	0.006866
19	0.052632	0.050553	0.008425	0.033702	0.003890	0.007228
18	0.055556	0.053361	0.008893	0.035574	0.004107	0.007629
16	0.062500	0.060031	0.010005	0.040020	0.004620	0.008583
14	0.071429	0.068607	0.011434	0.045738	0.005280	0.009809
12	0.083333	0.080041	0.013340	0.053361	0.006160	0.011444
11	0.090909	0.087317	0.014553	0.058212	0.006720	0.012484
10	0.100000	0.096049	0.016008	0.064033	0.007392	0.013733
9	0.111111	0.106721	0.017787	0.071147	0.008213	0.015259
8	0.125000	0.120061	0.020010	0.080041	0.009240	0.017166
7	0.142857	0.137213	0.022869	0.091475	0.010560	0.019618
6	0.166667	0.160082	0.026680	0.106721	0.012320	0.022888
5	0.20000	0.192098	0.032016	0.128065	0.014784	0.027466
4.5	0.222222	0.213442	0.035574	0.142295	0.016426	0.030518
4	0.250000	0.240123	0.040020	0.160082	0.018479	0.034332
3.5	0.285714	0.274426	0.045738	0.182951	0.021119	0.039237
3.25	0.307692	0.295536	0.049256	0.197024	0.022744	0.042255
3	0.333333	0.320164	0.053361	0.213442	0.024639	0.045776
2.875	0.347826	0.334084	0.055681	0.222722	0.025710	0.047767
2.75	0.363636	0.349269	0.058212	0.232846	0.026879	0.049938
2.625	0.380952	0.365901	0.060984	0.243934	0.028159	0.052316
2.5	0.400000	0.384196	0.064033	0.256131	0.029567	0.054932

Dimensions are in inches.

Allowances: Only Free Class and Medium Class bolts have an allowance. For nominal sizes of ¾ inch down to ¼ inch, the allowance is 30 per cent of the Medium Class bolt effective-diameter tolerance (0.3*T*); for sizes less than ¼ inch, the allowance for the ¼-inch size applies. Allowances are applied minus from the basic bolt dimensions; the tolerances are then applied to the reduced dimensions.

Table 3. British Standard Whitworth (BSW) and British Standard Fine (BSF) Screw Thread Series—Basic Dimensions *BS 84:1956 (obsolescent)*

Nominal Size, Inches	Threads per Inch	Pitch, Inches	Depth of Thread, Inches	Major Diameter, Inches	Effective Diameter, Inches	Minor Diameter, Inches	Area at Bottom of Thread, Sq. in.	Tap Drill Dia.
Coarse Thread Series (BSW)								
1/8[a]	40	0.02500	0.0160	0.1250	0.1090	0.9030	0.0068	2.55 mm
3/16	24	0.04167	0.0267	0.1875	0.1608	0.1341	0.0141	3.70 mm
1/4	20	0.05000	0.0320	0.2500	0.2180	0.1860	0.0272	5.10 mm
5/16	18	0.05556	0.0356	0.3125	0.2769	0.2413	0.0457	6.50 mm
3/8	16	0.06250	0.0400	0.3750	0.3350	0.2950	0.0683	7.90 mm
7/16	14	0.07143	0.0457	0.4375	0.3918	0.3461	0.0941	9.30 mm
1/2	12	0.08333	0.0534	0.5000	0.4466	0.3932	0.1214	10.50 mm
9/16[a]	12	0.08333	0.0534	0.5625	0.5091	0.4557	0.1631	12.10. mm
5/8	11	0.09091	0.0582	0.6250	0.5668	0.5086	0.2032	13.50 mm
11/16[a]	11	0.09091	0.0582	0.6875	0.6293	0.5711	0.2562	15.00 mm
3/4	10	0.10000	0.0640	0.7500	0.6860	0.6220	0.3039	16.25 mm
7/8	9	0.11111	0.0711	0.8750	0.8039	0.7328	0.4218	19.25 mm
1	8	0.12500	0.0800	1.0000	0.9200	0.8400	0.5542	22.00 mm
1 1/8	7	0.14286	0.0915	1.1250	1.0335	0.9420	0.6969	24.75 mm
1 1/4	7	0.14286	0.0915	1.2500	1.1585	1.0670	0.8942	28.00 mm
1 1/2	6	0.16667	0.1067	1.5000	1.3933	1.2866	1.3000	33.50 mm
1 3/4	5	0.20000	0.1281	1.7500	1.6219	1.4938	1.7530	39.00 mm
2	4.5	0.22222	0.1423	2.0000	1.8577	1.7154	2.3110	44.50 mm
2 1/4	4	0.25000	0.1601	2.2500	2.0899	1.9298	2.9250	
2 1/2	4	0.25000	0.1601	2.5000	2.3399	2.1798	3.7320	
2 3/4	3.5	0.28571	0.1830	2.7500	2.5670	2.3840	4.4640	Tap drill diameters shown in this column are recommended sizes listed in B.S. 1157:1975 and provide from 77 to 87% of full thread.
3	3.5	0.28571	0.1830	3.0000	2.8170	2.6340	5.4490	
3 1/4[a]	3.25	0.30769	0.1970	3.2500	3.0530	2.8560	6.4060	
3 1/2	3.25	0.30769	0.1970	3.5000	3.3030	3.1060	7.5770	
3 3/4[a]	3	0.33333	0.2134	3.7500	3.5366	3.3232	8.6740	
4	3	0.33333	0.2134	4.0000	3.7866	3.5732	10.0300	
4 1/2	2.875	0.34783	0.2227	4.5000	4.2773	4.0546	12.9100	
5	2.75	0.36364	0.2328	5.0000	4.7672	4.5344	16.1500	
5 1/2	2.625	0.38095	0.2439	5.5000	5.2561	5.0122	19.7300	
6	2.5	0.40000	0.2561	6.0000	5.7439	5.4878	23.6500	
Fine Thread Series (BSF)								
3/16[a,b]	32	0.03125	0.0200	0.1875	0.1675	0.1475	0.0171	4.00 mm
7/32[a]	28	0.03571	0.0229	0.2188	0.1959	0.1730	0.0235	4.60 mm
1/4	26	0.03846	0.0246	0.2500	0.2254	0.2008	0.0317	5.30 mm
9/32[a]	26	0.03846	0.0246	0.2812	0.2566	0.2320	0.0423	6.10 mm
5/16	22	0.04545	0.0291	0.3125	0.2834	0.2543	0.0508	6.80 mm
3/8	20	0.05000	0.0320	0.3750	0.3430	0.3110	0.0760	8.30 mm
7/16	18	0.05556	0.0 356	0.4375	0.4019	0.3363	0.1054	9.70 mm
1/2	16	0.06250	0.0400	0.5000	0.4600	0.4200	0.1385	11.10 mm
9/16	16	0.06250	0.0400	0.5625	0.5225	0.4825	0.1828	12.70 mm
5/8	14	0.07143	0.0457	0.6250	0.5793	0.5336	0.2236	14.00 mm
11/16[a]	14	0.07143	0.0457	0.6875	0.6418	0.5961	0.2791	15.50 mm
3/4	12	0.08333	0.0534	0.7500	0.6966	0.6432	0.3249	16.75 mm
7/8	11	0.09091	0.0582	0.8750	0.8168	0.7586	0.4520	19.75 mm
1	10	0.10000	0.0640	1.0000	0.9360	0.8720	0.5972	22.75 mm
1 1/8	9	0.11111	0.0711	1.1250	1.0539	0.9828	0.7586	25.50 mm
1 1/4	9	0.11111	0.0711	1.2500	1.1789	1.1078	0.9639	28.50 mm
1 3/8[a]	8	0.12500	0.0800	1.3750	1.2950	1.2150	1.1590	31.50 mm
1 1/2	8	0.12500	0.0800	1.5000	1.4200	1.3400	1.4100	34.50 mm
1 5/8[a]	8	0.12500	0.0800	1.6250	1.5450	1.4650	1.6860	
1 3/4	7	0.14286	0.0915	1.7500	1.6585	1.5670	1.9280	
2	7	0.14286	0.0915	2.0000	1.9085	1.8170	2.5930	Tap drill sizes listed in this column are recommended sizes shown in B.S. 1157:1975 and provide from 78 to 88% of full thread.
2 1/4	6	0.16667	0.1067	2.2500	2.1433	2.0366	3.2580	
2 1/2	6	0.16667	0.1067	2.5000	2.3933	2.2866	4.1060	
2 3/4	6	0.16667	0.1067	2.7500	2.6433	2.5366	5.0540	
3	5	0.20000	0.1281	3.0000	2.8719	2.7438	5.9130	
3 1/4	5	0.20000	0.1281	3.2500	3.1219	2.9938	7.0390	
3 1/2	4.5	0.22222	0.1423	3.5000	3.3577	3.2154	8.1200	
3 3/4	4.5	0.22222	0.1423	3.7500	3.6077	3.4654	9.4320	
4	4.5	0.22222	0.1423	4.0000	3.8577	3.7154	10.8400	
4 1/4	4	0.25000	0.1601	4.2500	4.0899	3.9298	12.1300	

[a] To be dispensed with wherever possible.

[b] The use of 2 BA threads is recommended.

British Association Standard Thread (BA) This screw thread system is recommended by the British Standards Institution for use in preference to the BSW and BSF systems for all screws smaller than ¼ inch except that the use of the "0" BA thread be discontinued in favor of the ¼-in. BSF It is further recommended that in the selection of sizes, preference be given to even numbered BA sizes. The thread form is shown by the diagram. .—

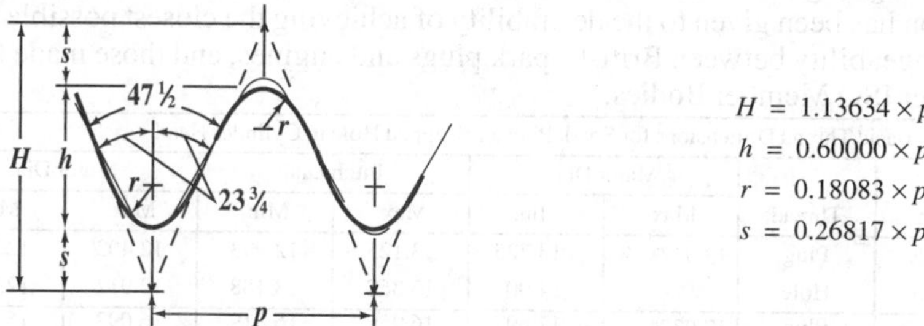

It is a symmetrical V-thread, of 47½ degree included angle, having its crests and roots rounded with equal radii, such that the basic depth of the thread is 0.6000 of the pitch. Where p = pitch of thread, H = depth of V-thread, h = depth of BA thread, r = radius at root and crest of thread, and s = root and crest truncation.

Table 4. British Association (BA) Standard Thread, Basic Dimensions

BS 93:1951 (obsolescent)

Designation Number	Pitch, mm	Depth of Thread, mm	Bolt and Nut Major Diameter, mm	Bolt and Nut Effective Diameter, mm	Bolt and Nut Minor Diameter, mm	Radius, mm	Threads per Inch (approx.)
0	1.0000	0.600	6.00	5.400	4.80	0.1808	25.4
1	0.9000	0.540	5.30	4.760	4.22	0.1627	28.2
2	0.8100	0.485	4.70	4.215	3.73	0.1465	31.4
3	0.7300	0.440	4.10	3.660	3.22	0.1320	34.8
4	0.6600	0.395	3.60	3.205	2.81	0.1193	38.5
5	0.5900	0.355	3.20	2.845	2.49	0.1067	43.0
6	0.5300	0.320	2.80	2.480	2.16	0.0958	47.9
7	0.4800	0.290	2.50	2.210	1.92	0.0868	52.9
8	0.4300	0.260	2.20	1.940	1.68	0.0778	59.1
9	0.3900	0.235	1.90	1.665	1.43	0.0705	65.1
10	0.3500	0.210	1.70	1.490	1.28	0.0633	72.6
11	0.3100	0.185	1.50	1.315	1.13	0.0561	82.0
12	0.2800	0.170	1.30	1.130	0.96	0.0506	90.7
13	0.2500	0.150	1.20	1.050	0.90	0.0452	102
14	0.2300	0.140	1.00	0.860	0.72	0.0416	110
15	0.2100	0.125	0.90	0.775	0.65	0.0380	121
16	0.1900	0.115	0.79	0.675	0.56	0.0344	134

Tolerances and Allowances: Two classes of bolts and one for nuts are provided: *Close Class bolts* are intended for precision parts subject to stress, no allowance being provided between maximum bolt and minimum nut sizes. *Normal Class bolts* are intended for general commercial production and general engineering use; for sizes 0 to 10 BA, an allowance of 0.025 mm is provided.

Table 5. Tolerance Formulas for British Association (BA) Screw Threads

	Class or Fit	Tolerance (+ for nuts, − for bolts) Major Dia.	Effective Dia.	Minor Dia.
Bolts	Close Class 0 to 10 BA incl.	$0.15p$ mm	$0.08p + 0.02$ mm	$0.16p + 0.04$ mm
	Normal Class 0 to 10 BA incl.	$0.20p$ mm	$0.10p + 0.025$ mm	$0.20p + 0.05$ mm
	Normal Class 11 to 16 BA incl.	$0.25p$ mm	$0.10p + 0.025$ mm	$0.20p + 0.05$ mm
Nuts	All Classes		$0.12p + 0.03$ mm	$0.375p$ mm

In these formulas, p = pitch in millimeters.

OTHER THREADS

British Standard for Spark Plugs BS 45:1972 (withdrawn).—This revised British Standard refers solely to spark plugs used in automobiles and industrial spark ignition internal combustion engines. The basic thread form is that of the ISO metric (see page 1827). In assigning tolerances to the threads of the spark plug and the tapped holes, full consideration has been given to the desirability of achieving the closest possible measure of interchangeability between British spark plugs and engines, and those made to the standards of other ISO Member Bodies.

Basic Thread Dimensions for Spark Plug and Tapped Hole in Cylinder Head

Nom. Size	Pitch	Thread	Major Dia.		Pitch Dia.		Minor Dia.	
			Max.	Min.	Max.	Min.	Max.	Min.
14	1.25	Plug	13.937[a]	13.725	13.125	12.993	12.402	12.181
14	1.25	Hole		14.00	13.368	13.188	12.912	12.647
18	1.5	Plug	17.933[a]	17.697	16.959	16.819	16.092	15.845
18	1.5	Hole		18.00	17.216	17.026	16.676	16.376

[a] Not specified

All dimensions are given in millimeters.

The tolerance grades for finished spark plugs and corresponding tapped holes in the cylinder head are: for 14 mm size, 6e for spark plugs and 6H for tapped holes which gives a minimum clearance of 0.063 mm; and for 18 mm size, 6e for spark plugs and 6H for tapped holes which gives a minimum clearance of 0.067 mm.

These minimum clearances are intended to prevent the possibility of seizure, as a result of combustion deposits on the bare threads, when removing the spark plugs and applies to both ferrous and non-ferrous materials. These clearances are also intended to enable spark plugs with threads in accordance with this standard to be fitted into existing holes.

S.A.E. Standard Threads for Spark Plugs

Size[a] Nom. × Pitch	Major Diameter		Pitch Diameter		Minor Diameter	
	Max.	Min.	Max.	Min.	Max.	Min.
Spark Plug Threads, mm (inches)						
M18 × 1.5	17.933	17.803	16.959	16.853	16.053	...
	(0.07060)	(0.7009)	(0.6677)	(0.6635)	(0.6320)	...
M14 × 1.25	13.868	13.741	13.104	12.997	12.339	...
	(0.5460)	(0.5410)	(0.5159)	(0.5117)	(0.4858)	...
M12 × 1.25	11.862	11.735	11.100	10.998	10.211	...
	(0.4670)	(0.4620)	(0.4370)	(0.4330)	(0.4020)	...
M10 × 1.0	9.974	9.794	9.324	9.212	8.747	...
	(0.3927)	(0.3856)	(0.3671)	(0.3627)	(0.3444)	...
Tapped Hole Threads, mm (inches)						
M18 × 1.5	...	18.039	17.153	17.026	16.426	16.266
	...	(0.7102)	(0.6753)	(0.6703)	(0.6467)	(0.6404)
M14 × 1.25	...	14.034	13.297	13.188	12.692	12.499
	...	(0.5525)	(0.5235)	(0.5192)	(0.4997)	(0.4921)
M12 × 1.25	...	12.000	11.242	11.188	10.559	10.366
	...	(0.4724)	(0.4426)	(0.4405)	(0.4157)	(0.4081)
M10 × 1.0	...	10.000	9.500	9.350	9.153	8.917
	...	(0.3937)	(0.3740)	(0.3681)	(0,3604)	(0.3511)

[a] M14 and M18 are preferred for new applications.

In order to keep the wear on the threading tools within permissible limits, the threads in the spark plug GO (ring) gage shall be truncated to the maximum minor diameter of the spark plug, and in the tapped hole GO (plug) gage to the minimum major diameter of the tapped hole. The plain plug gage for checking the minor diameter of the tapped hole shall be the minimum specified. The thread form is that of the ISO metric (see page 1827).

Table 1. ANSI Standard Hose Coupling Threads for NPSH, NH, and NHR Nipples and Coupling Swivels *ANSI/ASME B1.20.7-1991*

Nom. Size of Hose	Thds. per Inch	Thread Designation	Pitch	Basic Height of Thread	Nipple (External) Thread					Coupling (Internal) Thread				
					Major Dia.		Pitch Dia.		Minor Dia.	Minor Dia.		Pitch Dia.		Major Dia.
					Max.	Min.	Max.	Min.	Max.	Min.	Max.	Min.	Max.	Min.
½, ⅝, ¾	11.5	.75-11.5NH	.08696	.05648	1.0625	1.0455	1.0060	0.9975	0.9495	0.9595	0.9765	1.0160	1.0245	1.0725
½, ⅝, ¾	11.5	.75-11.5NHR	.08696	.05648	1.0520	1.0350	1.0100	0.9930	0.9495	0.9720	0.9930	1.0160	1.0280	1.0680
½	14	.5-14NPSH	.07143	.04639	0.8248	0.8108	0.7784	0.7714	0.7320	0.7395	0.7535	0.7859	0.7929	0.8323
¾	14	.75-14NPSH	.07143	.04639	1.0353	1.0213	0.9889	0.9819	0.9425	0.9500	0.9640	0.9964	1.0034	1.0428
1	11.5	1-11.5NPSH	.08696	.05648	1.2951	1.2781	1.2396	1.2301	1.1821	1.1921	1.2091	1.2486	1.2571	1.3051
1¼	11.5	1.25-11.5NPSH	.08696	.05648	1.6399	1.6229	1.5834	1.5749	1.5629	1.5369	1.5539	1.5934	1.6019	1.6499
1½	11.5	1.5-11.5 NPSH	.08696	.05648	1.8788	1.8618	1.8223	1.8138	1.7658	1.7758	1.7928	1.8323	1.8408	1.8888
2	11.5	2-11.5NPSH	.08696	.05648	2.3528	2.3358	2.2963	2.2878	2.2398	2.2498	2.2668	2.3063	2.3148	2.3628
2½	8	2.5-8NPSH	.12500	.08119	2.8434	2.8212	2.7622	2.7511	2.6810	2.6930	2.7152	2.7742	2.7853	2.8554
3	8	3-8NPSH	.12500	.08119	3.4697	3.4475	3.3885	3.3774	3.3073	3.3193	3.3415	3.4005	3.4116	3.4817
3½	8	3.5-8NPSH	.12500	.08119	3.9700	3.9478	3.8888	3.8777	3.8076	3.8196	3.8418	3.9008	3.9119	3.9820
4	8	4-8NPSH	.12500	.08119	4.4683	4.4461	4.3871	4.3760	4.3059	4.3179	4.3401	4.3991	4.4102	4.4803
4	6	4-6NH (SPL)	.16667	.10825	4.9082	4.8722	4.7999	4.7819	4.6916	4.7117	4.7477	4.8200	4.8380	4.9283

All dimensions are given in inches.

Dimensions given for the maximum minor diameter of the nipple are figured to the intersection of the worn tool arc with a centerline through crest and root. The minimum minor diameter of the nipple shall be that corresponding to a flat at the minor diameter of the minimum nipple equal to $\frac{1}{24}p$, and may be determined by subtracting $0.7939p$ from the minimum pitch diameter of the nipple. (See Fig. 1.)

Dimensions given for the minimum major diameter of the coupling correspond to the basic flat, $\frac{1}{8}p$, and the profile at the major diameter produced by a worn tool must not fall below the basic outline. The maximum major diameter of the coupling shall be that corresponding to a flat at the major diameter of the maximum coupling equal to $\frac{1}{24}p$ and may be determined by adding $0.7939p$ to the maximum pitch diameter of the coupling. (See Fig. 1.)

NH and NHR threads are used for garden hose applications. NPSH threads are used for steam, air and all other hose connections to be made up with standard pipe threads. NH (SPL) threads are used for marine applications.

Table 2. ANSI Standard Hose Coupling Screw Thread Lengths

ANSI/ASME B1.20.7-1991

Nom. Size of Hose	Thds. per Inch	I.D. of Nipple, *C*	Approx. O.D. of Ext. Thd.	Length of Nipple, *L*	Length of Pilot, *I*	Depth of Coupl., *H*	Coupl. Thd. Length, *T*	Approx. No. Thds. in Length *T*
1/2, 5/8, 3/4	11.5	25/32	1 1/16	9/16	1/8	17/32	3/8	4 1/4
1/2, 5/8, 3/4	11.5	25/32	1 1/16	9/16	1/8	17/32	3/8	4 1/4
1/2	14	17/32	13/16	1/2	1/8	15/32	5/16	4 1/4
3/4	14	25/32	1 1/32	9/16	1/8	17/32	3/8	5 1/4
1	11.5	1 1/32	1 9/32	9/16	5/32	17/32	3/8	4 1/4
1 1/4	11.5	1 9/32	1 5/8	5/8	5/32	19/32	15/32	5 1/2
1 1/2	11.5	1 17/32	1 7/8	5/8	5/32	19/32	15/32	5 1/2
2	11.5	2 1/32	2 11/32	3/4	3/16	23/32	19/32	6 3/4
2 1/2	8	2 17/32	2 27/32	1	1/4	15/16	11/16	5 1/2
3	8	3 1/32	3 15/32	1 1/8	1/4	1 1/16	13/16	6 1/2
3 1/2	8	3 17/32	3 31/32	1 1/8	1/4	1 1/16	13/16	6 1/2
4	8	4 1/32	4 15/32	1 1/8	1/4	1 1/16	13/16	6 1/2
4	6	4	4 29/32	1 1/8	5/16	1 1/16	3/4	4 1/2

All dimensions are given in inches. For thread designation see Table 1.

ANSI Standard Hose Coupling Screw Threads.—Threads for hose couplings, valves, and all other fittings used in direct connection with hose intended for domestic, industrial, and general service in sizes 1/2, 5/8, 3/4, 1, 1 1/4, 1 1/2, 2, 2 1/2, 3, 3 1/2, and 4 inches are covered by American National Standard ANSI/ASME B1.20.7-1991 These threads are designated as follows:

NH — Standard hose coupling threads of full form as produced by cutting or rolling.

NHR — Standard hose coupling threads for garden hose applications where the design utilizes thin walled material which is formed to the desired thread.

NPSH — Standard straight hose coupling thread series in sizes 1/2 to 4 inches for joining to American National Standard taper pipe threads using a gasket to seal the joint.

Thread dimensions are given in Table 1 and thread lengths in Table 2.

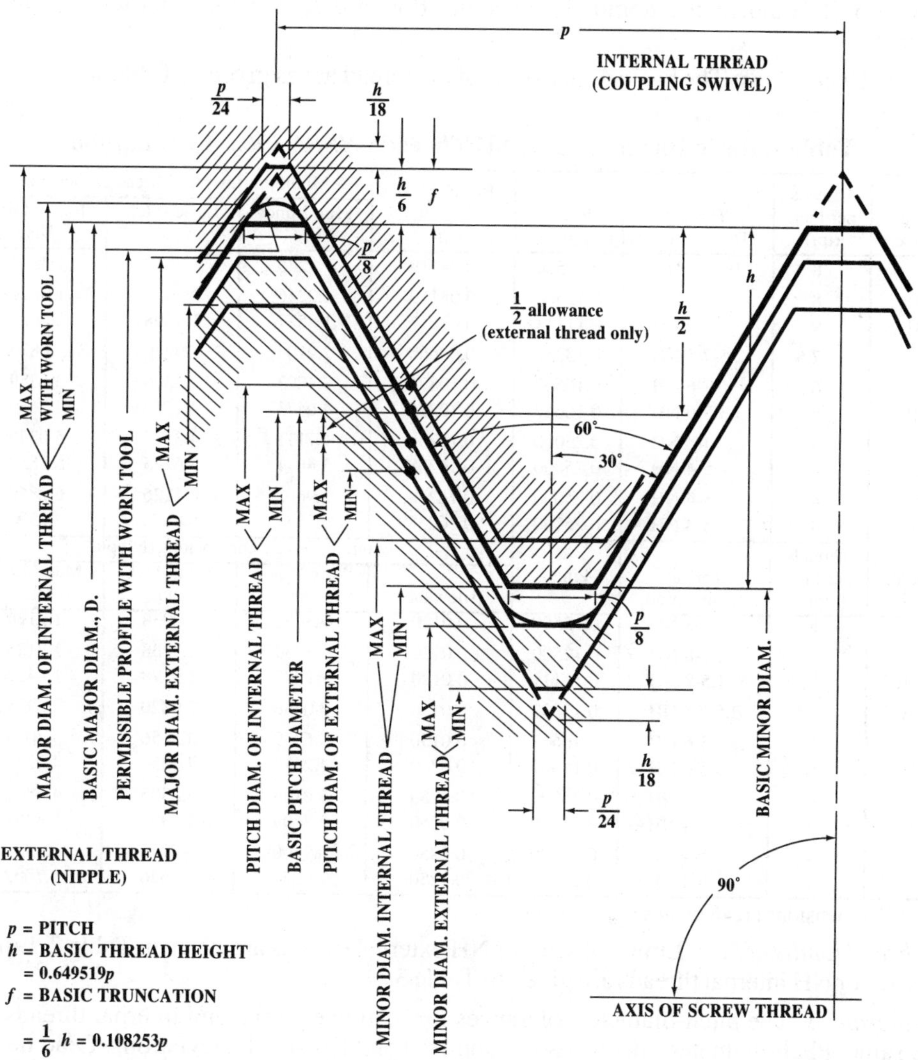

Fig. 1. Thread Form for ANSI Standard Hose Coupling Threads, NPSH, NH, and NHR. Heavy Line Shows Basic Size.

American National Fire Hose Connection Screw Thread.—This thread is specified in the National Fire Protection Association's Standard NFPA No. 194-1974. It covers the dimensions for screw thread connections for fire hose couplings, suction hose couplings, relay supply hose couplings, fire pump suctions, discharge valves, fire hydrants, nozzles, adaptors, reducers, caps, plugs, wyes, siamese connections, standpipe connections, and sprinkler connections.

Form of thread: The basic form of thread is as shown on page 1842. It has an included angle of 60 degrees and is truncated top and bottom. The flat at the root and crest of the basic thread form is equal to ⅛ (0.125) times the pitch in inches. The height of the thread is equal to 0.649519 times the pitch. The outer ends of both external and internal threads are terminated by the blunt start or "Higbee Cut" on full thread to avoid crossing and mutilation of thread.

Thread Designation: The thread is designated by specifying in sequence the nominal size of the connection, number of threads per inch followed by the thread symbol NH.

Thus, .75-8NH indicates a nominal size connection of 0.75 inch diameter with 8 threads per inch.

Basic Dimensions: The basic dimensions of the thread are as given in Table 1.

Table 1. Basic Dimensions of NH Threads *NFPA 1963–1993 Edition*

Nom. Size	Threads per Inch (tpi)	Thread Designation	Pitch, p	BasicThread Height, h	Minimum Internal Thread Dimensions		
					Min. Minor Dia.	Basic Pitch Dia.	BasicMajor Dia.
¾	8	0.75-8 NH	0.12500	0.08119	1.2246	1.3058	1.3870
1	8	1-8 NH	0.12500	0.08119	1.2246	1.3058	1.3870
1½	9	1.5-9 NH	0.11111	0.07217	1.8577	1.9298	2.0020
2½	7.5	2.5-7.5 NH	0.13333	0.08660	2.9104	2.9970	3.0836
3	6	3-6 NH	0.16667	0.10825	3.4223	3.5306	3.6389
3½	6	3.5-6 NH	0.16667	0.10825	4.0473	4.1556	4.2639
4	4	4-4 NH	0.25000	0.16238	4.7111	4.8735	5.0359
4½	4	4.5-4 NH	0.25000	0.16238	5.4611	5.6235	5.7859
5	4	5-4 NH	0.25000	0.16238	5.9602	6.1226	6.2850
6	4	6-4 NH	0.25000	0.16238	6.7252	6.8876	7.0500
Nom. Size	Threads per Inch (tpi)	Thread Designation	Pitch, p	External Thread Dimensions (Nipple)			
				Allowance	Max.Major Dia.	Max. Pitch Dia.	Max Minor Dia.
¾	8	0.75-8 NH	0.12500	0.0120	1.3750	1.2938	1.2126
1	8	1-8 NH	0.12500	0.0120	1.3750	1.2938	1.2126
1½	9	1.5-9 NH	0.11111	0.0120	1.9900	1.9178	1.8457
2½	7.5	2.5-7.5 NH	0.13333	0.0150	3.0686	2.9820	2.8954
3	6	3-6 NH	0.16667	0.0150	3.6239	3.5156	3.4073
3½	6	3.5-6 NH	0.16667	0.0200	4.2439	4.1356	4.0273
4	4	4-4 NH	0.25000	0.0250	5.0109	4.8485	4.6861
4½	4	4.5-4 NH	0.25000	0.0250	5.7609	5.5985	5.4361
5	4	5-4 NH	0.25000	0.0250	6.2600	6.0976	5.9352
6	4	6-4 NH	0.25000	0.0250	7.0250	6.8626	6.7002

All dimensions are in inches.

Thread Limits of Size: Limits of size for NH external threads are given in Table 2. Limits of size for NH internal threads are given in Table 3.

Tolerances: The pitch-diameter tolerances for mating external and internal threads are the same. Pitch-diameter tolerances include lead and half-angle deviations. Lead deviations consuming one-half of the pitch-diameter tolerance are 0.0032 inch for ¾-, 1-, and 1½-inch sizes; 0.0046 inch for 2½-inch size; 0.0052 inch for 3-, and 3½-inch sizes; and 0.0072 inch for 4-, 4½-, 5-, and 6-inch sizes. Half-angle deviations consuming one-half of the pitch-diameter tolerance are 1 degree, 42 minutes for ¾- and 1-inch sizes; 1 degree, 54 minutes for 1½-inch size; 2 degrees, 17 minutes for 2½-inch size; 2 degrees, 4 minutes for 3- and 3½-inch size; and 1 degree, 55 minutes for 4-, 4½-, 5-, and 6-inch sizes.

Tolerances for the external threads are

Major diameter tolerance = 2 × pitch-diameter tolerance

Minor diameter tolerance = pitch-diameter tolerance + $2h/9$

The minimum minor diameter of the external thread is such as to result in a flat equal to one-third of the $p/8$ basic flat, or $p/24$, at the root when the pitch diameter of the external thread is at its minimum value. The maximum minor diameter is basic, but may be such as results from the use of a worn or rounded threading tool. The maximum minor diameter is shown in the figure on page 1842 and is the diameter upon which the minor diameter tolerance formula shown above is based.

Tolerances for the internal threads are

Minor diameter tolerance = 2 × pitch-diameter tolerance

The minimum minor diameter of the internal thread is such as to result in a basic flat, $p/8$, at the crest when the pitch diameter of the thread is at its minimum value.

Major diameter tolerance = pitch-diameter tolerance - $2h/9$

Table 2. Limits of Size and Tolerances for NH External Threads (Nipples)
NFPA 1963, 1993 Edition

Nom. Size	Threads per Inch (tpi)	External Thread (Nipple)						
		Major Diameter			Pitch Diameter			Minor[a] Dia.
		Max.	Min.	Toler.	Max.	Min.	Toler.	Max.
¾	8	1.3750	1.3528	0.0222	1.2938	1.2827	0.0111	1.2126
1	8	1.3750	1.3528	0.0222	1.2938	1.2827	0.0111	1.2126
1½	9	1.9900	1.9678	0.0222	1.9178	1.9067	0.0111	1.8457
2½	7.5	3.0686	3.0366	0.0320	2.9820	2.9660	0.0160	2.8954
3	6	3.6239	3.5879	0.0360	3.5156	3.4976	0.0180	3.4073
3½	6	4.2439	4.2079	0.0360	4.1356	4.1176	0.0180	4.0273
4	4	5.0109	4.9609	0.0500	4.8485	4.8235	0.0250	4.6861
4½	4	5.7609	5.7109	0.0500	5.5985	5.5735	0.0250	5.4361
5	4	6.2600	6.2100	0.0500	6.0976	6.0726	0.0250	5.9352
6	4	7.0250	6.9750	0.0500	6.8626	6.8376	0.0250	6.7002

[a] Dimensions given for the maximum minor diameter of the nipple are figured to the intersection of the worn tool arc with a center line through crest and root. The minimum minor diameter of the nipple shall be that corresponding to a flat at the minor diameter of the minimum nipple equal to $p/24$ and may be determined by subtracting $11h/9$ (or $0.7939p$) from the minimum pitch diameter of the nipple.

All dimensions are in inches.

Table 3. Limits of Size and Tolerances for NH Internal Threads (Couplings)
NFPA 1963, 1993 Edition

Nom. Size	Threads per Inch (tpi)	Internal Thread (Coupling)						
		Major Diameter			Pitch Diameter			Minor[a] Dia.
		Min.	Max.	Toler.	Min.	Max.	Toler.	Min.
¾	8	1.2246	1.2468	0.0222	1.3058	1.3169	0.0111	1.3870
1	8	1.2246	1.2468	0.0222	1.3058	1.3169	0.0111	1.3870
1½	9	1.8577	1.8799	0.0222	1.9298	1.9409	0.0111	2.0020
2½	7.5	2.9104	2.9424	0.0320	2.9970	3.0130	0.0160	3.0836
3	6	3.4223	3.4583	0.0360	3.5306	3.5486	0.0180	3.6389
3½	6	4.0473	4.0833	0.0360	4.1556	4.1736	0.0180	4.2639
4	4	4.7111	4.7611	0.0500	4.8735	4.8985	0.0250	5.0359
4½	4	5.4611	5.5111	0.0500	5.6235	5.6485	0.0250	5.7859
5	4	5.9602	6.0102	0.0500	6.1226	6.1476	0.0250	6.2850
6	4	6.7252	6.7752	0.0500	6.8876	6.9126	0.0250	7.0500

[a] Dimensions for the minimum major diameter of the coupling correspond to the basic flat ($p/8$), and the profile at the major diameter produced by a worn tool must not fall below the basic outline. The maximum major diameter of the coupling shall be that corresponding to a flat at the major diameter of the maximum coupling equal to $p/24$ and may be determined by adding $11h/9$ (or $0.7939p$) to the maximum pitch diameter of the coupling.

All dimensions are in inches.

Gages and Gaging: Full information on gage dimensions and the use of gages in checking the NH thread are given in NFPA Standard No. 1963, 1993 Edition, published by the National Fire Protection Association, Batterymarch Park, Quincy, MA 02269.

The information and data taken from this standard are reproduced with the permission of the NFPA.

Rolled Threads for Screw Shells of Electric Sockets and Lamp Bases—American Standard

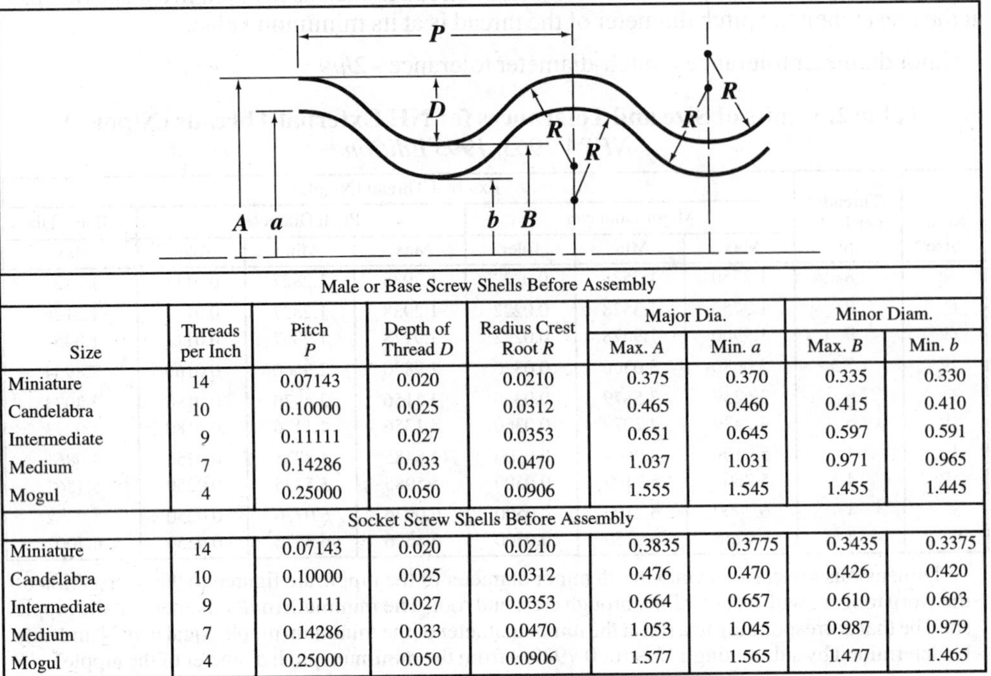

Size	Threads per Inch	Pitch P	Depth of Thread D	Radius Crest Root R	Major Dia. Max. A	Major Dia. Min. a	Minor Diam. Max. B	Minor Diam. Min. b
Male or Base Screw Shells Before Assembly								
Miniature	14	0.07143	0.020	0.0210	0.375	0.370	0.335	0.330
Candelabra	10	0.10000	0.025	0.0312	0.465	0.460	0.415	0.410
Intermediate	9	0.11111	0.027	0.0353	0.651	0.645	0.597	0.591
Medium	7	0.14286	0.033	0.0470	1.037	1.031	0.971	0.965
Mogul	4	0.25000	0.050	0.0906	1.555	1.545	1.455	1.445
Socket Screw Shells Before Assembly								
Miniature	14	0.07143	0.020	0.0210	0.3835	0.3775	0.3435	0.3375
Candelabra	10	0.10000	0.025	0.0312	0.476	0.470	0.426	0.420
Intermediate	9	0.11111	0.027	0.0353	0.664	0.657	0.610	0.603
Medium	7	0.14286	0.033	0.0470	1.053	1.045	0.987	0.979
Mogul	4	0.25000	0.050	0.0906	1.577	1.565	1.477	1.465

All dimensions are in inches.

Base Screw Shell Gage Tolerances: Threaded ring gages—"Go," Max. thread size to minus 0.0003 inch; "Not Go," Min. thread size to plus 0.0003 inch. Plain ring gages—"Go," Max. thread O.D. to minus 0.0002 inch; "Not Go," Min. thread O.D. to plus 0.0002 inch.

Socket Screw Shell Gages: Threaded plug gages—"Go," Min. thread size to plus 0.0003 inch; "Not Go," Max. thread size to minus 0.0003 inch. Plain plug gages—"Go," Min. minor dia. to plus 0.0002 inch; "Not Go," Max. minor dia. to minus 0.0002 inch.

Check Gages for Base Screw Shell Gages: Threaded plugs for checking threaded ring gages—"Go," Max. thread size to minus 0.0003 inch; "Not Go," Min. thread size to plus 0.0003 inch.

Instrument Makers' System.—The standard screw system of the Royal Microscopical Society of London, also known as the "Society Thread," is employed for microscope objectives and the nose pieces of the microscope into which these objectives screw. The form of the thread is the standard Whitworth form. The number of threads per inch is 36. The dimensions are as follows:

Male thread,	outside dia.,	max. 0.7982 inch,	min. 0.7952 inch;
	root dia.,	max. 0.7626 inch,	min. 0.7596 inch;
Female thread,	root of thread,	max. 0.7674 inch,	min. 0.7644 inch;
	top of thread,	max. 0.8030 inch,	min. 0.8000 inch.

Pipe Threads.—The types of threads used on pipe and pipe fittings may be classed according to their intended use: 1) threads that when assembled with a sealer will produce a pressure-tight joint; 2) threads that when assembled without a sealer will produce a pressure-tight joint; 3) threads that provide free- and loose-fitting mechanical joints without pressure tightness; and 4) threads that produce rigid mechanical joints without pressure tightness.

American National Standard pipe threads described in the following paragraphs provide taper and straight pipe threads for use in various combinations and with certain modifications to meet these specific needs.

American National Standard Taper Pipe Threads.—The basic dimensions of the ANSI Standard taper pipe thread are given in Table 4.

Form of Thread: The angle between the sides of the thread is 60 degrees when measured in an axial plane, and the line bisecting this angle is perpendicular to the axis. The depth of the truncated thread is based on factors entering into the manufacture of cutting tools and the making of tight joints and is given by the formulas in Table 4 or the data in Table 1 obtained from these formulas. Although the standard shows flat surfaces at the crest and root of the thread, some rounding may occur in commercial practice, and it is intended that the pipe threads of product shall be acceptable when crest and root of the tools or chasers lie within the limits shown in Table 1.

Table 1. Limits on Crest and Root of American National Standard External and Internal Taper Pipe Threads, NPT *ANSI/ASME B1.20.1-1983 (R1992)*

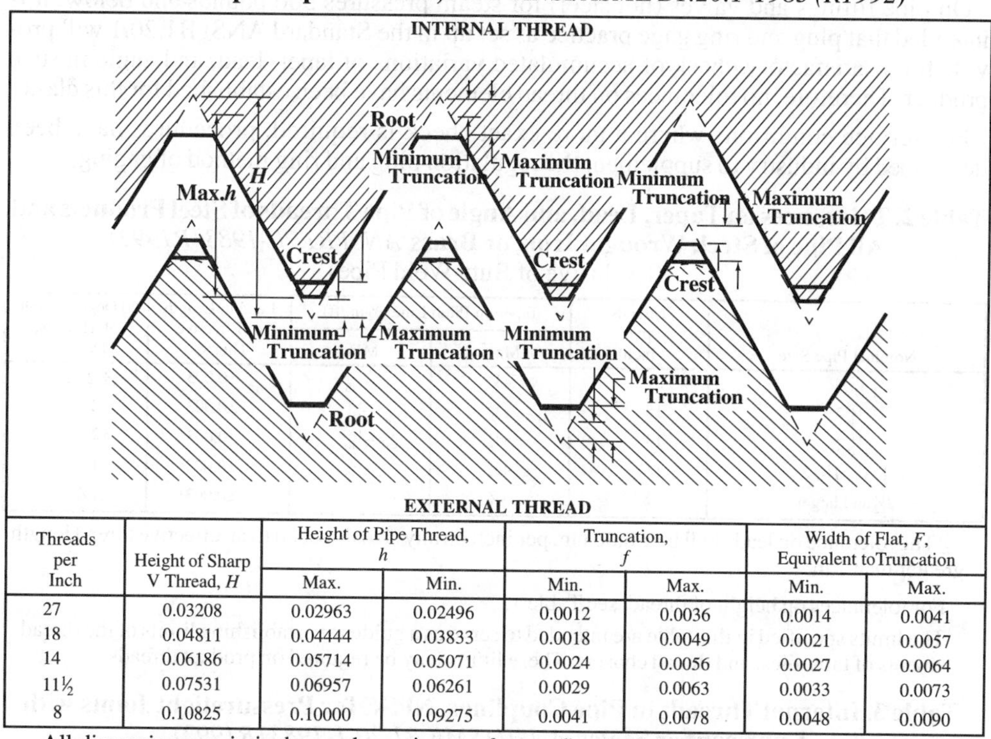

Threads per Inch	Height of Sharp V Thread, H	Height of Pipe Thread, h		Truncation, f		Width of Flat, F, Equivalent to Truncation	
		Max.	Min.	Min.	Max.	Min.	Max.
27	0.03208	0.02963	0.02496	0.0012	0.0036	0.0014	0.0041
18	0.04811	0.04444	0.03833	0.0018	0.0049	0.0021	0.0057
14	0.06186	0.05714	0.05071	0.0024	0.0056	0.0027	0.0064
11½	0.07531	0.06957	0.06261	0.0029	0.0063	0.0033	0.0073
8	0.10825	0.10000	0.09275	0.0041	0.0078	0.0048	0.0090

All dimensions are in inches and are given to four or five decimal places only to avoid errors in computations, not to indicate required precision.

Pitch Diameter Formulas: In the following formulas, which apply to the ANSI Standard taper pipe thread, E_0 = pitch diameter at end of pipe; E_1 = pitch diameter at the large end of the internal thread and at the gaging notch; D = outside diameter of pipe; L_1 = length of hand-tight or normal engagement between external and internal threads; L_2 = basic length of effective external taper thread; and p = pitch = 1 ÷ number of threads per inch.

$$E_0 = D - (0.05D + 1.1)p$$

$$E_1 = E_0 + 0.0625L_1$$

Thread Length: The formula for L_2 determines the length of the effective thread and includes approximately two usable threads that are slightly imperfect at the crest. The normal length of engagement, L_1, between external and internal taper threads, when assembled by hand, is controlled by the use of the gages.

$$L_2 = (0.80D + 6.8)p$$

Taper: The taper of the thread is 1 in 16, or 0.75 inch per foot, measured on the diameter and along the axis. The corresponding half-angle of taper or angle with the center line is 1 degree, 47 minutes.

Tolerances on Thread Elements.—The maximum allowable variation in the commercial product (manufacturing tolerance) is one turn large or small from the basic dimensions.

The permissible variations in thread elements on steel products and all pipe made of steel, wrought iron, or brass, exclusive of butt-weld pipe, are given in Table 2. This table is a guide for establishing the limits of the thread elements of taps, dies, and thread chasers. These limits may be required on product threads.

On pipe fittings and valves (not steel) for steam pressures 300 pounds and below, it is intended that plug and ring gage practice as set up in the Standard ANSI B1.20.1 will provide for a satisfactory check of accumulated variations of taper, lead, and angle in such product. Therefore, no tolerances on thread elements have been established for this class.

For service conditions where a more exact check is required, procedures have been developed by industry to supplement the regulation plug and ring method of gaging.

Table 2. Tolerances on Taper, Lead, and Angle of Pipe Threads of Steel Products and All Pipe of Steel, Wrought Iron, or Brass *ANSI B1.20-1983 (R1992)*
(Exclusive of Butt-Weld Pipe)

Nominal Pipe Size	Threads per Inch	Taper on Pitch Line (¾ in./ft)		Lead in Length of Effective Threads	60 Degree Angle of Threads, Degrees
		Max.	Min.		
1⁄16, 1⁄8	27	+1⁄8	−1⁄16	±0.003	± 2½
¼, 3⁄8	18	+1⁄8	−1⁄16	±0.003	±2
½, ¾	14	+1⁄8	−1⁄16	±0.003[a]	±2
1, 1¼, 1½, 2	11½	+1⁄8	−1⁄16	±0.003[a]	±1½
2½ and larger	8	+1⁄8	−1⁄16	±0.003[a]	±1½

[a] The tolerance on lead shall be ± 0.003 in. per inch on any size threaded to an effective thread length greater than 1 in.

For tolerances on height of thread, see Table 1.

The limits specified in this table are intended to serve as a guide for establishing limits of the thread elements of taps, dies, and thread chasers. These limits may be required on product threads.

Table 3. Internal Threads in Pipe Couplings, NPSC for Pressuretight Joints with Lubricant or Sealer *ANSI/ASME B1.20.1-1983 (R1992)*

Nom. Pipe-Size	Thds. per Inch	Minor[a] Dia.	Pitch Diameter[b]		Nom. Pipe	Thds. per Inch	Minor[a] Dia.	Pitch Diameter[b]	
		Min.	Min.	Max.			Min.	Min.	Max.
1⁄8	27	0.340	0.3701	0.3771	1½	11½	1.745	1.8142	1.8305
¼	18	0.442	0.4864	0.4968	2	11½	2.219	2.2881	2.3044
3⁄8	18	0.577	0.6218	0.6322	2½	8	2.650	2.7504	2.7739
½	14	0.715	0.7717	0.7851	3	8	3.277	3.3768	3.4002
¾	14	0.925	0.9822	0.9956	3½	8	3.777	3.8771	3.9005
1	11½	1.161	1.2305	1.2468	4	8	4.275	4.3754	4.3988
1¼	11½	1.506	1.5752	1.5915	…	…	…	…	…

[a] As the ANSI Standard Pipe Thread form is maintained, the major and minor diameters of the internal thread vary with the pitch diameter. All dimensions are given in inches.

[b] The actual pitch diameter of the straight tapped hole will be slightly smaller than the value given when gaged with a taper plug gage as called for in ANSI/ASME B1.20.1.

Table 4. Basic Dimensions, American National Standard Taper Pipe Threads, NPT
ANSI/ASME B1.20.1-1983 (R1992)

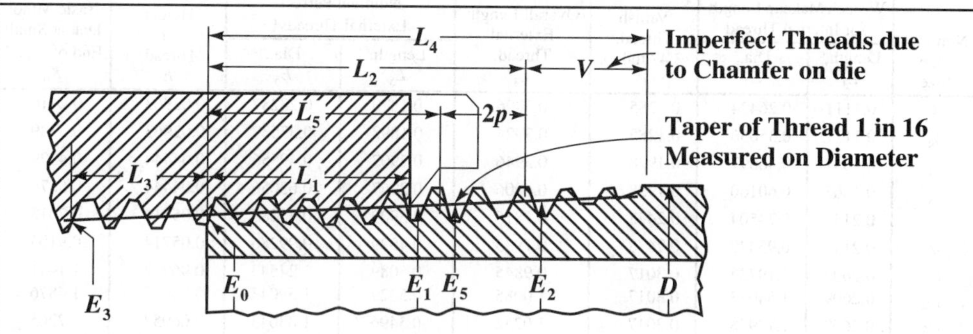

For all dimensions, see corresponding reference letter in table.

Angle between sides of thread is 60 degrees. Taper of thread, on diameter, is ¾ inch per foot. Angle of taper with center line is 1°47′.

The basic maximum thread height, h, of the truncated thread is 0.8 × pitch of thread. The crest and root are truncated a minimum of 0.033 × pitch for all pitches. For maximum depth of truncation, see Table 1.

Nominal Pipe Size	Outside Dia. of Pipe, D	Threads per Inch, n	Pitch of Thread, p	Pitch Diameter at Beginning of External Thread, E_0	Handtight Engagement Length,[b] L_1 In.	Handtight Engagement Dia.,[a] E_1	Effective Thread, Length,[c] L_2 In.	Effective Thread, Dia., E_2
1/16	0.3125	27	0.03704	0.27118	0.160	0.28118	0.2611	0.28750
1/8	0.405	27	0.03704	0.36351	0.1615	0.37360	0.2639	0.38000
¼	0.540	18	0.05556	0.47739	0.2278	0.49163	0.4018	0.50250
⅜	0.675	18	0.05556	0.61201	0.240	0.62701	0.4078	0.63750
½	0.840	14	0.07143	0.75843	0.320	0.77843	0.5337	0.79179
¾	1.050	14	0.07143	0.96768	0.339	0.98887	0.5457	1.00179
1	1.315	11½	0.08696	1.21363	0.400	1.23863	0.6828	1.25630
1¼	1.660	11½	0.08696	1.55713	0.420	1.58338	0.7068	1.60130
1½	1.900	11½	0.08696	1.79609	0.420	1.82234	0.7235	1.84130
2	2.375	11½	0.08696	2.26902	0.436	2.29627	0.7565	2.31630
2½	2.875	8	0.12500	2.71953	0.682	2.76216	1.1375	2.79062
3	3.500	8	0.12500	3.34062	0.766	3.38850	1.2000	3.41562
3½	4.000	8	0.12500	3.83750	0.821	3.88881	1.2500	3.91562
4	4.500	8	0.12500	4.33438	0.844	4.38712	1.3000	4.41562
5	5.563	8	0.12500	5.39073	0.937	5.44929	1.4063	5.47862
6	6.625	8	0.12500	6.44609	0.958	6.50597	1.5125	6.54062
8	8.625	8	0.12500	8.43359	1.063	8.50003	1.7125	8.54062
10	10.750	8	0.12500	10.54531	1.210	10.62094	1.9250	10.66562
12	12.750	8	0.12500	12.53281	1.360	12.61781	2.1250	12.66562
14 OD	14.000	8	0.12500	13.77500	1.562	13.87262	2.2500	13.91562
16 OD	16.000	8	0.12500	15.76250	1.812	15.87575	2.4500	15.91562
18 OD	18.000	8	0.12500	17.75000	2.000	17.87500	2.6500	17.91562
20 OD	20.000	8	0.12500	19.73750	2.125	19.87031	2.8500	19.91562
24 OD	24.000	8	0.12500	23.71250	2.375	23.86094	3.2500	23.91562

[a] Also pitch diameter at gaging notch (handtight plane).

[b] Also length of thin ring gage and length from gaging notch to small end of plug gage.

[c] Also length of plug gage.

Table 5. Basic Dimensions, American National Standard Taper Pipe Threads, NPT
ANSI/ASME B1.20.1-1983 (R1992)

Nominal Pipe Size	Wrench Makeup Length for Internal Thread: Length,[c] L_3	Dia., E_3	Vanish Thread, (3.47 thds.), V	Overall Length External Thread, L_4	Nominal Perfect External Threads[a]: Length, L_5	Dia., E_5	Height of Thread, h	Basic Minor Dia. at Small End of Pipe,[b] K_0
1/16	0.1111	0.26424	0.1285	0.3896	0.1870	0.28287	0.02963	0.2416
1/8	0.1111	0.35656	0.1285	0.3924	0.1898	0.37537	0.02963	0.3339
1/4	0.1667	0.46697	0.1928	0.5946	0.2907	0.49556	0.04444	0.4329
3/8	0.1667	0.60160	0.1928	0.6006	0.2967	0.63056	0.04444	0.5676
1/2	0.2143	0.74504	0.2478	0.7815	0.3909	0.78286	0.05714	0.7013
3/4	0.2143	0.95429	0.2478	0.7935	0.4029	0.99286	0.05714	0.9105
1	0.2609	1.19733	0.3017	0.9845	0.5089	1.24543	0.06957	1.1441
1 1/4	0.2609	1.54083	0.3017	1.0085	0.5329	1.59043	0.06957	1.4876
1 1/2	0.2609	1.77978	0.3017	1.0252	0.5496	1.83043	0.06957	1.7265
2	0.2609	2.25272	0.3017	1.0582	0.5826	2.30543	0.06957	2.1995
2 1/2	0.2500[d]	2.70391	0.4337	1.5712	0.8875	2.77500	0.100000	2.6195
3	0.2500[d]	3.32500	0.4337	1.6337	0.9500	3.40000	0.100000	3.2406
3 1/2	0.2500	3.82188	0.4337	1.6837	1.0000	3.90000	0.100000	3.7375
4	0.2500	4.31875	0.4337	1.7337	1.0500	4.40000	0.100000	4.2344
5	0.2500	5.37511	0.4337	1.8400	1.1563	5.46300	0.100000	5.2907
6	0.2500	6.43047	0.4337	1.9462	1.2625	6.52500	0.100000	6.3461
8	0.2500	8.41797	0.4337	2.1462	1.4625	8.52500	0.100000	8.3336
10	0.2500	10.52969	0.4337	2.3587	1.6750	10.65000	0.100000	10.4453
12	0.2500	12.51719	0.4337	2.5587	1.8750	12.65000	0.100000	12.4328
14 OD	0.2500	13.75938	0.4337	2.6837	2.0000	13.90000	0.100000	13.6750
16 OD	0.2500	15.74688	0.4337	2.8837	2.2000	15.90000	0.100000	15.6625
18 OD	0.2500	17.73438	0.4337	3.0837	2.4000	17.90000	0.100000	17.6500
20 OD	0.2500	19.72188	0.4337	3.2837	2.6000	19.90000	0.100000	19.6375
24 OD	0.2500	23.69688	0.4337	3.6837	3.0000	23.90000	0.100000	23.6125

[a] The length L_5 from the end of the pipe determines the plane beyond which the thread form is imperfect at the crest. The next two threads are perfect at the root. At this plane the cone formed by the crests of the thread intersects the cylinder forming the external surface of the pipe. $L_5 = L_2 - 2p$.

[b] Given as information for use in selecting tap drills.

[c] Three threads for 2-inch size and smaller; two threads for larger sizes.

[d] Military Specification MIL—P—7105 gives the wrench makeup as three threads for 3 in. and smaller. The E_3 dimensions are then as follows: Size 2½ in., 2.69609 and size 3 in., 3.31719.

All dimensions given in inches.

Increase in diameter per thread is equal to 0.0625/*n*.

The basic dimensions of the ANSI Standard Taper Pipe Thread are given in inches to four or five decimal places. While this implies a greater degree of precision than is ordinarily attained, these dimensions are the basis of gage dimensions and are so expressed for the purpose of eliminating errors in computations.

Engagement Between External and Internal Taper Threads.—The normal length of engagement between external and internal taper threads when screwed together handtight is shown as L_1 in Table 4. This length is controlled by the construction and use of the pipe thread gages. It is recognized that in special applications, such as flanges for high-pressure work, longer thread engagement is used, in which case the pitch diameter E_1 (Table 4) is maintained and the pitch diameter E_0 at the end of the pipe is proportionately smaller.

Railing Joint Taper Pipe Threads, NPTR.—Railing joints require a rigid mechanical thread joint with external and internal taper threads. The external thread is basically the same as the ANSI Standard Taper Pipe Thread, except that sizes ½ through 2 inches are shortened by 3 threads and sizes 2½ through 4 inches are shortened by 4 threads to permit the use of the larger end of the pipe thread. A recess in the fitting covers the last scratch or imperfect threads on the pipe.

Straight Pipe Threads in Pipe Couplings, NPSC.—Threads in pipe couplings made in accordance with the ANSI B1.20.1 specifications are straight (parallel) threads of the same thread form as the ANSI Standard Taper Pipe Thread. They are used to form pressuretight joints when assembled with an ANSI Standard external taper pipe thread and made up with lubricant or sealant. These joints are recommended for comparatively low pressures only.

Straight Pipe Threads for Mechanical Joints, NPSM, NPSL, and NPSH.—While external and internal taper pipe threads are recommended for pipe joints in practically every service, there are mechanical joints where straight pipe threads are used to advantage. Three types covered by ANSI B1.20.1 are:

Free-Fitting Mechanical Joints for Fixtures (External and Internal), NPSM: Standard iron, steel, and brass pipe are often used for special applications where there are no internal pressures. Where straight thread joints are required for mechanical assemblies, straight pipe threads are often found more suitable or convenient. Dimensions of these threads are given in Table 6.

Table 6. American National Standard Straight Pipe Threads for Mechanical Joints, NPSM and NPSL *ANSI/ASME B1.20.1-1983 (R1992)*

Nominal Pipe Size	Threads per Inch	External Thread					Internal Thread			
			Major Diameter		Pitch Diameter		Minor Diameter		Pitch Diameter	
		Allowance	Max.[a]	Min.	Max.	Min.	Min.[a]	Max.	Min.[b]	Max.
Free-fitting Mechanical Joints for Fixtures—NPSM										
⅛	27	0.0011	0.397	0.390	0.3725	0.3689	0.358	0.364	0.3736	0.3783
¼	18	0.0013	0.526	0.517	0.4903	0.4859	0.468	0.481	0.4916	0.4974
⅜	18	0.0014	0.662	0.653	0.6256	0.6211	0.603	0.612	0.6270	0.6329
½	14	0.0015	0.823	0.813	0.7769	0.7718	0.747	0.759	0.7784	0.7851
¾	14	0.0016	1.034	1.024	0.9873	0.9820	0.958	0.970	0.9889	0.9958
1	11½	0.0017	1.293	1.281	1.2369	1.2311	1.201	1.211	1.2386	1.2462
1¼	11½	0.0018	1.638	1.626	1.5816	1.5756	1.546	1.555	1.5834	1.5912
1½	11½	0.0018	1.877	1.865	1.8205	1.8144	1.785	1.794	1.8223	1.8302
2	11½	0.0019	2.351	2.339	2.2944	2.2882	2.259	2.268	2.2963	2.3044
2½	8	0.0022	2.841	2.826	2.7600	2.7526	2.708	2.727	2.7622	2.7720
3	8	0.0023	3.467	3.452	3.3862	3.3786	3.334	3.353	3.3885	3.3984
3½	8	0.0023	3.968	3.953	3.8865	3.8788	3.835	3.848	3.8888	3.8988
4	8	0.0023	4.466	4.451	4.3848	4.3771	4.333	4.346	4.3871	4.3971
5	8	0.0024	5.528	5.513	5.4469	5.4390	5.395	5.408	5.4493	5.4598
6	8	0.0024	6.585	6.570	6.5036	6.4955	6.452	6.464	6.5060	6.5165
Loose-fitting Mechanical Joints for Locknut Connections—NPSL										
⅛	27	…	0.409	…	0.3840	0.3805	0.362	…	0.3863	0.3898
¼	18	…	0.541	…	0.5038	0.4986	0.470	…	0.5073	0.5125
⅜	18	…	0.678	…	0.6409	0.6357	0.607	…	0.6444	0.6496
½	14	…	0.844	…	0.7963	0.7896	0.753	…	0.8008	0.8075
¾	14	…	1.054	…	1.0067	1.0000	0.964	…	1.0112	1.0179
1	11½	…	1.318	…	1.2604	1.2523	1.208	…	1.2658	1.2739
1¼	11½	…	1.663	…	1.6051	1.5970	1.553	…	1.6106	1.6187
1½	11½	…	1.902	…	1.8441	1.8360	1.792	…	1.8495	1.8576
2	11½	…	2.376	…	2.3180	2.3099	2.265	…	2.3234	2.3315
2½	8	…	2.877	…	2.7934	2.7817	2.718	…	2.8012	2.8129
3	8	…	3.503	…	3.4198	3.4081	3.344	…	3.4276	3.4393
3½	8	…	4.003	…	3.9201	3.9084	3.845	…	3.9279	3.9396
4	8	…	4.502	…	4.4184	4.4067	4.343	…	4.4262	4.4379
5	8	…	5.564	…	5.4805	5.4688	5.405	…	5.4884	5.5001
6	8	…	6.620	…	6.5372	6.5255	6.462	…	6.5450	6.5567
8	8	…	8.615	…	8.5313	8.5196	8.456	…	8.5391	8.5508
10	8	…	10.735	…	10.6522	10.6405	10.577	…	10.6600	10.6717
12	8	…	12.732	…	12.6491	12.6374	12.574	…	12.6569	12.6686

[a] As the ANSI Standard Straight Pipe Thread form of thread is maintained, the major and the minor diameters of the internal thread and the minor diameter of the external thread vary with the pitch diameter. The major diameter of the external thread is usually determined by the diameter of the pipe. These theoretical diameters result from adding the depth of the truncated thread ($0.666025 \times p$) to the maximum pitch diameters, and it should be understood that commercial pipe will not always have these maximum major diameters.

[b] This is the same as the pitch diameter at end of internal thread, E_1 *Basic.* (See Table 4.)

All dimensions are given in inches.

Notes for Free-fitting Fixture Threads:

The minor diameters of external threads and major diameters of internal threads are those as produced by commercial straight pipe dies and commercial ground straight pipe taps.

The major diameter of the external thread has been calculated on the basis of a truncation of $0.10825p$, and the minor diameter of the internal thread has been calculated on the basis of a truncation of $0.21651p$, to provide no interference at crest and root when product is gaged with gages made in accordance with the Standard.

Notes for Loose-fitting Locknut Threads:

The locknut thread is established on the basis of retaining the greatest possible amount of metal thickness between the bottom of the thread and the inside of the pipe.

In order that a locknut may fit loosely on the externally threaded part, an allowance equal to the "increase in pitch diameter per turn" is provided, with a tolerance of 1½ turns for both external and internal threads.

Loose-Fitting Mechanical Joints With Locknuts (External and Internal), NPSL: This thread is designed to produce a pipe thread having the largest diameter that it is possible to cut on standard pipe. The dimensions of these threads are given in Table 6. It will be noted that the maximum major diameter of the external thread is slightly greater than the nominal outside diameter of the pipe. The normal manufacturer's variation in pipe diameter provides for this increase.

Loose-Fitting Mechanical Joints for Hose Couplings (External and Internal), NPSH: Hose coupling joints are ordinarily made with straight internal and external loose-fitting threads. There are several standards of hose threads having various diameters and pitches. One of these is based on the ANSI Standard pipe thread and by the use of this thread series, it is possible to join small hose couplings in sizes ½ to 4 inches, inclusive, to ends of standard pipe having ANSI Standard External Pipe Threads, using a gasket to seal the joints. For the hose coupling thread dimensions see *ANSI Standard Hose Coupling Screw Threads* starting on page 1842.

Thread Designation and Notation.—American National Standard Pipe Threads are designated by specifying in sequence the nominal size, number of threads per inch, and the symbols for the thread series and form, as: ⅜—18 NPT. The symbol designations are as follows: NPT—American National Standard Taper Pipe Thread; NPTR—American National Standard Taper Pipe Thread for Railing Joints; NPSC—American National Standard Straight Pipe Thread for Couplings; NPSM—American National Standard Straight Pipe Thread for Free-fitting Mechanical Joints; NPSL—American National Standard Straight Pipe Thread for Loose-fitting Mechanical Joints with Locknuts; and NPSH—American National Standard Straight Pipe Thread for Hose Couplings.

American National Standard Dryseal Pipe Threads for Pressure-Tight Joints.— Dryseal pipe threads are based on the USA (American) pipe thread; however, they differ in that they are designed to seal pressure-tight joints without the necessity of using sealing compounds. To accomplish this, some modification of thread form and greater accuracy in manufacture is required. The roots of both the external and internal threads are truncated slightly more than the crests, i.e., roots have wider flats than crests so that metal-to-metal contact occurs at the crests and roots coincident with, or prior to, flank contact. Thus, as the threads are assembled by wrenching, the roots of the threads crush the sharper crests of the mating threads. This sealing action at both major and minor diameters tends to prevent spiral leakage and makes the joints pressure-tight without the necessity of using sealing compounds, provided that the threads are in accordance with standard specifications and tolerances and are not damaged by galling in assembly. The control of crest and root truncation is simplified by the use of properly designed threading tools. Also, it is desirable that both external and internal threads have full thread height for the length of hand engagement. Where not functionally objectionable, the use of a compatible lubricant or sealant is permissible to minimize the possibility of galling. This is desirable in assembling Dryseal pipe threads in refrigeration and other systems to effect a pressure-tight seal. The crest and root of Dryseal pipe threads may be slightly rounded, but are acceptable if they lie within the truncation limits given in Table 7.

Table 7. American National Standard Dryseal Pipe Threads—Limits on Crest and Root Truncation *ANSI B1.20.3-1976 (R1998)*

Threads Per Inch	Height of Sharp V Thread (H)	Truncation							
		Minimum				Maximum			
		At Crest		At Root		At Crest		At Root	
		Formula	Inch	Formula	Inch	Formula	Inch	Formula	Inch
27	0.03208	$0.047p$	0.0017	$0.094p$	0.0035	$0.094p$	0.0035	$0.140p$	0.0052
18	0.04811	$0.047p$	0.0026	$0.078p$	0.0043	$0.078p$	0.0043	$0.109p$	0.0061
14	0.06180	$0.036p$	0.0026	$0.060p$	0.0043	$0.060p$	0.0043	$0.085p$	0.0061
11½	0.07531	$0.040p$	0.0035	$0.060p$	0.0052	$0.060p$	0.0052	$0.090p$	0.0078
8	0.10825	$0.042p$	0.0052	$0.055p$	0.0069	$0.055p$	0.0069	$0.076p$	0.0095

All dimensions are given in inches. In the formulas, p = pitch.

Types of Dryseal Pipe Thread.—American National Standard ANSI B1.20.3-1976 (R1998) covers four types of standard Dryseal pipe threads:

NPTF—Dryseal USA (American) Standard Taper Pipe Thread

PTF-SAE SHORT—Dryseal SAE Short Taper Pipe Thread

NPSF—Dryseal USA (American) Standard Fuel Internal Straight Pipe Thread

NPSI—Dryseal USA (American) Standard Intermediate Internal Straight Pipe Thread

NPTF Threads: This type applies to both external and internal threads and is suitable for pipe joints in practically every type of service. Of all Dryseal pipe threads, NPTF external and internal threads mated are generally conceded to be superior for strength and seal since they have the longest length of thread and, theoretically, interference (sealing) occurs at every engaged thread root and crest. Use of tapered internal threads, such as NPTF or PTF-SAE SHORT in hard or brittle materials having thin sections will minimize the possibility of fracture.

There are two classes of NTPF threads. Class 1 threads are made to interfere (seal) at root and crest when mated, but inspection of crest and root truncation is not required. Consequently, Class 1 threads are intended for applications where close control of tooling is required for conformance of truncation or where sealing is accomplished by means of a sealant applied to the threads.

Class 2 threads are theoretically identical to those made to Class 1, however, inspection of root and crest truncation is required. Consequently, where a sealant is not used, there is more assurance of a pressure-tight seal for Class 2 threads than for Class 1 threads.

PTF-SAE SHORT Threads: External threads of this type conform in all respects with NPTF threads except that the thread length has been shortened by eliminating one thread from the small (entering) end. These threads are designed for applications where clearance is not sufficient for the full length of the NPTF threads or for economy of material where the full thread length is not necessary.

Internal threads of this type conform in all respects with NPTF threads, except that the thread length has been shortened by eliminating one thread from the large (entry) end. These threads are designed for thin materials where thickness is not sufficient for the full thread length of the NPTF threads or for economy in tapping where the full thread length is not necessary.

Pressure-tight joints without the use of lubricant or sealer can best be ensured where mating components are both threaded with NPTF threads. This should be considered before specifying PTF-SAE SHORT external or internal threads.

NPSF Threads: Threads of this type are straight (cylindrical) instead of tapered and are internal only. They are more economical to produce than tapered internal threads, but when assembled do not offer as strong a guarantee of sealing since root and crest interference will not occur for all threads. NPSF threads are generally used with soft or ductile materials which will tend to adjust at assembly to the taper of external threads, but may be used in hard or brittle materials where the section is thick.

NPSI Threads: Threads of this type are straight (cylindrical) instead of tapered, are internal only and are slightly larger in diameter than NPSF threads but have the same tolerance and thread length. They are more economical to produce than tapered threads and may be used in hard or brittle materials where the section is thick or where there is little expansion at assembly with external taper threads. As with NPSF threads, NPSI threads when assembled do not offer as strong a guarantee of sealing as do tapered internal threads.

For more complete specifications for production and acceptance of Dryseal pipe threads, see ANSI B1.20.3 (Inch) and ANSI B1.20.4 (Metric Translation), and for gaging and inspection, see ANSI Bl.20.5 (Inch) and ANSI B1.20.6M (Metric Translation).

Designation of Dryseal Pipe Threads: The standard Dryseal pipe threads are designated by specifying in sequence nominal size, thread series symbol, and class:

Examples: ⅛-27 NPTF-1

⅛-27 PTF-SAE SHORT

⅜-18 NPTF-1 AFTER PLATING

Table 8. Recommended Limitation of Assembly among the Various Types of Dryseal Threads

External Dryseal Thread		For Assembly with Internal Dryseal Thread	
Type	Description	Type	Description
1	NPTF (tapered), ext thd	1	NPTF (tapered), int thd
		2[a,b]	PTF-SAE SHORT (tapered), int thd
		3[a,c]	NPSF (straight), int thd
		4[a,c,d]	NPSI (straight), int thd
2[a,e]	PTF-SAE SHORT (tapered) ext thd	4	NPSI (straight), int thd
		1	NPTF (tapered), int thd

[a] Pressure-tight joints without the use of a sealant can best be ensured where both components are threaded with NPTF (full length threads), since theoretically interference (sealing) occurs at all threads, but there are two less threads engaged than for NPTF assemblies. When straight internal threads are used, there is interference only at one thread depending on ductility of materials.

[b] PTF-SAE SHORT internal threads are primarily intended for assembly with type 1-NPTF external threads. They are not designed for, and at extreme tolerance limits may not assemble with, type 2-PTF-SAE SHORT external threads.

[c] There is no external straight Dryseal thread.

[d] NPSI internal threads are primarily intended for assembly with type 2-PTF-SAE SHORT external threads but will also assemble with full length type 1 NPTF external threads.

[e] PTF-SAE SHORT external threads are primarily intended for assembly with type 4-NPSI internal threads but can also be used with type 1-NPTF internal threads. They are not designed for, and at extreme tolerance limits may not assemble with, type 2-PTF-SAE SHORT internal threads or type 3-NPSF internal threads.

An assembly with straight internal pipe threads and taper external pipe threads is frequently more advantageous than an all taper thread assembly, particularly in automotive and other allied industries where economy and rapid production are major considerations. Dryseal threads are not used in assemblies in which both components have straight pipe threads.

Table 9. Suggested Tap Drill Sizes for Internal Dryseal Pipe Threads

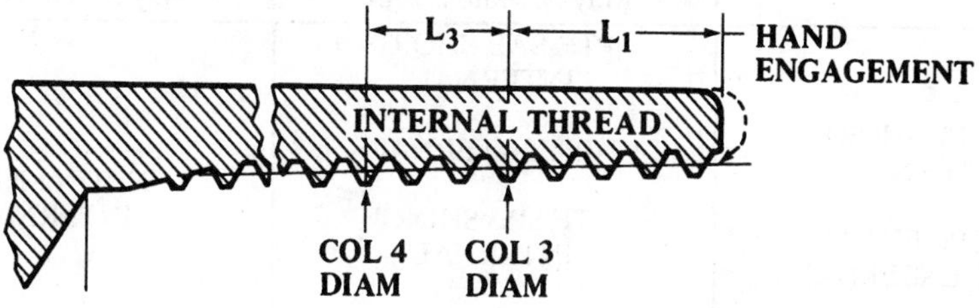

L3 L1 HAND ENGAGEMENT

INTERNAL THREAD

COL 4 DIAM COL 3 DIAM

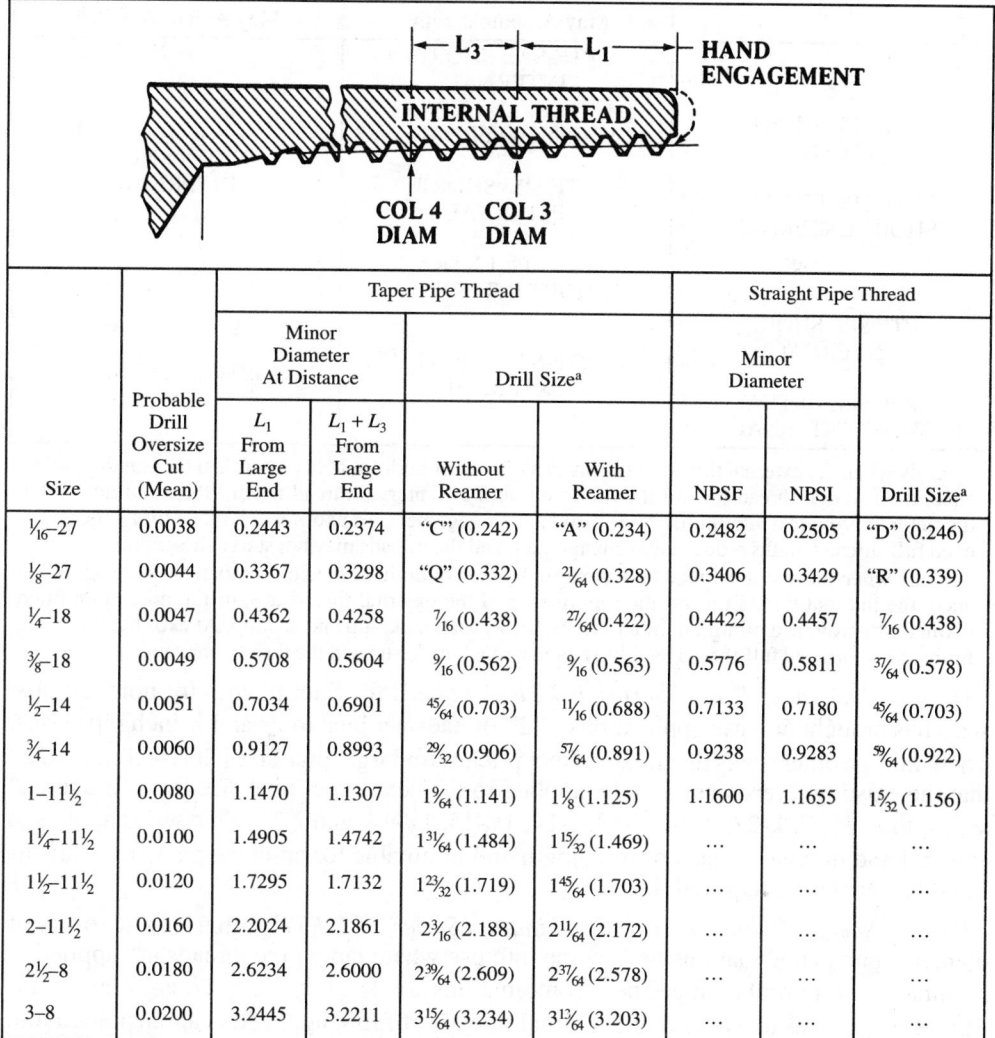

Size	Probable Drill Oversize Cut (Mean)	Taper Pipe Thread				Straight Pipe Thread		
		Minor Diameter At Distance		Drill Size[a]		Minor Diameter		Drill Size[a]
		L_1 From Large End	$L_1 + L_3$ From Large End	Without Reamer	With Reamer	NPSF	NPSI	
1/16–27	0.0038	0.2443	0.2374	"C" (0.242)	"A" (0.234)	0.2482	0.2505	"D" (0.246)
1/8–27	0.0044	0.3367	0.3298	"Q" (0.332)	21/64 (0.328)	0.3406	0.3429	"R" (0.339)
1/4–18	0.0047	0.4362	0.4258	7/16 (0.438)	27/64 (0.422)	0.4422	0.4457	7/16 (0.438)
3/8–18	0.0049	0.5708	0.5604	9/16 (0.562)	9/16 (0.563)	0.5776	0.5811	37/64 (0.578)
1/2–14	0.0051	0.7034	0.6901	45/64 (0.703)	11/16 (0.688)	0.7133	0.7180	45/64 (0.703)
3/4–14	0.0060	0.9127	0.8993	29/32 (0.906)	57/64 (0.891)	0.9238	0.9283	59/64 (0.922)
1–11 1/2	0.0080	1.1470	1.1307	1 9/64 (1.141)	1 1/8 (1.125)	1.1600	1.1655	1 5/32 (1.156)
1 1/4–11 1/2	0.0100	1.4905	1.4742	1 31/64 (1.484)	1 15/32 (1.469)	…	…	…
1 1/2–11 1/2	0.0120	1.7295	1.7132	1 23/32 (1.719)	1 45/64 (1.703)	…	…	…
2–11 1/2	0.0160	2.2024	2.1861	2 3/16 (2.188)	2 11/64 (2.172)	…	…	…
2 1/2–8	0.0180	2.6234	2.6000	2 39/64 (2.609)	2 37/64 (2.578)	…	…	…
3–8	0.0200	3.2445	3.2211	3 15/64 (3.234)	3 13/64 (3.203)	…	…	…

[a] Some drill sizes listed may not be standard drills.

All dimensions are given in inches.

Special Dryseal Threads.—Where design limitations, economy of material, permanent installation, or other limiting conditions prevail, consideration may be given to using a special Dryseal thread series.

Dryseal Special Short Taper Pipe Thread, PTF-SPL SHORT: Threads of this series conform in all respects to PTF-SAE SHORT threads except that the full thread length has been further shortened by eliminating one thread at the small end of internal threads or one thread at the large end of external threads.

Dryseal Special Extra Short Taper Pipe Thread, PTF-SPL EXTRA SHORT: Threads of this series conform in all respects to PTF-SAE SHORT threads except that the full thread length has been further shortened by eliminating two threads at the small end of internal threads or two threads at the large end of external threads.

Limitations of Assembly: Table 10 applies where Dryseal Special Short or Extra Short Taper Pipe Threads are to be assembled as special combinations.

Table 10. Assembly Limitations for Special Combinations of Dryseal Threads

	May Assemble with[a]	May Assemble with[b]
PTF SPL SHORT EXTERNAL PTF SPL EXTRA SHORT EXTERNAL	PTF-SAE SHORT INTERNAL NPSF INTERNAL PTF SPL SHORT INTERNAL PTF SPL EXTRA SHORT INTERNAL	NPTF or NPSI INTERNAL
PTF SPL SHORT INTERNAL PTF SPL EXTRA SHORT INTERNAL	PTF-SAE SHORT EXTERNAL	NPTF EXTERNAL

[a] Only when the external thread or the internal thread or both are held closer than the standard tolerance, the external thread toward the minimum and the internal thread toward the maximum pitch diameter to provide a minimum of one turn hand engagement. At extreme tolerance limits the shortened full-thread lengths reduce hand engagement and the threads may not start to assemble.

[b] Only when the internal thread or the external thread or both are held closer than the standard tolerance, the internal thread toward the minimum and the external thread toward the maximum pitch diameter to provide a minimum of two turns for wrench make-up and sealing. At extreme tolerance limits the shortened full-thread lengths reduce wrench make-up and the threads may not seal.

Dryseal Fine Taper Thread Series, F-PTF: The need for finer pitches for nominal pipe sizes has brought into use applications of 27 threads per inch to ¼- and ⅜-inch pipe sizes. There may be other needs that require finer pitches for larger pipe sizes. It is recommended that the existing threads per inch be applied to the next larger pipe size for a fine thread series, thus: ¼-27, ⅜-27, ½-18, ¾-18, 1-14, 1¼-14, 1½-14, and 2-14. This series applies to external and internal threads of full length and is suitable for applications where threads finer than NPTF are required.

Dryseal Special Diameter-Pitch Combination Series, SPL-PTF: Other applications of diameter-pitch combinations have come into use where taper pipe threads are applied to nominal size thin wall tubing. These combinations are: ½-27, ⅝-27, ¾-27, ⅞-27, and 1-27. This series applies to external and internal threads of full length and is applicable to thin wall nominal diameter outside tubing.

Designation of Special Dryseal Pipe Threads: The designations used for these special dryseal pipe threads are as follows:

⅛-27 PTF-SPL SHORT

⅛-27 PTF-SPL EXTRA SHORT

½-27 SPL PTF, OD 0.500

Note that in the last designation the OD of tubing is given.

British Standard Pipe Threads for Non-pressure-tight Joints The threads in BS 2779:1973—"Specifications for Pipe Threads where Pressure-tight Joints are not Made on the Threads" are Whitworth form parallel fastening threads that are generally used for fastening purposes such as the mechanical assembly of component parts of fittings, cocks and valves. They are not suitable where pressure-tight joints are made on the threads..—The crests of the basic Whitworth thread form may be truncated to certain limits of size given in the Standard except on internal threads, when they are likely to be assembled with external threads conforming to the requirements of BS 21 "British Standard Pipe Threads for Pressure-tight Joints" (see page 1857).

For external threads two classes of tolerance are provided and for internal, one class. The two classes of tolerance for external threads are Class A and Class B. For economy of manufacture the class B fit should be chosen whenever possible. The class A is reserved for those applications where the closer tolerance is essential. Class A tolerance is an entirely negative value, equivalent to the internal thread tolerance. Class B tolerance is an entirely negative value twice that of class A tolerance. Tables showing limits and dimensions are given in the Standard.

The thread series specified in this Standard shall be designated by the letter "G". A typical reference on a drawing might be "G½", for internal thread; "G½A", for external thread, class A: and "G ½B", for external thread, class B. Where no class reference is stated for external threads, that of class B will be assumed. The designation of truncated threads shall have the addition of the letter "T" to the designation, i.e., G ½T and G ½BT.

British Standard Pipe Threads (Non-pressure-tight Joints)—Metric and Inch Basic Sizes *BS 2779:1973*

Nominal Size, Inches	Threads per Inch[a]	Depth of Thread	Major Diameter	Pitch Diameter	Minor Diameter	Nominal Size, Inches	Threads per Inch[a]	Depth of Thread	Major Diameter	Pitch Diameter	Minor Diameter
1/16	28	0.581 *0.0229*	7.723 *0.3041*	7.142 *0.2812*	6.561 *0.2583*	1 3/4	11	1.479 *0.0582*	53.746 *2.1160*	52.267 *2.0578*	50.788 *1.9996*
1/8	28	0.581 *0.0229*	9.728 *0.3830*	9.147 *0.3601*	8.566 *0.3372*	2	11	1.479 *0.0582*	59.614 *2.3470*	58.135 *2.2888*	56.656 *2.2306*
1/4	19	0.856 *0.0337*	13.157 *0.5180*	12.301 *0.4843*	11.445 *0.4506*	2 1/4	11	1.479 *0.0582*	65.710 *2.5870*	64.231 *2.5288*	62.752 *2.4706*
3/8	19	0.856 *0.0337*	16.662 *0.6560*	15.806 *0.6223*	14.950 *0.5886*	2 1/2	11	1.479 *0.0582*	75.184 *2.9600*	73.705 *2.9018*	72.226 *2.8436*
1/2	14	1.162 *0.0457*	20.955 *0.8250*	19.793 *0.7793*	18.631 *0.7336*	2 3/4	11	1.479 *0.0582*	81.534 *3.2100*	80.055 *3.1518*	78.576 *3.0936*
5/8	14	1.162 *0.0457*	22.911 *0.9020*	21.749 *0.8563*	20.587 *0.8106*	3	11	1.479 *0.0582*	87.884 *3.4600*	86.405 *3.4018*	84.926 *3.3436*
3/4	14	1.162 *0.0457*	26.441 *1.0410*	25.279 *0.9953*	24.117 *0.9496*	3 1/2	11	1.479 *0.0582*	100.330 *3.9500*	98.851 *3.8918*	97.372 *3.8336*
7/8	14	1.162 *0.0457*	30.201 *1.1890*	29.039 *1.1433*	27.877 *1.0976*	4	11	1.479 *0.0582*	113.030 *4.4500*	111.551 *4.3918*	110.072 *4.3336*
1	11	1.479 *0.0582*	33.249 *1.3090*	31.770 *1.2508*	30.291 *1.1926*	4 1/2	11	1.479 *0.0582*	125.730 *4.9500*	124.251 *4.8918*	122.772 *4.8336*
1 1/8	11	1.479 *0.0582*	37.897 *1.4920*	36.418 *1.4338*	34.939 *1.3756*	5	11	1.479 *0.0582*	138.430 *5.4500*	136.951 *5.3918*	135.472 *5.3336*
1 1/4	11	1.479 *0.0582*	41.910 *1.6500*	40.431 *1.5918*	38.952 *1.5336*	5 1/2	11	1.479 *0.0582*	151.130 *5.9500*	149.651 *5.8918*	148.172 *5.8336*
1 1/2	11	1.479 *0.0582*	47.803 *1.8820*	46.324 *1.8238*	44.845 *1.7656*	6	11	1.479 *0.0582*	163.830 *6.4500*	162.351 *6.3918*	160.872 *6.3336*

[a] The thread pitches in millimeters are as follows: 0.907 for 28 threads per inch. 1.337 for 19 threads per inch, 1.814 for 14 threads per inch, and 2.309 for 11 threads per inch.

Each basic metric dimension is given in roman figures (nominal sizes excepted) and each basic inch dimension is shown in italics directly beneath it.

British Standard Pipe Threads for Pressure-tight Joints.—The threads in BS 21:1973—"Specification for Pipe Threads where Pressure-tight Joints are Made on the Threads" are based on the Whitworth thread form and are specified as: 1) *Jointing threads* These relate to pipe threads for joints made pressure-tight by the mating of the threads; they include taper external threads for assembly with either taper or parallel internal threads (parallel external pipe threads are not suitable as jointing threads); and 2) *Long-screw threads* These relate to parallel external pipe threads used for longscrews (connectors) specified in BS 1387 where a pressure-tight joint is achieved by the compression of a soft material onto the surface of the external thread by tightening a back nut against a socket.

British Standard External and Internal Pipe Threads (Pressure-tight Joints)—Metric and Inch Dimensions and Limits of Size *BS 21:1973*

Nominal Size	No. of Threads per Inch[a]	Basic Diameters at Gage Plane			Gage Length		Number of Useful Threads on Pipe for Basic Gage Length[b]	Tol., + and −, Gage Plane to Face of Int. Taper Thread	Tol., + and −, on Diameter of Parallel Int. Threads
		Major	Pitch	Minor	Basic	Tolerance (+ and −)			
1/16	28	7.723	7.142	6.561	(4⅜)	(1)	(7⅛)	(1¼)	0.071
		0.304	*0.2812*	*0.2583*	4.0	0.9	6.5	1.1	*0.0028*
⅛	28	9.728	9.147	8.566	(4⅜)	(1)	(7⅛)	(1¼)	0.071
		0.383	*0.3601*	*0.3372*	4.0	0.9	6.5	1.1	*0.0028*
¼	19	13.157	12.301	11.445	(4½)	(1)	(7¼)	(1¼)	0.104
		0.518	*0.4843*	*0.4506*	6.0	1.3	9.7	1.7	*0.0041*
⅜	19	16.662	15.806	14.950	(4¾)	(1)	(7½)	(1¼)	0.104
		0.656	0.6223	0.5886	6.4	1.3	10.1	1.7	0.0041
½	14	20.955	19.793	18.631	(4½)	(1)	(7¼)	(1¼)	0.142
		0.825	*0.7793*	*0.7336*	8.2	1.8	13.2	2.3	*0.0056*
¾	14	26.441	25.279	24.117	(5¼)	(1)	(8)	(1¼)	0.142
		1.041	*0.9953*	*0.9496*	9.5	1.8	14.5	2.3	*0.0056*
1	11	33.249	31.770	30.291	(4½)	(1)	(7¼)	(1¼)	0.180
		1.309	*1.2508*	*1.1926*	10.4	2.3	16.8	2.9	*0.0071*
1¼	11	41.910	40.431	38.952	(5½)	(1)	(8¼)	(1¼)	0.180
		1.650	*1.5918*	*1.5336*	12.7	2.3	19.1	2.9	*0.0071*
1½	11	47.803	46.324	44.845	(5½)	(1)	(8¼)	(1¼)	0.180
		1.882	*1.8238*	*1.7656*	12.7	2.3	19.1	2.9	*0.0071*
2	11	59.614	58.135	56.656	(6⅞)	(1)	(10⅛)	(1¼)	0.180
		2.347	*2.2888*	*2.2306*	15.9	2.3	23.4	2.9	*0.0071*
2½	11	75.184	73.705	72.226	(7 9/16)	(1½)	(11 9/16)	(1½)	0.216
		2.960	*2.9018*	*2.8436*	17.5	3.5	26.7	3.5	*0.0085*
3	11	87.884	86.405	84.926	(8 15/16)	(1½)	(12 15/16)	(1½)	0.216
		3.460	*3.4018*	*3.3436*	20.6	3.5	29.8	3.5	*0.0085*
4	11	113.030	111.551	110.072	(11)	(1½)	(15½)	(1½)	0.216
		4.450	*4.3918*	*4.3336*	25.4	3.5	35.8	3.5	*0.0085*
5	11	138.430	136.951	135.472	(12⅜)	(1½)	(17⅜)	(1½)	0.216
		5.450	*5.3918*	*5.3336*	28.6	3.5	40.1	3.5	*0.0085*
6	11	163.830	162.351	160.872	(12⅜)	(1½)	(17⅜)	(1½)	0.216
		6.450	*6.3918*	*6.3336*	28.6	3.5	40.1	3.5	*0.0085*

[a] In the Standard BS 21:1973 the thread pitches in millimeters are as follows: 0.907 for 28 threads per inch, 1.337 for 19 threads per inch, 1.814 for 14 threads per inch, and 2.309 for 11 threads per inch.

[b] This is the minimum number of useful threads on the pipe for the basic gage length; for the maximum and minimum gage lengths, the minimum numbers of useful threads are, respectively, greater and less by the amount of tolerance in the column to the left. The design of internally threaded parts shall make allowance for receiving pipe ends of up to the minimum number of useful threads corresponding to the maximum gage length; the minimum number of useful *internal* threads shall be no less than 80 per cent of the minimum number of useful external threads for the minimum gage length.

Each basic metric dimension is given in roman figures (nominal sizes excepted) and each basic inch dimension is shown in italics directly beneath it. Figures in ( ) are numbers of turns of thread with metric linear equivalents given beneath. Taper of taper thread is 1 in 16 on diameter.

MEASURING SCREW THREADS

Measuring Screw Threads

Pitch and Lead of Screw Threads.—The *pitch* of a screw thread is the distance from the center of one thread to the center of the next thread. This applies no matter whether the screw has a single, double, triple or quadruple thread. The *lead* of a screw thread is the distance the nut will move forward on the screw if it is turned around one full revolution. In a single-threaded screw, the pitch and lead are equal, because the nut would move forward the distance from one thread to the next, if turned around once. In a double-threaded screw, the nut will move forward two threads, or twice the pitch, so that in this case the lead equals twice the pitch. In a triple-threaded screw, the lead equals three times the pitch, and so on.

The word "pitch" is often, although improperly, used to denote the *number of threads per inch.* Screws are spoken of as having a 12-pitch thread, when twelve threads per inch is what is really meant. The number of threads per inch equals 1 divided by the pitch, or expressed as a formula:

$$\text{Number of threads per inch} = \frac{1}{\text{pitch}}$$

The pitch of a screw equals 1 divided by the number of threads per inch, or:

$$\text{Pitch} = \frac{1}{\text{number of threads per inch}}$$

If the number of threads per inch equals 16, the pitch = $\frac{1}{16}$. If the pitch equals 0.05, the number of threads equals $1 \div 0.05 = 20$. If the pitch is $\frac{2}{5}$ inch, the number of threads per inch equals $1 \div \frac{2}{5} = 2\frac{1}{2}$.

Confusion is often caused by the indefinite designation of multiple-thread screws (double, triple, quadruple, etc.). The expression, "four threads per inch, triple," for example, is not to be recommended. It means that the screw is cut with four triple threads or with twelve threads per inch, if the threads are counted by placing a scale alongside the screw. To cut this screw, the lathe would be geared to cut four threads per inch, but they would be cut only to the depth required for twelve threads per inch. The best expression, when a multiple-thread is to be cut, is to say, in this case, "$\frac{1}{4}$ inch lead, $\frac{1}{12}$ inch pitch, triple thread." For single-threaded screws, only the number of threads per inch and the form of the thread are specified. The word "single" is not required.

Measuring Screw Thread Pitch Diameters by Thread Micrometers.—As the pitch or angle diameter of a tap or screw is the most important dimension, it is necessary that the pitch diameter of screw threads be measured, in addition to the outside diameter.

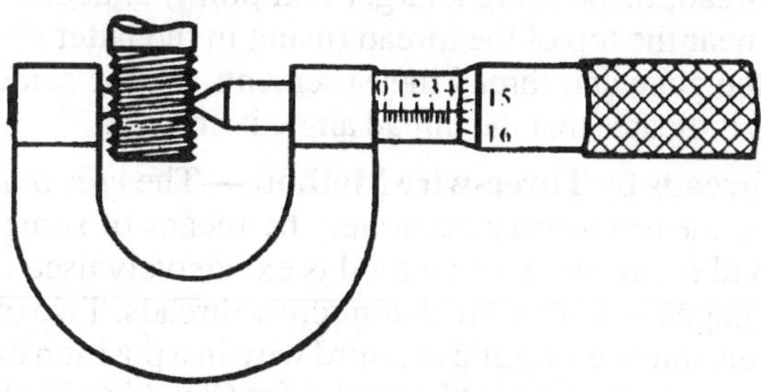

Fig. 1.

One method of measuring in the angle of a thread is by means of a special screw thread micrometer, as shown in the accompanying engraving, Fig. 1. The fixed anvil is W-shaped to engage two thread flanks, and the movable point is cone-shaped so as to enable it to enter the space between two threads, and at the same time be at liberty to revolve. The contact points are on the sides of the thread, as they necessarily must be in order that the pitch diameter may be determined. The cone-shaped point of the measuring screw is slightly rounded so that it will not bear in the bottom of the thread. There is also sufficient clearance at the bottom of the V-shaped anvil to prevent it from bearing on the top of the thread. The movable point is adapted to measuring all pitches, but the fixed anvil is limited in its capacity. To cover the whole range of pitches, from the finest to the coarsest, a number of fixed anvils are therefore required.

To find the theoretical pitch diameter, which is measured by the micrometer, subtract twice the addendum of the thread from the standard outside diameter. The addendum of the thread for the American and other standard threads is given in the section on screw thread systems.

Ball-point Micrometers.—If standard plug gages are available, it is not necessary to actually measure the pitch diameter, but merely to compare it with the standard gage. In this case, a ball-point micrometer, as shown in Fig. 2, may be employed. Two types of ball-point micrometers are ordinarily used. One is simply a regular plain micrometer with ball points made to slip over both measuring points. (See B, Fig. 2.) This makes a kind of combination plain and ball-point micrometer, the ball points being easily removed. These ball points, however, do not fit solidly on their seats, even if they are split, as shown, and are apt to cause errors in measurements. The best, and, in the long run, the cheapest, method is to use a regular micrometer arranged as shown at *A*. Drill and ream out both the end of the measuring screw or spindle and the anvil, and fit ball points into them as shown. Care should be taken to have the ball point in the spindle run true. The holes in the micrometer spindle and anvil and the shanks on the points are tapered to insure a good fit. The hole *H* in spindle *G* is provided so that the ball point can be easily driven out when a change for a larger or smaller size of ball point is required.

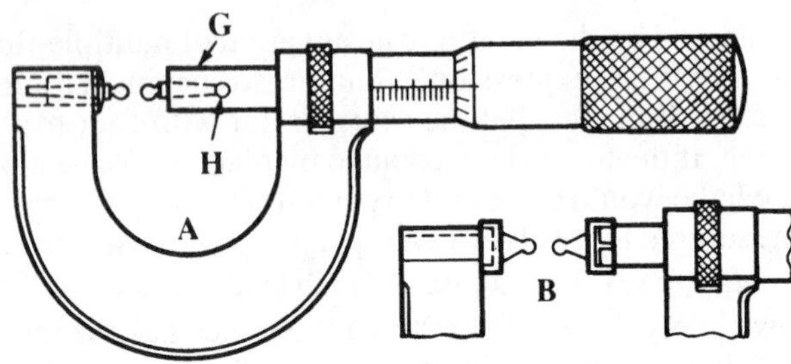

Fig. 2.

A ball-point micrometer may be used for comparing the *angle* of a screw thread, with that of a gage. This can be done by using different sizes of ball points, comparing the size first near the root of the thread, then (using a larger ball point) at about the point of the pitch diameter, and finally near the top of the thread (using in the latter case, of course, a much larger ball point). If the gage and thread measurements are the same at each of the three points referred to, this indicates that the thread angle is correct.

Measuring Screw Threads by Three-wire Method.—The *effective* or *pitch diameter* of a screw thread may be measured very accurately by means of some form of micrometer and three wires of equal diameter. This method is extensively used in checking the accuracy of threaded plug gages and other precision screw threads. Two of the wires are placed in contact with the thread on one side and the third wire in a position diametrically opposite as illustrated by the diagram, (see table "Formulas for Checking Pitch Diameters of Screw Threads") and the dimension over the wires is determined by means of a micrometer. An ordinary micrometer is commonly used but some form of "floating micrometer" is preferable, especially for measuring thread gages and other precision work. The floating micrometer is mounted upon a compound slide so that it can move freely in directions parallel or at right angles to the axis of the screw, which is held in a horizontal position between adjustable centers. With this arrangement the micrometer is held constantly at right angles to the axis of the screw so that only one wire on each side may be used instead of having two on one side and one on the other, as is necessary when using an ordinary micrometer. The pitch diameter may be determined accurately if the correct micrometer reading for wires of a given size is known.

Classes of Formulas for Three-Wire Measurement.—Various formulas have been established for checking the pitch diameters of screw threads by measurement over wires of known size. These formulas differ with regard to their simplicity or complexity and resulting accuracy. They also differ in that some show what measurement M over the wires should be to obtain a given pitch diameter E, whereas others show the value of the pitch diameter E for a given measurement M.

Formulas for Finding Measurement M: In using a formula for finding the value of measurement M, the required pitch diameter E is inserted in the formula. Then, in cutting or grinding a screw thread, the actual measurement M is made to conform to the calculated value of M. Formulas for finding measurement M may be modified so that the basic major or outside diameter is inserted in the formula instead of the pitch diameter; however, the pitch-diameter type of formula is preferable because the pitch diameter is a more important dimension than the major diameter.

Formulas for Finding Pitch Diameters E: Some formulas are arranged to show the value of the pitch diameter E when measurement M is known. Thus, the value of M is first determined by measurement and then is inserted in the formula for finding the corresponding pitch diameter E. This type of formula is useful for determining the pitch diameter of an existing thread gage or other screw thread in connection with inspection work. The formula for finding measurement M is more convenient to use in the shop or tool room in cutting or grinding new threads, because the pitch diameter is specified on the drawing and the problem is to find the value of measurement M for obtaining that pitch diameter.

General Classes of Screw Thread Profiles.—Thread profiles may be divided into three general classes or types as follows:

Screw Helicoid: Represented by a screw thread having a straight-line profile in the axial plane. Such a screw thread may be cut in a lathe by using a straight-sided single-point tool, provided the top surface lies in the axial plane.

Involute Helicoid: Represented either by a screw thread or a helical gear tooth having an involute profile in a plane perpendicular to the axis. A rolled screw thread, theoretically at least, is an exact involute helicoid.

Intermediate Profiles: An intermediate profile that lies somewhere between the screw helicoid and the involute helicoid will be formed on a screw thread either by milling or grinding with a straight-sided wheel set in alignment with the thread groove. The resulting form will approach closely the involute helicoid form. In milling or grinding a thread, the included cutter or wheel angle may either equal the standard thread angle (which is always measured in the axial plane) or the cutter or wheel angle may be reduced to approximate, at least, the thread angle in the normal plane. In practice, all these variations affect the three-wire measurement.

Accuracy of Formulas for Checking Pitch Diameters by Three-Wire Method.—The exact measurement M for a given pitch diameter depends upon the lead angle, the thread angle, and the profile or cross-sectional shape of the thread. As pointed out in the preceding paragraph, the profile depends upon the method of cutting or forming the thread. In a milled or ground thread, the profile is affected not only by the cutter or wheel angle, but also by the diameter of the cutter or wheel; hence, because of these variations, an absolutely exact and reasonably simple general formula for measurement M cannot be established; however, if the lead angle is low, as with a standard single-thread screw, and especially if the thread angle is high like a 60-degree thread, simple formulas that are not arranged to compensate for the lead angle are used ordinarily and meet most practical requirements, particularly in measuring 60-degree threads. If lead angles are large enough to greatly affect the result, as with most multiple threads (especially Acme or 29-degree worm threads), a formula should be used that compensates for the lead angle sufficiently to obtain the necessary accuracy.

The formulas that follow include 1) a very simple type in which the effect of the lead angle on measurement$M$ is entirely ignored. This simple formula usually is applicable to the measurement of 60-degree single-thread screws, except possibly when gage-making accuracy is required; 2) formulas that do include the effect of the lead angle but, nevertheless, are approximations and not always suitable for the higher lead angles when extreme accuracy is required; and 3) formulas for the higher lead angles and the most precise classes of work.

Where approximate formulas are applied consistently in the measurement of both thread plug gages and the thread "setting plugs" for ring gages, interchangeability might be secured, assuming that such approximate formulas were universally employed.

Wire Sizes for Checking Pitch Diameters of Screw Threads.—In checking screw threads by the 3-wire method, the general practice is to use measuring wires of the so-called "best size." The "best-size" wire is one that contacts at the pitch line or midslope of the thread because then the measurement of the pitch diameter is least affected by an error in the thread angle. In the following formula for determining approximately the "best-size" wire or the diameter for pitch-line contact, A = one-half included angle of thread in the axial plane.

$$\text{Best-size wire} = \frac{0.5 \text{ pitch}}{\cos A} = 0.5 \text{ pitch} \times \sec A$$

For 60-degree threads, this formula reduces to

$$\text{Best-size wire} = 0.57735 \times \text{pitch}$$

Diameters of Wires for Measuring American Standard and British Standard Whitworth Screw Threads

Threads per Inch	Pitch, Inch	Wire Diameters for American Standard Threads			Wire Diameters for Whitworth Standard Threads		
		Max.	Min.	Pitch-Line Contact	Max.	Min.	Pitch-Line Contact
4	0.2500	0.2250	0.1400	0.1443	0.1900	0.1350	0.1409
4½	0.2222	0.2000	0.1244	0.1283	0.1689	0.1200	0.1253
5	0.2000	0.1800	0.1120	0.1155	0.1520	0.1080	0.1127
5½	0.1818	0.1636	0.1018	0.1050	0.1382	0.0982	0.1025
6	0.1667	0.1500	0.0933	0.0962	0.1267	0.0900	0.0939
7	0.1428	0.1283	0.0800	0.0825	0.1086	0.0771	0.0805
8	0.1250	0.1125	0.0700	0.0722	0.0950	0.0675	0.0705
9	0.1111	0.1000	0.0622	0.0641	0.0844	0.0600	0.0626
10	0.1000	0.0900	0.0560	0.0577	0.0760	0.0540	0.0564
11	0.0909	0.0818	0.0509	0.0525	0.0691	0.0491	0.0512
12	0.0833	0.0750	0.0467	0.0481	0.0633	0.0450	0.0470
13	0.0769	0.0692	0.0431	0.0444	0.0585	0.0415	0.0434
14	0.0714	0.0643	0.0400	0.0412	0.0543	0.0386	0.0403
16	0.0625	0.0562	0.0350	0.0361	0.0475	0.0337	0.0352
18	0.0555	0.0500	0.0311	0.0321	0.0422	0.0300	0.0313
20	0.0500	0.0450	0.0280	0.0289	0.0380	0.0270	0.0282
22	0.0454	0.0409	0.0254	0.0262	0.0345	0.0245	0.0256
24	0.0417	0.0375	0.0233	0.0240	0.0317	0.0225	0.0235
28	0.0357	0.0321	0.0200	0.0206	0.0271	0.0193	0.0201
32	0.0312	0.0281	0.0175	0.0180	0.0237	0.0169	0.0176
36	0.0278	0.0250	0.0156	0.0160	0.0211	0.0150	0.0156
40	0.0250	0.0225	0.0140	0.0144	0.0190	0.0135	0.0141

These formulas are based upon a thread groove of zero lead angle because ordinary variations in the lead angle have little effect on the wire diameter and it is desirable to use one wire size for a given pitch regardless of the lead angle. A theoretically correct solution for finding the *exact* size for pitch-line contact involves the use of cumbersome indeterminate equations with solution by successive trials. The accompanying table gives the wire sizes for both American Standard (formerly, U.S. Standard) and the Whitworth Standard Threads. The following formulas for determining wire diameters do not give the extreme theoretical limits, but the smallest and largest practicable sizes. The diameters in the table are based upon these approximate formulas.

American Standard	Smallest wire diameter = 0.56 × pitch
	Largest wire diameter = 0.90 × pitch
	Diameter for pitch-line contact = 0.57735 × pitch
Whitworth	Smallest wire diameter = 0.54 × pitch
	Largest wire diameter = 0.76 × pitch
	Diameter for pitch-line contact = 0.56369 × pitch

Measuring Wire Accuracy.—A set of three measuring wires should have the same diameter within 0.0002 inch. To measure the pitch diameter of a screw-thread gage to an accuracy of 0.0001 inch by means of wires, it is necessary to know the wire diameters to 0.00002 inch. If the diameters of the wires are known only to an accuracy of 0.0001 inch, an accuracy better than 0.0003 inch in the measurement of pitch diameter cannot be expected. The wires should be accurately finished hardened steel cylinders of the maximum possible hardness without being brittle. The hardness should not be less than that corresponding to a Knoop indentation number of 630. A wire of this hardness can be cut with a file only with difficulty. The surface should not be rougher than the equivalent of a deviation of 3 microinches from a true cylindrical surface.

Measuring or Contact Pressure.—In measuring screw threads or screw-thread gages by the 3-wire method, variations in contact pressure will result in different readings. The effect of a variation in contact pressure in measuring threads of fine pitches is indicated by the difference in readings obtained with pressures of 2 and 5 pounds in checking a thread plug gage having 24 threads per inch. The reading over the wires with 5 pounds pressure was 0.00013 inch less than with 2 pounds pressure. For pitches finer than 20 threads per inch, a pressure of 16 ounces is recommended by the National Bureau of Standards, now National Institute of Standards and Technology (NIST). For pitches of 20 threads per inch and coarser, a pressure of 2 ½ pounds is recommended.

For Acme threads, the wire presses against the sides of the thread with a pressure of approximately twice that of the measuring instrument. To limit the tendency of the wires to wedge in between the sides of an Acme thread, it is recommended that pitch-diameter measurements be made at 1 pound on 8 threads per inch and finer, and at 2 ½ pounds for pitches coarser than 8 threads per inch.

Approximate Three-Wire Formulas That Do Not Compensate for Lead Angle.—A general formula in which the effect of lead angle is ignored is as follows (see accompanying notation used in formulas):

$$M = E - T\cot A + W(1 + \csc A) \quad (1)$$

This formula can be simplified for any given thread angle and pitch. To illustrate, because $T = 0.5P$, $M = E - 0.5P \cot 30° + W(1 + 2)$, for a 60-degree thread, such as the American Standard,

$$M = E - 0.86603P + 3W$$

The accompanying table contains these simplified formulas for different standard threads. Two formulas are given for each. The upper one is used when the measurement over wires, M, is known and the corresponding pitch diameter, E, is required; the lower formula gives the measurement M for a specified value of pitch diameter. These formulas are sufficiently accurate for checking practically all standard 60-degree single-thread screws because of the low lead angles, which vary from 1° 11′ to 4° 31′ in the American Standard Coarse-Thread Series.

Bureau of Standards (now NIST) General Formula.—Formula (2), which follows, compensates quite largely for the effect of the lead angle. It is from the National Bureau of Standards Handbook H 28 (1944), now FED-STD-H28. The formula, however, as here given has been arranged for finding the value of M (instead of E).

$$M = E - T\cot A + W(1 + \csc A + 0.5\tan^2 B\cos A\cot A) \tag{2}$$

This expression is also found in ANSI/ASME B1.2-1983 R1992. The Bureau of Standards uses Formula (2) in preference to Formula (1) when the value of $0.5W\tan^2 B\cos A\cot A$ exceeds 0.00015, with the larger lead angles. If this test is applied to American Standard 60-degree threads, it will show that Formula (1) is generally applicable; but for 29-degree Acme or worm threads, Formula (2) (or some other that includes the effect of lead angle) should be employed.

Notation Used in Formulas for Checking Pitch Diameters by Three-Wire Method

A = one-half included thread angle in the axial plane

A_n = one-half included thread angle in the normal plane or in plane perpendicular to sides of thread = one-half included angle of cutter when thread is milled ($\tan A_n = \tan A \times \cos B$). (Note: Included angle of milling cutter or grinding wheel may equal the nominal included angle of thread, or may be reduced to whatever normal angle is required to make the thread angle standard in the axial plane. In either case, A_n = one-half cutter angle.)

B = lead angle at pitch diameter = helix angle of thread as measured from a plane perpendicular to the axis. $\tan B = L \div 3.1416E$

D = basic major or outside diameter

E = pitch diameter (basic, maximum, or minimum) for which M is required, or pitch diameter corresponding to measurement M

F = angle required in Formulas (4b), (4d), and (4e)

G = angle required in Formula (4)

H = helix angle at pitch diameter and measured from axis = 90° − B or $\tan H = \cot B$

H_b = helix angle at R_b measured from axis

L = lead of thread = pitch P × number of threads S

M = dimension over wires

P = pitch = 1 ÷ number of threads per inch

R_b = radius required in Formulas (4) and (4e)

S = number of “starts” or threads on a multiple-threaded worm or screw

T = 0.5 P = width of thread in axial plane at diameter E

T_a = arc thickness on pitch cylinder in plane perpendicular to axis

W = wire or pin diameter

Formulas for Checking Pitch Diameters of Screw Threads

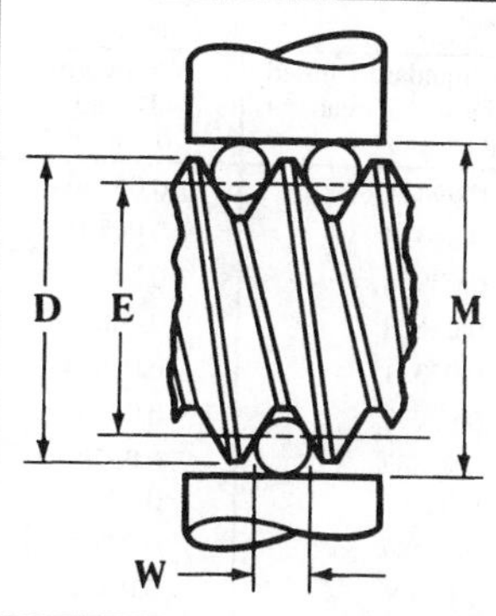

The formulas below do not compensate for the effect of the lead angle upon measurement M, but they are sufficiently accurate for checking standard single-thread screws unless exceptional accuracy is required. See accompanying information on effect of lead angle; also matter relating to measuring wire sizes, accuracy required for such wires, and contact or measuring pressure. The approximate best wire size for pitch-line contact may be obtained by the formula

$$W = 0.5 \times \text{pitch} \times \sec \tfrac{1}{2} \text{ included thread angle}$$

For 60-degree threads, $W = 0.57735 \times \text{pitch}$.

Form of Thread	Formulas for determining measurement M corresponding to correct pitch diameter and the pitch diameter E corresponding to a given measurement over wires.[a]
American National Standard Unified	When measurement M is known. $E = M + 0.86603P - 3W$ When pitch diameter E is used in formula. $M = E - 0.86603P + 3W$ The American Standard formerly was known as U.S. Standard.
British Standard Whitworth	When measurement M is known. $E = M + 0.9605P - 3.1657W$ When pitch diameter E is used in formula. $M = E - 0.9605P + 3.1657W$
British Association Standard	When measurement M is known. $E = M + 1.1363P - 3.4829W$ When pitch diameter E is used in formula. $M = E - 1.1363P + 3.4829W$
Lowenherz Thread	When measurement M is known. $E = M + P - 3.2359W$ When pitch diameter E is used in formula. $M = E - P + 3.2359W$
Sharp V-Thread	When measurement M is known. $E = M + 0.86603P - 3W$ When pitch diameter E is used in formula. $M = E - 0.86603P + 3W$
International Standard	Use the formula given above for the American National Standard Unified Thread.
Pipe Thread	See accompanying paragraph on *Buckingham Exact Involute Helicoid Formula Applied to Screw Threads.*
Acme and Worm Threads	See Buckingham Formulas page 1869; also *Three-wire Measurement of Acme and Stub Acme Thread Pitch Diameter.*
Buttress Form of Thread	Different forms of buttress threads are used. See paragraph on *Three-Wire Method Applied to Buttress Threads.*

[a] The wires must be lapped to a uniform diameter and it is very important to insert in the rule or formula the wire diameter as determined by precise means of measurement. Any error will be multiplied. See paragraph on Wire Sizes for Checking Pitch Diameters.

Values of Constants Used in Formulas for Measuring Pitch Diameters of Screws by the Three-wire System

No. of Threads per Inch	American Standard Unified and Sharp V-Thread 0.86603*P*	Whitworth Thread 0.9605*P*	No. of Threads per Inch	American Standard Unified and Sharp V-Thread 0.86603*P*	Whitworth Thread 0.9605*P*
2¼	0.38490	0.42689	18	0.04811	0.05336
2⅜	0.36464	0.40442	20	0.04330	0.04803
2½	0.34641	0.38420	22	0.03936	0.04366
2⅝	0.32992	0.36590	24	0.03608	0.04002
2¾	0.31492	0.34927	26	0.03331	0.03694
2⅞	0.30123	0.33409	28	0.03093	0.03430
3	0.28868	0.32017	30	0.02887	0.03202
3¼	0.26647	0.29554	32	0.02706	0.03002
3½	0.24744	0.27443	34	0.02547	0.02825
4	0.21651	0.24013	36	0.02406	0.02668
4½	0.19245	0.21344	38	0.02279	0.02528
5	0.17321	0.19210	40	0.02165	0.02401
5½	0.15746	0.17464	42	0.02062	0.02287
6	0.14434	0.16008	44	0.01968	0.02183
7	0.12372	0.13721	46	0.01883	0.02088
8	0.10825	0.12006	48	0.01804	0.02001
9	0.09623	0.10672	50	0.01732	0.01921
10	0.08660	0.09605	52	0.01665	0.01847
11	0.07873	0.08732	56	0.01546	0.01715
12	0.07217	0.08004	60	0.01443	0.01601
13	0.06662	0.07388	64	0.01353	0.01501
14	0.06186	0.06861	68	0.01274	0.01412
15	0.05774	0.06403	72	0.01203	0.01334
16	0.05413	0.06003	80	0.01083	0.01201

Constants Used for Measuring Pitch Diameters of Metric Screws by the Three-wire System

Pitch in mm	0.86603*P* in Inches	*W* in Inches	Pitch in mm	0.86603*P* in Inches	*W* in Inches	Pitch in mm	0.86603*P* in Inches	*W* in Inches
0.2	0.00682	0.00455	0.75	0.02557	0.01705	3.5	0.11933	0.07956
0.25	0.00852	0.00568	0.8	0.02728	0.01818	4	0.13638	0.09092
0.3	0.01023	0.00682	1	0.03410	0.02273	4.5	0.15343	0.10229
0.35	0.01193	0.00796	1.25	0.04262	0.02841	5	0.17048	0.11365
0.4	0.01364	0.00909	1.5	0.05114	0.03410	5.5	0.18753	0.12502
0.45	0.01534	0.01023	1.75	0.05967	0.03978	6	0.20457	0.13638
0.5	0.01705	0.01137	2	0.06819	0.04546	8	0.30686	0.18184
0.6	0.02046	0.01364	2.5	0.08524	0.05683	…	…	…
0.7	0.02387	0.01591	3	0.10229	0.06819	…	…	…

This table may be used for American National Standard Metric Threads. The formulas for American Standard Unified Threads on page 1865 are used. In the table above, the values of 0.86603*P* and *W* are in inches so that the values for *E* and *M* calculated from the formulas on page 1865 are also in inches.

Why Small Thread Angle Affects Accuracy of Three-Wire Measurement.—In measuring or checking Acme threads, or any others having a comparatively small thread angle *A*, it is particularly important to use a formula that compensates largely, if not entirely, for the effect of the lead angle, especially in all gage and precision work. The effect of the lead angle on the position of the wires and upon the resulting measurement *M* is much greater in a 29-degree thread than in a higher thread angle such, for example, as a 60-degree thread. This effect results from an increase in the cotangent of the thread angle as this angle becomes smaller. The reduction in the width of the thread groove in the normal plane due to the lead angle causes a wire of given size to rest higher in the groove of a thread having a small thread angle *A* (like a 29-degree thread) than in the groove of a thread with a larger angle (like a 60-degree American Standard).

Acme Threads: Three-wire measurements of high accuracy require the use of Formula (4). For most measurements, however, Formula (2) or (3) gives satisfactory results. The table on page 1872 lists suitable wire sizes for use in Formulas (2) and (4).

Dimensions Over Wires of Given Diameter for Checking Screw Threads of American National Form (U.S. Standard) and the V-Form

Dia. of Thread	No. of Threads per Inch	Wire Dia. Used	Dimension over Wires		Dia. of Thread	No. of Threads per Inch	Wire Dia. Used	Dimension over Wires	
			V-Thread	U.S. Thread				V-Thread	U.S. Thread
1/4	18	0.035	0.2588	0.2708	7/8	8	0.090	0.9285	0.9556
1/4	20	0.035	0.2684	0.2792	7/8	9	0.090	0.9525	0.9766
1/4	22	0.035	0.2763	0.2861	7/8	10	0.090	0.9718	0.9935
1/4	24	0.035	0.2828	0.2919	15/16	8	0.090	0.9910	1.0181
5/16	18	0.035	0.3213	0.3333	15/16	9	0.090	1.0150	1.0391
5/16	20	0.035	0.3309	0.3417	1	8	0.090	1.0535	1.0806
5/16	22	0.035	0.3388	0.3486	1	9	0.090	1.0775	1.1016
5/16	24	0.035	0.3453	0.3544	1 1/8	7	0.090	1.1476	1.1785
3/8	16	0.040	0.3867	0.4003	1 1/4	7	0.090	1.2726	1.3035
3/8	18	0.040	0.3988	0.4108	1 3/8	6	0.150	1.5363	1.5724
3/8	20	0.040	0.4084	0.4192	1 1/2	6	0.150	1.6613	1.6974
7/16	14	0.050	0.4638	0.4793	1 5/8	5 1/2	0.150	1.7601	1.7995
7/16	16	0.050	0.4792	0.4928	1 3/4	5	0.150	1.8536	1.8969
1/2	12	0.050	0.5057	0.5237	1 7/8	5	0.150	1.9786	2.0219
1/2	13	0.050	0.5168	0.5334	2	4 1/2	0.150	2.0651	2.1132
1/2	14	0.050	0.5263	0.5418	2 1/4	4 1/2	0.150	2.3151	2.3632
9/16	12	0.050	0.5682	0.5862	2 1/2	4	0.150	2.5170	2.5711
9/16	14	0.050	0.5888	0.6043	2 3/4	4	0.150	2.7670	2.28211
5/8	10	0.070	0.6618	0.6835	3	3 1/2	0.200	3.1051	3.1670
5/8	11	0.070	0.6775	0.6972	3 1/4	3 1/2	0.200	3.3551	3.4170
5/8	12	0.070	0.6907	0.7087	3 1/2	3 1/4	0.250	3.7171	3.7837
11/16	10	0.070	0.7243	0.7460	3 3/4	3	0.250	3.9226	3.9948
11/16	11	0.070	0.7400	0.7597	4	3	0.250	4.1726	4.2448
3/4	10	0.070	0.7868	0.8085	4 1/4	2 7/8	0.250	4.3975	4.4729
3/4	11	0.070	0.8025	0.8222	4 1/2	2 3/4	0.250	4.6202	4.6989
3/4	12	0.070	0.8157	0.8337	4 3/4	2 5/8	0.250	4.8402	4.9227
13/16	9	0.070	0.8300	0.8541	5	2 1/2	0.250	5.0572	5.1438
13/16	10	0.070	0.8493	0.8710	…	…	…	…	…

Buckingham Simplified Formula which Includes Effect of Lead Angle.—The Formula (3) which follows gives very accurate results for the lower lead angles in determining measurement *M*. However, if extreme accuracy is essential, it may be advisable to use the involute helicoid formulas as explained later.

$$M = E + W(1 + \sin A_n) \quad (3) \quad \text{where} \quad W = \frac{T \times \cos B}{\cos A_n} \quad (3a)$$

Theoretically correct equations for determining measurement *M* are complex and cumbersome to apply. Formula (3) combines simplicity with a degree of accuracy which meets all but the most exacting requirements, particularly for lead angles below 8 or 10 degrees and the higher thread angles. However, the wire diameter used in Formula (3) must conform to that obtained by Formula (3a) to permit a direct solution or one not involving indeterminate equations and successive trials.

Application of Buckingham Formula: In the application of Formula (3) to screw or worm threads, two general cases are to be considered.

Table for Measuring Whitworth Standard Threads by the Three-wire Method

Dia. of Thread	No. of Threads per Inch	Dia. of Wire Used	Dia. Measured over Wires	Dia. of Thread	No. of Threads per Inch	Dia. of Wire Used	Dia. Measured over Wires
1⁄8	40	0.018	0.1420	21⁄4	4	0.150	2.3247
3⁄16	24	0.030	0.2158	23⁄8	4	0.150	2.4497
1⁄4	20	0.035	0.2808	21⁄2	4	0.150	2.5747
5⁄16	18	0.040	0.3502	25⁄8	4	0.150	2.6997
3⁄8	16	0.040	0.4015	23⁄4	31⁄2	0.200	2.9257
7⁄16	14	0.050	0.4815	27⁄8	31⁄2	0.200	3.0507
1⁄2	12	0.050	0.5249	3	31⁄2	0.200	3.1757
9⁄16	12	0.050	0.5874	31⁄8	31⁄2	0.200	3.3007
5⁄8	11	0.070	0.7011	31⁄4	31⁄4	0.200	3.3905
11⁄16	11	0.070	0.7636	33⁄8	31⁄4	0.200	3.5155
3⁄4	10	0.070	0.8115	31⁄2	31⁄4	0.200	3.6405
13⁄16	10	0.070	0.8740	35⁄8	31⁄4	0.200	3.7655
7⁄8	9	0.070	0.9187	33⁄4	3	0.200	3.8495
15⁄16	9	0.070	0.9812	37⁄8	3	0.200	3.9745
1	8	0.090	1.0848	4	3	0.200	4.0995
11⁄16	8	0.090	1.1473	41⁄8	3	0.200	4.2245
11⁄8	7	0.090	1.1812	41⁄4	27⁄8	0.250	4.4846
13⁄16	7	0.090	1.2437	43⁄8	27⁄8	0.250	4.6096
11⁄4	7	0.090	1.3062	41⁄2	27⁄8	0.250	4.7346
15⁄16	7	0.090	1.3687	45⁄8	27⁄8	0.250	4.8596
13⁄8	6	0.120	1.4881	43⁄4	23⁄4	0.250	4.9593
17⁄16	6	0.120	1.5506	47⁄8	23⁄4	0.250	5.0843
11⁄2	6	0.120	1.6131	5	23⁄4	0.250	5.2093
19⁄16	6	0.120	1.6756	51⁄8	23⁄4	0.250	5.3343
15⁄8	5	0.120	1.6847	51⁄4	25⁄8	0.250	5.4316
111⁄16	5	0.120	1.7472	53⁄8	25⁄8	0.250	5.5566
13⁄4	5	0.120	1.8097	51⁄2	25⁄8	0.250	5.6816
113⁄16	5	0.120	1.8722	55⁄8	25⁄8	0.250	5.8066
17⁄8	41⁄2	0.150	1.9942	53⁄4	21⁄2	0.250	5.9011
115⁄16	41⁄2	0.150	2.0567	57⁄8	21⁄2	0.250	6.0261
2	41⁄2	0.150	2.1192	6	21⁄2	0.250	6.1511
21⁄8	41⁄2	0.150	2.2442	…	…	…	…

All dimensions are given in inches.

Case 1: The screw thread or worm is to be milled with a cutter having an included angle equal to the nominal or standard thread angle that is assumed to be the angle in the axial plane. For example, a 60-degree cutter is to be used for milling a thread. In this case, the thread angle in the plane of the axis will exceed 60 degrees by an amount increasing with the lead angle. This variation from the standard angle may be of little or no practical importance if the lead angle is small or if the mating nut (or teeth in worm gearing) is formed to suit the thread as milled.

Case 2: The screw thread or worm is to be milled with a cutter reduced to whatever normal angle is equivalent to the standard thread angle in the axial plane. For example, a 29-degree Acme thread is to be milled with a cutter having some angle smaller than 29 degrees (the reduction increasing with the lead angle) to make the thread angle standard in the plane of the axis. Theoretically, the milling cutter angle should always be corrected to suit the normal angle; but if the lead angle is small, such correction may be unnecessary.

If the thread is cut in a lathe to the standard angle as measured in the axial plane, Case 2 applies in determining the pin size W and the overall measurement M.

In solving all problems under Case 1, angle A_n used in Formulas (3) and (3a) equals one-half the included angle of the milling cutter.

When Case 2 applies, angle A_n for milled threads also equals one-half the included angle of the cutter, but the cutter angle is reduced and is determined as follows:

$$\tan A_n = \tan A \times \cos B$$

The included angle of the cutter or the normal included angle of the thread groove = $2A_n$. Examples 1 and 2, which follow, illustrate Cases 1 and 2.

Example 1(Case 1): Take, for example, an Acme screw thread that is milled with a cutter having an included angle of 29 degrees; consequently, the angle of the thread exceeds 29 degrees in the axial section.

The outside or major diameter is 3 inches; the pitch, ½ inch; the lead, 1 inch; the number of threads or "starts," 2. Find pin size *W* and measurement *M*.

Pitch diameter $E = 2.75$; $T = 0.25$; $L = 1.0$; $A_n = 14.50°$ tan $A_n = 0.258618$; sin A_n = 0.25038; and cos $A_n = 0.968148$.

$$\tan B = \frac{1.0}{3.1416 \times 2.75} = 0.115749 \qquad B = 6.6025°$$

$$W = \frac{0.25 \times 0.993368}{0.968148} = 0.25651 \text{ inch}$$

$$M = 2.75 + 0.25651 \times (1 + 0.25038) = 3.0707 \text{ inches}$$

Note: This value of *M* is only 0.0001 inch larger than that obtained by using the very accurate involute helicoid Formula (4) discussed on the following page.

Example 2(Case 2): A triple-threaded worm has a pitch diameter of 2.481 inches, pitch of 1.5 inches, lead of 4.5 inches, lead angle of 30 degrees, and nominal thread angle of 60 degrees in the axial plane. Milling cutter angle is to be reduced. $T = 0.75$ inch; cos B = 0.866025; and tan $A = 0.57735$. Again use Formula (3) to see if it is applicable.

Tan A_n = tan A × cos B = 0.57735 × 0.866025 = 0.5000; hence $A_n = 26.565°$, making the included cutter angle 53.13°. Cos $A_n = 0.89443$; sin $A_n = 0.44721$.

$$W = \frac{0.75 \times 0.866025}{0.89443} = 0.72618 \text{ inch}$$

$$M = 2.481 + 0.72618 \times (1 + 0.44721) = 3.532 \text{ inches}$$

Note: If the value of measurement *M* is determined by using the following Formula (4) it will be found that *M* = 3.515 + inches; hence the error equals 3.532 - 3.515 = 0.017 inch approximately, which indicates that Formula (3) is not accurate enough here. The application of this simpler Formula (3) will depend upon the lead angle and thread angle (as previously explained) and upon the class of work.

Buckingham Exact Involute Helicoid Formula Applied to Screw Threads.— When extreme accuracy is required in finding measurement *M* for obtaining a given pitch diameter, the equations that follow, although somewhat cumbersome to apply, have the merit of providing a direct and very accurate solution; consequently, they are preferable to the indeterminate equations and successive trial solutions heretofore employed when extreme precision is required. These equations are exact for involute helical gears and, consequently, give theoretically correct results when applied to a screw thread of the involute helicoidal form; they also give very close approximations for threads having intermediate profiles.

Helical Gear Equation Applied to Screw Thread Measurement: In applying the helical gear equations to a screw thread, use either the axial or normal thread angle and the lead angle of the helix. To keep the solution on a practical basis, either thread angle *A* or A_n, as the case may be, is assumed to equal the cutter angle of a milled thread. Actually, the profile of a milled thread will have some curvature in both axial and normal sections; hence angles A and A_n represent the angular approximations of these slightly curved profiles. The equations that follow give the values needed to solve the screw thread problem as a helical gear problem.

$$M = \frac{2R_b}{\cos G} + W \tag{4}$$

$$\tan F = \frac{\tan A}{\tan B} = \frac{\tan A_n}{\sin B} \quad (4a) \qquad R_b = \frac{E}{2}\cos F \quad (4b)$$

$$T_a = \frac{T}{\tan B} \quad (4c) \qquad \tan H_b = \cos F \times \tan H \quad (4d)$$

$$\operatorname{inv} G = \frac{T_a}{E} + \operatorname{inv} F + \frac{W}{2R_b \cos H_b} - \frac{\pi}{S} \tag{4e}$$

The tables of involute functions starting on page 98 provide values for angles from 14 to 51 degrees, used for gear calculations. The formula for involute functions on page 97 may be used to extend this table as required.

Example 3: To illustrate the application of Formula (4) and the supplementary formulas, assume that the number of starts $S = 6$; pitch diameter $E = 0.6250$; normal thread angle $A_n = 20°$; lead of thread $L = 0.864$ inch; $T = 0.072$; W = 0.07013 inch.

$$\tan B = \frac{L}{\pi E} = \frac{0.864}{1.9635} = 0.44003 \qquad B = 23.751°$$

$$\text{Helix angle } H = 90° - 23.751° = 66.249°$$

$$\tan F = \frac{\tan A_n}{\sin B} = \frac{0.36397}{0.40276} = 0.90369 \qquad F = 42.104°$$

$$R_b = \frac{E}{2}\cos F = \frac{0.6250}{2} \times 0.74193 = 0.23185$$

$$T_a = \frac{T}{\tan B} = \frac{0.072}{0.44003} = 0.16362$$

$$\tan H_b = \cos F \tan H = 0.74193 \times 2.27257 = 1.68609 \qquad H_b = 59.328°$$

The involute function of G is found next by Formula (4e).

$$\operatorname{inv} G = \frac{0.16362}{0.625} + 0.16884 + \frac{0.07013}{2 \times 0.23185 \times 0.51012} - \frac{3.1416}{6} = 0.20351$$

Since 0.20351 is outside the values for involute functions given in the tables on pages 98 through 101 use the formula for involute functions on page 97 to extend these tables as required. It will be found that 44 deg. 21 min. or 44.350 degrees is the angular equivalent of 0.20351; hence, $G = 44.350$ degrees.

$$M = \frac{2R_b}{\cos G} + W = \frac{2 \times 0.23185}{0.71508} + 0.07013 = 0.71859 \text{ inch}$$

Accuracy of Formulas (3) and (4) Compared.—With the involute helicoid Formula (4) any wire size that makes contact with the flanks of the thread may be used; however, in the preceding example, the wire diameter W was obtained by Formula (3a) in order to compare Formula (4) with (3) . If Example (3) is solved by Formula (3) , $M = 0.71912$; hence the difference between the values of M obtained with Formulas (3) and (4) equals 0.71912

- 0.71859 = 0.00053 inch. The included thread angle in this case is 40 degrees. If Formulas (3) and (4) are applied to a 29-degree thread, the difference in measurements M or the error resulting from the use of Formulas (3) will be larger. For example, with an Acme thread having a lead angle of about 34 degrees, the difference in values of M obtained by the two formulas equals 0.0008 inch.

Three-wire Measurement of Acme and Stub Acme Thread Pitch Diameter.—For single- and multiple-start Acme and Stub Acme threads having lead angles of less than 5 degrees, the approximate three-wire formula given on page 1863 and the best wire size taken from the table on page 1872 may be used.

Multiple-start Acme and Stub Acme threads commonly have a lead angle of greater than 5 degrees. For these, a direct determination of the actual pitch diameter is obtained by using the formula: $E = M - (C + c)$ in conjunction with the table on page 1873. To enter the table, the lead angle B of the thread to be measured must be known. It is found by the formula: $\tan B = L \div 3.1416E_1$ where L is the lead of the thread and E_1 is the nominal pitch diameter. The best wire size is now found by taking the value of w_1 as given in the table for lead angle B, with interpolation, and dividing it by the number of threads per inch. The value of $(C + c)_1$ given in the table for lead angle B is also divided by the number of threads per inch to get $(C + c)$. Using the best size wires, the actual measurement over wires M is made and the actual pitch diameter E found by using the formula: $E = M - (C + c)$.

Example: For a 5 tpi, 4-start Acme thread with a 13.952° lead angle, using three 0.10024-inch wires, $M = 1.1498$ inches, hence $E = 1.1498 - 0.1248 = 1.0250$ inches.

Under certain conditions, a wire may contact one thread flank at two points, and it is then advisable to substitute balls of the same diameter as the wires.

Checking Thickness of Acme Screw Threads.—In some instances it may be preferable to check the thread thickness instead of the pitch diameter, especially if there is a thread thickness tolerance.

A direct method, applicable to the larger pitches, is to use a vernier gear-tooth caliper for measuring the thickness in the *normal* plane of the thread. This measurement, for an American Standard General Purpose Acme thread, should be made at a distance below the *basic* outside diameter equal to $p/4$. The thickness at this basic pitch-line depth and in the axial plane should be $p/2 - 0.259 \times$ the pitch diameter allowance from the table on page 1800 with a tolerance of *minus* $0.259 \times$ the pitch diameter tolerance from the table on page 1801. The thickness in the normal plane or plane of measurement is equal to the thickness in the axial plane multiplied by the cosine of the helix angle. The helix angle may be determined from the formula:

$$\text{tangent of helix angle} = \text{lead of thread} \div (3.1416 \times \text{pitch diameter})$$

Three-Wire Method for Checking Thickness of Acme Threads.—The application of the 3-wire method of checking the thickness of an Acme screw thread is included in the Report of the National Screw Thread Commission. In applying the 3-wire method for checking thread thickness, the procedure is the same as in checking pitch diameter (see *Three-wire Measurement of Acme and Stub Acme Thread Pitch Diameter*), although a different formula is required. Assume that D = basic major diameter of screw; M = measurement over wires; W = diameter of wires; S = tangent of helix angle at pitch line; P = pitch; T = thread thickness at depth equal to $0.25P$.

$$T = 1.12931 \times P + 0.25862 \times (M - D) - W \times (1.29152 + 0.48407S^2)$$

This formula transposed to show the correct measurement M equivalent to a given required thread thickness is as follows:

$$M = D + \frac{W \times (1.29152 + 0.48407S^2) + T - 1.12931 \times P}{0.25862}$$

Wire Sizes for Three-Wire Measurement of Acme Threads with Lead Angles of Less than 5 Degrees

Threads per Inch	Best Size	Max.	Min.	Threads per Inch	Best Size	Max.	Min.
1	0.51645	0.65001	0.48726	5	0.10329	0.13000	0.09745
1⅓	0.38734	0.48751	0.36545	6	0.08608	0.10834	0.08121
1½	0.34430	0.43334	0.32484	8	0.06456	0.08125	0.06091
2	0.25822	0.32501	0.24363	10	0.05164	0.06500	0.04873
2½	0.20658	0.26001	0.19491	12	0.04304	0.05417	0.04061
3	0.17215	0.21667	0.16242	14	0.03689	0.04643	0.03480
4	0.12911	0.16250	0.12182	16	0.03228	0.04063	0.03045

Wire sizes are based upon zero helix angle. Best size = 0.51645 × pitch; maximum size = 0.650013 × pitch; minimum size = 0.487263 × pitch.

Example: An Acme General Purpose thread, Class 2G, has a 5-inch basic major diameter, 0.5-inch pitch, and 1-inch lead (double thread). Assume the wire size is 0.258 inch. Determine measurement M for a thread thickness T at the basic pitch line of 0.2454 inch. (T is the maximum thickness at the basic pitch line and equals 0.5P, the basic thickness, –0.259 × allowance from Table 4, page 1800.)

$$M = 5 + \frac{0.258 \times [1.29152 + 0.48407 \times (0.06701)^2] + 0.2454 - 1.12931 \times 0.5}{0.25862}$$

$$= 5.056 \text{ inches}$$

Testing Angle of Thread by Three-Wire Method.—The error in the angle of a thread may be determined by using sets of wires of two diameters, the measurement over the two sets of wires being followed by calculations to determine the amount of error, assuming that the angle cannot be tested by comparison with a standard plug gage, known to be correct. The diameter of the small wires for the American Standard thread is usually about 0.6 times the pitch and the diameter of the large wires, about 0.9 times the pitch. The total difference between the measurements over the large and small sets of wires is first determined. If the thread is an American Standard or any other form having an included angle of 60 degrees, the difference between the two measurements should equal three times the difference between the diameters of the wires used. Thus, if the wires are 0.116 and 0.076 inch in diameter, respectively, the difference equals 0.116 – 0.076 = 0.040 inch. Therefore, the difference between the micrometer readings for a standard angle of 60 degrees equals 3 × 0.040 = 0.120 inch for this example. If the angle is incorrect, the amount of error may be determined by the following formula, which applies to any thread regardless of angle:

$$\sin a = \frac{A}{B - A}$$

where A = difference in diameters of the large and small wires used

B = total difference between the measurements over the large and small wires

a = one-half the included thread angle

Example: The diameter of the large wires used for testing the angle of a thread is 0.116 inch and of the small wires 0.076 inch. The measurement over the two sets of wires shows a total difference of 0.122 inch instead of the correct difference, 0.120 inch, for a standard angle of 60 degrees when using the sizes of wires mentioned. The amount of error is determined as follows:

$$\sin a = \frac{0.040}{0.122 - 0.040} = \frac{0.040}{0.082} = 0.4878$$

A table of sines shows that this value (0.4878) is the sine of 29 degrees 12 minutes, approximately. Therefore, the angle of the thread is 58 degrees 24 minutes or 1 degree 36 minutes less than the standard angle.

Best Wire Diameters and Constants for Three-wire Measurement of Acme and Stub Acme Threads with Large Lead Angles—1–inch Axial Pitch

Lead angle, B, deg.	1-start threads		2-start threads		Lead angle, B, deg.	2-start threads		3-start threads	
	w_1	$(C+c)_1$	w_1	$(C+c)_1$		w_1	$(C+c)_1$	w_1	$(C+c)_1$
5.0	0.51450	0.64311	0.51443	0.64290	10.0	0.50864	0.63518	0.50847	0.63463
5.1	0.51442	0.64301	0.51435	0.64279	10.1	0.50849	0.63498	0.50381	0.63442
5.2	0.51435	0.64291	0.51427	0.64268	10.2	0.50834	0.63478	0.50815	0.63420
5.3	0.51427	0.64282	0.51418	0.64256	10.3	0.50818	0.63457	0.50800	0.63399
5.4	0.51419	0.64272	0.51410	0.64245	10.4	0.50802	0.63436	0.50784	0.63378
5.5	0.51411	0.64261	0.51401	0.64233	10.5	0.40786	0.63416	0.50768	0.63356
5.6	0.51403	0.64251	0.51393	0.64221	10.6	0.50771	0.63395	0.50751	0.63333
5.7	0.51395	0.64240	0.51384	0.64209	10.7	0.50755	0.63375	0.50735	0.63311
5.8	0.51386	0.64229	0.51375	0.64196	10.8	0.50739	0.53354	0.50718	0.63288
5.9	0.51377	0.64218	0.51366	0.64184	10.9	0.50723	0.63333	0.50701	0.63265
6.0	0.51368	0.64207	0.51356	0.64171	11.0	0.50707	0.63313	0.50684	0.63242
6.1	0.51359	0.64195	0.51346	0.64157	11.1	0.50691	0.63292	0.50667	0.63219
6.2	0.51350	0.64184	0.51336	0.64144	11.2	0.50674	0.63271	0.50649	0.63195
6.3	0.51340	0.64172	0.41327	0.64131	11.3	0.50658	0.63250	0.50632	0.63172
6.4	0.51330	0.64160	0.51317	0.64117	11.4	0.50641	0.63228	0.50615	0.63149
6.5	0.51320	0.64147	0.51306	0.64103	11.5	0.50623	0.63206	0.50597	0.63126
6.6	0.51310	0.64134	0.51296	0.64089	11.6	0.50606	0.63184	0.50579	0.63102
6.7	0.51300	0.64122	0.51285	0.64075	11.7	0.50589	0.63162	0.50561	0.63078
6.8	0.51290	0.64110	0.51275	0.64061	11.8	0.50571	0.63140	0.50544	0.63055
6.9	0.51280	0.64097	0.51264	0.64046	11.9	0.50553	0.63117	0.50526	0.63031
7.0	0.51270	0.64085	0.51254	0.64032	12.0	0.50535	0.63095	0.50507	0.63006
7.1	0.51259	0.64072	0.51243	0.64017	12.1	0.50517	0.63072	0.50488	0.62981
7.2	0.51249	0.64060	0.51232	0.64002	12.2	0.50500	0.63050	0.50470	0.62956
7.3	0.51238	0.64047	0.51221	0.63987	12.3	0.50482	0.63027	0.50451	0.62931
7.4	0.51227	0.64034	0.51209	0.63972	12.4	0.50464	0.63004	0.50432	0.62906
7.5	0.51217	0.64021	0.51198	0.63957	12.5	0.50445	0.62981	0.50413	0.62881
7.6	0.51206	0.64008	0.51186	0.63941	12.6	0.50427	0.62958	0.50394	0.62856
7.7	0.51196	0.63996	0.51174	0.63925	12.7	0.50408	0.62934	0.50375	0.62830
7.8	0.51186	0.63983	0.51162	0.63909	12.8	0.50389	0.62911	0.50356	0.62805
7.9	0.51175	0.63970	0.51150	0.63892	12.9	0.50371	0.62888	0.50336	0.62779
8.0	0.51164	0.63957	0.51138	0.63876	13.0	0.50352	0.62865		
8.1	0.51153	0.63944	0.51125	0.63859	13.1	0.50333	0.62841		
8.2	0.51142	0.63930	0.51113	0.63843	13.2	0.50313	0.62817		
8.3	0.51130	0.63916	0.51101	0.63827	13.3	0.50293	0.62792		
8.4	0.51118	0.63902	0.51088	0.63810	13.4	0.50274	0.62778		
8.5	0.51105	0.63887	0.51075	0.63793	13.5	0.50254	0.62743		
8.6	0.51093	0.63873	0.51062	0.63775	13.6	0.50234	0.62718		
8.7	0.51081	0.63859	0.51049	0.63758	13.7	0.50215	0.62694	For these 3-start thread values see table on following page.	
8.8	0.51069	0.63845	0.51035	0.63740	13.8	0.50195	0.62670		
8.9	0.51057	0.63831	0.51022	0.63722	13.9	0.50175	0.62645		
9.0	0.51044	0.63817	0.51008	0.63704	14.0	0.50155	0.62621		
9.1	0.51032	0.63802	0.50993	0.63685	14.1	0.50135	0.62596		
9.2	0.51019	0.63788	0.50979	0.63667	14.2	0.50115	0.62571		
9.3	0.51006	0.63774	0.50965	0.63649	14.3	0.50094	0.62546		
9.4	0.50993	0.63759	0.50951	0.63630	14.4	0.50073	0.62520		
9.5	0.50981	0.63744	0.50937	0.63612	14.5	0.50051	0.62494		
9.6	0.50968	0.63730	0.50922	0.63593	14.6	0.50030	0.62468		
9.7	0.50955	0.63715	0.50908	0.63574	14.7	0.50009	0.62442		
9.8	0.50941	0.63700	0.50893	0.63555	14.8	0.49988	0.62417		
9.9	0.50927	0.63685	0.50879	0.63537	14.9	0.49966	0.62391		
10.0	0.50913	0.63670	0.50864	0.63518	15.0	0.49945	0.62365		

All dimensions are in inches.

Values given for w_1 and $(C+c)_1$ in table are for 1-inch pitch axial threads. For other pitches, divide table values by number of threads per inch.

Courtesy of Van Keuren Co.

Best Wire Diameters and Constants for Three-wire Measurement of Acme and Stub Acme Threads with Large Lead Angles—1-inch Axial Pitch

Lead angle, B, deg.	3-start threads		4-start threads		Lead angle, B, deg.	3-start threads		4-start threads	
	w_1	$(C+c)_1$	w_1	$(C+c)_1$		w_1	$(C+c)_1$	w_1	$(C+c)_1$
13.0	0.50316	0.62752	0.50297	0.62694	18.0	0.49154	0.61250	0.49109	0.61109
13.1	0.50295	0.62725	0.50277	0.62667	18.1	0.49127	0.61216	0.49082	0.61073
13.2	0.50275	0.62699	0.50256	0.62639	18.2	0.49101	0.61182	0.49054	0.61037
13.3	0.50255	0.62672	0.50235	0.62611	18.3	0.49074	0.61148	0.49027	0.61001
13.4	0.50235	0.62646	0.50215	0.62583	18.4	0.49047	0.61114	0.48999	0.60964
13.5	0.50214	0.62619	0.50194	0.62555	18.5	0.49020	0.61080	0.48971	0.69928
13.6	0.50194	0.62592	0.50173	0.62526	18.6	0.48992	0.61045	0.48943	0.60981
13.7	0.50173	0.62564	0.50152	0.62498	18.7	0.48965	0.61011	0.48915	0.60854
13.8	0.50152	0.62537	0.50131	0.62469	18.8	0.48938	0.60976	0.48887	0.60817
13.9	0.50131	0.62509	0.50109	0.62440	18.9	0.48910	0.60941	0.48859	0.60780
14.0	0.50110	0.62481	0.50087	0.62411	19.0	0.48882	0.60906	0.48830	0.60742
14.1	0.50089	0.62453	0.50065	0.62381	19.1	0.48854	0.60871	0.48800	0.60704
14.2	0.50068	0.62425	0.50043	0.62351	19.2	0.48825	0.60835	0.48771	0.60666
14.3	0.50046	0.62397	0.50021	0.62321	19.3	0.48797	0.60799	0.48742	0.60628
14.4	0.50024	0.62368	0.49999	0.62291	19.4	0.48769	0.60764	0.48713	0.60590
14.5	0.50003	0.62340	0.49977	0.62262	19.5	0.48741	0.60729	0.48684	0.60552
14.6	0.49981	0.62312	0.49955	0.62232	19.6	0.48712	0.60693	0.48655	0.60514
14.7	0.49959	0.62883	0.49932	0.62202	19.7	0.48638	0.60657	0.48625	0.60475
14.8	0.49936	0.62253	0.49910	0.62172	19.8	0.48655	0.60621	0.48596	0.60437
14.9	0.49914	0.62224	0.49887	0.62141	19.9	0.48626	0.60585	0.48566	0.60398
15.0	0.49891	0.62195	0.49864	0.62110	20.0	0.48597	0.60549	0.48536	0.60359
15.1	0.49869	0.62166	0.49842	0.62080	20.1	…	…	0.48506	0.60320
15.2	0.49846	0.62137	0.49819	0.62049	20.2	…	…	0.48476	0.60281
15.3	0.49824	0.62108	0.49795	0.62017	20.3	…	…	0.48445	0.60241
15.4	0.42801	0.62078	0.49771	0.61985	20.4	…	…	0.48415	0.60202
15.5	0.49778	0.62048	0.49747	0.61953	20.5	…	…	0.48384	0.60162
15.6	0.49754	0.62017	0.49723	0.61921	20.6	…	…	0.48354	0.60123
15.7	0.49731	0.61987	0.49699	0.61889	20.7	…	…	0.48323	0.60083
15.8	0.49707	0.61956	0.49675	0.61857	20.8	…	…	0.48292	0.60042
15.9	0.49683	0.61926	0.49651	0.61825	20.9	…	…	0.48261	0.60002
16.0	0.49659	0.61895	0.49627	0.61793	21.0	…	…	0.48230	0.59961
16.1	0.49635	0.61864	0.49602	0.61760	21.1	…	…	0.48198	0.49920
16.2	0.49611	0.61833	0.49577	0.61727	21.2	…	…	0.481166	0.59879
16.3	0.49586	0.61801	0.49552	0.61694	21.3	…	…	0.48134	0.59838
16.4	0.49562	0.61770	0.49527	0.61661	21.4	…	…	0.48103	0.59797
16.5	0.49537	0.61738	0.49502	0.61628	21.5	…	…	0.48701	0.59756
16.6	0.49512	0.61706	0.49476	0.61594	21.6	…	…	0.48040	0.59715
16.7	0.49488	0.61675	0.49451	0.61560	21.7	…	…	0.48008	0.59674
16.8	0.40463	0.61643	0.49425	0.61526	21.8	…	…	0.47975	0.59632
16.9	0.49438	0.61611	0.49400	0.61492	21.9	…	…	0.47943	0.59590
17.0	0.49414	0.61580	0.49375	0.61458	22.0	…	…	0.47910	0.59548
17.1	0.49389	0.61548	0.49349	0.61424	22.1	…	…	0.47878	0.59507
17.2	0.49363	0.61515	0.49322	0.61389	22.2	…	…	0.47845	0.59465
17.3	0.49337	0.61482	0.49296	0.61354	22.3	…	…	0.47812	0.59422
17.4	0.49311	0.61449	0.49269	0.61319	22.4	…	…	0.47778	0.59379
17.5	0.49285	0.61416	0.49243	0.61284	22.5	…	…	0.47745	0.59336
17.6	0.49259	0.61383	0.49217	0.61250	22.6	…	…	0.47711	0.52993
17.7	0.49233	0.61350	0.49191	0.61215	22.7	…	…	0.47677	0.59250
17.8	0.49206	0.61316	0.49164	0.61180	22.8	…	…	0.47643	0.59207
17.9	0.49180	0.61283	0.49137	0.61144	22.9	…	…	0.47610	0.59164
…	…	…	…	…	23.0	…	…	0.47577	0.59121

All dimensions are in inches.

Values given for w_1 and $(C+c)_1$ in table are for 1-inch pitch axial threads. For other pitches divide table values by number of threads per inch.

Courtesy of Van Keuren Co.

Measuring Taper Screw Threads by Three-Wire Method.—When the 3-wire method is used in measuring a taper screw thread, the measurement is along a line that is not perpendicular to the axis of the screw thread, the inclination from the perpendicular equaling one-half the included angle of the taper. The formula that follows compensates for this inclination resulting from contact of the measuring instrument surfaces, with two wires on one side and one on the other. The taper thread is measured over the wires in the usual manner except that the single wire must be located in the thread at a point where the effective diameter is to be checked (as described more fully later). The formula shows the dimension equivalent to the correct pitch diameter at this given point. The general formula for taper screw threads follows:

$$M = \frac{E - (\cot a)/2N + W(1 + \csc a)}{\sec b}$$

where M = measurement over the 3 wires
E = pitch diameter
a = one-half the angle of the thread
N = number of threads per inch
W = diameter of wires; and
b = one-half the angle of taper.

This formula is not theoretically correct but it is accurate for screw threads having tapers of ¾ inch per foot or less. This general formula can be simplified for a given thread angle and taper. The simplified formula following (in which P = pitch) is for an American National Standard pipe thread:

$$M = \frac{E - (0.86603 \times P) + 3 \times W}{1.00049}$$

Standard pitch diameters for pipe threads will be found in the section "American Pipe Threads," which also shows the location, or distance, of this pitch diameter from the end of the pipe. In using the formula for finding dimension M over the wires,the single wire is placed in whatever part of the thread groove locates it at the point where the pitch diameter is to be checked. The wire must be accurately located at this point. The other wires are then placed on each side of the thread that is diametrically opposite the single wire. If the pipe thread is straight or without taper,

$$M = E - (0.86603 \times P) + 3 \times W$$

Application of Formula to Taper Pipe Threads: To illustrate the use of the formula for taper threads, assume that dimension M is required for an American Standard 3-inch pipe thread gage. Table 4 starting on page 1849 shows that the 3-inch size has 8 threads per inch, or a pitch of 0.125 inch, and a pitch diameter at the gaging notch of 3.3885 inches. Assume that the wire diameter is 0.07217 inch: Then when the pitch diameter is correct

$$M = \frac{3.3885 - (0.86603 \times 0.125) + 3 \times 0.07217}{1.00049} = 3.495 \text{ inches}$$

Pitch Diameter Equivalent to a Given Measurement Over the Wires: The formula following may be used to check the pitch diameter at any point along a tapering thread when measurement M over wires of a given diameter is known. In this formula, E = the effective or pitch diameter at the position occupied by the single wire. The formula is not theoretically correct but gives very accurate results when applied to tapers of ¾ inch per foot or less.

$$E = 1.00049 \times M + (0.86603 \times P) - 3 \times W$$

Example: Measurement M = 3.495 inches at the gaging notch of a 3-inch pipe thread and the wire diameter = 0.07217 inch. Then

$$E = 1.00049 \times 3.495 + (0.86603 \times 0.125) - 3 \times 0.07217 = 3.3885 \text{ inches}$$

Pitch Diameter at Any Point Along Taper Screw Thread: When the pitch diameter in any position along a tapering thread is known, the pitch diameter at any other position may be determined as follows:

Multiply the distance (measured along the axis) between the location of the known pitch diameter and the location of the required pitch diameter, by the taper per inch or by 0.0625 for American National Standard pipe threads. Add this product to the known diameter, if the required diameter is at a large part of the taper, or subtract if the required diameter is smaller.

Example: The pitch diameter of a 3-inch American National Standard pipe thread is 3.3885 at the gaging notch. Determine the pitch diameter at the small end. The table starting on page 1849 shows that the distance between the gaging notch and the small end of a 3-inch pipe is 0.77 inch. Hence the pitch diameter at the small end = 3.3885 − (0.77 × 0.0625) = 3.3404 inches.

Three-Wire Method Applied to Buttress Threads.—The angles of buttress threads vary somewhat, especially on the front or load-resisting side. Formula (1), which follows, may be applied to any angles required. In this formula, M = measurement over wires when *pitch diameter E* is correct; A = included angle of thread and thread groove; a = angle of front face or load-resisting side, measured from a line perpendicular to screw thread axis; P = pitch of thread; and W = wire diameter.

$$M = E - \left[\frac{P}{\tan a + \tan(A - a)}\right] + W\left[1 + \cos\left(\frac{A}{2} - a\right) \times \csc\frac{A}{2}\right] \quad (1)$$

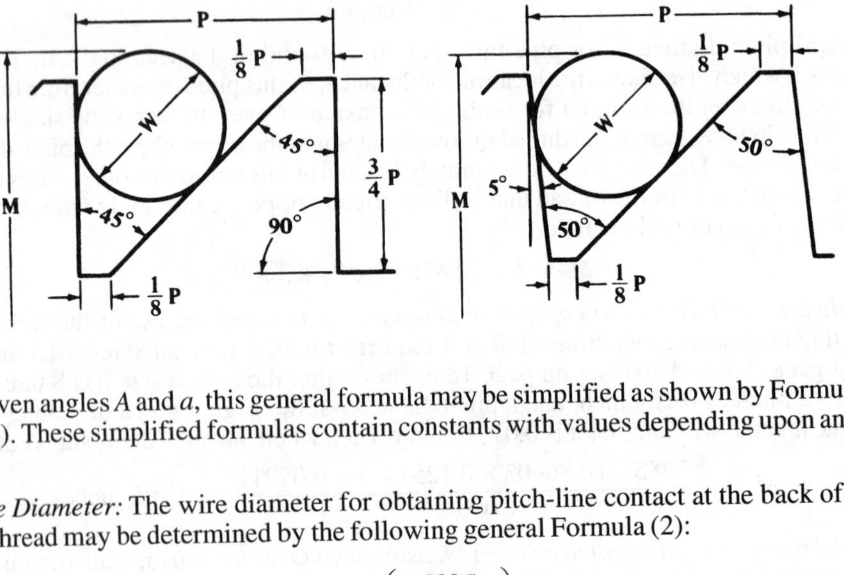

For given angles A and a, this general formula may be simplified as shown by Formulas (3) and (4). These simplified formulas contain constants with values depending upon angles A and a.

Wire Diameter: The wire diameter for obtaining pitch-line contact at the back of a buttress thread may be determined by the following general Formula (2):

$$W = P\left(\frac{\cos a}{1 + \cos A}\right) \quad (2)$$

45-Degree Buttress Thread: The buttress thread shown by the diagram at the left, has a front or load-resisting side that is perpendicular to the axis of the screw. Measurement M equivalent to a correct pitch diameter E may be determined by Formula (3):

$$M = E - P + (W \times 3.4142) \quad (3)$$

Wire diameter W for pitch-line contact at back of thread = 0.586 × pitch.

50-Degree Buttress Thread with Front-face Inclination of 5 Degrees: This buttress thread form is illustrated by the diagram at the right. Measurement M equivalent to the correct pitch diameter E may be determined by Formula (4):

$$M = E - (P \times 0.91955) + (W \times 3.2235) \tag{4}$$

Wire diameter W for pitch-line contact at back of thread = 0.606 × pitch. If the width of flat at crest and root = ⅛ × pitch, depth = 0.69 × pitch.

American National Standard Buttress Threads ANSI B1.9-1973: This buttress screw thread has an included thread angle of 52 degrees and a front face inclination of 7 degrees. Measurements M equivalent to a pitch diameter E may be determined by Formula (5):

$$M = E - 0.89064P + 3.15689W + c \tag{5}$$

The wire angle correction factor c is less than 0.0004 inch for recommended combinations of thread diameters and pitches and may be neglected. Use of wire diameter $W = 0.54147P$ is recommended.

Measurement of Pitch Diameter of Thread Ring Gages.—The application of direct methods of measurement to determine the pitch diameter of thread ring gages presents serious difficulties, particularly in securing proper contact pressure when a high degree of precision is required. The usual practice is to fit the ring gage to a master setting plug. When the thread ring gage is of correct lead, angle, and thread form, within close limits, this method is quite satisfactory and represents standard American practice. It is the only method available for small sizes of threads. For the larger sizes, various more or less satisfactory methods have been devised, but none of these have found wide application.

Screw Thread Gage Classification.—Screw thread gages are classified by their degree of accuracy, that is, by the amount of tolerance afforded the gage manufacturer and the wear allowance, if any.

There are also three classifications according to use: 1) Working gages for controlling production; 2) inspection gages for rejection or acceptance of the finished product; and 3) reference gages for determining the accuracy of the working and inspection gages.

American National Standard for Gages and Gaging for Unified Inch Screw Threads ANSI/ASME B1.2-1983 (R1991).—This standard covers gaging methods for conformance of Unified Screw threads and provides the essential specifications for applicable gages required for unified inch screw threads.

The standard includes the following gages for *Product Internal Thread:*

GO Working Thread Plug Gage for inspecting the maximum-material GO functional limit.

NOT GO (HI) Thread Plug Gage for inspecting the NOT GO (HI) functional diameter limit.

Thread Snap Gage—GO Segments or Rolls for inspecting the maximum-material GO functional limit.

Thread Snap Gage—NOT GO (HI) Segments or Rolls for inspecting the NOT GO (HI) functional diameter limit.

Thread Snap Gages—Minimum Material: Pitch Diameter Cone Type and Vee and Thread Groove Diameter Type for inspecting the minimum-material limit pitch diameter.

Thread-Setting Solid Ring Gage for setting internal thread indicating and snap gages.

Plain Plug, Snap, and Indicating Gages for checking the minor diameter of internal threads.

Snap and Indicating Gages for checking the major diameter of internal threads.

Functional Indicating Thread Gage for inspecting the maximum-material GO functional limit and size and the NOT GO (HI) functional diameter limit and size.

Minimum-Material Indicating Thread Gage for inspecting the minimum-material limit and size.

Indicating Runout Thread Gage for inspecting runout of the minor diameter to pitch diameter.

In addition to these gages for product internal threads, the Standard also covers differential gaging and such instruments as pitch micrometers, thread-measuring balls, optical comparator and toolmaker's microscope, profile tracing instrument, surface roughness measuring instrument, and roundness measuring equipment.

The Standard includes the following gages for *Product External Thread:*

GO Working Thread Ring Gage for inspecting the maximum-material GO functional limit.

NOT GO (LO) Thread Ring Gage for inspecting the NOT GO (LO) functional diameter limit.

Thread Snap Gage—GO Segments or Rolls for inspecting the maximum-material GO functional limit.

Thread Snap Gage—NOT GO (LO) Segments or Rolls for inspecting the NOT GO (LO) functional diameter limit.

Thread Snap Gages—Cone and Vee Type and Minimum Material Thread Groove Diameter Type for inspecting the minimum-material pitch diameter limit.

Plain Ring and Snap Gages for checking the major diameter.

Snap Gage for checking the minor diameter.

Functional Indicating Thread Gage for inspecting the maximum-material GO functional limit and size and the NOT GO (LO) functional diameter limit and size.

Minimum-Material Indicating Thread Gage for inspecting the minimum-material limit and size.

Indicating Runout Gage for inspecting the runout of the major diameter to the pitch diameter.

W Tolerance Thread-Setting Plug Gage for setting adjustable thread ring gages, checking solid thread ring gages, setting thread snap limit gages, and setting indicating thread gages.

Plain Check Plug Gage for Thread Ring Gage for verifying the minor diameter limits of thread ring gages after the thread rings have been properly set with the applicable thread-setting plug gages.

Indicating Plain Diameter Gage for checking the major diameter.

Indicating Gage for checking the minor diameter.

In addition to these gages for product external threads, the Standard also covers differential gaging and such instruments as thread micrometers, thread-measuring wires, optical comparator and toolmaker's microscope, profile tracing instrument, electromechanical lead tester, helical path attachment used with GO type thread indicating gage, helical path analyzer, surface roughness measuring equipment, and roundness measuring equipment.

The standard lists the following for use of Threaded and Plain Gages for verification of product internal threads:

Tolerance: Unless otherwise specified all thread gages which directly check the product thread shall be X tolerance for all classes.

GO Thread Plug Gages: GO thread plug gages must enter and pass through the full threaded length of the product freely. The GO thread plug gage is a cumulative check of all thread elements except the minor diameter.

NOT GO (HI) Thread Plug Gages: NOT GO (HI) thread plug gages when applied to the product internal thread may engage only the end threads (which may not be representative of the complete thread). Entering threads on product are incomplete and permit gage to start. Starting threads on NOT GO (HI) plugs are subject to greater wear than the remaining threads. Such wear in combination with the incomplete product threads permits further entry of the gage. NOT GO (HI) functional diameter is acceptable when the NOT GO (HI) thread plug gage applied to the product internal thread does not enter more than three complete turns. The gage should not be forced. Special requirements such as exceptionally thin or ductile material, small number of threads, etc., may necessitate modification of this practice.

GO and NOT GO Plain Plug Gages for Minor Diameter of Product Internal Thread: (Recommended in Class Z tolerance.) GO plain plug gages must completely enter and pass through the length of the product without force. NOT GO cylindrical plug gage must not enter.

The standard lists the following for use of Thread Gages for verification of product external threads:

GO Thread Ring Gages: Adjustable GO thread ring gages must be set to the applicable W tolerance setting plugs to assure they are within specified limits. The product thread must freely enter the GO thread ring gage for the entire length of the threaded portion. The GO thread ring gage is a cumulative check of all thread elements except the major diameter.

NOT GO (LO) Thread Ring Gages: NOT GO (LO) thread ring gages must be set to the applicable W tolerance setting plugs to assure that they are within specified limits. NOT GO (LO) thread ring gages when applied to the product external thread may engage only the end threads (which may not be representative of the complete product thread)

Starting threads on NOT GO (LO) rings are subject to greater wear than the remaining threads. Such wear in combination with the incomplete threads at the end of the product thread permit further entry in the gage. NOT GO (LO) functional diameter is acceptable when the NOT GO (LO) thread ring gage applied to the product external thread does not pass over the thread more than three complete turns. The gage should not be forced. Special requirements such as exceptionally thin or ductile material, small number of threads, etc., may necessitate modification of this practice.

GO and NOT GO Plain Ring and Snap Gages for Checking Major Diameter of Product External Thread: The GO gage must completely receive or pass over the major diameter of the product external thread to ensure that the major diameter does not exceed the maximum-material-limit. The NOT GO gage must not pass over the major diameter of the product external thread to ensure that the major diameter is not less than the minimum-material-limit.

Limitations concerning the use of gages are given in the standard as follows:

Product threads accepted by a gage of one type may be verified by other types. It is possible, however, that parts which are near either rejection limit may be accepted by one type and rejected by another. Also, it is possible for two individual limit gages of the same type to be at the opposite extremes of the gage tolerances permitted, and borderline product threads accepted by one gage could be rejected by another. For these reasons, a product screw thread is considered acceptable when it passes a test by any of the permissible gages in ANSI B1.3 for the gaging system that are within the tolerances.

Gaging large product external and internal threads equal to above 6.25-inch nominal size with plain and threaded plug and ring gages presents problems for technical and economic reasons. In these instances, verification may be based on use of modified snap or indicating gages or measurement of thread elements. Various types of gages or measuring devices in addition to those defined in the Standard are available and acceptable when properly correlated to this Standard. Producer and user should agree on the method and equipment used.

Thread Forms of Gages.—Thread forms of gages for product internal and external threads are given in Table 1. The Standard ANSI/ASME B1.2-1983 (R1991) also gives illustrations of the thread forms of truncated thread setting plug gages, the thread forms of full-form thread setting plug gages, the thread forms of solid thread setting ring gages, and an illustration that shows the chip groove and removal of partial thread.

Thread Gage Tolerances.—Gage tolerances of thread plug and ring gages, thread setting plugs, and setting rings for Unified screw threads, designated as W and X tolerances, are given in Table . W tolerances represent the highest commercial grade of accuracy and workmanship, and are specified for thread setting gages; X tolerances are larger than W tolerances and are used for product inspection gages. Tolerances for plain gages are given in Table 3.

Determining Size of Gages: The three-wire method of determining pitch diameter size of plug gages is recommended for gages covered by American National Standard B1.2, described in Appendix B of the 1983 issue of that Standard.

Size limit adjustments of thread ring and external thread snap gages are determined by their fit on their respective calibrated setting plugs. Indicating gages and thread gages for product external threads are controlled by reference to appropriate calibrated setting plugs.

Size limit adjustments of internal thread snap gages are determined by their fit on their respective calibrated setting rings. Indicating gages and other adjustable thread gages for product internal threads are controlled by reference to appropriate calibrated setting rings or by direct measuring methods.

Interpretation of Tolerances: Tolerances on lead, half-angle, and pitch diameter are variations which may be taken independently for each of these elements and may be taken to the extent allowed by respective tabulated dimensional limits. The tabulated tolerance on any one element must not be exceeded, even though variations in the other two elements are smaller than the respective tabulated tolerances.

Direction of Tolerance on Gages: At the maximum-material limit (GO), the dimensions of all gages used for final conformance gaging are to be within limits of size of the product thread. At the functional diameter limit, using NOT GO (HI and LO) thread gages, the standard practice is to have the gage tolerance within the limits of size of the product thread.

Formulas for Limits of Gages: Formulas for limits of American National Standard Gages for Unified screw threads are given in Table 5. Some constants which are required to determine gage dimensions are tabulated in Table 4.

Table 1. Thread Forms of Gages for Product Internal and External Threads

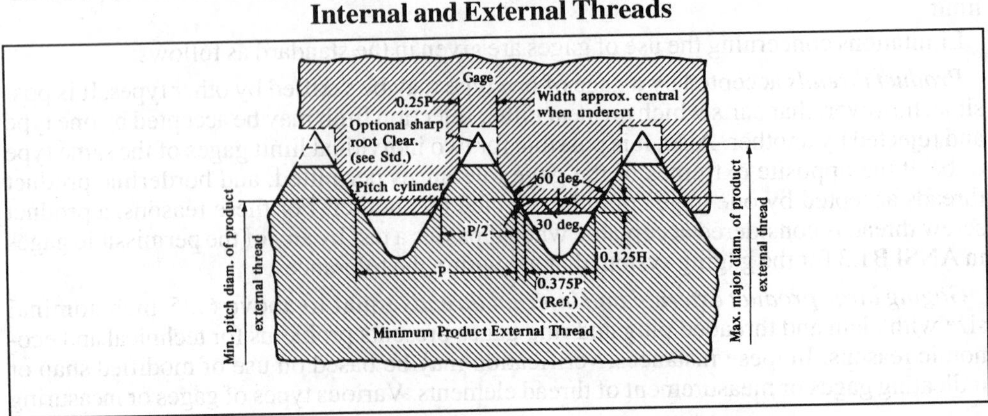

Table 1. *(Continued)* **Thread Forms of Gages for Product Internal and External Threads**

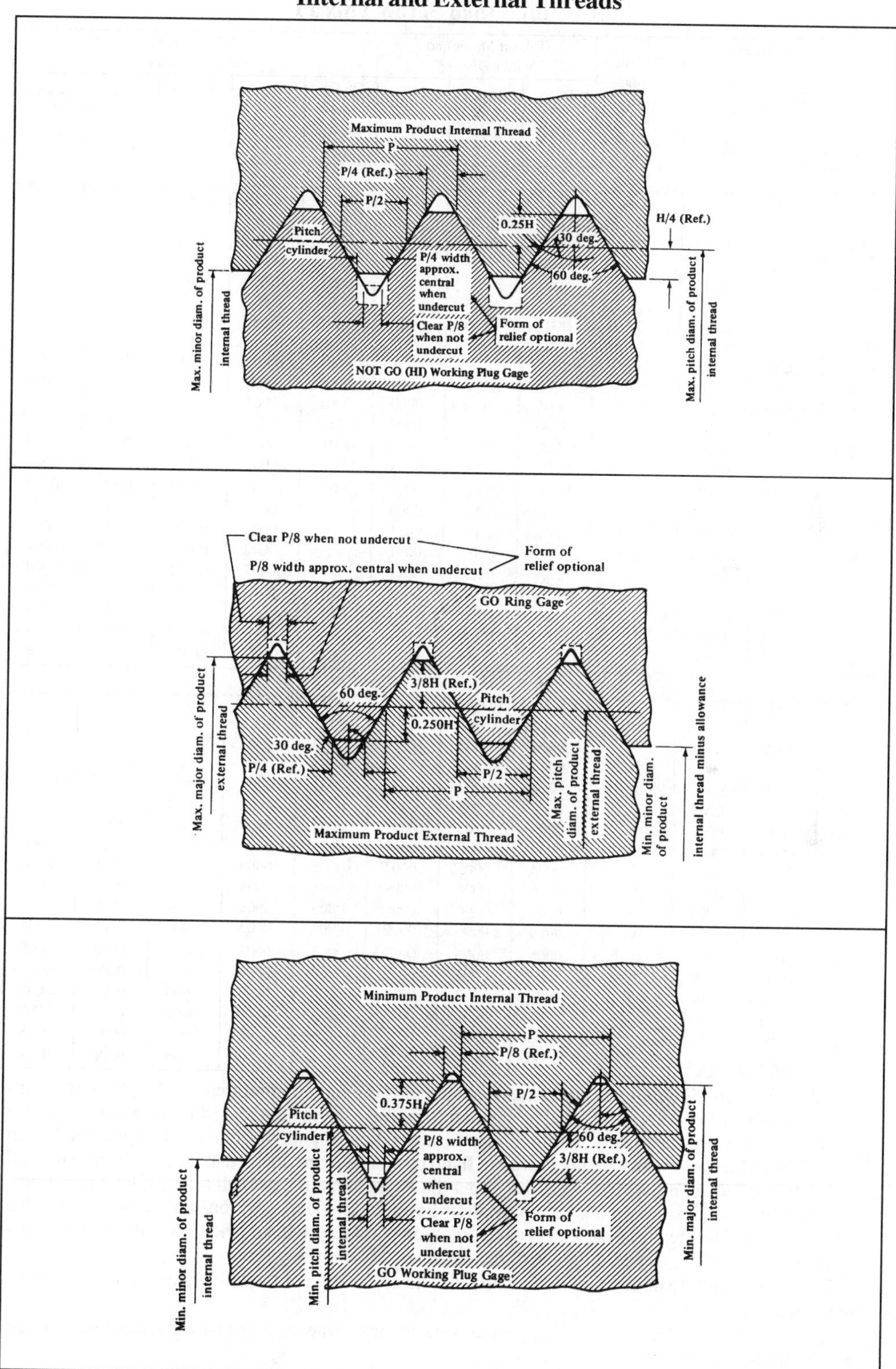

Table 2. American National Standard Tolerance for GO, HI, and LO Thread Gages for Unified Inch Screw Thread

Thds. per Inch	Tolerance on Lead[a] To & incl. ½ in. Dia.	Tolerance on Lead[a] Above ½ in. Dia.	Tol. on Thread Half-angle (±), minutes	Tol. on Major and Minor Diams.[b] To & incl. ½ in. Dia.	Tol. on Major and Minor Diams.[b] Above ½ to 4 in. Dia.	Tol. on Major and Minor Diams.[b] Above 4 in. Dia.	Tolerance on Pitch Diameter[b] To & incl. ½ in. Dia.	Tolerance on Pitch Diameter[b] Above ½ to 1½ in. Dia.	Tolerance on Pitch Diameter[b] Above 1½ to 4 in. Dia.	Tolerance on Pitch Diameter[b] Above 4 to 8 in. Dia.	Tolerance on Pitch Diameter[b] Above 8 to 12 in.[c] Dia.
W GAGES											
80, 72	.0001	.00015	20	.0003	.0003	…	.0001	.00015	…	…	…
64	.0001	.00015	20	.0003	.0004	…	.0001	.00015	…	…	…
56	.0001	.00015	20	.0003	.0004	…	.0001	.00015	.0002	…	…
48	.0001	.00015	18	.0003	.0004	…	.0001	.00015	.0002	…	…
44, 40	.0001	.00015	15	.0003	.0004	…	.0001	.00015	.0002	…	…
36	.0001	.00015	12	.0003	.0004	…	.0001	.00015	.0002	…	…
32	.0001	.00015	12	.0003	.0005	.0007	.0001	.00015	.0002	.00025	.0003
28, 27	.00015	.00015	8	.0005	.0005	.0007	.0001	.00015	.0002	.00025	.0003
24, 20	.00015	.00015	8	.0005	.0005	.0007	.0001	.00015	.0002	.00025	.0003
18	.00015	.00015	8	.0005	.0005	.0007	.0001	.00015	.0002	.00025	.0003
16	.00015	.00015	8	.0006	.0006	.0009	.0001	.0002	.00025	.0003	.0004
14, 13	.0002	.0002	6	.0006	.0006	.0009	.00015	.0002	.00025	.0003	.0004
12	.0002	.0002	6	.0006	.0006	.0009	.00015	.0002	.00025	.0003	.0004
11½	.0002	.0002	6	.0006	.0006	.0009	.00015	.0002	.00025	.0003	.0004
11	.0002	.0002	6	.0006	.0006	.0009	.00015	.0002	.00025	.0003	.0004
10	…	.00025	6	…	.0006	.0009	…	.0002	.0025	.0003	.0004
9	…	.00025	6	…	.0007	.0011	…	.0002	.00025	.0003	.0004
8	…	.00025	5	…	.0007	.0011	…	.0002	.00025	.0003	.0004
7	…	.0003	5	…	.0007	.0011	…	.0002	.00025	.0003	.0004
6	…	.0003	5	…	.0008	.0013	…	.0002	.00025	.0003	.0004
5	…	.0003	4	…	.0008	.0013	…	…	.00025	.0003	.0004
4½	…	.0003	4	…	.0008	.0013	…	…	.00025	.0003	.0004
4	…	.0003	4	…	.0009	.0015	…	…	.00025	.0003	.0004
X GAGES											
80, 72	.0002	.0002	30	.0003	.0003	…	.0002	.0002	…	…	…
64	.0002	.0002	30	.0004	.0004	…	.0002	.0002	…	…	…
56, 48	.0002	.0002	30	.0004	.0004	…	.0002	.0002	.0003	…	…
44, 40	.0002	.0002	20	.0004	.0004	…	.0002	.0002	.0003	…	…
36	.0002	.0002	20	.0004	.0004	…	.0002	.0002	.0003	…	…
32, 28	.0003	.0003	15	.0005	.0005	.0007	.0003	.0003	.0004	.0005	.0006
27, 24	.0003	.0003	15	.0005	.0005	.0007	.0003	.0003	.0004	.0005	.0006
20	.0003	.0003	15	.0005	.0005	.0007	.0003	.0003	.0004	.0005	.0006
18	.0003	.0003	10	.0005	.0005	.0007	.0003	.0003	.0004	.0005	.0006
16, 14	.0003	.0003	10	.0006	.0006	.0009	.0003	.0003	.0004	.0006	.0008
13, 12	.0003	.0003	10	.0006	.0006	.0009	.0003	.0003	.0004	.0006	.0008
11½	.0003	.0003	10	.0006	.0006	.0009	.0003	.0003	.0004	.0006	.0008
11, 10	.0003	.0003	10	.0006	.0006	.0009	.0003	.0003	.0004	.0006	.0008
9	.0003	.0003	10	.0007	.0007	.0011	.0003	.0003	.0004	.0006	.0008
8, 7	.0004	.0004	5	.0007	.0007	.0011	.0004	.0004	.0005	.0006	.0008
6	.0004	.0004	5	.0008	.0008	.0013	.0004	.0004	.0005	.0006	.0008
5, 4½	.0004	.0004	5	.0008	.0008	.0013	…	…	.0005	.0006	.0008
4	.0004	.0004	5	.0009	.0009	.0015	…	…	.0005	.0006	.0008

[a] Allowable variation in lead between any two threads not farther apart than the length of the standard gage as shown in ANSI B47.1. The tolerance on lead establishes the width of a zone, measured parallel to the axis of the thread, within which the actual helical path must lie for the specified length of the thread. Measurements are taken from a fixed reference point, located at the start of the first full thread, to a sufficient number of positions along the entire helix to detect all types of lead variations. The amounts that these positions vary from their basic (theoretical) positions are recorded with due respect to sign. The greatest variation in each direction (±) is selected, and the sum of their values, disregarding sign, must not exceed the tolerance limits specified for W gages.

[b] Tolerances apply to designated size of thread. The application of the tolerances is specified in the Standard.

[c] Above 12 in. the tolerance is directly proportional to the tolerance given in this column below, in the ratio of the diameter to 12 in.

All dimensions are given in inches unless otherwise specified.

Table 3. American National Standard Tolerances for Plain Cylindrical Gages *ANSI/ASME B1.2-1983 (R1991)*

Size Range		Tolerance Class[a]				
		XX	X	Y	Z	ZZ
Above	To and Including	Tolerance				
0.020	0.825	.00002	.00004	.00007	.00010	.00020
0.825	1.510	.00003	.00006	.00009	.00012	.00024
1.510	2.510	.00004	.00008	.00012	.00016	.00032
2.510	4.510	.00005	.00010	.00015	.00020	.00040
4.510	6.510	.000065	.00013	.00019	.00025	.00050
6.510	9.010	.00008	.00016	.00024	.00032	.00064
9.010	12.010	.00010	.00020	.00030	.00040	.00080

[a] Tolerances apply to actual diameter of plug or ring. Apply tolerances as specified in the Standard. Symbols XX, X, Y, Z, and ZZ are standard gage tolerance classes.

All dimensions are given in inches.

Table 4. Constants for Computing Thread Gage Dimensions *ANSI/ASME B1.2-1983 (R1991)*

Threads per Inch	Pitch, p	$0.060\sqrt[3]{p^2} + 0.017p$	$.05p$	$.087p$	Height of Sharp V-Thread, $H = .866025p$	$H/2 = .43301p$	$H/4 = .216506p$
80	.012500	.0034	.00063	.00109	.010825	.00541	.00271
72	.013889	.0037	.00069	.00122	.012028	.00601	.00301
64	.015625	.0040	.00078	.00136	.013532	.00677	.00338
56	.017857	.0044	.00089	.00155	.015465	.00773	.00387
48	.020833	.0049	.00104	.00181	.018042	.00902	.00451
44	.022727	.0052	.00114	.00198	.019682	.00984	.00492
40	.025000	.0056	.00125	.00218	.021651	.01083	.00541
36	.027778	.0060	.00139	.00242	.024056	.01203	.00601
32	.031250	.0065	.00156	.00272	.027063	.01353	.00677
28	.035714	.0071	.00179	.00311	.030929	.01546	.00773
27	.037037	.0073	.00185	.00322	.032075	.01604	.00802
24	.041667	.0079	.00208	.00361	.036084	.01804	.00902
20	.050000	.0090	.00250	.00435	.043301	.02165	.01083
18	.055556	.0097	.00278	.00483	.048113	.02406	.01203
16	.062500	.0105	.00313	.00544	0.54127	.02706	.01353
14	.071429	.0115	.00357	.00621	.061859	.03093	.01546
13	.076923	.0122	.00385	.00669	.066617	.03331	.01665
12	.083333	.0129	.00417	.00725	.072169	.03608	.01804
11½	.086957	.0133	.00435	.00757	.075307	.03765	.01883
11	.090909	.0137	.00451	.00791	.078730	.03936	.01968
10	.100000	.0146	.00500	.00870	.086603	.04330	.02165
9	.111111	.0158	.00556	.00967	.096225	.04811	.02406
8	.125000	.0171	.00625	.01088	.108253	.05413	.02706
7	.142857	.0188	.00714	.01243	.123718	.06186	.03093
6	.166667	.0210	.00833	.01450	.144338	.07217	.03608
5	.200000	.0239	.01000	.01740	.173205	.08660	.04330
4½	.222222	.0258	.01111	.01933	.192450	.09623	.04811
4	.250000	.0281	.01250	.02175	.216506	.10825	.05413

All dimensions are given in inches unless otherwise specified.

Table 5. Formulas for Limits of American National Standard Gages for Unified Inch Screw Threads *ANSI/ASME B1.2-1983 (R1991)*

No.	THREAD GAGES FOR EXTERNAL THREADS
1	GO Pitch Diameter = Maximum pitch diameter of external thread. Gage tolerance is *minus.*
2	GO Minor Diameter = Maximum pitch diameter of external thread minus $H/2$. Gage tolerance is *minus.*
3	NOT GO (LO) Pitch Diameter (for plus tolerance gage) = Minimum pitch diameter of external thread. Gage tolerance is *plus.*
4	NOT GO (LO) Minor Diameter = Minimum pitch diameter of external thread minus $H/4$. Gage tolerance is *plus.*
	PLAIN GAGES FOR MAJOR DIAMETER OF EXTERNAL THREADS
5	GO = Maximum major diameter of external thread. Gage tolerance is *minus.*
6	NOT GO = Minimum major diameter of external thread. Gage tolerance is *plus.*
	THREAD GAGES FOR INTERNAL THREADS
7	GO Major Diameter = Minimum major diameter of internal thread. Gage tolerance is *plus.*
8	GO Pitch Diameter = Minimum pitch diameter of internal thread. Gage tolerance is *plus.*
9	NOT GO (HI) Major Diameter = Maximum pitch diameter of internal thread plus $H/2$. Gage tolerance is *minus.*
10	NOT GO (HI) Pitch Diameter = Maximum pitch diameter of internal thread. Gage tolerance is *minus.*
	PLAIN GAGES FOR MINOR DIAMETER OF INTERNAL THREADS
11	GO = Minimum minor diameter of internal thread. Gage tolerance is *plus.*
12	NOT GO = Maximum minor diameter of internal thread. Gage tolerance is *minus.*
	FULL FORM AND TRUNCATED SETTING PLUGS
13	GO Major Diameter (Truncated Portion) = Maximum major diameter of external thread (= minimum major diameter of full portion of GO setting plug) minus $(0.060\sqrt[3]{p^2} + 0.017p)$. Gage tolerance is *minus.*
14	GO Major Diameter (Full Portion) = Maximum major diameter of external thread. Gage tolerance is *plus.*
15	GO Pitch Diameter = Maximum pitch diameter of external thread. Gage tolerance is *minus.*
16	[a]NOT GO (LO) Major Diameter (Truncated Portion) = Minimum pitch diameter of external thread plus $H/2$. Gage tolerance is *minus.*
17	NOT GO (LO) Major Diameter (Full Portion) = Maximum major diameter of external thread provided major diameter crest width shall not be less than 0.001 in. (0.0009 in. truncation). Apply W tolerance *plus* for maximum size except that for 0.001 in. crest width apply tolerance *minus*. For the 0.001 in. crest width, major diameter is equal to maximum major diameter of external thread plus 0.216506p minus the sum of external thread pitch diameter tolerance and 0.0017 in.
18	NOT GO (LO) Pitch Diameter = Minimum pitch diameter of external thread. Gage tolerance is *plus.*
	SOLID THREAD-SETTING RINGS FOR SNAP AND INDICATING GAGES
19	[b]GO Pitch Diameter = Minimum pitch diameter of internal thread. W gage tolerance is *plus.*
20	GO Minor Diameter = Minimum minor diameter of internal thread. W gage tolerance is *minus.*
21	[b]NOT GO (HI) Pitch Diameter = Maximum pitch diameter of internal thread. W gage tolerance is *minus.*
22	NOT GO (HI) Minor Diameter = Maximum minor diameter of internal thread. W gage tolerance is *minus.*

[a] Truncated portion is required when optional sharp root profile is used.

[b] Tolerances greater than W tolerance for pitch diameter are acceptable when internal indicating or snap gage can accommodate a greater tolerance and when agreed upon by supplier and user.

See data in Screw Thread Systems section for symbols and dimensions of Unified Screw Threads.

TAPPING AND THREAD CUTTING

Selection of Taps.—For most applications, a standard tap supplied by the manufacturer can be used, but some jobs may require special taps. A variety of standard taps can be obtained. In addition to specifying the size of the tap it is necessary to be able to select the one most suitable for the application at hand.

The elements of standard taps that are varied are: the number of flutes; the type of flute, whether straight, spiral pointed, or spiral fluted; the chamfer length; the relief of the land, if any; the tool steel used to make the tap; and the surface treatment of the tap.

Details regarding the nomenclature of tap elements are given in the section *TAPS AND THREADING DIES* starting on page 872, along with a listing of the standard sizes available.

Factors to consider in selecting a tap include: the method of tapping, by hand or by machine; the material to be tapped and its heat treatment; the length of thread, or depth of the tapped hole; the required tolerance or class of fit; and the production requirement and the type of machine to be used.

The diameter of the hole must also be considered, although this action is usually only a matter of design and the specification of the tap drill size.

Method of Tapping: The term *hand tap* is used for both hand and machine taps, and almost all taps can be applied by the hand or machine method. While any tap can be used for hand tapping, those having a concentric land without the relief are preferable. In hand tapping the tool is reversed periodically to break the chip, and the heel of the land of a tap with a concentric land (without relief) will cut the chip off cleanly or any portion of it that is attached to the work, whereas a tap with an eccentric or con-eccentric relief may leave a small burr that becomes wedged between the relieved portion of the land and the work. This wedging creates a pressure towards the cutting face of the tap that may cause it to chip; it tends to roughen the threads in the hole, and it increases the overall torque required to turn the tool. When tapping by machine, however, the tap is usually turned only in one direction until the operation is complete, and an eccentric or con-eccentric relief is often an advantage.

Chamfer Length: Three types of hand taps, used both for hand and machine tapping, are available, and they are distinguished from each other by the length of chamfer. *Taper taps* have a chamfer angle that reduces the height about 8–10 teeth; *plug taps* have a chamfer angle with 3–5 threads reduced in height; and *bottoming taps* have a chamfer angle with 1½ threads reduced in height. Since the teeth that are reduced in height do practically all the cutting, the chip load or chip thickness per tooth will be least for a taper tap, greater for a plug tap, and greatest for a bottoming tap.

For most through hole tapping applications it is necessary to use only a plug type tap, which is also most suitable for blind holes where the tap drill hole is deeper than the required thread. If the tap must bottom in a blind hole, the hole is usually threaded first with a plug tap and then finished with a bottoming tap to catch the last threads in the bottom of the hole. Taper taps are used on materials where the chip load per tooth must be kept to a minimum. However, taper taps should not be used on materials that have a strong tendency to work harden, such as the austenitic stainless steels.

Spiral Point Taps: Spiral point taps offer a special advantage when machine tapping through holes in ductile materials because they are designed to handle the long continuous chips that form and would otherwise cause a disposal problem. An angular gash is ground at the point or end of the tap along the face of the chamfered threads or lead teeth of the tap. This gash forms a left-hand helix in the flutes adjacent to the lead teeth which causes the chips to flow ahead of the tap and through the hole. The gash is usually formed to produce a rake angle on the cutting face that increases progressively toward the end of the tool. Since the flutes are used primarily to provide a passage for the cutting fluid, they are usually made narrower and shallower thereby strengthening the tool. For tapping thin workpieces short fluted spiral point taps are recommended. They have a spiral point gash along the cutting teeth; the remainder of the threaded portion of the tap has no flute. Most spiral pointed taps are of plug type; however, spiral point bottoming taps are also made.

Spiral Fluted Taps: Spiral fluted taps have a helical flute; the helix angle of the flute may be between 15 and 52 degrees and the hand of the helix is the same as that of the threads on the tap. The spiral flute and the rake that it forms on the cutting face of the tap combine to induce the chips to flow backward along the helix and out of the hole. Thus, they are ideally suited for tapping blind holes and they are available as plug and bottoming types. A higher spiral angle should be specified for tapping very ductile materials; when tapping harder materials, chipping at the cutting edge may result and the spiral angle must be reduced.

Holes having a pronounced interruption such as a groove or a keyway can be tapped with spiral fluted taps. The land bridges the interruption and allows the tap to cut relatively smoothly.

Serial Taps and Close Tolerance Threads: For tapping holes to close tolerances a set of serial taps is used.

They are usually available in sets of three: the No. 1 tap is undersize and is the first rougher; the No. 2 tap is of intermediate size and is the second rougher; and the No. 3 tap is used for finishing.

The different taps are identified by one, two, and three annular grooves in the shank adjacent to the square. For some applications involving finer pitches only two serial taps are required. Sets are also used to tap hard or tough materials having a high tensile strength, deep blind holes in normal materials, and large coarse threads. A set of more than three taps is sometimes required to produce threads of coarse pitch. Threads to some commercial tolerances, such as American Standard Unified 2B, or ISO Metric 6H, can be produced in one cut using a ground tap; sometimes even closer tolerances can be produced with a single tap. Ground taps are recommended for all close tolerance tapping operations. For much ordinary work, cut taps are satisfactory and more economical than ground taps.

Tap Steels: Most taps are made from high speed steel. The type of tool steel used is determined by the tap manufacturer and is usually satisfactory when correctly applied except in a few exceptional cases. Typical grades of high speed steel used to make taps are M-1, M-2, M-3, M-42, etc. Carbon tool steel taps are satisfactory where the operating temperature of the tap is low and where a high resistance to abrasion is not required as in some types of hand tapping.

Surface Treatment: The life of high speed steel taps can sometimes be increased significantly by treating the surface of the tap. A very common treatment is oxide coating, which forms a thin metallic oxide coating on the tap that has lubricity and is somewhat porous to absorb and retain oil. This coating reduces the friction between the tap and the work and it makes the surface virtually impervious to rust. It does not increase the hardness of the surface but it significantly reduces or prevents entirely galling, or the tendency of the work material to weld or stick to the cutting edge and to other areas on the tap with which it is in contact. For this reason oxide coated taps are recommended for metals that tend to gall and stick such as non-free cutting low carbon steels and soft copper. It is also useful for tapping other steels having higher strength properties.

Nitriding provides a very hard and wear resistant case on high speed steel. Nitrided taps are especially recommended for tapping plastics; they have also been used successfully on a variety of other materials including high strength high alloy steels. However, some caution must be used in specifying nitrided taps because the nitride case is very brittle and may have a tendency to chip.

Chrome plating has been used to increase the wear resistance of taps but its application has been limited because of the high cost and the danger of hydrogen embrittlement which can cause cracks to form in the tool. A flash plate of about .0001 in. or less in thickness is applied to the tap. Chrome-plated taps have been used successfully to tap a variety of ferrous and nonferrous materials including plastics, hard rubber, mild steel, and tool steel. Other surface treatments that have been used successfully to a limited extent are vapor blasting and liquid honing.

Rake Angle: For the majority of applications in both ferrous and nonferrous materials the rake angle machined on the tap by the manufacturer is satisfactory. This angle is approximately 5 to 7 degrees. In some instances it may be desirable to alter the rake angle of the tap to obtain beneficial results and Table 1 provides a guide that can be used. In selecting a rake angle from this table, consideration must be given to the size of the tap and the strength of the land. Most standard taps are made with a curved face with the rake angle measured as a chord between the crest and root of the thread. The resulting shape is called a hook angle.

Table 1. Tap Rake Angles for Tapping Different Materials

Material	Rake Angle, Degrees	Material	Rake Angle, Degrees
Cast Iron	0–3	Aluminum	8–20
Malleable Iron	5–8	Brass	2–7
Steel		Naval Brass	5–8
AISI 1100 Series	5–12	Phosphor Bronze	5–12
Low Carbon (up to .25 per cent)	5–12	Tobin Bronze	5–8
		Manganese Bronze	5–12
Medium Carbon, Annealed (.30 to .60 per cent)	5–10	Magnesium	10–20
		Monel	9–12
Heat Treated, 225–283 Brinell. (.30 to .60 per cent)	0–8	Copper	10–18
		Zinc Die Castings	10–15
High Carbon and High Speed	0–5	Plastic	
		Thermoplastic	5–8
Stainless	8–15	Thermosetting	0–3
Titanium	5–10	Hard Rubber	0–3

Cutting Speed.—The cutting speed for machine tapping is treated in detail on page 1046. It suffices to say here that many variables must be considered in selecting this cutting speed and any tabulation may have to be modified greatly. Where cutting speeds are mentioned in the following section, they are intended only to provide a guideline to show the possible range of speeds that could be used.

Tapping Specific Materials.—The work material has a great influence on the ease with which a hole can be tapped. For production work, in many instances, modified taps are recommended; however, for toolroom or short batch work, standard hand taps can be used on most jobs, providing reasonable care is taken when tapping. The following concerns the tapping of metallic materials; information on the tapping of plastics is given on page 599.

Low Carbon Steel (Less than 0.15% C): These steels are very soft and ductile resulting in a tendency for the work material to tear and to weld to the tap. They produce a continuous chip that is difficult to break and spiral pointed taps are recommended for tapping through holes; for blind holes a spiral fluted tap is recommended. To prevent galling and welding, a liberal application of a sulfur base or other suitable cutting fluid is essential and the selection of an oxide coated tap is very helpful.

Low Carbon Steels (0.15 to 0.30% C): The additional carbon in these steels is beneficial as it reduces the tendency to tear and to weld; their machinability is further improved by cold drawing. These steels present no serious problems in tapping provided a suitable cutting fluid is used. An oxide coated tap is recommended, particularly in the lower carbon range.

Medium Carbon Steels (0.30 to 0.60% C): These steels can be tapped without too much difficulty, although a lower cutting speed must be used in machine tapping. The cutting speed is dependent on the carbon content and the heat treatment. Steels that have a higher carbon content must be tapped more slowly, especially if the heat treatment has produced a pearlitic microstructure. The cutting speed and ease of tapping is significantly improved by heat treating to produce a spheroidized microstructure. A suitable cutting fluid must be used.

High Carbon Steels (More than 0.6% C): Usually these materials are tapped in the annealed or normalized condition although sometimes tapping is done after hardening and tempering to a hardness below 55 Rc. Recommendations for tapping after hardening and tempering are given under High Tensile Strength Steels. In the annealed and normalized condition these steels have a higher strength and are more abrasive than steels with a lower carbon content; thus, they are more difficult to tap. The microstructure resulting from the heat treatment has a significant effect on the ease of tapping and the tap life, a spheroidite structure being better in this respect than a pearlitic structure. The rake angle of the tap should not exceed 5 degrees and for the harder materials a concentric tap is recommended. The cutting speed is considerably lower for these steels and an activated sulfur-chlorinated cutting fluid is recommended.

Alloy Steels: This classification includes a wide variety of steels, each of which may be heat treated to have a wide range of properties. When annealed and normalized they are similar to medium to high carbon steels and usually can be tapped without difficulty, although for some alloy steels a lower tapping speed may be required. Standard taps can be used and for machine tapping a con-eccentric relief may be helpful. A suitable cutting fluid must be used.

High-Tensile Strength Steels: Any steel that must be tapped after being heat treated to a hardness range of 40–55 Rc is included in this classification. Low tap life and excessive tap breakage are characteristics of tapping these materials; those that have a high chromium content are particularly troublesome. Best results are obtained with taps that have concentric lands, a rake angle that is at or near zero degrees, and 6 to 8 chamfered threads on the end to reduce the chip load per tooth. The chamfer relief should be kept to a minimum. The load on the tap should be kept to a minimum by every possible means, including using the largest possible tap drill size; keeping the hole depth to a minimum; avoidance of bottoming holes; and, in the larger sizes, using fine instead of coarse pitches. Oxide coated taps are recommended although a nitrided tap can sometimes be used to reduce tap wear. An active sulfur-chlorinated oil is recommended as a cutting fluid and the tapping speed should not exceed about 10 feet per minute.

Stainless Steels: Ferritic and martensitic type stainless steels are somewhat like alloy steels that have a high chromium content, and they can be tapped in a similar manner, although a slightly slower cutting speed may have to be used. Standard rake angle oxide coated taps are recommended and a cutting fluid containing molybdenum disulphide is helpful to reduce the friction in tapping. Austenitic stainless steels are very difficult to tap because of their high resistance to cutting and their great tendency to work harden. A work-hardened layer is formed by a cutting edge of the tap and the depth of this layer depends on the severity of the cut and the sharpness of the tool. The next cutting edge must penetrate below the work-hardened layer, if it is to be able to cut. Therefore, the tap must be kept sharp and each succeeding cutting edge on the tool must penetrate below the work-hardened layer formed by the preceding cutting edge. For this reason, a taper tap should not be used, but rather a plug tap having 3–5 chamfered threads. To reduce the rubbing of the lands, an eccentric or con-eccentric relieved land should be used and a 10–15 degree rake angle is recommended. A tough continuous chip is formed that is difficult to break. To control this chip, spiral pointed taps are recommended for through holes and low-helix angle spiral fluted taps for blind holes. An oxide coating on the tap is very helpful and a sulfur-chlorinated mineral lard oil is recommended, although heavy duty soluble oils have also been used successfully.

Free Cutting Steels: There are large numbers of free cutting steels, including free cutting stainless steels, which are also called free machining steels. Sulfur, lead, or phosphorus are added to these steels to improve their machinability. Free machining steels are always easier to tap than their counterparts that do not have the free machining additives. Tool life is usually increased and a somewhat higher cutting speed can be used. The type of tap recommended depends on the particular type of free machining steel and the nature of the tapping operation; usually a standard tap can be used.

High Temperature Alloys: These are cobalt or nickel base nonferrous alloys that cut like austenitic stainless steel, but are often even more difficult to machine. The recommendations given for austenitic stainless steel also apply to tapping these alloys but the rake angle should be 0 to 10 degrees to strengthen the cutting edge. For most applications a nitrided tap or one made from M41, M42, M43, or M44 steel is recommended. The tapping speed is usually in the range of 5 to 10 feet per minute.

Titanium and Titanium Alloys: Titanium and its alloys have a low specific heat and a pronounced tendency to weld on to the tool material; therefore, oxide coated taps are recommended to minimize galling and welding. The rake angle of the tap should be from 6 to 10 degrees. To minimize the contact between the work and the tap an eccentric or con-eccentric relief land should be used. Taps having interrupted threads are sometimes helpful. Pure titanium is comparatively easy to tap but the alloys are very difficult. The cutting speed depends on the composition of the alloy and may vary from 40 to 10 feet per minute. Special cutting oils are recommended for tapping titanium.

Gray Cast Iron: The microstructure of gray cast iron can vary, even within a single casting, and compositions are used that vary in tensile strength from about 20,000 to 60,000 psi (160 to 250 Bhn). Thus, cast iron is not a single material, although in general it is not difficult to tap. The cutting speed may vary from 90 feet per minute for the softer grades to 30 feet per minute for the harder grades. The chip is discontinuous and straight fluted taps should be used for all applications. Oxide coated taps are helpful and gray cast iron can usually be tapped dry, although water soluble oils and chemical emulsions are sometimes used.

Malleable Cast Iron: Commercial malleable cast irons are also available having a rather wide range of properties, although within a single casting they tend to be quite uniform. They are relatively easy to tap and standard taps can be used. The cutting speed for ferritic cast irons is 60–90 feet per minute, for pearlitic malleable irons 40–50 feet per minute, and for martensitic malleable irons 30–35 feet per minute. A soluble oil cutting fluid is recommended except for martensitic malleable iron where a sulfur base oil may work better.

Ductile or Nodular Cast Iron: Several classes of nodular iron are used having a tensile strength varying from 60,000 to 120,000 psi. Moreover, the microstructure in a single casting and in castings produced at different times vary rather widely. The chips are easily controlled but have some tendency to weld to the faces and flanks of cutting tools. For this reason oxide coated taps are recommended. The cutting speed may vary from 15 fpm for the harder martensitic ductile irons to 60 fpm for the softer ferritic grades. A suitable cutting fluid should be used.

Aluminum: Aluminum and aluminum alloys are relatively soft materials that have little resistance to cutting. The danger in tapping these alloys is that the tap will ream the hole instead of cutting threads, or that it will cut a thread eccentric to the hole. For these reasons, extra care must be taken when aligning the tap and starting the thread. For production tapping a spiral pointed tap is recommended for through holes and a spiral fluted tap for blind holes; preferably these taps should have a 10 to 15 degree rake angle. A lead screw tapping machine is helpful in cutting accurate threads. A heavy duty soluble oil or a light base mineral oil should be used as a cutting fluid.

Copper Alloys: Most copper alloys are not difficult to tap, except beryllium copper and a few other hard alloys. Pure copper offers some difficulty because of its ductility and the ductile continuous chip formed, which can be difficult to control. However, with reasonable care and the use of medium heavy duty mineral lard oil it can be tapped successfully. Red brass, yellow brass, and similar alloys containing not more than 35 per cent zinc produce a continuous chip. While straight fluted taps can be used for hand tapping these alloys, machine tapping should be done with spiral pointed or spiral fluted taps for through and blind holes respectively. Naval brass, leaded brass, and cast brasses produce a discontinuous chip and a straight fluted tap can be used for machine tapping. These alloys exhibit a tendency to close in on the tap and sometimes an interrupted thread tap is used to reduce the resulting jamming effect. Beryllium copper and the silicon bronzes are the strongest of the copper alloys. Their strength combined with their ability to work harden can cause difficulties in tapping. For these alloys plug type taps should be used and the taps should be kept as sharp as possible. A medium or heavy duty water soluble oil is recommended as a cutting fluid.

Diameter of Tap Drill.—Tapping troubles are sometimes caused by tap drills that are too small in diameter. The tap drill should not be smaller than is necessary to give the required strength to the thread as even a very small decrease in the diameter of the drill will increase the torque required and the possibility of broken taps. Tests have shown that any increase in the percentage of full thread over 60 per cent does not significantly increase the strength of the thread. Often, a 55 to 60 per cent thread is satisfactory, although 75 per cent threads are commonly used to provide an extra measure of safety. The present thread specifications do not always allow the use of the smaller thread depths. However, the specification given on a part drawing must be adhered to and may require smaller minor diameters than might otherwise be recommended.

The depth of the thread in the tapped hole is dependent on the length of thread engagement and on the material. In general, when the engagement length is more than one and one-half times the nominal diameter a 50 or 55 per cent thread is satisfactory. Soft ductile materials may permit use of a slightly larger tapping hole than brittle materials such as gray cast iron.

It must be remembered that a twist drill is a roughing tool that may be expected to drill slightly oversize and that some variations in the size of the tapping holes are almost inevitable. When a closer control of the hole size is required it must be reamed. Reaming is recommended for the larger thread diameters and for some fine pitch threads.

For threads of Unified form (see *American National and Unified Screw Thread Forms* on page 1706) the selection of tap drills is covered in the following section, Factors Influencing Minor Diameter Tolerances of Tapped Holes and the hole size limits are given in Table 2. Tables 3 and 4 give tap drill sizes for American National Form threads based on 75 per cent of full thread depth. For smaller-size threads the use of slightly larger drills, if permissible, will reduce tap breakage. The selection of tap drills for these threads also may be based on the hole size limits given in Table 2 for Unified threads that take lengths of engagement into account.

Table 2. Recommended Hole Size Limits Before Tapping Unified Threads

Thread Size	Classes 1B and 2B								Class 3B							
	Length of Engagement (D = Nominal Size of Thread)															
	To and Including ⅓D		Above ⅓D to ⅔D		Above ⅔D to 1½D		Above 1½D to 3D		To and Including ⅓D		Above ⅓D to ⅔D		Above ⅔D to 1½D		Above 1½D to 3D	
	Recommended Hole Size Limits															
	Min[a]	Max	Min	Max	Min	Max[b]	Min	Max	Min[a]	Max	Min	Max	Min	Max[b]	Min	Max
0–80	0.0465	0.0500	0.0479	0.0514	0.0479	0.0514	0.0479	0.0514	0.0465	0.0500	0.0479	0.0514	0.0479	0.0514	0.0479	0.0514
1–64	0.0561	0.0599	0.0585	0.0623	0.0585	0.0623	0.0585	0.0623	0.0561	0.0599	0.0585	0.0623	0.0585	0.0623	0.0585	0.0623
1–72	0.0580	0.0613	0.0596	0.0629	0.0602	0.0635	0.0602	0.0635	0.0580	0.0613	0.0596	0.0629	0.0602	0.0635	0.0602	0.0635
2–56	0.0667	0.0705	0.0686	0.0724	0.0699	0.0737	0.0699	0.0737	0.0667	0.0705	0.0686	0.0724	0.0699	0.0737	0.0699	0.0737
2–64	0.0691	0.0724	0.0707	0.0740	0.0720	0.0753	0.0720	0.0753	0.0691	0.0724	0.0707	0.0740	0.0720	0.0753	0.0720	0.0753
3–48	0.0764	0.0804	0.0785	0.0825	0.0805	0.0845	0.0806	0.0846	0.0764	0.0804	0.0785	0.0825	0.0805	0.0845	0.0806	0.0846
3–56	0.0797	0.0831	0.0814	0.0848	0.0831	0.0865	0.0833	0.0867	0.0797	0.0831	0.0814	0.0848	0.0831	0.0865	0.0833	0.0867
4–40	0.0849	0.0894	0.0871	0.0916	0.0894	0.0939	0.0902	0.0947	0.0849	0.0894	0.0871	0.0916	0.0894	0.0939	0.0902	0.0947
4–48	0.0894	0.0931	0.0912	0.0949	0.0931	0.0968	0.0939	0.0976	0.0894	0.0931	0.0912	0.0949	0.0931	0.0968	0.0939	0.0976
5–40	0.0979	0.1020	0.1000	0.1041	0.1021	0.1062	0.1036	0.1077	0.0979	0.1020	0.1000	0.1041	0.1021	0.1062	0.1036	0.1077
5–44	0.1004	0.1042	0.1023	0.1060	0.1042	0.1079	0.1060	0.1097	0.1004	0.1042	0.1023	0.1060	0.1042	0.1079	0.1060	0.1097
6–32	0.104	0.109	0.106	0.112	0.109	0.114	0.112	0.117	0.1040	0.1091	0.1066	0.1115	0.1091	0.1140	0.1115	0.1164
6–40	0.111	0.115	0.113	0.117	0.115	0.119	0.117	0.121	0.1110	0.1148	0.1128	0.1167	0.1147	0.1186	0.1166	0.1205
8–32	0.130	0.134	0.132	0.137	0.134	0.139	0.137	0.141	0.1300	0.1345	0.1324	0.1367	0.1346	0.1389	0.1367	0.1410
8–36	0.134	0.138	0.136	0.140	0.138	0.142	0.140	0.144	0.1340	0.1377	0.1359	0.1397	0.1378	0.1416	0.1397	0.1435
10–24	0.145	0.150	0.148	0.154	0.150	0.156	0.152	0.159	0.1450	0.1502	0.1475	0.1528	0.1502	0.1555	0.1528	0.1581
10–32	0.156	0.160	0.158	0.162	0.160	0.164	0.162	0.166	0.1560	0.1601	0.1581	0.1621	0.1601	0.1641	0.1621	0.1661
12–24	0.171	0.176	0.174	0.179	0.176	0.181	0.178	0.184	0.1710	0.1758	0.1733	0.1782	0.1758	0.1807	0.1782	0.1831
12–28	0.177	0.182	0.179	0.184	0.182	0.186	0.184	0.188	0.1770	0.1815	0.1794	0.1836	0.1815	0.1857	0.1836	0.1878
12–32	0.182	0.186	0.184	0.188	0.186	0.190	0.188	0.192	0.1820	0.1858	0.1837	0.1877	0.1855	0.1895	0.1873	0.1913
¼–20	0.196	0.202	0.199	0.204	0.202	0.207	0.204	0.210	0.1960	0.2013	0.1986	0.2040	0.2013	0.2067	0.2040	0.2094
¼–28	0.211	0.216	0.213	0.218	0.216	0.220	0.218	0.222	0.2110	0.2152	0.2131	0.2171	0.2150	0.2190	0.2169	0.2209
¼–32	0.216	0.220	0.218	0.222	0.220	0.224	0.222	0.226	0.2160	0.2196	0.2172	0.2212	0.2189	0.2229	0.2206	0.2246
¼–36	0.220	0.224	0.221	0.225	0.224	0.226	0.225	0.228	0.2200	0.2243	0.2199	0.2243	0.2214	0.2258	0.2229	0.2273
5⁄16–18	0.252	0.259	0.255	0.262	0.259	0.265	0.262	0.268	0.2520	0.2577	0.2551	0.2604	0.2577	0.2630	0.2604	0.2657
5⁄16–24	0.267	0.272	0.270	0.275	0.272	0.277	0.275	0.280	0.2670	0.2714	0.2694	0.2734	0.2714	0.2754	0.2734	0.2774
5⁄16–32	0.279	0.283	0.281	0.285	0.283	0.286	0.285	0.289	0.2790	0.2817	0.2792	0.2832	0.2807	0.2847	0.2822	0.2862
5⁄16–36	0.282	0.286	0.284	0.288	0.285	0.289	0.287	0.291	0.2820	0.2863	0.2824	0.2863	0.2837	0.2877	0.2850	0.2890
⅜–16	0.307	0.314	0.311	0.318	0.314	0.321	0.318	0.325	0.3070	0.3127	0.3101	0.3155	0.3128	0.3182	0.3155	0.3209

Table 2. *(Continued)* **Recommended Hole Size Limits Before Tapping Unified Threads**

Thread Size	Classes 1B and 2B								Class 3B							
	Length of Engagement (D = Nominal Size of Thread)															
	To and Including ⅓D		Above ⅓D to ⅔D		Above ⅔D to 1½D		Above 1½D to 3D		To and Including ⅓D		Above ⅓D to ⅔D		Above ⅔D to 1½D		Above 1½D to 3D	
	Recommended Hole Size Limits															
	Min[a]	Max	Min	Max	Min	Max[b]	Min	Max	Min[a]	Max	Min	Max	Min	Max[b]	Min	Max
⅜–24	0.330	0.335	0.333	0.338	0.335	0.340	0.338	0.343	0.3300	0.3336	0.3314	0.3354	0.3332	0.3372	0.3351	0.3391
⅜–32	0.341	0.345	0.343	0.347	0.345	0.349	0.347	0.351	0.3410	0.3441	0.3415	0.3455	0.3429	0.3469	0.3444	0.3484
⅜–36	0.345	0.349	0.346	0.350	0.347	0.352	0.349	0.353	0.3450	0.3488	0.3449	0.3488	0.3461	0.3501	0.3474	0.3514
7/16–14	0.360	0.368	0.364	0.372	0.368	0.376	0.372	0.380	0.3600	0.3660	0.3630	0.3688	0.3659	0.3717	0.3688	0.3746
7/16–20	0.383	0.389	0.386	0.391	0.389	0.395	0.391	0.397	0.3830	0.3875	0.3855	0.3896	0.3875	0.3916	0.3896	0.3937
7/16–28	0.399	0.403	0.401	0.406	0.403	0.407	0.406	0.410	0.3990	0.4020	0.3995	0.4035	0.4011	0.4051	0.4017	0.4067
½–13	0.417	0.426	0.421	0.430	0.426	0.434	0.430	0.438	0.4170	0.4225	0.4196	0.4254	0.4226	0.4284	0.4255	0.4313
½–12	0.410	0.414	0.414	0.424	0.414	0.428	0.424	0.433	0.4100	0.4161	0.4129	0.4192	0.4160	0.4223	0.4192	0.4255
½–20	0.446	0.452	0.449	0.454	0.452	0.457	0.454	0.460	0.4460	0.4498	0.4477	0.4517	0.4497	0.4537	0.4516	0.4556
½–28	0.461	0.467	0.463	0.468	0.466	0.470	0.468	0.472	0.4610	0.4645	0.4620	0.4660	0.4636	0.4676	0.4652	0.4692
9/16–12	0.472	0.476	0.476	0.486	0.476	0.490	0.486	0.495	0.4720	0.4783	0.4753	0.4813	0.4783	0.4843	0.4813	0.4873
9/16–18	0.502	0.509	0.505	0.512	0.509	0.515	0.512	0.518	0.5020	0.5065	0.5045	0.5086	0.5065	0.5106	0.5086	0.5127
9/16–24	0.517	0.522	0.520	0.525	0.522	0.527	0.525	0.530	0.5170	0.5209	0.5186	0.5226	0.5204	0.5244	0.5221	0.5261
9/16–28	0.524	0.528	0.526	0.531	0.528	0.532	0.531	0.535	0.5240	0.5270	0.5245	0.5285	0.5261	0.5301	0.5277	0.5317
⅝–11	0.527	0.536	0.532	0.541	0.536	0.546	0.541	0.551	0.5270	0.5328	0.5298	0.5360	0.5329	0.5391	0.5360	0.5422
⅝–12	0.535	0.544	0.540	0.549	0.544	0.553	0.549	0.558	0.5350	0.5406	0.5377	0.5435	0.5405	0.5463	0.5434	0.5492
⅝–18	0.565	0.572	0.568	0.575	0.572	0.578	0.575	0.581	0.5650	0.5690	0.5670	0.5711	0.5690	0.5730	0.5711	0.5752
⅝–24	0.580	0.585	0.583	0.588	0.585	0.590	0.588	0.593	0.5800	0.5834	0.5811	0.5851	0.5829	0.5869	0.5846	0.5886
⅝–28	0.586	0.591	0.588	0.593	0.591	0.595	0.593	0.597	0.5860	0.5895	0.5870	0.5910	0.5886	0.5926	0.5902	0.5942
11/16–12	0.597	0.606	0.602	0.611	0.606	0.615	0.611	0.620	0.5970	0.6029	0.6001	0.6057	0.6029	0.6085	0.6057	0.6113
11/16–24	0.642	0.647	0.645	0.650	0.647	0.652	0.650	0.655	0.6420	0.6459	0.6436	0.6476	0.6454	0.6494	0.6471	0.6511
¾–10	0.642	0.653	0.647	0.658	0.653	0.663	0.658	0.668	0.6420	0.6481	0.6449	0.6513	0.6481	0.6545	0.6513	0.6577
¾–12	0.660	0.669	0.665	0.674	0.669	0.678	0.674	0.683	0.6600	0.6652	0.6626	0.6680	0.6653	0.6707	0.6680	0.6734
¾–16	0.682	0.689	0.686	0.693	0.689	0.696	0.693	0.700	0.6820	0.6866	0.6844	0.6887	0.6865	0.6908	0.6886	0.6929
¾–20	0.696	0.702	0.699	0.704	0.702	0.707	0.704	0.710	0.6960	0.6998	0.6977	0.7017	0.6997	0.7037	0.7016	0.7056
¾–28	0.711	0.716	0.713	0.718	0.716	0.720	0.718	0.722	0.7110	0.7145	0.7120	0.7160	0.7136	0.7176	0.7152	0.7192
13/16–12	0.722	0.731	0.727	0.736	0.731	0.740	0.736	0.745	0.7220	0.7276	0.7250	0.7303	0.7276	0.7329	0.7303	0.7356

Table 2. *(Continued)* **Recommended Hole Size Limits Before Tapping Unified Threads**

Thread Size	Classes 1B and 2B								Class 3B							
	Length of Engagement (D = Nominal Size of Thread)															
	To and Including ⅓D		Above ⅓D to ⅔D		Above ⅔D to 1½D		Above 1½D to 3D		To and Including ⅓D		Above ⅓D to ⅔D		Above ⅔D to 1½D		Above 1½D to 3D	
	Recommended Hole Size Limits															
	Min[a]	Max	Min	Max	Min	Max[b]	Min	Max	Min[a]	Max	Min	Max	Min	Max[b]	Min	Max
13/16–16	0.745	0.752	0.749	0.756	0.752	0.759	0.756	0.763	0.7450	0.7491	0.7469	0.7512	0.7490	0.7533	0.7511	0.7554
13/16–20	0.758	0.764	0.761	0.766	0.764	0.770	0.766	0.772	0.7580	0.7623	0.7602	0.7642	0.7622	0.7662	0.7641	0.7681
7/8–9	0.755	0.767	0.761	0.773	0.767	0.778	0.773	0.785	0.7550	0.7614	0.7580	0.7647	0.7614	0.7681	0.7647	0.7714
7/8–12	0.785	0.794	0.790	0.799	0.794	0.803	0.799	0.808	0.7850	0.7900	0.7874	0.7926	0.7900	0.7952	0.7926	0.7978
7/8–14	0.798	0.806	0.802	0.810	0.806	0.814	0.810	0.818	0.7980	0.8022	0.8000	0.8045	0.8023	0.8068	0.8045	0.8090
7/8–16	0.807	0.814	0.811	0.818	0.814	0.821	0.818	0.825	0.8070	0.8116	0.8094	0.8137	0.8115	0.8158	0.8136	0.8179
7/8–20	0.821	0.827	0.824	0.829	0.827	0.832	0.829	0.835	0.8210	0.8248	0.8227	0.8267	0.8247	0.8287	0.8266	0.8306
7/8–28	0.836	0.840	0.838	0.843	0.840	0.845	0.843	0.847	0.8360	0.8395	0.8370	0.8410	0.8386	0.8426	0.8402	0.8442
15/16–12	0.847	0.856	0.852	0.861	0.856	0.865	0.861	0.870	0.8470	0.8524	0.8499	0.8550	0.8524	0.8575	0.8550	0.8601
15/16–16	0.870	0.877	0.874	0.881	0.877	0.884	0.881	0.888	0.8700	0.8741	0.8719	0.8762	0.8740	0.8783	0.8761	0.8804
15/16–20	0.883	0.889	0.886	0.891	0.889	0.895	0.891	0.897	0.8830	0.8873	0.8852	0.8892	0.8872	0.8912	0.8891	0.8931
1–8	0.865	0.878	0.871	0.884	0.878	0.890	0.884	0.896	0.8650	0.8722	0.8684	0.8759	0.8722	0.8797	0.8760	0.8835
1–12	0.910	0.919	0.915	0.924	0.919	0.928	0.924	0.933	0.9100	0.9148	0.9123	0.9173	0.9148	0.9198	0.9173	0.9223
1–14	0.923	0.931	0.927	0.934	0.931	0.938	0.934	0.942	0.9230	0.9271	0.9249	0.9293	0.9271	0.9315	0.9293	0.9337
1–16	0.932	0.939	0.936	0.943	0.939	0.946	0.943	0.950	0.9320	0.9366	0.9344	0.9387	0.9365	0.9408	0.9386	0.9429
1–20	0.946	0.952	0.949	0.954	0.952	0.957	0.954	0.960	0.9460	0.9498	0.9477	0.9517	0.9497	0.9537	0.9516	0.9556
1–28	0.961	0.966	0.963	0.968	0.966	0.970	0.968	0.972	0.9610	0.9645	0.9620	0.9660	0.9636	0.9676	0.9652	0.9692
1 1/16–12	0.972	0.981	0.977	0.986	0.981	0.990	0.986	0.995	0.9720	0.9773	0.9748	0.9798	0.9773	0.9823	0.9798	0.9848
1 1/16–16	0.995	1.002	0.999	1.055	1.002	1.009	1.055	1.013	0.9950	0.9991	0.9969	1.0012	0.9990	1.0033	1.0011	1.0054
1 1/16–18	1.002	1.009	1.005	1.012	1.009	1.015	1.012	1.018	1.0020	1.0065	1.0044	1.0085	1.0064	1.0105	1.0085	1.0126
1 1/8–7	0.970	0.984	0.977	0.991	0.984	0.998	0.991	1.005	0.9700	0.9790	0.9747	0.9833	0.9789	0.9875	0.9832	0.9918
1 1/8–8	0.990	1.003	0.996	1.009	1.003	1.015	1.009	1.021	0.9900	0.9972	0.9934	1.0009	0.9972	1.0047	1.0010	1.0085
1 1/8–12	1.035	1.044	1.040	1.049	1.044	1.053	1.049	1.058	1.0350	1.0398	1.0373	1.0423	1.0398	1.0448	1.0423	1.0473
1 1/8–16	1.057	1.064	1.061	1.068	1.064	1.071	1.068	1.075	1.0570	1.0616	1.0594	1.0637	1.0615	1.0658	1.0636	1.0679
1 1/8–18	1.065	1.072	1.068	1.075	1.072	1.078	1.075	1.081	1.0650	1.0690	1.0669	1.0710	1.0689	1.0730	1.0710	1.0751
1 1/8–20	1.071	1.077	1.074	1.079	1.077	1.082	1.079	1.085	1.0710	1.0748	1.0727	1.0767	1.0747	1.0787	1.0766	1.0806
1 1/8–28	1.086	1.091	1.088	1.093	1.091	1.095	1.093	1.097	1.0860	1.0895	1.0870	1.0910	1.0886	1.0926	1.0902	1.0942
1 3/16–12	1.097	1.106	1.102	1.111	1.106	1.115	1.111	1.120	1.0970	1.1023	1.0998	1.1048	1.1023	1.1073	1.1048	1.1098

Table 2. *(Continued)* Recommended Hole Size Limits Before Tapping Unified Threads

Thread Size	Classes 1B and 2B								Class 3B							
	Length of Engagement (D = Nominal Size of Thread)															
	To and Including $\frac{1}{3}D$		Above $\frac{1}{3}D$ to $\frac{2}{3}D$		Above $\frac{2}{3}D$ to $1\frac{1}{2}D$		Above $1\frac{1}{2}D$ to $3D$		To and Including $\frac{1}{3}D$		Above $\frac{1}{3}D$ to $\frac{2}{3}D$		Above $\frac{2}{3}D$ to $1\frac{1}{2}D$		Above $1\frac{1}{2}D$ to $3D$	
	Recommended Hole Size Limits															
	Min[a]	Max	Min	Max	Min	Max[b]	Min	Max	Min[a]	Max	Min	Max	Min	Max[b]	Min	Max
1 3/16–16	1.120	1.127	1.124	1.131	1.127	1.134	1.131	1.138	1.1200	1.1241	1.1219	1.1262	1.1240	1.1283	1.1261	1.1304
1 3/16–18	1.127	1.134	1.130	1.137	1.134	1.140	1.137	1.143	1.1270	1.1315	1.1294	1.1335	1.1314	1.1355	1.1335	1.1376
1 1/4–7	1.095	1.109	1.102	1.116	1.109	1.123	1.116	1.130	1.0950	1.1040	1.0997	1.1083	1.1039	1.1125	1.1082	1.1168
1 1/4–8	1.115	1.128	1.121	1.134	1.128	1.140	1.134	1.146	1.1150	1.1222	1.1184	1.1259	1.1222	1.1297	1.1260	1.1335
1 1/4–12	1.160	1.169	1.165	1.174	1.169	1.178	1.174	1.183	1.1600	1.1648	1.1623	1.1673	1.1648	1.1698	1.1673	1.1723
1 1/4–16	1.182	1.189	1.186	1.193	1.189	1.196	1.193	1.200	1.1820	1.1866	1.1844	1.1887	1.1865	1.1908	1.1886	1.1929
1 1/4–18	1.190	1.197	1.193	1.200	1.197	1.203	1.200	1.206	1.1900	1.1940	1.1919	1.1960	1.1939	1.1980	1.1960	1.2001
1 1/4–20	1.196	1.202	1.199	1.204	1.202	1.207	1.204	1.210	1.1960	1.1998	1.1977	1.2017	1.1997	1.2037	1.2016	1.2056
1 5/16–12	1.222	1.231	1.227	1.236	1.231	1.240	1.236	1.245	1.2220	1.2273	1.2248	1.2298	1.2273	1.2323	1.2298	1.2348
1 5/16–16	1.245	1.252	1.249	1.256	1.252	1.259	1.256	1.263	1.2450	1.2491	1.2469	1.2512	1.2490	1.2533	1.2511	1.2554
1 5/16–18	1.252	1.259	1.256	1.262	1.259	1.265	1.262	1.268	1.2520	1.2565	1.2544	1.2585	1.2564	1.2605	1.2585	1.2626
1 3/8–6	1.195	1.210	1.203	1.221	1.210	1.225	1.221	1.239	1.1950	1.2046	1.1996	1.2096	1.2046	1.2146	1.2096	1.2196
1 3/8–8	1.240	1.253	1.246	1.259	1.253	1.265	1.259	1.271	1.2400	1.2472	1.2434	1.2509	1.2472	1.2547	1.2510	1.2585
1 3/8–12	1.285	1.294	1.290	1.299	1.294	1.303	1.299	1.308	1.2850	1.2898	1.2873	1.2923	1.2898	1.2948	1.2923	1.2973
1 3/8–16	1.307	1.314	1.311	1.318	1.314	1.321	1.318	1.325	1.3070	1.3116	1.3094	1.3137	1.3115	1.3158	1.3136	1.3179
1 3/8–18	1.315	1.322	1.318	1.325	1.322	1.328	1.325	1.331	1.3150	1.3190	1.3169	1.3210	1.3189	1.3230	1.3210	1.3251
1 7/16–12	1.347	1.354	1.350	1.361	1.354	1.365	1.361	1.370	1.3470	1.3523	1.3498	1.3548	1.3523	1.3573	1.3548	1.3598
1 7/16–16	1.370	1.377	1.374	1.381	1.377	1.384	1.381	1.388	1.3700	1.3741	1.3719	1.3762	1.3740	1.3783	1.3761	1.3804
1 7/16–18	1.377	1.384	1.380	1.387	1.384	1.390	1.387	1.393	1.3770	1.3815	1.3794	1.3835	1.3814	1.3855	1.3835	1.3876
1 1/2–6	1.320	1.335	1.328	1.346	1.335	1.350	1.346	1.364	1.3200	1.3296	1.3246	1.3346	1.3296	1.3396	1.3346	1.3446
1 1/2–8	1.365	1.378	1.371	1.384	1.378	1.390	1.384	1.396	1.3650	1.3722	1.3684	1.3759	1.3722	1.3797	1.3760	1.3835
1 1/2–12	1.410	1.419	1.4155	1.424	1.419	1.428	1.424	1.433	1.4100	1.4148	1.4123	1.4173	1.4148	1.4198	1.4173	1.4223
1 1/2–16	1.432	1.439	1.436	1.443	1.439	1.446	1.443	1.450	1.4320	1.4366	1.4344	1.4387	1.4365	1.4408	1.4386	1.4429
1 1/2–18	1.440	1.446	1.443	1.450	1.446	1.452	1.450	1.456	1.4400	1.4440	1.4419	1.4460	1.4439	1.4480	1.4460	1.4501
1 1/2–20	1.446	1.452	1.449	1.454	1.452	1.457	1.454	1.460	1.4460	1.4498	1.4477	1.4517	1.4497	1.4537	1.4516	1.4556
1 9/16–16	1.495	1.502	1.499	1.506	1.502	1.509	1.506	1.513	1.4950	1.4991	1.4969	1.5012	1.4990	1.5033	1.5011	1.5054
1 9/16–18	1.502	1.509	1.505	1.512	1.509	1.515	1.512	1.518	1.5020	1.5065	1.5044	1.5085	1.5064	1.5105	1.5085	1.5126

Table 2. *(Continued)* **Recommended Hole Size Limits Before Tapping Unified Threads**

Thread Size	Classes 1B and 2B								Class 3B							
	Length of Engagement (D = Nominal Size of Thread)															
	To and Including ⅓D		Above ⅓D to ⅔D		Above ⅔D to 1½D		Above 1½D to 3D		To and Including ⅓D		Above ⅓D to ⅔D		Above ⅔D to 1½D		Above 1½D to 3D	
	Recommended Hole Size Limits															
	Min[a]	Max	Min	Max	Min	Max[b]	Min	Max	Min[a]	Max	Min	Max	Min	Max[b]	Min	Max
1⅝–8	1.490	1.498	1.494	1.509	1.498	1.515	1.509	1.521	1.4900	1.4972	1.4934	1.5009	1.4972	1.5047	1.5010	1.5085
1⅝–12	1.535	1.544	1.540	1.549	1.544	1.553	1.549	1.558	1.5350	1.5398	1.5373	1.5423	1.5398	1.5448	1.5423	1.5473
1⅝–16	1.557	1.564	1.561	1.568	1.564	1.571	1.568	1.575	1.5570	1.5616	1.5594	1.5637	1.5615	1.5658	1.5636	1.5679
1⅝–18	1.565	1.572	1.568	1.575	1.572	1.578	1.575	1.581	1.5650	1.5690	1.5669	1.5710	1.5689	1.5730	1.5710	1.5751
1 11/16–16	1.620	1.627	1.624	1.631	1.627	1.634	1.631	1.638	1.6200	1.6241	1.6219	1.6262	1.6240	1.6283	1.6261	1.6304
1 11/16–18	1.627	1.634	1.630	1.637	1.634	1.640	1.637	1.643	1.6270	1.6315	1.6294	1.6335	1.6314	1.6355	1.6335	1.6376
1¾–5	1.534	1.551	1.543	1.560	1.551	1.568	1.560	1.577	1.5340	1.5455	1.5395	1.5515	1.5455	1.5575	1.5515	1.5635
1¾–8	1.615	1.628	1.621	1.634	1.628	1.640	1.634	1.646	1.6150	1.6222	1.6184	1.6259	1.6222	1.6297	1.6260	1.6335
1¾–12	1.660	1.669	1.665	1.674	1.669	1.678	1.674	1.683	1.6600	1.6648	1.6623	1.6673	1.6648	1.6698	1.6673	1.6723
1¾–16	1.682	1.689	1.686	1.693	1.689	1.696	1.693	1.700	1.6820	1.6866	1.6844	1.6887	1.6865	1.6908	1.6886	1.6929
1¾–20	1.696	1.702	1.699	1.704	1.702	1.707	1.704	1.710	1.6960	1.6998	1.6977	1.7017	1.6997	1.7037	1.7016	1.7056
1 13/16–16	1.745	1.752	1.749	1.756	1.752	1.759	1.756	1.763	1.7450	1.7491	1.7469	1.7512	1.7490	1.7533	1.7511	1.7554
1⅞–8	1.740	1.752	1.746	1.759	1.752	1.765	1.759	1.771	1.7400	1.7472	1.7434	1.7509	1.7472	1.7547	1.7510	1.7585
1⅞–12	1.785	1.794	1.790	1.799	1.794	1.803	1.799	1.808	1.7850	1.7898	1.7873	1.7923	1.7898	1.7948	1.7923	1.7973
1⅞–16	1.807	1.814	1.810	1.818	1.814	1.821	1.818	1.825	1.8070	1.8116	1.8094	1.8137	1.8115	1.8158	1.8136	1.1879
1 15/16–16	1.870	1.877	1.874	1.881	1.877	1.884	1.881	1.888	1.8700	1.8741	1.8719	1.8762	1.8740	1.8783	1.8761	1.8804
2–4½	1.759	1.777	1.768	1.786	1.777	1.795	1.786	1.804	1.7590	1.7727	1.7661	1.7794	1.7728	1.7861	1.7794	1.7927
2–8	1.865	1.878	1.871	1.884	1.878	1.890	1.884	1.896	1.8650	1.8722	1.8684	1.8759	1.8722	1.8797	1.8760	1.8835
2–12	1.910	1.919	1.915	1.924	1.919	1.928	1.924	1.933	1.9100	1.9148	1.9123	1.9173	1.9148	1.9198	1.9173	1.9223
2–16	1.932	1.939	1.936	1.943	1.939	1.946	1.943	1.950	1.9320	1.9366	1.9344	1.9387	1.9365	1.9408	1.9386	1.9429
2–20	1.946	1.952	1.949	1.954	1.952	1.957	1.954	1.960	1.9460	1.9498	1.9477	1.9517	1.9497	1.9537	1.9516	1.9556
2 1/16–16	1.995	2.002	2.000	2.006	2.002	2.009	2.006	2.012	1.9950	1.9991	1.9969	2.0012	1.9990	2.0033	2.0011	2.0054
2⅛–8	1.990	2.003	1.996	2.009	2.003	2.015	2.009	2.021	1.9900	1.9972	1.9934	2.0009	1.9972	2.0047	2.0010	2.0085
2⅛–12	2.035	2.044	2.040	2.049	2.044	2.053	2.049	2.058	2.0350	2.0398	2.0373	2.0423	2.0398	2.0448	2.0423	2.0473
2⅛–16	2.057	2.064	2.061	2.068	2.064	2.071	2.068	2.075	2.0570	2.0616	2.0594	2.0637	2.0615	2.0658	2.0636	2.0679
2 3/16–16	2.120	2.127	2.124	2.131	2.127	2.134	2.131	2.138	2.1200	2.1241	2.1219	2.1262	2.1240	2.1283	2.1261	2.1304
2¼–4½	2.009	2.027	2.018	2.036	2.027	2.045	2.036	2.054	2.0090	2.0227	2.0161	2.0294	2.0228	2.0361	2.0294	2.0427

Table 2. *(Continued)* **Recommended Hole Size Limits Before Tapping Unified Threads**

Thread Size	Classes 1B and 2B								Class 3B							
	Length of Engagement (D = Nominal Size of Thread)															
	To and Including ⅓D		Above ⅓D to ⅔D		Above ⅔D to 1½D		Above 1½D to 3D		To and Including ⅓D		Above ⅓D to ⅔D		Above ⅔D to 1½D		Above 1½D to 3D	
	Recommended Hole Size Limits															
	Min[a]	Max	Min	Max	Min	Max[b]	Min	Max	Min[a]	Max	Min	Max	Min	Max[b]	Min	Max
2¼–8	2.115	2.128	2.121	2.134	2.128	2.140	2.134	2.146	2.1150	2.1222	2.1184	2.1259	2.1222	2.1297	2.1260	2.1335
2¼–12	2.160	2.169	2.165	2.174	2.169	2.178	2.174	2.182	2.1600	2.1648	2.1623	2.1673	2.1648	2.1698	2.1673	2.1723
2¼–16	2.182	2.189	2.186	2.193	2.189	2.196	2.193	2.200	2.1820	2.1866	2.1844	2.1887	2.1865	2.1908	2.1886	2.1929
2¼–20	2.196	2.202	2.199	2.204	2.202	2.207	2.204	2.210	2.1960	2.1998	2.1977	2.2017	2.1997	2.2037	2.2016	2.2056
2⁵⁄₁₆–16	2.245	2.252	2.249	2.256	2.252	2.259	2.256	2.263	2.2450	2.2491	2.2469	2.2512	2.2490	2.2533	2.2511	2.2554
2⅜–12	2.285	2.294	2.290	2.299	2.294	2.303	2.299	2.308	2.2850	2.2898	2.2873	2.2923	2.2898	2.2948	2.2923	2.2973
2⅜–16	2.307	2.314	2.311	2.318	2.314	2.321	2.318	2.325	2.3070	2.3116	2.3094	2.3137	2.3115	2.3158	2.3136	2.3179
2⁷⁄₁₆–16	2.370	2.377	2.374	2.381	2.377	2.384	2.381	2.388	2.3700	2.3741	2.3719	2.3762	2.3740	2.3783	2.3761	2.3804
2½–4	2.229	2.248	2.238	2.258	2.248	2.267	2.258	2.277	2.2290	2.2444	2.2369	2.2519	2.2444	2.2594	2.2519	2.2669
2½–8	2.365	2.378	2.371	2.384	2.378	2.390	2.384	2.396	2.3650	2.3722	2.3684	2.3759	2.3722	2.3797	2.3760	2.3835
2½–12	2.410	2.419	2.415	2.424	2.419	2.428	2.424	2.433	2.4100	2.4148	2.4123	2.4173	2.4148	2.4198	2.4173	2.4223
2½–16	2.432	2.439	2.436	2.443	2.439	2.446	2.443	2.450	2.4320	2.4366	2.4344	2.4387	2.4365	2.4408	2.4386	2.4429
2½–20	2.446	2.452	2.449	2.454	2.452	2.457	2.454	2.460	2.4460	2.4498	2.4478	2.4517	2.4497	2.4537	2.4516	2.4556
2⅝–12	2.535	2.544	2.540	2.549	2.544	2.553	2.549	2.558	2.5350	2.5398	2.5373	2.5423	2.5398	2.5448	2.5423	2.5473
2⅝–16	2.557	2.564	2.561	2.568	2.564	2.571	2.568	2.575	2.5570	2.5616	2.5594	2.5637	2.5615	2.5658	2.5636	2.5679
2¾–4	2.479	2.498	2.489	2.508	2.498	2.517	2.508	2.527	2.4790	2.4944	2.4869	2.5019	2.4944	2.5094	2.5019	2.5169
2¾–8	2.615	2.628	2.621	2.634	2.628	2.640	2.634	2.644	2.6150	2.6222	2.6184	2.6259	2.6222	2.6297	2.6260	2.6335
2¾–12	2.660	2.669	2.665	2.674	2.669	2.678	2.674	2.683	2.6600	2.6648	2.6623	2.6673	2.6648	2.6698	2.6673	2.6723
2¾–16	2.682	2.689	2.686	2.693	2.689	2.696	2.693	2.700	2.6820	2.6866	2.6844	2.6887	2.6865	2.6908	2.6886	2.6929
2⅞–12	2.785	2.794	2.790	2.809	2.794	2.803	2.809	2.808	2.7850	2.7898	2.7873	2.7923	2.7898	2.7948	2.7923	2.7973
2⅞–16	2.807	2.814	2.811	2.818	2.814	2.821	2.818	2.825	2.8070	2.8116	2.8094	2.8137	2.8115	2.8158	2.8136	2.8179
3–4	2.729	2.748	2.739	2.758	2.748	2.767	2.758	2.777	2.7290	2.7444	2.7369	2.7519	2.7444	2.7594	2.7519	2.7669
3–8	2.865	2.878	2.871	2.884	2.878	2.890	2.884	2.896	2.8650	2.8722	2.8684	2.8759	2.8722	2.8797	2.8760	2.8835
3–12	2.910	2.919	2.915	2.924	2.919	2.928	2.924	2.933	2.9100	2.9148	2.9123	2.9173	2.9148	2.9198	2.9173	2.9223
3–16	2.932	2.939	2.936	2.943	2.939	2.946	2.943	2.950	2.9320	2.9366	2.9344	2.9387	2.9365	2.9408	2.9386	2.9429
3⅛–12	3.035	3.044	3.040	3.049	3.044	3.053	3.049	3.058	3.0350	3.0398	3.0373	3.0423	3.0398	3.0448	3.0423	3.0473
3⅛–16	3.057	3.064	3.061	3.068	3.064	3.071	3.068	3.075	3.0570	3.0616	3.0594	3.0637	3.0615	3.0658	3.0636	3.0679

Table 2. *(Continued)* **Recommended Hole Size Limits Before Tapping Unified Threads**

Thread Size	Classes 1B and 2B								Class 3B							
	Length of Engagement (D = Nominal Size of Thread)															
	To and Including ⅓D		Above ⅓D to ⅔D		Above ⅔D to 1½D		Above 1½D to 3D		To and Including ⅓D		Above ⅓D to ⅔D		Above ⅔D to 1½D		Above 1½D to 3D	
	Recommended Hole Size Limits															
	Min[a]	Max	Min	Max	Min	Max[b]	Min	Max	Min[a]	Max	Min	Max	Min	Max[b]	Min	Max
3¼–4	2.979	2.998	2.989	3.008	2.998	3.017	3.008	3.027	2.9790	2.9944	2.9869	3.0019	2.9944	3.0094	3.0019	3.0169
3¼–8	3.115	3.128	3.121	3.134	3.128	3.140	3.134	3.146	3.1150	3.1222	3.1184	3.1259	3.1222	3.1297	3.1260	3.1335
3¼–12	3.160	3.169	3.165	3.174	3.169	3.178	3.174	3.183	3.1600	3.1648	3.1623	3.1673	3.1648	3.1698	3.1673	3.1723
3¼–16	3.182	3.189	3.186	3.193	3.189	3.196	3.193	3.200	3.1820	3.1866	3.1844	3.1887	3.1865	3.1908	3.1886	3.1929
3⅜–12	3.285	3.294	3.290	3.299	3.294	3.303	3.299	3.299	3.2850	3.2898	3.2873	3.2923	3.2898	3.2948	3.2923	3.2973
3⅜–16	3.307	3.314	3.311	3.318	3.314	3.321	3.317	3.325	3.3070	3.3116	3.3094	3.3137	3.3115	3.3158	3.3136	3.3179
3½–4	3.229	3.248	3.239	3.258	3.248	3.267	3.258	3.277	3.2290	3.2444	3.2369	3.2519	3.2444	3.2594	3.2519	3.2669
3½–8	3.365	3.378	3.371	2.384	3.378	3.390	3.384	3.396	3.3650	3.3722	3.3684	3.3759	3.3722	3.3797	3.3760	3.3835
3½–12	3.410	3.419	3.415	3.424	3.419	3.428	3.424	3.433	3.4100	3.4148	3.4123	3.4173	3.4148	3.4198	3.4173	3.4223
3½–16	3.432	3.439	3.436	3.443	3.439	3.446	3.443	3.450	3.4320	3.4366	3.4344	3.4387	3.4365	3.4408	3.4386	3.4429
3⅝–12	3.535	3.544	3.544	3.549	3.544	3.553	3.549	3.553	3.5350	3.5398	3.5373	3.5423	3.5398	3.5448	3.5423	3.5473
3⅝–16	3.557	3.564	3.561	3.568	3.567	3.571	3.568	3.575	3.5570	3.5616	3.5594	3.5637	3.5615	3.5658	3.5636	3.5679
3¾–4	3.479	3.498	3.489	3.508	3.498	3.517	3.508	3.527	3.4790	3.4944	3.4869	3.5019	3.4944	3.5094	3.5019	3.5169
3¾–8	3.615	3.628	3.615	3.634	3.628	3.640	3.634	3.646	3.6150	3.6222	3.6184	3.6259	3.6222	3.6297	3.6260	3.6335
3¾–12	3.660	3.669	3.665	3.674	3.669	3.678	3.674	3.683	3.6600	3.6648	3.6623	3.6673	3.6648	3.6698	3.6673	3.6723
¾–16	3.682	3.689	3.686	3.693	3.689	3.696	3.693	3.700	3.6820	3.6866	3.6844	3.6887	3.6865	3.6908	3.6886	3.6929
3⅞–12	3.785	3.794	3.790	3.799	3.794	3.803	3.799	3.808	3.7850	3.7898	3.7873	3.7923	3.7898	3.7948	3.7923	3.7973
3⅞–16	3.807	3.814	3.811	3.818	3.814	3.821	3.818	3.825	3.8070	3.8116	3.8094	3.8137	3.8115	3.8158	3.8136	3.8179
4–4	3.729	3.748	3.739	3.758	3.748	3.767	3.758	3.777	3.7290	3.7444	3.7369	3.7519	3.7444	3.7594	3.7519	3.7669
4–8	3.865	3.878	3.871	3.884	3.878	3.890	3.884	3.896	3.8650	3.8722	3.8684	3.8759	3.8722	3.8797	3.8760	3.8835
4–12	3.910	3.919	3.915	3.924	3.919	3.928	3.924	3.933	3.9100	3.9148	3.9123	3.9173	3.9148	3.9198	3.9173	3.9223
4–16	3.932	3.939	3.936	3.943	3.939	3.946	3.943	3.950	3.9320	3.9366	3.9344	3.9387	3.9365	3.9408	3.9386	3.9429
4¼–4	3.979	3.998	3.989	4.008	3.998	4.017	4.008	4.027	3.9790	3.9944	3.9869	4.0019	3.9944	4.0094	4.0019	4.0169
4¼–8	4.115	4.128	4.121	4.134	4.128	4.140	4.134	4.146	4.1150	4.1222	4.1184	4.1259	4.1222	4.1297	4.1260	4.1335
4¼–12	4.160	4.169	4.165	4.174	4.169	4.178	4.174	4.183	4.1600	4.1648	4.1623	4.1673	4.1648	4.1698	4.1673	4.1723
4¼–16	4.182	4.189	4.186	4.193	4.189	4.196	4.193	4.200	4.1820	4.1866	4.1844	4.1887	4.1865	4.1908	4.1886	4.1929
4½–4	4.229	4.248	4.239	4.258	4.248	4.267	4.258	4.277	4.2290	4.2444	4.2369	4.2519	4.2444	4.2594	4.2519	4.2669

Table 2. *(Continued)* **Recommended Hole Size Limits Before Tapping Unified Threads**

Thread Size	Classes 1B and 2B								Class 3B							
	Length of Engagement (D = Nominal Size of Thread)															
	To and Including $\frac{1}{3}D$		Above $\frac{1}{3}D$ to $\frac{2}{3}D$		Above $\frac{2}{3}D$ to $1\frac{1}{2}D$		Above $1\frac{1}{2}D$ to $3D$		To and Including $\frac{1}{3}D$		Above $\frac{1}{3}D$ to $\frac{2}{3}D$		Above $\frac{2}{3}D$ to $1\frac{1}{2}D$		Above $1\frac{1}{2}D$ to $3D$	
	Recommended Hole Size Limits															
	Min[a]	Max	Min	Max	Min	Max[b]	Min	Max	Min[a]	Max	Min	Max	Min	Max[b]	Min	Max
4½–8	4.365	4.378	4.371	4.384	4.378	4.390	4.384	4.396	4.3650	4.3722	4.3684	4.3759	4.3722	4.3797	4.3760	4.3835
4½–12	4.410	4.419	4.419	4.424	4.419	4.428	4.424	4.433	4.4100	4.4148	4.4123	4.4173	4.4148	4.4198	4.4173	4.4223
4½–16	4.432	4.439	4.437	4.444	4.439	4.446	4.444	4.455	4.4320	4.4366	4.4344	4.4387	4.4365	4.4408	4.4386	4.4429
4¾–8	4.615	4.628	4.621	4.646	4.628	4.640	4.646	4.646	4.6150	4.6222	4.6184	4.6259	4.6222	4.6297	4.6260	4.6335
4¾–12	4.660	4.669	4.665	4.674	4.669	4.678	4.674	4.683	4.6600	4.6648	4.6623	4.6673	4.6648	4.6698	4.6673	4.6723
4¾–16	4.682	4.689	4.686	4.693	4.689	4.696	4.693	4.700	4.6820	4.6866	4.6844	4.6887	4.6865	4.6908	4.6886	4.6929
5–8	4.865	4.878	4.871	4.884	4.878	4.890	4.884	4.896	4.8650	4.8722	4.8684	4.8759	4.8722	4.8797	4.8760	4.8835
5–12	4.910	4.919	4.915	4.924	4.919	4.928	4.924	4.933	4.9100	4.9148	4.9123	4.9173	4.9148	4.9198	4.9173	4.9223
5–16	4.932	4.939	4.936	4.943	4.939	4.946	4.943	4.950	4.9320	4.9366	4.9344	4.9387	4.9365	4.9408	4.9386	4.9429
5¼–8	5.115	5.128	5.121	5.134	5.128	5.140	5.134	5.146	5.1150	5.1222	5.1184	5.1259	5.1222	5.1297	5.1260	5.1335
5¼–12	5.160	5.169	5.165	5.174	5.169	5.178	5.174	5.183	5.1600	5.1648	5.1623	5.1673	5.1648	5.1698	5.1673	5.1723
5¼–16	5.182	5.189	5.186	5.193	5.189	5.196	5.193	5.200	5.1820	5.1866	5.1844	5.1887	5.1865	5.1908	5.1886	5.1929
5½–8	5.365	5.378	5.371	5.384	5.378	5.390	5.384	5.396	5.3650	5.3722	5.3684	5.3759	5.3722	5.3797	5.3760	5.3835
5½–12	5.410	5.419	5.415	5.424	5.419	5.428	5.424	5.433	5.4100	5.4148	5.4123	5.4173	5.4148	5.4198	5.4173	5.4223
5½–16	5.432	5.439	5.436	5.442	5.439	5.446	5.442	5.450	5.4320	5.4366	5.4344	5.4387	5.4365	5.4408	5.4386	5.4429
5¾–8	5.615	5.628	5.621	5.634	5.628	5.640	5.634	5.646	5.6150	5.6222	5.6184	5.6259	5.6222	5.6297	5.6260	5.6335
5¾–12	5.660	5.669	5.665	5.674	5.669	5.678	5.674	5.683	5.6600	5.6648	5.6623	5.6673	5.6648	5.6698	5.6673	5.6723
5¾–16	5.682	5.689	5.686	5.693	5.689	5.696	5.693	5.700	5.6820	5.6866	5.6844	5.6887	5.6865	5.6908	5.6886	5.6929
6–8	5.865	5.878	5.871	5.896	5.878	5.890	5.896	5.896	5.8650	5.8722	5.8684	5.8759	5.8722	5.8797	5.8760	5.8835
6–12	5.910	5.919	5.915	5.924	5.919	5.928	5.924	5.933	5.9100	5.9148	5.9123	5.9173	5.9148	5.9198	5.9173	5.9223
6–16	5.932	5.939	5.935	5.943	5.939	5.946	5.943	5.950	5.9320	5.9366	5.9344	5.9387	5.9365	5.9408	5.9386	5.9429

[a] This is the minimum minor diameter specified in the thread tables, page 1716.

[b] This is the maximum minor diameter specified in the thread tables, page 1716.

All dimensions are in inches.

For basis of recommended hole size limits see accompanying text.

As an aid in selecting suitable drills, see the listing of American Standard drill sizes in the twist drill section. For amount of expected drill oversize, see page 857.

Table 3. Tap Drill Sizes for Threads of American National Form

Screw Thread		Commercial Tap Drills[a]		Screw Thread		Commercial Tap Drills[a]	
Outside Diam. Pitch	Root Diam.	Size or Number	Decimal Equiv.	Outside Diam. Pitch	Root Diam.	Size or Number	Decimal Equiv.
1/16–64	0.0422	3/64	0.0469	27	0.4519	15/32	0.4687
72	0.0445	3/64	0.0469	9/16–12	0.4542	31/64	0.4844
5/64–60	0.0563	1/16	0.0625	18	0.4903	33/64	0.5156
72	0.0601	52	0.0635	27	0.5144	17/32	0.5312
3/32–48	0.0667	49	0.0730	5/8–11	0.5069	17/32	0.5312
50	0.0678	49	0.0730	12	0.5168	35/64	0.5469
7/64–48	0.0823	43	0.0890	18	0.5528	37/64	0.5781
1/8–32	0.0844	3/32	0.0937	27	0.5769	19/32	0.5937
40	0.0925	38	0.1015	11/16–11	0.5694	19/32	0.5937
9/64–40	0.1081	32	0.1160	16	0.6063	5/8	0.6250
5/32–32	0.1157	1/8	0.1250	3/4–10	0.6201	21/32	0.6562
36	0.1202	30	0.1285	12	0.6418	43/64	0.6719
11/64–32	0.1313	9/64	0.1406	16	0.6688	11/16	0.6875
3/16–24	0.1334	26	0.1470	27	0.7019	23/32	0.7187
32	0.1469	22	0.1570	13/16–10	0.6826	23/32	0.7187
13/64–24	0.1490	20	0.1610	7/8–9	0.7307	49/64	0.7656
7/32–24	0.1646	16	0.1770	12	0.7668	51/64	0.7969
32	0.1782	12	0.1890	14	0.7822	13/16	0.8125
15/64–24	0.1806	10	0.1935	18	0.8028	53/64	0.8281
1/4–20	0.1850	7	0.2010	27	0.8269	27/32	0.8437
24	0.1959	4	0.2090	15/16–9	0.7932	53/64	0.8281
27	0.2019	3	0.2130	1–8	0.8376	7/8	0.8750
28	0.2036	3	0.2130	12	0.8918	59/64	0.9219
32	0.2094	7/32	0.2187	14	0.9072	15/16	0.9375
5/16–18	0.2403	F	0.2570	27	0.9519	31/32	0.9687
20	0.2476	17/64	0.2656	1 1/8–7	0.9394	63/64	0.9844
24	0.2584	I	0.2720	12	1.0168	1 3/64	1.0469
27	0.2644	J	0.2770	1 1/4–7	1.0644	1 7/64	1.1094
32	0.2719	9/32	0.2812	12	1.1418	1 11/64	1.1719
3/8–16	0.2938	5/16	0.3125	1 3/8–6	1.1585	1 7/32	1.2187
20	0.3100	21/64	0.3281	12	1.2668	1 19/64	1.2969
24	0.3209	Q	0.3320	1 1/2–6	1.2835	1 11/32	1.3437
27	0.3269	R	0.3390	12	1.3918	1 27/64	1.4219
7/16–14	0.3447	U	0.3680	1 5/8–5 1/2	1.3888	1 29/64	1.4531
20	0.3726	25/64	0.3906	1 3/4–5	1.4902	1 9/16	1.5625
24	0.3834	X	0.3970	1 7/8–5	1.6152	1 11/16	1.6875
27	0.3894	Y	0.4040	2–4 1/2	1.7113	1 25/32	1.7812
1/2–12	0.3918	27/64	0.4219	2 1/8–4 1/2	1.8363	1 29/32	1.9062
13	0.4001	27/64	0.4219	2 1/4–4 1/2	1.9613	2 1/32	2.0312
20	0.4351	29/64	0.4531	2 3/8–4	2.0502	2 1/8	2.1250
24	0.4459	29/64	0.4531	2 1/2–4	2.1752	2 1/4	2.2500

[a] These tap drill diameters allow approximately 75 per cent of a full thread to be produced. For small thread sizes in the first column, the use of drills to produce the larger hole sizes shown in Table 2 will reduce defects caused by tap problems and breakage.

Table 4. Tap Drills and Clearance Drills for Machine Screws with American National Thread Form

Size of Screw		No. of Threads per Inch	Tap Drills		Clearance Hole Drills			
					Close Fit		Free Fit	
No. or Diam.	Decimal Equiv.		Drill Size	Decimal Equiv.	Drill Size	Decimal Equiv.	Drill Size	Decimal Equiv.
0	.060	80	3⁄64	.0469	52	.0635	50	.0700
1	.073	64 72	53 53	.0595 .0595	48	.0760	46	.0810
2	.086	56 64	50 50	.0700 .0700	43	.0890	41	.0960
3	.099	48 56	47 45	.0785 .0820	37	.1040	35	.1100
4	.112	36[a] 40 48	44 43 42	.0860 .0890 .0935	32	.1160	30	.1285
5	.125	40 44	38 37	.1015 1040	30	.1285	29	.1360
6	.138	32 40	36 33	.1065 .1130	27	.1440	25	.1495
8	.164	32 36	29 29	.1360 .1360	18	.1695	16	.1770
10	.190	24 32	25 21	.1495 1590	9	.1960	7	.2010
12	.216	24 28	16 14	.1770 .1820	2	.2210	1	.2280
14	.242	20[a] 24[a]	10 7	.1935 .2010	D	.2460	F	.2570
1⁄4	.250	20 28	7 3	.2010 .2130	F	.2570	H	.2660
5⁄16	.3125	18 24	F I	.2570 .2720	P	.3230	Q	.3320
3⁄8	.375	16 24	5⁄16 Q	.3125 .3320	W	.3860	X	.3970
7⁄16	.4375	14 20	U 25⁄64	.3680 .3906	29⁄64	.4531	15⁄32	.4687
1⁄2	.500	13 20	27⁄64 29⁄64	.4219 .4531	33⁄64	.5156	17⁄32	.5312

[a] These screws are not in the American Standard but are from the former A.S.M.E. Standard.

The size of the tap drill hole for any desired percentage of full thread depth can be calculated by the formulas below. In these formulas the Per Cent Full Thread is expressed as a decimal; e.g., 75 per cent is expressed as .75. The tap drill size is the size nearest to the calculated hole size.

For American Unified Thread form:

$$\text{Hole Size} = \text{Basic Major Diameter} - \frac{1.08253 \times \text{Per Cent Full Thread}}{\text{Number of Threads per Inch}}$$

For ISO Metric threads (all dimensions in millimeters):

$$\text{Hole Size} = \text{Basic Major Diameter} - (1.08253 \times \text{Pitch} \times \text{Per Cent Full Thread})$$

The constant 1.08253 in the above equation represents $5H/8$ where H is the height of a sharp V-thread (see page 1706). (The pitch is taken to be 1.)

Factors Influencing Minor Diameter Tolerances of Tapped Holes.—As stated in the Unified screw thread standard, the principle practical factors that govern minor diameter tolerances of internal threads are tapping difficulties, particularly tap breakage in the small sizes, availability of standard drill sizes in the medium and large sizes, and depth (radial) of engagement. Depth of engagement is related to the stripping strength of the thread assembly, and thus also, to the length of engagement. It also has an influence on the tendency toward disengagement of the threads on one side when assembly is eccentric. The amount of possible eccentricity is one-half of the sum of the pitch diameter allowance and tolerances on both mating threads. For a given pitch, or height of thread, this sum increases with the diameter, and accordingly this factor would require a decrease in minor diameter tolerance with increase in diameter. However, such decrease in tolerance would often require the use of special drill sizes; therefore, to facilitate the use of standard drill sizes, for any given pitch the minor diameter tolerance for Unified thread classes 1B and 2B threads of ¼ inch diameter and larger is constant, in accordance with a formula given in the American Standard for Unified Screw Threads.

Effect of Length of Engagement of Minor Diameter Tolerances: There may be applications where the lengths of engagement of mating threads is relatively short or the combination of materials used for mating threads is such that the maximum minor diameter tolerance given in the Standard (based on a length of engagement equal to the nominal diameter) may not provide the desired strength of the fastening. Experience has shown that for lengths of engagement less than ⅔*D* (the minimum thickness of standard nuts) the minor diameter tolerance may be reduced without causing tapping difficulties. In other applications the length of engagement of mating threads may be long because of design considerations or the combination of materials used for mating threads. As the threads engaged increase in number, a shallower depth of engagement may be permitted and still develop stripping strength greater than the external thread breaking strength. Under these conditions the maximum tolerance given in the Standard should be increased to reduce the possibility of tapping difficulties. The following paragraphs indicate how the aforementioned considerations were taken into account in determining the minor diameter limits for various lengths of engagement given in Table 2.

Recommended Hole Sizes before Tapping.—Recommended hole size limits before threading to provide for optimum strength of fastenings and tapping conditions are shown in Table 2 for classes 1B, 2B, and 3B. The hole size limit before threading, and the tolerances between them, are derived from the minimum and maximum minor diameters of the internal thread given in the dimensional tables for Unified threads in the screw thread section using the following rules:

1) For lengths of engagement in the range to and including ⅓*D*, where *D* equals nominal diameter, the minimum hole size will be equal to the minimum minor diameter of the internal thread and the maximum hole size will be larger by one-half the minor diameter tolerance.

2) For the range from ⅓*D* to ⅔*D*, the minimum and maximum hole sizes will each be one quarter of the minor diameter tolerance larger than the corresponding limits for the length of engagement to and including ⅓*D*.

3) For the range from ⅔*D* to 1½*D* the minimum hole size will be larger than the minimum minor diameter of the internal thread by one-half the minor diameter tolerance and the maximum hole size will be equal to the maximum minor diameter.

4) For the range from 1½*D* to 3*D* the minimum and maximum hole sizes will each be one-quarter of the minor diameter tolerance of the internal thread larger than the corresponding limits for the ⅔*D* to 1½*D* length of engagement.

From the foregoing it will be seen that the difference between limits in each range is the same and equal to one-half of the minor diameter tolerance given in the Unified screw thread dimensional tables. This is a general rule, except that the minimum differences for sizes below ¼ inch are equal to the minor diameter tolerances calculated on the basis of lengths of engagement to and including ⅓*D*. Also, for lengths of engagement greater than ⅓*D* and for sizes ¼ inch and larger the values are adjusted so that the difference between limits is never less than 0.004 inch.

For diameter-pitch combinations other than those given in Table 2, the foregoing rules should be applied to the tolerances given in the dimensional tables in the screw thread section or the tolerances derived from the formulas given in the Standard to determine the hole size limits.

Selection of Tap Drills: In selecting standard drills to produce holes within the limits given in Table 2 it should be recognized that drills have a tendency to cut oversize. The material on page 857 may be used as a guide to the expected amount of oversize.

Table 5. Unified Miniature Screw Threads—Recommended Hole Size Limits Before Tapping

Thread Size		Internal Threads		Lengths of Engagement					
		Minor Diameter Limits		To and including ⅔D		Above ⅔D to 1½D		Above 1½D to 3D	
				Recommended Hole Size Limits					
	Pitch	Min	Max	Min	Max	Min	Max	Min	Max
Designation	mm	mm	mm	mm	mm	mm	mm	mm	mm
0.30 UNM	**0.080**	**0.217**	**0.254**	**0.226**	**0.240**	**0.236**	**0.254**	**0.245**	**0.264**
0.35 UNM	0.090	0.256	0.297	0.267	0.282	0.277	0.297	0.287	0.307
0.40 UNM	**0.100**	**0.296**	**0.340**	**0.307**	**0.324**	**0.318**	**0.340**	**0.329**	**0.351**
0.45 UNM	0.100	0.346	0.390	0.357	0.374	0.368	0.390	0.379	0.401
0.50 UNM	**0.125**	**0.370**	**0.422**	**0.383**	**0.402**	**0.396**	**0.422**	**0.409**	**0.435**
0.55 UNM	0.125	0.420	0.472	0.433	0.452	0.446	0.472	0.459	0.485
0.60 UNM	**0.150**	**0.444**	**0.504**	**0.459**	**0.482**	**0.474**	**0.504**	**0.489**	**0.519**
0.70 UNM	0.175	0.518	0.586	0.535	0.560	0.552	0.586	0.569	0.603
0.80 UNM	**0.200**	**0.592**	**0.668**	**0.611**	**0.640**	**0.630**	**0.668**	**0.649**	**0.687**
0.90 UNM	0.225	0.666	0.750	0.687	0.718	0.708	0.750	0.729	0.771
1.00 UNM	**0.250**	**0.740**	**0.832**	**0.763**	**0.798**	**0.786**	**0.832**	**0.809**	**0.855**
1.10 UNM	0.250	0.840	0.932	0.863	0.898	0.886	0.932	0.909	0.955
1.20 UNM	**0.250**	**0.940**	**1.032**	**0.963**	**0.998**	**0.986**	**1.032**	**1.009**	**1.055**
1.40 UNM	0.300	1.088	1.196	1.115	1.156	1.142	1.196	1.169	1.223
Designation	Thds. per in.	inch	inch	inch	inch	inch	inch	inch	inch
0.30 UNM	**318**	**0.0085**	**0.0100**	**0.0089**	**0.0095**	**0.0093**	**0.0100**	**0.0096**	**0.0104**
0.35 UNM	282	0.0101	0.0117	0.0105	0.0111	0.0109	0.0117	0.0113	0.0121
0.40 UNM	**254**	**0.0117**	**0.0134**	**0.0121**	**0.0127**	**0.0125**	**0.0134**	**0.0130**	**0.0138**
0.45 UNM	254	0.0136	0.0154	0.0141	0.0147	0.0145	0.0154	0.0149	0.0158
0.50 UNM	**203**	**0.0146**	**0.0166**	**0.0150**	**0.0158**	**0.0156**	**0.0166**	**0.0161**	**0.0171**
0.55 UNM	203	0.0165	0.0186	0.0170	0.0178	0.0176	0.0186	0.0181	0.0191
0.60 UNM	**169**	**0.0175**	**0.0198**	**0.0181**	**0.0190**	**0.0187**	**0.0198**	**0.0193**	**0.0204**
0.70 UNM	145	0.0204	0.0231	0.0211	0.0221	0.0217	0.0231	0.0224	0.0237
0.80 UNM	**127**	**0.0233**	**0.0263**	**0.0241**	**0.0252**	**0.0248**	**0.0263**	**0.0256**	**0.0270**
0.90 UNM	113	0.0262	0.0295	0.0270	0.0283	0.0279	0.0295	0.0287	0.0304
1.00 UNM	**102**	**0.0291**	**0.0327**	**0.0300**	**0.0314**	**0.0309**	**0.0327**	**0.0319**	**0.0337**
1.10 UNM	102	0.0331	0.0367	0.0340	0.0354	0.0349	0.0367	0.0358	0.0376
1.20 UNM	**102**	**0.0370**	**0.0406**	**0.0379**	**0.0393**	**0.0388**	**0.0406**	**0.0397**	**0.0415**
1.40 UNM	85	0.0428	0.0471	0.0439	0.0455	0.0450	0.0471	0.0460	0.0481

As an aid in selecting suitable drills, see the listing of American Standard drill sizes in the twist drill section. Thread sizes in heavy type are preferred sizes.

Hole Sizes for Tapping Unified Miniature Screw Threads.—Table 5 indicates the hole size limits recommended for tapping. These limits are derived from the internal thread minor diameter limits given in the American Standard for Unified Miniature Screw Threads ASA B1.10-1958 and are disposed so as to provide the optimum conditions for tapping. The maximum limits are based on providing a functionally adequate fastening for the most common applications, where the material of the externally threaded member is of a strength essentially equal to or greater than that of its mating part. In applications where, because of considerations other than the fastening, the screw is made of an appreciably weaker material, the use of smaller hole sizes is usually necessary to extend thread engagement to a greater depth on the external thread. Recommended minimum hole sizes are greater than the minimum limits of the minor diameters to allow for the spin-up developed in tapping.

In selecting drills to produce holes within the limits given in Table 5 it should be recognized that drills have a tendency to cut oversize. The material on page 857 may be used as a guide to the expected amount of oversize.

British Standard Tapping Drill Sizes for Screw and Pipe Threads.—British Standard BS 1157:1975 (1998) provides recommendations for tapping drill sizes for use with fluted taps for various ISO metric, Unified, British Standard fine, British Association, and British Standard Whitworth screw threads as well as British Standard parallel and taper pipe threads.

Table 6. British Standard Tapping Drill Sizes for ISO Metric Coarse Pitch Series Threads *BS 1157:1975 (1998)*

Nom. Size and Thread Diam.	Standard Drill Sizes[a]				Nom. Size and Thread Diam.	Standard Drill Sizes[a]			
	Recommended		Alternative			Recommended		Alternative	
	Size	Theoretical Radial Engagement with Ext. Thread (Per Cent)	Size	Theoretical Radial Engagement with Ext. Thread (Per Cent)		Size	Theoretical Radial Engagement with Ext. Thread (Per Cent)	Size	Theoretical Radial Engagement with Ext. Thread (Per Cent)
M 1	0.75	81.5	0.78	71.7	M 12	10.20	83.7	10.40	74.5[b]
M 1.1	0.85	81.5	0.88	71.7	M 14	12.00	81.5	12.20	73.4[b]
M 1.2	0.95	81.5	0.98	71.7	M 16	14.00	81.5	14.25	71.3[c]
M 1.4	1.10	81.5	1.15	67.9	M 18	15.50	81.5	15.75	73.4[c]
M 1.6	1.25	81.5	1.30	69.9	M 20	17.50	81.5	17.75	73.4[c]
M 1.8	1.45	81.5	1.50	69.9	M 22	19.50	81.5	19.75	73.4[c]
M 2	1.60	81.5	1.65	71.3	M 24	21.00	81.5	21.25	74.7[b]
M 2.2	1.75	81.5	1.80	72.5	M 27	24.00	81.5	24.25	74.7[b]
M 2.5	2.05	81.5	2.10	72.5	M 30	26.50	81.5	26.75	75.7[b]
M 3	2.50	81.5	2.55	73.4	M 33	29.50	81.5	29.75	75.7[b]
M 3.5	2.90	81.5	2.95	74.7	M 36	32.00	81.5	…	…
M 4	3.30	81.5	3.40	69.9[b]	M 39	35.00	81.5	…	…
M 4.5	3.70	86.8	3.80	76.1	M 42	37.50	81.5	…	…
M 5	4.20	81.5	4.30	71.3[b]	M 45	40.50	81.5	…	…
M 6	5.00	81.5	5.10	73.4	M 48	43.00	81.5	…	…
M 7	6.00	81.5	6.10	73.4	M 52	47.00	81.5	…	…
M 8	6.80	78.5	6.90	71.7[b]	M 56	50.50	81.5	…	…
M 9	7.80	78.5	7.90	71.7[b]	M 60	54.50	81.5	…	…
M 10	8.50	81.5	8.60	76.1	M 64	58.00	81.5	…	…
M 11	9.50	81.5	9.60	76.1	M 68	62.00	81.5	…	…

[a] These tapping drill sizes are for fluted taps only.

[b] For tolerance class 6H and 7H threads only.

[c] For tolerance class 7H threads only.

Drill sizes are given in millimeters.

In the accompanying Table 6, recommended and alternative drill sizes are given for producing holes for ISO metric coarse pitch series threads. These coarse pitch threads are suitable for the large majority of general-purpose applications, and the limits and tolerances for internal coarse threads are given in the table starting on page 1834. It should be noted that Table 6 is for fluted taps only since a fluteless tap will require for the same screw thread a different size of twist drill than will a fluted tap. When tapped, holes produced with drills of the recommended sizes provide for a theoretical radial engagement with the external thread of about 81 per cent in most cases. Holes produced with drills of the alternative sizes provide for a theoretical radial engagement with the external thread of about 70 to 75 per cent. In some cases, as indicated in Table 6, the alternative drill sizes are suitable only for medium (6H) or for free (7H) thread tolerance classes.

When relatively soft material is being tapped, there is a tendency for the metal to be squeezed down towards the root of the tap thread, and in such instances, the minor diameter of the tapped hole may become smaller than the diameter of the drill employed. Users may wish to choose different tapping drill sizes to overcome this problem or for special purposes, and reference can be made to the pages mentioned above to obtain the minor diameter limits for internal pitch series threads.

Reference should be made to this standard BS 1157:1975 (1998) for recommended tapping hole sizes for other types of British Standard screw threads and pipe threads.

Table 7. British Standard Metric Bolt and Screw Clearance Holes *BS 4186: 1967*

Nominal Thread Diameter	Clearance Hole Sizes			Nominal Thread Diameter	Clearance Hole Sizes		
	Close Fit Series	Medium Fit Series	Free Fit Series		Close Fit Series	Medium Fit Series	Free Fit Series
1.6	1.7	1.8	2.0	52.0	54.0	56.0	62.0
2.0	2.2	2.4	2.6	56.0	58.0	62.0	66.0
2.5	2.7	2.9	3.1	60.0	62.0	66.0	70.0
3.0	3.2	3.4	3.6	64.0	66.0	70.0	74.0
4.0	4.3	4.5	4.8	68.0	70.0	74.0	78.0
5.0	5.3	5.5	5.8	72.0	74.0	78.0	82.0
6.0	6.4	6.6	7.0	76.0	78.0	82.0	86.0
7.0	7.4	7.6	8.0	80.0	82.0	86.0	91.0
8.0	8.4	9.0	10.0	85.0	87.0	91.0	96.0
10.0	10.5	11.0	12.0	90.0	93.0	96.0	101.0
12.0	13.0	14.0	15.0	95.0	98.0	101.0	107.0
14.0	15.0	16.0	17.0	100.0	104.0	107.0	112.0
16.0	17.0	18.0	19.0	105.0	109.0	112.0	117.0
18.0	19.0	20.0	21.0	110.0	114.0	117.0	122.0
20.0	21.0	22.0	24.0	115.0	119.0	122.0	127.0
22.0	23.0	24.0	26.0	120.0	124.0	127.0	132.0
24.0	25.0	26.0	28.0	125.0	129.0	132.0	137.0
27.0	28.0	30.0	32.0	130.0	134.0	137.0	144.0
30.0	31.0	33.0	35.0	140.0	144.0	147.0	155.0
33.0	34.0	36.0	38.0	150.0	155.0	158.0	165.0
36.0	37.0	39.0	42.0	…	…	…	…
39.0	40.0	42.0	45.0	…	…	…	…
42.0	43.0	45.0	48.0	…	…	…	…
45.0	46.0	48.0	52.0	…	…	…	…
48.0	50.0	52.0	56.0	…	…	…	…

All dimensions are given in millimeters.

British Standard Clearance Holes for Metric Bolts and Screws.—The dimensions of the clearance holes specified in this British Standard BS 4186:1967 have been chosen in such a way as to require the use of the minimum number of drills. The recommendations cover three series of clearance holes, namely close fit (H 12), medium fit (H 13), and free fit (H 14) and are suitable for use with bolts and screws specified in the following metric British Standards: BS 3692, ISO metric precision hexagon bolts, screws, and nuts; BS 4168, Hexagon socket screws and wrench keys; BS 4183, Machine screws and machine screw nuts; and BS 4190, ISO metric black hexagon bolts, screws, and nuts. The sizes are in accordance with those given in ISO Recommendation R273, and the range has been extended up to 150 millimeters diameter in accordance with an addendum to that recommendation. The selection of clearance holes sizes to suit particular design requirements can of course be dependent upon many variable factors. It is however felt that the medium fit series should suit the majority of general purpose applications. In the Standard, limiting dimensions are given in a table which is included for reference purposes only, for use in instances where it may be desirable to specify tolerances.

To avoid any risk of interference with the radius under the head of bolts and screws, it is necessary to countersink slightly all recommended clearance holes in the close and medium fit series. Dimensional details for the radius under the head of fasteners made according to BS 3692 are given on page 1556; those for fasteners to BS 4168 are given on page 1614; those to BS 4183 are given on pages 1588 through 1592.

Cold Form Tapping.—Cold form taps do not have cutting edges or conventional flutes; the threads on the tap form the threads in the hole by displacing the metal in an extrusion or swaging process. The threads thus produced are stronger than conventionally cut threads because the grains in the metal are unbroken and the displaced metal is work hardened. The surface of the thread is burnished and has an excellent finish. Although chip problems are eliminated, cold form tapping does displace the metal surrounding the hole and countersinking or chamfering before tapping is recommended. Cold form tapping is not recommended if the wall thickness of the hole is less than two-thirds of the nominal diameter of the thread. If possible, blind holes should be drilled deep enough to permit a cold form tap having a four thread lead to be used as this will require less torque, produce less burr surrounding the hole, and give a greater tool life.

The operation requires 0 to 50 per cent more torque than conventional tapping, and the cold form tap will pick up its own lead when entering the hole; thus, conventional tapping machines and tapping heads can be used. Another advantage is the better tool life obtained. The best results are obtained by using a good lubricating oil instead of a conventional cutting oil.

The method can be applied only to relatively ductile metals, such as low-carbon steel, leaded steels, austenitic stainless steels, wrought aluminum, low-silicon aluminum die casting alloys, zinc die casting alloys, magnesium, copper, and ductile copper alloys. A higher than normal tapping speed can be used, sometimes by as much as 100 per cent.

Conventional tap drill sizes should not be used for cold form tapping because the metal is displaced to form the thread. The cold formed thread is stronger than the conventionally tapped thread, so the thread height can be reduced to 60 per cent without much loss of strength; however, the use of a 65 per cent thread is strongly recommended. The following formula is used to calculate the theoretical hole size for cold form tapping:

$$\text{Theoretical hole size} = \text{basic tap O.D.} - \frac{0.0068 \times \text{per cent of full thread}}{\text{threads per inch}}$$

The theoretical hole size and the tap drill sizes for American Unified threads are given in Table 8, and Table 9 lists drills for ISO metric threads. Sharp drills should be used to prevent cold working the walls of the hole, especially on metals that are prone to work hardening. Such damage may cause the torque to increase, possibly stopping the machine or breaking the tap. On materials that can be die cast, cold form tapping can be done in cored holes provided the correct core pin size is used. The core pins are slightly tapered, so the theoretical hole size should be at the position on the pin that corresponds to one-half of the required engagement length of the thread in the hole. The core pins should be designed to form a chamfer on the hole to accept the vertical extrusion.

Table 8. Theoretical and Tap Drill or Core Hole Sizes for Cold Form Tapping Unified Threads

Tap Size	Threads Per Inch	Percentage of Full Thread								
		75			65			55		
		Theor. Hole Size	Nearest Drill Size	Dec. Equiv.	Theor. Hole Size	Nearest Drill Size	Dec. Equiv.	Theor. Hole Size	Nearest Drill Size	Dec. Equiv.
0	80	0.0536	1.35 mm	0.0531	0.0545	…	…	0.0554	54	0.055
1	64	0.0650	1.65 mm	0.0650	0.0661	…	…	0.0672	51	0.0670
	72	0.0659	1.65 mm	0.0650	0.0669	1.7 mm	0.0669	0.0679	51	0.0670
2	56	0.0769	1.95 mm	0.0768	0.0781	5§64	0.0781	0.0794	2.0 mm	0.0787
	64	0.0780	5§64	0.0781	0.0791	2.0 mm	0.0787	0.0802	…	…
3	48	0.0884	2.25 mm	0.0886	0.0898	43	0.089	0.0913	2.3 mm	0.0906
	56	0.0889	43	0.089	0.0911	2.3 mm	0.0906	0.0924	2.35 mm	0.0925
4	40	0.0993	2.5 mm	0.0984	0.1010	39	0.0995	0.1028	2.6 mm	0.1024
	48	0.0104	38	0.1015	0.1028	2.6 mm	0.1024	0.1043	37	0.1040
5	40	0.1123	34	0.1110	0.1140	33	0.113	0.1158	32	0.1160
	44	0.1134	33	0.113	0.1150	2.9 mm	0.1142	0.1166	32	…
6	32	0.1221	3.1 mm	0.1220	0.1243	…	…	0.1264	3.2 mm	0.1260
	40	0.1253	1§8	0.1250	0.1270	3.2 mm	0.1260	0.1288	30	0.1285
8	32	0.1481	3.75 mm	0.1476	0.1503	25	0.1495	0.1524	24	0.1520
	36	0.1498	25	0.1495	0.1518	24	0.1520	0.1537	3.9 mm	0.1535
10	24	0.1688	…	…	0.1717	11§64	0.1719	0.1746	17	0.1730
	32	0.1741	17	0.1730	0.1763	…	…	0.1784	4.5 mm	0.1772
12	24	0.1948	10	0.1935	0.1977	5.0 mm	0.1968	0.2006	5.1 mm	0.2008
	28	0.1978	5.0 mm	0.1968	0.2003	8	0.1990	0.2028	…	…
1§4	20	0.2245	5.7 mm	0.2244	0.2280	1	0.2280	0.2315	…	…
	28	0.2318	…	…	0.2343	A	0.2340	0.2368	6.0 mm	0.2362
5§16	18	0.2842	7.2 mm	0.2835	0.2879	7.3 mm	0.2874	0.2917	7.4 mm	0.2913
	24	0.2912	7.4 mm	0.2913	0.2941	M	0.2950	0.2969	19§64	0.2969
3§8	16	0.3431	11§32	0.3437	0.3474	S	0.3480	0.3516	…	…
	24	0.3537	9.0 mm	0.3543	0.3566	…	…	0.3594	23§64	0.3594
7§16	14	0.4011	…	…	0.4059	13§32	0.4062	0.4108	…	…
	20	0.4120	Z	0.413	0.4154	…	…	0.4188	…	…
1§2	13	0.4608	…	…	0.4660	…	…	0.4712	12 mm	0.4724
	20	0.4745	…	…	0.4779	…	…	0.4813	…	…
9§16	12	0.5200	…	…	0.5257	…	…	0.5313	17§32	0.5312
	18	0.5342	13.5 mm	0.5315	0.5380	…	…	0.5417	…	…
5§8	11	0.5787	37§64	0.5781	0.5848	…	…	0.5910	15 mm	0.5906
	18	0.5976	19§32	0.5937	0.6004	…	…	0.6042	…	…
3§4	10	0.6990	…	…	0.7058	45§64	0.7031	0.7126	…	…
	16	0.7181	23§32	0.7187	0.7224	…	…	0.7266	…	…

Table 9. Tap Drill or Core Hole Sizes for Cold Form Tapping ISO Metric Threads

Nominal Size of Tap	Pitch	Recommended Tap Drill Size
1.6 mm	0.35 mm	1.45 mm
1.8 mm	0.35 mm	1.65 mm
2.0 mm	0.40 mm	1.8 mm
2.2 mm.	0.45 mm	2.0 mm
2.5 mm	0.45 mm	2.3 mm
3.0 mm	0.50 mm	2.8 mm[a]
3.5 mm	0.60 mm	3.2 mm
4.0 mm	0.70 mm	3.7 mm
4.5 mm	0.75 mm	4.2 mm[a]
5.0 mm	0.80 mm	4.6 mm
6.0 mm	1.00 mm	5.6 mm[a]
7.0 mm	1.00 mm	6.5 mm
8.0 mm	1.25 mm	7.4 mm
10.0 mm	1.50 mm	9.3 mm

[a] These diameters are the nearest stocked drill sizes and not the theoretical hole size, and may not produce 60 to 75 per cent full thread.

The sizes are calculated to provide 60 to 75 per cent of full thread.

Removing a Broken Tap.—Broken taps can be removed by electrodischarge machining (EDM), and this method is recommended when available. When an EDM machine is not available, broken taps may be removed by using a tap extractor, which has fingers that enter the flutes of the tap; the tap is backed out of the hole by turning the extractor with a wrench. Sometimes the injection of a small amount of a proprietary solvent into the hole will be helpful. A solvent can be made by diluting about one part nitric acid with five parts water. The action of the proprietary solvent or the diluted nitric acid on the steel loosens the tap so that it can be removed with pliers or with a tap extractor. The hole should be washed out afterwards so that the acid will not continue to work on the part. Another method is to add, by electric arc welding, additional metal to the shank of the broken tap, above the level of the hole. Care must be taken to prevent depositing metal on the threads in the tapped hole. After the shank has been built up, the head of a bolt or a nut is welded to it and then the tap may be backed out.

Tap Drills for Pipe Taps

Size of Tap	Drills for Briggs PipeTaps	Drills for Whitworth Pipe Taps	Size of Tap	Drills for Briggs Pipe Taps	Drills for Whitworth Pipe Taps	Size of Tap	Drills for Briggs Pipe Taps	Drills for Whitworth Pipe Taps
1/8	11/32	5/16	1 1/4	1 1/2	1 15/32	3 1/4	…	3 1/2
1/4	7/16	27/64	1 1/2	1 23/32	1 25/32	3 1/2	3 3/4	3 3/4
3/8	19/32	9/16	1 3/4	…	1 15/16	3 3/4	…	4
1/2	23/32	11/16	2	2 3/16	2 5/32	4	4 1/4	4 1/4
5/8	…	25/32	2 1/4	…	2 13/32	4 1/2	4 3/4	4 3/4
3/4	15/16	29/32	2 1/2	2 5/8	2 25/32	5	5 5/16	5 1/4
7/8	…	1 1/16	2 3/4	…	3 1/32	5 1/2	…	5 3/4
1	1 5/32	1 1/8	3	3 1/4	3 9/32	6	6 3/8	6 1/4

All dimensions are in inches.

To secure the best results, the hole should be reamed before tapping with a reamer having a taper of 3/4 inch per foot.

Power for Pipe Taps.—The power required for driving pipe taps is given in the following table, which includes nominal pipe tap sizes from 2 to 8 inches.

The holes to be tapped were reamed with standard pipe tap reamers before tapping. The horsepower recorded was read off just before the tap was reversed. The table gives the net horsepower, deductions being made for the power required to run the machine without a load. The material tapped was cast iron, except in two instances, where cast steel was tapped. It will be seen that nearly double the power is required for tapping cast steel. The power varies, of course, with the conditions. More power than that indicated in the table will be required if the cast iron is of a harder quality or if the taps are not properly relieved. The taps used in these experiments were of the inserted-blade type, the blades being made of high-speed steel.

Power Required for Pipe Taps

Nominal Tap Size	Rev. per Min.	Net H.P.	Thickness of Metal	Nominal Tap Size	Rev. per Min.	Net H.P.	Thickness of Metal
2	40	4.24	1⅛	3½	25.6	7.20	1¾
2½	40	5.15	1⅛	4	18	6.60	2
[a]2½	38.5	9.14	1⅛	5	18	7.70	2
3	40	5.75	1⅛	6	17.8	8.80	2
[a]3	38.5	9.70	1⅛	8	14	7.96	2½

[a] Tapping cast steel; other tests in cast iron.

Tap size and metal thickness are in inches.

High-Speed CNC Tapping.—Tapping speed depends on the type of material being cut, the type of cutting tool, the speed and rigidity of the machine, the rigidity of the part-holding fixture, and the proper use of coolants and cutting fluids. When tapping, each revolution of the tool feeds the tap a distance equal to the thread pitch. Both spindle speed and feed per revolution must be accurately controlled so that changes in spindle speed result in a corresponding change in feed rate. If the feed/rev is not right, a stripped thread or broken tap will result. NC/CNC machines equipped with the *synchronous tapping* feature are able to control the tap feed as a function of spindle speed. These machines can use rigid-type tap holders or automatic tapping attachments and are able to control depth very accurately. Older NC machines that are unable to reliably coordinate spindle speed and feed must use a tension-compression type tapping head that permits some variation of the spindle speed while still letting the tap feed at the required rate.

CNC machines capable of synchronous tapping accurately coordinate feed rate and rotational speed so that the tap advances at the correct rate regardless of the spindle speed. A canned tapping cycle (see Fixed Cycles in the Numerical Control section) usually controls the operation, and feed and speed are set by the machine operator or part programmer. Synchronized tapping requires reversing the tapping spindle twice for each hole tapped, once after finishing the cut and again at the end of the cycle. Because the rotating mass is fairly large (motor, spindle, chuck or tap holder, and tap), the acceleration and deceleration of the tap are rather slow and a lot of time is lost by this process. The frequent changes in cutting speed during the cut also accelerate tap wear and reduce tap life.

A self-reversing tapping attachment has a forward drive that rotates in the same direction as the machine spindle, a reverse drive that rotates in the opposite direction, and a neutral position in between the two. When a hole is tapped, the spindle feeds at a slightly slower rate than the tap to keep the forward drive engaged until the tap reaches the bottom of the hole. Through holes are tapped by feeding to the desired depth and then retracting the spindle, which engages the tapping-head reverse drive and backs the tap out of the hole—the spindle does not need to be reversed. For tapping blind holes, the spindle is fed to a depth equal to the thread depth minus the self-feed of the tapping attachment. When the spindle is retracted (without reversing), the tap continues to feed forward a short distance (the tapping head self-feed distance) before the reverse drive engages and reverse drives the tap out of the hole. The depth can be controlled to within about ¼ revolution of the tap. The tapping cycle normally used for the self-reversing tap attachment is a standard boring cycle with feed return and no dwell. A typical programming cycle is illustrated with a G85 block on page 1260. The inward feed is set to about 95 per cent of the normal tapping feed (i.e., 95 per cent of the pitch per revolution). Because the tap is lightweight, tap reversal is almost instantaneous and tapping speed is very fast compared with synchronous tapping.

Tapping speeds are usually given in surface feet per minute (sfm) or the equivalent feet per minute (fpm or ft/min), so a conversion is necessary to get the spindle speed in revolutions per minute. The tapping speed in rpm depends on the diameter of the tap, and is given by the following formula:

$$\text{rpm} = \frac{\text{sfm} \times 12}{d \times 3.14159} = \frac{\text{sfm} \times 3.82}{d}$$

where d is the nominal diameter of the tap in inches. As indicated previously, the feed in in/rev is equal to the thread pitch and is independent of the cutting speed. The feed rate in inches per minute is found by dividing the tapping speed in rpm by the number of threads per inch, or by multiplying the speed in rpm by the pitch or feed per revolution:

$$\text{feed rate (in/min)} = \frac{\text{rpm}}{\text{threads per inch}} = \text{rpm} \times \text{thread pitch} = \text{rpm} \times \text{feed/rev}$$

Example: If the recommended tapping speed for 1020 steel is given as 45 to 60 sfm, find the required spindle speed and feed rate for tapping a 1/4–20 UNF thread in 1020 steel.

Assuming that the machine being used is in good condition and rigid, and the tap is sharp, use the higher rate of 60 sfm and calculate the required spindle speed and feed rate as follows:

$$\text{speed} = \frac{60 \times 3.82}{0.25} = 916.8 \approx 920 \text{ rpm} \qquad \text{feed rate} = \frac{920}{20} = 46 \text{ in/min}$$

Coolant for Tapping.—Proper use of through-the-tap high-pressure coolant/lubricant can result in increased tap life, increased speed and feed, and more accurate threads. In most chip-cutting processes, cutting fluid is used primarily as a coolant, with lubrication being a secondary but important benefit. Tapping, however, requires a cutting fluid with lubricity as the primary property and coolant as a secondary benefit. Consequently, the typical blend of 5 per cent coolant concentrate to 95 per cent water is too low for best results. An increased percentage of concentrate in the blend helps the fluid to cling to the tap, providing better lubrication at the cutting interface. A method of increasing the tap lubrication qualities without changing the concentration of the primary fluid blend is to use a cutting fluid dispenser controlled by an M code different from that used to control the high-pressure flood coolant (for example, use an M08 code in addition to M07). The secondary coolant-delivery system applies a small amount of an edge-type cutting fluid (about a drop at a time) directly onto the tap-cutting surfaces providing the lubrication needed for cutting. The edge-type fluid applied in this way clings to the tap, increasing the lubrication effect and ensuring that the cutting fluid becomes directly involved in the cutting action at the shear zone.

High-pressure coolant fed through the tap is important in many high-volume tapping applications. The coolant is fed directly through the spindle or tool holder to the cutting zone, greatly improving the process of chip evacuation and resulting in better thread quality. High-pressure through-the-tap coolant flushes blind holes before the tap enters and can remove chips from the holes after tapping is finished. The flushing action prevents chip recutting by forcing chips through the flutes and back out of the hole, improving the surface of the thread and increasing tap life. By improving lubrication and reducing heat and friction, the use of high-pressure coolant may result in increased tap life up to five times that of conventional tapping and may permit speed and feed increases that reduce overall cycle time.

Combined Drilling and Tapping.—A special tool that drills and taps in one operation can save a lot of time by reducing setup and eliminating a secondary operation in some applications. A combination drill and tap can be used for through holes if the length of the fluted drill section is greater than the material thickness, but cannot be used for drilling and tapping blind holes because the tip (drill point) must cut completely through the material before the tapping section begins to cut threads. Drilling and tapping depths up to twice the tool diameter are typical. Determine the appropriate speed by starting the tool at the recommended speed for the tap size and material, and adjust the speed higher or lower to suit the application. Feed during tapping is dependent on the thread pitch. NC/CNC programs can use a fast drilling speed and a slower tapping speed to combine both operations into one and minimize cutting time.

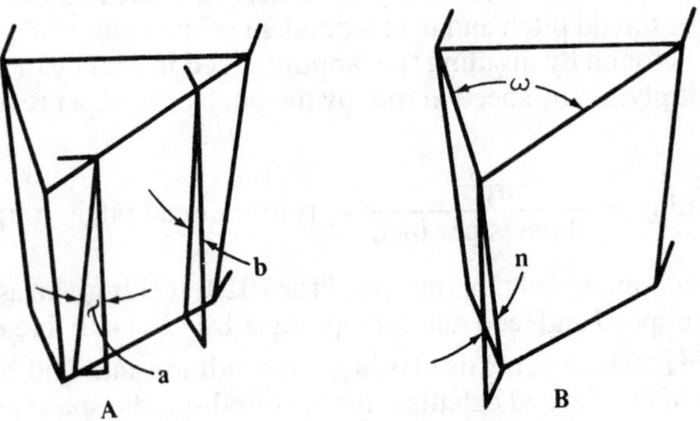

Two similar diagrams showing relationships of various relief angles of thread cutting tools

Relief Angles for Single-Point Thread Cutting Tools.—The surface finish on threads cut with single-point thread cutting tools is influenced by the relief angles on the tools. The leading and trailing cutting edges that form the sides of the thread, and the cutting edge at the nose of the tool must all be provided with an adequate amount of relief. Moreover, it is recommended that the effective relief angle, a_e, for all of these cutting edges be made equal, although the practice in some shops is to use slightly less relief at the trailing cutting edge. While too much relief may weaken the cutting edge, causing it to chip, an inadequate amount of relief will result in rough threads and in a shortened tool life. Other factors that influence the finish produced on threads include the following: the work material; the cutting speed; the cutting fluid used; the method used to cut the thread; and, the condition of the cutting edge.

Relief angles on single-point thread cutting tools are often specified on the basis of experience. While this method may give satisfactory results in many instances, better results can usually be obtained by calculating these angles, using the formulas provided further on. When special high helix angle threads are to be cut, the magnitude of the relief angles should always be calculated. These calculations are based on the effective relief angle, a_e; this is the angle between the flank of the tool and the sloping sides of the thread, measured in a direction parallel to the axis of the thread. Recommended values of this angle are 8 to 14 degrees for high speed steel tools, and 5 to 10 degrees for cemented carbide tools. The larger values are recommended for cutting threads on soft and gummy materials, and the smaller values are for the harder materials, which inherently take a better surface finish. Harder materials also require more support below the cutting edges, which is provided by using a smaller relief angle. These values are recommended for the relief angle below the cutting edge at the nose without any further modification. The angles below the leading and trailing side cutting edges are modified, using the formulas provided. The angles b and b' are the relief angles actually ground on the tool below the leading and trailing side cutting edges respectively; they are measured perpendicular to the side cutting edges. When designing or grinding the thread cutting tool, it is sometimes helpful to know the magnitude of the angle, n, for which a formula is provided. This angle would occur only in the event that the tool were ground to a sharp point. It is the angle of the edge formed by the intersection of the flank surfaces.

$$\tan\phi = \frac{\text{lead of thread}}{\pi K} \qquad \tan\phi' = \frac{\text{lead of thread}}{\pi D}$$

$$a = a_e + \phi$$

$$a' = a_e - \phi'$$

$$\tan b = \tan a \cos \tfrac{1}{2}\omega$$

$$\tan b' = \tan a' \cos \tfrac{1}{2}\omega$$

$$\tan n = \frac{\tan a - \tan a'}{2\tan \tfrac{1}{2}\omega}$$

where θ = helix angle of thread at minor diameter

θ' = helix angle of thread at major diameter

K = minor diameter of thread

D = major diameter of thread

a = side relief angle parallel to thread axis at leading edge of tool

a' = side relief angle parallel to thread axis at trailing edge of tool

a_e = effective relief angle

b = side relief angle perpendicular to leading edge of tool

b' = side relief angle perpendicular to trailing edge of tool

ω = included angle of thread cutting tool

n = nose angle resulting from intersection of flank surfaces

Example: Calculate the relief angles and the nose angle n for a single-point thread cutting tool that is to be used to cut a 1-inch diameter, 5-threads-per-inch, double Acme thread. The lead of this thread is $2 \times 0.200 = 0.400$ inch. The included angle ω of this thread is 29 degrees, the minor diameter K is 0.780 inch, and the effective relief angle a_e below all cutting edges is to be 10 degrees.

$$\tan\phi = \frac{\text{lead of thread}}{\pi K} = \frac{0.400}{\pi \times 0.780}$$

$$\phi = 9.27°(9°16')$$

$$\tan\phi' = \frac{\text{lead of thread}}{\pi D} = \frac{0.400}{\pi \times 1.000}$$

$$\phi' = 7.26°(7°15')$$

$$a = a_e + \phi = 10° + 9.27° = 19.27°$$

$$a' = a_e - \phi' = 10° - 7.26° = 2.74°$$

$$\tan b = \tan a \cos \tfrac{1}{2}\omega = \tan 19.27 \cos 14.5$$

$$b = 18.70°(18°42')$$

$$\tan b' = \tan a' \cos \tfrac{1}{2}\omega = \tan 2.74 \cos 14.5$$

$$b' = 2.65°(2°39')$$

$$\tan n = \frac{\tan a - \tan a'}{2\tan \tfrac{1}{2}\omega} = \frac{\tan 19.27 - \tan 2.74}{2\tan 14.5}$$

$$n = 30.26°(30°16')$$

Lathe Change Gears

Change Gears for Thread Cutting.—To determine the change gears to use for cutting a thread of given pitch, first find what number of threads per inch will be cut when gears of the same size are placed on the lead screw and spindle stud, either by trial or by referring to the index plate; then multiply this number, called the "lathe screw constant," by some trial number to obtain the number of teeth in the gear for the spindle stud, and multiply the threads per inch to be cut by the *same* trial number to obtain the number of teeth in the gear for the lead screw. Expressing this rule as a formula:

$$\frac{\text{Trial number} \times \text{lathe screw constant}}{\text{Trial number} \times \text{threads per inch to be cut}} = \frac{\text{teeth in gear on spindle stud}}{\text{teeth in gear on lead screw}}$$

For example, suppose the available change gears supplied with the lathe have 24, 28, 32, 36 teeth, etc., the number increasing by 4 up to 100, and that 10 threads per inch are to be cut in a lathe having a lathe screw constant of 6; then, if the screw constant is written as the numerator, the number of threads per inch to be cut as the denominator of a fraction, and both numerator and denominator are multiplied by some trial number, say, 4, it is found that gears having 24 and 40 teeth can be used. Thus:

$$\frac{6}{10} = \frac{6 \times 4}{10 \times 4} = \frac{24}{40}$$

The 24-tooth gear goes on the spindle stud and the 40-toothgear on the lead screw.

The lathe screw constant is, of course, equal to the number of threads per inch on the lead screw, provided the spindle stud and spindle are geared in the ratio of 1 to 1, which, however. is not always so.

Compound Gearing.—To find the change gears used in compound gearing, place the screw constant as the numerator and the number of threads per inch to be cut as the denominator of a fraction; resolve both numerator and denominator into two factors each, and multiply each "pair" of factors by the same number, until values are obtained representing suitable numbers of teeth for the change gears. (One factor in the numerator and one in the denominator make a "pair" of factors.)

Example:—1¾ threads per inch are to be cut in a lathe having a screw constant of 8; the available gears have 24, 28, 32, 36, 40 teeth. etc., increasing by 4 up to 100. Following the rule:

$$\frac{8}{1\frac{3}{4}} = \frac{2 \times 4}{1 \times 1\frac{3}{4}} = \frac{(2 \times 36) \times (4 \times 16)}{(1 \times 36) \times (1\frac{3}{4} \times 16)} = \frac{72 \times 64}{36 \times 28}$$

The gears having 72 and 64 teeth are the *driving* gears and those with 36 and 28 teeth are the *driven* gears.

Fractional Threads.—Sometimes the lead of a thread is given as a fraction of an inch instead of stating the number of threads per inch. For example, a thread may be required to be cut, having ⅜ inch lead. The expression "⅜ inch lead" should first be transformed to "number of threads per inch." The number of threads per inch (the thread being single) equals:

$$\frac{1}{\frac{3}{8}} = 1 \div \frac{3}{8} = \frac{8}{3} = 2\frac{2}{3}$$

To find the change gears to cut 2⅔ threads per inch in a lathe having a screw constant 8 and change gears ranging from 24 to 100 teeth, increasing in increments of 4, proceed as below:

$$\frac{8}{2\frac{2}{3}} = \frac{2 \times 4}{1 \times 2\frac{2}{3}} = \frac{(2 \times 36) \times (4 \times 24)}{(1 \times 36) \times (2\frac{2}{3} \times 24)} = \frac{72 \times 96}{36 \times 64}$$

Change Gears for Metric Pitches.—When screws are cut in accordance with the metric system, it is the usual practice to give the lead of the thread in millimeters, instead of the number of threads per unit of measurement. To find the change gears for cutting metric threads, when using a lathe having an inch lead screw, first determine the number of threads per inch corresponding to the given lead in millimeters. Suppose a thread of 3 millimeters lead is to be cut in a lathe having an inch lead screw and a screw constant of 6. As there are 25.4 millimeters per inch, the number of threads per inch will equal 25.4 ÷ 3. Place the screw constant as the numerator, and the number of threads per inch to be cut as the denominator:

$$\frac{6}{\frac{25.4}{3}} = 6 \div \frac{25.4}{3} = \frac{6 \times 3}{25.4}$$

The numerator and denominator of this fractional expression of the change gear ratio is next multiplied by some trial number to determine the size of the gears. The first whole number by which 25.4 can be multiplied so as to get a whole number as the result is 5. Thus, 25.4 × 5 = 127. Hence, one gear having 127 teeth is always used when cutting metric threads with an inch lead screw. The other gear required has 90 teeth. Thus:

$$\frac{6 \times 3 \times 5}{25.4 \times 5} = \frac{90}{127}$$

Therefore, the following rule can be used to find the change gears for cutting metric pitches with an inch lead screw:

Rule: Place the lathe screw constant multiplied by the lead of the required thread in millimeters multiplied by 5 as the numerator of the fraction and 127 as the denominator. The product of the numbers in the numerator equals the number of teeth for the spindle-stud gear, and 127 is the number of teeth for the lead-screw gear.

If the lathe has a metric pitch lead screw, and a screw having a given number of threads per inch is to be cut, first find the "metric screw constant" of the lathe or the lead of thread in millimeters that would be cut with change gears of equal size on the lead screw and spindle stud; then the method of determining the change gears is simply the reverse of the one already explained for cutting a metric thread with an inch lead screw.

Rule: To find the change gears for cutting inch threads with a metric lead screw, place 127 in the numerator and the threads per inch to be cut, multiplied by the metric screw constant multiplied by 5 in the denominator; 127 is the number of teeth on the spindle-stud gear and the product of the numbers in the denominator equals the number of teeth in the lead-screw gear.

Threads per Inch Obtained with a Given Combination of Gears.—To determine the number of threads per inch that will be obtained with a given combination of gearing, multiply the lathe screw constant by the number of teeth in the *driven* gear (or by the product of the numbers of teeth in both driven gears of compound gearing), and divide the product thus obtained by the number of teeth in the *driving* gear (or by the product of the two driving gears of a compound train). The quotient equals the number of threads per inch.

Change Gears for Fractional Ratios.—When gear ratios cannot be expressed exactly in whole numbers that are within the range of ordinary gearing, the combination of gearing required for the fractional ratio may be determined quite easily, often by the "cancellation method." To illustrate this method, assume that the speeds of two gears are to be in the ratio of 3.423 to 1. The number 3.423 is first changed to $^{3423}/_{1000}$ to clear it of decimals. Then, in order to secure a fraction that can be reduced, 3423 is changed to 3420;

$$\frac{3420}{1000} = \frac{342}{100} = \frac{3 \times 2 \times 57}{2 \times 50} = \frac{3 \times 57}{1 \times 50}$$

Then, multiplying $\frac{3}{1}$ by some trial number, say, 24, the following gear combination is obtained:

$$\frac{72}{24} \times \frac{57}{50} = \frac{4104}{1200} = \frac{3.42}{1}$$

As the desired ratio is 3.423 to I, there is an error of 0.003. When the ratios are comparatively simple, the cancellation method is not difficult and is frequently used; but by the logarithmic method to be described, more accurate results are usually possible.

Modifying the Quick-Change Gearbox Output.—On most modern lathes, the gear train connecting the headstock spindle with the lead screw contains a quick-change gearbox. Instead of using different change gears, it is only necessary to position the handles of the gearbox to adjust the speed ratio between the spindle and the lead screw in preparation for cutting a thread. However, a thread sometimes must be cut for which there is no quick-change gearbox setting. It is then necessary to modify the normal, or standard, gear ratio between the spindle and the gearbox by installing modifying change gears to replace the standard gears normally used. Metric and other odd pitch threads can be cut on lathes that have an inch thread lead screw and a quick-change gearbox having only settings for inch threads by using modifying-change gears in the gear train. Likewise, inch threads and other odd pitch threads can be cut on metric lead-screw lathes having a gearbox on which only metric thread settings can be made. Modifying-change gears also can be used for cutting odd pitch threads on lathes having a quick-change gearbox that has both inch and metric thread settings.

The sizes of the modifying-change gears can be calculated by formulas to be given later; they depend on the thread to be cut and on the setting of the quick-change gearbox. Many different sets of gears can be found for each thread to be cut. It is recommended that several calculations be made in order to find the set of gears that is most suitable for installation on the lathe. The modifying-change gear formulas that follow are based on the type of lead screw, i.e., whether the lead screw has inch or metric threads.

Metric Threads on Inch Lead-Screw Lathes: A 127-tooth translating gear must be used in the modifying-change gear train in order to be able to cut metric threads on inch lead-screw lathes. The formula for calculating the modifying change gears is:

$$\frac{5 \times \text{gearbox setting in thds/in.} \times \text{pitch in mm to be cut}}{127} = \frac{\text{driving gears}}{\text{driven gears}}$$

The numerator and denominator of this formula are multiplied by equal numbers, called trial numbers, to find the gears. If suitable gears cannot be found with one set, then another set of equal trial numbers is used. (Because these numbers are equal, such as 15/15 or 24/24, they are equal to the number one when thought of as a fraction; their inclusion has the effect of multiplying the formula by one, which does not change its value.) It is necessary to select the gearbox setting in threads per inch that must be used to cut the metric thread when using the gears calculated by the formula. One method is to select a quick-change gearbox setting that is close to the actual number of metric threads in a 1-inch length, called the equivalent threads per inch, which can be calculated by the following formula: Equivalent thds/in. = 25.4 ÷ pitch in millimeters to be cut.

Example: Select the quick-change gearbox setting and calculate the modifying change gears required to set up a lathe having an inch-thread lead screw in order to cut an M12 × 1.75 metric thread.

$$\text{Equivalent thds/in.} = \frac{25.4}{\text{pitch in mm to be cut}} = \frac{25.4}{1.75} = 1.45 \text{ (use 14 thds/in.)}$$

$$\frac{5 \times \text{gearbox setting in thds/in.} \times \text{pitch in mm to be cut}}{127} = \frac{5 \times 14 \times 1.75}{127}$$

$$= \frac{(24) \times 5 \times 14 \times 1.75}{(24) \times 127} = \frac{(5 \times 14) \times (24 \times 1.75)}{24 \times 127}$$

$$\frac{70 \times 42}{24 \times 127} = \frac{\text{driving gears}}{\text{driven gears}}$$

Odd Inch Pitch Threads: The calculation of the modifying change gears used for cutting odd pitch threads that are specified by their pitch in inches involves the sizes of the standard gears, which can be found by counting their teeth. Standard gears are those used to enable the lathe to cut the thread for which the gearbox setting is made; they are the gears that are normally used. The threads on worms used with worm gears are among the odd pitch threads that can be cut by this method. As before, it is usually advisable to calculate the actual number of threads per inch of the odd pitch thread and to select a gearbox setting that is close to this value. The following formula is used to calculate the modifying-change gears to cut odd inch pitch threads:

$$\frac{\text{Standard driving gear} \times \text{pitch to be cut in inches} \times \text{gearbox setting in thds/in.}}{\text{Standard driven gear}}$$

$$= \frac{\text{driving gears}}{\text{driven gears}}$$

Example: Select the quick-change gearbox setting and calculate the modifying change gears required to cut a thread having a pitch equal to 0.195 inch. The standard driving and driven gears both have 48 teeth. To find equivalent threads per inch:

$$\frac{\text{Thds}}{\text{in.}} = \frac{1}{\text{pitch}} = \frac{1}{0.195} = 5.13 \qquad \text{(use 5 thds/in.)}$$

$$\frac{\text{Standard driving gear} \times \text{pitch to be cut in inches} \times \text{gearbox setting in thds/in.}}{\text{Standard driven gear}}$$

$$= \frac{48 \times 0.195 \times 5}{48} = \frac{(1000) \times 0.195 \times 5}{(1000)} = \frac{195 \times 5}{500 \times 2} = \frac{39 \times 5}{100 \times 2} = \frac{39 \times 5 \times (8)}{50 \times 2 \times 2 \times (8)}$$

$$= \frac{39 \times 40}{50 \times 32} = \frac{\text{driving gears}}{\text{driven gears}}$$

It will be noted that in the second step above, 1000/1000 has been substituted for 48/48. This substitution does not change the ratio. The reason for this substitution is that 1000 × 0.195 = 195, a whole number. Actually, 200/200 might have been substituted because 200 × 0.195 = 39, also a whole number.

The procedure for calculating the modifying gears using the following formulas is the same as illustrated by the two previous examples.

Odd Threads per Inch on Inch Lead Screw Lathes:

$$\frac{\text{Standard driving gear} \times \text{gearbox setting in thds/in.}}{\text{Standard driven gear} \times \text{thds/in. to be cut}} = \frac{\text{driving gears}}{\text{driven gears}}$$

Inch Threads on Metric Lead Screw Lathes:

$$\frac{127}{5 \times \text{gearbox setting in mm pitch} \times \text{thds/in. to be cut}} = \frac{\text{driving gears}}{\text{driven gears}}$$

Odd Metric Pitch Threads on Metric Lead Screw Lathes:

$$\frac{\text{Standard driving gear} \times \text{mm pitch to be cut}}{\text{Standard driven gear} \times \text{gearbox setting in mm pitch}} = \frac{\text{driving gears}}{\text{driven gears}}$$

Finding Accurate Gear Ratios.—Tables included in the 23rd and earlier editions of this handbook furnished a series of logarithms of gear ratios as a quick means of finding ratios for all gear combinations having 15 to 120 teeth. The ratios thus determined could be factored into sets of 2, 4, 6, or any other even numbers of gears to provide a desired overall ratio.

Although the method of using logarithms of gear ratios provides results of suitable accuracy for many gear-ratio problems, it does not provide a systematic means of evaluating whether other, more accurate ratios are available. In critical applications, especially in the design of mechanisms using reduction gear trains, it may be desirable to find many or all possible ratios to meet a specified accuracy requirement. The methods best suited to such problems use *Continued Fractions* and *Conjugate Fractions* as explained starting on pages 13 and illustrated in the worked-out example on page 14 for a set of four change gears.

As an example, if an overall reduction of 0.31416 is required, a fraction must be found such that the factors of the numerator and denominator may be used to form a four-gear reduction train in which no gear has more than 120 teeth. By using the method of conjugate fractions discussed on page 14, the ratios listed above, and their factors are found to be successively closer approximations to the required overall gear ratio.

Ratio	Numerator Factors	Denominator	Error Factors
11/35	11	5×7	+0.00013
16/51	2×2×2×2	3×17	−0.00043
27/86	3×3×3	2×43	−0.00021
38/121	2×19	11×11	−0.00011
49/156	7×7	2×2×3×13	−0.00006
82/261	2×41	3×3×29	+0.00002
224/713	2×2×2×2×2×7	23×31	+0.000005
437/1391	19×23	13×107	+0.000002
721/2295	7×103	3×3×3×5×17	+0.000001
1360/4329	2×2×2×2395×17	3×3×13×53	+0.0000003
1715/5459	5×7×7×7	53×103	+0.0000001
3927/12500	3×7×11×17	2×2×5×5×5×5×5	0

Lathe Change-gears.—To calculate the change gears to cut any pitch on a lathe, the "constant" of the machine must be known. For any lathe, the ratio C:L = driver:driven gear, in which C = constant of machine and L = threads per inch.

For example, to find the change gears required to cut 1.7345 threads per inch on a lathe having a constant of 4, the formula:

$$\frac{C}{L} = \frac{4}{1.7345} = 2.306140$$

may be used. The method of conjugate fractions shown on page 14 will find the ratio, 113⁄49 = 2.306122, which is closer than any other having suitable factors. This ratio is in error by only 2.306140 − 2.306122 = 0.000018. Therefore, the driver should have 113 teeth and the driven gear 49 teeth.

Relieving Helical-Fluted Hobs.—Relieving hobs that have been fluted at right angles to the thread is another example of approximating a required change-gear ratio. The usual method is to change the angle of the helical flutes to agree with previously calculated change-gears. The ratio between the hob and the relieving attachment is expressed in the formula:

$$\frac{N}{(C \times \cos^2 \alpha)} = \frac{driver}{driven} \text{ gears}$$

and

$$\tan \alpha = \frac{P}{H_c}$$

in which: N = number of flutes in hob; α = helix angle of thread from plane perpendicular to axis; C = constant of relieving attachment; P = axial lead of hob; and H_c = hob pitch circumference, = 3.1416 times pitch diameter.

The constant of the relieving attachment is found on its index plate and is determined by the number of flutes that require equal gears on the change-gear studs. These values will vary with different makes of lathes.

For example, what four change-gears can be used to relieve a helical-fluted worm-gear hob, of 24 diametral pitch, six starts, 13 degrees, 41 minutes helix angle of thread, with eleven helical flutes, assuming a relieving attachment having a constant of 4 is to be used?

$$\frac{N}{(C \times \cos^2 \alpha)} = \frac{11}{(4 \times \cos^2 13°41')} = \frac{11}{(4 \times 0.944045)} = 2.913136$$

Using the conjugate fractions method discussed on page 14, the following ratios are found to provide factors that are successively closer approximations to the required change-gear ratio 2.913136.

Numerator/Denominator	*Ratio*	*Error*
67 × 78/(39 × 46)	2.913043	−0.000093
30 × 47/(22 × 22)	2.913223	+0.000087
80 × 26/(21 × 34)	2.913165	+0.000029
27 × 82/(20 × 38)	2.913158	+0.000021
55 × 75/(24 × 59)	2.913136	+0.0000004
74 × 92/(57 × 41)	2.913136	+0.00000005

THREAD ROLLING

Screw threads may be formed by rolling either by using some type of thread-rolling machine or by equipping an automatic screw machine or turret lathe with a suitable threading roll. If a thread-rolling machine is used, the unthreaded screw, bolt, or other "blank" is placed (either automatically or by hand) between dies having thread-shaped ridges that sink into the blank, and by displacing the metal, form a thread of the required shape and pitch. The thread-rolling process is applied where bolts, screws, studs, threaded rods, etc., are required in large quantities. Screw threads that are within the range of the rolling process may be produced more rapidly by this method than in any other way. Because of the cold-working action of the dies, the rolled thread is 10 to 20 per cent stronger than a cut or ground thread, and the increase may be much higher for fatigue resistance. Other advantages of the rolling process are that no stock is wasted in forming the thread, and the surface of a rolled thread is harder than that of a cut thread, thus increasing wear resistance.

Thread-Rolling Machine of Flat-Die Type.—One type of machine that is used extensively for thread rolling is equipped with a pair of flat or straight dies. One die is stationary and the other has a reciprocating movement when the machine is in use. The ridges on these dies, which form the screw thread, incline at an angle equal to the helix angle of the thread. In making dies for precision thread rolling, the threads may be formed either by milling and grinding after heat treatment, or by grinding "from the solid" after heat treating. A vitrified wheel is used.

In a thread-rolling machine, thread is formed in one passage of the work, which is inserted at one end of the dies, either by hand or automatically, and then rolls between the die faces until it is ejected at the opposite end. The relation between the position of the dies and a screw thread being rolled is such that the top of the thread-shaped ridge of one die, at the point of contact with the screw thread, is directly opposite the bottom of the thread groove in the other die at the point of contact. Some form of mechanism ensures starting the blank at the right time and square with the dies.

Thread-Rolling Machine of Cylindrical-Die Type.—With machines of this type, the blank is threaded while being rolled between two or three cylindrical dies (depending upon the type of machine) that are pressed into the blank at a rate of penetration adjusted to the hardness of the material, or wall thickness in threading operations on tubing or hollow parts. The dies have ground, or ground and lapped, threads and a pitch diameter that is a multiple of the pitch diameter of the thread to be rolled. As the dies are much larger in diameter than the work, a multiple thread is required to obtain the same lead angle as that of the work. The thread may be formed in one die revolution or even less, or several revolutions may be required (as in rolling hard materials) to obtain a gradual rate of penetration equivalent to that obtained with flat or straight dies if extended to a length of possibly 15 or 20 feet. Provisions for accurately adjusting or matching the thread rolls to bring them into proper alignment with each other are important features of these machines.

Two-Roll Type of Machine: With a two-roll type of machine, the work is rotated between two horizontal power-driven threading rolls and is supported by a hardened rest bar on the lower side. One roll is fed inward by hydraulic pressure to a depth that is governed automatically.

Three-Roll Type of Machine: With this machine, the blank to be threaded is held in a "floating position" while being rolled between three cylindrical dies that, through toggle arms, are moved inward at a predetermined rate of penetration until the required pitch diameter is obtained. The die movement is governed by a cam driven through change gears selected to give the required cycle of squeeze, dwell, and release.

Rate of Production.—Production rates in thread rolling depend upon the type of machine, the size of both machine and work, and whether the parts to be threaded are inserted by hand or automatically. A reciprocating flat die type of machine, applied to ordinary steels, may thread 30 or 40 parts per minute in diameters ranging from about ⅝ to 1⅛ inch, and 150 to 175 per minute in machine screw sizes from No. 10 (.190) to No. 6 (.138). In the case of heat-treated alloy steels in the usual hardness range of 26 to 32 Rockwell C, the production may be 30 or 40 per minute or less. With a cylindrical die type of machine, which is designed primarily for precision work and hard metals, 10 to 30 parts per minute are common production rates, the amount depending upon the hardness of material and allowable rate of die penetration per work revolution. These production rates are intended as a general guide only. The diameters of rolled threads usually range from the smallest machine screw sizes up to 1 or 1½ inches, depending upon the type and size of machine.

Precision Thread Rolling.—Both flat and cylindrical dies are used in aeronautical and other plants for precision work. With accurate dies and blank diameters held to close limits, it is practicable to produce rolled threads for American Standard Class 3 and Class 4 fits. The blank sizing may be by centerless grinding or by means of a die in conjunction with the heading operations. The blank should be round, and, as a general rule, the diameter tolerance should not exceed ½ to ⅔ the pitch diameter tolerance. The blank diameter should range from the correct size (which is close to the pitch diameter, but should be determined by actual trial), down to the allowable minimum, the tolerance being minus to insure a correct pitch diameter, even though the major diameter may vary slightly. Precision thread rolling has become an important method of threading alloy steel studs and other threaded parts, especially in aeronautical work where precision and high-fatigue resistance are required. Micrometer screws are also an outstanding example of precision thread rolling. This process has also been applied in tap making, although it is the general practice to finish rolled taps by grinding when the Class 3 and Class 4 fits are required.

Steels for Thread Rolling.—Steels vary from soft low-carbon types for ordinary screws and bolts, to nickel, nickel-chromium and molybdenum steels for aircraft studs, bolts, etc., or for any work requiring exceptional strength and fatigue resistance. Typical SAE alloy steels are No. 2330, 3135, 3140, 4027, 4042, 4640 and 6160. The hardness of these steels after heat-treatment usually ranges from 26 to 32 Rockwell C, with tensile strengths varying from 130,000 to 150,000 pounds per square inch. While harder materials might be rolled, grinding is more practicable when the hardness exceeds 40 Rockwell C. Thread rolling is applicable not only to a wide range of steels but for non-ferrous materials, especially if there is difficulty in cutting due to "tearing" the threads.

Diameter of Blank for Thread Rolling.—The diameter of the screw blank or cylindrical part upon which a thread is to be rolled should be less than the outside screw diameter by an amount that will just compensate for the metal that is displaced and raised above the original surface by the rolling process. The increase in diameter is approximately equal to the depth of one thread. While there are rules and formulas for determining blank diameters, it may be necessary to make slight changes in the calculated size in order to secure a well-formed thread. The blank diameter should be verified by trial, especially when rolling accurate screw threads. Some stock offers greater resistance to displacement than other stock, owing to the greater hardness or tenacity of the metal. The following figures may prove useful in establishing trial sizes. The blank diameters for screws varying from ¼ to ½ are from 0.002 to 0.0025 inch larger than the pitch diameter, and for screws varying from ½ to 1 inch or larger, the blank diameters are from 0.0025 to .003 inch larger than the pitch diameter. Blanks which are slightly less than the pitch diameter are intended for bolts, screws, etc., which are to have a comparatively free fit. Blanks for this class of work may vary from 0.002 to 0.003 inch less than the pitch diameter for screw thread sizes varying from ¼ to ½ inch, and from 0.003 to 0.005 inch less than the pitch diameter for sizes above ½ inch. If the screw threads are smaller than ¼ inch, the blanks are usually from 0.001 to 0.0015 inch less than the pitch diameter for ordinary grades of work.

Thread Rolling in Automatic Screw Machines.—Screw threads are sometimes rolled in automatic screw machines and turret lathes when the thread is behind a shoulder so that it cannot be cut with a die. In such cases, the advantage of rolling the thread is that a second operation is avoided. A circular roll is used for rolling threads in screw machines. The roll may be presented to the work either in a tangential direction or radially, either method producing a satisfactory thread. In the former case, the roll gradually comes into contact with the periphery of the work and completes the thread as it passes across the surface to be threaded. When the roll is held in a radial position, it is simply forced against one side until a complete thread is formed. The method of applying the roll may depend upon the relation between the threading operation and other machining operations. Thread rolling in automatic screw machines is generally applied only to brass and other relatively soft metals, owing to the difficulty of rolling threads in steel. Thread rolls made of chrome-nickel steel containing from 0.15 to 0.20 per cent of carbon have given fairly good results, however, when applied to steel. A 3 per cent nickel steel containing about 0.12 per cent carbon has also proved satisfactory for threading brass.

Factors Governing the Diameter of Thread Rolling.—The threading roll used in screw machines may be about the same diameter as the screw thread, but for sizes smaller than, say, ¾ inch, the roll diameter is some multiple of the thread diameter minus a slight amount to obtain a better rolling action. When the diameters of the thread and roll are practically the same, a single-threaded roll is used to form a single thread on the screw. If the diameter of the roll is made double that of the screw, in order to avoid using a small roll, then the roll must have a double thread. If the thread roll is three times the size of the screw thread, a triple thread is used, and so on. These multiple threads are necessary when the roll diameter is some multiple of the work, in order to obtain corresponding helix angles on the roll and work.

Diameter of Threading Roll.—The pitch diameter of a threading roll having a single thread is slightly less than the pitch diameter of the screw thread to be rolled, and in the case of multiple-thread rolls, the pitch diameter is not an exact multiple of the screw thread pitch diameter but is also reduced somewhat. The amount of reduction recommended by one screw machine manufacturer is given by the formula shown at the end of this paragraph. A description of the terms used in the formula is given as follows: D = pitch diameter of threading roll, d = pitch diameter of screw thread, N = number of single threads or "starts" on the roll (this number is selected with reference to diameter of roll desired), T = single depth of thread:

$$D = N\left(d - \frac{T}{2}\right) - T$$

Example: Find, by using above formula, the pitch diameter of a double-thread roll for rolling a ½-inch American standard screw thread. Pitch diameter d = 0.4500 inch and thread depth T = 0.0499 inch.

$$D = 2\left(0.4500 - \frac{0.0499}{2}\right) - 0.0499 = 0.8001 \text{ inch}$$

Kind of Thread on Roll and Its Shape.—The thread (or threads) on the roll should be left hand for rolling a right-hand thread, and *vice versa.* The roll should be wide enough to overlap the part to be threaded, provided there are clearance spaces at the ends, which should be formed if possible. The thread on the roll should be sharp on top for rolling an American (National) standard form of thread, so that less pressure will be required to displace the metal when rolling the thread. The bottom of the thread groove on the roll may also be left sharp or it may have a flat. If the bottom is sharp, the roll is sunk only far enough into the blank to form a thread having a flat top, assuming that the thread is the American form. The number of threads on the roll (whether double, triple, quadruple, etc.) is selected, as a rule, so that the diameter of the thread roll will be somewhere between 1¼ and 2¼ inches. In making a thread roll, the ends are beveled at an angle of 45 degrees, to prevent the threads on the ends of the roll from chipping. Precautions should be taken in hardening, because, if the sharp edges are burnt, the roll will be useless. Thread rolls are usually lapped after hardening, by holding them on an arbor in the lathe and using emery and oil on a piece of hard wood. To give good results a thread roll should fit closely in the holder. If the roll is made to fit loosely, it will mar the threads.

Application of Thread Roll.—The shape of the work and the character of the operations necessary to produce it, govern, to a large extent, the method employed in applying the thread roll. Some of the points to consider are as follows:

1) Diameter of the part to be threaded.
2) Location of the part to be threaded.
3) Length of the part to be threaded.
4) Relation that the thread rolling operation bears to the other operations.
5) Shape of the part to be threaded, whether straight, tapered or otherwise.
6) Method of applying the support.

When the diameter to be rolled is much smaller than the diameter of the shoulder preceding it, a cross-slide knurl-holder should be used. If the part to be threaded is not behind a shoulder, a holder on the swing principle should be used. When the work is long (greater in length than two-and-one-half times its diameter) a swing roll-holder should be employed, carrying a support. When the work can be cut off after the thread is rolled, a cross-slide roll-holder should be used. The method of applying the support to the work also governs to some extent the method of applying the thread roll. When no other tool is working at the same time as the thread roll, and when there is freedom from chips, the roll can be held more rigidly by passing it under instead of over the work. When passing the roll over the work, there is a tendency to raise the cross-slide. Where the part to be threaded is tapered, the roll can best be presented to the work by holding it in a cross-slide roll-holder.

Speeds and Feeds for Thread Rolling.—When the thread roll is made from high-carbon steel and used on brass, a surface speed as high as 200 feet per minute can be used. However, better results are obtained by using a lower speed than this. When the roll is held in a holder attached to the cross-slide, and is presented either tangentially or radially to the work, a considerably higher speed can be used than if it is held in a swing tool. This is due to the lack of rigidity in a holder of the swing type. The feeds to be used when a cross-slide roll-holder is used are given in the upper half of the table "Feeds for Thread Rolling;" the lower half of the table gives the feeds for thread rolling with swing tools. These feeds are applicable for rolling threads without a support, when the root diameter of the blank is not less than five times the double depth of the thread. When the root diameter is less than this, a support should be used. A support should also be used when the width of the roll is more than two-and-one-half times the smallest diameter of the piece to be rolled, irrespective of the pitch of the thread. When the smallest diameter of the piece to be rolled is much less than the root diameter of the thread, the smallest diameter should be taken as the deciding factor for the feed to be used.

Feeds for Thread Rolling

Root Diam. of Blank	Number of Threads per Inch													
	72	64	56	48	44	40	36	32	28	24	22	20	18	14
	Cross-slide Holders — Feed per Revolution in Inches													
1/8	0.0045	0.0040	0.0035	0.0030	0.0025	0.0020	0.0015	0.0010	.....	.....	.....	.....	.....	.....
3/16	0.0050	0.0045	0.0040	0.0035	0.0030	0.0025	0.0020	0.0015	0.0005	.....	.....	.....	.....	.....
1/4	0.0055	0.0050	0.0045	0.0040	0.0035	0.0030	0.0025	0.0020	0.0010	0.0005	0.0005	.....	.....	.....
5/16	0.0060	0.0055	0.0050	0.0045	0.0040	0.0035	0.0030	0.0025	0.0015	0.0010	0.0010	0.0005	0.0005	.....
3/8	0.0065	0.0060	0.0055	0.0050	0.0045	0.0040	0.0035	0.0030	0.0020	0.0015	0.0015	0.0010	0.0010	0.0005
7/16	0.0070	0.0065	0.0060	0.0055	0.0050	0.0045	0.0040	0.0035	0.0025	0.0020	0.0020	0.0015	0.0015	0.0010
1/2	0.0075	0.0070	0.0065	0.0060	0.0055	0.0050	0.0045	0.0040	0.0030	0.0025	0.0025	0.0020	0.0020	0.0015
5/8	0.0080	0.0075	0.0070	0.0065	0.0060	0.0055	0.0050	0.0045	0.0035	0.0030	0.0030	0.0025	0.0025	0.0020
3/4	0.0085	0.0080	0.0075	0.0070	0.0065	0.0060	0.0055	0.0050	0.0040	0.0035	0.0035	0.0030	0.0030	0.0025
7/8	0.0090	0.0085	0.0080	0.0075	0.0070	0.0065	0.0060	0.0055	0.0045	0.0040	0.0040	0.0035	0.0035	0.0030
1	0.0095	0.0090	0.0085	0.0080	0.0075	0.0070	0.0065	0.0060	0.0050	0.0045	0.0045	0.0040	0.0040	0.0035
Root Diam.	Swing Holders — Feed per Revolution in Inches													
1/8	0.0025	0.0020	0.0015	0.0010	0.0005	.....	.....	.....	.....	.....	.....	.....	.....	.....
3/16	0.0028	0.0025	0.0020	0.0015	0.0008	0.0005	.....	.....	.....	.....	.....	.....	.....	.....
1/4	0.0030	0.0030	0.0025	0.0020	0.0010	0.0010	0.0005	0.0005	0.0005	.....	.....	.....	.....	.....
5/16	0.0035	0.0035	0.0030	0.0025	0.0015	0.0015	0.0010	0.0010	0.0010	0.0005	.....	.....	.....	.....
3/8	0.0040	0.0040	0.0035	0.0030	0.0020	0.0020	0.0015	0.0015	0.0015	0.0010	0.0005	0.0005	0.0005	.....
7/16	0.0045	0.0045	0.0040	0.0035	0.0030	0.0025	0.0020	0.0020	0.0020	0.0015	0.0010	0.0010	0.0010	.....
1/2	0.0048	0.0048	0.0045	0.0040	0.0035	0.0030	0.0025	0.0025	0.0025	0.0020	0.0015	0.0015	0.0015	0.0005
5/8	0.0050	0.0050	0.0048	0.0043	0.0040	0.0035	0.0030	0.0030	0.0028	0.0025	0.0020	0.0020	0.0018	0.0010
3/4	0.0055	0.0052	0.0050	0.0045	0.0043	0.0040	0.0035	0.0035	0.0030	0.0028	0.0025	0.0022	0.0020	0.0013
7/8	0.0058	0.0055	0.0052	0.0048	0.0045	0.0043	0.0040	0.0038	0.0032	0.0030	0.0028	0.0025	0.0022	0.0015
1	0.0060	0.0058	0.0054	0.0050	0.0048	0.0047	0.0043	0.0040	0.0035	0.0032	0.0030	0.0028	0.0025	0.0018

THREAD GRINDING

Thread grinding is employed for precision tool and gage work and also in producing certain classes of threaded parts.

Thread grinding may be utilized 1) because of the accuracy and finish obtained; 2) hardness of material to be threaded; and 3) economy in grinding certain classes of screw threads when using modern machines, wheels, and thread-grinding oils.

In some cases pre-cut threads are finished by grinding; but usually, threads are ground "from the solid," being formed entirely by the grinding process. Examples of work include thread gages and taps of steel and tungsten carbide, hobs, worms, lead-screws, adjusting or traversing screws, alloy steel studs, etc. Grinding is applied to external, internal, straight, and tapering threads, and to various thread forms.

Accuracy Obtainable by Thread Grinding.— With single-edge or single-ribbed wheels it is possible to grind threads on gages to a degree of accuracy that requires but very little lapping to produce a so-called "master" thread gage. As far as lead is concerned, some thread grinding machine manufacturers guarantee to hold the lead within 0.0001 inch per inch of thread; and while it is not guaranteed that a higher degree of accuracy for lead is obtainable, it is known that threads have been ground to closer tolerances than this on the lead. Pitch diameter accuracies for either Class 3 or Class 4 fits are obtainable according to the grinding method used; with single-edge wheels, the thread angle can be ground to an accuracy of within two or three minutes in half the angle.

Wheels for Thread Grinding.— The wheels used for steel have an aluminous abrasive and, ordinarily, either a resinoid bond or a vitrified bond. The general rule is to use resinoid wheels when extreme tolerances are not required, and it is desirable to form the thread with a minimum number of passes, as in grinding threaded machine parts, such as studs, adjusting screws which are not calibrated, and for some classes of taps. *Resinoid wheels,* as a rule, will hold a fine edge longer than a vitrified wheel but they are more flexible and, consequently, less suitable for accurate work, especially when there is lateral grinding pressure that causes wheel deflection. *Vitrified wheels* are utilized for obtaining extreme accuracy in thread form and lead because they are very rigid and not easily deflected by side pressure in grinding. This rigidity is especially important in grinding pre-cut threads on such work as gages, taps and lead-screws. The progressive lead errors in long lead-screws, for example, might cause an increasing lateral pressure that would deflect a resinoid wheel. Vitrified wheels are also recommended for internal grinding.

Diamond Wheels: Diamond wheels set in a rubber or plastic bond are also used for thread grinding, especially for grinding threads in carbide materials and in other hardened alloys. Thread grinding is now being done successfully on a commercial basis on both taps and gages made from carbides. Gear hobs made from carbides have also been tested with successful results. Diamond wheels are dressed by means of silicon-carbide grinding wheels which travel past the diamond-wheel thread form at the angle required for the flanks of the thread to be ground. The action of the dressing wheels is, perhaps, best described as a "scrubbing" of the bond which holds the diamond grits. Obviously, the silicon-carbide wheels do not dress the diamonds, but they loosen the bond until the diamonds not wanted drop out.

Thread Grinding with Single-Edge Wheel.— With this type of wheel, the edge is trued to the cross-sectional shape of the thread groove. The wheel, when new, may have a diameter of 18 or 20 inches and, when grinding a thread, the wheel is inclined to align it with the thread groove. On some machines, lead variations are obtained by means of change-gears which transmit motion from the work-driving spindle to the lead-screw. Other machines are so designed that a lead-screw is selected to suit the lead of thread to be ground and transmits motion directly to the work-driving spindle.

Wheels with Edges for Roughing and Finishing.—The "three-ribbed" type of wheel has a roughing edge or rib which removes about two-thirds of the metal. This is followed by an intermediate rib which leaves about 0.005 inch for the third or finishing rib. The accuracy obtained with this triple-edge type compares with that of a single-edge wheel, which means that it may be used for the greatest accuracy obtainable in thread grinding.

When the accuracy required makes it necessary, this wheel can be inclined to the helix angle of the thread, the same as is the single-edge wheel.

The three-ribbed wheel is recommended not only for precision work but for grinding threads which are too long for the multi-ribbed wheel referred to later. It is also well adapted to tap grinding, because it is possible to dress a portion of the wheel adjacent to the finish rib for the purpose of grinding the outside diameter of the thread, as indicated in Fig. 1. Furthermore, the wheel can be dressed for grinding or relieving both crests and flanks at the same time.

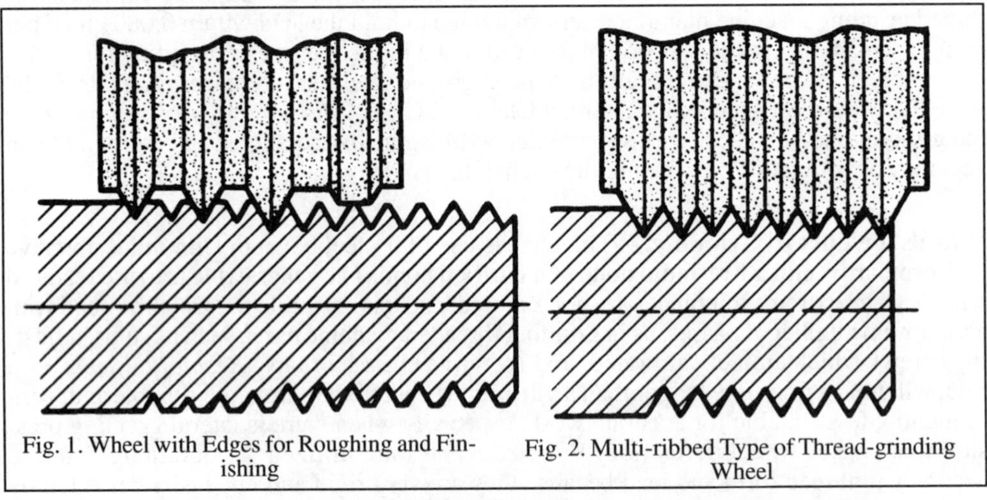

Fig. 1. Wheel with Edges for Roughing and Finishing

Fig. 2. Multi-ribbed Type of Thread-grinding Wheel

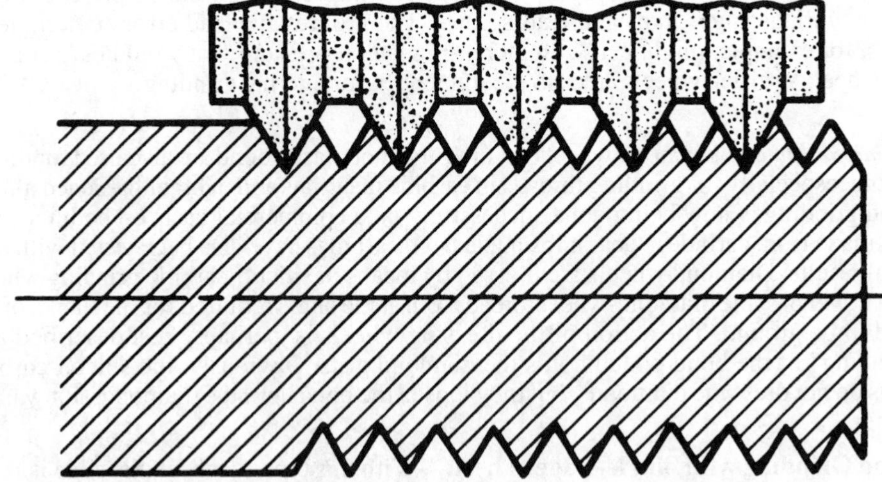

Fig. 3. Alternate-ribbed Wheel for Grinding the Finer Pitches

Multi-ribbed Wheels.—This type of wheel is employed when rapid production is more important than extreme accuracy, which means that it is intended primarily for the grinding of duplicate parts in manufacturing. A wheel 1¼ to 2 inches wide has formed upon its face a series of annular thread-shaped ridges (see Fig. 2); hence, if the length of the thread is not greater than the wheel width, a thread may be ground in one work revolution plus about one-half revolution for feeding in and withdrawing the wheel. The principle of operation is the same as that of thread milling with a multiple type cutter. This type of wheel is not inclined to the lead angle. To obtain a Class 3 fit, the lead angle should not exceed 4 degrees.

It is not practicable to use this form of wheel on thread pitches where the root is less than 0.007 inch wide, because of difficulties in wheel dressing. When this method can be applied, it is the fastest means known of producing threads in hardened materials. It is not recommended, however, that thread gages, taps, and work of this character be ground with multi-ribbed wheels. The single-ribbed wheel has a definite field for accurate, small-lot production.

It is necessary, in multi-ribbed grinding, to use more horsepower than is required for single-ribbed wheel grinding. Coarse threads, in particular, may require a wheel motor with two or three times more horsepower than would be necessary for grinding with a single-ribbed wheel.

Alternate-ribbed Wheel for Fine Pitches.—The spacing of ribs on this type of wheel (Fig. 3) equals twice the pitch, so that during the first revolution every other thread groove section is being ground; consequently, about two and one-half work revolutions are required for grinding a complete thread, but the better distribution of cooling oil and resulting increase in work speeds makes this wheel very efficient. This alternate-type of wheel is adapted for grinding threads of fine pitch. Since these wheels cannot be tipped to the helix angle of the thread, they are not recommended for anything closer than Class 3 fits. The "three-ribbed" wheels referred to in a previous paragraph are also made in the alternate type for the finer pitches.

Grinding Threads "from the Solid.".—The process of forming threads entirely by grinding, or without preliminary cutting, is applied both in the manufacture of certain classes of threaded parts and also in the production of precision tools, such as taps and thread gages. For example, in airplane engine manufacture, certain parts are heat-treated and then the threads are ground "from the solid," thus eliminating distortion. Minute cracks are sometimes found at the roots of threads that were cut and then hardened, or ground from the solid. Steel threads of coarse pitch that are to be surface hardened, may be rough threaded by cutting, then hardened and finally corrected by grinding. Many ground thread taps are produced by grinding from the solid after heat-treatment. Hardening high-speed steel taps before the thread is formed will make sure there are no narrow or delicate crests to interfere with the application of the high temperature required for uniform hardness and the best steel structure.

Number of Wheel Passes.—The number of cuts or passes for grinding from the solid depends upon the type of wheel and accuracy required. In general, threads of 12 or 14 per inch and finer may be ground in one pass of a single-edge wheel unless the "unwrapped" thread length is much greater than normal. Unwrapped length = pitch circumference × total number of thread turns, approximately. For example, a thread gage 1¼ inches long with 24 threads per inch would have an unwrapped length equal to 30 × pitch circumference. (If more convenient, outside circumference may be used instead of pitch circumference.) Assume that there are 6 or 7 feet of unwrapped length on a screw thread having 12 threads per inch. In this case, one pass might be sufficient for a Class 3 fit, whereas two passes might be recommended for a Class 4 fit. When two passes are required, too deep a roughing cut may break down the narrow edge of the wheel. To prevent this, try a roughing cut depth equal to about two-thirds the total thread depth, thus leaving one-third for the finishing cut.

Wheel and Work Rotation.—When a screw thread, on the side being ground, is moving *upward* or *against* the grinding wheel rotation, less heat is generated and the grinding operation is more efficient than when wheel and work are moving in the same direction on the grinding side; however, to avoid running a machine idle during its return stroke, many screw threads are ground during both the forward and return traversing movements, by reversing the work rotation at the end of the forward stroke. For this reason, thread grinders generally are equipped so that both forward and return work speeds may be changed; they may also be designed to accelerate the return movement when grinding in one direction only.

Wheel Speeds.—Wheel speeds should always be limited to the maximum specified on the wheel by the manufacturer. According to the American National Standard Safety Code, resinoid and vitrified wheels are limited to 12,000 surface feet per minute; however, according to Norton Co., the most efficient speeds are from 9,000 to 10,000 for resinoid wheels and 7,500 to 9,500 for vitrified wheels. Only tested wheels recommended by the wheel manufacturer should be used. After a suitable surface speed has been established, it should be maintained by increasing the rpm of the wheel, as the latter is reduced in diameter by wear.

Since thread grinding wheels work close to the limit of their stock-removing capacity, some adjustment of the wheel or work speed may be required to get the best results. If the wheel speed is too slow for a given job and excessive heat is generated, try an increase in speed, assuming that such increase is within the safety limits. If the wheel is too soft and the edge wears excessively, again an increase in wheel speed will give the effect of a harder wheel and result in better form-retaining qualities.

Work Speeds.—The work speed usually ranges from 3 to 10 feet per minute. In grinding with a comparatively heavy feed, and a mininum number of passes, the speed may not exceed 2½ or 3 feet per minute. If very light feeds are employed, as in grinding hardened high-speed steel, the work speed may be much higher than 3 feet per minute and should be determined by test. If excessive heat is generated by removing stock too rapidly, a work speed reduction is one remedy. If a wheel is working below its normal capacity, an increase in work speed would prevent dulling of the grains and reduce the tendency to heat or "burn" the work. An increase in work speed and reduction in feed may also be employed to prevent burning while grinding hardened steel.

Truing Grinding Wheels.—Thread grinding wheels are trued both to maintain the required thread form and also an efficient grinding surface. Thread grinders ordinarily are equipped with precision truing devices which function automatically. One type automatically dresses the wheel and also compensates for the slight amount removed in dressing, thus automatically maintaining size control of the work. While truing the wheel, a small amount of grinding oil should be used to reduce diamond wear. Light truing cuts are advisable, especially in truing resinoid wheels which may be deflected by excessive truing pressure. A master former for controlling the path followed by the truing diamond may require a modified profile to prevent distortion of the thread form, especially when the lead angles are comparatively large. Such modification usually is not required for 60-degree threads when the pitches for a given diameter are standard because then the resulting lead angles are less than 4½ degrees. In grinding Acme threads or 29-degree worm threads having lead angles greater than 4 or 5 degrees, modified formers may be required to prevent a bulge in the thread profile. The highest point of this bulge is approximately at the pitch line. A bulge of about 0.001 inch may be within allowable limits on some commercial worms but precision worms for gear hobbers, etc., require straight flanks in the axial plane.

Crushing Method: Thread grinding wheels are also dressed or formed by the crushing method, which is used in connection with some types of thread grinding machines. When this method is used, the annular ridge or ridges on the wheel are formed by a hardened steel cylindrical dresser or crusher. The crusher has a series of smooth annular ridges which are shaped and spaced like the thread that is to be ground. During the wheel dressing operation, the crusher is positively driven instead of the grinding wheel, and the ridges on the wheel face are formed by the rotating crusher being forced inward.

Wheel Hardness or Grade.—Wheel hardness or grade selection is based upon a compromise between efficient cutting and durability of the grinding edge. Grade selection depends on the bond and the character of the work. The following general recommendations are based upon Norton grading.

Vitrified wheels usually range from J to M, and resinoid wheels from R to U. For heat-treated screws or studs and the Unified Standard Thread, try the following. For 8 to 12 threads per inch, grade S resinoid wheel; for 14 to 20 threads per inch, grade T resinoid; for 24 threads per inch and finer, grades T or U resinoid. For high-speed steel taps 4 to 12 threads per inch, grade J vitrified or S resinoid; 14 to 20 threads per inch, grade K vitrified or T resinoid; 24 to 36 threads per inch, grade M vitrified or T resinoid.

Grain Size.—A thread grinding wheel usually operates close to its maximum stock-removing capacity, and the narrow edge which forms the root of the thread is the most vulnerable part. In grain selection, the general rule is to use the coarsest grained wheel that will hold its form while grinding a reasonable amount of work. Pitch of thread and quality of finish are two governing factors. Thus, to obtain an exceptionally fine finish, the grain size might be smaller than is needed to retain the edge profile. The usual grain sizes range from 120 to 150. For heat-treated screws and studs with Unified Standard Threads, 100 to 180 is the usual range. For precision screw threads of very fine pitch, the grain size may range from 220 to 320. For high-speed steel taps, the usual range is from 150 to 180 for Unified Standard Threads, and from 80 to 150 for pre-cut Acme threads.

Thread Grinding by Centerless Method.—Screw threads may be ground from the solid by the centerless method. A centerless thread grinder is similar in its operating principle to a centerless grinder designed for general work, in that it has a grinding wheel, a regulating or feed wheel (with speed adjustments), and a work-rest. Adjustments are provided to accommodate work of different sizes and for varying the rates of feed. The grinding wheel is a multi-ribbed type, being a series of annular ridges across the face. These ridges conform in pitch and profile with the thread to be ground. The grinding wheel is inclined to suit the helix or lead angle of the thread. In grinding threads on such work as socket type setscrews, the blanks are fed automatically and passed between the grinding and regulating wheels in a continuous stream. To illustrate production possibilities, hardened socket setscrews of ¼20 size may be ground from the solid at the rate of 60 to 70 per minute and with the wheel operating continuously for 8 hours without redressing. The lead errors of centerless ground screw threads may be limited to 0.0005 inch per inch or even less by reducing the production rate. The pitch diameter tolerances are within 0.0002 to 0.0003 inch of the basic size. The grain size for the wheel is selected with reference to the pitch of the thread, the following sizes being recommended: For 11 to 13 threads per inch, 150; for 16 threads per inch, 180; for 18 to 20 threads per inch, 220; for 24 to 28 threads per inch, 320; for 40 threads per inch, 400.

THREAD MILLING

Single-cutter Method.—Usually, when a single point cutter is used, the axis of the cutter is inclined an amount equal to the lead angle of the screw thread, in order to locate the cutter in line with the thread groove at the point where the cutting action takes place. Tangent of lead angle = lead of screw thread ÷ pitch circumference of screw.

The helical thread groove is generated by making as many turns around the workpiece diameter as there are pitches in the length of thread to be cut. For example, a 16-pitch thread, 1 inch long, would require 16 turns of the cutter around the work. The single cutter process is especially applicable to the milling of large screw threads of coarse pitch, and either single or multiple threads.

The cutter should revolve as fast as possible without dulling the cutting edges excessively, in order to mill a smooth thread and prevent the unevenness that would result with a slow-moving cutter, on account of the tooth spaces. As the cutter rotates, the part on which a thread is to be milled is also revolved, but at a very slow rate (a few inches per minute), since this rotation of the work is practically a feeding movement. The cutter is ordinarily set to the full depth of the thread groove and finishes a single thread in one passage, although deep threads of coarse pitch may require two or even three cuts. For fine pitches and short threads, the multiple-cutter method (described in the next paragraph) usually is preferable, because it is more rapid. The milling of taper screw threads may be done on a single-cutter type of machine by traversing the cutter laterally as it feeds along in a lengthwise direction, the same as when using a taper attachment on a lathe.

Multiple-cutter Method.—The multiple cutter for thread milling is practically a series of single cutters, although formed of one solid piece of steel, at least so far as the cutter proper is concerned. The rows of teeth do not lie in a helical path, like the teeth of a hob or tap, but they are annular or without lead. If the cutter had helical teeth the same as a gear hob, it would have to be geared to revolve in a certain fixed ratio with the screw being milled, but a cutter having annular teeth may rotate at any desired cutting speed, while the screw blank is rotated slowly to provide a suitable rate of feed. (The multiple thread milling cutters used are frequently called "hobs," but the term hob should be applied only to cutters having a helical row of teeth like a gear-cutting hob.)

The object in using a multiple cutter instead of a single cutter is to finish a screw thread complete in approximately one revolution of the work, a slight amount of over-travel being allowed to insure milling the thread to the full depth where the end of cut joins the starting point. The cutter which is at least one and one half or two threads or pitches wider than the thread to be milled, is fed in to the full thread depth and then either the cutter or screw blank is moved in a lengthwise direction a distance equal to the lead of the thread during one revolution of the work.

The multiple cutter is used for milling comparatively short threads and coarse, medium or fine pitches. The accompanying illustration shows typical examples of external and internal work for which the multiple-cutter type of thread milling has proved very efficient, although its usefulness is not confined to shoulder work and "blind" holes.

In using multiple cutters either for internal or external thread milling, the axis of the cutter is set parallel with the axis of the work, instead of inclining the cutter to suit the lead angle of the thread, as when using a single cutter. Theoretically, this is not the correct position for a cutter, since each cutting edge is revolving in a plane at right angles to the screw's axis while milling a thread groove of helical form. However, as a general rule, interference between the cutter and the thread, does not result a decided change in the standard thread form.

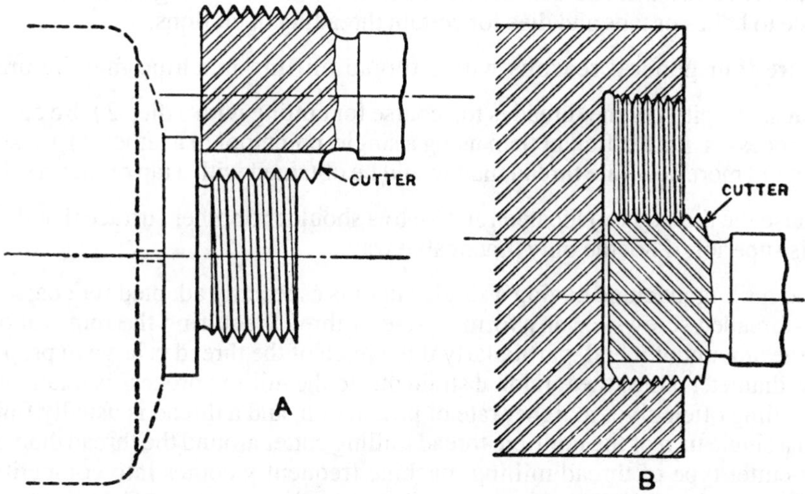

Examples of External and Internal Thread Milling with a Multiple Thread Milling Cutter

Usually the deviation is very slight and may be disregarded except when milling threads which incline considerably relative to the axis like a thread of multiple form and large lead angle. Multiple cutters are suitable for external threads having lead angles under 3½ degrees and for internal threads having lead angles under 2½ degrees. Threads which have steeper sides or smaller included angles than the American Standard or Whitworth forms have greater limitations on the maximum helix angle and may have to be milled with a single point cutter tilted to the helix angle, assuming that the milling process is preferable to other methods. For instance, in milling an Acme thread which has an included angle between the sides of 29 degrees, there might be considerable interference if a multiple cutter were used, unless the screw thread diameter were large enough in proportion to the pitch to prevent such interference. If an attempt were made to mill a square thread with a multiple cutter, the results would be unsatisfactory owing to the interference.

Interference between the cutter and work is more pronounced when milling internal threads, because the cutter does not clear itself so well. It is preferable to use as small a cutter as practicable, either for internal or external work, not only to avoid interference, but to reduce the strain on the driving mechanism. Some thread milling cutters, known as "topping cutters," are made for milling the outside diameter of the thread as well as the angular sides and root, but most are made non-tapping.

Planetary Method.—The planetary method of thread milling is similar in principle to planetary milling. The part to be threaded is held stationary and the thread milling cutter, while revolving about its own axis, is given a planetary movement around the work in order to mill the thread in one planetary revolution. The machine spindle and the cutter which is held by it is moved longitudinally for thread milling, an amount equal to the thread lead during one planetary revolution. This operation is applicable to both internal and external threads. Other advantages: Thread milling is frequently accompanied by milling operations on other adjoining surfaces, and may be performed with conventional and planetary methods. For example, a machine may be used for milling a screw thread and a concentric cylindrical surface simultaneously. When the milling operation begins, the cutter-spindle feeds the cutter in to the right depth and the planetary movement then begins, thus milling the thread and the cylindrical surface. Thin sharp starting edges are eliminated on threads milled by this method and the thread begins with a smooth gradual approach. One design of machine will mill internal and external threads simultaneously. These threads may be of the same hand or one may be right hand and the other left hand. The threads may also be either of the same pitch or of a different pitch, and either straight or tapered.

Classes of Work for Thread Milling Machines.—Thread milling machines are used in preference to lathes or taps and dies for certain threading operations.

There are four general reasons why a thread milling machine may be preferred:

1) Because the pitch of the thread is too coarse for cutting with a die; 2) because the milling process is more efficient than using a single-point tool in a lathe; 3) to secure a smoother and more accurate thread than would be obtained with a tap or die; and

4) because the thread is so located relative to a shoulder or other surface that the milling method is superior, if not the only practicable way.

A thread milling machine having a single cutter is especially adapted for coarse pitches, multiple-threaded screws, or any form or size of thread requiring the removal of a relatively large amount of metal, particularly if the pitch of the thread is large in proportion to the screw diameter, since the torsional strain due to the milling process is relatively small. Thread milling often gives a higher rate of production, and a thread is usually finished by means of a single turn of the multiple thread milling cutter around the thread diameter. The multiple-cutter type of thread milling machine frequently comes into competition with dies and taps, and especially self-opening dies and collapsing taps. The use of a multiple cutter is desirable when a thread must be cut close to a shoulder or to the bottom of a shallow recess, although the usefulness of the multiple cutter is not confined to shoulder work and "blind" holes.

Maximum Pitches of Die-cut Threads.—Dies of special design could be constructed for practically any pitch, if the screw blank were strong enough to resist the cutting strains and the size and cost of the die were immaterial; but, as a general rule, when the pitch is coarser than four or five threads per inch, the difficulty of cutting threads with dies increases rapidly, although in a few cases some dies are used successfully on screw threads having two or three threads per inch or less. Much depends upon the design of the die, the finish or smoothness required, and the relation between the pitch of the thread and the diameter of the screw. When the screw diameter is relatively small in proportion to the pitch, there may be considerable distortion due to the twisting strains set up when the thread is being cut. If the number of threads per inch is only one or two less than the standard number for a given diameter, a screw blank ordinarily will be strong enough to permit the use of a die.

Changing Pitch of Screw Thread Slightly.—A very slight change in the pitch of a screw thread may be necessary as, for example, when the pitch of a tap is increased a small amount to compensate for shrinkage in hardening. One method of obtaining slight variations in pitch is by means of a taper attachment. This attachment is set at an angle and the work is located at the same angle by adjusting the tailstock center. The result is that the tool follows an angular path relative to the movement of the carriage and, consequently, the pitch of the thread is increased slightly, the amount depending upon the angle to which the work and taper attachment are set. The cosine of this angle, for obtaining a given increase in pitch, equals the standard pitch (which would be obtained with the lathe used in the regular way) divided by the increased pitch necessary to compensate for shrinkage.

Example: If the pitch of a ¾-inch American standard screw is to be increased from 0.100 to 0.1005, the cosine of the angle to which the taper attachment and work should be set is found as follows:

$$\text{Cosine of required angle} = \frac{0.100}{0.1005} = 0.9950$$

which is the cosine of 5 degrees 45 minutes, nearly.

CHANGE GEARS FOR HELICAL MILLING

Lead of a Milling Machine.—If gears with an equal number of teeth are placed on the table feed-screw and the worm-gear stud, then the *lead of the milling machine* is the distance the table will travel while the index spindle makes one complete revolution. This distance is a constant used in figuring the change gears.

The lead of a helix or "spiral" is the distance, measured along the axis of the work, in which the helix makes one full turn around the work. The lead of the milling machine may, therefore, also be expressed as the lead of the helix that will be cut when gears with an equal number of teeth are placed on the feed-screw and the worm-gear stud, and an idler of suitable size is interposed between the gears.

Rule: To find the lead of a milling machine, place equal gears on the worm-gear stud and on the feed-screw, and multiply the number of revolutions made by the feed-screw to produce one revolution of the index head spindle, by the lead of the thread on the feed-screw. Expressing the rule given as a formula:

$$\begin{matrix}\text{lead of milling}\\\text{machine}\end{matrix} = \begin{matrix}\text{rev. of feed-screw for one}\\\text{revolution of index spindle}\\\text{with equal gears}\end{matrix} \times \begin{matrix}\text{lead of}\\\text{feed-screw}\end{matrix}$$

Assume that it is necessary to make 40 revolutions of the feed-screw to turn the index head spindle one complete revolution, when the gears are equal, and that the lead of the thread on the feed-screw of the milling machine is $\frac{1}{4}$ inch; then the lead of the machine equals $40 \times \frac{1}{4}$ inch = 10 inches.

Change Gears for Helical Milling.—To find the change gears to be used in the compound train of gears for helical milling, place the lead of the helix to be cut in the numerator and the lead of the milling machine in the denominator of a fraction; divide numerator and denominator into two factors each; and multiply each "pair" of factors by the *same* number until suitable numbers of teeth for the change gears are obtained. (One factor in the numerator and one in the denominator are considered as one "pair" in this calculation.)

Example: Assume that the lead of a machine is 10 inches, and that a helix having a 48-inch lead is to be cut. Following the method explained:

$$\frac{48}{10} = \frac{6 \times 8}{2 \times 5} = \frac{(6 \times 12) \times (8 \times 8)}{(2 \times 12) \times (5 \times 8)} = \frac{72 \times 64}{24 \times 40}$$

The gear having 72 teeth is placed on the worm-gear stud and meshes with the 24-tooth gear on the intermediate stud. On the same intermediate stud is then placed the gear having 64 teeth, which is driven by the gear having 40 teeth placed on the feed-screw. This makes the gears having 72 and 64 teeth the driven gears, and the gears having 24 and 40 teeth the driving gears. In general, for compound gearing, the following formula may be used:

$$\frac{\text{lead of helix to be cut}}{\text{lead of machine}} = \frac{\text{product of driven gears}}{\text{product of driving gears}}$$

Short-lead Milling.—If the lead to be milled is exceptionally short, the drive may be direct from the table feed-screw to the dividing head spindle to avoid excessive load on feed-screw and change-gears. If the table feed-screw has 4 threads per inch (usual standard), then

$$\text{Change-gear ratio} = \frac{\text{Lead to be milled}}{0.25} = \frac{\text{Driven gears}}{\text{Driving gears}}$$

For indexing, the number of teeth on the spindle change-gear should be some multiple of the number of divisions required, to permit indexing by disengaging and turning the gear.

Helix.—A helix is a curve generated by a point moving about a cylindrical surface (real or imaginary) at a constant rate in the direction of the cylinder's axis. The curvature of a screw thread is one common example of a helical curve.

Lead of Helix: The lead of a helix is the distance that it advances in an axial direction, in one complete turn about the cylindrical surface. To illustrate, the lead of a screw thread equals the distance that a thread advances in one turn; it also equals the distance that a nut would advance in one turn.

Development of Helix: If one turn of a helical curve were unrolled onto a plane surface (as shown by diagram), the helix would become a straight line forming the hypotenuse of a right angle triangle. The length of one side of this triangle would equal the circumference of the cylinder with which the helix coincides, and the length of the other side of the triangle would equal the lead of the helix.

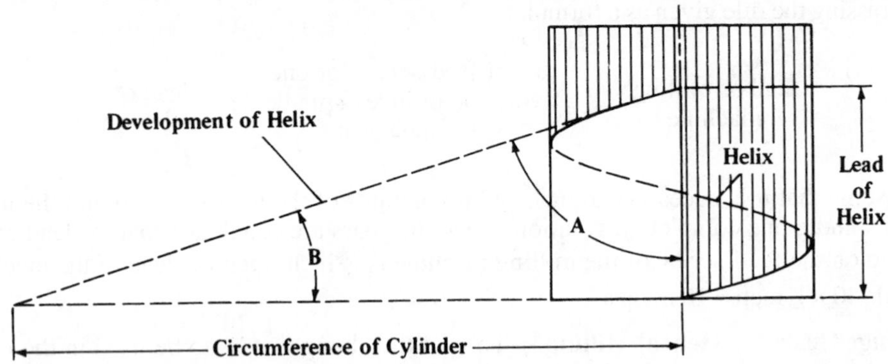

Helix Angles.—The triangular development of a helix has one angle *A* subtended by the circumference of the cylinder, and another angle *B* subtended by the lead of the helix. The term "helix angle" applies to angle *A*. For example, the helix angle of a helical gear, according to the general usage of the term, is always angle *A*, because this is the angle used in helical gear-designing formulas. Helix angle *A* would also be applied in milling the helical teeth of cutters, reamers, etc. Angle *A* of a gear or cutter tooth is a measure of its inclination relative to the axis of the gear or cutter.

Lead Angle: Angle *B* is applied to screw threads and worm threads and is referred to as the lead angle of the screw thread or worm. This angle *B* is a measure of the inclination of a screw thread from a plane that is perpendicular to the screw thread axis. Angle *B* is called the "lead angle" because it is subtended by the lead of the thread, and to distinguish it from the term "helix angle" as applied to helical gears.

Finding Helix Angle of Helical Gear: A helical gear tooth has an infinite number of helix angles, but the angle at the pitch diameter or mid-working depth is the one required in gear designing and gear cutting. This angle *A*, relative to the axis of the gear, is found as follows:

$$\text{tan helix angle} = \frac{3.1416 \times \text{pitch diameter of gear}}{\text{Lead of gear tooth}}$$

Finding Lead Angle of Screw Thread: The lead or helix angle at the pitch diameter of a screw thread usually is required when, for example, a thread milling cutter must be aligned with the thread. This angle measured from a plane perpendicular to the screw thread axis, is found as follows:

$$\text{tan lead angle} = \frac{\text{Lead of screw thread}}{3.1416 \times \text{pitch diameter of screw thread}}$$

Change Gears for Different Leads—0.670 Inch to 2.658 Inches

Lead in Inches	Driven	Driver	Driven	Driver	Lead in Inches	Driven	Driver	Driven	Driver	Lead in Inches	Driven	Driver	Driven	Driver
	Gear on Worm	First Gear on Stud	Second Gear on Stud	Gear on Screw		Gear on Worm	First Gear on Stud	Second Gear on Stud	Gear on Screw		Gear on Worm	First Gear on Stud	Second Gear on Stud	Gear on Screw
0.670	24	86	24	100	1.711	28	72	44	100	2.182	24	44	40	100
0.781	24	86	28	100	1.714	24	56	40	100	2.188	24	48	28	64
0.800	24	72	24	100	1.744	24	64	40	86	2.193	24	56	44	86
0.893	24	86	32	100	1.745	24	44	32	100	2.200	24	48	44	100
0.930	24	72	24	86	1.750	28	64	40	100	2.222	24	48	32	72
1.029	24	56	24	100	1.776	24	44	28	86	2.233	40	86	48	100
1.042	28	86	32	100	1.778	32	72	40	100	2.238	28	64	44	86
1.047	24	64	24	86	1.786	24	86	64	100	2.240	28	40	32	100
1.050	24	64	28	100	1.800	24	64	48	100	2.250	24	40	24	64
1.067	24	72	32	100	1.809	28	72	40	86	2.274	32	72	44	86
1.085	24	72	28	86	1.818	24	44	24	72	2.286	32	56	40	100
1.116	24	86	40	100	1.823	28	86	56	100	2.292	24	64	44	72
1.196	24	56	24	86	1.860	28	56	32	86	2.326	32	64	40	86
1.200	24	48	24	100	1.861	24	72	48	86	2.333	28	48	40	100
1.221	24	64	28	86	1.867	28	48	32	100	2.338	24	44	24	56
1.228	24	86	44	100	1.875	24	48	24	64	2.344	28	86	72	100
1.240	24	72	32	86	1.886	24	56	44	100	2.368	28	44	32	86
1.250	24	64	24	72	1.905	24	56	32	72	2.381	32	86	64	100
1.302	28	86	40	100	1.919	24	64	44	86	2.386	24	44	28	64
1.309	24	44	24	100	1.920	24	40	32	100	2.392	24	56	48	86
1.333	24	72	40	100	1.925	28	64	44	100	2.400	28	56	48	100
1.340	24	86	48	100	1.944	24	48	28	72	2.424	24	44	32	72
1.371	24	56	32	100	1.954	24	40	28	86	2.431	28	64	40	72
1.395	24	48	24	86	1.956	32	72	44	100	2.442	24	32	28	86
1.400	24	48	28	100	1.990	28	72	44	86	2.445	40	72	44	100
1.429	24	56	24	72	1.993	24	56	40	86	2.450	28	64	56	100
1.440	24	40	24	100	2.000	24	40	24	72	2.456	44	86	48	100
1.458	24	64	28	72	2.009	24	86	72	100	2.481	32	72	48	86
1.467	24	72	44	100	2.030	24	44	32	86	2.489	32	72	56	100
1.488	32	86	40	100	2.035	28	64	40	86	2.500	24	48	28	56
1.500	24	64	40	100	2.036	28	44	32	100	2.514	32	56	44	100
1.522	24	44	24	86	2.045	24	44	24	64	2.532	28	72	56	86
1.550	24	72	40	86	2.047	40	86	44	100	2.537	24	44	40	86
1.563	24	86	56	100	2.057	24	28	24	100	2.546	28	44	40	100
1.595	24	56	32	86	2.067	32	72	40	86	2.558	32	64	44	86
1.600	24	48	32	100	2.083	24	64	40	72	2.567	28	48	44	100
1.607	24	56	24	64	2.084	28	86	64	100	2.571	24	40	24	56
1.628	24	48	28	86	2.093	24	64	48	86	2.593	28	48	32	72
1.637	32	86	44	100	2.100	24	64	56	100	2.605	28	40	32	86
1.650	24	64	44	100	2.121	24	44	28	72	2.618	24	44	48	100
1.667	24	56	28	72	2.133	24	72	64	100	2.619	24	56	44	72
1.674	24	40	24	86	2.143	24	56	32	64	2.625	24	40	28	64
1.680	24	40	28	100	2.171	24	72	56	86	2.640	24	40	44	100
1.706	24	72	44	86	2.178	28	72	56	100	2.658	32	56	40	86

Change Gears for Different Leads—2.667 Inches to 4.040 Inches

Lead in Inches	Driven	Driver	Driven	Driver	Lead in Inches	Driven	Driver	Driven	Driver	Lead in Inches	Driven	Driver	Driven	Driver
	Gear on Worm	First Gear on Stud	Second Gear on Stud	Gear on Screw		Gear on Worm	First Gear on Stud	Second Gear on Stud	Gear on Screw		Gear on Worm	First Gear on Stud	Second Gear on Stud	Gear on Screw
2.667	40	72	48	100	3.140	24	86	72	64	3.588	72	56	24	86
2.674	28	64	44	72	3.143	40	56	44	100	3.600	72	48	24	100
2.678	24	56	40	64	3.150	28	100	72	64	3.618	56	72	40	86
2.679	32	86	72	100	3.175	32	56	40	72	3.636	24	44	32	48
2.700	24	64	72	100	3.182	28	44	32	64	3.637	48	44	24	72
2.713	28	48	40	86	3.189	32	56	48	86	3.646	40	48	28	64
2.727	24	44	32	64	3.190	24	86	64	56	3.655	40	56	44	86
2.743	24	56	64	100	3.198	40	64	44	86	3.657	64	56	32	100
2.750	40	64	44	100	3.200	28	100	64	56	3.663	72	64	28	86
2.778	32	64	40	72	3.214	24	56	48	64	3.667	40	48	44	100
2.791	28	56	48	86	3.225	24	100	86	64	3.673	24	28	24	56
2.800	24	24	28	100	3.241	28	48	40	72	3.684	44	86	72	100
2.812	24	32	24	64	3.256	24	24	28	86	3.686	86	56	24	100
2.828	28	44	32	72	3.267	28	48	56	100	3.704	32	48	40	72
2.843	40	72	44	86	3.273	24	40	24	44	3.721	24	24	32	86
2.845	32	72	64	100	3.275	44	86	64	100	3.733	48	72	56	100
2.849	28	64	56	86	3.281	24	32	28	64	3.750	24	32	24	48
2.857	24	48	32	56	3.300	44	64	48	100	3.763	86	64	28	100
2.865	44	86	56	100	3.308	32	72	64	86	3.771	44	56	48	100
2.867	86	72	24	100	3.333	32	64	48	72	3.772	24	28	44	100
2.880	24	40	48	100	3.345	28	100	86	72	3.799	56	48	28	86
2.894	28	72	64	86	3.349	40	86	72	100	3.809	24	28	32	72
2.909	32	44	40	100	3.360	56	40	24	100	3.810	64	56	24	72
2.917	24	64	56	72	3.383	32	44	40	86	3.818	24	40	28	44
2.924	32	56	44	86	3.403	28	64	56	72	3.819	40	64	44	72
2.933	44	72	48	100	3.409	24	44	40	64	3.822	86	72	32	100
2.934	32	48	44	100	3.411	32	48	44	86	3.837	24	32	44	86
2.946	24	56	44	64	3.422	44	72	56	100	3.840	64	40	24	100
2.960	28	44	40	86	3.428	24	40	32	56	3.850	44	64	56	100
2.977	40	86	64	100	3.429	40	28	24	100	3.876	24	72	100	86
2.984	28	48	44	86	3.438	24	48	44	64	3.889	32	64	56	72
3.000	24	40	28	56	3.488	40	64	48	86	3.896	24	44	40	56
3.030	24	44	40	72	3.491	64	44	24	100	3.907	56	40	24	86
3.044	24	44	48	86	3.492	32	56	44	72	3.911	44	72	64	100
3.055	28	44	48	100	3.500	40	64	56	100	3.920	28	40	56	100
3.056	32	64	44	72	3.520	32	40	44	100	3.927	72	44	24	100
3.070	24	40	44	86	3.535	28	44	40	72	3.929	32	56	44	64
3.080	28	40	44	100	3.552	56	44	24	86	3.977	28	44	40	64
3.086	24	56	72	100	3.556	40	72	64	100	3.979	44	72	56	86
3.101	40	72	48	86	3.564	56	44	28	100	3.987	24	28	40	86
3.111	28	40	32	72	3.565	28	48	44	72	4.000	24	40	32	48
3.117	24	44	32	56	3.571	24	48	40	56	4.011	28	48	44	64
3.125	28	56	40	64	3.572	48	86	64	100	4.019	72	86	48	100
3.126	48	86	56	100	3.582	44	40	28	86	4.040	32	44	40	72

Change Gears for Different Leads—4.059 Inches to 5.568 Inches

Lead in Inches	Driven: Gear on Worm	Driver: First Gear on Stud	Driven: Second Gear on Stud	Driver: Gear on Screw	Lead in Inches	Driven: Gear on Worm	Driver: First Gear on Stud	Driven: Second Gear on Stud	Driver: Gear on Screw	Lead in Inches	Driven: Gear on Worm	Driver: First Gear on Stud	Driven: Second Gear on Stud	Driver: Gear on Screw
4.059	32	44	48	86	4.567	72	44	24	86	5.105	28	48	56	64
4.060	64	44	24	86	4.572	40	56	64	100	5.116	44	24	24	86
4.070	28	32	40	86	4.582	72	44	28	100	5.119	86	56	24	72
4.073	64	44	28	100	4.583	44	64	48	72	5.120	64	40	32	100
4.074	32	48	44	72	4.584	32	48	44	64	5.133	56	48	44	100
4.091	24	44	48	64	4.651	40	24	24	86	5.134	44	24	28	100
4.093	32	40	44	86	4.655	64	44	32	100	5.142	72	56	40	100
4.114	48	28	24	100	4.667	28	40	32	48	5.143	24	28	24	40
4.125	24	40	44	64	4.675	24	28	24	44	5.156	44	32	24	64
4.135	40	72	64	86	4.687	40	32	24	64	5.160	86	40	24	100
4.144	56	44	28	86	4.688	56	86	72	100	5.168	100	72	32	86
4.167	28	48	40	56	4.691	86	44	24	100	5.185	28	24	32	72
4.186	72	64	32	86	4.714	44	40	24	56	5.186	64	48	28	72
4.200	48	64	56	100	4.736	64	44	28	86	5.195	32	44	40	56
4.242	28	44	32	48	4.762	40	28	24	72	5.209	100	64	24	72
4.253	64	56	32	86	4.773	24	32	28	44	5.210	64	40	28	86
4.264	40	48	44	86	4.778	86	72	40	100	5.226	86	64	28	72
4.267	64	48	32	100	4.784	72	56	32	86	5.233	72	64	40	86
4.278	28	40	44	72	4.785	48	28	24	86	5.236	72	44	32	100
4.286	24	28	24	48	4.800	48	24	24	100	5.238	44	28	24	72
4.300	86	56	28	100	4.813	44	40	28	64	5.250	24	32	28	40
4.320	72	40	24	100	4.821	72	56	24	64	5.256	86	72	44	100
4.341	48	72	56	86	4.849	32	44	48	72	5.280	48	40	44	100
4.342	64	48	28	86	4.861	40	32	28	72	5.303	28	44	40	48
4.361	100	64	24	86	4.884	48	64	56	86	5.316	40	28	32	86
4.363	24	40	32	44	4.889	32	40	44	72	5.328	72	44	28	86
4.364	40	44	48	100	4.898	24	28	32	56	5.333	40	24	32	100
4.365	40	56	44	72	4.900	56	32	28	100	5.347	44	64	56	72
4.375	24	24	28	64	4.911	40	56	44	64	5.348	44	32	28	72
4.386	24	28	44	86	4.914	86	56	32	100	5.357	40	28	24	64
4.400	24	24	44	100	4.950	56	44	28	72	5.358	64	86	72	100
4.444	64	56	28	72	4.961	64	48	32	86	5.375	86	64	40	100
4.465	64	40	24	86	4.978	56	72	64	100	5.400	72	32	24	100
4.466	48	40	32	86	4.984	100	56	24	86	5.413	64	44	32	86
4.477	44	32	28	86	5.000	24	24	28	56	5.426	40	24	28	86
4.479	86	64	24	72	5.017	86	48	28	100	5.427	40	48	56	86
4.480	56	40	32	100	5.023	72	40	24	86	5.444	56	40	28	72
4.500	72	64	40	100	5.029	44	28	32	100	5.455	48	44	28	56
4.522	100	72	28	86	5.040	72	40	28	100	5.469	40	32	28	64
4.537	56	48	28	72	5.074	40	44	48	86	5.473	86	44	28	100
4.545	24	44	40	48	5.080	64	56	32	72	5.486	64	28	24	100
4.546	28	44	40	56	5.088	100	64	28	86	5.500	44	40	24	48
4.548	44	72	64	86	5.091	56	44	40	100	5.556	40	24	24	72
4.558	56	40	28	86	5.093	40	48	44	72	5.568	56	44	28	64

Change Gears for Different Leads—5.581 Inches to 7.500 Inches

Lead in Inches	Driven	Driver	Driven	Driver	Lead in Inches	Driven	Driver	Driven	Driver	Lead in Inches	Driven	Driver	Driven	Driver
	Gear on Worm	First Gear on Stud	Second Gear on Stud	Gear on Screw		Gear on Worm	First Gear on Stud	Second Gear on Stud	Gear on Screw		Gear on Worm	First Gear on Stud	Second Gear on Stud	Gear on Screw
5.581	64	32	24	86	6.172	72	28	24	100	6.825	86	56	32	72
5.582	48	24	24	86	6.202	40	24	32	86	6.857	32	28	24	40
5.600	56	24	24	100	6.222	64	40	28	72	6.875	44	24	24	64
5.625	48	32	24	64	6.234	32	28	24	44	6.880	86	40	32	100
5.657	56	44	32	72	6.250	24	24	40	64	6.944	100	48	24	72
5.698	56	32	28	86	6.255	86	44	32	100	6.945	100	56	28	72
5.714	48	28	24	72	6.279	72	64	48	86	6.968	86	48	28	72
5.730	40	48	44	64	6.286	44	40	32	56	6.977	48	32	40	86
5.733	86	48	32	100	6.300	72	32	28	100	6.982	64	44	48	100
5.756	72	64	44	86	6.343	100	44	24	86	6.984	44	28	32	72
5.759	86	56	24	64	6.350	40	28	32	72	7.000	28	24	24	40
5.760	72	40	32	100	6.364	56	44	24	48	7.013	72	44	24	56
5.788	64	72	56	86	6.379	64	28	24	86	7.040	64	40	44	100
5.814	100	64	32	86	6.396	44	32	40	86	7.071	56	44	40	72
5.818	64	44	40	100	6.400	64	24	24	100	7.104	56	44	48	86
5.833	28	24	24	48	6.417	44	40	28	48	7.106	100	72	44	86
5.847	64	56	44	86	6.429	24	28	24	32	7.111	64	40	32	72
5.848	44	28	32	86	6.450	86	64	48	100	7.130	44	24	28	72
5.861	72	40	28	86	6.460	100	72	40	86	7.143	40	28	32	64
5.867	44	24	32	100	6.465	64	44	32	72	7.159	72	44	28	64
5.893	44	32	24	56	6.482	56	48	40	72	7.163	56	40	44	86
5.912	86	64	44	100	6.512	56	24	24	86	7.167	86	40	24	72
5.920	56	44	40	86	6.515	86	44	24	72	7.176	72	28	24	86
5.926	64	48	32	72	6.534	56	24	28	100	7.200	72	24	24	100
5.952	100	56	24	72	6.545	48	40	24	44	7.268	100	64	40	86
5.954	64	40	32	86	6.548	44	48	40	56	7.272	64	44	28	56
5.969	44	24	28	86	6.563	56	32	24	64	7.273	32	24	24	44
5.972	86	48	24	72	6.578	72	56	44	86	7.292	56	48	40	64
5.980	72	56	40	86	6.600	48	32	44	100	7.310	44	28	40	86
6.000	48	40	28	56	6.645	100	56	32	86	7.314	64	28	32	100
6.016	44	32	28	64	6.667	64	48	28	56	7.326	72	32	28	86
6.020	86	40	28	100	6.689	86	72	56	100	7.330	86	44	24	64
6.061	40	44	32	48	6.697	100	56	24	64	7.333	44	24	40	100
6.077	100	64	28	72	6.698	72	40	32	86	7.334	44	40	32	48
6.089	72	44	32	86	6.719	86	48	24	64	7.347	48	28	24	56
6.109	56	44	48	100	6.720	56	40	48	100	7.371	86	56	48	100
6.112	24	24	44	72	6.735	44	28	24	56	7.372	86	28	24	100
6.122	40	28	24	56	6.750	72	40	24	64	7.400	100	44	28	86
6.125	56	40	28	64	6.757	86	56	44	100	7.408	40	24	32	72
6.137	72	44	24	64	6.766	64	44	40	86	7.424	56	44	28	48
6.140	48	40	44	86	6.784	100	48	28	86	7.442	64	24	24	86
6.143	86	56	40	100	6.806	56	32	28	72	7.465	86	64	40	72
6.160	56	40	44	100	6.818	40	32	24	44	7.467	64	24	28	100
6.171	72	56	48	100	6.822	44	24	32	86	7.500	48	24	24	64

Change Gears for Different Leads—7.525 Inches to 9.598 Inches

Lead in Inches	Gear on Worm (Driven)	First Gear on Stud (Driver)	Second Gear on Stud (Driven)	Gear on Screw (Driver)	Lead in Inches	Gear on Worm (Driven)	First Gear on Stud (Driver)	Second Gear on Stud (Driven)	Gear on Screw (Driver)	Lead in Inches	Gear on Worm (Driven)	First Gear on Stud (Driver)	Second Gear on Stud (Driven)	Gear on Screw (Driver)
7.525	86	32	28	100	8.140	56	32	40	86	8.800	48	24	44	100
7.543	48	28	44	100	8.145	64	44	56	100	8.838	100	44	28	72
7.576	100	44	24	72	8.148	64	48	44	72	8.839	72	56	44	64
7.597	56	24	28	86	8.149	44	24	32	72	8.909	56	40	28	44
7.601	86	44	28	72	8.163	40	28	32	56	8.929	100	48	24	56
7.611	72	44	40	86	8.167	56	40	28	48	8.930	64	40	48	86
7.619	64	48	32	56	8.182	48	32	24	44	8.953	56	32	44	86
7.620	64	28	24	72	8.186	64	40	44	86	8.959	86	48	28	56
7.636	56	40	24	44	8.212	86	64	44	72	8.960	64	40	56	100
7.639	44	32	40	72	8.229	72	28	32	100	8.980	44	28	32	56
7.644	86	72	64	100	8.250	44	32	24	40	9.000	48	32	24	40
7.657	56	32	28	64	8.306	100	56	40	86	9.044	100	72	56	86
7.674	72	48	44	86	8.312	64	44	32	56	9.074	56	24	28	72
7.675	48	32	44	86	8.333	40	24	24	48	9.091	40	24	24	44
7.679	86	48	24	56	8.334	40	24	28	56	9.115	100	48	28	64
7.680	64	40	48	100	8.361	86	40	28	72	9.134	72	44	48	86
7.700	56	32	44	100	8.372	72	24	24	86	9.137	100	56	44	86
7.714	72	40	24	56	8.377	86	44	24	56	9.143	64	40	32	56
7.752	100	48	32	86	8.400	72	24	28	100	9.164	72	44	56	100
7.778	32	24	28	48	8.437	72	32	24	64	9.167	44	24	24	48
7.792	40	28	24	44	8.457	100	44	32	86	9.210	72	40	44	86
7.813	100	48	24	64	8.484	32	24	28	44	9.214	86	40	24	56
7.815	56	40	48	86	8.485	64	44	28	48	9.260	100	48	32	72
7.818	86	44	40	100	8.485	56	44	32	48	9.302	48	24	40	86
7.838	86	48	28	64	8.506	64	28	32	86	9.303	56	28	40	86
7.855	72	44	48	100	8.523	100	44	24	64	9.333	64	40	28	48
7.857	44	24	24	56	8.527	44	24	40	86	9.334	32	24	28	40
7.872	44	28	32	64	8.532	86	56	40	72	9.351	48	28	24	44
7.875	72	40	28	64	8.534	64	24	32	100	9.375	48	32	40	64
7.883	86	48	44	100	8.552	86	44	28	64	9.382	86	44	48	100
7.920	72	40	44	100	8.556	56	40	44	72	9.385	86	56	44	72
7.936	100	56	32	72	8.572	64	32	24	56	9.406	86	40	28	64
7.954	40	32	28	44	8.572	48	24	24	56	9.428	44	28	24	40
7.955	56	44	40	64	8.594	44	32	40	64	9.429	48	40	44	56
7.963	86	48	32	72	8.600	86	24	24	100	9.460	86	40	44	100
7.974	48	28	40	86	8.640	72	40	48	100	9.472	64	44	56	86
7.994	100	64	44	86	8.681	100	64	40	72	9.524	40	28	32	48
8.000	64	32	40	100	8.682	64	24	28	86	9.545	72	44	28	48
8.021	44	32	28	48	8.687	86	44	32	72	9.546	56	32	24	44
8.035	72	56	40	64	8.721	100	32	24	86	9.547	56	44	48	64
8.063	86	40	24	64	8.727	48	40	32	44	9.549	100	64	44	72
8.081	64	44	40	72	8.730	44	28	40	72	9.556	86	40	32	72
8.102	100	48	28	72	8.750	28	24	24	32	9.569	72	28	32	86
8.119	64	44	48	86	8.772	48	28	44	86	9.598	86	56	40	64

Change Gears for Different Leads—9.600 Inches to 12.375 Inches

Lead in Inches	Driven	Driver	Driven	Driver	Lead in Inches	Driven	Driver	Driven	Driver	Lead in Inches	Driven	Driver	Driven	Driver
	Gear on Worm	First Gear on Stud	Second Gear on Stud	Gear on Screw		Gear on Worm	First Gear on Stud	Second Gear on Stud	Gear on Screw		Gear on Worm	First Gear on Stud	Second Gear on Stud	Gear on Screw
9.600	72	24	32	100	10.370	64	24	28	72	11.314	72	28	44	100
9.625	44	32	28	40	10.371	64	48	56	72	11.363	100	44	24	48
9.643	72	32	24	56	10.390	40	28	32	44	11.401	86	44	28	48
9.675	86	64	72	100	10.417	100	32	24	72	11.429	32	24	24	28
9.690	100	48	40	86	10.419	64	40	56	86	11.454	72	40	28	44
9.697	64	48	32	44	10.451	86	32	28	72	11.459	44	24	40	64
9.723	40	24	28	48	10.467	72	32	40	86	11.467	86	24	32	100
9.741	100	44	24	56	10.473	72	44	64	100	11.512	72	32	44	86
9.768	72	48	56	86	10.476	44	24	32	56	11.518	86	28	24	64
9.773	86	44	24	48	10.477	48	28	44	72	11.520	72	40	64	100
9.778	64	40	44	72	10.500	56	32	24	40	11.574	100	48	40	72
9.796	64	28	24	56	10.558	86	56	44	64	11.629	100	24	24	86
9.818	72	40	24	44	10.571	100	44	40	86	11.638	64	40	32	44
9.822	44	32	40	56	10.606	56	44	40	48	11.667	56	24	24	48
9.828	86	28	32	100	10.631	64	28	40	86	11.688	72	44	40	56
9.844	72	32	28	64	10.655	72	44	56	86	11.695	64	28	44	86
9.900	72	32	44	100	10.659	100	48	44	86	11.719	100	32	24	64
9.921	100	56	40	72	10.667	64	40	48	72	11.721	72	40	56	86
9.923	64	24	32	86	10.694	44	24	28	48	11.728	86	40	24	44
9.943	100	44	28	64	10.713	40	28	24	32	11.733	64	24	44	100
9.954	86	48	40	72	10.714	48	32	40	56	11.757	86	32	28	64
9.967	100	56	48	86	10.750	86	40	24	48	11.785	72	48	44	56
9.968	100	28	24	86	10.800	72	32	48	100	11.786	44	28	24	32
10.000	56	28	24	48	10.853	56	24	40	86	11.825	86	32	44	100
10.033	86	24	28	100	10.859	86	44	40	72	11.905	100	28	24	72
10.046	72	40	48	86	10.909	72	44	32	48	11.938	56	24	44	86
10.057	64	28	44	100	10.913	100	56	44	72	11.944	86	24	24	72
10.078	86	32	24	64	10.937	56	32	40	64	11.960	72	28	40	86
10.080	72	40	56	100	10.945	86	44	56	100	12.000	48	24	24	40
10.101	100	44	32	72	10.949	86	48	44	72	12.031	56	32	44	64
10.159	64	28	32	72	10.972	64	28	48	100	12.040	86	40	56	100
10.175	100	32	28	86	11.000	44	24	24	40	12.121	40	24	32	44
10.182	64	40	28	44	11.021	72	28	24	56	12.153	100	32	28	72
10.186	44	24	40	72	11.057	86	56	72	100	12.178	72	44	64	86
10.209	56	24	28	64	11.111	40	24	32	48	12.216	86	44	40	64
10.228	72	44	40	64	11.137	56	32	28	44	12.222	44	24	32	48
10.233	48	24	44	86	11.160	100	56	40	64	12.245	48	28	40	56
10.238	86	28	24	72	11.163	72	24	32	86	12.250	56	32	28	40
10.267	56	24	44	100	11.169	86	44	32	56	12.272	72	32	24	44
10.286	48	28	24	40	11.198	86	48	40	64	12.277	100	56	44	64
10.312	48	32	44	64	11.200	56	24	48	100	12.286	86	28	40	100
10.313	72	48	44	64	11.225	44	28	40	56	12.318	86	48	44	64
10.320	86	40	48	100	11.250	72	24	24	64	12.343	72	28	48	100
10.336	100	72	64	86	11.313	64	44	56	72	12.375	72	40	44	64

Change Gears for Different Leads—12.403 Inches to 16.000 Inches

Lead in Inches	Driven: Gear on Worm	Driver: First Gear on Stud	Driven: Second Gear on Stud	Driver: Gear on Screw	Lead in Inches	Driven: Gear on Worm	Driver: First Gear on Stud	Driven: Second Gear on Stud	Driver: Gear on Screw	Lead in Inches	Driven: Gear on Worm	Driver: First Gear on Stud	Driven: Second Gear on Stud	Driver: Gear on Screw
12.403	64	24	40	86	13.438	86	24	24	64	14.668	44	24	32	40
12.444	64	40	56	72	13.469	48	28	44	56	14.694	72	28	32	56
12.468	64	28	24	44	13.500	72	32	24	40	14.743	86	28	48	100
12.500	40	24	24	32	13.514	86	28	44	100	14.780	86	40	44	64
12.542	86	40	28	48	13.566	100	24	28	86	14.800	100	44	56	86
12.508	86	44	64	100	13.611	56	24	28	48	14.815	64	24	40	72
12.558	72	32	48	86	13.636	48	32	40	44	14.849	56	24	28	44
12.571	64	40	44	56	13.643	64	24	44	86	14.880	100	48	40	56
12.572	44	28	32	40	13.650	86	28	32	72	14.884	64	28	56	86
12.600	72	32	56	100	13.672	100	32	28	64	14.931	86	32	40	72
12.627	100	44	40	72	13.682	86	40	28	44	14.933	64	24	56	100
12.686	100	44	48	86	13.713	64	40	48	56	14.950	100	56	72	86
12.698	64	28	40	72	13.715	64	28	24	40	15.000	48	24	24	32
12.727	64	32	28	44	13.750	44	24	24	32	15.050	86	32	56	100
12.728	56	24	24	44	13.760	86	40	64	100	15.150	100	44	32	48
12.732	100	48	44	72	13.889	100	24	24	72	15.151	100	44	48	72
12.758	64	28	48	86	13.933	86	48	56	72	15.202	86	44	56	72
12.791	100	40	44	86	13.935	86	24	28	72	15.238	64	28	48	72
12.798	86	48	40	56	13.953	72	24	40	86	15.239	64	28	32	48
12.800	64	28	56	100	13.960	86	44	40	56	15.272	56	40	48	44
12.834	56	40	44	48	13.968	64	28	44	72	15.278	44	24	40	48
12.857	72	28	32	64	14.000	56	24	24	40	15.279	100	40	44	72
12.858	48	28	24	32	14.025	72	44	48	56	15.306	100	28	24	56
12.900	86	32	48	100	14.026	72	28	24	44	15.349	72	24	44	86
12.963	56	24	40	72	14.063	72	32	40	64	15.357	86	28	24	48
12.987	100	44	32	56	14.071	86	44	72	100	15.429	72	40	48	56
13.020	100	48	40	64	14.078	86	48	44	56	15.469	72	32	44	64
13.024	56	24	48	86	14.142	72	40	44	56	15.480	86	40	72	100
13.030	86	44	32	48	14.204	100	44	40	64	15.504	100	48	64	86
13.062	64	28	32	56	14.260	56	24	44	72	15.556	64	32	56	72
13.082	100	64	72	86	14.286	40	24	24	28	15.584	48	28	40	44
13.090	72	40	32	44	14.318	72	32	28	44	15.625	100	24	24	64
13.096	44	28	40	48	14.319	72	44	56	64	15.636	86	40	32	44
13.125	72	32	28	48	14.322	100	48	44	64	15.677	86	32	28	48
13.139	86	40	44	72	14.333	86	40	32	48	15.714	44	24	24	28
13.157	72	28	44	86	14.352	72	28	48	86	15.750	72	32	28	40
13.163	86	28	24	56	14.400	72	24	48	100	15.767	86	24	44	100
13.200	72	24	44	100	14.536	100	32	40	86	15.873	100	56	64	72
13.258	100	44	28	48	14.545	64	24	24	44	15.874	100	28	32	72
13.289	100	28	32	86	14.583	56	32	40	48	15.909	100	40	28	44
13.333	64	24	24	48	14.584	40	24	28	32	15.925	86	48	64	72
13.393	100	56	48	64	14.651	72	32	56	86	15.926	86	24	32	72
13.396	72	40	64	86	14.659	86	44	48	64	15.989	100	32	44	86
13.437	86	32	28	56	14.667	64	40	44	48	16.000	64	24	24	40

Change Gears for Different Leads—16.042 Inches to 21.39 Inches

Lead in Inches	Driven	Driver	Driven	Driver	Lead in Inches	Driven	Driver	Driven	Driver	Lead in Inches	Driven	Driver	Driven	Driver
	Gear on Worm	First Gear on Stud	Second Gear on Stud	Gear on Screw		Gear on Worm	First Gear on Stud	Second Gear on Stud	Gear on Screw		Gear on Worm	First Gear on Stud	Second Gear on Stud	Gear on Screw
16.042	56	24	44	64	17.442	100	32	48	86	19.350	86	32	72	100
16.043	44	24	28	32	17.454	64	40	48	44	19.380	100	24	40	86
16.071	72	32	40	56	17.500	56	24	24	32	19.394	64	24	32	44
16.125	86	32	24	40	17.550	86	28	32	56	19.444	40	24	28	24
16.204	100	24	28	72	17.677	100	44	56	72	19.480	100	28	24	44
16.233	100	44	40	56	17.679	72	32	44	56	19.531	100	32	40	64
16.280	100	40	56	86	17.778	64	24	32	48	19.535	72	24	56	86
16.288	86	44	40	48	17.858	100	24	24	56	19.545	86	24	24	44
16.296	64	24	44	72	17.917	86	24	32	64	19.590	64	28	48	56
16.327	64	28	40	56	17.918	86	24	24	48	19.635	72	40	48	44
16.333	56	24	28	40	17.959	64	28	44	56	19.642	100	40	44	56
16.364	72	24	24	44	18.000	72	24	24	40	19.643	44	28	40	32
16.370	100	48	44	56	18.181	56	28	40	44	19.656	86	28	64	100
16.423	86	32	44	72	18.182	48	24	40	44	19.687	72	32	56	64
16.456	72	28	64	100	18.229	100	32	28	48	19.710	86	40	44	48
16.500	72	40	44	48	18.273	100	28	44	86	19.840	100	28	40	72
16.612	100	28	40	86	18.285	64	28	32	40	19.886	100	44	56	64
16.623	64	28	32	44	18.333	56	28	44	48	19.887	100	32	28	44
16.667	56	28	40	48	18.367	72	28	40	56	19.908	86	24	40	72
16.722	86	40	56	72	18.428	86	28	24	40	19.934	100	28	48	86
16.744	72	24	48	86	18.476	86	32	44	64	20.00	72	24	32	48
16.752	86	44	48	56	18.519	100	24	32	72	20.07	86	24	56	100
16.753	86	28	24	44	18.605	100	40	64	86	20.09	100	56	72	64
16.797	86	32	40	64	18.663	100	64	86	72	20.16	86	48	72	64
16.800	72	24	56	100	18.667	64	24	28	40	20.20	100	44	64	72
16.875	72	32	48	64	18.700	72	44	64	56	20.35	100	32	56	86
16.892	86	40	44	56	18.750	100	32	24	40	20.36	64	40	56	44
16.914	100	44	64	86	18.750	72	32	40	48	20.41	100	28	32	56
16.969	64	44	56	48	18.770	86	28	44	72	20.42	56	24	28	32
16.970	64	24	28	44	18.812	86	32	28	40	20.45	72	32	40	44
17.045	100	32	24	44	18.858	48	28	44	40	20.48	86	48	64	56
17.046	100	44	48	64	18.939	100	44	40	48	20.57	72	40	64	56
17.062	86	28	40	72	19.029	100	44	72	86	20.63	72	32	44	48
17.101	86	44	56	64	19.048	40	24	32	28	20.74	64	24	56	72
17.102	86	32	28	44	19.090	56	32	48	44	20.78	64	28	40	44
17.141	64	32	48	56	19.091	72	24	28	44	20.83	100	32	48	72
17.143	64	28	24	32	19.096	100	32	44	72	20.90	86	32	56	72
17.144	48	24	24	28	19.111	86	40	64	72	20.93	100	40	72	86
17.188	100	40	44	64	19.136	72	28	64	86	20.95	64	28	44	48
17.200	86	32	64	100	19.197	86	32	40	56	21.00	56	32	48	40
17.275	86	56	72	64	19.200	72	24	64	100	21.12	86	32	44	56
17.361	100	32	40	72	19.250	56	32	44	40	21.32	100	24	44	86
17.364	64	24	56	86	19.285	72	32	48	56	21.33	100	56	86	72
17.373	86	44	64	72	19.286	72	28	24	32	21.39	44	24	28	24

Change Gears for Different Leads—21.43 Inches to 32.09 Inches

Lead in Inches	Driven	Driver	Driven	Driver	Lead in Inches	Driven	Driver	Driven	Driver	Lead in Inches	Driven	Driver	Driven	Driver
	Gear on Worm	First Gear on Stud	Second Gear on Stud	Gear on Screw		Gear on Worm	First Gear on Stud	Second Gear on Stud	Gear on Screw		Gear on Worm	First Gear on Stud	Second Gear on Stud	Gear on Screw
21.43	100	40	48	56	24.88	100	72	86	48	28.05	72	28	48	44
21.48	100	32	44	64	24.93	64	28	48	44	28.06	100	28	44	56
21.50	86	24	24	40	25.00	72	24	40	48	28.13	100	40	72	64
21.82	72	44	64	48	25.08	86	24	28	40	28.15	86	28	44	48
21.88	100	40	56	64	25.09	86	40	56	48	28.29	72	28	44	40
21.90	86	24	44	72	25.13	86	44	72	56	28.41	100	32	40	44
21.94	86	28	40	56	25.14	64	28	44	40	28.57	100	56	64	40
21.99	86	44	72	64	25.45	64	44	56	32	28.64	72	44	56	32
22.00	64	32	44	40	25.46	100	24	44	72	28.65	100	32	44	48
22.04	72	28	48	56	25.51	100	28	40	56	28.67	86	40	64	48
22.11	86	28	72	100	25.57	100	64	72	44	29.09	64	24	48	44
22.22	100	40	64	72	25.60	86	28	40	48	29.17	100	40	56	48
22.34	86	44	64	56	25.67	56	24	44	40	29.22	100	56	72	44
22.40	86	32	40	48	25.71	72	24	48	56	29.32	86	48	72	44
22.50	72	24	48	64	25.72	72	24	24	28	29.34	64	24	44	40
22.73	100	24	24	44	25.80	86	24	72	100	29.39	72	28	64	56
22.80	86	48	56	44	25.97	100	44	64	56	29.56	86	32	44	40
22.86	64	24	24	28	26.04	100	32	40	48	29.76	100	28	40	48
22.91	72	44	56	40	26.06	86	44	64	48	29.86	100	40	86	72
22.92	100	40	44	48	26.16	100	32	72	86	29.90	100	28	72	86
22.93	86	24	64	100	26.18	72	40	64	44	30.00	56	28	48	32
23.04	86	56	72	48	26.19	44	24	40	28	30.23	86	32	72	64
23.14	100	24	40	72	26.25	72	32	56	48	30.30	100	48	64	44
23.26	100	32	64	86	26.33	86	28	48	56	30.48	64	24	32	28
23.33	64	32	56	48	26.52	100	44	56	48	30.54	100	44	86	64
23.38	72	28	40	44	26.58	100	28	64	86	30.56	44	24	40	24
23.44	100	48	72	64	26.67	64	28	56	48	30.61	100	28	48	56
23.45	86	40	48	44	26.79	100	48	72	56	30.71	86	24	48	56
23.52	86	32	56	64	26.88	86	28	56	64	30.72	86	24	24	28
23.57	72	28	44	48	27.00	72	32	48	40	30.86	72	28	48	40
23.81	100	48	64	56	27.13	100	24	56	86	31.01	100	24	64	86
23.89	86	32	64	72	27.15	100	44	86	72	31.11	64	24	56	48
24.00	64	40	72	48	27.22	56	24	28	24	31.25	100	28	56	64
24.13	86	28	44	56	27.27	100	40	48	44	31.27	86	40	64	44
24.19	86	40	72	64	27.30	86	28	64	72	31.35	86	32	56	48
24.24	64	24	40	44	27.34	100	32	56	64	31.36	86	24	28	32
24.31	100	32	56	72	27.36	86	40	56	44	31.43	64	28	44	32
24.43	86	32	40	44	27.43	64	28	48	40	31.50	72	32	56	40
24.44	44	24	32	24	27.50	56	32	44	28	31.75	100	72	64	28
24.54	72	32	48	44	27.64	86	40	72	56	31.82	100	44	56	40
24.55	100	32	44	56	27.78	100	32	64	72	31.85	86	24	64	72
24.57	86	40	64	56	27.87	86	24	56	72	31.99	100	56	86	48
24.64	86	24	44	64	27.92	86	28	40	44	32.00	64	28	56	40
24.75	72	32	44	40	28.00	100	64	86	48	32.09	56	24	44	32

Change Gears for Different Leads—32.14 Inches to 60.00 Inches

Lead in Inches	Driven	Driver	Driven	Driver	Lead in Inches	Driven	Driver	Driven	Driver	Lead in Inches	Driven	Driver	Driven	Driver
	Gear on Worm	First Gear on Stud	Second Gear on Stud	Gear on Screw		Gear on Worm	First Gear on Stud	Second Gear on Stud	Gear on Screw		Gear on Worm	First Gear on Stud	Second Gear on Stud	Gear on Screw
32.14	100	56	72	40	38.20	100	24	44	48	46.07	86	28	72	48
32.25	86	48	72	40	38.39	100	40	86	56	46.67	64	24	56	32
32.41	100	24	56	72	38.57	72	28	48	32	46.88	100	32	72	48
32.47	100	28	40	44	38.89	56	24	40	24	47.15	72	24	44	28
32.58	86	24	40	44	38.96	100	28	48	44	47.62	100	28	64	48
32.73	72	32	64	44	39.09	86	32	64	44	47.78	86	24	64	48
32.74	100	28	44	48	39.29	100	28	44	40	47.99	100	32	86	56
32.85	86	24	44	48	39.42	86	24	44	40	48.00	72	24	64	40
33.00	72	24	44	40	39.49	86	28	72	56	48.38	86	32	72	40
33.33	100	24	32	40	39.77	100	32	56	44	48.61	100	24	56	48
33.51	86	28	48	44	40.00	72	24	64	48	48.86	100	40	86	44
33.59	100	64	86	40	40.18	100	32	72	56	48.89	64	24	44	24
33.79	86	28	44	40	40.31	86	32	72	48	49.11	100	28	44	32
33.94	64	24	56	44	40.72	100	44	86	48	49.14	86	28	64	40
34.09	100	48	72	44	40.82	100	28	64	56	49.27	86	24	44	32
34.20	86	44	56	32	40.91	100	40	72	44	49.77	100	24	86	72
34.29	72	48	64	28	40.95	86	28	64	48	50.00	100	28	56	40
34.38	100	32	44	40	40.96	86	24	32	28	50.17	86	24	56	40
34.55	86	32	72	56	41.14	72	28	64	40	50.26	86	28	72	44
34.72	100	24	40	48	41.25	72	24	44	32	51.14	100	32	72	44
34.88	100	24	72	86	41.67	100	32	64	48	51.19	86	24	40	28
34.90	100	56	86	44	41.81	86	24	56	48	51.43	72	28	64	32
35.00	72	24	56	48	41.91	64	24	44	28	51.95	100	28	64	44
35.10	86	28	64	56	41.99	100	32	86	64	52.12	86	24	64	44
35.16	100	32	72	64	42.00	72	24	56	40	52.50	72	24	56	32
35.18	86	44	72	40	42.23	86	28	44	32	53.03	100	24	56	44
35.36	72	32	44	28	42.66	100	28	86	72	53.33	64	24	56	28
35.56	64	24	32	24	42.78	56	24	44	24	53.57	100	28	72	48
35.71	100	32	64	56	42.86	100	28	48	40	53.75	86	24	48	32
35.72	100	24	24	28	43.00	86	32	64	40	54.85	100	28	86	56
35.83	86	32	64	48	43.64	72	24	64	44	55.00	72	24	44	24
36.00	72	32	64	40	43.75	100	32	56	40	55.28	86	28	72	40
36.36	100	44	64	40	43.98	86	32	72	44	55.56	100	24	32	24
36.46	100	48	56	32	44.44	64	24	40	24	55.99	100	24	86	64
36.67	48	24	44	24	44.64	100	28	40	32	56.25	100	32	72	40
36.86	86	28	48	40	44.68	86	28	64	44	56.31	86	24	44	28
37.04	100	24	64	72	44.79	100	40	86	48	57.14	100	28	64	40
37.33	100	32	86	72	45.00	72	28	56	32	57.30	100	24	44	32
37.40	72	28	64	44	45.45	100	32	64	44	57.33	86	24	64	40
37.50	100	48	72	40	45.46	100	28	56	44	58.33	100	24	56	40
37.63	86	32	56	40	45.61	86	24	56	44	58.44	100	28	72	44
37.88	100	24	40	44	45.72	64	24	48	28	58.64	86	24	72	44
38.10	64	24	40	28	45.84	100	24	44	40	59.53	100	24	40	28
38.18	72	24	56	44	45.92	100	28	72	56	60.00	72	24	64	32

Lead of Helix for Given Helix Angle Relative to Axis, When Diameter = 1

Deg.	0′	6′	12′	18′	24′	30′	36′	42′	48′	54′	60′
0	Infin.	1800.001	899.997	599.994	449.993	359.992	299.990	257.130	224.986	199.983	179.982
1	179.982	163.616	149.978	138.438	128.545	119.973	112.471	105.851	99.967	94.702	89.964
2	89.964	85.676	81.778	78.219	74.956	71.954	69.183	66.617	64.235	62.016	59.945
3	59.945	58.008	56.191	54.485	52.879	51.365	49.934	48.581	47.299	46.082	44.927
4	44.927	43.827	42.780	41.782	40.829	39.918	39.046	38.212	37.412	36.645	35.909
5	35.909	35.201	34.520	33.866	33.235	32.627	32.040	31.475	30.928	30.400	29.890
6	29.890	29.397	28.919	28.456	28.008	27.573	27.152	26.743	26.346	25.961	25.586
7	25.586	25.222	24.868	24.524	24.189	23.863	23.545	23.236	22.934	22.640	22.354
8	22.354	22.074	21.801	21.535	21.275	21.021	20.773	20.530	20.293	20.062	19.835
9	19.835	19.614	19.397	19.185	18.977	18.773	18.574	18.379	18.188	18.000	17.817
10	17.817	17.637	17.460	17.287	17.117	16.950	16.787	16.626	16.469	16.314	16.162
11	16.162	16.013	15.866	15.722	15.581	15.441	15.305	15.170	15.038	14.908	14.780
12	14.780	14.654	14.530	14.409	14.289	14.171	14.055	13.940	13.828	13.717	13.608
13	13.608	13.500	13.394	13.290	13.187	13.086	12.986	12.887	12.790	12.695	12.600
14	12.600	12.507	12.415	12.325	12.237	12.148	12.061	11.975	11.890	11.807	11.725
15	11.725	11.643	11.563	11.484	11.405	11.328	11.252	11.177	11.102	11.029	10.956
16	10.956	10.884	10.813	10.743	10.674	10.606	10.538	10.471	10.405	10.340	10.276
17	10.276	10.212	10.149	10.086	10.025	9.964	9.904	9.844	9.785	9.727	9.669
18	9.669	9.612	9.555	9.499	9.444	9.389	9.335	9.281	9.228	9.176	9.124
19	9.124	9.072	9.021	8.971	8.921	8.872	8.823	8.774	8.726	8.679	8.631
20	8.631	8.585	8.539	8.493	8.447	8.403	8.358	8.314	8.270	8.227	8.184
21	8.184	8.142	8.099	8.058	8.016	7.975	7.935	7.894	7.855	7.815	7.776
22	7.776	7.737	7.698	7.660	7.622	7.584	7.547	7.510	7.474	7.437	7.401
23	7.401	7.365	7.330	7.295	7.260	7.225	7.191	7.157	7.123	7.089	7.056
24	7.056	7.023	6.990	6.958	6.926	6.894	6.862	6.830	6.799	6.768	6.737
25	6.737	6.707	6.676	6.646	6.617	6.586	6.557	6.528	6.499	6.470	6.441
26	6.441	6.413	6.385	6.357	6.329	6.300	6.274	6.246	6.219	6.192	6.166
27	6.166	6.139	6.113	6.087	6.061	6.035	6.009	5.984	5.959	5.933	5.908
28	5.908	5.884	5.859	5.835	5.810	5.786	5.762	5.738	5.715	5.691	5.668
29	5.668	5.644	5.621	5.598	5.575	5.553	5.530	5.508	5.486	5.463	5.441

Lead of Helix for Given Helix Angle Relative to Axis, When Diameter = 1 *(Continued)*

Deg.	0′	6′	12′	18′	24′	30′	36′	42′	48′	54′	60′
30	5.441	5.420	5.398	5.376	5.355	5.333	5.312	5.291	5.270	5.249	5.228
31	5.228	5.208	5.187	5.167	5.147	5.127	5.107	5.087	5.067	5.047	5.028
32	5.028	5.008	4.989	4.969	4.950	4.931	4.912	4.894	4.875	4.856	4.838
33	4.838	4.819	4.801	4.783	4.764	4.746	4.728	4.711	4.693	4.675	4.658
34	4.658	4.640	4.623	4.605	4.588	4.571	4.554	4.537	4.520	4.503	4.487
35	4.487	4.470	4.453	4.437	4.421	4.404	4.388	4.372	4.356	4.340	4.324
36	4.324	4.308	4.292	4.277	4.261	4.246	4.230	4.215	4.199	4.184	4.169
37	4.169	4.154	4.139	4.124	4.109	4.094	4.079	4.065	4.050	4.036	4.021
38	4.021	4.007	3.992	3.978	3.964	3.950	3.935	3.921	3.907	3.893	3.880
39	3.880	3.866	3.852	3.838	3.825	3.811	3.798	3.784	3.771	3.757	3.744
40	3.744	3.731	3.718	3.704	3.691	3.678	3.665	3.652	3.640	3.627	3.614
41	3.614	3.601	3.589	3.576	3.563	3.551	3.538	3.526	3.514	3.501	3.489
42	3.489	3.477	3.465	3.453	3.440	3.428	3.416	3.405	3.393	3.381	3.369
43	3.369	3.358	3.346	3.334	3.322	3.311	3.299	3.287	3.276	3.265	3.253
44	3.253	3.242	3.231	3.219	3.208	3.197	3.186	3.175	3.164	3.153	3.142
45	3.142	3.131	3.120	3.109	3.098	3.087	3.076	3.066	3.055	3.044	3.034
46	3.034	3.023	3.013	3.002	2.992	2.981	2.971	2.960	2.950	2.940	2.930
47	2.930	2.919	2.909	2.899	2.889	2.879	2.869	2.859	2.849	2.839	2.829
48	2.829	2.819	2.809	2.799	2.789	2.779	2.770	2.760	2.750	2.741	2.731
49	2.731	2.721	2.712	2.702	2.693	2.683	2.674	2.664	2.655	2.645	2.636
50	2.636	2.627	2.617	2.608	2.599	2.590	2.581	2.571	2.562	2.553	2.544
51	2.544	2.535	2.526	2.517	2.508	2.499	2.490	2.481	2.472	2.463	2.454
52	2.454	2.446	2.437	2.428	2.419	2.411	2.402	2.393	2.385	2.376	2.367
53	2.367	2.359	2.350	2.342	2.333	2.325	2.316	2.308	2.299	2.291	2.282
54	2.282	2.274	2.266	2.257	2.249	2.241	2.233	2.224	2.216	2.208	2.200
55	2.200	2.192	2.183	2.175	2.167	2.159	2.151	2.143	2.135	2.127	2.119
56	2.119	2.111	2.103	2.095	2.087	2.079	2.072	2.064	2.056	2.048	2.040
57	2.040	2.032	2.025	2.017	2.009	2.001	1.994	1.986	1.978	1.971	1.963
58	1.963	1.955	1.948	1.940	1.933	1.925	1.918	1.910	1.903	1.895	1.888
59	1.888	1.880	1.873	1.865	1.858	1.851	1.843	1.836	1.828	1.821	1.814

Lead of Helix for Given Helix Angle Relative to Axis, When Diameter = 1 *(Continued)*

Deg.	0′	6′	12′	18′	24′	30′	36′	42′	48′	54′	60′
60	1.814	1.806	1.799	1.792	1.785	1.777	1.770	1.763	1.756	1.749	1.741
61	1.741	1.734	1.727	1.720	1.713	1.706	1.699	1.692	1.685	1.677	1.670
62	1.670	1.663	1.656	1.649	1.642	1.635	1.628	1.621	1.615	1.608	1.601
63	1.601	1.594	1.587	1.580	1.573	1.566	1.559	1.553	1.546	1.539	1.532
64	1.532	1.525	1.519	1.512	1.505	1.498	1.492	1.485	1.478	1.472	1.465
65	1.465	1.458	1.452	1.445	1.438	1.432	1.425	1.418	1.412	1.405	1.399
66	1.399	1.392	1.386	1.379	1.372	1.366	1.359	1.353	1.346	1.340	1.334
67	1.334	1.327	1.321	1.314	1.308	1.301	1.295	1.288	1.282	1.276	1.269
68	1.269	1.263	1.257	1.250	1.244	1.237	1.231	1.225	1.219	1.212	1.206
69	1.206	1.200	1.193	1.187	1.181	1.175	1.168	1.162	1.156	1.150	1.143
70	1.143	1.137	1.131	1.125	1.119	1.112	1.106	1.100	1.094	1.088	1.082
71	1.082	1.076	1.069	1.063	1.057	1.051	1.045	1.039	1.033	1.027	1.021
72	1.021	1.015	1.009	1.003	0.997	0.991	0.985	0.978	0.972	0.966	0.960
73	0.960	0.954	0.948	0.943	0.937	0.931	0.925	0.919	0.913	0.907	0.901
74	0.901	0.895	0.889	0.883	0.877	0.871	0.865	0.859	0.854	0.848	0.842
75	0.842	0.836	0.830	0.824	0.818	0.812	0.807	0.801	0.795	0.789	0.783
76	0.783	0.777	0.772	0.766	0.760	0.754	0.748	0.743	0.737	0.731	0.725
77	0.725	0.720	0.714	0.708	0.702	0.696	0.691	0.685	0.679	0.673	0.668
78	0.668	0.662	0.656	0.651	0.645	0.639	0.633	0.628	0.622	0.616	0.611
79	0.611	0.605	0.599	0.594	0.588	0.582	0.577	0.571	0.565	0.560	0.554
80	0.554	0.548	0.543	0.537	0.531	0.526	0.520	0.514	0.509	0.503	0.498
81	0.498	0.492	0.486	0.481	0.475	0.469	0.464	0.458	0.453	0.447	0.441
82	0.441	0.436	0.430	0.425	0.419	0.414	0.408	0.402	0.397	0.391	0.386
83	0.386	0.380	0.375	0.369	0.363	0.358	0.352	0.347	0.341	0.336	0.330
84	0.330	0.325	0.319	0.314	0.308	0.302	0.297	0.291	0.286	0.280	0.275
85	0.275	0.269	0.264	0.258	0.253	0.247	0.242	0.236	0.231	0.225	0.220
86	0.220	0.214	0.209	0.203	0.198	0.192	0.187	0.181	0.176	0.170	0.165
87	0.165	0.159	0.154	0.148	0.143	0.137	0.132	0.126	0.121	0.115	0.110
88	0.110	0.104	0.099	0.093	0.088	0.082	0.077	0.071	0.066	0.060	0.055
89	0.055	0.049	0.044	0.038	0.033	0.027	0.022	0.016	0.011	0.005	0.000

Leads, Change Gears and Angles for Helical Milling

Lead of Helix, Inches	Change Gears: Gear on Worm	First Gear on Stud	Second Gear on Stud	Gear on Screw	Diameter of Work, Inches: ⅛	¼	⅜	½	⅝	¾	⅞	1	1¼	1½
					Approximate Angles for Milling Machine Table									
0.67	24	86	24	100	30¼	…	…	…	…	…	…	…	…	…
0.78	24	86	28	100	26	44½	…	…	…	…	…	…	…	…
0.89	24	86	32	100	23½	41	…	…	…	…	…	…	…	…
1.12	24	86	40	100	19	34½	…	…	…	…	…	…	…	…
1.34	24	86	48	100	16	30¼	41½	…	…	…	…	…	…	…
1.46	24	64	28	72	14¾	28	38½	…	…	…	…	…	…	…
1.56	24	86	56	100	13¾	26½	37	…	…	…	…	…	…	…
1.67	24	64	32	72	12¾	25	34¾	43¼	…	…	…	…	…	…
1.94	32	64	28	72	11¼	21¾	31	39	45	…	…	…	…	…
2.08	24	64	40	72	10¼	20½	29½	37	43¼	…	…	…	…	…
2.22	32	56	28	72	9¾	19¼	27½	35	41¼	…	…	…	…	…
2.50	24	64	48	72	8¾	17	25	32	38	43¼	…	…	…	…
2.78	40	56	28	72	8	15½	23	29½	35¼	40½	44¾	…	…	…
2.92	24	64	56	72	7½	15	21¾	28¼	34	39	43¼	…	…	…
3.24	40	48	28	72	6¾	13¼	19¾	25¾	31¼	36	40½	44¼	…	…
3.70	40	48	32	72	6	11¾	17½	23	28	32½	36½	40½	…	…
3.89	56	48	24	72	5½	11¼	16¾	22	26¾	31¼	35¼	39	…	…
4.17	40	72	48	64	5¼	10½	15¾	20½	25¼	29½	33½	37	43¼	…
4.46	48	40	32	86	4¾	9¾	14¾	19¼	23¾	27¾	31½	35	41½	…
4.86	40	64	56	72	4½	9	13½	17¾	22	25¾	29½	33	39	44¼
5.33	48	40	32	72	4	8¼	12¼	16½	20¼	23¾	27¼	30½	36½	41½
5.44	56	40	28	72	4	8	12	16	20	23½	26¾	30	36	41
6.12	56	40	28	64	3½	7¼	11	14½	17¾	21	24¼	27	33	37¾
6.22	56	40	32	72	3½	7	10¾	14¼	17½	20¾	23¾	26¾	32½	37¼
6.48	56	48	40	72	3¼	6¾	10¼	13½	16¾	20	23	25¾	31½	36¼
6.67	64	48	28	56	3¼	6½	10	13¼	16½	19½	22½	25¼	30¾	35¼
7.29	56	48	40	64	3	6¼	9¼	12¼	15	18	20½	23½	28½	33
7.41	64	48	40	72	3	6	9	12	14¾	17¾	20¼	22¾	28¼	32½
7.62	64	48	32	56	2¾	5¾	8¾	11½	14½	17¼	19¾	22¼	27½	32
8.33	48	32	40	72	2½	5¼	8	10½	13¼	15¾	18¼	20½	25½	29½
8.95	86	48	28	56	2½	5	7½	10	12½	14¾	17	19¼	24	28
9.33	56	40	48	72	2¼	4¾	7¼	9½	11¾	14	16¼	18½	23	27
9.52	64	48	40	56	2¼	4½	7	9¼	11½	13¾	16	18¼	22½	26½
10.29	72	40	32	56	2	4¼	6½	8¾	10¾	12¾	15	17¼	21	24¾
10.37	64	48	56	72	2	4¼	6½	8½	10½	12¾	14¾	17	20¾	24½
10.50	48	40	56	64	2	4¼	6¼	8½	10½	12½	14½	16¾	20½	24¼
10.67	64	40	48	72	2	4	6¼	8¼	10¼	12¼	14¼	16½	20¼	24
10.94	56	32	40	64	2	4	6	8¼	10¼	12	14	16¼	20	23½
11.11	64	32	40	72	2	4	6	8	10	11¾	13¾	16	19¾	23
11.66	56	32	48	72	1¾	3¾	5¾	7½	9½	11¼	13¼	15¼	18¾	22
12.00	72	40	32	48	1¾	3¾	5½	7¼	9¼	11	12¾	15	18¼	21½
13.12	56	32	48	64	1½	3½	5¼	6¾	8½	10¼	11¾	13½	16¾	20
13.33	56	28	48	72	1½	3¼	5	6½	8¼	10	11½	13¼	16½	19½
13.71	64	40	48	56	1½	3¼	4¾	6½	8	9¾	11¼	13	16	19
15.24	64	28	48	72	1½	3	4½	5¾	7¼	8¾	10¼	11¾	14½	17¼
15.56	64	32	56	72	1¼	2¾	4¼	5¾	7¼	8¾	10	11½	14¼	17
15.75	56	64	72	40	1¼	2¾	4¼	5½	7	8½	9¾	11¼	14	16¾
16.87	72	32	48	64	1¼	2½	4	5¼	6¾	7¾	9¼	10½	13¼	15¾
17.14	64	32	48	56	1¼	2½	4	5¼	6½	7¾	9	10¼	13	15½
18.75	72	32	40	48	1	2¼	3½	4¾	6	7¼	8¼	9½	12	14¼
19.29	72	32	48	56	1	2¼	3½	4 1/42	5¾	7	8	9¼	11½	13¾
19.59	64	28	48	56	1	2¼	3¼	4½	5¾	6¾	8	9¼	11½	13½
19.69	72	32	56	64	1	2¼	3¼	4½	5¾	6¾	8	9	11½	13½
21.43	72	24	40	56	1	2	3¼	4¼	5¼	6¼	7½	8½	10½	12½
22.50	72	28	56	64	1	2	3	4	5	6	7	8	10	12
23.33	64	32	56	48	1	2	3	4	5	5¾	6¾	7¾	9¾	11½
26.25	72	24	56	64	1	1¾	2¾	3½	4¼	5	6	7	8½	10¼
26.67	64	28	56	48	¾	1¾	2¾	3½	4¼	5	6	6¾	8½	10
28.00	64	32	56	40	¾	1¾	2½	3¼	4	4¾	5¾	6½	8	9½
30.86	72	28	48	40	¾	1½	2¼	3	3¾	4½	5	5¾	7¼	8¾

Leads, Change Gears and Angles for Helical Milling

Lead of Helix, Inches	Change Gears				Diameter of Work, Inches									
	Gear on Worm	First Gear on Stud	Second Gear on Stud	Gear on Screw	1¾	2	2¼	2½	2¾	3	3¼	3½	3¾	4
					Approximate Angles for Milling Machine Table									
6.12	56	40	28	64	42	…	…	…	…	…	…	…	…	…
6.22	56	40	32	72	41½	…	…	…	…	…	…	…	…	…
6.48	56	48	40	72	40¼	44¼	…	…	…	…	…	…	…	…
6.67	64	48	28	56	39½	43½	…	…	…	…	…	…	…	…
7.29	56	48	40	64	37	41	44¼	…	…	…	…	…	…	…
7.41	64	48	40	72	36½	40¼	43¾	…	…	…	…	…	…	…
7.62	64	48	32	56	36	39½	43	…	…	…	…	…	…	…
8.33	48	32	40	72	33½	37	40½	43½	…	…	…	…	…	…
8.95	86	48	28	56	31¾	35¼	38½	41¼	44	…	…	…	…	…
9.33	56	40	48	72	30½	34	37¼	40¼	43	…	…	…	…	…
9.52	64	48	40	56	30	33½	36½	39½	42¼	45	…	…	…	…
10.29	72	40	32	56	28¼	31½	34½	37½	40	42½	45	…	…	…
10.37	64	48	56	72	28	31¼	34¼	37¼	39¾	42¼	44¾	…	…	…
10.50	48	40	56	64	27¾	31	34	36¾	39½	42	44¼	…	…	…
10.67	64	40	48	72	27¼	30½	33½	36½	39	41½	43¾	…	…	…
10.94	56	32	40	64	26¾	30	33	35¾	38¼	40¾	43	…	…	…
11.11	64	32	40	72	26½	29½	32½	35¼	38	40¼	42½	44¾	…	…
11.66	56	32	48	72	25¼	28½	31¼	34	36½	39	41¼	43½	…	…
12.00	72	40	32	48	24¾	27¾	30½	33¼	35¾	38	40¼	42½	44¾	…
13.12	56	32	48	64	22¾	25¾	28¼	31	33¼	35¾	37¾	40	42	43¾
13.33	56	28	48	72	22½	25½	28	30½	33	35¼	37½	39½	41½	43¼
13.71	64	40	48	56	22	24¾	27¼	30	32¼	34½	36½	38¾	40¾	42½
15.24	64	28	48	72	20	22½	25	27¼	29½	31¾	34	35¾	37¾	39½
15.56	64	32	56	72	19½	22	24½	27	29	31¼	33¼	35¼	37	39
15.75	56	64	72	40	19¼	21¾	24¼	26½	28¾	31	33	35	36¾	38½
16.87	72	32	48	64	18¼	20½	22¾	25	27	29¼	31¼	33¼	35	36½
17.14	64	32	48	56	17¾	20¼	22¼	24¾	26¾	29	30¾	32¾	34½	36
18.75	72	32	40	48	16¼	18½	20¾	22¾	25	26¾	28½	30¼	32	33¾
19.29	72	32	48	56	16	18¼	20¼	22¼	24	26	28	29¾	31½	33
19.59	64	28	48	56	15¾	18	20	22	23¾	25¾	27½	29¼	31	32¾
19.69	72	32	56	64	15¾	17¾	20	21¾	23¾	25½	27½	29¼	31	32½
21.43	72	24	40	56	14½	16½	18½	20¼	22	23¾	25½	27¼	29	30¼
22.50	72	28	56	64	13¾	15¾	17½	19¼	21	22¾	24½	26	27¾	29¼
23.33	64	32	56	48	13¼	15¼	17	18¾	20¼	22	23½	25¼	27	28¼
26.25	72	24	56	64	12	13½	15	16¾	18¼	19¾	21¼	22¾	24¼	25½
26.67	64	28	56	48	11¾	13¼	14¾	16½	18	19½	21	22¼	23¾	25¼
28.00	64	32	56	40	11¼	12¾	14¼	15¾	17¼	18¾	20	21½	22¾	24
30.86	72	28	48	40	10	11½	13	14¼	15½	17	18½	19½	21	22
31.50	72	32	56	40	10	11¼	12¾	14	15¼	16½	18	19¼	20½	21¾
36.00	72	32	64	40	8¾	10	11	12¼	13½	14¾	16	17	18¼	19¼
41.14	72	28	64	40	7¾	8¾	9¾	10¾	11¾	13	14	15	16	17
45.00	72	28	56	32	7	8	9	10	11	11¾	12¾	13¾	14¾	15½
48.00	72	24	64	40	6½	7½	8½	9¼	10¼	11¼	12	13	13¾	14½
51.43	72	28	64	32	6	7	7¾	8¾	9½	10½	11¼	12	12¾	13¾
60.00	72	24	64	32	5¼	6	6¾	7½	8¼	9	9½	10¼	11	11¾
68.57	72	24	64	28	4¼	5¼	5¾	6½	7¼	8	8½	9	9¾	10¼

Helix Angle for Given Lead and Diameter.—The table on this and the preceding page gives helix angles (relative to axis) equivalent to a range of leads and diameters. The expression "Diameter of Work" at the top of the table might mean pitch diameter or outside diameter, depending upon the class of work. Assume, for example, that a plain milling cutter 4 inches in diameter is to have helical teeth and a helix angle of about 25 degrees is desired. The table shows that this angle will be obtained approximately by using change-gears that will give a lead of 26.67 inches. As the outside diameter of the cutter is 4 inches, the helix angle of 25¼ degrees is at the top of the teeth. The angles listed for different diameters are used in setting the table of a milling machine. In milling a right-hand helix (or cutter teeth that turn to the right as seen from the end of the cutter), swivel the right-hand end of the machine table toward the rear, and, inversely, for a left-hand helix, swivel the left-hand end of the table toward the rear. The angles in the table are based upon the following formula:

$$\text{cot helix angle relative to axis} = \frac{\text{lead of helix}}{3.1416 \times \text{diameter}}$$

Lead of Helix for Given Angle.—The lead of a helix or "spiral" for given angles measured with the axis of the work is given in the table, starting on page 1943, for a diameter of 1. For other diameters, lead equals the value found in the table multiplied by the given diameter. Suppose the angle is 55 degrees, and the diameter 5 inches; what would be the lead? By referring to the table starting on page 1943, it is found that the lead for a diameter of 1 and an angle of 55 degrees 0 minutes equals 2.200. Multiply this value by 5; 5 × 2.200 = 11 inches, which is the required lead. If the lead and diameter are given, and the angle is wanted, divide the given lead by the given diameter, thus obtaining the lead for a diameter equal to 1; then find the angle corresponding to this lead in the table. If the lead and angle are given, and the diameter is wanted, divide the lead by the value in the table for the angle.

Helix Angle for Given Lead and Pitch Radius.—To determine the helix angle for a helical gear, knowing the pitch radius and the lead, use the formula:

$$\tan\psi = 2\pi R/L$$

where ψ = helix angle
R = pitch radius of gear, and
L = lead of tooth

Example:

$$R = 3.000, L = 21.000, \tan\psi = (2 \times 3.1416 \times 3.000)/21.000 = 0.89760$$

$$\therefore \psi = 41.911 \text{ degrees}$$

Lead of Tooth Given Pitch Radius and Helix Angle.—To determine the lead of the tooth for a helical gear, given the helix angle and the pitch radius, the formula becomes: $L = 2\pi R/\tan\psi$.

$$\psi = 22.5^\circ, \qquad \therefore \tan\psi = 0.41421, \qquad R = 2.500.$$

$$L = \frac{2 \times 3.1416 \times 2.500}{0.41421} = 37.9228$$

Helix Angle and Lead, Given Normal DP and Numbers of Teeth.—When N_1 = number of teeth in pinion, N_2 = number of teeth in gear, P_n = normal diametral pitch, C = center distance, ψ = helix angle, L_1 = lead of pinion, and L_2 = lead ofgear, then:

$$\cos\psi = \frac{N_1 + N_2}{2P_nC}, \qquad L_1 = \frac{\pi N_1}{P_n \sin\psi}, \qquad L_2 = \frac{\pi N_2}{P_n \sin\psi}$$

$$P_n = 6, \quad N_1 = 18, \quad N_2 = 30, \quad C = 4.500$$

$$\cos\psi = \frac{18 + 30}{2 \times 6 \times 4.5} = 0.88889, \therefore \psi = 27.266^\circ, \text{ and } \sin\psi = 0.45812$$

$$L_1 = \frac{3.1416 \times 18}{6 \times 0.45812} = 20.5728, \text{ and } L_2 = \frac{3.1416 \times 30}{6 \times 0.45812} = 34.2880$$

SIMPLE, COMPOUND, DIFFERENTIAL, AND BLOCK INDEXING

Milling Machine Indexing.—Positioning a workpiece at a precise angle or interval of rotation for a machining operation is called indexing. A dividing head is a milling machine attachment that provides this fine control of rotational positioning through a combination of a crank-operated worm and worm gear, and one or more indexing plates with several circles of evenly spaced holes to measure partial turns of the worm crank. The indexing crank carries a movable indexing pin that can be inserted into and withdrawn from any of the holes in a given circle with an adjustment provided for changing the circle that the indexing pin tracks.

Hole Circles.—The Brown & Sharpe dividing head has three standard indexing plates, each with six circles of holes as listed in the table below.

Numbers of Holes in Brown & Sharpe Standard Indexing Plates

Plate Number	Numbers of Holes					
1	15	16	17	18	19	20
2	21	23	27	29	31	33
3	37	39	41	43	47	49

Dividing heads of Cincinnati Milling Machine design have two-sided, standard, and high-number plates with the numbers of holes shown in the following table.

Numbers of Holes in Cincinnati Milling Machine Standard Indexing Plates

Side	Standard Plate										
1	24	25	28	30	34	37	38	39	41	42	43
2	46	47	49	51	53	54	57	58	59	62	66
	High-Number Plates										
A	30	48	69	91	99	117	129	147	171	177	189
B	36	67	81	97	111	127	141	157	169	183	199
C	34	46	79	93	109	123	139	153	167	181	197
D	32	44	77	89	107	121	137	151	163	179	193
E	26	42	73	87	103	119	133	149	161	175	191
F	28	38	71	83	101	113	131	143	159	173	187

Some dividing heads provide for *Direct Indexing* through the attachment of a special indexing plate directly to the main spindle where a separate indexing pin engages indexing holes in the plate. The worm is disengaged from the worm gear during this quick method of indexing, which is mostly used for common, small-numbered divisions such as six, used in machining hexagonal forms for bolt heads and nuts, for instance.

Simple Indexing.—Also called *Plain Indexing* or *Indirect Indexing*, simple indexing is based on the ratio between the worm and the worm gear, which is usually, but not always, 40:1. The number of turns of the indexing crank needed for each indexing movement to produce a specified number of evenly spaced divisions is equal to the number of turns of the crank that produce exactly one full turn of the main spindle, divided by the specified number of divisions required for the workpiece. The accompanying tables provide data for the indexing movements to meet most division requirements, and include the simple indexing movements along with the more complex movements for divisions that are not available through simple indexing. The fractional entries in the tables are deliberately not reduced to lowest terms. Thus, the numerator represents the number of holes to be moved on the circle of holes specified by the denominator.

Setting up for an indexing job includes setting the sector arms to the fractional part of a turn required for each indexing movement to avoid the need to count holes each time. The current location of the indexing pin in the circle of holes to be used is always hole zero when counting the number of holes to be moved. The wormshaft hub carrying the dividing plate may also carry one or two sets of sector arms, each of which can be used to define two arcs of holes. As shown at the right in the drawing of a typical dividing head at the top of the Simple and Differential Indexing table on page 1979, these sector arms can make up an inner arc, A, and an outer arc, B. The inner arc is used most often, but some indexing movements require the use of the outer arc.

Example: With a worm/worm gear ratio of 40:1 making 35 divisions requires each indexing movement to be 40 ÷ 35 = 1 $\frac{1}{7}$ turns: one full turn of the indexing crank plus one-seventh of a full turn more. A full turn is easily achieved using any circle of holes, but to continue the indexing movement to completion for this example requires a circle in which the number of holes is evenly divisible by 7. The Brown & Sharpe dividing head has a 21-hole circle on plate 2 and a 49-hole circle on plate 3. Either circle could be used because 3/21 and 7/49 both equal 1/7th. The Cincinnati dividing head standard plate has a 28-hole circle on the first side and a 49-hole circle on the second side and again, either 4/28 or 7/49 could be used for the fractional part of a turn needed for 35 divisions. In selecting among equivalent indexing solutions, the one with the smallest number of holes in the fractional part of a turn is generally preferred (except that if an indexing plate with an alternate solution is already mounted on the dividing head, the alternate should be used to avoid switching indexing plates).

Compound Indexing.—Compound indexing is used to obtain divisions that are not available by simple indexing. Two simple indexing movements are used with different circles of holes on an indexing plate that is not bolted to the dividing head frame so that it is free to rotate on the worm shaft. A second, stationary indexing pin arrangement is clamped or otherwise fixed to the frame of the dividing head to hold the indexing plate in position except during the second portion of the compound indexing movement. If available, a double set of low-profile sector arms would improve the ease and reliability of this method. Sector arms for the innermost circle of an indexing movement should not reach as far as the outermost circle of the movement, and sector arms for the outermost circle should be full length. Positioning the outermost circle sector arms may have to wait until the indexing pin on the innermost circle is withdrawn, and may sometimes coincide with the position of that pin. The indexing pin on the crank is set to track the innermost of the two circles in the compound movement and the stationary indexing pin is set to track the outermost circle. Some divisions are only available using adjacent circles, so the intercircle spacing may become a constraining factor in the design or evaluation of a stationary pin arrangement.

The first part of the indexing movement is performed as in simple indexing by withdrawing the indexing pin on the crank arm from its hole in the indexing plate, rotating the crank to its next position, and reinserting the indexing pin in the new hole. For the second part of the movement, the stationary indexing pin is released from its hole in the indexing plate, and with the crank indexing pin seated in its hole, the crank is used to turn the crank arm and indexing plate together to the next position for reinserting the stationary pin into its new hole.

There are two possibilities for the separate movements in compound indexing: they may both be in the same direction of rotation, referred to as *positive compounding* and indicated in the table by a plus (+) sign between the two indexing movements, or they may be in opposite directions of rotation, referred to as *negative compounding* and indicated in the table by a minus (−) sign between the two indexing movements. In positive compounding, it does not matter whether the rotation is clockwise or counterclockwise, as long as it is the same throughout the job. In negative compounding, there will be one clockwise movement and one counterclockwise movement for each unit of the division. The mathematical difference is in whether the two fractional turns are to be added together or whether one is to be subtracted from the other. Operationally, this difference is important because of the backlash, or free play, between the worm and the worm gear of the dividing head. In positive compounding, this play is always taken up because the worm is turned continually in the same direction. In negative compounding, however, the direction of each turn is always opposite that of the previous turn, requiring each portion of each division to be started by backing off a few holes to allow the play to be taken up before the movement to the next position begins.

The table, Simple and Compound Indexing with Brown & Sharpe Plates, gives indexing movements for all divisions up to and including 250 with plain dividing heads of the Brown & Sharpe design. All the simple indexing movements, and many of the compound indexing movements, are exact for the divisions they provide. There remains a substantial number of divisions for which the indexing movements are approximate. For these divisions, the indexing movements shown come very close to the target number, but the price of getting close is increased length and complexity of the indexing movements. The table shows all divisions that can be obtained through simple indexing and all divisions for which exact compound indexing movements are available. Approximate movements are only used when it is necessary to obtain a division that would otherwise not be available. The approximate indexing movements usually involve multiple revolutions of the workpiece, with successive revolutions filling in spaces left during earlier turns.

Simple and Compound Indexing with Brown & Sharpe Plates

Number of Divisions	Whole Turns	Fractions of a Turn	Number of Divisions	Whole Turns	Fractions of a Turn	Number of Divisions	Whole Turns	Fractions of a Turn
2	20	…	15	2	26/39	33	1	7/33
3	13	5/15	16	2	8/16	34	1	3/17
3	13	7/21	17	2	6/17	35	1	3/21
3	13	13/39	18	2	4/18	35	1	7/49
4	10	…	18	2	6/27	36	1	2/18
5	8	…	19	2	2/19	36	1	3/27
6	6	10/15	20	2	…	37	1	3/37
6	6	14/21	21	1	19/21	38	1	1/19
6	6	26/39	22	1	27/33	39	1	1/39
7	5	15/21	23	1	17/23	40	1	…
8	5	…	24	1	10/15	41	…	40/41
9	4	8/18	24	1	14/21	42	…	20/21
9	4	12/27	24	1	26/39	43	…	40/43
10	4	…	25	1	9/15	44	…	30/33
11	3	21/33	26	1	21/39	45	…	16/18
12	3	5/15	27	1	13/27	45	…	24/27
12	3	7/21	28	1	9/21	46	…	20/23
12	3	13/39	29	1	11/29	47	…	40/47
13	3	3/39	30	1	5/15	48	…	15/18
14	2	18/21	30	1	7/21	49	…	40/49
14	2	42/49	30	1	13/39	50	…	12/15
15	2	10/15	31	1	9/31			
15	2	14/21	32	1	4/16			

Simple and Compound Indexing with Standard Brown &Sharpe Plates

Target Divisions	Indexing Movements	Work-piece Revolutions	Precise Number of Divisions	Diameter at Which Error = 0.001	Target Divisions	Indexing Movements	Work-piece Revolutions	Precise Number of Divisions	Diameter at Which Error = 0.001
51	10/15+2/17	1	51.00000	Exact	88	15/33	1	88.00000	Exact
52	30/39	1	52.00000	Exact	89	1 29/37 + 19/41	5	88.99971	96.58
53	26/29 + 19/31	2	52.99926	22.89	89	2 22/37 + 5/49	6	88.99980	138.50
53	14/43 + 4 45/47	7	52.99991	180.13	89	2 28/39 + 43/49	8	89.00015	194.65
53	5 43/47 + 43/49	9	53.00006	263.90	90	8/18	1	90.00000	Exact
54	20/27	1	54.00000	Exact	90	12/27	1	90.00000	Exact
55	24/33	1	55.00000	Exact	91	6/39 + 14/49	1	91.00000	Exact
56	15/21	1	56.00000	Exact	92	10/23	1	92.00000	Exact
56	35/49	1	56.00000	Exact	93	7/21+3/31	1	93.00000	Exact
57	5/15+7/19	1	57.00000	Exact	94	20/47	1	94.00000	Exact
58	20/29	1	58.00000	Exact	95	8/19	1	95.00000	Exact
59	18/37 + 9/47	1	58.99915	22.14	96	3/18 + 5/20	1	96.00000	Exact
59	4 2/43 + 1/47	6	59.00012	154.39	97	15/41 + 2/43	1	97.00138	22.45
59	5 15/37 + 3 20/49	13	58.99994	300.09	97	1 42/43 + 4/47	5	97.00024	128.66
60	10/15	1	60.00000	Exact	97	3 27/41 + 43/49	11	96.99989	281.37
60	14/21	1	60.00000	Exact	98	20/49	1	98.00000	Exact
60	26/39	1	60.00000	Exact	99	6/27 + 6/33	1	99.00000	Exact
61	2 3/43 + 26/47	4	60.99981	102.93	100	6/15	1	100.00000	Exact
61	3 42/47 + 2/49	6	60.99989	175.94	101	1 33/43 + 10/47	5	100.99950	64.33
61	4 31/41 + 2/49	8	61.00009	204.64	101	2 27/37 + 2/47	7	100.99979	154.99
62	20/31	1	62.00000	Exact	101	3 32/43 + 30/49	11	101.00011	295.10
63	11/21 + 3/27	1	63.00000	Exact	102	5/15 + 1/17	1	102.00000	Exact
64	10/16	1	64.00000	Exact	103	1 8/43 + 18/49	4	103.00031	107.31
65	24/39	1	65.00000	Exact	103	2 22/37 + 21/41	8	103.00021	154.52
66	20/33	1	66.00000	Exact	103	4 32/37 + 9/49	13	103.00011	300.09
67	29/37 + 16/39	2	66.99942	36.75	104	15/39	1	104.00000	Exact
67	2 27/41 + 16/49	5	67.00017	127.90	105	8/21	1	105.00000	Exact
67	4 20/43 + 2 5/49	11	67.00007	295.10	106	1 7/39 + 29/41	5	105.99934	50.90
68	10/17	1	68.00000	Exact	106	2 12/41 + 15/43	7	105.99957	78.57
69	14/21 – 2/23	1	69.00000	Exact	106	2 38/41 + 23/49	9	106.00029	115.11
70	12/21	1	70.00000	Exact	107	23/43+ 10/47	2	107.00199	17.15
70	28/49	1	70.00000	Exact	107	2 38/41 + 3/47	8	106.99983	196.28
71	35/37 + 32/43	3	71.00037	60.77	107	3 38/39 + 22/43	12	106.99987	256.23
71	2 34/41 + 27/49	6	70.99985	153.48	108	10/27	1	108.00000	Exact
71	4 25/39 + 2 28/41	13	70.99991	264.67	109	1 8/21 + 2/23	4	108.99859	24.60
72	10/18	1	72.00000	Exact	109	1 24/37 + 26/47	6	108.99974	132.85
72	15/27	1	72.00000	Exact	109	2 19/39 + 4/49	7	108.99980	170.32
73	5/43 + 48/49	2	73.00130	17.88	110	12/33	1	110.00000	Exact
73	2 19/43 + 14/47	5	72.99982	128.66	111	1/37 + 13/39	1	111.00000	Exact
73	2 28/47 + 3 48/49	12	73.00007	351.87	112	33/43 + 2 21/47	9	112.00123	28.95
74	20/37	1	74.00000	Exact	112	14/37 + 4 46/47	15	112.00086	41.52
75	8/15	1	75.00000	Exact	112	9 14/37 + 46/47	29	112.00044	80.26
76	10/19	1	76.00000	Exact	113	14/37 – 1/41	1	112.99814	19.32
77	9/21 + 3/33	1	77.00000	Exact	113	2 28/41 + 7/47	8	112.99982	196.28

Simple and Compound Indexing with Standard Brown &Sharpe Plates *(Continued)*

Target Divisions	Indexing Movements	Workpiece Revolutions	Precise Number of Divisions	Diameter at Which Error = 0.001	Target Divisions	Indexing Movements	Workpiece Revolutions	Precise Number of Divisions	Diameter at Which Error = 0.001
78	20/39	1	78.00000	Exact	113	4 20/37 + 3/49	13	113.00012	300.09
79	17/37 + 26/47	2	79.00057	44.28	114	10/15 – 6/19	1	114.00000	Exact
79	2 42/43 + 3/49	6	79.00016	160.96	115	8/23	1	115.00000	Exact
79	4 34/39 + 9/47	10	79.00011	233.38	116	10/29	1	116.00000	Exact
80	8/16	1	80.00000	Exact	117	1 16/41 + 15/47	5	117.00061	61.34
81	10/43 + 37/49	2	80.99952	53.65	117	7 1/47 – 9/49	20	117.00006	586.45
81	3 9/47 + 13/49	7	80.99987	205.26	118	1 8/39 + 24/49	5	117.99938	60.83
81	5 11/37 + 1 6/49	13	81.00009	300.09	118	30/41 + 2 15/47	9	117.99966	110.41
82	20/41	1	82.00000	Exact	119	15/43 + 31/47	3	118.99902	38.60
83	1 11/29 + 17/31	4	83.00058	45.79	119	2 4/23 + 17/33	8	119.00049	77.31
83	3 17/27 + 7/31	8	82.99969	85.26	119	3 31/37 + 25/47	13	118.99987	287.84
83	5 1/37 + 31/41	12	83.00011	231.78	120	5/15	1	120.00000	Exact
84	10/21	1	84.00000	Exact	120	7/21	1	120.00000	Exact
85	8/17	1	85.00000	Exact	120	13/39	1	120.00000	Exact
86	20/43	1	86.00000	Exact	121	8/37 + 38/49	3	121.00111	34.63
87	14/21 – 6/29	1	87.00000	Exact	121	1 4/43 + 2 10/47	10	120.99985	257.32
122	1 14/41 + 14/47	5	122.00063	61.34	157	22/41 + 2 38/49	13	157.00030	166.27
122	41/43 + 2 32/49	11	122.00026	147.55	158	1 4/39 + 8/49	5	157.99917	60.83
123	26/39 – 14/41	1	123.00000	Exact	158	1 29/39 + 23/43	9	158.00052	96.09
124	10/31	1	124.00000	Exact	159	14/37 + 27/43	4	159.00062	81.03
125	41/43 +16/49	4	124.99815	21.46	159	1 19/43 + 15/47	7	158.99972	180.13
125	2 3/43 + 8/47	7	125.00110	36.03	159	2 7/37 + 16/49	10	159.00022	230.84
125	3/41 + 3 21/47	11	125.00074	53.98	160	4/16	1	160.00000	Exact
126	2/21 + 6/27	1	126.00000	Exact	161	9/23 – 3/21	1	161.00000	Exact
127	2/39 + 42/47	3	126.99769	17.50	162	28/47 – 15/43	1	162.00401	12.87
127	2 6/37 + 2/47	7	127.00052	77.50	162	1 30/39 – 2/49	7	161.99818	28.39
127	2 23/39 + 12/49	9	127.00018	218.98	162	2 8/23 + 25/29	13	161.99907	55.20
128	5/16	1	128.00000	Exact	163	18/49 – 5/41	1	163.00203	25.58
129	13/39 – 1/43	1	129.00000	Exact	163	19/37 + 22/47	4	162.99941	88.57
130	12/39	1	130.00000	Exact	163	2 31/47 + 26/49	13	162.99986	381.20
131	5/37 + 8/47	1	130.99812	22.14	164	10/41	1	164.00000	Exact
131	4/37 + 1 18/43	5	131.00041	101.29	165	8/33	1	165.00000	Exact
131	2 27/43 + 20/47	10	130.99984	257.32	166	20/29 + 17/33	5	166.00173	30.46
132	10/33	1	132.00000	Exact	166	2 20/41 + 7/43	11	166.00043	123.46
133	1/37 + 27/47	2	133.00191	22.14	167	23/43 + 9/49	3	167.00132	40.24
133	12/31 + 17/33	3	133.00108	39.08	167	6/37 + 39/49	4	167.00058	92.34
133	1 23/29 + 19/31	8	133.00046	91.57	167	2 24/37 + 20/43	13	167.00040	131.67
134	4/29 + 25/33	3	134.00233	18.28	168	5/21	1	168.00000	Exact
134	1 13/43 + 37/47	7	133.99953	90.06	169	1/41 + 22/49	2	169.00105	51.16
134	3 27/47 + 15/49	13	134.00022	190.60	169	1 32/37 + 13/49	9	169.00052	103.88
135	8/27	1	135.00000	Exact	170	4/17	1	170.00000	Exact
136	5/17	1	136.00000	Exact	171	8/18 – 4/19	1	171.00000	Exact
137	9/37 + 31/49	3	137.00252	17.31	172	10/43	1	172.00000	Exact
137	11/41 + 1 38/49	7	136.99951	89.53	173	27/37 + 8/41	4	173.00071	77.26
137	17/43 + 2 40/49	11	137.00015	295.10	173	1 7/43 + 11/49	6	173.00034	160.96

Simple and Compound Indexing with Standard Brown &Sharpe Plates *(Continued)*

Target Divisions	Indexing Movements	Workpiece Revolutions	Precise Number of Divisions	Diameter at Which Error = 0.001	Target Divisions	Indexing Movements	Workpiece Revolutions	Precise Number of Divisions	Diameter at Which Error = 0.001
138	7/21 − 1/23	1	138.00000	Exact	174	7/21 − 3/29	1	174.00000	Exact
139	23/41 + 13/43	3	139.00131	33.67	175	3/37 + 26/43	3	174.99542	12.15
139	$1\frac{31}{39}$ + 9/41	7	139.00031	142.51	175	$1\frac{9}{37}$ + 5/39	6	174.99747	22.05
139	$3\frac{14}{43}$ + 6/47	12	138.99986	308.79	175	$2\frac{8}{41}$ + 15/47	11	175.00103	53.98
140	6/21	1	140.00000	Exact	176	$1\frac{14}{43}$ + 13/49	7	176.00239	23.47
141	29/47 − 13/39	1	141.00000	Exact	176	$2\frac{18}{37}$ + 22/47	13	175.99844	35.98
142	23/39 + 12/47	3	142.00129	35.01	177	6/37 + 3/47	1	176.99746	22.14
142	18/41 + $2\frac{31}{47}$	11	141.99967	134.94	177	$1\frac{17}{37}$ + 6/49	7	177.00139	40.40
143	13/37 + 20/41	3	143.00079	57.95	177	$2\frac{19}{47}$ + 4/49	11	176.99913	64.51
143	$1\frac{16}{27}$ + 20/31	8	143.00053	85.26	178	$1\frac{16}{39}$ + 7/43	7	177.99848	37.37
144	5/18	1	144.00000	Exact	178	$2\frac{11}{41}$ + 32/49	13	177.99966	166.27
145	8/29	1	145.00000	Exact	179	20/37 − 13/41	1	178.99705	19.31
146	16/41 − 5/43	1	146.00414	11.22	179	14/39 + 23/43	4	178.99933	85.41
146	3/37 + $1\frac{41}{49}$	7	145.99942	80.79	179	$1\frac{34}{47}$ + 36/49	11	179.00018	322.55
146	28/37 + $2\frac{33}{41}$	13	146.00037	125.55	180	4/18	1	180.00000	Exact
147	13/39 − 3/49	1	147.00000	Exact	180	6/27	1	180.00000	Exact
148	10/37	1	148.00000	Exact	181	20/37 + 6/49	3	180.99834	34.63
149	28/41 + 6/49	3	148.99876	38.37	181	39/41 + 28/47	7	180.99966	171.75
149	$1\frac{7}{39}$ + 7/43	5	149.00044	106.76	181	$2\frac{8}{39}$ + 21/47	12	180.99979	280.06
149	26/37 + $2\frac{37}{47}$	13	148.99984	287.84	182	3/39 + 7/49	1	182.00000	Exact
150	4/15	1	150.00000	Exact	183	8/29 + 5/31	2	183.00254	22.89
151	5/37 + 31/47	3	150.99855	33.21	183	1/43 + 40/47	4	182.99943	102.93
151	6/37 + 35/39	4	151.00065	73.49	183	$1\frac{24}{41}$ + 8/49	8	183.00028	204.64
151	$2\frac{21}{43}$ + 20/47	11	151.00017	283.05	184	5/23	1	184.00000	Exact
152	5/19	1	152.00000	Exact	185	8/37	1	185.00000	Exact
153	10/18 − 5/17	1	153.00000	Exact	186	17/31 − 7/21	1	186.00000	Exact
154	1/21 + 7/33	1	154.00000	Exact	187	19/37 + 5/39	3	186.99784	27.56
155	8/31	1	155.00000	Exact	187	21/23 + 10/27	6	187.00125	47.44
156	10/39	1	156.00000	Exact	187	$1\frac{38}{43}$ + 12/47	10	186.99977	257.32
157	18/47 − 5/39	1	157.00214	23.34	188	10/47	1	188.00000	Exact
157	22/47 + 27/49	4	157.00043	117.29	189	7/27 − 1/21	1	189.00000	Exact
190	4/19	1	190.00000	Exact	223	6/37 + 36/49	25	223.00123	57.71
191	1/21 + 18/31	3	191.00244	24.87	223	$1\frac{26}{37}$ + 38/47	14	222.99977	309.98
191	34/37 + 5/39	5	190.99934	91.86	223	$3\frac{11}{41}$ + 15/47	20	223.00014	490.71
191	28/39 + 45/47	8	190.99967	186.71	224	$1\frac{13}{37}$ + 11/43	9	223.99687	22.79
192	5/15 − 2/16	1	192.00000	Exact	224	$2\frac{16}{39}$ + 11/41	15	224.00187	38.17
193	5/37 + 34/49	4	193.00067	92.34	224	$3\frac{5}{43}$ + 13/47	19	223.99883	61.11
193	29/39 + 12/41	5	192.99940	101.80	225	1/15 + 2/18	1	225.00000	Exact
194	41/43 + 24/49	7	194.00197	31.30	226	28/37 + 5/39	5	225.99843	45.93
194	$1\frac{23}{37}$ + 11/47	9	194.00062	99.64	226	$1\frac{38}{39}$ + 16/49	13	225.99955	158.16
194	$2\frac{8}{47}$ + 25/49	13	193.99968	190.60	227	9/39 + 14/47	3	226.99690	23.34
195	8/39	1	195.00000	Exact	227	$1\frac{11}{37}$ + 25/39	11	227.00036	202.10
196	10/49	1	196.00000	Exact	227	$3\frac{3}{43}$ + 5/49	18	226.99985	482.89
197	17/37 − 10/39	1	196.99659	18.37	228	5/15 − 3/19	1	228.00000	Exact
197	19/39 + 5/41	3	197.00205	30.54	229	7/39 + 34/49	5	228.99940	121.66
197	$1\frac{39}{43}$ + 16/49	11	196.99958	147.55	229	$1\frac{19}{41}$ + 31/49	12	229.00024	306.95

Simple and Compound Indexing with Standard Brown &Sharpe Plates *(Continued)*

Target Divisions	Indexing Movements	Work-piece Revolutions	Precise Number of Divisions	Diameter at Which Error = 0.001	Target Divisions	Indexing Movements	Work-piece Revolutions	Precise Number of Divisions	Diameter at Which Error = 0.001
198	3/27 + 3/33	1	198.00000	Exact	229	2 35/41 + 20/43	19	229.00017	426.50
199	16/41 + 10/47	3	199.00172	36.80	230	4/23	1	230.00000	Exact
199	26/37 + 13/43	5	198.99937	101.29	231	3/21 + 1/33	1	231.00000	Exact
199	1 41/43 + 31/47	13	199.00019	334.52	232	5/29	1	232.00000	Exact
200	3/15	1	200.00000	Exact	233	2/37 + 31/49	4	232.99598	18.47
201	27/37 + 13/49	5	200.99778	28.85	233	21/37 + 26/41	7	233.00055	135.21
201	1 18/41 + 27/49	10	201.00050	127.90	233	1 23/37 + 41/43	15	232.99976	303.86
201	2 5/41 + 20/43	13	200.99978	291.81	234	8/41 + 31/47	5	234.00121	61.34
202	24/37 + 14/41	5	201.99734	24.14	234	2 17/43 + 24/47	17	233.99966	218.72
202	1/43 + 2 27/49	13	201.99853	43.59	235	8/47	1	235.00000	Exact
203	14/29 − 6/21	1	203.00000	Exact	236	22/37 + 29/49	7	236.00186	40.40
204	9/17 − 5/15	1	204.00000	Exact	236	2 30/43 + 9/49	17	236.00066	114.02
205	8/41	1	205.00000	Exact	237	1 7/39 + 7/41	8	236.99861	54.29
206	1/41 + 24/43	3	205.99805	33.67	237	1 26/37 + 6/39	11	236.99888	67.37
206	2 8/39 + 15/47	13	205.99957	151.70	237	12/47 + 1 46/49	13	236.99980	381.20
207	5/23 + 15/27	4	207.00000	Exact	238	7/37 + 28/43	5	237.99551	16.88
208	8/43 + 38/49	5	207.99605	16.77	238	1/43 + 1 23/47	9	237.99804	38.60
208	1 19/47 + 16/49	9	207.99799	32.99	238	2 17/39 + 4/47	15	238.00043	175.04
208	3 35/43 + 11/49	21	208.00094	70.42	239	1/37 + 12/39	2	239.00621	12.25
209	9/41 + 8/49	2	208.99870	51.16	239	32/39 + 9/49	6	238.99948	145.99
209	1 36/41 + 18/43	12	208.99975	269.37	239	2 3/41 + 26/43	16	239.00021	359.16
210	4/21	1	210.00000	Exact	240	3/18	1	240.00000	Exact
211	35/37 + 9/47	6	211.00101	66.42	241	4/39 + 17/43	3	241.00599	12.81
211	1 10/37 + 17/39	9	210.99919	82.68	241	26/41 + 17/47	6	241.00052	147.21
211	1 33/41 + 31/47	13	211.00021	318.96	241	1 25/37 + 35/43	15	240.99975	303.86
212	34/39 + 22/49	7	211.99683	21.29	242	4/37 + 19/49	3	242.00222	34.63
212	1 5/43 + 47/49	11	212.00091	73.77	242	1 37/39 + 26/49	15	242.00084	91.24
212	3 4/47 + 6/49	17	211.99946	124.62	242	2 22/39 + 39/43	21	241.99966	224.20
213	14/37 + 44/47	7	213.00087	77.50	243	22/37 + 3/47	4	243.00437	17.71
213	2 36/37 + 9/41	17	213.00021	328.36	243	32/41 + 2/47	5	243.00126	61.34
214	7/39 + 37/49	5	213.99776	30.41	243	29/41 + 46/49	10	242.99970	255.79
214	2 9/47 + 30/49	15	214.00031	219.92	244	36/39 + 11/49	7	243.99453	14.19
215	8/43	1	215.00000	Exact	244	1 10/37 + 8/39	9	244.00188	41.34
216	5/27	1	216.00000	Exact	244	1 28/37 + 2/43	11	244.00139	55.71
217	12/21 − 12/31	1	217.00000	Exact	245	8/49	1	245.00000	Exact
218	14/39 + 9/47	3	217.99802	35.01	246	13/39 − 7/41	1	246.00000	Exact
218	19/37 + 1 34/39	13	218.00116	59.71	247	17/37 + 21/41	6	247.00136	57.59
219	24/39 + 14/47	5	218.99642	19.45	247	15/43 + 1 5/49	14	247.00021	375.58
219	1 2/37 + 11/49	7	218.99914	80.79	248	5/31	1	248.00000	Exact
219	2 11/41 + 41/49	17	218.99968	217.42	249	20/37 + 5/49	4	248.99571	18.47
220	6/33	1	220.00000	Exact	249	10/37 + 1 46/47	14	249.00026	309.98
221	26/37 + 1/47	4	221.00079	88.57	249	4/43 + 2 47/49	19	249.00016	509.72
221	5/47 + 48/49	6	220.99960	175.94	250	1 8/41 + 12/49	9	249.99654	23.02
221	3 9/41 + 25/43	21	220.99985	471.39	250	2 2/43 + 33/49	17	250,00174	45.61
222	19/37 − 13/39	1	222.00000	Exact	250	3 16/47 + 48/49	27	249.99899	79.17

The greater spacing between successive machining operations may be used to advantage to spread out and reduce the effects of heat generation on the workpiece. The number of workpiece revolutions required by an approximation is shown in the table in the column to the right of the indexing movements. The table gives two or three choices for each division requiring approximate movements.

Two measures of the closeness of each approximation are provided to aid in the trade-off between complexity and precision. The first measure is the precise number of divisions that a set of indexing movements produces, offering a direct comparison of the degree of approximation. However, the difference between the precise number of divisions and the target number of divisions is angular in nature, so the error introduced by an approximation depends on the size of the circle being divided. The second measure of closeness reflects this characteristic by expressing the degree of approximation as the diameter at which the error is equal to 0.001. This second measure is unitless, so that taking the error as 0.001 inch means that the entries in that column are to be taken as diameters in inches, but the measure works as well with 0.001 centimeter and diameters in centimeters. The measure can also be used to calculate the error of approximation at a given diameter. Divide the given diameter by the value of the measure and multiply the result by 0.001 to determine the amount of error that using an approximation will introduce.

Example: A gear is to be cut with 127 teeth at 16 diametral pitch using a Brown & Sharpe plain dividing head. The indexing table gives three approximations for 127 divisions. The pitch diameter of a 16 DP gear with 127 teeth is about 7.9 inches, so the calculated error of approximation for the three choices would be about $(7.9 \div 17.5) \times 0.001 = 0.00045$ inch, $(7.9 \div 77.5) \times 0.001 = 0.00010$ inch, and $(7.9 \div 218.98) \times 0.001 = 0.000036$ inch. Considering the increased potential for operator error with longer indexing movements and such other factors as may be appropriate, assume that the first of the three approximations is selected. Plate 3 is mounted on the worm shaft of the dividing head but not bolted to the frame. A double set of sector arms is installed, if available; otherwise, the single pair of sector arms is installed. The indexing pin on the crank arm is set to track the 39-hole circle. The stationary indexing pin is installed and set to track the 47-hole circle. If only one pair of sector arms is used, it is used for the 42⁄47 movement and is set for 0-42 holes using the outer arc. Six holes should be showing in the inner arc on the 47-hole circle (the zero-hole, the 42-hole, and four extra holes). The second set of sector arms is set for 0-2 holes on the 39-hole circle using the inner arc (three holes showing). If there is no second pair of sector arms, this is a short enough movement to do freehand without adding much risk of error.

Angular Indexing.—The plain dividing head with a 40:1 gear ratio will rotate the main spindle and the workpiece 9 degrees for each full turn of the indexing crank, and therefore 1 degree for movements of 2⁄18 or 3⁄27 on Brown & Sharpe dividing heads and 6⁄54 on heads of Cincinnati design. To find the indexing movement for an angle, divide that angle, in degrees, by 9 to get the number of full turns and the remainder, if any. If the remainder, expressed in minutes, is evenly divisible by 36, 33.75, 30, 27, or 20, then the quotient is the number of holes to be moved on the 15-, 16-, 18-, 20-, or 27-hole circles, respectively, to obtain the fractional turn required (or evenly divisible by 22.5, 21.6, 18, 16.875, 15, 11.25, or 10 for the number of holes to be moved on the 24-, 25-, 30-, 32-, 36-, 48-, or 54-hole circles, respectively, for the standard and high number plates of a Cincinnati dividing head). If none of these divisions is even, it is not possible to index the angle (exactly) by this method.

Example: An angle of 61° 48′ is required. Expressed in degrees, this angle is 61.8°, which when divided by 9 equals 6 with a remainder of 7.8°, or 468′. Division of 468 by 20, 27, 30, 33.75, and 36 reveals an even division by 36, yielding 13. The indexing movement for 61° 48′ is six full turns plus 13 holes on the 15-hole circle.

Tables for Angular Indexing.—The table headed Angular Values of One-Hole Moves provides the angular movement obtained with a move of one hole in each of the indexing circles available on standard Brown & Sharpe and Cincinnati plates, for a selection of angles that can be approximated with simple indexing.

The table headed *Accurate Angular Indexing* provides the simple and compound indexing movements to obtain the full range of fractional turns with the standard indexing plates of both the Brown & Sharpe and Cincinnati dividing heads. Compound indexing movements depend on the presence of specific indexing circles on the same indexing plate, so some movements may not be available with plates of different configurations. To use the table to index an angle, first convert the angle to seconds and then divide the number of seconds in the angle by 32,400 (the number of seconds in 9 degrees, which is one full turn of the indexing crank). The whole-number portion of the quotient gives the number of full turns of the indexing crank, and the decimal fraction of the quotient gives the fractional turn required.

Angular Values of One-Hole Moves for B&S and Cincinnati Index Plates

Holes in Circle	Angle in Minutes	Holes in Circle	Angle in Minutes	Holes in Circle	Angle in Minutes
15	36.000	53	10.189	129	4.186
16	33.750	54	10.000	131	4.122
17	31.765	57	9.474	133	4.060
18	30.000	58	9.310	137	3.942
19	28.421	59	9.153	139	3.885
20	27.000	62	8.710	141	3.830
21	25.714	66	8.182	143	3.776
23	23.478	67	8.060	147	3.673
24	22.500	69	7.826	149	3.624
25	21.600	71	7.606	151	3.576
26	20.769	73	7.397	153	3.529
27	20.000	77	7.013	157	3.439
28	19.286	79	6.835	159	3.396
29	18.621	81	6.667	161	3.354
30	18.000	83	6.506	163	3.313
31	17.419	87	6.207	167	3.234
32	16.875	89	6.067	169	3.195
33	16.364	91	5.934	171	3.158
34	15.882	93	5.806	173	3.121
36	15.000	97	5.567	175	3.086
37	14.595	99	5.455	177	3.051
38	14.211	101	5.347	179	3.017
39	13.846	103	5.243	181	2.983
41	13.171	107	5.047	183	2.951
42	12.857	109	4.954	187	2.888
43	12.558	111	4.865	189	2.857
44	12.273	113	4.779	191	2.827
46	11.739	117	4.615	193	2.798
47	11.489	119	4.538	197	2.741
48	11.250	121	4.463	199	2.714
49	11.020	123	4.390	…	…
51	10.588	127	4.252	…	…

Accurate Angular Indexing

Part of a Turn	B&S, Becker, Hendey, K&T, & Rockford	Cincinnati and LeBlond	Part of a Turn	B&S, Becker, Hendey, K&T, & Rockford	Cincinnati and LeBlond
0.0010	12/49–10/41	15/51–17/58	0.0370	13/43–13/49	6/51–5/62
0.0020	24/49–20/41	23/51–22/49	0.0370	1/27	2/54
0.0030	8/23–10/29	7/39–6/34	0.0377	…	2/53
0.0040	1/41–1/49	4/66–3/53	0.0380	13/41–12/43	12/49–12/58
0.0050	4/39–4/41	3/24–3/25	0.0390	15/29–11/23	9/54–6/47
0.0060	9/29–7/23	18/51–17/49	0.0392	…	2/51
0.0070	11/31–8/23	5/46–6/59	0.0400	9/41–7/39	1/25
0.0080	2/41–2/49	10/49–10/51	0.0408	2/49	2/49
0.0090	1/23–1/29	8/24–12/37	0.0410	20/41–21/47	21/43–17/38
0.0100	8/39–8/41	6/24–6/25	0.0417	…	1/24
0.0110	6/39–7/49	11/59–10/57	0.0420	8/47–5/39	23/59–16/46
0.0120	9/47–7/39	15/49–15/51	0.0426	2/47	2/47
0.0130	2/33–1/21	3/54–2/47	0.0430	7/21–9/31	8/59–5/54
0.0140	19/47–16/41	10/46–12/59	0.0435	1/23	2/46
0.0150	8/29–6/23	9/24–9/25	0.0440	17/43–13/37	17/57–15/59
0.0152	…	1/66	0.0450	11/49–7/39	3/24–2/25
0.0160	18/49–13/37	20/49–20/51	0.0455	…	3/66
0.0161	…	1/62	0.0460	8/37–8/47	19/39–15/34
0.0169	…	1/59	0.0465	2/43	2/43
0.0170	15/41–15/43	9/49–11/66	0.0470	8/49–5/43	26/66–17/49
0.0172	…	1/58	0.0476	1/21	2/42
0.0175	…	1/57	0.0480	11/47–8/43	13/51–12/58
0.0180	9/39–10/47	11/42–10/41	0.0484	…	3/62
0.0185	…	1/54	0.0488	2/41	2/41
0.0189	…	1/53	0.0490	14/43–13/47	8/47–8/66
0.0190	7/37–8/47	6/49–6/58	0.0500	1/20	2/24–1/30
0.0196	…	1/51	0.0508	…	3/59
0.0200	16/39–16/41	3/30–2/25	0.0510	2/17–1/15	16/54–13/53
0.0204	1/49	1/49	0.0513	2/39	2/39
0.0210	3/43–2/41	15/46–18/59	0.0517	…	3/58
0.0213	1/47	1/47	0.0520	19/37–18/39	6/46–4/51
0.0217	…	1/46	0.0526	1/19	2/38
0.0220	12/37–13/43	22/59–20/57	0.0526	…	3/57
0.0230	4/37–4/47	3/47–2/49	0.0530	14/47–12/49	1/54+2/58
0.0233	1/43	1/43	0.0540	17/47–12/39	12/53–10/58
0.0238	…	1/42	0.0541	2/37	2/37
0.0240	18/47–14/39	23/58–19/51	0.0550	4/41–2/47	9/24–8/25
0.0244	1/41	1/41	0.0556	1/18	3/54
0.0250	2/16–2/20	2/30–1/24	0.0560	13/49–9/43	7/38–5/39
0.0256	1/39	1/39	0.0566	…	3/53
0.0260	4/33–2/21	3/46–2/51	0.0570	13/29–9/23	18/49–18/58
0.0263	…	1/38	0.0580	19/41–15/37	7/53–4/54
0.0270	3/23–3/29	6/53–5/58	0.0588	1/17	2/34
0.0270	1/37	1/37	0.0588	…	3/51
0.0280	17/43–18/49	20/46–24/59	0.0590	11/41–9/43	21/49–17/46
0.0290	11/37–11/41	4/49–3/57	0.0600	4/39–2/47	3/30–1/25
0.0294	…	1/34	0.0606	2/33	4/66
0.0300	2/39–1/47	7/25–6/24	0.0610	7/37–5/39	5/51–2/54
0.0303	1/33	2/66	0.0612	3/49	3/49
0.0310	13/39–13/43	13/34–13/37	0.0620	17/43–13/39	11/37–8/34
0.0320	11/37–13/49	8/37–7/38	0.0625	1/16	…
0.0323	1/31	2/62	0.0630	2/21–1/31	5/59–1/46
0.0330	11/49–9/47	11/49–9/47	0.0638	3/47	3/47

Accurate Angular Indexing *(Continued)*

Part of a Turn	B&S, Becker, Hendey, K&T, & Rockford	Cincinnati and LeBlond	Part of a Turn	B&S, Becker, Hendey, K&T, & Rockford	Cincinnati and LeBlond
0.0333	…	1/30	0.0640	23/49–15/37	10/51–7/53
0.0339	…	2/59	0.0645	2/31	4/62
0.0340	18/49–13/39	18/49–17/51	0.0650	5/43–2/39	11/25–9/24
0.0345	1/29	2/58	0.0652	…	3/46
0.0350	13/41–11/39	4/25–3/24	0.0660	22/49–18/47	22/49–18/47
0.0351	…	2/57	0.0667	1/15	2/30
0.0357	…	1/28	0.0670	5/39–3/49	19/49–17/53
0.0360	18/39–20/47	21/41–20/42	0.0678	…	4/59
0.0680	9/39–7/43	23/51–18/47	0.0980	…	5/51
0.0690	2/29	4/58	0.0990	1/29+2/31	20/47–16/49
0.0690	12/37–12/47	7/66–2/54	0.1000	2/20	3/30
0.0698	3/43	3/43	0.1010	6/27–4/33	18/59–10/49
0.0700	8/33–5/29	8/25–6/24	0.1017	…	6/59
0.0702	…	4/57	0.1020	6/47–1/39	11/54–6/59
0.0710	17/43–12/37	11/58–7/59	0.1020	5/49	5/49
0.0714	…	2/28	0.1026	4/39	4/39
0.0714	…	3/42	0.1030	21/49–14/43	18/57–10/47
0.0720	7/47–3/39	10/49–7/53	0.1034	…	6/58
0.0730	9/37–8/47	1/47+3/58	0.1035	3/29	…
0.0732	3/41	3/41	0.1040	21/41–20/49	15/62–8/58
0.0740	23/49–17/43	19/42–14/37	0.1050	11/41–8/49	12/25–9/24
0.0741	2/27	4/54	0.1053	2/19	4/38
0.0750	2/16–1/20	1/24+1/30	0.1053	…	6/57
0.0755	…	4/53	0.1060	1/27+2/29	2/54+4/58
0.0758	…	5/66	0.1061	…	7/66
0.0760	17/47–14/49	24/49–24/58	0.1064	5/47	5/47
0.0769	3/39	3/39	0.1070	10/37–8/49	2/51+4/59
0.0770	…	9/46–7/59	0.1071	…	3/28
0.0771	6/37–4/47	…	0.1080	1/23+2/31	24/53–20/58
0.0780	13/39–12/47	9/46–6/51	0.1081	4/37	4/37
0.0784	…	4/51	0.1087	…	5/46
0.0789	…	3/38	0.1090	8/47–3/49	25/58–19/59
0.0790	20/37–18/39	21/39–17/37	0.1100	2/41+3/49	9/25–6/24
0.0800	9/37–8/49	2/25	0.1110	15/49–8/41	32/59–22/51
0.0806	…	5/62	0.1111	3/27	6/54
0.0810	9/23–9/29	17/47–16/57	0.1111	2/18	…
0.0811	3/37	3/37	0.1120	23/41–22/49	23/47–20/53
0.0816	4/49	4/49	0.1129	…	7/62
0.0820	5/47–1/41	4/38–1/43	0.1130	13/41–10/49	1/49+5/54
0.0830	4/23–3/33	8/46–6/66	0.1132	…	6/53
0.0833	…	2/24	0.1140	7/43–2/41	22/58–13/49
0.0840	8/43–5/49	8/47–5/58	0.1150	13/31–7/23	6/25–3/24
0.0847	…	5/59	0.1160	6/29–3/33	14/53–8/54
0.0850	9/17–8/18	3/24–1/25	0.1163	5/43	5/43
0.0851	4/47	4/47	0.1170	5/33–1/29	5/59+2/62
0.0860	13/31–7/21	16/59–10/54	0.1176	2/17	4/34
0.0862	…	5/58	0.1176	…	6/51
0.0870	2/23	4/46	0.1180	6/23–3/21	27/59–18/53
0.0870	5/33–2/31	21/54–16/53	0.1186	…	7/59
0.0877	…	5/57	0.1190	10/29–7/31	4/47+2/59
0.0880	8/37–5/39	28/57–25/62	0.1190	…	5/42
0.0882	…	3/34	0.1200	8/39–4/47	3/25
0.0890	6/37–3/41	22/53–15/46	0.1207	…	7/58
0.0900	22/49–14/39	6/24–4/25	0.1210	21/43–18/49	26/53–17/46

Accurate Angular Indexing *(Continued)*

Part of a Turn	B&S, Becker, Hendey, K&T, & Rockford	Cincinnati and LeBlond	Part of a Turn	B&S, Becker, Hendey, K&T, & Rockford	Cincinnati and LeBlond
0.0909	3/33	6/66	0.1212	4/33	8/66
0.0910	23/49–14/37	21/51–17/53	0.1220	5/41	5/41
0.0920	4/31–1/27	4/34–1/39	0.1220	23/47–18/49	10/51–4/54
0.0926	…	5/54	0.1224	6/49	6/49
0.0930	15/29–14/33	5/34–2/37	0.1228	…	7/57
0.0930	4/43	4/43	0.1230	10/49–3/37	1/59+7/66
0.0940	23/47–17/43	15/49–14/66	0.1240	2/23+1/27	18/34–15/37
0.0943	…	5/53	0.1250	2/16	3/24
0.0950	5/43–1/47	9/24–7/25	0.1260	24/47–15/39	10/59–2/46
0.0952	2/21	4/42	0.1270	22/43–15/39	1/46+6/57
0.0960	11/39–8/43	11/39–8/43	0.1277	6/47	6/47
0.0968	3/31	6/62	0.1280	7/37–3/49	33/62–19/47
0.0970	22/43–17/41	12/59–5/47	0.1282	5/39	5/39
0.0976	4/41	4/41	0.1290	10/27–7/29	24/59–15/54
0.0980	7/27–5/31	16/47–19/66	0.1290	4/31	8/62
0.1296	…	7/54	0.1613	5/31	10/62
0.1300	10/43–4/39	6/24–3/25	0.1620	3/39+4/47	25/57–13/47
0.1304	3/23	6/46	0.1622	6/37	6/37
0.1310	3/43+3/49	8/49–2/62	0.1628	7/43	7/43
0.1316	…	5/38	0.1630	10/29–6/33	22/47–18/59
0.1320	11/47–5/49	11/47–5/49	0.1633	8/49	8/49
0.1321	…	7/53	0.1640	10/47–2/41	8/38–2/43
0.1330	8/29–3/21	2/37+3/38	0.1650	2/47+6/49	3/24+1/25
0.1333	2/15	4/30	0.1660	8/23–6/33	16/46–12/66
0.1340	7/17–5/18	19/53–11/49	0.1667	3/18	4/24
0.1350	10/43–4/41	9/24–6/25	0.1667	…	5/30
0.1351	5/37	5/37	0.1667	…	7/42
0.1356	…	8/59	0.1667	…	9/54
0.1360	5/16–3/17	11/47–5/51	0.1667	…	11/66
0.1364	…	9/66	0.1670	16/39–9/37	19/53–9/47
0.1370	3/41+3/47	10/47–5/66	0.1680	16/43–10/49	26/57–17/59
0.1373	…	7/51	0.1690	5/39+2/49	10/46–3/62
0.1379	4/29	8/58	0.1695	…	10/59
0.1380	23/47–13/37	18/39–11/34	0.1698	…	9/53
0.1390	28/49–16/37	16/38–11/39	0.1700	6/23–3/33	6/24–2/25
0.1395	6/43	6/43	0.1702	8/47	8/47
0.1400	8/49–1/43	1/25+3/30	0.1707	7/41	7/41
0.1404	…	8/57	0.1710	15/49–5/37	15/47–8/54
0.1410	11/47–4/43	12/66–2/49	0.1720	1/39+6/41	32/59–20/54
0.1420	8/49–1/47	21/57–12/53	0.1724	5/29	10/58
0.1429	…	4/28	0.1730	9/41–2/43	9/41–2/43
0.1429	3/21	6/42	0.1739	4/23	8/46
0.1429	7/49	7/49	0.1740	10/33–431	21/53–12/54
0.1430	10/47–3/43	24/51–19/58	0.1750	2/16+1/20	1/24+4/30
0.1440	19/43–14/47	20/49–14/53	0.1754	…	10/57
0.1450	9/47–2/43	15/24–12/25	0.1760	2/37+5/41	2/37+5/41
0.1452	…	9/62	0.1765	3/17	6/34
0.1460	26/49–15/39	2/47+6/58	0.1765	…	9/51
0.1463	6/41	6/41	0.1770	5/39+2/41	14/49–5/46
0.1470	4/21–1/23	16/41–9/37	0.1774	…	11/62
0.1471	…	5/34	0.1780	23/41–18/47	6/49+3/54
0.1480	4/19–1/16	9/37–4/42	0.1786	…	5/28
0.1481	4/27	8/54	0.1790	11/21–10/29	19/51–12/62
0.1489	7/47	7/47	0.1795	7/39	7/39

Accurate Angular Indexing *(Continued)*

Part of a Turn	B&S, Becker, Hendey, K&T, & Rockford	Cincinnati and LeBlond	Part of a Turn	B&S, Becker, Hendey, K&T, & Rockford	Cincinnati and LeBlond
0.1490	12/31−5/21	11/49−4/53	0.1800	11/47−2/37	2/25+3/30
0.1500	3/20	7/30−2/24	0.1810	11/37−5/43	18/38−12/41
0.1509	…	8/53	0.1818	6/33	12/66
0.1510	22/47−13/41	25/59−18/66	0.1820	9/39−2/41	21/46−14/51
0.1515	5/33	10/66	0.1830	5/17−2/18	33/58−22/57
0.1520	11/41−5/43	16/62	0.1837	9/49	9/49
0.1522	…	7/46	0.1840	8/31−2/27	19/62−6/49
0.1525	…	9/59	0.1842	…	7/38
0.1530	10/27−5/23	13/54−5/57	0.1850	4/37+3/39	14/25−9/24
0.1538	6/39	6/39	0.1852	5/27	10/54
0.1540	10/37−5/43	5/58+4/59	0.1860	1/29+5/33	18/47−13/66
0.1550	8/37−3/49	7/25−3/24	0.1860	8/43	8/43
0.1552	…	9/58	0.1864	…	11/59
0.1560	4/21−1/29	1/49+8/59	0.1870	16/49−6/43	24/57−11/47
0.1569	…	8/51	0.1875	3/16	…
0.1570	15/47−6/37	31/59−21/57	0.1880	12/29−7/31	30/49−28/66
0.1579	3/19	6/38	0.1887	…	10/53
0.1579	…	9/57	0.1890	13/29−7/27	23/58−11/53
0.1580	3/37+3/39	3/34+3/43	0.1892	7/37	7/37
0.1590	20/43−15/49	3/54+6/58	0.1897	…	11/58
0.1600	18/37−16/49	4/25	0.1900	10/43−2/47	11/25−6/24
0.1610	9/39−3/43	9/39−3/43	0.1905	4/21	8/42
0.1910	17/39−12/49	21/57−11/62	0.2222	4/18	…
0.1915	9/47	9/47	0.2222	6/27	12/54
0.1920	7/41+1/47	12/57−1/54	0.2230	15/31−6/23	11/46−1/62
0.1930	…	11/57	0.2240	5/41+5/49	15/38−7/41
0.1930	10/19−5/15	19/59−8/62	0.2241	…	13/58
0.1935	6/31	12/62	0.2245	11/49	11/49
0.1940	7/41+1/43	14/43−5/38	0.2250	2/16+2/20	3/24+3/30
0.1950	1/23+5/33	8/25−3/24	0.2258	7/31	14/62
0.1951	8/41	8/41	0.2260	7/39+2/43	2/49+10/54
0.1957	…	9/46	0.2264	…	12/53
0.1960	21/49−10/43	34/66−15/47	0.2270	7/27−1/31	14/54−2/62
0.1961	…	10/51	0.2273	…	15/66
0.1970	…	13/66	0.2280	11/39−2/37	23/49−14/58
0.1970	21/39−14/41	11/47−2/54	0.2281	…	13/57
0.1980	2/29+4/31	17/49−7/47	0.2290	18/39−10/43	29/51−18/53
0.1990	5/37+3/47	27/59−15/58	0.2300	7/37+2/49	12/25−6/24
0.2000	3/15	5/25	0.2308	9/39	9/39
0.2000	4/20	6/30	0.2310	28/49−16/47	26/59−13/62
0.2010	11/39−3/37	8/49+2/53	0.2320	20/41−11/43	26/57−13/58
0.2020	23/41−14/39	23/41−14/39	0.2326	10/43	10/43
0.2030	2/37+7/47	19/62−6/58	0.2330	27/47−14/41	25/62−8/47
0.2034	…	12/59	0.2333	…	7/30
0.2037	…	11/54	0.2340	24/41−13/37	2/54+13/66
0.2040	12/47−2/39	18/51−7/47	0.2340	11/47	11/47
0.2041	10/49	10/49	0.2350	11/27−5/29	9/25−3/24
0.2050	13/37−6/41	3/24+2/25	0.2353	4/17	8/34
0.2051	8/39	8/39	0.2353	…	12/51
0.2059	…	7/34	0.2360	10/39−1/49	29/66−12/59
0.2060	15/43−7/49	12/53−1/49	0.2368	…	9/38
0.2069	6/29	12/58	0.2370	23/37−15/39	23/37−15/39
0.2070	19/41−10/39	19/41−10/39	0.2373	…	14/59
0.2075	…	11/53	0.2380	2/43+9/47	34/57−19/53

Accurate Angular Indexing *(Continued)*

Part of a Turn	B&S, Becker, Hendey, K&T, & Rockford	Cincinnati and LeBlond	Part of a Turn	B&S, Becker, Hendey, K&T, & Rockford	Cincinnati and LeBlond
0.2080	15/31–8/29	13/58–1/62	0.2381	5/21	10/42
0.2083	…	5/24	0.2390	24/43–15/47	12/62+3/66
0.2090	16/33–8/29	8/46+2/57	0.2391	…	11/46
0.2093	9/43	9/43	0.2400	3/43+8/47	6/25
0.2097	…	13/62	0.2407	…	13/54
0.2100	22/37–15/39	6/24–1/25	0.2410	19/47–8/49	17/47–7/58
0.2105	4/19	8/38	0.2414	7/29	14/58
0.2105	…	12/57	0.2419	…	15/62
0.2110	22/41–14/43	22/41–14/43	0.2420	21/37–14/43	12/46–1/53
0.2120	2/27–4/29	4/54+8/58	0.2424	8/33	16/66
0.2121	7/33	14/66	0.2430	29/49–15/43	4/47+9/57
0.2128	10/47	10/47	0.2432	9/37	9/37
0.2130	23/49–10/39	2/30+6/41	0.2439	10/41	10/41
0.2140	20/37–16/49	12/51–1/47	0.2440	13/49–1/47	30/53–19/59
0.2143	…	6/28	0.2449	12/49	12/49
0.2143	…	9/42	0.2450	13/37–5/47	3/24+3/25
0.2150	11/43–2/49	9/24–4/25	0.2453	…	13/53
0.2157	…	11/51	0.2456	…	14/57
0.2160	2/23+4/31	25/51–17/62	0.2460	20/49–6/37	11/37–2/39
0.2162	8/37	8/37	0.2470	10/37–1/43	29/49–20/58
0.2170	11/41–2/39	28/59–17/66	0.2480	4/23+2/27	26/49–13/46
0.2174	5/23	10/46	0.2490	10/37–1/47	17/43–6/41
0.2180	3/31+4/33	21/59–8/58	0.2500	4/16	6/24
0.2190	11/23–7/27	3/47+9/58	0.2500	5/20	7/28
0.2195	9/41	9/41	0.2510	2/15+2/17	34/66–14/53
0.2200	4/41+6/49	3/25+3/30	0.2520	24/43–15/49	22/49–13/66
0.2203	…	13/59	0.2530	7/37+3/47	11/53+3/66
0.2210	18/49–6/41	21/47–14/62	0.2540	26/49–13/47	2/46+12/57
0.2220	25/41–19/49	7/51+5/59	0.2542	…	15/59
0.2549	…	13/51	0.2857	…	12/42
0.2550	4/21+2/31	9/24–3/25	0.2857	…	8/28
0.2553	12/47	12/47	0.2860	20/47–6/43	20/58–3/51
0.2558	11/43	11/43	0.2870	7/41+5/43	20/42–7/37
0.2560	13/33–4/29	9/47+4/62	0.2879	…	19/66
0.2564	10/39	10/39	0.2880	19/43–6/39	19/43–6/39
0.2570	20/39–11/43	8/53+7/66	0.2881	…	17/59
0.2576	…	17/66	0.2890	1/21+7/29	16/46–3/51
0.2580	15/29–7/27	24/54–11/59	0.2895	…	11/38
0.2581	8/31	16/62	0.2900	23/43–12/49	6/24+1/25
0.2586	…	15/58	0.2903	9/31	18/62
0.2590	24/49–9/39	16/42–5/41	0.2910	5/39+7/43	7/49+8/54
0.2593	7/27	14/54	0.2917	…	7/24
0.2600	20/43–8/39	4/25+3/30	0.2920	9/39+3/49	35/57–19/59
0.2609	6/23	12/46	0.2927	12/41	12/41
0.2610	15/33–6/31	5/53+9/54	0.2930	17/39–7/49	28/53–12/51
0.2619	…	11/42	0.2931	…	17/58
0.2620	18/37–11/49	16/51–3/58	0.2940	8/21–2/23	14/57+3/62
0.2630	13/41–2/37	13/46–1/51	0.2941	5/17	10/34
0.2632	5/19	10/38	0.2941	…	15/51
0.2632	…	15/57	0.2950	18/37–9/47	9/24–2/25
0.2640	22/47–10/49	22/47–10/49	0.2960	21/41–8/37	29/57–10/47
0.2642	…	14/53	0.2963	8/27	16/54
0.2647	…	9/34	0.2970	3/29+6/31	13/47+1/49
0.2650	8/37+2/41	15/24–9/25	0.2973	11/37	11/37

Accurate Angular Indexing *(Continued)*

Part of a Turn	B&S, Becker, Hendey, K&T, & Rockford	Cincinnati and LeBlond	Part of a Turn	B&S, Becker, Hendey, K&T, & Rockford	Cincinnati and LeBlond
0.2653	13/49	13/49	0.2979	14/47	14/47
0.2660	8/27−1/33	28/51−15/53	0.2980	11/21−7/31	12/37−1/38
0.2667	…	8/30	0.2982	…	17/57
0.2670	18/37−9/41	19/47−7/51	0.2990	19/43−7/49	19/43−4/28
0.2680	8/18−3/17	27/49−15/53	0.3000	6/20	9/30
0.2683	11/41	11/41	0.3010	1/41+13/47	19/54−3/59
0.2690	2/18+3/19	6/54+9/57	0.3019	…	16/53
0.2700	16/27−10/31	13/25−6/24	0.3020	7/29+2/33	23/62−4/58
0.2703	10/37	10/37	0.3023	13/43	13/43
0.2710	2/43+11/49	1/28+8/34	0.3030	15/39−4/49	25/57−8/59
0.2712	…	16/59	0.3030	10/33	20/66
0.2720	14/37−5/47	17/59−1/62	0.3040	16/31−7/33	33/59−12/47
0.2727	9/33	18/66	0.3043	7/23	14/46
0.2730	1/16+4/19	18/34−10/39	0.3050	1/31+9/33	15/24−8/25
0.2740	6/41+6/47	26/59−9/54	0.3051	…	18/59
0.2742	…	17/62	0.3060	17/31−8/33	33/54−18/59
0.2745	…	14/51	0.3061	15/49	15/49
0.2750	2/16+3/20	5/24+2/30	0.3065	…	19/62
0.2759	8/29	16/58	0.3070	18/37−7/39	1/53+17/59
0.2760	12/31−3/27	13/43−1/38	0.3077	12/39	12/39
0.2766	13/47	13/47	0.3080	5/41+8/43	16/49−1/54
0.2770	11/27−3/23	18/28−15/41	0.3090	1/43+14/49	19/30−12/37
0.2778	5/18	15/54	0.3095	…	13/42
0.2780	5/23+2/33	17/39−6/38	0.3100	16/37−6/49	14/25−6/24
0.2790	16/29−9/33	14/47−1/53	0.3103	9/29	18/58
0.2791	12/43	12/43	0.3110	3/39+11/47	19/46−5/49
0.2800	16/49−2/43	7/25	0.3120	8/21−2/29	2/49+16/59
0.2807	…	16/57	0.3125	5/16	…
0.2810	3/39+10/49	17/57−1/58	0.3130	9/37+3/43	4/24+6/41
0.2820	5/27+3/31	24/66−4/49	0.3137	…	16/51
0.2821	11/39	11/39	0.3140	12/27−3/23	14/47+1/62
0.2826	…	13/46	0.3148	…	17/54
0.2830	14/43−2/47	15/53	0.3150	26/41−15/47	21/24−14/25
0.2840	21/47−7/43	37/66−13/47	0.3158	…	12/38
0.2850	15/43−3/47	3/24+4/25	0.3158	…	18/57
0.2857	14/49	14/49	0.3160	6/37+6/39	6/34+6/43
0.3170	11/18−5/17	34/59−14/54	0.3485	…	23/66
0.3171	13/41	13/41	0.3488	15/43	15/43
0.3180	22/37−13/47	6/54+12/58	0.3490	11/29−1/33	2/47+19/62
0.3182	…	21/66	0.3500	7/20	6/24+3/30
0.3190	6/27+3/31	3/34+9/39	0.3509	…	20/57
0.3191	15/47	15/47	0.3510	13/27−3/23	25/53−7/58
0.3200	16/47−1/49	8/25	0.3514	13/37	13/37
0.3208	…	17/53	0.3519	…	19/54
0.3210	25/49−7/37	10/59+10/66	0.3520	4/37+10/41	24/62−2/57
0.3214	…	9/28	0.3529	6/17	12/34
0.3220	18/39−6/43	18/39−6/43	0.3529	…	18/51
0.3220	…	19/59	0.3530	4/37+12/49	31/59−10/58
0.3226	10/31	20/62	0.3540	14/37−1/41	22/59−1/53
0.3230	21/41−7/37	21/41−7/37	0.3548	11/31	22/62
0.3235	…	11/34	0.3550	10/43+6/49	12/25−3/24
0.3240	3/23+6/31	21/47−7/57	0.3559	…	21/59
0.3243	12/37	12/37	0.3560	5/41+11/47	12/49+6/54
0.3250	2/16+4/20	3/24+5/25	0.3570	20/43−4/37	20/43−4/37

Accurate Angular Indexing *(Continued)*

Part of a Turn	B&S, Becker, Hendey, K&T, & Rockford	Cincinnati and LeBlond	Part of a Turn	B&S, Becker, Hendey, K&T, & Rockford	Cincinnati and LeBlond
0.3256	14/43	14/43	0.3571	…	10/28
0.3260	21/49–4/39	23/59–3/47	0.3571	…	15/42
0.3261	…	15/46	0.3580	14/37–1/49	38/62–13/51
0.3265	16/49	16/49	0.3585	…	19/53
0.3270	24/49–7/43	17/58+2/59	0.3590	14/39	14/39
0.3276	…	19/58	0.3600	22/47–4/37	9/25
0.3280	26/41–15/49	15/51+2/59	0.3610	9/23–1/33	18/46–2/66
0.3290	5/21+3/33	23/43–7/34	0.3617	17/47	17/47
0.3300	4/47+12/49	6/24+2/25	0.3620	15/27–6/31	17/41–2/38
0.3310	17/31–5/23	23/59–3/51	0.3621	…	21/58
0.3320	28/43–15/47	30/59–9/51	0.3630	23/49–5/47	25/53–5/46
0.3330	7/43+8/47	36/51–22/59	0.3636	12/33	24/66
0.3333	5/15	8/24	0.3640	26/47–7/37	25/62–2/51
0.3333	6/18	10/30	0.3650	28/47–9/39	3/24+6/25
0.3333	7/21	13/39	0.3659	…	15/41
0.3333	9/27	14/42	0.3660	10/17–4/18	13/57+8/58
0.3333	11/33	17/51	0.3667	…	11/30
0.3333	13/39	18/54	0.3670	5/27+6/33	13/49+6/59
0.3333	…	19/57	0.3673	18/49	18/49
0.3333	…	22/66	0.3680	16/31–4/27	31/66–6/59
0.3340	7/41+8/49	29/47–15/53	0.3684	7/19	14/38
0.3350	21/37–10/43	9/24–1/25	0.3684	…	21/57
0.3360	28/41–17/49	9/46+8/57	0.3690	30/49–9/37	21/62+2/66
0.3370	2/39+14/49	33/57–15/62	0.3696	…	17/46
0.3380	10/23–3/31	25/62–3/46	0.3700	30/47–11/41	6/24+3/25
0.3387	…	21/62	0.3704	10/27	20/54
0.3390	19/49–2/41	20/59	0.3710	32/49–11/39	23/62
0.3396	…	18/53	0.3720	2/29+10/33	34/57–11/49
0.3400	25/49–8/47	6/25+3/30	0.3721	16/43	16/43
0.3404	16/47	16/47	0.3725	…	19/51
0.3410	12/27–3/29	22/46–7/51	0.3729	…	22/59
0.3415	14/41	14/41	0.3730	30/47–13/49	21/49–3/54
0.3420	4/21+5/33	25/62–3/49	0.3740	32/49–12/43	5/46+13/49
0.3421	…	13/38	0.3750	6/16	9/24
0.3430	13/23–6/27	37/57–15/49	0.3760	5/21+4/29	11/49+10/66
0.3440	2/39+12/41	14/54+5/59	0.3770	13/37+1/39	20/51–1/66
0.3448	…	20/58	0.3774	…	20/53
0.3450	2/23+8/31	15/24–7/25	0.3780	13/27–3/29	31/53–12/58
0.3460	18/41–4/43	18/41–4/43	0.3784	14/37	14/37
0.3469	17/49	17/49	0.3788	…	25/66
0.3470	7/31+4/33	7/38+7/43	0.3790	8/37+7/43	8/37+7/43
0.3478	8/23	16/46	0.3793	11/29	22/58
0.3480	20/33–8/31	31/59–11/62	0.3800	20/43–4/47	7/25+3/30
0.3810	8/21	16/42	0.4120	18/41–1/37	24/53–2/49
0.3810	19/47–1/43	8/51+13/58	0.4130	2/39+17/47	19/46
0.3820	25/49–5/39	40/62–15/57	0.4138	12/29	24/58
0.3824	…	13/34	0.4140	19/39–3/41	19/39–3/41
0.3830	18/47	18/47	0.4146	17/41	17/41
0.3830	27/43–12/49	37/58–13/51	0.4150	18/33–3/23	9/24+1/25
0.3840	27/43–10/41	27/43–10/41	0.4151	…	22/53
0.3846	15/39	15/39	0.4160	21/39–6/49	41/62–13/53
0.3850	15/37–1/49	15/24–6/25	0.4167	…	10/24
0.3860	1/37+14/39	22/57	0.4170	26/37–14/49	10/38+6/39
0.3870	3/17+4/19	16/39–1/43	0.4180	8/21+1/27	16/46+4/57

Accurate Angular Indexing *(Continued)*

Part of a Turn	B&S, Becker, Hendey, K&T, & Rockford	Cincinnati and LeBlond	Part of a Turn	B&S, Becker, Hendey, K&T, & Rockford	Cincinnati and LeBlond
0.3871	12/31	24/62	0.4186	18/43	18/43
0.3878	19/49	19/49	0.4190	18/39−2/47	31/46−13/51
0.3880	14/41+2/43	18/53+3/62	0.4194	13/31	26/62
0.3889	7/18	21/54	0.4200	24/49−3/43	8/25+3/30
0.3890	17/41−1/39	24/53−3/47	0.4210	1/39+17/43	23/49−3/62
0.3898	…	23/59	0.4211	8/19	16/38
0.3900	2/23+10/33	16/25−6/24	0.4211	…	24/57
0.3902	16/41	16/41	0.4220	15/29−2/21	3/41+15/43
0.3910	14/31−2/33	29/66−3/62	0.4230	24/43−5/37	20/54+3/57
0.3913	9/23	18/46	0.4237	…	25/59
0.3920	14/33−1/31	17/47+2/66	0.4240	4/27+8/29	41/62−14/59
0.3922	…	20/51	0.4242	14/33	28/66
0.3929	…	11/28	0.4250	6/16+1/20	7/24+4/30
0.3930	1/39+18/49	28/46−11/51	0.4255	20/47	20/47
0.3939	13/33	26/66	0.4259	…	23/54
0.3940	3/39+13/41	24/53−3/51	0.4260	27/41−10/43	28/57−3/46
0.3947	…	15/38	0.4270	27/39−13/49	29/59−4/62
0.3950	26/37−12/39	13/25−3/24	0.4280	12/43+7/47	33/57−8/53
0.3953	17/43	17/43	0.4286	9/21	12/28
0.3960	4/29+8/31	33/47−15/49	0.4286	21/49	18/42
0.3962	…	21/53	0.4286	…	21/49
0.3966	…	23/58	0.4290	30/47−9/43	21/51+1/58
0.3970	25/41−10/47	7/57+17/62	0.4300	22/43−4/49	17/25−6/24
0.3980	3/16+4/19	28/58−5/59	0.4310	11/39+7/47	25/58
0.3990	7/39+9/41	6/37+9/38	0.4314	…	22/51
0.4000	6/15	10/25	0.4320	4/23+8/31	28/62−1/51
0.4000	8/20	12/30	0.4324	16/37	16/37
0.4010	2/37+17/49	27/62−2/58	0.4330	5/37+14/47	26/42−8/43
0.4020	5/43+14/49	16/49+4/53	0.4333	…	13/30
0.4030	26/49−6/47	30/47−12/51	0.4340	5/31+9/33	23/53
0.4032	…	25/62	0.4348	10/23	20/46
0.4035	…	23/57	0.4350	21/31−8/33	14/25−3/24
0.4040	11/39+5/41	11/39+5/41	0.4355	…	27/62
0.4043	19/47	19/47	0.4359	…	17/39
0.4048	…	17/42	0.4360	6/31+8/33	42/59−16/58
0.4050	29/41−13/43	3/24+7/25	0.4370	27/39−12/47	31/49−9/46
0.4054	15/37	15/37	0.4375	7/16	…
0.4060	21/47−2/49	17/58+7/62	0.4380	13/27−1/23	24/57+1/59
0.4068	…	24/59	0.4386	…	25/57
0.4070	9/19−1/15	7/47+16/62	0.4390	18/43+1/49	34/59−7/51
0.4074	11/27	22/54	0.4390	18/41	18/41
0.4080	16/37−1/41	2/54+23/62	0.4394	…	29/66
0.4082	20/49	20/49	0.4400	8/41+12/49	11/25
0.4090	15/39+1/41	15/39+1/41	0.4407	…	26/59
0.4091	…	27/66	0.4410	10/37+7/41	10/37+7/41
0.4100	1/37+18/47	6/24+4/25	0.4412	…	15/34
0.4103	16/39	16/39	0.4419	19/43	19/43
0.4110	9/41+9/47	7/34+8/39	0.4420	18/33−3/29	34/62−5/47
0.4118	7/17	14/34	0.4430	4/39+16/47	20/51+3/59
0.4118	…	21/51	0.4440	9/41+11/49	14/51+10/59
0.4444	12/27	24/54	0.4737	…	27/57
0.4444	8/18	…	0.4740	9/37+9/39	9/34+9/43
0.4450	7/37+11/43	3/24+8/25	0.4746	…	28/59
0.4460	11/23−1/31	22/46−2/62	0.4750	6/16+2/20	9/24+3/30

Accurate Angular Indexing *(Continued)*

Part of a Turn	B&S, Becker, Hendey, K&T, & Rockford	Cincinnati and LeBlond	Part of a Turn	B&S, Becker, Hendey, K&T, & Rockford	Cincinnati and LeBlond
0.4468	21/47	21/47	0.4760	4/43+18/47	15/53+11/57
0.4470	6/21+5/31	14/49+10/62	0.4762	10/21	20/42
0.4474	…	17/38	0.4770	26/43−6/47	30/47−10/62
0.4480	10/41+10/49	27/41−8/38	0.4780	12/31+3/33	24/62+6/66
0.4483	13/29	26/58	0.4783	11/23	22/46
0.4490	22/49	22/49	0.4790	22/39−4/47	10/53+18/62
0.4490	20/39−3/47	14/57+12/59	0.4800	16/39+3/43	12/25
0.4500	9/20	6/24+6/30	0.4810	14/41+6/43	22/58+6/59
0.4510	…	23/51	0.4815	13/27	26/54
0.4510	5/15+2/17	42/62−12/53	0.4820	33/49−9/47	19/46+4/58
0.4516	14/31	28/62	0.4828	14/29	28/58
0.4520	14/39+4/43	4/49+20/54	0.4830	27/39−9/43	27/39−9/43
0.4524	…	19/42	0.4839	15/31	30/62
0.4528	…	24/53	0.4840	5/37+15/43	24/46−2/53
0.4530	3/23+10/31	16/59+12/66	0.4848	16/33	32/66
0.4540	14/27−2/31	1/54+27/62	0.4850	24/47−1/39	3/24+9/25
0.4545	15/33	30/66	0.4860	13/43+9/49	43/62−11/53
0.4550	25/47−3/39	9/24+2/25	0.4865	18/37	18/37
0.4560	9/37+10/47	17/62+12/66	0.4870	15/37+4/49	26/43−4/34
0.4561	…	26/57	0.4872	19/39	19/39
0.4565	…	21/46	0.4878	20/41	20/41
0.4570	4/37+15/43	27/39−8/34	0.4880	8/29+7/33	5/46+22/58
0.4576	…	27/59	0.4884	21/43	21/43
0.4580	27/49−4/43	20/53+5/62	0.4890	28/43−6/37	19/47+5/59
0.4583	…	11/24	0.4894	23/47	23/47
0.4590	35/49−12/47	18/51+7/66	0.4898	24/49	24/49
0.4595	17/37	17/37	0.4900	13/21−4/31	6/24+6/25
0.4600	16/41+3/43	9/25+3/30	0.4902	…	25/51
0.4610	13/39+6/47	22/34−8/43	0.4906	…	26/53
0.4615	18/39	18/39	0.4910	15/39+5/47	21/46+2/58
0.4620	15/47+7/49	36/62−7/59	0.4912	…	28/57
0.4630	…	25/54	0.4915	…	29/59
0.4630	…	10/46+14/57	0.4920	25/37−9/49	17/46+6/49
0.4631	9/21+1/29	…	0.4930	8/41+14/47	21/53+6/62
0.4634	19/41	19/41	0.4940	33/49−7/39	14/46+11/58
0.4640	21/43−1/41	32/58−5/57	0.4950	5/29+10/31	9/24+3/25
0.4643	…	13/28	0.4960	8/23+4/27	20/53+7/59
0.4650	21/37−4/39	15/24−4/25	0.4970	33/47−8/39	7/46+20/58
0.4651	20/43	20/43	0.4980	20/37−2/47	29/41−9/43
0.4655	…	27/58	0.4990	26/41−5/37	26/41−5/37
0.4660	13/41+7/47	31/47−12/62	0.5000	8/16	12/24
0.4667	7/15	14/30	0.5000	9/18	14/28
0.4670	19/37−2/43	25/34−11/41	0.5000	10/20	15/30
0.4677	…	29/62	0.5000	…	17/34
0.4680	11/27+2/33	3/49+24/59	0.5000	…	19/38
0.4681	22/47	22/47	0.5000	…	21/42
0.4690	8/23+4/33	35/49−13/53	0.5000	…	23/46
0.4694	23/49	23/49	0.5000	…	27/54
0.4697	…	31/66	0.5000	…	29/58
0.4700	19/29−5/27	18/25−6/24	0.5000	…	31/62
0.4706	8/17	16/34	0.5000	…	33/66
0.4706	…	24/51	0.5010	5/37+15/41	37/51−11/49
0.4710	12/39+8/49	12/47+11/51	0.5020	17/37+2/47	25/53+2/66
0.4717	…	25/53	0.5030	8/39+14/47	16/49+9/51

Accurate Angular Indexing *(Continued)*

Part of a Turn	B&S, Becker, Hendey, K&T, & Rockford	Cincinnati and LeBlond
0.4720	20/39−2/49	31/53−7/62
0.4730	6/39+15/47	29/59−1/54
0.4737	9/19	18/38
0.5070	33/47−8/41	28/46−6/59
0.5080	12/37+9/49	41/66−6/53
0.5085	…	30/59
0.5088	…	29/57
0.5090	24/39−5/47	27/51−1/49
0.5094	…	27/53
0.5098	…	26/51
0.5100	8/21+4/31	18/24−6/25
0.5102	25/49	25/49
0.5106	24/47	24/47
0.5110	6/37+15/43	26/49−1/51
0.5116	22/43	22/43
0.5120	21/29−7/33	45/66−9/53
0.5122	21/41	21/41
0.5128	20/39	20/39
0.5130	22/37−4/49	30/54−2/47
0.5135	19/37	19/37
0.5140	30/43−9/49	33/46−12/59
0.5150	1/39+23/47	16/25−3/24
0.5152	17/33	34/66
0.5160	28/43−5/37	37/59−6/54
0.5161	16/31	32/62
0.5170	13/39+9/49	9/49+17/51
0.5172	15/29	30/58
0.5180	9/47+16/49	31/41−10/42
0.5185	14/27	28/54
0.5190	27/41−6/43	6/49+23/58
0.5200	3/41+21/47	13/25
0.5210	17/39+4/47	41/59−8/46
0.5217	12/23	24/46
0.5220	19/31−3/33	14/47+13/58
0.5230	17/43+6/47	14/49+14/59
0.5238	11/21	22/42
0.5240	29/47−4/43	32/51−6/58
0.5250	6/16+3/20	7/24+7/30
0.5254	…	31/59
0.5260	28/37−9/39	26/46−2/51
0.5263	10/19	20/38
0.5263	…	30/57
0.5270	32/47−6/39	6/53+24/58
0.5280	19/39+2/49	35/59−3/46
0.5283	…	28/53
0.5290	18/37+2/47	30/53−2/54
0.5294	9/17	18/34
0.5294	…	27/51
0.5300	5/27+10/29	6/24+7/25
0.5303	…	35/66
0.5306	26/49	26/49
0.5310	15/23−4/33	24/37−4/34
0.5319	25/47	25/47
0.5320	7/27+9/33	5/51+23/53

Part of a Turn	B&S, Becker, Hendey, K&T, & Rockford	Cincinnati and LeBlond
0.5040	5/43+19/49	37/66−3/53
0.5050	21/31−5/29	15/24−3/25
0.5060	7/39+16/49	22/53+6/66
0.5370	…	29/54
0.5370	…	6/51+26/62
0.5371	17/41+6/49	…
0.5380	4/18+6/19	12/49+17/58
0.5385	21/39	21/39
0.5390	26/39−6/47	34/51−6/47
0.5400	25/41−3/43	6/25+9/30
0.5405	20/37	20/37
0.5410	12/47+14/49	2/38+21/43
0.5417	…	13/24
0.5420	4/43+22/49	7/46+23/59
0.5424	…	32/59
0.5430	33/43−11/49	22/54+8/59
0.5435	…	25/46
0.5439	…	31/57
0.5440	4/37+17/39	4/37+17/39
0.5450	3/39+22/47	15/24−2/25
0.5455	18/33	36/66
0.5460	13/27+2/31	2/34+19/39
0.5470	21/31−3/23	32/49−7/66
0.5472	…	29/53
0.5476	…	23/42
0.5480	12/41+12/47	25/39−4/43
0.5484	17/31	34/62
0.5490	10/15−2/17	8/47+25/66
0.5490	…	28/51
0.5500	11/20	6/24+9/30
0.5510	19/39+3/47	31/53−2/59
0.5510	…	27/49
0.5517	16/29	32/58
0.5520	31/41−10/49	29/46−4/51
0.5526	…	21/38
0.5530	15/21−5/31	28/54+2/58
0.5532	26/47	26/47
0.5540	12/23+1/31	12/53+19/58
0.5550	1/41+26/49	17/25−3/24
0.5556	10/18	30/54
0.5556	15/27	…
0.5560	32/41−11/49	35/47−10/53
0.5570	22/41+1/49	18/49+11/58
0.5580	3/29+15/33	7/53+23/54
0.5581	24/43	24/43
0.5588	…	19/34
0.5590	27/37−7/41	43/59−9/53
0.5593	…	33/59
0.5600	37/49−8/41	14/25
0.5606	…	37/66
0.5610	23/41	23/41
0.5610	4/23+12/31	5/51+25/54
0.5614	…	32/57
0.5620	1/23+14/27	9/34+11/37

Accurate Angular Indexing *(Continued)*

Part of a Turn	B&S, Becker, Hendey, K&T, & Rockford	Cincinnati and LeBlond	Part of a Turn	B&S, Becker, Hendey, K&T, & Rockford	Cincinnati and LeBlond
0.5323	…	33/62	0.5625	9/16	…
0.5330	18/37+2/43	16/46+10/54	0.5630	12/39+12/47	22/46+5/59
0.5333	8/15	16/30	0.5640	25/33−6/31	41/49−18/66
0.5340	28/41−7/47	37/51−9/47	0.5641	22/39	22/39
0.5345	…	31/58	0.5645	…	35/62
0.5349	23/43	23/43	0.5650	10/31+8/33	3/24+11/25
0.5350	16/37+4/39	9/24+4/25	0.5652	13/23	26/46
0.5357	…	15/28	0.5660	4/39+19/41	9/46+20/54
0.5360	1/41+22/43	5/49+23/53	0.5660	…	30/53
0.5366	22/41	22/41	0.5667	…	17/30
0.5670	33/47−5/37	25/47+2/57	0.5970	6/47+23/49	13/58+22/59
0.5676	21/37	21/37	0.5980	35/49−5/43	16/47+17/66
0.5680	23/31−4/23	21/47+8/66	0.5990	17/37+6/43	23/49+7/54
0.5686	…	29/51	0.6000	9/15	15/25
0.5690	…	33/58	0.6000	12/20	18/30
0.5690	28/39−7/47	42/59−7/49	0.6010	32/41−7/39	32/41−7/39
0.5700	21/43+4/49	6/24+8/25	0.6020	13/16−4/19	20/47+9/51
0.5710	9/43+17/47	39/57−6/53	0.6030	16/41+10/47	24/49+6/53
0.5714	12/21	16/28	0.6034	…	35/58
0.5714	28/49	24/42	0.6038	…	32/53
0.5714	…	28/49	0.6040	23/31−4/29	21/58+15/62
0.5720	31/43−7/47	40/58−6/51	0.6047	…	26/43
0.5730	12/39+13/49	23/57+10/59	0.6050	11/37+12/39	3/24+12/25
0.5740	14/41+10/43	23/37−2/42	0.6053	…	23/38
0.5741	…	31/54	0.6060	28/41−3/39	29/54+4/58
0.5745	27/47	27/47	0.6061	20/33	40/66
0.5750	3/15+6/16	9/24+6/30	0.6070	31/49−1/39	23/47+6/51
0.5758	19/33	38/66	0.6071	…	17/28
0.5760	21/29−4/27	24/49+5/58	0.6078	…	31/51
0.5763	…	34/59	0.6080	1/31+19/33	24/53+9/58
0.5770	5/37+19/43	32/46−7/59	0.6087	14/23	28/46
0.5780	2/21+14/29	25/49+4/59	0.6090	17/31+2/33	40/59−4/58
0.5789	11/19	22/38	0.6098	25/41	25/41
0.5789	…	33/57	0.6100	23/33−2/23	6/24+9/25
0.5790	26/43−1/39	38/62−2/59	0.6102	…	36/59
0.5800	3/43+25/49	12/25+3/30	0.6110	1/39+24/41	5/28+16/37
0.5806	18/31	36/62	0.6111	…	33/54
0.5810	21/39+2/47	18/53+14/58	0.6120	27/41−2/43	29/39−5/38
0.5814	25/43	25/43	0.6122	30/49	30/49
0.5820	6/21+8/27	23/38−1/43	0.6129	19/31	38/62
0.5830	11/37+14/49	8/46+27/66	0.6130	15/19−3/17	1/49+32/54
0.5833	…	14/24	0.6140	25/39−1/37	36/49−7/58
0.5840	18/39+6/49	13/57+21/59	0.6140	…	35/57
0.5849	…	31/53	0.6150	22/37+1/49	9/24+6/25
0.5850	3/23+15/33	15/24−1/25	0.6154	24/39	24/39
0.5854	24/41	24/41	0.6160	10/41+16/43	14/53+19/54
0.5860	20/39+3/41	17/54+16/59	0.6170	12/37+12/41	5/59+33/62
0.5862	17/29	34/58	0.6170	29/47	29/47
0.5870	…	27/46	0.6176	…	21/34
0.5870	30/47−2/39	37/53−6/54	0.6180	5/39+24/49	3/53+32/57
0.5880	1/37+23/41	28/57+6/62	0.6190	1/43+28/47	17/53+17/57
0.5882	10/17	20/34	0.6190	13/21	26/42
0.5882	…	30/51	0.6200	23/43+4/47	8/25+9/30
0.5890	32/41−9/47	8/46+22/53	0.6207	…	36/58

Accurate Angular Indexing *(Continued)*

Part of a Turn	B&S, Becker, Hendey, K&T, & Rockford	Cincinnati and LeBlond	Part of a Turn	B&S, Becker, Hendey, K&T, & Rockford	Cincinnati and LeBlond
0.5897	23/39	23/39	0.6210	29/37−7/43	6/46+26/53
0.5900	29/47−1/37	18/24−4/25	0.6212	…	41/66
0.5909	…	39/66	0.6216	23/37	23/37
0.5910	24/39−1/41	40/59−4/46	0.6220	14/27+3/29	15/53+20/59
0.5918	29/49	29/49	0.6226	…	33/53
0.5920	21/37+1/41	21/34−1/39	0.6230	24/37−1/39	24/37−1/39
0.5926	16/27	32/54	0.6240	5/29+14/31	4/49+32/59
0.5930	1/15+10/19	22/34−2/37	0.6250	10/16	15/24
0.5932	…	35/59	0.6260	12/43+17/49	21/46+10/59
0.5940	6/29+12/31	15/49+19/66	0.6270	17/47+13/49	24/46+6/57
0.5946	22/37	22/37	0.6271	…	37/59
0.5950	12/41+13/43	18/25−3/24	0.6275	…	32/51
0.5952	…	25/42	0.6279	27/43	27/43
0.5957	28/47	28/47	0.6280	23/33−2/29	28/47+2/62
0.5960	28/39−5/41	15/51+16/53	0.6290	11/39+17/49	12/54+24/59
0.5965	…	34/57	0.6290	…	39/62
0.5968	…	37/62	0.6296	17/27	34/54
0.6300	11/41+17/47	18/24−3/25	0.6610	…	39/59
0.6304	…	29/46	0.6613	…	41/62
0.6310	9/37+19/49	8/49+29/62	0.6620	13/23+3/31	36/53−1/58
0.6316	12/19	24/38	0.6630	35/49−2/39	11/39+16/42
0.6316	…	36/57	0.6640	13/41+17/49	33/51+1/59
0.6320	12/37+12/39	12/34+12/43	0.6650	16/37+10/43	15/24+1/25
0.6327	31/49	31/49	0.6660	34/41−8/49	21/51+15/59
0.6330	13/27+5/33	14/51+19/53	0.6667	10/15	16/24
0.6333	…	19/30	0.6667	12/18	20/30
0.6340	7/17+4/18	25/37−1/24	0.6667	14/21	26/39
0.6341	26/41	26/41	0.6667	18/27	28/42
0.6350	9/39+19/47	19/25−3/24	0.6667	22/33	34/51
0.6360	7/37+21/47	12/54+24/58	0.6667	26/39	36/54
0.6364	21/33	42/66	0.6667	…	38/57
0.6370	5/47+26/49	10/47+28/66	0.6667	…	44/66
0.6379	…	37/58	0.6670	24/41+4/49	5/51+33/58
0.6380	12/27+6/31	6/34+18/39	0.6680	14/41+16/49	11/47+23/53
0.6383	30/47	30/47	0.6690	5/23+14/31	10/46+28/62
0.6390	14/23+1/33	28/39−3/38	0.6700	37/49−4/47	18/24−2/25
0.6400	4/37+25/47	16/25	0.6710	9/21+8/33	15/47+19/54
0.6410	…	45/66−2/49	0.6720	15/41+15/49	7/54+32/59
0.6410	25/39	25/39	0.6724	…	39/58
0.6415	…	34/53	0.6730	7/43+25/49	42/57−3/47
0.6420	23/37+1/49	20/59+20/66	0.6735	33/49	33/49
0.6429	…	18/28	0.6739	…	31/46
0.6429	…	27/42	0.6740	4/39+28/49	21/53+15/54
0.6430	41/37−20/43	24/51+10/58	0.6744	29/43	29/43
0.6440	31/43−3/39	31/43−3/39	0.6750	10/16+1/20	9/24+9/30
0.6441	…	38/59	0.6757	25/37	25/37
0.6450	33/43−6/49	3/24+13/25	0.6760	20/23−6/31	43/62−1/57
0.6452	20/31	40/62	0.6765	…	23/34
0.6460	23/37+1/41	2/47+35/58	0.6770	7/37+20/41	26/53+11/59
0.6470	8/39+19/43	24/47+9/66	0.6774	21/31	42/62
0.6471	11/17	22/34	0.6780	…	40/59
0.6471	…	33/51	0.6780	21/39+6/43	6/49+30/54
0.6480	31/41−4/37	43/57−5/47	0.6786	…	19/28
0.6481	…	35/54	0.6790	7/37+24/49	19/51+19/62

Accurate Angular Indexing *(Continued)*

Part of a Turn	B&S, Becker, Hendey, K&T, & Rockford	Cincinnati and LeBlond	Part of a Turn	B&S, Becker, Hendey, K&T, & Rockford	Cincinnati and LeBlond
0.6486	24/37	24/37	0.6792	…	36/53
0.6490	3/23+14/27	8/30+13/34	0.6800	31/47+1/49	17/25
0.6491	…	37/57	0.6809	32/47	32/47
0.6500	13/20	6/24+12/30	0.6810	21/27−3/31	29/41−1/38
0.6510	18/29+1/33	25/59+15/66	0.6818	…	45/66
0.6512	…	28/43	0.6820	15/37+13/47	37/51−2/46
0.6515	…	43/66	0.6829	28/41	28/41
0.6520	8/31+13/33	46/59−6/47	0.6830	5/17+7/18	35/57+4/58
0.6522	15/23	30/46	0.6840	31/37−6/39	13/47+22/54
0.6530	24/31−4/33	22/54+14/57	0.6842	13/19	26/38
0.6531	32/49	32/49	0.6842	…	39/57
0.6540	23/41+4/43	34/58+4/59	0.6850	15/41+15/47	3/24+14/25
0.6550	23/31−2/23	9/24+7/25	0.6852	…	37/54
0.6552	19/29	38/58	0.6860	3/23+15/27	19/49+17/57
0.6560	29/41−2/39	23/24−13/43	0.6863	…	35/51
0.6570	10/23+6/27	20/46+12/54	0.6870	28/37−3/43	36/51−1/53
0.6579	…	25/38	0.6875	11/16	…
0.6580	10/21+6/33	20/34+3/43	0.6880	13/21+2/29	30/49+5/66
0.6585	27/41	27/41	0.6890	36/47−3/39	42/53−6/58
0.6590	15/27+3/29	3/54+35/58	0.6897	20/29	40/58
0.6596	31/47	31/47	0.6900	21/37+6/49	6/24+11/25
0.6600	8/47+24/49	9/25+9/30	0.6905	…	29/42
0.6604	…	35/53	0.6910	35/49−1/43	21/57+20/62
0.6610	2/41+30/49	34/57+4/62	0.6920	35/43−5/41	35/43−5/41
0.6923	27/39	27/39	0.7234	…	34/47
0.6930	19/37+7/39	19/37+7/39	0.7240	3/27+19/31	34/41−4/38
0.6935	…	43/62	0.7241	21/29	42/58
0.6939	34/49	34/49	0.7250	2/16+12/20	7/24+13/30
0.6940	10/23+7/27	14/38+14/43	0.7255	…	37/51
0.6949	…	41/59	0.7258	…	45/62
0.6950	24/33−1/31	9/24+8/25	0.7260	34/41−6/47	2/49+37/54
0.6957	16/23	32/46	0.7270	15/19−1/16	14/54+29/62
0.6960	15/31+7/33	32/47+1/66	0.7273	24/33	48/66
0.6970	…	46/66	0.7280	23/37+5/47	23/49+15/58
0.6970	24/39+4/49	24/51+12/53	0.7288	…	43/59
0.6977	30/43	30/43	0.7290	38/49−2/43	12/47+27/57
0.6980	22/29−2/33	4/47+38/62	0.7297	27/37	27/37
0.6981	…	37/53	0.7300	11/27+10/31	6/24+12/25
0.6990	34/47−1/41	14/58+27/59	0.7310	15/37+14/43	26/59+18/62
0.7000	14/20	21/30	0.7317	30/41	30/41
0.7010	24/43+7/49	28/47+6/57	0.7320	20/37+9/47	26/57+16/58
0.7018	…	40/57	0.7330	19/37+9/41	39/47−6/62
0.7020	10/21+7/31	7/37+20/39	0.7333	11/15	22/30
0.7021	33/47	33/47	0.7340	6/21+13/29	26/49+12/59
0.7027	26/37	26/37	0.7347	36/49	36/49
0.7030	25/31−3/29	23/58+19/62	0.7350	29/37−2/41	9/24+9/25
0.7037	19/27	38/54	0.7353	…	25/34
0.7040	8/37+20/41	47/59−5/54	0.7358	…	39/53
0.7050	19/37+9/47	15/24+2/25	0.7360	25/47+10/49	47/59−4/66
0.7059	12/17	24/34	0.7368	14/19	28/38
0.7059	…	36/51	0.7368	…	42/57
0.7060	18/37+9/41	38/58+3/59	0.7370	2/37+28/41	13/49+25/53
0.7069	…	41/58	0.7380	19/37+11/49	31/47+4/51
0.7070	6/39+26/47	7/30+18/38	0.7381	…	31/42

Accurate Angular Indexing *(Continued)*

Part of a Turn	B&S, Becker, Hendey, K&T, & Rockford	Cincinnati and LeBlond	Part of a Turn	B&S, Becker, Hendey, K&T, & Rockford	Cincinnati and LeBlond
0.7073	29/41	29/41	0.7390	6/31+18/33	12/62+36/66
0.7080	30/39−3/49	13/58+30/62	0.7391	17/23	34/46
0.7083	…	17/24	0.7400	8/39+23/43	6/25+15/30
0.7090	34/39−7/43	31/46+2/57	0.7407	20/27	40/54
0.7097	22/31	44/62	0.7410	9/39+25/49	49/57−7/59
0.7100	20/43+12/49	18/24−1/25	0.7414	…	43/58
0.7105	…	27/38	0.7419	23/31	46/62
0.7110	22/29−1/21	33/39−5/37	0.7420	17/39+15/49	28/51+11/57
0.7119	…	42/59	0.7424	…	49/66
0.7120	6/39+24/43	31/54+8/58	0.7430	19/39+11/43	1/53+42/58
0.7121	…	47/66	0.7436	…	29/39
0.7130	27/43+4/47	17/30+6/41	0.7440	4/29+20/33	27/49+11/57
0.7140	6/43+27/47	45/57−4/53	0.7442	32/43	32/43
0.7143	15/21	20/28	0.7447	35/47	35/47
0.7143	35/49	30/42	0.7450	17/21−2/31	15/24+3/25
0.7143	…	35/49	0.7451	…	38/51
0.7150	28/43+3/47	21/25−3/24	0.7458	…	44/59
0.7160	2/47+33/49	25/51+14/62	0.7460	13/47+23/49	2/59+47/66
0.7170	…	38/53	0.7470	23/41+8/43	29/49+9/58
0.7170	29/43+2/47	28/59+16/66	0.7480	19/43+15/49	10/53+33/59
0.7174	…	33/46	0.7490	13/15−2/17	36/47−1/59
0.7179	28/39	28/39	0.7500	12/16	18/24
0.7180	22/27−3/31	21/58+21/59	0.7500	15/20	21/28
0.7190	39/49−3/39	12/57+30/59	0.7510	11/23+9/33	39/53+1/66
0.7193	…	41/57	0.7520	19/23−2/27	39/57+4/59
0.7200	2/43+33/49	18/25	0.7530	23/39+8/49	11/53+36/66
0.7209	31/43	31/43	0.7540	6/37+29/49	25/46+12/57
0.7210	13/29+9/33	21/47+17/62	0.7544	…	43/57
0.7220	18/23−2/33	13/46+29/66	0.7547	…	40/53
0.7222	13/18	39/54	0.7550	24/37+5/47	21/24−3/25
0.7230	6/29+16/31	11/46+30/62	0.7551	37/49	37/49
0.7560	1/47+36/49	9/47+35/62	0.7860	17/37+16/49	49/58−3/51
0.7561	31/41	31/41	0.7870	10/39+26/49	30/37−1/42
0.7568	28/37	28/37	0.7872	37/47	37/47
0.7570	15/43+20/49	8/53+40/66	0.7879	26/33	52/66
0.7576	25/33	50/66	0.7880	6/39+26/41	45/51−5/53
0.7580	16/37+14/43	48/59−3/54	0.7890	19/41+14/43	37/49+2/59
0.7581	…	47/62	0.7895	…	30/38
0.7586	22/29	44/58	0.7895	…	45/57
0.7590	28/47+8/49	36/41−5/42	0.7900	15/37+15/39	18/24+1/25
0.7593	…	41/54	0.7903	…	49/62
0.7600	39/47−3/43	19/25	0.7907	34/43	34/43
0.7609	…	35/46	0.7910	8/29+17/33	7/49+35/54
0.7610	19/43+15/47	34/51+5/53	0.7917	…	19/24
0.7619	16/21	32/42	0.7920	8/29+16/31	4/47+41/58
0.7620	38/47−2/43	16/51+26/58	0.7925	…	42/53
0.7627	…	45/59	0.7930	10/39+22/41	7/47+38/59
0.7630	14/37+15/39	36/46−1/51	0.7931	23/29	46/58
0.7632	…	29/38	0.7940	28/43+7/49	14/57+34/62
0.7640	29/39+1/49	27/57+18/62	0.7941	…	27/34
0.7647	13/17	26/34	0.7949	31/39	31/39
0.7647	…	39/51	0.7950	24/37+6/41	21/24−2/25
0.7650	16/27+5/29	3/24+16/25	0.7959	39/49	39/49
0.7660	36/47	36/47	0.7960	2/39+35/47	18/37+13/42

Accurate Angular Indexing *(Continued)*

Part of a Turn	B&S, Becker, Hendey, K&T, & Rockford	Cincinnati and LeBlond	Part of a Turn	B&S, Becker, Hendey, K&T, & Rockford	Cincinnati and LeBlond
0.7660	13/37+17/41	4/37+25/38	0.7963	…	43/54
0.7667	…	23/30	0.7966	…	47/59
0.7670	14/41+20/47	31/38–2/41	0.7970	40/47–2/37	13/53+32/58
0.7674	33/43	33/43	0.7980	14/39+18/41	33/51+8/53
0.7680	21/41+11/43	21/41+11/43	0.7990	3/37+28/39	10/28+19/43
0.7690	16/47+21/49	15/54+28/57	0.8000	12/15	20/25
0.7692	30/39	30/39	0.8000	16/20	24/30
0.7700	14/23+5/31	6/24+13/25	0.8010	32/37–3/47	10/47+30/51
0.7710	21/39+10/43	48/59–2/47	0.8020	27/31–2/29	54/62–4/58
0.7719	…	44/57	0.8030	18/39+14/41	18/39+14/41
0.7720	2/37+28/39	2/37+28/39	0.8030	…	53/66
0.7727	…	51/66	0.8039	…	41/51
0.7730	20/27+1/31	1/34+29/39	0.8040	10/43+28/49	32/49+8/53
0.7736	…	41/53	0.8043	…	37/46
0.7740	32/39–2/43	32/39–2/43	0.8049	33/41	33/41
0.7742	24/31	48/62	0.8050	22/23–5/33	3/24+17/25
0.7750	6/16+8/20	9/24+12/30	0.8060	34/41–1/43	13/47+27/51
0.7755	38/49	38/49	0.8065	25/31	50/62
0.7759	…	45/58	0.8070	5/15+9/19	15/58+34/62
0.7760	36/41–5/49	5/51+40/59	0.8070	…	46/57
0.7770	6/23+16/31	47/59–1/51	0.8080	19/21–3/31	22/39+10/41
0.7778	14/18	42/54	0.8085	38/47	38/47
0.7778	21/27	…	0.8090	22/39+12/49	28/53+16/57
0.7780	16/41+19/49	41/47–5/53	0.8095	…	34/42
0.7790	6/41+31/49	49/57–5/62	0.8100	33/43+2/47	6/24+14/25
0.7797	…	46/59	0.8103	…	47/58
0.7800	28/37+1/43	17/25+3/30	0.8108	30/37	30/37
0.7805	32/41	32/41	0.8110	7/27+16/29	34/53+10/59
0.7810	12/23+7/27	17/57+28/58	0.8113	…	43/53
0.7820	28/31–4/33	45/49–9/66	0.8120	17/29+7/31	34/58+14/62
0.7826	18/23	36/46	0.8125	13/16	…
0.7830	13/31+12/33	23/46+15/53	0.8130	6/43+33/49	16/24+6/41
0.7838	29/37	29/37	0.8136	…	48/59
0.7840	21/23–4/31	34/47+4/66	0.8140	35/43	35/43
0.7843	…	40/51	0.8140	28/33–1/29	14/47+32/62
0.7850	32/43+2/49	15/24+4/25	0.8148	22/27	44/54
0.7857	…	22/28	0.8150	22/47+17/49	9/24+11/25
0.7857	…	33/42	0.8158	…	31/38
0.8160	5/37+32/47	23/34+6/43	0.8478	…	39/46
0.8163	40/49	40/49	0.8480	30/41+5/43	31/59+20/62
0.8170	12/17+2/18	13/54+34/59	0.8485	28/33	56/66
0.8180	30/39+2/41	33/54+12/58	0.8490	13/41+25/47	2/47+50/62
0.8182	27/33	54/66	0.8491	…	45/53
0.8190	26/37+5/43	20/34+9/39	0.8500	17/20	18/24+3/30
0.8200	28/39+5/49	8/25+15/30	0.8510	5/21+19/31	22/37+10/39
0.8205	32/39	32/39	0.8511	40/47	40/47
0.8210	10/21+10/29	10/59+43/66	0.8519	23/27	46/54
0.8214	…	23/28	0.8520	1/16+15/19	55/62–2/57
0.8220	18/41+18/47	11/57+39/62	0.8529	…	29/34
0.8226	…	51/62	0.8530	17/21+1/23	19/58+31/59
0.8230	34/39–2/41	4/49+43/58	0.8537	35/41	35/41
0.8235	14/17	28/34	0.8540	15/39+23/49	54/62–1/59
0.8235	…	42/51	0.8548	…	53/62
0.8240	15/37+18/43	19/53+27/58	0.8550	19/37+14/41	9/24+12/25

Accurate Angular Indexing *(Continued)*

Part of a Turn	B&S, Becker, Hendey, K&T, & Rockford	Cincinnati and LeBlond	Part of a Turn	B&S, Becker, Hendey, K&T, & Rockford	Cincinnati and LeBlond
0.8246	…	47/57	0.8560	24/43+14/47	37/53+9/57
0.8250	6/16+9/20	7/24+16/30	0.8570	3/43+37/47	51/57−2/53
0.8260	4/31+23/33	39/62+13/66	0.8571	18/21	24/28
0.8261	19/23	38/46	0.8571	42/49	36/42
0.8270	32/41+2/43	46/58+2/59	0.8571	…	42/49
0.8276	24/29	48/58	0.8580	25/43+13/47	38/51+7/62
0.8280	35/41−1/39	35/41−1/39	0.8590	11/27+14/31	22/54+28/62
0.8290	5/37+34/49	10/34+23/43	0.8596	…	49/57
0.8293	34/41	34/41	0.8600	1/43+41/49	9/25+15/30
0.8298	39/47	39/47	0.8605	37/43	37/43
0.8300	17/23+3/33	18/24+2/25	0.8610	16/37+21/49	18/46+31/66
0.8302	…	44/53	0.8620	13/37+24/47	17/38+17/41
0.8305	…	49/59	0.8621	25/29	50/58
0.8310	34/39−2/49	39/49+2/57	0.8627	…	44/51
0.8320	27/43+10/49	27/53+20/62	0.8630	38/41−3/47	18/46+25/53
0.8330	9/37+23/39	42/47−4/66	0.8636	…	57/66
0.8333	15/18	20/24	0.8640	11/16+3/17	56/62−2/51
0.8333	…	25/30	0.8644	…	51/59
0.8333	…	35/42	0.8649	32/37	32/37
0.8333	…	45/54	0.8650	4/41+33/43	15/24+6/25
0.8333	…	55/66	0.8660	10/17+5/18	13/57+37/58
0.8340	15/23+6/33	20/38+12/39	0.8667	13/15	26/30
0.8350	43/49−2/47	21/24−1/25	0.8670	3/21+21/29	28/54+23/66
0.8360	2/41+37/47	32/46+8/57	0.8679	…	46/53
0.8367	41/49	41/49	0.8680	36/47+5/49	53/59−2/66
0.8370	19/29+6/33	33/57+16/62	0.8684	…	33/38
0.8372	…	36/43	0.8690	40/43−3/49	39/47+2/51
0.8378	…	31/37	0.8696	20/23	40/46
0.8380	7/39+27/41	20/46+25/62	0.8700	4/39+33/43	18/24+3/25
0.8387	26/31	52/62	0.8704	…	47/54
0.8390	30/39+3/43	3/49+42/54	0.8710	27/31	54/62
0.8400	19/37+16/49	21/25	0.8710	17/27+7/29	31/51+15/57
0.8410	23/43+15/49	44/51−1/46	0.8718	34/39	34/39
0.8420	34/37−3/39	46/49−6/62	0.8720	30/37+3/49	26/58+25/59
0.8421	16/19	32/38	0.8723	41/47	41/47
0.8421	…	48/57	0.8730	15/39+21/43	21/49+24/54
0.8430	6/37+32/47	9/47+43/66	0.8740	15/39+23/47	5/53+46/59
0.8431	…	43/51	0.8750	14/16	21/24
0.8440	17/21+1/29	41/54+5/59	0.8760	21/23−1/27	48/57+2/59
0.8448	…	49/58	0.8770	3/37+39/49	37/51+10/66
0.8450	29/37+3/49	3/24+18/25	0.8772	…	50/57
0.8460	27/37+5/43	22/54+25/57	0.8776	43/49	43/49
0.8462	33/39	33/39	0.8780	24/47+18/49	31/53+17/58
0.8470	5/23+17/27	26/38+7/43	0.8780	36/41	36/41
0.8475	…	50/59	0.8788	…	58/66
0.8790	22/43+18/49	52/58−1/57	0.9110	31/37+3/41	31/37+3/41
0.8793	…	51/58	0.9118	…	31/34
0.8800	31/39+4/47	22/25	0.9120	26/37+9/43	52/43−11/37
0.8810	…	37/42	0.9123	…	52/57
0.8810	19/29+7/31	8/51+42/58	0.9130	2/31+28/33	35/62+23/66
0.8814	…	52/59	0.9130	21/23	42/46
0.8820	20/37+14/41	42/57+9/62	0.9138	…	53/58
0.8824	15/17	30/34	0.9140	7/21+18/31	42/47+1/49
0.8824	…	45/51	0.9149	43/47	43/47

Accurate Angular Indexing *(Continued)*

Part of a Turn	B&S, Becker, Hendey, K&T, & Rockford	Cincinnati and LeBlond	Part of a Turn	B&S, Becker, Hendey, K&T, & Rockford	Cincinnati and LeBlond
0.8830	27/41+11/49	7/51+44/59	0.9150	8/17+8/18	21/24+1/25
0.8837	38/43	38/43	0.9153	…	54/59
0.8840	23/29+3/33	37/47+6/62	0.9160	35/43+5/49	40/53+10/62
0.8850	7/23+18/31	3/24+19/25	0.9167	…	22/24
0.8860	38/41−2/49	38/59+15/62	0.9170	19/23+3/33	29/38+6/39
0.8868	…	47/53	0.9180	1/41+42/47	39/46+4/57
0.8870	28/41+10/49	34/51+13/59	0.9184	45/49	45/49
0.8871	…	55/62	0.9189	34/37	34/37
0.8880	18/41+22/49	28/51+20/59	0.9190	14/23+9/29	8/46+38/51
0.8889	16/18	48/54	0.9194	…	57/62
0.8889	24/27	…	0.9200	28/37+8/49	23/25
0.8890	8/41+34/49	53/57−2/49	0.9210	17/37+18/39	23/49+28/62
0.8900	39/41−3/49	6/24+16/25	0.9211	…	35/38
0.8910	29/41+9/49	52/57−1/47	0.9216	…	47/51
0.8913	…	41/46	0.9220	26/39+12/47	10/34+27/43
0.8919	33/37	33/37	0.9229	31/37+4/47	…
0.8920	19/41+21/49	17/47+35/66	0.9230	…	29/54+22/57
0.8929	…	25/28	0.9231	36/39	36/39
0.8930	27/37+8/49	5/46+40/51	0.9240	15/41+24/43	45/59+10/62
0.8936	42/47	42/47	0.9242	…	61/66
0.8939	…	59/66	0.9245	…	49/53
0.8940	27/29−1/27	28/49+20/62	0.9250	14/16+1/20	7/24+19/30
0.8947	17/19	34/38	0.9259	25/27	50/54
0.8947	…	51/57	0.9260	17/43+26/49	2/51+47/53
0.8950	30/41+8/49	9/24+13/25	0.9268	38/41	38/41
0.8960	20/41+20/49	8/47+45/62	0.9270	28/37+8/47	29/59+27/62
0.8966	26/29	52/58	0.9280	3/39+40/47	47/57+6/58
0.8970	14/43+28/49	7/57+48/62	0.9286	…	26/28
0.8974	35/39	35/39	0.9286	…	39/42
0.8980	44/49	44/49	0.9290	12/37+26/43	16/53+37/59
0.8980	1/39+41/47	28/57+24/59	0.9298	…	53/57
0.8983	…	53/59	0.9300	5/29+25/33	6/24+17/25
0.8990	8/39+34/49	42/51+4/53	0.9302	40/43	40/43
0.9000	18/20	27/30	0.9310	25/37+12/47	7/30+30/43
0.9010	28/29−2/31	27/58+27/62	0.9310	27/29	54/58
0.9020	20/27+5/31	46/51	0.9320	30/39+7/43	59/62+1/51
0.9020	…	29/53+22/62	0.9322	…	55/59
0.9024	37/41	37/41	0.9330	34/39+3/49	5/42+35/43
0.9030	17/41+21/43	17/41+21/43	0.9333	14/15	28/30
0.9032	28/31	56/62	0.9340	1/37+39/43	56/59−1/66
0.9040	17/41+23/47	7/53+44/57	0.9348	…	43/46
0.9048	19/21	38/42	0.9350	2/39+38/43	9/24+14/25
0.9050	38/43+1/47	15/24+7/25	0.9355	29/31	58/62
0.9057	…	48/53	0.9360	15/37+26/49	13/58+42/59
0.9060	17/43+24/47	17/58+38/62	0.9362	44/47	44/47
0.9070	39/43	39/43	0.9370	19/21+1/31	29/53+23/59
0.9070	14/29+14/33	7/47+47/62	0.9375	15/16	…
0.9074	…	49/54	0.9380	13/39+26/43	21/46+26/54
0.9080	1/27+27/31	29/54+23/62	0.9388	46/49	46/49
0.9090	14/37+26/49	8/53+47/62	0.9390	30/37+5/39	30/37+5/39
0.9091	30/33	60/66	0.9394	31/33	62/66
0.9100	14/39+27/49	18/24+4/25	0.9400	35/39+2/47	16/25+9/30
0.9410	30/41+9/43	25/47+27/66	0.9670	9/47+38/49	8/34+30/41
0.9412	16/17	32/34	0.9677	30/31	60/62

Accurate Angular Indexing *(Continued)*

Part of a Turn	B&S, Becker, Hendey, K&T, & Rockford	Cincinnati and LeBlond	Part of a Turn	B&S, Becker, Hendey, K&T, & Rockford	Cincinnati and LeBlond
0.9412	…	48/51	0.9680	26/37+13/49	2/46+49/53
0.9420	15/37+22/41	42/47+3/62	0.9690	26/39+13/43	5/51+54/62
0.9430	9/23+16/29	12/49+37/53	0.9697	32/33	64/66
0.9434	…	50/53	0.9700	12/23+13/29	6/24+18/25
0.9440	9/43+36/49	26/46+25/66	0.9706	…	33/34
0.9444	17/18	51/54	0.9710	26/37+11/41	14/46+34/51
0.9450	37/41+2/47	15/24+8/25	0.9720	26/43+18/49	31/53+24/62
0.9459	35/37	35/37	0.9730	36/37	36/37
0.9460	12/39+30/47	22/46+29/62	0.9730	20/23+3/29	26/54+29/59
0.9470	33/47+12/49	14/49+41/62	0.9737	…	37/38
0.9474	18/19	36/38	0.9740	16/21+7/33	26/34+9/43
0.9474	…	54/57	0.9744	38/39	38/39
0.9480	18/37+18/39	11/38+27/41	0.9750	10/16+7/20	13/24+13/30
0.9483	…	55/58	0.9756	40/41	40/41
0.9487	37/39	37/39	0.9760	14/39+29/47	10/49+44/57
0.9490	1/15+15/17	13/47+39/58	0.9762	…	41/42
0.9492	…	56/59	0.9767	42/43	42/43
0.9500	19/20	6/24+21/30	0.9770	33/37+4/47	30/47+21/62
0.9510	29/43+13/47	41/53+11/62	0.9780	25/37+13/43	25/37+13/43
0.9512	39/41	39/41	0.9783	…	45/46
0.9516	…	59/62	0.9787	46/47	46/47
0.9520	8/43+36/47	30/53+22/57	0.9790	13/23+12/29	10/53+49/62
0.9524	20/21	40/42	0.9796	48/49	48/49
0.9530	5/43+41/49	16/59+45/66	0.9800	23/39+16/41	17/25+9/30
0.9535	41/43	41/43	0.9804	…	50/51
0.9540	29/37+8/47	28/54+27/62	0.9810	22/43+23/49	51/58+6/59
0.9545	…	63/66	0.9811	…	52/53
0.9550	7/39+38/49	21/24+2/25	0.9815	…	53/54
0.9560	13/37+26/43	13/37+26/43	0.9820	30/39+10/47	19/46+33/58
0.9565	22/23	44/46	0.9825	…	56/57
0.9570	14/21+9/31	21/47+25/49	0.9828	…	57/58
0.9574	45/47	45/47	0.9830	26/41+15/43	45/57+12/62
0.9580	5/39+39/47	20/53+36/62	0.9831	…	58/59
0.9583	…	23/24	0.9839	…	61/62
0.9590	21/41+21/47	18/51+40/66	0.9840	13/37+31/49	1/46+51/53
0.9592	47/49	47/49	0.9848	…	65/66
0.9600	7/39+32/41	24/25	0.9850	6/23+21/29	15/24+9/25
0.9608	…	49/51	0.9860	16/23+9/31	42/53+12/62
0.9610	11/23+14/29	5/34+35/43	0.9870	15/21+9/33	13/34+26/43
0.9620	28/41+12/43	52/59+5/62	0.9880	7/39+38/47	5/46+51/58
0.9623	…	51/53	0.9890	33/39+7/49	20/39+20/42
0.9630	26/27	52/54	0.9900	10/29+20/31	18/24+6/25
0.9630	30/43+13/49	1/51+50/53	0.9910	22/23+1/29	21/46+31/58
0.9640	21/39+20/47	52/57+3/58	0.9920	39/41+2/49	40/53+14/59
0.9643	…	27/28	0.9930	8/23+20/31	21/53+37/62
0.9649	…	55/57	0.9940	7/23+20/29	14/46+40/58
0.9650	11/39+28/41	3/24+21/25	0.9950	35/39+4/41	21/24+3/25
0.9655	28/29	56/58	0.9960	40/41+1/49	43/46+3/49
0.9660	13/39+31/49	15/39+25/43	0.9970	15/23+10/29	30/46+20/58
0.9661	…	57/59	0.9980	20/41+25/49	12/51+45/59
0.9667	…	29/30	0.9990	10/41+37/49	6/51+52/59

Use the table to locate the indexing movement for the decimal fraction nearest to the decimal fraction of the quotient for which there is an entry in the column for the dividing head to be used. If the decimal fraction of the quotient is close to the midpoint between two table entries, calculate the mathematical value of the two indexing movements to more decimal places to make the closeness determination.

Example: Movement through an angle of 31° 27′ 50″ is required. Expressed in seconds, this angle 113270″, which, divided by 32,400, equals 3.495987. The indexing movement is three full turns of the crank plus a fractional turn of 0.495987. The nearest table entry is for 0.4960, which requires a compound indexing movement of 8 holes on the 23-hole circle plus 4 holes on the 27-hole circle in the same direction. Checking the value of these movements shows that $\frac{8}{23} + \frac{4}{27} = 0.347826 + 0.148148 = 0.495974$, which, multiplied by 32,400, = 16,069.56, or 4° 27′ 49.56″ from the fractional turn. Adding the 27° from three full turns gives a total movement of 31° 27′ 49.56″.

Approximate Indexing for Small Angles.—To find *approximate* indexing movements for small angles, such as the remainder from the method discussed in *Angular Indexing* starting on page 1956, on a dividing head with a 40:1 worm-gear ratio, divide 540 by the number of minutes in the angle, and then divide the number of holes in each of the available indexing circles by this quotient. The result that is closest to a whole number is the best approximation of the angle for a simple indexing movement and is the number of holes to be moved in the corresponding circle of holes. If the angle is greater than 9 degrees, the whole number will be greater than the number of holes in the circle, indicating that one or more full turns of the crank are required. Dividing by the number of holes in the indicated circle of holes will reduce the required indexing movement to the number of full turns, and the remainder will be the number of holes to be moved for the fractional turn. If the angle is less than about 11 minutes, it cannot be indexed by simple indexing with standard B & S plates (the corresponding angle for standard plates on a Cincinnati head is about 8 minutes, and for Cincinnati high number plates, 2.7 minutes).

Example: An angle of 7° 25′ is to be indexed. Expressed in minutes, it is 445′ and 540 divided by 445 equals 1.213483. The indexing circles available on standard B & S plates are 15, 16, 17, 18, 19, 20, 21, 23, 27, 29, 31, 33, 37, 39, 41, 43, 47, and 49. Each of these numbers is divided by 1.213483 and the closest to a whole number is found to be 17 ÷ 1.213483 = 14.00926. The best approximation for a simple indexing movement to obtain 7° 25′ is 14 holes on the 17-hole circle.

Differential Indexing.—This method is the same, in principle, as compound indexing, but differs from the latter in that the index plate is rotated by suitable gearing that connects it to the spiral-head spindle. This rotation or differential motion of the index plate takes place when the crank is turned, the plate moving either in the same direction as the crank or opposite to it, as may be required. The result is that the *actual* movement of the crank, at every indexing, is either greater or less than its movement with relation to the index plate. The differential method makes it possible to obtain almost any division by using only one circle of holes for that division and turning the index crank in one direction, as with plain indexing.

The gears to use for turning the index plate the required amount (when gears are required) are shown by the table Simple and Differential Indexing with Brown & Sharpe Plates, which shows what divisions can be obtained by plain indexing, and when it is necessary to use gears and the differential system. For example, if 50 divisions are required, the 20-hole index circle is used and the crank is moved 16 holes, but no gears are required. For 51 divisions, a 24-tooth gear is placed on the wormshaft and a 48-tooth gear on the spindle. These two gears are connected by two idler gears having 24 and 44 teeth, respectively.

To illustrate the principle of differential indexing, suppose a dividing head is to be geared for 271 divisions. The table calls for a gear on the wormshaft having 56 teeth, a spindle gear with 72 teeth, and a 24-tooth idler to rotate the index plate in the same direction as the crank. The sector arms should be set to give the crank a movement of 3 holes in the 21-hole circle. If the spindle and the index plate were not connected through gearing, 280 divisions would be obtained by successively moving the crank 3 holes in the 21-hole circle, but the gears cause the index plate to turn in the same direction as the crank at such a rate that, when 271 indexings have been made, the work is turned one complete revolution. Therefore, we have 271 divisions instead of 280, the number being reduced because the total movement of the crank, for each indexing, is equal to the movement relative to the index plate, *plus* the movement of the plate itself when, as here, the crank and plate rotate in the same direction.

If they were rotated in opposite directions, the crank would have a total movement equal to the amount it turned relative to the plate, *minus* the plate's movement. Sometimes it is necessary to use compound gearing to move the index plate the required amount for each turn of the crank. The differential method cannot be used in connection with helical or spiral milling because the spiral head is then geared to the leadscrew of the machine.

Finding Ratio of Gearing for Differential Indexing.—To find the ratio of gearing for differential indexing, first select some approximate number A of divisions either greater or less than the required number N. For example, if the required number N is 67, the approximate number A might be 70. Then, if 40 turns of the index crank are required for 1 revolution of the spindle, the gearing ratio $R = (A - N) \times 40/A$. If the approximate number A is less than N, the formula is the same as above except that $A - N$ is replaced by $N - A$.

Example: Find the gearing ratio and indexing movement for 67 divisions.

$$\text{If } A = 70, \text{ gearing ratio } = (70 - 67) \times \frac{40}{70} = \frac{12}{7} = \frac{\text{gear on spindle (driver)}}{\text{gear on worm (driven)}}$$

The fraction 12⁄7 is raised to obtain a numerator and a denominator to match gears that are available. For example, 12⁄7 = 48⁄28.

Various combinations of gearing and index circles are possible for a given number of divisions. The index numbers and gear combinations in the accompanying table apply to a given series of index circles and gear-tooth numbers. The approximate number A on which any combination is based may be determined by dividing 40 by the fraction representing the indexing movement. For example, the approximate number used for 109 divisions equals $40 \div 6/16$, or $40 \times 16/6 = 106\tfrac{2}{3}$. If this approximate number is inserted in the preceding formula, it will be found that the gear ratio is 7⁄8, as shown in the table.

Second Method of Determining Gear Ratio.—In illustrating a somewhat different method of obtaining the gear ratio, 67 divisions will again be used. If 70 is selected as the approximate number, then 40⁄70 = 4⁄7 or 12⁄21 turn of the index crank will be required. If the crank is indexed four-sevenths of a turn, sixty-seven times, it will make $4/7 \times 67 = 38\tfrac{2}{7}$ revolutions. This number is $1\tfrac{5}{7}$ turns less than the 40 required for one revolution of the work (indicating that the gearing should be arranged to rotate the index plate in the same direction as the index crank to increase the indexing movement). Hence the gear ratio $1\tfrac{5}{7} =$ 12⁄7.

To Find the Indexing Movement.—The indexing movement is represented by the fraction $40/A$. For example, if 70 is the approximate number A used in calculating the gear ratio for 67 divisions, then, to find the required movement of the index crank, reduce 40⁄70 to any fraction of equal value and having as denominator any number equal to the number of holes available in an index circle.

$$\text{To illustrate, } \frac{40}{70} = \frac{4}{7} = \frac{12}{21} = \frac{\text{number of holes indexed}}{\text{number of holes in index circle}}$$

Use of Idler Gears.—In differential indexing, idler gears are used to rotate the index plate in the same direction as the index crank, thus *increasing* the resulting indexing movement, or to rotate the index plate in the opposite direction, thus *reducing* the resulting indexing movement.

Example 1: If the approximate number A is *greater* than the required number of divisions N, simple gearing will require one idler, and compound gearing, no idler. Index plate and crank rotate in the same direction.

Example 2: If the approximate number A is *less* than the required number of divisions N, simple gearing requires two idlers, and compound gearing, one idler. Index plate and crank rotate in opposite directions.

When Compound Gearing Is Required.—It is sometimes necessary, as shown in the table, to use a train of four gears to obtain the required ratio with the gear-tooth numbers that are available.

Example: Find the gear combination and indexing movement for 99 divisions, assuming that an approximate number A of 100 is used.

$$\text{Ratio} = (100 - 99) \times \frac{40}{100} = \frac{4}{10} = \frac{4 \times 1}{5 \times 2} = \frac{32}{40} \times \frac{28}{56}$$

The final numbers here represent available gear sizes. The gears having 32 and 28 teeth are the drivers (gear on spindle and first gear on stud), and gears having 40 and 56 teeth are driven (second gear on stud and gear on wormshaft). The indexing movement is represented by the fraction $^{40}/_{100}$, which is reduced to $^{8}/_{20}$, the 20-hole index circle being used here.

Example: Determine the gear combination to use for indexing 53 divisions. If 56 is used as an approximate number (possibly after one or more trial solutions to find an approximate number and resulting gear ratio coinciding with available gears):

$$\text{Gearing ratio} = (56 - 53) \times \frac{40}{56} = \frac{15}{7} = \frac{3 \times 5}{1 \times 7} = \frac{72 \times 40}{24 \times 56}$$

The tooth numbers above the line here represent *gear on spindle* and *first gear on stud*. The tooth numbers below the line represent *second gear on stud* and *gear on wormshaft*.

$$\text{Indexing movement} = \frac{40}{56} = \frac{5}{7} = \frac{5 \times 7}{7 \times 7} = \frac{35 \text{ holes}}{49\text{-hole circle}}$$

To Check the Number of Divisions Obtained with a Given Gear Ratio and Index Movement.—Invert the fraction representing the indexing movement. Let C = this inverted fraction and R = gearing ratio.

Example 1: If simple gearing with one idler, or compound gearing with no idler, is used: number of divisions $N = 40C - RC$.

For instance, if the gear ratio is $^{12}/_{7}$, there is simple gearing and one idler, and the indexing movement is $^{12}/_{21}$, making the inverted fraction C, $^{21}/_{12}$; find the number of divisions N.

$$N = \left(40 \times \frac{21}{12}\right) - \left(\frac{12}{7} \times \frac{21}{12}\right) = 70 - \frac{21}{7} = 67$$

Example 2: If simple gearing with two idlers, or compound gearing with one idler, is used: number of divisions $N = 40C + RC$.

For instance, if the gear ratio is $^{7}/_{8}$, two idlers are used with simple gearing, and the indexing movement is 6 holes in the 16-hole circle, then number of divisions:

$$N = \left(40 \times \frac{16}{6}\right) + \left(\frac{7}{8} \times \frac{16}{6}\right) = 109$$

Simple and Differential Indexing - Browne & Sharpe Milling Machines

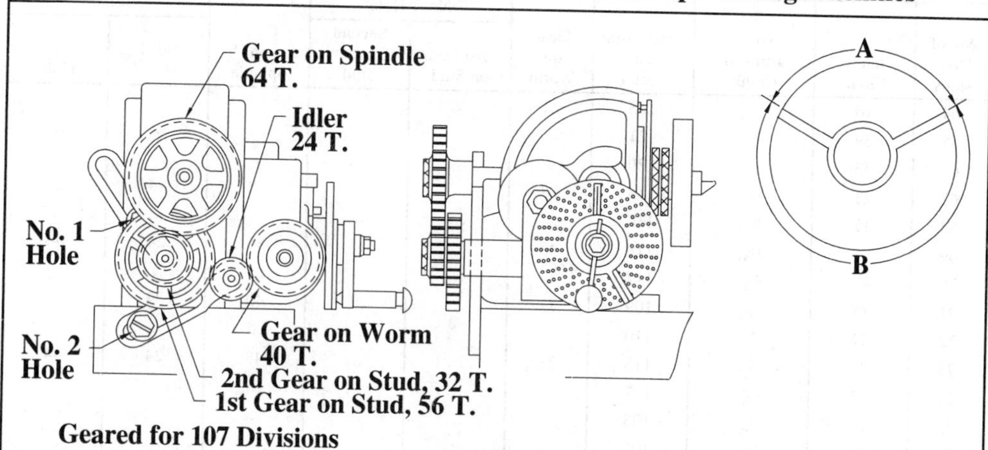

Geared for 107 Divisions

Note: Graduations in table indicate setting for sector arms when index crank moves through arc *A*, except figures marked *, when crank moves through arc *B*.

No. of Div.	Index Circle	No. of Turns of Crank	Graduation on Sector	No. of Div.	Index Circle	No. of Turns of Crank	Graduation on Sector	Gear on Worm	No. 1 Hole: First Gear on Stud	No. 1 Hole: Second Gear on Stud	Gear on Spindle	Idlers: No. 1 Hole	Idlers: No. 2 Hole
2	Any	20	…	33	33	$1\frac{7}{33}$	41						
3	39	$13\frac{13}{39}$	65	34	17	$1\frac{3}{17}$	33						
4	Any	10	…	35	49	$1\frac{7}{49}$	26						
5	Any	8	…	36	27	$1\frac{3}{27}$	21						
6	39	$6\frac{26}{39}$	132	37	37	$1\frac{3}{37}$	15						
7	49	$5\frac{35}{49}$	140	38	19	$1\frac{1}{19}$	9						
8	Any	5	…	39	39	$1\frac{1}{39}$	3						
9	27	$4\frac{12}{27}$	88	40	Any	1	…						
10	Any	4	…	41	41	$\frac{40}{41}$	3*						
11	33	$3\frac{21}{33}$	126	42	21	$\frac{20}{21}$	9*						
12	39	$3\frac{13}{39}$	65	43	43	$\frac{40}{43}$	12*						
13	39	$3\frac{3}{39}$	14	44	33	$\frac{30}{33}$	17*						
14	49	$2\frac{42}{49}$	169	45	27	$\frac{24}{27}$	21*						
15	39	$2\frac{20}{39}$	132	46	23	$\frac{20}{23}$	172						
16	20	$2\frac{10}{20}$	98	47	47	$\frac{40}{47}$	168						
17	17	$2\frac{6}{17}$	69	48	18	$\frac{15}{18}$	165						
18	27	$2\frac{6}{27}$	43	49	49	$\frac{40}{49}$	161						
19	19	$2\frac{2}{19}$	19	50	20	$\frac{16}{20}$	158						
20	Any	2	…	51	17	$\frac{14}{17}$	33*	24	…	…	48	24	44
21	21	$1\frac{19}{21}$	18*	52	39	$\frac{30}{39}$	152	…	…	…	…	…	…
22	33	$1\frac{27}{33}$	161	53	49	$\frac{35}{49}$	140	56	40	24	72	…	…
23	23	$1\frac{17}{23}$	147	54	27	$\frac{20}{27}$	147	…	…	…	…	…	…
24	39	$1\frac{26}{39}$	132	55	33	$\frac{24}{33}$	144	…	…	…	…	…	…
25	20	$1\frac{12}{20}$	118	56	49	$\frac{35}{49}$	140	…	…	…	…	…	…
26	39	$1\frac{21}{39}$	106	57	21	$\frac{15}{21}$	142	56	…	…	40	24	44
27	27	$1\frac{13}{27}$	95	58	29	$\frac{20}{29}$	136	…	…	…	…	…	…
28	49	$1\frac{21}{49}$	83	59	39	$\frac{26}{39}$	132	48	…	…	32	44	…
29	29	$1\frac{11}{29}$	75	60	39	$\frac{26}{39}$	132	…	…	…	…	…	…
30	39	$1\frac{13}{39}$	65	61	39	$\frac{26}{39}$	132	48	…	…	32	24	44
31	31	$1\frac{9}{31}$	56	62	31	$\frac{20}{31}$	127	…	…	…	…	…	…
32	20	$1\frac{5}{20}$	48	63	39	$\frac{26}{39}$	132	24	…	…	48	24	44

Differential Indexing

Certain divisions such as 51, 53, 57, etc., require the use of differential indexing. In differential indexing, change gears are used to transmit motion from the main spindle of the dividing head to the index plate, which turns (either in the same direction as the index plate or in the opposite direction) whatever amount is required to obtain the correct indexing movement.

The numbers in the columns below represent numbers of teeth for the change gears necessary to give the divisions required. Where no numbers are shown simple indexing, which does not require change gears, is used.

Differential Gears

Simple and Differential Indexing - Browne & Sharpe Milling Machines

No. of Divisions	Index Circle	No. of Turns of Crank	Graduation on Sector	Gear on Worm	No. 1 Hole		Gear on Spindle	Idlers	
					First Gear on Stud	Second Gear on Stud		No. 1 Hole	No. 2 Hole
64	16	10/16	123	…	…	…	…	…	…
65	39	24/39	121	…	…	…	…	…	…
66	33	20/33	120	…	…	…	…	…	…
67	21	12/21	113	28	…	…	48	44	…
68	17	10/17	116	…	…	…	…	…	…
69	20	12/20	118	40	…	…	56	24	44
70	49	28/49	112	…	…	…	…	…	…
71	18	10/18	109	72	…	…	40	24	…
72	27	15/27	110	…	…	…	…	…	…
73	21	12/21	113	28	…	…	48	24	44
74	37	20/37	107	…	…	…	…	…	…
75	15	8/15	105	…	…	…	…	…	…
76	19	10/19	103	…	…	…	…	…	…
77	20	10/20	98	32	…	…	48	44	…
78	39	20/39	101	…	…	…	…	…	…
79	20	10/20	98	48	…	…	24	44	…
80	20	10/20	98	…	…	…	…	…	…
81	20	10/20	98	48	…	…	24	24	44
82	41	20/41	96	…	…	…	…	…	…
83	20	10/20	98	32	…	…	48	24	44
84	21	10/21	94	…	…	…	…	…	…
85	17	8/17	92	…	…	…	…	…	…
86	43	20/48	91	…	…	…	…	…	…
87	15	7/15	92	40	…	…	24	24	44
88	33	15/33	89	…	…	…	…	…	…
89	18	8/18	87	72	…	…	32	44	…
90	27	12/27	88	…	…	…	…	…	…
91	39	18/39	91	24	…	…	48	24	44
92	23	10/23	86	…	…	…	…	…	…
93	18	8/18	87	24	…	…	32	24	44
94	47	20/47	83	…	…	…	…	…	…
95	19	8/19	82	…	…	…	…	…	…
96	21	9/21	85	28	…	…	32	24	44
97	20	8/20	78	40	…	…	48	44	…
98	49	20/49	79	…	…	…	…	…	…
99	20	8/20	78	56	28	40	32	…	…
100	20	8/20	78	…	…	…	…	…	…
101	20	8/20	78	72	24	40	48	…	24
102	20	8/20	78	40	…	…	32	24	44
103	20	8/20	78	40	…	…	48	24	44
104	39	15/39	75	…	…	…	…	…	…
105	21	8/21	75	…	…	…	…	…	…
106	43	16/43	73	86	24	24	48	…	…
107	20	8/20	78	40	56	32	64	…	24
108	27	10/27	73	…	…	…	…	…	…
109	16	6/16	73	32	…	…	28	24	44
110	33	12/33	71	…	…	…	…	…	…
111	39	13/39	65	24	…	…	72	32	…
112	39	13/39	65	24	…	…	64	44	…
113	39	13/39	65	24	…	…	56	44	…
114	39	13/39	65	24	…	…	48	44	…

Simple and Differential Indexing - Browne & Sharpe Milling Machines *(Continued)*

No. of Divisions	Index Circle	No. of Turns of Crank	Graduation on Sector	Gear on Worm	No. 1 Hole		Gear on Spindle	Idlers	
					First Gear on Stud	Second Gear on Stud		No. 1 Hole	No. 2 Hole
115	23	8/23	68	…	…	…	…	…	…
116	29	10/29	68	…	…	…	…	…	…
117	39	13/39	65	24	…	…	24	56	…
118	39	13/39	65	48	…	…	32	44	…
119	39	13/39	65	72	…	…	24	44	…
120	39	13/39	65	…	…	…	…	…	…
121	39	13/39	65	72	…	…	24	24	44
122	39	13/39	65	48	…	…	32	24	44
123	39	13/39	65	24	…	…	24	24	44
124	31	10/31	63	…	…	…	…	…	…
125	39	13/39	65	24	…	…	40	24	44
126	39	13/39	65	24	…	…	48	24	44
127	39	13/39	65	24	…	…	56	24	44
128	16	5/16	61	…	…	…	…	…	…
129	39	13/39	65	24	…	…	72	24	44
130	39	12/39	60	…	…	…	…	…	…
131	20	6/20	58	40	…	…	28	44	…
132	33	10/33	59	…	…	…	…	…	…
133	21	6/21	56	24	…	…	48	44	…
134	21	6/21	56	28	…	…	48	44	…
135	27	8/27	58	…	…	…	…	…	…
136	17	5/17	57	…	…	…	…	…	…
137	21	6/21	56	28	…	…	24	56	…
138	21	6/21	56	56	…	…	32	44	…
139	21	6/21	56	56	32	48	24	…	…
140	49	14/49	55	…	…	…	…	…	…
141	18	5/18	54	48	…	…	40	44	…
142	21	6/21	56	56	…	…	32	24	44
143	21	6/21	56	28	…	…	24	24	44
144	18	5/18	54	…	…	…	…	…	…
145	29	8/29	54	…	…	…	…	…	…
146	21	6/21	56	28	…	…	48	24	44
147	21	6/21	56	24	…	…	48	24	44
148	37	10/37	53	…	…	…	…	…	…
149	21	6/21	56	28	…	…	72	24	44
150	15	4/15	52	…	…	…	…	…	…
151	20	5/20	48	32	…	…	72	44	…
152	19	5/19	51	…	…	…	…	…	…
153	20	5/20	48	32	…	…	56	44	…
154	20	5/20	48	32	…	…	48	44	…
155	31	8/31	50	…	…	…	…	…	…
156	39	10/39	50	…	…	…	…	…	…
157	20	5/20	48	32	…	…	24	56	…
158	20	5/20	48	48	…	…	24	44	…
159	20	5/20	48	64	32	56	28	…	…
160	20	5/20	48	…	…	…	…	…	…
161	20	5/20	48	64	32	56	28	…	24
162	20	5/20	48	48	…	…	24	24	44
163	20	5/20	48	32	…	…	24	24	44
164	41	10/41	47	…	…	…	…	…	…
165	33	8/33	47	…	…	…	…	…	…

Simple and Differential Indexing - Browne & Sharpe Milling Machines *(Continued)*

No. of Divisions	Index Circle	No. of Turns of Crank	Graduation on Sector	Gear on Worm	No. 1 Hole		Gear on Spindle	Idlers	
					First Gear on Stud	Second Gear on Stud		No. 1 Hole	No. 2 Hole
166	20	5/20	48	32	…	…	48	24	44
167	20	5/20	48	32	…	…	56	24	44
168	21	5/21	47	…	…	…	…	…	…
169	20	5/20	48	32	…	…	72	24	44
170	17	4/17	45	…	…	…	…	…	…
171	21	5/21	47	56	…	…	40	24	44
172	43	10/43	44	…	…	…	…	…	…
173	18	4/18	43	72	56	32	64	…	…
174	18	4/18	43	24	…	…	32	56	…
175	18	4/18	43	72	40	32	64	…	…
176	18	4/18	43	72	24	24	64	…	…
177	18	4/18	43	72	…	…	48	24	…
178	18	4/18	43	72	…	…	32	44	…
179	18	4/18	43	72	24	48	32	…	…
180	18	4/18	43	…	…	…	…	…	…
181	18	4/18	43	72	24	48	32	…	24
182	18	4/18	43	72	…	…	32	24	44
183	18	4/18	43	48	…	…	32	24	44
184	23	5/23	42	…	…	…	…	…	…
185	37	8/37	42	…	…	…	…	…	…
186	18	4/18	43	48	…	…	64	24	44
187	18	4/18	43	72	48	24	56	…	24
188	47	10/47	40	…	…	…	…	…	…
189	18	4/18	43	32	…	…	64	24	44
190	19	4/19	40	…	…	…	…	…	…
191	20	4/20	38	40	…	…	72	24	…
192	20	4/20	38	40	…	…	64	44	…
193	20	4/20	38	40	…	…	56	44	…
194	20	4/20	38	40	…	…	48	44	…
195	39	8/39	39	…	…	…	…	…	…
196	49	10/49	38	…	…	…	…	…	…
197	20	4/20	38	40	…	…	24	56	…
198	20	4/20	38	56	28	40	32	…	…
199	20	4/20	38	100	40	64	32	…	…
200	20	4/20	38	…	…	…	…	…	…
201	20	4/20	38	72	24	40	24	…	24
202	20	4/20	38	72	24	40	48	…	24
203	20	4/20	38	40	…	…	24	24	44
204	20	4/20	38	40	…	…	32	24	44
204	20	4/20	38	40	…	…	32	24	44
205	41	8/41	37	…	…	…	…	…	…
206	20	4/20	38	40	…	…	48	24	44
207	20	4/20	38	40	…	…	56	24	44
208	20	4/20	38	40	…	…	64	24	44
209	20	4/20	38	40	…	…	72	24	44
210	21	4/21	37	…	…	…	…	…	…
211	16	3/16	36	64	…	…	28	44	…
212	43	8/43	35	86	24	24	48	…	…
213	27	5/27	36	72	…	…	40	44	…
214	20	4/20	38	40	56	32	64	…	24
215	43	8/43	35	…	…	…	…	…	…

Simple and Differential Indexing - Browne & Sharpe Milling Machines *(Continued)*

No. of Divisions	Index Circle	No. of Turns of Crank	Graduation on Sector	Gear on Worm	No. 1 Hole		Gear on Spindle	Idlers	
					First Gear on Stud	Second Gear on Stud		No. 1 Hole	No. 2 Hole
216	27	5/27	36	…	…	…	…	…	…
217	21	4/21	37	48	…	…	64	24	44
218	16	3/16	36	64	…	…	56	24	44
219	21	4/21	37	28	…	…	48	24	44
220	33	6/33	35	…	…	…	…	…	…
221	17	3/17	33	24	…	…	24	56	…
222	18	3/18	32	24	…	…	72	44	…
223	43	8/43	35	86	8	24	64	…	24
224	18	3/18	32	24	…	…	64	44	…
225	27	5/27	36	24	…	…	40	24	44
226	18	3/18	32	24	…	…	56	44	…
227	49	8/49	30	56	64	28	72	…	…
228	18	3/18	32	24	…	…	48	44	…
229	18	3/18	32	24	…	…	44	48	…
230	23	4/23	34	…	…	…	…	…	…
231	18	3/18	32	32	…	…	48	44	…
232	29	5/29	33	…	…	…	…	…	…
233	18	3/18	32	48	…	…	56	44	…
234	18	3/18	32	24	…	…	24	56	…
235	47	8/47	32	…	…	…	…	…	…
236	18	3/18	32	48	…	…	32	44	…
237	18	3/18	32	48	…	…	24	44	…
238	18	3/18	32	72	…	…	24	44	…
239	18	3/18	32	72	24	64	32	…	…
240	18	3/18	32	…	…	…	…	…	…
241	18	3/18	32	72	24	64	32	…	24
242	18	3/18	32	72	…	…	24	24	44
243	18	3/18	32	64	…	…	32	24	44
244	18	3/18	32	48	…	…	32	24	44
245	49	8/49	30	…	…	…	…	…	…
246	18	3/18	32	24	…	…	24	24	44
247	18	3/18	32	48	…	…	56	24	44
248	31	5/31	31	…	…	…	…	…	…
249	18	3/18	32	32	…	…	48	24	44
250	18	3/18	32	24	…	…	40	24	44
251	18	3/18	32	48	44	32	64	…	24
252	18	3/18	32	24	…	…	48	24	44
253	33	5/33	29	24	…	…	40	56	…
254	18	3/18	32	24	…	…	56	24	44
255	18	3/18	32	48	40	24	72	…	24
256	18	3/18	32	24	…	…	64	24	44
257	49	8/49	30	56	48	28	64	…	24
258	43	7/43	31	32	…	…	64	24	44
259	21	3/21	28	24	…	…	72	44	…
260	39	6/39	29	…	…	…	…	…	…
261	29	4/29	26	48	64	24	72	…	…
262	20	3/20	28	40	…	…	28	44	…
263	49	8/49	30	56	64	28	72	…	24
264	33	5/33	29	…	…	…	…	…	…
265	21	3/21	28	56	40	24	72	…	…
266	21	3/21	28	32	…	…	64	44	…

Simple and Differential Indexing - Browne & Sharpe Milling Machines *(Continued)*

No. of Divisions	Index Circle	No. of Turns of Crank	Graduation on Sector	Gear on Worm	No. 1 Hole		Gear on Spindle	Idlers	
					First Gear on Stud	Second Gear on Stud		No. 1 Hole	No. 2 Hole
267	27	$\frac{4}{27}$	28	72	…	…	32	44	…
268	21	$\frac{3}{21}$	28	28	…	…	48	44	…
269	20	$\frac{3}{20}$	28	64	32	40	28	…	24
270	27	$\frac{4}{27}$	28	…	…	…	…	…	…
271	21	$\frac{3}{21}$	28	56	24	24	72	…	…
272	21	$\frac{3}{21}$	28	56	…	…	64	24	…
273	21	$\frac{3}{21}$	28	24	…	…	24	56	…
274	21	$\frac{3}{21}$	28	56	…	…	48	44	…
275	21	$\frac{3}{21}$	28	56	…	…	40	44	…
276	21	$\frac{3}{21}$	28	56	…	…	32	44	…
277	21	$\frac{3}{21}$	28	56	…	…	24	44	…
278	21	$\frac{3}{21}$	28	56	32	48	24	…	…
279	27	$\frac{4}{27}$	28	24	…	…	32	24	44
280	49	$\frac{7}{49}$	26	…	…	…	…	…	…
281	21	$\frac{3}{21}$	28	72	24	56	24	…	24
282	43	$\frac{6}{43}$	26	86	24	24	56	…	…
283	21	$\frac{3}{21}$	28	56	…	…	24	24	44
284	21	$\frac{3}{21}$	28	56	…	…	32	24	44
285	21	$\frac{3}{21}$	28	56	…	…	40	24	44
286	21	$\frac{3}{21}$	28	56	…	…	48	24	44
287	21	$\frac{3}{21}$	28	24	…	…	24	24	44
288	21	$\frac{3}{21}$	28	28	…	…	32	24	44
289	21	$\frac{3}{21}$	28	56	24	24	72	…	24
290	29	$\frac{4}{29}$	26	…	…	…	…	…	…
291	15	$\frac{2}{15}$	25	40	…	…	48	44	…
292	21	$\frac{3}{21}$	28	28	…	…	48	24	44
293	15	$\frac{2}{15}$	25	48	32	40	56	…	…
294	21	$\frac{3}{21}$	28	24	…	…	48	24	44
295	15	$\frac{2}{15}$	25	48	…	…	32	44	…
296	37	$\frac{5}{37}$	26	…	…	…	…	…	…
297	33	$\frac{4}{33}$	23	28	48	24	56	…	…
298	21	$\frac{3}{21}$	28	28	…	…	72	24	44
299	23	$\frac{3}{23}$	25	24	…	…	24	56	…
300	15	$\frac{2}{15}$	25	…	…	…	…	…	…
301	43	$\frac{6}{43}$	26	24	…	…	48	24	44
302	16	$\frac{2}{16}$	24	32	…	…	72	24	…
303	15	$\frac{2}{15}$	25	72	24	40	48	…	24
304	16	$\frac{2}{16}$	24	24	…	…	48	44	…
305	15	$\frac{2}{15}$	25	48	…	…	32	24	44
306	15	$\frac{2}{15}$	25	40	…	…	32	24	44
307	15	$\frac{2}{15}$	25	72	48	40	56	…	24
308	16	$\frac{2}{16}$	24	32	…	…	48	44	…
309	15	$\frac{2}{15}$	25	40	…	…	48	24	44
310	31	$\frac{4}{31}$	24	…	…	…	…	…	…
311	16	$\frac{2}{16}$	24	64	24	24	72	…	…
312	39	$\frac{5}{39}$	24	…	…	…	…	…	…
313	16	$\frac{2}{16}$	24	32	…	…	28	56	…
314	16	$\frac{2}{16}$	24	32	…	…	24	56	…
315	16	$\frac{2}{16}$	24	64	…	…	40	24	…
316	16	$\frac{2}{16}$	24	64	…	…	32	44	…
317	16	$\frac{2}{16}$	24	64	…	…	24	44	…

Simple and Differential Indexing - Browne & Sharpe Milling Machines *(Continued)*

No. of Divisions	Index Circle	No. of Turns of Crank	Graduation on Sector	Gear on Worm	No. 1 Hole		Gear on Spindle	Idlers	
					First Gear on Stud	Second Gear on Stud		No. 1 Hole	No. 2 Hole
318	16	2/16	24	56	28	48	24	…	…
319	29	4/29	26	48	64	24	72	…	24
320	16	2/16	24	…	…	…	…	…	…
321	16	2/16	24	72	24	64	24	…	24
322	23	3/23	25	32	…	…	64	24	44
323	16	2/16	24	64	…	…	24	24	44
324	16	2/16	24	64	…	…	32	24	44
325	16	2/16	24	64	…	…	40	24	44
326	16	2/16	24	32	…	…	24	24	44
327	16	2/16	24	32	…	…	28	24	44
328	41	5/41	23	…	…	…	…	…	…
329	16	2/16	24	64	24	24	72	…	24
330	33	4/33	23	…	…	…	…	…	…
331	16	2/16	24	64	44	24	48	…	24
332	16	2/16	24	32	…	…	48	24	44
333	18	2/18	21	24	…	…	72	44	…
334	16	2/16	24	32	…	…	56	24	44
335	33	4/33	23	72	48	44	40	…	24
336	16	2/16	24	32	…	…	64	24	44
337	43	5/43	21	86	40	32	56	…	…
338	16	2/16	24	32	…	…	72	24	44
339	18	2/18	21	24	…	…	56	44	…
340	17	2/17	22	…	…	…	…	…	…
341	43	5/43	21	86	24	32	40	…	…
342	18	2/18	21	32	…	…	64	44	…
343	15	2/15	25	40	64	24	86	…	24
344	43	5/43	21	…	…	…	…	…	…
345	18	2/18	21	24	…	…	40	56	…
346	18	2/18	21	72	56	32	64	…	…
347	43	5/43	21	86	24	32	40	…	24
348	18	2/18	21	24	…	…	32	56	…
349	18	2/18	21	72	44	24	48	…	…
350	18	2/18	21	72	40	32	64	…	…
351	18	2/18	21	24	…	…	24	56	…
352	18	2/18	21	72	24	24	64	…	…
353	18	2/18	21	72	24	24	56	…	…
354	18	2/18	21	72	…	…	48	24	…
355	18	2/18	21	72	…	…	40	24	…
356	18	2/18	21	72	…	…	32	24	…
357	18	2/18	21	72	…	…	24	44	…
358	18	2/18	21	72	32	48	24	…	…
359	43	5/43	21	86	48	32	100	…	24
360	18	2/18	21	…	…	…	…	…	…
361	19	2/19	19	32	…	…	64	44	…
362	18	2/18	21	72	28	56	32	…	24
363	18	2/18	21	72	…	…	24	24	44
364	18	2/18	21	72	…	…	32	24	44

Notes: On B & S numbers 1, 1½, and 2 machines, number 2 hole is in the machine table. On numbers 3 and 4 machines, number 2 hole is in the head.

Indexing Movements for Standard Index Plate — Cincinnati Milling Machine

The standard index plate indexes all numbers up to and including 60; all even numbers and those divisible by 5 up to 120; and all divisions listed below up to 400. This plate is drilled on both sides, and has holes as follows:

First side: 24, 25, 28, 30, 34, 37, 38, 39, 41, 42, 43.

Second side: 46, 47, 49, 51, 53, 54, 57, 58, 59, 62, 66.

No. of Divisions	Index Plate Circle	No. of Turns	No. of Holes	No. of Divisions	Index Plate Circle	No. of Holes	No. of Divisions	Index Plate Circle	No. of Holes	No. of Divisions	Index Plate Circle	No. of Holes
2	Any	20	…	44	66	60	104	39	15	205	41	8
3	24	13	8	45	54	48	105	42	16	210	42	8
4	Any	10	…	46	46	40	106	53	20	212	53	10
5	Any	8	…	47	47	40	108	54	20	215	43	8
6	24	6	16	48	24	20	110	66	24	216	54	10
7	28	5	20	49	49	40	112	28	10	220	66	12
8	Any	5	…	50	25	20	114	57	20	224	28	5
9	54	4	24	51	51	40	115	46	16	228	57	10
10	Any	4	…	52	39	30	116	58	20	230	46	8
11	66	3	42	53	53	40	118	59	20	232	58	10
12	24	3	8	54	54	40	120	66	22	235	47	8
13	39	3	3	55	66	48	124	62	20	236	59	10
14	49	2	42	56	28	20	125	25	8	240	66	11
15	24	2	16	57	57	40	130	39	12	245	49	8
16	24	2	12	58	58	40	132	66	20	248	62	10
17	34	2	12	59	59	40	135	54	16	250	25	4
18	54	2	12	60	42	28	136	34	10	255	51	8
19	38	2	4	62	62	40	140	28	8	260	39	6
20	Any	2	…	64	24	15	144	54	15	264	66	10
21	42	1	38	65	39	24	145	58	16	270	54	8
22	66	1	54	66	66	40	148	37	10	272	34	5
23	46	1	34	68	34	20	150	30	8	280	28	4
24	24	1	16	70	28	16	152	38	10	290	58	8
25	25	1	15	72	54	30	155	62	16	296	37	5
26	39	1	21	74	37	20	156	39	10	300	30	4
27	54	1	26	75	30	16	160	28	7	304	38	5
28	42	1	18	76	38	20	164	41	10	310	62	8
29	58	1	22	78	39	20	165	66	16	312	39	5
30	24	1	8	80	34	17	168	42	10	320	24	3
31	62	1	18	82	41	20	170	34	8	328	41	5
32	28	1	7	84	42	20	172	43	10	330	66	8
33	66	1	14	85	34	16	176	66	15	336	42	5
34	34	1	6	86	43	20	180	54	12	340	34	4
35	28	1	4	88	66	30	184	46	10	344	43	5
36	54	1	6	90	54	24	185	37	8	360	54	6
37	37	1	3	92	46	20	188	47	10	368	46	5
38	38	1	2	94	47	20	190	38	8	370	37	4
39	39	1	1	95	38	16	192	24	5	376	47	5
40	Any	1	…	96	24	10	195	39	8	380	38	4
41	41	…	40	98	49	20	196	49	10	390	39	4
42	42	…	40	100	25	10	200	30	6	392	49	5
43	43	…	40	102	51	20	204	51	10	400	30	3

Indexing Movements for High Numbers — Cincinnati Milling Machine

This set of 3 index plates indexes all numbers up to and including 200; all even numbers and those divisible by 5 up to and including 400. The plates are drilled on each side, making six sides *A*, *B*, *C*, *D*, *E* and *F*.

Example:—It is required to index 35 divisions. The preferred side is *F*, since this requires the least number of holes; but should one of plates *D*, *A* or *E* be in place, either can be used, thus avoiding the changing of plates.

No. of Divisions	Side	Circle	Turns	Holes	No. of Divisions	Side	Circle	Turns	Holes	No. of Divisions	Side	Circle	Turns	Holes
2	Any	Any	20	....	15	*C*	93	2	62	28	*D*	77	1	33
3	*A*	30	13	10	15	*F*	159	2	106	28	*A*	91	1	39
3	*B*	36	13	12	16	*E*	26	2	13	29	*E*	87	1	33
3	*E*	42	13	14	16	*F*	28	2	14	30	*A*	30	1	10
3	*C*	93	13	31	16	*A*	30	2	15	30	*B*	36	1	12
3	*F*	159	13	53	16	*D*	32	2	16	30	*E*	42	1	14
4	Any	Any	10	....	16	*C*	34	2	17	30	*C*	93	1	31
5	Any	Any	8	....	16	*B*	36	2	18	30	*F*	159	1	53
6	*A*	30	6	20	17	*C*	34	2	12	31	*C*	93	1	27
6	*B*	36	6	24	17	*E*	119	2	42	32	*F*	28	1	7
6	*E*	42	6	28	17	*C*	153	2	54	32	*D*	32	1	8
6	*C*	93	6	62	17	*F*	187	2	66	32	*B*	36	1	9
6	*F*	159	6	106	18	*B*	36	2	8	32	*A*	48	1	12
7	*F*	28	5	20	18	*A*	99	2	22	33	*A*	99	1	21
7	*E*	42	5	30	18	*C*	153	2	34	34	*C*	34	1	6
7	*D*	77	5	55	19	*F*	38	2	4	34	*E*	119	1	21
7	*A*	91	5	65	19	*E*	133	2	14	34	*F*	187	1	33
8	Any	Any	5	....	19	*A*	171	2	18	35	*F*	28	1	4
9	*B*	36	4	16	20	Any	Any	2	....	35	*D*	77	1	11
9	*A*	99	4	44	21	*E*	42	1	38	35	*A*	91	1	13
9	*C*	153	4	68	21	*A*	147	1	133	35	*E*	119	1	17
10	Any	Any	4	....	22	*D*	44	1	36	36	*B*	36	1	4
11	*D*	44	3	28	22	*A*	99	1	81	36	*A*	99	1	11
11	*A*	99	3	63	22	*F*	143	1	117	36	*C*	153	1	17
11	*F*	143	3	91	23	*C*	46	1	34	37	*B*	111	1	9
12	*A*	30	3	10	23	*A*	69	1	51	38	*F*	38	1	2
12	*B*	36	3	12	23	*E*	161	1	119	38	*E*	133	1	7
12	*E*	42	3	14	24	*A*	30	1	20	38	*A*	171	1	9
12	*C*	93	3	31	24	*B*	36	1	24	39	*A*	117	1	3
12	*F*	159	3	53	24	*E*	42	1	28	40	Any	Any	1	....
13	*E*	26	3	2	24	*C*	93	1	62	41	*C*	123	....	120
13	*A*	91	3	7	24	*F*	159	1	106	42	*E*	42	....	40
13	*F*	143	3	11	25	*A*	30	1	18	42	*A*	147	....	140
13	*B*	169	3	13	25	*E*	175	1	105	43	*A*	129	....	120
14	*F*	28	2	24	26	*F*	26	1	14	44	*D*	44	....	40
14	*E*	42	2	36	26	*A*	91	1	49	44	*A*	99	....	90
14	*D*	77	2	66	26	*B*	169	1	91	44	*F*	143	....	130
14	*A*	91	2	78	27	*B*	81	1	39	45	*B*	36	....	32
15	*A*	30	2	20	27	*A*	189	1	91	45	*A*	99	....	88
15	*B*	36	2	24	28	*F*	28	1	12	45	*C*	153	....	136
15	*E*	42	2	28	28	*E*	42	1	18	46	*C*	46	....	40

Indexing Movements for High Numbers - Cincinnati Milling Machine

No. of Division	Side	Circle	Holes	No. of Divisions	Side	Circle	Holes	No. of Division	Side	Circle	Holes
46	*A*	69	60	70	*E*	119	68	96	*B*	36	15
46	*E*	161	140	71	*F*	71	40	96	*A*	48	20
47	*B*	141	120	72	*B*	36	20	97	*B*	97	40
48	*A*	30	25	72	*A*	117	65	98	*A*	147	60
48	*B*	36	30	72	*C*	153	85	99	*A*	99	40
49	*A*	147	120	73	*E*	73	40	100	*A*	30	12
50	*A*	30	24	74	*B*	111	60	100	*E*	175	70
50	*E*	175	140	75	*A*	30	16	101	*F*	101	40
51	*C*	153	120	76	*F*	38	20	102	*C*	153	60
52	*E*	26	20	76	*E*	133	70	103	*E*	103	40
52	*A*	91	70	76	*A*	171	90	104	*E*	26	10
52	*F*	143	110	77	*D*	77	40	104	*A*	91	35
52	*B*	169	130	78	*A*	117	60	104	*F*	143	55
53	*F*	159	120	79	*C*	79	40	104	*B*	169	65
54	*B*	81	60	80	*E*	26	13	105	*E*	42	16
54	*A*	189	140	80	*F*	28	14	105	*A*	147	56
55	*D*	44	32	80	*A*	30	15	106	*F*	159	60
55	*F*	143	104	80	*D*	32	16	107	*D*	107	40
56	*F*	28	20	80	*C*	34	17	108	*B*	81	30
56	*E*	42	30	80	*B*	36	18	108	*A*	189	70
56	*D*	77	55	80	*E*	42	21	109	*C*	109	40
56	*A*	91	65	81	*B*	81	40	110	*D*	44	16
57	*A*	171	120	82	*C*	123	60	110	*A*	99	36
58	*E*	87	60	83	*F*	83	40	110	*F*	143	52
59	*A*	177	120	84	*E*	42	20	111	*B*	111	40
60	*A*	30	20	84	*A*	147	70	112	*F*	28	10
60	*B*	36	24	85	*C*	34	16	112	*E*	42	15
60	*E*	42	28	85	*E*	119	56	113	*F*	113	40
60	*F*	159	106	85	*F*	187	88	114	*A*	171	60
61	*B*	183	120	86	*A*	129	60	115	*C*	46	16
62	*C*	93	60	87	*E*	87	40	115	*A*	69	24
63	*A*	189	120	88	*D*	44	20	115	*E*	161	56
64	*D*	32	20	88	*A*	99	45	116	*E*	87	30
64	*A*	48	30	88	*F*	143	65	117	*A*	117	40
65	*E*	26	16	89	*D*	89	40	118	*A*	177	60
65	*A*	91	56	90	*B*	36	16	119	*E*	119	40
65	*F*	143	88	90	*A*	99	44	120	*A*	30	10
65	*B*	169	104	90	*C*	153	68	120	*B*	36	12
66	*A*	99	60	91	*A*	91	40	120	*E*	42	14
67	*B*	67	40	92	*C*	46	20	120	*C*	93	31
68	*C*	34	20	92	*A*	69	30	120	*F*	159	53
68	*E*	119	70	92	*E*	161	70	121	*D*	121	40
68	*F*	187	110	93	*C*	93	40	122	*B*	183	60
69	*A*	69	40	94	*B*	141	60	123	*C*	123	40
70	*F*	28	16	95	*F*	38	16	124	*C*	93	30
70	*D*	42	24	95	*E*	133	56	125	*E*	175	56
70	*A*	91	52	95	*A*	171	72	126	*A*	189	60
127	*B*	127	40	160	*A*	48	12	198	*A*	99	20
128	*D*	32	10	161	*E*	161	40	199	*B*	199	40
128	*A*	48	15	162	*B*	81	20	200	*A*	30	6
129	*A*	129	40	163	*D*	163	40	200	*E*	175	35
130	*E*	26	8	164	*C*	123	30	202	*F*	101	20
130	*A*	91	28	165	*A*	99	24	204	*C*	153	30
130	*F*	143	44	166	*F*	83	20	205	*C*	123	24
130	*B*	169	52	167	*C*	167	40	206	*E*	103	20
131	*F*	131	40	168	*E*	42	10	208	*E*	26	5
132	*A*	99	30	168	*A*	147	35	210	*E*	42	8
133	*E*	133	40	169	*B*	169	40	210	*A*	147	28
134	*B*	67	20	170	*C*	34	8	212	*F*	159	30
135	*B*	81	24	170	*E*	119	28	214	*D*	107	20
135	*A*	189	56	170	*F*	187	44	215	*A*	129	24
136	*C*	34	10	171	*A*	171	40	216	*B*	81	15
136	*E*	119	35	172	*A*	129	30	216	*A*	189	35

Indexing Movements for High Numbers - Cincinnati Milling Machine *(Continued)*

No. of Division	Side	Circle	Holes	No. of Divisions	Side	Circle	Holes	No. of Division	Side	Circle	Holes
137	*D*	137	40	173	*F*	173	40	218	*C*	109	20
138	*A*	69	20	174	*E*	87	20	220	*D*	44	8
139	*C*	139	40	175	*E*	175	40	220	*A*	99	18
140	*F*	28	8	176	*D*	44	10	220	*F*	143	26
140	*E*	42	12	177	*A*	177	40	222	*B*	111	20
140	*D*	77	22	178	*D*	89	20	224	*F*	28	5
140	*A*	91	26	179	*D*	179	40	226	*F*	113	20
141	*B*	141	40	180	*B*	36	8	228	*A*	171	30
142	*F*	71	20	180	*A*	99	22	230	*C*	46	8
143	*F*	143	40	180	*C*	153	34	230	*A*	69	12
144	*B*	36	10	181	*C*	181	40	230	*E*	161	28
145	*E*	87	24	182	*A*	91	20	232	*E*	87	15
146	*E*	73	20	183	*B*	183	40	234	*A*	117	20
147	*A*	147	40	184	*C*	46	10	235	*B*	141	24
148	*B*	111	30	184	*A*	69	15	236	*A*	177	30
149	*E*	149	40	184	*E*	161	35	238	*E*	119	20
150	*A*	30	8	185	*B*	111	24	240	*A*	30	5
151	*D*	151	40	186	*C*	93	20	240	*B*	36	6
152	*F*	38	10	187	*F*	187	40	240	*E*	42	7
152	*E*	133	35	188	*B*	141	30	240	*A*	48	8
152	*A*	171	45	189	*A*	189	40	242	*D*	121	20
153	*C*	153	40	190	*F*	38	8	244	*B*	183	30
154	*D*	77	20	190	*E*	133	28	245	*A*	147	24
155	*C*	93	24	190	*A*	171	36	246	*C*	123	20
156	*A*	117	30	191	*E*	191	40	248	*C*	93	15
157	*B*	157	40	192	*A*	48	10	250	*E*	175	28
158	*C*	79	20	193	*D*	193	40	252	*A*	189	30
159	*F*	159	40	194	*B*	97	20	254	*B*	127	20
160	*F*	28	7	195	*A*	117	24	255	*C*	153	24
160	*D*	32	8	196	*A*	147	30	256	*D*	32	5
160	*B*	36	9	197	*C*	197	40	258	*A*	129	20
260	*E*	26	4	304	*F*	38	5	354	*A*	177	20
260	*A*	91	14	305	*B*	183	24	355	*F*	71	8
260	*F*	143	22	306	*C*	153	20	356	*D*	89	10
260	*B*	169	26	308	*D*	77	10	358	*D*	179	20
262	*F*	131	20	310	*C*	93	12	360	*B*	36	4
264	*A*	99	15	312	*A*	117	15	360	*A*	99	11
265	*F*	159	24	314	*B*	157	20	360	*C*	153	17
266	*E*	133	20	315	*A*	189	24	362	*C*	181	20
268	*B*	67	10	316	*C*	79	10	364	*A*	91	10
270	*B*	81	12	318	*F*	159	20	365	*E*	73	8
270	*A*	189	28	320	*D*	32	4	366	*B*	183	20
272	*C*	34	5	320	*A*	48	6	368	*C*	46	5
274	*D*	137	20	322	*E*	161	20	370	*B*	111	12
276	*A*	69	10	324	*B*	81	10	372	*C*	93	10
278	*C*	139	20	326	*D*	163	20	374	*F*	187	20
280	*F*	28	4	328	*C*	123	15	376	*B*	141	15
280	*E*	42	6	330	*A*	99	12	378	*A*	189	20
280	*D*	77	11	332	*F*	83	10	380	*F*	38	4
280	*A*	91	13	334	*C*	167	20	380	*E*	133	14
282	*B*	141	20	335	*B*	67	8	380	*A*	171	18
284	*F*	71	10	336	*E*	42	5	382	*E*	191	20
285	*A*	171	24	338	*B*	169	20	384	*A*	48	5
286	*F*	143	20	340	*C*	34	4	385	*D*	77	8
288	*B*	36	5	340	*E*	119	14	386	*D*	193	20
290	*E*	87	12	340	*F*	187	22	388	*B*	97	10
292	*E*	73	10	342	*A*	171	20	390	*A*	117	12
294	*A*	147	20	344	*A*	129	15	392	*A*	147	15
295	*A*	177	24	345	*A*	69	8	394	*C*	197	20
296	*B*	111	15	346	*F*	173	20	395	*C*	79	8
298	*E*	149	20	348	*E*	87	10	396	*A*	99	10
300	*A*	30	4	350	*E*	175	20	398	*B*	199	20
302	*D*	151	20	352	*D*	44	5	400	*A*	30	3

Indexing Tables.—Indexing tables are usually circular, with a flat, T-slotted table, 12 to 24 in. in diameter, to which workpieces can be clamped. The flat table surface may be horizontal, universal, or angularly adjustable. The table can be turned continuously through 360° about an axis normal to the surface. Rotation is through a worm drive with a graduated scale, and a means of angular readout is provided. Indexed locations to 0.25° with accuracy of ±0.1 second can be obtained from mechanical means, or greater accuracy from an autocollimator or sine-angle attachment built into the base, or under numerical control. Provision is made for locking the table at any angular position while a machining operation is being performed.

Power for rotation of the table during machining can be transmitted, as with a dividing head, for cutting a continuous, spiral scroll, for instance. The indexing table is usually more rigid and can be used with larger workpieces than the dividing head.

Block or Multiple Indexing for Gear Cutting.—With the block system of indexing, numbers of teeth are indexed at one time, instead of cutting the teeth consecutively, and the gear is revolved several times before all the teeth are finished. For example, when cutting a gear having 25 teeth, the indexing mechanism is geared to index four teeth at once (see table) and the first time around, six widely separated tooth spaces are cut. The second time around, the cutter is one tooth behind the spaces originally milled. On the third indexing, the cutter has dropped back another tooth, and the gear in question is thus finished by indexing it through four cycles.

The various combinations of change gears to use for block or multiple indexing are given in the accompanying table. The advantage claimed for block indexing is that the heat generated by the cutter (especially when cutting cast iron gears of coarse pitch) is distributed more evenly about the rim and is dissipated to a greater extent, thus avoiding distortion due to local heating and permitting higher speeds and feeds to be used.

The table gives values for use with Brown & Sharpe automatic gear cutting machines, but the gears for any other machine equipped with a similar indexing mechanism can be calculated easily. Assume, for example, that a gear cutter requires the following change gears for indexing a certain number of teeth: driving gears having 20 and 30 teeth, respectively, and driven gears having 50 and 60 teeth.

Then if it is desired to cut, for instance, every fifth tooth, multiply the fractions 20/60 and 30/50 by 5. Then $20/60 \times 30/50 \times 5/1 = 1/1$. In this instance, the blank could be divided so that every fifth space was cut, by using gears of equal size. The number of teeth in the gear and the number of teeth indexed in each block must not have a common factor.

Block or Multiple Indexing for Gear Cutting

Number of Teeth to be Cut	Number Indexed at Once	1st Driver	1st Follower	2nd Driver	2nd Follower	Turns of Locking Disk	Number of Teeth to be Cut	Number Indexed at Once	1st Driver	1st Follower	2nd Driver	2nd Follower	Turns of Locking Disk
25	4	100	50	72	30	4	46	5	100	30	90	92	4
26	3	100	50	90	52	4	47	5	100	30	90	94	4
27	2	100	50	60	54	4	48	5	100	30	90	96	4
28	3	100	50	90	56	4	49	5	100	30	90	98	4
29	3	100	50	90	58	4	50	7	100	50	84	40	4
30	7	100	30	84	40	4	51	4	100	30	96	68	2
31	3	100	50	90	62	4	52	5	100	30	90	52	2
32	3	100	50	90	64	4	54	5	100	30	90	54	2
33	4	100	50	80	44	4	55	4	100	50	96	44	2
34	3	100	50	90	68	4	56	5	100	30	90	56	2
35	4	100	50	96	56	4	57	4	100	30	96	76	2
36	5	100	48	80	40	4	58	5	100	30	90	58	2
37	5	100	30	90	74	4	60	7	100	30	84	40	2
38	5	100	30	90	76	4	62	5	100	30	90	62	2
39	5	100	30	90	78	4	63	5	100	30	80	56	2
40	3	100	50	90	80	4	64	5	100	30	90	64	2

Block or Multiple Indexing for Gear Cutting *(Continued)*

Number of Teeth to be Cut	Number Indexed at Once	1st Driver	1st Follower	2nd Driver	2nd Follower	Turns of Locking Disk	Number of Teeth to be Cut	Number Indexed at Once	1st Driver	1st Follower	2nd Driver	2nd Follower	Turns of Locking Disk
41	5	100	30	90	82	4	65	4	100	50	96	52	2
42	5	100	30	90	84	4	66	5	100	44	80	40	2
43	5	100	30	90	86	4	67	5	100	30	90	67	2
44	5	100	30	90	88	4	68	5	100	30	90	68	2
45	7	100	50	70	30	4	69	5	100	46	80	40	2
70	3	100	50	90	70	2	132	5	100	88	80	40	2
72	5	100	30	90	72	2	133	4	100	70	96	76	2
74	5	100	30	90	74	2	134	5	100	60	90	67	2
75	7	100	30	84	50	2	135	7	100	50	84	54	2
76	5	100	30	90	76	2	136	5	100	60	90	68	2
77	4	100	70	96	44	2	138	5	100	92	80	40	2
78	5	100	30	90	78	2	140	3	50	50	90	70	2
80	3	100	50	90	80	2	141	5	100	94	80	40	2
81	7	100	30	84	52	2	143	6	90	66	96	52	2
82	5	100	30	90	82	2	144	5	100	60	90	72	2
84	5	100	30	90	84	2	145	6	100	50	72	58	2
85	4	100	50	96	68	2	147	5	100	98	80	40	2
86	5	100	30	90	86	2	148	5	100	60	90	74	2
87	7	100	30	84	58	2	150	7	100	60	84	50	2
88	5	100	30	90	88	2	152	5	100	60	90	76	2
90	7	100	30	70	50	2	153	5	100	68	80	60	2
91	3	100	70	72	52	2	154	5	100	56	72	66	2
92	5	100	30	90	92	2	155	6	100	50	72	62	2
93	7	100	30	84	62	2	156	5	100	60	90	78	2
94	5	100	30	90	94	2	160	7	100	50	84	64	2
95	4	100	50	96	76	2	161	5	100	70	60	46	2
96	5	100	30	90	96	2	162	7	100	60	84	52	2
98	5	100	30	90	98	2	164	5	100	60	90	82	2
99	10	100	30	80	44	2	165	7	100	50	84	66	2
100	7	100	50	84	40	2	168	5	100	60	90	84	2
102	5	100	30	60	68	2	169	6	96	52	90	78	2
104	5	100	60	90	52	2	170	7	100	50	84	68	2
105	4	100	70	96	60	2	171	5	70	42	80	76	2
108	7	100	30	70	60	2	172	5	100	60	90	86	2
110	7	100	50	84	44	2	174	7	100	60	84	58	2
111	5	100	74	80	40	2	175	8	100	50	96	70	2
112	5	100	60	90	56	2	176	5	100	60	90	88	2
114	7	100	30	84	76	2	180	7	100	60	70	50	2
115	8	100	50	96	46	2	182	9	90	56	96	52	2
116	5	100	60	90	58	2	184	5	100	60	90	92	2
117	8	100	30	96	78	2	185	6	100	50	72	74	2
119	3	100	70	72	68	2	186	7	100	60	84	62	2
120	7	100	50	70	40	2	187	5	100	44	48	68	2
121	4	60	66	96	44	2	188	5	100	60	90	94	2
123	7	100	30	84	82	2	189	5	100	60	80	84	2
124	5	100	60	90	62	2	190	7	100	50	84	76	2
125	7	100	50	84	50	2	192	5	100	60	90	96	2
126	5	100	50	50	42	2	195	7	100	50	84	78	2
128	5	100	60	90	64	2	196	5	100	60	90	98	2
129	7	100	30	84	86	2	198	7	100	50	70	66	2
130	7	100	50	84	52	2	200	7	60	60	84	40	2

Indexing Movements for 60-Tooth Worm-Wheel Dividing Head

Divisions	Index Circle	No. of Turns	No. of Holes	Divisions	Index Circle	No. of Turns	No. of Holes	Divisions	Index Circle	No. of Holes	Divisions	Index Circle	No. of Holes
2	Any	30	..	50	60	1	12	98	49	30	146	73	30
3	Any	20	..	51	17	1	3	99	33	20	147	49	20
4	Any	15	..	52	26	1	4	100	60	36	148	37	15
5	Any	12	..	53	53	1	7	101	101	60	149	149	60
6	Any	10	..	54	27	1	3	102	17	10	150	60	24
7	21	8	12	55	33	1	3	103	103	60	151	151	60
8	26	7	13	56	28	1	2	104	26	15	152	76	30
9	21	6	14	57	19	1	1	105	21	12	153	51	20
10	Any	6	..	58	29	1	1	106	53	30	154	77	30
11	33	5	15	59	59	1	1	107	107	60	155	31	12
12	Any	5	..	60	Any	1	..	108	27	15	156	26	10
13	26	4	16	61	61	..	60	109	109	60	157	157	60
14	21	4	6	62	31	..	30	110	33	18	158	79	30
15	Any	4	..	63	21	..	20	111	37	20	159	53	20
16	28	3	21	64	32	..	30	112	28	15	160	32	12
17	17	3	9	65	26	..	24	113	113	60	161	161	60
18	21	3	7	66	33	..	30	114	19	10	162	27	10
19	19	3	3	67	67	..	60	115	23	12	163	163	60
20	Any	3	..	68	17	..	15	116	29	15	164	41	15
21	21	2	18	69	23	..	20	117	39	20	165	33	12
22	33	2	24	70	21	..	18	118	59	30	166	83	30
23	23	2	14	71	71	..	60	119	119	60	167	167	60
24	26	2	13	72	60	..	50	120	26	13	168	28	10
25	60	2	24	73	73	..	60	121	121	60	169	169	60
26	26	2	8	74	37	..	30	122	61	30	170	17	6
27	27	2	6	75	60	..	48	123	41	20	171	57	20
28	21	2	3	76	19	..	15	124	31	15	172	43	15
29	29	2	2	77	77	..	60	125	100	48	173	173	60
30	Any	2	..	78	26	..	20	126	21	10	174	29	10
31	31	1	29	79	79	..	60	127	127	60	175	35	12
32	32	1	28	80	28	..	21	128	32	15	176	44	15
33	33	1	27	81	27	..	20	129	43	20	177	59	20
34	17	1	13	82	41	..	30	130	26	12	178	89	30
35	21	1	15	83	83	..	60	131	131	60	179	179	60
36	21	1	14	84	21	..	15	132	33	15	180	21	7
37	37	1	23	85	17	..	12	133	133	60	181	181	60
38	19	1	11	86	43	..	30	134	67	30	182	91	30
39	26	1	14	87	29	..	20	135	27	12	183	61	20
40	26	1	13	88	44	..	30	136	68	30	184	46	15
41	41	1	19	89	89	..	60	137	137	60	185	37	12
42	21	1	9	90	21	..	14	138	23	10	186	31	10
43	43	1	17	91	91	..	60	139	139	60	187	187	60
44	33	1	12	92	23	..	15	140	21	9	188	47	15
45	21	1	7	93	31	..	20	141	47	20	189	63	20
46	23	1	7	94	47	..	30	142	71	30	190	19	6
47	47	1	13	95	19	..	12	143	143	60	191	191	60
48	28	1	7	96	32	..	20	144	60	25	192	32	10
49	49	1	11	97	97	..	60	145	29	12	193	193	60

Linear Indexing for Rack Cutting.—When racks are cut on a milling machine, two general methods of linear indexing are used. One is by using the graduated dial on the feed-screw and the other is by using an indexing attachment. The accompanying table shows the indexing movements when the first method is employed. This table applies to milling machines having feed-screws with the usual lead of ¼ inch and 250 dial graduations each equivalent to 0.001 inch of table movement.

$$\text{Actual rotation of feed-screw} = \frac{\text{Linear pitch of rack}}{\text{Lead of feed-screw}}$$

Multiply *decimal* part of turn (obtained by above formula) by 250, to obtain dial reading for fractional part of indexing movement, assuming that dial has 250 graduations.

Linear Indexing Movements for Cutting Rack Teeth on a Milling Machine These movements are for table feed-screws having the usual lead of ¼ inch

Pitch of Rack Teeth		Indexing, Movement		Pitch of Rack Teeth		Indexing, Movement	
Diametral Pitch	Linear or Circular	No. of Whole Turns	No. of 0.001 Inch Divisions	Diametral Pitch	Linear or Circular	No. of Whole Turns	No. of 0.001 Inch Divisions
2	1.5708	6	70.8	12	0.2618	1	11.8
2¼	1.3963	5	146.3	13	0.2417	0	241.7
2½	1.2566	5	6.6	14	0.2244	0	224.4
2¾	1.1424	4	142.4	15	0.2094	0	208.4
3	1.0472	4	47.2	16	0.1963	0	196.3
3½	0.8976	3	147.6	17	0.1848	0	184.8
4	0.7854	3	35.4	18	0.1745	0	174.8
5	0.6283	2	128.3	19	0.1653	0	165.3
6	0.5263	2	23.6	20	0.1571	0	157.1
7	0.4488	1	198.8	22	0.1428	0	142.8
8	0.3927	1	142.7	24	0.1309	0	130.9
9	0.3491	1	99.1	26	0.1208	0	120.8
10	0.3142	1	64.2	28	0.1122	0	112.2
11	0.2856	1	35.6	30	0.1047	0	104.7

Note: The linear pitch of the rack equals the circular pitch of gear or pinion which is to mesh with the rack. The table gives both standard diametral pitches and their equivalent linear or circular pitches.

Example: Find indexing movement for cutting rack to mesh with a pinion of 10 diametral pitch.

Indexing movement equals 1 whole turn of feed-screw plus 64.2 thousandths or divisions on feed-screw dial. The feed-screw may be turned this fractional amount by setting dial back to its zero position for each indexing (without backward movement of feed-screw), or, if preferred, 64.2 (in this example) may be added to each successive dial position as shown below.

Dial reading for second position = 64.2 × 2 = 128.4 (complete movement = 1 turn × 64.2 additional divisions by turning feed-screw until dial reading is 128.4).

Third dial position = 64.2 × 3 = 192.6 (complete movement = 1 turn + 64.2 additional divisions by turning until dial reading is 192.6).

Fourth position = 64.2 × 4 − 250 = 6.8 (1 turn + 64.2 additional divisions by turning feed-screw until dial reading is 6.8 divisions past the zero mark); or, to simplify operation, set dial back to zero for fourth indexing (without moving feed-screw) and then repeat settings for the three previous indexings or whatever number can be made before making a complete turn of the dial.

Contour Milling.—Changing the direction of a linear milling operation by a specific angle requires a linear offset before changing the angle of cut. This compensates for the radius of the milling cutters, as illustrated in Figs. 1a and 1b.

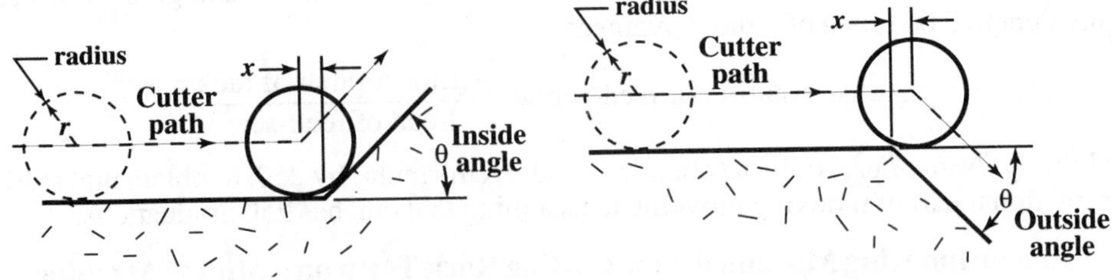

Fig. 1a. Inside Milling   Fig. 1b. Outside Milling

For inside cuts the offset is subtracted from the point at which the cutting direction changes (Fig. 1a), and for outside cuts the offset is added to the point at which the cutting direction changes (Fig. 1b). The formula for the offset is

$$x = rM$$

where x = offset distance; r = radius of the milling cutter; and, M = the multiplication factor ($M = \tan \theta/2$). The value of M for certain angles can be found in Table 1.

Table 1. Offset Multiplication Factors

Deg°	*M*	Deg°	*M*	Deg°	*M*	Deg°	*M*	Deg°	*M*
1°	0.00873	19°	0.16734	37°	0.33460	55°	0.52057	73°	0.73996
2°	0.01746	20°	0.17633	38°	0.34433	56°	0.53171	74°	0.75355
3°	0.02619	21°	0.18534	39°	0.35412	57°	0.54296	75°	0.76733
4°	0.03492	22°	0.19438	40°	0.36397	58°	0.55431	76°	0.78129
5°	0.04366	23°	0.20345	41°	0.37388	59°	0.56577	77°	0.79544
6°	0.05241	24°	0.21256	42°	0.38386	60°	0.57735	78°	0.80978
7°	0.06116	25°	0.22169	43°	0.39391	61°	0.58905	79°	0.82434
8°	0.06993	26°	0.23087	44°	0.40403	62°	0.60086	80°	0.83910
9°	0.07870	27°	0.24008	45°	0.41421	63°	0.61280	81°	0.85408
10°	0.08749	28°	0.24933	46°	0.42447	64°	0.62487	82°	0.86929
11°	0.09629	29°	0.25862	47°	0.43481	65°	0.63707	83°	0.88473
12°	0.10510	30°	0.26795	48°	0.44523	66°	0.64941	84°	0.90040
13°	0.11394	31°	0.27732	49°	0.45573	67°	0.66189	85°	0.91633
14°	0.12278	32°	0.28675	50°	0.46631	68°	0.67451	86°	0.93252
15°	0.13165	33°	0.29621	51°	0.47698	69°	0.68728	87°	0.94896
16°	0.14054	34°	0.30573	52°	0.48773	70°	0.70021	88°	0.96569
17°	0.14945	35°	0.31530	53°	0.49858	71°	0.71329	89°	0.98270
18°	0.15838	36°	0.32492	54°	0.50953	72°	0.72654	90°	1.00000

Multiply factor *M* by the tool radius *r* to determine the offset dimension *x*.

GEARS, SPLINES, AND CAMS

GEARS AND GEARING

GEARS AND GEARING (cont.)

INTERNAL GEARING

HYPOID AND BEVEL GEARING

WORM GEARING

HELICAL GEARING

PLANETARY GEARING

RATCHET GEARING

MODULE SYSTEM

CHECKING GEAR SIZES

GEAR MATERIALS

GEARS, SPLINES, AND CAMS

SPLINES AND SERRATIONS

SPLINES AND SERRATIONS (cont.)

CAMS AND CAM DESIGN

GEARS AND GEARING

External spur gears are cylindrical gears with straight teeth cut parallel to the axes. Gears transmit drive between parallel shafts. Tooth loads produce no axial thrust. Excellent at moderate speeds but tend to be noisy at high speeds. Shafts rotate in opposite directions.

Internal spur gears provide compact drive arrangements for transmitting motion between parallel shafts rotating in the same direction.

Helical gears are cylindrical gears with teeth cut at an angle to the axes. Provide rive between shafts rotating in opposite directions, with superior load carrying capacity and quietness than spur gears. Tooth loads produce axial thrust.

Crossed helical gears are helical gears that mesh together on non-parallel axes.

Straight bevel gears have teeth that are radial toward the apex and are of conical form. Designed to operate on intersecting axes, bevel gears are used to connect two shafts on intersecting axes. The angle between the shafts equals the angle between the two axes of the meshing teeth. End thrust developed under load tends to separate the gears.

Spiral bevel gears have curved oblique teeth that contact each other smoothly and gradually from one end of a tooth to the other. Meshing is similar to that of straight bevel gears but is smoother and quieter in use. Left hand spiral teeth incline away from the axis in an anti-clockwise direction looking on small end of pinion or face of gear, right-hand teeth incline away from axis in clockwise direction. The hand of spiral of the pinion is always opposite to that of the gear and is used to identify the hand of the gear pair. Used to connect two shafts on intersecting axes as with straight bevel gears. The spiral angle does not affect the smoothness and quietness of operation or the efficiency but does affect the direction of the thrust loads created. A left-hand spiral pinion driving clockwise when viewed from the large end of the pinion creates an axial thrust that tends to move the pinion out of mesh.

Zerol bevel gears have curved teeth lying in the same general direction as straight bevel teeth but should be considered to be spiral bevel gears with zero spiral angle.

Hypoid bevel gears are a cross between spiral bevel gears and worm gears. The axes of hypoid bevel gears are non-intersecting and non-parallel. The distance between the axes is called the offset. The offset permits higher ratios of reduction than is practicable with other bevel gears. Hypoid bevel gears have curved oblique teeth on which contact begins gradually and continues smoothly from one end of the tooth to the other.

Worm gears are used to transmit motion between shafts at right angles, that do not lie in a common plane and sometimes to connect shafts at other angles. Worm gears have line tooth contact and are used for power transmission, but the higher the ratio the lower the efficiency.

Definitions of Gear Terms.—The following terms are commonly applied to the various classes of gears:

Active face width is the dimension of the tooth face width that makes contact with a mating gear.

Addendum is the radial or perpendicular distance between the pitch circle and the top of the tooth.

Arc of action is the arc of the pitch circle through which a tooth travels from the first point of contact with the mating tooth to the point where contact ceases.

Arc of approach is the arc of the pitch circle through which a tooth travels from the first point of contact with the mating tooth to the pitch point.

Arc of recession is the arc of the pitch circle through which a tooth travels from its contact with a mating tooth at the pitch point until contact ceases.

Axial pitch is the distance parallel to the axis between corresponding sides of adjacent teeth.

Axial plane is the plane that contains the two axes in a pair of gears. In a single gear the axial plane is any plane containing the axis and any given point.

Axial thickness is the distance parallel to the axis between two pitch line elements of the same tooth.

Backlash is the shortest distance between the non-driving surfaces of adjacent teeth when the working flanks are in contact.

Base circle is the circle from which the involute tooth curve is generated or developed.

Base helix angle is the angle at the base cylinder of an involute gear that the tooth makes with the gear axis.

Base pitch is the circular pitch taken on the circumference of the base circles, or the distance along the line of action between two successive and corresponding involute tooth profiles. The *normal base pitch* is the base pitch in the normal plane and the *axial base pitch* is the base pitch in the axial plane.

Base tooth thickness is the distance on the base circle in the plane of rotation between involutes of the same pitch.

Bottom land is the surface of the gear between the flanks of adjacent teeth.

Center distance is the shortest distance between the non-intersecting axes of mating gears, or between the parallel axes of spur gears and parallel helical gears, or the crossed axes of crossed helical gears or worm gears.

Central plane is the plane perpendicular to the gear axis in a worm gear, which contains the common perpendicular of the gear and the worm axes. In the usual arrangement with the axes at right angles, it contains the worm axis.

Chordal addendum is the radial distance from the circular thickness chord to the top of the tooth, or the height from the top of the tooth to the chord subtending the circular thickness arc.

Chordal thickness is the length of the chord subtended by the circular thickness arc. The dimension obtained when a gear tooth caliper is used to measure the tooth thickness at the pitch circle.

Circular pitch is the distance on the circumference of the pitch circle, in the plane of rotation, between corresponding points of adjacent teeth. The length of the arc of the pitch circle between the centers or other corresponding points of adjacent teeth.

Circular thickness is the thickness of the tooth on the pitch circle in the plane of rotation, or the length of arc between the two sides of a gear tooth measured on the pitch circle.

Clearance is the radial distance between the top of a tooth and the bottom of a mating tooth space, or the amount by which the dedendum in a given gear exceeds the addendum of its mating gear.

Contact diameter is the smallest diameter on a gear tooth with which the mating gear makes contact.

Contact ratio is the ratio of the arc of action in the plane of rotation to the circular pitch, and is sometimes thought of as the average number of teeth in contact. This ratio is obtained most directly as the ratio of the length of action to the base pitch.

Contact ratio – face is the ratio of the face advance to the circular pitch in helical gears.

Contact ratio – total is the ratio of the sum of the arc of action and the face advance to the circular pitch.

Contact stress is the maximum compressive stress within the contact area between mating gear tooth profiles. Also called the Hertz stress.

Cycloid is the curve formed by the path of a point on a circle as it rolls along a straight line. When such a circle rolls along the outside of another circle the curve is called an *epicycloid*, and when it rolls along the inside of another circle it is called a *hypocycloid*. These curves are used in defining the former American Standard composite Tooth Form.

Dedendum is the radial or perpendicular distance between the pitch circle and the bottom of the tooth space.

Diametral pitch is the ratio of the number of teeth to the number of inches in the pitch diameter in the plane of rotation, or the number of gear teeth to each inch of pitch diameter. Normal diametral pitch is the diametral pitch as calculated in the normal plane, or the diametral pitch divided by the cosine of the helix angle.

Efficiency is the torque ratio of a gear set divided by its gear ratio.

Equivalent pitch radius is the radius of curvature of the pitch surface at the pitch point in a plane normal to the pitch line element.

Face advance is the distance on the pitch circle that a gear tooth travels from the time pitch point contact is made at one end of the tooth until pitch point contact is made at the other end.

Fillet radius is the radius of the concave portion of the tooth profile where it joins the bottom of the tooth space.

Fillet stress is the maximum tensile stress in the gear tooth fillet.

Flank of tooth is the surface between the pitch circle and the bottom land, including the gear tooth fillet.

Gear ratio is the ratio between the numbers of teeth in mating gears.

Helical overlap is the effective face width of a helical gear divided by the gear axial pitch.

Helix angle is the angle that a helical gear tooth makes with the gear axis at the pitch circle, unless specified otherwise.

Hertz stress, see *Contact stress*.

Highest point of single tooth contact (HPSTC) is the largest diameter on a spur gear at which a single tooth is in contact with the mating gear.

Interference is the contact between mating teeth at some point other than along the line of action.

Internal diameter is the diameter of a circle that coincides with the tops of the teeth of an internal gear.

Internal gear is a gear with teeth on the inner cylindrical surface.

Involute is the curve generally used as the profile of gear teeth. The curve is the path of a point on a straight line as it rolls along a convex base curve, usually a circle.

Land: The top land is the top surface of a gear tooth and the *bottom land* is the surface of the gear between the fillets of adjacent teeth.

Lead is the axial advance of the helix in one complete turn, or the distance along its own axis on one revolution if the gear were free to move axially.

Length of action is the distance on an involute line of action through which the point of contact moves during the action of the tooth profile.

Line of action is the portion of the common tangent to the base cylinders along which contact between mating involute teeth occurs.

Lowest point of single tooth contact (LPSTC) is the smallest diameter on a spur gear at which a single tooth is in contact with its mating gear. Gear set contact stress is determined with a load placed on the pinion at this point.

Module is the ratio of the pitch diameter to the number of teeth, normally the ratio of pitch diameter in mm to the number of teeth. Module in the inch system is the ratio of the pitch diameter in inches to the number of teeth.

Normal plane is a plane normal to the tooth surfaces at a point of contact and perpendicular to the pitch plane.

Number of teeth is the number of teeth contained in a gear.

Outside diameter is the diameter of the circle that contains the tops of the teeth of external gears.

Pitch is the distance between similar, equally-spaced tooth surfaces in a given direction along a given curve or line.

Pitch circle is the circle through the pitch point having its center at the gear axis.

Pitch diameter is the diameter of the pitch circle. The operating pitch diameter is the pitch diameter at which the gear operates.

Pitch plane is the plane parallel to the axial plane and tangent to the pitch surfaces in any pair of gears. In a single gear, the pitch plane may be any plane tangent to the pitch surfaces.

Pitch point is the intersection between the axes of the line of centers and the line of action.

Plane of rotation is any plane perpendicular to a gear axis.

Pressure angle is the angle between a tooth profile and a radial line at its pitch point. In involute teeth, the pressure angle is often described as the angle between the line of action and the line tangent to the pitch circle. *Standard pressure angles* are established in connection with standard tooth proportions. A given pair of involute profiles will transmit smooth motion at the same velocity ratio when the center distance is changed. Changes in center distance in gear design and gear manufacturing operations may cause changes in pitch diameter, pitch and pressure angle in the same gears under different conditions. Unless otherwise specified, the pressure angle is the *standard pressure angle at the standard pitch diameter*. The *operating pressure angle* is determined by the center distance at which a pair of gears operate. In oblique teeth such as helical and spiral designs, the pressure angle is specified in the transverse, normal or axial planes.

Principle reference planes are pitch plane, axial plane and transverse plane, all intersecting at a point and mutually perpendicular.

Rack: A rack is a gear with teeth spaced along a straight line, suitable for straight line motion. A basic rack is a rack that is adopted as the basis of a system of interchangeable gears. Standard gear tooth dimensions are often illustrated on an outline of a basic rack.

Roll angle is the angle subtended at the center of a base circle from the origin of an involute to the point of tangency of a point on a straight line from any point on the same involute. The radian measure of this angle is the tangent of the pressure angle of the point on the involute.

Root diameter is the diameter of the circle that contains the roots or bottoms of the tooth spaces.

Tangent plane is a plane tangent to the tooth surfaces at a point or line of contact.

Tip relief is an arbitrary modification of a tooth profile where a small amount of material is removed from the involute face of the tooth surface near the tip of the gear tooth.

Tooth face is the surface between the pitch line element and the tooth tip.

Tooth surface is the total tooth area including the flank of the tooth and the tooth face.

Total face width is the dimensional width of a gear blank and may exceed the effective face width as with a double-helical gear where the total face width includes any distance separating the right-hand and left-hand helical gear teeth.

Transverse plane is a plane that is perpendicular to the axial plane and to the pitch plane. In gears with parallel axes, the transverse plane and the plane of rotation coincide.

Trochoid is the curve formed by the path of a point on the extension of a radius of a circle as it rolls along a curve or line. A trochoid is also the curve formed by the path of a point on a perpendicular to a straight line as the straight line rolls along the convex side of a base curve. By the first definition, a trochoid is derived from the *cycloid*, by the second definition it is derived from the *involute*.

True involute form diameter is the smallest diameter on the tooth at which the point of tangency of the involute tooth profile exists. Usually this position is the point of tangency of the involute tooth profile and the fillet curve, and is often referred to as the TIF diameter.

Undercut is a condition in generated gear teeth when any part of the fillet curve lies inside a line drawn at a tangent to the working profile at its lowest point. Undercut may be introduced deliberately to facilitate shaving operations, as in pre-shaving.

Whole depth is the total depth of a tooth space, equal to the addendum plus the dedendum and equal to the working depth plus clearance.

Working depth is the depth of engagement of two gears, or the sum of their addendums. The standard working distance is the depth to which a tooth extends into the tooth space of a mating gear when the center distance is standard.

Definitions of gear terms are given in AGMA Standards 112.05, 115.01, and 116.01 entitled "Terms, Definitions, Symbols and Abbreviations," "Reference Information—Basic Gear Geometry," and "Glossary—Terms Used in Gearing," respectively; obtainable from American Gear Manufacturers Assn., 1500 King. St., Alexandria, VA 22314.

Comparative Sizes and Shape of Gear Teeth

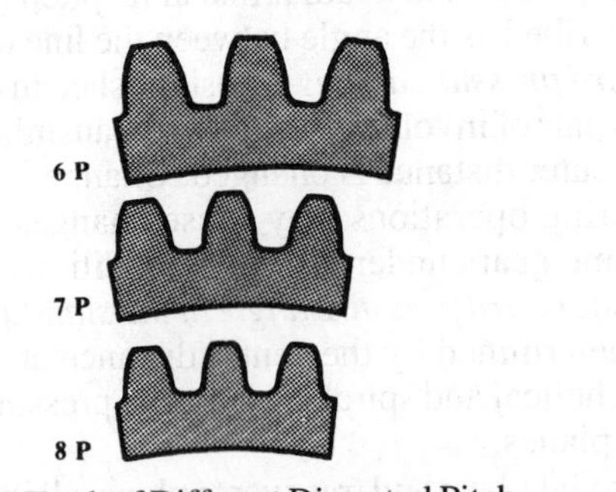

Gear Teeth of Different Diametral Pitch

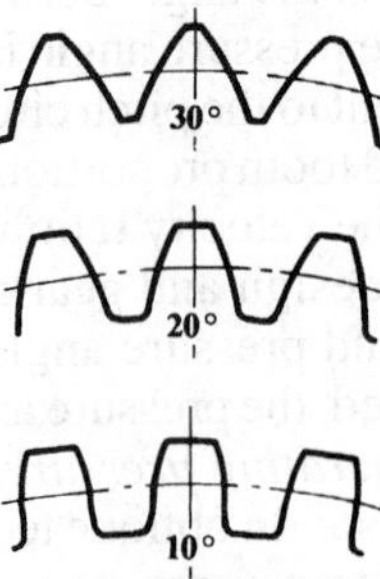

Shapes of Gear Teeth of Different Pressure Angles

Nomenclature of Gear Teeth

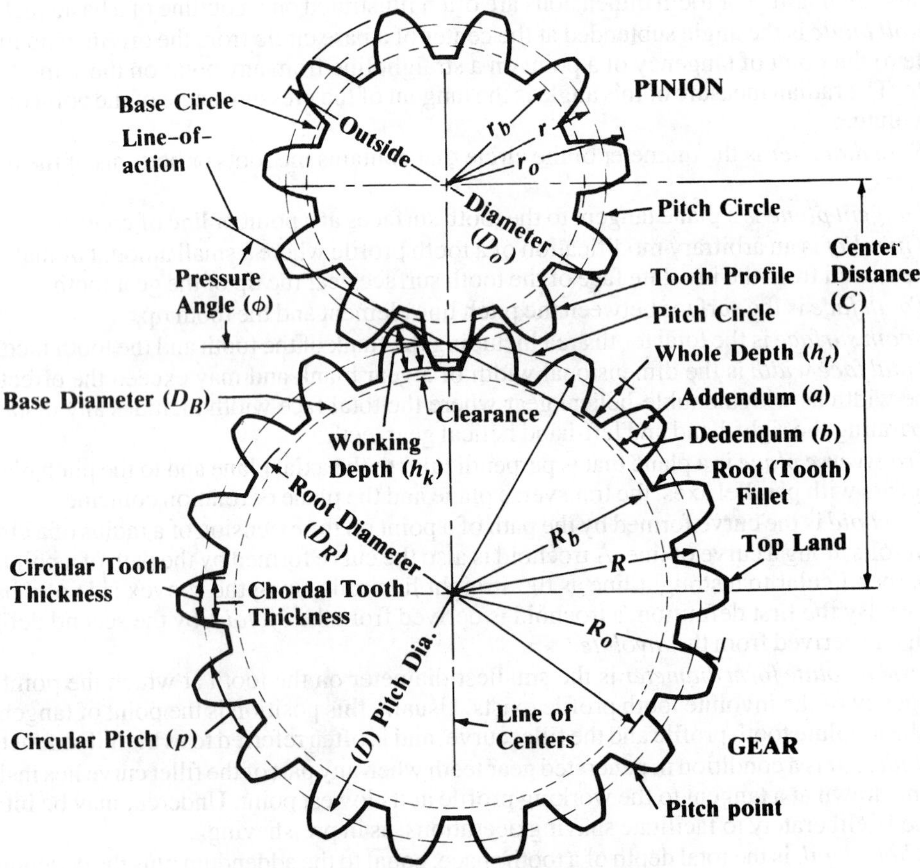

Terms Used in Gear Geometry from Table 1 on page 2004

Properties of the Involute Curve.—The involute curve is used almost exclusively for gear-tooth profiles, because of the following important properties.

1) The form or shape of an involute curve depends upon the diameter of the base circle from which it is derived. (If a taut line were unwound from the circumference of a circle—the *base circle* of the involute—the end of that line or any point on the unwound portion, would describe an involute curve.)

2) If a gear tooth of involute curvature acts against the involute tooth of a mating gear while rotating at a uniform rate, the angular motion of the driven gear will also be uniform, even though the center-to-center distance is varied.

3) The relative rate of motion between driving and driven gears having involute tooth curves is established by the diameters of their base circles.

4) Contact between intermeshing involute teeth on a driving and driven gear is along a straight line that is tangent to the two base circles of these gears. This is the *line of action.*

5) The point where the line of action intersects the common center-line of the mating involute gears, establishes the radii of the pitch circles of these gears; hence true pitch circle diameters are affected by a change in the center distance. (Pitch diameters obtained by dividing the number of teeth by the diametral pitch apply when the center distance equals the total number of teeth on both gears divided by twice the diametral pitch.)

6) The pitch diameters of mating involute gears are directly proportional to the diameters of their respective base circles; thus, if the base circle of one mating gear is three times as large as the other, the pitch circle diameters will be in the same ratio.

7) The angle between the line of action and a line perpendicular to the common center-line of mating gears, is the *pressure angle*; hence the pressure angle is affected by any change in the center distance.

8) When an involute curve acts against a straight line (as in the case of an involute pinion acting against straight-sided rack teeth), the straight line is tangent to the involute and perpendicular to its line of action.

9) The pressure angle, in the case of an involute pinion acting against straight-sided rack teeth, is the angle between the line of action and the line of the rack's motion. If the involute pinion rotates at a uniform rate, movement of the rack will also be uniform.

Nomenclature:

ϕ = Pressure Angle
a = Addendum a_G = Addendum of Gear a_P = Addendum of Pinion
b = Dedendum
c = Clearance
C = Center Distance
D = Pitch Diameter D_G = Pitch Diameter of Gear D_P = Pitch Diameter of Pinion
D_B = Base Circle Diameter D_O = Outside Diameter D_R = Root Diameter
F = Face Width
h_k = Working Depth of Tooth h_t = Whole Depth of Tooth
m_G = Gear Ratio
N = Number of Teeth N_G = Number of Teeth in Gear N_P = Number of Teeth in Pinion
p = Circular Pitch P = Diametral Pitch

Diametral and Circular Pitch Systems.—Gear tooth system standards are established by specifying the tooth proportions of the basic rack. The diametral pitch system is applied to most of the gearing produced in the United States. If gear teeth are larger than about one diametral pitch, it is common practice to use the circular pitch system. The circular pitch system is also applied to cast gearing and it is commonly used in connection with the design and manufacture of worm gearing.

Pitch Diameters Obtained with Diametral Pitch System.—The diametral pitch system is arranged to provide a series of standard tooth sizes, the principle being similar to the standardization of screw thread pitches. Inasmuch as there must be a whole number of teeth on each gear, the increase in pitch diameter per tooth varies according to the pitch. For example, the pitch diameter of a gear having, say, 20 teeth of 4 diametral pitch, will be 5 inches; 21 teeth, 5¼ inches; and so on, the increase in diameter for each additional tooth being equal to ¼ inch for 4 diametral pitch. Similarly, for 2 diametral pitch the variations for successive numbers of teeth would equal ½ inch, and for 10 diametral pitch the variations would equal $\frac{1}{10}$ inch, etc. Where a given center distance must be maintained and no standard diametral pitch can be used, gears should be designed with reference to the gear set center distance procedure discussed in *Gears for Given Center Distance and Ratio* starting on page 2012.

Table 1. Formulas for Dimensions of Standard Spur Gears

To Find	Formula	
Base Circle Diameter	$D_B = D\cos\phi$	(1)
Circular Pitch	$p = \frac{3.1416D}{N}$	(2a)
	$p = \frac{3.1416}{P}$	(2b)
Center Distance	$C = \frac{N_P(m_G+1)}{2P}$	(3a)
	$C = \frac{D_P+D_G}{2}$	(3b)
	$C = \frac{N_G+N_P}{2P}$	(3c)
	$C = \frac{(N_G+N_P)p}{6.2832}$	(3d)
Diametral Pitch	$P = \frac{3.1416}{p}$	(4a)
	$P = \frac{N}{D}$	(4b)
	$P = \frac{N_P(m_G+1)}{2C}$	(4c)
Gear Ratio	$m_G = \frac{N_G}{N_P}$	(5)

To Find	Formula	
Number of Teeth	$N = P \times D$	(6a)
	$N = \frac{3.1416D}{p}$	(6b)
Outside Diameter (Full-depth Teeth)	$D_O = \frac{N+2}{P}$	(7a)
	$D_O = \frac{(N+2)p}{3.1416}$	(7b)
Outside Diameter (American Standard Stub Teeth)	$D_O = \frac{N+1.6}{P}$	(8a)
	$D_O = \frac{(N+1.6)p}{3.1416}$	(8b)
Outside Diameter	$D_O = D + 2a$	(9)
Pitch Diameter	$D = \frac{N}{P}$	(10a)
	$D = \frac{Np}{3.1416}$	(10b)
Root Diameter[a]	$D_R = D - 2b$	(11)
Whole Depth	$a + b$	(12)
Working Depth	$a_G + a_P$	(13)

[a] See also formulas in Tables 2 and 4 on pages 2004 and 2008.

Table 2. Formulas for Tooth Parts, 20-and 25-degree Involute Full-depth Teeth ANSI Coarse Pitch Spur Gear Tooth Forms *ANSI B6.1-1968, (R1974)*

To Find	Diametral Pitch, *P*, Known	Circular Pitch, *p*, Known
Addendum	$a = 1.000 \div P$	$a = 0.3183 \times p$
Dedendum (Preferred)	$b = 1.250 \div P$	$b = 0.3979 \times p$
(Shaved or Ground Teeth)[a]	$b = 1.350 \div P$	$b = 0.4297 \times p$
Working Depth	$h_k = 2.000 \div P$	$h_k = 0.6366 \times p$

Table 2. *(Continued)* **Formulas for Tooth Parts, 20-and 25-degree Involute Full-depth Teeth ANSI Coarse Pitch Spur Gear Tooth Forms** *ANSI B6.1-1968 (R1974)*

To Find	Diametral Pitch, P, Known	Circular Pitch, p, Known
Whole Depth (Preferred)	$h_t = 2.250 \div P$	$h_t = 0.7162 \times p$
(Shaved or Ground Teeth)	$h_t = 2.350 \div P$	$h_t = 0.7480 \times p$
Clearance (Preferred)[b]	$c = 0.250 \div P$	$c = 0.0796 \times p$
(Shaved or Ground Teeth)	$c = 0.350 \div P$	$c = 0.1114 \times p$
Fillet Radius (Rack)[c]	$r_f = 0.300 \div P$	$r_f = 0.0955 \times p$
Pitch Diameter	$D = N \div P$	$D = 0.3183 \times N_p$
Outside Diameter	$D_O = (N + 2) \div P$	$D_O = 0.3183 \times (N + 2)p$
Root Diameter (Preferred)	$D_R = (N - 2.5) \div P$	$D_R = 0.3183 \times (N - 2.5)p$
(Shaved or Ground Teeth)	$D_R = (N - 2.7) \div P$	$D_R = 0.3183 \times (N - 2.7)p$
Circular Thickness—Basic	$t = 1.5708 \div P$	$t = p \div 2$

[a] When gears are preshave cut on a gear shaper the dedendum will usually need to be increased to 1.40/*P* to allow for the higher fillet trochoid produced by the shaper cutter. This is of particular importance on gears of few teeth or if the gear blank configuration requires the use of a small diameter shaper cutter, in which case the dedendum may need to be increased to as much as 1.45/*P*. This should be avoided on highly loaded gears where the consequently reduced *J* factor will increase gear tooth stress excessively.

[b] A minimum clearance of 0.157/*P* may be used for the basic 20-degree and 25-degree pressure angle rack in the case of shallow root sections and use of existing hobs or cutters. However, whenever less than standard clearance is used, the location of the TIF diameter should be determined by the method shown in *True Involute Form Diameter* starting on page 2030. The TIF diameter must be less than the Contact Diameter determined by the method shown on page 2028.

[c] The fillet radius of the basic rack should not exceed 0.235/*P* for a 20-degree pressure angle rack or 0.270/*P* for a 25-degree pressure angle rack for a clearance of 0.157/*P*. The basic rack fillet radius must be *reduced* for teeth with a 25-degree pressure angle having a clearance in excess of 0.250/*P*.

American National Standard and Former American Standard Gear Tooth Forms *ANSI B6.1-1968, (R1974)* and *ASA B6.1-1932*

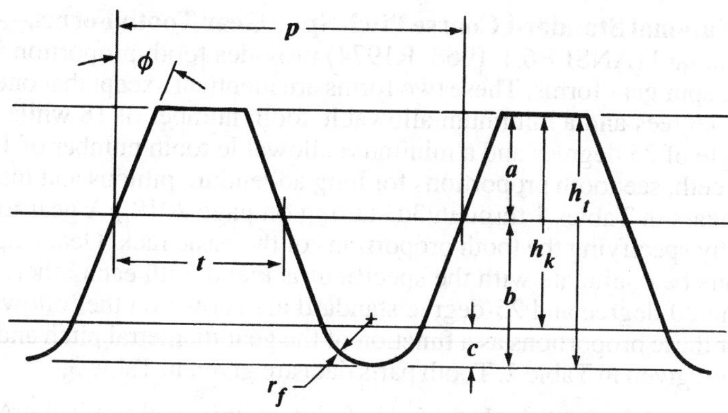

a = addendum
b = dedendum
c = clearance
h_k = working depth
h_t = whole depth
p = circular pitch
r_f = fillet radius of basic rack
t = circular tooth thickness — basic
ϕ = pressure angle

Basic Rack of the 20-Degree and 25-Degree Full-Depth Involute Systems

American National Standard and Former American Standard Gear Tooth Forms
ANSI B6.1-1968, R1974 **and** *ASA B6.1-1932 (Continued)*

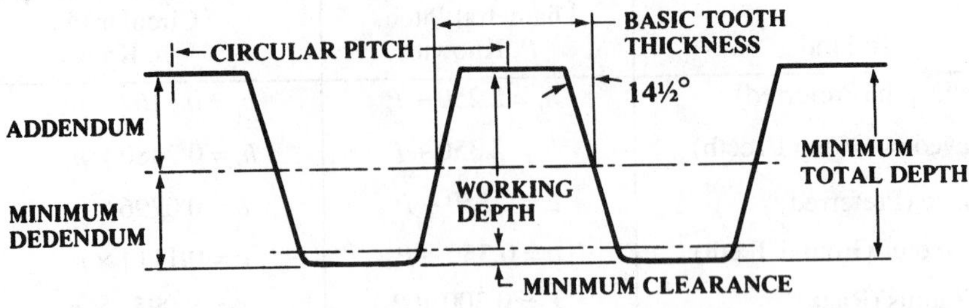

Basic Rack of the 14½-Degree Full-Depth Involute System

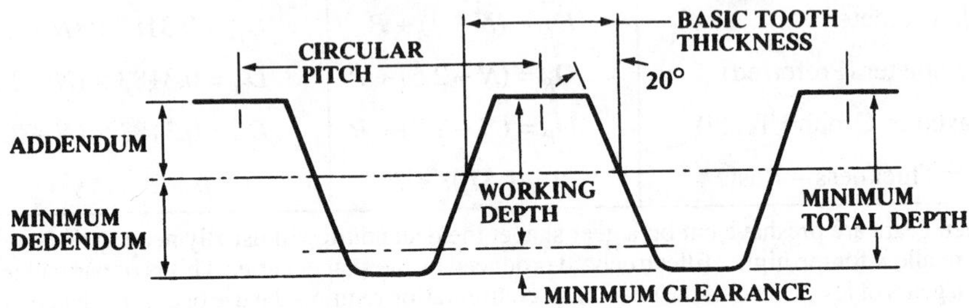

Basic Rack of the 20-Degree Stub Involute System

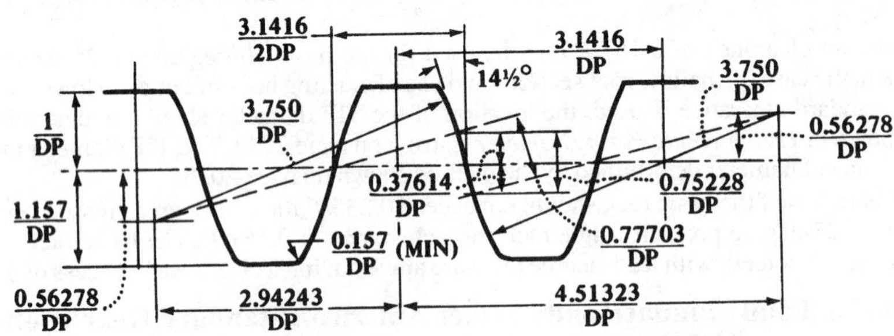

Approximation of Basic Rack for the 14½-Degree Composite System

American National Standard Coarse Pitch Spur Gear Tooth Forms.—The American National Standard (ANSI B6.1-1968, R1974) provides tooth proportion information on two involute spur gear forms. These two forms are identical except that one has a pressure angle of 20 degrees and a minimum allowable tooth number of 18 while the other has a pressure angle of 25 degrees and a minimum allowable tooth number of 12. (For pinions with fewer teeth, see tooth proportions for long addendum pinions and their mating short addendum gears in Tables 1 through 3d starting on page 2019.) A gear tooth standard is established by specifying the tooth proportions of the basic rack. Gears made to this standard will thus be conjugate with the specified rack and with each other. The basic rack forms for the 20-degree and 25-degree standard are shown on the following page; basic formulas for these proportions as a function of the gear diametral pitch and also of the circular pitch are given in Table 2. Tooth parts data are given in Table 3.

In recent years the established standard of almost universal use is the ANSI 20-degree standard spur gear form. It provides a gear with good strength and without fillet undercut in pinions of as few as eighteen teeth. Some more recent applications have required a tooth form of even greater strength and fewer teeth than eighteen. This requirement has stimulated the establishment of the ANSI 25-degree standard. This 25-degree form will give greater tooth strength than the 20-degree standard, will provide pinions of as few as twelve teeth without fillet undercut and will provide a lower contact compressive stress for greater gear set surface durability.

Table 3. Gear Tooth Parts for American National Standard Coarse Pitch 20- and 25-Degree Pressure Angle Gears

Dia. Pitch	Circ. Pitch	Stand. Addend.[a]	Stand. Dedend.	Spec. Dedend.[b]	Min. Dedend.	Stand. F. Rad.	Min. F. Rad.
P	p	a	b	b	b	r_f	r_f
0.3142	10.	3.1831	3.9789	4.2972	3.6828	0.9549	0.4997
0.3307	9.5	3.0239	3.7799	4.0823	3.4987	0.9072	0.4748
0.3491	9.	2.8648	3.5810	3.8675	3.3146	0.8594	0.4498
0.3696	8.5	2.7056	3.3820	3.6526	3.1304	0.8117	0.4248
0.3927	8.	2.5465	3.1831	3.4377	2.9463	0.7639	0.3998
0.4189	7.5	2.3873	2.9842	3.2229	2.7621	0.7162	0.3748
0.4488	7.	2.2282	2.7852	3.0080	2.5780	0.6685	0.3498
0.4833	6.5	2.0690	2.5863	2.7932	2.3938	0.6207	0.3248
0.5236	6.	1.9099	2.3873	2.5783	2.2097	0.5730	0.2998
0.5712	5.5	1.7507	2.1884	2.3635	2.0256	0.5252	0.2749
0.6283	5.	1.5915	1.9894	2.1486	1.8414	0.4775	0.2499
0.6981	4.5	1.4324	1.7905	1.9337	1.6573	0.4297	0.2249
0.7854	4.	1.2732	1.5915	1.7189	1.4731	0.3820	0.1999
0.8976	3.5	1.1141	1.3926	1.5040	1.2890	0.3342	0.1749
1.	3.1416	1.0000	1.2500	1.3500	1.1570	0.3000	0.1570
1.25	2.5133	0.8000	1.0000	1.0800	0.9256	0.2400	0.1256
1.5	2.0944	0.6667	0.8333	0.9000	0.7713	0.2000	0.1047
1.75	1.7952	0.5714	0.7143	0.7714	0.6611	0.1714	0.0897
2.	1.5708	0.5000	0.6250	0.6750	0.5785	0.1500	0.0785
2.25	1.3963	0.4444	0.5556	0.6000	0.5142	0.1333	0.0698
2.5	1.2566	0.4000	0.5000	0.5400	0.4628	0.1200	0.0628
2.75	1.1424	0.3636	0.4545	0.4909	0.4207	0.1091	0.0571
3.	1.0472	0.3333	0.4167	0.4500	0.3857	0.1000	0.0523
3.25	0.9666	0.3077	0.3846	0.4154	0.3560	0.0923	0.0483
3.5	0.8976	0.2857	0.3571	0.3857	0.3306	0.0857	0.0449
3.75	0.8378	0.2667	0.3333	0.3600	0.3085	0.0800	0.0419
4.	0.7854	0.2500	0.3125	0.3375	0.2893	0.0750	0.0392
4.5	0.6981	0.2222	0.2778	0.3000	0.2571	0.0667	0.0349
5.	0.6283	0.2000	0.2500	0.2700	0.2314	0.0600	0.0314
5.5	0.5712	0.1818	0.2273	0.2455	0.2104	0.0545	0.0285
6.	0.5236	0.1667	0.2083	0.2250	0.1928	0.0500	0.0262
6.5	0.4833	0.1538	0.1923	0.2077	0.1780	0.0462	0.0242
7.	0.4488	0.1429	0.1786	0.1929	0.1653	0.0429	0.0224
7.5	0.4189	0.1333	0.1667	0.1800	0.1543	0.0400	0.0209
8.	0.3927	0.1250	0.1563	0.1687	0.1446	0.0375	0.0196
8.5	0.3696	0.1176	0.1471	0.1588	0.1361	0.0353	0.0185
9.	0.3491	0.1111	0.1389	0.1500	0.1286	0.0333	0.0174
9.5	0.3307	0.1053	0.1316	0.1421	0.1218	0.0316	0.0165
10.	0.3142	0.1000	0.1250	0.1350	0.1157	0.0300	0.0157
11.	0.2856	0.0909	0.1136	0.1227	0.1052	0.0273	0.0143
12.	0.2618	0.0833	0.1042	0.1125	0.0964	0.0250	0.0131
13.	0.2417	0.0769	0.0962	0.1038	0.0890	0.0231	0.0121
14.	0.2244	0.0714	0.0893	0.0964	0.0826	0.0214	0.0112
15.	0.2094	0.0667	0.0833	0.0900	0.0771	0.0200	0.0105
16.	0.1963	0.0625	0.0781	0.0844	0.0723	0.0188	0.0098
17.	0.1848	0.0588	0.0735	0.0794	0.0681	0.0176	0.0092
18.	0.1745	0.0556	0.0694	0.0750	0.0643	0.0167	0.0087
19.	0.1653	0.0526	0.0658	0.0711	0.0609	0.0158	0.0083
20.	0.1571	0.0500	0.0625	0.0675	0.0579	0.0150	0.0079

[a] When using equal addendums on pinion and gear the minimum number of teeth on the pinion is 18 and the minimum total number of teeth in the pair is 36 for 20-degree full depth involute tooth form and 12 and 24, respectively, for 25-degree full depth tooth form.

[b] The dedendum in this column is used when the gear tooth is shaved. It allows for the higher fillet cut by a protuberance hob.

The working depth is equal to twice the addendum.

The whole depth is equal to the addendum plus the dedendum.

Table 4. Tooth Proportions for Fine-Pitch Involute Spur and Helical Gears of 14½-, 20-, and 25-Degree Pressure Angle *ANSI B6.7-1977*

Item	Spur	Helical
Addendum, a	$\dfrac{1.000}{P}$	$\dfrac{1.000}{P_n}$
Dedendum, b	$\dfrac{1.200}{P} + 0.002\text{ (min.)}$	$\dfrac{1.200}{P_n} + 0.002\text{ (min.)}$
Working Depth, h_k	$\dfrac{2.000}{P}$	$\dfrac{2.000}{P_n}$
Whole Depth, h_t	$\dfrac{2.200}{P} + 0.002\text{ (min.)}$	$\dfrac{2.200}{P_n} + 0.002\text{ (min.)}$
Clearance, c (Standard)	$\dfrac{0.200}{P} + 0.002\text{ (min.)}$	$\dfrac{0.200}{P_n} + 0.002\text{ (min.)}$
(Shaved or Ground Teeth)	$\dfrac{0.350}{P} + 0.002\text{ (min.)}$	$\dfrac{0.350}{P_n} + 0.002\text{ (min.)}$
Tooth Thickness, t At Pitch Diameter	$t = \dfrac{1.5708}{P}$	$t_n = \dfrac{1.5708}{P_n}$
Circular Pitch, p	$p = \dfrac{\pi D}{N}$ or $\dfrac{\pi d}{n}$ or $\dfrac{\pi}{P}$	$p_n = \dfrac{\pi}{P_n}$
Pitch Diameter Pinion, d	$\dfrac{n}{P}$	$\dfrac{n}{P_n \cos \psi}$
Gear, D	$\dfrac{N}{P}$	$\dfrac{N}{P_n \cos \psi}$
Outside Diameter Pinion, d_o	$\dfrac{n+2}{P}$	$\dfrac{1}{P_n}\left(\dfrac{n}{\cos \psi} + 2\right)$
Gear, D_o	$\dfrac{N+2}{P}$	$\dfrac{1}{P_n}\left(\dfrac{N}{\cos \psi} + 2\right)$
Center Distance, C	$\dfrac{N+n}{2P}$	$\dfrac{N+n}{2P_n \cos \psi}$

All dimensions are in inches.
P = Transverse Diametral Pitch
P_n = Normal Diametral Pitch
t_n = Normal Tooth Thickness at Pitch Diameter
p_n = Normal Circular Pitch
ψ = Helix Angle
n = Number of pinion teeth
N = Number of gear teeth

American National Standard Tooth Proportions for Fine-Pitch Involute Spur and Helical Gears.—The proportions of spur gears in this Standard (ANSI B6.7-1977) follow closely ANSI B6.1-1968, R1974, "Tooth Proportions for Coarse-Pitch Involute Spur Gears." The main difference between fine-pitch and coarse-pitch gears is the greater clearance specified for fine-pitch gears. The increased clearance provides for any foreign material that may tend to accumulate at the bottoms of the teeth and also the relatively larger fillet radius resulting from proportionately greater wear on the tips of fine-pitch cutting tools.

Pressure Angle: The standard pressure angle for fine-pitch gears is 20 degrees and is recommended for most applications. For helical gears this pressure angle applies in the *nor-*

mal plane. In certain cases, notably sintered or molded gears, or in gearing where greatest strength and wear resistance are desired, a 25-degree pressure angle may be required. However, pressure angles greater than 20 degrees tend to require use of generating tools having very narrow point widths, and higher pressure angles require closer control of center distance when backlash requirements are critical.

In those cases where consideration of angular position or backlash is critical and both pinion and gear contain relatively large numbers of teeth, a 14½-degree pressure angle may be desirable. In general, pressure angles less than 20 degrees require greater amounts of tooth modification to avoid undercutting problems and are limited to larger total numbers of teeth in pinion and gear when operating at a standard center distance. Information Sheet B in the Standard provides tooth proportions for both 14½- and 25-degree pressure angle fine-pitch gears. Table 4 provides tooth proportions for fine-pitch spur and helical gears with 14½-, 20-, and 25-degree pressure angles, and Table 5 provides tooth parts.

Diametral Pitches: Diametral pitches preferred are: 20, 24, 32, 40, 48, 64, 72, 80, 96, and 120.

Table 5. American National Standard Fine Pitch Standard Gear Tooth Parts— 14½-, 20-, and 25-Degree Pressure Angles

Diametral Pitch	Circular Pitch	Circular Thickness	Standard Addend.	Standard Dedend.	Special Dedend.[a]
P	*p*	*t*	*a*	*b*	*b*
20	0.1571	0.0785	0.0500	0.0620	0.0695
24	0.1309	0.0654	0.0417	0.0520	0.0582
32	0.0982	0.0491	0.0313	0.0395	0.0442
40	0.0785	0.0393	0.0250	0.0320	0.0358
48	0.0654	0.0327	0.0208	0.0270	0.0301
64	0.0491	0.0245	0.0156	0.0208	0.0231
72	0.0436	0.0218	0.0139	0.0187	0.0208
80	0.0393	0.0196	0.0125	0.0170	0.0189
96	0.0327	0.0164	0.0104	0.0145	0.0161
120	0.0262	0.0131	0.0083	0.0120	0.0132

[a] Based upon clearance for shaved or ground teeth.

The working depth is equal to twice the addendum. The whole depth is equal to the addendum plus the dedendum. For minimum number of teeth see page 2027.

Other American Spur Gear Standards.—An appended information sheet in the American National Standard ANSI B6.1-1968, R1974 provides tooth proportion information for three spur gear forms with the notice that they are "not recommended for new designs." These forms are therefore considered to be obsolescent but the information is given on their proportions because they have been used widely in the past. These forms are the 14½-degree full depth form, the 20-degree stub involute form and the 14½-degree composite form which were covered in the former American Standard (ASA B6.1-1932). The basic rack for the 14½-degree full depth form is shown on page 2005; basic formulas for these proportions are given in Table 6.

Table 6. Formulas for Tooth Parts—Former American Standard Spur Gear Tooth Forms *ASA B6.1-1932*

To Find	Diametral Pitch, *P* Known	Circular Pitch, *p* Known
14½-Degree Involute Full-depth Teeth		
Addendum	$a = 1.000 \div P$	$a = 0.3183 \times p$
Minimum Dedendum	$b = 1.157 \div P$	$b = 0.3683 \times p$
Working Depth	$h_k = 2.000 \div P$	$h_k = 0.6366 \times p$
Minimum Whole Depth	$h_t = 2.157 \div P$	$h_t = 0.6866 \times p$
Basic Tooth Thickness on Pitch Line	$t = 1.5708 \div P$	$t = 0.500 \times p$
Minimum Clearance	$c = 0.157 \div P$	$c = 0.050 \times p$
20-Degree Involute Stub Teeth		
Addendum	$a = 0.800 \div P$	$a = 0.2546 \times p$
Minimum Dedendum	$b = 1.000 \div P$	$b = 0.3183 \times p$
Working Depth	$h_k = 1.600 \div P$	$h_k = 0.5092 \times p$
Minimum Whole Depth	$h_t = 1.800 \div P$	$h_t = 0.5729 \times p$
Basic Tooth Thickness on Pitch Line	$t = 1.5708 \div P$	$t = 0.500 \times p$
Minimum Clearance	$c = 0.200 \div P$	$c = 0.0637 \times p$

Note: Radius of fillet equals 1⅓ × clearance for 14½-degree full-depth teeth and 1½ × clearance for 20-degree full-depth teeth.
Note: A suitable working tolerance should be considered in connection with all minimum recommendations.

Fellows Stub Tooth.—The system of stub gear teeth introduced by the Fellows Gear Shaper Co. is based upon the use of two diametral pitches. One diametral pitch, say, 8, is used as the basis for obtaining the dimensions for the addendum and dedendum, while another diametral pitch, say, 6, is used for obtaining the dimensions of the thickness of the tooth, the number of teeth, and the pitch diameter. Teeth made according to this system are designated as 6⁄8 pitch, 12⁄14 pitch, etc., the numerator in this fraction indicating the pitch determining the thickness of the tooth and the number of teeth, and the denominator, the pitch determining the depth of the tooth. The clearance is made greater than in the ordinary gear-tooth system and equals 0.25 ÷ denominator of the diametral pitch. The pressure angle is 20 degrees.

This type of stub gear tooth is now used infrequently. For information as to the tooth part dimensions see 18th and earlier editions of Machinery's Handbook.

Basic Gear Dimensions.—The basic dimensions for all involute spur gears may be obtained using the formulas shown in Table 1. This table is used in conjunction with Table 3 to obtain dimensions for coarse pitch gears and Table 5 to obtain dimensions for fine pitch standard spur gears. To obtain the dimensions of gears that are specified at a standard circular pitch, the equivalent diametral pitch is first calculated by using the formula in Table 1. If the required number of teeth in the pinion (N_p) is less than the minimum specified in either Table 3 or Table 5, whichever is applicable, the gears must be proportioned by the long and short addendum method shown on page 2022.

Formulas for Outside and Root Diameters of Spur Gears that are Finish-hobbed, Shaped, or Pre-shaved

Notation

D = Pitch Diameter
D_O = Outside Diameter
D_R = Root Diameter
P = Diametral Pitch
a = Standard Addendum
b = Standard Minimum Dedendum
b_s = Standard Dedendum
b_{ps} = Dedendum for Pre-shaving

14½-, 20-, And 25-degree Involute Full-depth Teeth (19P and coarser)[a]

$$D_O = D + 2a = \frac{N}{P} + \left(2 \times \frac{1}{P}\right)$$

$$D_R = D - 2b = \frac{N}{P} - \left(2 \times \frac{1.157}{P}\right) \quad \text{(Hobbed)}^{b}$$

$$D_R = D - 2b_s = \frac{N}{P} - \left(2 \times \frac{1.25}{P}\right) \quad \text{(Shaped)}^{c}$$

$$D_R = D - 2b_{ps} = \frac{N}{P} - \left(2 \times \frac{1.35}{P}\right) \quad \text{(Pre-shaved)}^{d}$$

$$D_R = D - 2b_{ps} = \frac{N}{P} - \left(2 \times \frac{1.40}{P}\right) \quad \text{(Pre-shaved)}^{e}$$

20-degree Involute Fine-pitch Full-depth Teeth (20P and finer)

$$D_O = D + 2a = \frac{N}{P} + \left(2 \times \frac{1}{P}\right)$$

$$D_R = D - 2b = \frac{N}{P} - 2\left(\frac{1.2}{P} + 0.002\right) \quad \text{(Hobbed or Shaped)}^{f}$$

$$D_R = D - 2b_{ps} = \frac{N}{P} - 2\left(\frac{1.35}{P} + 0.002\right) \quad \text{(Pre-shaved)}^{g}$$

20-degree Involute Stub Teeth[a]

$$D_O = D + 2a = \frac{N}{P} + \left(2 \times \frac{0.8}{P}\right)$$

$$D_R = D - 2b = \frac{N}{P} - \left(2 \times \frac{1}{P}\right) \quad \text{(Hobbed)}$$

$$D_R = D - 2b_{ps} = \frac{N}{P} - \left(2 \times \frac{1.35}{P}\right) \quad \text{(Pre-shaved)}$$

[a] 14½-degree full-depth and 20-degree stub teeth are not recommended for new designs.

[b] According to ANSI B6.1-1968 a minimum clearance of 0.157/P may be used for the basic 20-degree and 25-degree pressure angle rack in the case of shallow root sections and the use of existing hobs and cutters.

[c] According to ANSI B6.1-1968 the preferred clearance is 0.250/P.

[d] According to ANSI B6.1-1968 the clearance for teeth which are shaved or ground is 0.350/P.

[e] When gears are preshave cut on a gear shaper the dedendum will usually need to be increased to 1.40/P to allow for the higher fillet trochoid produced by the shaper cutter; this is of particular importance on gears of few teeth or if the gear blank configuration requires the use of a small diameter shaper cutter, in which case the dedendum may need to be increased to as much as 1.45/P. This should be avoided on highly loaded gears where the consequently reduced J factor will increase gear tooth stress excessively.

[f] According to ANSI B6.7-1967 the standard clearance is 0.200/P + 0.002 (min.).

[g] According to ANSI B6.7-1967 the clearance for shaved or ground teeth is 0.350/P + 0.002 (min.).

Gears for Given Center Distance and Ratio.—When it is necessary to use a pair of gears of given ratio at a specified center distance C_1, it may be found that no gears of standard diametral pitch will satisfy the center distance requirement. Gears of standard diametral pitch P may need to be redesigned to operate at other than their standard pitch diameter D and standard pressure angle ϕ. The diametral pitch P_1 at which these gears will operate is

$$P_1 = \frac{N_P + N_G}{2C_1} \quad (1)$$

where N_p = number of teeth in pinion

N_G = number of teeth in gear

and their operating pressure angle ϕ_1 is

$$\phi_1 = \arccos\left(\frac{P_1}{P}\cos\phi\right) \quad (2)$$

Thus although the pair of gears are cut to a diametral pitch P and a pressure angle ϕ, they operate as standard gears of diametral pitch $P1$ and pressure angle ϕ_1. The pitch P and pressure angle ϕ should be chosen so that ϕ_1 lies between about 18 and 25 degrees.

The operating pitch diameters of the pinion D_{p1} and of the gear D_{G1} are

$$D_{P1} = \frac{N_P}{P_1} \quad (3a) \qquad \text{and} \qquad D_{G1} = \frac{N_G}{P_1} \quad (3b)$$

The base diameters of the pinion D_{PB1} and of the gear D_{GB1} are

$$D_{PB1} = D_{P1}\cos\phi_1 \quad (4a) \qquad \text{and} \qquad D_{GB1} = D_{G1}\cos\phi_1 \quad (4b)$$

The basic tooth thickness, t_1, at the operating pitch diameter for both pinion and gear is

$$t_1 = \frac{1.5708}{P_1} \quad (5)$$

The root diameters of the pinion D_{PR1} and gear D_{GR1} and the corresponding outside diameters D_{PO1} and D_{GO1} are not standard because each gear is to be cut with a cutter that is not standard for the operating pitch diameters D_{P1} and D_{G1}.

The root diameters are

$$D_{PR_1} = \frac{N_P}{P} - 2b_{P1} \quad (6a) \qquad \text{and} \qquad D_{GR_1} = \frac{N_G}{P} - 2b_{G1} \quad (6b)$$

where

$$b_{P1} = b_c - \left(\frac{t_{P2} - 1.5708/P}{2\tan\phi}\right) \quad (7a)$$

and

$$b_{G1} = b_c - \frac{t_{G2} - 1.5708/P}{2\tan\phi} \quad (7b)$$

where b_c is the hob or cutter addendum for the pinion and gear.

The tooth thicknesses of the pinion t_{P2} and the gear t_{G2} are

$$t_{P2} = \frac{N_P}{P}\left(\frac{1.5708}{N_P} + \mathrm{inv}\,\phi_1 - \mathrm{inv}\,\phi\right) \quad (8a)$$

$$t_{G2} = \frac{N_G}{P}\left(\frac{1.5708}{N_G} + \text{inv}\,\phi_1 - \text{inv}\,\phi\right) \quad (8b)$$

The outside diameter of the pinion D_{PO} and the gear D_{GO} are

$$D_{PO} = 2 \times C_1 - D_{GR1} - 2(b_c - 1/P) \quad (9a)$$

and

$$D_{GO} = 2 \times C_1 - D_{PR1} - 2(b_c - 1/P) \quad (9b)$$

Example: Design gears of 8 diametral pitch, 20-degree pressure angle, and 28 and 88 teeth to operate at 7.50-inch center distance. The gears are to be cut with a hob of 0.169-inch addendum.

$$P_1 = \frac{28 + 88}{2 \times 7.50} = 7.7333 \quad (1)$$

$$\phi_1 = \arccos\left(\frac{7.7333}{8} \times 0.93969\right) = 24.719° \quad (2)$$

$$D_{P1} = \frac{28}{7.7333} = 3.6207 \text{ in.} \quad (3a)$$

and

$$D_{G1} = \frac{88}{7.7333} = 11.3794 \text{ in.} \quad (3b)$$

$$D_{PB1} = 3.6207 \times 0.90837 = 3.2889 \text{ in.} \quad (4a)$$

and

$$D_{GB1} = 11.3794 \times 0.90837 = 10.3367 \text{ in.} \quad (4b)$$

$$t_1 = \frac{1.5708}{7.7333} = 0.20312 \text{ in.} \quad (5)$$

$$D_{PR1} = \frac{28}{8} - 2 \times 0.1016 = 3.2968 \text{ in.} \quad (6a)$$

and

$$D_{GR1} = \frac{88}{8} - 2 \times (-0.0428) = 11.0856 \text{ in.} \quad (6b)$$

$$b_{P1} = 0.169 - \left(\frac{0.2454 - 1.5708/8}{2 \times 0.36397}\right) = 0.1016 \text{ in.} \quad (7a)$$

$$b_{G1} = 0.169 - \left(\frac{0.3505 - 1.5708/8}{2 \times 0.36397}\right) = -0.0428 \text{ in.} \quad (7b)$$

$$t_{P2} = \frac{28}{8}\left(\frac{1.5708}{28} + 0.028922 - 0.014904\right) = 0.2454 \text{ in.} \quad (8a)$$

$$t_{G2} = \frac{88}{8}\left(\frac{1.5708}{88} + 0.028922 - 0.014904\right) - 0.3505 \text{ in.} \quad (8b)$$

$$D_{PO1} = 2 \times 7.50 - 11.0856 - 2(0.169 - 1/8) = 3.8264 \text{ in.} \quad (9a)$$

$$D_{GO1} = 2 \times 7.50 - 3.2968 - 2(0.169 - 1/8) = 11.6152 \text{ in.} \quad (9b)$$

Tooth Thickness Allowance for Shaving.—Proper stock allowance is important for good results in shaving operations. If too much stock is left for shaving, the life of the shaving tool is reduced and, in addition, shaving time is increased. The following figures represent the amount of stock to be left on the teeth for removal by shaving under average conditions: For diametral pitches of 2 to 4, a thickness of 0.003 to 0.004 inch (one-half on each side of the tooth); for 5 to 6 diametral pitch, 0.0025 to 0.0035 inch; for 7 to 10 diametral pitch, 0.002 to 0.003 inch; for 11 to 14 diametral pitch, 0.0015 to 0.0020 inch; for 16 to 18 diametral pitch, 0.001 to 0.002 inch; for 20 to 48 diametral pitch, 0.0005 to 0.0015 inch; and for 52 to 72 diametral pitch, 0.0003 to 0.0007 inch.

The thickness of the gear teeth may be measured in several ways to determine the amount of stock left on the sides of the teeth to be removed by shaving. If it is necessary to measure the tooth thickness during the preshaving operation while the gear is in the gear shaper or hobbing machine, a gear tooth caliper or pins would be employed. Caliper methods of measuring gear teeth are explained in detail on page 2020 for measurements over single teeth, and on page 2109 for measurements over two or more teeth.

When the preshaved gear can be removed from the machine for checking, the center distance method may be employed. In this method, the preshaved gear is meshed without backlash with a gear of standard tooth thickness and the increase in center distance over standard is noted. The amount of total tooth thickness over standard that is left on the preshaved gear can then be determined by the formula: $t_2 = 2 \tan \phi \times d$, where t_2 = amount that the total thickness of the tooth exceeds the standard thickness, ϕ = pressure angle, and d = amount that the center distance between the two gears exceeds the standard center distance.

Circular Pitch for Given Center Distance and Ratio.—When it is necessary to use a pair of gears of given ratio at a specified center distance, it may be found that no gears of standard diametral pitch will satisfy the center distance requirement. Hence, circular pitch gears may be selected. To find the required circular pitch p, when the center distance C and total number of teeth N in both gears are known, use the following formula:

$$p = \frac{C \times 6.2832}{N}$$

Example: A pair of gears having a ratio of 3 is to be used at a center distance of 10.230 inches. If one gear has 60 teeth and the other 20, what must be their circular pitch?

$$p = \frac{10.230 \times 6.2832}{60 + 20} = 0.8035 \text{ inch}$$

Circular Thickness of Tooth when Outside Diameter is Standard.—For a full-depth or stub tooth gear of standard outside diameter, the tooth thickness on the pitch circle (circular thickness or arc thickness) is found by the following formula:

$$t = \frac{1.5708}{P}$$

where t = circular thickness and P = diametral pitch. In the case of Fellows stub tooth gears the diametral pitch used is the numerator of the pitch fraction (for example, 6 if the pitch is 6⁄8).

Example 1: Find the tooth thickness on the pitch circle of a 14½-degree full-depth tooth of 12 diametral pitch.

$$t = \frac{1.5708}{12} = 0.1309 \text{ inch}$$

Example 2: Find the tooth thickness on the tooth circle of a 20-degree full-depth involute tooth having a diametral pitch of 5.

$$t = \frac{1.5708}{5} = 0.31416, \text{ say } 0.3142 \text{ inch}$$

The tooth thickness on the pitch circle can be determined very accurately by means of measurement over wires which are located in tooth spaces that are diametrically opposite or as nearly diametrically opposite as possible. Where measurement over wires is not feasible, the circular or arc tooth thickness can be used in determining the chordal thickness which is the dimension measured with a gear tooth caliper.

Circular Thickness of Tooth when Outside Diameter has been Enlarged.—When the outside diameter of a small pinion is not standard but is enlarged to avoid undercut and to improve tooth action, the teeth are located farther out radially relative to the standard pitch diameter and consequently the circular tooth thickness at the standard pitch diameter is increased. To find this increased arc thickness the following formula is used, where t = tooth thickness; e = amount outside diameter is increased over standard; ϕ = pressure angle; and p = circular pitch at the standard pitch diameter.

$$t = \frac{p}{2} + e\tan\phi$$

Example: The outside diameter of a pinion having 10 teeth of 5 diametral pitch and a pressure angle of 14½ degrees is to be increased by 0.2746 inch. The circular pitch equivalent to 5 diametral pitch is 0.6283 inch. Find the arc tooth thickness at the standard pitch diameter.

$$t = \frac{0.6283}{2} + (0.2746 \times \tan 14\tfrac{1}{2}^\circ)$$

$$t = 0.3142 + (0.2746 \times 0.25862) = 0.3852 \text{ inch}$$

Circular Thickness of Tooth when Outside Diameter has been Reduced.—If the outside diameter of a gear is reduced, as is frequently done to maintain the standard center distance when the outside diameter of the mating pinion is increased, the circular thickness of the gear teeth at the standard pitch diameter will be reduced.This decreased circular thickness can be found by the following formula where t = circular thickness at the standard pitch diameter; e = amount outside diameter is reduced under standard; ϕ = pressure angle; and p = circular pitch.

$$t = \frac{p}{2} - e\tan\phi$$

Example: The outside diameter of a gear having a pressure angle of 14½ degrees is to be reduced by 0.2746 inch or an amount equal to the increase in diameter of its mating pinion. The circular pitch is 0.6283 inch. Determine the circular tooth thickness at the standard pitch diameter.

$$t = \frac{0.6283}{2} - (0.2746 \times \tan 14\tfrac{1}{2}^\circ)$$

$$t = 0.3142 - (0.2746 \times 0.25862) = 0.2432 \text{ inch}$$

Chordal Thickness of Tooth when Outside Diameter is Standard.—To find the chordal or straight line thickness of a gear tooth the following formula can be used where t_c = chordal thickness; D = pitch diameter; and N = number of teeth.

$$t_c = D\sin\left(\frac{90^\circ}{N}\right)$$

Example: A pinion has 15 teeth of 3 diametral pitch; the pitch diameter is equal to 15 ÷ 3 or 5 inches. Find the chordal thickness at the standard pitch diameter.

$$t_c = 5\sin\left(\frac{90^\circ}{15}\right) = 5\sin 6^\circ = 5 \times 0.10453 = 0.5226 \text{ inch}$$

Chordal Thicknesses and Chordal Addenda of Milled, Full-depth Gear Teeth and of Gear Milling Cutters

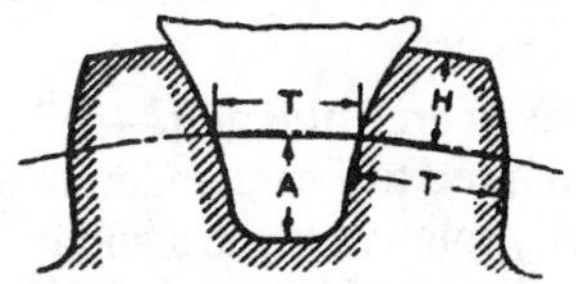

T = chordal thickness of gear tooth and cutter tooth at pitch line;
H = chordal addendum for full-depth gear tooth;
A = chordal addendum of cutter 5 (2.157 4 diametral pitch) 2 H 5 (0.6866 3 circular pitch) 2 H.

Diametral Pitch	Dimension	Number of Gear Cutter, and Corresponding Number of Teeth							
		No. 1 135 Teeth	No. 2 55 Teeth	No. 3 35 Teeth	No. 4 26 Teeth	No. 5 21 Teeth	No. 6 17 Teeth	No. 7 14 Teeth	No. 8 12 Teeth
1	T	1.5707	1.5706	1.5702	1.5698	1.5694	1.5686	1.5675	1.5663
	H	1.0047	1.0112	1.0176	1.0237	1.0294	1.0362	1.0440	1.0514
1½	T	1.0471	1.0470	1.0468	1.0465	1.0462	1.0457	1.0450	1.0442
	H	0.6698	0.6741	0.6784	0.6824	0.6862	0.6908	0.6960	0.7009
2	T	0.7853	0.7853	0.7851	0.7849	0.7847	0.7843	0.7837	0.7831
	H	0.5023	0.5056	0.5088	0.5118	0.5147	0.5181	0.5220	0.5257
2½	T	0.6283	0.6282	0.6281	0.6279	0.6277	0.6274	0.6270	0.6265
	H	0.4018	0.4044	0.4070	0.4094	0.4117	0.4144	0.4176	0.4205
3	T	0.5235	0.5235	0.5234	0.5232	0.5231	0.5228	0.5225	0.5221
	H	0.3349	0.3370	0.3392	0.3412	0.3431	0.3454	0.3480	0.3504
3½	T	0.4487	0.4487	0.4486	0.4485	0.4484	0.4481	0.4478	0.4475
	H	0.2870	0.2889	0.2907	0.2919	0.2935	0.2954	0.2977	0.3004
4	T	0.3926	0.3926	0.3926	0.3924	0.3923	0.3921	0.3919	0.3915
	H	0.2511	0.2528	0.2544	0.2559	0.2573	0.2590	0.2610	0.2628
5	T	0.3141	0.3141	0.3140	0.3139	0.3138	0.3137	0.3135	0.3132
	H	0.2009	0.2022	0.2035	0.2047	0.2058	0.2072	0.2088	0.2102
6	T	0.2618	0.2617	0.2617	0.2616	0.2615	0.2614	0.2612	0.2610
	H	0.1674	0.1685	0.1696	0.1706	0.1715	0.1727	0.1740	0.1752
7	T	0.2244	0.2243	0.2243	0.2242	0.2242	0.2240	0.2239	0.2237
	H	0.1435	0.1444	0.1453	0.1462	0.1470	0.1480	0.1491	0.1502
8	T	0.1963	0.1963	0.1962	0.1962	0.1961	0.1960	0.1959	0.1958
	H	0.1255	0.1264	0.1272	0.1279	0.1286	0.1295	0.1305	0.1314
9	T	0.1745	0.1745	0.1744	0.1744	0.1743	0.1743	0.1741	0.1740
	H	0.1116	0.1123	0.1130	0.1137	0.1143	0.1151	0.1160	0.1168
10	T	0.1570	0.1570	0.1570	0.1569	0.1569	0.1568	0.1567	0.1566
	H	0.1004	0.1011	0.1017	0.1023	0.1029	0.1036	0.1044	0.1051
11	T	0.1428	0.1428	0.1427	0.1427	0.1426	0.1426	0.1425	0.1424
	H	0.0913	0.0919	0.0925	0.0930	0.0935	0.0942	0.0949	0.0955
12	T	0.1309	0.1309	0.1308	0.1308	0.1308	0.1307	0.1306	0.1305
	H	0.0837	0.0842	0.0848	0.0853	0.0857	0.0863	0.0870	0.0876
14	T	0.1122	0.1122	0.1121	0.1121	0.1121	0.1120	0.1119	0.1118
	H	0.0717	0.0722	0.0726	0.0731	0.0735	0.0740	0.0745	0.0751
16	T	0.0981	0.0981	0.0981	0.0981	0.0980	0.0980	0.0979	0.0979
	H	0.0628	0.0632	0.0636	0.0639	0.0643	0.0647	0.0652	0.0657
18	T	0.0872	0.0872	0.0872	0.0872	0.0872	0.0871	0.0870	0.0870
	H	0.0558	0.0561	0.0565	0.0568	0.0571	0.0575	0.0580	0.0584
20	T	0.0785	0.0785	0.0785	0.0785	0.0784	0.0784	0.0783	0.0783
	H	0.0502	0.0505	0.0508	0.0511	0.0514	0.0518	0.0522	0.0525

Chordal Thicknesses and Chordal Addenda of Milled, Full-depth Gear Teeth and of Gear Milling Cutters

Circular Pitch	Dimension	Number of Gear Cutter, and Corresponding Number of Teeth							
		No. 1 135 Teeth	No. 2 55 Teeth	No. 3 35 Teeth	No. 4 26 Teeth	No. 5 21 Teeth	No. 6 17 Teeth	No. 7 14 Teeth	No. 8 12 Teeth
¼	T	0.1250	0.1250	0.1249	0.1249	0.1249	0.1248	0.1247	0.1246
	H	0.0799	0.0804	0.0809	0.0814	0.0819	0.0824	0.0830	0.0836
5⁄16	T	0.1562	0.1562	0.1562	0.1561	0.1561	0.1560	0.1559	0.1558
	H	0.0999	0.1006	0.1012	0.1018	0.1023	0.1030	0.1038	0.1045
⅜	T	0.1875	0.1875	0.1874	0.1873	0.1873	0.1872	0.1871	0.1870
	H	0.1199	0.1207	0.1214	0.1221	0.1228	0.1236	0.1245	0.1254
7⁄16	T	0.2187	0.2187	0.2186	0.2186	0.2185	0.2184	0.2183	0.2181
	H	0.1399	0.1408	0.1416	0.1425	0.1433	0.1443	0.1453	0.1464
½	T	0.2500	0.2500	0.2499	0.2498	0.2498	0.2496	0.2495	0.2493
	H	0.1599	0.1609	0.1619	0.1629	0.1638	0.1649	0.1661	0.1673
9⁄16	T	0.2812	0.2812	0.2811	0.2810	0.2810	0.2808	0.2806	0.2804
	H	0.1799	0.1810	0.1821	0.1832	0.1842	0.1855	0.1868	0.1882
⅝	T	0.3125	0.3125	0.3123	0.3123	0.3122	0.3120	0.3118	0.3116
	H	0.1998	0.2012	0.2023	0.2036	0.2047	0.2061	0.2076	0.2091
11⁄16	T	0.3437	0.3437	0.3436	0.3435	0.3434	0.3432	0.3430	0.3427
	H	0.2198	0.2213	0.2226	0.2239	0.2252	0.2267	0.2283	0.2300
¾	T	0.3750	0.3750	0.3748	0.3747	0.3747	0.3744	0.3742	0.3740
	H	0.2398	0.2414	0.2428	0.2443	0.2457	0.2473	0.2491	0.2509
13⁄16	T	0.4062	0.4062	0.4060	0.4059	0.4059	0.4056	0.4054	0.4050
	H	0.2598	0.2615	0.2631	0.2647	0.2661	0.2679	0.2699	0.2718
⅞	T	0.4375	0.4375	0.4373	0.4372	0.4371	0.4368	0.4366	0.4362
	H	0.2798	0.2816	0.2833	0.2850	0.2866	0.2885	0.2906	0.2927
15⁄16	T	0.4687	0.4687	0.4685	0.4684	0.4683	0.4680	0.4678	0.4674
	H	0.2998	0.3018	0.3035	0.3054	0.3071	0.3092	0.3114	0.3137
1	T	0.5000	0.5000	0.4998	0.4997	0.4996	0.4993	0.4990	0.4986
	H	0.3198	0.3219	0.3238	0.3258	0.3276	0.3298	0.3322	0.3346
1⅛	T	0.5625	0.5625	0.5623	0.5621	0.5620	0.5617	0.5613	0.5610
	H	0.3597	0.3621	0.3642	0.3665	0.3685	0.3710	0.3737	0.3764
1¼	T	0.6250	0.6250	0.6247	0.6246	0.6245	0.6241	0.6237	0.6232
	H	0.3997	0.4023	0.4047	0.4072	0.4095	0.4122	0.4152	0.4182
1⅜	T	0.6875	0.6875	0.6872	0.6870	0.6869	0.6865	0.6861	0.6856
	H	0.4397	0.4426	0.4452	0.4479	0.4504	0.4534	0.4567	0.4600
1½	T	0.7500	0.7500	0.7497	0.7495	0.7494	0.7489	0.7485	0.7480
	H	0.4797	0.4828	0.4857	0.4887	0.4914	0.4947	0.4983	0.5019
1¾	T	0.8750	0.8750	0.8746	0.8744	0.8743	0.8737	0.8732	0.8726
	H	0.5596	0.5633	0.5666	0.5701	0.5733	0.5771	0.5813	0.5855
2	T	1.0000	1.0000	0.9996	0.9994	0.9992	0.9986	0.9980	0.9972
	H	0.6396	0.6438	0.6476	0.6516	0.6552	0.6596	0.6644	0.6692
2¼	T	1.1250	1.1250	1.1246	1.1242	1.1240	1.1234	1.1226	1.1220
	H	0.7195	0.7242	0.7285	0.7330	0.7371	0.7420	0.7474	0.7528
2½	T	1.2500	1.2500	1.2494	1.2492	1.2490	1.2482	1.2474	1.2464
	H	0.7995	0.8047	0.8095	0.8145	0.8190	0.8245	0.8305	0.8365
3	T	1.5000	1.5000	1.4994	1.4990	1.4990	1.4978	1.4970	1.4960
	H	0.9594	0.9657	0.9714	0.9774	0.9828	0.9894	0.9966	1.0038

Chordal Thickness of Tooth when Outside Diameter is Special.—When the outside diameter is larger or smaller than standard the chordal thickness at the standard pitch diameter is found by the following formula where t_c = chordal thickness at the standard pitch diameter D; t = circular thickness at the standard pitch diameter of the enlarged pinion or reduced gear being measured.

$$t_c = t - \frac{t^3}{6 \times D^2}$$

Example 1: The outside diameter of a pinion having 10 teeth of 5 diametral pitch has been *enlarged* by 0.2746 inch. This enlargement has increased the circular tooth thickness at the standard pitch diameter (as determined by the formula previously given) to 0.3852 inch. Find the equivalent chordal thickness.

$$t_c = 0.3852 - \frac{0.385^3}{6 \times 2^2} = 0.3852 - 0.0024 = 0.3828 \text{ inch}$$

(The error introduced by rounding the circular thickness to three significant figures before cubing it only affects the fifth decimal place in the result.)

Example 2: A gear having 30 teeth is to mesh with the pinion in Example 1 and is *reduced* so that the circular tooth thickness at the standard pitch diameter is 0.2432 inch. Find the equivalent chordal thickness.

$$t_c = 0.2432 - \frac{0.243^3}{6 \times 6^2} = 0.2432 - 0.00007 = 0.2431 \text{ inch}$$

Chordal Addendum.—In measuring the chordal thickness, the vertical scale of a gear tooth caliper is set to the chordal or "corrected" addendum to locate the caliper jaws at the pitch line (see *Method of setting a gear tooth caliper* on page 2021). The simplified formula which follows may be used in determining the chordal addendum either when the addendum is standard for full-depth or stub teeth or when the addendum is either longer or shorter than standard as in case of an enlarged pinion or a gear which is to mesh with an enlarged pinion and has a reduced addendum to maintain the standard center distance. If a_c = chordal addendum; a = addendum; and t = circular thickness of tooth at pitch diameter D; then,

$$a_c = a + \frac{t^2}{4D}$$

Example 1: The outside diameter of an 8 diametral pitch 14-tooth pinion with 20-degree full-depth teeth is to be increased by using an enlarged addendum of 1.234 ÷ 8 = 0.1542 inch (see Table 1 on page 2019). The basic tooth thickness of the enlarged pinion is 1.741 ÷ 8 = 0.2176 inch. What is the chordal addendum?

$$\text{Chordal addendum} = 0.1542 + \frac{0.2176^2}{4 \times (14 \div 8)} = 0.1610 \text{ inch}$$

Example 2: The outside diameter of a 14½-degree pinion having 12 teeth of 2 diametral pitch is to be enlarged 0.624 inch to avoid undercut (see Table 2 on page 2019), thus increasing the addendum from 0.5000 to 0.8120 inch and the arc thickness at the pitch line from 0.7854 to 0.9467 inch. Then,

$$\text{Chordal addendum of pinion} = 0.8120 + \frac{0.9467^2}{4 \times (12 \div 2)} = 0.8493 \text{ inch}$$

Table 1. Addendums and Tooth Thicknesses for Coarse-Pitch Long-Addendum Pinions and their Mating Short-Addendum Gears—20- and 25-degree Pressure Angles *ANSI B6.1-1968 (R1974)*

Number of Teeth in Pinion	Addendum		Basic Tooth Thickness		Number of Teeth in Gear
	Pinion	Gear	Pinion	Gear	
N_P	a_P	a_G	t_P	t_G	N_G (min)
20-Degree Involute Full Depth Tooth Form (Less than 20 Diametral Pitch)					
10	1.468	.532	1.912	1.230	25
11	1.409	.591	1.868	1.273	24
12	1.351	.649	1.826	1.315	23
13	1.292	.708	1.783	1.358	22
14	1.234	.766	1.741	1.400	21
15	1.175	.825	1.698	1.443	20
16	1.117	.883	1.656	1.486	19
17	1.058	.942	1.613	1.529	18
25-Degree Involute Full Depth Tooth Form (Less than 20 Diametral Pitch)					
10	1.184	.816	1.742	1.399	15
11	1.095	.905	1.659	1.482	14

All values are for 1 diametral pitch. For any other sizes of teeth all linear dimensions should be divided by the diametral pitch. Basic tooth thicknesses do not include an allowance for backlash.

Table 2. Enlarged Pinion and Reduced Gear Dimensions to Avoid Interference Coarse Pitch 14½-degree Involute Full Depth Teeth

Number of Pinion Teeth	Changes in Pinion and Gear Diameters	Circular Tooth Thickness		Min. No. of Teeth in Mating Gear	
		Pinion	Mating Gear	To Avoid Undercut	For Full Involute Action
10	1.3731	1.9259	1.2157	54	27
11	1.3104	1.9097	1.2319	53	27
12	1.2477	1.8935	1.2481	52	28
13	1.1850	1.8773	1.2643	51	28
14	1.1223	1.8611	1.2805	50	28
15	1.0597	1.8449	1.2967	49	28
16	0.9970	1.8286	1.3130	48	28
17	0.9343	1.8124	1.3292	47	28
18	0.8716	1.7962	1.3454	46	28
19	0.8089	1.7800	1.3616	45	28
20	0.7462	1.7638	1.3778	44	28
21	0.6835	1.7476	1.3940	43	28
22	0.6208	1.7314	1.4102	42	27
23	0.5581	1.7151	1.4265	41	27
24	0.4954	1.6989	1.4427	40	27
25	0.4328	1.6827	1.4589	39	26
26	0.3701	1.6665	1.4751	38	26
27	0.3074	1.6503	1.4913	37	26
28	0.2447	1.6341	1.5075	36	25
29	0.1820	1.6179	1.5237	35	25
30	0.1193	1.6017	1.5399	34	24
31	0.0566	1.5854	1.5562	33	24

All dimensions are given in inches and are for 1 diametral pitch. For other pitches divide tabular values by desired diametral pitch.

Add to the standard outside diameter of the pinion the amount given in the second column of the table divided by the desired diametral pitch, and (to maintain standard center distance) subtract the same amount from the outside diameter of the mating gear. Long addendum pinions will mesh with standard gears, but the center distance will be greater than standard.

Example 3: The outside diameter of the mating gear for the pinion in Example 3 is to be reduced 0.624 inch. The gear has 60 teeth and the addendum is reduced from 0.5000 to 0.1881 inch (to maintain the standard center distance), thus reducing the arc thickness to 0.6240 inch. Then,

$$\text{Chordal addendum of gear} = 0.1881 + \frac{0.6240^2}{4 \times (60 \div 2)} = 0.1913 \text{ inch}$$

When a gear addendum is reduced as much as the mating pinion addendum is enlarged, the minimum number of gear teeth required to prevent undercutting depends upon the enlargement of the mating pinion. To illustrate, if a 14½-degree pinion with 13 teeth is enlarged 1.185 inches, then the reduced mating gear should have a minimum of 51 teeth to avoid undercut (see Table 2 on page 2019).

Tables for Chordal Thicknesses and Chordal Addenda of Milled, Full-depth Teeth.—Two convenient tables for checking gears with milled, full-depth teeth are given on pages 2016 and 2017. The first shows chordal thicknesses and chordal addenda for the lowest number of teeth cut by gear cutters Nos. 1 through 8, and for the commonly used diametral pitches. The second gives similar data for commonly used circular pitches. In each case the data shown are accurate for the number of gear teeth indicated, but are approximate for other numbers of teeth within the range of the cutter under which they appear in the table. For the higher diametral pitches and lower circular pitches, the error introduced by using the data for any tooth number within the range of the cuuter under which it appears is comparatively small. The chordal thicknesses and chordal addenda for gear cutters Nos. 1 through 8 of the more commonly used diametral and circular pitches can be obtained from the table and formulas on pages 2016 and 2017.

Caliper Measurement of Gear Tooth.—In cutting gear teeth, the general practice is to adjust the cutter or hob until it grazes the outside diameter of the blank; the cutter is then sunk to the total depth of the tooth space plus whatever slight additional amount may be required to provide the necessary play or backlash between the teeth. (For recommendations concerning backlash and excess depth of cut required, see *Backlash* starting on page 2036.) If the outside diameter of the gear blank is correct, the tooth thickness should also be correct after the cutter has been sunk to the depth required for a given pitch and backlash. However, it is advisable to check the tooth thickness by measuring it, and the vernier gear-tooth caliper (see following illustration) is commonly used in measuring the thickness.

The vertical scale of this caliper is set so that when it rests upon the top of the tooth as shown, the lower ends of the caliper jaws will be at the height of the pitch circle; the horizontal scale then shows the chordal thickness of the tooth at this point. If the gear is being cut on a milling machine or with the type of gear-cutting machine employing a formed milling cutter, the tooth thickness is checked by first taking a trial cut for a short distance at one side of the blank; then the gear blank is indexed for the next space and another cut is taken far enough to mill the full outline of the tooth. The tooth thickness is then measured.

Before the gear-tooth caliper can be used, it is necessary to determine the correct chordal thickness and also the chordal addendum (or "corrected addendum" as it is sometimes called). The vertical scale is set to the chordal addendum, thus locating the ends of the jaws at the height of the pitch circle. The rules or formulas to use in determining the chordal thickness and chordal addendum will depend upon the outside diameter of the gear; for example, if the outside diameter of a small pinion is enlarged to avoid undercut and improve the tooth action, this must be taken into account in figuring the chordal thickness and chordal addendum as shown by the accompanying rules. The detail of a gear tooth included with the gear-tooth caliper illustration, represents the chordal thickness T, the addendum S, and the chordal addendum H. For the caliper measurements over two or more teeth see *Checking Spur Gear Size by Chordal Measurement Over Two or More Teeth* starting on page 2109.

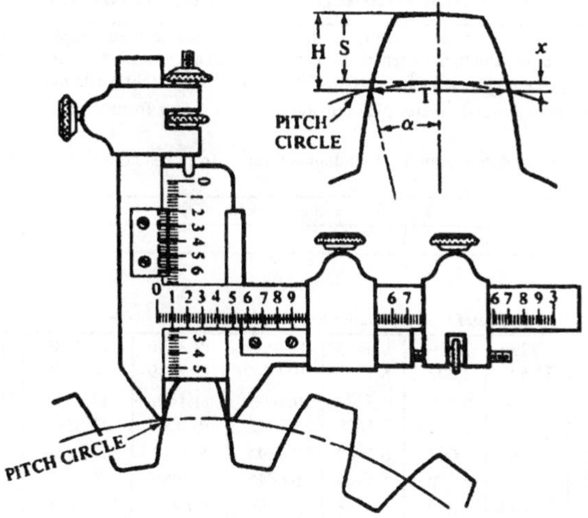

Method of setting a gear tooth caliper

Selection of Involute Gear Milling Cutter for a Given Diametral Pitch and Number of Teeth.—When gear teeth are cut by using formed milling cutters, the cutter must be selected to suit both the pitch and the number of teeth, because the shapes of the tooth spaces vary according to the number of teeth. For instance, the tooth spaces of a small pinion are not of the same shape as the spaces of a large gear of equal pitch. Theoretically, there should be a different formed cutter for every tooth number, but such refinement is unnecessary in practice. The involute formed cutters commonly used are made in series of eight cutters for each diametral pitch (see *Series of Involute, Finishing Gear Milling Cutters for Each Pitch*). The shape of each cutter in this series is correct for a certain number of teeth only, but it can be used for other numbers within the limits given. For instance, a No. 6 cutter may be used for gears having from 17 to 20 teeth, but the tooth outline is correct only for 17 teeth or the lowest number in the range, which is also true of the other cutters listed. When this cutter is used for a gear having, say, 19 teeth, too much material is removed from the upper surfaces of the teeth, although the gear meets ordinary requirements. When greater accuracy of tooth shape is desired to ensure smoother or quieter operation, an intermediate series of cutters having half-numbers may be used provided the number of gear teeth is between the number listed for the regular cutters (see *Series of Involute, Finishing Gear Milling Cutters for Each Pitch*).

Involute gear milling cutters are designed to cut a composite tooth form, the center portion being a true involute while the top and bottom portions are cycloidal. This composite form is necessary to prevent tooth interference when milled mating gears are meshed with each other. Because of their composite form, milled gears will not mate satisfactorily enough for high grade work with those of generated, full-involute form. Composite form hobs are available, however, which will produce generated gears that mesh with those cut by gear milling cutters.

Metric Module Gear Cutters: The accompanying table for selecting the cutter number to be used to cut a given number of teeth may be used also to select metric module gear cutters except that the numbers are designated in reverse order. For example, cutter No. 1, in the metric module system, is used for 12–13 teeth, cutter No. 2 for 14–16 teeth, etc.

Circular Pitch in Gears—Pitch Diameters, Outside Diameters, and Root Diameters

For any particular circular pitch and number of teeth, use the table as shown in the example to find the pitch diameter, outside diameter, and root diameter. *Example:* Pitch diameter for 57 teeth of 6-inch circular pitch = 10 × pitch diameter given under factor for 5 teeth plus pitch diameter given under factor for 7 teeth. (10 × 9.5493) + 13.3690 = 108.862 inches.

Outside diameter of gear equals pitch diameter plus outside diameter factor from next-to-last column in table = 108.862 + 3.8197 = 112.682 inches.

Root diameter of gear equals pitch diameter minus root diameter factor from last column in table = 108.862 − 4.4194 = 104.443 inches.

Circular Pitch in Inches	Factor for Number of Teeth									Outside Dia. Factor	Root Diameter Factor
	1	2	3	4	5	6	7	8	9		
	Pitch Diameter Corresponding to Factor for Number of Teeth										
6	1.9099	3.8197	5.7296	7.6394	9.5493	11.4591	13.3690	15.2788	17.1887	3.8197	4.4194
$5\frac{1}{2}$	1.7507	3.5014	5.2521	7.0028	8.7535	10.5042	12.2549	14.0056	15.7563	3.5014	4.0511
5	1.5915	3.1831	4.7746	6.3662	7.9577	9.5493	11.1408	12.7324	14.3239	3.1831	3.6828
$4\frac{1}{2}$	1.4324	2.8648	4.2972	5.7296	7.1620	8.5943	10.0267	11.4591	12.8915	2.8648	3.3146
4	1.2732	2.5465	3.8197	5.0929	6.3662	7.6394	8.9127	10.1859	11.4591	2.5465	2.9463
$3\frac{1}{2}$	1.1141	2.2282	3.3422	4.4563	5.5704	6.6845	7.7986	8.9127	10.0267	2.2282	2.5780
3	0.9549	1.9099	2.8648	3.8197	4.7746	5.7296	6.6845	7.6394	8.5943	1.9099	2.2097
$2\frac{1}{2}$	0.7958	1.5915	2.3873	3.1831	3.9789	4.7746	5.5704	6.3662	7.1620	1.5915	1.8414
2	0.6366	1.2732	1.9099	2.5465	3.1831	3.8197	4.4563	5.0929	5.7296	1.2732	1.4731
$1\frac{7}{8}$	0.5968	1.1937	1.7905	2.3873	2.9841	3.5810	4.1778	4.7746	5.3715	1.1937	1.3811
$1\frac{3}{4}$	0.5570	1.1141	1.6711	2.2282	2.7852	3.3422	3.8993	4.4563	5.0134	1.1141	1.2890
$1\frac{5}{8}$	0.5173	1.0345	1.5518	2.0690	2.5863	3.1035	3.6208	4.1380	4.6553	1.0345	1.1969
$1\frac{1}{2}$	0.4775	0.9549	1.4324	1.9099	2.3873	2.8648	3.3422	3.8197	4.2972	0.9549	1.1049
$1\frac{7}{16}$	0.4576	0.9151	1.3727	1.8303	2.2878	2.7454	3.2030	3.6606	4.1181	0.9151	1.0588
$1\frac{3}{8}$	0.4377	0.8754	1.3130	1.7507	2.1884	2.6261	3.0637	3.5014	3.9391	0.8754	1.0128
$1\frac{5}{16}$	0.4178	0.8356	1.2533	1.6711	2.0889	2.5067	2.9245	3.3422	3.7600	0.8356	0.9667
$1\frac{1}{4}$	0.3979	0.7958	1.1937	1.5915	1.9894	2.3873	2.7852	3.1831	3.5810	0.7958	0.9207
$1\frac{3}{16}$	0.3780	0.7560	1.1340	1.5120	1.8900	2.2680	2.6459	3.0239	3.4019	0.7560	0.8747
$1\frac{1}{8}$	0.3581	0.7162	1.0743	1.4324	1.7905	2.1486	2.5067	2.8648	3.2229	0.7162	0.8286
$1\frac{1}{16}$	0.3382	0.6764	1.0146	1.3528	1.6910	2.0292	2.3674	2.7056	3.0438	0.6764	0.7826
1	0.3183	0.6366	0.9549	1.2732	1.5915	1.9099	2.2282	2.5465	2.8648	0.6366	0.7366
$\frac{15}{16}$	0.2984	0.5968	0.8952	1.1937	1.4921	1.7905	2.0889	2.3873	2.6857	0.5968	0.6905
$\frac{7}{8}$	0.2785	0.5570	0.8356	1.1141	1.3926	1.6711	1.9496	2.2282	2.5067	0.5570	O.6445
$\frac{13}{16}$	0.2586	0.5173	0.7759	1.0345	1.2931	1.5518	1.8104	2.0690	2.3276	0.5173	0.5985
$\frac{3}{4}$	0.2387	0.4475	0.7162	0.9549	1.1937	1.4324	1.6711	1.9099	2.1486	0.4775	0.5524
$\frac{11}{16}$	0.2188	0.4377	0.6565	0.8754	1.0942	1.3130	1.5319	1.7507	1.9695	0.4377	0.5064
$\frac{2}{3}$	0.2122	0.4244	0.6366	0.8488	1.0610	1.2732	1.4854	1.6977	1.9099	0.4244	0.4910
$\frac{5}{8}$	0.1989	0.3979	0.5968	0.7958	0.9947	1.1937	1.3926	1.5915	1.7905	0.3979	0.4604
$\frac{9}{16}$	0.1790	0.3581	0.5371	0.7162	0.8952	1.0743	1.2533	1.4324	1.6114	0.3581	0.4143
$\frac{1}{2}$	0.1592	0.3183	0.4775	0.6366	0.7958	0.9549	1.1141	1.2732	1.4324	0.3183	0.3683
$\frac{7}{16}$	0.1393	0.2785	0.4178	0.5570	0.6963	0.8356	0.9748	1.1141	1.2533	0.2785	0.3222
$\frac{3}{8}$	0.1194	0.2387	0.3581	0.4775	0.5968	0.7162	0.8356	0.9549	1.0743	0.2387	0.2762
$\frac{1}{3}$	0.1061	0.2122	0.3183	0.4244	0.5305	0.6366	0.7427	0.8488	0.9549	0.2122	0.2455
$\frac{5}{16}$	0.0995	0.1989	0.2984	0.3979	0.4974	0.5968	0.6963	0.7958	0.8952	0.1989	0.2302
$\frac{1}{4}$	0.0796	0.1592	0.2387	0.3183	0.3979	0.4775	0.5570	0.6366	0.7162	0.1592	0.1841
$\frac{3}{16}$	0.0597	0.1194	0.1790	0.2387	0.2984	0.3581	0.4178	0.4775	0.5371	0.1194	0.1381
$\frac{1}{8}$	0.0398	0.0796	0.1194	0.1592	0.1989	0.2387	0.2785	0.3183	0.3581	0.0796	0.0921
$\frac{1}{16}$	0.0199	0.0398	0.0597	0.0796	0.0995	0.1194	0.1393	0.1592	0.1790	0.0398	0.0460

Increasing Pinion Diameter to Avoid Undercut or Interference.—On coarse-pitch pinions with small numbers of teeth (10 to 17 for 20-degree and 10 and 11 for 25-degree pressure angle involute tooth forms) undercutting of the tooth profile or fillet interference with the tip of the mating gear can be avoided by making certain changes from the standard tooth proportions that are specified in Table on page 2007. These changes consist essentially in increasing the addendum and hence the outside diameter of the pinion and decreasing the addendum and hence the outside diameter of the mating gear. These changes in outside diameters of pinion and gear do not change the velocity ratio or the procedures in cutting the teeth on a hobbing machine or generating type of shaper or planer.

Data in Table 1 on page 2019 are taken from ANSI Standard B6.1-1968, reaffirmed 1974, and show for 20-degree and 25-degree full-depth standard tooth forms, respectively, the addendums and tooth thicknesses for long addendum pinions and their mating short addendum gears when the number of teeth in the pinion is as given. Similar data for former standard 14½-degree full-depth teeth (20 diametral pitch and coarser) are given in Table 2 on page 2019.

Example: A 14-tooth, 20-degree pressure angle pinion of 6 diametral pitch is to be enlarged. What will be the outside diameters of the pinion and a 60-tooth mating gear? If the mating gear is to have the minimum number of teeth to avoid undercut, what will be its outside diameter?

$$D_o(\text{ pinion}) = \frac{N_P}{P} + 2a = \frac{14}{6} + 2\left(\frac{1.234}{6}\right) = 2.745 \text{ inches}$$

$$D_o(\text{ gear}) = \frac{N_G}{P} + 2a = \frac{60}{6} + 2\left(\frac{0.766}{6}\right) = 10.255 \text{ inches}$$

For a mating gear with minimum number of teeth to avoid undercut:

$$D_o(\text{ gear}) = \frac{N_G}{P} + 2a = \frac{21}{6} + 2\left(\frac{0.766}{6}\right) = 3.755 \text{ inches}$$

Series of Involute, Finishing Gear Milling Cutters for Each Pitch

Number of Cutter	Will cut Gears from	Number of Cutter	Will cut Gears from
1	135 teeth to a rack	5	21 to 25 teeth
2	55 to 134 teeth	6	17 to 20 teeth
3	35 to 54 teeth	7	14 to 16 teeth
4	26 to 34 teeth	8	12 to 13 teeth
The regular cutters listed above are used ordinarily. The cutters listed below (an intermediate series having half numbers) may be used when greater accuracy of tooth shape is essential in cases where the number of teeth is between the numbers for which the regular cutters are intended.			
Number of Cutter	Will cut Gears from	Number of Cutter	Will cut Gears from
1½	80 to 134 teeth	5½	19 to 20 teeth
2½	42 to 54 teeth	6½	15 to 16 teeth
3½	30 to 34 teeth	7½	13 teeth
4½	23 to 25 teeth	…	…

Roughing cutters are made with No. 1 form only. Dimensions of roughing and finishing cutters are given on page 791. Dimensions of cutters for bevel gears are given on page 792.

Enlarged Fine-Pitch Pinions: American Standard ANSI B6.7–1977, Information Sheet A provides a different system for 20-degree pressure angle pinion enlargement than is used for coarse-pitch gears. Pinions with 11 through 23 teeth (9 through 14 teeth for 25-degree pressure angle) are enlarged so that a standard tooth thickness rack with addendum $1.05/P$ will start contact 5° of roll above the base circle radius. The use of $1.05/P$ for the addendum allows for center distance variation and eccentricity of the mating gear outside diameter; the 5° roll angle avoids the fabrication of the involute in the troublesome area near the base circle.

Pinions with less than 11 teeth (9 teeth for 25-degree pressure angle) are enlarged to the extent that the highest point of undercut coincides with the start of contact with the standard rack described previously. The height of undercut considered is that produced by a sharp-cornered 120 pitch hob. Pinions with less than 13 teeth (11 teeth for 25-degree pressure angle) are truncated to provide a top land of $0.275/P$. Data for enlarged pinions may be found in Tables 3a, 3b, 3c, and 3d.

Table 3a. Increase in Dedendum, Δ for 20-, and 25-Degree Pressure Angle Fine-Pitch Enlarged Pinions and Reduced Gears *ANSI B6. 7-1977*

Diametral Pitch, *P*	Δ	Diametral Pitch, *P*	Δ	Diametral Pitch, *P*	Δ	Diametral Pitch, *P*	Δ	Diametral Pitch, *P*	Δ
20	0.0000	32	0.0007	48	0.0012	72	0.0015	96	0.0016
24	0.0004	40	0.0010	64	0.0015	80	0.0015	120	0.0017

Δ = increase in standard dedendum to provide increased clearance. See footnote to Table 3d.

Table 3b. Dimensions Required when Using Enlarged, Fine-pitch, 14½-Degree Pressure Angle Pinions *ANSI B6.7-1977*, **Information Sheet B**

Enlarged Pinion			Standard Center-distance System (Long and Short Addendum)				Enlarged Center-distance System		
			Reduced Mating Gear				Enlarged Pinion Mating with St'd. Gear	Two Equal Enlarged Mating Pinions[a]	
							Increase over St'd. Center Distance		
No. of Teeth *n*	Outside Diame-ter	Cir. Tooth Thickness at Standard Pitch Dia.	Decrease in Standard Outside Dia.[b]	Cir. Tooth Thickness at Standard Pitch Dia.	Recommended Minimum No. of Teeth *N*	Contact Ratio, *n* Mating with *N*			Contact Ratio of Two Equal Enlarged Mating Pin-ions
10	13.3731	1.9259	1.3731	1.2157	54	1.831	0.6866	1.3732	1.053
11	14.3104	1.9097	1.3104	1.2319	53	1.847	0.6552	1.3104	1.088
12	15.2477	1.8935	1.2477	1.2481	52	1.860	0.6239	1.2477	1.121
13	16.1850	1.8773	1.1850	1.2643	51	1.873	0.5925	1.1850	1.154
14	17.1223	1.8611	1.1223	1.2805	50	1.885	0.5612	1.2223	1.186
15	18.0597	1.8448	1.0597	1.2967	49	1.896	0.5299	1.0597	1.217
16	18.9970	1.8286	0.9970	1.3130	48	1.906	0.4985	0.9970	1.248
17	19.9343	1.8124	0.9343	1.3292	47	1.914	0.4672	0.9343	1.278
18	20.8716	1.7962	0.8716	1.3454	46	1.922	0.4358	0.8716	1.307
19	21.8089	1.7800	0.8089	1.3616	45	1.929	0.4045	0.8089	1.336
20	22.7462	1.7638	0.7462	1.3778	44	1.936	0.3731	0.7462	1.364
21	23.6835	1.7476	0.6835	1.3940	43	1.942	0.3418	0.6835	1.392
22	24.6208	1.7314	0.6208	1.4102	42	1.948	0.3104	0.6208	1.419
23	25.5581	1.7151	0.5581	1.4265	41	1.952	0.2791	0.5581	1.446
24	26.4954	1.6989	0.4954	1.4427	40	1.956	0.2477	0.4954	1.472
25	27.4328	1.6827	0.4328	1.4589	39	1.960	0.2164	0.4328	1.498
26	28.3701	1.6665	0.3701	1.4751	38	1.963	0.1851	0.3701	1.524
27	29.3074	1.6503	0.3074	1.4913	37	1.965	0.1537	0.3074	1.549
28	30.2447	1.6341	0.2448	1.5075	36	1.967	0.1224	0.2448	1.573
29	31.1820	1.6179	0.1820	1.5237	35	1.969	0.0910	0.1820	1.598
30	32.1193	1.6017	0.1193	1.5399	34	1.970	0.0597	0.1193	1.622
31	33.0566	1.5854	0.0566	1.5562	33	1.971	0.0283	0.0566	1.646

[a] If enlarged mating pinions are of unequal size, the center distance is increased by an amount equal to one-half the sum of their increase over standard outside diameters. Data in this column are not given in the standard.

[b] To maintain standard center distance when using an enlarged pinion, the mating gear diameter must be decreased by the amount of the pinion enlargement.

All dimensions are given in inches and are for 1 diametral pitch. For other pitches divide tabulated dimensions by the diametrical pitch.

Table 3c. Tooth Proportions Recommended for Enlarging Fine-Pitch Pinions of 20-Degree Pressure Angle—20 Diametral Pitch and Finer
ANSI B6.7-1977

Enlarged Pinion Dimensions					Enlarged C.D. System Pinion Mating with Standard Gear		Standard Center Distance (Long and Short Addendums) Reduced Gear Dimensions				
Number of Teeth,[a] n	Outside Diameter, D_{oP}	Addendum, a_P	Basic Tooth Thickness, t_P	Dedendum Based on 20 Pitch,[b] b_P	Contact Ratio Two Equal Pinions	Contact Ratio with a 24-Tooth Gear	Addendum, a_G	Basic Tooth Thickness, t_G	Dedendum Based on 20 Pitch,[b] b_G	Recommended Minimum No. of Teeth, N	Contact Ratio n Mating with N
7	10.0102	1.5051	2.14114	0.4565	0.697	1.003	0.2165	1.00045	2.0235	42	1.079
8	11.0250	1.5125	2.09854	0.5150	0.792	1.075	0.2750	1.04305	1.9650	40	1.162
9	12.0305	1.5152	2.05594	0.5735	0.893	1.152	0.3335	1.08565	1.9065	39	1.251
10	13.0279	1.5140	2.01355	0.6321	0.982	1.211	0.3921	1.12824	1.8479	38	1.312
11	14.0304	1.5152	1.97937	0.6787	1.068	1.268	0.4387	1.16222	1.8013	37	1.371
12	15.0296	1.5148	1.94703	0.7232	1.151	1.322	0.4832	1.19456	1.7568	36	1.427
13	15.9448	1.4724	1.91469	0.7676	1.193	1.353	0.5276	1.22690	1.7124	35	1.457
14	16.8560	1.4280	1.88235	0.8120	1.232	1.381	0.5720	1.25924	1.6680	34	1.483
15	17.7671	1.3836	1.85001	0.8564	1.270	1.408	0.6164	1.29158	1.6236	33	1.507
16	18.6782	1.3391	1.81766	0.9009	1.323	1.434	0.6609	1.32393	1.5791	32	1.528
17	19.5894	1.2947	1.78532	0.9453	1.347	1.458	0.7053	1.35627	1.5347	31	1.546
18	20.5006	1.2503	1.75298	0.9897	1.385	1.482	0.7497	1.38861	1.4903	30	1.561
19	21.4116	1.2058	1.72064	1.0342	1.423	1.505	0.7942	1.42095	1.4458	29	1.574
20	22.3228	1.1614	1.68839	1.0786	1.461	1.527	0.8386	1.45320	1.4014	28	1.584
21	23.2340	1.1170	1.65595	1.1230	1.498	1.548	0.8830	1.48564	1.3570	27	1.592
22	24.1450	1.0725	1.62361	1.1675	1.536	1.568	0.9275	1.51798	1.3125	26	1.598
23	25.0561	1.0281	1.59127	1.2119	1.574	1.588	0.9719	1.55032	1.2681	25	1.601
24	26.0000	1.0000	1.57080	1.2400	1.602	1.602	1.0000	1.57080	1.2400	24	1.602

[a] Caution should be exercised in the use of pinions above the horizontal lines. They should be checked for suitability, particularly in the areas of contact ratio (less than 1.2 is not recommended), center distance, clearance, and tooth strength.

[b] The actual dedendum is calculated by dividing the values in this column by the desired diametral pitch and then adding to the result an amount Δ found in Table 3a. As an example, a 20-degree pressure angle 7-tooth pinion meshing with a 42-tooth gear would have, for 24 diametral pitch, a dedendum of $0.4565 \div 24 + 0.0004 = 0.0194$. The 42-tooth gear would have a dedendum of $2.0235 \div 24 + 0.004 = 0.0847$ inch.

All dimensions are given in inches.

Table 3d. Tooth Proportions Recommended for Enlarging Fine-Pitch Pinions of 20-Degree Pressure Angle—20 Diametral Pitch and Finer

ANSI B6.7-1977

Enlarged Pinion Dimensions					Enlarged C.D. System Pinion Mating with Standard Gear		Standard Center Distance (Long and Short Addendums) Reduced Gear Dimensions				
Number of Teeth,[a] n	Outside Diameter, D_{oP}	Addendum, a_P	Basic Tooth Thickness, t_P	Dedendum Based on 20 Pitch, Dedendum Based on 20 Pitch, [b] b_P	Contact Ratio Two Equal Pinions	Contact Ratio with a 15-Tooth Gear	Addendum, a_G	Basic Tooth Thickness, t_G	Dedendum Based on 20 Pitch, [b] b_G	Recommended Minimum No. of Teeth, N	Contact Ratio n Mating with N
6	8.7645	1.3822	2.18362	0.5829	0.696	0.954	0.3429	0.95797	1.8971	24	1.030
7	9.7253	1.3626	2.10029	0.6722	0.800	1.026	0.4322	1.04130	1.8078	23	1.108
8	10.6735	1.3368	2.01701	0.7616	0.904	1.094	0.5216	1.12459	1.7184	22	1.177
9	11.6203	1.3102	1.94110	0.8427	1.003	1.156	0.6029	1.20048	1.6371	20	1.234
10	12.5691	1.2846	1.87345	0.9155	1.095	1.211	0.6755	1.26814	1.5645	19	1.282
11	13.5039	1.2520	1.80579	0.9880	1.183	1.261	0.7480	1.33581	1.4920	18	1.322
12	14.3588	1.1794	1.73813	1.0606	1.231	1.290	0.8206	1.40346	1.4194	17	1.337
13	15.2138	1.1069	1.67047	1.1331	1.279	1.317	0.8931	1.47112	1.3469	16	1.347
14	16.0686	1.0343	1.60281	1.2057	1.328	1.343	0.9657	1.53878	1.2743	15	1.352
15	17.0000	1.0000	1.57030	1.2400	1.358	1.358	1.0000	1.57080	1.2400	15	1.358

[a] Caution should be exercised in the use of pinions above the horizontal lines. They should be checked for suitability, particularly in the areas of contact ratio (less than 1.2 is not recommended), center distance, clearance, and tooth strength.

[b] The actual dedendum is calculated by dividing the values in this column by the desired diametral pitch and then adding to the result an amount Δ found in Table 3a. As an example, a 20-degree pressure angle 7-tooth pinion meshing with a 42-tooth gear would have, for 24 diametral pitch, a dedendum of $0.4565 \div 24 + 0.0004 = 0.0194$. The 42-tooth gear would have a dedendum of $2.0235 \div 24 + 0.004 = 0.0847$ inch.

All dimensions are given in inches.

All values are for 1 diametral pitch. For any other sizes of teeth, all linear dimensions should be divided by the diametral pitch.

Note: The tables in the ANSI B6.7-1977 standard also specify Form Diameter, Roll Angle to Form Diameter, and Top Land. These are not shown here. The top land is in no case less than 0.275/*P*. The form diameters and the roll angles to form diameter shown in the Standard are the values which should be met with a standard hob when generating the tooth thicknesses shown in the tables. These form diameters provides more than enough length of involute profile for any mating gear smaller than a rack. However,since these form diameters are based on gear tooth generation using standard hobs, they should impose little or no hardship on manufacture except in cases of the most critical quality levels. In such cases, form diameter specifications and master gear design should be based upon actual mating conditions.

Minimum Number of Teeth to Avoid Undercutting by Hob.—The data in the above tables give tooth proportions for low numbers of teeth to avoid interference between the gear tooth tip and the pinion tooth flank. Consideration must also be given to possible undercutting of the pinion tooth flank by the hob used to cut the pinion. The minimum number of teeth N_{min} of standard proportion that may be cut without undercut is:

$N_{min} = 2P\csc^2\phi\,[a_H - r_t(1 - \sin\phi)]$ where: a_H = cutter addendum; r_t = radius at cutter tip or corners; ϕ = cutter pressure angle; and P = diametral pitch.

Gear to Mesh with Enlarged Pinion.—Data in the fifth column of Table 2 show minimum number of teeth in a mating gear which can be cut with hob or rack type cutter without undercut, when outside diameter of gear has been reduced an amount equal to the pinion enlargement to retain the standard center distance. To calculate N for the gear, insert addendum a of enlarged mating pinion in the formula $N = 2a \times \csc^2\phi$.

Example: A gear is to mesh with a 24-tooth pinion of 1 diametral pitch which has been enlarged 0.4954 inch, as shown by the table. The pressure angle is 14½ degrees. Find minimum number of teeth N for reduced gear.

$$\text{Pinion addendum} = 1 + (0.4954 \div 2) = 1.2477$$

$$\text{Hence,}\quad N = 2 \times 1.2477 \times 15.95 = 39.8 \text{ (use 40)}$$

In the case of fine pitch gears with reduced outside diameters, the recommended minimum numbers of teeth given in Tables 3b, 3c, and 3d, are somewhat more than the minimum numbers required to prevent undercutting and are based upon studies made by the *American Gear Manufacturers Association*.

Standard Center-distance System for Enlarged Pinions.—In this system, sometimes referred to as "long and short addendums," the center distance is made standard for the numbers of teeth in pinion and gear. The outside diameter of the gear is decreased by the same amount that the outside of the pinion is enlarged.

The advantages of this system are: 1) No change in center distance or ratio is required; 2) The operating pressure angle remains standard; and 3) A slightly greater contact ratio is obtained than when the center distance is increased.

The disadvantages are 1) The gears as well as the pinion must be changed from standard dimensions; 2) Pinions having fewer than the minimum number of teeth to avoid undercut cannot be satisfactorily meshed together; and 3) In most cases where gear trains include idler gears, the standard center-distance system cannot be used.

Enlarged Center-distance System for Enlarged Pinions.—If an enlarged pinion is meshed with another enlarged pinion or with a gear of standard outside diameter, the center distance must be increased. For fine-pitch gears, it is usually satisfactory to increase the center distance by an amount equal to one-half of the enlargements (see eighth column of Table 3b). This is an approximation as theoretically there is a slight increase in backlash.

The advantages of this system are: 1) Only the pinions need be changed from the standard dimensions; 2) Pinions having fewer than 18 teeth may engage other pinions in this range; 3) The pinion tooth, which is the weaker member, is made stronger by the enlargement; and 4) The tooth contact stress, which controls gear durability, is lowered by being moved away from the pinion base circle.

The disadvantages are: 1) Center distances must be enlarged over the standard; 2) The operating pressure angle increases slightly with different combinations of pinions and gears, which is usually not important; and 3) The contact ratio is slightly smaller than that obtained with the standard center-distance system.

This consideration is of minor importance as in the worst case the loss is approximately only 6 per cent.

Enlarged Pinions Meshing without Backlash: When two enlarged pinions are to mesh without backlash, their center distance will be greater than the standard and less than that for the enlarged center-distance system. This center distance may be calculated by the formulas given in the following section.

Center Distance at Which Modified Mating Spur Gears Will Mesh with No Backlash.—When the tooth thickness of one or both of a pair of mating spur gears has been increased or decreased from the standard value ($\pi \div 2P$), the center distance at which they will mesh tightly (without backlash) may be calculated from the following formulas:

$$\text{inv}\,\phi_1 = \text{inv}\,\phi_1 + \frac{P(t+T)-\pi}{n+N}$$

$$C = \frac{n+N}{2P}$$

$$C_1 = \frac{\cos\phi}{\cos\phi_1} \times C$$

In these formulas, P = diametral pitch; n = number of teeth in pinion; N = number of teeth in gear; t and T are the actual tooth thicknesses of the pinion and gear, respectively, on their standard pitch circles; inv ϕ = involute function of standard pressure angle of gears; C = standard center distance for the gears; C_1 = center distance at which the gears mesh without backlash; and inv ϕ_1 = involute function of operating pressure angle when gears are meshed tightly at center distance C_1.

Example: Calculate the center distance for no backlash when an enlarged 10-tooth pinion of 100 diametral pitch and 20-degree pressure angle is meshed with a standard 30-tooth gear, the circular tooth thickness of the pinion and gear, respectively, being 0.01873 and 0.015708 inch.

$$\text{inv}\,\phi_1 = \text{inv}\,20° + \frac{100(0.01873+0.015708)-\pi}{(10+30)}$$

From the table of involute functions, inv 20-degrees = 0.014904. Therefore,

$$\text{inv}\,\phi_1 = 0.014904 + \frac{0.34438-0.31416}{4} = 0.022459$$

$$\phi_1 = 22°49' \text{ from page 99}$$

$$C = \frac{n+N}{2P} = \frac{10+30}{2\times 100} = 0.2000 \text{ inch}$$

$$C_1 = \frac{\cos 20°}{\cos 22°49'} \times 0.2000 = \frac{0.93969}{0.92175} \times 0.2000 = 0.2039 \text{ inch}$$

Contact Diameter.—For two meshing gears it is important to know the contact diameter of each. A first gear with number of teeth, n, and outside diameter, d_0, meshes at a standard center distance with a second gear with number of teeth, N, and outside diameter, D_0; both gears have a diametral pitch, P, and pressure angle, ϕ, a, A, b, and B are unnamed angles used only in the calculations. The contact diameter, d_c, is found by a three-step calculation that can be done by hand using a trigonometric table and a logarithmic table or a desk calculator. Slide rule calculation is not recommended because it is not accurate enough to give good results. The three-step formulas to find the contact diameter, d_c, of the first gear are:

$$\cos A = \frac{N\cos\phi}{D_o \times P} \qquad (1)$$

$$\tan b = \tan\phi - \frac{N}{n}(\tan A - \tan\phi) \qquad (2)$$

$$d_c = \frac{n\cos\phi}{P\cos b} \qquad (3)$$

Similarly the three-step formulas to find the contact diameter, D_c, of the second gear are:

$$\cos a = \frac{n\cos\phi}{d_o \times P} \qquad (4)$$

$$\tan B = \tan\phi - \frac{n}{N}(\tan a - \tan\phi) \qquad (5)$$

$$D_c = \frac{N\cos\phi}{P\cos B} \qquad (6)$$

Contact Ratio.—The contact ratio of a pair of mating spur gears must be well over 1.0 to assure a smooth transfer of load from one pair of teeth to the next pair as the two gears rotate under load. Because of a reduction in contact ratio due to such factors as tooth deflection, tooth spacing errors, tooth tip breakage, and outside diameter and center distance tolerances, the contact ratio of gears for power transmission as a general rule should not be less than about 1.4. A contact ratio of as low as 1.15 may be used in extreme cases, provided the tolerance effects mentioned above are accounted for in the calculation. The formula for determining the contact ratio, m_f, using the nomenclature in the previous section is:

$$m_f = \frac{N}{6.28318}(\tan A - \tan B) \qquad (7a)$$

or

$$m_f = \frac{N}{6.28318}(\tan a - \tan b) \qquad (7b)$$

or

$$m_f = \frac{\sqrt{R_0^2 - R_B^2} + \sqrt{r_0^2 - r_B^2} - C\sin\theta}{P\cos\theta} \qquad (7c)$$

where R_0 = outside radius of first gear; R_B = = base radius of first gear ; r_0 = outside radius of second gear; r_B = base radius of second gear; C = center distance; I = pressure angle; and, p = circular pitch.

Both formulas Equations (7a) and Equations (7b) should give the same answer. It is good practice to use both formulas as a check on the previous calculations.

Lowest Point of Single Tooth Contact.—This diameter on the pinion (sometimes referred to as LPSTQ is used to find the maximum contact compressive stress (sometimes called the Hertz Stress) of a pair of mating spur gears. The two-step formulas for determining this pinion diameter, d_L, using the same nomenclature as in the previous sections with c and C as unnamed angles used only in the calculations are:

$$\tan c = \tan a - \frac{6.28318}{n} \qquad (8)$$

$$d_L = \frac{n\cos\phi}{P\cos c} \qquad (9)$$

In some cases it is necessary to have a plot of the compressive stress over the whole cycle of contact; in this case the LPSTC for the gear is required also. The similar two-step formulas for this gear diameter are:

$$\tan C = \tan A - \frac{6.28318}{N} \qquad (10)$$

$$D_L = \frac{N\cos\phi}{P\cos C} \tag{11}$$

Maximum Hob Tip Radius.—The standard gear tooth proportions given by the formulas in Table 2 on page 2004 provide a specified size for the rack fillet radius in the general form of (a constant) × (pitch). For any given standard this constant may vary up to a maximum which it is geometrically impossible to exceed; this maximum constant, r_c (max), is found by the formula:

$$r_c\,(\max) = \frac{0.785398\cos\phi - b\sin\phi}{1 - \sin\phi} \tag{12}$$

where b is the similar constant in the specified formula for the gear dedendum. The hob tip radius of any standard hob to finish cut any standard gear may vary from zero up to this limiting value.

Undercut Limit for Hobbed Involute Gears.—It is well to avoid designing and specifying gears that will have a hobbed trochoidal fillet that undercuts the involute gear tooth profile. This should be avoided because it may cause the involute profile to be cut away up to a point above the required contact diameter with the mating gear so that involute action is lost and the contact ratio reduced to a level that may be too low for proper conjugate action. An undercut fillet will also weaken the beam strength and thus raise the fillet tensile stress of the gear tooth. To assure that the hobbed gear tooth will not have an undercut fillet, the following formula must be satisfied:

$$\frac{b - r_c}{\sin\phi} + r_c \leq 0.5n\sin\phi \tag{13}$$

where b is the dedendum constant; r_c is the hob or rack tip radius constant; n is the number of teeth in the gear; and ϕ is the gear and hob pressure angle. If the gear is not standard or the hob does not roll at the gear pitch diameter, this formula can not be applied and the determination of the expected existence of undercut becomes a considerably more complicated procedure.

Highest Point of Single Tooth Contact.—This diameter is used to place the maximum operating load for the determination of the gear tooth fillet stress. The two-step formulas for determining this diameter, d_H, of the pinion using the same nomenclature as in the previous sections with d and D as unnamed angles used only in the calculations are:

$$\tan d = \tan b + \frac{6.28318}{n} \tag{14}$$

$$d_H = \frac{n\cos\phi}{P\cos d} \tag{15}$$

Similarly for the gear:

$$\tan D = \tan B + \frac{6.28318}{N} \tag{16}$$

$$D_H = \frac{N\cos\phi}{P\cos D} \tag{17}$$

True Involute Form Diameter.—The point on the gear tooth at which the fillet and the involute profile are tangent to each other should be determined to assure that it lies at a smaller diameter than the required contact diameter with the mating gear. If the TIF diameter is larger than the contact diameter, then fillet interference will occur with severe damage to the gear tooth profile and rough action of the gear set. This two-step calculation is made by using the following two formulas with e and E as unnamed angles used only in the calculations:

$$\tan e = \tan\phi - \frac{4}{n}\left(\frac{b - r_c}{\sin 2\phi} + \frac{r_c}{2\cos\phi}\right) \tag{18}$$

$$d_{TIF} = \frac{n\cos\phi}{P\cos e} \tag{19}$$

As in the previous sections, ϕ is the pressure angle of the gear; n is the number of teeth in the pinion; b is the dedendum constant, r_c is the rack or hob tip radius constant, P is the gear diametral pitch and d_{TIF} is the true involute form diameter.

Similarly, for the mating gear:

$$\tan E = \tan\phi - \frac{4}{N}\left(\frac{b - r_c}{\sin 2\phi} + \frac{r_c}{2\cos\phi}\right) \tag{20}$$

$$D_{TIF} = \frac{N\cos\phi}{P\cos E} \tag{21}$$

where N is the number of teeth in this mating gear and D_{TIF} is the true involute form diameter.

Profile Checker Settings.—The actual tooth profile tolerance will need to be determined on high performance gears that operate either at high unit loads or at high pitch-line velocity. This is done on an involute checker, a machine which requires two settings, the gear base radius and the roll angle in degrees to significant points on the involute. From the smallest diameter outward these significant points are: TIF, Contact Diameter, LPSTC, Pitch Diameter, HPSTC, and Outside Diameter.

The base radius is:

$$R_b = \frac{N\cos\phi}{2P} \tag{22}$$

The roll angle, in degrees, at any point is equal to the tangent of the pressure angle at that point multiplied by 57.2958. The following table shows the tangents to be used at each significant diameter.

Significant Point on Tooth Profile	Pinion	Gear	For Computation
TIF	tan e	tan E	(See Formulas (18) & (20))
Contact Dia.	tan b	tan B	(See Formulas (2) & (5))
LPSTC	tan c	tan C	(See Formulas (8) & (10))
Pitch Dia.	tan ϕ	tan ϕ	(ϕ = Pressure angle)
HPSTC	tan d	tan D	(See Formulas (14) & (16))
Outside Dia.	tan a	tan A	(See Formulas (4) & (1))

Example: Find the significant diameters, contact ratio and hob tip radius for a 10-diametral pitch, 23-tooth, 20-degree pressure angle pinion of 2.5-inch outside diameter if it is to mesh with a 31-tooth gear of 3.3-inch outside diameter.

Thus: $n = 23$

$d_O = 2.5$

$P = 10$

$N = 31$

$D_O = 3.3$

$\phi = 20°$

1) Pinion contact diameter, d_c

$$\cos A = \frac{31 \times 0.93969}{3.3 \times 10} \tag{1}$$
$$= 0.88274 \qquad A = 28°1'30''$$

$$\tan b = 0.36397 - \frac{31}{23}(0.53227 - 0.36397) \tag{2}$$
$$= 0.13713 \qquad b = 7°48'26''$$

$$d_c = \frac{23 \times 0.93969}{10 \times 0.99073} \tag{3}$$
$$= 2.1815 \text{ inches}$$

2) Gear contact diameter, D_c

$$\cos a = \frac{23 \times 0.93963}{2.5 \times 10} \tag{4}$$
$$= 0.86452 \qquad a = 30°10'20''$$

$$\tan B = 0.36397 - \frac{23}{31}(0.58136 - 0.36937) \tag{5}$$
$$= 0.20267 \qquad B = 11°27'26''$$

$$D_c = \frac{31 \times 0.93969}{10 \times 0.98000} \tag{6}$$
$$= 2.9725 \text{ inches}$$

3) Contact ratio, m_f

$$m_f = \frac{31}{6.28318}(0.53227 - 0.20267) \tag{7a}$$
$$= 1.626$$
$$m_f = \frac{23}{6.28318}(0.58136 - 0.13713) \tag{7b}$$
$$= 1.626$$

4) Pinion LPSTC, d_L

$$\tan c = 0.58136 - \frac{6.28318}{23} \tag{8}$$
$$= 0.30818 \qquad c = 17°7'41''$$

$$d_L = \frac{23 \times 0.93969}{10 \times 0.95565} \tag{9}$$
$$= 2.2616 \text{ inches}$$

5) Gear LPSTC, D_L

$$\tan C = 0.53227 - \frac{6.28318}{31} \tag{10}$$
$$= 0.32959 \qquad C = 18°14'30''$$

$$D_L = \frac{31 \times 0.93969}{10 \times 0.94974} \tag{11}$$
$$= 3.0672 \text{ inches}$$

6) Maximum permissible hob tip radius, r_c (max). The dedendum factor is 1.25.

$$r_c\,(\max) = \frac{0.785398 \times 0.93969 - 1.25 \times 0.34202}{1 - 0.34202} \qquad (12)$$
$$= 0.4719 \text{ inch}$$

7) If the hob tip radius r_c is 0.30, determine if the pinion involute is undercut.

$$\frac{1.25 - 0.30}{0.34202} + 0.30 \le 0.5 \times 23 \times 0.34202 \qquad (13)$$
$$3.0776 < 3.9332$$

8) therefore there is no involute undercut.

9) Pinion HPSTC, D_H

$$\tan d = 0.13713 + \frac{6.28318}{23} \qquad (14)$$
$$= 0.41031 \qquad d = 22°18'32''$$

$$d_H = \frac{23 \times 0.93969}{10 \times 0.92515} \qquad (15)$$
$$= 2.3362 \text{ inches}$$

10) Gear HPSTC, D_H

$$\tan D = 0.20267 + \frac{6.28318}{31} \qquad (16)$$
$$= 0.40535 \qquad D = 22°3'55''$$

$$D_H = \frac{31 \times 0.93969}{10 \times 0.92676} \qquad (17)$$
$$= 3.1433 \text{ inches}$$

11) Pinion TIF diameter, d_{TIF}

$$\tan e = 0.36397 - \frac{4}{23}\left(\frac{1.25 - 0.30}{0.64279} + \frac{0.30}{2 \times 0.93969}\right) \qquad (18)$$
$$= 0.07917 \qquad e = 4°31'36''$$

$$d_{TIF} = \frac{23 \times 0.93969}{10 \times 0.99688} \qquad (19)$$
$$= 2.1681 \text{ inches}$$

12) Gear TIF diameter, D_{TIF}

$$\tan E = 0.36397 - \frac{4}{31}\left(\frac{1.25 - 0.30}{0.64279} + \frac{0.30}{2 \times 0.93969}\right) \qquad (20)$$
$$= 0.15267 \qquad E = 8°40'50''$$

$$D_{TIF} = \frac{31 \times 0.93969}{10 \times 0.98855} = 2.9468 \text{ inches} \qquad (21)$$

Gear Blanks for Fine-pitch Gears.—The accuracy to which gears can be produced is considerably affected by the design of the gear blank and the accuracy to which the various surfaces of the blank are machined. The following recommendations should not be regarded as inflexible rules, but rather as minimum average requirements for gear-blank quality compatible with the expected quality class of the finished gear.

Design of Gear Blanks: The accuracy to which gears can be produced is affected by the design of the blank, so the following points of design should be noted: 1) Gears designed with a hole should have the hole large enough that the blank can be adequately supported during machining of the teeth and yet not so large as to cause distortion; 2) Face widths should be wide enough, in proportion to outside diameters, to avoid springing and to permit obtaining flatness in important surfaces; 3) Short bore lengths should be avoided wherever possible. It is feasible, however, to machine relatively thin blanks in stacks, provided the surfaces are flat and parallel to each other; 4) Where gear blanks with hubs are to be designed, attention should be given to the wall sections of the hubs. Too thin a section will not permit proper clamping of the blank during machining operations and may also affect proper mounting of the gear; and 5) Where pinions or gears integral with their shafts are to be designed, deflection of the shaft can be minimized by having the shaft length and shaft diameter well proportioned to the gear or pinion diameter. The foregoing general principles may also be useful when applied to blanks for coarser pitch gears.

Specifying Spur and Helical Gear Data on Drawings.—The data that may be shown on drawings of spur and helical gears falls into three groups: The first group consists of data basic to the design of the gear; the second group consists of data used in manufacturing and inspection; and the third group consists of engineering reference data. The accompanying table may be used as a checklist for the various data which may be placed on gear drawings and the sequence in which they should appear.

Explanation of Terms Used in Gear Specifications: 1) Number of teeth is the number of teeth in 360 deg of gear circumference. In a sector gear, both the actual number of teeth in the sector and the theoretical number of teeth in 360 deg should be given.

2) Diametral pitch is the ratio of the number of teeth in the gear to the number of inches in the standard pitch diameter. It is used in this standard as a nominal specification of tooth size.

a) Normal diametral pitch is the diametral pitch in the normal plane.

b) Transverse diametral pitch is the diametral pitch in the transverse plane.

c) Module is the ratio of the number of teeth in the gear to the number of mm in the standard pitch diameter.

d) Normal module is the module measured in the normal plane.

e) Transverse module is the module measured in the transverse plane.

3) Pressure angle is the angle between the gear tooth profile and a radial line at the pitch point. It is used in this standard to specify the pressure angle of the basic rack used in defining the gear tooth profile.

a) Normal pressure angle is the pressure angle in the normal plane.

b) Transverse pressure angle is the pressure angle in the transverse plane.

4) Helix angle is the angle between the pitch helix and an element of the pitch cylinder, unless otherwise specified.

a) Hand of helix is the direction in which the teeth twist as they recede from an observer along the axis. A right hand helix twists clockwise and a left hand helix twists counterclockwise.

5) Standard pitch diameter is the diameter of the pitch circle. It equals the number of teeth divided by the transverse diametral pitch.

6) Tooth form may be specified as standard addendum, long addendum, short addendum, modified involute or special. If a modified involute or special tooth form is required, a detailed view should be shown on the drawing. If a special tooth form is specified, roll angles must be supplied (see page 2031).

7) Addendum is the radial distance between the standard pitch circle and the outside circle. The actual value depends on the specification of outside diameter.

8) Whole depth is the total radial depth of the tooth space. The actual value is dependent on the specification of outside diameter and root diameter.

9) Maximum calculated circular thickness on the standard pitch circle is the tooth thickness which will provide the desired minimum backlash when the gear is assembled in mesh with its mate on minimum center distance. Control may best be exerted by testing in tight mesh with a master which integrates all errors in the several teeth in mesh through the arc of action as explained on page 2042. This value is independent of the effect of runout.

a) Maximum calculated *normal* circular thickness is the circular tooth thickness in the normal plane which satisfies requirements explained in (9).

10) Gear testing radius is the distance from its axis of rotation to the standard pitch line of a standard master when in intimate contact under recommended pressure on a variable-center-distance running gage. Maximum testing radius should be calculated to provide the maximum circular tooth thickness specified in (9) when checked as explained on page 2042. This value is affected by the runout of the gear. Tolerance on testing radius must be equal to or greater than the total composite error permitted by the quality class specified in (11).

11) Quality class is specified for convenience when talking or writing about the accuracy of the gear.

12) Maximum total composite error, and (13). Maximum tooth-to-tooth composite error. Actual tolerance values (12 and 13) permitted by the quality class (11) are specified in inches to provide machine operator or inspector with tolerances required to inspect the gear.

13) Testing pressure recommendations are given on page 2042. Incorrect testing pressure will result in incorrect measurement of testing radius.

14) Master specifications by tool or code number may be required to call for the use of a special master gear when tooth thickness deviates excessively from standard.

15) Measurement over two 0.xxxx diameter pins may be specified to assist the manufacturing department in determining size at machine for setup only.

16) Outside diameter is usually shown on the drawing of the gear together with other blank dimensions so that it will not be necessary for machine operators to search gear tooth data for this dimension. Since outside diameter is also frequently used in the manufacture and inspection of the teeth, it may be included in the data block with other tooth specifications if preferred. To permit use of topping hobs for cutting gears on which the tooth thickness has been modified from standard, the outside diameter should be related to the specified gear testing radius (10).

17) Maximum root diameter is specified to assure adequate clearance for the outside diameter of the mating gear. This dimension is usually considered acceptable if the gear is checked with a master and meets specifications (10) through (13).

18) Active profile diameter of a gear is the smallest diameter at which the mating gear tooth profile can make contact. Because of difficulties involved in checking, this specification is not recommended for gears finer than 48 pitch.

19) Surface roughness on active profile surfaces may be specified in microinches to be checked by instrument up to about 32 pitch, or by visual comparison in the finer pitch ranges. It is difficult to determine accurately the surface roughness of fine pitch gears. For many commercial applications surface roughness may be considered acceptable on gears which meet the maximum tooth-to-tooth-error specification (13).

20) Mating gear part number may be shown as a convenient reference. If the gear is used in several applications, all mating gears may be listed but usual practice is to record this information in a reference file.

21) Number of teeth in mating gear, and (23). Minimum operating center distance. This information is often specified to eliminate the necessity of getting prints of the mating gear and assemblies for checking the design specifications, interference, backlash, determination of master gear specification, and acceptance or rejection of gears made out of tolerance.

Data for Spur and Helical Gear Drawings

Type of Data	Min. Spur Gear Data	Min. Helical Gear-Data	Add'l Optional Data	Item Number[a]	Data[a]
Basic Specifications	X	X		1	Number of teeth
	X			2	Diametral pitch or module
		X		2a	Normal diametral pitch or module
			X	2b	Transverse diametral pitch or module
	X			3	Pressure angle
		X		3a	Normal pressure angle
			X	3b	Transverse pressure angle
		X		4	Helix angle
		X		4a	Hand of helix
	X	X		5	Standard pitch diameter
	X	X		6	Tooth form
			X	7	Addendum
			X	8	Whole depth
	X			9	Max. calc. circular thickness on std. pitch circle
		X		9a	Max. calc. normal circular thickness on std.pitch circle
Manufacturing and Inspection			X	10	Roll angles
	X	X		11	A.G.M.A. quality class
	X	X		12	Max. total composite error
	X	X		13	Max. tooth-to-tooth composite error
			X	14	Testing pressure (Ounces)
	X	X		15	Master specification
			X	16	Meas. over two .xxxx dia. pins (For setup only)
	X	X		17	Outside diameter (Preferably shown on drawing of gear)
			X	18	Max. root diameter
			X	19	Active profile diameter
			X	20	Surface roughness of active profile
Engineering Reference			X	21	Mating gear part number
			X	22	Number of teeth in mating gear
			X	23	Minimum operating center distance

[a] An item-by-item explanation of the terms used in this table is given beginning on page 2034.

Backlash

In general, backlash in gears is play between mating teeth. For purposes of measurement and calculation, backlash is defined as the amount by which a tooth space exceeds the thickness of an engaging tooth. It does not include the effect of center-distance changes of the mountings and variations in bearings. When not otherwise specified, numerical values of backlash are understood to be given on the pitch circles. The general purpose of backlash is to prevent gears from jamming together and making contact on both sides of their teeth simultaneously. Lack of backlash may cause noise, overloading, overheating of the-gears and bearings, and even seizing and failure.

Excessive backlash is objectionable, particularly if the drive is frequently reversing, or if there is an overrunning load as in cam drives. On the other hand, specification of an unnecessarily small amount of backlash allowance will increase the cost of gears, because errors in runout, pitch, profile, and mounting must be held correspondingly smaller. Backlash does not affect involute action and usually is not detrimental to proper gear action.

Determining Proper Amount of Backlash.—In specifying proper backlash and tolerances for a pair of gears, the most important factor is probably the maximum permissible amount of runout in both gear and pinion (or worm). Next are the allowable errors in profile, pitch, tooth thickness, and helix angle. Backlash between a pair of gears will vary as successive teeth make contact because of the effect of composite tooth errors, particularly runout, and errors in the gear center distances and bearings.

Other important considerations are speed and space for lubricant film. Slow-moving gears, in general, require the least backlash. Fast-moving fine-pitch gears are usually lubricated with relatively light oil, but if there is insufficient clearance for an oil film, and particularly if oil trapped at the root of the teeth cannot escape, heat and excessive tooth loading will occur.

Heat is a factor because gears may operate warmer, and, therefore, expand more, than the housings. The heat may result from oil churning or from frictional losses between the teeth, at bearings or oil seals, or from external causes. Moreover, for the same temperature rise, the material of the gears—for example, bronze and aluminum—may expand more than the material of the housings, usually steel or cast iron.

The higher the helix angle or spiral angle, the more transverse backlash is required for a given normal backlash. The transverse backlash is equal to the normal backlash divided by the cosine of the helix angle.

In designs employing normal pressure angles higher than 20 degrees, special consideration must be given to backlash, because more backlash is required on the pitch circles to obtain a given amount of backlash in a direction normal to the tooth profiles.

Errors in boring the gear housings, both in center distance and alignment, are of extreme importance in determining allowance to obtain the backlash desired. The same is true in the mounting of the gears, which is affected by the type and adjustment of bearings, and similar factors. Other influences in backlash specification are heat treatment subsequent to cutting the teeth, lapping operations, need for recutting, and reduction of tooth thickness through normal wear.

Minimum backlash is necessary for timing, indexing, gun-sighting, and certain instrument gear trains. If the operating speed is very low and the necessary precautions are taken in the manufacture of such gear trains, the backlash may be held to extremely small limits. However, the specification of "zero backlash," so commonly stipulated for gears of this nature, usually involves special and expensive techniques, and is difficult to obtain.

Table 1. AGMA Recommended Backlash Range for Coarse-Pitch Spur, Helical and Herringbone Gearing

Center Distance (Inches)	Normal Diametral Pitches				
	0.5–1.99	2–3.49	3.5–5.99	6–9.99	10–19.99
	Backlash, Normal Plane, Inches[a]				
Up to 5					.005–.015
Over 5 to 10				.010–.020	.010–.020
Over 10 to 20			.020–.030	.015–.025	.010–.020
Over 20 to 30		.030–.040	.025–.030	.020–.030	
Over 30 to 40	.040–.060	.035–.045	.030–.040	.025–.035	
Over 40 to 50	.050–.070	.040–.055	.035–.050	.030–.040	
Over 50 to 80	.060–.080	.045–.065	.040–.060		
Over 80 to 100	.070–.095	.050–.080			
Over 100 to 120	.080–.110				

[a] Suggested backlash, on nominal centers, measured after rotating to the point of closest engagement. For helical and herringbone gears, divide above values by the cosine of the helix angle to obtain the transverse backlash.

The above backlash tolerances contain allowance for gear expansion due to differential in the operating temperature of the gearing and their supporting structure. The values may be used where the operating temperatures are up to 70 deg F higher than the ambient temperature.

For most gearing applications the recommended backlash ranges will provide proper running clearance between engaging teeth of mating gears. Deviation below the minimum or above the maximum values shown, which do not affect operational use of the gearing, should not be cause for rejection.

Definite backlash tolerances on coarse-pitch gearing are to be considered binding on the gear manufacturer only when agreed upon in writing.

Some applications may require less backlash than shown in the above table. In such cases the amount and tolerance should be by agreement between manufacturer and purchaser.

Recommended Backlash: In the following tables American Gear Manufacturers Association recommendations for backlash ranges for various kinds of gears are given.* For purposes of measurement and calculation, backlash is defined as the amount by which a tooth space exceeds the thickness of an engaging tooth. When not otherwise specified, numerical values of backlash are understood to be measured at the tightest point of mesh on the pitch circle in a direction normal to the tooth surface when the gears are mounted in their specified position.

Coarse-Pitch Gears: Table 1 gives the recommended backlash range for coarse-pitch spur, helical and herringbone gearing. Because backlash for helical and herringbone gears is more conveniently measured in the normal plane, Table 1 has been prepared to show backlash in the normal plane for coarse-pitch helical and herringbone gearing and in the transverse plane for spur gears. To obtain backlash in the transverse plane for helical and herringbone gears, divide the normal plane backlash in Table 1 by the cosine of the helix angle.

Table 2. AGMA Recommended Backlash Range for Bevel and Hypoid Gears

Diametral Pitch	Normal Backlash, Inch: Quality Numbers 7 through 13	Normal Backlash, Inch: Quality Numbers 3 through 6	Diametral Pitch	Normal Backlash, Inch: Quality Numbers 7 through 13	Normal Backlash, Inch: Quality Numbers 3 through 6
1.00 to 1.25	0.020–0.030	0.045–0.065	5.00 to 6.00	0.005–0.007	0.006–0.013
1.25 to 1.50	0.018–0.026	0.035–0.055	6.00 to 8.00	0.004–0.006	0.005–0.010
1.50 to 1.75	0.016–0.022	0.025–0.045	8.00 to 10.00	0.003–0.005	0.004–0.008
1.75 to 2.00	0.014–0.018	0.020–0.040	10.00 to 16.00	0.002–0.004	0.003–0.005
2.00 to 2.50	0.012–0.016	0.020–0.030	16.00 to 20.00	0.001–0.003	0.002–0.004
2.50 to 3.00	0.010–0.013	0.015–0.025	20 to 50	0.000–0.002	0.000–0.002
3.00 to 3.50	0.008–0.011	0.012–0.022	50 to 80	0.000–0.001	0.000–0.001
3.50 to 4.00	0.007–0.009	0.010–0.020	80 and finer	0.000–0.0007	0.000–0.0007
4.00 to 5.00	0.006–0.008	0.008–0.016	…	…	…

Measured at tightest point of mesh

The backlash tolerances given in this table contain allowances for gear expansion due to a differential in the operating temperature of the gearing and their supporting structure. The values may be used where the operating temperature is up to 70 degrees F. higher than the ambient temperature. These backlash values will provide proper running clearances for most gear applications.

The following important factors must be considered in establishing backlash tolerances:

V) Center distance tolerance; W) Parallelism of gear axes; X) Side runout or wobble;

Y) Tooth thickness tolerance; Z) Pitch line runout tolerance; AA) Profile tolerance;

AB) Pitch tolerance; AC) Lead tolerance; AD) Types of bearings and subsequent wear;

AE) Deflection under load; AF) Gear tooth wear; AG) Pitch line velocity; AH) Lubrication requirements; and AI) Thermal expansion of gears and housing.

A tight mesh may result in objectionable gear sound, increased power losses, overheating, rupture of the lubricant film, overloaded bearings and premature gear failure. However, it is recognized that there are some gearing applications where a tight mesh (zero backlash) may be required.

Specifying unnecessarily close backlash tolerances will increase the cost of the gearing. It is obvious from the above summary that the desired amount of backlash is difficult to evaluate. It is, therefore, recommended that when a designer, user or purchaser includes a reference to backlash in a gearing specification and drawing, consultation be arranged with the manufacturer.

* Extracted from Gear Classification Manual, AGMA 390.03 with permission of the publisher, the American Gear Manufacturers Association, 1500 King St., Alexandria, VA 22314.

Bevel and Hypoid Gears: Table 2 gives similar backlash range values for bevel and hypoid gears. These are values based upon average conditions for general purpose gearing, but may require modification to meet specific needs.

Backlash on bevel and hypoid gears can be controlled to some extent by axial adjustment of the gears during assembly. However, due to the fact that actual adjustment of a bevel or hypoid gear in its mounting will alter the amount of backlash, it is imperative that the amount of backlash cut into the gears during manufacture is not excessive. Bevel and hypoid gears must always be capable of operation without interference when adjusted for zero backlash. This requirement is imposed by the fact that a failure of the axial thrust bearing might permit the gears to operate under this condition. Therefore, bevel and hypoid gears should never be designed to operate with normal backlash in excess of $0.080/P$ where P is diametral pitch.

Fine-Pitch Gears: Table 3 gives similar backlash range values for fine-pitch spur, helical and herringbone gearing.

Providing Backlash.—In order to obtain the amount of backlash desired, it is necessary to decrease tooth thicknesses. However, because of manufacturing and assembling inaccuracies not only in the gears but also in other parts, the allowances made on tooth thickness almost always must exceed the desired amount of backlash. Since the amounts of these allowances depend on the closeness of control exercised on all manufacturing operations, no general recommendations for them can be given.

It is customary to make half of the allowance for backlash on the tooth thickness of each gear of a pair, although there are exceptions. For example, on pinions having very low numbers of teeth it is desirable to provide all of the allowance on the mating gear, so as not to weaken the pinion teeth. In worm gearing, ordinary practice is to provide all of the allowance on the worm which is usually made of a material stronger than that of the worm gear.

In some instances the backlash allowance is provided in the cutter, and the cutter is then operated at the standard tooth depth. In still other cases, backlash is obtained by setting the distance between two tools for cutting the two sides of the teeth, as in straight bevel gears, or by taking side cuts, or by changing the center distance between the gears in their mountings. In spur and helical gearing, backlash allowance is usually obtained by sinking the cutter deeper into the blank than the standard depth. The accompanying table gives the excess depth of cut for various pressure angles.

Excess Depth of Cut E to Provide Backlash Allowance

Distribution of Backlash	Pressure Angle ϕ, Degrees				
	14½	17½	20	25	30
Excess Depth of Cut E to Obtain Circular Backlash B[a]					
All on One Gear	$1.93B$	$1.59B$	$1.37B$	$1.07B$	$0.87B$
One-half on Each Gear	$0.97B$	$0.79B$	$0.69B$	$0.54B$	$0.43B$
Excess Depth of Cut E to Obtain Backlash B_b Normal to Tooth Profile[b]					
All on One Gear	$2.00B_b$	$1.66B_b$	$1.46B_b$	$1.18B_b$	$1.99B_b$
One-half on Each Gear	$1.00B_b$	$0.83B_b$	$0.73B_b$	$0.59B_b$	$0.50B_b$

[a] Circular backlash is the amount by which the width of a tooth space is greater than the thickness of the engaging tooth on the pitch circles. As described in pages 2036 and 2040 this is what is meant by backlash unless otherwise specified.

[b] Backlash measured normal to the tooth profile by inserting a feeler gage between meshing teeth; to convert to circular backlash, $B = B_b \div \cos\phi$.

Control of Backlash Allowances in Production.—Measurement of the tooth thickness of gears is perhaps the simplest way of controlling backlash allowances in production.

There are several ways in which this may be done including: 1) chordal thickness measurements as described on page 2018; 2) caliper measurements over two or more teeth as described on page 2109; and 3) measurements over wires.

In this last method, first the theoretical measurement over wires when the backlash allowance is zero is determined by the method described on page 2094; then the amount this measurement must be reduced to obtain a desired backlash allowance is taken from the table on page 2108.

It should be understood, as explained in the section *Measurement of Backlash* that merely making tooth thickness allowances will not guarantee the amount of backlash in the ready-to-run assembly of two or more gears. Manufacturing limitations will introduce such gear errors as runout, pitch error, profile error, and lead error, and gear-housing errors in both center distance and alignment. All of these make the backlash of the assembled gears different from that indicated by tooth thickness measurements on the individual gears.

Measurement of Backlash.—Backlash is commonly measured by holding one gear of a pair stationary and rocking the other back and forth. The movement is registered by a dial indicator having its pointer or finger in a plane of rotation at or near the pitch diameter and in a direction parallel to a tangent to the pitch circle of the moving gear. If the direction of measurement is normal to the teeth, or other than as specified above, it is recommended that readings be converted to the plane of rotation and in a tangent direction at or near the pitch diameter, for purposes of standardization and comparison.

In spur gears, parallel helical gears, and bevel gears, it is immaterial whether the pinion or gear is held stationary for the test. In crossed helical and hypoid gears, readings may vary according to which member is stationary; hence, it is customary to hold the pinion stationary and measure on the gear.

In some instances, backlash is measured by thickness gages or feelers. A similar method utilizes a soft lead wire inserted between the teeth as they pass through mesh. In both methods, it is likewise recommended that readings be converted to the plane of rotation and in a tangent direction at or near the pitch diameter, taking into account the normal pressure angle, and the helix angle or spiral angle of the teeth.

Sometimes backlash in parallel helical or herringbone gears is checked by holding the gear stationary, and moving the pinion axially back and forth, readings being taken on the face or shaft of the pinion, and converted to the plane of rotation by calculation. Another method consists of meshing a pair of gears tightly together on centers and observing the variation from the specified center distance. Such readings should also be converted to the plane of rotation and in a tangent direction at or near the pitch diameter for the reasons previously given.

Measurements of backlash may vary in the same pair of gears, depending on accuracy of manufacturing and assembling. Incorrect tooth profiles will cause a change of backlash at different phases of the tooth action. Eccentricity may cause a substantial difference between maximum and minimum backlash at different positions around the gears. In stating amounts of backlash, it should always be remembered that merely making allowances on tooth thickness does not guarantee the minimum amount of backlash that will exist in assembled gears.

Fine-Pitch Gears: The measurement of backlash of fine-pitch gears, when assembled, cannot be made in the same manner and by the same techniques employed for gears of coarser pitches. In the very fine pitches, it is virtually impossible to use indicating devices for measuring backlash. Sometimes a toolmaker's microscope is used for this purpose to good advantage on very small mechanisms.

Another means of measuring backlash in fine-pitch gears is to attach a beam to one of the shafts and measure the angular displacement in inches when one member is held stationary. The ratio of the length of the beam to the nominal pitch radius of the gear or pinion to which the beam is attached gives the approximate ratio of indicator reading to circular backlash. Because of the limited means of measuring backlash between a pair of fine-pitch gears, gear centers and tooth thickness of the gears when cut must be held to very close limits. Tooth thickness of fine-pitch spur and helical gears can best be checked on a variable-center-distance fixture using a master gear. When checked in this manner, tooth thickness change = 2 × center distance change × tangent of transverse pressure angle, approximately.

Control of Backlash in Assemblies.—Provision is often made for adjusting one gear relative to the other, thereby affording complete control over backlash at initial assembly and throughout the life of the gears. Such practice is most common in bevel gearing. It is fairly common in spur and helical gearing when the application permits slight changes between shaft centers. It is practical in worm gearing only for single thread worms with low lead angles. Otherwise faulty contact results.

Another method of controlling backlash quite common in bevel gears and less common in spur and helical gears is to match the high and low spots of the runout gears of one to one ratio and mark the engaging teeth at the point where the runout of one gear cancels the runout of the mating gear.

Table 3. AGMA Backlash Allowance and Tolerance for Fine-Pitch Spur, Helical and Herringbone Gearing

Backlash Designation	Normal Diametral Pitch Range	Tooth Thinning to Obtain Backlash[a]		Resulting Approximate Backlash (per Mesh) Normal Plane[b] Inch
		Allowance, per Gear, Inch	Tolerance, per Gear, Inch	
A	20 thru 45	.002	0 to .002	.004 to .008
	46 thru 70	.0015	0 to .002	.003 to .007
	71 thru 90	.001	0 to .00175	.002 to .0055
	91 thru 200	.00075	0 to .00075	.0015 to .003
B	20 thru 60	.001	0 to .001	.002 to .004
	61 thru 120	.00075	0 to .00075	.0015 to .003
	121 thru 200	.0005	0 to .0005	.001 to .002
C	20 thru 60	.0005	0 to .0005	.001 to .002
	61 thru 120	.00035	0 to .0004	.0007 to .0015
	121 thru 200	.0002	0 to .0 to .0003	.0004 to .001
D	20 thru 60	.0025	0 to .00025	.0005 to .001
	61 thru 120	.0002	0 to .0002	.0004 to .0008
	121 thru 200	.0001	0 to .0001	.0002 to .0004
E	20 thru 60		0 to .00025	0 to .0005
	61 thru 120		0 to .0002	0 to .0004
	121 thru 200	Zero[c]	0 to .0001	0 to .0002

[a] These dimensions are shown primarily for the benefit of the gear manufacturer and represent the amount that the thickness of teeth should be reduced in the pinion and gear below the standard calculated value, to provide for backlash in the mesh. In some cases, particularly with pinions involving small numbers of teeth, it may be desirable to provide for total backlash by thinning the teeth in the gear member only by twice the allowance value shown in column (3). In this case both members will have the tolerance shown in column (4). In some cases, particularly in meshes with a small number of teeth, backlash may be achieved by an increase in basic center at distance. In such cases, neither member is reduced by the allowance shown in column (3).

[b] These dimensions indicate the approximate backlash that will occur in a mesh in which each of the mating pairs of gears have the teeth thinned by the amount referred to in Note 1, and are meshed on theoretical centers.

[c] Backlash in gear sets can also be achieved by increasing the center distance above nominal and using the teeth at standard tooth thickness. Class E backlash designation infers gear sets operating under these conditions.

Backlash in gears is the play between mating tooth surfaces. For purposes of measurement and calculation, backlash is defined as the amount by which a tooth space exceeds the thickness of an engaging tooth. When not otherwise specified, numerical values of backlash are understood to be measured at the tightest point of mesh on the pitch circle in a direction normal to the tooth surface when the gears are mounted in their specified position.

Allowance is the basic amount that a tooth is thinned from basic calculated circular tooth thickness to obtain the required backlash class.

Tolerance is the total permissible variation in the circular thickness of the teeth.

Angular Backlash in Gears.—When the backlash on the pitch circles of a meshing pair of gears is known, the angular backlash or angular play corresponding to this backlash may be computed from the following formulas.

$$\theta_D = \frac{6875B}{D} \text{ minutes} \qquad \theta_d = \frac{6875B}{d} \text{ minutes}$$

In these formulas, B = backlash between gears, in inches; D = pitch diameter of larger gear, in inches; d = pitch diameter of smaller gear, in inches; θ_D = angular backlash or angular movement of larger gear in minutes when smaller gear is held fixed and larger gear rocked back and forth; and θ_d = angular backlash or angular movement of smaller gear, in minutes, when the larger gear is held fixed and the smaller gear rocked back and forth.

Inspection of Gears.—Perhaps the most widely used method of determining relative accuracy in a gear is to rotate the gear through at least one complete revolution in intimate contact with a master gear of known accuracy. The gear to be tested and the master gear are mounted on a variable-center-distance fixture and the resulting radial displacements or changes in center distance during rotation of the gear are measured by a suitable device. Except for the effect of backlash, this so-called "composite check" approximates the action of the gear under operating conditions and gives the combined effect of the following errors: runout; pitch error; tooth-thickness variation; profile error; and lateral runout (sometimes called wobble).

Tooth-to-Tooth Composite Error, illustrated below, is the error that shows up as flicker on the indicator of a variable-center-distance fixture as the gear being tested is rotated from tooth to tooth in intimate contact with the master gear. Such flicker shows the combined or composite effect of circular pitch error, tooth-thickness variation, and profile error.

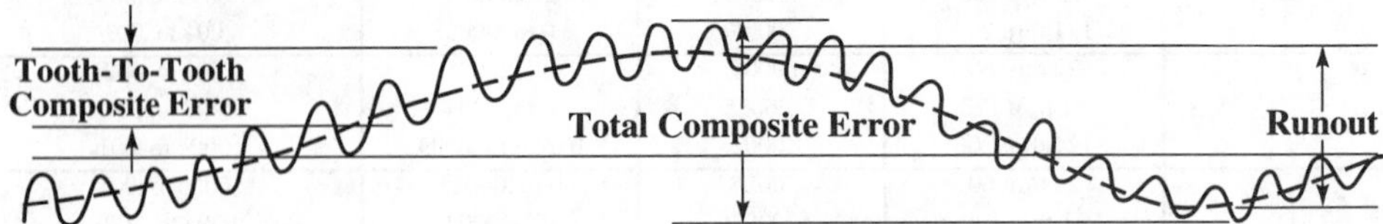

Diagram Showing Nature of Composite Errors

Total Composite Error, shown above, is made up of runout, wobble, and the tooth-to-tooth composite error; it is the total center-distance displacement read on the indicating device of the testing fixture, as shown in the accompanying diagram.

Pressure for Composite Checking of Fine-Pitch Gears.—In using a variable-center-distance fixture, excessive pressure on fine-pitch gears of narrow face width will result in incorrect readings due to deflection of the teeth. Based on tests, the following checking pressures are recommended for gears of 0.100-inch face width: 20 to 29 diametral pitch, 28 ounces; 30 to 39 pitch, 24 ounces; 40 to 49 pitch, 20 ounces; 50 to 59 pitch, 16 ounces; 60 to 79 pitch, 12 ounces; 80 to 99 pitch, 8 ounces, 100 to 149 pitch, 4 ounces; and 150 and finer pitches, 2 ounces, minimum. These recommended checking pressures are based on the use of antifriction mountings for the movable head of the checking fixture and include the pressure of the indicating device. For face widths less than 0.100 inch, the recommended pressures should be reduced proportionately; for larger widths, no increase is necessary although the force may be increased safely in the proper proportion.

British Standard for Spur and Helical Gears

British Standard For Spur And Helical Gears.—BS 436: Part 1: 1967: Spur and Helical Gears, Basic Rack Form, Pitches and Accuracy for Diametral Pitch Series, now has sections concerned with basic requirements for general tooth form, standard pitches, accuracy and accuracy testing procedures, and the showing of this information on engineering drawings to make sure that the gear manufacturer receives the required data. The latest form of the standard complies with ISO agreements. The standard pitches are in accordance with ISO R54, and the basic rack form and its modifications are in accordance with the ISO R53 "Basic Rack of Cylindrical Gears for General Engineering and for Heavy Engineering Standard".

Five grades of gear accuracy in previous versions are replaced by grades 3 to 12 of the draft ISO Standard. Grades 1 and 2 cover master gears that are not dealt with here. BS 436: Part 1: 1967 is a companion to the following British Standards:

BS 235 "Gears for Traction"

BS 545 "Bevel Gears (Machine Cut)"

BS 721 "Worm Gearing"

BS 821 "Iron Castings for Gears and Gear Blanks (Ordinary, Medium and High Grade)"

BS 978 "Fine Pitch Gears"Part 1, "Involute, Spur and Helical Gears"; Part 2, "Cycloidal Gears" (with addendum 1, PD 3376: "Double Circular Arc Type Gears."; Part 3, "Bevel Gears"

BS 1807 "Gears for Turbines and Similar Drives" Part 1, "Accuracy" Part 2, "Tooth Form and Pitches"

BS 2519 "Glossary of Terms for Toothed Gearing"

BS 3027 "Dimensions for Worm Gear Units"

BS 3696 "Master Gears"

Part 1 of BS 436 applies to external and internal involute spur and helical gears on parallel shafts and having normal diametral pitch of 20 or coarser. The basic rack and tooth form are specified, also first and second preference standard pitches and fundamental tolerances that determine the grades of gear accuracy, and requirements for terminology and notation. These requirements include: center distance a; reference circle diameter d, for pinion d_1 and wheel d_2; tip diameter d_a for pinion d_{a1} and wheel d_{a2}; center distance modification coefficient γ; face width b for pinion b_1 and wheel b_2; addendum modification coefficient x; for pinion x_1 and wheel x_2; length of arc l; diametral pitch P_t; normal diametral pitch p_n; transverse pitch p_t; number of teeth z, for pinion z_1 and wheel z_2; helix angle at reference cylinder β; pressure angle at reference cylinder α; normal pressure angle at reference cylinder α_n; transverse pressure angle at reference cylinder α_t; and transverse pressure angle, working, α_{tw}.

The basic rack tooth profile has a pressure angle of 20°.

The Standard permits the total tooth depth to be varied within 2.25 to 2.40, so that the root clearance can be increased within the limits of 0.25 to 0.040 to allow for variations in manufacturing processes; and the root radius can be varied within the limits of 0.25 to 0.39. Tip relief can be varied within the limits shown at the right in the illustration.

Standard normal diametral pitches P_n, BS 436 Part 1:1967, are in accordance with ISO R54. The preferred series, rather than the second choice, should be used where possible.

Preferred normal diametral pitches for spur and helical gears (second choices in parentheses) are: 20 (18), 16 (14), 12 (11), 10 (9), 8 (7), 6 (5.5), 5 (4.5), 4 (3.5), 3 (2.75), 2.5 (2.25), 2 (1.75), 1.5, 1.25, and 1.

Information to be Given on Drawings: British Standard BS 308, "Engineering Drawing Practice", specifies data to be included on drawings of spur and helical gears. For all gears the data should include: number of teeth, normal diametral pitch, basic rack tooth form, axial pitch, tooth profile modifications, blank diameter, reference circle diameter, and helix angle at reference cylinder (0° for straight spur gears), tooth thickness at reference cylinder, grade of gear, drawing number of mating gear, working center distance, and backlash.

For single helical gears, the above data should be supplemented with hand and lead of the tooth helix; and for double helical gears, with the hand in relation to a specific part of the face width and the lead of tooth helix.

Inspection instructions also should be included, care being taken to avoid conflicting requirements for accuracy of individual elements, and single- and dual-flank testing. Supplementary data covering specific design, manufacturing and inspection requirements or limitations may also be needed, together with other dimensions and their tolerances, material, heat treatment, hardness, case depth, surface texture, protective finishes, and drawing scale.

Addendum Modification to Involute Spur and Helical Gears.—The British Standards Institute guide PD 6457:1970 contains certain design recommendations aimed at making it possible to use standard cutting tools for some sizes of gears. Essentially, the guide covers addendum modification and includes formulas for both English and metric units.

Addendum Modification is an enlargement or reduction of gear tooth dimensions that results from displacement of the reference plane of the generating rack from its normal position. The displacement is represented by the coefficient X, X1 , or X2 , where X is the equivalent dimension for gears of unit module or diametral pitch. The addendum modification establishes a datum tooth thickness at the reference circle of the gear but does not necessarily establish the height of either the reference addendum or the working addendum. In any pair of gears, the datum tooth thicknesses are those that always give zero backlash at the meshing center distance. Normal practice requires allowances for backlash for all unmodified gears.

Taking full advantage of the adaptability of the involute system allows various tooth design features to be obtained. Addendum modification has the following applications: avoiding undercut tooth profiles; achieving optimum tooth proportions and control of the proportion of receding to approaching contact; adapting a gear pair to a predetermined center distance without recourse to non-standard pitches; and permitting use of a range of working pressure angles using standard geometry tools.

BS 436, Part 3:1986 "Spur and Helical Gears".—This part provides methods for calculating contact and root bending stresses for metal involute gears, and is somewhat similar to the ANSI/AGMA Standard for calculating stresses in pairs of involute spur or helical gears. Stress factors covered in the British Standard include the following:

Tangential Force is the nominal force for contact and bending stresses.

Zone Factor accounts for the influence of tooth flank curvature at the pitch point on Hertzian stress.

Contact Ratio Factor takes account of the load-sharing influence of the transverse contact ratio and the overlap ratio on the specific loading.

Elasticity Factor takes into account the influence of the modulus of elasticity of the material and of Poisson's ratio on the Hertzian stress.

Basic Endurance Limit for contact makes allowance for the surface hardness.

Material Quality covers the quality of the material used.

Lubricant Influence, Roughness, and Speed The lubricant viscosity, surface roughness and pitch line speed affect the lubricant film thickness, which in turn, affects the Hertzian stresses.

Work Hardening Factor accounts for the increase in surface durability due to the meshing action.

Size Factor covers the possible influences of size on the material quality and its response to manufacturing processes.

Life Factor accounts for the increase in permissible stresses when the number of stress cycles is less than the endurance life.

Application Factor allows for load fluctuations from the mean load or loads in the load histogram caused by sources external to the gearing.

Dynamic Factor allows for load fluctuations arising from contact conditions at the gear mesh.

Load Distribution accounts for the increase in local load due to maldistribution of load across the face of the gear tooth caused by deflections, alignment tolerances and helix modifications.

Minimum Demanded and Actual Safety Factor The minimum demanded safety factor is agreed between the supplier and the purchaser. The actual safety factor is calculated.

Geometry Factors allow for the influence of the tooth form, the effect of the fillet and the helix angle on the nominal bending stress for the application of load at the highest point of single pair tooth contact.

Sensitivity Factor allows for the sensitivity of the gear material to the presence of notches such as the root fillet.

Surface Condition Factor accounts for reduction of the endurance limit due to flaws in the material and the surface roughness of the tooth root fillets.

ISO TC/600.—The ISO TC/600 Standard is similar to BS 436, Part 3:1986, but is far more comprehensive. For general gear design, the ISO Standard provides a complicated method of arriving at a conclusion similar to that reached by the less complex British Standard. Factors additional to the above that are included in the ISO Standard include the following

Application Factor takes account of dynamic overloads from sources external to the gearing.

Dynamic Factor allows for internally generated dynamic loads caused by vibrations of the pinion and wheel against each other.

Load Distribution makes allowance for the effects of non-uniform distribution of load across the face width, depending on the mesh alignment error of the loaded gear pair and the mesh stiffness.

Transverse Load Distribution Factor takes into account the effect of the load distribution on gear tooth contact stresses.

Gear Tooth Stiffness Constants are defined as the load needed to deform one or several meshing gear teeth having 1 mm face width, by an amount of 1 μm (0.00004 in).

Allowable Contact Stress is the permissible Hertzian pressure on the gear tooth face.

Minimum demanded and Calculated Safety Factors The minimum demanded safety factor is agreed between the supplier and the customer. The calculated safety factor is the actual safety factor of the gear pair.

Zone Factor accounts for the influence on the Hertzian pressure of the tooth flank curvature at the pitch point.

Elasticity Factor takes account of the influence of the material properties such as the modulus of elasticity and Poisson's ratio.

Contact Ratio Factor accounts for the influence of the transverse contact ratio and the overlap ratio on the specific surface load of the gears.

Helix Angle Factor makes allowance for the influence of the helix angle on the surface durability.

Endurance Limit is the limit of repeated Hertzian stresses that can be permanently endured by a given material

Life Factor takes account of a higher permissible Hertzian stress if only limited durability is demanded.

Lubrication Film Factor The film of lubricant between the tooth flanks influences the surface load capacity. Factors include the oil viscosity, pitch line velocity and roughness of the tooth flanks.

Work Hardening Factor takes account of the increase in surface durability due to meshing a steel wheel with a hardened pinion having smooth tooth surfaces.

Coefficient of Friction The mean value of the local coefficient of friction depends on the lubricant, surface roughness, the lay of surface irregularities, material properties of the tooth flanks, and the force and size of tangential velocities.

Bulk Temperature Thermal Flash Factor is dependent on the moduli of elasticity and thermal contact coefficients of pinion and wheel materials and the geometry of the line of action.

Welding Factor Accounts for different tooth materials and heat treatments.

Geometrical Factor is defined as a function of the gear ratio and the dimensionless parameter on the line of action.

Integral Temperature Criterion The integral temperature of the gears depends on the lubricant viscosity and tendency toward cuffing and scoring of the gear materials.

Examination of the above factors shows the similarity in the approach of the British and the ISO Standards to that of the ANSI/AGMA Standards. Slight variations in the methods used to calculate the factors will result in different allowable stress figures. Experimental work using some of the stressing formulas has shown wide variations and designers must continue to rely on experience to arrive at satisfactory results.

Standards Nomenclature

All standards are referenced and identified throughout this book by an alphanumeric prefix which designates the organization that administered the development work on the standard, and followed by a standards number.

All standards are reviewed by the relevant committees at regular time intervals, as specified by the overseeing standards organization, to determine whether the standard should be confirmed (reissued without changes other than correction of typographical errors), updated, or removed from service.

The following is for example use only. ANSI B18.8.2-1984, R1994 is a standard for Taper, Dowel, Straight, Grooved, and Spring Pins. ANSI refers to the American National Standards Institute that is responsible for overseeing the development or approval of the standard, and B18.8.2 is the number of the standard. The first date, 1984, indicates the year in which the standard was issued, and the sequence R1994 indicates that this standard was reviewed and reaffirmed in that 1994. The current designation of the standard, ANSI/ASME B18.8.2-1995, indicates that it was revised in 1995; it is ANSI approved; and, ASME (American Society of Mechanical Engineers) was the standards body responsible for development of the standard. This standard is sometimes also designated ASME B18.8.2-1995.

ISO (International Organization for Standardization) standards use a slightly different format, for example, ISO 5127-1:1983. The entire ISO reference number consists of a prefix ISO, a serial number, and the year of publication.

Aside from the content, ISO standards differ from American National standards in that they often smaller focused documents, which in turn reference other standards or other parts of the same standard. Unlike the numbering scheme used by ANSI, ISO standards related to a particular topic often do not carry sequential numbers nor are they in consecutive series.

British Standards Institute standards use the following format: BS 1361: 1971 (1986). The first part is the organization prefix BS, followed by the reference number and the date of issue. The number in parenthesis is the date that the standard was most recently reconfirmed. British Standards may also be designated *withdrawn* (no longer to be used) and *obsolescent* (going out of use, but may be used for servicing older equipment).

Organization	Web Address	Organization	Web Address
ISO (International Organization for Standardization)	www.iso.ch	JIS (Japanese Industrial Standards)	www.jisc.org
IEC (International Electrotechnical Commission)	www.iec.ch	ASME (American Society of Mechanical Engineers)	www.asme.org
ANSI (American National Standards Institute)	www.ansi.org	SAE (Society of Automotive Engineers)	www.sae.org
BSI (British Standards Institute)	www.bsi-inc.org	SME (Society of Manufacturing Engineers)	www.sme.org

INTERNAL GEARING

Internal Spur Gears.—An internal gear may be proportioned like a standard spur gear turned "outside in" or with addendum and dedendum in reverse positions; however, to avoid interference or improve the tooth form and action, the internal diameter of the gear should be increased and the outside diameter of the mating pinion is also made larger than the size based upon standard or conventional tooth proportions. The extent of these enlargements will be illustrated by means of examples given following table, *Rules for Internal Gears—20-degree Full-Depth Teeth*. The 20-degree involute full-depth tooth form is recommended for internal gears; the 20-degree stub tooth and the 14½-degree full-depth tooth are also used.

Methods of Cutting Internal Gears.—Internal spur gears are cut by methods similar in principle to those employed for external spur gears.

They may be cut by one of the following methods: 1) By a generating process, as when using a Fellows gear shaper; 2) by using a formed cutter and milling the teeth; 3) by planing, using a machine of the template or form-copying type (especially applicable to gears of large pitch); and 4) by using a formed tool that reproduces its shape and is given a planing action either on a slotting or a planing type of machine.

Internal gears frequently have a web at one side that limits the amount of clearance space at the ends of the teeth. Such gears may be cut readily on a gear shaper. The most practical method of cutting very large internal gears is on a planer of the form-copying type. A regular spur gear planer is equipped with a special tool holder for locating the tool in the position required for cutting internal teeth.

Formed Cutters for Internal Gears.—When formed cutters are used, a special cutter usually is desirable, because the tooth spaces of an internal gear are not the same shape as the tooth spaces of external gearing having the same pitch and number of teeth. This difference is because an internal gear is a spur gear "turned outside in." According to one rule, the standard No. 1 cutter for external gearing may be used for internal gears of 4 diametral pitch and finer, when there are 60 or more teeth. This No. 1 cutter, as applied to external gearing, is intended for all gears having from 135 teeth to a rack. The finer the pitch and the larger the number of teeth, the better the results obtained with a No. 1 cutter. The standard No. 1 cutter is considered satisfactory for jobbing work, and usually when the number of gears to be cut does not warrant obtaining a special cutter, although the use of the No. 1 cutter is not practicable when the number of teeth in the pinion is large in proportion to the number of teeth in the internal gear.

Arc Thickness of Internal Gear Tooth.—*Rule:* If internal diameter of an internal gear is enlarged as determined by Rules 1 and 2 for Internal Diameters (see *Rules for Internal Gears—20-degree Full-Depth Teeth*), the arc tooth thickness at the pitch circle equals 1.3888 divided by the diametral pitch, assuming a pressure angle of 20 degrees.

Arc Thickness of Pinion Tooth.—*Rule:* If the pinion for an internal gear is larger than conventional size (see Outside Diameter of Pinion for Internal Gear, under *Rules for Internal Gears—20-degree Full-Depth Teeth*), then the arc tooth thickness on the pitch circle equals 1.7528 divided by the diametral pitch, assuming a pressure angle of 20 degrees.

Note: For chordal thickness and chordal addendum, see rules and formulas for spur gears.

Relative Sizes of Internal Gear and Pinion.—If a pinion is too large or too near the size of its mating internal gear, serious interference or modification of the tooth shape may occur.

Rule: For internal gears having a 20-degree pressure angle and full-depth teeth, the difference between the numbers of teeth in gear and pinion should not be less than 12. For teeth of stub form, the smallest difference should be 7 or 8 teeth. For a pressure angle of 14½ degrees, the difference in tooth numbers should not be less than 15.

Rules for Internal Gears—20-degree Full-Depth Teeth

To Find	Rule
Pitch Diameter	*Rule:* To find the pitch diameter of an internal gear, divide the number of internal gear teeth by the diametral pitch. The pitch diameter of the mating pinion also equals the number of pinion teeth divided by the diametral pitch, the same as for external spur gears.
Internal Diameter (Enlarged to Avoid Interference)	*Rule* 1: For internal gears to mesh with pinions having 16 teeth or more, subtract 1.2 from the number of teeth and divide the remainder by the diametral pitch. *Example:* An internal gear has 72 teeth of 6 diametral pitch and the mating pinion has 18 teeth; then $\text{Internal diameter} = \frac{72 - 1.2}{6} = 11.8 \text{ inches}$ *Rule* 2: If circular pitch is used, subtract 1.2 from the number of internal gear teeth, multiply the remainder by the circular pitch, and divide the product by 3.1416.
Internal Diameter (Based upon Spur Gear Reversed)	*Rule:* If the internal gear is to be designed to conform to a spur gear turned outside in, subtract 2 from the number of teeth and divide the remainder by the diametral pitch to find the internal diameter. *Example:* (Same as Example above.) $\text{Internal diameter} = \frac{72 - 2}{6} = 11.666 \text{ inches}$
Outside Diameter of Pinion for Internal Gear	*Note:* If the internal gearing is to be proportioned like standard spur gearing, use the rule or formula previously given for spur gears in determining the outside diameter. The rule and formula following apply to a pinion that is enlarged and intended to mesh with an internal gear enlarged as determined by the preceding Rules 1 and 2 above. *Rule:* For pinions having 16 teeth or more, add 2.5 to the number of pinion teeth and divide by the diametral pitch. *Example* 1: A pinion for driving an internal gear is to have 18 teeth (full depth) of 6 diametral pitch; then $\text{Outside diameter} = \frac{18 + 2.5}{6} = 3.416 \text{ inches}$ By using the rule for external spur gears, the outside diameter = 3.333 inches.
Center Distance	*Rule:* Subtract the number of pinion teeth from the number of internal gear teeth and divide the remainder by two times the diametral pitch.
Tooth Thickness	See paragraphs, *Arc Thickness of Internal Gear Tooth* and *Effect of Diameter of Cutting on Profile and Pressure Angle of Worms*, on previous page.

Hypoid Gears

Hypoid gears are offset and in effect, are spiral gears whose axes do not intersect but are staggered by an amount decided by the application. Due to the offset, contact between the teeth of the two gears does not occur along a surface line of the cones as it does with spiral bevels having intersecting axes, but along a curve in space inclined to the surface line. The basic solids of the hypoid gear members are not cones, as in spiral bevels, but are hyperboloids of revolution which cannot be projected into the common plane of ordinary flat gears, thus the name hypoid. The visualization of hypoid gears is based on an imaginary flat gear which is a substitute for the theoretically correct helical surface. If certain rules are observed during the calculations to fix the gear dimensions, the errors that result from the use of an imaginary flat gear as an approximation are negligible.

The staggered axes result in meshing conditions that are beneficial to the strength and running properties of the gear teeth. A uniform sliding action takes place between the teeth, not only in the direction of the tooth profile but also longitudinally, producing ideal conditions for movement of lubricants. With spiral gears, great differences in sliding motion arise over various portions of the tooth surface, creating vibration and noise. Hypoid gears are almost free from the problems of differences in these sliding motions and the teeth also have larger curvature radii in the direction of the profile. Surface pressures are thus reduced so that there is less wear and quieter operation.

The teeth of hypoid gears are 1.5 to 2 times stronger than those of spiral bevel gears of the same dimensions, made from the same material. Certain limits must be imposed on the dimensions of hypoid gear teeth so that their proportions can be calculated in the same way as they are for spiral bevel gears. The offset must not be larger than 1/7th of the ring gear outer diameter, and the tooth ratio must not be much less than 4 to 1. Within these limits, the tooth proportions can be calculated in the same way as for spiral bevel gears and the radius of lengthwise curvature can be assumed in such a way that the normal module is a maximum at the center of the tooth face width to produce stabilized tooth bearings.

If the offset is larger or the ratio is smaller than specified above, a tooth form must be selected that is better adapted to the modified meshing conditions. In particular, the curvature of the tooth length curve must be determined with other points in view. The limits are only guidelines since it is impossible to account for all other factors involved, including the pitch line speed of the gears, lubrication, loads, design of shafts and bearings, and the general conditions of operation.1

Of the three different designs of hypoid bevel gears now available, the most widely used, especially in the automobile industry, is the Gleason system. Two other hypoid gear systems have been introduced by Oerlikon (Swiss) and Klingelnberg (German). All three methods use the involute gear form, but they have teeth with differing curvatures, produced by the cutting method. Teeth in the Gleason system are arc shaped and their depth tapers. Both the European systems are designed to combine rolling with the sideways motion of the teeth and use a constant tooth depth. Oerlikon uses an epicycloidal tooth form and Klingelnberg uses a true involute form.

With their circular arcuate tooth face curves, Gleason hypoid gears are produced with multi-bladed face milling cutters. The gear blank is rolled relative to the rotating cutter to make one inter-tooth groove, then the cutter is withdrawn and returned to its starting position while the blank is indexed into the position for cutting the next tooth. Both roughing and finishing cutters are kept parallel to the tooth root lines, which are at an angle to the gear pitch line. Depending on this angularity, plus the spiral angle, a correction factor must be calculated for both the leading and trailing faces of the gear tooth.

In operation, the convex faces of the teeth on one gear always bear on the concave faces of the teeth on the mating gear. For correct meshing between the pinion and gear wheel, the spiral angles should not vary over the full face width. The tooth form generated is a logarithmic spiral and, as a compromise, the cutter radius is made equal to the mean radius of a corresponding logarithmic spiral.

The involute tooth face curves of the Klingelnberg system gears have constant-pitch teeth cut by (usually) a single-start taper hob. The machine is set up to rotate both the cutter and the gear blank at the correct relative speeds. The surface of the hob is set tangential to a circle radius, which is the gear base circle, from which all the parallel involute curves are struck. To keep the hob size within reasonable dimensions, the cone must lie a minimum distance within the teeth and this requirement governs the size of the module.

Both the module and the tooth depth are constant over the full face width and the spiral angle varies. The cutting speed variations, especially with regard to crown wheels, over the cone surface of the hob, make it difficult to produce a uniform surface finish on the teeth, so a finishing cut is usually made with a truncated hob which is tilted to produce the required amount of crowning automatically, for correct tooth marking and finishing. The dependence of the module, spiral angle and other features on the base circle radius, and the need for suitable hob proportions restrict the gear dimensions and the system cannot be used for gears with a low or zero angle. However, gears can be cut with a large root radius giving teeth of high strength. The favorable geometry of the tooth form gives quieter running and tolerance of inaccuracies in assembly.

Teeth of gears made by the Oerlikon system have elongated epicycloidal form, produced with a face-type rotating cutter. Both the cutter and the gear blank rotate continuously, with no indexing. The cutter head has separate groups of cutters for roughing, outside cutting and inside cutting so that tooth roots and flanks are cut simultaneously, but the feed is divided into two stages. As stresses are released during cutting, there is some distortion of the blank and this distortion will usually be worse for a hollow crown wheel than for a solid pinion.

All the heavy cuts are taken during the first stages of machining with the Oerlikon system and the second stage is used to finish the tooth profile accurately, so distortion effects are minimized. As with the Klingelnberg process, the Oerlikon system produces a variation in spiral angle and module over the width of the face, but unlike the Klingelnberg method, the tooth length curve is cycloidal. It is claimed that, under load, the tilting force in an Oerlikon gear set acts at a point 0.4 times the distance from the small diameter end of the gear and not in the mid-tooth position as in other gear systems, so that the radius is obviously smaller and the tilting moment is reduced, resulting in lower loading of the bearings.

Gears cut by the Oerlikon system have tooth markings of different shape than gears cut by other systems, showing that more of the face width of the Oerlikon tooth is involved in the load-bearing pattern. Thus, the surface loading is spread over a greater area and becomes lighter at the points of contact.

Bevel Gearing

Types of Bevel Gears.—Bevel gears are conical gears, that is, gears in the shape of cones, and are used to connect shafts having intersecting axes. Hypoid gears are similar in general form to bevel gears, but operate on axes that are offset. With few exceptions, most bevel gears may be classified as being either of the straight-tooth type or of the curved-tooth type. The latter type includes spiral bevels, Zerol bevels, and hypoid gears. The following is a brief description of the distinguishing characteristics of the different types of bevel gears.

Straight Bevel Gears: The teeth of this most commonly used type of bevel gear are straight but their sides are tapered so that they would intersect the axis at a common point called the pitch cone apex if extended inward. The face cone elements of most straight bevel gears, however, are now made parallel to the root cone elements of the mating gear to obtain uniform clearance along the length of the teeth. The face cone elements of such gears, therefore, would intersect the axis at a point inside the pitch cone. Straight bevel gears are the easiest to calculate and are economical to produce.

Straight bevel gear teeth may be generated for full-length contact or for localized contact. The latter are slightly convex in a lengthwise direction so that some adjustment of the gears during assembly is possible and small displacements due to load deflections can occur without undesirable load concentration on the ends of the teeth. This slight lengthwise rounding of the tooth sides need not be computed in the design but is taken care of automatically in the cutting operation on the newer types of bevel gear generators.

Zerol Bevel Gears: The teeth of Zerol bevel gears are curved but lie in the same general direction as the teeth of straight bevel gears. They may be thought of as spiral bevel gears of zero spiral angle and are manufactured on the same machines as spiral bevel gears. The face cone elements of Zerol bevel gears do not pass through the pitch cone apex but instead are approximately parallel to the root cone elements of the mating gear to provide uniform tooth clearance. The root cone elements also do not pass through the pitch cone apex because of the manner in which these gears are cut. Zerol bevel gears are used in place of straight bevel gears when generating equipment of the spiral type but not the straight type is available, and may be used when hardened bevel gears of high accuracy (produced by grinding) are required.

Spiral Bevel Gears: Spiral bevel gears have curved oblique teeth on which contact begins gradually and continues smoothly from end to end. They mesh with a rolling contact similar to straight bevel gears. As a result of their overlapping tooth action, however, spiral bevel gears will transmit motion more smoothly than straight bevel or Zerol bevel gears, reducing noise and vibration that become especially noticeable at high speeds.

One of the advantages associated with spiral bevel gears is the complete control of the localized tooth contact. By making a slight change in the radii of curvature of the mating tooth surfaces, the amount of surface over which tooth contact takes place can be changed to suit the specific requirements of each job. Localized tooth contact promotes smooth, quiet running spiral bevel gears, and permits some mounting deflections without concentrating the load dangerously near either end of the tooth. Permissible deflections established by experience are given under the heading *Mountings for Bevel Gears*.

Because their tooth surfaces can be ground, spiral bevel gears have a definite advantage in applications requiring hardened gears of high accuracy. The bottoms of the tooth spaces and the tooth profiles may be ground simultaneously, resulting in a smooth blending of the tooth profile, the tooth fillet, and the bottom of the tooth space. This feature is important from a strength standpoint because it eliminates cutter marks and other surface interruptions that frequently result in stress concentrations.

Hypoid Gears: In general appearance, hypoid gears resemble spiral bevel gears, except that the axis of the pinion is offset relative to the gear axis. If there is sufficient offset, the shafts may pass one another thus permitting the use of a compact straddle mounting on the gear and pinion. Whereas a spiral bevel pinion has equal pressure angles and symmetrical profile curvatures on both sides of the teeth, a hypoid pinion properly conjugate to a mating gear having equal pressure angles on both sides of the teeth must have nonsymmetrical profile curvatures for proper tooth action. In addition, to obtain equal arcs of motion for both sides of the teeth, it is necessary to use unequal pressure angles on hypoid pinions. Hypoid gears are usually designed so that the pinion has a larger spiral angle than the gear. The advantage of such a design is that the pinion diameter is increased and is stronger than a corresponding spiral bevel pinion. This diameter increment permits the use of comparatively high ratios without the pinion becoming too small to allow a bore or shank of adequate size. The sliding action along the lengthwise direction of their teeth in hypoid gears is a function of the difference in the spiral angles on the gear and pinion. This sliding effect makes such gears even smoother running than spiral bevel gears. Grinding of hypoid gears can be accomplished on the same machines used for grinding spiral bevel and Zerol bevel gears.

Applications of Bevel and Hypoid Gears.—Bevel and hypoid gears may be used to transmit power between shafts at practically any angle and speed. The particular type of gearing best suited for a specific job, however, depends on the mountings and the operating conditions.

Straight and Zerol Bevel Gears: For peripheral speeds up to 1000 feet per minute, where maximum smoothness and quietness are not the primary consideration, straight and Zerol bevel gears are recommended. For such applications, plain bearings may be used for radial and axial loads, although the use of antifriction bearings is always preferable. Plain bearings permit a more compact and less expensive design, which is one reason why straight and Zerol bevel gears are much used in differentials. This type of bevel gearing is the simplest to calculate and set up for cutting, and is ideal for small lots where fixed charges must be kept to a minimum.

Zerol bevel gears are recommended in place of straight bevel gears where hardened gears of high accuracy are required, because Zerol gears may be ground; and when only spiral-type equipment is available for cutting bevel gears.

Spiral Bevel and Hypoid Gears: Spiral bevel and hypoid gears are recommended for applications where peripheral speeds exceed 1000 feet per minute or 1000 revolutions per minute. In many instances, they may be used to advantage at lower speeds, particularly where extreme smoothness and quietness are desired. For peripheral speeds above 8000 feet per minute, ground gears should be used.

For large reduction ratios the use of spiral and hypoid gears will reduce the overall size of the installation because the continuous pitch line contact of these gears makes it practical to obtain smooth performance with a smaller number of teeth in the pinion than is possible with straight or Zerol bevel gears.

Hypoid gears are recommended for industrial applications: when maximum smoothness of operation is desired; for high reduction ratios where compactness of design, smoothness of operation, and maximum pinion strength are important; and for nonintersecting shafts.

Bevel and hypoid gears may be used for both speed-reducing and speed-increasing drives. In speed-increasing drives, however, the ratio should be kept as low as possible and the pinion mounted on antifriction bearings; otherwise bearing friction will cause the drive to lock.

Notes on the Design of Bevel Gear Blanks.—The quality of any finished gear is dependent, to a large degree, on the design and accuracy of the gear blank. A number of factors that affect manufacturing economy as well as performance must be considered.

A gear blank should be designed to avoid localized stresses and serious deflections within itself. Sufficient thickness of metal should be provided under the roots of gear teeth to give them proper support. As a general rule, the amount of metal under the root should equal the whole depth of the tooth; this metal depth should be maintained under the small ends of the teeth as well as under the middle. On webless-type ring gears, the minimum stock between the root line and the bottom of tap drill holes should be one-third the tooth depth. For heavily loaded gears, a preliminary analysis of the direction and magnitude of the forces is helpful in the design of both the gear and its mounting. Rigidity is also necessary for proper chucking when cutting the teeth. For this reason, bores, hubs, and other locating surfaces must be in proper proportion to the diameter and pitch of the gear. Small bores, thin webs, or any condition that necessitates excessive overhang in cutting should be avoided.

Other factors to be considered are the ease of machining and, in gears that are to be hardened, proper design to ensure the best hardening conditions. It is desirable to provide a locating surface of generous size on the backs of gears. This surface should be machined or ground square with the bore and is used both for locating the gear axially in assembly and for holding it when the teeth are cut. The front clamping surface must, of course, be flat and parallel to the back surface. In connection with cutting the teeth on Zerol bevel, spiral bevel, and hypoid gears, clearance must be provided for face-mill type cutters; front and rear hubs should not intersect the extended root line of the gear or they will interfere with the path of the cutter. In addition, there must be enough room in the front of the gear for the clamp nut that holds the gear on the arbor, or in the chuck, while cutting the teeth. The same considerations must be given to straight bevel gears that are to be generated using a circular-type cutter instead of reciprocating tools.

Mountings for Bevel Gears.—Rigid mountings should be provided for bevel gears to keep the displacements of the gears under operating loads within recommended limits. To align gears properly, care should be taken to ensure accurately machined mountings, properly fitted keys, and couplings that run true and square.

As a result of deflection tests on gears and their mountings, and having observed these same units in service, the *Gleason Works* recommends that the following allowable deflections be used for gears from 6 to 15 inches in diameter: neither the pinion nor the gear should lift or depress more than 0.003 inch at the center of the face width; the pinion should not yield axially more than 0.003 inch in either direction; and the gear should not yield axially more than 0.003 inch in either direction on 1 to 1 ratio gears (miter gears), or near miters, or more than 0.010 inch away from the pinion on higher ratios.

When deflections exceed these limits, additional problems are involved in obtaining satisfactory gears. It becomes necessary to narrow and shorten the tooth contacts to suit the more flexible mounting. These changes decrease the bearing area, raise the unit tooth pressure, and reduce the number of teeth in contact, resulting in increased noise and the danger of surface failure as well as tooth breakage.

Spiral bevel and hypoid gears in general should be mounted on antifriction bearings in an oil-tight case. Designs for a given set of conditions may use plain bearings for radial and thrust loads, maintaining gears in satisfactory alignment is usually more easily accomplished with ball or roller bearings.

Bearing Spacing and Shaft Stiffness: Bearing spacing and shaft stiffness are extremely important if gear deflections are to be minimized. For both straddle mounted and overhung mounted gears the spread between bearings should never be less than 70 per cent of the pitch diameter of the gear. On overhung mounted gears the spread should be at least 2½ times the overhang and, in addition, the shaft diameter should be equal to or preferably greater than the overhang to provide sufficient shaft stiffness. When two spiral bevel or hypoid gears are mounted on the same shaft, the axial thrust should be taken at one place only and near the gear where the greater thrust is developed. Provision should be made for adjusting both the gear and pinion axially in assembly. Details on how this may be accomplished are given in the *Gleason Works* booklet, "Assembling Bevel Gears."

Cutting Bevel Gear Teeth.—A correctly formed bevel gear tooth has the same sectional shape throughout its length, but on a uniformly diminishing scale from the large to the small end. The only way to obtain this correct form is by using a generating type of bevel gear cutting machine. This accounts, in part, for the extensive use of generating type gear cutting equipment in the production of bevel gears.

Bevel gears too large to be cut by generating equipment (100 inches or over in diameter) may be produced by a form-copying type of gear planer. With this method, a template or former is used to mechanically guide a single cutting tool in the proper path to cut the profile of the teeth. Since the tooth profile produced by this method is dependent on the contour of the template used, it is possible to produce tooth profiles to suit a variety of requirements.

Although generating methods are to be preferred, there are still some cases where straight bevel gears are produced by milling. Milled gears cannot be produced with the accuracy of generated gears and generally are not suitable for use in high-speed applications or where angular motion must be transmitted with a high degree of accuracy. Milled gears are used chiefly as replacement gears in certain applications, and gears which are subsequently to be finished on generating type equipment are sometimes roughed out by milling. Formulas and methods used for the cutting of bevel gears are given in the latter part of this section.

In producing gears by generating methods, the tooth curvature is generated from a straight-sided cutter or tool having an angle equal to the required pressure angle. This tool represents the side of a crown gear tooth. The teeth of a true involute crown gear, however, have sides which are very slightly curved. If the curvature of the cutting tool conforms to that of the involute crown gear, an involute form of bevel gear tooth will be obtained. The use of a straight-sided tool is more practical and results in a very slight change of tooth shape to what is known as the "octoid" form. Both the octoid and involute forms of bevel gear tooth give theoretically correct action.

Bevel gear teeth, like those for spur gears, differ as to pressure angle and tooth proportions. The whole depth and the addendum at the large end of the tooth may be the same as for a spur gear of equal pitch. Most bevel gears, however, both of the straight tooth and spiral-bevel types, have lengthened pinion addendums and shortened gear addendums as in the case of some spur gears, the amount of departure from equal addendums varying with the ratio of gearing. Long addendums on the pinion are used principally to avoid undercut and to increase tooth strength. In addition, where long and short addendums are used, the tooth thickness of the gear is decreased and that of the pinion increased to provide a better balance of strength. See the Gleason Works System for straight and spiral bevel gears and also the British Standard.

Nomenclature for Bevel Gears.—The accompanying diagram, *Bevel Gear Nomenclature*, illustrates various angles and dimensions referred to in describing bevel gears. In connection with the face angles shown in the diagram, it should be noted that the face cones are made parallel to the root cones of the mating gears to provide uniform clearance along the length of the teeth.

American Standard for Bevel Gears.—American Standard ANSI/AGMA 2005-B88, Design Manual for Bevel Gears, replaces AGMA Standards 202.03, 208.03, 209.04, and 330.01, and provides standards for design of straight, zerol, and spiral bevel gears and hypoid gears with information on fabrication, inspection, and mounting. The information covers preliminary design, drawing formats, materials, rating, strength, inspection, lubrication, mountings, and assembly. Blanks for standard taper, uniform depth, duplex taper, and tilted root designs are included so that the material applies to users of Gleason, Klingelnberg, and Oerlikon gear cutting machines.

Formulas for Dimensions of Milled Bevel Gears.—As explained earlier, most bevel gears are produced by generating methods. Even so, there are applications for which it may be desired to cut a pair of mating bevel gears by using rotary formed milling cutters. Examples of such applications include replacement gears for certain types of equipment and gears for use in experimental developments.

The tooth proportions of milled bevel gears differ in some respects from those of generated gears, the principal difference being that for milled bevel gears the tooth thicknesses of pinion and gear are made equal, and the addendum and dedendum of the pinion are respectively the same as those of the gear. The rules and formulas in the accompanying table may be used to calculate the dimensions of milled bevel gears with shafts at a right angle, an acute angle, and an obtuse angle.

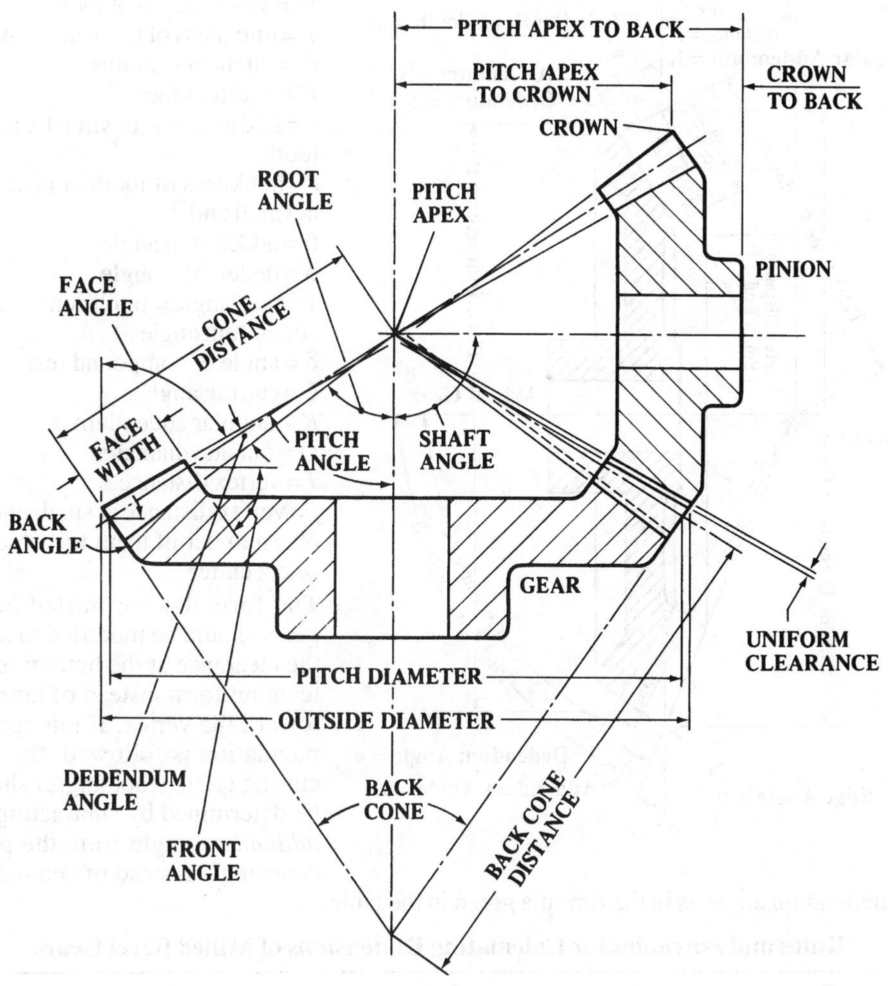

Bevel Gear Nomenclature

In the accompanying diagram and list of notations, the various terms and symbols applied to milled bevel gears are as indicated.

N = number of teeth
P = diametral pitch
p = circular pitch
α = pitch cone angle and edge angle
Σ = angle between shafts
D = pitch diameter
S = addendum
$S+A$ = dedendum (A = clearance)

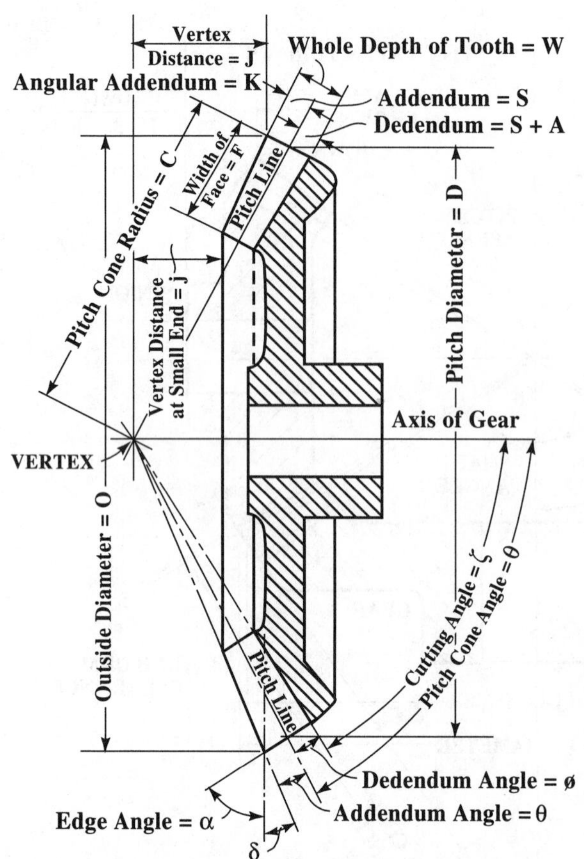

W = whole depth of tooth
T = thickness of tooth at pitch line
C = pitch cone radius
F = width of face
s = addendum at small end of tooth
t = thickness of tooth at pitch line at small end
θ = addendum angle
ϕ = dedendum angle
γ = face angle = pitch cone angle + addendum angle
δ = angle of compound rest
ζ = cutting angle
K = angular addendum
O = outside diameter
J = vertex distance
j = vertex distance at small end
N' = number of teeth for which to select cutter

The formulas for milled bevel gears should be modified to make the clearance at the bottom of the teeth uniform instead of tapering toward the vertex. If this recommendation is followed, then the cutting angle (root angle) should be determined by subtracting the *addendum* angle from the pitch cone angle instead of subtracting the dedendum angle as in the formula given in the table.

Rules and Formulas for Calculating Dimensions of Milled Bevel Gears

To Find	Rule	Formula
Pitch Cone Angle of Pinion	Divide the sine of the shaft angle by the sum of the cosine of the shaft angle and the quotient obtained by dividing the number of teeth in the gear by the number of teeth in the pinion; this gives the tangent. *Note:* For shaft angles greater than 90° the cosine is negative.	$\tan\alpha P = \frac{\sin\Sigma}{\frac{N_G}{N_P} + \cos\Sigma}$ For 90° shaft angle, $\tan\alpha P = \frac{N_P}{N_G}$
Pitch Cone Angle of Gear	Subtract the pitch cone angle of the pinion from the shaft angle.	$\alpha G = \Sigma - \alpha P$
Pitch Diameter	Divide the number of teeth by the diametral pitch.	$D = N \div P$

Rules and Formulas for Calculating Dimensions of Milled Bevel Gears *(Continued)*

	To Find	Rule	Formula
These dimensions are the same for both gear and pinion.	Addendum	Divide 1 by the diametral pitch.	$S = 1 \div P$
	Dedendum	Divide 1.157 by the diametral pitch.	$S + A = 1.157 \div P$
	Whole Depth of Tooth	Divide 2.157 by the diametral pitch.	$W = 2.157 \div P$
	Thickness of Tooth at Pitch Line	Divide 1.571 by the diametral pitch.	$T = 1.571 \div P$
	Pitch Cone Radius	Divide the pitch diameter by twice the sine of the pitch cone angle.	$C = \frac{D}{2 \times \sin\alpha}$
	Addendum of Small End of Tooth	Subtract the width of face from the pitch cone radius, divide the remainder by the pitch cone radius and multiply by the addendum.	$s = S \times \frac{C - F}{C}$
	Thickness of Tooth at Pitch Line at Small End	Subtract the width of face from the pitch cone radius, divide the remainder by the pitch cone radius and multiply by the thickness of the tooth at pitch line.	$t = T \times \frac{C - F}{C}$
	Addendum Angle	Divide the addendum by the pitch cone radius to get the tangent.	$\tan\theta = \frac{S}{C}$
	Dedendum Angle	Divide the dedendum by the pitch cone radius to get the tangent.	$\tan\phi = \frac{S + A}{C}$
	Face Width (Max.)	Divide the pitch cone radius by 3 or divide 8 by the diametral pitch, whichever gives the smaller value.	$F = \frac{C}{3}$ or $F = \frac{8}{P}$
	Circular Pitch	Divide 3.1416 by the diametral pitch.	$\rho = 3.1416 \div P$
Face Angle		Add the addendum angle to the pitch cone angle	$\gamma = \alpha + \theta$
Compound Rest Angle for Turning Blank		Subtract both the pitch cone angle and the addendum angle from 90 degrees.	$\delta = 90° - \alpha - \theta$
Cutting Angle		Subtract the dedendum angle from the pitch cone angle.	$\zeta = \alpha - \phi$
Angular Addendum		Multiply the addendum by the cosine of the pitch cone angle.	$K = S \times \cos\alpha$
Outside Diameter		Add twice the angular addendum to the pitch diameter.	$O = D + 2K$
Vertex or Apex Distance		Multiply one-half the outside diameter by the cotangent of the face angle.	$J = \frac{O}{2} \times \cot\gamma$
Vertex Distance at Small End of Tooth		Subtract the width of face from the pitch cone radius; divide the remainder by the pitch cone radius and multiply by the apex distance.	$j = J \times \frac{C - F}{C}$
Number of Teeth for which to Select Cutter		Divide the number of teeth by the cosine of the pitch cone angle.	$N' = \frac{N}{\cos\alpha}$

Numbers of Formed Cutters Used to Mill Teeth in Mating Bevel Gear and Pinion with Shafts at Right Angles

Number of Teeth in Gear	Number of Teeth in Pinion																
	12	13	14	15	16	17	18	19	20	21	22	23	24	25	26	27	28
12	7-7	…	…	…	…	…	…	…	…	…	…	…	…	…	…	…	…
13	6-7	6-6	…	…	…	…	…	…	…	…	…	…	…	…	…	…	…
14	5-7	6-6	6-6	…	…	…	…	…	…	…	…	…	…	…	…	…	…
15	5-7	5-6	5-6	5-5	…	…	…	…	…	…	…	…	…	…	…	…	…
16	4-7	5-7	5-6	5-6	5-5	…	…	…	…	…	…	…	…	…	…	…	…
17	4-7	4-7	4-6	5-6	5-5	5-5	…	…	…	…	…	…	…	…	…	…	…
18	4-7	4-7	4-6	4-6	4-5	4-5	5-5	…	…	…	…	…	…	…	…	…	…
19	3-7	4-7	4-6	4-6	4-6	4-5	4-5	4-4	…	…	…	…	…	…	…	…	…
20	3-7	3-7	4-6	4-6	4-6	4-5	4-5	4-4	4-4	…	…	…	…	…	…	…	…
21	3-8	3-7	3-7	3-6	4-6	4-5	4-5	4-5	4-4	4-4	…	…	…	…	…	…	…
22	3-8	3-7	3-7	3-6	3-6	3-5	4-5	4-5	4-4	4-4	4-4	…	…	…	…	…	…
23	3-8	3-7	3-7	3-6	3-6	3-5	3-5	3-5	3-4	4-4	4-4	4-4	…	…	…	…	…
24	3-8	3-7	3-7	3-6	3-6	3-6	3-5	3-5	3-4	3-4	3-4	4-4	4-4	…	…	…	…
25	2-8	2-7	3-7	3-6	3-6	3-6	3-5	3-5	3-5	3-4	3-4	3-4	4-4	3-3	…	…	…
26	2-8	2-7	3-7	3-6	3-6	3-6	3-5	3-5	3-5	3-4	3-4	3-4	3-4	3-3	3-3	…	…
27	2-8	2-7	2-7	2-6	3-6	3-6	3-5	3-5	3-5	3-4	3-4	3-4	3-4	3-4	3-3	3-3	…
28	2-8	2-7	2-7	2-6	2-6	3-6	3-5	3-5	3-5	3-4	3-4	3-4	3-4	3-4	3-3	3-3	3-3
29	2-8	2-7	2-7	2-7	2-6	2-6	3-5	3-5	3-5	3-4	3-4	3-4	3-4	3-4	3-3	3-3	3-3
30	2-8	2-7	2-7	2-7	2-6	2-6	2-5	2-5	3-5	3-5	3-4	3-4	3-4	3-4	3-4	3-3	3-3
31	2-8	2-7	2-7	2-7	2-6	2-6	2-6	2-5	2-5	2-5	3-4	3-4	3-4	3-4	3-4	3-3	3-3
32	2-8	2-7	2-7	2-7	2-6	2-6	2-6	2-5	2-5	2-5	2-4	2-4	3-4	3-4	3-4	3-3	3-3
33	2-8	2-8	2-7	2-7	2-6	2-6	2-6	2-5	2-5	2-5	2-4	2-4	2-4	3-4	3-4	3-4	3-3
34	2-8	2-8	2-7	2-7	2-6	2-6	2-6	2-5	2-5	2-5	2-4	2-4	2-4	2-4	2-4	3-4	3-3
35	2-8	2-8	2-7	2-7	2-6	2-6	2-6	2-5	2-5	2-5	2-4	2-4	2-4	2-4	2-4	2-4	2-3
36	2-8	2-8	2-7	2-7	2-6	2-6	2-6	2-5	2-5	2-5	2-5	2-4	2-4	2-4	2-4	2-4	2-3
37	2-8	2-8	2-7	2-7	2-6	2-6	2-6	2-5	2-5	2-5	2-5	2-4	2-4	2-4	2-4	2-4	2-3
38	2-8	2-8	2-7	2-7	2-6	2-6	2-6	2-5	2-5	2-5	2-5	2-4	2-4	2-4	2-4	2-4	2-4
39	2-8	2-8	2-7	2-7	2-6	2-6	2-6	2-5	2-5	2-5	2-5	2-4	2-4	2-4	2-4	2-4	2-4
40	1-8	2-8	2-7	2-7	2-6	2-6	2-6	2-5	2-5	2-5	2-5	2-4	2-4	2-4	2-4	2-4	2-4
41	1-8	1-8	2-7	2-7	2-6	2-6	2-6	2-6	2-5	2-5	2-5	2-4	2-4	2-4	2-4	2-4	2-4
42	1-8	1-8	2-7	2-7	2-6	2-6	2-6	2-6	2-5	2-5	2-5	2-5	2-4	2-4	2-4	2-4	2-4
43	1-8	1-8	1-7	2-7	2-6	2-6	2-6	2-6	2-5	2-5	2-5	2-5	2-4	2-4	2-4	2-4	2-4
44	1-8	1-8	1-7	1-7	2-6	2-6	2-6	2-6	2-5	2-5	2-5	2-5	2-4	2-4	2-4	2-4	2-4
45	1-8	1-8	1-7	1-7	1-6	2-6	2-6	2-6	2-5	2-5	2-5	2-5	2-4	2-4	2-4	2-4	2-4
46	1-8	1-8	1-7	1-7	1-7	2-6	2-6	2-6	2-5	2-5	2-5	2-5	2-4	2-4	2-4	2-4	2-4
47	1-8	1-8	1-7	1-7	1-7	1-6	2-6	2-6	2-5	2-5	2-5	2-5	2-4	2-4	2-4	2-4	2-4
48	1-8	1-8	1-7	1-7	1-7	1-6	1-6	2-6	2-5	2-5	2-5	2-5	2-4	2-4	2-4	2-4	2-4
49	1-8	1-8	1-7	1-7	1-7	1-6	1-6	1-6	2-5	2-5	2-5	2-5	2-4	2-4	2-4	2-4	2-4
50	1-8	1-8	1-7	1-7	1-7	1-6	1-6	1-6	2-5	2-5	2-5	2-5	2-4	2-4	2-4	2-4	2-4
51	1-8	1-8	1-7	1-7	1-7	1-6	1-6	1-6	1-5	2-5	2-5	2-5	2-4	2-4	2-4	2-4	2-4
52	1-8	1-8	1-7	1-7	1-7	1-6	1-6	1-6	1-5	1-5	2-5	2-5	2-4	2-4	2-4	2-4	2-4
53	1-8	1-8	1-7	1-7	1-7	1-6	1-6	1-6	1-5	1-5	1-5	2-5	2-4	2-4	2-4	2-4	2-4
54	1-8	1-8	1-7	1-7	1-7	1-6	1-6	1-6	1-5	1-5	1-5	1-5	2-4	2-4	2-4	2-4	2-4
55	1-8	1-8	1-7	1-7	1-7	1-6	1-6	1-6	1-5	1-5	1-5	1-5	1-4	2-4	2-4	2-4	2-4

Numbers of Formed Cutters Used to Mill Teeth in Mating Bevel Gear and Pinion with Shafts at Right Angles *(Continued)*

Number of Teeth in Gear	Number of Teeth in Pinion																
	12	13	14	15	16	17	18	19	20	21	22	23	24	25	26	27	28
56	1-8	1-8	1-7	1-7	1-6	1-6	1-6	1-6	1-5	1-5	1-5	1-5	1-4	1-4	2-4	2-4	2-4
57	1-8	1-8	1-7	1-7	1-6	1-6	1-6	1-6	1-5	1-5	1-5	1-5	1-4	1-4	1-4	2-4	2-4
58	1-8	1-8	1-7	1-7	1-6	1-6	1-6	1-6	1-5	1-5	1-5	1-5	1-4	1-4	1-4	1-4	2-4
59	1-8	1-8	1-7	1-7	1-6	1-6	1-6	1-6	1-5	1-5	1-5	1-5	1-5	1-4	1-4	1-4	1-4
60	1-8	1-8	1-7	1-7	1-6	1-6	1-6	1-6	1-5	1-5	1-5	1-5	1-5	1-4	1-4	1-4	1-4
61	1-8	1-8	1-7	1-7	1-6	1-6	1-6	1-6	1-5	1-5	1-5	1-5	1-5	1-4	1-4	1-4	1-4
62	1-8	1-8	1-7	1-7	1-6	1-6	1-6	1-6	1-5	1-5	1-5	1-5	1-5	1-4	1-4	1-4	1-4
63	1-8	1-8	1-7	1-7	1-6	1-6	1-6	1-6	1-5	1-5	1-5	1-5	1-5	1-4	1-4	1-4	1-4
64	1-8	1-8	1-7	1-7	1-6	1-6	1-6	1-6	1-5	1-5	1-5	1-5	1-5	1-4	1-4	1-4	1-4
65	1-8	1-8	1-7	1-7	1-7	1-6	1-6	1-6	1-6	1-5	1-5	1-5	1-5	1-4	1-4	1-4	1-4
66	1-8	1-8	1-7	1-7	1-7	1-6	1-6	1-6	1-6	1-5	1-5	1-5	1-5	1-4	1-4	1-4	1-4
67	1-8	1-8	1-7	1-7	1-7	1-6	1-6	1-6	1-6	1-5	1-5	1-5	1-5	1-4	1-4	1-4	1-4
68	1-8	1-8	1-7	1-7	1-7	1-6	1-6	1-6	1-6	1-5	1-5	1-5	1-5	1-4	1-4	1-4	1-4
69	1-8	1-8	1-7	1-7	1-7	1-6	1-6	1-6	1-6	1-5	1-5	1-5	1-5	1-4	1-4	1-4	1-4
70	1-8	1-8	1-7	1-7	1-7	1-6	1-6	1-6	1-6	1-5	1-5	1-5	1-5	1-4	1-4	1-4	1-4
71	1-8	1-8	1-7	1-7	1-7	1-6	1-6	1-6	1-6	1-5	1-5	1-5	1-5	1-4	1-4	1-4	1-4
72	1-8	1-8	1-7	1-7	1-7	1-6	1-6	1-6	1-6	1-5	1-5	1-5	1-5	1-4	1-4	1-4	1-4
73	1-8	1-8	1-7	1-7	1-7	1-6	1-6	1-6	1-6	1-5	1-5	1-5	1-5	1-4	1-4	1-4	1-4
74	1-8	1-8	1-7	1-7	1-7	1-6	1-6	1-6	1-6	1-5	1-5	1-5	1-5	1-4	1-4	1-4	1-4
75	1-8	1-8	1-7	1-7	1-7	1-6	1-6	1-6	1-6	1-5	1-5	1-5	1-5	1-4	1-4	1-4	1-4
76	1-8	1-8	1-7	1-7	1-7	1-6	1-6	1-6	1-6	1-5	1-5	1-5	1-5	1-4	1-4	1-4	1-4
77	1-8	1-8	1-7	1-7	1-7	1-6	1-6	1-6	1-6	1-5	1-5	1-5	1-5	1-4	1-4	1-4	1-4
78	1-8	1-8	1-7	1-7	1-7	1-6	1-6	1-6	1-6	1-5	1-5	1-5	1-5	1-4	1-4	1-4	1-4
79	1-8	1-8	1-7	1-7	1-7	1-6	1-6	1-6	1-6	1-5	1-5	1-5	1-5	1-4	1-4	1-4	1-4
80	1-8	1-8	1-7	1-7	1-7	1-6	1-6	1-6	1-6	1-5	1-5	1-5	1-5	1-4	1-4	1-4	1-4
81	1-8	1-8	1-7	1-7	1-7	1-6	1-6	1-6	1-6	1-5	1-5	1-5	1-5	1-4	1-4	1-4	1-4
82	1-8	1-8	1-7	1-7	1-7	1-6	1-6	1-6	1-6	1-5	1-5	1-5	1-5	1-4	1-4	1-4	1-4
83	1-8	1-8	1-7	1-7	1-7	1-6	1-6	1-6	1-6	1-5	1-5	1-5	1-5	1-4	1-4	1-4	1-4
84	1-8	1-8	1-7	1-7	1-7	1-6	1-6	1-6	1-6	1-5	1-5	1-5	1-5	1-4	1-4	1-4	1-4
85	1-8	1-8	1-7	1-7	1-7	1-6	1-6	1-6	1-6	1-5	1-5	1-5	1-5	1-4	1-4	1-4	1-4
86	1-8	1-8	1-7	1-7	1-7	1-6	1-6	1-6	1-6	1-5	1-5	1-5	1-5	1-4	1-4	1-4	1-4
87	1-8	1-8	1-7	1-7	1-7	1-6	1-6	1-6	1-6	1-5	1-5	1-5	1-5	1-4	1-4	1-4	1-4
88	1-8	1-8	1-7	1-7	1-7	1-6	1-6	1-6	1-6	1-5	1-5	1-5	1-5	1-4	1-4	1-4	1-4
89	1-8	1-8	1-7	1-7	1-7	1-6	1-6	1-6	1-6	1-5	1-5	1-5	1-5	1-4	1-4	1-4	1-4
90	1-8	1-8	1-7	1-7	1-7	1-6	1-6	1-6	1-6	1-5	1-5	1-5	1-5	1-4	1-4	1-4	1-4
91	1-8	1-8	1-7	1-7	1-7	1-6	1-6	1-6	1-6	1-5	1-5	1-5	1-5	1-4	1-4	1-4	1-4
92	1-8	1-8	1-7	1-7	1-7	1-6	1-6	1-6	1-6	1-5	1-5	1-5	1-5	1-4	1-4	1-4	1-4
93	1-8	1-8	1-7	1-7	1-7	1-6	1-6	1-6	1-6	1-5	1-5	1-5	1-5	1-4	1-4	1-4	1-4
94	1-8	1-8	1-7	1-7	1-7	1-6	1-6	1-6	1-6	1-5	1-5	1-5	1-5	1-4	1-4	1-4	1-4
95	1-8	1-8	1-7	1-7	1-7	1-6	1-6	1-6	1-6	1-5	1-5	1-5	1-5	1-4	1-4	1-4	1-4
96	1-8	1-8	1-7	1-7	1-7	1-6	1-6	1-6	1-6	1-5	1-5	1-5	1-5	1-4	1-4	1-4	1-4
97	1-8	1-8	1-7	1-7	1-7	1-6	1-6	1-6	1-6	1-5	1-5	1-5	1-5	1-4	1-4	1-4	1-4
98	1-8	1-8	1-7	1-7	1-7	1-6	1-6	1-6	1-6	1-5	1-5	1-5	1-5	1-4	1-4	1-4	1-4
99	1-8	1-8	1-7	1-7	1-7	1-6	1-6	1-6	1-6	1-5	1-5	1-5	1-5	1-4	1-4	1-4	1-4
100	1-8	1-8	1-7	1-7	1-7	1-6	1-6	1-6	1-6	1-5	1-5	1-5	1-5	1-4	1-4	1-4	1-4

Number of cutter for gear given first, followed by number for pinion. See text, page 2060

Selecting Formed Cutters for Milling Bevel Gears.—For milling 14½-degree pressure angle bevel gears, the standard cutter series furnished by manufacturers of formed milling cutters is commonly used. There are 8 cutters in the series for each diametral pitch to cover the full range from a 12-tooth pinion to a crown gear. The difference between formed cutters used for milling spur gears and those used for bevel gears is that bevel gear cutters are thinner because they must pass through the narrow tooth space at the small end of the bevel gear; otherwise the shape of the cutter and hence, the cutter number, are the same.

To select the proper number of cutter to be used when a bevel gear is to be milled, it is necessary, first, to compute what is called the "Number of Teeth, N' for which to Select Cutter." This number of teeth can then be used to select the proper number of bevel gear cutter from the spur gear milling cutter table on page 2023. The value of N' may be computed using the last formula in the table on page 2056.

Example 1: What numbers of cutters are required for a pair of bevel gears of 4 diametral pitch and 70 degree shaft angle if the gear has 50 teeth and the pinion 20 teeth?

The pitch cone angle of the pinion is determined by using the first formula in the table on page 2056:

$$\tan\alpha_P = \frac{\sin\Sigma}{\dfrac{N_G}{N_P} + \cos\Sigma} = \frac{\sin 70°}{\dfrac{50}{20} + \cos 70°} = 0.33064;\ \alpha_P = 18°18'$$

The pitch cone angle of the gear is determined from the second formula in the table on page 2056:

$$\alpha_G = \Sigma - \alpha_P = 70° - 18°18' = 51°42'$$

The numbers of teeth N′ for which to select the cutters for the gear and pinion may now be determined from the last formula in the table on page 2056:

$$N'\text{ for the pinion} = \frac{N_P}{\cos\alpha_P} = \frac{20}{\cos 18°18'} = 21.1 \approx 21\text{ teeth}$$

$$N'\text{ for the gear} = \frac{N_G}{\cos\alpha_G} = \frac{50}{\cos 51°42'} = 80.7 \approx 81\text{ teeth}$$

From the table on page 2023 the numbers of the cutters for pinion and gear are found to be, respectively, 5 and 2.

Example 2: Required the cutters for a pair of bevel gears where the gear has 24 teeth and the pinion 12 teeth. The shaft angle is 90 degrees. As in the first example, the formulas given in the table on page 2056 will be used.

$$\tan\alpha_P = N_P \div N_G = 12 \div 24 = 0.5000\text{ and }\alpha_P = 26°34'$$

$$\alpha_G = \Sigma - \alpha_P = 90° - 26°34' = 63°26'$$

$$N'\text{ for pinion} = 12 \div \cos 26°34' = 13.4 \approx 13\text{ teeth}$$

$$N'\text{ for gear} = 24 \div \cos 63°26' = 53.6 \approx 54\text{ teeth}$$

And from the table on page 2023 the cutters for pinion and gear are found to be, respectively, 8 and 3.

Use of Table for Selecting Formed Cutters for Milling Bevel Gears.—The table beginning on page 2058 gives the numbers of cutters to use for milling various numbers of teeth in the gear and pinion. The table applies only to bevel gears with axes at right angles. Thus, in Example 2 given above, the numbers of the cutters could have been obtained directly by entering the table with the actual numbers of teeth in the gear, 24, and the pinion, 12.

Offset of Cutter for Milling Bevel Gears.—When milling bevel gears with a rotary formed cutter, it is necessary to take two cuts through each tooth space with the gear blank slightly off center, first on one side and then on the other, to obtain a tooth of approximately the correct form. The gear blank is also rotated proportionately to obtain the proper tooth thickness at the large and small ends. The amount that the gear blank or cutter should be offset from the central position can be determined quite accurately by the use of the table *Factors for Obtaining Offset for Milling Bevel Gears* in conjunction with the following rule: Find the factor in the table corresponding to the number of cutter used and to the ratio of the pitch cone radius to the face width; then divide this factor by the diametral pitch and subtract the result from half the thickness of the cutter at the pitch line.

Factors for Obtaining Offset for Milling Bevel Gears

No. of Cutter	Ratio of Pitch Cone Radius to Width of Face $\left(\frac{C}{F}\right)$												
	$\frac{3}{1}$	$\frac{3\frac{1}{4}}{1}$	$\frac{3\frac{1}{2}}{1}$	$\frac{3\frac{3}{4}}{1}$	$\frac{4}{1}$	$\frac{4\frac{1}{4}}{1}$	$\frac{4\frac{1}{2}}{1}$	$\frac{4\frac{3}{4}}{1}$	$\frac{5}{1}$	$\frac{5\frac{1}{2}}{1}$	$\frac{6}{1}$	$\frac{7}{1}$	$\frac{8}{1}$
1	0.254	0.254	0.255	0.256	0.257	0.257	0.257	0.258	0.258	0.259	0.260	0.262	0.264
2	0.266	0.268	0.271	0.272	0.273	0.274	0.274	0.275	0.277	0.279	0.280	0.283	0.284
3	0.266	0.268	0.271	0.273	0.275	0.278	0.280	0.282	0.283	0.286	0.287	0.290	0.292
4	0.275	0.280	0.285	0.287	0.291	0.293	0.296	0.298	0.298	0.302	0.305	0.308	0.311
5	0.280	0.285	0.290	0.293	0.295	0.296	0.298	0.300	0.302	0.307	0.309	0.313	0.315
6	0.311	0.318	0.323	0.328	0.330	0.334	0.337	0.340	0.343	0.348	0.352	0.356	0.362
7	0.289	0.298	0.308	0.316	0.324	0.329	0.334	0.338	0.343	0.350	0.360	0.370	0.376
8	0.275	0.286	0.296	0.309	0.319	0.331	0.338	0.344	0.352	0.361	0.368	0.380	0.386

Note.—For obtaining offset by above table, use formula:

$$\text{Offset} = \frac{T}{2} - \frac{\text{factor from table}}{P}$$

P = diametral pitch of gear to be cut
T = thickness of cutter used, measured at pitch line

To illustrate, what would be the amount of offset for a bevel gear having 24 teeth, 6 diametral pitch, 30-degree pitch cone angle and 1¼-inch face or tooth length? In order to obtain a factor from the table, the ratio of the pitch cone radius to the face width must be determined. The pitch cone radius equals the pitch diameter divided by twice the sine of the pitch cone angle $= 4 \div (2 \times 0.5) = 4$ inches. As the face width is 1.25, the ratio is $4 \div 1.25$ or about 3¼ to 1. The factor in the table for this ratio is 0.280 with a No. 4 cutter, which would be the cutter number for this particular gear. The thickness of the cutter at the pitch line is measured by using a vernier gear tooth caliper. The depth $S + A$ (see Fig. 1; S = addendum; A = clearance) at which to take the measurement equals 1.157 divided by the diametral pitch; thus, $1.157 \div 6 = 0.1928$ inch. The cutter thickness at this depth will vary with different cutters and even with the same cutter as it is ground away, because formed bevel gear cutters are commonly provided with side relief. Assuming that the thickness is 0.1745 inch, and substituting the values in the formula given, we have:

$$\text{Offset} = \frac{0.1745}{2} = \frac{0.280}{6} = 0.0406 \text{ inch}$$

Adjusting the Gear Blank for Milling.—After the offset is determined, the blank is adjusted laterally by this amount, and the tooth spaces are milled around the blank. After having milled one side of each tooth to the proper dimensions, the blank is set over in the opposite direction the same amount from a position central with the cutter, and is rotated to line up the cutter with a tooth space at the small end. A trial cut is then taken, which will leave the tooth being milled a little too thick, provided the cutter is thin enough—as it should be—to pass through the small end of the tooth space of the finished gear. This trial tooth is made the proper thickness by rotating the blank toward the cutter. To test the amount of offset, measure the tooth thickness (with a vernier caliper) at the large and small ends. The caliper should be set so that the addendum at the small end is in proper proportion to the addendum at the large end; that is, in the ratio, $(C - F)/C$ (see Fig. 1).

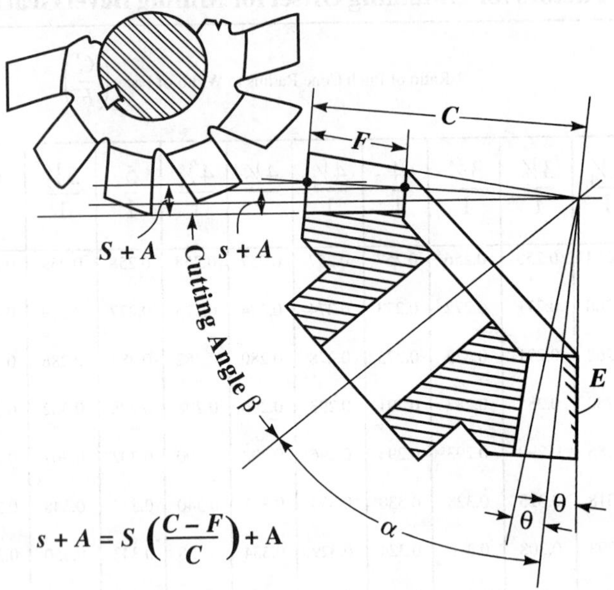

Fig. 1.

In taking these measurements, if the thicknesses at both ends (which should be in this same ratio) are too great, rotate the tooth toward the cutter and take trial cuts until the proper thickness at either the large or small end is obtained. If the large end of the tooth is the right thickness and the small end too thick, the blank was offset too much; inversely, if the small end is correct and the large end too thick, the blank was not set enough off center, and, either way, its position should be changed accordingly. The formula and table previously referred to will enable a properly turned blank to be set accurately enough for general work. The dividing head should be set to the cutting angle β (see Fig. 1), which is found by subtracting the addendum angle θ from the pitch cone angle α. After a bevel gear is cut by the method described, the sides of the teeth at the small end should be filed as indicated by the shade lines at *E*; that is, by filing off a triangular area from the point of the tooth at the large end to the point at the small end, thence down to the pitch line and back diagonally to a point at the large end.

Typical Steels Used for Bevel Gear Applications

SAE or AISI No.	Type of Steel	Preliminary Heat Treatment	Brinell Hardness Number	ASTM Grain Size	Remarks
Carburizing Steels					
		Purchase Specifications			
1024	Manganese	Normalize			Low Alloy — oil quench limited to thin sections
2512	Nickel Alloy	Normalize — Anneal	163–228	5–8	Aircraft quality
3310 3312X	Nickel-Chromium	Normalize, then heat to 1450°F, cool in furnace. Reheat to1170°F — cool in air	163–228	5–8	Used for maximum resistance to wear and fatigue
4028	Molybdenum	Normalize	163–217		Low Alloy
4615 4620	Nickel-Molybdenum	Normalize — 1700°F–1750°F	163–217	5–8	Good machining qualities. Well adapted to direct quench — gives tough core with minimum distortion
4815 4820	Nickel-Molybdenum	Normalize	163–241	5–8	For aircraft and heavily loaded service
5120	Chromium	Normalize	163–217	5–8	
8615 8620 8715 8720	Chromium-Nickel-Molybdenum	Normalize — cool at hammer	163–217	5–8	Used as an alternate for 4620
Oil Hardening and Flame Hardening Steels					
1141	Sulfurized free-cutting carbon steel	Normalize Heat-treated	179–228 255–269	5 or Coarser	Free-cutting steel used for unhardened gears, oil-treated gears, and for gears to be surface hardened where stresses are low
4140 4640	Chromium-Molybdenum Nickel-Molybdenum	For oil hardening, Normalize — Anneal For surface hardening, Normalize, reheat, quench, and draw	179–212 235–269 269–302 302–341		Used for heat-treated, oil-hardened, and surface-hardened gears. Machine qualities of 4640 are superior to 4140, and it is the preferred steel for flame hardening
6145	Chromium-Vanadium	Normalize—reheat, quench, and draw	235–269 269–302 302–341		Fair machining qualities. Used for surface hardened gears when 4640 is not available
8640 8739	Chromium-Nickel-Molybdenum	Same as for 4640			Used as an alternate for 4640
Nitriding Steels					
Nitralloy H & G	Special Alloy	Anneal	163–192		Normal hardness range for cutting is 20–28 Rockwell C

Circular Thickness, Chordal Thickness, and Chordal Addendum of Milled Bevel Gear Teeth.—In the formulas that follow, T = circular tooth thickness on pitch circle at large end of tooth; t = circular thickness at small end; T_c and t_c = chordal thickness at large and small ends, respectively; S_c and s_c = chordal addendum at large and small ends, respectively; D = pitch diameter at large end; and C, F, P, S, s, and α are as defined on page 2054.

$$T = \frac{1.5708}{P} \qquad T_c = T - \frac{T^3}{6D^2} \qquad S_c = S + \frac{T^2 \cos\alpha}{4D}$$

$$t = \frac{T(C-F)}{C} \qquad t_c = t - \frac{t^3}{6(D-2F\sin\alpha)^2} \qquad s_c = s + \frac{t^2\cos\alpha}{4(D-2F\sin\alpha)}$$

Worm Gearing

Worm Gearing.—Worm gearing may be divided into two general classes, fine-pitch worm gearing, and coarse-pitch worm gearing. Fine-pitch worm gearing is segregated from coarse-pitch worm gearing for the following reasons:

1) Fine-pitch worms and wormgears are used largely to transmit motion rather than power. Tooth strength except at the coarser end of the fine-pitch range is seldom an important factor; durability and accuracy, as they affect the transmission of uniform angular motion, are of greater importance.

2) Housing constructions and lubricating methods are, in general, quite different for fine-pitch worm gearing.

3) Because fine-pitch worms and wormgears are so small, profile deviations and tooth bearings cannot be measured with the same accuracy as can those of coarse pitches.

4) Equipment generally available for cutting fine-pitch wormgears has restrictions which limit the diameter, the lead range, the degree of accuracy attainable, and the kind of tooth bearing obtainable.

5) Special consideration must be given to top lands in fine-pitch hardened worms and wormgear-cutting tools.

6) Interchangeability and high production are important factors in fine-pitch worm gearing; individual matching of the worm to the gear, as often practiced with coarse-pitch precision worms, is impractical in the case of fine-pitch worm drives.

American Standard Design for Fine-pitch Worm Gearing (ANSI B6.9-1977).—This standard is intended as a design procedure for fine-pitch worms and wormgears having axes at right angles. It covers cylindrical worms with helical threads, and wormgears hobbed for fully conjugate tooth surfaces. It does not cover helical gears used as wormgears.

Hobs: The hob for producing the gear is a duplicate of the mating worm with regard to tooth profile, number of threads, and lead. The hob differs from the worm principally in that the outside diameter of the hob is larger to allow for resharpening and to provide bottom clearance in the wormgear.

Pitches: Eight standard axial pitches have been established to provide adequate coverage of the pitch range normally required: 0.030, 0.040, 0.050, 0.065, 0.080, 0.100, 0.130, and 0.160 inch.

Axial pitch is used as a basis for this design standard because: 1) Axial pitch establishes lead which is a basic dimension in the production and inspection of worms; 2) the axial pitch of the worm is equal to the circular pitch of the gear in the central plane; and 3) only one set of change gears or one master lead cam is required for a given lead, regardless of lead angle, on commonly-used worm-producing equipment.

Table 1. Formulas for Proportions of American Standard Fine-pitch Worms and Wormgears *ANSI B6.9-1977*

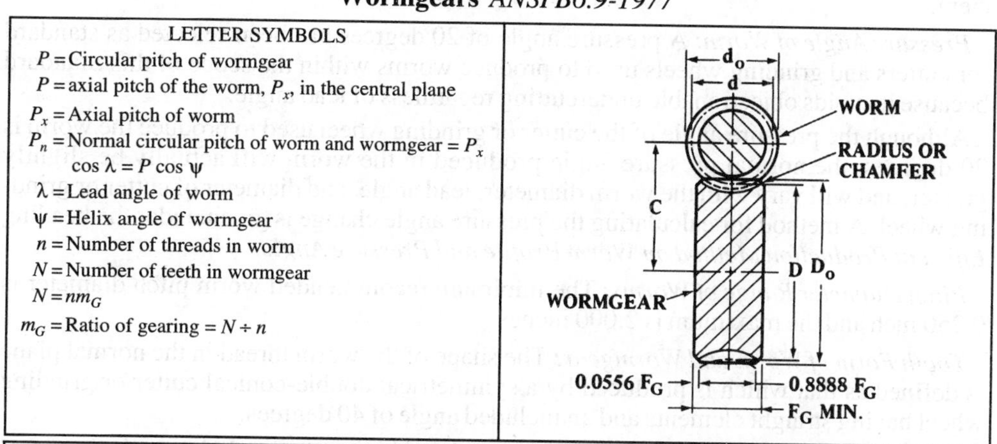

LETTER SYMBOLS

P = Circular pitch of wormgear

P = axial pitch of the worm, P_x, in the central plane

P_x = Axial pitch of worm

P_n = Normal circular pitch of worm and wormgear = $P_x \cos\lambda = P\cos\psi$

λ = Lead angle of worm

ψ = Helix angle of wormgear

n = Number of threads in worm

N = Number of teeth in wormgear

$N = nm_G$

m_G = Ratio of gearing = $N \div n$

Item	Formula	Item	Formula
WORM DIMENSIONS		WORMGEAR DIMENSIONS[a]	
Lead	$l = nP_x$	Pitch Diameter	$D = NP \div \pi = N\Pi_\xi \div \pi$
Pitch Diameter	$d = l \div (\pi \tan\lambda)$	Outside Diameter	$D_o = 2C - d + 2a$
Outside Diameter	$d_o = d + 2a$	Face Width	$F_{G\min} = 1.125 \times \sqrt{(d_o + 2c)^2 - (d_o - 4a)^2}$
Safe Minimum Length of Threaded Portion of Worm[b]	$F_W = \sqrt{D_o^2 - D^2}$		
DIMENSIONS FOR BOTH WORM AND WORMGEAR			
Addendum	$a = 0.3183P_n$	Tooth thickness	$t_n = 0.5P_n$
Whole Depth	$h_t = 0.7003P_n + 0.002$	Approximate normal pressure angle[c]	$\phi_n = 20$ degrees
Working Depth	$h_k = 0.6366P_n$		
Clearance	$c = h_t - h_k$	Center distance	$C = 0.5\,(d + D)$

[a] Current practice for fine-pitch worm gearing does not require the use of throated blanks. This results in the much simpler blank shown in the diagram which is quite similar to that for a spur or helical gear. The slight loss in contact resulting from the use of non-throated blanks has little effect on the load-carrying capacity of fine-pitch worm gears. It is sometimes desirable to use topping hobs for producing wormgears in which the size relation between the outside and pitch diameters must be closely controlled. In such cases the blank is made slightly larger than D_o by an amount (usually from 0.010 to 0.020) depending on the pitch. Topped wormgears will appear to have a small throat which is the result of the hobbing operation. For all intents and purposes, the throating is negligible and a blank so made is not to be considered as being a throated blank.

[b] This formula allows a sufficient length for fine-pitch worms.

[c] As stated in the text on page 2066, the actual pressure angle will be slightly greater due to the manufacturing process.

All dimensions in inches unless otherwise indicated.

Lead Angles: Fifteen standard lead angles have been established to provide adequate coverage: 0.5, 1, 1.5, 2, 3, 4, 5, 7, 9, 11, 14, 17, 21, 25, and 30 degrees.

This series of lead angles has been standardized to: 1) Minimize tooling; 2) permit obtaining geometric similarity between worms of different axial pitch by keeping the same lead angle; and 3) take into account the production distribution found in fine-pitch worm gearing applications.

For example, most fine-pitch worms have either one or two threads. This requires smaller increments at the low end of the lead angle series. For the less frequently used thread numbers, proportionately greater increments at the high end of the lead angle series are sufficient.

Pressure Angle of Worm: A pressure angle of 20 degrees has been selected as standard for cutters and grinding wheels used to produce worms within the scope of this Standard because it avoids objectionable undercutting regardless of lead angle.

Although the pressure angle of the cutter or grinding wheel used to produce the worm is 20 degrees, the normal pressure angle produced in the worm will actually be slightly greater, and will vary with the worm diameter, lead angle, and diameter of cutter or grinding wheel. A method for calculating the pressure angle change is given under the heading *Effect of Production Method on Worm Profile and Pressure Angle*.

Pitch Diameter Range of Worms: The minimum recommended worm pitch diameter is 0.250 inch and the maximum is 2.000 inches.

Tooth Form of Worm and Wormgear: The shape of the worm thread in the normal plane is defined as that which is produced by a symmetrical double-conical cutter or grinding wheel having straight elements and an included angle of 40 degrees.

Because worms and wormgears are closely related to their method of manufacture, it is impossible to specify clearly the tooth form of the wormgear without referring to the mating worm. For this reason, worm specifications should include the method of manufacture and the diameter of cutter or grinding wheel used. Similarly, for determining the shape of the generating tool, information about the method of producing the worm threads must be given to the manufacturer if the tools are to be designed correctly.

The worm profile will be a curve that departs from a straight line by varying amounts, depending on the worm diameter, lead angle, and the cutter or grinding wheel diameter. A method for calculating this deviation is given in the Standard. The tooth form of the wormgear is understood to be made fully conjugate to the mating worm thread.

Effect of Diameter of Cutting on Profile and Pressure Angle of Worms

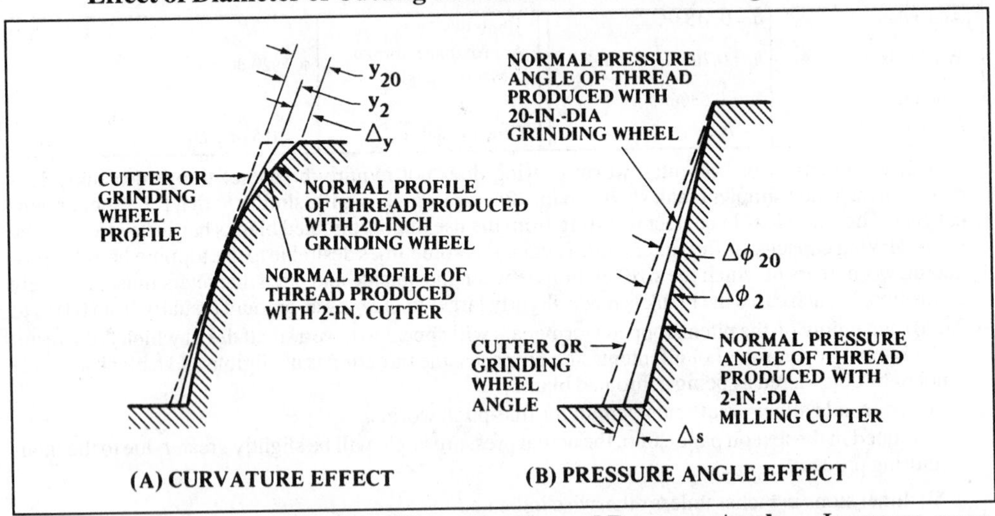

Effect of Production Method on Worm Profile and Pressure Angle.—In worm gearing, tooth bearing is usually used as the means of judging tooth profile accuracy since direct profile measurements on fine-pitch worms or wormgears is not practical. According to AGMA 370.01, Design Manual for Fine-Pitch Gearing, a minimum of 50 per cent initial area of contact is suitable for most fine-pitch worm gearing, although in some cases, such as when the load fluctuates widely, a more restricted initial area of contact may be desirable.

Except where single-pointed lathe tools, end mills, or cutters of special shape are used in the manufacture of worms, the pressure angle and profile produced by the cutter are different from those of the cutter itself. The amounts of these differences depend on several factors, namely, diameter and lead angle of the worm, thickness and depth of the worm thread, and diameter of the cutter or grinding wheel. The accompanying diagram shows the curvature and pressure angle effects produced in the worm by cutters and grinding wheels, and how the amount of variation in worm profile and pressure angle is influenced by the diameter of the cutting tool used.

Materials for Worm Gearing.—Worm gearing, especially for power transmission, should have steel worms and phosphor bronze wormgears. This combination is used extensively. The worms should be hardened and ground to obtain accuracy and a smooth finish.

The phosphor bronze wormgears should contain from 10 to 12 per cent of tin. The S.A.E. phosphor gear bronze (No. 65) contains 88–90% copper, 10–12% tin, 0.50% lead, 0.50% zinc (but with a maximum total lead, zinc and nickel content of 1.0 per cent), phosphorous 0.10–0.30%, aluminum 0.005%. The S.A.E. nickel phosphor gear bronze (No. 65 + Ni) contains 87% copper, 11% tin, 2% nickel and 0.2% phosphorous.

Single-thread Worms.—The ratio of the worm speed to the wormgear speed may range from 1.5 or even less up to 100 or more. Worm gearing having high ratios are not very efficient as transmitters of power; nevertheless high as well as low ratios often are required. Since the ratio equals the number of wormgear teeth divided by the number of threads or "starts" on the worm, single-thread worms are used to obtain a high ratio. As a general rule, a ratio of 50 is about the maximum recommended for a single worm and wormgear combination, although ratios up to 100 or higher are possible. When a high ratio is required, it may be preferable to use, in combination, two sets of worm gearing of the multi-thread type in preference to one set of the single-thread type in order to obtain the same total reduction and a higher combined efficiency.

Single-thread worms are comparatively inefficient because of the effect of the low lead angle; consequently, single-thread worms are not used when the primary purpose is to transmit power as efficiently as possible but they may be employed either when a large speed reduction with one set of gearing is necessary, or possibly as a means of adjustment, especially if "mechanical advantage" or self-locking are important factors.

Multi-thread Worms.—When worm gearing is designed primarily for transmitting power efficiently, the lead angle of the worm should be as high as is consistent with other requirements and preferably between, say, 25 or 30 and 45 degrees. This means that the worm must be multi-threaded. To obtain a given ratio, some number of wormgear teeth divided by some number of worm threads must equal the ratio. Thus, if the ratio is 6, combinations such as the following might be used:

$$\frac{24}{4}, \frac{30}{5}, \frac{36}{6}, \frac{42}{7}$$

The numerators represent the number of wormgear teeth and the denominators, the number of worm threads or "starts." The number of wormgear teeth may not be an exact multiple of the number of threads on a multi-thread worm in order to obtain a "hunting tooth" action.

Number of Threads or "Starts" on Worm: The number of threads on the worm ordinarily varies from one to six or eight, depending upon the ratio of the gearing. As the ratio is increased, the number of worm threads is reduced, as a general rule. In some cases, however, the higher of two ratios may also have a larger number of threads. For example, a ratio of $6\frac{1}{5}$ would have 5 threads whereas a ratio of $6\frac{5}{6}$ would have 6 threads. Whenever the ratio is fractional, the number of threads on the worm equals the denominator of the fractional part of the ratio.

HELICAL GEARING

Basic Rules and Formulas for Helical Gear Calculations.—The rules and formulas in the following table and elsewhere in this article are basic to helical gear calculations. The notation used in the formulas is: P_n = normal diametral pitch of cutter; D = pitch diameter; N = number of teeth; α = helix angle; γ = center angle or angle between shafts; C = center distance; N' = number of teeth for which to select a formed cutter for milled teeth; L = lead of tooth helix; S = addendum; W = whole depth; T_n = normal tooth thickness at pitch line; and O = outside diameter.

Rules and Formulas for Helical Gear Calculations

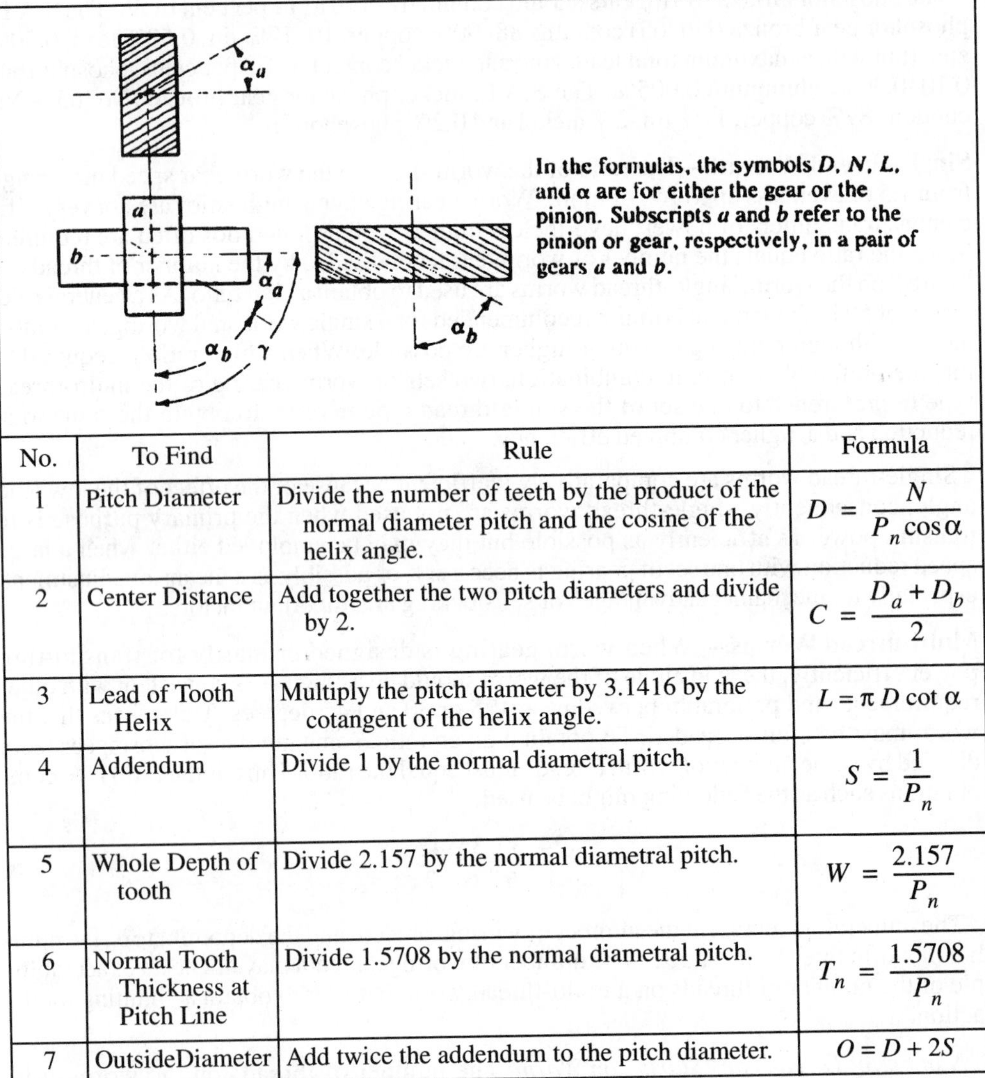

In the formulas, the symbols *D, N, L,* and α are for either the gear or the pinion. Subscripts *a* and *b* refer to the pinion or gear, respectively, in a pair of gears *a* and *b*.

No.	To Find	Rule	Formula
1	Pitch Diameter	Divide the number of teeth by the product of the normal diameter pitch and the cosine of the helix angle.	$D = \dfrac{N}{P_n \cos\alpha}$
2	Center Distance	Add together the two pitch diameters and divide by 2.	$C = \dfrac{D_a + D_b}{2}$
3	Lead of Tooth Helix	Multiply the pitch diameter by 3.1416 by the cotangent of the helix angle.	$L = \pi D \cot\alpha$
4	Addendum	Divide 1 by the normal diametral pitch.	$S = \dfrac{1}{P_n}$
5	Whole Depth of tooth	Divide 2.157 by the normal diametral pitch.	$W = \dfrac{2.157}{P_n}$
6	Normal Tooth Thickness at Pitch Line	Divide 1.5708 by the normal diametral pitch.	$T_n = \dfrac{1.5708}{P_n}$
7	OutsideDiameter	Add twice the addendum to the pitch diameter.	$O = D + 2S$

Determining Direction of Thrust.—The first step in helical gear design is to determine the desired direction of the thrust. When the direction of the thrust has been determined and the relative positions of the driver and driven gears are known, then the direction of helix (right- or left-hand) may be found from the accompanying thrust diagrams, *Directions of rotation and resulting thrust for parallel shaft and 90 degree shaft angle helical gears*. The diagrams show the directions of rotation and the resulting thrust for parallel-shaft and 90-

degree shaft angle helical gears. The thrust bearings are located so as to take the thrust caused by the tooth loads. The direction of the thrust depends on the direction of the helix, the relative positions of driver and driven gears, and the direction of rotation. The thrust may be changed to the opposite direction by changing any one of the three conditions, namely, by changing the hand of the helix, by reversing the direction of rotation, or by exchanging of driver and driven gear positions.

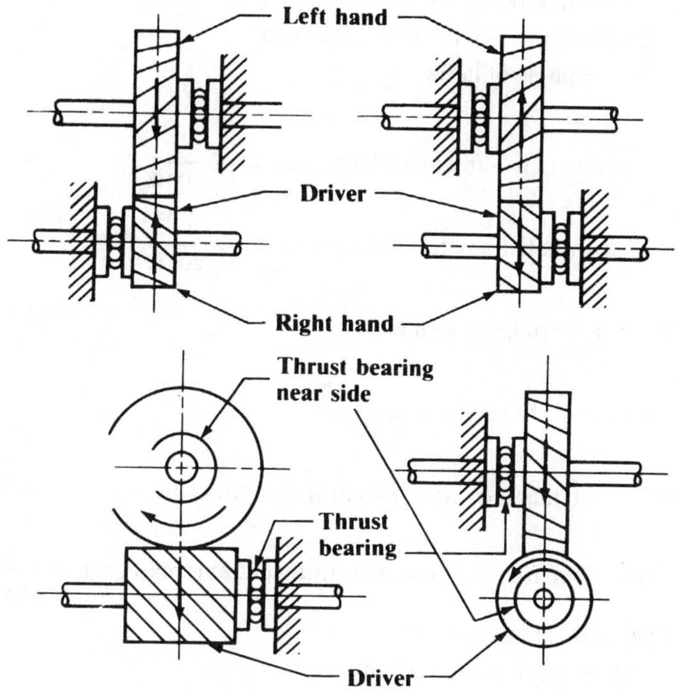

Directions of rotation and resulting thrust for parallel shaft and 90 degree shaft angle helical gears

Determining Helix Angles.—The following rules should be observed for helical gears with shafts at any given angle. If each helix angle is less than the shaft angle, then the sum of the helix angles of the two gears will equal the angle between the shafts, and the helix angle is of the same hand for both gears; if the helix angle of one of the gears is larger than the shaft angle, then the difference between the helix angles of the two gears will be equal to the shaft angle, and the gears will be of opposite hand.

Pitch of Cutter to be Used.—The thickness of the cutter at the pitchline for cutting helical gears should equal one-half the *normal* circular pitch. The normal pitch varies with the helix angle, hence, the helix angle must be considered when selecting a cutter. The cutter should be of the same pitch as the *normal* diametral pitch of the gear. This normal pitch is found by dividing the transverse diametral pitch of the gear by the cosine of the helix angle. To illustrate, if the pitch diameter of a helical gear is 6.718 and there are 38 teeth having a helix angle of 45 degrees, the transverse diametral pitch equals 38 divided by 6.718 = 5.656; then the normal diametral pitch equals 5.656 divided by 0.707 = 8. A cutter, then, of 8 diametral pitch is the one to use for this particular gear.

Helical gears should preferably be cut on a generating-type gear cutting machine such as a hobber or shaper. Milling machines are used in some shops when hobbers or shapers are not available or when single, replacement gears are being made. In such instances, the pitch of the formed cutter used in milling a helical gear must not only conform to the normal diametral pitch of the gear, but the cutter number must also be determined. See *Selecting Cutter for Milling Helical Gears* starting on page 2077.

1. Shafts Parallel, Center Distance Approximate.—Given or assumed:

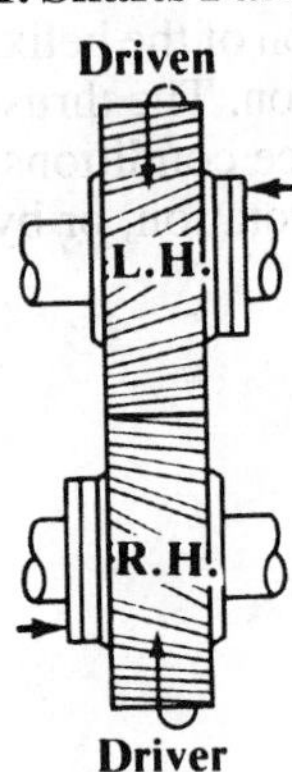

1) Position of gear having right- or left-hand helix, depending upon rotation and direction in which thrust is to be received
2) C_a = approximate center distance
3) P_n = normal diametral pitch
4) N = number of teeth in large gear
5) n = number of teeth in small gear
6) α = angle of helix

To find:

1) D = pitch diameter of large gear = $\dfrac{N}{P_n \cos\alpha}$

2) d = pitch diameter of small gear = $\dfrac{n}{P_n \cos\alpha}$

3) O = outside diameter of large gear = $D + \dfrac{2}{P_n}$

4) o = outside diameter of small gear = $d + \dfrac{2}{P_n}$

5) T = number of teeth marked on formed milling cutter (large gear) = $\dfrac{N}{\cos^3\alpha}$

6) t = number of teeth marked on formed milling cutter (small gear) = $\dfrac{n}{\cos^3\alpha}$

7) L = lead of helix on large gear = $\pi D \cot\alpha$
8) l = lead of helix on small gear = $\pi d \cot\alpha$
9) C = center distance (if not right, vary α) = $\frac{1}{2}(D + d)$

Example

Given or assumed:
1) See illustration 2) C_a = 17 inches 3) $P_n = 2$ 4) $N = 48$ 5) $n = 20$ 6) $\alpha = 20$
To find:

$$1)\ D = \frac{N}{P_n \cos\alpha} = \frac{48}{2 \times 0.9397} = 25.541 \text{ inches}$$

$$2)\ d = \frac{n}{P_n \cos\alpha} = \frac{20}{2 \times 0.9397} = 10.642 \text{ inches}$$

$$3)\ O = \frac{2}{P_n} = 25.541 + \frac{2}{2} = 26.541 \text{ inches}$$

$$4)\ o = d + \frac{2}{P_n} = 10.642 + \frac{2}{2} = 11.642 \text{ inches}$$

$$5)\ T = \frac{N}{\cos^3\alpha} = \frac{48}{(0.9397)^3} = 57.8\text{, say 58 teeth}$$

$$6)\ t = \frac{n}{\cos^3\alpha} = \frac{20}{(0.9397)^3} = 24.1\text{, say 24 teeth}$$

$$7)\ L = \pi D \cot\alpha = 3.1416 \times 25.541 \times 2.747 = 220.42 \text{ inches}$$

8) $l = \pi d \cot\alpha = 3.1416 \times 10.642 \times 2.747 = 91.84$ inches

9) $C = \frac{1}{2}(D + d) = \frac{1}{2}(25.541 + 10.642) = 18.091$ inches

2. Shafts Parallel, Center Distance Exact.—Given or assumed:

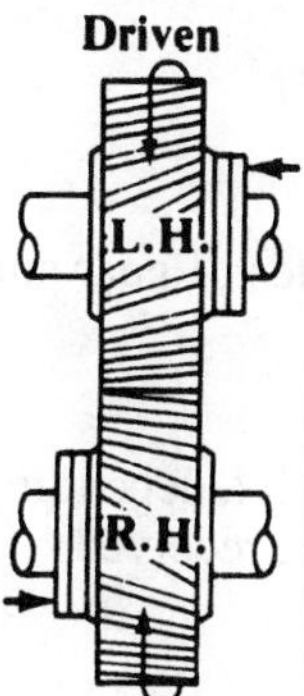

1) Position of gear having right- or left-hand helix, depending upon rotation and direction in which thrust is to be received

2) C = exact center distance

3) P_n = normal diametral pitch (pitch of cutter)

4) N = number of teeth in large gear

5) n = number of teeth in small gear

To find:

1) $\cos\alpha = \dfrac{N + n}{2P_nC}$

2) D = pitch diameter of large gear $= \dfrac{N}{P_n \cos\alpha}$

3) d = pitch diameter of small gear $= \dfrac{n}{P_n \cos\alpha}$

4) O = outside diameter of large gear $= D + \dfrac{2}{P_n}$

5) o = outside diameter of small gear $= d + \dfrac{2}{P_n}$

6) T = number of teeth marked on formed milling cutter (large gear) $= \dfrac{N}{\cos^3\alpha}$

7) t = number of teeth marked on formed milling cutter (small gear) $= \dfrac{n}{\cos^3\alpha}$

8) L = lead of helix (large gear) $= \pi D \cot\alpha$

9) l = lead of helix (small gear) $= \pi d \cot\alpha$

Example

Given or assumed:

1) See illustration 2) $C = 18.75$ inches 3) $P_n = 4$ 4) $N = 96$ 5) $n = 48$

To find:

1) $\cos\alpha = \dfrac{N + n}{2P_nC} = \dfrac{96 + 48}{2 \times 4 \times 18.75} = 0.96$, or $\alpha = 16°16'$

2) $D = \dfrac{N}{P_n \cos\alpha} = \dfrac{96}{4 \times 0.96} = 25$ inches

3) $d = \dfrac{n}{P_n \cos\alpha} = \dfrac{48}{4 \times 0.96} = 12.5$ inches

4) $O = D + \dfrac{2}{P_n} = 25 + \dfrac{2}{4} = 25.5$ inches

5) $o = d + \dfrac{2}{P_n} = 12.5 + \dfrac{2}{4} = 13$ inches

6) $T = \frac{N}{\cos^3 \alpha} = \frac{96}{(0.96)^3} = 108$ teeth

7) $t = \frac{n}{\cos^3 \alpha} = \frac{48}{(0.96)^3} = 54$ teeth

8) $L = \pi D \cot \alpha = 3.1416 \times 25 \times 3.427 = 269.15$ inches

9) $l = \pi d \cot \alpha = 3.1416 \times 12.5 \times 3.427 = 134.57$ inches

3. Shafts at Right Angles, Center Distance Approx.—Sum of helix angles of gear and pinion must equal 90 degrees.

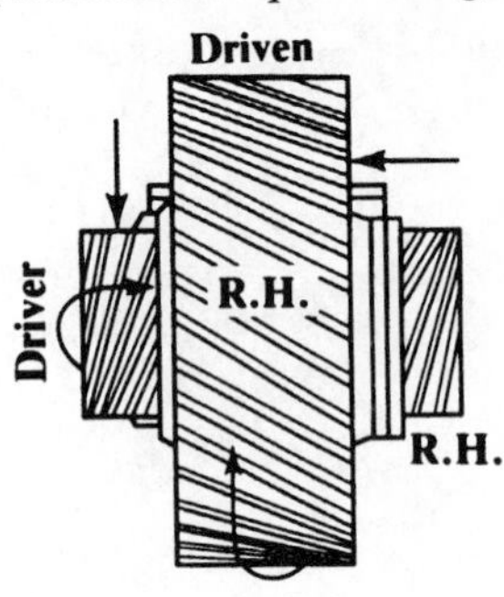

Given or assumed:

1) Position of gear having right- or left-hand helix, depending on rotation and direction in which thrust is to be received

2) C_a= approximate center distance

3) P_n = normal diametral pitch (pitch of cutter)

4) R = ratio of gear to pinion size

5) n = number of teeth in pinion = $\frac{1.41 C_a P_n}{R + 1}$ for 45 degrees;

and $\frac{2 C_a P_n \cos \alpha \cos \beta}{R \cos \beta + \cos \alpha}$ for any angle

6) N = number of teeth in gear = nR

7) α = angle of helix of gear

8) β = angle of helix of pinion

To find:

A) When helix angles are 45 degrees,

1) D = pitch diameter of gear = $\frac{N}{0.70711 P_n}$

2) d = pitch diameter of pinion = $\frac{n}{0.70711 P_n}$

3) O = outside diameter of gear = $D + \frac{2}{P_n}$

4) o = outside diameter of pinion = $d + \frac{2}{P_n}$

5) T = number of formed cutter (gear) = $\frac{N}{0.353}$

6) t = number of formed cutter (pinion) = $\frac{n}{0.353}$

7) L = lead of helix of gear = πD

8) l = lead of helix of pinion = πd

9) C = center distance (exact) = $\frac{D + d}{2}$

B) When helix angles are other than 45 degrees

1) $D = \dfrac{N}{P_n \cos\alpha}$ 2) $d = \dfrac{n}{P_n \cos\beta}$ 3) $T = \dfrac{N}{\cos^3\alpha}$

4) $t = \dfrac{n}{\cos^3\beta}$ 5) $L = \pi D \cot\alpha$ 6) $l = \pi d \cot\beta$

Example

Given or assumed:

1) See illustration 2) $C_a = 3.2$ inches 3) $P_n = 10$ 4) $R = 1.5$

5) $n = \dfrac{1.41 C_a P_n}{R+1} = \dfrac{1.41 \times 3.2 \times 10}{1.5 + 1} =$ say 18 teeth

6) $N = nR = 18 \times 1.5 = 27$ teeth

7) $\alpha = 45$ degrees

8) $\beta = 45$ degrees

To find:

1) $D = \dfrac{N}{0.70711 P_n} = \dfrac{27}{0.70711 \times 10} = 3.818$ inches

2) $d = \dfrac{n}{0.70711 P_n} = \dfrac{18}{0.70711 \times 10} = 2.545$ inches

3) $O = D + \dfrac{2}{P_n} = 3.818 + \dfrac{2}{10} = 4.018$ inches

4) $o = d + \dfrac{2}{P_n} = 2.545 + \dfrac{2}{10} = 2.745$ inches

5) $T = \dfrac{N}{0.353} = \dfrac{27}{0.353} = 76.5$, say 76 teeth

6) $t = \dfrac{n}{0.353} = \dfrac{18}{0.353} = 51$ teeth

7) $L = \pi D = 3.1416 \times 3.818 = 12$ inches

8) $l = \pi d = 3.1416 \times 2.545 = 8$ inches

9) $C = \dfrac{D+d}{2} = \dfrac{3.818 + 2.545}{2} = 3.182$ inches

4A. Shafts at Right Angles, Center Distance Exact.—Gears have same direction of helix. Sum of the helix angles will equal 90 degrees.

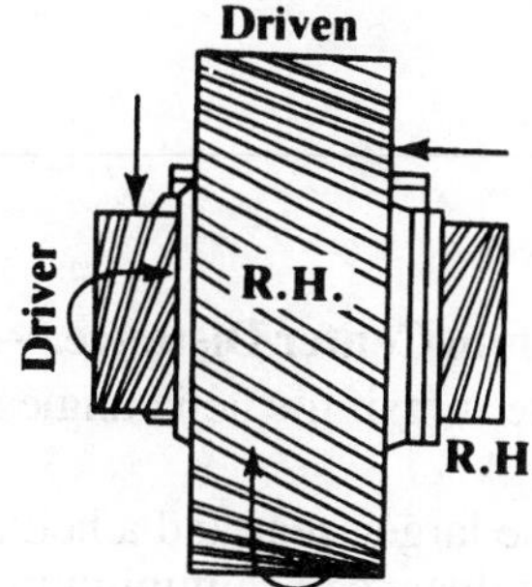

Given or assumed:

1) Position of gear having right- or left-hand helix depending on rotation and direction in which thrust is to be received

2) P_n = normal diametral pitch (pitch of cutter)

3) R = ratio of number of teeth in large gear to number of teeth in small gear

4) α_a = approximate helix angle of large gear

5) C = exact center distance

To find:

1) n = number of teeth in small gear nearest = $2\ CP_n \sin\alpha_a \div 1 + R\tan\alpha_a$

2) N = number of teeth in large gear = Rn

3) α = exact helix angle of large gear, found by trial from $R\sec\alpha + \operatorname{cosec}\alpha = 2\ CP_n \div n$

4) β = exact helix angle of small gear = $90° - \alpha$

5) D = pitch diameter of large gear = $\dfrac{N}{P_n \cos\alpha}$

6) d = pitch diameter of small gear = $\dfrac{n}{P_n \cos\beta}$

7) O = outside diameter of large gear = $D + \dfrac{2}{P_n}$

8) o = outside diameter of small gear = $d + \dfrac{2}{P_n}$

9) N' and n' = numbers of teeth marked on cuttters for large and small gears (see page 2077)

10) L = lead of helix on large gear = $\pi D\cot\alpha$

11) l = lead of helix on small gear = $\pi d\cot\beta$

Example

Given or assumed:

1) See illustration 2) $P_n = 8$ 3) $R = 3$ 4) $\alpha_a = 45$ degrees 5) $C = 10$ in

To find:

$$1)\ n = \frac{2CP_n \sin\alpha_a}{1 + R\tan\alpha_a} = \frac{2 \times 10 \times 8 \times 0.70711}{1 + 3} = 28.25\text{, say 28 teeth}$$

$$2)\ N = Rn = 3 \times 28 = 84\text{ teeth}$$

$$3)\ R\sec\alpha + \operatorname{cosec}\alpha = \frac{2CP_n}{n} = \frac{2 \times 10 \times 8}{28} = 5.714\text{, or } \alpha = 46°6'$$

$$4)\ \beta = 90° - \alpha = 90° - 46°6' = 43°54'$$

$$5)\ D = \frac{N}{P_n \cos\alpha} = \frac{84}{8 \times 0.6934} = 15.143\text{ inches}$$

$$6)\ d = \frac{n}{P_n \cos\beta} = \frac{28}{8 \times 0.72055} = 4.857\text{ inches}$$

$$7)\ O = D + \frac{2}{P_n} = 15.143 + 0.25 = 15.393\text{ inches}$$

$$8)\ o = d + \frac{2}{P_n} = 4.857 + 0.25 = 5.107\text{ inches}$$

9) $N' = 275$; $n' = 94$ (see page 2077)

10) $L = \pi D\cot\alpha = 3.1416 \times 15.143 \times 0.96232 = 45.78$ inches

11) $l = \pi d\cot\beta = 3.1416 \times 4.857 \times 1.0392 = 15.857$ inches

4B. Shafts at Right Angles, Any Ratio, Helix Angle for Minimum Center Distance.— Diagram similar to 4A. Gears have same direction of helix. The sum of the helix angles will equal 90 degrees.

For any given ratio of gearing R there is a helix angle α for the larger gear and a helix angle $\beta = 90° - \alpha$ for the smaller gear that will make the center distance C a minimum.

Helix angle α is found from the formula cot $\alpha = R^{1/3}$. As an example, using the data found in Case 4A, helix angles α and β for minimum center distance would be: cot $\alpha = R^{1/3} = 1.4422$; $\alpha = 34°44'$ and $\beta = 90° - 34°44' = 55°16'$. Using these helix angles, $D = 12.777$; $d = 6.143$; and $C = 9.460$ from the formulas for D and d given under Case 4A.

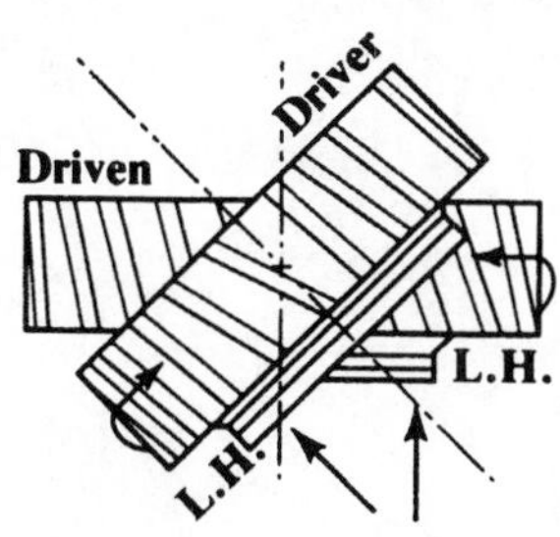

5. Shafts at Any Angle, Center Distance Approx.— The sum of the helix angles of the two gears equals the shaft angle, and the gears are of the same hand, if each angle is less than the shaft angle. The difference between the helix angles equals the shaft angle, and the gears are of opposite hand, if either angle is greater than the shaft angle.

Given or assumed:

1) Hand of helix, depending on rotation and direction in which thrust is to be received

2) C_a = center distance

3) P_n = normal diametral pitch (pitch of cutter)

4) R = ratio of gear to pinion = $\frac{N}{n}$

5) α = angle of helix, gear

6) β = angle of helix, pinion

7) n = number of teeth in pinion nearest $\frac{2C_aP_n\cos\alpha\cos\beta}{R\cos\beta + \cos\alpha}$

for any angle, and $\frac{2C_aP_n\cos\alpha}{R+1}$ when both angles are equal

8) N = number of teeth in gear = Rn

To find:

1) D = pitch diameter of gear = $\frac{N}{P_n\cos\alpha}$

2) d = pitch diameter of pinion = $\frac{n}{P_n\cos\beta}$

3) O = outside diameter of gear = $D + \frac{2}{P_n}$

4) o = outside diameter of pinion = $d + \frac{2}{P_n}$

5) T = number of teeth marked on cutter for gear = $\frac{N}{\cos^3\alpha}$

6) t = number of teeth marked on cutter for pinion = $\frac{n}{\cos^3\beta}$

7) L = lead of helix on gear = $\pi D\cot\alpha$

8) l = lead of helix on pinion = $\pi d\cot\beta$

9) C = actual center distance = $\frac{D+d}{2}$

Example

Given or assumed (angle of shafts, 60 degrees):

1) See illustration 2) C_a = 12 inches 3) $P_n = 8$

4) $R = 4$ 5) $\alpha = 30$ degrees 6) $\beta = 30$ degrees

7) $n = \dfrac{2C_a P_n \cos\alpha}{R+1} = \dfrac{2 \times 12 \times 8 \times 0.86603}{4+1} = 33$ teeth

8) $N = 4 \times 33 = 132$ teeth

To find:

1) $D = \dfrac{N}{P_n \cos\alpha} = \dfrac{132}{8 \times 0.86603} = 19.052$ inches

2) $d = \dfrac{n}{P_n \cos\beta} = \dfrac{33}{8 \times 0.86603} = 4.763$ inches

3) $O = D + \dfrac{2}{P_n} = 19.052 + \dfrac{2}{8} = 19.302$ inches

4) $o = d + \dfrac{2}{P_n} = 4.763 + \dfrac{2}{8} = 5.010000000000000003$ inches

5) $T = \dfrac{N}{\cos^3\alpha} = \dfrac{132}{0.65} = 203$ teeth

6) $t = \dfrac{n}{\cos^3\beta} = \dfrac{33}{0.65} = 51$ teeth

7) $L = \pi D \cot\alpha = \pi \times 19.052 \times 1.732 = 103.66$ inches

8) $l = \pi d \cot\beta = \pi \times 4.763 \times 1.732 = 25.92$ inches

9) $C = \dfrac{D+d}{2} = \dfrac{19.052 + 4.763}{2} = 11.9075$ inches

6. Shafts at Any Angle, Center Distance Exact.—The sum of the helix angles of the two gears equals the shaft angle, and the gears are of the same hand, if each angle is less than the shaft angle. The difference between the helix angles equals the shaft angle, and the gears are of opposite hand, if either angle is greater than the shaft angle.

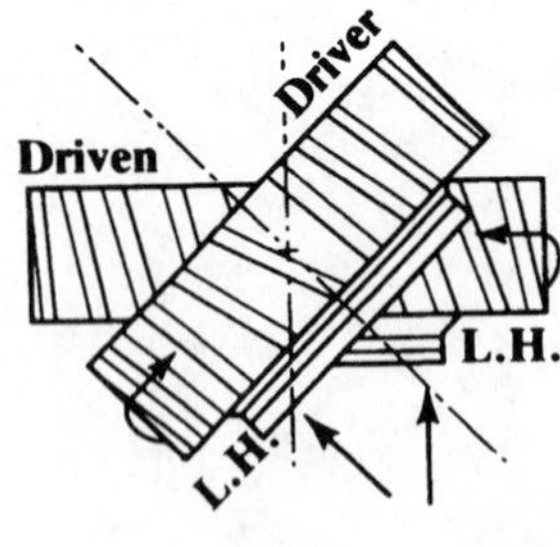

Given or assumed:

1) Hand of helix, depending on rotation and direction in which thrust is to be received

2) C = center distance

3) P_n = normal diametral pitch (pitch of cutter)

4) α_a = approximate helix angle of gear

5) β_a = approximate helix angle of pinion

6) R = ratio of gear to pinion size = $\dfrac{N}{n}$

7) n = number of pinion teeth nearest $\dfrac{2CP_n \cos\alpha_a \cos\beta_a}{R\cos\beta_a + \cos\alpha_a}$

8) N = number of gear teeth = Rn

To find:

1) α and β, exact helix angles, found by trial from $R\sec\alpha + \sec\beta = \dfrac{2CP_n}{n}$

2) D = pitch diameter of gear = $\dfrac{N}{P_n \cos\alpha}$

3) d = pitch diameter of pinion = $\dfrac{n}{P_n \cos\beta}$

4) O = outside diameter of gear = $D + \dfrac{2}{P_n}$

5) o = outside diameter of pinion = $d + \dfrac{2}{P_n}$

6) N' = number of teeth marked on formed cutter for gear (see below)

7) n' = number of teeth marked on formed cutter for pinion (see below)

8) L = lead of helix on gear = $\pi D \cot\alpha$

9) l = lead of helix on pinion = $\pi d \cot\beta$

Selecting Cutter for Milling Helical Gears.—The proper milling cutter to use for *spur* gears depends on the pitch of the teeth and also upon the number of teeth as explained on page 2021 but a cutter for milling helical gears is not selected with reference to the actual number of teeth in the gear, as in spur gearing, but rather with reference to a calculated number N' that takes into account the effect on the tooth profile of lead angle, normal diametral pitch, and cutter diameter.

In the helical gearing examples starting on page 2070 the number of teeth N' on which to base the selection of the cutter has been determined using the approximate formula $N' = N \div \cos^3\alpha$ or $N' = N\sec^3\alpha$, where N = the actual number of teeth in the helical gear and α = the helix angle. However, the use of this formula may, where a combination of high helix angle and low tooth number is involved, result in the selection of a higher number of cutter than should actually be used for greatest accuracy. This condition is most likely to occur when the aforementioned formula is used to calculate N' for gears of high helix angle and low number of teeth.

To avoid the possibility of error in choice of cutter number, the following formula, which gives theoretically correct results for all combinations of helix angle and tooth numbers, is to be preferred:

$$N' = N\sec^3\alpha + P_n D_c \tan^2\alpha \quad (1)$$

where: N' = number of teeth on which to base selection of cutter number from table on page 2023; N = actual number of teeth in helical gear; α = helix angle; P_n = normal diametral pitch of gear and cutter; and D_c = pitch diameter of cutter.

Factors for Selecting Cutters for Milling Helical Gears

Helix Angle, α	K	K′	Helix Angle, α	K	K′	Helix Angle, α	K	K′	Helix Angle, α	K	K′
0	1.000	0	16	1.127	0.082	32	1.640	0.390	48	3.336	1.233
1	1.001	0	17	1.145	0.093	33	1.695	0.422	49	3.540	1.323
2	1.002	0.001	18	1.163	0.106	34	1.755	0.455	50	3.767	1.420
3	1.004	0.003	19	1.182	0.119	35	1.819	0.490	51	4.012	1.525
4	1.007	0.005	20	1.204	0.132	36	1.889	0.528	52	4.284	1.638
5	1.011	0.008	21	1.228	0.147	37	1.963	0.568	53	4.586	1.761
6	1.016	0.011	22	1.254	0.163	38	2.044	0.610	54	4.925	1.894
7	1.022	0.015	23	1.282	0.180	39	2.130	0.656	55	5.295	2.039
8	1.030	0.020	24	1.312	0.198	40	2.225	0.704	56	5.710	2.198
9	1.038	0.025	25	1.344	0.217	41	2.326	0.756	57	6.190	2.371
10	1.047	0.031	26	1.377	0.238	42	2.436	0.811	58	6.720	2.561
11	1.057	0.038	27	1.414	0.260	43	2.557	0.870	59	7.321	2.770
12	1.068	0.045	28	1.454	0.283	44	2.687	0.933	60	8.000	3.000
13	1.080	0.053	29	1.495	0.307	45	2.828	1	61	8.780	3.254
14	1.094	0.062	30	1.540	0.333	46	2.983	1.072	62	9.658	3.537
15	1.110	0.072	31	1.588	0.361	47	3.152	1.150	63	10.687	3.852

$K = 1 \div \cos^3 \alpha = \sec^3 \alpha; K' = \tan^2 \alpha$

Outside and Pitch Diameters of Standard Involute-form Milling Cutters

Normal Diametral Pitch, P_n	Outside Dia., D_o	Pitch Dia., D_c	$Q = P_nD_c$	Normal Diametral Pitch, P_n	Outside Dia., D_o	Pitch Dia., D_c	$Q = P_nD_c$	Normal Diametral Pitch, P_n	Outside Dia., D_o	Pitch Dia., D_c	$Q = P_nD_c$
1	8.500	6.18	6.18	6	3.125	2.76	16.56	20	2.000	1.89	37.80
1¼	7.750	5.70	7.12	7	2.875	2.54	17.78	24	1.750	1.65	39.60
1½	7.000	5.46	8.19	8	2.875	2.61	20.88	28	1.750	1.67	46.76
1¾	6.500	5.04	8.82	9	2.750	2.50	22.50	32	1.750	1.68	53.76
2	5.750	4.60	9.20	10	2.375	2.14	21.40	36	1.750	1.69	60.84
2½	5.750	4.83	12.08	12	2.250	2.06	24.72	40	1.750	1.70	68.00
3	4.750	3.98	11.94	14	2.125	1.96	27.44	48	1.750	1.70	81.60
4	4.250	3.67	14.68	16	2.125	1.98	31.68	..	…	…	…
5	3.750	3.29	16.45	18	2.000	1.87	33.66	..	…	…	…

Pitch diameters shown in the table are computed from the formula: $D_c = D_o - 2(1.57 \div P_n)$. This same formula may be used to compute the pitch diameter of a non-standard outside diameter cutter when the normal diametral pitch P_n and the outside diameter D_o are known.

To simplify calculations, Formula (1) may be written as follows:

$$N' = NK + QK' \quad (2)$$

In this formula, K, K' and Q are constants obtained from the tables on page 2078.

Example: Helix angle = 30 degrees; number of teeth in helical gear = 15; and normal diametral pitch = 20. From the tables on page 2077 K, K', and Q are, respectively, 1.540, 0.333, and 37.80.

$$N' = (15 \times 1.540) + (37.80 \times 0.333) = 23.10 + 12.60$$
$$= 35.70, \text{ say, } 36$$

Hence, from page 2023 select a number 3 cutter. Had the approximate formula been used, then a number 5 cutter would have been selected on the basis of $N' = 23$.

Milling the Helical Teeth.—The teeth of a helical gear are proportioned from the normal pitch and not the circular pitch. The whole depth of the tooth can be found by dividing 2.157 by the normal diametral pitch of the gear, which corresponds to the pitch of the cutter. The thickness of the tooth at the pitch line equals 1.571 divided by the normal diametral pitch. After a tooth space has been milled, the cutter should be prevented from dragging through it when being returned for another cut. This can be done by lowering the blank slightly, or by stopping the machine and turning the cutter to such a position that the teeth will not touch the work. If the gear has teeth coarser than 10 or 12 diametral pitch, it is well to take a roughing and a finishing cut. When pressing a helical gear blank on the arbor, it should be remembered that it is more likely to slip when being milled than a spur gear, because the pressure of the cut, being at an angle, tends to rotate the blank on the arbor.

Angular Position of Table: When cutting a helical gear on a milling machine, the table is set to the helix angle of the gear. If the lead of the helical gear is known, but not the helix angle, the helix angle is determined by multiplying the pitch diameter of the gear by 3.1416 and dividing this product by the lead; the result is the tangent of the lead angle which may be obtained from trigonometric tables or a calculator.

American National Standard Fine-Pitch Teeth For Helical Gears.—This Standard, ANSI B6.7-1977, provides a 20-degree tooth form for both spur and helical gears of 20 diametral pitch and finer. Formulas for tooth parts are given on page 2008.

Enlargement of Helical Pinions of 20-Degree Normal Pressure Angle: Formula (4) and the accompanying graph are based on the use of hobs having sharp corners at their top lands. Pinions cut by shaper cutters may not require as much modification as indicated by Formula (4) or the graph. The number 2.1 appearing in (4) results from the use of a standard tooth thickness rack having an addendum of $1.05/P_n$ which will start contact at a roll angle 5 degrees above the base radius. The roll angle of 5 degrees is also reflected in Formula (4).

To avoid undercutting of the teeth and to provide more favorable contact conditions near the base of the tooth, it is recommended that helical pinions with less than 24 teeth be enlarged in accordance with the following graph and formulas. As with enlarged spur pinions, when an enlarged helical pinion is used it is necessary either to reduce the diameter of the mating gear or to increase the center distance. In the formulas that follow, ϕ_n = normal pressure angle; ϕ_t = transverse pressure angle; ψ = helix angle of pinion; P_n = normal diametral pitch; P_t = transverse diametral pitch; d = pitch diameter of pinion; d_o = outside diameter of enlarged pinion, K_h = enlargement for full depth pinions of 1 normal diametral pitch; and n = number of teeth in pinion.

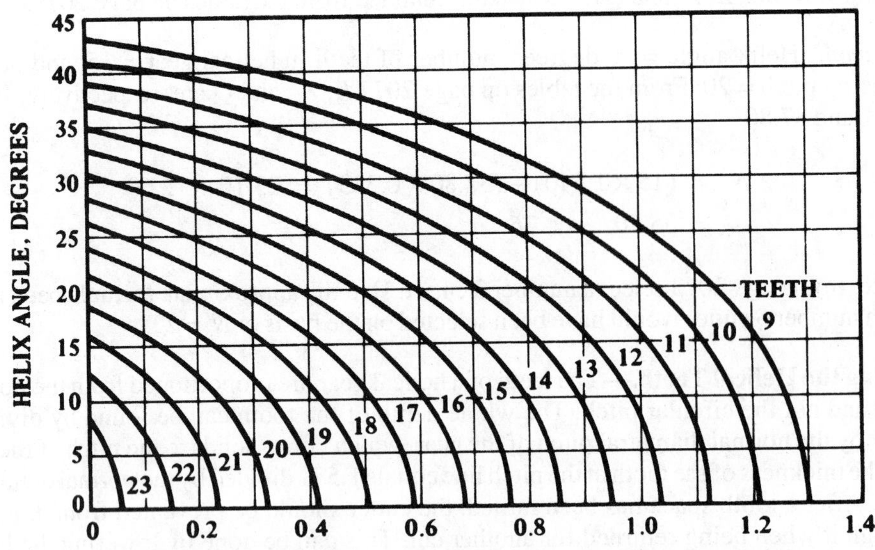

ENLARGEMENT K_h, IN INCHES, FOR 1 DIAMETRAL PITCH, 20° PRESSURE ANGLE PINIONS

To eliminate the need for making the calculations indicated in Formulas (3) and (4), the accompanying graph may be used to obtain the value of K_h directly for full-depth pinions of 20-degree normal pressure angle.

$$P_t = P_n \cos\psi \quad (1)$$

$$d = n \div P_t \quad (2)$$

$$\tan\psi_t = \tan\phi_n \div \cos\psi \quad (3)$$

$$K_h = 2.1 - \frac{n}{\cos\psi}(\sin\phi_t - \cos\phi_t \tan 5°)\sin\phi_t \quad (4)$$

$$d_o = d + \frac{2 + K_h}{P_n} \quad (5)$$

Example: Find the outside diameter of a helical pinion having 12 teeth, 32 normal diametral pitch, 20-degree pressure angle, and 18-degree helix angle.

$$P_t = P_n \cos\psi = 32\cos 18° = 32 \times 0.95106 = 30.4339$$

$$d = n \div P_t = 12 \div 30.4339 = 0.3943 \text{ inch}$$

$$K_h = 0.851 \text{ (from graph)}$$

$$d_o = 0.3943 + \frac{2 + 0.851}{32} = 0.4834$$

Center Distance at Which Modified Mating Helical Gears Will Mesh with no Backlash.—If the helical pinion in the previous example on page 2080 had been made to standard dimensions, that is, not enlarged, and was in tight mesh with a standard 24-tooth mating gear, the center distance for tight mesh could be calculated from the formula on page 2008:

$$C = \frac{n + N}{2P_n \cos\psi} = \frac{12 + 24}{2 \times 32 \times \cos 18°} = 0.5914 \text{ inch} \quad (1)$$

However, if the pinion is enlarged as in the example and meshed with the same standard 24-tooth gear, then the center distance for tight mesh will be increased. To calculate the new center distance, the following formulas and calculations are required:

First, calculate the transverse pressure angle ϕ_t using Formula (2):

$$\tan\phi_t = \tan\phi_n \div \cos\psi = \tan 20^\circ \div \cos 18^\circ = 0.38270 \qquad (2)$$

and from a calculator the angle ϕ_t is found to be 20° 56′ 30″. In the table on page 98, inv ϕ_t is found to be 0.017196, and the cosine from a calculator as 0.93394.

Next, using Formula (3), calculate the pressure angle ϕ at which the gears are in tight mesh:

$$\text{inv}\,\phi = \text{inv}\,\phi_t + \frac{(t_{nP} + t_{nG}) - \pi}{n + N} \qquad (3)$$

In this formula, the value for t_{nP} for 1 diametral pitch is that found in Table 3c on page 2025, for a 12-tooth pinion, in the fourth column: 1.94703. The value of t_{nG} for 1 diametral pitch for a standard gear is always 1.5708.

$$\text{inv}\,\phi = 0.017196 + \frac{(1.94703 + 1.5708) - \pi}{12 + 24} = 0.027647$$

From the table on page 99, or a calculator, 0.027647 is the involute of 24° 22′ 7″ and the cosine corresponding to this angle is 0.91091.

Finally, using Formula (4), the center distance for tight mesh, C' is found:

$$C' = \frac{C\cos\phi_t}{\cos\phi} = \frac{0.5914 \times 0.93394}{0.91091} = 0.606 \text{ inch} \qquad (4)$$

Change-gears for Helical Gear Hobbing.—If a gear-hobbing machine is not equipped with a differential, there is a fixed relation between the index and feed gears and it is necessary to compensate for even slight errors in the index gear ratio, to avoid excessive lead errors. This may be done readily (as shown by the example to follow) by modifying the ratio of the feed gears slightly, thus offsetting the index gear error and making very accurate leads possible.

Machine Without Differential: The formulas which follow may be applied in computing the index gear ratio.

R = index-gear ratio
L = lead of gear, inches
F = feed per gear revolution, inch
K = machine constant
T = number of threads on hob
N = number of teeth on gear
P_n = normal diametral pitch
P_{nc} = normal circular pitch
A = helix angle, relative to axis
M = feed gear constant

$$R = \frac{L \div F}{(L \div F) \pm 1} \times \frac{KT}{N} = \frac{L}{L \pm F} \times \frac{KT}{N} = \frac{\text{Driving gear sizes}}{\text{Driven gear sizes}} \qquad (1)$$

Use minus (−) sign in Formulas (1) and (2) when gear and hob are the same "hand" and plus (+) sign when they are of opposite hand; when *climb* hobbing is to be used, reverse this rule.

$$R = \frac{KT}{N \pm \dfrac{P_n \times \sin A \times F}{\pi}} = \frac{KT}{N \pm \dfrac{\sin A \times F}{P_{nc}}} \qquad (2)$$

$$\text{Ratio of feed gears} = \frac{F}{M} \qquad F = \frac{L(NR-KT)}{NR} \tag{3}$$

$$L = \frac{FNR}{(NR-KT)} = \text{lead obtained with available index and feed gears} \tag{4}$$

Note: If gear and hob are of opposite hand, then in Formulas (3) and (4) change $(NR-KT)$ to $(KT-NR)$. This change is also made if gear and hob are of same hand but *climb* hobbing is used.

Example: A right-hand helical gear with 48 teeth of 10 normal diametral pitch, has a lead of 44.0894 inches. The feed is to be 0.035 inch, with whatever slight adjustment may be necessary to compensate for the error in available index gears. $K = 30$ and $M = 0.075$. A single-thread right-hand hob is to be used.

$$R = \frac{44.0894}{44.0894 - 0.035} \times \frac{30 \times 1}{48} = 0.62549654$$

Using the method of *Conjugate Fractions* beginning on page 14, several suitable ratios close to 0.62549654 were found. One of these, $(34 \times 53)/(43 \times 67) = 0.625477264839$ will be used as the index ratio. Other usable ratios and their decimal values were found to be as follows:

$$\frac{32 \times 38}{27 \times 72} = 0.6255144 \qquad \frac{27 \times 42}{42 \times 37} = 0.62548263$$

$$\frac{44 \times 29}{34 \times 60} = 0.6254902 \qquad \frac{26 \times 97}{96 \times 42} = 0.62549603$$

$$\frac{20 \times 41}{23 \times 57} = 0.62547674$$

Index ratio error = 0.62549654 − 0.62547726 = 0.00001928.

Now use Formula (3) to find slight change required in rate of feed. This change compensates sufficiently for the error in available index gears.

Change in Feed Rate: Insert in Formula (3) obtainable index ratio.

$$F = \frac{44.0894 \times (48 \times 0.62547726 - 30)}{48 \times 0.62547726} = 0.0336417$$

$$\text{Modified feed gear ratio} = \frac{F}{M} = \frac{0.0336417}{0.075} = 0.448556$$

$$\text{Log } 0.448556 = \bar{1}.651817 \qquad \text{log of reciprocal} = 0.348183$$

To find close approximation to modified feed gear ratio, proceed as in finding suitable gears for index ratio, thus obtaining $\frac{106}{71} \times \frac{112}{75}$. Inverting, modified feed gear ratio = $\frac{71}{106} \times \frac{75}{112} = 0.448534$.

Modified feed F = obtainable modified feed ratio × M = 0.448534 × 0.075 = 0.03364 inch. If the feed rate is not modified, even a small error in the index gear ratio may result in an excessive lead error.

Checking Accuracy of Lead: The modified feed and obtainable index ratio are inserted in Formula (4). Desired lead = 44.0894 inches. Lead obtained = 44.087196 inches; hence the computed error = 44.0894 − 44.087196 = 0.002204 inch or about 0.00005 inch per inch of lead.

Machine with Differential: If a machine is equipped with a differential, the *lead gears* are computed in order to obtain the required helix angle and lead. The instructions of the hobbing machine manufacturer should be followed in computing the lead gears, because the ratio formula is affected by the location of the differential gears. If these gears are *ahead* of the index gears, the lead gear ratio is not affected by a change in the number of teeth to be cut (see Formula (5)); hence, the same lead gears are used when, for example, a gear and pinion are cut on the same machine. In the formulas which follow, the notation is the same as previously given, with these exceptions: R_d = lead gear ratio for machine with differential; P_a = axial or linear pitch of helical gear = distance from center of one tooth to center of next tooth measured parallel to gear axis = total lead $L \div$ number of teeth N.

$$R_d = \frac{P_a \times T}{K} = \frac{L \times T}{N \times K} = \frac{\pi \times \operatorname{cosec} A \times T}{P_n \times K} = \frac{\text{Driven gear sizes}}{\text{Driving gear sizes}} \qquad (5)$$

The number of hob threads T is included in the formula because double-thread hobs are used sometimes, especially for roughing in order to reduce the hobbing time. Lead gears having a ratio sufficiently close to the required ratio may be determined by using the table of gear ratio logarithms as previously described in connection with the non-differential type of machine. When using a machine equipped with a differential, the effect of a lead-gear ratio error upon the lead of the gear is small in comparison with the effect of an index gear error when using a non-differential type of machine. The lead obtained with a given or obtainable lead gear ratio may be determined by the following formula: $L = (R_d NK) \div T$. In this formula, R_d represents the ratio obtained with available gears. If the given lead is 44.0894 inches, as in the preceding example, then the desired ratio as obtained with Formula (5) would be 0.9185292 if $K = 1$. Assume that the lead gears selected by using logs of ratios have a ratio of 0.9184704; then this ratio error of 0.0000588 would result in a computed lead error of only 0.000065 inch per inch.

Formula (5), as mentioned, applies to machines having the differential located *ahead* of the index gears. If the differential is located after the index gears, it is necessary to change lead gears whenever the index gears are changed for hobbing a different number of teeth, as indicated by the following formula which gives the lead gear ratio. In this formula, D = pitch diameter.

$$R_d = \frac{L \times T}{K} = \frac{D \times \pi \times T}{K \times \tan A} = \frac{\text{Driven gear sizes}}{\text{Driving gear sizes}} \qquad (6)$$

General Remarks on Helical Gear Hobbing.—In cutting teeth having large angles, it is desirable to have the direction of helix of the hob the same as the direction of helix of the gear, or in other words, the gear and the hob of the same "hand." Then the direction of the cut will come against the movement of the blank. At ordinary angles, however, one hob will cut both right- and left-hand gears. In setting up the hobbing machine for helical gears, care should be taken to see that the vertical feed does not trip until the machine has been stopped or the hob has fed down past the finished gear.

Herringbone Gears

Double helical or herringbone gears are commonly used in parallel-shaft transmissions, especially when a smooth, continuous action (due to the gradual overlapping engagement of the teeth) is essential, as in high-speed drives where the pitch-line velocity may range from about 1000 to 3000 feet per minute in commercial gearing and up to 12,000 feet per minute or higher in more specialized installations. These relatively high speeds are encountered in marine reduction gears, in certain speed-reducing and speed-increasing units, and in various other transmissions, particularly in connection with steam turbine and electric motor drives.

General Classes of Helical Gear Problems.—There are two general classes of problems. In one, the problem is to design gears capable of transmitting a given amount of power at a given speed, safely and without excessive wear; hence, the required proportions must be determined. In the second, the proportions and speed are known and the power-transmitting capacity is required. The first is the more difficult and the more common problem.

Causes of Herringbone Gear Failures.—Where failure occurs in a herringbone gear transmission, it is rarely due to tooth breakage but usually to excessive wear or sub-surface failures, such as pitting and spalling; hence, it is common practice to base the design of such gears upon durability, or upon tooth pressures which are within the allowable limits for wear. In this connection, it seems to have been well established by tests of both spur gears and herringbone gears, that there is a critical surface pressure value for teeth having given physical properties and coefficient of friction. According to these tests, pressures above the critical value result in rapid wear and a short gear life, whereas when pressures are below the critical, wear is negligible. The yield point or endurance limit of the material marks the critical loading point, and in practical designing a reasonable factor of safety would, of course, be employed.

Planetary Gearing

Planetary or epicyclic gearing provides means of obtaining a compact design of transmission, with driving and driven shafts in line, and a large speed reduction when required. Typical arrangements of planetary gearing are shown by the following diagrams which are accompanied by speed ratio formulas. When planetary gears are arranged as shown by Fig. 5, 6, 9 and 12, the speed of the follower relative to the driver is increased, whereas Fig. 7, 8, 10, and 11 illustrate speed-reducing mechanisms.

Direction of Rotation.—In using the following formulas, if the final result is preceded by a minus sign (negative), this indicates that the driver and follower will rotate in opposite directions; otherwise, both will rotate in the same direction.

Compound Drive.—The formulas accompanying Figs. 19 through 22 are for obtaining the speed ratios when there are *two* driving members rotating at different speeds. For example, in Fig. 19, the central shaft with its attached link is one driver. The internal gear z, instead of being fixed, is also rotated. In Fig. 22, if $z = 24$, $B = 60$ and $S = 3\frac{1}{2}$, with both drivers rotating in the same direction, then $F = 0$, thus indicating, in this case, the point where a larger value of S will reverse follower rotation.

Planetary Bevel Gears.—Two forms of planetary gears of the bevel type are shown in Fig. 23 and 24. The planet gear in Fig. 23 rotates about a fixed bevel gear at the center of which is the driven shaft. Fig. 24 illustrates the Humpage reduction gear. This is sometimes referred to as cone-pulley back-gearing because of its use within the cone pulleys of certain types of machine tools.

Ratios of Planetary or Epicyclic Gearing

D = rotation of *driver* per revolution of follower or driven member

F = rotation of *follower* or driven member per revolution of driver. (In Figs. 1 through 4 F = rotation of planet type follower about its axis.)

A = size of driving gear (use either number of teeth or pitch diameter). Note: When follower derives its motion both from A and from a secondary driving member, A = size of *initial* driving gear, and formula gives speed relationship between A and follower.

B = size of *driven gear or follower* (use either pitch diameter or number of teeth)

C = size of *fixed gear* (use either pitch diameter or number of teeth)

x = size of *planet gear* as shown by diagram (use either pitch diameter or number of teeth)

y = size of *planet gear* as shown by diagram (use either pitch diameter or number of teeth)

z = size of secondary or *auxiliary driving gear*, when follower derives its motion from two driving members

S = rotation of *secondary driver*, per revolution of *initial driver*. S is negative when secondary and initial drivers rotate in opposite directions. (Formulas in which S is used, give speed relationship between follower and the initial driver.)

Note: In all cases, if D is known, $F = 1 \div D$, or, if F is known, $D = 1 \div F$.

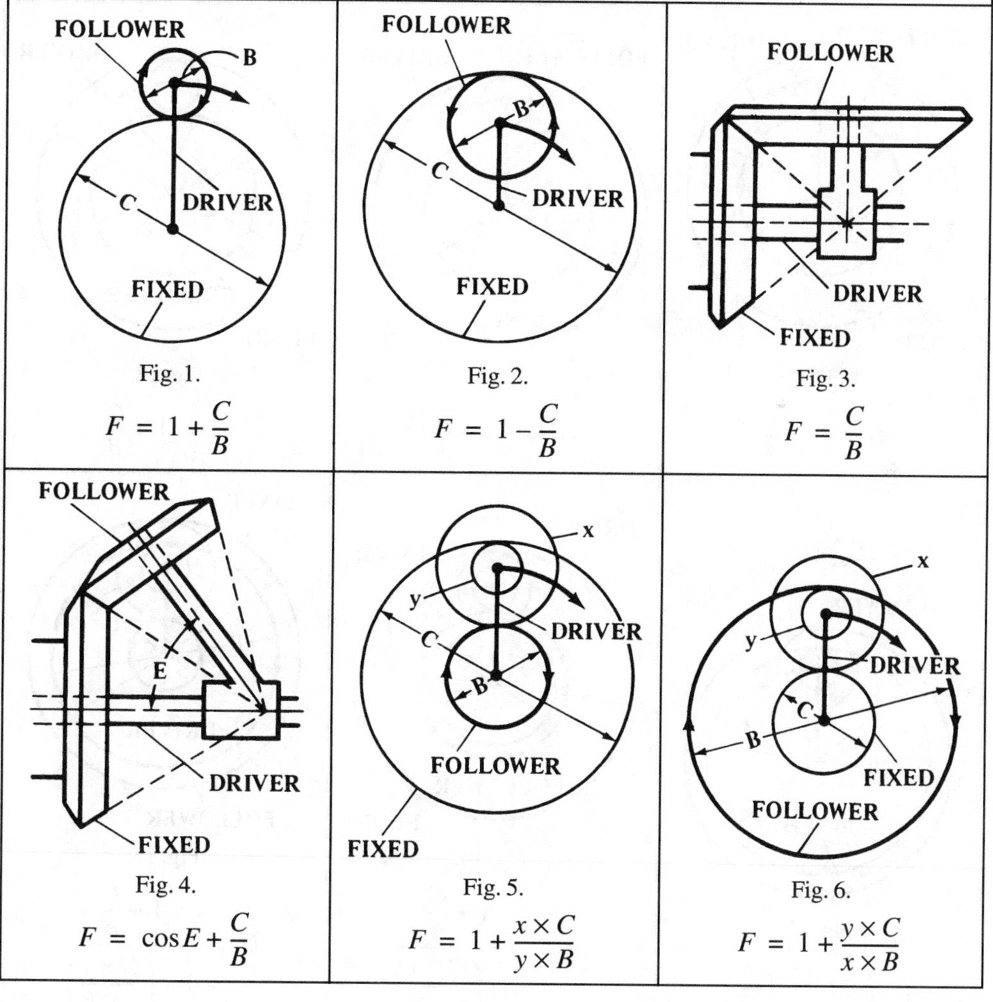

Fig. 1. $F = 1 + \frac{C}{B}$

Fig. 2. $F = 1 - \frac{C}{B}$

Fig. 3. $F = \frac{C}{B}$

Fig. 4. $F = \cos E + \frac{C}{B}$

Fig. 5. $F = 1 + \frac{x \times C}{y \times B}$

Fig. 6. $F = 1 + \frac{y \times C}{x \times B}$

Ratios of Planetary or Epicyclic Gearing *(Continued)*

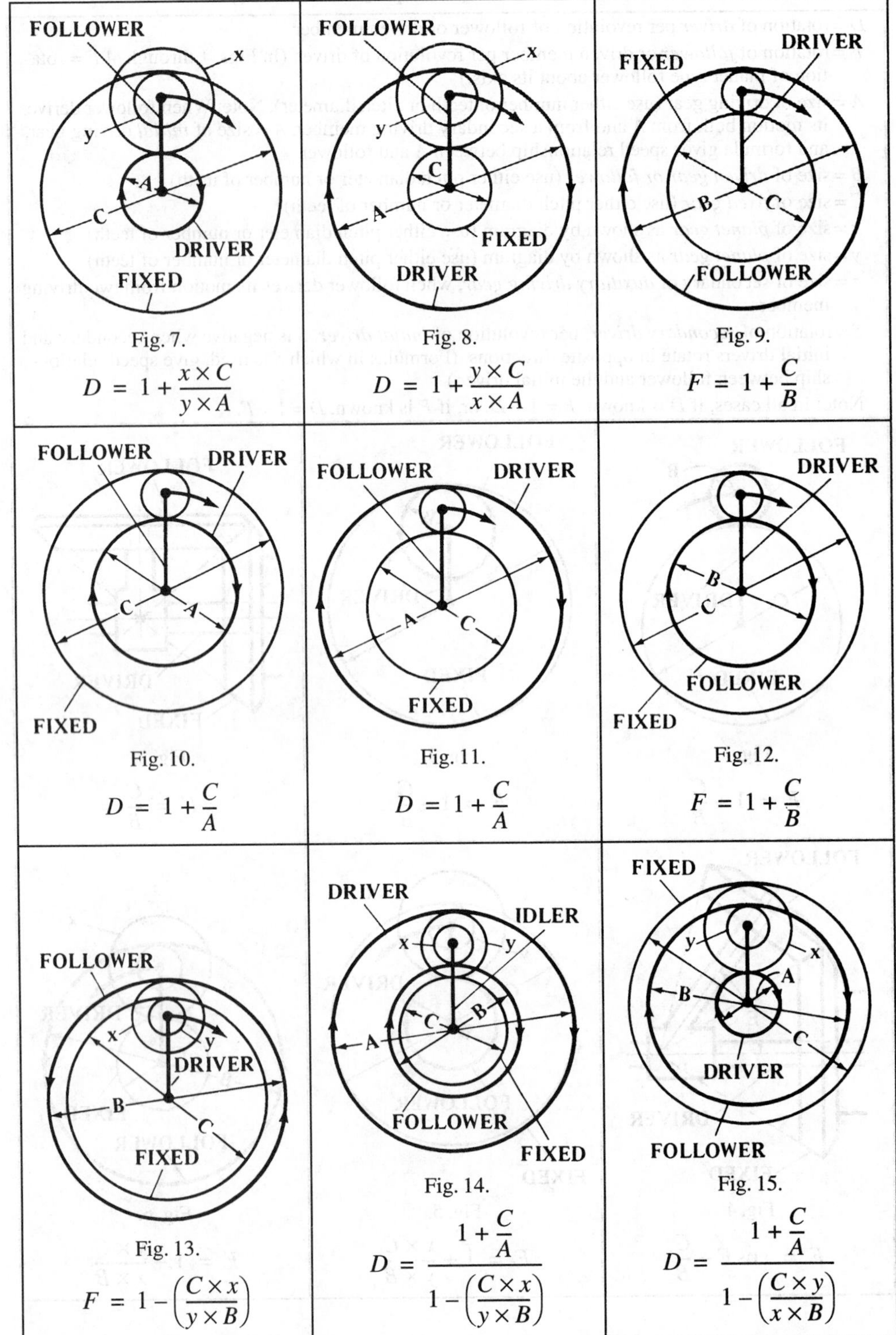

Fig. 7.

$$D = 1 + \frac{x \times C}{y \times A}$$

Fig. 8.

$$D = 1 + \frac{y \times C}{x \times A}$$

Fig. 9.

$$F = 1 + \frac{C}{B}$$

Fig. 10.

$$D = 1 + \frac{C}{A}$$

Fig. 11.

$$D = 1 + \frac{C}{A}$$

Fig. 12.

$$F = 1 + \frac{C}{B}$$

Fig. 13.

$$F = 1 - \left(\frac{C \times x}{y \times B}\right)$$

Fig. 14.

$$D = \frac{1 + \frac{C}{A}}{1 - \left(\frac{C \times x}{y \times B}\right)}$$

Fig. 15.

$$D = \frac{1 + \frac{C}{A}}{1 - \left(\frac{C \times y}{x \times B}\right)}$$

Ratios of Planetary or Epicyclic Gearing *(Continued)*

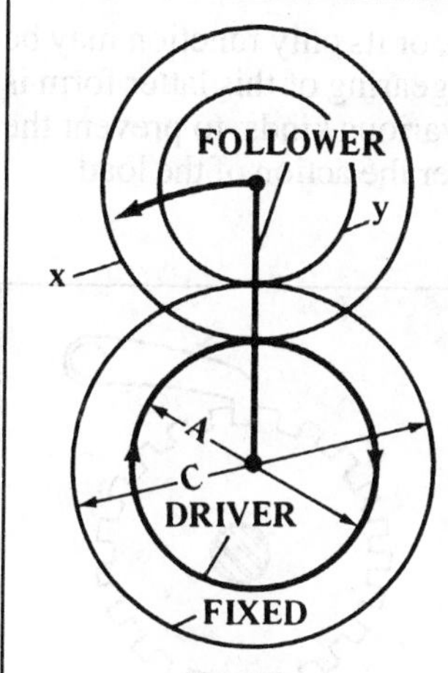

Fig. 16.

$$D = 1 - \left(\frac{C \times x}{y \times A}\right)$$

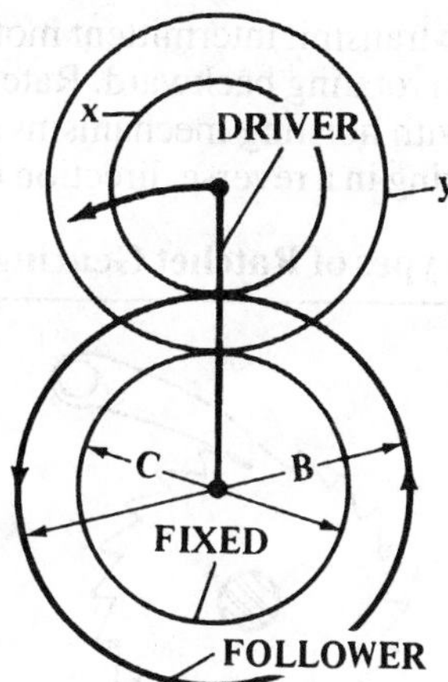

Fig. 17.

$$F = 1 - \left(\frac{C \times x}{y \times B}\right)$$

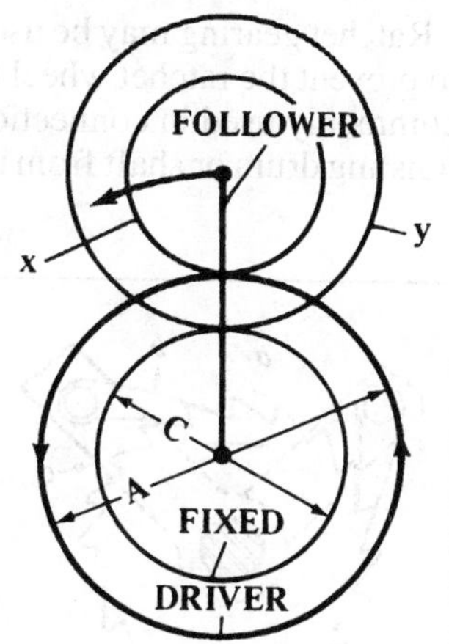

Fig. 18.

$$D = 1 - \left(\frac{C \times x}{y \times A}\right)$$

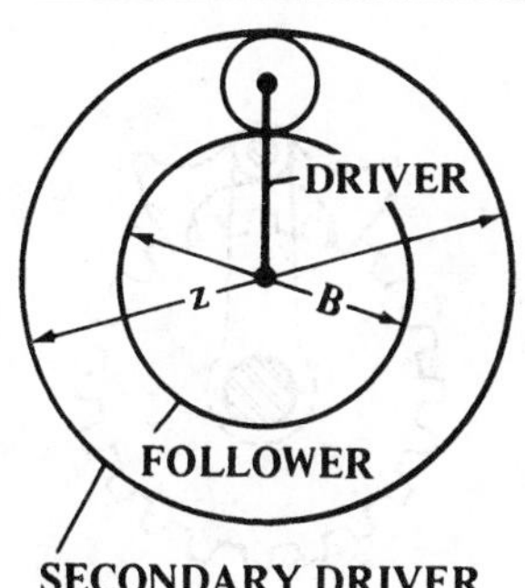

Fig. 19.

$$F = 1 + \frac{z \times (1 - S)}{B}$$

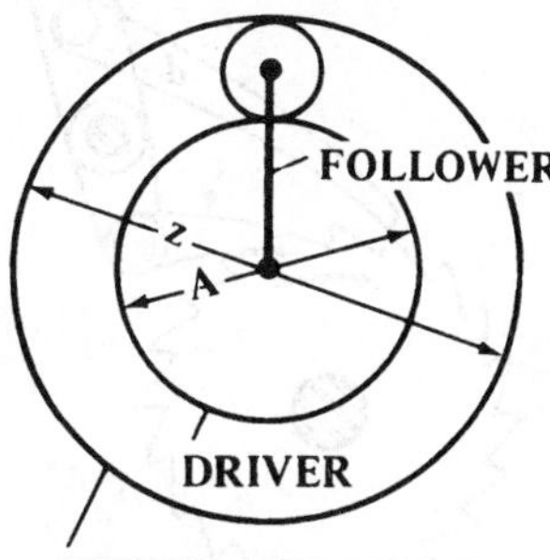

Fig. 20.

$$D = \frac{A + z}{A + (S \times z)}$$

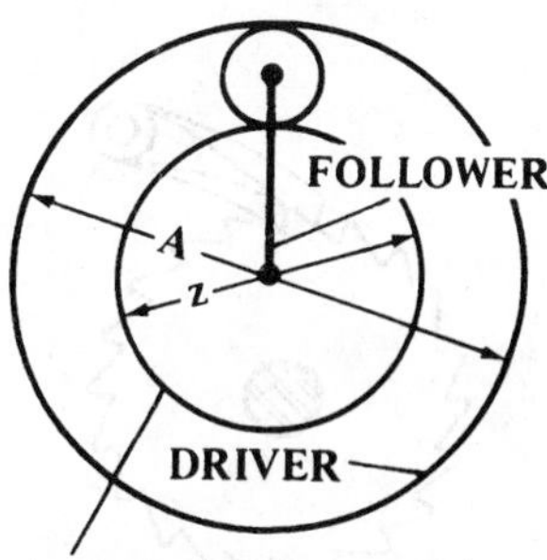

Fig. 21.

$$D = \frac{A + z}{A + (S \times z)}$$

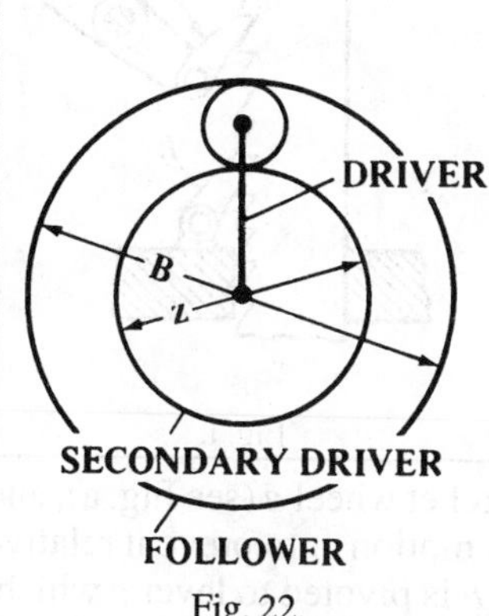

Fig. 22.

$$F = 1 + \frac{z \times (1 - S)}{B}$$

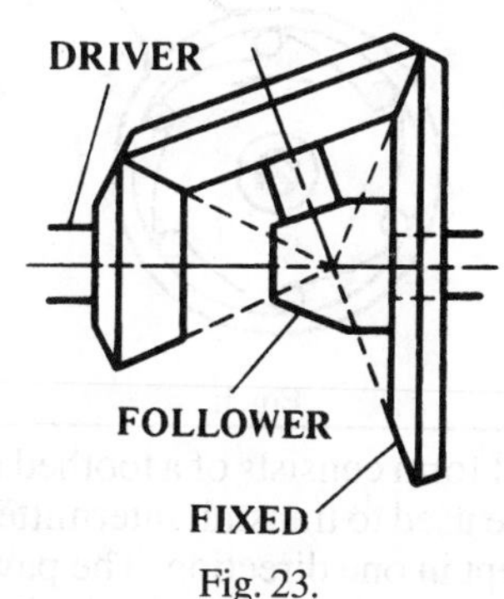

Fig. 23.

$$D = 1 + \frac{C}{A}$$

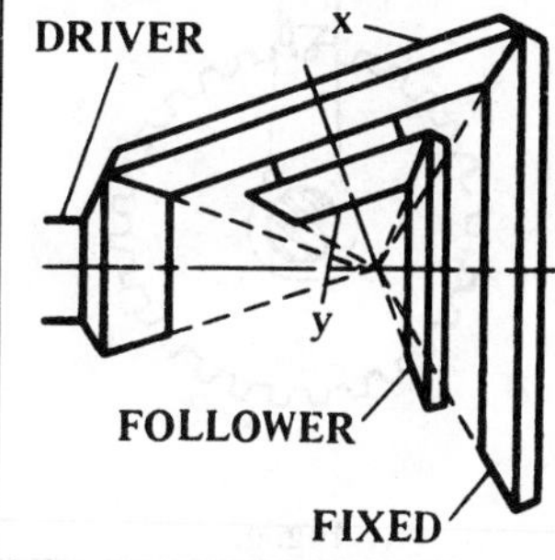

Fig. 24.

$$D = \frac{1 + \dfrac{C}{A}}{1 - \left(\dfrac{C \times y}{x \times B}\right)}$$

Ratchet Gearing

Ratchet gearing may be used to transmit intermittent motion, or its only function may be to prevent the ratchet wheel from rotating backward. Ratchet gearing of this latter form is commonly used in connection with hoisting mechanisms of various kinds, to prevent the hoisting drum or shaft from rotating in a reverse direction under the action of the load.

Types of Ratchet Gearing

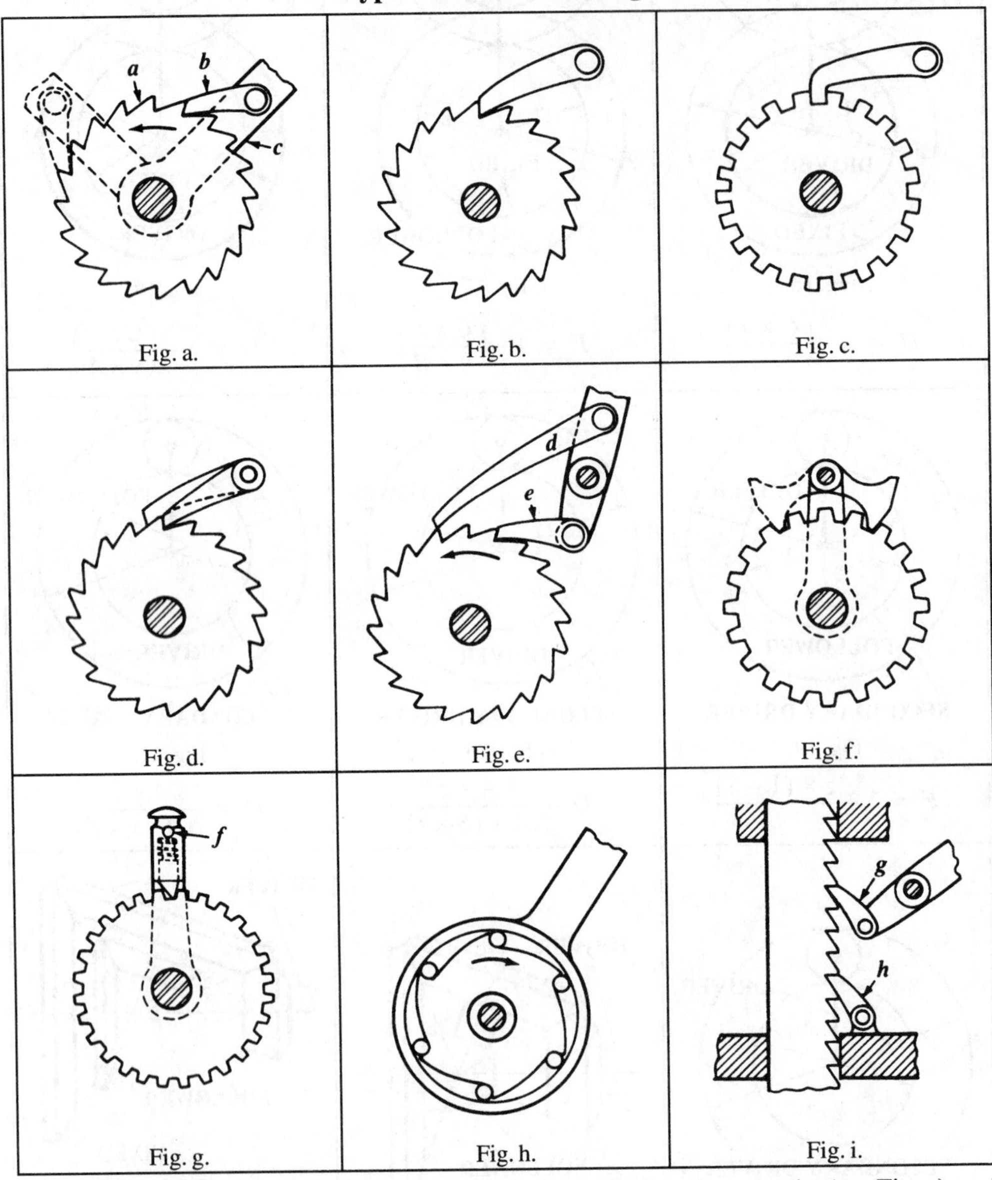

Fig. a. Fig. b. Fig. c. Fig. d. Fig. e. Fig. f. Fig. g. Fig. h. Fig. i.

Ratchet gearing in its simplest form consists of a toothed ratchet wheel *a* (see Fig. a), and a pawl or detent *b*, and it may be used to transmit intermittent motion or to prevent relative motion between two parts except in one direction. The pawl *b* is pivoted to lever *c* which, when given an oscillating movement, imparts an intermittent rotary movement to ratchet wheel *a*. Fig. b illustrates another application of the ordinary ratchet and pawl mechanism. In this instance, the pawl is pivoted to a stationary member and its only function is to prevent the ratchet wheel from rotating backward. With the stationary design, illustrated at

Fig. c, the pawl prevents the ratchet wheel from rotating in either direction, so long as it is in engagement with the wheel.

The principle of *multiple-pawl ratchet gearing* is illustrated at Fig. d, which shows the use of two pawls. One of these pawls is longer than the other, by an amount equal to one-half the pitch of the ratchet-wheel teeth, so that the practical effect is that of reducing the pitch one-half. By placing a number of driving pawls side byside and proportioning their lengths according to the pitch of the teeth, a very fine feed can be obtained with a ratchet wheel of comparatively coarse pitch.

This method of obtaining a fine feed from relatively coarse-pitch ratchets may be preferable to the use of single ratchets of fine pitch which, although providing the feed required, may have considerably weaker teeth.

The type of ratchet gearing shown at Fig. e is sometimes employed to impart a rotary movement to the ratchet wheel for both the forward and backward motions of the lever to which the two pawls are attached.

A simple form of *reversing ratchet* is illustrated at Fig. f. The teeth of the wheel are so shaped that either side may be used for driving by simply changing the position of the double-ended pawl, as indicated by the full and dotted lines.

Another form of reversible ratchet gearing for shapers is illustrated at Fig. g. The pawl, in this case, instead of being a pivoted latch, is in the form of a plunger which is free to move in the direction of its axis, but is normally held into engagement with the ratchet wheel by a small spring. When the pawl is lifted and turned one-half revolution, the driving face then engages the opposite sides of the teeth and the ratchet wheel is given an intermittent rotary motion in the opposite direction.

The *frictional type* of ratchet gearing differs from the designs previously referred to, in that there is no positive engagement between the driving and driven members of the ratchet mechanism, the motion being transmitted by frictional resistance. One type of frictional ratchet gearing is illustrated at Fig. h. Rollers or balls are placed between the ratchet wheel and an outer ring which, when turned in one direction, causes the rollers or balls to wedge between the wheel and ring as they move up the inclined edges of the teeth.

Fig. i illustrates one method of utilizing ratchet gearing for moving the driven member in a straight line, as in the case of a lifting jack. The pawl g is pivoted to the operating lever of the jack and does the lifting, whereas the pawl h holds the load while the lifting pawl g is being returned preparatory to another lifting movement.

Shape of Ratchet Wheel Teeth.—When designing ratchet gearing, it is important to so shape the teeth that the pawl will remain in engagement when a load is applied. The faces of the teeth which engage the end of the pawl should be in such relation with the center of the pawl pivot that a line perpendicular to the face of the engaging tooth will pass somewhere between the center of the ratchet wheel and the center of the pivot about which the pawl swings. This is true if the pawl *pushes* the ratchet wheel, or if the ratchet wheel *pushes* the pawl. However, if the pawl *pulls* the ratchet wheel or if the ratchet wheel *pulls* the pawl, the perpendicular from the face of the ratchet teeth should fall outside the pawl pivot center. Ratchet teeth may be either cut by a milling cutter having the correct angle, or hobbed in a gear-hobbing machine by the use of a special hob.

Pitch of Ratchet Wheel Teeth.—The pitch of ratchet wheels used for holding suspended loads may be calculated by the following formula, in which P = circular pitch, in inches, measured at the outside circumference; M = turning moment acting upon the ratchet wheel shaft, in inch-pounds; L = length of tooth face, in inches (thickness of ratchet gear); S = safe stress (for steel, 2500 pounds per square inch when subjected to shock, and 4000 pounds per square inch when not subjected to shock); N = number of teeth in ratchet wheel; F = a factor the value of which is 50 for ratchet gears with 12 teeth or less, 35 for gears having from 12 to 20 teeth, and 20 for gears having over 20 teeth:

$$P = \sqrt{\frac{FM}{LSN}}$$

This formula has been used in the calculation of ratchet gears for crane design.

Gear Design Based upon Module System.—The *module* of a gear is equal to the pitch diameter divided by the number of teeth, whereas *diametral pitch* is equal to the number of teeth divided by the pitch diameter. The module system (see accompanying table and diagram) is in general use in countries that have adopted the metric system; hence, the term module is usually understood to mean the pitch diameter *in millimeters* divided by the number of teeth. The module system, however, may also be based on inch measurements and then it is known as the English module to avoid confusion with the metric module. Module is an actual dimension, whereas diametral pitch is only a ratio. Thus, if the pitch diameter of a gear is 50 millimeters and the number of teeth 25, the module is 2, which means that there are 2 millimeters of pitch diameter for each tooth. The table *Tooth Dimensions Based Upon Module System* shows the relation among module, diametral pitch, and circular pitch.

German Standard Tooth Form for Spur and Bevel Gears *DIN 867*

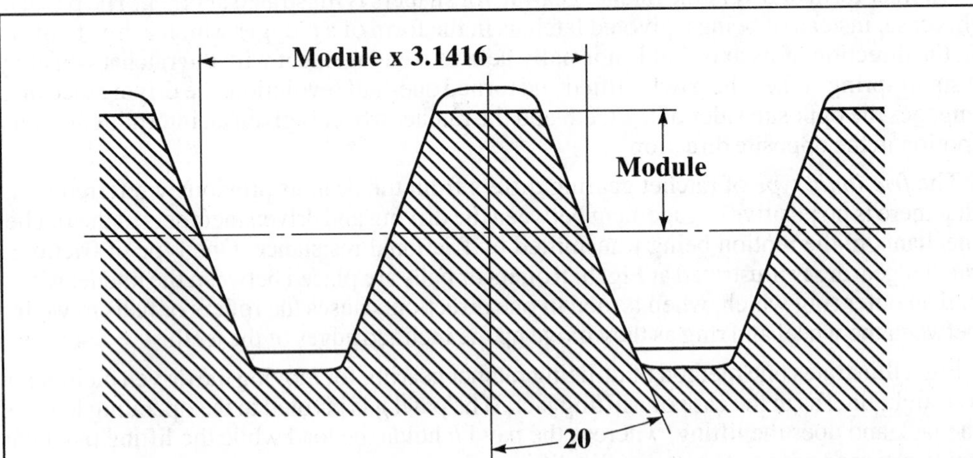

The flanks or sides are straight (involute system) and the pressure angle is 20 degrees. The shape of the root clearance space and the amount of clearance depend upon the method of cutting and special requirements. The amount of clearance may vary from 0.1 × module to 0.3 × module.

To Find	Module Known	Circular Pitch Known
Addendum	Equals module	0.31823 × Circular pitch
Dedendum	1.157 × module* 1.167 × module**	0.3683 × Circular pitch* 0.3714 × Circular pitch**
Working Depth	2 × module	0.6366 × Circular pitch
Total Depth	2.157 × module* 2.167 × module**	0.6866 × Circular pitch* 0.6898 × Circulate pitch**
Tooth Thickness on Pitch Line	1.5708 × module	0.5 × Circular pitch

Formulas for dedendum and total depth, marked (*) are used when clearance equals 0.157 × module. Formulas marked (**) are used when clearance equals one-sixth module. It is common practice among American cutter manufacturers to make the clearance of metric or module cutters equal to 0.157 × module.

Tooth Dimensions Based Upon Module System

Module, *DIN* Standard Series	Equivalent Diametral Pitch	Circular Pitch		Addendum, Millimeters	Dedendum, Millimeters[a]	Whole Depth,[a] Millimeters	Whole Depth,[b] Millimeters
		Millimeters	Inches				
0.3	84.667	0.943	0.0371	0.30	0.35	0.650	0.647
0.4	63.500	1.257	0.0495	0.40	0.467	0.867	0.863
0.5	50.800	1.571	0.0618	0.50	0.583	1.083	1.079
0.6	42.333	1.885	0.0742	0.60	0.700	1.300	1.294
0.7	36.286	2.199	0.0865	0.70	0.817	1.517	1.510
0.8	31.750	2.513	0.0989	0.80	0.933	1.733	1.726
0.9	28.222	2.827	0.1113	0.90	1.050	1.950	1.941
1	25.400	3.142	0.1237	1.00	1.167	2.167	2.157
1.25	20.320	3.927	0.1546	1.25	1.458	2.708	2.697
1.5	16.933	4.712	0.1855	1.50	1.750	3.250	3.236
1.75	14.514	5.498	0.2164	1.75	2.042	3.792	3.774
2	12.700	6.283	0.2474	2.00	2.333	4.333	4.314
2.25	11.289	7.069	0.2783	2.25	2.625	4.875	4.853
2.5	10.160	7.854	0.3092	2.50	2.917	5.417	5.392
2.75	9.236	8.639	0.3401	2.75	3.208	5.958	5.932
3	8.466	9.425	0.3711	3.00	3.500	6.500	6.471
3.25	7.815	10.210	0.4020	3.25	3.791	7.041	7.010
3.5	7.257	10.996	0.4329	3.50	4.083	7.583	7.550
3.75	6.773	11.781	0.4638	3.75	4.375	8.125	8.089
4	6.350	12.566	0.4947	4.00	4.666	8.666	8.628
4.5	5.644	14.137	0.5566	4.50	5.25	9.750	9.707
5	5.080	15.708	0.6184	5.00	5.833	10.833	10.785
5.5	4.618	17.279	0.6803	5.50	6.416	11.916	11.864
6	4.233	18.850	0.7421	6.00	7.000	13.000	12.942
6.5	3.908	20.420	0.8035	6.50	7.583	14.083	14.021
7	3.628	21.991	0.8658	7.	8.166	15.166	15.099
8	3.175	25.132	0.9895	8.	9.333	17.333	17.256
9	2.822	28.274	1.1132	9.	10.499	19.499	19.413
10	2.540	31.416	1.2368	10.	11.666	21.666	21.571
11	2.309	34.558	1.3606	11.	12.833	23.833	23.728
12	2.117	37.699	1.4843	12.	14.000	26.000	25.884
13	1.954	40.841	1.6079	13.	15.166	28.166	28.041
14	1.814	43.982	1.7317	14.	16.332	30.332	30.198
15	1.693	47.124	1.8541	15.	17.499	32.499	32.355
16	1.587	50.266	1.9790	16.	18.666	34.666	34.512
18	1.411	56.549	2.2263	18.	21.000	39.000	38.826
20	1.270	62.832	2.4737	20.	23.332	43.332	43.142
22	1.155	69.115	2.7210	22.	25.665	47.665	47.454
24	1.058	75.398	2.9685	24.	28.000	52.000	51.768
27	0.941	84.823	3.339	27.	31.498	58.498	58.239
30	0.847	94.248	3.711	30.	35.000	65.000	64.713
33	0.770	103.673	4.082	33.	38.498	71.498	71.181
36	0.706	113.097	4.453	36.	41.998	77.998	77.652
39	0.651	122.522	4.824	39.	45.497	84.497	84.123
42	0.605	131.947	5.195	42.	48.997	90.997	90.594
45	0.564	141.372	5.566	45.	52.497	97.497	97.065
50	0.508	157.080	6.184	50.	58.330	108.330	107.855
55	0.462	172.788	6.803	55.	64.163	119.163	118.635
60	0.423	188.496	7.421	60.	69.996	129.996	129.426
65	0.391	204.204	8.040	65.	75.829	140.829	140.205
70	0.363	219.911	8.658	70.	81.662	151.662	150.997
75	0.339	235.619	9.276	75.	87.495	162.495	161.775

[a] Dedendum and total depth when clearance = 0.1666 × module, or one-sixth module.

[b] Total depth equivalent to American standard full-depth teeth. (Clearance = 0.157 × module.)

Rules for Module System of Gearing

To Find	Rule
Metric Module	*Rule 1:* To find the metric module, divide the pitch diameter in millimeters by the number of teeth. *Example 1:* The pitch diameter of a gear is 200 millimeters and the number of teeth, 40; then $\text{Module} = \frac{200}{40} = 5$ *Rule 2:* Multiply circular pitch in millimeters by 0.3183. *Example 2:* (Same as Example 1. Circular pitch of this gear equals 15.708 millimeters.) $\text{Module} = 15.708 \times 0.3183 = 5$ *Rule 3:* Divide outside diameter in millimeters by the number of teeth plus 2.
English Module	*Note:* The module system is usually applied when gear dimensions are expressed in millimeters, but module may also be based on inch measurements. *Rule:* To find the English module, divide pitch diameter in inches by the number of teeth. *Example:* A gear has 48 teeth and a pitch diameter of 12 inches. $\text{Module} = \frac{12}{48} = \frac{1}{4}$ module or 4 diametral pitch
Metric Module Equivalent to Diametral Pitch	*Rule:* To find the metric module equivalent to a given diametral pitch, divide 25.4 by the diametral pitch. *Example:* Determine metric module equivalent to 10 diameteral pitch. $\text{Equivalent module} = \frac{25.4}{10} = 2.54$ *Note:* The nearest standard module is 2.5.
Diametral Pitch Equivalent to Metric Module	*Rule:* To find the diametral pitch equivalent to a given module, divide 25.4 by the module. (25.4 = number of millimeters per inch.) *Example:* The module is 12; determine equivalent diametral pitch. $\text{Equivalent diametral pitch} = \frac{25.4}{12} = 2.117$ *Note:* A diametral pitch of 2 is the nearest *standard* equivalent.
Pitch Diameter	*Rule:* Multiply number of teeth by module. *Example:* The metric module is 8 and the gear has 40 teeth; then $D = 40 \times 8 = 320 \text{ millimeters} = 12.598 \text{ inches}$
OutsideDiameter	*Rule:* Add 2 to the number of teeth and multiply sum by the module. *Example:* A gear has 40 teeth and module is 6. Find outside or blank diameter. $\text{Outside diameter} = (40 + 2) \times 6 = 252 \text{ millimeters}$

For tooth dimensions, see table *Tooth Dimensions Based Upon Module System*; also formulas in *German Standard Tooth Form for Spur and Bevel Gears DIN 867*.

Equivalent Diametral Pitches, Circular Pitches, and Metric Modules

Commonly Used Pitches and Modules in Bold Type

Diametral Pitch	Circular Pitch, Inches	Module Millimeters	Diametral Pitch	Circular Pitch, Inches	Module Millimeters	Diametral Pitch	Circular Pitch, Inches	Module Millimeters
1/2	6.2832	50.8000	2.2848	**1 3/8**	11.1170	10.0531	**5/16**	2.5266
0.5080	6.1842	**50**	2.3091	1.3605	**11**	10.1600	0.3092	**2 1/2**
0.5236	**6**	48.5104	**2 1/2**	1.2566	10.1600	**11**	0.2856	2.3091
0.5644	5.5658	**45**	2.5133	**1 1/4**	10.1063	**12**	0.2618	2.1167
0.5712	**5 1/2**	44.4679	2.5400	1.2368	**10**	12.5664	**1/4**	2.0213
0.6283	**5**	40.4253	**2 3/4**	1.1424	9.2364	12.7000	0.2474	**2**
0.6350	4.9474	**40**	2.7925	**1 1/8**	9.0957	**13**	0.2417	1.9538
0.6981	**4 1/2**	36.3828	2.8222	1.1132	**9**	**14**	0.2244	1.8143
0.7257	4.3290	**35**	**3**	1.0472	8.4667	**15**	0.2094	1.6933
3/4	4.1888	33.8667	3.1416	**1**	8.0851	**16**	0.1963	1.5875
0.7854	**4**	32.3403	3.1750	0.9895	**8**	16.7552	**3/16**	1.5160
0.8378	**3 3/4**	30.3190	3.3510	**15/16**	7.5797	16.9333	0.1855	**1 1/2**
0.8467	3.7105	**30**	**3 1/2**	0.8976	7.2571	**17**	0.1848	1.4941
0.8976	**3 1/2**	28.2977	3.5904	**7/8**	7.0744	**18**	0.1745	1.4111
0.9666	**3 1/4**	26.2765	3.6286	0.8658	**7**	**19**	0.1653	1.3368
1	3.1416	25.4000	3.8666	**13/16**	6.5691	**20**	0.1571	1.2700
1.0160	3.0921	**25**	3.9078	0.8040	**6 1/2**	**22**	0.1428	1.1545
1.0472	**3**	24.2552	**4**	0.7854	6.3500	**24**	0.1309	1.0583
1.1424	**2 3/4**	22.2339	4.1888	**3/4**	6.0638	**25**	0.1257	1.0160
1 1/4	2.5133	20.3200	4.2333	0.7421	**6**	25.1328	**1/8**	1.0106
1.2566	**2 1/2**	20.2127	4.5696	**11/16**	5.5585	25.4000	0.1237	**1**
1.2700	2.4737	**20**	4.6182	0.6803	**5 1/2**	**26**	0.1208	0.9769
1.3963	**2 1/4**	18.1914	**5**	0.6283	5.0800	**28**	0.1122	0.9071
1.4111	2.2263	**18**	5.0265	**5/8**	5.0532	**30**	0.1047	0.8467
1 1/2	2.0944	16.9333	5.0800	0.6184	**5**	**32**	0.0982	0.7937
1.5708	**2**	16.1701	5.5851	**9/16**	4.5478	**34**	0.0924	0.7470
1.5875	1.9790	**16**	5.6443	0.5566	**4 1/2**	**36**	0.0873	0.7056
1.6755	**1 7/8**	15.1595	**6**	0.5236	4.2333	**38**	0.0827	0.6684
1.6933	1.8553	**15**	6.2832	**1/2**	4.0425	**40**	0.0785	0.6350
1 3/4	1.7952	14.5143	6.3500	0.4947	**4**	**42**	0.0748	0.6048
1.7952	**1 3/4**	14.1489	**7**	0.4488	3.6286	**44**	0.0714	0.5773
1.8143	1.7316	**14**	7.1808	**7/16**	3.5372	**46**	0.0683	0.5522
1.9333	**1 5/8**	13.1382	7.2571	0.4329	**3 1/2**	**48**	0.0654	0.5292
1.9538	1.6079	**13**	**8**	0.3927	3.1750	**50**	0.0628	0.5080
2	1.5708	12.7000	8.3776	**3/8**	3.0319	50.2656	**1/16**	0.5053
2.0944	**1 1/2**	12.1276	8.4667	0.3711	**3**	50.8000	0.0618	**1/2**
2.1167	1.4842	**12**	**9**	0.3491	2.8222	**56**	0.0561	0.4536
2 1/4	1.3963	11.2889	**10**	0.3142	2.5400	**60**	0.0524	0.4233

The module of a gear is the pitch diameter divided by the number of teeth. The module may be expressed in any units; but when no units are stated, it is understood to be in millimeters. The metric module, therefore, equals the pitch diameter in millimeters divided by the number of teeth. To find the metric module equivalent to a given diametral pitch, divide 25.4 by the diametral pitch. To find the diametral pitch equivalent to a given module, divide 25.4 by the module. (25.4 = number of millimeters per inch.)

CHECKING GEAR SIZES

Checking Gear Size by Measurement Over Wires or Pins

The wire or pin method of checking gear sizes is accurate, easily applied, and especially useful in shops with limited inspection equipment. Two cylindrical wires or pins of predetermined diameter are placed in diametrically opposite tooth spaces (see diagram). If the gear has an odd number of teeth, the wires are located as nearly opposite as possible, as shown by the diagram at the right. The overall measurement *M* is checked by using any sufficiently accurate method of measurement. The value of measurement *M* when the pitch diameter is correct can be determined easily and quickly by means of the calculated values in the accompanying tables.

Measurements for Checking External Spur Gears when Wire Diameter Equals 1.728 Divided by Diametral Pitch.—Tables 1 and 2 give measurements *M*, in inches, for checking the pitch diameters of external spur gears of 1 diametral pitch. For any other diametral pitch, divide the measurement given in the table by whatever diametral pitch is required. The result shows what measurement *M* should be when the pitch diameter is correct *and there is no allowance for backlash.* The procedure for obtaining a given amount of backlash will be explained later. Tables 1 through 4 inclusive are based on wire sizes conforming to the Van Keuren standard. For external spur gears, the wire size equals 1.728 divided by the diametral pitch. The wire diameters for various diametral pitches will be found in the left-hand section of Table 5.

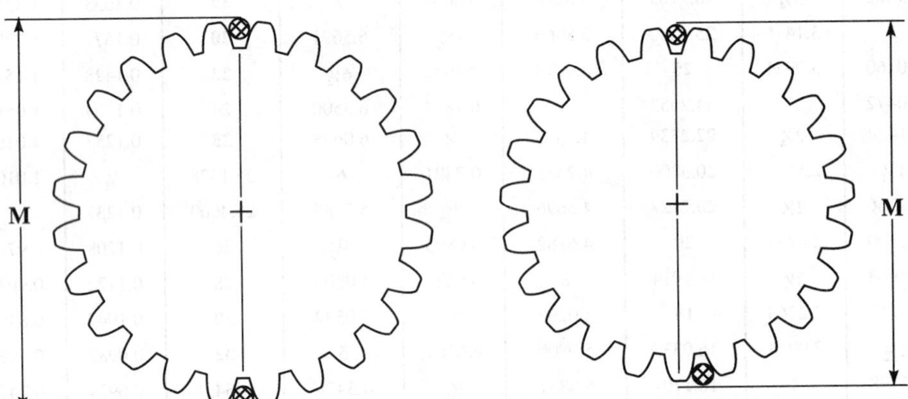

Even Number of Teeth: Table 1 is for even numbers of teeth. To illustrate the use of the table, assume that a spur gear has 32 teeth of 4 diametral pitch and a pressure angle of 20 degrees. Table 1 shows that the measurement for 1 diametral pitch is 34.4130; hence, for 4 diametral pitch, the measurement equals 34.4130 ÷ 4 = 8.6032 inches. This dimension is the measurement over the wires when the pitch diameter is correct, provided there is no allowance for backlash. The wire diameter here equals 1.728 ÷ 4 = 0.432 inch (Table 5).

Measurement for even numbers of teeth above 170 and not in Table 1 may be determined as shown by the following example: Assume that number of teeth = 240 and pressure angle = 14½ degrees; then, for 1 diametral pitch, figure at left of decimal point = given No. of teeth + 2 = 240 + 2 = 242. Figure at right of decimal point lies between decimal values given in table for 200 teeth and 300 teeth and is obtained by interpolation. Thus, 240 − 200 = 40 (change to 0.40); 0.5395 − 0.5321 = 0.0074 = difference between decimal values for 300 and 200 teeth; hence, decimal required = 0.5321 + (0.40 × 0.0074) = 0.53506. Total dimension = 242.53506 divided by the diametral pitch required.

Odd Number of Teeth: Table 2 is for odd numbers of teeth. Measurement for odd numbers above 171 and not in Table 2 may be determined as shown by the following example: Assume that number of teeth = 335 and pressure angle = 20 degrees; then, for 1 diametral pitch, figure at left of decimal point = given No. of teeth + 2 = 335 + 2 = 337. Figure at right of decimal point lies between decimal values given in table for 301 and 401 teeth. Thus, 335 − 301 = 34 (change to 0.34); 0.4565 − 0.4538 = 0.0027; hence, decimal required = 0.4538 + (0.34 × 0.0027) = 0.4547. Total dimension = 337.4547.

Table 1. Checking External Spur Gear Sizes by Measurement Over Wires

EVEN NUMBERS OF TEETH

Dimensions in table are for 1 diametral pitch and Van Keuren standard wire sizes. For any other diametral pitch, divide dimension in table by given pitch.

$$\text{Wire or pin diameter} = \frac{1.728}{\text{Diametral Pitch}}$$

No. of Teeth	Pressure Angle				
	14½°	17½°	20°	25°	30°
6	8.2846	8.2927	8.3032	8.3340	8.3759
8	10.3160	10.3196	10.3271	10.3533	10.3919
10	12.3399	12.3396	12.3445	12.3667	12.4028
12	14.3590	14.3552	14.3578	14.3768	14.4108
14	16.3746	16.3677	16.3683	16.3846	16.4169
16	18.3877	18.3780	18.3768	18.3908	18.4217
18	20.3989	20.3866	20.3840	20.3959	20.4256
20	22.4087	22.3940	22.3900	22.4002	22.4288
22	24.4172	24.4004	24.3952	24.4038	24.4315
24	26.4247	26.4060	26.3997	26.4069	26.4339
26	28.4314	28.4110	28.4036	28.4096	28.4358
28	30.4374	30.4154	30.4071	30.4120	30.4376
30	32.4429	32.4193	32.4102	32.4141	32.4391
32	34.4478	34.4228	34.4130	34.4159	34.4405
34	36.4523	36.4260	36.4155	36.4176	36.4417
36	38.4565	38.4290	38.4178	38.4191	38.4428
38	40.4603	40.4317	40.4198	40.4205	40.4438
40	42.4638	42.4341	42.4217	42.4217	42.4447
42	44.4671	44.4364	44.4234	44.4228	44.4455
44	46.4701	46.4385	46.4250	46.4239	46.4463
46	48.4729	48.4404	48.4265	48.4248	48.4470
48	50.4756	50.4422	50.4279	50.4257	50.4476
50	52.4781	52.4439	52.4292	52.4265	52.4482
52	54.4804	54.4454	54.4304	54.4273	54.4487
54	56.4826	56.4469	56.4315	56.4280	56.4492
56	58.4847	58.4483	58.4325	58.4287	58.4497
58	60.4866	60.4496	60.4335	60.4293	60.4501
60	62.4884	62.4509	62.4344	62.4299	62.4506
62	64.4902	64.4520	64.4352	64.4304	64.4510
64	66.4918	66.4531	66.4361	66.4309	66.4513
66	68.4933	68.4542	68.4369	68.4314	68.4517
68	70.4948	70.4552	70.4376	70.4319	70.4520
70	72.4963	72.4561	72.4383	72.4323	72.4523
72	74.4977	74.4570	74.4390	74.4327	74.4526
74	76.4990	76.4578	76.4396	76.4331	76.4529
76	78.5002	78.4586	78.4402	78.4335	78.4532
78	80.5014	80.4594	80.4408	80.4339	80.4534
80	82.5026	82.4601	82.4413	82.4342	82.4536
82	84.5037	84.4608	84.4418	84.4345	84.4538
84	86.5047	86.4615	86.4423	86.4348	86.4540
86	88.5057	88.4621	88.4428	88.4351	88.4542
88	90.5067	90.4627	90.4433	90.4354	90.4544
90	92.5076	92.4633	92.4437	92.4357	92.4546
92	94.5085	94.4639	94.4441	94.4359	94.4548

Table 1. *(Continued)* **Checking External Spur Gear Sizes by Measurement Over**

EVEN NUMBERS OF TEETH

Dimensions in table are for 1 diametral pitch and Van Keuren standard wire sizes. For any other diametral pitch, divide dimension in table by given pitch.

$$\text{Wire or pin diameter} = \frac{1.728}{\text{Diametral Pitch}}$$

No. of Teeth	Pressure Angle				
	14½°	17½°	20°	25°	30°
94	96.5094	96.4644	96.4445	96.4362	96.4550
96	98.5102	98.4649	98.4449	98.4364	98.4552
98	100.5110	100.4655	100.4453	100.4367	100.4554
100	102.5118	102.4660	102.4456	102.4369	102.4555
102	104.5125	104.4665	104.4460	104.4370	104.4557
104	106.5132	106.4669	106.4463	106.4372	106.4558
106	108.5139	108.4673	108.4466	108.4374	108.4560
108	110.5146	110.4678	110.4469	110.4376	110.4561
110	112.5152	112.4682	112.4472	112.4378	112.4562
112	114.5159	114.4686	114.4475	114.4380	114.4563
114	116.5165	116.4690	116.4478	116.4382	116.4564
116	118.5171	118.4693	118.4481	118.4384	118.4565
118	120.5177	120.4697	120.4484	120.4385	120.4566
120	122.5182	122.4701	122.4486	122.4387	122.4567
122	124.5188	124.4704	124.4489	124.4388	124.4568
124	126.5193	126.4708	126.4491	126.4390	126.4569
126	128.5198	128.4711	128.4493	128.4391	128.4570
128	130.5203	130.4714	130.4496	130.4393	130.4571
130	132.5208	132.4717	132.4498	132.4394	132.4572
132	134.5213	134.4720	134.4500	134.4395	134.4573
134	136.5217	136.4723	136.4502	136.4397	136.4574
136	138.5221	138.4725	138.4504	138.4398	138.4575
138	140.5226	140.4728	140.4506	140.4399	140.4576
140	142.5230	142.4730	142.4508	142.4400	142.4577
142	144.5234	144.4733	144.4510	144.4401	144.4578
144	146.5238	146.4736	146.4512	146.4402	146.4578
146	148.5242	148.4738	148.4513	148.4403	148.4579
148	150.5246	150.4740	150.4515	150.4404	150.4580
150	152.5250	152.4742	152.4516	152.4405	152.4580
152	154.5254	154.4745	154.4518	154.4406	154.4581
154	156.5257	156.4747	156.4520	156.4407	156.4581
156	158.5261	158.4749	158.4521	158.4408	158.4582
158	160.5264	160.4751	160.4523	160.4409	160.4582
160	162.5267	162.4753	162.4524	162.4410	162.4583
162	164.5270	164.4755	164.4526	164.4411	164.4584
164	166.5273	166.4757	166.4527	166.4411	166.4584
166	168.5276	168.4759	168.4528	168.4412	168.4585
168	170.5279	170.4760	170.4529	170.4413	170.4585
170	172.5282	172.4761	172.4531	172.4414	172.4586
180	182.5297	182.4771	182.4537	182.4418	182.4589
190	192.5310	192.4780	192.4542	192.4421	192.4591
200	202.5321	202.4786	202.4548	202.4424	202.4593
300	302.5395	302.4831	302.4579	302.4443	302.4606
400	402.5434	402.4854	402.4596	402.4453	402.4613
500	502.5458	502.4868	502.4606	502.4458	502.4619

Table 2. Checking External Spur Gear Sizes by Measurement Over Wires

ODD NUMBERS OF TEETH

Dimensions in table are for 1 diametral pitch and Van Keuren standard wire sizes. For any other diametral pitch, divide dimension in table by given pitch.

$$\text{Wire or pin diameter} = \frac{1.728}{\text{Diametral Pitch}}$$

No. of Teeth	Pressure Angle				
	14½°	17½°	20°	25°	30°
7	9.1116	9.1172	9.1260	9.1536	9.1928
9	11.1829	11.1844	11.1905	11.2142	11.2509
11	13.2317	13.2296	13.2332	13.2536	13.2882
13	15.2677	15.2617	15.2639	15.2814	15.3142
15	17.2957	17.2873	17.2871	17.3021	17.3329
17	19.3182	19.3072	19.3053	19.3181	19.3482
19	21.3368	21.3233	21.3200	21.3310	21.3600
21	23.3524	23.3368	23.3321	23.3415	23.3696
23	25.3658	25.3481	25.3423	25.3502	25.3775
25	27.3774	27.3579	27.3511	27.3576	27.3842
27	29.3876	29.3664	29.3586	29.3640	29.3899
29	31.3966	31.3738	31.3652	31.3695	31.3948
31	33.4047	33.3804	33.3710	33.3743	33.3991
33	35.4119	35.3863	35.3761	35.3786	35.4029
35	37.4185	37.3916	37.3807	37.3824	37.4063
37	39.4245	39.3964	39.3849	39.3858	39.4094
39	41.4299	41.4007	41.3886	41.3889	41.4120
41	43.4348	43.4047	43.3920	43.3917	43.4145
43	45.4394	45.4083	45.3951	45.3942	45.4168
45	47.4437	47.4116	47.3980	47.3965	47.4188
47	49.4477	49.4147	49.4007	49.3986	49.4206
49	51.4514	51.4175	51.4031	51.4006	51.4223
51	53.4547	53.4202	53.4053	53.4024	53.4239
53	55.4579	55.4227	55.4074	55.4041	55.4254
55	57.4609	57.4249	57.4093	57.4056	57.4267
57	59.4637	59.4271	59.4111	59.4071	59.4280
59	61.4664	61.4291	61.4128	61.4084	61.4292
61	63.4689	63.4310	63.4144	63.4097	63.4303
63	65.4712	65.4328	65.4159	65.4109	65.4313
65	67.4734	67.4344	67.4173	67.4120	67.4323
67	69.4755	69.4360	69.4186	69.4130	69.4332
69	71.4775	71.4375	71.4198	71.4140	71.4341
71	73.4795	73.4389	73.4210	73.4150	73.4349
73	75.4813	75.4403	75.4221	75.4159	75.4357
75	77.4830	77.4416	77.4232	77.4167	77.4364
77	79.4847	79.4428	79.4242	79.4175	79.4371
79	81.4863	81.4440	81.4252	81.4183	81.4378
81	83.4877	83.4451	83.4262	83.4190	83.4384
83	85.4892	85.4462	85.4271	85.4196	85.4390
85	87.4906	87.4472	87.4279	87.4203	87.4395
87	89.4919	89.4481	89.4287	89.4209	89.4400
89	91.4932	91.4490	91.4295	91.4215	91.4405
91	93.4944	93.4499	93.4303	93.4221	93.4410
93	95.4956	95.4508	95.4310	95.4227	95.4415
95	97.4967	97.4516	97.4317	97.4232	97.4420

Table 2. *(Continued)* **Checking External Spur Gear Sizes by Measurement Over**

ODD NUMBERS OF TEETH

Dimensions in table are for 1 diametral pitch and Van Keuren standard wire sizes. For any other diametral pitch, divide dimension in table by given pitch.

$$\text{Wire or pin diameter} = \frac{1.728}{\text{Diametral Pitch}}$$

No. of Teeth	Pressure Angle				
	14½°	17½°	20°	25°	30°
97	99.4978	99.4524	99.4323	99.4237	99.4424
99	101.4988	101.4532	101.4329	101.4242	101.4428
101	103.4998	103.4540	103.4335	103.4247	103.4432
103	105.5008	105.4546	105.4341	105.4252	105.4436
105	107.5017	107.4553	107.4346	107.4256	107.4440
107	109.5026	109.4559	109.4352	109.4260	109.4443
109	111.5035	111.4566	111.4357	111.4264	111.4447
111	113.5044	113.4572	113.4362	113.4268	113.4450
113	115.5052	115.4578	115.4367	115.4272	115.4453
115	117.5060	117.4584	117.4372	117.4275	117.4456
117	119.5068	119.4589	119.4376	119.4279	119.4459
119	121.5075	121.4594	121.4380	121.4282	121.4462
121	123.5082	123.4599	123.4384	123.4285	123.4465
123	125.5089	125.4604	125.4388	125.4288	125.4468
125	127.5096	127.4609	127.4392	127.4291	127.4471
127	129.5103	129.4614	129.4396	129.4294	129.4473
129	131.5109	131.4619	131.4400	131.4297	131.4476
131	133.5115	133.4623	133.4404	133.4300	133.4478
133	135.5121	135.4628	135.4408	135.4302	135.4480
135	137.5127	137.4632	137.4411	137.4305	137.4483
137	139.5133	139.4636	139.4414	139.4307	139.4485
139	141.5139	141.4640	141.4418	141.4310	141.4487
141	143.5144	143.4644	143.4421	143.4312	143.4489
143	145.5149	145.4648	145.4424	145.4315	145.4491
145	147.5154	147.4651	147.4427	147.4317	147.4493
147	149.5159	149.4655	149.4430	149.4319	149.4495
149	151.5164	151.4658	151.4433	151.4321	151.4497
151	153.5169	153.4661	153.4435	153.4323	153.4498
153	155.5174	155.4665	155.4438	155.4325	155.4500
155	157.5179	157.4668	157.4440	157.4327	157.4502
157	159.5183	159.4671	159.4443	159.4329	159.4504
159	161.5188	161.4674	161.4445	161.4331	161.4505
161	163.5192	163.4677	163.4448	163.4333	163.4507
163	165.5196	165.4680	165.4450	165.4335	165.4508
165	167.5200	167.4683	167.4453	167.4337	167.4510
167	169.5204	169.4686	169.4455	169.4338	169.4511
169	171.5208	171.4688	171.4457	171.4340	171.4513
171	173.5212	173.4691	173.4459	173.4342	173.4514
181	183.5230	183.4704	183.4469	183.4350	183.4520
191	193.5246	193.4715	193.4478	193.4357	193.4526
201	203.5260	203.4725	203.4487	203.4363	203.4532
301	303.5355	303.4790	303.4538	303.4402	303.4565
401	403.5404	403.4823	403.4565	403.4422	403.4582
501	503.5433	503.4843	503.4581	503.4434	503.4592

Table 3. Checking Internal Spur Gear Sizes by Measurement Between Wires

EVEN NUMBERS OF TEETH

Dimensions in table are for 1 diametral pitch and Van Keuren standard wire sizes. For any other diametral pitch, divide dimension in table by given pitch.

$$\text{Wire or pin diameter} = \frac{1.44}{\text{Diametral Pitch}}$$

No. of Teeth	Pressure Angle				
	14½°	17½°	20°	25°	30°
10	8.8337	8.7383	8.6617	8.5209	8.3966
12	10.8394	10.7404	10.6623	10.5210	10.3973
14	12.8438	12.7419	12.6627	12.5210	12.3978
16	14.8474	14.7431	14.6630	14.5210	14.3982
18	16.8504	16.7441	16.6633	16.5210	16.3985
20	18.8529	18.7449	18.6635	18.5211	18.3987
22	20.8550	20.7456	20.6636	20.5211	20.3989
24	22.8569	22.7462	22.6638	22.5211	22.3991
26	24.8585	24.7467	24.6639	24.5211	24.3992
28	26.8599	26.7471	26.6640	26.5211	26.3993
30	28.8612	28.7475	28.6641	28.5211	28.3994
32	30.8623	30.7478	30.6642	30.5211	30.3995
34	32.8633	32.7481	32.6642	32.5211	32.3995
36	34.8642	34.7483	34.6643	34.5212	34.3996
38	36.8650	36.7486	36.6642	36.5212	36.3996
40	38.8658	38.7488	38.6644	38.5212	38.3997
42	40.8665	40.7490	40.6644	40.5212	40.3997
44	42.8672	42.7492	42.6645	42.5212	42.3998
46	44.8678	44.7493	44.6645	44.5212	44.3998
48	46.8683	46.7495	46.6646	46.5212	46.3999
50	48.8688	48.7496	48.6646	48.5212	48.3999
52	50.8692	50.7497	50.6646	50.5212	50.3999
54	52.8697	52.7499	52.6647	52.5212	52.4000
56	54.8701	54.7500	54.6647	54.5212	54.4000
58	56.8705	56.7501	56.6648	56.5212	56.4001
60	58.8709	58.7502	58.6648	58.5212	58.4001
62	60.8712	60.7503	60.6648	60.5212	60.4001
64	62.8715	62.7504	62.6648	62.5212	62.4001
66	64.8718	64.7505	64.6649	64.5212	64.4001
68	66.8721	66.7505	66.6649	66.5212	66.4001
70	68.8724	68.7506	68.6649	68.5212	68.4001
72	70.8727	70.7507	70.6649	70.5212	70.4002
74	72.8729	72.7507	72.6649	72.5212	72.4002
76	74.8731	74.7508	74.6649	74.5212	74.4002
78	76.8734	76.7509	76.6649	76.5212	76.4002
80	78.8736	78.7509	78.6649	78.5212	78.4002
82	80.8738	80.7510	80.6649	80.5212	80.4002
84	82.8740	82.7510	82.6649	82.5212	82.4002
86	84.8742	84.7511	84.6650	84.5212	84.4002
88	86.8743	86.7511	86.6650	86.5212	86.4003
90	88.8745	88.7512	88.6650	88.5212	88.4003
92	90.8747	90.7512	90.6650	90.5212	90.4003
94	92.8749	92.7513	92.6650	92.5212	92.4003
96	94.8750	94.7513	94.6650	94.5212	94.4003

Table 3. ***(Continued)*** **Checking Internal Spur Gear Sizes by Measurement Between**

EVEN NUMBERS OF TEETH

Dimensions in table are for 1 diametral pitch and Van Keuren standard wire sizes. For any other diametral pitch, divide dimension in table by given pitch.

$$\text{Wire or pin diameter} = \frac{1.44}{\text{Diametral Pitch}}$$

No. of Teeth	Pressure Angle				
	14½°	17½°	20°	25°	30°
98	96.8752	96.7513	96.6650	96.5212	96.4003
100	98.8753	98.7514	98.6650	98.5212	98.4003
102	100.8754	100.7514	100.6650	100.5212	100.4003
104	102.8756	102.7514	102.6650	102.5212	102.4003
106	104.8757	104.7515	104.6650	104.5212	104.4003
108	106.8758	106.7515	106.6650	106.5212	106.4003
110	108.8759	108.7515	108.6651	108.5212	108.4004
112	110.8760	110.7516	110.6651	110.5212	110.4004
114	112.8761	112.7516	112.6651	112.5212	112.4004
116	114.8762	114.7516	114.6651	114.5212	114.4004
118	116.8763	116.7516	116.6651	116.5212	116.4004
120	118.8764	118.7517	118.6651	118.5212	118.4004
122	120.8765	120.7517	120.6651	120.5212	120.4004
124	122.8766	122.7517	122.6651	122.5212	122.4004
126	124.8767	124.7517	124.6651	124.5212	124.4004
128	126.8768	126.7518	126.6651	126.5212	126.4004
130	128.8769	128.7518	128.6652	128.5212	128.4004
132	130.8769	130.7518	130.6652	130.5212	130.4004
134	132.8770	132.7518	132.6652	132.5212	132.4004
136	134.8771	134.7519	134.6652	134.5212	134.4004
138	136.8772	136.7519	136.6652	136.5212	136.4004
140	138.8773	138.7519	138.6652	138.5212	138.4004
142	140.8773	140.7519	140.6652	140.5212	140.4004
144	142.8774	142.7519	142.6652	142.5212	142.4004
146	144.8774	144.7520	144.6652	144.5212	144.4004
148	146.8775	146.7520	146.6652	146.5212	146.4004
150	148.8775	148.7520	148.6652	148.5212	148.4005
152	150.8776	150.7520	150.6652	150.5212	150.4005
154	152.8776	152.7520	152.6652	152.5212	152.4005
156	154.8777	154.7520	154.6652	154.5212	154.4005
158	156.8778	156.7520	156.6652	156.5212	156.4005
160	158.8778	158.7520	158.6652	158.5212	158.4005
162	160.8779	160.7520	160.6652	160.5212	160.4005
164	162.8779	162.7521	162.6652	162.5212	162.4005
166	164.8780	164.7521	164.6652	164.5212	164.4005
168	166.8780	166.7521	166.6652	166.5212	166.4005
170	168.8781	168.7521	168.6652	168.5212	168.4005
180	178.8783	178.7522	178.6652	178.5212	178.4005
190	188.8785	188.7522	188.6652	188.5212	188.4005
200	198.8788	198.7523	198.6652	198.5212	198.4005
300	298.8795	298.7525	298.6654	298.5212	298.4005
400	398.8803	398.7527	398.6654	398.5212	398.4006
500	498.8810	498.7528	498.6654	498.5212	498.4006

Table 4. Checking Internal Spur Gear Sizes by Measurement Between Wires

ODD NUMBERS OF TEETH

Dimensions in table are for 1 diametral pitch and Van Keuren standard wire sizes. For any other diametral pitch, divide dimensions in table by given pitch.

$$\text{Wire or pin diameter} = \frac{1.44}{\text{Diametral Pitch}}$$

No. of Teeth	Pressure Angle				
	14½°	17½°	20°	25°	30°
7	5.6393	5.5537	5.4823	5.3462	5.2232
9	7.6894	7.5976	7.5230	7.3847	7.2618
11	9.7219	9.6256	9.5490	9.4094	9.2867
13	11.7449	11.6451	11.5669	11.4265	11.3040
15	13.7620	13.6594	13.5801	13.4391	13.3167
17	15.7752	15.6703	15.5902	15.4487	15.3265
19	17.7858	17.6790	17.5981	17.4563	17.3343
21	19.7945	19.6860	19.6045	19.4625	19.3405
23	21.8017	21.6918	21.6099	21.4676	21.3457
25	23.8078	23.6967	23.6143	23.4719	23.3501
27	25.8130	25.7009	25.6181	25.4755	25.3538
29	27.8176	27.7045	27.6214	27.4787	27.3571
31	29.8216	29.7076	29.6242	29.4814	29.3599
33	31.8251	31.7104	31.6267	31.4838	31.3623
35	33.8282	33.7128	33.6289	33.4860	33.3645
37	35.8311	35.7150	35.6310	35.4879	35.3665
39	37.8336	37.7169	37.6327	37.4896	37.3682
41	39.8359	39.7187	39.6343	39.4911	39.3698
43	41.8380	41.7203	41.6357	41.4925	41.3712
45	43.8399	43.7217	43.6371	43.4938	43.3725
47	45.8416	45.7231	45.6383	45.4950	45.3737
49	47.8432	47.7243	47.6394	47.4960	47.3748
51	49.8447	49.7254	49.6404	49.4970	49.3758
53	51.8461	51.7265	51.6414	51.4979	51.3768
55	53.8474	53.7274	53.6422	53.4988	53.3776
57	55.8486	55.7283	55.6431	55.4996	55.3784
59	57.8497	57.7292	57.6438	57.5003	57.3792
61	59.8508	59.7300	59.6445	59.5010	59.3799
63	61.8517	61.7307	61.6452	61.5016	61.3806
65	63.8526	63.7314	63.6458	63.5022	63.3812
67	65.8535	65.7320	65.6464	65.5028	65.3818
69	67.8543	67.7327	67.6469	67.5033	67.3823
71	69.8551	69.7332	69.6475	69.5038	69.3828
73	71.8558	71.7338	71.6480	71.5043	71.3833
75	73.8565	73.7343	73.6484	73.5048	73.3838
77	75.8572	75.7348	75.6489	75.5052	75.3842
79	77.8573	77.7352	77.6493	77.5056	77.3846
81	79.8584	79.7357	79.6497	79.5060	79.3850
83	81.8590	81.7361	81.6501	81.5064	81.3854
85	83.8595	83.7365	83.6505	83.5067	83.3858
87	85.8600	85.7369	85.6508	85.5071	85.3861
89	87.8605	87.7373	87.6511	87.5074	87.3864
91	89.8610	89.7376	89.6514	89.5077	89.3867
93	91.8614	91.7379	91.6517	91.5080	91.3870
95	93.8619	93.7383	93.6520	93.5082	93.3873
97	95.8623	95.7386	95.6523	95.5085	95.3876
99	97.8627	97.7389	97.6526	97.5088	97.3879
101	99.8631	99.7391	99.6528	99.5090	99.3881
103	101.8635	101.7394	101.6531	101.5093	101.3883
105	103.8638	103.7397	103.6533	103.5095	103.3886
107	105.8642	105.7399	105.6535	105.5097	105.3888
109	107.8645	107.7402	107.6537	107.5099	107.3890
111	109.8648	109.7404	109.6539	109.5101	109.3893
113	111.8651	111.7406	111.6541	111.5103	111.3895
115	113.8654	113.7409	113.6543	113.5105	113.3897
117	115.8657	115.7411	115.6545	115.5107	115.3899

Table 4. ***(Continued)*** **Checking Internal Spur Gear Sizes by Measurement Between**

ODD NUMBERS OF TEETH

Dimensions in table are for 1 diametral pitch and Van Keuren standard wire sizes. For any other diametral pitch, divide dimensions in table by given pitch.

$$\text{Wire or pin diameter} = \frac{1.44}{\text{Diametral Pitch}}$$

No. of Teeth	Pressure Angle				
	14½°	17½°	20°	25°	30°
119	117.8660	117.7413	117.6547	117.5109	117.3900
121	119.8662	119.7415	119.6548	119.5110	119.3902
123	121.8663	121.7417	121.6550	121.5112	121.3904
125	123.8668	123.7418	123.6552	123.5114	123.3905
127	125.8670	125.7420	125.6554	125.5115	125.3907
129	127.8672	127.7422	127.6556	127.5117	127.3908
131	129.8675	129.7424	129.6557	129.5118	129.3910
133	131.8677	131.7425	131.6559	131.5120	131.3911
135	133.8679	133.7427	133.6560	133.5121	133.3913
137	135.8681	135.7428	135.6561	135.5123	135.3914
139	137.8683	137.7430	137.6563	137.5124	137.3916
141	139.8685	139.7431	139.6564	139.5125	139.3917
143	141.8687	141.7433	141.6565	141.5126	141.3918
145	143.8689	143.7434	143.6566	143.5127	143.3919
147	145.8691	145.7436	145.6568	145.5128	145.3920
149	147.8693	147.7437	147.6569	147.5130	147.3922
151	149.8694	149.7438	149.6570	149.5131	149.3923
153	151.8696	151.7439	151.6571	151.5132	151.3924
155	153.8698	153.7441	153.6572	153.5133	153.3925
157	155.8699	155.7442	155.6573	155.5134	155.3926
159	157.8701	157.7443	157.6574	157.5135	157.3927
161	159.8702	159.7444	159.6575	159.5136	159.3928
163	161.8704	161.7445	161.6576	161.5137	161.3929
165	163.8705	163.7446	163.6577	163.5138	163.3930
167	165.8707	165.7447	165.6578	165.5139	165.3931
169	167.8708	167.7448	167.6579	167.5139	167.3932
171	169.8710	169.7449	169.6580	169.5140	169.3933
181	179.8717	179.7453	179.6584	179.5144	179.3937
191	189.8721	189.7458	189.6588	189.5148	189.3940
201	199.8727	199.7461	199.6591	199.5151	199.3944
301	299.8759	299.7485	299.6612	299.5171	299.3965
401	399.8776	399.7496	399.6623	399.5182	399.3975
501	499.8786	499.7504	499.6629	499.5188	499.3981

Table 5. Van Keuren Wire Diameters for Gears

External Gears Wire Dia. = 1.728 ÷ D.P.				Internal Gears Wire Dia. = 1.44 ÷ D.P.			
D.P.	Dia.	D.P.	Dia.	D.P.	Dia.	D.P.	Dia.
2	0.86400	16	0.10800	2	0.72000	16	0.09000
2½	0.69120	18	0.09600	2½	0.57600	18	0.08000
3	0.57600	20	0.08640	3	0.48000	20	0.07200
4	0.43200	22	0.07855	4	0.36000	22	0.06545
5	0.34560	24	0.07200	5	0.28800	24	0.06000
6	0.28800	28	0.06171	6	0.24000	28	0.05143
7	0.24686	32	0.05400	7	0.20571	32	0.04500
8	0.21600	36	0.04800	8	0.18000	36	0.04000
9	0.19200	40	0.04320	9	0.16000	40	0.03600
10	0.17280	48	0.03600	10	0.14400	48	0.03000
11	0.15709	64	0.02700	11	0.13091	64	0.02250
12	0.14400	72	0.02400	12	0.12000	72	0.02000
14	0.12343	80	0.02160	14	0.10286	80	0.01800

Measurements for Checking Internal Gears when Wire Diameter Equals 1.44 Divided by Diametral Pitch.—Tables 3 and 4 give measurements between wires for checking internal gears of 1 diametral pitch. For any other diametral pitch, divide the measurement given in the table by the diametral pitch required. These measurements are based upon the Van Keuren standard wire size, which, for internal spur gears, equals 1.44 divided by the diametral pitch (see Table 5).

Even Number of Teeth: For an even number of teeth above 170 and not in Table 3, proceed as shown by the following example: Assume that the number of teeth = 380 and pressure angle is 14½ degrees; then, for 1 diametral pitch, figure at left of decimal point = given number of teeth – 2 = 380 – 2 = 378. Figure at right of decimal point lies between decimal values given in table for 300 and 400 teeth and is obtained by interpolation. Thus, 380 – 300 = 80 (change to 0.80); 0.8803 – 0.8795 = 0.0008; hence, decimal required = 0.8795 + (0.80 × 0.0008) 0.88014. Total dimension = 378.88014.

Odd Number of Teeth: Table 4 is for internal gears having odd numbers of teeth. For tooth numbers above 171 and not in the table, proceed as shown by the following example: Assume that number of teeth = 337 and pressure angle is 14½ degrees; then, for 1 diametral pitch, figure at left of decimal point = given No. of teeth – 2 = 337 – 2 = 335. Figure at right of decimal point lies between decimal values given in table for 301 and 401 teeth and is obtained by interpolation. Thus, 337 – 301 = 36 (change to 0.36); 0.8776 – 0.8759 = 0.0017; hence, decimal required = 0.8759 + (0.36 × 0.0017) = 0.8765. Total dimension = 335.8765.

Measurements for Checking External Spur Gears when Wire Diameter Equals 1.68 Divided by Diametral Pitch.—Tables 7 and 8 give measurements M, in inches, for checking the pitch diameters of external spur gears of 1 diametral pitch. For any other diametral pitch, divide the measurement given in the table by whatever diametral pitch is required. The result shows what measurement M should be when the pitch diameter is correct and there is no allowance for backlash. The procedure for checking for a given amount of backlash when the diameter of the measuring wires equals 1.68 divided by the diametral pitch is explained under a subsequent heading. Tables 7 and 8 are based upon wire sizes equal to 1.68 divided by the diametral pitch. The corresponding wire diameters for various diametral pitches are given in Table 6.

Table 6. Wire Diameters for Spur and Helical Gears Based upon 1.68 Constant

Diametral or Normal Diametral Pitch	Wire Diameter	Diametral or Normal Diametral Pitch	Wire Diameter	Diametral or Normal Diametral Pitch	Wire Diameter	Diametral or Normal Diametral Pitch	Wire Diameter
2	0.840	8	0.210	18	0.09333	40	0.042
2½	0.672	9	0.18666	20	0.084	48	0.035
3	0.560	10	0.168	22	0.07636	64	0.02625
4	0.420	11	0.15273	24	0.070	72	0.02333
5	0.336	12	0.140	28	0.060	80	0.021
6	0.280	14	0.120	32	0.0525	…	…
7	0.240	16	0.105	36	0.04667	…	…

Pin diameter = 1.68 ÷ diametral pitch for spur gears and 1.68 ÷ normal diametral pitch for helical gears.

To find measurement M of an external spur gear using wire sizes equal to 1.68 inches divided by the diametral pitch, the same method is followed in using Tables 7 and 8 as that outlined for Tables 1 and 2.

Table 7. Checking External Spur Gear Sizes by Measurement Over Wires

EVEN NUMBERS OF TEETH

Dimensions in table are for 1 diametral pitch and 1.68-inch series wire sizes (a Van Keuren standard). For any other diametral pitch, divide dimension in table by given pitch.

$$\text{Wire or pin diameter} = \frac{1.68}{\text{Diametral Pitch}}$$

No. of Teeth	Pressure Angle				
	14½°	17½°	20°	25°	30°
6	8.1298	8.1442	8.1600	8.2003	8.2504
8	10.1535	10.1647	10.1783	10.2155	10.2633
10	12.1712	12.1796	12.1914	12.2260	12.2722
12	14.1851	14.1910	14.2013	14.2338	14.2785
14	16.1964	16.2001	16.2091	16.2397	16.2833
16	18.2058	18.2076	18.2154	18.2445	18.2871
18	20.2137	20.2138	20.2205	20.2483	20.2902
20	22.2205	22.2190	22.2249	22.2515	22.2927
22	24.2265	24.2235	24.2286	24.2542	24.2949
24	26.2317	26.2275	26.2318	26.2566	26.2967
26	28.2363	28.2309	28.2346	28.2586	28.2982
28	30.2404	30.2339	30.2371	30.2603	30.2996
30	32.2441	32.2367	32.2392	32.2619	32.3008
32	34.2475	34.2391	34.2412	34.2632	34.3017
34	36.2505	36.2413	36.2430	36.2644	36.3026
36	38.2533	38.2433	38.2445	38.2655	38.3035
38	40.2558	40.2451	40.2460	40.2666	40.3044
40	42.2582	42.2468	42.2473	42.2675	42.3051
42	44.2604	44.2483	44.2485	44.2683	44.3057
44	46.2624	46.2497	46.2496	46.2690	46.3063
46	48.2642	48.2510	48.2506	48.2697	48.3068
48	50.2660	50.2522	50.2516	50.2704	50.3073
50	52.2676	52.2534	52.2525	52.2710	52.3078
52	54.2691	54.2545	54.2533	54.2716	54.3082
54	56.2705	56.2555	56.2541	56.2721	56.3086
56	58.2719	58.2564	58.2548	58.2726	58.3089
58	60.2731	60.2572	60.2555	60.2730	60.3093
60	62.2743	62.2580	62.2561	62.2735	62.3096
62	64.2755	64.2587	64.2567	64.2739	64.3099
64	66.2765	66.2594	66.2572	66.2742	66.3102
66	68.2775	68.2601	68.2577	68.2746	68.3104
68	70.2785	70.2608	70.2582	70.2749	70.3107
70	72.2794	72.2615	72.2587	72.2752	72.3109
72	74.2803	74.2620	74.2591	74.2755	74.3111
74	76.2811	76.2625	76.2596	76.2758	76.3113
76	78.2819	78.2631	78.2600	78.2761	78.3115
78	80.2827	80.2636	80.2604	80.2763	80.3117
80	82.2834	82.2641	82.2607	82.2766	82.3119
82	84.2841	84.2646	84.2611	84.2768	84.3121
84	86.2847	86.2650	86.2614	86.2771	86.3123
86	88.2854	88.2655	88.2617	88.2773	88.3124
88	90.2860	90.2659	90.2620	90.2775	90.3126
90	92.2866	92.2662	92.2624	92.2777	92.3127
92	94.2872	94.2666	94.2626	94.2779	94.3129
94	96.2877	96.2670	96.2629	96.2780	96.3130
96	98.2882	98.2673	98.2632	98.2782	98.3131
98	100.2887	100.2677	100.2635	100.2784	100.3132
100	102.2892	102.2680	102.2638	102.2785	102.3134
102	104.2897	104.2683	104.2640	104.2787	104.3135
104	106.2901	106.2685	106.2642	106.2788	106.3136
106	108.2905	108.2688	108.2644	108.2789	108.3137

Table 7. *(Continued)* **Checking External Spur Gear Sizes by Measurement Over**

EVEN NUMBERS OF TEETH

Dimensions in table are for 1 diametral pitch and 1.68-inch series wire sizes (a Van Keuren standard). For any other diametral pitch, divide dimension in table by given pitch.

$$\text{Wire or pin diameter} = \frac{1.68}{\text{Diametral Pitch}}$$

No. of Teeth	Pressure Angle				
	14½°	17½°	20°	25°	30°
108	110.2910	110.2691	110.2645	110.2791	110.3138
110	112.2914	112.2694	112.2647	112.2792	112.3139
112	114.2918	114.2696	114.2649	114.2793	114.3140
114	116.2921	116.2699	116.2651	116.2794	116.3141
116	118.2925	118.2701	118.2653	118.2795	118.3142
118	120.2929	120.2703	120.2655	120.2797	120.3142
120	122.2932	122.2706	122.2656	122.2798	122.3143
122	124.2936	124.2708	124.2658	124.2799	124.3144
124	126.2939	126.2710	126.2660	126.2800	126.3145
126	128.2941	128.2712	128.2661	128.2801	128.3146
128	130.2945	130.2714	130.2663	130.2802	130.3146
130	132.2948	132.2716	132.2664	132.2803	132.3147
132	134.2951	134.2718	134.2666	134.2804	134.3147
134	136.2954	136.2720	136.2667	136.2805	136.3148
136	138.2957	138.2722	138.2669	138.2806	138.3149
138	140.2960	140.2724	140.2670	140.2807	140.3149
140	142.2962	142.2725	142.2671	142.2808	142.3150
142	144.2965	144.2727	144.2672	144.2808	144.3151
144	146.2967	146.2729	146.2674	146.2809	146.3151
146	148.2970	148.2730	148.2675	148.2810	148.3152
148	150.2972	150.2732	150.2676	150.2811	150.3152
150	152.2974	152.2733	152.2677	152.2812	152.3153
152	154.2977	154.2735	154.2678	154.2812	154.3153
154	156.2979	156.2736	156.2679	156.2813	156.3154
156	158.2981	158.2737	158.2680	158.2813	158.3155
158	160.2983	160.2739	160.2681	160.2814	160.3155
160	162.2985	162.2740	162.2682	162.2815	162.3155
162	164.2987	164.2741	164.2683	164.2815	164.3156
164	166.2989	166.2742	166.2684	166.2816	166.3156
166	168.2990	168.2744	168.2685	168.2816	168.3157
168	170.2992	170.2745	170.2686	170.2817	170.3157
170	172.2994	172.2746	172.2687	172.2818	172.3158
180	182.3003	182.2752	182.2691	182.2820	182.3160
190	192.3011	192.2757	192.2694	192.2823	192.3161
200	202.3018	202.2761	202.2698	202.2825	202.3163
300	302.3063	302.2790	302.2719	302.2839	302.3173
400	402.3087	402.2804	402.2730	402.2845	402.3178
500	502.3101	502.2813	502.2736	502.2850	502.3181

Allowance for Backlash: Tables 1, 2, 7, and 8 give measurements over wires when the pitch diameters are correct and there is no allowance for backlash or play between meshing teeth. Backlash is obtained by cutting the teeth somewhat deeper than standard, thus reducing the thickness. Usually, the teeth of both mating gears are reduced in thickness an amount equal to one-half of the total backlash desired. However, if the pinion is small, it is common practice to reduce the gear teeth the full amount of backlash and the pinion is made to standard size. The changes in measurements M over wires, for obtaining backlash in external spur gears, are listed in Table 9.

Table 8. Checking External Spur Gear Sizes by Measurement Over Wires

ODD NUMBERS OF TEETH

Dimensions in table are for 1 diametral pitch and 1.68-inch series wire sizes (a Van Keuren standard). For any other diametral pitch, divide dimension in table by given pitch.

$$\text{Wire or pin diameter} = \frac{1.68}{\text{Diametral Pitch}}$$

No. of Teeth	Pressure Angle				
	14½°	17 ½°	20°	25°	30°
5	6.8485	6.8639	6.8800	6.9202	6.9691
7	8.9555	8.9679	8.9822	9.0199	9.0675
9	11.0189	11.0285	11.0410	11.0762	11.1224
11	13.0615	13.0686	13.0795	13.1126	13.1575
13	15.0925	15.0973	15.1068	15.1381	15.1819
15	17.1163	17.1190	17.1273	17.1570	17.1998
17	19.1351	19.1360	19.1432	19.1716	19.2136
19	21.1505	21.1498	21.1561	21.1832	21.2245
21	23.1634	23.1611	23.1665	23.1926	23.2334
23	25.1743	25.1707	25.1754	25.2005	25.2408
25	27.1836	27.1788	27.1828	27.2071	27.2469
27	29.1918	29.1859	29.1892	29.2128	29.2522
29	31.1990	31.1920	31.1948	31.2177	31.2568
31	33.2053	33.1974	33.1997	33.2220	33.2607
33	35.2110	35.2021	35.2041	35.2258	35.2642
35	37.2161	37.2065	37.2079	37.2292	37.2674
37	39.2208	39.2104	39.2115	39.2323	39.2702
39	41.2249	41.2138	41.2147	41.2349	41.2726
41	43.2287	43.2170	43.2174	43.2374	43.2749
43	45.2323	45.2199	45.2200	45.2396	45.2769
45	47.2355	47.2226	47.2224	47.2417	47.2788
47	49.2385	49.2251	49.2246	49.2435	49.2805
49	51.2413	51.2273	51.2266	51.2452	51.2820
51	53.2439	53.2294	53.2284	53.2468	53.2835
53	55.2463	55.2313	55.2302	55.2483	55.2848
55	57.2485	57.2331	57.2318	57.2497	57.2861
57	59.2506	59.2348	59.2333	59.2509	59.2872
59	61.2526	61.2363	61.2347	61.2521	61.2883
61	63.2545	63.2378	63.2360	63.2532	63.2893
63	65.2562	65.2392	65.2372	65.2543	65.2902
65	67.2579	67.2406	67.2383	67.2553	67.2911
67	69.2594	69.2419	69.2394	69.2562	69.2920
69	71.2609	71.2431	71.2405	71.2571	71.2928
71	73.2623	73.2442	73.2414	73.2579	73.2935
73	75.2636	75.2452	75.2423	75.2586	75.2942
75	77.2649	77.2462	77.2432	77.2594	77.2949
77	79.2661	79.2472	79.2440	79.2601	79.2955
79	81.2673	81.2481	81.2448	81.2607	81.2961
81	83.2684	83.2490	83.2456	83.2614	83.2967
83	85.2694	85.2498	85.2463	85.2620	85.2972
85	87.2704	87.2506	87.2470	87.2625	87.2977
87	89.2714	89.2514	89.2476	89.2631	89.2982
89	91.2723	91.2521	91.2482	91.2636	91.2987
91	93.2732	93.2528	93.2489	93.2641	93.2991
93	95.2741	95.2534	95.2494	95.2646	95.2996
95	97.2749	97.2541	97.2500	97.2650	97.3000
97	99.2757	99.2547	99.2506	99.2655	99.3004

Table 8. ***(Continued)*** **Checking External Spur Gear Sizes by Measurement Over**

ODD NUMBERS OF TEETH

Dimensions in table are for 1 diametral pitch and 1.68-inch series wire sizes (a Van Keuren standard). For any other diametral pitch, divide dimension in table by given pitch.

$$\text{Wire or pin diameter} = \frac{1.68}{\text{Diametral Pitch}}$$

No. of Teeth	Pressure Angle				
	14½°	17 ½°	20°	25°	30°
99	101.2764	101.2553	101.2511	101.2659	101.3008
101	103.2771	103.2558	103.2516	103.2663	103.3011
103	105.2778	105.2563	105.2520	105.2667	105.3015
105	107.2785	107.2568	107.2525	107.2671	107.3018
107	109.2791	109.2573	109.2529	109.2674	109.3021
109	111.2798	111.2578	111.2533	111.2678	111.3024
111	113.2804	113.2583	113.2537	113.2681	113.3027
113	115.2809	115.2588	115.2541	115.2684	115.3030
115	117.2815	117.2592	117.2544	117.2687	117.3033
117	119.2821	119.2596	119.2548	119.2690	119.3036
119	121.2826	121.2601	121.2552	121.2693	121.3038
121	123.2831	123.2605	123.2555	123.2696	123.3041
123	125.2836	125.2608	125.2558	125.2699	125.3043
125	127.2841	127.2612	127.2562	127.2702	127.3046
127	129.2846	129.2615	129.2565	129.2704	129.3048
129	131.2851	131.2619	131.2568	131.2707	131.3050
131	133.2855	133.2622	133.2571	133.2709	133.3053
133	135.2859	135.2626	135.2574	135.2712	135.3055
135	137.2863	137.2629	137.2577	137.2714	137.3057
137	139.2867	139.3632	139.2579	139.2716	139.3059
139	141.2871	141.2635	141.2582	141.2718	141.3060
141	143.2875	143.2638	143.2584	143.2720	143.3062
143	145.2879	145.2641	145.2587	145.2722	145.3064
145	147.2883	147.2644	147.2589	147.2724	147.3066
147	149.2887	149.2647	149.2591	149.2726	149.3068
149	151.2890	151.2649	151.2594	151.2728	151.3069
151	153.2893	153.2652	153.2596	153.2730	153.3071
153	155.2897	155.2654	155.2598	155.2732	155.3073
155	157.2900	157.2657	157.2600	157.2733	157.3074
157	159.2903	159.2659	159.2602	159.2735	159.3076
159	161.2906	161.2661	161.2604	161.2736	161.3077
161	163.2909	163.2663	163.2606	163.2738	163.3078
163	165.2912	165.2665	165.2608	165.2740	165.3080
165	167.2915	167.2668	167.2610	167.2741	167.3081
167	169.2917	169.2670	169.2611	169.2743	169.3083
169	171.2920	171.2672	171.2613	171.2744	171.3084
171	173.2922	173.2674	173.2615	173.2746	173.3085
181	183.2936	183.2684	183.2623	183.2752	183.3091
191	193.2947	193.2692	193.2630	193.2758	193.3097
201	203.2957	203.2700	203.2636	203.2764	203.3101
301	303.3022	303.2749	303.2678	303.2798	303.3132
401	403.3056	403.2774	403.2699	403.2815	403.3147
501	503.3076	503.2789	503.2711	503.2825	503.3156

Table 9. Backlash Allowances for External and Internal Spur Gears

External Gears: For each 0.001 inch reduction in pitch-line tooth thickness, *reduce* measurement over wires obtained from Tables 1, 2, 7, or 8 by the amount shown below.

Internal Gears: For each 0.001 inch reduction in pitch-line tooth thickness, *increase* measurement between wires obtained from Tables 3 or 4 by the amounts shown below.

Backlash on pitch line equals double tooth thickness reduction when teeth of *both* mating gears are reduced. If teeth of *one* gear only are reduced, backlash on pitch line equals amount of reduction.

Example: For a 30-tooth, 10-diametral pitch, 20-degree pressure angle, external gear the measurement over wires from Table 1 is 32.4102 ÷ 10. For a backlash of 0.002 this measurement must be reduced by 2 × 0.0024 to 3.2362 or (3.2410 – 0.0048).

No. of Teeth	14½°		17½°		20°		25°		30°	
	Ext.	Int.	Ext.	Int.	Ext.	Int.	Ext.	Int.	Ext.	Int.
5	.0019	.0024	.0018	.0024	.0017	.0023	.0015	.0021	.0013	.0019
10	.0024	.0029	.0022	.0027	.0020	.0026	.0017	.0022	.0015	.0018
20	.0028	.0032	.0025	.0029	.0023	.0027	.0019	.0022	.0016	.0018
30	.0030	.0034	.0026	.0030	.0024	.0027	.0020	.0022	.0016	.0018
40	.0031	.0035	.0027	.0030	.0025	.0027	.0020	.0022	.0017	.0018
50	.0032	.0036	.0028	.0031	.0025	.0027	.0020	.0022	.0017	.0018
100	.0035	.0037	.0030	.0031	.0026	.0027	.0021	.0022	.0017	.0017
200	.0036	.0038	.0031	.0031	.0027	.0027	.0021	.0022	.0017	.0017

Measurements for Checking Helical Gears using Wires or Balls.—Helical gears may be checked for size by using one wire, or ball; two wires, or balls; and three wires, depending on the case at hand. Three wires may be used for measurement of either even or odd tooth numbers provided that the face width and helix angle of the gear permit the arrangement of two wires in adjacent tooth spaces on one side of the gear and a third wire on the opposite side. The wires should be held between flat, parallel plates. The measurement between these plates, and perpendicular to the gear axis, will be the same for both even and odd numbers of teeth because the axial displacement of the wires with the odd numbers of teeth does not affect the perpendicular measurement between the plates. The calculation of measurements over three wires is the same as described for measurements over two wires for even numbers of teeth.

Measurements over One Wire or One Ball for Even or Odd Numbers of Teeth: This measurement is calculated by the method for measurement over two wires for even numbers of teeth and the result divided by two to obtain the measurement from over the wire or ball to the center of the gear mounted on an arbor.

Measurement over Two Wires or Two Balls for Even Numbers of Teeth: The measurement over two wires (or two balls kept in the same plane by holding them against a surface parallel to the face of the gear) is calculated as follows: First, calculate the pitch diameter of the helical gear from the formula D = Number of teeth divided by the product of the normal diametral pitch and the cosine of the helix angle, $D = N \div (P_n \times \cos\psi)$. Next, calculate the number of teeth, N_e, there would be in a spur gear for it to have the same tooth curvature as the helical gear has in the normal plane: $N_e = N/\cos^3\psi$. Next, refer to Table 7 for spur gears with even tooth numbers and find, by interpolation, the *decimal* value of the constant for this number of teeth under the given *normal* pressure angle. Finally, add 2 to this decimal value and divide the sum by the normal diametral pitch P_n. The result of this calculation, added to the pitch diameter D, is the measurement over two wires or balls.

Example: A helical gear has 32 teeth of 6 normal diametral pitch, 20 degree pressure angle, and 23 degree helix angle. Determine the measurement over two wires, M, without allowance for backlash.

$D = 32 \div 6 \times \cos 23° = 5.7939$; $N_e = 32 \div \cos^3 23° = 41.027$; and in Table 7, fourth column, the decimal part of the measurement for 40 teeth is .2473 and that for 42 teeth is .2485. The decimal part for 41.027 teeth is, by interpolation, $\frac{(41.027-40)}{(42-40)} \times (.2485-.2473)+.2473$ $=0.2479$; $(0.2479+2) \div 6 = 0.3747$; and $M = 0.3747 + 5.7939 = 6.1686$.

This measurement over wires or balls is based upon the use of $1.68/P_n$ wires or balls. If measurements over $1.728/P_n$ diameter wires or balls are preferred, use Table 1 to find the decimal part described above instead of Table 7.

Measurement over Two Wires or Two Balls for Odd Numbers of Teeth: The procedure is similar to that for two wire or two ball measurement for even tooth numbers except that a correction is made in the final M value to account for the wires or balls not being diametrically opposite by one-half tooth interval. In addition, care must be taken to ensure that the balls or wires are kept in a plane of the gear's rotation as described previously.

Example: A helical gear has 13 teeth of 8 normal diametral pitch, 14½ degree pressure angle, and 45 degree helix angle. Determine measurement M without allowance for backlash based upon the use of $1.728/P_n$ balls or wires.

As before, $D = 13/8 \times \cos 45° = 2.2981$; $N_e = 13/\cos^3 45° = 36.770$; and in the second column of Table 1 the *decimal* part of the measurement for 36 teeth is .4565 and that for 38 teeth is .4603. The decimal part for 36.770 teeth is, by interpolation, $\frac{(36.770-36)}{(38-36)} \times$ $(.4603 - .4565) + .4565 = 0.4580$; $(0.4580 + 2)/8 = 0.3073$; and $M = 0.3073 + 2.2981 =$ 2.6054. This measurement is correct for three-wire measurements but, for two balls or wires held in the plane of rotation of the gear, M must be corrected as follows:

$$M \text{ corrected} = (M - \text{Ball Diam.}) \times \cos(90°/N) + \text{Ball Diam.}$$
$$= (2.6054 - 1.728/8) \times \cos(90°/13) + 1.728/8 = 2.5880$$

Checking Spur Gear Size by Chordal Measurement Over Two or More Teeth.— Another method of checking gear sizes, that is generally available, is illustrated by the diagram accompanying Table 10. A vernier caliper is used to measure the distance M over two or more teeth. The diagram illustrates the measurement over two teeth (or with one intervening tooth space), but three or more teeth might be included, depending upon the pitch. The jaws of the caliper are merely held in contact with the sides or profiles of the teeth and perpendicular to the axis of the gear. Measurement M for involute teeth of the correct size is determined as follows

General Formula for Checking External and Internal Spur Gears by Measurement Over Wires: The following formulas may be used for pressure angles or wire sizes not covered by the tables. In these formulas, M = measurement *over* wires for external gears or measurement *between* wires for internal gears; D = pitch diameter; T = arc tooth thickness on pitch circle; W = wire diameter; N = number of gear teeth; A = pressure angle of gear; a = angle, the cosine of which is required in Formulas (2) and (3).

First determine the involute function of angle a (inv a); then the corresponding angle a is found by referring to the tables of involute functions beginning on page 98,

$$\text{inv}\, a = \text{inv}\, A \pm \frac{T}{D} \pm \frac{W}{D\cos A} \mp \frac{\pi}{N} \qquad (1)$$

$$\text{For even numbers of teeth, } M = \frac{D\cos A}{\cos a} \pm W \qquad (2)$$

$$\text{For odd numbers of teeth, } M = \left(\frac{D\cos A}{\cos a}\right)\left(\cos\frac{90°}{N}\right) \pm W \qquad (3)$$

Note: In Formulas (1), (2), and (3), use the upper sign for *external* and the lower sign for *internal* gears wherever a ± or ∓ appears in the formulas.

Table 10. Chordal Measurements over Spur Gear Teeth of 1 Diametral Pitch

Find value of M under pressure angle and opposite number of teeth; divide M by diametral pitch of gear to be measured and then subtract one-half total backlash to obtain a measurement M equivalent to given pitch and backlash. The number of teeth to gage or measure over is shown by Table 11.

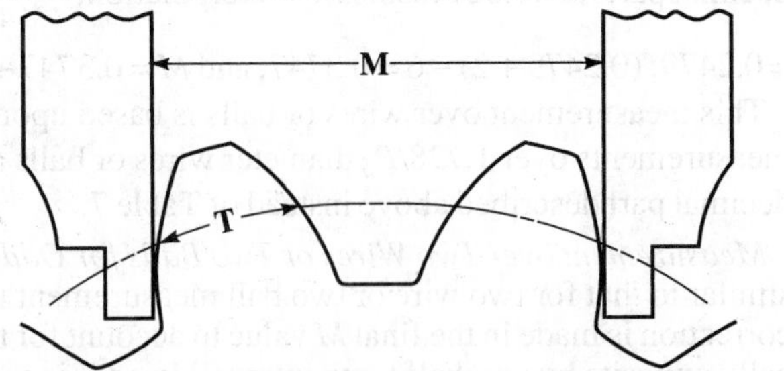

Number of Gear Teeth	M in Inches for 1 D.P.	Number of Gear Teeth	M in Inches for 1 D.P.	Number of Gear Teeth	M in Inches for 1 D.P.	Number of Gear Teeth	M in Inches for 1 D.P.
Pressure Angle, 14½ Degrees							
12	4.6267	37	7.8024	62	14.0197	87	20.2370
13	4.6321	38	10.8493	63	17.0666	88	23.2838
14	4.6374	39	10.8547	64	17.0720	89	23.2892
15	4.6428	40	10.8601	65	17.0773	90	23.2946
16	4.6482	41	10.8654	66	17.0827	91	23.2999
17	4.6536	42	10.8708	67	17.0881	92	23.3053
18	4.6589	43	10.8762	68	17.0934	93	23.3107
19	7.7058	44	10.8815	69	17.0988	94	23.3160
20	7.7112	45	10.8869	70	17.1042	95	23.3214
21	7.7166	46	10.8923	71	17.1095	96	23.3268
22	7.7219	47	10.8976	72	17.1149	97	23.3322
23	7.7273	48	10.9030	73	17.1203	98	23.3375
24	7.7326	49	10.9084	74	17.1256	99	23.3429
25	7.7380	50	10.9137	75	17.1310	100	23.3483
26	7.7434	51	13.9606	76	20.1779	101	26.3952
27	7.7488	52	13.9660	77	20.1833	102	26.4005
28	7.7541	53	13.9714	78	20.1886	103	26.4059
29	7.7595	54	13.9767	79	20.1940	104	26.4113
30	7.7649	55	13.9821	80	20.1994	105	26.4166
31	7.7702	56	13.9875	81	20.2047	106	26.4220
32	7.7756	57	13.9929	82	20.2101	107	26.4274
33	7.7810	58	13.9982	83	20.2155	108	26.4327
34	7.7683	59	14.0036	84	20.2208	109	26.4381
35	7.7917	60	14.0090	85	20.2262	110	26.4435
36	7.7971	61	14.0143	86	20.2316	…	…
Pressure Angle, 20 Degrees							
12	4.5963	30	10.7526	48	16.9090	66	23.0653
13	4.6103	31	10.7666	49	16.9230	67	23.0793
14	4.6243	32	10.7806	50	16.9370	68	23.0933
15	4.6383	33	10.7946	51	16.9510	69	23.1073
16	4.6523	34	10.8086	52	16.9650	70	23.1214
17	4.6663	35	10.8226	53	16.9790	71	23.1354
18	4.6803	36	10.8366	54	16.9930	72	23.1494
19	7.6464	37	13.8028	55	19.9591	73	26.1155
20	7.6604	38	13.8168	56	19.9731	74	26.1295
21	7.6744	39	13.8307	57	19.9872	75	26.1435
22	7.6884	40	13.8447	58	20.0012	76	26.1575
23	7.7024	41	13.8587	59	20.0152	77	26.1715
24	7.7165	42	13.8727	60	20.0292	78	26.1855
25	7.7305	43	13.8867	61	20.0432	79	26.1995
26	7.7445	44	13.9007	62	20.0572	80	26.2135
27	7.7585	45	13.9147	63	20.0712	81	26.2275
28	10.7246	46	16.8810	64	23.0373	…	…
29	10.7386	47	16.8950	65	23.0513	…	…

Table for Determining the Chordal Dimension: Table 10 gives the chordal dimensions for one diametral pitch when measuring over the number of teeth indicated in Table 11. To obtain any chordal dimension, it is simply necessary to divide chord M in the table (opposite the given number of teeth) by the diametral pitch of the gear to be measured and then subtract from the quotient one-half the total backlash between the mating pair of gears. In cases where a small pinion is used with a large gear and all of the backlash is to be obtained by reducing the gear teeth, the total amount of backlash is subtracted from the chordal dimension of the gear and nothing from the chordal dimension of the pinion. The application of the tables will be illustrated by an example.

Table 11. Number of Teeth Included in Chordal Measurement

Tooth Range for 14½° Pressure Angle	Tooth Range for 20° Pressure Angle	Number of Teeth to Gage Over	Tooth Range for 14½° Pressure Angle	Tooth Range for 20° Pressure Angle	Number of Teeth to Gage Over
12 to 18	12 to 18	2	63 to 75	46 to 54	6
19 to 37	19 to 27	3	76 to 87	55 to 63	7
38 to 50	28 to 36	4	88 to 100	64 to 72	8
51 to 62	37 to 45	5	101 to 110	73 to 81	9

This table shows the number of teeth to be included between the jaws of the vernier caliper in measuring dimension M as explained in connection with Table 10.

Example: Determine the chordal dimension for checking the size of a gear having 30 teeth of 5 diametral pitch and a pressure angle of 20 degrees. A total backlash of 0.008 inch is to be obtained by reducing equally the teeth of both mating gears.

Table 10 shows that the chordal distance for 30 teeth of one diametral pitch and a pressure angle of 20 degrees is 10.7526 inches; one-half of the backlash equals 0.004 inch; hence,

$$\text{Chordal dimension} = \frac{10.7526}{5} - 0.004 = 2.1465 \text{ inches}$$

Table 11 shows that this is the chordal dimension when the vernier caliper spans four teeth, this being the number of teeth to gage over whenever gears of 20-degree pressure angle have any number of teeth from 28 to 36, inclusive.

If it is considered necessary to leave enough stock on the gear teeth for a shaving or finishing cut, this allowance is simply added to the chordal dimension of the finished teeth to obtain the required measurement over the teeth for the roughing operation. It may be advisable to place this chordal dimension for rough machining on the detail drawing.

Formula for Chordal Dimension *M*.—The required measurement M over spur gear teeth may be obtained by the following formula in which R = pitch radius of gear, A = pressure angle, T = tooth thickness along pitch circle, N = number of gear teeth, S = number of tooth *spaces* between caliper jaws, F = a factor depending on the pressure angle = 0.01109 for 14½°; = 0.01973 for 17½°; = 0.0298 for 20°; = 0.04303 for 22½°; = 0.05995 for 25°. This factor F equals twice the involute function of the pressure angle.

$$M = R \times \cos A \times \left(\frac{T}{R} + \frac{6.2832 \times S}{N} + F\right)$$

Example: A spur gear has 30 teeth of 6 diametral pitch and a pressure angle of 14½ degrees. Determine measurement M over three teeth, there being two intervening tooth spaces.

The pitch radius = 2½ inches, the arc tooth thickness equivalent to 6 diametral pitch is 0.2618 inch (if no allowance is made for backlash) and factor F for 14½ degrees = 0.01109 inch.

$$M = 2.5 \times 0.96815 \times \left(\frac{0.2618}{2.5} + \frac{6.2832 \times 2}{30} + 0.01109 \right) = 1.2941 \text{ inches}$$

Checking Enlarged Pinions by Measuring Over Pins or Wires.—When the teeth of small spur gears or pinions would be undercut if generated by an unmodified straight-sided rack cutter or hob, it is common practice to make the outside diameter larger than standard. The amount of increase in outside diameter varies with the pressure angle and number of teeth, as shown by Table 1 on page 2019. The teeth are always cut to standard depth on a generating type of machine such as a gear hobber or gear shaper; and because the number of teeth and pitch are not changed, the pitch diameter also remains unchanged. The tooth thickness on the pitch circle, however, is increased and wire sizes suitable for standard gears are not large enough to extend above the tops of these enlarged gears or pinions; hence, the Van Keuren wire size recommended for these enlarged pinions equals 1.92 ÷ diametral pitch. Table 12 gives measurements over wires of this size, for checking full-depth involute gears of 1 diametral pitch. For any other pitch, merely divide the measurement given in the table by the diametral pitch. Table 12 applies to pinions that have been enlarged by the same amounts as given in tables 1 and 2, starting on page 2019. These enlarged pinions will mesh with standard gears; but if the standard center distance is to be maintained, reduce the gear diameter below the standard size by as much as the pinion diameter is increased.

Table 12. Checking Enlarged Spur Pinions by Measurement Over Wires

Measurements over wires are given in table for 1 diametral pitch. For any other diametral pitch, divide measurement in table by given pitch. Wire size equals 1.92 ÷ diametral pitch.

Number of Teeth	Outside or Major Diameter (Note 1)	Circular Tooth Thickness (Note 2)	Measurement Over Wires	Number of Teeth	Outside or Major Diameter (Note 1)	Circular Tooth Thickness (Note 2)	Measurement Over Wires
14½-degree full-depth involute teeth:				20-degree full-depth involute teeth:			
10	13.3731	1.9259	13.6186	10	12.936	1.912	13.5039
11	14.3104	1.9097	14.4966	11	13.818	1.868	14.3299
12	15.2477	1.8935	15.6290	12	14.702	1.826	15.4086
13	16.1850	1.8773	16.5211	13	15.584	1.783	16.2473
14	17.1223	1.8611	17.6244	14	16.468	1.741	17.2933
15	18.0597	1.8449	18.5260	15	17.350	1.698	18.1383
16	18.9970	1.8286	19.6075	16	18.234	1.656	19.1596
17	19.9343	1.8124	20.5156	17	19.116	1.613	20.0080
18	20.8716	1.7962	21.5806				
19	21.8089	1.7800	22.4934				
20	22.7462	1.7638	23.5451				
21	23.6835	1.7476	24.4611				
22	24.6208	1.7314	25.5018				
23	25.5581	1.7151	26.4201				
24	26.4954	1.6989	27.4515				
25	27.4328	1.6827	28.3718				
26	28.3701	1.6665	29.3952				
27	29.3074	1.6503	30.3168				
28	30.2447	1.6341	31.3333				
29	31.1820	1.6179	32.2558				
30	32.1193	1.6017	33.2661				
31	33.0566	1.5854	34.1889				

Note 1: These enlargements, which are to improve the tooth form and avoid undercut, conform to those given in tables 1 and 2, starting on page 2019 where data will be found on the minimum number of teeth in the mating gear.

Note 2: The circular or arc thickness is at the standard pitch diameter. The corresponding chordal thickness may be found as follows: Multiply arc thickness by 90 and then divide product by 3.1416 × pitch radius; find sine of angle thus obtained and multiply it by pitch diameter.

GEAR MATERIALS

Classification of Gear Steels.—Gear steels may be divided into two general classes — the plain carbon and the alloy steels. Alloy steels are used to some extent in the industrial field, but heat-treated plain carbon steels are far more common. The use of untreated alloy steels for gears is seldom, if ever, justified, and then, only when heat-treating facilities are lacking. The points to be considered in determining whether to use heat-treated plain carbon steels or heat-treated alloy steels are: Does the service condition or design require the superior characteristics of the alloy steels, or, if alloy steels are not required, will the advantages to be derived offset the additional cost? For most applications, plain carbon steels, heat-treated to obtain the best of their qualities for the service intended, are satisfactory and quite economical. The advantages obtained from using heat-treated alloy steels in place of heat-treated plain carbon steels are as follows:

1) Increased surface hardness and depth of hardness penetration for the same carbon content and quench.

2) Ability to obtain the same surface hardness with a less drastic quench and, in the case of some of the alloys, a lower quenching temperature, thus giving less distortion.

3) Increased toughness, as indicated by the higher values of yield point, elongation, and reduction of area.

4) Finer grain size, with the resulting higher impact toughness and increased wear resistance.

5) In the case of some of the alloys, better machining qualities or the possibility of machining at higher hardnesses.

Use of Casehardening Steels.—Each of the two general classes of gear steels may be further subdivided as follows: 1) Casehardening steels; 2) full-hardening steels; and 3) steels that are heat-treated and drawn to a hardness that will permit machining.

The first two — casehardening and full-hardening steels — are interchangeable for some kinds of service, and the choice is often a matter of personal opinion. Casehardening steels with their extremely hard, fine-grained (when properly treated) case and comparatively soft and ductile core are generally used when resistance to wear is desired. Casehardening alloy steels have a fairly tough core, but not as tough as that of the full-hardening steels. In order to realize the greatest benefits from the core properties, casehardened steels should be double-quenched. This is particularly true of the alloy steels, because the benefits derived from their use seldom justify the additional expense, unless the core is refined and toughened by a second quench. The penalty that must be paid for the additional refinement is increased distortion, which may be excessive if the shape or design does not lend itself to the casehardening process.

Use of "Thru-Hardening" Steels.—Thru-hardening steels are used when great strength, high endurance limit, toughness, and resistance to shock are required. These qualities are governed by the kind of steel and treatment used. Fairly high surface hardnesses are obtainable in this group, though not so high as those of the casehardening steels. For that reason, the resistance to wear is not so great as might be obtained, but when wear resistance combined with great strength and toughness is required, this type of steel is superior to the others. Thru-hardening steels become distorted to some extent when hardened, the amount depending upon the steel and quenching medium used. For that reason, thru-hardening steels are not suitable for high-speed gearing where noise is a factor, or for gearing where accuracy is of paramount importance, except, of course, in cases where grinding of the teeth is practicable. The medium and high-carbon percentages require an oil quench, but a water quench may be necessary for the lower carbon contents, in order to obtain the highest physical properties and hardness. The distortion, however, will be greater with the water quench.

Heat-Treatment that Permits Machining.—When the grinding of gear teeth is not practicable and a high degree of accuracy is required, hardened steels may be drawn or tempered to a hardness that will permit the cutting of the teeth. This treatment gives a highly refined structure, great toughness, and, in spite of the low hardness, excellent wearing qualities. The lower strength is somewhat compensated for by the elimination of the increment loads due to the impacts which are caused by inaccuracies. When steels that have a low degree of hardness penetration from surface to core are treated in this manner, the design cannot be based on the physical properties corresponding to the hardness at the surface. Since the physical properties are determined by the hardness, the drop in hardness from surface to core will give lower physical properties at the root of the tooth, where the stress is greatest. The quenching medium may be either oil, water, or brine, depending on the steel used and hardness penetration desired. The amount of distortion, of course, is immaterial, because the machining is done after heat-treating.

Making Pinion Harder than Gear to Equalize Wear.—Beneficial results from a wear standpoint are obtained by making the pinion harder than the gear. The pinion, having a lesser number of teeth than the gear, naturally does more work per tooth, and the differential in hardness between the pinion and the gear (the amount being dependent on the ratio) serves to equalize the rate of wear. The harder pinion teeth correct the errors in the gear teeth to some extent by the initial wear and then seem to burnish the teeth of the gear and increase its ability to withstand wear by the greater hardness due to the cold-working of the surface. In applications where the gear ratio is high and there are no severe shock loads, a casehardened pinion running with an oil-treated gear, treated to a Brinell hardness at which the teeth may be cut after treating, is an excellent combination. The pinion, being relatively small, is distorted but little, and distortion in the gear is circumvented by cutting the teeth after treatment.

Forged and Rolled Carbon Steels for Gears.—These compositions cover steel for gears in three groups, according to heat treatment, as follows: D) case-hardened gears;

E) unhardened gears, not heat treated after machining; and F) hardened and tempered gears.

Forged and rolled carbon gear steels are purchased on the basis of the requirements as to chemical composition specified in Table 1. Class N steel will normally be ordered in ten point carbon ranges within these limits. Requirements as to physical properties have been omitted, but when they are called for the requirements as to carbon shall be omitted. The steels may be made by either or both the open hearth and electric furnace processes.

Table 1. Compositions of Forged and Rolled Carbon Steels for Gears

Heat Treatment	Class	Carbon	Manganese	Phosphorus	Sulfur
Case-hardened	C	0.15–0.25	0.40–0.70	0.045 max	0.055 max
Untreated	N	0.25–0.50	0.50–0.80	0.045 max	0.055 max
Hardened (or untreated)	H	0.40–0.50	0.40–0.70	0.045 max	0.055 max

Forged and Rolled Alloy Steels for Gears.—These compositions cover alloy steel for gears, in two classes according to heat treatment, as follows: G) casehardened gears; and

H) hardened and tempered gears.

Forged and rolled alloy gear steels are purchased on the basis of the requirements as to chemical composition specified in Table 2. Requirements as to physical properties have been omitted. The steel shall be made by either or both the open hearth and electric furnace process.

Table 2. Compositions of Forged and Rolled Alloy Steels for Gears

Steel Specification	Chemical Composition[a]					
	C	Mn	Si	Ni	Cr	Mo
AISI 4130	0.28–0.30	0.40–0.60	0.20–0.35	…	0.80–1.1	0.15–0.25
AISI 4140	0.38–0.43	0.75–1.0	0.20–0.35	…	0.80–1.1	0.15–0.25
AISI 4340	0.38–0.43	0.60–0.80	0.20–0.35	1.65–2.0	0.70–90	0.20–0.30
AISI 4615	0.13–0.18	0.45–0.65	0.20–0.35	1.65–2.0	…	0.20–0.30
AISI 4620	0.17–0.22	0.45–0.65	0.20–0.35	1.65–2.0	…	0.20–0.30
AISI 8615	0.13–0.18	0.70–0.90	0.20–0.35	0.40–0.70	0.40–0.60	0.15–0.25
AISI 8620	0.18–0.23	0.70–0.90	0.20–0.35	0.40–0.70	0.40–0.60	0.15–0.25
AISI 9310	0.08–0.13	0.45–0.65	0.20–0.35	3.0–3.5	1.0–1.4	0.08–0.15
Nitralloy						
Type N[b]	0.20–0.27	0.40–0.70	0.20–0.40	3.2–3.8	1.0–1.3	0.20–0.30
135 Mod.[b]	0.38–0.45	0.40–0.70	0.20–0.40	…	1.4–1.8	0.30–0.45

[a] C = carbon; Mn = manganese; Si = silicon; Ni = nickel; Cr = chromium, and Mo = molybdenum.
[b] Both Nitralloy alloys contain aluminum 0.85–1.2%

Steel Castings for Gears.—It is recommended that steel castings for cut gears be purchased on the basis of chemical analysis and that only two types of analysis be used, one for case-hardened gears and the other for both untreated gears and those which are to be hardened and tempered. The steel is to be made by the open hearth, crucible, or electric furnace processes. The converter process is not recognized. Sufficient risers must be provided to secure soundness and freedom from undue segregation. Risers should not be broken off the unannealed castings by force. Where risers are cut off with a torch, the cut should be at least one-half inch above the surface of the castings, and the remaining metal removed by chipping, grinding, or other noninjurious method.

Steel for use in gears should conform to the requirements for chemical composition indicated in Table 3. All steel castings for gears must be thoroughly normalized or annealed, using such temperature and time as will entirely eliminate the characteristic structure of unannealed castings.

Table 3. Compositions of Cast Steels for Gears

Steel Specification	Chemical Composition[a]			
	C	Mn	Si	
SAE-0022	0.12–0.22	0.50–0.90	0.60 Max.	May be carburized
SAE-0050	0.40–0.50	0.50–0.90	0.80 Max.	Hardenable 210–250

[a] C = carbon; Mn = manganese; and Si = silicon.

Effect of Alloying Metals on Gear Steels.—The effect of the various alloying elements on steel are here summarized to assist in deciding on the particular kind of alloy steel to use for specific purposes. The characteristics outlined apply only to heat-treated steels. When the effect of the addition of an alloying element is stated, it is understood that reference is made to alloy steels of a given carbon content, compared with a plain carbon steel of the same carbon content.

Nickel: The addition of nickel tends to increase the hardness and strength, with but little sacrifice of ductility. The hardness penetration is somewhat greater than that of plain carbon steels. Use of nickel as an alloying element lowers the critical points and produces less distortion, due to the lower quenching temperature. The nickel steels of the case-hardening group carburize more slowly, but the grain growth is less.

Chromium: Chromium increases the hardness and strength over that obtained by the use of nickel, though the loss of ductility is greater. Chromium refines the grain and imparts a greater depth of hardness. Chromium steels have a high degree of wear resistance and are easily machined in spite of the fine grain.

Manganese: When present in sufficient amounts to warrant the use of the term alloy, the addition of manganese is very effective. It gives greater strength than nickel and a higher degree of toughness than chromium. Owing to its susceptibility to cold-working, it is likely to flow under severe unit pressures. Up to the present time, it has never been used to any great extent for heat-treated gears, but is now receiving an increasing amount of attention.

Vanadium: Vanadium has a similar effect to that of manganese—increasing the hardness, strength, and toughness. The loss of ductility is somewhat more than that due to manganese, but the hardness penetration is greater than for any of the other alloying elements. Owing to the extremely fine-grained structure, the impact strength is high; but vanadium tends to make machining difficult.

Molybdenum: Molybdenum has the property of increasing the strength without affecting the ductility. For the same hardness, steels containing molybdenum are more ductile than any other alloy steels, and having nearly the same strength, are tougher; in spite of the increased toughness, the presence of molybdenum does not make machining more difficult. In fact, such steels can be machined at a higher hardness than any of the other alloy steels. The impact strength is nearly as great as that of the vanadium steels.

Chrome-Nickel Steels: The combination of the two alloying elements chromium and nickel adds the beneficial qualities of both. The high degree of ductility present in nickel steels is complemented by the high strength, finer grain size, deep hardening, and wear-resistant properties imparted by the addition of chromium. The increased toughness makes these steels more difficult to machine than the plain carbon steels, and they are more difficult to heat treat. The distortion increases with the amount of chromium and nickel.

Chrome-Vanadium Steels: Chrome-vanadium steels have practically the same tensile properties as the chrome-nickel steels, but the hardening power, impact strength, and wear resistance are increased by the finer grain size. They are difficult to machine and become distorted more easily than the other alloy steels.

Chrome-Molybdenum Steels: This group has the same qualities as the straight molybdenum steels, but the hardening depth and wear resistance are increased by the addition of chromium. This steel is very easily heat treated and machined.

Nickel-Molybdenum Steels: Nickel-molybdenum steels have qualities similar to chrome-molybdenum steel. The toughness is said to be greater, but the steel is somewhat more difficult to machine.

Sintered Materials.—For high production of low and moderately loaded gears, significant production cost savings may be effected by the use of a sintered metal powder. With this material, the gear is formed in a die under high pressure and then sintered in a furnace. The primary cost saving comes from the great reduction in labor cost of machining the gear teeth and other gear blank surfaces. The volume of production must be high enough to amortize the cost of the die and the gear blank must be of such a configuration that it may be formed and readily ejected from the die.

Steels for Industrial Gearing

Material Specification	Hardness		Typical Heat Treatment, Characteristics, and Uses
	Case Rc	Core Bhn	
Case-Hardening Steels			
AISI 1020			Carburize, harden, temper at 350°F.
AISI 1116	55–60	160–230	For gears that must be wear-resistant. Normalizedmaterial is easily machined. Core is ductile but has little strength.
AISI 4130			Harden, temper at 900°F, Nitride.
AISI 4140	50–55	270–370	For parts requiring greater wear resistance than that of through-hardened steels but cannot tolerate the distortion of carburizing. Case is shallow, core is tough.
AISI 4615			Carburize, harden, temper at 350°F.
AISI 4620 AISI 8615	55–60	170–260	For gears requiring high fatigue resistance and strength. The 86xx series has better machinability.
AISI 8620	55–60	200–300	The 20 point steels are used for coarser teeth.
AISI 9310	58–63	250–350	Carburize, harden, temper at 300°F. Primarily for aerospace gears that are highly loaded and operate at high pitch line velocity and for other gears requiring high reliability under extreme operating conditions. This material is not used at high temperatures.
Nitralloy N and Type 135 Mod. (15-N)	90–94	300–370	Harden, temper at 1200°F, Nitride. For gears requiring high strength and wear resistance that cannot tolerate the distortion of thecarburizing process or that operate at high temperatures. Gear teeth are usually finished before nitriding. Care must be exercised in running nitrided gears together to avoid crazing of case-hardened surfaces.
Through-Hardening Steels			
AISI 1045 AISI 1140	24–40	…	Harden and temper to required hardness. Oil quench for lower hardness and water quench for higher hardness. For gears of medium and large size requiring moderate strength and wear resistance. Gears that must have consistent, solid sections to withstand quenching.
AISI 4140			Harden (oil quench), temper to required hardness.
AISI 4340	24–40	…	For gears requiring high strength and wear resistance, and high shock loading resistance. Use 41xx series for moderate sections and 43xx series for heavy sections. Gears must have consistent, solid sections to withstand quenching.

Bronze and Brass Gear Castings.—These specifications cover nonferrous metals for spur, bevel, and worm gears, bushings and flanges for composition gears. This material shall be purchased on the basis of chemical composition. The alloys may be made by any approved method.

Spur and Bevel Gears: For spur and bevel gears, hard cast bronze is recommended (ASTM B-10-18; SAE No. 62; and the well-known 88-10-2 mixture) with the following limits as to composition: Copper, 86 to 89; tin, 9 to 11; zinc, 1 to 3; lead (max), 0.20; iron (max), 0.06 per cent. Good castings made from this bronze should have the following minimum physical characteristics: Ultimate strength, 30,000 pounds per square inch; yield point, 15,000 pounds per square inch; elongation in 2 inches, 14 per cent.

Worm Gears: For bronze worm gears, two alternative analyses of phosphor bronze are recommended, SAE No. 65 and No. 63.

SAE No. 65 (called phosphor gear bronze) has the following composition: Copper, 88 to 90; tin, 10 to 12; phosphorus, 0.1 to 0.3; lead, zinc, and impurities (max) 0.5 per cent.

Good castings made of this alloy should have the following minimum physical characteristics: Ultimate strength, 35,000 pounds per square inch; yield point, 20,000 pounds per square inch; elongation in 2 inches, 10 per cent.

The composition of SAE No. 63 (called leaded gun metal) follows: copper, 86 to 89; tin, 9 to 11; lead, 1 to 2.5; phosphorus (max), 0.25; zinc and impurities (max), 0.50 per cent.

Good castings made of this alloy should have the following minimum physical characteristics: Ultimate strength, 30,000 pounds per square inch; yield point, 12,000 pounds per square inch; elongation in 2 inches, 10 per cent.

These alloys, especially No. 65, are adapted to chilling for hardness and refinement of grain. No. 65 is to be preferred for use with worms of great hardness and fine accuracy. No. 63 is to be preferred for use with unhardened worms.

Gear Bushings: For bronze bushings for gears, SAE No. 64 is recommended of the following analysis: copper, 78.5 to 81.5; tin, 9 to 11; lead, 9 to 11; phosphorus, 0.05 to 0.25; zinc (max), 0.75; other impurities (max), 0.25 per cent. Good castings of this alloy should have the following minimum physical characteristics: Ultimate strength, 25,000 pounds per square inch; yield point, 12,000 pounds per square inch; elongation in 2 inches, 8 per cent.

Flanges for Composition Pinions: For brass flanges for composition pinions ASTM B-30-32T, and SAE No. 40 are recommended. This is a good cast red brass of sufficient strength and hardness to take its share of load and wear when the design is such that the flanges mesh with the mating gear. The composition is as follows: copper, 83 to 86; tin, 4.5 to 5.5; lead, 4.5 to 5.5; zinc, 4.5 to 5.5; iron (max) 0.35; antimony (max), 0.25 per cent; aluminum, none. Good castings made from this alloy should have the following minimum physical characteristics: ultimate strength, 27,000 pounds per square inch; yield point, 12,000 pounds per square inch; elongation in 2 inches, 16 per cent.

Materials for Worm Gearing.—The Hamilton Gear & Machine Co. conducted an extensive series of tests on a variety of materials that might be used for worm gears, to ascertain which material is the most suitable. According to these tests chill-cast nickel-phosphor-bronze ranks first in resistance to wear and deformation. This bronze is composed of approximately 87.5 per cent copper, 11 per cent tin, 1.5 per cent nickel, with from 0.1 to 0.2 per cent phosphorus. The worms used in these tests were made from SAE-2315, 3½ per cent nickel steel, case-hardened, ground, and polished. The Shore scleroscope hardness of the worms was between 80 and 90. This nickel alloy steel was adopted after numerous tests of a variety of steels, because it provided the necessary strength, together with the degree of hardness required.

The material that showed up second best in these tests was a No. 65 SAE bronze. Navy bronze (88-10-2) containing 2 per cent zinc, with no phosphorus, and not chilled, performed satisfactorily at speeds of 600 revolutions per minute, but was not sufficiently strong at lower speeds. Red brass (85-5-5) proved slightly better at from 1500 to 1800 revolutions per minute, but would bend at lower speeds, before it would show actual wear.

Non-metallic Gearing.—Non-metallic or composition gearing is used primarily where quietness of operation at high speed is the first consideration. Non-metallic materials are also applied very generally to timing gears and numerous other classes of gearing. Rawhide was used originally for non-metallic gears, but other materials have been introduced that have important advantages. These later materials are sold by different firms under various trade names, such as Micarta, Textolite, Formica, Dilecto, Spauldite, Phenolite, Fibroc, Fabroil, Synthane, Celoron, etc. Most of these gear materials consist of layers of canvas or other material that is impregnated with plastics and forced together under hydraulic pressure, which, in conjunction with the application of heat, forms a dense rigid mass.

Although phenol resin gears in general are resilient, they are self-supporting and require no side plates or shrouds unless subjected to a heavy starting torque. The phenol resinoid element protects these gears from vermin and rodents.

The non-metallic gear materials referred to are generally assumed to have the power-transmitting capacity of cast iron. Although the tensile strength may be considerably less than that of cast iron, the resiliency of these materials enables them to withstand impact and abrasion to a degree that might result in excessive wear of cast-iron teeth. Thus, composition gearing of impregnated canvas has often proved to be more durable than cast iron.

Application of Non-metallic Gears.—The most effective field of use for these non-metallic materials is for high-speed duty. At low speeds, when the starting torque may be high, or when the load may fluctuate widely, or when high shock loads may be encountered, these non-metallic materials do not always prove satisfactory. In general, non-metallic materials should not be used for pitch-line velocities below 600 feet per minute.

Tooth Form: The best tooth form for non-metallic materials is the 20-degree stub-tooth system. When only a single pair of gears is involved and the center distance can be varied, the best results will be obtained by making the non-metallic driving pinion of all-addendum form, and the driven metal gear with standard tooth proportions. Such a drive will carry from 50 to 75 per cent greater loads than one of standard tooth proportions.

Material for Mating Gear: For durability under load, the use of hardened steel (over 400 Brinell) for the mating metal gear appears to give the best results. A good second choice for the material of the mating member is cast iron. The use of brass, bronze, or soft steel (under 400 Brinell) as a material for the mating member of phenolic laminated gears leads to excessive abrasive wear.

Power-Transmitting Capacity of Non-metallic Gears.—The characteristics of gears made of phenolic laminated materials are so different from those of metal gears that they should be considered in a class by themselves. Because of the low modulus of elasticity, most of the effects of small errors in tooth form and spacing are absorbed at the tooth surfaces by the elastic deformation, and have but little effect on the strength of the gears.

If S = safe working stress for a given velocity
S_s = allowable static stress
V = pitch-line velocity in feet per minute

then, according to the recommended practice of the American Gear Manufacturers' Association,

$$S = S_s \times \left(\frac{150}{200 + V} + 0.25\right)$$

The value of S_s for phenolic laminated materials is given as 6000 pounds per square inch. The accompanying table gives the safe working stresses S for different pitch-line velocities. When the value of S is known, the horsepower capacity is determined by substituting the value of S for S_s in the appropriate equations in the section on power-transmitting capacity of plastics gears starting on page 601.

Safe Working Stresses for Non-metallic Gears

Pitch-Line Velocity, Feet per Minute, *V*	Safe Working Stresses	Pitch-Line Velocity, Feet per Minute, *V*	Safe Working Stresses	Pitch-Line Velocity, Feet per Minute, *V*	Safe Working Stresses
600	2625	1800	1950	4000	1714
700	2500	2000	1909	4500	1691
800	2400	2200	1875	5000	1673
900	2318	2400	1846	5500	1653
1000	2250	2600	1821	6000	1645
1200	2143	2800	1800	6500	1634
1400	2063	3000	1781	7000	1622
1600	2000	3500	1743	7500	1617

The tensile strength of the phenolic laminated materials used for gears is slightly less than that of cast iron. These materials are far softer than any metal, and the modulus of elasticity is about one-thirtieth that of steel. In other words, if the tooth load on a steel gear that causes a deformation of 0.001 inch were applied to the tooth of a similar gear made of phenolic laminated material, the tooth of the non-metallic gear would be deformed about $\frac{1}{32}$ inch. Under these conditions, several things will happen. With all gears, regardless of the theoretical duration of contact, one tooth only will carry the load until the load is sufficient to deform the tooth the amount of the error that may be present. On metal gears, when the tooth has been deformed the amount of the error, the stresses set up in the materials may approach or exceed the elastic limit of the material. Hence, for standard tooth forms and those generated from standard basic racks, it is dangerous to calculate their strength as very much greater than that which can safely be carried on a single tooth. On gears made of phenolic laminated materials, on the other hand, the teeth will be deformed the amount of this normal error without setting up any appreciable stresses in the material, so that the load is actually supported by several teeth.

All materials have their own peculiar and distinct characteristics, so that under certain specific conditions, each material has a field of its own where it is superior to any other. Such fields may overlap to some extent, and only in such overlapping fields are different materials directly competitive. For example, steel is more or less ductile, has a high tensile strength, and a high modulus of elasticity. Cast iron, on the other hand, is not ductile, has a low tensile strength, but a high compressive strength, and a low modulus of elasticity. Hence, when stiffness and high tensile strength are essential, steel is far superior to cast iron. On the other hand, when these two characteristics are unimportant, but high compressive strength and a moderate amount of elasticity are essential, cast iron is superior to steel.

Preferred Pitch for Non-metallic Gears.—The pitch of the gear or pinion should bear a reasonable relation either to the horsepower or speed or to the applied torque, as shown by the accompanying table. The upper half of this table is based upon horsepower transmitted at a given pitch-line velocity. The lower half gives the torque in pounds-feet or the torque at a 1-foot radius. This torque *T* for any given horsepower and speed can be obtained from the following formula:

$$T = \frac{5252 \times \text{hp}}{\text{rpm}}$$

Bore Sizes for Non-metallic Gears.—For plain phenolic laminated pinions, that is, pinions without metal end plates, a drive fit of 0.001 inch per inch of shaft diameter should be used. For shafts above 2.5 inches in diameter, the fit should be constant at 0.0025 to 0.003 inch. When metal reinforcing end plates are used, the drive fit should conform to the same standards as used for metal.

Preferred Pitches for Non-metallic Gears

Diametral Pitch for Given Horsepower and Pitch Line Velocities			
Horsepower Transmitted	Pitch Line Velocity up to 1000 Feet per Minute	Pitch Line Velocity from 1000 to 2000 Feet per Minute	Pitch Line Velocity over 2000 Feet per Minute
¼–1	8–10	10–12	12–16
1–2	7–8	8–10	10–12
2–3	6–7	7–8	8–10
3–7½	5–6	6–7	7–8
7½–10	4–5	5–6	6–7
10–15	3–4	4–5	5–6
15–25	2½–3	3–4	4–5
25–60	2–2½	2½–3	3–4
60–100	1¾–2	2–2½	2½–3
100–150	1½–1¾	1¾–2	2–2½

Torque in Pounds-feet for Given Diametral Pitch					
Diametral Pitch	Torque in Pounds-feet		Diametral Pitch	Torque in Pounds-feet	
	Minimum	Maximum		Minimum	Maximum
16	1	2	4	50	100
12	2	4	3	100	200
10	4	8	2½	200	450
8	8	15	2	450	900
6	15	30	1½	900	1800
5	30	50	1	1800	3500

These preferred pitches are applicable both to rawhide and the phenolic laminated types of materials.

The root diameter of a pinion of phenolic laminated type should be such that the minimum distance from the edge of the keyway to the root diameter will be at least equal to the depth of tooth.

Keyway Stresses for Non-metallic Gears.—The keyway stress should not exceed 3000 pounds per square inch on a plain phenolic laminated gear or pinion. The keyway stress is calculated by the formula

$$S = \frac{33{,}000 \times \text{hp}}{V \times A}$$

where S = unit stress in pounds per square inch

hp = horsepower transmitted

V = peripheral speed of shaft in feet per minute; and

A = square inch area of keyway in pinion (length × height)

If the keyway stress formula is expressed in terms of shaft radius r and revolutions per minute, it will read

$$S = \frac{63{,}000 \times \text{hp}}{\text{rpm} \times r \times A}$$

When the design is such that the keyway stresses exceed 3000 pounds, metal reinforcing end plates may be used. Such end plates should not extend beyond the root diameter of the teeth. The distance from the outer edge of the retaining bolt to the root diameter of the teeth shall not be less than a full tooth depth. The use of drive keys should be avoided, but if required, metal end plates should be used on the pinion to take the wedging action of the key.

For phenolic laminated pinions, the face of the mating gear should be the same or slightly greater than the pinion face.

Invention of Gear Teeth.—The invention of gear teeth represents a gradual evolution from gearing of primitive form. The earliest evidence we have of an investigation of the problem of *uniform motion* from toothed gearing and the successful solution of that problem dates from the time of Olaf Roemer, the celebrated Danish astronomer, who, in the year 1674, proposed the epicycloidal form to obtain uniform motion. Evidently Robert Willis, professor at the University of Cambridge, was the first to make a practical application of the epicycloidal curve so as to provide for an interchangeable series of gears. Willis gives credit to Camus for conceiving the idea of interchangeable gears, but claims for himself its first application. The involute tooth was suggested as a theory by early scientists and mathematicians, but it remained for Willis to present it in a practical form. Perhaps the earliest conception of the application of this form of teeth to gears was by Philippe de Lahire, a Frenchman, who considered it, in theory, equally suitable with the epicycloidal for tooth outlines. This was about 1695 and not long after Roemer had first demonstrated the epicycloidal form. The applicability of the involute had been further elucidated by Leonard Euler, a Swiss mathematician, born at Basel, 1707, who is credited by Willis with being the first to suggest it. Willis devised the Willis odontograph for laying out involute teeth.

A pressure angle of 14½ degrees was selected for three different reasons. First, because the sine of 14½ degrees is nearly ¼, making it convenient in calculation; second, because this angle coincided closely with the pressure angle resulting from the usual construction of epicycloidal gear teeth; third, because the angle of the straight-sided involute rack is the same as the 29-degree worm thread.

Calculating Replacement-Gear Dimensions from Simple Measurements.—The following tables provide formulas with which to calculate the dimensions needed to produce replacement spur, bevel, and helical gears when only the number of teeth, the outside diameter, and the tooth depth of the gear to be replaced are known.

For helical gears, exact helix angles can be obtained by the following procedure.

1) Using a common protractor, measure the approximate helix angle *A* at the approximate pitch line.

2) Place sample or its mating gear on the arbor of a gear hobbing machine.

3) Calculate the index and lead gears differentially for the angle obtained by the measurements, and set up the machine as though a gear is to be cut.

4) Attach a dial indicator on an adjustable arm to the vertical swivel head, with the indicator plunger in a plane perpendicular to the gear axis and in contact with the tooth face. Contact may be anywhere between the top and the root of the tooth.

5) With the power shut off, engage the starting lever and traverse the indicator plunger axially by means of the handwheel.

6) If angle *A* is correct, the indicator plunger will not move as it traverses the face width of the gear. If it does move from 0, note the amount. Divide the amount of movement by the width of the gear to obtain the tangent of the angle by which to correct angle *A*, plus or minus, depending on the direction of indicator movement.

Spur Gears

Tooth Form and Pressure Angle	Diametral Pitch P	Pitch Diameter D	Circular Pitch P_c	Outside Diameter O	Addendum J
American Standard 14½-and 20-degree full depth	$\frac{N+2}{O}$	$\frac{N}{P}$	$\frac{3.1416}{P}$	$\frac{N+2}{P}$	$\frac{1}{P}$
American Standard 20-degree stub	$\frac{N+1.6}{O}$	$\frac{N}{P}$	$\frac{3.1416}{P}$	$\frac{N+1.6}{P}$	$\frac{0.8}{P}$
Fellows 20-degree stub	Note	$\frac{N}{P_N}$	$\frac{3.1416}{P_N}$	$\frac{N}{P_N}+\frac{2}{P_D}$	$\frac{1}{P_D}$

Tooth Form and Pressure Angle	Dedendum K	Whole Tooth Depth W	Clearance $K-J$	Tooth Thickness on Pitch Circle
American Standard 14½-and 20-degree full depth	$\frac{1.157}{P}$	$\frac{2.157}{P}$	$\frac{0.157}{P}$	$\frac{1.5708}{P}$
American Standard 20-degree stub	$\frac{1}{P}$	$\frac{1.8}{P}$	$\frac{0.2}{P}$	$\frac{1.5708}{P}$
Fellows 20-degree stub	$\frac{1.25}{P_D}$	$\frac{2.25}{P_D}$	$\frac{0.25}{P_D}$	$\frac{1.5708}{P_N}$

N = number of teeth.

In the Fellows stub-tooth system, P_N = diametral pitch in numerator of stub-tooth designation and is used to determine circular pitch and number of teeth, and P_D = diametral pitch in the diameter of stub-tooth designation and is used to determine tooth depth.

Tooth Form and Pressure Angle	Tangent of Pitch Cone Angle of Gear, tan A	Tangent of PitchCone Angle of Pinion, tan a	Diametral Pitch of Both Gear and Pinion, P[a]	Outside Diameter of Gear, O, or Pinion, o
American Standard 14½- and 20-degree full depth	$\frac{N_G}{N_P}$	$\frac{N_P}{N_G}$	$\frac{N_a+2\cos A}{O}$ or $\frac{N_P+2\cos a}{o}$	$\frac{N_a+2\cos A}{P}$ or $\frac{N_P+2\cos a}{P}$

Tooth Form and Pressure Angle	Tangent of Pitch Cone Angle of Gear, tan A	Tangent of PitchCone Angle of Pinion, tan a	Diametral Pitch of Both Gear and Pinion, P[a]	Outside Diameter of Gear, O, or Pinion, o
American Standard 20-degree stub	$\frac{N_G}{N_P}$	$\frac{N_P}{N_G}$	$\frac{N_a + 1.6\cos A}{O}$ or $\frac{N_P + 1.6\cos a}{o}$	$\frac{N_a + 1.6\cos A}{P}$ or $\frac{N_P + 1.6\cos a}{P}$
Fellows 20-degree stub	$\frac{N_G}{N_P}$	$\frac{N_P}{N_G}$	…	$\frac{N_G}{P_N} + \frac{2\cos A}{P_D}$ or $\frac{N_P}{P_N} + \frac{2\cos a}{P_D}$

[a] These values are the same for both gear and pinion.

Tooth Form and Pressure Angle	Pitch-Cone Radius or Cone Distance, E[a]	Tangent of Addendum Angle[a]	Tangent of Dedendum Angle[a]	Cosine of Pitch-Cone Angle of Gear, cos A[a]
American Standard 14½- and 20-degree full depth	$\frac{D}{2\sin A}$ or $\frac{d}{2\sin a}$	$\frac{2\sin A}{N_a}$ or $\frac{2\sin a}{N_P}$	$\frac{2.314\sin A}{N_G}$ or $\frac{2.314\sin a}{N_P}$	$\frac{(P \times O) - N_G}{2}$
American Standard 20-degree stub	$\frac{D}{2\sin A}$ or $\frac{d}{2\sin a}$	$\frac{1.6\sin A}{N_G}$ or $\frac{1.6\sin a}{N_P}$	$\frac{2\sin A}{N_G}$ or $\frac{2\sin a}{N_P}$	$\frac{(P \times O) - N_G}{1.6}$
Fellows 20-degree stub	$\frac{D}{2\sin A}$ or $\frac{d}{2\sin a}$	$\frac{2P_N\sin A}{N_G \times P_D}$ or $\frac{2P_N\sin a}{N_P \times P_D}$	$\frac{2.5P_N\sin A}{N_G \times P_D}$ or $\frac{2.5P_N\sin a}{N_P \times P_D}$	$\frac{P_D[(O \times P_N) - N_G]}{2P_N}$

[a] The same formulas apply to the pinion, substituting N_P for N_G and o for O.

N_G = number of teeth in gear; N_P = number of teeth in pinion; O = outside diameter of gear; o = outside diameter of pinion; D = pitch diameter of gear = $N_G \div P$; d = pitch diameter of pinion = $N_P \div P$; P_c = circular pitch; J = addendum; K = dedendum; W = whole depth.

See footnote in *Spur Gears* table for meaning of P_N and P_D. The tooth thickness on the pitch circle is found by means of the formulas in the last column under spur gears.

Helical Gears

Tooth Form and Pressure Angle	Normal Diametral Pitch P_N	Diametral Pitch P	Outside Diameter of Blank O	Pitch Diameter D
American Standard 14½- and 20-degree full depth	$\frac{N+2\cos A}{O\times\cos A}$ or $\frac{P}{\cos A}$	$P_N\cos A$ or $\frac{N+2\cos A}{O}$	$\frac{N+2\cos A}{P_N\cos A}$ or $\frac{N+2\cos A}{P}$	$\frac{N}{P_N\cos A}$ or $\frac{N}{P}$
American Standard 20-degree stub	$\frac{N+1.6\cos A}{O\times\cos A}$ or $\frac{P}{\cos A}$	$P_n\cos A$ or $\frac{N+1.6\cos A}{O}$	$\frac{N+1.6\cos A}{P_N\cos A}$ or $\frac{N+1.6\cos A}{P}$	$\frac{N}{P_N\cos A}$ or $\frac{N}{P}$
Fellows 20-degree stub	…	…	$\frac{N}{(P_N)_N\cos A}+\frac{2}{(P_N)_D}$	$\frac{N}{(P_N)_N\cos A}$

Tooth Form and Pressure Angle	Cosine of Helix Angle A	Addendum	Dedendum	Whole Depth
American Standard 14½- and 20-degree full depth	$\frac{P}{P_N}$ or $\frac{N}{O\times P_N-2}$	$\frac{1}{P_N}$ or $\frac{\cos A}{P}$	$\frac{1.157}{P_N}$ or $\frac{1.157\cos A}{P}$	$\frac{2.157}{P_N}$ or $\frac{2.157\cos A}{P}$
American Standard 20-degree stub	$\frac{P}{P_N}$ or $\frac{N}{O\times P_N-1.6}$	$\frac{0.8}{P_N}$ or $\frac{0.8\cos A}{P}$	$\frac{1}{P_N}$ or $\frac{\cos A}{P}$	$\frac{1.8}{P_N}$ or $\frac{1.8\cos A}{P}$
Fellows 20-degree stub	$\frac{N}{(P_N)_N\left(O-\frac{2}{(P_N)_D}\right)}$	$\frac{1}{(P_N)_D}$	$\frac{1.25}{(P_N)_D}$	$\frac{2.25}{(P_N)_D}$

P_N = normal diametral pitch; = normal diametral pitch of cutter or hob used to cut teeth

P = diametral pitch
O = outside diameter of blank
D = pitch diameter
A = helix angle
N = number of teeth

$(P_N)_N$ = normal diametral pitch in numerator of stub-tooth designation, which determines thickness of tooth and number of teeth

$(P_N)_D$ = normal diametral pitch in denominator of stub-tooth designation, which determines the addendum, dedendum, and whole depth

SPLINES AND SERRATIONS

A splined shaft is one having a series of parallel keys formed integrally with the shaft and mating with corresponding grooves cut in a hub or fitting; this arrangement is in contrast to a shaft having a series of keys or feathers fitted into slots cut into the shaft. The latter construction weakens the shaft to a considerable degree because of the slots cut into it and consequently, reduces its torque-transmitting capacity.

Splined shafts are most generally used in three types of applications: 1) for coupling shafts when relatively heavy torques are to be transmitted without slippage; 2) for transmitting power to slidably-mounted or permanently-fixed gears, pulleys, and other rotating members; and 3) for attaching parts that may require removal for indexing or change in angular position.

Splines having straight-sided teeth have been used in many applications (see SAE Parallel Side Splines for Soft Broached Holes in Fittings); however, the use of splines with teeth of involute profile has steadily increased since 1) involute spline couplings have greater torque-transmitting capacity than any other type; 2) they can be produced by the same techniques and equipment as is used to cut gears; and 3) they have a self-centering action under load even when there is backlash between mating members.

Involute Splines

American National Standard Involute Splines*.—These splines or multiple keys are similar in form to internal and external involute gears. The general practice is to form the external splines either by hobbing, rolling, or on a gear shaper, and internal splines either by broaching or on a gear shaper. The internal spline is held to basic dimensions and the external spline is varied to control the fit. Involute splines have maximum strength at the base, can be accurately spaced and are self-centering, thus equalizing the bearing and stresses, and they can be measured and fitted accurately.

In American National Standard ANSI B92.1-1970 (R 1993), many features of the 1960 standard are retained; plus the addition of three tolerance classes, for a total of four. The term "involute serration," formerly applied to involute splines with 45-degree pressure angle, has been deleted and the standard now includes involute splines with 30-, 37.5-, and 45-degree pressure angles. Tables for these splines have been rearranged accordingly. The term "serration" will no longer apply to splines covered by this Standard.

The Standard has only one fit class for all side fit splines; the former Class 2 fit. Class 1 fit has been deleted because of its infrequent use. The major diameter of the flat root side fit spline has been changed and a tolerance applied to include the range of the 1950 and the 1960 standards. The interchangeability limitations with splines made to previous standards are given later in the section entitled "Interchangeability."

There have been no tolerance nor fit changes to the major diameter fit section.

The Standard recognizes the fact that proper assembly between mating splines is dependent only on the spline being within effective specifications from the tip of the tooth to the form diameter. Therefore, on side fit splines, the internal spline major diameter now is shown as a maximum dimension and the external spline minor diameter is shown as a minimum dimension. The minimum internal major diameter and the maximum external minor diameter must clear the specified form diameter and thus do not need any additional control.

The spline specification tables now include a greater number of tolerance level selections. These tolerance classes were added for greater selection to suit end product needs. The selections differ only in the tolerance as applied to space widthand tooth thickness.

* See American National Standard ANSI B92.2M-1980 (R1989), Metric Module Involute Splines; also see page 2148.

The tolerance class used in ASA B5.15-1960 is the basis and is now designated as tolerance Class 5. The new tolerance classes are based on the following formulas:

Tolerance Class 4 = Tolerance Class 5 × 0.71

Tolerance Class 6 = Tolerance Class 5 × 1.40

Tolerance Class 7 = Tolerance Class 5 × 2.00

All dimensions listed in this standard are for the finished part. Therefore, any compensation that must be made for operations that take place during processing, such as heat treatment, must be taken into account when selecting the tolerance level for manufacturing.

The standard has the same internal minimum effective space width and external maximum effective tooth thickness for all tolerance classes and has two types of fit. For tooth side fits, the minimum effective space width and the maximum effective tooth thickness are of equal value. This basic concept makes it possible to have interchangeable assembly between mating splines where they are made to this standard regardless of the tolerance class of the individual members. A tolerance class "mix" of mating members is thus allowed, which often is an advantage where one member is considerably less difficult to produce than its mate, and the "average" tolerance applied to the two units is such that it satisfies the design need. For instance, assigning a Class 5 tolerance to one member and Class 7 to its mate will provide an assembly tolerance in the Class 6 range. The maximum effective tooth thickness is less than the minimum effective space width for major diameter fits to allow for eccentricity variations.

In the event the fit as provided in this standard does not satisfy a particular design need and a specific amount of effective clearance or press fit is desired, the change should be made only to the external spline by a reduction or an increase in effective tooth thickness and a like change in actual tooth thickness. The minimum effective space width, in this standard, is always basic. The basic minimum effective space width should always be retained when special designs are derived from the concept of this standard.

Terms Applied to Involute Splines.—The following definitions of involute spline terms, here listed in alphabetical order, are given in the American National Standard. Some of these terms are illustrated in the diagram in Tables 6.

Active Spline Length (L_a) is the length of spline that contacts the mating spline. On sliding splines, it exceeds the length of engagement.

Actual Space Width (s) is the circular width on the pitch circle of any single space considering an infinitely thin increment of axial spline length.

Actual Tooth Thickness (t) is the circular thickness on the pitch circle of any single tooth considering an infinitely thin increment of axial spline length.

Alignment Variation is the variation of the effective spline axis with respect to the reference axis (see Fig. 1c).

Base Circle is the circle from which involute spline tooth profiles are constructed.

Base Diameter (D_b) is the diameter of the base circle.

Basic Space Width is the basic space width for 30-degree pressure angle splines; half the circular pitch. The basic space width for 37.5- and 45-degree pressure angle splines, however, is greater than half the circular pitch. The teeth are proportioned so that the external tooth, at its base, has about the same thickness as the internal tooth at the form diameter. This proportioning results in greater minor diameters than those of comparable involute splines of 30-degree pressure angle.

Circular Pitch (p) is the distance along the pitch circle between corresponding points of adjacent spline teeth.

Depth of Engagement is the radial distance from the minor circle of the internal spline to the major circle of the external spline, minus corner clearance and/or chamfer depth.

Diametral Pitch (P) is the number of spline teeth per inch of pitch diameter. The diametral pitch determines the circular pitch and the basic space width or tooth thickness. In conjunction with the number of teeth, it also determines the pitch diameter. (See also Pitch.)

Effective Clearance (c_v) is the effective space width of the internal spline minus the effective tooth thickness of the mating external spline.

Effective Space Width (S_v) of an internal spline is equal to the circular tooth thickness on the pitch circle of an imaginary perfect external spline that would fit the internal spline without looseness or interference considering engagement of the entire axial length of the spline. The minimum effective space width of the internal spline is always basic, as shown in Table 3. Fit variations may be obtained by adjusting the tooth thickness of the external spline.

Three types of involute spline variations

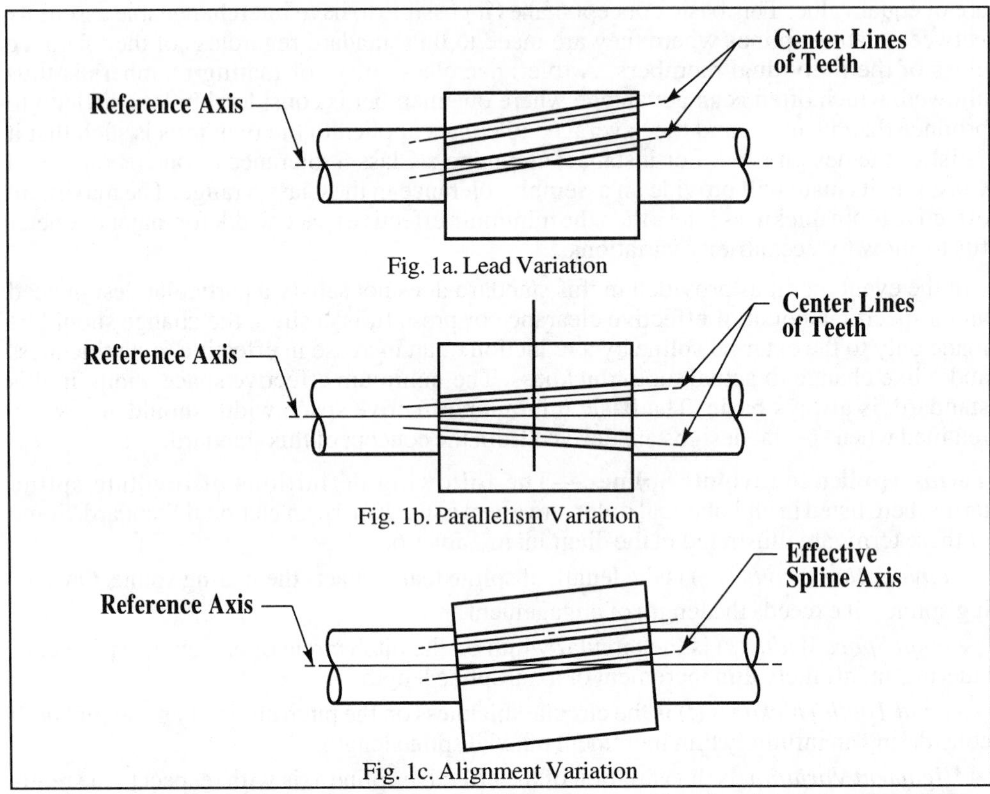

Fig. 1a. Lead Variation

Fig. 1b. Parallelism Variation

Fig. 1c. Alignment Variation

Effective Tooth Thickness (t_v) of an external spline is equal to the circular space width on the pitch circle of an imaginary perfect internal spline that would fit the external spline without looseness or interference, considering engagement of the entire axial length of the spline.

Effective Variation is the accumulated effect of the spline variations on the fit with the mating part.

External Spline is a spline formed on the outer surface of a cylinder.

Fillet is the concave portion of the tooth profile that joins the sides to the bottom of the space.

Fillet Root Splines are those in which a single fillet in the general form of an arc joins the sides of adjacent teeth.

Flat Root Splines are those in which fillets join the arcs of major or minor circles to the tooth sides.

Form Circle is the circle which defines the deepest points of involute form control of the tooth profile. This circle along with the tooth tip circle (or start of chamfer circle) determines the limits of tooth profile requiring control. It is located near the major circle on the internal spline and near the minor circle on the external spline.

Form Clearance (c_F) is the radial depth of involute profile beyond the depth of engagement with the mating part. It allows for looseness between mating splines and for eccentricities between the minor circle (internal), the major circle (external), and their respective pitch circles.

Form Diameter (D_{Fe}, D_{Fi}) the diameter of the form circle.

Internal Spline is a spline formed on the inner surface of a cylinder.

Involute Spline is one having teeth with involute profiles.

Lead Variation is the variation of the direction of the spline tooth from its intended direction parallel to the reference axis, also including parallelism and alignment variations (see Fig. 1a). *Note:* Straight (nonhelical) splines have an infinite lead.

Length of Engagement (L_q) is the axial length of contact between mating splines.

Machining Tolerance (m) is the permissible variation in actual space width or actual tooth thickness.

Major Circle is the circle formed by the outermost surface of the spline. It is the outside circle (tooth tip circle) of the external spline or the root circle of the internal spline.

Major Diameter (D_o, D_{ri}) is the diameter of the major circle.

Minor Circle is the circle formed by the innermost surface of the spline. It is the root circle of the external spline or the inside circle (tooth tip circle) of the internal spline.

Minor Diameter (D_{re}, D_i) is the diameter of the minor circle.

Nominal Clearance is the actual space width of an internal spline minus the actual tooth thickness of the mating external spline. It does not define the fit between mating members, because of the effect of variations.

Out of Roundness is the variation of the spline from a true circular configuration.

Parallelism Variation is the variation of parallelism of a single spline tooth with respect to any other single spline tooth (see Fig. 1b).

Pitch (P/P_s) is a combination number of a one-to-two ratio indicating the spline proportions; the upper or first number is the diametral pitch, the lower or second number is the stub pitch and denotes, as that fractional part of an inch, the basic radial length of engagement, both above and below the pitch circle.

Pitch Circle is the reference circle from which all transverse spline tooth dimensions are constructed.

Pitch Diameter (D) is the diameter of the pitch circle.

Pitch Point is the intersection of the spline tooth profile with the pitch circle.

Pressure Angle (ϕ) is the angle between a line tangent to an involute and a radial line through the point of tangency. Unless otherwise specified, it is the standard pressure angle.

Profile Variation is any variation from the specified tooth profile normal to the flank.

Spline is a machine element consisting of integral keys (spline teeth) or keyways (spaces) equally spaced around a circle or portion thereof.

Standard (Main) Pressure Angle (ϕ_D) is the pressure angle at the specified pitch diameter.

Stub Pitch (P_s) is a number used to denote the radial distance from the pitch circle to the major circle of the external spline and from the pitch circle to the minor circleof the internal spline. The stub pitch for splines in this standard is twice the diametral pitch.

Total Index Variation is the greatest difference in any two teeth (adjacent or otherwise) between the actual and the perfect spacing of the tooth profiles.

Total Tolerance ($m + \lambda$) is the machining tolerance plus the variation allowance.

Variation Allowance (λ) is the permissible effective variation.

Tooth Proportions.—There are 17 pitches: 2.5/5, 3/6, 4/8,5/10, 6/12, 8/16, 10/20, 12/24, 16/32, 20/40, 24/48, 32/64, 40/80, 48/96, 64/128, 80/160, and 128/256. The numerator in this fractional designation is known as the diametral pitch and controls the pitch diameter; the denominator, which is always double the numerator, is known as the stub pitch and controls the tooth depth. For convenience in calculation, only the numerator is used in the formulas given and is designated as *P*. Diametral pitch, as in gears, means the number of teeth per inch of pitch diameter.

Table 1 shows the symbols and Table 2 the formulas for basic tooth dimensions of involute spline teeth of various pitches. Basic dimensions are given in Table 3.

Table 1. American National Standard Involute Spline Symbols
ANSI B92.1-1970, R1993

Symbol	Meaning	Symbol	Meaning
c_v	effective clearance	M_i	measurement between pins, internal spline
c_F	form clearance		
D	pitch diameter	N	number of teeth
D_b	base diameter	P	diametral pitch
D_{ci}	pin contact diameter, internal spline	P_s	stub pitch
		p	circular pitch
D_{ce}	pin contact diameter, external spline	r_f	fillet radius
		s	actual space width, circular
D_{Fe}	form diameter, external spline	s_v	effective space width, circular
D_{Fi}	form diameter, internal spline	s_c	allowable compressive stress, psi
D_i	minor diameter, internal spline	s_s	allowable shear stress, psi
D_o	major diameter, external spline	t	actual tooth thickness, circular
D_{re}	minor diameter, external spline (root)	t_v	effective tooth thickness, circular
		λ	variation allowance
D_{ri}	major diameter, internal spline (root)	$\in$	involute roll angle
		ϕ	pressure angle
d_e	diameter of measuring pin for external spline	ϕ_D	standard pressure angle
		ϕ_{ci}	pressure angle at pin contact diameter, internal spline
d_i	diameter of measuring pin for internal spline		
		ϕ_{ce}	pressure angle at pin contact diameter, external spline
K_e	change factor for external spline		
K_i	change factor for internal spline	ϕ_i	pressure angle at pin center, internal spline
L	spline length		
L_a	active spline length	ϕ_e	pressure angle at pin center, external spline
L_g	length of engagement		
m	machining tolerance	ϕ_F	pressure angle at form diameter
M_e	measurement over pins, external spline		

Table 2. Formulas for Basic Dimensions *ANSI B92.1-1970, R1993*

Term	Symbol	Formula				
		30 deg ϕ_D			37.5 deg ϕ_D	45 deg ϕ_D
		Flat Root Side Fit	Flat Root Major Dia Fit	Fillet Root Side Fit	Fillet Root Side Fit	Fillet Root Side Fit
		2.5/5–32/64 Pitch	3/6–16/32 Pitch	2.5/5–48/96 Pitch	2.5/5–48/96 Pitch	10/20–128/256 Pitch
Stub Pitch	P_s	$2P$	$2P$	$2P$	$2P$	$2P$
Pitch Diameter	D	$\frac{N}{P}$	$\frac{N}{P}$	$\frac{N}{P}$	$\frac{N}{P}$	$\frac{N}{P}$
Base Diameter	D_b	$D \cos \phi_D$	$D \cos \phi_D$	$D \cos \phi_D$	$D \cos \phi_D$	$D \cos \phi_D$
Circular Pitch	p	$\frac{\pi}{P}$	$\frac{\pi}{P}$	$\frac{\pi}{P}$	$\frac{\pi}{P}$	$\frac{\pi}{P}$
Minimum Effective Space Width	s_v	$\frac{\pi}{2P}$	$\frac{\pi}{2P}$	$\frac{\pi}{2P}$	$\frac{0.5\pi + 0.1}{P}$	$\frac{0.5\pi + 0.2}{P}$
Major Diameter, Internal	D_{ri}	$\frac{N + 1.35}{P}$	$\frac{N + 1}{P}$	$\frac{N + 1.8}{P}$	$\frac{N + 1.6}{P}$	$\frac{N + 1.4}{P}$
Major Diameter, External	D_o	$\frac{N + 1}{P}$	$\frac{N + 1}{P}$	$\frac{N + 1}{P}$	$\frac{N + 1}{P}$	$\frac{N + 1}{P}$
Minor Diameter, Internal	D_i	$\frac{N - 1}{P}$	$\frac{N - 1}{P}$	$\frac{N - 1}{P}$	$\frac{N - 0.8}{P}$	$\frac{N - 0.6}{P}$

Table 2. *(Continued)* **Formulas for Basic Dimensions** *ANSI B92.1-1970, R1993*

Term		Symbol	Formula				
			30 deg ϕ_D			37.5 deg ϕ_D	45 deg ϕ_D
			Flat Root Side Fit	Flat Root Major Dia Fit	Fillet Root Side Fit	Fillet Root Side Fit	Fillet Root Side Fit
			2.5/5–32/64 Pitch	3/6–16/32 Pitch	2.5/5–48/96 Pitch	2.5/5–48/96 Pitch	10/20–128/256 Pitch
Minor Dia. Ext.	2.5/5 thru 12/24 pitch	D_{re}	$\frac{N-1.35}{P}$		$\frac{N-1.8}{P}$	$\frac{N-1.3}{P}$	…
	16/32 pitch and finer				$\frac{N-2}{P}$		
	10/20 16/32 pitch and finer				…		$\frac{N-1}{P}$
Form Diameter, Internal		D_{Fi}	$\frac{N+1}{P}+2cF$	$\frac{N+0.8}{P}-0.004+2cF$	$\frac{N+1}{P}+2cF$	$\frac{N+1}{P}+2cF$	$\frac{N+1}{P}+2cF$
Form Diameter, External		D_{Fe}	$\frac{N-1}{P}-2cF$	$\frac{N-1}{P}-2cF$	$\frac{N-1}{P}-2cF$	$\frac{N-0.8}{P}-2cF$	$\frac{N-0.6}{P}-2cF$
Form Clearance (Radial)		c_F	0.001 D, with max of 0.010, min of 0.002				
π = 3.1415927. NOTE: All spline specification table dimensions in the standard are derived from these basic formulas by application of tolerances.							

Table 3. Basic Dimensions for Involute Splines *ANSI B92.1-1970, R1993*

Pitch, P/P_s	Circular Pitch, p	Min Effective Space Width (BASIC), S_v min			Pitch, P/P_s	Circular Pitch, p	Min Effective Space Width (BASIC), S_v min		
		30 deg φ	37.5 deg φ	45 deg φ			30 deg φ	37.5 deg φ	45 deg φ
2.5/5	1.2566	0.6283	0.6683	…	20/40	0.1571	0.0785	0.0835	0.0885
3/6	1.0472	0.5236	0.5569	…	24/48	0.1309	0.0654	0.0696	0.0738
4/8	0.7854	0.3927	0.4177	…	32/64	0.0982	0.0491	0.0522	0.0553
5/10	0.6283	0.3142	0.3342	…	40/80	0.0785	0.0393	0.0418	0.0443
6/12	0.5236	0.2618	0.2785	…	48/96	0.0654	0.0327	0.0348	0.0369
8/16	0.3927	0.1963	0.2088	…	64/128	0.0491	…	…	0.0277
10/20	0.3142	0.1571	0.1671	0.1771	80/160	0.0393	…	…	0.0221
12/24	0.2618	0.1309	0.1392	0.1476	128/256	0.0246	…	…	0.0138
16/32	0.1963	0.0982	0.1044	0.1107	…	…	…	…	…

Tooth Numbers.—The American National Standard covers involute splines having tooth numbers ranging from 6 to 60 with a 30- or 37.5-degree pressure angle and from 6 to 100 with a 45-degree pressure angle. In selecting the number of teeth for a given spline application, it is well to keep in mind that there are no advantages to be gained by using odd numbers of teeth and that the diameters of splines with odd tooth numbers, particularly internal splines, are troublesome to measure with pins since no two tooth spaces are diametrically opposite each other.

Types and Classes of Involute Spline Fits.—Two types of fits are covered by the American National Standard for involute splines, the side fit, and the major diameter fit. Dimensional data for flat root side fit, flat root major diameter fit, and fillet root side fit splines are tabulated in this standard for 30-degree pressure angle splines; but for only the fillet root side fit for 37.5- and 45-degree pressure angle splines.

Side Fit: In the side fit, the mating members contact only on the sides of the teeth; major and minor diameters are clearance dimensions. The tooth sides act as drivers and centralize the mating splines.

Major Diameter Fit: Mating parts for this fit contact at the major diameter for centralizing. The sides of the teeth act as drivers. The minor diameters are clearance dimensions.

The major diameter fit provides a minimum effective clearance that will allow for contact and location at the major diameter with a minimum amount of location or centralizing effect by the sides of the teeth. The major diameter fit has only one space width and tooth thickness tolerance which is the same as side fit Class 5.

A fillet root may be specified for an external spline, even though it is otherwise designed to the flat root side fit or major diameter fit standard. An internal spline with a fillet root can be used only for the side fit.

Classes of Tolerances.—This standard includes four classes of tolerances on space width and tooth thickness. This has been done to provide a range of tolerances for selection to suit a design need. The classes are variations of the former single tolerance which is now Class 5 and are based on the formulas shown in the footnote of Table 4. All tolerance classes have the same minimum effective space width and maximum effective tooth thickness limits so that a mix of classes between mating parts is possible.

Table 4. Maximum Tolerances for Space Width and Tooth Thickness of Tolerance Class 5 Splines *ANSI B92.1-1970, R1993*
(Values shown in ten thousandths; 20 = 0.0020)

No. of Teeth	Pitch, P/P_s							
	2.5/5 and 3/6	4/8 and 5/10	6/12 and 8/16	10/20 and 12/24	16/32 and 20/40	24/48 thru 48/96	64/128 and 80/160	128/256
N	Machining Tolerance, *m*							
10	15.8	14.5	12.5	12.0	11.7	11.7	9.6	9.5
20	17.6	16.0	14.0	13.0	12.4	12.4	10.2	10.0
30	18.4	17.5	15.5	14.0	13.1	13.1	10.8	10.5
40	21.8	19.0	17.0	15.0	13.8	13.8	11.4	—
50	23.0	20.5	18.5	16.0	14.5	14.5	—	—
60	24.8	22.0	20.0	17.0	15.2	15.2	—	—
70	—	—	—	18.0	15.9	15.9	—	—
80	—	—	—	19.0	16.6	16.6	—	—
90	—	—	—	20.0	17.3	17.3	—	—
100	—	—	—	21.0	18.0	18.0	—	—
N	Variation Allowance, λ							
10	23.5	20.3	17.0	15.7	14.2	12.2	11.0	9.8
20	27.0	22.6	19.0	17.4	15.4	13.4	12.0	10.6
30	30.5	24.9	21.0	19.1	16.6	14.6	13.0	11.4
40	34.0	27.2	23.0	21.6	17.8	15.8	14.0	—
50	37.5	29.5	25.0	22.5	19.0	17.0	—	—
60	41.0	31.8	27.0	24.2	20.2	18.2	—	—
70	—	—	—	25.9	21.4	19.4	—	—
80	—	—	—	27.6	22.6	20.6	—	—
90	—	—	—	29.3	23.8	21.8	—	—
100	—	—	—	31.0	25.0	23.0	—	—
N	Total Index Variation							
10	20	17	15	15	14	12	11	10
20	24	20	18	17	15	13	12	11
30	28	22	20	19	16	15	14	13
40	32	25	22	20	18	16	15	—
50	36	27	25	22	19	17	—	—
60	40	30	27	24	20	18	—	—
70	—	—	—	26	21	20	—	—
80	—	—	—	28	22	21	—	—
90	—	—	—	29	24	23	—	—
100	—	—	—	31	25	24	—	—
N	Profile Variation							
All	+7 −10	+6 −8	+5 −7	+4 −6	+3 −5	+2 −4	+2 −4	+2 −4

Lead Variation												
L_g, in.	0.3	0.5	1	2	3	4	5	6	7	8	9	10
Variation	2	3	4	5	6	7	8	9	10	11	12	13

For other tolerance classes: Class 4 = 0.71 × Tabulated value
Class 5 = As tabulated in table
Class 6 = 1.40 × Tabulated value
Class 7 = 2.00 × Tabulated value

Fillets and Chamfers.—Spline teeth may have either a flat root or a rounded fillet root.

Flat Root Splines: are suitable for most applications. The fillet that joins the sides to the bottom of the tooth space, if generated, has a varying radius of curvature. Specification of this fillet is usually not required. It is controlled by the form diameter, which is the diameter at the deepest point of the desired true involute form (sometimes designated as TIF).

When flat root splines are used for heavily loaded couplings that are not suitable for fillet root spline application, it may be desirable to minimize the stress concentration in the flat root type by specifying an approximate radius for the fillet.

Because internal splines are stronger than external splines due to their broad bases and high pressure angles at the major diameter, broaches for flat root internal splines are normally made with the involute profile extending to the major diameter.

Fillet Root Splines: are recommended for heavy loads because the larger fillets provided reduce the stress concentrations. The curvature along any generated fillet varies and cannot be specified by a radius of any given value.

External splines may be produced by generating with a pinion-type shaper cutter or with a hob, or by cutting with no generating motion using a tool formed to the contour of a tooth space. External splines are also made by cold forming and are usually of the fillet root design. Internal splines are usually produced by broaching, by form cutting, or by generating with a shaper cutter. Even when full-tip radius tools are used, each of these cutting methods produces a fillet contour with individual characteristics. Generated spline fillets are curves related to the prolate epicycloid for external splines and the prolate hypocycloid for internal splines. These fillets have a minimum radius of curvature at the point where the fillet is tangent to the external spline minor diameter circle or the internal spline major diameter circle and a rapidly increasing radius of curvature up to the point where the fillet comes tangent to the involute profile.

Chamfers and Corner Clearance: In major diameter fits, it is always necessary to provide corner clearance at the major diameter of the spline coupling. This clearance is usually effected by providing a chamfer on the top corners of the external member. This method may not be possible or feasible because of the following:

A) If the external member is roll formed by plastic deformation, a chamfer cannot be provided by the process.

B) A semitopping cutter may not be available.

C) When cutting external splines with small numbers of teeth, a semitopping cutter may reduce the width of the top land to a prohibitive point.

In such conditions, the corner clearance can be provided on the internal spline, as shown in Fig. 2.

When this option is used, the form diameter may fall in the protuberance area.

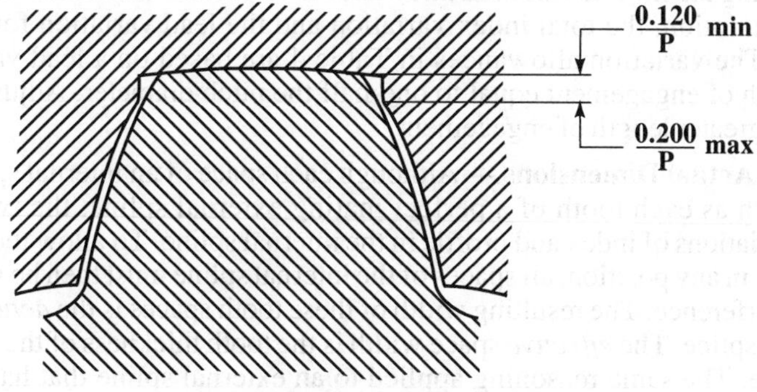

Fig. 2. Internal corner clearance.

Spline Variations.—The maximum allowable variations for involute splines are listed in Table 4.

Profile Variation: The reference profile, from which variations occur, passes through the point used to determine the actual space width or tooth thickness. This is either the pitch point or the contact point of the standard measuring pins.

Profile variation is positive in the direction of the space and negative in the direction of the tooth. Profile variations may occur at any point on the profile for establishing effective fits and are shown in Table 4.

Lead Variations: The lead tolerance for the total spline length applies also to any portion thereof unless otherwise specified.

Out of Roundness: This condition may appear merely as a result of index and profile variations given in Table 4 and requires no further allowance. However, heat treatment and deflection of thin sections may cause out of roundness, which increases index and profile variations. Tolerances for such conditions depend on many variables and are therefore not tabulated. Additional tooth and/or space width tolerance must allow for such conditions.

Eccentricity: Eccentricity of major and minor diameters in relation to the effective diameter of side fit splines should not cause contact beyond the form diameters of the mating splines, even under conditions of maximum effective clearance. This standard does not establish specific tolerances.

Eccentricity of major diameters in relation to the effective diameters of major diameter fit splines should be absorbed within the maximum material limits established by the tolerances on major diameter and effective space width or effective tooth thickness.

If the alignment of mating splines is affected by eccentricity of locating surfaces relative to each other and/or the splines, it may be necessary to decrease the effective and actual tooth thickness of the external splines in order to maintain the desired fit condition. This standard does not include allowances for eccentric location.

Effect of Spline Variations.—Spline variations can be classified as index variations, profile variations, or lead variations.

Index Variations: These variations cause the clearance to vary from one set of mating tooth sides to another. Because the fit depends on the areas with minimum clearance, index variations reduce the effective clearance.

Profile Variations: Positive profile variations affect the fit by reducing effective clearance. Negative profile variations do not affect the fit but reduce the contact area.

Lead Variations: These variations will cause clearance variations and therefore reduce the effective clearance.

Variation Allowance: The effect of individual spline variations on the fit (effective variation) is less than their total, because areas of more than minimum clearance can be altered without changing the fit. The variation allowance is 60 percent of the sum of twice the positive profile variation, the total index variation and the lead variation for the length of engagement. The variation allowances in Table 4 are based on a lead variation for an assumed length of engagement equal to one-half the pitch diameter. Adjustment may be required for a greater length of engagement.

Effective and Actual Dimensions.—Although each space of an internal spline may have the same width as each tooth of a perfect mating external spline, the two may not fit because of variations of index and profile in the internal spline. To allow the perfect external spline to fit in any position, all spaces of the internal spline must then be widened by the amount of interference. The resulting width of these tooth spaces is the *actual* space width of the internal spline. The *effective* space width is the tooth thickness of the perfect mating external spline. The same reasoning applied to an external spline that has variations of index and profile when mated with a perfect internal spline leads to the concept of effective tooth thickness, which exceeds the actual tooth thickness by the amount of the effective variation.

The effective space width of the internal spline minus the effective tooth thickness of the external spline is the effective clearance and defines the fit of the mating parts. (This statement is strictly true only if high points of mating parts come into contact.) Positive effective clearance represents looseness or backlash. Negative effective clearance represents tightness or interference.

Space Width and Tooth Thickness Limits.—The variation of actual space width and actual tooth thickness within the machining tolerance causes corresponding variations of effective dimensions, so that there are four limit dimensions for each component part.

These variations are shown diagrammatically in Table 5.

Table 5. Specification Guide for Space Width and Tooth Thickness
ANSI B92.1-1970, R1993

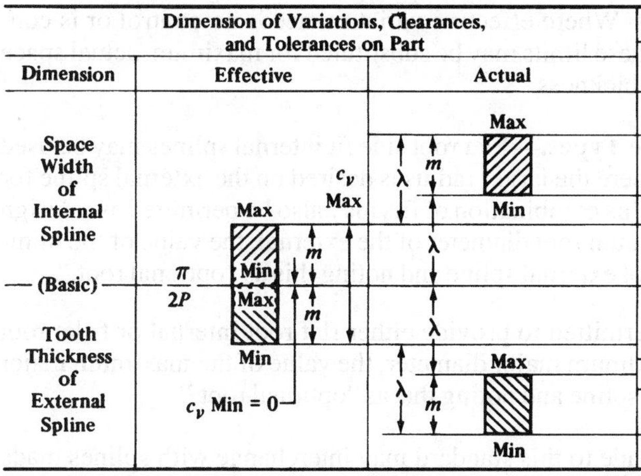

Dimension	Dimension of Variations, Clearances, and Tolerances on Part: Effective	Actual	Dimensioning Method: Standard	Alternatives: A	B
Space Width of Internal Spline		Max	Required	Required	Ref.
	c_v Max (λ, m)	Min (λ, m)	Ref.	Ref.	Ref.
	Max (m)	(λ)	Ref.	Required	Required
— (Basic) —	$\frac{\pi}{2P}$ Min / Max		Required	Required	Required
Tooth Thickness of External Spline	Min (m)	Max (λ)	Ref.	Required	Required
	c_v Min = 0	(λ, m)	Ref.	Ref.	Ref.
		Min	Required	Required	Ref.

The minimum effective space width is always basic. The maximum effective tooth thickness is the same as the minimum effective space width except for the major diameter fit. The major diameter fit maximum effective tooth thickness is less than the minimum effective space width by an amount that allows for eccentricity between the effective spline and the major diameter. The permissible variation of the effective clearance is divided between the internal and external splines to arrive at the maximum effective space width and the minimum effective tooth thickness. Limits for the actual space width and actual tooth thickness are constructed from suitable variation allowances.

Use of Effective and Actual Dimensions.—Each of the four dimensions for space width and tooth thickness shown in Table 5 has a definite function.

Minimum Effective Space Width and Maximum Effective Tooth Thickness: These dimensions control the minimum effective clearance, and must always be specified.

Minimum Actual Space Width and Maximum Actual Tooth Thickness: These dimensions cannot be used for acceptance or rejection of parts. If the actual space width is less than the minimum without causing the effective space width to be undersized, or if the actual tooth thickness is more than the maximum without causing the effective tooth thickness to be oversized, the effective variation is less than anticipated; such parts are desirable and not defective. The specification of these dimensions as processing reference dimensions is optional. They are also used to analyze undersize effective space width or oversize effective tooth thickness conditions to determine whether or not these conditions are caused by excessive effective variation.

Maximum Actual Space Width and Minimum Actual Tooth Thickness: These dimensions control machining tolerance and limit the effective variation. The spread between these dimensions, reduced by the effective variation of the internal and external spline, is the maximum effective clearance. Where the effective variation obtained in machining is appreciably less than the variation allowance, these dimensions must be adjusted in order to maintain the desired fit.

Maximum Effective Space Width and Minimum Effective Tooth Thickness: These dimensions define the maximum effective clearance but they do not limit the effective variation. They may be used, in addition to the maximum actual space width and minimum actual tooth thickness, to prevent the increase of maximum effective clearance due to reduction of effective variations. The notation "inspection optional" may be added where maximum effective clearance is an assembly requirement, but does not need absolute control. It will indicate, without necessarily adding inspection time and equipment, that the actual space width of the internal spline must be held below the maximum, or the actual tooth thickness of the external spline above the minimum, if machining methods result in less than the allowable variations. Where effective variation needs no control or is controlled by laboratory inspection, these limits may be substituted for maximum actual space width and minimum actual tooth thickness.

Combinations of Involute Spline Types.—Flat root side fit internal splines may be used with fillet root external splines where the larger radius is desired on the external spline for control of stress concentrations. This combination of fits may also be permitted as a design option by specifying for the minimum root diameter of the external, the value of the minimum root diameter of the fillet root external spline and noting this as "optional root."

A design option may also be permitted to provide either flat root internal or fillet root internal by specifying for the maximum major diameter, the value of the maximum major diameter of the fillet root internal spline and noting this as "optional root."

Interchangeability.—Splines made to this standard may interchange with splines made to older standards. Exceptions are listed below.

External Splines: These external splines will mate with older internal splines as follows:

Year	Major Dia. Fit	Flat Root Side Fit	Fillet Root Side Fit
1946	Yes	No (A)[a]	No (A)
1950[b]	Yes (B)	Yes (B)	Yes (C)
1950[c]	Yes (B)	No (A)	Yes (C)
1957 SAE	Yes	No (A)	Yes (C)
1960	Yes	No (A)	Yes (C)

[a] For exceptions A, B, C, see the paragraph on *Exceptions* that follows.
[b] Full dedendum.
[c] Short dedendum.

Internal Splines: These will mate with older external splines as follows:

Year	Major Dia. Fit	Flat Root Side Fit	Fillet Root Side Fit
1946	No (D)[a]	No (E)	No (D)
1950	Yes (F)	Yes	Yes (C)
1957 SAE	Yes (G)	Yes	Yes
1960	Yes (G)	Yes	Yes

[a] For exceptions C, D, E, F, G, see the paragraph on *Exceptions* that follows.

Table 6. Spline Terms, Symbols, and Drawing Data, 30-Degree Pressure Angle, Flat Root Side Fit *ANSI B92.1-1970, R1993*

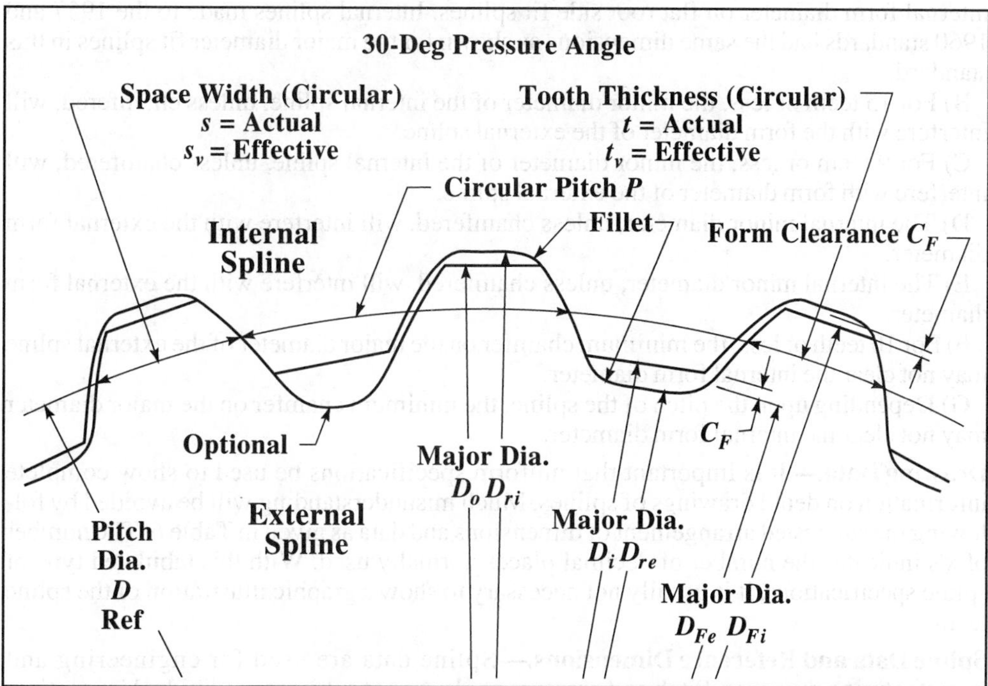

The fit shown is used in restricted areas (as with tubular parts with wall thickness too small to permit use of fillet roots, and to allow hobbing closer to shoulders, etc.) and for economy (when hobbing, shaping, etc., and using shorter broaches for the internal member).

Press fits are not tabulated because their design depends on the degree of tightness desired and must allow for such factors as the shape of the blank, wall thickness, materila, hardness, thermal expansion, etc. Close tolerances or selective size grouping may be required to limit fit variations.

Drawing Data

Internal Involute Spline Data		External Involute Spline Data	
Flat Root Side Fit		Flat Root Side Fit	
Number of Teeth	xx	Number of Teeth	xx
Pitch	xx/xx	Pitch	xx/xx
Pressure Angle	30°	Pressure Angle	30°
Base Diameter	x.xxxxxx Ref	Base Diameter	x.xxxxxx Ref
Pitch Diameter	x.xxxxxx Ref	Pitch Diameter	x.xxxxxx Ref
Major Diameter	x.xxx max	Major Diameter	x.xxx/x.xxx
Form Diameter	x.xxx	Form Diameter	x.xxx
Minor Diameter	x.xxx/x.xxx	Minor Diameter	x.xxx min
Circular Space Width		Circular Tooth Thickness	
Max Actual	x.xxxx	Max Effective	x.xxxx
Min Effective	x.xxxx	Min Actual	x.xxxx
The following information may be added as required:		The following information may be added as required:	
Max Measurement Between Pins	x.xxx Ref	Min Measurement Over Pins	x.xxxx Ref
Pin Diameter	x.xxxx	Pin Diameter	x.xxxx

The above drawing data are required for the spline specifications. The standard system is shown; for alternate systems, see Table 5. Number of x's indicates number of decimal places normally used.

Exceptions:

A) The external major diameter, unless chamfered or reduced, may interfere with the internal form diameter on flat root side fit splines. Internal splines made to the 1957 and 1960 standards had the same dimensions as shown for the major diameter fit splines in this standard.

B) For 15 teeth or less, the minor diameter of the internal spline, unless chamfered, will interfere with the form diameter of the external spline.

C) For 9 teeth or less, the minor diameter of the internal spline, unless chamfered, will interfere with form diameter of the external spline.

D) The internal minor diameter, unless chamfered, will interfere with the external form diameter.

E) The internal minor diameter, unless chamfered, will interfere with the external form diameter.

F) For 10 teeth or less, the minimum chamfer on the major diameter of the external spline may not clear the internal form diameter.

G) Depending upon the pitch of the spline, the minimum chamfer on the major diameter may not clear the internal form diameter.

Drawing Data.—It is important that uniform specifications be used to show complete information on detail drawings of splines. Much misunderstanding will be avoided by following the suggested arrangement of dimensions and data as given in Table 6. The number of x's indicates the number of decimal places normally used. With this tabulated type of spline specifications, it is usually not necessary to show a graphic illustration of the spline teeth.

Spline Data and Reference Dimensions.—Spline data are used for engineering and manufacturing purposes. Pitch and pressure angle are not subject to individual inspection.

As used in this standard, *reference* is an added notation or modifier to a dimension, specification, or note when that dimension, specification, or note is:

1) Repeated for drawing clarification.

2) Needed to define a nonfeature datum or basis from which a form or feature is generated.

3) Needed to define a nonfeature dimension from which other specifications or dimensions are developed.

4) Needed to define a nonfeature dimension at which toleranced sizes of a feature are specified.

5) Needed to define a nonfeature dimension from which control tolerances or sizes are developed or added as useful information.

Any dimension, specification, or note that is noted "REF" should not be used as a criterion for part acceptance or rejection.

Estimating Key and Spline Sizes and Lengths.—Fig. 1 may be used to estimate the size of American Standard involute splines required to transmit a given torque. It also may be used to find the outside diameter of shafts used with single keys. After the size of the shaft is found, the proportions of the key can be determined from Table 1 on page 2342.

Curve A is for flexible splines with teeth hardened to Rockwell C 55–65. For these splines, lengths are generally made equal to or somewhat greater than the pitch diameter for diameters below 1¼ inches; on larger diameters, the length is generally one-third to two-thirds the pitch diameter. Curve A also applies for a single key used as a fixed coupling, the length of the key being one to one and one-quarter times the shaft diameter. The stress in the shaft, neglecting stress concentration at the keyway, is about 7500 pounds per square inch. See also *Effect of Keyways on Shaft Strength* starting on page 283.

Curve B represents high-capacity single keys used as fixed couplings for stresses of 9500 pounds per square inch, neglecting stress concentration. Key-length is one to one and one-quarter times shaft diameter and both shaft and key are of moderately hard heat-treated steel. This type of connection is commonly used to key commercial flexible couplings to motor or generator shafts.

Curve C is for multiple-key fixed splines with lengths of three-quarters to one and one-quarter times pitch diameter and shaft hardness of 200–300 BHN.

Curve D is for high-capacity splines with lengths one-half to one times the pitch diameter. Hardnesses up to Rockwell C 58 are common and in aircraft applications the shaft is generally hollow to reduce weight.

Curve E represents a solid shaft with 65,000 pounds per square inch shear stress. For hollow shafts with inside diameter equal to three-quarters of the outside diameter the shear stress would be 95,000 pounds per square inch.

Length of Splines: Fixed splines with lengths of one-third the pitch diameter will have the same shear strength as the shaft, assuming uniform loading of the teeth; however, errors in spacing of teeth result in only half the teeth being fully loaded. Therefore, for balanced strength of teeth and shaft the length should be two-thirds the pitch diameter. If weight is not important, however, this may be increased to equal the pitch diameter. In the case of flexible splines, long lengths do not contribute to load carrying capacity when there is misalignment to be accommodated. Maximum effective length for flexible splines may be approximated from Fig. 2.

Formulas for Torque Capacity of Involute Splines.—The formulas for torque capacity of 30-degree involute splines given in the following paragraphs are derived largely from an article "When Splines Need Stress Control" by D. W. Dudley, *Product Engineering*, Dec. 23, 1957.

In the formulas that follow the symbols used are as defined on page 2130 with the following additions: D_h = inside diameter of hollow shaft, inches; K_a = application factor from Table 1; K_m = load distribution factor from Table 2; K_f = fatigue life factor from Table 3; K_w = wear life factor from Table 4; L_e = maximum effective length from Fig. 2, to be used in stress formulas even though the actual length may be greater; T = transmitted torque, pound-inches. For fixed splines without helix modification, the effective length L_e should never exceed $5000\,D^{3.5} \div T$.

Table 1. Spline Application Factors, K_a

Power Source	Type of Load			
	Uniform (Generator-Fan)	Light Shock (Oscillating Pumps, etc.)	Intermittent Shock (Actuating Pumps, etc.)	Heavy Shock (Punches, Shears, etc.)
	Application Factor, K_a			
Uniform (Turbine, Motor)	1.0	1.2	1.5	1.8
Light Shock (Hydraulic Motor)	1.2	1.3	1.8	2.1
Medium Shock (Internal Combustion, Engine)	2.0	2.2	2.4	2.8

Table 2. Load Distribution Factors, K_m, for Misalignment of Flexible Splines

Misalignment, inches per inch	Load Distribution Factor, K_m[a]			
	½-in. Face Width	1-in. Face Width	2-in. Face Width	4-in. Face Width
0.001	1	1	1	1 ½
0.002	1	1	1 ½	2
0.004	1	1 ½	2	2 ½
0.008	1 ½	2	2 ½	3

[a] For fixed splines, K_m=1.

For fixed splines, $K_m = 1$.

Table 3. Fatigue-Life Factors, K_f, for Splines

No. of Torque Cycles[a]	Fatigue-Life Factor, K_f	
	Unidirectional	Fully-reversed
1,000	1.8	1.8
10,000	1.0	1.0
100,000	0.5	0.4
1,000,000	0.4	0.3
10,000,000	0.3	0.2

[a] A torque cycle consists of one start and one stop, not the number of revolutions.

Table 4. Wear Life Factors, K_w, for Flexible Splines

Number of Revolutions of Spline	Life Factor, K_w	Number of Revolutions of Spline	Life Factor, K_w
10,000	4.0	100,000,000	1.0
100,000	2.8	1,000,000,000	0.7
1,000,000	2.0	10,000,000,000	0.5
10,000,000	1.4	…	…

Wear life factors, unlike fatigue life factors given in Table 3, are based on the total number of revolutions of the spline, since each revolution of a flexible spline results in a complete cycle of rocking motion which contributes to spline wear.

Definitions: A *fixed* spline is one which is either shrink fitted or loosely fitted but piloted with rings at each end to prevent rocking of the spline which results in small axial movements that cause wear. A *flexible* spline permits some rocking motion such as occurs when the shafts are not perfectly aligned. This flexing or rocking motion causes axial movement and consequently wear of the teeth. Straight-toothed flexible splines can accommodate only small angular misalignments (less than 1 deg.) before wear becomes a serious problem. For greater amounts of misalignment (up to about 5 deg.), crowned splines are preferable to reduce wear and end-loading of the teeth.

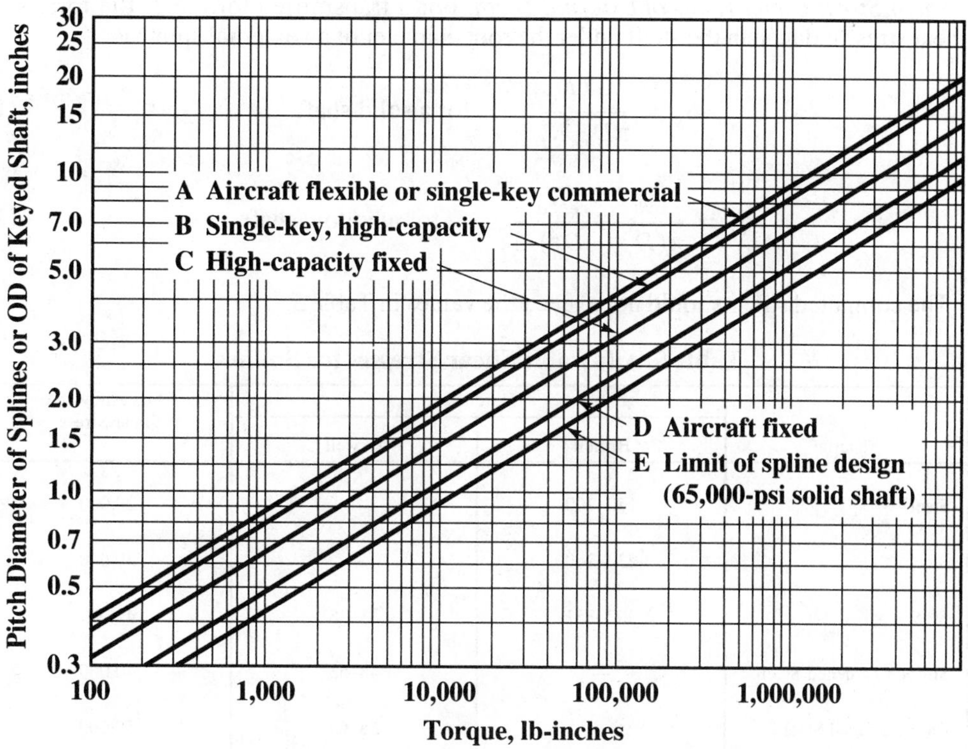

Fig. 1. Chart for Estimating Involute Spline Size Based on Diameter-Torque Relationships

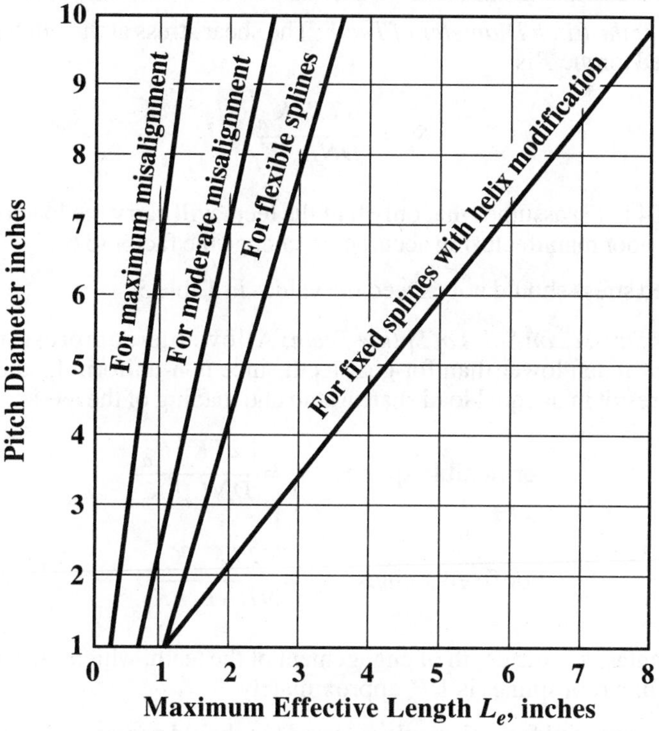

Fig. 2. Maximum Effective Length for Fixed and Flexible Splines

Shear Stress Under Roots of External Teeth: For a transmitted torque T, the torsional shear stress induced in the shaft under the root diameter of an external spline is:

$$S_s = \frac{16TK_a}{\pi D_{re}^3 K_f} \quad \text{for a solid shaft} \qquad (1)$$

$$S_s = \frac{16TD_{re}K_a}{\pi(D_{re}^4 - D_h^4)K_f} \quad \text{for a hollow shaft} \qquad (2)$$

The computed stress should not exceed the values in Table 5.

Table 5. Allowable Shear Stresses for Splines

Material	Hardness		Max. Allowable Shear Stress, psi
	Brinell	Rockwell C	
Steel	160–200	—	20,000
Steel	230–260	—	30,000
Steel	302–351	33–38	40,000
Surface-hardened Steel	—	48–53	40,000
Case-hardened Steel	—	58–63	50,000
Through-hardened Steel (Aircraft Quality)	—	42–46	45,000

Shear Stress at the Pitch Diameter of Teeth: The shear stress at the pitch line of the teeth for a transmitted torque T is:

$$S_s = \frac{4TK_aK_m}{DNL_etK_f} \qquad (3)$$

The factor of 4 in (3) assumes that only half the teeth will carry the load because of spacing errors. For poor manufacturing accuracies, change the factor to 6.

The computed stress should not exceed the values in Table 5.

Compressive Stresses on Sides of Spline Teeth: Allowable compressive stresses on splines are very much lower than for gear teeth since non-uniform load distribution and misalignment result in unequal load sharing and end loading of the teeth.

$$\text{For flexible splines,} \quad S_c = \frac{2TK_mK_a}{DNL_ehK_w} \qquad (4)$$

$$\text{For fixed splines,} \quad S_c = \frac{2TK_mK_a}{9DNL_ehK_f} \qquad (5)$$

In these formulas, h is the depth of engagement of the teeth, which for flat root splines is $0.9/P$ and for fillet root splines is $1/P$, approximately.

The stresses computed from Formulas (4) and (5) should not exceed the values in Table 6.

Table 6. Allowable Compressive Stresses for Splines

Material	Hardness		Max. Allowable Compressive Stress, psi	
	Brinell	Rockwell C	Straight	Crowned
Steel	160–200	—	1,500	6,000
Steel	230–260	—	2,000	8,000
Steel	302–351	33–38	3,000	12,000
Surface-hardened Steel	—	48–53	4,000	16,000
Case-hardened Steel	—	58–63	5,000	20,000

Bursting Stresses on Splines: Internal splines may burst due to three kinds of tensile stress: 1) tensile stress due to the radial component of the transmitted load; 2) centrifugal tensile stress; and 3) tensile stress due to the tangential force at the pitch line causing bending of the teeth.

$$\text{Radial load tensile stress, } S_1 = \frac{T \tan\phi}{\pi D t_w L} \tag{6}$$

where t_w = wall thickness of internal spline = outside diameter of spline sleeve minus spline major diameter, all divided by 2. L = full length of spline.

$$\text{Centrifugal tensile stress, } S_2 = \frac{1.656 \times (\text{rpm})^2 (D_{oi}^2 + 0.212 D_{ri}^2)}{1,000,000} \tag{7}$$

where D_{oi} = outside diameter of spline sleeve.

$$\text{Beam loading tensile stress, } S_3 = \frac{4T}{D^2 L_e Y} \tag{8}$$

In this equation, Y is the Lewis form factor obtained from a tooth layout. For internal splines of 30-deg. pressure angle a value of $Y = 1.5$ is a satisfactory estimate. The factor 4 in (8) assumes that only half the teeth are carrying the load.

The total tensile stress tending to burst the rim of the external member is: $S_t = [K_a K_m (S_1 + S_3) + S_2]/K_f$; and should be less than those in Table 7.

Table 7. Allowable Tensile Stresses for Splines

Material	Hardness		Max. Allowable Stress, psi
	Brinell	Rockwell C	
Steel	160–200	—	22,000
Steel	230–260	—	32,000
Steel	302–351	33–38	45,000
Surface-hardened Steel	—	48–53	45,000
Case-hardened Steel	—	58–63	55,000
Through-hardened Steel	—	42–46	50,000

Crowned Splines for Large Misalignments.—As mentioned on page 2142, crowned splines can accommodate misalignments of up to about 5 degrees. Crowned splineshave considerably less capacity than straight splines of the same size if both are operating with precise alignment. However, when large misalignments exist, the crowned spline has greater capacity.

American Standard tooth forms may be used for crowned external members so that they may be mated with straight internal members of Standard form.

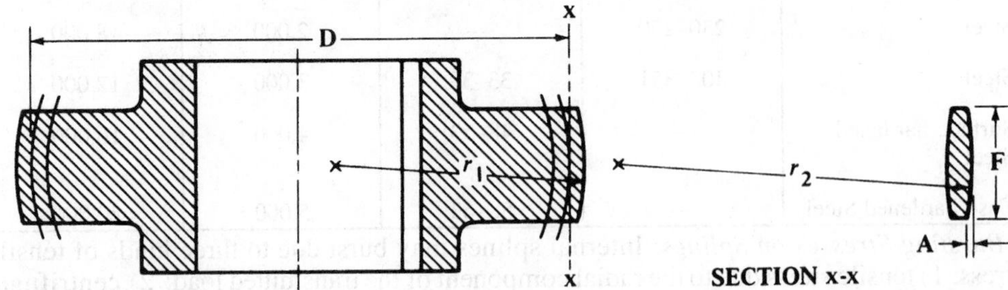

The accompanying diagram of a crowned spline shows the radius of the crown r_1; the radius of curvature of the crowned tooth, r_2; the pitch diameter of the spline, D; the face width, F; and the relief or crown height A at the ends of the teeth. The crown height A should always be made somewhat greater than one-half the face width multiplied by the tangent of the misalignment angle. For a crown height A, the approximate radius of curvature r_2 is $F^2 \div 8A$, and $r_1 = r_2 \tan \phi$, where ϕ is the pressure angle of the spline.

For a torque T, the compressive stress on the teeth is:

$$S_c = 2290\sqrt{2T \div DNhr_2};$$

and should be less than the value in Table 6.

Fretting Damage to Splines and Other Machine Elements.—Fretting is wear that occurs when cyclic loading, such as vibration, causes two surfaces in intimate contact to undergo small oscillatory motions with respect to each other. During fretting, high points or asperities of the mating surfaces adhere to each other and small particles are pulled out, leaving minute, shallow pits and a powdery debris. In steel parts exposed to air, the metallic debris oxidizes rapidly and forms a red, rustlike powder or sludge; hence, the coined designation "fretting corrosion."

Fretting is mechanical in origin and has been observed in most materials, including those that do not oxidize, such as gold, platinum, and nonmetallics; hence, the corrosion accompanying fretting of steel parts is a secondary factor.

Fretting can occur in the operation of machinery subject to motion or vibration or both. It can destroy close fits; the debris may clog moving parts; and fatigue failure may be accelerated because stress levels to initiate fatigue in fretted parts are much lower than for undamaged material. Sites for fretting damage include interference fits; splined, bolted, keyed, pinned, and riveted joints; between wires in wire rope; flexible shafts and tubes; between leaves in leaf springs; friction clamps; small amplitude-of-oscillation bearings; and electrical contacts.

Vibration or cyclic loadings are the main causes of fretting. If these factors cannot be eliminated, greater clamping force may reduce movement but, if not effective, may actually worsen the damage. Lubrication may delay the onset of damage; hard plating or surface hardening methods may be effective, not by reducing fretting, but by increasing the fatigue strength of the material. Plating soft materials having inherent lubricity onto contacting surfaces is effective until the plating wears through.

Involute Spline Inspection Methods.—Spline gages are used for routine inspection of production parts.

Analytical inspection, which is the measurement of individual dimensions and variations, may be required:

A) To supplement inspection by gages, for example, where NOT GO composite gages are used in place of NOT GO sector gages and variations must be controlled.

B) To evaluate parts rejected by gages.

C) For prototype parts or short runs where spline gages are not used.

D) To supplement inspection by gages where each individual variation must be restrained from assuming too great a portion of the tolerance between the minimum material actual and the maximum material effective dimensions.

Inspection with Gages.— A variety of gages is used in the inspection of involute splines.

Types of Gages: A composite spline gage has a full complement of teeth. A sector spline gage has two diametrically opposite groups of teeth. A sector plug gage with only two teeth per sector is also known as a "paddle gage." A sector ring gage with only two teeth per sector is also known as a "snap ring gage." A progressive gage is a gage consisting of two or more adjacent sections with different inspection functions. Progressive GO gages are physical combinations of GO gage members that check consecutively first one feature or one group of features, then their relationship to other features. GO and NOT GO gages may also be combined physically to form a progressive gage.

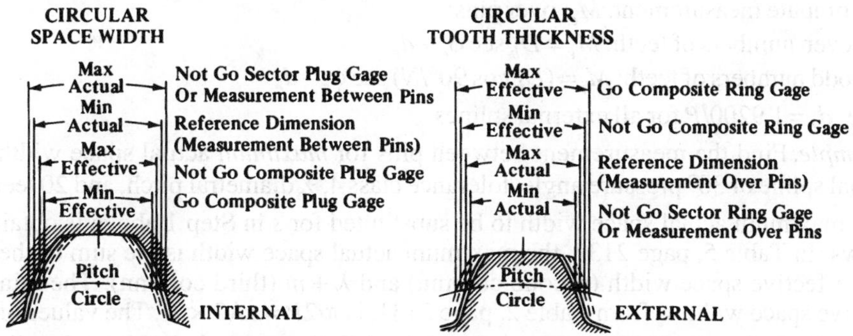

Fig. 3. Space width and tooth-thickness inspection.

GO and NOT GO Gages: GO gages are used to inspect maximum material conditions (maximum external, minimum internal dimensions). They may be used to inspect an individual dimension or the relationship between two or more functional dimensions. They control the minimum looseness or maximum interference.

NOT GO gages are used to inspect minimum material conditions (minimum external, maximum internal dimensions), thereby controlling the maximum looseness or minimum interference. Unless otherwise agreed upon, a product is acceptable only if the NOT GO gage does not enter or go on the part. A NOT GO gage can be used to inspect only one dimension. An attempt at simultaneous NOT GO inspection of more than one dimension could result in failure of such a gage to enter or go on (acceptance of part), even though all but one of the dimensions were outside product limits. In the event all dimensions are outside the limits, their relationship could be such as to allow acceptance.

Effective and Actual Dimensions: The effective space width and tooth thickness are inspected by means of an accurate mating member in the form of a composite spline gage.

The actual space width and tooth thickness are inspected with sector plug and ring gages, or by measurements with pins.

Measurements with Pins.— The actual space width of internal splines, and the actual tooth thickness of external splines, may be measured with pins. These measurements do not determine the fit between mating parts, but may be used as part of the analytic inspection of splines to evaluate the effective space width or effective tooth thickness by approximation.

Formulas for 2-Pin Measurement: For measurement *between* pins of internal splines using the symbols given on page 2130:

1) Find involute of pressure angle at pin center:

$$\operatorname{inv}\phi_i = s/D + \operatorname{inv}\phi_d - d_i/D_b$$

2) Find the value of ϕ_i, in degrees, in the involute function tables beginning on page 19. Find sec ϕ_i = 1/cosine ϕ_i in the trig tables, pages 16 through 18, using interpolation to obtain higher accuracy.

3) Compute measurement, M_i, between pins:

For even numbers of teeth: $M_i = D_b \sec\phi_i - d_i$

For odd numbers of teeth: $M_i = (D_b \cos 90°/N) \sec\phi_i - d_i$

where: $d_i = 1.7280/P$ for 30° and 37.5° standard pressure angle (ϕ_D) splines

$d_i = 1.9200/P$ for 45° pressure angle splines

For measurement *over* pins of external splines:

1) Find involute of pressure angle at pin center:

$$\operatorname{inv}\phi_e = t/D + \operatorname{inv}\phi_D + d_e/D_b - \pi/N$$

2) Find the value of ϕ_e and sec ϕ_e from the involute function tables beginning on page 19.

3) Compute measurement, M_e, over pins:

For even numbers of teeth: $M_e = D_b \sec\phi_e + d_e$

For odd numbers of teeth: $M_i = (D_b \cos 90°/N) \sec\phi_e - d_e$

where $d_e = 1.9200/P$ for all external splines

Example: Find the measurement between pins for *maximum* actual space width of an internal spline of 30° pressure angle, tolerance class 4, 3/6 diametral pitch, and 20 teeth.

The maximum actual space width to be substituted for *s* in Step 1 above is obtained as follows: In Table 5, page 2137, the maximum actual space width is the sum of the minimum effective space width (second column) and $\lambda + m$ (third column). The minimum effective space width s_v from Table 2, page 2131, is $\pi/2P = \pi/(2 \times 3)$. The values of λ and *m* from Table 4, page 2134, are, for a class 4 fit, 3/6 diametral pitch, 20-tooth spline: $\lambda = 0.0027 \times 0.71 = 0.00192$; and $m = 0.00176 \times 0.71 = 0.00125$, so that $s = 0.52360 + 0.00192 + 0.00125 = 0.52677$.

Other values required for Step 1 are:

$D = N/P = 20/3 = 6.66666$

$inv\phi_D$ = inv 30° = 0.053751 from a calculator

$d_i = 1.7280/3 = 0.57600$

$D_b = D \cos\phi_D = 6.66666 \times 0.86603 = 5.77353$

The computation is made as follows:

1) inv $\phi_i = 0.52677/6.66666 + 0.053751 - 0.57600/5.77353 = 0.03300$

2) From a calculator, $\phi_i = 25°46.18'$ and sec $\phi_i = 1.11044$

3) $M_i = 5.77353 \times 1.11044 - 0.57600 = 5.8352$ inches

American National Standard Metric Module Splines.—ANSI B92.2M-1980 (R1989) is the American National Standards Institute version of the International Standards Organization involute spline standard. It is not a "soft metric" conversion of any previous, inch-based, standard,* and splines made to this hard metric version are not intended for use with components made to the B92.1 or other, previous standards. The ISO 4156 Standard from

* A "soft" conversion is one in which dimensions in inches, when multiplied by 25.4 will, after being appropriately rounded off, provide equivalent dimensions in millimeters. In a "hard" system the tools of production, such as hobs, do not bear a usable relation to the tools in another system; i.e., a 10 diametral pitch hob calculates to be equal to a 2.54 module hob in the metric module system, a hob that does not exist in the metric standard.

which this one is derived is the result of a cooperative effort between the ANSI B92 committee and other members of the ISO/TC 14-2 involute spline committee.

Many of the features of the previous standard, ANSI B92.1-1970 (R1993), have been retained such as: 30-, 37.5-, and 45-degree pressure angles; flat root and fillet root side fits; the four tolerance classes 4, 5, 6, and 7; tables for a single class of fit; and the effective fit concept.

Among the major differences are: use of modules of from 0.25 through 10 mm in place of diametral pitch; dimensions in millimeters instead of inches; the "basic rack"; removal of the major diameter fit; and use of ISO symbols in place of those used previously. Also, provision is made for calculating three defined clearance fits.

The Standard recognizes that proper assembly between mating splines is dependent only on the spline being within effective specifications from the tip of the tooth to the form diameter. Therefore, the internal spline major diameter is shown as a maximum dimension and the external spline minor diameter is shown as a minimum dimension. The minimum internal major diameter and the maximum external minor diameter must clear the specified form diameter and thus require no additional control. All dimensions are for the finished part; any compensation that must be made for operations that take place during processing, such as heat treatment, must be considered when selecting the tolerance level for manufacturing.

The Standard provides the same internal minimum effective space width and external maximum effective tooth thickness for all tolerance classes. This basic concept makes possible interchangeable assembly between mating splines regardless of the tolerance class of the individual members, and permits a tolerance class "mix" of mating members. This arrangement is often an advantage when one member is considerably less difficult to produce than its mate, and the "average" tolerance applied to the two units is such that it satisfies the design need. For example, by specifying Class 5 tolerance for one member and Class 7 for its mate, an assembly tolerance in the Class 6 range is provided.

If a fit given in this Standard does not satisfy a particular design need, and a specific clearance or press fit is desired, the change shall be made only to the external spline by a reduction of, or an increase in, the effective tooth thickness and a like change in the actual tooth thickness. The minimum effective space width is always basic and this basic width should always be retained when special designs are derived from the concept of this Standard.

Spline Terms and Definitions: The spline terms and definitions given for American National Standard ANSI B92.1-1970 (R1993) described in the preceding section, may be used in regard to ANSI B92.2M-1980 (R1989). The 1980 Standard utilizes ISO symbols in place of those used in the 1970 Standard; these differences are shown in Table 1.

Table 1. Comparison of Symbols Used in *ANSI B92.2M-1980 (R1989)* **and Those in** *ANSI B92.1-1970, R1993*

Symbol			Symbol		
B92.2M	B92.1	Meaning of Symbol	B92.2M	B92.1	Meaning of Symbol
c	…	theoretical clearance	m	…	module
c_v	c_v	effective clearance	…	P	diametral pitch
c_F	c_F	form clearance	…	P_s	stub pitch = $2P$
D	D	pitch diameter	P_b	…	base pitch
DB	D_b	base diameter	p	p	circular pitch
d_{ce}	D_{ce}	pin contact diameter, external spline	π	π	3.141592654
d_{ci}	D_{ci}	pin contact diameter, internal spline	rfe	r_f	fillet rad., ext. spline
DEE	D_o	major diam., ext. spline	rfi	r_f	fillet rad., int. spline
DEI	D_{ri}	major diam., int. spline	E_{bsc}	s_v min	basic circular space width
DFE	D_{Fe}	form diam., ext. spline	E_{max}	s	max. actual circular space width
DFI	D_{Fi}	form diam., int. spline	E_{min}	s	min. actual circular space width
DIE	D_{re}	minor diam., ext. spline	EV	s_v	effective circular space width
DII	D_i	minor diam., int. spline	S_{bsc}	t_v max	basic circular tooth thickness
DRE	d_e	pin diam., ext. spline	S_{max}	t	max. actual circular tooth thick.
DRI	d_i	pin diam., int. spline	S_{min}	t	min. actual circular tooth thick.
h_s	…	see Figs. 1a, 1b, 1c, and 1d	SV	t_v	effective circular tooth thick.
λ	λ	effective variation	α	ϕ	pressure angle
INV α	…	involute $\alpha = \tan\alpha - \text{arc}\ \alpha$	α_D	ϕ_D	standard pressure angle
KE	K_e	change factor, ext. spline	α_{ci}	ϕ_{ci}	press. angle at pin contact diameter, internal spline
KI	K_i	change factor, int. spline	α_{ce}	ϕ_{ce}	press. angle at pin contact diameter, external spline
g	L	spline length	α_i	ϕ_i	press. angle at pin center, internal spline
g_w	…	active spline length	α_e	ϕ_e	press. angle at pin center, external spline
$g\gamma$	…	length of engagement	α_{Fe}	ϕ_F	press. angle at form diameter, external spline
T	m	machining tolerance	α_{Fi}	ϕ_F	press. angle at form diameter, internal spline
MRE	M_e	meas. over 2 pins, ext. spline	es	…	ext. spline cir. tooth thick.modification for required fit class=c_v min (Table 3)
MRI	M_i	meas. bet. 2 pins, int. spline	h, f, e, or d	…	tooth thick, size modifiers (called fundamental deviation in ISO R286), Table 3
Z	N	number of teeth	H	…	space width size modifier (called fundamental deviation in ISO R286), Table 3

Dimensions and Tolerances: Dimensions and tolerances of splines made to the 1980 Standard may be calculated using the formulas given in Table 2. These formulas are for metric module splines in the range of from 0.25 to 10 mm metric module of side-fit design and having pressure angles of 30-, 37.5-, and 45-degrees. The standard modules in the system are: 0.25; 0.5; 0.75; 1; 1.25; 1.5; 1.75; 2; 2.5; 3; 4; 5; 6; 8; and 10. The range of from 0.5 to 10 module applies to all splines except 45-degree fillet root splines; for these, the range of from 0.25 to 2.5 module applies.

Table 2. Formulas for Dimensions and Tolerances for All Fit Classes—Metric Module Involute Splines

Term	Symbol	Formula			
		30-Degree Flat Root	30-Degree Fillet Root	37.5-Degree Fillet Root	45-Degree Fillet Root
		0.5 to 10 module	0.5 to 10 module	0.5 to 10 module	0.25 to 2.5 module
Pitch Diameter	D	mZ			
Base Diameter	DB	$mZ \cos \alpha_D$			
Circular Pitch	p	πm			
Base Pitch	p_b	$\pi m \cos \alpha_D$			
Tooth Thick Mod	es	According to selected fit class, H/h, H/f, H/e, or H/d (see Table 3)			
Min Maj. Diam. Int	DEI min	$m(Z+1.5)$	$m(Z+1.8)$	$m(Z+1.4)$	$m(Z+1.2)$
Max Maj Diam. Int.	DEI max	DEI min $+ (T+\lambda)/\tan \alpha_D$ (see Note 1)			
Form Diam, Int.	DFI	$m(Z+1)+2c_F$	$m(Z+1)+2c_F$	$m(Z+0.9)+2c_F$	$m(Z+0.8)+2c_F$
Min Minor Diam, Int	DII min	$DFE + 2c_F$ (see Note 2)			
Max Minor Diam, Int	DII max	DII min $+ (0.2m^{0.667} - 0.01m^{-0.5})$[a]			
Cir Space Width,					
Basic	E_{bsc}	$0.5\pi m$			
Min Effective	EV min	$0.5\pi m$			
Max Actual	E max	EV min $+ (T+\lambda)$ for classes 4, 5, 6, and 7 (see Table 4 for $T+\lambda$)			
Min Actual	E min	EV min $+ \lambda$ (see text on page 2153 for λ)			
Max Effective	EV max	E max $- \lambda$ (see text on page 2153 for λ)			
Max Major Diam, Ext	DEE max	$m(Z+1) - es/\tan \alpha_D$[b]	$m(Z+1) - es/\tan \alpha_D$[b]	$m(Z+0.9) - es/\tan \alpha_D$[b]	$m(Z+0.8) - es/\tan \alpha_D$[b]
Min Major Diam. Ext	DEE min	DEE max $- (0.2m^{0.667} - 0.01m^{-0.5})$[a]			
Form Diam, External	DFE	$2 \times \sqrt{(0.5DB)^2 + \left[0.5D\sin\alpha_D - \frac{h_s + \left(\frac{0.5es}{\tan\alpha_D}\right)}{\sin\alpha_D}\right]^2}$			
Max Minor Diam, Ext	DIE max	$m(Z-1.5) - es/\tan \alpha_D$[b]	$m(Z-1.8) - es/\tan \alpha_D$[b]	$m(Z-1.4) - es/\tan \alpha_D$[b]	$m(Z-1.2) - es/\tan \alpha_D$[b]

Table 2. *(Continued)* **Formulas for Dimensions and Tolerances for All Fit Classes—Metric Module Involute Splines**

Term	Symbol	Formula			
		30-Degree Flat Root	30-Degree Fillet Root	37.5-Degree Fillet Root	45-Degree Fillet Root
		0.5 to 10 module	0.5 to 10 module	0.5 to 10 module	0.25 to 2.5 module
Min Minor Diam, Ext	*DIE* min	DIE max − $(T + \lambda)/\tan \alpha_D$ (see Note 1)			
Cir Tooth Thick,					
Basic	S_{bsc}	$0.5\pi m$			
Max Effective	*SV* max	$S_{bsc} - es$			
Min Actual	*S* min	*SV* max − $(T + \lambda)$ for classes 4, 5, 6, and 7 (see Table 4 for $T + \lambda$)			
Max Actual	*S* max	*SV* max − λ (see text on page 2153 for λ)			
Min Effective	*SV* min	*S* min + λ (see text on page 2153 for λ)			
Total Tolerance on Circular Space Width or Tooth Thickness	$(T + \lambda)$	See formulas in Table 4			
Machining Tolerance on Circular Space Width or Tooth Thickness	T	$T = (T + \lambda)$ from Table 4 − λ from text on page 2153.			
Effective Variation Allowed on Circular Space Width or Tooth Thickness	λ	See text on page 2153.			
Form Clearance	c_F	$0.1m$			
Rack Dimension	h_s	$0.6m$(see Fig. 1a)	$0.6m$(see Fig. 1b)	$0.55m$(see Fig. 1c)	$0.5m$(see Fig. 1d)

[a] Values of $(0.2m^{0.667} - 0.01m^{-0.5})$ are as follows: for 10 module, 0.93; for 8 module, 0.80; for 6 module, 0.66; for 5 module, 0.58; for 4 module, 0.50; for 3 module, 0.41; for 2.5 module, 0.36; for 2 module, 0.31; for 1.75 module, 0.28; for 1.5 module, 0.25; for 1.25 module, 0.22; for 1 module, 0.19; for 0.75 module, 0.15; for 0.5 module, 0.11; and for 0.25 module, 0.06.

[b] See Table 6 for values of $es/\tan \alpha_D$.

Note 1: Use $(T + \lambda)$ for class 7 from Table 4.

Note 2: For all types of fit, always use the *DFE* value corresponding to the H/h fit.

Fit Classes: Four classes of side fit splines are provided: spline fit class H/h having a minimum effective clearance, $c_v = es = 0$; classes H/f, H/e, and H/d having tooth thickness modifications, *es*, of f, e, and d, respectively, to provide progressively greater effective clearance c_v, The tooth thickness modifications h, f, e, and d in Table 3 are fundamental deviations selected from ISO R286, "ISO System of Limits and Fits." They are applied to the external spline by shifting the tooth thickness total tolerance below the basic tooth thickness by the amount of the tooth thickness modification to provide a prescribed minimum effective clearance c_v.

Table 3. Tooth Thickness Modification, *es*, for Selected Spline Fit Classes

Pitch Diameter in mm, *D*	External Splines[a] Selected Fit Class d	e	f	h	Pitch Diameter in mm, *D*	External Splines[a] Selected Fit Class d	e	f	h
	Tooth Thickness Modification (Reduction) Relative to Basic Tooth Thickness at Pitch Diameter, *es*, in mm					Tooth Thickness Modification (Reduction) Relative to Basic Tooth Thickness at Pitch Diameter, *es*, in mm			
≤ 3	0.020	0.014	0.006	0	> 120 to 180	0.145	0.085	0.043	0
> 3 to 6	0.030	0.020	0.010	0	> 180 to 250	0.170	0.100	0.050	0
> 6 to 10	**0.040**	0.025	0.013	0	> 250 to 315	0.190	0.110	0.056	0
> 10 to 18	0.050	0.032	0.016	0	> 315 to 400	0.210	**0.125**	**0.062**	0
> 18 to 30	0.065	0.040	0.020	0	> 400 to 500	0.230	0.135	**0.068**	0
> 30 to 50	0.080	0.050	0.025	0	> 500 to 630	0.260	0.145	**0.076**	0
> 50 to 80	0.100	0.060	0.030	0	> 630 to 800	0.290	0.160	**0.080**	0
> 80 to 120	0.120	**0.072**	0.036	0	> 800 to 1000	0.320	**0.170**	**0.086**	0

[a] Internal splines are fit class H and have space width modification from basic space width equal to zero; thus, an H/h fit class has effective clearance $c_v = 0$.

Note: The values listed in this table are taken from ISO R286 and have been computed on the basis of the geometrical mean of the size ranges shown. Values in **boldface** type do not comply with any documented rule for rounding but are those used by ISO R286; they are used in this table to comply with established international practice.

Basic Rack Profiles: The basic rack profile for the standard pressure angle splines are shown in see Fig. 1a, 1b, 1c, and 1d. The dimensions shown are for maximum material condition and for fit class H/h.

Spline Machining Tolerances and Variations.—The total tolerance $(T + \lambda)$, Table 4, is the sum of Effective Variation, λ, and a Machining Tolerance, T.

Table 4. Space Width and Tooth Thickness Total Tolerance, $(T + \lambda)$, in Millimeters

Spline Tolerance Class	Formula for Total Tolerance, $(T+\lambda)$	Spline Tolerance Class	Formula for Total Tolerance, $(T+\lambda)$	In these formulas, i* and i** are tolerance units based upon pitch diameter and tooth thickness, respectively:
4	10i* + 40i**	6	25i* + 100i**	$i^* = 0.001(0.45\sqrt[3]{D} + 0.001D)$ for $D \leq 500$mm
5	16i* + 64i**	7	40i* + 160i**	$= 0.001(0.004D + 2.1)$ for $D > 500$mm
				$i^{**} = 0.001(0.45\sqrt[3]{S_{bsc}} + 0.001S_{bsc})$

Effective Variation: The effective variation, λ, is the combined effect that total index variation, positive profile variation, and tooth alignment variation has on the effective fit of mating involute splines. The effect of the individual variations is less than the sum of the allowable variations because areas of more than minimum clearance can have profile, tooth alignment, or index variations without changing the fit. It is also unlikely that these variations would occur in their maximum amounts simultaneously on the same spline. For this reason, total index variation, total profile variation, and tooth alignment variation are used to calculate the combined effect by the following formula:

$$\lambda = 0.6\sqrt{(F_p)^2 + (f_f)^2 + (F_\beta)^2} \text{ millimeters}$$

The above variation is based upon a length of engagement equal to one-half the pitch diameter of the spline; adjustment of λ may be required for a greater length of engagement. Formulas for values of F_p, f_f, and F_β used in the above formula are given in Table 5.

Table 5. Formulas for F_p, f_f, and F_β used to calculate λ

Spline Toler-ance Class	Total Index Variation, in mm, F_p	Total Profile Variation, in mm, f_f	Total Lead Variation, in mm, F_β
4	$0.001(2.5\sqrt{mZ\pi/2}+6.3)$	$0.001\,[1.6m(1+0.0125Z)+10]$	$0.001(0.8\sqrt{g}+4)$
5	$0.001(3.55\sqrt{mZ\pi/2}+9)$	$0.001\,[2.5m(1+0.0125Z)+16]$	$0.001(1.0\sqrt{g}+5)$
6	$0.001(5\sqrt{mZ\pi/2}+12.5)$	$0.001\,[4m(1+0.0125Z)+25]$	$0.001(1.25\sqrt{g}+6.3)$
7	$0.001(7.1\sqrt{mZ\pi/2}+18)$	$0.001\,[6.3m(1+0.0125Z)+40]$	$0.001(2\sqrt{g}+10)$

g = length of spline in millimeters.

Table 6. Reduction, $es/\tan\alpha_D$, of External Spline Major and Minor Diameters Required for Selected Fit Classes

Pitch Diameter D in mm	Standard Pressure Angle, in Degrees									
	30	37.5	45	30	37.5	45	30	37.5	45	All
	Classes of Fit									
	d			e			f			h
	$es/\tan\alpha_D$ in millimeters									
≤3	0.035	0.026	0.020	0.024	0.018	0.014	0.010	0.008	0.006	0
>3 to 6	0.052	0.039	0.030	0.035	0.026	0.020	0.017	0.013	0.010	0
> 6 to 10	0.069	0.052	0.040	0.043	0.033	0.025	0.023	0.017	0.013	0
> 10 to 18	0.087	0.065	0.050	0.055	0.042	0.032	0.028	0.021	0.016	0
> 18 to 30	0.113	0.085	0.065	0.069	0.052	0.040	0.035	0.026	0.020	0
> 30 to 50	0.139	0.104	0.080	0.087	0.065	0.050	0.043	0.033	0.025	0
> 50 to 80	0.173	0.130	0.100	0.104	0.078	0.060	0.052	0.039	0.030	0
> 80 to 120	0.208	0.156	0.120	0.125	0.094	0.072	0.062	0.047	0.036	0
> 120 to 180	0.251	0.189	0.145	0.147	0.111	0.085	0.074	0.056	0.043	0
> 180 to 250	0.294	0.222	0.170	0.173	0.130	0.100	0.087	0.065	0.050	0
> 250 to 315	0.329	0.248	0.190	0.191	0.143	0.110	0.097	0.073	0.056	0
> 315 to 400	0.364	0.274	0.210	0.217	0.163	0.125	0.107	0.081	0.062	0
> 400 to 500	0.398	0.300	0.230	0.234	0.176	0.135	0.118	0.089	0.068	0
> 500 to 630	0.450	0.339	0.260	0.251	0.189	0.145	0.132	0.099	0.076	0
> 630 to 800	0.502	0.378	0.290	0.277	0.209	0.160	0.139	0.104	0.080	0
> 800 to 1000	0.554	0.417	0.320	0.294	0.222	0.170	0.149	0.112	0.086	0

These values are used with the applicable formulas in Table 2.

Machining Tolerance: A value for machining tolerance may be obtained by subtracting the effective variation, λ, from the total tolerance ($T + \lambda$). Design requirements or specific processes used in spline manufacture may require a different amount of machining tolerance in relation to the total tolerance.

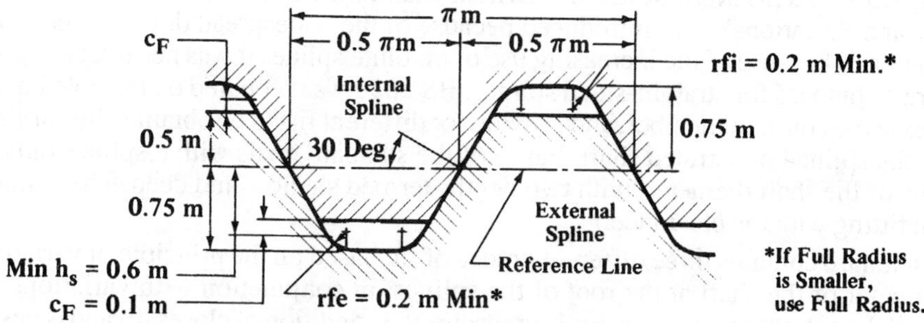

Fig. 1a. Profile of Basic Rack for 30° Flat Root Spline

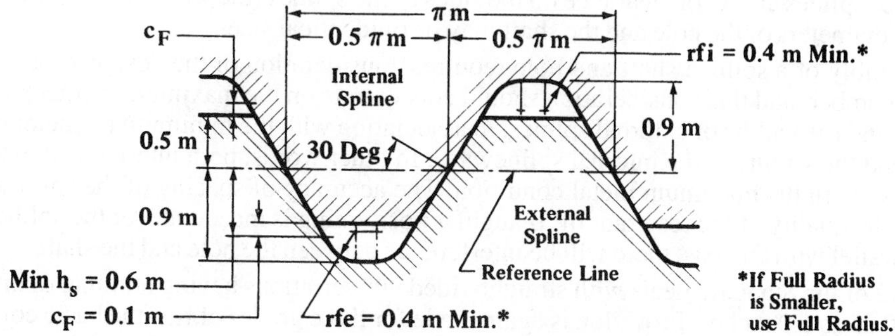

Fig. 1b. Profile of Basic Rack for 30° Fillet Root Spline

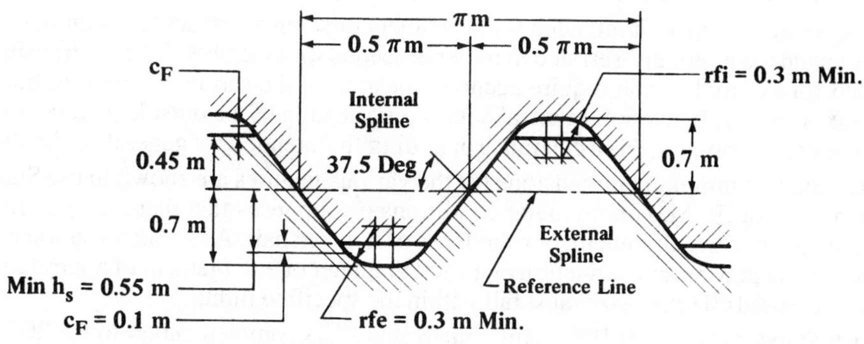

Fig. 1c. Profile of Basic Rack for 37.5° Fillet Root Spline

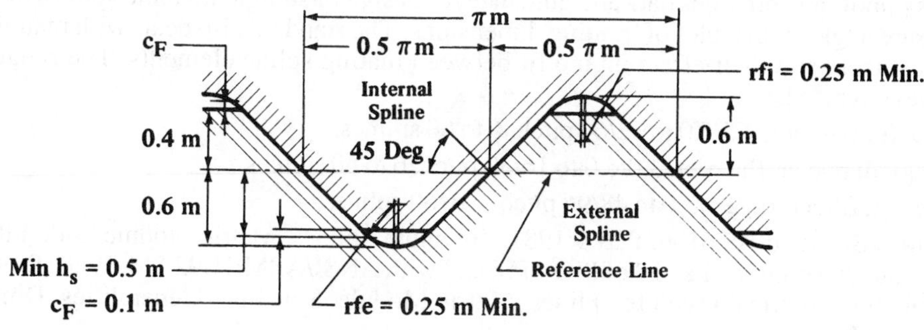

Fig. 1d. Profile of Basic Rack for 45° Fillet Root Spline

British Standard Striaght Splines.—British Standard BS 2059:1953, "Straight-sided Splines and Serrations", was introduced because of the widespread development and use of splines and because of the increasing use of involute splines it was necessary to provide a separate standard for straight-sided splines. BS 2059 was prepared on the hole basis, the hole being the constant member, and provide for different fits to be obtained by varying the size of the splined or serrated shaft. Part 1 of the standard deals with 6 splines only, irrespective of the shaft diameter, with two depths termed shallow and deep. The splines are bottom fitting with top clearance.

The standard contains three different grades of fit, based on the principle of variations in the diameter of the shaft at the root of the splines, in conjunction with variations in the widths of the splines themselves. Fit 1 represents the condition of closest fit and is designed for minimum backlash. Fit 2 has a positive allowance and is designed for ease of assembly, and Fit 3 has a larger positive allowance for applications that can accept such clearances. all these splines allow for clearance on the sides of the splines (the widths), but in Fit 1, the minor diameters of the hole and the shaft may be of identical size.

Assembly of a splined shaft and hole requires consideration of the designed profile of each member, and this consideration should concentrate on the maximum diameter of the shafts and the widths of external splines, in association with the minimum diameter of the hole and the widths of the internal splineways. In other words, both internal and external splines are in the maximum metal condition. The accuracy of spacing of the splines will affect the quality of the resultant fit. If angular positioning is inaccurate, or the splines are not parallel with the axis, there will be interference between the hole and the shaft.

Part 2 of the Standard deals with straight-sided 90° serrations having nominal diameters from 0.25 to 6.0 inches. Provision is again made for three grades of fits, the basic constant being the serrated hole size. Variations in the fits of these serrations is obtained by varying the sizes of the serrations on the shaft, and the fits are related to flank bearing, the depth of engagement being constant for each size and allowing positive clearance at crest and root.

Fit 1 is an interference fit intended for permanent or semi-permanent ass emblies. Heating to expand the internally-serrated member is needed for assembly. Fit 2 is a transition fit intended for assemblies that require accurate location of the serrated members, but must allow disassembly. In maximum metal conditions, heating of the outside member may be needed for assembly. Fit. 3 is a clearance or sliding fit, intended for general applications.

Maximum and minimum dimensions for the various features are shown in the Standard for each class of fit. Maximum metal conditions presupposes that there are no errors of form such as spacing, alignment, or roundness of hole or shaft. Any compensation needed for such errors may require reduction of a shaft diameter or enlargement of a serrated bore, but the measured effective size must fall within the specified limits.

British Standard BS 3550:1963, "Involute Splines", is complementary to BS 2059, and the basic dimensions of all the sizes of splines are the same as those in the ANSI/ASME B5.15-1960, for major diameter fit and side fit. The British Standard uses the same terms and symbols and provides data and guidance for design of straight involute splines of 30° pressure angle, with tables of limiting dimensions. The standard also deals with manufacturing errors and their effect on the fit between mating spline elements. The range of splines covered is:

Side fit, flat root, 2.5/5.0 to 32/64 pitch, 6 to 60 splines.

Major diameter, flat root, 3.0/6.0 to 16/32 pitch, 6 to 60 splines.

Side fit, fillet root, 2.5/5.0 to 48/96 pitch, 6 to 60 splines.

British Standard BS 6186, Part 1:1981, "Involute Splines, Metric Module, Side Fit" is identical with sections 1 and 2 of ISO 4156 and with ANSI/ASME B92.2M-1980 (R1989) "Straight Cylindrical Involute Splines, Metric Module, Side Fit – Generalities, Dimensions and Inspection".

Table 1. S.A.E. Standard Splined Fittings

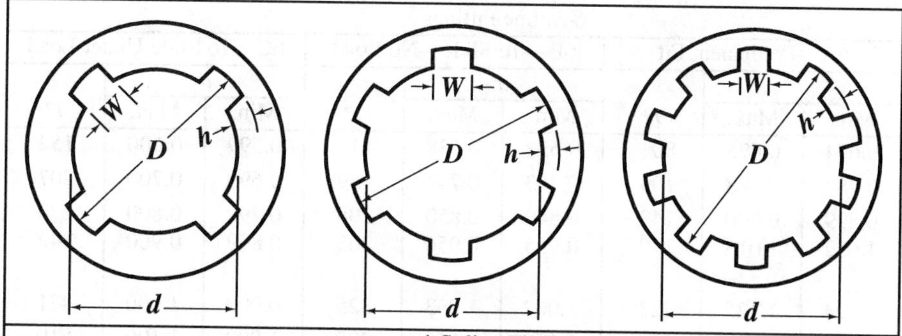

Nom. Diam	4-Spline Fittings: For All Fits, *D* Min.	*D* Max.	*W* Min.	*W* Max.	4A—Permanent Fit, *d* Min.	*d* Max.	*h* Min.	*h* Max.	T^a
¾	0.749	0.750	0.179	0.181	0.636	0.637	0.055	0.056	78
⅞	0.874	0.875	0.209	0.211	0.743	0.744	0.065	0.066	107
1	0.999	1.000	0.239	0.241	0.849	0.850	0.074	0.075	139
1⅛	1.124	1.125	0.269	0.271	0.955	0.956	0.083	0.084	175
1¼	1.249	1.250	0.299	0.301	1.061	1.062	0.093	0.094	217
1⅜	1.374	1.375	0.329	0.331	1.168	1.169	0.102	0.103	262
1½	1.499	1.500	0.359	0.361	1.274	1.275	0.111	0.112	311
1⅝	1.624	1.625	0.389	0.391	1.380	1.381	0.121	0.122	367
1¾	1.749	1.750	0.420	0.422	1.486	1.487	0.130	0.131	424
2	1.998	2.000	0.479	0.482	1.698	1.700	0.148	0.150	555
2¼	2.248	2.250	0.539	0.542	1.910	1.912	0.167	0.169	703
2½	2.498	2.500	0.599	0.602	2.123	2.125	0.185	0.187	865
3	2.998	3.000	0.720	0.723	2.548	2.550	0.223	0.225	1249

Nom. Diam.	4-Spline Fittings: 4B—To Slide—No Load, *d* Min.	*d* Max.	*h* Min.	*h* Max.	T^a	6-Spline Fittings: For All Fits, *D* Min.	*D* Max.	*W* Min.	*W* Max.
¾	0.561	0.562	0.093	0.094	123	0.749	0.750	0.186	0.188
⅞	0.655	0.656	0.108	0.109	167	0.874	0.875	0.217	0.219
1	0.749	0.750	0.124	0.125	219	0.999	1.000	0.248	0.250
1⅛	0.843	0.844	0.140	0.141	277	1.124	1.125	0.279	0.281
1¼	0.936	0.937	0.155	0.156	341	1.249	1.250	0.311	0.313
1⅜	1.030	1.031	0.171	0.172	414	1.374	1.375	0.342	0.344
1½	1.124	1.125	0.186	0.187	491	1.499	1.500	0.373	0.375
1⅝	1.218	1.219	0.202	0.203	577	1.624	1.625	0.404	0.406
1¾	1.311	1.312	0.218	0.219	670	1.749	1.750	0.436	0.438
2	1.498	1.500	0.248	0.250	875	1.998	2.000	0.497	0.500
2¼	1.685	1.687	0.279	0.281	1106	2.248	2.250	0.560	0.563
2½	1.873	1.875	0.310	0.312	1365	2.498	2.500	0.622	0.625
3	2.248	2.250	0.373	0.375	1969	2.998	3.000	0.747	0.750

[a] See note at end of Table 4.

Table 2. S.A.E. Standard Splined Fittings

6-Spline Fittings									
Nom. Diam.	6A—Permanent Fit			6B—To Slide—No Load			6C—To Slide Under Load		
	d			*d*			*d*		
	Min.	Max.	*T*[a]	Min.	Max.	*T*[a]	Min.	Max.	*T*[a]
¾	0.674	0.675	80	0.637	0.638	117	0.599	0.600	152
⅞	0.787	0.788	109	0.743	0.744	159	0.699	0.700	207
1	0.899	0.900	143	0.849	0.850	208	0.799	0.800	270
1⅛	1.012	1.013	180	0.955	0.956	263	0.899	0.900	342
1¼	1.124	1.125	223	1.062	1.063	325	0.999	1.000	421
1⅜	1.237	1.238	269	1.168	1.169	393	1.099	1.100	510
1½	1.349	1.350	321	1.274	1.275	468	1.199	1.200	608
1⅝	1.462	1.463	376	1.380	1.381	550	1.299	1.300	713
1¾	1.574	1.575	436	1.487	1.488	637	1.399	1.400	827
2	1.798	1.800	570	1.698	1.700	833	1.598	1.600	1080
2¼	2.023	2.025	721	1.911	1.913	1052	1.798	1.800	1367
2½	2.248	2.250	891	2.123	2.125	1300	1.998	2.000	1688
3	2.698	2.700	1283	2.548	2.550	1873	2.398	2.400	2430

[a] See note at end of Table 4.

10-Spline Fittings							
Nom. Diam.	For All Fits				10A—Permanent Fit		
	D		*W*		*d*		
	Min.	Max.	Min.	Max.	Min.	Max.	*T*[a]
¾	0.749	0.750	0.115	0.117	0.682	0.683	120
⅞	0.874	0.875	0.135	0.137	0.795	0.796	165
1	0.999	1.000	0.154	0.156	0.909	0.910	215
1⅛	1.124	1.125	0.174	0.176	1.023	1.024	271
1¼	1.249	1.250	0.193	0.195	1.137	1.138	336
1⅜	1.374	1.375	0.213	0.215	1.250	1.251	406
1½	1.499	1.500	0.232	0.234	1.364	1.365	483
1⅝	1.624	1.625	0.252	0.254	1.478	1.479	566
1¾	1.749	1.750	0.271	0.273	1.592	1.593	658
2	1.998	2.000	0.309	0.312	1.818	1.820	860
2¼	2.248	2.250	0.348	0.351	2.046	2.048	1088
2½	2.498	2.500	0.387	0.390	2.273	2.275	1343
3	2.998	3.000	0.465	0.468	2.728	2.730	1934
3½	3.497	3.500	0.543	0.546	3.182	3.185	2632
4	3.997	4.000	0.621	0.624	3.637	3.640	3438
4½	4.497	4.500	0.699	0.702	4.092	4.095	4351
5	4.997	5.000	0.777	0.780	4.547	4.550	5371
5½	5.497	5.500	0.855	0.858	5.002	5.005	6500
6	5.997	6.000	0.933	0.936	5.457	5.460	7735

Table 3. S.A.E. Standard Splined Fittings

Nom. Diam.	10-Spline Fittings					
	10B—To Slide—No Load			10C—To Slide Under Load		
	d			*d*		
	Min.	Max.	*T*[a]	Min.	Max.	*T*[a]
¾	0.644	0.645	183	0.607	0.608	241
⅞	0.752	0.753	248	0.708	0.709	329
1	0.859	0.860	326	0.809	0.810	430
1⅛	0.967	0.968	412	0.910	0.911	545
1¼	1.074	1.075	508	1.012	1.013	672
1⅜	1.182	1.183	614	1.113	1.114	813
1½	1.289	1.290	732	1.214	1.215	967
1⅝	1.397	1.398	860	1.315	1.316	1135
1¾	1.504	1.505	997	1.417	1.418	1316
2	1.718	1.720	1302	1.618	1.620	1720
2¼	1.933	1.935	1647	1.821	1.823	2176
2½	2.148	2.150	2034	2.023	2.025	2688
3	2.578	2.580	2929	2.428	2.430	3869
3½	3.007	3.010	3987	2.832	2.835	5266
4	3.437	3.440	5208	3.237	3.240	6878
4½	3.867	3.870	6591	3.642	3.645	8705
5	4.297	4.300	8137	4.047	4.050	10746
5½	4.727	4.730	9846	4.452	4.455	13003
6	5.157	5.160	11718	4.857	4.860	15475

[a] See note at end of Table 4.

Nom. Diam.	16-Spline Fittings						
	For All Fits				16A—Permanent Fit		
	D		*W*		*d*		
	Min.	Max.	Min.	Max.	Min.	Max.	*T*[a]
2	1.997	2.000	0.193	0.196	1.817	1.820	1375
2½	2.497	2.500	0.242	0.245	2.273	2.275	2149
3	2.997	3.000	0.291	0.294	2.727	2.730	3094
3½	3.497	3.500	0.340	0.343	3.182	3.185	4212
4	3.997	4.000	0.389	0.392	3.637	3.640	5501
4½	4.497	4.500	0.438	0.441	4.092	4.095	6962
5	4.997	5.000	0.487	0.490	4.547	4.550	8595
5½	5.497	5.500	0.536	0.539	5.002	5.005	10395
6	5.997	6.000	0.585	0.588	5.457	5.460	12377

Table 4. S.A.E. Standard Splined Fittings

Nom. Diam.	16-Spline Fittings					
	16B—To Slide—No Load			16C—To Slide Under Load		
	d			*d*		
	Min.	Max.	*T*[a]	Min.	Max.	*T*[a]
2	1.717	1.720	2083	1.617	1.620	2751
2½	2.147	2.150	3255	2.022	2.025	4299
3	2.577	2.580	4687	2.427	2.430	6190
3½	3.007	3.010	6378	2.832	2.835	8426
4	3.437	3.440	8333	3.237	3.240	11005
4½	3.867	3.870	10546	3.642	3.645	13928
5	4.297	4.300	13020	4.047	4.050	17195
5½	4.727	4.730	15754	4.452	4.455	20806
6	5.157	5.160	18749	4.857	4.860	24760

[a] *Torque Capacity of Spline Fittings:* The torque capacities of the different spline fittings are given in the columns headed "*T*." The torque capacity, per inch of bearing length at 1000 pounds pressure per square inch on the sides of the spline, may be determined by the following formula, in which T = torque capacity in inch-pounds per inch of length, N = number of splines, R = mean radius or radial distance from center of hole to center of spline, h = depth of spline: $T = 1000NRh$

Table 5. Formulas for Determining Dimensions of S.A.E. Standard Splines

No. of Splines	*W* For All Fits	*A* Permanent Fit		*B* To Slide Without Load		*C* To Slide Under Load	
		h	*d*	*h*	*d*	*h*	*d*
Four	0.241*D*[a]	0.075*D*	0.850*D*	0.125*D*	0.750*D*	…	…
Six	0.250*D*	0.050*D*	0.900*D*	0.075*D*	0.850*D*	0.100*D*	0.800*D*
Ten	0.156*D*	0.045*D*	0.910*D*	0.070*D*	0.860*D*	0.095*D*	0.810*D*
Sixteen	0.098*D*	0.045*D*	0.910*D*	0.070*D*	0.860*D*	0.095*D*	0.810*D*

[a] Four splines for fits *A* and *B* only.

The formulas in the table above give the maximum dimensions for *W*, *h*, and *d*, as listed in Tables 1 through 4 inclusive.

S.A.E. Standard Spline Fittings.—The S.A.E. spline fittings (Tables 1 through 4 inclusive) have become an established standard for many applications in the agricultural, automotive, machine tool, and other industries. The dimensions given, in inches, apply only to soft broached holes. The tolerances given may be readily maintained by usual broaching methods. The tolerances selected for the large and small diameters may depend upon whether the fit between the mating part, as finally made, is on the large or the small diameter. The other diameter, which is designed for clearance, may have a larger manufactured tolerance. If the final fit between the parts is on the sides of the spline only, larger tolerances are permissible for both the large and small diameters. The spline should not be more than 0.006 inch per foot out of parallel with respect to the shaft axis. No allowance is made for corner radii to obtain clearance. Radii at the corners of the spline should not exceed 0.015 inch.

Polygon-Type Shaft Connections Involute-form and straight-sided splines are used for both fixed and sliding connections between machine members such as shafts and gears. Polygon-type connections, so called because they resemble regular polygons but with curved sides, may be used similarly. German DIN Standards 32711 and 32712 include data for three- and four-sided metric polygon connections. Data for 11

of the sizes shown in those Standards, but converted to inch dimensions by Stoffel Polygon Systems, are given in the accompanying table..— Dimensions of Three- and Four-Sided Polygon-type Shaft Connections

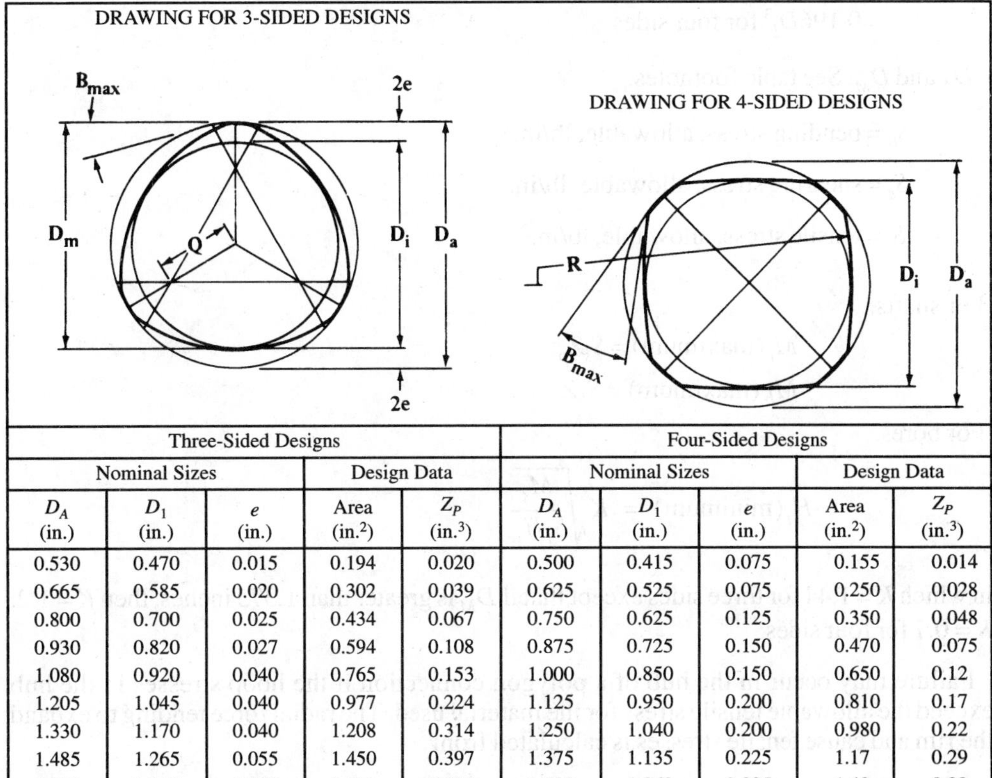

Three-Sided Designs					Four-Sided Designs				
Nominal Sizes			Design Data		Nominal Sizes			Design Data	
D_A (in.)	D_1 (in.)	e (in.)	Area (in.²)	Z_P (in.³)	D_A (in.)	D_1 (in.)	e (in.)	Area (in.²)	Z_P (in.³)
0.530	0.470	0.015	0.194	0.020	0.500	0.415	0.075	0.155	0.014
0.665	0.585	0.020	0.302	0.039	0.625	0.525	0.075	0.250	0.028
0.800	0.700	0.025	0.434	0.067	0.750	0.625	0.125	0.350	0.048
0.930	0.820	0.027	0.594	0.108	0.875	0.725	0.150	0.470	0.075
1.080	0.920	0.040	0.765	0.153	1.000	0.850	0.150	0.650	0.12
1.205	1.045	0.040	0.977	0.224	1.125	0.950	0.200	0.810	0.17
1.330	1.170	0.040	1.208	0.314	1.250	1.040	0.200	0.980	0.22
1.485	1.265	0.055	1.450	0.397	1.375	1.135	0.225	1.17	0.29
1.610	1.390	0.055	1.732	0.527	1.500	1.260	0.225	1.43	0.39
1.870	1.630	0.060	2.378	0.850	1.750	1.480	0.250	1.94	0.64
2.140	1.860	0.070	3.090	1.260	2.000	1.700	0.250	2.60	0.92

Dimensions Q and R shown on the diagrams are approximate and used only for drafting purposes: $Q \approx 7.5e$; $R \approx D_1/2 + 16e$.

Dimension $D_M = D_1 + 2e$. Pressure angle B_{max} is approximately $344e/D_M$ degrees for three sides, and $299e/D_M$ degrees for four sides.

Tolerances: ISO H7 tolerances apply to bore dimensions. For shafts, g6 tolerances apply for sliding fits; k7 tolerances for tight fits.

Choosing Between Three- and Four-Sided Designs: Three-sided designs are best for applications in which no relative movement between mating components is allowed while torque is transmitted. If a hub is to slide on a shaft while under torque, four-sided designs, which have larger pressure angles B_{max} than those of three-sided designs, are better suited to sliding even though the axial force needed to move the sliding member is approximately 50 percent greater than for comparable involute spline connections.

Strength of Polygon Connections: In the formulas that follow,

H_w = hub width, inches

H_t = hub wall thickness, inches

M_b = bending moment, lb-inch

M_t = torque, lb-inch

Z = section modulus, bending, in.³

$= 0.098 D_M^4 / D_A$ for three sides

$= 0.15 D_I^3$ for four sides

Z_P = polar section modulus, torsion, in.^{3}

$= 0.196 D_M^4 / D_A$ for three sides

$= 0.196 D_I^3$ for four sides

D_A and D_M. See table footnotes.

S_b = bending stress, allowable, lb/in.^{2}

S_s = shearing stress, allowable, lb/in.^{2}

S_t = tensile stress, allowable, lb/in.^{2}

For shafts,

$$M_t \text{ (maximum)} = S_s Z_p;$$
$$M_b \text{ (maximum)} = S_b Z$$

For bores,

$$H_t(\text{minimum}) = K\sqrt{\frac{M_t}{S_t H_w}}$$

in which $K = 1.44$ for three sides except that if D_M is greater than 1.375 inches, then $K = 1.2$; $K = 0.7$ for four sides.

Failure may occur in the hub of a polygon connection if the hoop stresses in the hub exceed the allowable tensile stress for the material used. The radial force tending to expand the rim and cause tensile stresses is calculated from

$$\text{Radial Force, lb} = \frac{2M_t}{D_I n \tan(B_{max} + 11.3)}$$

This radial force acting at n points may be used to calculate the tensile stress in the hub wall using formulas from strength of materials.

Manufacturing: Polygon shaft profiles may be produced using conventional machining processes such as hobbing, shaping, contour milling, copy turning, and numerically controlled milling and grinding. Bores are produced using broaches, spark erosion, gear shapers with generating cutters of appropriate form, and, in some instances, internal grinders of special design. Regardless of the production methods used, points on both of the mating profiles may be calculated from the following equations:

$$X = (D_I/2 + e)\cos\alpha - e\cos n\alpha \;\cos\;\alpha - ne\;\sin$$
$$Y = (D_I/2 + e)\sin\alpha - e\cos n\alpha \;\sin\;\alpha + ne\;\sin$$

In these equations, α is the angle of rotation of the workpiece from any selected reference position; n is the number of polygon sides, either 3 or 4; D_I is the diameter of the inscribed circle shown on the diagram in the table; and e is the dimension shown on the diagram in the table and which may be used as a setting on special polygon grinding machines. The value of e determines the shape of the profile. A value of 0, for example, results in a circular shaft having a diameter of D_I. The values of e in the table were selected arbitrarily to provide suitable proportions for the sizes shown.

CAMS AND CAM DESIGN

Classes of Cams.—Cams may, in general, be divided into two classes: uniform motion cams and accelerated motion cams. The uniform motion cam moves the follower at the same rate of speed from the beginning to the end of the stroke; but as the movement is started from zero to the full speed of the uniform motion and stops in the same abrupt way, there is a distinct shock at the beginning and end of the stroke, if the movement is at all rapid. In machinery working at a high rate of speed, therefore, it is important that cams are so constructed that sudden shocks are avoided when starting the motion or when reversing the direction of motion of the follower.

The uniformly accelerated motion cam is suitable for moderate speeds, but it has the disadvantage of sudden changes in acceleration at the beginning, middle and end of the stroke. A cycloidal motion curve cam produces no abrupt changes in acceleration and is often used in high-speed machinery because it results in low noise, vibration and wear. The cycloidal motion displacement curve is so called because it can be generated from a cycloid which is the locus of a point of a circle rolling on a straight line.*

Cam Follower Systems.—The three most used cam and follower systems are radial and offset translating roller follower, Figs. 1a and 1b; and the swinging roller follower, Fig. 1c. When the cam rotates, it imparts a translating motion to the roller followers in Figs. 1a and 1b and a swinging motion to the roller follower in Fig. 1c. The motionof the follower is, of course, dependent on the shape of the cam; and the following section on displacement diagrams explains how a favorable motion is obtained so that the cam can rotate at high speed without shock.

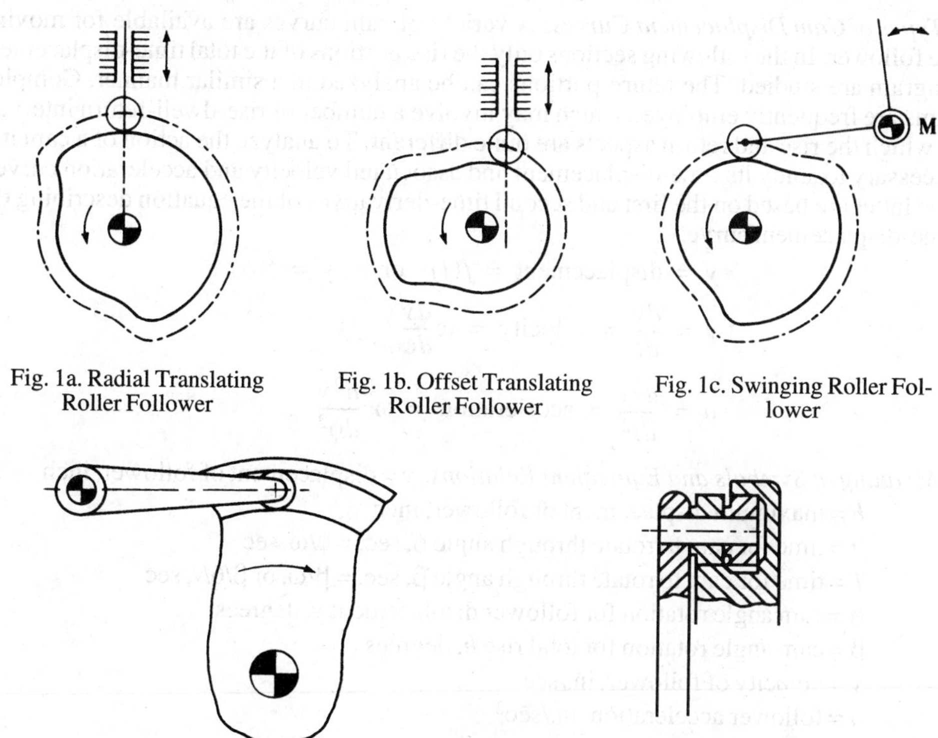

Fig. 1a. Radial Translating Roller Follower

Fig. 1b. Offset Translating Roller Follower

Fig. 1c. Swinging Roller Follower

Fig. 2a. Closed-Track Cam

Fig. 2b. Closed-Track Cam With Two Rollers

The arrangements in Figs. 1a, 1b, and 1c show open-track cams. In Figs. 2a and 2b the roller is forced to move in a closed track. Open-track cams build smaller than closed-track

* Jensen, P. W., *Cam Design and Manufacture,* Industrial Press Inc.

cams but, in general, springs are necessary to keep the roller in contact with the cam at all times. Closed-track cams do not require a spring and have the advantage of positive drive throughout the rise and return cycle. The positive drive is sometimes required as in the case where a broken spring would cause serious damage to a machine.

Displacement Diagrams.—Design of a cam begins with the displacement diagram. A simple displacement diagram is shown in Fig. 3. One cycle means one whole revolution of the cam; i.e., one cycle represents 360°. The horizontal distances T_1, T_2, T_3, T_4 are expressed in units of time (seconds); or radians or degrees. The vertical distance, *h*, represents the maximum "rise" or stroke of the follower.

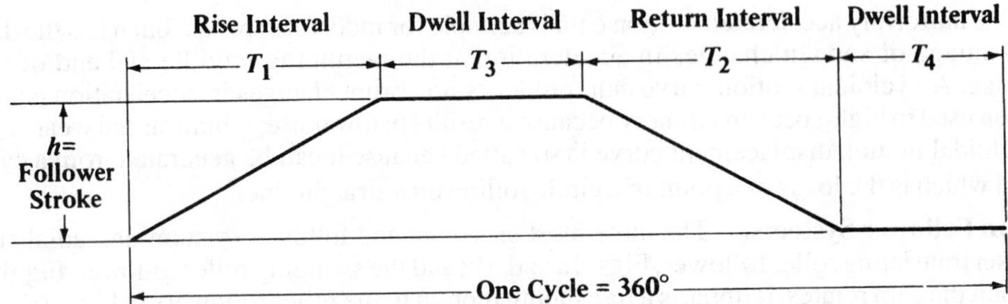

Fig. 3. A Simple Displacement Diagram

The displacement diagram of Fig. 3 is not a very favorable one because the motion from rest (the horizontal lines) to constant velocity takes place instantaneously and this means that accelerations become infinitely large at these transition points.

Types of Cam Displacement Curves: A variety of cam curves are available for moving the follower. In the following sections only the rise portions of the total time-displacement diagram are studied. The return portions can be analyzed in a similar manner. Complex cams are frequently employed which may involve a number of rise-dwell-return intervals in which the rise and return aspects are quite different. To analyze the action of a cam it is necessary to study its time-displacement and associated velocity and acceleration curves. The latter are based on the first and second time-derivatives of the equation describing the time-displacement curve:

$$y = \text{displacement} = f(t) \quad \text{or} \quad y = f(\phi)$$

$$v = \frac{dy}{dt} = \text{velocity} = \omega\frac{dy}{d\phi}$$

$$a = \frac{d^2y}{dt^2} = \text{acceleration} = \omega^2\frac{d^2y}{d\phi^2}$$

Meaning of Symbols and Equivalent Relations: y = displacement of follower, inch

h = maximum displacement of follower, inch

t = time for cam to rotate through angle ϕ, sec, = ϕ/ω, sec

T = time for cam to rotate through angle β, sec, = β/ω, or $\beta/6N$, sec

ϕ = cam angle rotation for follower displacement y, degrees

β = cam angle rotation for total rise h, degrees

v = velocity of follower, in./sec

a = follower acceleration, in./sec^2

$t/T = \phi/\beta$

N = cam speed, rpm

ω = angular velocity of cam, degrees/sec = $\beta/T = \phi/t = d\phi/dt = 6N$

ω_R = angular velocity of cam, radians/sec = $\pi\omega/180$

W = effective weight, lbs

g = gravitational constant = 386 in./sec^2

$f(t)$ = means a function of t

$f(\phi)$ = means a function of ϕ

R_{min} = minimum radius to the cam pitch curve, inch

R_{max} = maximum radius to the cam pitch curve, inch

r_f = radius of cam follower roller, inch

ρ = radius of curvature of cam pitch curve (path of center of roller follower), inch

R_c = radius of curvature of actual cam surface, in., = $\rho - r_f$ for convex surface; = $\rho + r_f$ for concave surface.

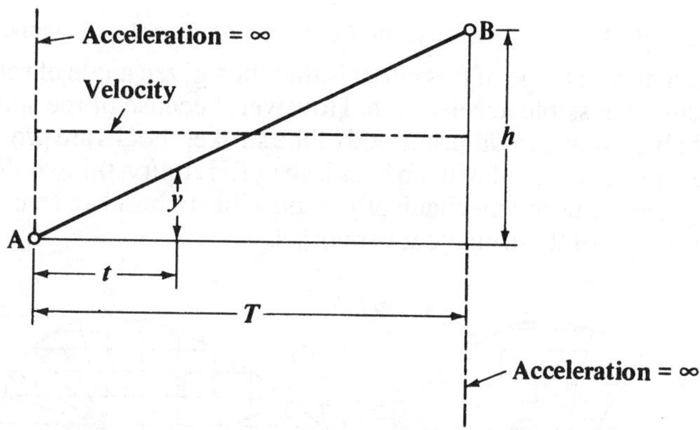

Fig. 4. Cam Displacement, Velocity, and Acceleration Curves for Constant Velocity Motion

Four displacement curves are of the greatest utility in cam design.

1. Constant-Velocity Motion: (Fig. 4)

$y = h\frac{t}{T}$ or $y = \frac{h\phi}{\beta}$	(1a)	
$v = \frac{dy}{dt} = \frac{h}{T}$ or $v = \frac{h\omega}{\beta}$	(1b)	} $0 < t < T$
$a = \frac{d^2y}{dt^2} = 0^*$	(1c)	

* Except at $t = 0$ and $t = T$ where the acceleration is theoretically infinite.

This motion and its disadvantages were mentioned previously. While in the unaltered form shown it is rarely used except in very crude devices, nevertheless, the advantage of uniform velocity is an important one and by modifying the start and finish of the follower stroke this form of cam motion can be utilized. Such modification is explained in the section Displacement Diagram Synthesis.

2. Parabolic Motion: (Fig. 5)

For $0 \le t \le T/2$ and $0 \le \phi \le \beta/2$		For $T/2 \le t \le T$ and $\beta/2 \le \phi \le \beta$	
$y = 2h(t/T)^2 = 2h(\phi/\beta)^2$	(2a)	$y = h[1 - 2(1 - t/T)^2] = h[1 - 2(1 - \phi/\beta)^2]$	(2d)
$v = 4ht/T^2 = 4h\omega\phi/\beta^2$	(2b)	$v = 4h/T(1 - t/T) = (4h\omega/\beta)(1 - \phi/\beta)$	(2e)
$a = 4h/T^2 = 4h(\omega/\beta)^2$	(2c)	$a = -4h/T^2 = -4h(\omega/\beta)^2$	(2f)

Examination of the above formulas shows that the velocity is zero when $t = 0$ and $y = 0$; and when $t = T$ and $y = h$.

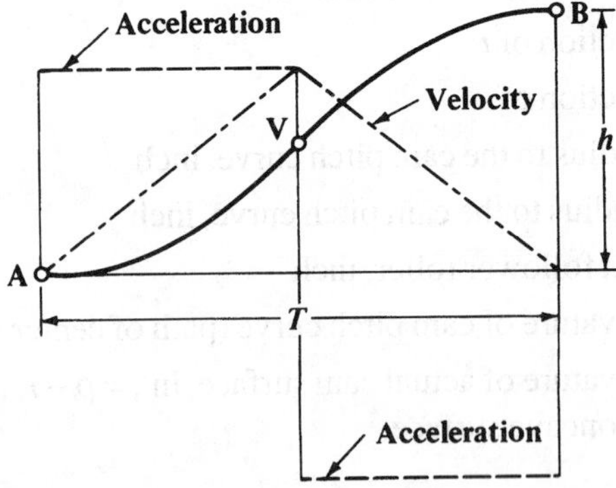

Fig. 5. Cam Displacement, Velocity, and Acceleration Curves for Parabolic Motion

The most important advantage of this curve is that for a given angle of rotation and rise it produces the smallest possible acceleration. However, because of the sudden changes in acceleration at the beginning, middle, and end of the stroke, shocks are produced. If the follower system were perfectly rigid with no backlash or flexibility, this would be of little significance. But such systems are mechanically impossible to build and a certain amount of impact is caused at each of these changeover points.

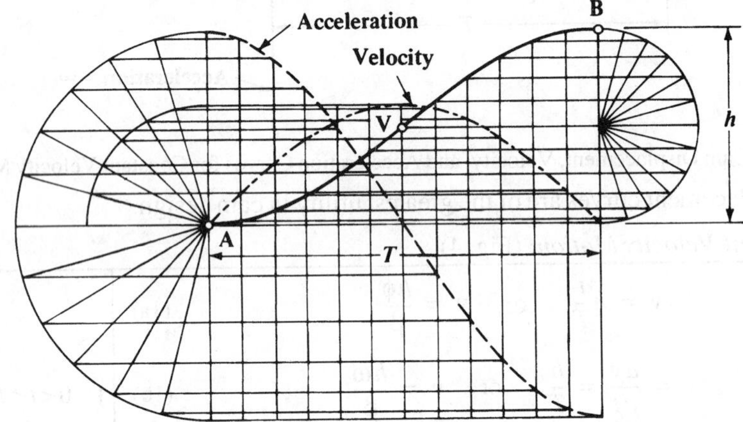

Fig. 6. Cam Displacement, Velocity, and Acceleration Curves for Simple Harmonic Motion

3. Simple Harmonic Motion: (Fig. 6)

$$y = \frac{h}{2}\left[1 - \cos\left(\frac{180°t}{T}\right)\right] \quad \text{or} \quad y = \frac{h}{2}\left[1 - \cos\left(\frac{180°\phi}{\beta}\right)\right] \quad (3a)$$

$$v = \frac{h}{2} \cdot \frac{\pi}{T}\sin\left(\frac{180°t}{T}\right) \quad \text{or} \quad v = \frac{h}{2} \cdot \frac{\pi\omega}{\beta}\sin\left(\frac{180°\phi}{\beta}\right) \quad (3b) \quad \} \quad 0 \le t \le T$$

$$a = \frac{h}{2} \cdot \frac{\pi^2}{T^2}\cos\left(\frac{180°t}{T}\right) \quad \text{or} \quad a = \frac{h}{2} \cdot \left(\frac{\pi\omega}{\beta}\right)^2 \cos\left(\frac{180°\phi}{\beta}\right) \quad (3c)$$

Smoothness in velocity and acceleration during the stroke is the advantage inherent in this curve. However, the instantaneous changes in acceleration at the beginning and end of the stroke tend to cause vibration, noise, and wear. As can be seen from Fig. 6, the maximum acceleration values occur at the ends of the stroke. Thus, if inertia loads are to be overcome by the follower, the resulting forces cause stresses in the members. These forces are in many cases much larger than the externally applied loads.

4. Cycloidal Motion: (Fig. 7)

$$y = h\left[\frac{t}{T} - \frac{1}{2\pi}\sin\left(\frac{360°t}{T}\right)\right] \quad \text{or} \quad y = h\left[\frac{\phi}{\beta} - \frac{1}{2\pi}\sin\left(\frac{360°\phi}{\beta}\right)\right] \quad (4a)$$

$$v = \frac{h}{T}\left[1 - \cos\left(\frac{360°t}{T}\right)\right] \quad \text{or} \quad v = \frac{h\omega}{\beta}\left[1 - \cos\left(\frac{360°\phi}{\beta}\right)\right] \quad (4b) \quad \Bigg\} \quad 0 \le i \le T$$

$$a = \frac{2\pi h}{T^2}\sin\left(\frac{360°t}{T}\right) \quad \text{or} \quad a = \frac{2\pi h\omega^2}{\beta^2}\sin\left(\frac{360°\phi}{\beta}\right) \quad (4c)$$

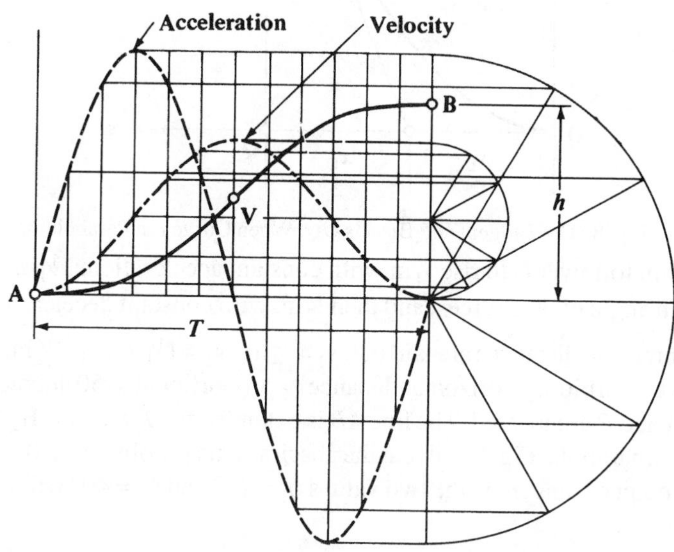

Fig. 7. Cam Displacement, Velocity, and Acceleration Curves for Cycloidal Motion

This time-displacement curve has excellent acceleration characteristics; there are no abrupt changes in its associated acceleration curve. The maximum value of the acceleration of the follower for a given rise and time is somewhat higher than that of the simple harmonic motion curve. In spite of this, the cycloidal curve is used often as a basis for designing cams for high-speed machinery because it results in low levels of noise, vibration, and wear.

Displacement Diagram Synthesis.—The straight-line graph shown in Fig. 3 has the important advantage of uniform velocity. This is so desirable that many cams based on this graph are used. To avoid impact at the beginning and end of the stroke, a modification is introduced at these points. There are many different types of modifications possible, ranging from a simple circular arc to much more complicated curves. One of the better curves used for this purpose is the parabolic curve given by Equation (2a). As seen from the derived time graphs, this curve causes the follower to begin a stroke with zero velocity but having a finite and constant acceleration. We must accept the necessity of acceleration, but effort should be made to hold it to a minimum.

Matching of Constant Velocity and Parabolic Motion Curves: By matching a parabolic cam curve to the beginning and end of a straight-line cam displacement diagram it is possible to reduce the acceleration from infinity to a finite constant value to avoid impact loads. As illustrated in Fig. 8, it can be shown that for any parabola the vertex of which is at O, the tangent to the curve at the point P intersects the line OQ at its midpoint. This means that the tangent at P represents the velocity of the follower at time X_0 as shown in Fig. 8. Since the tangent also represents the velocity of the follower over the constant velocity portion of the stroke, the transition from rest to the maximum velocity is accomplished with smoothness.

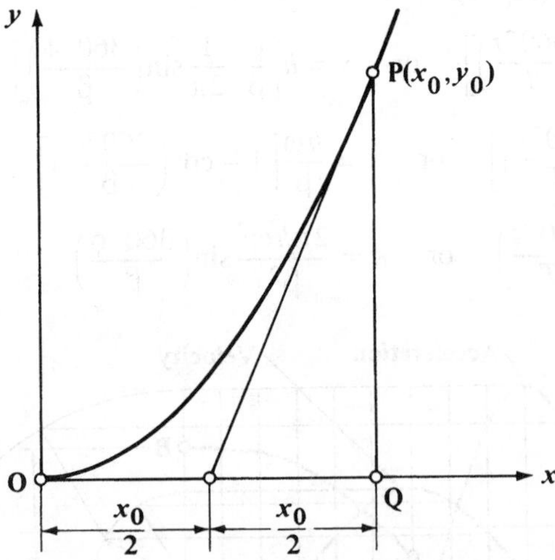

Fig. 8. The Tangent at P Bisects OQ, When Curve is a Parabola

Example: A cam follower is to rise ¼ in. with constant acceleration; 1¼ in. with constant velocity, over an angle of 50 degrees; and then ½ in. with constant deceleration.

In Fig. 9 the three rise distances are laid out, $y_1 = ¼$ in., $y_2 = 1¼$ in., $y_3 = ½$ in., and horizontals drawn. Next, an arbitrary horizontal distance ϕ_2 proportional to 50 degrees is measured off and points A and B are located. The line AB is extended to M_1 and M_2. By remembering that a tangent to a parabola, Fig. 8, will cut the abscissa axis at point $(X_0/2,0)$ where X_0 is the abscissa of the point of tangency, the two values $\phi_1 = 20°$ and $\phi_3 = 40°$ will be found. Analytically,

$$\frac{M_1E}{\phi_2} = \frac{y_1}{y_2} \qquad \frac{\frac{1}{2}\phi_1}{50°} = \frac{0.25}{1.25} \qquad \therefore \phi_1 = 20°$$

$$\frac{FM_2}{\phi_2} = \frac{y_3}{y_2} \qquad \frac{\frac{1}{2}\phi_3}{50°} = \frac{0.50}{1.25} \qquad \therefore \phi_3 = 40°$$

In Fig. 9, the portions of the parabola have been drawn in; the details of this operation are as follows:

Assume that accuracy to the nearest thousandth of one inch is desired, and it is decided to plot values for every 5 degrees of cam rotation.

The formula for the acceleration portion of the parabolic curve is:

$$y = \frac{2h}{T^2}t^2 = 2h\left(\frac{\phi}{\beta}\right)^2 \tag{5}$$

Two different parabolas are involved in this example; one for accelerating the follower during a cam rotation of 20 degrees, the other for decelerating it in 40 degrees, these two being tangent, to opposite ends of the same line AB.

In Fig. 9 only the first half of a complete acceleration-deceleration parabolic curve is used to blend with the left end of the straight line AB. Therefore, in using the Formula (5) substitute $2y_1$ for h and $2\phi_1$ for β so that

$$y = \frac{2h\phi^2}{\beta^2} = \frac{(2)(2y_1)}{(2\phi_1)^2}\phi^2$$

For the right end of the straight line *AB*, the calculations are similar but, in using Formula (5), calculated *y* values are *subtracted* from the *total* rise of the cam $(y_1 + y_2 + y_3)$ to obtain the follower displacement.

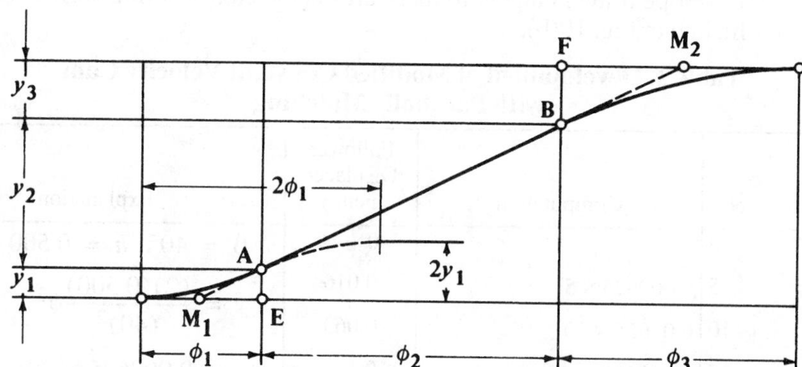

Fig. 9. Matching a Parabola at Each End of Straight Line Displacement Curve *AB* to Provide More Acceptable Acceleration and Deceleration

Table 1 shows the computations and resulting values for the cam displacement diagram described. The calculations are shown in detail so that if equations are programmed for a digital computer, the results can be verified easily. Obviously, the intermediate points are not needed to draw the straight line, but when the cam profile is later to be drawn or cut, these values will be needed since they are to be measured on radial lines.

The matching procedure when using cycloidal motion is exactly the same as for parabolic motion, because parabolic and cycloidal motion have the same maximum velocity for equal rise (or return) and lift angle (or return angle).

Cam Profile Determination.—In the cam constructions that follow an artificial device called an *inversion* is used. This represents a mental concept which is very helpful in performing the graphical work. The construction of a cam profile requires the drawing of many positions of the cam with the follower in each case in its related location. However, instead of revolving the cam, it is assumed that the follower rotates around the *fixed* cam. It requires the drawing of many follower positions, but since this is done more or less diagrammatically, it is relatively simple.

As part of the inversion process, the direction of rotation is important. In order to preserve the correct sequence of events, the artificial rotation of the follower must be the reverse of the cam's prescribed rotation. Thus, in Fig. 10 the cam rotation is counterclockwise, whereas the artificial rotation of the follower is clockwise.

Radial Translating Roller Follower: The time-displacement diagram for a camwith a radial translating roller follower is shown in Fig. 10(a). This diagram is read from left to right as follows: For 100 degrees of cam shaft rotation the follower rises *h* inches (*AB*), dwells in its upper position for 20 degrees (*BC*), returns over 180 degrees (*CD*), and finally dwells in its lowest position for 60 degrees (*DE*). Then the entire cycle is repeated.

Fig. 10(b) shows the cam construction layout with the cam pitch curve as a dot and dash line. To locate a point on this curve, take a point on the displacement curve, as 6′ at the 6o-degree position, and project this horizontally to point 6″ on the 0-degree position of the cam construction diagram. Using the center of cam rotation, an arc is struck from point 6″ to intercept the 60-degree position radial line which gives point 6‴ on the cam pitch curve. It will be seen that the smaller circle in the cam construction layout has a radius R_{min} equal to the smallest distance from the center of cam rotation to the pitch curve and, similarly, the larger circle has a radius R_{max} equal to the largest distance to the pitch curve. Thus, the difference in radii of these two circles is equal to the maximum rise h of the follower.

The cam pitch curve is also the actual profile or working surface when a knife-edged follower is used. To get the profile or working surface for a cam with a roller follower, a series of arcs with centers on the pitch curve and radii equal to the radius of the roller are drawn and the inner envelope drawn tangent to these arcs is the cam working surface or profile shown as a solid line in Fig. 10(b).

Table 1. Development of Modified Constant Velocity Cam with Parabolic Matching

Rise Angle	ϕ Degrees	Computation	Follower Displacement y	Explanation
	0	0	0	$\beta = 40°,\ h = 0.500$
	5	0.000625×5^2	0.016	$y = \frac{(2)(0.500)}{(40)^2}\phi^2$
$\phi_1 =$ 20°	{ 10	0.00625×10^2	0.063	
	15	0.000625×15^2	0.141	$= 0.000625\phi^2$
	20	0.000625×20^2	0.250	
	25		0.375	
	30		0.500	
	35		0.625	
	40		0.750	
$\phi_2 =$ 50°	{ 45		0.875	1.250 in. divided into 10 uniform divisions
	50		1.000	
	55		1.125	
	60		1.250	
	65		1.375	
	70		1.500	
	75	$2.000 - (0.0003125 \times 35^2)$	1.617	$\beta = 80°,\ h = 1.000$
	80	$2.000 - (0.0003125 \times 30^2)$	1.617	$y = 2 - \frac{(2)(1.000)}{(80°)^2}(110° - \phi)^2$
	85	$2.000 - (0.0003125 \times 25^2)$	1.805	
	90	$2.000 - (0.0003125 \times 20^2)$	1.875	$= 2 - 0.0003125(110° - \phi)^2$
$\phi_3 =$ 40°	{ 95	$2.000 - (0.0003125 \times 15^2)$	1.930	
	100	$2.000 - (0.0003125 \times 10^2)$	1.969	
	105	$2.000 - (0.0003125 \times 5^2$	1.992	
	110	$2.000 - (0.0003125 \times 0^2)$	2.000	
Since the deceleration portion of a parabolic cam is the same shape as the acceleration portion, but inverted, Formula (5) may be used to calculate the y values by substituting $2y_3$ for h and for β and the result subtracted from the total rise $(y_1 + y_2 + y_3)$ to obtain the follower displacement.				

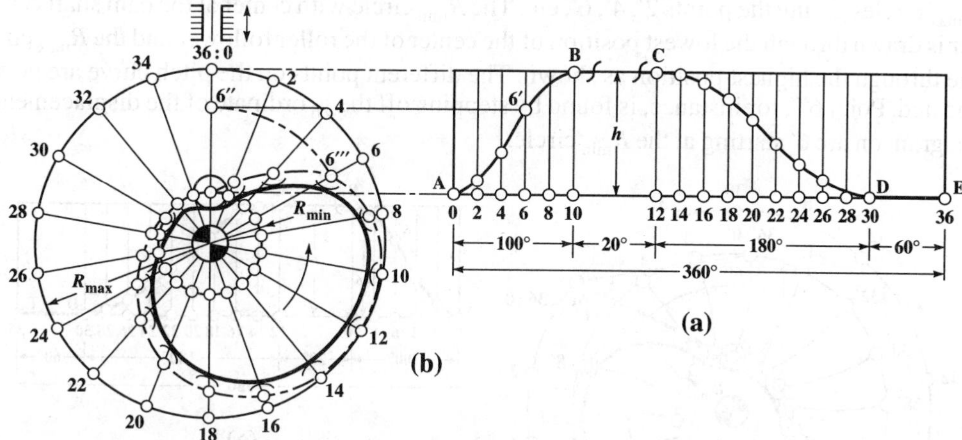

Fig. 10. (a) Time-Displacement Diagram for Cam to be Laid Out; (b) Construction of Contour of Cam With Radial Translating Roller Follower

Offset Translating Roller Follower: Given the time-displacement diagram Fig. 11(a) and an offset follower. The construction of the cam in this case is very similar to the foregoing case and is shown in Fig. 11(b). In this construction it will be noted that the angular position lines are not drawn radially from the cam shaft center but tangent to a circle having a radius equal to the amount of offset of the center line of the cam follower from the center of the cam shaft. For counterclockwise rotation of the cam, points 6′, 6″, and 6‴ are located in succession as indicated.

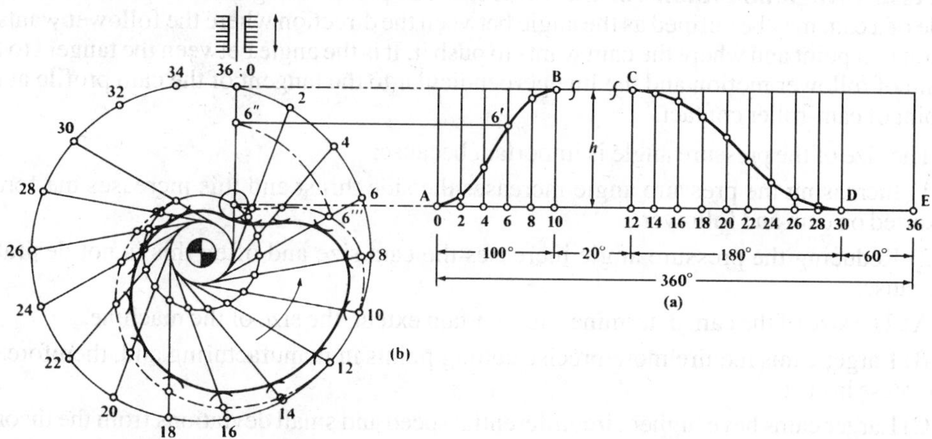

Fig. 11. (a) Time-Displacement Diagram for Cam to be Laid Out; (b) Construction of Contour of Cam With Offset Translating Roller Follower

Swinging Roller Follower: Given the time-displacement diagram Fig. 12(a) and the length of the swinging follower arm L_f, it is required that the displacement of the follower center along the circular arc that it describes be equal to the corresponding displacements in the time-displacement diagram. If ϕ_0 is known, the displacement h of Fig. 12(a) would be found from the formula $h = \pi\phi_0 L_f/180°$: otherwise the maximum rise h of the follower is stepped off on the arc drawn with M as a center and starting at a point on the R_{min} circle. Point M is the actual position of the pivot center of the swinging follower with respect to the cam shaft center. It is again required that the rotation of the cam be counterclockwise and therefore M is considered to have been rotated clockwise around the cam shaft center, whereby the points 2, 4, 6, etc., are obtained as shown in Fig. 12(b). Around each of the pivot points, 2, 4, 6, etc., circular arcs whose radii equal L_f are drawn between the R_{min} and

R_{max} circles giving the points 2′, 4′, 6′, etc. The R_{min} circle with center at the cam shaft center is drawn through the lowest position of the center of the roller follower and the R_{max} circle through the highest position as shown. The different points on the pitch curve are now located. Point 6‴, for instance, is found by stepping off the y_6 ordinate of the displacement diagram on arc 6′ starting at the R_{min} circle.

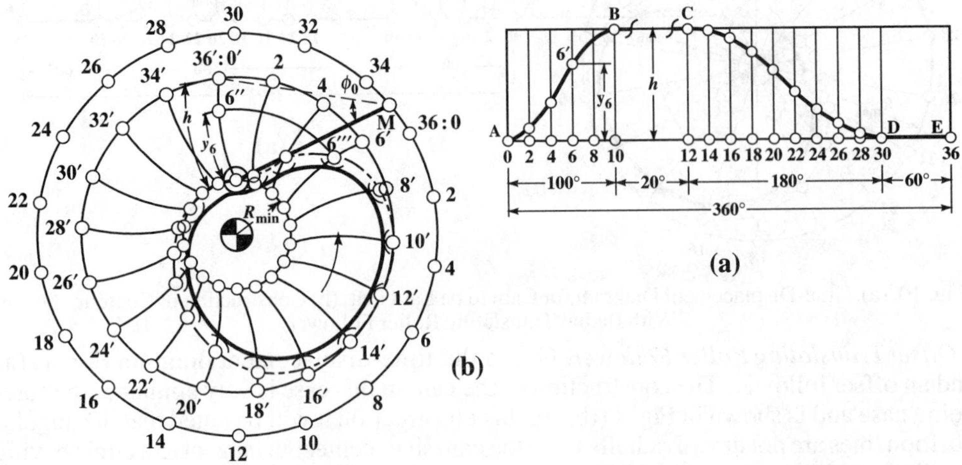

Fig. 12. (a) Time-Displacement Diagram for Cam to be Laid Out; (b) Construction of Contour of Cam With Swinging Roller Follower

Pressure Angle and Radius of Curvature.—The pressure angle at any point on the profile of a cam may be defined as the angle between the direction where the follower wants to go at that point and where the cam wants to push it. It is the angle between the tangent to the path of follower motion and the line perpendicular to the tangent of the cam profile at the point of cam-roller contact.

The size of the pressure angle is important because:

1) Increasing the pressure angle increases the side thrust and this increases the forces exerted on cam and follower.

2) Reducing the pressure angle increases the cam size and often this is not desirable because:

A) The size of the cam determines, to a certain extent, the size of the machine.

B) Larger cams require more precise cutting points in manufacturing and, therefore, an increase in cost.

C) Larger cams have higher circumferential speed and small deviations from the theoretical path of the follower cause additional acceleration, the size of which increases with the square of the cam size.

D) Larger cams mean more revolving weight and in high-speed machines this leads to increased vibrations in the machine.

E) The inertia of a large cam may interfere with quick starting and stopping.

The maximum pressure angle α_m should, in general, be kept at or below 30 degrees for translating-type followers and at or below 45 degrees for swinging-type followers.These values are on the conservative side and in many cases may be increased considerably, but beyond these limits trouble could develop and an analysis is necessary.

In the following, graphical methods are described by which a cam mechanism can be designed with translating or swinging roller followers having specified maximum pressure angles for rise and return. These methods are applicable to any kind of time-displacement diagram.

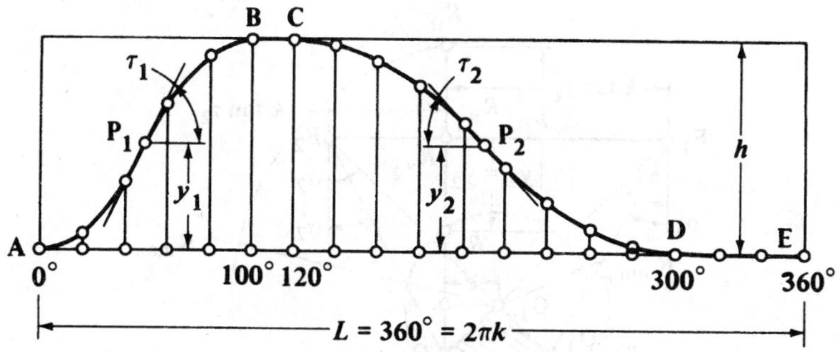

Fig. 13. Displacement Diagram

Determination of Cam Size for a Radial or an Offset Translating Follower.—Fig. 13 shows a time-displacement diagram. The maximum displacement is preferably made to scale, but the length of the abscissa, L, can be chosen arbitrarily. The distance L from 0 to 360 degrees is measured and is set equal to $2\pi k$ from which

$$k = \frac{L}{2\pi}$$

k is calculated and laid out as length E to M in Fig. 14.

In Fig. 13 the two points P_1 and P_2 having the maximum angles of slope, τ_1, and τ_2, are located by inspection. In this example y_1 and y_2 are of equal length.

Angles τ_1 and τ_2 are laid out as shown in Fig. 14, and the points of intersection with a perpendicular to EM at M determine Q_1 and Q_2. The measured distances

$$MQ_1 = k\tan\tau_1 \quad \text{and} \quad MQ_2 = k\tan\tau_2$$

are laid out in Fig. 15, which is constructed as follows:

Draw a vertical line R_uR_o of length h equal to the stroke of the roller follower, R_u being the lowest position and R_o the highest position of the center of the roller follower. From R_u lay out $R_uR_{y1} = y_1$ and $R_uR_{y2} = y_2$; these are equal lengths in this example. Next, if the rotation of the cam is counterclockwise, lay out $k \tan \tau_1$, to the left, $k \tan \tau_2$ to the right from points R_{y1} and R_{y2}, respectively, R_{y1} and R_{y2} being the same point in this case.

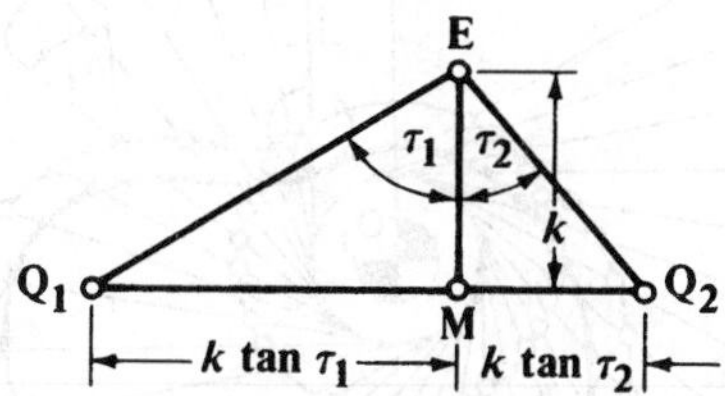

Fig. 14. Construction to Find $k \tan \tau_1$ and $k \tan \tau_2$

The specified maximum pressure angle α_1 is laid out at E_1 as shown, and a ray (line) E_1F_1 is determined. Any point on this ray chosen as the cam shaft center will proportion the cam so that the pressure angle at a point on the cam profile corresponding to point P_1, of the displacement diagram will be exactly α_1.

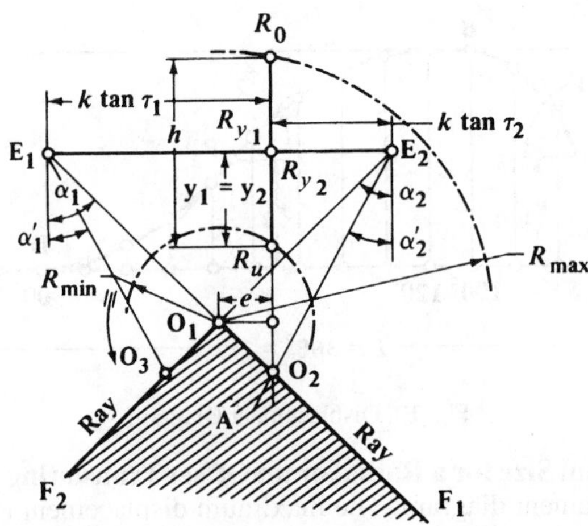

Fig. 15. Finding Proportions of Cam; Offset Translating Follower

The angle α_2 is laid out at E_2 as shown, and another ray E_2F_2 is determined. Similarly, any point on this ray chosen as the cam shaft center will proportion the cam so that the pressure angle at a point on the cam profile corresponding to point P_2 of the displacement diagram will be exactly α_2.

Any point chosen within the cross-hatched area A as the cam center will yield a cam whose pressure angles at points corresponding to P_1 and P_2 will not exceed the specified values α_1 and α_2 respectively. If O_1 is chosen as the cam shaft center, the pressure angles on the cam profile corresponding to points P_1 and P_2 are exactly α_1 and α_2, respectively. Selection of point O_1 also yields the smallest possible cam for the given requirements and requires an offset follower in which e is the offset distance.

If O_2 is chosen as the cam shaft center, a radial translating follower is obtained (zero offset). In that case, the pressure angle α_1 for the rise is unchanged, whereas the pressure angle for the return is changed from α_2 to α'_2. That is, the pressure angle on the return stroke is reduced at the point P_2. If point O_3 had been selected, then α_2 would remain unchanged but α_1 would be decreased and the offset, e, increased.

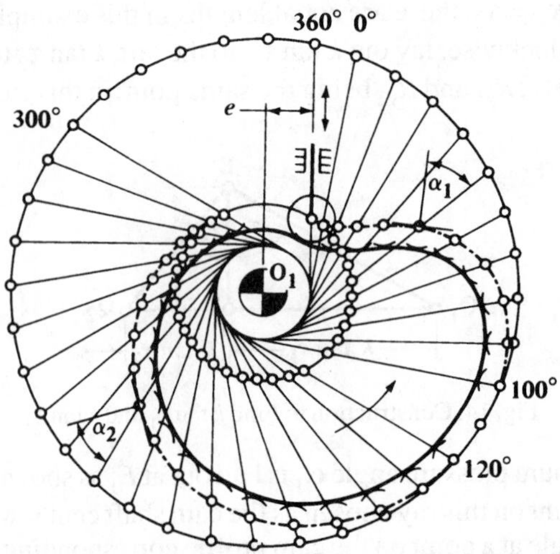

Fig. 16. Construction of Cam Contour; Offset Translating Follower

Fig. 16 shows the shape of the cam when O_1 from Fig. 15 is chosen as the cam shaft center, and it is seen that the pressure angle at a point on the cam profile corresponding to point P_1 is α_1 and at a point corresponding to point P_2 is α_2.

In the foregoing, a cam mechanism has been so proportioned that the pressure angles α_1 and α_2 at points on the cam corresponding to points P_1 and P_2 were obtained. Even though P_1 and P_2 are the points of greatest slope on the displacement diagram, the pressure angles produced at some other points on the actual cam may be slightly greater.

However, if the pressure angles α_1 and α_2 are not to be exceeded at any point—i.e., they are to be maximum pressure angles—then P_1 and P_2 must be selected to be at the locations where these maximum pressure angles occur. If these locations are not known, then the graphical procedure described must be repeated, letting P_1 take various positions on the curve for rise (AB) and P_2 various positions on the return curve (CD) and then setting R_{min} equal to the largest of the values determined from the various positions.

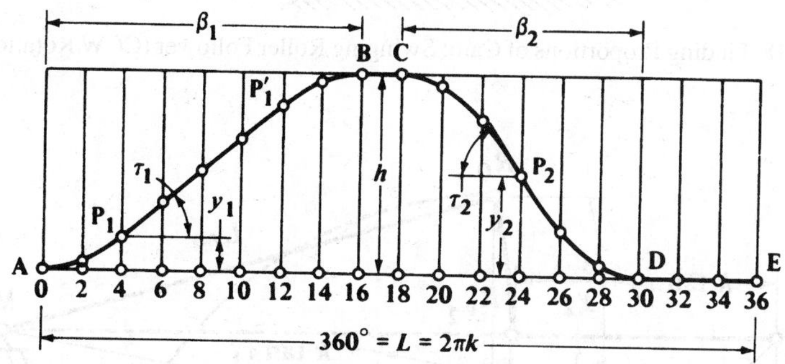

Fig. 17. Displacement Diagram

Determination of Cam Size for Swinging Roller Follower.—The proportioning of a cam with swinging roller follower having specific pressure angles at selected points follows the same procedure as that for a translating follower.

Example: Given the diagram for the roller displacement along its circular arc, Fig. 17 with $h = 1.95$ in., the periods of rise and fall, respectively, $\beta_1 = 160°$ and $\beta_2 = 120°$, the length of the swinging follower arm $L_f = 3.52$ in., rotation of the cam away from pivot point M, and pressure angles $\alpha_1 = \alpha_2 = 45°$ (corresponding to the points P_1 and P_2 in the displacement diagram). Find the cam proportions.

Solution: Distances k tan τ_1 and k tan τ_2 are determined as in the previous example, Fig. 14. In Fig. 18, R_{y1} is determined by making the distance $R_uR_{y1} = y_1$ along the arc R_uR_o and R_{y2} by making $R_uR_{y2} = y_2$. The arc $R_uR_o = h$ and R_u indicates thelowest position of the center of the swinging roller follower and R_o the highest position.

Because the cam (i.e., the surface of the cam as it passes under the follower roller) rotates away from pivot point M, k tan τ_1 is laid out away from M, that is, from R_{y1} to E_1 and k tan τ_2 is laid out toward M from R_{y2} to E_2. Angle α_1 at E_1 determines one ray and α_2 at E_2 another ray, which together subtend an area A having the property that if the cam shaft center is chosen inside this area, the pressure angles at the points of the cam corresponding to P_1 and P_2 in the displacement diagram will not exceed the given values α_1 and α_2, respectively. If the cam shaft center is chosen on the ray drawn from E_1 at an angle $\alpha_1 = 45°$, the pressure angle α_1 on the cam profile corresponding to point P_1 will be exactly 45°, and if chosen on the ray from E_2, the pressure angle α_2 corresponding to P_2 will be exactly 45°. If another point, O_2 for example, is chosen as the cam shaft center, the pressure angle corresponding to P_1 will be α'_1 and that corresponding to P_2 will be α_2.

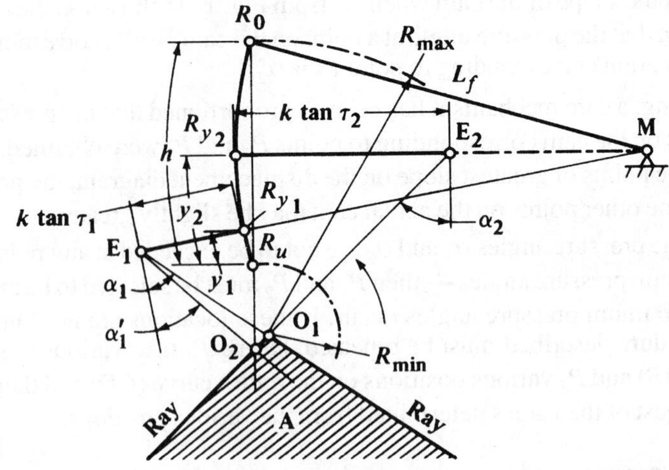

Fig. 18. Finding Proportions of Cam; Swinging Roller Follower (CCW Rotation)

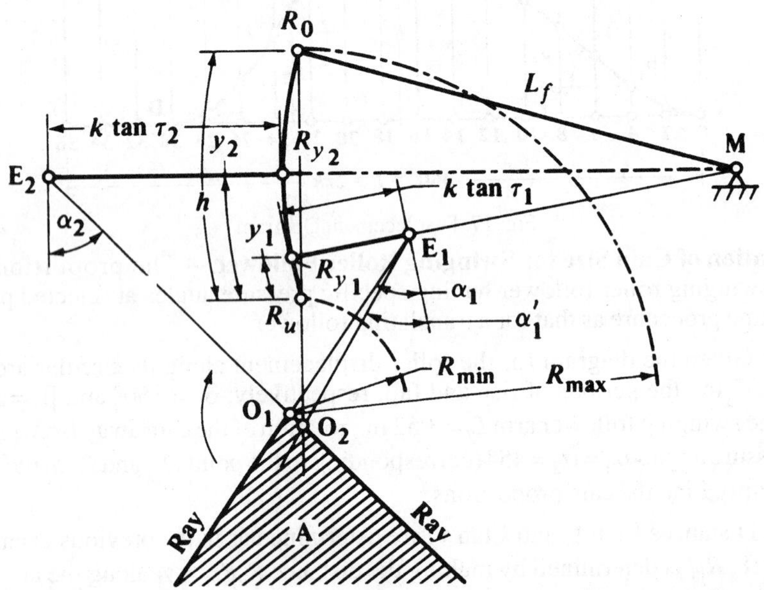

Fig. 19. Finding Proportions of Cam; Swinging Roller Follower (CW Rotation)

Fig. 19 shows the construction for rotation toward pivot point M (clockwise rotation of the cam in this case). The layout of the cam curve is made in a manner similar to that shown previously in Fig. 12.

In this example, the cam mechanism was so proportioned that the pressure angles at certain points (corresponding to P_1 and P_2) do not exceed certain specified values (namely α_1 and α_2).

To make sure that the pressure angle at *no point* along the cam profile exceeds the specified value, the previous procedure should be repeated for a series of points along the profile.

Formulas for Calculating Pressure Angles.—The graphical methods described previously are useful because they permit layout and measurement of pressure angles and radii of curvature of *any* cam profile. For cams of complicated profiles, and especially if the profile cannot be represented by a simple formula, the graphical method may be the only practical solution. However, for some of the standard cam profiles utilizing *radial* translating roller followers, the following formulas may be used to determine key cam dimensions before laying out the cam. These formulas enable the designer to specify the maximum pressure angle (usually 30° or less) and, using the specified value, to calculate the minimum cam size that will satisfy the requirement.

The following symbols are in addition to those starting on page 2164.

α_{max} = specified maximum pressure angle, degrees

$R_{\alpha max}$ = radius from cam center to point on pitch curve where α_{max} is located, inches

ϕ_p = rise angle, in degrees, corresponding to α_{max} and $R_{\alpha\,max}$

α = pressure angle at any selected point, degrees

R_α = radius from cam center to pitch curve at α, inches

ϕ = rise angle, in degrees, corresponding to α and R_α

For Uniform Velocity Motion: $\alpha = \arctan\left[\frac{180°h}{\pi\beta R_\alpha}\right]$ at radius R_α to the pitch curve (6a)

$$\alpha_{max} = \arctan\left[\frac{180°h}{\pi\beta R_{min}}\right] \text{ at radius } R_{min} \text{ of the pitch curve } (\phi=0°). \qquad (6b)$$

If α_{max} is specified, then the minimum radius to the lowest point on the pitch curve, R_{min}, is:

$$R_{min} = \frac{180°h}{\pi\beta\tan\alpha_{max}} \text{ which corresponds to } \phi=0°. \qquad (6c)$$

For Parabolic Motion: $\alpha = \arctan\left[\frac{720°h\phi}{\pi\beta^2 R_\alpha}\right]$ at radius R_α to the pitch curve at angle ϕ, where $0 \le \phi \le \beta/2$

(7a)

$\alpha = \arctan\left[\frac{720°h(1-\phi/\beta)}{\pi\beta R_\alpha}\right]$ at radius R_α to the pitch curve at angle ϕ, where $\beta/2 \le \phi \le \beta$.

$$\alpha = \arctan\left[\frac{360°h}{\pi\beta R_\alpha}\right] \text{ which occurs at } \phi=\beta/2 \text{ and } R_\alpha=R_{min}+h/2 \qquad (7b)$$

If α_{max} is specified, then the minimum radius to the lowest point of the pitch curve is:

$$R_{min} = \left[\frac{360°h}{\pi\beta\tan\alpha_{max}} - \frac{h}{2}\right] \text{ which corresponds to } \phi=0°. \qquad (7c)$$

For Simple Harmonic Motion: $\alpha = \arctan\left[\frac{90°h}{\beta R_\alpha}\sin\left(\frac{180°\phi}{\beta}\right)\right]$ at radius R_α to the pitch curve at angle ϕ

(8a)

$\phi_p = \left(\frac{\beta}{180°}\right)\left[\operatorname{arccot}\left(\frac{\beta}{180^\circ}\tan\alpha_{max}\right)\right]$ = value of ϕ where specified pressure angle α_{max} occurs

(8b)

$$R_{\alpha max} = \frac{h[\sin(180°\phi_p/\beta)]^2}{2\cos(180°\phi_p/\beta)} \text{ at point where } \alpha=\alpha_{max} \text{ and } \phi=\phi_{max}$$

(8c)

$$R_{min} = R_{\alpha\ max} - \frac{h}{2}\left[1 - \cos\left(\frac{180°\phi_p}{\beta}\right)\right] \quad (8d)$$

For Cycloidal Motion: $\alpha = \arctan\left[\frac{180°}{\pi\beta R_\alpha}\left[1 - \cos\left(\frac{360°\phi}{\beta}\right)\right]\right]$ at radius R_α to the pitch curve at angle ϕ (9a)

$$\phi_p = \frac{\beta}{180°}\left[\text{arccot}\left(\frac{\beta\tan\alpha_{max}}{360°}\right)\right]$$ ϕ_p=value of ϕ where specified pressure angle α_{max} occurs (9b)

$$R_{\alpha max} = \frac{h}{2\pi}\frac{[1 - \cos(360°\phi_p/\beta)]^2}{\sin(360°\phi_p/\beta)}$$ at point where $\alpha=\alpha_{max}$ and $\phi=\phi_p$ (9c)

$$R_{min} = R_{\alpha\ max} - h\left[\frac{\phi_p}{\beta} - \frac{1}{2\pi}\sin\left(\frac{360°\phi_p}{\beta}\right)\right] \quad (9d)$$

Radius of Curvature.—The minimum radius of curvature of a cam should be kept as large as possible (1) to prevent undercutting of the convex portion of the cam and (2) to prevent too high surface stresses. Figs. 20(a), (b) and (c) illustrate how undercutting occurs.

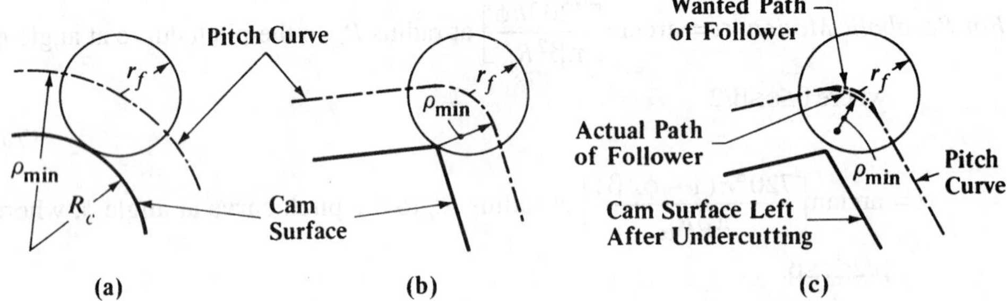

Fig. 20. (a) No Undercutting. (b) Sharp Corner on Cam. (c) Undercutting

In Fig. 20(a) the radius of curvature of the path of the follower is ρ_{min} and the cam will at that point have a radius of curvature $R_c = \rho_{min} - r_f$.

In Fig. 20(b) $\rho_{min} = r_f$ and $R_c = 0$. Therefore, the actual cam will have a sharp corner which in most cases will result in too high surface stresses.

In Fig. 20(c) is shown the case where $\rho_{min} < r_f$. This case is not possible because undercutting will occur and the actual motion of the roller follower will deviate from the desired one as shown.

Undercutting cannot occur at the *concave* portion of the cam profile (working surface), but caution should be exerted in not making the radius of curvature equal to the radius of the roller follower. This condition would occur if there is a cusp on the displacement diagram which, of course, should always be avoided. To enable milling or grinding of *concave* portions of a cam profile, the radius of curvature of concave portions of the cam, $R_c = \rho_{min} + r_f$, must be larger than the radius of the cutter to be used.

The radius of curvature is used in calculating surface stresses (see following section), and may be determined by measurement on the cam layout or, in the case of radial translating followers, may be calculated using the formulas that follow. Although these formulas are exact for radial followers, they may be used for offset and swinging followers to obtain an approximation.

Based upon polar coordinates, the radius of curvature is:

$$\rho = \frac{\left[r^2 + \left(\frac{dr}{d\phi}\right)^2\right]^{3/2}}{r^2 + 2\left(\frac{dr}{d\phi}\right)^2 - r\left(\frac{d^2r}{d\phi^2}\right)} \qquad (10)$$

*Positive values (+) indicate convex curve; negative values (−), concave.

In Equation (10), $r = (R_{min} + y)$, where R_{min} is the smallest radius to the pitch curve (see Fig. 12) and y is the displacement of the follower from its lowest position given in terms of ϕ, the angle of cam rotation. The following formulas for r, $dr/d\phi$, and $d^2r/d\phi^2$ may be substituted into Equation (10) to calculate the radius of curvature at any point of the cam pitch curve; however, to determine the possibility of undercutting of the convex portion of the cam, it is the minimum radius of curvature on the convex portion, ρ_{min}, that is needed. The minimum radius of curvature occurs, generally, at the point of maximum *negative* acceleration.

Parabolic motion:

$$r = R_{min} + h - 2h\left(1 - \frac{\phi}{\beta}\right)^2 \qquad (11a)$$

$$\frac{dr}{d\phi} = \frac{720°h}{\pi\beta}\left(1 - \frac{\phi}{\beta}\right) \qquad (11b) \qquad \frac{\beta}{2} \le \phi \le \beta$$

$$\frac{d^2r}{d\phi^2} = \frac{-4(180°)^2h}{\pi^2\beta^2} \qquad (11c)$$

These equations are for the deceleration portion of the curve as explained in the footnote to Table 1.

The minimum radius of curvature can occur at either $\phi = \beta/2$ or at $\phi = \beta$, depending on the magnitudes of h, R_{min}, and β. Therefore, to determine which is the case, make two calculations using Formula (10), one for $\phi = \beta/2$, and the other for $\phi = \beta$.

Simple harmonic motion:

$$r = R_{min} + \frac{h}{2}\left[1 - \cos\left(\frac{180°\phi}{\beta}\right)\right] \qquad (12a)$$

$$\frac{dr}{d\phi} = \frac{180°h}{2\beta}\sin\left(\frac{180°\phi}{\beta}\right) \qquad (12b) \quad 0 \le \phi \le \beta$$

$$\frac{d^2r}{d\phi^2} = \frac{(180°)^2h}{2\beta^2}\cos\left(\frac{180°\phi}{\beta}\right) \qquad (12c)$$

The minimum radius of curvature can occur at either $\phi = \beta/2$ or at $\phi = \beta$, depending on the magnitudes of h, R_{min}, and β. Therefore, to determine which is the case, make two calculations using Formula (10), one for $\phi = \beta/2$, and the other for $\phi = \beta$.

Cycloidal motion:

$$r = R_{min} + h\left[\frac{\phi}{\beta} - \frac{1}{2\pi}\sin\left(\frac{360°\phi}{\beta}\right)\right] \qquad (13a)$$

$$\frac{dr}{d\phi} = \frac{180°h}{\pi\beta}\left[1 - \cos\left(\frac{360°\phi}{\beta}\right)\right] \qquad (13b) \quad 0 \le \phi \le \beta$$

$$\frac{d^2r}{d\phi^2} = \frac{2(180°)^2h}{\pi\beta^2}\sin\left(\frac{360°\phi}{\beta}\right) \qquad (13c)$$

$$\rho_{min} = \frac{[(R_{min}+0.91h)^2+(180°h/\pi\beta)^2]^{3/2}}{(R_{min}+0.91h)^2+2(180°h/\pi\beta)^2+(R_{min}+0.91h)[2(180°)^2h/\pi\beta^2]} \qquad (13d)$$

(ρ_{min} occurs near $\phi = 0.75\beta$.)

Example: Given $h = 1$ in., $R_{min} = 2.9$ in., and $\beta = 60°$. Find ρ_{min} for parabolic motion, simple harmonic motion, and cycloidal motion.

Solution: $\rho_{min} = 2.02$ in. for parabolic motion, from Equation (10)
$\rho_{min} = 1.8$ in. for simple harmonic motion, from Equation (10)
$\rho_{min} = 1.6$ in. for cycloidal motion, from Equation (13d)

The value of ρ_{min} on *any* cam may also be obtained by measurement on the layout of the cam using a compass.

Cam Forces, Contact Stresses, and Materials.—After a cam and follower configuration has been determined, the forces acting on the cam may be calculated or otherwise determined. Next, the stresses at the cam surface are calculated and suitable materials to withstand the stress are selected. If the calculated maximum stress is too great, it will be necessary to change the cam design.

Such changes may include: 1) increasing the cam size to decrease pressure angle and increase the radius of curvature; 2) changing to an offset or swinging follower to reduce the pressure angle; 3) reducing the cam rotation speed to reduce inertia forces; 4) increasing the cam rise angle, β, during which the rise, h, occurs; 5) increasing the thickness of the cam, provided that deflections of the follower are small enough to maintain uniform loading across the width of the cam; and 6) using a more suitable cam curve or modifying the cam curve at critical points.

Although parabolic motion seems to be the best with respect to minimizing the calculated maximum acceleration and, therefore, also the maximum acceleration forces, nevertheless, in the case of high speed cams, cycloidal motion yields the lower maximum acceleration forces. Thus, it can be shown that owing to the sudden change in acceleration (called *jerk* or *pulse*) in the case of parabolic motion, the actual forces acting on the cam are doubled and sometimes even tripled at high speed, whereas with cycloidal motion, owing to the gradually changing acceleration, the actual dynamic forces are only slightly higher than the theoretical. Therefore, the calculated force due to acceleration should be multiplied by at least a factor of 2 for parabolic and 1.05 for cycloidal motion to provide an allowance for the load-increasing effects of elasticity and backlash.

The main factors influencing cam forces are: 1) displacement and cam speed (forces due to acceleration); 2) dynamic forces due to backlash and flexibility; 3) linkage dimensions which affect weight and weight distribution; 4) pressure angle and friction forces; and 5) spring forces.

The main factors influencing stresses in cams are: 1) radius of curvature for cam and roller; and 2) materials.

Acceleration Forces: The formula for the force acting on a translating body given an acceleration a is:

$$R = \frac{Wa}{g} = \frac{Wa}{386} \qquad (14)$$

In this formula, g = 386 inches/second squared, a = acceleration of W in inches/second squared; R = resultant of all the external forces (except friction) acting on the weight W. For cam analysis purposes, W, in pounds, consists of the weight of the follower, a portion of the weight of the return spring (⅓), and the weight of the members of the external mechanism against which the follower pushes, for example, the weight of a piston:

$$W = W_f + \tfrac{1}{3} W_s + W_e \tag{15}$$

where W = equivalent single weight; W_f = follower weight; W_s = spring weight; and W_e = external weight, all in pounds.

Spring Forces: The return spring, K_s, shown in Fig. 21a must be strong enough to hold the follower against the cam at all times. At high cam speeds the main force attempting to separate the follower from the cam surface is the acceleration force R at the point of maximum *negative* acceleration. Thus, at that point the spring must exert a force F_s,

$$F_s = R - W_f - F_e - F_f \tag{16}$$

where F_e = external force resisting motion of follower, and F_f = friction force from follower guide bushings and other sources.

When the follower is at its lowest position (R_{min} in Fig. 21a), it is usual practice to have the spring provide some estimated preload to account for "set" that takes place in a spring after repeated use and to prevent roller sliding at the start of movement.

The required spring constant, K_s, in pounds per inch of spring deflection is:

$$K_s = \frac{F_s - \text{preload}}{y_a} \tag{17}$$

where y_a = rise of cam from R_{min} to height at which maximum negative acceleration takes place.

The force, F_y, that the spring exerts at any height y above R_{min} is:

$$F_y = yK_s + \text{preload} \tag{18}$$

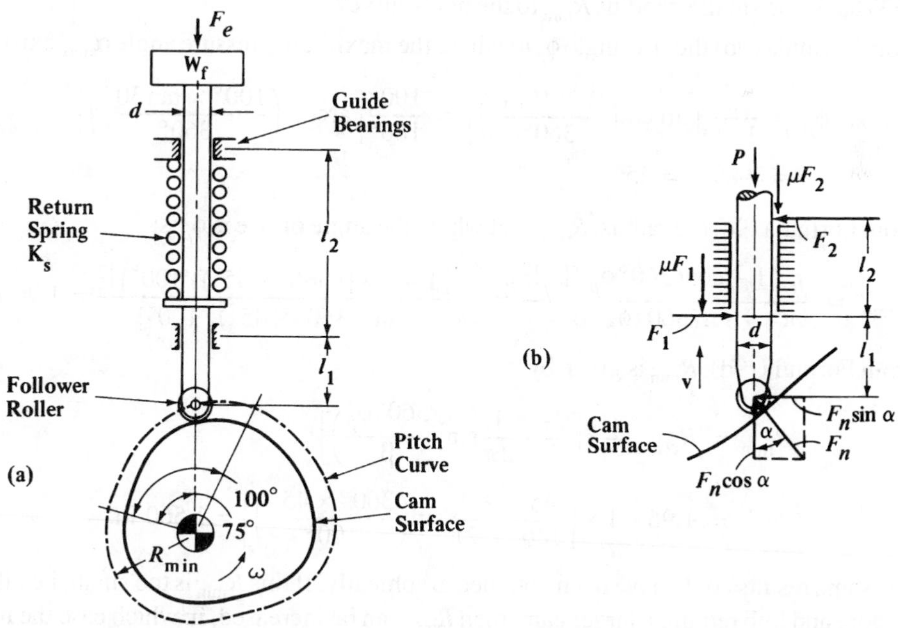

Fig. 21. (a) Radial Translating Follower and Cam System (b) Force Acting on a Translating Follower

Pressure Angle and Friction Forces: As shown in Fig. 21b, the pressure angle of the cam causes a sideways component $F_n \sin\alpha$ which produces friction forces μF_1 and μF_2 in the guide bushing. If the follower rod is too flexible, bending of the follower will increase these friction forces. The effect of the friction forces and the pressure angle are taken into account in the formula,

$$F_n = \frac{P}{\cos\alpha - \frac{\mu\sin\alpha}{l_2}(2l_1 + l_2 - \mu d)} \tag{19a}$$

where μ = coefficient of friction in bushing; l_1, l_2, and d are as shown in Fig. 21; and P = the sum of all the forces acting down against the upward motion of the follower (acceleration force + spring force + follower weight + external force)

$$P = \frac{W \times a}{386} + (yK_s + \text{preload}) + W_f + F_e \tag{19b}$$

Cam Torque: The follower pressing against the cam causes resisting torques during the rise period and assisting torques during the return period. The maximum value of the resisting torque determines the cam drive requirements. Instantaneous torque values may be calculated from

$$T_o = \frac{30vF_n\cos\alpha}{\pi N} = (R_{min} + y)F_n\sin\alpha \tag{20}$$

in which T_o = instantaneous torque in pound-inches.

Exampleof Force Analysis: A radial translating follower system is shown in Fig. 21a. The follower is moved with cycloidal motion over a distance of 1 in. and an angle of lift β = 100°. Cam speed N = 900 rpm. The weight of the follower mass, W_f, is 2 pounds. Both the spring weight W_s and the external weight W_e are negligible. The follower stem diameter is 0.75 in., l_1 = 1.5 in., l_2 = 4 in., coefficient of friction μ = 0.05, external force F_e = 10 lbs, and the pressure angle is not to exceed 30°.

(a) What is the smallest radius R_{min} to the pitch curve?

From Formula (9b) the rise angle ϕ_p to where the maximum pressure angle α_{max} exists is:

$$\phi_p = \frac{\beta}{180°}\left[\text{arccot}\left(\frac{\beta\tan\alpha_{max}}{360°}\right)\right] = \frac{100°}{180°}\left[\text{arccot}\left(\frac{100° \times \tan 30°}{360°}\right)\right]$$
$$= 44.94° = 45°$$

From Formula (9c) the radius, $R_{\alpha\,max}$, at which the angle of rise is ϕ_p is:

$$R_{\alpha max} = \frac{h}{2\pi}\frac{[1-\cos(360°\phi_p/\beta)]^2}{\sin(360°\phi_p/\beta)} = \frac{1}{2\pi}\frac{[1-\cos[(360° \times 45°)/100°]]^2}{\sin[(360° \times 45°)/100°]} = 1.96 \text{ in.}$$

From Formula (9d), R_{min} is given by

$$R_{min} = R_{\alpha max} - h\left[\frac{\phi_p}{\beta} - \frac{1}{2\pi}\sin\left(\frac{360°\phi_p}{\beta}\right)\right]$$
$$= 1.96 - 1 \times \left[\frac{45°}{100°} - \frac{1}{2\pi}\sin\left(\frac{360° \times 45°}{100°}\right)\right] = 1.560 \text{ in.}$$

The same results could have been obtained graphically. If this R_{min} is too small, i.e., if the cam bore and hub require a larger cam, then R_{min} can be increased, in which case the maximum pressure angle will be less than 30°.

(b) If the return spring K_s is specified to provide a preload of 36 lbs when the follower is at R_{min}, what is the spring constant required to hold the follower on the cam throughout the cycle?

The follower tends to leave the cam at the point of maximum *negative* acceleration. Fig. 7 shows this to be at $\phi = \frac{3}{4}\beta = 75°$.

From Formula (4c),

$$a = \frac{2\pi h\omega^2}{\beta^2}\sin\left(\frac{360°\phi}{\beta}\right) = \frac{2\pi \times 1 \times (6 \times 900)^2}{(100°)^2}\sin\left(\frac{360° \times 75°}{100°}\right) = -18{,}300 \text{ in./sec}^2$$

From Formulas (14) and (15),

$$R = \frac{Wa}{386} = \frac{(W_f + \frac{1}{3}W_s + W_e)a}{386} = \frac{(2 + 0 + 0)(-18{,}300\ )}{386} = 95 \text{ lbs (upward)}$$

Using Formula (16) to determine the spring force F_s to hold the follower on the cam,

$$F_s = R - W_f - F_e - F_f$$

as stated on page 2180, the value of R in the above formula should be multiplied by 1.05 for cycloidal motion to provide a factor of safety for dynamic pulses. Thus,

$$F_s = 1.05R - W_f - F_e - F_f = 1.05 \times 95 - 2 - 10 - 0 = 88 \text{ lbs (downward)}$$

The spring constant from Formula (17) is:

$$K_s = \frac{F_s - \text{preload}}{y_a} = \frac{88 - 36}{y_a}$$

and, from Formula (4a) y_a is:

$$y_a = h\left[\frac{\phi}{\beta} - \frac{1}{2\pi}\sin\left(\frac{360°\phi}{\beta}\right)\right] = 1 \times \left[\frac{75°}{100°} - \frac{1}{2\pi}\sin\left(\frac{360° \times 75°}{100°}\right)\right] = 0.909 \text{ in.}$$

so that $K_s = (88 - 36)/0.909 = 57$ lb/in.

(c) At the point where the pressure angle α_{max} is 30° ($\phi = 45°$) the rise of the follower is 1.96 − 1.56 = 0.40 in. What is the normal force, F_n, on the cam? From Formulas (19a) and (19b)

$$F_n = \frac{Wa/386 + yK_s + \text{preload} + W_f + F_e}{\cos\alpha - \dfrac{\mu\sin\alpha}{l_2}(2l_1 + l_2 - \mu d)}$$

using $\phi = 45°$, $h = 1$ in., $\beta = 100°$, and $\omega = 6 \times 900$ in Formula (4c) gives $a = 5660$ in./sec². So that, with $W = 2$ lbs, $y = 0.4$, $K_s = 57$, preload = 36 lbs, $W_f = 2$ lbs, $F_e = 10$ lbs, $\alpha = 30°$, $\mu = 0.05$, $l_1 = 1.5$, $l_2 = 4$, and $d = 0.75$,

$$F_n = \frac{(2 \times 5660)/386 + 0.4 \times 57 + 36 + 2}{\cos 30° - \dfrac{0.05 \times \sin 30°}{4}(2 \times 1.5 + 4 - 0.05 \times 0.75)} = 110 \text{ lbs}$$

Note: If the coefficient of friction had been assumed to be 0, then $F_n = 104$; on the other hand, if the follower is too flexible, so that sidewise bending occurs causing jamming in the bushing, the coefficient of friction may increase to, say, 0.5, in which case the calculated F_n = 200 lbs.

(d) Assuming that in the manufacture of this cam that an error or "bump" resulting from a chattermark or as a result of poor blending occurred, and that this "bump" rose to a height of 0.001 in. in a 1° rise of the cam in the vicinity of $\phi = 45°$. What effect would this bump have on the acceleration force R?

One formula that may be used to calculate the change in acceleration caused by such a cam error is:

$$\Delta a = \pm 2e\left(\frac{6N}{\Delta\phi}\right)^2 \tag{21}$$

where Δa = change in acceleration,

e = error in inches,

$\Delta\phi$ = width of error in degrees. The plus (+) sign is used for a "bump" and the minus (−) sign for a dent or hollow in the surface

For $e = 0.001$, $\Delta\phi = 1°$, and $N = 900$ rpm,

$$\Delta a = +2 \times 0.001\left(\frac{6 \times 900}{1°}\right)^2 = 58{,}320 \text{ in./sec}^2$$

which is 10 times the acceleration calculated for a perfect cam and would cause sufficient force F_n to damage the cam surface. On high speed cams, therefore, accuracy is of considerable importance.

(e) What is the cam torque at $\phi = 45°$?

From Formula (20),

$$T_o = (R_{\min} + y)F_n \sin\alpha$$
$$= (1.56 + 0.4) \times 110 \times \sin 30° = 108 \text{ in.-lbs}$$

(f) What is the radius of curvature at $\phi = 45°$?

From Formula (10),

$$\rho = \frac{\left[r^2 + \left(\frac{dr}{d\phi}\right)^2\right]^{3/2}}{r^2 + 2\left(\frac{dr}{d\phi}\right)^2 - r\left(\frac{d^2r}{d\phi^2}\right)}$$

$$r = R_{\min} + y = 1.56 + 0.4 = 1.96$$

From Formula (13b),

$$\frac{dr}{d\phi} = \frac{180°h}{\pi\beta}\left[1 - \cos\left(\frac{360°\phi}{\beta}\right)\right] = \frac{180° \times 1}{\pi \times 100°}\left[1 - \cos\left(\frac{360° \times 45°}{100°}\right)\right]$$
$$= 1.12$$

From Formula (13c),

$$\frac{d^2r}{d\phi^2} = \frac{2(180°)^2h}{\pi\beta^2}\sin\left(\frac{360°\phi}{\beta}\right) = \frac{2 \times (180°)^2 \times 1}{\pi \times (100°)^2}\sin\left(\frac{360° \times 45°}{100°}\right)$$
$$= 0.64$$

$$\rho = \frac{[(1.96)^2 + (1.12)^2]^{3/2}}{(1.96)^2 + 2(1.12)^2 - 1.96 \times 0.64} = 2.26 \text{ in.}$$

Calculation of Contact Stresses.—When a roller follower is loaded against a cam, the compressive stress developed at the surface of contact may be calculated from

$$S_c = 2290\sqrt{\frac{F_n}{b}\left(\frac{1}{r_f} \pm \frac{1}{R_c}\right)} \tag{22}$$

for a steel roller against a steel cam. For a steel roller on a cast iron cam, use 1850 instead of 2290 in Equation (22).

S_c = maximum calculated compressive stress, psi

F_n = normal load, lb

b = width of cam, inch

R_c = radius of curvature of cam surface, inch

r_f = radius of roller follower, inch

The plus sign in (21) is used in calculating the maximum compressive stress when the roller is in contact with the convex portion of the cam profile and the minus sign is used when the roller is in contact with the concave portion. When the roller is in contact with the straight (flat) portion of the cam profile, $R_c = \infty$ and $1/R_c = 0$. In practice, the greatest compressive stress is most apt to occur when the roller is in contact with that part of the cam profile which is convex and has the smallest radius of curvature.

Example: Given the previous cam example, the radius of the roller r_f = 0.25 in., the convex radius of the cam R_c = (2.26–0.25) in., the width of contact b = 0.3 in., and the normal load F_n = 110 lbs. Find the maximum surface compressive stress. From (21),

$$S_c = 2290\sqrt{\frac{110}{0.3}\left(\frac{1}{0.25} + \frac{1}{2.01}\right)} = 93{,}000 \text{ psi}$$

This calculated stress should be less than the allowable stress for the material selected from Table 2.

Cam Materials: In considering materials for cams it is difficult to select any single material as being the best for every application. Often the choice is based on custom or the machinability of the material rather than its strength. However, the failure of a cam or roller is commonly due to fatigue, so that an important factor to be considered is the limiting wear load, which depends on the surface endurance limits of the materials used and the relative hardnesses of the mating surfaces.

Table 2. Cam Materials

Cam Materials for Use with Roller of Hardened Steel	Maximum Allowable Compressive Stress, psi
Gray-iron casting, ASTM A 48-48, Class 20, 160–190 Bhn, phosphate-coated	58,000
Gray-iron casting, ASTM A 339-51T, Grade 20, 140–160 Bhn	51,000
Nodular-iron casting, ASTM A 339-51T, Grade 80-60-03, 207–241Bhn	72,000
Gray-iron casting, ASTM A 48–48, Class 30, 200–220 Bhn	65,000
Gray-iron casting, ASTM A 48–48, Class 35, 225–225 Bhn	78,000
Gray-iron casting, ASTM A 48-48, Class 30, heat treated (Austempered), 225–300 Bhn	90,000
SAE 1020 steel, 130–150 Bhn	82,000
SAE 4150 steel, heat treated to 270–300 Bhn, phosphate coated	20,000
SAE 4150 steel, heat treated to 270–300 Bhn	188,000
SAE 1020 steel, carburized to 0.045 in. depth of case, 50–58 Rc	226,000
SAE 1340 steel, induction hardened to 45–55 Rc	198,000
SAE 4340 steel, induction hardened to 50–55 Rc	226,000

Based on United Shoe Machinery Corp. data by Guy J. Talbourdet.

In Table 2 are given maximum permissible compressive stresses (surface endurance limits) for various cam materials when in contact with a roller of hardened steel. The stress values shown are based on 100,000,000 cycles or repetitions of stress for pure rolling. Where the repetitions of stress are considerably greater than 100,000,000, where there is appreciable misalignment, or where there is sliding, more conservative stress figures must be used.

Layout of Cylinder Cams.—In Fig. 22 is shown the development of a uniformly accelerated motion cam curve laid out on the surface of a cylindrical cam. This development is necessary for finding the projection on the cylindrical surface, as shown at *KL*. To construct the developed curve, first divide the base circle of the cylinder into, say, twelve equal parts. Set off these parts along line *ag*. Only one-half of the layout has been shown, as the other half is constructed in the same manner, except that the curve is here falling instead of rising. Divide line *aH* into the same number of divisions as the half circle, the divisions being in the proportion 1 : 3 : 5 : 5 : 3 : 1. Draw horizontal lines from these division points and vertical lines from *a*, *b*, *c*, etc. The intersections between the two sets of lines are points on the developed cam curve. These points are transferred to the cylindrical surface at the left by projection in the usual manner.

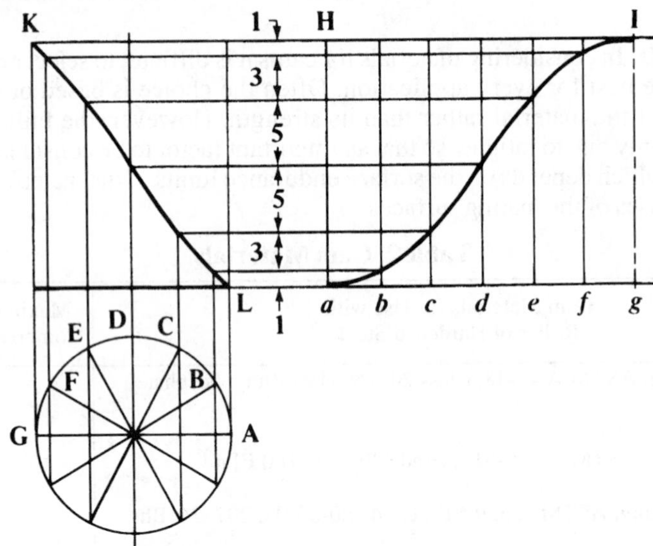

Fig. 22. Development of Cylindrical Cam

Shape of Rolls for Cylinder Cams.—The rolls for cylindrical cams working in a groove in the cam should be conical rather than cylindrical in shape, in order that they may rotate freely and without excessive friction. Fig. 23(a) shows a straight roll and groove, the action of which is faulty because of the varying surface speed at the top and bottom of the groove. Fig. 23(b) shows a roll with curved surface. For heavy work, however, the small bearing area is quickly worn down and the roll presses a groove into the side of the cam as well, thus destroying the accuracy of the movement and creating backlash. Fig. 23(c) shows the conical shape which permits a true rolling action in the groove. The amount of taper depends on the angle of spiral of the cam groove. As this angle, as a rule, is not constant for the whole movement, the roll and groove should be designed to meet the requirements on that section of the cam where the heaviest duty is performed. Frequently the cam groove is of a nearly even spiral angle for a considerable length. The method for determining the angle of the roll and groove to work correctly during the important part of the cycle is as follows:

In Fig. 23(d), *b* is the circumferential distance on the surface of the cam that includes the section of the groove for which correct rolling action is required. The throw of the cam for this circumferential movement is *a*. Line *OU* is the development of the movement of the cam roll during the given part of the cycle, and c is the movement corresponding to b, but on a circle the diameter of which is equal to that of the cam at the bottom of the groove. With the same throw a as before, the line OV will be the development of the cam at the bottom of the groove. OU then is the length of the helix traveled by the top of the roll, while OV is the travel at the bottom of the groove. If, then, the top width and bottom width of the groove be made proportional to OU and OV, the groove will be properly proportioned.

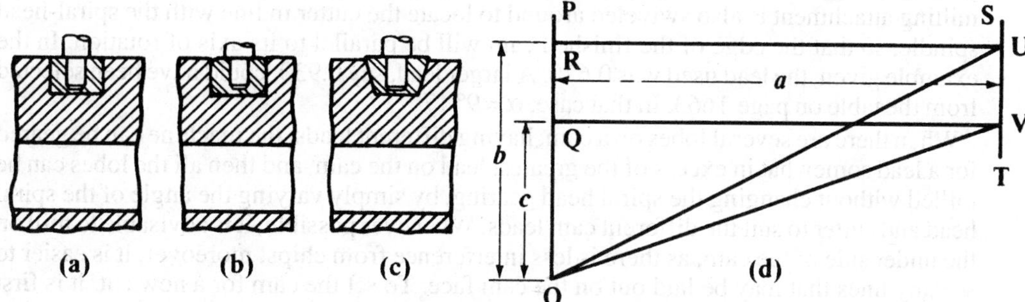

Fig. 23. Shape of Rolls for Cylinder Cams

Cam Milling.—Plate cams having a constant rise, such as are used on automatic screw machines, can be cut in a universal milling machine, with the spiral head set at an angle α, as shown by the illustration. When the spiral head is set vertical, the "lead" of the cam (or its rise for one complete revolution) is the same as the lead for which the machine is geared; but when the spiral head and cutter are inclined, any lead or rise of the cam can be obtained, provided it is less than the lead for which the machine is geared, that is, less than the forward feed of the table for one turn of the spiral-head spindle. The cam lead, then, can be varied within certain limits by simply changing the inclination α of the spiral head and cutter. In the following formulas for determining this angle of inclination, for a given rise of cam and with the machine geared for a lead, L, selected from the tables beginning on page 1933, let

α = angle to which index head and milling attachment are set from horizontal as shown in the accompanying diagram

r = rise of cam in given part of circumference

L = spiral lead for which milling machine is geared

ϕ = angle in which rise is required, expressed in degrees

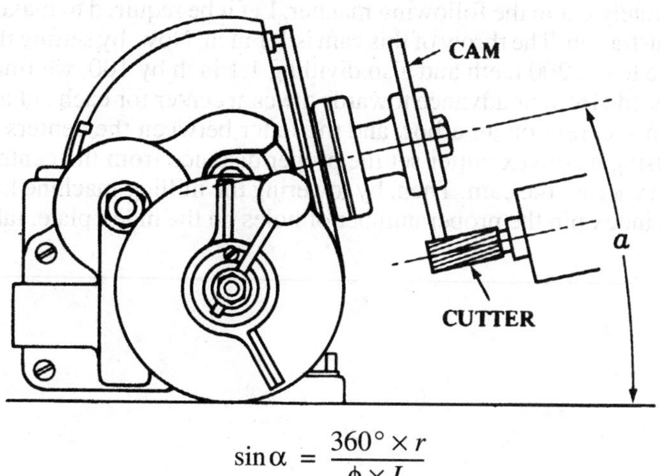

$$\sin\alpha = \frac{360° \times r}{\phi \times L}$$

For example, suppose a cam is to be milled having a rise of 0.125 inch in 300 degrees and that the machine is geared for the smallest possible lead, or 0.670 inch; then:

$$\sin\alpha = \frac{360° \times 0.125}{300° \times 0.670} = 0.2239$$

which is the sine of 12° 56′. Therefore, to secure a rise of 0.125 inch with the machine geared for 0.670 inch lead, the spiral head is elevated to an angle of 12° 56′ and the vertical milling attachment is also swiveled around to locate the cutter in line with the spiral-head spindle, so that the edge of the finished cam will be parallel to its axis of rotation. In the example given, the lead used was 0.670. A larger lead, say 0.930, could have been selected from the table on page 1663. In that case, $\alpha = 9° 17'$.

When there are several lobes on a cam, having different leads, the machine can be geared for a lead somewhat in excess of the greatest lead on the cam, and then all the lobes can be milled without changing the spiral head gearing, by simply varying the angle of the spiral head and cutter to suit the different cam leads. Whenever possible, it is advisable to mill on the under side of the cam, as there is less interference from chips; moreover, it is easier to see any lines that may be laid out on the cam face. To set the cam for a new cut, it is first turned back by operating the handle of the table feed screw, after which the index crank is disengaged from the plate and turned the required amount.

Simple Method for Cutting Uniform Motion Cams.—Some cams are laid out with dividers, machined and filed to the line; but for a cam that must advance a certain

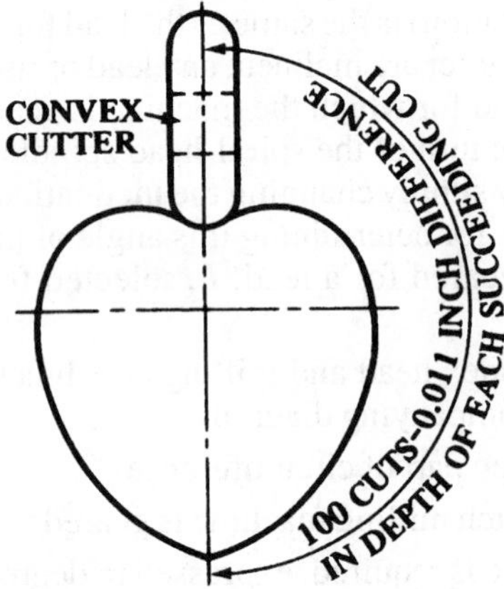

number of thousandths per revolution of spindle this method is not accurate. Cams are easily and accurately cut in the following manner. Let it be required to make the heart cam shown in the illustration. The throw of this cam is 1.1 inch. Now, by setting the index on the milling machine to cut 200 teeth and also dividing 1.1 inch by 100, we find that we have 0.011 inch to recede from or advance towards the cam center for each cut across the cam. Placing the cam securely on an arbor, and the latter between the centers of the milling machine, and using a convex cutter set the proper distance from the center of the arbor, make the first cut across the cam. Then, by lowering the milling machine knee 0.011 inch and turning the index pin the proper number of holes on the index plate, take the next cut and so on.

MACHINE ELEMENTS

PLAIN BEARINGS

THRUST BEARINGS

PLAIN BEARING MATERIALS

BALL, ROLLER, AND NEEDLE BEARINGS

MACHINE ELEMENTS

STANDARD METAL BALLS

LUBRICANTS AND LUBRICATION

COUPLINGS AND CLUTCHES

FRICTION BRAKES

KEYS AND KEYSEATS

FLEXIBLE BELTS AND SHEAVES

FLEXIBLE BELTS AND SHEAVES (cont.)

TRANSMISSION CHAINS

TRANSMISSION CHAINS (cont.)

STANDARDS FOR ELECTRIC MOTORS

STANDARDS FOR ELECTRIC MOTORS (cont.)

ADHESIVES AND SEALANTS

MOTION CONTROL

MOTION CONTROL (cont.)

O-RINGS

ROLLED STEEL SECTIONS, WIRE AND SHEET-METAL GAGES

PIPE AND PIPE FITTINGS

PLAIN BEARINGS

On the following pages are given data and procedures for designing full-film or hydrodynamically lubricated bearings of the journal and thrust types. However, before proceeding to these design methods, it is thought useful to first review those bearing aspects concerning the types of bearings available; lubricants and lubrication methods; hardness and surface finish; machining methods; seals; retainers; and typical length-to-diameter ratios for various applications.

The following paragraphs preceding the design sections provide guidance in these matters and suggest modifications in allowable loads when other than full-film operating conditions exist in a bearing.

Classes of Plain Bearings.—Bearings that provide sliding contact between mating surfaces fall into three general classes: *radial bearings* that support rotating shafts or journals; *thrust bearings* that support axial loads on rotating members; and *guide* or *slipper bearings* that guide moving parts in a straight line. Radial sliding bearings, more commonly called sleeve bearings, may be of several types, the most usual being the plain full journal bearing, which has 360-degree contact with its mating journal, and the partial journal bearing, which has less than 180-degree contact. This latter type is used when the load direction is constant and has the advantages of simplicity, ease of lubrication, and reduced frictional loss.

The relative motions between the parts of plain bearings may take place: 1) As pure sliding without the benefit of a liquid or gaseous lubricating medium between the moving surfaces such as with the dry operation of nylon or Teflon; 2) with hydrodynamic lubrication in which a wedge or film buildup of lubricating medium is produced, with either whole or partial separation of the bearing surfaces; 3) with hydrostatic lubrication in which a lubricating medium is introduced under pressure between the mating surfaces causing a force opposite to the applied load and a lifting or separation of these surfaces; and 4) with a hybrid form or combination of hydrodynamic and hydrostatic lubrication.

Listed below are some of the advantages and disadvantages of sliding contact (plain) bearings as compared with rolling contact (antifriction) bearings.

Advantages: 1) Require less space; 2) are quieter in operation; 3) are lower in cost, particularly in high-volume production; 4) have greater rigidity; and 5) their life is generally not limited by fatigue.

Disadvantages: 1) Have higher frictional properties resulting in higher power consumption; 2) are more susceptible to damage from foreign material in lubrication system; 3) have more stringent lubrication requirements; and 4) are more susceptible to damage from interrupted lubrication supply.

Types of Journal Bearings.—Many types of journal bearing configurations have been developed; some of these are shown in Fig. 1.

Circumferential-groove bearings, Fig. 1(a), have an oil groove extending circumferentially around the bearing. The oil is maintained under pressure in the groove. The groove divides the bearing into two shorter bearings that tend to run at a slightly greater eccentricity. However, the advantage in terms of stability is slight, and this design is most commonly used in reciprocating-load main and connecting-rod bearings because of the uniformity of oil distribution.

Short cylindrical bearings are a better solution than the circumferential-groove bearing for high-speed, low-load service. Often the bearing can be shortened enough to increase the unit loading to a substantial value, causing the shaft to ride at a position of substantial eccentricity in the bearing. Experience has shown that instability rarely results when the shaft eccentricity is greater than 0.6. Very short bearings are not often used for this type of application, because they do not provide a high temporary rotating-load capacity in the event some unbalance should be created in the rotor during service.

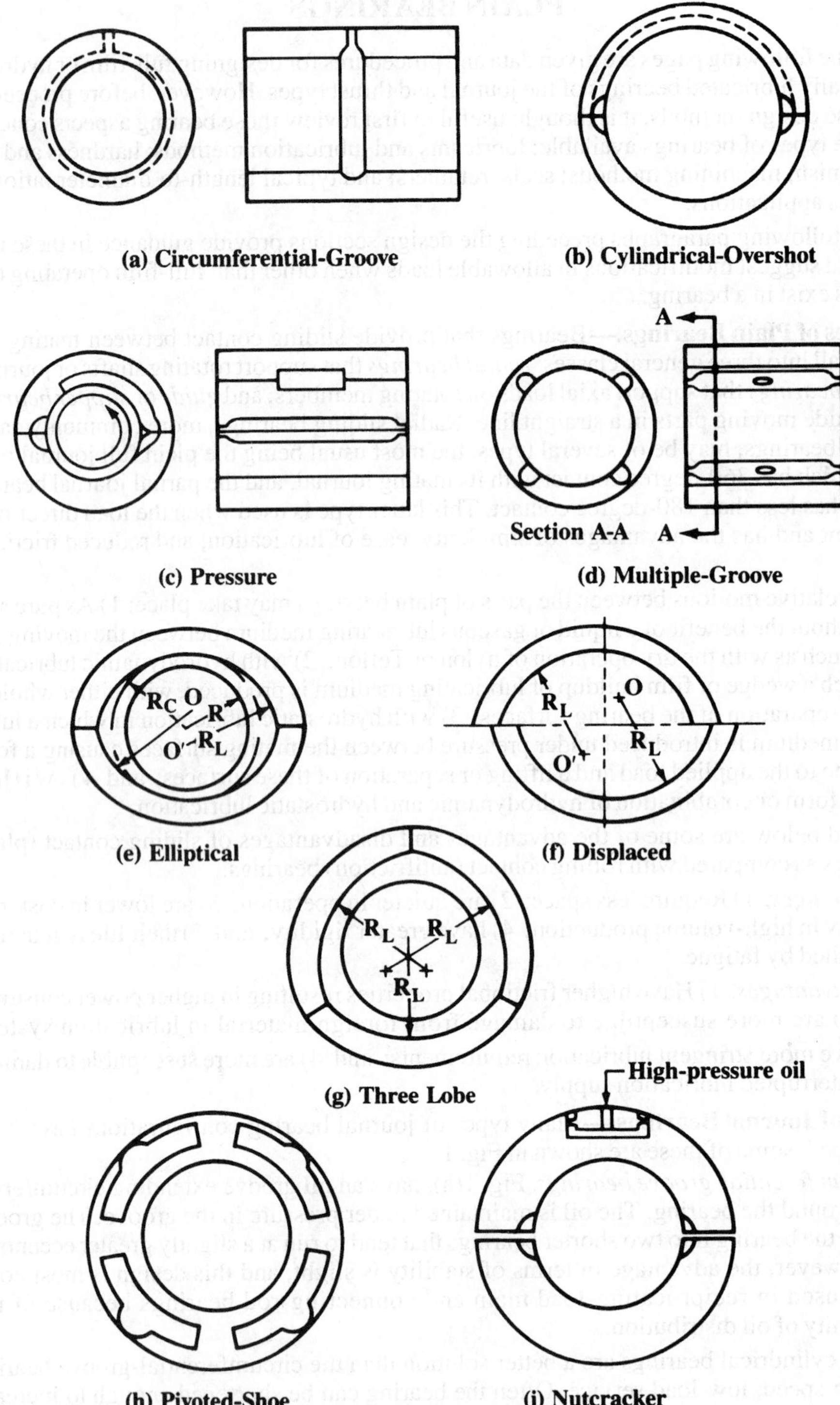

Fig. 1. Typical shapes of several types of pressure-fed bearings.

Cylindrical-overshot bearings,: Fig. 1(b), are used where surface speeds of 10,000 fpm or more exist, and where additional oil flow is desired to maintain a reasonable bearing temperature. This bearing has a wide circumferential groove extending from one axial oil groove to the other over the upper half of the bearing. Oil is usually admitted to the trailing-edge oil groove. An inlet orifice is used to control the oil flow. Cooler operation results from the elimination of shearing action over a large section of the upper half of the bearing and, to a great extent, from the additional flow of cool oil over the top half of the bearing.

Pressure bearings,: Fig. 1(c), employ a groove over the top half of the bearing. The groove terminates at a sharp dam about 45 degrees beyond the vertical in the direction of shaft rotation. Oil is pumped into this groove by shear action from the rotation of the shaft and is then stopped by the dam. In high-speed operation, this situation creates a high oil pressure over the upper half of the bearing. The pressure created in the oil groove and surrounding upper half of the bearing increases the load on the lower half of the bearing. This self-generated load increases the shaft eccentricity. If the eccentricity is increased to 0.6 or greater, stable operation under high-speed, low-load conditions can result. The central oil groove can be extended around the lower half of the bearing, further increasing the effective loading. This design has one primary disadvantage: Dirt in the oil will tend to abrade the sharp edge of the dam and impair ability to create high pressures.

Multiple-groove bearings,: Fig. 1(d), are sometimes used to provide increased oil flow. The interruptions in the oil film also appear to give this bearing some merit as a stable design.

Elliptical bearings,: Fig. 1(e), are not truly elliptical, but are formed from two sections of a cylinder. This two-piece bearing has a large clearance in the direction of the split and a smaller clearance in the load direction at right angles to the split. At light loads, the shaft runs eccentric to both halves of the bearing, and hence, the elliptical bearing has a higher oil flow than the corresponding cylindrical bearing. Thus, the elliptical bearing will run cooler and will be more stable than a cylindrical bearing.

Elliptical-overshot bearings: (not shown) are elliptical bearings in which the upper half is relieved by a wide oil groove connecting the axial oil grooves. They are analogous to cylindrical-overshot bearings.

Displaced elliptical bearings,: Fig. 1(f), shift the centers of the two bearing arcs in both a horizontal and a vertical direction. This design has greater stiffness than a cylindrical bearing, in both horizontal and vertical directions, with substantially higher oil flow. It has not been extensively used, but offers the prospect of high stability and cool operation.

Three-lobe bearings,: Fig. 1(g), are made up in cross section of three circular arcs. They are most effective as antioil whip bearings when the centers of curvature of each of the three lobes lie well outside the clearance circle that the shaft center can describe within the bearing. Three axial oil-feed grooves are used. It is a more difficult design to manufacture, because it is almost necessary to make it in three parts instead of two. The bore is machined with shims between each of the three parts. The shims are removed after machining is completed.

Pivoted-shoe bearings,: Fig. 1(h), are one of the most stable bearings. The bearing surface is divided into three or more segments, each of which is pivoted at the center. In operation, each shoe tilts to form a wedge-shaped oil film, thus creating a force tending to push the shaft toward the center of the bearing. For single-direction rotation, the shoes are sometimes pivoted near one end and forced toward the shaft by springs.

Nutcracker bearings,: Fig. 1(i), consist of two cylindrical half-bearings. The upper half-bearing is free to move in a vertical direction and is forced toward the shaft by a hydraulic cylinder. External oil pressure may be used to create load on the upper half of the bearing through the hydraulic cylinder. Or the high-pressure oil may be obtained from the lower half of the bearing by tapping a hole into the high-pressure oil film, thus creating a self-loading bearing. Either type can increase bearing eccentricity to the point where stable operation can be achieved.

Hydrostatic Bearings.—Hydrostatic bearings are used when operating conditions require full film lubrication that cannot be developed hydrodynamically. The hydrostatically lubricated bearing, either thrust or radial, is supplied with lubricant under pressure from an external source. Some advantages of the hydrostatic bearing over bearings of other types are: low friction; high load capacity; high reliability; high stiffness; and long life.

Hydrostatic bearings are used successfully in many applications including machine tools, rolling mills, and other heavily loaded slow-moving machinery. However, specialized techniques, including a thorough understanding of hydraulic components external to the bearing package is required. The designer is cautioned against use of this type of bearing without a full knowledge of all aspects of the problem. Determination of the operating performance of hydrostatic bearings is a specialized area of the lubrication field and is described in specialized reference books.

Design.—The design of a sliding bearing is generally accomplished in one of two ways:

1) a bearing operating under similar conditions is used as a model or basis from which the new bearing is designed; and 2) in the absence of any previous experience with similar bearings in similar environments, certain assumptions concerning operating conditions and requirements are made and a tentative design prepared based on general design parameters or rules of thumb. Detailed lubrication analysis is then performed to establish design and operating details and requirements.

Modes of Bearing Operation.—The load-carrying ability of a sliding bearing depends upon the kind of fluid film that is formed between its moving surfaces. The formation of this film is dependent, in part, on the design of the bearing and, in part, on the speed of rotation. The bearing has three modes or regions of operation designated as *full-film, mixed-film,* and *boundary* lubrication with effects on bearing friction, as shown in Fig. 2.

In terms of physical bearing operation these three modes may be further described as follows:

1) Full-film, or hydrodynamic, lubrication produces a complete physical separation of the sliding surfaces resulting in low friction and long wear-free service life.

To promote full-film lubrication in hydrodynamic operation, the following parameters should be satisfied: 1) Lubricant selected has the correct viscosity for the proposed operation; 2) proper lubricant flow rates are maintained; 3) proper design methods and considerations have been utilized; and 4) surface velocity in excess of 25 feet per minute is maintained.

When full-film lubrication is achieved, a coefficient of friction between 0.001 and 0.005 can be expected.

2) Mixed-film lubrication is a mode of operation between the full-film and boundary modes. With this mode, there is a partial separation of the sliding surfaces by the lubricant film; however, as in boundary lubrication, limitations on surface speed and wear will result. With this type of lubrication, a surface velocity in excess of 10 feet per minute is required with resulting coefficients of friction of 0.02 to 0.08.

3) Boundary lubrication takes place when the sliding surfaces are rubbing together with only an extremely thin film of lubricant present. This type of operation is acceptable only in applications with oscillating or slow rotary motion. In complete boundary lubrication, the oscillatory or rotary motion is usually less than 10 feet per minute with resulting coefficients of friction of 0.08 to 0.14. These bearings are usually grease lubricated or periodically oil lubricated.

In starting up and accelerating to its operating point, a journal bearing passes through all three modes of operation. At rest, the journal and bearing are in contact, and thus when starting, the operation is in the boundary lubrication region. As the shaft begins to rotate more rapidly and the hydrodynamic film starts to build up, bearing operation enters the region of mixed-film lubrication. When design speeds and loads are reached, the hydrodynamic action in a properly designed bearing will promote full-film lubrication.

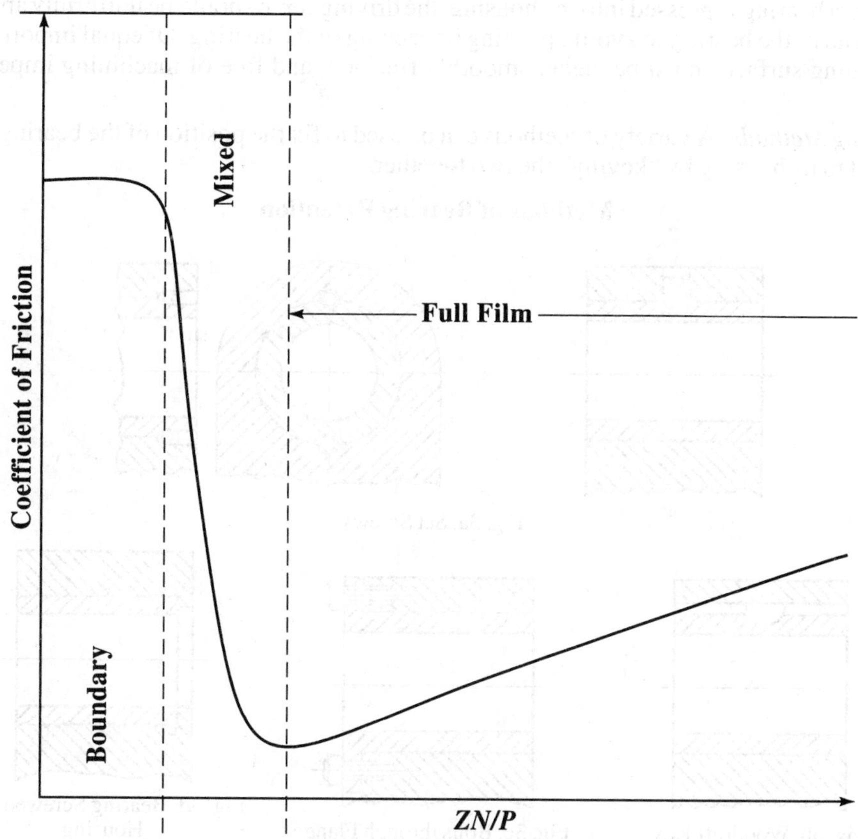

Fig. 2. Three modes of bearing operation.

Methods of Retaining Bearings.—Several methods are available to ensure that a bearing remains in place within a housing. Which method to use depends upon the particular application but requires first that the unit lends itself to convenient assembly and disassembly; additionally, the bearing wall should be of uniform thickness to avoid introduction of weak points in the construction that may lead to elastic or thermal distortion.

Press or Shrink Fit: One common and satisfactory technique for retaining the bearing is to press or shrink the bearing in the housing with an interference fit. This method permits the use of bearings having uniform wall thickness over the entire bearing length.

Standard bushings with finished inside and outside diameters are available in sizes up to approximately 5 inches inside diameter. Stock bushings are commonly provided 0.002 to 0.003 inch over nominal on outside diameter sizes of 3 inches or less. For diameters greater than 3 inches, outside diameters are 0.003 to 0.005 inch over nominal. Because these tolerances are built into standard bushings, the amount of press fit is controlled by the housing-bore size.

As a result of a press or shrink fit, the bore of the bearing material "closes in" by some amount. In general, this diameter decrease is approximately 70 to 100 per cent of the amount of the interference fit. Any attempt to accurately predict the amount of reduction, in an effort to avoid final clearance machining, should be avoided.

Shrink fits may be accomplished by chilling the bearing in a mixture of dry ice and alcohol, or in liquid air. These methods are easier than heating the housing and are preferred. Dry ice in alcohol has a temperature of −110 degrees F and liquid air boils at −310 degrees F.

When a bearing is pressed into the housing, the driving force should be uniformly applied to the end of the bearing to avoid upsetting or peening of the bearing. Of equal importance, the mating surfaces must be clean, smoothly finished, and free of machining imperfections.

Keying Methods: A variety of methods can be used to fix the position of the bearing with respect to its housing by "keying" the two together.

Methods of Bearing Retention

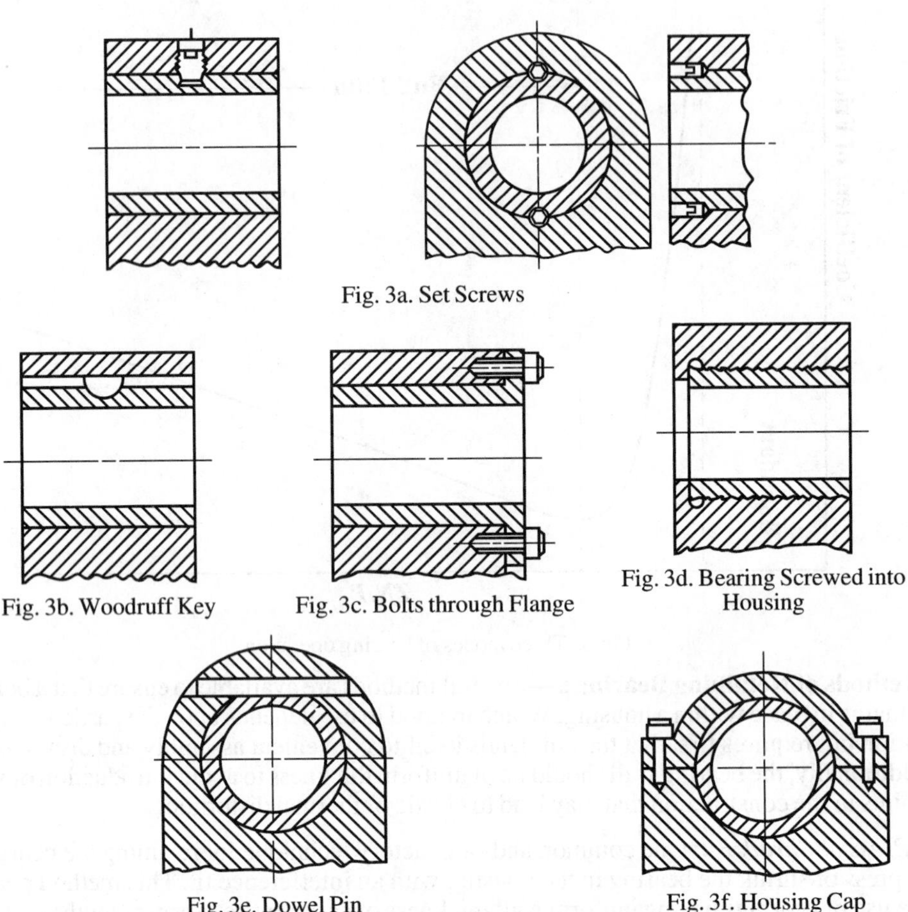

Fig. 3a. Set Screws

Fig. 3b. Woodruff Key

Fig. 3c. Bolts through Flange

Fig. 3d. Bearing Screwed into Housing

Fig. 3e. Dowel Pin

Fig. 3f. Housing Cap

Possible keying methods are shown in Figs. 3a through 3f including: D) set screws; E) Woodruff keys; F) bolted bearing flanges; G) threaded bearings; H) dowel pins; and I) housing caps.

Factors to be considered when selecting one of these methods are as follows:

1) Maintaining uniform wall thickness of the bearing material, if possible, especially in the load-carrying region of the bearing.

2) Providing as much contact area as possible between bearing and housing. Mating surfaces should be clean, smooth, and free from imperfections to facilitate heat transfer.

3) Preventing any local deformation of the bearing that might result from the keying method. Machining after keying is recommended.

4) Considering the possibility of bearing distortion resulting from the effect of temperature changes on the particular keying method.

Methods of Sealing.—In applications where lubricants or process fluids are utilized in operation, provision must be made normally to prevent leakage to other areas. This provision is made by the use of static and dynamic type sealing devices. In general, three terms are used to describe the devices used for sealing:

Seal: A means of preventing migration of fluids, gases, or particles across a joint or opening in a container.

Packing: A dynamic seal, used where some form of relative motion occurs between rigid members of an assembly.

Gaskets: A static seal, used where there is no relative motion between joined parts.

Two major functions must be achieved by all sealing applications: prevent escape of fluid; and prevent migration of foreign matter from the outside.

The first determination in selecting the proper seal is whether the application is static or dynamic. To meet the requirements of a static application there must be no relative motion between the joining parts or between the seal and the mating part. If there is any relative motion, the application must be considered dynamic, and the seal selected accordingly.

Dynamic sealing requires control of fluids leaking between parts with relative motion. Two primary methods are used to this end: positive contact or rubbing seals; and controlled clearance noncontact seals.

Positive Contact or Rubbing Seals: These seals are used where positive containment of liquids or gases is required, or where the seal area is continuously flooded. If properly selected and applied, contact seals can provide zero leakage for most fluids. However, because they are sensitive to temperature, pressure, and speed, improper application can result in early failure. These seals are applicable to rotating and reciprocating shafts. In many assemblies, positive-contact seals are available as off-the-shelf items. In other instances, they are custom-designed to the special demands of a particular application. Custom design is offered by many seal manufacturers and, for extreme cases, probably offers the best solution to the sealing problem.

Controlled Clearance Noncontact Seals: Representative of the controlled-clearance seals, which includes all seals in which there is no rubbing contact between the rotating and stationary members, are throttling bushings and labyrinths. Both types operate by fluid-throttling action in narrow annular or radial passages.

Clearance seals are frictionless and very insensitive to temperature and speed. They are chiefly effective as devices for limiting leakage rather than stopping it completely. Although they are employed as primary seals in many applications, the clearance seal also finds use as auxiliary protection in contact-seal applications. These seals are usually designed into the equipment by the designer himself, and they can take on many different forms.

Advantages of this seal are that friction is kept to an absolute minimum and there is no wear or distortion during the life of the equipment. However, there are two significant disadvantages: The seal has limited use when leakage rates are critical; and it becomes quite costly as the configuration becomes elaborate.

Static Seals: Static seals such as gaskets. "O" rings, and molded packings cover very broad ranges of both design and materials.

Some of the typical types are as follows: 1) Molded packings:A. lip type, B. squeeze-molded; 2) simple compression packings; 3) diaphragm seals; 4) nonmetallic gaskets; 5) "O" rings;; and 6) metallic gaskets and "O" rings..

Data on "O" rings are found starting on page 2482.

Detailed design information for specific products should be obtained directly from manufacturers.

Hardness and Surface Finish.—Even in well-lubricated full-film sleeve bearings, momentary contact between journal and bearing may occur under such conditions as starting, stopping, or overloading. In mixed-film and boundary-film lubricated sleeve bearings, continuous metal-to-metal contact occurs. Hence, to allow for any necessary wearing-in, the journal is usually made harder than the bearing material. This arrangement allows the effects of scoring or wearing to take place on the bearing, which is more easily replaced, rather than on the more expensive shaft. As a general rule, recommended Brinell (Bhn) hardness of the journal is at least 100 points harder than the bearing material.

The softer cast bronzes used for bearings are those with high lead content and very little tin. Such bronzes give adequate service in boundary-and mixed-film applications where full advantage is taken of their excellent "bearing" characteristics.

High-tin, low-lead content cast bronzes are the harder bronzes and these have high ulimate load-carrying capacity: higher journal hardnesses are required with these bearing bronzes. Aluminum bronze, for example, requires a journal hardness in the range of 550 to 600 Bhn.

In general, harder bearing materials require better alignment and more reliable lubrication to minimize local heat generation if and when the journal touches the shaft. Also, abrasives that find their way into the bearing are a problem for the harder bearing materials and greater care should be taken to exclude them.

Surface Finish: Whether bearing operation is complete boundary, mixed film, or fluid film, surface finishes of the journal and bearing must receive careful attention. In applications where operation is hydrodynamic or full-film, peak surface variations should be less than the expected minimum film thickness; otherwise, peaks on the journal surface will contact peaks on the bearing surface, with resulting high friction and temperature rise. Ranges of surface roughness obtained by various finishing methods are: boring, broaching, and reaming, 32 to 64 microinches, rms; grinding, 16 to 64 microinches, rms; and fine grinding, 4 to 16 microinches, rms.

In general, the better surface finishes are required for full-film bearings operating at high eccentricity ratios because full-film lubrication must be maintained with small clearances, and metal-to-metal contact must be avoided. Also, the harder the material, the better the surface finish required. For boundary- and mixed-film applications, surface finish requirements may be somewhat relaxed because bearing wear-in will in time smooth the surfaces.

Fig. 4 is a general guide to the ranges required for bearing and journal surface finishes. Selecting a particular surface finish in each range can be simplified by observing the general rule that smoother finishes are required for the harder materials, for high loads, and for high speeds.

Machining.—The methods most commonly used in finishing journal bearing bores are boring, broaching, reaming, and burnishing.

Broaching is a rapid finishing method providing good size and alignment control when adequate piloting is possible. Soft babbitt materials are particularly compatible with the broaching method. A third finishing method, reaming, facilitates good size and alignment control when piloting is utilized. Reaming can be accomplished both manually or by machine, the machine method being preferred. Burnishing is a fast sizing operation that gives good alignment control, but does not give as good size control as the cutting methods. It is not recommended for soft materials such as babbitt. Burnishing has an ironing effect that gives added seating of the bushing outside diameter in the housing bore; consequently, it is often used for this purpose, especially on a $\frac{1}{32}$-inch wall bushing, even if a further sizing operation is to be used subsequently.

Boring of journal bearings provides the best concentricity, alignment, and size control and is the finishing method of choice when close tolerances and clearances are desirable.

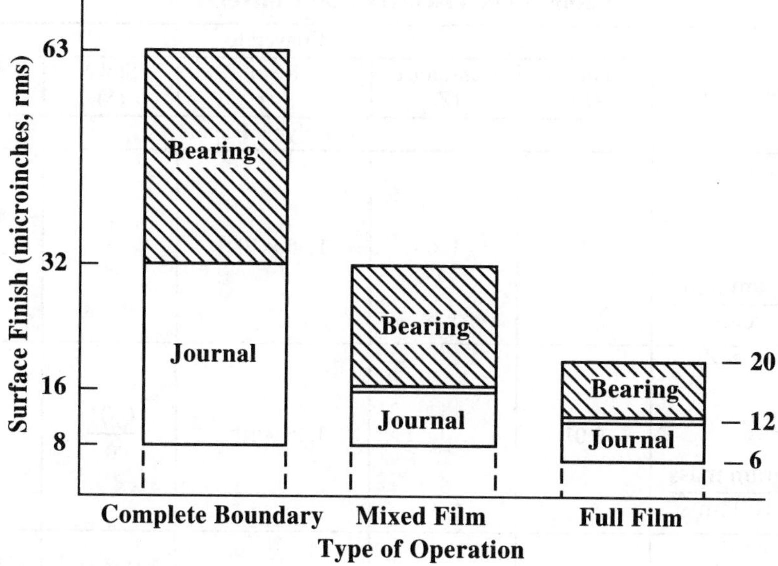

Fig. 4. Recommended ranges of surface finish for the three types of sleeve bearing operations.

Methods of Lubrication.—There are numerous ways to supply lubricant to bearings. The more common of these are described in the following.

Pressure lubrication,: in which an abundance of oil is fed to the bearing from a central groove, single or multiple holes, or axial grooves, is effective and efficient. The moving oil assists in flushing dirt from the bearing and helps keep the bearing cool. In fact, it removes heat faster than other lubricating methods and, therefore, permits thinner oil films and unimpaired load capacities. The oil-supply pressure needed for bushings carrying the basic load is directly proportional to the shaft speed, but for most installations, 50 psi will be adequate.

Splash fed: applies to a variety of intermittently lubricated bushings. It includes everything from bearings spattered with oil from the action of other moving parts to bearings regularly dipped in oil. Like oil bath lubrication, splash feeding is practical when the housing can be made oiltight and when the moving parts do not churn the oil. The fluctuating nature of the load and the intermittent oil supply in splash fed applications requires the designer to use experience and judgment when determining the probable load capacity of bearings lubricated in this way.

Oil bath lubrication,: in which the bushing is submerged in oil, is the most reliable of all methods except pressure lubrication. It is practical if the housing can be made oil tight, and if the shaft speed is not so great as to cause excessive churning of the oil.

Oil ring lubrication,: in which oil is supplied to the bearing by a ring in contact with the shaft, will, within reasonable limits, bring enough oil to the bearing to maintain hydrodynamic lubrication. If the shaft speed is too low, little oil will follow the ring to the bearing; and, if the speed is too high, the ring speed will not keep pace with the shaft. Also, a ring revolving at high speed will lose oil by centrifugal force. For best results, the peripheral speed of the shaft should be between 200 and 2000 feet per minute. Safe load to achieve hydrodynamic lubrication should be one-half of that for pressure fed bearings. Unless the load is light, hydrodynamic lubrication is doubtful. The safe load, then, to achieve hydrodynamic lubrication, should be one-quarter of that of pressure fed bearings.

Table 1. Oil Viscosity Unit Conversion

Convert from	Convert to				
	Poise (P)	Centipoise (Z)	Reyn (μ)	Stoke (S)	Centistoke (ν)
	Multiplying Factors				
Poise (P) $\frac{\text{dyne-s}}{\text{cm}^2}$ or $\frac{\text{gram mass}}{\text{cm-s}}$	1	100	1.45×10^{-5}	$\frac{1}{\rho}$	$\frac{100}{\rho}$
Centipoise (Z) $\frac{\text{dyne-s}}{100\ \text{cm}^2}$ or $\frac{\text{gram mass}}{100\text{cm-s}}$	0.01	1	1.45×10^{-7}	$\frac{0.01}{\rho}$	$\frac{1}{\rho}$
Reyn (μ) $\frac{\text{lb force-s}}{\text{in}^2}$	6.9×10^4	6.9×10^4	1	$\frac{6.9 \times 10^4}{\rho}$	$\frac{6.9 \times 10^6}{\rho}$
Stoke (S) $\frac{\text{cm}^2}{\text{s}}$	ρ	$100\ \rho$	$1.45 \times 10^{-5}\ \rho$	1	100
Centistoke (ν) $\frac{\text{cm}^2}{100\ \text{s}}$	$0.01\ \rho$	ρ	$1.45 \times 10^{-7}\ \rho$	0.01	1

ρ = Specific gravity of the oil.

To convert from a value in the "Convert from" column to a value in a "Convert to" column, multiply the "Convert from" column by the figure in the intersecting block, e.g. to change from Centipoise to Reyn, multiply Centipoise value by 1.45×10^{-7}.

Wick or waste pack lubrication: delivers oil to a bushing by the capillary action of a wick or waste pack; the amount delivered is proportional to the size of the wick or pack.

Lubricants: The value of an oil as a lubricant depends mainly on its film-forming capacity, that is, its capability to maintain a film of oil between the bearing surfaces. The film-forming capacity depends to a large extent on the viscosity of the oil, but this should not be understood to mean that oil of the highest viscosity is always the most suitable lubricant. For practical reasons, an oil of the lowest viscosity that will retain an unbroken oil film between the bearing surfaces is the most suitable for purposes of lubrication. A higher viscosity than that necessary to maintain the oil film results in a waste of power due to the expenditure of energy necessary to overcome the internal friction of the oil itself.

Fig. 5 provides representative values of viscosity in centipoises for SAE mineral oils. Table 1 is provided as a means of converting viscosities of other units to centipoises.

Grease: packed in a cavity surrounding the bushing is less adequate than an oil system, but it has the advantage of being more or less permanent. Although hydrodynamic lubrication is possible under certain very favorable circumstances, boundary lubrication is the usual state.

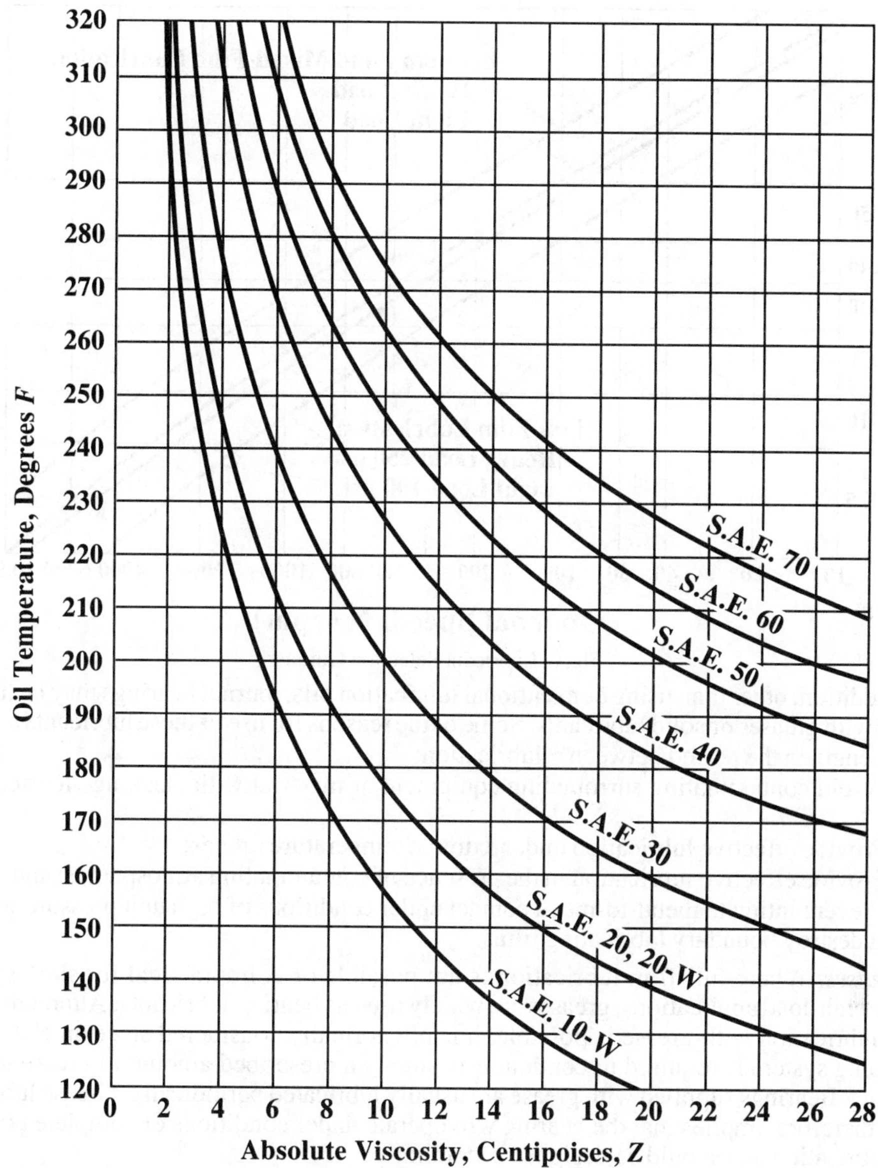

Fig. 5. Viscosity vs. temperature—SAE oils.

Lubricant Selection.— In selecting lubricants for journal bearing operation, several factors must be considered: 1) Type of operation (full, mixed, or boundary film) anticipated;

2) Surface speed; and 3) Bearing loading.

Fig. 6 combines these factors and facilitates general selection of the proper lubricant viscosity range.

As an example of using these curves, consider a lightly loaded bearing operating at 2000 rpm. At the bottom of the figure, locate 2000 rpm and move vertically to intersect the light-load full-film lubrication curve, which indicates an SAE 5 oil.

As a general rule-of-thumb, heavier oils are recommended for high loads and lighter oils for high speeds.

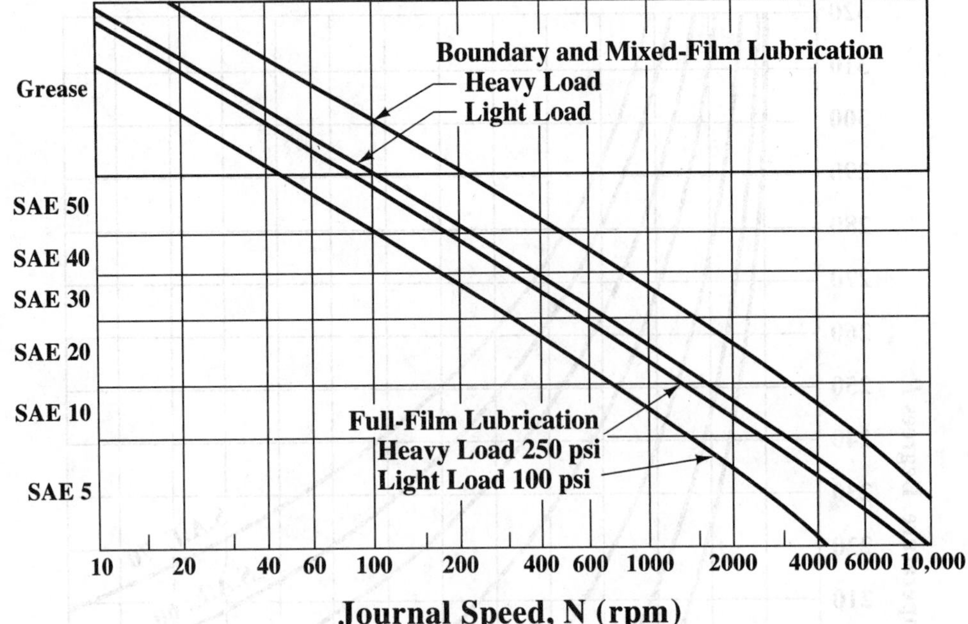

Fig. 6. Lubricant Selection Guide

In addition, other than using conventional lubrication oils, journal bearings may be lubricated with greases or solid lubricants. Some of the reasons for use of these lubricants are to:

1) Lengthen the period between relubrication:

2) Avoid contaminating surrounding equipment or material with "leaking" lubricating oil;

3) Provide effective lubrication under extreme temperature ranges;

4) Provide effective lubrication in the presence of contaminating atmospheres; and

5) Prevent intimate metal-to-metal contact under conditions of high unit pressure which might destroy boundary lubricating films.

Greases: Where full-film lubrication is not possible or is impractical for slow-speed fairly high-load applications, greases are widely used as bearing lubricants. Although full-film lubrication with grease is possible, it is not normally considered since an elaborate pumping system is required to continuously supply a prescribed amount of grease to the bearing. Bearings supplied with grease are usually lubricated periodically. Grease lubrication, therefore, implies that the bearing will operate under conditions of complete boundary lubrication and should be designed accordingly.

Lubricating greases are essentially a combination of a mineral lubricating oil and a thickening agent, which is usually a metallic soap. When suitably mixed, they make excellent bearing lubricants. There are many different types of greases which, in general, may be classified according to the soap base used. Information on commonly used greases is shown in Table 2.

Synthetic greases are composed of normal types of soaps but use synthetic hydrocarbons instead of normal mineral oils. They are available in a range of consistencies in both water-soluble and insoluble types. Synthetic greases can accommodate a wide range of variation in operating temperature; however, recommendations on special-purpose greases should be obtained from the lubricant manufacturer.

Application of grease is accomplished by one of several techniques depending upon grease consistency. These classifications are shown in Table 3 along with typical methods of application. Grooves for grease are generally greater in width, up to 1.5 times, than for oil.

Table 2. Commonly Used Greases and Solid Lubricants

Type	Operating Temperature, Degrees F	Load	Comments
Greases			
Calcium or lime soap	160	Moderate	…
Sodium soap	300	Wide	For wide speed range
Aluminum soap	180	Moderate	…
Lithium soap	300	Moderate	Good low temperature
Barium soap	350	Wide	…
Solid Lubricants			
Graphite	1000	Wide	…
Molybdenum disulfide	−100 to 750	Wide	…

Table 3. NLGI Consistency Numbers

NLGI Consistency No.	Consistency of Grease	Typical Method of Application
0	Semifluid	Brush or gun
1	Very soft	Pin-type cup or gun
2	Soft	Pressure gun or centralized pressure system
3	Light cup grease	Pressure gun or centralized pressure system
4	Medium cup grease	Pressure gun or centralized pressure system
5	Heavy cup grease	Pressure gun or hand
6	Block grease	Hand, cut to fit

NLGI is National Lubricating Grease Institute

Coefficients of friction for grease-lubricated bearings range from 0.08 to 0.16, depending upon consistency of the grease, frequency of lubrication, and type of grease. An average value of 0.12 may be used for design purposes.

Solid Lubricants: The need for effective high-temperature lubricants led to the development of several solid lubricants. Essentially, solid lubricants may be described as low-shear-strength solid materials. Their function within a bronze bearing is to act as an intermediary material between sliding surfaces. Since these solids have very low shear strength, they shear more readily than the bearing material and thereby allow relative motion. So long as solid lubricant remains between the moving surfaces, effective lubrication is provided and friction and wear are reduced to acceptable levels.

Solid lubricants provide the most effective boundary films in terms of reduced friction, wear, and transfer of metal from one sliding component to the other. However, there is a significant deterioration in these desirable properties as the operating temperature of the boundary film approaches the melting point of the solid film. At this temperature the friction may increase by a factor of 5 to 10 and the rate of metal transfer may increase by as much as 1000. What occurs is that the molecules of the lubricant lose their orientation to the surface that exists when the lubricant is solid. As the temperature further increases, additional deterioration sets in with the friction increasing by some additional small amount but the transfer of metal accelerates by an additional factor of 20 or more. The final effect of too high temperature is the same as metal-to-metal contact without benefit of lubricant. These changes, which are due to the physical state of the lubricant, are reversed when cooling takes place.

The effects just described also partially explain why fatty acid lubricants are superior to paraffin base lubricants. The fatty acid lubricants react chemically with the metallic surfaces to form a metallic soap that has a higher melting point than the lubricant itself, the result being that the breakdown temperature of the film, now in the form of a metallic soap is raised so that it acts more like a solid film lubricant than a fluid film lubricant.

Journal or Sleeve Bearings

Although this type of bearing may take many shapes and forms, there are always three basic components: journal or shaft, bushing or bearing, and lubricant. Fig. 7 shows these components with the nomenclature generally used to describe a journal bearing: W = applied load, N = revolution, e = eccentricity of journal center to bearing center, θ = attitude angle, which is the angle between the applied load and the point of minimum film thickness, d = diameter of the shaft, c_d = bearing clearance, $d + c_d$ = diameter of the bearing and h_o = minimum film thickness.

Grooving and Oil Feeding.—Grooving in a journal bearing has two purposes: 1) to establish and maintain an efficient film of lubricant between the bearing moving surfaces and; and 2) to provide adequate bearing cooling.

The obvious and only practical location for introducing lubricant to the bearing is in a region of low pressure. A typical pressure profile of a bearing is shown by Fig. 8. The arrow W shows the applied load. Typical grooving configurations used for journal bearings are shown in Figs. 10a through 10e.

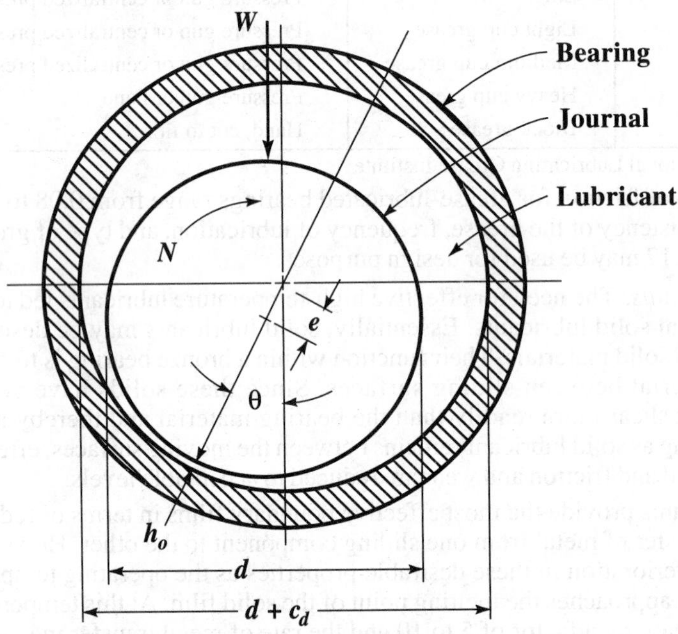

Fig. 7. Basic components of a journal bearing.

Heat Radiating Capacity.—In a self-contained lubrication system for a journal bearing, the heat generated by bearing friction must be removed to prevent continued temperature rise to an unsatisfactory level. The heat-radiating capacity H_R of the bearing in foot-pounds per minute may be calculated from the formula $H_R = Ld\,Ct_R$ in which C is a constant determined by O. Lasche, and t_R is temperature rise in degrees Fahrenheit.

Values for the product Ct_R may be found from the curves in Fig. 9 for various values of bearing temperature rise t_R and for three operating conditions. In this equation, L = total length of the bearing in inches and d = bearing diameter in inches.

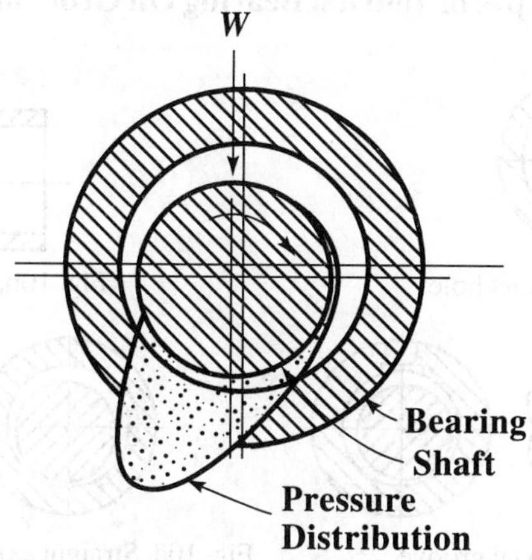

Fig. 8. Typical pressure profile of journal bearing.

Journal Bearing Design Notation.—The symbols used in the following step-by-step procedure for lubrication analysis and design of a plain sleeve or journal bearing are as follows:

c = specific heat of lubricant, Btu/lb/degree F

c_d = diametral clearance, inches

C_n = bearing capacity number

d = journal diameter, inches

e = eccentricity, inches

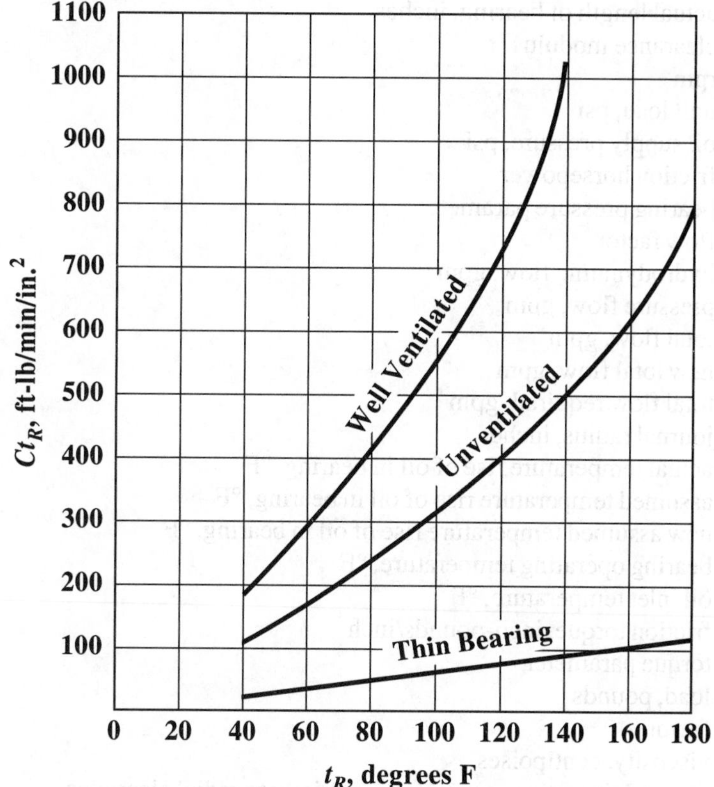

Fig. 9. Heat-radiating capacity factor, Ct_R, vs. bearing temperature rise, t_R—journal bearings.

Types of Journal Bearing Oil Grooving

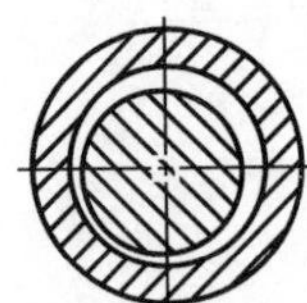

Fig. 10a. Single inlet hole

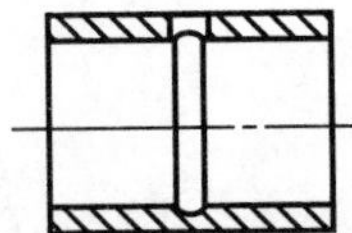

Fig. 10b. Circular groove

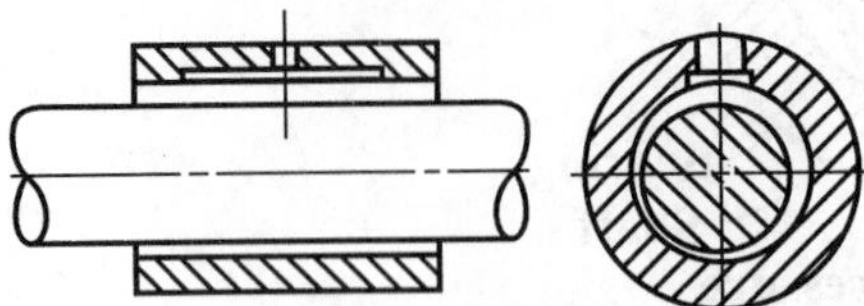

Fig. 10c. Straight axial groove

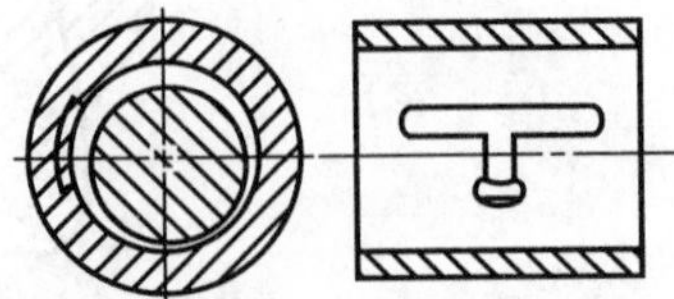

Fig. 10d. Straight axial groove with feeder groove

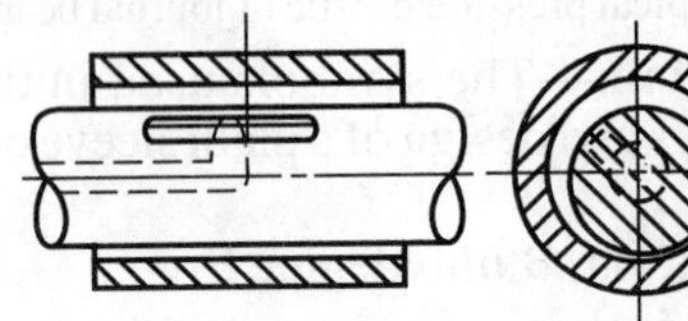

Fig. 10e. Straight axial groove in shaft

h_o = minimum film thickness, inch
K = constants
l = bearing length as defined in Fig. 11, inches
L = actual length of bearing, inches
m = clearance modulus
N = rpm
p_b = unit load, psi
p_s = oil supply pressure, psi
P_f = friction horsepower
P' = bearing pressure parameter
q = flow factor
Q_1 = hydrodynamic flow, gpm
$Q2$ = pressure flow, gpm
Q = total flow, gpm
Q_{new} = new total flow, gpm
Q_R = total flow required, gpm
r = journal radius, inches
Δt = actual temperature rise of oil in bearing, °F
Δt_a = assumed temperature rise of oil in bearing, °F
Δt_{new} = new assumed temperature rise of oil in bearing, °F
t_b = bearing operating temperature, °F
t_{in} = oil inlet temperature, °F
T_f = friction torque, inch-pounds/inch
T' = torque parameter
W = load, pounds
X = factor
Z = viscosity, centipoises
$\in$ = eccentricity ratio — ratio of eccentricity to radial clearance
α = oil density, lbs/inch^3

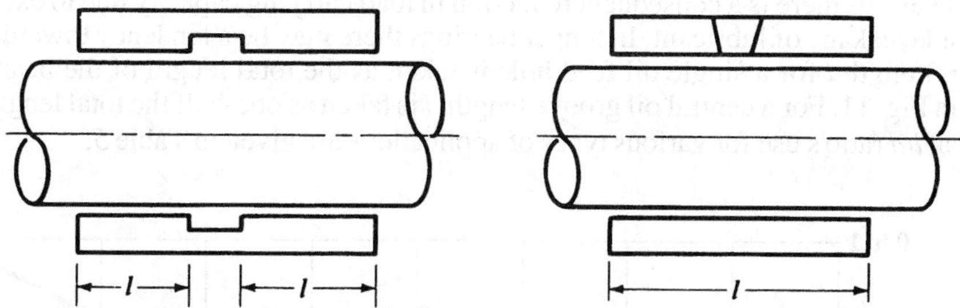

Fig. 11. Length, l, of bearing for circular groove type (left) and single inlet hole type (right).

Journal Bearing Lubrication Analysis.—The following procedure leads to a complete lubrication analysis which forms the basis for the bearing design.

1) *Diameter of bearing d.* This is usually determined by considering strength and/or deflection requirements for the shaft using principles of strength of materials.

2) *Length of bearing L.* This is determined by an assumed l/d ratio in which l may or may not be equal to the overall length, L (See Step 6). Bearing pressure and the possibility of edge loading due to shaft deflection and misalignment are factors to be considered. In general, shaft misalignment resulting from location tolerances and/or shaft deflections should be maintained below 0.0003 inch per inch of length.

3) *Bearing pressure* p_b. The unit load in pound per square inch is calculated from the formula:

$$p_b = \frac{W}{Kld}$$

where $K = 1$ for single oil hole

$K = 2$ for central groove

W = load, pounds

l = bearing length as defined in Fig. 11, inches

d = journal diameter, inches

4) Typical unit loads in service are shown in Table 4. These pressures can be used as a safe guide in selection. However, if space limitations impose a higher limit of loading, the complete lubrication analysis and evaluation of material properties will determine acceptability.

Diametral clearance c_d. This is selected on a trial basis from Fig. 12 which shows suggested diametral clearance ranges for various shaft sizes and for two speed ranges. These are *hot* or *operating* clearances so that thermal expansion of journal and bearing to these temperatures must be taken into consideration in establishing machining dimensions. The optimum operating clearance should be determined on the basis of a complete lubrication analysis (See paragraph following Step 23).

Clearance modulus m. This is calculated from the formula:

$$m = \frac{c_d}{d}$$

Length to diameter ratio l/d. This is usually between 1 and 2; however, with the modern trend toward higher speeds and more compact units, lower ratios down to 0.3 are used. In shorter bearings there is a consequent reduction in load carrying capacity due to excessive end or side leakage of lubricant. In longer bearings there may be a tendency towards edge loading. Length l for a single oil feed hole is taken as the total length of the bearing as shown in Fig. 11. For a central oil groove length, l is taken as one-half the total length.

Typical l/d ratio's use for various types of applications are given in Table 5.

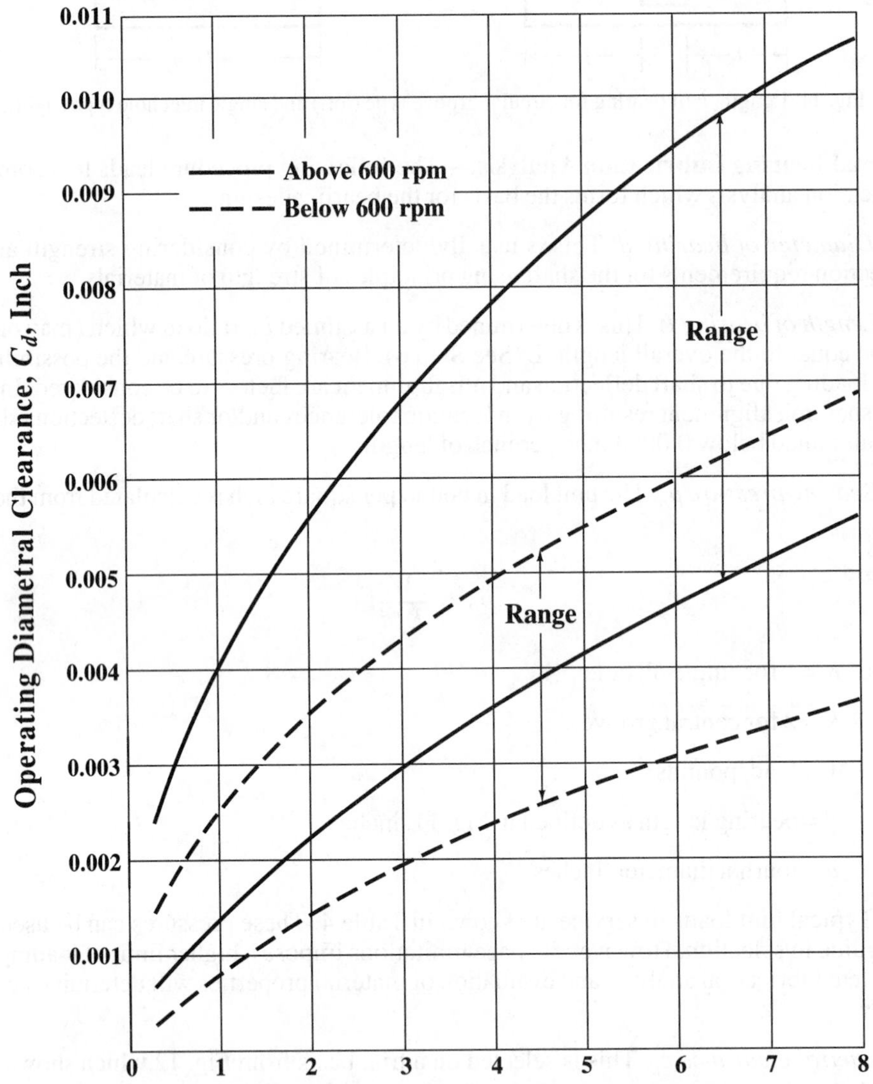

Fig. 12. Operating Diametral Clearance C_d vs. Journal Diameter d.

5) *Assumed operating temperature* t_b. A temperature rise of the lubricant as it passes through the bearing is assumed and the consequent operating temperature in degrees F is calculated from the formula:

$$t_b = t_{in} + \Delta t_a$$

where t_{in} = inlet temperature of oil in °F

Δt_a = assumed temperature rise of oil in bearing in °F

6) An initial assumption of 20°F is usually made.

7) *Viscosity of lubricant Z.* The viscosity in centipoises at the assumed bearing operating temperature is found from the curve in Fig. 5 which shows the viscosity of SAE grade oils versus temperature.

Bearing pressure parameter P′. This value is required to find the eccentricity ratio and is calculated from the formula:

$$P' = \frac{6.9(1000m)^2 p_b}{ZN}$$

where N = rpm

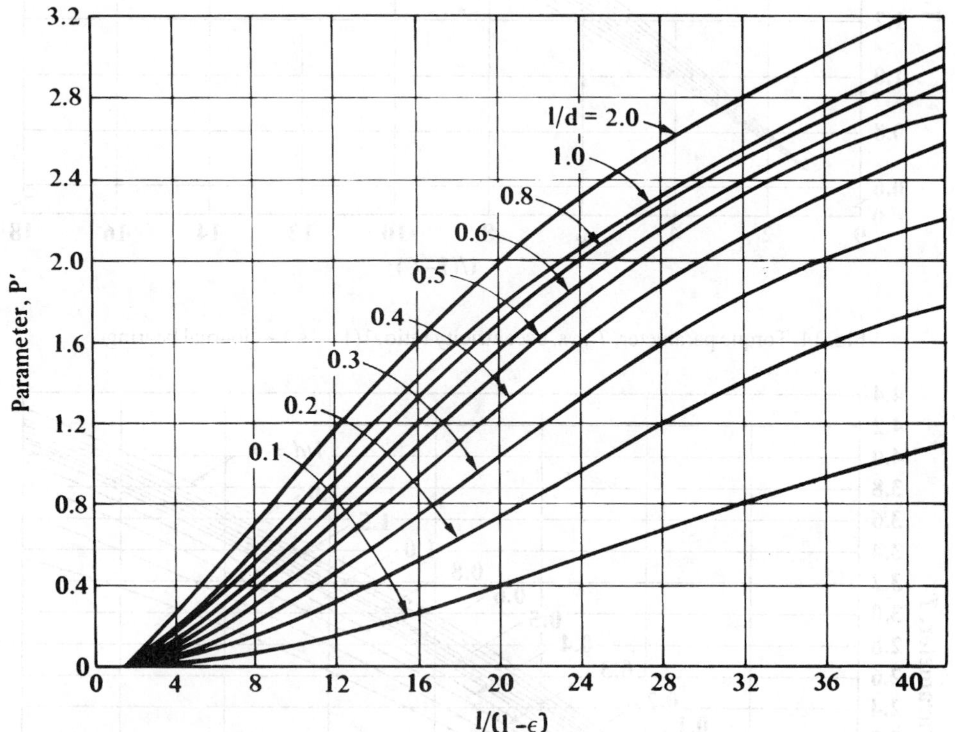

Fig. 13. Bearing parameter. P', vs. eccentricity ratio. $1/(1 - \epsilon)$—journal bearings.

8) *Eccentricity ratio* ϵ. Using P' and l/d, the value of $1/(1 - \epsilon)$ is determined from Fig. 13 and from this ϵ can be determined.

9) *Torque parameter T′.* This value is obtained from Fig. 14 or Fig. 15 using $1/(1 - \epsilon)$ and l/d.

Friction torque T. This value is calculated from the formula:

$$T = \frac{T' r^2 ZN}{6900(1000m)}$$

where r = journal radius, inches

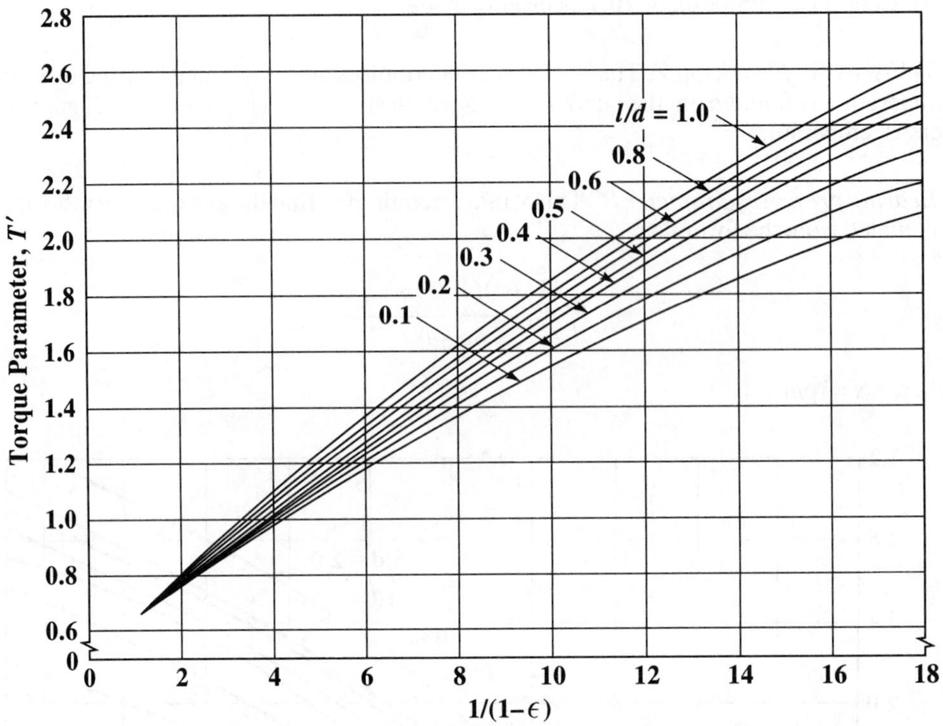

Fig. 14. Torque parameter, T', vs. eccentricity ratio, $1(1 - \epsilon)$ — journal bearings.

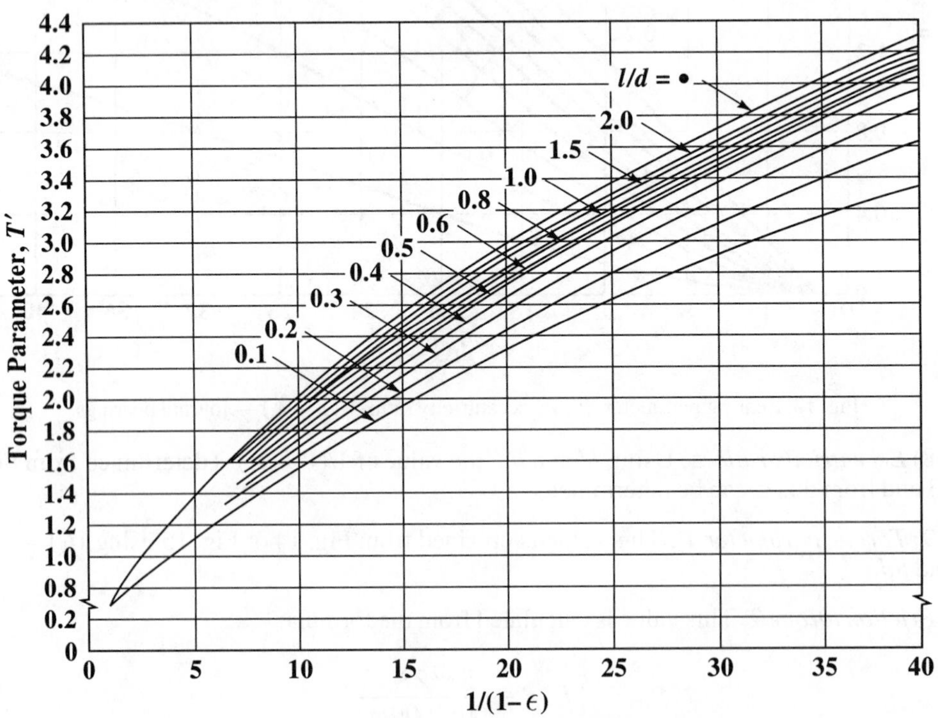

Fig. 15. Torque parameter, T', vs eccentricity ratio, $1/(1 - \epsilon)$ — journal bearings.

10) *Friction horsepower* P_f. This value is calculated from the formula:

$$P_f = \frac{KTNl}{63{,}000}$$

where $K = 1$ for single oil hole, 2 for central groove.

Factor X. This factor is used in the calculation of the lubricant flow and can either be obtained from Table 6 or calculated from the formula:

$$X = 0.1837/\alpha c$$

where α = oil density in pounds per cubic inch

c = specific heat of lubricant in Btu/lb/°F

Total flow of lubricant required Q_R. This is calculated from the formula:

$$Q_R = \frac{X(P_f)}{\Delta t_a}$$

Bearing capacity number C_n. This value is needed to obtain the flow factor and is calculated from the formula:

$$C_n = \left(\frac{l}{d}\right)^2 / 60P'$$

Flow factor q. This value is obtained from the curve in Fig. 16.

Hydrodynamic flow of lubricant Q_1. This flow in gallons per minute is calculated from the formula:

$$Q_1 = \frac{Nlc_d qd}{294}$$

Pressure flow of lubricant Q_2. This flow in gallons per minute is calculated from the formula:

$$Q_2 = \frac{Kp_s c_d^3 d(1 + 1.5\epsilon^2)}{Zl}$$

where $K = 1.64 \times 10^5$ for single oil hole

$K = 2.35 \times 10^5$ for central groove

p_s = oil supply pressure

Total flow of lubricant Q. This value is obtained by adding the hydrodynamic flow and the pressure flow.

$$Q = Q_1 + Q_2$$

Bearing temperature rise Δt. This temperature rise in degrees F is obtained from the formula:

$$\Delta t = \frac{X(P_f)}{Q}$$

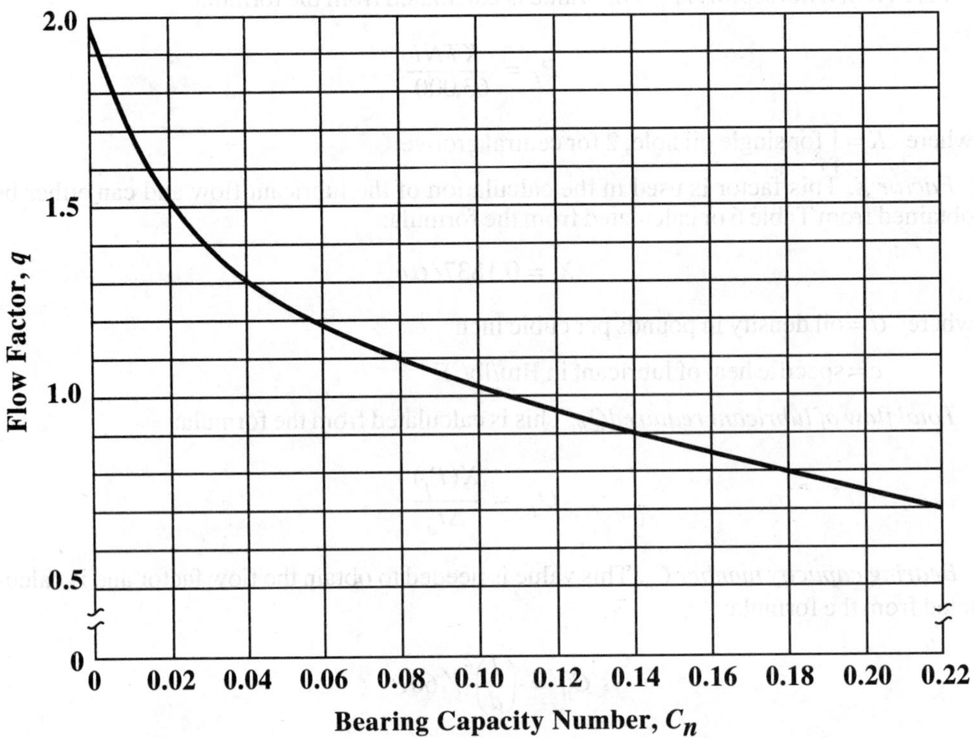

Fig. 16. Flow factor, q, vs. bearing capacity number, C_n—journal bearings.

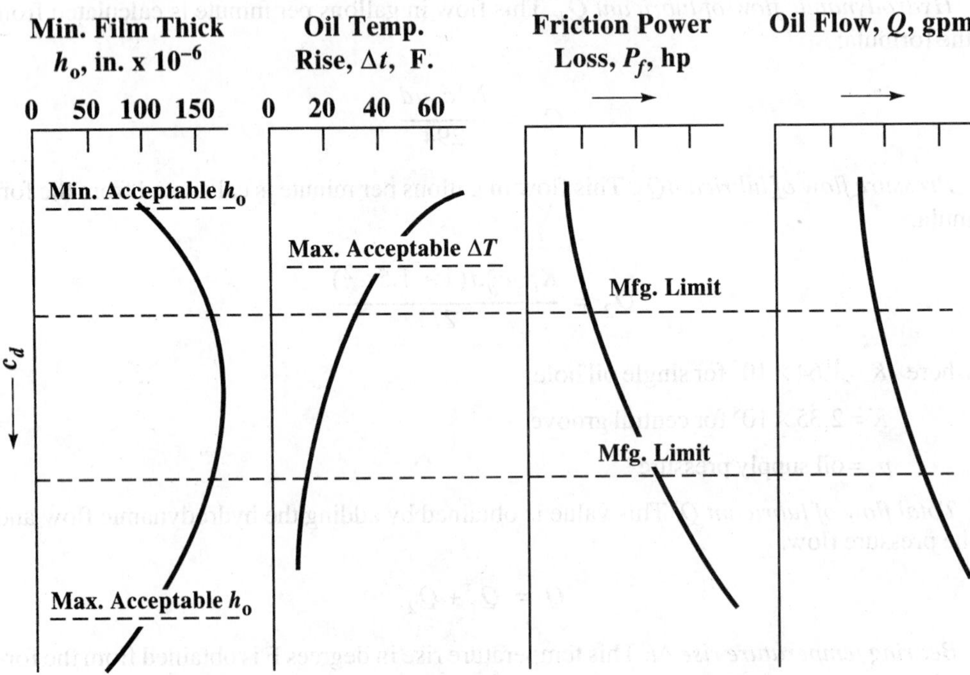

Fig. 17. Example of lubrication analysis curves for journal bearing.

11) *Comparison of actual and assumed temperature rises.* At this point if Δt_a and Δt differ by more than 5 degrees F, Steps 7 through 22 are repeated using a Δt_{new} halfway between the former Δt_a and Δt.

Minimum film thickness h_o. When Step 22 has been satisfied, the minimum film thickness in inches is calculated from the formula: $h_o = \frac{1}{2}C_d(1 - \epsilon)$.

A new diametral clearance c_d is now assumed and Steps 5 through 23 are repeated. When this repetition has been done for a sufficient number of values for c_d, the full lubrication study is plotted as shown in Fig. 17. From this chart a working range of diametral clearance can be determined that optimizes film thickness, differential temperature, friction horsepower and oil flow.

Table 4. Allowable Sleeve Bearing Pressures for Various Classes of Bearings

Types of Bearing or Kind of Service	Pressure, Lbs. per Sq. In.	Types of Bearing or Kind of Service	Pressure, Lbs. per Sq. In.
Electric Motor & Generator		Diesel Engine	
Bearings (General)	100–200	Rod	1000–2000
Turbine & Reduction		Wrist Pins	1800–2000
Gears	100–250	Automotive,	
Heavy Line Shafting	100–150	Main Bearings	500–700
Locomotive Axles	300–350	Rod Bearings	1500–2500
Light Line Shafting	15–35	Centrifugal Pumps	80–100
Diesel Engine, Main	800–1500	Aircraft Rod Bearings	700–3000

These pressures in pounds per square inch of area equal to length times diameter are intended as a general guide only. The allowable unit pressure depends upon operating conditions, especially in regard to lubrication, design of bearings, workmanship, velocity, and nature of load.

Table 5. Representative *l/d* Ratios

Type of Service	*l/d*	Type of Service	*l/d*
Gasoline and diesel engine		Light shafting	2.5 to 3.5
main bearings and crankpins	0.3 to 1.0	Heavy shafting	2.0 to 3.0
Generators and motors	1.2 to 2.5	Steam engine	
Turbogenerators	0.8 to 1.5	Main bearings	1.5 to 2.5
Machine tools	2.0 to 3.0	Crank and wrist pins	1.0 to 1.3

Table 6. *X* Factor vs. Temperature of Mineral Oils

Temperature	*X* Factor
100	12.9
150	12.4
200	12.1
250	11.8
300	11.5

Use of Lubrication Analysis.—Once the lubrication analysis has been completed and plotted as shown in Fig. 17, the following steps lead to the optimum bearing design, taking into consideration both basic operating requirements and requirements peculiar to the application.

1) Examine the curve (Fig. 17)

1) for minimum film thickness and determine the acceptable range of diametral clearance, c_d, based on a) a minimum of 200×10^{-6} inches for small bearings under 1 inch diameter; b) a minimum of 500×10^{-6} inches for bearings from 1 to 4 inches diameter; 1) and

a) a minimum of 750×10^{-6} inches for larger bearings. More conservative designs would increase these requirements.

2) Determine the minimum acceptable c_d based on a maximum Δt of 40°F from the oil temperature rise curve (Fig. 17).

3) If there are no requirements for maintaining low friction horsepower and oil flow, the possible limits of diametral clearance are now defined.

4) The required manufacturing tolerances can now be placed within this band to optimize h_o as shown by Fig. 17.

5) If oil flow and power loss are a consideration, the manufacturing tolerances may then be shifted, within the range permitted by the requirements for h_o and Δt.

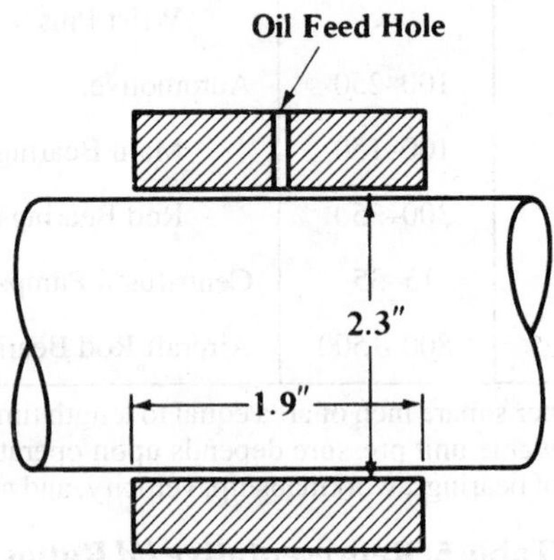

Fig. 18. Full journal bearing example design.

A full journal bearing, Fig. 18, 2.3 inches in diameter and 1.9 inches long is to carry a load of 6000 pounds at 4800 rpm, using SAE 30 oil supplied at 200°F through a single oil hole at 30 psi. Determine the operating characteristics of this bearing as a function of diametral clearance.

1) *Diameter of bearing.* Given as 2.3 inches.

2) *Length of bearing.* Given as 1.9 inches.

3) *Bearing pressure.*

$$p_b = \frac{6000}{1 \times 1.9 \times 2.3} = 1372 \text{ lbs. per sq. in.}$$

4) *Diametral clearance.* Assume c_d is equal to 0.003 inch from Fig. 12 for first calculation.

5) *Clearance modulus.*

$$m = \frac{0.003}{2.3} = 0.0013 \text{ inch}$$

6) *Length-to-diameter ratio.*

$$\frac{l}{d} = \frac{1.9}{2.3} = 0.83$$

7) *Assumed operating temperature.* If the temperature rise Δt_a is assumed to be 20°F,

$$t_b = 200 + 20 = 220°F$$

8) *Viscosity of lubricant.* From Fig. 5, $Z = 7.7$ centipoises

9) *Bearing-pressure parameter.*

$$P' = \frac{6.9 \times 1.3^2 \times 1372}{7.7 \times 4800} = 0.43$$

10) *Eccentricity ratio.* From Fig. 13, $\frac{1}{1-\epsilon} = 6.8$ and $\epsilon = 0.85$

11) *Torque parameter.* From Fig. 14, $T' = 1.46$

12) *Friction torque.*

$$T_f = \frac{1.46 \times 1.15^2 \times 7.7 \times 4800}{6900 \times 1.3} = 7.96 \text{ inch-pounds per inch}$$

13) *Friction horsepower.*

$$P_f = \frac{1 \times 7.96 \times 4800 \times 1.9}{63{,}000} = 1.15 \text{ horsepower}$$

14) *Factor X.* From Table 6, $X = 12$, approximately

15) *Total flow of lubricant required.*

$$Q_R = \frac{12 \times 1.15}{20} = 0.69 \text{ gallon per minute}$$

16) *Bearing-capacity number.*

$$C_n = \frac{0.83^2}{60 \times 0.43} = 0.027$$

17) *Flow factor.* From Fig. 16, $q = 1.43$

18) *Actual hydrodynamic flow of lubricant.*

$$Q_1 = \frac{4800 \times 1.9 \times 0.003 \times 1.43 \times 2.3}{294} = 0.306 \text{ gallon per minute}$$

19) *Actual pressure flow of lubricant.*

$$Q_2 = \frac{1.64 \times 10^5 \times 30 \times 0.003^3 \times 2.3 \times (1 + 1.5 \times 0.85^2)}{7.7 \times 1.9} = 0.044 \text{gallon per min}$$

20) *Actual total flow of lubricant.*

$$Q = 0.306 + 0.044 = 0.350 \text{ gallon per minute}$$

21) *Actual bearing-temperature rise.*

$$\Delta t = \frac{12 \times 1.15}{0.350} = 39.4°F$$

22) *Comparison of actual and assumed temperature rises.* Because Δt_a and Δt differ by more than 5°F, a new Δt_a, midway between these two, of 30°F is assumed and Steps 7 through 22 are repeated.

7a. *Assumed operating temperature.*

$$t_b = 200 + 30 = 230°F$$

8a. *Viscosity of lubricant.* From Fig. 5, $Z = 6.8$ centipoises

9a. *Bearing-pressure parameter.*

$$P' = \frac{6.9 \times 1.3^2 \times 1372}{6.8 \times 4800} = 0.49$$

10a. *Eccentricity ratio.* From Fig. 13,

$$\frac{1}{1-\epsilon} = 7.4$$

and $\in = 0.86$

11a. *Torque parameter.* From Fig. 14, $T' = 1.53$

12a. *Friction torque.*

$$T_f = \frac{1.53 \times 1.15^2 \times 6.8 \times 4800}{6900 \times 1.3} = 7.36 \text{ inch-pounds per inch}$$

13a. *Friction horsepower.*

$$P_f = \frac{1 \times 7.36 \times 4800 \times 1.9}{63{,}000} = 1.07 \text{ horsepower}$$

14a. *Factor X.* From Table 6, $X = 11.9$ approximately

15a. *Total flow of lubricant required.*

$$Q_R = \frac{11.9 \times 1.07}{30} = 0.42 \text{ gallon per minute}$$

16a. *Bearing-capacity number.*

$$C_n = \frac{0.83^2}{60 \times 0.49} = 0.023$$

17a. *Flow factor.* From Fig. 16, $q = 1.48$

18a. *Actual hydrodynamic flow of lubricant.*

$$Q_1 = \frac{4800 \times 1.9 \times 0.003 \times 1.48 \times 2.3}{294} = 0.317 \text{ gallon per minute}$$

19a. *Pressure flow.*

$$Q_2 = \frac{1.64 \times 10^5 \times 30 \times 0.003^3 \times 2.3 \times (1 + 1.5 \times 0.86^2)}{6.8 \times 1.9} = 0.050 \text{ gallon per minute}$$

20a. *Actual flow of lubricant.*

$$Q_{\text{new}} = 0.317 + 0.050 = 0.367 \text{ gallon per minute}$$

21a. *Actual bearing-temperature rise.*

$$\Delta t = \frac{11.9 \times 1.06}{0.367} = 34.4°\text{F}$$

22a. *Comparison of actual and assumed temperature rises.* Now Δt and Δt_a are within 5 degrees F.

23) *Minimum film thickness.*

$$h_o = \frac{0.003}{2}(1 - 0.86) = 0.00021 \text{ inch}$$

This analysis may now be repeated for other values of c_d determined from Fig. 12 and a complete lubrication analysis performed and plotted as shown in Fig. 17. An operating range for c_d can then be determined to optimize minimum clearance, friction horsepower loss, lubricant flow, and temperature rise.

THRUST BEARINGS

As the name implies, thrust bearings are used either to absorb axial shaft loads or to position shafts axially. Brief descriptions of the normal designs for these bearings follow with approximate design methods for each. The generally accepted load ranges for these types of bearings are given in Table 7 and the schematic configurations are shown in Fig. 19.

The parallel or flat plate thrust bearing is probably the most frequently used type. It is the simplest and lowest in cost of those considered; however, it is also the least capable of absorbing load, as can be seen from Table 7. It is most generally used as a positioning device where loads are either light or occasional.

The step bearing, like the parallel plate, is also a relatively simple design. This type of bearing will accept the normal range of thrust loads and lends itself to low-cost, high-volume production. However, this type of bearing becomes sensitive to alignment as its size increases.

The tapered land thrust bearing, as shown in Table 7, is capable of high load capacity. Where the step bearing is generally used for small sizes, the tapered land type can be used in larger sizes. However, it is more costly to manufacture and does require good alignment as size is increased.

The tilting pad or Kingsbury thrust bearing (as it is commonly referred to) is also capable of high thrust capacity. Because of its construction it is more costly, but it has the inherent advantage of being able to absorb significant amounts of misalignment.

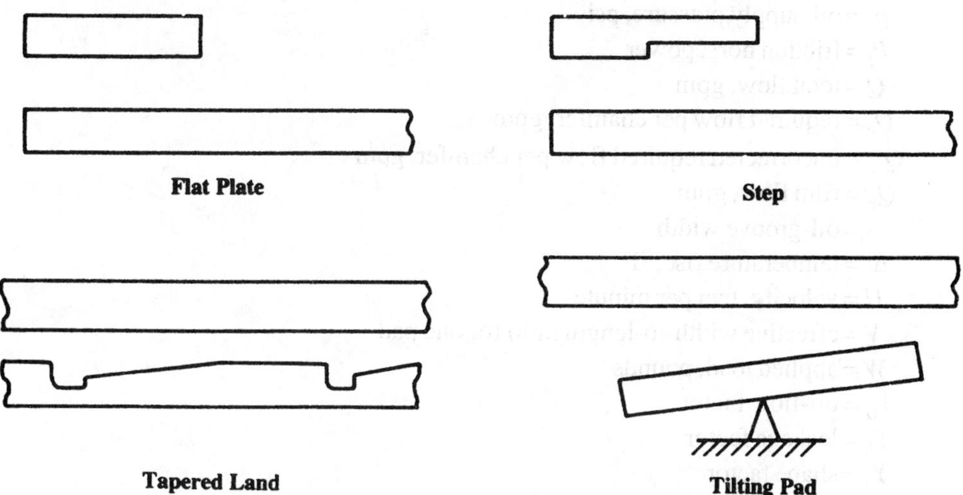

Fig. 19. Types of thrust bearings.

Table 7. Thrust Bearing Loads*

Type	Normal Unit Loads, Lb per Sq. In.	Maximum Unit Loads, Lb per Sq. In.
Parallel surface	<75	<150
Step	200	500
Tapered land	200	500
Tilting pad	200	500

Thrust Bearing Design Notation.—The symbols used in the design procedures that follow for flat plate, step, tapered land, and tilting pad thrust bearings are as follows:

a = radial width of pad, inches

* Reproduced with permission from Wilcock and Booser, *Bearing Design and Applications,* McGraw-Hill Book Co., Copyright © 1957.

b = circumferential length of pad at pitch line, inches
b_2 = pad step length
B = circumference of pitch circle, inches
c = specific heat of oil, Btu/gal/°F
D = diameter, inches
e = depth of step, inch
f = coefficient of friction
g = depth of 45° chamfer, inches
h = film thickness, inch
i = number of pads
J = power loss coefficient
K = film thickness factor
K_g = fraction of circumference occupied by the pads; usually, 0.8
l = length of chamfer, inches
M = horsepower per square inch
N = revolutions per minute
O = operating number
p = bearing unit load, psi
p_s = oil-supply pressure, psi
P_f = friction horsepower
Q = total flow, gpm
Q_c = required flow per chamfer, gpm
$Q^o{}_c$ = uncorrected required flow per chamfer, gpm
Q_F = film flow, gpm
s = oil-groove width
Δt = temperature rise, °F
U = velocity, feet per minute
V = effective width-to-length ratio for one pad
W = applied load, pounds
Y_G = oil-flow factor
Y_L = leakage factor
Y_S = shape factor
Z = viscosity, centipoises
α = dimensionless film-thickness factor
δ = taper
ξ = kinetic energy correction factor

Note: In the following, subscript 1 denotes inside diameter and subscript 2 denotes outside diameter. Subscript *i* denotes inlet and subscript *o* denotes outlet.

Flat Plate Thrust Bearing Design.—The following steps define the performance of a flat plate thrust bearing, one section of which is shown in Fig. 20. Although each bearing section is wedge shaped, as shown below right, for the purposes of design calculation, it is considered to be a rectangle with a length b equal to the circumferential length along the pitch line of the section being considered, and a width a equal to the difference in the external and internal radii.

General Parameters: X) From Table 7, the maximum unit load is between 75 and 100 pounds per square inch; and Y) The outside diameter is usually between 1.5 and 2.5 times the inside diameter.

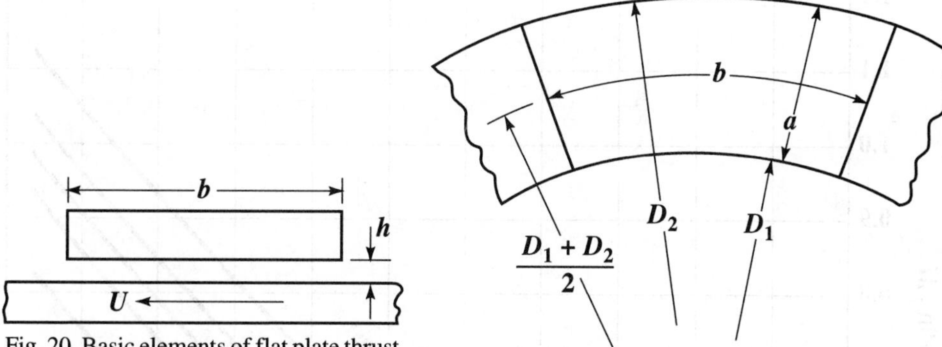

Fig. 20. Basic elements of flat plate thrust bearing.*

Basic elements of flat plate thrust bearing.*

1) *Inside diameter, D_1*. Determined by shaft size and clearance.

2) *Outside diameter, D_2*. Calculated by the formula

$$D_2 = \left(\frac{4W}{\pi K_g p} + D_1^2\right)^{1/2}$$

where W = applied load, pounds

K_g = fraction of circumference occupied by pads; usually, 0.8

p = bearing unit load, psi

3) *Radial pad width, a*. Equal to one-half the difference between the inside and outside diameters.

$$a = \frac{D_2 - D_1}{2}$$

4) *Pitch line circumference, B*. Found from the pitch diameter.

$$B = \pi(D_2 - a)$$

Number of pads, i. Assume an oil groove width, *s*. If the length of pad is assumed to be optimum, i.e., equal to its width,

$$i_{app} = \frac{B}{a + s}$$

Take *i* as nearest even number.

5) *Length of pad, b*. If number of pads and oil groove width are known,

$$b = \frac{B - (i \times s)}{i}$$

6) *Actual unit load, p*. Calculated in pounds per square inch based on pad dimensions.

$$p = \frac{W}{iab}$$

7) *Pitch line velocity, U*. Found in feet per minute from

$$U = \frac{BN}{12}$$

where N = rpm

8) *Friction power loss, P_f*. Friction power loss is difficult to calculate for this type of bearing because there is no theoretical method of determining the operating film thickness. However, a good approximation can be made using Fig. 21. From this curve, the value of *M*, horsepower loss per square inch of bearing surface, can be obtained. The total power loss, P_f, is then calculated from

$$P_f = iabM$$

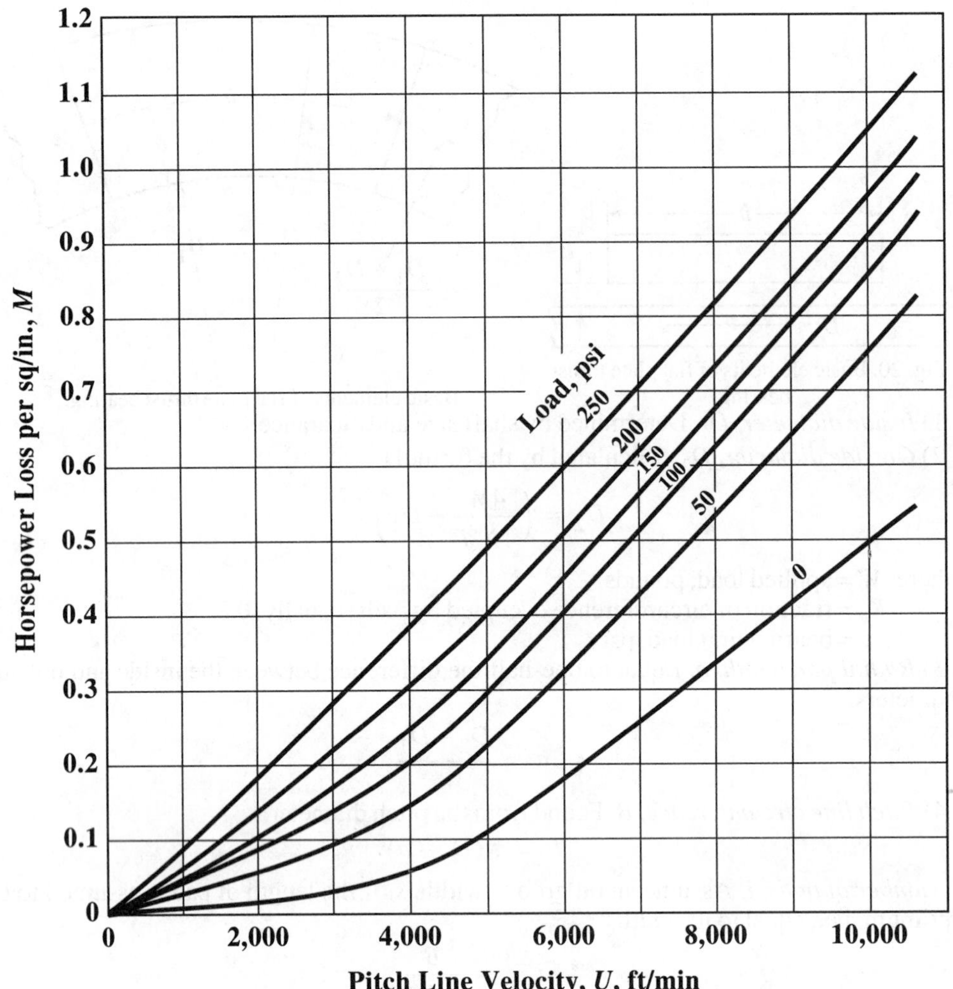

Fig. 21. Friction power loss, *M*, vs. peripheral speed, *U* — thrust bearings.*

9) *Oil flow required, Q*. May be estimated in gallons per minute for a given temperature rise from

$$Q = \frac{42.4 P_f}{c \Delta t}$$

where c = specific heat of oil in Btu/gal/°F

Δt = temperature rise of the oil in °F

Note: A Δt of 50°F is an acceptable maximum.

Because there is no theoretical method of predicting the minimum film thickness in this type of bearing, only an approximation, based on experience, of the film flow can be made. For this reason and based on practical experience, it is desirable to have a minimum of one-half of the desired oil flow pass through the chamfer.

10) *Film flow, Q_F*. Calculated in gallons per minute from

$$Q_F = \frac{(1.5)(10^5) i V h^3 p_s}{Z_2}$$

where V = effective width-to-length ratio for one pad, a/b

* See footnote on page page 2219.

Z_2 = oil viscosity at outlet temperature
h = film thickness

Note: Because h cannot be calculated, use $h = 0.002$ inch.

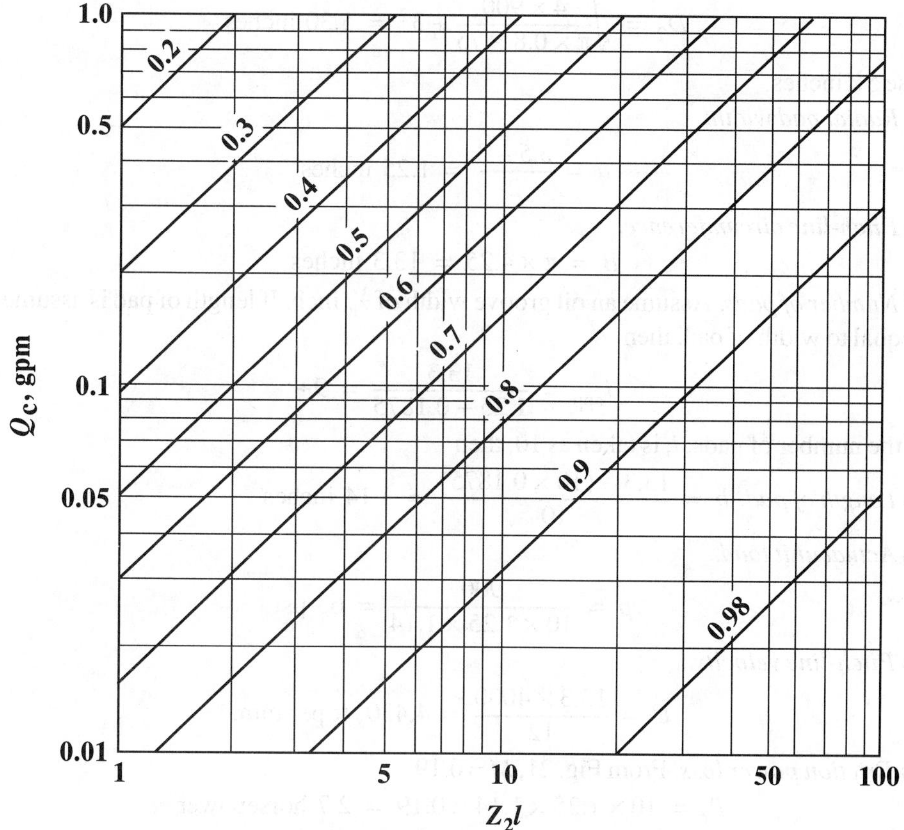

Fig. 22. Kinetic energy correction factor, ξ—thrust bearings.*

11) *Required flow per chamfer, Q_c.* Readily found from the formula

$$Q_c = \frac{Q}{i}$$

12) *Kinetic energy correction factor, ξ.* Found by assuming a chamfer length l and entering Fig. 22 with a value $Z_2 l$ and Q_c.

13) *Uncorrected required flow per chamfer, $Q^0{}_c$.* Found from the formula

$$Q_c^0 = \frac{Q_c}{\xi}$$

14) *Depth of chamfer, g.* Found from the formula

$$g = \sqrt[4]{\frac{Q_c^0 l Z_2}{4.74 \times 10^4 p_s}}$$

Example:—Design a flat plate thrust bearing to carry 900 pounds load at 4000 rpm using an SAE 10 oil with a specific heat of 3.5 Btu/gal/°F at 120°F and 30-psi inlet conditions. The shaft is 2¾ inches in diameter and the temperature rise is not to exceed 40°F. Fig. 23 shows the final design of this bearing.

* See footnote on page page 2219.

1) *Inside diameter.* Assumed to be 3 inches to clear shaft.

2) *Outside diameter.* Assuming a unit bearing load of 75 pounds per square inch from Table 7,

$$D_2 = \sqrt{\frac{4 \times 900}{\pi \times 0.8 \times 75} + 3^2} = 5.30 \text{ inches}$$

Use 5½ inches.

3) *Radial pad width.*

$$a = \frac{5.5 - 3}{2} = 1.25 \text{ inches}$$

4) *Pitch-line circumference.*

$$B = \pi \times 4.25 = 13.3 \text{ inches}$$

5) *Number of pads.* Assume an oil groove width of 3⁄16 inch. If length of pad is assumed to be equal to width of pad, then

$$i_{app} = \frac{13.3}{1.25 + 0.1875} = 9+$$

If the number of pads, i, is taken as 10, then

6) *Length of pad.* $b = \dfrac{13.3 - (10 \times 0.1875)}{10} = 1.14$ inches

7) *Actual unit load.*

$$p = \frac{900}{10 \times 1.25 \times 1.14} = 63 \text{ psi}$$

8) *Pitch-line velocity.*

$$U = \frac{13.3 \times 4000}{12} = 4{,}430 \text{ ft per min.}$$

9) *Friction power loss.* From Fig. 21, $M = 0.19$

$$P_f = 10 \times 1.25 \times 1.14 \times 0.19 = 2.7 \text{ horsepower}$$

10) *Oil flow required.*

$$Q = \frac{42.4 \times 2.7}{3.5 \times 40} = 0.82 \text{ gallon per minute}$$

(Assuming a temperature rise of 40°F—the maximum allowable according to the given condition—then the assumed operating temperature will be 120°F + 40°F = 160°F and the oil viscosity Z_2 is found from Fig. 5 to be 9.6 centipoises.)

11) *Film flow.*

$$Q_F = \frac{1.5 \times 10^5 \times 10 \times 1 \times 0.002^3 \times 30}{9.6} = 0.038 \text{ gpm}$$

Because 0.038 gpm is a very small part of the required flow of 0.82 gpm, the bulk of the flow must be carried through the chamfers.

12) *Required flow per chamfer.* Assume that all the oil flow is to be carried through the chamfers.

$$Q_c = \frac{0.82}{10} = 0.082 \text{ gpm}$$

13) *Kinetic energy correction factor.* If l, the length of chamfer is made ⅛ inch, then $Z_2 l$ = 9.6 × ⅛ = 1.2. Entering Fig. 22 with this value and $Q_c = 0.082$,

$$\xi = 0.44$$

14) *Uncorrected required oil flow per chamfer.*

$$Q_c^0 = \frac{0.082}{0.44} = 0.186 \text{ gpm}$$

15) *Depth of chamfer.*

$$g = \sqrt[4]{\frac{0.186 \times 0.125 \times 9.6}{4.74 \times 10^4 \times 30}}$$

$$g = 0.02 \text{ inch}$$

A schematic drawing of this bearing is shown in Fig. 23.

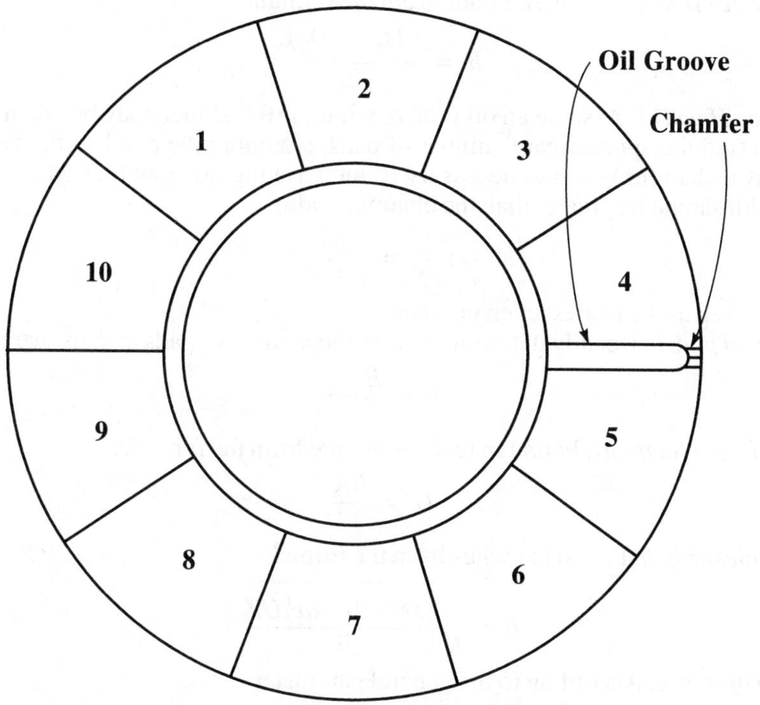

Fig. 23. Flat plate thrust bearing example design.*

Step Thrust Bearing Design.—The following steps define the performance of a step thrust bearing, one section of which is shown in Fig. 24.

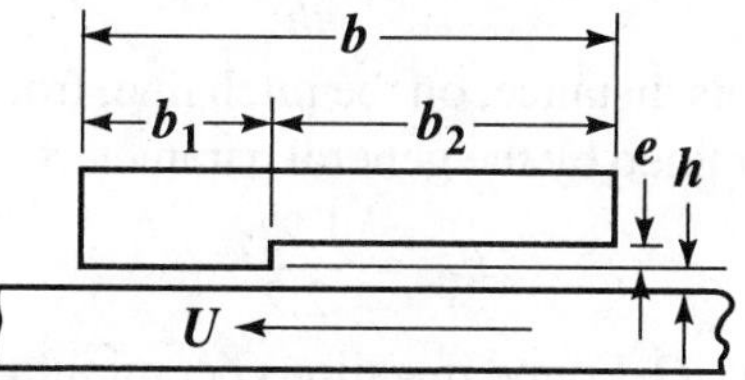

Fig. 24. Basic elements of step thrust bearing.*

Although each bearing section is wedge shaped, as shown at the right in Fig. 24, for the purposes of design calculation it is considered to be a rectangle with a length b equal to the circumferential length along the pitch line of the section being considered, and a width a equal to the difference in the external and internal radii.

General Parameters: For optimum proportions, $a = b$, $b_2 = 1.2b_1$, and $e = 0.7h$.

1) *Internal diameter, D_1.* An internal diameter is assumed that is sufficient to clear the shaft.

* See footnote on page page 2219.

2) *External diameter, D_2.* A unit bearing pressure is assumed from Table 7 and the external diameter is then found from the formula

$$D_2 = \sqrt{\frac{4W}{\pi K_g p} + D_1^2}$$

3) *Radial pad width, a.* Equal to the difference between the external and internal radii.

$$a = \frac{D_2 - D_1}{2}$$

4) *Pitch-line circumference, B.* Found from the formula

$$B = \frac{\pi(D_1 + D_2)}{2}$$

5) *Number of pads, i.* Assume an oil groove width, *s* (0.062 inch may be taken as a minimum), and find the approximate number of pads, assuming the pad length is equal to *a*. Note that if a chamfer is found necessary to increase the oil flow (see Step 13), the oil groove width should be greater than the chamfer width.

$$i_{app} = \frac{B}{a + s}$$

Then *i* is taken as the nearest even number.

6) *Length of pad, b.* Readily determined from the number of pads and groove width.

$$b = \frac{B}{i} - s$$

7) *Pitch-line velocity, U.* Found in feet per minute from the formula

$$U = \frac{BN}{12}$$

8) *Film thickness, h.* Found in inches from the formula

$$h = \sqrt{\frac{2.09\times10^{-9} ia^3 UZ}{W}}$$

9) *Depth of step, e.* According to the general parameter

$$e = 0.7h$$

10) *Friction power loss, P_f.* Found from the formula

$$P_f = \frac{7.35\times10^{-13} ia^2 U^2 Z}{h}$$

11) *Pad step length, b_2.* This distance, on the pitch line, from the leading edge of the pad to the step in inches is determined by the general parameters

$$b_2 = \frac{1.2b}{2.2}$$

12) *Hydrodynamic oil flow, Q.* Found in gallons per minute from the formula

$$Q = 6.65\times10^{-4} iahU$$

13) *Temperature rise, Δt.* Found in degrees F from the formula

$$\Delta t = \frac{42.4P_f}{cQ}$$

If the flow is insufficient, as indicated by too high a temperature rise, chamfers can be added to provide adequate flow as in Steps 12–15 of the flat plate thrust bearing design.

Example: Design a step thrust bearing for positioning a ⅞-inch diameter shaft operating with a 25-pound thrust load and a speed of 5,000 rpm. The lubricating oil has a viscosity of

25 centipoises at the operating temperature of 160 deg. F and has a specific heat of 3.4 Btu per gal. per deg. F.

1) *Internal diameter.* Assumed to be 1 inch to clear the shaft.

2) *External diameter.* Because the example is a positioning bearing with low total load, unit load will be negligible and the external diameter is not established by using the formula given in Step 2 of the procedure, but a convenient size is taken to give the desired overall bearing proportions.

$$D_2 = 3 \text{ inches}$$

3) *Radial pad width.*

$$a = \frac{3-1}{2} = 1 \text{ inch}$$

4) *Pitch-line circumference.*

$$B = \frac{\pi(3+1)}{2} = 6.28 \text{ inches}$$

5) *Number of pads.* Assuming a minimum groove width of 0.062 inch,

$$i_{app} = \frac{6.28}{1+0.062} = 5.9$$

Take $i = 6$.

6) *Length of pad.*

$$b = \frac{6.28}{6} - 0.062 = 0.985$$

7) *Pitch-line velocity.*

$$U = \frac{6.28 \times 5{,}000}{12} = 2{,}620 \text{ fpm}$$

8) *Film thickness.*

$$h = \sqrt{\frac{2.09\times10^{-9} \times 6 \times 1^3 \times 2{,}620 \times 25}{25}} = 0.0057 \text{ inch}$$

9) *Depth of step.*

$$e = 0.7 \times 0.0057 = 0.004 \text{ inch}$$

10) *Power loss.*

$$P_f = \frac{7.35\times10^{-13} \times 6 \times 1^2 \times 2{,}620^2 \times 25}{0.0057} = 0.133 \text{ hp}$$

11) *Pad step length.*

$$b_2 = \frac{1.2 \times 0.985}{2.2} = 0.537 \text{ inch}$$

12) *Total hydrodynamic oil flow.*

$$Q = 6.65\times10^{-4} \times 6 \times 1 \times 0.0057 \times 2{,}620 = 0.060 \text{ gpm}$$

13) *Temperature rise.*

$$\Delta t = \frac{42.4 \times 0.133}{3.4 \times 0.060} = 28° \text{ F}$$

Tapered Land Thrust Bearing Design.—The following steps define the performance of a tapered land thrust bearing, one section of which is shown in Fig. 25. Although each bearing section is wedge shaped, as shown in Fig. 25, right, for the purposes of design calculation, it is considered to be a rectangle with a length *b* equal to the circumferential length along the pitch line of the section being considered and a width *a* equal to the difference in the external and internal radii.

General Parameters: Usually, the taper extends to only 80 per cent of the pad length with the remainder being flat, thus: $b_2 = 0.8b$ and $b_1 = 0.2b$.

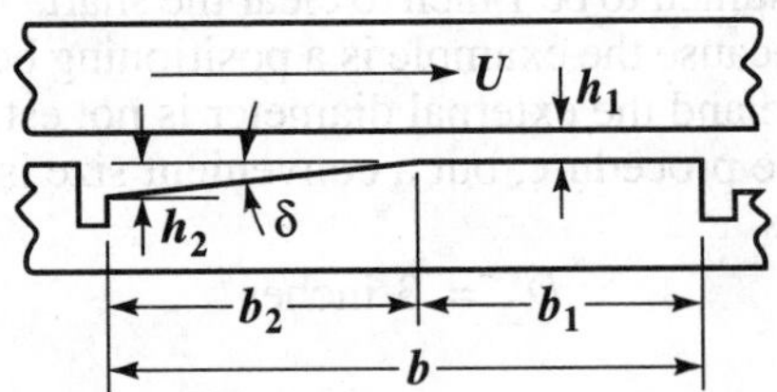

Fig. 25. Basic elements of tapered land thrust bearing.*

1) *Inside diameter, D_1.* Determined by shaft size and clearance.

2) *Outside diameter, D_2.* Calculated by the formula

$$D_2 = \left(\frac{4W}{\pi K_g P_a} + D_1^2\right)^{1/2}$$

where $K_g = 0.8$ or 0.9 and W = applied load, pounds
P_a = assumed unit load from Table 7, page 2219

3) *Radial pad width, a.* Equal to one-half the difference between the inside and outside diameters.

$$a = \frac{D_2 - D_1}{2}$$

4) *Pitch-line circumference, B.* Found from the mean diameter:

$$B = \frac{\pi(D_1 + D_2)}{2}$$

5) *Number of pads, i.* Assume an oil groove width, *s*, and find the approximate number of pads, assuming the pad length is equal to *a*.

$$i_{app} = \frac{B}{a + s}$$

Then *i* is taken as the nearest even number.

6) *Length of pad, b.* Readily determined because the number of pads and groove width are known.

$$b = \frac{B - is}{i}$$

7) *Taper values, δ_1 and δ_2.* Can be taken from Table 8.

8) *Actual bearing unit load, p.* Calculated in pounds per square inch from the formula

$$p = \frac{W}{iab}$$

9) *Pitch-line velocity, U.* Found in feet per minute at the pitch circle from the formula

$$U = \frac{BN}{12}$$

where N = rpm

10) *Oil leakage factor, Y_L.* Found either from Fig. 26 which shows curves for Y_L as functions of the pad width a and length of land *b* or from the formula

$$Y_L = \frac{b}{1 + (\pi^2 b^2 / 12a^2)}$$

11) *Film thickness factor, K.* Calculated using the formula

* See footnote on page page 2219.

$$K = \frac{5.75 \times 10^6 p}{UY_L Z}$$

12) *Minimum film thickness, h.* Using the value of K just determined and the selected taper values δ_1 and δ_2, h is found from Fig. 27. In general, h should be 0.001 inch for small bearings and 0.002 inch for larger and high-speed bearings.

13) *Friction power loss, P_f.* Using the film thickness h, the coefficient J can be obtained from Fig. 28. The friction loss in horsepower is then calculated from the formula

$$P_f = 8.79 \times 10^{-13} iabJU^2 Z$$

14) *Required oil flow, Q.* May be estimated in gallons per minute for a given temperature rise Δ_t from the formula

$$Q = \frac{42.4 P_f}{c \Delta t}$$

where c = specific heat of the oil in Btu/gal/°F

Note: A Δt of 50°F is an acceptable maximum.

15) *Shape factor, Y_s.* Needed to compute the actual oil flow and calculated from the formula

$$Y_S = \frac{8ab}{D_2^2 - D_1^2}$$

16) *Oil flow factor, Y_G.* Found from Fig. 29 using Y_s and D_1/D_2.

17) *Actual oil film flow, Q_F.* The amount of oil in gallons per minute that the bearing film will pass is calculated from the formula

$$Q_F = \frac{8.9 \times 10^{-4} i \delta_2 D_2^3 N Y_G Y_S^2}{D_2 - D_1}$$

18) If the flow is insufficient, the tapers can be increased or chamfers calculated to provide adequate flow, as in Steps 12–15 of the flat plate thrust bearing design procedure.

Example: Design a tapered land thrust bearing for 70,000 pounds at 3600 rpm. The shaft diameter is 6.5 inches. The oil inlet temperature is 110°F at 20 psi.

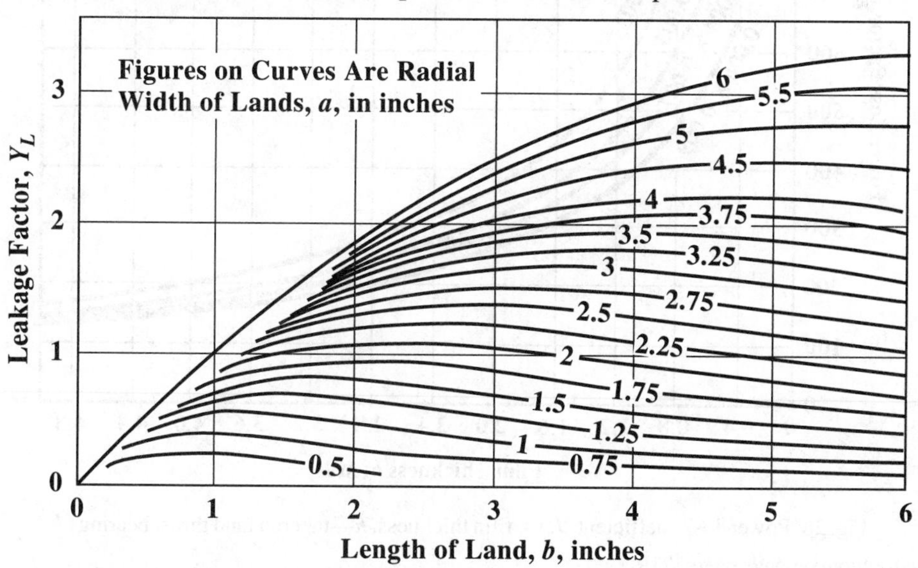

Fig. 26. Leakage factor, Y_L, vs. pad dimensions a and b—tapered land thrust bearings.*

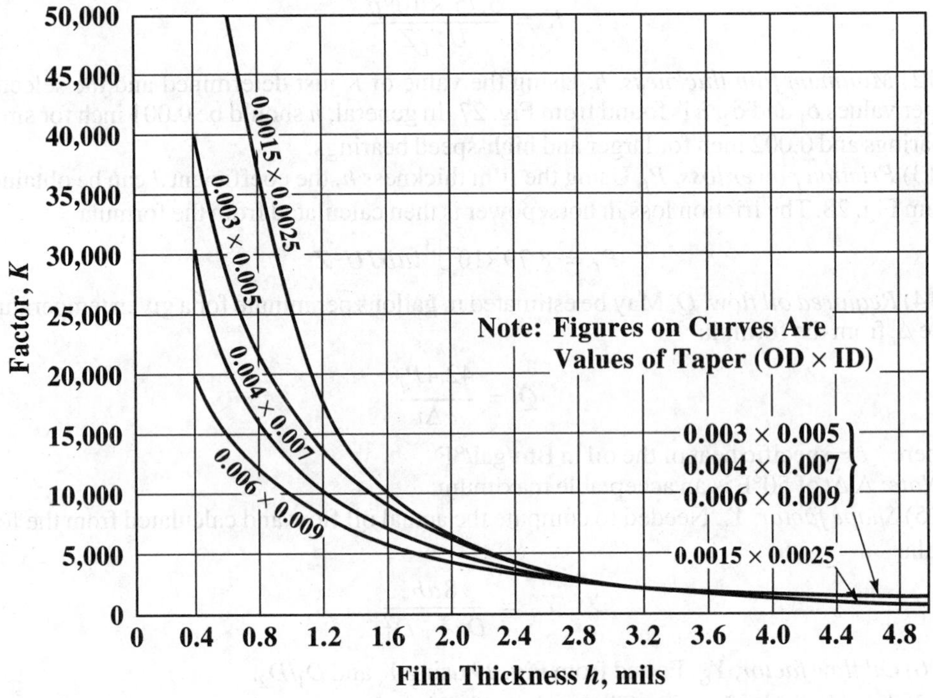

Fig. 27. Thickness, *h*, vs. factor *K*—tapered land thrust bearings.*

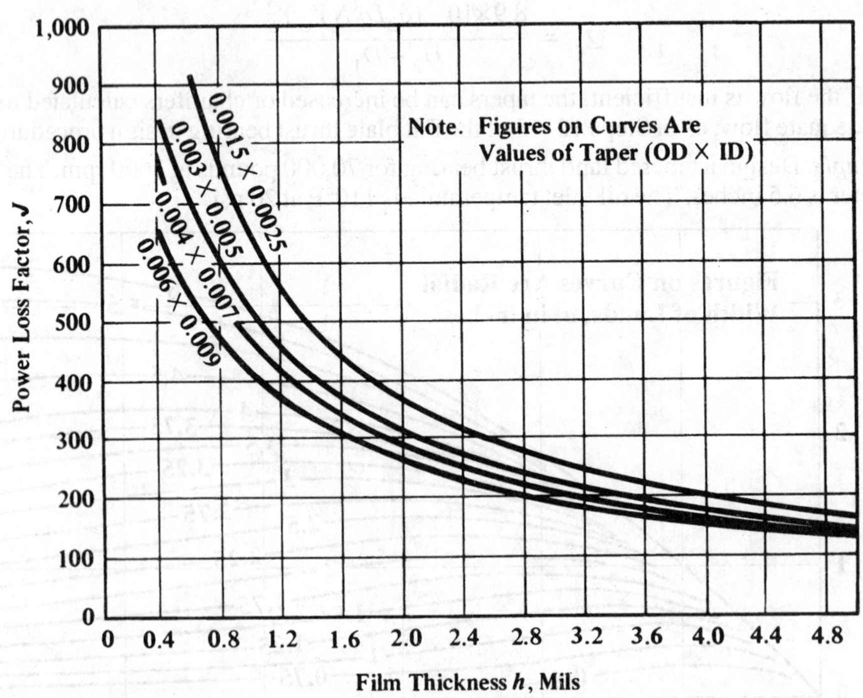

Fig. 28. Power-loss coefficient. *J*, vs. film thickness, *h*—tapered land thrust bearings.*

* See footnote on page page 2219.

* See footnote on page page 2219.

A maximum temperature rise of 50°F is acceptable and results in a viscosity of 18 centipoises. Use values of $K_g = 0.9$ and $c = 3.5$ Btu/gal/°F.

1) *Internal diameter.* Assume $D_1 = 7$ inches to clear shaft.

2) *External diameter.* Assume a unit bearing load p_a of 400 pounds per square inch from Table 7, then

$$D_2 = \sqrt{\frac{4 \times 70{,}000}{3.14 \times 0.9 \times 400} + 7^2} = 17.2 \text{ inches}$$

Round off to 17 inches.

3) *Radial pad width.*

$$a = \frac{17 - 7}{2} = 5 \text{ inches}$$

4) *Pitch-line circumference.*

$$B = \frac{3.14(17 + 7)}{2} = 37.7 \text{ inches}$$

5) *Number of pads.* Assume groove width of 0.5 inch, then

$$i_{\text{app}} = \frac{37.7}{5 + 0.5} = 6.85$$

Take $i = 6$.

6) *Length of pad.*

$$b = \frac{37.7 - 6 \times 0.5}{6} = 5.78 \text{ inches}$$

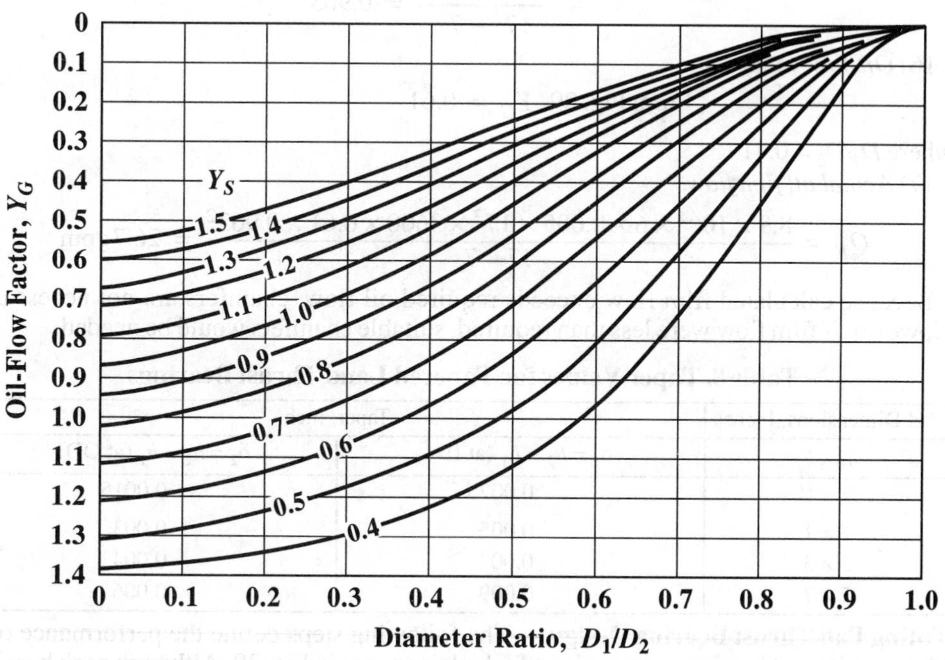

Fig. 29. Oil-flow factor, Y_G, vs. diameter ratio D_1/D_2—tapered land bearings.*

7) *Taper values.* Interpolate in Table 8 to obtain

$$\delta_1 = 0.008 \text{ inch} \quad \text{and} \quad \delta_2 = 0.005 \text{ inch}$$

8) *Actual bearing unit load.*

* See footnote on page page 2219.

$$p = \frac{70{,}000}{6 \times 5 \times 5.78} = 404 \text{ psi}$$

9) *Pitch-line velocity.*

$$U = \frac{37.7 \times 3600}{12} = 11{,}300 \text{ ft per min}$$

10) *Oil leakage factor.*

From Fig. 26, $Y_L = 2.75$

11) *Film-thickness factor.*

$$K = \frac{5.75 \times 10^6 \times 404}{11,300 \times 2.75 \times 18} = 4150$$

12) *Minimum film thickness.*

From Fig. 27, $h = 2.2$ mils

13) *Friction power loss.* From Fig. 28, $J = 260$, then

$$P_f = 8.79 \times 10^{-13} \times 6 \times 5 \times 5.78 \times 260 \times 11{,}300^2 \times 18 = 91 \text{ hp}$$

14) *Required oil flow.*

$$Q = \frac{42.4 \times 91}{3.5 \times 50} = 22.0 \text{ gpm}$$

See footnote on page 2219.

15) *Shape factor.*

$$Y_S = \frac{8 \times 5 \times 5.78}{17^2 - 7^2} = 0.963$$

16) *Oil-flow factor.*

From Fig. 29, $Y_G = 0.61$

where $D_1/D_2 = 0.41$

17) *Actual oil film flow.*

$$Q_F = \frac{8.9 \times 10^{-4} \times 6 \times 0.005 \times 17^3 \times 3600 \times 0.61 \times 0.963^2}{17 - 7} = 26.7 \text{gpm}$$

Because calculated film flow exceeds required oil flow, chamfers are not necessary. However, if film flow were less than required, suitable chamfers would be needed.

Table 8. Taper Values for Tapered Land Thrust Bearings

Pad Dimensions, Inches	Taper, Inch	
$a \times b$	$\delta_1 = h_2 - h_1$ (at ID)	$\delta_2 = h_2 - h_1$ (at OD)
½ × ½	0.0025	0.0015
1 × 1	0.005	0.003
3 × 3	0.007	0.004
7 × 7	0.009	0.006

Tilting Pad Thrust Bearing Design.—The following steps define the performance of a tilting pad thrust bearing, one section of which is shown in Fig. 30. Although each bearing section is wedge shaped, as shown at the right below, for the purposes of design calculation, it is considered to be a rectangle with a length b equal to the circumferential length along the pitch line of the section being considered and a width a equal to the difference in the external and internal radii, as shown at left in Fig. 30. The location of the pivot shown in Fig. 30 is optimum. If shaft rotation in both directions is required, however, the pivot must be at the midpoint, which results in little or no detrimental effect on the performance.

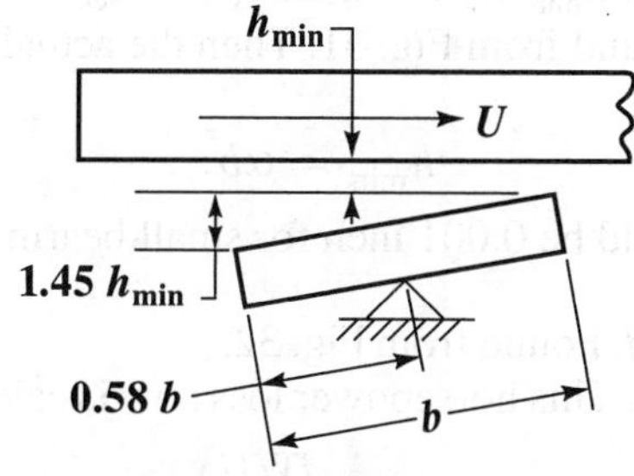

Fig. 30. Basic elements of tilting pad thrust bearing.*

1) *Inside diameter, D_1*. Determined by shaft size and clearance.

2) *Outside diameter, D_2*. Calculated from the formula

$$D_2 = \left(\frac{4W}{\pi K_g p} + D_1^2\right)^{1/2}$$

where W = applied load, pounds
K_g = 0.8
p = unit load from Table 7

3) *Radial pad width, a.* Equal to one-half the difference between the inside and outside diameters:

$$a = \frac{D_2 - D_1}{2}$$

4) *Pitch-line circumference, B.* Found from the mean diameter:

$$B = \pi\left(\frac{D_1 + D_2}{2}\right)$$

5) *Number of pads, i.* The number of pads may be estimated from the formula

$$i = \frac{BK_g}{a}$$

Select the nearest even number.

6) *Length of pad, b.* Found from the formula

$$b \cong \frac{BK_g}{i}$$

7) *Pitch-line velocity, U.* Calculated in feet per minute from the formula

$$U = \frac{BN}{12}$$

8) *Bearing unit load, p.* Calculated from the formula

$$p = \frac{W}{iab}$$

9) *Operating number, O.* Calculated from the formula

$$O = \frac{1.45 \times 10^{-7} Z_2 U}{5pb}$$

10) where Z_2 = viscosity of oil at outlet temperature (inlet temperature plus assumed temperature rise through the bearing).

* See footnote on page page 2219.

11) *Minimum film thickness,* h_{min}. By using the operating number, the value of α = dimensionless film thickness is found from Fig. 31. Then the actual minimum film thickness is calculated from the formula:

$$h_{min} = \alpha b$$

In general, this value should be 0.001 inch for small bearings and 0.002 inch for larger and high-speed bearings.

12) *Coefficient of friction, f.* Found from Fig. 32.

13) *Friction power loss,* P_f. This horsepower loss now is calculated by the formula

$$P_f = \frac{fWU}{33,000}$$

14) *Actual oil flow, Q.* This flow over the pad in gallons per minute is calculated from the formula

$$Q = 0.0591\alpha iabU$$

15) *Temperature rise,* Δt. Found from the formula

$$\Delta t = 0.0217\frac{fp}{\alpha c}$$

16) where c = specific heat of oil in Btu/gal/°F

If the flow is insufficient, as indicated by too high a temperature rise, chamfers can be added to provide adequate flow, as in Steps 12–15 of the flat plate thrust bearing design.

Example: Design a tilting pad thrust bearing for 70,000 pounds thrust at 3600 rpm. The shaft diameter is 6.5 inches and a maximum OD of 15 inches is available. The oil inlet temperature is 110°F and the supply pressure is 20 pounds per square inch. A maximum temperature rise of 50°F is acceptable and results in a viscosity of 18 centipoises. Use a value of 3.5 Btu/gal/°F for c.

1) *Inside diameter.* Assume D_1, = 7 inches to clear shaft.

2) *Outside diameter.* Given maximum D_2 = 15 inches.

3) *Radial pad width.*

$$a = \frac{15-7}{2} = 4 \text{ inches}$$

4) *Pitch-line circumference.*

$$B = \pi\left(\frac{7+15}{2}\right) = 34.6 \text{ inches}$$

5) *Number of pads.*

$$i = \frac{34.6 \times 0.8}{4} = 6.9$$

Select 6 pads: $i = 6$.

6) *Length of pad.*

$$b = \frac{34.6 \times 0.8}{6} = 4.61 \text{ inches}$$

Make $b = 4.75$ inches.

7) *Pitch-line velocity.*

$$U = \frac{34.6 \times 3600}{12} = 10,400 \text{ ft/min}$$

8) *Bearing unit load.*

$$p = \frac{70,000}{6 \times 4 \times 4.75} = 614 \text{ psi}$$

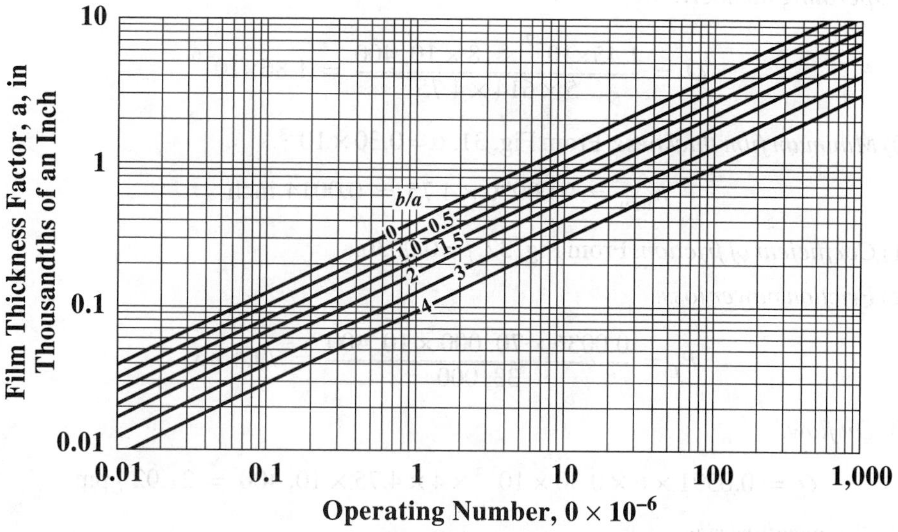

Fig. 31. Dimensionless minimum film thickness, α, vs. operating number, *O*—tilting pad thrust bearings.*

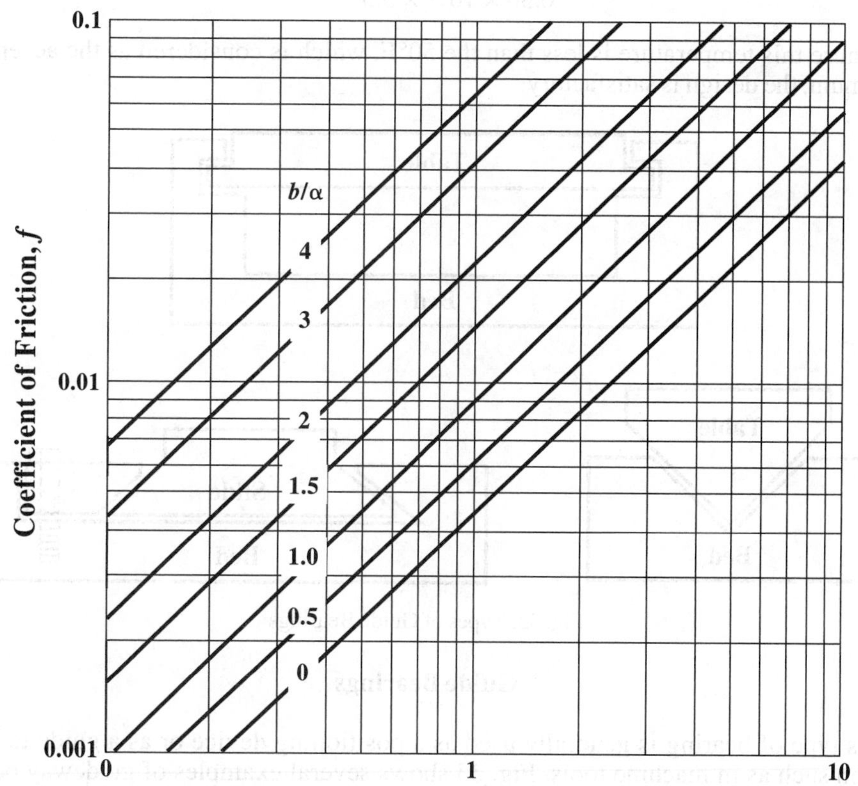

Fig. 32. Coefficient of friction *f* vs. dimensionless film thickness α for tilting pad thrust bearings with optimum pivot location.†

* See footnote on page page 2219.

† See footnote on page page 2219.

9) *Operating number.*

$$O = \frac{1.45\times10^{-7}\times 18\times 10,400}{5\times 614\times 4.75} = 1.86\times10^{-6}$$

10) *Minimum film thickness.* From Fig. 31, $\alpha = 0.30\times10^{-3}$.

$$h_{min} = 0.00030\times 4.75 = 0.0014 \text{ inch}$$

11) *Coefficient of friction.* From Fig. 32, $f = 0.0036$.

12) *Friction power loss.*

$$P_f = \frac{0.0036\times 70,000\times 10,400}{33,000} = 79.4 \text{ hp}$$

13) *Oil flow.*

$$Q = 0.0591\times 6\times 0.30\times 10^{-3}\times 4\times 4.75\times 10,400 = 21.02 \text{ gpm}$$

14) *Temperature rise.*

$$\Delta t = \frac{0.0217\times 0.0036\times 614}{0.30\times 10^{-3}\times 3.5} = 45.7^{\circ}\text{F}$$

Because this temperature is less than the 50°F, which is considered as the acceptable maximum, the design is satisfactory.

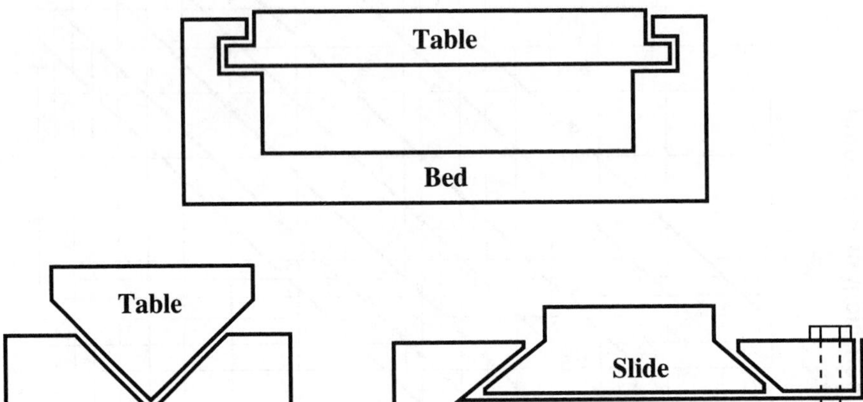

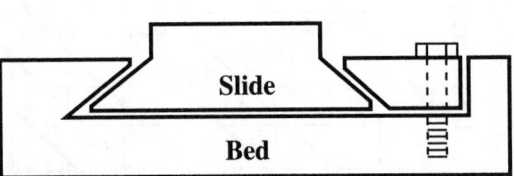

Fig. 33. Types of Guide Bearings

Guide Bearings

This type of bearing is generally used as a positioning device or as a guide to linear motion such as in machine tools. Fig. 33 shows several examples of guideway bearing designs. It is normal for this type of bearing to operate in the boundary lubrication region with either dry, dry film such as molybdenum disulfide (MoS_2) or tetrafluorethylene (TFE), grease, oil, or gaseous lubrication. Hydrostatic lubrication is often used to improve performance, reduce wear, and increase stability. This type of design uses pumps to supply air or gas under pressure to pockets designed to produce a bearing film and maintain complete separation of the sliding surfaces.

PLAIN BEARING MATERIALS

Materials used for sliding bearings cover a wide range of metals and nonmetals. To make the optimum selection requires a complete analysis of the specific application. The important general categories are: Babbitts, alkali-hardened lead, cadmium alloys, copper lead, aluminum bronze, silver, sintered metals, plastics, wood, rubber, and carbon graphite.

Properties of Bearing Materials.—For a material to be used as a plain bearing, it must possess certain physical and chemical properties that permit it to operate properly. If a material does not possess all of these characteristics to some degree, it will not function long as a bearing. It should be noted, however, that few, if any, materials are outstanding in all these characteristics. Therefore, the selection of the optimum bearing material for a given application is at best a compromise to secure the most desirable combination of properties required for that particular usage.

The seven properties generally acknowledged to be the most significant are: 1) Fatigue resistance; 2) Embeddability; 3) Compatibility; 4) Conformability; 5) Thermal conductivity; 6) Corrosion resistance; and 7) Load capacity.

These properties are described as follows:

1) *Fatigue resistance* is the ability of the bearing lining material to withstand repeated applications of stress and strain without cracking, flaking, or being destroyed by some other means.

2) *Embeddability* is the ability of the bearing lining material to absorb or embed within itself any of the larger of the small dirt particles present in a lubrication system. Poor embeddability permits particles circulating around the bearing to score both the bearing surface and the journal or shaft. Good embeddability will permit these particles to be trapped and forced into the bearing surface and out of the way where they can do no harm.

3) *Compatibility or antiscoring tendencies* permit the shaft and bearing to "get along" with each other. It is the ability to resist galling or seizing under conditions of metal-to-metal contact such as at startup. This characteristic is most truly a bearing property, because contact between the bearing and shaft in good designs occurs only at startup.

4) *Conformability* is defined as malleability or as the ability of the bearing material to creep or flow slightly under load, as in the initial stages of running, to permit the shaft and bearing contours to conform with each other or to compensate for nonuniform loading caused by misalignment.

5) *High thermal conductivity* is required to absorb and carry away the heat generated in the bearing. This conductivity is most important, not in removing frictional heat generated in the oil film, but in preventing seizures due to hot spots caused by local asperity breakthroughs or foreign particles.

6) *Corrosion resistance* is required to resist attack by organic acids that are sometimes formed in oils at operating conditions.

7) *Load capacity or strength* is the ability of the material to withstand the hydrodynamic pressures exerted upon it during operation.

Babbitt or White Metal Alloys.—Many different bearing metal compositions are referred to as babbitt metals. The exact composition of the original babbitt metal is not known; however, the ingredients were probably tin, copper, and antimony in approximately the following percentages: 89.3, 3.6, and 7.1. Tin and lead-base babbitts are probably the best known of all bearing materials. With their excellent embeddability and compatibility characteristics under boundary lubrication, babbitt bearings are used in a wide range of applications including household appliances, automobile and diesel engines, railroad cars, electric motors, generators, steam and gas turbines, and industrial and marine gear units.

Table 9. Bearing and Bushing Alloys—Composition, Forms, Characteristics, and Applications *SAE General Information*

SAE No.and Alloy Grouping		Nominal Composition, Per cent	Form of Use (1), Characteristics (2), and Applications (3)
Sn-Base Alloys	11	{Sn, 87.5; Sb, 6.75; Cu,5.75	(1) Cast on steel, bronze, or brass backs, or directly in the bearing housing. (2) Soft, corrosion-resistant with moderate fatigue resistance. (3) Main and connecting-rod bearings; motor bushings. Operates with either hard or soft journal.
	12	{ Sn, 89; Sb,7.5;Cu, 3.5	
Pb-Base Alloys	13	{ Pb, 84; Sb, 10; Sn, 6	(1) SAE 13 and 14 are cast on steel, bronze, or brass, or in the bearing housing; SAE 15 is cast on steel; and SAE 16 is cast into and on a porous sintered matrix, usually copper-nickel bonded to steel. (2) Soft, moderately fatigue-resistant, corrosion-resistant. (3) Main and connecting-rod bearings. Operates with hard or soft journal with good finish.
	14	{ Pb, 75; Sb, 15; Sn, 10	
	15	{Pb, 83; Sb,15 ;Sn,14 ; As,1	
	16	{ Pb, 92; Sb, 3.5;Sn, 4.5	
Pb-Sn Overlays	19	{ Pb, 90; Sn, 10	(1) Electrodeposited as a thin layer on copper-lead or silver bearings faces. (2) Soft, corrosion-resistant. Bearings so coated run satisfactorily against soft shafts throughout the life of the coating. (3) Heavy-duty, high-speed main and connecting-rod bearings.
	190	{ Pb, 93; Sn, 7	
Cu-Pb Alloys	49	{ Cu, 76; Pb, 24	(1) Cast or sintered on steel back with the exception of SAE 481, which is cast on steel back only. (2) Moderately hard. Somewhat subject to oil corrosion. Some oils minimize this; protection with overlay may be desirable. Fatigue resistance good to fairly good. Listed in order of decreasing hardness and fatigue resistance. (3) Main and connecting-rod bearings. The higher lead alloys can be used unplated against a soft shaft, although an overlay is helpful. The lower lead alloys may be used against a hard shaft, or with an overlay against a soft one.
	48	{ Cu, 70; Pb, 30	
	480	{ Cu, 65; Pb, 35	
	481	{ Cu, 60; Pb, 40	
Cu-Pb-Sn-Alloys	482	{ Cu, 67; Pb, 28; Sn, 5	(1) Steel-backed and lined with a structure combining sintered copper alloy matrix with corrosion-resistant lead alloy. (2) Moderately hard. Corrosion resistance improved over copper-leads of equal lead content without tin. Fatigue resistance fairly good. Listed in order of decreasing hardness and fatigue resistance. (3) Main and connecting-rod bearings. Generally used without overlay. SAE 484 and 485 may be used with hard or soft shaft, and a hardened or cast shaft is recommended for SAE 482.
	484	{ Cu, 55; Pb, 42; Sn, 3	
	485	{ Cu, 46; Pb, 51; Sn, 3	
Al-Base Alloys	770	{ Al, 91.75; Sn, 6.25; Cu, 1; Ni, 1	(1) SAE 770 cast in permanent molds; work-hardened to improve physical properties. SAE 780 and 782 usually bonded to steel back but is procurable in strip form without steel backing. SAE 781 usually bonded to steel back but can be produced as castings or wrought strip without steel back. (2) Hard, extremely fatigue-resistant, resistant to oil corrosion. (3) Main and connecting-rod bearings. Generally used with suitable overlay. SAE 781 and 782 also used for bushings and thrust bearings with or without overlay.
	780	{ Al, 91; Sn, 6; Si, 1.5; Cu, 1; Ni, 0.5	
	781	{ Al, 95; Si, 4; Cd, 1	
	782	{ Al, 95; Cu, 1;Ni, 1; Cd, 3	
Other Cu-Base Alloys	795	{ Cu, 90; Zn, 9.5; Sn, 0.5	(1) Wrought solid bronze, (2) Hard, strong, good fatigue resistance, (3) Intermediate-load oscillating motion such as tie-rods and brake shafts.
	791	{Cu, 88; Zn,4;Sn,4; Pb, 4	(1) SAE 791, wrought solid bronze; SAE 793, cast on steel back; SAE 798, sintered on steel back. (2) General-purpose bearing material, good shock and load capacity. Resistant to high temperatures. Hard shaft desirable. Less score-resistant than higher lead alloys. (3) Medium to high loads. Transmission bushings and thrust washers. SAE 791 also used for piston pin and 793 and 798 for chassis bushings.
	793	{ Cu, 84; Pb, 8; Sn, 4; Zn, 4	
	798	{Cu, 84; Pb, 8; Sn,4; Zn, 4	
Other Cu-Base Alloys	792	{ Cu, 80; Sn, 10; Pb, 10	(1) SAE 792, cast on steel back, SAE 797, sintered on steel back. (2) Has maximum shock and load-carrying capacity of conventional cast bearing alloys; hard, both fatigue- and corrosion-resistant. Hard shaft desirable. (3) Heavy loads with oscillating or rotating motion. Used for piston pins, steering knuckles, differential axles, thrust washers, and wear plates.
	797	{ Cu, 80; Sn, 10; Pb, 10	
	794	{ Cu, 73.5; Pb, 23; Sn, 3.5	(1) SAE 794, cast on steel back; SAE 799, sintered on steel back. (2) Higher lead content gives improved surface action for higher speeds but results in somewhat less corrosion resistance. (3) Intermediate load application for both oscillating and rotating shafts, that is, rocker-arm bushings, transmissions, and farm implements.
	799	{ Cu, 73.5,; Pb, 23; Sn, 3.5	

Table 10. White Metal Bearing Alloys—Composition and Properties *ASTM B23-83, reapproved 1988*

ASTM Alloy Number	Nominal Composition, Per Cent				Compressive Yield Point,[a] psi		Ultimate Compressive Strength,[b] psi		Brinell Hardness[c]		Melting Point °F	Proper Pouring Temperature, °F
	Sn	Sb	Pb	Cu	68 °F	212 °F	68 °F	212 °F	68 °F	212 °F		
1	91.0	4.5	…	4.5	4400	2650	12,850	6950	17.0	8.0	433	825
2	89.0	7.5	…	3.5	6100	3000	14,900	8700	24.5	12.0	466	795
3	83.33	8.33	…	8.33	6600	3150	17,600	9900	27.0	14.5	464	915
4	75.0	12.0	10.0	3.0	5550	2150	16,150	6900	24.5	12.0	363	710
5	65.0	15.0	18.0	2.0	5050	2150	15,050	6750	22.5	10.0	358	690
6	20.0	15.0	63.5	1.5	3800	2050	14,550	8050	21.0	10.5	358	655
7[d]	10.0	15.0	bal.	…	3550	1600	15,650	6150	22.5	10.5	464	640
8[d]	5.0	15.0	bal.	…	3400	1750	15,600	6150	20.0	9.5	459	645
10	2.0	15.0	83.0	…	3350	1850	15,450	5750	17.5	9.0	468	630
11	…	15.0	85.0	…	3050	1400	12,800	5100	15.0	7.0	471	630
12	…	10.0	90.0	…	2800	1250	12,900	5100	14.5	6.5	473	625
15[e]	1.0	16.0	bal.	0.5	…	…	…	…	21.0	13.0	479	662
16	10.0	12.5	77.0	0.5	…	…	…	…	27.5	13.6	471	620
19	5.0	9.0	86.0	…	…	…	15,600	6100	17.7	8.0	462	620

[a] The values for yield point were taken from stress-strain curves at the deformation of 0.125 per cent reduction of gage.

[b] The ultimate strength values were taken as the unit load necessary to produce a deformation of 25 per cent of the length of the specimen.

[c] These values are the average Brinell number of three impressions on each alloy using a 10-mm ball and a 500-kg load applied for 30 seconds.

[d] Also nominal arsenic, 0.45 per cent.

[e] Also nominal arsenic, 1 per cent.

Data for ASTM alloys 1, 2, 3, 7, 8, and 15 appear in the Appendix of ASTM B23-83; the data for alloys 4, 5, 6, 10, 11, 12, 16, and 19 are given in ASTM B23-49. All values are for reference purposes only.

The compression test specimens were cylinders 1.5 inches in length and 0.5 inch in diameter, machined from chill castings 2 inches in length and 0.75 inch in diameter. The Brinell tests were made on the bottom face of parallel machined specimens cast in a 2-inch diameter by 0.625-inch deep steel mold at room temperature.

Both the Society of Automotive Engineers and American Society for Testing and Materials have classified white metal bearing alloys. Tables 9 and 10 give compositions and properties or characteristics for the two classifications.

In small bushings for fractional-horsepower motors and in automotive engine bearings, the babbitt is generally used as a thin coating over a flat steel strip. After forming oil distribution grooves and drilling required holes, the strip is cut to size, then rolled and shaped into the finished bearing. These bearings are available for shaft diameters from 0.5 to 5 inches. Strip bearings are turned out by the millions yearly in highly automated factories and offer an excellent combination of low cost with good bearing properties.

For larger bearings in heavy-duty equipment, a thicker babbitt is cast on a rigid backing of steel or cast iron. Chemical and electrolytic cleaning of the bearing shell, thorough rinsing, tinning, and then centrifugal casting of the babbitt are desirable for sound bonding of the babbitt to the bearing shell. After machining, the babbitt layer is usually $\frac{1}{2}$ to $\frac{1}{4}$ inch thick.

Compared to other bearing materials, babbitts generally have lower load-carrying capacity and fatigue strength, are a little higher in cost, and require a more complicated design. Also, their strength decreases rapidly with increasing temperature. These shortcomings can be avoided by using an intermediate layer of high-strength, fatigue-resistant material that is placed between a steel backing and a thin babbitt surface layer. Such composite bearings frequently eliminate any need for using alternate materials having poorer bearing characteristics.

Tin babbitt is composed of 80 to 90 per cent tin to which is added about 3 to 8 per cent copper and 4 to 14 per cent antimony. An increase in copper or antimony produces increased hardness and tensile strength and decreased ductility. However, if the percentages of these alloys are increased above those shown in Table 10, the resulting alloy will have decreased fatigue resistance. These alloys have very little tendency to cause wear to their journals because of their ability to embed dirt. They resist the corrosive effects of acids, are not prone to oil-film failure, and are easily bonded and cast. Two drawbacks are encountered from use of these alloys because they have low fatigue resistance and their hardness and strength drop appreciably at low temperatures.

Lead babbitt compositions generally range from 10 to 15 per cent antimony and up to 10 per cent tin in combination with the lead. Like tin-base babbitts, these alloys have little tendency to cause wear to their journals, embed dirt well, resist the corrosive effects of acids, are not prone to oil-film failure and are easily bonded and cast. Their chief disadvantages when compared with tin-base alloys are a rather lower strength and a susceptibility to corrosion.

Cadmium Base.—Cadmium alloy bearings have a greater resistance to fatigue than babbitt bearings, but their use is very limited due to their poor corrosion resistance. These alloys contain 1 to 15 per cent nickel, or 0.4 to 0.75 per cent copper, and 0.5 to 2.0 per cent silver. Their prime attribute is their high-temperature capability. The load-carrying capacity and relative basic bearing properties are shown in Tables 11 and 12.

Table 11. Properties of Bearing Alloys

Material	Recommended Shaft Hardness, Brinell	Load-Carrying Capacity, psi	Maximum Operating Temp., °F
Tin-Base Babbitt	150 or less	800–1500	300
Lead-Base Babbitt	150 or less	800–1200	300
Cadmium Base	200–250	1200–2000	500
Copper-Lead	300	1500–2500	350
Tin-Bronze	300–400	4000+	500+
Lead-Bronze	300	3000–4500	450–500
Aluminum	300	4000+	225–300
Silver-Overplate	300	4000+	500
Trimetal-Overplate	230 or less	2000–4000+	225–300

Table 12. Bearing Characteristics Ratings

Material	Compatibility	Conformability and Embeddability	Corrosion Resistance	Fatigue Strength
Tin-Base Babbitt	1	1	1	5
Lead-Base Babbitt	1	1	3	5
Cadmium Base	1	2	5	4
Copper-Lead	2	2	5	3
Tin-Bronze	3	5	2	1
Lead-Bronze	3	4	4	2
Aluminum	5	3	1	2
Silver Overplate	2	3	1	1
Trimetal-Overplate	1	2	2	3

Note: 1 is best; 5 is worst.

Copper-Lead.—Copper-lead bearings are a binary mixture of copper and lead containing from 20 to 40 per cent lead. Lead is practically insoluble in copper, so a cast microstructure consists of lead pockets in a copper matrix. A steel backing is commonly used with this material and high volume is achieved either by continuous casting or by powder metallurgy techniques. This material is very often used with an overplate such as lead-tin and lead-tin-copper to increase basic bearing properties. Tables 11 and 12 provide comparisons of material properties.

The combination of good fatigue strength, high-load capacity, and high-temperature performance has resulted in extensive use of this material for heavy-duty main and connecting-rod bearings as well as moderate-load and speed applications in turbines and electric motors.

Leaded Bronze and Tin-Bronze.—Leaded and tin-bronzes contain up to 25 per cent lead or approximately 10 per cent tin, respectively. Cast leaded bronze bearings offer good compatibility, excellent casting, and easy machining characteristics, low cost, good structural properties and high-load capacity, usefulness as a single material that requires neither a separate overlay nor a steel backing. Bronzes are available in standard bar stock, sand or permanent molds, investment, centrifugal or continuous casting. Leaded bronzes have better compatibility than tin-bronzes because the spheroids of lead smear over the bearing surface under conditions of inadequate lubrication. These alloys are generally a first choice at intermediate loads and speeds. Tables 11 and 12 provide comparisons of basic bearing properties of these materials.

Aluminum.—Aluminum bearings are either cast solid aluminum, aluminum with a steel backing, or aluminum with a suitable overlay. The aluminum is usually alloyed with small amounts of tin, silicon, cadmium, nickel, or copper, as shown in Table 9. An aluminum bearing alloy with 20 to 30 per cent tin alloy and up to 3 per cent copper has shown promise as a substitute for bronzes in some industrial applications.

These bearings are best suited for operation with hard journals. Owing to the high thermal expansion of the metal (resulting in diametral contraction when it is confined as a bearing in a rigid housing), large clearances are required, which tend to make the bearing noisy, especially on starting. Overlays of lead-tin, lead, or lead-tin-copper may be applied to aluminum bearings to facilitate their use with soft shafts.

Aluminum alloys are available with properties specifically designed for bearing applications, such as high load-carrying capacity, fatigue strength, and thermal conductivity, in addition to excellent corrosion resistance and low cost.

Silver.—Silver bearings were developed for and have an excellent record in heavy-duty applications such as aircraft master rod and diesel engine main bearings. Silver has a higher fatigue rating than any of the other bearing materials; the steel backing used with this material may show evidence of fatigue before the silver. The advent of overlays, or more commonly called overplates, made it possible for silver to be used as a bearing material. Silver by itself does not possess any of the desirable bearing qualities except high fatigue resistance and high thermal conductivity. The overlays such as lead, lead-tin, or lead-indium improve the embeddability and antiscoring properties of silver. The relative basic properties of this material, when used as an overplate, are shown in Tables 11 and 12.

Cast Iron.—Cast iron is an inexpensive bearing material capable of operation at light loads and low speeds, i.e., up to 130 ft/min and 150 lb/in.^{2} These bearings must be well lubricated and have a rather large clearance so as to avoid scoring from particles torn from the cast iron that ride between bearing and journal. A journal hardness of between 150 and 250 Brinell has been found to be best when using cast-iron bearings.

Porous Metals.—Porous metal self-lubricating bearings are usually made by sintering metals such as plain or leaded bronze, iron, and stainless steel. The sintering produces a spongelike structure capable of absorbing fairly large quantities of oil, usually 10–35 per cent of the total volume. These bearings are used where lubrication supply is difficult, inadequate, or infrequent. This type of bearing should be flooded from time to time to resaturate the material. Another use of these porous materials is to meter a small quantity of oil to the bearings such as in drip feed systems. The general design operating characteristics of this class of materials are shown in Table 13.

Table 13. Application Limits — Sintered Metal and Nonmetallic Bearings

Bearing Material	Load Capacity (psi)	Maximum Temperature (°F)	Surface Speed, V_{max} (max. fpm)	PV Limit P = psi load V = surface ft/min
Acetal	1000	180	1000	3000
Graphite (dry)	600	750	2500	15,000
Graphite (lubricated)	600	750	2500	150,000
Nylon, Polycarbonate	1000	200	1000	3000
Nylon composite	…	400	…	16,000
Phenolics	6000	200	2500	15,000
Porous bronze	4500	160	1500	50,000
Porous iron	8000	160	800	50,000
Porous metals	4000–8000	150	1500	50,000
Virgin Teflon (TFE)	500	500	50	1000
Reinforced Teflon	2500	500	1000	10,000–15,000
TFE fabric	60,000	500	150	25,000
Rubber	50	150	4000	15,000
Maple & Lignum Vitae	2000	150	2000	15,000

Table 14 gives the chemical compositions, permissible loads, interference fits, and running clearances of bronze-base and iron-base metal-powder sintered bearings that are specified in the ASTM specifications for oil-impregnated metal-powder sintered bearings (B438-83a and B439-83).

Plastics Bearings.—Plastics are finding increased use as bearing materials because of their resistance to corrosion, quiet operation, ability to be molded into many configurations, and their excellent compatibility, which minimizes or eliminates the need for lubrication. Many plastics are capable of operating as bearings, especially phenolic, tetrafluoroethylene (TFE), and polyamide (nylon) resins. The general application limits for these materials are shown in Table 13.

Laminated Phenolics: These composite materials consist of cotton fabric, asbestos, or other fillers bonded with phenolic resin. They have excellent compatibility with various fluids as well as strength and shock resistance. However, precautions must be taken to maintain adequate bearing cooling because the thermal conductivity of these materials is low.

Nylon: This material has the widest use for small, lightly loaded applications. It has low frictional properties and requires no lubrication.

Teflon: This material, with its exceptional low coefficient of friction, self-lubricating characteristics, resistance to attack by almost any chemicals, and its wide temperature range, is one of the most interesting of the plastics for bearing use. High cost combined with low load capacity cause Teflon to be selected mostly in modified form, where other less expensive materials have proved inadequate for design requirements.

Bearings made of laminated phenolics, nylon, or Teflon are all unaffected by acids and alkalies except if highly concentrated and therefore can be used with lubricants containing dilute acids or alkalies. Water is used to lubricate most phenolic laminate bearings but oil, grease, and emulsions of grease and water are also used. Water and oil are used as lubricants for nylon and Teflon bearings. Almost all types of plastic bearings absorb water and oil to some extent. In some the dimensional change caused by the absorption may be as much as three per cent in one direction. This means that bearings have to be treated before use so that proper clearances will be kept. This may be done by boiling in water, for water lubricated bearings. Boiling in water makes bearings swell the maximum amount. Clearances for phenolic bearings are kept at about 0.001 inch per inch of diameter on treated bearings. Partially lubricated or dry nylon bearings are given a clearance of 0.004 to 0.006 inches for a one-inch diameter bearing.

Table 14. Copper- and Iron-Base Sintered Bearings (Oil Impregnated) — *ASTM B438-83a (R1989), B439-83 (R1989), and Appendices*

Chemical Requirements								
Alloying Elements[a]	Percentage Composition							
	Copper-Base Bearings				Iron-Base Bearings			
	Grade 1		Grade 2		Grades			
	Class A	Class B	Class A	Class B	1	2	3	4
Cu	87.5–99.5	87.5–90.5	87.5–90.5	87.5–90.5	…	…	7.0–11.0	18.0–22.0
Sn	9.5–10.5	9.5–10.5	9.5–10.5	9.5–10.5	…	…	…	…
Graphite	0.1 max.	1.75 max.	0.1 max.	1.75 max.	…	…	…	…
Pb	…	…	2.0–4.0	2.0–4.0	…	…	…	…
Fe	1.0 max.	1.0 max.	1.0 max.	1.0 max.	96.25 min.	95.9 min.	Balance[b]	Balance[b]
Comb. C[c]	…	…	…	…	0.25 max.	0.25–0.60	…	…
Si, max.	…	…	…	…	0.3	0.3	…	…
Al, max.	…	…	…	…	0.2	0.2	…	…
Others	0.5 max.	0.5 max.	1.0 max.	1.0 max.	3.0 max.	3.0 max.	3.0 max.	3.0 max.

[a] Abbreviations used for the alloying elements are as follows: Cu, copper; Fe, iron; Sn, tin; Pb, lead; Zn, zinc; Ni, nickel; Sb, antimony; Si, silicon; Al, aluminum; and C, carbon.

[b] Total of iron plus copper shall be 97 per cent, minimum.

[c] Combined carbon (on basis of iron only) may be a metallographic estimate of the carbon in the iron.

Permissible Loads						
Copper-Base Bearings				Iron-Base Bearings		
	Grades 1 & 2					
	Type 1	Type 2	Types 3 & 4		Grades 1 & 2	Grades 3 & 4
Shaft Velocity, fpm	Max. Load, psi			Shaft Velocity, fpm	Max. Load, psi	
Slow and intermittent	3200	4000	4000	Slow and intermittent	3600	8000
25	2000	2000	2000	25	1800	3000
50 to 100	500	500	550	50 to 100	450	700
Over 100 to 150	365	325	365	Over 100 to 150	300	400
Over 150 to 200	280	250	280	Over 150 to 200	225	300
Over 200		a	a	a	Over 200	a

[a] For shaft velocities over 200 fpm, the permissible loads may be calculated as follows: $P = 50{,}000/V$; where P = safe load, psi of projected area; and V = shaft velocity, fpm. With a shaft velocity of less than 50 fpm and a permissible load greater than 1,000 psi, an extreme pressure lubricant should be used; with heat dissipation and removal techniques, higher PV ratings can be obtained.

Clearances						
Press-Fit Clearances			Running Clearances[a]			
Copper- and Iron-Base			Copper-Base		Iron-Base	
Bearing OD	Min.	Max.	Shaft Size	Min. Clearance	Shaft Size	Min. Clearance
Up to 0.760	0.001	0.003	Up to 0.250	0.0003	Up to 0.760	0.0005
0.761–1.510	0.0015	0.004	0.250–0.760	0.0005	0.761–1.510	0.001
1.511–2.510	0.002	0.005	0.760–1.510	0.0010	1.511–2.510	0.0015
2.511–3.010	0.002	0.006	1.510–2.510	0.0015	Over 2.510	0.002
Over 3.010	0.002	0.007	Over 2.510	0.0020		

[a] Only minimum recommended clearances are listed. It is assumed that ground steel shafting will be used and that all bearings will be oil-impregnated.

Table 15. Copper- and Iron-Base Sintered Bearings (Oil Impregnated) —
ASTM B438-83a (R1989), B439-83 (R1989), and Appendices

Commercial Dimensional Tolerances[a,b]							
Diameter Tolerance		Length Tolerance		Diameter Tolerance		Length Tolerance	
Copper Base				Iron Base			
Inside or Outside Diameter	Total Diameter Tolerances	Length	Total Length Tolerances	Inside or Outside Diameter	Total Diameter Tolerances	Length	Total Length Tolerances
Up to 1	0.001	Up to 1.5	0.01	Up to 0.760	−0.001	Up to 1.495	0.01
1 to 1.5	0.0015	1.5 to 3	0.01	0.761–1.510	−0.0015	1.496–1.990	0.02
1.5 to 2	0.002	3 to 4.5	0.02	1.511–2.510	−0.002	1.991–2.990	0.02
2 to 2.5	0.0025	…	…	2.511–3.010	−0.003	2.991–4.985	0.03
2.5 to 3	0.003	…	…	3.011–4.010	−0.005	…	…
…	…	…	…	4.011–5.010	−0.005	…	…
…	…	…	…	5.011–6.010	−0.006	…	…

[a] For copper-base bearings with 4-to-1 maximum-length-diameter ratio and a 24-to-1 maximum-length-to-wall-thickness ratio; bearings with greater ratios are not covered here.

[b] For iron-base bearings with a 3-to-1 maximum-length-to-inside diameter ration and a 20-to-1 maximum-length-to-wall-thickness ratio; bearings with greater ratios are not covered here.

Concentricity Tolerance[a,b,a]		
Iron Base		
Outside Diameter	Max. Wall Thickness	Concentricity Tolerance
Up to 1.510	Up to 0.355	0.003
1.511 to 2.010	Up to 0.505	0.004
2.011 to 4.010	Up to 1.010	0.005
4.011 to 5.010	Up to 1.510	0.006
5.011 to 6.010	Up to 2.010	0.007
Copper Base		
Outside Diameter	Length	Concentricity Tolerance
Up to 1	0 to 1	0.00.
	1 to 2	0.004
	2 to 3	0.005
1 to 2	0 to 1	0.004
	1 to 2	0.005
	2 to 3	0.006
2 to 3	0 to 1	0.005
	1 to 2	0.006
	2 to 3	0.007

[a] Total indicator reading.

Table 16. Copper- and Iron-Base Sintered Bearings (Oil Impregnated) — *ASTM B438-83a (R1989), B439-83 (R1989), and Appendices*

Diameter Range	Flange and Thrust Bearings, Diameter, and Thickness Tolerances[a]			
	Flange Diameter Tolerance		Flange Thickness Tolerance	
	Standard	Special	Standard	Special
0 to 1½	±0.005	±0.0025	±0.005	±0.0025
Over 1½ to 3	±0.010	±0.005	±0.010	±0.007
Over 3 to 6	±0.025	±0.010	±0.015	±0.010
Diameter Range	Parallelism on Faces, max.			
	Copper Base		Iron Base	
	Standard	Special	Standard	Special
0 to 1½	0.003	0.002	0.005	0.003
Over 1½ to 3	0.004	0.003	0.007	0.005
Over 3 to 6	0.005	0.004	0.010	0.007

[a] Standard and special tolerances are specified for diameters, thicknesses, and parallelism. Special tolerances should not be specified unless required because they require additional or secondary operations and, therefore, are costlier. Thrust bearings (¼ inch thickness, max.) have a standard thickness tolerance of ±0.005 inch and a special thickness tolerance of ±0.0025 inch for all diameters.

All dimensions in inches except where otherwise noted.

Wood: Bearings made from such woods as lignum vitae, rock maple, or oak offer self-lubricating properties, low cost, and clean operation. However, they have frequently been displaced in recent years by various plastics, rubber and sintered-metal bearings. General applications are shown in Table 10.

Rubber: Rubber bearings give excellent performance on propeller shafts and rudders of ships, hydraulic turbines, pumps, sand and gravel washers, dredges and other industrial equipment that handle water or slurries. The resilience of rubber helps to isolate vibration and provide quiet operation, allows running with relatively large clearances and helps to compensate for misalignment. In these beatings a fluted rubber structure is supported by a metal shell. The flutes or scallops in the rubber form a series of grooves through which lubricant or, as generally used, water and foreign material such as sand may pass through the beating.

Carbon-Graphite.—Bearings of molded and machined carbon-graphite are used where regular maintenance and lubrication cannot be given. They are dimensionally stable over a wide range of temperatures, may be lubricated if desired, and are not affected by chemicals. These bearings may be used up to temperatures of 700 to 750 degrees F. in air or 1200 degrees F. in a non-oxidizing atmosphere, and generally are operated at a maximum load of 20 pounds per square inch. In some instances a metal or metal alloy is added to the carbon-graphite composition to improve such properties as compressive strength and density. The temperature limitation depends upon the melting point of the metal or alloy and the maximum load is generally 350 pounds per square inch when used with no lubrication or 600 pounds per square inch when used with lubrication.

Normal running clearances for both types of carbon-graphite bearings used with steel shafts and operating at a temperature of less than 200 degrees F. are as follows: 0.001 inch for bearings of 0.187 to 0.500-inch inside diameter, 0.002 inch for bearings of 0.501 to 1.000-inch inside diameter, 0.003 inch for bearings of 1.001 to 1.250-inch inside diameter, 0.004 inch for bearings of 1.251 to 1.500-inch diameter, and 0.005 inch for bearings of 1.501 to 2.000-inch inside diameter. Speeds depend upon too many variables to list specifically so it can only be stated here that high loads require a low number of rpm and low loads permit a high number of rpm. Smooth journals are necessary in these bearings as rough ones tend to abrade the bearings quickly. Cast iron and hard chromium-plate steel shafts of 400 Brinell and over, and phosphor-bronze shafts over 135 Brinell are recommended.

BALL, ROLLER, AND NEEDLE BEARINGS

Rolling contact bearings substitute a rolling element, ball or roller, for a hydrodynamic or hydrostatic fluid film to carry an impressed load without wear and with reduced friction. Because of their greatly reduced starting friction, when compared to the conventional journal bearing, they have acquired the common designation of "anti-friction" bearings. Although normally made with hardened rolling elements and races, and usually utilizing a separator to space the rolling elements and reduce friction, many variations are in use throughout the mechanical and electrical industries. The most common anti-friction bearing application is that of the deep-groove ball bearing with ribbon-type separator and sealed-grease lubrication used to support a shaft with radial and thrust loads in rotating equipment. This shielded or sealed bearing has become a standard and commonplace item ordered from a supplier's catalogue in much the same manner as nuts and bolts. Because of the simple design approach and the elimination of a separate lubrication system or device, this bearing is found in as many installations as the wick-fed or impregnated porous plain bushing.

Currently, a number of manufacturers produce a complete range of ball and roller bearings in a fully interchangeable series with standard dimensions, tolerances and fits as specified in Anti-Friction Bearing Manufacturers Association (AFBMA) Standards. Except for deep-groove ball bearings, performance standards are not so well defined and sizing and selection must be done in close conformance with the specific manufacturer's catalogue requirements. In general, desired functional features should be carefully gone over with the vendor's representatives.

Rolling contact bearings are made to high standards of accuracy and with close metallurgical control. Balls and rollers are normally held to diametral tolerances of .0001 inch or less within one bearing and are often used as "gage" blocks in routine toolroom operations. This accuracy is essential to the performance and durability of rolling-contact bearings and in limiting runout, providing proper radial and axial clearances, and ensuring smoothness of operation.

Because of their low friction, both starting and running, rolling-contact bearings are utilized to reduce the complexity of many systems that normally function with journal bearings. Aside from this advantage and that of precise radial and axial location of rotating elements, however, they also are desirable because of their reduced lubrication requirements and their ability to function during brief interruptions in normal lubrication.

In applying rolling-contact bearings it is well to appreciate that their life is limited by the fatigue life of the material from which they are made and is modified by the lubricant used. In rolling-contact fatigue, precise relationships among life, load, and design characteristics are not predictable, but a statistical function described as the "probability of survival" is used to relate them according to equations recommended by the AFBMA. Deviations from these formulas result when certain extremes in applications such as speed, deflection, temperature, lubrication, and internal geometry must be dealt with.

Types of Anti-friction Bearings.—The general types are usually determined by the shape of the rolling element, but many variations have been developed that apply conventional elements in unique ways. Thus it is well to know that special bearings can be procured with races adapted to specific applications, although this is not practical for other than high volume configurations or where the requirements cannot be met in a more economical manner. "Special" races are appreciably more expensive. Quite often, in such situations, races are made to incorporate other functions of the mechanism, or are "submerged" in the surrounding structure, with the rolling elements supported by a shaft or housing that has been hardened and finished in a suitable manner. Typical anti-friction bearing types are shown in Table 1.

Types of Ball Bearings.—Most types of ball bearings originate from three basic designs: the single-row radial, the single-row angular contact, and the double-row angular contact.

Table 1. Types of Rolling Element Bearings and Their Symbols

BALL BEARINGS, SINGLE ROW, RADIAL CONTACT			
Symbol	Description	Symbol	Description
BC	Non-filling slot assembly 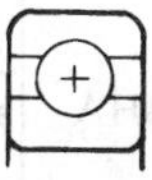	BH	Non-separable counter-bore assembly 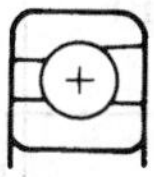
BL	Filling slot assembly 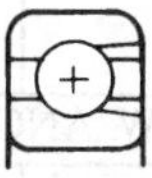	BM	Separable assembly 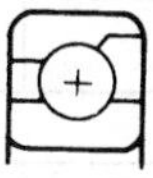
BALL BEARINGS, SINGLE ROW, ANGULAR CONTACT[a]			
Symbol	Description	Symbol	Description
BN	Non-separable Nominal contact angle: from above 10° to and including 22° 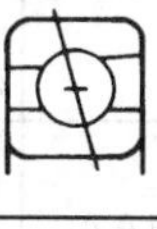	BAS	Separable inner ring Nominal contact angle: from above 22° to and including 32° 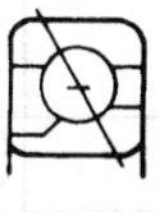
BNS	Separable outer ring Nominal contact angle: from above 10° to and including 22° 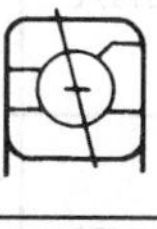	BT	Non-separable Nominal contact angle: from above 32° to and including 45° 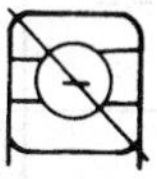
BNT	Separable inner ring Nominal contact angle: from above 10° to and including 22° 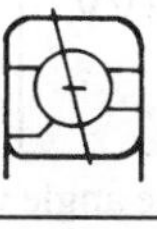	BY	Two-piece outer ring 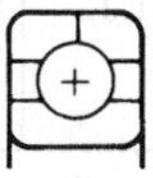
BA	Non-separable Nominal contact angle: from above 22° to and including 32° 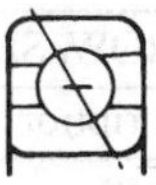	BZ	Two-piece inner ring 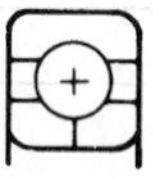
BALL BEARINGS, SINGLE ROW, RADIAL CONTACT, SPHERICAL OUTSIDE SURFACE			
Symbol	Description	Symbol	Description
BCA	Non-filling slot assembly 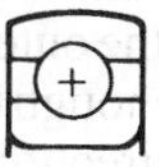	BLA	Filling slot assembly 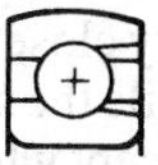

[a] A line through the ball contact points forms an acute angle with a perpendicular to the bearing axis.

Single-row Radial, Non-filling Slot: This is probably the most widely used ball bearing and is employed in many modified forms. It is also known as the "Conrad" type or "Deep-groove" type. It is a symmetrical unit capable of taking combined radial and thrust loads in which the thrust component is relatively high, but is not intended for pure thrust loads, however. Because this type is not self-aligning, accurate alignment between shaft and housing bore is required.

Single-row Radial, Filling Slot: This type is designed primarily to carry radial loads. Bearings of this type are assembled with as many balls as can be introduced by eccentric displacement of the rings, as in the non-filling slot type, and then several more balls are inserted through the loading slot, aided by a slight spreading of the rings and heat expansion of the outer ring, if necessary. This type of bearing will take a certain degree of thrust when in combination with a radial load but is not recommended where thrust loads exceed 60 per cent of the radial load.

BALL BEARINGS, DOUBLE ROW, RADIAL CONTACT			
Symbol	Description	Symbol	Description
BF	Filling slot assembly 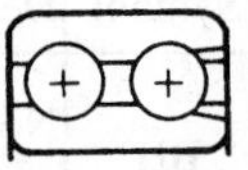	BHA	Non-separable two-piece outer ring 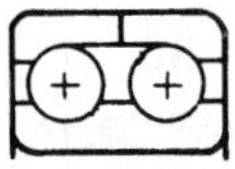
BK	Non-filling slot assembly 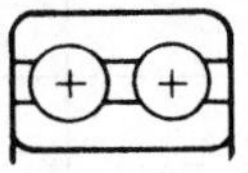		

BALL BEARINGS, DOUBLE ROW, ANGULAR CONTACT[a]			
Symbol	Description	Symbol	Description
BD	Filling slot assembly Vertex of contact angles inside bearing 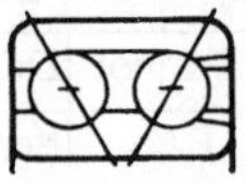	BG	Non-filling slot assembly Vertex of contact angles outside bearing 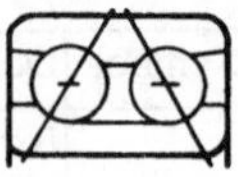
BE	Filling slot assembly Vertex of contact angles outside bearing 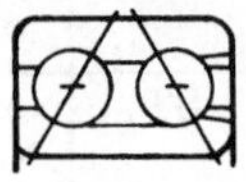	BAA	Non-separable Vertex of contact angles inside bearing Two-piece outer ring 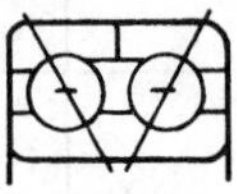
BJ	Non-filling slot assembly Vertex of contact angles inside bearing 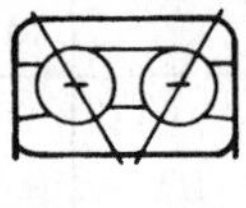	BVV	Separable Vertex of contact angles outside bearing Two-piece inner ring 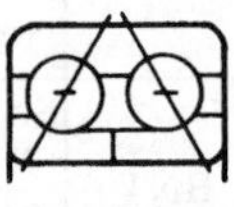

[a] A line through the ball contact points forms an acute angle with a perpendicular to the bearing axis.

BALL BEARINGS, DOUBLE ROW, SELF-ALIGNING[a]	
Symbol	Description
BS	Raceway of outer ring spherical 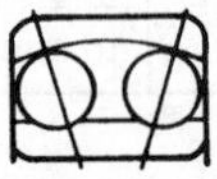

Single-row Angular-contact: This type is designed for combined radial and thrust loads where the thrust component may be large and axial deflection must be confined within very close limits. A high shoulder on one side of the outer ring is provided to take the thrust, while the shoulder on the other side is only high enough to make the bearing non-separable. Except where used for a pure thrust load in one direction, this type is applied either in pairs (duplex) or one at each end of the shaft, opposed.

Double-row Bearings: These are, in effect, two single-row angular-contact bearings built as a unit with the internal fit between balls and raceway fixed at the time of bearing assembly. This fit is therefore not dependent upon mounting methods for internal rigidity. These bearings usually have a known amount of internal preload built in for maximum resistance to deflection under combined loads with thrust from either direction. Thus, with balls and races under compression before an external load is applied, due to this internal preload, the bearings are very effective for radial loads where bearing deflection must be minimized.

Other Types: Modifications of these basic types provide arrangements for self-sealing, location by snap ring, shielding, etc., but the fundamentals of mounting are not changed. A special type is the *self-aligning* ball bearing which can be used to compensate for an appreciable degree of misalignment between shaft and housing due to shaft deflections, mounting inaccuracies, or other causes commonly encountered. With a single row of balls, alignment is provided by a spherical outer surface on the outer ring; with a double row of balls, alignment is provided by a spherical raceway on the outer ring. Bearings in the wide series have a considerable amount of thrust capacity.

CYLINDRICAL ROLLER BEARING, SINGLE ROW, NON-LOCATING TYPE			
Symbol	Description	Symbol	Description
RU	Inner ring without ribs Double-ribbed outer-ring Inner ring separable 	RNS	Double-ribbed inner ring Outer ring without ribs Outer ring separable Spherical outside surface 
RUP	Inner ring without ribs Double-ribbed outer ring with oneloose rib Both rings separable 	RAB	Inner ring without ribs Single-ribbedouter ring Both rings separable 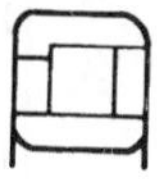
RUA	Inner ring without ribs Double-ribbed outer ring Inner ring separable Spherical outside surface 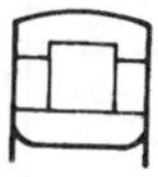	RM	Inner ring without ribs Rollers located by cage, end-rings or internal snap rings recesses in outer ring Inner ring separable 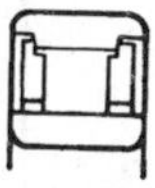
RN	Double-ribbed inner ring Outer ring without ribs Outer ring separable 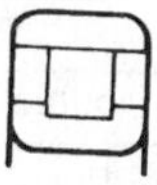	RNU	Inner ring without ribs Outer ring without ribs Both rings separable 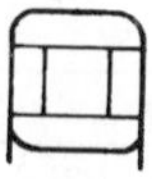
CYLINDRICAL ROLLER BEARINGS, SINGLE ROW, ONE-DIRECTION-LOCATING TYPE			
Symbol	Description	Symbol	Description
RR	Single-ribbed inner-ring Outer ring with two internal snap rings Inner ring separable 	RF	Double-ribbed inner ring Single-ribbed outer ring Outer ring separable 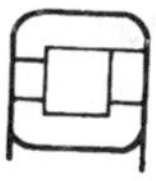
RJ	Single-ribbedinner ring Double-ribbed outer ring Inner ring separable 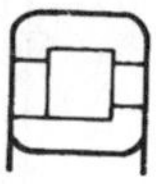	RS	Single-ribbed inner ring Outer ring with one rib and one internal snap ring Inner ring separable 
RJP	Single-ribbed inner ring Double-ribbed outer ring with one loose rib Both rings separable 	RAA	Single-ribbed inner ring Single-ribbed outer ring Both rings separable 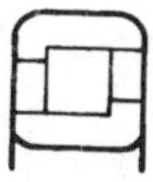

CYLINDRICAL ROLLER BEARINGS, SINGLE ROW, TWO-DIRECTION-LOCATING TYPE

Symbol	Description	Symbol	Description
RK	Double-ribbed inner ring Outer ring with two internal snap rings Non-separable	 RY	Double-ribbed inner ring Outer ring with one rib and one internal snap ring Non-separable 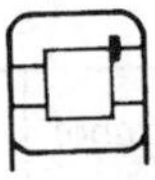
RC	Doble-ribbed inner ring Double-ribbed outer ring Non-separable	 RCS	Double-ribbed inner ring Double-ribbed outer ring Non-separable Spherical outside surface 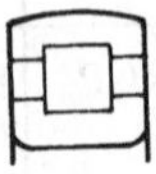
RG	Inner ring, with one rib and one snap ring Double-ribbed outer ring Non-separable 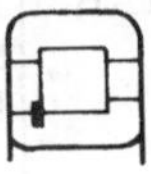		
RP	Double-ribbed inner ring Double-ribbed outer ring with one loose rib Outer ring separable	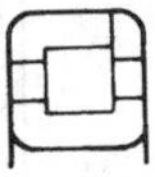 RT	Double-ribbed inner ring with one loose rib Double-ribbed outer ring Inner ring separable 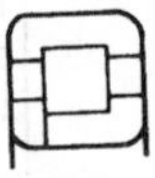

CYLINDRICAL ROLLER BEARINGS

Double Row Non-Locating Type		Double Row Two-Direction-Locating Type	
Symbol	Description	Symbol	Description
RA	Inner ring without ribs Three integral ribs on outer ring Inner ring separable	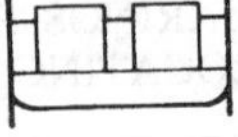 RB	Three integral ribs on inner ring Outer ring without ribs, with two internal snap rings Non-separable 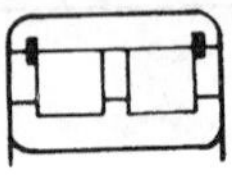
		Multi-Row Non-Locating Type	
RD	Three integral ribs on inner ring Outer ring without ribs Outer ring separable	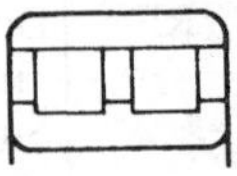 Symbol	Description
RE	Inner ring without ribs Outer rings without ribs, with two internal snap rings Inner ring separable	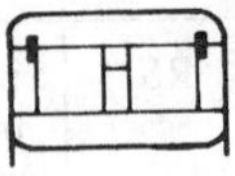 RV	Inner ring without ribs Double-ribbed outer ring (loose ribs) Both rings separable 

SELF-ALIGNING ROLLER BEARINGS, DOUBLE ROW

Symbol	Description	Symbol	Description
SD	Three integral ribs on inner ring Raceway of outer ring spherical	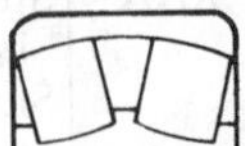 SL	Raceway of outer ring spherical Rollers guided by the cage Two integral ribs on inner ring 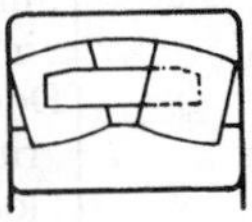

SELF-ALIGNING ROLLER BEARINGS, DOUBLE ROW			
Symbol	Description	Symbol	Description
SE	Raceway of outer ring spherical Rollers guided by separate center guide ring in outer ring		SELF-ALIGNING ROLLER BEARINGS SINGLE ROW
		Symbol	Description
SW	Raceway of inner ring spherical	SR	Inner ring with ribs Raceway of outer ring spherical Radial contact
SC	Raceway of outer ring spherical Rollers guided by separate axially floating guide ring on inner ring	SA	Raceway of outer ring spherical Angular contact
		SB	Raceway of inner ring spherical Angular contact
THRUST BALL BEARINGS			
Symbol	Description	Symbol	Description
TA TB[a]	Single direction, grooved raceways, flat seats	TDA	Double direction, washers with grooved raceways, flat seats
TBF[a]	Single direction, flat washers, flat seats		
THRUST ROLLER BEARINGS			
Symbol	Description	Symbol	Description
TS	Single direction, aligning flat seats, spherical rollers	TPC[a]	Single direction, flat seats, flat races, outside band, cylindrical rollers
TP	Single direction, flat seats, cylindrical rollers	TR[a]	Single direction, flat races, aligning seat with aligning washer, cylindrical rollers

[a] Inch dimensioned only.

Types of Roller Bearings.—Types of roller bearings are distinguished by the design of rollers and raceways to handle axial, combined axial and thrust, or thrust loads.

Cylindrical Roller: These bearings have solid or helically wound hollow cylindrical rollers. The free ring may have a restraining flange to provide some restraint to endwise movement in one direction or may be without a flange so that the bearing rings may be displaced axially with respect to each other. Either rolls or roller path on the races may be slightly crowned to prevent edge loading under slight shaft misalignment. Low friction makes this type suitable for relatively high speeds.

Barrel Roller: These bearings have rollers that are barrel-shaped and symmetrical. They are furnished in both single- and double-row mountings. As with cylindrical roller bearings, the single-row mounting type has a low thrust capacity, but angular mounting of rolls in the double-row type permits its use for combined axial and thrust loads.

Spherical Roller: These bearings are usually furnished in a double-row, self-aligning mounting. Both rows of rollers have a common spherical outer raceway. The rollers are barrel-shaped with one end smaller than the other to provide a small thrust to keep the rollers in contact with the center guide flange. This type of roller bearing has a high radial and thrust load carrying capacity with the ability to maintain this capacity under some degree of misalignment of shaft and bearing housing.

Tapered Roller: In this type, straight tapered rollers are held in accurate alignment by means of a guide flange on the inner ring. The basic characteristic of these bearings is that the apexes of the tapered working surfaces of both rollers and races, if extended, would coincide on the bearing axis. These bearings are separable. They have a high radial and thrust carrying capacity.

TAPERED ROLLER BEARINGS — INCH			
Symbol	Description	Symbol	Description
TS	Single row	 TDI	Two row, double-cone single cups 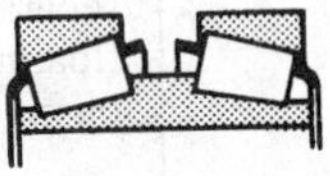
TDO	Two row, double-cup single-cone adjustable	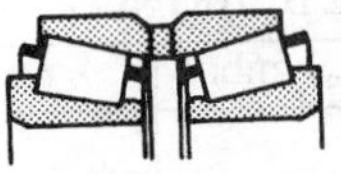 TNA	Two row, double-cup single cone nonadjustable 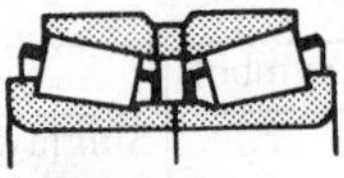
TQD, TQI	Four row, cup adjusted	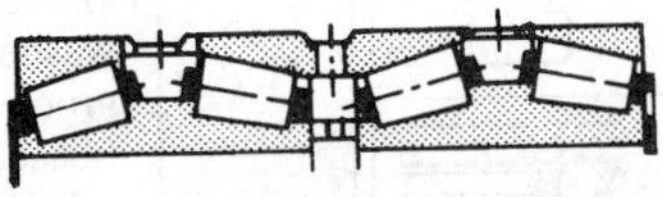	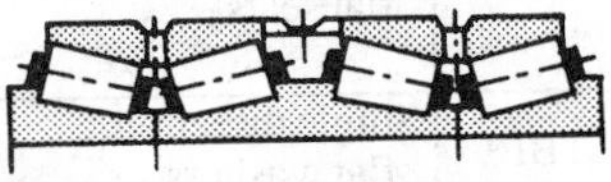

TAPERED ROLLER BEARINGS — METRIC			
Symbol	Description	Symbol	Description
TS	Single row, straight bore	 TSF	Single row, straight bore, flanged cup 
TDO	Double row, straight bore, two single cones, one double cup with lubrication hole and groove	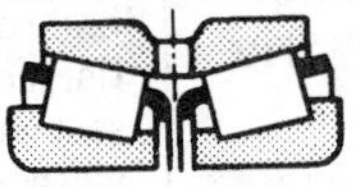 2TS	Double row, straight bore, two single cones, two single cups 

THRUST TAPERED ROLLER BARINGS	
Symbol	Description
TT	Thrust bearings 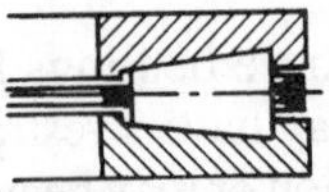

Types of Ball and Roller Thrust Bearings.—Are designed to take thrust loads alone or in combination with radial loads.

One-direction Ball Thrust: These bearings consist of a shaft ring and a flat or spherical housing ring with a single row of balls between. They are capable of carrying pure thrust loads in one direction only. They cannot carry any radial load.

Two-direction Ball Thrust: These bearings consist of a shaft ring with a ball groove in either side, two sets of balls, and two housing rings so arranged that thrust loads in either direction can be supported. No radial loads can be carried.

Spherical Roller Thrust: This type is similar in design to the radial spherical roller bearing except that it has a much larger contact angle. The rollers are barrel shaped with one end smaller than the other. This type of bearing has a very high thrust load carrying capacity and can also carry radial loads.

Tapered Roller Thrust: In this type the rollers are tapered and several different arrangements of housing and shaft are used.

Roller Thrust: In this type the rollers are straight and several different arrangements of housing and shaft are used.

NEEDLE ROLLER BEARINGS, DRAWN CUP			
Symbol[a]	Description	Symbol[a]	Description
NIB NB	Needle roller bearing, full complement, drawn cup, without inner ring	NIYM NYM	Needle roller bearing, full complement, rollers retained by lubricant, drawn cup, closed end, without inner ring
NIBM NBM	Needle roller bearing, full complement, drawn cup, closed end, without inner ring	NIH NH	Needle roller bearing, with cage, drawn cup, without inner ring
NIY NY	Needle roller bearing, full complement, rollers retained by lubricant, drawn cup, without inner ring	NIHM NHM	Needle roller bearing, with cage, drawn cup, closed end, without inner ring
NEEDLE ROLLER BEARINGS		NEEDLE ROLLER AND CAGE ASSEMBLIES	
Symbol[a]	Description	Symbol[a]	Description
NIA NA	Needle roller bearing, with cage, machined ring lubrication hole and groove in OD, without inner ring	NIM NM	Needle roller and cage assembly
NEEDLE ROLLER BEARING INNER RINGS			
Symbol[a]	Description		
NIR NR	Needle roller bearing inner ring, lubrication hole and groove in bore	Machined Ring Needle Roller Bearings Type NIA may be used with inch dimensioned inner rings, Type NIR, and Type NA may be used with metric dimensional inner rings, Type NR.	

[a] Symbols with I, as NIB, are inch-dimensioned, and those without the I, as NB, are metric dimensioned.

Types of Needle Bearings.—Needle bearings are characterized by their relatively small size rollers, usually not above ¼ inch in diameter, and a relatively high ratio of length to diameter, usually ranging from about 3 to 1 and 10 to 1. Another feature that is characteristic of several types of needle bearings is the absence of a cage or separator for retaining the individual rollers. Needle bearings may be divided into three classes: loose-roller, outer race and retained roller, and non-separable units.

Loose-roller: This type of bearing has no integral races or retaining members, the needles being located directly between the shaft and the outer bearing bore. Usually both shaft and outer bore bearing surfaces are hardened and retaining members that have smooth unbroken surfaces are provided to prevent endwise movement. Compactness and high radial load capacity are features of this type.

Outer Race and Retained Roller: There are two types of outer race and retained roller bearings. In the *Drawn Shell* type, the needle rollers are enclosed by a hardened shell that acts as a retaining member and as a hardened outer race. The needles roll directly on the shaft, the bearing surface of which should be hardened. The capacity for given roller length and shaft diameter is about two-thirds that of the loose roller type. It is mounted in the housing with a press fit.

In the *Machined Race* type, the outer race consists of a heavy machined member. Various modifications of this type provide heavy ends or faces for end location of the needle rollers, or open end construction with end washers for roller retention, or a cage that maintains alignment of the rollers and is itself held in place by retaining rings. An auxiliary outer member with spherical seat that holds the outer race may be provided for self-alignment. This type is applicable where split housings occur or where a press fit of the bearing into the housing is not possible.

Non-separable: This type consists of a non-separable unit of outer race, rollers and inner race. These bearings are used where high static or oscillating motion loads are expected as in certain aircraft components and where both outer and inner races are necessary.

Special or Unconventional Types.—Rolling contact bearings have been developed for many highly specialized applications. They may be constructed of non-corrosive materials, non-magnetic materials, plastics, ceramics, and even wood. Although the materials are chosen to adapt more conventional configurations to difficult applications or environments, even greater ingenuity has been applied in utilizing rolling contact for solving particular problems. Thus, linear or recirculating bearings are available to provide low friction, accurate location, and simplified lubrication features to such applications as machine ways, axial motion devices, jack-screws, steering linkages, collets, and chucks. This type of bearing utilizes the "full-complement" style of loading the rolling elements between "races" or ways without a cage and with each element advancing by the action of "races" in the loaded areas and by contact with the adjacent element in the unloaded areas. The "races" may not be cylindrical or bodies of revolution but plane surfaces, with suitable interruptions to free the rolling elements so that they can follow a return trough or slot back to the entry-point at the start of the "race" contact area. Combinations of radial and thrust bearings are available for the user with special requirements.

Plastics Bearings.—A more recent development has been the use of Acetal resin rollers and balls in applications where abrasive, corrosive and difficult-to-lubricate conditions exist. Although these bearings do not have the load carrying capacity nor the low friction factor of their hard steel counterparts, they do offer freedom from indentation, wear, and corrosion, while at the same time providing significant weight savings.

Of additional value are: 1) their resistance to indentation from shock loads or oscillation; and 2) their self-lubricating properties.

Usually these bearings are not available from stock, but must be designed and produced in accordance with the data made available by the plastics processor.

Pillow Block and Flanged Housing Bearings.—Of great interest to the shop man and particularly adaptable to "line-shafting" applications are a series of ball and roller bearings supplied with their own housings, adapters, and seals. Often called pre-mounted bearings, they come with a wide variety of flange mountings permitting location on faces parallel to or perpendicular to the shaft axis.

Inner races can be mounted directly on ground shafts, or can be adapter-mounted to "drill-rod" or to commercial shafting. For installations sensitive to imbalance and vibration, the use of accurately ground shaft seats is recommended.

Most pillow block designs incorporate self-aligning types of bearings so they do not require the precision mountings utilized with more normal bearing installations.

Conventional Bearing Materials.—Most rolling contact bearings are made with all load carrying members of full hard steel, either through- or case-hardened. For greater reliability this material is controlled and selected for cleanliness and alloying practices in conformity with rigid specifications in order to reduce anomalies and inclusions that could limit the useful fatigue life. Magnaflux inspection is employed to ensure that elements are free from both material defects and cracks. Likewise, a light etch is employed between rough and finish grinding to allow detection of burns due to heavy stock removal and associated decarburization in finished pieces.

Cage Materials.—Standard bearings are normally made with cages of free-machining brass or low carbon sulfurized steel. In high-speed applications or where lubrication may be intermittent or marginal, special materials may be employed. Iron-silicon-bronze, laminated phenolics, silver-plating, over-lays, solid-film baked-on coatings, carbon-graphite inserts, and, in extreme cases, sintered or even impregnated materials are used in separators.

Commercial bearings usually rely on stamped steel with or without a phosphate treatment; some low cost varieties are found with snap-in plastic or metallic cages.

So long as lubrication is adequate and speeds are both reasonable and steady, the materials and design of the cage are of secondary importance when compared with those of the rolling elements and their contacts with the races. In spite of this tolerance, a good portion of all rolling bearing failures encountered can be traced to cage failures resulting from inadequate lubrication. It can never be overemphasized that *no bearing can be designed to run continuously without lubrication!*

Standard Method of Bearing Designation.—The Anti-Friction Bearing Manufacturers Association has adopted a standard identification code that provides a specific designation for each different ball, roller, and needle bearing. Thus, for any given bearing, a uniform designation is provided for manufacturer and user alike, so that the confusion of different company designations can be avoided.

In this identification code there is a "basic number" for each bearing that consists of three elements: a one- to four-digit number to indicate the size of the bore in numbers of millimeters (metric series); a two- or three-letter symbol to indicate the type of bearing; and a two-digit number to identify the dimension series to which the bearing belongs.

In addition to this "basic number" other numbers and letters are added to designate type of tolerance, cage, lubrication, fit up, ring modification, addition of shields, seals, mounting accessories, etc. Thus, a complete designating symbol might be *50BC02JPXE0A10,* for example. The basic number is *50BC02* and the remainder is the supplementary number. For a radial bearing, this latter consists of up to four letters to indicate modification of design, one or two digits to indicate internal fit and tolerances, a letter to indicate lubricants and preservatives, and up to three digits to indicate special requirements.

For a thrust bearing the supplementary number would consist of two letters to indicate modifications of design, one digit to indicate tolerances, one letter to indicate lubricants and preservatives, and up to three digits to indicate special requirements.

For a needle bearing the supplementary number would consist of up to three letters indicating cage material or integral seal information or whether the outer ring has a crowned outside surface and one letter to indicate lubricants or preservatives.

Dimension Series: Annular ball, cylindrical roller, and self-aligning roller bearings are made in a series of different outside diameters for every given bore diameter and in a series of different widths for every given outside diameter. Thus, each of these bearings belongs to a dimension series that is designated by a two-digit number such as or, 23, 93, etc. The first digit (8, 0, 1, 2, 3, 4, 5, 6 and 9) indicates the *width series* and the second digit (7, 8, 9, 0, 1, 2, 3, and 4) the *diameter series* to which the bearing belongs. Similar types of identification codes are used for ball and roller thrust bearings and needle roller bearings.

Table 1. ABEC-1 and RBEC-1 Tolerance Limits for Metric Ball and Roller Bearings *ANSI/ABMA 20-1987*

Basic Inner Ring Bore Diameter, d		V_{dp},[a] max			Δ_{dmp}[b]		K_{ia}[c]
mm		Diameter Series					
Over	Incl.	7,8,9	0,1	2,3,4	High	Low	max
2.5	10	10	8	6	0	−8	10
10	18	10	8	6	0	−8	10
18	30	13	10	8	0	−10	13
30	50	15	12	9	0	−12	15
50	80	19	19	11	0	−15	20
80	120	25	25	15	0	−20	25
120	180	31	31	19	0	−25	30
180	250	38	38	23	0	−30	40
250	315	44	44	26	0	−35	50
315	400	50	50	30	0	−40	60

[a] Bore diameter variation in a single radial plane.
[b] Single plane mean bore diameter deviation from basic. (For a basically tapered bore, Δ_{dmp} refers only to the theoretical small end of the bore.)
[c] Radial runout of assembled bearing inner ring.

Basic Outer Ring Outside Outerside Diameter, D		V_{Dp},[a] max				Δ_{Dmp}[b]		K_{ea}[c]
		Open Bearings			Capped Bearings[d]			
mm		Diameter Series						
Over	Incl.	7,8,9	0,1	2,3,4	2,3,4	High	Low	max
6	18	10	8	6	10	0	−8	15
18	30	12	9	7	12	0	−9	15
30	50	14	11	8	16	0	−11	20
50	80	16	13	10	20	0	−13	25
80	120	19	19	11	26	0	−15	35
120	150	23	23	14	30	0	−18	40
150	180	31	31	19	38	0	−25	45
180	250	38	38	23	…	0	−30	50
250	315	44	44	26	…	0	−35	60
315	400	50	50	50	…	0	−40	70

[a] Outside diameter variation in a single radial plane. Applies before mounting and after removal of internal or external snap ring.
[b] Single plane mean outside diameter deviation from basic.
[c] Radial runout of assembled bearing outer ring.
[d] No values have been established for diameters series 7, 8, 9, 0, and 1.

Width Tolerances									
d		Δ_{Bs}[a]			d		Δ_{Bs}[a]		
mm		All	Normal	Modified[b]	mm		All	Normal	Modified[b]
Over	Incl.	High	Low		Over	Incl.	High	Low	
2.5	10	0	−120	−250	80	120	0	−200	−380
10	18	0	−120	−250	120	180	0	−250	−500
18	30	0	−120	−250	180	250	0	−300	−500
30	50	0	−120	−250	250	315	0	−350	−500
50	80	0	−150	−380	315	400	0	−400	−630

[a] Single inner ring width deviation from basic. Δ_{Cs} (single outer ring width deviation from basic) is identical to Δ_{Bs} of inner ring of same bearing.
[b] Refers to the rings of single bearings made for paired or stack mounting.

All units are micrometers, unless otherwise indicated.For sizes beyond range of this table, see Standard.This table does not cover tapered roller bearings.

Table 2. ABEC-3 AND RBEC-3 Tolerance Limits for Metric Ball and Roller Bearings *ANSI/ABMA 20-1987*

Basic Inner Ring Bore Diameter, d		V_{dp},[a] max			Δ_{dmp}[b]		K_{ia}[c]
mm		Diameter Series					
Over	Incl.	7, 8, 9	0, 1	2, 3, 4	High	Low	max
2.5	10	9	7	5	0	−7	6
10	18	9	7	5	0	−7	7
18	30	10	8	6	0	−8	8
30	50	13	10	8	0	−10	10
50	80	15	15	9	0	−12	10
80	120	19	19	11	0	−15	13
120	180	23	23	14	0	−18	18
180	250	28	28	17	0	−22	20
250	315	31	31	19	0	−25	25
315	400	38	38	23	0	−30	30

[a] Bore diameter variation in a single radial plane.

[b] Single plane mean bore diameter deviation from basic. (For a basically tapered bore, Δ_{dmp} refers only to the theoretical small end of the bore.)

[c] Radial runout of assembled bearing inner ring.

Basic Outer Ring Outside Outerside Diameter, D		V_{Dp},[a] max				Δ_{Dmp}[b]		K_{ea}[c]
		Open Bearings			Capped Bearings[d]			
mm		Diameter Series						
Over	Incl.	7,8,9	0,1	2,3,4	2,3,4	High	Low	max
6	18	9	7	5	9	0	−7	8
18	30	10	8	6	10	0	−8	9
30	50	11	9	7	13	0	−9	10
50	80	14	11	8	16	0	−11	13
80	120	16	16	10	20	0	−13	18
120	150	19	19	11	25	0	−15	20
150	180	23	23	14	30	0	−18	23
180	250	25	25	15	…	0	−20	25
250	315	31	31	19	…	0	−25	30
315	400	35	35	21	…	0	−28	35

[a] Outside diameter variation in a single radial plane. Applies before mounting and after removal of internal or external snap ring.

[b] Single plane mean outside diameter deviation from basic.

[c] Radial runout of assembled bearing outer ring.

[d] No values have been established for diameter series 7, 8, 9, 0, and 1.

Width Tolerances									
d		Δ_{Bs}[a]			d		Δ_{Bs}[a]		
mm		All	Normal	Modified[b]	mm		All	Normal	Modified[b]
Over	Incl.	High	Low		Over	Incl.	High	Low	
2.5	10	0	−120	−250	80	120	0	−200	−380
10	18	0	−120	−250	120	180	0	−250	−500
18	30	0	−120	−250	180	250	0	−300	−500
30	50	0	−120	−250	250	315	0	−350	−500
50	80	0	−150	−380	315	400	0	−400	−630

[a] Single inner ring width deviation from basic. Δ_{Cs} (single outer ring width deviation from basic) is identical to Δ_{Bs} of inner ring of same bearing.

[b] Refers to the rings of single bearings made for paired or stack mounting.

All units are micrometers, unless otherwise indicated.For sizes beyond range of this table, see Standard.This table does not cover tapered roller bearings.

Table 3. ABEC-5 and RBEC-5 Tolerance Limits for Metric Ball and Roller Bearings ***ANSI/ABMA 20-1987***

INNER RING													
Inner Ring BoreBasic Dia.,d		V_{dp},[a] max		Δ_{dmp}[b]		Radial Runout K_{ia}	Ref. Face Runout with Bore S_d	Axial Runout of Assembled Bearing with Inner Ring S_{ia}[c]	Width				
mm		Diameter Series							Δ_{Bs}[d]			V_{Bs}[e]
Over	Incl.	7, 8, 9	0, 1, 2, 3, 4	High	Low	max	max	max	All	Normal	Modified[f]	max
2.5	10	5	4	0	−5	4	7	7	0	−40	−250	5
10	18	5	4	0	−5	4	7	7	0	−80	−250	5
18	30	6	5	0	−6	4	8	8	0	−120	−250	5
30	50	8	6	0	−8	5	8	8	0	−120	−250	5
50	80	9	7	0	−9	5	8	8	0	−150	−250	6
80	120	10	8	0	−10	6	9	9	0	−200	−380	7
120	180	13	10	0	−13	8	10	10	0	−250	−380	8
180	250	15	12	0	−15	10	11	13	0	−300	−500	10

[a] Bore (V_{dp}) and outside diameter (V_{Dp}) variation in a single radial plane.

[b] Single plane mean bore (Δ_{dmp}) and outside diameter (Δ_{Dmp}) deviation from basic. (For a basically tapered bore, Δ_{dmp} refers only to the theoretical small end of the bore.)

[c] Applies to groove-type ball bearings only.

[d] Single bore (Δ_{Bs}) and outer ring (Δ_{Cs}) width variation.

[e] Inner (V_{Bs}) and outer (V_{Cs}) ring width deviation from basic.

[f] Applies to the rings of single bearings made for paired or stack mounting.

OUTER RING											
Basic Outer Ring Outside Dia., D		V_{Dp},[aa] max		Δ_{Dmp}[b]		Radial Runout K_{ea}	Outside Cylindrical Surface Runout with Outer Ring Ref. Face S_D	Axial Runout of Assembled Bearing with Outer Ring S_{ea}	Width		
mm		Diameter Series							Δ_{Cs}[d]		V_{Cs}[e]
Over	Incl.	7, 8, 9	0, 1, 2, 3, 4	High	Low	max	max	max	High	Low	max
6	18	5	4	0	−5	5	8	8	Identical to Δ_{Bs} of inner ring of same bearing		5
18	30	6	5	0	−6	6	8	8			5
30	50	7	5	0	−7	7	8	8			5
50	80	9	7	0	−9	8	8	10			6
80	120	10	8	0	−10	10	9	11			8
120	150	11	8	0	−11	11	10	13			8
150	180	13	10	0	−13	13	10	14			8
180	250	15	11	0	−15	15	11	15			10

[a] No values have been established for capped bearings.

All units are micrometers, unless otherwise indicated.For sizes beyond range of this table, see Standard.This table does not cover instrument bearings and tapered roller bearings.

Table 4. ABEC-7 Tolerance Limits for Metric Ball and Roller Bearings *ANSI/ABMA 20-1987*

INNER RING																
Inner Ring Bore Basic Dia., d		V_{dp},[a] max		Δ_{dmp}[b]		Δ_{ds}[c]		Radial Runout K_{ia}	Ref. Face Runout with Bore S_d	Axial Runout of AssembledBearing with Inner Ring S_{ia}[d]	Width					
mm		Diameter Series									Δ_{Bs}[e]				V_{Bs}[f]	
Over	Incl.	7, 8, 9	0, 1, 2, 3, 4	High	Low	High	Low	max	max	max	All	Normal	Modified[g]		max	
2.5	10	4	3	0	−4	0	−4	2.5	3	3	0	−40	−250		2.5	
10	18	4	3	0	−4	0	−4	2.5	3	3	0	−80	−250		2.5	
18	30	5	4	0	−5	0	−5	3	4	4	0	−120	−250		2.5	
30	50	6	5	0	−6	0	−6	4	4	4	0	−120	−250		3	
50	80	7	5	0	−7	0	−7	4	5	5	0	−150	−250		4	
80	120	8	6	0	−8	0	−8	5	5	5	0	−200	−380		4	
120	180	10	8	0	−10	0	−10	6	6	7	0	−250	−380		5	
180	250	12	9	0	−12	0	−12	8	7	8	0	−300	−500		6	

[a] Bore (V_{dp}) and outside diameter (V_{Dp}) variation in a single radial plane.

[b] Single plane mean bore (Δ_{dmp}) and outside diameter (Δ_{Dmp}) deviation from basic. (For a basically tapered bore, Δ_{dmp} refers only to the theoretical small end of the bore.)

[c] Single bore (Δ_{ds}) and outside diameter (Δ_{Ds}) deviations from basic. These deviations apply to diameter series 0, 1, 2, 3, and 4 only.

[d] Applies to groove-type ball bearings only.

[e] Single bore (Δ_{Bs}) and outer ring (Δ_{Cs}) width deviation from basic.

[f] Inner (V_{Bs}) and outer (V_{Cs}) ring width variation.

[g] Applies to the rings of single bearings made for paired or stack mounting.

OUTER RING														
Basic Outer Ring Outside Dia., D		V_{Dp},[aa] max		Δ_{Dmp}[b]		Δ_{Ds}[c]		Radial Runout K_{ea}	Outside Cylindrical Surface Runout Outer Ring Ref. Face S_D	Axial Runout of Assembled Bearing Outer Ring S_{ea}[d]	Width		
mm		Diameter Series									Δ_{Cs}[e]		V_{Cs}[f]
Over	Incl.	7, 8, 9	0, 1, 2, 3, 4	High	low	High	Low	max	max	max	High	Low	max
6	18	4	3	0	−4	0	−4	3	4	5	Identical to Δ_{Bs} of inner ring of same bearing		2.5
18	30	5	4	0	−5	0	−5	4	4	5			2.5
30	50	6	5	0	−6	0	−6	5	4	5			2.5
50	80	7	5	0	−7	0	−7	5	4	5			3
80	120	8	6	0	−8	0	−8	6	5	6			4
120	150	9	7	0	−9	0	−9	7	5	7			5
150	180	10	8	0	−10	0	−10	8	5	8			5
180	250	11	8	0	−11	0	−11	19	7	10			7

[a] No values have been established for capped bearings.

All units are micrometers, unless otherwise indicated. For sizes beyond range of this table, see Standard. This table does not cover instrument bearings.

Table 5. ABEC-9 Tolerance Limits for Metric Ball and Roller Bearing *ANSI/ABMA 20-1987*

INNER RING

Inner Ring Bore Basic Dia., d mm		V_{dp},[a] max	Δ_{dmp}[b]		Δ_{ds}[c]		Radial Runout K_{ia}	Ref. Face Runout with Bore S_d	Axial Runout of Assembled Bearing with Inner Ring S_{ia}[d]	Width Δ_{Bs}[e]		Width V_{Bs}[f]
Over	Incl.	max	High	Low	High	Low	max	max	max	High	Low	max
2.5	10	2.5	0	−2.5	0	−2.5	1.5	1.5	1.5	0	−40	1.5
10	18	2.5	0	−2.5	0	−2.5	1.5	1.5	1.5	0	−80	1.5
18	30	2.5	0	−2.5	0	−2.5	2.5	1.5	2.5	0	−120	1.5
30	50	2.5	0	−2.5	0	−2.5	2.5	1.5	2.5	0	−120	1.5
50	80	4	0	−4	0	−4	2.5	1.5	2.5	0	−150	1.5
50	80	4	0	−4	0	−4	2.5	1.5	2.5	0	−150	1.5
80	120	5	0	−5	0	−5	2.5	2.5	2.5	0	−200	2.5
120	150	7	0	−7	0	−7	2.5	2.5	2.5	0	−250	2.5
150	180	7	0	−7	0	−7	5	4	5	0	−300	4
180	250	8	0	−8	0	−8	5	5	5	0	−350	5

[a] Bore (V_{dp}) and outside diameter (V_{Dp}) variation in a single radial plane.

[b] Single plane mean bore (Δ_{dmp}) and outside diameter (Δ_{Dmp}) deviation from basic. (For a basically tapered bore, Δ_{dmp} refers to the theoretical small end of the bore.)

[c] Single bore diameter (Δ_{ds}) and outside diameter (Δ_{Ds}) deviation from basic.

[d] Applies to groove-type ball bearings only.

[e] Single bore (Δ_{Bs}) and outer ring (Δ_{Cs}) width variation from basic.

[f] Inner (V_{Bs}) and outer (V_{Cs}) ring width variation.

OUTER RING

Basic Outside Diameter, D mm		V_{Dp}[aa]	Δ_{Dmp}[b]		Δ_{Ds}[c]		Radial Runout K_{ea}	Outside Cylindrical Surface Runout with Outer Ring S_D	Axial Runout of Assembled Bearing with Outer Ring S_{ea}	Width Δ_{Cs}[e]		Width V_{Cs}[f]
Over	Incl.	max	High	Low	High	Low	max	max	max	High	Low	max
6	18	2.5	0	−2.5	0	−2.5	1.5	1.5	1.5	Identical to Δ_{Bs} of inner ring of same bearing		1.5
18	30	4	0	−4	0	−4	2.5	1.5	2.5			1.5
30	50	4	0	−4	0	−4	2.5	1.5	2.5			1.5
50	80	4	0	−4	0	−4	4	1.5	4			1.5
80	120	5	0	−5	0	−5	5	2.5	5			1.5
120	150	5	0	−5	0	−5	5	2.5	5			1.5
150	180	7	0	−7	0	−7	5	2.5	5			2.5
180	250	8	0	−8	0	−8	7	4	7			4
250	315	8	0	−8	0	−8	7	5	7			5

[a] No values have been established for capped bearings.

All units are micrometers, unless otherwise indicated.For sizes beyond range of this table, see Standard.This table does not cover instrument bearings.

Bearing Tolerances.—In order to provide standards of precision for proper application of ball or roller bearings in all types of equipment, five classes of tolerances have been established by the Anti-Friction Bearing Manufacturers Association for ball bearings, three for cylindrical roller bearings and one for spherical roller bearings. These tolerances are given in Tables 1, 2, 3, 4, and 5. They are designated as ABEC-1, ABEC-3, ABEC-5, ABEC-7 and ABEC-9 for ball bearings, the ABEC-9 being the most precise, RBEC-1, RBEC-3, and RBEC-5 for roller bearings. In general, bearings to specifications closer than ABEC-1 or RBEC-1 are required because of the need for very precise fits on shaft or housing, to reduce eccentricity or runout of shaft or supported part, or to permit operation at very high speeds. All five classes include tolerances for bore, outside diameter, ring width, and radial runouts of inner and outer rings. ABEC-5, ABEC-7 and ABEC-9 provide added tolerances for parallelism of sides, side runout and groove parallelism with sides.

Thrust Bearings: Anti-Friction Bearing Manufacturers Association and American National Standard tolerance limits for metric single direction thrust ball and roller bearings are given in Table 6. Tolerance limits for single direction thrust ball bearings, inch dimensioned are given in Table 7, and for cylindrical thrust roller bearings, inch dimensioned in Table 8.

Table 6. AFBMA and American National Standard Tolerance Limits for Metric Single Direction Thrust Ball (Type TA) and Roller Type (Type TS) Bearings ***ANSI/ABMA 24.1-1989***

Bore Dia.of Shaft Washer, d mm		Δd_{mp}[a]		S_i, S_e[b]	ΔT_s[c]			Outside Dia. of Housing Washer, D mm		ΔD_{mp}[a]	
Over	Incl.	High	Low	Max	Max	Min Type TA	MinType TS	Over	Incl.	High	Low
18	30	0	−10	10	20	−250	…	10	18	0	−11
30	**50**	**0**	**−12**	10	**20**	−250	−300	18	30	0	−13
50	**80**	**0**	**−15**	10	**20**	−300	−400	30	50	0	−16
80	**120**	**0**	**−20**	15	**25**	−300	−400	**50**	**80**	**0**	**−19**
120	**180**	**0**	**−25**	15	**25**	−400	−500	**80**	**120**	**0**	**−22**
180	**250**	**0**	**−30**	20	**30**	−400	−500	**120**	**180**	**0**	**−25**
250	**315**	**0**	**−35**	25	**40**	−400	−700	**180**	**250**	**0**	**−30**
315	**400**	**0**	**−40**	30	**40**	−500	−700	**250**	**315**	**0**	**−35**
400	**500**	**0**	**−45**	30	**50**	−500	−900	**315**	**400**	**0**	**−40**
500	**630**	**0**	**−50**	35	**60**	−600	−1200	**400**	**500**	**0**	**−45**

[a] Single plane mean bore diameter deviation of central shaft washer (Δd_{mp}) and outside diameter (ΔD_{mp}) variation.

[b] Raceway parallelism with the face, housing-mounted (S_e) and boremounted (S_i) race or washer.

[c] Deviation of the actual bearing height.

All dimensions in micrometers, unless otherwise indicated.

Tolerances are for normal tolerance class only. For sizes beyond the range of this table and for other tolerance class values, see Standard. All entries apply to type TA bearings; boldface entries also apply to type TS bearings.

Table 7. Tolerance Limits for Single Direction Ball Thrust Bearings—Inch Design ***ANSI/ABMA 24.2-1998***

Bore Diameter[a] d, Inches		Single Plane Mean Bore Dia. Variation, d, Inch		Outside Diameter D, Inches		Single Plane Mean O.D. Variation, D, Inch	
Over	Incl.	High	Low	Over	Incl.	High	Low
0	6.7500	+0.005	0	0	5.3125	+0	−0.002
6.7500	20.0000	+0.010	0	5.3125	17.3750	+0	−0.003
…	…	…	…	17.3750	39.3701	+0	−0.004

[a] Bore tolerance limits are: For bore diameters over 0 to 1.8125 inches, inclusive, +0.005, −0.005; over 1.8125 to 12.000 inches, inclusive, +0.010, −0.010; over 12.000 to 20.000, inclusive, +0.0150, −0.0150.

Table 8. Tolerance Limits for Cylindrical Roller Thrust Bearings—Inch Design *ANSI/ABMA 24.2-1998*

Basic Bore Dia., d		Δd_{mp} [a]		ΔT_s [b]		Basic Outside dia., D		ΔD_{mp} [c]	
Over	Incl.	Low	High	High	Low	Over	Incl.	High	Low
EXTRA LIGHT SERIES—TYPE TP									
0	0.9375	+.0040	+.0060	+.0050	−.0050	0	4.7188	+0	−.0030
0.9375	1.9375	+.0050	+.0070	+.0050	−.0050	4.7188	5.2188	+0	−.0030
1.9375	3.0000	+.0060	+.0080	+.0050	−.0050	…	…	…	…
3.0000	3.5000	+.0080	+.0100	−.0100	−.0100	…	…	…	…

[a] Single plane mean bore diameter deviation.
[b] Deviation of the actual bearing height, single direction bearing.
[c] Single plane mean outside diameter deviation.

Basic Bore Diameter, d		Δd_{mp} [a]		Basic Outside Diameter, D		Outside Dia., D Tolerance Limits		Basic Bore Diameter, d		ΔT_s	
Over	Inc.	High	Low	Over	Incl.	High	Low	Over	Incl.	High	Low
LIGHT SERIES—TYPE TP											
0	1.1870	+0	−.0005	0	2.8750	+.0005	−0	0	2.0000	+0	−.006
1.1870	1.3750	+0	−.0006	2.8750	3.3750	+.0007	−0	2.0000	3.0000	+0	−.008
1.3750	1.5620	+0	−.0007	3.3750	3.7500	+.0009	−0	3.0000	6.0000	+0	−.010
1.5620	1.7500	+0	−.0008	3.7500	4.1250	+.0011	−0	6.0000	10.0000	+0	−.015
1.7500	1.9370	+0	−.0009	4.1250	4.7180	+.0013	−0	10.0000	18.0000	+0	−.020
1.9370	2.1250	+0	−.0010	4.7180	5.2180	+.0015	−0	18.0000	30.0000	+0	−.025
2.1250	2.5000	+0	−.0011	…	…	…	…	…	…	…	…
2.2500	3.0000	+0	−.0012	…	…	…	…	…	…	…	…
3.0000	3.5000	+0	−.0013	…	…	…	…	…	…	…	…
HEAVY SERIES—TYPE TP											
2.0000	3.0000	+0	−.0010	5.0000	10.0000	+.0015	−0	0	2.000	+0	−.006
3.0000	3.5000	+0	−.0012	10.0000	18.0000	+.0020	−0	2.000	3.000	+0	−.008
3.5000	9.0000	+0	−.0015	18.0000	26.0000	+.0025	−0	3.000	6.000	+0	−.010
9.0000	12.0000	+0	−.0018	26.0000	34.0000	+.0030	−0	6.000	10.000	+0	−.015
12.0000	18.0000	+0	−.0020	34.0000	44.0000	+.0040	−0	10.000	18.000	+0	−.020
18.0000	22.0000	+0	−.0025	…	…	…	…	18.000	30.000	+0	−.025
22.0000	30.0000	+0	−.003	…	…	…	…	…	…	…	…
TYPE TPC											
0	2.0156	+.010	−0	2.5000	4.0000	+.005	−.005	0	2.0156	+0	−.008
2.0156	3.0156	+.010	−.020	4.0000	6.0000	+.006	−.006	2.0156	3.0156	+0	−.010
3.0156	6.0156	+.015	−.020	6.0000	10.0000	+.010	−.010	3.0156	6.0156	+0	−.015
6.0156	10.1560	+.015	−.050	10.0000	18.0000	+.012	−.012	6.0156	10.1560	+0	−.020

All dimensions are in inches.

For Type TR bearings, see Standard.

Only one class of tolerance limits is established for metric thrust bearings.

Radial Needle Roller Bearings: Tolerance limits for needle roller bearings, drawn cup, without inner ring, inch types NIB, NIBM, NIY, NIYM, NIH, and NIHM are given in Table 9 and for metric types NB, NBM, NY, NYM, NH and NHM in Table 10. Standard tolerance limits for needle roller bearings, with cage, machined ring, without inner ring, inch type NIA are given in Table 11 and for needle roller bearings inner rings, inch type NIR in Table 12.

Table 9. AFBMA and American National Standard Tolerance Limits for Needle Roller Bearings, Drawn Cup, Without Inner Ring — Inch Types NIB, NIBM, NIY, NIYM, NIH, and NIHM *ANSI/ABMA 18.2-1982 (R1993)*

Ring Gage Bore Diameter[a]			Basic Bore Diameter under Needle Rollers, F_w		Allowable Deviation from F_w[a]		Allowable Deviation from Width, B	
Basic Outside Diameter, D Inch		Deviation from D Inch	Inch		Inch		Inch	
Over	Incl.		Over	Incl.	Low	High	High	Low
0.1875	0.9375	+0.0005	0.1875	0.6875	+0.0015	+0.0024	+0	−0.0100
0.9375	4.0000	−0.0005	0.6875	1.2500	+0.0005	+0.0014	+0	−0.0100
			1.2500	1.3750	+0.0005	+0.0015	+0	−0.0100
			1.3750	1.6250	+0.0005	+0.0016	+0	−0.0100
			1.6250	1.8750	+0.0005	+0.0017	+0	−0.0100
			1.8750	2.0000	+0.0006	+0.0018	+0	−0.0100
			2.0000	2.5000	+0.0006	+0.0020	+0	−0.0100
			2.5000	3.5000	+0.0010	+0.0024	+0	−0.0100

For fitting and mounting practice see Table 18.

[a] The bore diameter under needle rollers can be measured only when bearing is pressed into a ring gage, which rounds and sizes the bearing.

Table 10. AFBMA and American National Standard Tolerance Limits for Needle Roller Bearings, Drawn Cup, Without Inner Ring — Metric Types NB, NBM, NY, NYM, NH, and NHM *ANSI/ABMA 18.1-1982 (R1994)*

Ring Gage Bore Diameter[a]			Basic Bore Diameter underNeedle Rollers, F_w		Allowable Deviation from F_w[a]		Allowable Deviation from Width, B	
Basic Outside Diameter, D mm		Deviation from D Micrometers	mm		Micrometers		Micrometers	
Over	Incl.		Over	Incl.	Low	High	High	Low
6	10	−16	3	6	+10	+28	+0	−250
10	18	−20	6	10	+13	+31	+0	−250
30	50	−28	18	30	+20	+41	+0	−250
50	80	−33	30	50	+25	+50	+0	−250
…	…	…	50	70	+30	+60	+0	−250

[a] The bore diameter under needle rollers can be measured only when bearing is pressed into a ring gage, which rounds and sizes the bearing.

For fitting and mounting practice, see Table 18.

Table 11. AFBMA and American National Standard Tolerance Limits for Needle Roller Bearings, With Cage, Machined Ring, Without Inner Ring—Inch Type NIA *ANSI/ABMA 18.2-1982 (R1993)*

Basic Outside Diameter, D		Allowable Deviation From D of Single Mean Diameter, D_{mp}		Basic Bore Diameter under Needle Rollers, F_w		Allowable Deviation from F_w		Allowable Deviation from Width, B	
Inch		Inch		Inch		Inch		Inch	
Over	Incl.	High	Low	Over	Incl.	Low	High	High	Low
0.7500	2.0000	+0	−0.0005	0.3150	0.7087	+0.0008	+0.0017	+0	−0.0050
2.0000	3.2500	+0	−0.0006	0.7087	1.1811	+0.0009	+0.0018	+0	−0.0050
3.2500	4.7500	+0	−0.0008	1.1811	1.6535	+0.0010	+0.0019	+0	−0.0050
4.7500	7.2500	+0	−0.0010	1.6535	1.9685	+0.0010	+0.0020	+0	−0.0050
				1.9685	2.7559	+0.0011	+0.0021	+0	−0.0050
7.2500	10.2500	+0	−0.0012	2.7559	3.1496	+0.0011	+0.0023	+0	−0.0050
10.2500	11.1250	+0	−0.0014	3.1496	4.0157	+0.0012	+0.0024	+0	−0.0050
…	…	…	…	4.0157	4.7244	+0.0012	+0.0026	+0	−0.0050
…	…	…	…	4.7244	6.2992	+0.0013	+0.0027	+0	−0.0050
…	…	…	…	6.2992	7.0866	+0.0013	+0.0029	+0	−0.0050
…	…	…	…	7.0866	7.8740	+0.0014	+0.0030	+0	−0.0050
…	…	…	…	7.8740	9.2520	+0.0014	+0.0032	+0	−0.0050

For fitting and mounting practice, see Table 19.

Table 12. AFBMA and American National Standard Tolerance Limits for Needle Roller Bearing Inner Rings—Inch Type NIR *ANSI/ABMA 18.2-1982 (R1993)*

Basic Outside Diameter, *F*		Allowable Deviation From *F* of Single Mean Diameter, F_{mp}		Basic Bore Diameter *d*		Allowable Deviation from *d* of Single Mean Diameter, d_{mp}		Allowable Deviation from Width, *B*	
Inch		Inch		Inch		Inch		Inch	
Over	Incl.	High	Low	Over	Incl.	High	Low	High	Low
0.3937	0.7087	−0.0005	−0.0009	0.3125	0.7500	+0	−0.0004	+0.0100	+0.0050
0.7087	1.0236	−0.0007	−0.0012	0.7500	2.0000	+0	−0.0005	+0.0100	+0.0050
1.0236	1.1811	−0.0009	−0.0014	2.0000	3.2500	+0	−0.0006	+0.0100	+0.0050
1.1811	1.3780	−0.0009	−0.0015	3.2500	4.2500	+0	−0.0008	+0.0100	+0.0050
1.3780	1.9685	−0.0010	−0.0016	4.2500	4.7500	+0	−0.0008	+0.0150	+0.0100
1.9685	3.1496	−0.0011	−0.0018	4.7500	7.0000	+0	−0.0010	+0.0150	+0.0100
3.1496	3.9370	−0.0013	−0.0022	7.0000	8.0000	+0	−0.0012	+0.0150	+0.0100
3.9370	4.7244	−0.0015	−0.0024	…	…	…	…	…	…
4.7244	5.5118	−0.0015	−0.0025	…	…	…	…	…	…
5.5118	7.0866	−0.0017	−0.0027	…	…	…	…	…	…
7.0866	8.2677	−0.0019	−0.0031	…	…	…	…	…	…
8.2677	9.2520	−0.0020	−0.0032	…	…	…	…	…	…

For fitting and mounting practice, see Table 20.

Metric Radial Ball and Roller Bearing Shaft and Housing Fits.—To select the proper fits, it is necessary to consider the type and extent of the load, bearing type, and certain other design and performance requirements.

The required shaft and housing fits are indicated in Tables 13 and 14. The terms "Light," "Normal," and "Heavy" loads refer to radial loads that are generally within the following limits, with some overlap (*C* being the Basic Load Rating computed in accordance with AFBMA-ANSI Standards):

Bearing Type	Radial Load		
	Light	Normal	Heavy
Ball	Up to 0.075*C*	From 0.075*C* to 0.15*C*	Over 0.15*C*
Cylindrical Roller	Up to 0.075*C*	From 0.075*C* to 0.2*C*	Over 0.15*C*
Spherical Roller	Up to 0.075*C*	From 0.070*C* to 0.25*C*	Over 0.15*C*

Shaft Fits: Table 13 indicates the initial approach to shaft fit selection. Note that for most normal applications where the shaft rotates and the radial load direction is constant, an interference fit should be used. Also, the heavier the load, the greater is the required interference. For stationary shaft conditions and constant radial load direction, the inner ring may be moderately loose on the shaft.

For pure thrust (axial) loading, heavy interference fits are not necessary; only a moderately loose to tight fit is needed.

The upper part of Table 15 shows how the shaft diameters for various ANSI shaft limit classifications deviate from the basic bore diameters.

Table 16 gives metric values for the shaft diameter and housing bore tolerance limits given in Table 15.

The lower parts of Tables 15 and 16 show how housing bores for various ANSI hole limit classifications deviate from the basic shaft outside diameters.

Table 13. Selection of Shaft Tolerance Classifications for Metric Radial Ball and Roller Bearings of ABEC-1 and RBEC-1 Tolerance Classes *ANSI/ABMA 7-1995*

Operating Conditions			Nominal Shaft Diameter						Tolerance Symbol[a]
			Ball Bearings		Cylindrical Roller Bearings		Spherical Roller Bearings		
			mm	Inch	mm	Inch	mm	Inch	
Inner ring stationary in relation to the direction of the load.	All loads	Inner ring has to be easily displaceable	All diameters	All diameters	All diameters	All diameters	All diameters	All diameters	g6
		Inner ring does not have to be easily displaceable	All diameters	All diameters	All diameters	All diameters	All diameters	All diameters	h6
Direction of load indeterminate or the inner ring rotating in relation to the direction of the load.	Radial load: LIGHT		≤18 >18	≤0.71 >0.71	≤40 (40)-140 (140)–320 (320)–500 >500	≤1.57 (1.57)–5.51 (5.51)–12.6 (126)–19.7 >19.7	≤40 (40)–100 (100)–320 (320)–500 >500	≤1.57 (1.57)–3.94 (3.94)–12.6 (126)–19.7 >19.7	h5 j6[b] k6[b] m6[b] n6 p6
	NORMAL		≤18 >18	≤0.71 >0.71	≤40 (40)–100 (100)–140 (140)–320 (320)–500 >500	≤1.57 (1.57)–3.94 (3.94)–5.51 (5.51)–12.6 (12.6)–19.7 >19.7	≤40 (40)–65 (65)–100 (100)–140 (140)–280 (280)–500 >500	≤1.57 (1.57)–2.56 (2.56)–3.94 (3.94)–5.51 (5.51)–11.0 (11.0)–19.7 >19.7	j5 k5 m5 m6 n6 p6 r6 r7
	HEAVY		(18)–100 >100	(0.71)–3.94 >3.94	≤40 (40)-65 (65)–140 (140)–200 (200)–500 >500	≤1.57 (1.57)–2.56 (2.56)–5.51 (5.51)–7.87 (7.87)–19.7 >19.7	≤40 (40)–65 (65)–100 (100)–140 (140)–200 >200	≤1.57 (1.57)–2.56 (2.56)–3.94 (3.94)–5.51 (5.51)–7.87 >7.87	k5 m5 m6[b] n6[b] p6[b] r6[b] r7[b]
Pure Thrust Load			All diams.	All diams.	Consult Bearing Manufacturer				j6

[a] For solid steel shafts. For hollow or nonferrous shafts, tighter fits may be needed.

[b] When greater accuracy is required, use j5, k5, and m5 instead of j6, k6, and m6, respectively.

Numerical values are given in Tables 15 and 16.

Table 14. Selection of Housing Tolerance Classifications for Metric Radial Ball and Roller Bearings of ABEC-1 and RBEC-1 Tolerance Classes

Design and Operating Conditions				Tolerance Classification[a]
Rotational Conditions	Loading	Outer Ring Axial Displacement Limitations	Other Conditions	
Outer ring stationary in relation to load direction	Light Normal and Heavy	Outer ring must be easily axially displaceable	Heat input through shaft	G7
			Housing split axially	H7[b]
			Housing not split axially	H6[b]
	Shock with temporary complete unloading	Transitional Range[c]		J6[b]
Load direction is indeterminate	Light and normal			
	Normal and Heavy		split housing not recommended	K6[b]
	Heavy Shock			M6[b]
Outer ring rotating in relation to load direction	Light			
	Normal and Heavy	Outer ring need not be axially displaceable		N6[b]
	Heavy		Thin wall housing not split	P6[b]

[a] For cast iron or steel housings. For housings of nonferrous alloys tighter fits may be needed.

[b] Where wider tolerances are permissible, use tolerance classifications P7, N7, M7, K7, J7, and H7, in place of P6, N6, M6, K6, J6, and H6, respectively.

[c] The tolerance zones are such that the outer ring may be either tight or loose in the housing.

Table 15. AFBMA and American National Standard Shaft Diameter and Housing Bore Tolerance Limits ***ANSI/ABMA 7-1995***

Inches		mm		Allowable Deviations of Shaft Diameter from Basic Bore Diameter, Inch												
Over	Incl.	Over	Incl.													
Base Bore Diameter				g6	h6	h5	j5	j6	k5	k6	m5	m6	n6	p6	r6	r7
0.2362	0.3937	6	10	−.0002 −.0006	0 −.0004	0 −.0002	+.0002 −.0001	+.0003 −.0001	+.0003 0		+.0005 +.0002					
0.3937	0.7087	10	18	−.0002 −.0007	0 −.0004	0 −.0003	+.0002 −.0001	+.0003 −.0001	+.0004 0		+.0006 +.0003					
0.7087	1.1811	18	30	−.0003 −.0008	0 −.0005		+.0002 −.0002	+.0004 −.0002	+.0004 +.0001		+.0007 +.0003					
1.1811	1.9685	30	50	−.0004 −.0010	0 −.0006		+.0002 −.0002	+.0004 −.0002	+.0005 +.0001	+.0007 +.0001	+.0008 +.0004	+.0010 +.0004				
1.9685	3.1496	50	80	−.0004 −.0011	0 −.0007		+.0002 −.0003	+.0005 −.0003	+.0006 +.0001	+.0008 +.0001	+.0009 +.0004	+.0012 +.0004	+.0018 +.0009			
3.1496	4.7244	80	120	−.0005 −.0013	0 −.0009		+.0002 −.0004	+.0005 −.0004	+.0007 +.0001	+.0010 +.0001	+.0011 +.0005	+.0014 +.0005	+.0019 +.0010	+.0023 +.0015		
				Allowable Deviations of Housing Bore from Basic Outside Diameter of Shaft, Inch												
Basic Outside Diameter				G7	H7	H6	J7	J6	K6	K7	M6	M7	N6	N7	P6	P7
0.7087	1.1811	18	30	+.0003 +.0011	0 +.0008	0 +.0005	−.0004 +.0005	−.0002 +.0003	−.0004 +.0001	−.0006 +.0002	−.0007 +.0002	−.0008 0	−.0009 −.0004	−.0011 −.0003	−.0012 −.0007	−.0014 −.0006
1.1811	1.9685	30	50	+.0004 +.0013	0 +.0010	0 +.0006	−.0004 +.0006	−.0002 +.0004	−.0005 +.0001	−.0007 +.0003	−.0008 −.0002	−.0010 0	−.0011 −.0005	−.0013 −.0003	−.0015 −.0008	−.0017 −.0007
1.9685	3.1496	50	80	+.0004 +.0016	0 +.0012	0 +.0007	−.0005 +.0007	−.0002 +.0005	−.0006 +.0002	−.0008 +.0004	−.0009 −.0002	−.0012 0	−.0013 −.0006	−.0015 −.0004	−.0018 −.0010	−.0020 −.0008
3.1496	4.7244	80	120	+.0005 +.0019	0 +.0014	0 +.0009	−.0005 +.0009	−.0002 +.0006	−.0007 +.0002	−.0010 +.0004	−.0011 −.0002	−.0014 0	−.0015 −.0006	−.0018 −.0004	−.0020 −.0012	−.0023 −.0009
4.7244	7.0866	120	180	+.0006 +.0021	0 +.0016	0 +.0010	−.0006 +.0010	−.0003 +.0007	−.0008 +.0002	−.0011 +.0005	−.0013 −.0003	−.0016 0	−.0018 −.0008	−.0020 −.0005	−.0024 −.0014	−.0027 −.0011
7.0866	9.8425	180	250	+.0006 +.0024	0 +.0018	0 +.0011	−.0006 +.0012	−.0003 +.0009	−.0009 +.0002	−.0013 +.0005	−.0015 −.0003	−.0018 0	−.0020 −.0009	−.0024 −.0006	−.0028 −.0016	−.0031 −.0013

Based on ANSI B4.1-1967 (R1994) Preferred Limits and Fits for Cylindrical Parts. Symbols g6, h6, etc., are shaft and G7, H7, etc., hole limits designations. For larger diameters and metric values see AFBMA Standard 7.

Table 16. AFBMA and American National Standard Shaft Diameter and Housing Bore Tolerance Limits *ANSI/ABMA 7-1995*

Allowable Deviations of Shaft Diameter from Basic Bore Diameter, mm																
Inches		mm														
Over	Incl.	Over	Incl.													
Base Bore Diameter				g6	h6	h5	j5	j6	k5	k6	m5	m6	n6	p6	r6	r7
0.2362	0.3937	6	10	–.005 –.014	0 –.009	0 –.006	+.004 –.002	+.007 –.002	+.007 –.001		+.012 +.006					
0.3937	0.7087	10	18	–.006 –.017	0 –.011	0 –.008	+.005 –.003	+.008 –.003	+.009 +.001		+.015 +.007					
0.7087	1.1811	18	30	–.007 –.020	0 –.013		+.005 –.004	+.009 –.004	+.011 +.002		+.017 +.008					
1.1811	1.9685	30	50	–.009 –.025	0 –.016		+.006 –.005	+.011 –.005	+.013 +.002	+.018 +.002	+.020 +.009	+.025 +.009				
1.9685	3.1496	50	80	–.010 –.029	0 –.019		+.006 –.007	+.012 –.007	+.015 +.002	+.021 +.002	+.024 +.011	+.030 +.011	+.039 +.020			
3.1496	4.7244	80	120	–.012 –.034	0 –.022		+.006 –.009	+.013 –.009	+.018 +.003	+.025 +.003	+.028 +.013	+.035 +.013	+.045 +.023	+.059 +.037		
Allowable Deviations of Housing Bore from Basic Outside Diameter of Shaft, mm																
Basic Outside Diameter				G7	H7	H6	J7	J6	K6	K7	M6	M7	N6	N7	P6	P7
.7086	1.1811	18	30	+.007 +.028	0 +.021	0 +.013	–.009 +.012	–.005 +.008	–.011 +.002	–.015 +.006	–.017 –.004	–.021 0	–.024 –.011	–.028 –.007	–.031 –.018	–.035 –.014
1.1811	1.9685	30	50	+.009 +.034	0 +.025	0 +.016	–.011 +.014	–.006 +.010	–.013 +.003	–.018 +.007	–.020 –.004	–.025 0	–.028 –.012	–.033 –.008	–.037 –.021	–.042 –.017
1.9685	3.1496	50	80	+.010 +.040	0 +.030	0 +.019	–.012 +.018	–.006 +.013	–.015 +.004	–.021 +.009	–.024 –.005	–.030 0	–.033 –.014	–.039 –.009	–.045 –.026	–.051 –.021
3.1496	4.7244	80	120	+.012 +.047	0 +.035	0 +.022	–.013 +.022	–.006 +.016	–.018 +.004	–.025 +.010	–.028 –.006	–.035 0	–.038 –.016	–.045 –.010	–.052 –.030	–.059 –.024
4.7244	7.0866	120	180	+.014 +.054	0 +.040	0 +.025	–.014 +.026	–.007 +.018	–.021 +.004	–.028 +.012	–.033 –.008	–.040 0	–.045 –.020	–.052 –.012	–.061 –.036	–.068 –.028
7.0866	9.8425	180	250	+.015 +.061	0 +.046	0 +.029	–.016 +.030	–.007 +.022	–.024 +.005	–.033 +.013	–.037 –.008	–.046 0	–.051 –.022	–.060 –.014	–.070 –.041	–.079 –.033

Based on ANSI B4.1-1967 (R1994) Preferred Limits and Fits for Cylindrical Parts. Symbols g6, h6, etc., are shaft and G7, H7, etc., hole limits designations. For larger diameters and metric values see AFBMA Standard 7.

Design and Installation Considerations.—Interference fitting will reduce bearing radial internal clearance, so it is recommended that prospective users consult bearing manufacturers to make certain that the required bearings are correctly specified to satisfy all mounting, environmental and other operating conditions and requirements. This check is particularly necessary where heat sources in associated parts may further diminish bearing clearances in operation.

Standard values of radial internal clearances of radial bearings are listed in AFBMA-ANSI Standard 20.

Allowance for Axial Displacement.—Consideration should be given to axial displacement of bearing components owing to thermal expansion or contraction of associated parts. Displacement may be accommodated either by the internal construction of the bearing or by allowing one of the bearing rings to be axially displace-able. For unusual applications consult bearing manufacturers.

Needle Roller Bearing Fitting and Mounting Practice.—The tolerance limits required for shaft and housing seat diameters for needle roller beatings with inner and outer rings as well as limits for raceway diameters where inner or outer rings or both are omitted and rollers operate directly upon these surfaces are given in Tables 17 through 20, inclusive. Unusual design and operating conditions may require a departure from these practices. In such cases, bearing manufacturers should be consulted.

Needle Roller Bearings, Drawn Cup: These beatings without inner ring, Types NIB, NB, NIBM, NBM, NIY, NY, NIYM, NYM, NIH, NH, NIHM, NHM, and Inner Rings, Type NIR depend on the housings into which they are pressed for their size and shape. Therefore, the housings must not only have the proper bore dimensions but also must have sufficient strength. Tables 17 and 18, show the bore tolerance limits for rigid housings such as those made from cast iron or steel of heavy radial section equal to or greater than the ring gage section given in AFBMA Standard 4, 1984. The bearing manufacturers should be consulted for recommendations if the housings must be of lower strength materials such as aluminum or even of steel of thin radial section. The shape of the housing bores should be such that when the mean bore diameter of a housing is measured in each of several radial planes, the maximum difference between these mean diameters should not exceed 0.0005 inch (0.013 mm) or one-half the housing bore tolerance limit, if smaller. Also, the radial deviation from circular form should not exceed 0.00025 inch (0.006 mm). The housing bore surface finish should not exceed 125 micro-inches (3.2 micrometers) arithmetical average.

Table 17. AFBMA and American National Standard Tolerance Limits for Shaft Raceway and Housing Bore Diameters—Needle Roller Bearings, Drawn Cup, Without Inner Ring, Inch Types NIB, NIBM, NIY, NIYM, NIH, and NIHM

ANSI/ABMA 18.2-1982 (R1993)

Basic Bore Diameter under Needle Rollers, F_w		Shaft Raceway Diameter[a] Allowable Deviation from F_w		Basic Outside Diameter, D		Housing Bore Diameter[a] Allowable Deviation from D	
Inch		Inch		Inch		Inch	
Over	Incl.	High	Low	Over	Incl.	Low	High
OUTER RING STATIONARY RELATIVE TO LOAD							
0.1875	1.8750	+0	−0.0005	0.3750	4.0000	−0.0005	+0.0005
1.8750	3.5000	+0	−0.0006	…	…	…	…
OUTER RING ROTATING RELATIVE TO LOAD							
0.1875	1.8750	−0.0005	−0.0010	0.3750	4.0000	−0.0010	+0
1.8750	3.5000	−0.0005	−0.0011	…	…	…	…

[a] See text for additional requirements.

For bearing tolerances, see Table 9.

Table 18. AFBMA and American National Standard Tolerance Limits for Shaft Raceway and Housing Bore Diameters—Needle Roller Bearings, Drawn Cup, Without Inner Ring, Metric Types NB, NBM, NY, NYM, NH, and NHM
ANSI/ABMA 18.1-1982 (R1994)

Basic Bore Diameter Under Needle Rollers, F_w				Shaft Raceway Diameter-[a]Allowable Deviation from F_w		Basic Outside Diameter, D				Housing Bore Diameter-[a]Allowable Deviation from D	
OUTER RING STATIONARY RELATIVE TO LOAD											
mm		Inch		ANSI h6, Inch		mm		Inch		ANSI N7, Inch	
Over	Incl.	Over	Incl.	High	Low	Over	Incl.	Over	Incl.	Low	High
3	6	0.1181	0.2362	+0	−0.0003	6	10	0.2362	0.3937	−0.0007	−0.0002
6	10	0.2362	0.3937	+0	−0.0004	10	18	0.3937	0.7087	−0.0009	−0.0002
10	18	0.3937	0.7087	+0	−0.0004	18	30	0.7087	1.1811	−0.0011	−0.0003
18	30	0.7087	1.1811	+0	−0.0005	30	50	1.1811	1.9685	−0.0013	−0.0003
30	50	1.1811	1.9685	+0	−0.0006	50	80	1.9685	3.1496	−0.0015	−0.0004
50	80	1.9685	3.1496	+0	−0.0007	…	…	…	…	…	…
OUTER RING ROTATING RELATIVE TO LOAD											
mm		Inch		ANSI f6, Inch		mm		Inch		ANSI R7, Inch	
Over	Incl.	Over	Incl.	High	Low	Over	Incl.	Over	Incl.	Low	High
3	6	0.1181	0.2362	−0.0004	−0.0007	6	10	0.2362	0.3937	−0.0011	−0.0005
6	10	0.2362	0.3937	−0.0005	−0.0009	10	18	0.3937	0.7087	−0.0013	−0.0006
10	18	0.3937	0.7087	−0.0006	−0.0011	18	30	0.7087	1.1811	−0.0016	−0.0008
18	30	0.7087	1.1811	−0.0008	−0.0013	30	50	1.1811	1.9685	−0.0020	−0.0010
30	50	1.1811	1.9685	−0.0010	−0.0016	50	65	1.9685	2.5591	−0.0024	−0.0012
50	80	1.9685	3.1496	−0.0012	−0.0019	65	80	2.5591	3.1496	−0.0024	−0.0013

For bearing tolerances, see Table 10.

Table 19. AFBMA and American National Standard Tolerance Limits for Shaft Raceway and Housing Bore Diameters—Needle Roller Bearings, With Cage, Machined Ring, Without Inner Ring, Inch Type NIA
ANSI/ABMA 18.2-1982 (R1993)

Basic Bore Diameter under Needle Rollers, F_w		Shaft Raceway Diameter[a] Allowable Deviation from F_w		Basic Outside Diameter, D		Housing Bore Diameter[a] Allowable Deviation from D	
OUTER RING STATIONARY RELATIVE TO LOAD							
Inch		ANSI h6, Inch		Inch		ANSI H7, Inch	
Over	Incl.	High	Low	Over	Incl.	Low	High
0.2362	0.3937	+0	−0.0004	0.3937	0.7087	+0	+0.0007
0.3937	0.7087	+0	−0.0004	0.7087	1.1811	+0	+0.0008
0.7087	1.1811	+0	−0.0005	1.1811	1.9685	+0	+0.0010
1.1811	1.9685	+0	−0.0006	1.9685	3.1496	+0	+0.0012
1.9685	3.1496	+0	−0.0007	3.1496	4.7244	+0	+0.0014
3.1496	4.7244	+0	−0.0009	4.7244	7.0866	+0	+0.0016
4.7244	7.0866	+0	−0.0010	7.0866	9.8425	+0	+0.0018
7.0866	9.8425	+0	−0.0011	9.8425	12.4016	+0	+0.0020
OUTER RING ROTATING RELATIVE TO LOAD							
Inch		ANSI f6, Inch		Inch		ANSI N7, Inch	
Over	Incl.	High	Low	Over	Incl.	Low	High
0.2362	0.3937	−0.0005	−0.0009	0.3937	0.7087	−0.0009	−0.0002
0.3937	0.7087	−0.0006	−0.0011	0.7087	1.1811	−0.0011	−0.0003
0.7087	1.1811	−0.0008	−0.0013	1.1811	1.9685	−0.0013	−0.0003
1.1811	1.9685	−0.0010	−0.0016	1.9685	3.1496	−0.0015	−0.0004
1.9685	3.1496	−0.0012	−0.0019	3.1496	4.7244	−0.0018	−0.0004
3.1496	4.7244	−0.0014	−0.0023	4.7244	7.0866	−0.0020	−0.0005
4.7244	7.0866	−0.0016	−0.0027	7.0866	9.8425	−0.0024	−0.0006
7.0866	9.8425	−0.0020	−0.0031	9.8425	11.2205	−0.0026	−0.0006

[a] See text for additional requirements.

For bearing tolerances, see Table 11.

Table 20. AFBMA and American National Standard Tolerance Limits for Shaft Diameters—Needle Roller Bearing Inner Rings, Inch Type NIR (Used with Bearing Type NIA) ***ANSI/ABMA 18.2-1982 (R1993)***

Basic Bore, *d*		Shaft Diameter[a]			
		Shaft Rotating Relative to Load, Outer Ring Stationary Relative to Load Allowable Deviation from *d*		Shaft Stationary Relative to Load, Outer Ring Rotating Relative to Load Allowable Deviation from *d*	
Inch		ANSI m5, Inch		ANSI g6, Inch	
Over	Incl.	High	Low	High	Low
0.2362	0.3937	+0.0005	+0.0002	−0.0002	−0.0006
0.3937	0.7087	+0.0006	+0.0003	−0.0002	−0.0007
0.7087	1.1811	+0.0007	+0.0003	−0.0003	−0.0008
1.1811	1.9685	+0.0008	+0.0004	−0.0004	−0.0010
1.9685	3.1496	+0.0009	+0.0004	−0.0004	−0.0011
3.1496	4.7244	+0.0011	+0.0005	−0.0005	−0.0013
4.7244	7.0866	+0.0013	+0.0006	−0.0006	−0.0015
7.0866	9.8425	+0.0015	+0.0007	−0.0006	−0.0017

[a] See text for additional requirements.

For inner ring tolerance limits, see Table 12.

Most needle roller bearings do not use inner rings, but operate directly on the surfaces of shafts. When shafts are used as inner raceways, they should be made of bearing quality steel hardened to Rockwell C 58 minimum. Tables 14 and 18 show the shaft raceway tolerance limits and Table 20 shows the shaft seat tolerance limits when inner rings are used. However, whether the shaft surfaces are used as inner raceways or as seats for inner rings, the mean outside diameter of the shaft surface in each of several radial planes should be determined. The difference between these mean diameters should not exceed 0.0003 inch (0.008 mm) or one-half the diameter tolerance limit, if smaller. The radial deviation from circular form should not exceed 0.0001 inch (0.0025 mm), for diameters up to and including 1 in. (25.4 mm). Above one inch the allowable deviation is 0.0001 times the shaft diameter. The surface finish should not exceed 16 micro-inches (0.4 micrometer) arithmetical average. The housing bore and shaft diameter tolerance limits depend upon whether the load rotates relative to the shaft or the housing.

Needle Roller Bearing With Cage, Machined Ring, Without Inner Ring: The following covers needle roller bearings Type NIA and inner rings Type NIR. The shape of the housing bores should be such that when the mean bore diameter of a housing is measured in each of several radial planes, the maximum difference between these mean diameters does not exceed 0.0005 inch (0.013 mm) or one-half the housing bore tolerance limit, if smaller. Also, the radial deviation from circular form should not exceed 0.00025 inch (0.006 mm). The housing bore surface finish should not exceed 125 micro-inches (3.2 micrometers) arithmetical average. Table 20 shows the housing bore tolerance limits.

When shafts are used as inner raceways their requirements are the same as those given above for Needle Roller Bearings, Drawn Cup. Table 19 shows the shaft raceway tolerance limits and Table 20 shows the shaft seat tolerance limits when inner rings are used.

Needle Roller and Cage Assemblies, Types NIM and NM: For information concerning boundary dimensions, tolerance limits, and fitting and mounting practice, reference should be made to ANSI/ABMA 18.1-1982 (R1994) and ANSI/ABMA 18.2-1982 (R1993).

Bearing Mounting Practice.—Because of their inherent design and material rigidity, rolling contact bearings must be mounted with careful control of their alignment and runout. Medium-speed or slower (400,000 *DN* values or less where *D* is the bearing bore in millimeters and N is the beating speed in revolutions per minute), and medium to light load (C/P values of 7 or greater where C is the beating specific dynamic capacity in pounds and P is the average beating load in pounds) applications can endure misalignments equivalent to those acceptable for high-capacity, precision journal beatings utilizing hard bearing materials such as silver, copper-lead, or aluminum. In no case, however, should the maximum shaft deflection exceed .001 inch per inch for well-crowned roller bearings, and .003 inch per inch for deep-groove ball-beatings. Except for self-aligning ball-bearings and spherical or barrel roller bearings, all other types require shaft alignments with deflections no greater than .0002 inch per inch. With preloaded ball bearings, this same limit is recommended as a maximum. Close-clearance tapered bearings or thrust beatings of most types require the same shaft alignment also.

Of major importance for all bearings requiring good reliability, is the location of the races on the shaft and in the housing.

Assembly methods must insure: 1) that the faces are square, before the cavity is closed; 2) that the cover face is square to the shoulder and pulled in evenly; and 3) that it will be located by a face parallel to it when finally seated against the housing.

These requirements are shown in the accompanying figure. In applications not controlled by automatic tooling with closely controlled fixtures and bolt torquing mechanisms, races should be checked for squareness by sweeping with a dial indicator mounted as shown below. For commercial applications with moderate life and reliability requirements, outer race runouts should be held to .0005 inch per inch of radius and inner race runout to .0004 inch per inch of radius. In preloaded and precision applications, these tolerances must be cut in half. In regard to the question of alignment, it must be recognized that rolling-contact bearings, being made of fully-hardened steel, do not wear in as may certain journal bearings when carefully applied and initially operated. Likewise, rolling contact bearings absorb relatively little deflection when loaded to C/P values of 6 or less. At such stress levels the rolling element-race deformation is generally not over .0002 inch. Consequently, proper mounting and control of shaft deflections are imperative for reliable bearing performance. Aside from inadequate lubrication, these factors are the most frequent causes of premature bearing failures.

Mountings for Precision and Quiet-running Applications.—In applications of rolling-element bearings where vibration or smoothness of operation is critical, special precautions must be taken to eliminate those conditions which can serve to initiate radial and axial motions. These exciting forces can result in shaft excursions which are in resonance with shaft or housing components over a range of frequencies from well below shaft speed to as much as 100 times above it. The more sensitive the configuration, the greater is the need for precision bearings and mountings to be used.

Precision bearings are normally made to much closer tolerances than standard and therefore benefit from better finishing techniques. Special inspection operations are required, however, to provide races and rolling elements with smoothness and runouts compatible with the needs of the application. Similarly, shafts and housings must be carefully controlled.

Among the important elements to be controlled are shaft, race, and housing roundness; squareness of faces, diameters, shoulders, and rolling paths. Though not readily appreciated, grinding chatter, lobular and compensating out-of-roundness, waviness, and flats of less than .0005 inch deviation from the average or mean diameter can cause significant roughness. To detect these and insure the selection of good pieces, three-point electronic indicator inspection must be made. For ultra-precise or quiet applications, pieces are often checked on a "Talyrond" or a similar continuous recording instrument capable of measuring to within a few millionths of an inch. Though this may seem extreme, it has been found that shaft deformities will be reflected through inner races shrunk onto them. Similarly, tight-fit outer races pick up significant deviations in housings. In many instrument and in missile guidance applications, such deviations and deformities may have to be limited to less than .00002 inch.

In most of these precision applications, bearings are used with rolling elements controlled to less than 5 millionths of an inch deviation from roundness and within the same range for diameter.

Special attention is required both in housing design and in assembly of the bearing to shaft and housing. Housing response to axial excursions forced by bearing wobble (which in itself is a result of out-of-square mounting) has been found to be a major source of small electric and other rotating equipment noise and howl. Stiffer, more massive housings and careful alignment of bearing races can make significant improvements in applications where noise or vibration has been found to be objectionable.

Commercial Application Alignment Tolerances

1. **Housing Face Runout** — Square to shaft center within .0004 inch/inch of radius full indicator-eading.

2. **Outer Race Face Runout** — Square to shaft center within .0004 inch/inch of radius full indicator reading and complementary to the housing runout (not opposed).

3. **Inner Race Face Runout** — Square to shaft center within .0003 inch/inch of radius full indicator reading.

4. and 5. **Cover and Closure Mounting Face Parallelism** — Parallel within .001.

6. **Housing Mounting Face Parallelism** — Parallel within .001

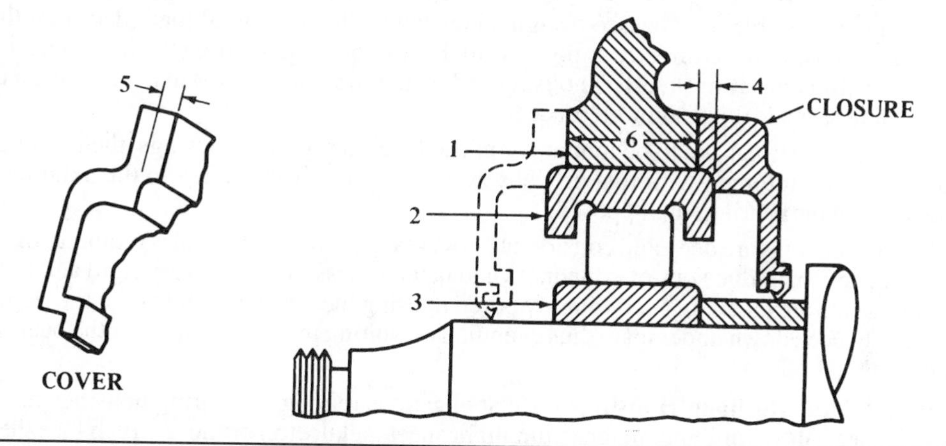

Squareness and Alignment.—In addition to the limits for roundness and wall variation of the races and their supports, squareness of end faces and shoulders must be closely controlled. Tolerances of .0001 inch full indicator reading per inch of diameter are normally required for end faces and shoulders, with appropriately selected limits for fillet eccentricities. The latter must also fall within specified limits for radii tolerances to prevent interference and the resulting cocking of the race. Reference should be made to the bearing dimension tables which list corner radii for typical bearings. Shoulders must also be of a sufficient height to insure proper support for the races, since they are of hardened steel and are less capable of absorbing shock loads and abuse. The general subject of squareness and alignment is of primary importance to the life of rolling element bearings.

The following recommendations for shaft and housing design are given by the New Departure Division of General Motors Corporation:*

"As a rule, there is little trouble experienced with inaccuracies in shafts. Bearings seats and locating shoulders are turned and ground to size with the shaft held on centers and, with ordinary care, there is small chance for serious out-of-roundness or taper. Shaft shoul-

* New Departure Handbook. Vol. II — 1951.

ders should present sufficient surface in contact with the bearing face to assure positive and accurate location.

"Where an undercut must be made for wheel runout in grinding a bearing seat, care should be exercised that no sharp corners are left, for it is at such points that fatigue is most likely to result in shaft breakage. It is best to undercut as little as possible and to have the undercut end in a fillet instead of a sharp corner.

"Where clamping nuts are to be used, it is important to cut the threads as true and square as possible in order to insure even pressure at all points on the bearing inner ring faces when the nuts are set up tight. It is also important not to cut threads so far into the bearing seat as to leave part of the inner ring unsupported or carried on the threads. Excessive deflection is usually the result of improperly designed or undersized machine parts. With a weak shaft, it is possible to seriously affect bearing operation through misalignment due to shaft deflection. Where shafts are comparatively long, the diameter between bearings must be great enough to properly resist bending. In general, the use of more than two bearings on a single shaft should be avoided, owing to the difficulty of securing accurate alignment. With bearings mounted close to each other, this can result in extremely heavy bearing loads.

"Design is as important as careful machining in construction of accurate bearing housings. There should be plenty of metal in the wall sections and large, thin areas should be avoided as much as possible, since they are likely to permit deflection of the boring tool when the housing is being finish-machined.

"Wherever possible, it is best to design a housing so that the radial load placed on the bearing is transmitted as directly as possible to the wall or rib supporting the housing. Diaphragm walls connecting an offset housing to the main wall or side of a machine are apt to deflect unless made thick and well braced.

"When two bearings are to be mounted opposed, but in separate housings, the housings should be so reinforced with fins or webs as to prevent deflection due to the axial load under which the bearings are opposed.

"Where housings are deep and considerable overhang of the boring tool is required, there is a tendency to produce out-of-roundness and taper, unless the tool is very rigid and light finishing cuts are taken. In a too roughly bored housing there is a possibility for the ridges of metal to peen down under load, thus eventually resulting in too loose a fit for the bearing outer ring."

Soft Metal and Resilient Housings.—In applications relying on bearing housings made of soft materials (aluminum, magnesium, light sheet metal, etc.) or those which lose their fit because of differential thermal expansion, outer race mounting must be approached in a cautious manner. Of first importance is the determination of the possible consequences of race loosening and turning. In conjunction with this, the type of loading must be considered for it may serve to magnify the effect of race loosening. It must be remembered that generally, balancing processes do not insure zero unbalance at operating speeds, but rather an "acceptable" maximum. This force exerted by the rotating element on the outer race can initiate a precession which will aggravate the race loosening problem by causing further attrition through wear, pounding, and abrasion. Since this force is generally of an order greater than the friction forces in effect between the outer race, housing, and closures (retaining nuts also), no foolproof method can be recommended for securing outer races in housings which deform significantly under load or after appreciable service wear. Though many such "fixes" are offered, the only sure solution is to press the race into a housing of sufficient stiffness with the heaviest fit consistent with the installed and operating clearances. In many cases, inserts, or liners of cast iron or steel are provided to maintain the desired fit and increase useful life of both bearing and housing.

Quiet or Vibration-free Mountings.—In seeming contradiction is the approach to bearing mountings in which all shaft or rotating element excursions must be isolated from the frame, housing, or supporting structure. Here bearing outer races are often supported on elastomeric or metallic springs. Fundamentally, this is an isolation problem and must be approached with caution to insure solution of the primary bearing objective — location and restraint of the rotating body, as well as the reduction or elimination of the dynamic problem. Again, the danger of skidding rolling elements must be considered and reference to the resident engineers or sales engineers of the numerous bearing companies is recommended, as this problem generally develops requirements for special, or non-catalog-type bearings.

General Mounting Precautions.—Since the last operations involving the bearing application — mounting and closing — have such important effects on bearing performance, durability, and reliability, it must be cautioned that more bearings are abused or "killed" in this early stage of their life than wear out or "die" under conditions for which they were designed. Hammer and chisel "mechanics" invariably handle bearings as though no blow could be too hard, no dirt too abrasive, and no misalignment of any consequence. Proper tools, fixtures, and techniques are a must for rolling bearing application, and it is the responsibility of the design engineer to provide for this in his design, advisory notes, mounting instructions, and service manuals. Nicks, dents, scores, scratches, corrosion staining, and dirt must be avoided if reliability, long life, and smooth running are to be expected of rolling bearings. All manufacturers have pertinent service instructions available for the bearing user. These should be followed for best performance. In a later section, methods for inspecting bearings and descriptions of most common bearing deficiencies will be given.

Seating Fits for Bearings.—Anti-Friction Bearing Manufacturers Association (AFBMA) standard shaft and housing bearing seat tolerances are given in Tables 12 through 17, inclusive.

Clamping and Retaining Methods.—Various methods of clamping bearings to prevent axial movement on the shaft are employed, one of the most common being a nut screwed on the end of the shaft and held in place by a tongued lock washer (see Table 21). The shaft thread for the clamping nut (see Table 22) should be cut in accurate relation to bearing seats and shoulders if bearing stresses are to be avoided. The threads used are of American National Form, Class 3; special diameters and data for these are given in Tables 23 and 24. Where somewhat closer than average accuracy is required, the washers and locknut faces may be obtained ground for closer alignment with the threads. For a high degree of accuracy the shaft threads are ground and a more precise clamping means is employed. Where a bearing inner ring is to be clamped, it is important to provide a sufficiently high shoulder on the shaft to locate the bearing positively and accurately. If the difference between bearing bore and maximum shaft diameter gives a low shoulder which would enter the corner of the radius of the bearing, a shoulder ring that extends above the shoulder and well into the shaft corner is employed. A shoulder ring with snap wire fitting into a groove in the shaft is sometimes used where no locating shaft shoulder is present. A snap ring fitting into a groove is frequently employed to prevent endwise movement of the bearing away from the locating shoulder where tight clamping is not required. Such a retaining ring should not be used where a slot in the shaft surface might lead to fatigue failure. Snap rings are also used to locate the outer bearing ring in the housing. Dimensions of snap rings used for this latter purpose are given in AFBMA and ANSI standards.

Table 21. AFBMA Standard Lockwashers (Series W-00) for Ball Bearings and Cylindrical and Spherical Roller Bearings and (Series TW-100) for Tapered Roller Bearings. Inch Design.

W-00 through W-16

TW-100 through TW-116

Type W No.	Q	Type TW No.	Q	Tangs		Key								Bore R		Diameter		Dia. Over Tangs. Max.	
					Width[a]	Project.[a]	Width S		X		X'								
				No.	T	V	Min.	Max.	Min.	Max.	Min.	Max.		Min.	Max.	E	Tol.	B	B'
W-00	.032	TW-100	.032	9	.120	.031	.110	.120	.334	.359	.334	.359		.406	0.421	0.625	+.015	0.875	0.891
W-01	.032	TW-101	.032	9	.120	.031	.110	.120	.412	.437	.412	.437		.484	.499	0.719	+.015	1.016	1.031
W-02	.032	TW-102	.048	11	.120	.031	.110	.120	.529	.554	.513	.538		.601	.616	0.813	+.015	1.156	1.156
W-03	.032	TW-103	.048	11	.120	.031	.110	.120	.607	.632	.591	.616		.679	.694	0.938	+.015	1.328	1.344
W-04	.032	TW-104	.048	11	.166	.031	.156	.176	.729	.754	.713	.738		.801	.816	1.125	+.015	1.531	1.563
W-05	.040	TW-105	.052	13	.166	.047	.156	.176	.909	.939	.897	.927		.989	1.009	1.281	+.015	1.719	1.703
W-06	.040	TW-106	.052	13	.166	.047	.156	.176	1.093	1.128	1.081	1.116		1.193	1.213	1.500	+.015	1.922	1.953
		TW-065	.052	15	.166	…	.156	.176	…	…	1.221	1.256		1.333	1.353	1.813	+.015	…	2.234
W-07	.040	TW-107	.052	15	.166	.047	.156	.176	1.296	1.331	1.284	1.319		1.396	1.416	1.813	+.015	2.250	2.250
W-08	.048	TW-108	.062	15	.234	.047	.250	.290	1.475	1.510	1.461	1.496		1.583	1.603	2.000	+.030	2.469	2.484
W-09	.048	TW-109	.062	17	.234	.062	.250	.290	1.684	1.724	1.670	1.710		1.792	1.817	2.281	+.030	2.734	2.719
W-10	.048	TW-110	.062	17	.234	.062	.250	.290	1.884	1.924	1.870	1.910		1.992	2.017	2.438	+.030	2.922	2.922
W-11	.053	TW-111	.062	17	.234	.062	.250	.290	2.069	2.109	2.060	2.100		2.182	2.207	2.656	±.030	3.109	3.094
W-12	.053	TW-112	.072	17	.234	.062	.250	.290	2.267	2.307	2.248	2.288		2.400	2.425	2.844	+.030	3.344	3.328
W-13	.053	TW-113	.072	19	.234	.062	.250	.290	2.455	2.495	2.436	2.476		2.588	2.613	3.063	+.030	3.578	3.563
W-14	.053	TW-114	.072	19	.234	.094	.250	.290	2.658	2.698	2.639	2.679		2.791	2.816	3.313	+.030	3.828	3.813
W-15	.062	TW-115	.085	19	.328	.094	.250	.290	2.831	2.876	2.808	2.853		2.973	3.003	3.563	+.030	4.109	4.047
W-16	.062	TW-116	.085	19	.328	.094	.313	.353	3.035	3.080	3.012	3.057		3.177	3.207	3.844	+.030	4.375	4.391

[a] *Tolerances:* On width, T,–.010 inch for Types W-00 to W-03 and TW-100 to TW-103; –.020 inch for W-04 to W-07 and TW-104 to TW-107; –.030 inch for all others shown. On Projection V, +.031 inch for all sizes up through W-13 and TW-113; +.062 inch for all others shown.

All dimensions in inches. For dimensions in millimeters, multiply inch values by 25.4 and round result to two decimal places.

Data for sizes larger than shown are given in ANSI/AFBMA Standard 8.2-1991.

Table 22. AFBMA Standard Locknuts (Series N-00) for Ball Bearings and Cylindrical and Spherical Roller Bearings and (Series TN-00) for Tapered Roller Bearings. Inch Design.

Runout and parallelism of faces measured on a tight fitting threaded arbor.

N-00 to N-06 = .002 Max.
N-07 to AN-15 = .004 Max.

TN-065 to TAN-15 = .002 Max.

Surface Finish Note

TN-065 to TN-11, 100μ in., max.
TN-12 to TAN-15, 120μ in., max.

G H C

N-00 through AN-15
TN-065 through TAN-15

D 50° d E Threads (see note)
60° Max.
45° Min.

BB & RB Nut No.	TRB Nut No.	Thds. per Inch	Thread Minor Deam.		Thread Pitch Dia.		Thd. Major Dia. d	Outside Dia. C	Face Dia. E		Slot dimension: Width G		Slot dimension: Height H	Thickness D	
			Min.	Max.	Min.	Max.	Min.	Max.	Min.	Max.	Min.	Max.	Max.	Min.	Max.
N-00	—	32	0.3572	0.3606	0.3707	0.3733	0.391	0.755	.605	.625	.120	.130	.073	.209	.229
N-01	—	32	0.4352	0.4386	0.4487	0.4513	0.469	0.880	.699	.719	.120	.130	.073	.303	.323
N-02	—	32	0.5522	0.5556	0.5657	0.5687	0.586	1.005	.793	.813	.120	.130	.104	.303	.323
N-03	—	32	0.6302	0.6336	0.6437	0.6467	0.664	1.130	.918	.938	.120	.130	.104	.334	.354
N-04	—	32	0.7472	0.7506	0.7607	0.7641	0.781	1.380	1.105	1.125	.178	.198	.104	.365	.385
N-05	—	32	0.9352	0.9386	0.9487	0.9521	0.969	1.568	1.261	1.281	.178	.198	.104	.396	.416
N-06	—	18	1.1129	1.1189	1.1369	1.1409	1.173	1.755	1.480	1.500	.178	.198	.104	.396	.416
	TN-065	18	1.2524	1.2584	1.2764	1.2804	1.312	2.068	1.793	1.813	.178	.198	.104	.428	.448
N-07	TN-07	18	1.3159	1.3219	1.3399	1.3439	1.376	2.068	1.793	1.813	.178	.198	.104	.428	.448
N-08	TN-08	18	1.5029	1.5089	1.5269	1.5314	1.563	2.255	1.980	2.000	.240	.260	.104	.428	.448
N-09	TN-09	18	1.7069	1.7129	1.7309	1.7354	1.767	2.536	2.261	2.281	.240	.260	.104	.428	.448
N-10	TN-10	18	1.9069	1.9129	1.9309	1.9354	1.967	2.693	2.418	2.438	.240	.260	.104	.490	.510
N-11	TN-11	18	2.0969	2.1029	2.1209	2.1260	2.157	2.974	2.636	2.656	.240	.260	.135	.490	.510
N-12	TN-12	18	2.2999	2.3059	2.3239	2.3290	2.360	3.161	2.824	2.844	.240	.260	.135	.521	.541
N-13	TN-13	18	2.4879	2.4949	2.5119	2.5170	2.548	3.380	3.043	3.063	.240	.260	.135	.553	.573
N-14	TN-14	18	2.6909	2.6969	2.7149	2.7200	2.751	3.630	3.283	3.313	.240	.260	.135	.553	.573
AN-15	TAN-15	12	2.8428	2.8518	2.8789	2.8843	2.933	3.880	3.533	3.563	.360	.385	.135	.584	.604

All dimensions in inches. For dimensions in millimeters, multiply inch values, except thread diameters, by 25.4 and round result to two decimal places.

Threads are American National form, Class 3.

Typical steels for locknuts are: AISI, C1015, C1018, C1020, C1025, C1035, C1117, C1118, C1212, C1213, and C1215. Minimum hardness, tensile strength, yield strength and elongation are given in ANSI/ABMA 8.2-1991 which also lists larger sizes of locknuts.

Table 23. AFBMA Standard for Shafts for Locknuts (series N-00) for Ball Bearings and Cylindrical and Spherical Roller Bearings. Inch Design.

Locknut Number	Bearing Bore	V_2	Threads[a]					Relief		Keyway		
			No. per inch	Major Dia.	Pitch Dia.	Minor Dia.	Length L	Dia. A	Width W	Depth H	Width S	M
		Max.		Max.	Max.	Max.	Max.	Max.	Max.	Min.	Min.	Min.
N-00	0.3937	0.312	32	0.391	0.3707	0.3527	0.297	0.3421	0.078	0.062	0.125	0.094
N-01	0.4724	0.406	32	0.469	0.4487	0.4307	0.391	0.4201	0.078	0.062	0.125	0.094
N-02	0.5906	0.500	32	0.586	0.5657	0.5477	0.391	0.5371	0.078	0.078	0.125	0.094
N-03	0.6693	0.562	32	0.664	0.6437	0.6257	0.422	0.6151	0.078	0.078	0.125	0.094
N-04	0.7874	0.719	32	0.781	0.7607	0.7427	0.453	0.7321	0.078	0.078	0.188	0.094
N-05	0.9843	0.875	32	0.969	0.9487	0.9307	0.484	0.9201	0.078	0.094	0.188	0.125
N-06	1.1811	1.062	18	1.173	1.1369	1.1048	0.484	1.0942	0.109	0.094	0.188	0.125
N-07	1.3780	1.250	18	1.376	1.3399	1.3078	0.516	1.2972	0.109	0.094	0.188	0.125
N-08	1.5748	1.469	18	1.563	1.5269	1.4948	0.547	1.4842	0.109	0.094	0.312	0.125
N-09	1.7717	1.688	18	1.767	1.7309	1.6988	0.547	1.6882	0.141	0.094	0.312	0.156
N-10	1.9685	1.875	18	1.967	1.9309	1.8988	0.609	1.8882	0.141	0.094	0.312	0.156
N-11	2.1654	2.062	18	2.157	2.1209	2.0888	0.609	2.0782	0.141	0.125	0.312	0.156
N-12	2.3622	2.250	18	2.360	2.3239	2.2918	0.641	2.2812	0.141	0.125	0.312	0.156
N-13	2.5591	2.438	18	2.548	2.5119	2.4798	0.672	2.4692	0.141	0.125	0.312	0.156
N-14	2.7559	2.625	18	2.751	2.7149	2.6828	0.672	2.6722	0.141	0.125	0.312	0.250
AN-15	2.9528	2.781	12	2.933	2.8789	2.8308	0.703	2.8095	0.172	0.125	0.312	0.250
AN-16	3.1496	3.000	12	3.137	3.0829	3.0348	0.703	3.0135	0.172	0.125	0.375	0.250

[a] Threads are American National form Class 3.

All dimensions in inches. For dimensions in millimeters, multiply inch values, except thread diameters, by 25.4 and round result to two decimal places. See footnote to Table 24 for material other than sttel.For sizes larger than shown, see ANSI/ABMA 8.2-1991.

Table 24. AFBMA Standard for Shafts for Tapered Roller Bearing Locknuts. Inch Design.

CLAMPED MOUNTING

ADJUSTABLE MOUNTING

Locknut Number	Bearing Bore	V_2	Threads[a]						Relief		Keyway			
			No. per inch	Major Dia.	Pitch Dia.	Minor Dia.	Length L_1	Length L_2	Dia. A	Width W	Depth H	Width S	M	U
		Max.		Max.	Max.	Max.	Max.	Max.	Max.	Max.	Max.	Min.	Min.	Min.
N-00	0.3937	0.312	32	0.391	0.3707	0.3527	0.609	0.391	0.3421	0.078	0.094	0.125	0.094	0.469
N-01	0.4724	0.406	32	0.469	0.4487	0.4307	0.797	0.484	0.4201	0.078	0.094	0.125	0.094	0.562
N-02	0.5906	0.500	32	0.586	0.5657	0.5477	0.828	0.516	0.5371	0.078	0.094	0.125	0.094	0.594
N-03	0.6693	0.562	32	0.664	0.6437	0.6257	0.891	0.547	0.6151	0.078	0.094	0.125	0.094	0.625
N-04	0.7874	0.703	32	0.781	0.7607	0.7427	0.922	0.547	0.7321	0.078	0.078	0.125	0.094	0.625
N-05	0.9843	0.875	32	0.969	0.9487	0.9307	1.016	0.609	0.9201	0.078	0.094	0.188	0.094	0.625
N-06	1.1811	1.062	18	1.173	1.1369	1.1048	1.016	0.609	1.0942	0.078	0.125	0.188	0.125	0.719
TN-065	1.3750	1.188	18	1.312	1.2764	1.2443	1.078	0.641	1.2337	0.109	0.125	0.188	0.125	0.719
TN-07	1.3780	1.250	18	1.376	1.3399	1.3078	1.078	0.641	1.2972	0.109	0.125	0.188	0.125	0.750
TN-08	1.5748	1.438	18	1.563	1.5269	1.4948	1.078	0.641	1.4842	0.109	0.125	0.188	0.125	0.750
TN-09	1.7717	1.656	18	1.767	1.7309	1.6988	1.078	0.641	1.6882	0.109	0.125	0.312	0.125	0.750
TN-10	1.9685	1.859	18	1.967	1.9309	1.8988	1.203	0.703	1.882	0.141	0.125	0.312	0.156	0.781
TN-11	2.1654	2.047	18	2.157	2.1209	2.0888	1.203	0.703	2.0782	0.141	0.125	0.312	0.156	0.844
TN-12	2.3622	2.250	18	2.360	2.3239	2.2918	1.297	0.766	2.2812	0.141	0.125	0.312	0.156	0.844
TN-13	2.5591	2.422	18	2.548	2.5119	2.4798	1.359	0.797	2.4692	0.141	0.156	0.312	0.156	0.906
TN-14	2.7559	2.625	18	2.751	2.7149	2.6828	1.359	0.797	2.6722	0.141	0.156	0.312	0.156	0.938
TAN-15	2.9528	2.781	12	2.933	2.8789	2.8308	1.422	0.828	2.8095	0.172	0.156	0.312	0.250	1.000
TAN-16	3.1496	3.000	12	3.137	3.0829	3.0348	1.422	0.828	3.0135	0.172	0.188	0.312	0.250	1.031
											0.188	0.375	0.250	1.031

[a] Threads are American National form Class 3.

All dimensions in inches. For dimensions in millimeters, multiply inch values, except thread diameters, by 25.4 and round results to two decimal places. These data apply to steel. When either the nut or the shaft is made of stainless steel, aluminum, or other material having a tendency to seize, it is recommended that the maximum thread diameter of the shaft, both major and pitch, be reduced by 20 per cent of the pitch diameter tolerance listed in the Standard. For sizes larger than shown, see ANSI/ABMA 8.2-1991.

Bearing Closures.—Shields, seals, labyrinths, and slingers are employed to retain the lubricant in the bearing and to prevent the entry of dirt, moisture, or other harmful substances. The type selected for a given application depends upon the lubricant, shaft, speed, and the atmospheric conditions in which the unit is to operate. The shields or seals may be located in the bearing itself. Shields differ from seals in that they are attached to one bearing race but there is a definite clearance between the shield and the other, usually the inner, race. When a shielded bearing is placed in a housing in which the grease space has been filled, the bearing in running will tend to expel excess grease past the shields or to accept grease from the housing when the amount in the bearing itself is low.

Seals of leather, rubber, cork, felt, or plastic composition may be used. Since they must bear against the rotating member, excessive pressure should be avoided and some lubricant must be allowed to flow into the area of contact in order to prevent seizing and burning of the seal and scoring of the rotating member. Some seals are made up in the form of cartridges which can be pressed into the end of the bearing housing.

Leather seals may be used over a wide range of speeds. Although lubricant is best retained with a leather cupped inward toward the bearing, this arrangement is not suitable at high speeds due to danger of burning the leather. At high speeds where abrasive dust is present, the seal should be arranged with the leather cupped outward to lead some lubricant into the contact area. Only light pressure of leather against the shaft should be maintained.

Bearing Fits.—The slipping or creeping of a bearing ring on a rotating shaft or in a rotating housing occurs when the fit of the ring on the shaft or in the housing is loose. Such slipping or creeping action may cause rapid wear of both shaft and bearing ring when the surfaces are dry and highly loaded. To prevent this action the bearing is customarily mounted with the rotating ring a press fit and the stationary ring a push fit, the tightness or looseness depending upon the service intended. Thus, where shock or vibratory loads are to be encountered, fits should be made somewhat tighter than for ordinary service. The stationary ring, if correctly fitted, is allowed to creep very slowly so that prolonged stressing of one part of the raceway is avoided.

To facilitate the assembly of a bearing on a shaft it may become necessary to expand the inner ring by heating. This should be done in clean oil or in a temperature-controlled furnace at a temperature of between 200 and 250°F. The utmost care must be used to make sure that the temperature does not exceed 250°F. as overheating will tend to reduce the hardness of the rings. Prelubricated bearings should not be mounted by this method.

Friction Losses in Rolling Element Bearings.—The static and kinematic torques of rolling element bearings are generally small and in many applications are not significant. Bearing torque is a measure of the frictional resistance of the bearing to rotation and is the sum of three components: the torque due to the applied load; the torque due to viscous forces in lubricated rolling element bearings; and the torque due to roller end motions, for example, thrust loads against flanges. The friction or torque data may be used to calculate power absorption or heat generation within the bearing and can be utilized in efficiency or system-cooling studies.

Empirical equations have been developed for each of the torque components. These equations are influenced by such factors as bearing load, lubrication environment, and bearing design parameters. These design parameters include sliding friction from contact between the rolling elements and separator surfaces or between adjacent rolling elements; rolling friction from material deformations during the passage of the rolling elements over the race path; skidding or sliding of the Hertzian contact; and windage friction as a function of speed.

Starting or breakaway torques are also of interest in some situations. Breakaway torques tend to be between 1.5 and 1.8 times the running or kinetic torques.

When evaluating the torque requirements of a system under design, it should be noted that other components of the bearing package, such as seals and closures, can increase the overall system torque significantly. Seal torques have been shown to vary from a fraction of the bearing torque to several times that torque. In addition, the torque values given can vary significantly when load, speed of rotation, temperature, or lubrication are outside normal ranges.

For small instrument bearings friction torque has implications more critical than for larger types of bearings. These bearings have three operating friction torques to consider: starting torque, normal running torque, and peak running torque. These torque levels may vary between manufacturers and among lots from a given manufacturer.

Instrument bearings are even more critically dependent on design features — radial play, retainer type, and race conformity — than larger bearings. Typical starting torque values for small bearings are given in the accompanying table, extracted from the New Departure General Catalog.

Finally, if accurate control of friction torque is critical to a particular application, tests of the selected bearings should be conducted to evaluate performance.

Starting Torque — ABEC7

Bearing Bore (in.)	Max. Starting Torque (g cm)	Thrust Load (g)	Minimum Radial Play Range (inches)	
			High Carbon Chrome Steel and All Miniatures	Stainless Steel Except Miniatures
0.125	0.10	75	0.0003–0.0005	—
	0.14	75	0.0002–0.0004	0.0004–0.0006
	0.18	75	0.0001–0.0003	0.0003–0.0005
	0.22	75	0.0001–0.0003	0.0001–0.0003
0.1875–0.312	0.40	400	0.0005–0.0008	—
	0.45	400	0.0004–0.0006	0.0005–0.0008
	0.50	400	0.0003–0.0005	0.0003–0.0005
	0.63	400	0.0001–0.0003	0.0002–0.0004
0.375	0.50	400	0.0005–0.0008	0.0008–0.0011
	0.63	400	0.0004–0.0006	0.0005–0.0008
	0.75	400	0.0003–0.0005	0.0004–0.0006
	0.95	400	0.0002–0.0004	0.0003–0.0005

Selection of Ball and Roller Bearings.—As compared with sleeve bearings, ball and roller bearings offer the following advantages: 1) Starting friction is low; 2) Less axial space is required; 3) Relatively accurate shaft alignment can be maintained; 4) Both radial and axial loads can be carried by certain types; 5) Angle of load application is not restricted; 6) Replacement is relatively easy; 7) Comparatively heavy overloads can be carried momentarily; 8) Lubrication is simple; and 9) Design and application can be made with the assistance of bearing supplier engineers.

In selecting a ball or roller bearing for a specific application five choices must be made:

1) the bearing series; 2) the type of bearing; 3) the size of bearing; 4) the method of lubrication; and 5) the type of mounting.

Naturally these considerations are modified or affected by the anticipated operating conditions, expected life, cost, and overhaul philosophy.

It is well to review the possible history of the bearing and its function in the machine it will be applied to, thus: 1) Will it be expected to endure removal and reapplication?; 2) Must it be free from maintenance attention during its useful life?; 3) Can wear of the housing or shaft be tolerated during the overhaul period?; 4) Must it be adjustable to take up wear, or to change shaft location?; 5) How accurately can the load spectrum be estimated? and; and 6) Will it be relatively free from abuse in operation?.

Though many cautions could be pointed out, it should always be remembered that inadequate design approaches limit the utilization of rolling element bearings, reduce customer satisfaction, and reduce reliability. Time spent in this stage of design is the most rewarding effort of the bearing engineer, and here again he can depend on the bearing manufacturers' field organization for assistance.

Type: Where loads are low, ball bearings are usually less expensive than roller bearings in terms of unit-carrying capacity. Where loads are high, the reverse is usually true.

For a purely radial load, almost any type of radial bearing can be used, the actual choice being determined by other factors. To support a combination of thrust and radial loads, several types of bearings may be considered. If the thrust load component is large, it may be most economical to provide a separate thrust bearing. When a separate thrust bearing cannot be used due to high speed, lack of space, or other factors, the following types may be considered: angular contact ball bearing, deep groove ball bearing without filling slot, tapered roller bearing with steep contact angle, and self-aligning bearing of the wide type. If movement or deflection in an axial direction must be held to a minimum, then a separate thrust bearing or a preloaded bearing capable of taking considerable thrust load is required. To minimize deflection due to a moment in an axial plane, a rigid bearing such as a double row angular contact type with outwardly converging load lines is required. In such cases, the resulting stresses must be taken into consideration in determining the proper size of the bearing.

For shock loads or heavy loads of short duration, roller bearings are usually preferred.

Special bearing designs may be required where accelerations are usually high as in planetary or crank motions.

Where the problem of excessive shaft deflection or misalignment between shaft and housing is present, a self-aligning type of bearing may be a satisfactory solution.

It should be kept in mind that a great deal of difficulty can be avoided if standard types of bearings are used in preference to special designs, wherever possible.

Size: The size of bearing required for a given application is determined by the loads that are to be carried and, in some cases, by the amount of rigidity that is necessary to limit deflection to some specified amount.

The forces to which a bearing will be subjected can be calculated by the laws of engineering mechanics from the known loads, power, operating pressure, etc. Where loads are irregular, varying, or of unknown magnitude, it may be difficult to determine the actual forces. In such cases, empirical determination of such forces, based on extensive experience in bearing design, may be needed to attack the problem successfully. Where such experience is lacking, the bearing manufacturer should be consulted or the services of a bearing expert obtained.

If a ball or roller bearing is to be subjected to a combination of radial and thrust loads, an *equivalent radial load* is computed in the case of radial or angular type bearings and an *equivalent thrust load* is computed in the case of thrust bearings.

Method of Lubrication.—If speeds are high, relubrication difficult, the shaft angle other than horizontal, the application environment incompatible with normal lubrication, leakage cannot be tolerated; if other elements of the mechanism establish the lubrication requirements, bearing selection must be made with these criteria as controlling influences. Modern bearing types cover a wide selection of lubrication means. Though the most popular type is the "cartridge" type of sealed grease ball bearing, many applications have requirements which dictate against them. Often, operating environments may subject bearings to temperatures too high for seals utilized in the more popular designs. If minute leakage or the accumulation of traces of dirt at seal lips cannot be tolerated by the application (as in baking industry machinery), then the selections of bearings must be made with other sealing and lubrication systems in mind.

High shaft speeds generally dictate bearing selection based on the need for cooling, the suppression of churning or aeration of conventional lubricants, and most important of all, the inherent speed limitations of certain bearing types. An example of the latter is the effect of cage design and of the roller-end thrust-flange contact on the lubrication requirements in commercial taper roller bearings, which limit the speed they can endure and the thrust load they can carry. Reference to the manufacturers' catalog and application-design manuals is recommended before making bearing selections.

Type of Mounting.—Many bearing installations are complicated because the best adapted type was not selected. Similarly, performance, reliability, and maintenance operations are restricted because the mounting was not thoroughly considered. There is no universally adaptable bearing for all needs. Careful reviews of the machine requirements should be made before designs are implemented. In many cases complicated machining, redundant shaft and housings, and use of an oversize bearing can be eliminated if the proper bearing in a well-thought-out mounting is chosen.

Advantage should be taken of the many race variations available in "standard" series of bearings. Puller grooves, tapered sleeves, ranged outer races, split races, fully demountable rolling-element and cage assemblies, flexible mountings, hydraulic removal features, relubrication holes and grooves, and many other innovations are available beyond the obvious advantages which are inherent in the basic bearing types.

Radial and Axial Clearance.—In designing the bearing mounting, a major consideration is to provide running clearances consistent with the requirements of the application. Race fits must be expected to absorb some of the original bearing clearance so that allowance should be made for approximately 80 per cent of the actual interference showing up in the diameter of the race. This will increase for heavy, stiff housings or for extra light series races shrunk onto solid shafts, while light metal housings (aluminum, magnesium, or sheet metal) and tubular shafts with wall sections less than the race wall thickness will cause a lesser change in the race diameter.

Where the application will impose heat losses through housing or shaft, or where a temperature differential may be expected, allowances must be made in the proper direction to insure proper operating clearance. Some compromises are required in applications where the indicated modification cannot be fully accommodated without endangering the bearing performance at lower speeds, during starting, or under lower temperature conditions than anticipated. Some leeway can be relied on with ball bearings since they can run with moderate preloads (.0005 inch, max.) without affecting bearing life or temperature rise. Roller bearings, however, have a lesser tolerance for preloading, and must be carefully controlled to avoid overheating and resulting self-destruction.

In all critical applications axial and radial clearances should be checked with feeler gages or dial indicators to insure mounted clearances within tolerances established by the design engineer. Since chips, scores, race misalignment, shaft or housing denting, housing distortion, end cover (closure) off-squareness, and mismatch of rotor and housing axial dimensions can rob the bearing of clearance, careful checks of running clearance is recommended.

For precision applications, taper-sleeve mountings, opposed ball or tapered-roller bearings with adjustable or shimmed closures are employed to provide careful control of radial and/or axial clearances. This practice requires skill and experience as well as the initial assistance of the bearing manufacturer's field engineer.

Tapered bore bearings are often used in applications such as these, again requiring careful and well worked-out assembly procedures. They can be assembled on either tapered shafts or on adapter sleeves. Advancement of the inner race over the tapered shaft can be done either by controlled heating (to expand the race as required) or by the use of a hydraulic jack. The adapter sleeve is supplied with a lock-nut which is used to advance the race on the tapered sleeve. With the heavier fits normally required to effect the clearance changes compatible with such mountings, hydraulic removal devices are normally recommended.

For the conventional application, with standard fits, clearances provided in the standard bearing are suitable for normal operation. To insure that the design conditions are "normal," a careful review of the application requirements, environments, operating speed range, anticipated abuses, and design parameters must be made.

General Bearing Handling Precautions.—To insure that rolling element bearings are capable of achieving their design life and that they perform without objectionable noise, temperature rise, or shaft excursions, the following precautions are recommended:

1) Use the best bearing available for the application, consistent with the value of the application. Remember, the cost of the best bearing is generally small compared to the replacement costs of the rotating components that can be destroyed if a bearing fails or malfunctions.

2) If questions arise in designing the bearing application, seek out the assistance of the bearing manufacturer's representative.

3) Handle bearings with care, keeping them in the sealed, original container until ready to use.

4) Follow the manufacturer's instructions in handling and assembling the bearings.

5) Work with clean tools, clean dry hands, and in clean surroundings.

6) Do not wash or wipe bearings prior to installation unless special instructions or requirements have been established to do so.

7) Place unwrapped bearings on clean paper and keep them similarly covered until applied, if they cannot be kept in the original container.

8) Don't use wooden mallets, brittle or chipped tools, or dirty fixtures and tools in mounting bearings.

9) Don't spin uncleaned bearings, nor spin *any* bearing with an air blast.

10) Use care not to scratch or nick bearings.

11) Don't strike or press on race flanges.

12) Use adapters for mounting which provide uniform steady pressure rather than hammering on a drift or sleeve.

13) Insure that races are started onto shafts and into housings evenly so as to prevent cocking.

14) Inspect shafts and housings before mounting beating to insure that proper fits will be maintained.

15) When removing beatings, clean housings, covers, and shafts before exposing the bearings. All dirt can be considered an abrasive, dangerous to the reuse of any rolling bearing.

16) Treat used beatings, which may be reused, as new ones.

17) Protect dismantled bearings from dirt and moisture.

18) Use clean, lint-free rags if bearings are wiped.

19) Wrap beatings in clean, oil-proof paper when not in use.

20) Use clean filtered, water-free Stoddard's solvent or flushing oil to clean bearings.

21) In heating beatings for mounting onto shafts, follow manufacturer's instructions.

22) In assembling bearings onto shafts *never* strike the outer race, or press on it to force the inner race. Apply the pressure on the inner race only. In dismantling follow the same precautions.

23) Do not press, strike, or otherwise force the seal or shield on factory-sealed beatings.

Bearing Failures, Deficiencies, and Their Origins.— The general classifications of failures and deficiencies requiting bearing removal are:

1) Overheating a) Inadequate or insufficient lubrication; b) Excessive lubrication; c) Grease liquefaction or aeration; d) Oil foaming; e) Abrasive or corrosive action due to contaminants in beating; f) Distortion of housing due to warping, or out-of-round; g) Seal rubbing or failure; h) Inadequate or blocked scavenge oil passages; i) Inadequate beating-clearance or bearing-preload; j) Race turning; k) Cage wear; 1) and a) Shaft expansion — loss of bearing or seal clearance..

2) Vibration a) Dirt or chips in bearing; b) Fatigued race or rolling elements; c) Race turning; d) Rotor unbalance; e) Out-of-round shaft; f) Race misalignment; g) Housing resonance; h) Cage wear; i) Flats on races or rolling elements; j) Excessive clearance; k) Corrosion; l) False-brinelling or indentation of races; m) Electrical discharge (similar to corrosion effects); n) Mixed rolling element diameters; 3) and a) Out-of-square rolling paths in races.

4) Turning on shaft a) Growth of race due to overheating; b) Fretting wear; c) Improper initial fit; d) Excessive shaft deflection; e) Initially coarse shaft finish; 5) and a) Seal rub on inner race.

6) Binding of the shaft a) Lubricant breakdown; b) Contamination by abrasive or corrosive matter; c) Housing distortion or out-of-round pinching bearing; d) Uneven shimming of housing with loss of clearance; e) Tight rubbing seals; f) Preloaded beatings; g) Cocked races; h) Loss of clearance due to excessive tightening of adapter; i) Thermal expansion of shaft or housing; 7) and a) Cage failure.

8) Noisy bearing a) Lubrication breakdown, inadequate lubrication, stiff grease; b) Contamination; c) Pinched beating; d) Seal rubbing; e) Loss of clearance and preloading; f) Bearing slipping on shaft or in housing; g) Flatted roller or ball; h) Brinelling due to assembly abuse, handling, or shock loads; i) Variation in size of rolling elements; j) Out-of-round or lobular shaft; k) Housing bore waviness; 9) and a) Chips or scores under beating race seat.

10) Displaced shaft a) Bearing wear; b) Improper housing or closure assembly; c) Overheated and shifted bearing; d) Inadequate shaft or housing shoulder; e) Lubrication and cage failure permitting rolling elements to bunch; f) Loosened retainer nut or adapter; g) Excessive heat application in assembling inner race, causing growth and shifting on shaft; 11) and a) Housing pounding out.

12) Lubricant leakage a) Overfilling of lubricant; b) Grease churning due to use of too soft a consistency; c) Grease deterioration due to excessive operating temperature; d) Operating life longer than grease life (grease breakdown, aeration, and purging); e) Seal wear; f) Wrong shaft attitude (bearing seals designed for horizontal mounting only); g) Seal failure; h) Clogged breather; i) Oil foaming due to churning or air flow through housing; j) Gasket (O-ring) failure or misapplication; k) Porous housing or closure; 13) and a) Lubricator set at wrong flow rate.

Load Ratings and Fatigue Life

Ball and Roller Bearing Life.— The performance of ball and roller bearings is a function of many variables. These include the bearing design, the characteristics of the material from which the bearings are made, the way in which they are manufactured, as well as many variables associated with their application. The only sure way to establish the satisfactory operation of a bearing selected for a specific application is by actual performance in the application. As this is often impractical, another basis is required to estimate the suitability of a particular bearing for a given application. Two factors are taken into consideration: the beating fatigue life, and its ability to withstand static loading.

Life Criterion: Even if a ball or roller bearing is properly mounted, adequately lubricated, protected from foreign matter and not subjected to extreme operating conditions, it can ultimately fatigue. Under ideal conditions, the repeated stresses developed in the contact areas between the balls or rollers and the raceways eventually can result in the fatigue of the material which manifests itself with the spalling of the load-carrying surfaces. In most applications the fatigue life is the maximum useful life of a bearing.

Static Load Criterion: A static load is a load acting on a non-rotating bearing. Permanent deformations appear in balls or rollers and raceways under a static load of moderate magnitude and increase gradually with increasing load. The permissible static load is, therefore, dependent upon the permissible magnitude of permanent deformation. It has been found that for ball and roller bearings suitably manufactured from hardened alloy steel, deformations occurring under maximum contact stress of 4,000 megapascals (580,000 pounds per square inch) acting at the center of contact (in the case of roller beatings, of a uniformly loaded roller) do not greatly impair smoothness or friction. Depending on requirements for smoothness of operation, friction, or sound level, higher or lower static load limits may be tolerated.

Ball Bearing Types Covered.—AFBMA and American National Standard ANSI/ABMA 9-1990 sets forth the method of determining ball bearing Rating Life and Static Load Rating and covers the following types:

1) *Radial, deep groove and angular contact ball bearings* whose inner ring race-ways have a cross-sectional radius not larger than 52 percent of the ball diameter and whose outer ring raceways have a cross-sectional radius not larger than 53 percent of the ball diameter.

2) *Radial, self-aligning ball bearings* whose inner ring raceways have cross-sectional radii not larger than 53 percent of the ball diameter.

3) *Thrust ball bearings* whose washer raceways have cross-sectional radii not larger than 54 percent of the ball diameter.

4) *Double row, radial and angular contact ball bearings* and double direction thrust ball bearings are presumed to be symmetrical.

Limitations for Ball Bearings.—The following limitations apply:

1) *Truncated contact area.* This standard* may not be safely applied to ball bearings subjected to loading which causes the contact area of the ball with the raceway to be truncated by the raceway shoulder. This limitation depends strongly on details of bearing design which are not standardized.

2) *Material.* This standard applies only to ball bearings fabricated from hardened good quality steel.

3) *Types.* The f_c factors specified in the basic load rating formulas are valid only for those ball bearing types specified above.

4) *Lubrication.* The Rating Life calculated according to this standard is based on the assumption that the bearing is adequately lubricated. The determination of adequate lubrication depends upon the bearing application.

5) *Ring support and alignment.* The Rating Life calculated according to this standard assumes that the bearing inner and outer rings are rigidly supported and the inner and outer ring axes are properly aligned.

6) *Internal clearance.* The radial ball bearing Rating Life calculated according to this standard is based on the assumption that only a nominal interior clearance occurs in the mounted bearing at operating speed, load and temperature.

7) *High speed effects.* The Rating Life calculated according to this standard does not account for high speed effects such as ball centrifugal forces and gyroscopic moments. These effects tend to diminish fatigue life. Analytical evaluation of these effects frequently

* All references to "standard" are to AFBMA and American National Standard "Load Ratings and Fatigue Life for Ball Bearings" ANSI/ABMA 9-1990.

requires the use of high speed digital computation devices and hence is not covered in the standard.

8) *Groove radii.* If groove radii are smaller than those specified in the bearing types covered, the ability of a bearing to resist fatigue is not improved: however, it is diminished by the use of larger radii.

Ball Bearing Rating Life.—According to the Anti-Friction Bearing Manufacturers Association standards the Rating Life L_{10} of a group of apparently identical ball bearings is the life in millions of revolutions that 90 percent of the group will complete or exceed. For a single bearing, L_{10} also refers to the life associated with 90 percent reliability.

Radial and Angular Contact Ball Bearings: The magnitude of the Rating Life L_{10} in millions of revolutions, for a radial or angular contact ball bearing application is given by the formula:

$$L_{10} = \left(\frac{C}{P}\right)^3 \qquad (1)$$

where C = basic load rating, newtons (pounds). See Formulas (2), (3a) and (3b)

P = equivalent radial load, newtons (pounds). See Formula (4)

Table 25. Values of f_c for Radial and Angular Contact Ball Bearings

$\frac{D\cos\alpha}{d_m}$	Single Row Radial Contact; Single and Double Row Angular Contact, Groove Type[a]		Double Row Radial Contact Groove Type		Self-Aligning	
	Values of f_c					
	Metric[b]	Inch[c]	Metric[b]	Inch[c]	Metric[b]	Inch[c]
0.05	46.7	3550	44.2	3360	17.3	1310
0.06	49.1	3730	46.5	3530	18.6	1420
0.07	51.1	3880	48.4	3680	19.9	1510
0.08	52.8	4020	50.0	3810	21.1	1600
0.09	54.3	4130	51.4	3900	22.3	1690
0.10	55.5	4220	52.6	4000	23.4	1770
0.12	57.5	4370	54.5	4140	25.6	1940
0.14	58.8	4470	55.7	4230	27.7	2100
0.16	59.6	4530	56.5	4290	29.7	2260
0.18	59.9	4550	56.8	4310	31.7	2410
0.20	59.9	4550	56.8	4310	33.5	2550
0.22	59.6	4530	56.5	4290	35.2	2680
0.24	59.0	4480	55.9	4250	36.8	2790
0.26	58.2	4420	55.1	4190	38.2	2910
0.28	57.1	4340	54.1	4110	39.4	3000
0.30	56.0	4250	53.0	4030	40.3	3060
0.32	54.6	4160	51.8	3950	40.9	3110
0.34	53.2	4050	50.4	3840	41.2	3130
0.36	51.7	3930	48.9	3730	41.3	3140
0.38	50.0	3800	47.4	3610	41.0	3110
0.40	48.4	3670	45.8	3480	40.4	3070

[a] A. When calculating the basic load rating for a unit consisting of two similar, single row, radial contact ball bearings, in a duplex mounting, the pair is considered as one, double row, radial contact ball bearing.
B. When calculating the basic load rating for a unit consisting of two, similar, single row, angular contact ball bearings in a duplex mounting, "face-to-face" or "back-to-back," the pair is considered as one, double row, angular contact ball bearing.
C. When calculating the basic load rating for a unit consisting of two or more similar, single angular contact ball bearings mounted "in tandem," properly manufactured and mounted for equal load distribution, the rating of the combination is the number of bearings to the 0.7 power times the rating of a single row ball bearing. If the unit may be treated as a number of individually interchangeable single row bearings, this footnote "C" does not apply.

[b] **Use to obtain C in newtons when D is given in mm.**

[c] Use to obtain C in pounds when D is given in inches.

Table 26. Values of X and Y for Computing Equivalent Radial Load P of Radial and Angular Contact Ball Bearings

Contact Angle, α	Table Entering Factors[a]				Single Row Bearings[b] $\frac{F_a}{F_r} > e$		Double Row Bearings $\frac{F_a}{F_r} \leq e$		Double Row Bearings $\frac{F_a}{F_r} > e$	
RADIAL CONTACT GROOVE BEARINGS										
	F_a/C_o	F_a/iZD^2 Metric Units	F_a/iZD^2 Inch Units	e	X	Y	X	Y	X	Y
0°	0.014	0.172	25	0.19	0.56	2.30	1	0	0.56	2.30
	0.028	0.345	50	0.22		1.99				1.99
	0.056	0.689	100	0.26		1.71				1.71
	0.084	1.03	150	0.28		1.56				1.55
	0.11	1.38	200	0.30		1.45				1.45
	0.17	2.07	300	0.34		1.31				1.31
	0.28	3.45	500	0.38		1.15				1.15
	0.42	5.17	750	0.42		1.04				1.04
	0.56	6.89	1000	0.44		1.00				1.00
ANGULAR CONTACT GROOVE BEARINGS										
	iF_a/C_o	F_a/ZD^2 Metric Units	F_a/ZD^2 Inch Units	e	X	Y	X	Y	X	Y
5°	0.014	0.172	25	0.23	For this type use the X, Y, and e values applicable to single row radial contact bearings		1	2.78	0.78	3.74
	0.028	0.345	50	0.26				2.40		3.23
	0.056	0.689	100	0.30				2.07		2.78
	0.085	1.03	150	0.34				1.87		2.52
	0.11	1.38	200	0.36				1.75		2.36
	0.17	2.07	300	0.40				1.58		2.13
	0.28	3.45	500	0.45				1.39		1.87
	0.42	5.17	750	0.50				1.26		1.69
	0.56	6.89	1000	0.52				1.21		1.63
10°	0.014	0.172	25	0.29	0.46	1.88	1	2.18	0.75	3.06
	0.029	0.345	50	0.32		1.71		1.98		2.78
	0.057	0.689	100	0.36		1.52		1.76		2.47
	0.086	1.03	150	0.38		1.41		1.63		2.20
	0.11	1.38	200	0.40		1.34		1.55		2.18
	0.17	2.07	300	0.44		1.23		1.42		2.00
	0.29	3.45	500	0.49		1.10		1.27		1.79
	0.43	5.17	750	0.54		1.01		1.17		1.64
	0.57	6.89	1000	0.54		1.00		1.16		1.63

Table 26. *(Continued)* **Values of *X* and *Y* for Computing Equivalent Radial Load *P* of Radial and Angular Contact Ball Bearings**

	0.015	0.172	25	0.38		1.47		1.65		2.39
	0.029	0.345	50	0.40		1.40		1.57		2.28
	0.058	0.689	100	0.43		1.30		1.46		2.11
	0.087	1.03	150	0.46		1.23		1.38		2.00
15°	0.12	1.38	200	0.47	0.44	1.19	1	1.34	0.72	1.93
	0.17	2.07	300	0.50		1.12		1.26		1.82
	0.29	3.45	500	0.55		1.02		1.14		1.66
	0.44	5.17	750	0.56		1.00		1.12		1.63
	0.58	6.89	1000	0.56		1.00		1.12		1.63
20°	…	…	…	0.57	0.43	1.00	1	1.09	0.70	1.63
25°	…	…	…	0.68	0.41	0.87	1	0.92	0.67	1.41
30°	…	…	…	0.80	0.39	0.76	1	0.78	0.63	1.24
35°	…	…	…	0.95	0.37	0.66	1	0.66	0.60	1.07
40°	…	…	…	1.14	0.35	0.57	1	0.55	0.57	0.98
Self-aligning Ball Bearings				1.5 tan α	0.40	0.4 cot α	1	0.42 cot α	0.65	0.65 cot α

[a] Symbol definitions are given on the following page.

[b] For single row bearings when $F_a/F_r \leq e$, use $X = 1$, $Y = 0$. Two similar, single row, angular contact ball bearings mounted face-to-face or back-to-back are considered as one double row, angular contact bearing.

Values of *X*, *Y*, and *e* for a load or contact angle other than shown are obtained by linear interpolation. Values of *X*, *Y*, and *e* do not apply to filling slot bearings for applications in which ball-raceway contact areas project substantially into the filling slot under load. Symbol Definitions: F_a is the applied axial load in newtons (pounds); C_o is the static load rating in newtons (pounds) of the bearing under consideration and is found by Formula (20); *i* is the number of rows of balls in the bearing; *Z* is the number of balls per row in a radial or angular contact bearing or the number of balls in a single row, single direction thrust bearing; *D* is the ball diameter in millimeters (inches); and F_r is the applied radial load in newtons (pounds).

For radial and angular contact ball bearings with balls not larger than 25.4 mm (1 inch) in diameter, *C* is found by the formula:

$$C = f_c(i\cos\alpha)^{0.7}Z^{2/3}D^{1.8} \qquad (2)$$

and with balls larger than 25.4 mm (1 inch) in diameter *C* is found by the formula:

$$C = 3.647f_c(i\cos\alpha)^{0.7}Z^{2/3}D^{1.4} \quad \text{(metric)} \qquad (3a)$$

$$C = f_c(i\cos\alpha)^{0.7}Z^{2/3}D^{1.4} \text{ (inch)} \qquad (3b)$$

where f_c = a factor which depends on the geometry of the bearing components, the accuracy to which the various bearing parts are made and the material. Values of f_c, are given in Table 25

i = number of rows of balls in the bearing

α = nominal contact angle, degrees

Z = number of balls per row in a radial or angular contact bearing

D = ball diameter, mm (inches)

The magnitude of the equivalent radial load, *P*, in newtons (pounds) for radial and angular contact ball bearings, under combined constant radial and constant thrust loads is given by the formula:

$$P = XF_r + YF_a \qquad (4)$$

where F_r = the applied radial load in newtons (pounds)

F_a = the applied axial load in newtons (pounds)

X = radial load factor as given in Table 28

Y = axial load factor as given in Table 28

Thrust Ball Bearings: The magnitude of the Rating Life L_{10} in millions of revolutions for a thrust ball bearing application is given by the formula:

$$L_{10} = \left(\frac{C_a}{P_a}\right)^3 \qquad (5)$$

where C_a = the basic load rating, newtons (pounds). See Formulas (6) to (10)

P_a = equivalent thrust load, newtons (pounds). See Formula (11)

For single row, single and double direction, thrust ball bearing with balls not larger than 25.4 mm (1 inch) in diameter, C_a is found by the formulas:

$$\text{for } \alpha = 90 \text{ degrees,} \quad C_a = f_c Z^{2/3} D^{1.8} \qquad (6)$$

$$\text{for } \alpha \neq 90 \text{ degrees,} \quad C_a = f_c (\cos\alpha)^{0.7} Z^{2/3} D^{1.8} \tan\alpha \qquad (7)$$

and with balls larger than 25.4 mm (1 inch) in diameter, C_a is found by the formulas:

$$\text{for } \alpha = 90 \text{ degrees,} \quad C_a = 3.647 f_c Z^{2/3} D^{1.4} \quad \text{(metric)} \qquad (8a)$$

$$C_a = f_c Z^{2/3} D^{1.4} \quad \text{(inch)} \qquad (8b)$$

$$\text{for } \alpha \neq 90 \text{ degrees,} \quad C_a = 3.647 f_c (\cos\alpha)^{0.7} Z^{2/3} D^{1.4} \tan\alpha \quad \text{(metric)} \qquad (9a)$$

$$C_a = f_c (\cos\alpha)^{0.7} Z^{2/3} D^{1.4} \tan\alpha \quad \text{(inch)} \qquad (9b)$$

where f_c = a factor which depends on the geometry of the bearing components, the accuracy to which the various bearing parts are made, and the material. Values of f_c are given in Table 27

Z = number of balls per row in a single row, single direction thrust ball bearing

D = ball diameter, mm (inches)

α = nominal contact angle, degrees

Table 27. Values of f_c for Thrust Ball Bearings

$\frac{D}{d_m}$	α = 90°		$D \cos \alpha$	α = 45°		α = 60°		α = 75°	
	Metric[a]	Inch[b]		Metric[a]	Inch[b]	Metric[a]	Inch[b]	Metric[a]	Inch[b]
0.01	36.7	2790	0.01	42.1	3200	39.2	2970	37.3	2840
0.02	45.2	3430	0.02	51.7	3930	48.1	3650	45.9	3490
0.03	51.1	3880	0.03	58.2	4430	54.2	4120	51.7	3930
0.04	55.7	4230	0.04	63.3	4810	58.9	4470	56.1	4260
0.05	59.5	4520	0.05	67.3	5110	62.6	4760	59.7	4540
0.06	62.9	4780	0.06	70.7	5360	65.8	4990	62.7	4760
0.07	65.8	5000	0.07	73.5	5580	68.4	5190	65.2	4950
0.08	68.5	5210	0.08	75.9	5770	70.7	5360	67.3	5120
0.09	71.0	5390	0.09	78.0	5920	72.6	5510	69.2	5250
0.10	73.3	5570	0.10	79.7	6050	74.2	5630	70.7	5370
0.12	77.4	5880	0.12	82.3	6260	76.6	5830	…	…
0.14	81.1	6160	0.14	84.1	6390	78.3	5950	…	…
0.16	84.4	6410	0.16	85.1	6470	79.2	6020	…	…
0.18	87.4	6640	0.18	85.5	6500	79.6	6050	…	…
0.20	90.2	6854	0.20	85.4	6490	79.5	6040	…	…
0.22	92.8	7060	0.22	84.9	6450	…	…	…	…
0.24	95.3	7240	0.24	84.0	6380	…	…	…	…
0.26	97.6	7410	0.26	82.8	6290	…	…	…	…
0.28	99.8	7600	0.28	81.3	6180	…	…	…	…
0.30	101.9	7750	0.30	79.6	6040	…	…	…	…
0.32	103.9	7900	…	…	…	…	…	…	…
0.34	105.8	8050	…	…	…	…	…	…	…

[a] **Use to obtain C_a in newtons when D is given in mm.**

[b] Use to obtain C_a in pounds when D is given in inches.

For thrust ball bearings with two or more rows of similar balls carrying loads in the same direction, the basic load rating, C_a, in newtons (pounds) is found by the formula:

$$C_a = (Z_1 + Z_2 + \dots Z_n)\left[\left(\frac{Z_1}{C_{a1}}\right)^{10/3} + \left(\frac{Z_2}{C_{a2}}\right)^{10/3} + \dots \left(\frac{Z_n}{C_{an}}\right)^{10/3}\right]^{-0.3} \quad (10)$$

where $Z_1, Z_2 \dots Z_n$ = number of balls in respective rows of a single-direction multi-row thrust ball bearing

$C_{a1}, C_{a2} \dots C_{an}$ = basic load rating per row of a single-direction, multi-row thrust ball bearing, each calculated as a single-row bearing with $Z_1, Z_2 \dots Z_n$ balls, respectively

The magnitude of the equivalent thrust load, P_a, in newtons (pounds) for thrust ball bearings with $\alpha \neq 90$ degrees under combined constant thrust and constant radial loads is found by the formula:

$$P_a = XF_r + YF_a \quad (11)$$

where F_r = the applied radial load in newtons (pounds)
F_a = the applied axial load in newtons (pounds)
X = radial load factor as given in Table 28
Y = axial load factor as given in Table 28

Table 28. Values of *X* and *Y* for Computing Equivalent Thrust Load P_a for Thrust Ball Bearings

Contact Angle		Single Direction Bearings		Double Direction Bearings			
		$\frac{F_a}{F_r} > e$		$\frac{F_a}{F_r} \le e$		$\frac{F_a}{F_r} > e$	
α	e	X	Y	X	Y	X	Y
45°	1.25	0.66	1	1.18	0.59	0.66	1
60°	2.17	0.92	1	1.90	0.54	0.92	1
75°	4.67	1.66	1	3.89	0.52	1.66	1

For $\alpha = 90°$, $F_r = 0$ and $Y = 1$.

Roller Bearing Types Covered.—This standard* applies to *cylindrical, tapered and self-aligning radial and thrust roller bearings* and to *needle roller bearings.* These bearings are presumed to be within the size ranges shown in the AFBMA dimensional standards, of good quality and produced in accordance with good manufacturing practice.

Roller bearings vary considerably in design and execution. Since small differences in relative shape of contacting surfaces may account for distinct differences in load carrying ability, this standard does not attempt to cover all design variations, rather it applies to basic roller bearing designs.

The following limitations apply:

1) *Truncated contact area.* This standard may not be safely applied to roller bearings subjected to application conditions which cause the contact area of the roller with the raceway to be severely truncated by the edge of the raceway or roller.

2) *Stress concentrations.* A cylindrical, tapered or self-aligning roller bearing must be expected to have a basic load rating less than that obtained using a value of f_c taken from Table 29 or Table 30 if, under load, a stress concentration is present in some part of the roller-raceway contact. Such stress concentrations occur in the center of nominal point contacts, at the contact extremities for line contacts and at inadequately blended junctions of a rolling surface profile. Stress concentrations can also occur if the rollers are not accu-

* All references to "standard" are to AFBMA and American National Standard "Load Ratings and Fatigue Life for Roller Bearings" ANSI/AFBMA Std 11–1990.

rately guided such as in bearings without cages and bearings not having rigid integral flanges. Values of f_c given in Tables 29 and 30 are based upon bearings manufactured to achieve optimized contact. For no bearing type or execution will the factor f_c be greater than that obtained in Tables 29 and 30.

3) *Material.* This standard applies only to roller bearings fabricated from hardened, good quality steel.

4) *Lubrication.* Rating Life calculated according to this standard is based on the assumption that the bearing is adequately lubricated. Determination of adequate lubrication depends upon the bearing application.

5) *Ring support and alignment.* Rating Life calculated according to this standard assumes that the bearing inner and outer rings are rigidly supported, and that the inner and outer ring axes are properly aligned.

6) *Internal clearance.* Radial roller bearing Rating Life calculated according to this standard is based on the assumption that only a nominal internal clearance occurs in the mounted bearing at operating speed, load, and temperature.

7) *High speed effects.* The Rating Life calculated according to this standard does not account for high speed effects such as roller centrifugal forces and gyroscopic moments: These effects tend to diminish fatigue life. Analytical evaluation of these effects frequently requires the use of high speed digital computation devices and hence, cannot be included.

Table 29. Values of f_c for Radial Roller Bearings

$\frac{D\cos\alpha}{d_m}$	f_c Metric[a]	f_c Inch[b]	$\frac{D\cos\alpha}{d_m}$	f_c Metric[a]	f_c Inch[b]	$\frac{D\cos\alpha}{d_m}$	f_c Metric[a]	f_c Inch[b]
0.01	52.1	4680	0.18	88.8	7980	0.35	79.5	7140
0.02	60.8	5460	0.19	88.8	7980	0.36	78.6	7060
0.03	66.5	5970	0.20	88.7	7970	0.37	77.6	6970
0.04	70.7	6350	0.21	88.5	7950	0.38	76.7	6890
0.05	74.1	6660	0.22	88.2	7920	0.39	75.7	6800
0.06	76.9	6910	0.23	87.9	7890	0.40	74.6	6700
0.07	79.2	7120	0.24	87.5	7850	0.41	73.6	6610
0.08	81.2	7290	0.25	87.0	7810	0.42	72.5	6510
0.09	82.8	7440	0.26	86.4	7760	0.43	71.4	6420
0.10	84.2	7570	0.27	85.8	7710	0.44	70.3	6320
0.11	85.4	7670	0.28	85.2	7650	0.45	69.2	6220
0.12	86.4	7760	0.29	84.5	7590	0.46	68.1	6120
0.13	87.1	7830	0.30	83.8	7520	0.47	67.0	6010
0.14	87.7	7880	0.31	83.0	7450	0.48	65.8	5910
0.15	88.2	7920	0.32	82.2	7380	0.49	64.6	5810
0.16	88.5	7950	0.33	81.3	7300	0.50	63.5	5700
0.17	88.7	7970	0.34	80.4	7230	…	…	…

[a] For $\alpha = 0°$, $F_a = 0$ and $X = 1$.

[b] Use to obtain C in pounds when l_{eff} and D are given in inches.

Table 30. Values of f_c for Thrust Roller Bearings

$\frac{D\cos\alpha}{d_m}$	$45° < \alpha < 60°$ f_c Metric[a]	$45° < \alpha < 60°$ f_c Inch[b]	$60° < \alpha < 75°$ f_c Metric[a]	$60° < \alpha < 75°$ f_c Inch[b]	$75° \leq \alpha < 90°$ f_c Metric[a]	$75° \leq \alpha < 90°$ f_c Inch[b]	$\alpha = 90°$ $\frac{D}{d_m}$	$\alpha = 90°$ f_c Metric[a]	$\alpha = 90°$ f_c Inch[b]
0.01	109.7	9840	107.1	9610	105.6	9470	0.01	105.4	9500
0.02	127.8	11460	124.7	11180	123.0	11030	0.02	122.9	11000
0.03	139.5	12510	136.2	12220	134.3	12050	0.03	134.5	12100
0.04	148.3	13300	144.7	12980	142.8	12810	0.04	143.4	12800
0.05	155.2	13920	151.5	13590	149.4	13400	0.05	150.7	13200
0.06	160.9	14430	157.0	14080	154.9	13890	0.06	156.9	14100

Table 30. *(Continued)* **Values of f_c for Thrust Roller Bearings**

$\frac{D\cos\alpha}{d_m}$	45° < α < 60°		60° < α < 75°		75° ≤ α < 90°		$\frac{D}{d_m}$	α = 90°	
	f_c							f_c	
	Metric[a]	Inch[b]	Metric[a]	Inch[b]	Metric[a]	Inch[b]		Metric[a]	Inch[b]
0.07	165.6	14850	161.6	14490	159.4	14300	0.07	162.4	14500
0.08	169.5	15200	165.5	14840	163.2	14640	0.08	167.2	15100
0.09	172.8	15500	168.7	15130	166.4	14930	0.09	171.7	15400
0.10	175.5	15740	171.4	15370	169.0	15160	0.10	175.7	15900
0.12	179.7	16120	175.4	15730	173.0	15520	0.12	183.0	16300
0.14	182.3	16350	177.9	15960	175.5	15740	0.14	189.4	17000
0.16	183.7	16480	179.3	16080	…	…	0.16	195.1	17500
0.18	184.1	16510	179.7	16120	…	…	0.18	200.3	18000
0.20	183.7	16480	179.3	16080	…	…	0.20	205.0	18500
0.22	182.6	16380	…	…	…	…	0.22	209.4	18800
0.24	180.9	16230	…	…	…	…	0.24	213.5	19100
0.26	178.7	16030	…	…	…	…	0.26	217.3	19600
0.28	…	…	…	…	…	…	0.28	220.9	19900
0.30	…	…	…	…	…	…	0.30	224.3	20100

[a] **Use to obtain C_a in newtons when l_{eff} and D are given in mm.**

[b] Use to obtain C_a in pounds when l_{eff} and D are given in inches.

Roller Bearing Rating Life.—The Rating Life L_{10} of a group of apparently identical roller bearings is the life in millions of revolutions that 90 percent of the group will complete or exceed. For a single bearing, L_{10} also refers to the life associated with 90 percent reliability.

Radial Roller Bearings: The magnitude of the Rating Life, L_{10}, in millions of revolutions, for a radial roller bearing application is given by the formula:

$$L_{10} = \left(\frac{C}{P}\right)^{10/3} \tag{12}$$

where C = the basic load rating in newtons (pounds), see Formula (13); and, P = equivalent radial load in newtons (pounds), see Formula (14).

For radial roller bearings, C is found by the formula:

$$C = f_c(il_{eff}\cos\alpha)^{7/9}Z^{3/4}D^{29/27} \tag{13}$$

where f_c = a factor which depends on the geometry of the bearing components, the accuracy to which the various bearing parts are made, and the material. Maximum values of f_c are given in Table 29

i = number of rows of rollers in the bearing

l_{eff} = effective length, mm (inches) α = nominal contact angle, degrees

Z = number of rollers per row in a radial roller bearing

D = roller diameter, mm (inches) (mean diameter for a tapered roller, major diameter for a spherical roller)

When rollers are longer than 2.5D, a reduction in the f_c value must be anticipated. In this case, the bearing manufacturer may be expected to establish load ratings accordingly.

In applications where rollers operate directly on a shaft surface or a housing surface, such a surface must be equivalent in all respects to the raceway it replaces to achieve the basic load rating of the bearing.

When calculating the basic load rating for a unit consisting of two or more similar single-row bearings mounted "in tandem," properly manufactured and mounted for equal load distribution, the rating of the combination is the number of bearings to the 7/9 power times the rating of a single-row bearing. If, for some technical reason, the unit may be treated as a number of individually interchangeable single-row bearings, this consideration does not apply.

The magnitude of the equivalent radial load, P, in newtons (pounds), for radial roller bearings, under combined constant radial and constant thrust loads is given by the formula:

$$P = XF_r + YF_a \qquad (14)$$

where F_r = the applied radial load in newtons (pounds)

F_a = the applied axial load in newtons (pounds)

X = radial load factor as given in Table 31

Y = axial load factor as given in Table 31

Typical Bearing Life for Various Design Applications

Uses	Design life in hours	Uses	Design life in hours
Agricultural equipment	3000 – 6000	Gearing units	
Aircraft equipment	500 – 2000	Automotive	600 – 5000
Automotive		Multipurpose	8000 – 15000
Race car	500 – 800	Machine tools	20000
Light motor cycle	600 – 1200	Rail Vehicles	15000 – 25000
Heavy motor cycle	1000 – 2000	Heavy rolling mill	> 50000
Light cars	1000 – 2000	Machines	
Heavy cars	1500 – 2500	Beater mills	20000 – 30000
Light trucks	1500 – 2500	Briquette presses	20000 – 30000
Heavy trucks	2000 – 2500	Grinding spindles	1000 – 2000
Buses	2000 – 5000	Machine tools	10000 – 30000
Electrical		Mining machinery	4000 – 15000
Household appliances	1000 – 2000	Paper machines	50000 – 80000
Motors ≤ ½ hp	1000 – 2000	Rolling mills	
Motors ≤ 3 hp	8000 – 10000	Small cold mills	5000 – 6000
Motors, medium	10000 – 15000	Large multipurpose mills	8000 – 10000
Motors, large	20000 – 30000	Rail vehicle axle	
Elevator cables sheaves	40000 – 60000	Mining cars	5000
Mine ventillation fans	40000 – 50000	Motor rail cars	16000 – 20000
Propeller thrust bearings	15000 – 25000	Open–pit mining cars	20000 – 25000
Propeller shaft bearings	> 80000	Streetcars	20000 – 25000
Gear drives		Passenger cars	26000
Boat gearing units	3000 – 5000	Freight cars	35000
Gear drives	> 50000	Locomotive outer bearings	20000 – 25000
Ship gear drives	20000 – 30000	Locomotive inner bearings	30000 – 40000
Machinery for 8 hour service which are not always fully utilized	14000 – 20000	Machinery for short or intermittent opearation where service interruption is of minor importance	4000 – 8000
Machinery for 8 hour service which are fully utilized	20000 – 30000	Machinery for intermittent service where reliable opeara-tion is of great importance	8000 – 14000
Machinery for continuous 24 hour service	50000 – 60000	Instruments and apparatus in frequent use	0 – 500

Table 31. Values of *X* and *Y* for Computing Equivalent Radial Load *P* for Radial Roller Bearing

	$\frac{F_a}{F_r} \le e$		$\frac{F_a}{F_r} > e$	
Bearing Type	*X*	*Y*	*X*	*Y*
Self-Aligning and Tapered Roller Bearings[a] $\alpha \neq 0°$	Single Row Bearings			
	1	0	0.4	0.4 cot α
	Double Row Bearings[a]			
	1	0.45 cot α	0.67	0.67 cot α

[a] For $\alpha = 0°$, $F_a = 0$ and $X = 1$.

$e = 1.5 \tan \alpha$

Roller bearings are generally designed to achieve optimized contact; however, they usually support loads other than the loading at which optimized contact is maintained. The 10⁄3 exponent in Rating Life Formulas (12) and (15) was selected to yield satisfactory Rating Life estimates for a broad spectrum from light to heavy loading. When loading exceeds that which develops optimized contact, e.g., loading greater than $C/4$ to $C/2$ or $C_a/4$ to $C_a/2$, the user should consult the bearing manufacturer to establish the adequacy of the Rating Life formulas for the particular application.

Thrust Roller Bearings: The magnitude of the Rating Life, L_{10}, in millions of revolutions for a thrust roller bearing application is given by the formula:

$$L_{10} = \left(\frac{C_a}{P_a}\right)^{10/3} \qquad (15)$$

where C_a = basic load rating, newtons (pounds). See Formulas (16) to (18)

P_a = equivalent thrust load, newtons (pounds). See Formula (19)

For single row, single and double direction, thrust roller bearings, the magnitude of the basic load rating, C_a, in newtons (pounds), is found by the formulas:

$$\text{for } \alpha = 90°,\ C_a = f_c l_{eff}^{7/9} Z^{3/4} D^{29/27} \qquad (16)$$

$$\text{for } \alpha \neq 90°,\ C_a = f_c (l_{eff} \cos\alpha)^{7/9} Z^{3/4} D^{29/27} \tan\alpha \qquad (17)$$

where f_c = a factor which depends on the geometry of the bearing components, the accuracy to which the various parts are made, and the material. Values of f_c are given in Table 30

l_{eff} = effective length, mm (inches)

Z = number of rollers in a single row, single direction, thrust roller bearing

D = roller diameter, mm (inches) (mean diameter for a tapered roller, major diameter for a spherical roller)

α = nominal contact angle, degrees

For thrust roller bearings with two or more rows of rollers carrying loads in the same direction the magnitude of C_a is found by the formula:

$$C_a = (Z_1 l_{eff1} + Z_2 l_{eff2} \ldots Z_n l_{effn}) \left\{ \left[\frac{Z_1 l_{eff1}}{C_{a1}}\right]^{9/2} + \left[\frac{Z_2 l_{eff2}}{C_{a2}}\right]^{9/2} + \ldots \left[\frac{Z_n l_{effn}}{C_{an}}\right]^{9/2} \right\}^{-2/9} \qquad (18)$$

Where $Z_1, Z_2 . . . Z_n$ = the number of rollers in respective rows of a single direction, multi-row bearing

$C_{a1}, C_{a2} .... C_{an}$ = the basic load rating per row of a single direction, multi-row, thrust roller bearing, each calculated as a single row bearing with $Z_1, Z_2 ... Z_n$ rollers respectively

$l_{eff1}, l_{eff2} ... l_{effn}$ = effective length, mm (inches), or rollers in the respective rows

In applications where rollers operate directly on a surface supplied by the user, such a surface must be equivalent in all respects to the washer raceway it replaces to achieve the basic load rating of the bearing.

In case the bearing is so designed that several rollers are located on a common axis, these rollers are considered as one roller of a length equal to the total effective length of contact of the several rollers. Rollers as defined above, or portions thereof which contact the same washer-raceway area, belong to one row.

When the ratio of the individual roller effective length to the pitch diameter (at which this roller operates) is too large, a reduction of the f_c value must be anticipated due to excessive slip in the roller-raceway contact.

When calculating the basic load rating for a unit consisting of two or more similar single row bearings mounted "in tandem," properly manufactured and mounted for equal load distribution, the rating of the combination is defined by Formula (18). If, for some technical reason, the unit may be treated as a number of individually interchangeable single-row bearings, this consideration does not apply.

The magnitude of the equivalent thrust load, P_a, in pounds, for thrust roller bearings with α not equal to 90 degrees under combined constant thrust and constant radial loads is given by the formula:

$$P_a = XF_r + YF_a \qquad (19)$$

where F_r = applied radial load, newtons (pounds)

F_a = applied axial load, newtons (pounds)

X = radial load factor as given in Table 32

Y = axial load factor as given in Table 32

Table 32. Values of X and Y for Computing Equivalent Thrust Load P_a for Thrust Roller Bearings

Bearing Type	Single Direction Bearings		Double Direction Bearings			
	$\frac{F_a}{F_r} > e$		$\frac{F_a}{F_r} \le e$		$\frac{F_a}{F_r} > e$	
	X	Y	X	Y	X	Y
Self-Aligning Tapered Thrust Roller Bearings[a] $\alpha \neq 0$	tan α	1	1.5 tan α	0.67	tan α	1

[a] For $\alpha = 90°$, $F_r = 0$ and $Y = 1$.

$e = 1.5 \tan \alpha$

Life Adjustment Factors.—In certain applications of ball or roller bearings it is desirable to specify life for a reliability other than 90 per cent. In other cases the bearings may be fabricated from special bearing steels such as vacuum-degassed and vacuum-melted steels, and improved processing techniques. Finally, application conditions may indicate other than normal lubrication, load distribution, or temperature. For such conditions a series of life adjustment factors may be applied to the fatigue life formula. This is fully explained in AFBMA and American National Standard "Load Ratings and Fatigue Life for Ball Bearings"ANSI/AFBMA Std 9–1990 and AFBMA and American National Standard "Load Ratings and Fatigue Life for Roller Bearings"ANSI/AFBMA Std 11–1990. In addition to consulting these standards it may be advantageous to also obtain information from the bearing manufacturer.

Life Adjustment Factor for Reliability.—For certain applications, it is desirable to specify life for a reliability greater than 90 per cent which is the basis of the Rating Life.

To determine the bearing life of ball or roller bearings for reliability greater than 90 per cent, the Rating Life must be adjusted by a factor a_1 such that $L_n = a_1 L_{10}$. For a reliability of 95 per cent, designated as L_5, the life adjustment factor a_1 is 0.62; for 96 per cent, L_4, a_1 is 0.53; for 97 per cent, L_3, a_1 is 0.44; for 98 per cent, L_2, a_1 is 0.33; and for 99 per cent, L_1, a_1 is 0.21.

Life Adjustment Factor for Material.—For certain types of ball or roller bearings which incorporate improved materials and processing, the Rating Life can be adjusted by a factor a_2 such that $L_{10}' = a_2 L_{10}$. Factor a_2 depends upon steel analysis, metallurgical processes, forming methods, heat treatment, and manufacturing methods in general. Ball and roller bearings fabricated from consumable vacuum remelted steels and certain other special analysis steels, have demonstrated extraordinarily long endurance. These steels are of exceptionally high quality, and bearings fabricated from these are usually considered special manufacture. Generally, a_2 values for such steels can be obtained from the bearing manufacturer. However, all of the specified limitations and qualifications for the application of the Rating Life formulas still apply.

Life Adjustment Factor for Application Condition.—Application conditions which affect ball or roller bearing life include: 1) lubrication; 2) load distribution (including effects of clearance, misalignment, housing and shaft stiffness, type of loading, and thermal gradients); and 3) temperature.

Items 2 and 3 require special analytical and experimental techniques, therefore the user should consult the bearing manufacturer for evaluations and recommendations.

Operating conditions where the factor a_3 might be less than 1 include: D) exceptionally low values ofNd_m (rpm times pitch diameter, in mm); e.g.,$Nd_m < 10{,}000$; E) lubricant viscosity at less than 70 SSU for ball bearings and 100 SSU for roller bearings at operating temperature; and F) excessively high operating temperatures.

When a_3 is less than 1 it may not be assumed that the deficiency in lubrication can be overcome by using an improved steel. When this factor is applied, $L_{10}' = a_3 L_{10}$.

In most ball and roller bearing applications, lubrication is required to separate the rolling surfaces, i.e., rollers and raceways, to reduce the retainer-roller and retainer-land friction and sometimes to act as a coolant to remove heat generated by the bearing.

Factor Combinations.—A fatigue life formula embodying the foregoing life adjustment factors is $L_{10}' = a_1 a_2 a_3 L_{10}$. Indiscriminate application of the life adjustment factors in this formula may lead to serious overestimation of bearing endurance, since fatigue life is only one criterion for bearing selection. Care must be exercised to select bearings which are of sufficient size for the application.

Ball Bearing Static Load Rating.—For ball bearings suitably manufactured from hardened alloy steels, the static radial load rating is that uniformly distributed static radial bearing load which produces a maximum contact stress of 4,000 megapascals (580,000 pounds per square inch). In the case of a single row, angular contact ball bearing, the static radial load rating refers to the radial component of that load which causes a purely radial displacement of the bearing rings in relation to each other. The static axial load rating is that uniformly distributed static centric axial load which produces a maximum contact stress of 4,000 megapascals (580,000 pounds per square inch).

Radial and Angular Contact Groove Ball Bearings: The magnitude of the static load rating C_o in newtons (pounds) for radial ball bearings is found by the formula:

$$C_o = f_o i Z D^2 \cos\alpha \quad (20)$$

where f_o = a factor for different kinds of ball bearings given in Table 33

i = number of rows of balls in bearing

Z = number of balls per row

D = ball diameter, mm (inches)

α = nominal contact angle, degrees

This formula applies to bearings with a cross sectional raceway groove radius not larger than 0.52 D in radial and angular contact groove ball bearing inner rings and 0.53 D in radial and angular contact groove ball bearing outer rings and self-aligning ball bearing inner rings.

The load carrying ability of a ball bearing is not necessarily increased by the use of a smaller groove radius but is reduced by the use of a larger radius than those indicated above.

Radial or Angular Contact Ball Bearing Combinations: The basic static load rating for two similar single row radial or angular contact ball bearings mounted side by side on the same shaft such that they operate as a unit (duplex mounting) in "back-to-back" or "face-to-face" arrangement is two times the rating of one single row bearing.

The basic static radial load rating for two or more single row radial or angular contact ball bearings mounted side by side on the same shaft such that they operate as a unit (duplex or stack mounting) in "tandem" arrangement, properly manufactured and mounted for equal load distribution, is the number of bearings times the rating of one single row bearing.

Thrust Ball Bearings: The magnitude of the static load rating C_{oa} for thrust ball bearings is found by the formula:

$$C_{oa} = f_o Z D^2 \cos\alpha \quad (21)$$

where f_o = a factor given in Table 33

Z = number of balls carrying the load in one direction

D = ball diameter, mm (inches)

α = nominal contact angle, degrees

This formula applies to thrust ball bearings with a cross sectional raceway radius not larger than 0.54 D. The load carrying ability of a bearing is not necessarily increased by use of a smaller radius, but is reduced by use of a larger radius.

Roller Bearing Static Load Rating: For roller bearings suitably manufactured from hardened alloy steels, the static radial load rating is that uniformly distributed static radial bearing load which produces a maximum contact stress of 4,000 megapascals (580,000 pounds per square inch) acting at the center of contact of the most heavily loaded rolling element. The static axial load rating is that uniformly distributed static centric axial load which produces a maximum contact stress of 4,000 megapascals (580,000 pounds per square inch) acting at the center of contact of each rolling element.

Table 33. Values of f_o for Calculating Static Load Rating for Ball Bearings

$\frac{D\cos\alpha}{d_m}$	Radial and Angular Contact Groove Type		Radial Self-Aligning		Thrust	
	Metric[a]	Inch[b]	Metric[a]	Inch[b]	Metric[a]	Inch[b]
0.00	12.7	1850	1.3	187	51.9	7730
0.01	13.0	1880	1.3	191	52.6	7620
0.02	13.2	1920	1.3	195	51.7	7500
0.03	13.5	1960	1.4	198	50.9	7380
0.04	13.7	1990	1.4	202	50.2	7280
0.05	14.0	2030	1.4	206	49.6	7190
0.06	14.3	2070	1.5	210	48.9	7090
0.07	14.5	2100	1.5	214	48.3	7000
0.08	14.7	2140	1.5	218	47.6	6900
0.09	14.5	2110	1.5	222	46.9	6800
0.10	14.3	2080	1.6	226	46.4	6730
0.11	14.1	2050	1.6	231	45.9	6660
0.12	13.9	2020	1.6	235	45.5	6590
0.13	13.6	1980	1.7	239	44.7	6480
0.14	13.4	1950	1.7	243	44.0	6380
0.15	13.2	1920	1.7	247	43.3	6280
0.16	13.0	1890	1.7	252	42.6	6180
0.17	12.7	1850	1.8	256	41.9	6070
0.18	12.5	1820	1.8	261	41.2	5970
0.19	12.3	1790	1.8	265	40.4	5860
0.20	12.1	1760	1.9	269	39.7	5760
0.21	11.9	1730	1.9	274	39.0	5650
0.22	11.6	1690	1.9	278	38.3	5550
0.23	11.4	1660	2.0	283	37.5	5440
0.24	11.2	1630	2.0	288	37.0	5360
0.25	11.0	1600	2.0	293	36.4	5280
0.26	10.8	1570	2.1	297	35.8	5190
0.27	10.6	1540	2.1	302	35.0	5080
0.28	10.4	1510	2.1	307	34.4	4980
0.29	10.3	1490	2.1	311	33.7	4890
0.30	10.1	1460	2.2	316	33.2	4810
0.31	9.9	1440	2.2	321	32.7	4740
0.32	9.7	1410	2.3	326	32.0	4640
0.33	9.5	1380	2.3	331	31.2	4530
0.34	9.3	1350	2.3	336	30.5	4420
0.35	9.1	1320	2.4	341	30.0	4350
0.36	8.9	1290	2.4	346	29.5	4270
0.37	8.7	1260	2.4	351	28.8	4170
0.38	8.5	1240	2.5	356	28.0	4060
0.39	8.3	1210	2.5	361	27.2	3950
0.40	8.1	1180	2.5	367	26.8	3880
0.41	8.0	1160	2.6	372	26.2	3800
0.42	7.8	1130	2.6	377	25.7	3720
0.43	7.6	1100	2.6	383	25.1	3640
0.44	7.4	1080	2.7	388	24.6	3560
0.45	7.2	1050	2.7	393	24.0	3480
0.46	7.1	1030	2.8	399	23.5	3400
0.47	6.9	1000	2.8	404	22.9	3320
0.48	6.7	977	2.8	410	22.4	3240
0.49	6.6	952	2.9	415	21.8	3160
0.50	6.4	927	2.9	421	21.2	3080

[a] Use to obtain C_o or C_{oa} in newtons when D is given in mm.

[b] Use to obtain C_o or C_{oa} in pounds when D is given in inches.

Note: Based on modulus of elasticity = 2.07×10^5 megapascals (30×10^6 pounds per square inch) and Poisson's ratio = 0.3.

Radial Roller Bearings: The magnitude of the static load rating C_o in newtons (pounds) for radial roller bearings is found by the formulas:

$$C_o = 44\left(1 - \frac{D\cos\alpha}{d_m}\right)iZl_{eff}D\cos\alpha \qquad \text{(metric)} \qquad (22a)$$

$$C_o = 6430\left(1 - \frac{D\cos\alpha}{d_m}\right)iZl_{eff}D\cos\alpha \qquad \text{(inch)} \qquad (22b)$$

where D = roller diameter, mm (inches); mean diameter for a tapered roller and major diameter for a spherical roller

d_m = mean pitch diameter of the roller complement, mm (inches)

i = number of rows of rollers in bearing

Z = number of rollers per row

l_{eff} = effective length, mm (inches); overall roller length minus roller chamfers or minus grinding undercuts at the ring where contact is shortest

α = nominal contact angle, degrees

Radial Roller Bearing Combinations: The static load rating for two similar single row roller bearings mounted side by side on the same shaft such that they operate as a unit is two times the rating of one single row bearing.

The static radial load rating for two or more similar single row roller bearings mounted side by side on the same shaft such that they operate as a unit (duplex or stack mounting) in "tandem" arrangement, properly manufactured and mounted for equal load distribution, is the number of bearings times the rating of one single row bearing.

Thrust Roller Bearings: The magnitude of the static load rating C_{oa} in newtons (pounds) for thrust roller bearings is found by the formulas:

$$C_{oa} = 220\left(1 - \frac{D\cos\alpha}{d_m}\right)Zl_{eff}D\sin\alpha \qquad \text{(metric)} \qquad (23a)$$

$$C_{oa} = 32150\left(1 - \frac{D\cos\alpha}{d_m}\right)Zl_{eff}D\sin\alpha \qquad \text{(inch)} \qquad (23b)$$

where the symbol definitions are the same as for Formulas (22a) and (22b).

Thrust Roller Bearing Combination: The static axial load rating for two or more similar single direction thrust roller bearings mounted side by side on the same shaft such that they operate as a unit (duplex or stack mounting) in "tandem" arrangement, properly manufactured and mounted for equal load distribution, is the number of bearings times the rating of one single direction bearing. The accuracy of this formula decreases in the case of single direction bearings when $F_r > 0.44\ F_a \cot\alpha$ where F_r is the applied radial load in newtons (pounds) and F_a is the applied axial load in newtons (pounds).

Ball Bearing Static Equivalent Load.—For ball bearings the static equivalent radial load is that calculated static radial load which produces a maximum contact stress equal in magnitude to the maximum contact stress in the actual condition of loading. The static equivalent axial load is that calculated static centric axial load which produces a maximum contact stress equal in magnitude to the maximum contact stress in the actual condition of loading.

Radial and Angular Contact Ball Bearings: The magnitude of the static equivalent radial load P_o in newtons (pounds) for radial and angular contact ball bearings under combined thrust and radial loads is the greater of:

$$P_o = X_oF_r + Y_oF_a \qquad (24)$$

$$P_o = F_r \qquad (25)$$

where X_o = radial load factor given in Table 34

Y_o = axial load factor given in Table 34

F_r = applied radial load, newtons (pounds)

F_a = applied axial load, newtons (pounds)

Table 34. Values of X_o and Y_o for Computing Static Equivalent Radial Load P_o of Ball Bearings

Contact Angle	Single Row Bearings[a]		Double Row Bearings	
	X_o	Y_o[b]	X_o	Y_o[b]
RADIAL CONTACT GROOVE BEARINGS[c,a]				
α = 0°	0.6	0.5	0.6	0.5
ANGULAR CONTACT GROOVE BEARINGS				
α = 15°	0.5	0.47	1	0.94
α = 20°	0.5	0.42	1	0.84
α = 25°	0.5	0.38	1	0.76
α = 30°	0.5	0.33	1	0.66
α = 35°	0.5	0.29	1	0.58
α = 40°	0.5	0.26	1	0.52
SELF-ALIGNING BEARINGS				
…	0.5	0.22 cot α	1	0.44 cot α

[a] P_o is always ≥ F_r.

[b] Values of Y_o for intermediate contact angles are obtained by linear interpolation.

[c] Permissible maximum value of F_a/C_o (where F_a is applied axial load and C_o is static radial load rating) depends on the bearing design (groove depth and internal clearance).

Thrust Ball Bearings: The magnitude of the static equivalent axial load P_{oa} in newtons (pounds) for thrust ball bearings with contact angle $\alpha \neq 90°$ under combined radial and thrust loads is found by the formula:

$$P_{oa} = F_a + 2.3F_r \tan\alpha \qquad (26)$$

where the symbol definitions are the same as for Formulas (24) and (25). This formula is valid for all load directions in the case of double direction ball bearings. For single direction ball bearings, it is valid where $F_r/F_a \leq 0.44 \cot\alpha$ and gives a satisfactory but less conservative value of P_{oa} for F_r/F_a up to 0.67 cot α.

Thrust ball bearings with α = 90° can support axial loads only. The static equivalent load for this type of bearing is $P_{oa} = F_a$.

Roller Bearing Static Equivalent Load.—The static equivalent radial load for roller bearings is that calculated, static radial load which produces a maximum contact stress acting at the center of contact of a uniformly loaded rolling element equal in magnitude to the maximum contact stress in the actual condition of loading. The static equivalent axial load is that calculated, static centric axial load which produces a maximum contact stress acting at the center of contact of a uniformly loaded rolling element equal in magnitude to the maximum contact stress in the actual condition of loading.

Radial Roller Bearings: The magnitude of the static equivalent radial load P_o in newtons (pounds) for radial roller bearings under combined radial and thrust loads is the greater of:

$$P_o = X_oF_r + Y_oF_a \tag{27}$$

$$P_o = F_r \tag{28}$$

where X_o = radial factor given in Table 35

Y_o = axial factor given in Table 35

F_r = applied radial load, newtons (pounds)

F_a = applied axial load, newtons (pounds)

Table 35. Values of X_o and Y_o for Computing Static Equivalent Radial Load P_o for Self-Aligning and Tapered Roller Bearings

Bearing Type	Single Row[a]		Double Row	
	X_o	Y_o	X_o	Y_o
Self-Aligningand Tapered $\alpha \neq 0$	0.5	0.22 cot α	1	0.44 cot α

[a] P_o is always $\geq F_r$.

The static equivalent radial load for radial roller bearings with α = 0° and subjected to radial load only is $P_{or} = F_r$.

Note: The ability of radial roller bearings with α = 0° to support axial loads varies considerably with bearing design and execution. The bearing user should therefore consult the bearing manufacturer for recommendations regarding the evaluation of equivalent load in cases where bearings with α = 0° are subjected to axial load.

Radial Roller Bearing Combinations: When calculating the static equivalent radial load for two similar single row angular contact roller bearings mounted side by side on the same shaft such that they operate as a unit (duplex mounting) in "back-to-back" or "face-to-face" arrangement, use the X_o and Y_o values for a double row bearing and the F_r and F_a values for the total loads on the arrangement.

When calculating the static equivalent radial load for two or more similar single row angular contact roller bearings mounted side by side on the same shaft such that they operate as a unit (duplex or stack mounting) in "tandem" arrangement, use the X_o and Y_o values for a single row bearing and the F_r and F_a values for the total loads on the arrangement.

Thrust Roller Bearings: The magnitude of the static equivalent axial load P_{oa} in newtons (pounds) for thrust roller bearings with contact angle $\alpha \neq 90°$, under combined radial and thrust loads is found by the formula:

$$P_{oa} = F_a + 2.3F_r \tan\alpha \tag{29}$$

where F_a = applied axial load, newtons (pounds)

F_r = applied radial load, newtons (pounds)

α = nominal contact angle, degrees

The accuracy of this formula decreases for single direction thrust roller bearings when $F_r > 0.44\, F_a \cot\alpha$.

Thrust Roller Bearing Combinations: When calculating the static equivalent axial load for two or more thrust roller bearings mounted side by side on the same shaft such that they operate as a unit (duplex or stack mounting) in "tandem" arrangement, use the F_r and F_a values for the total loads acting on the arrangement.

STANDARD METAL BALLS

Standard Metal Balls.— American National Standard ANSI/AFBMA Std 10-1989 provides information for the user of metal balls permitting them to be described readily and accurately. It also covers certain measurable characteristics affecting ball quality.

On the following pages, tables taken from this Standard cover standard balls for bearings and other purposes by type of material, grade, and size range; preferred ball sizes; ball hardness corrections for curvature; various tolerances, marking increments, and maximum surface roughnesses by grades; total hardness ranges for various materials; and minimum case depths for carbon steel balls. The numbers of balls per pound and per kilogram for ferrous and nonferrous metals are also shown.

Definitions and Symbols.— The following definitions and symbols apply to American National Standard metal balls.

Nominal Ball Diameter, D_w: The diameter value that is used for the general identification of a ball size, e.g., ¼ inch, 6 mm, etc.

Single Diameter of a Ball, D_{ws}: The distance between two parallel planes tangent to the surface of a ball.

Mean Diameter of a Ball, D_{wm}: The arithmetical mean of the largest and smallest single diameters of a ball.

Ball Diameter Variation, V_{Dws}: The difference between the largest and smallest single diameters of one ball.

Deviation from Spherical Form, ΔR_w: The greatest radial distance in any radial plane between a sphere circumscribed around the ball surface and any point on the ball surface.

Lot: A definite quantity of balls manufactured under conditions that are presumed uniform, considered and identified as an entirety.

Lot Mean Diameter, D_{wmL}: The arithmetical mean of the mean diameter of the largest ball and that of the smallest ball in the lot.

Lot Diameter Variation, V_{DwL}: The difference between the mean diameter of the largest ball and that of the smallest ball in the lot.

Nominal Ball Diameter Tolerance: The maximum allowable deviation of any ball lot mean diameter from the Nominal Ball Diameter.

Container Marking Increment: The Standard unit steps in millionths of an inch or in micrometers used to express the Specific Diameter.

Specific Diameter: The amount by which the lot mean diameter (D_{wmL}) differs from the nominal diameter (D_w), accurate to the container marking increment for that grade; the specific diameter should be marked on the unit container.

Ball Gage Deviation, ΔS: The difference between the lot mean diameter and the sum of the nominal mean diameter and the ball gage.

Surface Roughness, R_a: Surface roughness consists of all those irregularities that form surface relief and are conventionally defined within the area where deviations of form and waviness are eliminated. (See Handbook Surface Texture Section.)

Ordering Specifications.— Unless otherwise agreed between producer and user, orders for metal balls should provide the following information: quantity, material, nominal ball diameter, grade, and ball gage. A *ball grade* embodies a specific combination of dimensional form, and surface roughness tolerances. A *ball gage(s)* is the prescribed small amount, expressed with the proper algebraic sign, by which the lot mean diameter (arithmetic mean of the mean diameters of the largest and smallest balls in the lot) should differ from the nominal diameter, this amount being one of an established series of amounts as shown in the table below. The 0 ball gage is commonly referred to as "OK".

Preferred Ball Gages for Grades 3 to 200

Grade	Ball Gages (in 0.0001-in. units)			Ball Gages (in 1μm units)		
	Minus	OK	Plus	Minus	OK	Plus
3, 5	− 3 − 2 − 1	0	+ 1 + 2 + 3	− 8 − 7 − 6 − 5 − 4 − 3 − 2 − 1	0	+ 1 + 2 + 3 + 4 + 5 + 6 + 7 + 8
10, 16	− 4 − 3 − 2 − 1	0	+ 1 + 2 + 3 + 4	− 10 − 8 − 6 − 4 − 2	0	+ 2 + 4 + 6 + 8 + 10
24	− 5 − 4 − 3 − 2 − 1	0	+ 1 + 2 + 3 + 4 + 5	− 12 − 10 − 8 − 6 − 4 − 2	0	+ 2 + 4 + 6 + 8 + 10 + 12
48	− 6 − 4 − 2	0	+ 2 + 4 + 6	− 16 − 12 − 8 − 4	0	+ 4 + 8 + 12 + 16
100		0			0	
200		0			0	

Table 1. AFBMA Standard Balls — Tolerances for Individual Balls and for Lots of Balls

Grade	Allowable Ball Diameter Variation	Allowable Deviation from Spherical Form	Maximum Surface Roughness R_a	Allowable Lot Diameter Variation	Nominal Ball Diameter Tolerance (±)	Container Marking Increments
	For Individual Balls			For Lots of Balls		
	Millionths of an Inch					
3	3	3	0.5	5	[a]	10
5	5	5	0.8	10	[a]	10
10	10	10	1	20	[a]	10
16	16	16	1	32	[a]	10
24	24	24	2	48	[a]	10
48	48	48	3	96	[a]	50
100	100	100	5	200	500	[a]
200	200	200	8	400	1000	[a]
500	500	500	[a]	1000	2000	[a]
1000	1000	1000	[a]	2000	5000	[a]
	Micrometers					
3	0.08	0.08	0.012	0.13	[a]	0.25
5	0.13	0.13	0.02	0.25	[a]	0.25
10	0.25	0.25	0.025	0.5	[a]	0.25
16	0.4	0.4	0.025	0.8	[a]	0.25
24	0.6	0.6	0.05	1.2	[a]	0.25
48	1.2	1.2	0.08	2.4	[a]	1.25
100	2.5	2.5	0.125	5	12.5	[a]
200	5	5	0.2	10	25	[a]
500	13	13	[a]	25	50	[a]
1000	25	25	[a]	50	125	[a]

[a] Not applicable.

Allowable ball gage (see text) deviation is for Grade 3: + 0.000030, – 0.000030 inch (+0.75, – 0.75 μm); for Grades 5, 10, and 16: + 0.000050, – 0.000040 inch (+ 1.25, – 1 μm); and for Grade 24: + 0.000100, – 0.000100 inch (+ 2.5, – 2.5 μm). Other grades not given.

Examples: A typical order, in inch units, might read as follows: 80,000 pieces, chrome alloy steel, ¼-inch Nominal Diameter, Grade 16, and Ball Gage to be –0.0002 inch.

A typical order, in metric units, might read as follows: 80,000 pieces, chrome alloy steel, 6 mm Nominal Diameter, Grade 16, and Ball Gage to be –4 μm.

Package Marking: The ball manufacturer or supplier will identify packages containing each lot with information provided on the orders, as given above. In addition, the specific diameter of the contents shall be stated. Container marking increments are listed in Table 1.

Examples: Balls supplied to the order of the first of the previous examples would, if perfect size, be D_{wmL} = 0.249800 inch. In Grade 16 these balls would be acceptable with D_{wmL} from 0.249760 to 0.249850 inch. If they actually measured 0.249823 (which would be rounded off to 0.249820), each package would be marked: 5,000 Balls, Chrome Alloy Steel, 1⁄4″ Nominal Diameter, Grade 16, −0.0002 inch Ball Gage, and −0.000180 inch Specific Diameter.

Balls supplied to the order of the second of the two previous examples would, if perfect size, be D_{wmL} = 5.99600 mm. In Grade 16 these balls would be acceptable with a D_{wmL} from 5.99500 to 5.99725 mm. If they actually measured 5.99627 mm (which would be rounded off to 5.99625 mm), each package would be marked: 5,000 Balls, Chrome Alloy Steel, 6 mm Nominal Diameter, Grade 16, −4 μm Ball Gage, and −3.75 μm Specific Diameter.

Table 2. AFBMA Standard Balls — Typical Nominal Size Ranges by Material and Grade

Steel Balls[a]			
		Size Range[b]	
Material	Grade	Inch	mm
Chrome Alloy	3	1⁄32–1	0.8–25
	5,10, 16,24	1⁄64–1 1⁄2	0.3–38
	48, 100, 200, 500	1⁄32–2 7⁄8	0.8–75
	1000	3⁄8–4 1⁄2	10–115
AISI M-50	3 5,10,16	1⁄32–1⁄2	0.8–12
	24,48	1⁄32–1 5⁄8	0.8–40
Corrosion Resisting Hardened	3,5,10,16	1⁄64–3⁄4	0.3–19
	24	1⁄32–1	0.8–25
	48	1⁄32–2	0.8–50
	100,200	1⁄32–4 1⁄2	0.8–115
Corrosion-Resisting Unhardened	100,200, 500	1⁄16–3⁄4	1.5–19
Carbon Steel[c]	100,200, 500, 1000	1⁄16–1 1⁄2	1.5–38
Silicon Molybdenum	200	1⁄4–1 1⁄8	6.5–28

Non-Ferrous Balls[a]			
		Size Range[b]	
Material Grade	Grade	Inch	mm
Aluminum	200	1⁄16–1	1.5–25
Aluminum Bronze	200	13⁄16–4	20–100
Brass	100,200, 500, 1000	1⁄16–3⁄4	1.5–19
Bronze	200,500, 1000	1⁄16–3⁄4	1.5–19
Monel Metal 400	100,200, 500	1⁄16–3⁄4	1.5–19
K-Monel Metal 500	100	1⁄16–3⁄4	1.5–19
	200	1⁄16–1 11⁄16	1.5–45
Tungsten Carbide	5	3⁄64–1⁄2	1.2–12
	10	3⁄64–3⁄4	1.2–19
	16	3⁄64–1	1.2–25
	24	3⁄64–1 1⁄4	1.2–32

[a] For hardness rages see Table 3.

[b] For tolerances see Table 1.

[c] For minimum case depths refer to the Standard.

Table 3. AFBMA Standard Balls—Typical Hardness Ranges

Material	Common Standard	SAE Unified Number	Rockwell Value[a,b]
Steel—			
Alloy tool	AISI/SAE M50	T-11350	60–65 "C"[c,d]
Carbon[e]	AISI/SAE 1008	G-10080	60 Minimum "C"[b]
	AISI/SAE 1013	G-10130	60 Minimum "C"[b]
	AISI/SAE 1018	G-10180	60 Minimum "C"[b]
AISI/SAE 1022	G-10220	60 Minimum "C"[b]	
Chrome alloy	AISI/SAE E52100	G-5298660	60–67 "C"[c,d]
	AISI/SAE E51100	G-51986	60-67 "C"[c,d]
Corrosion-resisting			
	AISI/SAE 440C	S-44004	58–65 "C"[f,d]
hardened	AISI/SAE 440B	S-44003	55–62 "C"[f,d]
	AISI/SAE 420	S-42000	52 Minimum "C"[f,d]
	AISI/SAE 410	S-41000	97 "B"; 41 "C"[f,d]
	AISI/SAE 329	S-32900	45 Minimum "C"[f,d]
Corrosion-resisting			
unhardened	AISI/SAE 3025	S-30200	25–39 "C"[d,g]
	AISI/SAE 304	S-304000	25–39 "C"[d,g]
	AISI/SAE 305	S-305000	25–39 "C"[d,g]
	AISI/SAE 316	S-31600	25–39 "C"[d,g]
	AISI/SAE 430	S-43000	48–63 "A"[d]
Silicon molybdenum	AISI/SAE S2	T-41902	52–60 "C"[c]
Aluminum	AA-2017	A-92017	54–72 "B"
Aluminium bronze	CDA-624	C-62400	94–98 "B"
	CDA-630	C-63000	94–98 "B"
Brass	CDA-260	C-26000	75–87 "B"
Bronze	CDA-464	C-46400	75–98 "B"
Monel 400	AMS-4730	N-04400	85–95 "B"
Monel K-500	QA-N-286	N-05500	24 Minimum "C"
Tungsten carbide	JIC Carbide Classification	…	84-91.5 "A"

[a] Rockwell Hardness Tests shall be conducted on parallel flats in accordance with ASTM Standard E-18 unless otherwise specified.

[b] Hardness readings taken on spherical surfaces are subject to the corrections shown in Table 5. Hardness readings for carbon steel balls smaller than 5 mm (¼ inch) shall be taken by the microhardness method (detailed in ANSI/AFBMA Std 10-1989) or as agreed between manufacturer and purchaser.

[c] Hardness of balls in any one lot shall be within 3 points on Rockwell C scale.

[d] When microhardness method (see ANSI/AFBMA Std 10-1989 is used, the Rockwell hardness values given are converted to DPH in accordance with ASTM Standard E 140, "Standard Hardness Conversion Tables for Metals."

[e] Choice of carbon steels shown to be at ball manufacturer's option.

[f] Hardness of balls in any one lot shall be within 4 points on Rockwell C scale.

[g] Annealed hardness of 75-90 "B" is available when specified.

For complete details as to material requirements, quality specifications, quality assurance provisions, and methods of hardness testing, reference should be made to the Standard.

Table 4. Preferred Ball Sizes

Nominal Ball Sizes Metric	Diameter mm	Diameter Inches	Nominal Ball Sizes Inch	Nominal Ball Sizes Metric	Diameter mmb	Diameter Inches	Nominal Ball Sizes Inch
0.3	0.300 00	0.011 810			0.793 75	0.031 250	1/32
	0.396 88	0.015 625	1/64	0.8	0.800 00	0.031 496	
0.4	0.400 00	0.015 750		1	1.000 00	0.039 370	
0.5	0.500 00	0.019 680			1.190 63	0.046 875	3/64
	0.508 00	0.020 000	0.020	1.2	1.200 00	0.047 240	
0.6	0.600 00	0.023 620		1.5	1.500 00	0.059 060	
	0.635 00	0.025 000	0.025		1.587 50	0.062 500	1/16
0.7	0.700 00	0.027 560			1.984 38	0.078 125	5/64
2	2.000 00	0.078 740		21	21.000 000	0.826 770	
	2.381 25	0.093 750	3/32		21.431 25	0.843 750	27/32
2.5	2.500 00	0.098 420		22	22.000 00	0.866 140	
	2.778 00	0.109 375	7/64		22.225 00	0.875 000	7/8
3	3.000 00	0.118 110		23	23.000 00	0.905 510	
	3.175 00	0.125 000	1/8		23.018 75	0.906 250	29/32
3.5	3.500 00	0.137 800			23.812 50	0.937 500	15/16
	3.571 87	0.140 625	9/64	24	24.000 00	0.944 880	
	3.968 75	0.156 250	5/32		24.606 25	0.968 750	31/32
4	4.000 00	0.157 480		25	25.000 00	0.984 250	
	4.365 63	0.171 875	11/64		25.400 00	1.000 000	1
4.5	4.500 00	0.177 160		26	26.000 00	1.023 620	
	4.762 50	0.187 500	3/16		26.987 50	1.062 500	1 1/16
5	5.000 00	0.196 850		28	28.000 00	1.102 360	
5.5	5.500 00	0.216 540			28.575 00	1.125 000	1 1/8
	5.556 25	0.218 750	7/32	30	30.000 00	1.181 100	
	5.953 12	0.234 375	15/64		30.162 50	1.187 500	1 3/16
6	6.000 00	0.236 220			31.750 00	1.250 000	1 1/4
	6.350 00	0.250 000	1/4	32	32.000 00	1.259 840	
6.5	6.500 00	0.255 900			33.337 50	1.312 500	1 5/16
	6.746 88	0.265 625	17/64	34	34.000 00	1.338 580	
7	7.000 00	0.275 590			34.925 00	1.375 000	1 3/8
	7.143 75	0.281 250	9/32	35	35.000 00	1.377 950	
7.5	7.500 00	0.295 280		36	36.000 00	1.417 320	
	7.540 63	0.296 875	19/64		36.512 50	1.437 500	1 7/16
	7.937 50	0.312 500	5/16	38	38.000 00	1.496 060	
8	8.000 00	0.314 960			38.100 00	1.500 000	1 1/2
8.5	8.500 00	0.334 640			39.687 50	1.562 500	1 9/16
	8.731 25	0.343 750	11/32	40	40.000 00	1.574 800	
9	9.000 00	0.354 330			41.275 00	1.625 000	1 5/8
	9.128 12	0.359 375	23/64		42.862 50	1.687 500	1 11/16
	9.525 00	0.375 000	3/8		44.450 00	1.750 000	1 3/4
	9.921 87	0.390 625	25/64	45	45.000 00	1.771 650	
10	10.000 00	0.393 700			46.037 50	1.812 500	1 13/16
	10.318 75	0.406 250	13/32		47.625 00	1.875 000	1 7/8
11	11.000 00	0.433 070			49.212 50	1.937 500	1 15/16
	11.112 50	0.437 500	7/16	50	50.000 00	1.968 500	
11.5	11.500 00	0.452 756			50.800 00	2.000 000	2
	11.509 38	0.453 125	29/64		53.975 00	2.125 000	2 1/8

Table 4. *(Continued)* **Preferred Ball Sizes**

Nominal Ball Sizes Metric	Diameter mm	Diameter Inches	Nominal Ball Sizes Inch	Nominal Ball Sizes Metric	Diameter mmb	Diameter Inches	Nominal Ball Sizes Inch
	11.906 25	0.468 750	15⁄32	55	55.000 00	2.165 354	
12	12.000 00	0.472 440			57.150 00	2.250 000	2¼
	12.303 12	0.484 375	31⁄64	60	60.000 00	2.362 205	
	12.700 00	0.500 000	½		60.325 00	2.375 00	2⅜
13	13.000 00	0.511 810			63.500 00	2.500 000	2½
	13.493 75	0.531 250	17⁄32	65	65.000 00	2.559 055	
14	14.000 00	0.551 180			66.675 00	2.625 000	2⅝
	14.287 50	0.562 500	9⁄16		69.850 00	2.750 000	2¾
15	15.000 00	0.590 550			73.025 00	2.875 000	2⅞
	15.081 25	0.593 750	19⁄32		76.200 00	3.000 000	3
	15.875 00	0.625 000	⅝		79.375 00	3.125 000	3 ⅛
16	16.000 00	0.629 920			82.550 00	3.250 000	3¼
	16.668 75	0.656 250	21⁄32		85.725 00	3.375 00	3⅜
17	17.000 00	0.669 290			88.900 00	3.500 000	3½
	17.462 50	0.687 500	11⁄16		92.075 00	3.625 000	3⅝
18	18.000 00	0.708 660			95.250 00	3.750 000	3¾
	18.256 25	0.718 750	23⁄32		98.425 00	3.875 000	3⅞
19	19.000 00	0.748 030			101.600 00	4.000 000	4
	19.050 00	0.750 000	¾		104.775 00	4.125 000	4⅛
	19.843 75	0.781 250	25⁄32		107.950 00	4.250 000	4¼
20	20.000 00	0.787 400			111.125 00	4.375 000	4⅜
	20.637 50	0.812 500	13⁄16		114.300 00	4.500 000	4½

Table 5. Ball Hardness Corrections for Curvatures

Hardness Reading, Rockwell C	Ball Diameters, Inch						
	¼	5⁄16	⅜	½	⅝	¾	1
	Correction—Rockwell C						
20	12.1	9.3	7.7	6.1	4.9	4.1	3.1
25	11.0	8.4	7.0	5.5	4.4	3.7	2.7
30	9.8	7.5	6.2	4.9	3.9	3.2	2.4
35	8.6	6.6	5.5	4.3	3.4	2.8	2.1
40	7.5	5.7	4.7	3.6	2.9	2.4	1.7
45	6.3	4.9	4.0	3.0	2.4	1.9	1.4
50	5.2	4.0	3.2	2.4	1.9	1.5	1.1
55	4.1	3.1	2.5	1.8	1.4	1.1	0.8
60	2.9	2.2	1.8	1.2	0.9	0.7	0.4
65	1.8	1.3	1.0	0.5	0.3	0.2	0.1
20	12.8	9.3	7.6	6.6	5.2	4.0	3.2
25	11.7	8.4	6.9	5.9	4.6	3.5	2.8
30	10.5	7.5	6.1	5.2	4.1	3.1	2.4
35	9.4	6.6	5.4	4.6	3.6	2.7	2.1
40	8.0	5.7	4.5	3.8	3.0	2.2	1.8
45	6.7	4.9	3.8	3.2	2.5	1.8	1.4
50	5.5	4.0	3.0	2.6	2.0	1.4	1.1
55	4.3	3.1	2.3	1.9	1.5	1.0	0.8
60	3.0	2.2	1.7	1.2	1.0	0.6	0.4
65	1.9	1.3	0.9	0.6	0.4	0.2	0.1

Corrections to be added to Rockwell C readings obtained on spherical surfaces of chrome alloy steel, corrosion resisting hardened and unhardened steel, and carbon steel balls. For other ball sizes and hardness readings, interpolate between correction values shown.

Table 6. Number of Metal Balls per Pound

Nom. Dia.,[a] Inches	Material Density, Pounds per Cubic Inch												
	.101	.274	.277	.279	.283	.284	.286	.288	.301	.304	.306	.319	.540
1⁄32	620 000	228 000	226 000	224 000	221 000	220 000	219 000	217 000	208 000	206 000	205 000	196 000	116 000
1⁄16	77 500	28 600	28 200	28 000	27 600	27 500	27 400	27 200	26 000	25 700	25 600	24 500	14 500
3⁄32	22 900	8 460	8 370	8 310	8 190	8 160	8 100	8 050	7 700	7 620	7 570	7 270	4 290
1⁄8	9 680	3 570	3 530	3 500	3 460	3 440	3 420	3 400	3 250	3 220	3 200	3 070	1 810
5⁄32	4 960	1 830	1 810	1 790	1 770	1 760	1 750	1 740	1 660	1 650	1 640	1 570	927
3⁄16	2 870	1 060	1 050	1 040	1 020	1 020	1 010	1 010	963	953	947	908	537
7⁄32	1 810	666	659	654	645	642	638	634	606	600	596	572	338
1⁄4	1 210	446	441	438	432	430	427	424	406	402	399	383	226
9⁄32	850	313	310	308	303	302	300	298	285	282	281	269	159
5⁄16	620	228	226	224	221	220	219	217	208	206	205	196	116
11⁄32	466	172.	170.	169.	166.	166.	164.	163.	156.	155.	154.	147.	87.1
3⁄8	359	132.	131.	130.	128.	128.	127.	126.	120.	119.	118.	114.	67.1
13⁄32	282	104.	103.	102.	101.	100.	99.6	98.9	94.6	93.7	93.1	89.3	52.8
7⁄16	226	83.2	82.3	81.7	80.6	80.3	79.7	79.2	75.8	75.0	74.5	71.5	42.2
15⁄32	184	67.7	66.9	66.5	65.5	65.3	64.8	64.4	61.6	61.0	60.6	58.1	34.3
1⁄2	151.	55.8	55.2	54.8	54.0	53.8	53.4	53.1	50.8	50.3	49.9	47.9	28.3
17⁄32	126.	46.5	46.0	45.7	45.0	44.9	44.5	44.2	42.3	41.9	41.6	39.9	23.6
9⁄16	106.	39.2	38.7	38.5	37.9	37.8	37.5	37.3	35.7	35.3	35.1	33.6	19.9
19⁄32	90.3	33.3	32.9	32.7	32.2	32.1	31.9	31.7	30.3	30.0	29.8	28.6	16.9
5⁄8	77.5	28.6	28.2	28.0	27.6	27.5	27.4	27.2	26.0	25.7	25.6	24.5	14.5
21⁄32	66.9	24.7	24.4	24.2	23.9	23.8	23.6	23.5	22.5	22.2	22.1	21.2	12.5
11⁄16	58.2	21.5	21.2	21.1	20.8	20.7	20.6	20.4	19.5	19.3	19.2	18.4	10.9
23⁄32	50.9	18.8	18.6	18.4	18.2	18.1	18.0	17.9	17.1	16.9	16.8	16.1	9.53
3⁄4	44.8	16.5	16.3	16.2	16.0	15.9	15.8	15.7	15.0	14.9	14.8	14.2	8.38
25⁄32	39.7	14.6	14.5	14.4	14.2	14.1	14.0	13.9	13.3	13.2	13.1	12.6	7.42
13⁄16	35.3	13.0	12.9	12.8	12.6	12.5	12.5	12.4	11.8	11.7	11.6	11.2	6.59
27⁄32	31.5	11.6	11.5	11.4	11.2	11.2	11.1	11.0	10.6	10.5	10.4	9.97	5.89
7⁄8	28.2	10.4	10.3	10.2	10.1	10.0	9.97	9.90	9.47	9.38	9.32	8.94	5.28
29⁄32	25.4	9.37	9.26	9.20	9.07	9.04	8.97	8.91	8.53	8.44	8.39	8.04	4.75
15⁄16	22.9	8.46	8.37	8.31	8.19	8.16	8.10	8.05	7.70	7.62	7.57	7.27	4.29
31⁄32	20.8	7.67	7.58	7.53	7.42	7.40	7.35	7.29	6.98	6.91	6.87	6.59	3.89
1	18.9	6.97	6.89	6.85	6.75	6.72	6.68	6.63	6.35	6.28	6.24	5.99	3.54

[a] For sizes above 1 in. diameter, use the following formula: No. balls per pound = 1.91 ÷ [(nom. dia., in.)^3 × (material density, lbs. per cubic in.)].

Ball material densities in pounds per cubic inch: aluminum .101; aluminum bronze .274; corrosion resisting hardened steel .277; AISI M-50 and silicon molybdenum steels .279; chrome alloy steel .283; carbon steel .284; AISI 302 corrosion resisting unhardened steel .286; AISI 316 corrosion resisting unhardened steel .288; bronze .304; brass and K-Monel metal .306; Monel metal .319; and tungsten carbide .540.

Table 7. Number of Metal Balls per Kilogram

Nom.Dia.,[a] mm	Material Density, Grams per Cubic Centimeter											
	2.796	7.584	7.667	7.723	7.833	7.861	7.916	7.972	8.332	8.415	8.470	8.830	14.947
0.3	25 300 000	9 330 000	9 230 000	9 160 000	9 030 000	9 000 000	8 940 000	8 870 000	8 490 000	8 410 000	8 350 000	8 010 000	4 730 000
0.4	10 670 000	3 930 000	3 890 000	3 860 000	3 810 000	3 800 000	3 770 000	3 740 000	3 580 000	3 550 000	3 520 000	3 380 000	2 000 000
0.5	5 470 000	2 010 000	1 990 000	1 980 000	1 950 000	1 940 000	1 930 000	1 920 000	1 830 000	1 820 000	1 800 000	1 730 000	1 020 000
0.7	1 990 000	734 000	726 000	721 000	711 000	708 000	703 000	698 000	668 000	662 000	657 000	631 000	373 000
0.8	1 330 000	492 000	487 000	483 000	476 000	475 000	471 000	468 000	448 000	443 000	440 000	422 000	250 000
1.0	683 000	252 000	249 000	247 000	244 000	243 000	241 000	240 000	229 000	227 000	225 000	216 000	128 000
1.2	395 000	146 000	144 000	143 000	141 000	141 000	140 000	139 000	133 000	131 000	130 000	125 000	73 900
1.5	202 000	74 600	73 800	73 300	72 200	72 000	71 500	71 000	67 900	67 200	66 800	64 100	37 900
2.0	85 400	31 500	31 100	30 900	30 500	30 400	30 200	29 900	28 700	28 400	28 200	27 000	16 000
2.5	43 700	16 100	15 900	15 800	15 600	15 500	15 400	15 300	14 700	14 500	14 400	13 800	8 180
3.0	25 300	9 330	9 230	9 160	9 030	9 000	8 940	8 870	8 490	8 410	8 350	8 010	4 730
3.5	15 900	5 870	5 810	5 770	5 690	5 670	5 630	5 590	5 350	5 290	5 260	5 040	2 980
4.0	10 700	3 930	3 890	3 860	3 810	3 800	3 770	3 740	3 580	3 550	3 520	3 380	2 000
4.5	7 500	2 760	2 730	2 710	2 680	2 670	2 650	2 630	2 520	2 490	2 470	2 370	1 400
5.0	5 470	5 010	1 990	1 980	1 950	1 940	1 930	1 920	1 830	1 820	1 800	1 730	1 020
5.5	4 110	1 510	1 500	1 490	1 470	1 460	1 450	1 440	1 380	1 360	1 360	1 300	768
6.0	3 160	1 170	1 150	1 140	1 130	1 120	1 120	1 110	1 060	1 050	1 040	1 000	592
6.5	2 490	917	907	901	888	885	878	872	835	826	821	788	465
7.0	1 990	734	726	721	711	708	703	698	668	662	657	631	373
7.5	1 620	597	590	586	578	576	572	568	543	538	534	513	303
8.0	1 330	492	487	483	476	475	471	468	448	443	440	422	250
8.5	1 110	410	406	403	397	396	393	390	373	370	367	352	208
9.0	937	345	342	339	334	333	331	329	314	311	309	297	175
10.0	683	252	249	247	244	243	241	240	229	227	225	216	128
11.0	513.0	189.0	187.0	186.0	183.0	183.0	181.0	180.0	172.0	171.0	169.0	163.0	96.0
11.5	449.0	166.0	164.0	163.0	160.0	160.0	159.0	158.0	151.0	149.0	148.0	142.0	84.0
12.0	395.0	146.0	144.0	143.0	141.0	141.0	140.0	139.0	133.0	131.0	130.0	125.0	73.9
13.0	311.0	115.0	113.0	113.0	111.0	111.0	110.0	109.0	104.0	103.0	103.0	98.5	58.2
14.0	249.0	91.8	90.8	90.1	88.9	88.5	87.9	87.3	83.5	82.7	82.2	78.8	46.6
15.0	202.0	74.6	73.8	73.3	72.2	72.0	71.5	71.0	67.9	67.2	66.8	64.1	37.9
16.0	167.0	61.5	60.8	60.4	59.5	59.3	58.9	58.5	56.0	55.4	55.1	52.8	31.2
17.0	139.0	51.3	50.7	50.3	49.6	49.5	49.1	48.8	46.7	46.2	45.9	44.0	26.0

[a] For sizes above 17 mm diameter, use the following formula: No. balls per kilogram = 1,910,000 ÷ [(nom. dia., mm)^3 × (material density, grams per cu. cm)].

Ball material densities in grams per cubic centimeter: aluminum, 2.796; aluminum bronze, 7.584; corrosion-resisting hardened steel, 7.677; AISI M-50 and silicon molybdenum steel, 7.723; chrome alloy steel, 7.833; carbon steel, 7.861; AISI 302 corrosion-resisting unhardened steel, 7.916; AISI 316 corrosion-resisting unhardened steel, 7.972; bronze, 8.415; brass and K-Monel metal, 8.470; Monel metal, 8.830; tungsten carbide, 14.947.

LUBRICANTS AND LUBRICATION

A lubricant is used for one or more of the following purposes: to reduce friction; to prevent wear; to prevent adhesion; to aid in distributing the load; to cool the moving elements; and to prevent corrosion.

The range of materials used as lubricants has been greatly broadened over the years, so that in addition to oils and greases, many plastics and solids and even gases are now being applied in this role. The only limitations on many of these materials are their ability to replenish themselves, to dissipate frictional heat, their reaction to high environmental temperatures, and their stability in combined environments. Because of the wide selection of lubricating materials available, great care is advisable in choosing the material and the method of application. The following types of lubricants are available: petroleum fluids, synthetic fluids, greases, solid films, working fluids, gases, plastics, animal fat, metallic and mineral films, and vegetable oils.

Lubricating Oils.—The most versatile and best-known lubricant is mineral oil. When applied in well-designed applications that provide for the limitations of both mechanical and hydraulic elements, oil is recognized as the most reliable lubricant. Concurrently, it is offered in a wide selection of stocks, carefully developed to meet the requirements of the specific application.

Lubricating oils are seldom marketed without additives blended for a narrow range of applications. These "additive packages" are developed for particular applications, so it is advisable to consult the sales-engineering representatives of a reputable petroleum company on the proper selection for the conditions under consideration. The following are the most common types of additives: wear preventive, oxidation inhibitor, rust inhibitor, detergent-dispersant, viscosity index improver, defoaming agent, and pour-point depressant.

A more recent development in the field of additives is a series of organic compounds that leave no ash when heated to a temperature high enough to evaporate or burn off the base oil. Initially produced for internal-combustion-engine applications these additives have found ready acceptance in those other applications where metallic or mineral trace elements would promote catalytic, corrosive, deposition, or degradation effects on mechanism materials.

Additives usually are not stable over the entire temperature and shear-rate ranges considered acceptable for the base stock oil application. Because of this problem, additive type oils must be carefully monitored to ensure that they are not continued in service after their principal capabilities have been diminished or depleted. Of primary importance in this regard is the action of the detergent-dispersant additives that function so well to reduce and control degradation products that would otherwise deposit on the operating parts and oil cavity walls. Because the materials cause the oil to carry a higher than normal amount of the breakdown products in a fine suspension, they may cause an accelerated deposition rate or foaming when they have been depleted or degenerated by thermal or contamination action. Ingestion of water by condensation or leaking can cause markedly harmful effects.

Viscosity index improvers serve to modify oils so that their change in viscosity is reduced over the operating temperature range. These materials may be used to improve both a heavy or a light oil; however, the original stock will tend to revert to its natural state when the additive has been depleted or degraded due to exposure to high temperatures or to the high shear rates normally encountered in the load-carrying zones of bearings and gears. In heavy-duty installations, it is generally advisable to select a heavier or a more highly refined oil (and one that is generally more costly) rather than to rely on a less stable viscosity-index-improvement product. Viscosity-index-improved oils are generally used in applications where the shear rate is well below 1,000,000 reciprocal seconds, as determined by the following formula:

$$\text{Shear rate}(s^{-1}) = \frac{DN}{60t}$$

where D is the journal diameter in inches, N is the journal speed in rpm, and t is the film thickness in inches.

Types of Oils.— Aside from being aware of the many additives available to satisfy particular application requirements and improve the performance of fluids, the designer must also be acquainted with the wide variety of oils, natural and synthetic, which are also available. Each oil has its own special features that make it suitable for specific applications and limit its utility in others. Though a complete description of each oil and its application feasibility cannot be given here, reference to major petroleum and chemical company sales engineers will provide full descriptions and sound recommendations. In some applications, however, it must be accepted that the interrelation of many variables, including shear rate, load, and temperature variations, prohibit precise recommendations or predictions of fluid durability and performance. Thus, prototype and rig testing are often required to ensure the final selection of the most satisfactory fluid.

The following table lists the major classifications and properties of available commercial petroleum oils.

Properties of Commercial Petroleum Oils and Their Applications

Group A				Group B			
Type	Viscosity, Centistokes 100°F	Viscosity, Centistokes 210°F	Density, g/cc at 60°F	Type	Viscosity, Centistokes 100°F	Viscosity, Centistokes 210°F	Density, g/cc at 60°F
SAE 10 W	41	6.0	0.870	General Purpose	22	3.9	0.880
SAE 20 W	71	8.5	0.885		44	6.0	0.898
SAE 30	114	11.2	0.890		66	7.0	0.915
SAE 40	173	14.5	0.890		110	9.9	0.915
SAE 50	270	19.5	0.900		200	15.5	0.890
Group C				Group D			
SAE 75	47	7.0	0.930, approx.	Turbine			
SAE 80	69	8.0		Light	32	5.5	0.871
SAE 90	285	20.5		Medium	65	8.1	0.876
SAE 140	725	34.0		Heavy	99	10.7	0.885
SAE 250	1,220	47.0					
Group E				Group F			
Aviation	5	1.5	0.858	Aviation	76	9.3	0.875
	10	2.5	0.864		268	20.0	0.891
					369	25.0	0.892

Group A. Automotive. With increased additives, diesel and marine reciprocating engines.
Group B. Gear trains and transmissions. With E. P. additives, hypoid gears.
Group C. Machine tools and other industrial applications.
Group D. Marine propulsion and stationary power turbines.
Group E. Turbojet engines.
Group F. Reciprocating engines.

Viscosity.— As noted before, fluids used as lubricants are generally categorized by their viscosity at 100 and 210 deg. F. Absolute viscosity is defined as a fluid's resistance to shear or motion—its internal friction in other words. This property is described in several ways, but basically it is the force required to move a plane surface of unit area with unit speed parallel to a second plane and at unit distance from it. In the metric system, the unit of viscosity is called the "poise" and in the English system is called the "reyn." One reyn is equal to 68,950 poises. One poise is the viscosity of a fluid, such that one dyne force is required to move a surface of one square centimeter with a speed of one centimeter per second, the distance between surfaces being one centimeter. The range of kinematic viscosity for a series of typical fluids is shown in the table on page 2312. Kinematic viscosity is related directly to the flow time of a fluid through the viscosimeter capillary. By multiplying the kinematic viscosity by the density of the fluid at the test temperature, one can determine the absolute viscosity. Because, in the metric system, the mass density is equal to the specific gravity, the conversion from kinematic to absolute viscosity is generally made in this system and then converted to English units where required. The densities of typical lubricating fluids with comparable viscosities at 100 deg. F and 210 deg. F are shown in this same table.

The following conversion table may be found helpful.

Multiply	By	To Get
Centipoises, Z, $\frac{\text{dyne-s}}{100\ \text{cm}^2}$	1.45×10^{-7}	Reyns, μ, $\frac{\text{lb force-s}}{\text{in.}^2}$
Centistokes, ν, $\frac{\text{cm}^2}{100\ \text{s}}$	Density in g/cc	Centipoises, Z, $\frac{\text{dyne-s}}{100\ \text{cm}^2}$
Saybolt Universal Seconds, t_s	$0.22t_s - \frac{180}{t_s}$	Centistokes, ν, $\frac{\text{cm}^2}{100\ \text{s}}$

Finding Specific Gravity of Oils at Different Temperatures.—The standard practice in the oil industry is to obtain a measure of specific gravity at 60 deg. F on an arbitrary scale, in degrees API, as specified by the American Petroleum Institute. As an example, API gravity, ρ_{API}, may be expressed as 27.5 degrees at 60 deg. F.

The relation between gravity in API degrees and specific gravity (grams of mass per cubic centimeter) at 60 deg. F, ρ_{60}, is

$$\rho_{60} = \frac{141.5}{131.5 + \rho_{API}}$$

The specific gravity, ρ_T, at some other temperature, T, is found from the equation

$$\rho_T = \rho_{60} - 0.00035(T - 60)$$

Normal values of specific gravity for sleeve-bearing lubricants range from 0.75 to 0.95 at 60 deg. F. If the API rating is not known, an assumed value of 0.85 may be used.

Application of Lubricating Oils.—In the selection and application of lubricating oils, careful attention must be given to the temperature in the critical operating area and its effect on oil properties. Analysis of each application should be made with detailed attention given to cooling, friction losses, shear rates, and contaminants.

Many oil selections are found to result in excessive operating temperatures because of a viscosity that is initially too high, which raises the friction losses. As a general rule, the lightest-weight oil that can carry the maximum load should be used. Where it is felt that the load carrying capacity is borderline, lubricity improvers may be employed rather than an arbitrarily higher viscosity fluid. It is well to remember that in many mechanisms the thicker fluid may increase friction losses sufficiently to lower the operating viscosity into the range provided by an initially lighter fluid. In such situations also, improved cooling, such as may be accomplished by increasing the oil flow, can improve the fluid properties in the load zone.

Similar improvements can be accomplished in many gear trains and other mechanisms by reducing churning and aeration through improved scavenging, direction of oil jets, and elimination of obstacles to the flow of the fluid. Many devices, such as journal bearings, are extremely sensitive to the effects of cooling flow and can be improved by greater flow rates with a lighter fluid. In other cases it is well to remember that the load carrying capacity of a petroleum oil is affected by pressure, shear rate, and bearing surface finish as well as initial viscosity and therefore these must be considered in the selection of the fluid. Detailed explanation of these factors is not within the scope of this text; however the technical representatives of the petroleum companies can supply practical guides for most applications.

Other factors to consider in the selection of an oil include the following:

1) Compatibility with system materials
2) Water absorption properties
3) Break-in requirements
4) Detergent requirements

5) Corrosion protection
6) Low temperature properties
7) Foaming tendencies
8) Boundary lubrication properties
9) Oxidation resistance (high temperature properties)
10) Viscosity/temperature stability (Viscosity Temperature Index).

Generally, the factors listed above are those which are usually modified by additives as described earlier. Since additives are used in limited amounts in most petroleum products, blended oils are not as durable as the base stock and must therefore be used in carefully worked-out systems. Maintenance procedures must be established to monitor the oil so that it may be replaced when the effect of the additive is noted or expected to degrade. In large systems supervised by a lubricating engineer, sampling and associated laboratory analysis can be relied on, while in customer-maintained systems as in automobiles and reciprocating engines, the design engineer must specify a safe replacement period which takes into account any variation in type of service or utilization.

Some large systems, such as turbine-power units, have complete oil systems which are designed to filter, cool, monitor, meter, and replenish the oil automatically. In such facilities, much larger oil quantities are used and they are maintained by regularly assigned lubricating personnel. Here reliance is placed on conservatively chosen fluids with the expectation that they will endure many months or even years of service.

Centralized Lubrication Systems.— Various forms of centralized lubrication systems are used to simplify and render more efficient the task of lubricating machines. In general, a central reservoir provides the supply of oil, which is conveyed to each bearing either through individual lines of tubing or through a single line of tubing that has branches extending to each of the different bearings. Oil is pumped into the lines either manually by a single movement of a lever or handle, or automatically by mechanical drive from some revolving shaft or other part of the machine. In either case, all bearings in the central system are lubricated simultaneously. Centralized force-feed lubrication is adaptable to various classes of machine tools such as lathes, planers, and milling machines and to many other types of machines. It permits the use of a lighter grade of oil, especially where complete coverage of the moving parts is assured.

Gravity Lubrication Systems.— Gravity systems of lubrication usually consist of a small number of distributing centers or manifolds from which oil is taken by piping as directly as possible to the various surfaces to be lubricated, each bearing point having its own independent pipe and set of connections. The aim of the gravity system, as of all lubrication systems, is to provide a reliable means of supplying the bearing surfaces with the proper amount of lubricating oil. The means employed to maintain this steady supply of oil include drip feeds, wick feeds, and the wiping type of oiler. Most manifolds are adapted to use either or both drip and wick feeds.

Drip-feed Lubricators: A drip feed consists of a simple cup or manifold mounted in a convenient position for filling and connected by a pipe or duct to each bearing to be oiled. The rate of feed in each pipe is regulated by a needle or conical valve. A loose-fitting cover is usually fitted to the manifold in order to prevent cinders or other foreign matter from becoming mixed with the oil. When a cylinder or other chamber operating under pressure is to be lubricated, the oil-cup takes the form of a lubricator having a tight-fitting screw cover and a valve in the oil line. To fill a lubricator of this kind, it is only necessary to close the valve and unscrew the cover.

Operation of Wick Feeds: For a wick feed, the siphoning effect of strands of worsted yarn is employed. The worsted wicks give a regular and reliable supply of oil and at the same time act as filters and strainers. A wick composed of the proper number of strands is fitted into each oil-tube. In order to insure using the proper sizes of wicks, a study should be made of the oil requirements of each installation, and the number of strands necessary to meet the demands of bearings at different rates of speed should be determined. When the necessary data have been obtained, a table should be prepared showing the size of wick or the number of strands to be used for each bearing of the machine.

Oil-conducting Capacity of Wicks: With the oil level maintained at a point $\frac{3}{8}$ to $\frac{3}{4}$ inch below the top of an oil-tube, each strand of a clean worsted yarn will carry slightly more than one drop of oil a minute. A twenty-four-strand wick will feed approximately thirty drops a minute, which is ordinarily sufficient for operating a large bearing at high speed. The wicks should be removed from the oil-tubes when the machinery is idle. If left in place, they will continue to deliver oil to the bearings until the supply in the cup is exhausted, thus wasting a considerable quantity of oil, as well as flooding the bearing. When bearings require an extra supply of oil temporarily, it may be supplied by dipping the wicks or by pouring oil down the tubes from an oil-can or, in the case of drip feeds, by opening the needle valves. When equipment that has remained idle for some time is to be started up, the wicks should be dipped and the moving parts oiled by hand to insure an ample initial supply of oil. The oil should be kept at about the same level in the cup, as otherwise the rate of flow will be affected. Wicks should be lifted periodically to prevent dirt accumulations at the ends from obstructing the flow of oil.

How Lubricating Wicks are Made: Wicks for lubricating purposes are made by cutting worsted yarn into lengths about twice the height of the top of the oil-tube above the bottom of the oil-cup, plus 4 inches. Half the required number of strands are then assembled and doubled over a piece of soft copper wire, laid across the middle of the strands. The free ends are then caught together by a small piece of folded sheet lead, and the copper wire twisted together throughout its length. The lead serves to hold the lower end of the wick in place, and the wire assists in forcing the other end of the wick several inches into the tube. When the wicks are removed, the free end of the copper wire may be hooked over the tube end to indicate which tube the wick belongs to. Dirt from the oil causes the wick to become gummy and to lose its filtering effect. Wicks that have thus become clogged with dirt should be cleaned or replaced by new ones. The cleaning is done by boiling the wicks in soda water and then rinsing them thoroughly to remove all traces of the soda. Oil-pipes are sometimes fitted with openings through which the flow of oil can be observed. In some installations, a short glass tube is substituted for such an opening.

Wiper-type Lubricating Systems: Wiper-type lubricators are used for out-of-the-way oscillating parts. A wiper consists of an oil-cup with a central blade or plate extending above the cup, and is attached to a moving part. A strip of fibrous material fed with oil from a source of supply is placed on a stationary part in such a position that the cup in its motion scrapes along the fibrous material and wipes off the oil, which then passes to the bearing surfaces.

Oil manifolds, cups, and pipes should be cleaned occasionally with steam conducted through a hose or with boiling soda water. When soda water is used, the pipes should be disconnected, so that no soda water can reach the bearings.

Oil Mist Systems.— A very effective system for both lubricating and cooling many elements which require a limited quantity of fluid is found in a device which generates a mist of oil, separates out the denser and larger (wet) oil particles, and then distributes the mist through a piping or conduit system. The mist is delivered into the bearing, gear, or lubricated element cavity through a condensing or spray nozzle, which also serves to meter the flow. In applications which do not encounter low temperatures or which permit the use of visual devices to monitor the accumulation of solid oil, oil mist devices offer advantages in providing cooling, clean lubricant, pressurized cavities which prevent entrance of contaminants, efficient application of limited lubricant quantities, and near-automatic performance. These devices are supplied with fluid reservoirs holding from a few ounces up to several gallons of oil and with accommodations for either accepting shop air or working from a self-contained compressor powered by electricity. With proper control of the fluid temperature, these units can atomize and dispense most motor and many gear oils.

Lubricating Greases.—In many applications, fluid lubricants cannot be used because of the difficulty of retention, relubrication, or the danger of churning. To satisfy these and other requirements such as simplification, greases are applied. These formulations are usually petroleum oils thickened by dispersions of soap, but may consist of synthetic oils with soap or inorganic thickeners, or oil with silaceous dispersions. In all cases, the thickener, which must be carefully prepared and mixed with the fluid, is used to immobilize the oil, serving as a storehouse from which the oil bleeds at a slow rate. Though the thickener very often has lubricating properties itself, the oil bleeding from the bulk of the grease is the determining lubricating function. Thus, it has been shown that when the oil has been depleted to the level of 50 per cent of the total weight of the grease, the lubricating ability of the material is no longer reliable. In some applications requiring an initially softer and wetter material, however, this level may be as high as 60 per cent.

Grease Consistency Classifications.—To classify greases as to mobility and oil content, they are divided into Grades by the NLGI (National Lubricating Grease Institute). These grades, ranging from 0, the softest, up through 6, the stiffest, are determined by testing in a penetrometer, with the depth of penetration of a specific cone and weight being the controlling criterion. To insure proper averaging of specimen resistance to the cone, most specifications include a requirement that the specimen be worked in a sieve-like device before being packed into the penetrometer cup for the penetration test. Since many greases exhibit thixotropic properties (they soften with working, as they often do in an application with agitation of the bulk of the grease by the working elements or accelerations), this penetration of the worked specimen should be used as a guide to compare the material to the original manufactured condition of it and other greases, rather than to the exact condition in which it will be found in the application. Conversely, many greases are found to stiffen when exposed to high shear rates at moderate loads as in automatic grease dispensing equipment. The application of a grease, therefore must be determined by a carefully planned cut-and-try procedure. Most often this is done by the original equipment manufacturer with the aid of the petroleum company representatives, but in many cases it is advisable to include the bearing engineer as well. In this general area it is well to remember that shock loads, axial or thrust movement within or on the grease cavity can cause the grease to contact the moving parts and initiate softening due to the shearing or working thus induced. To limit this action, grease-lubricated bearing assemblies often utilize dams or dividers to keep the bulk of the grease contained and unchanged by this working. Successful application of a grease depends however, on a relatively small amount of mobile lubricant (the oil bled out of the bulk) to replenish that small amount of lubricant in the element to be lubricated. If the space between the bulk of the mobile grease and the bearing is too large, then a critical delay period (which will be regulated by the grease bleed rate and the temperature at which it is held) will ensue before lubricant in the element can be resupplied. Since most lubricants undergo some attrition due to thermal degradation, evaporation, shearing, or decomposition in the bearing area to which applied, this delay can be fatal.

To prevent this from leading to failure, grease is normally applied so that the material in the cavity contacts the bearing in the lower quadrants, insuring that the excess originally packed into it impinges on the material in the reservoir. With the proper selection of a grease which does not slump excessively, and a reservoir construction to prevent churning, the initial action of the bearing when started into operation will be to purge itself of excess grease, and to establish a flow path for bleed oil to enter the bearing. For this purpose, most greases selected will be of a grade 2 or 3 consistency, falling into the "channelling" variety or designation.

Types of Grease.—Greases are made with a variety of soaps and are chosen for many particular characteristics. Most popular today, however, are the lithium, or soda-soap grease and the modified-clay thickened materials. For high temperature applications (250 deg. F. and above) certain finely divided dyes and other synthetic thickeners are applied. For all-around use the lithium soap greases are best for moderate temperature applications (up to 225 deg. F.) while a number of soda-soap greases have been found to work well up to 285 deg. F. Since the major suppliers offer a number of different formulations for these temperature ranges it is recommended that the user contact the engineering representatives of a reputable petroleum company before choosing a grease. Greases also vary in volatility and viscosity according to the oil used. Since the former will affect the useful life of the bulk applied to the bearing and the latter will affect the load carrying capacity of the grease, they must both be considered in selecting a grease.

For application to certain gears and slow-speed journal bearings, a variety of greases are thickened with carbon, graphite, molybdenum disulfide, lead, or zinc oxide. Some of these materials are likewise used to inhibit fretting corrosion or wear in sliding or oscillating mechanisms and in screw or thread applications. One material used as a "gear grease" is a residual asphaltic compound which is known as a "Crater Compound." Being extremely stiff and having an extreme temperature-viscosity relationship, its application must also be made with careful consideration of its limitations and only after careful evaluation in the actual application. Its oxidation resistance is limited and its low mobility in winter temperature ranges make it a material to be used with care. However, it is used extensively in the railroad industry and in other applications where containment and application of lubricants is difficult. In such conditions its ability to adhere to gear and chain contact surfaces far outweighs its limitations and in some extremes it is "painted" onto the elements at regular intervals.

Temperature Effects on Grease Life.—Since most grease applications are made where long life is important and relubrication is not too practical, operating temperatures must be carefully considered and controlled. Being a hydro-carbon, and normally susceptible to oxidation, grease is subject to the general rule that: Above a critical threshold temperature, each 15- to 18-deg. F. rise in temperature reduces the oxidation life of the lubricant by half. For this reason, it is vital that all elements affecting the operating temperature of the application be considered, correlated, and controlled. With sealed-for-life bearings, in particular, grease life must be determined for representative bearings and limits must be established for all subsequent applications.

Most satisfactory control can be established by measuring bearing temperature rise during a controlled test, at a consistent measuring point or location. Once a base line and limiting range are determined, all deviating bearings should be dismantled, inspected, and reassembled with fresh lubricant for retest. In this manner mavericks or faulty assemblies will be ferreted out and the reliability of the application established. Generally, a well lubricated grease packed bearing will have a temperature rise above ambient, as measured at the outer race, of from 10 to 50 deg. F. In applications where heat is introduced into the bearing through the shaft or housing, a temperature rise must be added to that of the frame or shaft temperature.

In bearing applications care must be taken not to fill the cavity too full. The bearing should have a practical quantity of grease worked into it with the rolling elements thoroughly coated and the cage covered, but the housing (cap and cover) should be no more than 75 per cent filled; with softer greases, this should be no more than 50 per cent. Excessive packing is evidenced by overheating, churning, aerating, and eventual purging with final failure due to insufficient lubrication. In grease lubrication, *never* add a bit more for good luck — hold to the prescribed amount and determine this with care on a number of representative assemblies.

Relubricating with Grease.—In some applications, sealed-grease methods are not applicable and addition of grease at regular intervals is required. Where this is recommended by the manufacturer of the equipment, or where the method has been worked out as part of a development program, the procedure must be carefully followed. *First,* use the proper lubricant — the same as recommended by the manufacturer or as originally applied (grease performance can be drastically impaired if contaminated with another lubricant). *Second,* clean the lubrication fitting thoroughly with materials which will not affect the mechanism or penetrate into the grease cavity. *Third,* remove the cap (and if applicable, the drain or purge plug). *Fourth,* clean and inspect the drain or scavenge cavity. *Fifth,* weigh the grease gun or calibrate it to determine delivery rate. *Sixth,* apply the directed quantity or fill until grease is detected coming out the drain or purge hole. *Seventh,* operate the mechanism with the drain open so that excess grease is purged. *Last,* continue to operate the mechanism while determining the temperature rise and insure that it is within limits. Where there is access to a laboratory, samples of the purged material may be analyzed to determine the deterioration of the lubricant and to search for foreign material which may be evidence of contamination or of bearing failure.

Normally, with modern types of grease and bearings, lubrication need only be considered at overhaul periods or over intervals of three to ten years.

Solid Film Lubricants.—Solids such as graphite, molybdenum disulfide, polytetrafluoroethylene, lead, babbit, silver, or metallic oxides are used to provide dry film lubrication in high-load, slow-speed or oscillating load conditions. Though most are employed in conjunction with fluid or grease lubricants, they are often applied as the primary or sole lubricant where their inherent limitations are acceptable. Of foremost importance is their inability to carry away heat. Second, they cannot replenish themselves, though they generally do lay down an oriented film on the contacting interface. Third, they are relatively immobile and must be bonded to the substrate by a carrier, by plating, fusing, or by chemical or thermal deposition.

Though these materials do not provide the low coefficient of friction associated with fluid lubrication, they do provide coefficients in the range of 0.4 down to 0.02, depending on the method of application and the material against which they rub. Polytetrafluoroethylene, in normal atmospheres and after establishing a film on both surfaces has been found to exhibit a coefficient of friction down to 0.02. However, this material is subject to cold flow and must be supported by a filler or on a matrix to continue its function. Since it can now be cemented in thin sheets and is often supplied with a fine glass fiber filler, it is practical in a number of installations where the speed and load do not combine to melt the bond or cause the material to sublime.

Bonded films of molybdenum disulfide, using various resins and ceramic combinations as binders, are deposited over phosphate treated steel, aluminum, or other metals with good success. Since its action produces a gradual wear of the lubricant, its life is limited by the thickness which can be applied (not over a thousandth or two in the conventional application). In most applications this is adequate if the material is used to promote break-in, prevent galling or pick-up, and to reduce fretting or abrasion in contacts otherwise impossible to separate.

In all applications of solid film lubricants, the performance of the film is limited by the care and preparation of the surface to which they are applied. If they can't adhere properly, they cannot perform, coming off in flakes and often jamming under flexible components. The best advice is to seek the assistance of the supplier's field engineer and set up a close control of the surface preparation and solid film application procedure. It should be noted that the functions of a good solid film lubricant cannot overcome the need for better surface finishing. Contacting surfaces should be smooth and flat to insure long life and minimum friction forces. Generally, surfaces should be finished to no more than 24 micro-inches AA with wariness no greater than 0.00002 inch.

Anti-friction Bearing Lubrication.—The limiting factors in bearing lubrication are the load and the linear velocity of the centers of the balls or rollers. Since these are difficult to evaluate, a speed factor which consists of the inner race bore diameter × RPM is used as a criterion. This factor will be referred to as S_i where the bore diameter is in inches and S_m where it is in millimeters.

In order to be suitable for use in anti-friction bearings, grease must have the following properties:

1) Freedom from chemically or mechanically active ingredients such as uncombined metals or oxides, and similar mineral or solid contaminants.

2) The slightest possible tendency of change in consistency, such as thickening, separation of oil, evaporation or hardening.

3) A melting point considerably higher than the operating temperatures.

The choice of lubricating oils is easier. They are more uniform in their characteristics and if resistant to oxidation, gumming and evaporation, can be selected primarily with regard to a suitable viscosity.

Grease Lubrication: Anti-friction bearings are normally grease lubricated, both because grease is much easier than oil to retain in the housing over a long period and because it acts to some extent as a seal against the entry of dirt and other contaminants into the bearings. For almost all applications, a No. 2 soda-base grease or a mixed-base grease with up to 5 per cent calcium soap to give a smoother consistency, blended with an oil of around 250 to 300 SSU (Saybolt Universal Seconds) at 100 degrees F. is suitable. In cases where speeds are high, say S_i is 5000 or over, a grease made with an oil of about 150 SSU at 100 degrees F. may be more suitable especially if temperatures are also high. In many cases where bearings are exposed to large quantities of water, it has been found that a standard soda-base ball-bearing grease, although classed as water soluble gives better results than water-insoluble types. Greases are available that will give satisfactory lubrication over a temperature range of −40 degrees to +250 degrees F.

Conservative grease renewal periods will be found in the accompanying chart. Grease should not be allowed to remain in a bearing for longer than 48 months or if the service is very light and temperatures low, 60 months, irrespective of the number of hours' operation during that period as separation of the oil from the soap and oxidation continue whether the bearing is in operation or not.

Before renewing the grease in a hand-packed bearing, the bearing assembly should be removed and washed in clean kerosene, degreasing fluid or other solvent. As soon as the bearing is quite clean it should be washed at once in clean light mineral oil, preferably rust-inhibited. The bearing should *not be spun* before or while it is being oiled. Caustic solutions may be used if the old grease is hard and difficult to remove, but the best method is to soak the bearing for a few hours in light mineral oil, preferably warmed to about 130 degrees F., and then wash in cleaning fluid as described above. The use of chlorinated solvents is best avoided.

When replacing the grease, it should be forced with the fingers between the balls or rollers, dismantling the bearing, if convenient. The available space inside the bearing should be filled completely and the bearing then spun by hand. Any grease thrown out should be wiped off. The space on each side of the bearing in the housing should be not more than half-filled. Too much grease will result in considerable churning, high bearing temperatures and the possibility of early failure. Unlike any other kind of bearing, anti-friction bearings more often give trouble due to over-rather than to under-lubrication.

Grease is usually not very suitable for speed factors over 12,000 for S_i or 300,000 for S_m (although successful applications have been made up to an S_i of 50,000) or for temperatures much over 210 degrees F., 300 degrees F. being the extreme practical upper limit, even if synthetics are used. For temperatures above 210 degrees F., the grease renewal periods are very short.

Oil Lubrication: Oil lubrication is usually adopted when speeds and temperatures are high or when it is desired to adopt a central oil supply for the machine as a whole. Oil for anti-friction bearing lubrication should be well refined with high film strength and good resistance to oxidation and good corrosion protection. Anti-oxidation additives do no harm but are not really necessary at temperatures below about 200 degrees F. Anti-corrosion additives are always desirable. The accompanying table gives recommended viscosities of oil for ball bearing lubrication other than by an air-distributed oil mist. Within a given temperature and speed range, an oil towards the lighter end of the grade should be used, if convenient, as speeds increase. Roller bearings usually require an oil one grade heavier than do ball bearings for a given speed and temperature range. Cooled oil is sometimes circulated through an anti-friction bearing to carry off excess heat resulting from high speeds and heavy loads.

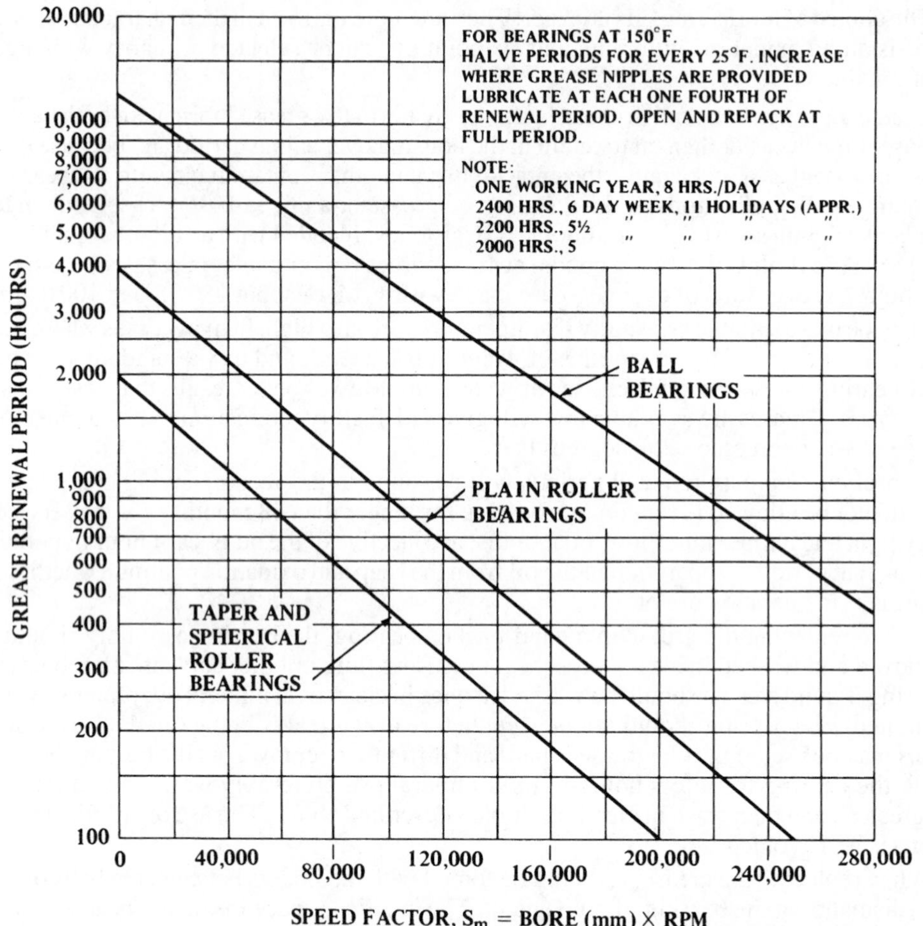

Oil Viscosities and Temperature Ranges for Ball Bearing Lubrication

Maximum Temperature Range Degrees F.	Optimum Temperature Range, Degrees F.	Speed Factor, S_i[a] Under 1000	Over 1000
		Viscosity	
− 40 to + 100	− 40 to − 10	80 to 90 SSU[b]	70 to 80 SSU[b]
− 10 to + 100	− 10 to + 30	100 to 115 SSU[b]	80 to 100 SSU[b]
+ 30 to + 150	+ 30 to + 150	SAE 20	SAE 10
+ 30 to + 200	+ 150 to + 200	SAE 40	SAE 30
+ 50 to + 300	+ 200 to + 300	SAE 70	SAE 60

[a] Inner race bore diameter (inches) × RPM.

[b] At 100 deg. F.

Not applicable to air-distributed oil mist lubrication.

Aerodynamic Lubrication

A natural extension of hydrodynamic lubrication consists in using air or some other gas as the lubricant. The viscosity of air is 1,000 times smaller than that of a very thin mineral oil. Consequently, the viscous resistance to motion is very much less. However, the distance of nearest approach, i.e. the closest distance between the shaft and the bearing is also correspondingly smaller, so that special precautions must be taken.

To obtain full benefit from such aerodynamic lubrication, the surfaces must have a very fine finish, the alignment must be very good, the speeds must be high and the loading relatively low. If all these conditions are fulfilled extremely successful bearing system can be made to run at very low coefficients of friction. They may also operate at very high temperatures since chemical degradation of the lubricant need not occur. Furthermore, if air is used as the lubricant, it costs nothing. This type of lubrication mechanism is very important for oil-free compressors and gas turbines. Another area of growing application for aerodynamic bearings is in data recording heads for computers. Air is used as the lubricant for the recording heads which are designed to be separated from the magnetic recording disc by a thin air film. The need for high recording densities in magnetic discs necessitates the smallest possible air film thickness between the head and disc. A typical thickness is around 1μm.

The analysis of aerodynamic bearings is very similar to liquid hydrodynamic bearings. The main difference, however, is that the gas compressibility is now a distinctive feature and has to be incorporated into the analysis.

Elastohydrodynamic Lubrication.—In the arrangement of the shaft and bearing it is usually assumed that the surfaces are perfectly rigid and retain their geometric shape during operation. However, a question might be posed: what is the situation if the geometry or mechanical properties of the materials are such that appreciable elastic deformation of the surfaces occurs? Suppose a steel shaft rests on a rubber block. It deforms the block elastically and provides an approximation to a half-bearing (see Figure 1 a).

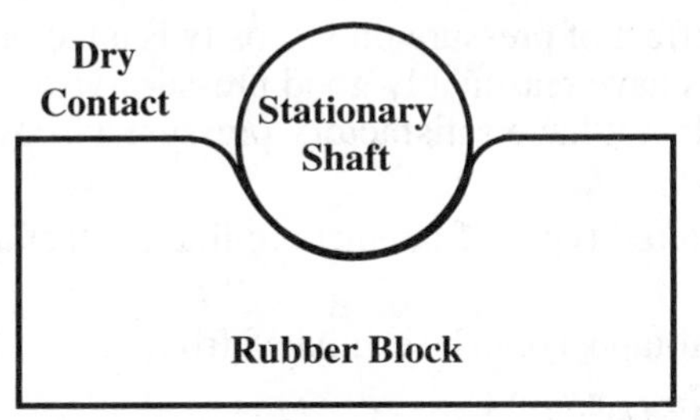

Fig. 1a.

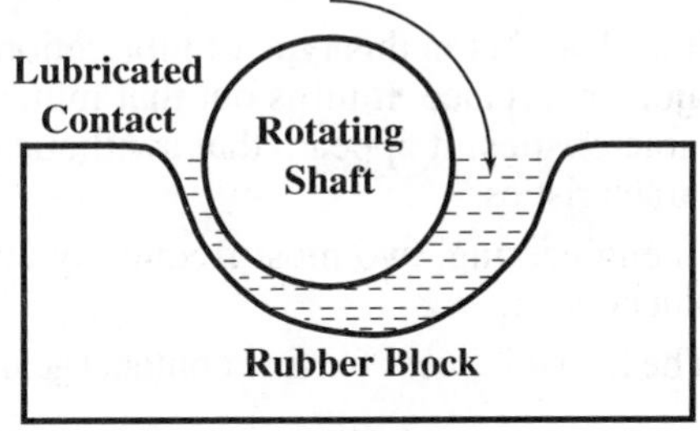

Fig. 1b.

If a lubricant is applied to the system it will be dragged into the interface and, if the conditions are right, it will form a hydrodynamic film. However, the pressures developed in the oil film will now have to match up with the elastic stresses in the rubber. In fact the shape of the rubber will be changed as indicated in Figure 1 b.

This type of lubrication is known as elastohydrodynamic lubrication. It occurs between rubber seals and shafts. It also occurs, rather surprisingly, in the contact between a windshield wiper blade and a windshield in the presence of rain. The geometry of the deformable member, its elastic properties, the load, the speed and the viscosity of the liquid and its dependence on the contact pressure are all important factors in the operation of elastohydrodynamic lubrication.

With conventional journals and bearings the average pressure over the bearing is of the order of 7×10^{-6} N/rn^2. With elastohydrodynamic bearings using a material such as rubber the pressures are perhaps 10 to 20 times smaller. At the other end of pressure spectrum, for instance in gear teeth, contact pressures of the order of 700×10^6 N/in^2 may easily be reached. Because the metals used for gears are very hard this may still be within the range of elastic deformation. With careful alignment of the engaging gear teeth and appropriate surface finish, gears can in fact run successfully under these conditions using an ordinary mineral oil as the lubricant. If the thickness of the elastohydrodynamic film formed at such pressures is calculated it will be found that it is less than an atomic diameter. Sincc even the smoothest metal surfaces are far rougher than this (a millionth of an inch is about 100 atomic diameters) it seems hard to understand why lubrication is effective in these circumstances.

The explanation was first provided by A.N.Grubin in 1949 and a little later (1958) by A.W.Crook. With most mineral oils the application of a high pressure can lead to an enormous increase in viscosity. For example, at a pressure of 700×10^6 N/rn^2 the viscosity may be increased 10,000-fold. The oil entering the gap between the gear teeth is trapped between the surfaces and at the high pressures existing in the contact region behaves virtually like a solid separating layer. This process explains why many mechanisms in engineering practice operate under much severer conditions than the classical theory would allow.

This type of elastohydrodynamic lubrication becomes apparent only when the film thickness is less than about 0.25 to 1 μm. To be exploited successfully it implies that the surfaces must be very smooth and very carefully aligned. If these conditions are met systems such as gears or cams and tappets can operate effectively at very high contact pressures without any metallic contact occurring. The coefficient of friction depends on the load, contact geometry, speed, etc., but generally it lies between about $\mu = 0.01$ at the lightest pressures and $\mu = 0.1$ at the highest pressures. The great success of elastohydrodynamic theory in explaining effective lubrication at very high contact pressures also raises a problem that has not yet been satisfactorily resolved: why do lubricants ever fail, since the harder they are squeezed the harder it is to extrude them? It is possible that high temperature flashes are responsible; alternatively the high rates of shear can actually fracture the lubricant film since when it is trapped between the surfaces it is, instantaneously, more like a wax than an oil.

It is clear that in this type of lubrication the effect of pressure on viscosity is a factor of major importance. It turns out that mineral oils have reasonably good pressure-viscosity characteristics. It appears that synthetic oils do not have satisfactory pressure-viscosity characteristics.

In engineering, two most frequently encountered types of contact are line contact and point contact.

The film thickness for line contact (gears, cam-tappet) can be estimated from:

$$h_o = 2.65 \frac{\alpha^{0.54} (\eta_o U)^{0.7} R_e^{0.43}}{w^{0.13} E_e^{0.03}}$$

In the case of point contact (ball bearings), the film thickness is given by:

$$h_o = 0.84 \alpha \eta_o U^{0.74} 0.41 R_e \left(\frac{E_e}{W}\right)^{0.074}$$

In the above equations the symbols used are defined as:

α = the pressure-viscosity coefficient. A typical value for a mineral oil is 1.8x1 0-8 m^2/N

ν = the viscosity of the lubricant at atmospheric pressure Ns/m^2

U = the entraining surface velocity, $U = (U_A + U_B)/2$ m/s, where the subscripts A and B refer to the velocities of bodies 'A' and 'B' respectively.

W = the load on the contact, N

w = the load per unit width of line contact, N/m

E_O = the reduced Young's modulus $\frac{1}{E_e} = \frac{1}{2}\left(\frac{1-\upsilon_A^2}{E_A} + \frac{1-\upsilon_B^2}{E_B}\right)$ N/m² where 'ν_A and ν_B are the Poisson's ratios of the contacting bodies 'A' and 'B' respectively; E_A and E_B are the Young's moduli of the contacting bodies 'A' and 'B' respectively.

R_e = - is the reduced radius of curvature (meters) and is given by different equations for different contact configurations.

In ball bearings (see Figure 2) the reduced radius is given by:

- contact between the ball and inner race: $R_e = \frac{rR_1}{R_1 + r}$
- contact between the ball and outer race: $R_e = \frac{r(R_1 + 2r)}{R_1 + r}$

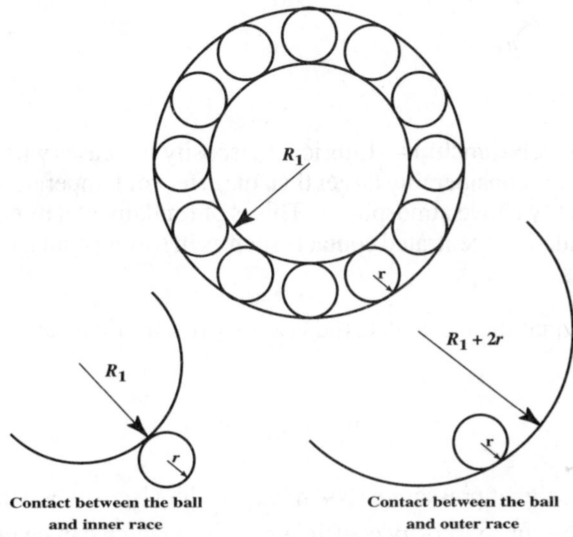

Fig. 2.

For involute gears it can readily be shown that the contact at a distance s from the pitch point can be represented by two cylinders of radii $R_{1,2}$ sinψ + s rotating with the angular velocity of the wheels (see Figure 3). In the expression below R_1 or R_2 represent pitch radii of the wheels and ψ is the pressure angle. Thus,

$$R_e = \frac{(R_1 \sin\psi + s)(R_2 \sin\psi + s)}{(R_1 + R_2)\sin\psi}$$

The thickness of the film developed in the contact zone between smooth surfaces must be related to the topography of the actual surfaces. The most commonly used parameter for this purpose is the specific film thickness defined as the ratio of the minimum film thickness for smooth surfaces (given by the above equations) to the roughness parameter of the contacting surfaces.

$$\lambda = \frac{h_o}{\sqrt{R_{m1}^2 + R_{m2}^2}}$$

where $R_m = 1.11R_a$ is the root-mean-square height of surface asperities, and R_a is the centre-line-average height of surface asperities.

If λ is greater than 3 then it is usually assumed that there is full separation of contacting bodies by an elastohydrodynamic film.

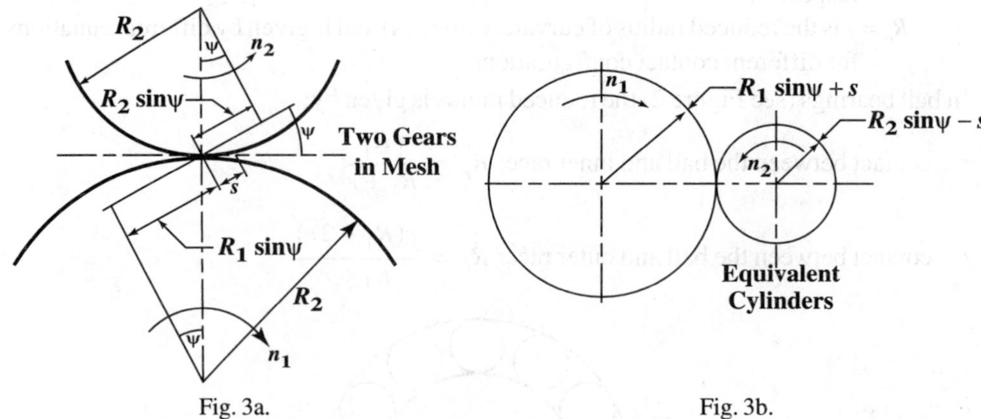

Fig. 3a. Fig. 3b.

Viscosity-pressure relationship.—Lubricant viscosity increases with pressure. For most lubricants this effect is considerably larger than the effect of temperature or shear when the pressure is appreciably above atmospheric. This is of fundamental importance in the lubrication of highly loaded concentrated contacts such as in rolling contact bearings, gears and cam-tappet systems.

The best known equation to calculate the viscosity of a lubricant at moderate pressures is the Barus equation.

$$\eta_p = \eta_o e^{\alpha p}$$

where η is the viscosity at pressure p (Ns/m^2), η_0 is the viscosity at atmospheric pressure (Ns/in^2), α is the pressure-viscosity coefficient (m^2/N) which can be obtained by plotting the natural logarithm of dynamic viscosity η measured at pressure p. The slope of the graph is α and p is the pressure of concern (N/m^2).

Values of dynamic viscosity η and pressure-viscosity coefficient α for most commonly used lubricants are given in Table 1.

Table 1. Dynamic Viscosity η and Pressure-viscosity Coefficient α for Lubricants

Lubricant	Dynamic viscosity η measured at atmospheric pressure and room temperature $\eta \times 10^{-3}$ Ns/m^2	Pressure-viscosity coefficient α measured at room temperature $\alpha \times 10^{-3}$ m^2/N
Light machine oil	45	28
Heavy machine oil	153	23.7
Cylinder oil	810	34
Spindle oil	18.6	20
Medicinal whale oil	107	29.5
Castor oil	360	15.9
Glycerol (glycerine)	535	5.9

COUPLINGS AND CLUTCHES

Connecting Shafts.—For couplings to transmit up to about 150 horsepower, simple flange-type couplings of appropriate size, as shown in the table, are commonly used. The design shown is known as a safety flange coupling because the bolt heads and nuts are shrouded by the flange, but such couplings today are normally shielded by a sheet metal or other cover.

Safety Flange Couplings

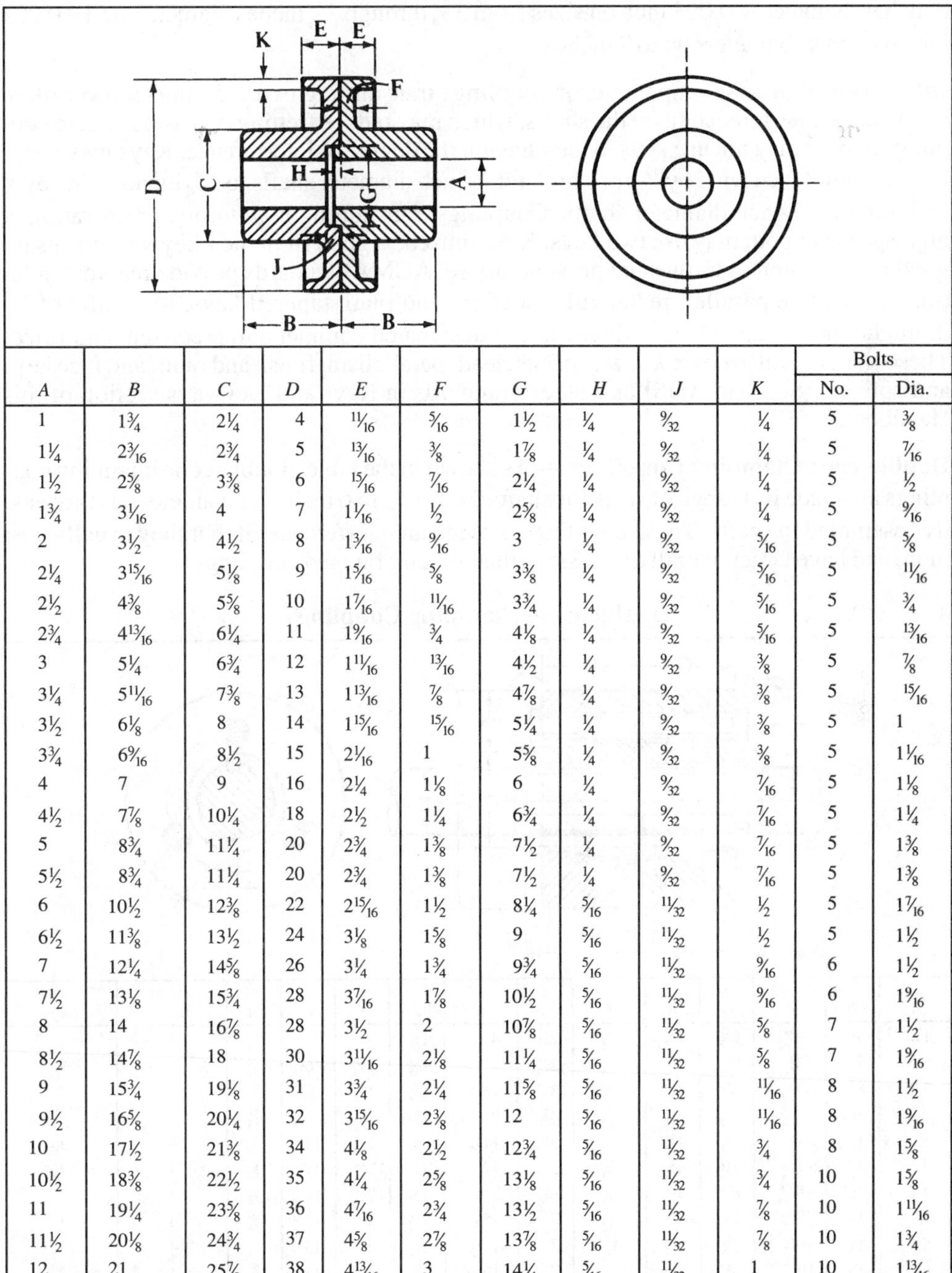

A	B	C	D	E	F	G	H	J	K	Bolts No.	Bolts Dia.
1	1 3/4	2 1/4	4	11/16	5/16	1 1/2	1/4	9/32	1/4	5	3/8
1 1/4	2 3/16	2 3/4	5	13/16	3/8	1 7/8	1/4	9/32	1/4	5	7/16
1 1/2	2 5/8	3 3/8	6	15/16	7/16	2 1/4	1/4	9/32	1/4	5	1/2
1 3/4	3 1/16	4	7	1 1/16	1/2	2 5/8	1/4	9/32	1/4	5	9/16
2	3 1/2	4 1/2	8	1 3/16	9/16	3	1/4	9/32	5/16	5	5/8
2 1/4	3 15/16	5 1/8	9	1 5/16	5/8	3 3/8	1/4	9/32	5/16	5	11/16
2 1/2	4 3/8	5 5/8	10	1 7/16	11/16	3 3/4	1/4	9/32	5/16	5	3/4
2 3/4	4 13/16	6 1/4	11	1 9/16	3/4	4 1/8	1/4	9/32	5/16	5	13/16
3	5 1/4	6 3/4	12	1 11/16	13/16	4 1/2	1/4	9/32	3/8	5	7/8
3 1/4	5 11/16	7 3/8	13	1 13/16	7/8	4 7/8	1/4	9/32	3/8	5	15/16
3 1/2	6 1/8	8	14	1 15/16	15/16	5 1/4	1/4	9/32	3/8	5	1
3 3/4	6 9/16	8 1/2	15	2 1/16	1	5 5/8	1/4	9/32	3/8	5	1 1/16
4	7	9	16	2 1/4	1 1/8	6	1/4	9/32	7/16	5	1 1/8
4 1/2	7 7/8	10 1/4	18	2 1/2	1 1/4	6 3/4	1/4	9/32	7/16	5	1 1/4
5	8 3/4	11 1/4	20	2 3/4	1 3/8	7 1/2	1/4	9/32	7/16	5	1 3/8
5 1/2	8 3/4	11 1/4	20	2 3/4	1 3/8	7 1/2	1/4	9/32	7/16	5	1 3/8
6	10 1/2	12 3/8	22	2 15/16	1 1/2	8 1/4	5/16	11/32	1/2	5	1 7/16
6 1/2	11 3/8	13 1/2	24	3 1/8	1 5/8	9	5/16	11/32	1/2	5	1 1/2
7	12 1/4	14 5/8	26	3 1/4	1 3/4	9 3/4	5/16	11/32	9/16	6	1 1/2
7 1/2	13 1/8	15 3/4	28	3 7/16	1 7/8	10 1/2	5/16	11/32	9/16	6	1 9/16
8	14	16 7/8	28	3 1/2	2	10 7/8	5/16	11/32	5/8	7	1 1/2
8 1/2	14 7/8	18	30	3 11/16	2 1/8	11 1/4	5/16	11/32	5/8	7	1 9/16
9	15 3/4	19 1/8	31	3 3/4	2 1/4	11 5/8	5/16	11/32	11/16	8	1 1/2
9 1/2	16 5/8	20 1/4	32	3 15/16	2 3/8	12	5/16	11/32	11/16	8	1 9/16
10	17 1/2	21 3/8	34	4 1/8	2 1/2	12 3/4	5/16	11/32	3/4	8	1 5/8
10 1/2	18 3/8	22 1/2	35	4 1/4	2 5/8	13 1/8	5/16	11/32	3/4	10	1 5/8
11	19 1/4	23 5/8	36	4 7/16	2 3/4	13 1/2	5/16	11/32	7/8	10	1 11/16
11 1/2	20 1/8	24 3/4	37	4 5/8	2 7/8	13 7/8	5/16	11/32	7/8	10	1 3/4
12	21	25 7/8	38	4 13/16	3	14 1/4	5/16	11/32	1	10	1 13/16

For small sizes and low power applications, a setscrew may provide the connection between the hub and the shaft, but higher power usually requires a key and perhaps two setscrews, one of them above the key. A flat on the shaft and some means of locking the setscrew(s) in position are advisable. In the AGMA Class I and II fits the shaft tolerances are –0.0005 inch from ½ to 1 ½ inches diameter and -0.001 inch on larger diameters up to 7 inches.

Class I coupling bore tolerances are + 0.001 inch up to 1 ½ inches diameter, then + 0.0015 inch to 7 inches diameter. Class II coupling bore tolerances are + 0.002 inch on sizes up to 3 inches diameter, + 0.003 inch on sizes from 3 ¼ through 3¾ inches diameter, and + 0.004 inch on larger diameters up to 7 inches.

Interference Fits.—Components of couplings transmitting over 150 horsepower often are made an interference fit on the shafts, which may reduce fretting corrosion. These couplings may or may not use keys, depending on the degree of interference. Keys may range in size from ⅛ inch wide by 1/16 inch high for ½-inch diameter shafts to 1 ¾ inches wide by ⅞ inch high for 7-inch diameter shafts. Couplings transmitting high torque or operating at high speeds or both may use two keys. Keys must be a good fit in their keyways to ensure good transmission of torque and prevent failure. AGMA standards provide recommendations for square parallel, rectangular section, and plain tapered keys, for shafts of 5/16 through 7 inches diameter, in three classes designated commercial, precision, and fitted. These standards also cover keyway offset, lead, parallelism, finish and radii, and face keys and splines. (See also ANSI and other Standards in Keys and Keyways section of this Handbook.)

Double-cone Clamping Couplings.—As shown in the table, double-cone clamping couplings are made in a range of sizes for shafts from 1 7/16 to 6 inches in diameter, and are easily assembled to shafts. These couplings provide an interference fit, but they usually cost more and have larger overall dimensions than regular flanged couplings.

Double-cone Clamping Couplings

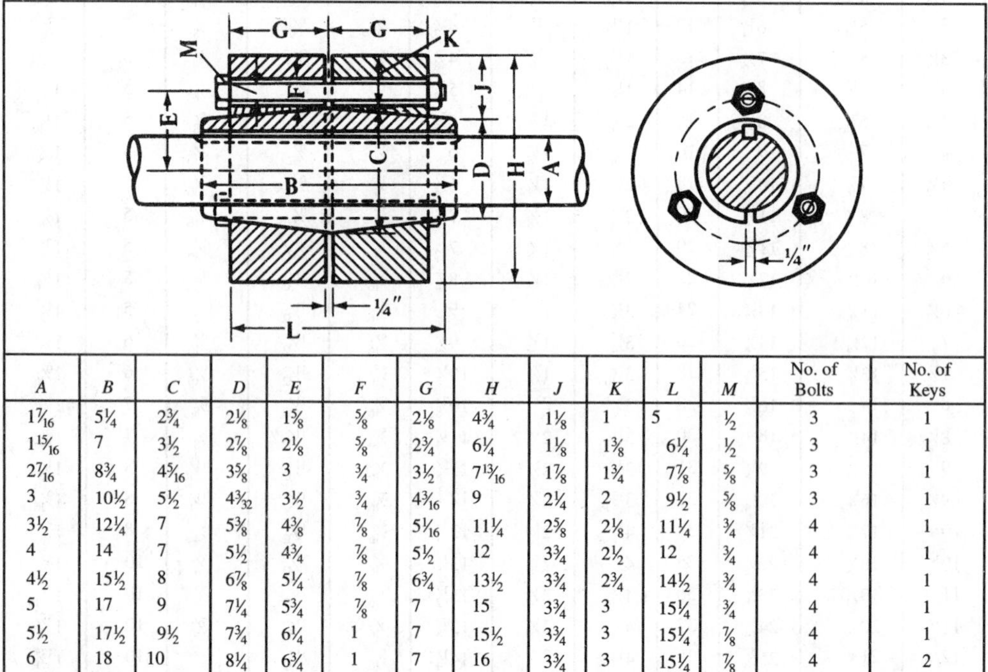

A	B	C	D	E	F	G	H	J	K	L	M	No. of Bolts	No. of Keys
1 7/16	5¼	2¾	2⅛	1⅝	⅝	2⅛	4¾	1⅛	1	5	½	3	1
1 15/16	7	3½	2⅞	2⅛	⅝	2¾	6¼	1⅛	1⅜	6¼	½	3	1
2 7/16	8¾	4 5/16	3⅝	3	¾	3½	7 13/16	1⅞	1¾	7⅞	⅝	3	1
3	10½	5½	4 3/32	3½	¾	4 3/16	9	2¼	2	9½	⅝	3	1
3½	12¼	7	5⅜	4⅜	⅞	5 1/16	11¼	2⅝	2⅛	11¼	¾	4	1
4	14	7	5½	4¾	⅞	5½	12	3¾	2½	12	¾	4	1
4½	15½	8	6⅞	5¼	⅞	6¾	13½	3¾	2¾	14½	¾	4	1
5	17	9	7¼	5¾	⅞	7	15	3¾	3	15¼	¾	4	1
5½	17½	9½	7¾	6¼	1	7	15½	3¾	3	15¼	⅞	4	1
6	18	10	8¼	6¾	1	7	16	3¾	3	15¼	⅞	4	2

Flexible Couplings.—Shafts that are out of alignment laterally or angularly can be connected by any of several designs of flexible couplings. Such couplings also permit some degree of axial movement in one or both shafts. Some couplings use disks or diaphragms to transmit the torque. Another simple form of flexible coupling consists of two flanges connected by links or endless belts made of leather or other strong, pliable material. Alternatively, the flanges may have projections that engage spacers of molded rubber or other flexible materials that accommodate uneven motion between the shafts. More highly developed flexible couplings use toothed flanges engaged by correspondingly toothed elements, permitting relative movement. These couplings require lubrication unless one or more of the elements is made of a self-lubricating material. Other couplings use diaphragms or bellows that can flex to accommodate relative movement between the shafts.

The Universal Joint.—This form of coupling, originally known as a Cardan or Hooke's coupling, is used for connecting two shafts the axes of which are not in line with each other, but which merely intersect at a point. There are many different designs of universal joints or couplings, which are based on the principle embodied in the original design. One well-known type is shown by the accompanying diagram.

As a rule, a universal joint does not work well if the angle α (see illustration) is more than 45 degrees, and the angle should preferably be limited to about 20 degrees or 25 degrees, excepting when the speed of rotation is slow and little power is transmitted.

Variation in Angular Velocity of Driven Shaft: Owing to the angularity between two shafts connected by a universal joint, there is a variation in the angular velocity of one shaft during a single revolution, and because of this, the use of universal couplings is sometimes prohibited. Thus, the angular velocity of the driven shaft will not be the same at all points of the revolution as the angular velocity of the driving shaft. In other words, if the driving shaft moves with a uniform motion, then the driven shaft will have a variable motion and, therefore, the universal joint should not be used when absolute uniformity of motion is essential for the driven shaft.

Determining Maximum and Minimum Velocities: If shaft *A* (see diagram) runs at a constant speed, shaft *B* revolves at maximum speed when shaft A occupies the position shown in the illustration, and the minimum speed of shaft *B* occurs when the fork of the driving shaft *A* has turned 90 degrees from the position illustrated. The maximum speed of the driven shaft may be obtained by multiplying the speed of the driving shaft by the secant of angle α. The minimum speed of the driven shaft equals the speed of the driver multiplied by cosine α. Thus, if the driver rotates at a constant speed of 100 revolutions per minute and the shaft angle is 25 degrees, the maximum speed of the driven shaft is at a rate equal to $1.1034 \times 100 = 110.34$ R.P.M. The minimum speed rate equals $0.9063 \times 100 = 90.63$; hence, the extreme variation equals $110.34 - 90.63 = 19.71$ R.P.M.

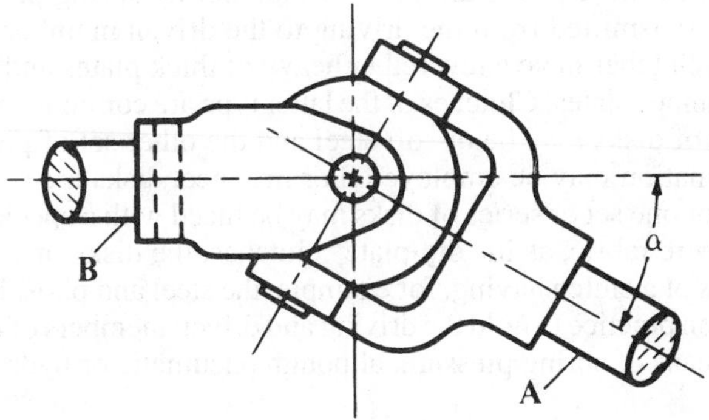

Use of Intermediate Shaft between Two Universal Joints.—The lack of uniformity in the speed of the driven shaft resulting from the use of a universal coupling, as previously explained, is objectionable for some forms of mechanisms. This variation may be avoided if the two shafts are connected with an intermediate shaft and two universal joints, provided the latter are properly arranged or located. Two conditions are necessary to obtain a constant speed ratio between the driving and driven shafts. First, the shafts must make the same angle with the intermediate shaft; second, the universal joint forks (assuming that the fork design is employed) on the intermediate shaft must be placed relatively so that when the plane of the fork at the left end coincides with the center lines of the intermediate shaft and the shaft attached to the left-hand coupling, the plane of the right-hand fork must also coincide with the center lines of the intermediate shaft and the shaft attached to the right-hand coupling; therefore the driving and the driven shafts may be placed in a variety of positions. One of the most common arrangements is with the driving and driven shafts parallel. The forks on the intermediate shafts should then be placed in the same plane.

This intermediate connecting shaft is frequently made telescoping, and then the driving and driven shafts can be moved independently of each other within certain limits in longitudinal and lateral directions. The telescoping intermediate shaft consists of a rod which enters a sleeve and is provided with a suitable spline, to prevent rotation between the rod and sleeve and permit a sliding movement. This arrangement is applied to various machine tools.

Knuckle Joints.—Movement at the joint between two rods may be provided by knuckle joints, for which typical proportions are seen in the table *Proportions of Knuckle Joints* that follows.

Friction Clutches.—Clutches which transmit motion from the driving to the driven member by the friction between the engaging surfaces are built in many different designs, although practically all of them can be classified under four general types, namely, conical clutches; radially expanding clutches; contracting-band clutches; and friction disk clutches in single and multiple types. There are many modifications of these general classes, some of which combine the features of different types. The proportions of various sizes of cone clutches are given in the table "Cast-iron Friction Clutches." The multicone friction clutch is a further development of the cone clutch. Instead of having a single cone-shaped surface, there is a series of concentric conical rings which engage annular grooves formed by corresponding rings on the opposite clutch member. The internal-expanding type is provided with shoes which are forced outward against an enclosing drum by the action of levers connecting with a collar free to slide along the shaft. The engaging shoes are commonly lined with wood or other material to increase the coefficient of friction. Disk clutches are based on the principle of multiple-plane friction, and use alternating plates or disks so arranged that one set engages with an outside cylindrical case and the other set with the shaft. When these plates are pressed together by spring pressure, or by other means, motion is transmitted from the driving to the driven members connected to the clutch. Some disk clutches have a few rather heavy or thick plates and others a relatively large number of thinner plates. Clutches of the latter type are common in automobile transmissions. One set of disks may be of soft steel and the other set of phosphor-bronze, or some other combination may be employed. For instance, disks are sometimes provided with cork inserts, or one set or series of disks may be faced with a special friction material such as asbestos-wire fabric, as in "dry plate" clutches, the disks of which are not lubricated like the disks of a clutch having, for example, the steel and phosphor-bronze combination. It is common practice to hold the driving and driven members of friction clutches in engagement by means of spring pressure, although pneumatic or hydraulic pressure may be employed.

Proportions of Knuckle Joints

For sizes not given below:

$a = 1.2D$	$h = 2D$
$b = 1.1D$	$i = 0.5D$
$c = 1.2D$	$j = 0.25D$
$e = 0.75D$	$k = 0.5D$
$f = 0.6D$	$l = 1.5D$
$g = 1.5D$	

D	*a*	*b*	*c*	*e*	*f*	*g*	*h*	*i*	*j*	*k*	*l*
1/2	5/8	9/16	5/8	3/8	5/16	3/4	1	1/4	1/8	1/4	3/4
3/4	7/8	3/4	7/8	9/16	7/16	1 1/8	1 1/2	3/8	3/16	3/8	1 1/8
1	1 1/4	1 1/8	1 1/4	3/4	5/8	1 1/2	2	1/2	1/4	1/2	1 1/2
1 1/4	1 1/2	1 3/8	1 1/2	15/16	3/4	1 7/8	2 1/2	5/8	5/16	5/8	1 7/8
1 1/2	1 3/4	1 5/8	1 3/4	1 1/8	7/8	2 1/4	3	3/4	3/8	3/4	2 1/4
1 3/4	2 1/8	2	2 1/8	1 5/16	1 1/16	2 5/8	3 1/2	7/8	7/16	7/8	2 5/8
2	2 3/8	2 1/4	2 3/8	1 1/2	1 3/16	3	4	1	1/2	1	3
2 1/4	2 3/4	2 1/2	2 3/4	1 11/16	1 3/8	3 3/8	4 1/2	1 1/8	9/16	1 1/8	3 3/8
2 1/2	3	2 3/4	3	1 7/8	1 1/2	3 3/4	5	1 1/4	5/8	1 1/4	3 3/4
2 3/4	3 1/4	3	3 1/4	2 1/16	1 5/8	4 1/8	5 1/2	1 3/8	11/16	1 3/8	4 1/8
3	3 5/8	3 1/4	3 5/8	2 1/4	1 13/16	4 1/2	6	1 1/2	3/4	1 1/2	4 1/2
3 1/4	4	3 3/8	4	2 7/16	2	4 7/8	6 1/2	1 5/8	13/16	1 5/8	4 7/8
3 1/2	4 1/4	3 7/8	4 1/4	2 5/8	2 1/8	5 1/4	7	1 3/4	7/8	1 3/4	5 1/4
3 3/4	4 1/2	4 1/8	4 1/2	2 13/16	2 1/4	5 5/8	7 1/2	1 7/8	15/16	1 7/8	5 5/8
4	4 3/4	4 3/8	4 3/4	3	2 3/8	6	8	2	1	2	6
4 1/4	5 1/8	4 3/4	5 1/8	3 3/16	2 9/16	6 3/8	8 1/2	2 1/8	1 1/16	2 1/8	6 3/8
4 1/2	5 1/2	5	5 1/2	3 3/8	2 3/4	6 3/4	9	2 1/4	1 1/8	2 1/4	6 3/4
4 3/4	5 3/4	5 1/4	5 3/4	3 9/16	2 7/8	7 1/8	9 1/2	2 3/8	1 3/16	2 3/8	7 1/8
5	6	5 1/2	6	3 3/4	3	7 1/2	10	2 1/2	1 1/4	2 1/2	7 1/2

Power Transmitting Capacity of Friction Clutches.—When selecting a clutch for a given class of service, it is advisable to consider any overloads that may be encountered and base the power transmitting capacity of the clutch upon such overloads. When the load varies or is subject to frequent release or engagement, the clutch capacity should be greater than the actual amount of power transmitted. If the power is derived from a gas or gasoline engine, the horsepower rating of the clutch should be 75 or 100 per cent greater than that of the engine.

Power Transmitted by Disk Clutches.—The approximate amount of power that a disk clutch will transmit may be determined from the following formula, in which H = horsepower transmitted by the clutch; μ = coefficient of friction; r = mean radius of engaging surfaces; F = axial force in pounds (spring pressure) holding disks in contact; N = number of frictional surfaces; S = speed of shaft in revolutions per minute:

$$H = \frac{\mu rFNS}{63{,}000}$$

Cast-iron Friction Clutches

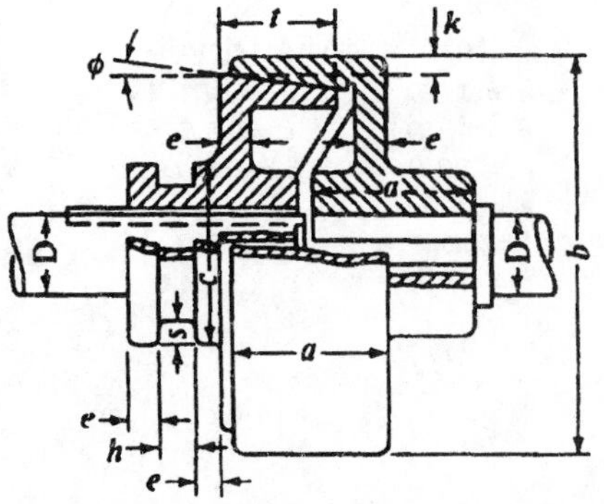

For sizes not given below:

$a = 2D$
$b = 4 \text{ to } 8D$
$c = 2\frac{1}{4}D$
$t = 1\frac{1}{2}D$
$e = \frac{3}{8}D$
$h = \frac{1}{2}D$
$s = \frac{5}{16}D$, nearly
$k = \frac{1}{4}D$

Note:— The angle ϕ of the cone may be from 4 to 10 degrees

D	*a*	*b*	*c*	*t*	*e*	*h*	*s*	*k*
1	2	4 – 8	2 1/4	1 1/2	3/8	1/2	5/16	1/4
1 1/4	2 1/2	5–10	2 7/8	1 7/8	1/2	5/8	3/8	5/16
1 1/2	3	6–12	3 3/8	2 1/4	5/8	3/4	1/2	3/8
1 3/4	3 1/2	7–14	4	2 5/8	5/8	7/8	5/8	7/16
2	4	8–16	4 1/2	3	3/4	1	5/8	1/2
2 1/4	4 1/2	9–18	5	3 3/8	7/8	1 1/8	5/8	9/16
2 1/2	5	10–20	5 5/8	3 3/4	1	1 1/4	3/4	5/8
2 3/4	5 1/2	11–22	6 1/4	4 1/8	1	1 3/8	7/8	11/16
3	6	12–24	6 3/4	4 1/2	1 1/8	1 1/2	7/8	3/4
3 1/4	6 1/2	13–26	7 3/8	4 7/8	1 1/4	1 5/8	1	13/16
3 1/2	7	14–28	7 7/8	5 1/4	1 3/8	1 3/4	1	7/8
3 3/4	7 1/2	15–30	8 1/2	5 5/8	1 3/8	1 7/8	1 1/4	15/16
4	8	16–32	9	6	1 1/2	2	1 1/4	1
4 1/4	8 1/2	17–34	9 1/2	6 3/8	1 5/8	2 1/8	1 3/8	1 1/16
4 1/2	9	18–36	10 1/4	6 3/4	1 3/4	2 1/4	1 3/8	1 1/8
4 3/4	9 1/2	19–38	10 3/4	7 1/8	1 3/4	2 3/8	1 1/2	1 3/16
5	10	20–40	11 1/4	7 1/2	1 7/8	2 1/2	1 1/2	1 1/4
5 1/4	10 1/2	21–42	11 3/4	7 7/8	2	2 5/8	1 5/8	1 5/16
5 1/2	11	22–44	12 3/8	8 1/4	2	2 3/4	1 3/4	1 3/8
5 3/4	11 1/2	23–46	13	8 5/8	2 1/4	2 7/8	1 3/4	1 7/16
6	12	24–48	13 1/2	9	2 1/4	3	1 7/8	1 1/2

Frictional Coefficients for Clutch Calculations.—While the frictional coefficients used by designers of clutches differ somewhat and depend upon variable factors, the following values may be used in clutch calculations: For greasy leather on cast iron about 0.20 or 0.25, leather on metal that is quite oily 0.15; metal and cork on oily metal 0.32; the same on dry metal 0.35; metal on dry metal 0.15; disk clutches having lubricated surfaces 0.10.

Formulas for Cone Clutches.—In cone clutch design, different formulas have been developed for determining the horsepower transmitted. These formulas, at first sight, do not seem to agree, there being a variation due to the fact that in some of the formulas the friction clutch surfaces are assumed to engage without slip, whereas, in others, some allowance is made for slip. The following formulas include both of these conditions:

H.P. = horsepower transmitted

N = revolutions per minute
r = mean radius of friction cone, in inches
r_1 = large radius of friction cone, in inches
r_2 = small radius of friction cone, in inches
R_1 = outside radius of leather band, in inches
R_2 = inside radius of leather band, in inches
V = velocity of a point at distance r from the center, in feet per minute
F = tangential force acting at radius r, in pounds
P_n = total normal force between cone surfaces, in pounds
P_s = spring force, in pounds
α = angle of clutch surface with axis of shaft = 7 to 13 degrees
β = included angle of clutch leather, when developed, in degrees
f = coefficient of friction = 0.20 to 0.25 for greasy leather on iron
p = allowable pressure per square inch of leather band = 7 to 8 pounds
W = width of clutch leather, in inches

$$R_1 = \frac{r_1}{\sin\alpha} \qquad R_2 = \frac{r_2}{\sin\alpha}$$

DEVELOPMENT OF CLUTCH LEATHER

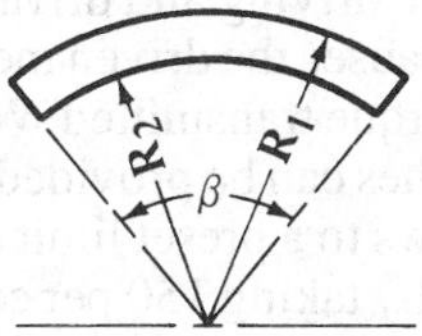

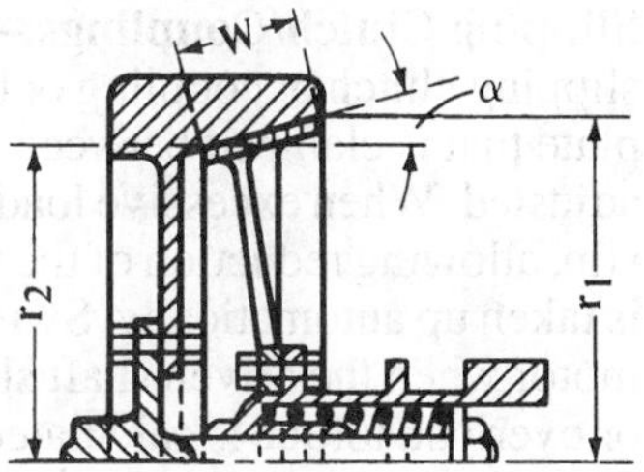

$$\beta = \sin\alpha \times 360 \qquad r = \frac{r_1 + r_2}{2}$$

$$V = \frac{2\pi rN}{12}$$

$$F = \frac{\text{HP} \times 33{,}000}{V} \qquad W = \frac{P_n}{2\pi rp} \qquad \text{HP} = \frac{P_n frN}{63{,}025}$$

For engagement with some slip:

$$P_n = \frac{P_s}{\sin\alpha} \qquad P_s = \frac{\text{HP} \times 63{,}025 \sin\alpha}{frN}$$

For engagement without slip:

$$P_n = \frac{P_s}{\sin\alpha + f\cos\alpha} \qquad P_s = \frac{\text{HP} \times 63{,}025(\sin\alpha + f\cos\alpha)}{frN}$$

Angle of Cone.—If the angle of the conical surface of the cone type of clutch is too small, it may be difficult to release the clutch on account of the wedging effect, whereas, if the angle is too large, excessive spring force will be required to prevent slipping. The minimum angle for a leather-faced cone is about 8 or 9 degrees and the maximum angle about 13 degrees. An angle of 12½ degrees appears to be the most common and is generally considered good practice. These angles are given with relation to the clutch axis and are one-half the included angle.

Magnetic Clutches.—Many disk and other clutches are operated electromagnetically with the magnetic force used only to move the friction disk(s) and the clutch disk(s) into or out of engagement against spring or other pressure. On the other hand, in a magnetic particle clutch, transmission of power is accomplished by magnetizing a quantity of metal particles enclosed between the driving and the driven components. forming a bond between them. Such clutches can be controlled to provide either a rigid coupling or uniform slip, useful in wire drawing and manufacture of cables.

Another type of magnetic clutch uses eddy currents induced in the input member which interact with the field in the output rotor. Torque transmitted is proportional to the coil current, so precise control of torque is provided. A third type of magnetic clutch relies on the hysteresis loss between magnetic fields generated by a coil in an input drum and a close-fitting cup on the output shaft, to transmit torque. Torque transmitted with this type of clutch also is proportional to coil current, so close control is possible.

Permanent-magnet types of clutches also are available, in which the engagement force is exerted by permanent magnets when the electrical supply to the disengagement coils is cut off. These types of clutches have capacities up to five times the torque-to-weight ratio of spring-operated clutches. In addition, if the controls are so arranged as to permit the coil polarity to be reversed instead of being cut off, the combined permanent magnet and electromagnetic forces can transmit even greater torque.

Centrifugal and Free-wheeling Clutches.—Centrifugal clutches have driving members that expand outward to engage a surrounding drum when speed is sufficient to generate centrifugal force. Free-wheeling clutches are made in many different designs and use balls, cams or sprags, ratchets, and fluids to transmit motion from one member to the other. These types of clutches are designed to transmit torque in only one direction and to take up the drive with various degrees of gradualness up to instantaneously.

Slipping Clutch/Couplings.—Where high shock loads are likely to be experienced, a slipping clutch or coupling or both should be used. The most common design uses a clutch plate that is clamped between the driving and driven plates by spring pressure that can be adjusted. When excessive load causes the driven member to slow, the clutch plate surfaces slip, allowing reduction of the torque transmitted. When the overload is removed, the drive is taken up automatically. Switches can be provided to cut off current supply to the driving motor when the driven shaft slows to a preset limit or to signal a warning or both. The slip or overload torque is calculated by taking 150 per cent of the normal running torque.

Wrapped-spring Clutches.—For certain applications, a simple steel spring sized so that its internal diameter is a snug fit on both driving and driven shafts will transmit adequate torque in one direction. The tightness of grip of the spring on the shafts increases as the torque transmitted increases. Disengagement can be effected by slight rotation of the spring, through a projecting tang, using electrical or mechanical means, to wind up the spring to a larger internal diameter, allowing one of the shafts to run free within the spring.

Normal running torque T_r in lb-ft = (required horsepower × 5250) ÷ rpm. For heavy shock load applications, multiply by a 200 per cent or greater overload factor. (See Motors, factors governing selection.)

The clutch starting torque T_c, in lb-ft, required to accelerate a given inertia in a specific time is calculated from the formula:

$$T_c = \frac{WR^2 \times \Delta N}{308t}$$

where WR^2 = total inertia encountered by clutch in lb-ft^2 (W = weight and R = radius of gyration of rotating part)

ΔN = final rpm — initial rpm; 308 = constant (see Motors, factors governing selection)

t = time to required speed in seconds

Example: If the inertia is 80 lb-ft^2, and the speed of the driven shaft is to be increased from 0 to 1500 rpm in 3 seconds, find the clutch starting torque in lb-ft.

$$T_c = \frac{80 \times 1500}{308 \times 3} = 130 \text{ lb-ft}$$

The heat E, in BTU, generated in one engagement of a clutch can be calculated from the formula:

$$E = \frac{T_c \times WR^2 \times (N_1^2 - N_2^2)}{(T_c - T_1) \times 4.7 \times 10^6}$$

where: WR^2 = total inertia encountered by clutch in lb-ft.^{2}

N_1 = final rpm

N_2 = initial rpm

Tc = clutch torque in lb-ft

T_1 = torque load in lb-ft

Example: Calculate the heat generated for each engagement under the conditions cited for the first example.

$$E = \frac{130 \times 80 \times (1500)^2}{(130 - 10) \times 4.7 \times 10^6} = 41.5 \text{ BTU}$$

The preferred location for a clutch is on the high- rather than on the low-speed shaft because a smaller-capacity unit, of lower cost and with more rapid dissipation of heat, can be used. However, the heat generated may also be more because of the greater slippage at higher speeds, and the clutch may have a shorter life. For light-duty applications, such as to a machine tool, where cutting occurs after the spindle has reached operating speed, the calculated torque should be multiplied by a safety factor of 1.5 to arrive at the capacity of the clutch to be used. Heavy-duty applications such as frequent starting of a heavily loaded vibratory-finishing barrel require a safety factor of 3 or more.

Positive Clutches.—When the driving and driven members of a clutch are connected by the engagement of interlocking teeth or projecting lugs, the clutch is said to be "positive" to distinguish it from the type in which the power is transmitted by frictional contact. The positive clutch is employed when a sudden starting action is not objectionable and when the inertia of the driven parts is relatively small. The various forms of positive clutches differ merely in the angle or shape of the engaging surfaces. The least positive form is one having planes of engagement which incline backward, with respect to the direction of motion. The tendency of such a clutch is to disengage under load, in which case it must be held in position by axial pressure.

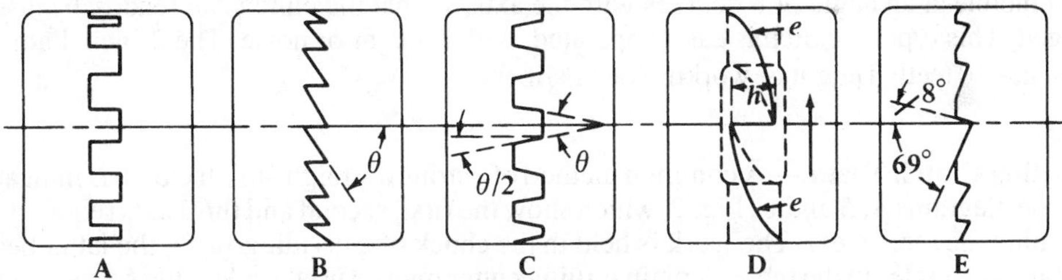

Fig. 1. Types of Clutch Teeth

This pressure may be regulated to perform normal duty, permitting the clutch to slip and disengage when over-loaded. Positive clutches, with the engaging planes parallel to the axis of rotation, are held together to obviate the tendency to jar out of engagement, but they provide no safety feature against over-load. So-called "under-cut" clutches engage more tightly the heavier the load, and are designed to be disengaged only when free from load. The teeth of positive clutches are made in a variety of forms, a few of the more common styles being shown in Fig. 1. Clutch *A* is a straight-toothed type, and *B* has angular or saw-shaped teeth. The driving member of the former can be rotated in either direction: the latter is adapted to the transmission of motion in one direction only, but is more readily engaged. The angle θ of the cutter for a saw-tooth clutch *B* is ordinarily 60 degrees. Clutch *C* is similar to *A*, except that the sides of the teeth are inclined to facilitate engagement and disengagement. Teeth of this shape are sometimes used when a clutch is required to run in either direction without backlash. Angle θ is varied to suit requirements and should not exceed 16 or 18 degrees. The straight-tooth clutch A is also modified to make the teeth engage more readily, by rounding the corners of the teeth at the top and bottom. Clutch *D* (commonly called a "spiral-jaw" clutch) differs from *B* in that the surfaces *e* are helicoidal. The driving member of this clutch can transmit motion in only one direction.

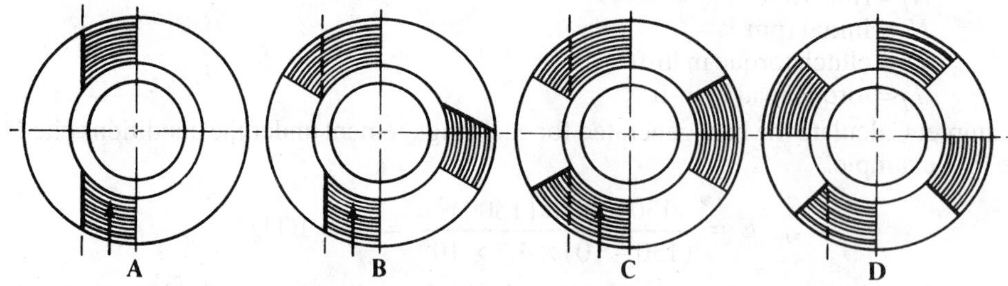

Fig. 2. Diagrammatic View Showing Method of Cutting Clutch Teeth

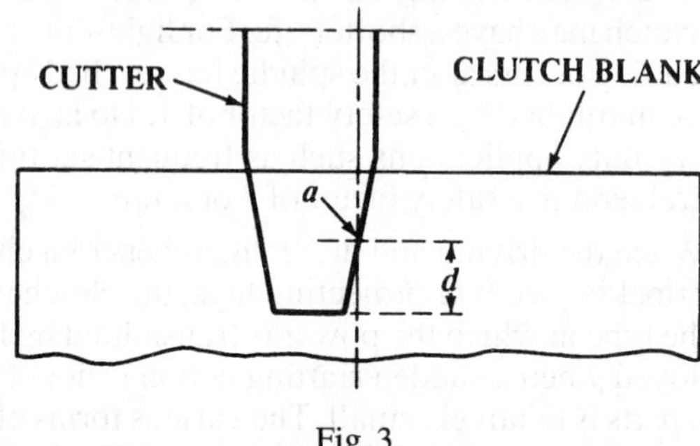

Fig. 3.

Clutches of this type are known as right- and left-hand, the former driving when turning to the right, as indicated by the arrow in the illustration. Clutch *E* is the form used on the back-shaft of the Brown & Sharpe automatic screw machines. The faces of the teeth are radial and incline at an angle of 8 degrees with the axis, so that the clutch can readily be disengaged. This type of clutch is easily operated, with little jar or noise. The 2-inch diameter size has 10 teeth. Height of working face, ⅛ inch.

Cutting Clutch Teeth.—A common method of cutting a straight-tooth clutch is indicated by the diagrams *A*, *B* and *C*, Fig. 2, which show the first, second and third cuts required for forming the three teeth. The work is held in the chuck of a dividing-head, the latter being set at right angles to the table. A plain milling cutter may be used (unless the corners of the teeth are rounded), the side of the cutter being set to exactly coincide with the center-line. When the number of teeth in the clutch is odd, the cut can be taken clear across the blank as shown, thus finishing the sides of two teeth with one passage of the cutter. When the number of teeth is even, as at *D*, it is necessary to mill all the teeth on one side and then set the cutter for finishing the opposite side. Therefore, clutches of this type commonly have an odd number of teeth. The maximum width of the cutter depends upon the width of the space at the narrow ends of the teeth. If the cutter must be quite narrow in order to pass the narrow ends, some stock may be left in the tooth spaces, which must be removed by a separate cut. If the tooth is of the modified form shown at *C*, Fig. 1, the cutter should be set as indicated in Fig. 3; that is, so that a point a on the cutter at a radial distance *d* equal to one-half the depth of the clutch teeth lies in a radial plane. When it is important to eliminate all backlash, point *a* is sometimes located at a radial distance *d* equal to six-tenths of the depth of the tooth, in order to leave clearance spaces at the bottoms of the teeth; the two clutch members will then fit together tightly. Clutches of this type must be held in mesh.

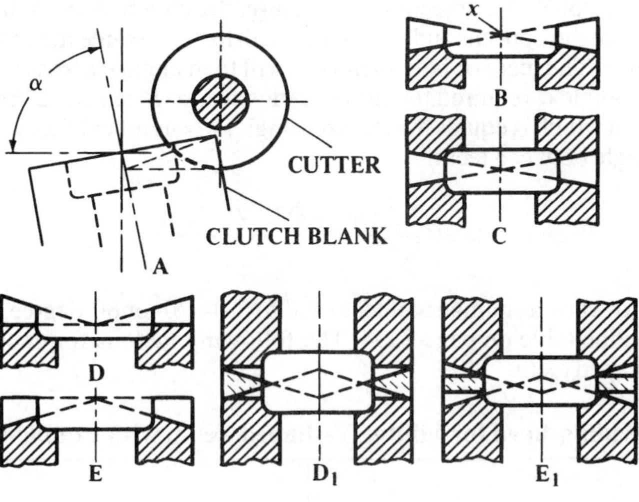

Fig. 4.

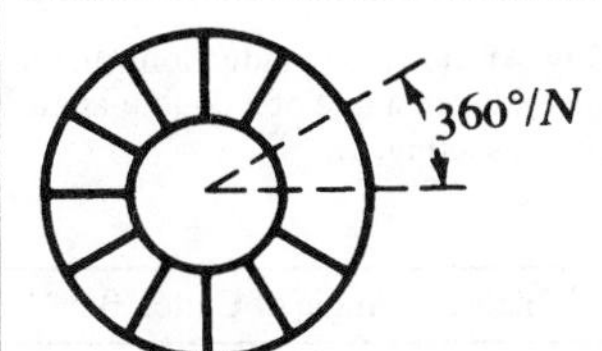

$$\cos\alpha = \frac{\tan(360°/N) \times \cot\theta}{2}$$

α is the angle shown in Fig. 4 and is the angle shown by the graduations on the dividing head. θ is the included angle of a single cutter, see Fig. 1.

No. of Teeth, *N*	Angle of Single-angle Cutter, θ			No. of Teeth, *N*	Angle of Single-angle Cutter, θ		
	60°	70°	80°		60°	70°	80°
	Dividing Head Angle, α				Dividing Head Angle, α		
5	27° 19.2′	55° 56.3′	74° 15.4′	18	83° 58.1′	86° 12.1′	88° 9.67′
6	60	71 37.6	81 13	19	84 18.8	86 25.1	88 15.9
7	68 46.7	76 48.5	83 39.2	20	84 37.1	86 36.6	88 21.5
8	73 13.3	79 30.9	84 56.5	21	84 53.5	86 46.9	88 26.5
9	75 58.9	81 13	85 45.4	22	85 8.26	86 56.2	88 31
10	77 53.6	82 24.1	86 19.6	23	85 21.6	87 4.63	88 35.1
11	79 18.5	83 17	86 45.1	24	85 33.8	87 12.3	88 38.8
12	80 24.4	83 58.1	87 4.94	25	85 45	87 19.3	88 42.2
13	81 17.1	84 31.1	87 20.9	26	85 55.2	87 25.7	88 45.3
14	82 .536	84 58.3	87 34	27	86 4.61	87 31.7	88 48.2
15	82 36.9	85 21.2	87 45	28	86 13.3	87 37.2	88 50.8
16	83 7.95	85 40.6	87 54.4	29	86 21.4	87 42.3	88 53.3
17	83 34.7	85 57.4	88 2.56	30	86 28.9	87 47	88 55.6

Cutting Saw-tooth Clutches: When milling clutches having angular teeth as shown at *B*, Fig. 1, the axis of the clutch blank should be inclined a certain angle α as shown at *A* in Fig. 4. If the teeth were milled with the blank vertical, the tops of the teeth would incline towards the center as at *D*, whereas, if the blank were set to such an angle that the tops of the teeth were square with the axis, the bottoms would incline upwards as at *E*. In either case, the two clutch members would not mesh completely: the engagement of the teeth cut as shown at *D* and *E* would be as indicated at D_1 and E_1 respectively. As will be seen, when the outer points of the teeth at D_1 are at the bottom of the grooves in the opposite member, the inner ends are not together, the contact area being represented by the dotted lines. At E_1 the inner ends of the teeth strike first and spaces are left between the teeth around the outside of the clutch. To overcome this objectionable feature, the clutch teeth should be cut as indicated at *B*, or so that the bottoms and tops of the teeth have the same inclination, converging at a central point *x*. The teeth of both members will then engage across the entire width as shown at *C*. The angle α required for cutting a clutch as at *B* can be determined by the following formula in which α equals the required angle, *N* = number of teeth, θ = cutter angle, and 360°/*N* = angle between teeth:

$$\cos\alpha = \frac{\tan(360°/N) \times \cot\theta}{2}$$

The angles α for various numbers of teeth and for 60-, 70- or 80-degree single-angle cutters are given in the table on page 2334. The following table is for double-angle cutters used to cut V-shaped teeth.

Angle of Dividing-head for Milling V-shaped Teeth with Double-angle Cutter

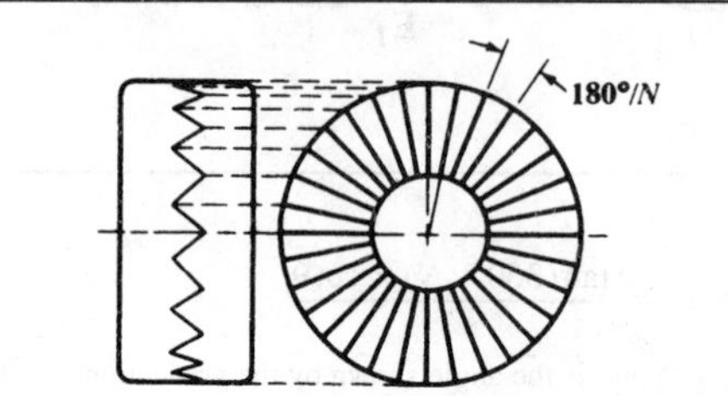

$$\cos\alpha = \frac{\tan(180°/N) \times \cot(\theta/2)}{2}$$

This is the angle (α, Fig. 4) shown by graduations on the dividing-head. θ is the included angle of a double-angle cutter, see Fig. 1.

No. of Teeth, *N*	Included Angle of Cutter, θ				No. of Teeth, *N*	Included Angle of Cutter, θ			
	60°		90°			60°		90°	
	Dividing Head Angle, α					Dividing Head Angle, α			
10	73°	39.4′	80°	39′	31	84°	56.9′	87°	5.13′
11	75	16.1	81	33.5	32	85	6.42	87	10.6
12	76	34.9	82	18	33	85	15.4	87	15.8
13	77	40.5	82	55.3	34	85	23.8	87	20.7
14	78	36	83	26.8	35	85	31.8	87	25.2
15	79	23.6	83	54	36	85	39.3	87	29.6
16	80	4.83	84	17.5	37	85	46.4	87	33.7
17	80	41	84	38.2	38	85	53.1	87	37.5
18	81	13	84	56.5	39	85	59.5	87	41.2
19	81	41.5	85	12.8	40	86	5.51	87	44.7
20	82	6.97	85	27.5	41	86	11.3	87	48
21	82	30	85	40.7	42	86	16.7	87	51.2
22	82	50.8	85	52.6	43	86	22	87	54.2
23	83	9.82	86	3.56	44	86	26.9	87	57
24	83	27.2	86	13.5	45	86	31.7	87	59.8
25	83	43.1	86	22.7	46	86	36.2	88	2.4
26	83	57.8	26	31.2	47	86	40.6	88	4.91
27	84	11.4	86	39	48	86	44.8	88	7.32
28	84	24	86	46.2	49	86	48.8	88	9.63
29	84	35.7	86	53	50	86	52.6	88	11.8
30	84	46.7	86	59.3	51	86	56.3	88	14

The angles given in the table above are applicable to the milling of V-shaped grooves in brackets, etc., which must have toothed surfaces to prevent the two members from turning relative to each other, except when unclamped for angular adjustment

FRICTION BRAKES

Formulas for Band Brakes.—In any band brake, such as shown in Fig. 1, in the tabulation of formulas, where the brake wheel rotates in a clockwise direction, the tension in that part of the band marked x equals $P\frac{1}{e^{\mu\theta}-1}$

The tension in that part marked y equals $P\frac{e^{\mu\theta}}{e^{\mu\theta}-1}$.

P = tangential force in pounds at rim of brake wheel

e = base of natural logarithms = 2.71828

μ = coefficient of friction between the brake band and the brake wheel

θ = angle of contact of the brake band with the brake wheel expressed in

radians (one radian $= \frac{180\text{ deg.}}{\pi\text{ radians}} = 57.296\frac{\text{deg.}}{\text{radian}}$).

For simplicity in the formulas presented, the tensions at x and y (Fig. 1) are denoted by T_1 and T_2 respectively, for clockwise rotation. When the direction of the rotation is reversed, the tension in x equals T_2, and the tension in y equals T_1, which is the reverse of the tension in the clockwise direction.

The value of the expression $e^{\mu\theta}$ in these formulas may be most easily found by using a hand-held calculator of the scientific type; that is, one capable of raising 2.71828 to the power $\mu\theta$. The following example outlines the steps in the calculations.

Table of Values of $e^{\mu\theta}$

Proportion of Contact to Whole Circumference, $\frac{\theta}{2\pi}$	Steel Band on Cast Iron, $\mu = 0.18$	Leather Belt on			
		Wood	Cast Iron		
		Slightly Greasy; $\mu = 0.47$	Very Greasy; $\mu = 0.12$	Slightly Greasy; $\mu = 0.28$	Damp; $\mu = 0.38$
0.1	1.12	1.34	1.08	1.19	1.27
0.2	1.25	1.81	1.16	1.42	1.61
0.3	1.40	2.43	1.25	1.69	2.05
0.4	1.57	3.26	1.35	2.02	2.60
0.425	1.62	3.51	1.38	2.11	2.76
0.45	1.66	3.78	1.40	2.21	2.93
0.475	1.71	4.07	1.43	2.31	3.11
0.5	1.76	4.38	1.46	2.41	3.30
0.525	1.81	4.71	1.49	2.52	3.50
0.55	1.86	5.07	1.51	2.63	3.72
0.6	1.97	5.88	1.57	2.81	4.19
0.7	2.21	7.90	1.66	3.43	5.32
0.8	2.47	10.60	1.83	4.09	6.75
0.9	2.77	14.30	1.97	4.87	8.57
1.0	3.10	19.20	2.12	5.81	10.90

Formulas for Simple and Differential Band Brakes

F = force in pounds at end of brake handle; P = tangential force in pounds at rim of brake wheel; e = base of natural logarithms = 2.71828; μ = coefficient of friction between the brake band and the brake wheel; θ = angle of contact of the brake band with the brake wheel, expressed in radians (one radian = 57.296 degrees).

$$T_1 = P\frac{1}{e^{\mu\theta}-1} \qquad T_2 = P\frac{e^{\mu\theta}}{e^{\mu\theta}-1}$$

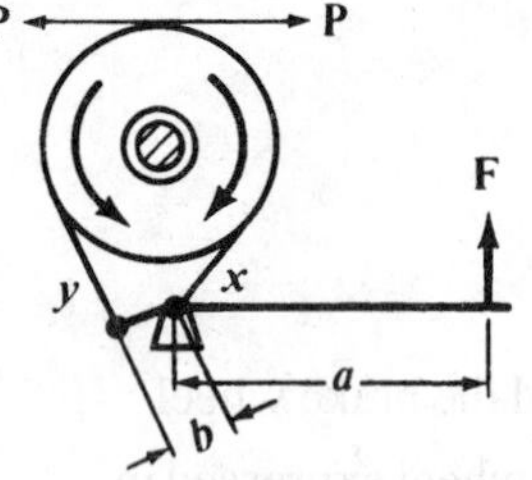

Fig. 1.

Simple band brake.
For clockwise rotation:

$$F = \frac{bT_2}{a} = \frac{Pb}{a}\left(\frac{e^{\mu\theta}}{e^{\mu\theta}-1}\right)$$

For counter clockwise rotation:

$$F = \frac{bT_1}{a} = \frac{Pb}{a}\left(\frac{1}{e^{\mu\theta}-1}\right)$$

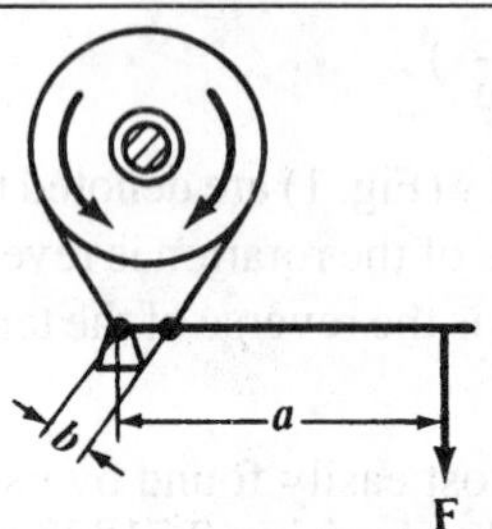

Fig. 2.

Simple band brake.
For clockwise rotation:

$$F = \frac{bT_1}{a} = \frac{Pb}{a}\left(\frac{1}{e^{\mu\theta}-1}\right)$$

For counter clockwise rotation:

$$F = \frac{bT_2}{a} = \frac{Pb}{a}\left(\frac{e^{\mu\theta}}{e^{\mu\theta}-1}\right)$$

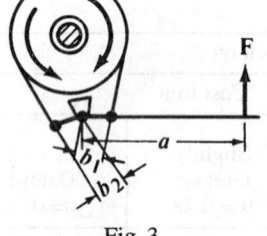

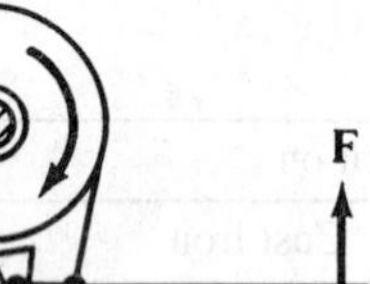

Fig. 3.

Differential band brake.
For clockwise rotation:

$$F = \frac{b_2T_2 - b_1T_1}{a} = \frac{P}{a}\left(\frac{b_2e^{\mu\theta} - b_1}{e^{\mu\theta}-1}\right)$$

For counter clockwise rotation:

$$F = \frac{b_2T_1 - b_1T_2}{a} = \frac{P}{a}\left(\frac{b_2 - b_1e^{\mu\theta}}{e^{\mu\theta}-1}\right)$$

In this case, if b_2 is equal to, or less than, $b_1e^{\mu\theta}$, the force F will be 0 or negative and the band brake works automatically.

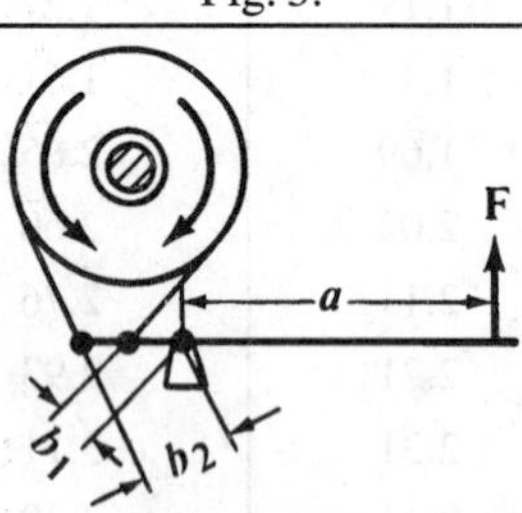

Fig. 4.

Differential band brake.
For clockwise rotation:

$$F = \frac{b_2T_2 + b_1T_1}{a} = \frac{P}{a}\left(\frac{b_2e^{\mu\theta} + b_1}{e^{\mu\theta}-1}\right)$$

For counter clockwise rotation:

$$F = \frac{b_1T_2 + b_2T_1}{a} = \frac{P}{a}\left(\frac{b_1e^{\mu\theta} + b_2}{e^{\mu\theta}-1}\right)$$

If $b_2 = b_1$, both of the above formulas reduce to $F = \frac{Pb_1}{a}\left(\frac{e^{\mu\theta}+1}{e^{\mu\theta}-1}\right)$. In this case, the same force F is required for rotation in either direction.

In a band brake of the type in Fig. 1, dimension $a = 24$ inches, and $b = 4$ inches; force $P = 100$ pounds; coefficient $\mu = 0.2$, and angle of contact = 240 degrees, or

$$\theta = \frac{240}{180} \times \pi = 4.18$$

The rotation is clockwise. Find force F required.

$$F = \frac{Pb}{a}\left(\frac{e^{\mu\theta}}{e^{\mu\theta} - 1}\right)$$

$$= \frac{100 \times 4}{24}\left(\frac{2.71828^{0.2 \times 4.18}}{2.71828^{0.2 \times 4.18} - 1}\right)$$

$$= \frac{400}{24} \times \frac{2.71828^{0.836}}{2.71828^{0.836} - 1}$$

$$= 16.66 \times \frac{2.31}{2.31 - 1} = 29.4$$

If a hand-held calculator is not used, determining the value of $e^{\mu\theta}$ is rather tedious, and the table on page 2337 will save calculations.

Coefficient of Friction in Brakes.—The coefficients of friction that may be assumed for friction brake calculations are as follows: Iron on iron, 0.25 to 0.3 leather on iron, 0.3; cork on iron, 0.35. Values somewhat lower than these should be assumed when the velocities exceed 400 feet per minute at the beginning of the braking operation.

For brakes where wooden brake blocks are used on iron drums, poplar has proved the best brake-block material. The best material for the brake drum is wrought iron. Poplar gives a high coefficient of friction, and is little affected by oil. The average coefficient of friction for poplar brake blocks and wrought-iron drums is 0.6; for poplar on cast iron, 0.35 for oak on wrought iron, 0.5; for oak on cast iron, 0.3; for beech on wrought iron, 0.5; for beech on cast iron, 0.3; for elm on wrought iron, 0.6; and for elm on cast iron, 0.35. The objection to elm is that the friction decreases rapidly if the friction surfaces are oily. The coefficient of friction for elm and wrought iron, if oily, is less than 0.4.

Calculating Horsepower from Dynamometer Tests.—When a dynamometer is arranged for measuring the horsepower transmitted by a shaft, as indicated by the diagrammatic view in the illustration on page 2340, the horsepower may be obtained by the formula:

$$\text{HP} = \frac{2\pi LPN}{33{,}000}$$

in which H.P. = horsepower transmitted; N = number of revolutions per minute; L = distance (as shown in illustration) from center of pulley to point of action of weight P, in feet; P = weight hung on brake arm or read on scale.

By adopting a length of brake arm equal to 5 feet 3 inches, the formula may be reduced to the simple form:

$$\text{HP} = \frac{NP}{1000}$$

If a length of brake arm equal to 2 feet 7 1⁄2 inches is adopted as a standard, the formula takes the form:

$$\text{HP} = \frac{NP}{2000}$$

The *transmission* type of dynamometer measures the power by transmitting it through the mechanism of the dynamometer from the apparatus in which it is generated, or to the apparatus in which it is to be utilized. Dynamometers known as *indicators* operate by simultaneously measuring the pressure and volume of a confined fluid. This type may be used for the measurement of the power generated by steam or gas engines or absorbed by refrigerating machinery, air compressors, or pumps. An electrical dynamometer is for measuring the power of an electric current, based on the mutual action of currents flowing in two coils. It consists principally of one fixed and one movable coil, which, in the normal position, are at right angles to each other. Both coils are connected in series, and, when a current traverses the coils, the fields produced are at right angles; hence, the coils tend to take up a parallel position. The movable coil with an attached pointer will be deflected, the deflection measuring directly the electric current.

Formulas for Block Brakes

<table>
<tr><td colspan="2">F = force in pounds at end of brake handle;
P = tangential force in pounds at rim of brake wheel;
μ = coefficient of friction between the brake block and brake wheel.</td></tr>
<tr><td>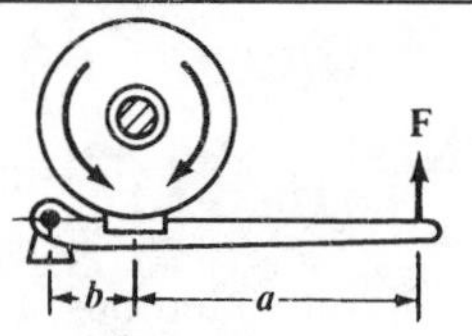
Fig. 1.</td><td>Block brake.
For rotation in either direction:
$$F = P\frac{b}{a+b} \times \frac{1}{\mu} = \frac{Pb}{a+b}\left(\frac{1}{\mu}\right)$$</td></tr>
<tr><td>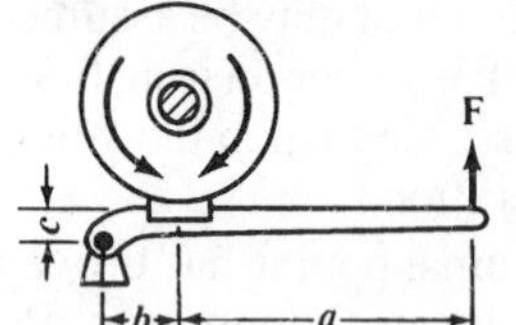
Fig. 2.</td><td>Block brake.
For clockwise rotation:
$$F = \frac{\frac{Pb}{\mu} - Pc}{a+b} = \frac{Pb}{a+b}\left(\frac{1}{\mu} - \frac{c}{b}\right)$$
For counter clockwise rotation:
$$F = \frac{\frac{Pb}{\mu} + Pc}{a+b} = \frac{Pb}{a+b}\left(\frac{1}{\mu} + \frac{c}{b}\right)$$</td></tr>
<tr><td>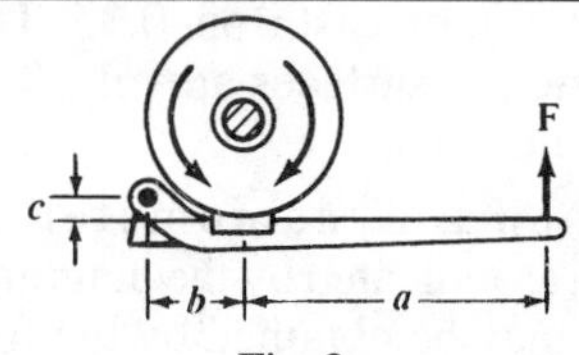
Fig. 3.</td><td>Block brake.
For clockwise rotation:
$$F = \frac{\frac{Pb}{\mu} + Pc}{a+b} = \frac{Pb}{a+b}\left(\frac{1}{\mu} + \frac{c}{b}\right)$$
For counter clockwise rotation:
$$F = \frac{\frac{Pb}{\mu} - Pc}{a+b} = \frac{Pb}{a+b}\left(\frac{1}{\mu} - \frac{c}{b}\right)$$</td></tr>
<tr><td>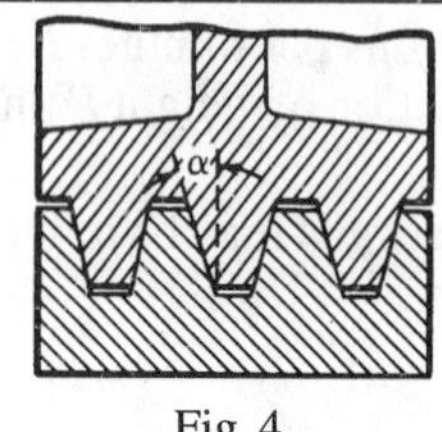
Fig. 4.</td><td>The brake wheel and friction block of the block brake are often grooved as shown in Fig. 4. In this case, substitute for μ in the above equations the value $\frac{\mu}{\sin\alpha + \mu\cos\alpha}$ where α is one-half the angle included by the facts of the grooves.</td></tr>
</table>

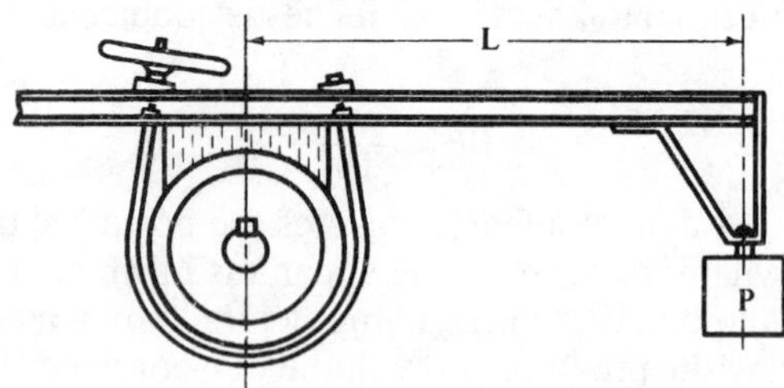

Friction Wheels for Power Transmission

When a rotating member is driven intermittently and the rate of driving does not need to be positive, friction wheels are frequently used, especially when the amount of power to be transmitted is comparatively small. The driven wheels in a pair of friction disks should always be made of a harder material than the driving wheels, so that if the driven wheel should be held stationary by the load, while the driving wheel revolves under its own pressure, a flat spot may not be rapidly worn on the driven wheel. The driven wheels, therefore, are usually made of iron, while the driving wheels are made of or covered with, rubber, paper, leather, wood or fiber. The safe working force per inch of face width of contact for various materials are as follows: Straw fiber, 150; leather fiber, 240; tarred fiber, 240; leather, 150; wood, 100 to 150; paper, 150. Coefficients of friction for different combinations of materials are given in the following table. Smaller values should be used for exceptionally high speeds, or when the transmission must be started while under load.

Horsepower of Friction Wheels.—Let D = diameter of friction wheel in inches; N = Number of revolutions per minute; W = width of face in inches; f = coefficient of friction; P = force in pounds, per inch width of face. Then:

$$\text{H.P.} = \frac{3.1416 \times D \times N \times P \times W \times f}{33{,}000 \times 12}$$

Assume

$$\frac{3.1416 \times P \times f}{33{,}000 \times 12} = C$$

then,

for $P = 100$ and $f = 0.20$, $C = 0.00016$

for $P = 150$ and $f = 0.20$, $C = 0.00024$

for $P = 200$ and $f = 0.20$, $C = 0.00032$

Working Values of Coefficient of Friction

Materials	Coefficient of Friction
Straw fiber and cast iron	0.26
Straw fiber and aluminum	0.27
Leather fiber and cast iron	0.31
Leather fiber and aluminum	0.30
Tarred fiber and cast iron	0.15
Paper and cast iron	0.20
Tarred fiber and aluminum	0.18
Leather and cast iron	0.14
Leather and aluminum	0.22
Leather and typemetal	0.25
Wood and metal	0.25

The horsepower transmitted is then:

$$\text{HP} = D \times N \times W \times C$$

Example:—Find the horsepower transmitted by a pair of friction wheels; the diameter of the driving wheel is 10 inches, and it revolves at 200 revolutions per minute. The width of the wheel is 2 inches. The force per inch width of face is 150 pounds, and the coefficient of friction 0.20.

$$\text{HP} = 10 \times 200 \times 2 \times 0.00024 = 0.96 \text{ horsepower}$$

Horsepower Which May be Transmitted by Means of a Clean Paper Friction Wheel of One-inch Face when Run Under a Force of 150 Pounds (Rockwood Mfg. Co.)

Dia. of Friction Wheel	Revolutions per Minute										
	25	50	75	100	150	200	300	400	600	800	1000
4	0.023	0.047	0.071	0.095	0.142	0.190	0.285	0.380	0.571	0.76	0.95
6	0.035	0.071	0.107	0.142	0.214	0.285	0.428	0.571	0.856	1.14	1.42
8	0.047	0.095	0.142	0.190	0.285	0.380	0.571	0.761	1.142	1.52	1.90
10	0.059	0.119	0.178	0.238	0.357	0.476	0.714	0.952	1.428	1.90	2.38
14	0.083	0.166	0.249	0.333	0.499	0.666	0.999	1.332	1.999	2.66	3.33
16	0.095	0.190	0.285	0.380	0.571	0.761	1.142	1.523	2.284	3.04	3.80
18	0.107	0.214	0.321	0.428	0.642	0.856	1.285	1.713	2.570	3.42	4.28
24	0.142	0.285	0.428	0.571	0.856	1.142	1.713	2.284	3.427	4.56	5.71
30	0.178	0.357	0.535	0.714	1.071	1.428	2.142	2.856	4.284	5.71	7.14
36	0.214	0.428	0.642	0.856	1.285	1.713	2.570	3.427	5.140	6.85	8.56
42	0.249	0.499	0.749	0.999	1.499	1.999	2.998	3.998	5.997	7.99	9.99
48	0.285	0.571	0.856	1.142	1.713	2.284	3.427	4.569	6.854	9.13	11.42
50	0.297	0.595	0.892	1.190	1.785	2.380	3.570	4.760	7.140	9.52	11.90

KEYS AND KEYSEATS

ANSI Standard Keys and Keyseats.—American National Standard, B17.1 Keys and Keyseats, based on current industry practice, was approved in 1967, and reaffirmed in 1989. This standard establishes a uniform relationship between shaft sizes and key sizes for parallel and taper keys as shown in Table 1. Other data in this standard are given in Tables 2 and 3 through 7. The sizes and tolerances shown are for single key applications only.

The following definitions are given in the standard:

Key: A demountable machinery part which, when assembled into keyseats, provides a positive means for transmitting torque between the shaft and hub.

Keyseat: An axially located rectangular groove in a shaft or hub.

This standard recognizes that there are two classes of stock for parallel keys used by industry. One is a close, plus toleranced key stock and the other is a broad, negative toleranced bar stock. Based on the use of two types of stock, two classes of fit are shown:

Class 1: A clearance or metal-to-metal side fit obtained by using bar stock keys and keyseat tolerances as given in Table 4. This is a relatively free fit and applies only to parallel keys.

Class 2: A side fit, with possible interference or clearance, obtained by using key stock and keyseat tolerances as given in Table 4. This is a relatively tight fit.

Class 3: This is an interference side fit and is not tabulated in Table 4 since the degree of interference has not been standardized. However, it is suggested that the top and bottom fit range given under Class 2 in Table 4, for parallel keys be used.

Table 1. Key Size Versus Shaft Diameter *ANSI B17.1-1967 (R1998)*

Nominal Shaft Diameter		Nominal Key Size			Normal Keyseat Depth	
			Height, *H*		*H*/2	
Over	To (Incl.)	Width, *W*	Square	Rectangular	Square	Rectangular
5/16	7/16	3/32	3/32	…	3/64	…
7/16	9/16	1/8	1/8	3/32	1/16	3/64
9/16	7/8	3/16	3/16	1/8	3/32	1/16
7/8	1 1/4	1/4	1/4	3/16	1/8	3/32
1 1/4	1 3/8	5/16	5/16	1/4	5/32	1/8
1 3/8	1 3/4	3/8	3/8	1/4	3/16	1/8
1 3/4	2 1/4	1/2	1/2	3/8	1/4	3/16
2 1/4	2 3/4	5/8	5/8	7/16	5/16	7/32
2 3/4	3 1/4	3/4	3/4	1/2	3/8	1/4
3 1/4	3 3/4	7/8	7/8	5/8	7/16	5/16
3 3/4	4 1/2	1	1	3/4	1/2	3/8
4 1/2	5 1/2	1 1/4	1 1/4	7/8	5/8	7/16
5 1/2	6 1/2	1 1/2	1 1/2	1	3/4	1/2
Square Keys preferred for shaft diameters above this line; rectangular keys, below						
6 1/2	7 1/2	1 3/4	1 3/4	1 1/2[a]	7/8	3/4
7 1/2	9	2	2	1 1/2	1	3/4
9	11	2 1/2	2 1/2	1 3/4	1 1/4	7/8

[a] Some key standards show 1 1/4 inches; preferred height is 1 1/2 inches.

All dimensions are given in inches. For larger shaft sizes, see *ANSI Standard Woodruff Keys and Keyseats.*

Key Size vs. Shaft Diameter: Shaft diameters are listed in Table 1 for identification of various key sizes and are not intended to establish shaft dimensions, tolerances or selections. For a stepped shaft, the size of a key is determined by the diameter of the shaft at the point of location of the key. Up through 6½-inch diameter shafts square keys are preferred; rectangular keys are preferred for larger shafts.

If special considerations dictate the use of a keyseat in the hub shallower than the preferred nominal depth shown, it is recommended that the tabulated preferred nominal standard keyseat always be used in the shaft.

Keyseat Alignment Tolerances: A tolerance of 0.010 inch, max is provided for offset (due to parallel displacement of keyseat centerline from centerline of shaft or bore) of keyseats in shaft and bore. The following tolerances for maximum lead (due to angular displacement of keyseat centerline from centerline of shaft or bore and measured at right angles to the shaft or bore centerline) of keyseats in shaft and bore are specified: 0.002 inch for keyseat length up to and including 4 inches; 0.0005 inch per inch of length for keyseat lengths above 4 inches to and including 10 inches; and 0.005 inch for keyseat lengths above 10 inches. For the effect of keyways on shaft strength, see *Effect of Keyways on Shaft Strength* on page 283.

Table 2. Depth Control Values *S* and *T* for Shaft and Hub

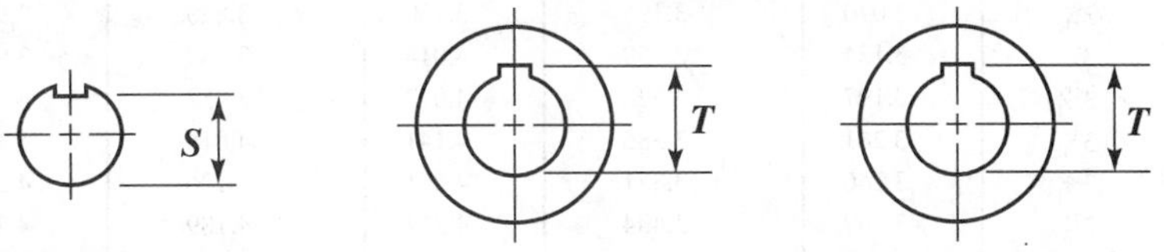

Nominal Shaft Diameter	Parallel and Taper		Parallel		Taper	
	Square	Rectangular	Square	Rectangular	Square	Rectangular
	S	*S*	*T*	*T*	*T*	*T*
½	0.430	0.445	0.560	0.544	0.535	0.519
9⁄16	0.493	0.509	0.623	0.607	0.598	0.582
5⁄8	0.517	0.548	0.709	0.678	0.684	0.653
11⁄16	0.581	0.612	0.773	0.742	0.748	0.717
¾	0.644	0.676	0.837	0.806	0.812	0.781
13⁄16	0.708	0.739	0.900	0.869	0.875	0.844
7⁄8	0.771	0.802	0.964	0.932	0.939	0.907
15⁄16	0.796	0.827	1.051	1.019	1.026	0.994
1	0.859	0.890	1.114	1.083	1.089	1.058
1 1⁄16	0.923	0.954	1.178	1.146	1.153	1.121
1⅛	0.986	1.017	1.241	1.210	1.216	1.185
1 3⁄16	1.049	1.080	1.304	1.273	1.279	1.248
1¼	1.112	1.144	1.367	1.336	1.342	1.311
1 5⁄16	1.137	1.169	1.455	1.424	1.430	1.399
1⅜	1.201	1.232	1.518	1.487	1.493	1.462
1 7⁄16	1.225	1.288	1.605	1.543	1.580	1.518
1½	1.289	1.351	1.669	1.606	1.644	1.581
1 9⁄16	1.352	1.415	1.732	1.670	1.707	1.645
1⅝	1.416	1.478	1.796	1.733	1.771	1.708
1 11⁄16	1.479	1.541	1.859	1.796	1.834	1.771
1¾	1.542	1.605	1.922	1.860	1.897	1.835
1 13⁄16	1.527	1.590	2.032	1.970	2.007	1.945
1⅞	1.591	1.654	2.096	2.034	2.071	2.009
1 15⁄16	1.655	1.717	2.160	2.097	2.135	2.072
2	1.718	1.781	2.223	2.161	2.198	2.136
2 1⁄16	1.782	1.844	2.287	2.224	2.262	2.199
2⅛	1.845	1.908	2.350	2.288	2.325	2.263
2 3⁄16	1.909	1.971	2.414	2.351	2.389	2.326
2¼	1.972	2.034	2.477	2.414	2.452	2.389
2 5⁄16	1.957	2.051	2.587	2.493	2.562	2.468
2⅜	2.021	2.114	2.651	2.557	2.626	2.532

Table 2. *(Continued)* **Depth Control Values *S* and *T* for Shaft and Hub**

2 7/16	2.084	2.178	2.714	2.621	2.689	2.596
2 1/2	2.148	2.242	2.778	2.684	2.753	2.659
2 9/16	2.211	2.305	2.841	2.748	2.816	2.723
2 5/8	2.275	2.369	2.905	2.811	2.880	2.786
2 11/16	2.338	2.432	2.968	2.874	2.943	2.849
2 3/4	2.402	2.495	3.032	2.938	3.007	2.913
2 13/16	2.387	2.512	3.142	3.017	3.117	2.992
2 7/8	2.450	2.575	3.205	3.080	3.180	3.055
2 15/16	2.514	2.639	3.269	3.144	3.244	3.119
3	2.577	2.702	3.332	3.207	3.307	3.182
3 1/16	2.641	2.766	3.396	3.271	3.371	3.246
3 1/8	2.704	2.829	3.459	3.334	3.434	3.309
3 3/16	2.768	2.893	3.523	3.398	3.498	3.373
3 1/4	2.831	2.956	3.586	3.461	3.561	3.436
3 5/16	2.816	2.941	3.696	3.571	3.671	3.546
3 3/8	2.880	3.005	3.760	3.635	3.735	3.610
3 7/16	2.943	3.068	3.823	3.698	3.798	3.673
3 1/2	3.007	3.132	3.887	3.762	3.862	3.737
3 9/16	3.070	3.195	3.950	3.825	3.925	3.800
3 5/8	3.134	3.259	4.014	3.889	3.989	3.864
3 11/16	3.197	3.322	4.077	3.952	4.052	3.927
3 3/4	3.261	3.386	4.141	4.016	4.116	3.991
3 13/16	3.246	3.371	4.251	4.126	4.226	4.101
3 7/8	3.309	3.434	4.314	4.189	4.289	4.164
3 15/16	3.373	3.498	4.378	4.253	4.353	4.228
4	3.436	3.561	4.441	4.316	4.416	4.291
4 3/16	3.627	3.752	4.632	4.507	4.607	4.482
4 1/4	3.690	3.815	4.695	4.570	4.670	4.545
4 3/8	3.817	3.942	4.822	4.697	4.797	4.672
4 7/16	3.880	4.005	4.885	4.760	4.860	4.735
4 1/2	3.944	4.069	4.949	4.824	4.924	4.799
4 3/4	4.041	4.229	5.296	5.109	5.271	5.084
4 7/8	4.169	4.356	5.424	5.236	5.399	5.211
4 15/16	4.232	4.422	5.487	5.300	5.462	5.275
5	4.296	4.483	5.551	5.363	5.526	5.338
5 3/16	4.486	4.674	5.741	5.554	5.716	5.529
5 1/4	4.550	4.737	5.805	5.617	5.780	5.592
5 7/16	4.740	4.927	5.995	5.807	5.970	5.782
5 1/2	4.803	4.991	6.058	5.871	6.033	5.846
5 3/4	4.900	5.150	6.405	6.155	6.380	6.130
5 15/16	5.091	5.341	6.596	6.346	6.571	6.321
6	5.155	5.405	6.660	6.410	6.635	6.385
6 1/4	5.409	5.659	6.914	6.664	6.889	6.639
6 1/2	5.662	5.912	7.167	6.917	7.142	6.892
6 3/4	5.760	[a]5.885	7.515	[a]7.390	7.490	[a]7.365
7	6.014	[a]6.139	7.769	[a]7.644	7.744	[a]7.619
7 1/4	6.268	[a]6.393	8.023	[a]7.898	7.998	[a]7.873
7 1/2	6.521	[a]6.646	8.276	[a]8.151	8.251	[a]8.126
7 3/4	6.619	6.869	8.624	8.374	8.599	8.349
8	6.873	7.123	8.878	8.628	8.853	8.603
9	7.887	8.137	9.892	9.642	9.867	9.617
10	8.591	8.966	11.096	10.721	11.071	10.696
11	9.606	9.981	12.111	11.736	12.086	11.711
12	10.309	10.809	13.314	12.814	13.289	12.789
13	11.325	11.825	14.330	13.830	14.305	13.805
14	12.028	12.528	15.533	15.033	15.508	15.008
15	13.043	13.543	16.548	16.048	16.523	16.023

[a] 1 3/4 × 1 1/2 inch key.

All dimensions are given in inches. See Table 4 for tolerances.

Table 3. ANSI Standard Plain and Gib Head Keys *ANSI B17.1-1967 (R1998)*

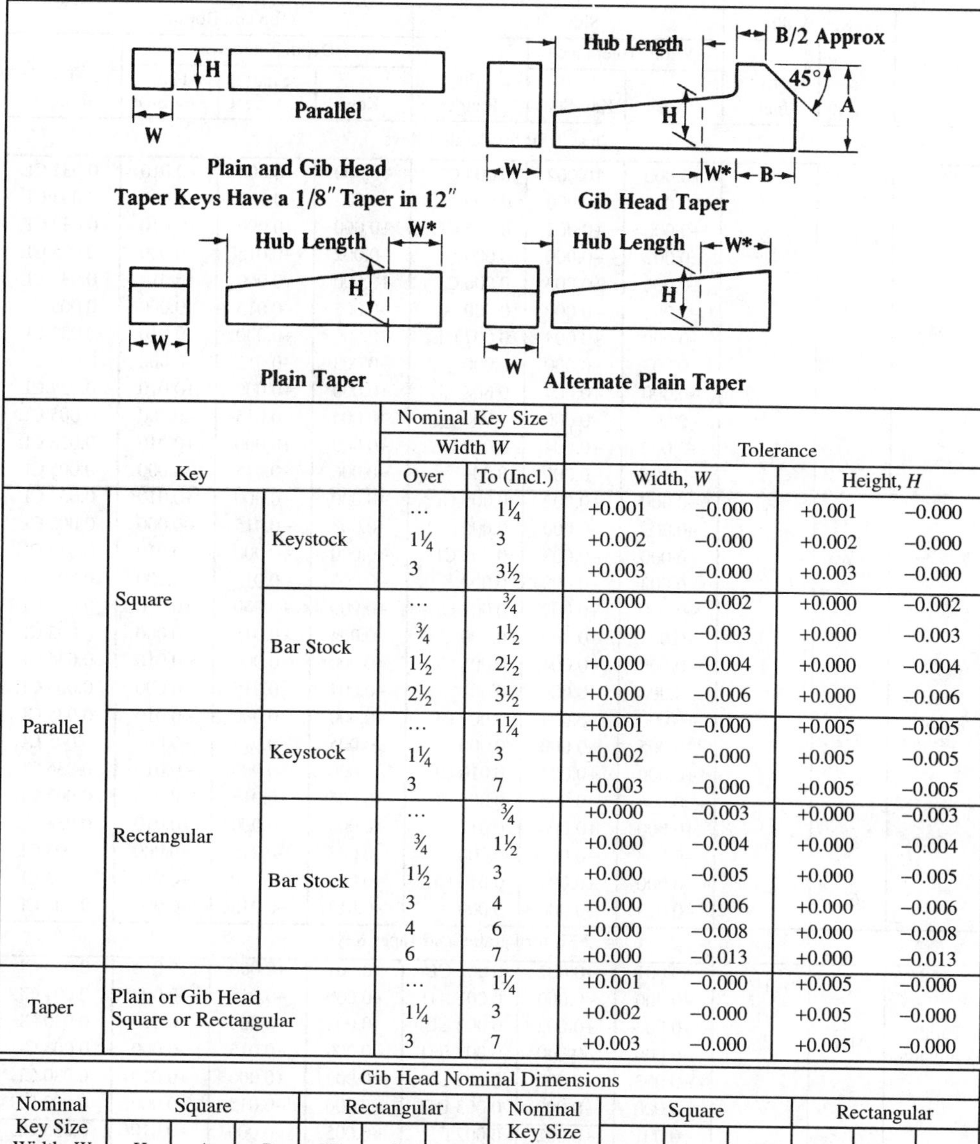

Key			Nominal Key Size Width *W* Over	To (Incl.)	Tolerance Width, *W*		Height, *H*	
Parallel	Square	Keystock	…	1 1/4	+0.001	−0.000	+0.001	−0.000
			1 1/4	3	+0.002	−0.000	+0.002	−0.000
			3	3 1/2	+0.003	−0.000	+0.003	−0.000
		Bar Stock	…	3/4	+0.000	−0.002	+0.000	−0.002
			3/4	1 1/2	+0.000	−0.003	+0.000	−0.003
			1 1/2	2 1/2	+0.000	−0.004	+0.000	−0.004
			2 1/2	3 1/2	+0.000	−0.006	+0.000	−0.006
	Rectangular	Keystock	…	1 1/4	+0.001	−0.000	+0.005	−0.005
			1 1/4	3	+0.002	−0.000	+0.005	−0.005
			3	7	+0.003	−0.000	+0.005	−0.005
		Bar Stock	…	3/4	+0.000	−0.003	+0.000	−0.003
			3/4	1 1/2	+0.000	−0.004	+0.000	−0.004
			1 1/2	3	+0.000	−0.005	+0.000	−0.005
			3	4	+0.000	−0.006	+0.000	−0.006
			4	6	+0.000	−0.008	+0.000	−0.008
			6	7	+0.000	−0.013	+0.000	−0.013
Taper	Plain or Gib Head Square or Rectangular		…	1 1/4	+0.001	−0.000	+0.005	−0.000
			1 1/4	3	+0.002	−0.000	+0.005	−0.000
			3	7	+0.003	−0.000	+0.005	−0.000

Gib Head Nominal Dimensions

Nominal Key Size Width, *W*	Square *H*	Square *A*	Square *B*	Rectangular *H*	Rectangular *A*	Rectangular *B*	Nominal Key Size Width, *W*	Square *H*	Square *A*	Square *B*	Rectangular *H*	Rectangular *A*	Rectangular *B*
1/8	1/8	1/4	1/4	3/32	3/16	1/8	1	1	1 5/8	1 1/8	3/4	1 1/4	7/8
3/16	3/16	5/16	5/16	1/8	1/4	1/4	1 1/4	1 1/4	2	1 7/16	7/8	1 3/8	1
1/4	1/4	7/16	3/8	3/16	5/16	5/16	1 1/2	1 1/2	2 3/8	1 3/4	1	1 5/8	1 1/8
5/16	5/16	1/2	7/16	1/4	7/16	3/8	1 3/4	1 3/4	2 3/4	2	1 1/2	2 3/8	1 3/4
3/8	3/8	5/8	1/2	1/4	7/16	3/8	2	2	3 1/2	2 1/4	1 1/2	2 3/8	1 3/4
1/2	1/2	7/8	5/8	3/8	5/8	1/2	2 1/2	2 1/2	4	3	1 3/4	2 3/4	2
5/8	5/8	1	3/4	7/16	3/4	9/16	3	3	5	3 1/2	2	3 1/2	2 1/4
3/4	3/4	1 1/4	7/8	1/2	7/8	5/8	3 1/2	3 1/2	6	4	2 1/2	4	3
7/8	7/8	1 3/8	1	5/8	1	3/4	…	…	…	…	…	…	…

All dimensions are given in inches.

*For locating position of dimension *H*. Tolerance does not apply.

For larger sizes the following relationships are suggested as guides for establishing *A* and *B*: A = 1.8*H* and *B* = 1.2*H*.

Table 4. ANSI Standard Fits for Parallel and Taper Keys *ANSI B17.1-1967 (R1998)*

Type of Key	Key Width		Side Fit			Top and Bottom Fit			
			Width Tolerance			Depth Tolerance			
	Over	To (Incl.)	Key	Key-Seat	Fit Range[a]	Key	Shaft Key-Seat	Hub Key-Seat	Fit Range[a]
Class 1 Fit for Parallel Keys									
Square	…	½	+0.000 –0.002	+0.002 –0.000	0.004 CL 0.000	+0.000 –0.002	+0.000 –0.015	+0.010 –0.000	0.032 CL 0.005 CL
	½	¾	+0.000 –0.002	+0.003 –0.000	0.005 CL 0.000	+0.000 –0.002	+0.000 –0.015	+0.010 –0.000	0.032 CL 0.005 CL
	¾	1	+0.000 –0.003	+0.003 –0.000	0.006 CL 0.000	+0.000 –0.003	+0.000 –0.015	+0.010 –0.000	0.033 CL 0.005 CL
	1	1½	+0.000 –0.003	+0.004 –0.000	0.007 CL 0.000	+0.000 –0.003	+0.000 –0.015	+0.010 –0.000	0.033 CL 0.005 CL
	1½	2½	+0.000 –0.004	+0.004 –0.000	0.008 CL 0.000	+0.000 –0.004	+0.000 –0.015	+0.010 –0.000	0.034 CL 0.005 CL
	2½	3½	+0.000 –0.006	+0.004 –0.000	0.010 CL 0.000	+0.000 –0.006	+0.000 –0.015	+0.010 –0.000	0.036 CL 0.005 CL
Rectan-gular	…	½	+0.000 –0.003	+0.002 –0.000	0.005 CL 0.000	+0.000 –0.003	+0.000 –0.015	+0.010 –0.000	0.033 CL 0.005 CL
	½	¾	+0.000 –0.003	+0.003 –0.000	0.006 CL 0.000	+0.000 –0.003	+0.000 –0.015	+0.010 –0.000	0.033 CL 0.005 CL
	¾	1	+0.000 –0.004	+0.003 –0.000	0.007 CL 0.000	+0.000 –0.004	+0.000 –0.015	+0.010 –0.000	0.034 CL 0.005 CL
	1	1½	+0.000 –0.004	+0.004 –0.000	0.008 CL 0.000	+0.000 –0.004	+0.000 –0.015	+0.010 –0.000	0.034 CL 0.005 CL
	1½	3	+0.000 –0.005	+0.004 –0.000	0.009 CL 0.000	+0.000 –0.005	+0.000 –0.015	+0.010 –0.000	0.035 CL 0.005 CL
	3	4	+0.000 –0.006	+0.004 –0.000	0.010 CL 0.000	+0.000 –0.006	+0.000 –0.015	+0.010 –0.000	0.036 CL 0.005 CL
	4	6	+0.000 –0.008	+0.004 –0.000	0.012 CL 0.000	+0.000 –0.008	+0.000 –0.015	+0.010 –0.000	0.038 CL 0.005 CL
	6	7	+0.000 –0.013	+0.004 –0.000	0.017 CL 0.000	+0.000 –0.013	+0.000 –0.015	+0.010 –0.000	0.043 CL 0.005 CL
Class 2 Fit for Parallel and Taper Keys									
Parallel Square	…	1¼	+0.001 –0.000	+0.002 –0.000	0.002 CL 0.001 INT	+0.001 –0.000	+0.000 –0.015	+0.010 –0.000	0.030 CL 0.004 CL
	1¼	3	+0.002 –0.000	+0.002 –0.000	0.002 CL 0.002 INT	+0.002 –0.000	+0.000 –0.015	+0.010 –0.000	0.030 CL 0.003 CL
	3	3½	+0.003 –0.000	+0.002 –0.000	0.002 CL 0.003 INT	+0.003 –0.000	+0.000 –0.015	+0.010 –0.000	0.030 CL 0.002 CL
Parallel Rectan-gular	…	1¼	+0.001 –0.000	+0.002 –0.000	0.002 CL 0.001 INT	+0.005 –0.005	+0.000 –0.015	+0.010 –0.000	0.035 CL 0.000 CL
	1¼	3	+0.002 –0.000	+0.002 –0.000	0.002 CL 0.002 INT	+0.005 –0.005	+0.000 –0.015	+0.010 –0.000	0.035 CL 0.000 CL
	3	7	+0.003 –0.000	+0.002 –0.000	0.002 CL 0.003 INT	+0.005 –0.005	+0.000 –0.015	+0.010 –0.000	0.035 CL 0.000 CL
Taper	…	1¼	+0.001 –0.000	+0.002 –0.000	0.002 CL 0.001 INT	+0.005 –0.000	+0.000 –0.015	+0.010 –0.000	0.005 CL 0.025 INT
	1¼	3	+0.002 –0.000	+0.002 –0.000	0.002 CL 0.002 INT	+0.005 –0.000	+0.000 –0.015	+0.010 –0.000	0.005 CL 0.025 INT
	3	[b]	+0.003 –0.000	+0.002 –0.000	0.002 CL 0.003 INT	+0.005 –0.000	+0.000 –0.015	+0.010 –0.000	0.005 CL 0.025 INT

[a] Limits of variation. CL = Clearance; INT = Interference.

[b] To (Incl.) 3½-inch Square and 7-inch Rectangular key widths.

All dimensions are given in inches. See also text on page 2342.

Table 5. Suggested Keyseat Fillet Radius and Key Chamfer
ANSI B17.1-1967 (R1998)

Keyseat Depth, *H*/2		Fillet Radius	45 deg. Chamfer	Keyseat Depth, *H*2		Fillet Radius	45 deg. Chamfer
Over	To (Incl.)			Over	To (Incl.)		
1/8	1/4	1/32	3/64	7/8	1 1/4	3/16	7/32
1/4	1/2	1/16	5/64	1 1/4	1 3/4	1/4	9/32
1/2	7/8	1/8	5/32	1 3/4	2 1/2	3/8	13/32

All dimensions are given in inches.

Table 6. ANSI Standard Keyseat Tolerances for Electric Motor and Generator Shaft Extensions *ANSI B17.1-1967 (R1998)*

Keyseat Width		Width Tolerance	Depth Tolerance
Over	To (Incl.)		
…	1/4	+0.001 −0.001	+0.000 −0.015
1/4	3/4	+0.000 −0.002	+0.000 −0.015
3/4	1 1/4	+0.000 −0.003	+0.000 −0.015

All dimensions are given in inches.

Table 7. Set Screws for Use Over Keys *ANSI B17.1-1967 (R1998)*

Nom. Shaft Diam.		Nom. Key Width	Set Screw Diam.	Nom. Shaft Diam.		Nom. Key Width	Set Screw Diam.
Over	To (Incl.)			Over	To (Incl.)		
5/16	7/16	3/32	No. 10	2 1/4	2 3/4	5/8	1/2
7/16	9/16	1/8	No. 10	2 3/4	3 1/4	3/4	5/8
9/16	7/8	3/16	1/4	3 1/4	3 3/4	7/8	3/4
7/8	1 1/4	1/4	5/16	3 3/4	4 1/2	1	3/4
1 1/4	1 3/8	5/16	3/8	4 1/2	5 1/2	1 1/4	7/8
1 3/8	1 3/4	3/8	3/8	5 1/2	6 1/2	1 1/2	1
1 3/4	2 1/4	1/2	1/2	…	…	…	…

All dimensions are given in inches.

These set screw diameter selections are offered as a guide but their use should be dependent upon design considerations.

ANSI Standard Woodruff Keys and Keyseats.—American National Standard B17.2 was approved in 1967, and reaffirmed in 1990. Data from this standard are shown in Tables 8, 9, and 10.

Table 8. ANSI Standard Woodruff Keys *ANSI B17.2-1967 (R1998)*

+0.000 −0.010 *B*; +0.001 −0.000 *W*; *E*; *C*; *F*; Break corners 0.020 max *R*; **Full radius type**

+0.000 −0.010 *B*; +0.001 −0.000 *W*; *E*; *D*; *F*; Break corners 0.020 max *R*; **Flat bottom type**

Key No.	Nominal Key Size *W* X *B*	Actual Length *F* +0.000 −0.010	Height of Key *C* Max.	*C* Min.	*D* Max.	*D* Min.	Distance Below Center *E*
202	1/16 × 1/4	0.248	0.109	0.104	0.109	0.104	1/64
202.5	1/16 × 5/16	0.311	0.140	0.135	0.140	0.135	1/64
302.5	3/32 × 5/16	0.311	0.140	0.135	0.140	0.135	1/64
203	1/16 × 3/8	0.374	0.172	0.167	0.172	0.167	1/64
303	3/32 × 3/8	0.374	0.172	0.167	0.172	0.167	1/64
403	1/8 × 3/8	0.374	0.172	0.167	0.172	0.167	1/64
204	1/16 × 1/2	0.491	0.203	0.198	0.194	0.188	3/64
304	3/32 × 1/2	0.491	0.203	0.198	0.194	0.188	3/64
404	1/8 × 1/2	0.491	0.203	0.198	0.194	0.188	3/64
305	3/32 × 5/8	0.612	0.250	0.245	0.240	0.234	1/16
405	1/8 × 5/8	0.612	0.250	0.245	0.240	0.234	1/16
505	5/32 × 5/8	0.612	0.250	0.245	0.240	0.234	1/16
605	3/16 × 5/8	0.612	0.250	0.245	0.240	0.234	1/16
406	1/8 × 3/4	0.740	0.313	0.308	0.303	0.297	1/16
506	5/32 × 3/4	0.740	0.313	0.308	0.303	0.297	1/16
606	3/16 × 3/4	0.740	0.313	0.308	0.303	0.297	1/16
806	1/4 × 3/4	0.740	0.313	0.308	0.303	0.297	1/16
507	5/32 × 7/8	0.866	0.375	0.370	0.365	0.359	1/16
607	3/16 × 7/8	0.866	0.375	0.370	0.365	0.359	1/16
707	7/32 × 7/8	0.866	0.375	0.370	0.365	0.359	1/16
807	1/4 × 7/8	0.866	0.375	0.370	0.365	0.359	1/16
608	3/16 × 1	0.992	0.438	0.433	0.428	0.422	1/16
708	7/32 × 1	0.992	0.438	0.433	0.428	0.422	1/16
808	1/4 × 1	0.992	0.438	0.433	0.428	0.422	1/16
1008	5/16 × 1	0.992	0.438	0.433	0.428	0.422	1/16
1208	3/8 × 1	0.992	0.438	0.433	0.428	0.422	1/16
609	3/16 × 1 1/8	1.114	0.484	0.479	0.475	0.469	5/64
709	7/32 × 1 1/8	1.114	0.484	0.479	0.475	0.469	5/64
809	1/4 × 1 1/8	1.114	0.484	0.479	0.475	0.469	5/64
1009	5/16 × 1 1/8	1.114	0.484	0.479	0.475	0.469	5/64
610	3/16 × 1 1/4	1.240	0.547	0.542	0.537	0.531	5/64
710	7/32 × 1 1/4	1.240	0.547	0.542	0.537	0.531	5/64
810	1/4 × 1 1/4	1.240	0.547	0.542	0.537	0.531	5/64
1010	5/16 × 1 1/4	1.240	0.547	0.542	0.537	0.531	5/64
1210	3/8 × 1 1/4	1.240	0.547	0.542	0.537	0.531	5/64
811	1/4 × 1 3/8	1.362	0.594	0.589	0.584	0.578	3/32
1011	5/16 × 1 3/8	1.362	0.594	0.589	0.584	0.578	3/32
1211	3/8 × 1 3/8	1.362	0.594	0.589	0.584	0.578	3/32
812	1/4 × 1 1/2	1.484	0.641	0.636	0.631	0.625	7/64
1012	5/16 × 1 1/2	1.484	0.641	0.636	0.631	0.625	7/64
1212	3/8 × 1 1/2	1.484	0.641	0.636	0.631	0.625	7/64

All dimensions are given in inches.

The Key numbers indicate normal key dimensions. The last two digits give the nominal diameter *B* in eighths of an inch and the digits preceding the last two give the nominal width *W* in thirty-seconds of an inch.

Table 9. ANSI Standard Woodruff Keys ***ANSI B17.2-1967 (R1998)***

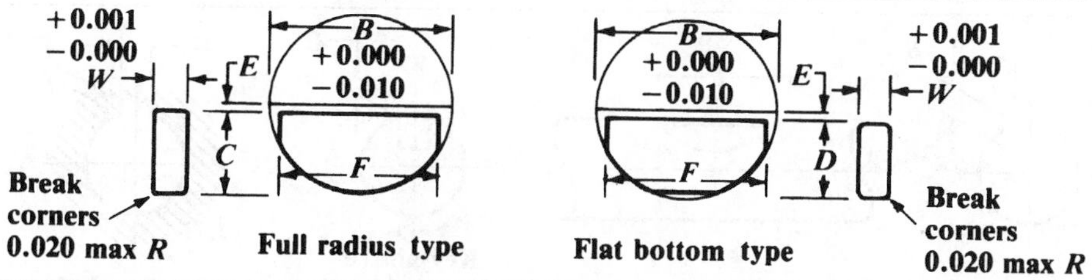

+0.001 −0.000 W; E; B +0.000 −0.010; C; F; Break corners 0.020 max R; **Full radius type**

B +0.000 −0.010; E; +0.001 −0.000 W; D; F; **Flat bottom type**; Break corners 0.020 max R

Key No.	Nominal Key Size $W \times B$	Actual Length F +0.000 −0.010	Height of Key C Max.	Height of Key C Min.	Height of Key D Max.	Height of Key D Min.	Distance Below Center E
617-1	$\frac{3}{16} \times 2\frac{1}{8}$	1.380	0.406	0.401	0.396	0.390	$\frac{21}{32}$
817-1	$\frac{1}{4} \times 2\frac{1}{8}$	1.380	0.406	0.401	0.396	0.390	$\frac{21}{32}$
1017-1	$\frac{5}{16} \times 2\frac{1}{8}$	1.380	0.406	0.401	0.396	0.390	$\frac{21}{32}$
1217-1	$\frac{3}{8} \times 2\frac{1}{8}$	1.380	0.406	0.401	0.396	0.390	$\frac{21}{32}$
617	$\frac{3}{16} \times 2\frac{1}{8}$	1.723	0.531	0.526	0.521	0.515	$\frac{17}{32}$
817	$\frac{1}{4} \times 2\frac{1}{8}$	1.723	0.531	0.526	0.521	0.515	$\frac{17}{32}$
1017	$\frac{5}{16} \times 2\frac{1}{8}$	1.723	0.531	0.526	0.521	0.515	$\frac{17}{32}$
1217	$\frac{3}{8} \times 2\frac{1}{8}$	1.723	0.531	0.526	0.521	0.515	$\frac{17}{32}$
822-1	$\frac{1}{4} \times 2\frac{3}{4}$	2.000	0.594	0.589	0.584	0.578	$\frac{25}{32}$
1022-1	$\frac{5}{16} \times 2\frac{3}{4}$	2.000	0.594	0.589	0.584	0.578	$\frac{25}{32}$
1222-1	$\frac{3}{8} \times 2\frac{3}{4}$	2.000	0.594	0.589	0.584	0.578	$\frac{25}{32}$
1422-1	$\frac{7}{16} \times 2\frac{3}{4}$	2.000	0.594	0.589	0.584	0.578	$\frac{25}{32}$
1622-1	$\frac{1}{2} \times 2\frac{3}{4}$	2.000	0.594	0.589	0.584	0.578	$\frac{25}{32}$
822	$\frac{1}{4} \times 2\frac{3}{4}$	2.317	0.750	0.745	0.740	0.734	$\frac{5}{8}$
1022	$\frac{5}{16} \times 2\frac{3}{4}$	2.317	0.750	0.745	0.740	0.734	$\frac{5}{8}$
1222	$\frac{3}{8} \times 2\frac{3}{4}$	2.317	0.750	0.745	0.740	0.734	$\frac{5}{8}$
1422	$\frac{7}{16} \times 2\frac{3}{4}$	2.317	0.750	0.745	0.740	0.734	$\frac{5}{8}$
1622	$\frac{1}{2} \times 2\frac{3}{4}$	2.317	0.750	0.745	0.740	0.734	$\frac{5}{8}$
1228	$\frac{3}{8} \times 3\frac{1}{2}$	2.880	0.938	0.933	0.928	0.922	$\frac{13}{16}$
1428	$\frac{7}{16} \times 3\frac{1}{2}$	2.880	0.938	0.933	0.928	0.922	$\frac{13}{16}$
1628	$\frac{1}{2} \times 3\frac{1}{2}$	2.880	0.938	0.933	0.928	0.922	$\frac{13}{16}$
1828	$\frac{9}{16} \times 3\frac{1}{2}$	2.880	0.938	0.933	0.928	0.922	$\frac{13}{16}$
2028	$\frac{5}{8} \times 3\frac{1}{2}$	2.880	0.938	0.933	0.928	0.922	$\frac{13}{16}$
2228	$\frac{11}{16} \times 3\frac{1}{2}$	2.880	0.938	0.933	0.928	0.922	$\frac{13}{16}$
2428	$\frac{3}{4} \times 3\frac{1}{2}$	2.880	0.938	0.933	0.928	0.922	$\frac{13}{16}$

All dimensions are given in inches.

The key numbers indicate nominal key dimensions. The last two digits give the nominal diameter *B* in eighths of an inch and the digits preceding the last two give the nominal width *W* in thirty-seconds of an inch.

The key numbers with the −1 designation, while representing the nominal key size have a shorter length *F* and due to a greater distance below center *E* are less in height than the keys of the same number without the −1 designation.

Table 10. ANSI Keyseat Dimensions for Woodruff

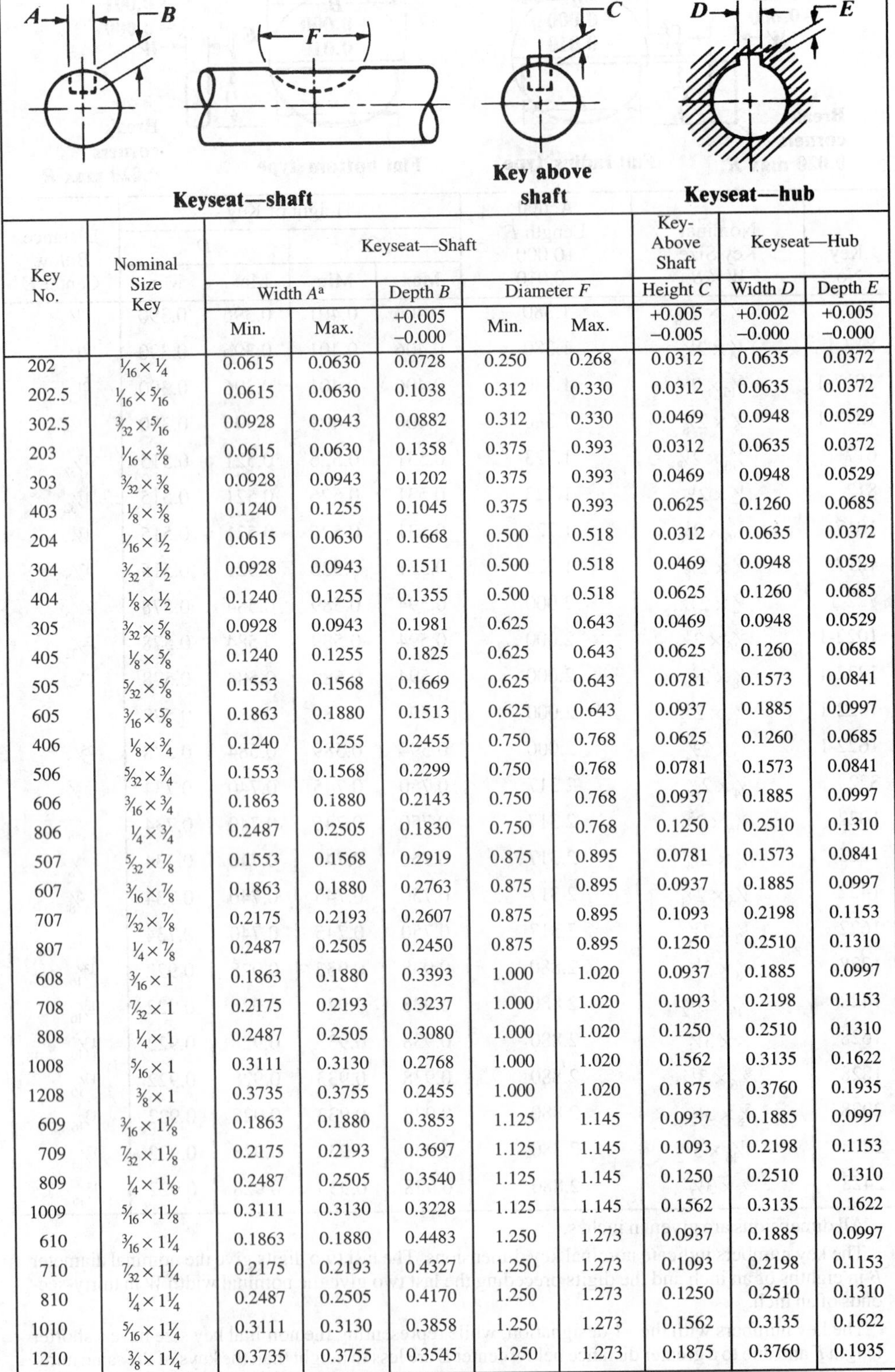

Key No.	Nominal Size Key	Keyseat—Shaft Width A[a] Min.	Keyseat—Shaft Width A[a] Max.	Keyseat—Shaft Depth B +0.005 −0.000	Keyseat—Shaft Diameter F Min.	Keyseat—Shaft Diameter F Max.	Key Above Shaft Height C +0.005 −0.005	Keyseat—Hub Width D +0.002 −0.000	Keyseat—Hub Depth E +0.005 −0.000
202	$1/16 \times 1/4$	0.0615	0.0630	0.0728	0.250	0.268	0.0312	0.0635	0.0372
202.5	$1/16 \times 5/16$	0.0615	0.0630	0.1038	0.312	0.330	0.0312	0.0635	0.0372
302.5	$3/32 \times 5/16$	0.0928	0.0943	0.0882	0.312	0.330	0.0469	0.0948	0.0529
203	$1/16 \times 3/8$	0.0615	0.0630	0.1358	0.375	0.393	0.0312	0.0635	0.0372
303	$3/32 \times 3/8$	0.0928	0.0943	0.1202	0.375	0.393	0.0469	0.0948	0.0529
403	$1/8 \times 3/8$	0.1240	0.1255	0.1045	0.375	0.393	0.0625	0.1260	0.0685
204	$1/16 \times 1/2$	0.0615	0.0630	0.1668	0.500	0.518	0.0312	0.0635	0.0372
304	$3/32 \times 1/2$	0.0928	0.0943	0.1511	0.500	0.518	0.0469	0.0948	0.0529
404	$1/8 \times 1/2$	0.1240	0.1255	0.1355	0.500	0.518	0.0625	0.1260	0.0685
305	$3/32 \times 5/8$	0.0928	0.0943	0.1981	0.625	0.643	0.0469	0.0948	0.0529
405	$1/8 \times 5/8$	0.1240	0.1255	0.1825	0.625	0.643	0.0625	0.1260	0.0685
505	$5/32 \times 5/8$	0.1553	0.1568	0.1669	0.625	0.643	0.0781	0.1573	0.0841
605	$3/16 \times 5/8$	0.1863	0.1880	0.1513	0.625	0.643	0.0937	0.1885	0.0997
406	$1/8 \times 3/4$	0.1240	0.1255	0.2455	0.750	0.768	0.0625	0.1260	0.0685
506	$5/32 \times 3/4$	0.1553	0.1568	0.2299	0.750	0.768	0.0781	0.1573	0.0841
606	$3/16 \times 3/4$	0.1863	0.1880	0.2143	0.750	0.768	0.0937	0.1885	0.0997
806	$1/4 \times 3/4$	0.2487	0.2505	0.1830	0.750	0.768	0.1250	0.2510	0.1310
507	$5/32 \times 7/8$	0.1553	0.1568	0.2919	0.875	0.895	0.0781	0.1573	0.0841
607	$3/16 \times 7/8$	0.1863	0.1880	0.2763	0.875	0.895	0.0937	0.1885	0.0997
707	$7/32 \times 7/8$	0.2175	0.2193	0.2607	0.875	0.895	0.1093	0.2198	0.1153
807	$1/4 \times 7/8$	0.2487	0.2505	0.2450	0.875	0.895	0.1250	0.2510	0.1310
608	$3/16 \times 1$	0.1863	0.1880	0.3393	1.000	1.020	0.0937	0.1885	0.0997
708	$7/32 \times 1$	0.2175	0.2193	0.3237	1.000	1.020	0.1093	0.2198	0.1153
808	$1/4 \times 1$	0.2487	0.2505	0.3080	1.000	1.020	0.1250	0.2510	0.1310
1008	$5/16 \times 1$	0.3111	0.3130	0.2768	1.000	1.020	0.1562	0.3135	0.1622
1208	$3/8 \times 1$	0.3735	0.3755	0.2455	1.000	1.020	0.1875	0.3760	0.1935
609	$3/16 \times 1\,1/8$	0.1863	0.1880	0.3853	1.125	1.145	0.0937	0.1885	0.0997
709	$7/32 \times 1\,1/8$	0.2175	0.2193	0.3697	1.125	1.145	0.1093	0.2198	0.1153
809	$1/4 \times 1\,1/8$	0.2487	0.2505	0.3540	1.125	1.145	0.1250	0.2510	0.1310
1009	$5/16 \times 1\,1/8$	0.3111	0.3130	0.3228	1.125	1.145	0.1562	0.3135	0.1622
610	$3/16 \times 1\,1/4$	0.1863	0.1880	0.4483	1.250	1.273	0.0937	0.1885	0.0997
710	$7/32 \times 1\,1/4$	0.2175	0.2193	0.4327	1.250	1.273	0.1093	0.2198	0.1153
810	$1/4 \times 1\,1/4$	0.2487	0.2505	0.4170	1.250	1.273	0.1250	0.2510	0.1310
1010	$5/16 \times 1\,1/4$	0.3111	0.3130	0.3858	1.250	1.273	0.1562	0.3135	0.1622
1210	$3/8 \times 1\,1/4$	0.3735	0.3755	0.3545	1.250	1.273	0.1875	0.3760	0.1935

Table 10. *(Continued)* **ANSI Keyseat Dimensions for Woodruff**

811	$\frac{1}{4}\times 1\frac{3}{8}$	0.2487	0.2505	0.4640	1.375	1.398	0.1250	0.2510	0.1310
1011	$\frac{5}{16}\times 1\frac{3}{8}$	0.3111	0.3130	0.4328	1.375	1.398	0.1562	0.3135	0.1622
1211	$\frac{3}{8}\times 1\frac{3}{8}$	0.3735	0.3755	0.4015	1.375	1.398	0.1875	0.3760	0.1935
812	$\frac{1}{4}\times 1\frac{1}{2}$	0.2487	0.2505	0.5110	1.500	1.523	0.1250	0.2510	0.1310
1012	$\frac{5}{16}\times 1\frac{1}{2}$	0.3111	0.3130	0.4798	1.500	1.523	0.1562	0.3135	0.1622
1212	$\frac{3}{8}\times 1\frac{1}{2}$	0.3735	0.3755	0.4485	1.500	1.523	0.1875	0.3760	0.1935
617-1	$\frac{3}{16}\times 2\frac{1}{8}$	0.1863	0.1880	0.3073	2.125	2.160	0.0937	0.1885	0.0997
817-1	$\frac{1}{4}\times 2\frac{1}{8}$	0.2487	0.2505	0.2760	2.125	2.160	0.1250	0.2510	0.1310
1017-1	$\frac{5}{16}\times 2\frac{1}{8}$	0.3111	0.3130	0.2448	2.125	2.160	0.1562	0.3135	0.1622
1217-1	$\frac{3}{8}\times 2\frac{1}{8}$	0.3735	0.3755	0.2135	2.125	2.160	0.1875	0.3760	0.1935
617	$\frac{3}{16}\times 2\frac{1}{8}$	0.1863	0.1880	0.4323	2.125	2.160	0.0937	0.1885	0.0997
817	$\frac{1}{4}\times 2\frac{1}{8}$	0.2487	0.2505	0.4010	2.125	2.160	0.1250	0.2510	0.1310
1017	$\frac{5}{16}\times 2\frac{1}{8}$	0.3111	0.3130	0.3698	2.125	2.160	0.1562	0.3135	0.1622
1217	$\frac{3}{8}\times 2\frac{1}{8}$	0.3735	0.3755	0.3385	2.125	2.160	0.1875	0.3760	0.1935
822-1	$\frac{1}{4}\times 2\frac{3}{4}$	0.2487	0.2505	0.4640	2.750	2.785	0.1250	0.2510	0.1310
1022-1	$\frac{5}{16}\times 2\frac{3}{4}$	0.3111	0.3130	0.4328	2.750	2.785	0.1562	0.3135	0.1622
1222-1	$\frac{3}{8}\times 2\frac{3}{4}$	0.3735	0.3755	0.4015	2.750	2.785	0.1875	0.3760	0.1935
1422-1	$\frac{7}{16}\times 2\frac{3}{4}$	0.4360	0.4380	0.3703	2.750	2.785	0.2187	0.4385	0.2247
1622-1	$\frac{1}{2}\times 2\frac{3}{4}$	0.4985	0.5005	0.3390	2.750	2.785	0.2500	0.5010	0.2560
822	$\frac{1}{4}\times 2\frac{3}{4}$	0.2487	0.2505	0.6200	2.750	2.785	0.1250	0.2510	0.1310
1022	$\frac{5}{16}\times 2\frac{3}{4}$	0.3111	0.3130	0.5888	2.750	2.785	0.1562	0.3135	0.1622
1222	$\frac{3}{8}\times 2\frac{3}{4}$	0.3735	0.3755	0.5575	2.750	2.785	0.1875	0.3760	0.1935
1422	$\frac{7}{16}\times 2\frac{3}{4}$	0.4360	0.4380	0.5263	2.750	2.785	0.2187	0.4385	0.2247
1622	$\frac{1}{2}\times 2\frac{3}{4}$	0.4985	0.5005	0.4950	2.750	2.785	0.2500	0.5010	0.2560
1228	$\frac{3}{8}\times 3\frac{1}{2}$	0.3735	0.3755	0.7455	3.500	3.535	0.1875	0.3760	0.1935
1428	$\frac{7}{16}\times 3\frac{1}{2}$	0.4360	0.4380	0.7143	3.500	3.535	0.2187	0.4385	0.2247
1628	$\frac{1}{2}\times 3\frac{1}{2}$	0.4985	0.5005	0.6830	3.500	3.535	0.2500	0.5010	0.2560
1828	$\frac{9}{16}\times 3\frac{1}{2}$	0.5610	0.5630	0.6518	3.500	3.535	0.2812	0.5635	0.2872
2028	$\frac{5}{8}\times 3\frac{1}{2}$	0.6235	0.6255	0.6205	3.500	3.535	0.3125	0.6260	0.3185
2228	$\frac{11}{16}\times 3\frac{1}{2}$	0.6860	0.6880	0.5893	3.500	3.535	0.3437	0.6885	0.3497
2428	$\frac{3}{4}\times 3\frac{1}{2}$	0.7485	0.7505	0.5580	3.500	3.535	0.3750	0.7510	0.3810

[a] These Width *A* values were set with the maximum keyseat (shaft) width as that figure which will receive a key with the greatest amount of looseness consistent with assuring the key's sticking in the keyseat (shaft). Minimum keyseat width is that figure permitting the largest shaft distortion acceptable when assembling maximum key in minimum keyseat. Dimensions *A*, *B*, *C*, *D* are taken at side intersection.

All dimensions are given in inches.

The following definitions are given in this standard:

Woodruff Key: A Remountable machinery part which, when assembled into key-seats, provides a positive means for transmitting torque between the shaft and hub.

Woodruff Key Number: An identification number by which the size of key may be readily determined.

Woodruff Keyseat—Shaft: The circular pocket in which the key is retained.

Woodruff Keyseat—Hub: An axially located rectangular groove in a hub. (This has been referred to as a keyway.)

Woodruff Keyseat Milling Cutter: An arbor type or shank type milling cutter normally used for milling Woodruff keyseats in shafts.

Taper Shaft Ends with Slotted Nuts *SAE Standard*

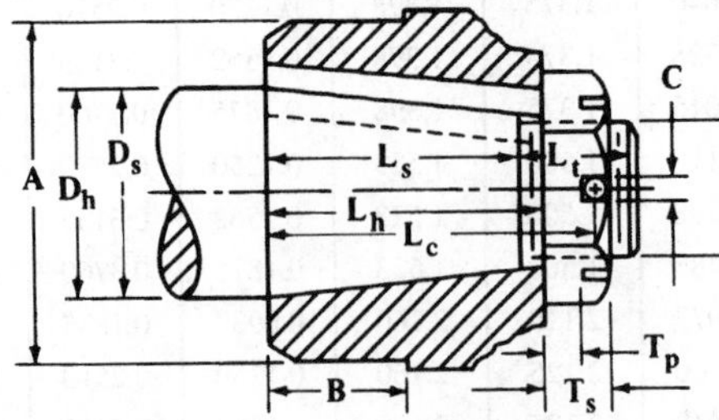

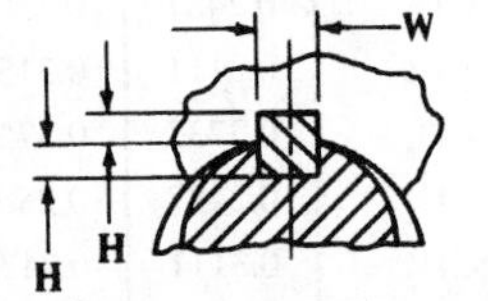

TAPER PER FOOT = $1.500^{\pm 0.002}$ IN.
H IS MEASURED PERPENDICULAR TO KEY.
C, COTTER-PIN HOLE HAS CENTERLINE DISPLACED 90° FROM KEYWAY CENTER

Nom. Diam.	Diam. of Shaft, D_s		Diam. of Hole, D_h		L_c	L_s	L_h	L_t	T_s	T_p	Nut Width, Flats
	Max.	Min.	Max.	Min.							
1/4	0.250	0.249	0.248	0.247	9/16	5/16	3/8	5/16	7/32	9/64	5/16
3/8	0.375	0.374	0.373	0.372	47/64	7/16	1/2	23/64	17/64	3/16	1/2
1/2	0.500	0.499	0.498	0.497	63/64	11/16	3/4	23/64	17/64	3/16	1/2
5/8	0.625	0.624	0.623	0.622	1 3/32	11/16	3/4	17/32	7/16	1/4	3/4
3/4	0.750	0.749	0.748	0.747	1 11/32	15/16	1	17/32	7/16	1/4	3/4
7/8	0.875	0.874	0.873	0.872	1 11/16	1 1/8	1 1/4	11/16	1/2	5/16	15/16
1	1.001	0.999	0.997	0.995	1 15/16	1 3/8	1 1/2	11/16	1/2	5/16	1 1/16
1 1/8	1.126	1.124	1.122	1.120	1 15/16	1 3/8	1 1/2	11/16	1/2	5/16	1 1/4
1 1/4	1.251	1.249	1.247	1.245	1 15/16	1 3/8	1 1/2	11/16	1/2	5/16	1 7/16
1 3/8	1.376	1.374	1.372	1.370	2 7/16	1 7/8	2	11/16	1/2	5/16	1 7/16
1 1/2	1.501	1.499	1.497	1.495	2 7/16	1 7/8	2	11/16	1/2	5/16	1 7/16
1 5/8	1.626	1.624	1.622	1.620	2 13/16	2 1/8	2 1/4	13/16	5/8	7/16	2 3/16
1 3/4	1.751	1.749	1.747	1.745	2 13/16	2 1/8	2 1/4	13/16	5/8	7/16	2 3/16
1 7/8	1.876	1.874	1.872	1.870	3 1/16	2 3/8	2 1/2	13/16	5/8	7/16	2 3/16
2	2.001	1.999	1.997	1.995	3 9/16	2 7/8	3	13/16	5/8	7/16	2 3/16
2 1/4	2.252	2.248	2.245	2.242	3 9/16	2 7/8	3	13/16	5/8	7/16	2 3/8
2 1/2	2.502	2.498	2.495	2.492	4 9/32	3 3/8	3 1/2	1 1/4	1	5/8	3 1/8
2 3/4	2.752	2.748	2.745	2.742	4 9/32	3 3/8	3 1/2	1 1/4	1	5/8	3 1/8
3	3.002	2.998	2.995	2.992	2 25/32	3 7/8	4	1 1/4	1	5/8	3 1/8
3 1/4	3.252	3.248	3.245	3.242	5 1/32	4 1/8	4 1/4	1 1/4	1	5/8	3 1/8
3 1/2	3.502	3.498	3.495	3.492	5 7/16	4 3/8	4 1/2	1 3/8	1 1/8	3/4	3 7/8
4	4.002	3.998	3.995	3.992	6 7/16	5 3/8	5 1/2	1 3/8	1 1/8	3/4	3 7/8

Nom. Diam.	D_t	Thds. per Inch	Keyway				Square Key		A	B	C
			W		H						
			Max.	Min.	Max.	Min.	Max.	Min.			
1/4	#10	40	0.0625	.0615	.037	.033	0.0635	0.0625	1/2	3/16	5/64
3/8	5/16	32	0.0937	.0927	.053	.049	0.0947	0.0937	11/16	1/4	5/64
1/2	5/16	32	0.1250	.1240	.069	.065	0.1260	0.1250	7/8	3/8	5/64
5/8	1/2	28	0.1562	.1552	.084	.080	0.1572	0.1562	1 1/16	3/8	1/8
3/4	1/2	28	0.1875	.1865	.100	.096	0.1885	0.1875	1 1/4	5/8	1/8
7/8	5/8	24	0.2500	.2490	.131	.127	0.2510	0.2500	1 1/2	3/4	5/32
1	3/4	20	0.2500	.2490	.131	.127	0.2510	0.2500	1 3/4	7/8	5/32
1 1/8	7/8	20	0.3125	.3115	.162	.158	0.3135	0.3125	2	7/8	5/32
1 1/4	1	20	0.3125	.3115	.162	.158	0.3135	0.3125	2 1/8	7/8	5/32
1 3/8	1	20	0.3750	.3740	.194	.190	0.3760	0.3750	2 1/4	1	5/32
1 1/2	1	20	0.3750	.3740	.194	.190	0.3760	0.3750	2 1/2	1	5/32
1 5/8	1 1/4	18	0.4375	.4365	.225	.221	0.4385	0.4375	2 3/4	1 1/4	5/32
1 3/4	1 1/4	18	0.4375	.4365	.225	.221	0.4385	0.4375	3	1 1/4	5/32
1 7/8	1 1/4	18	0.4375	.4365	.225	.221	0.4385	0.4375	3 1/8	1 1/4	5/32
2	1 1/4	18	0.5000	.4990	.256	.252	0.5010	0.5000	3 1/4	1 1/2	5/32
2 1/4	1 1/2	18	0.5625	.5610	.287	.283	0.5640	0.5625	3 1/2	1 1/2	5/32
2 1/2	2	16	0.6250	.6235	.319	.315	0.6265	0.6250	4	1 3/4	7/32
2 3/4	2	16	0.6875	.6860	.350	.346	0.6890	0.6875	4 3/8	1 3/4	7/32
3	2	16	0.7500	.7485	.381	.377	0.7515	0.7500	4 3/4	2	7/32
3 1/4	2	16	0.7500	.7485	.381	.377	0.7515	0.7500	5	2 1/8	7/32
3 1/2	2 1/2	16	0.8750	.8735	.444	.440	0.8765	0.8750	5 1/2	2 1/4	9/32
4	2 1/2	16	1.0000	.9985	.506	.502	1.0015	1.0000	6 1/4	2 3/4	9/32

All dimensions in inches except where otherwise noted.

Chamfered Keys and Filleted Keyseats.—In general practice, chamfered keys and filleted keyseats are not used. However, it is recognized that fillets in keyseats decrease stress concentration at corners. When used, fillet radii should be as large as possible without causing excessive bearing stresses due to reduced contact area between the key and its mating parts. Keys must be chamfered or rounded to clear fillet radii. Values in Table 5 assume general conditions and should be used only as a guide when critical stresses are encountered.

Table 11. Finding Depth of Keyseat and Distance from Top of Key to Bottom of Shaft

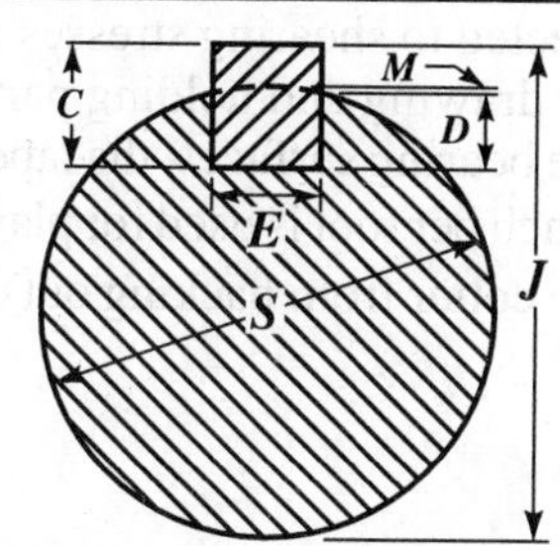

For milling keyseats, the total depth to feed cutter in from outside of shaft to bottom of keyseat is $M + D$, where D is depth of keyseat.

For checking an assembled key and shaft, caliper measurement J between top of key and bottom of shaft is used.

$$J = S - (M + D) + C$$

where C is depth of key. For Woodruff keys, dimensions C and D can be found in Tables 8 through 10. Assuming shaft diameter S is normal size, the tolerance on dimension J for Woodruff keys in keyslots are + 0.000, –0.010 inch.

Dia. of Shaft S, Inches	Width of Keyseat, E														
	1/16	3/32	1/8	5/32	3/16	7/32	1/4	5/16	3/8	7/16	1/2	9/16	5/8	11/16	3/4
	Dimension M, Inch														
0.3125	.0032	…	…	…	…	…	…	…	…	…	…	…	…	…	…
0.3437	.0029	.0065	…	…	…	…	…	…	…	…	…	…	…	…	…
0.3750	.0026	.0060	.0107	…	…	…	…	…	…	…	…	…	…	…	…
0.4060	.0024	.0055	.0099	…	…	…	…	…	…	…	…	…	…	…	…
0.4375	.0022	.0051	.0091	…	…	…	…	…	…	…	…	…	…	…	…
0.4687	.0021	.0047	.0085	.0134	…	…	…	…	…	…	…	…	…	…	…
0.5000	.0020	.0044	.0079	.0125	…	…	…	…	…	…	…	…	…	…	…
0.5625	…	.0039	.0070	.0111	.0161	…	…	…	…	…	…	…	…	…	…
0.6250	…	.0035	.0063	.0099	.0144	.0198	…	…	…	…	…	…	…	…	…
0.6875	…	.0032	.0057	.0090	.0130	.0179	.0235	…	…	…	…	…	…	…	…
0.7500	…	.0029	.0052	.0082	.0119	.0163	.0214	.0341	…	…	…	…	…	…	…
0.8125	…	.0027	.0048	.0076	.0110	.0150	.0197	.0312	…	…	…	…	…	…	…
0.8750	…	.0025	.0045	.0070	.0102	.0139	.0182	.0288	…	…	…	…	…	…	…
0.9375	…	…	.0042	.0066	.0095	.0129	.0170	.0263	.0391	…	…	…	…	…	…
1.0000	…	…	.0039	.0061	.0089	.0121	.0159	.0250	.0365	…	…	…	…	…	…
1.0625	…	…	.0037	.0058	.0083	.0114	.0149	.0235	.0342	…	…	…	…	…	…
1.1250	…	…	.0035	.0055	.0079	.0107	.0141	.0221	.0322	.0443	…	…	…	…	…
1.1875	…	…	.0033	.0052	.0074	.0102	.0133	.0209	.0304	.0418	…	…	…	…	…
1.2500	…	…	.0031	.0049	.0071	.0097	.0126	.0198	.0288	.0395	…	…	…	…	…
1.3750	…	…	…	.0045	.0064	.0088	.0115	.0180	.0261	.0357	.0471	…	…	…	…
1.5000	…	…	…	.0041	.0059	.0080	.0105	.0165	.0238	.0326	.0429	…	…	…	…
1.6250	…	…	…	.0038	.0054	.0074	.0097	.0152	.0219	.0300	.0394	.0502	…	…	…
1.7500	…	…	…	…	.0050	.0069	.0090	.0141	.0203	.0278	.0365	.0464	…	…	…
1.8750	…	…	…	…	.0047	.0064	.0084	.0131	.0189	.0259	.0340	.0432	.0536	…	…
2.0000	…	…	…	…	.0044	.0060	.0078	.0123	.0177	.0242	.0318	.0404	.0501	…	…
2.1250	…	…	…	…	…	.0056	.0074	.0116	.0167	.0228	.0298	.0379	.0470	.0572	.0684
2.2500	…	…	…	…	…	…	.0070	.0109	.0157	.0215	.0281	.0357	.0443	.0538	.0643
2.3750	…	…	…	…	…	…	…	.0103	.0149	.0203	.0266	.0338	.0419	.0509	.0608
2.5000	…	…	…	…	…	…	…	…	.0141	.0193	.0253	.0321	.0397	.0482	.0576
2.6250	…	…	…	…	…	…	…	…	.0135	.0184	.0240	.0305	.0377	.0457	.0547
2.7500	…	…	…	…	…	…	…	…	…	.0175	.0229	.0291	.0360	.0437	.0521
2.8750	…	…	…	…	…	…	…	…	…	.0168	.0219	.0278	.0344	.0417	.0498
3.0000	…	…	…	…	…	…	…	…	…	…	.0210	.0266	.0329	.0399	.0476

Depths for Milling Keyseats.—The above table has been compiled to facilitate the accurate milling of keyseats. This table gives the distance M (see illustration accompanying table) between the top of the shaft and a line passing through the upper corners or edges of the keyseat. Dimension M is calculated by the formula: $M = \frac{1}{2}(S - \sqrt{S^2 - E^2})$ where S is diameter of shaft, and E is width of keyseat. A simple approximate formula that gives M to within 0.001 inch is $M = E_2 \div 4S$.

Cotters.—A cotter is a form of key that is used to connect rods, etc., that are subjected either to tension or compression or both, the cotter being subjected to shearing stresses at two transverse cross-sections. When taper cotters are used for drawing and holding parts together, if the cotter is held in place by the friction between the bearing surfaces, the taper should not be too great. Ordinarily a taper varying from ¼ to ½ inch per foot is used for plain cotters. When a set-screw or other device is used to prevent the cotter from backing out of its slot, the taper may vary from 1 ½ to 2 inches per foot.

British Keys and Keyways

British Standard Metric Keys and Keyways.—This British Standard, BS 4235:Part 1:1972 (1986), covers square and rectangular parallel keys and keyways, and square and rectangular taper keys and keyways. Plain and gib-head taper keys are specified. There are three classes of fit for the square and rectangular parallel keys and keyways, designated free, normal, and close. A *free fit* is applied when the application requires the hub of an assembly to slide over the key; a *normal fit* is employed when the key is to be inserted in the keyway with the minimum amount of fitting, as may be required in mass-production assembly work; and a *close fit* is applied when accurate fitting of the key is required under maximum material conditions, which may involve selection of components.

The Standard does not provide for misalignment or offset greater than can be accommodated within the dimensional tolerances. If an assembly is to be heavily stressed, a check should be made to ensure that the cumulative effect of misalignment or offset, or both, does not prevent satisfactory bearing on the key. Radii and chamfers are not normally provided on keybar and keys as supplied, but they can be produced during manufacture by agreement between the user and supplier.

Unless otherwise specified, keys in compliance with this Standard are manufactured from steel made to BS 970 having a tensile strength of not less than 550 MN/m² in the finished condition. BS 970, Part 1, lists the following steels and maximum section sizes, respectively, that meet this tensile strength requirement: 070M20, 25 × 14 mm; 070M26, 36 × 20 mm; 080M30, 90 × 45 mm; and 080M40, 100 × 50 mm.

At the time of publication of this Standard, the demand for metric keys was not sufficient to enable standard ranges of lengths to be established. The lengths given in the accompanying table are those shown as standard in ISO Recommendations R773: 1969, "Rectangular or Square Parallel Keys and their Corresponding Keyways (Dimensions in Millimeters)," and R 774: 1969, "Taper Keys and their Corresponding Keyways—with or without Gib Head (Dimensions in Millimeters)."

Tables 1 through 4 on the following pages cover the dimensions and tolerances of square and rectangular keys and keyways, and square and rectangular taper keys and keyways.

Table 1. British Standard Metric Keyways for Square and Rectangular Parallel Keys,

Section x-x

Enlarged Detail of Key and Keyways

Shaft		Key	Keyway											
Nonimal Diameter d			Width, b						Depth				Radius r	
				Free Fit		Normal Fit		Close Fit	Shaft t_1		Hub t_2			
Over	Up to and Incl.	Size, $b \times h$	Nom.	Shaft (H9)	Hub (D10)	Shaft (N9)	Hub (J_s9)[a]	Shaft and Hub (P9)	Nom.	Tol.	Nom.	Tol.	Max.	Min.
				Tolerances										
Keyways for Square Parallel Keys														
6	8	2 × 2	2	+0.025	+0.060	−0.004	+0.012	−0.006	1.2		1		0.16	0.08
8	10	3 × 3	3	0	+0.020	−0.029	−0.012	−0.031	1.8		1.4		0.16	0.08
10	12	4 × 4	4						2.5	+0.1 0	1.8	+0.1 0	0.16	0.08
12	17	5 × 5	5	+0.030 0	+0.078 +0.030	0 −0.030	+0.015 −0.015	−0.012 −0.042	3		2.3		0.25	0.16
17	22	6 × 6	6						3.5		2.8		0.25	0.16

Table 1. *(Continued)* **British Standard Metric Keyways for Square and Rectangular Parallel Keys,**

22	30	8 × 7	8	+0.036	+0.098	0	+0.018	−0.015	4		3.3		0.25	0.16
30	38	10 × 8	10	0	+0.040	−0.036	−0.018	−0.051	5		3.3		0.40	0.25
38	44	12 × 8	12						5		3.3		0.40	0.25
44	50	14 × 9	14	+0.043	+0.120	0	+0.021	−0.018	5.5		3.8		0.40	0.25
50	58	16 × 10	16		+0.050	−0.043	−0.021	−0.061	6		4.3		0.40	0.25
58	65	18 × 11	18						7	+0.2 0	4.4	+0.2 0	0.40	0.25
65	75	20 × 12	20						7.5		4.9		0.60	0.40
75	85	22 × 14	22	+0.052	+0.149	0	+0.026	−0.022	9		5.4		0.60	0.40
85	95	25 × 14	25		+0.065	−0.052	−0.026	−0.074	9		5.4		0.60	0.40
95	110	28 × 16	28						10		6.4		0.60	0.40
110	130	32 × 18	32						11		7.4		0.60	0.40
130	150	36 × 20	36						12		8.4		1.00	0.70
150	170	40 × 22	40	+0.062	+0.180	0	+0.031	−0.026	13		9.4		1.00	0.70
170	200	45 × 25	45		−0.080	−0.062	−0.031	−0.088	15		10.4		1.00	0.70
200	230	50 × 28	50						17		11.4		1.00	0.70
230	260	56 × 32	56						20	+0.3 0	12.4	+0.3 0	1.60	1.20
260	290	63 × 32	63	+0.074	+0.220	0	+0.037	−0.032	20		12.4		1.60	1.20
290	330	70 × 36	70		+0.100	−0.074	−0.037	−0.106	22		14.4		1.60	1.20
330	380	80 × 40	80						25		15.4		2.50	2.00
380	440	90 × 45	90	+0.087	+0.260	0	+0.043	−0.037	28		17.4		2.50	2.00
440	500	100 × 50	100	0	+0.120	−0.087	−0.043	−0.124	31		19.5		2.50	2.00

[a] Tolerance limits J_s9 are quoted from BS 4500, "ISO Limits and Fits," to three significant figures.

All dimensions in millimeters.

Table 2. British Standard Metric Keyways for Square and Rectangular Taper Keys, *BS 4235:Part 1:1972 (1986)*

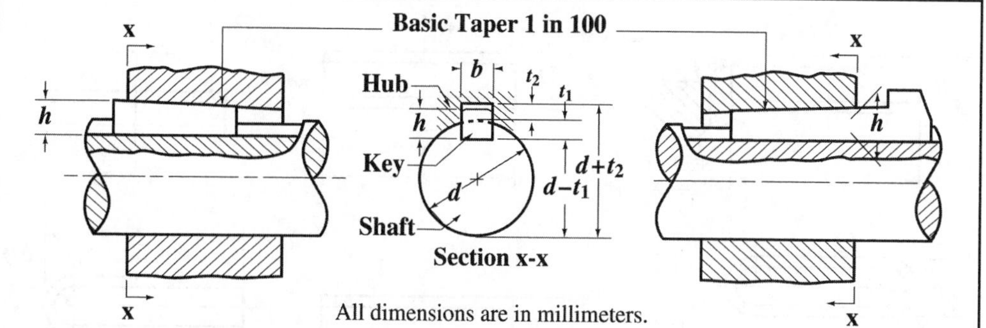

Basic Taper 1 in 100

All dimensions are in millimeters.

Shaft Nominal Diameter d Over	Shaft Nominal Diameter d Up to and Incl.	Key Size, $b \times h$	Keyway Width b, Shaft and Hub Nom.	Keyway Width b, Shaft and Hub Tol. (D10)	Depth Shaft t_1 Nom.	Depth Shaft t_1 Tol.	Depth Hub t_2 Nom.	Depth Hub t_2 Tol.	Corner Radius of Keyway Max.	Corner Radius of Keyway Min.
Keyways for Square Taper Keys										
6	8	2 × 2	2	} + 0.060 +0.020	1.2	} +0.1 0	0.5	} +0.1 0	0.16	0.08
8	10	3 × 3	3		1.8		0.9		0.16	0.08
10	12	4 × 4	4	} +0.078 +0.030	2.5		1.2		0.16	0.08
12	17	5 × 5	5		3		1.7		0.25	0.16
17	22	6 × 6	6		3.5	+ 0.2 0	2.2	+ 0.2 0	0.25	0.16
Keyways for Rectangular Taper Keys										
22	30	8 × 7	8	} + 0.098 +0.040	4	} +0.2 0	2.4	} +0.2 0	0.25	0.16
30	38	10 × 8	10		5		2.4		0.40	0.25
38	44	12 × 8	12	} +0.120 +0.050	5		2.4		0.40	0.25
44	50	14 × 9	14		5.5		2.9		0.40	0.25
50	58	16 × 10	16		6		3.4		0.40	0.25
58	65	18 × 11	18		7		3.4		0.40	0.25
65	75	20 × 12	20	} +0.149 +0.065	7.5		3.9		0.60	0.40
75	85	22 × 14	22		9		4.4		0.60	0.40
85	95	25 × 14	25		9		4.4		0.60	0.40
95	110	28 × 16	28		10		5.4		0.60	0.40
110	130	32 × 18	32	} +0.180 +0.080	11		6.4		0.60	0.40
130	150	36 × 20	36		12	} +0.3 0	7.1	} +0.3 0	1.00	0.70
150	170	40 × 22	40		13		8.1		1.00	0.70
170	200	45 × 25	45		15		9.1		1.00	0.70
200	230	50 × 28	50		17		10.1		1.00	0.70
230	260	56 × 32	56	} +0.220 +0.120	20		11.1		1.60	1.20
260	290	63 × 32	63		20		11.1		1.60	1.20
290	330	70 × 36	70		22		13.1		1.60	1.20
330	380	80 × 40	80		25		14.1		2.50	2.00
380	440	90 × 45	90	} + 0.260 +0.120	28		16.1		2.50	2.00
440	500	100 × 50	100		31		18.1		2.50	2.00

Table 3. British Standard Metric Square and Rectangular Parallel Keys. ***BS 4235:Part 1:1972 (1986)***

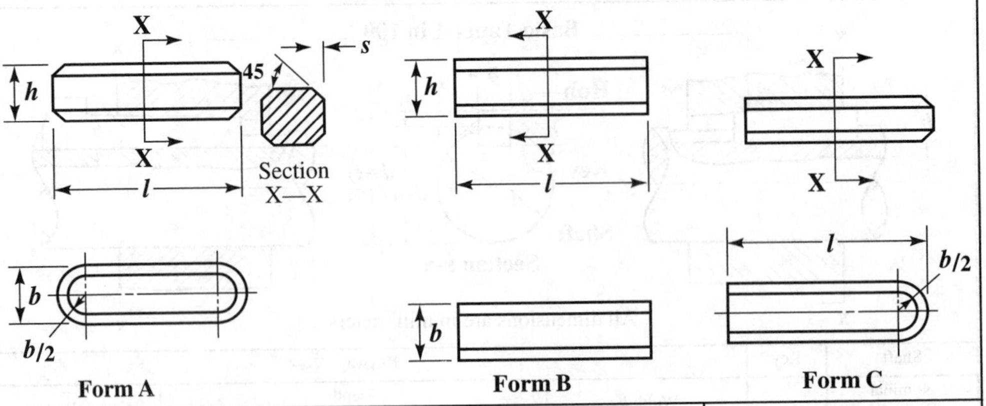

Width, b		Thickness, h		Chamfer, s		Length Range, l	
Nom.	Tol.[a]	Nom.	Tol.[a]	Min.	Max.	From	To
Square Parallel Keys							
2	0 / −0.025	2	0 / −0.025	0.16	0.25	6	20
3		3		0.16	0.25	6	36
4	0 / −0.030	4	0 / −0.030	0.16	0.25	8	45
5		5		0.25	0.40	10	56
6		6		0.25	0.40	14	70
Rectangular Parallel Keys							
8	0 / −0.036	7	0 / −0.090	0.25	0.40	18	90
10		8		0.40	0.60	22	110
12	0 / −0.043	8		0.40	0.60	28	140
14		9		0.40	0.60	36	160
16		10		0.40	0.60	45	180
18		11	0 / −0.110	0.40	0.60	50	200
20	0 / −0.052	12		0.60	0.80	56	220
22		14		0.60	0.80	63	250
25		14		0.60	0.80	70	280
28		16		0.60	0.80	80	320
32	0 / −0.062	18		0.60	0.80	90	360
36		20	0 / −0.130	1.00	1.20	100	400
40		22		1.00	1.20	…	…
45		25		1.00	1.20	…	…
50		28		1.00	1.20	…	…
56	0 / −0.074	32	0 / −0.160	1.60	2.00	…	…
63		32		1.60	2.00	…	…
70		36		1.60	2.00	…	…
80		40		2.50	3.00	…	…
90	0 / −0.087	45		2.50	3.00	…	…
100		50		2.50	3.00	…	…

[a] The tolerance on the width and thickness of square taper keys is h9, and on the width and thickness of rectangular keys, h9 and h11, respectively, in accordance with ISO metric limits and fits. All dimensions in millimeters.

Table 4. British Standard Metric Square and Rectangular Taper Keys
BS 4235:Part 1:1972 (1986)

Width b		Thickness h		Chamfer s		Length Range l		Gib head h_1	Radius r
Nom.	Tol.[a]	Nom.	Tol.[a]	Min.	Max.	From	To	Nom.	Nom.
Square Taper Keys									
2	0	2	0	0.16	0.25	6	20	…	…
3	−0.025	3	−0.025	0.16	0.25	6	36	…	…
4		4		0.16	0.25	8	45	7	0.25
5	0	5	0	0.25	0.40	10	56	8	0.25
6	−0.030	6	−0.030	0.25	0.40	14	70	10	0.25
Rectangular Taper Keys									
8	0	7		0.25	0.40	18	90	11	1.5
10	−0.036	8	0	0.40	0.60	22	110	12	1.5
12		8	−0.090	0.40	0.60	28	140	12	1.5
14	0	9		0.40	0.60	36	160	14	1.5
16	−0.043	10		0.40	0.60	45	180	16	3.2
18		11		0.40	0.60	50	200	18	3.2
20		12		0.60	0.80	56	220	20	3.2
22	0	14	0	0.60	0.80	63	250	22	3.2
25	−0.052	14	−0.110	0.60	0.80	70	280	22	3.2
28		16		0.60	0.80	80	320	25	3.2
32		18		0.60	0.80	90	360	28	6.4
36	0	20		1.00	1.20	100	400	32	6.4
40	−0.062	22	0	1.00	1.20	…	…	36	6.4
45		25	−0.130	1.00	1.20	…	…	40	6.4
50		28		1.00	1.20	…	…	45	6.4
56		32		1.60	2.00	…	…	50	9.5
63	0	32		1.60	2.00	…	…	50	9.5
70	−0.074	36	0	1.60	2.00	…	…	56	9.5
80		40	−0.160	2.50	3.00	…	…	63	9.5
90	0	45		2.50	3.00	…	…	70	9.5
100	−0.087	50		2.50	3.00	…	…	80	9.5

[a] The tolerance on the width and thickness of square taper keys is h9, and on the width and thickness of rectangular taper keys, h9 and h11 respectively, in accordance with ISO metric limits and fits. Does not apply to gib head dimensions.

British Standard Keys and Keyways: Tables 1 through 6 from BS 46:Part 1:1958 (1985) (obsolescent) provide data for rectangular parallel keys and keyways, square parallel keys and keyways, plain and gib head rectangular taper keys and key-ways, plain and gib head square taper keys and keyways, and Woodruff keys and keyways.

Parallel Keys: These keys are used for transmitting unidirectional torques in transmissions not subject to heavy starting loads and where periodic withdrawal or sliding of the hub member may be required. In many instances, particularly couplings, a gib-head cannot be accommodated, and there is insufficient room to drift out the key from behind. It is then necessary to withdraw the component over the key and a parallel key is essential. Parallel square and rectangular keys are normally side fitting with top clearance and are usually retained in the shaft rather more securely than in the hub. The rectangular key is the general-purpose key for shafts greater than 1 inch in diameter; the square key is intended for use with shafts up to and including 1-inch diameter or for shafts up to 6-inch diameter where it is desirable to have a greater key depth than is provided by rectangular keys. In stepped shafts, the larger diameters are usually required by considerations other than torque, e.g., resistance to bending. Where components such as fans, gears, impellers, etc., are attached to the larger shaft diameter, the use of a key smaller than standard for that diameter may be permissible. As this results in unequal disposition of the key in the shaft and its related hub, the dimensions *H* and *h* must be recalculated to maintain the *T*/2 relationship.

British Standard Preferred Lengths of Metric Keys *BS 4235:Part 1:1972 (1986)*

Length	Type of key				Length	Type of key			
	Sq.	Rect.	Sq. Taper	Rect. Taper		Sq.	Rect.	Sq. Taper	Rect. Taper
6	x		x		63	x	x	x	x
8	x		x		70	x	x	x	x
10	x		x		80		x		x
12	x		x		90		x		x
14	x		x		100		x		x
16	x		x		110		x		x
18	x	x	x	x	125		x		x
20	x	x	x	x	140		x		x
22	x	x	x	x	160		x		x
25	x	x	x	x	180		x		x
28	x	x	x	x	200		x		x
32	x	x	x	x	220		x		x
36	x	x	x	x	250		x		x
40	x	x	x	x	280		x		x
45	x	x	x	x	320		x		x
50	x	x	x	x	360		x		x
56	x	x	x	x	400		x		x

Taper Keys: These keys are used for transmitting heavy unidirectional, reversing, or vibrating torques and in applications where periodic withdrawal of the key may be necessary. Taper keys are usually top fitting, but may be top and side fitting where required, and the keyway in the hub should then have the same width value as the keyway in the shaft. Taper keys of rectangular section are used for general purposes and are of less depth than square keys; square sections are for use with shafts up to and including 1-inch diameter or for shafts up to 6-inch diameter where it is desirable to have greater key depth.

Woodruff Keys: These keys are used for light applications or the angular location of associated parts on tapered shaft ends. They are not recommended for other applications, but if so used, corner radii in the shaft and hub keyways are advisable to reduce stress concentration.

Dimensions and Tolerances for British Parallel and Taper Keys and Keyways: Dimensions and tolerances for key and keyway widths given in Tables 1, 2, 3, and 4 are based on the width of key *W* and provide a fitting allowance. The fitting allowance is designed to permit an interference between the key and the shaft keyway and a slightly easier condition between the key and the hub keyway. In shrink and heavy force fits, it may be found necessary to depart from the width and depth tolerances specified. Any variation in the width of the keyway should be such that the greatest width is at the end from which the key enters and any variation in the depth of the keyway should be such that the greatest depth is at the end from which the key enters.

Keys and keybar normally are not chamfered or radiused as supplied, but this may be done at the time of fitting. Radii and chamfers are given in Tables 1, 2, 3, and 4. Corner radii are recommended for keyways to alleviate stress concentration.

Table 1. British Standard Rectangular Parallel Keys, Keyways, and Keybars *B.S. 46: Part I: 1958*

Diameter of Shaft		Key					Keyway in Shaft				Keyway in Hub				Nominal Keyway Radius, r[a]	Keybar			
Over	Up to and Including	Size $W \times T$	Width, W		Thickness, T		Width W_s		Depth H		Width W_h		Depth h			Width W		Thickness T	
			Max.	Min.	Max.	Min.	Min.	Max.	Min.	Max.	Min.	Max.	Min.	Max.		Max.	Min.	Max.	Min.
1	1¼	5/16 × 1/4	0.314	0.312	0.253	0.250	0.311	0.312	0.146	0.152	0.312	0.313	0.112	0.118	0.010	0.314	0.312	0.253	0.250
1¼	1½	3/8 × 1/4	0.377	0.375	0.253	0.250	0.374	0.375	0.150	0.156	0.375	0.376	0.108	0.114	0.010	0.377	0.375	0.253	0.250
1½	1¾	7/16 × 5/16	0.440	0.438	0.315	0.312	0.437	0.438	0.186	0.192	0.438	0.439	0.135	0.141	0.020	0.440	0.438	0.315	0.312
1¾	2	1/2 × 5/16	0.502	0.500	0.315	0.312	0.499	0.500	0.190	0.196	0.500	0.501	0.131	0.137	0.020	0.502	0.500	0.315	0.312
2	2½	5/8 × 7/16	0.627	0.625	0.441	0.438	0.624	0.625	0.260	0.266	0.625	0.626	0.185	0.191	0.020	0.627	0.625	0.441	0.438
2½	3	3/4 × 1/2	0.752	0.750	0.503	0.500	0.749	0.750	0.299	0.305	0.750	0.751	0.209	0.215	0.020	0.752	0.750	0.503	0.500
3	3½	7/8 × 5/8	0.877	0.875	0.629	0.625	0.874	0.875	0.370	0.376	0.875	0.876	0.264	0.270	0.062	0.877	0.875	0.629	0.625
3½	4	1 × 3/4	1.003	1.000	0.754	0.750	0.999	1.000	0.441	0.447	1.000	1.001	0.318	0.324	0.062	1.003	1.000	0.754	0.750
4	5	1¼ × 7/8	1.253	1.250	0.879	0.875	1.248	1.250	0.518	0.524	1.250	1.252	0.366	0.372	0.062	1.253	1.250	0.879	0.875
5	6	1½ × 1	1.504	1.500	1.006	1.000	1.498	1.500	0.599	0.605	1.500	1.502	0.412	0.418	0.062	1.504	1.500	1.006	1.000
6	7	1½ × 1¼	1.754	1.750	1.256	1.250	1.748	1.750	1.740	0.746	1.750	1.752	0.526	0.532	0.125				
7	8	2 × 1⅜	2.005	2.000	1.381	1.375	1.998	2.000	0.818	0.824	2.000	2.002	0.573	0.579	0.125				
8	9	2¼ × 1½	2.255	2.250	1.506	1.500	2.248	2.250	0.897	0.905	2.250	2.252	0.619	0.627	0.125	Bright keybar is not normally available in sections larger than the above.			
9	10	2½ × 1⅝	2.505	2.500	1.631	1.625	2.498	2.500	0.975	0.983	2.500	2.502	0.666	0.674	0.187				
10	11	2¾ × 1⅞	2.755	2.750	1.881	1.875	2.748	2.750	1.114	1.122	2.750	2.752	0.777	0.785	0.187				
11	12	3 × 2	3.006	3.000	2.008	2.000	2.998	3.000	1.195	1.203	3.000	3.002	0.823	0.831	0.187				
12	13	3¼ × 2⅛	3.256	3.250	2.133	2.125	3.248	3.250	1.273	1.281	3.250	3.252	0.870	0.878	0.187				
13	14	3½ × 2⅜	3.506	3.500	2.383	2.375	3.498	3.500	1.413	1.421	3.500	3.502	0.980	0.988	0.250				
14	15	3¾ × 2½	3.756	3.750	2.508	2.500	3.748	3.750	1.492	1.502	3.750	3.752	1.026	1.036	0.250	All dimensions in inches.			
15	16	4 × 2⅝	4.008	4.000	2.633	2.625	3.998	4.000	1.571	1.581	4.000	4.002	1.072	1.082	0.250				
16	17	4¼ × 2⅞	4.258	4.250	2.883	2.875	4.248	4.250	1.711	1.721	4.250	4.252	1.182	1.192	0.312				
17	18	4½ × 3	4.508	4.500	3.010	3.000	4.498	4.500	1.791	1.801	4.500	4.502	1.229	1.239	0.312				
18	19	4¼ × 3⅛	4.758	4.750	3.135	3.125	4.748	4.750	1.868	1.878	4.750	4.752	1.277	1.287	0.312				
19	20	5 × 3⅜	5.008	5.000	3.385	3.375	4.998	5.000	2.010	2.020	5.000	5.002	1.385	1.395	0.312				

[a] The key chamfer shall be the minimum to clear the keyway radius. Nominal values are given.

Table 2. British Standard Square Parallel Keys, Keyways, and Keybars *B.S. 46: Part I: 1958*

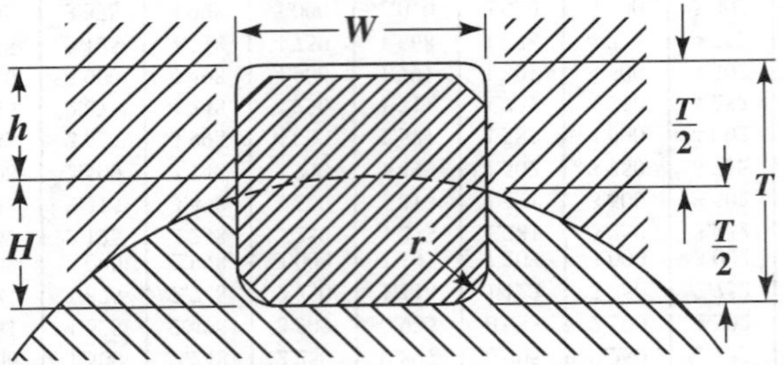

Diameter of Shaft			Key		Keyway in Shaft				Keyway in Hub				Nominal Keyway Radius, r^a	Bright Keybar	
			Width, W and Thickness, T		Width, W_s		Depth, H		Width, W_h		Depth, h			Width, W and Thickness, T	
Over	Up to and Including	Size, $W \times T$	Max.	Min.	Min.	Max.	Min.	Max.	Min.	Max.	Min.	Max.		Max.	Min.
1/4	1/2	1/8 × 1/8	0.127	0.125	0.124	0.125	0.072	0.078	0.125	0.126	0.060	0.066	0.010	0.127	0.125
1/2	3/4	3/16 × 3/16	0.190	0.188	0.187	0.188	0.107	0.113	0.188	0.189	0.088	0.094	0.010	0.190	0.188
3/4	1	1/4 × 1/4	0.252	0.250	0.249	0.250	0.142	0.148	0.250	0.251	0.115	0.121	0.010	0.252	0.250
1	1 1/4	5/16 × 5/16	0.314	0.312	0.311	0.312	0.177	0.183	0.312	0.313	0.142	0.148	0.010	0.314	0.312
1 1/4	1 1/2	3/8 × 3/8	0.377	0.375	0.374	0.375	0.213	0.219	0.375	0.376	0.169	0.175	0.010	0.377	0.375
1 1/2	1 3/4	7/16 × 7/16	0.440	0.438	0.437	0.438	0.248	0.254	0.438	0.439	0.197	0.203	0.020	0.440	0.438
1 3/4	2	1/2 × 1/2	0.502	0.500	0.499	0.500	0.283	0.289	0.500	0.501	0.224	0.230	0.020	0.502	0.500
2	2 1/2	5/8 × 5/8	0.627	0.625	0.624	0.625	0.354	0.360	0.625	0.626	0.278	0.284	0.020	0.627	0.625
2 1/2	3	3/4 × 3/4	0.752	0.750	0.749	0.750	0.424	0.430	0.750	0.751	0.333	0.339	0.020	0.752	0.750
3	3 1/2	7/8 × 7/8	0.877	0.875	0.874	0.875	0.495	0.501	0.875	0.876	0.387	0.393	0.062	0.877	0.875
3 1/2	4	1 × 1	1.003	1.000	0.999	1.000	0.566	0.572	1.000	1.001	0.442	0.448	0.062	1.003	1.000
4	5	1 1/4 × 1 1/4	1.253	1.250	1.248	1.250	0.707	0.713	1.250	1.252	0.551	0.557	0.062	1.253	1.250
5	6	1 1/2 × 1 1/2	1.504	1.500	1.498	1.500	0.848	0.854	1.500	1.502	0.661	0.667	0.062	1.504	1.500

[a] The key chamfer shall be the minimum to clear the keyway radius. Nominal values are given. All dimensions in inches.

Table 3. British Standard Rectangular Taper Keys and Keyways, Gib-head and Plain *B.S. 46: Part 1: 1958*

h
h
H
r
Section at Deep End of Keyway in Hub

Alternative Design Showing a Parallel Extension with a Drilled Hole To Facilitate Extraction
Taper 1 in 100
T
Plain Taper Key

A
D
B
C
45
R
T
Taper 1 in 100
All dimensions in inches
Gib-Head Key

Diameter of Shaft		Key					Keyway in Shaft and Hub								Nominal Keyway Radius, r[a]	Gib-head[b]				Radius, R
Over	Up to and Including	Size, $W \times T$	Width, W		Thickness, T		Keyway in Shaft Width, W_s		Keyway in Hub Width, W_h		Depth in Shaft, H		Depth in Hub at Deep End of Keyway, h			A	B	C	D	
			Max.	Min.	Max.	Min.	Min.	Max.	Min.	Max.	Min.	Max.	Min.	Max.						
1	1 1/4	5/16 × 1/4	0.314	0.312	0.254	0.249	0.311	0.312	0.312	0.313	0.146	0.152	0.090	0.096	0.010	3/8	7/16	1/4	0.3	1/16
1 1/4	1 1/2	3/8 × 1/4	0.377	0.375	0.254	0.249	0.374	0.375	0.375	0.376	0.150	0.156	0.086	0.092	0.010	7/16	7/16	9/32	0.3	1/16
1 1/2	1 3/4	7/16 × 5/16	0.440	0.438	0.316	0.311	0.437	0.438	0.438	0.439	0.186	0.192	0.112	0.118	0.020	1/2	9/16	5/16	0.4	1/16
1 3/4	2	1/2 × 5/16	0.502	0.500	0.316	0.311	0.499	0.500	0.500	0.0.501	0.190	0.196	0.108	0.114	0.020	9/16	5/8	3/8	0.4	1/16
2	2 1/2	5/8 × 7/16	0.627	0.625	0.442	0.437	0.624	0.625	0.625	0.626	0.260	0.266	0.162	0.168	0.020	11/16	3/4	7/16	0.5	1/8
2 1/2	3	3/4 × 1/2	0.752	0.750	0.504	0.499	0.749	0.750	0.750	0.751	0.299	0.305	0.185	0.191	0.020	13/16	7/8	17/32	0.5	1/8
3	3 1/2	7/8 × 5/8	0.877	0.875	0.630	0.624	0.874	0.875	0.875	0.876	0.370	0.376	0.239	0.245	0.062	15/16	1	21/32	0.6	1/8
3 1/2	4	1 × 3/4	1.003	1.000	0.755	0.749	0.999	1.000	1.000	1.001	0.441	0.447	0.293	0.299	0.062	1 1/16	1 1/4	23/32	0.6	1/8
4	5	1 1/4 × 7/8	1.253	1.250	0.880	0.874	1.248	1.250	1.250	1252	0.518	0.524	0.340	0.346	0.062	1 5/16	1 1/2	27/32	0.7	1/4
5	6	1 1/2 × 1	1.504	1.500	1.007	0.999	1.498	1.500	1.500	1.502	0.599	0.605	0.384	0.390	0.062	1 9/16	1 5/8	1 1/32	0.7	1/4
6	7	1 3/4 × 1 1/4	1.754	1.750	1.257	1.249	1.748	1.750	1.750	1.752	0.740	0.746	0.493	0.499	0.125	1 13/16	2	1 7/32	0.8	1/4
7	8	2 × 1 3/8	2.005	2.000	1.382	1.374	1.998	2.000	2.000	2.002	0.818	0.824	0.539	0.545	0.125	2 1/16	2 1/4	1 13/32	0.8	1/4
8	9	2 1/4 × 1 1/2	2.255	2.250	1.509	1.499	2.248	2.250	2.250	2.252	0.897	0.905	0.581	0.589	0.125	2 5/16	2 1/2	1 9/16	0.9	3/8
9	10	2 1/2 × 1 5/8	2.505	2.500	1.634	1.624	2.498	2.500	2.500	2.502	0.975	0.983	0.628	0.636	0.187	2 9/16	2 3/4	1 11/16	0.9	3/8
10	11	2 3/4 × 1 7/8	2.755	2.750	1.884	1.874	2.748	2.750	2.750	2.752	1.114	1.122	0.738	0.746	0.187	2 13/16	3	1 15/16	1.0	3/8
11	12	3 × 2	3.006	3.000	2.014	1.999	2.998	3.000	3.000	3.002	1.195	1.203	0.782	0.790	0.187	3 1/16	3 1/4	2 1/16	1.0	3/8

[a] The key chamfer shall be the minimum to clear the keyway radius. Nominal values shall be given.
[b] Dimensions A, B, C, D, and R pertain to gib-head keys only.
All dimensions in inches.

Table 4. British Standard Square Taper Keys and Keyways, Gib-head or Plain *B.S. 46: Part I: 1958*

Section at Deep End of Keyway in Hub

Alternative Design Showing a Parallel Extension with a Drilled Hole To Facilitate Extraction

Taper 1 in 100

Plain Taper Key

All dimensions in inches

Taper 1 in 100

Gib-Head Key

Diameter of Shaft		Key					Keyway in Shaft and Hub								Nominal Keyway Radius, r[a]	Gib-head[b]				
							Keyway in Shaft		Keyway in Hub											
Over	Up to and Including	Size $W \times T$	Width, W		Thickness, T		Width W_s		Width W_h		Depth in Shaft, H		Depth in Hub at Deep End of Keyway, h			A	B	C	D	Radius, R
			Max.	Min.	Max.	Min.	Min.	Max.	Min.	Max.	Min.	Max.	Min.	Max.						
1/4	1/2	1/8 × 1/8	0.127	0.125	0.129	0.124	0.124	0.125	0.125	0.126	0.072	0.078	0.039	0.045	0.010	3/16	1/4	5/32	0.1	1/32
1/2	3/4	3/16 × 3/16	0.190	0.188	0.192	0.187	0.187	0.188	0.188	0.189	0.107	0.113	0.067	0.073	0.010	1/4	3/8	7/32	0.2	1/32
3/4	1	1/4 × 1/4	0.252	0.250	0.254	0.249	0.249	0.250	0.250	0.251	0.142	0.148	0.094	0.100	0.010	5/16	7/16	9/32	0.2	1/16
1	1 1/4	5/16 × 5/16	0.314	0.312	0.316	0.311	0.311	0.312	0.312	0.313	0.177	0.183	0.121	0.127	0.010	3/8	9/16	11/32	0.3	1/16
1 1/4	1 1/2	3/8 × 3/8	0.377	0.375	0.379	0.374	0.374	0.375	0.375	0.376	0.213	0.219	0.148	0.154	0.010	7/16	5/8	13/32	0.3	1/16
1 1/2	1 3/4	7/16 × 7/16	0.440	0.438	0.442	0.437	0.437	0.438	0.438	0.439	0.248	0.254	0.175	0.181	0.020	1/2	3/4	15/32	0.4	1/16
1 3/4	2	1/2 × 1/2	0.502	0.500	0.504	0.499	0.499	0.500	0.500	0.501	0.283	0.289	0.202	0.208	0.020	9/16	7/8	17/32	0.4	1/16
2	2 1/2	5/8 × 5/8	0.627	0.625	0.630	0.624	0.624	0.625	0.625	0.626	0.354	0.360	0.256	0.262	0.020	11/16	1	21/32	0.5	1/8
2 1/2	3	3/4 × 3/4	0.752	0.750	0.755	0.749	0.749	0.750	0.750	0.751	0.424	0.430	0.310	0.316	0.020	13/16	1 1/4	25/32	0.5	1/8
3	3 1/2	7/8 × 7/8	0.877	0.875	0.880	0.874	0.874	0.875	0.875	0.876	0.495	0.501	0.364	0.370	0.062	15/16	1 3/8	20/32	0.6	1/8
3 1/2	4	1 × 1	1.003	1.000	1.007	0.999	0.999	1.000	1.000	1.001	0.566	0.572	0.418	0.424	0.062	1 1/16	1 5/8	1 1/32	0.6	1/8
4	5	1 1/4 × 1 1/4	1.253	1.250	1.257	1.249	1.248	1.250	1.250	1.252	0.707	0.713	0.526	0.532	0.062	1 5/16	2	1 9/32	0.7	1/4
5	6	1 1/2 × 1 1/2	1.504	1.500	1.509	1.499	1.498	1.500	1.500	1.502	0.848	0.854	0.635	0.641	0.062	1 9/16	2 1/2	1 17/32	0.7	1/4

[a] The key chamfer shall be the minimum to clear the keyway radius. Nominal values shall be given.

[b] Dimensions A, B, C, D, and R pertain to gib-head keys only.

All dimensions in inches.

Dimensions and Tolerances of British Woodruff Keys and Keyways.—Dimensions and tolerances are shown in Table 5. An optional alternative design of the Woodruff key that differs from the normal form in its depth is given in the illustration accompanying the table. The method of designating British Woodruff Keys is the same as the American method explained in the footnote on page 2348.

Table 5. British Standard Woodruff Keys and Keyways *BS 6: Part 1: 1958*

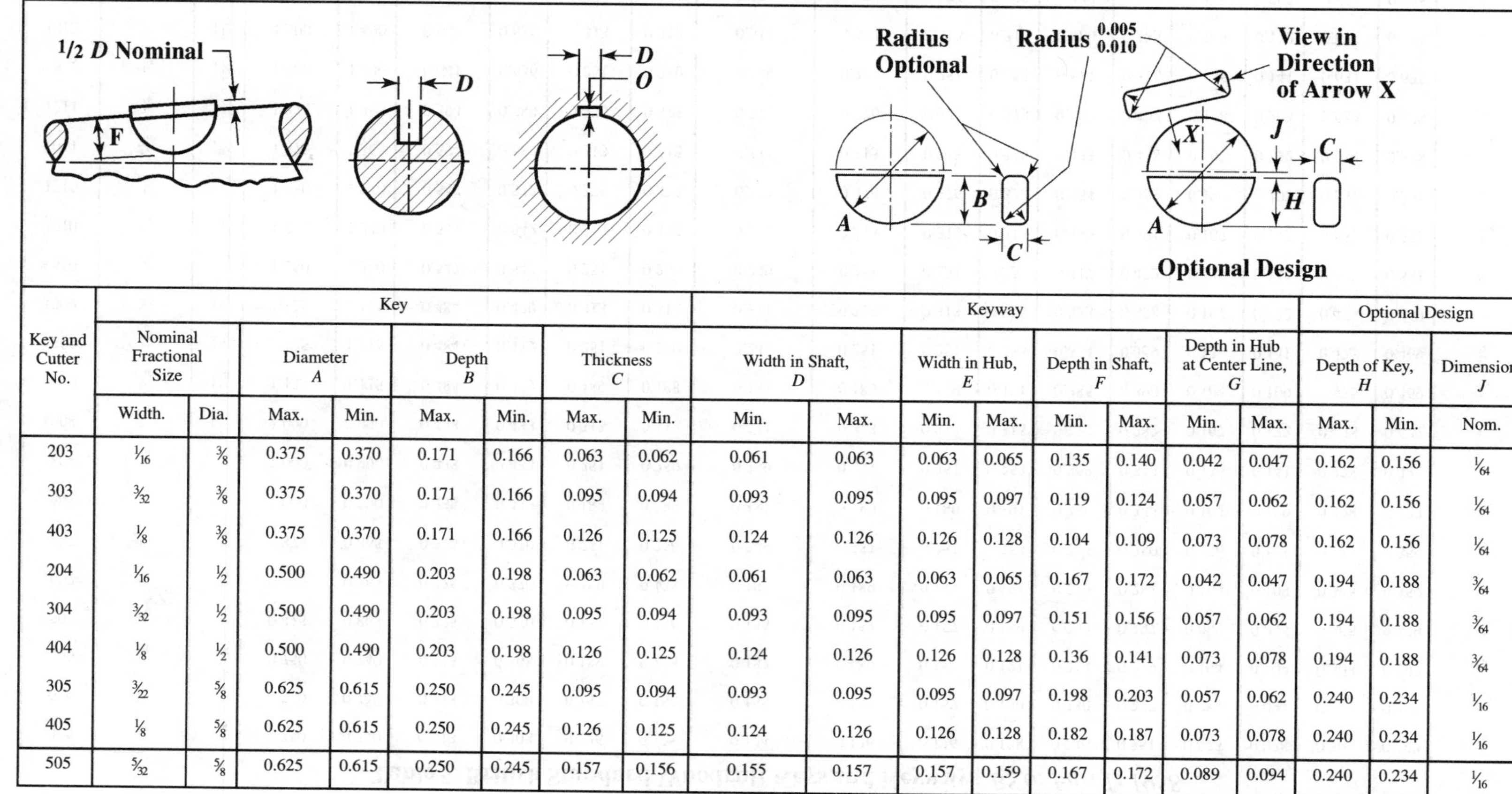

Key and Cutter No.	Key								Keyway								Optional Design		
	Nominal Fractional Size		Diameter *A*		Depth *B*		Thickness *C*		Width in Shaft, *D*		Width in Hub, *E*		Depth in Shaft, *F*		Depth in Hub at Center Line, *G*		Depth of Key, *H*		Dimension, *J*
	Width.	Dia.	Max.	Min.	Max.	Min.	Max.	Min.	Min.	Max.	Min.	Max.	Min.	Max.	Min.	Max.	Max.	Min.	Nom.
203	1/16	3/8	0.375	0.370	0.171	0.166	0.063	0.062	0.061	0.063	0.063	0.065	0.135	0.140	0.042	0.047	0.162	0.156	1/64
303	3/32	3/8	0.375	0.370	0.171	0.166	0.095	0.094	0.093	0.095	0.095	0.097	0.119	0.124	0.057	0.062	0.162	0.156	1/64
403	1/8	3/8	0.375	0.370	0.171	0.166	0.126	0.125	0.124	0.126	0.126	0.128	0.104	0.109	0.073	0.078	0.162	0.156	1/64
204	1/16	1/2	0.500	0.490	0.203	0.198	0.063	0.062	0.061	0.063	0.063	0.065	0.167	0.172	0.042	0.047	0.194	0.188	3/64
304	3/32	1/2	0.500	0.490	0.203	0.198	0.095	0.094	0.093	0.095	0.095	0.097	0.151	0.156	0.057	0.062	0.194	0.188	3/64
404	1/8	1/2	0.500	0.490	0.203	0.198	0.126	0.125	0.124	0.126	0.126	0.128	0.136	0.141	0.073	0.078	0.194	0.188	3/64
305	3/32	5/8	0.625	0.615	0.250	0.245	0.095	0.094	0.093	0.095	0.095	0.097	0.198	0.203	0.057	0.062	0.240	0.234	1/16
405	1/8	5/8	0.625	0.615	0.250	0.245	0.126	0.125	0.124	0.126	0.126	0.128	0.182	0.187	0.073	0.078	0.240	0.234	1/16
505	5/32	5/8	0.625	0.615	0.250	0.245	0.157	0.156	0.155	0.157	0.157	0.159	0.167	0.172	0.089	0.094	0.240	0.234	1/16

Table 5. British Standard Woodruff Keys and Keyways *BS 6: Part 1: 1958*

406	1/8	3/4	0.750	0.740	0.313	0.308	0.126	0.125	0.124	0.126	0.126	0.128	0.246	0.251	0.073	0.078	0.303	0.297	1/16
506	5/22	3/4	0.750	0.740	0.313	0.308	0.157	0.156	0.155	0.157	0.157	0.159	0.230	0.235	0.089	0.094	0.303	0.297	1/16
606	3/16	3/4	0.750	0.740	0.313	0.308	0.189	0.188	0.187	0.189	0.189	0.191	0.214	0.219	0.104	0.109	0.303	0.297	1/16
507	5/32	7/8	0.875	0.865	0.375	0.370	0.157	0.156	0.155	0.157	0.157	0.159	0.292	0.297	0.089	0.094	0.365	0.359	1/16
607	3/16	7/8	0.875	0.865	0.375	0.370	0.189	0.188	0.187	0.189	0.189	0.191	0.276	0.281	0.104	0.109	0.365	0.359	1/16
807	1/4	7/8	0.875	0.865	0.375	0.370	0.251	0.250	0.249	0.251	0.251	0.253	0.245	0.250	0.136	0.141	0.365	0.359	1/16
608	3/16	1	1.000	0.990	0.438	0.433	0.189	0.188	0.187	0.189	0.189	0.191	0.339	0.344	0.104	0.109	0.428	0.422	1/16
808	1/4	1	1.000	0.990	0.438	0.433	0.251	0.250	0.249	0.251	0.251	0.253	0.308	0.313	0.136	0.141	0.428	0.422	1/16
1008	5/16	1	1.000	0.990	0.438	0.433	0.313	0.312	0.311	0.313	0.313	0.315	0.277	0.282	0.167	0.172	0.428	0.422	1/16
609	3/16	1 1/8	1.125	1.115	0.484	0.479	0.189	0.188	0.187	0.189	0.189	0.191	0.385	0.390	0.104	0.109	0.475	0.469	5/64
809	1/4	1 1/8	1.125	1.115	0.484	0.479	0.251	0.250	0.249	0.251	0.251	0.253	0.354	0.359	0.136	0.141	0.475	0.469	5/64
1009	5/16	1 1/8	1.125	1.115	0.484	0.479	0.313	0.312	0.311	0.313	0.313	0.315	0.323	0.328	0.167	0.172	0.475	0.469	5/64
810	1/4	1 1/4	1.250	1.240	0.547	0.542	0.251	0.250	0.249	0.251	0.251	0.253	0.417	0.422	0.136	0.141	0.537	0.531	5/64
1010	5/16	1 1/4	1.250	1.240	0.547	0.542	0.313	0.312	0.311	0.313	0.313	0.315	0.386	0.391	0.167	0.172	0.537	0.531	5/64
1210	3/8	1 1/4	1.250	1.240	0.547	0.542	0.376	0.375	0.374	0.376	0.376	0.378	0.354	0.359	0.198	0.203	0.537	0.531	5/64
1011	5/16	1 3/8	1.375	1.365	0.594	0.589	0.313	0.312	0.311	0.313	0.313	0.315	0.433	0.438	0.167	0.172	0.584	0.578	3/32
1211	3/8	1 3/8	1.375	1.365	0.594	0.589	0.376	0.375	0.374	0.376	0.376	0.378	0.402	0.407	0.198	0.203	0.584	0.578	3/32
812	1/4	1 1/2	1.500	1.490	0.641	0.636	0.251	0.250	0.249	0.251	0.251	0.253	0.511	0.516	0.136	0.141	0.631	0.625	7/64
1012	5/16	1 1/2	1.500	1.490	0.641	0.636	0.313	0.312	0.311	0.313	0.313	0.315	0.480	0.485	0.167	0.172	0.631	0.625	7/64
1212	3/8	1 1/2	1.500	1.490	0.641	0.636	0.376	0.375	0.374	0.376	0.376	0.378	0.448	0.453	0.198	0.203	0.631	0.625	7/64

All dimensions are in inches.

Table 6. British Preferred Lengths of Plain (Parallel or Taper) and Gib-head Keys, Rectangular and Square Section *BS 46:Part 1:1958 (1985) Appendix*

Plain Key Size $W \times T$	Overall Length, L														
	¾	1	1¼	1½	1¾	2	2¼	2½	2¾	3	3½	4	4½	5	6
⅛ × ⅛	X	X													
3/16 × 3/16	X	X	X	X	X	X									
¼ × ¼	X	X	X	X	X	X	X	X	X	X					
5/16 × ¼	X	X	X	X	X	X	X	X	X	X					
5/16 × 5/16	X	X	X	X	X	X	X	X	X	X					
⅜ × ¼		X	X	X	X	X	X	X	X	X	X	X			
⅜ × ⅜		X	X	X	X	X	X	X	X	X	X	X			
7/16 × 5/16				X	X	X	X	X	X	X	X	X			
7/16 × 7/16					X	X	X	X	X	X	X	X			
½ × 5/16					X	X	X	X	X	X	X	X	X	X	
½ × ½						X	X	X	X	X	X	X	X	X	
⅝ × 7/16								X	X		X	X	X	X	
⅝ × ⅝								X		X	X	X	X	X	X
¾ × ½										X	X	X	X	X	X
¾ × ¾											X	X	X	X	X
⅞ × ⅝											X	X	X	X	X

Gib-head Key Size, $W \times T$	Overall Length, L																
	1½	1¾	2	2¼	2½	2¾	3	3½	4	4½	5	5½	6	6½	7	7½	8
3/16 × 3/16	X	X	X	X	X		X										
¼ × ¼	X	X	X	X	X	X	X	X	X								
5/16 × ¼			X	X	X	X	X	X	X								
5/16 × 5/16			X	X	X	X	X	X	X	X							
⅜ × ¼			X	X	X	X	X	X	X	X	X						
⅜ × ⅜			X	X	X	X	X	X	X	X	X	X	X				
7/16 × 5/16					X		X	X	X	X	X	X	X				
7/16 × 7/16					X		X	X	X	X	X	X	X				
½ × 5/16					X		X	X	X	X	X	X	X	X			
½ × ½					X		X	X	X	X	X	X	X	X			
⅝ × 7/16							X		X	X	X	X	X	X	X		
⅝ × ⅝									X	X	X	X	X	X	X		X
¾ × ½									X	X	X	X	X	X	X	X	X
¾ × ¾									X		X		X		X		X
⅞ × ⅝											X		X	X	X	X	X
⅞ × ⅞													X		X		X
1 × ¾													X		X	X	X
1 × 1													X		X		X

All dimension are in inches

Flat Belting

Flat belting was originally made from leather because it was the most durable material available and could easily be cut and joined to make a driving belt suitable for use with cylindrical or domed pulleys. This type of belting was popular because it could be used to transmit high torques over long distances and it was employed in factories to drive many small machines from a large common power source such as a steam engine. As electric motors became smaller, more efficient, and more powerful, and new types of belts and chains were made possible by modern materials and manufacturing processes, flat belts fell out of favor. Flat belts are still used for some drive purposes, but leather has been replaced by other natural and synthetic materials such as urethanes, which can be reinforced by high-strength polyamide or steel fabrics to provide properties such as resistance to stretching. The high modulus of elasticity in these flat belts eliminates the need for periodic retensioning that is usually necessary with V-belts.

Driving belts can be given a coating of an elastomer with a high coefficient of friction, to enable belts to grip pulleys without the degree of tension common with earlier materials. Urethanes are commonly used for driving belts where high resistance to abrasion is required, and will also resist attack by chemical solvents of most kinds. Flat belts having good resistance to high temperatures are also available. Typical properties of polyurethane belts include tensile strength up to 40,000 psi, depending on reinforcement type and Shore hardness of 85 to 95. Most polyurethane belts are installed under tension. The amount of the tension varies with the belt cross-section, being greater for belts of small section. Belt tension can be measured by marking lines 10 in. apart on an installed belt, then applying tension until the separation increases by the desired percentage. For 2 per cent tension, the lines on the tensioned belt would be 10.2 in. apart. Mechanical failure may result when belt tensioning is excessive, and 2 to 2.5 per cent elongation should be regarded as the limit.

Flat belts offer high load capacities and are capable of transmitting power over long distances, maintaining relative rotational direction, can operate without lubricants, and are generally inexpensive to maintain or replace when worn. Flat belt systems will operate with little maintenance and only periodic adjustment. Because they transmit motion by friction, flat belts have the ability to slip under excessive loads, providing a fail-safe action to guard against malfunctions. This advantage is offset by the problem that friction drives can both slip and creep so that they do not offer exact, consistent velocity ratios nor precision timing between input and output shafts. Flat belts can be made to any desired length, being joined by reliable chemical bonding processes.

Increasing centrifugal force has less effect on the load-carrying capacity of flat belts at high speeds than it has on V-belts, for instance. The low thickness of a flat belt, compared with a V-belt, places its center of gravity near the pulley surface. Flat belts therefore may be run at surface speeds of up to 16,000 or even 20,000 ft/min (81.28 and 101.6 m/s), although ideal speeds are in the range of 3,000 to 10,000 ft/min (15.25 to 50.8 m/s). Elastomeric drive surfaces on flat belts have eliminated the need for belt dressings that were often needed to keep leather belts in place. These surface coatings can also contain antistatic materials. Belt pulley wear and noise are low with flat belts shock and vibration are damped, and efficiency is generally greater than 98 per cent compared with 96 per cent for V-belts.

Driving belt load capacities can be calculated from torque $T = F(d/2)$ and horsepower $HP = T \times rpm/396{,}000$, where T is the torque in in.-lb, F is the force transmitted in lb, and d is the pulley diameter in in. Pulley width is usually about 10 per cent larger than the belt, and for good tracking, pulleys are often crowned by 0.012 to 0.10 in. for diameters in the range of 1.5 to 80 in.

Before a belt specification is written, the system should be checked for excessive startup and shut-down loads, which sometimes are more than 10 per cent above operating conditions. In overcoming such loads, the belt will transmit considerably more force than during normal operation. Large starting and stopping forces will also shorten belt life unless they are taken into account during the design stage.

Belt speed plays an important role in the amount of load a friction drive system can transmit. Higher speeds will require higher preloads (increased belt tension) to compensate for the higher centrifugal force. In positive drive (toothed belt) systems, higher speeds generate dynamic forces caused by unavoidable tolerance errors that may result in increased tooth or pin stresses and shorter belt life.

Pulley Diameters and Drive Ratios: Minimum pulley diameters determined by belt manufacturers are based on the minimum radius that a belt can wrap around a pulley without stressing the load-carrying members. For positive drive systems, minimum pulley diameters are also determined by the minimum number of teeth that must be engaged with the sprocket to guarantee the operating load.

Diameters of driving and driven pulleys determine the velocity ratio of the input relative to the output shaft and are derived from the following formulas: for all belt systems, velocity ratio $V = D_{pi}/D_{po}$, and for positive (toothed) drive systems, velocity ratio $V = N_i/N_o$, where D_{pi} is the pitch diameter of the driving pulley, D_{po} is the pitch diameter of the driven pulley, N_i is the number of teeth on the driving pulley, and N_o is the number of teeth on the driven pulley. For most drive systems, a velocity ratio of 8:1 is the largest that should be attempted with a single reduction drive, and 6:1 is a reasonable maximum.

Wrap Angles and Center-to-Center Distances: The radial distance for which the belt is in contact with the pulley surface, or the number of teeth in engagement for positive drive belts, is called the wrap angle. Belt and sprocket combinations should be chosen to ensure a wrap angle of about 120° around the smaller pulley. The wrap angle should not be less than 90°, especially with positive drive belts, because if too few teeth are in engagement, the belt may jump a tooth or pin and timing or synchronization may be lost.

For flat belts, the minimum allowable center-to-center distance (CD) for any belt-and-sprocket combination should be chosen to ensure a minimum wrap angle around the smaller pulley. For high-velocity systems, a good rule of thumb is a minimum CD equal to the sum of the pitch diameter of the larger sprocket and one-half the pitch diameter of the smaller sprocket. This formula ensures a minimum wrap angle of approximately 120°, which is generally sufficient for friction drives and will ensure that positive drive belts do not jump teeth.

Pulley Center Distances and Belt Lengths: Maximum center distances of pulleys should be about 15 to 20 times the pitch diameter of the smaller pulley. Greater spacing requires tight control of the belt tension because a small amount of stretch will cause a large drop in tension. Constant belt tension can be obtained by application of an adjustable tensioning pulley applied to the slack side of the belt. Friction drive systems using flat belts require much more tension than positive drive belt systems.

Belt length can be calculated from: $L = 2C + \pi(D_2 + D_1)/2 + (D_2 - D_1)^2/4C$ for friction drives, and length $L = 2C + \pi(D_2 + D_1)/2 + (D_2 + D_1)^2/4C$ for crossed belt friction belt drives, where C is the center distance, D_1 is the pitch diameter of the small pulley, and D_2 is the pitch diameter of the large pulley. For serrated belt drives, the length determined by use of these equations should be divided by the serration pitch. The belt length must then be adjusted to provide a whole number of serrations.

Calculating Diameters and Speeds of Pulleys

Pulley Diameters and Speeds.—If D = diameter of driving pulley, d = diameter of driven pulley, S = speed of driving pulley, and s = speed of driven pulley:

$$D = \frac{d \times s}{S}, \qquad d = \frac{D \times S}{s}, \qquad S = \frac{d \times s}{D}, \quad \text{and} \quad s = \frac{D \times S}{d}$$

Example 1: If the diameter of the driving pulley D is 24 inches, its speed is 100 rpm, and the driven pulley is to run at 600 rpm, the diameter of the driven pulley, $d = 24 \times 100/600 =$ 4 inches.

Example 2: If the diameter of the driven pulley d is 36 inches, its required speed is to be 150 rpm, and the speed of the driving pulley is to be 600 rpm, the diameter of the driving pulley $D = 36 \times 150/600 = 9$ inches.

Example 3: If the diameter of the driven pulley d is 4 inches, its required speed is 800 rpm, and the diameter of the driving pulley D is 26 inches, the speed of the driving pulley $= 4 \times 800/26 = 123$ rpm.

Example 4: If the diameter of the driving pulley D is 15 inches and its speed is 180 rpm, and the diameter of the driven pulley d is 9 inches, then the speed of the driven pulley = 15 $\times$ 180/9 = 300 rpm.

Pulley Diameters in Compound Drive.—If speeds of driving and driven pulleys, A, B, C, and D (see illustration) are known, the first step in finding their diameters is to form a fraction with the driving pulley speed as the numerator and the driven pulley speed as the, denominator, and then reduce this fraction to its lowest terms. Resolve the numerator and the denominator into two pairs of factors (a pair being one factor in the numerator and one in the denominator) and, if necessary, multiply each pair by a trial number that will give pulleys of suitable diameters.

Example: If the speed of pulley A is 260 rpm and the required speed of pulley D is 720 rpm, find the diameters of the four pulleys. Reduced to its lowest terms, the fraction 260/720 = 13/36, which represents the required speed ratio. Resolve this ratio 13/36 into two factors:

$$\frac{13}{36} = \frac{1 \times 13}{2 \times 18}$$

Multiply by trial numbers 12 and 1 to get:

$$\frac{(1 \times 12) \times (13 \times 1)}{(2 \times 12) \times (18 \times 1)} = \frac{12 \times 13}{24 \times 18}$$

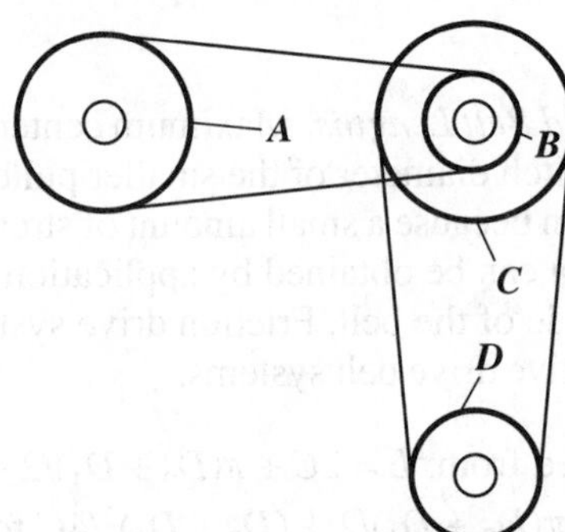

Compound Drive with Four Pulleys.

The values 12 and 13 in the numerator represent the diameters of the *driven* pulleys, B and D, and the values 24 and 18 in the denominator represent the diameters of the *driving* pulleys, A and C, as shown in the illustration.

Speed of Driven Pulley in Compound Drive.—If diameters of pulleys *A*, *B*, *C*, and *D* (see illustration above), and speed of pulley *A* are known, the speed of the driven pulley *D* is found from:

$$\frac{\text{driving pulley diameter}}{\text{driven pulley diameter}} \times \frac{\text{driving pulley diameter}}{\text{driven pulley diameter}} \times \text{speed of first driving pulley}$$

Example: If the diameters of driving pulleys *A* and *C* are 18 and 24 inches, diameters of driven pulleys *B* and *D* are 12 and 13 inches, and the speed of driving pulley *A* is 260 rpm, speed of driven pulley

$$D = \frac{18 \times 24}{12 \times 13} \times 260 = 720 \text{ rpm}$$

Length of Belt Traversing Three Pulleys.—The length *L* of a belt traversing three pulleys, as shown in the diagram below, and touching them on one side only, can be found by the following formula.

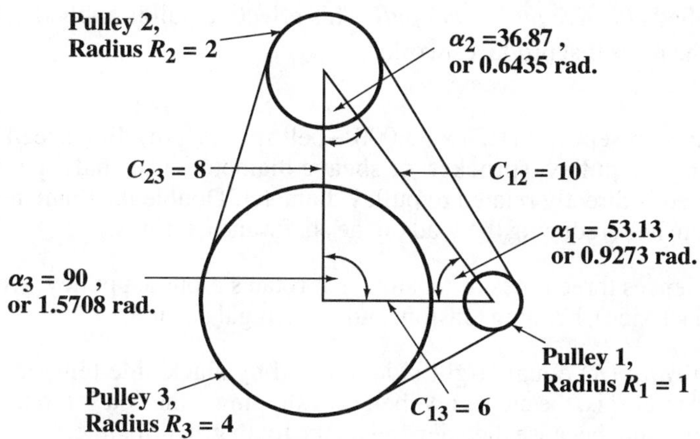

Flat Belt Traversing Three Pulleys.

Referring to the diagram, R_1, R_2, and R_3 are the radii of the three pulleys; C_{12}, C_{13}, and C_{23} are the center distances; and α_1, α_2, and α_3 are the angles, in radians, of the triangle formed by the center distances. Then:

$$L = C_{12} + C_{13} + C_{23} + \frac{1}{2}\left[\frac{(R_2 - R_1)^2}{C_{12}} + \frac{(R_3 - R_1)^2}{C_{13}} + \frac{(R_3 - R_2)^2}{C_{23}}\right]$$
$$+\pi(R_1 + R_2 + R_3) - (\alpha_1 R_1 + \alpha_2 R_2 + \alpha_3 R_3)$$

For example: assume $R_1 = 1$, $R_2 = 2$, $R_3 = 4$, $C_{12} = 10$, $C_{13} = 6$, $C_{23} = 8$, $\alpha_1 = 53.13$ degrees or 0.9273 radian, $\alpha_2 = 36.87$ degrees or 0.6435 radian, and $\alpha_3 = 90$ degrees or 1.5708 radians. Then:

$$L = 10 + 6 + 8 + \frac{1}{2}\left[\frac{(2-1)^2}{10} + \frac{(4-1)^2}{6} + \frac{(4-2)^2}{8}\right]$$
$$+\pi(1 + 2 + 4) - 0.9273 \times 1 + 0.6435 \times 2 + 1.5708 \times 4$$
$$= 24 + 1.05 + 21.9911 - 8.4975 = 38.5436$$

FLEXIBLE BELTS AND SHEAVES

Flexible belt drives are used in industrial power transmission applications, especially when the speeds of the driver and driven shafts must be different or when shafts must be widely separated. The trend toward higher speed prime movers and the need to achieve a slower, useful driven speed are additional factors favoring the use of belts. Belts have numerous advantages over other means of power transmission; these advantages include overall economy, cleanliness, no need for lubrication, lower maintenance costs, easy installation, dampening of shock loads, and the abilities to be used for clutching and variable speed power transmission between widely spaced shafts.

Power Transmitted By Belts.—With belt drives, the force that produces work acts on the rim of a pulley or sheave and causes it to rotate. Since a belt on a drive must be tight enough to prevent slip, there is a belt pull on both sides of a driven wheel. When a drive is stationary or operating with no power transmitted, the pulls on both sides of the driven wheel are equal. When the drive is transmitting power, however, the pulls are not the same. There is a tight side tension T_T and a slack side tension, T_S. The difference between these two pulls $(T_T - T_S)$ is called *effective pull* or *net pull*. This effective pull is applied at the rim of the pulley and is the force that produces work.

Net pull equals horsepower (HP) × 33,000 ÷ belt speed (fpm). Belt speed in fpm can be set by changing the pulley, sprocket, or sheave diameter. The shaft speeds remain the same. Belt speed is directly related to pulley diameter. Double the diameter and the total belt pull is cut in half, reducing the load on the shafts and bearings.

A belt experiences three types of tension as it rotates around a pulley: working tension (tight side – slack side), bending tension, and centrifugal tension.

The *tension ratio* (*R*) equals tight side divided by slack side tension (measured in pounds). The larger *R* is, the closer a V-belt is to slipping—the belt is too loose. (Synchronous belts do not slip, because they depend on the tooth grip principle.)

In addition to working tension (tight side – slack side), two other tensions are developed in a belt when it is operating on a drive. *Bending tension* T_B occurs when the belt bends around the pulley. One part of the belt is in tension and the other is in compression, so compressive stresses also occur. The amount of tension depends on the belt's construction and the pulley diameter. *Centrifugal tension* (T_C) occurs as the belt rotates around the drive and is calculated by $T_C = MV^2$, where T_C is centrifugal tension in pounds, *M* is a constant dependent on the belt's weight, and *V* is the belt velocity in feet per minute. Neither the bending nor centrifugal tensions are imposed on the pulley, shaft, or bearing—only on the belt.

Combining these three types of tension results in *peak tension* which is important in determining the degree of performance or belt life: $T_{peak} = T_T + T_B + T_C$.

Measuring the Effective Length.—The effective length of a V-belt is determined by placing the belt on a measuring device having two equal diameter sheaves with standard groove dimensions. The shaft of one of the sheaves is fixed. A specified measuring tension is applied to the housing for the shaft of the other sheave, moving it along a graduated scale. The belt is rotated around the sheaves at least two revolutions of the belt to seat it properly in the sheave grooves and to divide the total tension equally between the two strands of the belt.

The effective length of the belt is obtained by adding the effective (outside) circumference of one of the measuring sheaves to twice the center distance. Synchronous belts are measured in a similar manner.

The following sections cover common belts used in industrial applications for power transmission and specified in Rubber Manufacturers Association (RMA), Mechanical Power Transmission Association (MPTA), and The Rubber Association of Canada (RAC) standards. The information presented does not apply to automotive or agricultural drives, for which other standards exist. The belts covered in this section are Narrow, Classical, Double, and Light-Duty V-Belts, V-Ribbed Belts, Variable-Speed Belts, 60 deg V-Belts, and Synchronous (Timing) Belts.

Narrow V-Belts ANSI/RMA IP-22.—Narrow V-belts serve the same applications as multiple, classical V-belts, but allow for a lighter, more compact drive. Three basic cross sections—3V and 3VX, 5V and 5VX, and 8V—are provided, as shown in Fig. 1. The 3VX and 5VX are molded, notched V-belts that have greater power capacity than conventional belts. Narrow V-belts are specified by cross section and effective length and have top widths ranging from ⅜ to 1 in.

Narrow V-belts usually provide substantial weight and space savings over classical belts. Some narrow belts can transmit up to three times the horsepower of conventional belts in the same drive space, or the same horsepower in one-third to one-half the space. These belts are designed to operate in multiples and are also available in the joined configuration.

Belt Cross Sections: Nominal dimensions of the three cross sections are given in Fig. 1.

Belt Size Designation: Narrow V-belt sizes are identified by a standard belt number. The first figure of this number followed by the letter V denotes the belt cross section. An X following the V indicates a notched cross section. The remaining figures show the effective belt length in tenths of an inch. For example, the number 5VX1400 designates a notched V-belt with a 5V cross section and an effective length of 140.0 in. Standard effective lengths of narrow V-belts are shown in Table 1.

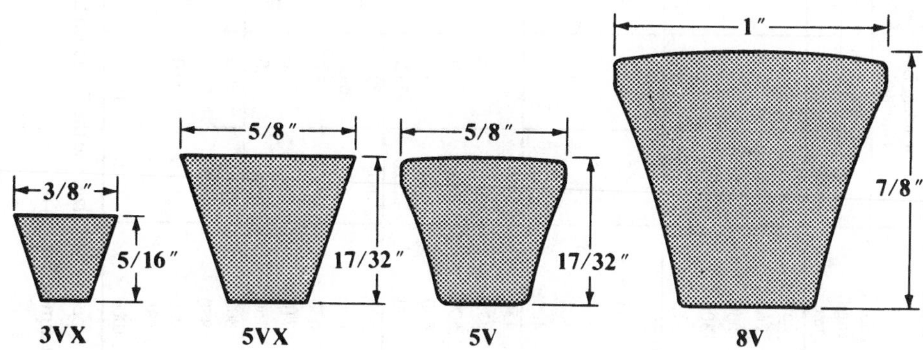

Fig. 1. Nominal Narrow V-Belt Dimensions

Table 1. Narrow V-Belt Standard Effective Lengths *ANSI/RMA IP-22, 1983*

Standard Length Designation[a]	Standard Effective Outside Length, Cross Section			Permissible Deviation from Standard Length	Matching Limits for One Set	Standard Length Designation[a]	Standard Effective Outside Length, Cross Section			Permissible Deviation from Standard Length	Matching Limits for One Set
	3V	5V	8V				3V	5V	8V		
250	25.0	…	…	±0.3	0.15	1060	106.0	106.0	106.0	±0.6	0.30
265	26.5	…	…	±0.3	0.15	1120	112.0	112.0	112.0	±0.6	0.30
280	28.0	…	…	±0.3	0.15	1180	118.0	118.0	118.0	±0.6	0.30
300	30.0	…	…	±0.3	0.15	1250	125.0	125.0	125.0	±0.6	0.30
315	31.5	…	…	±0.3	0.15	1320	132.0	132.0	132.0	±0.6	0.30
335	33.5	…	…	±0.3	0.15	1400	140.0	140.0	140.0	±0.6	0.30
355	35.5	…	…	±0.3	0.15	1500	…	150.0	150.0	±0.8	0.30
375	37.5	…	…	±0.3	0.15	1600	…	160.0	160.0	±0.8	0.45
400	40.0	…	…	±0.3	0.15	1700	…	170.0	170.0	±0.8	0.45
425	42.5	…	…	±0.3	0.15	1800	…	180.0	180.0	±0.8	0.45
450	45.0	…	…	±0.3	0.15	1900	…	190.0	190.0	±0.8	0.45
475	47.5	…	…	±0.3	0.15	2000	…	200.0	200.0	±0.8	0.45
500	50.0	50.0	…	±0.3	0.15	2120	…	212.0	212.0	±0.8	0.45
530	53.0	53.0	…	±0.4	0.15	2240	…	224.0	224.0	±0.8	0.45
560	56.0	56.0	…	±0.4	0.15	2360	…	236.0	236.0	±0.8	0.45
600	60.0	60.0	…	±0.4	0.15	2500	…	250.0	250.0	±0.8	0.45
630	63.0	63.0	…	±0.4	0.15	2650	…	265.0	265.0	±0.8	0.60
670	67.0	67.0	…	±0.4	0.30	2800	…	280.0	280.0	±0.8	0.60
710	71.0	71.0	…	±0.4	0.30	3000	…	300.0	300.0	±0.8	0.60
750	75.0	75.0	…	±0.4	0.30	3150	…	315.0	315.0	±1.0	0.60
800	80.0	80.0	…	±0.4	0.30	3350	…	335.0	335.0	±1.0	0.60
850	85.0	85.0	…	±0.5	0.30	3550	…	355.0	355.0	±1.0	0.60
900	90.0	90.0	…	±0.5	0.30	3750	…	…	375.0	±1.0	0.60
950	95.0	95.0	…	±0.5	0.30	4000	…	…	400.0	±1.0	0.75
1000	100.0	100.0	100.0	±0.5	0.30	4250	…	…	425.0	±1.2	0.75

[a] To specify belt size, use the Standard Length Designation prefixed by the cross section, for example, 5 V850.

All dimensions in inches.

Table 2. Narrow V-Belt Standard Sheave and Groove Dimensions *ANSI/RMA IP-22, 1983*

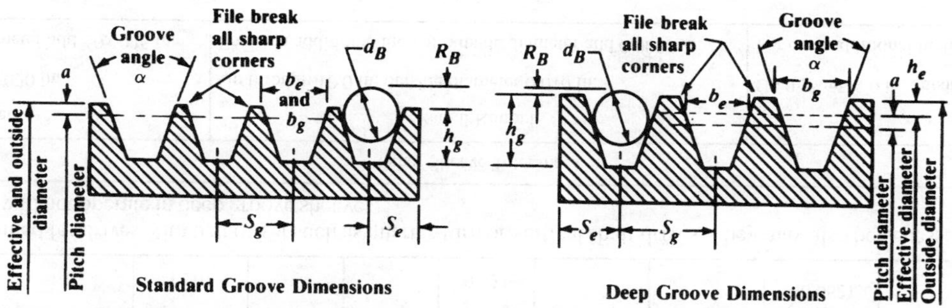

Standard Groove Dimensions

Deep Groove Dimensions

Face Width of Standard and Deep Groove Sheaves = $s_g (N_g - 1) + 2S_e$, where N_g = number of grooves

Cross Section	Standard Groove Outside Diameter	Standard Groove Dimensions: Groove Angle, α, ±0.25 deg	b_g ±0.005	b_e (Ref)	h_g (Min)	R_B (Min)	d_B ±0.0005	S_g[a] ±0.015	S_e	Design Factors: Min Recommended OD	$2a$
3V	Up through 3.49	36	0.350	0.350	0.340	0.181	0.3438	0.406	0.344 (+0.099, −0.031)	2.65	0.050
	Over 6.00 up to and including 12.00	40				0.186					
	Over 9.99 up to and including 16.00	40	0.600	0.600	0.590	0.332	0.5938	0.688	0.500 +0.125, −0.047)	7.10	0.100
	Over 16.00	42				0.336					
8V	Up through 15.99	38	1.000	1.000	0.990	0.575	1.0000	1.125	0.750 (+0.250, −0.062)	12.50	0.200
	Over 15.99 up to and including 22.40	40				0.580					
	Over 22.40	42				0.585					

[a] Summation of the deviations from S_g for all grooves in any one sheave should not exceed ±0.031 in. The variations in pitch diameter between the grooves in any one sheave must be within the following limits: Up through 19.9 in. outside diameter and up through 6 grooves—0.010 in. (add 0.0005 in. for each additional groove). 20.0 in. and over on outside diameter and up through 10 grooves—0.015 in. (add 0.0005 in. for each additional groove). This variation can be obtained by measuring the distance across two measuring balls or rods placed in the grooves diametrically opposite each other. Comparing this "diameter over balls or rods" measurement between grooves will give the variation in pitch diameter.

Table 2. *(Continued)* **Narrow V-Belt Standard Sheave and Groove Dimensions ANSI/RMA IP-22 (1983)**

Cross Section	Deep Groove Outside Diameter	Deep Groove Dimensions[a]								Design Factors		
		Groove Angle, α, ±0.25 deg	b_g ±0.005	b_e (Ref)	h_g (Min)	R_B (Min)	d_B ±0.0005	S_g[a] ±0.015	S_e	Min Recommended OD	$2a$	$2h_e$
3V	Up through 3.71	36	0.421			0.070						
	Over 3.71 up to and including 6.22	38	0.425			0.073						
	Over 6.22 up to and including 12.22	40	0.429	0.350	0.449	0.076	0.3438	0.500	0.375 (+0.094, −0.031)	2.87	0.050	0.218
	Over 12.22	42	0.434			0.078						
5V	Up through 10.31	38	0.710			0.168						
	Over 10.31 up to and including 16.32	40	0.716	0.600	0.750	0.172	0.5938	0.812	0.562 (+0.125, − 0.047)	7.42	0.100	0.320
	Over 16.32	42	0.723			0.175						
8V	Up through 16.51	38	1.180			0.312						
	Over 16.51 up to and including 22.92	40	1.191	1.000	1.252	0.316	1.0000	1.312	0.844 (+0.250, −0.062)	13.02	0.200	0.524
	Over 22.92	42	1.201			0.321						

[a] Deep groove sheaves are intended for drives with belt offset such as quarter-turn or vertical shaft drives. They may also be necessary where oscillations in the center distance may occur. Joined belts will not operate in deep groove sheaves.

Other Sheave Tolerances		
Outside Diameter	Radial Runout[a]	Axial Runout[a]
Up through 8.0 in. outside diameter ±0.020 in.	Up through 10.0 in. outside diameter 0.010 in.	Up through 5.0 in. outside diameter 0.005 in.
For each additional inch of outside diameter add ±0.0025 in.	For each additional inch of outside diameter add 0.0005 in.	For each additional inch of outside diameter add 0.001 in.

[a] Total indicator reading.
All dimensions in inches.

Sheave Dimensions: Groove angles and dimensions for sheaves and face widths of sheaves for multiple belt drives are given in Table 2, along with various tolerance values. Standard sheave outside diameters are given in Table 3.

Table 3. Standard Sheave Outside Diameters *ANSI/RMA IP-22, 1983*

3V			5V			8V		
Nom	Min	Max	Nom	Min	Max	Nom	Min	Max
2.65	2.638	2.680	7.10	7.087	7.200	12.50	12.402	12.600
2.80	2.795	2.840	7.50	7.480	7.600	13.20	13.189	13.400
3.00	2.953	3.000	8.00	7.874	8.000	14.00	13.976	14.200
3.15	3.150	3.200	8.50	8.346	8.480	15.00	14.764	15.000
3.35	3.346	3.400	9.00	8.819	8.960	16.00	15.748	16.000
3.55	3.543	3.600	9.25	9.291	9.440	17.00	16.732	17.000
3.65	3.642	3.700	9.75	9.567	9.720	18.00	17.717	18.000
4.00	3.937	4.000	10.00	9.843	10.000	19.00	18.701	19.000
4.12	4.055	4.120	10.30	10.157	10.320	20.00	19.685	20.000
4.50	4.409	4.480	10.60	10.433	10.600	21.20	20.866	21.200
4.75	4.646	4.720	10.90	10.709	10.880	22.40	22.047	22.400
5.00	4.921	5.000	11.20	11.024	11.200	23.60	23.622	24.000
5.30	5.197	5.280	11.80	11.811	12.000	24.80	24.803	25.200
5.60	5.512	5.600	12.50	12.402	12.600	30.00	29.528	30.000
6.00	5.906	6.000	13.20	13.189	13.400	31.50	31.496	32.000
6.30	6.299	6.400	14.00	13.976	14.200	35.50	35.433	36.000
6.50	6.496	6.600	15.00	14.764	15.000	40.00	39.370	40.000
6.90	6.890	7.000	16.00	15.748	16.000	44.50	44.094	44.800
8.00	7.874	8.000	18.70	18.701	19.000	50.00	49.213	50.000
10.00	9.843	10.000	20.00	19.685	20.000	52.00	51.969	52.800
10.60	10.433	10.600	21.20	20.866	21.200	63.00	62.992	64.000
12.50	12.402	12.600	23.60	23.622	24.000	71.00	70.866	72.000
14.00	13.976	14.200	25.00	24.803	25.200	79.00	78.740	80.000
16.00	15.748	16.000	28.00	27.953	28.400	99.00	98.425	100.000
19.00	18.701	19.000	31.50	31.496	32.000	…	…	…
20.00	19.685	20.000	37.50	37.402	38.000	…	…	…
25.00	24.803	25.200	40.00	39.370	40.000	…	…	…
31.50	31.496	32.000	44.50	44.094	44.800	…	…	…
33.50	33.465	34.000	50.00	49.213	50.000	…	…	…
…	…	…	63.00	62.992	64.000	…	…	…
…	…	…	71.00	70.866	72.000	…	…	…

All dimensions in inches. The nominal diameters were selected from R40 and R80 preferred numbers (see page 19).

Cross Section Selection: The chart (Fig. 2, on page 2379) is a guide to the V-belt cross section to use for any combination of design horsepower and speed of the faster shaft. When the intersection of the design horsepower and speed of the faster shaft falls near a line between two areas on the chart, it is advisable to investigate the possibilities in both areas. Special circumstances (such as space limitations) may lead to a choice of belt cross section different from that indicated in the chart.

Horsepower Ratings: The horsepower ratings of narrow V-belts can be calculated using the following formula:

$$\mathrm{HP} = d_p r[K_1 - K_2/d_p - K_3(d_p r)^2 - K_4 \log(d_p r)] + K_{SR} r$$

where d_p = the pitch diameter of the small sheave, in.; r = rpm of the faster shaft divided by 1000; K_{SR}, speed ratio correction factor, and K_1, K_2, K_3, and K_4, cross section parameters, are listed in the accompanying tables. This formula gives the basic horsepower rating, corrected for the speed ratio. To obtain the horsepower per belt for an arc of contact other than 180° and for belts shorter or longer than average length, multiply the horsepower obtained from this formula by the length correction factor (Table 4) and the arc of contact correction factor (Table 5).

Table 4. Length Correction Factors

Standard Length Designation	Cross Section 3V	5V	8V	Standard Length Designation	Cross Section 3V	5V	8V
250	0.83			1180	1.12	0.99	0.89
265	0.84			1250	1.13	1.00	0.90
280	0.85			1320	1.14	1.01	0.91
300	0.86			1400	1.15	1.02	0.92
315	0.87			1500		1.03	0.93
335	0.88			1600		1.04	0.94
355	0.89			1700		1.05	0.94
375	0.90			1800		1.06	0.95
400	0.92			1900		1.07	0.96
425	0.93			2000		1.08	0.97
450	0.94			2120		1.09	0.98
475	0.95			2240		1.09	0.98
500	0.96	0.85		2360		1.10	0.99
530	0.97	0.86		2500		1.11	1.00
560	0.98	0.87		2650		1.12	1.01
600	0.99	0.88		2800		1.13	1.02
630	1.00	0.89		3000		1.14	1.03
670	1.01	0.90		3150		1.15	1.03
710	1.02	0.91		3350		1.16	1.04
750	1.03	0.92		3550		1.17	1.05
800	1.04	0.93		3750			1.06
850	1.06	0.94		4000			1.07
900	1.07	0.95		4250			1.08
950	1.08	0.96		4500			1.09
1000	1.09	0.96	0.87	4750			1.09
1060	1.10	0.97	0.88	5000			1.10
1120	1.11	0.98	0.88	…			…

Table 5. Arc of Contact Correction Factors

$\frac{D_e - d_e}{C}$	Arc of Contact, θ, on Small Sheave (deg)	Correction Factor	$\frac{D_e - d_e}{C}$	Arc of Contact, θ, on Small Sheave (deg)	Correction Factor
0.00	180	1.00	0.80	133	0.87
0.10	174	0.99	0.90	127	0.85
0.20	169	0.97	1.00	120	0.82
0.30	163	0.96	1.10	113	0.80
0.40	157	0.94	1.20	106	0.77
0.50	151	0.93	1.30	99	0.73
0.60	145	0.91	1.40	91	0.70
0.70	139	0.89	1.50	83	0.65

Speed Ratio Correction Factors

Speed Ratio[a] Range	K_{SR} Cross Section 3VX	5VX	Speed Ratio[a] Range	K_{SR} Cross Section 5V	8V
1.00–1.01	0.0000	0.0000	1.00–1.01	0.0000	0.0000
1.02–1.03	0.0157	0.0801	1.02–1.05	0.0963	0.4690
1.04–1.06	0.0315	0.1600	1.06–1.11	0.2623	1.2780
1.07–1.09	0.0471	0.2398	1.12–1.18	0.4572	2.2276
1.10–1.13	0.0629	0.3201	1.19–1.26	0.6223	3.0321
1.14–1.18	0.0786	0.4001	1.27–1.38	0.7542	3.6747
1.19–1.25	0.0944	0.4804	1.39–1.57	0.8833	4.3038
1.26–1.35	0.1101	0.5603	1.58–1.94	0.9941	4.8438
1.36–1.57	0.1259	0.6405	1.95–3.38	1.0830	5.2767
Over 1.57	0.1416	0.7202	Over 3.38	1.1471	5.5892

[a] D_p/d_p, where D_p (d_p) is the pitch diameter of the large (small) sheave.

Cross Section Correction Factors

Cross Section	K_1	K_2	K_3	K_4
3VX	1.1691	1.5295	1.5229×10^{-4}	0.15960
5VX	3.3038	7.7810	3.6432×10^{-4}	0.43343
5V	3.3140	10.123	5.8758×10^{-4}	0.46527
8V	8.6628	49.323	1.5804×10^{-3}	1.1669

Number of Belts: The number of belts required for an application is obtained by dividing the design horsepower by the corrected horsepower rating for one belt.

Minimum Sheave Size: The recommended minimum sheave size depends on the rpm of the faster shaft. Minimum sheave diameters for each belt cross-section are listed in Table 3.

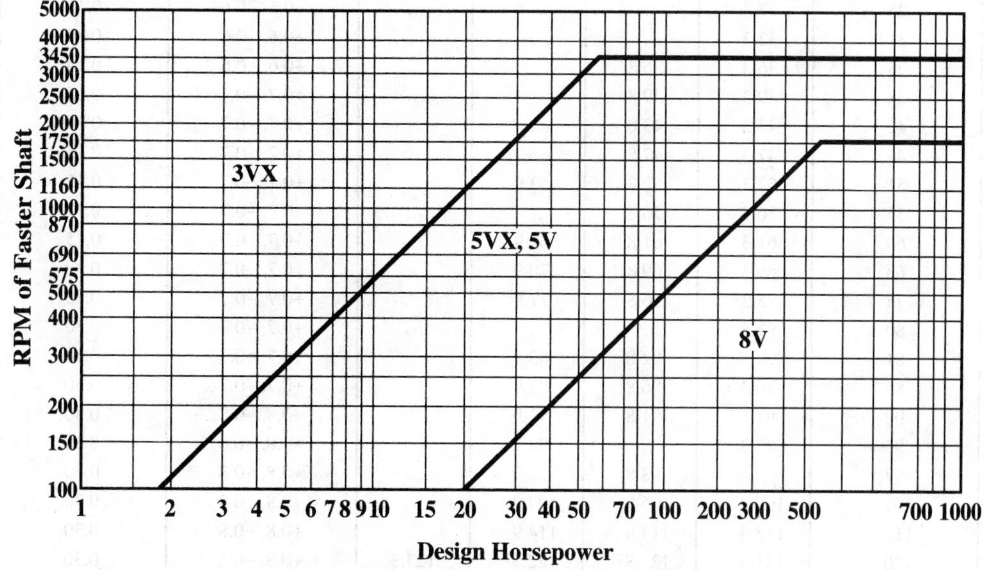

Fig. 2. Selection of Narrow V-Belt Cross Section

Arc of contact on the small sheave may be determined by the formulas.

Exact formula:

$$\text{Arc of Contact (deg)} = 2\cos^{-1}\left(\frac{D_e - d_e}{2C}\right)$$

Approximate formula:

$$\text{Arc of Contact (deg)} = 180 - \frac{(D_e - d_e)60}{C}$$

where: D_e = Effective diameter of large sheave, inch

d_e = Effective diameter of small sheave, inch

C = Center distance, inch

Classical V-Belts ANSI/RMA IP-20.—Classical V-belts are most commonly used in heavy-duty applications and include these standard cross sections: A, AX, B, BX, C, CX, D, and DX (Fig. 3, page 2383). Top widths range from ½ to 1¼ in. and are specified by cross section and nominal length. Classical belts can be teamed in multiples of two or more. These multiple drives can transmit up to several hundred horsepower continuously and absorb reasonable shock loads.

Belt Cross Sections: Nominal dimensions of the four cross sections are given in Fig. 3.

Belt Size Designation: Classical V-belt sizes are identified by a standard belt number consisting of a letter-numeral combination. The letter identifies the cross section; the numeral identifies the length as shown in Table 6. For example, A60 indicates an A cross section and a standard length designation of 60. An X following the section letter designation indicates a molded notch cross section, for example, AX60.

Table 6. Classical V-Belt Standard Datum Length *ANSI/RMA IP-20, 1988*

Standard Length Designation[a]	Standard Datum lengths				Permissible Deviations from Std. Datum Length	Matching Limits for One Set
	Cross Section					
	A, AX	B, BX	C, CX	D		
26	27.3	…	…	…	+0.6, −0.6	0.15
31	32.3	…	…	…	+0.6, −0.6	0.15
35	36.3	36.8	…	…	+0.6, −0.6	0.15
38	39.3	39.8	…	…	+0.7, −0.7	0.15
42	43.3	43.8	…	…	+0.7, −0.7	0.15
46	47.3	47.8	…	…	+0.7, −0.7	0.15
51	52.3	52.8	53.9	…	+0.7, −0.7	0.15
55	56.3	56.8	…	…	+0.7, −0.7	0.15
60	61.3	61.8	62.9	…	+0.7, −0.7	0.15
68	69.3	69.8	70.9	…	+0.7, −0.7	0.30
75	75.3	76.8	77.9	…	+0.7, −0.7	0.30
80	81.3	…	…	…	+0.7, −0.7	0.30
81	…	82.8	83.9	…	+0.7, −0.7	0.30
85	86.3	86.8	87.9	…	+0.7, −0.7	0.30
90	91.3	91.8	92.9	…	+0.8, −0.8	0.30
96	97.3	…	98.9	…	+0.8, −0.8	0.30
97	…	98.8	…	…	+0.8, −0.8	0.30
105	106.3	106.8	107.9	…	+0.8, −0.8	0.30
112	113.3	113.8	114.9	…	+0.8, −0.8	0.30
120	121.3	121.8	122.9	123.3	+0.8, −0.8	0.30
128	129.3	129.8	130.9	131.3	+0.8, −0.8	0.30
144	…	145.8	146.9	147.3	+0.8, −0.8	0.30
158	…	159.8	160.9	161.3	+1.0, −1.0	0.45
173	…	174.8	175.9	176.3	+1.0, −1.0	0.45
180	…	181.8	182.9	183.3	+1.0, −1.0	0.45
195	…	196.8	197.9	198.3	+1.1, −1.1	0.45
210	…	211.8	212.9	213.3	+1.1, −1.1	0.45
240	…	240.3	240.9	240.8	+1.3, −1.3	0.45
270	…	270.3	270.9	270.8	+1.6, −1.6	0.60
300	…	300.3	300.0	300.8	+1.6, −1.6	0.60
330	…	…	330.9	330.8	+2.0, −2.0	0.60
360	…	…	380.9	360.8	+2.0, −2.0	0.60
540	…	…	…	540.8	+3.3, −3.3	0.90
390	…	…	390.9	390.8	+2.0, −2.0	0.75
420	…	…	420.9	420.8	+3.3, −3.3	0.75
480	…	…	…	480.8	+3.3, −3.3	0.75
600	…	…	…	600.8	+3.3, −3.3	0.90
660	…	…	…	660.8	+3.3, −3.3	0.90

[a] To specify belt size use the Standard Length Designation prefixed by the letter indicating the cross section, e.g., B90.

All dimensions in inches.

Table 7. Classical V-Belt Sheave and Groove Dimensions *ANSI/RMA IP-20, 1988*

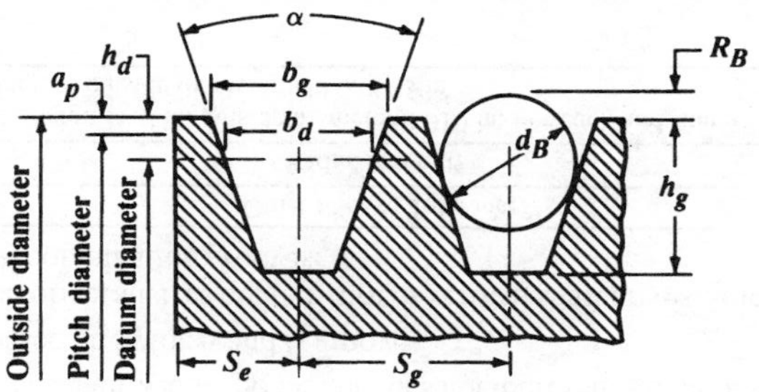

Face Width of Standard and Deep Groove Sheaves = $S_g (N_g - 1) + 2S_e$, where N_g = number of grooves

Cross Section		Datum[a] Diameter Range	α Groove Angle ±0.33°	b_d Ref	b_g		h_g Min	$2h_d$	R_B Min	d_B ±0.0005	S_g[b] ±0.025	S_e		Min Recom. Datum Diameter	$2a_p$
			Standard Groove Dimensions											Design Factors	
A, AX		Through 5.4	34	0.418	0.494	± 0.005	0.460	0.250	0.148	0.4375	0.625	0.375	+ 0.090	A 3.0	0
		Over 5.4	38		0.504				0.149	(7/16)			− 0.062	AX 2.2	
B, BX		Through 7.0	34	0.530	0.637	± 0.006	0.550	0.350	0.189	0.5625	0.750	0.500	+ 0.120	B 5.4	0
		Over 7.0	38		0.650				0.190	(9/16)			− 0.065	BX 4.0	
Com-bina-tion	A, AX Belt	Through 7.4[c]	34	0.508[d]	0.612	±0.006	0.612	0.634[e]	0.230	0.5625	0.750	0.500	+0.120	A 3.6[c]	0.37
		Over 7.4	38		0.625			0.602[e]	0.226	(9/16)			−0.065	AX 2.8	
	B, BX Belt	Through 7.4[c]	34		0.612	±0.006		0.333[e]	0.230					B 5.7[c]	−0.01
		Over 7.4	38		0.625			0.334[e]	0.226					BX 4.3	
C, CX		Through 7.99	34	0.757	0.879	±0.007	0.750	0.400	0.274	0.7812	1.000	0.688	+0.160	C 9.0	0
		Over 7.99 to and incl. 12.0	36		0.887				0.276	(25/32)			−0.070	CX 6.8	
		Over 12.0	38		0.895				0.277						
D		Through 12.99	34	1.076	1.259	± 0.008	1.020	0.600	0.410	1.1250	1.438	0.875	+ 0.220	13.0	0
		Over 12.99 to and incl. 17.0	36		1.271				0.410	(1 1/8)			− 0.080		
		Over 17.0	38		1.283				0.411						

Table 7. *(Continued)* **Classical V-Belt Sheave and Groove Dimensions** *ANSI/RMA IP-20, 1988*

		Deep Groove Dimensions[f]									Design Factors	
Cross Section	Datum[a] Dia. Range	α Groove Angle ± 0.33°	b_g Ref	b_g	h_g Min	$2h_d$ Ref	R_B Min	d_B ± 0.0005	S_g[b] ± 0.025	S_e	Min Rec. Datum Diameter	$2a_p$
B, BX	Through 7.0	34	0.530	0.747 ± 0.006	0.730	0.710	0.007	0.5625 (9⁄16)	0.875	0.562 + 0.120 − 0.065	B 5.4	0.36
	Over 7.0	38		0.774			0.008				BX 4.0	
C, CX	Through 7.99	34	0.757	1.066	1.055	1.010	– 0.035	0.7812 (25⁄32)	1.250	0.812 + 0.160 − 0.070	C 9.0	0.61
	Over 7.99 to and incl. 12.0	36		1.085 ± 0.007			– 0.032				CX 6.8	
	Over 12.0	38		1.105			−0.031					
D	Through 12.99	34	1.076	1.513	1.435	1.430	−0.010	1.1250 (1⅛)	1.750	1.062 +0.220 −0.080	13.0	0.83
	Over 12.99 to and incl. 17.0	36		1.514 ±0.008			−0.009					
	Over 17.0	38		1.569			−0.008					

[a] The A/AX, B/BX combination groove should be used when deep grooves are required for A or AX belts.

[b] Summation of the deviations from S_g for all grooves in any one sheave should not exceed ±0.050 in. The variation in datum diameter between the grooves in any one sheave must be within the following limits: Through 19.9 in. outside diameter and through 6 grooves: 0.010 in. (add 0.0005 in. for each additional groove). 20.0 in. and over on outside diameter and through 10 grooves: 0.015 in. (add 0.0005 in. for each additional groove). This variation can be obtained by measuring the distance across two measuring balls or rods placed diametrically opposite each other in a groove. Comparing this "diameter over balls or rods" measurement between grooves will give the variation in datum diameter.

[c] Diameters shown for combination grooves are outside diameters. A specific datum diameter does not exist for either A or B belts in combination grooves.

[d] The b_d value shown for combination grooves is the "constant width" point, but does not represent a datum width for either A or B belts ($2h_d = 0.340$ ref).

[e] $2h_d$ values for combination grooves are calculated based on b_d for A and B grooves.

[f] Deep groove sheaves are intended for drives with belt offset such as quarter-turn or vertical shaft drives. Joined belts will not operate in deep groove sheaves. Also, A and AX joined belts will not operate in A/AX and B/BX combination grooves.

Other Sheave Tolerances		
Outside Diameter	Radial Runout[a]	Axial Runout[a]
Through 8.0 in. outside diameter ±0.020 in. For each additional inch of outside diameter add ±0.005 in.	Through 10.0 in. outside diameter 0.010 in. For each additional inch of outside diameter add 0.0005 in.	Through 5.0 in. outside diameter 0.005 in. For each additional inch of outside diameter add 0.001 in.

[a] Total indicator readings.

A, AX & B, BX Combin. All dimensions in inches.

Sheave Dimensions: Groove angles and dimensions for sheaves and the face widths of sheaves for multiple belt drives are given in Table 7, along with various tolerance values.

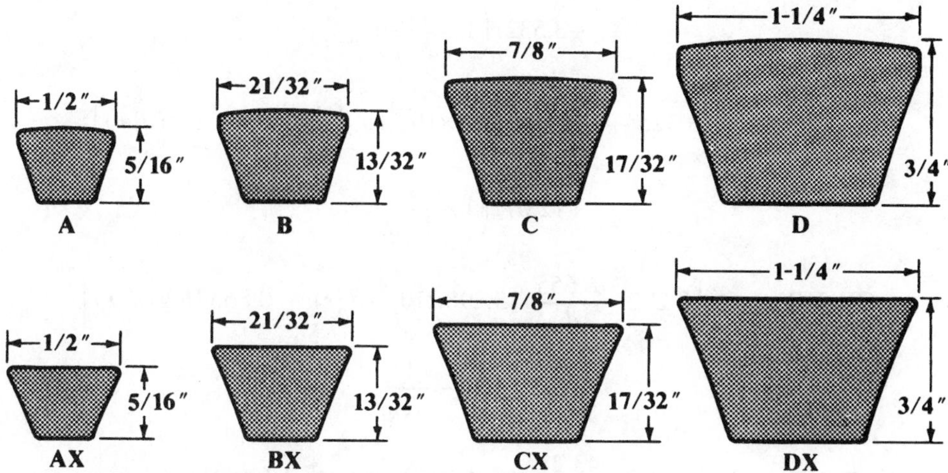

Fig. 3. Classical V-Belt Cross Sections

Cross Section Selection: Use the chart (Fig. 4) as a guide to the Classical V-belt cross section for any combination of design horsepower and speed of the faster shaft. When the intersection of the design horsepower and speed of the faster shaft falls near a line between two areas on the chart, the possibilities in both areas should be investigated. Special circumstances (such as space limitations) may lead to a choice of belt cross section different from that indicated in the chart.

Horsepower Ratings: The horsepower rating formulas for classical V-belts are:

$$\mathbf{A:}\text{HP} = d_p r\left[1.004 - \frac{1.652}{d_p} - 1.547\times10^{-4}(d_p r)^2 - 0.2126\log(d_p r)\right] + 1.652r\left(1 - \frac{1}{K_{SR}}\right)$$

$$\mathbf{AX:}\text{HP} = d_p r\left[1.462 - \frac{2.239}{d_p} - 2.198\times10^{-4}(d_p r)^2 - 0.4238\log(d_p r)\right] + 2.239r\left(1 - \frac{1}{K_{SR}}\right)$$

$$\mathbf{B:}\text{HP} = d_p r\left[1.769 - \frac{4.372}{d_p} - 3.081\times10^{-4}(d_p r)^2 - 0.3658\log(d_p r)\right] + 4.372r\left(1 - \frac{1}{K_{SR}}\right)$$

$$\mathbf{BX}{:}\mathrm{HP} = d_p r\left[2.051 - \frac{3.532}{d_p} - 3.097\times10^{-4}(d_p r)^2 - 0.5735\log(d_p r)\right] + 3.532r\left(1 - \frac{1}{K_{SR}}\right)$$

$$\mathbf{C}{:}\mathrm{HP} = d_p r\left[3.325 - \frac{12.07}{d_p} - 5.828\times10^{-4}(d_p r)^2 - 0.6886\log(d_p r)\right] + 12.07r\left(1 - \frac{1}{K_{SR}}\right)$$

$$\mathbf{CX}{:}\mathrm{HP} = d_p r\left[3.272 - \frac{6.655}{d_p} - 5.298\times10^{-4}(d_p r)^2 - 0.8637\log(d_p r)\right] + 6.655r\left(1 - \frac{1}{K_{SR}}\right)$$

$$\mathbf{D}{:}\mathrm{HP} = d_p r\left[7.160 - \frac{43.21}{d_p} - 1.384\times10^{-3}(d_p r)^2 - 1.454\log(d_p r)\right] + 43.21r\left(1 - \frac{1}{K_{SR}}\right)$$

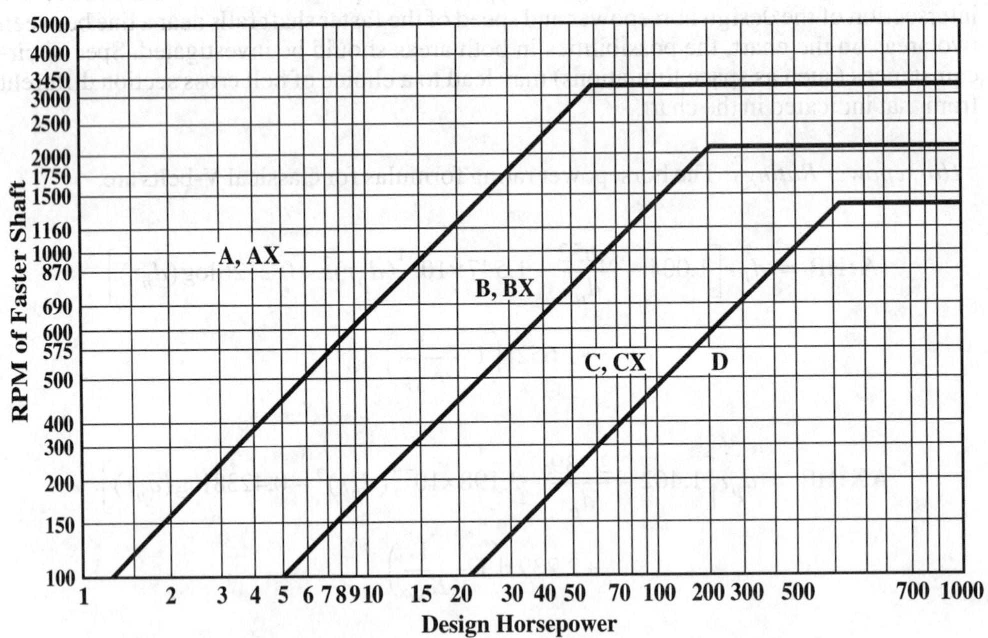

Fig. 4. Selection of Classic V-Belt Cross Sections

In these equations, d_p = pitch diameter of small sheave, in.; r = rpm of the faster shaft divided by 1000; K_{SR} = speed ratio factor given in the accompanying table. These formulas give the basic horsepower rating, corrected for the speed ratio. To obtain the horsepower per belt for an arc of contact other than 180 degrees and for belts shorter or longer than average length, multiply the horsepower obtained from these formulas by the length correction factor (Table 8) and the arc of contact correction factor (Table 9).

Table 8. Length Correction Factors

Std. Length Designation	Cross Section			
	A, AX	B, BX	C, CX	D
26	0.78	…	…	…
31	0.82	…	…	…
35	0.85	0.80	…	…
38	0.87	0.82	…	…
42	0.89	0.84	…	…
46	0.91	0.86	…	…
51	0.93	0.88	0.80	…
55	0.95	0.89	…	…
60	0.97	0.91	0.83	…
68	1.00	0.94	0.85	…
75	1.02	0.96	0.87	…
80	1.04	…	…	…
81	…	0.98	0.89	…
85	1.05	0.99	0.90	…
90	1.07	1.00	0.91	…
96	1.08	…	0.92	…
97	…	1.02	…	…
105	1.10	1.03	0.94	…
112	1.12	1.05	0.95	…
120	1.13	1.06	0.96	0.88
128	1.15	1.08	0.98	0.89
144	…	1.10	1.00	0.91
158	…	1.12	1.02	0.93
173	…	1.14	1.04	0.94
180	…	1.15	1.05	0.95
195	…	1.17	1.08	0.96
210	…	1.18	1.07	0.98
240	…	1.22	1.10	1.00
270	…	1.24	1.13	1.02
300	…	1.27	1.15	1.04
330	…	…	1.17	1.06
360	…	…	1.18	1.07
390	…	…	1.20	1.09
420	…	…	1.21	1.10
480	…	…	…	1.13
540	…	…	…	1.15
600	…	…	…	1.17
660	…	…	…	1.18

Table 9. Arc of Contact Correction Factors

$\frac{D_d - d_d}{C}$	Arc of Contact, θ, Small Sheave (deg)	Correction Factor		$\frac{D_d - d_d}{C}$	Arc of Contact, θ Small Sheave (deg)	Correction Factor	
		V-V	V-Flat[a]			V-V	V-Flat[a]
0.00	180	1.00	0.75	0.80	133	0.87	0.85
0.10	174	0.99	0.76	0.90	127	0.85	0.85
0.20	169	0.97	0.78	1.00	120	0.82	0.82
0.30	163	0.96	0.79	1.10	113	0.80	0.80
0.40	157	0.94	0.80	1.20	106	0.77	0.77
0.50	151	0.93	0.81	1.30	99	0.73	0.73
0.60	145	0.91	0.83	1.40	91	0.70	0.70
0.70	139	0.89	0.84	1.50	83	0.65	0.65

[a] A V-flat drive is one using a small sheave and a large diameter flat pulley.

Speed Ratio Correction Factors

Speed Ratio[a] Range	K_{SR}	Speed Ratio[a] Range	K_{SR}
1.00–1.01	1.0000	1.15–1.20	1.0586
1.02–1.04	1.0112	1.21–1.27	1.0711
1.05–1.07	1.0226	1.28–1.39	1.0840
1.08–1.10	1.0344	1.40–1.64	1.0972
1.11–1.14	1.0463	Over 1.64	1.1106

[a] D_p/d_p, where D_p (d_p) is the pitch diameter of the large (small) sheave.

Arc of contact on the small sheave may be determined by the formulas.

Exact formula: Arc of Contact (deg) $= 2\cos^{-1}\left(\frac{D_d - d_d}{2C}\right)$

Approximate formula: Arc of Contact (deg) $= 180 - \left(\frac{(D_d - d_d)60}{C}\right)$

where D_d = Datum diameter of large sheave or flat pulley, inch; d_d = Datum diameter of small sheave, inch; and, C = Center distance, inch.

Number of Belts: The number of belts required for an application is obtained by dividing the design horsepower by the corrected horsepower rating for one belt.

Minimum Sheave Size: The recommended minimum sheave size depends on the rpm of the faster shaft. Minimum groove diameters for each belt cross section are listed in Table 11.

Double V-Belts ANSI/RMA IP-21.—Double V-belts or hexagonal belts are used when power input or takeoff is required on both sides of the belt. Designed for use on "serpentine" drives, which consist of sheaves rotating in opposite directions, the belts are available in AA, BB, CC, and DD cross sections and operate in standard classical sheaves. They are specified by cross section and nominal length.

Belt Cross Sections: Nominal dimensions of the four cross sections are given in Fig. 5.

Belt Size Designation: Double V-belt sizes are identified by a standard belt number, consisting of a letter-numeral combination. The letters identify the cross section; the numbers identify length as shown in Column 1 of Table 10. For example, AA51 indicates an AA cross section and a standard length designation of 51.

Table 10. Double V-Belt Standard Effective Lengths *ANSI/RMA IP-21, 1984*

Standard Length Designation[a]	Standard Effetive Length Cross Section				Permissible Deviation from Standard Effective Length	Matching Limits for One Set
	AA	BB	CC	DD		
51	53.1	53.9	…	…	±0.7	0.15
55	…	57.9	…	…	±0.7	0.15
60	62.1	62.9	…	…	±0.7	0.15
68	70.1	70.9	…	…	±0.7	0.30
75	77.1	77.9	…	…	±0.7	0.30
80	82.1	…	…	…	±0.7	0.30
81	…	83.9	85.2	…	±0.7	0.30
85	87.1	87.9	89.2	…	±0.7	0.30
90	92.1	92.9	94.2	…	±0.8	0.30
96	98.1	…	100.2	…	±0.8	0.30
97	…	99.9	…	…	±0.8	0.30
105	107.1	107.9	109.2	…	±0.8	0.30
112	114.1	114.9	116.2	…	±0.8	0.30
120	122.1	122.9	124.2	125.2	±0.8	0.30
128	130.1	130.9	132.2	133.2	±0.8	0.30
144	…	146.9	148.2	149.2	±0.8	0.30
158	…	160.9	162.2	163.2	±1.0	0.45
173	…	175.9	177.2	178.2	±1.0	0.45
180	…	182.9	184.2	185.2	±1.0	0.45
195	…	197.9	199.2	200.2	±1.1	0.45
210	…	212.9	214.2	215.2	±1.1	0.45
240	…	241.4	242.2	242.7	±1.3	0.45
270	…	271.4	272.2	272.7	±1.6	0.60
300	…	301.4	302.2	302.7	±1.6	0.60
330	…	…	332.2	332.7	±2.0	0.60
360	…	…	362.2	362.7	±2.0	0.60

[a] To specify belt size use the Standard Length Designation prefixed by the letters indicating cross section; for example, BB90.

All dimensions in inches.

Sheave Dimensions: Groove angles and dimensions for sheaves and face widths of sheaves for multiple belt drives are given in Table 11, along with various tolerance values.

Table 11. Double V-Belt Sheave and Groove Dimensions *ANSI/RMP IP-21, 1984*

Face Width of Standard and Deep Groove Sheaves = $S_g(N_g - 1) + 2S_e$, where N_g = number of grooves

			Standard Groove Dimensions								Drive Design Factors	
Cross Section	Outside Diameter Range	Groove Angle, α ±0.33°	b_g		h_g (Min.)	$2h_d$	R_B (Min.)	d_B ±0.0005	S_g[a] ±0.025	S_e	Min. Recomm. Outside Diam.	$2a_p$[b]
AA	Up through 5.65	34	0.494	±0.005	0.460	0.250	0.148	0.4375	0.625	0.375 +0.090	3.25	0.0
	Over 5.65	38	0.504				0.149	(7⁄16)		−0.062		
BB	Up through 7.35	34	0.637	±0.006	0.550	0.350	0.189	0.5625	0.750	0.500 +0.120	5.75	0.0
	Over 7.35	38	0.650				0.190	(9⁄16)		−0.065		
AA-BB	Up through 7.35	34	0.612	±0.006	0.612	A = 0.750	0.230	0.5625	0.750	0.500 +0.120	A = 3.620	A = 0.370
	Over 7.35	38	0.625			B = 0.350	0.226	(9⁄16)		−0.065	B = 5.680	B = −0.070
CC	Up through 8.39	34	0.879	±0.007	0.750	0.400	0.274	0.7812	1.000	0.688 +0.160	9.4	0.0
	Over 8.39 up to and including 12.40	36	0.887				0.276	(25⁄32)		−0.070		
	Over 12.40	38	0.895				0.277					
DD	Up through 13.59	34	1.259	±0.008	1.020	0.600	0.410	1.1250	1.438	0.875 +0.220	13.6	0.0
	Over 13.59 up to and including 17.60	36	1.271				0.410	(1⅛)		−0.080		
	Over 17.60	38	1.283				0.411					

Table 11. *(Continued)* **Double V-Belt Sheave and Groove Dimensions** *ANSI/RMP IP-21, 1984*

		Deep Groove Dimensions[c]										Drive Design Factors	
Cross Section	Outside Diameter Range	Groove Angle, α ±0.33°	b_g		h_g (Min.)	$2h_d$	R_B (Min.)	d_B ±0.0005	S_g[a] ±0.025	S_e		Minimum Recommended Outside Diameter	$2a_p$
AA	Up through 5.96	34	0.589	±0.005	0.615	0.560	−0.009	0.4375 (7⁄16)	0.750	0.438	+0.090 −0.062	3.56	0.310
	Over 5.96	38	0.611				−0.008						
BB	Up through 7.71	34	0.747	±0.006	0.730	0.710	+0.007	0.5625 (9⁄16)	0.875	0.562	+0.120 −0.065	6.11	0.360
	Over 7.71	38	0.774				+0.008						
CC	Up through 9.00	34	1.066	±0.007	1.055	1.010	−0.035	0.7812 (25⁄32)	1.250	0.812	+0.160 −0.070	10.01	0.610
	Over 9.00 up to and including 13.01	36	1.085				−0.032						
	Over 13.01	38	1.105				−0.031						
DD	Up through 14.42	34	1.513	±0.008	1.435	1.430	−0.010	1.1250 (1⅛)	1.750	1.062	+0.220 −0.080	14.43	0.830
	Over 14.42 up to and including 18.43	36	1.541				−0.009						
	Over 18.43	38	1.569				−0.008						

[a] Summation of the deviations from S_g for all grooves in any one sheave shall not exceed ±0.050 in. The variation in pitch diameter between the grooves in any one sheave must be within the following limits: Up through 19.9 in. outside diameter and up through 6 grooves: 0.010 in. (add 0.005 in. for each additional groove). 20.0 in. and over on outside diameter and up through 10 grooves: 0.015 in. (add 0.0005 in. for each additional groove). This variation can be obtained easily by measuring the distance across two measuring balls or rods placed diametrically opposite each other in a groove. Comparing this "diameter over balls or rods" measurement between grooves will give the variation in pitch diameter.

[b] The a_p values shown for the A/B combination sheaves are the geometrically derived values. These values may be different from those shown in manufacturer's catalogs.

[c] Deep groove sheaves are intended for drives with belt offset such as quarter-turn or vertical shaft drives.

Other Sheave Tolerances		
Outside Diameter	Radial Runout[a]	Axial Runout[a]
Up through 4.0 in. outside diameter ±0.020 in. For each additional inch of outside diameter add ±0.005 in.	Up through 10.0 in. outside diameter ±0.010 in. For each additional inch of outside diameter add 0.0005 in.	Up through 5.0 in. outside diameter 0.005 in. For each additional inch of outside diameter add 0.001 in.

[a] Total indicator reading.

All dimensions in inches.

Cross Section Selection: Use the chart (Fig. 5) as a guide to the double V-belt cross section for any combination of design horsepower and speed of the faster shaft. When the intersection of the design horsepower and speed of the faster shaft falls near a line between two areas on the chart, it is best to investigate the possibilities in both areas. Special circumstances (such as space limitations) may lead to a choice of belt cross section different from that indicated in the chart.

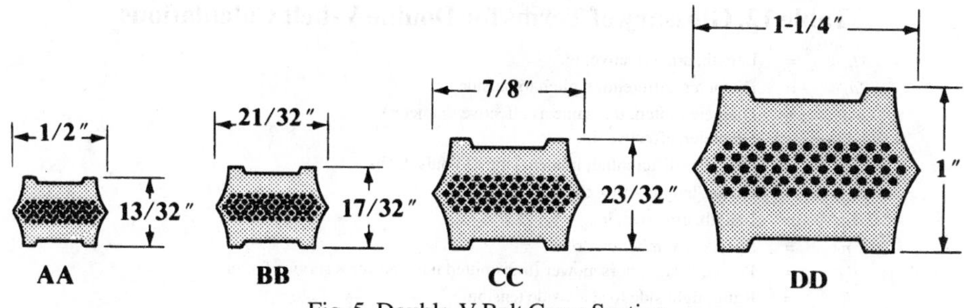

Fig. 5. Double-V Belt Cross Section

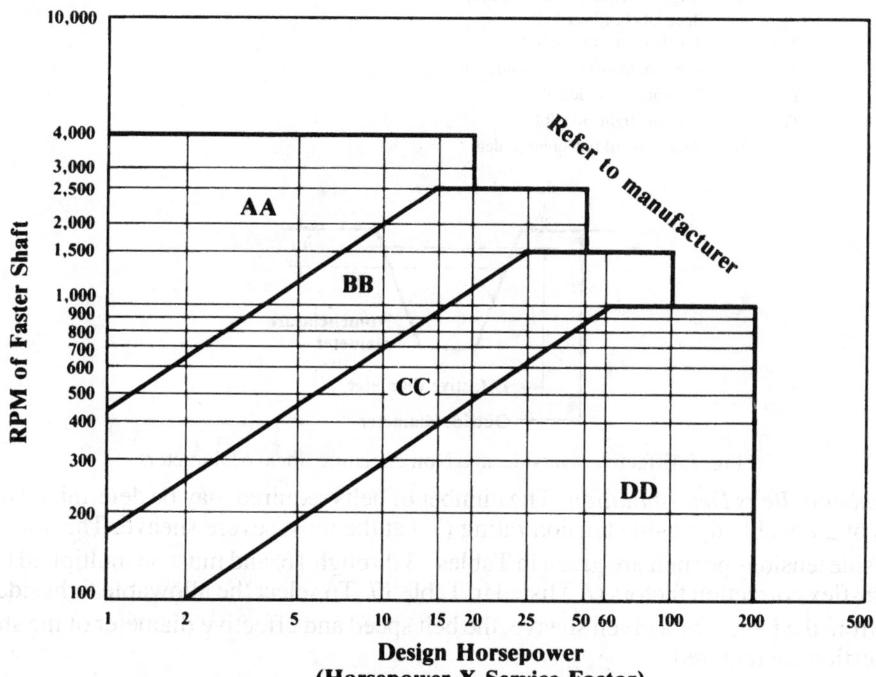

Fig. 6. Selection of Double V-Belt Cross Section

Effective Diameter Determination: Fig. 6 shows the relationship of effective diameter, outside diameter, and nomenclature diameter. Nomenclature diameter is used when ordering sheaves for double V-belt drives. The effective diameter is determined as follows:

$$\text{Effective diameter} = \text{Nomenclature diameter} + 2h_d - 2a_p$$

The values of $2h_d$ and $2a_p$ are given in Table 11.

Double V-belt Length Determination: The effective belt length of a specific drive may be determined by making a scaled layout of the drive. Draw the sheaves in terms of their effective diameters and in the position when a new belt is applied and first brought to driving tension. Next, measure the tangents and calculate the effective arc length (AL_e) of each sheave (see Table 12 for a glossary of terms):

$$AL_e = \frac{d_e\theta}{115}$$

The effective length of the belt will then be the sum of the tangents and the connecting arc lengths. Manufacturers may be consulted for mathematical calculation of effective belt length for specific drive applications.

Table 12. Glossary of Terms for Double V-belt Calculations

AL_e	=	Length, arc, effective, in.
$2a_p$	=	Diameter, differential, pitch to outside, in.
d	=	Diameter, pitch, in. (same as effective diameter)
d_e	=	Diameter, effective, in.
$2h_d$	=	Diameter differential, nomenclature to outside, in.
K_f	=	Factor, length – flex correction
L_e	=	Length, effective, in.
n	=	Sheaves, number on drive
P_d	=	Power, design, horsepower (transmitted horsepower × service factor)
R	=	Ratio, tight side to slack side tension
$R/(R-1)$	=	Factor, tension ratio
r	=	Angular velocity, faster shaft, rpm/1000
S	=	Speed, belt, fpm/1000
T_e	=	Tension, effective pull, lbf
T_r	=	Tension, allowable tight side, lbf
T_S	=	Tension, slack side, lbf
T_T	=	Tension, tight side, lbf
θ	=	Angle, arc of belt contact, deg

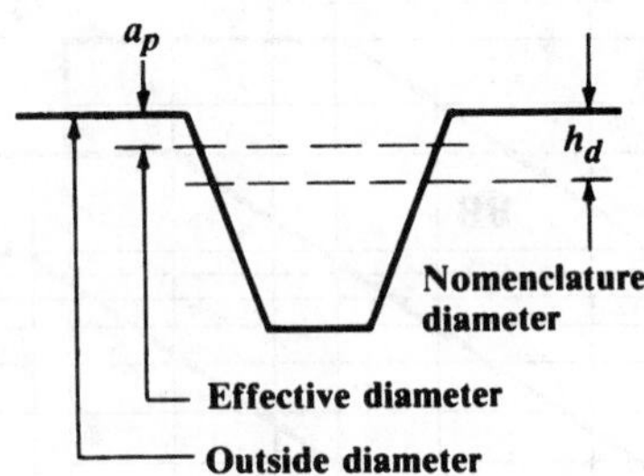

Fig. 7. Effective, Outside, and Nomenclature Sheave Diameters

Number of Belts Determination: The number of belts required may be determined on the basis of allowable tight side tension rating (T_r) at the most severe sheave. The allowable tight side tensions per belt are given in Tables 13 through 16, and must be multiplied by the length-flex correction factors (K_f) listed in Table 17. To select the allowable tight side tension from the tables for a given sheave, the belt speed and effective diameter of the sheave in question are required.

Double V-Belt Drive Design Method: The fourteen drive design steps are as follows:

1) Number the sheaves starting from the driver in the opposite direction to belt rotation; include the idlers.

2) Select the proper service factor for each loaded driven unit.

3) Multiply the horsepower requirement for each loaded driven sheave by the corresponding service factor. This is the design horsepower at each sheave.

4) Calculate driver design horsepower. This hp is equal to the sum of all the driven design horsepowers.

5) Calculate belt speed (S) in thousands of feet per minute: $S = rd/3.820$.

6) Calculate effective tension (T_e) for each loaded sheave: $T_e = 33P_d/S$.

7) Determine minimum $R/(R-1)$ for each loaded sheave from Table 18 using the arc of contact determined from the drive layout.

8) In most drives, slippage will occur first at the driver sheave. Assume this to be true and calculate T_T and T_S for the driver: $T_T = T_e\,[R/(R-1)]$ and $T_s = T_T - T_e$. Use $R/(R-1)$ from Step 7 and T_e from Step 6 for the driver sheave.

9) Starting with the first driven sheave, determine T_T and T_S for each segment of the drive. The T_T for the driver becomes T_S for that sheave and is equal to $T_T - T_e$. Proceed around the drive in like manner.

10) Calculate actual $R/(R-1)$ for each sheave using: $R/(R-1) = T_T/T_e = T_T/(T_T - T_S)$. The T_T and T_S values are for those determined in Step 9. If these values are equal to or greater than those determined in Step 7, the assumption that slippage will first occur at the driver is correct and the next two steps are not necessary. If the value is less, the assumption was not correct, so proceed with Step 11.

11) Take the sheave where the actual value $R/(R-1)$ (Step 10) is less than the minimum, as determined in Step 7, and calculate a new T_T and T_S for this sheave using the minimum $R/(R-1)$ as determined in Step 7: $T_T = T_e\,[R/(R-1)]$ and $T_S = T_T - T_e$.

12) Start with this sheave and recalculate the tension in each segment of the drive as in Step 9.

13) The length-flex factor (K_f) is taken from Table 17. Before using this table, calculate the value of L_e/n. Be sure to use the appropriate belt cross-section column when selecting the correction factor.

14) Beginning with the driver sheave, determine the number of belts (N_b) needed to satisfy the conditions at each loaded sheave using: $N_b = T_T/T_rK_f$. Note: T_T is tight side tension as determined in Step 9 or 11 and 12. T_r is allowable tight side tension as shown in Tables 15-18. K_f is the length-flex correction factor from Table 17. The sheave that requires the largest number of belts is the number of belts required for the drive. Any fraction of a belt should be treated as a whole belt.

Table 13. Allowable Tight Side Tension for an AA Section

Belt Speed (fpm)	Sheave Effective Diameter (in.)							
	3.0	3.5	4.0	4.5	5.0	5.5	6.0	6.5
200	30	46	57	66	73	79	83	88
400	23	38	49	58	65	71	76	80
600	18	33	44	53	60	66	71	75
800	14	30	41	50	57	63	67	72
1000	12	27	38	47	54	60	65	69
1200	9	24	36	45	52	57	62	66
1400	7	22	34	42	49	55	60	64
1600	5	20	32	40	47	53	58	62
1800	3	18	30	38	46	51	56	60
2000	1	16	28	37	44	50	54	58
2200	…	15	26	35	42	48	53	57
2400	…	13	24	33	40	46	51	55
2600	…	11	23	31	39	44	49	53
2800	…	9	21	30	37	43	47	51
3000	…	8	19	28	35	41	46	50
3200	…	6	17	26	33	39	44	48
3400	…	4	16	24	31	37	42	46
3600	…	2	14	23	30	35	40	44
3800	…	1	12	21	28	34	38	43
4000	…	…	10	19	26	32	37	41
4200	…	…	8	17	24	30	35	39
4400	…	…	6	15	22	28	33	37
4600	…	…	4	13	20	26	31	35
4800	…	…	2	11	18	24	29	33
5000	…	…	…	9	16	22	27	31
5200	…	…	…	7	14	20	24	28
5400	…	…	…	4	12	17	22	26
5600	…	…	…	2	9	15	20	24
5800	…	…	…	…	7	13	18	22

The allowable tight side tension must be evaluated for each sheave in the system (see Step 14). Values must be corrected by K_f from Table 17.

Table 14. Allowable Tight Side Tension for a BB Section

Belt Speed (fpm)	Sheave Effective Diameter (in.)								
	5.0	5.5	6.0	6.5	7.0	7.5	8.0	8.5	9.0
200	81	93	103	111	119	125	130	135	140
400	69	81	91	99	107	113	118	123	128
600	61	74	84	92	99	106	111	116	121
800	56	68	78	87	94	101	106	111	115
1000	52	64	74	83	90	96	102	107	111
1200	48	60	71	79	86	93	98	103	107
1400	45	57	67	76	83	89	95	100	104
1600	42	54	64	73	80	86	92	97	101
1800	39	51	61	70	77	84	89	94	98
2000	36	49	59	67	74	81	86	91	96
2200	34	46	56	64	72	78	84	89	93
2400	31	43	53	62	69	75	81	86	90
2600	29	41	51	59	67	73	78	83	88
2800	26	38	48	57	64	70	76	81	85
3000	23	35	45	54	61	68	73	78	82
3200	21	33	43	51	59	65	70	75	80
3400	18	30	40	49	56	62	68	73	77
3600	15	27	37	46	53	59	65	70	74
3800	12	24	35	43	50	57	62	67	71
4000	9	22	32	40	47	54	59	64	69
4200	7	19	29	37	45	51	56	61	66
4400	4	16	26	34	42	48	53	58	63
4600	1	13	23	31	39	45	50	55	60
4800	…	10	20	28	35	42	47	52	57
5000	…	6	16	25	32	39	44	49	53
5200	…	3	13	22	29	35	41	46	50
5400	…	…	10	18	26	32	38	42	47
5600	…	…	6	15	22	29	34	39	43
5800	…	…	3	11	19	25	31	36	40

The allowable tight side tension must be evaluated for each sheave in the system (see Step 14). Values must be corrected by K_f from Table 17.

Table 15. Allowable Tight Side Tension for a CC Section

Belt Speed (fpm)	Sheave Effective Diameter (in.)								
	7.0	8.0	9.0	10.0	11.0	12.0	13.0	14.0	15.0
200	121	158	186	207	228	244	257	268	278
400	99	135	164	187	206	221	234	246	256
600	85	122	151	173	192	208	221	232	242
800	75	112	141	164	182	198	211	222	232
1000	67	104	133	155	174	190	203	214	224
1200	60	97	126	149	167	183	196	207	217
1400	54	91	120	142	161	177	190	201	211
1600	48	85	114	137	155	171	184	196	205
1800	43	80	108	131	150	166	179	190	200
2000	38	75	103	126	145	160	174	185	195
2200	33	70	98	121	140	155	169	180	190
2400	28	65	93	116	135	150	164	175	185
2600	23	60	88	111	130	145	159	170	180
2800	18	55	83	106	125	140	154	165	175
3000	13	50	78	101	120	135	149	160	170
3200	8	45	73	96	115	130	144	155	165
3400	3	39	68	91	110	125	138	150	160
3600	…	34	63	86	104	120	133	145	154
3800	…	29	58	80	99	115	128	139	149
4000	…	24	52	75	94	109	123	134	144
4200	…	18	47	70	88	104	117	128	138
4400	…	12	41	64	83	98	112	123	133
4600	…	7	35	58	77	93	106	117	127
4800	…	1	29	52	71	87	100	111	121
5000	…	…	23	46	65	81	94	105	115
5200	…	…	17	40	59	75	88	99	109
5400	…	…	11	34	53	68	81	93	103
5600	…	…	5	27	46	62	75	86	96
5800	…	…	…	21	40	55	68	80	90

The allowable tight side tension must be evaluated for each sheave in the system (see Step 14). Values must be corrected by K_f from Table 17.

Table 16. Allowable Tight Side Tension for a DD Section

Belt Speed (fpm)	Sheave Effective Diameter (in.)								
	12,0	13.0	14.0	15.0	16.0	17.0	18.0	19.0	20.0
200	243	293	336	373	405	434	459	482	503
400	195	245	288	325	358	386	412	434	455
600	167	217	259	297	329	358	383	406	426
800	146	196	239	276	308	337	362	385	405
1000	129	179	222	259	291	320	345	368	389
1200	114	164	207	244	277	305	331	353	374
1400	101	151	194	231	263	292	318	340	361
1600	89	139	182	219	251	280	305	328	349
1800	78	128	170	207	240	269	294	317	337
2000	67	117	159	196	229	258	283	306	326
2200	56	106	149	186	218	247	272	295	316
2400	45	95	138	175	208	236	262	284	305
2600	35	85	128	165	197	226	251	274	294
2800	24	74	117	154	187	215	241	263	284
3000	14	64	106	144	176	205	230	253	273
3200	3	53	96	133	165	194	219	242	263
3400	…	42	85	122	155	183	209	231	252
3600	…	31	74	111	144	172	198	220	241
3800	…	20	63	100	132	161	186	209	230
4000	…	9	51	89	121	150	175	198	218
4200	…	…	40	77	109	138	163	186	207
4400	…	…	28	65	97	126	152	174	195
4600	…	…	16	53	85	114	139	162	183
4800	…	…	3	40	73	102	127	150	170
5000	…	…	…	28	60	89	114	137	158
5200	…	…	…	15	47	76	101	124	145
5400	…	…	…	1	34	62	88	111	131
5600	…	…	…	…	20	49	74	97	118
5800	…	…	…	…	6	35	60	83	104

The allowable tight side tension must be evaluated for each sheave in the system (see Step 14). Values must be corrected by K_f from Table 17.

Table 17. Length-Flex Correction Factors K_f

$\frac{L_e}{n}$	Belt Cross Section				$\frac{L_e}{n}$	Belt Cross Section			
	AA	BB	CC	DD		AA	BB	CC	DD
10	0.64	0.58	…	…	70	…	1.03	0.95	0.91
15	0.74	0.68	…	…	80	…	1.06	0.98	0.94
20	0.82	0.74	0.68	…	90	…	1.09	1.00	0.96
25	0.87	0.79	0.73	0.70	100	…	1.11	1.03	0.99
30	0.92	0.84	0.77	0.74	110	…	…	1.05	1.00
35	0.96	0.87	0.80	0.77	120	…	…	1.06	1.02
40	0.99	0.90	0.83	0.80	130	…	…	1.08	1.04
45	1.02	0.93	0.86	0.82	140	…	…	1.10	1.05
50	1.05	0.95	0.88	0.84	150	…	…	1.11	1.07
60	…	0.99	0.92	0.88	…	…	…	…	…

Minimum Sheave Size: The recommended minimum sheave size depends on the rpm of the faster shaft. Minimum sheave diameters for each cross-section belt are listed in Table 11.

Tension Ratings: The tension rating formulas are:

$$\text{AA} \quad T_r = 118.5 - \frac{318.2}{d} - 0.8380S^2 - 25.76\log S$$

$$\text{BB} \quad T_r = 186.3 - \frac{665.1}{d} - 1.269S^2 - 39.02\log S$$

$$\text{CC} \quad T_r = 363.9 - \frac{2060}{d} - 2.400S^2 - 73.77\log S$$

$$\text{DD} \quad T_r = 783.1 - \frac{7790}{d} - 5.078S^2 - 156.1\log S$$

where T_r = The allowable tight side tension for a double-V belt drive, lbf (not corrected for tension ratio or length-flex correction factor)

d = Pitch diameter of small sheave, inch

S = Belt speed, fpm/1000

Table 18. Tension Ratio/Arc of Contact Factors

Arc of Contact, θ (deg.)	Design $\frac{R}{R-1}$	Arc of Contact, θ(deg.)	Design $\frac{R}{R-1}$
300	1.07	170	1.28
290	1.08	160	1.31
280	1.09	150	1.35
270	1.10	140	1.40
260	1.11	130	1.46
250	1.12	120	1.52
240	1.13	110	1.60
230	1.15	100	1.69
220	1.16	90	1.81
210	1.18	80	1.96
200	1.20	70	2.15
190	1.22	60	2.41
180	1.25	50	2.77

Light Duty V-Belts ANSI/RMA IP-23.—Light duty V-belts are typically used with fractional horsepower motors or small engines, and are designed primarily for fractional horsepower service. These belts are intended and specifically designed for use with small diameter sheaves and drives of loads and service requirements that are within the capacity of a single belt.

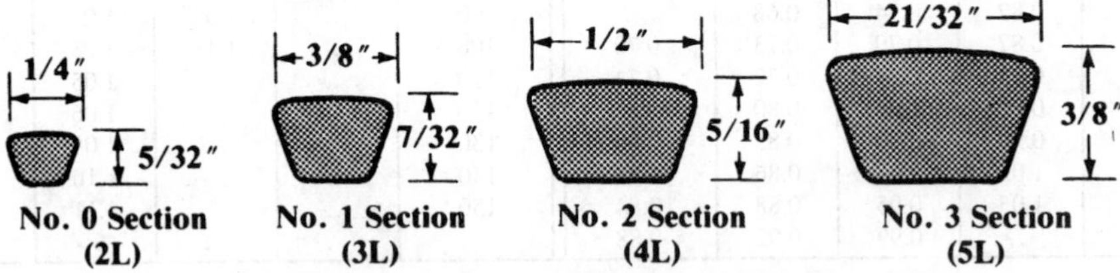

Fig. 8. Light Duty V-Belt Cross Sections

The four belt cross sections and sheave groove sizes are 2L, 3L, 4L, and 5L. The 2L is generally used only by OEMs and is not covered in the RMA standards.

Belt Cross Sections.—Nominal dimensions of the four cross sections are given in Fig. 8.

Belt Size Designation.—V-belt sizes are identified by a standard belt number, consisting of a letter-numeral combination. The first number and letter identify the cross section; the remaining numbers identify length as shown in Table 19. For example, a 3L520 belt has a 3L cross section and a length of 52.0 in.

Table 19. Light Duty V-Belt Standard Dimensions *ANSI/RMA IP-23, 1968*

Standard Effective Outside Length (in.)				Permissible Deviation From Standard Effective Length (in.)	Standard Effective Outside Length (in.)				Permissible Deviation From Standard Effective Length (in.)
Cross Section					Cross Section				
2L	3L	4L	5L		2L	3L	4L	5L	
8	…	…	…	+0.12, −0.38	…	…	53	53	+0.25, −0.62
9	…	…	…	+0.12, −0.38	…	54	54	54	+0.25, −0.62
10	…	…	…	+0.12, −0.38	…	…	55	55	+0.25, −0.62
11	…	…	…	+0.12, −0.38	…	56	56	56	+0.25, −0.62
12	…	…	…	+0.12, −0.38	…	…	57	57	+0.25, −0.62
13	…	…	…	+0.12, −0.38	…	58	58	58	+0.25, −0.62
14	14	…	…	+0.12, −0.38	…	…	59	59	+0.25, −0.62
15	15	…	…	+0.12, −0.38	…	60	60	60	+0.25, −0.62
16	16	…	…	+0.12, −0.38	…	…	61	61	+0.31, −0.69
17	17	…	…	+0.12, −0.38	…	…	62	62	+0.31, −0.69
18	18	18	…	+0.12, −0.38	…	…	63	63	+0.31, −0.69
19	19	19	…	+0.12, −0.38	…	…	64	64	+0.31, −0.69
20	20	20	…	+0.12, −0.38	…	…	65	65	+0.31, −0.69
…	21	21	…	+0.25, −0.62	…	…	66	66	+0.31, −0.69
…	22	22	…	+0.25, −0.62	…	…	67	67	+0.31, −0.69
…	23	23	…	+0.25, −0.62	…	…	68	68	+0.31, −0.69
…	24	24	…	+0.25, −0.62	…	…	69	69	+0.31, −0.69
…	25	25	25	+0.25, −0.62	…	…	70	70	+0.31, −0.69
…	26	26	26	+0.25, −0.62	…	…	71	71	+0.31, −0.69
…	27	27	27	+0.25, −0.62	…	…	72	72	+0.31, −0.69
…	28	28	28	+0.25, −0.62	…	…	73	73	+0.31, −0.69
…	29	29	29	+0.25, −0.62	…	…	74	74	+0.31, −0.69
…	30	30	30	+0.25, −0.62	…	…	75	75	+0.31, −0.69
…	31	31	31	+0.25, −0.62	…	…	76	76	+0.31, −0.69
…	32	32	32	+0.25, −0.62	…	…	77	77	+0.31, −0.69
…	33	33	33	+0.25, −0.62	…	…	78	78	+0.31, −0.69
…	34	34	34	+0.25, −0.62	…	…	79	79	+0.31, −0.69
…	35	35	35	+0.25, −0.62	…	…	80	80	+0.62, −0.88
…	36	36	36	+0.25, −0.62	…	…	82	82	+0.62, −0.88
…	37	37	37	+0.25, −0.62	…	…	84	84	+0.62, −0.88
…	38	38	38	+0.25, −0.62	…	…	86	86	+0.62, −0.88
…	39	39	39	+0.25, −0.62	…	…	88	88	+0.62, −0.88
…	40	40	40	+0.25, −0.62	…	…	90	90	+0.62, −0.88
…	41	41	41	+0.25, −0.62	…	…	92	92	+0.62, −0.88
…	42	42	42	+0.25, −0.62	…	…	94	94	+0.62, −0.88
…	43	43	43	+0.25, −0.62	…	…	96	96	+0.62, −0.88
…	44	44	44	+0.25, −0.62	…	…	98	98	+0.62, −0.88
…	45	45	45	+0.25, −0.62	…	…	100	100	+0.62, −0.88
…	46	46	46	+0.25, −0.62	…	…	…	…	…
…	47	47	47	+0.25, −0.62	…	…	…	…	…
…	48	48	48	+0.25, −0.62					
…	49	49	49	+0.25, −0.62	…	…	…	…	…
…	50	50	50	+0.25, −0.62	…	…	…	…	…
…	…	51	51	+0.25, −0.62	…	…	…	…	…
…	52	52	52	+0.25, −0.62	…	…	…	…	…

All dimensions in inches.

Sheave Dimensions: Groove angles and dimensions for sheaves and various sheave tolerances are given in Table 20.

Table 20. Light Duty V-Belt Sheave and Groove Dimensions
ANSI/RMA IP-23, 1968

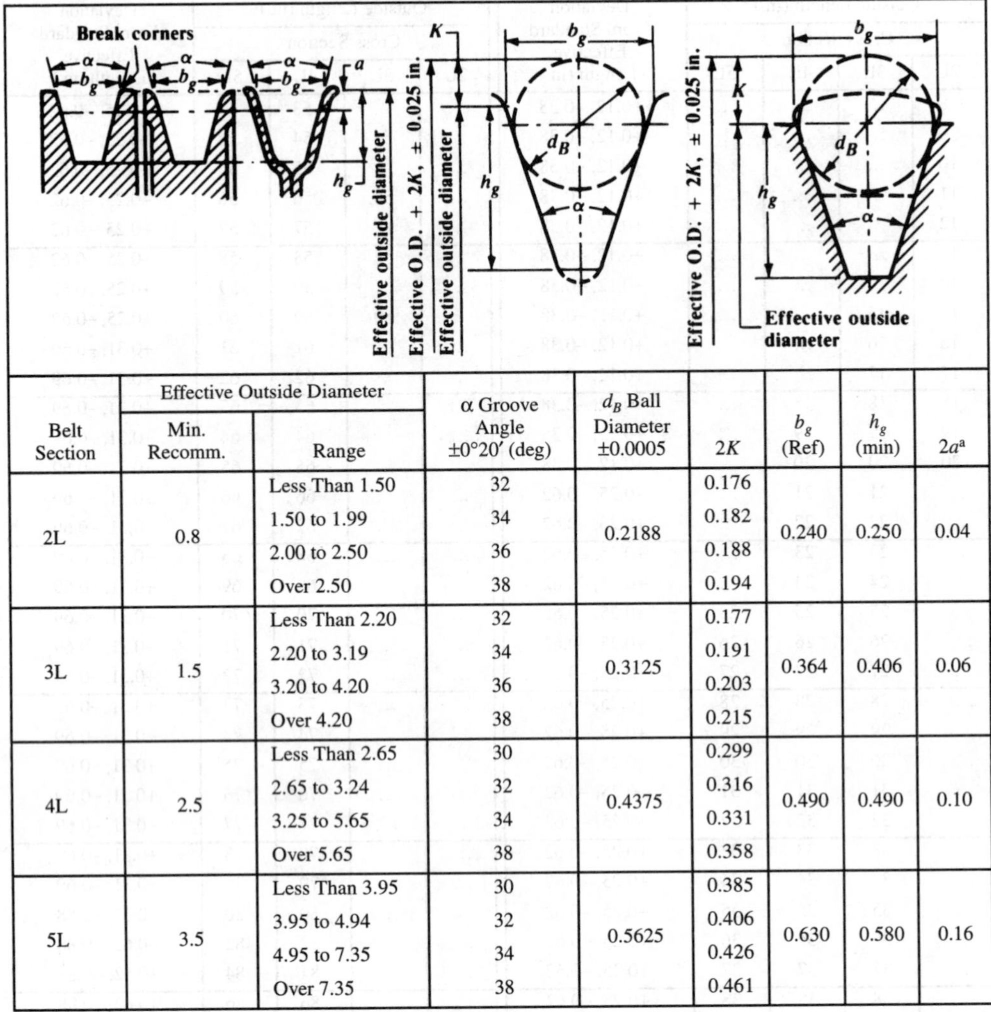

Belt Section	Effective Outside Diameter: Min. Recomm.	Effective Outside Diameter: Range	α Groove Angle ±0°20′ (deg)	d_B Ball Diameter ±0.0005	2K	b_g (Ref)	h_g (min)	$2a$[a]
2L	0.8	Less Than 1.50	32	0.2188	0.176	0.240	0.250	0.04
		1.50 to 1.99	34		0.182			
		2.00 to 2.50	36		0.188			
		Over 2.50	38		0.194			
3L	1.5	Less Than 2.20	32	0.3125	0.177	0.364	0.406	0.06
		2.20 to 3.19	34		0.191			
		3.20 to 4.20	36		0.203			
		Over 4.20	38		0.215			
4L	2.5	Less Than 2.65	30	0.4375	0.299	0.490	0.490	0.10
		2.65 to 3.24	32		0.316			
		3.25 to 5.65	34		0.331			
		Over 5.65	38		0.358			
5L	3.5	Less Than 3.95	30	0.5625	0.385	0.630	0.580	0.16
		3.95 to 4.94	32		0.406			
		4.95 to 7.35	34		0.426			
		Over 7.35	38		0.461			

[a] The diameter used in calculating speed ratio and belt speed is obtained by subtracting the $2a$ value from the Effective Outside Diameter of the sheave.

Other Sheave Tolerances		
Outside Diameters	Outside Diameter Eccentricity[a]	Groove Side Wobble & Runout[a]
For outside diameters under 6.0 in. ±0.015 in.	For outside diameters 10.0 in. and under 0.010 in.	For outside diameters 20.0 in. and under 0.0015 in. per inch of outside diameter.
For outside diameters 6.0 to 12.0 in. ±0.020 in.	For each additional inch of outside diameter, add 0.0005 in.	For each additional inch of outside diameter, add 0.0005 in.
For outside diameters over 12.0 in. ±0.040 in.		

[a] Total indicator reading.

All dimensions in inches.

Horsepower Ratings: The horsepower ratings for light duty V-belts can be calculated from the following formulas:

$$\text{3L} \quad \text{HP} = r\left(\frac{0.2164d^{0.91}}{r^{0.09}} - 0.2324 - 0.0001396r^2d^3\right)$$

$$\text{4L} \quad \text{HP} = r\left(\frac{0.4666d^{0.91}}{r^{0.09}} - 0.7231 - 0.0002286r^2d^3\right)$$

$$\text{5L} \quad \text{HP} = r\left(\frac{0.7748d^{0.91}}{r^{0.09}} - 1.727 - 0.0003641r^2d^3\right)$$

where $d = d_0 - 2a$; d_0 = effective outside diameter of small sheave, in.; r = rpm of the faster shaft divided by 1000. The corrected horsepower rating is obtained by dividing the horsepower rating by the combined correction factor (Table 21), which accounts for drive geometry and service factor requirements.

Table 21. Combined Correction Factors

Type of Driven Unit	Speed Ratio	
	Less than 1.5	1.5 and Over
Fans and blowers	1.0	0.9
Domestic laundry machines	1.1	1.0
Centrifugal pumps	1.1	1.0
Generators	1.2	1.1
Rotary compressors	1.2	1.1
Machine tools	1.3	1.2
Reciprocating pumps	1.4	1.3
Reciprocating compressors	1.4	1.3
Wood working machines	1.4	1.3

V-Ribbed Belts ANSI/RMA IP-26.— V-ribbed belts are a cross between flat belts and V-belts. The belt is basically flat with V-shaped ribs projecting from the bottom, which guide the belt and provide greater stability than that found in a flat belt. The ribs operate in grooved sheaves.

V-ribbed belts do not have the wedging action of a V-belt and thus operate at higher tensions. This design provides excellent performance in high-speed and serpentine applications, and in drives that utilize small diameter sheaves. The V-ribbed belt comes in five cross sections: H, J, K, L, and M, specified by effective length, cross section and number of ribs.

Belt Cross Sections: Nominal dimensions of the five cross sections are given in Table 22.

Table 22. Nominal Dimensions of V-Ribbed Belt Cross Sections
ANSI/RMA IP-26, 1977

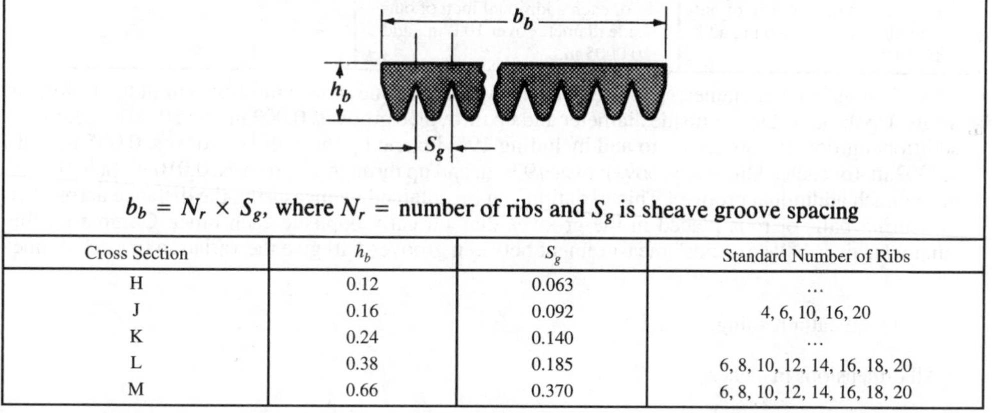

$b_b = N_r \times S_g$, where N_r = number of ribs and S_g is sheave groove spacing

Cross Section	h_b	S_g	Standard Number of Ribs
H	0.12	0.063	…
J	0.16	0.092	4, 6, 10, 16, 20
K	0.24	0.140	…
L	0.38	0.185	6, 8, 10, 12, 14, 16, 18, 20
M	0.66	0.370	6, 8, 10, 12, 14, 16, 18, 20

All dimensions in inches.

Table 23. V-Ribbed Belt Sheave and Groove Dimensions ***ANSI / RMA IP-26, 1977***

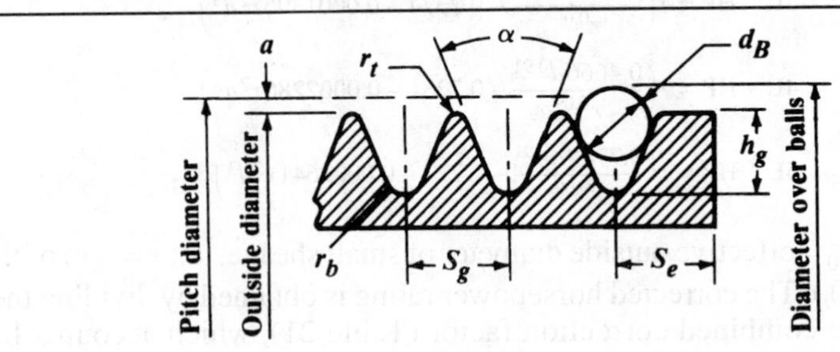

Face width = $S_e (N_g - 1) + 2S_e$, where N_g is number of grooves

Cross Section	Minimum Recommended Outside Diameter	α Groove Angle ±0.25 (deg)	S_g[a]	r_t +0.005, −0.000	2a	r_b	h_g (min)	d_B ±0.0005	S_e
H	0.50	40	0.063 ±0.001	0.005	0.020	0.013 +0.000 −0.005	0.041	0.0469	0.080 +0.020 −0.010
J	0.80	40	0.092 ±0.001	0.008	0.030	0.015 +0.000 −0.005	0.071	0.0625	0.125 +0.030 −0.015
K	1.50	40	0.140 ±0.002	0.010	0.038	0.020 +0.000 −0.005	0.122	0.1093	0.125 +0.050 −0.000
L	3.00	40	0.185 ±0.002	0.015	0.058	0.015 +0.000 −0.005	0.183	0.1406	0.375 +0.075 −0.030
M	7.00	40	0.370 ±0.003	0.030	0.116	0.030 +0.000 −0.010	0.377	0.2812	0.500 +0.100 −0.040

[a] Summation of the deviations from S_g for all grooves in any one sheave shall not exceed ±0.010 in.

Other Sheave Tolerances[a]		
Outside Diameter	**Radial Runout[b]**	**Axial Runout[b]**
Up through 2.9 in. outside diameter ±0.010 in.	Up through 2.9 in. outside diameter 0.005 in.	0.001 in. per inch of outside diameter
Over 2.9 in. to and including 8.0 in. outside diameter ±0.020 in.	Over 2.9 in. to and including 10.0 in. outside diameter 0.010 in.	
For each additional inch of outside diameter over 8.0 in., add ±0.0025 in.	For each additional inch of outside diameter over 10.0 in., add 0.0005 in.	

[a] Variations in pitch diameter between the grooves in any one sheave must be within the following limits: Up through 2.9 in. outside diameter and up through 6 grooves, 0.002 in. (add 0.001 in. for each additional groove); over 2.9 in. to and including 19.9 in. and up through 10 grooves, 0.005 in. (add 0.0002 in. for each additional groove); over 19.9 in. and up through 10 grooves, 0.010 in. (add 0.0005 in. for each additional groove). This variation can be obtained by measuring the distance across two measuring balls or rods placed in the grooves diametrically opposite each other. Comparing this "diameter-over-balls or -rods" measurement between grooves will give the variation in pitch diameter.

[b] Total indicator reading.

All dimensions in inches

elt Size Designation: Belt sizes are identified by a standard belt number, which consists of belt effective length to the nearest tenth of an inch, a letter designating cross section, and the number of ribs. For example, 540L6 signifies a 54.0 in. effective length, L belt, six ribs wide.

Sheave Dimensions.: Groove angles and dimensions for sheaves and face widths of sheaves for multiple belt drives are given in Table 23, along with various tolerance values.

Cross Section Selection.: Use the chart (Fig. 9) as a guide to the V-ribbed belt cross section for any combination of design horsepower and speed of the faster shaft. When the intersection of the design horsepower and speed of the faster shaft falls near a line between two areas on the chart, the possibilities in both areas should be explored. Special circumstances (such as space limitations) may lead to a choice of belt cross section different from that indicated in the chart. H and K cross sections are not included because of their specialized use. Belt manufacturers should be contacted for specific data.

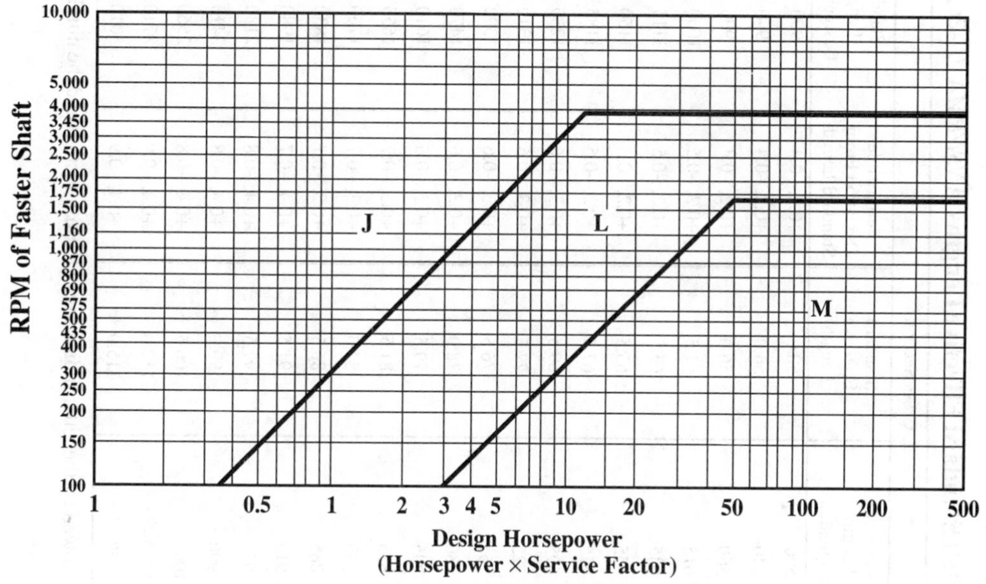

Fig. 9. Selection of V-Ribbed Belt Cross Section

Horsepower Ratings.: The horsepower rating formulas are:

$$\mathbf{J:}\mathrm{HP} = d_p r\left[\frac{0.1240}{(d_p r)^{0.09}} - \frac{0.08663}{d_p} - 0.2318\times10^{-4}(d_p r)^2\right] + 0.08663 r\left[1 - \frac{1}{K_{SR}}\right]$$

$$\mathbf{L:}\mathrm{HP} = d_p r\left[\frac{0.5761}{(d_p r)^{0.09}} - \frac{0.8987}{d_p} - 1.018\times10^{-4}(d_p r)^2\right] + 0.8987 r\left[1 - \frac{1}{K_{SR}}\right]$$

$$\mathbf{M:}\mathrm{HP} = d_p r\left[\frac{1.975}{(d_p r)^{0.09}} - \frac{6.597}{d_p} - 3.922\times10^{-4}(d_p r)^2\right] + 6.597 r\left[1 - \frac{1}{K_{SR}}\right]$$

In these equations, d_p = pitch diameter of the small sheave, in.; r = rpm of the faster shaft divided by 1000; K_{SR} = speed ratio factor given in the accompanying table. These formulas give the maximum horsepower per rib recommended, corrected for the speed ratio. To obtain the horsepower per rib for an arc of contact other than 180 degrees, and for belts longer or shorter than the average length, multiply the horsepower obtained from these formulas by the length correction factor (Table 25) and the arc of contact correction factor (Table 26).

Table 24. V-Ribbed Belt Standard Effective Lengths *ANSI/RMA IP-26, 1977*

J Cross Section			L Cross Section			M Cross Section		
Standard Length Designation[a]	Standard Effective Length	Permissible Deviation From Standard Length	Standard Length Designation[a]	Standard Effective Length	Permissible Deviation From Standard Length	Standard Length Designation[a]	Standard Effective Length	Permissible Deviation From Standard Length
180	18.0	+0.2, −0.2	500	50.0	+0.2, −0.4	900	90.0	+0.4, −0.7
190	19.0	+0.2, −0.2	540	54.0	+0.2, −0.4	940	94.0	+0.4, −0.8
200	20.0	+0.2, −0.2	560	56.0	+0.2, −0.4	990	99.0	+0.4, −0.8
220	22.0	+0.2, −0.2	615	61.5	+0.2, −0.5	1060	106.0	+0.4, −0.8
240	24.0	+0.2, −0.2	635	63.5	+0.2, −0.5	1115	111.5	+0.4, −0.9
260	26.0	+0.2, −0.2	655	65.5	+0.2, −0.5	1150	115.0	+0.4, −0.9
280	28.0	+0.2, −0.2	675	67.5	+0.3, −0.6	1185	118.5	+0.4, −0.9
300	30.0	+0.2, −0.3	695	69.5	+0.3, −0.6	1230	123.0	+0.4, −1.0
320	32.0	+0.2, −0.3	725	72.5	+0.3, −0.6	1310	131.0	+0.5, −1.1
340	34.0	+0.2, −0.3	765	76.5	+0.3, −0.6	1390	139.0	+0.5, −1.1
360	36.0	+0.2, −0.3	780	78.0	+0.3, −0.6	1470	147.0	+0.6, −1.2
380	38.0	+0.2, −0.3	795	79.5	+0.3, −0.6	1610	161.0	+0.6, −1.2
400	40.0	+0.2, −0.4	815	81.5	+0.3, −0.7	1650	165.0	+0.6, −1.3
430	43.0	+0.2, −0.4	840	84.0	+0.3, −0.7	1760	176.0	+0.7, −1.4
460	46.0	+0.2, −0.4	865	86.5	+0.3, −0.7	1830	183.0	+0.7, −1.4
490	49.0	+0.2, −0.4	915	91.5	+0.4, −0.7	1980	198.0	+0.8, −1.6
520	52.0	+0.2, −0.4	975	97.5	+0.4, −0.8	2130	213.0	+0.8, −1.6
550	55.0	+0.2, −0.4	990	99.0	+0.4, −0.8	2410	241.0	+0.9, −1.6
580	58.0	+0.2, −0.5	1065	106.5	+0.4, −0.8	2560	256.0	+1.0, −1.8
610	61.0	+0.2, −0.5	1120	112.0	+0.4, −0.9	2710	271.0	+1.1, −2.2
650	65.0	+0.2, −0.5	1150	115.0	+0.4, −0.9	3010	301.0	+1.2, −2.4

[a] To specify belt size, use the standard length designation, followed by the letter indicating belt cross section and the number of ribs desired. For example: 865L10. All dimensions in inches.

Table 25. Length Correction Factors

Std. Length Designation	Cross Section J	Cross Section L	Cross Section M	Std. Length Designation	Cross Section J	Cross Section L	Cross Section M
180	0.83	…	…	1230	…	1.08	0.94
200	0.85	…	…	1310	…	1.10	0.96
240	0.89	…	…	1470	…	1.12	0.098
280	0.92	…	…	1610	…	1.14	1.00
320	0.95	…	…	1830	…	1.17	1.03
360	0.98	…	…	1980	…	1.19	1.05
400	1.00	…	…	2130	…	1.21	1.06
440	1.02	…	…	2410	…	1.24	1.09
500	1.05	0.89	…	2710	…	…	1.12
550	1.07	0.91	…	3010	…	…	1.14
610	1.09	0.93	…	3310	…	…	1.16
690	1.12	0.96	…	3610	…	…	1.18
780	1.16	0.98	…	3910	…	…	1.20
910	1.18	1.02	0.88	4210	…	…	1.22
940	1.19	1.02	0.89	4810	…	…	1.25
990	1.20	1.04	0.90	5410	…	…	1.28
1060	…	1.05	0.91	6000	…	…	1.30
1150	…	1.07	0.93	…	…	…	…

Table 26. Arc of Contact Correction Factors

$\frac{D_o - d_o}{C}$	Arc of Contact, θ, on Small Sheave, (deg)	Correction Factor
0.00	180	1.00
0.10	174	0.98
0.20	169	0.97
0.30	163	0.95
0.40	157	0.94
0.50	151	0.92
0.60	145	0.90
0.70	139	0.88
0.80	133	0.85
0.90	127	0.83
1.00	120	0.80
1.10	113	0.77
1.20	106	0.74
1.30	99	0.71
1.40	91	0.67
1.50	83	0.63

Number of Ribs: The number of ribs required for an application is obtained by dividing the design horsepower by the corrected horsepower rating for one rib.

Arc of contact on the small sheave may be determined by the following formulas:

Exact Formula:

$$\text{Arc of Contact (deg)} = 2\cos^{-1}\left(\frac{D_o - d_o}{2C}\right)$$

Approximate Formula:

$$\text{Arc of Contact (deg)} = 180 - \frac{(D_o - d_o)60}{C}$$

where D_o = Effective outside diameter of large sheave, in; d_o = Effective outside diameter of small sheave, in; and, C = Center distance, inch.

Speed Ratio Correction Factors

Speed Ratio[a]	K_{SR}
1.00 to and incl. 1.10	1.0000
Over 1.01 to and incl. 1.04	1.0136
Over 1.04 to and incl. 1.08	1.0276
Over 1.08 to and incl. 1.12	1.0419
Over 1.12 to and incl. 1.18	1.0567
Over 1.18 to and incl. 1.24	1.0719
Over 1.24 to and incl. 1.34	1.0875
Over 1.34 to and incl. 1.51	1.1036
Over 1.51 to and incl. 1.99	1.1202
Over 1.99	1.1373

[a] D_p/d_p, where D_p (d_p) is the pitch diameter of the large (small) sheave.

Variable Speed Belts ANSI, RMA IP-25.—For drives that require more speed variation than can be obtained with conventional industrial V-belts, standard-line variable-speed drives are available. These drives use special wide, thin belts. Package units of standard-line variable-speed belts and sheaves, combined with the motor and output gearbox are available in ranges from approximately ½ through 100 horsepower.

The speed ranges of variable-speed drives can be much greater than those drives using classical V-belts. Speed ranges up to 10:1 can be obtained on lower horsepower units.

This section covers 12 variable speed belt cross sections and sheave groove sizes designed 1422V, 1922V, 2322V 1926V, 2926V, 3226V, 2530V, 3230V, 4430V, 4036V, 4436V, and 4836V. The industry supplies many other sizes that are not listed in this section.

Belt Cross Sections and Lengths: Nominal dimensions of the 12 cross sections are given in Table 27, and lengths in Table 28.

Table 27. Normal Variable-Speed Belt Dimensions ***ANSI/RMA IP-25, 1982***

Cross Section	b_b	h_b	h_b/b_b	Cross Section	b_b	h_b	h_b/b_b
1422V	0.88	0.31	0.35	2530V	1.56	0.59	0.38
1922V	1.19	0.38	0.32	3230V	2.00	0.62	0.31
2322V	1.44	0.44	0.31	4430V	2.75	0.69	0.25
1926V	1.19	0.44	0.37	4036V	2.50	0.69	0.28
2926V	1.81	0.50	0.28	4436V	2.75	0.72	0.26
3226V	2.0	0.53	0.27	4836V	3.00	0.75	0.25

All dimensions in inches.

Table 28. Variable-Speed V-Belt Standard Belt Lengths *ANSI/RMA IP-25, 1982*

Standard Pitch Length Designation	Standard Effective Lengths Cross Section												Permissible Deviations From Standard Length
	1422V	1922V	2322V	1926V	2926V	3226V	2530V	3230V	4430V	4036V	4436V	4836V	
315	32.1	…	…	…	…	…	…	…	…	…	…	…	±0.7
335	34.1	…	…	…	…	…	…	…	…	…	…	…	±0.7
355	36.1	36.2	…	36.3	…	…	…	…	…	…	…	…	±0.7
375	38.1	38.2	…	38.3	…	…	…	…	…	…	…	…	±0.7
400	40.6	40.7	40.8	40.8	…	…	…	…	…	…	…	…	±0.7
425	43.1	43.2	43.3	43.3	…	…	…	…	…	…	…	…	±0.8
450	45.6	45.7	45.8	45.8	…	…	…	…	…	…	…	…	±0.8
475	48.1	48.2	48.3	48.3	…	…	…	…	…	…	…	…	±0.8
500	50.6	50.7	50.8	50.8	…	…	50.9	…	…	…	…	…	±0.8
530	53.6	53.7	53.8	53.8	53.9	…	53.9	…	…	…	…	…	±0.8
560	56.6	56.7	56.8	56.8	56.9	56.9	56.9	57.1	57.3	57.3	57.3	57.4	±0.9
600	60.6	60.7	60.8	60.8	60.9	60.9	60.9	61.1	61.3	61.3	61.3	61.4	±0.9
630	63.6	63.7	63.8	63.8	63.9	63.9	63.9	64.1	64.3	64.3	64.3	64.4	±0.9
670	67.6	67.7	67.8	67.8	67.9	67.9	67.9	68.1	68.3	68.3	68.3	68.4	±0.9
710	71.6	71.7	71.8	71.8	71.9	71.9	71.9	72.1	72.3	72.3	72.3	72.4	±0.9
750	75.6	75.7	75.8	75.8	75.9	75.9	75.9	76.1	76.3	76.3	76.3	76.4	±1.0
800	…	80.7	80.8	80.8	80.9	80.9	80.9	81.1	81.3	81.3	81.3	81.4	±1.0
850	…	85.7	85.8	85.8	85.9	85.9	85.9	86.1	86.3	86.3	86.3	86.4	±1.1
900	…	90.7	90.8	90.8	90.9	90.9	90.9	91.1	91.3	91.3	91.3	91.4	±1.1
950	…	95.7	95.8	95.8	95.9	95.9	95.9	96.1	96.3	96.3	96.3	96.4	±1.1
1000	…	100.7	100.8	100.8	100.9	100.9	100.9	101.1	101.3	101.3	101.3	101.4	±1.2
1060	…	106.7	106.8	106.8	106.9	106.9	106.9	107.1	107.3	107.3	107.3	107.4	±1.2
1120	…	112.7	112.8	112.8	112.9	112.9	112.9	113.1	113.3	113.3	113.3	113.4	±1.2
1180	…	118.7	118.8	118.8	118.9	118.9	118.9	119.1	119.3	119.3	119.3	119.4	±1.3
1250	…	…	…	…	125.9	125.9	125.9	126.1	126.3	126.3	126.3	126.4	±1.3
1320	…	…	…	…	…	132.9	…	133.1	133.3	133.3	133.3	133.4	±1.3

All dimensions in inches.

The lengths given in this table are not necessarily available from all manufacturers. Availability should be investigated prior to design commitment.

Table 29. Variable-Speed Sheave and Groove Dimensions

	Standard Groove Dimensions									Drive Design Factors			
	Variable					Companion							
Cross Section	α Groove Angle ±0.67 (deg)	b_g[a] Closed +0.000 −0.030	b_{go} Open Max	h_{gv} Min	S_g ±0.03	α Groove Angle ±0.33 (deg)	b_g ±0.010	h_g Min	S_g ±0.03	Min. Recomm. Pitch Diameter	$2a$	$2av$ Max	CL Min
1422V	22	0.875	1.63	2.33	1.82	22	0.875	0.500	1.82	2.0	0.20	3.88	0.08
1922V	22	1.188	2.23	3.14	2.42	22	1.188	0.562	2.42	3.0	0.22	5.36	0.08
2322V	22	1.438	2.71	3.78	2.89	22	1.438	0.625	2.89	3.5	0.25	6.52	0.08
1926V	26	1.188	2.17	2.65	2.36	26	1.188	0.625	2.36	3.0	0.25	4.26	0.08
2926V	26	1.812	3.39	4.00	3.58	26	1.812	0.750	3.58	3.5	0.30	6.84	0.08
3226V	26	2.000	3.75	4.41	3.96	26	2.000	0.781	3.96	4.0	0.30	7.60	0.08
2530V	30	1.562	2.81	3.01	2.98	30	1.562	0.844	2.98	4.0	0.30	4.64	0.10
3230V	30	2.000	3.67	3.83	3.85	30	2.000	0.875	3.85	4.5	0.35	6.22	0.10
4430V	30	2.750	5.13	5.23	5.38	30	2.750	0.938	5.38	5.0	0.40	8.88	0.10
4036V	36	2.500	4.55	3.95	4.80	36	2.500	0.938	4.80	4.5	0.40	6.32	0.10
4436V	36	2.750	5.03	4.33	5.30	36	2.750	0.969	5.30	5.0	0.40	7.02	0.10
4836V	36	3.000	5.51	4.72	5.76	36	3.000	1.000	5.76	6.0	0.45	7.74	0.10

[a] The effective width (b_e), a reference dimension, is the same as the ideal top width of closed variable-speed sheave (b_g) and the ideal top width of the companion sheave (b_g).

Other Sheave Tolerances					
Outside Diameter		Radial Runout[a]		Axial Runout[a]	
Up through 4.0 in. outside diameter	±0.020 in.	Up through 10.0 in. outside diameter	0.010 in.	Up through 5.0 in. outside diameter	0.005 in.
For each additional inch of outside diameter add ±0.005 in.		For each additional inch of outside diameter add 0.0005 in.		For each additional inch of outside diameter add 0.001 in.	

[a] Total indicator reading.

Surface Finish			
Machined Surface Area	Max Surface Roughness Height, R_a (AA) (μ in.)	Machined Surface Area	Max Surface Roughness Height, R_a (AA) (μ in.)
V-Sheave groove sidewalls	125	Straight bores with 0.002 in. or less total tolerance	125
Rim edges and ID, Hub ends and OD	500	Taper and straight bores with total tolerance over 0.002 in.	250

All dimensions in inches, except where noted.

Belt Size Designation: Variable-speed belt sizes are identified by a standard belt number. The first two digits denote the belt top width in sixteenths of an inch; the third and fourth digits indicate the angle of the groove in which the belt is designed to operate. The letter V (for variable) follows the first four digits. The digits after the V indicate the pitch length to the nearest 0.1 in. For example, 1422V450 is a belt of ⅞ in. (14⁄16 in.) nominal top width designed to operate in a sheave of 22 degree groove angle and having a pitch length of 45.0 in.

Sheave Groove Data: A variable speed sheave is an assembly of movable parts, designed to permit one or both flanges of the sheave to be moved axially causing a radial movement of the variable speed belt in the sheave groove. This radial movement permits stepless speed variation within the physical limits of the sheave and the belt. A companion sheave may be a solid sheave having a constant diameter and groove profile or another variable sheave. Variable speed sheave designs should conform to the dimensions in Table 29 and Fig. 10. The included angle of the sheaves, top width, and clearance are boundary dimensions. Groove angles and dimensions of companion sheaves should conform to Table 29 and Fig. 11. Various tolerance values are also given in Table 29.

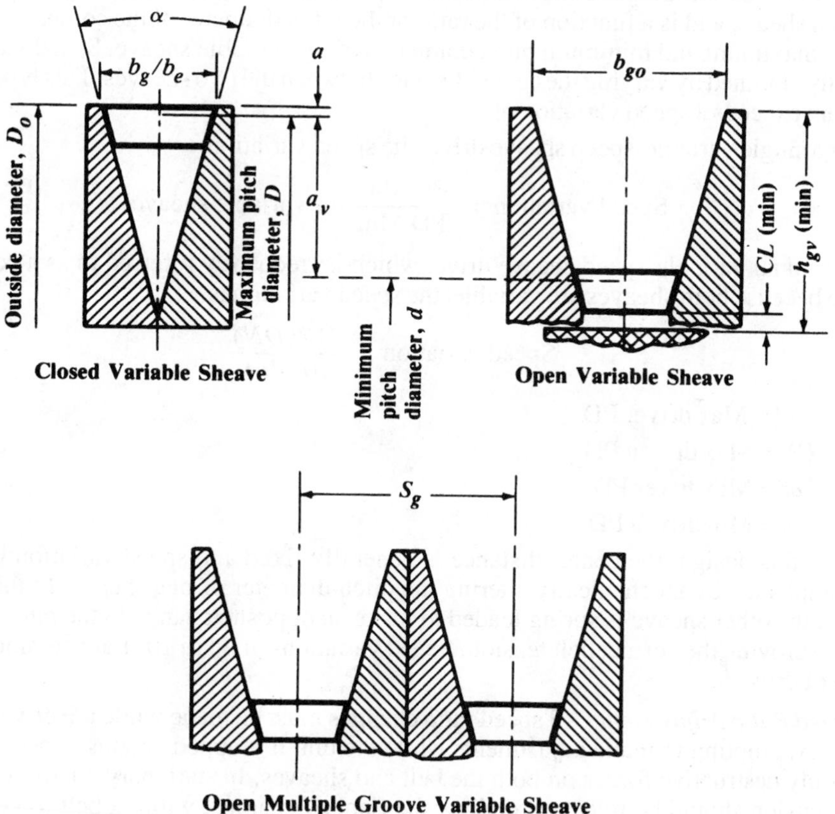

Fig. 10. Variable Sheaves

Variable-Speed Drive Design: Variable-speed belts are designed to operate in sheaves that are an assembly of movable parts. The sheave design permits one or both flanges of the sheave to be moved axially, causing a radial movement of the variable-speed belt in the sheave groove. The result is a stepless speed variation within the physical limits of the sheave and the variable-speed belt. Therefore, besides transmitting power, variable-speed belt drives provide speed variation.

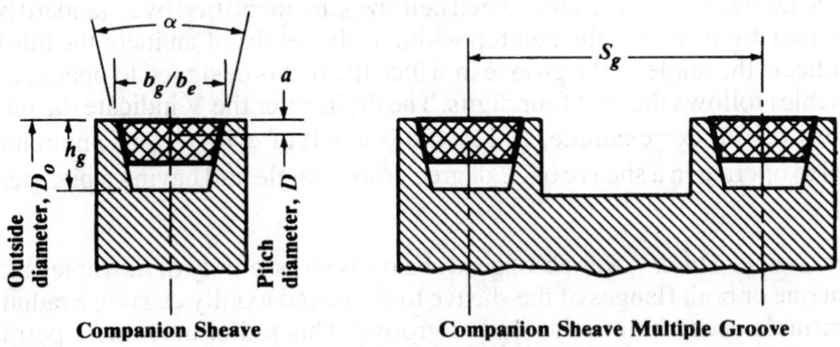

Fig. 11. Companion Sheaves

The factors that determine the amount of pitch diameter change on variable-speed sheaves are belt top width, belt thickness, and sheave angle. This pitch diameter change, combined with the selected operating pitch diameters for a sheave, determines the possible speed variation.

The range of output speeds from a variable-speed sheave drive is established by the companion sheave and is a function of the ratio of the pitch diameter of the companion sheave to the maximum and minimum pitch diameters of the variable sheave. Speed variation is usually obtained by varying the center distance between the two sheaves. This type of drive seldom exceeds a speed variation of 3:1.

For a single variable-speed sheave drive, the speed variation

$$\text{Speed variation} = \frac{\text{PD Max}}{\text{PD Min}} \text{ (of variable sheave)}$$

For a dual variable-speed sheave drive, which is frequently referred to as a compound drive because both sheaves are variable, the speed variation is

$$\text{Speed variation} = \frac{DR(DN)}{dr(dn)}$$

where DR = Max driver PD

DN = Max driven PD

dr = Min driver PD

dn = Min driven PD

With this design, the center distance is generally fixed and speed variation is usually accomplished by mechanically altering the pitch diameter of one sheave. In this type of drive, the other sheave is spring loaded to make an opposite change in the pitch diameter and to provide the correct belt tension. Speed variations of up to 10: 1 are common on this type of drive.

Speed Ratio Adjustment: All speed ratio changes must be made while the drives are running. Attempting to make adjustments while the unit is stopped creates unnecessary and possibly destructive forces on both the belt and sheaves. In stationary control drives, the belt tension should be released to allow the flanges to adjust without belt force interference.

Cross Section Selection: Selection of a variable speed belt cross section is based on the drive design horsepower and speed variation. Table 29 shows the maximum pitch diameter variation (2*av*) that each cross section can attain.

Horsepower Ratings: The general horsepower formulas for variable-speed belts are:

$$1422\text{V HP} = d_p r\left[0.4907(d_p r)^{-0.09} - \frac{0.8378}{d_p} - 0.000337(d_p r)^2\right] + 0.8378r\left(1 - \frac{1}{K_{SR}}\right)$$

$$1922\,\text{VHP} = d_p r\left[0.8502(d_p r)^{-0.09} - \frac{1.453}{d_p} - 0.000538(d_p r)^2\right] + 1.453r\left(1 - \frac{1}{K_{SR}}\right)$$

$$2322\,\text{VHP} = d_p r\left[1.189(d_p r)^{-0.09} - \frac{2.356}{d_p} - 0.000777(d_p r)^2\right] + 2.356r\left(1 - \frac{1}{K_{SR}}\right)$$

$$1926\,\text{VHP} = d_p r\left[1.046(d_p r)^{-0.09} - \frac{1.833}{d_p} - 0.000589(d_p r)^2\right] + 1.833r\left(1 - \frac{1}{K_{SR}}\right)$$

$$2926\,\text{VHP} = d_p r\left[1.769(d_p r)^{-0.09} - \frac{4.189}{d_p} - 0.001059(d_p r)^2\right] + 4.189r\left(1 - \frac{1}{K_{SR}}\right)$$

$$3226\,\text{VHP} = d_p r\left[2.073(d_p r)^{-0.09} - \frac{5.236}{d_p} - 0.001217(d_p r)^2\right] + 5.236r\left(1 - \frac{1}{K_{SR}}\right)$$

$$2530\,\text{VHP} = d_p r\left[2.395(d_p r)^{-0.09} - \frac{6.912}{d_p} - 0.001148(d_p r)^2\right] + 6.912r\left(1 - \frac{1}{K_{SR}}\right)$$

$$3230\,\text{VHP} = d_p r\left[2.806(d_p r)^{-0.09} - \frac{7.854}{d_p} - 0.001520(d_p r)^2\right] + 7.854r\left(1 - \frac{1}{K_{SR}}\right)$$

$$4430\,\text{VHP} = d_p r\left[3.454(d_p r)^{-0.09} - \frac{7.854}{d_p} - 0.002196(d_p r)^2\right] + 9.818r\left(1 - \frac{1}{K_{SR}}\right)$$

$$4036\,\text{VHP} = d_p r\left[3.566(d_p r)^{-0.09} - \frac{9.687}{d_p} - 0.002060(d_p r)^2\right] + 9.687r\left(1 - \frac{1}{K_{SR}}\right)$$

$$4436\,\text{VHP} = d_p r\left[4.041(d_p r)^{-0.09} - \frac{11.519}{d_p} - 0.002297(d_p r)^2\right] + 11.519r\left(1 - \frac{1}{K_{SR}}\right)$$

$$4836\,\text{VHP} = d_p r\left[4.564(d_p r)^{-0.09} - \frac{13.614}{d_p} - 0.002634(d_p r)^2\right] + 13.614r\left(1 - \frac{1}{K_{SR}}\right)$$

In these equations, d_p = pitch diameter of small sheave, in.; r = rpm of faster shaft divided by 1000; K_{SR} = speed ratio factor given in the accompanying table. These formulas give the basic horsepower rating, corrected for the speed ratio. To obtain the horsepower for arcs of contact other than 180 degrees and for belts longer or shorter than average length, multiply the horsepower obtained from these formulas by the arc of contact correction factor (Table 30) and the length correction factor (Table 31).

Table 30. Arc of Contact Correction Factors

$\frac{D-d}{C}$	Arc of Contact, θ, on Small Sheave, (deg)	Correction Factor	$\frac{D-d}{C}$	Arc of Contact, θ, on Small Sheave, (deg)	Correction Factor
0.00	180	1.00	0.80	0.80	0.87
0.10	174	0.99	0.90	0.90	0.85
0.20	169	0.97	1.00	1.00	0.82
0.30	163	0.96	1.10	1.10	0.80
0.40	157	0.94	1.20	1.20	0.77
0.50	151	0.93	1.30	1.30	0.73
0.60	145	0.91	1.40	1.40	0.70
0.70	139	0.89	1.50	1.50	0.65

Table 31. Length Correction Factors

Standard Pitch Length Designation	Cross Section											
	1422V	1922V	2322V	1926V	2926V	3226V	2530V	3230V	4430V	4036V	4436V	4836V
315	0.93	…	…	…	…	…	…	…	…	…	…	…
335	0.94	…	…	…	…	…	…	…	…	…	…	…
355	0.95	0.90	…	0.90	…	…	…	…	…	…	…	…
375	0.96	0.91	…	0.91	…	…	…	…	…	…	…	…
400	0.97	0.92	0.90	0.92	…	…	…	…	…	…	…	…
425	0.98	0.93	0.91	0.93	…	…	…	…	…	…	…	…
450	0.99	0.94	0.92	0.94	…	…	…	…	…	…	…	…
475	1.00	0.95	0.93	0.95	…	…	…	…	…	…	…	…
500	1.01	0.95	0.94	0.95	…	…	0.90	…	…	…	…	…
530	1.02	0.96	0.95	0.96	0.92	…	0.92	…	…	…	…	…
560	1.03	0.97	0.96	0.97	0.93	0.92	0.93	0.91	0.90	0.91	0.91	0.92
600	1.04	0.98	0.97	0.98	0.94	0.93	0.94	0.93	0.92	0.93	0.92	0.93
630	1.05	0.99	0.98	0.99	0.95	0.94	0.95	0.94	0.93	0.94	0.93	0.94
670	1.06	1.00	0.99	1.00	0.97	0.95	0.96	0.95	0.94	0.95	0.95	0.95
710	1.07	1.01	1.00	1.01	0.98	0.96	0.98	0.96	0.96	0.96	0.96	0.96
750	1.08	1.02	1.01	1.02	0.99	0.98	0.99	0.97	0.97	0.97	0.97	0.98
800	…	1.03	1.02	1.03	1.00	0.99	1.00	0.99	0.99	0.99	0.99	0.99
850	…	1.04	1.03	1.04	1.01	1.00	1.01	1.00	1.00	1.00	1.00	1.00
900	…	1.05	1.04	1.05	1.02	1.01	1.02	1.01	1.01	1.01	1.01	1.01
950	…	1.06	1.05	1.06	1.03	1.02	1.04	1.02	1.03	1.02	1.02	1.02
1000	…	1.07	1.06	1.07	1.04	1.03	1.05	1.03	1.04	1.03	1.04	1.03
1060	…	1.08	1.07	1.07	1.06	1.04	1.06	1.05	1.06	1.05	1.05	1.04
1120	…	1.09	1.08	1.08	1.07	1.06	1.07	1.06	1.07	1.06	1.06	1.06
1180	…	1.09	1.09	1.09	1.08	1.07	1.08	1.07	1.08	1.07	1.07	1.07
1250	…	…	…	…	1.09	1.08	1.10	1.08	1.10	1.08	1.09	1.08
1320	…	…	…	…	…	1.09	…	1.09	1.11	1.09	1.10	1.09

Rim Speed: The material and design selected for sheaves must be capable of withstanding the high rim speeds that may occur in variable-speed drives. The rim speed is calculated as follows: Rim speed (fpm) = $(\pi/12)(D_o)$ (rpm).

60 Degree V-Belts.—60 degree V-belts are ideal for compact drives. Their 60 degree angle and ribbed top are specifically designed for long life on small diameter sheaves. These belts offer extremely smooth operation at high speeds (in excess of 10,000 rpm) and can be used on drives with high speed ratios. They are available in 3M, 5M, 7M, and 11M (3, 5, 7, 11 mm) cross sections (top widths) and are commonly found in the joined configuration, which provides extra stability and improved performance. They are specified by cross section and nominal length; for example, a 5M315 designation indicates a belt having a 5 mm cross section and an effective length of 315 mm.

Speed Ratio Correction Factors

Speed Ratio[a]	K_{SR}	Speed Ratio[a]	K_{SR}
1.00–1.01	1.0000	1.19–1.24	1.0719
1.02–1.04	1.0136	1.25–1.34	1.0875
1.05–1.08	1.0276	1.35–1.51	1.1036
1.09–1.12	1.0419	1.52–1.99	1.1202
1.13–1.18	1.0567	2.0 and over	1.1373

[a] D_p/d_p, where D_p (d_p) is the pitch diameter of the large (small) sheave.

Arc of contact on the small sheave may be determined by the formulas:

Exact Formula: Arc of Contact (deg) $= 2\cos^{-1}\left(\frac{D-d}{2C}\right)$

Approximate Formula:

$$\text{Arc of Contact (deg)} = 180 - \frac{(D-d)60}{C}$$

where D = Pitch diameter of large sheave or flat pulley, inch

d = Pitch diameter of small sheave, inch

C = Center distance, inch

Industry standards have not yet been published for 60 degree V-belts. Therefore, belt manufacturers should be contacted for specific applications, specifications, and additional information.

Synchronous Belts ANSI/RMA IP-24.—Synchronous belts are also known as timing or positive-drive belts. These belts have evenly spaced teeth on their surfaces, which mesh with teeth on pulleys or sprockets to produce a positive, no-slip transmission of power. Such designs should not be confused with molded notched V-belts, which transmit power by means of the wedging action of the V-shape.

Synchronous belts are used where driven shaft speeds must be synchronized to the rotation of the driver shaft and to eliminate the noise and maintenance problems of chain drives.

Standard Timing Belts: Conventional trapezoidal, or rectangular tooth, timing belts come in six cross sections, which relate to the pitch of the belt. Pitch is the distance from center to center of the teeth. The six basic cross sections or pitches are MXL (mini extra light), XL (extra light), L (light), H (heavy), XH (extra heavy), and XXH (double extra heavy) (Fig. 12). Belts are specified by pitch length, cross section (pitch), and width.

Double-sided timing belts have identical teeth on both sides of the belt and are used where synchronization is required from each belt face. They are available in XL, L, and H cross sections.

Size Designations: Synchronous belt sizes are identified by a standard number. The first digits specify the belt length to 0.1 in. followed by the belt section (pitch) designation. The digits following the belt section designation represent the nominal belt width times 100. For example, an L section belt 30.000 in. pitch length and 0.75 in. in width would be specified as a 300L075 synchronous belt.

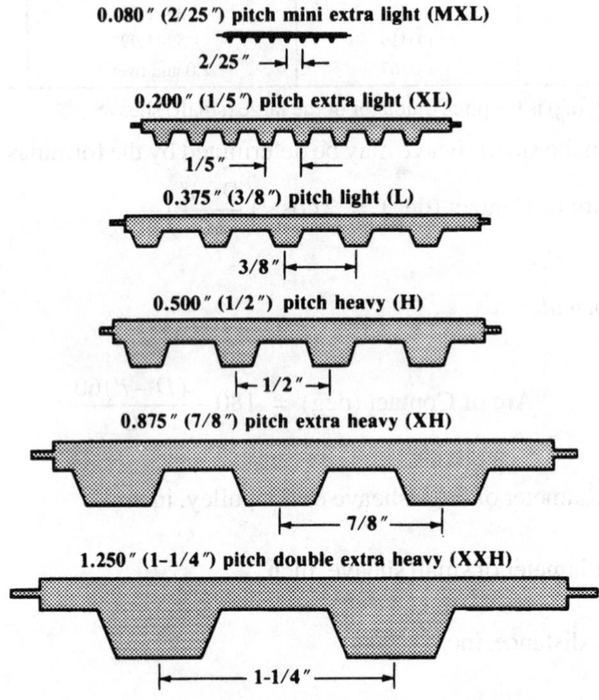

Fig. 12. Standard Synchronous Belt Sections

The RMA nomenclature for double-sided belts is the same as for single-sided belts with the addition of the prefix "D" in front of the belt section. However, some manufacturers use their own designation system for double-sided belts.

Standard Sections: Belt sections are specified in terms of pitch. Table 33 gives the Standard Belt Sections and their corresponding pitches.

Pitch Lengths: Standard belt pitch lengths, belt length designations, and numbers of teeth are shown in Table 34. Belt length tolerances are also given in this table; these tolerances apply to all belt sections and represent the total manufacturing tolerance on belt length.

Table 32. Synchronous Belt Standard Pulley and Flange Dimensions
ANSI/RMA IP-24, 1983

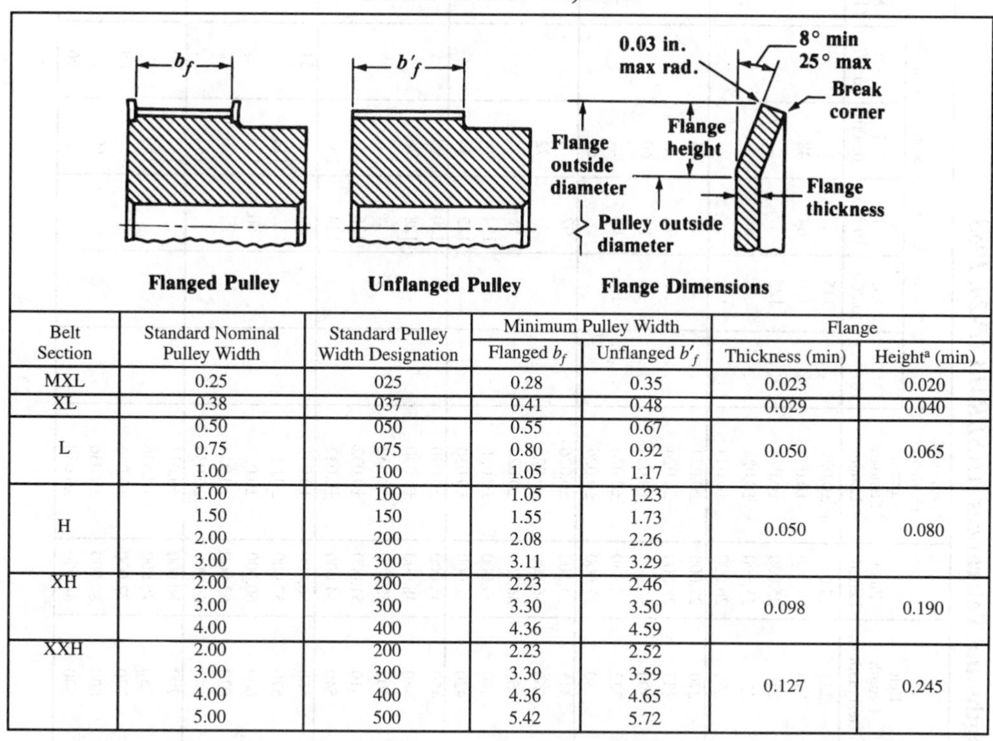

Belt Section	Standard Nominal Pulley Width	Standard Pulley Width Designation	Minimum Pulley Width		Flange	
			Flanged b_f	Unflanged b'_f	Thickness (min)	Height[a] (min)
MXL	0.25	025	0.28	0.35	0.023	0.020
XL	0.38	037	0.41	0.48	0.029	0.040
L	0.50	050	0.55	0.67	0.050	0.065
	0.75	075	0.80	0.92		
	1.00	100	1.05	1.17		
H	1.00	100	1.05	1.23	0.050	0.080
	1.50	150	1.55	1.73		
	2.00	200	2.08	2.26		
	3.00	300	3.11	3.29		
XH	2.00	200	2.23	2.46	0.098	0.190
	3.00	300	3.30	3.50		
	4.00	400	4.36	4.59		
XXH	2.00	200	2.23	2.52	0.127	0.245
	3.00	300	3.30	3.59		
	4.00	400	4.36	4.65		
	5.00	500	5.42	5.72		

Nominal Tooth Dimensions: Table 33 shows the nominal tooth dimensions for each of the standard belt sections. Tooth dimensions for single- and double-sided belts are identical.

Table 33. Synchronous Belt Nominal Tooth and Section Dimensions
ANSI/RMP IP-24, 1983

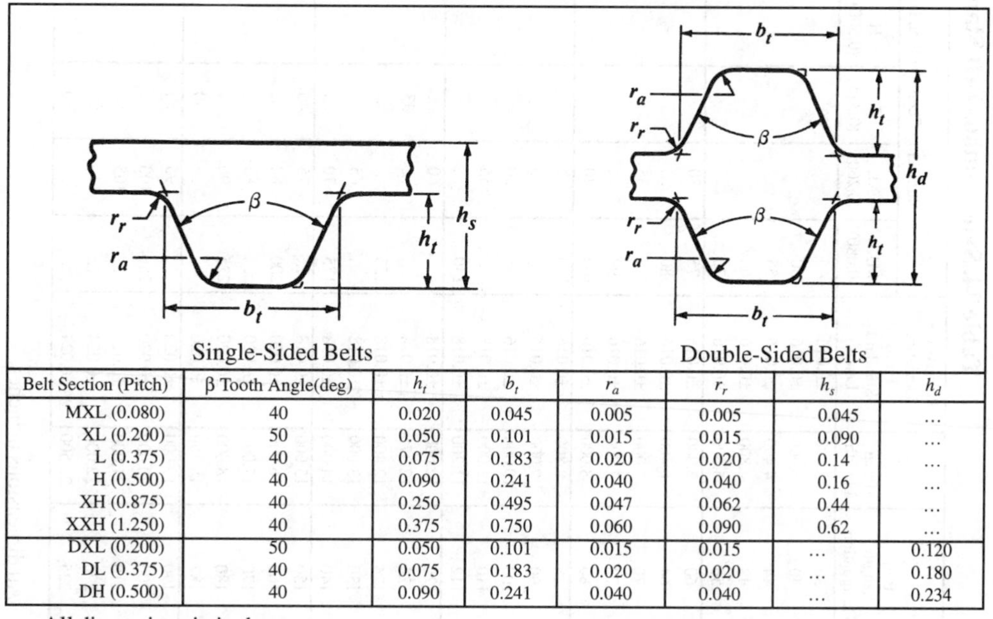

Belt Section (Pitch)	β Tooth Angle(deg)	h_t	b_t	r_a	r_r	h_s	h_d
MXL (0.080)	40	0.020	0.045	0.005	0.005	0.045	…
XL (0.200)	50	0.050	0.101	0.015	0.015	0.090	…
L (0.375)	40	0.075	0.183	0.020	0.020	0.14	…
H (0.500)	40	0.090	0.241	0.040	0.040	0.16	…
XH (0.875)	40	0.250	0.495	0.047	0.062	0.44	…
XXH (1.250)	40	0.375	0.750	0.060	0.090	0.62	…
DXL (0.200)	50	0.050	0.101	0.015	0.015	…	0.120
DL (0.375)	40	0.075	0.183	0.020	0.020	…	0.180
DH (0.500)	40	0.090	0.241	0.040	0.040	…	0.234

All dimensions in inches.

Table 34. Synchronous Belt Standard Pitch Lengths and Tolerances *ANSI/RMA IP-24, 1983*

Belt Length Designation	Pitch Length	Permissible Deviation From Standard Length	Number of Teeth for Standard Lengths						Belt Length Designation	Pitch Length	Permissible Deviation From Standard Length	Number of Teeth for Standard Lengths					
			MXL (0.080)	XL (0.200)	L (0.375)	H (0.500)	XH (0.875)	XXH (1.250)				MXL (0.080)	XL (0.200)	L (0.375)	H (0.500)	XH (0.875)	XXH (1.250)
36	3.600	±0.016	45						230	23.000	±0.024	…	115	…	…		
40	4.000	±0.016	50						240	24.000	±0.024	…	120	64	48		
44	4.400	±0.016	55						250	25.000	±0.024	…	125	…	…		
48	4.800	±0.016	60						255	25.500	±0.024	…	…	68	…		
56	5.600	±0.016	70						260	26.000	±0.024	…	130	…	…		
60	6.000	±0.016	75	30					270	27.000	±0.024	…	…	72	54		
64	6.400	±0.016	80	…					285	28.500	±0.024	…	…	76	…		
70	7.000	±0.016	…	35					300	30.000	±0.024	…	…	80	60		
72	7.200	±0.016	90	…					322	32.250	±0.026	…	…	86	…		
80	8.000	±0.016	100	40					330	33.000	±0.026	…	…	…	66		
88	8.800	±0.016	110	…					345	34.500	±0.026			92	…		
90	9.000	±0.016	…	45					360	36.000	±0.026			…	72		
100	10.000	±0.016	125	50					367	36.750	±0.026			98	…		
110	11.000	±0.018	…	55					390	39.000	±0.026			104	78		
112	11.200	±0.018	140	…					420	42.000	±0.030			112	84		
120	12.000	±0.018	…	60	…				450	45.000	±0.030			120	90	…	
124	12.375	±0.018	…	…	33				480	48.000	±0.030			128	96	…	
124	12.400	±0.018	155	…	…				507	50.750	±0.032			…	…	58	
130	13.000	±0.018	…	65	…				510	51.000	±0.032			136	102	…	
140	14.000	±0.018	175	70	…				540	54.000	±0.032			144	108	…	
150	15.000	±0.018	…	75	40				560	56.000	±0.032			…	…	64	
160	16.000	±0.020	200	80	…				570	57.000	±0.032			…	114	…	
170	17.000	±0.020	…	85	…				600	60.000	±0.032			160	120	…	
180	18.000	±0.020	225	90	…				630	63.000	±0.034			…	126	72	
187	18.750	±0.020	…	…	50				660	66.000	±0.034			…	132	…	
190	19.000	±0.020	…	95	…				700	70.000	±0.034				140	80	56
200	20.000	±0.020	250	100	…				750	75.000	±0.036				150	…	…
210	21.000	±0.024	…	105	56				770	77.000	±0.036				…	88	…
220	22.000	±0.024	…	110	…				800	80.000	±0.036				160	…	64
225	22.500	±0.024	…	…	60				840	84.000	±0.038				…	96	…

All dimensions in inches.

Widths.: Standard belt widths, width designations, and width tolerances are shown in Table 35.

Table 35. Synchronous Belt Standard Widths and Tolerances
ANSI/RMA IP-24, 1983

Belt Section	Standard Belt Widths		Tolerances on Width for Belt Pitch Lengths		
	Designation	Dimensions	Up to and including 33 in.	Over 33 in. up to and including 66 in.	Over 66 in.
MXL (0.080)	012 019 025	0.12 0.19 0.25	+0.02 −0.03	…	…
XL (0.200)	025 037	0.25 0.38	+0.02 −0.03	…	…
L (0.375)	050 075 100	0.50 0.75 1.00	+0.03 −0.03	+0.03 −0.05	…
H (0.500)	075 100 150	0.75 1.00 1.50	+0.03 −0.03	+0.03 −0.05	+0.03 −0.05
	200	2.00	+0.03 −0.05	+0.05 −0.05	+0.05 −0.06
	300	3.00	+0.05 −0.06	+0.06 −0.06	+0.06 −0.08
XH (0.875)	200 300 400	2.00 3.00 4.00	…	+0.19 −0.19	+0.19 −0.19
XXH (1.250)	200 300 400 500	2.00 3.00 4.00 5.00	…	…	+0.19 −0.19

Length Determination.: The pitch length of a synchronous belt is determined by placing the belt on a measuring fixture having two pulleys of equal diameter, a method of applying force, and a means of measuring the center distance between the two pulleys. The position of one of the two pulleys is fixed and the other is movable along a graduated scale.

Synchronous Belt Pulley Diameters: Table 36 lists the standard pulley diameters by belt section (pitch). Fig. 13 defines the pitch, pitch diameter, outside diameter and pitch line differential.

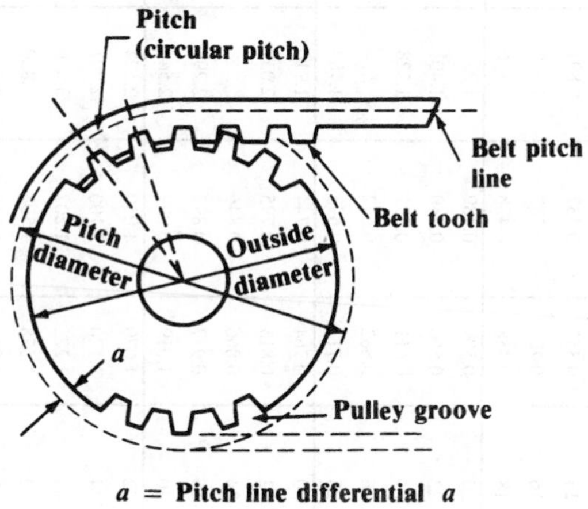

Fig. 13. Synchronous Belt Pulley Dimensions

Table 36. Synchronous Belt Standard Pulley Diameters *ANSI/RMA IP-24, 1983*

Number of Grooves	Belt Section											
	MXL (0.080)		XL (0.200)		L (0.375)		H (0.500)		XH (0.875)		XXH (1.250)	
	Diameters		Diameters		Diameters		Diameters		Diameters		Diameters	
	Pitch	Outside	Pitch	Outside	Pitch	Outside	Pitch	Outside	Pitch	Outside	Pitch	Outside
10	0.255	0.235	0.637	0.617	1.194[a]	1.164	…	…	…	…	…	…
12	0.306	0.286	0.764	0.744	1.432[a]	1.402	…	…	…	…	…	…
14	0.357	0.337	0.891	0.871	1.671	1.641	2.228[a]	2.174	…	…	…	…
16	0.407	0.387	1.019	0.999	1.910	1.880	2.546	2.492	…	…	…	…
18	0.458	0.438	1.146	1.126	2.149	2.119	2.865	2.811	5.013	4.903	7.162	7.042
20	0.509	0.489	1.273	1.253	2.387	2.357	3.183	3.129	5.570	5.460	7.958	7.838
22	0.560	0.540	1.401	1.381	2.626	2.596	3.501	3.447	6.127	6.017	8.754	8.634
24	0.611	0.591	1.528	1.508	2.865	2.835	3.820	3.766	6.685	6.575	9.549	9.429
26	0.662	0.642	…	…	3.104	3.074	4.138	4.084	7.242	7.132	10.345	10.225
28	0.713	0.693	1.783	1.763	3.342	3.312	4.456	4.402	7.799	7.689	…	…
30	0.764	0.744	1.910	1.890	3.581	3.551	4.775	4.721	8.356	8.246	11.937	11.817
32	0.815	0.795	2.037	2.017	3.820	3.790	5.093	5.039	8.913	8.803	…	…
34	0.866	0.846	…	…	…	…	…	…	…	…	13.528	13.408
36	0.917	0.897	2.292	2.272	4.297	4.267	5.730	5.676	…	…	…	…
40	1.019	0.999	2.546	2.526	4.775	4.745	6.366	6.312	11.141	11.031	15.915	15.795
42	1.070	1.050	2.674	2.654	…	…	…	…	…	…	…	…
44	1.120	1.100	2.801	2.781	5.252	5.222	7.003	6.949	…	…	…	…
48	1.222	1.202	3.056	3.036	5.730	5.700	7.639	7.585	13.369	13.259	19.099	18.979
60	1.528	1.508	3.820	3.800	7.162	7.132	9.549	9.495	16.711	16.601	23.873	23.753
72	1.833	1.813	4.584	4.564	8.594	8.564	11.459	11.405	20.054	19.944	28.648	28.528
84	…	…	…	…	10.027	9.997	13.369	13.315	23.396	23.286	…	…
90	…	…	…	…	…	…	…	…	…	…	35.810	35.690
96	…	…	…	…	…	…	15.279	15.225	26.738	26.628	…	…
120	…	…	…	…	…	…	19.099	19.045	33.423	33.313	…	…

All dimensions in inches.

* Usually not available in all widths — consult supplier.

Widths: Standard pulley widths for each belt section are shown in Table 32. The nominal pulley width is specified in terms of the maximum standard belt width the pulley will accommodate. The minimum pulley width, whether flanged or unflanged, is also shown in Table 32, along with flange dimensions and various pulley tolerances.

Pulley Size Designation: Synchronous belt pulleys are designated by the number of grooves, the belt section, and a number representing 100 times the nominal width. For example, a 30 groove L section pulley with a nominal width of 0.75 in. would be designated by 30L075. Pulley tolerances are shown in Table 37.

Table 37. Pulley Tolerances (All Sections)

Outside Diameter Range	Outside Diameter Tolerance	Pitch to Pitch Tolerance	
		Adjacent Grooves	Accumulative Over 90 Degrees
Up thru 1.000	+0.002 −0.000	±0.001	±0.003
Over 1.000 to and including 2.000	+0.003 −0.000	±0.001	±0.004
Over 2.000 to and including 4.000	+0.004 −0.000	±0.001	±0.005
Over 4.000 to and including 7.000	+0.005 −0.000	±0.001	±0.005
Over 7.000 to and including 12.000	+0.006 −0.000	±0.001	±0.006
Over 12.000 to and including 20.000	+0.007 −0.000	±0.001	±0.007
Over 20.000	+0.008 −0.000	±0.001	±0.008

Radial Runout[a]	Axial Runout[b]
For outside diameters 8.0 in. and under 0.005 in. For each additional inch of outside diameter add 0.0005 in.	For outside diameters 1.0 in. and under 0.001 in. For each additional inch of outside diameter up through 10.0 in., add 0.001 in. For each additional inch of outside diameter over 10.0 in., add 0.0005 in.

[a] Flange outside diameter equals pulley outside diameter plus twice flange height.
[b] Total indicator reading.

All dimensions in inches.

Cross Section Selection: The chart (Fig. 14) may be used as a guide to the selection of a synchronous belt for any combination of design horsepower and speed of the faster shaft. When the intersection of the design horsepower and speed of the faster shaft falls near a line between two areas on the chart, the possibilities in both areas should be explored. Special circumstances (such as space limitations) may result in selection of a belt cross section different from that indicated in the chart. Belt manufacturers should be contacted for specific data.

Torque Ratings: It is customary to use torque load requirements rather than horsepower load when designing drives using the small pitch MXL section belts. These belts operate on small diameters resulting in relatively low belt speeds, so torque is essentially constant for all rpm. The torque rating formulas for MXL sections are:

$$Q_r = d[1.13 - 1.38\times10^{-3} d^2] \text{ for belt width } = 0.12 \text{ in.}$$

$$Q_r = d[1.88 - 2.30\times10^{-3} d^2] \text{ for belt width } = 0.19 \text{ in.}$$

$$Q_r = d[2.63 - 3.21\times10^{-3} d^2] \text{ for belt width } = 0.25 \text{ in.}$$

where Q_r = the maximum torque rating (lbf-in.) for a belt of specified width having six or more teeth in mesh and a pulley surface speed of 6500 fpm or less. Torque ratings for drives with less than six teeth in mesh must be corrected as shown in Table 38. d = pitch diameter of smaller pulley, in.

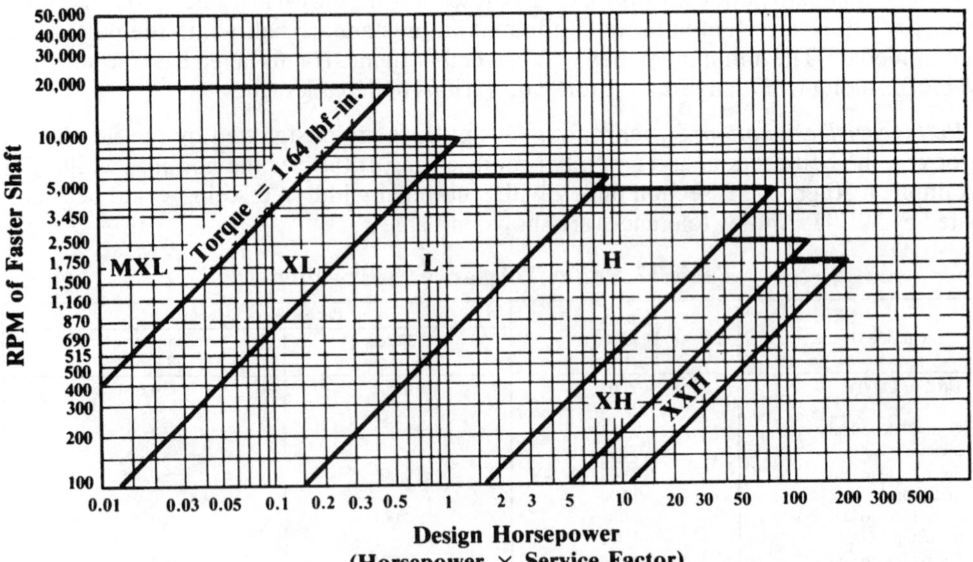

Fig. 14. Selection of Synchronous Belt Cross Section

Table 38. Teeth in Mesh Factor

Teeth in Mesh	Factor K_z	Teeth in Mesh	Factor K_z
6 or more	1.00	3	0.40
5	0.80	2	0.20
4	0.60		

Horsepower Rating Formulas: The horsepower rating formulas for synchronous belts, other than the MLX section, are determined from the following formulas, where the number in parentheses is the belt width in inches.

$$XL(0.38)HP = dr[0.0916 - 7.07\times10^{-5}(dr)^2]$$

$$L(1.00)HP = dr[0.436 - 3.01\times10^{-4}(dr)^2]$$

$$H(3.00)HP = dr[3.73 - 1.41\times10^{-3}(dr)^2]$$

$$XH(4.00)HP = dr[7.21 - 4.68\times10^{-3}(dr)^2]$$

$$XXH(5.00)HP = dr[11.4 - 7.81\times10^{-3}(dr)^2]$$

where HP = the maximum horsepower rating recommended for the specified standard belt width having six or more teeth in mesh and a pulley surface speed of 6500 fpm or less. Horsepower ratings for drives with less than six teeth in mesh must be corrected as shown in Table 38. d = Pitch diameter of smaller pulley, in. r = rpm of faster shaft divided by 1000. Total horsepower ratings are the same for double-sided as for single-sided belts. Contact manufacturers for percentage of horsepower available for each side of the belt.

Finding the Required Belt Width: The belt width should not exceed the small pulley diameter or excessive side thrust will result.

Torque Rating Method (MXL Section): Divide the design torque by the teeth in mesh factor to obtain the corrected design torque. Compare the corrected design torque with the torque rating given in Table 39 for the pulley diameter being considered. Select the narrowest belt width that has a torque rating equal to or greater than the corrected design torque.

Table 39. Torque Rating for MXL Section (0.080 in. Pitch)

Belt Width, (in.)	Rated Torque (lbf-in.) for Small Pulley (Number of Grooves and Pitch Diameter, in.)									
	10MXL 0.255	12MXL 0.306	14MXL 0.357	16MXL 0.407	18MXL 0.458	20MXL 0.509	22MXL 0.560	24MXL 0.611	28MXL 0.713	30MXL 0.764
0.12	0.29	0.35	0.40	0.46	0.52	0.57	0.63	0.69	0.81	0.86
0.19	0.48	0.58	0.67	0.77	0.86	0.96	1.05	1.15	1.34	1.44
0.25	0.67	0.80	0.94	1.07	1.20	1.34	1.47	1.61	1.87	2.01

Horsepower Rating Method (XL, L, H, XH, and XXH Sections): Multiply the horsepower rating for the widest standard belt of the selected section by the teeth in mesh factor to obtain the corrected horsepower rating. Divide the design horsepower by the corrected horsepower rating to obtain the required belt width factor. Compare the required belt width factor with those shown in Table 40. Select the narrowest belt width that has a width factor equal to or greater than the required belt width factor.

Table 40. Belt Width Factor

Belt Section	Belt Width (in.)											
	0.12	0.19	0.25	0.38	0.50	0.75	1.00	1.50	2.00	3.00	4.00	5.00
MXL (0.080)	0.43	0.73	1.00									
XL (0.200)			0.62	1.00								
L (0.375)					0.45	0.72	1.00					
H (0.500)						0.21	0.29	0.45	0.63	1.00		
XH (0.875)									0.45	0.72	1.00	
XXH (1.250)									0.35	0.56	0.78	1.00

Drive Selection: Information on design and selection of synchronous belt drives is available in engineering manuals published by belt manufacturers. Manufacturers should be consulted on such matters as preferred stock sizes, desirable speeds, center distances, etc.

Minimum Pulley Size: The recommended minimum pulley size depends on the rpm of the faster shaft. Minimum sheave diameters for each cross-section belt are listed in Table 36.

Selection of Flanged Pulleys: To determine when to use flanged pulleys, consider the following conditions:

1) On all two-pulley drives, the minimum flanging requirements are two flanges on one pulley, or one flange on each pulley on opposite sides.

2) On drives where the center distance is more than eight times the diameter of the small pulley, both pulleys should be flanged on both sides.

3) On vertical shaft drives, one pulley should be flanged on both sides and other pulleys in the system should be flanged on the bottom side only.

4) On drives with more than two pulleys, the minimum flanging requirements are two flanges on every other pulley, or one flange on every pulley, alternating sides around the system.

Service Factors: Service factors for V-belts are listed in Table 41 and for synchronous belts in Table 42.

Belt Storage and Handling.—To achieve maximum belt performance, proper belt storage procedures should always be practiced. If belts are not stored properly, their performance can be adversely affected. Four key rules are:

1) Do not store belts on floors unless they are protected by appropriate packaging.

2) Do not store belts near windows where the belts may be exposed to direct sunlight or moisture.

3) Do not store belts near electrical devices that may generate ozone (transformers, electric motors, etc.).

4) Do not store belts in areas where solvents or chemicals may be present in the atmosphere.

Table 41. Service Factors for V-Belts

Driving Unit	*AC Motors:* Normal Torque, Squirrel Cage, Synchronous and Split Phase. *DC Motors:* Shunt Wound. *Engines:* Multiple Cylinder Internal Combustion.		
Types of Driven Machines	Intermittent Service (3–5 hours daily or seasonal)	Normal Service (8–10 hours daily)	Continuous Service (16–24 hours daily)
Agitators for liquids; Blowers and exhausters; Centrifugal pumps & compressors; Fans up to 10 horsepower; Light duty conveyors	1.1	1.2	1.3
Belt conveyors for sand, grain, etc.; Dough mixers; Fans over 10 horsepower; Generators; Line shafts; Laundry machinery; Machine tools; Punches, presses, shears; Printing machinery; Positive displacement rotary pumps; Revolving and vibrating screens	1.2	1.3	1.4
Brick machinery; Bucket elevators; Exciters; Piston compressors; Conveyors (drag, pan, screw); Hammer mills; Paper mill beaters; Piston pumps; Positive displacement blowers; Pulverizers; Saw mill and woodworking machinery; Textile machinery	1.4	1.5	1.6
Crushers (gyratory, jaw, roll); Mills (ball, rod, tube); Hoists; Rubber calendars, extruders, mills	1.5	1.6	1.8

Driving Unit	*AC Motors:* High Torque, High Slip, Repulsion-Induction, Single Phase, Series Wound, Slip Ring. *DC Motors:* Series Wound, Compound Wound. *Engines:* Single Cylinder Internal Combustion. *Line Shafts, Clutches*		
Types of Driven Machines	Intermittent Service (3–5 hours daily or seasonal)	Normal Service (8–10 hours daily)	Continuous Service (16–24 hours daily)
Agitators for liquids; Blowers and exhausters; Centrifugal pumps & compressors; Fans up to 10 horsepower; Light duty conveyors	1.1	1.2	1.3
Belt conveyors for sand, grain, etc.; Dough mixers; Fans over 10 horsepower; Generators; Line shafts; Laundry machinery; Machine tools; Punches, presses, shears; Printing machinery; Positive displacement rotary pumps; Revolving and vibrating screens	1.2	1.3	1.4
Brick machinery; Bucket elevators; Exciters; Piston compressors; Conveyors (drag, pan, screw); Hammer mills; Paper mill beaters; Piston pumps; Positive displacement blowers; Pulverizers; Saw mill and woodworking machinery; Textile machinery	1.4	1.5	1.6
Crushers (gyratory, jaw, roll); Mills (ball, rod, tube); Hoists; Rubber calendars, extruders, mills	1.5	1.6	1.8

The machines listed above are representative samples only. Select the group listed above whose load characteristics most closely approximate those of the machine being considered.

Belts should be stored in a cool, dry environment. When stacked on shelves, the stacks should be short enough to avoid excess weight on the bottom belts, which may cause distortion. When stored in containers, the container size and contents should be sufficiently limited to avoid distortion.

V-Belts: A common method is to hang the belts on pegs or pin racks. Very long belts stored this way should use sufficiently large pins or crescent shaped "saddles" to prevent their weight from causing distortion.

Joined V-belts, Synchronous Belts, V-Ribbed Belts: Like V-belts, these belts may be stored on pins or saddles with precautions taken to avoid distortion. However, belts of this type up to approximately 120 in. are normally shipped in a "nested" configuration and should be stored in the same manner. Nests are formed by laying a belt on its side on a flat surface and placing as many belts inside the first belt as possible without undue force. When the nests are tight and are stacked with each rotated 180° from the one below, they may be stacked without damage.

Belts of this type over 120 in. may be "rolled up" and tied for shipment. These rolls may be stacked for easy storage. Care should be taken to avoid small bend radii which could damage the belts.

Table 42. Service Factors for Synchronous Belt Drives

Driving Units	*AC Motors:* Normal Torque, Squirrel Cage, Synchronous and Split Phase. *DC Motors:* Shunt Wound. *Engines:* Multiple Cylinder Internal Combustion.		
Types of Driven Machines	Intermittent Service (3–5 hours daily or seasonal)	Normal Service (8–10 hours daily)	Continuous Service (16–24 hours daily)
Display, Dispensing, Projection, Medical equipment; Instrumentation; Measuring devices	1.0	1.2	1.4
Appliances, sweepers, sewing machines; Office equipment; Wood lathes, band saws	1.2	1.4	1.6
Conveyors: belt, light package, oven, screens, drums, conical	1.3	1.5	1.7
Agitators for liquids; Dough mixers; Drill presses, lathes; Screw machines, jointers; Circular saws, planes; Laundry, Paper, Printing machinery	1.4	1.6	1.8
Agitators for semiliquids; Brick machinery (except pug mills); Conveyor belt: ore, coal, sand; Line shafts; Machine tools: grinder, shaper, boring mill, milling machines; Pumps: centrifugal, gear, rotary	1.5	1.7	1.9
Conveyor: apron, pan, bucket, elevator; Extractors, washers; Fans, blowers; centifugal, induced draft exhausters; Generators & exciters; Hoists, elevators; Rubber calenders, mills, extruders; Saw mill, Textile machinery inc. looms, spinning frames, twisters	1.6	1.8	2.0
Centrifuges; Conveyors: flight, screw; Hammer mills; Paper pulpers	1.7	1.9	2.1
Brick & clay pug mills; Fans, blowers, propeller mine fans, positive blowers	1.8	2.0	2.2

Driving Units	*AC Motors:* High Torque, High Slip, Repulsion-Induction, Single Phase Series Wound and Slip Ring. *DC Motors:* Series Wound and Compound Wound. *Engines:* Single Cylinder Internal Combustion. *Line Shafts. Clutches.*		
Types of Driven Machines	Intermittent Service (3–5 hours daily or seasonal)	Normal Service (8–10 hours daily)	Continuous Service (16–24 hours daily)
Display, Dispensing, Projection, Medical equipment; Instrumentation; Measuring devices	1.2	1.4	1.6
Appliances, sweepers, sewing machines; Office equipment; Wood lathes, band saws	1.4	1.6	1.8
Conveyors: belt, light package, oven, screens, drums, conical	1.5	1.7	1.9
Agitators for liquids; Dough mixers; Drill presses, lathes; Screw machines, jointers; Circular saws, planes; Laundry, Paper, Printing machinery	1.6	1.8	2.0
Agitators for semiliquids; Brick machinery (except pug mills); Conveyor belt: ore, coal, sand; Line shafts; Machine tools:grinder, shaper, boring mill, milling machines; Pumps: centrifugal, gear, rotary	1.7	1.9	2.1
Conveyor: apron, pan, bucket, elevator; Extractors, washers; Fans, blowers; centifugal, induced draft exhausters; Generators & exciters; Hoists, elevators; Rubber calenders, mills, extruders; Saw mill, Textile machinery inc. looms, spinning frames, twisters	1.8	2.0	2.2
Centrifuges; Conveyors: flight, screw; Hammer mills; Paper pulpers	1.9	2.1	2.3
Brick & clay pug mills; Fans, blowers, propeller mine fans, positive blowers	2.0	2.2	2.4

Synchronous belts will not slip, and therefore must be belted for the highest loadings anticipated in the system. A minimum service factor of 2.0 is recommended for equipment subject to chocking.

Variable Speed Belts: Variable speed belts are more sensitive to distortion than most other belts, and should not be hung from pins or racks but stored on shelves in the sleeves in which they are shipped.

SAE Standard V-Belts.—The data for V-belts and pulleys shown in Table 43 cover nine sizes, three of which—0.250, 0.315, and 0.440—were added in 1977 to conform to existing practice. This standard was reaffirmed in 1987.

Table 43. SAE V-Belt and Pulley Dimensions

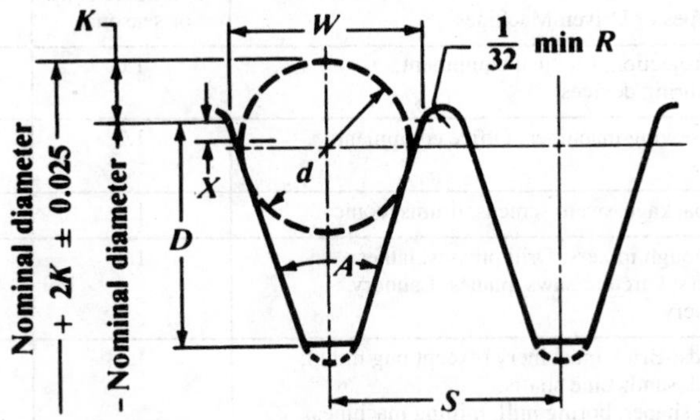

SAE Size	Recommended Min. Eff Dia[a]	*A* Groove Angle (deg) ±0.5	*W* Eff. Groove Width	*D* Groove Depth Min	*d* Ball or Rod Dia (±0.0005)	2*K* 2 × Ball Extension	2*X*[b]	*S* Groove[c] Spacing (±0.015)
0.250	2.25	36	0.248	0.276	0.2188	0.164	0.04	0.315
0.315	2.25	36	0.315	0.354	0.2812	0.222	0.05	0.413
0.380	2.40	36	0.380	0.433	0.3125	0.154	0.06	0.541
0.440	2.75	36	0.441	0.512	0.3750	0.231	0.07	0.591
0.500	3.00	36	0.500	0.551	0.4375	0.314	0.08	0.661
11/16	3.00	34	0.597	0.551	0.500	0.258	0.00	0.778
	Over 4.00	36				0.280		
	Over 6.00	38				0.302		
¾	3.00	34	0.660	0.630	0.5625	0.328	0.02	0.841
	Over 4.00	36				0.352		
	Over 6.00	38				0.374		
⅞	3.50	34	0.785	0.709	0.6875	0.472	0.04	0.966
	Over 4.50	36				0.496		
	Over 6.00	38				0.520		
1	4.00	34	0.910	0.827	0.8125	0.616	0.06	1.091
	Over 6.00	36				0.642		
	Over 8.00	38				0.666		

[a] Pulley effective diameters below those recommended should be used with caution, because power transmission and belt life may be reduced.

[b] The *X* dimension is radial; 2*X* is to be subtracted from the effective diameter to obtain "pitch diameter" for speed ratio calculations.

[c] These values are intended for adjacent grooves of the same effective width (*W*). Choice of pulley manufacture or belt design parameter may justify variance from these values. The *S* dimension should be the same on all multiple groove pulleys in a drive using matched belts.

All dimensions in inches.

V-belts are produced in a variety of constructions in a basic trapezoidal shape and are to be dimensioned in such a way that they are functional in pulleys dimensioned as described in the standard. Standard belt lengths are in increments of ½ inch up to and including 80 inches. Standard lengths above 80 inches up to and including 100 inches are in increments of 1 inch, without fractions. Standard belt length tolerances are based on the center distance and are as follows: For belt lengths of 50 inches or less, ± 0.12 inch; over 50 to 60 inches, inclusive, ± 0.16 inch; over 60 to 80 inches, inclusive, ± 0.19; and over 80 to 100 inches, inclusive, ± 0.22.

TRANSMISSION CHAINS

Types of Chains

In addition to the standard roller and inverted tooth types, a wide variety of drive chains of different construction is available. Such chains are manufactured to various degrees of precision ranging from unfinished castings or forgings to chains having certain machined parts. Practically all of these chains as well as standard roller chains can be equipped with attachments to fit them for conveyor use. A few such types are briefly described in the following paragraphs. Detailed information about them can be obtained from the manufacturers.

Types of chains.—*Detachable Chains:* The links of this type of chain, which are identical, are easily detachable. Each has a hook-shaped end in which the bar of the adjacent link articulates. These chains are available in malleable iron or pressed steel. The chief advantage is the ease with which any link can be removed.

Cast Roller Chains: Cast roller chains are constructed, wholly or partly, of cast metal parts and are available in various styles. In general the rollers and side bars are accurately made castings without machine finish. The links are usually connected by means of forged pins secured by nuts or cotters. Such chains are used for slow speeds and moderate loads, or where the precision of standard roller chains is not required.

Pintle Chains: Unlike the roller chain, the pintle chain is composed of hollow-cored cylinders cast or forged integrally with two offset side bars and each link identical. The links are joined by pins inserted in holes in the ends of the side bars and through the cored holes in the adjacent links. Lugs prevent turning of the pins in the side bars ensuring articulation of the chain between the pin and the cored cylinder.

Standard Roller Transmission Chains

A roller chain is made up of two kinds of links: roller links and pin links alternately spaced throughout the length of the chain as shown in Table 1.

Roller chains are manufactured in several types, each designed for the particular service required. All roller chains are so constructed that the rollers are evenly spaced throughout the chain. The outstanding advantage of this type of chain is the ability of the rollers to rotate when contacting the teeth of the sprocket. Two arrangements of roller chains are in common use: the single-strand type and the multiple-strand type. In the latter type, two or more chains are joined side by side by means of common pins which maintain the alignment of the rollers in the different strands.

Types of roller chains.—*Standard roller chains* are manufactured to the specifications in the American National Standard for precision power transmission roller chains, attachments, and sprockets ANSI/ASME B29.1M-1993 and, where indicated, the data in the subsequent tables have been taken from this standard. These roller chains and sprockets are commonly used for the transmission of power in industrial machinery, machine tools, motor trucks, motorcycles, tractors, and similar applications. In tabulating the dimensional information in ANSI/ASME B29.1M, customary inch-pound units were used. Metric (SI) units are given in separate tabulations in the Standard.

Nonstandard roller chains, developed individually by various manufacturers prior to the adoption of the ANSI standard, are similar in form and construction to standard roller chains but do not conform dimensionally to standard chains. Some sizes are still available from the originating manufacturers for replacement on existing equipment. They are not recommended for new installations, since their manufacture is being discontinued as rapidly as possible.

Table 1. ANSI Nomenclature for Roller Chain Parts ***ANSI/ASME B29.1M-1993***

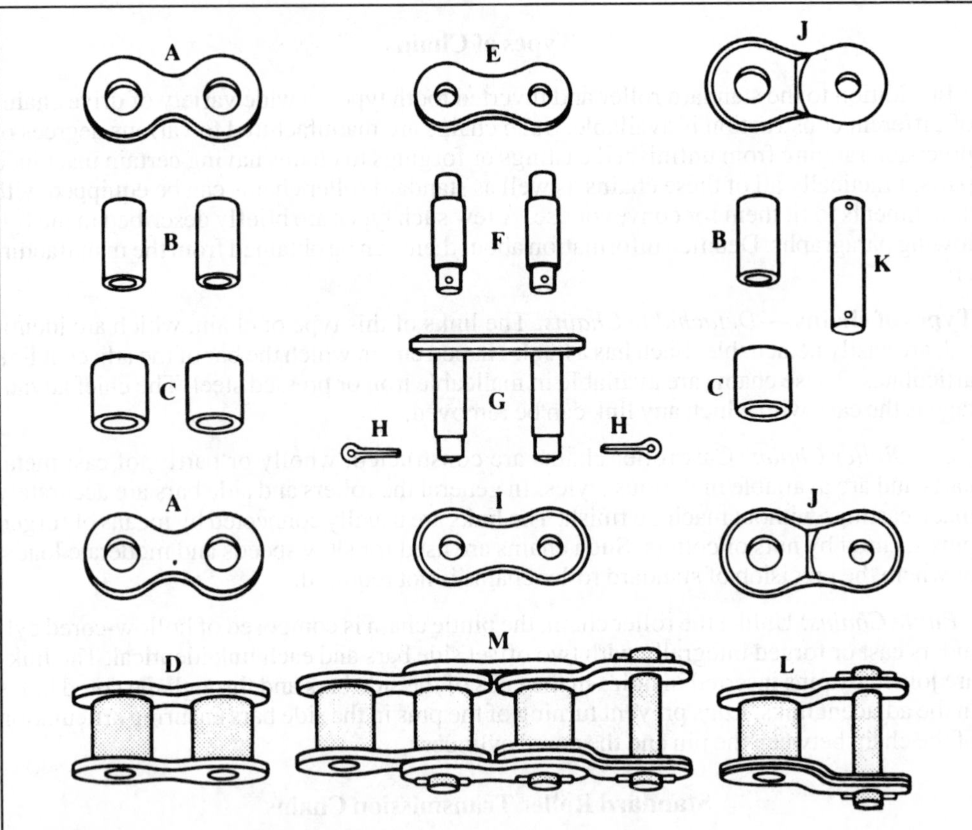

Roller Link D. — An inside link consisting of two inside plates, two bushings, and two rollers.

Pin Link G and E. — An outside link consisting of two pin-link plates assembled with two pins.

Inside Plate A. — One of the plates forming the tension members of a roller link.

Pin Link Plate E. — One of the plates forming the tension members of a pin link.

Pin F. — A stud articulating within a bushing of an inside link and secured at its ends by the pin-link plates.

Bushing B. — A cylindrical bearing in which the pin turns.

Roller C. — A ring or thimble which turns over a bushing.

Assembled Pins G. — Two pins assembled with one pin-link plate.

Connecting-Link G and I. — A pin link having one side plate detachable.

Connecting-Link Plate I. — The detachable pin-link plate belonging to a connecting link. It is retained by cotter pins or by a one-piece spring clip (not shown).

Connecting Link Assembly M. — A unit designed to connect two roller links.

Offset Link L. — A link consisting of two offset plates assembled with a bushing and roller at one end and an offset link pin at the other.

Offset Plate J. — One of the plates forming the tension members of the offset link.

Offset Link Pin K. — A pin used in offset links.

Standard double-pitch roller chains are like standard roller chains, except that their link plates have twice the pitch of the corresponding standard-pitch chain. Their design conforms to specifications in the ANSI Standard for double-pitch power transmission roller chains and sprockets ANSI/ASME B29.3M-1994. They are especially useful for low speeds, moderate loads, or long center distances.

Transmission Roller Chain

Standard Roller Chain Nomenclature, Dimensions and Loads.—Standard nomenclature for roller chain parts are given in Table 1. Dimensions for Standard Series roller chain are given in Table 2.

Table 2. ANSI Roller Chain Dimensions ***ASME/ANSI B29.1M-1986***

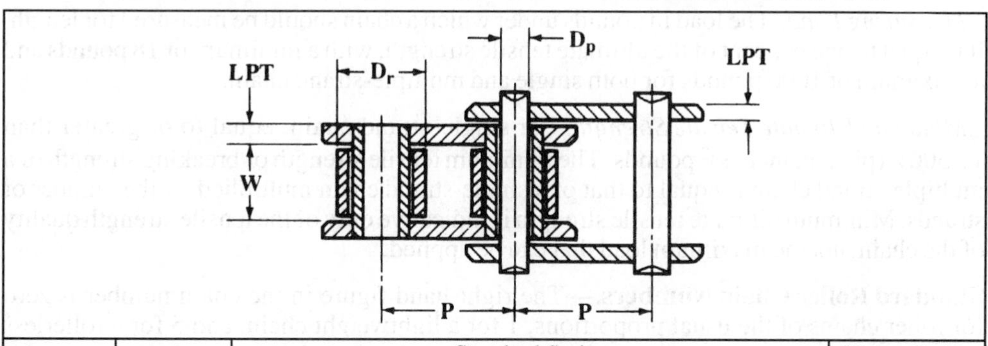

Pitch P	Max. Roller Diameter D_r	Standard Series					Heavy Series
		Standard Chain No.	Width W	Pin Diameter D_p	Thickness of Link Plates *LPT*	Measuring Load,† Lb.	Thickness of Link Plates *LPT*
0.250	[a]0.130	25	0.125	0.0905	0.030	18	…
0.375	[a]0.200	35	0.188	0.141	0.050	18	…
0.500	0.306	41	0.250	0.141	0.050	18	…
0.500	0.312	40	0.312	0.156	0.060	31	…
0.625	0.400	50	0.375	0.200	0.080	49	…
0.750	0.469	60	0.500	0.234	0.094	70	0.125
1.000	0.625	80	0.625	0.312	0.125	125	0.156
1.250	0.750	100	0.750	0.375	0.156	195	0.187
1.500	0.875	120	1.000	0.437	0.187	281	0.219
1.750	1.000	140	1.000	0.500	0.219	383	0.250
2.000	1.125	160	1.250	0.562	0.250	500	0.281
2.250	1.406	180	1.406	0.687	0.281	633	0.312
2.500	1.562	200	1.500	0.781	0.312	781	0.375
3.000	1.875	240	1.875	0.937	0.375	1000	0.500

[a] Bushing diameter. This size chain has no rollers.

All dimensions are in inches.

Roller Diameters D_r are approximately $\frac{5}{8}P$.

The *width W* is defined as the distance between the link plates. It is approximately $\frac{5}{8}$ of the chain pitch.

Pin Diameters D_p are approximately $\frac{5}{16}P$ or $\frac{1}{2}$ of the roller diameter.

Thickness LPT of Inside and Outside Link Plates for the standard series is approximately $\frac{1}{8}P$.

Thickness of Link Plates for the heavy series of any pitch is approximately that of the next larger pitch Standard Series chain.

Maximum Height of Roller Link Plates = 0.95 *P*.

Maximum Height of Pin Link Plates = 0.82 *P*.

Maximum Pin Diameter = nominal pin diameter + 0.0005 inch.

Minimum Hole in Bushing = nominal pin diameter + 0.0015 inch.

Maximum Width of Roller Link = nominal width of chain + (2.12 × nominal link plate thickness.)

Minimum Distance between Pin Link Plates = maximum width of roller link + 0.002 inch.

Chain Pitch: Distance in inches between centers of adjacent joint members. Other dimensions are proportional to the pitch.

Tolerances for Chain Length: New chains, under standard measuring load, must not be underlength. Overlength tolerance is 0.001/(pitch in inches)^2 + 0.015 inch per foot. Length measurements are to be taken over a length of at least 12 inches.

Measuring Load: The load in pounds under which a chain should be measured for length. It is equal to one per cent of the ultimate tensile strength, with a minimum of 18 pounds and a maximum of 1000 pounds for both single and multiple-strand chain.

Minimum Ultimate Tensile Strength: For single-strand chain, equal to or greater than 12,500 × (pitch in inches)^2 pounds. The minimum tensile strength or breaking strength of a multiple-strand chain is equal to that of a single-strand chain multiplied by the number of strands. Minimum ultimate tensile strength is indicative only of the tensile strength quality of the chain, not the maximum load that can be applied.

Standard Roller Chain Numbers.—The right-hand figure in the chain number is zero for roller chains of the usual proportions, 1 for a lightweight chain, and 5 for a rollerless bushing chain. The numbers to the left of the right-hand figure denote the number of ⅛ inches in the pitch. The letter *H* following the chain number denotes the heavy series; thus the number 80 *H* denotes a 1-inch pitch heavy chain. The hyphenated number 2 suffixed to the chain number denotes a double strand, 3 a triple strand, 4 a quadruple strand chain and so on.

Heavy Series: These chains, made in ¾-inch and larger pitches, have thicker link plates than those of the regular standard. Their value is only in the acceptance of higher loads at lower speeds.

Light-weight Machinery Chain: This chain is designated as No. 41. It is ½ inch pitch; ¼ inch wide; has 0.306-inch diameter rollers and a 0.141-inch pin diameter. The minimum ultimate tensile strength is 1500 pounds.

Multiple-strand Chain: This is essentially an assembly of two or more single-strand chains placed side by side with pins that extend through the entire width to maintain alignment of the different strands.

Types of Sprockets.—Four different designs or types of roller-chain sprockets are shown by the sectional views, Fig. 1. Type *A* is a plain plate; type *B* has a hub on one side only; type *C*, a hub on both sides; and type *D*, a detachable hub. Also used are shear pin and slip clutch sprockets designed to prevent damage to the drive or to other equipment caused by overloads or stalling.

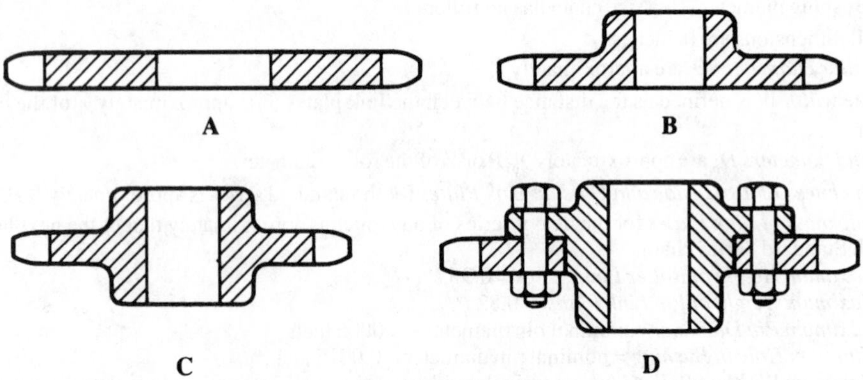

Fig. 1. Types of Sprockets

Attachments.—Modifications to standard chain components to adapt the chain for use in conveying, elevating, and timing operations are known as "attachments." The components commonly modified are: 1) the link plates, which are provided with extended lugs which may be straight or bent; and 2) the chain pins, which are extended in length so as to project substantially beyond the outer surface of the pin link plates.

Hole diameters, thicknesses, hole locations and offset dimensions for straight link and bent link plate extensions and lengths and diameters of extended pins are given in Table 3.

Table 3. Straight and Bent Link Plate Extensions and Extended Pin Dimensions *ANSI/ASME B29.1M-1993*

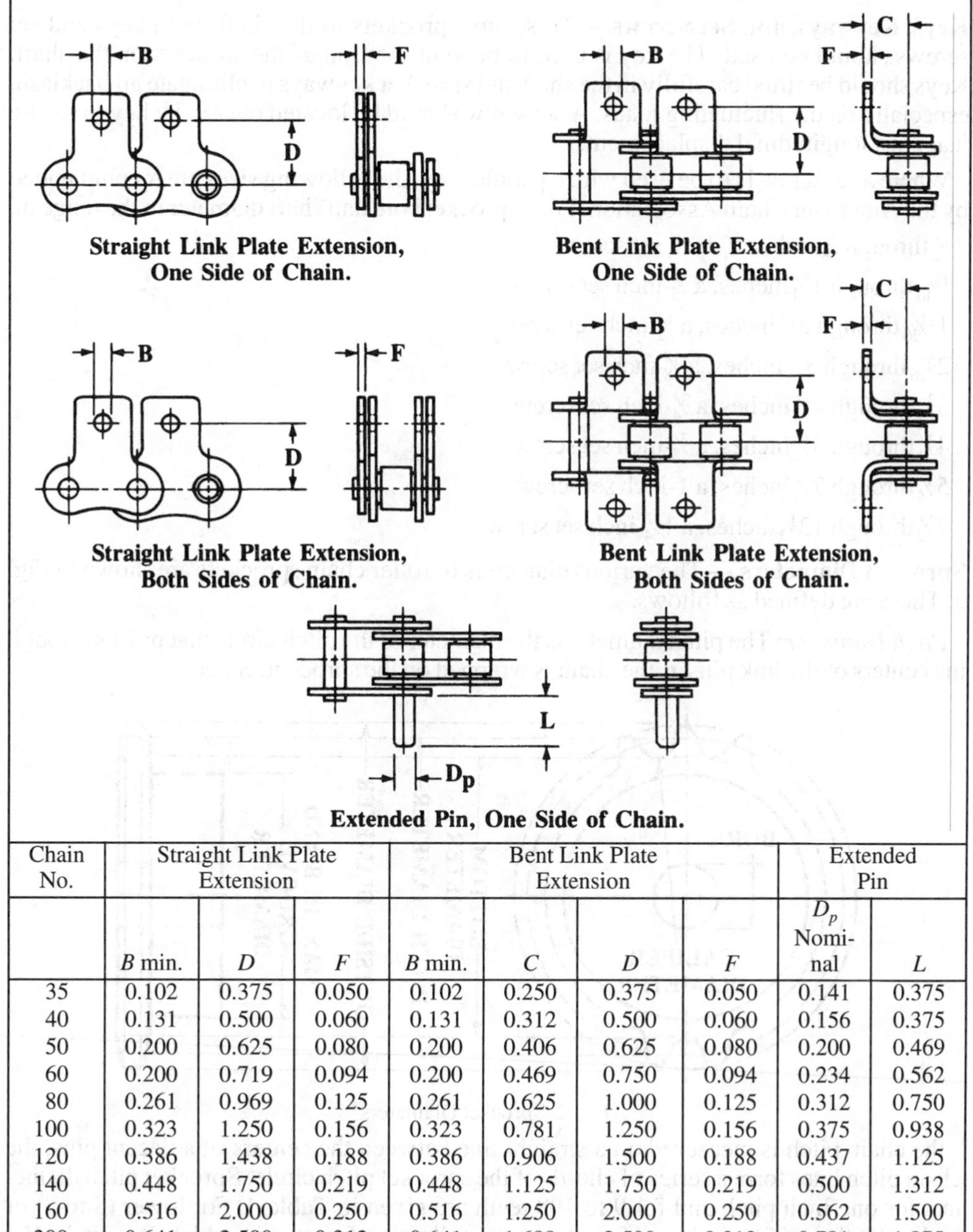

Straight Link Plate Extension, One Side of Chain.

Bent Link Plate Extension, One Side of Chain.

Straight Link Plate Extension, Both Sides of Chain.

Bent Link Plate Extension, Both Sides of Chain.

Extended Pin, One Side of Chain.

Chain No.	Straight Link Plate Extension			Bent Link Plate Extension				Extended Pin	
	B min.	*D*	*F*	*B* min.	*C*	*D*	*F*	D_p Nominal	*L*
35	0.102	0.375	0.050	0.102	0.250	0.375	0.050	0.141	0.375
40	0.131	0.500	0.060	0.131	0.312	0.500	0.060	0.156	0.375
50	0.200	0.625	0.080	0.200	0.406	0.625	0.080	0.200	0.469
60	0.200	0.719	0.094	0.200	0.469	0.750	0.094	0.234	0.562
80	0.261	0.969	0.125	0.261	0.625	1.000	0.125	0.312	0.750
100	0.323	1.250	0.156	0.323	0.781	1.250	0.156	0.375	0.938
120	0.386	1.438	0.188	0.386	0.906	1.500	0.188	0.437	1.125
140	0.448	1.750	0.219	0.448	1.125	1.750	0.219	0.500	1.312
160	0.516	2.000	0.250	0.516	1.250	2.000	0.250	0.562	1.500
200	0.641	2.500	0.312	0.641	1.688	2.500	0.312	0.781	1.875

All dimensions are in inches.

Sprocket Classes.—The American National Standard ANSI/ASME B29.1M-1993 provides for two classes of sprockets designated as Commercial and Precision. The selection of either is a matter of drive application judgment. The usual moderate to slow speed commercial drive is adequately served by Commercial sprockets. Where extreme high speed in combination with high load is involved, or where the drive involves fixed centers, critical timing, or register problems, or close clearance with outside interference, then the use of Precision sprockets may be more appropriate.

As a general guide, drives requiring Type A or Type B lubrication (see page 2443) would be served by Commercial sprockets. Drives requiring Type C lubrication may require Precision sprockets; the manufacturer should be consulted.

Keys, Keyways, and Set Screws.—To secure sprockets to the shaft, both keys and set screws should be used. The key is used to prevent rotation of the sprocket on the shaft. Keys should be fitted carefully in the shaft and sprocket keyways to eliminate all backlash, especially on the fluctuating loads. A set screw should be located over a flat key to secure it against longitudinal displacement.

Where a set screw is to be used with a parallel key, the following sizes are recommended by the American Chain Association. For a sprocket bore and shaft diameter in the range of

$\frac{1}{2}$ through $\frac{7}{8}$ inch, a $\frac{1}{4}$-inch set screw

$\frac{15}{16}$ through $1\frac{3}{4}$ inches, a $\frac{3}{8}$-inch set screw

$1\frac{13}{16}$ through $2\frac{1}{4}$ inches, a $\frac{1}{2}$-inch set screw

$2\frac{5}{16}$ through $3\frac{1}{4}$ inches, a $\frac{5}{8}$-inch set screw

$3\frac{3}{8}$ through $4\frac{1}{2}$ inches, a $\frac{3}{4}$-inch set screw

$4\frac{3}{4}$ through $5\frac{1}{2}$ inches, a $\frac{7}{8}$-inch set screw

$5\frac{3}{4}$ through $7\frac{3}{8}$ inches, a 1-inch set screw

$7\frac{1}{2}$ through $12\frac{1}{2}$ inches, a $1\frac{1}{4}$-inch set screw

Sprocket Diameters.—The various diameters of roller chain sprockets are shown in Fig. 2. These are defined as follows.

Pitch Diameter: The pitch diameter is the diameter of the pitch circle that passes through the centers of the link pins as the chain is wrapped on the sprocket. Since

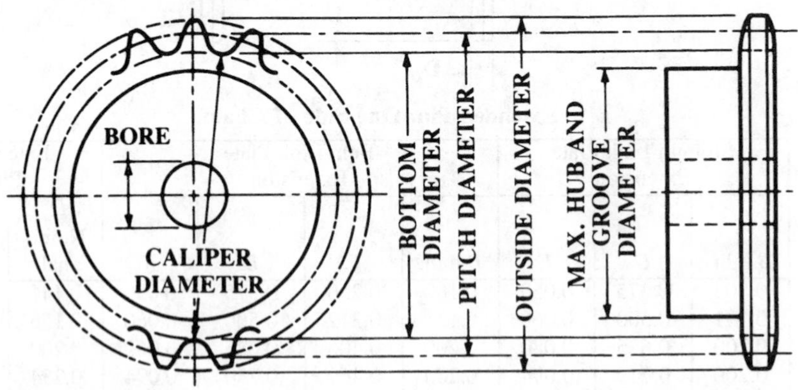

Fig. 2. Sprocket Diameters

the chain pitch is measured on a straight line between the centers of adjacent pins, the chain pitch lines form a series of chords of the sprocket pitch circle. Sprocket pitch diameters for one-inch pitch and for 9 to 108 teeth are given in Table 4. For lower (5 to 8) or higher (109 to 200) numbers of teeth use the following formula in which P = pitch, N = number of teeth: Pitch Diameter = $P \div \sin(180° \div N)$.

Table 4. ANSI Roller Chain Sprocket Diameters *ANSI/ASME B29.1M-1993*

These diameters and caliper factors apply only to chain of 1-inch pitch. For any other pitch, multiply the values given below by the pitch.

Caliper Dia. (even teeth) = Pitch Diameter − Roller Dia.

Caliper Dia. (odd teeth) = Caliper factor × Pitch − Roller Dia.

See Table 5 for tolerances on Caliper Diameters.

No. Teeth[a]	Pitch Diameter	Outside Diameter		Caliper Factor	No. Teeth[a]	Pitch Diameter	Outside Diameter		Caliper Factor
		Turned	Topping Hob Cut				Turned	Topping Hob Cut	
9	2.9238	3.348	3.364	2.8794	59	18.7892	19.363	19.361	18.7825
10	3.2361	3.678	3.676		60	19.1073	19.681	19.680	
11	3.5495	4.006	3.990	3.5133	61	19.4255	20.000	19.998	19.4190
12	3.8637	4.332	4.352		62	19.7437	20.318	20.316	
13	4.1786	4.657	4.666	4.1481	63	20.0618	20.637	20.634	20.0556
14	4.4940	4.981	4.982		64	20.3800	20.956	20.952	
15	4.8097	5.304	5.298	4.7834	65	20.6982	21.274	21.270	20.6921
16	5.1258	5.627	5.614		66	21.0164	21.593	21.588	
17	5.4422	5.949	5.930	5.4190	67	21.3346	21.911	21.907	21.3287
18	5.7588	6.271	6.292		68	21.6528	22.230	22.225	
19	6.0755	6.593	6.609	6.0548	69	21.9710	22.548	22.543	21.9653
20	6.3924	6.914	6.926		70	22.2892	22.867	22.861	
21	6.7095	7.235	7.243	6.6907	71	22.6074	23.185	23.179	22.6018
22	7.0267	7.555	7.560		72	22.9256	23.504	23.498	
23	7.3439	7.876	7.877	7.3268	73	23.2438	23.822	23.816	23.2384
24	7.6613	8.196	8.195		74	23.5620	24.141	24.134	
25	7.9787	8.516	8.512	7.9630	75	23.8802	24.459	24.452	23.8750
26	8.2962	8.836	8.829		76	24.1984	24.778	24.770	
27	8.6138	9.156	9.147	8.5992	77	24.5166	25.096	25.089	24.5116
28	8.9314	9.475	9.465		78	24.8349	25.415	25.407	
29	9.2491	9.795	9.782	9.2355	79	25.1531	25.733	25.725	25.1481
30	9.5668	10.114	10.100		80	25.4713	26.052	26.043	
31	9.8845	10.434	10.418	9.8718	81	25.7896	26.370	26.362	25.7847
32	10.2023	10.753	10.736		82	26.1078	26.689	26.680	
33	10.5201	11.073	11.053	10.5082	83	26.4260	27.007	26.998	26.4213
34	10.8379	11.392	11.371		84	26.7443	27.326	27.316	
35	11.1558	11.711	11.728	11.1446	85	27.0625	27.644	27.635	27.0579
36	11.4737	12.030	12.046		86	27.3807	27.962	27.953	
37	11.7916	12.349	12.364	11.7810	87	27.6990	28.281	28.271	27.6945
38	12.1095	12.668	12.682		88	28.0172	28.599	28.589	
39	12.4275	12.987	13.000	12.4174	89	28.3354	28.918	28.907	28.3310
40	12.7455	13.306	13.318		90	28.6537	29.236	29.226	
41	13.0635	13.625	13.636	13.0539	91	28.9719	29.555	29.544	28.9676
42	13.3815	13.944	13.954		92	29.2902	29.873	29.862	
43	13.6995	14.263	14.272	13.6904	93	29.6084	30.192	30.180	29.6042
44	14.0175	14.582	14.590		94	29.9267	30.510	30.499	
45	14.3355	14.901	14.908	14.3269	95	30.2449	30.828	30.817	30.2408
46	14.6535	15.219	15.226		96	30.5632	31.147	31.135	
47	14.9717	15.538	15.544	14.9634	97	30.8815	31.465	31.454	30.8774
48	15.2898	15.857	15.862		98	31.1997	31.784	31.772	
49	15.6079	16.176	16.180	15.5999	99	31.5180	32.102	32.090	31.5140
50	15.9260	16.495	16.498		100	31.8362	32.421	32.408	
51	16.2441	16.813	16.816	16.2364	101	32.1545	32.739	32.727	32.1506
52	16.5622	17.132	17.134		102	32.4727	33.057	33.045	
53	16.8803	17.451	17.452	16.8729	103	32.7910	33.376	33.363	32.7872
54	17.1984	17.769	17.770		104	33.1093	33.694	33.681	
55	17.5165	18.088	18.089	17.5094	105	33.4275	34.013	34.000	33.4238
56	17.8347	18.407	18.407		106	33.7458	34.331	34.318	
57	18.1528	18.725	18.725	18.1459	107	34.0641	34.649	34.636	34.0604
58	18.4710	19.044	19.043		108	34.3823	34.968	34.954	

[a] For 5 – 8 and 109–200 teeth see text, pages 2426, 2428.

Bottom Diameter: The bottom diameter is the diameter of a circle tangent to the curve (called the seating curve) at the bottom of the tooth gap. It equals the pitch diameter minus the diameter of the roller.

Caliper Diameter: The caliper diameter is the same as the bottom diameter for a sprocket with an even number of teeth. For a sprocket with an odd number of teeth, it is defined as the distance from the bottom of one tooth gap to that of the nearest opposite tooth gap. The caliper diameter for an even tooth sprocket is equal to pitch diameter–roller diameter. The caliper diameter for an odd tooth sprocket is equal to caliper factor–roller diameter. Here, the caliper factor = $PD[\cos(90° \div N)]$, where PD = pitch diameter and N = number of teeth. Caliper factors for 1-in. pitch and sprockets having 9–108 teeth are given in Table 4. For other tooth numbers use above formula. Caliper diameter tolerances are minus only and are equal to $0.002P\sqrt{N} + 0.006$ inch for the Commercial sprockets and $0.001P\sqrt{N} + 0.003$ inch for Precision sprockets. Tolerances are given in Table 5.

Table 5. Minus Tolerances on the Caliper Diameters of Precision Sprockets
ANSI/ASME B29.1M-1993

	Number of Teeth				
Pitch	Up to 15	16–24	25–35	36–48	49–63
0.250	0.004	0.004	0.004	0.005	0.005
0.375	0.004	0.004	0.004	0.005	0.005
0.500	0.004	0.005	0.0055	0.006	0.0065
0.625	0.005	0.0055	0.006	0.007	0.008
0.750	0.005	0.006	0.007	0.008	0.009
1.000	0.006	0.007	0.008	0.009	0.010
1.250	0.007	0.008	0.009	0.010	0.012
1.500	0.007	0.009	0.0105	0.012	0.013
1.750	0.008	0.010	0.012	0.013	0.015
2.000	0.009	0.011	0.013	0.015	0.017
2.250	0.010	0.012	0.014	0.016	0.018
2.500	0.010	0.013	0.015	0.018	0.020
3.000	0.012	0.015	0.018	0.021	0.024
	Number of Teeth				
Pitch	64–80	81–99	100–120	121–143	144 up
0.250	0.005	0.005	0.006	0.006	0.006
0.375	0.006	0.006	0.006	0.007	0.007
0.500	0.007	0.0075	0.008	0.0085	0.009
0.625	0.009	0.009	0.009	0.010	0.011
0.750	0.010	0.010	0.011	0.012	0.013
1.000	0.011	0.012	0.013	0.014	0.015
1.250	0.013	0.014	0.016	0.017	0.018
1.500	0.015	0.016	0.018	0.019	0.021
1.750	0.017	0.019	0.020	0.022	0.024
2.000	0.019	0.021	0.023	0.025	0.027
2.250	0.021	0.023	0.025	0.028	0.030
2.500	0.023	0.025	0.028	0.030	0.033
3.000	0.027	0.030	0.033	0.036	0.039

Minus tolerances for Commercial sprockets are twice those shown in this table.

Outside Diameter: OD is the diameter over the tips of teeth. Sprocket ODs for 1-in. pitch and 9–108 teeth are given in Table 4. For other tooth numbers the OD may be determined by the following formulas in which O = approximate OD; P = pitch of chain; N = number of sprocket teeth: $O = P\,[0.6 + \cot(180° \div N)]$, for turned sprocket; O = pitch diameter − roller diameter + 2 × whole depth of topping hob cut, for topping hob cut sprocket.*

Table 6. American National Standard Roller Chain Sprocket Flange Thickness and Tooth Section Profile Dimension *ANSI/ASME B29.1M-1993*

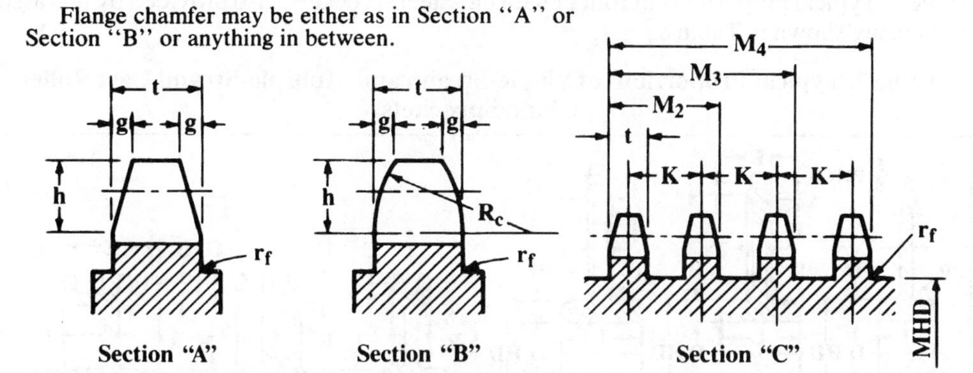

Flange chamfer may be either as in Section "A" or Section "B" or anything in between.

Section "A" Section "B" Section "C"

Sprocket Flange Thickness											
		Maximum Sprocket Flange Thickness, *t*			Minus Tolerance on *t*		Tolerance on *M*		Max. Variation of *t* on Each Flange	
Std. Chain No.	Width of Chain, *W*	Single	Double & Triple	Quad. & Over	Commercial	Precision	Commercial Plus or Minus	Precision Minus Only	Commercial	Precision
25	0.125	0.110	0.106	0.096	0.021	0.007	0.007	0.007	0.021	0.004
35	0.188	0.169	0.163	0.150	0.027	0.008	0.008	0.008	0.027	0.004
41	0.250	0.226	…	…	0.032	0.009	…	…	0.032	0.004
40	0.312	0.284	0.275	0.256	0.035	0.009	0.009	0.009	0.035	0.004
50	0.375	0.343	0.332	0.310	0.036	0.010	0.010	0.010	0.036	0.005
60	0.500	0.459	0.444	0.418	0.036	0.011	0.011	0.011	0.036	0.006
80	0.625	0.575	0.556	0.526	0.040	0.012	0.012	0.012	0.040	0.006
100	0.750	0.0692	0.669	0.633	0.046	0.014	0.014	0.014	0.046	0.007
120	1.000	0.924	0.894	0.848	0.057	0.016	0.016	0.016	0.057	0.008
140	1.000	0.924	0.894	0.848	0.057	0.016	0.016	0.016	0.057	0.008
160	1.250	1.156	1.119	1.063	0.062	0.018	0.018	0.018	0.062	0.009
180	1.406	1.302	1.259	1.198	0.068	0.020	0.020	0.020	0.068	0.010
200	1.500	1.389	1.344	1.278	0.072	0.021	0.021	0.021	0.072	0.010
240	1.875	1.738	1.682	1.602	0.087	0.025	0.025	0.025	0.087	0.012

Sprocket Tooth Section Profile Dimensions						
					Transverse Pitch *K*	
Std. Chain No.	Chain Pitch *P*	Depth of Chamfer *h*	Width of Chamfer *g*	Minimum Radius R_c	Standard Series	Heavy Series
25	0.250	0.125	0.031	0.265	0.252	…
35	0.375	0.188	0.047	0.398	0.399	…
41	0.500	0.250	0.062	0.531	…	…
40	0.500	0.250	0.062	0.531	0.566	…
50	0.625	0.312	0.078	0.664	0.713	…
60	0.750	0.375	0.094	0.796	0.897	1.028
80	1.000	0.500	0.125	1.062	1.153	1.283
100	1.250	0.625	0.156	1.327	1.408	1.539
120	1.500	0.750	0.188	1.593	1.789	1.924
140	1.750	0.875	0.219	1.858	1.924	2.055
160	2.000	1.000	0.250	2.124	2.305	2.437
180	2.250	1.125	0.281	2.392	2.592	2.723
200	2.500	1.250	0.312	2.654	2.817	3.083
240	3.000	1.500	0.375	3.187	3.458	3.985

All dimensions are in inches. $r_f$max = 0.04 *P* for max. hub diameter.

* This dimension was added in 1984 as a desirable goal for the future. It should in no way obsolete existing tools or sprockets. The whole depth *WD* is found from the formula: $WD = \frac{1}{2}D_r + P[0.3 - \frac{1}{2}\tan(90 \text{ deg} \div N_a)]$, where N_a is the intermediate number of teeth for the topping hob. For teeth range 5, N_a = 5; 6, 6; 7–8, 7.47; 9–11, 9.9; 12–17, 14.07; 18–34, 23.54; 35 and over, 56.

Proportions of Sprockets.—Typical proportions of single-strand and multiple-strand cast roller chain sprockets, as provided by the American Chain Association, are shown in Table 7. Typical proportions of roller chain bar-steel sprockets, also provided by this association, are shown in Table 8.

Table 7. Typical Proportions of Single-Strand and Multiple-Strand Cast Roller Chain Sprockets

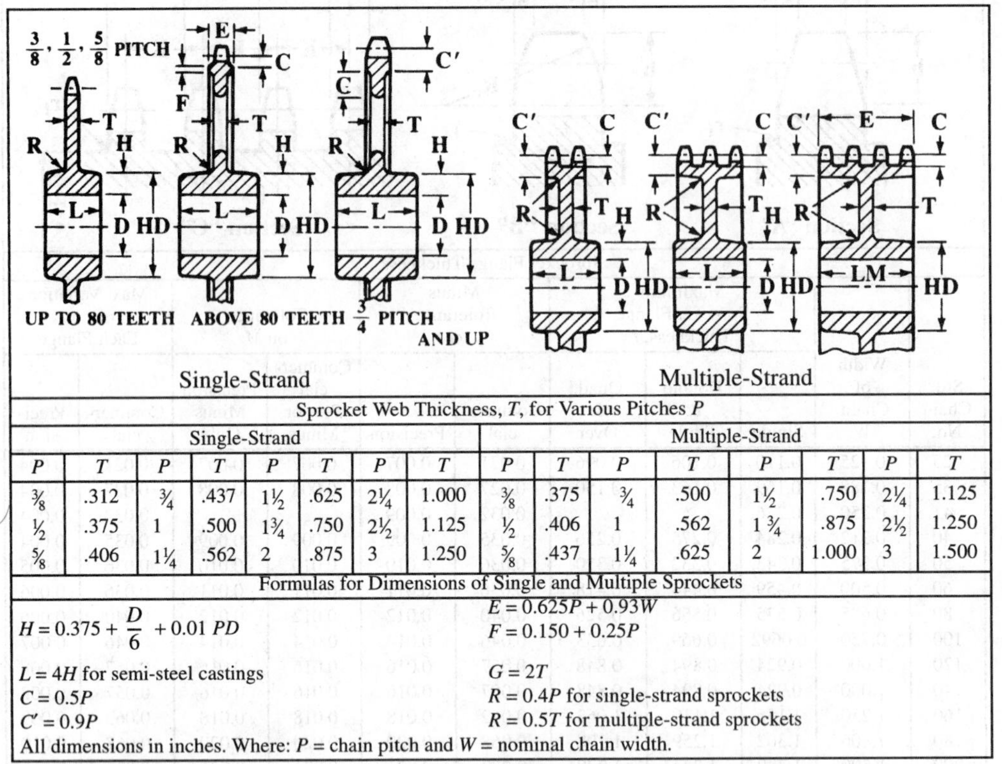

Sprocket Web Thickness, T, for Various Pitches P															
Single-Strand								Multiple-Strand							
P	T	P	T	P	T	P	T	P	T	P	T	P	T	P	T
3/8	.312	3/4	.437	1½	.625	2¼	1.000	3/8	.375	3/4	.500	1½	.750	2¼	1.125
1/2	.375	1	.500	1¾	.750	2½	1.125	1/2	.406	1	.562	1¾	.875	2½	1.250
5/8	.406	1¼	.562	2	.875	3	1.250	5/8	.437	1¼	.625	2	1.000	3	1.500

Formulas for Dimensions of Single and Multiple Sprockets

$H = 0.375 + \frac{D}{6} + 0.01\ PD$

$E = 0.625P + 0.93W$

$F = 0.150 + 0.25P$

$L = 4H$ for semi-steel castings

$G = 2T$

$C = 0.5P$

$R = 0.4P$ for single-strand sprockets

$C' = 0.9P$

$R = 0.5T$ for multiple-strand sprockets

All dimensions in inches. Where: P = chain pitch and W = nominal chain width.

Table 8. Typical Proportions of Roller Chain Bar-steel Sprockets

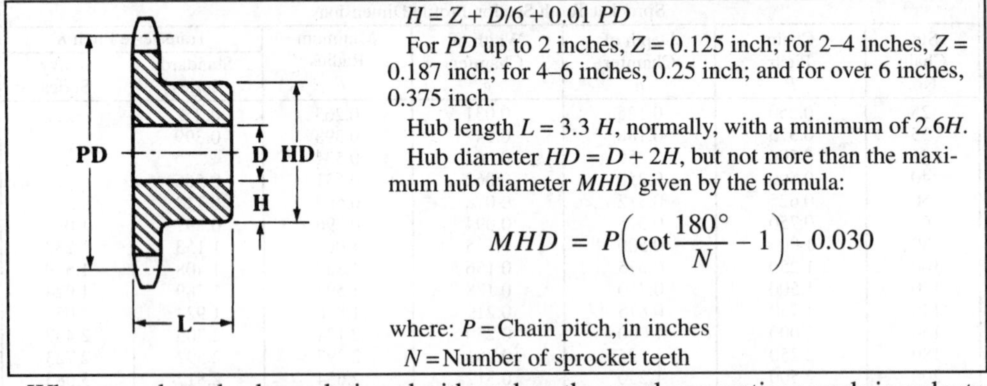

$H = Z + D/6 + 0.01\ PD$

For PD up to 2 inches, Z = 0.125 inch; for 2–4 inches, Z = 0.187 inch; for 4–6 inches, 0.25 inch; and for over 6 inches, 0.375 inch.

Hub length $L = 3.3\ H$, normally, with a minimum of $2.6H$.

Hub diameter $HD = D + 2H$, but not more than the maximum hub diameter MHD given by the formula:

$$MHD = P\left(\cot\frac{180°}{N} - 1\right) - 0.030$$

where: P = Chain pitch, in inches

N = Number of sprocket teeth

When sprocket wheels are designed with spokes, the usual assumptions made in order to determine suitable proportions are as follows: 1) That the maximum torque load acting on a sprocket is the chain tensile strength times the sprocket pitch radius; 2) That the torque load is equally divided between the arms by the rim; and 3) That each arm acts as a cantilever beam.

The arms are generally elliptical in cross section, with the major axis twice the minor axis.

Selection of Chain and Sprockets.—The smallest applicable pitch of roller chain is desirable for quiet operation and high speed. The horsepower capacity varies with the chain pitch as shown in Table . However, short pitch with high working load can often be obtained by the use of multiple-strand chain.

The small sprocket selected must be large enough to accommodate the shaft. Table 10 gives maximum bore and hub diameters consistent with commercial practice for sprockets with up to 25 teeth.

After selecting the small sprocket, the number of teeth in the larger sprocket is determined by the desired ratio of the shaft speed. Overemphasis on the exactness in the speed ratio may result in a cumbersome and expensive installation. In most cases, satisfactory operation can be obtained with a minor change in speed of one or both shafts.

Table 9. Horsepower Ratings for Roller Chain–1986

To properly use this table the following factors must be taken into consideration:
1) Service factors
2) Multiple Strand Factors
3) Lubrication

Service Factors: See Table 15.

Multiple Strand Factors: For two strands, the multiple strand factor is 1.7; for three strands, it is 2.5; and for four strands, it is 3.3.

Lubrication:

Required type of lubrication is indicated at the bottom of each roller chain size section of the table. For a description of each type of lubrication, see page 2443.

Type A — Manual or Drip Lubrication
Type B — Bath or Disc Lubrication
Type C — Oil Stream Lubrication

To find the required horsepower table rating, use the following formula:

$$\text{Required hp Table Rating} = \frac{\text{hp to be Transmitted} \times \text{Service Factor}}{\text{Multiple-Strand Factor}}$$

	No. of Teeth Small Spkt.	Revolutions per Minute — Small Sprocket[a]												
		50	100	300	500	700	900	1200	1500	1800	2100	2500	3000	3500
		Horsepower Rating												
1/4-inch Pitch Standard Single-Strand Roller Chain — No. 25	11	0.03	0.05	0.14	0.23	0.31	0.39	0.50	0.62	0.73	0.83	0.98	1.15	1.32
	12	0.03	0.06	0.16	0.25	0.34	0.43	0.55	0.68	0.80	0.92	1.07	1.26	1.45
	13	0.04	0.06	0.17	0.27	0.37	0.47	0.60	0.74	0.87	1.00	1.17	1.38	1.58
	14	0.04	0.07	0.19	0.30	0.40	0.50	0.65	0.80	0.94	1.08	1.27	1.49	1.71
	15	0.04	0.08	0.20	0.32	0.43	0.54	0.70	0.86	1.01	1.17	1.36	1.61	1.85
	16	0.04	0.08	0.22	0.34	0.47	0.58	0.76	0.92	1.09	1.25	1.46	1.72	1.98
	17	0.05	0.09	0.23	0.37	0.50	0.62	0.81	0.99	1.16	1.33	1.56	1.84	2.11
	18	0.05	0.09	0.25	0.39	0.53	0.66	0.86	1.05	1.24	1.42	1.66	1.96	2.25
	19	0.05	0.10	0.26	0.41	0.56	0.70	0.91	1.11	1.31	1.50	1.76	2.07	2.38
	20	0.06	0.10	0.28	0.44	0.59	0.74	0.96	1.17	1.38	1.59	1.86	2.19	2.52
	21	0.06	0.11	0.29	0.46	0.62	0.78	1.01	1.24	1.46	1.68	1.96	2.31	2.66
	22	0.06	0.11	0.31	0.48	0.66	0.82	1.07	1.30	1.53	1.76	2.06	2.43	2.79
	23	0.06	0.12	0.32	0.51	0.69	0.86	1.12	1.37	1.61	1.85	2.16	2.55	2.93
	24	0.07	0.13	0.34	0.53	0.72	0.90	1.17	1.43	1.69	1.94	2.27	2.67	3.07
	25	0.07	0.13	0.35	0.56	0.75	0.94	1.22	1.50	1.76	2.02	2.37	2.79	3.21
	26	0.07	0.14	0.37	0.58	0.79	0.98	1.28	1.56	1.84	2.11	2.47	2.91	3.34
	28	0.08	0.15	0.40	0.63	0.85	1.07	1.38	1.69	1.99	2.29	2.68	3.15	3.62
	30	0.08	0.16	0.43	0.68	0.92	1.15	1.49	1.82	2.15	2.46	2.88	3.40	3.90
	32	0.09	0.17	0.46	0.73	0.98	1.23	1.60	1.95	2.30	2.64	3.09	3.64	4.18
	35	0.10	0.19	0.51	0.80	1.08	1.36	1.76	2.15	2.53	2.91	3.41	4.01	4.61
	40	0.12	0.22	0.58	0.92	1.25	1.57	2.03	2.48	2.93	3.36	3.93	4.64	5.32
	45	0.13	0.25	0.66	1.05	1.42	1.78	2.31	2.82	3.32	3.82	4.47	5.26	6.05
		Type A				Type B								

Table 9. *(Continued)* **Horsepower Ratings for Roller Chain–1986**

$\frac{3}{8}$-inch Pitch Standard Single-Strand Roller Chain — No. 35

No. of Teeth Small Spkt.	Revolutions per Minute — Small Sprocket[a]												
	50	100	300	500	700	900	1200	1500	1800	2100	2500	3000	3500
	Horsepower Rating												
11	0.10	0.18	0.49	0.77	1.05	1.31	1.70	2.08	2.45	2.82	3.30	2.94	2.33
12	0.11	0.20	0.54	0.85	1.15	1.44	1.87	2.29	2.70	3.10	3.62	3.35	2.66
13	0.12	0.22	0.59	0.93	1.26	1.57	2.04	2.49	2.94	3.38	3.95	3.77	3.00
14	0.13	0.24	0.63	1.01	1.36	1.71	2.21	2.70	3.18	3.66	4.28	4.22	3.35
15	0.14	0.25	0.68	1.08	1.47	1.84	2.38	2.91	3.43	3.94	4.61	4.68	3.71
16	0.15	0.27	0.73	1.16	1.57	1.97	2.55	3.12	3.68	4.22	4.94	5.15	4.09
17	0.16	0.29	0.78	1.24	1.68	2.10	2.73	3.33	3.93	4.51	5.28	5.64	4.48
18	0.17	0.31	0.83	1.32	1.78	2.24	2.90	3.54	4.18	4.80	5.61	6.15	4.88
19	0.18	0.33	0.88	1.40	1.89	2.37	3.07	3.76	4.43	5.09	5.95	6.67	5.29
20	0.19	0.35	0.93	1.48	2.00	2.51	3.25	3.97	4.68	5.38	6.29	7.20	5.72
21	0.20	0.37	0.98	1.56	2.11	2.64	3.42	4.19	4.93	5.67	6.63	7.75	6.15
22	0.21	0.38	1.03	1.64	2.22	2.78	3.60	4.40	5.19	5.96	6.97	8.21	6.59
23	0.22	0.40	1.08	1.72	2.33	2.92	3.78	4.62	5.44	6.25	7.31	8.62	7.05
24	0.23	0.42	1.14	1.80	2.44	3.05	3.96	4.84	5.70	6.55	7.66	9.02	7.51
25	0.24	0.44	1.19	1.88	2.55	3.19	4.13	5.05	5.95	6.84	8.00	9.43	7.99
26	0.25	0.46	1.24	1.96	2.66	3.33	4.31	5.27	6.21	7.14	8.35	9.84	8.47
28	0.27	0.50	1.34	2.12	2.88	3.61	4.67	5.71	6.73	7.73	9.05	10.7	9.47
30	0.29	0.54	1.45	2.29	3.10	3.89	5.03	6.15	7.25	8.33	9.74	11.5	10.5
32	0.31	0.58	1.55	2.45	3.32	4.17	5.40	6.60	7.77	8.93	10.4	12.3	11.6
35	0.34	0.64	1.71	2.70	3.66	4.59	5.95	7.27	8.56	9.84	11.5	13.6	13.2
40	0.39	0.73	1.97	3.12	4.23	5.30	6.87	8.40	9.89	11.4	13.3	15.7	16.2
45	0.45	0.83	2.24	3.55	4.80	6.02	7.80	9.53	11.2	12.9	15.1	17.8	19.3
	Type A				Type B						Type C		

$\frac{1}{2}$-inch Pitch Standard Single-Strand Roller Chain — No. 40

No. of Teeth Small Spkt.	Revolutions per Minute — Small Sprocket[a]												
	50	100	200	300	400	500	700	900	1000	1200	1400	1600	1800
	Horsepower Rating												
11	0.23	0.43	0.80	1.16	1.50	1.83	2.48	3.11	3.42	4.03	4.63	5.22	4.66
12	0.25	0.47	0.88	1.27	1.65	2.01	2.73	3.42	3.76	4.43	5.09	5.74	5.31
13	0.28	0.52	0.96	1.39	1.80	2.20	2.97	3.73	4.10	4.83	5.55	6.26	5.99
14	0.30	0.56	1.04	1.50	1.95	2.38	3.22	4.04	4.44	5.23	6.01	6.78	6.70
15	0.32	0.60	1.12	1.62	2.10	2.56	3.47	4.35	4.78	5.64	6.47	7.30	7.43
16	0.35	0.65	1.20	1.74	2.25	2.75	3.72	4.66	5.13	6.04	6.94	7.83	8.18
17	0.37	0.69	1.29	1.85	2.40	2.93	3.97	4.98	5.48	6.45	7.41	8.36	8.96
18	0.39	0.73	1.37	1.97	2.55	3.12	4.22	5.30	5.82	6.86	7.88	8.89	9.76
19	0.42	0.78	1.45	2.09	2.71	3.31	4.48	5.62	6.17	7.27	8.36	9.42	10.5
20	0.44	0.82	1.53	2.21	2.86	3.50	4.73	5.94	6.53	7.69	8.83	9.96	11.1
21	0.46	0.87	1.62	2.33	3.02	3.69	4.99	6.26	6.88	8.11	9.31	10.5	11.7
22	0.49	0.91	1.70	2.45	3.17	3.88	5.25	6.58	7.23	8.52	9.79	11.0	12.3
23	0.51	0.96	1.78	2.57	3.33	4.07	5.51	6.90	7.59	8.94	10.3	11.6	12.9
24	0.54	1.00	1.87	2.69	3.48	4.26	5.76	7.23	7.95	9.36	10.8	12.1	13.5
25	0.56	1.05	1.95	2.81	3.64	4.45	6.02	7.55	8.30	9.78	11.2	12.7	14.1
26	0.58	1.09	2.04	2.93	3.80	4.64	6.28	7.88	8.66	10.2	11.7	13.2	14.7
28	0.63	1.18	2.20	3.18	4.11	5.03	6.81	8.54	9.39	11.1	12.7	14.3	15.9
30	0.68	1.27	2.38	3.42	4.43	5.42	7.33	9.20	10.1	11.9	13.7	15.4	17.2
32	0.73	1.36	2.55	3.67	4.75	5.81	7.86	9.86	10.8	12.8	14.7	16.5	18.4
35	0.81	1.50	2.81	4.04	5.24	6.40	8.66	10.9	11.9	14.1	16.2	18.2	20.3
40	0.93	1.74	3.24	4.67	6.05	7.39	10.0	12.5	13.8	16.3	18.7	21.1	23.4
45	1.06	1.97	3.68	5.30	6.87	8.40	11.4	14.2	15.7	18.5	21.2	23.9	26.6
	Type A					Type B						Type C	

Table 9. *(Continued)* **Horsepower Ratings for Roller Chain–1986**

½-inch Pitch Light Weight Machinery Roller Chain — No. 41

No. of Teeth Small Spkt.	Revolutions per Minute — Small Sprocket[a]												
	10	25	50	100	200	300	400	500	700	900	1000	1200	1400
	Horsepower Rating												
11	0.03	0.07	0.13	0.24	0.44	0.64	0.82	1.01	1.37	1.71	1.88	1.71	1.36
12	0.03	0.07	0.14	0.26	0.49	0.70	0.91	1.11	1.50	1.88	2.07	1.95	1.55
13	0.04	0.08	0.15	0.28	0.53	0.76	0.99	1.21	1.63	2.05	2.25	2.20	1.75
14	0.04	0.09	0.16	0.31	0.57	0.83	1.07	1.31	1.77	2.22	2.44	2.46	1.95
15	0.04	0.09	0.18	0.33	0.62	0.89	1.15	1.41	1.91	2.39	2.63	2.73	2.17
16	0.04	0.10	0.19	0.36	0.66	0.95	1.24	1.51	2.05	2.57	2.82	3.01	2.39
17	0.05	0.11	0.20	0.38	0.71	1.02	1.32	1.61	2.18	2.74	3.01	3.29	2.61
18	0.05	0.12	0.22	0.40	0.75	1.08	1.40	1.72	2.32	2.91	3.20	3.59	2.85
19	0.05	0.12	0.23	0.43	0.80	1.15	1.49	1.82	2.46	3.09	3.40	3.89	3.09
20	0.06	0.13	0.24	0.45	0.84	1.21	1.57	1.92	2.60	3.26	3.59	4.20	3.33
21	0.06	0.14	0.26	0.48	0.89	1.28	1.66	2.03	2.74	3.44	3.78	4.46	3.59
22	0.06	0.14	0.27	0.50	0.93	1.35	1.74	2.13	2.89	3.62	3.98	4.69	3.85
23	0.06	0.15	0.28	0.53	0.98	1.41	1.83	2.24	3.03	3.80	4.17	4.92	4.11
24	0.07	0.16	0.29	0.55	1.03	1.48	1.92	2.34	3.17	3.97	4.37	5.15	4.38
25	0.07	0.17	0.31	0.57	1.07	1.55	2.00	2.45	3.31	4.15	4.57	5.38	4.66
26	0.07	0.17	0.32	0.60	1.12	1.61	2.09	2.55	3.46	4.33	4.76	5.61	4.94
28	0.08	0.19	0.35	0.65	1.21	1.75	2.26	2.77	3.74	4.69	5.16	6.08	5.52
30	0.08	0.20	0.38	0.70	1.31	1.88	2.44	2.98	4.03	5.06	5.56	6.55	6.13
32	0.09	0.22	0.40	0.75	1.40	2.02	2.61	3.20	4.33	5.42	5.96	7.03	6.75
35	0.10	0.24	0.44	0.83	1.54	2.22	2.88	3.52	4.76	5.97	6.57	7.74	7.72
40	0.12	0.27	0.51	0.96	1.78	2.57	3.33	4.07	5.50	6.90	7.59	8.94	9.43
45	0.14	0.31	0.58	1.08	2.02	2.92	3.78	4.62	6.25	7.84	8.62	10.2	11.3
	Type A				Type B							Type C	

⅝-inch Pitch Standard Single-Strand Roller Chain — No. 50

No. of Teeth Small Spkt.	Revolutions per Minute — Small Sprocket[a]												
	25	50	100	200	300	400	500	700	900	1000	1200	1400	1600
	Horsepower Rating												
11	0.24	0.45	0.84	1.56	2.25	2.92	3.57	4.83	6.06	6.66	7.85	8.13	6.65
12	0.26	0.49	0.92	1.72	2.47	3.21	3.92	5.31	6.65	7.31	8.62	9.26	7.58
13	0.29	0.54	1.00	1.87	2.70	3.50	4.27	5.78	7.25	7.97	9.40	10.4	8.55
14	0.31	0.58	1.09	2.03	2.92	3.79	4.63	6.27	7.86	8.64	10.2	11.7	9.55
15	0.34	0.63	1.17	2.19	3.15	4.08	4.99	6.75	8.47	9.31	11.0	12.6	10.6
16	0.36	0.67	1.26	2.34	3.38	4.37	5.35	7.24	9.08	9.98	11.8	13.5	11.7
17	0.39	0.72	1.34	2.50	3.61	4.67	5.71	7.73	9.69	10.7	12.6	14.4	12.8
18	0.41	0.76	1.43	2.66	3.83	4.97	6.07	8.22	10.3	11.3	13.4	15.3	13.9
19	0.43	0.81	1.51	2.82	4.07	5.27	6.44	8.72	10.9	12.0	14.2	16.3	15.1
20	0.46	0.86	1.60	2.98	4.30	5.57	6.80	9.21	11.5	12.7	15.0	17.2	16.3
21	0.48	0.90	1.69	3.14	4.53	5.87	7.17	9.71	12.2	13.4	15.8	18.1	17.6
22	0.51	0.95	1.77	3.31	4.76	6.17	7.54	10.2	12.8	14.1	16.6	19.1	18.8
23	0.53	1.00	1.86	3.47	5.00	6.47	7.91	10.7	13.4	14.8	17.4	20.0	20.1
24	0.56	1.04	1.95	3.63	5.23	6.78	8.29	11.2	14.1	15.5	18.2	20.9	21.4
25	0.58	1.09	2.03	3.80	5.47	7.08	8.66	11.7	14.7	16.2	19.0	21.9	22.8
26	0.61	1.14	2.12	3.96	5.70	7.39	9.03	12.2	15.3	16.9	19.9	22.8	24.2
28	0.66	1.23	2.30	4.29	6.18	8.01	9.79	13.2	16.6	18.3	21.5	24.7	27.0
30	0.71	1.33	2.48	4.62	6.66	8.63	10.5	14.3	17.9	19.7	23.2	26.6	30.0
32	0.76	1.42	2.66	4.96	7.14	9.25	11.3	15.3	19.2	21.1	24.9	28.6	32.2
35	0.84	1.57	2.93	5.46	7.86	10.2	12.5	16.9	21.1	23.2	27.4	31.5	35.5
40	0.97	1.81	3.38	6.31	9.08	11.8	14.4	19.5	24.4	26.8	31.6	36.3	41.0
45	1.10	2.06	3.84	7.16	10.3	13.4	16.3	22.1	27.7	30.5	35.9	41.3	46.5
	Type A			Type B					Type C				

Table 9. *(Continued)* **Horsepower Ratings for Roller Chain–1986**

3/4-inch Pitch Standard Single-Strand Roller Chain — No. 60

No. of Teeth Small Spkt.	Revolutions per Minute — Small Sprocket[a]												
	25	50	100	150	200	300	400	500	600	700	800	900	1000
	Horsepower Rating												
11	0.41	0.77	1.44	2.07	2.69	3.87	5.02	6.13	7.23	8.30	9.36	10.4	11.4
12	0.45	0.85	1.58	2.28	2.95	4.25	5.51	6.74	7.94	9.12	10.3	11.4	12.6
13	0.50	0.92	1.73	2.49	3.22	4.64	6.01	7.34	8.65	9.94	11.2	12.5	13.7
14	0.54	1.00	1.87	2.69	3.49	5.02	6.51	7.96	9.37	10.8	12.1	13.5	14.8
15	0.58	1.08	2.01	2.90	3.76	5.41	7.01	8.57	10.1	11.6	13.1	14.5	16.0
16	0.62	1.16	2.16	3.11	4.03	5.80	7.52	9.19	10.8	12.4	14.0	15.6	17.1
17	0.66	1.24	2.31	3.32	4.30	6.20	8.03	9.81	11.6	13.3	15.0	16.7	18.3
18	0.70	1.31	2.45	3.53	4.58	6.59	8.54	10.4	12.3	14.1	15.9	17.7	19.5
19	0.75	1.39	2.60	3.74	4.85	6.99	9.05	11.1	13.0	15.0	16.9	18.8	20.6
20	0.79	1.47	2.75	3.96	5.13	7.38	9.57	11.7	13.8	15.8	17.9	19.8	21.8
21	0.83	1.55	2.90	4.17	5.40	7.78	10.1	12.3	14.5	16.7	18.8	20.9	23.0
22	0.87	1.63	3.05	4.39	5.68	8.19	10.6	13.0	15.3	17.5	19.8	22.0	24.2
23	0.92	1.71	3.19	4.60	5.96	8.59	11.1	13.6	16.0	18.4	20.8	23.1	25.4
24	0.96	1.79	3.35	4.82	6.24	8.99	11.6	14.2	16.8	19.3	21.7	24.2	26.6
25	1.00	1.87	3.50	5.04	6.52	9.40	12.2	14.9	17.5	20.1	22.7	25.3	27.8
26	1.05	1.95	3.65	5.25	6.81	9.80	12.7	15.5	18.3	21.0	23.7	26.4	29.0
28	1.13	2.12	3.95	5.69	7.37	10.6	13.8	16.8	19.8	22.8	25.7	28.5	31.4
30	1.22	2.28	4.26	6.13	7.94	11.4	14.8	18.1	21.4	24.5	27.7	30.8	33.8
32	1.31	2.45	4.56	6.57	8.52	12.3	15.9	19.4	22.9	26.3	29.7	33.0	36.3
35	1.44	2.69	5.03	7.24	9.38	13.5	17.5	21.4	25.2	29.0	32.7	36.3	39.9
40	1.67	3.11	5.81	8.37	10.8	15.6	20.2	24.7	29.1	33.5	37.7	42.0	46.1
45	1.89	3.53	6.60	9.50	12.3	17.7	23.0	28.1	33.1	38.0	42.9	47.7	52.4
	Type A		Type B							Type C			

1-inch Pitch Standard Single-Strand Roller Chain — No. 80

No. of Teeth Small Spkt.	Revolutions per Minute — Small Sprocket[a]												
	25	50	100	150	200	300	400	500	600	700	800	900	1000
	Horsepower Ratings												
11	0.97	1.80	3.36	4.84	6.28	9.04	11.7	14.3	16.9	19.4	21.9	23.0	19.6
12	1.06	1.98	3.69	5.32	6.89	9.93	12.9	15.7	18.5	21.3	24.0	26.2	22.3
13	1.16	2.16	4.03	5.80	7.52	10.8	14.0	17.1	20.2	23.2	26.2	29.1	25.2
14	1.25	2.34	4.36	6.29	8.14	11.7	15.2	18.6	21.9	25.1	28.4	31.5	28.2
15	1.35	2.52	4.70	6.77	8.77	12.6	16.4	20.0	23.6	27.1	30.6	34.0	31.2
16	1.45	2.70	5.04	7.26	9.41	13.5	17.6	21.5	25.3	29.0	32.8	36.4	34.4
17	1.55	2.88	5.38	7.75	10.0	14.5	18.7	22.9	27.0	31.0	35.0	38.9	37.7
18	1.64	3.07	5.72	8.25	10.7	15.4	19.9	24.4	28.7	33.0	37.2	41.4	41.1
19	1.74	3.25	6.07	8.74	11.3	16.3	21.1	25.8	30.4	35.0	39.4	43.8	44.5
20	1.84	3.44	6.41	9.24	12.0	17.2	22.3	27.3	32.2	37.0	41.7	46.3	48.1
21	1.94	3.62	6.76	9.74	12.6	18.2	23.5	28.8	33.9	39.0	43.9	48.9	51.7
22	2.04	3.81	7.11	10.2	13.3	19.1	24.8	30.3	35.7	41.0	46.2	51.4	55.5
23	2.14	4.00	7.46	10.7	13.9	20.1	26.0	31.8	37.4	43.0	48.5	53.9	59.3
24	2.24	4.19	7.81	11.3	14.6	21.0	27.2	33.2	39.2	45.0	50.8	56.4	62.0
25	2.34	4.37	8.16	11.8	15.2	21.9	28.4	34.7	40.9	47.0	53.0	59.0	64.8
26	2.45	4.56	8.52	12.3	15.9	22.9	29.7	36.2	42.7	49.1	55.3	61.5	67.6
28	2.65	4.94	9.23	13.3	17.2	24.8	32.1	39.3	46.3	53.2	59.9	66.7	73.3
30	2.85	5.33	9.94	14.3	18.5	26.7	34.6	42.3	49.9	57.3	64.6	71.8	78.9
32	3.06	5.71	10.7	15.3	19.9	28.6	37.1	45.4	53.5	61.4	69.2	77.0	84.6
35	3.37	6.29	11.7	16.9	21.9	31.6	40.9	50.0	58.9	67.6	76.3	84.8	93.3
40	3.89	7.27	13.6	19.5	25.3	36.4	47.2	57.7	68.0	78.1	88.1	99.0	108
45	4.42	8.25	15.4	22.2	28.7	41.4	53.6	65.6	77.2	88.7	100	111	122
	Type A	Type B						Type C					

Table 9. ***(Continued)*** **Horsepower Ratings for Roller Chain–1986**

1¼-inch Pitch Standard Single-Strand Roller Chain — No. 100

No. of Teeth Small Spkt.	Revolutions per Minute — Small Sprocket[a]												
	10	25	50	100	150	200	300	400	500	600	700	800	900
	Horsepower Rating												
11	0.81	1.85	3.45	6.44	9.28	12.0	17.3	22.4	27.4	32.3	37.1	32.8	27.5
12	0.89	2.03	3.79	7.08	10.2	13.2	19.0	24.6	30.1	35.5	40.8	37.3	31.3
13	0.97	2.22	4.13	7.72	11.1	14.4	20.7	26.9	32.8	38.7	44.5	42.1	35.3
14	1.05	2.40	4.48	8.36	12.0	15.6	22.5	29.1	35.6	41.9	48.2	47.0	39.4
15	1.13	2.59	4.83	9.01	13.0	16.8	24.2	31.4	38.3	45.2	51.9	52.2	43.7
16	1.22	2.77	5.17	9.66	13.9	18.0	26.0	33.6	41.1	48.4	55.6	57.5	48.2
17	1.30	2.96	5.52	10.3	14.8	19.2	27.7	35.9	43.9	51.7	59.4	63.0	52.8
18	1.38	3.15	5.88	11.0	15.8	20.5	29.5	38.2	46.7	55.0	63.2	68.6	57.5
19	1.46	3.34	6.23	11.6	16.7	21.7	31.2	40.5	49.5	58.3	67.0	74.4	62.3
20	1.55	3.53	6.58	12.3	17.7	22.9	33.0	42.8	52.3	61.6	70.8	79.8	67.3
21	1.63	3.72	6.94	13.0	18.7	24.2	34.8	45.1	55.1	65.0	74.6	84.2	72.4
22	1.71	3.91	7.30	13.6	19.6	25.4	36.6	47.4	58.0	68.3	78.5	88.5	77.7
23	1.80	4.10	7.66	14.3	20.6	26.7	38.4	49.8	60.8	71.7	82.3	92.8	83.0
24	1.88	4.30	8.02	15.0	21.5	27.9	40.2	52.1	63.7	75.0	86.2	97.2	88.5
25	1.97	4.49	8.38	15.6	22.5	29.2	42.0	54.4	66.6	78.4	90.1	102	94.1
26	2.05	4.68	8.74	16.3	23.5	30.4	43.8	56.8	69.4	81.8	94.0	106	99.8
28	2.22	5.07	9.47	17.7	25.5	33.0	47.5	61.5	75.2	88.6	102	115	112
30	2.40	5.47	10.2	19.0	27.4	35.5	51.2	66.3	81.0	95.5	110	124	124
32	2.57	5.86	10.9	20.4	29.4	38.1	54.9	71.1	86.9	102	118	133	136
35	2.83	6.46	12.0	22.5	32.4	42.0	60.4	78.3	95.7	113	130	146	156
40	3.27	7.46	13.9	26.0	37.4	48.5	69.8	90.4	111	130	150	169	188
45	3.71	8.47	15.8	29.5	42.5	55.0	79.3	103	126	148	170	192	213
	Type A		Type B					Type C					

1½-inch Pitch Standard Single-Strand Roller Chain — No. 120

No. of Teeth Small Spkt.	Revolutions per Minute — Small Sprocket[a]												
	10	25	50	100	150	200	300	400	500	600	700	800	900
	Horsepower Rating												
11	1.37	3.12	5.83	10.9	15.7	20.3	29.2	37.9	46.3	54.6	46.3	37.9	31.8
12	1.50	3.43	6.40	11.9	17.2	22.3	32.1	41.6	50.9	59.9	52.8	43.2	36.2
13	1.64	3.74	6.98	13.0	18.8	24.3	35.0	45.4	55.5	65.3	59.5	48.7	40.8
14	1.78	4.05	7.56	14.1	20.3	26.3	37.9	49.1	60.1	70.8	66.5	54.4	45.6
15	1.91	4.37	8.15	15.2	21.9	28.4	40.9	53.0	64.7	76.3	73.8	60.4	50.6
16	2.05	4.68	8.74	16.3	23.5	30.4	43.8	56.8	69.4	81.8	81.3	66.5	55.7
17	2.19	5.00	9.33	17.4	25.1	32.5	46.8	60.6	74.1	87.3	89.0	72.8	61.0
18	2.33	5.32	9.92	18.5	26.7	34.6	49.8	64.5	78.8	92.9	97.0	79.4	66.5
19	2.47	5.64	10.5	19.6	28.3	36.6	52.8	68.4	83.6	98.5	105	86.1	72.1
20	2.61	5.96	11.1	20.7	29.9	38.7	55.8	72.2	88.3	104	114	92.9	77.9
21	2.75	6.28	11.7	21.9	31.5	40.8	58.8	76.2	93.1	110	122	100	83.8
22	2.90	6.60	12.3	23.0	33.1	42.9	61.8	80.1	97.9	115	131	107	89.9
23	3.04	6.93	12.9	24.1	34.8	45.0	64.9	84.0	103	121	139	115	96.1
24	3.18	7.25	13.5	25.3	36.4	47.1	67.9	88.0	108	127	146	122	102
25	3.32	7.58	14.1	26.4	38.0	49.3	71.0	91.9	112	132	152	130	109
26	3.47	7.91	14.8	27.5	39.7	51.4	74.0	95.9	117	138	159	138	115
28	3.76	8.57	16.0	29.8	43.0	55.7	80.2	104	127	150	172	154	129
30	4.05	9.23	17.2	32.1	46.3	60.0	86.4	112	137	161	185	171	143
32	4.34	9.90	18.5	34.5	49.6	64.3	92.6	120	147	173	199	188	158
35	4.78	10.9	20.3	38.0	54.7	70.9	102	132	162	190	219	215	180
40	5.52	12.6	23.5	43.9	63.2	81.8	118	153	187	220	253	…	…
45	6.27	14.3	26.7	49.8	71.7	92.9	134	173	212	250	287	…	…
	Type A	Type B					Type C						

[a] For lower or higher rpm, larger chain sizes, and rpm above 3500, see B29.1M-1993.
For use of table see page 2431.

Table 10. Recommended Roller Chain Sprocket Maximum Bore and Hub Diameters

No. of Teeth	Roller Chain Pitch 3/8		1/2		5/8		3/4		1	
	Max. Bore	Max. Hub Dia.	Max. Bore	Max. Hub Dia.	Max. Bore	Max. Hub Dia.	Max. Bore	Max. Hub Dia.	Max. Bore	Max. Hub Dia.
11	19/32	55/64	25/32	1 11/64	31/32	1 15/32	1 1/4	1 49/64	1 5/8	2 3/8
12	5/8	63/64	7/8	1 21/64	1 5/32	1 43/64	1 9/32	2 1/64	1 25/32	2 45/64
13	3/4	1 7/64	1	1 1/2	1 9/32	1 7/8	1 1/2	2 1/4	2	3 1/64
14	27/32	1 15/64	1 5/32	1 21/32	1 5/16	2 5/64	1 3/4	2 1/2	2 9/32	3 11/32
15	7/8	1 23/64	1 1/4	1 13/16	1 17/32	2 9/32	1 25/32	2 3/4	2 13/32	3 43/64
16	31/32	1 15/32	1 9/32	1 63/64	1 11/16	2 31/64	1 31/32	2 63/64	2 23/32	3 63/64
17	1 3/32	1 19/32	1 3/8	2 9/64	1 25/32	2 11/16	2 7/32	3 7/32	2 13/16	4 5/16
18	1 7/32	1 23/32	1 17/32	2 19/64	1 7/8	2 57/64	2 9/32	3 15/32	3 1/8	4 41/64
19	1 1/4	1 27/32	1 11/16	2 29/64	2 1/16	3 5/64	2 7/16	3 45/64	3 5/16	4 61/64
20	1 9/32	1 61/64	1 25/32	2 5/8	2 1/4	3 9/32	2 11/16	3 61/64	3 1/2	5 9/32
21	1 5/16	2 5/64	1 25/32	2 25/32	2 9/32	3 31/64	2 13/16	4 3/16	3 3/4	5 19/32
22	1 7/16	2 13/16	1 15/16	2 15/16	2 7/16	3 11/16	2 15/16	4 7/16	3 7/8	5 59/64
23	1 9/16	2 5/16	2 3/32	3 3/32	2 5/8	3 57/64	3 1/8	4 43/64	4 3/16	6 15/64
24	1 11/16	2 7/16	2 1/4	3 17/64	2 13/16	4 5/64	3 1/4	4 29/32	4 9/16	6 9/16
25	1 3/4	2 9/16	2 9/32	3 27/64	2 27/32	4 9/32	3 3/8	5 5/32	4 11/16	6 7/8

No. of Teeth	Roller Chain Pitch 1 1/4		1 1/2		1 3/4		2		2 1/2	
	Max. Bore	Max. Hub Dia.	Max. Bore	Max. Hub Dia.	Max. Bore	Max. Hub Dia.	Max. Bore	Max. Hub Dia.	Max. Bore	Max. Hub Dia.
11	1 31/32	2 31/32	2 5/16	3 37/64	2 13/16	4 11/64	3 9/32	4 25/32	3 15/16	5 63/64
12	2 9/32	3 3/8	2 3/4	4 1/16	3 1/4	4 3/4	3 5/8	5 27/64	4 23/32	6 51/64
13	2 17/32	3 25/32	3 1/16	4 35/64	3 9/16	5 5/16	4 1/16	6 5/64	5 3/32	7 39/64
14	2 11/16	4 3/16	3 5/16	5 1/32	3 7/8	5 7/8	4 11/16	6 23/32	5 23/32	8 27/64
15	3 3/32	4 19/32	3 3/4	5 33/64	4 7/16	6 29/64	4 7/8	7 3/8	6 1/4	9 7/32
16	3 9/32	5	4	6	4 11/16	7 1/64	5 1/2	8 1/64	7	10 1/32
17	3 21/32	5 13/32	4 15/32	6 31/64	5 1/16	7 37/64	5 11/16	8 21/32	7 7/16	10 27/32
18	3 25/32	5 51/64	4 21/32	6 31/32	5 5/8	8 9/64	6 1/4	9 5/16	8 1/8	11 41/64
19	4 3/16	6 13/64	4 15/16	7 29/64	5 11/16	8 45/64	6 7/8	9 61/64	9	12 7/16
20	4 19/32	6 39/64	5 7/16	7 15/16	6 1/4	9 17/64	7	10 19/32	9 3/4	13 1/4
21	4 11/16	7	5 11/16	8 27/64	6 13/16	9 53/64	7 3/4	11 15/64	10	14 3/64
22	4 7/8	7 13/32	5 7/8	8 57/64	7 1/4	10 25/64	8 3/8	11 7/8	10 7/8	14 27/32
23	5 5/16	7 13/16	6 3/8	9 3/8	7 7/16	10 15/16	9	12 33/64	11 5/8	15 21/32
24	5 11/16	8 13/64	6 13/16	9 55/64	8	11 1/2	9 5/8	13 5/32	13	16 29/64
25	5 23/32	8 39/64	7 1/4	10 11/32	8 9/16	12 1/16	10 1/4	13 51/64	13 1/2	17 1/4

All dimensions in inches.

For standard key dimensions see pages 2342 through 2343.

Source: American Chain Association.

Center Distance between Sprockets.—The center-to-center distance between sprockets, as a general rule, should not be less than 1 1/2 times the diameter of the larger sprocket and not less than thirty times the pitch nor more than about 50 times the pitch, although much depends upon the speed and other conditions. A center distance equivalent to 80 pitches may be considered an approved maximum. Very long center distances result in catenary tension in the chain. If roller-chain drives are designed correctly, the center-to-center distance for some transmissions may be so short that the sprocket teeth nearly touch each other, assuming that the load is not too great and the number of teeth is not too small. To avoid interference of the sprocket teeth, the center distance must, of course, be somewhat greater than one-half the sum of the outside diameters of the sprockets. The chain should extend around at least 120 degrees of the pinion circumference, and this minimum amount of contact is obtained for all center distances provided the ratio is less than 3½ to 1. Other things being equal, a fairly long chain is recommended in preference to the shortest one allowed by the sprocket diameters, because the rate of chain elongation due to natural wear is inversely proportional to the length, and also because the greater elasticity of the longer strand tends to absorb irregularities of motion and to decrease the effect of shocks.

If possible, the center distance should be adjustable in order to take care of slack due to elongation from wear and this range of adjustment should be at least one and one-half pitches. A little slack is desirable as it allows the chain links to take the best position on the sprocket teeth and reduces the wear on the bearings. Too much sag or an excessive distance between the sprockets may cause the chain to whip up and down — a condition detrimental to smooth running and very destructive to the chain. The sprockets should run in a vertical plane, the sprocket axes being approximately horizontal, unless an idler is used on the slack side to keep the chain in position. The most satisfactory results are obtained when the slack side of the chain is on the bottom.

Center Distance for a Given Chain Length.—When the distance between the driving and driven sprockets can be varied to suit the length of the chain, this center distance for a tight chain may be determined by the following formula, in which c = center-to-center distance in inches; L = chain length in pitches; P = pitch of chain; N = number of teeth in large sprocket; n = number of teeth in small sprocket.

$$c = \frac{P}{8}\left(2L - N - n + \sqrt{(2L - N - n)^2 - 0.810(N - n)^2}\right)$$

This formula is approximate, but the error is less than the variation in the length of the best chains. The length L in pitches should be an even number for a roller chain, so that the use of an offset connecting link will not be necessary.

Idler Sprockets.—When sprockets have a fixed center distance or are non-adjustable, it may be advisable to use an idler sprocket for taking up the slack. The idler should preferably be placed against the slack side between the two strands of the chain. When a sprocket is applied to the tight side of the chain to reduce vibration, it should be on the lower side and so located that the chain will run in a straight line between the two main sprockets. A sprocket will wear excessively if the number of teeth is too small and the speed too high, because there is impact between the teeth and rollers even though the idler carries practically no load.

Length of Driving Chain.—The total length of a block chain should be given in multiples of the pitch, whereas for a roller chain, the length should be in multiples of twice the pitch, because the ends must be connected with an outside and inside link. The length of a chain can be calculated accurately enough for ordinary practice by the use of the following formula, in which L = chain length in pitches; C = center distance in pitches; N = number of teeth in large sprocket; n = number of teeth in small sprocket:

$$L = 2C + \frac{N}{2} + \frac{n}{2} + \left(\frac{N - n}{2\pi}\right)^2 \times \frac{1}{C}$$

Table 11. ANSI Sprocket Tooth Form for Roller Chain ***ANSI/ASME B29.1M-1993***

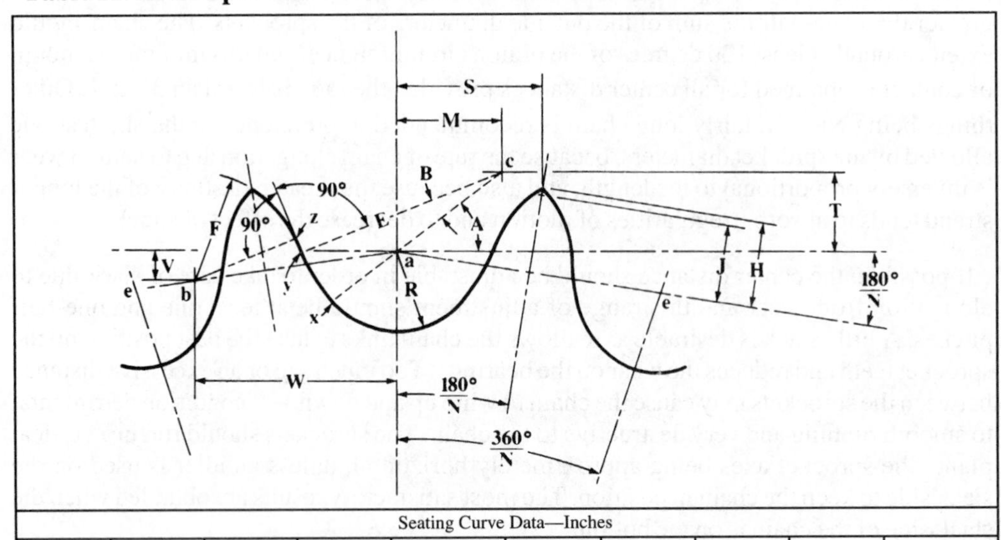

Seating Curve Data—Inches

P	D_r	Min. R	Min. D_s	D_s Tol.[a]	P	D_r	Min. R	Min. D_s	D_s Tol.[a]
0.250	0.130	0.0670	0.134	0.0055	1.250	0.750	0.3785	0.757	0.0070
0.375	0.200	0.1020	0.204	0.0055	1.500	0.875	0.4410	0.882	0.0075
0.500	0.306	0.1585	0.317	0.0060	1.750	1.000	0.5040	1.008	0.0080
0.500	0.312	0.1585	0.317	0.0060	2.000	1.125	0.5670	1.134	0.0085
0.625	0.400	0.2025	0.405	0.0060	2.250	1.406	0.7080	1.416	0.0090
0.750	0.469	0.2370	0.474	0.0065	2.500	1.562	0.7870	1.573	0.0095
1.000	0.625	0.3155	0.631	0.0070	3.000	1.875	0.9435	1.887	0.0105

[a] Plus tolerance only.

$P =$ pitch (ae)

$N =$ number of teeth; $D_r =$ nominal roller diameter

$D_s =$ seating curve diameter $= 1.005\,D_r + 0.003$ (in inches)

$R = \frac{1}{2} D_s$ D_s has only plus tolerance

$A = 35° + (60° \div N)$; $B = 18° - (56° \div N)$; $ac = 0.8\,D_r$

$M = 0.8\,D_r \cos(35° + (60° \div N))$

$T = 0.8\,D_r \sin(35° + (60° \div N))$

$E = 1.3025\,D_r + 0.0015$ (in inches)

Chord $xy = (2.605\,D_r + 0.003) \sin 9° - (28° \div N))$ (in inches)

$yz = D_r\,[1.4 \sin(17° - (64° \div N)) - 0.8 \sin(18° - (56° \div N))]$

Length of a line between a and $b = 1.4\,D_r$

$W = 1.4\,D_r \cos(180° \div N)$; $V = 1.4\,D_r \sin(180° \div N)$

$F = D_r\,[0.8 \cos(18° - (56° \div N)) + 1.4 \cos(17° - (64° \div N)) - 1.3025] - 0.0015$ inch

$H = \sqrt{F^2 - (1.4 D_r - 0.5 P)^2}$

$S = 0.5\,P \cos(180° \div N) + H \sin(180° \div N)$

Approximate O.D. of sprocket when J is $0.3\,P = P\,[0.6 + \cot(180° \div N)]$

O.D. of sprocket when tooth is pointed $+ P \cot(180° \div N) + \cos(180° \div N)\,(D_s - D_r) + 2H$

Pressure angle for new chain $= xab = 35° - (120° \div N)$

Minimum pressure angle $= xab - B = 17° - (64° \div N)$;
Average pressure angle $= 26° - (92° \div N)$

Table 12. Standard Hob Design for Roller Chain Sprockets

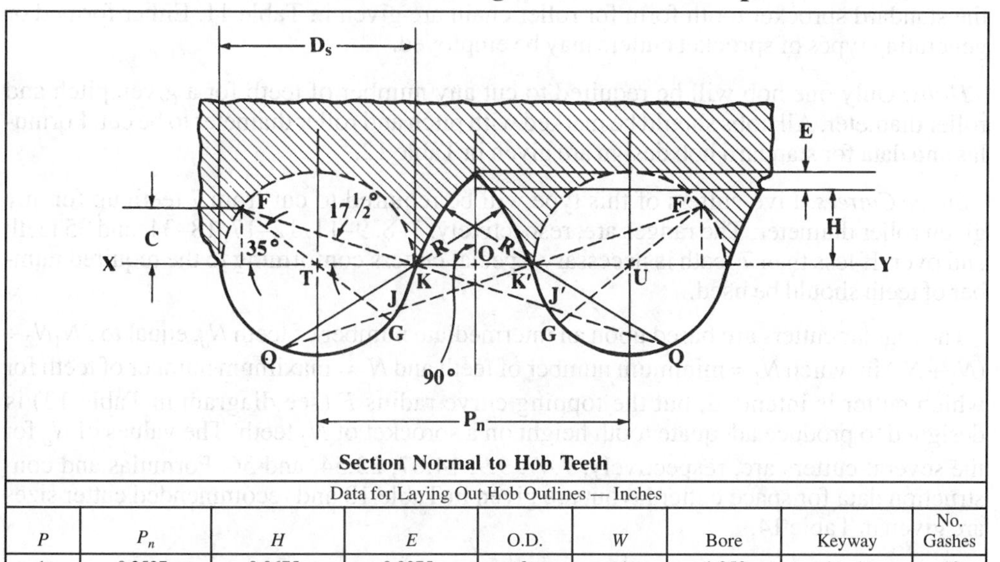

Section Normal to Hob Teeth

Data for Laying Out Hob Outlines — Inches								
P	P_n	*H*	*E*	O.D.	*W*	Bore	Keyway	No. Gashes
1/4	0.2527	0.0675	0.0075	2 5/8	2 1/2	1.250	1/4 × 1/8	13
3/8	0.379	0.101	0.012	3 1/8	2 1/2	1.250	1/4 × 1/8	13
1/2	0.506	0.135	0.015	3 3/8	2 1/2	1.250	1/4 × 1/8	12
5/8	0.632	0.170	0.018	3 5/8	2 1/2	1.250	11/4 × 1/8	12
3/4	0.759	0.202	0.023	3 3/4	2 7/8	1.250	1/4 × 1/8	11
1	1.011	0.270	0.030	4 3/8	3 3/4	1.250	1/4 × 1/8	11
1 1/4	1.264	0.337	0.038	4 3/4	4 1/2	1.250	1/4 × 1/8	10
1 1/2	1.517	0.405	0.045	5 3/8	5 1/4	1.250	1/4 × 1/8	10
1 3/4	1.770	0.472	0.053	6 3/8	6	1.500	3/8 × 3/16	9
2	2.022	0.540	0.060	6 7/8	6 3/4	1.500	3/8 × 3/16	9
2 1/4	2.275	0.607	0.068	8	8 1/2	1.750	3/8 × 3/16	8
2 1/2	2.528	0.675	0.075	8 5/8	9 3/8	1.750	3/8 × 3/16	8
3	3.033	0.810	0.090	9 3/4	11 1/4	2.000	1/2 × 3/6	8

Hobs designed for a given roller diameter (D_r) and chain pitch (P) will cut any number of teeth.

P = Pitch of Chain

P_n = Normal Pitch of Hob = 1.011 P inches

D_s = Minimum Diameter of Seating Curve = 1.005 D_r + 0.003 inches

F = Radius Center for Arc GK; $TO = OU = P_n \div 2$

$H = 0.27\,P$; $E = 0.03\,P$ = Radius of Fillet Circle

Q is located on line passing through F and J. Point J is intersection of line XY with circle of diameter D_s. R is found by trial and the arc of this radius is tangent to arc KG at K and to fillet radius.

OD = Outside Diameter = 1.7 (Bore + D_r + 0.7 P) approx.

D_h = Pitch Diameter = $OD - D_s$; M = Helix Angle; $\sin M = P_n \div \pi D_h$

L = Lead = $P_n \div \cos M$; W = Width = Not less than 2 × Bore, or 6 D_r, or 3.2 P

To the length obtained by this formula, add enough to make a whole number (and for a roller chain, an even number) of pitches. If a roller chain has an odd number of pitches, it will be necessary to use an offset connecting link.

Another formula for obtaining chain length in which D = distance between centers of shafts; R = pitch radius of large sprocket; r = pitch radius of small sprocket: N = number of teeth in large sprocket; n = number of teeth in small sprocket; P = pitch of chain and sprockets; and l = required chain length in inches, is:

$$l = \frac{180° + 2\alpha}{360°}NP + \frac{180° - 2\alpha}{360°}nP + 2D\cos\alpha \qquad \text{where } \sin\alpha = \frac{R-r}{D}$$

Cutting Standard Sprocket Tooth Form.—The proportions and seating curve data for the standard sprocket tooth form for roller chain are given in Table 11. Either formed or generating types of sprocket cutters may be employed.

Hobs: Only one hob will be required to cut any number of teeth for a given pitch and roller diameter. All hobs should be marked with pitch and roller diameter to be cut. Formulas and data for standard hob design are given in Table 12.

Space Cutters: Five cutters of this type will be required to cut from 7 teeth up for any given roller diameter. The ranges are, respectively, 7–8, 9–11, 12–17, 18–34, and 35 teeth and over. If less than 7 teeth is necessary, special cutters conforming to the required number of teeth should be used.

The regular cutters are based upon an intermediate number of teeth N_a, equal to $2N_1N_2 \div (N_1 + N_2)$ in which N_1 = minimum number of teeth and N_2 = maximum number of teeth for which cutter is intended; but the topping curve radius F (see diagram in Table 13) is designed to produce adequate tooth height on a sprocket of N_2 teeth. The values of N_a for the several cutters are, respectively, 7.47, 9.9, 14.07, 23.54, and 56. Formulas and construction data for space cutter layout are given in Table 13 and recommended cutter sizes are given in Table 14.

Table 13. Standard Space Cutters for Roller-Chain Sprockets

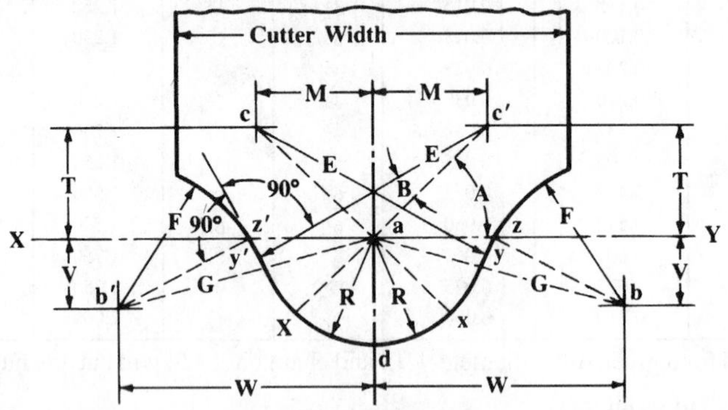

Data for Laying Out Space Cutter

Range of Teeth	*M*	*T*	*W*	*V*
7–8	0.5848 D_r	0.5459 D_r	1.2790 D_r	0.5694 D_r
9–11	0.6032 D_r	0.5255 D_r	1.3302 D_r	0.4365 D_r
12–17	0.6194 D_r	0.5063 D_r	1.3694 D_r	0.2911 D_r
18–34	0.6343 D_r	0.4875 D_r	1.3947 D_r	0.1220 D_r
35 up	0.6466 D_r	0.4710 D_r	1.4000 D_r	0
Range of Teeth	***F***	**Chord *xy***	***yz***	**Angle *Yab***
7–8	0.8686 D_r – 0.0015	0.2384 D_r + 0.0003	0.0618 D_r	24°
9–11	0.8554 D_r – 0.0015	0.2800 D_r + 0.0003	0.0853 D_r	18° 10′
12–17	0.8364 D_r – 0.0015	0.3181 D_r + 0.0004	0.1269 D_r	12°
18–34	0.8073 D_r – 0.0015	0.3540 D_r + 0.0004	0.1922 D_r	5°
35 up	0.7857 D_r – 0.0015	0.3850 D_r + 0.0004	0.2235 D_r	0°

E (same for all ranges) = 1.3025 D_r + 0.0015; *G* (same for all ranges) = 1.4 D_r

See Table 14 for recommended cutter sizes.

Angle *Yab* is equal to 180° ÷ *N* when the cutter is made for a specific number of teeth. For the design of cutters covering a range of teeth, angle *Yab* was determined by layout to ensure chain roller clearance and to avoid pointed teeth on the larger sprockets of each range. It has values as given below for cutters covering the range of teeth shown. The following formulas are for cutters covering the standard ranges of teeth where N_a equals intermediate values given on page 2440.

$$W = 1.4D_r \cos Yab \qquad V = 1.4D_r \sin Yab$$

$$yz = D_r\left[1.4\sin\left(17° + \frac{116°}{N_a} - Yab\right) - 0.8\sin\left(18° - \frac{56°}{N_a}\right)\right]$$

$$F = D_r\left[0.8\cos\left(18° - \frac{56°}{N_a}\right) + 1.4\cos\left(17° + \frac{116°}{N_a} - Yab\right) - 1.3025\right] - 0.0015 \text{ in.}$$

For other points, use the value of N_a for N in the standard formulas in Table 11.

Table 14. Recommended Space Cutter Sizes for Roller-Chain Sprockets

		Number of Teeth					
		6	7–8	9–11	12–17	18–34	35 up
Pitch	Roller Dia.	Cutter Diameter (Minimum)					
0.250	0.130	2.75	2.75	2.75	2.75	2.75	2.75
0.375	0.200	2.75	2.75	2.75	2.75	2.75	2.75
0.500	0.312	3.00	3.00	3.12	3.12	3.12	3.12
0.625	0.400	3.12	3.12	3.25	3.25	3.25	3.25
0.725	0.469	3.25	3.25	3.38	3.38	3.38	3.38
1.000	0.625	3.88	4.00	4.12	4.12	4.25	4.25
1.250	0.750	4.25	4.38	4.50	4.50	4.62	4.62
1.500	0.875	4.38	4.50	4.62	4.62	4.75	4.75
1.750	1.000	5.00	5.12	5.25	5.38	5.50	5.50
2.000	1.125	5.38	5.50	5.62	5.75	5.88	5.88
2.250	1.406	5.88	6.00	6.25	6.38	6.50	6.50
2.500	1.563	6.38	6.62	6.75	6.88	7.00	7.12
3.000	1.875	7.50	7.75	7.88	8.00	8.00	8.25
Pitch	Roller Dia.	Cutter Width (Minimum)					
0.250	0.130	0.31	0.31	0.31	0.31	0.28	0.28
0.375	0.200	0.47	0.47	0.47	0.44	0.44	0.41
0.500	0.312	0.75	0.75	0.75	0.75	0.72	0.69
0.625	0.400	0.75	0.75	0.75	0.75	0.72	0.69
0.750	0.469	0.91	0.91	0.91	0.88	0.84	0.81
1.000	0.625	1.50	1.50	1.47	1.47	1.41	1.34
1.250	0.750	1.81	1.81	1.78	1.75	1.69	1.62
1.500	0.875	1.81	1.81	1.78	1.75	1.69	1.62
1.750	1.000	2.09	2.09	2.06	2.03	1.97	1.88
2.000	1.125	2.41	2.41	2.38	2.31	2.25	2.16
2.250	1.406	2.69	2.69	2.66	2.59	2.47	2.41
2.500	1.563	3.00	3.00	2.94	2.91	2.75	2.69
3.000	1.875	3.59	3.59	3.53	3.47	3.34	3.22

Where the same roller diameter is commonly used with chains of two different pitches it is recommended that stock cutters be made wide enough to cut sprockets for both chains.

Marking of Cutters.— All cutters are to be marked, giving pitch, roller diameter and range of teeth to be cut.

Bores for Sprocket Cutters (recommended practice) are approximately as calculated from the formula:

$$\text{Bore} = 0.7\sqrt{(\text{Width of Cutter} + \text{Roller Diameter} + 0.7 \text{ Pitch})}$$

and are equal to 1 inch for ¼- through ¾-inch pitches; 1¼ inches for 1- through 1½-inch for 1¾- through 2¼-inch pitches; 1¾ inches for 2½-inch pitch; and 2 inches for 3-inch pitch.

Minimum Outside Diameters of Space Cutters for 35 teeth and over (recommended practice) are approximately as calculated from the formula:

$$\text{Outside Diameter} = 1.2(\text{Bore} + \text{Roller Diameter} + 0.7 \text{ Pitch}) + 1 \text{ in.}$$

Shaper Cutters: Only one will be required to cut any number of teeth for a given pitch and roller diameter. The manufacturer should be referred to for information concerning the cutter form design to be used.

Sprocket Manufacture.—Cast sprockets have cut teeth, and the rim, hub face, and bore are machined. The smaller sprockets are generally cut from steel bar stock, and are finished all over. Sprockets are often made from forgings or forged bars. The extent of finishing depends on the particular specifications that are applicable. Many sprockets are made by welding a steel hub to a steel plate. This process produces a one-piece sprocket of desired proportions and one that can be heat-treated.

Sprocket Materials.—For large sprockets, cast iron is commonly used, especially in drives with large speed ratios, since the teeth of the larger sprocket are subjected to fewer chain engagements in a given time. For severe service, cast steel or steel plate is preferred.

The smaller sprockets of a drive are usually made of steel. With this material the body of the sprocket can be heat-treated to produce toughness for shock resistance, and the tooth surfaces can be hardened to resist wear.

Stainless steel or bronze may be used for corrosion resistance, and Formica, nylon or other suitable plastic materials for special applications.

Roller Chain Drive Ratings.—In 1961, under auspices of The American Sprocket Chain Manufacturers Association (now called American Chain Association), a joint research program was begun to study pin-bushing interaction at high speeds and to gain further data on the phenomenon of chain joint galling among other research areas. These studies have shown that a separating film of lubricant is formed in chain joints in a manner similar to that found in journal bearings. These developments appear in ANSI/ASME B29.1M-1993, and are contained in Table . The ratings shown in Table are below the galling range.

The horsepower ratings in Table 15 apply to lubricated, single-pitch, single-strand roller chains, both ANSI Standard and Heavy series. To obtain ratings of multiple-strand chains, a multiple-strand factor is applied.

The ratings in Table 15 are based upon: 1) A service factor of 1.; 2) A chain length of approximately 100 pitches.; 3) Use of recommended lubrication methods.; and 4) A drive arrangement where two aligned sprockets are mounted on parallel shafts in a horizontal plane..

Under these conditions, approximately 15,000 hours of service life at full load operation may be expected.

Table 15. Roller Chain Drive Service Factors

Type of Driven Load	Type of Input Power		
	Internal Combustion Engine with Hydraulic Drive	Electric Motor or Turbine	Internal Combustion Engine with Mechanical Drive
Smooth	1.0	1.0	1.2
Moderate Shock	1.2	1.3	1.4
Heavy Shock	1.4	1.5	1.7

Substantial increases in rated speed loads can be utilized, as when a service life of less than 15,000 hours is satisfactory, or when full load operation is encountered only during a portion of the required service life. Chain manufacturers should be consulted for assistance with any special application requirements.

The horsepower ratings shown in Table relate to the speed of the smaller sprocket and drive selections are made on this basis, whether the drive is speed reducing or speed increasing. Drives with more than two sprockets, idlers, composite duty cycles, or other unusual conditions often require special consideration. Where quietness or extra smooth operation are of special importance, small-pitch chain operating over large diameter sprockets will minimize noise and vibration.

When making drive selection, consideration is given to the loads imposed on the chain by the type of input power and the type of equipment to be driven. Service factors are used to compensate for these loads and the *required* horsepower rating of the chain is determined by the following formula:

$$\text{Required hp Table Rating} = \frac{\text{hp to be Transmitted} \times \text{Service Factor}}{\text{Multiple-Strand Factor}}$$

Service Factors: The service factors in Table 15 are for normal chain loading. For unusual or extremely severe operating conditions not shown in this table, it is desirable to use larger service factors.

Multiple-Strand Factors: The horsepower ratings for multiple-strand chains equal single-strand ratings multiplied by these factors: for two strands, a factor of 1.7; for three strands, 2.5; and and for four strands, 3.3.

Lubrication.—It has been shown that a separating wedge of fluid lubricant is formed in operating chain joints much like that formed in journal bearings. Therefore, fluid lubricant must be applied to ensure an oil supply to the joints and minimize metal-to-metal contact. If supplied in sufficient volume, lubrication also provides effective cooling and impact damping at higher speeds. For this reason, it is important that lubrication recommendations be followed. *The ratings in* Table *apply only to drives lubricated in the manner specified in this table*.

Chain drives should be protected against dirt and moisture and the oil supply kept free of contamination. Periodic oil change is desirable. A good grade of non-detergent petroleum base oil is recommended. Heavy oils and greases are generally too stiff to enter and fill the chain joints. The following lubricant viscosities are recommended: For temperatures of 20° to 40°F, use SAE 20 lubricant; for 40° to 100°, use SAE 30; for 100° to 120°, use SAE 40; and for 120° to 140°, use SAE 50.

There are three basic types of lubrication for roller chain drives. The recommended type shown in Table as Type A, Type B, or Type C is influenced by the chain speed and the amount of power transmitted. These are *minimum* lubrication requirements and the use of a better type (for example, Type C instead of Type B) is acceptable and may be beneficial. Chain life can vary appreciably depending upon the way the drive is lubricated. The better the chain lubrication, the longer the chain life. For this reason, it is important that the lubrication recommendations be followed when using the ratings given in Table . The types of lubrication are as follows:

Type A — Manual or Drip Lubrication: In manual lubrication, oil is applied copiously with a brush or spout can at least once every eight hours of operation. Volume and frequency should be sufficient to prevent overheating of the chain or discoloration of the chain joints. In drip lubrication, oil drops from a drip lubricator are directed between the link plate edges. The volume and frequency should be sufficient to prevent discoloration of the lubricant in the chain joints. Precautions must be taken against misdirection of the drops by windage.

Type B — Bath or Disc Lubrication: In bath lubrication, the lower strand of the chain runs through a sump of oil in the drive housing. The oil level should reach the pitch line of the chain at its lowest point while operating. In disc lubrication, the chain operates above the oil level. The disc picks up oil from the sump and deposits it onto the chain, usually by means of a trough. The diameter of the disc should be such as to produce rim speeds of between 600 and 8000 feet per minute.

Type C — Oil Stream Lubrication: The lubricant is usually supplied by a circulating pump capable of supplying each chain drive with a continuous stream of oil. The oil should be applied inside the chain loop evenly across the chain width, and directed at the slack strand.

The chain manufacturer should be consulted when it appears desirable to use a type of lubricant other than that recommended.

Installation and Alignment.—Sprockets should have the tooth form, thickness, profile, and diameters conforming to ASME/ANSI B29.1M. For maximum service life small sprockets operating at moderate to high speeds, or near the rated horsepower, should have hardened teeth. Normally, large sprockets should not exceed 120 teeth.

In general a center distance of 30 to 50 chain pitches is most desirable. The distance between sprocket centers should provide at least a 120 degree chain wrap on the smaller sprocket. Drives may be installed with either adjustable or fixed center distances. Adjustable centers simplify the control of chain slack. Sufficient housing clearance must always be provided for the chain slack to obtain full chain life.

Accurate alignment of shafts and sprocket tooth faces provides uniform distribution of the load across the entire chain width and contributes substantially to optimum drive life.

Shafting, bearings, and foundations should be suitable to maintain the initial alignment. Periodic maintenance should include an inspection of alignment.

Example of Roller Chain Drive Design Procedure.—The selection of a roller chain and sprockets for a specific design requirement is best accomplished by a systematic step-by-step procedure such as is used in the following example.

Example: Select a roller chain drive to transmit 10 horsepower from a countershaft to the main shaft of a wire drawing machine. The countershaft is $1\frac{15}{16}$-inches diameter and operates at 1000 rpm. The main shaft is also $1\frac{15}{16}$-inches diameter and must operate between 378 and 382 rpm. Shaft centers, once established, are fixed and by initial calculations must be approximately $22\frac{1}{2}$ inches. The load on the main shaft is uneven and presents "peaks," which place it in the heavy shock load category. The input power is supplied by an electric motor. The driving head is fully enclosed and all parts are lubricated from a central system.

Step 1. Service Factor: From Table 15 the service factor for heavy shock load and an electric motor drive is 1.5.

Step 2. Design Horsepower: The horsepower upon which the chain selection is based (design horsepower) is equal to the specified horsepower multiplied by the service factor, $10 \times 1.5 = 15$ hp.

Step 3. Chain Pitch and Small Sprocket Size for Single-Strand Drive: In Table under 1000 rpm, a $\frac{5}{8}$-inch pitch chain with a 24-tooth sprocket or a $\frac{3}{4}$-inch pitch chain with a 15-tooth sprocket are possible choices.

Step 4. Check of Chain Pitch and Sprocket Selection: From Table 10 it is seen that only the 24-tooth sprocket in Step 3 can be bored to fit the $1\frac{15}{16}$-inch diameter main shaft. In Table a $\frac{5}{8}$-pitch chain at a small sprocket speed of 1000 rpm is rated at 15.5 hp for a 24-tooth sprocket.

Step 5. Selection of Large Sprocket: Since the driver is to operate at 1000 rpm and the driven at a minimum of 378 rpm, the speed ratio 1000/378 = 2.646. Therefore the large sprocket should have $24 \times 2.646 = 63.5$ (use 63) teeth.

This combination of 24 and 63 teeth will produce a main drive shaft speed of 381 rpm which is within the limitation of 378 to 382 rpm established in the original specification.

Step 6. Computation of Chain Length: Since the 24- and 63-tooth sprockets are to be placed on $22\frac{1}{2}$-inch centers, the chain length is determined from the formula:

$$L = 2C + \frac{N}{2} + \frac{n}{2} + \left(\frac{N-n}{2\pi}\right)^2 \times \frac{1}{C}$$

where L = chain length in pitches; C = shaft center distance in pitches; N = number of teeth in large sprocket; and n = number of teeth in small sprocket.

$$L = 2 \times 36 + \frac{63 + 24}{2} + \left(\frac{63 - 24}{6.28}\right)^2 \times \frac{1}{36} = 116.57 \text{ pitches}$$

Step 7. Correction of Center Distance: Since the chain is to couple at a whole number of pitches, 116 pitches will be used and the center distance recomputed based on this figure using the formula on page 2437 where c is the center distance in inches and P is the pitch.

$$c = \frac{P}{8}(2L - N - n + \sqrt{(2L - N - n)^2 - 0.810(N - n)^2})$$

$$c = \frac{5}{64}(2 \times 116 - 63 - 24 + \sqrt{(2 \times 116 - 63 - 24)^2 - 0.810(63 - 24)^2})$$

$$c = \frac{5}{64}(145 + 140.69) = 22.32 \text{ inches, say } 22\frac{3}{8} \text{ inches}$$

STANDARDS FOR ELECTRIC MOTORS

Classes of NEMA Standards.—National Electrical Manufacturers Association Standards, available from the Association at 2101 L Street, NW, Washington, DC 20037, are of two classes: 1) *NEMA Standard,* which relates to a product commercially standardized and subject to repetitive manufacture, which standard has been approved by at least 90 per cent of the members of the Subdivision eligible to vote thereon; and 2) *Suggested Standard for Future Design,* which may not have been regularly applied to a commercial product, but which suggests a sound engineering approach to future development and has been approved by at least two-thirds of the members of the Subdivision eligible to vote thereon.

Authorized Engineering Information consists of explanatory data and other engineering information of an informative character not falling within the classification of NEMA Standard or Suggested Standard for Future Design.

Mounting Dimensions and Frame Sizes for Electric Motors.—Dimensions for foot-mounted electric motors as standardized in the United States by the National Electrical Manufacturers Association (NEMA) include the spacing of bolt holes in the feet of the motor, the distance from the bottom of the feet to the center-line of the motor shaft, the size of the conduit, the length and diameter of shaft, and other dimensions likely to be required by designers or manufacturers of motor-driven equipment. The Standard provides dimensions for face-mounted and flange-mounted motors by means of standard motor frame numbers.

Standard dimensions also are given where the motor is to be mounted upon a belt-tightening base or upon rails.

The NEMA standards also prescribe lettering for dimension drawings, mounting and terminal housing locations and dimensions, symbols and terminal connections, and provision for grounding of field wiring. In addition, the standards give recommended knock-out and clearance hole dimensions; tolerances on shaft extension diameters and keyseats; methods of measuring shaft run-out and eccentricity, also face runout of mounting surfaces; and tolerances of face-mounted and flanged-mounted motors.

Design Letters of Polyphase Integral-horsepower Motors.—Designs A, B, C, and D motors are squirrel-cage motors designed to withstand full voltage starting and developing locked-rotor torque and breakdown torque, drawing locked-rotor current, and having a slip as specified below:

Design A: Locked-rotor torque as shown in Table 2, breakdown torque as shown in Table 3, locked-rotor current higher than the values shown in Table 1, and a slip at rated load of less than 5 per cent. Motors with 10 or more poles may have a slightly greater slip.

Table 1. NEMA Standard Locked-rotor Current of 3-phase 60-hertz Integral-horsepower Squirrel-cage Induction Motors Rated at 230 Volts

Horse-power	Locked-rotor Current, Amps.	Design Letters	Horse-power	Locked-rotor Current, Amps.	Design Letters	Horse-power	Locked-rotor Current, Amps.	Design Letters
½	20	B, D	7½	127	B, C, D	50	725	B, C, D
¾	25	B, D	10	162	B, C, D	60	870	B, C, D
1	30	B, D	15	232	B, C, D	75	1085	B, C, D
1½	40	B, D	20	290	B, C, D	100	1450	B, C, D
2	50	B, D	25	365	B, C, D	125	1815	B, C, D
3	64	B, C, D	30	435	B, C, D	150	2170	B, C, D
5	92	B, C, D	40	580	B, C, D	200	2900	B, C

Note: The locked-rotor current of a motor is the steady-state current taken from the line with the rotor locked and with rated voltage and frequency applied to the motor.

For motors designed for voltages other than 230 volts, the locked-rotor current is inversely proportional to the voltages. For motors larger than 200 hp, see NEMA Standard MG 1-12.34.

Table 2. NEMA Standard Locked-rotor Torque of Single-speed Polyphase 60- and 50-hertz Squirrel-cage Integral-horsepower Motors with Continuous Ratings

Hp	Designs A and B							Design C		
	Synchronous Speed, rpm									
60 hertz	3600	1800	1200	900	720	600	514	1800	1200	900
50 hertz	3000	1500	1000	750	…	…	…	1500	1000	750
	Percent of Full-load Torque[a]									
½	…	…	…	140	140	115	110	…	…	…
¾	…	…	175	135	135	115	110	…	…	…
1	…	275	170	135	135	115	110	…	…	…
1½	175	250	165	130	130	115	110	…	…	…
2	170	235	160	130	125	115	110	…	…	…
3	160	215	155	130	125	115	110	…	250	225
5	150	185	150	130	125	115	110	250	250	225
7½	140	175	150	125	120	115	110	250	225	200
10	135	165	150	125	120	115	110	250	225	200
15	130	160	140	125	120	115	110	225	200	200
20	130	150	135	125	120	115	110	200 for all sizes above 15 hp.		
25	130	150	135	125	120	115	110			
30	130	150	135	125	120	115	110			
40	125	140	135	125	120	115	110			
50	120	140	135	125	120	115	110			
60	120	140	135	125	120	115	110			
75	105	140	135	125	120	115	110	For Design D motors, see footnote.		
100	105	125	125	125	120	115	110			
125	100	110	125	120	115	115	110			
150	100	110	120	120	115	115	…			
200	100	100	120	120	115	…	…			

[a] These values represent the upper limit of application for these motors.

Note: The locked-rotor torque of a motor is the minimum torque which it will develop at rest for all angular positions of the rotor, with rated voltage applied at rated frequency.

The locked-rotor torque of Design D, 60- and 50-hertz 4-, 6-, and 8-pole single-speed, polyphase squirrel-cage motors rated 150 hp and smaller, with rated voltage and frequency applied is 275 per cent of full-load torque, which represents the upper limit of application for these motors.

For motors larger than 200 hp, see NEMA Standard MG 1-12.37.

Table 3. NEMA Standard Breakdown Torque of Single-speed Polyphase Squirrel-cage, Integral-horsepower Motors with Continuous Ratings

Horsepower	Synchronous Speed, rpm						
60 hertz	3600	1800	1200	900	720	600	514
50 hertz	3000	1500	1000	750	…	…	…
	Per Cent of Full Load Torque						
	Designs A and B[a]						
½	…	…	…	225	200	200	200
¾	…	…	275	220	200	200	200
1	…	300	265	215	200	200	200
1½	250	280	250	210	200	200	200
2	240	270	240	210	200	200	200
3	230	250	230	205	200	200	200
5	215	225	215	205	200	200	200
7½	200	215	205	200	200	200	200
10–125, incl.	200	200	200	200	200	200	200
150	200	200	200	200	200	200	…
200	200	200	200	200	200	…	…
	Design C						
3	…	…	225	200	…	…	…
5	…	200	200	200	…	…	…
7½–200, incl.	…	190	190	190	…	…	…

[a] Design A values are in excess of those shown.

These values represent the upper limit of the range of application for these motors. For above 200 hp, see NEMA Standard MG1-12.38.

Design B: Locked-rotor torque as shown in Table 2, breakdown torque as shown in Table 3, locked-rotor current not exceeding that in Table 1, and a slip at rated load of less than 5 per cent. Motors with 10 or more poles may have a slightly greater slip.

Design C: Locked-rotor torque for special high-torque applications up to values shown in Table 2, breakdown torque up to values shown in Table 3, locked-rotor current not exceeding values shown in Table 1 and a slip at rated load of less than 5 per cent.

Design D: Locked-rotor torque as indicated in Table 2, locked-rotor current not greater than that shown in Table 1 and a slip at rated load of 5 per cent or more.

Torque and Current Definitions.—The definitions which follow have been adopted as standard by the National Electrical Manufacturers Association.

Locked-Rotor or Static Torque: The locked-rotor torque of a motor is the minimum torque which it will develop at rest for all angular positions of the rotor, with rated voltage applied at rated frequency.

Breakdown Torque: The breakdown torque of a motor is the maximum torque which the motor will develop, with rated voltage applied at rated frequency, without an abrupt drop in speed (see Table 4).

Full-Load Torque: The full-load torque of a motor is the torque necessary to produce its rated horsepower at full load speed. In pounds at 1-foot radius, it is equal to the horsepower times 5252 divided by the full-load speed.

Pull-Out Torque: The pull-out torque of a synchronous motor is the maximum sustained torque which the motor will develop at synchronous speed with rated voltage applied at rated frequency and with normal excitation.

Pull-In Torque: The pull-in torque of a synchronous motor is the maximum constant torque under which the motor will pull its connected inertia load into synchronism at rated voltage and frequency, when its field excitation is applied.

Pull-Up Torque: The pull-up torque of an alternating current motor is the minimum torque developed by the motor during the period of acceleration from rest to the speed at which breakdown torque occurs. For motors which do not have a definite breakdown torque, the pull-up torque is the minimum torque developed up to rated speed.

Locked Rotor Current: The locked rotor current of a motor is the steady-state current taken from the line with the rotor locked and with rated voltage (and rated frequency in the case of alternating-current motors) applied to the motor.

Table 4. NEMA Standard Breakdown Torque of Polyphase Wound-rotor Motors with Continuous Ratings — 60- and 50-hertz

Horsepower	Speed, rpm			Horsepower	Speed, rpm		
	1800	1200	900		1800	1200	900
	Per cent of Full-load Torque				Per cent of Full-load Torque		
1	…	…	250	7½	275	250	225
1½	…	…	250	10	275	250	225
2	275	275	250	15	250	225	225
3	275	275	250	20–200, incl.	225	225	225
5	275	275	250	…	…	…	…

These values represent the upper limit of the range of application for these motors.

Standard Direction of Motor Rotation.—The standard direction of rotation for all non-reversing direct-current motors, all alternating-current single-phase motors, all synchronous motors, and all universal motors, is *counterclockwise* when facing that end of the motor opposite the drive.

This rule does not apply to two- and three-phase induction motors, as in most applications the phase sequence of the power lines is rarely known.

Motor Types According to Variability of Speed.—Five types of motors classified according to variability of speed are:

Constant-speed Motors: In this type of motor the normal operating speed is constant or practically constant; for example, a synchronous motor, an induction motor with small slip, or a direct-current shunt-wound motor.

Varying-speed Motor: In this type of motor, the speed varies with the load, ordinarily decreasing when the load increases; such as a series-wound or repulsion motor.

Adjustable-speed Motor: In this type of motor, the speed can be varied gradually over a considerable range, but when once adjusted remains practically unaffected by the load; such as a direct-current shunt-wound motor with field resistance control designed for a considerable range of speed adjustment.

The base speed of an adjustable-speed motor is the lowest rated speed obtained at rated load and rated voltage at the temperature rise specified in the rating.

Adjustable Varying-speed Motor: This type of motor is one in which the speed can be adjusted gradually, but when once adjusted for a given load will vary in considerable degree with the change in load; such as a direct-current compound-wound motor adjusted by field control or a wound-rotor induction motor with rheostatic speed control.

Multispeed Motor: This type of motor is one which can be operated at any one of two or more definite speeds, each being practically independent of the load; such as a direct-current motor with two armature windings or an induction motor with windings capable of various pole groupings. In the case of multispeed permanent-split capacitor and shaded pole motors, the speeds are dependent upon the load.

Pull-up Torque.—NEMA Standard pull up torques for single-speed, polyphase, squirrel-cage integral-horsepower motors, Designs A and B, with continuous ratings and with rated voltage and frequency applied are as follows: When the locked-rotor torque given in Table 2 is 110 per cent or less, the pull-up torque is 90 per cent of the locked-rotor torque; when the locked-rotor torque is greater than 110 per cent but less than 145 per cent, the pull-up torque is 100 per cent of full-load torque; and when the locked-rotor torque is 145 per cent or more, the pull-up torque is 70 per cent of the locked-rotor torque.

For Design C motors, with rated voltage and frequency applied, the pull-up torque is not less than 70 per cent of the locked-rotor torque as given in Table 2.

Types and Characteristics of Electric Motors

Types of Direct-Current Motors.—Direct-current motors may be grouped into three general classes: series-wound; shunt-wound; and compound-wound.

In the *series-wound motor* the field windings, which are fixed in the stator frame, and the armature windings, which are placed around the rotor, are connected in series so that all current passing through the armature also passes through the field. In the *shunt-wound motor,* both armature and field are connected across the main power supply so that the armature and field currents are separate. In the *compound-wound motor,* both series and shunt field windings are provided and these may be connected so that the currents in both are flowing in the same direction, called *cumulative compounding,* or so that the currents in each are flowing in opposite directions, called *differential compounding.*

Characteristics of Series-wound Direct-Current Motors.—In the series-wound motor, any increase in load results in more current passing through the armature and the field windings. As the field is strengthened by this increased current, the motor speed decreases. Conversely, as the load is decreased the field is weakened and the speed increases and at very light loads may become excessive. For this reason, series-wound direct-current motors are usually directly connected or geared to the load to prevent "runaway." (A series-wound motor, designated as series-shunt wound, is sometimes provided with a light shunt field winding to prevent dangerously high speeds at light loads.) The increase in armature current with increasing load produces increased torque, so that the series-wound motor is particularly suited to heavy starting duty and where severe overloads may be expected. Its speed may be adjusted by means of a variable resistance placed in series with the motor, but due to variation with load, the speed cannot be held at any constant value. This variation of speed with load becomes greater as the speed is reduced. Series-wound motors are used where the load is practically constant and can easily be controlled by hand. They are usually limited to traction and lifting service.

Shunt-wound Direct-Current Motors.—In the shunt-wound motor, the strength of the field is not affected appreciably by change in the load, so that a fairly constant speed (about 10 to 12 per cent drop from no load to full load speed) is obtainable. This type of motor may be used for the operation of machines requiring an approximately constant speed and imposing low starting torque and light overload on the motor.

The shunt-wound motor becomes an adjustable-speed motor by means of field control or by armature control. If a variable resistance is placed in the field circuit, the amount of current in the field windings and hence the speed of rotation can be controlled. As the speed increases, the torque decreases proportionately, resulting in nearly constant horsepower. A speed range of 6 to 1 is possible using field control, but 4 to 1 is more common. Speed regulation is somewhat greater than in the constant-speed shunt-wound motors, ranging from about 15 to 22 per cent. If a variable resistance is placed in the armature circuit, the voltage applied to the armature can be reduced and hence the speed of rotation can be reduced over a range of about 2 to 1. With armature control, speed regulation becomes poorer as speed is decreased, and is about 100 per cent for a 2 to 1 speed range. Since the current through the field remains unchanged, the torque remains constant.

Machine Tool Applications: The adjustable-speed shunt-wound motors are useful on larger machines of the boring mill, lathe, and planer type and are particularly adapted to spindle drives because constant horsepower characteristics permit heavy cuts at low speed and light or finishing cuts at high speed. They have long been used for planer drives because they can provide an adjustable low speed for the cutting stroke and a high speed for the return stroke. Their application has been limited, however, to plants in which direct-current power is available.

Adjustable-voltage Shunt-wound Motor Drive.—More extensive use of the shunt-wound motor has been made possible by a combination drive that includes a means of converting alternating current to direct current. This conversion may be effected by a self-contained unit consisting of a separately excited direct-current generator driven by a constant speed alternating-current motor connected to the regular alternating-current line, or by an electronic rectifier with suitable controls connected to the regular alternating-current supply lines. The latter has the advantage of causing no vibration when mounted directly on the machine tool, an important factor in certain types of grinders.

In this type of adjustable-speed, shunt-wound motor drive, speed control is effected by varying the voltage applied to the armature while supplying constant voltage to the field. In addition to providing for the adjustment of the voltage supplied by the conversion unit to the armature of the shunt-wound motor, the amount of current passing through the motor field may also be controlled. In fact, a single control may be provided to vary the motor speed from minimum to base speed (speed of the motor at full load with rated voltage on armature and field) by varying the voltage applied to the armature and from base speed to maximum speed by varying the current flowing through the field. When so controlled, the motor operates at constant torque up to base speed and at constant horsepower above base speed.

Speed Range: Speed ranges of at least 20 to 1 below base speed and 4 or 5 to 1 above base speed (a total range of 100 to 1, or more) are obtainable as compared with about 2 to 1 below normal speed and 3 or 4 to 1 above normal speed for the conventional type of control. Speed regulation may be as great as 25 per cent at high speeds. Special electronic controls, when used with this type shunt motor drive, make possible maintenance of motor speeds with as little variation as ½ to 1 per cent of full load speed from full load to no load over a line voltage variation of ± 10 per cent and over any normal variation in motor temperature and ambient temperature.

Applications: These direct-current, adjustable-voltage drives, as they are sometimes called, have been applied successfully to such machine tools as planers, milling machines, boring mills and lathes, as well as to other industrial machines where wide, stepless speed control, uniform speed under all operating conditions, constant torque acceleration and adaptability to automatic operation are required.

Compound-wound Motors.—In the compound-wound motor, the speed variation due to load changes is much less than in the series-wound motor, but greater than in the shunt-wound motor (ranging up to 25 per cent from full load to no load). It has a greater starting torque than the shunt-wound motor, is able to withstand heavier overloads, but has a narrower adjustable speed range. Standard motors of this type have a cumulative-compound winding, the differential-compound winding being limited to special applications. They are used where the starting load is very heavy or where the load changes suddenly and violently as with reciprocating pumps, printing presses and punch presses.

Types of Polyphase Alternating-Current Motors.—The most widely used polyphase motors are of the induction type. The *"squirrel cage" induction motor* consists of a wound stator which is connected to an external source of alternating-current power and a laminated steel core rotor with a number of heavy aluminum or copper conductors set into the core around its periphery and parallel to its axis. These conductors are connected together at each end of the rotor by a heavy ring, which provides closed paths for the currents induced in the rotor to circulate. The rotor bars form, in effect, a "squirrel cage" from which the motor takes its name.

Wound-rotor type of *Induction motor:* This type has in addition to a squirrel cage, a series of coils set into the rotor which are connected through slip-rings to external variable resistors. By varying the resistance of the wound-rotor circuits, the amount of current flowing in these circuits and hence the speed of the motor can be controlled. Since the rotor of an induction motor is not connected to the power supply, the motor is said to operate by transfer action and is analogous to a transformer with a short-circuited secondary that is free to rotate. Induction motors are built with a wide range of speed and torque characteristics which are discussed under "Operating Characteristics of Squirrel-cage Induction Motors."

Synchronous Motor: The other type of polyphase alternating-current motor used industrially is the *synchronous motor.* In contrast to the induction motor, the rotor of the synchronous motor is connected to a direct-current supply which provides a field that rotates in step with the alternating-current field in the stator. After having been brought up to synchronous speed, which is governed by the frequency of the power supply and the number of poles in the rotor, the synchronous motor operates at this constant speed throughout its entire load range.

Operating Characteristics of Squirrel-cage Induction Motors.—In general, squirrel-cage induction motors are simple in design and construction and offer rugged service. They are essentially constant-speed motors, their speed changing very little with load and not being subject to adjustment. They are used for a wide range of industrial applications calling for integral horsepower ratings. According to the NEMA (National Electrical Manufacturers Association) Standards, there are four classes of squirrel-cage induction motors designated respectively as *A*, *B*, *C*, and *D*.

Design A motors are not commonly used since Design *B* has similar characteristics with the advantage of lower starting current.

Design B: motors may be designated as a general purpose type suitable for the majority of polyphase alternating-current applications such as blowers, compressors, drill presses, grinders, hammer mills, lathes, planers, polishers, saws, screw machines, shakers, stokers, etc. The starting torque at 1800 R.P.M. is 250 to 275 per cent of full load torque for 3 H.P. and below; for 5 H.P. to 75 H. P. ratings the starting torque ranges from 185 to 150 per cent of full load torque. They have low starting current requirements, usually no more than 5 to 6 times full load current and can be started at full voltage. Their slip (difference between synchronous speed and actual speed at rated load) is relatively low.

Design C: motors have high starting torque (up to 250 per cent of full load torque) but low starting current. They can be started at full voltage. Slip at rated load is relatively low. They are used for compressors requiring a loaded start, heavy conveyors, reciprocating pumps and other applications requiring high starting torque.

Design D: motors have high slip at rated load, that is, the motor speed drops off appreciably as the load increases, permitting use of the stored energy of a flywheel. They provide heavy starting torque, up to 275 per cent of full load torque, are quiet in operation and have relatively low starting current. Applications are for impact, shock and other high peak loads or flywheel drives such as trains, elevators, hoists, punch and drawing presses, shears, etc.

Design F: motors are no longer standard. They had low starting torque, about 125 per cent of full-load torque, and low starting current. They were used to drive machines which required infrequent starting at no load or at very light load.

Multiple-Speed Induction Motors.—This type has a number of windings in the stator so arranged and connected that the number of effective poles and hence the speed can be changed. These motors are for the same types of starting conditions as the conventional squirrel-cage induction motors and are available in designs that provide constant horsepower at all rated speeds and in designs that provide constant torque at all rated speeds.

Typical speed combinations obtainable in these motors are 600, 900, 1200 and 1800 R.P.M.; 450, 600, 900 and 1200 R.P.M.; and 600, 720, 900 and 1200 R.P.M.

Where a gradual change in speed is called for, a wound rotor may be provided in addition to the multiple stator windings.

Wound-Rotor Induction Motors.—These motors are designed for applications where extremely low starting current with high starting torque are called for, such as in blowers, conveyors, compressors, fans and pumps. They may be employed for adjustable-varying speed service where the speed range does not extend below 50 per cent of synchronous speed, as for steel plate-forming rolls, printing presses, cranes, blowers, stokers, lathes and milling machines of certain types. The speed regulation of a wound rotor induction motor ranges from 5 to 10 per cent at maximum speed and from 18 to 30 per cent at low speed. They are also employed for reversing service as in cranes, gates, hoists and elevators.

High-Frequency Induction Motors.—This type is used in conjunction with frequency changers when very high speeds are desired, as on grinders, drills, routers, portable tools or woodworking machinery. These motors have an advantage over the series-wound or universal type of high speed motor in that they operate at a relatively constant speed over the entire load range. A motor-generator set, a two-unit frequency converter or a single unit inductor frequency converter may be used to supply three-phase power at the frequency required. The single unit frequency converter may be obtained for delivering any one of a number of frequencies ranging from 360 to 2160 cycles and it is self-driven and self-excited from the general polyphase power supply.

Synchronous Motors.—These are widely used in electric timing devices; to drive machines that must operate in synchronism; and also to operate compressors, rolling mills, crushers which are started without load, paper mill screens, shredders, vacuum pumps and motor-generator sets. Synchronous motors have an inherently high power factor and are often employed to make corrections for the low power factor of other types of motors on the same system.

Types of Single-Phase Alternating-Current Motors.—Most of the single-phase alternating-current motors are basically induction motors distinguished by different arrangements for starting. (A single-phase induction motor with only a squirrel-cage rotor has no starting torque.) In the *capacitor-start* single-phase motor, an auxiliary winding in the stator is connected in series with a capacitor and a centrifugal switch. During the starting and accelerating period the motor operates as a two-phase induction motor. At about two-thirds full-load speed, the auxiliary circuit is disconnected by the switch and the motor then runs as a single-phase induction motor. In the *capacitor-start, capacitor-run* motor, the auxiliary circuit is arranged to provide high effective capacity for high starting torque and to remain connected to the line but with reduced capacity during the running period. In the *single-value capacitor* or *capacitor split-phase* motor, a relatively small continuously-rated capacitor is permanently connected in one of the two stator windings and the motor both starts and runs like a two-phase motor.

In the *repulsion-start* single-phase motor, a drum-wound rotor circuit is connected to a commutator with a pair of short-circuited brushes set so that the magnetic axis of the rotor winding is inclined to the magnetic axis of the stator winding. The current flowing in this rotor circuit reacts with the field to produce starting and accelerating torques. At about two-thirds full load speed the brushes are lifted, the commutator is short circuited and the motor runs as a single-phase squirrel-cage motor. The *repulsion* motor employs a repulsion winding on the rotor for both starting and running. The *repulsion-induction* motor has an outer winding on the rotor acting as a repulsion winding and an inner squirrel-cage winding. As the motor comes up to speed, the induced rotor current partially shifts from the repulsion winding to the squirrel-cage winding and the motor runs partly as an induction motor.

In the *split-phase* motor, an auxiliary winding in the stator is used for starting with either a resistance connected in series with the auxiliary winding (*resistance-start*) or a reactor in series with the main winding (*reactor-start*).

The *series-wound* single-phase motor has a rotor winding in series with the stator winding as in the series-wound direct-current motor. Since this motor may also be operated on direct current, it is called a *universal* motor.

Characteristics of Single-Phase Alternating-Current Motors.—Single-phase motors are used in sizes up to about 7½ horsepower for heavy starting duty chiefly in home and commercial appliances for which polyphase power is not available. The *capacitor-start* motor is available in normal starting torque designs for such applications as centrifugal pumps, fans, and blowers and in high-starting torque designs for reciprocating compressors, pumps, loaded conveyors, or belts. The *capacitor-start, capacitor-run* motor is exceptionally quiet in operation when loaded to at least 50 per cent of capacity. It is available in low-torque designs for fans and centrifugal pumps and in high-torque designs for applications similar to those of the capacitor-start motor.

The *capacitor split-phase* motor requires the least maintenance of all single-phase motors, but has very low starting torque. Its high maximum torque makes it potentially useful in floor sanders or in grinders where momentary overloads due to excessive cutting pressure are experienced. It is also used for slow-speed direct connected fans.

The *repulsion-start, induction-run* motor has higher starting torque than the capacitor motors, although for the same current, the capacitor motors have equivalent pull-up and maximum torque. Electrical and mechanical noise and the extra maintenance sometimes required are disadvantages. These motors are used for compressors, conveyors and stokers starting under full load. The *repulsion-induction* motor has relatively high starting torque and low starting current. It also has a smooth speed-torque curve with no break and a greater ability to withstand long accelerating periods than capacitor type motors. It is particularly suitable for severe starting and accelerating duty and for high inertia loads such as laundry extractors. Brush noise is, however, continuous.

The *repulsion* motor has no limiting synchronous speed and the speed changes with the load. At certain loads, slight changes in load cause wide changes in speed. A brush shifting arrangement may be provided to adjust the speed which may have a range of 4 to 1 if full rated constant torque is applied but a decreasing range as the torque falls below this value. This type of motor may be reversed by shifting the brushes beyond the neutral point. These motors are suitable for machines requiring constant-torque and adjustable speed.

The *split-phase* and *universal* motors are limited to about ⅓ H.P. ratings and are used chiefly for small appliance and office machine applications.

Motors with Built-in Speed Reducers.—Electric motors having built-in speed-changing units are compact and the design of these motorized speed reducers tends to improve the appearance of the machines which they drive. There are several types of these speed reducers; they may be classified according to whether they are equipped with worm gearing, a regular gear train with parallel shafts, or planetary gearing.

The claims made for the worm gearing type of reduction unit are that the drive is quiet in operation and well adapted for use where the slow-speed shaft must be at right angles to the motor shaft and where a high speed ratio is essential.

For very low speeds, the double reduction worm gearing units are suitable. In these units two sets of worm gearing form the gear train, and both the slow-speed shaft and the armature shaft are parallel. The intermediate worm gear shaft can be built to extend from the housing, if required, so as to make two countershaft speeds available on the same unit.

In the parallel-shaft type of speed reducer, the slow-speed shaft is parallel with the armature shaft. The slow-speed shaft is rotated by a pinion on the armature shaft, this pinion meshing with a larger gear on the slow-speed shaft.

Geared motors having built-in speed-changing units are available with constant-mesh change gears for varying the speed ratio.

Planetary gearing permits a large speed reduction with few parts; hence, it is well adapted for geared-head motor units where economy and compactness are essential. The slow-speed shaft is in line with the armature shaft.

Factors Governing Motor Selection

Speed, Horsepower, Torque and Inertia Requirements.—Where more than one speed or a range of speeds are called for, one of the following types of motors may be selected, depending upon other requirements: For direct-current, the standard shunt-wound motor with field control has a 2 to 1 range in some designs; the adjustable speed motor may have a range of from 3 to 1 up to 6 to 1; the shunt motor with adjustable voltage supply has a range up to 20 to 1 or more below base speed and 4 or 5 to 1 above base speed, making a total range of up to 100 to 1 or more. For polyphase alternating current, multi-speed squirrel-cage induction motors have 2, 3 or 4 fixed speeds; the wound-rotor motor has a 2 to 1 range. The two-speed wound-rotor motor has a 4 to 1 range. The brush-shifting shunt motor has a 4 to 1 range. The brush-shifting series motor has a 3 to 1 range; and the squirrel-cage motor with a variable-frequency supply has a very wide range. For single-phase alternating current, the brush-shifting repulsion motor has a 2½ to 1 range; the capacitor motor with tapped winding has a 2 to 1 range and the multi-speed capacitor motor has 2 or 3 fixed speeds. Speed regulation (variation in speed from no load to full load) is greatest with motors having series field windings and entirely absent with synchronous motors.

Horsepower: Where the load to be carried by the motor is not constant but follows a definite cycle, a horsepower-time curve enables the peak horsepower to be determined as well as the root-mean-square-average horsepower, which indicates the proper motor rating from a heating standpoint. Where the load is maintained at a constant value for a period of from 15 minutes to 2 hours depending on the size, the horsepower rating required will usually not be less than this constant value. When selecting the size of an induction motor, it should be kept in mind that this type of motor operates at maximum efficiency when it is loaded to full capacity. Where operation is to be at several speeds, the horsepower requirement for each speed should be considered.

Torque: Starting torque requirements may vary from 10 per cent of full load to 250 per cent of full load torque depending upon the type of machine being driven. Starting torque may vary for a given machine because of frequency of start, temperature, type and amount of lubricant, etc., and such variables should be taken into account. The motor torque supplied to the machine must be well above that required by the driven machine at all points up to full speed. The greater the excess torque, the more rapid the acceleration. The approximate time required for acceleration from rest to full speed is given by the formula:

$$\text{Time} = \frac{N \times WR^2}{T_a \times 308} \text{ seconds}$$

where N = Full load speed in R.P.M.

T_a = Torque = average foot-pounds available for acceleration.

WR^2 = Inertia of rotating part in pounds feet squared (W = weight and R = radius of gyration of rotating part).

308 = Combined constant converting minutes into seconds, weight into mass and radius into circumference.

If the time required for acceleration is greater than 20 seconds, special motors or starters may be required to avoid overheating.

The running torque T_r is found by the formula:

$$T_r = \frac{5250 \times \text{HP}}{N} \text{ foot pounds}$$

where *H.P.* = Horsepower being supplied to the driven machine

N = Running speed in R.P.M.

5250 = Combined constant converting horsepower to foot-pounds per minute and work per revolution into torque.

The peak horsepower determines the maximum torque required by the driven machine and the motor must have a maximum running torque in excess of this value.

Inertia: The inertia or flywheel effect of the rotating parts of a driven machine will, if large, appreciably affect the accelerating time and, hence, the amount of heating in the motor. If synchronous motors are used, the inertia (WR^2) of both the motor rotor and the rotating parts of the machine must be known since the pull-in torque (torque required to bring the driven machine up to synchronous speed) varies approximately as the square root of the total inertia of motor and load.

Space Limitations in Motor Selection.—If the motor is to become an integral part of the machine which it drives and space is at a premium, a partial motor may be called for. A complete motor is one made up of a stator, a rotor, a shaft, and two end shields with bearings. A *partial motor* is without one or more of these elements. One common type is furnished without drive-end end shield and bearing and is directly connected to the end or side of the machine which it drives, such as the headstock of a lathe. A so-called *shaftless type of motor* is supplied without shaft, end shields or bearings and is intended for built-in application in such units as multiple drilling machines, precision grinders, deep well pumps, compressors and hoists where the rotor is actually made a part of the driven machine. Where a partial motor is used, however, proper ventilation, mounting, alignment and bearings must be arranged for by the designer of the machine to which it is applied.

Sometimes it is possible to use a motor having a smaller frame size and wound with Class *B* insulation, permitting it to be subjected to a higher temperature rise than the larger-frame Class *A* insulated motor having the same horsepower rating.

Temperatures.—The applicability of a given motor is limited not only by its load starting and carrying ability, but also by the temperature which it reaches under load. Motors are given temperature ratings which are based upon the type of insulation (Class A or Class B are the most common) used in their construction and their type of frame (open, semienclosed, or enclosed).

Insulating Materials: Class A materials are: cotton, silk, paper, and similar organic materials when either impregnated or immersed in a liquid dielectric; molded and laminated materials with cellulose filler, phenolic resins, and other resins of similar properties;

films and sheets of cellulose acetate and other cellulose derivatives of similar properties;

and varnishes (enamel) as applied to conductors.

Class B insulating materials are: materials or combinations of materials such as mica, glass fiber, asbestos, etc., with suitable bonding substances. Other materials shown capable of operation at Class B temperatures may be included.

Ambient Temperature and Allowable Temperature Rise: Normal ambient temperature is taken to be 40°C (104°F). For open general-purpose motors with Class A insulation, the normal temperature rise on which the performance guarantees are based is 40°C (104°F).

Motors with Class A insulation having protected, semiprotected, drip-proof, or splash-proof, or drip-proof protected enclosures have a 50°C (122°F) rise rating.

Motors with Class A insulation and having totally enclosed, fan-cooled, explosion-proof, waterproof, dust-tight, submersible, or dust-explosion-proof enclosures have a 55°C (131°F) rise rating.

Motors with Class B insulation are permissible for total temperatures up to 110 degrees C (230°F) for open motors and 115°C (239°F) for enclosed motors.

Motors Exposed to Injurious Conditions.—Where motors are to be used in locations imposing unusual operating conditions, the manufacturer should be consulted, especially where any of the following conditions apply: exposure to chemical fumes; operation in damp places; operation at speeds in excess of specified overspeed; exposure to combustible or explosive dust; exposure to gritty or conducting dust; exposure to lint; exposure to steam; operation in poorly ventilated rooms; operation in pits, or where entirely enclosed in boxes; exposure to inflammable or explosive gases; exposure to temperatures below 10°C (50°F); exposure to oil vapor; exposure to salt air; exposure to abnormal shock or vibration from external sources; where the departure from rated voltage is excessive; and or where the alternating-current supply voltage is unbalanced.

Improved insulating materials and processes and greater mechanical protection against falling materials and liquids make it possible to use general-purpose motors in many locations where special-purpose motors were previously considered necessary. *Splash-proof motors* having well-protected ventilated openings and specially treated windings are used where they are to be subjected to falling and splashing water or are to be washed down as with a hose. Where climatic conditions are not severe, this type of motor is also successfully used in unprotected outdoor installations.

If the surrounding atmosphere carries abnormal quantities of metallic, abrasive, or non-explosive dust or acid or alkali fumes, a *totally enclosed fan-cooled motor* may be called for. In this type, the motor proper is completely enclosed but air is blown through an outer shell that completely or partially surrounds the inner case. If the dust in the atmosphere tends to pack or solidify and close the air passages of open splash-proof or totally enclosed fan-cooled motors, *totally enclosed (nonventilated) motors* are used. This type, which is limited to low horsepower ratings, is also used for outdoor service in mild or severe climates.

Table 1. Characteristics and Applications of D.C. Motors, 1–300 hp

Type	Starting Duty	Maximum Momentary Running Torque	Speed Regulation	Speed Control[a]	Applications
Shunt-wound, constant-speed	Medium starting torque. Varies with voltage supplied to armature, and is limited by starting resistor to 125 to 200% full-load torque	125 to 200%. Limited by commutation	8 to 12%	Basic speed to 200% basic speed by field control	Drives where starting requirements are not severe. Use constant-speed or adjustable-speed, depending on speed required. Centrifugal pumps, fans, blowers, conveyors, elevators, wood- and metalworking machines
Shunt-wound, adjustable speed	Medium starting torque. Varies with voltage supplied to armature, and is limited by starting resistor to 125 to 200% full-load torque	125 to 200%. Limited by commutation	10 to 20%, increases with weak fields	Basic speed to 60% basic speed (lower for some ratings) by field control	Drives where starting requirements are not severe. Use constant-speed or adjustable-speed, depending on speed required. Centrifugal pumps, fans, blowers, conveyors, elevators, wood- and metalworking machines
Shunt-wound, adjustable voltage control	Medium starting torque. Varies with voltage supplied to armature, and is limited by starting resistor to 125 to 200% full-load torque	125 to 200%. Limited by commutation	Up to 25%. Less than 5% obtainable with special rotating regulator	Basic speed to 2% basic speed and basic speed to 200% basic speed	Drives where wide, stepless speed control, uniform speed, constant-torque acceleration and adaptability to automatic operation are required. Planers, milling machines, boring machines, lathes, etc.
Compound wound, constant-speed	Heavy starting torque, Limited by starting resistor to 130 to 260% of full-load torque	130 to 260%. Limited by commutation	Standard compounding 25%. Depends on amount of series winding	Basic speed to 125% basic speed by field control	Drives requiring high starting torque and fairly constant speed. Pulsating loads. Shears, bending rolls, pumps, conveyors, crushers, etc.
Series-wound, varying-speed	Very heavy starting torque. Limited to 300 to 350% full-load torque	300 to 350%. Limited by commutation	Very high. Infinite no-load speed	From zero to maximum speed, depending on control and load	Drives where very high starting torque is required and speed can be regulated. Cranes, hoists, gates, bridges, car dumpers, etc.

[a] Minimum speed below basic speed by armature control limited by heating.

Table 2. Characteristics and Applications of Polyphase AC Motors

Polyphase Type	Ratings hp	Speed Regulation	Speed Control	Starting Torque	Breakdown Torque	Applications
General-purpose squirrel cage, normal stg current, normal stg torque. Design B	0.5 to 200	Less than 5%	None, except multi-speed types, designed for two to four fixed speeds	100 to 250% of full-load	200 to 300% of full-load	Constant-speed service where starting torque is not excessive. Fans, blowers, rotary compressors, centrifugal pumps, woodworking machines, machine tools, line shafts
Full-voltage starting, high stg torque, normal stg current, squirrel-cage, Design C	3 to 150	Less than 5%	None except multi-speed types, designed for two to four fixed speeds	200 to 250% of full-load	190 to 225% of full-load	Constant-speed service where fairly high starting torque is required at infrequent intervals with starting current of about 500% full-load. Reciprocating pumps and compressors, conveyors, crushers, pulverizers, agitators, etc.
Full-voltage starting, high stg-torque, high-slip squirrel cage, Design D	0.5 to 150	Drops about 7 to 12% from no load to full load	None, except multi-speed types, designed for two to four fixed speeds	275% of full-load depending on speed and rotor resistance	275% of full-load Will usually not stall until loaded to its maximum torque, which occurs at standstill	Constant-speed service and high-starting torque if starting not too frequent, and for taking high-peak loads with or without flywheels. Punch presses, die stamping, shears, bulldozers, bailers, hoists, cranes, elevators, etc.
Wound-rotor, external-resistance starting	0.5 to several thousand	With rotor rings short-circuited drops about 3% for large to 5% for small sizes	Speed can be reduced to 50% of normal by rotor resistance. Speed varies inversely as the load	Up to 300% depending on external resistance in rotor circuit and how distributed	200% when rotor slip rings are short circulated	Where high-starting torque with low-starting current or where limited speed control is required. Fans, centrifugal and plunger pumps, compressors, conveyors, hoists, cranes, ball mills, gate hoists, etc.
Synchronous	25 to several thousand	Constant	None, except special motors designed for two fixed speeds	40% for slow speed to 160% for medium speed 80% p-f designs. Special high-torque designs	Pull-out torque of unity-p-f motors 170%; 80%-p-f motors 225%. Special designs up to 300%	For constant-speed service, direct connection to slow-speed machines and where power-factor correction is required.

In addition to these special-purpose motors, there are two types of *explosion-proof motors* designed for hazardous locations. One type is for operation in hazardous dust locations (Class II, Group *G* of the National Electrical Code) and the other is for atmospheres containing explosive vapors and fumes classified as Class I, Group *D* (gasoline, naphtha, alcohols, acetone, lacquer-solvent vapors, natural gas).

Electric Motor Maintenance

Electric Motor Inspection Schedule.—Frequency and thoroughness of inspection depend upon such factors as 1) importance of the motor in the production scheme; 2) percentage of days the motor operates; 3) nature of service; and 4) winding conditions.

The following schedules, recommended by the General Electric Company, and covering both AC and DC motors are based on average conditions in so far as duty and dirt are concerned.

Weekly Inspection.—1) *Surroundings.* Check to see if the windings are exposed to any dripping water, acid or alcoholic fumes; also, check for any unusual amount of dust, chips, or lint on or about the motor. See if any boards, covers, canvas, etc., have been misplaced that might interfere with the motor ventilation or jam moving parts.

2) *Lubrication of sleeve-bearing motors.* In sleeve-bearing motors check oil level, if a gage is used, and fill to the specified line. If the journal diameter is less than 2 inches, the motor should be stopped before checking the oil level. For special lubricating systems, such as wool-packed, forced lubrication, flood and disk lubrication, follow instruction book. Oil should be added to bearing housing only when motor is at rest. A check should be made to see if oil is creeping along the shaft toward windings where it may harm the insulation.

3) *Mechanical condition.* Note any unusual noise that may be caused by metal-to-metal contact or any odor as from scorching insulation varnish.

4) *Ball or roller bearings.* Feel ball- or roller-bearing housings for evidence of vibration, and listen for any unusual noise. Inspect for creepage of grease on inside of motor.

5) *Commutators and brushes.* Check brushes and commutator for sparking. If the motor is on cyclic duty it should be observed through several cycles. Note color and surface condition of the commutator. A stable copper oxide-carbon film (as distinguished from a pure copper surface) on the commutator is an essential requirement for good commutation. Such a film may vary in color all the way from copper to straw, chocolate to black. It should be clean and smooth and have a high polish. All brushes should be checked for wear and pigtail connections for looseness. The commutator surface may be cleaned by using a piece of dry canvas or other hard, nonlinting material that is wound around and securely fastened to a wooden stick, and held against the rotating commutator.

6) *Rotors and armatures.* The air gap on sleeve-bearing motors should be checked, especially if they have been recently overhauled. After installing new bearings, make sure that the average reading is within 10 per cent, provided reading should be less than 0.020 inch. Check air passages through punchings and make sure they are free of foreign matter.

7) *Windings.* If necessary clean windings by suction or mild blowing. After making sure that the motor is dead, wipe off windings with dry cloth, note evidence of moisture, and see if any water has accumulated in the bottom of frame. Check if any oil or grease has worked its way up to the rotor or armature windings. Clean with carbon tetrachloride in a well-ventilated room.

8) *General.* This is a good time to check the belt, gears, flexible couplings, chain, and sprockets for excessive wear or improper location. The motor starting should be checked to make sure that it comes up to proper speed each time power is applied.

Monthly or Bimonthly Inspection.—1) *Windings.* Check shunt, series, and commutating field windings for tightness. Try to move field spools on the poles, as drying out may have caused some play. If this condition exists, a service shop should be consulted. Check motor cable connections for tightness.

2) *Brushes.* Check brushes in holders for fit and free play. Check the brush-spring pressure. Tighten brush studs in holders to take up slack from drying out of washers, making sure that studs are not displaced, particularly on DC motors. Replace brushes that are worn down almost to the brush rivet, examine brush faces for chipped toes or heels, and for heat cracks. Damaged brushes should be replaced immediately.

3) *Commutators.* Examine commutator surface for high bars and high mica, or evidence of scratches or roughness. See that the risers are clean and have not been damaged.

4) *Ball or roller bearings.* On hard-driven, 24-hour service ball- or roller-bearing motors, purge out old grease through drain hole and apply new grease. Check to make sure grease or oil is not leaking out of the bearing housing. If any leakage is present, correct the condition before continuing to operate.

5) *Sleeve bearings.* Check sleeve bearings for wear, including end-play bearing surfaces. Clean out oil wells if there is evidence of dirt or sludge. Flush with lighter oil before refilling.

6) *Enclosed gears.* For motors with enclosed gears, open drain plug and check oil flow for presence of metal scale, sand, or water. If condition of oil is bad, drain, flush, and refill as directed. Rock rotor to see if slack or backlash is increasing.

7) *Loads.* Check loads for changed conditions, bad adjustment, poor handling, or control.

8) *Couplings and other drive details.* Note if belt-tightening adjustment is all used up. Shorten belt if this condition exists. See if belt runs steadily and close to inside (motor edge) of pulley. Chain should be checked for evidence of wear and stretch. Clean inside of chain housing. Check chain-lubricating system. Note inclination of slanting base to make sure it does not cause oil rings to rub on housing.

Annual or Biannual Inspection.—1) *Windings.* Check insulation resistance by using either a megohmmeter or a voltmeter having a resistance of about 100 ohms per volt. Check insulation surfaces for dry cracks and other evidence of need for coatings of insulating material. Clean surfaces and ventilating passages thoroughly if inspection shows accumulation of dust. Check for mold or water standing in frame to determine if windings need to be dried out, varnished, and baked.

2) *Air gap and bearings.* Check air gap to make sure that average reading is within 10 per cent, provided reading should be less than 0.020 inch. All bearings, ball, roller, and sleeve should be thoroughly checked and defective ones replaced. Waste-packed and wick-oiled bearings should have waste or wicks renewed, if they have become glazed or filled with metal or dirt, making sure that new waste bears well against shaft.

3) *Rotors (squirrel-cage).* Check squirrel-cage rotors for broken or loose bars and evidence of local heating. If fan blades are not cast in place, check for loose blades. Look for marks on rotor surface indicating foreign matter in air gap or a worn bearing.

4) *Rotors (wound).* Clean wound rotors thoroughly around collector rings, washers, and connections. Tighten connections if necessary. If rings are rough, spotted, or eccentric, refer to service shop for refinishing. See that all top sticks or wedges are tight. If any are loose, refer to service shop.

5) *Armatures.* Clean all armature air passages thoroughly if any are obstructed. Look for oil or grease creeping along shaft, checking back to bearing. Check commutator for surface condition, high bars, high mica, or eccentricity. If necessary, remachine the commutator to secure a smooth fresh surface.

6) *Loads.* Read load on motor with instruments at no load, full load, or through an entire cycle, as a check on the mechanical condition of the driven machine.

ADHESIVES AND SEALANTS

By strict definition, an adhesive is any substance that fastens or bonds materials to be joined (adherends) by means of surface attachment. The bond durability depends on the strength of the adhesive to the substrate (adhesion) and the strength within the adhesive (cohesion). Besides bonding a joint, an adhesive may serve as a seal against foreign matter. When an adhesive performs both bonding and sealing functions, it is usually referred to as an adhesive sealant. Joining materials with adhesives offers significant benefits compared with mechanical methods of uniting two materials.

Among these benefits are that an adhesive distributes a load over an area rather than concentrating it at a point, resulting in a more even distribution of stresses. The adhesive bonded joint is therefore more resistant to flexural and vibrational stresses than, for example, a bolted, riveted, or welded joint. Another benefit is that an adhesive forms a seal as well as a bond. This seal prevents the corrosion that may occur with dissimilar metals, such as aluminum and magnesium, or mechanically fastened joints, by providing a dielectric insulation between the substrates. An adhesive also joins irregularly shaped surfaces more easily than does a mechanical fastener. Other benefits include negligible weight addition and virtually no change to part dimensions or geometry.

Most adhesives are available in liquids, gels, pastes, and tape forms. The growing variety of adhesives available can make the selection of the proper adhesive or sealant a challenging experience. In addition to the technical requirements of the adhesive, time and costs are also important considerations. Proper choice of an adhesive is based on knowledge of the suitability of the adhesive or sealant for the particular substrates. Appropriate surface preparation, curing parameters, and matching the strength and durability characteristics of the adhesive to its intended use are essential. The performance of an adhesive-bonded joint depends on a wide range of these factors, many of them quite complex. Adhesive suppliers can usually offer essential expertise in the area of appropriate selection.

Adhesives can be classified as structural or nonstructural. In general, an adhesive can be considered structural when it is capable of supporting heavy loads; nonstructural when it cannot support such loads. Many adhesives and sealants, under various brand names, may be available for a particular bonding application. It is always advisable to check the adhesive manufacturers' information before making an adhesive sealant selection. Also, testing under end-use conditions is always suggested to help ensure bonded or sealed joints meet or exceed expected performance requirements.

Though not meant to be all-inclusive, the following information correlates the features of some successful adhesive compositions available in the marketplace.

Bonding Adhesives

Reactive-type bonding adhesives are applied as liquids and react (cure) to solids under appropriate conditions. The cured adhesive is either a thermosetting or thermoplastic polymer. These adhesives are supplied as two-component no-mix, two-component mix, and one-component no-mix types, which are discussed in the following paragraphs.

Two-Component No-Mix Adhesives

Types of Adhesives.—*Anaerobic (Urethane Methacrylate Ester) Structural Adhesives:* Anaerobic structural adhesives are mixtures of acrylic esters that remain liquid when exposed to air but harden when confined between metal substrates. These adhesives can be used for large numbers of industrial purposes where high reliability of bond joints is required. Benefits include: no mixing is required (no pot-life or waste problems), flexible/durable bonds are made that withstand thermal cycling, have excellent resistance to solvents and severe environments, and rapid cure at room temperatures (eliminating expensive ovens). The adhesives are easily dispensed with automatic equipment. An activator is usually required to be present on one surface to initiate the cure for these adhesives.

Applications for these adhesives include bonding of metals, magnets (ferrites), glass, thermosetting plastics, ceramics, and stone.

Acrylic Adhesives: Acrylic adhesives are composed of a polyurethane polymer backbone with acrylate end groups. They can be formulated to cure through heat or the use of an activator applied to the substrate surface, but many industrial acrylic adhesives are cured by light. Light-cured adhesives are used in applications where the bond geometry allows light to reach the adhesive and the production rate is high enough to justify the capital expense of a light source. Benefits include: no mixing is required (no pot-life or waste problems); formulations cure (solidify) with activator, heat, or light; the adhesive will bond to a variety of substrates, including metal and most thermoplastics; and tough and durable bonds are produced with a typical resistance to the effects of temperatures up to 180°C. Typical applications include automobile body parts (steel stiffeners), assemblies subjected to paint-baking cycles, speaker magnets to pole plates, and bonding of motor magnets, sheet steel, and many other structural applications. Other applications include bonding glass, sheet metal, magnets (ferrite), thermosetting and thermoplastic plastics, wood, ceramics, and stone.

Two-Component Mix Adhesives

Types of Adhesives.—*Epoxy Adhesives:* Two-component epoxy adhesives are well-established adhesives that offer many benefits in manufacturing. The reactive components of these adhesives are separated prior to use, so they usually have a good shelf life without refrigeration. Polymerization begins upon mixing, and a thermoset polymer is formed. Epoxy adhesives cure to form thermosetting polymers made up of a base side with the polymer resin and a second part containing the catalyst. The main benefit of these systems is that the depth of cure is unlimited. As a result, large volume can be filled for work such as potting, without the cure being limited by the need for access to an external influence such as moisture or light to activate the curing process.

For consistent adhesive performance, it is important that the mix ratio remain constant to eliminate variations in adhesive performance. Epoxies can be handled automatically, but the equipment involves initial and maintenance costs. Alternatively, adhesive components can be mixed by hand. However, this approach involves labor costs and the potential for human error. The major disadvantage of epoxies is that they tend to be very rigid and consequently have low peel strength. This lack of peel strength is less of a problem when bonding metal to metal than it is when bonding flexible substrates such as plastics.

Applications of epoxy adhesives include bonding, potting, and coating of metals, bonding of glass, rigid plastics, ceramics, wood, and stone.

Polyurethane Adhesives: Like epoxies, polyurethane adhesives are available as two-part systems or as one-component frozen premixes. They are also available as one-part moisture-cured systems. Polyurethane adhesives can provide a wide variety of physical properties. Their flexibility is greater than that of most epoxies. Coupled with the high cohesive strength, this flexibility provides a tough polymer able to achieve better peel strength and lower flexural modulus than most epoxy systems. This superior peel resistance allows use of polyurethanes in applications that require high flexibility. Polyurethanes bond very well to a variety of substrates, though a primer may be needed to prepare the substrate surface. These primers are moisture-reactive and require several hours to react sufficiently for the parts to be used. Such a time requirement may cause a production bottleneck if the bond-strength requirements are such that a primer is needed.

Applications for polyurethane adhesives include bonding of metals, glass, rubber, thermosetting and thermoplastic plastics, and wood.

One-Component No-Mix Adhesives

Types of Adhesives.—*Light-Curable Adhesives:* Light-curing systems use a unique curing mechanism. The adhesives contain photoinitiators that absorb light energy and dissociate to form radicals. These radicals then initiate the polymerization of the polymers, oligomers, and monomers in the adhesive. The photoinitiator acts as a chemical solar cell, converting the light energy into chemical energy for the curing process. Typically, these systems are formulated for use with ultraviolet light sources. However, newer products have been formulated for use with visible light sources.

One of the biggest benefits that light-curing adhesives offer to the manufacturer is the elimination of the work time to work-in-progress trade-off, which is embodied in most adhesive systems. With light-curing systems, the user can take as much time as needed to position the part without fear of the adhesive curing. Upon exposure to the appropriate light source, the adhesive then can be fully cured in less than 1 minute, minimizing the costs associated with work in progress. Adhesives that utilize light as the curing mechanism are often one-part systems with good shelf life, which makes them even more attractive for manufacturing use.

Applications for light-curable adhesives include bonding of glass, and glass to metal, tacking of wires, surface coating, thin-film encapsulation, clear substrate bonding, and potting of components,

Cyanoacrylate Adhesives (Instant Adhesives): Cyanoacrylates or instant adhesives are often called Superglue™. Cyanoacrylates are one-part adhesives that cure rapidly, as a result of the presence of surface moisture, to form high-strength bonds, when confined between two substrates. Cyanoacrylates have excellent adhesion to many substrates, including most plastics and they achieve fixture strength in seconds and full strength within 24 hours. These qualities make cyanoacrylates suitable for use in automated production environments. They are available in viscosities ranging from water-thin liquids to thixotropic gels.

Because cyanoacrylates are a relatively mature adhesive family, a wide variety of specialty formulations is now available to help the user address difficult assembly problems. One of the best examples is the availability of polyolefin primers, which allow users to obtain high bond strengths on difficult-to-bond plastics such as polyethylene and polypropylene. One common drawback of cyanoacrylates is that they form a very rigid polymer matrix, resulting in very low peel strengths. To address this problem, formulations have been developed that are rubber-toughened. Although the rubber toughening improves the peel strength of the system to some extent, peel strength remains a weak point for this system, and, therefore, cyanoacrylates are poor candidates for joint designs that require high peel resistance. In manufacturing environments with low relative humidity, the cure of the cyanoacrylate can be significantly retarded.

This problem can be addressed in one of two ways. One approach is to use accelerators that deposit active species on the surface to initiate the cure of the product. The other approach is to use specialty cyanoacrylate formulations that have been engineered to be surface-insensitive. These formulations can cure rapidly even on dry or slightly acidic surfaces.

Applications for cyanoacrylate adhesives include bonding of thermoplastic and thermosetting plastics, rubber, metals, wood, and leather, also strain relief of wires.

Hot-Melt Adhesives: Hot-melt adhesives are widely used in assembly applications. In general, hot-melt adhesives permit fixturing speeds that are much faster than can be achieved with water- or solvent-based adhesives. Usually supplied in solid form, hot-melt adhesives liquify when exposed to elevated temperatures. After application, they cool quickly, solidifying and forming a bond between two mating substrates. Hot-melt adhesives have been used successfully for a wide variety of adherends and can greatly reduce both the need for clamping and the length of time for curing. Some drawbacks with hot-melt adhesives are their tendency to string during dispensing and relatively low-temperature resistance.

Applications for hot-melt adhesives are bonding of fabrics, wood, paper, plastics, and cardboard.

Rubber-Based Solvent Cements: Rubber-based solvent cements are adhesives made by combining one or more rubbers or elastomers in a solvent. These solutions are further modified with additives to improve the tack or stickiness, the degree of peel strength, flexibility, and the viscosity or body. Rubber-based adhesives are used in a wide variety of applications such as contact adhesive for plastics laminates like counter tops, cabinets, desks, and tables. Solvent-based rubber cements have also been the mainstay of the shoe and leather industry for many years.

Applications for rubber-based solvent cements include bonding of plastics laminates, wood, paper, carpeting, fabrics, and leather.

Moisture-Cured Polyurethane Adhesives: Like heat-curing systems, moisture-cured polyurethanes have the advantage of a very simple curing process. These adhesives start to cure when moisture from the atmosphere diffuses into the adhesive and initiates the polymerization process. In general, these systems will cure when the relative humidity is above 25 per cent, and the rate of cure will increase as the relative humidity increases.

The dependence of these systems on the permeation of moisture through the polymer is the source of their most significant process limitations. As a result of this dependence, depth of cure is limited to between 0.25 and 0.5 in. (6.35 and 12.7 mm). Typical cure times are in the range of 12 to 72 hours. The biggest use for these systems is for windshield bonding in automobile bodies.

Applications for moisture-cured polyurethane adhesives include bonding of metals, glass, rubber, thermosetting and thermoplastic plastics, and wood.

Retaining Compounds

The term *retaining compounds* is used to describe adhesives used in circumferential assemblies joined by inserting one part into the other. In general, retaining compounds are anaerobic adhesives composed of mixtures of acrylic esters that remain liquid when exposed to air but harden when confined between cylindrical machine components. A typical example is a bearing held in an electric motor housing with a retaining compound. The first retaining compounds were launched in 1963, and the reaction among users of bearings was very strong because these retaining compounds enabled buyers of new bearings to salvage worn housings and minimize their scrap rate.

The use of retaining compounds has many benefits, including elimination of bulk needed for high friction forces, ability to produce more accurate assemblies and to augment or replace press fits, increased strength in heavy press fits, and reduction of machining costs. Use of these compounds also helps in dissipating heat through assembly, and eliminating distortion when installing drill bushings, fretting corrosion and backlash in keys and splines, and bearing seizure during operation.

The major advantages of retaining compounds for structural assemblies are that they require less severe machining tolerances and no securing of parts. Components are assembled quickly and cleanly, and they transmit high forces and torques, including dynamic forces. Retaining compounds also seal, insulate, and prevent micromovements so that neither fretting corrosion nor stress corrosion occurs. The adhesive joint can be taken apart easily after heating above 450°F (230°C) for a specified time.

Applications for retaining compounds include mounting of bearings in housings or on shafts, avoiding distortion of precision tooling and machines, mounting of rotors on shafts, inserting drill jig bushings, retaining cylinder linings, holding oil filter tubes in castings, retaining engine-core plugs, restoring accuracy to worn machine tools, and eliminating keys and set screws.

Threadlocking

The term *threadlocker* is used to describe adhesives used in threaded assemblies for locking the threaded fasteners by filling the spaces between the nut and bolt threads with a hard, dense material that prevents loosening. In general, thread-lockers are anaerobic adhesives comprising mixtures of acrylic esters that remain liquid when exposed to air but harden when confined between threaded components. A typical example is a mounting bolt on a motor or a pump. Threadlocker strengths range from very low strength (removable) to high strength (permanent).

It is important that the total length of the thread is coated and that there is no restriction to the curing of the threadlocker material. (Certain oils or cleaning systems can impede or even completely prevent the adhesive from curing by anaerobic reaction.) The liquid threadlocker may be applied by hand or with special dispensing devices. Proper coating (wetting) of a thread is dependent on the size of the thread, the viscosity of the adhesive, and the geometry of the parts. With blind-hole threads, it is essential that the adhesive be applied all the way to the bottom of the threaded hole. The quantity must be such that after assembly, the displaced adhesive fills the whole length of the thread.

Some threadlocking products cured by anaerobic reaction have a positive influence on the coefficient of friction in the thread. The values are comparable with those of oiled bolts. Prestress and installation torque therefore can be defined exactly. This property allows threadlocking products cured by anaerobic reaction to be integrated into automated production lines using existing assembly equipment. The use of thread-lockers has many benefits including ability to lock and seal all popular bolt and nut sizes with all industrial finishes, and to replace mechanical locking devices. The adhesive can seal against most industrial fluids and will lubricate threads so that the proper clamp load is obtained. The materials also provide vibration-resistant joints that require handtool dismantling for servicing, prevent rusting of threads, and cure (solidify) without cracking or shrinking.

The range of applications includes such uses as locking and sealing nuts on hydraulic pistons, screws on vacuum cleaner bell housings, track bolts on bulldozers, hydraulic-line fittings, screws on typewriters, oil-pressure switch assembly, screws on carburetors, rocker nuts, machinery driving keys, and on construction equipment.

Sealants

The primary role of a sealant composition is the prevention of leakage from or access by dust, fluids, and other materials to assembly structures. Acceptable leak rates can range from a slight drip to bubbletight to molecular diffusion through the base materials. Equipment users in the industrial market want trouble-free operation, but it is not always practical to specify zero leak rates. Factors influencing acceptable leak rates are toxicity, product or environmental contamination, combustibility, economics, and personnel considerations. All types of fluid seals perform the same basic function: they seal the process fluid (gas, liquid, or vapor) and keep it where it belongs. A general term for these assembly approaches is gasketing. Many products are being manufactured that are capable of sealing a variety of substrates.

Types of Sealants.—*Anaerobic Formed-in-Place Gasketing Materials:* Mechanical assemblies that require the joining of metal-to-metal flange surfaces have long been designed with prefabricated, precut materials required to seal the imperfect surfaces of the assembly. Numerous gasket materials that have been used to seal these assemblies include paper, cork, asbestos, wood, metals, dressings, and even plastics. Fluid seals are divided into static and dynamic systems, depending on whether or not the parts move in relationship to each other. Flanges are classed as static systems, although they may be moved relative to each other by vibration, temperature, and/or pressure changes, shocks, and impacts.

The term *anaerobic formed-in-place gasketing* is used to describe sealants that are used in flanged assemblies to compensate for surface imperfections of metal-to-metal components by filling the space between the substrates with a flexible, nonrunning material. In general, anaerobic formed-in-place gaskets are sealants made up of mixtures of acrylic esters that remain liquid when exposed to air but harden when confined between components. A typical example is sealing two halves of a split crankcase.

The use of anaerobic formed-in-place gaskets has many benefits, including the ability to seal all surface imperfections, allow true metal-to-metal contact, eliminate compression set and fastener loosening, and add structural strength to assemblies. These gaskets also help improve torque transmission between bolted flange joints, eliminate bolt retorquing needed with conventional gaskets, permit use of smaller fasteners and lighter flanges, and provide for easy disassembly and cleaning.

Applications in which formed-in-place gasketing can be used to produce leakproof joints include pipe flanges, split crankcases, pumps, compressors, power takeoff covers, and axle covers. These types of gaskets may also be used for repairing damaged conventional gaskets and for coating soft gaskets.

Silicone Rubber Formed-in-Place Gasketing: Another type of formed-in-place gasket uses room-temperature vulcanizing (RTV) silicone rubbers. These materials are one-component sealants that cure on exposure to atmospheric moisture. They have excellent properties for vehicle use such as flexibility, low volatility, good adhesion, and high resistance to most automotive fluids. The materials will also withstand temperatures up to 600°F (320°C) for intermittent operation.

RTV silicones are best suited for fairly thick section (gap) gasketing applications where flange flexing is greatest. In the form of a very thin film, for a rigid metal-to-metal seal, the cured elastomer may abrade and eventually fail under continual flange movement. The RTV silicone rubber does not unitize the assembly, and it requires relatively clean, oil-free surfaces for sufficient adhesion and leakproof seals.

Because of the silicone's basic polymeric structure, RTV silicone elastomers have several inherent characteristics that make them useful in a wide variety of applications. These properties include outstanding thermal stability at temperatures from 400 to 600°F (204 to 320°C), and good low-temperature flexibility at −85 to −165°F (−65 to −115°C). The material forms an instant seal, as is required of all liquid gaskets, and will fill large gaps up to 0.250 in. (6.35 mm) for stamped metal parts and flanges. The rubber also has good stability in ultraviolet light and excellent weathering resistance.

Applications for formed-in-place RTV silicones in the automotive field are valve, camshaft and rocker covers, manual transmission (gearbox) flanges, oil pans, sealing panels, rear axle housings, timing chain covers, and window plates. The materials are also used on oven doors and flues.

Tapered Pipe-thread Sealing

Thread sealants are used to prevent leakage of gases and liquids from pipe joints. All joints of this type are considered to be dynamic because of vibration, changing pressures, or changing temperatures.

Several types of sealants are used on pipe threads including noncuring pipe dopes, which are one of the oldest methods of sealing the spiral leak paths of threaded joints. In general, pipe dopes are pastes made from oils and various fillers. They lubricate joints and jam threads but provide no locking advantage. They also squeeze out under pressure, and have poor solvent resistance. Noncuring pipe dopes are not suitable for use on straight threads.

Another alternative is solvent-drying pipe dopes, which are an older method of sealing tapered threaded joints. These types of sealant offer the advantages of providing lubrication and orifice jamming and they also extrude less easily than noncuring pipe dopes. One disadvantage is that they shrink during cure as the solvents evaporate and fittings must be retorqued to minimize voids. These materials generally lock the threaded joint together by friction. A third type of sealer is the trapped elastomer supplied in the form of a thin tape incorporating polytetrafluorethylene (PTFE). This tape gives a good initial seal and resists chemical attack, and is one of the only materials used for sealing systems that will seal against oxygen gas.

Some other advantages of PTFE are that it acts as a lubricant, allows for high torquing, and has a good resistance to various solvents. Some disadvantages are that it may not provide a true seal between the two threaded surfaces, and it lubricates in the off direction, so it may allow fittings to loosen. In dynamic joints, tape may allow creep, resulting in leakage over time. The lubrication effect may allow overtightening, which can add stress or lead to breakage. Tape also may be banned in some hydraulic systems due to shredding, which may cause clogging of key orifices.

Anaerobic Pipe Sealants.—*Anaerobic Pipe Sealants:* The term *anaerobic pipe sealants* is used to describe anaerobic sealants used in tapered threaded assemblies for sealing and locking threaded joints. Sealing and locking are accomplished by filling the space between the threads with the sealant. In general, these pipe sealants are anaerobic adhesives consisting of mixtures of acrylic esters that remain liquid when exposed to air but harden when confined between threaded components to form an insoluble tough plastics. The strength of anaerobic pipe sealants is between that of elastomers and yielding metal.

Clamp loads need be only tight enough to prevent separation in use. Because they develop strength by curing after they are in place, these sealants are generally forgiving of tolerances, tool marks, and slight misalignment. These sealants are formulated for use on metal substrates. If the materials are used on plastics, an activator or primer should be used to prepare the surfaces.

Among the advantages of these anaerobic sealers are that they lubricate during assembly, they seal regardless of assembly torque, and they make seals that correspond with the burst rating of the pipe. They also provide controlled disassembly torque, do not cure outside the joint, and are easily dispensed on the production line. These sealants also have the lowest cost per sealed fitting. Among the disadvantages are that the materials are not suitable for oxygen service, for use with strong oxidizing agents, or for use at temperatures above 200°C. The sealants also are typically not suitable for diameters over M80 (approximately 3 inches).

The many influences faced by pipe joints during service should be known and understood at the design stage, when sealants are selected. Sealants must be chosen for reliability and long-term quality. Tapered pipe threads must remain leak-free under the severest vibration and chemical attack, also under heat and pressure surges.

Applications of aerobic sealants are found in industrial plant fluid power systems, the textile industry, chemical processing, utilities and power generation facilities, petroleum refining, and in marine, automotive, and industrial equipment. The materials are also used in the pulp and paper industries, in gas compression and distribution, and in waste-treatment facilities.

MOTION CONTROL

The most important factor in the manufacture of accurately machined components is the control of motion, whatever power source is used. For all practical purposes, motion control is accomplished by electrical or electronic circuits, energizing or deenergizing actuators such as electric motors or solenoid valves connected to hydraulic or pneumatic cylinders or motors. The accuracy with which a machine tool slide, for example, may be brought to a required position, time after time, controls the dimensions of the part being machined. This accuracy is governed by the design of the motion control system in use.

There is a large variety of control systems, with power outputs from milliwatts to megawatts, and they are used for many purposes besides motion control. Such a system may control a mechanical positioning unit, which may be linear or rotary, its velocity, acceleration, or combinations of these motion parameters. A control system may also be used to set voltage, tension, and other manufacturing process variables and to actuate various types of solenoid-operated valves. The main factors governing design of control systems are whether they are to be open- or closed-loop; what kinds and amounts of power are available; and the function requirements.

Factors governing selection of control systems are listed in Table 1.

Table 1. Control System Application Factors

Type of System	Nature of required control motion, i.e., position, velocity, acceleration
Accuracy	Controlled output versus input
Mechanical Load	Viscous friction, coulomb friction, starting friction, load inertia
Impact Loads	Hitting mechanical stops and load disturbances
Ratings	Torque or force, and speed
Torque	Peak instantaneous torque
Duty Cycle	Load response, torque level, and duration and effect on thermal response
Ambient Temperature	Relation to duty cycle and internal temperature rise, and to the effect of temperature on the sensor
Speed of Response	Time to reach commanded condition. Usually defined by a response to a stepped command
Frequency Response	Output to input ratio versus frequency, for varying frequency and specified constant input amplitude. Usually expressed in decibels
No-Load Speed	Frequently applies to maximum kinetic energy and to impact on stops; avoiding overspeeding
Backdriving	With power off, can the load drive the motor? Is a fail-safe brake required? Can the load backdrive with power on without damage to the control electronics? (Electric motor acting as a generator)
Power Source	Range of voltage and frequency within which the system must work. Effect of line transients
Environmental Conditions	Range of nonoperating and operating conditions, reliability and serviceability, scheduled maintenance

Open-Loop Systems.—The term open-loop typically describes use of a rheostat or variable resistance to vary the input voltage and thereby adjust the speed of an electric motor, a low-accuracy control method because there is no output sensor to measure the performance. However, use of stepper motors (see Table 2, and page 2473) in open-loop systems can make them very accurate. Shafts of stepper motors are turned through a fixed angle for every electrical pulse transmitted to them. The maximum pulse rate can be high, and the shaft can be coupled with step-down gear drives to form inexpensive, precise drive units with wide speed ranges. Although average speed with stepper motors is exact, speed modulation can occur at low pulse rates and drives can incur serious resonance problems.

Table 2. Control Motor Types

AC Motors	Induction motors, simplest, lowest cost, most rugged, can work directly off the ac line or through an inexpensive, efficient, and compact thyristor controller. Useful in fan and other drives where power increases rapidly with speed as well as in simple speed regulation. Ac motors are larger than comparable permanent-magnet motors
Two-Phase Induction Motors	Often used as control motors in small electromechanical control systems. Power outputs range from a few milliwatts to tens of watts
Split-Field Series Motors	Work on both ac and dc. Feature high starting torque, low cost, uniform power output over a wide speed range, and are easily reversed with a single-pole three-position switch. Very easy to use with electric limit switches for controlling angle of travel
Permanent-Magnet Motors	Operate on dc, with high power output and high efficiency. The most powerful units use rare-earth magnets and are more expensive than conventional types. Lower-cost ferrite magnets are much less expensive and require higher gear-reduction ratios, but at their higher rated speeds are very efficient
Brushless DC Motors	Use electrical commutation and may be applied as simple drive motors or as four-quadrant control motors. The absence of brushes for commutation ensures high reliability and low electromagnetic interference
Stepper Motors	Index through a fixed angle for each input pulse so that speed is in exact proportion to pulse rate and the travel angle increases uniformly with the number of pulses. Proper application in systems with backlash and load inertia requires special care
Wound-Field DC Motors	For subfractional to integral horsepower applications where size is not significant. Cost is moderate because permanent magnets are not required. Depending on the windings, output characteristics can be adjusted for specific applications

Open-loop systems are only as accurate as the input versus output requirement can be calibrated, including the effects of changes in line voltage, temperature, and other operating conditions.

Closed-Loop Systems.—Table 3 shows some parameters and characteristics of closed-loop systems, and a simple example of such a system is shown below. A command may be input by a human operator, it may be derived from another piece of system equipment, or it may be generated by a computer. Generally, the command is in the form of an electrical signal. The system response is converted by the output sensor to a compatible, scaled electrical signal that may be compared with the input command, the difference constituting an error signal. It is usually required that the error be small, so it is amplified and applied to an appropriate driving unit. The driver may take many forms, but for motion control it is usually a motor.

The amplified error voltage drives the motor to correct the error. If the input command is constant, the system is a closed-loop regulator.

Closed-loop systems use feedback sensors that measure system output and give instructions to the power drive components, based on the measured values. A typical closed-loop speed control, for instance, uses a tachometer as a feedback sensor and will correct automatically for differences between the tachometer output and the commanded speed. All motion control systems require careful design to achieve good practical performance. Closed-loop systems generally cost more than open-loop systems because of the extra cost of the tachometer or transducer used for output measurement. Faster response components also increase cost.

Table 3. Closed-Loop System Parameters and Characteristics

Step Response	The response of the system to a step change in the input command. The response to a large step, which can saturate the system amplifier, is different from the response to a small nonsaturating step. Initial overshoots may not be permissible in some types of equipment
Frequency Response	System response to a specified small-amplitude sinusoidal command where frequency is varied over the range of interest. The response is in decibels (dB), where dB = 20 $\log_{10}$(output/input). This characteristic determines whether the system is responsive enough to meet requirements
Bandwidth	The effective range of input frequencies within which the control system responds well. The bandwidth is often described by the point where the frequency response is down by three decibels. Bandwidth is usually defined in Hz (cycles/sec) or $\omega = 2\pi \times$ Hz (radians/sec)
Loading	The torque required to drive the load and the load inertia. The amplifier must supply enough power to meet acceleration as well as output power requirements. If the load is nonlinear, its effect on error must be within specifications. Behavior may vary considerably, depending on whether the load aids or opposes motor torque, as in a hoist
Output Stiffness	A measure of the system's response to load disturbances. Dynamic stiffness measures the system's response to a rapidly varying load
Resonant Peaks	Can show up in frequency-response testing as sharp (undamped) resonances. These resonances cannot be tolerated in the normal frequency range of the control system because they can lead to oscillation and vibration
No load, or maximum speed and maximum torque	Can be controlled by voltage or current limiting in the electronic amplifier. A slip clutch can also be used for torque limiting, particularly to avoid impact damage

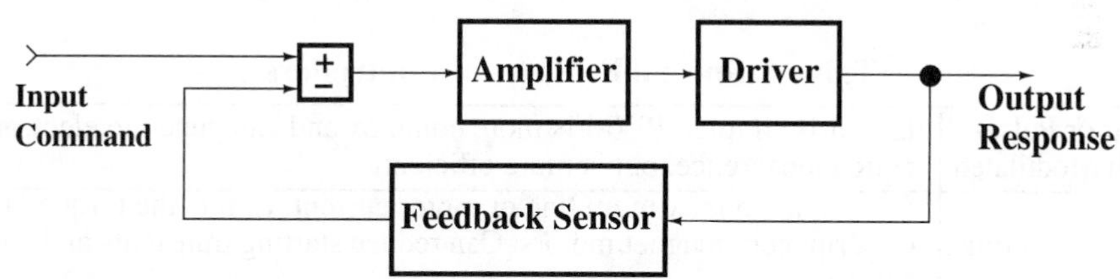

Fig. 1. General Arrangement of a Closed-Loop Control System

Accuracy of closed-loop systems is directly related to the accuracy of the sensor, so that choosing between open-loop and closed-loop controls may mean choosing between low price and consistent, accurate repeatability. In the closed-loop arrangement in Fig. 1, the sensor output is compared with the input command and the difference is amplified and applied to the motor to produce a correction. When the amplifier gain is high (the difference is greatly enlarged), even a small error will generate a correction. However, a high gain can lead to an unstable system due to inherent delays between the electrical inputs and outputs, especially with the motor.

Response accuracy depends not only on the precision of the feedback sensor and the gain of the amplifier, but also on the rate at which the command signal changes. The ability of the control system to follow rapidly changing inputs is naturally limited by the maximum motor speed and acceleration.

Amplified corrections cannot be applied to the motor instantaneously, and the motor does not respond immediately. Overshoots and oscillations can occur and the system must be adjusted or tuned to obtain acceptable performance. This adjustment is called damping the system response. Table 4 lists a variety of methods of damping, some of which require specialized knowledge.

Table 4. Means of Damping System Response

Network Damping	Included in the electrical portion of the closed loop. The networks adjust amplitude and phase to minimize control system feedback oscillations. Notch networks are used to reduce gain at specific frequencies to avoid mechanical resonance oscillations
Tachometer Damping	Feedback proportional to output velocity is added to the error signal for system stabilization
Magnetic Damping	Viscous or inertial dampers on the motor rear shaft extension for closed-loop stabilization. Similar dampers use silicone fluid instead of magnetic means to provide damping
Nonlinear Damping	Used for special characteristics. Inverse error damping provides low damping for large errors, permitting fast slewing toward zero and very stable operation at zero. Other nonlinearities meet specific needs, for example, coulomb friction damping works well in canceling backlash oscillations
Damping Algorithms	With information on output position or velocity, or both, sampled data may be used with appropriate algorithms to set motor voltage for an optimum system response

The best damping methods permit high error amplification and accuracy, combined with the desired degree of stability. Whatever form the output takes, it is converted by the output sensor to an electrical signal of compatible form that can be compared with the input command. The error signal thus generated is amplified before being applied to the driving unit.

Drive Power.—Power for the control system often depends on what is available and may vary from single- and three-phase ac 60 or 400 Hz, through dc and other types. Portable or mobile equipment is usually battery-powered dc or an engine-driven electrical generator. Hydraulic and pneumatic power may also be available. Cost is often the deciding factor in the choice.

Table 5. Special Features of Controllers

Linear or Pulse-Width Modulated	Linear is simpler, PWM is more complex and can generate electromagnetic interference, but is more efficient
Current Limiting	Sets limits to maximum line or motor current. Limits the torque output of permanent magnet motors. Can reduce starting transients and current surges
Voltage Limiting	Sets limits to maximum motor speed. Permits more uniform motor performance over a wide range of line voltages
Energy Absorption	Ability of the controller to absorb energy from a dc motor drive, back-driven by the load
EMI Filtering	Especially important when high electrical gain is required, as in thermocouple circuits, for example
Isolation	Of input and output, sometimes using optoisolators, or transformers, when input and output circuits require a high degree of isolation

Control Function.—The function of the control is usually set by the designer of the equipment and needs careful definition because it is the basis for the overall design. For instance, in positioning a machine tool table, such aspects as speed of movement and permissible variations in speed, accuracy of positioning, repeatability, and overshoot are among dozens of factors that must be considered. Some special features of controllers are listed in Table 5. Complex electromechanical systems require more knowledge of design and debugging than are needed for strictly mechanical systems.

Electromechanical Control Systems.—Wiring is the simplest way to connect components, so electromechanical controls are more versatile than pure hydraulic or pneumatic controls. The key to this versatility is often in the controller, the fundamental characteristic of which is its power output. The power output must be compatible with motor and load requirements. Changes to computer chips or software can usually change system performance to suit the application.

When driving a dc motor, for instance, the controller must supply sufficient power to match load requirements as well as motor operating losses, at minimum line voltage and maximum ambient temperature. The system's wiring must not be greatly sensitive to transient or steady-state electrical interference, and power lines must be separated from control signal lines, or appropriately shielded and isolated to avoid cross-coupling. Main lines to the controller must often include electrical interference filters so that the control system does not affect the power source, which may influence other equipment connected to the same source. For instance, an abruptly applied step command can be smoothed out so that heavy motor inrush currents are avoided. The penalty is a corresponding delay in response.

Use of current limiting units in a controller will not only set limits to line currents, but will also limit motor torque. Electronic torque limiting can frequently avoid the need for mechanical torque limiting. An example of the latter is using a slip clutch to avoid damage due to overtravel, the impact of which usually includes the kinetic energy of the moving machine elements. In many geared systems, most of the kinetic energy is in the motor. Voltage limiting is less useful than current limiting but may be needed to isolate the motor from voltage transients on the power line, to prevent overspeeding, as well as to protect electronic components.

Mechanical Stiffness.—When output motion must respond to a rapidly changing input command, the control system must have a wide bandwidth. Where the load mass (in linear motion systems) or the polar moment of inertia (in rotary systems) is high, there is a possibility of resonant oscillations. For the most stable and reliable systems, with a defined load, a high system mechanical stiffness is preferred. To attain this stiffness requires strengthening shafts, preloading bearings, and minimizing free play or backlash. In the best-performing systems, motor and load are coupled without intervening compliant members. Even tightly bolted couplings can introduce compliant oscillations resulting from extremely minute slippages caused by the load motions.

Backlash is a factor in the effective compliance of any coupling but has little effect on the resonant frequency because little energy is exchanged as the load is moved through the backlash region. However, even in the absence of significant torsional resonance, a high-gain control system can "buzz" in the backlash region. Friction is often sufficient to eliminate this small-amplitude, high-frequency component.

The difficulty with direct-drive control systems lies in matching motor to load. Most electric motors deliver rated power at higher speeds than are required by the driven load, so that load power must be delivered by the direct-drive motor operating at a slow and relatively inefficient speed. Shaft power at low speed involves a correspondingly high torque, which requires a large motor and a high-power controller. Motor copper loss (heating) is high in delivering the high motor torque. However, direct-drive motors provide maximum load velocity and acceleration, and can position massive loads within seconds of arc (rotational) or tenths of thousandths of an inch (linear) under dynamic conditions.

Where performance requirements are moderate, the required load torque can be traded off against speed by using a speed-changing transmission, typically, a gear train. The transmission effectively matches the best operating region of the motor to the required operating region of the load, and both motor and controller can be much smaller than would be needed for a comparable direct drive.

Torsional Vibration.—Control system instabilities can result from insufficient stiffness between the motor and the inertia of the driven load. The behavior of such a system is similar to that of a torsional pendulum, easily excited by commanded motions of the control system. If frictional losses are moderate to low, sustained oscillations will occur. In spite of the complex dynamics of the closed-loop system, the resonant frequency, as for a torsional pendulum, is given to a high degree of accuracy by the formula:

$$f_n = \frac{1}{2\pi} \times \sqrt{\frac{K}{J_L}}$$

where f_n is in hertz, K is torsional stiffness in in.-lb/rad, and J_L is load inertia in in.-lb-sec²/rad. If this resonant frequency falls within the bandwidth of the control system, self-sustained oscillations are likely to occur. These oscillations are often overlooked by control systems analysts because they do not appear in simple control systems, and they are very difficult to correct.

Friction inherently reduces the oscillation by dissipating the energy in the system inertia. If there is backlash between motor and load, coulomb friction (opposing motion but independent of speed) is especially effective in damping out the oscillation. However, the required friction for satisfactory damping can be excessive, introducing positioning error and adding to motor (and controller) power requirements. Friction also varies with operating conditions and time.

The most common method of eliminating torsional oscillation is to introduce a filter in the error channel of the control system to shape the gain characteristic as a function of frequency. If the torsional resonance is within the required system bandwidth, little can be done except stiffening the mechanical system and increasing the resonant frequency. If the filter reduces the gain within the required bandwidth, it will reduce performance. This method will work only if the natural resonance is above the minimum required performance bandwidth.

The simplest shaping network is the notch network (Table 4, network damping), which, in effect, is a band-rejection filter that sharply reduces gain at the notch frequency. By locating the notch frequency so as to balance out the torsional resonance peak, the oscillation can be eliminated. Where there are several modes of oscillation, several filter networks can be connected in series.

Electric Motors.—Electric motors for control systems must suit the application. Motors used in open-loop systems (excluding stepper motors) need not respond quickly to input command changes. Where the command is set by a human, response times of hundreds of milliseconds to several seconds may be acceptable. Slow response does not lead to the instabilities that time delays can introduce into closed-loop systems.

Closed-loop systems need motors with fast response, of which the best are permanent-magnet dc units, used where wide bandwidth, efficient operation, and high power output are required. Table 2 lists some types of control motors and their characteristics. An important feature of high-performance, permanent-magnet motors using high-energy, rare-earth magnets is that their maximum torque output capacity can be 10 to 20 or more times higher than their rated torque. In intermittent or low-duty-cycle applications, very high torque loads can be driven by a given motor. However, when rare-earth magnets (samarium cobalt or neodymium) are not used, peak torque capability may be limited by the possibility of demagnetization. Rare-earth magnets are relatively expensive, so it is important to verify peak torque capabilities for lower-cost motors that may use weaker Alnico or ferrite magnets.

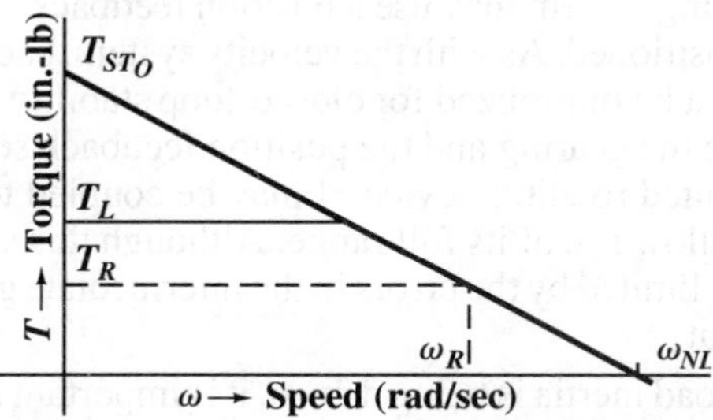

Fig. 2. Idealized Control Motor Characteristics for a Consistent Set of Units

Duty-cycle calculations are an aspect of thermal analysis that are well understood and are not covered here. Motor manufacturers usually supply information on thermal characteristics including thermal time constants and temperature rise per watt of internal power dissipation.

Characteristics of permanent-magnet motors are defined with fair accuracy by relatively few parameters. The most important characteristics are: D_M motor damping in lb-in.-sec/rad; J_M motor inertia in lb-in.-sec$_2$/rad; and R winding resistance in ohms. Fig. 1 shows other control motor characteristics, T_{STo} stall torque with no current limiting; T_L maximum torque with current limiting; ω_{NL} no-load speed; ω_R rated speed. Other derived motor parameters include V rated voltage in volts; $I_{STo} = V/R$ current in amperes at stall with no current limiting; I_L ampere limit, adjusted in amplifier; I_R rated current; $K_T = T_{STo}/I_{STo}$ torque constant in in.-lb/ampere; $K_E = V/\omega_{NL}$ voltage constant in volt/rad/sec; $K_M = K_T/\sqrt{R}$, torque per square root of winding resistance; $D_M = T_{STo}/\omega_{NL}$ motor damping in in.-lb/rad/sec; and $T_M = J_M/D_M$ motor mechanical time constant in seconds.

Stepper Motors.—In a stepper motor, power is applied to a wound stator, causing the brushless rotor to change position to correspond with the internal magnetic field. The rotor maintains its position relative to the internal magnetic field at all times. In its most common mode of operation, the stepper motor is energized by an electronic controller whose current output to the motor windings defines the position of the internally generated magnetic field. Applying a command pulse to the controller will change the motor currents to reposition the rotor. A series of pulses, accompanied by a direction command, will cause rotation in uniformly spaced steps in the specified direction.

If the pulses are applied at a sufficiently high frequency, the rotor will be carried along with the system's inertia and will rotate relatively uniformly but with a modulated velocity. At the other extreme, the response to a single pulse will be a step followed by an overshoot and a decaying oscillation. Where the application cannot permit the oscillation, damping can be included in the controller.

Stepper motors are often preferred because positions of the rotor are known from the number of pulses and the step size. An initial index point is required as an output position reference, and care is required in the electronic circuits to avoid introducing random pulses that will cause false positions. As a minimum, the output index point on an appropriate shaft can verify the step count during operation.

Gearing.—In a closed-loop system, gearing may be used to couple a high-speed, low-torque motor to a lower-speed, higher-torque load. The gearing must meet requirements for accuracy, strength, and reliability to suit the application. In addition, the closed loop requires minimum backlash at the point where the feedback sensor is coupled. In a velocity-controlled system, the feedback sensor is a tachometer that is usually coupled directly to the rotor shaft. Backlash between motor and tachometer, as well as torsional compliance, must be minimized for stable operation of a high-performance system. Units combining motor and tachometer on a single shaft can usually be purchased as an assembly.

By contrast, a positioning system may use a position feedback sensor that is closely coupled to the shaft being positioned. As with the velocity system, backlash between the motor and feedback sensor must be minimized for closed-loop stability. Antibacklash gearing is frequently used between the gearing and the position feedback sensor. When the position feedback sensor is a limited rotation device, it may be coupled to a gear that turns faster than the output gear to allow use of its full range. Although this step-up gearing enhances it, accuracy is ultimately limited by the errors in the intermediate gearing between the position sensor and the output.

When an appreciable load inertia is being driven, it is important that the mechanical stiffness between the position sensor coupling point and the load be high enough to avoid natural torsional resonances in the passband.

Feedback Transducers.—Controlled variables are measured by feedback transducers and are the key to accuracy in operation of closed-loop systems. When the accuracy of a carefully designed control system approaches the accuracy of the feedback transducer, the need for precision in the other system components is reduced.

Transducers may measure the quantity being controlled in digital or analog form, and are available for many different parameters such as pressure and temperature, as well as distance traveled or degrees of rotation. Machine designers generally need to measure and control linear or rotary motion, velocity, position, and sometimes acceleration. Although some transducers are nonlinear, a linear relationship between the measured variable and the (usually electrical) output is most common.

Output characteristics of an analog linear-position transducer are shown in Fig. 2. By dividing errors into components, accuracy can be increased by external adjustments, and slope error and zero offsets are easily trimmed in. Nonlinearity is controlled by the manufacturer. In Fig. 2 are seen the discrete error components that can be distinguished because of the ease with which they can be canceled out individually by external adjustments. The most common compensation is for zero-position alignment, so that when the machine has been set to the start position for a sequence, the transducer can be positioned to read zero output. Alternatively, with all components in fixed positions, a small voltage can be inserted in series with the transducer output for a very accurate alignment of mechanical and electrical zeros. This method helps in canceling long-term drift, particularly in the mechanical elements.

The second most common adjustment of a position transducer is of its output gradient, that is, transducer output volts per degree. Depending on the type of analog transducer, it is usually possible to add a small adjustment to the electrical input, to introduce a proportional change in output gradient. As with the zero-position adjustment, the gradient may be set very accurately initially and during periodic maintenance. The remaining errors shown in Fig. 2, such as intrinsic nonlinearity or nonconformity, result from limitations in design and manufacture of the transducer.

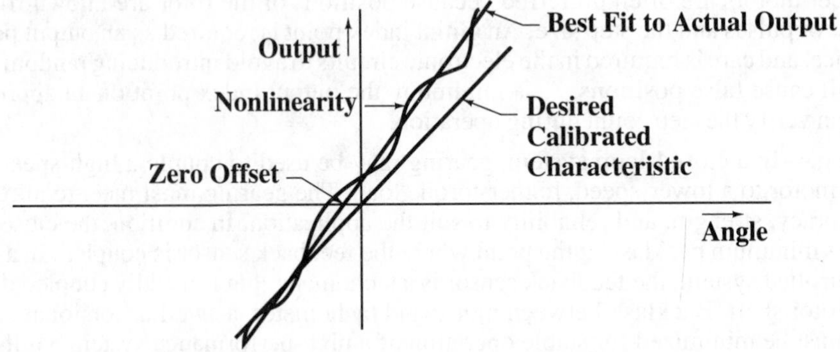

Fig. 3. Output Characteristics of a General Linear Position Transducer

Greater accuracy can be achieved in computer-controlled systems by using the computer to cancel out transducer errors. The system's mechanical values and corresponding transducer values are stored in a lookup table in the computer and referred to as necessary. Accuracies approaching the inherent repeatability and stability of the system can thus be secured. If necessary, recalibration can be performed at frequent intervals.

Analog Transducers.—The simplest analog position transducer is the resistance potentiometer, the resistance element in which is usually a deposited-film rather than a wire-wound type. Very stable resistance elements based on conductive plastics, with resolution to a few microinches and operating lives in the 100 million rotations, are available, capable of working in severe environments with high vibrations and shock and at temperatures of 150 to 200°C. Accuracies of a few hundredths, and stability of thousandths, of a per cent, can be obtained from these units by trimming the plastics resistance element as a function of angle.

Performance of resistance potentiometers deteriorates when they operate at high speeds, and prolonged operation at speeds above 10 rpm causes excessive wear and increasing output noise. An alternative to the resistance potentiometer is the variable differential transformer, which uses electrical coupling between ac magnetic elements to measure angular or linear motion without sliding contacts. These units have unlimited resolution with accuracy comparable to the best resistance potentiometers but are more expensive and require compatible electronic circuits.

A variable differential transformer needs ac energization, so an ac source is required. A precision demodulator is frequently used to change the ac output to dc. Sometimes the ac output is balanced against an ac command signal whose input is derived from the same ac source. In dealing with ac signals, phase-angle matching and an accurate amplitude-scale factor are required for proper operation. Temperature compensation also may be required, primarily due to changes in resistance of the copper windings. Transducer manufacturers will supply full sets of compatible electronic controls.

Synchros and Resolvers.—Synchros and resolvers are transducers that are widely used for sensing of angles at accuracies down to 10 to 20 arc-seconds. More typically, and at much lower cost, their accuracies are 1 to 2 arc-minutes. Cost is further reduced when accuracies of 0.1 degree or higher are acceptable.

Synchros used as angle-position transducers are made as brush types with slip rings and in brushless types. These units can rotate continuously at high speeds, the operating life of brushless designs being limited only by the bearing life. Synchros have symmetrical three-wire stator windings that facilitate transmission of angle data over long distances (thousands of feet). Such a system is also highly immune to noise and coupled signals. Practically the only trimming required for very long line systems is matching the line-to-line capacitances.

Because synchros can rotate continuously, they can be used in multispeed arrangements, where, for example, full-scale system travel may be represented by 36 or 64 full rotations. When reduced by gearing to a single, full-scale turn, a synchro's electrical inaccuracy is the typical 0.1° error divided by 36 or 64 or whatever gear ratio is used. This error is insignificant compared with the error of the gearing coupling the high-speed synchro and the single speed (1 rotation for full scale) output shaft. The accuracy is dependable and stable, using standard synchros and gearing.

Hydraulic and Pneumatic Systems

In Fig. 1 is shown a schematic of a hydraulic cylinder and the relationships between force and area that govern all hydraulic systems. Hydraulic actuators that drive the load may be cylinders or motors, depending on whether linear or rotary motion is required. The load must be defined by its torque–speed characteristics and inertia, and a suitable hydraulic actuator selected before the remaining system components can be chosen. Fluid under pressure and suitable valves are needed to control motion. Both single- and double-acting hydraulic cylinders are available, and the latter type is seen in Fig. 1.

Pressure can be traded off against velocity, if desired, by placing a different effective area at each side of the piston. The same pressure on a smaller area will move the piston at a higher speed but lower force for a given rate of fluid delivery. The cylinder shown in Fig. 1 can drive loads in either direction. The simple formulas of plane geometry relate cylinder areas, force, fluid flow, and rate of movement. Other configurations can develop equal forces and speeds in both directions.

The rotary equivalent of the cylinder is the hydraulic motor, which is defined by the fluid displacement required to turn the output shaft through one revolution, by the output torque, and by the load requirements of torque and speed. Output torque is proportional to fluid pressure, which can be as high as safety permits. Output speed is defined by the number of gallons per minute supplied to the motor. As an example, if 231 cu. in. = 1 gallon, an input of 6 gallons/min (gpm) with a 5-cu. in. displacement gives a mean speed of $6 \times 231/5 = 277$ rpm. The motor torque must be defined by lb-in. per 100 $lb_f/in.^2$ (typically) from which the required pressure can be determined. Various motor types are available.

Hydraulic Pumps.—The most-used hydraulic pump is the positive-displacement type, which delivers a fixed amount of fluid for every cycle. These pumps are also called hydrostatic because they deliver energy by static pressure rather than by the kinetic energy of a moving fluid. Positive-displacement pumps are rated by the gpm delivered at a stated speed and by the maximum pressure, which are the key parameters defining the power capacity of the hydraulic actuator. Delivered gpm are reduced under load due to leakage, and the reduction is described by the volumetric efficiency, which is the ratio of actual to theoretical output.

Hydraulic Fluids.—The hydraulic fluid is the basic means of transmitting power, and it also provides lubrication and cooling when passed through a heat exchanger. The fluid must be minimally compressible to avoid springiness and delay in response. The total system inertia reacts with fluid compliance to generate a resonant frequency, much as inertia and mechanical compliance react in an electromechanical system. Compliance must be low enough that resonances do not occur in the active bandwidth of the servomechanism, and that unacceptable transients do not occur under shock loads. Seal friction and fluid viscosity tend to damp out resonant vibrations. Shock-absorbing limit stops or cushions are usually located at the travel limits to minimize transient impact forces.

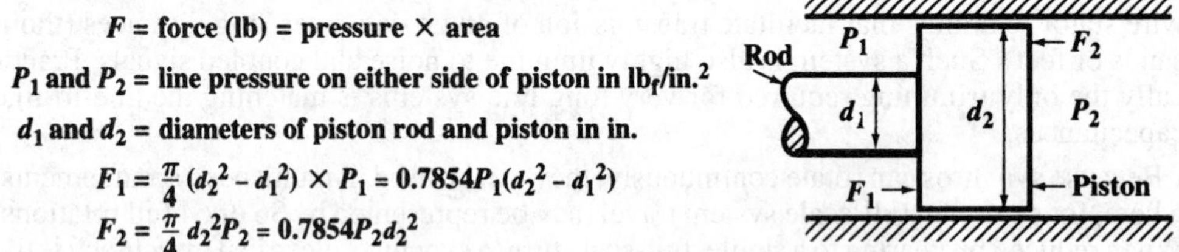

F = force (lb) = pressure × area
P_1 and P_2 = line pressure on either side of piston in $lb_f/in.^2$
d_1 and d_2 = diameters of piston rod and piston in in.
$F_1 = \frac{\pi}{4}(d_2^2 - d_1^2) \times P_1 = 0.7854P_1(d_2^2 - d_1^2)$
$F_2 = \frac{\pi}{4} d_2^2 P_2 = 0.7854P_2 d_2^2$
P_1
F_2
Rod
d_1
d_2
P_2
F_1
Piston

Fig. 1. Elementary Hydraulic Force/Area Formulas

Hydraulic fluids with special additives for lubrication minimize wear between moving parts. An auxiliary function is prevention of corrosion and pitting. Hydraulic fluids must also be compatible with gaskets, seals, and other nonmetallic materials.

Viscosity is another critical parameter of hydraulic fluids as high viscosity means high resistance to fluid flow with a corresponding power loss and heating of the fluid, pressure drop in the hydraulic lines, difficulty in removing bubbles, and sometimes overdamped operation. Unfortunately, viscosity falls very rapidly with increasing temperature, which can lead to reduction of the lubrication properties and excessive wear as well as increasing leakage. For hydraulic actuators operating at very low temperatures, the fluid pour point is important. Below this temperature, the hydraulic fluid will not flow. Design guidelines similar to those used with linear or rotating bearings are applicable in these conditions. Fire-resistant fluids are available for use in certain conditions such as in die casting, where furnaces containing molten metal are often located near hydraulic systems.

A problem with hydraulic systems that is absent in electromechanical systems is that of dirt, air bubbles, and contaminants in the fluid. Enclosed systems are designed to keep out contaminants, but the main problem is with the reservoir or fluid storage unit. A suitable sealer must be used in the reservoir to prevent corrosion and a filter should be used during filling. Atmospheric pressure is required on the fluid surface in the reservoir except where a pressurized reservoir is used. Additional components include coarse and fine filters to remove contaminants and these filters may be rated to remove micron sized particles (1 micron = 0.00004 in.).

Very fine filters are sometimes used in high-pressure lines, where dirt might interfere with the operation of sensitive valves. Where a high-performance pump is used, a fine filter is a requirement. Usually, only coarse filters are used on fluid inlet lines because fine filters might introduce excessive pressure drop.

Aside from the reservoir used for hydraulic fluid storage, line connections, fittings, and couplings are needed. Expansion of these components under pressure increases the mechanical compliance of the system, reducing the frequencies of any resonances and possibly interfering with the response of wide-band systems.

Formulas relating fluid flow and mechanical power follow. These formulas supplement the general force, torque, speed, and power formulas of mechanical systems.

$$F = P \times A$$

$$A = 0.7854 \times d^2$$

$$hp = 0.000583q \times \text{pressure in lb}_f/\text{in.}^2$$

1 gallon of fluid flow/min at 1 $\text{lb}_f/\text{in.}^2$ pressure = 0.000582 hp.

For rotary outputs,

$$hp = \text{torque} \times \text{rpm}/63{,}025$$

where torque is in lb-in. (Theoretical hp output must be multiplied by the efficiency of the hydraulic circuits to determine actual output.)

In the preceding equations,

P = pressure in $\text{lb}_f/\text{in.}^2$

A = piston area in in.^{2}

F = force in lb

q = fluid flow in gallons/min

d = piston diameter in inches

Hydraulic and Pneumatic Control Systems.—Control systems for hydraulic and pneumatic circuits are more mature than those for electromechanical systems because they have been developed over many more years. Hydraulic components are available at moderate prices from many sources. Although their design is complex, application and servicing of these systems are usually more straightforward than with electromechanical systems.

Electromechanical and hydraulic/pneumatic systems may be analyzed by similar means. The mathematical requirements for accuracy and stability are analogous, as are most performance features, although nonlinearities are caused by different physical attributes. Nonlinear friction, backlash, and voltage and current limiting are common to both types of system, but hydraulic/pneumatic systems also have the behavior characteristics of fluid-driven systems such as thermal effects and fluid flow dynamics including turbulence, leakage caused by imperfect seals, and contamination.

Both control types require overhead equipment that does not affect performance but adds to overall cost and complexity. For instance, electromechanical systems require electrical power sources and power control components, voltage regulators, fuses and circuit breakers, relays and switches, connectors, wiring and related devices. Hydraulic/pneumatic systems require fluid stored under pressure, motor-driven pumps or compressors, valves, pressure regulators/limiters, piping and fasteners, as well as hydraulic/pneumatic motors and cylinders. Frequently, the optimum system is selected on the basis of overhead equipment already available.

Electromechanical systems are generally slower and heavier than hydraulic systems and less suited to controlling heavy loads. The bandwidths of hydraulic control systems can respond to input signals of well over 100 Hz as easily as an electromechanical system can respond to, say, 10 to 20 Hz. Hydraulic systems can drive very high torque loads without intermediate transmissions such as the gear trains often used with electromechanical systems. Also, hydraulic/pneumatic systems using servo valves and piston/cylinder arrangements are inherently suited to linear motion operation, whereas electromechanical controls based on conventional electrical machines are more naturally suited to driving rotational loads.

Until recently, electromechanical systems were limited to system bandwidths of about 10 Hz, with power outputs of a few hundred watts. However, their capabilities have now been sharply extended through the use of rare-earth motor magnets having much higher energies than earlier designs. Similarly, semiconductor power components deliver much higher output power at lower prices than earlier equipment. Electromechanical control systems are now suited to applications of more than 100 hp with bandwidths up to 40 Hz and sometimes up to 100 Hz.

Although much depends on the specific design, the edge in reliability, even for high-power, fast-response needs, is shifting toward electromechanical systems. Basically, there are more things that can go wrong in hydraulic/pneumatic systems, as indicated by the shift to more electrical systems in aircraft.

Hydraulic Control Systems.—Using essentially incompressible fluid, hydraulic systems are suited to a wide range of applications, whereas pneumatic power is generally limited to simpler uses. In Fig. 2 are shown the essential features of a simple linear hydraulic control system and a comparable system for driving a rotating load.

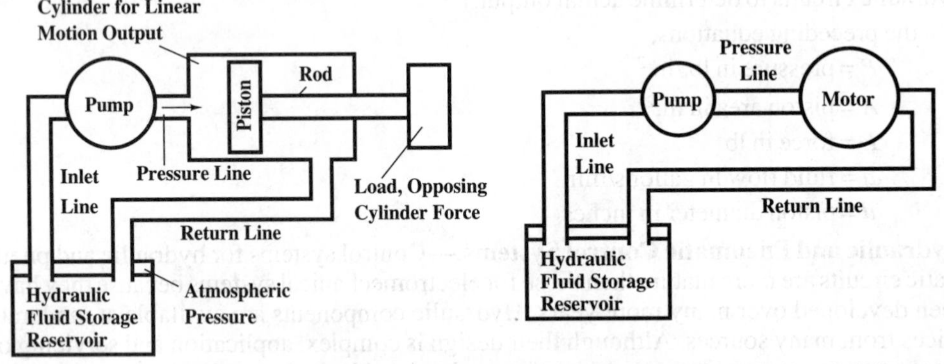

Fig. 2. (left) A Simple Linear Hydraulic Control System in Which the Load Force Returns the Piston and (right) a Comparable System for Driving a Rotating Load

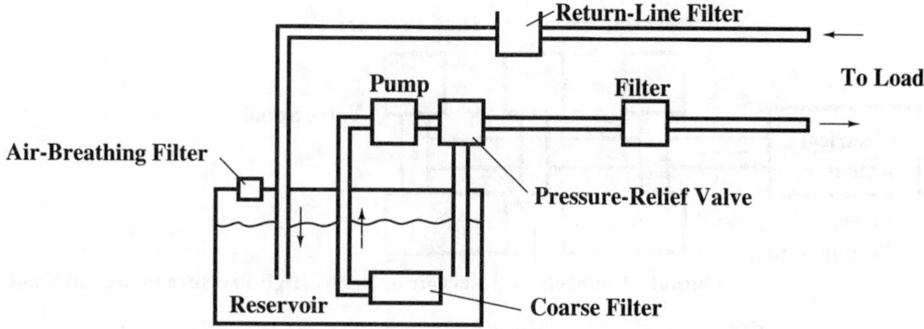

Fig. 3. Some of the Auxiliary Components Used in a Practical Hydraulic System

Hydraulic controls of the type shown have fast response and very high load capacities. In a linear actuator, for example, each $lb_f/in.^2$ of system pressure acts against the area of the piston to generate the force applied. Hydraulic pressures of up to 3000 $lb_f/in.^2$ are readily obtained from hydraulic pumps, so that cylinders can exert forces of hundreds of tons without the need for speed-reducing transmission systems to increase the force. The hydraulic fluid distributes heat, so it helps cool the system.

Systems similar to those in Fig. 2 can be operated in open- or closed-loop modes. Open-loop operation can be controlled by programming units that initiate each step by operating relays, limit switches, solenoid valves, and other components to generate the forces over the required travel ranges. Auxiliary components are used to ensure safe operation and make such systems flexible and reliable, as shown in Fig. 3.

In the simplest mode, whether open- or closed-loop, hydraulic system operation may be discontinuous or proportional. Discontinuous operation, sometimes called bang-bang, or on–off, works well, is widely used in low- to medium-accuracy systems, and is easy to maintain. In this closed-loop mode, accuracy is limited; if the response to error is set too high, the system will oscillate between on–off modes, with average output at about the desired value. This oscillation, however, can be noisy, introduces system transients, and may cause rapid wear of system components.

Another factor to be considered in on–off systems is the shock caused by sudden opening and closing of high-pressure valves, which introduce transient pulses in the fluid flow and can cause high stresses in components. These problems can be addressed by the use of pressure-limiting relief valves and other units.

Proportional Control Systems.—Where the highest accuracy is required, perhaps in two directions, and with aiding or opposing forces or torques, a more sophisticated proportional control, closed-loop system is preferred. As shown in Fig. 4, the amplifier and electric servomotor used in electromechanical closed-loop systems is replaced in the closed-loop hydraulic system by an electronically controlled servo-valve. In its simplest form, the valve uses a linear motor to position the spool that determines the flow path for the hydraulic fluid. In some designs, the linear motor may be driven by a solenoid against a bias spring on the value spool. In other arrangements, the motor may be a bidirectional unit that permits a fluid flow depending on the polarity and amplitude of the voltage supplied to the motor.

Such designs can be used in proportional control systems to achieve smooth operation and minimum nonlinearities, and will give the maximum accuracy required by the best machine tool applications. Where very high power must be controlled, use is often made of a two-stage valve in which the output from the first stage is used to drive the second-stage valve, as shown in Fig. 5.

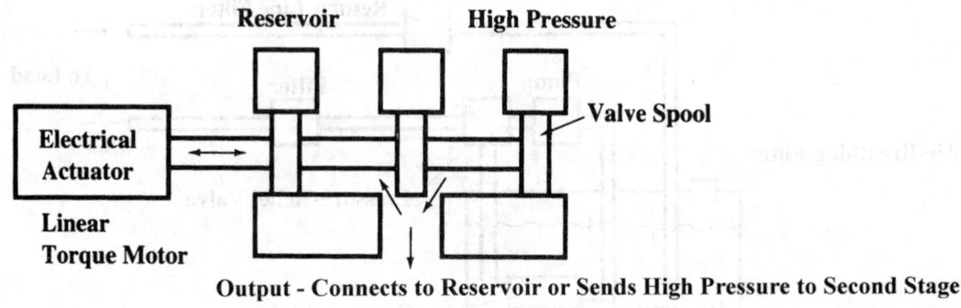

Fig. 4.

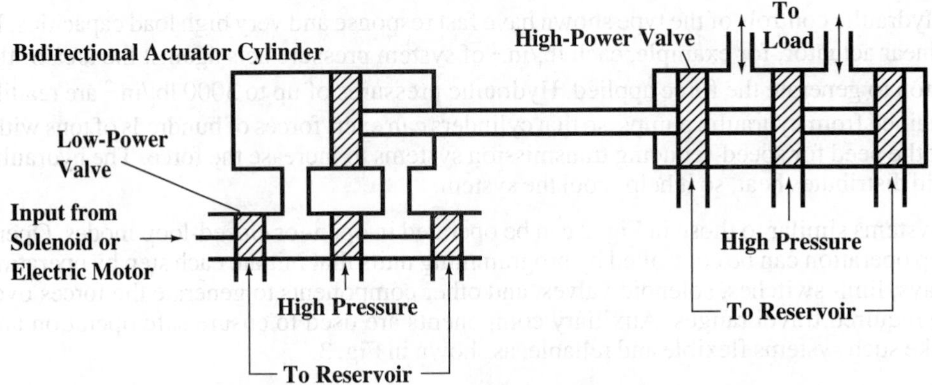

Fig. 5. Two-Stage Valve for Large-Power Control from a Low-Power Input

Electronic Controls.—An error-sensing electronic amplifier drives the solenoid motor of Fig. 5, which provides automatic output correction in a closed-loop system. The input is an ideal place to introduce electrical control features, adding greatly to the versatility of the control system. The electronic amplifier can provide the necessary driving power using pulse-width modulation as required, for minimum heating. The output can respond to signals in the low-microvolt range.

A major decision is whether to use analog or digital control. Although analog units are simple, they are much less versatile than their digital counterparts. Digital systems can be readjusted for total travel, speed, and acceleration by simple reprogramming. Use of appropriate feedback sensors can match accuracy to any production requirement, and a single digital system can be easily adapted to a great variety of similar applications. This adaptability is an important cost-saving feature for moderate-sized production runs. Modern microprocessors can integrate the operation of sets of systems.

Because nonlinearities and small incremental motions are easy to implement, digital systems are capable of very smooth acceleration, which avoids damaging shocks and induced leaks, and enhances reliability so that seals and hose connections last longer. The accuracy of digital control systems depends on transducer availability, and a full range of such devices has been developed and is now available.

Other features of digital controls are their capacity for self-calibration, easy digital readout, and periodic self-compensation. For example, it is easy to incorporate backlash compensation. Inaccuracies can be corrected by using lookup tables that may themselves be updated as necessary. Digital outputs can be used as part of an inspection plan, to indicate need for tool changing, adjustment or sharpening, or for automatic record keeping. Despite continuing improvements in analog systems, digital control of hydraulic systems is favored in large plants.

Pneumatic Systems.—Hydraulic systems transmit power by means of the flow of an essentially incompressible fluid. Pneumatic systems use a highly compressible gas. For this reason, a pneumatic system is slower in responding to loads, especially sudden output loads, than a hydraulic system. Similarly, torque or force requires time and output motion to build up. Response to sudden output loads shows initial overshoot. Much more complex networks or other damping means are required to develop stable response in closed-loop systems. On the other hand, there are no harmful shock waves analogous to the transients that can occur in hydraulic systems, and pneumatic system components last comparatively longer.

Notwithstanding their performance deficiencies, pneumatic systems have numerous desirable features. Pneumatic systems avoid some fire hazards compared with the most preferred hydraulic fluids. Air can be vented to the atmosphere so a flow line only is needed, reducing the complexity, cost, and weight of the overall system. Pneumatic lines, couplings, and fittings are lighter than their hydraulic counterparts, often a significant advantage. The gaseous medium also is lighter than hydraulic fluid, and pneumatic systems are usually easier to clean, assemble, and generally maintain. Fluid viscosity and its temperature variations are virtually negligible with pneumatic systems.

Among drawbacks with pneumatics are that lubrication must be carefully designed in, and more power is needed to achieve a desired pressure when the fluid medium is a compressible gas. Gas under high pressure can cause an explosion if its storage tank is damaged, so storage must have substantial safety margins. Gas compressibility makes pneumatic systems 1 or 2 orders of magnitude slower than hydraulic systems.

The low stiffness of pneumatic systems is another indicator of the long response time. Resonances occur between the compressible gas and equivalent system inertias at lower frequencies. Even the relatively low speed of sound in connecting lines contributes to response delay, adding to the difficulty of closed-loop stabilization. Fortunately, it is possible to construct pneumatic analogs to electrical networks to simplify stabilization at the exact point of the delays. Such pneumatic stabilizing means are commercially available and are important elements of closed-loop pneumatic control systems.

In contrast with hydraulic systems, where speed may be controlled by varying pump output, pneumatic system control is almost exclusively by valves, which control the flow from a pneumatic accumulator or pressure source. The pressure is maintained between limits by an intermittently operated pump. Low-pressure outlet ports must be large enough to accommodate the high volume of the expanded gas. In Fig. 6 is shown a simplified system for closed-loop position control applied to an air cylinder, in which static accuracy is controlled by the position sensor. Proper design requires a good theoretical analysis and attention to practical design if good, stable, closed-loop response is to be achieved.

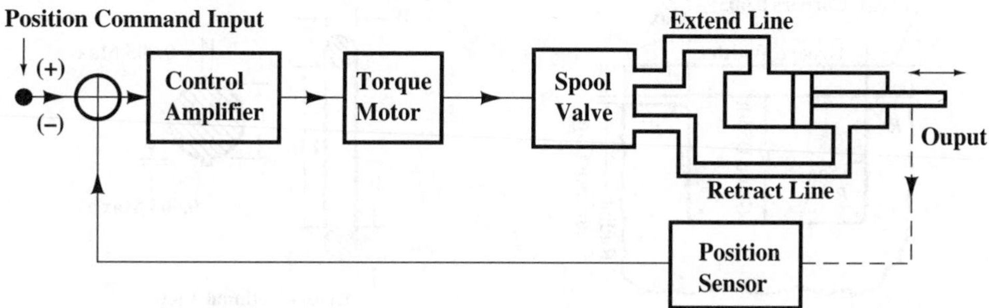

Fig. 6. A Pneumatic Closed-Loop Linear Control System

O-RINGS

An O-ring is a one-piece molded elastomeric seal with a circular cross-section that seals by distortion of its resilient elastic compound. Dimensions of O-rings are given in ANSI/SAE AS568A, Aerospace Size Standard for O-rings. The standard ring sizes have been assigned identifying dash numbers that, in conjunction with the compound (ring material), completely specifies the ring. Although the ring sizes are standardized, ANSI/SAE AS568A does not cover the compounds used in making the rings; thus, different manufacturers will use different designations to identify various ring compounds. For example, 230-8307 represents a standard O-ring of size 230 (2.484 in. ID by 0.139 in. width) made with compound 8307, a general-purpose nitrile compound. O-ring material properties are discussed at the end of this section.

When properly installed in a groove, an O-ring is normally slightly deformed so that the naturally round cross-section is squeezed diametrically out of round prior to the application of pressure. This compression ensures that under static conditions, the ring is in contact with the inner and outer walls enclosing it, with the resiliency of the rubber providing a zero-pressure seal. When pressure is applied, it tends to force the O-ring across the groove, causing the ring to further deform and flow up to the fluid passage and seal it against leakage, as in Fig. 1(a). As additional pressure is applied, the O-ring deforms into a D shape, as in Fig. 1(b). If the clearance gap between the sealing surface and the groove corners is too large or if the pressure exceeds the deformation limits of the O-ring material (compound), the O-ring will extrude into the clearance gap, reducing the effective life of the seal. For very low-pressure static applications, the effectiveness of the seal can be improved by using a softer durometer compound or by increasing the initial squeeze on the ring, but at higher pressures, the additional squeeze may reduce the ring's dynamic sealing ability, increase friction, and shorten ring life.

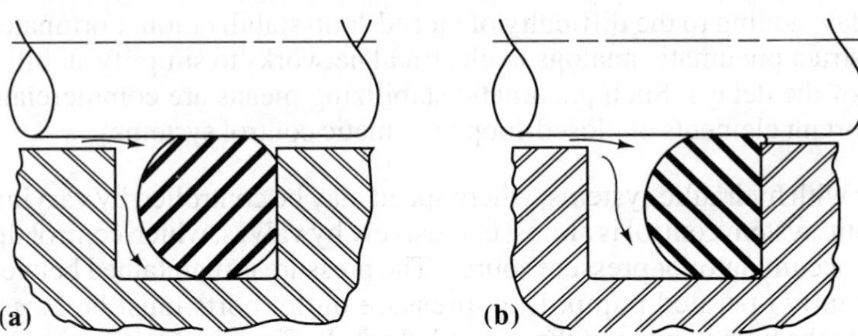

Fig. 1.

The initial diametral squeeze of the ring is very important in the success of an O-ring application. The squeeze is the difference between the ring width *W* and the gland depth

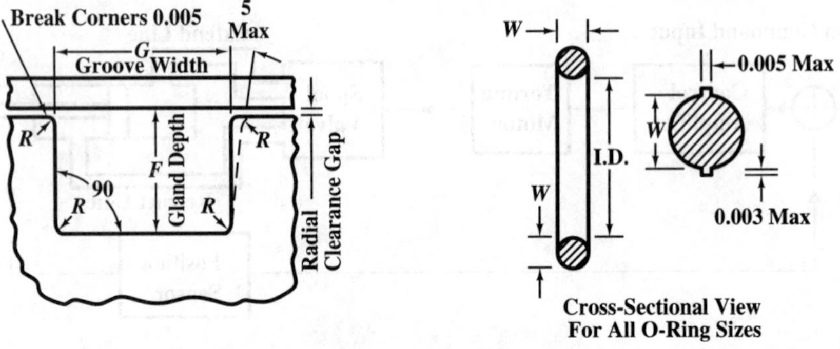

Fig. 2. Groove and Ring Details

F (Fig. 2) and has a great effect on the sealing ability and life of an O-ring application. The ideal squeeze varies according to the ring cross-section, with the average being about 20 per cent, i.e., the ring's cross-section *W* is about 20 per cent greater than the gland depth *F* (groove depth plus clearance gap). The groove width is normally about 1.5 times larger than the ring width *W*. When installed, an O-ring compresses slightly and distorts into the free space within the groove. Additional expansion or swelling may also occur due to contact of the ring with fluid or heat. The groove must be large enough to accommodate the maximum expansion of the ring or the ring may extrude into the clearance gap or rupture the assembly. In a dynamic application, the extruded ring material will quickly wear and fray, severely limiting seal life.

To prevent O-ring extrusion or to correct an O-ring application, reduce the clearance gap by modifying the dimensions of the system, reduce the system operating pressure, install antiextrusion backup rings in the groove with the O-ring, as in Fig. 3, or use a harder O-ring compound. A harder compound may result in higher friction and a greater tendency of the seal to leak at low pressures. Backup rings, frequently made of leather, Teflon, metal, phenolic, hard rubber, and other hard materials, prevent extrusion and nibbling where large clearance gaps and high pressure are necessary.

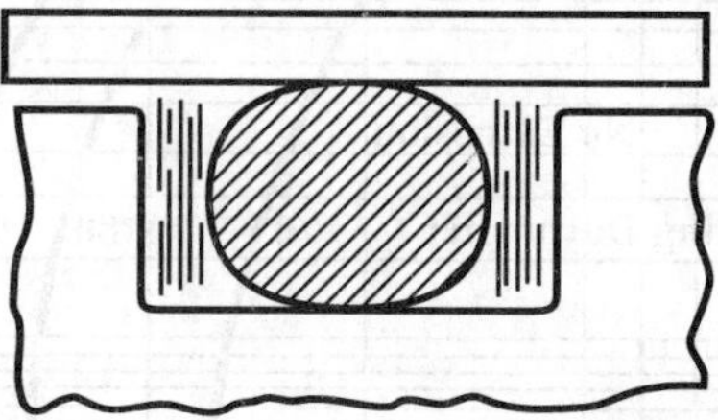

Fig. 3. Preferred Use of Backup Washers

The most effective and reliable sealing is generally provided by using the diametrical clearances given in manufacturers' literature. However, the information in Table 1 may be used to estimate the gland depth (groove depth plus radial clearance) required in O-ring applications. The radial clearance used (radial clearance equals one-half the diametral clearance) also depends on the system pressure, the ring compound and hardness, and specific details of the application.

Table 1. Gland Depth for O-Ring Applications

Standard O-Ring Cross-Sectional Diameter (in.)	Gland Depth (in.)	
	Reciprocating Seals	Static Seals
0.070	0.055 to 0.057	0.050 to 0.052
0.103	0.088 to 0.090	0.081 to 0.083
0.139	0.121 to 0.123	0.111 to 0.113
0.210	0.185 to 0.188	0.170 to 0.173
0.275	0.237 to 0.240	0.226 to 0.229

Source: Auburn Manufacturing Co. When possible, use manufacturer recommendations for clearance gaps and groove depth.

Fig. 4 indicates conditions where O-ring seals may be used, depending on the fluid pressure and the O-ring hardness. If the conditions of use fall to the right of the curve, extrusion of the O-ring into the surrounding clearance gap will occur, greatly reducing the life of the ring. If conditions fall to the left of the curve, no extrusion of the ring will occur, and the ring may be used under these conditions. For example, in an O-ring application with a 0.004-in. diametral clearance and 2500-psi pressure, extrusion will occur with a 70 durometer O-ring but not with an 80 durometer O-ring. As the graph indicates, high-pressure applications require lower clearances and harder O-rings for effective sealing.

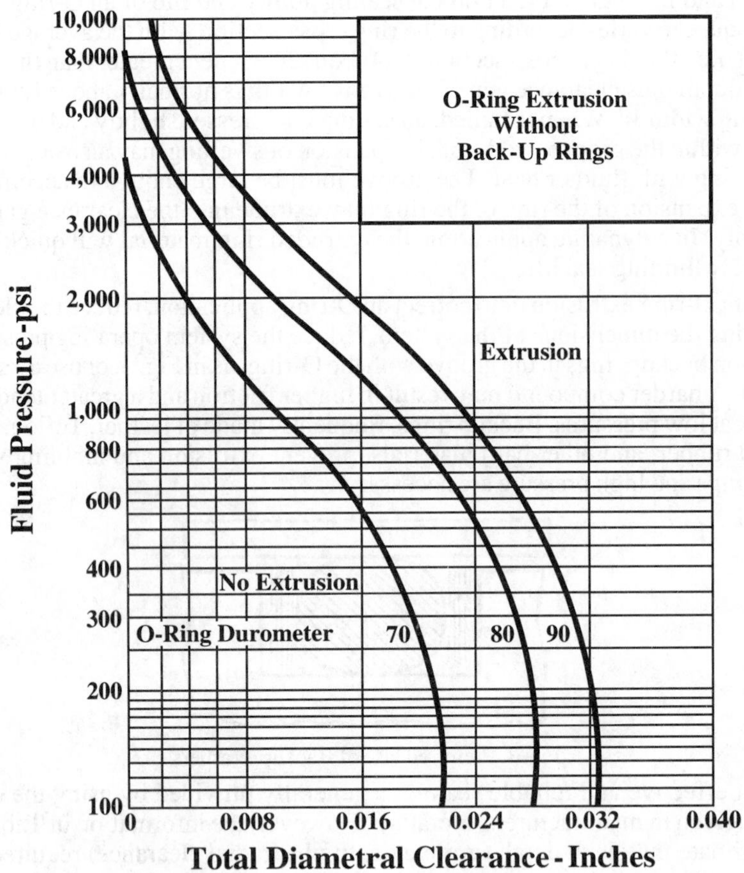

Fig. 4. Extrusion Potential of O-Rings as a Function of Hardness and Clearance

Recommended groove width, clearance dimensions, and bottom-of-groove radius for O-ring numbers up to 475 (25.940-in. ID by 0.275-in. width) can be found using Table in conjunction with Fig. 5. In general, except for ring cross-sections smaller than 1/16 in., the groove width is approximately 1.5W, where W is the ring cross-sectional diameter. Straight-sided grooves are best for preventing extrusion of the ring or nibbling; however, for low-pressure applications (less than 1500 psi) sloped sides with an angle up to 5° can be used to simplify machining of the groove. The groove surfaces should be free of burrs, nicks, or scratches. For static seals (i.e., no contact between the O-ring and any moving parts), the groove surfaces should have a maximum roughness of 32 to 63 µin. rms for liquid-sealing applications and 16 to 32 µin. rms for gaseous-sealing applications. In dynamic seals, relative motion exists between the O-ring and one or more parts and the maximum groove surface roughness should be 8 to 16 µin. rms for sliding contact applications (reciprocating seals, for example) and 16 to 32 µin. rms for rotary contact applications (rotating and oscillating seals).

In dynamic seal applications, the roughness of surfaces in contact with O-rings (bores, pistons, and shafts, for example) should be 8 to 16 µin. rms, without longitudinal or circumferential scratches. Surface finishes of less than 5 µin. rms are too smooth to give a good seal life because they wipe too clean, causing the ring to wear against the housing in the absence of a lubricating film. The best-quality surfaces are honed, burnished, or hard chromium plated. Soft and stringy metals such as aluminum, brass, bronze, Monel, or free machining stainless steel should not be used in contact with moving seals. In static applications, O-ring contacting surfaces should have a maximum surface roughness of 64 to 125 µin. rms.

Table 2. Diametral Clearance and Groove Sizes for O-Ring Applications

ANSI/SAE AS568 Number	Tolerances A	Tolerances B	Diametral Clearance, D: Reciprocating & Static Seals	Diametral Clearance, D: Rotary Seals	Groove Width, G: Backup Rigs None	Groove Width, G: Backup Rigs One	Groove Width, G: Backup Rigs Two	Bottom of Groove Radius, R
001	+0.001 −0.000	+0.000 −0.001	0.002 to 0.004	0.012 to 0.016	0.063			0.005 to 0.015
002					0.073			
003					0.083			
004 to 012					0.094	0.149	0.207	
013 to 050	+0.002 −0.000	+0.000 −0.002	0.002 to 0.005	0.016 to 0.020				
102 to 129					0.141	0.183	0.245	
130 to 178			0.002 to 0.006					
201 to 284					0.188	0.235	0.304	0.010 to 0.025
309 to 395			0.003 to 0.007		0.281	0.334	0.424	
425 to 475	+0.003 −0.000	+0.000 −0.003	0.004 to 0.010		0.375	0.475	0.579	0.020 to 0.035

Source: Auburn Manufacturing Co. All dimensions are in inches. Clearances listed are minimum and maximum values; standard groove widths may be reduced by about 10 per cent for use with ring compounds that free swell less than 15 per cent. Dimension *A* is the ID of any surface contacted by the outside circumference of the ring; *B* is the OD of any surface contacted by the inside circumference of the ring.

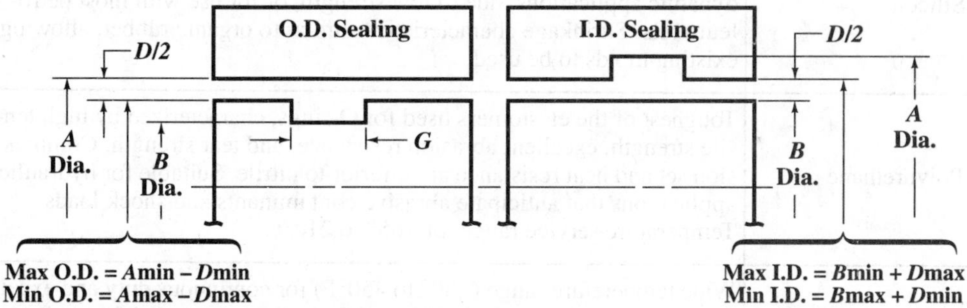

Fig. 5. Installation Data for Use with Table . Max and Min are maximum and minimum piston and bore diameters for O.D. and I.D., respectively.

The preferred bore materials are steel and cast iron, and pistons should be softer than the bore to avoid scratching them. The bore sections should be thick enough to resist expansion and contraction under pressure so that the radial clearance gap remains constant, reducing the chance of damage to the O-ring by extrusion and nibbling. Some compatibility problems may occur when O-rings are used with plastics parts because certain compounding ingredients may attack the plastics, causing crazing of the plastics surface.

O-rings are frequently used as driving belts in round bottom or V-grooves with light tension for low-power drive elements. Special compounds are available with high resistance to stress relaxation and fatigue for these applications. Best service is obtained in drive belt applications when the initial belt tension is between 80 and 200 psi and the initial installed stretch is between 8 and 25 per cent of the circumferential length. Most of the compounds used for drive belts operate best between 10 and 15 per cent stretch, although polyurethane has good service life when stretched as much as 20 to 25 per cent.

Table 3. Typical O-Ring Compounds

Nitrile	General-purpose compound for use with most petroleum oils, greases, gasoline, alcohols and glycols, LP gases, propane and butane fuels. Also for food service to resist vegetable and animal fats. Effective temperature range is about −40° to 250°F. Excellent compression set, tear and abrasion resistance, but poor resistance to ozone, sunlight and weather. Higher-temperature nitrile compounds with similar properties are also available.
Hydrogenated Nitrile	Similar to general-purpose nitrile compounds with improved high-temperature performance, resistance to aging, and petroleum product compatibility.
Polychloroprene (Neoprene)	General-purpose compound with low compression set and good resistance to elevated temperatures. Good resistance to sunlight, ozone, and weathering, and fair oil resistance. Frequently used for refrigerator gases such as Freon. Effective temperature range is about −40° to 250°F.
Ethylene Propylene	General-purpose compound with excellent resistance to polar fluids such as water, steam, ketones, and phosphate esters, and brake fluids, but not resistant to petroleum oils and solvents. Excellent resistance to ozone and flexing. Recommended for belt-drive applications. Continuous duty service in temperatures up to 250°F.
Silicon	Widest temperature range (−150° to 500°F) and best low-temperature flexibility of all elastomeric compounds. Not recommended for dynamic applications, due to low strength, or for use with most petroleum oils. Shrinkage characteristics similar to organic rubber, allowing existing molds to be used.
Polyurethane	Toughest of the elastomers used for O-rings, characterized by high tensile strength, excellent abrasion resistance, and tear strength. Compression set and heat resistance are inferior to nitrile. Suitable for hydraulic applications that anticipate abrasive contaminants and shock loads. Temperature service range of −65° to 212°F.
Fluorosilicone	Wide temperature range (−80° to 450°F) for continuous duty and excellent resistance to petroleum oils and fuels. Recommended for static applications only, due to limited strength and low abrasion resistance.
Polyacrylate	Heat resistance better than nitrile compounds, but inferior low temperature, compression set, and water resistance. Often used in power steering and transmission applications due to excellent resistance to oil, automatic transmission fluids, oxidation, and flex cracking. Temperature service range of −20° to 300°F.
Fluorocarbon (Viton)	General-purpose compound suitable for applications requiring resistance to aromatic or halogenated solvents or to high temperatures (−20° to 500°F with limited service to 600°F). Outstanding resistance to blended aromatic fuels, straight aromatics, and halogenated hydrocarbons and other petroleum products. Good resistance to strong acids (temperature range in acids (−20° to 250°F), but not effective for use with very hot water, steam, and brake fluids.

Ring Materials.—Thousands of O-ring compounds have been formulated for specific applications. Some of the most common types of compounds and their typical applications are given in Table 3. The Shore A durometer is the standard instrument used for measuring the hardness of elastomeric compounds. The softest O-rings are 50 and 60 Shore A and stretch more easily, exhibit lower breakout friction, seal better on rough surfaces, and need less clamping pressure than harder rings. For a given squeeze, the higher the durometer hardness of a ring, the greater the associated friction because a greater compressive force is exerted by hard rings than soft rings.

The most widely used rings are medium-hard O-rings with 70 Shore A hardness, which have the best wear resistance and frictional properties for running seals. Applications that involve oscillating or rotary motion frequently use 80 Shore A materials. Rings with a hardness above 85 Shore A often leak more because of less effective wiping action. These harder rings have a greater resistance to extrusion, but for small sizes may break easily during installation. O-ring hardness varies inversely with temperature, but when used for continuous service at high temperatures, the hardness may eventually increase after an initial softening of the compound.

O-ring compounds have thermal coefficients of expansion in the range of 7 to 20 times that of metal components, so shrinkage or expansion with temperature change can pose problems of leakage past the seal at low temperatures and excessive pressures at high temperatures when a ring is installed in a tight-fitting groove. Likewise, when an O-ring is immersed in a fluid, the compound usually absorbs some of the fluid and consequently increases in volume. Manufacturer's data give volumetric increase data for compounds completely immersed in various fluids. For confined rings (those with only a portion of the ring exposed to fluid), the size increase may be considerably lower than for rings completely immersed in fluid. Certain fluids can also cause ring shrinkage during "idle" periods, i.e., when the seal has a chance to dry out. If this shrinkage is more than 3 to 4 per cent, the seal may leak.

Excessive swelling due to fluid contact and high temperatures softens all compounds approximately 20 to 30 Shore A points from room temperature values and designs should anticipate the expected operating conditions. At low temperatures, swelling may be beneficial because fluid absorption may make the seal more flexible. However, the combination of low temperature and low pressure makes a seal particularly difficult to maintain. A soft compound should be used to provide a resilient seal at low temperatures. Below −65°F, only compounds formulated with silicone are useful; other compounds are simply too stiff, especially for use with air and other gases.

Compression set is another material property and a very important sealing factor. It is a measure of the shape memory of the material, that is, the ability to regain shape after being deformed. Compression set is a ratio, expressed as a percentage, of the unrecovered to original thickness of an O-ring compressed for a specified period of time between two heated plates and then released. O-rings with excessive compressive set will fail to maintain a good seal because, over time, the ring will be unable to exert the necessary compressive force (squeeze) on the enclosing walls. Swelling of the ring due to fluid contact tends to increase the squeeze and may partially compensate for the loss due to compression set. Generally, compression set varies by compound and ring cross-sectional diameter, and increases with the operating temperature.

ROLLED STEEL SECTIONS, WIRE AND SHEET-METAL GAGES

Rolled Steel Sections

Lengths of Angles Bent to Circular Shape.—To calculate the length of an angle-iron used either inside or outside of a tank or smokestack, the following table of constants may be used: Assume, for example, that a stand-pipe, 20 feet inside diameter, is provided with a 3 by 3 by $\frac{3}{8}$ inch angle-iron on the inside at the top. The circumference of a circle 20 feet in diameter is 754 inches. From the table of constants, find the constant for a 3 by 3 by $\frac{3}{8}$ inch angle-iron, which is 4.319. The length of the angle then is $754 - 4.319 = 749.681$ inches. Should the angle be on the outside, add the constant instead of subtracting it; thus, $754 + 4.319 = 758.319$ inches.

Size of Angle	Const.	Size of Angle	Const.	Size of Angle	Const.
$\frac{1}{4} \times 2 \times 2$	2.879	$\frac{5}{16} \times 3 \times 3$	4.123	$\frac{1}{2} \times 5 \times 5$	6.804
$\frac{5}{16} \times 2 \times 2$	3.076	$\frac{3}{8} \times 3 \times 3$	4.319	$\frac{3}{8} \times 6 \times 6$	7.461
$\frac{3}{8} \times 2 \times 2$	3.272	$\frac{1}{2} \times 3 \times 3$	4.711	$\frac{1}{2} \times 6 \times 6$	7.854
$\frac{1}{4} \times 2\frac{1}{2} \times 2\frac{1}{2}$	3.403	$\frac{3}{8} \times 3\frac{1}{2} \times 3\frac{1}{2}$	4.843	$\frac{3}{4} \times 6 \times 6$	8.639
$\frac{5}{16} \times 2\frac{1}{2} \times 2\frac{1}{2}$	3.600	$\frac{1}{2} \times 3\frac{1}{2} \times 3\frac{1}{2}$	5.235	$\frac{1}{2} \times 8 \times 8$	9.949
$\frac{3}{8} \times 2\frac{1}{2} \times 2\frac{1}{2}$	3.796	$\frac{3}{8} \times 4 \times 4$	5.366	$\frac{3}{4} \times 8 \times 8$	10.734
$\frac{1}{2} \times 2\frac{1}{2} \times 2\frac{1}{2}$	4.188	$\frac{1}{2} \times 4 \times 4$	5.758	$1 \times 8 \times 8$	11.520
$\frac{1}{4} \times 3 \times 3$	3.926	$\frac{3}{8} \times 5 \times 5$	6.414	…	…

Standard Designations of Rolled Steel Shapes.—Through a joint effort, the American Iron and Steel Institute (AISI) and the American Institute of Steel Construction (AISC) have changed most of the designations for their hot-rolled structural steel shapes. The present designations, standard for steel producing and fabricating industries, should be used when designing, detailing, and ordering steel. The accompanying table compares the present designations with the previous descriptions.

Hot-Rolled Structural Steel Shape Designations (AISI and AISC)

Present Designation	Type of Shape	Previous Designation
W 24 × 76	W shape	24 WF 76
W 14 × 26	W shape	14 B 26
S 24 × 100	S shape	24 I 100
M 8 × 18.5	M shape	8 M 18.5
M 10 × 9	M shape	10 JR 9.0
M 8 × 34.3	M shape	8 × 8 M 34.3
C 12 × 20.7	American Standard Channel	12 [20.7
MC 12 × 45	Miscellaneous Channel	12 × 4 [45.0
MC 12 × 10.6	Miscellaneous Channel	12 JR [10.6
HP 14 × 73	HP shape	14 BP 73
L 6 × 6 × $\frac{3}{4}$	Equal Leg Angle	∠6 × 6 × $\frac{3}{4}$
L 6 × 4 × $\frac{5}{8}$	Unequal Leg Angle	∠6 × 4 × $\frac{5}{8}$
WT 12 × 38	Structural Tee cut from W shape	ST 12 WF 38
WT 7 × 13	Structural Tee cut from W shape	ST 7 B 13
St 12 × 50	Structural Tee cut from S shape	ST 12 I 50
MT 4 × 9.25	Structural Tee cut from M shape	ST 4 M 9.25
MT 5 × 4.5	Structural Tee cut from M shape	ST 5 JR 4.5
MT 4 × 17.15	Structural Tee cut from M shape	ST 4 M 17.15
PL $\frac{1}{2}$ × 18	Plate	PL 18 × $\frac{1}{2}$
Bar 1	Square Bar	Bar 1
Bar 1 $\frac{1}{4}$Ø	Round Bar	Bar 1$\frac{1}{4}$Ø
Bar 2$\frac{1}{2}$ × $\frac{1}{2}$	Flat Bar	Bar 2$\frac{1}{2}$ × $\frac{1}{2}$
Pipe 4 Std.	Pipe	Pipe 4 Std.
Pipe 4 X-Strong	Pipe	Pipe 4 X-Strong
Pipe 4 XX-Strong	Pipe	Pipe 4 XX-Strong
TS 4 × 4 × .375	Structural Tubing: Square	Tube 4 × 4 × .375
TS 5 × 3 × .375	Structural Tubing: Rectangular	Tube 5 × 3 × .375
TS 3 OD × .250	Structural Tubing: Circular	Tube 3 OD × .250

Data taken from the "Manual of Steel Construction," 8th Edition, 1980, with permission of the American Institute of Steel Construction.

Steel Wide-Flange Sections—1

Wide-flange sections are designated, in order, by a section letter, nominal depth of the member in inches, and the nominal weight in pounds per foot; thus:

W 18 × 64

indicates a wide-flange section having a nominal depth of 18 inches, and a nominal weight per foot of 64 pounds. Actual geometry for each section can be obtained from the values below.

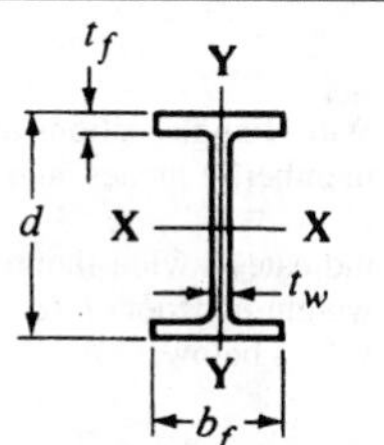

Designation	Area, A	Depth, d	Flange Width, b_f	Flange Thickness, t_f	Web Thickness, t_w	Axis X-X I	Axis X-X S	Axis X-X r	Axis Y-Y I	Axis Y-Y S	Axis Y-Y r
	in.^{2}	in.	in.	in.	in.	in.^{4}	in.^{3}	in.	in.^{4}	in.^{3}	in.
[a]W 27 × 178	52.3	27.81	14.085	1.190	0.725	6990	502	11.6	555	78.8	3.26
× 161	47.4	27.59	14.020	1.080	0.660	6280	455	11.5	497	70.9	3.24
× 146	42.9	27.38	13.965	0.975	0.605	5630	411	11.4	443	63.5	3.21
× 114	33.5	27.29	10.070	0.930	0.570	4090	299	11.0	159	31.5	2.18
× 102	30.0	27.09	10.015	0.830	0.515	3620	267	11.0	139	27.8	2.15
× 94	27.7	26.92	9.990	0.745	0.490	3270	243	10.9	124	24.8	2.12
× 84	24.8	26.71	9.960	0.640	0.460	2850	213	10.7	106	21.2	2.07
W 24 × 162	47.7	25.00	12.955	1.220	0.705	5170	414	10.4	443	68.4	3.05
× 146	43.0	24.74	12.900	1.090	0.650	4580	371	10.3	391	60.5	3.01
× 131	38.5	24.48	12.855	0.960	0.605	4020	329	10.2	340	53.0	2.97
× 117	34.4	24.26	12.800	0.850	0.550	3540	291	10.1	297	46.5	2.94
× 104	30.6	24.06	12.750	0.750	0.500	3100	258	10.1	259	40.7	2.91
× 94	27.7	24.31	9.065	0.875	0.515	2700	222	9.87	109	24.0	1.98
× 84	24.7	24.10	9.020	0.770	0.470	2370	196	9.79	94.4	20.9	1.95
× 76	22.4	23.92	8.990	0.680	0.440	2100	176	9.69	82.5	18.4	1.92
× 68	20.1	23.73	8.965	0.585	0.415	1830	154	9.55	70.4	15.7	1.87
× 62	18.2	23.74	7.040	0.590	0.430	1550	131	9.23	34.5	9.80	1.38
× 55	16.2	23.57	7.005	0.505	0.395	1350	114	9.11	29.1	8.30	1.34
W 21 × 147	43.2	22.06	12.510	1.150	0.720	3630	329	9.17	376	60.1	2.95
× 132	38.8	21.83	12.440	1.035	0.650	3220	295	9.12	333	53.5	2.93
× 122	35.9	21.68	12.390	0.960	0.600	2960	273	9.09	305	49.2	2.92
× 111	32.7	21.51	12.340	0.875	0.550	2670	249	9.05	274	44.5	2.90
× 101	29.8	21.36	12.290	0.800	0.500	2420	227	9.02	248	40.3	2.89
× 93	27.3	21.62	8.420	0.930	0.580	2070	192	8.70	92.9	22.1	1.84
× 83	24.3	21.43	8.355	0.835	0.515	1830	171	8.67	81.4	19.5	1.83
× 73	21.5	21.24	8.295	0.740	0.455	1600	151	8.64	70.6	17.0	1.81
× 68	20.0	21.13	8.270	0.685	0.430	1480	140	8.60	64.7	15.7	1.80
× 62	18.3	20.99	8.240	0.615	0.400	1330	127	8.54	57.5	13.9	1.77
× 57	16.7	21.06	6.555	0.650	0.405	1170	111	8.36	30.6	9.35	1.35
× 50	14.7	20.83	6.530	0.535	0.380	984	94.5	8.18	24.9	7.64	1.30
× 44	13.0	20.66	6.500	0.450	0.350	843	81.6	8.06	20.7	6.36	1.26
W 18 × 119	35.1	18.97	11.265	1.060	0.655	2190	231	7.90	253	44.9	2.69
× 106	31.1	18.73	11.200	0.940	0.590	1910	204	7.84	220	39.4	2.66
× 97	28.5	18.59	11.145	0.870	0.535	1750	188	7.82	201	36.1	2.65
× 86	25.3	18.39	11.090	0.770	0.480	1530	166	7.77	175	31.6	2.63
× 76	22.3	18.21	11.035	0.680	0.425	1330	146	7.73	152	27.6	2.61
× 71	20.8	18.47	7.635	0.810	0.495	1170	127	7.50	60.3	15.8	1.70
× 65	19.1	18.35	7.590	0.750	0.450	1070	117	7.49	54.8	14.4	1.69
× 60	17.6	18.24	7.555	0.695	0.415	984	108	7.47	50.1	13.3	1.69
× 55	16.2	18.11	7.530	0.630	0.390	890	98.3	7.41	44.9	11.9	1.67
× 50	14.7	17.99	7.495	0.570	0.355	800	88.9	7.38	40.1	10.7	1.65
× 46	13.5	18.06	6.060	0.605	0.360	712	78.8	7.25	22.5	7.43	1.29
× 40	11.8	17.90	6.015	0.525	0.315	612	68.4	7.21	19.1	6.35	1.27
× 35	10.3	17.70	6.000	0.425	0.300	510	57.6	7.04	15.3	5.12	1.22

[a] Consult the AISC Manual, noted above, for W steel shapes having nominal depths greater than 27 in.

Symbols: I = moment of inertia; *S* = section modulus; *r* = radius of gyration.

Data taken from the "Manual of Steel Construction," 8th Edition, 1980, with permission of the American Institute of Steel Construction.

Steel Wide-Flange Sections–2

Wide-flange sections are designated, in order, by a section letter, nominal depth of the member in inches, and the nominal weight in pounds per foot; thus:

W 16 × 78

indicates a wide-flange section having a nominal depth of 16 inches, and a nominal weight per foot of 78 pounds. Actual geometry for each section can be obtained from the values below.

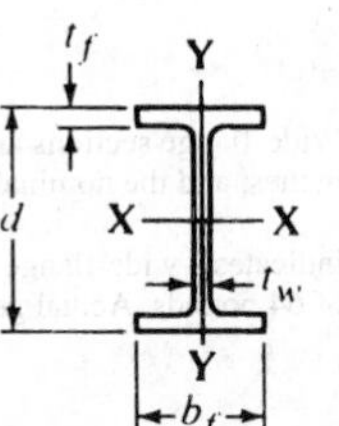

			Flange			Axis X-X			Axis Y-Y		
	Area, A	Depth, d	Width, b_f	Thickness, t_f	Web Thickness, t_w	I	S	r	I	S	r
Designation	in.^{2}	in.	in.	in.	in.	in.^{4}	in.^{3}	in.	in.^{4}	in.^{3}	in.
W 16 × 100	29.4	16.97	10.425	0.985	0.585	1490	175	7.10	186	35.7	2.51
× 89	26.2	16.75	10.365	0.875	0.525	1300	155	7.05	163	31.4	2.49
× 77	22.6	16.52	10.295	0.760	0.455	1110	134	7.00	138	26.9	2.47
× 67	19.7	16.33	10.235	0.665	0.395	954	117	6.96	119	23.2	2.46
× 57	16.8	16.43	7.120	0.715	0.430	758	92.2	6.72	43.1	12.1	1.60
× 50	14.7	16.26	7.070	0.630	0.380	659	81.0	6.68	37.2	10.5	1.59
× 45	13.3	16.13	7.035	0.565	0.345	586	72.7	6.65	32.8	9.34	1.57
× 40	11.8	16.01	6.995	0.505	0.305	518	64.7	6.63	28.9	8.25	1.57
× 36	10.6	15.86	6.985	0.430	0.295	448	56.5	6.51	24.5	7.00	1.52
× 31	9.12	15.88	5.525	0.440	0.275	375	47.2	6.41	12.4	4.49	1.17
× 26	7.68	15.69	5.500	0.345	0.250	301	38.4	6.26	9.59	3.49	1.12
W 14 × 730	215.0	22.42	17.890	4.910	3.070	14300	1280	8.17	4720	527	4.69
× 665	196.0	21.64	17.650	4.520	2.830	12400	1150	7.98	4170	472	4.62
× 605	178.0	20.92	17.415	4.160	2.595	10800	1040	7.80	3680	423	4.55
× 550	162.0	20.24	17.200	3.820	2.380	9430	931	7.63	3250	378	4.49
× 500	147.0	19.60	17.010	3.500	2.190	8210	838	7.48	2880	339	4.43
× 455	134.0	19.02	16.835	3.210	2.015	7190	756	7.33	2560	304	4.38
× 426	125.0	18.67	16.695	3.035	1.875	6600	707	7.26	2360	283	4.34
× 398	117.0	18.29	16.590	2.845	1.770	6000	656	7.16	2170	262	4.31
× 370	109.0	17.92	16.475	2.660	1.655	5440	607	7.07	1990	241	4.27
× 342	101.0	17.54	16.360	2.470	1.540	4900	559	6.98	1810	221	4.24
× 311	91.4	17.12	16.230	2.260	1.410	4330	506	6.88	1610	199	4.20
× 283	83.3	16.74	16.110	2.070	1.290	3840	459	6.79	1440	179	4.17
× 257	75.6	16.38	15.995	1.890	1.175	3400	415	6.71	1290	161	4.13
× 233	68.5	16.04	15.890	1.720	1.070	3010	375	6.63	1150	145	4.10
× 211	62.0	15.72	15.800	1.560	0.980	2660	338	6.55	1030	130	4.07
× 193	56.8	15.48	15.710	1.440	0.890	2400	310	6.50	931	119	4.05
× 176	51.8	15.22	15.650	1.310	0.830	2140	281	6.43	838	107	4.02
× 159	46.7	14.98	15.565	1.190	0.745	1900	254	6.38	748	96.2	4.00
× 145	42.7	14.78	15.500	1.090	0.680	1710	232	6.33	677	87.3	3.98

Symbols: I = moment of inertia; S = section modulus; r = radius of gyration.

Data taken from the "Manual of Steel Construction," 8th Edition, 1980, with permission of the American Institute of Steel Construction.

Steel Wide-Flange Sections—3

Wide-flange sections are designated, in order, by a section letter, nominal depth of the member in inches, and the nominal weight in pounds per foot; thus:

W 14 × 38

indicates a wide-flange section having a nominal depth of 14 inches, and a nominal weight per foot of 38 pounds. Actual geometry for each section can be obtained from the values below.

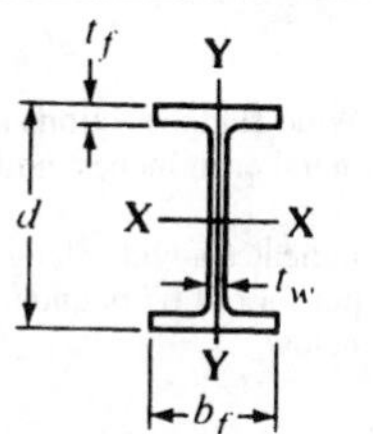

			Flange		Web	Axis X-X			Axis Y-Y		
	Area, A	Depth, d	Width, b_f	Thickness, t_f	Thickness, t_w	I	S	r	I	S	r
Designation	in.²	in.	in.	in.	in.	in.⁴	in.³	in.	in.⁴	in.³	in.
W 14 × 132	38.8	14.66	14.725	1.030	0.645	1530	209	6.28	548	74.5	3.76
× 120	35.3	14.48	14.670	0.940	0.590	1380	190	6.24	495	67.5	3.74
× 109	32.0	14.32	14.605	0.860	0.525	1240	173	6.22	447	61.2	3.73
× 99	29.1	14.16	14.565	0.780	0.485	1110	157	6.17	402	55.2	3.71
× 90	26.5	14.02	14.520	0.710	0.440	999	143	6.14	362	49.9	3.70
× 82	24.1	14.31	10.130	0.855	0.510	882	123	6.05	148	29.3	2.48
× 74	21.8	14.17	10.070	0.785	0.450	796	112	6.04	134	26.6	2.48
×68	20.0	14.04	10.035	0.720	0.415	723	103	6.01	121	24.2	2.46
× 61	17.9	13.89	9.995	0.645	0.375	640	92.2	5.98	107	21.5	2.45
× 53	15.6	13.92	8.060	0.660	0.370	541	77.8	5.89	57.7	14.3	1.92
× 48	14.1	13.79	8.030	0.595	0.340	485	70.3	5.85	51.4	12.8	1.91
× 43	12.6	13.66	7.995	0.530	0.305	428	62.7	5.82	45.2	11.3	1.89
× 38	11.2	14.10	6.770	0.515	0.310	385	54.6	5.87	26.7	7.88	1.55
× 34	10.0	13.98	6.745	0.455	0.285	340	48.6	5.83	23.3	6.91	1.53
× 30	8.85	13.84	6.730	0.385	0.270	291	42.0	5.73	19.6	5.82	1.49
× 26	7.69	13.91	5.025	0.420	0.255	245	35.3	5.65	8.91	3.54	1.08
× 22	6.49	13.74	5.000	0.335	0.230	199	29.0	5.54	7.00	2.80	1.04
W 12 × 336	98.8	16.82	13.385	2.955	1.775	4060	483	6.41	1190	177	3.47
× 305	89.6	16.32	13.235	2.705	1.625	3550	435	6.29	1050	159	3.42
× 279	81.9	15.85	13.140	2.470	1.530	3110	393	6.16	937	143	3.38
× 252	74.1	15.41	13.005	2.250	1.395	2720	353	6.06	828	127	3.34
× 230	67.7	15.05	12.895	2.070	1.285	2420	321	5.97	742	115	3.31
× 210	61.8	14.71	12.790	1.900	1.180	2140	292	5.89	664	104	3.28
× 190	55.8	14.38	12.670	1.735	1.060	1890	263	5.82	589	93.0	3.25
× 170	50.0	14.03	12.570	1.560	0.960	1650	235	5.74	517	82.3	3.22
× 152	44.7	13.71	12.480	1.400	0.870	1430	209	5.66	454	72.8	3.19
× 136	39.9	13.41	12.400	1.250	0.790	1240	186	5.58	398	64.2	3.16
× 120	35.3	13.12	12.320	1.105	0.710	1070	163	5.51	345	56.0	3.13
× 106	31.2	12.89	12.220	0.990	0.610	933	145	5.47	301	49.3	3.11
× 96	28.2	12.71	12.160	0.900	0.550	833	131	5.44	270	44.4	3.09
× 87	25.6	12.53	12.125	0.810	0.515	740	118	5.38	241	39.7	3.07
× 79	23.2	12.38	12.080	0.735	0.470	662	107	5.34	216	35.8	3.05
× 72	21.1	12.25	12.040	0.670	0.430	597	97.4	5.31	195	32.4	3.04
× 65	19.1	12.12	12.000	0.605	0.390	533	87.9	5.28	174	29.1	3.02
× 58	17.0	12.19	10.010	0.640	0.360	475	78.0	5.28	107	21.4	2.51
× 53	15.6	12.06	9.995	0.575	0.345	425	70.6	5.23	95.8	19.2	2.48
× 50	14.7	12.19	8.080	0.640	0.370	394	64.7	5.18	56.3	13.9	1.96
× 45	13.2	12.06	8.045	0.575	0.335	350	58.1	5.15	50.0	12.4	1.94
× 40	11.8	11.94	8.005	0.515	0.295	310	51.9	5.13	44.1	11.0	1.93
× 35	10.3	12.50	6.560	0.520	0.300	285	45.6	5.25	24.5	7.47	1.54
× 30	8.79	12.34	6.520	0.440	0.260	238	38.6	5.21	20.3	6.24	1.52
× 26	7.65	12.22	6.490	0.380	0.230	204	33.4	5.17	17.3	5.34	1.51
× 22	6.48	12.31	4.030	0.425	0.260	156	25.4	4.91	4.66	2.31	0.847
× 19	5.57	12.16	4.005	0.350	0.235	130	21.3	4.82	3.76	1.88	0.822
× 16	4.71	11.99	3.990	0.265	0.220	103	17.1	4.67	2.82	1.41	0.773
× 14	4.16	11.91	3.970	0.225	0.200	88.6	14.9	4.62	2.36	1.19	0.753

Data taken from the "Manual of Steel Construction," 8th Edition, 1980, with permission of the American Institute of Steel Construction.

Steel Wide-Flange Sections—4

Wide-flange sections are designated, in order, by a section letter, nominal depth of the member in inches, and the nominal weight in pounds per foot; thus:

W 8 × 67

indicates a wide-flange section having a nominal depth of 8 inches, and a nominal weight per foot of 67 pounds. Actual geometry for each section can be obtained from the values below.

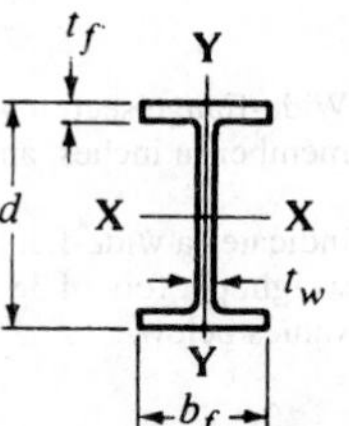

			Flange		Web	Axis X-X			Axis Y-Y		
Designation	Area, A in.^{2}	Depth, d in.	Width, b_f in.	Thick-ness, t_f in.	Thick-ness, t_w in.	I in.^{4}	S in.^{3}	r in.	I in.^{4}	S in.^{3}	r in.
W 10 × 112	32.9	11.36	10.415	1.250	0.755	716	126	4.66	236	45.3	2.68
× 100	29.4	11.10	10.340	1.120	0.680	623	112	4.60	207	40.0	2.65
× 88	25.9	10.84	10.265	0.990	0.605	534	98.5	4.54	179	34.8	2.63
× 77	22.6	10.60	10.190	0.870	0.530	455	85.9	4.49	154	30.1	2.60
× 68	20.0	10.40	10.130	0.770	0.470	394	75.7	4.44	134	26.4	2.59
× 60	17.6	10.22	10.080	0.680	0.420	341	66.7	4.39	116	23.0	2.57
× 54	15.8	10.09	10.030	0.615	0.370	303	60.0	4.37	103	20.6	2.56
× 49	14.4	9.98	10.000	0.560	0.340	272	54.6	4.35	93.4	18.7	2.54
× 45	13.3	10.10	8.020	0.620	0.350	248	49.1	4.32	53.4	13.3	2.01
× 39	11.5	9.92	7.985	0.530	0.315	209	42.1	4.27	45.0	11.3	1.98
× 33	9.71	9.73	7.960	0.435	0.290	170	35.0	4.19	36.6	9.20	1.94
× 30	8.84	10.47	5.810	0.510	0.300	170	32.4	4.38	16.7	5.75	1.37
× 26	7.61	10.33	5.770	0.440	0.260	144	27.9	4.35	14.1	4.89	1.36
× 22	6.49	10.17	5.750	0.360	0.240	118	23.2	4.27	11.4	3.97	1.33
× 19	5.62	10.24	4.020	0.395	0.250	96.3	18.8	4.14	4.29	2.14	0.874
× 17	4.99	10.11	4.010	0.330	0.240	81.9	16.2	4.05	3.56	1.78	0.844
× 15	4.41	9.99	4.000	0.270	0.230	68.9	13.8	3.95	2.89	1.45	0.810
× 12	3.54	9.87	3.960	0.210	0.190	53.8	10.9	3.90	2.18	1.10	0.785
W 8 × 67	19.7	9.00	8.280	0.935	0.570	272	60.4	3.72	88.6	21.4	2.12
× 58	17.1	8.75	8.220	0.810	0.510	228	52.0	3.65	75.1	18.3	2.10
× 48	14.1	8.50	8.110	0.685	0.400	184	43.3	3.61	60.9	15.0	2.08
× 40	11.7	8.25	8.070	0.560	0.360	146	35.5	3.53	49.1	12.2	2.04
× 35	10.3	8.12	8.020	0.495	0.310	127	31.2	3.51	42.6	10.6	2.03
× 31	9.13	8.00	7.995	0.435	0.285	110	27.5	3.47	37.1	9.27	2.02
× 28	8.25	8.06	6.535	0.465	0.285	98.0	24.3	3.45	21.7	6.63	1.62
× 24	7.08	7.93	6.495	0.400	0.245	82.8	20.9	3.42	18.3	5.63	1.61
× 21	6.16	8.28	5.270	0.400	0.250	75.3	18.2	3.49	9.77	3.71	1.26
× 18	5.26	8.14	5.250	0.330	0.230	61.9	15.2	3.43	7.97	3.04	1.23
× 15	4.44	8.11	4.015	0.315	0.245	48.0	11.8	3.29	3.41	1.70	0.876
× 13	3.84	7.99	4.000	0.255	0.230	39.6	9.91	3.21	2.73	1.37	0.843
× 10	2.96	7.89	3.940	0.205	0.170	30.8	7.81	3.22	2.09	1.06	0.841
W 6 × 25	7.34	6.38	6.080	0.455	0.320	53.4	16.7	2.70	17.1	5.61	1.52
× 20	5.87	6.20	6.020	0.365	0.260	41.4	13.4	2.66	13.3	4.41	1.50
× 16	4.74	6.28	4.030	0.405	0.260	32.1	10.2	2.60	4.43	2.20	0.966
× 15	4.43	5.99	5.990	0.260	0.230	29.1	9.72	2.56	9.32	3.11	1.46
× 12	3.55	6.03	4.000	0.280	0.230	22.1	7.31	2.49	2.99	1.50	0.918
× 9	2.68	5.90	3.940	0.215	0.170	16.4	5.56	2.47	2.19	1.11	0.905
W 5 × 19	5.54	5.15	5.030	0.430	0.270	26.2	10.2	2.17	9.13	3.63	1.28
× 16	4.68	5.01	5.000	0.360	0.240	21.3	8.51	2.13	7.51	3.00	1.27
W 4 × 13	3.83	4.16	4.060	0.345	0.280	11.3	5.46	1.72	3.86	1.90	1.00

Symbols: I = moment of inertia; S = section modulus; r = radius of gyration.

Data taken from the "Manual of Steel Construction," 8th Edition, 1980, with permission of the American Institute of Steel Construction.

Steel S Sections

"S" is the section symbol for "I" Beams. S shapes are designated, in order, by their section letter, actual depth in inches, and nominal weight in pounds per foot. Thus:

S 5 × 14.75

indicates an S shape (or I beam) having a depth of 5 inches and a nominal weight of 14.75 pounds per foot.

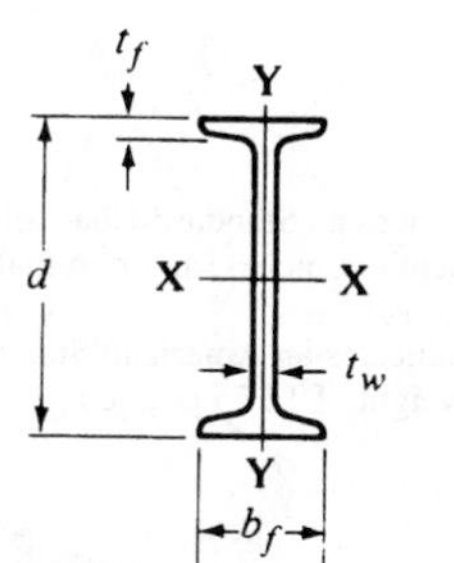

Designation	Area A	Depth, d	Flange Width, b_f	Flange Thickness, t_f	Web Thickness, t_w	Axis X-X I	Axis X-X S	Axis X-X r	Axis Y-Y I	Axis Y-Y S	Axis Y-Y r
	in.^{2}	in.	in.	in.	in.	in.^{4}	in.^{3}	in.	in.^{4}	in.^{3}	in.
S 24 × 121	35.6	24.50	8.050	1.090	0.800	3160	258	9.43	83.3	20.7	1.53
× 106	31.2	24.50	7.870	1.090	0.620	2940	240	9.71	77.1	19.6	1.57
×100	29.3	24.00	7.245	0.870	0.745	2390	199	9.02	47.7	13.2	1.27
× 90	26.5	24.00	7.125	0.870	0.625	2250	187	9.21	44.9	12.6	1.30
× 80	23.5	24.00	7.000	0.870	0.500	2100	175	9.47	42.2	12.1	1.34
S 20 × 96	28.2	20.30	7.200	0.920	0.800	1670	165	7.71	50.2	13.9	1.33
× 86	25.3	20.30	7.060	0.920	0.660	1580	155	7.89	46.8	13.3	1.36
× 75	22.0	20.00	6.385	0.795	0.635	1280	128	7.62	29.8	9.32	1.16
× 66	19.4	20.00	6.255	0.795	0.505	1190	119	7.83	27.7	8.85	1.19
S 18 × 70	20.6	18.00	6.251	0.691	0.711	926	103	6.71	24.1	7.72	1.08
× 54.7	16.1	18.00	6.001	0.691	0.461	804	89.4	7.07	20.8	6.94	1.14
S 15 × 50	14.7	15.00	5.640	0.622	0.550	486	64.8	5.75	15.7	5.57	1.03
× 42.9	12.6	15.00	5.501	0.622	0.411	447	59.6	5.95	14.4	5.23	1.07
S 12 × 50	14.7	12.00	5.477	0.659	0.687	305	50.8	4.55	15.7	5.74	1.03
× 40.8	12.0	12.00	5.252	0.659	0.462	272	45.4	4.77	13.6	5.16	1.06
× 35	10.3	12.00	5.078	0.544	0.428	229	38.2	4.72	9.87	3.89	0.980
× 31.8	9.35	12.00	5.000	0.544	0.350	218	36.4	4.83	9.36	3.74	1.00
S 10 × 35	10.3	10.00	4.944	0.491	0.594	147	29.4	3.78	8.36	3.38	0.901
× 25.4	7.46	10.00	4.661	0.491	0.311	124	24.7	4.07	6.79	2.91	0.954
S 8 × 23	6.77	8.00	4.171	0.426	0.441	64.9	16.2	3.10	4.31	2.07	0.798
× 18.4	5.41	8.00	4.001	0.426	0.271	57.6	14.4	3.26	3.73	1.86	0.831
S 7 × 20	5.88	7.00	3.860	0.392	0.450	42.4	12.1	2.69	3.17	1.64	0.734
× 15.3	4.50	7.00	3.662	0.392	0.252	36.7	10.5	2.86	2.64	1.44	0.766
S 6 × 17.25	5.07	6.00	3.565	0.359	0.465	26.3	8.77	2.28	2.31	1.30	0.675
× 12.5	3.67	6.00	3.332	0.359	0.232	22.1	7.37	2.45	1.82	1.09	0.705
S 5 × 14.75	4.34	5.00	3.284	0.326	0.494	15.2	6.09	1.87	1.67	1.01	0.620
× 10	2.94	5.00	3.004	0.326	0.214	12.3	4.92	2.05	1.22	0.809	0.643
S 4 × 9.5	2.79	4.00	2.796	0.293	0.326	6.79	3.39	1.56	0.903	0.646	0.569
× 7.7	2.26	4.00	2.663	0.293	0.193	6.08	3.04	1.64	0.764	0.574	0.581
S 3 × 7.5	2.21	3.00	2.509	0.260	0.349	2.93	1.95	1.15	0.586	0.468	0.516
× 5.7	1.67	3.00	2.330	0.260	0.170	2.52	1.68	1.23	0.455	0.390	0.522

Data taken from the "Manual of Steel Construction," 8th Edition, 1980, with permission of the American Institute of Steel Construction.

American Standard Steel Channels

American Standard Channels are designated, in order, by a section letter, actual depth in inches, and nominal weight per foot in pounds. Thus:

C 7 × 14.75

indicates an American Standard Channel with a depth of 7 inches and a nominal weight of 14.75 pounds per foot.

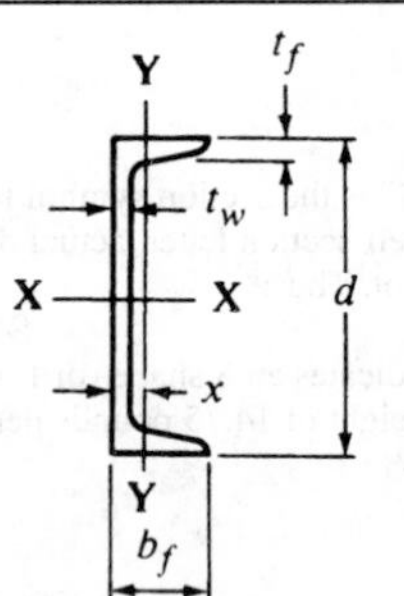

			Flange		Web	Axis X-X			Axis Y-Y			
	Area, A	Depth, d	Width, b_f	Thickness, t_f	Thickness, t_w	I	S	r	I	S	r	x
Designation	in.^{2}	in.	in.	in.	in.	in.^{4}	in.^{3}	in.	in.^{4}	in.^{3}	in.	in.
C 15 × 50	14.7	15.00	3.716	0.650	0.716	404	53.8	5.24	11.0	3.78	0.867	0.798
× 40	11.8	15.00	3.520	0.650	0.520	349	46.5	5.44	9.23	3.37	0.886	0.777
× 33.9	9.96	15.00	3.400	0.650	0.400	315	42.0	5.62	8.13	3.11	0904	0.787
C 12 × 30	8.82	12.00	3.170	0.501	0.510	162	27.0	4.29	5.14	2.06	0.763	0.674
× 25	7.35	12.00	3.047	0.501	0.387	144	24.1	4.43	4.47	1.88	0.780	0.674
× 20.7	6.09	12.00	2.942	0.501	0.282	129	21.5	4.61	3.88	1.73	0.799	0.698
C 10 × 30	8.82	10.00	3.033	0.436	0.673	103	20.7	3.42	3.94	1.65	0.669	0.649
× 25	7.35	10.00	2.886	0.436	0.526	91.2	18.2	3.52	3.36	1.48	0.676	0.617
× 20	5.88	10.00	2.739	0.436	0.379	78.9	15.8	3.66	2.81	1.32	0.692	0.606
× 15.3	4.49	10.00	2.600	0.436	0.240	67.4	13.5	3.87	2.28	1.16	0.713	0.634
C 9 × 20	5.88	9.00	2.648	0.413	0.448	60.9	13.5	3.22	2.42	1.17	0.642	0.583
× 15	4.41	9.00	2.485	0.413	0.285	51.0	11.3	3.40	1.93	1.01	0.661	0.586
× 13.4	3.94	9.00	2.433	0.413	0.233	47.9	10.6	3.48	1.76	0.962	0.669	0.601
C 8 × 18.75	5.51	8.00	2.527	0.390	0.487	44.0	11.0	2.82	1.98	1.01	0.599	0.565
× 13.75	4.04	8.00	2.343	0.390	0.303	36.1	9.03	2.99	1.53	0.854	0.615	0.553
× 11.5	3.38	8.00	2.260	0.390	0.220	32.6	8.14	3.11	1.32	0.781	0.625	0.571
C 7 × 14.75	4.33	7.00	2.299	0.366	0.419	27.2	7.78	2.51	1.38	0.779	0.564	0.532
× 12.25	3.60	7.00	2.194	0.366	0.314	24.2	6.93	2.60	1.17	0.703	0.571	0.525
× 9.8	2.87	7.00	2.090	0.366	0.210	21.3	6.08	2.72	0.968	0.625	0.581	0.540
C 6 × 13	3.83	6.00	2.157	0.343	0.437	17.4	5.80	2.13	1.05	0.642	0.525	0.514
× 10.5	3.09	6.00	2.034	0.343	0.314	15.2	5.06	2.22	0.866	0.564	0.529	0.499
× 8.2	2.40	6.00	1.920	0.343	0.200	13.1	4.38	2.34	0.693	0.492	0.537	0.511
C 5 × 9	2.64	5.00	1.885	0.320	0.325	8.90	3.56	1.83	0.632	0.450	0.489	0.478
× 6.7	1.97	5.00	1.750	0.320	0.190	7.49	3.00	1.95	0.479	0.378	0.493	0.484
C 4 × 7.25	2.13	4.00	1.721	0.296	0.321	4.59	2.29	1.47	0.433	0.343	0.450	0.459
× 5.4	1.59	4.00	1.584	0.296	0.184	3.85	1.93	1.56	0.319	0.283	0.449	0.457
C 3 × 6	1.76	3.00	1.596	0.273	0.356	2.07	1.38	1.08	0.305	0.268	0.416	0.455
× 5	1.47	3.00	1.498	0.273	0.258	1.85	1.24	1.12	0.247	0.233	0.410	0.438
× 4.1	1.21	3.00	1.410	0.273	0.170	1.66	1.10	1.17	0.197	0.202	0.404	0.436

Symbols: I = moment of inertia; S = section modulus; r = radius of gyration; x = distance from center of gravity of section to outer face of structural shape.

Data taken from the "Manual of Steel Construction," 8th Edition, 1980, with permission of the American Institute of Steel Construction.

Steel Angles with Equal Legs

These angles are commonly designated by section symbol, width of each leg, and thickness, thus:

L 3 × 3 × ¼

indicates a 3 × 3-inch angle of ¼-inch thickness.

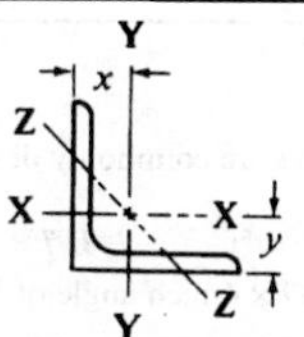

Size	Thickness	Weight per Foot	Area	Axis X-X & Y-Y			Z-Z
				I	r	x or y	r
in.	in.	lb.	in.^{2}	in.^{4}	in.	in.	in.
8 × 8	1⅛	56.9	16.7	98.0	2.42	2.41	1.56
	1	51.0	15.0	89.0	2.44	2.37	1.56
	⅞	45.0	13.2	79.6	2.45	2.32	1.57
	¾	38.9	11.4	69.7	2.47	2.28	1.58
	⅝	32.7	9.61	59.4	2.49	2.23	1.58
	9/16	29.6	8.68	54.1	2.50	2.21	1.59
	½	26.4	7.75	48.6	2.50	2.19	1.59
6 × 6	1	37.4	11.00	35.5	1.80	1.86	1.17
	⅞	33.1	9.73	31.9	1.81	1.82	1.17
	¾	28.7	8.44	28.2	1.83	1.78	1.17
	⅝	24.2	7.11	24.2	1.84	1.73	1.18
	9/16	21.9	6.43	22.1	1.85	1.71	1.18
	½	19.6	5.75	19.9	1.86	1.68	1.18
	7/16	17.2	5.06	17.7	1.87	1.66	1.19
	⅜	14.9	4.36	15.4	1.88	1.64	1.19
	5/16	12.4	3.65	13.0	1.89	1.62	1.20
5 × 5	⅞	27.2	7.98	17.8	1.49	1.57	.973
	¾	23.6	6.94	15.7	1.51	1.52	.975
	⅝	20.0	5.86	13.6	1.52	1.48	.978
	½	16.2	4.75	11.3	1.54	1.43	.983
	7/16	14.3	4.18	10.0	1.55	1.41	.986
	⅜	12.3	3.61	8.74	1.56	1.39	.990
	5/16	10.3	3.03	7.42	1.57	1.37	.994
4 × 4	¾	18.5	5.44	7.67	1.19	1.27	.778
	⅝	15.7	4.61	6.66	1.20	1.23	.779
	½	12.8	3.75	5.56	1.22	1.18	.782
	7/16	11.3	3.31	4.97	1.23	1.16	.785
	⅜	9.8	2.86	4.36	1.23	1.14	.788
	5/16	8.2	2.40	3.71	1.24	1.12	.791
	¼	6.6	1.94	3.04	1.25	1.09	.795
3½ × 3½	½	11.1	3.25	3.64	1.06	1.06	.683
	7/16	9.8	2.87	3.26	1.07	1.04	.684
	⅜	8.5	2.48	2.87	1.07	1.01	.687
	5/16	7.2	2.09	2.45	1.08	.990	.690
	¼	5.8	1.69	2.01	1.09	.968	.694
3 × 3	½	9.4	2.75	2.22	.898	.932	.584
	7/16	8.3	2.43	1.99	.905	.910	.585
	⅜	7.2	2.11	1.76	.913	.888	.587
	5/16	6.1	1.78	1.51	.922	.865	.589
	¼	4.9	1.44	1.24	.930	.842	.592
	3/16	3.71	1.09	.962	.939	.820	.596
2½ × 2½	½	7.7	2.25	1.23	.739	.806	.487
	⅜	5.9	1.73	.984	.753	.762	.487
	5/16	5.0	1.46	.849	.761	.740	.489
	¼	4.1	1.19	.703	.769	.717	.491
	3/16	3.07	.902	.547	.778	.694	.495
2 × 2	⅜	4.7	1.36	.479	.594	.636	.389
	5/16	3.92	1.15	.416	.601	.614	.390
	¼	3.19	.938	.348	.609	.592	.391
	3/16	2.44	.715	.272	.617	.569	.394
	⅛	1.65	.484	.190	.626	.546	.398

Data taken from the "Manual of Steel Construction," 8th Edition, 1980, with permission of the American Institute of Steel Construction.

Steel Angles with Unequal Legs

These angles are commonly designated by section symbol, width of each leg, and thickness, thus:

L 7 × 4 × ½

indicates a 7 × 4-inch angle of ½-inch thickness.

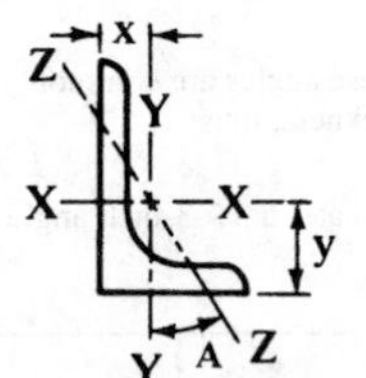

Size	Thickness	Weight per Ft.	Area	Axis X-X				Axis Y-Y			Axis Z-Z		
				I	*S*	*r*	*y*	*I*	*S*	*r*	*x*	*r*	Tan *A*
in.	in.	lb.	in.^{2}	in.^{4}	in.^{3}	in.	in.	in.^{4}	in.^{3}	in.	in.	in.	
9 × 4	⅝	26.3	7.73	64.9	11.5	2.90	3.36	8.32	2.65	1.04	.858	.847	.216
	9⁄16	23.8	7.00	59.1	10.4	2.91	3.33	7.63	2.41	1.04	.834	.850	.218
	½	21.3	6.25	53.2	9.34	2.92	3.31	6.92	2.17	1.05	.810	.854	.220
8 × 6	1	44.2	13.0	80.8	15.1	2.49	2.65	38.8	8.92	1.73	1.65	1.28	.543
	⅞	39.1	11.5	72.3	13.4	2.51	2.61	34.9	7.94	1.74	1.61	1.28	.547
	¾	33.8	9.94	63.4	11.7	2.53	2.56	30.7	6.92	1.76	1.56	1.29	.551
	⅝	28.5	8.36	54.1	9.87	2.54	2.52	26.3	5.88	1.77	1.52	1.29	.554
	9⁄16	7	7.56	49.3	8.95	2.55	2.50	24.0	5.34	1.78	1.50	1.30	.556
	½	23.0	6.75	44.3	8.02	2.56	2.47	21.7	4.79	1.79	1.47	1.30	.558
	7⁄16	20.2	5.93	39.2	7.07	2.57	2.45	19.3	4.23	1.80	1.45	1.31	.560
8 × 4	1	37.4	11.0	69.6	14.1	2.52	3.05	11.6	3.94	1.03	1.05	.846	.247
	¾	28.7	8.44	54.9	10.9	2.55	2.95	9.36	3.07	1.05	.953	.852	.258
	9⁄16	21.9	6.43	42.8	8.35	2.58	2.88	7.43	2.38	1.07	.882	.861	.265
	½	19.6	5.75	38.5	7.49	2.59	2.86	6.74	2.15	1.08	.859	.865	.267
7 × 4	¾	26.2	7.69	37.8	8.42	2.22	2.51	9.05	3.03	1.09	1.01	.860	.324
	⅝	22.1	6.48	32.4	7.14	2.24	2.46	7.84	2.58	1.10	.963	.865	.329
	½	17.9	5.25	26.7	5.81	2.25	2.42	6.53	2.12	1.11	.917	.872	.335
	⅜	13.6	3.98	20.6	4.44	2.27	2.37	5.10	1.63	1.13	.870	.880	.340
6 × 4	⅞	27.2	7.98	27.7	7.15	1.86	2.12	9.75	3.39	1.11	1.12	.857	.421
	¾	23.6	6.94	24.5	6.25	1.88	2.08	8.68	2.97	1.12	1.08	.860	.428
	⅝	20.0	5.86	21.1	5.31	1.90	2.03	7.52	2.54	1.13	1.03	.864	.435
	9⁄16	18.1	5.31	19.3	4.83	1.90	2.01	6.91	2.31	1.14	1.01	.866	.438
	½	16.2	4.75	17.4	4.33	1.91	1.99	6.27	2.08	1.15	.987	.870	.440
	7⁄16	14.3	4.18	15.5	3.83	1.92	1.96	5.60	1.85	1.16	.964	.873	.443
	⅜	12.3	3.61	13.5	3.32	1.93	1.94	4.90	1.60	1.17	.941	.877	.446
	5⁄16	10.3	3.03	11.4	2.79	1.94	1.92	4.18	1.35	1.17	.918	.882	.448
6 × 3½	½	15.3	4.50	16.6	4.24	1.92	2.08	4.25	1.59	.972	.833	.759	.344
	⅜	11.7	3.42	12.9	3.24	1.94	2.04	3.34	1.23	.988	.787	.676	.350
	5⁄16	9.8	2.87	10.9	2.73	1.95	2.01	2.85	1.04	.996	.763	.772	.352
5 × 3½	¾	19.8	5.81	13.9	4.28	1.55	1.75	5.55	2.22	.977	.996	.748	.464
	⅝	16.8	4.92	12.0	3.65	1.56	1.70	4.83	1.90	.991	.951	.751	.472
	½	13.6	4.00	9.99	2.99	1.58	1.66	4.05	1.56	1.01	.906	.755	.479
	7⁄16	12.0	3.53	8.90	2.64	1.59	1.63	3.63	1.39	1.01	.883	.758	.482
	⅜	10.4	3.05	7.78	2.29	1.60	1.61	3.18	1.21	1.02	.861	.762	.486
	5⁄16	8.7	2.56	6.60	1.94	1.61	1.59	2.72	1.02	1.03	.838	.766	.489
	¼	7.0	2.06	5.39	1.57	1.62	1.56	2.23	.830	1.04	.814	.770	.492
5 × 3	⅝	15.7	4.61	11.4	3.55	1.57	1.80	3.06	1.39	.815	.796	.644	.349

Steel Angles with Unequal Legs *(Continued)*

5 × 3	½	12.8	3.75	9.45	2.91	1.59	1.75	2.58	1.15	.829	.750	.648	.357
	7/16	11.3	3.31	8.43	2.58	1.60	1.73	2.32	1.02	.837	.727	.651	.361
	⅜	9.8	2.86	7.37	2.24	1.61	1.70	2.04	.888	.845	.704	.654	.364
	5/16	8.2	2.40	6.26	1.89	1.61	1.68	1.75	.753	.853	.681	.658	.368
	¼	6.6	1.94	5.11	1.53	1.62	1.66	1.44	.614	.861	.657	.663	.371
4 × 3½	⅝	14.7	4.30	6.37	2.35	1.22	1.29	4.52	1.84	1.03	1.04	.719	.745
	½	11.9	3.50	5.32	1.94	1.23	1.25	3.79	1.52	1.04	1.00	.722	.750
	7/16	10.6	3.09	4.76	1.72	1.24	1.23	3.40	1.35	1.05	.978	.724	.753
	⅜	9.1	2.67	4.18	1.49	1.25	1.21	2.95	1.17	1.06	.955	.727	.755
	5/16	7.7	2.25	3.56	1.26	1.26	1.18	2.55	.994	1.07	.932	.730	.757
	¼	6.2	1.81	2.91	1.03	1.27	1.16	2.09	.808	1.07	.909	.734	.759
4 × 3	⅝	13.6	3.98	6.03	2.30	1.23	1.37	2.87	1.35	.849	.871	.637	.534
	½	11.1	3.25	5.05	1.89	1.25	1.33	2.42	1.12	.864	.827	.639	.543
	7/16	9.8	2.87	4.52	1.68	1.25	1.30	2.18	.992	.871	.804	.641	.547
	⅜	8.5	2.48	3.96	1.46	1.26	1.28	1.92	.866	.879	.782	.644	.551
	5/16	7.2	2.09	3.38	1.23	1.27	1.26	1.65	.734	.887	.759	.647	.554
	¼	5.8	1.69	2.77	1.00	1.28	1.24	1.36	.599	.896	.736	.651	.558
3½ × 3	½	10.2	3.00	3.45	1.45	1.07	1.13	2.33	1.10	.881	.875	.621	.714
	7/16	9.1	2.65	3.10	1.29	1.08	1.10	2.09	.975	.889	.853	.622	.718
	⅜	7.9	2.30	2.72	1.13	1.09	1.08	1.85	.851	.897	.830	.625	.721
	5/16	6.6	1.93	2.33	.954	1.10	1.06	1.58	.722	.905	.808	.627	.724
	¼	5.4	1.56	1.91	.776	1.11	1.04	1.30	.589	.914	.785	.631	.727
3½ × 2½	½	9.4	2.75	3.24	1.41	1.09	1.20	1.36	.760	.704	.705	.534	.486
	7/16	8.3	2.43	2.91	1.26	1.09	1.18	1.23	.677	.711	.682	.535	.491
	⅜	7.2	2.11	2.56	1.09	1.10	1.16	1.09	.592	.719	.660	.537	.496
	5/16	6.1	1.78	2.19	.927	1.11	1.14	.939	.504	.727	.637	.540	.501
	¼	4.9	1.44	1.80	.755	1.12	1.11	.777	.412	.735	.614	.544	.506
3 × 2½	½	8.5	2.50	2.08	1.04	.913	1.00	1.30	.744	.722	.750	.520	.667
	7/16	7.6	2.21	1.88	.928	.920	.978	1.18	.664	.729	.728	.521	.672
	⅜	6.6	1.92	1.66	.810	.928	.956	1.04	.581	.736	.706	.522	.676
	5/16	5.6	1.62	1.42	.688	.937	.933	.898	.494	.744	.683	.525	.680
	¼	4.5	1.31	1.17	.561	.945	.911	.743	.404	.753	.661	.528	.684
	3/16	3.39	.996	.907	.430	.954	.888	.577	.310	.761	.638	.533	.688
3 × 2	½	7.7	2.25	1.92	1.00	.924	1.08	.672	.474	.546	.583	.428	.414
	7/16	6.8	2.00	1.73	.894	.932	1.06	.609	.424	.553	.561	.429	.421
	⅜	5.9	1.73	1.53	.781	.940	1.04	.543	.371	.559	.539	.430	.428
	5/16	5.0	1.46	1.32	.664	.948	1.02	.740	.317	.567	.516	.432	.435
	¼	4.1	1.19	1.09	.542	.957	.993	.392	.260	.574	.493	.435	.440
	3/16	3.07	.902	.842	.415	.966	.970	.307	.200	.583	.470	.439	.446
2½ × 2	⅜	5.3	1.55	.912	.547	.768	.831	.514	.363	.577	.581	.420	.614
	5/16	4.5	1.31	.788	.466	.776	.809	.446	.310	.584	.559	.422	.620
	¼	3.62	1.06	.654	.381	.784	.787	.372	.254	.592	.537	.424	.626
	3/16	2.75	.809	.509	.293	.793	.764	.291	.196	.600	.514	.427	.631

Symbols: I = moment of inertia; *S* = section modulus; *r* = radius of gyration; *x* = distance from center of gravity of section to outer face of structural shape.

Data taken from the "Manual of Steel Construction," 8th Edition, 1980, with permission of the American Institute of Steel Construction.

Aluminum Association Standard Structural Shapes

I-BEAMS — CHANNELS

Depth	Width	Weight per Foot	Area	Flange Thickness	Web Thickness	Fillet Radius	Axis X-X I	Axis X-X S	Axis X-X r	Axis Y-Y I	Axis Y-Y S	Axis Y-Y r	Axis Y-Y x
in.	in.	lb.	in.^{2}	in.	in.	in.	in.^{4}	in.^{3}	in.	in.^{4}	in.^{3}	in.	in.
I-BEAMS													
3.00	2.50	1.637	1.392	0.20	0.13	0.25	2.24	1.49	1.27	0.52	0.42	0.61	…
3.00	2.50	2.030	1.726	0.26	0.15	0.25	2.71	1.81	1.25	0.68	0.54	0.63	…
4.00	3.00	2.311	1.965	0.23	0.15	0.25	5.62	2.81	1.69	1.04	0.69	0.73	…
4.00	3.00	2.793	2.375	0.29	0.17	0.25	6.71	3.36	1.68	1.31	0.87	0.74	…
5.00	3.50	3.700	3.146	0.32	0.19	0.30	13.94	5.58	2.11	2.29	1.31	0.85	…
6.00	4.00	4.030	3.427	0.29	0.19	0.30	21.99	7.33	2.53	3.10	1.55	0.95	…
6.00	4.00	4.692	3.990	0.35	0.21	0.30	25.50	8.50	2.53	3.74	1.87	0.97	…
7.00	4.50	5.800	4.932	0.38	0.23	0.30	42.89	12.25	2.95	5.78	2.57	1.08	…
8.00	5.00	6.181	5.256	0.35	0.23	0.30	59.69	14.92	3.37	7.30	2.92	1.18	…
8.00	5.00	7.023	5.972	0.41	0.25	0.30	67.78	16.94	3.37	8.55	3.42	1.20	…
9.00	5.50	8.361	7.110	0.44	0.27	0.30	102.02	22.67	3.79	12.22	4.44	1.31	…
10.00	6.00	8.646	7.352	0.41	0.25	0.40	132.09	26.42	4.24	14.78	4.93	1.42	…
10.00	6.00	10.286	8.747	0.50	0.29	0.40	155.79	31.16	4.22	18.03	6.01	1.44	…
12.00	7.00	11.672	9.925	0.47	0.29	0.40	255.57	42.60	5.07	26.90	7.69	1.65	…
12.00	7.00	14.292	12.153	0.62	0.31	0.40	317.33	52.89	5.11	35.48	10.14	1.71	…
CHANNELS													
2.00	1.00	0.577	0.491	0.13	0.13	0.10	0.288	0.288	0.766	0.045	0.064	0.303	0.298
2.00	1.25	1.071	0.911	0.26	0.17	0.15	0.546	0.546	0.774	0.139	0.178	0.391	0.471
3.00	1.50	1.135	0.965	0.20	0.13	0.25	1.41	0.94	1.21	0.22	0.22	0.47	0.49
3.00	1.75	1.597	1.358	0.26	0.17	0.25	1.97	1.31	1.20	0.42	0.37	0.55	0.62
4.00	2.00	1.738	1.478	0.23	0.15	0.25	3.91	1.95	1.63	0.60	0.45	0.64	0.65
4.00	2.25	2.331	1.982	0.29	0.19	0.25	5.21	2.60	1.62	1.02	0.69	0.72	0.78
5.00	2.25	2.212	1.881	0.26	0.15	0.30	7.88	3.15	2.05	0.98	0.64	0.72	0.73
5.00	2.75	3.089	2.627	0.32	0.19	0.30	11.14	4.45	2.06	2.05	1.14	0.88	0.95
6.00	2.50	2.834	2.410	0.29	0.17	0.30	14.35	4.78	2.44	1.53	0.90	0.80	0.79
6.00	3.25	4.030	3.427	0.35	0.21	0.30	21.04	7.01	2.48	3.76	1.76	1.05	1.12
7.00	2.75	3.205	2.725	0.29	0.17	0.30	22.09	6.31	2.85	2.10	1.10	0.88	0.84
7.00	3.50	4.715	4.009	0.38	0.21	0.30	33.79	9.65	2.90	5.13	2.23	1.13	1.20
8.00	3.00	4.147	3.526	0.35	0.19	0.30	37.40	9.35	3.26	3.25	1.57	0.96	0.93
8.00	3.75	5.789	4.923	0.471	0.25	0.35	52.69	13.17	3.27	7.13	2.82	1.20	1.22
9.00	3.25	4.983	4.237	0.35	0.23	0.35	54.41	12.09	3.58	4.40	1.89	1.02	0.93
9.00	4.00	6.970	5.927	0.44	0.29	0.35	78.31	17.40	3.63	9.61	3.49	1.27	1.25
10.00	3.50	6.136	5.218	0.41	0.25	0.35	83.22	16.64	3.99	6.33	2.56	1.10	1.02
10.00	4.25	8.360	7.109	0.50	0.31	0.40	116.15	23.23	4.04	13.02	4.47	1.35	1.34
12.00	4.00	8.274	7.036	0.47	0.29	0.40	159.76	26.63	4.77	11.03	3.86	1.25	1.14
12.00	5.00	11.822	10.053	0.62	0.35	0.45	239.69	39.95	4.88	25.74	7.60	1.60	1.61

Structural sections are available in 6061-T6 aluminum alloy. Data supplied by The Aluminum Association.

Wire and Sheet-metal Gages

The thicknesses of sheet metals and the diameters of wires conform to various gaging systems. These gage sizes are indicated by numbers, and the following tables give the decimal equivalents of the different gage numbers. Much confusion has resulted from the use of gage numbers, and in ordering materials it is preferable to give the exact dimensions in decimal fractions of an inch. While the dimensions thus specified should conform to the gage ordinarily used for a given class of material, any error in the specification due, for example, to the use of a table having "rounded off" or approximate equivalents, will be apparent to the manufacturer at the time the order is placed. Furthermore, the decimal method of indicating wire diameters and sheet metal thicknesses has the advantage of being self-explanatory, whereas arbitrary gage numbers are not. The decimal system of indicating gage sizes is now being used quite generally, and gage numbers are gradually being discarded. Unfortunately, there is considerable variation in the use of different gages. For example, a gage ordinarily used for copper, brass and other non-ferrous materials, may at times be used for steel, and vice versa. The gages specified in the following are the ones ordinarily employed for the materials mentioned, but there are some minor exceptions and variations in the different industries.

Wire Gages.—The wire gage system used by practically all of the steel producers in the United States is known by the name Steel Wire Gage or to distinguish it from the Standard Wire Gage (S.W.G.) used in Great Britain it is called the United States Steel Wire Gage. It is the same as the Washburn and Moen, American Steel and Wire Company, and Roebling Wire Gages. The name has the official sanction of the Bureau of Standards at Washington but is not legally effective. The only wire gage which has been recognized in Acts of Congress is the Birmingham Gage (also known as Stub's Iron Wire). The Birmingham Gage is, however, nearly obsolete in both the United States and Great Britain, where it originated. Copper and aluminum wires are specified in decimal fractions. They were formerly universally specified in the United States by the American or Brown & Sharpe Wire Gage. Music spring steel wire, one of the highest quality wires of several types used for mechanical springs, is specified by the piano or music wire gage.

In Great Britain one wire gage has been legalized. This is called the Standard Wire Gage (S.W.G.), formerly called Imperial Wire Gage.

Gages for Rods.—Steel wire rod sizes are designated by fractional or decimal parts of an inch and by the gage numbers of the United States Steel Wire Gage. Copper and aluminum rods are specified by decimal fractions and fractions. Drill rod may be specified in decimal fractions but in the carbon and alloy tool steel grades may also be specified in the Stub's Steel Wire Gage and in the high-speed steel drill rod grade may be specified by the Morse Twist Drill Gage (Manufacturers' Standard Gage for Twist Drills). For gage numbers with corresponding decimal equivalents see the tables of American Standard Straight Shank Twist Drills.

Gages for Wall Thicknesses of Tubing.—At one time the Birmingham or Stub's Iron Wire Gage was used to specify the wall thickness of the following classes of tubing: seamless brass, seamless copper, seamless steel, and aluminum. The Brown & Sharpe Wire Gage was used for brazed brass and brazed copper tubing. Wall thicknesses are now specified by decimal parts of an inch but the wall thickness of steel pressure tubes and steel mechanical tubing may be specified by the Birmingham or Stub's Iron Wire Gage. In Great Britain the Standard Wire Gage (S.W.G.) is used to specify the wall thickness of some kinds of steel tubes.

Table 1. Wire Gages in Approximate Decimals of an Inch

No. of Wire Gage	American Wire or Brown & Sharpe Gage	Steel Wire Gage (U.S.)[a]	British Standard Wire Gage (Imperial Wire Gage)	Music or Piano Wire Gage	Birmingham or Stub's Iron Wire Gage	Stub's Steel Wire Gage	No. of Wire Gage	Stub's Steel Wire Gage
7/0	…	0.4900	0.5000	…	…	…	51	0.066
6/0	0.5800	0.4615	0.4640	0.004	…	…	52	0.063
5/0	0.5165	0.4305	0.4320	0.005	0.5000	…	53	0.058
4/0	0.4600	0.3938	0.4000	0.006	0.4540	…	54	0.055
3/0	0.4096	0.3625	0.3720	0.007	0.4250	…	55	0.050
2/0	0.3648	0.3310	0.3480	0.008	0.3800	…	56	0.045
1/0	0.3249	0.3065	0.3240	0.009	0.3400	…	57	0.042
1	0.2893	0.2830	0.3000	0.010	0.3000	0.227	58	0.041
2	0.2576	0.2625	0.2760	0.011	0.2840	0.219	59	0.040
3	0.2294	0.2437	0.2520	0.012	0.2590	0.212	60	0.039
4	0.2043	0.2253	0.2320	0.013	0.2380	0.207	61	0.038
5	0.1819	0.2070	0.2120	0.014	0.2200	0.204	62	0.037
6	0.1620	0.1920	0.1920	0.016	0.2030	0.201	63	0.036
7	0.1443	0.1770	0.1760	0.018	0.1800	0.199	64	0.035
8	0.1285	0.1620	0.1600	0.020	0.1650	0.197	65	0.033
9	0.1144	0.1483	0.1440	0.022	0.1480	0.194	66	0.032
10	0.1019	0.1350	0.1280	0.024	0.1340	0.191	67	0.031
11	0.0907	0.1205	0.1160	0.026	0.1200	0.188	68	0.030
12	0.0808	0.1055	0.1040	0.029	0.1090	0.185	69	0.029
13	0.0720	0.0915	0.0920	0.031	0.0950	0.182	70	0.027
14	0.0641	0.0800	0.0800	0.033	0.0830	0.180	71	0.026
15	0.0571	0.0720	0.0720	0.035	0.0720	0.178	72	0.024
16	0.0508	0.0625	0.0640	0.037	0.0650	0.175	73	0.023
17	0.0453	0.0540	0.0560	0.039	0.0580	0.172	74	0.022
18	0.0403	0.0475	0.0480	0.041	0.0490	0.168	75	0.020
19	0.0359	0.0410	0.0400	0.043	0.0420	0.164	76	0.018
20	0.0320	0.0348	0.0360	0.045	0.0350	0.161	77	0.016
21	0.0285	0.0318	0.0320	0.047	0.0320	0.157	78	0.015
22	0.0253	0.0286	0.0280	0.049	0.0280	0.155	79	0.014
23	0.0226	0.0258	0.0240	0.051	0.0250	0.153	80	0.013
24	0.0201	0.0230	0.0220	0.055	0.0220	0.151	…	…
25	0.0179	0.0204	0.0200	0.059	0.0200	0.148	…	…
26	0.0159	0.0181	0.0180	0.063	0.0180	0.146	…	…
27	0.0142	0.0173	0.0164	0.067	0.0160	0.143	…	…
28	0.0126	0.0162	0.0149	0.071	0.0140	0.139	…	…
29	0.0113	0.0150	0.0136	0.075	0.0130	0.134	…	…
30	0.0100	0.0140	0.0124	0.080	0.0120	0.127	…	…
31	0.00893	0.0132	0.0116	0.085	0.0100	0.120	…	…
32	0.00795	0.0128	0.0108	0.090	0.0090	0.115	…	…
33	0.00708	0.0118	0.0100	0.095	0.0080	0.112	…	…
34	0.00630	0.0104	0.0092	0.100	0.0070	0.110	…	…
35	0.00561	0.0095	0.0084	0.106	0.0050	0.108	…	…
36	0.00500	0.0090	0.0076	0.112	0.0040	0.106	…	…
37	0.00445	0.0085	0.0068	0.118	…	0.103	…	…
38	0.00396	0.0080	0.0060	0.124	…	0.101	…	…
39	0.00353	0.0075	0.0052	0.130	…	0.099	…	…
40	0.00314	0.0070	0.0048	0.138	…	0.097	…	…
41	0.00280	0.0066	0.0044	0.146	…	0.095	…	…
42	0.00249	0.0062	0.0040	0.154	…	0.092	…	…
43	0.00222	0.0060	0.0036	0.162	…	0.088	…	…
44	0.00198	0.0058	0.0032	0.170	…	0.085	…	…
45	0.00176	0.0055	0.0028	0.180	…	0.081	…	…
46	0.00157	0.0052	0.0024	…	…	0.079	…	…
47	0.00140	0.0050	0.0020	…	…	0.077	…	…
48	0.00124	0.0048	0.0016	…	…	0.075	…	…
49	0.00111	0.0046	0.0012	…	…	0.072	…	…
50	0.00099	0.0044	0.0010	…	…	0.069	…	…

[a] Also known as Washburn and Moen, American Steel and Wire Co. and Roebling Wire Gages. A greater selection of sizes is available and is specified by what are known as split gage numbers. They can be recognized by ½ fractions which follow the gage number; i.e., 4½. The decimal equivalents of split gage numbers are in the Steel Products Manual entitled: *Wire and Rods, Carbon Steel* published by the American Iron and Steel Institute, Washington, DC.

Sheet-Metal Gages.—Thicknesses of steel sheets are based upon a weight of 41.82 pounds per square foot per inch of thickness, which is known as the Manufacturers' Standard Gage for Sheet Steel. This gage differs from the older United States Standard Gage for iron and steel sheets and plates, established by Congress in 1893, based upon a weight of 40 pounds per square foot per inch of thickness which is the weight of wrought-iron plate.

Thicknesses of aluminum, copper, and copper-base alloys were formerly designated by the American or Brown & Sharpe Wire Gage but now are specified in decimals or fractions of an inch. American National Standard B32.1-1952 (R1988) entitled Preferred Thicknesses for Uncoated Thin Flat Metals (see accompanying Table 2) gives thicknesses that are based on the 20- and 40-series of preferred numbers in American National Standard Preferred Numbers — ANSI Z17.1 (see Handbook page 19) and are applicable to uncoated, thin, flat metals and alloys. Each number of the 20-series is approximately 12 percent greater than the next smaller one and each number of the 40-series is approximately 6 percent greater than the next smaller one.

Table 2. Preferred Thicknesses for Uncoated Metals and Alloys—Under 0.250 Inch in Thickness *ANSI B32.1-1952 (R1994)*

Preferred Thickness, Inches							
Based on 20-Series	Based on 40-Series	Based on 20-Series	Based on 40-Series	Based on 20-Series	Based on 40-Series	Based on 20-Series	Based on 40-Series
…	0.236	0.100	0.100	…	0.042	0.018	0.018
0.224	0.224	…	0.095	0.040	0.040	…	0.017
…	0.212	0.090	0.090	…	0.038	0.016	0.016
0.200	0.200	…	0.085	0.036	0.036	…	0.015
…	0.190	0.080	0.080	…	0.034	0.014	0.014
0.180	0.180	…	0.075	0.032	0.032	…	0.013
…	0.170	0.071	0.071	…	0.030	0.012	0.012
0.160	0.160	…	0.067	0.028	0.028	0.011	0.011
…	0.150	0.063	0.063	…	0.026	0.010	0.010
0.140	0.140	…	0.060	0.025	0.025	0.009	0.009
…	0.132	0.056	0.056	…	0.024	0.008	0.008
0.125	0.125	…	0.053	0.022	0.022	0.007	0.007
…	0.118	0.050	0.050	…	0.021	0.006	0.006
0.112	0.112	…	0.048	0.020	0.020	0.005	0.005
…	0.106	0.045	0.045	…	0.019	0.004	0.004

The American National Standard ANSI B32.1-1952 (R1994) lists preferred thicknesses that are based on the 20- and 40-series of preferred numbers and states that those based on the 40-series should provide adequate coverage. However, where intermediate thicknesses are required, the Standard recommends that thicknesses be based on the 80-series of preferred numbers (see Handbook page 19).

Thicknesses for copper and copper-base alloy flat products below ¼ inch thick are specified by the 20-series of American National Standard Preferred Numbers given in ANSI B32.1. Although the table in ANSI B32.1 gives only the 20- and 40-series of numbers, it states that when intermediate thicknesses are required they should be selected from thicknesses based on the 80-series of numbers (see Handbook page 19).

Zinc sheets are usually ordered by specifying decimal thickness although a zinc gage exists and is shown in Table 3.

Table 3. Sheet-Metal Gages in Approximate Decimals of an Inch

Gage No.	Steel Gage	B.G.[a]	Galva-nized Sheet	Zinc Gage	Gage No.	Steel Gage	B.G.[a]	Galvanized Sheet	Zinc Gage
15/0	…	1.000	…	…	20	0.0359	0.0392	0.0396	0.070
14/0	…	0.9583	…	…	21	0.0329	0.0349	0.0366	0.080
13/0	…	0.9167	…	…	22	0.0299	0.03125	0.0336	0.090
12/0	…	0.8750	…	…	23	0.0269	0.02782	0.0306	0.100
11/0	…	0.8333	…	…	24	0.0239	0.02476	0.0276	0.125
10/0	…	0.7917	…	…	25	0.0209	0.02204	0.0247	…
9/0	…	0.7500	…	…	26	0.0179	0.01961	0.0217	…
8/0	…	0.7083	…	…	27	0.0164	0.01745	0.0202	…
7/0	…	0.6666	…	…	28	0.0149	0.01562	0.0187	…
6/0	…	0.6250	…	…	29	0.0135	0.01390	0.0172	…
5/0	…	0.5883	…	…	30	0.0120	0.01230	0.0157	…
4/0	…	0.5416	…	…	31	0.0105	0.01100	0.0142	…
3/0	…	0.5000	…	…	32	0.0097	0.00980	0.0134	…
2/0	…	0.4452	…	…	33	0.0090	0.00870	…	…
1/0	…	0.3964	…	…	34	0.0082	0.00770	…	…
1	…	0.3532	…	…	35	0.0075	0.00690	…	…
2	…	0.3147	…	…	36	0.0067	0.00610	…	…
3	0.2391	0.2804	…	0.006	37	0.0064	0.00540	…	…
4	0.2242	0.2500	…	0.008	38	0.0060	0.00480	…	…
5	0.2092	0.2225	…	0.010	39	…	0.00430	…	…
6	0.1943	0.1981	…	0.012	40	…	0.00386	…	…
7	0.1793	0.1764	…	0.014	41	…	0.00343	…	…
8	0.1644	0.1570	0.1681	0.016	42	…	0.00306	…	…
9	0.1495	0.1398	0.1532	0.018	43	…	0.00272	…	…
10	0.1345	0.1250	0.1382	0.020	44	…	0.00242	…	…
11	0.1196	0.1113	0.1233	0.024	45	…	0.00215	…	…
12	0.1046	0.0991	0.1084	0.028	46	…	0.00192	…	…
13	.0897	0.0882	0.0934	0.032	47	…	0.00170	…	…
14	0.0747	0.0785	0.0785	0.036	48	…	0.00152	…	…
15	0.0673	0.0699	0.0710	0.040	49	…	0.00135	…	…
16	0.0598	0.0625	0.0635	0.045	50	…	0.00120	…	…
17	0.0538	0.0556	0.0575	0.050	51	…	0.00107	…	…
18	.0478	0.0495	0.0516	0.055	52	…	0.00095	…	…
19	0.0418	0.0440	0.0456	0.060	…	…	…	…	…

[a] B.G. is the Birmingham Gage for sheets and hoops.

The United States Standard Gage (not shown above) for iron and steel sheets and plates was established by Congress in 1893 and was primarily a *weight* gage rather than a thickness gage. The equivalent thicknesses were derived from the weight of wrought iron. The weight per cubic foot was taken at 480 pounds, thus making the weight of a plate 12 inches square and 1 inch thick, 40 pounds. In converting weight to equivalent thickness, gage tables formerly published contained thicknesses equivalent to the basic weights just mentioned. For example, a No. 3 U.S. gage represents a wrought-iron plate having a weight of 10 pounds per square foot; hence, if the weight per square foot per inch thick is 40 pounds, the plate thickness for a No. 3 gage = 10 ÷ 40 = 0.25 inch, which was the original thickness equivalent for this gage number. Because this and the other thickness equivalents were derived from the weight of wrought iron, they are not correct for steel.

Most sheet-metal products in Great Britain are specified by the British Standard Wire Gage (Imperial Wire Gage). Black iron and steel sheet and hooping, and galvanized flat and corrugated steel sheet, however, are specified by the Birmingham Gage (B.G.), which was legalized in 1914. This Birmingham Gage should not be confused with the Birmingham or Stub's Iron Wire Gage mentioned previously.

Metric Sizes for Flat Metal Products.—American National Standard B32.3 M-1984 establishes a preferred series of metric thicknesses, widths, and lengths for flat metal products of rectangular cross section; the thickness and width values are also applicable to base metals that may be coated in later operations. Table 4 lists the preferred thicknesses. Whenever possible, the Preferred Thickness values should be used, with the Second or Third Preference chosen only if no suitable Preferred size is available. Since not all metals and grades are produced in each of the sizes given in Table 4, producers or distributors should be consulted to determine a particular product and size combination's availability.

Table 4. Preferred Metric Thicknesses for All Flat Metal Products
ANSI/ASME B32.3M-1984

Preferred Thickness	Second Preference	Third Preference	Preferred Thickness	Second Preference	Third Preference
0.050	…	…	…	…	3.6
0.060	…	…	…	3.8	…
0.080	…	…	4.0	…	…
0.10	…	…	…	4.2	…
0.12	…	…	…	4.5	…
…	0.14	…	…	4.8	…
0.16	…	…	5.0	…	…
…	0.18	…	…	5.5	…
0.20	…	…	6.0	…	…
…	0.22	…	…	…	6.5
0.25	…	…	…	7.0	…
…	0.28	…	…	…	7.5
0.30	…	…	8.0	…	…
…	0.35	…	…	9.0	…
0.40	…	…	10	…	…
…	0.45	…	…	11	…
0.50	…	…	12	…	…
…	0.55	…	…	14	…
0.60	…	…	16	…	…
…	0.65	…	…	18	…
…	0.70	…	20	…	…
…	…	0.75	…	22	…
0.80	…	…	25	…	…
…	…	0.85	…	28	…
…	0.90	…	30	…	…
…	…	0.95	…	32	…
1.0	…	…	35	…	…
…	…	1.05	…	38	…
…	1.1	…	40	…	…
1.2	…	…	…	45	…
…	…	1.3	50	…	…
…	1.4	…	…	55	…
…	…	1.5	60	…	…
1.6	…	…	…	70	…
…	…	1.7	80	…	…
…	1.8	…	…	90	…
…	…	1.9	100	…	…
2.0	…	…	…	110	…
…	…	2.1	120	…	…
…	2.2	…	…	130	…
…	…	2.4	140	…	…
2.5	…	…	…	150	…
…	…	2.6	160	…	…
…	2.8	…	180	…	…
3.0	…	…	200	…	…
…	3.2	…	250	…	…
…	…	3.4	300	…	…
3.5	…	…			

All dimensions are in millimeters.

PIPE AND PIPE FITTINGS

Wrought Steel Pipe.—ANSI/ASME B36.10M-1995 covers dimensions of welded and seamless wrought steel pipe, for high or low temperatures or pressures.

The word *pipe* as distinguished from *tube* is used to apply to tubular products of dimensions commonly used for pipelines and piping systems. Pipe dimensions of sizes 12 inches and smaller have outside diameters numerically larger than the corresponding nominal sizes whereas outside diameters of tubes are identical to nominal sizes.

Size: The size of all pipe is identified by the nominal pipe size. The manufacture of pipe in the nominal sizes of ⅛ inch to 12 inches, inclusive, is based on a standardized outside diameter (OD). This OD was originally selected so that pipe with a standard OD and having a wall thickness which was typical of the period would have an inside diameter (ID) approximately equal to the nominal size. Although there is now no such relation between the existing standard thicknesses, ODs and nominal sizes, these nominal sizes and standard ODs continue in use as "standard."

The manufacture of pipe in nominal sizes of 14-inch OD and larger proceeds on the basis of an OD corresponding to the nominal size.

Weight: The nominal weights of steel pipe are calculated values and are tabulated in Table 1. They are based on the following formula:

$$W_{pe} = 10.68(D-t)t$$

where W_{pe} = nominal plain end weight to the nearest 0.01 lb/ft.
D = outside diameter to the nearest 0.001 inch
t = specified wall thickness rounded to the nearest 0.001 inch

Wall thickness: The nominal wall thicknesses are given in Table 1 which also indicates the wall thicknesses in API Standard 5L.

The wall thickness designations "Standard," "Extra-Strong," and "Double Extra-Strong" have been commercially used designations for many years. The Schedule Numbers were subsequently added as a convenient designation for use in ordering pipe. "Standard" and Schedule 40 are identical for nominal pipe sizes up to 10 inches, inclusive. All larger sizes of "Standard" have ⅜-inch wall thickness. "Extra-Strong" and Schedule 80 are identical for nominal pipe sizes up to 8 inch, inclusive. All larger sizes of "Extra-Strong" have ½-inch-wall thickness.

Wall Thickness Selection: When the selection of wall thickness depends primarily on capacity to resist internal pressure under given conditions, the designer shall compute the exact value of wall thickness suitable for conditions for which the pipe is required as prescribed in the "ASME Boiler and Pressure Vessel Code," "ANSI B31 Code for Pressure Piping," or other similar codes, whichever governs the construction. A thickness can then be selected from Table 1 to suit the value computed to fulfill the conditions for which the pipe is desired.

Metric Weights and Mass: Standard SI metric dimensions in millimeters for outside diameters and wall thicknesses may be found by multiplying the inch dimensions by 25.4. Outside diameters converted from those shown in Table 1 should be rounded to the nearest 0.1 mm and wall thicknesses to the nearest 0.01 mm.

The following formula may be used to calculate the SI metric plain end mass in kg/m using the converted metric diameters and thicknesses:

$$W_{pe} = 0.02466(D-t)t$$

where W_{pe} = nominal plain end mass rounded to the nearest 0.01 kg/m.
D = outside diameter to the nearest 0.1 mm for sizes shown in Table 1.
t = specified wall thickness rounded to the nearest 0.01 mm.

Table 1. American National Standard Weights and Dimensions of Welded and Seamless Wrought Steel Pipe *ANSI/ASME B36.10M-1995*

Nom. Size and (O.D.), in.	Wall Thick., in.	Plain End Wgt., lb/ft	Identification		
			Sch. No.	Other	
⅛ (0.405)	0.057	0.21	30	…	…
	0.068	0.24	40	5L	STD
	0.095	0.31	80	5L	XS
¼ (0.540)	0.073	0.36	30	…	…
	0.088	0.42	40	5L	STD
	0.119	0.54	80	5L	XS
⅜ (0.675)	0.073	0.47	30	…	…
	0.091	0.57	40	5L	STD
	0.126	0.74	80	5L	XS
½ (0.840)	0.095	0.76	30	…	…
	0.109	0.85	40	5L	STD
	0.147	1.09	80	5L	XS
	0.188	1.31	160	…	…
	0.294	1.71	…	5L	XXS
¾ (1.050)	0.095	0.97	30	…	…
	0.113	1.13	40	5L	STD
	0.154	1.47	80	5L	XS
	0.219	1.94	160	…	…
	0.308	2.44	…	5L	XXS
1 (1.315)	0.114	1.46	30	…	…
	0.133	1.68	40	5L	STD
	0.179	2.17	80	5L	XS
	0.250	2.84	160	…	…
	0.358	3.66	…	5L	XXS
1¼ (1.660)	0.117	1.93	30	…	…
	0.140	2.27	40	5L	STD
	0.191	3.00	80	5L	XS
	0.250	3.76	160	…	…
	0.382	5.21	…	5L	XXS
1½ (1.900)	0.125	2.37	30	…	…
	0.145	2.72	40	5L	STD
	0.200	3.63	80	5L	XS
	0.281	4.86	160	…	…
	0.400	6.41	…	5L	XXS
2 (2.375)	0.083	2.03	…	5L	…
	0.109	2.64	…	5L	…
	0.125	3.00	…	5L	…
	0.141	3.36	…	5L	…
	0.154	3.65	40	5L	STD
	0.172	4.05	…	5L	…
	0.188	4.39	…	5L	…
	0.218	5.02	80	5L	XS
	0.250	5.67	…	5L	…
	0.281	6.28	…	…	…
	0.344	7.46	160	…	…
	0.436	9.03	…	…	XXS
2½ (2.875)	0.083	2.47	…	5L	…
	0.109	3.22	…	5L	…
	0.125	3.67	…	5L	…
	0.141	4.12	…	5L	…
	0.156	4.53	…	5L	…
	0.172	4.97	…	5L	…
	0.188	5.40	…	5L	…
	0.203	5.79	40	5L	STD
	0.216	6.13	…	5L	…
	0.250	7.01	…	5L	…
	0.276	7.66	80	5L	XS
	0.375	10.01	160	…	…
	0.552	13.69	…	5L	XXS
3 (3.500)	0.083	3.03	…	5L	…
	0.109	3.95	…	5L	…
	0.125	4.51	…	5L	…
3 (3.500)	0.141	5.06	…	5L	…
	0.156	5.57	…	5L	…
	0.172	6.11	…	5L	…
	0.188	6.65	…	5L	…
	0.216	7.58	40	5L	STD
	0.250	8.68	…	5L	…
	0.281	9.66	…	5L	…
	0.300	10.25	80	5L	XS
	0.438	14.32	160	…	…
	0.600	18.58	…	5L	XXS
3½ (4.000)	0.083	3.47	…	5L	…
	0.109	4.53	…	5L	…
	0.125	5.17	…	5L	…
	0.141	5.81	…	5L	…
	0.156	6.40	…	5L	…
	0.172	7.03	…	5L	…
	0.188	7.65	…	5L	…
	0.226	9.11	40	5L	STD
	0.250	10.01	…	5L	…
	0.281	11.16	…	5L	…
	0.318	12.50	80	5L	XS
4 (4.500)	0.083	3.92	…	5L	…
	0.109	5.11	…	5L	…
	0.125	5.84	…	5L	…
	0.141	6.56	…	5L	…
	0.156	7.24	…	5L	…
	0.172	7.95	…	5L	…
	0.188	8.66	…	5L	…
	0.203	9.32	…	5L	…
	0.219	10.01	…	5L	…
	0.237	10.79	40	5L	STD
	0.250	11.35	…	5L	…
	0.281	12.66	…	5L	…
	0.312	13.96	…	5L	…
	0.337	14.98	80	5L	XS
	0.438	19.00	120	5L	…
	0.531	22.51	160	5L	…
	0.674	27.54	…	5L	XXS
5 (5.563)	0.083	4.86	…	5L	…
	0.125	7.26	…	5L	…
	0.156	9.01	…	5L	…
	0.188	10.79	…	5L	…
	0.219	12.50	…	5L	…
	0.258	14.62	40	5L	STD
	0.281	15.85	…	5L	…
	0.312	17.50	…	5L	…
	0.344	19.17	…	5L	…
	0.375	20.78	80	5L	XS
	0.500	27.04	120	5L	…
	0.625	32.96	160	5L	…
	0.750	38.55	…	5L	XXS
6 (6.625)	0.083	5.80	…	5L	…
	0.109	7.59	…	5L	…
	0.125	8.68	…	5L	…
	0.141	9.76	…	5L	…
	0.156	10.78	…	5L	…
	0.172	11.85	…	5L	…

Table 1. *(Continued)* **American National Standard Weights and Dimensions of Welded and Seamless Wrought Steel Pipe** *ANSI/ASME B36.10M-1995*

Nom. Size and (O.D.), in.	Wall Thick., in.	Plain End Wgt., lb/ft	Identification		
			Sch. No.		Other
6 (6.625)	0.188	12.92	…	5L	…
	0.203	13.92	…	5L	…
	0.219	14.98	…	5L	…
	0.250	17.02	…	5L	…
	0.280	18.97	40	5L	STD
	0.312	21.04	…	5L	…
	0.344	23.08	…	5L	…
	0.375	25.03	…	5L	…
	0.432	28.57	80	5L	XS
	0.500	32.71	…	5L	…
	0.562	36.39	120	5L	…
	0.625	40.05	…	5L	…
	0.719	45.35	160	5L	…
	0.750	47.06	…	5L	…
	0.864	53.16	…	5L	XXS
	0.875	53.73	…	5L	…
8 (8.625)	0.125	11.35	…	5L	…
	0.156	14.11	…	5L	…
	0.188	16.94	…	5L	…
	0.203	18.26	…	5L	…
	0.219	19.66	…	5L	…
	0.250	22.36	20	5L	…
	0.277	24.70	30	5L	…
	0.312	27.70	…	5L	…
	0.322	28.55	40	5L	STD
	0.344	30.42	…	5L	…
	0.375	33.04	…	5L	…
	0.406	35.64	60	…	…
	0.438	38.30	…	5L	…
	0.500	43.39	80	5L	XS
	0.562	48.40	…	5L	…
	0.594	50.95	100	…	…
	0.625	53.40	…	5L	…
	0.719	60.71	120	5L	…
	0.750	63.08	…	5L	…
	0.812	67.76	140	5L	…
	0.875	72.42	…	5L	XXS
	0.906	74.69	160	…	…
	1.000	81.44	…	5L	…
10 (10.750)	0.156	17.65	…	5L	…
	0.188	21.21	…	5L	…
	0.203	22.87	…	5L	…
	0.219	24.63	…	5L	…
	0.250	28.04	20	5L	…
	0.279	31.20	…	5L	…
	0.307	34.24	30	5L	…
	0.344	38.23	…	5L	…
	0.365	40.48	40	5L	STD
	0.438	48.24	…	5L	…
	0.500	54.74	60	5L	XS
	0.562	61.15	…	5L	…
	0.594	64.43	80	…	…
	0.625	67.58	…	5L	…
	0.719	77.03	100	5L	…
	0.812	86.18	…	5L	…
	0.844	89.29	120	…	…
	0.875	92.28	…	5L	…
	0.938	98.30	…	5L	…
	1.000	104.13	140	5L	XXS
10 (10.750)	1.125	115.64	160	…	…
	1.250	126.83	…	5L	…
12 (12.750)	0.172	23.11	…	5L	…
	0.188	25.22	…	5L	…
	0.203	27.20	…	…	…
	0.219	29.31	…	5L	…
	0.250	33.38	20	5L	…
	0.281	37.42	…	5L	…
	0.312	41.45	…	5L	…
	0.330	43.77	30	5L	…
	0.344	45.58	…	5L	…
	0.375	49.56	…	5L	STD
	0.406	53.52	40	5L	…
	0.438	57.59	…	5L	…
	0.500	65.42	…	5L	XS
	0.562	73.15	60	5L	…
	0.625	80.93	…	5L	…
	0.688	88.63	80	5L	…
	0.750	96.12	…	5L	…
	0.812	103.53	…	5L	…
	0.844	107.32	100	…	…
	0.875	110.97	…	5L	…
	0.938	118.33	…	5L	…
	1.000	125.49	120	5L	XXS
	1.062	132.57	…	5L	…
	1.125	139.67	140	5L	…
	1.250	153.53	…	5L	…
	1.312	160.27	160	5L	…
14 (14.000)	0.188	27.73	…	5L	…
	0.203	29.91	…	5L	…
	0.210	30.93	…	5L	…
	0.219	32.23	…	5L	…
	0.250	36.71	10	5L	…
	0.281	41.17	…	5L	…
	0.312	45.61	20	5L	…
	0.344	50.17	…	5L	…
	0.375	54.57	30	5L	STD
	0.406	58.94	…	5L	…
	0.438	63.44	40	5L	…
	0.469	67.78	…	5L	…
	0.500	72.09	…	5L	XS
	0.562	80.66	…	5L	…
	0.594	85.05	60	…	…
	0.625	89.28	…	5L	…
	0.688	97.81	…	5L	…
	0.750	106.13	80	5L	…
	0.812	114.37	…	5L	…
	0.875	122.65	…	5L	…
	0.938	130.85	100	5L	…
	1.000	138.84	…	5L	…
	1.062	146.74	…	5L	…
	1.094	150.79	120	…	…
	1.125	154.69	…	5L	…
	1.250	170.21	140	5L	…
	1.406	189.11	160	…	…
	2.000	256.32	…	…	…
	2.125	269.50	…	…	…
	2.200	277.25	…	…	…
	2.500	307.05	…	…	…

Table 2. Properties of American National Standard Schedule 40 Welded and Seamless Wrought Steel Pipe

Diameter, Inches			Wall Thickness, Inches	Cross-Sectional Area of Metal	Weight per Foot, Pounds		Capacity per Foot of Length		Length of Pipe in Feet to Contain		Properties of Sections		
Nominal	Actual Inside	Actual Outside			Of Pipe	Of Water in Pipe	In Cubic Inches	In Gallons	One Cubic Foot	One Gallon	Moment of Inertia	Radius of Gyration	Section Modulus
⅛	0.269	0.405	0.068	0.072	0.24	0.025	0.682	0.003	2532.	338.7	0.00106	0.122	0.00525
¼	0.364	0.540	0.088	0.125	0.42	0.045	1.249	0.005	1384.	185.0	0.00331	0.163	0.01227
⅜	0.493	0.675	0.091	0.167	0.57	0.083	2.291	0.010	754.4	100.8	0.00729	0.209	0.02160
½	0.622	0.840	0.109	0.250	0.85	0.132	3.646	0.016	473.9	63.35	0.01709	0.261	0.4070
¾	0.824	1.050	0.113	0.333	1.13	0.231	6.399	0.028	270.0	36.10	0.03704	0.334	0.07055
1	1.049	1.315	0.133	0.494	1.68	0.374	10.37	0.045	166.6	22.27	0.08734	0.421	0.1328
1¼	1.380	1.660	0.140	0.669	2.27	0.648	17.95	0.078	96.28	12.87	0.1947	0.539	0.2346
1½	1.610	1.900	0.145	0.799	2.72	0.882	24.43	0.106	70.73	9.456	0.3099	0.623	0.3262
2	2.067	2.375	0.154	1.075	3.65	1.454	40.27	0.174	42.91	5.737	0.6658	0.787	0.5607
2½	2.469	2.875	0.203	1.704	5.79	2.074	57.45	0.249	30.08	4.021	1.530	0.947	1.064
3	3.068	3.500	0.216	2.228	7.58	3.202	88.71	0.384	19.48	2.604	3.017	1.163	1.724
3½	3.548	4.000	0.226	2.680	9.11	4.283	118.6	0.514	14.56	1.947	4.788	1.337	2.394
4	4.026	4.500	0.237	3.174	10.79	5.515	152.8	0.661	11.31	1.512	7.233	1.510	3.215
5	5.047	5.563	0.258	4.300	14.62	8.666	240.1	1.04	7.198	0.9622	15.16	1.878	5.451
6	6.065	6.625	0.280	5.581	18.97	12.52	346.7	1.50	4.984	0.6663	28.14	2.245	8.496
8	7.981	8.625	0.322	8.399	28.55	21.67	600.3	2.60	2.878	0.3848	72.49	2.938	16.81
10	10.020	10.750	0.365	11.91	40.48	34.16	946.3	4.10	1.826	0.2441	160.7	3.674	29.91
12	11.938	12.750	0.406	15.74	53.52	48.49	1343.	5.81	1.286	0.1720	300.2	4.364	47.09
16	15.000	16.000	0.500	24.35	82.77	76.55	2121.	9.18	0.8149	0.1089	732.0	5.484	91.50
18	16.876	18.000	0.562	30.79	104.7	96.90	2684.	11.62	0.6438	0.0861	1172.	6.168	130.2
20	18.812	20.000	0.594	36.21	123.1	120.4	3335.	14.44	0.5181	0.0693	1706.	6.864	170.6
24	22.624	24.000	0.688	50.39	171.3	174.1	4824.	20.88	0.3582	0.0479	3426.	8.246	285.5
32	30.624	32.000	0.688	67.68	230.1	319.1	8839.	38.26	0.1955	0.0261	8299.	11.07	518.7

Note: Torsional section modulus equals twice section modulus.

Table 3. Properties of American National Standard Schedule 80 Welded and Seamless Wrought Steel Pipe

Diameter, Inches			Wall Thickness, Inches	Cross-Sectional Area of Metal	Weight per Foot, Pounds		Capacity per Foot of Length		Length of Pipe in Feet to Contain		Properties of Sections		
Nominal	Actual Inside	Actual Outside			Of Pipe	Of Water in Pipe	In Cubic Inches	In Gallons	One Cubic Foot	One Gallon	Moment of Inertia	Radius of Gyration	Section Modulus
⅛	0.215	0.405	0.095	0.093	0.315	0.016	0.436	0.0019	3966.	530.2	0.00122	0.115	0.00600
¼	0.302	0.540	0.119	0.157	0.537	0.031	0.860	0.0037	2010.	268.7	0.00377	0.155	0.01395
⅜	0.423	0.675	0.126	0.217	0.739	0.061	1.686	0.0073	1025.	137.0	0.00862	0.199	0.02554
½	0.546	0.840	0.147	0.320	1.088	0.101	2.810	0.0122	615.0	82.22	0.02008	0.250	0.04780
¾	0.742	1.050	0.154	0.433	1.474	0.187	5.189	0.0225	333.0	44.52	0.04479	0.321	0.08531
1	0.957	1.315	0.179	0.639	2.172	0.312	8.632	0.0374	200.2	26.76	0.1056	0.407	0.1606
1¼	1.278	1.660	0.191	0.881	2.997	0.556	15.39	0.0667	112.3	15.01	0.2418	0.524	0.2913
1½	1.500	1.900	0.200	1.068	3.631	0.766	21.21	0.0918	81.49	10.89	0.3912	0.605	0.4118
2	1.939	2.375	0.218	1.477	5.022	1.279	35.43	0.1534	48.77	6.519	0.8680	0.766	0.7309
2½	2.323	2.875	0.276	2.254	7.661	1.836	50.86	0.2202	33.98	4.542	1.924	0.924	1.339
3	2.900	3.500	0.300	3.016	10.25	2.861	79.26	0.3431	21.80	2.914	3.895	1.136	2.225
3½	3.364	4.000	0.318	3.678	12.50	3.850	106.7	0.4617	16.20	2.166	6.280	1.307	3.140
4	3.826	4.500	0.337	4.407	14.98	4.980	138.0	0.5972	12.53	1.674	9.611	1.477	4.272
5	4.813	5.563	0.375	6.112	20.78	7.882	218.3	0.9451	7.915	1.058	20.67	1.839	7.432
6	5.761	6.625	0.432	8.405	28.57	11.29	312.8	1.354	5.524	0.738	40.49	2.195	12.22
8	7.625	8.625	0.500	12.76	43.39	19.78	548.0	2.372	3.153	0.422	105.7	2.878	24.52
10	9.562	10.750	0.594	18.95	64.42	31.11	861.7	3.730	2.005	0.268	245.2	3.597	45.62
12	11.374	12.750	0.688	26.07	88.63	44.02	1219.	5.278	1.417	0.189	475.7	4.271	74.62
14	12.500	14.000	0.750	31.22	106.1	53.16	1473.	6.375	1.173	0.157	687.4	4.692	98.19
16	14.312	16.000	0.844	40.19	136.6	69.69	1931.	8.357	0.895	0.120	1158.	5.366	144.7
18	16.124	18.000	0.938	50.28	170.9	88.46	2450.	10.61	0.705	0.094	1835.	6.041	203.9
20	17.938	20.000	1.031	61.44	208.9	109.5	3033.	13.13	0.570	0.076	2772.	6.716	277.2
22	19.750	22.000	1.125	73.78	250.8	132.7	3676.	15.91	0.470	0.063	4031.	7.391	366.4

Note: Torsional section modulus equals twice section modulus.

Volume of Flow at 1 Foot Per-Minute Velocity in Pipe and Tube

Nominal Dia., Inches	Schedule 40 Pipe			Schedule 80 Pipe			Type K Copper Tube			Type L Copper Tube		
	Cu. Ft. per Minute	Gallons per Minute	Pounds 60 F Water per Min.	Cu. Ft. per Minute	Gallons per Minute	Pounds 60 F Water per Min.	Cu. Ft. per Minute	Gallons per Minute	Pounds 60 F Water per Min.	Cu. Ft. per Minute	Gallons per Minute	Pounds 60 F Water per Min.
⅛	0.0004	0.003	0.025	0.0003	0.002	0.016	0.0002	0.0014	0.012	0.0002	0.002	0.014
¼	0.0007	0.005	0.044	0.0005	0.004	0.031	0.0005	0.0039	0.033	0.0005	0.004	0.034
⅜	0.0013	0.010	0.081	0.0010	0.007	0.061	0.0009	0.0066	0.055	0.0010	0.008	0.063
½	0.0021	0.016	0.132	0.0016	0.012	0.102	0.0015	0.0113	0.094	0.0016	0.012	0.101
¾	0.0037	0.028	0.232	0.0030	0.025	0.213	0.0030	0.0267	0.189	0.0034	0.025	0.210
1	0.0062	0.046	0.387	0.0050	0.037	0.312	0.0054	0.0404	0.338	0.0057	0.043	0.358
1¼	0.0104	0.078	0.649	0.0088	0.067	0.555	0.0085	0.0632	0.53	0.0087	0.065	0.545
1½	0.0141	0.106	0.882	0.0123	0.092	0.765	0.0196	0.1465	1.22	0.0124	0.093	0.770
2	0.0233	0.174	1.454	0.0206	0.154	1.280	0.0209	0.1565	1.31	0.0215	0.161	1.34
2½	0.0332	0.248	2.073	0.0294	0.220	1.830	0.0323	0.2418	2.02	0.0331	0.248	2.07
3	0.0514	0.383	3.201	0.0460	0.344	2.870	0.0461	0.3446	2.88	0.0473	0.354	2.96
3½	0.0682	0.513	4.287	0.0617	0.458	3.720	0.0625	0.4675	3.91	0.0640	0.479	4.00
4	0.0884	0.660	5.516	0.0800	0.597	4.970	0.0811	0.6068	5.07	0.0841	0.622	5.20
5	0.1390	1.040	8.674	0.1260	0.947	7.940	0.1259	0.9415	7.87	0.1296	0.969	8.10
6	0.2010	1.500	12.52	0.1820	1.355	11.300	0.1797	1.3440	11.2	0.1862	1.393	11.6
8	0.3480	2.600	21.68	0.3180	2.380	19.800	0.3135	2.3446	19.6	0.3253	2.434	20.3
10	0.5476	4.10	34.18	0.5560	4.165	31.130	0.4867	3.4405	30.4	0.5050	3.777	21.6
12	0.7773	5.81	48.52	0.7060	5.280	44.040	0.6978	5.2194	43.6	0.7291	5.454	45.6
14	0.9396	7.03	58.65	0.8520	6.380	53.180	—	—	—	—	—	—
16	1.227	9.18	76.60	1.1170	8.360	69.730	—	—	—	—	—	—
18	1.553	11.62	96.95	1.4180	10.610	88.500	—	—	—	—	—	—
20	1.931	14.44	120.5	1.7550	13.130	109.510	—	—	—	—	—	—

To obtain volume of flow at any other velocity, multiply values in table by velocity in feet per minute.

Plastics Pipe.—Shortly after World War II, plastics pipe became an acceptable substitute, under certain service conditions, for other piping materials. Now, however, plastics pipe is specified on the basis of its own special capabilities and limitations. The largest volume of application has been for water piping systems.

Besides being light in weight, plastics pipe performs well in resisting deterioration from corrosive or caustic fluids. Even if the fluid borne is harmless, the chemical resistance of plastics pipe offers protection against a harmful exterior environment, such as when buried in a corrosive soil.

Generally, plastics pipe is limited by its temperature and pressure capacities. The higher the operating pressure of the pipe system, the less will be its temperature capability. The reverse is true, also. Since it is formed from organic resins, plastics pipe will burn. For various piping compositions, ignition temperatures vary from 700° to 800°F (370° to 430°C).

The following are accepted methods for joining plastics pipe:

Solvent Welding is usually accomplished by brushing a solvent cement on the end of the length of pipe and into the socket end of a fitting or the flange of the next pipe section. A chemical weld then joins and seals the pipe after connection.

Threading: is a procedure not recommended for thin-walled plastics pipe or for specific grades of plastics. During connection of thicker-walled pipe, strap wrenches are used to avoid damaging and weakening the plastics.

Heat Fusion: involves the use of heated air and plastics filler rods to weld plastics pipe assemblies. A properly welded joint can have a tensile strength equal to 90 percent that of the pipe material.

Elastomeric Sealing: is used with bell-end piping. It is a recommended procedure for large diameter piping and for underground installations. The joints are set quickly and have good pressure capabilities.

Table 1. Dimensions and Weights of Thermoplastics Pipe

Nominal Pipe Size		Outside Diameter		Schedule 40				Schedule 80			
				Nom. Wall Thickness		Nominal Weight		Nom. Wall Thickness		Nominal Weight	
in.	cm	in.	cm	in.	cm	lb/100′	kg/m	in.	cm	lb/100′	kg/m
1/8	0.3	0.405	1.03	0.072	0.18	3.27	0.05	0.101	0.256	4.18	0.06
1/4	0.6	0.540	1.37	0.093	0.24	5.66	0.08	0.126	0.320	7.10	0.11
3/8	1.0	0.675	1.71	0.096	0.24	7.57	0.11	0.134	0.340	9.87	0.15
1/2	1.3	0.840	2.13	0.116	0.295	11.4	0.17	0.156	0.396	14.5	0.22
3/4	2.0	1.050	2.67	0.120	0.305	15.2	0.23	0.163	0.414	19.7	0.29
1	2.5	1.315	3.34	0.141	0.358	22.5	0.33	0.190	0.483	29.1	0.43
1 1/4	3.2	1.660	4.22	0.148	0.376	30.5	0.45	0.202	0.513	40.1	0.60
1 1/2	3.8	1.900	4.83	0.154	0.391	36.6	0.54	0.212	0.538	48.7	0.72
2	5.1	2.375	6.03	0.163	0.414	49.1	0.73	0.231	0.587	67.4	1.00
2 1/2	6.4	2.875	7.30	0.215	0.546	77.9	1.16	0.293	0.744	103	1.5
3	7.6	3.500	8.89	0.229	0.582	102	1.5	0.318	0.808	138	2.1
3 1/2	8.9	4.000	10.16	0.240	0.610	123	1.8	0.337	0.856	168	2.5
4	10.2	4.500	11.43	0.251	0.638	145	2.2	0.357	0.907	201	3.0
5	12.7	5.563	14.13	0.273	0.693	197	2.9	0.398	1.011	280	4.2
6	15.2	6.625	16.83	0.297	0.754	256	3.8	0.458	1.163	385	5.7
8	20.3	8.625	21.91	0.341	0.866	385	5.7	0.530	1.346	584	8.7
10	25.4	10.75	27.31	0.387	0.983	546	8.1	0.629	1.598	867	12.9
12	30.5	12.75	32.39	0.430	1.09	722	10.7	0.728	1.849	1192	17.7

The nominal weights of plastics pipe given in this table are based on an empirically chosen material density of 1.00 g/cm^3. The nominal unit weight for a specific plastics pipe formulation can be obtained by multiplying the weight values from the table by the density in g/cm^3 or by the specific gravity of the particular platics composition.

The following are ranges of density factors for various plastics pipe materials: PE, 0.93 to 0.96; PVC, 1.35 to 1.40; CPVC, 1.55; ABS, 1.04 to 1.08; SR, 1.05; PB, 0.91 to 0.92; and PP, 0.91. For meanings of abbreviations see Table 2.

Information supplied by the Plastics Pipe Institute.

Insert Fitting: is particularly useful for PE and PB pipe. For joining pipe sections, insert fittings are pushed into the pipe and secured by stainless steel clamps.

Transition Fitting: involves specially designed connectors to join plastic pipe with other materials, such as cast iron, steel, copper, clay, and concrete.

Plastic pipe can be specified by means of Schedules 40, 80, and 120, which conform dimensionally to metal pipe, or through a Standard Dimension Ratio (SDR). The SDR is a rounded value obtained by dividing the average outside diameter of the pipe by the wall thickness. Within an individual SDR series of pipe, pressure ratings are uniform, regardless of pipe diameter.

Table 1 provides the weights and dimensions for Schedule 40 and 80 thermoplastic pipe, Table 2 gives properties of plastics pipe, Table 3 gives maximum non-shock operating pressures for several varieties of Schedule 40 and 80 plastics pipe at 73°F, and Table 4 gives correction factors to pressure ratings for elevated temperatures.

Table 2. General Properties and Uses of Plastic Pipe

Plastic Pipe Material	Properties	Common Uses	Operating Temperature[a]		Joining Methods
			With Pressure	Without Pressure	
ABS (Acrylonitrilebutadiene styrene)	Rigid; excellent impact strength at low temperatures; maintains rigidity at higher temperatures.	Water, Drain, Waste, Vent, Sewage.	100°F (38°C)	180°F (82°C)	Solvent cement, Threading, Transition fitting.
PE (Polyethylene)	Flexible; excellent impact strength; good performance at low temperatures.	Water, Gas, Chemical, Irrigation.	100°F (38°C)	180°F (82°C)	Heat fusion, Insert and Transition fitting.
PVC (Polyvinylchloride)	Rigid; fire self-extinguishing; high impact and tensile strength.	Water, Gas, Sewage, Industrial process, Irrigation.	100°F (38°C)	180°F (82°C)	Solvent cement, Elastomeric seal, Mechanical coupling, Transition fitting.
CPVC (Chlorinated polyvinyl chloride)	Rigid; fire self-extinguishing; high impact and tensile strength.	Hot and cold water, Chemical.	180°F (82°C) at 100 psig (690kPa) for SDR-11		Solvent cement, Threading, Mechanical coupling, Transition fitting.
PB (Polybutylene)	Flexible; good performance at elevated temperatures.	Water, Gas, Irrigation.	180°F (82°C)	200°F (93°C)	Insert fitting, Heat fusion, Transition fitting.
PP (Polypropylene)	Rigid; very light; high chemical resistance, particularly to sulfur-bearing compounds.	Chemical waste and processing.	100°F (38°C)	180°F (82°C)	Mechanical coupling, Heat fusion, Threading.
SR (Styrene rubber plastic)	Rigid; moderate chemical resistance; fair impact strength.	Drainage, Septic fields.	150°F (66°C)	…	Solvent cement, Transition fitting, Elastomeric seal.

[a] The operating temperatures shows are general guide points. For specific operating temperature and pressure data for various grades of the types of plastic pipe given, please consult the pipe manufacturer or the Plastics Pipe Institute.

From information supplied by the Plastics Pipe Institute.

Table 3. Maximum Nonshock Operating Pressure (psi) for Thermoplastic Piping at 73°F

Nominal Pipe Size (inch)	Schedule 40		Schedule 80						
	PVC & CPVC (Socket End)	ABS	PVC & CPVC		Polypropylene		PVDF		ABS
			Socket End	Threaded End	Thermo-seal Joint	Threaded End[a]	Thermoseal Joint	Threaded End	
½	600	476	850	420	410	20	580	290	678
¾	480	385	690	340	330	20	470	230	550
1	450	360	630	320	310	20	430	210	504
1¼	370	294	520	260	260	20	…	…	416
1½	330	264	470	240	230	20	326	160	376
2	280	222	400	200	200	…	270	140	323
2½	300	243	420	210	…	20	…	…	340
3	260	211	370	190	160	20	250	NR	297
4	220	177	320	160	140	NR	220	NR	259
6	180	141	280	NR	…	…	190	NR	222
8	160	…	250[b]	NR	…	…	…	…	…
10	140	…	230	NR	…	…	…	…	…
12	130	…	230	NR	…	…	…	…	…

[a] Recommended for intermittent drainage pressure not exceeding 20 psi.

[b] 8-inch CPVC Tee, 90° Ell, and 45° Ell are rated at half the pressure shown.

ABS pressures refer to unthreaded pipe only.

For service at higher temperature, multiply the pressure obtained from this table by the correction factor from Table 2.

NR is not recommended.

Table 4. Temperature-Correction Factors for Thermoplastic Piping Operating Pressures

Operating Temperature, °F	Pipe Material			
	PVC	CPVC	Polypropylene	PVDF
70	1	1	1	1
80	0.90	0.96	0.97	0.95
90	0.75	0.92	0.91	0.87
100	0.62	0.85	0.85	0.80
110	0.50	0.77	0.80	0.75
115	0.45	0.74	0.77	0.71
120	0.40	0.70	0.75	0.68
125	0.35	0.66	0.71	0.66
130	0.30	0.62	0.68	0.62
140	0.22	0.55	0.65	0.58
150	NR	0.47	0.57	0.52
160	NR	0.40	0.50	0.49
170	NR	0.32	0.26	0.45
180	NR	0.25	[a]	0.42
200	NR	0.18	NR	0.36
210	NR	0.15	NR	0.33
240	NR	NR	NR	0.25
280	NR	NR	NR	0.18

[a] Recommended for intermittent drainage pressure not exceeding 20 psi.

NR = not recommended.

For more detailed information concerning the properties of a particular plastic pipe formulation, consult the pipe manufacturer or The Plastics Pipe Institute, 355 Lexington Ave., New York, N.Y. 10017.

MEASURING UNITS

THERMAL ENERGY OR HEAT

Thermometer Scales.—There are two thermometer scales in general use: the Fahrenheit (F), which is used in the United States and in other countries still using the English system of units, and the Celsius (C) or Centigrade used throughout the rest of the world.

In the Fahrenheit thermometer, the freezing point of water is marked at 32 degrees on the scale and the boiling point, at atmospheric pressure, at 212 degrees. The distance between these two points is divided into 180 degrees. On the Celsius scale, the freezing point of water is at 0 degrees and the boiling point at 100 degrees. The following formulas may be used for converting temperatures given on any one of the scales to the other scale:

$$\text{Degrees Fahrenheit} = \frac{9 \times \text{degrees C}}{5} + 32$$

$$\text{Degrees Celsius} = \frac{5 \times (\text{degrees F} - 32)}{9}$$

Tables on the pages that follow can be used to convert degrees Celsius into degrees Fahrenheit or vice versa. In the event that the conversions are not covered in the tables, use those applicable portions of the formulas given above for converting.

Absolute Temperature and Absolute Zero.—A point has been determined on the thermometer scale, by theoretical considerations, that is called the absolute zero and beyond which a further decrease in temperature is inconceivable. This point is located at −273.2 degrees Celsius or −459.7 degrees F. A temperature reckoned from this point, instead of from the zero on the ordinary thermometers, is called absolute temperature. Absolute temperature in degrees C is known as "degrees Kelvin" or the "Kelvin scale" (K) and absolute temperature in degrees F is known as "degrees Rankine" or the "Rankine scale" (R).

$$\text{Degrees Kelvin} = \text{degrees C} + 273.2$$

$$\text{Degrees Rankine} = \text{degrees F} + 459.7$$

Measures of the Quantity of Thermal Energy.—The unit of quantity of thermal energy used in the United States is the British thermal unit, which is the quantity of heat or thermal energy required to raise the temperature of one pound of pure water one degree F. (American National Standard abbreviation, Btu; conventional British symbol, B.Th.U.) The French thermal unit, or *kilogram calorie,* is the quantity of heat or thermal energy required to raise the temperature of one kilogram of pure water one degree C. One kilogram calorie = 3.968 British thermal units = 1000 gram calories. The number of foot-pounds of mechanical energy equivalent to one British thermal unit is called the *mechanical equivalent of heat,* and equals 778 foot-pounds.

In the modern metric or SI system of units, the unit for thermal energy is the *joule* (J); a commonly used multiple being the kilojoule (kJ), or 1000 joules. See page 2520 for an explanation of the SI System. One kilojoule = 0.9478 Btu. Also in the SI System, the *watt* (W), equal to joule per second (J/s), is used for power, where one watt = 3.412 Btu per hour.

Fahrenheit-Celsius (Centigrade) Conversion.—A simple way to convert a Fahrenheit temperature reading into a Celsius temperature reading or vice versa is to enter the accompanying table in the center or boldface column of figures. These figures refer to the temperature in either Fahrenheit or Celsius degrees. If it is desired to convert from Fahrenheit to Celsius degrees, consider the center column as a table of Fahrenheit temperatures and read the corresponding Celsius temperature in the column at the left. If it is desired to convert from Celsius to Fahrenheit degrees, consider the center column as a table of Celsius values, and read the corresponding Fahrenheit temperature on the right.

Interpolation Factors

deg C		deg F	deg C		deg F
0.56	**1**	1.8	3.33	**6**	10.8
1.11	**2**	3.6	3.89	**7**	12.6
1.67	**3**	5.4	4.44	**8**	14.4
2.22	**4**	7.2	5.00	**9**	16.2
2.78	**5**	9.0	5.56	**10**	18.0

Interpolation factors are given for use with that portion of the table in which the center column advances in increments of 10. To illustrate, suppose it is desired to find the Fahrenheit equivalent of 314 degrees C. The equivalent of 310 degrees C, found in the body of the main table, is seen to be 590.0 degrees F. The Fahrenheit equivalent of a 4-degree C difference is seen to be 7.2, as read in the table of interpolating factors. The answer is the sum or 597.2 degrees F.

Fahrenheit-Celsius (Centigrade) Conversion Table

deg C		deg F	deg C		deg F	deg C		deg F	deg C		deg F
−273	**−459.4**	...	−101	**−150**	−238	−8.3	**17**	62.6	9.4	**49**	120.2
−268	**−450**	...	−96	**−140**	−220	−7.8	**18**	64.4	10.0	**50**	122.0
−262	**−440**	...	−90	**−130**	−202	−7.2	**19**	66.2	10.6	**51**	123.8
−257	**−430**	...	−84	**−120**	−184	−6.7	**20**	68.0	11.1	**52**	125.6
−251	**−420**	...	−79	**−110**	−166	−6.1	**21**	69.8	11.7	**53**	127.4
−246	**−410**	...	−73	**−100**	−148	−5.6	**22**	71.6	12.2	**54**	129.2
−240	**−400**	...	−68	**−90**	−130	−5.0	**23**	73.4	12.8	**55**	131.0
−234	**−390**	...	−62	**−80**	−112	−4.4	**24**	75.2	13.3	**56**	132.8
−229	**−380**	...	−57	**−70**	−94	−3.9	**25**	77.0	13.9	**57**	134.6
−223	**−370**	...	−51	**−60**	−76	−3.3	**26**	78.8	14.4	**58**	136.4
−218	**−360**	...	−46	**−50**	−58	−2.8	**27**	80.6	15.0	**59**	138.2
−212	**−350**	...	−40	**−40**	−40	−2.2	**28**	82.4	15.6	**60**	140.0
−207	**−340**	...	−34	**−30**	−22	−1.7	**29**	84.2	16.1	**61**	141.8
−201	**−330**	...	−29	**−20**	−4	−1.1	**30**	86.0	16.7	**62**	143.6
−196	**−320**	...	−23	**−10**	14	−0.6	**31**	87.8	17.2	**63**	145.4
−190	**−310**	...	−17.8	**0**	32–	0–	**32**	89.6	17.8	**64**	147.2
−184	**−300**	...	−17.2	**1**	33.8	0.6	**33**	91.4	18.3	**65**	149.0
−179	**−290**	...	−16.7	**2**	35.6	1.1	**34**	93.2	18.9	**66**	150.8
−173	**−280**	...	−16.1	**3**	37.4	1.7	**35**	95.0	19.4	**67**	152.6
−169	**−273**	−459.4	−15.6	**4**	39.2	2.2	**36**	96.8	20.0	**68**	154.4
−168	**−270**	−454	−15.0	**5**	41.0	2.7	**37**	98.6	20.6	**69**	156.2
−162	**−260**	−436	−14.4	**6**	42.8	3.3	**38**	100.4	21.1	**70**	158.0
−157	**−250**	−418	−13.9	**7**	44.6	3.9	**39**	102.2	21.7	**71**	159.8
−151	**−240**	−400	−13.3	**8**	46.4	4.4	**40**	104.0	22.2	**72**	161.6
−146	**−230**	−382	−12.8	**9**	48.2	5.0	**41**	105.8	22.8	**73**	163.4
−140	**−220**	−364	−12.2	**10**	50.0	5.6	**42**	107.6	23.3	**74**	165.2
−134	**−210**	−346	−11.7	**11**	51.8	6.1	**43**	109.4	23.9	**75**	167.0
−129	**−200**	−328	−11.1	**12**	53.6	6.7	**44**	111.2	24.4	**76**	168.8
−123	**−190**	−310	−10.6	**13**	55.4	7.2	**45**	113.0	25.0	**77**	170.6
−118	**−180**	−292	−10.0	**14**	57.2	7.8	**46**	114.8	25.6	**78**	172.4
−112	**−170**	−274	−9.4	**15**	59.0	8.3	**47**	116.6	26.1	**79**	174.2
−107	**−160**	−256	−8.9	**16**	60.8	8.9	**48**	118.4	26.7	**80**	176.0
27.2	**81**	177.8	58.3	**137**	278.6	89.4	**193**	379.4	304.4	**580**	1076
27.8	**82**	179.6	58.9	**138**	280.4	90.0	**194**	381.2	310.0	**590**	1094
28.3	**83**	181.4	59.4	**139**	282.2	90.6	**195**	383.0	315.6	**600**	1112
28.9	**84**	183.2	60.0	**140**	284.0	91.1	**196**	384.8	321.1	**610**	1130
29.4	**85**	185.0	60.6	**141**	285.8	91.7	**197**	386.6	326.7	**620**	1148
30.0	**86**	186.8	61.1	**142**	287.6	92.2	**198**	388.4	332.2	**630**	1166
30.6	**87**	188.6	61.7	**143**	289.4	92.8	**199**	390.2	337.8	**640**	1184

Fahrenheit-Celsius (Centigrade) Conversion Table *(Continued)*

31.1	**88**	190.4	62.2	**144**	291.2	93.3	**200**	392.0	343.3	**650**	1202
31.7	**89**	192.2	62.8	**145**	293.0	93.9	**201**	393.8	348.9	**660**	1220
32.2	**90**	194.0	63.3	**146**	294.8	94.4	**202**	395.6	354.4	**670**	1238
32.8	**91**	195.8	63.9	**147**	296.6	95.0	**203**	397.4	360.0	**680**	1256
33.3	**92**	197.6	64.4	**148**	298.4	95.6	**204**	399.2	365.6	**690**	1274
33.9	**93**	199.4	65.0	**149**	300.2	96.1	**205**	401.0	371.1	**700**	1292
34.4	**94**	201.2	65.6	**150**	302.0	96.7	**206**	402.8	376.7	**710**	1310
35.0	**95**	203.0	66.1	**151**	303.8	97.2	**207**	404.6	382.2	**720**	1328
35.6	**96**	204.8	66.7	**152**	305.6	97.8	**208**	406.4	387.8	**730**	1346
36.1	**97**	206.6	67.2	**153**	307.4	98.3	**209**	408.2	393.3	**740**	1364
36.7	**98**	208.4	67.8	**154**	309.2	98.9	**210**	410.0	398.9	**750**	1382
37.2	**99**	210.2	68.3	**155**	311.0	99.4	**211**	411.8	404.4	**760**	1400
37.8	**100**	212.0	68.9	**156**	312.8	100.0	**212**	413.6	410.0	**770**	1418
38.3	**101**	213.8	69.4	**157**	314.6	104.4	**220**	428.0	415.6	**780**	1436
38.9	**102**	215.6	70.0	**158**	316.4	110.0	**230**	446.0	421.1	**790**	1454
39.4	**103**	217.4	70.6	**159**	318.2	115.6	**240**	464.0	426.7	**800**	1472
40.0	**104**	219.2	71.1	**160**	320.0	121.1	**250**	482.0	432.2	**810**	1490
40.6	**105**	221.0	71.7	**161**	321.8	126.7	**260**	500.0	437.8	**820**	1508
41.1	**106**	222.8	72.2	**162**	323.6	132.2	**270**	518.0	443.3	**830**	1526
41.7	**107**	224.6	72.8	**163**	325.4	137.8	**280**	536.0	448.9	**840**	1544
42.2	**108**	226.4	73.3	**164**	327.2	143.3	**290**	554.0	454.4	**850**	1562
42.8	**109**	228.2	73.9	**165**	329.0	148.9	**300**	572.0	460.0	**860**	1580
43.3	**110**	230.0	74.4	**166**	330.8	154.4	**310**	590.0	465.6	**870**	1598
43.9	**111**	231.8	75.0	**167**	332.6	160.0	**320**	608.0	471.1	**880**	1616
44.4	**112**	233.6	75.6	**168**	334.4	165.6	**330**	626.0	476.7	**890**	1634
45.0	**113**	235.4	76.1	**169**	336.2	171.1	**340**	644.0	482.2	**900**	1652
45.6	**114**	237.2	76.7	**170**	338.0	176.7	**350**	662.0	487.8	**910**	1670
46.1	**115**	239.0	77.2	**171**	339.8	182.2	**360**	680.0	493.3	**920**	1688
46.7	**116**	240.8	77.8	**172**	341.6	187.8	**370**	698.0	498.9	**930**	1706
47.2	**117**	242.6	78.3	**173**	343.4	193.3	**380**	716.0	504.4	**940**	1724
47.8	**118**	244.4	78.9	**174**	345.2	198.9	**390**	734.0	510.0	**950**	1742
48.3	**119**	246.2	79.4	**175**	347.0	204.4	**400**	752.0	515.6	**960**	1760
48.9	**120**	248.0	80.0	**176**	348.8	210.0	**410**	770.0	521.1	**970**	1778
49.4	**121**	249.8	80.6	**177**	350.6	215.6	**420**	788	526.7	**980**	1796
50.0	**122**	251.6	81.1	**178**	352.4	221.1	**430**	806	532.2	**990**	1814
50.6	**123**	253.4	81.7	**179**	354.2	226.7	**440**	824	537.8	**1000**	1832
51.1	**124**	255.2	82.2	**180**	356.0	232.2	**450**	842	565.6	**1050**	1922
51.7	**125**	257.0	82.8	**181**	357.8	237.8	**460**	860	593.3	**1100**	2012
52.2	**126**	258.8	83.3	**182**	359.6	243.3	**470**	878	621.1	**1150**	2102
52.8	**127**	260.6	83.9	**183**	361.4	248.9	**480**	896	648.9	**1200**	2192
53.3	**128**	262.4	84.4	**184**	363.2	254.4	**490**	914	676.7	**1250**	2282
53.9	**129**	264.2	85.0	**185**	365.0	260.0	**500**	932	704.4	**1300**	2372
54.4	**130**	266.0	85.6	**186**	366.8	265.6	**510**	950	732.2	**1350**	2462
55.0	**131**	267.8	86.1	**187**	368.6	271.1	**520**	968	760.0	**1400**	2552
55.6	**132**	269.6	86.7	**188**	370.4	276.7	**530**	986	787.8	**1450**	2642
56.1	**133**	271.4	87.2	**189**	372.2	282.2	**540**	1004	815.6	**1500**	2732
56.7	**134**	273.2	87.8	**190**	374.0	287.8	**550**	1022	1093.9	**2000**	3632
57.2	**135**	275.0	88.3	**191**	375.8	293.3	**560**	1040	1648.9	**3000**	5432
57.8	**136**	276.8	88.9	**192**	377.6	298.9	**570**	1058	2760.0	**5000**	9032

Above 1000 in the center column, the table increases in increments of 50. To convert 1462 degrees F to Celsius, for instance, add to the Celsius equivalent of 1400 degrees F ten times the interpolation factor for 6 and the interpolation factor for 2 or 760.0 + 33.3 + 1.11, which equals 794.4.

WEIGHTS AND MEASURES

Measures of Length

1mile = 1760 yards = 5280 feet
1yard = 3 feet = 36 inches *1foot* = 12 inches
1mil = 0.001 inch
1fathom = 2 yards = 6 feet *1rod* = 5.5 yards = 16.5 feet.
1hand = 4 inches *1span* = 9 inches
1micro-inch = one millionth inch or 0.000001 inch (1 micrometer or micron one millionth meter = 0.00003937 inch.)

Surveyor's Measure

1mile = 8 furlongs = 80 chains
1furlong = 10 chains = 220 yards
1chain = 4 rods = 22 yards = 66 feet = 100 links
1link = 7.92 inches

Nautical Measure

1league = 3 nautical miles
1nauticalmile = 6076.11549 feet = 1.1508 statute miles
1knot = nautical unit of speed = 1 nautical mile per hour
One degree at the equator = 60 nautical miles = 69.047 statute miles 360 degrees = 21,600 nautical miles = 24,856.8 statute miles = circumference at equator

Square Measure

1squaremile = 640 acres = 6400 square chains
1acre = 10 square chains = 4840 square yards = 43,560 square feet. An acre is equal to a square, the side of which is 208.7 feet.
1squarechain = 16 square rods = 484 square yards = 4356 square feet.
1squarerod = 30.25 square yards = 272.25 square feet = 625 square links
1squareyard = 9 square feet
1squarefoot = 144 square inches

Measure used for Diameters and Areas of Electric Wires

1circularinch = area of circle 1 inch in diameter = 0.7854 square inch
1circularinch = 1,000,000 circular mils
1squareinch = 1.2732 circular inch = 1,273,239 circular mils
1circularmil = the area of a circle 0.001 inch in diameter

Cubic Measure

1cubicyard = 27 cubic feet
1cubicfoot = 1728 cubic inches

The following measures are also used for wood and masonry:

1cordofwood = $4 \times 4 \times 8$ feet = 128 cubic feet
1perchofmasonry = $16\frac{1}{2} \times 1\frac{1}{2} \times 1$ foot = $24\frac{3}{4}$ cubic feet

Shipping Measure

For measuring entire internal capacity of a vessel:

1 register ton = 100 cubic feet

For measurement of cargo:

Approximately 40 cubic feet of merchandise is considered a shipping ton, unless that bulk would weigh more than 2000 pounds, in which case the freight charge may be based upon weight

40 cubic feet = 32.143 U.S. bushels = 31.16 Imperial bushels

Dry Measure

1 bushel (U.S. or Winchester struck bushel) = 1.2445 cubic feet = 2150.42 cubic inches

1 bushel = 4 pecks = 32 quarts = 64 pints

1 peck = 8 quarts = 16 pints

1 quart = 2 pints

1 heaped bushel = $1\frac{1}{4}$ struck bushel

1 cubic foot = 0.8036 struck bushel

1 British Imperial bushel = 8 Imperial gallons = 1.2837 cubic feet = 2218.19 cubic inches

Liquid Measure

1 U.S. gallon = 0.1337 cubic foot = 231 cubic inches = 4 quarts = 8 pints

1 quart = 2 pints = 8 gills *1 pint* = 4 gills

1 British Imperial gallon = 1.2009

1 U.S. gallon = 277.42 cubic inches

1 cubic foot = 7.48 U.S. gallons

Old Liquid Measure

1 barrel = $31\frac{1}{2}$ gallons *1 hogshead* = 2 barrels = 63 gallons

1 pipe or *butt* = 2 hogsheads = 4 barrels = 126 gallons

1 tierce = 42 gallons *1 puncheon* = 2 tierces = 84 gallons

1 tun = 2 pipes = 3 puncheons

Apothecaries' Fluid Measure

1 U.S. fluid ounce = 8 drachms = 1.805 cubic inch = $\frac{1}{128}$ U.S. gallon

1 fluid drachm = 60 minims

1 British fluid ounce = 1.732 cubic inch

Avoirdupois or Commercial Weight

1 gross or long ton = 2240 pounds *1 net or short ton* = 2000 pounds

1 pound = 16 ounces = 7000 grains

1 ounce = 16 drachms = 437.5 grains

The following measures for weight are now seldom used in the United States:

1 hundred-weight = 4 quarters = 112 pounds (1 gross or long ton = 20 hundred-weights); 1 quarter = 28 pounds; 1 stone = 14 pounds; 1 quintal = 100 pounds

Troy Weight, used for Weighing Gold and Silver

1 pound = 12 ounces = 5760 grains

1 ounce = 20 pennyweights = 480 grains

1 pennyweight = 24 grains

1 carat (used in weighing diamonds) = 3.086 grains

1 grain Troy = 1 grain avoirdupois = 1 grain apothecaries' weight

Apothecaries' Weight

1pound = 12 ounces = 5760 grains
1ounce = 8 drachms = 480 grains
1drachm = 3 scruples = 60 grains
1scruple = 20 grains

Measures of Pressure

1 pound per square inch = 144 pounds per square foot = 0.068 atmosphere = 2.042 inches of mercury at 62° F. = 27.7 inches of water at 62° F. = 2.31 feet of water at 62° F

1atmosphere = 30 inches of mercury at 62° F. = 14.7 pounds per square inch = 2116.3 pounds per square foot = 33.95 feet of water at 62° F

1 foot of water at 62° F. = 62.355 pounds per square foot = 0.433 pound per square inch

1 inch of mercury at 62° F. = 1.132 foot of water = 13.58 inches of water = 0.491 pound per square inch

Miscellaneous

1 great gross = 12 gross = 144 dozen
1quire = 24 sheets
1gross = 12 dozen = 144 units
1ream = 20 quires = 480 sheets
1dozen = 12 units
1 ream printing paper = 500 sheets
1score = 20 units

Decimal Equivalents of Fractions of an Inch

1/64	0.015 625	11/32	0.343 75	43/64	0.671 875
1/32	0.031 25	23/64	0.359 375	11/16	0.687 5
3/64	0.046 875	3/8	0.375	45/64	0.703 125
1/16	0.062 5	25/64	0.390 625	23/32	0.718 75
5/64	0.078 125	13/32	0.406 25	47/64	0.734 375
3/32	0.093 75	27/64	0.421 875	3/4	0.750
7/64	0.109 375	7/16	0.437 5	49/64	0.765 625
1/8	0.125	29/64	0.453 125	25/32	0.781 25
9/64	0.140 625	15/32	0.468 75	51/64	0.796 875
5/32	0.156 25	31/64	0.484 375	13/16	0.812 5
11/64	0.171 875	1/2	0.500	53/64	0.828 125
3/16	0.187 5	33/64	0.515 625	27/32	0.843 75
13/64	0.203 125	17/32	0.531 25	55/64	0.859 375
7/32	0.218 75	35/64	0.546 875	7/8	0.875
15/64	0.234 375	9/16	0.562 5	57/64	0.890 625
1/4	0.250	37/64	0.578 125	29/32	0.906 25
17/64	0.265 625	19/32	0.593 75	59/64	0.921 875
9/32	0.281 25	39/64	0.609 375	15/16	0.937 5
19/64	0.296 875	5/8	0.625	61/64	0.953 125
5/16	0.312 5	41/64	0.640 625	31/32	0.968 75
21/64	0.328 125	21/32	0.656 25	63/64	0.984 375

METRIC SYSTEMS OF MEASUREMENT

A metric system of measurement was first established in France in the years following the French Revolution, and various systems of metric units have been developed since that time. All metric unit systems are based, at least in part, on the International Metric Standards, which are the meter and kilogram, or decimal multiples or submultiples of these standards.

In 1795, a metric system called the centimeter-gram-second (cgs) system was proposed, and was adopted in France in 1799. In 1873, the British Association for the Advancement of Science recommended the use of the cgs system, and since then it has been widely used in all branches of science throughout the world. From the base units in the cgs system are derived the following:

$$\begin{aligned} \textit{Unit of velocity} &= 1 \text{ centimeter per second} \\ \textit{Acceleration due to gravity (at Paris)} &= 981 \text{ centimeters per second per second} \\ \textit{Unit of force} &= 1 \text{ dyne} = \tfrac{1}{981} \text{ gram} \\ \textit{Unit of work} &= 1 \text{ erg} = 1 \text{ dyne-centimeter} \\ \textit{Unit of power} &= 1 \text{ watt} = 10{,}000{,}000 \text{ ergs per second} \end{aligned}$$

Another metric system called the MKS (meter-kilogram-second) system of units was proposed by Professor G. Giorgi in 1902. In 1935, the International Electro-technical Commission (IEC) accepted his recommendation that this system of units of mechanics should be linked with the electromagnetic units by the adoption of a fourth base unit. In 1950, the IEC adopted the ampere, the unit of electric current, as the fourth unit, and the MKSA system thus came into being.

A gravitational system of metric units, known as the technical system, is based on the meter, the kilogram as a force, and the second. It has been widely used in engineering. Because the standard of force is defined as the weight of the mass of the standard kilogram, the fundamental unit of force varies due to the difference in gravitational pull at different locations around the earth. By international agreement, a standard value for acceleration due to gravity was chosen (9.81 meters per second squared) that for all practical measurements is approximately the same as the local value at the point of measurement.

The International System of Units (SI).—The Conference Generale des Poids et Mesures (CGPM), which is the body responsible for all international matters concerning the metric system, adopted in 1954, a rationalized and coherent system of units, based on the four MKSA units (see above), and including the *kelvin* as the unit of temperature and the *candela* as the unit of luminous intensity. In 1960, the CGPM formally named this system the Système International d'Unites, for which the abbreviation is SI in all languages. In 1971, the 14th CGPM adopted a seventh base unit, the *mole,* which is the unit of quantity ("amount of substance").

In the period since the first metric system was established in France toward the end of the 18th century, most of the countries of the world have adopted a metric system. At the present time, most of the industrially advanced metric-using countries are changing from their traditional metric system to SI. Those countries that are currently changing or considering change from the English system of measurement to metric have the advantage that they can convert directly to the modernized system. The United Kingdom, which can be said to have led the now worldwide move to change from the English system, went straight to SI.

The use of SI units instead of the traditional metric units has little effect on everyday life or trade. The units of linear measurement, mass, volume, and time remain the same, viz. meter, kilogram, liter, and second.

The SI, like the traditional metric system, is based on decimal arithmetic. For each physical quantity, units of different sizes are formed by multiplying or dividing a single base value by powers of 10. Thus, changes can be made very simply by adding zeros or shifting decimal points. For example, the meter is the basic unit of length; the kilometer is a multiple (1000 meters); and the millimeter is a sub-multiple (one-thousandth of a meter).

In the older metric systems, the simplicity of a series of units linked by powers of ten is an advantage for plain quantities such as length, but this simplicity is lost as soon as more complex units are encountered. For example, in different branches of science and engineering, energy may appear as the erg, the calorie, the kilogram-meter, the liter-atmosphere, or the horsepower-hour. In contrast, the SI provides only one basic unit for each physical quantity, and universality is thus achieved.

As mentioned before, there are seven base units, which are for the basic quantities of length, mass, time, electric current, thermodynamic temperature, amount of substance, and luminous intensity, expressed as the meter (m), the kilogram (kg), the second (s), the ampere (A), the kelvin (K), the mole (mol), and the candela (cd). The units are defined in the accompanying table.

The SI is a coherent system. A system is said to be coherent if the product or quotient of any two unit quantities in the system is the unit of the resultant quantity. For example, in a coherent system in which the foot is the unit of length, the square foot is the unit of area, whereas the acre is not.

Other physical quantities are derived from the base units. For example, the unit of velocity is the meter per second (m/s), which is a combination of the base units of length and time. The unit of acceleration is the meter per second squared (m/s^2). By applying Newton's second law of motion—force is proportional to mass multiplied by acceleration—the unit of force is obtained that is the kilogram-meter per second squared ($kg\text{-}m/s^2$). This unit is known as the newton, or N. Work, or force times distance is the kilogram-meter squared per second squared ($kg\text{-}m^2/s^2$), which is the joule (1 joule = 1 newton-meter), and energy is also expressed in these terms. The abbreviation for joule is J. Power or work per unit time is the kilogram-meter squared per second cubed ($kg\text{-}m^2/s^3$), which is the watt (1 watt = 1 joule per second = 1 newton-meter per second). The abbreviation for watt is W. The term horsepower is not used in the SI and is replaced by the watt, which together with multiples and submultiples—kilowatt and milliwatt, for example—is the same unit as that used in electrical work.

The use of the newton as the unit of force is of particular interest to engineers. In practical work using the English or traditional metric systems of measurements, it is a common practice to apply weight units as force units. Thus, the unit of force in those systems is that force that when applied to unit mass produces an acceleration *g* rather than unit acceleration. The value of gravitational acceleration *g* varies around the earth, and thus the weight of a given mass also varies. In an effort to account for this minor error, the kilogram-force and pound-force were introduced, which are defined as the forces due to "standard gravity" acting on bodies of one kilogram or one pound mass, respectively. The standard gravitational acceleration is taken as 9.80665 meters per second squared or 32.174 feet per second squared. The newton is defined as "that force, which when applied to a body having a mass of one kilogram, gives it an acceleration of one meter per second squared." It is independent of *g*. As a result, the factor *g* disappears from a wide range of formulas in dynamics. However, in some formulas in statics, where the weight of a body is important rather than its mass, *g* does appear where it was formerly absent (the weight of a mass of *W* kilograms is equal to a force of *Wg* newtons, where *g* = approximately 9.81 meters per second squared). Details concerning the use of SI units in mechanics calculations are given on page 116 and throughout the Mechanics section in this Handbook. The use of SI units in strength of materials calculations is covered in the section on that subject.

Decimal multiples and sub-multiples of the SI units are formed by means of the prefixes given in the following table, which represent the numerical factors shown.

Factors and Prefixes for Forming Decimal Multiples and Sub-multiples of the SI Units

Factor by which the unit is multiplied	Prefix	Symbol	Factor by which the unit is multiplied	Prefix	Symbol
10^{12}	tera	T	10^{-2}	centi	c
10^{9}	giga	G	10^{-3}	milli	m
10^{6}	mega	M	10^{-6}	micro	µ
10^{3}	kilo	k	10^{-9}	nano	n
10^{2}	hecto	h	10^{-12}	pico	p
10	deka	da	10^{-15}	femto	f
10^{-1}	deci	d	10^{-18}	atto	a

Standard of Length.—In 1866 the United States, by act of Congress, passed a law making legal the meter, the only measure of length that has been legalized by the United States Government. The United States yard is defined by the relation: 1 yard = $^{3600}/_{3937}$ meter. The legal equivalent of the meter for commercial purposes was fixed as 39.37 inches, by law, in July, 1866, and experience having shown that this value was exact within the error of observation, the United States Office of Standard Weights and Measures was, in 1893, authorized to derive the yard from the meter by the use of this relation. The United States prototype meters Nos. 27 and 21 were received from the International Bureau of Weights and Measures in 1889. Meter No. 27, sealed in its metal case, is preserved in a fireproof vault at the Bureau of Standards.

Comparisons made prior to 1893 indicated that the relation of the yard to the meter, fixed by the Act of 1866, was by chance the exact relation between the international meter and the British imperial yard, within the error of observation. A subsequent comparison made between the standards just mentioned indicates that the legal relation adopted by Congress is in error 0.0001 inch; but, in view of the fact that certain comparisons made by the English Standards Office between the imperial yard and its authentic copies show variations as great if not greater than this, it cannot be said with certainty that there is a difference between the imperial yard of Great Britain and the United States yard derived from the meter. The bronze yard No. 11, which was an exact copy of the British imperial yard both in form and material, had shown changes when compared with the imperial yard in 1876 and 1888, which could not reasonably be said to be entirely due to changes in Bronze No. 11. On the other hand, the new meters represented the most advanced ideas of standards, and it therefore seemed that greater stability as well as higher accuracy would be secured by accepting the international meter as a fundamental standard of length.

For more information on SI practice, the reader is referred to the following publications:

Metric Practice Guide, published by the American Society for Testing and Materials, 1916 Race St., Philadelphia, PA 19103.

ISO International Standard 1000. This publication covers the rules for use of SI units, their multiples and sub-multiples. It can be obtained from the American National Standards Institute 11 West 42nd Street, New York, NY 10036.

The International System of Units, Special Publication 330 of the National Bureau of Standards—available from the Superintendent of Documents, U.S. Government Printing Office, Washington, DC 20402.

Traditional Metric Measures

Measures of Length

10millimeters(mm) = 1 centimeter (cm).

10centimeters = 1 decimeter (dm).

10decimeters = 1 meter (m).

1000meters = 1 kilometer (km.)

Square Measure

100 square millimeters (mm^2) = 1 square centimeter (cm^2).
100 square centimeters = 1 square decimeter (dm^2).
100 square decimeters = 1 square meter (m^2).

Surveyor's Square Measure

100 square meters (m^2) = 1 are (a).
100 ares = 1 hectare (ha).
100 hectares = 1 square kilometer (km^2).

Cubic Measure

1000 cubic millimeters (mm^3 = 1 cubic centimeter (cm^3).
1000 cubic centimeters = 1 cubic decimeter ($dm^{3)}$.
100 cubic decimeters = 1 cubic meter (m^3).

Dry and Liquid Measure

10 milliliters (ml) = 1 centiliter (cl).
10 centiliters = 1 deciliter (dl).
10 deciliters = 1 liter (l).
100 liters = 1 hectoliter (hl).
1 liter = 1 cubic decimeter = the volume of 1 kilogram of pure water at a temperature of 39.2 degrees F.

Measures of Weight

10 milligrams (mg) = 1 centigram (cg).
10 centigrams = 1 decigram (dg).
10 decigrams = 1 gram (g).
10 grams = 1 dekagram (dag).
10 dekagrams = 1 hectogram (hg).
10 hectograms = 1 kilogram (kg).
1000 kilograms = 1 (metric) ton (t).

Table 1. International System (SI) Units

PHYSICAL QUANTITY	NAME OF UNIT	UNIT SYMBOL	DEFINITION
Basic SI Units			
Length	metre	m	Distance traveled by light in vacuo during 1/299, 792, 458 of a second.
Mass	kilogram	kg	Mass of the international prototype which is in the custody of the Bureau International des Poids et Mesures (BIPM) at Sèvres, near Paris.
Time	second	s	The duration of 9,192,631,770 periods of the radiation corresponding to the transition between the two hyperfine levels of the ground state of the cesium-133 atom.
Electric Current	ampere	A	The constant current which, if maintained in two parallel rectilinear conductors of infinite length, of negligible circular cross section, and placed at a distance of one metre apart in a vacuum, would produce between these conductors a force equal to 2×10^{-7} N/m length.
Thermodynamic Temperature	degree kelvin	K	The fraction 1/273.16 of the thermodynamic temperature of the triple point of water.
Amount of Substance	mole	mol	The amount of substance of a system which contains as many elementary entities as there are atoms in 0.012 kilogram of carbon 12.
Luminous Intensity	candela	cd	Luminous intensity, in the perpendicular direction, of a surface of 1/600,000 square metre of a black body at the temperature of freezing platinum under a pressure of 101,325 newtons per square metre.

Table 1. International System (SI) Units *(Continued)*

PHYSICAL QUANTITY	NAME OF UNIT	UNIT SYMBOL	DEFINITION
SI Units Having Special Names			
Force	newton	$N = kg{\cdot}m/s^2$	That force which, when applied to a body having a mass of one kilogram, gives it an acceleration of one metre per second squared.
Work, Energy, Quantity of Heat	joule	$J = N{\cdot}m$	The work done when the point of application of a force of one newton is displaced through a distance of one metre in the direction of the force.
Electric Charge	coulomb	$C = A{\cdot}s$	The quantity of electricity transported in one second by a current of one ampere.
Electric Potential	volt	$V = W/A$	The difference of potential between two points of a conducting wire carrying a constant current of one ampere, when the power dissipated between these points is equal to one watt.
Electric Capacitance	farad	$F = C/V$	The capacitance of a capacitor between the plates of which there appears a difference of potential of one volt when it is charged by a quantity of electricity equal to one coulomb.
Electric Resistance	ohm	$\Omega = V/A$	The resistance between two points of a conductor when a constant difference of potential of one volt, applied between these two points, produces in this conductor a current of one ampere, this conductor not being the source of any electromotive force.
Magnetic Flux	weber	$Wb = V\ s$	The flux which, linking a circuit of one turn produces in it an electromotive force of one volt as it is reduced to zero at a uniform rate in one second.
Inductance	henry	$H = V\ s/A$	The inductance of a closed circuit in which an electromotive force of one volt is produced when the electric current in the circuit varies uniformly at the rate of one ampere per second.
Luminous Flux	lumen	$1m = cd\ sr$	The flux emitted within a unit solid angle of one steradian by a point source having a uniform intensity of one candela.
Illumination	lux	$lx = lm/m^2$	An illumination of one lumen per square metre.

Table 2. International System (SI) Units with Complex Names

PHYSICAL QUANTITY	SI UNIT	UNIT SYMBOL
SI Units Having Complex Names		
Area	square metre	m^2
Volume	cubic metre	m^3
Frequency	hertz[a]	Hz
Density (Mass Density)	kilogram per cubic metre	kg/m^3
Velocity	metre per second	m/s
Angular Velocity	radian per second	rad/s
Acceleration	metre per second squared	m/s^2
Angular Acceleration	radian per second squared	rad/s^2
Pressure	pascal[b]	Pa
Surface Tension	newton per metre	N/m
Dynamic Viscosity	newton second per metre squared	$N\ s/m^2$
Kinematic Viscosity Diffusion Coefficient }	metre squared per second	m^2/s
Thermal Conductivity	watt per metre degree Kelvin	W/(m °K)
Electric Field Strength	volt per metre	V/m
Magnetic Flux Density	tesla[c]	T
Magnetic Field Strength	ampere per metre	A/m
Luminance	candela per square metre	cd/m^2

[a] Hz = cycle/second.

[b] Pa = newton/metre^2.

[c] T = weber/metre^2.

Metric System

In the table of conversion factors that follows, the symbols of SI units, multiples and submultiples are given in parantheses in the right hand column.

Table 1. Metric Conversion Factors

Multiply	By	To Obtain
	Length	
centimeter	0.03280840	foot
centimeter	0.3937008	inch
fathom	1.8288[a]	meter (m)
foot	0.3048[a]	meter (m)
foot	30.48[a]	centimeter (cm)
foot	304.8[a]	millimeter (mm)
inch	0.0254[a]	meter (m)
inch	2.54[a]	centimeter (cm)
inch	25.4[a]	millimeter (mm)
kilometer	0.6213712	mile [U.S. statute]
meter	39.37008	inch
meter	0.5468066	fathom
meter	3.280840	foot
meter	0.1988388	rod
meter	1.093613	yard
meter	0.0006213712	mile [U. S. statute]
microinch	0.0254[a]	micrometer [micron] (μm)
micrometer [micron]	39.37008	microinch
mile [U. S. statute]	1609.344[a]	meter (m)
mile [U. S. statute]	1.609344[a]	kilometer (km)
millimeter	0.003280840	foot
millimeter	0.03937008	inch
rod	5.0292[a]	meter (m)
yard	0.9144[a]	meter (m)
	Area	
acre	4046.856	meter^2 (m^2)
acre	0.4046856	hectare
centimeter^2	0.1550003	inch^2
centimeter^2	0.001076391	foot^2
foot^2	0.09290304[a]	meter^2 (m^2)
foot^2	929.0304[a]	centimeter^2 (cm^2)
foot^2	92,903.04[a]	millimeter^2 (mm^2)
hectare	2.471054	acre
inch^2	645.16[a]	millimetere^2 (mm^2)
inch^2	6.4516[a]	centimeter^2 (cm^2)
inch^2	0.00064516[a]	meter^2 (m^2)
meter^2	1550.003	inch^2
meter^2	10.763910	foot^2
meter^2	1.195990	yard^2
meter^2	0.0002471054	acre

Table 1. *(Continued)* **Metric Conversion Factors**

Multiply	By	To Obtain
mile^2	2.5900	kilometer^2
millimeter^2	0.00001076391	foot^2
millimeter^2	0.001550003	inch^2
yard^2	0.8361274	meter^2 (m^2)
Volume (including Capacity)		
centimeter^3	0.06102376	inch^3
foot^3	28.31685	litre
foot^3	28.31685	litre
gallon [U.K. liquid]	0.004546092	meter^3 (m^3)
gallon [U.K. liquid]	4.546092	litre
gallon [U. S. liquid]	0.003785412	meter^3 (m^3)
gallon [U.S. liquid]	3.785412	litre
inch^3	16,387.06	millimeter^3 (mm^3)
inch^3	16.38706	centimeter^3 (cm^3)
inch^3	0.00001638706	meter^3 (m^3)
litre	0.001[a]	meter^3 (m^3)
litre	0.2199692	gallon [U. K. liquid]
litre	0.2641720	gallon [U. S. liquid]
litre	0.03531466	foot^3
meter^3	219.9692	gallon [U. K. liquid]
meter^3	264.1720	gallon [U. S. liquid]
meter^3	35.31466	foot^3
meter^3	1.307951	yard^3
meter^3	1000.[a]	litre
meter^3	61,023.76	inch^3
millimeter^3	0.00006102376	inch^3
quart[U.S. Liquid]	0.946	litre
quart[U.K. Liquid]	1.136	litre
yard^3	0.7645549	meter^3 (m^3)
Velocity, Acceleration, and Flow		
centimeter/second	1.968504	foot/minute
centimeter/second	0.03280840	foot/second
centimeter/minute	0.3937008	inch/minute
foot/hour	0.00008466667	meter/second (m/s)
foot/hour	0.00508[a]	meter/minute
foot/hour	0.3048[a]	meter/hour
foot/minute	0.508[a]	centimeter/second
foot/minute	18.288[a]	meter/hour
foot/minute	0.3048[a]	meter/minute
foot/minute	0.00508[a]	meter/second (m/s)
foot/second	30.48[a]	centimeter/second
foot/second	18.288[a]	meter/minute
foot/second	0.3048[a]	meter/second (m/s)
foot/second^2	0.3048[a]	meter/second^2 (m/s^2)
foot^3/minute	28.31685	litre/minute

Table 1. *(Continued)* **Metric Conversion Factors**

Multiply	By	To Obtain
foot^3/minute	0.0004719474	meter^3/second (m^3/s)
gallon [U. S. liquid]/min.	0.003785412	meter^3/minute
gallon [U. S. liquid]/min.	0.00006309020	meter^3/second (m^3/s)
gallon [U. S. liquid]/min.	0.06309020	litre/second
gallon [U. S. liquid]/min.	3.785412	litre/minute
gallon [U. K. liquid]/min.	0.004546092	meter^3/minute
gallon [U. K. liquid]/min.	0.00007576820	meter^3/second (m^3/s)
inch/minute	25.4[a]	millimeter/minute
inch/minute	2.54[a]	centimeter/minute
inch/minute	0.0254[a]	meter/minute
inch/second^2	0.0254[a]	meter/second^2 (m/s^2)
kilometer/hour	0.6213712	mile/hour [U. S. statute]
litre/minute	0.03531466	foot^3/minute
litre/minute	0.2641720	gallon [U.S. liquid]/minute
litre/second	15.85032	gallon [U. S. liquid]/minute
mile/hour	1.609344[a]	kilometer/hour
millimeter/minute	0.03937008	inch/minute
meter/second	11,811.02	foot/hour
meter/second	196.8504	foot/minute
meter/second	3.280840	foot/second
meter/second^2	3.280840	foot/second^2
meter/second^2	39.37008	inch/second^2
meter/minute	3.280840	foot/minute
meter/minute	0.05468067	foot/second
meter/minute	39.37008	inch/minute
meter/hour	3.280840	foot/hour
meter/hour	0.05468067	foot/minute
meter^3/second	2118.880	foot^3/minute
meter^3/second	13,198.15	gallon [U. K. liquid]/minute
meter^3/second	15,850.32	gallon [U. S. liquid]/minute
meter^3/minute	219.9692	gallon [U. K. liquid]/minute
meter^3/minute	264.1720	gallon [U. S. liquid]/minute
	Mass and Density	
grain [$\frac{1}{7000}$ 1b avoirdupois]	0.06479891	gram (g)
gram	15.43236	grain
gram	0.001[a]	kilogram (kg)
gram	0.03527397	ounce [avoirdupois]
gram	0.03215074	ounce [troy]
gram/centimeter^3	0.03612730	pound/inch^3
hundredweight [long]	50.80235	kilogram (kg)
hundredweight [short]	45.35924	kilogram (kg)
kilogram	1000.[a]	gram (g)
kilogram	35.27397	ounce [avoirdupois]
kilogram	32.15074	ounce [troy]

Table 1. *(Continued)* **Metric Conversion Factors**

Multiply	By	To Obtain
kilogram	2.204622	pound [avoirdupois]
kilogram	0.06852178	slug
kilogram	0.0009842064	ton [long]
kilogram	0.001102311	ton [short]
kilogram	0.001[a]	ton [metric]
kilogram	0.001[a]	tonne
kilogram	0.01968413	hundredweight [long]
kilogram	0.02204622	hundredweight [short]
kilogram/meter^3	0.06242797	pound/foot^3
kilogram/meter^3	0.01002242	pound/gallon [U. K. liquid]
kilogram/meter^3	0.008345406	pound/gallon [U. S. liquid]
ounce [avoirdupois]	28.34952	gram (g)
ounce [avoirdupois]	0.02834952	kilogram (kg)
ounce [troy]	31.10348	gram (g)
ounce [troy]	0.03110348	kilogram (kg)
pound [avoirdupois]	0.4535924	kilogram (kg)
pound/foot^3	16.01846	kilogram/meter^3 (kg/m^3)
pound/inch^3	27.67990	gram/centimeter^3 (g/cm^3)
pound/gal [U. S. liquid]	119.8264	kilogram/meter^3 (kg/m^3)
pound/gal [U. K. liquid]	99.77633	kilogram/meter^3 (kg/m^3)
slug	14.59390	kilogram (kg)
ton [long 2240 lb]	1016.047	kilogram (kg)
ton [short 2000 lb]	907.1847	kilogram (kg)
ton [metric]	1000.[a]	kilogram (kg)
ton [Metric]	0.9842	ton[long 2240 lb]
ton [Metric]	1.1023	ton[short 2000 lb]
tonne	1000.[a]	kilogram (kg)
	Force and Force/Length	
dyne	0.00001[a]	newton (N)
kilogram-force	9.806650[a]	newton (N)
kilopond	9.806650[a]	newton (N)
newton	0.1019716	kilogram-force
newton	0.1019716	kilopond
newton	0.2248089	pound-force
newton	100,000.[a]	dyne
newton	7.23301	poundal
newton	3.596942	ounce-force
newton/meter	0.005710148	pound/inch
newton/meter	0.06852178	pound/foot
ounce-force	0.2780139	newton (N)
pound-force	4.448222	newton (N)
poundal	0.1382550	newton (N)
pound/inch	175.1268	newton/meter (N/m)
pound/foot	14.59390	newton/meter (N/m)

Table 1. *(Continued)* **Metric Conversion Factors**

Multiply	By	To Obtain
Bending Moment or Torque		
dyne-centimeter	0.0000001[a]	newton-meter (N · m)
kilogram-meter	9.806650[a]	newton-meter (N · m)
ounce-inch	7.061552	newton-millimeter
ounce-inch	0.007061552	newton-meter (N · m)
newton-meter	0.7375621	pound-foot
newton-meter	10,000,000.[a]	dyne-centimeter
newton-meter	0.1019716	kilogram-meter
newton-meter	141.6119	ounce-inch
newton-millimeter	0.1416119	ounce-inch
pound-foot	1.355818	newton-meter (N · m)
Moment Of Inertia and Section Modulus		
moment of inertia [kg · m^2]	23.73036	pound-foot^2
moment of inertia [kg · m^2]	3417.171	pound-inch^2
moment of inertia [lb · ft^2]	0.04214011	kilogram-meter^2 (kg · m^2)
moment of inertia [lb · inch^2]	0.0002926397	kilogram-meter^2 (kg · m^2)
moment of section [foot^4]	0.008630975	meter^4 (m^4)
moment of section [inch^4]	41.62314	centimeter^4
moment of section [meter^4]	115.8618	foot^4
moment of section [centimeter^4]	0.02402510	inch^4
section modulus [foot^3]	0.02831685	meter^3 (m^3)
section modulus [inch^3]	0.00001638706	meter^3 (m^3)
section modulus [meter^3]	35.31466	foot^3
section modulus [meter^3]	61,023.76	inch^3
Momentum		
kilogram-meter/second	7.233011	pound-foot/second
kilogram-meter/second	86.79614	pound-inch/second
pound-foot/second	0.1382550	kilogram-meter/second (kg · m/s)
pound-inch/second	0.01152125	kilogram-meter/second (kg · m/s)
Pressure and Stress		
atmosphere [14.6959 lb/inch^2]	101,325.	pascal (Pa)
bar	100,000.[a]	pascal (Pa)
bar	14.50377	pound/inch^2
bar	100,000.[a]	newton/meter^2 (N/m^2)
hectobar	0.6474898	ton [long]/inch^2
kilogram/centimeter^2	14.22334	pound/inch^2
kilogram/meter^2	9.806650[a]	newton/meter^2 (N/m^2)
kilogram/meter^2	9.806650[a]	pascal (Pa)
kilogram/meter^2	0.2048161	pound/foot^2
kilonewton/meter^2	0.1450377	pound/inch^2
newton/centimeter^2	1.450377	pound/inch^2
newton/meter^2	0.00001[a]	bar
newton/meter^2	1.0[a]	pascal (Pa)

Table 1. *(Continued)* **Metric Conversion Factors**

Multiply	By	To Obtain
newton/meter^2	0.0001450377	pound/inch^2
newton/meter^2	0.1019716	kilogram/meter^2
newton/millimeter^2	145.0377	pound/inch^2
pascal	0.00000986923	atmosphere
pascal	0.00001[a]	bar
pascal	0.1019716	kilogram/meter^2
pascal	1.0[a]	newton/meter^2 (N/m^2)
pascal	0.02088543	pound/foot^2
pascal	0.0001450377	pound/inch^2
pound/foot^2	4.882429	kilogram/meter^2
pound/foot^2	47.88026	pascal (Pa)
pound/inch^2	0.06894757	bar
pound/inch^2	0.07030697	kilogram/centimeter^2
pound/inch^2	0.6894757	newton/centimeter^2
pound/inch^2	6.894757	kilonewton/meter^2
pound/inch^2	6894.757	newton/meter^2 (N/m^2)
pound/inch^2	0.006894757	newton/millimeter^2 (N/mm^2)
pound/inch^2	6894.757	pascal (Pa)
ton [long]/inch^2	1.544426	hectobar
	Energy and Work	
Btu [International Table]	1055.056	joule (J)
Btu [mean]	1055.87	joule (J)
calorie [mean]	4.19002	joule (J)
foot-pound	1.355818	joule (J)
foot-poundal	0.04214011	joule (J)
joule	0.0009478170	Btu [International Table]
joule	0.0009470863	Btu [mean]
joule	0.2386623	calorie [mean]
joule	0.7375621	foot-pound
joule	23.73036	foot-poundal
joule	0.9998180	joule [International U. S.]
joule	0.9999830	joule [U. S. legal, 1948]
joule [International U. S.]	1.000182	joule (J)
joule [U. S. legal, 1948]	1.000017	joule (J)
joule	.0002777778	watt-hour
watt-hour	3600.[a]	joule (J)
	Power	
Btu [International Table]/hour	0.2930711	watt (W)
foot-pound/hour	0.0003766161	watt (W)
foot-pound/minute	0.02259697	watt (W)
horsepower [550 ft-lb/s]	0.7456999	kilowatt (kW)
horsepower [550 ft-lb/s]	745.6999	watt (W)
horsepower [electric]	746.[a]	watt (W)
horsepower [metric]	735.499	watt (W)
horsepower [U. K.]	745.70	watt (W)
kilowatt	1.341022	horsepower [550 ft-lb/s]

Table 1. *(Continued)* **Metric Conversion Factors**

Multiply	By	To Obtain
watt	2655.224	foot-pound⁄hour
watt	44.25372	foot-pound⁄minute
watt	0.001341022	horsepower [550 ft-lb⁄s]
watt	0.001340483	horsepower [electric]
watt	0.001359621	horsepower [metric]
watt	0.001341022	horsepower [U. K.]
watt	3.412141	Btu [International Table]⁄hour
	Viscosity	
centipoise	0.001[a]	pascal-second (Pa · s)
centistoke	0.000001[a]	meter²⁄second (m²⁄s)
meter²⁄second	1,000,000.[a]	centistoke
meter²⁄second	10,000.[a]	stoke
pascal-second	1000.[a]	centipoise
pascal-second	10.[a]	poise
poise	0.1[a]	pascal-second (Pa · s)
stoke	0.0001[a]	meter²⁄second (m²⁄s)
	Temperature	
To Convert From	To	Use Formula
temperature Celsius, t_C	temperature Kelvin, t_K	$t_K = t_C + 273.15$
temperature Fahrenheit, t_F	temperature Kelvin, t_K	$t_K = (t_F + 459.67)/1.8$
temperature Celsius, t_C	temperature Fahrenheit, t_F	$t_F = 1.8\ t_C + 32$
temperature Fahrenheit, t_F	temperature Celsius, t_C	$t_C = (t_F - 32)/1.8$
temperature Kelvin, t_K	temperature Celsius, t_C	$t_C = t_K - 273.15$
temperature Kelvin, t_K	temperature Fahrenheit, t_F	$t_F = 1.8\ t_K - 459.67$
temperature Kelvin, t_K	temperature Rankine, t_R	$t_R = 9/5\ t_K$
temperature Rankine, t_R	temperature Kelvin, t_K	$t_K = 5/9\ t_R$

[a] Where an asterisk is shown, the figure is exact.
Symbols of SI units, multiples and sub-multiples are given in parentheses in the right-hand column

Metric and English Equivalents

Use of Conversion Tables.—On this and following pages, tables are given that permit conversion from English to metric units and vice versa over a wide range of values. Where the desired value cannot be obtained directly from these tables, a simple addition of two or more values taken directly from the table will suffice as shown in the following examples:

Example 1: Find the millimeter equivalent of 0.4476 inch.

.4	in	=	10.16000 mm
.04	in	=	1.01600 mm
.007	in	=	.17780 mm
.0006	in	=	.01524 mm
.4476	in	=	11.36904 mm

Example 2: Find the inch equivalent of 84.9 mm.

80.	mm	=	3.14961 in.
4.	mm	=	0.15748 in.
0.9	mm	=	0.03543 in.
84.9	mm	=	3.34252 in.

Miscellaneous Conversion Factors (English Units)

Multiply	By	To Obtain
atmospheres	29.92	inches of mercury (32 deg. F.)
atmospheres	14.70	pounds/inch^2
British thermal units/hour	12.96	foot-pounds/minute
circular mils	0.7854	square mils
feet of water (60 deg. F.)	0.8843	inches of mercury (60 deg. F.)
feet of water (60 deg. F.)	0.4331	pounds/inch^2
feet/minute	0.01136	miles/hour
foot-pounds/second	0.07716	British thermal units/minute
gallons (U.S.) of water (60 deg. F.)	8.337	pounds of water (60 deg. F.)
gallons (U.S.)/second	8.021	feet^3/minute
inches of mercury (32 deg. F.)	0.03342	atmospheres
inches of mercury (60 deg. F.)	1.131	feet of water (60 deg. F.)
inches of mercury (60 deg. F.)	0.4898	pounds/inch^2
inches of water (60 deg. F.)	0.03609	pounds/inch^2
knots (International)	1.151	miles (statute)/hour
miles/hour	88	feet/minute
miles (statute)/hour	0.8690	knots (International)
ounces (avoirdupois)	0.9115	ounces (troy)
ounces (troy)	1.097	ounces (avoirdupois)
ounces (troy)	0.06857	pounds (avoirdupois)
pounds (avoirdupois)	14.58	ounces (troy)
pounds of water (60 deg. F.)	0.01603	feet^3
pounds of water (60 deg. F.)	0.1199	gallons (U.S.)
pounds/inch^2	0.06805	atmospheres
pounds/inch^2	2.309	feet of water (60 deg. F.)
pounds/inch^2	2.042	inches of mercury (60 deg. F.)
pounds/inch^2	27.71	inches of water (60 deg. F.)
square mils	1.273	circular mils

Ohm's Law.—The following figure represents basic electrical relationships. This chart has been formatted in such a way that each variable has been related to the other three variables. This figure is simply for reference.

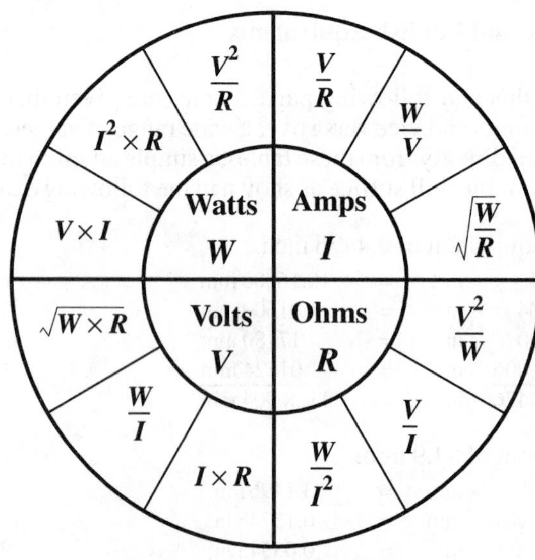

Key to variables:
V = Voltage (Volts)
R = Resistance (Ohms)
I = Current (Amps)
W = Power (Watts)

Circular Model of Electrical Relations

Binary Multiples.—The International Electrotechnical Commission has assigned the following prefixes to represent exponential binary multiples. This avoids confusion with standard SI decimal prefixes when representing powers of 2, as in bits and bytes.

Symbol	Name	Binary Power	Symbol	Name	Binary Power
Ki	kibi	2^{10}	Ti	tebi	2^{40}
Mi	mebi	2^{20}	Pi	pebi	2^{50}
Gi	gibi	2^{30}	Ei	exbi	2^{60}

Example 1: 2 Ki = $2 \times 2^{10} = 2 \times 1{,}024 = 2{,}048$. This does *not* equal 2 K = $2 \times 10^3 = 2{,}000$.

Example 2: 1 mebibyte = 1×2^{20} = 1,048,576 bytes. Again this does *not* equal 1 megabyte = 1×10^6 = 1,000,000 bytes, a value that is often confused with 1,048,576 bytes.

Linear Measure

1kilometer = 0.6214 mile
1meter = 39.37 inches = 3.2808 feet = 1.0936 yards
1centimeter = 0.3937 inch
1millimeter = 0.03937 inch
1mile = 1.609 kilometers
1yard = 0.9144 meter
1foot = 0.3048 meter = 304.8 millimeters
1inch = 2.54 centimeters = 25.4 millimeters

Table 2. Inch to Millimeters and Millimeters to Inch Convesion Table

Inches To Millimeters

in.	mm	in.	mm	in.	mm	in.	mm	in.	mm	in.	mm
10	254.00000	1	25.40000	0.1	2.54000	.01	0.25400	0.001	0.02540	0.0001	0.00254
20	508.00000	2	50.80000	0.2	5.08000	.02	0.50800	0.002	0.05080	0.0002	0.00508
30	762.00000	3	76.20000	0.3	7.62000	.03	0.76200	0.003	0.07620	0.0003	0.00762
40	1,016.00000	4	101.60000	0.4	10.16000	.04	1.01600	0.004	0.10160	0.0004	0.01016
50	1,270.00000	5	127.00000	0.5	12.70000	.05	1.27000	0.005	0.12700	0.0005	0.01270
60	1,524.00000	6	152.40000	0.6	15.24000	.06	1.52400	0.006	0.15240	0.0006	0.01524
70	1,778.00000	7	177.80000	0.7	17.78000	.07	1.77800	0.007	0.17780	0.0007	0.01778
80	2,032.00000	8	203.20000	0.8	20.32000	.08	2.03200	0.008	0.20320	0.0008	0.02032
90	2,286.00000	9	228.60000	0.9	22.86000	.09	2.2860	0.009	0.22860	0.0009	0.02286
100	2,540.00000	10	254.00000	1.0	25.40000	.10	2.54000	0.010	0.25400	0.0010	0.02540

Millimeters To Inches

mm	in.	mm	in.	mm	in	mm	in.	mm	in.	mm	in.
100	3.93701	10	0.39370	1	0.03937	0.1	0.00394	0.01	.000039	0.001	0.00004
200	7.87402	20	0.78740	2	0.07874	0.2	0.00787	0.02	.00079	0.002	0.00008
300	11.81102	30	1.18110	3	0.11811	0.3	0.01181	0.03	.00118	0.003	0.00012
400	15.74803	40	1.57480	4	0.15748	0.4	0.01575	0.04	.00157	0.004	0.00016
500	19.68504	50	1.96850	5	0.19685	0.5	0.01969	0.05	.00197	0.005	0.00020
600	23.62205	60	2.36220	6	0.23622	0.6	0.02362	0.06	.00236	0.006	0.00024
700	27.55906	70	2.75591	7	0.27559	0.7	0.02756	0.07	.00276	0.007	0.00028
800	31.49606	80	3.14961	8	0.31496	0.8	0.03150	0.08	.00315	0.008	0.00031
900	35.43307	90	3.54331	9	0.35433	0.9	0.03543	0.09	.00354	0.009	0.00035
1,000	39.37008	100	3.93701	10	0.39370	1.0	0.03937	0.10	.00394	0.010	0.00039

For inches to centimeters, shift decimal point in mm column one place to left and read centimeters, thus:

40 in. = 1016 mm = 101.6 cm

For centimeters to inches, shift decimal point of centimeter value one place to right and enter mm column, thus:

70 cm = 700 mm = 27.55906 inches

Decimals of an Inch to Millimeters
(Based on 1 inch = 25.4 millimeters, exactly)

Inches	0.000	0.001	0.002	0.003	0.004	0.005	0.006	0.007	0.008	0.009
	Millimeters									
0.000	…	0.0254	0.0508	0.0762	0.1016	0.1270	0.1524	0.1778	0.2032	0.2286
0.010	0.2540	0.2794	0.3048	0.3302	0.3556	0.3810	0.4064	0.4318	0.4572	0.4826
0.020	0.5080	0.5334	0.5588	0.5842	0.6096	0.6350	0.6604	0.6858	0.7112	0.7366
0.030	0.7620	0.7874	0.8128	0.8382	0.8636	0.8890	0.9144	0.9398	0.9652	0.9906
0.040	1.0160	1.0414	1.0668	1.0922	1.1176	1.1430	1.1684	1.1938	1.2192	1.2446
0.050	1.2700	1.2954	1.3208	1.3462	1.3716	1.3970	1.4224	1.4478	1.4732	1.4986
0.060	1.5240	1.5494	1.5748	1.6002	1.6256	1.6510	1.6764	1.7018	1.7272	1.7526
0.070	1.7780	1.8034	1.8288	1.8542	1.8796	1.9050	1.9304	1.9558	1.9812	2.0066
0.080	2.0320	2.0574	2.0828	2.1082	2.1336	2.1590	2.1844	2.2098	2.2352	2.2606
0.090	2.2860	2.3114	2.3368	2.3622	2.3876	2.4130	2.4384	2.4638	2.4892	2.5146
0.100	2.5400	2.5654	2.5908	2.6162	2.6416	2.6670	2.6924	2.7178	2.7432	2.7686
0.110	2.7940	2.8194	2.8448	2.8702	2.8956	2.9210	2.9464	2.9718	2.9972	3.0226
0.120	3.0480	3.0734	3.0988	3.1242	3.1496	3.1750	3.2004	3.2258	3.2512	3.2766
0.130	3.3020	3.3274	3.3528	3.3782	3.4036	3.4290	3.4544	3.4798	3.5052	3.5306
0.140	3.5560	3.5814	3.6068	3.6322	3.6576	3.6830	3.7084	3.7338	3.7592	3.7846
0.150	3.8100	3.8354	3.8608	3.8862	3.9116	3.9370	3.9624	3.9878	4.0132	4.0386
0.160	4.0640	4.0894	4.1148	4.1402	4.1656	4.1910	4.2164	4.2418	4.2672	4.2926
0.170	4.3180	4.3434	4.3688	4.3942	4.4196	4.4450	4.4704	4.4958	4.5212	4.5466
0.180	4.5720	4.5974	4.6228	4.6482	4.6736	4.6990	4.7244	4.7498	4.7752	4.8006
0.190	4.8260	4.8514	4.8768	4.9022	4.9276	4.9530	4.9784	5.0038	5.0292	5.0546
0.200	5.0800	5.1054	5.1308	5.1562	5.1816	5.2070	5.2324	5.2578	5.2832	5.3086
0.210	5.3340	5.3594	5.3848	5.4102	5.4356	5.4610	5.4864	5.5118	5.5372	5.5626
0.220	5.5880	5.6134	5.6388	5.6642	5.6896	5.7150	5.7404	5.7658	5.7912	5.8166
0.230	5.8420	5.8674	5.8928	5.9182	5.9436	5.9690	5.9944	6.0198	6.0452	6.0706
0.240	6.0960	6.1214	6.1468	6.1722	6.1976	6.2230	6.2484	6.2738	6.2992	6.3246
0.250	6.3500	6.3754	6.4008	6.4262	6.4516	6.4770	6.5024	6.5278	6.5532	6.5786
0.260	6.6040	6.6294	6.6548	6.6802	6.7056	6.7310	6.7564	6.7818	6.8072	6.8326
0.270	6.8580	6.8834	6.9088	6.9342	6.9596	6.9850	7.0104	7.0358	7.0612	7.0866
0.280	7.1120	7.1374	7.1628	7.1882	7.2136	7.2390	7.2644	7.2898	7.3152	7.3406
0.290	7.3660	7.3914	7.4168	7.4422	7.4676	7.4930	7.5184	7.5438	7.5692	7.5946
0.300	7.6200	7.6454	7.6708	7.6962	7.7216	7.7470	7.7724	7.7978	7.8232	7.8486
0.310	7.8740	7.8994	7.9248	7.9502	7.9756	8.0010	8.0264	8.0518	8.0772	8.1026
0.320	8.1280	8.1534	8.1788	8.2042	8.2296	8.2550	8.2804	8.3058	8.3312	8.3566
0.330	8.3820	8.4074	8.4328	8.4582	8.4836	8.5090	8.5344	8.5598	8.5852	8.6106
0.340	8.6360	8.6614	8.6868	8.7122	8.7376	8.7630	8.7884	8.8138	8.8392	8.8646
0.350	8.8900	8.9154	8.9408	8.9662	8.9916	9.0170	9.0424	9.0678	9.0932	9.1186
0.360	9.1440	9.1694	9.1948	9.2202	9.2456	9.2710	9.2964	9.3218	9.3472	9.3726
0.370	9.3980	9.4234	9.4488	9.4742	9.4996	9.5250	9.5504	9.5758	9.6012	9.6266
0.380	9.6520	9.6774	9.7028	9.7282	9.7536	9.7790	9.8044	9.8298	9.8552	9.8806
0.390	9.9060	9.9314	9.9568	9.9822	10.0076	10.0330	10.0584	10.0838	10.1092	10.1346
0.400	10.1600	10.1854	10.2108	10.2362	10.2616	10.2870	10.3124	10.3378	10.3632	10.3886
0.410	10.4140	10.4394	10.4648	10.4902	10.5156	10.5410	10.5664	10.5918	10.6172	10.6426
0.420	10.6680	10.6934	10.7188	10.7442	10.7696	10.7950	10.8204	10.8458	10.8712	10.8966
0.430	10.9220	10.9474	10.9728	10.9982	11.0236	11.0490	11.0744	11.0998	11.1252	11.1506
0.440	11.1760	11.2014	11.2268	11.2522	11.2776	11.3030	11.3284	11.3538	11.3792	11.4046
0.450	11.4300	11.4554	11.4808	11.5062	11.5316	11.5570	11.5824	11.6078	11.6332	11.6586
0.460	11.6840	11.7094	11.7348	11.7602	11.7856	11.8110	11.8364	11.8618	11.8872	11.9126
0.470	11.9380	11.9634	11.9888	12.0142	12.0396	12.0650	12.0904	12.1158	12.1412	12.1666
0.480	12.1920	12.2174	12.2428	12.2682	12.2936	12.3190	12.3444	12.3698	12.3952	12.4206
0.490	12.4460	12.4714	12.4968	12.5222	12.5476	12.5730	12.5984	12.6238	12.6492	12.6746
0.500	12.7000	12.7254	12.7508	12.7762	12.8016	12.8270	12.8524	12.8778	12.9032	12.9286

Decimals of an Inch to Millimeters *(Continued)*

(Based on 1 inch = 25.4 millimeters, exactly)

Inches	0.000	0.001	0.002	0.003	0.004	0.005	0.006	0.007	0.008	0.009
	Millimeters									
0.510	12.9540	12.9794	13.0048	13.0302	13.0556	13.0810	13.1064	13.1318	13.1572	13.1826
0.520	13.2080	13.2334	13.2588	13.2842	13.3096	13.3350	13.3604	13.3858	13.4112	13.4366
0.530	13.4620	13.4874	13.5128	13.5382	13.5636	13.5890	13.6144	13.6398	13.6652	13.6906
0.540	13.7160	13.7414	13.7668	13.7922	13.8176	13.8430	13.8684	13.8938	13.9192	13.9446
0.550	13.9700	13.9954	14.0208	14.0462	14.0716	14.0970	14.1224	14.1478	14.1732	14.1986
0.560	14.2240	14.2494	14.2748	14.3002	14.3256	14.3510	14.3764	14.4018	14.4272	14.4526
0.570	14.4780	14.5034	14.5288	14.5542	14.5796	14.6050	14.6304	14.6558	14.6812	14.7066
0.580	14.7320	14.7574	14.7828	14.8082	14.8336	14.8590	14.8844	14.9098	14.9352	14.9606
0.590	14.9860	15.0114	15.0368	15.0622	15.0876	15.1130	15.1384	15.1638	15.1892	15.2146
0.600	15.2400	15.2654	15.2908	15.3162	15.3416	15.3670	15.3924	15.4178	15.4432	15.4686
0.610	15.4940	15.5194	15.5448	15.5702	15.5956	15.6210	15.6464	15.6718	15.6972	15.7226
0.620	15.7480	15.7734	15.7988	15.8242	15.8496	15.8750	15.9004	15.9258	15.9512	15.9766
0.630	16.0020	16.0274	16.0528	16.0782	16.1036	16.1290	16.1544	16.1798	16.2052	16.2306
0.640	16.2560	16.2814	16.3068	16.3322	16.3576	16.3830	16.4084	16.4338	16.4592	16.4846
0.650	16.5100	16.5354	16.5608	16.5862	16.6116	16.6370	16.6624	16.6878	16.7132	16.7386
0.660	16.7640	16.7894	16.8148	16.8402	16.8656	16.8910	16.9164	16.9418	16.9672	16.9926
0.670	17.0180	17.0434	17.0688	17.0942	17.1196	17.1450	17.1704	17.1958	17.2212	17.2466
0.680	17.2720	17.2974	17.3228	17.3482	17.3736	17.3990	17.4244	17.4498	17.4752	17.5006
0.690	17.5260	17.5514	17.5768	17.6022	17.6276	17.6530	17.6784	17.7038	17.7292	17.7546
0.700	17.7800	17.8054	17.8308	17.8562	17.8816	17.9070	17.9324	17.9578	17.9832	18.0086
0.710	18.0340	18.0594	18.0848	18.1102	18.1356	18.1610	18.1864	18.2118	18.2372	18.2626
0.720	18.2880	18.3134	18.3388	18.3642	18.3896	18.4150	18.4404	18.4658	18.4912	18.5166
0.730	18.5420	18.5674	18.5928	18.6182	18.6436	18.6690	18.6944	18.7198	18.7452	18.7706
0.740	18.7960	18.8214	18.8468	18.8722	18.8976	18.9230	18.9484	18.9738	18.9992	19.0246
0.750	19.0500	19.0754	19.1008	19.1262	19.1516	19.1770	19.2024	19.2278	19.2532	19.2786
0.760	19.3040	19.3294	19.3548	19.3802	19.4056	19.4310	19.4564	19.4818	19.5072	19.5326
0.770	19.5580	19.5834	19.6088	19.6342	19.6596	19.6850	19.7104	19.7358	19.7612	19.7866
0.780	19.8120	19.8374	19.8628	19.8882	19.9136	19.9390	19.9644	19.9898	20.0152	20.0406
0.790	20.0660	20.0914	20.1168	20.1422	20.1676	20.1930	20.2184	20.2438	20.2692	20.2946
0.800	20.3200	20.3454	20.3708	20.3962	20.4216	20.4470	20.4724	20.4978	20.5232	20.5486
0.810	20.5740	20.5994	20.6248	20.6502	20.6756	20.7010	20.7264	20.7518	20.7772	20.8026
0.820	20.8280	20.8534	20.8788	20.9042	20.9296	20.9550	20.9804	21.0058	21.0312	21.0566
0.830	21.0820	21.1074	21.1328	21.1582	21.1836	21.2090	21.2344	21.2598	21.2852	21.3106
0.840	21.3360	21.3614	21.3868	21.4122	21.4376	21.4630	21.4884	21.5138	21.5392	21.5646
0.850	21.5900	21.6154	21.6408	21.6662	21.6916	21.7170	21.7424	21.7678	21.7932	21.8186
0.860	21.8440	21.8694	21.8948	21.9202	21.9456	21.9710	21.9964	22.0218	22.0472	22.0726
0.870	22.0980	22.1234	22.1488	22.1742	22.1996	22.2250	22.2504	22.2758	22.3012	22.3266
0.880	22.3520	22.3774	22.4028	22.4282	22.4536	22.4790	22.5044	22.5298	22.5552	22.5806
0.890	22.6060	22.6314	22.6568	22.6822	22.7076	22.7330	22.7584	22.7838	22.8092	22.8346
0.900	22.8600	22.8854	22.9108	22.9362	22.9616	22.9870	23.0124	23.0378	23.0632	23.0886
0.910	23.1140	23.1394	23.1648	23.1902	23.2156	23.2410	23.2664	23.2918	23.3172	23.3426
0.920	23.3680	23.3934	23.4188	23.4442	23.4696	23.4950	23.5204	23.5458	23.5712	23.5966
0.930	23.6220	23.6474	23.6728	23.6982	23.7236	23.7490	23.7744	23.7998	23.8252	23.8506
0.940	23.8760	23.9014	23.9268	23.9522	23.9776	24.0030	24.0284	24.0538	24.0792	24.1046
0.950	24.1300	24.1554	24.1808	24.2062	24.2316	24.2570	24.2824	24.3078	24.3332	24.3586
0.960	24.3840	24.4094	24.4348	24.4602	24.4856	24.5110	24.5364	24.5618	24.5872	24.6126
0.970	24.6380	24.6634	24.6888	24.7142	24.7396	24.7650	24.7904	24.8158	24.8412	24.8666
0.980	24.8920	24.9174	24.9428	24.9682	24.9936	25.0190	25.0444	25.0698	25.0952	25.1206
0.990	25.1460	25.1714	25.1968	25.2222	25.2476	25.2730	25.2984	25.3238	25.3492	25.3746
1.000	25.4000	…	…	…	…	…	…	…	…	…

Use previous table to obtain whole inch equivalents to add to decimal equivalents above. All values given in this table are exact; figures to the right of the last place figures are all zeros.

Millimeters to Inches
(Based on 1 inch = 25.4 millimeters, exactly)

Millimeters	0	1	2	3	4	5	6	7	8	9
	Inches									
0	…	0.03937	0.07874	0.11811	0.15748	0.19685	0.23622	0.27559	0.31496	0.35433
10	0.39370	0.43307	0.47244	0.51181	0.55118	0.59055	0.62992	0.66929	0.70866	0.74803
20	0.78740	0.82677	0.86614	0.90551	0.94488	0.98425	1.02362	1.06299	1.10236	1.14173
30	1.18110	1.22047	1.25984	1.29921	1.33858	1.37795	1.41732	1.45669	1.49606	1.53543
40	1.57480	1.61417	1.65354	1.69291	1.73228	1.77165	1.81102	1.85039	1.88976	1.92913
50	1.96850	2.00787	2.04724	2.08661	2.12598	2.16535	2.20472	2.24409	2.28346	2.32283
60	2.36220	2.40157	2.44094	2.48031	2.51969	2.55906	2.59843	2.63780	2.67717	2.71654
70	2.75591	2.79528	2.83465	2.87402	2.91339	2.95276	2.99213	3.03150	3.07087	3.11024
80	3.14961	3.18898	3.22835	3.26772	3.30709	3.34646	3.38583	3.42520	3.46457	3.50394
90	3.54331	3.58268	3.62205	3.66142	3.70079	3.74016	3.77953	3.81890	3.85827	3.89764
100	3.93701	3.97638	4.01575	4.05512	4.09449	4.13386	4.17323	4.21260	4.25197	4.29134
110	4.33071	4.37008	4.40945	4.44882	4.48819	4.52756	4.56693	4.60630	4.64567	4.68504
120	4.72441	4.76378	4.80315	4.84252	4.88189	4.92126	4.96063	5.00000	5.03937	5.07874
130	5.11811	5.15748	5.19685	5.23622	5.27559	5.31496	5.35433	5.39370	5.43307	5.47244
140	5.51181	5.55118	5.59055	5.62992	5.66929	5.70866	5.74803	5.78740	5.82677	5.86614
150	5.90551	5.94488	5.98425	6.02362	6.06299	6.10236	6.14173	6.18110	6.22047	6.25984
160	6.29921	6.33858	6.37795	6.41732	6.45669	6.49606	6.53543	6.57480	6.61417	6.65354
170	6.69291	6.73228	6.77165	6.81102	6.85039	6.88976	6.92913	6.96850	7.00787	7.04724
180	7.08661	7.12598	7.16535	7.20472	7.24409	7.28346	7.32283	7.36220	7.40157	7.44094
190	7.48031	7.51969	7.55906	7.59843	7.63780	7.67717	7.71654	7.75591	7.79528	7.83465
200	7.87402	7.91339	7.95276	7.99213	8.03150	8.07087	8.11024	8.14961	8.18898	8.22835
210	8.26772	8.30709	8.34646	8.38583	8.42520	8.46457	8.50394	8.54331	8.58268	8.62205
220	8.66142	8.70079	8.74016	8.77953	8.81890	8.85827	8.89764	8.93701	8.97638	9.01575
230	9.05512	9.09449	9.13386	9.17323	9.21260	9.25197	9.29134	9.33071	9.37008	9.40945
240	9.44882	9.48819	9.52756	9.56693	9.60630	9.64567	9.68504	9.72441	9.76378	9.80315
250	9.84252	9.88189	9.92126	9.96063	10.0000	10.0394	10.0787	10.1181	10.1575	10.1969
260	10.2362	10.2756	10.3150	10.3543	10.3937	10.4331	10.4724	10.5118	10.5512	10.5906
270	10.6299	10.6693	10.7087	10.7480	10.7874	10.8268	10.8661	10.9055	10.9449	10.9843
280	11.0236	11.0630	11.1024	11.1417	11.1811	11.2205	11.2598	11.2992	11.3386	11.3780
290	11.4173	11.4567	11.4961	11.5354	11.5748	11.6142	11.6535	11.6929	11.7323	11.7717
300	11.8110	11.8504	11.8898	11.9291	11.9685	12.0079	12.0472	12.0866	12.1260	12.1654
310	12.2047	12.2441	12.2835	12.3228	12.3622	12.4016	12.4409	12.4803	12.5197	12.5591
320	12.5984	12.6378	12.6772	12.7165	12.7559	12.7953	12.8346	12.8740	12.9134	12.9528
330	12.9921	13.0315	13.0709	13.1102	13.1496	13.1890	13.2283	13.2677	13.3071	13.3465
340	13.3858	13.4252	13.4646	13.5039	13.5433	13.5827	13.6220	13.6614	13.7008	13.7402
350	13.7795	13.8189	13.8583	13.8976	13.9370	13.9764	14.0157	14.0551	14.0945	14.1339
360	14.1732	14.2126	14.2520	14.2913	14.3307	14.3701	14.4094	14.4488	14.4882	14.5276
370	14.5669	14.6063	14.6457	14.6850	14.7244	14.7638	14.8031	14.8425	14.8819	14.9213
380	14.9606	15.0000	15.0394	15.0787	15.1181	15.1575	15.1969	15.2362	15.2756	15.3150
390	15.3543	15.3937	15.4331	15.4724	15.5118	15.5512	15.5906	15.6299	15.6693	15.7087
400	15.7480	15.7874	15.8268	15.8661	15.9055	15.9449	15.9843	16.0236	16.0630	16.1024
410	16.1417	16.1811	16.2205	16.2598	16.2992	16.3386	16.3780	16.4173	16.4567	16.4961
420	16.5354	16.5748	16.6142	16.6535	16.6929	16.7323	16.7717	16.8110	16.8504	16.8898
430	16.9291	16.9685	17.0079	17.0472	17.0866	17.1260	17.1654	17.2047	17.2441	17.2835
440	17.3228	17.3622	17.4016	17.4409	17.4803	17.5197	17.5591	17.5984	17.6378	17.6772
450	17.7165	17.7559	17.7953	17.8346	17.8740	17.9134	17.9528	17.9921	18.0315	18.0709
460	18.1102	18.1496	18.1890	18.2283	18.2677	18.3071	18.3465	18.3858	18.4252	18.4646
470	18.5039	18.5433	18.5827	18.6220	18.6614	18.7008	18.7402	18.7795	18.8189	18.8583
480	18.8976	18.9370	18.9764	19.0157	19.0551	19.0945	19.1339	19.1732	19.2126	19.2520
490	19.2913	19.3307	19.3701	19.4094	19.4488	19.4882	19.5276	19.5669	19.6063	19.6457

Millimeters to Inches *(Continued)*
(Based on 1 inch = 25.4 millimeters, exactly)

Millimeters	0	1	2	3	4	5	6	7	8	9
	Inches									
500	19.6850	19.7244	19.7638	19.8031	19.8425	19.8819	19.9213	19.9606	20.0000	20.0394
510	20.0787	20.1181	20.1575	20.1969	20.2362	20.2756	20.3150	20.3543	20.3937	20.4331
520	20.4724	20.5118	20.5512	20.5906	20.6299	20.6693	20.7087	20.7480	20.7874	20.8268
530	20.8661	20.9055	20.9449	20.9843	21.0236	21.0630	21.1024	21.1417	21.1811	21.2205
540	21.2598	21.2992	21.3386	21.3780	21.4173	21.4567	21.4961	21.5354	21.5748	21.6142
550	21.6535	21.6929	21.7323	21.7717	21.8110	21.8504	21.8898	21.9291	21.9685	22.0079
560	22.0472	22.0866	22.1260	22.1654	22.2047	22.2441	22.2835	22.3228	22.3622	22.4016
570	22.4409	22.4803	22.5197	22.5591	22.5984	22.6378	22.6772	22.7165	22.7559	22.7953
580	22.8346	22.8740	22.9134	22.9528	22.9921	23.0315	23.0709	23.1102	23.1496	23.1890
590	23.2283	23.2677	23.3071	23.3465	23.3858	23.4252	23.4646	23.5039	23.5433	23.5827
600	23.6220	23.6614	23.7008	23.7402	23.7795	23.8189	23.8583	23.8976	23.9370	23.9764
610	24.0157	24.0551	24.0945	24.1339	24.1732	24.2126	24.2520	24.2913	24.3307	24.3701
620	24.4094	24.4488	24.4882	24.5276	24.5669	24.6063	24.6457	24.6850	24.7244	24.7638
630	24.8031	24.8425	24.8819	24.9213	24.9606	25.0000	25.0394	25.0787	25.1181	25.1575
640	25.1969	25.2362	25.2756	25.3150	25.3543	25.3937	25.4331	25.4724	25.5118	25.5512
650	25.5906	25.6299	25.6693	25.7087	25.7480	25.7874	25.8268	25.8661	25.9055	25.9449
660	25.9843	26.0236	26.0630	26.1024	26.1417	26.1811	26.2205	26.2598	26.2992	26.3386
670	26.3780	26.4173	26.4567	26.4961	26.5354	26.5748	26.6142	26.6535	26.6929	26.7323
680	26.7717	26.8110	26.8504	26.8898	26.9291	26.9685	27.0079	27.0472	27.0866	27.1260
690	27.1654	27.2047	27.2441	27.2835	27.3228	27.3622	27.4016	27.4409	27.4803	27.5197
700	27.5591	27.5984	27.6378	27.6772	27.7165	27.7559	27.7953	27.8346	27.8740	27.9134
710	27.9528	27.9921	28.0315	28.0709	28.1102	28.1496	28.1890	28.2283	28.2677	28.3071
720	28.3465	28.3858	28.4252	28.4646	28.5039	28.5433	28.5827	28.6220	28.6614	28.7008
730	28.7402	28.7795	28.8189	28.8583	28.8976	28.9370	28.9764	29.0157	29.0551	29.0945
740	29.1339	29.1732	29.2126	29.2520	29.2913	29.3307	29.3701	29.4094	29.4488	29.4882
750	29.5276	29.5669	29.6063	29.6457	29.6850	29.7244	29.7638	29.8031	29.8425	29.8819
760	29.9213	29.9606	30.0000	30.0394	30.0787	30.1181	30.1575	30.1969	30.2362	30.2756
770	30.3150	30.3543	30.3937	30.4331	30.4724	30.5118	30.5512	30.5906	30.6299	30.6693
780	30.7087	30.7480	30.7874	30.8268	30.8661	30.9055	30.949	30.9843	31.0236	31.0630
790	31.1024	31.1417	31.1811	31.2205	31.2598	31.2992	31.3386	31.3780	31.4173	31.4567
800	31.4961	31.5354	31.5748	31.6142	31.6535	31.6929	31.7323	31.7717	31.8110	31.8504
810	31.8898	31.9291	31.9685	32.0079	32.0472	32.0866	32.1260	32.1654	32.2047	32.2441
820	32.2835	32.3228	32.3622	32.4016	32.4409	32.4803	32.5197	32.5591	32.5984	32.6378
830	32.6772	32.7165	32.7559	32.7953	32.8346	32.8740	32.9134	32.9528	32.9921	33.0315
840	33.0709	33.1102	33.1496	33.1890	33.2283	33.2677	33.3071	33.3465	33.3858	33.4252
850	33.4646	33.5039	33.5433	33.5827	33.6220	33.6614	33.7008	33.7402	33.7795	33.8189
860	33.8583	33.8976	33.9370	33.9764	34.0157	34.0551	34.0945	34.1339	34.1732	34.2126
870	34.2520	34.2913	34.3307	34.3701	34.4094	34.4488	34.4882	34.5276	34.5669	34.6063
880	34.6457	34.6850	34.7244	34.7638	34.8031	34.8425	34.8819	34.9213	34.9606	35.0000
890	35.0394	35.0787	35.1181	35.1575	35.1969	35.2362	35.2756	35.3150	35.3543	35.3937
900	35.4331	35.4724	35.5118	35.5512	35.5906	35.6299	35.6693	35.7087	35.7480	35.7874
910	35.8268	35.8661	35.9055	35.9449	35.9843	36.0236	36.0630	36.1024	36.1417	36.1811
920	36.2205	36.2598	36.2992	36.3386	36.3780	36.4173	36.4567	36.4961	36.5354	36.5748
930	36.6142	36.6535	36.6929	36.7323	36.7717	36.8110	36.8504	36.8898	36.9291	36.9685
940	37.0079	37.0472	37.0866	37.1260	37.1654	37.2047	37.2441	37.2835	37.3228	37.3622
950	37.4016	37.409	37.4803	37.5197	37.5591	37.5984	37.6378	37.6772	37.7165	37.7559
960	37.7953	37.8346	37.8740	37.9134	37.9528	37.9921	38.0315	38.0709	38.1102	38.1496
970	38.1800	38.2283	38.2677	38.3071	38.3465	38.3858	38.4252	38.4646	38.5039	38.5433
980	38.5827	38.6220	38.6614	38.7008	38.7402	38.7795	38.8189	38.8583	38.8976	38.9370
990	38.9764	39.0157	39.0551	39.0945	39.1339	39.1732	39.2126	39.2520	39.2913	39.3307
1000	39.3701	…	…	…	…	…	…	…	…	…

Fractional Inch — Millimeter and Foot — Millimeter Conversation Tables (Based on 1 inch = 25.4 millimeters, exactly)

Fractional Inch to Millimeters

in.	mm	in.	mm	in.	mm	in.	mm
1/64	0.397	17/64	6.747	33/64	13.097	49/64	19.447
1/32	0.794	9/32	7.144	17/32	13.494	25/32	19.844
3/64	1.191	19/64	7.541	35/64	13.891	51/64	20.241
1/16	1.588	5/16	7.938	9/16	14.288	13/16	20.638
5/64	1.984	21/64	8.334	37/64	14.684	53/64	21.034
3/32	2.381	11/32	8.731	19/32	15.081	27/32	21.431
7/64	2.778	23/64	9.128	39/64	15.478	55/64	21.828
1/8	3.175	3/8	9.525	5/8	15.875	7/8	22.225
9/64	3.572	25/64	9.922	41/64	16.272	57/64	22.622
5/32	3.969	13/32	10.319	21/32	16.669	29/32	23.019
11/64	4.366	27/64	10.716	43/64	17.066	59/64	23.416
3/16	4.762	7/16	11.112	11/16	17.462	15/16	23.812
13/64	5.159	29/64	11.509	45/64	17.859	61/64	24.209
7/32	5.556	15/32	11.906	23/32	18.256	31/32	24.606
15/64	5.953	31/64	12.303	47/64	18.653	63/64	25.003
1/4	6.350	1/2	12.700	3/4	19.050	1	25.400

Inches to Millimeters

in.	mm	in.	mm	in.	mm	in.	mm	in.	mm	in.	mm
1	25.4	3	76.2	5	127.0	7	177.8	9	228.6	11	279.4
2	50.8	4	101.6	6	152.4	8	203.2	10	254.0	12	304.8

Feet to Millimeters

ft	mm	ft	mm	ft	mm	ft	mm	ft	mm
100	30,480	10	3,048	1	304.8	0.1	30.48	0.01	3.048
200	60,960	20	6,096	2	609.6	0.2	60.96	0.02	6.096
300	91,440	30	9,144	3	914.4	0.3	91.44	0.03	9.144
400	121,920	40	12,192	4	1,219.2	0.4	121.92	0.04	12.192
500	152,400	50	15,240	5	1,524.0	0.5	152.400	0.05	15.240
600	182,880	60	18,288	6	1,828.8	0.6	182.88	0.06	18.288
700	213,360	70	21,336	7	2,133.6	0.7	213.36	0.07	21.336
800	243,840	80	24,384	8	2,438.4	0.8	243.84	0.08	24.384
900	274,320	90	27,432	9	2,743.2	0.9	274.32	0.09	27.432
1,000	304,800	100	30,480	10	3,048.0	1.0	304.80	0.10	30.480

Example 1: Find millimeter equivalent of 293 feet, 5 47/64 inches.

200 ft	=	600,960.	mm
90 ft	=	27,432.	mm
3 ft	=	914.4	mm
5 in.	=	127.0	mm
47/64 in.	=	18.653	mm
293 ft 5 47/64 in.	=	89,452.053	mm

Example 2: Find millimeter equivalent of 71.86 feet.

70.	ft	=	21,336.	mm
1.	ft	=	304.8	mm
.80	ft	=	243.84	mm
.06	ft	=	18.288	mm
71.86	ft	=	21,902.928	mm

Microinches to Micrometers (microns)
(Based on 1 microinch = 0.0254 micrometers, exactly)

Micro-inches	0	1	2	3	4	5	6	7	8	9
	Micrometers (microns)									
0	…	0.025	0.051	0.076	0.102	0.127	0.152	0.178	0.203	0.229
10	0.254	0.279	0.305	0.330	0.356	0.381	0.406	0.432	0.457	0.483
20	0.508	0.533	0.559	0.584	0.610	0.635	0.660	0.686	0.711	0.737
30	0.762	0.787	0.813	0.838	0.864	0.889	0.914	0.940	0.965	0.991
40	1.016	1.041	1.067	1.092	1.118	1.143	1.168	1.194	1.219	1.245
50	1.270	1.295	1.321	1.346	1.372	1.397	1.422	1.448	1.473	1.499
60	1.524	1.549	1.575	1.600	1.626	1.651	1.676	1.702	1.727	1.753
70	1.778	1.803	1.829	1.854	1.880	1.905	1.930	1.956	1.981	2.007
80	2.032	2.057	2.083	2.108	2.134	2.159	2.184	2.210	2.235	2.261
90	2.286	2.311	2.337	2.362	2.388	2.413	2.438	2.464	2.489	2.515
100	2.540	2.565	2.591	2.616	2.642	2.667	2.692	2.718	2.743	2.769
110	2.794	2.819	2.845	2.870	2.896	2.921	2.946	2.972	2.997	3.023
120	3.048	3.073	3.099	3.124	3.150	3.175	3.200	3.226	3.251	3.277
130	3.302	3.327	3.353	3.378	3.404	3.429	3.454	3.480	3.505	3.531
140	3.556	3.581	3.607	3.632	3.658	3.683	3.708	3.734	3.759	3.785
150	3.810	3.835	3.861	3.886	3.912	3.937	3.962	3.988	4.013	4.039
160	4.064	4.089	4.115	4.140	4.166	4.191	4.216	4.242	4.267	4.293
170	4.318	4.343	4.369	4.394	4.420	4.445	4.470	4.496	4.521	4.547
180	4.572	4.597	4.623	4.648	4.674	4.699	4.724	4.750	4.775	4.801
190	4.826	4.851	4.877	4.902	4.928	4.953	4.978	5.004	5.029	5.055
200	5.080	5.105	5.131	5.156	5.182	5.207	5.232	5.258	5.283	5.309
210	5.334	5.359	5.385	5.410	5.436	5.461	5.486	5.512	5.537	5.563
220	5.588	5.613	5.639	5.664	5.690	5.715	5.740	5.766	5.791	5.817
230	5.842	5.867	5.893	5.918	5.944	5.969	5.994	6.020	6.045	6.071
240	6.096	6.121	6.147	6.172	6.198	6.223	6.248	6.274	6.299	6.325
250	6.350	6.375	6.401	6.426	6.452	6.477	6.502	6.528	6.553	6.579
260	6.604	6.629	6.655	6.680	6.706	6.731	6.756	6.782	6.807	6.833
270	6.858	6.883	6.909	6.934	6.960	6.985	7.010	7.036	7.061	7.087
280	7.112	7.137	7.163	7.188	7.214	7.239	7.264	7.290	7.315	7.341
290	7.366	7.391	7.417	7.442	7.468	7.493	7.518	7.544	7.569	7.595

The following short table permits conversion of microinches to micrometers for ranges higher than in the main table given above. Appropriate quantities chosen from both tables are simply added to obtain the higher converted value:

μin.	μm	μin.	μm	μin.	μm	μin.	μm	μin.	μm
300	7.620	900	22.860	1500	38.100	2100	53.340	2700	68.580
600	15.240	1200	30.480	1800	45.720	2400	60.960	3000	76.200

Example: Convert 1375 μin. to μm:

From above table:	1200 μin.	=	30.480 μm
From main table:	175 μin.	=	4.445 μm
	1375 μin.	=	34.925 μm

Micrometers (microns) to Microinches (Based on 1 microinch = 0.0254 micrometers, exactly)

Micrometers Microns	0	0.01	0.02	0.03	0.04	0.05	0.06	0.07	0.08	0.09
	Microinches									
0	…	0.4	0.8	1.2	1.6	2.0	2.4	2.8	3.1	3.5
0.10	3.9	4.3	4.7	5.1	5.5	5.9	6.3	6.7	7.1	7.5
0.20	7.9	8.3	8.7	9.1	9.4	9.8	10.2	10.6	11.0	11.4
0.30	11.8	12.2	12.6	13.0	13.4	13.8	14.2	14.6	15.0	15.4
0.40	15.7	16.1	16.5	16.9	17.3	17.7	18.1	18.5	18.9	19.3
0.50	19.7	20.1	20.5	20.9	21.3	21.7	22.0	22.4	22.8	23.2
0.60	23.6	24.0	24.4	24.8	25.2	25.6	26.0	26.4	26.8	27.2
0.70	27.6	28.0	28.3	28.7	29.1	29.5	29.9	30.3	30.7	31.1
0.80	31.5	31.9	32.3	32.7	33.1	33.5	33.9	34.3	34.6	35.0
0.90	35.4	35.8	36.2	36.6	37.0	37.4	37.8	38.2	38.6	39.0
1.00	39.4	39.8	40.2	40.6	40.9	41.3	41.7	42.1	42.5	42.9
1.10	43.3	43.7	44.1	44.5	44.9	45.3	45.7	46.1	46.5	46.9
1.20	47.2	47.6	48.0	48.4	48.8	49.2	49.6	50.0	50.4	50.8
1.30	51.2	51.6	52.0	52.4	52.8	53.1	53.5	53.9	54.3	54.7
1.40	55.1	55.5	55.9	56.3	56.7	57.1	57.5	57.9	58.3	58.7
1.50	59.1	59.4	59.8	60.2	60.6	61.0	61.4	61.8	62.2	62.6
1.60	63.0	63.4	63.8	64.2	64.6	65.0	65.4	65.7	66.1	66.5
1.70	66.9	67.3	67.7	68.1	68.5	68.9	69.3	69.7	70.1	70.5
1.80	70.9	71.3	71.7	72.0	72.4	72.8	73.2	73.6	74.0	74.4
1.90	74.8	75.2	75.6	76.0	76.4	76.8	7.2	7.6	78.0	78.3
2.00	78.7	79.1	79.5	79.9	80.3	80.7	81.1	81.5	81.9	82.3
2.10	82.7	83.1	83.5	83.9	84.3	84.6	85.0	85.4	85.8	86.2
2.20	86.6	87.0	87.4	87.8	88.2	88.6	89.0	89.4	89.8	90.2
2.30	90.6	90.9	91.3	91.7	92.1	92.5	92.9	93.3	93.7	94.1
2.40	94.5	94.9	95.3	95.7	96.1	96.5	96.9	97.2	97.6	98.0
2.50	98.4	98.8	99.2	99.6	100.0	100.4	100.8	101.2	101.6	102.0
2.60	102.4	102.8	103.1	103.5	103.9	104.3	104.7	105.1	105.5	105.9
2.70	106.3	106.7	107.1	107.5	107.9	108.3	108.7	109.1	109.4	109.8
2.80	110.2	110.6	111.0	111.4	111.8	112.2	112.6	113.0	113.4	113.8
2.90	114.2	114.6	115.0	115.4	115.7	116.1	116.5	116.9	117.3	117.7
3.00	118.1	118.5	118.9	119.3	119.7	120.1	120.5	120.9	121.3	121.7
3.10	122.0	122.4	122.8	123.2	123.6	124.0	124.4	124.8	125.2	125.6
3.20	126.0	126.4	126.8	127.2	127.6	128.0	128.3	128.7	129.1	129.5
3.30	129.9	130.3	130.7	131.1	131.5	131.9	132.3	132.7	133.1	133.5
3.40	133.9	134.3	134.6	135.0	135.4	135.8	136.2	136.6	137.0	137.4
3.50	137.8	138.2	138.6	139.0	139.4	139.8	140.2	140.6	140.9	141.3
3.60	141.7	142.1	142.5	142.9	143.3	143.7	144.1	144.5	144.9	145.3
3.70	145.7	146.1	146.5	146.9	147.2	147.6	148.0	148.4	148.8	149.2
3.80	149.6	150.0	150.4	150.8	151.2	151.6	152.0	152.4	152.8	153.1
3.90	153.5	153.9	154.3	154.7	155.1	155.5	155.9	156.3	156.7	157.1
4.00	157.5	157.9	158.3	158.7	159.1	159.4	159.8	160.2	160.6	161.0
4.10	161.4	161.8	162.2	162.6	163.0	163.4	163.8	164.2	164.6	165.0
4.20	165.4	165.7	166.1	166.5	166.9	167.3	167.7	168.1	168.5	168.9
4.30	169.3	169.7	170.1	170.5	170.9	171.3	171.7	172.0	172.4	172.8
4.40	173.2	173.6	174.0	174.4	174.8	175.2	175.6	176.0	176.4	176.8
4.50	177.2	177.6	178.0	178.3	178.7	179.1	179.5	179.9	180.3	180.7
4.60	181.1	181.5	181.9	182.3	182.7	183.1	183.5	183.9	184.3	184.6
4.70	185.0	185.4	185.8	186.2	186.6	187.0	187.4	187.8	188.2	188.6
4.80	189.0	189.4	189.8	190.2	190.6	190.9	191.3	191.7	192.1	192.5
4.90	192.9	193.3	193.7	194.1	194.5	194.9	195.3	195.7	196.1	196.5
5.00	196.9	197.2	197.6	198.0	198.4	198.8	199.2	199.6	200.0	200.4

The table given below can be used with the preceding main table to obtain higher converted values, simply by adding appropriate quantities chosen from each table:

μm	μin.	μm	μin.	μm	μin.	μm	μin.	μm	μin.
10	393.7	20	787.4	30	1,181.1	40	1,574.8	50	1,968.5
15	590.6	25	984.3	35	1,378.0	45	1,771.7	55	2,165.4

Example: Convert 23.55 μm to μin.:

From above table:	20.00 μm	=	787.4 μin.
From main table:	3.55 μm	=	139.8 μin.
	23.55 μm	=	927.2 μin.

Foot — Meter and Mile — Kilometer Conversion Tables (Based on 1 foot = 0.3048 meter, exactly)

Feet to Meters (1 ft = 0.3048 m, exactly)									
feet	meters	feet	meters	feet	meters	feet	meters	feet	meters
100	30.480	10	3.048	1	0.305	0.1	0.030	0.01	0.003
200	60.960	20	6.096	2	0.610	0.2	0.061	0.02	0.006
300	91.440	30	9.144	3	0.914	0.3	0.091	0.03	0.009
400	121.920	40	12.192	4	1.219	0.4	0.122	0.04	0.012
500	152.400	50	15.240	5	1.524	0.5	0.152	0.05	0.015
600	182.880	60	18.288	6	1.829	0.6	0.183	0.06	0.018
700	213.360	70	21.336	7	2.134	0.7	0.213	0.07	0.021
800	243.840	80	24.384	8	2.438	0.8	0.244	0.08	0.024
1,000	304.800	100	30.480	10	3.048	1.0	0.305	0.10	0.030

Meters to Feet (1 m = 3.280840 ft)									
meters	feet	meters	feet	meters	feet	meters	feet	meters	feet
100	328.084	10	32.808	1	3.281	0.1	0.328	0.01	0.033
200	656.168	20	65.617	2	6.562	0.2	0.656	0.02	0.066
300	984.252	30	98.425	3	9.843	0.3	0.984	0.03	0.098
400	1,312.336	40	131.234	4	13.123	0.4	1.312	0.04	0.131
500	1,640.420	50	164.042	5	16.404	0.5	1.640	0.05	0.164
600	1,968.504	60	196.850	6	19.685	0.6	1.969	0.06	0.197
700	2,296.588	70	229.659	7	22.966	0.7	2.297	0.07	0.230
800	2,624.672	80	262.467	8	26.247	0.8	2.625	0.08	0.262
900	2,952.756	90	295.276	9	29.528	0.9	2.953	0.09	0.295
1,000	3,280.840	100	328.084	10	32.808	1.0	3.281	0.10	0.328

Miles to Kilometers (1 mile = 1.609344 km, exactly)									
miles	km	miles	km	miles	km	miles	km	miles	km
1,000	1,609.34	100	160.93	10	16.09	1	1.61	0.1	0.16
2,000	3,218.69	200	321.87	20	32.19	2	3.22	0.2	0.32
3,000	4,828.03	300	482.80	30	48.28	3	4.83	0.3	0.48
4,000	6,437.38	400	643.74	40	64.37	4	6.44	0.4	0.64
5,000	8,046.72	500	804.67	50	80.47	5	8.05	0.5	0.80
6,000	9,656.06	600	965.61	60	96.56	6	9.66	0.6	0.97
7,000	11,265.41	700	1,126.54	70	112.65	7	11.27	0.7	1.13
8,000	12,874.75	800	1,287.48	80	128.75	8	12.87	0.8	1.29
9,000	14,484.10	900	1,448.41	90	144.84	9	14.48	0.9	1.45
10,000	16,093.44	1,000	1,609.34	100	160.93	10	16.09	1.0	1.61

Kilometers to Miles (1 km = 0.6213712 mile)									
km	miles	km	miles	km	miles	km	miles	km	miles
1,000	621.37	100	62.14	10	6.21	1	0.62	0.1	0.06
2,000	1,242.74	200	124.27	20	12.43	2	1.24	0.2	0.12
3,000	1,864.11	300	186.41	30	18.64	3	1.86	0.3	0.19
4,000	2,485.48	400	248.55	40	24.85	4	2.49	0.4	0.25
5,000	3,106.86	500	310.69	50	31.07	5	3.11	0.5	0.31
6,000	3,728.23	600	372.82	60	37.28	6	3.73	0.6	0.37
7,000	4,349.60	700	434.96	70	43.50	7	4.35	0.7	0.43
8,000	4,970.97	800	497.10	80	49.71	8	4.97	0.8	0.50
9,000	5,592.34	900	559.23	90	55.92	9	5.59	0.9	0.56
10,000	6,213.71	1,000	621.37	100	62.14	10	6.21	1.0	0.62

Square Measure

1 square kilometer = 0.3861 square mile = 247.1 acres

1 hectare = 2.471 acres = 107,639 square feet

1 are = 0.0247 acre = 1076.4 square feet

1 square meter = 10.764 square feet = 1.196 square yards

1 square centimeter = 0.155 square inch

1 square millimeter = 0.00155 square inch

1 square mile = 2.5899 square kilometers

1 acre = 0.4047 hectare = 40.47 ares

1 square yard = 0.836 square meter

1 square foot = 0.0929 square meter = 929 square centimeters

1 square inch = 6.452 square centimeters = 645.2 square millimeters

Square Inch — Square Centimeter and Square Foot — Square Meter Conversion Tables (Based on 1 inch = 2.54 centimeters, exactly)

in.^{2}	cm^2	in.^{2}	cm^2	in.^{2}	cm^2	in.^{2}	cm^2	in.^{2}	cm^2
Square Inches to Square Centimeters (1 in.^{2} = 6.4516 cm^2, exactly)									
100	645.16	10	64.52	1	6.45	0.1	0.65	0.01	0.06
200	1,290.32	20	129.03	2	12.90	0.2	1.29	0.02	0.13
300	1,935.48	30	193.55	3	19.35	0.3	1.94	0.03	0.19
400	2,580.64	40	258.06	4	25.81	0.4	2.58	0.04	0.26
500	3,225.80	50	322.58	5	32.26	0.5	3.23	0.05	0.32
600	30,870.96	60	387.10	6	38.71	0.6	3.87	0.06	0.39
700	4,516.12	70	451.61	7	45.16	0.7	4.52	0.07	0.45
800	5,161.28	80	516.13	8	51.61	0.8	5.16	0.08	0.52
900	5,806.44	90	580.64	9	58.06	0.9	5.81	0.09	0.58
1,000	6,451.60	100	645.16	10	64.52	1.0	6.45	0.10	0.65
Square Centimeters to Square Inches (1 cm^2 = 0.1550003 in.^{2})									
cm^2	in.^{2}	cm^2	in.^{2}	cm^2	in.^{2}	cm^2	in.^{2}	cm^2	in.^{2}
100	15.500	10	1.550	1	0.155	0.1	0.016	0.01	0.002
200	31,000	20	3.100	2	0.310	0.2	0.031	0.02	0.003
300	46.500	30	4.650	3	0.465	0.3	0.047	0.03	0.005
400	62.000	40	6.200	4	0.620	0.4	0.062	0.04	0.006
500	77.500	50	7.750	5	0.75	0.5	0.078	0.05	0.008
600	93.000	60	9.300	6	0.930	0.6	0.093	0.06	0.009
700	108.500	70	10.850	7	1.085	0.7	0.109	0.07	0.011
800	124.000	80	12.400	8	1.240	0.8	0.124	0.08	0.012
900	139.500	90	13.950	9	1.395	0.9	0.140	0.09	0.014
1,000	155.000	100	15.500	10	1.550	1.0	0.155	0.10	0.016
Square Feet to Square Meters (1 ft^2 = 0.09290304 m^2, exactly)									
ft^2	m^2	ft^2	m^2	ft^2	m^2	ft^2	m^2	ft^2	m^2
1,000	92.903	100	9.290	10	0.929	1	0.093	0.1	0.009
2,000	185.806	200	18.581	20	1.858	2	0.186	0.2	0.019
3,000	278.709	300	27.871	30	2.787	3	0.279	0.3	0.028
4,000	371.612	400	37.161	40	3.716	4	0.372	0.4	0.037
5,000	464.515	500	46.452	50	4.645	5	0.465	0.5	0.046
6,000	557.418	600	55.742	60	5.574	6	0.557	0.6	0.056
7,000	650.321	700	65.032	70	6.503	7	0.650	0.7	0.065
8,000	743.224	800	74.322	80	7.432	8	0.743	0.8	0.074
9,000	836.127	900	83.613	90	8.361	9	0.836	0.9	0.084
10,000	929.030	1,000	92.903	100	9.290	10	0.929	1.0	0.093
Square Meters to Square Feet (1 m^2 = 10.76391 ft^2)									
m^2	ft^2	m^2	ft^2	m^2	ft^2	m^2	ft^2	m^2	ft^2
100	1,076.39	10	107.64	1	10.76	0.1	1.08	0.01	0.11
200	2,152.78	20	215.28	2	21.53	0.2	2.15	0.02	0.22
300	3,229.17	30	322.92	3	32.29	0.3	3.23	0.03	0.32
400	4,305.56	40	430.56	4	43.06	0.4	4.31	0.04	0.43
500	5,381.96	50	538.20	5	53.82	0.5	5.38	0.05	0.54
600	6,458.35	60	645.83	6	64.58	0.6	6.46	0.06	0.65
700	7,534.74	70	753.47	7	75.35	0.7	7.53	0.07	0.75
800	8,611.13	80	861.11	8	86.11	0.8	8.61	0.08	0.86
900	9,687.52	90	968.75	9	96.88	0.9	9.69	0.09	0.97
1,000	10,763.91	100	1,076.39	10	107.64	1.0	10.76	0.10	1.08

Cubic Measure

1 cubic meter = 35.315 cubic feet = 1.308 cubic yards.

1 cubic meter = 264.2 U.S. gallons.

1 cubic centimeter = 0.061 cubic inch.

1 liter (cubic decimeter) = 0.0353 cubic foot = 61.023 cubic inches.

1 liter = 0.2642 U.S. gallon = 1.0567 U.S. quarts.

1 cubic yard = 0.7646 cubic meter.

1 cubic foot = 0.02832 cubic meter = 28.317 liters.

1 cubic inch = 16.38706 cubic centimeters.

1 U.S. gallon = 3.785 liters.

1 U.S. quart = 0.946 liter.

Cubic Inch — Cubic Centimeter and Cubic Foot — Cubic Meter Conversion Tables (Based on 1 inch = 2.54 centimeters, exactly)

Cubic Inches to Cubic Centimeters ($1 \text{ in.}^3 = 16.38706 \text{ cm}^3$)

in.³	cm³	in.³	cm³	in.³	cm³	in.³	cm³	in.³	cm³
100	1,638.71	10	163.87	1	16.39	0.1	1.64	0.01	0.16
200	3,277.41	20	327.74	2	32.77	0.2	3.28	0.02	0.33
300	4,916.12	30	491.61	3	49.16	0.3	4.92	0.03	0.49
400	6,554.82	40	655.48	4	65.55	0.4	6.55	0.04	0.66
500	8,193.53	50	819.35	5	81.94	0.5	8.19	0.05	0.82
600	9,832.24	60	983.22	6	98.32	0.6	9.83	0.06	0.98
700	11,470.94	70	1,147.09	7	114.71	0.7	11.47	0.07	1.15
800	13,109.65	80	1,310.96	8	131.10	0.8	13.11	0.08	1.31
900	14,748.35	90	1,474.84	9	147.48	0.9	14.75	0.09	1.47
1,000	16,387.06	100	1,638.71	10	163.87	1.0	16.39	0.10	1.64

Cubic Centimeres to Cubic Inches ($1 \text{ cm}^3 = 0.06102376 \text{ in.}^3$)

cm³	in.³	cm³	in³	cm³	in.³	cm³	in³	cm³	in³
1,000	61.024	100	6.102	10	0.610	1	0.061	0.1	0.006
2,000	122.048	200	12.205	20	1.220	2	0.122	0.2	0.012
3,000	183.071	300	18.307	30	1,831	3	0.183	0.3	0.018
4,000	244.095	400	24.410	40	2.441	4	0.244	0.4	0.024
5,000	305.119	500	30.512	50	3.051	5	0.305	0.5	0.031
6,000	366.143	600	36.614	60	3.661	6	0.366	0.6	0.037
7,000	427.166	700	42.717	70	4.272	7	0.427	0.7	0.043
8,000	488.190	800	48.819	80	4.882	8	0.488	0.8	0.049
9,000	549.214	900	54.921	90	5.492	9	0.549	0.9	0.055
10,000	610.238	1,000	61.024	100	6.102	10	0.610	1.0	0.061

Cubic Feet to Cubic Meters ($1 \text{ ft}^3 = 0.02831685 \text{ m}^3$)

ft³	m³	ft³	m³	ft³	m³	ft³	m³	ft³	m³
1,000	28.317	100	2.832	10	0.283	1	0.028	0.1	0.003
2,000	56.634	200	5.663	20	0.566	2	0.057	0.2	0.006
3,000	84.951	300	8.495	30	0.850	3	0.085	0.3	0.008
4,000	113.267	400	11.327	40	1.133	4	0.113	0.4	0.011
5,000	141.584	500	14.158	50	1.416	5	0.142	0.5	0.014
6,000	169.901	600	16.990	60	1.699	6	0.170	0.6	0.017
7,000	198.218	700	19.822	70	1.982	7	0.198	0.7	0.020
8,000	226.535	800	22.653	80	2.265	8	0.227	0.8	0.023
9,000	254.852	900	25.485	90	2.549	9	0.255	0.9	0.025
10,000	283.168	1,000	28.317	100	2.832	10	0.283	1.0	0.028

Cubic Meters to Cubic Feet ($1 \text{ m}^3 = 35.31466 \text{ ft}^3$)

m³	ft³	m³	ft³	m³	ft³	m³	ft³	m³	ft³
100	3,531.47	10	353.15	1	35.31	0.1	3.53	0.01	0.35
200	7,062.93	20	706.29	2	70.63	0.2	7.06	0.02	0.71
300	10,594.40	30	1,059.44	3	105.94	0.3	10.59	0.03	1.06
400	14,125.86	40	4,412.59	4	141.26	0.4	14.13	0.04	1.41
500	17,657.33	50	1,756.73	5	176.57	0.5	17.66	0.05	1.77
600	21,188.80	60	2,118.88	6	211.89	0.6	21.19	0.06	2.12
700	24,720.26	70	2,472.03	7	247.20	0.7	24.72	0.07	2.47
800	28,251.73	80	2,825.17	8	282.52	0.8	28.25	0.08	2.83
900	31,783.19	90	3,178.32	9	317.83	0.9	31.78	0.09	3.18
1,000	35,314.66	100	3,531.47	10	353.15	1.0	35.311	0.10	3.53

Acre—Hectare and U.K. Gallon—Liter Conversion Tables

Acres to Hectares
(1 acre = 0.4046856 hectare)

acres	0	10	20	30	40	50	60	70	80	90
	hectares									
0	…	4.047	8.094	12.141	16.187	20.234	24.281	28.328	32.375	36.422
100	40.469	44.515	48.562	52.609	56.656	60.703	64.750	68.797	72.843	76.890
200	80.937	84.984	89.031	93.078	97.125	101.171	105.218	109.265	113.312	117.359
300	121.406	125.453	129.499	133.546	137.593	141.640	145.687	149.734	153.781	157.827
400	161.874	165.921	169.968	174.015	178.062	182.109	186.155	190.202	194.249	198.296
500	202.343	206.390	240.437	214.483	218.530	222.577	226.624	230.671	234.718	238.765
600	242.811	246.858	250.905	254.952	258.999	263.046	267.092	271.139	275.186	279.233
700	283.280	287.327	291.374	295.420	299.467	303.514	307.561	311.608	315.655	319.702
800	323.748	327.795	331.842	335.889	339.936	343.983	348.030	352.076	356.123	360.170
900	364.217	368.264	372.311	376.358	380.404	384.451	388.498	392.545	396.592	400.639
1000	404.686	…	…	…	…	…	…	…	…	…

Hectares to Acres
(1 hectare = 2.471054 acres)

hect-ares	0	10	20	30	40	50	60	70	80	90
	acres									
0	…	24.71	49.42	74.13	98.84	123.55	148.26	172.97	197.68	222.39
100	247.11	271.82	296.53	321.24	345.95	370.66	395.37	420.08	444.79	469.50
200	494.21	518.92	543.63	568.34	593.05	617.76	642.47	667.18	691.90	716.61
300	741.32	766.03	790.74	815.45	840.16	864.87	889.58	914.29	939.00	963.71
400	988.42	1013.13	1037.84	1062.55	1087.26	1111.97	1136.68	1161.40	1186.11	1210.82
500	1235.53	1260.24	1284.95	1309.66	1334.37	1359.08	1383.79	1408.50	1433.21	1457.92
600	1482.63	1507.34	1532.05	1556.76	1581.47	1606.19	1630.90	1655.61	1680.32	1705.03
700	1729.74	1754.45	1779.16	1803.87	1828.58	1853.29	1878.00	1902.71	1927.42	1952.13
800	1976.84	2001.55	2026.26	2050.97	2075.69	2100.40	2125.11	2149.82	2174.53	2199.24
900	2223.95	2248.66	2273.37	2298.08	2322.79	2347.50	2372.21	2396.92	2421.63	2446.34
1000	2471.05	…	…	…	…	…	…	…	…	…

U.K. Gallons to Liters
(1 U.K. gallon = 4.546092 liters)

Imp. gals	0	1	2	3	4	5	6	7	8	9
	liters									
0	…	4.546	9.092	13.638	18.184	22.730	27.277	31.823	36.369	40.915
10	45.461	50.007	54.553	59.099	63.645	68.191	72.737	77.284	81.830	86.376
20	90.922	95.468	100.014	104.560	109.106	113.652	118.198	122.744	127.291	131.837
30	136.383	140.929	145.475	150.021	154.567	159.113	163.659	168.205	172.751	177.298
40	181.844	186.390	190.936	195.482	200.028	204.574	209.120	213.666	218.212	222.759
50	227.305	231.851	236.397	240.943	245.489	250.035	254.581	259.127	263.673	268.219
60	272.766	277.312	281.858	286.404	290.950	295.496	300.042	304.588	309.134	313.680
70	318.226	322.773	327.319	331.865	336.411	340.957	345.503	350.049	354.595	359.141
80	363.687	368.233	372.780	377.326	381.872	386.418	390.964	395.510	400.056	404.602
90	409.148	413.694	418.240	422.787	427.333	431.879	436.425	440.971	445.517	450.063
100	454.609	459.155	463.701	468.247	472.794	477.340	481.886	486.432	490.978	495.524

Liters to U.K. Gallons
(1 liter = 0.2199692 U.K. gallons)

liters	0	1	2	3	4	5	6	7	8	9
	Imperial gallons									
0	…	0.220	0.440	0.660	0.880	1.100	1.320	1.540	1.760	1.980
10	2.200	2.420	2.640	2.860	3.080	3.300	3.520	3.739	3.959	4.179
20	4.399	4.619	4.839	5.059	5.279	5.499	5.719	5.939	6.159	6.379
30	6.599	6.819	7.039	7.259	7.479	7.699	7.919	8.139	8.359	8.579
40	8.799	9.019	9.239	9.459	9.679	9.899	10.119	10.339	10.559	10.778
50	10.998	11.218	11.438	11.658	11.878	12.098	12.318	12.538	12.758	12.978
60	13.198	13.418	13.638	13.858	14.078	14.298	14.518	14.738	14.958	15.178
70	15.398	15.618	15.838	16.058	16.278	16.498	16.718	16.938	17.158	17.378
80	17.598	17.818	18.037	18.257	18.477	18.697	18.917	19.137	19.357	19.577
90	19.797	20.017	20.237	20.457	20.677	20.897	21.117	21.337	21.557	21.777
100	21.997	22.217	22.437	22.657	22.877	23.097	23.317	23.537	23.757	23.977

Cubic Foot—Liter and Gallon—Liter Conversion Tables (Based on 1 liter = 1000 cubic centimeters)

Cubic Feet to Liters
(1 ft^3 = 28.31685 liters)

ft^3	liters	ft^3	liters	ft^3	liters	ft^3	liters	ft^3	liters
100	2,831.68	10	283.17	1	28.32	0.1	2.83	0.01	0.28
200	5,663.37	20	566.34	2	56.63	0.2	5.66	0.02	0.57
300	8,495.06	30	849.51	3	84.95	0.3	8.50	0.03	0.85
400	11,326.74	40	1,132.67	4	113.27	0.4	11.33	0.04	1.13
500	14,158.42	50	1,415.84	5	141.58	0.5	14.16	0.05	1.42
600	16,990.11	60	1,699.01	6	169.90	0.6	16.99	0.06	1.70
700	19,821.80	70	1,982.18	7	198.22	0.7	19.82	0.07	1.98
800	22,653.48	80	2,263.35	8	226.53	0.8	22.65	0.08	2.27
900	25,485.16	90	2,548.52	9	254.85	0.9	25.49	0.09	2.55
1,000	28,316.85	100	2,831.68	10	283.17	1.0	28.32	0.10	2.83

Liters to Cubic Feet
(1 liter = 0.03531466 ft^3)

liters	ft^3	liters	ft^3	liters	ft^3	liters	ft^3	liters	ft^3
1,000	35.315	100	3.531	10	0.353	1	0.035	0.1	0.004
2,000	70.629	200	7.063	20	0.706	2	0.071	0.2	0.007
3,000	105.944	300	10.594	30	1.059	3	0.106	0.3	0.011
4,000	141.259	400	14.126	40	1.413	4	0.141	0.4	0.014
5,000	176.573	500	17.657	50	1.766	5	0.177	0.5	0.018
6,000	211.888	600	21.189	60	2.119	6	0.212	0.6	0.021
7,000	247.203	700	24.720	70	2.472	7	0.247	0.7	0.025
8,000	282.517	800	28.252	80	2.825	8	0.283	0.8	0.028
9,000	317.832	900	31.783	90	3.178	9	0.318	0.9	0.032
10,000	353.147	1,000	35.315	100	3.531	10	0.353	1.0	0.035

U.S. Gallons to Liters
(1 U.S. gallon = 3785412 liters)

gals	liters	gals	liters	gals	liters	gals	liters	gals	liters
1,000	3,785.41	100	378.54	10	37.85	1	3.79	0.1	0.38
2,000	7,570.82	200	757.08	20	75.71	2	7.57	0.2	0.76
3,000	11,356.24	300	1,135.62	30	113.56	3	11.36	0.3	1.14
4,000	15,141.65	400	1,514.16	40	151.42	4	15.14	0.4	1.51
5,000	18,927.06	500	1,892.71	50	189.27	5	18.93	0.5	1.89
6,000	22,712.47	600	2,271.25	60	227.12	6	22.71	0.6	2.27
7,000	26,497.88	700	2,649.79	70	264.98	7	26.50	0.7	2.65
8,000	30,283.30	800	3,028.33	80	302.83	8	30.28	0.8	3.03
9,000	34,068.71	900	3,406.87	90	340.69	9	34.07	0.9	3.41
10,000	37.854.12	1,000	3.785.41	100	378.54	10	37.85	1.0	3.79

Liters to U.S. Gallons
(1 liter = 0.2641720 U.S. gallon)

liters	gals	liters	gals	liters	gals	liters	gals	liters	gals
1,000	264.17	100	26.42	10	2.64	1	0.26	0.1	0.03
2,000	528.34	200	52.83	20	5.28	2	0.53	0.2	0.05
3,000	792.52	300	79.25	30	7.93	3	0.79	0.3	0.08
4,000	1,056.69	400	105.67	40	10.57	4	1.06	0.4	0.11
5,000	1,320.86	500	132.09	50	13.21	5	1.32	0.5	0.13
6,000	1,585.03	600	158.50	60	15.85	6	1.59	0.6	0.16
7,000	1,849.20	700	184.92	70	18.49	7	1.85	0.7	0.18
8,000	2,113.38	800	211.34	80	21.13	8	2.11	0.8	0.21
9,000	2,377.55	900	237.75	90	23.78	9	2.38	0.9	0.24
10,000	2,641.72	1,000	264.17	100	26.42	10	2.64	1.0	0.26

Weight

1 metric ton = 0.9842 ton (of 2240 pounds) = 2204.6 pounds.
1 kilogram = 2.2046 pounds = 35.274 ounces avoirdupois.
1 gram = 0.03215 ounce troy = 0.03527 ounce avoirdupois.
1 gram = 15.432 grains.
1 ton (of 2240 pounds) = 1.016 metric ton = 1016 kilograms.
1 pound = 0.4536 kilogram = 453.6 grams.
1 ounce avoirdupois = 28.35 grams.
1 ounce troy = 31.103 grams.
1 grain = 0.0648 gram.
1 kilogram per square millimeter = 1422.32 pounds per square inch.
1 kilogram per square centimeter = 14.223 pounds per square inch.
1 kilogram-meter = 7.233 foot-pounds.
1 pound per square inch = 0.0703 kilogram per square centimeter.
1 calorie (kilogram calorie) = 3.968 Btu (British thermal unit).

Pound—Kilogram and Ounce—Gram Conversion Tables

Pounds to Kilograms (1 pound = 0.4535924 kilogram)									
lb	kg	lb	kg	lb	kg	lb	kg	lb	kg
1,000	453.59	100	45.36	10	4.54	1	0.45	0.1	0.05
2,000	907.18	200	90.72	20	9.07	2	0.91	0.2	0.09
3,000	1,360.78	300	136.08	30	13.61	3	1.36	0.3	0.14
4,000	1,814.37	400	181.44	40	18.14	4	1.81	0.4	0.18
5,000	2,267.96	500	226.80	50	22.68	5	2.27	0.5	0.23
6,000	2,721.55	600	272.16	60	27.22	6	2.72	0.6	0.27
7,000	3,175.15	700	317.51	70	31.75	7	3.18	0.7	0.32
8,000	3,628.74	800	362.87	80	36.29	8	3.63	0.8	0.36
9,000	4,082.33	900	408.23	90	40.82	9	4.08	0.9	0.41
10,000	4,535.92	1,000	453.59	100	45.36	10	4.54	1.0	0.45
Kilograms to Pounds (1 kilogram = 2.204622 pounds)									
kg	lb	kg	lb	kg	lb	kg	lb	kg	lb
1,000	2,204.62	100	220.46	10	22.05	1	2.20	0.1	0.22
2,000	4,409.24	200	440.92	20	44.09	2	4.41	0.2	0.44
3,000	6,613.87	300	661.39	30	66.14	3	6.61	0.3	0.66
4,000	8,818.49	400	881.85	40	88.18	4	8.82	0.4	0.88
5,000	11,023.11	500	1,102.31	50	110.23	5	11.02	0.5	1.10
6,000	13,227.73	600	1,322.77	60	132.28	6	13.23	0.6	1.32
7,000	15,432.35	700	1,543.24	70	154.32	7	15.43	0.7	1.54
8,000	17,636.98	800	1,763.70	80	176.37	8	17.64	0.8	1.76
9,000	19,841.60	900	1,984.16	90	198.42	9	19.84	0.9	1.98
10,000	22,046.22	1,000	2,204.62	100	220.46	10	22.05	1.0	2.20
Ounces to Grams (1 ounce = 28.34952 grams)									
oz	g	oz	g	oz	g	oz	g	oz	g
10	283.50	1	28.35	0.1	2.83	0.01	0.28	0.001	0.03
20	566.99	2	56.70	0.2	5.67	0.02	0.57	0.002	0.06
30	850.49	3	85.05	0.3	8.50	0.03	0.85	0.003	0.09
40	1,133.98	4	113.40	0.4	11.34	0.04	1.13	0.004	0.11
50	1,417.48	5	141.75	0.5	14.17	0.05	1.42	0.005	0.14
60	1,700.97	6	170.10	0.6	17.01	0.06	1.70	0.006	0.17
70	1,984.47	7	198.45	0.7	19.84	0.07	1.98	0.007	0.20
80	2,267.96	8	226.80	0.8	22.68	0.08	2.27	0.008	0.23
90	2,551.46	9	255.15	0.9	25.51	0.09	2.55	0.009	0.26
100	2,834.95	10	283.50	1.0	28.35	0.10	2.83	0.010	0.28

Pound—Kilogram and Ounce—Gram Conversion Tables *(Continued)*

Grams to Ounces (1 gram = 0.03527397 ounce)									
g	oz	g	oz	g	oz	g	oz	g	oz
100	3.527	10	0.353	1	0.035	0.1	0.004	0.01	0.000
200	7.055	20	0.705	2	0.071	0.2	0.007	0.02	0.001
300	10.582	30	1.058	3	0.106	0.3	0.011	0.03	0.001
400	14.110	40	1.411	4	0.141	0.4	0.014	0.04	0.001
500	17.637	50	1.764	5	0.176	0.5	0.018	0.05	0.002
600	21.164	60	2.116	6	0.212	0.6	0.021	0.06	0.002
700	24.692	70	2.469	7	0.247	0.7	0.025	0.07	0.002
800	28.219	80	2.822	8	0.282	0.8	0.028	0.08	0.003
900	31.747	90	3.175	9	0.317	0.9	0.032	0.09	0.003
1,000	35.274	100	3.527	10	0.353	1.0	0.035	0.10	0.004

Pounds per Square Inch—Kilograms per Square Centimeter and Pounds per Square Foot—Kilograms per Square Meter Conversion Tables

Pounds per Square Inch to Kilograms per Square Centimeter (1 lb/in.^{2} = 0.07030697 kg/cm^2)									
lb/in.^{2}	kg/cm^2	lb/in.^{2}	kg/cm^2	lb/in.^{2}	kg/cm^2	lb/in.^{2}	kg/cm^2	lb/in.^{2}	kg/cm^2
1,000	70.307	100	7.031	10	0.703	1	0.070	0.1	0.007
2,000	140.614	200	14.061	20	1.406	2	0.141	0.2	0.014
3,000	210.921	300	21.092	30	2.109	3	0.211	0.3	0.021
4,000	281.228	400	28.123	40	2.812	4	0.281	0.4	0.028
5,000	351.535	500	35.153	50	3.515	5	0.352	0.5	0.035
6,000	421.842	600	42.184	60	4.218	6	0.422	0.6	0.042
7,000	492.149	700	49.215	70	4.921	7	0.492	0.7	0.049
8,000	562.456	800	56.246	80	5.625	8	0.562	0.8	0.056
9,000	632.763	900	63.276	90	6.328	9	0.633	0.9	0.063
10,000	703.070	1,000	70.307	100	7.031	10	0.703	1.0	0.070
Kilogram per Square Centimeter to Pounds per Square Inch (1 kg/cm^2 = 14.22334 lb/in.^{2})									
kg/cm^2	lb/in.^{2}	kg/cm^2	lb/in.^{2}	kg/cm^2	lb/in.^{2}	kg/cm^2	lb/in.^{2}	kg/cm^2	lb/in.^{2}
100	1,422.33	10	142.23	1	14.22	0.1	1.42	0.01	0.14
200	2,844.67	20	284.47	2	28.45	0.2	2.84	0.02	0.28
300	4,267.00	30	426.70	3	42.67	0.3	4.27	0.03	0.43
400	5,689.34	40	568.93	4	56.89	0.4	5.69	0.04	0.57
500	7,111.67	50	711.17	5	71.12	0.5	7.11	0.05	0.71
600	8,534.00	60	853.40	6	85.34	0.6	8.53	0.06	0.85
700	9,956.34	70	995.63	7	99.56	0.7	9.96	0.07	1.00
800	11,378.67	80	1,137.87	8	113.79	0.8	11.38	0.08	1.14
900	12,801.01	90	1,280.10	9	128.01	0.9	12.80	0.09	1.28
1,000	14,223.34	100	1,422.33	10	142.23	1.0	14.22	0.10	1.42
Pounds per Square Foot to Kilograms per Square Meter (1 lb/ft^2 = 4.882429 kg/m^2)									
lb/ft^2	kg/m^2	lb/ft^2	kg/m^2	lb/ft^2	kg/m^2	lb/ft^2	kg/m^2	lb/ft^2	kg/m^2
1,000	4,882.43	100	488.24	10	48.82	1	4.88	0.1	0.49
2,000	9,764.86	200	976.49	20	97.65	2	9.76	0.2	0.98
3,000	14,647.29	300	1,464.73	30	146.47	3	14.65	0.3	1.46
4,000	19,529.72	400	1,952.97	40	195.30	4	19.53	0.4	1.95
5,000	24,412.14	500	2,441.21	50	244.12	5	24.41	0.5	2.44
6,000	29,294.57	600	2,929.46	60	292.95	6	29.29	0.6	2.93
7,000	34,177.00	700	3,417.70	70	341.77	7	34.18	0.7	3.42
8,000	39,059.43	800	3,905.94	80	390.59	8	39.06	0.8	3.91
9,000	43,941.86	900	4,394.19	90	439.42	9	43.94	0.9	4.39
10,000	48,824.28	10,000	4,882.43	100	488.24	10	48.82	1.0	4.88

Pounds per Square Inch—Kilograms per Square Centimeter and Pounds per Square Foot—Kilograms per Square Meter Conversion Tables*(Continued)*

Kilograms per Square Meter to Pounds per Square Foot (1 kg/m² = 0.2048161 lb/ft²)									
kg/m²	lb/ft²	kg/m²	lb/ft²	kg/m²	lb/ft²	kg/m²	lb/ft²	kg/m²	lb/ft²
1,000	204.82	100	20.48	10	2.05	1	0.20	0.1	0.02
2,000	409.63	200	40.96	20	4.10	2	0.41	0.2	0.04
3,000	614.45	300	61.44	30	6.14	3	0.61	0.3	0.06
4,000	819.26	400	81.93	40	8.19	4	0.82	0.4	0.08
5,000	1,024.08	500	102.41	50	10.24	5	1.02	0.5	0.10
6,000	1,228.90	600	122.89	60	12.29	6	1.23	0.6	0.12
7,000	1,433.71	700	143.37	70	14.34	7	1.43	0.7	0.14
8,000	1,638.53	800	163.85	80	16.39	8	1.64	0.8	0.16
9,000	1,843.34	900	184.33	90	18.43	9	1.84	0.9	0.18
10,000	2,048.16	1,000	204.82	100	20.48	10	2.05	1.0	0.20

Pounds per Square Inch—Kilopscals Conversion Table

Pounds per Square Inch to Kilopascals (1 lb/in² = 6.894757 kPa)										
lb/in.²	0	1	2	3	4	5	6	7	8	9
	kilopascals									
0	…	6.895	13.790	20.684	27.579	34.474	41.369	48.263	55.158	62.053
10	68.948	75.842	82.737	89.632	96.527	103.421	110.316	117.211	124.106	131.000
20	137.895	144.790	151.685	158.579	165.474	172.369	179.264	186.158	193.053	199.948
30	206.843	213.737	220.632	227.527	234.422	241.316	248.211	255.106	262.001	268.896
40	275.790	282.685	289.580	296.475	303.369	310.264	317.159	324.054	330.948	337.843
50	344.738	351.633	358.527	365.422	372.317	379.212	386.106	393.001	399.896	406.791
60	413.685	420.580	427.475	434.370	441.264	448.159	455.054	461.949	468.843	475.738
70	482.633	489.528	496.423	503.317	510.212	517.107	524.002	530.896	537.791	544.686
80	551.581	558.475	565.370	572.265	579.160	586.054	592.949	599.844	606.739	613.633
90	620.528	627.423	634.318	641.212	648.107	655.002	661.897	668.791	675.686	682.581
100	689.476	696.370	703.265	710.160	717.055	723.949	730.844	737.739	744.634	751.529
Kilopascals to Pounds Per Square Inch (1 kPa = 0.1450377 lb/in²)										
kPa	0	1	2	3	4	5	6	7	8	9
	lb/in.²									
0	…	0.145	0.290	0.435	0.580	0.725	0.870	1.015	1.160	1.305
10	1.450	1.595	1.740	1.885	2.031	2.176	2.321	2.466	2.611	2.756
20	2.901	3.046	3.191	3.336	3.481	3.626	3.771	3.916	4.061	4.206
30	4.351	4.496	4.641	4.786	4.931	5.076	5.221	5.366	5.511	5.656
40	5.802	5.947	6.092	6.237	6.382	6.527	6.672	6.817	6.962	7.107
50	7.252	7.397	7.542	7.687	7.832	7.977	8.122	8.267	8.412	8.557
60	8.702	8.847	8.992	9.137	9.282	9.427	9.572	9.718	9.863	10.008
70	10.153	10.298	10.443	10.588	10.733	10.878	11.023	11.168	11.313	11.458
80	11.603	11.748	11.893	12.038	12.183	12.328	12.473	12.618	12.763	12.908
90	13.053	13.198	13.343	13.489	13.634	13.779	13.924	14.069	14.214	14.359
100	14.504	14.649	14.794	14.939	15.084	15.229	15.374	15.519	15.664	15.809

Note: 1 kilopascal = 1 kilonewton/meter².

Pounds per Cubic Inch—Grams per Cubic Centimeter and Pounds per Cubic Foot—Kilograms per Cubic Meter Conversion Tables

Pounds per Cubic Inch to Grams per Cubic Centimeter (1 lb/in.³ = 27.67990 g/cm³)									
lb/in.³	g/cm³	lb/in.³	g/cm³	lb/in.³	g/cm³	lb/in³	g/cm³	lb/in.³	g/cm³
100	2,767.99	10	276.80	1	27.68	0.1	2.77	0.01	0.28
200	5,535.98	20	553.60	2	55.36	0.2	5.54	0.02	0.55
300	8,303.97	30	830.40	3	83.04	0.3	8.30	0.03	0.83
400	11,071.96	40	1,107.20	4	110.72	0.4	11.07	0.04	1.11
500	13,839.95	50	1,384.00	5	138.40	0.5	13.84	0.05	1.38
600	16,607.94	60	1,660.79	6	166.08	0.6	16.61	0.06	1.66
700	19,375.93	70	1,937.59	7	193.76	0.7	19.38	0.07	1.94
800	22,143.92	80	2,214.39	8	221.44	0.8	22.14	0.08	2.21
900	24,911.91	90	2,491.19	9	249.12	0.9	24.91	0.09	2.49
1,000	27,679.90	100	2,767.99	10	276.80	1.0	27.68	0.10	2.77

Grams per Cubic Centimeter to Pounds per Cubic Inch (1 g/cm³ = 0.03612730 lb/in.³)									
g/cm³	lb/in.³	g/cm³	lb/in.³	g/cm³	lb/in.³	g/cm³	lb/in.³	g/cm³	lb/in.³
1,000	36.127	100	3.613	10	0.361	1	0.036	0.1	0.004
2,000	72.255	200	7.225	20	0.723	2	0.072	0.2	0.007
3,000	108.382	300	10.838	30	1.084	3	0.108	0.3	0.011
4,000	144.509	400	14.451	40	1.445	4	0.145	0.4	0.014
5,000	180.636	500	18.064	50	1.806	5	0.181	0.5	0.018
6,000	216.764	600	21.676	60	2.168	6	0.217	0.6	0.022
7,000	252.891	700	25.289	70	2.529	7	0.253	0.7	0.025
8,000	289.018	800	28.902	80	2.890	8	0.289	0.8	0.029
9,000	325.146	900	32.515	90	3.251	9	0.325	0.9	0.033
10,000	361.273	1,000	36.127	100	3.613	10	0.361	1.0	0.036

Pounds per Cubic Foot to Kilograms per Cubic Meter (1 lb/ft³ = 16.01846 kg/m³)									
lb/lb/ft³	lb/kg/m³	lb/lb/ft³	lb/kg/m³	lb/lb/ft³	lb/kg/m³	lb/lb/ft³	lb/kg/m³	lb/lb/ft³	lb/kg/m³
100	1,601.85	10	160.18	1	16.02	0.1	1.60	0.01	0.16
200	3,203.69	20	320.37	2	32.04	0.2	3.20	0.02	0.32
300	4,805.54	30	480.55	3	48.06	0.3	4.81	0.03	0.48
400	6,407.38	40	640.74	4	64.07	0.4	6.41	0.04	0.64
500	8,009.23	50	800.92	5	80.09	0.5	8.01	0.05	0.80
600	9,611.08	60	961.11	6	96.11	0.6	9.61	0.06	0.96
700	11,212.92	70	1,121.29	7	112.13	0.7	11.21	0.07	1.12
800	12,814.77	80	1,281.48	8	128.15	0.8	12.81	0.08	1.28
900	14,416.61	90	1,441.66	9	144.17	0.9	14.42	0.09	1.44
1,000	16,018.46	100	1,601.85	10	160.18	1.0	16.02	0.10	1.60

Kilograms per Cubic Meter to Pounds per Cubic Foot (1 kg/m³ =0.06242797 lb/ft³)									
kg/m³	lb/ft³	kg/m³	lb/ft³	kg/m³	lb/ft³	kg/m³	lb/ft³	kg/m³	lb/ft³
1,000	62.428	100	6.243	10	0.624	1	0.062	0.1	0.006
2,000	124.856	200	12.486	20	1.249	2	0.125	0.2	0.012
3,000	187.284	300	18.728	30	1.873	3	0.187	0.3	0.019
4,000	249.712	400	24.971	40	2.497	4	0.250	0.4	0.025
5,000	312.140	500	31.214	50	3.121	5	0.312	0.5	0.031
6,000	374.568	600	37.457	60	3.746	6	0.375	0.6	0.037
7,000	436.996	700	43.700	70	4.370	7	0.437	0.7	0.044
8,000	499.424	800	49.942	80	4.994	8	0.499	0.8	0.050
9,000	561.852	900	56.185	90	5.619	9	0.562	0.9	0.056
10,000	624.280	1,000	62.428	100	6.243	10	0.624	1.0	0.062

Pounds-force to Newtons Conversion Table

Pounds-Force to Newtons (1 pound-force = 4.448222 newtons.)										
lbf	0	1	2	3	4	5	6	7	8	9
	newtons									
0	…	4.448	8.896	13.345	17.793	22.241	26.689	31.138	35.586	40.034
10	44.482	48.930	53.379	57.827	62.275	66.723	71.172	75.620	80.068	84.516
20	88.964	93.413	97.861	102.309	106.757	111.206	115.654	120.102	124.550	128.998
30	133.447	137.895	142.343	146.791	151.240	155.688	160.136	164.584	169.032	173.481
40	177.929	182.377	186.825	191.274	195.722	200.170	204.618	209.066	213.515	217.963
50	222.411	226.859	231.308	235.756	240.204	244.652	249.100	253.549	257.997	262.445
60	266.893	271.342	275.790	280.238	284.686	289.134	293.583	298.031	302.479	306.927
70	311.376	315.824	320.272	324.720	329.168	333.617	338.065	342.513	346.961	351.410
80	355.858	360.306	364.754	369.202	373.651	378.099	382.547	386.995	391.444	395.892
90	400.340	404.788	409.236	413.685	418.133	422.581	427.029	431.478	435.926	440.374
100	444.822	449.270	453.719	458.167	462.615	467.063	471.512	475.960	480.408	484.856

Newtons to Pounds-Force (1 newton = 0.2248089 pound-force.)										
N	0	1	2	3	4	5	6	7	8	9
	pounds-force									
0	…	0.22481	0.44962	0.67443	0.89924	1.12404	1.34885	1.57366	1.79847	2.02328
10	2.24809	2.47290	2.69771	2.92252	3.14732	3.37213	3.59694	3.82175	4.04656	4.27137
20	4.49618	4.72099	4.94580	5.17060	5.39541	5.62022	5.84503	6.06984	6.29465	6.51946
30	6.74427	6.96908	7.19388	7.41869	7.64350	7.86831	8.09312	8.31793	8.54274	8.76755
40	8.99236	9.21716	9.44197	9.66678	9.89159	10.1164	10.3412	10.5660	10.7908	11.0156
50	11.2404	11.4653	11.6901	11.9149	12.1397	12.3645	12.5893	12.8141	13.0389	13.2637
60	13.4885	13.7133	13.9382	14.1630	14.3878	14.6126	14.8374	15.0622	15.2870	15.5118
70	15.7366	15.9614	16.1862	16.4110	16.6359	16.8607	17.0855	17.3103	17.5351	17.7599
80	17.9847	18.2095	18.4343	18.6591	18.8839	19.1088	19.3336	19.5584	19.7832	20.0080
90	20.2328	20.4576	20.6824	20.9072	21.1320	21.3568	21.5817	21.8065	22.0313	22.2561
100	22.4809	22.7057	22.9305	23.1553	23.3801	23.6049	23.8297	24.0546	24.2794	24.5042

Pound-Inches to Newton-Meters Conversion Table

Pound-Inches to Newton-Meters (1 pound-force-inch = 0.1129848 newton-meter)									
lbf-in.	N·m	lbf-in.	N·m	lbf-in.	N·m	lbf-in.	N·m	lbf-in.	N·m
100	11.298	10	1.130	1	0.113	0.1	0.011	0.01	0.001
200	22.597	20	2.260	2	0.226	0.2	0.023	0.02	0.002
300	33.895	30	3.390	3	0.339	0.3	0.034	0.03	0.003
400	45.194	40	4.519	4	0.452	0.4	0.045	0.04	0.005
500	56.492	50	5.649	5	0.565	0.5	0.056	0.05	0.006
600	67.791	60	6.779	6	0.678	0.6	0.068	0.06	0.007
700	79.089	70	7.909	7	0.791	0.7	0.079	0.07	0.008
800	90.388	80	9.039	8	0.904	0.8	0.090	0.08	0.009
900	101.686	90	10.169	9	1.017	0.9	0.102	0.09	0.010
1000	112.985	100	11.298	10	1.130	1.0	0.113	0.10	0.011

Pound-Inches to Newton-Meters Conversion Table *(Continued)*

Newton-Meters to Pound-Inches (1 newton meter = 8.850748 pound-force-inches)									
N·m	lbf-in.	N·m	lbf-in.	N·m	lbf-in.	N·m	lbf-in.	N·m	lbf-in.
100	885.07	10	88.51	1	8.85	0.1	0.89	0.01	0.09
200	1770.15	20	177.01	2	17.70	0.2	1.77	0.02	0.18
300	2655.22	30	265.52	3	26.55	0.3	2.66	0.03	0.27
400	3540.30	40	354.03	4	35.40	0.4	3.54	0.04	0.35
500	4425.37	50	442.54	5	44.25	0.5	4.43	0.05	0.44
600	5310.45	60	531.04	6	53.10	0.6	5.31	0.06	0.53
700	6195.52	40	619.55	7	61.96	0.7	6.20	0.07	0.62
800	7080.60	80	708.06	8	70.81	0.8	7.08	0.08	0.71
900	7965.67	90	796.57	9	79.66	0.9	7.97	0.09	0.80
1000	8850.75	100	885.07	10	88.51	1.0	8.85	0.10	0.89

Power and Heat Equivalents

1horsepower-hour = 0.746 kilowatt-hour = 1,980,000 foot-pounds = 2545 Btu (British thermal units) = 2.64 pounds of water evaporated at 212°F = 17 pounds of water raised from 62° to 212°F.

1kilowatt-hour = 100 watt-hours = 1.34 horsepower-hour = 2,655,200 foot-pounds = 3,600,000 joules = 3415 Btu = 3.54 pounds of water evaporated at 212°F = 22.8 pounds of water raised from 62° to 212°F.

1horsepower = 746 watts = 0.746 kilowatt = 33,000 foot-pounds per minute = 550 foot-pounds per second = 2545 Btu per hour = 42.4 Btu per minute = 0.71 Btu per second = 2.64 lbs. of water evaporated per hour at 212°F.

1kilowatt = 1000 watts = 1.34 horsepower = 2,654,200 foot-pounds per hour = 44,200 foot-pounds per minute = 737 foot-pounds per second = 3415 Btu per hour = 57 Btu per minute = 0.95 Btu per second = 3.54 pounds of water evaporated per hour at 212°F.

1watt = 1 joule per second = 0.00134 horsepower = 0.001 kilowatt = 3.42 Btu per hour = 44.22 foot-pounds per minute = 0.74 foot-pounds per second = 0.0035 pound of water evaporated per hour at 212°F.

1 Btu (British thermal unit) = 1052 watt-seconds = 778 foot-pounds = 0.252 kilogram-calorie = 0.000292 kilowatt-hour = 0.000393, horse-power-hour = 0.00104 pound of water evaporated at 212°F.

1foot-pound = 1.36 joules = 0.000000377 kilowatt-hour = 0.00129 Btu = 0.0000005 horsepower-hour.

1joule = 1 watt-second = 0.00000078 kilowatt-hour = 0.00095 Btu = 0.74 foot-pound.

British Thermal Unit — Foot-Pound and Horsepower — Kilowatt Conversion Tables

British Thermal Units to Foot-Pounds (1 Btu = 778.26 ft.-lb.)[a]									
Btu	Ft.-lb.	Btu	Ft.-lb.	Btu	Ft.-lb.	Btu	Ft.-lb.	Btu	Ft.-lb.
100	77,826	10	7,783	1	778	0.1	78	0.01	8
200	155,652	20	15,565	2	1,557	0.2	156	0.02	16
300	233,478	30	23,348	3	2,335	0.3	233	0.03	23
400	311,304	40	31,130	4	3,113	0.4	311	0.04	31
500	389,130	50	38,913	5	3,891	0.5	389	0.05	39
600	466,956	60	46,696	6	4,670	0.6	467	0.06	47
700	544,782	70	54,478	7	5,448	0.7	545	0.07	54
800	622,608	80	62,261	8	6,226	0.8	623	0.08	62
900	700,434	90	70,043	9	7,004	0.9	700	0.09	70
1,000	778,260	100	77,826	10	7,783	1.0	778	0.10	78

Foot-Pounds to British Thermal Units (1 ft.-lb. = 0.00128492 Btu)[a]									
Ft.-lb.	Btu	Ft.-lb.	Btu	Ft.-lb.	Btu	Ft.-lb.	Btu	Ft.-lb.	Btu
10,000	12.849	1,000	1.285	100	0.128	10	0.013	1	0.001
20,000	25.698	2,000	2.570	200	0.257	20	0.026	2	0.003
30,000	38.548	3,000	3.855	300	0.385	30	0.039	3	0.004
40,000	51.397	4,000	5.140	400	0.514	40	0.051	4	0.005
50,000	64.246	5,000	6.425	500	0.642	50	0.064	5	0.006
60,000	77.095	6,000	7.710	600	0.771	60	0.077	6	0.008
70,000	89.944	7,000	8.994	700	0.899	70	0.090	7	0.009
80,000	102.794	8,000	10.279	800	1.028	80	0.103	8	0.010
90,000	115.643	9,000	11.564	900	1.156	90	0.116	9	0.012
100,000	128.492	10,000	12.849	1,000	1.285	100	0.128	10	0.013

Horsepower to Kilowatts (1 hp = 0.7456999 kW)[b]									
hp	kW	hp	kW	hp	kW	hp	kW	hp	kW
1,000	745.7	100	74.6	10	7.5	1	0.7	0.1	0.07
2,000	1,491.4	200	149.1	20	14.9	2	1.5	0.2	0.15
3,000	2,237.1	300	223.7	30	22.4	3	2.2	0.3	0.22
4,000	2,982.8	400	298.3	40	29.8	4	3.0	0.4	0.30
5,000	3,728.5	500	372.8	50	37.3	5	3.7	0.5	0.37
6,000	4,474.2	600	447.4	60	44.7	6	4.5	0.6	0.45
7,000	5,219.9	700	522.0	70	52.2	7	5.2	0.7	0.52
8,000	5,965.6	800	596.6	80	59.7	8	6.0	0.8	0.60
9,000	6,711.3	900	671.1	90	67.1	9	6.7	0.9	0.67
10,000	7,457.0	1,000	745.7	100	74.6	10	7.5	1.0	0.75

Kilowatts to Horsepower (1 kW = 1.341022 hp)[b]									
kW	hp	kW	hp	kW	hp	kW	hp	kW	hp
1,000	1,341.0	100	134.1	10	13.4	1	1.3	0.1	0.13
2,000	2,682.0	200	268.2	20	26.8	2	2.7	0.2	0.27
3,000	4,023.1	300	402.3	30	40.2	3	4.0	0.3	0.40
4,000	5,364.1	400	536.4	40	53.6	4	5.4	0.4	0.54
5,000	6,705.1	500	670.5	50	67.1	5	6.7	0.5	0.67
7,000	9,387.2	700	938.7	70	93.9	7	9.4	0.7	0.94
8,000	10,728.2	800	1,072.8	80	107.3	8	10.7	0.8	1.07
9,000	12,069.2	900	1,206.9	90	120.7	9	12.1	0.9	1.21
10,000	13,410.2	1,000	1,341.0	100	134.1	10	13.4	1.0	1.34

[a] Conversion factor defined by International Steam Table Conference, 1929.

[b] Based on 1 horsepower = 550 foot-pounds per second.

British Thermal Unit — Kilojoule and Foot-Pound — Joule Conversion Tables

British Thermal Units to Kilojoules (1 Btu = 1055.056 joules.)										
Btu	0	100	200	300	400	500	600	700	800	900
	kilojoules									
0	…	105.51	211.01	316.52	422.02	527.53	633.03	738.54	844.04	949.55
1000	1055.06	1160.56	1266.07	1371.57	1477.08	1582.58	1688.09	1793.60	1899.10	2004.61
2000	2110.11	2215.62	2321.12	2426.63	2532.13	2637.64	2743.15	2848.65	2954.16	3059.66
3000	3165.17	3270.67	3376.18	3481.68	3587.19	3692.70	3798.20	3903.71	4009.21	4114.72
4000	4220.22	4325.73	4431.24	4536.74	4642.25	4747.75	4853.26	4958.76	5064.27	5169.77
5000	5275.28	5380.79	5486.29	5591.80	5697.30	5802.81	5908.31	6013.82	6119.32	6224.83
6000	6330.34	6435.84	6541.35	6646.85	6752.36	6857.86	6963.37	7068.88	7174.38	7279.89
7000	7385.39	7490.90	7596.40	7701.91	7807.41	7912.92	8018.43	8123.93	8229.44	8334.94
8000	8440.45	8545.95	8651.46	8756.96	8862.47	8967.98	9073.48	9178.99	9284.49	9390.00
9000	9495.50	9601.01	9706.52	9812.02	9917.53	10023.0	10128.5	10234.0	10339.5	10445.1
10000	10550.6	…	…	…	…	…	…	…	…	…

Kilojoules to British Thermal Units 1 joule = 0.0009478170 Btu.)										
kJ	0	100	200	300	400	500	600	700	800	900
	British Thermal Units									
0	…	94.78	189.56	284.35	379.13	473.91	568.69	663.47	758.25	853.04
1000	947.82	1042.60	1137.38	1232.16	1326.94	1421.73	1516.51	1611.29	1706.07	1800.85
2000	1895.63	1990.42	2085.20	2179.98	2274.76	2369.54	2464.32	2559.11	2653.89	2748.67
3000	2843.45	2938.23	3033.01	3127.80	3222.58	3317.36	3412.14	3506.92	3601.70	3696.49
4000	3791.27	3886.05	3980.83	4075.61	4170.39	4265.18	4359.96	4454.74	4549.52	4644.30
5000	4739.08	4833.87	4928.65	5023.43	5118.21	5212.99	5307.78	5402.56	5497.34	5592.12
6000	5686.90	5781.68	5876.47	5971.25	6066.03	6160.81	6255.59	6350.37	6445.16	6539.94
7000	6634.72	6729.50	6824.28	6919.06	7013.85	7108.63	7203.41	7298.19	7392.97	7487.75
8000	7582.54	7677.32	7772.10	7866.88	7961.66	8056.44	8151.23	8246.01	8340.79	8435.57
9000	8530.35	8625.13	8719.92	8814.70	8909.48	9004.26	9099.04	9193.82	9288.61	9383.39
10000	9478.17	…	…	…	…	…	…	…	…	…

Foot-Pounds to Joules (1 foot-pound = 1.355818 joules)										
ft-lb	0	1	2	3	4	5	6	7	8	9
	joules									
0	…	1.356	2.712	4.067	5.423	6.779	8.135	9.491	10.847	12.202
10	13.558	14.914	16.270	17.626	18.981	20.337	21.693	23.049	24.405	25.761
20	27.116	28.472	29.828	31.184	32.540	33.895	35.251	36.607	37.963	39.319
30	40.675	42.030	43.386	44.742	46.098	47.454	48.809	50.165	51.521	52.877
40	54.233	55.589	56.944	58.300	59.656	61.012	62.368	63.723	65.079	66.435
50	67.791	69.147	70.503	71.858	73.214	74.570	75.926	77.282	78.637	79.993
60	81.349	82.705	84.061	85.417	86.772	88.128	89.484	90.840	92.196	93.551
70	94.907	96.263	97.619	98.975	100.331	101.686	103.042	104.398	105.754	107.110
80	108.465	109.821	111.177	112.533	113.889	115.245	116.600	117.956	119.312	120.668
90	122.024	123.379	124.735	126.091	127.447	128.803	130.159	131.514	132.870	134.226
100	135.582	136.938	138.293	139.649	141.005	142.361	143.717	145.073	146.428	147.784

Joules to Foot-Pounds (1 joule = 0.7375621 foot-pound)										
J	0	1	2	3	4	5	6	7	8	9
	foot-pounds									
0	…	0.7376	1.4751	2.2127	2.9502	3.6878	4.4254	5.1629	5.9005	6.6381
10	7.3756	8.1132	8.8507	9.5883	10.3259	11.0634	11.8010	12.5386	13.2761	14.0137
20	14.7512	15.4888	16.2264	16.9639	17.7015	18.4391	19.1766	19.4142	20.6517	21.3893
30	22.1269	22.8644	23.6020	24.3395	25.0771	25.8147	26.5522	27.2898	28.0274	28.7649
40	29.5025	30.2400	30.9776	31.7152	32.4527	33.1903	33.9279	34.6654	35.4030	36.1405
50	36.8781	37.6157	38.3532	39.0908	39.8284	40.5659	41.3035	42.0410	42.7786	43.5162
60	44.2537	44.9913	45.7289	46.4664	47.2040	47.9415	48.6791	49.4167	50.1542	50.8918
70	51.6293	52.3669	53.1045	53.8420	54.5796	55.3172	56.0547	56.7923	57.5298	58.2674
80	59.0050	59.7425	60.4801	61.2177	61.9552	62.6928	63.4303	64.1679	64.9055	65.6430
90	66.3806	67.1182	67.8557	68.5933	69.3308	70.0684	70.8060	71.5435	72.2811	73.0186
100	73.7562	74.4938	75.2313	75.9689	76.7065	77.4440	78.1816	78.9191	79.6567	80.3943

Greek Letters and Standard Abbreviations

The Greek letters are frequently used in mathematical expressions and formulas. The Greek alphabet is given below.

Α	α	Alpha	Η	η	Eta	Ν	ν	Nu	Τ	τ	Tau
Β	β	Beta	Θ	ϑ θ	Theta	Ξ	ξ	Xi	Υ	υ	Upsilon
Γ	γ	Gamma	Ι	ι	Iota	Ο	ο	Omicron	Φ	φ	Phi
Δ	δ	Delta	Κ	κ	Kappa	Π	π	Pi	Χ	χ	Chi
Ε	ε	Epsilon	Λ	λ	Lambda	Ρ	ρ	Rho	Ψ	ψ	Psi
Ζ	ζ	Zeta	Μ	μ	Mu	Σ	σ ς	Sigma	Ω	ω	Omega

American National Standard Abbreviations for Scientific and Engineering Terms

ANSI Y1.1-1972, (R 1984)

Absolute	abs	Decibel	dB
Alternating current	ac	Degree	deg or°
Ampere	amp	Degree Centigrade	°C
Ampere-hour	amp hr	Degree Fahrenheit	°F
Angstrom unit	A	Degree Kelvin	K
Antilogarithm	antilog	Diameter	dia
Arithmetical average	aa	Direct current	dc
Atmosphere	atm	Dozen	doz
Atomic weight	at wt	Dram	dr
Avoirdupois	avdp	Efficiency	eff
Barometer	baro	Electric	elec
Board feet (feet board measure)	fbm	Electromotive force	emf
Boiler pressure	bopress	Elevation	el
Boiling point	bp	Engine	eng
Brinell hardness number	Bhn	Engineer	engr
British thermal unit	Btu or B	Engineering	engrg
Bushel	bu	Equation	eq
Calorie	cal	External	ext
Candle	cd	Fluid	fl
Center to center	c to c	Foot	ft
Centimeter	cm	Foot-candle	fc
Centimeter-gram-second (system)	cgs	Foot-Lambert	fL or fl
Chemical	chem	Foot per minute	fpm
Chemically pure	cp	Foot per second	fps
Circular	circ	Foot-pound	ft lb
Circular mil	cmil	Foot-pound-second (system)	fps
Coefficient	coef	Free on board	fob
Cologarithm	colog	Freezing point	fp
Concentrate	conc	Frequency	freq
Conductivity	cndct	Fusion point	fnpt
Constant	const	Gallon	gal
Cord	cd	Gallon per minute	gpm
Cosecant	csc	Gallon per second	gps
Cosine	cos	Grain	gr
Cost, insurance, and freight	cif	Gram	g
Cotangent	ctn	Greatest common divisor	gcd
Counter electromotive force	cemf		
Cubic	cu	High pressure	hp
Cubic centimeter	cm^3 or cc	Horsepower	hp
Cubic foot	ft^3 or cu ft	Horsepower-hour	hp hr
Cubic feet per second	ft^3 or cfs	Hour	h or hr
Cubic inch	in^3 or cu in	Hyperbolic cosine	cosh
Cubic meter	m^3 or cu m	Hyperbolic sine	sinh
Cubic millimeter	mm^3 or cumm	Hyperbolic tangent	tanh
Cubic yard	yd^3 or cu yd	pound-force foot	$lb_f \cdot ft$ or lb ft
Current density	cd	pound-force inch	$lb_f \cdot in$ or lb in
Cylinder	cyl	Inch	in
Indicated horsepower-hour	iph	Inch per second	in/s or ips

American National Standard Abbreviations for Scientific and Engineering Terms
ANSI Y1.1-1972, R 1984 (Continued)

Intermediate pressure	ip	Inch-pound	in lb
Internal	intl	pound-force per square foot	lb_f/ft^2 or psf
Kilovolt-ampere/hour	KVA-h or kVah	pound-force per square inch	lb_f/in^2 or psi
Kilowatt-hour meter	kwhm	pound per horsepower	lb/hp or php
Latitude	lat	Power factor	pf
Least common multiple	lcm	Quart	qt
Liquid	liq	Reactive volt-ampere meter	rva
Logarithm (common)	log	Revolution per minute	r/min or rpm
Logarithm (natural)	ln	Revolution per second	r/s or rps
Low pressure	lp	Root mean square	rms
Lumen per watt	lm/W or lpw	Round	rnd
Magnetomotive force	mmf	Secant	sec
Mathematics (ical)	math	Second	s or sec
Maximum	max	Sine	sin
Mean effective pressure	mep	Specific gravity	sp gr
Melting point	mp	Specific heat	sp ht
Meter	m	Square	sq
Meter-kilogram-second	mks	Square centimeter	cm^2 or sq cm
Microfarad	μF	Square foot	ft^2 or sq ft
Mile	mi	Square inch	in^2 or sq in
Mile per hour	mi/h or mph	Square kilometer	km^2 or sq km
Milliampere	m/A	Square root of mean square	rms
Minimum	min	Standard	std
Molecular weight	mol wt	Tangent	tan
Molecule	mo	Temperature	temp
National Electrical Code	NEC	Tensile strength	ts
Ounce	oz	Versed sine	vers
Ounce-inch	oz in	Volt	V
Pennyweight	dwt	Watt	W
Pint	pt	Watthour	Wh
Potential	pot	Week	wk
Potential difference	pd	Weight	wt
Pound	lb	Yard	yd

Alternative abbreviations conforming to the practice of the International Electrotechnical Commission.

Ampere	A	Kilovolt-ampere	kVA	Microfarad	μF	Milliampere	mA
Ampere-hour	Ah	Kilowatt	kW	Microwatt	μW	Volt	V
Coulomb	C			Milliampere	mA	Volt-ampere	VA
Farad	F	Kilowatthour	kWh	Millifarad	mF	Volt-coulomb	VC
Henry	H	Megawatt	MW	Millihenry	mH	Watt	W
Joule	J	Megohm	Mω	Millivolt	mV	Watthour	Wh
Kilovolt	kV	Microampere	μA	Ohm	ω	Volt	VA

Only the most commonly used terms have been included. These forms are recommended for those whose familiarity with the terms used makes possible a maximum of abbreviations. For others, less contracted combinations made up from this list may be used. For example, the list gives the abbreviation of the term "feet per second" as "fps." To some, however, ft per sec will be more easily understood.

Abbreviations should be used sparingly and only where their meaning will be clear. If there is any doubt, then spell out the term or unit of measurement.

The following points are good practice when preparing engineering documentation. Terms denoting units of measurement should be abbreviated in text only when preceded by the amounts indicated in numerals: "several inches," "one inch," "12 in." A sentence should not begin with a numeral followed by an abbreviation. The use of conventional signs for abbreviations in text should be avoided: use "lb," not "#" or "in," not ″.

Symbols for the chemical elements are listed in the table on page 363.

A

B

C

D

E

F

G

I

J

K

L

M

N

Q

R

S

T

U

V

W

Y

Z

NOTES

NOTES